NOBEL PRIZES

YEAR	PHYSICS	CHEMISTRY	PHYSIOLOGY OR MEDICINE
1940	(No award)	(No award)	(No award)
1941	(No award)	(No award)	(No award)
1942	(No award)	(No award)	
1943	Otto Stern	George C. de Hevesy	
1944	Isidor Isaac Rabi	Otto Hahn	
1945	Wolfgang Pauli	Artturi I. Virtanen	Sir Alexander Fleming Ernst Boris Chain Lord Howard W. Florey
1946	Percy W. Bridgman	James B. Sumner John H. Northrop Wendell Stanley	Hermann J. Muller
1947	Sir Edward Appleton	Sir Robert Robinson	Carl F. Cori Gerty T. Cori Bernardo A. Houssay
1948	Patrick M. Blackett	Arne Tiselius	Paul H. Müller
1949	Hideki Yukawa	William F. Giauque	Walter R. Hess Antonio Egas Moniz
1950	Cecil F. Powell	Otto P. Diels Kurt Alder	Philip S. Hench Edward C. Kendall Tadeus Reichstein
1951	Sir John D. Cockcroft Ernest T. Walton	Edwin M. McMillan Glenn T. Seaborg	Max Theiler
1952	Felix Bloch Edward Purcell	Archer J. Martin Richard L. Synge	Selman A. Waksman
1953	Frits Zernike	Hermann Staudinger	Fritz A. Lipmann Sir Hans A. Krebs
1954	Max Born Walther Bothe	Linus Carl Pauling	John F. Enders Frederick C. Robbins Thomas H. Weller
1955	Willis Lamb, Jr. Polykarp Kusch	Vincent du Vigneaud	Axel Hugo Theorell
1956	William Shockley Walter H. Brattain John Bardeen	Sir Cyril Hinshelwood Nikolai N. Semenov	Dickinson W. Richards André F. Cournaand Werner Forssmann
1957	Tsung-Dao Lee Chen Ning Yang	Sir Alexander Todd	Daniel Bovet
1958	Pavel A. Cherenkov Igor Tamm Ilya M. Frank	Frederick Sanger	Joshua Lederberg George W. Beadle Edward L. Tatum
1959	Emilio Segrè Owen Chamberlain	Jaroslav Heyrovsky	Severo Ochoa Arthur Kornberg
1960	Donald A. Glaser	Willard F. Libby	Sir Frank M. Burnet Peter B. Medawar
1961	Robert Hofstadter Rudolf Mössbauer	Melvin Calvin	Georg von Bekesy
1962	Lev D. Landau	Max F. Perutz John C. Kendrew	James C. Watson Francis H. Crick Maurice H. Wilkins
1963	J. Hans D. Jensen Maria Goeppert-Mayer Eugene Paul Wigner	Giulio Natta Karl Ziegler	Sir John C. Eccles Alan L. Hodgkin Andrew F. Huxley
1964	Charles H. Townes Nikolai G. Basov Aleksandr M. Prokhorov	Dorothy C. Hodgkin	Konrad Bloch Feodor Lynen
1965	Julian Schwinger Richard P. Feynman Sin-itiro Tomonaga	Robert B. Woodward	Francois Jacob Andre M. Lwoff Jacques L. Monod
1966	Alfred Kastler	Robert S. Mulliken	Charles B. Huggins Francis P. Rous
1967	Hans A. Bethe	Manfred Eigen Ronald G. Norrish George Porter	Haldan K. Hartline George Wald Ragnar A. Granit

WORLD WHO'S WHO
IN SCIENCE

A BIOGRAPHICAL DICTIONARY OF NOTABLE SCIENTISTS
FROM ANTIQUITY TO THE PRESENT

A Component Volume of
The Marquis Biographical Library

FIRST EDITION

Editor
ALLEN G. DEBUS
Professor of the History of Science
University of Chicago

Associate Editors
RONALD S. CALINGER   EDWARD J. COLLINS

Managing Editor
STEPHEN J. KENNEDY

MARQUIS-WHO'S WHO
INCORPORATED

The A. N. Marquis Company—Founded 1897

Marquis-Who's Who Building
200 East Ohio Street
Chicago, Illinois 60611

Standard Book Number 8379-1001-3

LIBRARY OF CONGRESS CATALOG CARD NUMBER 68-56149

PRINTED AND BOUND AT HANNIBAL, MISSOURI, BY
WESTERN PUBLISHING COMPANY

TABLE OF CONTENTS

PREFACE

The importance of science and technology in our society has been generally accepted only since the close of the second World War. Prior to that time the average man—at least in the United States—looked with wonder on the inventions of men such as Edison and Morse, but paid little attention to research professors who devoted their lives to the laboratory and turned out a stream of published papers which often seemed unintelligible and certainly seemed uninteresting. John Q. Public pictured the scientist as a distant creature who lived in his ivory tower and had no interest in or concern for the needs of mankind. Most men would have agreed that Marconi—the inventor—was a benefactor of society, but for the same people the scientist was of a different breed. Few knew this better than radio comedians who were assured of a laugh by the mere mention of Einstein and the theory of relativity.

Perhaps no one event was more influential in changing this attitude than the bombing of Hiroshima. Suddenly the scientist assumed a new significance, and his new status has grown ever more substantial over the past two decades. During this time we have been subjected to an unceasing series of developments in all areas of the sciences which have with increasing regularity pushed more mundane matters from the headlines. Nor are scientists still pictured as men divorced of all interest in the world about them. A recent study by the University of Chicago shows that scientists have risen in the public esteem from eighth place in 1947 to third place in 1963. The same study gave the even more dramatic finding that the status of the "nuclear physicist" had risen from fifteenth place to tie "scientist" for third place during the same period.[*] The curricula of our major educational institutions reflect this change. Ambitious new programs in the history of science, the sociology of science, and the philosophy of science—subjects of interest to only a handful of scholars a generation ago—testify to a growing concern with one of the most potent factors shaping our civilization.

Recognizing the very real need for a new reference guide to prominent scientists and their achievements, the A. N. Marquis Company, publishers of *Who's Who in America*, began planning the present *World Who's Who in Science* in 1964. The editors were convinced from the start that a single volume of approximately thirty thousand names covering all major scientists from antiquity to the present could be prepared if the biographical sketches were compiled in the concise and familiar style associated with *Who's Who in America*. These short biographies were to indicate both the scientific contribution of the individual and his vital statistics. Of course the physical and the biological sciences were to be treated in detail, but at the same time the scope of the work was conceived to be broader than this.

Subjects which no longer form an integral role in "science" (e.g. alchemy and astrology) were to be covered here provided that they were thought to be a legitimate path to truth in nature in earlier periods. Also, to a lesser extent, it was decided to include the social sciences (i.e., geography, anthropology, archaeology, psychology, sociology, economics). In short, a different sort of scientific reference work was planned from any which had been prepared earlier. It was to be international in scope and useful to the historian of science as well as to those who needed information on living scientists. These initial goals have been kept in mind throughout the intervening years of preparation.

It is one thing to plan such a book and quite another one to complete it. The problem is not a new one. Two centuries ago Diderot wrote that

> The most important consideration, however, and one that lends added weight to the previous ones, is that an encyclopedia, like a dictionary, must be begun, carried forward and completed within a certain period of time . . . you must shorten your labors by multiplying the number of your helpers, an expedient that is not, indeed, without its disadvantages . . .[†]

The *Encyclopédie ou Dictionnaire Raisonné des Sciences, des Arts et des Métiers* appeared in seventeen volumes of text and another eleven volumes of plates between the years 1751 and 1772. *World Who's Who in Science* surely does not claim to be a project of comparable magnitude, but the problem seen by Diderot is common to all those who engage themselves in the production of research works. If, on the one hand, an advisory board aided us in the selection of names to be included in the volume, on the other hand, it was essential—in order to complete the work within a reasonable period of time—to turn to a small army of researchers to uncover the needed data on the chosen individuals. Here we relied on well over one hundred researchers. Of these a high percentage were graduate students at Chicago area universities who worked part time on this project while completing their academic requirements for their

---

[*]Robert Hodge, Paul Siegel, and Peter Rossi, "Occupational Prestige," *American Journal of Sociology,* LXX (Nov. 1964), pp. 290-292.

[†]Denis Diderot, "Encyclopédie," in *Encyclopédie ou Dictionnaire Raisonné des Sciences, des Arts et des Métiers* (Paris: Chez Briasson, David, Le Breton, Durand, 1755) V, pp. 636, 637.

degrees. We tried to locate individuals with different language, science, and historical specialties—and to put them to work in areas where they had special competence.

After the name lists had been compiled and a research staff assembled, the remaining mechanics of the operation appeared to be relatively simple and straightforward. Information on living scientists was to be obtained by mailing them questionnaires on which they were requested to supply the needed data on their careers and their scientific contribution. Historical names were assigned to members of the research staff who were to seek out the required information in standard histories of science and biographical dictionaries.

However, plans seldom materialize as smoothly as one would wish, and this volume proved to be no exception. It is, of course, possible to make up lists of prominent scientists. For contemporary names one may turn to directories of scientific societies, lists of Nobel—and other—prize winners, and catalogs of the world's major educational institutions. Similarly there are a host of other guides such as *American Men of Science, Who's Who in Soviet Science and Technology,* and the *World Nuclear Directory.* These, and other works like them, are useful for pinpointing scientists of importance. Yet, these directories concentrate on the general research interests of the scientists rather than their actual scientific contributions. Our task, therefore, with living scientists, was to search the guides we had at hand for likely candidates and then to mail questionnaires directly to them for further information. The data returned were reviewed by the research staff and a final decision made (sketches prepared from data supplied or approved by the individuals themselves are marked with an asterisk). "Must" names who did not reply were researched in the same fashion as deceased scientists.

At all times we attempted to maintain scientific contribution (or influence) as our prime standard for acceptance or rejection. With historical names there was little difficulty since the work of these men had already been assessed by historians of science. But in many cases this type of information is difficult to obtain for our contemporaries. Scientists who replied to our mailings often gave only their areas of research interest when asked to remark on their scientific contribution. One scientist replied—with considerable justification—that only his professional colleagues could properly assess the value of his work.

Surely no reference work can do more than reflect its sources, and in those cases where we had no more to guide us than research interests we occasionally turned to other factors. Among these we considered position in the scientific community. When we considered this as a major factor we made the assumption that the scientist's research had been the ultimate reason for his prominence. We would emphasize, however, that the editorial staff tried to avoid this type of listing. The sketches of this sort which were included result from our belief that the biographical information alone might be valuable for those who will consult this volume.

The determination of contribution was far less difficult for historical names. The careful study of standard histories of science resulted in a relatively rapid compilation of a historical "must" list. And yet, if biographical data are easy to locate for living notables, they are generally more difficult to find for non-living scientists. J. C. Poggendorff's *Biographisch-literarisches Handwörterbuch zur Geschichte der exacten Wissenschaften,* the *Neue Deutsche Biographie,* the *Dictionnaire Biographie Française,* the *Dictionary of National Biography* and other standard biographical collections are helpful, but for scientists of less than major importance career information was frequently difficult to find. We might generalize that for living persons we started with biographical data and then had to find the scientific contribution; while, in contrast, for historical names we began with the subject's contribution and then had to find the necessary biographical material to complete the sketch.

Apart from these general considerations there were other problems which had to be solved. Some of these were the result of the size of our research staff and one of the most persistent has been that of spelling. Historians of science have employed considerable latitude in their choice of the proper spelling of names. One might use the Latin form of a name while another might choose to translate the author's name into its normal vulgar form. Since our researchers were assigned different books to work on, they very naturally accepted the spellings they encountered. However, filing clerks often did not spot the identity of these different names. The result would be two or more sketches of the same scientist. As an example, we did not note until the arrival of page proof that we had on the same page biographical sketches of "Baldwin" and "Balduinus"—both the same person.

An allied problem is that of Islamic scientists who were known to Western Europe through a Latinized form of their names. Here, if the individual was particularly well known to Western scholars we generally placed the sketch under the Latin form of the name with an appropriate cross reference under the Arabic form. Where the work of the individual was not as well known in the West, and in those cases where he is best known today through the Arabic form of his name, we chose the other alternative. With Arabic names we generally followed the spelling and alphabetization employed by George Sarton in his *Introduction to the History of Science.*

In accordance with normal Marquis style we generally followed alphabetical order in cases where names included "De", "D'", "La", "Van", "Von", and other similar prefixes. Exceptions were made in cases of individuals who are most commonly referred to without the prefix. Thus van Helmont will be found here under "H", and de Fourcroy under "F". Among the most difficult cases to resolve are those individuals with compound Latin names. An excellent example is a sixteenth-seventeenth century Portuguese physician whom we found referred to by authorities as José Rodrigo da Castro, Etienne Rodriguez de Castro, Estaran Roderigo da Castro, Estevan Rodrigues de Castro and Stephanus Rodericus Castrensis. The same authorities indexed him under "R", "D" and "C". By the time we were ready to send copy to the printer we found that we had completed several sketches on him which were located under different letters. Our final decision was to list him under "R" as "Rodrigues de Castro, Estevan." With Rodrigues de Castro and in other similar cases we attempted to include enough cross references to make the sketch easy to find for those who might be more familiar with another spelling than the one we chose. The point of all this is not an unimportant one. It is unlikely that we have eliminated all duplicate sketches in this volume even though we have tried to do so. We can only hope that those who use this volume will call to our attention any remaining duplications.

Another persistent problem was that of translation. Although we have had on our staff specialists in German, French, Latin, Arabic, Russian and Polish, even the best translators slip up on occasion. While reading page proof we noted in the Richard Assman sketch the rather surprising statement, "probably coined the term weather." On checking the original source we found that he was the "Begründer der Z/s 'Das Wetter'." This error was caught, but others may have slipped past. Again, accommodating European scientists sometimes filled out their data cards in English. There were times when their very literal translations were most difficult for us to interpret correctly. And we might mention here also that our original sketches were occasionally modified within our own organization. Professional sketch writers took our copy and reduced it to a standard style. This, of course was to be desired and it seldom caused any problems, but at times statements of contribution were inadvertently altered incorrectly.

Finally, we would wish to inject a note of caution to those who might be tempted to use this volume as the basis of a statistical analysis of the world's scientists. Only in a most general way would the results of such an analysis be valid. The sources upon which this volume is ultimately based are most complete for the United States and the countries of Western Europe. Accordingly, the highest percentage of individuals included here are from these geographical areas—even though we attempted to give as broad coverage to Eastern Europe, Central and South America, Africa and the Orient as we could with the material available to us.

Similarly we would warn against the interpretation of the length of the sketches as a valid guide to our estimate of the importance of the included scientists. In most cases the length of a sketch indicates little more than the fact that we had more information on hand for one man than another. Living scientists have been able to supply us with much detailed information on their lives and to reproduce this requires many lines of type. Far less space may have been given to important scientists of the past on whom we have less solid information in regard to their careers. Here a seeming lack of consistency should rather be ascribed to our desire to give as much pertinent information as we have access to in our files.

If we have been candid at this point about possible errors in *World Who's Who in Science* our intention has not been to disparage our efforts. We have, through a series of checks and counterchecks, tried to eliminate as many errors as possible and it is our hope that few remain. Nevertheless, we feel that it would be unrealistic to expect the first edition of a research volume of this magnitude to be totally devoid of error even though we might wish it to be so. We might add that because all Marquis books are continuing publications, the significance of *World Who's Who in Science* can only increase in the future. A permanent research staff will have as its object the review of the present volume and the collection of data for the planned revised edition. It is our hope that readers and reviewers will bring to our attention any errors which may be contained within these pages. They will be corrected in later editions.

It is our belief that this volume will fill a much needed gap on our research shelves. Here will be found approximately thirty thousand sketches of prominent scientists from antiquity to the present time. Nearly half of these are historical. Most of the biographical sketches include information on the families, the careers, and the scientific interests and contributions of the individuals. A high percentage of the living scientists have supplied us with their own data and these sketches may be considered definitive. Similarly, the sketches of a high percentage of the most important historical names were prepared by professional historians of science. We can only hope that the value of the volume will justify the thousands of hours of research which have gone into its preparation.

Allen G. Debus
August 16, 1968

ADVISORY COUNCIL

ACKNOWLEDGEMENTS

The editors of *World Who's Who in Science* offer their utmost thanks to Nicholas Chako, professor of mathematics at Queen's College, New York, who from the very inception of this book has unselfishly given of his time in offering valuable suggestions, and who has used his good offices to personally obtain essential data on many leading European scientists, without which this book would not be complete.

Grateful acknowledgement is also accorded to the many individuals and organizations who offered suggestions, nominated biographees, and supplied data, including Pierre Rouard, professor and dean, Faculty of Sciences, Université D'Aix-Marseille, France; Mircea Abagiu, second secretary, Embassy of the Socialist Republic of Rumania, Washington, D.C.; and Ljubomir Georgiev Iliev, Science secretary general and director, Mathematics Institute, Bulgarian Academy of Sciences.

Without the enthusiastic dedication of the members of the Science Project and the Marquis staff, the objectives of this book could not have been realized. Effort beyond the call of duty was unstintingly given. Everyone at Marquis contributed, but there would have been no *World Who's Who in Science* without the following: Bonnie D. Fors, assistant to the editors and senior Science researcher; Nicholas J. Malinski, Jr., and Marilyn Prado, medical researchers; Samuel Adetunji, Sofia Biernadski, John Calfee, Brunilda Debus, Jane Dietrich, Diana Duggan, Marie Feigl, Peter Gran, Susan F. Greenberg, Richard Gudell, Eric Larsen, Chari H. McNally, Lorry Plagenz, Susan J. Pollett, Will Provine, Melissa Reed, Sarah Rosenbloom, Edward S. Ross, P. P. G. Sarath, and Jim Albert Zatte, Science researchers and proofreaders; Don B. T. Freeman and Oscar B. Treiman, Assistants to the President; John K. Kloster, Jr., Senior Vice-President and Treasurer; Robert P. Balsitis, Vice-President and Comptroller; Elizabeth Pedersen, Vice-President for Administrative Services; Robert A. Irwin, Director of Research and Development, and Barbara Koppel, his staff artist; Betty Cahill, Vice-President for Production; Dolores Smith, Director of Compilation; Madelyn Corbett Jacobs, Associate Director, and Joyce Myers Fancher, Assistant Director; Mary Fran Sheldon and Betty Sammons Connor, Directors of Sketchwriting; Rosmarie Chrapkowski, Associate Director, and Ann Lewis, Joen Kinnan, Diane Skinner and Jean Bassan, Sketchwriters; Barbara Wardell Galgoul, Editorial Director, and her Assistant, Lois Heuser; and Lucille Lewis, Director of Order and Billing, who made available names and data on black scientists. The project was under the general direction of Kenneth N. Anglemire, President, The A. N. Marquis Company, Inc.

The Editors

SELECTED BIBLIOGRAPHY*

Allgemeine Deutsche Biographie. 56 vols., Leipzig, 1875-1912.

American Men of Science. Ed. Jacques Cattell Press, 11th edn., New York, 1967.

Asimov, Isaac. *Asimov's Biographical Encyclopedia of Science and Technology.* Garden City, New York, 1965.

Cajori, Florian. *A History of Physics.* New York, 1929.

Chamber's Biographical Dictionary. Ed. J. O. Thorne, New York, 1962.

Dampier, William. *A History of Science.* 4th edn., Cambridge, Eng., 1948.

Darmstaedter, Ludwig. *Handbuch zur Geschichte der Naturwissenschaften und der Technik.* Berlin, 1908.

Daumas, Maurice. *Histoire de la Science.* Paris, 1957.

Dictionary of American Biography. Ed. Allen Johnson, *et al.,* 10 vols. and supplements, New York, 1927-58.

The Dictionary of National Biography. Ed. Leslie Stephen, *et al.,* 21 vols. and supplements, Oxford, 1908-59.

Dictionnaire de Biographie Française. Ed. J. Balteau, *et al.,* 9 vols. and supplements, Paris, 1932-66.

Ferguson, John. *Bibliotheca Chemica.* 2 vols., London, 1954.

Garrison, Fielding H. *An Introduction to the History of Medicine.* Philadelphia, 1929.

Needham, Joseph. *Science and Civilization in China.* 4 vols., Cambridge, Eng., 1954-62.

Neue Deutsche Biographie. Ed. Historischen Kommission of the Bayerischen Akademie der Wissenschaften, 7 vols. to date, Berlin, 1953-present.

Neugebauer, Otto. *The Exact Sciences in Antiquity.* New York: Harper Torchbooks, 1962.

Nordenskiöld, Eric. *The History of Biology.* New York, 1928.

The Oxford Classical Dictionary. Ed. Max Cary, *et al.,* Oxford, 1949.

Pannekoek, A. *A History of Astronomy.* New York, 1961.

Partington, J. R. *A History of Chemistry.* vols. 2-4, London, 1961-64.

Poggendorff, J. O. *Biographisch-literarisches Handwörterbuch zur Geschichte der exacten Wissenschaften.* 7 vols., Leipzig and Berlin, 1863-1962.

Sarton, George. *A History of Science.* 2 vols., Cambridge, Mass., 1952-59.

Sarton, George. *Introduction to the History of Science.* 3 vols., Baltimore, 1927-48.

Scott, J. F. *A History of Mathematics.* London, 1960.

Singer, Charles. *A History of Biology.* New York, 1950.

Singer, Charles, E. J. Holmyard, A. R. Hall, T. I. Williams, *et al. A History of Technology,* 5 vols., Oxford, 1954-58.

Struve, Otto, and Velta Zebergs. *Astronomy of the 20th Century,* New York, 1962.

Taton, René. *Histoire Générale des Sciences.* 4 vols. in 3 tomes, Paris, 1957-64.

Thorndike, Lynn. *A History of Magic and Experimental Science.* 8 vols., New York, 1923-58.

Turkevich, John. *Soviet Men of Science.* Princeton, N.J., 1963.

Zischka, Gert A. *Allgemeines Gelehrten-Lexikon.* Stuttgart, 1961.

*The purpose of this very brief bibliography is not to provide a comprehensive guide to the biographical literature on living scientists or to the history of science. Rather it is a list of a few sources which were found to be of value in the compilation of this book and which the editors feel might be of use to those readers who desire further information.

A FEDERAL COURT HOLDS THAT "WHO'S WHO" MEANS MARQUIS

This permanent injunction by a Federal court upholds the long-standing contention of Marquis-Who's Who, Inc. "that the words 'Who's Who' have acquired a secondary meaning and are a mark by which publications of the Plaintiff (Marquis-Who's Who) are distinguished from all other publications of the same class." Further, this action protects biographees against exploitation of their listings in our "Who's Whos." Following is the substance of the injunction:

UNITED STATES DISTRICT COURT
MIDDLE DISTRICT OF FLORIDA—TAMPA DIVISION

MARQUIS-WHO'S WHO, INC.,
a Delaware corporation, Plaintiff,

vs.

THE NATIONAL REGISTER
OF WHO'S WHO, INC., a Florida
corporation,   Defendant

No. 67-240-Civ. T.

FILED

July 25, 1967

Tampa, Fla.

PERMANENT INJUNCTION

5. That through the expenditure of many thousands of dollars in the advertising, publication and sale of said directories the words "Who's Who" in the title of a book of biographical reference have come to mean in the minds of the general public that such book is being or will be published by Plaintiff, and that the words "Who's Who" have acquired a secondary meaning and are a mark by which publications of the Plaintiff are distinguished from all other publications of the same class.

6. That the Defendant was incorporated on or about March 7, 1967, taking the similar name and trade-mark of "The National Register of Who's Who, Inc.", and that commencing on or about May 20, 1967, the Defendant mailed out, prior to the issuance of the Preliminary Injunction, large numbers of envelopes and business reply cards in the name of "The National Register of Who's Who, Inc."; that in this solicitation defendant described the Register as a list of individuals who have achieved notable prominence in their business or community.

7. That the mail solicitation campaign heretofore conducted by the Defendant under the name and designation of "The National Register of Who's Who, Inc." has caused confusion among the public as tending to show that the said Register is a product of Plaintiff, and, unless permanently restrained by this court, will cause serious and irreparable injury to the Plaintiff. It is therefore, upon consideration,

ORDERED, ADJUDGED and DECREED:

1. That the Stipulation and Agreement entered into between Plaintiff and Defendant for the entry of a Permanent Injunction in this cause be and is hereby approved.

2. That the Defendant and its officers, agents and employees be, and they are hereby, permanently enjoined and restrained from using or purporting to authorize, or aiding, abetting or cooperating with others to use the name and trademark "The National Register of Who's Who, Inc." and the words "Who's Who" alone or in combination with other words, or any other name or colorable imitation of or name similar to Plaintiff's registered name and the other of plaintiff's registered trade-marks using the words "Who's Who" . . .

DONE and ORDERED at Tampa, Florida, this 25th day of July, 1967.

Joseph P. Lieb
United States District Judge

FOR YOUR INFORMATION AND ASSURANCE:

Marquis-Who's Who, Incorporated, publishes only these titles

Who's Who in America	Who's Who in the South and Southwest
World Who's Who in Commerce and Industry	Who's Who in the West
Who's Who of American Women	Directory of Medical Specialists
Who Was Who in America	World Who's Who in Science
Who's Who in the East	Who's Who In The Methodist Church
Who's Who in the Midwest	Columbia College Alumni Register

AN IMPORTANT NOTICE

This work, like all the standard Marquis biographical reference publications, is *not* merely an alphabetical arrangement of biographical data. And it is not a mere compilation of the names of those responding to mailings to lists obtainable from mailing-list concerns and other sources. When there is no rigorous enforcement of a selective standard which assures inclusion of the listings necessary to coverage of basic reference requirements, no reference standing, completeness, or worth can result.

Its compilation has been so patterned that this work will serve usefully and effectively in private or public libraries as a sound biographical reference tool. Biographees have been painstakingly searched out who fall under the unique measures of reference incidence assembled at large expense by Marquis-Who's Who, Incorporated, to meet its standards for selection resulting from 70 years of biographical reference experience.

Exploitation in any manner whatsoever of the names herein, thus selected and under which the biographical sketches appear alphabetically, is beyond the fair uses intended in the sale of the work, and in certain obvious circumstances, if damage follows to the publisher from a misappropriation of his time and expense expended in their selection, may bring legal action.

In respect to those selected for inclusion in the work, the great majority of whom have by making biographical data available cooperated in assuring the completeness vital to actual reference use, this notice is directed toward fending off on their behalf use of their names as canvassing and mailing lists; in respect to others, its purpose is obvious.

MARQUIS BIOGRAPHICAL LIBRARY

WORLD WHO'S WHO IN SCIENCE

WHO'S WHO IN AMERICA

WHO WAS WHO IN AMERICA
HISTORICAL VOLUME (1607-1896)
VOLUME I (1897-1942)
VOLUME II (1943-1950)
VOLUME III (1951-1960)
VOLUME IV (1961-1968)

WORLD WHO'S WHO IN COMMERCE AND INDUSTRY

WHO'S WHO OF AMERICAN WOMEN

WHO'S WHO IN THE EAST

WHO'S WHO IN THE MIDWEST

WHO'S WHO IN THE WEST

WHO'S WHO IN THE SOUTH AND SOUTHWEST

DIRECTORY OF MEDICAL SPECIALISTS

TABLE OF ABBREVIATIONS

*Signifies that data were supplied and/or approved by the biographee.

A.A., Associate in Arts..
A.A.A.S., American Association for the Advancement of Science.
AAC, Army Air Corps.
A. and M., Agricultural and Mechanical.
AAF, Army Air Force.
A.A.H.P.E.R., American Association for Health, Physical Education, and Recreation.
A.B., Bachelor of Arts.
ABC, American Broadcasting Company.
AC, Air Corps.
AC, alternating current.
acad., academy; academic.
A.C.P., American College of Physicians.
A.C.S., American College of Surgeons.
A.D., *anno Domini.*
a.d.c., aide-de-camp.
adj., adjutant; adjunct.
adminstr., administrator.
adminstrn., administration.
adminstrv., administrative.
adv., advocate; advisory.
A.E., Agricultural Engineer.
AEC, Atomic Energy Commission.
aero., aeronautical, aeronautic.
AFB, Air Force Base.
A.F.D., Doctor of Fine Arts.
agr., agriculture.
agrl., agricultural.
agt., agent.
A.H., *anno Hegirae.*
A.I.A., American Institute of Architects.
AID, Agency for International Development.
A.I.M., American Institute of Management.
Ala., Alabama.
A.L.A., American Library Association.
Alta., Alberta.
Am., American, America.
A.M., *ante meridiam* (before noon).
A.M., Master of Arts.
AM, amplitude modulation.
A.M.A., American Medical Association.
Am. Inst. E.E., American Institute of Electrical Engineers.
Am. Soc. C.E., American Society of Civil Engineers.
Am. Soc. M.E., American Society of Mechanical Engineers.
anat., anatomical.
ann., annual.
anthrop., anthropological.
apptd., appointed.
apt., apartment.
A.R.C., American Red Cross.
archeol., archeological.
archtl., architectural.
Ariz., Arizona.
Ark., Arkansas.
arty., artillery.
assn., association.
asso., associate; associated.
ASSR, Autonomous Soviet Socialist Republic.
asst., assistant.
astron., astronomical
astrophys., astrophysical.

atty., attorney.
at. wt., atomic weight.
Av., Avenue.

b., born.
B., Bachelor.
B.A., Bachelor of Arts.
B.Agr., Bachelor of Agriculture.
Balt., Baltimore.
B.A.S., Bachelor of Agricultural Science.
B.B.A., Bachelor of Business Administration.
BBC, British Broadcasting Corp.
B.C., before Christ; British Columbia.
B.C.E., Bachelor of Civil Engineering.
B.Chir., Bachelor of Surgery.
B.C.L., Bachelor of Civil Law.
bd., board.
B.Di., Bachelor of Didactics.
B.E., Bachelor of Education.
B.E.E., Bachelor of Electrical Engineering.
bibliog., bibliographical.
biog., biographical.
biol., biological.
Bklyn., Brooklyn.
B.L., Bachelor of Letters.
bldg., building.
Blvd., Boulevard.
bot., botanical.
B.Pd., Bachelor of Pedagogy.
B.Py., Bachelor of Pedagogy.
br., branch.
Brit., British; Britannica.
Bro., Brother.
B.S., Bachelor of Science.
B.S.A., Bachelor of Agricultural Science.
B.S.D., Bachelor of Didactic Science.
bull., bulletin.
bur., bureau.
bus., business.
B.W.I., British West Indies.

C., centigrade.
Cal., California.
C.Am., Central America.
CAA, Civil Aeronautics Administrn.
CAB, Civil Aeronautics Board.
Can., Canada.
Cantab., Of or pertaining to Cambridge University, England.
CBS, Columbia Broadcasting System.
C.E., Civil Engineer, Corps of Engineers.
CERN., Conseil Européen pour la Recherche Nucléaire.
ch., church.
Ch.D., Doctor of Chemistry.
chem., chemical.
Chem. E., Chemical Engineer.
Chgo., Chicago.
Chirurg., Chirurgical.
chmn., chairman.
chpt., chapter.
CIA, Central Intelligence Agency.
Cin., Cincinnati.
Cleve., Cleveland.
climatol., climatological.
clin., clinical.
clk., clerk.
C.M., Master in Surgery.
Co., Company, County.
C. of C., Chamber of Commerce.

coll., college.
Colo., Colorado.
com., committee.
comd., commanded.
comdg., commanding.
comdr., commander.
comdt., commandant.
commd., commissioned.
comml., commercial.
commn., commission.
commr., commissioner.
condr., conductor.
conf., conference.
Conn., Connecticut.
cons., consulting, consultant.
consol., consolidated.
constl., constitutional.
constn., constitution.
constrn., construction.
contbd., contributed.
contbg., contributing.
contbn., contribution.
contbr., contributor.
corp., corporation.
corr., correspondent; corresponding; correspondence.
C.P.H., Certificate of Public Health.
C.S., Christian Science.
C.S.B., Bachelor of Christian Science.
C.S.D., Doctor of Christian Science.
ct., court.
cyclo., cyclopedia.
C.Z., Canal Zone.

d., daughter.
D., Doctor.
D.Agr., Doctor of Agriculture.
dau., daughter.
DC, direct current.
D.C., District of Columbia.
D.C.L., Doctor of Civil Law.
D.C.S., Doctor of Commercial Science.
D.D., Doctor of Divinity.
D.D.S., Doctor of Dental Surgery.
DDT, Dichloro-diphenyl-trichloro-ethane.
dec., deceased.
Def., Defense.
Del., Delaware.
del., delegate.
D.Eng., Doctor of Engineering.
dep., deputy.
dept., department.
dermatol., dermatological.
desc., descendant.
devel., development.
dir., director.
dist., district.
distbg., distributing.
distbn., distribution.
distbr., distributor.
div., division; divinity; divorce proceedings.
D.Litt., Doctor of Literature.
D.M.D., Doctor of Medical Dentistry.
D.M.S., Doctor of Medical Science.
D.O., Doctor of Osteopathy.
D.P.H., Diploma in Public Health.
Dr., Doctor, Drive.
Dr.P.H., Doctor of Public Health; Doctor of Public Hygiene.
D.Sc., Doctor of Science
D.T.M., Doctor of Tropical Medicine.

D.V.M., Doctor of Veterinary Medicine.
D.V.S., Doctor of Veterinary Surgery.

E., East.
eccles., ecclesiastical.
ecol., ecological.
econ., economic.
ECOSOC, Economic and Social Council (of the UN).
ed., educated; edited; editor.
E.D., Doctor of Engineering.
Ed.B., Bachelor of Education.
Ed.D., Doctor of Education.
edit., edition.
Ed.M., Master of Education.
edn., education.
ednl., educational.
E.E., Electrical Engineer.
e.g., *exempli gratia* (for example).
Egyptol., Egyptological.
elec., electrical.
electrochem., electrochemical.
electrophys., electrophysical.
E.M., Engineer of Mines.
ency., encyclopaedia.
Eng., England.
engr., engineer.
engring., engineering.
entomol., entomological.
ethnol., ethnological.
exam., examination; examining.
exec., executive.
exhbn., exhibition.
expdn., expedition.
expn., exposition.
expt., experiment.
exptl., experimental.

F., Fahrenheit; Fellow.
FAA, Federal Aviation Agency
FAO, Food and Agriculture Organization (of the UN).
FCC, Federal Communications Commission.
FCDA, Federal Civil Defense Administration.
FDA, Food and Drug Administration.
F.F., Forest Engineer.
Fed., Federal.
Fedn., Federation.
Fgn., Foreign.
Fla., Florida.
Found., Foundation.
F.R.A.S., Fellow Royal Astronomical Society.
F.R.C.P., Fellow Royal College of Physicians (England).
F.R.C.S., Fellow Royal College of Surgeons (England).
F.R.G.S., Fellow Royal Geographical Society.
Ft., Fort.

Ga., Georgia.
gastroent., gastroenterological.
gen., general.
geneal., genealogical.
geod., geodetic.
geog., geographical; geographic.
geol., geological.
geophys., geophysical.
gov., governor.
govt., government.
govtl., governmental.
grad., graduated; graduate.

Gt., Great.
gynecol., gynecological.

h.c., *honoris causa.*
H.I., Hawaiian Islands.
hist., historical.
homeo., homeopathic.
hon., honorary; honorable.
hort., horticultural.
hosp., hospital.
Hwy., Highway.
hydrog., hydrographic.

Ia., Iowa.
IAEA, International Atomic Energy Agency.
IBM, International Business Machines Corp.
ICBM, Intercontinental ballistic missile.
Ida., Idaho.
I.E.E.E., Institute of Electrical and Electronic Engineers.
IGY, International Geophysical Year.
Ill., Illinois.
illus., illustrated.
Inc., Incorporated.
Ind., Indiana.
ind., independent.
Indpls., Indianapolis.
indsl., industrial.
ins., insurance.
insp., inspector.
inst., institute.
instl., institutional.
instn., institution.
instr., instructor.
instrn., instruction.
internat., international.
intro., introduction.
IQ, intelligence quotient.
I.R.E., Institute of Radio Engineers.
IUPAC, International Union of Pure and Applied Chemistry.

J.B., Jurum Baccalaureus.
J.C.B., Juris Canonici Bachelor.
J.C.L., Juris Canonici Lector.
J.D., Doctor of Jurisprudence.
Jour., journal.
jr., junior.
J.S.D., Doctor of Juristic Science.
jud., judicial.
J.U.D., Juris Utriusque Doctor; Doctor of Both (Canon and Civil) Laws.

K., Kelvin.
Ky., Kentucky.

La., Louisiana.
lab., laboratory.
lang., language.
laryngol., laryngological.
lectr., lecturer.
L.H.D., Doctor of Humane Letters.
L.I., Long Island.
lit., literary; literature.
Litt.D., Doctor of Letters.
LL.D., Doctor of Laws.
log., logarithm.
L.R.C.P., Licentiate Royal Coll. Physicians.
L.R.C.S., Licentiate Royal Coll. Surgeons.
Ltd., Limited.

m., married.
M., Master.
M.A., Master of Arts.
mag., magazine.
M.Agr., Master of Agriculture.
maj., major.
Man., Manitoba.
Mass., Massachusetts.
math., mathematical.
M.B., Bachelor of Medicine.
M.B.A., Master of Business Administration.
MBS, Mutual Broadcasting System.
M.C.E., Master of Civil Engineering.
mcht., merchant.
Md., Maryland.
M.D., Doctor of Medicine.
M.D.V., Doctor of Veterinary Medicine.
Me., Maine.
M.E., Mechanical Engineer.
mech., mechanical.
med., medical.
M.E.E., Master of Electrical Engineering.
mem., member.
Meml., Memorial.
met., metropolitan.
metall., metallurgical.
Met.E., Metallurgical Engineer
meteorol., meteorological.
metrol., metrological.
M.F., Master of Forestry.
mfg., manufacturing.
mfr., manufacturer.
mgmt., management.
mgr., manager.
Mich., Michigan.
micros., microscopical.
mil., military.
Milw., Milwaukee.
mineral., mineralogical.
Minn., Minnesota.
M.I.T., Massachusetts Institute of Technology.
M.L., Master of Laws.
M.L.D., Magister Legnum Diplomatic.
M.Litt., Master of Literature.
M.M.E., Master of Mechanical Engineering.
mng., managing.
Mo., Missouri.
Mont., Montana.
M.P., Member of Parliament.
M.Pd., Master of Pedagogy.
M.P.H., Master of Public Health.
M.P.L., Master of Patent Law.
Mpls., Minneapolis.
MS, manuscript.
M.S., Master of Science.
M.Sc., Master of Science.
M.S.F., Master of Science of Forestry.
Mt., Mount.
mus., museum; musical.
mycol., mycological.

N., North.
NACA, National Advisory Committee for Aeronautics.
N.Am., North America.
NASA, National Aeronautics and Space Administration.
nat., national.
NATO, North Atlantic Trenty Organization.
nav., navigation.

N.B., New Brunswick.
NBC, National Broadcasting Company.
N.C., North Carolina.
N.Dak., North Dakota.
NDRC, National Defense Research Committee.
N.E., Northeast.
N.E.A., National Education Association.
Nebr., Nebraska.
neurol., neurological.
Nev., Nevada.
New Eng., New England
N.H., New Hampshire.
NIH, National Institutes of Health.
N.J., New Jersey.
N.M., New Mexico.
No., Northern.
nr., near.
NRC, National Research Council.
n.s., new style.
N.S., Nova Scotia.
NSC, National Security Council.
NSF, National Science Foundation.
N.T., New Testament.
numis., numismatic.
N.W., Northwest.
N.Y., New York.
N.Y.C., New York City.

O., Ohio.
OAS, Organization of American States.
obs., observatory.
obstet., obstetrical.
OECD, Organization European Cooperation and Development.
OEEC, Organization European Economic Cooperation.
ofcl., official.
Okla., Oklahoma.
Ont., Ontario.
ophthal., ophthalmological.
Ore., Oregon.
orgn., organization.
ornithol., ornithological.
o.s., old style.
OSRD, Office of Scientific Research and Development.
OSS, Office of Strategic Services.
osteo., osteopathic.
O.T., Old Testament.
otol., otological.
otolaryn., otolaryngological.
Oxon., Of or pertaining to Oxford University, Eng.

Pa., Pennsylvania.
paleontol., paleontological.
path., pathological.
Pd.B., Bachelor of Pedagogy.
Pd.D., Doctor of Pedagogy.
Pd.M., Master of Pedagogy.
Pe.B., Bachelor of Pediatrics.
P.E.I., Prince Edward Island.
pharm., pharmaceutical.
Pharm.D., Doctor of Pharmacy.
Pharm.M., Master of Pharmacy.
Ph.B., Bachelor of Philosophy.
Ph.C., Pharmaceutical Chemist.
Ph.D., Doctor of Philosophy.
Ph.G., Graduate in Pharmacy.
Phila., Philadelphia.
philol., philological.
philos., philosophical.
photog., photographic.

phys., physical.
physiol., physiological.
P.I., Philippine Islands.
Pitts., Pittsburgh.
Pkwy., Parkway.
Pl., Place.
P.O., Post Office.
polit., political.
poly., polytechnic; polytechnical.
pomol., pomological.
P.R., Puerto Rico.
prep., preparatory.
pres., president.
presdl., presidential.
prin., principal.
proc., proceedings.
prodn., production.
prof., professor.
profl., professional.
propr., proprietor.
psychiat., psychiatric.
psychol., psychological.
pub., public; publisher; publishing; published.
publ., publication.

quar., quarterly.
Que., Quebec (province).

radiol., radiological.
RCA, Radio Corporation of America.
Rd., Road.
rec., recording.
rehab., rehabilitation.
rep., representative.
ret., retired.
Rev., Reverend.
rev., review, revised.
rhinol., rhinological.
R.I., Rhode Island.
R.N., Registered Nurse.
röntgenol., röntgenological.
R.R., Railroad.
RSFSR, Russican Soviet Federated Socialist Republic.
Ry., Railway.

s., son.
S., South.
S.Am., South America.
san., sanitary.
Sask., Saskatchewan.
S.B., Bachelor of Science.
S.C., South Carolina.
Sc.B., Bachelor of Science.
Sc.D., Doctor of Science.
sch., school.
sci., science; scientific.
S.Dak., South Dakota.
S.E., Southeast.
SEATO, Southeast Asia Treaty Organization.
SEC, Securities and Exchange Commn.
sec., secretary.
sect., section.
seismol., seismological.
S.J., Society of Jesus (Jesuit).
S.J.D., Doctor Juristic Science.
S.M., Master of Science.
So., Southern
soc., society.
sociol., sociological.
spl., special.
splty., specialty.
Sq., Square.

sr., senior.
S.S., Steamship.
SSR, Soviet Socialist Republic.
St., Saint; Street.
sta., station.
statis., statistical.
supr., supervisor.
supt., superintendent.
surg., surgical.
S.W., Southwest.

Tb (or TB), Tuberculosis.
tchr., teacher.
tech., technical; technology.
technol., technological.
temp., temporary.
Tenn., Tennessee.
Ter., Territory.
theol., theological.
tng., training.
TNT, trinitrotoluene.
topog., topographical.
trans., transactions; transferred.
transl., translation.
treas., treasurer.
TV, Television.
TVA, Tennessee Valley Authority.
Twp., Township.

U., University.
UAR, United Arab Republic.
UHF, ultra high frequency.
U.K., United Kingdom.
UN, United Nations.
UNESCO, United Nations Educational, Scientific and Cultural Organization.
UNICEF, United Nations International Childrens Emergency Fund.
univ., university.
U.P.I., United Press International.
urol., urological.
U.S., United States.
U.S.A., United States of America.
USIA, United States Information Agency.
USIS, United States Information Service.
USPHS, United States Public Health Service.
USSR, Union of Soviet Socialist Republics.

VA, Veterans Administration.
Va., Virginia.
vet., veteran; veterinary.
VHF, very high frequency.
V.I., Virgin Islands.
vice pres., vice president.
vis., visiting.
vol., volunteer; volume.
v.p., vice president.
vs., versus.
Vt., Vermont.

W., West.
Wash., Washington (state).
WHO, World Health Organization (of the UN).
W.I., West Indies.
Wis., Wisconsin.
wt., weight.
W.Va., West Virginia.
Wyo., Wyoming.

yr., year.

zoöl., zoölogical.

World Who's Who
in Science

FROM ANTIQUITY TO THE PRESENT

First Edition

BIOGRAPHIES

A

AACH, Hans Gunther, German biologist; b. Oldenburg, Germany, Oct. 2, 1919; s. Emil and Kathe (Schwarting) A.; student Berlin U., 1945-48; Sc.D., Göttingen (Germany) U., 1952; m. Margret Althüser, Apr. 1, 1960; children—Til, Helmut. Asst., Max Planck Inst., Tübingen, Germany, 1952-58; asst., dozent Cologne (Germany) U., 1958-61; research fellow virology dept. U. Cal. at Berkeley, 1962; prof. botany Tech. U., Aachen, Germany, 1962—, dir. Bot. Inst., 1963—. Mem. Deutsche Botanische Gesellschaft, Bot. Soc. Am., Scandinavian Soc. for Plant Physiology. Research, publs. on steering of algae cultures, 1952, vernalization and photoperiodism of plants, 1953-56, mutation of plant viruses and phages, 1957-—. Home: 8 Am Hang, 5105 Laurensberg, Germany. Office: Botanisches Institut, 51 Aachen, Germany.*

AAES-JORGENSEN, Erik, Danish biochemist; b. Sdr. Felding, Denmark, Mar. 30, 1918; s. Jorgen David and Helen Margrethe (Aaes) Jorgensen; mag. scient. in biochemistry U. Copenhagen (Denmark), 1948, dr.phil., 1954; m. Yrsa Aggis Holst Hansen, Dec. 28, 1945; children—Britta, Jorgen, Torben, Erling. With dept. biochemistry and nutrition Poly Inst., Copenhagen, 1948—, docent, 1959—; research fellow U. Minn., Hormel Inst., Austin, 1956-58. Cons. Danish Fat Research Inst., 1963—. Mem. Am. Oil Chemistry Soc., Am. Inst. Nutrition, Danish Biol. Soc., Danish Biochem. Soc., Danish Chem. Soc. Author: (with H. Dam.) Biokemi, I, and II, 1950; Grundrids af Biokemi og Ernaering, 1955; contbg. author: Laerebog i Biokemi, 1965-66; also articles. Editorial bd. Jour. Atherosclerosis Research, 1960, editor European sector, 1962-63; editorial bd. Dansk Kemi, 1961—. Research on deficiencies of essential fatty acids and vitamin E, nutritive effect and metabolism of isomers of unsaturated fatty acids. Home: 43 Holmeparken, Virum, Denmark. Office: Danmarks Tekniske Hojskole, Afdeling for Biokemi og Ernaering, Ostervoldgarde 10, Trp. L2, Copenhagen K, Denmark.*

AARON OF ALEXANDRIA, Greek physician, philosopher; probably flourished 610-41; author number of writings on Greek medicine under title Pandectae medicinae (a 30-section Greek med. ency., contained description of small pox, was translated into Syriac, Arabic).

AARONS, Jules, Am. space scientist; b. N.Y.C., Oct. 3, 1921; s. Joseph and Sadie (Gold) A.; B.S., Coll. City N.Y., 1942; M.A., Boston U., 1950; Ph.D., U. Paris, 1954; m. Jeanette Lampert, June 21, 1944; children—Herbert G., Philip E. With USAF Cambridge Research Labs., Bedford, Mass., 1946-—, chief radio astronomy br., 1956-—, coordinator joint satellite studies group, 1959-—. Recipient Loeser award Cambridge Research Labs., 1964. Mem. I.E.-E.E., Am. Geophys, Union, A.A.A.S., Am. Astron. Soc. Author: Radio Astronomical and Satellite Studies of the Atmosphere, 1962; Solar System Radio Astronomy, 1965; also articles. Discovered radiation at protongyro-frequency; studies of irregularities in earth's atmosphere; solar eclipse. Home: 46 Kingswood Rd., Newton, Mass. 02166. Office: USAF Cambridge Research Labs., Bedford, Mass. 01730.*

AASER, Carl S., Norwegian veterinarian; b. Oslo, Norway, Aug. 8, 1887; s. Peter and Louise A.; ed. Coll. Vet. and Agrl. Studies, Oslo U., Ph.D.; m. Alfhild Gerg, May 10, 1917; children—Siri, Hans Peter. Asst. vet. lab., Copenhagen, 1913-14; chief asst. pathology dept. Oslo U., 1914-27, prof. alimentary hygiene and legal medicine, 1938-53, prof. legal medicine Vet. Coll., 1953-59; chief veterinarian town of Oslo, 1927-38; mem Internat Office Epizooty, 1948-55 Mem Norwegian (gold medal), Swedish (corr.) Veterinary assns., Acad. Sci. Oslo. Recipient gold medal Soc. for Protection of Animals, Denmark, Finland, Norway, Sweden. Research, numerous publs. on pathology, alimentary hygiene, legal vet. medicine. Address: Norske Videnskaps-Akademi, Drammensveien 78, Oslo, Norway.

AASTED FRANDSEN, Victor, Danish physician; b. Copenhagen, Denmark, Sept. 21, 1923; s. Christian and Ebba (Esmann) A-F.; M.D., U. Copenhagen, 1951, Dr.med., 1963; m. Anne Lykke, Nov. 7, 1953; children—Nina, Finn, Tine. Research fellow lab. exptl. medicine Broussais Hosp., Paris, France, 1951-52; intern dept. dermatology, pediatrics and medicine Univ., Frederiksberg hosps., Copenhagen, 1953-55; asst. hormone dept. Statens Serum Inst., Copenhagen, 1955-—. Recipient Gold medal U. Copenhagen, 1949. Author: The Excretion of Oestriol in Normal Human Pregnancy, 1963. Research, publs. on estriol excretion in pregnancy and its estimation; site of prodn. of estriol. Home: 22 A Kroyersveg, Klampenborg, Denmark. Office: Statens Serum Inst., Copenhagen, Denmark.*

ABABEI, Leonid, Rumanian biochemist; b. Zapodeni, Rumania, Jan. 23, 1930; s. Pavel and Elena (Teodoras) A.; student Inst. Medicine, Gorkii, USSR, 1954; M.D., Humboldt U., Berlin, Germany, 1963; m. Elena Vîlcoci, Feb. 5, 1963; children—Adriana, Iuliana. Med. sci. asst. Bucharest, Rumania, 1954-58; sci. investigator Humboldt U. Biochem. Inst., 1958-61, 63-65; sci. investigator Inst. Normal and Path. Physiology, Bucharest, 1961-63; dir. biochem. lab., prof. biochemistry Med. Sch., Institut-Med., Jassy, Rumania, 1965-—. Mem. Physiologische Gesellshaft (Berlin), Fedn. European Biochem. Socs., Biochem. Soc. Rumania, N.Y. Acad. Sci., Pan Am. Med. Assn. Research and publs. on glutaminase activity in erythrocytes; described enzymes NADP-Lactate oxidoreduc-tase, coenzyme independent-LDH, regulatory mechanisms at level of LDH (crabtree effect). Home: Bucuresti-str. Biserica Amzei 8, et.I.Ap.5, Rumania. Office: Iasi Str. Săulescu. 4-Biochem. Labor, Rumania.*

ABAILARD, Peter, see Abélard, Peter.

ABALOS, Jorge Washington, Argentinian zoologist; b. La Plata, Argentina; s. Gabriel and Pilar (Urieta) A.; Doctor honoris causa in biologic scis., Universidad Nacional de Tucumán, Argentina, 1950; m. Leonie Albaca, July 12, 1947; children—Jorge Eduardo, Iván, Gabriel Alberto. Entomologist, Instituto de Medicina Regional, Universidad Nacional de Tucumán, 1953, prof. biology, 1954; entomologist Dirección de Lucha contra la Enfermedad de Chagas, 1953; dir. Inst. Poisonous Animals, Santiago del Estero, Argentina, 1957; tech. investigator poisonous animals Nat. Microbiol. Inst., 1958; prof. forest zoology Faculty Forest Engring., 1959 ; titular prof. invcrte brate zoology Nat. U. Córdoba (Argentina), 1966-—. Author: (with P. Wygodzinsky) Las Triatominae Argentina, 1951; Quésabe usted de Viboras?, 1964. Research, publs. in med. zoology. Home: 427 Gral. Paz, Córdoba. Office: 299 Vélez Sársfield, Córdoba, Argentina.*

ABANO, Pietro d', see d'Abano, Pietro.

ABAS-ZADE, Abaskuli Agabalaogly, Russian physicist, methodologist; b. 1906; grad. Azerbaijan U., 1930; D. Physico-Math. Sci. Former dep. dir. Azerbaijan Acad. Pedagogical Sci.; head chair gen. and theoretical physics Azerbaijan Pedagogical Inst. Mem. Russian Soviet Federative Socialist Republic Acad. Pedagogical Sci. Author: Thermodynamics, 1948, also author works on teaching physics in secondary and higher ednl. establishments. Editor: Indsl. Tng. Jour. (in Azerbaijan). Research in thermodynamics and molecular physics of liquids and gases. Address: Gosudarstvenny Pedagogichesky Institute, Ulitsa Shaumyana 39, Baku, Azerbaijan, SSR.

ABAUZIT, Firmin, natural philosopher; b. Uzes, France, Nov. 11, 1679; educated in Geneva, Switzerland; friend of Newton, Bayle, Rousseau, Voltaire; declined patronage of William III of Eng., also professorship of philosophy at Geneva, Switzerland; accepted lifetime position as city librarian, Geneva; author theol., archeol. treatises, also history of Geneva; detected error in 1st edit. Newton's Principia; defended Newton against Castel; known for his scepticism. Died Geneva, March 20, 1767.

ABBATI, Pietro, mathematician; b. 1768; one of those who reached fundamental results in theory of groups of finite order; gave complete proof of Lagrange's theorem. Died 1842.

ABBATT, John Dilworth, med. radiobiologist; b. Poulton-le-Fylde, Eng., May 13, 1923; s. Geoffrey Peile and Millicent (Jackson) A.; M.B., Ch.B., D.M.R., U. Edinburgh; m. Ainslie Ann Goer Ferguson, Nov. 19, 1955; children—Sarah C. D., Joanna L. P., Jonathan P. D. External staff Med. Research Council, Brit. Postgrad. Med. Sch., Hammersmith, London, 1950-57; radiobiologist, chief med. officer Gen. Elec. Co. Ltd., Erith, Eng., 1957-64; asst. chief radiation protection div. Dept. Nat. Health and Welfare, Ottawa, Ont., Can., 1966——. Lectr. U. London, 1954-62; cons. Internat. Atomic Energy Agy., Vienna, WHO, Geneva, 1962——. Fellow Zool. Soc. London; mem. Brit. Inst. Radiology, Brit. Radiation Research Soc., Brit. Photobiology Group. Author: (with DeHevesy and Forssberg) Advances in Radiobiology, 1957; (with Lakey and Mathias) Protection Against Radiation, 1961; also articles. Research in understanding mechanism of acute radiation syndrome in man; devel. early applications of radioactive materials combined with drugs as potentiators in humans, especially human radiation epidemiological work. Home: 33 Commanche Dr., Ottawa 5. Office: Radiation Protection Div., Dept. Nat. Health and Welfare, Brookfield Rd., Ottawa 8, Ont., Can.*

ABBATUCCI, Jacques Séverin, French physician; b. Bastia, France, Feb. 22, 1923; s. Jean Charles and Madeleine (Forcioli-Conti) A.; Doctorat in Medecine 1954, Diplôme d'Electro-radiologie, 1957; m. Yvette, Lamperier, Dec. 23, 1952; children—Marie-Gabrielle, Marie-Laure, Marie-Helène, Séverin, Pierre. Chief telecobalt and radiotherapy service Centre Regional anticanereux de Caen, Normandy, France, 1963——. Mem. Société Francaise pour l'Etude du Cancer, Societe d'Electro-Radiologie, Groupe Cooperateur de Radiobiologie Clinique. Author: (with J. Bloquel, J. P. Viallaneix, O. Blin) Techniques de Télécobalthérapie, 1966; also numerous articles. Cancer research in clin. and therapeutic field; research on radiobiology of exptl. tumors. Home: 6 Fosses Saint Julien, Caen 14. Office: Centre Anti-Cancereux, Caen 14, France.*

ABBE, Cleveland, Am. meteorologist; b. N.Y.C., Dec. 3, 1838; s. George Waldo and Charlotte (Colgate) A.; A.B., Coll. City N.Y., 1857, A.M., 1860, Ph.D., 1891; S.B., Harvard, 1864; LL.D., U. Mich., 1888, U. Glasgow, 1896; studied astronomy with Brünnow, of Ann Arbor, Mich., 1858-59, under B.A. Gould, Cambridge, Mass., 1860-64; m. Frances Martha Neal, May 10, 1870 (dec. July, 1908); m. 2d, Margaret A. Percival, Apr. 12, 1909; children—Cleveland A., Jr., William and Truman A. Asst. prof. Mich. Agrl. Coll., 1859; tutor, U. Mich., 1859-60; aid in U. S. Coast and Geod. Survey, 1860-64; guest at Nicholas Central Obs., Poulkova, nr. St. Petersburg, Russia, 1864-66; aid U. S. Naval Obs., 1867-68; dir. Cincinnati Obs. 1868-73; meteorologist in the U. S. Signal Service, 1871-91, U. S. Weather Bur., from 1891; editor Monthly Weather Review, 1873 and 1892-1915; editor Bulletin Mount Weather Observatory, 1909-13; prof. meteorology, Columbian (now George Washington) U., 1886-1905; lectr. meteorology Johns Hopkins, 1896-1914. Mem. Nat. Acad. Scis. Author: Report on the Total Solar Eclipse of July 29, 1878, 1881; Report on Standard Time, 1879, which started the agitation that resulted in the modern standard hour meredians from Greenwich; Meteorological Apparatus and Methods, 1887; Preliminary Studies in Storm and Weather Prediction; 1889; The Mechanics of the Earth's Atmosphere, Vol. I, 1891, Vol. II, 1909; The Altitude of the Aurora, 1896; Physical Basis of Long-Range Forecasting, 1902; plus nearly 300 sci. papers. Inaugurated daily weather report for Cin. C. of C., which led U. S. Govt. to take up similar reporting, 1869. Died Chevy Chase, Md., Oct. 28, 1916.

ABBE, Ernst Karl, physicist, optician; b. Eisenach, Germany, Jan. 23, 1840; s. Georg Adam and Elisabeth Christina (Barchfeldt) A.; student univs. Jena and Göttingen (both Germany), 1857-61; m. Elisabeth Snell, 1871; 2 daus. Asst. lectr. on mechanics, exptl. physics U. Jena, from 1870, apptd. asso. prof., 1873, prof., 1879, dir. obs., 1877-1901; partner optical works Carl Zeiss, owner, from 1888. At his plant improved conditions of labor with a noncontributory pension fund and a discharge compensation fund. Established Carl Zeiss Foundation for scientific research and social betterment, 1889. Author: Gesammelte Abhandlungen, 5 vols., 1904-40; writings in 3 vols., 1903-06. Constructed (Abbe) refractometer to determine refraction index of liquids, 1871; devel. laws of image of non-luminous objects, thereby laying found. for exact theory of microscope, also introduced lighting system adequate even for strongest microscope objective, 1872; built 1st system with homogeneous immersion for microscope, 1878; perfected Fizeau's interference method, 1884; introduced projection oculars for microphotographic purposes, also described system of mirror and lenses known as Abbe's condenser, 1886; built image reversal system meeting all requirements for telescope, 1895; invented achromatic microscope objective; showed that wave aspect of light restricts validity of geometrical objects; used wave theory to demonstrate limits of optical instruments due to interference and diffraction; defined power of resolution. Died Jena, Jan. 14, 1905.

ABBE, Robert, Am. surgeon; b. N.Y.C., Apr. 13, 1851; s. George Waldo and Charlotte (Colgate) A.; A.B., Coll. City of N.Y., 1870; M.D., Coll. Phys. and Surg., Columbia, 1874; m. Mrs. Catherine Amory Palmer, Nov. 14, 1891. Instr. English, drawing and geometry, Coll. City of N.Y., 1870-72; attending surgeon out-patient dept. N.Y. Hosp., 1877-84; prof. didactic surgery, Woman's Med. Coll., 2 yrs.; surgeon to St. Luke's Hosp., from 1884, N.Y. Cancer Hosp., from 1893; attending surgeon N.Y. Babies' Hosp., 1892-97; prof. surgery N.Y. Post-Grad. Med. Sch., 1889-97. Developed catgut rings used for supporting ends of intestines in operations for intestinal anastomosis, 1889; developed string method treatment for esophageal stricture, 1893; demonstrated that beta rays (not gamma rays) are effective in cell destruction, 1914; 1st American surgeon to use radium to treat cancer; pioneer in using X-rays for detection of kidney stones. Died Mar. 7, 1928.

ABBES, Michel José Louis, French surgeon; b. Béziers, France, June 29, 1936; s. Marcel and Claudine (Berger) A.; grad. in medicine Faculté de Medecine, Toulouse, 1956, Doctorat en Medicine, 1961; agrégé in cancerology; m. Janine Paschetta, Aug. 6, 1957; children—Veronique, Sophie. Became asst. Centre de Lutte contre le cancer, Nice, France, 1961; became head surgeon Service du Centre du cancer de Nice, 1964; named head diagnosis consultations Centre des Rumeurs, 1961; instr. Ecole des Infirmières de la Croix Rouge et des Assistantes Sociales Nice. Mem. French Surgery Assn., Internat. Coll. Surgeons, Nice Soc. Surgery and Medicine, French Anti-Cancer League. Research and publs. on metastic spread of cancer via lymphatic system, diagnostic radiology, intra-arterial chemotherapy of tumors. Home: 42 Rue Verdi, Nice 06. Office: 36 Av. Voie Romaine, Nice, France.*

ABBOT, Charles Greeley, Am. astrophysicist; b. Wilton, N.H., May 31, 1872; s. Harris and Caroline (Greeley) A.; B.S., Mass. Inst. Tech., 1894, M.S., 1895; D.Sc., U. Melbourne, 1914; LL.D., U. Toronto (Ont., Can.), 1933; D.Sc., Case Inst. Tech., 1930, George Washington U., 1937; m. Lillian Moore, Oct. 13, 1897 (dec. June 1944); m. 2d, Virginia Johnston, June 9, 1954. With Smithsonian Instn., Washington, 1895——, dir. obs., 1907-44, asst. sec., 1918-28, sec., 1928-44, research asso., 1944——. Mem. Am. Philos. Soc., Am. Acad. Arts and Scis. (Rumford medal 1916), Nat. Acad. Scis. (Draper medal 1910), A.A.-A.S. Author: Annuals Smithsonian Astrophysical Observatory, vol. 1, 1900, vol. 2, 1908, vol. 3, 1913, vol. 4, 1922, vol. 5, 1932, vol. 6, 1942; The Sun, 1911; The Sun and the Welfare of Man, 1929; Every Day Mysteries, 1926; The Earth and the Stars, 1958; Adventures in a World of Science, 1958; also numerous articles. Research on solar spectrum absorption; lines; diameters, and radiation bright stars; observed variations in solar radiation from sea level to 14,500 feet elevation, its effect upon climactic conditions on earth, and its relation to the sunspot cycle; determined solar constant; discovered harmonic family periods in sun and weather; made useful long range forecasts of weather; invented many research instruments. Home: 4409 Beechwood Rd., Hyattsville, Md. 20782. Office: Smithsonian Instn., Washington 25.

ABBOTT, Charles Conrad, Am. naturalist; b. Trenton, N.J., June 4, 1843; s. Timothy and Susan (Conrad) A.; M.D., U. Pa., 1865; m. Julia Boggs Olden, Feb. 13, 1867. Made large collection of archaeol. specimens, now in Peabody Mus., Cambridge, Mass., where he was asst., 1876-89. Made 2d large archaeol. collection of Del. Valley for Princeton, 1901-07. Author: The Stone Age in New Jersey, 1876; Primitive Industry, 1881; A Naturalist's Rambles About Home, 1884; Upland and Meadow, 1886; Wasteland Wanderings, 1887; Archaeological Explorations in the Valley of the Delaware, 1894; Archaeologia Nova Caesarea, 1907, 1908, 1909; Ten Years' Diggings in Lanapè Land, 1912. Demonstrated existence of man in Delaware River Valley during glacial and subsequent prehistoric periods. Died July 27, 1919.

ABBOTT, Clinton Gilbert, ornithologist; b. Liverpool, Eng., Apr. 17, 1881 (parents Am. citizens); s. Lewis Lowe and Grace (Van Dusen) A.; came to U. S. 1897; A.B., Columbia, 1903, postgrad. Cornell U., 1914-15; m. Dorothy Clarke, May 18, 1915; children —Dorothea Van Dusen (Mrs. Hal G. Evarts, Jr.). Lois Virginia (Mrs. Peter D. Whitney), Lucia Grace. Lectr. on ornithology and travel Bd. Edn., N.Y.; confidential sec., editor, Conservation Comm., State of N.Y., 1918-21; in charge pub. edn. San Diego Natural History Mus., 1921-22, dir., from 1922. Fellow San Diego Soc. Natural History (pres. 1923-25); mem. Linnaean Soc. N.Y. (sec. 1901-10, v.p. 1910-14). Am. Ornithologists' Union, Cooper Ornith. Club (pres. so. div. 1934), Am. Soc. Mammalogists, Western Soc. Naturalists, Nat. Audubon Soc. Internat. Com. for Wildlife Protection, Phi Beta Kappa. Author: The Home Life of the Osprey, 1911. Contbr. to sci. journs. Died Mar. 5, 1946.

ABBOTT, Donald Putnam, Am. zoologist; b. Chgo., Oct. 14, 1920; s. Donald Putnam and Marion (Dummer) A.; B.A., U. Hawaii, 1941; M.A., U. Cal. at Berkeley, 1948, Ph.D., 1950; m. Isabella Aiona, Mar. 3, 1943; 1 dau., Ann Kaiue. Instr., U. Hawaii, 1942-43; faculty Stanford, 1950——, prof. zoology, 1963——, asst. dir. Hopkins Marine Sta., Pacific Grove, Cal., 1962——. Mem. Pacific Sci. Bd. Ifaluk Atoll Expdn., 1953, Hawaii-P.I.-Sulu Sea Expdn., 1957, Galapagos Internat. Sci. project, 1964; chief sci. Te Vega Indian Ocean Expdn., 1964. Fellow A.A.A.S., Cal. Acad. Scis.; mem. Am. Soc. Zoologists, Soc. for Study Evolution, Ecol. Soc. Am., Soc. Systematic Zoology. Author: (with others) Intertidal Invertebrates of the Central California Coast, 1954; (with M. Bates) Coral Island, 1958; also articles. Research in ecology of coral atolls, reprodn., devel. structure and classification of ascidians, hydrobiology of Monterey Bay, Cal. Home: 210 Asilomar Blvd., Pacific Grove, Cal. 93950.*

ABBOTT, E. V., Am. plant pathologist; b. Ashland, Ore., July 9, 1899; s. Charles E. and Johanna (Arneson) A.; B.S., Ore. State U., 1922; M.S., Ia. State U., 1923 Ph.D., 1925; m. Elizabeth Oldham, Sept. 4, 1934; 1 dau., Ruth Elizabeth (Mrs. Keith H. Remy). Asst. plant pathologist La. Agr. Exptl. Sta., 1925-26; asst. soil bacteriologist La. Agr. Expt. Sta., 1926-27; plant pathologist Estacion Exptl. Agricola de la Molina, Lima, Peru, 1927-30; plant pathologist U. S. Sugar Cane Field Sta., Houma, La., 1930-66, supt. 1950-66. Recipient Superior Service award U. S. Dept. Agr., 1957. Mem. Am. Soc. Sugarcane Technologists (past pres., hon. life mem.). Author: Sugar-Cane Diseases of the World. vol. I (with J. P. Martin), 1961, vol. II (with C. G. Hughes, C. A. Wismer), 1964; also numerous articles. Research on diseases of sugar cane. Home: 605 Morninside Dr., Houma 70360. Office: U. S. Sugar Cane Field Sta., Houma, La. 70360.*

ABBOTT, Frederick Wallace, Am. physician; b. Dover, N.H., Mar. 5, 1861; s. Sylvester and Elizabeth Graves (Wortman) A.; A.B., U. of America, 1883; attended Med. Sch. of Me. (mem. dept. Bowdoin Coll.), 1884, 1885; M.D. Eclectic Med. Coll. of Me.; 1886; H.F.B.S., 1893; A.M., Taylor U. 1901; Ph.D., Nat. Normal U., 1901; LL.D., Potomac U., 1905; D.P.H., Eclectic Med. U., Kansas City, Mo., 1917; m. Sylvina Apphia Emery, Sept. 2, 1886. Practiced in Taunton, Mass., from 1886; lectr. on physiology and hygiene Merrimack County Acad., 1892; asso. editor Mass. Med. Jour., 1894-1904, Am. Med. Jour., 1906——; sr. censor Eclectic Med. Coll. City of N.Y.; prof. eugenics, Eclectic Med. U., Kansas City, Mo., from 1913; prof. med. history, med. ethics, Middlesex Coll. Medicine and Surgery, cons. physician Middlesex Hosp., from 1916. Pres. Mass. Eclectic Med. Soc., 1894, N.E. Eclectic Med. Assn., 1900, Am. Eclectic Materia Medica Assn., 1905-07, Boston Dist. Eclectic Med. Soc., 1910; v.p. Am. Anti-Tb League. 1907. Academician of Toulouse, France; life-mem. and medalist of 1st class (gold) Italian Acad. Physics and Chemistry. Author: Limitation of the Family, 1891; The Education of Youth Upon Matters Sexual, 1895. Died June 19, 1919.

ABBOTT, George (Alonzo), Am. chemist and toxicologist; born in Alma, Illinois, July 7, 1874; son of John Baughman and Harriet (Stuart) A.; B.S., De Pauw U., 1895, A.M. pro merito, 1896; Ph.D., Mass. Inst. Tech., 1908; LL.D. (Honoris Causa) U. of North Dakota, 1951; Sc.D. (hon.), N.D. State College, 1962; married Ruth Ware, June 15, 1910; children—Marian Ware (Mrs. Russell Huxsol), Stuart Ware. Instr. in chemistry Evansville (Ind.) High Sch., 1896-99, Duluth, Minn., 1899-1900, Indianapolis, 1900-09; asst. prof. and chemist, expt. sta., State Coll., Fargo, N.D., 1909-10; prof. chemistry and head of dept., N.D., 1910-47, prof. of chemistry since 1948; exchange lecturer U. of Manitoba, 1912; guest prof. City Coll., N.Y. City, 1940. Spl. water consultant Nat. Resources Bd., 1937, consultant since 1910; expert toxicologist and court witness in coroners' cases and murder trials. State del. 8th Internat. Congress Applied Chemistry, 1912. Fellow Am. Inst. Chemists, Ind. Acad. Sci.; mem. Am. Assn. Univ. Profs., Am. Chem. Soc. (mem. senate on chem. edn.; charter mem. and 1st chmn. Red River Valley sect. 1948), N.D. Acad. Sci. (charter mem.; sec. since 1910), S.D. Acad. Sci. (acad. lecturer); Polytechnic Soc. (charter mem. and a founder), Phi Beta Kappa (charter member N.D. Alpha chapter; elected chapter pres. 3 times); Sigma Xi (past pres. chapter), Contbg. editor: Jour. Chem. Edn. 1924-40. Author: The First Fifty Years—a History of North Dakota Academy of Science, 1958; also bulls. and articles. Investigations in basic and applied chemistry; studies of Lianite, Leonardite, soils and waters; surveyed ground waters; research on ionization relations of ortho and pyro phosphoric acids, occurrence of selenium in soils and waters, reaction of sugars with zirconyl salts, humic acids, colloidal graphite lubrication. Home: 505 Hamline St., Grand Forks, N.D.*

ABBOTT, Lynn DeForrest, Jr., Am. biochemist; b. Ithaca, New York, Nov. 23, 1913; s. Lynn DeForrest and Olive (Ruth) A.; B.S. Wayne State U. 1936, M.S., 1937; Ph.D., U. Mich. 1940; m. Hester May Easton, Aug. 24, 1940; children—Lynn DeForrest III, James Easton, William Lewis. Faculty biochemistry Med. Coll. Va., Richmond, 1940——, prof., 1956-—, chmn. dept., 1962——. Mem. Am. Soc. Biol. Chemists, Am. Chem. Soc. (chmn. Va. sect. 1950-51), Soc. Exptl. Biology and Medicine, A.A.A.S., Va. Acad. Sci. (J. Shelton Horsley Research award 1954), Am. Soc. Clin. Nutrition, Sigma Xi (pres. Med. Coll. Va. chpt. 1961-62). Research, publs. on hemoglobin synthesis; biochem. edn.; amino acid metabolism. Home: 607 Horsepen Rd., Richmond, Va. 23229.*

ABBOTT, Osler Almon, Am. surgeon; b. Hamilton, Ont., Can., June 6, 1912; s. H. P. Almon and Rachel (Gwyn) A.; came to U. S., 1914, naturalized, 1921; A.B., Princeton, 1933; M.D., Johns Hopkins, 1937; m. Sarah Kendall Yancey, June 24, 1939; children—William O. A., John Y., Edward K., Caroline G. Practice medicine, specializing in thoracic surgery, Atlanta, 1945—; faculty Sch. Medicine Emory U., 1945—, prof. thoracic surgery, 1961—, chief thoracic and cardiovascular surgery U. Clinic, 1952—; chief thoracic and cardiovascular surgery Emory U. Hosp., Henrietta Egleston Hosp. for Children, Grady, Meml. Hosp., 1945—. Sr. cons. 3d Army, Atlanta Va. hosps. Recipient Shipley Surg. medal So. Surg. Assn., 1962, Susan and Theodore Cummings Humanitarian award Am. Coll. Cardiology, 1962, 64. Diplomate Am. Bd. Surgery, Bd. Thoracic Surgery. Fellow A.C.S., Am. Coll. Cardiology (past pres.), Am. Coll. Chest Physicians (regent), S.E. Surg. Congress, So. Surg. Assn. (pres. elect), Hellenic Cardiol. Soc. (Greece); mem. Am. Assn. Thoracic Surgery, Soc. Thoracic Surgeons, So. Thoracic Surg. Assn., Soc. U. Surgeons, Am. Thoracic Soc., Am. Bronchoesophagalogical Assn., Am. Fedn. Clin. Research, Am. Med. Writers Assn., Am. Heart Assn., A.M.A., So., Pan Am. med. assns., N.Y. Acad. Scis., Turkish Med. Soc., Internat. Platform Assn., Sigma Xi, Alpha Omega Alpha. Contbr. chpts. to Bronchopulmonary Diseases, 1957; Clinical Cardiopulmonary Physiology, 1960. Contbr. numerous articles to profl. jours. Research in esophageal surgery, cerebral air embolism, pulmonary emphysema, cardiovascular surgery. Home: 3037 W. Pine Valley Rd. N.W., Atlanta 30305. Office: Emory U. Clinic, Atlanta 30322.*

ABBOTT, Percival William Henry, English mathematician; b. June 7, 1869; s. J. S. Abbott; B.A., London (Eng.) U.; m. Isobel M. Lucy; 1 dau. Math. master Poly. Sch., London, head math. dept., 1895-1919, headmaster, 1919-34. Mem. Tchrs.' Registration Council, chmn. tech. sect., 1912-26; leader tech. panel Burnham Com., 1919-22. Mem. Math. Assn. (hon. sec. teaching com.), Assn. Tchrs. in Tech. Institutions (hon. sec., pres.), Inc. Assn. Headmasters. Author: Mathematical Tables and Formulae, 1918; Numerical Trigonometry, 1918; Exercises in Arithmetic and Mensuration, 1913; National Certificate Mathematics, 3 vols., 1938; Teach yourself Trigonometry, 1940; Teach yourself Mechanics, 1941; Teach yourself Algebra, 1942; Teach yourself Calculus, 1946; Teach yourself Geometry, 1947. Editor Tech. Coll. Series, also The Tech. Jour. Died Jan. 29, 1954.

ABBOTT, William Osler, Am. physician; b. Phila., 1902; devised (with T. G. Miller) double-lumen intestinal drainage tube for relief of distention (Miller-Abbott tube), 1934, (with Arthur Joy Rawson) double-barreled gastroenterostomy tube for use in postoperative care (Abbott-Rawson tube), 1937. Died 1943.

ABBOUD, Francois Mitry, physiologist, physician; born Cairo, Egypt, Jan. 5, 1931; s. Mitry Y. and Asma (Habac) A.; student U. Cairo, 1948-52; M.B., B.Ch., Ein Chams U., 1955; m. Doris Evelyn Khal, June 5, 1955; children—Mary Agnese, Susan Marie, Nancy Louise, Anthony Lawrence. Came to U. S., 1955, naturalized, 1963. Am. Heart Assn. research fellow cardiovascular labs. Marquette U., 1958-60; Am. Heart Assn. advanced fellow U. Iowa, Iowa City, 1960-62, asst. prof., 1961-65, asso. prof. medicine, 1965—; attending physician VA Hosp., Iowa City, 1963—; U. Iowa Hosps., 1961—. European Traveling fellow French Govt., 1948; NIH grantee, awardee, 1962—. Mem. Am. Soc. Clin. Investigation, Central Soc. for Clin. Research, Soc. Exptl. Biology and Medicine, Am. Physiol. Soc., A.C.P., A.M.A., Am. Fedn. Clin. Research (councillor); Am. Assn. U. Profs., Sigma Xi, Research, publs. in cardiovascular physiology on vascular reactivity and autonomic pharmacology in man and animals. Home: 1911 Rochester Ct., Iowa City 52240.*

ABD AL-MALIK AL-SHIRAZI (Abu al-Husain Abd al-Malik ibn Muhammad al-Shirazi), Muslim mathematician, astronomer; flourished 2d half of 12th century. Wrote summary of Apollonios' treatise on conics in Arabic; also an abridgment of Almagest.

ABD AL-RAHMAN, Umar al-Sufi (Abu al-Husayn), astronomer; b. Rayy, Persia, Dec. 8, 903; s. Omar; attended vizier Abu 'l-Fadl ben al-'Amid, Isfahan, Persia, 948-49; at court 'Adud al-Dawla, 960-61, became court astronomer (made silver globe preserved in library Fatimid Palace, Cairo, Egypt). Author: Suwar al-Kawakid al-Thabita (known under various titles, described fixed stars according to Ptolemaic and Arabic systems, illus. by drawings traced from celestial globe), circa 965; handbook of astronomy and astrology; treatise on use of astrolabe. Died May 25, 986.

ABDEL-RASSOUL, Ahmed, Egyptian radiochemist; b. Alexandria, Egypt, Mar. 4, 1929; s. Ahmed Abdel-Rassoul and Raoufa (Ghoneim) Gadalla; B.Sc., Alexandria U., 1949; M.Sc., 1955; Dr.rer.nat., Johannes Gutenberg U., Mainz, Germany, 1960; Degree in Research Devel., Nat. Inst. Mgmt. Devel., 1964. Demonstrator chemistry U. Alexandria, 1949; research worker Max-Planck/Otto Hahn Institut für Radiochemie, Mainz, 1956-60; lectr. nuclear chemistry UAR Atomic Energy Establishment, Cairo, 1960, asst. prof., 1964—; vis. research fellow Institutt for Atomenergi, Kjeller, Norway, 1961; staff Radioisotope Prodn. Lab., Inchass, Egypt, 1962-64. Mem. Egyptian Chem. Soc. Research, publs. on prodn. of a series of radionuclides in high chem. and radiochem. purity for med. applications; devel. analytical schemes based on nuclear activiation to solve med. and biochem. problems, particularly disposition of anti-Bilharzial antimonial drugs, purity control of reactor materials; scintillation nuclear spectrometric analytical investigations for successive and other induced nuclear reactions. Home: 19 Abdel-Aziz Fahmy/Heliopolis, Cairo. Office: Nuclear Chemistry Dept., UAR Atomic Energy, P.O. B, Cairo, Egypt.*

ABDEL-WAHAB, Mohamed Fathy, Egyptian biochemist; b. Samatout, Egypt, Mar. 27, 1925; s. Abdel-Wahab Ahmed Toni; B.Sc. with honors, Faculty Sci., Cairo U., 1951, M.Sc., 1955, D.B.A., 1952, D.Sc., 1965; Ph.D., Vienna (Austria) U., 1958; m. May 18, 1952; children—Sannaa, Ahmed, Wafaa, Hatem. Mgr. printing co., 1942-47; demonstrator biochemistry dept. Ein-Shams U., Faculty Medicine, Cairo, Egypt, 1951-56, lectr. chemistry radioisotope dept., 1958-63, asso. prof. radiochemistry, 1963—; mem. mission U. Vienna, 1956-58. Expert radiochemistry Middle Eastern Regional Radioisotope Centre for Arab Countries in cooperation with IAEA, Cairo; expert, cons. in drug research. Recipient State prize, 1962; Decoration Sci. and Art, Pres. Nasser, 1961. Mem. Chem. Soc. Cairo, Syndicate Sci. Author: The Peaceful Uses of Cobalt-60. Founder sch. for use radioisotopes in biochem. problems, 1958; pioneered preparation of labelled compounds and hormones using simple, cheap method; introduced new techniques for screening of new drugs using radioisotopes. Home: Kinawi House, Pyramid St., Giza, Egypt. Office: M.E.R.R. Centre, Sh. Malaeb El-Gamaa, Dokki, Cairo, Egypt.*

ABDERHALDEN, Emil, Swiss physiologist, biochemist; born Oberuzwil, Switzerland, Mar. 9, 1877; son of Nikolaus and Anna Barbara (Stamm) A.; student of Emil Fischer; M.D., U. Basel (Switzerland), 1902; Ph.D. (hon.), U. Halle (Germany); M.D.V. (hon.), U. Zurich (Switzerland); m. Margarethe Barth, 1909; children—Rudolph, Klaus, 3 daus., including Barbara. Became asst. to Emil Fischer, Berlin, Germany, 1902; apptd. prof. physiology Vet. U. Berlin, 1908; prof. physiol. chemistry and physiology U. Halle, 1911-45; guest prof. physiol. chemistry U. Zurich, 1946-47. Mem. German Acad. Scientists (pres.), numerous other assns. Author: Lehrbuch der physiologischen Chemie, 1906; Biochemische Handlexikon, 1910-25; Handbuch der biochemischen und biologischen Arbeitsmethoden, 1920-39; Lehrbuch der Physiologie, 1925; numerous papers on vitamins, hormones, related subjects. Devised test for pregnancy known as Abderhalden or Abderhalden-Fauser reaction, 1912; isolated (with A. Weil) norleucine, 1913; continued Fischer's work preparing esters of amino acids; investigated metabolism of foods; studied albumin analysis and synthesis. Died Zurich, Aug. 5, 1950.

ABDULLAEV, Khabib Mukhamedovich, Russian geologist; b. Aug. 31, 1912; grad. Central Asian Indsl. Inst., Tashkent, 1935. Mem. staff Central Asian Indsl. Inst., 1940-41; posts with Party and govt., Uzbek, 1941-48, 55-56; attended 20th Internat. Congress Geologists, Mexico, 1956. Recipient Lenin prize, 1956. Mem. Uzbek Acad. Sci. (v.p., 1947-52, pres., 1956), bur. mem. Dept. Geol. and Geog. Sci., USSR Acad. Sci. (del. to India). Author: Geology of the Scheelitic Skarns of Central Asia, 1947; The Genetic Relationship of Mineralization to Granitoid Intrucions, 1954; Dikes and Mineralizations, 1957. Research on theory of formation of ore and magna deposits; also role of granitoids in post-magnetic ore-formation.

ABDULLAH, Abdul Jabbar, meteorologist; b. Qualat-Saleh, Iraq, Nov. 14, 1911; s. Shaikh Abdullah Sam and Nofa Roomi; B.A. with distinction in Math., Am. U., Beirut; Sc.D. in Meteorology (univ. scholar), Mass. Inst. Tech.; m. Kismet Anaisi Fayyadh, Sept. 7, 1943; children—Sana, Sinan, Heitham, Thabit. Tchr. physics, math. secondary schs., Iraq, 1934-40; asst. meteorologist Airport, Bassrah, Iraq, 1940-44; faculty Mass. Inst. Tech., 1946-49, Coll. Edn., Baghdad, Iraq, 1949-52, 55-58, N.Y. U., 1952-55; mem. founding council U. Baghdad, Iraq, 1958-59, pres. U., 1959-63, emeritus, 1963—. Research scientist Nat. Center for Atmospheric Research, Boulder, Colo., 1965—. Recipient Rafidain medal Govt. Iraq, 1956. Mem. Am. Instrument Soc., Royal Meteorol. Soc., Am. Geophys. Union, A.A.A.S., Iraq Soc. Physics and Math., Internat. Geophys. and Geodes. Soc., Internat. Soc. Limnology, Sigma Xi. Author: (Arabic) A Glossary of Meteorology, 1942; The Measurement of Upper Winds, 1943; A Text-Book of General Physics, 1950; Theory of Sound, 1954; also articles. Translator into Arabic: Introduction to Atomic Physics (Semat). Contbns. to gen. theory of atmospheric waves and dispersion of energy in atmosphere, theory of atmospheric vortices, theory of pressure jumps of atmosphere and relation to cyclogenesis and squall lines, theory of atmospheric solitary waves. Home: 97B 11/21 Hussam-Eddin, Baghdad, Iraq. Office: Nat. Center for Atmospheric Research, Boulder, Colo. 80301.*

ABDULLAH, Mohammad, entomologist-zoologist; b. Allahabad, India, July 8, 1938; s. Sabir and Razia (Qasim-Ali) H.; B.Sc., U. Allahabad, 1954; M.Sc., Aligarh Muslim U., India, 1956; M.S., U. Ill., 1959; Ph.D., U. Reading (Eng.), 1965; m. Abida Ebrahim, Dec. 4, 1964. Research asst. U. Karadri, Pakistan, 1956-57; with U. Ill., Urbana, 1957-62, instr., 1961-62; postdoctorate fellow dept. entomology Macdonald Coll., McGill U., Montreal, Que., Can. Postdoctorate fellow NRC Council Can., 1965-66. Fellow Royal Entomol. Soc. London, Zool. Soc., Zool. Soc. London. Author: The Natural Classification of Anthicidae; also articles. Research on natural classification and relationship of various beetles, discovery of fossils, beetle genera in Eastern Prussia. Home: 14/13 III C Nazimabad, Karachi 18, Pakistan.*

ABE, Matazo, Japanese microbiologist; b. nr. Niigata City, Japan, Oct. 30, 1909; s. Takazo and Tase (Takai) A.; B.S., Tokyo Imperial U., 1938, Ph.D., 1946; m. Hatsuko Shiga, July 23, 1937; children—Michiko (Mrs. Shozo Matano), Matanobu. With research labs. Takeda Chem. Industries Ltd., Osaka, 1938-47, head applied microbiology dept., 1960-64; chief sect. Inst. for Fermentation, Osaka, 1947-60; prof. indsl. microbiology, faculty agr. Tokyo U. Edn., 1964—. Recipient Outstanding Sci. award Japanese Assn. Agrl. Sci. Socs., 1947, Suzuki award, 1947. Mem. Agrl. Chem. Soc. Japan (councilor 1947—, dir. 1959-61), Soc. Fermentation Tech. Japan (councilor 1964—). Numerous publs. on investigation of morphology, physiology and ecology of parasitic and saprophytic ergot fungi; devel. of Abe's culture medium; chemistry, biochemistry of ergot alkaloids.

ABE, Shoo (1st name Teruto, posthumous name Tomonoshin), Japanese botanist; b. Morioka, Iwate Prefecture, Japan, 1650; studied medicine and botany, Chekiang Province, South China. Lived in China, 18 years, then returned to Yedo, Japan; entered Shogunate Govt. service in program to train botanists and promote Chinese pharmacology, 1729; gathered specimens of med. herbs in mountains of Kai Province, Yamanashi Prefecture, 1722; made important specimen discoveries in Tōhoku and Hokkaidō, 1729; opened Med. Herb Garden for exptl. research in pharmacology, Konyamachi, Yedo. Author books including: Saiyakushisetsu, Honzō Kōmoku Ruikō. Died 1753.

ABEGG, Richard, chemist; b. Danzig, Poland, Jan. 9, 1869; s. Wilhelm A. and Margarethe (Friedenthal) A.; student univs. Kiel, Tubingen, Berlin (Germany), from 1886; degree in phys. chemistry, 1891. Became pvt. asst. lectr., dept. head. Chemistry Inst., Wroclaw, Poland, 1899; apptd. asso. prof. U. Wroclaw, 1900, prof. Tech. U. Wroclaw, 1909. Author: Die Theorie der elektrolytischen Dissociation, Sammlungen und Chemischer und Chem-Technischer Vortrage, 1903; Die Elektroaffinitat, ein neues Prinzip der chemischen Systematik, 1904-25; (with F. Auerbach) Handbuch der anorganischen Chemie, 1905; (with F. Auerbach and R. Luther) Messungen Elektromotorischer Krafte Galvanischer Kette mit wassrigen Elektrolyte, abhandlungen D. Bunsen Geschaft, 1911. Editor Zeitschrift für Elektrokemie, 1901-10. Worked in phys. and electro-chemistry on osmotic pressures, freezing point depressions in dilute solutions, dielectric constant of ice, speed of translation of ions and chem. equilibrium; tested Faraday's Law; studied effect on chem. valence of new electronic view of atom; developed (with Bodländer) theory of secondary valence. Died Koszalin, Poland, Apr. 3, 1910.

ABEILLE, Louis Paul, French economist; b. Toulouse, France, June 2, 1719; 1 son, Jean-Louis; lawyer at parliament of Bret; insp. gen. French manufactures; sec.-gen., consul Bur. Commerce, France. Fellow Royal Soc., 1762; mem. Soc. Agr., Macquer. Author: Table raisonnée de ordonnances, 1757; Principes sur la liberté du commerce des grains, 1768. Patron agr. socs.; encouraged Ami Argant (lamp inventor); responsible for Corps d'observations de la Societé d'agriculture . . . , 1761. Died July 28, 1807.

ABEL, Clarke, Brit. naturalist; b. England, circa 1780; ed. for med. profession. Joined H.M.S. Alceste at Spithead, 1816, went to China, amassed large collections there; lost collections in shipwreck, except for small portion he had given to Sir George Stauton; his plants were described by Robert Brown (who named genus Abelia in his honor); described orangoutang, boa, made observations on geology of the Cape; physician to Lord Amherst, gov.-gen. of India. Fellow Royal Soc., 1819, Linnean Soc., Geol. Soc. London; mem. Asiatic Soc., Med. and Phys. Soc. Calcutta. Author: Narrative of a Journey in the Interior of China, 1816-17, pub. 1818. Died Nov. 24, 1826.

ABEL, Edward William, English chemist; b. Kenfig Hill, Wales, U.K., Dec. 3, 1931; s. Sydney John and Donna (Grabham) A.; B.Sc., U. Coll., Cardiff, Wales, 1952; postgrad. No. Poly.; Ph.D., Imperial Coll., London, Eng., 1957, D.Sc., 1965; m. Margaret Rosina Edwards, Aug. 6, 1960; children—Edward Christopher, Julia Margaret. Lectr. inorganic chemistry U. Bristol (Eng.), 1959-67, reader, 1967—. Mem. Chem. Soc., Faraday Soc. Research, publs. on organometallic chemistry of main group and transitional elements. Home: 29 Grove Av., Bristol 9, Eng.*

ABEL, Sir Frederick Augustus, English chemist; b. Woolwich, Eng., July 17, 1827; s. John Leopold and Louisa (Hopkins) A.; student Royal Poly. Instn.; student Royal Coll. Chemistry (under Hoffman), 1845-46; D.C.L. (hon.), Oxford (Eng.) U., 1883; D.Sc. (hon.), Cambridge (Eng.) U., 1888; m. Sarah Selina, 1854 (died 1888); m. 2d Guileitta de la Feuillade, 1889 (died 1892). Apptd. asst. Royal Coll. Chemistry, 1846; became demonstrator chemistry St. Bartholemew's Hosp., 1851; named lectr. chemistry Royal Mil. Acad., Woolwich, 1852, ordnance chemist, 1854; became chemist War Dept., Woolwich, 1856; chief authority on explosives, Woolwich, 1854-88. Mem. Ordnance Select Com., Woolwich; pres. Explosives Com., 1888-1902. Created knight, 1883. Recipient Royal medal, 1887, Bessemer Gold medal, 1897; Albert medal, 1891, Telford medal Instn. Civil Engrs., 1897. Fellow Royal Soc., 1860; mem. Chem. Soc. (pres. 1875-77), Soc. Chem. Industry (pres. 1883), Iron and Steel Inst. (pres. 1891), Brit. Assn. (pres. 1890). Author works on Explosives. Developed high explosives and smokeless gunpowder; studied (with Nobel) relation between speed of projectile, size and form of powder grain, and behavior of black powder when fired; investigated guncotton, nitroglycerine, dust and mine explosions, hardening of steel; invented Abel open-test for petroleum, 1868, Abel close-test, 1879; invented (with James Dewar) cordite, 1889. Died London, Sept. 6, 1902.

ABEL, John Jacob, Am. physiol. chemist; b. Cleve., May 19, 1857; s. George M. and Mary (Becker) A.; Ph.B., U. Mich., 1883, A.M., 1903, Sc.D., 1912; Sc.D., U. Pitts., 1915; LL.D., U. Cambridge (Eng.), 1920; Sc.D., Harvard, 1925, Yale U., 1927; hon. M.D., John Casimir U., Lwow, Poland, 1926; LL.D., U. Aberdeen, 1932; postgrad. Johns Hopkins, 1883-84, chemistry and medicine at Leipzig, Strassburg, Heidelberg, Vienna, Berne, Würsburg and Berlin, 1884-91; M.D., Strassburg, 1888; m. Mary W. Hinman, July 10, 1883. Lectr., prof. materia medica and therapeutics U. Mich., 1891-93; prof. pharmacology Johns Hopkins, 1893-1932; prof. emeritus pharmacology, and dir. Lab. for Endocrine Research, from 1932. Founder, editor Jour. of Pharmacology and Exptl. Therapeutics, 1909-32. Recipient Research Corp. prize, 1925; first award of lectureship Kober Found., Am. Assn. Physicians, 1925; Willard Gibbs medal Chgo. sect. Am. Chem. Soc., 1926; gold medal, Soc. of Apothecaries, London, 1928; Philip A. Conné medal N.Y. Chemists' Club, 1932; Kober medal, 1934. Fellow Royal Soc., 1938. Known for investigations chem. composition of animal tissues and fluids; did outstanding research on suprarenal glands; investigated toxic and therapeutical action of many substances; isolated epinephrine (adrenalin), 1897; (with L. G. Rowntree) studied action of the phthaleins, 1909; devised plasmapheresis (method of plasma removal), 1914; 1st to obtain insulin in crystalline form, 1926. Died Baltimore, Md., May 26, 1938.

ABEL, Niels Henrik, Norwegian mathematician; b. Finnoy, Norway, Aug. 25, 1802; s. Anne Marie Simonsen; studied on state pension Christiana U., Oslo, Norway, 1821-22, degree; visited Germany and France on state pension, 1825; contbd. to making success of Crelle's Math. Jour.; returned to Norway, destitute and ill with Tb, 1827, taught at Christiana U.; offered post as prof. math. U. Berlin (Germany), 1829, but offer did not reach Norway until after his death. Author: Mémoire sur une propriété générale d'une classe trè-étendue de fonctions transcendentes (his masterpiece, written in Freiburg, Germany), 1826; also several articles in Crelle's Jour. Proved impossibility of algebraic solution of quintic equation, 1824; responsible for vast devel. of algebraic geometry; provided new rigor in his work on equation theory, function theory, and power series; was one of leading analysts of his time; best known work dealt with algebraic approach to elliptical functions, a new theory generalizing trigonometric functions; discovered several types of equations, including those later called Abelian equations; proved there is no gen. radical solution for 5th degree and higher degree equations, 1824. Died Froland, Norway, Apr. 6, 1829.

ABEL, Othenio Lothar Franz Anton Louis, paleobiologist; b. Vienna, Austria, June 20, 1875; s. Lothar Paul F. K. and Mathilde Franziska Antonia (Schneider) A.; Ph.D., U. Vienna, 1899; hon. degrees U. Capetown (S. Africa), Athens (Greece); m. Friedericke Dengg, Feb. 18, 1901; children—Wolfgang, Elfriede. Employee State Inst. Geology, Vienna, 1900-07; became asst. to prof. E. Suess, U. Vienna, 1898, asso. prof., 1907, prof., 1917-34; prof. U. Göttingen (Germany), 1935-40. Recipient Bigsby Gold Medal Geol. Soc. London, 1911, Elliot Gold medal Acad. Scis. Washington, 1921. Mem. acads. Göttingen, Halle (Germany), Leningrad (Russia), Upsala (Sweden), Vienna, others. Author: Grundzüge der Paläobiologie der Wirbeltiere, 1912; Vorzeitliche Säugetiere, 1914; Paläobiologie der Cephalopoden, 1916; Stäume der Wirbeltiere, 1919; Lehrbuch der Paläozologie, 1920; Die Methoden der paläobiologischen Forschung, 1921; Geschichte und Methode der Rekonstruktion vorzeitlicher Wirbeltiere, 1924; Amerikafahrs, 1926; Die Stellung des Menschen im Rahmen der Wirbeltiere, 1931. Research on mollusca, 1916; founded a paleobiol. inst.; instrumental in establishing paleobiology as ind. sci. Died Pichlam, Mondsee, July 4, 1946.

ABÉLARD, Peter, French philosopher and theologian; b. Le Pallet, nr. Nantes, France, 1079; student William of Champeaux at Notre Dame; student Anselm of Laon's theology sch., circa 1114; m. Héloise; 1 son, Astrolabe. Tchr. at Melun, also Corbeil; tchr. logic, theology in Paris (Mt. St. Geneviève), later at Notre Dame; monk Abby St. Denis; hermit, Nogent-sur-Seine, where he established a school (The Paraclete); abbot St. Gildas de Rhuys, Brittany, from 1125; tchr. of Peter Lombard, John of Salisbury, others; returned to Mt. St. Geneviève, 1136; condemned by Council of Sens, 1141, and in Rome, 1142; opposition led by Bernard of Clairvaux. Author: Theologia summa boni; Theologia christiana; Theologia scholarium; Tracitus de unitate et trinitate; Apologia; Sic et non (exposition of contradictions in scriptures and work of Ch. fathers); Exposition in hexaemeron; Expositio in epistolam ad Romanes; Sermons; Scito te ipsum; Judaeum et Christianum; Dialectica; De intellectibus; Sententiae; commentaries on Aristotle, Porphyry, Boethius; Love Letters to Héloise; Letters to St. Bernard; Historia calamitatum; various poems. Attempted to use sic et non method to resolve contradictory statements; adopted a position of moderate realism in medieval scholastic debates; appealed to logic and reason to solve problems; argued that faith and reason are distinct and different; emphasized the method of Aristotle's dialectic in one of earliest and most systematic attempts to apply reason to Christian theology; helped establish the Scholastic manner of philosophizing. Died Chalon-sur-Saône, France, Apr. 21, 1142.

ABELIN, Richard, German veterinarian; b. Munich, Germany, June 24, 1891; s. Karl and Sofia (Buhmann) A.; doctorate in vet. medicine U. Munich; m. Elisabeth Brockhoff, 1922; children—Rolf, Dieter, Werner, Helga. Practiced vet. medicine, specializing in disorders of reproduction, 1919-35, also from 1945; asso. prof. vet. medicine U. Munich, 1935-45, emeritus, from 1956. Introduced artificial insemination in Bavaria; studied bovine trichomoniasis. Address: Stauffenbergstrasse 7, 8 Munich 13, West Germany.

ABELL, George Ogden, Am. astronomer; b. Los Angeles, Mar. 1, 1927; s. Theodore Curtis and Annamarie (Ogden) A.; B.S., Cal. Inst. Tech., 1951, M.S., 1952, Ph.D., 1956; m. Lois Everson, June 16, 1951; children—Anthony Alan, Jonathan Edward. Observer, Nat. Geog. Soc.-Palomar Obs. Sky Survey, 1953-56; lectr. Griffith Obs., 1953-60; faculty U. Cal., Los Angeles, 1956—, prof., since 1967—; ques Max-Planck-Institut für Physik und Astrophysik, Munich, Germany, 1965-66; guest investigator Mt. Wilson and Palomar Obs., 1958—. Cons. Space Tech. Labs., 1958-59, Jet Propulsion Lab., 1962-66, Douglas Aircraft, 1964-66; vis. prof. Am. Astron. Soc., 1962—. Mem. Internat. Astron. Union, Am. Astron. Soc. (com. chmn.), Astron. Soc. Pacific (dir.), A.A.A.S., Sigma Xi. Author: Exploration of the Universe, 1964. Contbr. numerous articles to profl. jours. Discovered numerous star clusters, planetary nebulae, clusters of galaxies, three comets; investigated distbn. of rich clusters of galaxies, properties of rich clusters of galaxies, properties of 86 old planetary nebulae, origin of planetary nebulae. Home: 4244 Woodcliff Rd., Sherman Oaks, Cal. 91403. Office: Dept. Astronomy, U. Cal., Los Angeles 90024.*

ABELL, Irvin, Am. surgeon; b. Lebanon, Ky., Sept. 13, 1876; s. Irvin and Sarah Silesia (Rogers) A.; A.M., St. Mary's Coll., Ky., 1894; M.D., Louisville Med. Coll., 1897; postgrad. U. Berlin (Germany), 1898; Sc.D., U. of Louisville, 1937; D.Sc., Georgetown U., 1939, Manhattan Coll., 1939; LL.D., Marquette U., 1939; Sc.D., U. of Ky., 1942; LL.D., U. of Cin., 1943; m. Carrie Harting, Oct. 19, 1907; children—Irvin, William, Rogers (dec.), Spalding. Practiced in Louisville, Ky., from 1900; prof. surgery U. Louisville, 1904-47; vis. surgeon, Louisville Public Hosp., St. Joseph Infirmary; cons. surgeon, Children's Free Hosp., Kosair Hosp. for Crippled Children. Recipient Laetare medal U. Notre Dame, 1938. Trustee U. Louisville. Fellow A.C.S. (pres. 1945-46), Am. Surg. Assn.; hon. fellow Royal Surgeons, Eng., 1947; member A.M.A. (pres. 1938-39), Ky. Med. Assn. (pres. 1927), So. Med. Assn. (pres. 1933), So. Surg. Assn. (pres. 1926), Am. Urol. Assn., Am. Gastroenterological Assn. (pres. 1939-40), Assn. Mil. Surgs. (pres. 1945-46), Southeastern Surg. Congress (pres. 1937), Phi Chi, Alpha Omega Alpha, Phi Kappa Phi. Died Aug. 28, 1949.

ABELL, Richard Gurley, psychoanalyst, psychiatrist; b. Phila., Jan. 24, 1904; s. Edward W. and Bertha (Halsey) A.; A.B., Swarthmore Coll., 1926; M.A., U. Pa., 1930, Ph.D., 1934, M.D., 1947; certificate in Psychoanalysis, William Alanson White Inst. Psychoanalysis, 1957; m. Ellen Yerzley, Apr. 9, 1928 (div.); 1 dau., Margaret Lessing (Mrs. Frederick Ditmars); m. 2d, Corlis Wilber, June 15, 1962. Instr. zoology U. Pa., 1928-30, instr. anatomy Med. Sch., 1930-36, asst. prof., 1936-47; interne Bryn Mawr (Pa.) Hosp., 1947-48; resident in psychiatry, USPHS fellow N.Y. State Psychiatric Inst., 1948-49; resident in psychiatry Bellevue Hosp., N.Y.C., 1949-51; resident VA Mental Hygiene Clinic, N.Y.C., 1949-51, psychiatrist, 1951-52; individual practice psychoanalysis and psychiatry, N.Y.C., 1952—; psychoanalyst, low cost psychoanalytic service William Alanson White Inst. N.Y.C., 1955-60, mem. faculty, 1964—; dir. group psychotherapy Roosevelt Hosp., N.Y.C., 1966—. Lectr. in mental hygiene, cons. psychiatrist Barnard Coll., Columbia, 1961—; faculty New Sch. for Social Research, N.Y.C., 1964; mem. corp. Marine Biol. Lab., Woods Hole, Mass., 1945—; sec. Am. Found. for Psychoanalysis and Psychoanalysis in Groups, 1961—. John Lockwood Meml. fellow Swarthmore Coll., 1932-33. Fellow Acad. Psychoanalysis; mem. William Alanson White Psychoanalytic Soc. (sec. 1964-65), Am. Psychiat. Assn., A.M.A., N.Y. State Med. Soc., Assn. Med. Group Psychoanalysts (pres. 1960-62), Am. Group Therapy Assn., N.Y. Soc. Clin. Psychiatrists, Sigma Xi. Contbr. articles on histophysiology, personality devel., technics of psychoanalytic therapy, application of modern learning theory to the psychoanalytic process, psychoanalysis in groups to tech. jours. Conducted studies of living mammalian arterioles, capillaries and venules with high powers of microscope, during anaphylactic and traumatic shock, exptl. hypertension, studies of etiology and treatment neuroses. Home: Address: 11 E. 68th St., N.Y.C. 10021.

ABELMANN, Walter H., physician; born Frankfort/Main, Germany, May 16, 1921; s. Arthur and Else (Weill) A.; came to U.S., 1939, naturalized, 1944; A.B. magna cum laude, Harvard, 1943; M.D., U. Rochester, 1946; m. Rena J. White, June 8, 1958; children—Karen S., Nancy A., Ruth E., Arthur W., Charles H. Practice medicine, specializing in cardiology, Boston, 1955—; research fellow Thorndike Lab. Boston City Hosp., 1951-55, asst. physician, 1955—; reasearch fellow Am. Heart Assn., 1953-55, established investigator, 1955-58; asso. in medicine Med. Sch., Harvard, 1955-58, asst. prof. medicine, 1958-64, asso. clin. prof., 1964—; asst. cardiologist Childrens Hosp. Med. Center, 1961—; cons. cardiology Mt. Auburn Hosp., Cambridge, Mass., 1957—, West Roxbury VA Hosp., 1964—, Cambridge City Hosp., 1966—. Fellow A.C.P.; mem. A.A.A.S., Am. Fedn. Clin. Research, Am. Heart Assn., Am. Soc. Clin. Investigation, Assn. U. Cardiologists, New Eng. Cardiovascular Soc. (past pres.), N.Y. Acad. Sci., Phi Beta Kappa. Contbr. numerous articles to profl. jours. Studies of cardiovascular system in health and disease with emphasis on clin. and diagnostic aspects. Home: 7 Moon Hill Rd., Lexington, Mass. 02173. Office: 818 Harrison Av., Boston 02118.*

ABELOOS, Marcel Auguste, French biologist; b. Alfortville, France, Jan. 13, 1901; ed. École normale supérieure; Dr. ès scis., agrégé in natural scis.; m. Renée Parize, Sept. 30, 1925; 1 son, Daniel. Asst., Sorbonne, 1926-30, head of work Faculty Scis., Caen, France, 1930-33; prof. Faculty Scis., Rennes, France, 1933-38, Poitiers, France, 1938-48, Marseille, France, 1948—. Author: Croissance et régénération des planaires; La régénération et la morphogénèse; Biologie animale; La croissance; Les métamorphoses. Home: 4, av. J.-Siegfried, Marseille 9, France.

ABELSON, Philip Hauge, Am. phys. chemist; b. Tacoma, Apr. 27, 1913; s. Ole Andrew and Ellen (Hauge) A.; B.S. in Chemistry, Wash. State Coll., 1933, M.S. in Physics, 1935; Ph.D. in Nuclear Physics, U. Cal. at Berkeley, 1939; D.Sc. (hon.), Yale, 1946; m. Neva Martin, Dec. 30, 1936; 1 dau., Ellen Hauge. Asst. physicist Carnegie Inst. Washington, 1939-41, chmn. biophysics sect, dept. terrestial magnetism, 1946-53, dir. Geophys. Lab., 1953—; asso. physicist Naval Research Lab., Washington, 1941-42, physicist, 1942-44, sr. physicist, 1944-45, prin. physicist, 1945, civilian in charge br. at Navy Yard, Phila., 1944-45. Recipient, Navy Distinguished Civilian Service medal, 1945, ann. award phys. scis. Washington Acad. Scis., 1950, Distinguished Alumnus award Washington State U., 1962, Hillebrand award Chem. Soc. Washington, 1962. Mem. Nat. Acad. Scis., Am. Acad. Arts and Scis., Am. Phys. Soc., Am. Chem. Soc., Geol. Soc. Am., Am. Geophys. Union, Am. Philos. Soc., Brit. Biochem. Soc., Phi Beta Kappa, Sigma Xi. Editor: Researches in Geochemistry, 1959; (with J. A. Peoples, Jr.) Jour. Geophys. Research, 1959—, Science, 1962—. Co-initiator (with E. M. McMillan) of the study of transuranium elements, 1940; co-discoverer (with E. M. McMillan) Neptunium, 1940; advocate of the use of thermal diffusion of uranium hexafluoride to obtain purer uranium; instrumental in obtaining nuclear fission. Home: 32 Woodale Rd., Phila. 19118. Office: 2801 Upton St., N.W., Washington 20008.*

ABENGUEFIT (or Ibn al-Wafid), Arabian physician; b. Toledo, Spain, 997; apptd. vizier, Toledo. Author: Liber de medicamentis simplicibus (book on materia medica based on Dioscorides and Galen). Recommended dietetic measures; when drugs were needed, recommended simplest ones; devised method of investigating action of drugs; wrote on balneotherapy. Died circa 1075.

ABERCROMBIE, John, Scottish physician; b. Aberdeen, Scotland, Oct. 10, 1780; s. George Abercrombie; M.D., Edinburgh (Scotland), 1803; postgrad., London, 1803-04; M.D. (hon.) Oxford (Eng.) U., 1835. Practiced medicine, Edinburgh, from 1804; became 1 of chief cons. physicians in Scotland, 1821; apptd. physician in ordinary to King in Scotland. Fellow Coll. Physicians. Author: Pathological and Practical Researches on Diseases of the Brain and Spinal Cord, 1818; Pathological and Practical Researches on Diseases of the Stomach, The Intestinal Canal, the Liver, and the other Viscera of the Abdomen, 1828; Inquiry on the Intellectual Faculties,

1830; Philosophy of Moral Sentiments, 1833; papers on pathology. Forerunner physiol. and path. psychology. Died Edinburgh, Nov. 14, 1844.

ABERCROMBIE, Michael, English biologist; b. Ryton, Eng., Aug. 14, 1912; s. Lascelles and Catherine (Gwatkin) A.; B.A., U. Oxford, 1934, M.A., B.Sc., 1936; m. Minnnie Louie Johnson, July 19, 1939; 1 son, Nicholas. Jr. research fellow Queen's Coll., Oxford U., 1937-40, Beit meml. fellow, 1940-43; asst. lectr. U. Birmingham, 1943-45; faculty U. Coll. London, 1945—, Jodrell prof. zoology,1962—. Fellow Royal Soc., 1958; member Internat. Inst. Embryology, Soc. Exptl. Biology. Author: (with C. J. Hickman, M. L. Johnson) Dictionary of Biology, 1951. Co-editor Advances in Morphogenesis, 1961—, editor Jour. Embryology and Exptl. Morphology, 1952-61. Publs. on quantitative analysis of cellular events in injured nerves; formation of connective tissue in wound healing; cell arrangement in tissue cultures. Home: 14 Park Dr., London N.W. 11, Eng.*

ABERLE, David Friend, Am. anthropologist; b. St. Paul, Nov. 23, 1918; s. David Winfield and Lisette (Friend) A.; A.B., Harvard, 1940; postgrad. U. N.M.; Ph.D., Columbia, 1950; postgrad. Harvard; m. Eleanor Kathleen Gough, Sept. 5, 1955; 1 son, Stephen Daniel. Instr., Harvard, 1947-50, research fellow Sch. Pub. Health, 1948-50; vis. asso. prof. Page Sch., Johns Hopkins, 1950-52; asso. prof. U. Mich., 1952-58, prof., 1958-60; fellow Center for Advanced Study in Behavioral Scis., Stanford, Cal., 1955-56; Simon vis. prof., hon. research fellow dept. social anthropology Manchester (Eng.) U., 1960-61; prof., chmn. dept. anthropology Brandeis U., Waltham, Mass., 1961-63; prof. U. Ore., Eugene, 1963-67; prof. University of Brit. Columbia, 1967—. Fellow Am. Anthrop. Assn., Am. Sociol. Assn., Royal Anthrop. Inst. Gt. Britain and Ireland; mem. Am. Ethnol. Soc., Soc. for Applied Anthropology, Phi Beta Kappa. Author: The Psychological Analysis of a Hopi Life History, 1951; The Kinship System of the Kalmuk Mongols, 1953; Navaho and Ute Peyotism: a Chronological and Distributional Study (with Omer C. Stewart), 1957; Chahar and Dagor Mongol Bureaucratic Administration: 1912-1945, 1962; The Peyote Religion Among the Navaho, 1966; also articles. Office: Dept. Anthropology and Sociology, U. B.C., Vancouver 8, B.C., Can.*

ABERNETHY, John, English physician, surgeon; b. London, Apr. 3, 1764; s. John Abernethy; ed. Wolverhampton Grammar Sch.; m. Anne Threfall, 1800; became asst. surgeon, St. Bartholomew's, 1787, surgeon, 1815, founder Sch. Medicine. Fellow Royal Soc., 1796. Author: Surgical Observations on Tumours, 1804; Essay on the Constitutional Origin of Local Diseases, 1806; Surgical Works, 1811. Defended John Hunter's theory of life, extended his operation for cure of aneurysm by tying external iliac artery, 1796 (known as Abernethy's operation); studied relation of local diseases to certain digestive system disorders; opposed excessive use of trephine. Died Enfield, Eng., Apr. 28, 1831.

ABERT, John James, Am. topog. engr.; b. Shepherdstown, Va., Sept. 17, 1788; s. John and Margarita (Meng) A.; grad. U. S. Mil. Acad., 1811; m. Ellen Matlack Stretch, Jan. 5, 1812, 6 children including James William, Silvanus Thayer, William Stretch. Asst. to chief clk. War Office, 1811-14; admitted to D.C. bar, 1813; volunteer in D.C. Militia, 1814; served in battle of Bladensburg; apptd. maj. Topog. Engrs., U. S. Army, 1814, assisted in geodetic surveys Atlantic coast, 1816-18; made topog. surveys for river and harbor improvements, canals and defense; brevetted lt. col., 1824; in charge of Topog. Bur., War Dept., 1829-31, chief Topog. Bur. (after its creation as separate br. of War Dept.), 1831, 34-61; commr. for Indian affairs, 1832-34; promoted to col. when Topog. Engrs. became a staff corps of army, 1838; founder, dir. Nat. Inst. Sci., Washington, D.C.; mem. Geog. Soc. of Paris; mem. bd. visitors U. S. mil. Acad., 1842. Died Washington, Jan. 27, 1863.

ABETTI, Antonio, Italian astronomer; b. Gorizia, June 19, 1846; ed. at Pisa. m. Giovanna Colbachini; 1 son, Giorgio. Employed in obs. at Pisa; later became dir. Arcetri Obs., Florence. An authority on the minor planets. Died Feb. 20, 1928.

ABETTI, Giorgio, Italian astronomer; b. Padua, Italy, Oct. 5, 1882; s. Antonio and Giovanna (Colbachini) A.; ed. univs. Rome, Padua; doctorate in phys. scis.; m. Anna Garino, Apr. 20, 1920; 1 son, Pier Antonio. Prof., U. Florence (Italy); pres. Nat. Inst. Optics, Florence; dir. Arcetri Obs., Florence. Recipient Gold medal Ministry Pub. Instrn., Silver medal Italian Geog. Soc. Mem. Socio Nazionale, Accademia dei Lincei, Royal Astronomy Soc., Royal Soc. Edinburgh. Author: The Sun; Storia dell'Astronomia; Stars and Planets. Research in astrophysics and astron. divulgation. Home: Largo Enrico Fermi 7, Florence 50125. Office: Nat. Inst. of Optics, Florence 50125, Italy.

ABHYANKER, Shreeram Shankar, mathematician; b. Ujjain, India, July 22, 1930; s. Shankar Keshav and Uma (Tamhankar) A.; student Victoria Coll., Gwalior, India, 1947-49; B.Sc., Inst. Sci., Bombay, India, 1951; A.M., Harvard, 1952, Ph.D., 1955; m. Yvonne Margit Kraft, June 5, 1959. Came to U. S., 1951. Research instr. math. Columbia, 1955-56, vis. asst. prof., 1956-57; asst. prof. math. Cornell U., Ithaca, N.Y., 1957-59; asso. prof. math. Johns Hopkins U., 1959-63; prof. math. Purdue U., Lafayette, Ind., 1963—. Mem. Am. Math. Soc. Author: Ramification Theoretic Methods in Algebraic Geometry, 1959; Local Analytic Geometry, 1964. Contbr. research papers to math. jours. Home: 325 University St., Lafayette, Ind.*

ABICH, Otto Wilhelm Hermann von, see von Abich.

ABICH, Rudolph Adam, German chemist; maintained validity of phlogiston theory; performed expts. on calcination of zinc in closed vessels which were important in criticizing theory that phlogiston is same as inflammable air. Died Schöningen, Germany, 1809.

ABIGDOR, Abraham, see Abraham ben Meshullam ben Solomon Abigdor.

ABILDGAARD, Peter Christian, Danish physician, naturalist; b. Copenhagen, 1740; ed. vet. sch., Lyon. Founder, dir. Vet. Coll., Copenhagen; active in founding a soc. of natural history, Copenhagen; sec. Acad. Scis. Author several works on medicine, zoology, mineralogy. Gave description of Megatherium at same time as Cuvier, 1796; 1st to examine mineral cryolite (chrysolith, brought from Greenland), named it kryolith. Died Jan. 11, 1801.

ABNEY, William de Wiveleslie, English chemist; b. Derby, Eng., July 24, 1843; s. Edward Henry and Catharine (Strutt) A.; educated Woolwich Royal Military Academy (graduated 1861); married Agnes Mathilda Smith, 1864 (died 1888); 1 son, 2 daughters; m. 2d, Mary Louisa Meade, 1890; 1 dau. Lt., Royal Engrs., 1861, advanced through grades to capt., 1873, ret., 1881; became instr. chemistry and photography Sch. Mil. Engring., Chatham, Eng., 1871, head sch., 1874; became mem. sci. and art dept. (later incorporated in Bd. Edn.), South Kensington, Eng., 1877, asst. dir. sci., 1884, dir., 1893, prin. asst. sec. Bd. Edn., 1899, ret., 1903, became sci. advisor; mem. adv. council for edn. War Office. Fellow Royal Soc., 1876 (Rumford medal 1882); mem. Royal Photog. Soc. (pres. 1892-94, 96, 1903, 05), Royal Astron. Soc. (pres. 1893-95), Phys. Soc. (pres. 1895-97). Author: Photography with Emulsions; Instruction on Photography, 1870; Treatise on Photography, 1875; Thebes and Its Five Great Temples, 1876; Instantaneous Photography, 1895; Trichromatic Theory of Colour, 1914; (with C. D. Cunningham) The Pioneers of the Alps, 1888; numerous articles. Worked in photog. chemistry, color photography and printing, spectro-photography, stellar photometry, color analysis, color vision; produced rapid gelatine emulsion (made instantaneous photography possible), 1878-79; introduced gelatino-citro-chloride emulsion process (forerunner of modern printing out paper); investigated alkaline photographic image, 1877; introduced hydroquinone (new developing agt.), 1880; devised method for measuring and classifying colors; redrew 3 color sensation curves of Young and Hemholtz; made photog. plates sensitive to red and infrared, 1880; was able to photograph infrared solar spectrum as far as 11,000 angstroms. Died Folkestone, Eng., Dec. 3, 1920.

ABOOD, Leo George, Am. neurochemist; b. Erie, Pa., Jan. 15, 1922; s. George E. and Sarah (Muffet) A.; B.S., Ohio State U., 1943; Ph.D., U. Chgo., 1950; m. Lois Wuchner, Sept. 12, 1947; children—George T., Mary Ellen. Instr. physiology U. Chgo., 1950-52; faculty U. Ill. Coll. of Medicine, Chgo., 1952-65, prof. neurophysiology and biochemistry, 1963-65; prof. Center for Brain Research, also dept. biochemistry U. Rochester (N.Y.), 1965—. Mem. program project com. NIH, 1963—; bd. sci. counsellors Nat. Inst. Neurol. Diseases and Blindness, 1965—; sci. adv. bd. Am. Schizophrenic Found. Mem. Am. Physiol. Soc., Am. Chem. Soc., Am. Coll. Neuropsychopharmacology, Soc. for Exptl. Biology and Medicine. Mem. editorial bds. Jour. Medicinal Chemistry, 1966—, Jour. Neuropsychiatry, 1957—. Research, numerous publs. in chemistry of nervous system, chemistry and action of hallucinogenic drugs and chemistry of nerve conduction. Home: 45 Crandon Way, Rochester, N.Y. 14618.*

ABRAGAM, Anatole, physicist; b. Moscow, USSR, Dec. 15, 1914; s. Simon and Anna (Maimin) A.; student Sorbonne, Paris, France, 1933-37; D.Phil., Oxford (Eng.) U., 1950; m. Suzanne Lequesme, Oct. 21, 1944. Brit. Council fellow Oxford U., 1948-50; research fellow Harvard, 1952-53; with French Atomic Energy, Paris, 1950—, head nuclear and solid state dept., 1959-65, dir. physics, 1965—; prof. Collège de France, Paris, 1960—. Decorated Legion d'Honneur, officer Ordre Nat. du Mérite. Mem. French Phys. Soc. (pres. elect 1967—). Author: The Principles of Nuclear Magnetism, 1961; also numerous articles. Research on structure of hyperfine structure in solid state, theory of perturbed angular correlations, theory of spin temperature, dynamic nuclear polarization; invention of polarized targets for nuclear reactions. Home: 33 Croulebarbe Paris. Office: Collège de France, Paris, France.*

ABRAHAM, Ambrus, Hungarian neurohistologist; b. Tusnád, Csik, Hungary, Nov. 20, 1893; s. István and Teréz (Hóródy) A.; Ph.D., U. Budapest, 1922. With Pázmány Péter U. Budapest, 1917-19, asst. to zool. and comparative Anatom. Inst., 1919-34; 1st asst., leader zool. dept. Tchrs. Tng. Coll., Szeged, Hungary, 1934-40; prof. gen. zoology and biology, dir. gen. zool. and biol. inst. Horthy Miklós U. Szeged, 1940—. Recipient Kossuth prize Govt. Hungary, 1952. Mem. Hungarian Biol. Soc. (pres. 1958-61), Hungarian Acad. Scis., Royal Soc. Medicine, Acad. Zoology in India, European Soc. Comparative Endocrinology, Commn. Comp. Neuroanatomy. Author (with others) Anatómia-Élettan/Anatomy-Physiology, 1958; Die mikroskopische Innervation des Herzens und der Blutgefässe von Vertebraten, 1964; Összehasonlító Allatszervezettan/Comparative Study of the Animal Organism, 1964; The Microscopic Innervation of the Heart and Blood Vessels in Vertebrates, Including Man, 1967. Research publs. on peripheral and central nervous systems in animals and man; cardiovascular system, sensory organs, interoreceptors, interneuronal synapses, neuroendocrine system. Home: Somogyi N. 3, Szeged, Hungary. Office: Táncsics M. 2, Szeged, Hungary.*

ABRAHAM, Edward Penley, English biochemist; b. Southampton, Eng., June 10, 1913; s. Albert Penley and Mary (Hearn) A.; M.A., D.Phil., Queen's Coll., Oxford, 1941; grad. with 1st class honors, Sch. Natural Sci., Oxford, 1936; m. Asbjorg Harung, Nov. 1, 1939; 1 son, Michael. Rockefeller Found. Travelling fellow, Stockholm, Sweden, 1938-39, U. Cal. at Berkeley, 1947; sr. research officer Sir William Dunn Sch. Pathology, Oxford, 1948—, fellow Lincoln Coll., Oxford, 1948—, prof. chem. pathology, Oxford, 1964—. Ciba lectr. Rutgers U., 1957; vis. lectr. U. Sydney (Australia), 1960; Rennebohm lectr. U. Wis., 1967. Fellow Royal Soc., 1958; mem. Biochem. Soc. (past mem. editorial bd.), Chem. Soc., Soc. for Gen. Microbiology. Author: (with others) Antibiotics, 1949; Biochemistry of Some Peptide and Steroid Antibiotics, 1957; also numerous articles; contbg. author: General Pathology, 1960. Research (with H. W. Florey, E. Chain, others) leading to isolation of penicillin, determination of its structure and its introduction to medicine, 1939-45; discovered (with G. G. F. Newton) cephalosporin C which led to introduction of cephalosporins to medicine; 1950-60; studied isolation, structure and biol. properties of other antibiotics and various peptides with biol. activity. Home: Badger's Wood, Bedwells Heath, Boars Hill, Oxford, Eng.*

ABRAHAM, Henri, physicist, inventor; b. Paris, France, 1868; student Ecole Normale Supérieure, 1868; D.Sc., 1892; prof. Sorbonne, U. Paris, 1912-43. Invented microgalvanometer or electrostatic voltmeter (Abraham's voltmeter); introduced deForest triode into France; invented (with E. Bloch) multivibrator, 1918. Died 1943.

ABRAHAM, Max, physicist; b. Danzig, Poland, Mar. 26, 1875; became asst. Max Planck Inst., 1897; prof., Göttingen, Germany, 1900-09, Milan, Italy, 1910-14, Munich, Germany, also Stuttgart, Germany, after 1919. Author: Prinzipien der Dynamik des Elektrons, 1903; Lichtdruck auf einen bewegten Spiegel und das Gesetz der schwarzen Stahlung, 1904; Theorie der Elektriztät, 1904-05. Refined Maxwell's theory of electricity; helped develop wireless telegraphy in theory and practice, circa 1900; stated calculations of J. J. Thompson in his model of atom, 1903; introduced notion of electromagnetic moment; showed that mass of electron approached infinity when its speed approached that of light. Died Munich, Sept. 16, 1922.

ABRAHAM, Phineas Simon, dermatologist; b. Falmouth, Jamaica; s. Phineas Abraham; B.Sc., Univ. Coll., London, Eng.; M.A., Trinity Coll, Dublin, Ireland; M.D.; ed. also Royal Coll. Sci., Dublin, St. Bartholomew's Hosp., Ecole de Medecine, Paris, France, Clausthal, Germany; m. Ellen Chard; 1 dau. Curator, Mus., Royal Coll. Surgeons, Dublin, 1879, later mem. ct. examiners; lectr. physiology, histology Westminster Hosp. Med. Sch., from 1885; asso. Royal Coll. Sci., Ireland; cons. dermatologist W. London Hosp.; lectr. dermatology W. London Post Grad. Coll. Med. sec. Nat. Leprosy Fund, 1889; represented Eng., Internat. Lepra Conf., Berlin, Germany, 1897. Fellow Royal Coll. Surgeons Ireland, Royal Soc. Medicine; mem. W. London Medico-Chirurg. Soc. (pres. 1910-11), Brit. Med. Assn. (pres. dermatol. sect., London meeting 1910), Irish Med. Schs. and Grads. Assn. (pres.), New London Dermatol. Soc. (pres.), Royal Acad. Medicine Ireland (founder), Dermatol. Soc. Gt. Britain and Ireland (founder). Author: Jour of the Leprosy Investigation Com.; contbr. articles to sci. publs. Died Feb. 23, 1921.

ABRAHAM BAR HIYYA, mathematician, astronomer; b. Barcelona, Spain, 1070. Author: Chochmat ha-Schiur, 1116; Cheschbon ha-Ibbur, 1123; Cheschbon Mahalach ha-Kochabim. Pioneered spread of Arabic sci. (begun by Spanish Jews) to Western world; his math. and philos. writings played important role in devel. of Western sci. Died Provence, France, 1136.

ABRAHAM BEN MESHULLAM BEN SOLOMON ABIGDOR, physician, sci. writer; b. Aries, France, 1350 or 51; student Montpellier, France, circa 1367-79. Practiced medicine, Arles, until 1399 or later. Translator from Latin to Hebrew, Mabo Bi-Melekeh (Introductorium in practicam pro provectis in theorica, collection of recipes derived from Ibn Sina's Qanum by Bernard Alberti of Montpellier, 1358), Almazori (abridged transl. Gerard of Cremona's transl. 9th book of Kitab al-Mansuri of Al-Razi, with Almansoris liber nonus cum expositione Geraldi de Solo).

ABRAHAM IBN DAVID (Abraham ibn Daud Halevi) Spanish biologist, astronomer; b. Toledo, Spain, circa 1110. Attempted to reconcile Judaism with Aristotelianism. Author: Book of Tradition; Book of the Sublime or Highest Doctrine; also astron. work, 1180. First strict Aristotelian among Jews; believed no bounds should be set to sci. Died 1175.

ABRAHAM OF ARAGON, Jewish physician, oculist; b. Aaragon, Spain; flourished 1253; practiced medicine, Languedoc, France; said to have contbd. to flow of Arabic knowledge into Europe.

ABRAHAM OF TOLEDO (or Abraham Alfaquin, Alfaker), Spanish physician. Physician to King Alfonso (VIII) el Sabio, employed by king to translate sci. works from Arabic to Spanish. Jewish religion. Translator treatise of Ibn Al-Haitham on configuration of universe, treatise of al-Zarqali on constrn. and use (chiefly astrological) of his astrolabe, 70th chpt. of Quran, Surat al-mu'àrij (Chpt. of the Ascents). Died Toledo, Spain, 1231 or 1239.

ABRAHAMS, Sidney Cyril, crystallographer; b. London, Eng., May 28, 1924; s. A. Harry and Freda Abrahams; B.Sc., U. Glasgow, 1946, Ph.D., 1949, D.Sc., 1957; m. Rhoda Banks, May 1, 1950; children—David Mark, Peter Brian, Jennifer Anne. Came to U. S., 1948, naturalized, 1963. Research fellow U. Minn., 1949-50; mem. staff Mass. Inst. Tech., 1950-54; fellow U. Glasgow, Scotland, 1954-57; mem. tech. staff Bell Tel. Labs., Murray Hill, N.J., 1957—; guest scientist Brookhaven Nat. Lab., 1957—. Mem. U. S. Nat. Com. for crystallography, 1966—. Mem. Am. Crystallographic Assn. (editor Transactions 1965, v.p. 1967—), Am. Phys. Soc., Internat. Union Crystallography (commn. crystallographic apparatus 1963—), Chem. Soc. London, Sigma Xi. Mem. editorial bd. Rev. Sci. Instruments, 1963-66. Studies, numerous publs. on magnetic and crystal structure of inorganic materials using x-ray and neutron diffraction; relation of phys. properties to structure; applications of automation to this field. Home: 30 Fox Run. Office: Bell Tel. Labs., Murray Hill, N.J. 07971.*

ABRAHAMSON, Edwin William, Am. chemist; b. Medina, N.Y., Jan. 6, 1922; s. Edwin W. and Doris (Reeves) A.; B.A., U. Buffalo, 1944; M.A., Columbia U., 1948; Ph.D., Syracuse U., 1952; m. Corinne Manghi, June 19, 1948; children—Carla Maria, Kirsten. Research chemist Manhattan Project, 1944-46; instr. chemistry Colgate U., 1948-50; research asso. Mass. Inst. Tech., 1952-53; asst. prof. Syracuse U. Coll. Forestry, 1955-59; asso. prof. chemistry Case Inst. Tech., Cleve., 1959-66; prof. chemistry Case Western Res. U., Cleve., 1967—. Mem. Am. Chem. Soc., Am. Inst. Chemists, Am. Phys. Soc., Am. Inst. Biol. Chemists, Sigma Xi. Research and publs. on chemistry of vision, spectroscopic basis of photochemistry, photosynthesis. Home: 3051 Scarborough Rd., Cleveland Heights, O. 44118. Office: Case Western Res. U., Cleve. 44106.*

ABRAMOVITCH, Rudolph Abraham, chemist; b. Alexandria, Egypt, July 19, 1930; s. Lazare W. and Elise (Litinsky) A.; B.Sc. with 1st class honors, Alexandria U., 1950; B.Sc. with 1st class honors, U. London (Eng.), 1950, Ph.D., 1953, D.Sc., 1964; m. Liliane Esther Guetta, July 8, 1952; children—Michael, Danny. Sr. tchr. sci. and math. Menasce High Sch., Alexandria, 1946-51, prin., 1948-50; Med. Research council postdoctoral fellow U. Exeter (Devon, Eng.), 1954; research scientist Weizmann Inst., Israel, 1955; Imperial Chems. Industry Research fellow, U. London, King's Coll., 1956-57; faculty U. Sask., Saskatoon, Can., 1957-68, prof. chemistry, 1964-68; prof. chemistry U. Ala., University, 1968—. Cons. Warner Lambert Research Inst., Morris Plains, N.J., 1966—. Fellow Chem. Inst. Can., Chem. Soc. London; mem. Am. Chem. Soc. (sr.). Research, numerous publs. on mechanism and orientation in nucleophilic and free radical aromatic substitution, mechanism and stereochemistry of nucleophilic addition reactions, formation and properties of carbenes and nitrenes, synthesis and properties of heterocyclic nitrogen compounds. Home: U. Ala., University, Ala. 35486.*

ABRAMS, Albert, Am. physician; b. San Francisco, Dec. 8, 1863; s. Marcus and Rachel (Leavey) A.; M.D., U. Heidelberg, 1882; A.M., Portland U., 1892 (LL.D.); post-grad. London, Berlin, Vienna, Paris; m. Jeanne Roth, Nov. 25, 1897; 2d, Blanche Schwabacher, Sept. 28, 1915. Prof. pathology Cooper Med. Coll., 1893-98; pres. Emanuel Polyclinic, from 1904. Author: Synopsis of Morbid Renal Secretions, 1892; Manual of Clinical Diagnosis, 1894; Consumption—Its Causes and Prevention, 1895; Scattered Leaves of a Physician's Diary, 1900, Diseases of Heart, 1901; Nervous Breakdown, 1901; Hygiene, in A System of Physiologic Therapeutics, 1901; Diseases of the Lung, 1905; Self-Poisoning; Diagnostic-Therapeutics, 1909; Spinal Therapeutics, 1909; Spondylotherapy, in Reference Handbook Medical Sciences, 1917. Discoverer Abram's Reflex of the lungs, Abram's heart reflex and Electronic Reactions of Abrams. Died San Francisco, Jan. 13, 1924.

ABRAMS, Herbert LeRoy, Am. radiologist; b. N.Y.C., Aug. 16, 1920; s. Morris and Freda (Sugarman) A.; B.A., Cornell U., 1941; M.D., State U. Medicine, N.Y., 1946; m. Marilyn Spitz, Mar. 23, 1943; children—Nancy (Mrs. Harlan Jacobson), John. Practice medicine, specializing in radiology, Stanford, Cal., 1959-67; faculty Sch. Medicine Stanford, 1951-67, dir. div. diagnostic roentgenology, 1961-67, prof. radiology, 1962-67; Cook prof., chmn. dept. radiology, Harvard, 1967—. Radiation study sect. NIH, 1962-66; cons. Survey on Renovascular Hypertension, Nat. Heart Inst.; cons. to hosps., profl. socs.; lectr., vis. faculty numerous univs., profl. socs. Recipient Phi Epsilon Pi award for scholastic achievement; Nat. Cancer Inst. fellow, 1950; Spl. Research fellow Nat. Heart Inst., 1960; Malcolm Rogers Meml. lectr. Wis. Heart Assn., 1963; David M. Gould Meml. lectr. Johns Hopkins, 1964. Diplomate Am. Bd. Radiology. Mem. Assn. U. Radiologists, Am. Coll. Radiology, A.A.A.S., Am. Assn. Cancer Research Western Soc. Clin. Research, A.M.A. Author: (with H. S. Kaplan) Angiocardiographic Interpretation in Congenital Heart Disease, 1956; (with others) Congenital Heart Disease, 1965. Editor: Angiography, 1961; Investigative Radiology. Research, publs. on cardiovascular disease and radiology. Home: 294 Buckminster Rd., Brookline, Mass. 02146. Office: Harvard Med. Sch., Boston.*

ABRAMS, Richard, Am. biochemist; b. Chgo., Sept. 19, 1917; s. David and Matilda (Hornstein) A.; B.S., U. Chgo., 1938, Ph.D., 1941; m. Thelma E. Peterson, Oct. 31, 1947; children—Peter Arnold, Erika Karen, Kersti Elida, Lauren Jan. Staff, U. Chgo., 1941-46, group leader Manhattan Project, 1942-46, asst. prof. Inst. Radiobiology and Biophysics, 1947-51; fellow Donner Found., Karolinska Inst., Stockholm, Sweden, 1946-47; asso. dir. Research Inst., Montefiore Hosp., Pitts., 1951-58, dir., 1958-65; faculty dept. biochemistry and nutrition U. Pitts. Grad. Sch. Pub. Health, 1959—, prof., head dept., 1965—. Cons. neurochemistry VA labs., Pitts., 1958-64; mem. com. on leukemia and cancer Health Research and Service Found., 1964—; study sect. mem. NIH, 1966—. Fellow A.A.A.S.; mem. Am. Soc. Biol. Chemists, Am. Chem. Soc., Biochem. Soc. (London, Eng.), N.Y. Acad. Scis. Research, numerous publs. on elucidation of biosynthetic pathways leading to nucleic acids in animal cells, control mechanisms for genetic repliciation in mamalian cells. Home: 301 Hillcrest Av., Pitts. 15237.*

ABRAMSON, Arthur Simon, physician; b. Montreal, Que., Can., June 4, 1912; s. Jacob Joseph and Dora (Rosenthal) A.; B.Sc., McGill U., 1933, M.D., C.M.; 1937; m. Ruth Mary Rumsey, Aug. 1, 1956; 1 son, Daniel Rumsey. Came to U. S., 1937, naturalized, 1942. Asst. chief phys. medicine and rehab. service VA Hosp., Bronx, N.Y., 1948-50, chief, 1950-55, cons., 1955—, area med. cons. VA, 1955—, cons. VA hosps., 1956—; asst. clin. prof. N.Y. U. Coll. Medicine, 1950-52; spl. lectr. Columbia, 1950-59; clin. prof. phys. medicine and rehab. N.Y. Med. Coll., 1952-55; vis. physician Met. Hosp., Welfare Island, N.Y., 1952-55, Bird S. Coler Meml. Hosp. and Home, Welfare Island, 1952-55; attending physician Flower and Fifth Av. hosps., N.Y.C., 1952-55; prof., chmn. dept. rehab. medicine Yeshiva U. Albert Einstein Coll. Medicine, N.Y.C., 1955—; dir., vis. physician phys. medicine and rehab. Bronx Municipal Hosp. Center, 1955—; med. dir., div. phys. therapy, vis. prof. Ithaca (N.Y.) Coll., 1959—; cons. Beth Abraham Hosp., Kessler Inst. for Rehab., Jewish Chronic Disease Hosp., Bklyn., Misericordia Hosp., N.Y.C. Mem. President's Com. on Employment Physically Handicapped, 1956—, Gov. Rockefeller's Council on Rehab., 1959—; mem. nat. adv. council Vocational Rehab. Adminstrn., 1962; vis. lectr. Loyola U. Stritch Med. Sch., 1962, others; mem. Commn. on Edn. in Phys. Medicine and Rehab., 1964. Recipient President's Trophy for Handicapped Man of Year, 1956, award in medicine N.Y. Philanthropic League, 1956, award Assn. Rehab. Coordinators and Dirs., 1958, others. Office Vocational Rehab. Spl. research fellow Polio Inst., Copenhagen, Denmark, 1961-62. Diplomate Am. Bd. Phys. Medicine and Rehab. (vice chmn. 1962-68). Fellow A.C.P., N.Y. Acad. Medicine; mem. (chmn. com. on prostheses and orthoses 1958-59, dir., Gold Key award 1966), Am. Acad. Phys. Medicine and Rehab. N.Y. Soc. Phys. Medicine, and Rehab. (pres. 1955), N.Y. Acad. Scis., Israel Assn. Phys. Therapists (hon.), Internat. Soc. for Rehab. Disabled, Arthritis and Rheumatism Found. N.Y., A.A.A.S., Am. Rheumatism Assn., Amvets (nat. surgeon gen. 1955-56), Paralyzed Vets. Am. (med. cons.). Research, publs. on spastic urinary bladder, metabolism of paralyses. Home: Hawthorne Way, Hartsdale, N.Y., 10530. Office: Albert Einstein Coll. Medicine, 1300 Morris Park Av., N.Y.C. 10461.*

ABRAMSON, David Irvin, Am. physician; b. N.Y., Oct. 14, 1905; s. Aaron and Anna (Oschrin) A.; student Coll. City N.Y., 1922-23, Columbia, 1923-24; M.D., L.I. Coll. Medicine, 1929; m. Louise Felson, Aug. 17, 1940; children—Julie Syril, Marian Beth. Intern Bushwick Hosp., Bklyn., 1929-30; pvt. practice peripheral vascular disease, Chgo., 1946—; instr. physiology L.I. Coll. Medicine, 1930-36; dir. cardiovascular research May Inst. Med. Research, Cin., 1938-42; asst. prof. medicine U. Ill., 1946-54, asso. prof., 1954-55, prof. dept. medicine, head dept. phys. medicine and rehab., 1955—; cons. Hines and West Side VA Hosp.; attending physician Michael Reese, Mt. Sinai hosps., Chgo., 1954—. Diplomate Am. Bd. Internal Medicine, 1946. Fellow A.C.P., Am. Heart Assn.; mem. Am. Physiol. Soc., Soc. Exptl. Biology and Medicine, Central Soc. Clin. Research, Am. Soc. Clin. Investigation, Chgo. Medical Society, Chicago Society for Internal Medicine, American Congress Physical Medicine and Rehabilitation, Sigma Xi. Author: Vascular Responses in Extremities of Main in Health and Disease, 1944; Diagnosis and Treatment of Peripheral Vascular Disorders, 1956; Circulation in the Extremities, 1967. Editor: Blood Vessels and Lymphatics, 1962. Contbr. articles med. jours. Research on cardiovascular disease, human blood flow in various normal and abnormal conditions. Home: 916 N. Oak Park Av., Oak Park, Ill. Office: 8 S. Michigan Av., Chgo. 60603.*

ABRAMSON, Harold Alexander, Am. psychiatrist; b. N.Y.C., Nov. 27, 1899; s. Samuel and Rose (Richard) A.; A.B., Columbia, 1920, M.D., 1923; m. Barbara H. Smith, June 26, 1933, (div. 1952); children—Alexandra M. (Mrs. Atilla Orhun), Harold Alexander II, Barbara Howland, Howland Wilson; m. 2d, Virginia T. Wildman, 1955. Asst. inorganic chemistry Columbia, 1918-20; asst. prof. physiology, 1935-59; Libman travelling fellow U. Coll., London, 1925-26; practice medicine, specializing in psychiatry, N.Y.C., 1925—; NRC fellow Kaiser-Wilhelm Inst. for Phys. Chemistry and Electrochemistry, Berlin, Germany, 1926-28; instr. medicine Johns Hopkins, 1928-29; instr. biochem. scis. Harvard, 1929-31; asso. in bacteriology and immunology Cornell U., 1934-35; chief allergy clinic Mt. Sinai Hosp., N.Y.C., 1948-58; chief psychobiology sect. Biol. Lab., L.I., N.Y., 1952-61; dir. psychiat. research South Oaks Psychiat. Hosp., Amityville, N.Y., 1956—. Cons. research psychiatry State Hosp., Central Islip, N.Y.; cons. Community Hosp., Glen Cove, N.Y., Huntington and Seaview Hosp., N.Y.; v.p. Asthmatic Children's Found. of N.Y.; sci., ednl. council Allergy Found. Am., 1958—. Fellow Am. Psychiat. Assn.; mem. Am. Psychosomatic Soc., Soc. Exptl. Biology and Medicine, N.Y. Acad. Medicine, Am. Coll. Allergists, A.A.A.S., N.Y. County Med. Soc., N.Y. Acad. Sci., Am. Acad. Allergists, Am. Coll. Chest Physicians, Am. Physiol. Soc., Soc. Biol. Chemists Author: Electrokinetic Phenomena, 1934; Electrophoresis of Proteins, 1942; The Patient Speaks, 1956. Editor: Dimensional Analysis for Students of Medicine, 1950; Problems of Consciousness, 5 vols., 1951-53; Somatic and Psychiatric Treatment of Asthma, 1952; Neuropharmacology, 5 vols., 1955-59; The Use of LSD in Psychotherapy, 1960; The use of LSD in Psychotherapy and Alcoholism, 1967; Editor-in-chief Asthma Research, 1963—. Research, publs. measured electric charge of blood cells; initiated and directed penicillin aerosol therapy for lungs; pioneer electrophoretic fractionation of ragweed and other pollen extracts; investigated LSD psychosis and its use in therapy; blocking of LSD by brain extract and other compounds; psychoanalytic theories of eczema and asthma. Home: Cold Spring Harbor, N.Y. Office: 133 E. 58th St., N.Y.C. 10022.*

ABRAMSON, Hyman Norman, Am. mech. engr.; b. San Antonio, Mar. 4, 1926; s. Nathan and Pearl (Westerman) A.; B.S., Stanford, 1950, M.S., 1951; Ph.D., U. Tex., 1956; m. Idelle R. Ringel, Apr. 20, 1947; children—Phillip D., Mark D. Research asst. Stanford, 1948-51; project analytical engr. Chance Vought Aircraft Co., Dallas, 1951-52; asso. prof. aero. engring. Tex. A. and M. Coll., 1952-56; dir. dept. mech. scis. S. W. Research Inst., San Antonio, 1956—. Cons. coms. Nat. Acad. Scis., NRC, govt. agys. Author: Dynamics of Airplanes, 1958. Asso. editor Applied Mechanics Rev., 1954—. Research, publs. in aeros. and astronautics especially rocket dynamics and motions of liquid propellants; also, dynamics of ships. Home: 1511 Spanish Oak Dr., San Antonio 78213. Office: 8500 Culebra Rd., San Antonio 78206.*

ABRIA, Jérémie-Joseph-Benoit, French physicist; b. Limoges, France, Mar. 19, 1811; ed. L'Ecole Normale, 1831-34, D.Sc., 1837. Tchr., Coll. of Limoges, 1834-36; prof. physics Faculty of Scis., Bordeaux, dean faculty, 1845-86, hon. dean, 1886. Decorated chevalier, officer Legion of Honor. Mem. Société philomathique, Acad. Scis. Bordeaux, Regional Sci. Assn., French Acad. Scis., Soc. Phys. and Natural Scis. (founder). Research in electricity and optics; proved intensity of induced current is proportional to that of current of inductor. Died Bordeaux, France, Apr. 4, 1892.

ABRIKOSOV (or ABRIKOSOFF), Aleksei Ivanovich, Russian pathologist; b. Russia, 1875; prof. 1st Med. Inst., Moscow; named to chair Russian Soc. Pathologists; Recipient Stalin prize, 1942; named hero of socialist labor. Mem. USSR Acad. Med. Scis. (dir. Inst. Normal and Path. Morphology), Moscow Soc. Anatomists and Pathologists (hon. pres.), Acad. Scis. USSR, Ukrainian Soc. Pathologists (hon.), Internat. Assn. Clin. Pathologists, London. Author: Pathological-Anatomical Changes in Early Stages of Consumption, 1904; Pathological Anatomy of Sympathetic Ganglia, 1921; New Myogenic Tumour, 1925; Myoblastenmyori, 1931; Allergic Changes of Blood Vessels, Fundamentals of General Pathological Anatomy, 1933. Described Abrikosov's tumor of striated muscle made of cell groups resembling primitive myoblasts.

ABSE, David Wilfred, clinical psychiatrist, psychoanalyst; b. Cardiff, Wales, U.K., Mar. 15, 1915; s. Rudolph and Kate (Shepherd) A.; B.Sc., U. Wales, 1935, M.D., 1948; M.B., B.Ch., Welsh Nat. Sch.

Medicine, 1938; Diploma in Psychol. Medicine, U. London, 1940; m. Elizabeth Smith, July 26, 1960; 2 children—Edward and Nathan Abse. Came to the United States, in 1951, naturalized in 1956. Clin. psychiatrist Westminster Hosp., London, Eng., 1942-43; dep. med. supt. Monmouthshire Mental Hosp., Abergavenny, Wales, 1947-48; chief asst. psychiatrist Med. Sch., Charing Cross Hosp., London, 1949-51; Lectr. in psychology City Coll., London, 1950-51; clin. dir. Dorothea Dix State Hosp., Raleigh, N.C., 1952-53; clin. asst. prof. dept. psychiatry U. N.C., Chapel Hill, N.C., 1952-53, dir. inpatient service, asso. prof., 1953-57, dir. postgrad. edn. in psychiatry, 1957-59, prof., 1958-62; prof. dept. neurology and psychiatry, U. Va., Charlottesville, 1962—. Fellow Brit. Psychol. Soc., Royal Soc. Medicine, A.A.A.S.; mem., N.Y. Acad. Sci. Author: Diagnosis of Hysteria, 1950; Marriage Counseling in Medical Practice, 1964; Hysteria and Related Mental Disorders, 1966. Research on psychodynamics of shock therapy, effects of psychotropic drugs, psychoanalysis of hysteria. Home: 1852 Winston Rd., Charlottesville, Va. 22901.*

ABSOLON, Karel Bedrich, physician; b. Brno, Czechoslovakia, Mar. 21, 1926; s. Karel and Valerie (Hauska-Minkusiewicz) A.; B.S., Gymnasium, Brno, 1944; B.M., Masaryk U., Brno, 1948; M.D., Yale, 1952; M.S., U. Minn., 1963, Ph.D., 1963; m. Mary Joan Bendix, Oct. 9, 1954; children—Mary Therese, Charles Frederick, John Bendix, Peter Henry, Martha Jane. Naturalized, 1954. Practice medicine, specializing in surgery, Mpls., 1952-65, Amarillo, Tex., 1965—; research fellow USPHS, U. Minn., Med. Sch., 1958-63, asst. prof. surgery, 1963-66; cardiovascular surgeon St. Anthony Hosp., Amarillo, 1966—. Recipient Mosby Thesis award Yale, 1952. Mem. A.A.A.S., A.M.A., History of Medicine Soc., Soc. Biology and Medicine. Research, numerous publs. on first successful human liver transplantation; immunosuppressive effects of lymphoresis; hemicorporectomy feasibility of coronary endarterectomy; history of medicine, surg. schs. Home: 1523 Crockett St. Office: 2714 W. 10th St., Amarillo, Tex. 79102.*

ABT, Helmut A., astronomer; b. Helmstedt, Germany, May 26, 1925; s. Karl Max and Margaret (Siemon) A.; B.S. in Math., Northwestern U., 1946, M.S. in Physics, 1948; Ph.D. in Astronomy, Cal. Inst. Tech., 1952. Jr. research astronomer Lick Obs., U. Cal., 1952-53; asst. prof. Yerkes Obs., U. Chgo., 1953-59; asso. astronomer Kitt Peak Nat. Obs., Tucson, 1959-62, astronomer, 1962—. Vis. prof. NSF-Am. Astron. Soc., 1959-62. Mem. Am., Royal astron. socs., Astron. Soc. Pacific (dir. 1964—, pres. 1966—). Research on spectroscopic binary stars, stellar rotation, pulsating stars, magnetic stars, shock waves in stars. Home: 3819 E. 3d St., Tucson 85716. Office: Box 4130, 950 N. Cherry Av., Tucson 85717.*

ABT, Isaac Arthur, Am. pediatrician; b. Wilmington, Ill., Dec. 18, 1867; s. Levi and Henrietta (Hart) A.; completed preliminary med. course Johns Hopkins, 1889; M.D., Chgo. Med. Coll., 1891; interne Michael Reese Hosp., Chgo., 1891-93; postgrad. in Vienna and Berlin, 1893-94; Sc.D., Northwestern U., 1931; m. Lena Rosenberg, Aug. 20, 1897; children—Arthur F., Lawrence E. Prof. diseases of children Northwestern U. Woman's Med. Sch., 1897-1901; asso. prof. diseases of children Rush Med. Coll., 1902-08; prof. diseases of children Northwestern U. Med. Sch., 1909-42, prof. emeritus, from 1942; cons. physician diseases of children Children's Meml. Hosp., Sarah Morris Children's Hosp., St. Luke's Hosp.; attending physician diseases of children Passavant Hosp. Hon mem. Deutsche Gesellschaft für Kinderheilkunde; mem. Am. Pediatric Soc. (pres. 1926-27), Am. Acad. Pediatrics (pres. 1930-31), A.M.A., (Distinguished Service award 1948), Central States Pediatric Soc., Am. Assn. Teachers of Diseases of Children (pres. 1922), Children's Hosp. Assn. Am., Washington Med. and Surg. Soc. (hon.), Alpha Omega Alpha (hon. Minn. Chpt.), Chevalier of Legion of Honor (France). Author: The Baby's Food, 1917, The Baby Doctor, 1944; also monographs. Editor vol. on Pediatrics in the Practical Medicine Series; also editor, A System of Pediatrics. Died Nov. 22, 1955.

ABU AL-FARAG (Bar-Hebraeus), physician, astronomer; b. Malatiya, Syria, 1225 or 26; s. Aaron (Ahron); studied Melitene, Antioch, Tripolis. Became monk at age 17, Jacobite bishop 3 years later, Bishop of Lakabhin in 1247, of Aleppo, 1253, and made Maphrian, 1264; a Christian encyclopedist who wrote in Syrian on wide range of subjects, including history, sci. and medicine. Author: Mahtebhanut Zabhne: A History of Syria (written to teach countrymen of their culture and past, divided into two parts: Syria Chronicle, Eccles. Chronicle); The Ascension of the Spirit (essay on astronomy); also an abridgement of Muhammad al-Ghafiqi's Simple Drugs. Died 1286.

ABU AL-HASAN AHMED, Arab physician; flourished 2d half 10th century; s. Muhammad al-Tabari; physician to Buwayid rukn al-Dawla; author book on Hippocratic treatments (10 vols., contains 1st definite identification of acarus scabiei).

ABU AL-SALT (abu al-Salt Umaiya ibn 'abd al-Aziz ibn abi al-Salt al-Andalust), Muslim physician, mathematician, astronomer; b. Denia, Spain, 1067; author: Kitab al-adwiya al-mufrada (treatise on simple drugs); Tagwim al-dhihn (Rectification of the Understanding, treatise on logic); Risala fi-l-'amal bi-l-istarlab (treatise on astrolabe). Died Mahdiya, Tunis, 1134.

ABU BARAKAT (or ABU'L BARAKAT), philosopher; b. Balad, nr. Mosul, 1077; of a Jewish family, converted to Islam at an advanced age; pupil of Abu 'l-Hasan Sa 'id; practiced medicine, Baghdad; rival of Christian physician Ibn-al-Tilmidh; friend and tchr. of Ishak, son of Abraham ben Ezra. Author: Kitab al-Mu 'Tabar (attempt at a systematic philosophy, covers logic, naturalia, metaphysics); commentary on Ecclesiastes; also credited with Ris Ala fi Sabab Zuhur al-Kawakib Layl wa-Khafa 'iha Nahar. Accepted a Platonic physics, psychology from Shifa, a priori knowledge; believed in infinite space, a priori apperception of time, close connection of time with being, corpuscular nature of earth, multiple faculties of soul; denied Aristotelian distinction of intellect and soul; identified prime matter and the body; wrote on movement of projectiles, expressed 1st recorded instance of idea of acceleration from a constant force; influenced al-Razi; originated a crisis in Muslim philos. speculation. Died Baghdad, 1164.

ABU BISR, physician; b. Syria, flourished middle 10th century; Nestorian physician, contemporary of al-Farabi; contbd. to devel. Aristotelian studies (wrote commentaries in Arabic and Syriac on all then-known works of Aristotle); mentor of Abu Zakariya Yahya.

ABU GAFAR ABDALIAH AL-MANSUR, geometer, astronomer; b. Bagdad, Iraq; flourished 872-73; worked (with al-Hassan) in geometry and astronomy, built obs. in Baghdad, wrote numerous works, patronized by 3 sons of Musa ben Sakir. Author: Book by the Three Brothers about Geometry; works on math and mech. problems, including constrn. of automatic mus. instrument.

ABU KAMIL SHUDJA, mathematician; b. 850. Author: al-Tara'if (on integral solutions of indeterminate equations); Algebra (work on which his fame rests, extant in Latin and Hebrew); books on astrological, math. and related subjects, including On Augmenting and Diminishing, and On the Two Error; On the Pentagon and Decagon (treatise on algebraic methods applied to geometry in manner superior to that of al-Kharizmi, used later by Leonard of Pisa in his Practica geometriae). Died 930.

ABULADZE, Kalenik Sardinovich, Russian physiologist; b. 1895; D. Med. Sci., 1950 Sr. asso., head Pavlov physiology dept. Lab. Pathology of Higher Nervous Activity, Leningrad Inst. Exptl. Medicine, USSR Acad. Med. Sci., 1932—. Mem. USSR Acad. Med. Sci. (corr.). Author: (monographs) Study of the Reflex Activity of the Salivary and Lachrymal Glands, 1953; Functions of Twin Organs, 1961. Address: Inst. of Exptl. Medicine, USSR Acad. Med. Sci., Solyanka 14, Leningrad, USSR.

ABUL WAFA (Abul al-Wafa, al Buzjani) (also Abul Wefa, or Albuzjani, or al-Hasib), mathematician, astronomer; b. Buzshan, Chorassan, Persia, 940; worked in obs. built by Saraf-ed-Daula, son of Emir Adud-ed-Daula, Bagdad. Author: Commentaries on Euclid; Diophantos al-Khwarizmi (all lost); also astron. tables, book of applied geometry (both adaptations from original); Zyal-wahid (Practical Arithematic); Kitab al handasa (Book of Applied Geometry); Theories of the Moon. Translated and popularized Almagest (Ptolemy). Important in devel. of trigonometry; 1st to show generality of sine theorem relative to spherical triangles; computed sine tables; introduced secant and cosecant, understood relationship between 6 trigonometric lines used to define them; solved biquadratic equations geometrically, some geometric problems using one opening of compass; constructed square equivalent to other squares; proposed and studied problems of cube and 4th power roots; in his work on lunar theory, used trigonometric lines, tangents and cotangents for the 1st time; believed to have discovered 2d large deviation in moon's orbit, 980 (discovery originally thought to have been made by Tycho Brahe); built 1st wall quadrant for observing stars. Died Bagdad, 998.

ABU MA'SHAR, see Albumasar.

ABU UTMAN, physician, mathematician; b. Damascus, Syria, 1st half of 10th century. Named dir. all. hosps. in Bagdad by Vizir Ali ben Isa, 914. Translator 10th book of Elements (Euclid); also works by Aristotle and Galen.

ABU YAHYA AL-BATRIQ, Arab translator; one of 1st translators employed by al-Mansur; translated works of Galen, Hippocrates, Ptolemy (Opus quadripartitum). Died circa 800.

ACCASCINA, Filippo, Italian chemist; b. Palermo, Italy, May 23, 1919; s. Giuseppe and Margherith (Morreale) A.; doctorate, U. Palermo, 1940; m. Giovanna Carnesi, Feb. 4, 1945; children—Giuseppe, Giorgio. Faculty, U. Rome, 1944-51, asso. prof., 1950-60; prof. phys. chemistry Palermo U., 1961—; research staff electrochemistry Brown U., Providence, 1954-55. Mem. Am. Chem. Soc., Chem. Soc., Società Chimica Italiana. Author: (with R. M. Fuoss) Conducibilita elettrolitica, 1959; Electrolytic Conductance, 1960. Research on theory of elec. conductance of solutions, interaction in solution between ions and between ions solvent molecules, absorption of ultrasonics in liquid mixtures and kinetic of hydrolysis. Home: 10 Giardino, Palermo, Italy.*

ACCETTA, Giulio, mathematician; b. Francavilla, Calabre; nominated corr. Maraldi, 1751; prof. math. U. Turin (Italy); Augustinian monk; mem. French Acad. Scis. Author: Gli elementi di Euclide . . . , 1753. Died Turin, Sept. 25, 1752.

ACCORNERO, Ferdinando, Italian psychiatrist; b. Genoa, Italy, Mar. 26, 1910; s. d'Anselmo and Irma (Battaglia) A.; M.D., U. Rome, m. Maria Figari, Jan. 4, 1936; 1 child, Neri. Prof., Clinic Mental and Nervous Illness; prof. pediatric neuropsychiatry, staff aide Neuropsychiat. Clinic; under-dir., prof. classes for specialists in neuropsychiatry U. Rome; dir. Castello della Quiete sanitorium. Mem. Italian Soc. Physchiatry, Italian Soc. Neurology, Metapsichica. Author: L'istopatologia del sistema nervoso; L'Organizzazione del lovore intellettuale; also others. Home: via Anapo 7. Office: via del Babuine 29, Rome, Italy.

ACCUM, Friedrich Christian, chemist; b. Bückeburg, Germany, Mar. 29, 1769; s. Christian and Judith (La Motte) A.; m. Mary Ann Simpson, May 10, 1798. Went to London, Eng., 1793; lectr. on chemistry and physics Surrey Inst., 1803; engr. London Chartered Gaslight & Coke Co., from 1810; librarian Royal Instn.; prof. Gewerbe-Institut, Berlin, Germany, 1822-38. Author: System of Theoretical and Practical Chemistry, 1803; Essay on Analysis of Minerals, 1804; Manual of Analytical Mineralogy, 1808; Crystallography, 1813; Practical Treatise on Gas Light, 1815; Chemical Amusement, 1817; Description of the Process of Manufacturing Coal Gas, 1819; Chalybeate Spring at Thetford, 1819; Adulterations of Food and Culinary Poisons, 1820; Brewing, 1820; Art of Making Wine, 1820; Culinary Chemistry, 1821; Art of Making Wholesome Bread, 1821; Chemische und physikalische Beobachtungen der Baumaterialen, 2 vols., 1826. Studied coal gas; helped found gas industry; introduced (with Ackermann) gas lighting to English towns; worked on chemistry of foods; found that swimming bladder of fresh carp contains ordinary air. Died Berlin, June 28, 1838.

ACH, Narziss Kasper, German psychologist; b. Ermershausen, Germany, Oct. 29, 1871; s. Michael Joseph and Margarete (Burger) A.; M.D., Ph.D., 1902; research with Müller at Göttingen (Germany), 1900, with Külpe at Würzburg (Germany), 1904; m. Maria Mez, Aug. 17, 1911; 4 sons, 2 daus. Apptd. pvt. docent, Göttingen, 1902, prof., 1922; apptd. pvt. docent, Marburg, Germany, 1904, Berlin, 1906; became prof., Königsberg, Germany, 1907. Author: Über die Willenstätigkeit und das Denken, 1905; Über den Willensakt und das Temperament, 1910; Über die Begriffsbildung, 1921; Analyse des Willens, 1935; numerous articles. Mem. of Külpe's Wurzburg sch.; formulated many of Watt's results; worked on problems of action and thought; invented term systematic exptl. introspection; probably invented term determining tendency; invented term Bewusstheit (awareness), to refer to vague or intangible conscious content, neither image nor sensation; worked on imageless thought; emphasized that some thinking is not sensory; defended method of self-observation; created Ach-Ducker law of spl. determination; invented chronotyper (tool for exptl. psychology). Died Munich, Germany, July 25, 1946.

ACHARD, Emile-Charles, French physician; b. Paris, July 24, 1860; prof. Faculty Medicine, U. Paris; gen. sec. Acad. Medicine; mem. French Acad. Scis. Studied pathogenic bacteria; discovered (with Bensaude) paratyphoid bacilli, 1896; pioneer in serodiagnosis in use of agglutination phenomenon to diagnose paratyphoid fever; conducted inconclusive experiments on Tb of joints following injury, 1889; described (with Joseph Thiers) Achard-Thiers syndrome of hirsutism in female asso. with diabetes, 1921. Died Versailles, France, Aug. 8, 1944.

ACHARD, Franz Karl, German chemist, physicist; b. Berlin, Germany, Apr. 28, 1753; s. Guilleaume and Marie E. H. (Rouppert) A.; m. Caroline Repper, 1776; 1 son; m. 2d, Henriette Koeppen; 1 son, 2 daus. Became aide to Andreas Sigismund Marggraf (1st successful producer beet sugar) at Acad. Scis., 1776; apptd. dir. physics class Acad. Scis., Berlin, 1782; founded 1st beet sugar factory in Germany (became sch. for promoting and teaching beet sugar prodn. 1812), Kunern estate, Silesia, 1801. Author: Vorlesung über Experimentalphysik, 4 vols., 1791-92; Die europische Zuckerfabrikat, aus Runkelrüben, 3 vols., 1809. Improved German tobaccos (on commn. Frederick the Great); 1st to prepare platinum crucible, 1784; 1st to describe the phenomenon that liquids begin boiling only at higher temperature than normal boiling points, 1785; one of 1st to suspect that thermal conductivity is proportional to elec. conductivity; pioneered use of Marggraf's discovery of sugar in beetroots. Died Kunern, Silesia, Apr. 20, 1821.

ACHARIUS, Erik, Swedish physician, botanist; b. Gefle, Sweden, Oct. 10, 1757; student Linneaus, U. Uppsala (Sweden), 1773; M.D., Lund (Sweden), 1782; practiced medicine, Landskrona, later at Wadstena, Sweden, 1789-1819; apptd. prof. botany Wadstena Acad., 1801; illus. plates in Trans. Acad. Scis., Stockholm, Sweden; mem. Acad. Botany, Stockholm. Author: Lichenographiae Suecicae prodromus, 1798; Methodus qua ommnes detectos Lichenes secundum organa carpomorpha ad genera . . . , 1803; Lichenographia universalis, 1804; Synopsis methodica Lichenum, 1814. Studied cryptogams; did fundamental work on lichens (divided them into 40 genera, 800 species). Died Wadstena, Aug. 14, 1819.

ACHENWALL, Gottfried, German statistician; b. Elbing, Germany, Oct. 20, 1719; s. Gottfried and Elisabeth (Zachert) A.; student U. Jena, from 1738, U. Halle, from 1740 (both Germany); Magister, U. Leipzig (Germany), 1746; m. Sofie Eleonore Walther, 1752; m. 2d, Wilhelmine Luise Moser, 1754; m. 3d, Maria Jäger, 1763. Lectr. law, polit. history, statistics U. Marburg, Germany, from 1746; asso. prof. Faculty Philosophy, also prof. Faculty Law, U. Göttingen (Germany), 1748-72. Author: Staatsver Fassung der heutigen vornehmsten europaischen Reiche und Volkerim Grundrisse (used statis. method to describe constns. and econ. conditions of leading countries), 1752. Introduced word Statistik; considered father of statistics, his work led to popularized use of statistics in Germany. Died Göttingen, May 1, 1772.

ACHESON, Edward Goodrich, Am. inventor; b. Washington, Pa., Mar. 9, 1856; s. William and Sarah D. (Ruplé) A.; acad. edn. Bellefonte, Pa.; Sc.D., U. Pitts., 1909; m. Margaret Maher, 1884; children —Veronica Belle, Edward Goodrich, Raymond Maher, Sarah Ruth, George Wilson, John Huyler, Margaret Irene, Jean Ellen, Howard Archibald. Draftsman in the experimental department of Thomas A. Edison Company, 1880-81; installed electric lights in Europe for two years; employed by Westinghouse in Pittsburgh, 1886-89. Recipient Rumford medal Am. Acad. Arts and Scis., 1908; Perkin Research medal, 1910; grand prix, Paris Expn., 1900, St. Louis Expn., 1904. Officer Royal Order of Polar Star (Sweden). Hon. mem. Russian Imperial Tech. Soc., Swedish Tech. Soc.; mem. Electrochem Soc. (pres.; established Acheson biennial prize 1928, received Acheson medal 1929), Am. Inst. Chem. Engrs. (v.p.). Inventor of carborundum, siloxicon, Egyptianized clay, Aquadag, Oildag, and a method of making graphite. Died N.Y.C., July 6, 1931.

ACHESON, Louis Kruzan, Jr., Am. physicist; b. Brazil, Ind., Apr. 2, 1926; s. Louis Kruzan and Ruth (Morrison) A.; B.S. in Elec. Engring., Case Inst. Tech., 1946; Ph.D. in Physics, Mass. Inst. Tech., 1950; m. Hyla Armstrong Cook, July 12, 1958; children—Mary Ruth, William Louis. Mem. tech. staff Hughes Aircraft Co., Culver City, Cal., 1950-54, sr. staff engr., 1954-58, sr. scientist, 1959——. Mem. tech. staff Inst. for Def. Analyses, Washington, 1958-59 (on leave from Hughes). Mem. Am. Inst. Aeros. and Astronautics, Am. Phys. Soc., Am. Geophys. Union, A.A.A.S., N.Y. Acad. Sci., Sigma Xi, Tau Beta Pi, Eta Kappa Nu, Theta Tau. Research on electron scattering by atomic nuclei; contbns. to devel. airborne interceptor fire control systems for def. against bombers; devel. systems for def. against ballistic missiles; devel. mil. satellite systems. Home: 17721 Marcello Pl., Encino, Cal. 91316. Office: care Hughes Aircraft, Space System Div., El Segundo, Cal.*

ACHESON, Roy Malcolm, epidemiologist; b. Belfast, Ireland, Aug. 18, 1921; s. Malcolm King and Dorothy (Rennoldson) A.; B.A., U. Dublin, 1945, M.A., 1949, Sc.D., 1962; M.A., U. Oxford, 1951, B.M.B.Ch., 1951, D.M., 1954; M.A. (hon.), Yale, 1964; m. Fiona Marigo O'Brien, Mar. 16, 1950; children Malcolm O'Brien, Vincent Rennoldson, Marigo Fiona. Rockefeller Travelling fellow Western Res. U., Harvard, 1955-56; lectr. in social medicine U. Dublin, 1955-59; sr. Lectr., then reader in social and preventive medicine U. London, 1959-62; asso. prof. epidemiology and medicine Yale, 1962-64, prof., 1964——. Cons. on med. edn., East Pakistan, 1964; cons. Atomic Bomb Casualty Commn., 1965, Pan Am. Health Orgn., Latin Am., 1965, Cal. State med. edn., 1966-68. Mem. expert com. on methods in epidemiology of chronic disease WHO, 1966, expert panel on health statistics, 1966——. Ernest Hart Meml. fellow Brit. Med. Assn., 1954; Radcliffe travelling fellow U. Coll., Oxford, 1955-57; fellow Trinity Coll., Dublin, 1957-59, Jonathan Edwards Coll., Yale, 1966. Mem. Internat. Epidemiol. Assn., Am. Pub. Health Assn., Am. Heart Assn., Am. Assn. Phys. Anthropology. Editor: Comparability in International Epidemiology, 1965. Research, numerous publs. on phys. growth of children, Tb, epidemiology of common cardiovascular diseases, cancer and arthritis, med. care and edn. Home: 211 Highland St., New Haven 06511.*

ACHILLINI, Alessandro, Italian anatomist, physician; b. Bologna, Italy, 1463; student in Paris; became tchr., Bologna, 1485; prof. medicine and philosophy, Padua, Italy, 1506-09; Author: Corporis humani anatomia, 1516; Anatomia annotationes, 1520. One of 1st in modern times to dissect human body; pioneer in intestinal research; credited with discovery of malleus and incus of middle ear, also trochlear nerves; believed to be 1st to recognize function of olfactory nerves. Died Bologna, 1512.

ACHUCARRO, Nicolás, neurohistologist, neuropathologist; b. Bilbao, Spain, 1881; ed. Madrid, Paris, Munich, Florence; M.D., Madrid, 1906. Worked in lab. Govt. Hosp. for Insane (now St. Elizabeth's Hosp.), Washington, 1908. Devised silver-tannin stain (Achúcarro's stain) for impregnating connective tissue, especially neuroglia, 1911; his process of silver impregnation led Hortega to discovery of microglia; delineated nature of various cell types in human cerebral cortex, 1910; did important work on the pineal body, rabies, gen. paresis (1911), brain tumors, including ganglioneuroma of central nervous system (1913), Alzheimer's Disease (1914). Died 1918.

ACKER, Charles Ernest, Am. inventor, mfr.; b. Bourbon, Ind., Mar. 19, 1868; s. William James and Mercia (Grant) A.; Ph.B., Cornell, 1888; m. Alice Reynolds Beal, Apr. 26, 1892. Elec. engr., Chgo., 1888-93; Recipient Elliott Cresson gold medal Franklin Inst.; other medals. Originator Acker process of mfg. caustic soda by electrolysis of molten salt, and built works at Niagara Falls; originator chem. and electro chem. processes; granted many patents. Died Oct. 18, 1920.

ACKER, Ludwig, German food chemist; b. Mannheim, Germany, Sept. 17, 1913; s. Ludwig and Susanna (Betz) A.; Dipl.-Chemiker, U. Heidelberg (Germany), 1939, Dr.rer.nat., 1943; Food chemist, U. Frankfort/Main (Germany), 1947; m. Josefine Bechtum, May 19, 1945; children—Barbara, Rolf-Dieter. Food chemist Lebensmitteluntersuchungsamt, Frankfort/Main, 1948-58; privatdozent U. Frankfort/Main, 1953-59; prof. U. Giessen (Germany), 1959; prof. U. Münster (Germany), 1959——, dir. Institut für Lebensmittelchemie, 1965——. Mem. Bundesgesundheitsrat, 1967. Mem. Gesellschaft-Deutscher Chemiker, Deutsche Gesellschaft für Fettwissenschaft, Deutsche Gesellschaft für Ernährung. Research, publs. on relations between relative humidity and enzymic reactions in foods during storage, composition of lipids in cereal starches and wheat flours, analytical problems in field of foods; discovered phosopholipid splitting enzymes in cereals and studied their role in different foods, especially during brewing. Home: 3 Lühnstiege, Münster, Westfalen, Deutsche Bundesrepublik.*

ACKER, Thomas Stephen, Am. biologist; b. Cleve., July 21, 1929; s. Frederick N. and Margaret (Zurlinden) A.; A.B., Loyola U., Chgo., 1952, Licentiate of Philosophy, 1954; Ph.D. in Biology, Stanford, 1961; S.T.B., West Baden (Ind.) U., 1964. Ordained priest Jesuit Order, Roman Catholic Ch., 1963; research asso. NIH, West Baden U., 1963-65; asst. prof. biology John Carroll U., Cleve., 1965-67; asst. prof. biology University of Detroit, 1967——. Mem. N.Y. Acad. Sci. (life), Entomol. Soc. Am., Pan-Pacific Entomology Club, Am. Inst. Biol. Sci., Am. Soc. Zoologists, Sigma Xi. Author pamphlet: Three Dimensions to Birth Control, 1964; also articles. Developed an evolutionary and morphological terminology for insect genitalia, growth and devel. insect internal and external sex organs, physiol. regulation and normalization human female cycles. Address: Dept. Biology, U. Detroit, Detroit 48221.

ACKERL, Franz, Austrian mathematician; b. Vienna, Austria, May 19, 1901; s. Franz and Anna (Mattausch) A.; ed. sch. mathematics U. Vienna; D. Engring.; m. Maria Fontana, June 26, 1926; children —Maria, Annelies. With U. Vienna, from 1923, charge of classes, 1928; apptd. prof. geodesy, 1934, prof. geodesy and photogrammetry, 1946, chairman, 1952. Mem. pres. Austrian Soc. Photogrammetry. Author: Geodäsie und Photogrammetrie, 2 vols., 1951-56; also numerous articles. Home: Vienna 18, Gersthoferstrasse 28. Office: Vienna 19, Peter Jordanstrasse 82, Austria.

ACKERMAN, Edward Augustus, Am. geographer; b. Post Falls, Ida., Dec. 5, 1911; s. August and Augusta (Anderson) A.; B.A., Harvard, 1934, M.A., 1936, Ph.D., 1939; m. Adrienne Aymard Desjardins, Sept. 24, 1949; children—Helen Augusta, Francis Edward, Julia Hardy, Justin Kemp, Elizabeth Chantal. Prof. geography U. Chgo., 1948-55; dir. water resources program Resources For The Future, Inc., 1954-58; dep. exec. officer Carnegie Instn., Washington, 1958-60, exec. officer, 1960——. Chmn. Nat. Acad. Scis.—NRC ad hoc com. geography, 1963-65; dir. Planning Found. Am., 1965——; mem. U. S. Nat. Com. Internat. Biol. Program, 1965; mem. Dept. Interior Potomac River Planning Task Force, 1965; cons. govt. agys. Mem. Assn. Am. Geographers, Phi Beta Kappa, Sigma Xi. Author: New England's Fishing Industry, 1941; A report on Japanese Natural Resources, 1948; (with others) Ten Rivers in America's Future, 1950; (with J. Russell Whitaker) American Resources: Their Management and Conservation, 1951; Japan's Natural Resources and Their Relation to Japan's Economic Future, 1953; (with George O. G. Lof) Technology in American Water Development, 1959; The Impact of New Techniques on Integrated Multiple-Purpose Water Development, 1960. Office: 1530 P St. N.W., Washington 20005.*

ACKERMAN, Eugene, Am. biophysicist; b. Bklyn., N.Y., July 8, 1920; s. Saul Benton and Dorothy (Salwen) A.; B.A., Swarthmore Coll., 1941; Sc.M., Brown U., 1943; Ph.D., U. Wis., 1949; postdoctoral student U. Pa., 1949-51, fellow 1957-58; m. Dorothy Hopkirk, June 5, 1943; children—Francis H., Emmanuel T., Amy R. Instr., Brown U., 1943; from asst. prof. to prof. biophysics Pa. State U., 1951-60; mem. faculty Mayo Grad. Sch. Medicine, U. Minn., 1960——, prof. biophysics, 1965——; staff cons. biophysics Mayo Found. and Mayo Clinic, 1960-67; Hill Family Found. prof. biomedical computing, professor biometry U. Minn., Mpls., 1967——; dir. computer facility Mayo Found., 1964-65. Cons. bioacoustics USAF, 1957-62; mem. epidemiology and biometry tng. com. NIH, 1963-67, spl. study sect. ultrasonic applications, 1965——. Research grantee Am. Cancer Soc., 1953-56, NSF, 1958-64, NIH, 1954——. Mem. A.A.-A.S., Biophys. Soc., Am. Physiol. Soc., N.Y. Acad. Scis., Minn. Acad. Sci., Assn. Computing Machinery, Soc. Social Responsibility Sci., I.E.E.E., Phi Beta Kappa, Sigma Xi. Author: Biophysical Science, 1962; also articles, tech. reports, chpts. in books. Research in enzyme thermodynamics, altitude physiology, bioacoustics, biomath., computer applications, compartmental analysis. Home: 140 14th St. N.E., Rochester 55901. Office: Mayo Clinic, Rochester, Minn. 55901.

ACKERMAN, Gustave Adolph, Am. histologist, anatomist; b. Columbus, O., Nov. 23, 1927; s. Gustave Adolph and Ethel (Huffman) A.; A.B., Ohio State U., 1948, M.S., 1949, Ph.D., 1954, M.D., 1954; m. Ellen Bomar Harris, Sept. 1, 1956; children—Carlyle Adolph, Sue Ellen. Faculty, Ohio State U., 1957——, prof. anatomy, 1964——. Recipient Borden award, 1954; Lederle Med. Faculty award, 1961-63. Mem. Am. Assn. Anatomists, Histochem. Soc., Am. Soc. Hematology, Am. Soc. Cell Biology, Electron Microscope Soc. Am., N.Y. Acad. Scis., Sigma Xi. Contbr. chpts. to books, also numerous articles to profl. jours. Research, publs. on structure, chemistry various blood cells and blood forming organs, immunological role of thymus and bursa of Fabricius, morphological, histochemical studies leukemic blood cells. Home: 629 Sanbridge Circle E., Worthington, O. 43085. Office: 1645 Neil Av., Columbus, O. 43210.*

ACKERMAN, Lauren Vedder, Am. physician, educator; b. Auburn, N.Y., Mar. 12, 1905; s. John Walter and Bertha (Vedder) A.; A.B., Hamilton Coll., 1927, D.Sc., 1962; M.D., U. Rochester, 1932; m. Elizabeth Fitts, May 7, 1938; children—John, Gretchen, Jennifer, Alison. Pathologist, med. dir. Ellis Fischel State Cancer Hosp., Columbia, Mo., 1940-48; faculty Washington U. Sch. Medicine, St. Louis, 1948——, prof. surg. pathology and pathology, 1952——. Cons. Armed Forces Inst. Pathology, 1948——; mem. subcom. on oncology NRC, 1955——; mem. com. on tumor nomenclature Unio International Contre de Cancer, 1955——. Mem. Am. Assn. Pathologists and Bacteriologists, A.M.A., Internat. Acad. Pathology, Alpha Omega Alpha. Author: Cancer Diagnosis Treatment and Prognosis, 1947; Surgical Pathology, 1953. Research, numerous publs. on cancer. Home: 1353 Green Tree Lane, St. Louis 63122. Office: 600 S. Kingshighway, St. Louis 63110.*

ACKERMAN, Norman Bernard, Am. surgeon; b. N.Y.C., Nov. 27, 1930; s. Louis and Anne (Mukasie) A.; A.B., Harvard, 1952; M.D., U. Pa., 1956; Ph.D. in Surgery U. Minn., 1964; m. Anne Linda Gross, June 14, 1953; children—Sara Jean, Beth Leslie, Amy Susan, Jane Ellen. Med. fellow surgery U. Minn., 1957-64; asso. prof. surgery Boston U. Sch. Medicine, 1967——; asst. vis. surgeon Univ. Hosp., Boston, 1965——; cons. surgery Boston and Providence VA Hosps., 1965——; faculty research asso. Am. Cancer Soc., 1966——. Recipient Mary Ellis Bell award, 1956. Nat. Cancer Inst. Research fellow, 1960-63. Fellow A.C.S.; mem. Soc. U. Surgeons, Am. Fedn. Clin. Research, Soc. Nuclear Medicine. Contbr. numerous articles to med. jours. Devised new diagnostic tests for cancer with radioisotopes. Home: 54 Old Colony Rd., Chestnut Hill, Mass. 02167. Office: 750 Harrison Av., Boston 02118.*

ACKERMANN, Conrad Theodor, German physician; b. Wismar, Germany, Sept. 17, 1825; ed. U. Griefswald (Germany), U. Wurzburg (Germany), U. Prague (Czechoslovakia); M.D., U. Rostock (Germany), 1852. Apptd. privatdocent U. Roscock, 1856, asso. prof., 1859, prof., 1865; founder Inst. Path. anatomy and Exptl. Pathology, 1865. Author: Beobachtungen über einige physiologischen Emitka, 1856; Anweisung zur Erkenntniss und Behandlung der wichtigsten äusseren Verletzungen und inneren Krankheiten auf Seeschiffen, 1869; Über die Wirkunger der Digitalis, 1873; Über die Schädeleformität bei der Encephalocele cogenita, 1882. Described angle at skull base characteristic of kyphosis, 1882 (known as Ackermann's angle). Died 1896.

ACKERMANN, Dankwart, German chemist; b. Halle-S, Germany, Nov. 11, 1878; s. Theodor and Mathilde (Fritzsche) A.; ed. Rostock, Munich, Fribourg univs.; M.D.; doctor of natural sciences, Honoris causa; m. Marianne von Frey; 3 children. Agrégé, 1908-1953; prof. physiol. chemistry U. Wurtzbourg, from 1922. Mem. Soc. Physical Medicine Wurtzbourg (hon.), Soc. Physiol. Chemistry, Leopoldina, Bayer-Verdienstorder acads. Discovered 14 biogenous amines, including histamine. Address: Seinsheimstrasse 12, Würzburg, West Germany.

ACKERMANN, Johann Christian Gottlieb, German physician; b. Zeulenroda, Germany, Feb. 1756; s. Johann Samuel and Eva Rosine (Steinmuller) A.; student Univ. Jena (Germany); M.D., Univ. Gottingen, 1775; m. Eleanore Friedrike von Wolfersdorff; 9 children; Privatedocent Univ. Halle (Germany); 1775-78, practiced medicine, Zeulenroda, 1778-86; apptd. prof. chemistry U. Altdorf (Switzerland), 1786, prof. pathology, also dir. hosp., 1794. Author: Serenus

Samonicus, 1786; Sextus Placidus Papyriensis, 1788; Lucius Apuleius, 1788; Regimen sanitatus Salerni, 1790; Institutiones historiae medicinae, 1792; Institutiones therapiae generalis, 2 vols., 1794-95; Handbuch der Kriegsarzneikunde . . . , 2 vols., 1794-95; Hand und Hülfsbuch fur Feldärzte, 1797; Pathologische-praktische Abhandlung über die Blähungen, 1800. Pioneer in army hygiene. Died Altdorf, Mar. 9, 1801.

ACKERMANN, Weston Wilbur, Am. virologist; b. New Orleans, June 15, 1925; s. R. Herbert and Isabelle (Trepagnier) A.; B.S., Tulane U., 1944, M.S., 1945; Ph.D., U. Tex., 1947. Postdoctoral fellow U. Tex., U. Wis. Faculty, U. Mich., Ann Arbor, 1949—, prof. epidemiology, 1957—. Recipient Eli Lilly award in microbiology, 1957. Research on biochem. and dynamic aspects of viral infections at cellular level. Home: 662 Hampstead Lane, Ann Arbor, Mich.

ACKERMANN, William Carl, Am. hydrologist; b. Sheboygan, Wis., Oct. 7, 1913; s. William H. and Frances E. (Shirmer) A.; student Lawrence Coll., 1930-32; B.S. in Civil Engring. with honors, U. Wis., 1935; m. Margaret A. Kreysell, May 6, 1942; children—William Charles, Nancy Ann (Mrs. W. David Price), Arthur John. Field engr. Kimberly-Clark Paper Corp., Neenah, Wis., 1935; head hydrology sect. TVA, Knoxville, 1935-54; head watershed hydrology sect. U. S. Dept. Agr., Washington, 1954-56; chief Ill. Water Survey, Champaign, 1956—; tech. asst. Exec. Office Pres., Washington, 1963-64; prof. civil engring. U. Ill., Urbana, 1958—. Recipient Distinguished Service award U. Wis., 1964. Mem. Am. Geophys. Union (pres. 1966-68), Am. Soc. C.E., Am. Water Works Assn., Sigma Xi, Chi Epsilon, Tau Beta Pi. Research numerous publs. on hydrologic cycle on exptl. watersheds, meteorology, precipitation, stream flow, ground water, evaporation, sedimentation. Home: 701 Hamilton Dr., Champaign, Ill. 61820. Office: Ill. Water Survey, Box 232, Urbana, Ill. 61801.*

ACKERSON, Clifton Walter, Am. biochemist; b. Elbow Lake, Minn., Dec. 29, 1896; s. Jonathan Francis and Christine (Olson) A.; B.S., U. Minn., 1921, M.S., 1922; Ph.D., U. Neb., 1926; m. Harriet Wooster Smith, June 27, 1925; children—Kathleen Wilmot (Mrs. Frank Osler Foote), Elizabeth Christine (Mrs. Dwight Harold Houseman), Harriet Jean (Mrs. James Vaclav Potmesil). Asst. in agrl. biochemistry U. Minn., 1920-22; faculty agrl. chemistry U. Neb., Lincoln, 1922—, chmn. dept., 1946-53, prof. biochemistry, 1953-65, prof. emeritus, 1965—; cons. animal nutrition; spl. writer for nutrition jours., feed mags. Mem. Am. Chem. Soc., Poultry Sci. Assn., Sigma Xi, Phi Lambda Upsilon, Gamma Sigma Delta, Alpha Chi Sigma, Tau Kappa Epsilon. Research, numerous publs. on the vitamin, mineral protein and amino acid requirements of poultry; pioneer on vitamin D requirement of chicks as opposed to that of rats; nitrate poisoning problems in forages and feeds. Home: 3707 Holdrege St., Lincoln, Neb. 68503.*

ACKERT, James Edward, Am. zoologist, parasitologist; b. Woosung, Ill., Aug. 31, 1879; s. Abram and Eva (Nowell) A.; grad. No. Ill. U., 1903; A.B., U. Ill., 1909, A.M., 1911, Ph.D., 1913; m. Florence Mae Tanner, Aug. 15, 1914; 1 dau., Jane (Mrs. John Trafton Fleetwood) A.; Prin. high sch., Algonquin, Ill., 1903-07; fellow zoology U. Ill., 1911-13; faculty Kan. State U., 1913—, prof., 1918-50, emeritus prof., 1950—, head dept. zoology, 1944-45, parasitologist expt. sta., 1913-50, head grad. study, 1923-31, dean grad. sch., 1931-45, dean emeritus, 1945—; mem. expdn. for hookworm control Internat. Health Bd., Rockefeller Found., Island of Trinidad, 1921. Vis. lectr. to Brazil and Chile, 1953. Recipient Centennial Distinguished Service award Kan. State U., 1963, Agr. award, Gamma Sigma Delta, 1964, Distinguished award No. Ill. U., 1965. Fellow A.A.A.S.; mem. Soc. Naturalists, Am. Soc. Zoologists, Am. Micros. Soc. (past pres., hon. mem.), Am. Soc. Parasitologists (emeritus, past pres.), Soc. Tropical Medicine and Hygiene, Kan. Acad. Sci. (past pres., life mem.), U. S. Livestock San. Assn., Sigma Xi, Phi Kappa Phi, Gamma Alpha (past nat. sec.), Alpha Zeta, Delta Phi. Author: The Graduate Student of Kansas State University, 1868-1945; Laboratory Manual of Parasitology, 1920; also numerous articles, book revs. Raised 1st healthy chickens free of rickets, 1923; 1st exptl. evidence of effect of parasitic worm on its host; studied effect of vitamin deficiency on resistance to ascarid worms (leading to incorporation of adequate amounts of vitamin A and B in comml. feeds and diets); found effect of inheritance, proteins and minerals on ascarid resistance. Home: 1923 Leavenworth St., Manhattan, Kan. 66502.*

ACKMAN, Robert George, Canadian chemist; b. Dorchester, N.B., Can., Sept. 27, 1927; s. Robert Kenneth and Charlotte (Ward) A.; B.A., U. Toronto (Ont., Can.), 1946-50; M.Sc., Dalhousie U., (Halifax, N.S., Can.), 1952; D.I.C., Ph.D., Imperial Coll., U. London, 1956; m. Catherine Isobel McKinnon, Sept. 11, 1957; children—Elizabeth Kathleen, Margaret Louise. Staff, Halifax Lab., Fisheries Research Bd. Can., 1950-53, 56—, asst. dir., 1966—; research asso. chemistry dept. Dalhousie U., 1965—. Vice chmn. Canadian Com. on Fats and Oils, 1965—. Mem. Chem. Inst. Can. Contbr. numerous articles to tech. jours. Developed gas-liquid chromatographic techniques applicable to analysis fatty acids and esters; research on fatty acid composition, marine oils and lipids. Home: 60 Sinclair St., Dartmouth, N.S. Office: P.O. Box 429, Halifax, N.S., Can.*

ACLAND, Henry Wentworth, English physician; b. Killerton, Eng.; Aug. 23, 1815; s. Thomas Dyke and Lydia Elizabeth (Hoare) A.; B.A., Oxford (Eng.) U., 1840, M.A., 1842, M.B., 1846, M.D., 1848; M.D. (hon.), LL.D., Dublin (Ireland) U.; D.C.L. (hon.) U. Durham (Eng.), m. Sarah Cotton, July 14, 1846; 7 sons, 1 dau. Elected fellow All Soul's Coll., Oxford, 1840; Lee's reader anatomy Christ Ch., Oxford, 1845-58; apptd. Radcliffe librarian, Aldrichian prof. clin. medicine Oxford U., 1851, Regius prof. medicine, 1858-94, master Ewelme Hosp. Created baronet, 1890. Mem. Gen Med. Council, 1858-87, pres., 1874-87. Fellow Royal Soc., 1847, Royal Coll. Phys. (Harveian orator 1865), Brit. Med. Assn. (pres. 1868); mem. various fgn. med. and philos. socs. Author: Feigned Insanity, 1844; Remarks on the Extension of Education at the University of Oxford, 1848; Synopsis of the Physiological Series in the Christ Church Museum, arranged for the use of students after the plan of the Hunterian Collection, 1854; Memoir of the Cholera at Oxford in the year 1854, with considerations suggested by the Epidemic, Maps and Plans, 1856; Notes on Drainage, with especial reference to the Sewers and Swamps of the Upper Thames, 1857; The Oxford Museum, 1859; The Harveian Oration, 1865; Health in the Village, 1884; Village Health and Village Life, 1884. Tried to introduce biology and chemistry into ordinary curriculum at Oxford; studied relations between practice of medicine and state of the natural sciences. Died Oxford, Oct. 16, 1900.

ACOSTA, Cristobal (Christopher), botanist, surgeon; b. Mozambique, 1515; lived in Portuguese colonies; researched and collected med. plants and drugs, Goa, 1658; later practiced surgery, Burgos, Spain. Author: Tratado de las drogas y medicinas de las Indias orientales, con sus plantas debuscadas al vivo (on plants and herbs of E. Indies; 1st European work on plants of Goa). Died Burgos, 1580.

ACOSTA, José de see de Acosta, José.

ACOSTA, Virgilio, physicist; born Camaguey, Cuba, June 14, 1916; s. Virgilio and Belen (Martinez) A.; B.S., Las Villas Inst., Havana, Cuba, 1936; D.Sc., U. Havana, Cuba, 1942; m. Idalia Perales, Aug. 8, 1948; children—Idalia Marie, Virgilio T., Helen Elizabeth. Came to U. S., 1960, naturalized, 1967. Prof. physics La Vibora Inst., Havana, 1942-60; prof. physics, applied math. Riemann Inst., Havana, 1948-54; prof. physics U. Villanueva, Havana, Cuba, 1950-60, head physics dept., asst. dean dept. sci., tech., 1951-60; research prof. physics Catholic U. Am., 1960-65, part-time, 1965—; asso. prof. physics U. S. Naval Acad., 1965—; cons. Hydrospace Research Corp., Rockville, Md., 1963-64. Mem. Am. Phys. Soc., Am. Geophys. Union, Am. Assn. Physics Tchrs., Sigma Xi, Sigma Pi Sigma. Author: Introductory Physics 1966; Introductory Elementary Physics, 1951; College Physics, 2 Vols., 4th edn., 1954; Experiments in Physics, 1956. Contbr. articles in field to sci. jours. Research on weak magnetic fields, leading to an electronic magnetometer; in neutrino detection with Cosmic Ray Group. Inventor of equipment for electrolysis and recombination of hydrogen and oxygen. Home: 3509 Jeffry St., Wheaton, Md. 20906. Office: Dept. Sci., U. S. Naval Acad., Annapolis, Md. 21402.*

ACOSTA-SISON, Honoria, physician; b. Philippines, Dec. 30, 1888; d. Mariano and Paslora (Dizon) Acosta; student Drexel Inst., 1904-05, Woman's Med. Coll. Pa., 1905-09; m. Antonio P. Sison, Oct. 10, 1910; children—Antonio, Honora (Mrs. Rogue), Pastora. First asst. obstetrics St. Paul's Hosp., Manila, Philippines, 1912-14; faculty U. Philippines, 1914—, emeritus prof., 1955—, also acting head obstetrics. Cited as Foremost Filipino Woman Physician, U. Philippines Med. Assn., 1951; recipient Presdl. medal for med. research, 1951; cited by Bd. regents U. Philippines, 1963. Fellow Philippine Obstet. and Gynecol. Soc.; mem. Philippine Acad. Sci. and Humanities (founding), Philippine Med. Assn., Obstet. and Gynecol. Soc. (hon. mem., 1st pres.), Coll. Surgeons (charter), Nat. Research Council (editorial bd. Acta Medica Philippina), Women's Med. Assn. (hon.), Coll. Medicine U. Philippines (hon. alumna), Pi Gamma Mu, Phi Kappa Phi. Research and numerous publs. on eclampsia, choriocarcinoma, disease states in pregnancy, difficulties in parturition; obstet. and gynecol. practices in various countries; sex edn. in Philippines. Office: U. Philippines, Manila, Philippines.*

ACQUAVIVA, Sabino Samele, Italian sociologist; b. Padua, Italy, Apr. 29, 1927; s. Vito and Francesca (Ricci) A.; Laurea, Padua U., 1951; m. Eugenia Gaudenzio, Aug. 30, 1954; 1 son, Francesco. Prof. sociology Padua U., 1959—, U. Cá Foscari, Venice, 1964—. Mem. sci. com. Institut Européen des Hautes Etudes Internationales, Nice (France) U., 1966. Mem. Internat. Honor Social Sci. Soc., Accademia Tiberina, 1964, Accademia dei 500, 1964, Accademia di Abruzzo, 1966. Author: Sociologia Dinamica, 1957; Automazione e nuova classe, 1958, 64; Der Untergang des Heiligen in der industriellen Gesellschaft, 1964; La scelta illusoria, 1965; La logica nelle scienze umane, 1964; also articles. Dir. rev. Sociologia Religiosa, 1958; dir. Centro di sociologia religiosa, 1958. Research on sociology of religion and politics, logic and methodology of social scis., problems of indsl. soc. Address: 16 Altinate, Padua, Italy.*

ACREE, Solomon Farley, Am. chemist; b. McGregor, Tex., Dec. 18, 1875; s. George Wren and Elizabeth Virginia (Grimes) A.; B.S., U. Tex., 1896, M.S., 1897; Ph.D., U. Chgo., 1902; postgrad. U. Berlin, 1903-04; Johnston scholar Johns Hopkins, 1904-06; m. Ruby Jarvis Tiller, May 1, 1917; children—Ruby Jane de Haven, George Wren. Asst. in chemistry U. Tex., 1896-97; fellow, asst. in chemistry U. Chgo., 1898-1901; asst. prof. chemistry U. Utah, 1901-04; asst. and asso. prof. Johns Hopkins, 1906-14; prof. forest chemistry U. Wis., 1914-17, prof., chief of chem. sect., Forest Products Lab., 1914-17; prof. forest chemistry Syracuse U., 1917-19; v.p. Internat. Chem. Products Co. and pres. Grahame Chem. Co., 1920-24; prof. phys. organic chemistry George Washington U., 1924-26; prin. chemist U. S. Bur. Standards, Washington, from 1927, chief of sect. Fiber Structures and Hydrogen Ion Measurements, from 1931. Fellow A.A.A.S., Texas Acad. Sci., Chem. Soc. (London); mem. Am., German chem. socs., Phi Beta Kappa, Sigma Xi. Writer on graphs and theory of errors, physical organic chemistry, reactions of ions and molecules; also articles; research on utilization of agrl. and lumber waste; pH standards. Died Oct. 23, 1957.

ACREL, Olaf, Swedish surgeon; b. Stockholm, Sweden, Nov. 26, 1717; ed. U. Upsala, U. Stockholm, Sweden; U. Strasbourg, U. Paris, France (under Jean-Louis Petit); surgeon-in-chief French Army, 1743-44; apptd. surgeon-in-chief Seraphins Hosp., 1752, prof. surgery, 1755; became gen. dir Swedish hosps., 1776; made a noble, 1780; Acad. Scis. Stockholm (pres.), Acad. Surgery Paris. Author: Utfoerlig Foerklaring om friska saars egenskaper, 1745; Kirurgiska Händelser, anmärkte och samlade uti K. Lazarethet och annorstädes, 1759; Historia tumorum variorum circa carpum et in vola manus obvenientium, 1779. Reformed med. teaching, improved orgn. of hosps.; known as father of Swedish surgery. Died Stockholm, May 28, 1806.

ACRON, physician; b. Agrigentum, flourished 480-430 B.C.; disciple of Empedocles; important mem. Sicilian med. sch.; recommended building large fires in streets and fumigating houses during plague in Athens, Greece, 473 B.C. (resulted in reduced contagion and mortality according to Plutarch); distinguished between different currents of air, from their mixtures drew conclusions about man's state of health; an empiricist.

ACRONIUS, Johannes, astronomer; b. Akkrum, probably 1520; s. Bernhard Acronius; student, Basel, Switzerland; M.D., 1564; became prof. math., 1547, prof. logic, 1549. Author: De motu terrae; De sphaera; De astrolabiis et annuli astronomici confectione; Prognosticon astronomicum. Died Basel, Oct. 18, 1564.

ACZEL, Janos Dezso, mathematician; b. Budapest, Hungary, Dec. 26, 1924; s. Dezso and Iren (Adler) A.; M.A., U. Budapest, 1947, Ph.D., 1947; Habil, Hungarian Acad. Sci., 1952, D.Sc., 1957; m. Susan Kende, Dec. 14, 1946; children—Catherine, Julie. Faculty, U. Szeged, Hungary, 1948-50, Inst. Tech., Miskolc, 1950-52, U. Debrecen, Hungary, 1952-65; vis. prof. U. Fla., Gainesville, 1963-64, U. Koln, Germany, 1965, U. Giessen, 1966; prof. math. U. Waterloo, Ont., Can., 1965—. Recipient M. Beke award J. Bolyai Math. Soc., 1961. Mem. Am. Math. Soc., Osterreichische Mathematische Gesellschaft. Author: (with S. Golab) Funktionalgleichungen der Theorie der geometrischen Objekte, 1960; Ein Blick auf Funktionalgleichungen und ihre Anwendungen, 1962; Lectures on Functional Equations and their Applications, 1966; Vorlesungen uber Funktionalgleichungen und ihre Anwendungen, 1961; also articles. Initiated modern theory of functional equations; gave gen. theorems and applications in geometry, probability, and information theory; theories of mean values and measurement, ordered and continuous groups, semigroups, quasi-groups, and nets. Home: 66 Westmount S., Waterloo, Ont., Can.*

ADA, Gordon Leslie, Australian biochemist; b. Sydney, Australia, Dec. 6, 1922; s. William Leslie and Erica Maud (Flower) A.; B.Sc., U. Sydney, 1943, M.Sc., 1946; D.Sc., 1960; m. Jean Macpherson, 1946; children—Ian Douglas, Andrew Leslie, Louise Margaret, Neil Ross. Research officer Red Cross Blood Transfusion Service, Sydney, 1943, Commonwealth Serum Labs., Melbourne, 1944-46, div. biophysics Nat. Inst. Med. Research, London, 1946-48, biophysics unit Walter and Eliza Hall Inst., Melbourne, 1948-60; vis. specialist virus research unit Molteno Inst., Cambridge, Eng., 1956, div. immunology Nat. Inst. Med. Research, London, 1964; head biochem. and biophysics unit Walter and Eliza Hall Inst., Melbourne, 1965—. Fellow Australian Acad. Sci.; mem. Australian Biochem. Soc. (pres. 1966-68). Research, numerous publs. on chem. and phys. properties of viruses, isolation and crystallization of bacterial enzymes, biol. fate of purified antigens, formation of specific antibody. Home: 44 Victoria St., Macleod, Victoria 3085.

Office: Royal Melbourne Hosp. Post Office, Victoria, Australia 3050.*

ADAIR, Fred Lyman, Am. physician; b. Anamosa, Ia., July 28, 1877; grad. U. Minn., 1898, M.A., 1908; M.D., Rush Med. Coll., 1901; postgrad. Charité Hosp., Berlin, Germany, 1908-09. Intern Michael Reese Hosp., Chgo.; gen. practice medicine Mpls., 1903-08, specialist obstetrics, from 1908; faculty U. Minn., 1905-29; prof., chmn. dept. obstetrics, gynecology U. Chgo., from 1929; chief staff Lying-in-Hosp., Chgo. Decorated Croix Civique (Belgium), 1919. Author: Obstetric Medicine; Obstetrics and Gynecology; Fetal and Neonatal Death. Promoted pub. edn. and influenced legislation regarding mothers and infants.

ADAIR, John, Scottish cartographer, surveyor; flourished 1683; commd. by Privy Council to survey Scottish shires, 1683. Fellow Royal Soc., 1688; pub. charts of Scottish coasts, 1703; many maps survive including Sea Coast and Islands of Scotland. Died circa 1722.

ADAIR, Robert K(emp), Am. physicist; b. Ft. Wayne, Ind., Aug. 14, 1924; s. Robert Cleland and Margaret (Weigman) A.; Ph.B., U. Wis., 1947, Ph.D., 1951; m. Eleanor Reed, June 21, 1952; children—Douglas McVeigh, Margaret Guthrie, James Cleland. Instr., U. Wis., 1950-53; physicist Brookhaven Nat. Lab., 1953-58; mem. faculty Yale, 1958—, prof. physics, 1961—, chmn. dept., 1967—. Guggenheim fellow, 1954; Ford Found. fellow, 1962-63; Sloane Found. fellow, 1962-63. Fellow Am. Phys. Soc. Author: (with Earle C. Fowler) Strange Particles, 1963. Asso. editor Phys. Rev., 1962-65. Research in measurement of neutron cross sects., scattering of polarized neutrons, exptl. evidence for optical model of nucleus, charge independence of nuclear reactions, devel. hydrogen bubble chambers, spins of elementary particles, resonances of elementary particles, K-meson reactions and decays, searches for elementary triplets-quarks; proton-proton scattering. Home: 88 Killdeer Rd., Hamden, Conn. Office: J. W. Gibbs Lab., Yale Univ., New Haven.*

ADAM OF CREMONA, med. writer; flourished 1225; wrote treatise in 3 vols. (on diet sleep, blood-letting, sea sickness, care of feet, religious purpose of a crusade, other topics). Author: Regimen iter argentium vel peregrinantium (composed for Frederick II, on hygiene of a crusading army).

ADAM, Neil Kensington, English phys. chemist; b. Cambridge, Eng.; s. James and Adela Marion (Kensington) A.; B.A., Trinity Coll., Cambridge U., 1913, M.A., 1917, Sc.D., 1928; m. Winifred Wright, June 24, 1916; children—Jean Alison (Mrs. John Bernard Martin Adams) (dec.), Arthur Mylne. Chemist, Royal Naval Airships, 1914-17; fellow Trinity Coll., Cambridge U., 1915-21; Sorby Research fellow U. Sheffield (Eng.), 1921-29; hon. research asso. U. Coll., London, Eng., 1930-36, lectr., 1936-37; prof. chemistry U. Southampton (Eng.), 1937-57. Fellow Royal Soc., 1935, Royal Inst. Chemistry; mem. Chem. Soc., Faraday Soc. (hon. life), Am. Chem. Soc. Author: The Physics and Chemistry of Surfaces, 1930; Physical Chemistry, 1954; also numerous articles. Continued Langmuir's investigation of surface films; research in surface chemistry, especially unimolecular surface films on water, 1920-39, wetting, and detergency. Home: 95 Highfield Lane, Southampton, Eng.*

ADAMANTIDIS, Demètre, veterinarian; b. Kilkis, Greece, Sept. 1, 1917; s. Stav and Eugenie (Campar) A.; ed. U. Paris; D.V.M.; m. Colette Horicks, Feb. 21, 1946; children—Philippe, Nicole. Practiced vet. medicine, Belgium; dir. vet. services Ruanda-Urundi; tech. dir. Diamond Labs. Internat. S.A. Mem. Union vétérinaire belge, Société vétérinaire royal du Brabant. Author: Organisation et exploitation des élevages porcins ANGB; Monographie pastorale de Ruanda-Urundi; Elevage, érosion, protection de la nature; Les zoonoses africainees; also others. Home: 44 A, av. de Foestraets, Uccle (Bruxelles 18). Office: 54-56, Honeniersstraat, Malines, Belgique.

ADAMCZEWSKI, Ignacy, Polish physicist; b. Warsaw, Poland, Jan. 25, 1907; s. Lucas and Eleonora (Stroncynska) A.; Mgr.Phil., U. Warsaw, 1932, Dr. Phys., 1936; m. Lopinska Janina, June 21, 1950; 1 son, Christopher. Staff, U. Warsaw, 1932-45; reader U. Lodz (Poland), 1945; faculty, head dept. physics Tech. U., Gdansk, Poland, 1945—, prof., 1965—, dean chem. faculty, 1945—; prof. Acad. Medicine, Gdansk, 1964—. Decorated Officers Cross of Polonia Restituta, Gold Cross of Merit. Mem. Polish Phys. Soc. (pres.), Med. Physics Soc., Gdansk Sci. Soc. Author: Physics for Physicians, 1955; Health Protection against Ionizing Radiation, 1959; Biophysics, 1967; Ionization and Conductivity in Liquid Dielectrics, 1965 (French, English edits. 1968); also numerous articles. Research in physics of dielectrics, health physics, radiation dosimetry, application of radioisotopes, biophysics, Home: 5 Fahrenheita, Gdansk, Poland.*

ADAMCZYK, Bogdan, Polish physicist, logopathologist; b. Lublin, Poland, Aug. 4, 1930; s. Adolf and Helena (Adamiak) A.; magister exptl. physics Maria Curie-Sklodowska U., Lublin, 1955, d. physics, 1963; m. Barbara Wilczynska, Oct. 28, 1954; 1 dau., Monika. Staff, Lab. Exptl. Physics, Maria Curie-Sklodowska U., 1952—, sci. asst., 1953-63, sci. adjunct, 1963—; logopathologist Lublin Dept. Health, 1957—; bursar Netherlands Ministry Edn. Lab. for Mass Separation, Amsterdam, 1965; logopathologist, cons. Psychiat. Clinic, U. Leyden (Netherlands), 1965. Recipient award Ministry Communication, 1963. Mem. Phys. Soc. Poland, Logopedical Soc. Poland, Psychol. Soc. Poland. Research, publs. on new method of speech remedy using artificial echo, echo telephone system for correction of stuttering, application mass spectrometer of analysis of gases excreted by human skin. Home: 25 Nadbystrzycka, Lublin, Poland.*

ADAMEK, Karel Václav, Czechoslovakian ethnographer, archaeologist; b. Hlinsko, Czechoslovakia, Sept. 19, 1868; s. Karel A. and Terezie A. (Stulikova) A.; ed. Charles U., Prague, Czechoslovakia. Ethnographic, topographic, historic researcher Czech museums, Hlinsko, 1893, Kutná Hora, 1898, Prague, 1908; co-founder E. Bohemian Nat. Mus., Chrudim, also Bohemian Archaeologic Soc., Prague; alderman, Hlinsko, 1911-18. Mem. Royal Soc. Letters and Scis. of Bohemia, Prague, Czech. Acad. (topog. com.). Author books and papers on history, geography, topography, statistics, ethnography.

ADAMETZ, Leopold, zoologist; b. Brünn, Austria-Hungary, Nov. 11, 1861; studied at Vienna (Austria), Leipzig (Germany), Pasteur Inst., Paris, France; m. Emma Freiin v. Ruepprecht, Mar. 10, 1894; 1 son, 1 dau. Became prof. extraordinary U. Cracow (Poland), 1891, named prof., 1894; joined faculty Coll. Agr., Vienna, 1898. Mem. Viennese Acad. Sci. Author: Untersuchungen über den niederen Pilzed ackerbrume, 1886; Saccharomyces lactis, . . . , 1886; Über der Aufgaben der modlinen Tierproduktionslehr, 1901; Lehrbuch der allgemeine Tierzucht, 1926. Research in animal breeding, microbiology, animal diseases from Asia to Europe. Died Vienna, Jan. 27, 1941.

ADAMI, Enrico, Italian pharmacologist; b. Milan, Italy, Sept. 10, 1906; s. Matteo and Zenobia (Camilletti-Perotti) A.; M.D., U. Milan, 1931, Libera Docenza in Pharmacology, 1938; m. Johanna Kiesswetter, Jan. 6, 1939; children—Enrica (Mrs. Cesare Corradini), Carlo, Giuseppe, Francesco. Asst., Sanatory Legnano (Italy), 1933-34; faculty U. Milan, 1933-43, prof. pharmacology Faculty Vet. Medicine, 1943-54; sci. dir. Istituto De Angeli, Milan, 1948—. Mem. Societa Italiana di Farmacol. (v.p. 1965—), Soc. Ital. di Scienze Farmaceutische; soc. Ital. Gastroenterologia, Biometric Soc., European Soc. for Study Drug Toxicity. Author: Farmacologia e Farmacoterapia, 1960 (5th edit.), Manuale di Farmaccologia Veterinaria, 1956 (3d edit.); also numerous articles. Research on pharmacology of anti-inflammatory agts., pharmacological synergisms, anti-ulcer drugs. Home: 8 Passaggio Centrale, Milan. Office: 15 Via Serio, Milan, Italy.*

ADAMI, John George, pathologist; b. Ashton-on-Mersey, Eng., Jan. 12, 1862; s. John George and Sarah Ann (Leech) A.; ed. Old Trafford Sch., Owens Coll.; entered Christ's Coll., Cambridge (Eng.) U., 1880; m. Mary Stuart Cantlie, 1894; m. 2d, Marie Wilkinson, 1922; 1 son, 1 dau. Worked under Roux, Metchinkoff, and Pasteur at Pasteur Inst. in Paris, France, 1890; apptd. to chair of pathology and bacteriology McGill U., Montreal, Que., Can., 1892; asst. dir. med. services to Canadian Expeditionary Force (World War I); apptd. vice chancellor U. Liverpool, 1919. Fellow Royal Soc., 1905; pres. Canadian Assn. for prevention Tuberculosis, 1911, Assn. Am. Physicians, 1912. Author: (with A. G. Nicholls) Principles of Pathology, 2 vols., 1908-09; (with John Macrae) Textbook of Pathology, 1912; The War Story of the Canadian Army Medical Corps, 1918. Worked on pub. health in Can., child welfare, Tb prevention; responsible for standardization of Wassermann technique (syphilis test) in Brit. labs; offered hypothesis on heredity with particular reference to inheritance of acquired characteristics (similar to Erlich's side-chain theory of immunity), 1901. Died Ruthin, Eng., Aug. 29, 1926.

ADAMIECKI, Karol, Polish management scientist; b. Dabrowa Gornicza, Poland, Mar. 18, 1866; engineering degree, St. Petersburg, 1891. Engr., Bank Smelting Works, Dabrowa Gornicza, 1891-99; dept. head, Hartman Smelting Works, Lugansk, from 1899; tech. dir., mills, Ekaterinoslaw, to 1905; consulting engr., Poland and Russia, 1906-18; dir. smelting works, Ostrow, 1906; managing dir., The Ceramic Works Ltd., Korwinow, 1907-11; lectr., Warsaw Polytechnic, 1919-22; 1st prof., Chair of Industrial Organization and Management, Warsaw Polytechnic, 1922-33. Co-founder, 1st Chairman of Board, 1st Dir., Polish Institute of Scientific Management, Warsaw. Awarded Commodore Cross, Order of Polonia Restituta; Order of White Lion, Czech.; Gold Medal, Int. Comm. for Sci. Management; hon. mem., Masaryk Acad. of Arts, Czech.; Acad. of Technical Science, Warsaw; pres. of exec. comm., Assn. of Organizing Engineers, Poland; vice-pres., International Committee for Scientific Management. Author: Harmonizacja Pracy, 1948. Investigated problems of rolling mill output; designed construction installations, ceramic furnaces; studied socio-economic effects of rationalization; wage payment; engineer's role in industry; humanization of work; developed theory of harmonization in management planning and control of production teamwork. Died May 16, 1933.

ADAMKIEWICZ, Albert, pathologist; b. Zerkow, Poland, Aug. 11, 1850; ed. Konigsberg, Germany, Wroclaw, Poland, Würzburg, Germany; recipient degree, 1872; became asst. Physiol. Inst., Königsberg, 1873; apptd. prof. pathology U. Cracow (Poland), 1878. Author: La secretion des seurs, fonctions des nerfs bilatérauxsymetriques, 1878; Les vaisseaux sanguins de la moelle épinière chez l'homme, 1881-82. Described color reaction for albumin using sulfuric and glacial acetic acids (Adamkiewicz test), 1875; described demilunes (of Adamkiewicz), or crescent-shaped cells under neurilemma of medullated nerve fibers, 1885. Died Vienna, Austria, 1921.

ADAMKIEWICZ, Vincent Witold, educator; b. Poland, Nov. 29, 1924; s. Georges and Zofia (Rogala-Lewicka) A.; Dip. Agriculture Czernichow U., Poland, 1943, B.S., 1949; M.Sc. Pharmacology, U. Bristol (Eng.), 1950; Ph.D., U. Montreal, Can., 1954; m. Lidia Maria Gowor, July 11, 1953; children—Paul, Mark, Thomas, Michael and Mathias (twins). Agriculturist, Poland 1943; research asso. sch. vet. med. U. Bristol, 1950; asst. prof. U. St. Francis Xavier, N.S., Can., 1951; research chemist F. W. Horner Ltd., pharmaceuticals, Montreal, Que., 1955; faculty U. Montreal Sch. Medicine, 1956—, professor immuno-physiology, 1967—. Mem. Am., Canadian physiol. socs., Can. Soc. Immunology, N.Y. Acad. Sci., Assn. des Physiologistes de Langue Francaise, Histamine Soc. Research, publs. on physiologic aspects of hypersensitivity, allergy; anaphylactoid allergies and their transference. Home: 5507 Côte St. Antoine, Montreal 28. Office: 2900 Blvd. Mont Royal, Montreal 26, Que., Can.*

ADAMS, Andrew Leith, zoologist; b. 1827; s. Francis Adams. Army surgeon, 1848, surgeon-maj., 1861, reported on Maltese cholera epidemic, 1865, retired from army, 1873; became professor zoology College Sciences, Dublin, Ireland, 1873, professor natural history Queen's College, Cork, Ireland, 1878. Fellow Geol. Soc., Royal Soc., 1872. Author: Wanderings of a Naturalist in India; The Western Himalayas and Cashemere, 1867; Notes of a Naturalist in the Nile Valley and Malta, 1870; Field and Forest Rambles, with Notes and Observations on the Natural History of Eastern Canada, 1873; Monographs of the British Fossil Elephants, 1877. Made valuable observations on fossil elephant remains. Died 1882.

ADAMS, Brooks, Am. social theorist; b. Quincy, Mass., June 24, 1848; s. Charles Francis and Abigail Brown (Brooks) A.; A.B., Harvard, 1870; Harvard Law Sch., 1 yr.; m. Evelyn Davis, Sept. 7, 1899. Admitted to bar, 1873; practiced at Boston, 1873-81; lectr. Boston U. Sch. of Law, 1904-11. Mem. Nat. Inst. Arts and Letters. Author: The Emancipation of Massachusetts, 1887; The Law of Civilization and Decay, 1896; America's Economic Supremacy, 1900; The New Empire, 1902; two chapters in Centralization and the Law, 1905; Railways as Public Agents, 1910; The Theory of Social Revolutions, 1913; Emancipation of Massachusetts—The Dream and the Reality, 1919. Theorized that human socs. differ according to their sources of energy; formulated law of civilization and decay (civilizations rise and decline according to comml. growth); predicted in 1900 the decline of Brit. Empire and France, with USSR and U. S. as only potential great powers remaining. Died Feb. 13, 1927.

ADAMS, Charles Baker, Am. geologist, naturalist; b. Dorchester, Mass., Jan. 11, 1814; s. Charles J. Adams; grad. Amherst, 1834; m. Mary Holmes, Feb. 1839. Asst. in Geol. Survey of N.Y., 1836; tutor, lectr. geology Amherst, 1837, prof. natural history, astronomy, 1847-53; prof. chemistry, natural history Middlebury (Vt.) Coll., 1838-47; state geologist Vt., 1845-48. Author: Contributions to Conchology, 1849-52; Catalogue of Shells Collected at Panama, 1852. Research on Am. conchology; wrote classic studies of Antillean and Panamic mollusks; grouped animal species to aid understanding of relationships. Died Jan. 18, 1853.

ADAMS, Charles Christopher, Am. ecologist; b. Clinton, Ill., July 23, 1873; s. William Henry Harrison and Hannah Westfall (Concklin) A.; B.S., Ill. Wesleyan U, 1895; M.S., Harvard, 1899. Ph.D., U. Chgo., 1908; Sc.D., Ill. Wesleyan U., 1920; m. Alice Luthera Norton, Oct. 3, 1908 (dec. Sept. 1931); 1 dau., Harriet Dyer. Asst. in biology Ill. Wesleyan U., 1895-96; asst. entomologist Ill. State Lab. of Natural History, 1896-98; curator U. Museum, U. Mich., 1903-06; dir. Cin. Soc. Natural History 1906-07; asso., animal ecology U. Ill., 1908-14; asst. prof. forest zoölogy N.Y. State Coll. Forestry. Syracuse U., 1914-16, prof., 1916; dir. Roosevelt Wild Life Forest Expt. Sta., 1919-26; N.Y. State Mus., 1926-43. Mem. Am. Soc. Naturalists, Assn. Am. Geographers, Ecol. Soc. Am. (pres. 1923), Am. Soc. Mammalogists, Brit. Ecol. Soc., A.A.A.S., History Sci. Soc., Sigma Xi. Author: Guide to the Study of Animal Ecology, 1913; also articles on migration, evolution, geog. distbn., forestry and food supply, human ecology and its application to society. Died May 22, 1955.

ADAMS, Colin Wallace Maitland, English pathologist, histochemist; b. Westcliff, Essex, Eng., Feb. 17, 1928; s. Sidney Ewart and Gladys (Keddie) A.; B.A., Christ's Coll., Cambridge, Eng., 1949, M.A., M.B., B.Chir., 1952, M.D., 1960, D.Sc., 1967; m. Anne Brownhill, May 16, 1953; children—Richard James, Maitland Adams. House officer, pathology asst., Freedom Research fellow London Hosp., 1952-57; faculty Guy's Hosp. Med. Sch., London, 1958—, Sir William Dunn prof. pathology, 1965—; vis. scientist NIH, Bethesda, Md., 1960-61. Hon. cons. pathologist Guy's Hosp., since 1965—. Recipient Sir Lionel Whitby medal, 1959-60. Mem. Royal Coll. Physicians London, Coll. Pathologists London, Path. Soc., Histochem. Soc., (U. S.), Royal Micros. Soc., Royal Soc. Medicine. Author: Neurohistochemistry, 1965; Vascular Histochemistry, 1967; also numerous articles. Research on lipid penetration into arteries, cause of various arterial degenerations, cellular changes and lipid disorders of demyelinating diseases, including multiple sclerosis, pituitary gland; tech. advances in histochem. demonstration of lipids, proteins and enzymes. Home: The Knoll, Rayleigh, Essex, Eng. Office: Dept. Pathology, Guy's Hosp. Med. Sch., London, S.E.1, Eng.*

ADAMS, Comfort Avery, Am. elec. engr.; b. Cleve., Nov. 1, 1868; s. Comfort Avery and Katherine Emily (Peticolas) A.; S.B., Case Sch. Applied Sci., 1890, E.E., 1905, Dr. Engring., 1925; hon. Dr. Engring., Lehigh U., 1938; postgrad. math. and physics, Harvard, 1891-93; m. Elizabeth Challis Parsons, June 21, 1894; children—John, Clayton Comfort. Asst. in physics, Case Sch. Applied Sci., Cleve., 1886-90; designing engr., 1890-91; instr., 1891-95, asst. prof., 1896-1905, prof. elec. engring., 1906-16, Lawrence prof. engring., Harvard U., 1914-36, Gordon McKay prof. elec. engring., 1935-36, emeritus, from 1936; dean Harvard Engring. Sch., 1919; chmn. div. engring. NRC, 1919-21; cons. engr. Am. Tool & Machine Co., Boston, The Okonite Co., Okonite Callender Cable Co., Babcock and Wilcox Co., Gen. Electric Co., The Budd Co. Mem. internat. awards jury (dept. electricity), St. Louis Expn., 1904. Recipient Miller medal for conspicuous contbns. to art of welding, 1929; Lamme medal, 1940, Edison medal, 1956 (both Am. Inst. E.E.). Fellow Am. Acad. Arts and Sciences, Am. Inst. E.E. (pres. 1918-19), Am. Soc. M.E., A.A.A.S.; mem. Nat. Acad. Scis., Instn. Elec. Engrs., Société Française 'des Electriciens, Am. Soc. C.E., Soc. Promotion Engring. Edn., Am. Phys. Soc., Am. Welding Soc. (pres. 1919-20), Am. Bur. Welding (dir. 1919-36), Engring. Found., Am. Engring. Standards Com. (chmn. 1918-20). Author: Dynamo Design Schedules; also articles. Contbd. to theory and design of alternating current elec. machinery. Died Feb. 21, 1958.

ADAMS, David Hemsley, English biochemist; b. Oxford, Eng., June 1, 1923; s. Ernest and Ruth (Markwick) A.; B.Sc., U. London, 1943, B.Sc. in Chemistry, 1945; D.Phil., U. Oxford, 1948; D.Sc., U. London, 1965; m. Agatha Mary Anne, Dec. 6, 1958; children—John Nicholas, Catherine Anne. Mem. cancer research dept. London Hosp. Med. Coll., London, 1948-61; Brit. Empire Cancer Campaign Research fellow, Can., 1954-55; sr. vis. scientist Stanford Research Inst., Menlo Park, Cal., 1962-63; lectr. dept. biochemistry Inst. Psychiatry, Denmark Hill, London, 1963-66; biochemist demyelinating diseases unit Med. Research Council, Newcastle-on-Tyne, Eng., 1966—; lectr. biochemistry U. London, 1963-66; hon. lectr. U. Newcastle-on-Tyne, 1966—. Fellow Royal Inst. Chemistry; mem. Biochem. Soc., Brit. Soc. for Cell Biology. Research, publs. on host-tumor relationships, cancer chemotherapy, tumor cell viruses, nucleic acid synthesis in developing cerebral cortex and amino acid incorporation by developing cerebral cortex. Home: 132 Runnymede Rd., Ponteland. Office: Demyelinating Diseases Research Unit, Med. Research Council, Newcastle-on-Tyne, 2, Eng.*

ADAMS, Donald Keith, Am. psychologist; b. Orrville, O., Mar. 6, 1902; s. John Furnan and Elizabeth (Taylor) A.; A.B., Pa. State U., 1923; A.M., Harvard, 1925; Ph.D., Yale, 1927; m. Naomi Quinter Holsopple, June 27, 1927; 1 dau., Nancy Quinter (Mrs. Robert A. Malone). Instr., Wesleyan U., 1928-29; faculty Duke, 1931—, prof., 1946—, chmn. dept. 1946-51; mem. Salzburg (Austria) Sem., 1950. Sterling fellow Yale, 1927-28; Nat. Research fellow, Berlin, 1929-31; Guggenheim fellow, Geneva, Switzerland, Swarthmore, Pa., 1937-38; Fulbright Sr. fellow Seewiesen, W. Germany, 1958-59. Home: 1214 W. Cornwallis Rd., Durham, N.C. 27705.*

ADAMS, Edwin Plimpton, physicist; b. Prague, Jan. 23, 1878; s. Edwin Augustus and Caroline Amelia (Plimpton) A.; B.S., Beloit Coll., 1899, Sc.D., 1931; M.S., Harvard, 1901, Ph.D., 1904; studied univs. of Berlin, Göttingen, Trinity Coll., Cambridge U.; Prof. physics Princeton, 1906-41. Mem. Am. Phys. Soc., Am. Math. Soc., Am. Philos. Soc., Beta Theta Pi. Editor: Smithsonian Mathematical Formulae, 1922. Died Dec. 31, 1956.

ADAMS, Elijah, Am. biochemist; b. Buffalo, Jan. 14, 1918; s. Joseph and Matilda (Berkman) A.; B.A., Johns Hopkins, 1938; M.D., U. Rochester, 1942; m. Blanche Macoff, June 27, 1943; children—Margaret, Janet, Joseph, James. House staff medicine Strong Meml. Hosp., 1942-43, Grace-New Haven Hosp., 1946-47, 48-49; fellow physiol. chemistry Yale, 1947-48; fellow biochemistry U. Cal., Los Angeles, 1949-50; commd. corps USPHS, 1950-55; vis. scientist U. Utah, 1950-52; staff mem. NIH, 1952-55; faculty N.Y. U., 1955-58; prof., head pharmacology St. Louis U. Sch. Medicine, 1958-63; prof., head biochemistry U. Md. Sch. Medicine, Balt., 1963—. Mem. Am. Chem. Soc., Am. Soc. Biol. Chemists. Research, numerous publs. on enzymes, amino acid metabolism in mammals and bacteria. Home: 2405 Ken Oak Rd., Balt. 21209. Office: 660 W. Redwood St., Balt. 21201.*

ADAMS, Ernst, engr., mathematician; b. Essen, Feb. 2, 1928; s. Werner and Margaret A.; ed. Technische Hochschule, Braunschweig, and Darmstadt; degree in math., 1953; D. Eng., 1956; m. Adelheid Beckert, Sept. 3, 1960; children—Bernard, Georg, Ulrike. Collaborator lab. scis., Braunschweig, 1955-57; dept. chief NASA, U. S., 1957-63; prof. sch. of mechanics U. Alabama, 1961-63; dep. sci. research, Fribourg-Br., 1963-66, U. Karlsruhe, 1966—. Mem. Div. Sci. Soc. Contbg. author chapters, articles. Home: Thüringerstrasse 14, 7501 Grünwettersbach, Germany. Office: University of Karlsruhe, 75 Karlsruhe, Germany.*

ADAMS, Francis, Scottish physician, classical scholar; b. Lumphanan, Aberdeenshire, Scotland, Mar. 13, 1796; s. James Adams; M.A., King's Coll. Aberdeen, Scotland, M.D. (hon.), 1856; LL.D., U. Glasgow (Scotland), 1846. Student medicine; m. Miss Shaw; at least 2 sons. Practiced medicine Banchory Ternan, Scotland. Mem. Coll. Surgeons (London). Translator: The Seven Books of Paulus Aegineta (translated from Greek with commentary), 1844-47; The Genuine Works of Hippocrates, 1849 (only complete English translation); The Extant works of Aretaeus, the Cappadocian, 1856. Contbr. med. memoirs, articles to profl. jours. Did more in area of Greek medicine than had been done by any Brit. scholar for a century and a half. Died Feb. 26, 1861.

ADAMS, Frank Dawson, geologist; b. Montreal, Que., Can., Sept. 17, 1859; ed. McGill U., Yale, Johns Hopkins, Heidelberg (Germany) U.; m. Marie Stuart Finley, 1892 (dec. 1937). Became mem. staff Geol. Survey Can., 1880; apptd. lectr. in geology McGill U., Montreal, 1889, later dean faculty applied sci., vice prin., Logan prof. geology. Recipient Wollaston Gold medal Geol. Soc. London. Fellow Royal Soc., 1907, also Canada; mem. Geol. Soc. Am., Instn. Mining and Metallurgy Great Britain. Author: History of the Birth and Development of the Geological Sciences, 1938; History of Christ Church, Montreal. Research, publs. on problems of metamorphism and older crystalline rocks of earth's crust. Died 1942.

ADAMS, Gail Dayton, Am. radiol. physicist; b. Cleve., Jan. 27, 1918; s. Gail D. and Edna (Baker) A.; B.S. magna cum laude, Case Inst. Tech., 1940; M.S., U. Ill., 1942, Ph.D., 1943; m. Lucille A. W. Pyne, May 26, 1942 (div.); children—Kenneth C., Frances L.; m. 2d, Helen L. Newman, Oct. 1, 1966. Research physicist U. of Ill., Urbana, 1943-48, research asst. prof., 1948-50; clin. asso. prof. physics U. Cal. at San Francisco, 1951-52, prof., 1952-65, asso. dir. radiol. lab., 1952-65; prof. radiation physics U. Okla. Med. Center, Oklahoma City, 1965—. Cons. USPHS, VA and USN hosps. Diplomate Am. Bd. Radiology, Am. Bd. Health Physics. Fellow Am. Phys. Soc., Am. Coll. Radiologists; mem. A.A.A.S., Am. Assn. Physicists in Medicine (pres. 1958-60), Radiation Research Soc. (sec.-treas. 1963—), Cal. Wildlife Fedn. (dir. 1958-64, treas. 1962-64), Am. Coll. Radiology, Health Physics Soc., Radiol. Soc. N.Am., Sigma Xi, Tau Beta Pi, Phi Kappa Phi, Pi Mu Epsilon. Research in betatron devel. and application, 1940-50, 70 Mev synchrotron devel. and application to cancer therapy, 1951—; patentee in betatron devel. and radiation dosimetry. Home 6913 Edgewater Dr., Oklahoma City 73116.*

ADAMS, George (the Elder), Brit. engr.; b. Eng.; children—George, Dudley; math. instrument maker to George III. Author: Micrographia illustrata, 1746; The Description and Use of a New Seaquadrant for taking the altitude of the sun from the visible horizon, 1748; The Description and Use of the Universal Trigonometrical Octant, 1753; Treatise describing and explaining the construction and use of new celestial and terrestrial globes, 1766. Widely known as maker of celestial and terrestial globes; proposed artificial horizon (sea instrument), 1748; made high-accuracy spiral slide rules, circa 1748. Died 1773.

ADAMS, George (the Younger), Brit. engr.; b. Eng., 1750; s. George Adams; succeeded father as math. instrument maker to George III. Author: An Essay on Electricity, to which is added an Essay on Magnetism, 1784; Essays on the Microscope, 1787; An Essay on Vision, 1789; Astronomical and Geographical Essays, 1790; A Short Dissertation on the Barometer, 1790; Geometrical and Graphical Essays, 1790; Lectures on Natural and Experimental Philosophy, 1794; numerous elementary sci. works which comprised a course of instrn.; works on use of math. instruments. Constructed light microscope, 1787; invented elec. apparatus. Died Southampton, Eng., Aug. 4, 1795.

ADAMS, George Plimpton, Jr., Am. economist; b. Berkeley, Cal., Apr. 27, 1909; s. George P. and Mary (Woodle) A.; A.B., Harvard, 1929; Ph.D., U. Cal. at Berkeley, 1940; m. Evelyn Yonker, June 13, 1934. Instr., Robert Coll., Istanbul. Turkey, 1929-32; pvt. tchr., Berkeley, Cal., 1932-38; staff Brookings Instn., Washington, 1938-40; faculty Cornell U., Ithaca, N.Y., 1940—, prof., 1950—, chmn. econs. dept., 1950-62; economist U. S. Dept. State, Washington, 1944-46. Mem. Am. Econ. Assn., Am. Assn. U. Profs., Phi Beta Kappa. Author: Wartime Price Control, 1942; Competitive Economic Systems, 1955; also articles. Research on history of econ. thought and its relevance to problems of contemporary econ. policy, structure and performance of contemporary econ. systems. Home: 309 Roat St., Ithaca, N.Y. 14850.*

ADAMS, Henry, cons. engr.; b. Mar. 24, 1846; s. J. H. Adams; ed. King's Coll., London, City London Coll.; m. twice; 6 sons, 3 daus. Outdoor mgr. Sir W. G. Armstrong & Co., 1865-77; prof. engring. City London Coll. for 35 years. Chief examiner (engring.) Bd. Edn., 1905-11; extensive lectr. Fellow Royal Instn. Chartered Surveyors, Royal Inst. Brit. Architects, Royal San. Inst. (chmn. examiners); mem. Instn. Civil Engrs., Instn. Mech. Engrs., Civil and Mech. Engring. Soc. (pres. 1889), Soc. Engrs. (pres. 1890), City London Coll. Sci. Soc. (pres. 1900), Instn. San. Engrs. (pres. 1908), Instn. Engrs.-in-Charge (pres. 1910), Instn. Structural Engrs. (pres. 1914-16), Soc. Architects (chmn. examiners, mem. council). Author: Some Reminiscences, 1926; many standard texts; numerous tech. books. Inventor Adams Vortex Blast Pipe for Locos. Died Aug. 13, 1935.

ADAMS, James Alexander, Brit. physician; b. Glasgow, Scotland, 1857; M.D., Glasgow U.; 1 son, 2 daus. Practiced medicine, specializing in surgery; demonstrator anatomy, lectr., examiner; Glasgow U.; sr. surgeon Erskine Limbless Hosp.; cons. surgeon Royal Infirmary, Glasgow. Mem. Gen. Med. Council, London, Eng.; pres. Royal Faculty Physicians and Surgeons. Author: Anatomical Relation Between Disease of Brain and Disease of Ear; A New Operation for Uterine Displacements; Pylorectomy; other publs. Devised operation to correct retroversion of uterus by shortening round ligaments, 1882; made 1st x-ray diagnosis of renal calculus, 1897. Died Dec. 28, 1930.

ADAMS, John, Brit. surgeon; b. Eng., 1806; studied under Andrews at London Hosp. Demonstrator in anatomy London Hosp. Sch., 1828, also asst. surgeon London Hosp.; later prof. anatomy and physiology and titular surgeon. Author: The Anatomy and Diseases of the Prostate Gland, 1851. Credited with being 1st to distinguish between benign hypertrophy and carcinoma of prostate, 1851. Died London, Eng., 1877.

ADAMS, John Couch, English astronomer; b. Laneast, Eng., June 5, 1819; s. Thomas and Tabitha Knill (Grylls) A.; grad. Cambridge U., 1843; m. Eliza Bruce, 1863. Became fellow St. John's Coll., Cambridge, 1843, life fellow Pembroke Coll., 1853; prof. math. St. Andrew's U., Aberdeen, Scotland, 1858-59; Lowndean prof. astronomy and geometry Cambridge U., 1859-92, dir. obs., 1861. Recipient Copley medal, 1848. Fellow Royal Soc., 1849; mem. Royal Astron. Soc. (pres. 1851-53, 74-76, gold medal for research on theory of secular acceleration of moon's mean motion 1866), French Acad. Scis. (corr.). Author: An explanation of the observed irregularities in the motion of Uranus, 1846; Scientific Papers, 2 vols. 1896-1900; memoires presented to Royal Astron. Soc. Discovered (independently of Leverrier) planet Neptune, 1841-45; studied motion of moon's perigee, 1853; determined cometlike orbit of Leonid swarm of meteors, 1866; researched terrestrial magnetism, gravitational astronomy. Died Cambridge, Eng., Jan. 21, 1892.

ADAMS, John Emery, Am. geologist; b. Salon, Ia., June 5, 1899; s. Harry Delvie and Virginia (Bacon) A.; B.A., U. of Ia., 1922, M.S., 1923; grad. study, U. of Chicago, 1923-24, U. of Wis., 1925, U. of Tex., 1926-27; m. Margaret MacLaughlin, July 25, 1926; 1 dau., Mary Ann (Mrs. A. B. Plunkett). Began career as geologist at the Roxana Petroleum Corp., St. Louis, 1923; biology instr., Tex. A. and M. Coll., 1925-26; asso. Texas Bur. of Econ. Geology, Austin, Tex., 1926-27; geologist, later sr. geologist Cal. Co. and Standard Oil Co. of Tex., 1927-64; geological consultant, Midland, Texas, 1964—. Mem. Am. Geophys. Union, Am. Assn. Petroleum Geologists (hon. mem.; pres. 1953), Geol. Soc. Am. (councilor, 1945-47), Soc. Econ. Paleontol. and Mineralogy, West Tex. Geol. Soc. (hon. life mem.; v.p. 1931, pres. 1940), Soc. Ind. Earth Scientists, Soc. Econ. Paleontologists and Mineralogists (hon. life), Sigma Xi. Research on origin, migration and accumulation of petroleum, limestone deposition, saline residues in evaporite lagoons, origin of dolomite, sediment-starved basins, structural deformation in large basins, geology of Permian Basin province in W. Tex. and eastern N.M. Home: 1107 N. Ainslee. Office: Petroleum Life Bldg., Midland, Tex. 79704.*

ADAMS, John Frank, English mathematician; b. London, Eng., Nov. 5, 1930; s. William F. and Jean M. (Baines) A.; B.A., Trinity Coll., Cambridge, Eng., 1952, Ph.D., 1955, M.A., 1956; m. Grace Rhoda Carty, Dec. 12, 1953; children—Alice Grace, Adrian John, Lucy Jane, Katy Ruth. Fellow Trinity Coll., 1955-58; jr. lectr. Oxford (Eng.) U., 1955-56; mem. Inst. for Advanced Study, Princeton, 1957-58; fellow Trinity Hall, asst. lectr. Cambridge U., 1958-61; faculty U. Manchester (Eng.), 1962—, prof. math.,

1964—. Fellow Royal Soc., 1964; mem. London Math. Soc., Am. Math. Soc. Author: Stable Homotopy Theory, 1964; also articles. Research in algebraic topology. Home: Bracklyn, Priory Rd., Bowdon, Altrincham, Cheshire, Eng. Office: Dept. Math. U. Manchester, Manchester 13, Eng.*

ADAMS, John M., Am. physician, educator; b. Mpls., June 7, 1905; s. Paul and Olive (Marshall) A.; B.S., Princeton, 1929; M.D., Columbia, 1933, Ph.D., U. Minn., 1943; m. Carolyn Gaston, Mar. 24, 1934; children—John M., Herbert G., William M. Practice medicine specializing in pediatrics, with Nicollet Clinic, Mpls., 1937-42; asst. prof. pediatrics, U. Minn., Mpls., 1942-46, asso. prof., 1946-50; prof., chmn. dept. pediatrics U. Cal. Med. Sch., Los Angeles, 1950-64, prof., 1950—; dir. Am. Bd. Pediatrics. Mem. Am. Pediatric Soc., Soc. for Pediatric Research, Soc. for Exptl. Biology and Medicine, Soc. for Exptl. Pathology. Author: Newer Virus Diseases, 1960; Viruses and Colds, 1967. Research, publs. in field of infectious diseases. Home: 415 24th St., Santa Monica, Cal. 90402. Office: U. Cal. at Los Angeles Med. Center, Cal. 90024.*

ADAMS, John Wagstaff, Am. geologist; b. N.Y.C., Sept. 17, 1915; s. John F. and Margaret (Wagstaff) A.; Geol. Engr., Colo. Sch. Mines, 1941; postgrad. Columbia; m. Ruth Ann Lisco, Sept. 17, 1942; children—Virginia Ann, John Garrison. Student engr. N.J. Zinc Co., Franklin, N.J. and Hanover, N.M., 1942-43; mem. U. S. Geol. Survey, Pegmatite investigations, S.D., 1943-45; mining geologist U. S. Vanadium Corp., Am. Zinc, Lead & Smelting Co., Bishop, Cal. and Ouray, Colo., 1946-47; geologist U. S. Geol. Survey, N.M., N.H., Colo., 1947-58; mineralogist Colo. Sch. Mines Research Found., Inc., Golden, 1958; commodity geologist for rare earths U. S. Geol. Survey, Denver, 1958—. Fellow Mineral. Soc. Am.; mem. Geol. Soc. Am., Soc. Econ. Geologists, Colo. Sci. Soc., Mineral. Assn. Can. Research, publs. on mineralogy and paragenetic relations of minerals in uranium vein deposits, wall-rock control of uranium deposition, visible region absorption spectra of rare-earth minerals. Home: 705 Garland St., Lakewood, Colo. 80215. Office: Federal Center, Bldg. 25, Denver 80225.*

ADAMS, Joseph Henry, Am. inventor; b. Bklyn., 1867; student Coll. Phys. and Surg., Columbia, until 1889; D.Sc. (hon.), U. Miami, 1928; m. Helen Mar Howell, 1889; children—Ray, Alfred, Mrs. Charles W. Beeching, Mrs. Alvah Davison. Author 4 Harpers Books for Boys, 1900-1910. Perfected oil-cracking process used to manufacture gasoline, 1897-1917. Died Feb. 1941.

ADAMS, Leason Heberling, Am. phys. chemist; b. Cherryvale, Kan., Jan. 16, 1887; s. William Bardon and Katherine (Heberling) A.; B.S., U. Ill., 1906; Sc.D., Tufts U., 1941; m. Jeannette Blaisdell, Jan. 25, 1908 (dec.); children—Leason B., William M., Madeline J. (Mrs. Kinsey Jones), Ralston H.; m. 2d, Freda R. Ostraw, Aug. 14, 1956. Chemist, Morris & Co., Chgo., 1906, Mo. Pacific R.R. Co., St. Louis, 1907-08, tech. br. U. S. Geol. Survey, 1908-10; phys. chemist Geophys. Lab., Carnegie Instn. Washington, 1910-36, dir., 1936-52; vis. prof. U. Cal. at Los Angeles, 1956—; chief div. 1 Office Sci. Research and Devel., 1941-46. Recipient Longsteth medal Franklin Inst., 1924; William Bowie medal Am. Geophys. Union, 1950; Medal for merit U. S. Govt., 1946. Fellow Royal Astron., Soc. (Gt. Britain); mem. Nat. Acad. Scis., Sigma Xi. Research, publs. on materials under high pressure, interior earth; co-inventor process for annealing glass. Home: 8700 Rayborn Rd., Bethesda, Md. 20034.*

ADAMS, Leonard Caldwell, Am. elec. engr.; b. Saluda, S.C., Nov. 7, 1921; s. James P. and Amelia (Minnick) A.; B.S. in Elec. Engring., Clemson U., 1943; M.S., Okla. State U., 1950; Ph.D., U. Fla., 1953; m. Rachel Lorean Adams, June 18, 1945; children—James Phillip, Richard C., Ann S. Faculty, Clemson U., 1946-58, prof., 1956-58, dir. engring. research, 1959-61; prof., head elec. engring. dept. La. State U., Baton Rouge, 1961—; exptl. physicist E.I. du Pont de Nemours & Co., 1958-59. Engring. tng. cons. So. Bell Tel. & Tel., 1956-58, 65. So. fellow, 1949-50. Mem. Am. Soc. Elec. Engring. (chmn. B.R. sect. 1966-67), I.E.E.E., Baton Rouge Council Engring. and Sci. Socs. Home: 12123 N. Lakeview Dr., Baton Rouge 70810.*

ADAMS, Mildred, Am. chemist; Manchester, N.H., July 21, 1899; d. James Edward and Grace (Gibson) A.; B.A., magna cum laude, Smith Coll., 1921; M.A., Columbia, 1923, Ph.D., 1927. Carnegie Research asst. Columbia, 1923-29; instr. Centenary Jr. Coll., Hackettstown, N.J., 1929-30; asst. prof. Mayo Found., U. Minn., also cons. in clin. metabolism Mayo Clinic, Rochester, Minn., 1930-36; biochemist NIH, Bethesda, Md., 1936-42; v.p. in charge research Takamine Labs., Clifton, N.J., 1942-49; chief exptl. nutrition lab. human nutrition research div. Agrl. Research Service, U. S. Dept. Agr., 1950—. Fellow A.A.A.S., N.Y. Acad. Scis.; mem. Am. Chem. Soc., Am. Soc. Biol. Chemists, Soc. for Exptl. Biology and Medicine, Inst. Nutrition, Phi Beta Kappa, Sigma Xi. Contbr. numerous articles to tech. jours., chpts. to books. Research on amylases, muscular weakness disorders, penicillin prodn., basic nutritional studies using rat. Home: 10101 Pierce Dr., Silver Spring, Md. 20901. Office: U. S. Dept. Agr. Human Nutrition Div., Agrl. Research Service, Beltsville, Md. 20705.*

ADAMS, Norman Ilsley, Jr., Am. physicist; b. Winthrop, Mass., Sept. 20, 1895; s. Norman Ilsley and Mabel E. (George) A.; B.A., Yale, 1917, Ph.D., 1923; m. Genevieve A. Sloan, July 28, 1926; children—Norman Ilsley 3d, Harry Bell. Engr., Am. Tel.-&Tel. Co., 1923-24; faculty Yale, 1925—, prof. physics, 1944-64, prof. emeritus, 1964—; vis. prof. U. Ida., 1964-65, U. Del., 1965-66; cons. engr. radiobroadcasting, 1927—. Fellow Am. Phys. Soc.; mem. Phi Beta Kappa, Sigma Xi, Gamma Alpha. Author: (with L. Page) Principles of Electricity, 1931; Electrodynamics, 1940; also articles. Research in electricity and magnetism, electronics, radioactivity and thermodynamics. Home: 870 Prospect St., New Haven 06511.*

ADAMS, Raymond DeLacy, Am. physician, educator; b. Portland, Ore., Feb. 13, 1911; s. William Henry and Eva Mable (Morris) A.; B.A., U. Ore., 1932, M.A. in Psychology, 1933; M.D., Duke, 1936; m. Margaret Elinor Clark, May 12, 1933; children—Mary Elinor, John William, Carol Ann, Sarah Ellen. Rockefeller fellow in neurology and psychiatry Harvard, Boston, 1938-40, faculty neurology, 1941-53, Bullard prof. neuropathology, 1954—; Rockefeller fellow in neurology and psychiatry Yale, 1940-41; chief neurology service Mass. Gen Hosp., 1951—. Mem. Am. Neurol. Assn., Am. Assn. Neuropathologists, Am. Acad. Neurology, A.C.P., Assn. for Research and Mental Disease, World Fedn. Neurology, German Assn. Neuropathologists, Soc. French Neurology, Assn. Brit. Neurologists, Swiss Neurol. Soc., Alpha Omega Alpha. Author: (with Merritt and Solomon) Neurosyphilis, 1946; (with Denny-Brown and Pearson) Diseases of Muscle, 1952. Editor: Principles of Internal Medicine, 1954. Contbr. numerous articles and essays to sci. publs. Research on vascular diseases of the brain, nutritional, metabolic, and alcoholic diseases of the nervous system, and diseases of the skeletal muscle. Home: 320 Adams St., Milton, Mass. 02186. Office: Mass. Gen. Hospital, Boston 02114.*

ADAMS, Raymond Voiles, Am. physicist, educator; b. Eureka, Kan., Oct. 5, 1920; s. Raymond Voiles and Gladys Mae (Grove) A.; B.S. in Physics, Kan. State U., 1941; Ph.D. in Physics, Cal. Inst. Tech., 1948. Research fellow Cal. Inst. Tech., 1947-49; asst. prof. Yale, 1949-52; asso. prof. Mt. Holyoke Coll., South Hadley, Mass., 1952-55; asso. prof. Wayne State U., Detroit, 1955-59, prof., 1959-60, chmn. dept. physics, 1959-60; prof., chmn. dept. physics Cal. State Coll., Fullerton, 1960—. Recipient Naval Ordnance Devel. award, 1945. Mem. Am. Phys. Soc., Sigma Xi, Sigma Nu, Phi Kappa Phi. Research on partical physics and cosmic radiation. Home: P.O. Box 3177, Fullerton, Cal. 92631.*

ADAMS, Richard Newbold, Am. anthropologist; b. Ann Arbor, Mich., Aug. 4, 1924; s. Randolph Greenfield and Helen Constance (Spiller) A.; A.B., U. Mich., 1947; M.A., Yale, 1949, Ph.D., 1951; m. Betty Virginia Hannstein, Nov. 4, 1951; children—Walter Randolph, Tani Marilena, Gina Constance. Ethnologist, Inst. Social Anthropology, Smithsonian Instn., Guatemala City, 1950-51; specialist grantee State Dept. 1951-52; scientist WHO, Guatemala City, 1953-56; prof. sociology and anthropology Mich. State U., 1956-62; vis. prof. anthropology U. Cal. at Berkeley, 1960-61; prof. Anthropology, asst. dir. Inst. Latin Am. Studies, Univ. of Texas, Austin, 1962—, also chmn. dept. of anthropology. Served to lt. (j.g.), USNR, 1943-46. Fellow Am. Anthrop. Soc., A.A.A.S.; mem. Latin Am. Studies Association (pres. 1967-68), Soc. Applied Anthropology (pres. 1962-63), Am. Sociol. Assn., Am. Ethnol. Soc., Sigma Xi. Author: Cultural Surveys of Panama-Nicaragua-Guatemala-El Salvador-Honduras, 1957; A Community in the Andes, 1959; Second Sowing: Power and Secondary Development in Latin America, published in 1967. Co-author: United States University Cooperation in Latin America, 1960; Introducción a la Antropologia aplicada, 1964. Co-editor: Human Organization Research, 1960; Contemporary Cultures and Societies of Latin America, 1964. Research in Latin Am. sociology, analysis of complex cultures, social and applied anthropology, human soc. and energy. Address: Inst. Latin Am. Studies, U. Texas, Austin 12, Tex.

ADAMS, Robert, Irish physician; born Dublin, Ireland, 1791; B.A., Dublin U., 1814, M.A., Dublin U., 1832, M.D., Dublin U., 1842; later studied in Europe. Became surgeon at Jervis Street Hospital, Dublin, later at Richmond Hosp.; founder (with others), lectr. Richmond (name later changed to Carmichael) Sch. Medicine; apptd. surgeon to Queen in Ireland, 1861; became Regius prof. surgery Dublin U., 1861. Fellow Royal Coll. Surgeons Ireland (pres. 3 times). Author: Treatise on Rheumatic Gout, or Chronic Rheumatic Arthritis of all the Joints (classic work, contains original research), 1857. Gave classic picture of heart block characterized by syncope with or without convulsive seizures (Stokes-Adams syndrome or Spens syndrome), 1826. Died Jan. 13, 1875.

ADAMS, Robert McCormick, Jr., Am. anthropologist; b. Chgo., Ill., July 23, 1926; s. Robert McCormick and Janet (Lawrence) A.; Ph.B., U. Chgo., 1947, M.A., 1952, Ph.D., 1956; m. Ruth Salzman Skinner, July 24, 1953; 1 dau., Megan. Archaeol. field tng. in Jarmo, Iraq, 1950-51, Yucatan, Mexico, 1953; field studies history irrigation and urban settlement, Iraq and Iran, 1956—; reconnaissance and excavation ancient Mayan settlement patterns, Chiapas, Mexico, 1958-61; mem. faculty U. Chgo. and staff Oriental Inst., 1955—, asso. prof., 1961-64, prof., 1964—, dir. inst., 1962—. Fellow A.A.A.S., Am. Anthrop. Assn.; mem. Soc. Am. Archaeology, German Archaeol. Inst., Sigma Xi. Author: Land Behind Baghdad, 1965. Editor: (with C. H. Kraeling) City Invincible: A Symposium on Urbanization and Cultural Development in the Ancient Near East, 1960; The Evolution of Urban Society, 1966. Intensive, regionally oriented studies of long-term patterns of settlement and land-use in Middle East; comparisons of early growth of civilization in Middle East and New World. Home: 5201 S. Kimbark Av., Chgo. 60615.*

ADAMS, Roger, Am. chemist; b. Boston, Jan. 2, 1889; s. Austin Winslow and Lydia (Curtis) A.; A.B., Harvard, 1909, A.M., 1910, Ph.D., 1912, Sc.D., 1945; Sc.D., Bklyn. Poly. Inst., 1935, Northwestern U., 1942, U. Rochester, 1943, U. Pa., 1947, Yale, 1948, Drexel Inst. Tech., 1955, U. Ill., 1957, Bridgeport U., 1960; LL.D., U. Mich., 1954; m. Lucile Wheeler, Aug. 29, 1918; 1 dau., Lucile (Mrs. William E. Ranz). Faculty, U. Ill., Urbana, 1916—, prof., head dept. chemistry, chem. engring., 1926-54, research prof., 1954-57, prof. emeritus, 1957—. Bd. dirs. Chemurgic Council, 1934-61; v.p. Internat. Union Pure and Applied Chemistry, 1951-55; trustee Sloan-Kettering Inst. Cancer Research Com. Sci. Policy, 1954—; mem. sci. adv. bd. Sugar Research Found., 1959—. Recipient Cresson Medal (Franklin Inst.), 1944, Davy Medal (Royal Soc.), 1945, Priestley Medal (Am. Chem. Soc.), 1946, Franklin Medal (Franklin Inst.), 1960, Nat. Medal of Sci. (U. S. govt.), 1964, numerous other awards from profl. orgns., 1927-64. Hon. fellow Chem. Soc. London, Société Chimique de France, Swiss Chem. Soc.; mem. Am. Chem. Soc. (past pres., dir., chmn. bd.), A.A.A.S. (past pres., chmn. bd.), Am Inst. Chemists, Harvey Soc., German, Polish (hon.), Spanish (hon.), Japan (hon.), Argentine (hon.), chem. socs., Am. Acad. Arts and Scis., Am. Philos. Soc., Nat. Acad. Scis., Phi Beta Kappa, Sigma Xi, many others. Author: ((with J. R. Johnson) Laboratory Experiments in Organic Chemistry, 1928-63. Research, numerous publs. on catalysts, various chemotherapeutic agts., local anesthetics, structure of gossypol, alkali insoluble phenols, active principle of marihuana and synthetics with similar activity, stereochemistry, natural product structures; 1st to produce butyn, procaine in U. S. Home: 603 W. Michigan Av., Urbana, Ill. 61801.*

ADAMS, Walter Sydney, astronomer; b. Antioch, North Syria, Dec. 20, 1876; s. Lucien Harper and Dora (Francis) A.; A.B., Dartmouth, 1898; D.Sc., 1913; A.M., U. Chgo., 1900, U. Munich, 1901; LL.D., Pomona, 1926; D.Sc., Columbia, 1926, U. So. Cal., 1930, U. Chgo., 1945, Princeton, 1947; m. Lillian M. Wickham, June 2, 1910 (dec. 1920); m. 2d, Adeline L. Miller, June 15, 1922; children—Edmund Miller, John Francis. Asst., Yerkes Obs., 1901-03, instr. 1903-04; asst. astronomer Mt. Wilson Obs., Carnegie Instn., Pasadena, Cal., 1904-09, astronomer, from 1909, acting dir., 1910-11, asst. dir., 1913-23, dir., 1923-46; research asso. Cal. Inst. Tech., 1947-48. Mem. Am. Astron. Soc. (pres. 1931-34), Astron. Soc. Pacific (pres. 1923), Royal Acad. Scis. of Uppsala, Astron. Union (v.p. 1935-38), Am. Philos. Soc.; fgn. mem. Royal Soc. London, Royal Astron. Soc. (London), Inst. France (Acad. Scis.), Royal Swedish Acad. Scis. Recipient Gold medal Royal Astron. Soc., 1917, Draper medal Nat. Acad. Scis., 1918; Janssen medal Société Astronomique de France, 1926; Bruce medal Astron. Soc. Pacific, 1928; Janssen medal Paris Acad. Scis., 1935. Deduced a star's luminosity from its spectrum, 1914, also calculated its distance by stellar parallax method of comparing luminosity to apparent brightness; deduced that companion star to Sirius must be white dwarf in order to confirm Einstein's theory of gen. relativity in regard to red line shift, 1915, found shift, 1925; applied method to Venus and found it rich in carbon dioxide, 1932; studied rotation of sun, spectra of sunspots, gases in interstellar space, radial velocities; made spectroscopic studies of sun at different solar latitudes, also of stars. Died Pasadena, Cal., May 11, 1956.

ADAMS, William Bridges, English inventor; m. Sarah Flower, 1834; mfr. ry. plant at works in Bow, Eng. Author: English Pleasure Carriages, 1837; Railway and Permanent Way, 1854; Roads and Rails, 1862. Invented fish-joint for ry. rails; improved rolling stock; reduced weight of early locomotives; patentee improvements in common road carriages, ship propulsion, guns, wood-carving and other machines. Died Broadstairs, Eng., 1872.

ADAMS, William Elias, Am. surgeon, educator; b. Nichols, Ia., May 1, 1902; s. Frank A. and Alvina (Mills) A.; M.D., B.S., State U. Ia., 1926; m. Huberta M. Livingstone, June 9, 1928; 1 dau., Diana I. L. Faculty surgery U. Chgo., 1933—, prof., 1947-54, Raymond prof. surgery, 1954—, chmn. dept. surgery, 1959—. Hon. prof. surgery U. Guadalajara (Mexico), 1955, U. Madrid (Spain), 1956. Fellow Kansas City Acad. Medicine (hon.); mem. A.M.A., A.C.S., Am. Assn. for Thoracic Surgery, Am. Coll. Chest Physicians (2d v.p.), Soc. U. Surgeons, Chgo. Surg. Soc., Ill.

Med. Soc. (chmn. bd. trustees), Am., Central surg. assns., Internat. Congress on Smoking and Health (treas.), Soc. Clin. Surgery, Société Internationale de Chirurgie, Sigma Xi, Alpha Kappa Kappa; hon. mem. Soc. Cancerology (Guadalajara), Guadalajara Surg. Soc., Alpha Omega Alpha. Research, publs. on gen. and thoracic surgery; a method of maintaining adequate circulation in animals in cardiac arrest, with future applicability to humans. Home: 5805 Dorchester Av., Chgo. 60637.*

ADAMS, William Grylls, English physicist; b. Laneast, Eng., Feb. 16, 1836; ed. St. John's Coll., Cambridge; M.A., D.Sc.; married Mary Dingle, 1869; 3 sons, 1 daughter. Professor natural philosophy and astronomy King's College, London, to 1906; fellow St. John's College. Fellow Royal Soc., 1872, Geol. Soc.; member Math. Soc. London; pres. Soc. Elec. Engrs.; pres., v.p. Phys. Soc. Publs. on simultaneous magnetic disturbances, action of light on selenium, polariscopes, equipotential curves and surfaces, alternate current machine, dynamo-machines. Died Apr. 10, 1915.

ADAMS, William Mansfield, Am. geophysicist; b. Kissimmee, Fla., Feb. 19, 1932; s. Shirah Devoy and Olive (Goding) A.; A.B., U. Chgo., 1951; B.A., U. Cal. at Berkeley, 1953; M.S., St. Louis U., 1955, Ph.D., 1957; M.B.A., Santa Clara U., 1964; m. Roberta Kay Blackwell, Feb. 19, 1955; children—William Mansfield, Jonathan Blackwell, Christopher Daniel. Staff, Shell Oil Co., 1953, Stanolind Oil Co., 1953, Western Geophys. Co., Los Angeles, 1954; tech. officer Govt. Can., Ottawa, Ont., 1956; chief geophysicist Geotech. Corp., Dallas, 1957-58; tech. dir. Lawrence Radiation Lab., U. Cal., Livermore, 1959-62; pres. Planetary Scis. Inc., Santa Clara, Cal., 1962—; prof. U. Hawaii, Honolulu, 1964—. Gulf fellow, 1955, Fulbright grantee, 1956-57. Fellow Geol. Soc. Am. (mem. com. on engring. seismology 1962—; mem. Am. Geophys. Union, Soc. Exptl. Geophysics, European Soc. Exptl. Geophysics, Seismol. Soc. Am. (editor Bull. 1962-65), Sigma Xi. Author: Earthquakes, An Introduction to Observational Seismology, 1964. Contbr. articles to tech. jours. Home: 509 University Av. Office: 2525 Correa Av., Honolulu.*

ADAMS, William Ritchie, Am. forester; b. Phila., Feb. 17, 1902; s. Wakeman Ritchie and Florence (Douglass) A.; B.S., Syracuse U., 1926; M.S., U. Vt., 1928; Ph.D., Yale, 1935; m. Nelle Swisher Alexander, June 21, 1930. With U. Vt., Burlington, 1926—, prof. forestry, 1948—, chmn. forestry dept., 1947-67; with Vt. Agrl. Experiment Sta., Burlington, 1926—, forester, 1947—. Chmn. gov.'s Com. Forest Mgmt. Resource and Devel. Study, 1963; mem. Gov.'s Maple Industry Com., 1964—; mem. forestry com. N.Y.-Vt. Interstate Commn. on the Lake Champlain Basin, 1956. Mem. Soc. Am. Foresters, Am. Forestry Assn., Sigma Xi, Theta Chi (nat. treas. 1962—, Alumni award, 1962), Alpha Zeta (Distinguished Service award, 1955), Gamma Alpha. Research and numerous publs. on tree physiology and forest tree growth, competition effects on pine growth, stand thinning effects, effect of soil temperature on pine root devel., comparative growth of introduced and native conifers. Home: 36 Overlake Park, Burlington, Vt. 05401.*

ADAMS, William Sprague, Am. physician; b. Sodus, N.Y., May 28, 1919; s. Ephraim Crawford Brown and Bessie (Granger) A.; student Cornell U., 1936-39; M.D., U. Rochester, 1943; m. Esther Driver Stratton, June 19, 1947; children—Thomas G., Mary B., Nancy Q., Esther S. Instr. medicine U. Rochester, 1947-48, research fellow in medicine, 1946-47; faculty U. Cal., Los Angeles, 1948—, prof. medicine, 1958—; adminstv. sect. chief, asst. in research VA Hosp., Los Angeles, 1949-50. Adv. com. on research on pathogenesis cancer Am. Cancer Soc., 1956-60; mem. cancer research tng. com. USPHS, 1960-64. Mem. A.M.A., A.C.P., Western Soc. for Clin. Research, Western Assn. Physicians, N.Y. Acad. Scis., A.A.A.S. Research on purine and pyrimidine biochemistry in man. Home: 206 24th St., Santa Monica, Cal. Office: Dept. Medicine, U. Cal. at Los Angeles Center for Health Services, Los Angeles 90024.*

ADAMS, Wright Rowe, Am. physician, educator; b. Sheridan, Ill., June 14, 1903; s. Harry H. and Bessie (Law) A.; B.S., U. Ill., 1925, M.D., 1929; m. Ruth Chatfield, Oct. 8, 1927; children—Karen Louise, Judson Chatfield. Faculty medicine U. Chgo., 1934—, prof., 1949—, asso. dean div. biol. scis., 1947—, chmn. dept. medicine, 1949-61, chief staff U. Chgo. Hosps. and Clinics, 1961—. Trustee Am. Optical Co., Southbridge, Mass. Diplomate Am. Bd. Internal Medicine (chmn. 1961-63). Fellow A.C.P. (gov. No. Ill. 1957—); mem. A.M.A. (chmn. sect. internal medicine 1962-63), Am. (award of merit 1966, dir., exec. com. 1957—, chmn. council clin. cardiology), Chgo. (pres. 1957-59) heart assns., Assn. Profs. Medicine, Am. Soc. Clin. Investigation, Inst. Medicine Chgo., Chgo. Soc. Internal Medicine, Am. Physiol. Soc., Assn. Am. Physicians, Alpha Omega Alpha, Nu Sigma Nu. Editor: Pulmonary Circulation, 1959; editorial bd. Circulation, 1959-64. Research, publs. on output, size of heart, interpretation of electrocardiogram, effect of altitude on circulation. Home: 5755 Harper Av., Chgo. 60637.*

ADAMSON, Arthur Wilson, Am. chemist, educator; b. Shanghai, China, Aug. 15, 1919 (parents Am. citizens); s. Arthur Quintin and Ethel (Rhoda) A.; B.S. with honors, U. Cal. at Berkeley, 1940; Ph.D. in Chemistry, U. Chgo., 1944; m. Virginia Louise Dillman, Mar. 24, 1942; children—Carol Ann (Mrs. Trenton Parker), Janet Louise, Jean Elizabeth. Research asso. plutonium project Oak Ridge Nat. Lab., 1944-46; faculty chemistry U. So. Cal., Los Angeles, 1946—, prof., 1953—, Unilever prof. Bristol (Eng.) U., 1965-66. Fellow Am. Inst. Chemists, A.A.A.S.; mem. Am. Chem. Soc. (chmn. So. Cal. sect. 1964), Am. Assn. U. Profs. (past pres. U. So. Cal. chpt.). Author: Physical Chemistry of Surfaces, 1960; Understanding Physical Chemistry, 1964. Tech. work, numerous publs. in coordination and surface chemistry. Home: 4400 Via Pavion, Palos Verdes Estates, Cal. Office: Dept. Chemistry, U. So. Cal., Los Angeles 90007.*

ADANSON, Michel, French botanist; b. Aix-en-Provence, France, Apr. 7, 1727; student Coll. Sainte-Barbe, Coll. du Plessis; visited Azores and Canary Islands, 1748; made voyage to Senegal, 1949-54; Fellow Royal Soc., 1761; mem. French Acad. Scis., 1750. Author: Histoire naturelle du Sénégal, 1757; Les Familles naturelles des plantes, 1763; Traité de physiologie végétale; Traité de Botanique rurale. Devised new system of plant and animal classification based on natural succession as indicated by their resemblances (work interrupted by French Revolution, never completed); described flora of Senegal; baobab genus (monkey-bread tree) named Adansonia in his honor. Died Paris, France, Aug. 3, 1806.

ADDINALL, Carl Rupert, chemist; b. Dewsbury, Eng., Dec. 1, 1890; s. Robert and Henrietta (Parker) A.; came to U. S., 1923, naturalized, 1928; B.S. Harvard, 1925, M.A., 1926, Ph.D. in Organic Chemistry, 1930; m. Anna Maria Josephson, Sept. 20, 1947. Instr. chemistry Radcliffe Coll., 1926-27; instr. Harvard, 1928-30; research chemist Merck & Co., Inc., Rahway, N.J., 1930-34, asst. dir. research and devel., 1934-44, dir. tech. information dept., 1945-48, fgn. sci. adv., 1948-56, cons. 1956-62. Sect. editor Chemical Abstracts, 1960, chem. cons., translator, abstractor, 1962—; cons. Dept. Commerce, Washington, 1946-47. Recipient Bowdoin medal Harvard, 1927. Fellow N.Y. Acad. Sci., Chem. Soc., A.A.A.S. Editor: Merck Index, 1960. Author: Story of Vitamin B, 1939; Story of Amino Acids, 1939. Home: 746 Belvidere Av., Westfield, N.J. 07090.*

ADDISON, Christopher, English anatomist; b. Hogsthorpe, Eng., June 19, 1869; s. Robert and Susan Addison; M.D., B.S., U. London; D.Sc. (hon.) Cambridge U.; D.C.L., Oxford U.; LL.D., Sheffield (Eng.) U.; m. Isobel Gray, 1902 (dec. 1934), 2 sons, 2 daus.; m. 2d, Dorothy Low, 1937. Prof. anatomy Univ. Coll., Sheffield; lectr. anatomy St. Bartholomew's Hosp.; mem. faculty medicine, also chmn. bd. intermediate med. studies, mem. bd. human anatomy and morphology U. London; Hunterian prof., examiner in anatomy, univs. Cambridge and London. Mem. Parliament, 1910-22, 29-31, 34-45; parliamentary sec. Bd. Edn., 1914-15, Office Munitions, 1915-16; minister munitions, 1916-17; minister in charge reconstruction, 1917; pres. local govt. bd., 1919; 1st minister health, 1919-21; minister without portfolio, 1921; parliamentary sec. Ministry Agrl., 1929-30; minister agl. and fisheries, 1930-31; sec. state for commonwealth relations, 1945-47; leader House Lords, 1945-51; chmn. Med. Research Council, 1948; paymaster-gen., 1948-49; lord privy seal, 1947-51; lord pres. council, 1951. Created baron, 1937, viscount, 1945. Fellow Royal Coll. Surgeons Eng.; mem. Anat. Soc. Great Britain and Ireland (sec.). Author: The Betrayal of the Slums, 1922; Politics from Within, 2 vols., 1924; Practical Socialism, 2 vols., 1926; Four and a Half Years, 2 vols., 1934; A Policy for British Agriculture, 1939; On the Topographical Anatomy of the Pancreas and Adjoining Viscera; (with J. W. Jennings) With the Abyssinians in Somaliland. Editor Quar. Med. Jour.; Ellis's Demonstrations of Anatomy, 12th edit. Described some points, planes, lines in topography of thorax and abdomen. Died Dec. 11, 1951.

ADDISON, Cyril Clifford, English inorganic chemist; b. Penrith, Eng., Nov. 28, 1913; s. Edward Thomas and Olive (Clifford) A.; B.Sc., U. Durham (Eng.), 1934, Ph.D., 1937, D.Sc., 1947; m. Marjorie Whineray Thompson, July 28, 1939; children—Clifford William Edward, Helen Marjorie. Sci. officer Brit. Launderers Research Assn., 1936-38; lectr. Harris Inst., Preston, Eng., 1938-39; with Ministry Supply, 1939-45; staff Chem. Def. Research Establishment, 1945; faculty U. Nottingham (Eng.), 1966—, prof. inorganic chemistry, 1960—. Fellow Royal Inst. Chemistry (past mem. council, v.p. 1965—), Inst. Physics; mem. Chem. Soc. (past mem. council). Research, numerous publs. on chemistry of non-aqueous solvents, particularly liquid nitrogen oxides and liquid alkali metals. Office: Dept. Chemistry, The Univ., Nottingham, Eng.*

ADDISON, Thomas, English physician; b. Long Benton, Eng., Apr. 1793; s. Joseph Addison; M.D., U. Edinburgh, 1815; Practiced medicine, London; house surgeon Lock Hosp., London; apptd. asst. physician Guy's Hosp., 1824, lectr. on materia medica, 1827, physician, also lectr. (concurrently with Bright) on medicine, 1837; Author: Observations of the Disorders of Females connected with Uterine Irritation, 1830; Constitutional and Local Effects of Disorders of the Renal Glands, 1855; Essay on Disease of the Suprarenal Capsules, 1855; (with John Morgan) Essay upon the Operation of Poisonous Agents, 1829; (with Richard Bright) Elements of Practice of Medicine, 1839; pneumonia research, 1837, 43, and Tb. Described true keloid and Alibert's keloid, 1854; divided anemia into 2 types, pernicious and Addison's disease (prog. destruction of suprarenal cortex), 1855; described inflammation of appendix vermiformis. Died Lanercost, Eng., June 29, 1860.

ADDISON, William, physician; b. Eng., 1802; Fellow Royal Soc., 1846. Gave 1st known description of leukocytosis (under another name), also credited with 1st identification of diapedesis, 1843. Died 1881.

ADEL, Arthur, Am. astrophysicist; b. Bklyn., Nov. 22, 1908; s. Morris and Jennie (Schrieber) A.; A.B., U. Mich., 1931; Ph.D., 1933; m. Catharine Emilia Backus, Sept. 11, 1935. Research asso. Lowell Obs., Flagstaff, Ariz., 1933-42; asst. prof. physics U. Mich., Ann Arbor, 1942-46, asst. prof. astronomy, 1946-48; prof. physics Ariz. State Coll., Flagstaff, 1948—, dir. Atmospheric Research Obs. Fellow Am. Phys. Soc.; mem. Am. Astron. Soc., Am. Meteorol. Soc., Phi Beta Kappa, Sigma Xi. Important works include discovery of atmospheric nitrous oxide, 1939, preparation first prismatic and grating maps of infrared solar-telluric spectra, 1941, discovery atmospheric window, 16-24 microns, 1942, periodic ozone phenomena in atmosphere, 1959. Home: 610 N. Bertrand St., Flagstaff, Ariz. 86002.*

ADELARD OF BATH (or Aethelard), English philosopher; b. Bath, Eng., circa 1090; studied in Tours, Laon, France, also in Spain; became tchr., Laon, also Paris; traveled across Europe, Asia Minor, No. Africa. Author: Perdifficiles quaestiones naturales (summary of Arabic sci.), printed 15th century; De eodem et diverso (written before 1116, attempts a solution to the problems of realism and nominalism). Mathematical algorist; made 1st Latin translation of Euclid's elements; one of earliest translators from Arabic into Latin; introduced Muslim trigonometry into Latin-reading world with transl. of al-Khwarizimi's tables of astron. calculations and sines; used Arabic numbers (established later by Fibonacci). Died circa 1150.

ADELMANN, Howard Bernhardt, Am. embryologist; b. Buffalo, N.Y., May 8, 1898; s. Charles Michael and Louise Henrietta (Kohler) A.; A.B., Cornell U., 1920, A.M., 1922, Ph.D., 1924; student U. Freiburg (Germany), 1927; Sc.D. honoris causa, Ohio State U., 1962. Asst. histology and embryology Cornell U., 1919-21, instr., 1921-25, asst. prof., 1925-37, prof., 1937-66, prof. emeritus Cornell, 1967—; chmn. dept. zoology, 1944-59, faculty rep. bd. trustees 1947-51. NRC fellow biol. sci., 1927-28. Recipient Order Star Italian Solidarity (Italy), 1962. Fellow Institut Internat. d'Embryologie (Amsterdam); mem. Am. Assn. Anatomists, Am. Soc. Zoologists, Hist. Sci. Soc. (Pfizer Award, 1967), Am. Assn. History Medicine (recipient of William H. Welch medal 1967), The Bibliog. Soc. (London), Phi Beta Kappa, Sigma Xi. Author: The Embryological Treatises of Hieronymus Fabricius (Crofts prize Cornell U. Press); Marcello Malpighi and the Evolution of Embryology, 5 vols., 1966. Asso. editor: Jour. Morphology, 1948-51. Contbr. sci. papers to profl. jours. Studies of the embryonic devel. of head, mesoderm, eyes. Home: 410 Columbia St., Ithaca, N.Y.*

ADELON, Nicolas-Philibert, French physician; b. Dijon, France, Aug. 20, 1782; M.D., Paris, 1809; became prof. legal medicine, l'Ecole de Médecine, 1827; mem. Acad. Médicine. Founder (with others) jour. Annales publiques d'hygiene et de médecine légale. Author: Anatomie Physiologique du Cerveau d'après le Systeme de Gall, 1818; Traité de la Physiologie de l'Homme, 1823-24; contbg. author: Biographie universelle; Dictionnaire de médecine, 1821. Died Paris, France, July 19, 1862.

ADELSON, Bernard Henry, Am. physician; b. Tampa, Fla., Mar. 16, 1920; s. Edward H. and Esther (Madesman) A.; A.A., U. Fla., 1938; B.S., Northwestern U., 1942, Ph.D., 1946, M.D., 1950; m. Martha Ann Stein, June 13, 1950; children—Duffie, Edward, David. Asst. prof. internal medicine Northwestern U. Med. Sch., Chgo., 1960—; practice medicine specializing in internal medicine, Winnetka, Ill., 1957—; attending physician Evanston (Ill.) Hosp., 1957—, dir. artificial kidney unit, 1960—; asso. attending physician Cook County Hosp., Chgo., 1960—; cons. physician Municipal Tb Sanatarium, Chgo., 1960—. Mem. A.A.A.S., A.M.A., Sigma Xi. Investigations with artificial kidney. Home: 595 Lincoln St., Glencoe, Ill. 60022. Office: 750 Green Bay St., Winnetka, Ill. 60093.*

ADELSON, Daniel, Am. psychologist; born New York City, Sept. 19, 1918; son of Louis Hirsch and Esther (Edelson) A.; B.A., Coll. City N.Y., 1940; M.A., New Sch. for Social Research, 1953; Ph.D., Columbia, 1957; m. Suzanne Blanche Maricq, July 23, 1948; children—Isabelle Rachel, Ann Deborah, Mark David. Team officer Children's Center, UNRRA, Germany, 1946, dep. chief vol. agy. div., 1946-48; research asso., social psychologist Altro Health and Rehab. Services, N.Y.C., 1953-54, 55-56; social sci. asso. Jewish Bd. Guardians, N.Y.C., 1954-55; research cons. Child Welfare League Am., N.Y.C., 1956; prin. investigator, dir. project drug study with schizophrenics Cal. Dept. Mental Hygiene, Berkeley,

1957-64, asst. dir. Center for Tng. in Community Psychiatry, 1961—. Lectr. psychology Sch. Nursing, U. Cal. at San Francisco, 1960—, research asso. Inst. Human Devel., Berkeley, 1960—; staff U. Cal. Med. Center, San Francisco, 1965—, dir. community mental health program, 1966—, asso. prof. psychology in residence Sch. Nursing, 1967—. Mem. Am., Western psychol. assns., A.A.A.S. Research, publs. on treatment schizophrenia, attitudes of identification and self-acceptance, drug efficacy. Home: 26 Alvarado Rd., Berkeley, Cal. 94705. Office: U. Cal. Sch. Nursing, San Francisco Med. Center, San Francisco 94122.*

ADELSON, Edward, Am. physician, educator; b. Boston, Sept. 14, 1923; s. Abraham and Rebecca (Friedman) A.; B.S. magna cum laude, Tufts U., 1943; M.D. magna cum laude, 1947; m. Lois Potts, Nov. 7, 1954; children—Andrew Jay, Nancy Robin, Stephen Gary. Damon Runyan Clin. research fellow in hematology New Eng. Med. Center, Boston, 1951-53; hematology research Walter Reed Army Inst., Washington, 1953-55, cons., 1955—; pvt. practice medicine specializing in internal medicine and hematology, Washington, 1955—; dir. blood research lab., attending physician George Washington U. Med. Sch., Washington, 1955-67, asso. clin. prof. medicine, 1961—; sr. attending physician Washington Hosp. Center, 1964—; attending physician D.C. Gen. Hosp., 1955—; cons. in hematology Mt. Alto VA, Martinsburg (W.Va.) VA, Andrews AFB, St. Elizabeth's hosps. Mem. Am., Internat. socs. hematology, Am. Fedn. Clin. Research, Soc. Nuclear Medicine, D.C. Med. Soc., A.M.A., Phi Beta Kappa, Alpha Omega Alpha. Research, publs. on blood platelets and blood coagulation. Home: 7020 Richard Dr., Bethesda, Md. 20034. Office: 1100 22d St. N.W., Washington 20037.*

ADER, Clément, French aero. pioneer, inventor; b. Muret, France, Apr. 2, 1841; chief of constrn. Dept. Pub. Engring.; helped build Toulouse-Rayonne Ry., 1862; became builder and technician, Paris, France, 1876. Author: La Primière étape de l'aviation militaire, 1907; L'Aviation militaire, 1909. Sometimes known as father of aviation; built machine which made 1st airplane flight, 1890; flew 300 meters, 1897; coined French word avion (airplane); inventor rubber tires for velocipede, 1866, microphone, 1876; installed 1st telephone system in Paris, 1878. Died Muret, 1923.

ADER, Robert, Am. psychologist; b. N.Y.C., Feb. 20, 1932; s. Nathan and Mae (Levine) A.; B.S., Tulane U., 1953; Ph.D., Cornell U., 1957; m. Gayle Simon, June 2, 1957; children—Deborah Nan, Janet Lynn, Norine Lee, Leslie Allison. Faculty, U. Rochester Sch. Medicine and Dentistry, 1957—, research asso. prof. psychiatry, 1964—. Recipient USPHS Research Career Devel. award, 1964—. Mem. Am., Eastern psychol. assns., Am. Psychosomatic Soc., Animal Behavior Soc., Soc. for Psychophysiol. Research, Psychonomic Soc., Sigma Xi. Research, publs. on psychosomatic phenomena in animals, elucidation of role of early life experiences on adult susceptibility to disease. Home: 31 Bradford Rd., Pittsford, N.Y. 14534. Office: U. Rochester Med. Center, Rochester, N.Y. 14620.*

ADERHOLD, Rudolf Ferdinand Theodor, German biologist; b. Bad Frankenhausen, Germany, Feb. 12, 1865; s. Friedrich Oskar and Magdalena A. F. (Picht) A.; ed. Berlin, also Jena, Germany; m. Anna Elise Clementine Haccius, May 28, 1895; 1 son. Employee, Inst. for fruit culture and wine growing, Geisenheim; apptd. head bot. dept., research sta. Pomol. Inst., Proskau, Silesia, 1893; became head bot. lab. Kaiserliches Gesundheitamt, Berlin, 1901; became dir. Biol. Inst. Agrl. and Forestry, 1905. Research in mycology, applied physiology and pathology of plants, especially fruit trees. Died Berlin-Dahlem, Germany, Mar. 17, 1907.

ADES, Harlow Whiting, Am. neurophysiologist; b. Rockford, Ill., Dec. 31, 1911; s. Arthur William and Mary (Thayer) A.; B.S., U. Ill., 1934, M.S., 1935; Ph.D., 1938; m. Claire Frenkel McMillan, July 24, 1963; children—Helen May (Mrs. Myron Butler), Elizabeth Anne (Mrs. Ralph K. Park), William Edward, Susan. Faculty, U. Ill., 1934-38, Johns Hopkins, 1938-39, Emory U., 1939-54, Northwestern U., 1947-48; prof. anatomy U. Tex. Southwestern Med. Sch., 1954-57; head neurol. scis. div. research dept. U.S. Naval Sch. Aviation Medicine, Pensacola, Fla., 1957—; prof. elec. engring. and physiology; biophysics U. Ill., Urbana, 1965—. Cons. VA hosps. Atlanta, McKinney, Tex., USN, U. Ill.; mem. Navy-Air Force Project on Biol. Effects of Noise-Exploratory, 1953; mem. Armed Forces-NRC Com. on Hearing, Bioacoustics, 1955, NASA Research Adv. Com. on Biotech. Mem. Am. Physiol. Soc., Am. Assn. Anatomists, Am. Neurol. Assn., Am. Acad. Neurology, So. E.E.G. Soc., Aerospace Med. Assn., A.A.A.S., Sigma Xi. Research, publs. on structure, function, pathology of auditory and vestibular sytems. Home: 508 E. Sunnycrest Ct. Office: Dept. Elec. Engring., U. Ill., Urbana, Ill. 61801.*

ADET, Pierre August, chemist; b. Nevers, France, May 18, 1763. Named sec. to 1st commr. to Santo Domingo, later chief of colony adminstrn.; became physician to Ministry of Marine, also adjunct minister; apptd. minister plenipotentiary to U. S., 1795; elected mem. French Senate, 1809. Author: Méthode de nomenclature chimique, 1787; a reply to Priestley's Considerations on Phlogiston, 1797; Leçons élémentaires de chimie, 1804. Editor Annales de chimie et de physique. Theorized that inflammable air was hydrogen in charcoal; defended chem. nomenclature introduced by Lavoisier and others against criticism of De la Metherie; devised (with Hassenfratz) symbols to accompany new nomenclature, 1787. Responsible for the diffusion of Priestley's ideas in France. Died 1832 or 1834.

ADEY, William Ross, physiologist, educator; b. Adelaide, Australia, Jan. 31, 1922; s. William James and Constance (Weston) A.; M.B., B.S., U. Adelaide, 1943, M.D. (Nuffield Found. fellow in medicine), 1949; postgrad. U. Oxford (Eng.); m. Alwynne Sydney Morris, June 24, 1944; children—David John, Susan Ruth, Geoffrey Robert. Lectr., sr. lectr., reader in anatomy U. Adelaide, 1946-53; sr. lectr. in anatomy U. Melbourne (Australia), 1954-56; prof. anatomy and physiology U. Cal. at Los Angeles, 1957—, dir. Space Biology Lab., Brain Research Inst., 1961—. Mem. cybernetics panel Pres.'s Sci. Adv. Com., 1963—; cons. USAF Sci. Adv. Bd., 1960—, NASA, 1961—, VA, 1957—; mem. Nat. Research Resources Adv. Com. NIH, 1962—. Royal Soc. London and Nuffield Found. fellow, 1956; recipient C. J. Herrick award Am. Assn. Anatomists, 1963. Mem. I.E.E.E. (nat. adv. com. 1961—), Am. Electroencephalographic Soc. (council). Research, publs. on brain mechanisms in learning and memory, techniques for measurement and study of brain functions. Home: 7127 Dume Dr., Malibu, Cal. 90265. Office: Dept. Anatomy, U. Cal. at Los Angeles Health Scis. Center, Los Angeles 90024.*

ADIERNA, Giovanni Battista, see Hodierna, Giovanni Battista.

ADKINS, Dorothy Christina, Am. psychologist; b. Atlanta, Ohio, Apr. 6, 1912; d. George Hoadley and Pearl (James) A.; B.Sc., Ohio State U., 1931, Ph.D., 1937. Asst. examiner U. Chgo., 1936-38, research asso., 1938-40; chief, research and test constrn. U. S. Social Security Bd., 1940-44; policy cons., chief test devel. U. S. Civil Service Comn., 1944-48; prof. psychology U. N.C., Chapel Hill, 1948-65, chmn. dept., 1950-60; prof., researcher Coll. Edn., U. Hawaii, Honolulu, 1965—, also dir. Head Start Evaluation and Research Center. Fellow Am. Psychol. Assn. (recording sec. 1949-52, dir. 1955-58); mem. Psychometric Soc. (pres. 1949-50), Inter-Am. Soc. Psychology, Am. Edn. Research Assn., Sigma Xi. Author: Construction and Analysis of Achievement Tests, 1947; (with S. B. Lyerly) Factor Analysis of Reasoning Tests, 1952; Test Construction, 1960; Statistics: an Introduction for Students in the Behavioral Sciences, 1964; also articles. Asst. mng. editor Psychometrika, 1937-50, mng. editor, 1950-56, council of editors, 1956—. Research in test devel. and statis. item analysis methods. Home: 921 Kealaolu Av., Honolulu 96815.*

ADKINS, Homer Burton, Am. chemist; b. Newport, O., Jan. 16, 1892; s. Alvin and Emily (Middeswart) A.; grad. Denison U., 1915, D. Sc. (hon.), 1938; M.S., Ohio State U., 1916, Ph.D., 1918. Research chemist War Dept., 1918; instr. organic chemistry Ohio State U., 1918-19; research chemist E. I. Du Pont De Nemours & Co., summer 1919; prof. U. Wis., 1919-49; with Bakelite Corp., summers 1924, 26; adminstr., research dir. NDRC, also OSRD, 1942-45. Recipient Medal of Merit, U. S. Govt. Mem. Nat. Acad. Scis., Chem. Soc. London, Chemists Club N.Y., Am. Assn. Univ. Profs. Author: Reactions of Hydrogen, 1937. Co-author: Practice of Organic Chemistry; Elementary Organic Chemistry. Research in causation of organic chem. reactions, relations of catalysts, reactivity and structure of organic compounds, reaction of hydrogen with organic compounds, oxidation potential; developed high pressure hydrogenation as tool for use in synthetic chemistry; gave 1st practical methods of preparing compounds, especially catalytic hydrogenation of esters. Died 1949.

ADKINS, Walter Scott, Am. geologist; b. Dec. 24, 1890; s. Richard Byron and Elizabeth (Thomas) A.; B.S., U. Tenn., 1910, grad. asst., 1910-ll; postgrad. Columbia, 1911-12, U. Paris (France), 1925-26; m. Mary Grace Muse. Asst. prof. biology Tex. Christian U., Fort Worth, 1912-16; tchr. anatomy U. Ill. Med Sch., 1916-18, Baylor U. Med. Sch., Dallas, 1918-19; mem. staff bur. econ. geol. U. Tex., 1919-21, 1926-34, became cons., 1951; with Compañia Mexicana de Petroleo, Tampico, Mexico, 1921-25, Shell Oil Co., 1934-50. Guggenheim fellow Europe, also Brit. Mus. Natural History, 1931-32. Mem. Geol. Soc. Am., Am. Assn. Petroleum Geologists, Paleontol. Soc., Soc. Econ. Paleontologists and Mineralogists, Sociedad Geologica Mexicana, Société Géologique de France, Tex. Acad. Sci. Contbd. articles to profl. jours. Authority on cretaceous of Tex. and S.W.; contbd. to Tex. Mesozoic geology. Died Sept. 22, 1956.

ADKINSON, Burton Wilbur, Am. geographer; b. Everson, Wash., Mar. 5, 1909; s. Jason Hope and Clara (Warriner) A.; B.A., U. Wash., 1936, M.A., 1939; Ph.D., Clark U., 1942; m. Margaret Louise Klock, Sept. 10, 1942; children—Karen Louise, Margaret Jane. Asst. chief, map div. Library of Congress, Washington, 1945-47, chief, 1947-48, asst. dir. reference dept., 1948-49, dir., 1949-57; head Office of Sci. Information Service, NSF, Washington, since 1957—. Fellow A.A.A.S., member of Assn. Am. Geographers, Internat. Fedn. for Documentation (past pres.), Am. Documentation Inst., Spl. Libraries Assn., A.L.A. Home: 5907 Welborn Dr. N.W., Washington 20016. Office: 1800 G St. N.W., Washington 20550.*

ADKISSON, Perry Lee, Am. entomologist; b. Hickman, Ark., Mar. 11, 1929; s. Robert Louis and Imogene (Perry) A.; B.S. in Agr., U. Ark., 1950, M.S. in Agronomy, 1954; Ph.D. in Entomology, Kan. State U., 1956; postgrad. (fellow) Harvard; m. Frances Rozelle, Dec. 29, 1956; a dau., Jean Amanda. Asst. prof. U. Mo., 1956-58; faculty Tex. A. and M. U., College Station, 1958—, prof. entomology, 1963—. Recipient Faculty Distinguished Achievement award for research Tex. A. and M. U., 1965. Mem. A.A.A.S., Entomol. Soc. Am. (recipient J. Everett Bussart award 1967), Kansas Entomol. Soc., Sigma Xi. Research, numerous publs. on cotton insect control, photoperiodic control insect diapause. Home: 305 W. Brookside St., Bryan, Tex. 77803. Office: Dept, Entomology, Tex. A. and M. U., College Station, Tex. 77843.*

ADLER, Alfred, psychologist; b. Vienna, Austria, Feb. 7, 1870; s. Leopold and Pauline Adler; M.D., Vienna U., 1895; LL.D.; m. Raissa Epstein, 1898; 1 son, 3 daus. Worked in Vienna Gen. Hosp. and Polyclinic, 1895-97; gen. med. practitioner, nerve specialist, Vienna, 1897-1927; joined Freud's discussion group (became 1st psychoanalytic soc.), 1907, broke with Freud and began Soc. for Free Psychoanalysis (later Soc. for Individual Psychology), 1911; organized (with others) Child Guidance Centers to prevent neurosis and delinquency in childhood, at 30 schs. in Vienna, from 1912; founded Jour. Individual Psychology, 1914; attached to Austrian Army, Vienna, also Cracow, 1914-18; lectr. Pedagogical Inst. City of Vienna, 1924; lectr. Columbia U., N.Y.C., 1927, later at Med. Centre, N.Y.C.; clin. dir. Mariahilfer Ambulatorium, Vienna, 1928; became vis. prof. med. psychology, cons. psychologist L.I. Coll. Medicine, 1932; founded Jour. Indivdual Psychology in U. S. A., 1935. Author: Individual Psychology; Practice and Theory of Individual Psychology, 1920, transl., 1927; The Science of Living, 1929; Education of Children, 1930; The Pattern of Life, 1930; What Life Should Mean to You, 1931; Social Interest: a Challenge to Mankind, 1937; Understanding Human Nature, transl. 1946; Problem Child, transl. 1963; The Education of the Individual; also numerous books in German on individual psychology, edn., religion, homosexuality. Founder sci. of individual psychology; theorized (early in career) that failure of individual to adapt to organic weakness results in emotional disturbance, later turned his attention to instinct; introduced concept of the aggressive drive (subordinating primary drives); introduced concept that need for superiority and power is primary driving force of personality, differing from Freud's emphasis on sexual libido; introduced terms inferiority complex (admitted inferiority) and compensation (effort to overcome inferiority); believed it is the therapist's task to interpret how the patient deceives himself in regard to this life style; saw behavior disorders as overcompensation for deficiencies (either phys. or environmental); thus concentrated more on overt personality problems than on the unconscious in his therapy; his theories on aggression and compensation foreshadowed many later theories in psychoanalysis and had widespread influence on devel. of psychoanalysis. Died Aberdeen, Scotland, May 28, 1937.

ADLER, Erich, chemist; b. Frankenreuth, Germany, Oct. 24, 1905; s. Max and Elfriede (Putzker) A.; Dr.Ing., Technische Hochschule, Munich, Germany, 1931; m. Anne-marie Pawel, Aug. 19, 1935; children—Eva Marianne (Mrs. Allan Garellick), Rolf Mikael. Research asst. Techn. Hochschule, Munich, 1931-32, City Hosp. Lab., Mannheim, Germany, 1932-33; research asst. Inst. Biochemistry, U. Stockholm (Sweden), 1933-38, sr. research asst. Inst. Organic Chemistry, 1938-44; research asso. Forest Products Research Lab., Stockholm, 1944-52; asst. prof. organic chemistry Royal Inst. Tech., Stockholm, 1946-52; prof. organic chemistry Chalmers U. Tech., Göteborg, Sweden, 1952—, head chemistry dept., 1962-66. Decorated Riddare av Nordstjärne-Orden. Mem. Kungl. Vetenskaps-och Vitterhetssamhället, Göteborg, Royal Swedish Acad. Engring. Scis. Research, numerous publs. in biochemistry including enzymes involved in dehydrogenation processes, organic chemistry, wood chemistry, chemistry of phenol-formaldehyde resins, synthesis of estrogenic substances, oxidation reactions with phenols, structure and reactions of lignin. Home: 11, Raketgatan, S-41320 Göteborg, Sweden.*

ADLER, Felix T., physicist; born Zurich, Switzerland, Jan. 5, 1915; s. Friedrich Wolfgang and Catherine (Germanschskaya) A.; M.S. in Math., U. Zurich, 1936, Ph.D. in Physics, 1939; m. Donatella G. Baroncini, Aug. 17, 1959. Came to U. S., 1940, naturalized, 1949. Mem. Inst. for Advanced Study, Princeton, N.J., 1941-42; vis. asst. prof. physics U. Wis., Madison, 1942-45, 47-50; sci. officer Minister of Supply, Montreal, Can. 1943-46; asso. prof. physics U. Colo., Boulder, 1946; asso. prof. physics Carnegie Inst. Tech., Pitts., 1950-56; mem.

staff Gen. Atomic, San Diego, 1956-58, cons., 1959-61; prof. physics and nuclear engring. U. Ill., Urbana, 1958——; cons. Oak Ridge Nat Lab., 1955-57, U. Cal. Radiation Lab., Livermore, 1960-61, Los Alamos, (N.M.) Sci. Lab., 1961——, Argonne (Ill.) Nat. Lab. 1959——, Westinghouse Astro Nuclear lab., 1963-65. Recipient Carnegie award for teaching, 1954. Fellow Am. Phys. Soc.; mem. Am. Nuclear Soc. Asso. editor Am. Jour. Physics, 1949-52. Home: 20 G. H. Baker Dr., Urbana, Ill. 61801.*

ADLER, Friedrich, German archaeologist, architect; born Berlin, Germany, Oct. 15, 1827; son of Johann Karl and Maria Louise Dorothea Adler; educated at Berlin Architectural Academy; married Christiane Dorothea Pauline Kohler, Oct. 21, 1854; 5 daughters; married 2d, Karolina D. Trendlenburg; 1 dau. Became tchr. Berlin Constrn. Acad., 1859, docent, 1861, professor, 1863, consulting architect to Prussian minister of public works, 1877-1903. Member Royal Prussian Academy Art, German Archaeol. Inst. Author: Ma. Backsteinbauten der preussischen Staates, 1859-89; Baugeschichtlischer Forschungen, 2 vols., 1870-79; Das Mausoleum zu Halikarnass, 1900; (with E. Curtius) Olympia, 5 vols., 1890-97; Zur Kunstgeschichte, 1906. Synthesized classical forms of architecture and medieval principles of construction; made archeological expeditions to Olympia, Italy, Asia Minor, Palestine; archeol. work foreshadowed that of H. Schliemann. Died Berlin, Sept. 15, 1908.

ADLER, Gottlieb, physicist; b. Steken, Bohemia (now Czechoslovakia), Mar. 7, 1860; s. Josef and Franziska (Pichler) A.; student U. Vienna (Austria), 1877-81, degree, 1882; postgrad. U. Berlin (Germany), 1881-82. Gymnasium tchr., 1882-83; became lecturing prof., 1885; apptd. asso. prof. physics U. Vienna, 1893. Author: Über die Energie und den Zwangszustand im elektrostatischen Felde, 1885. Research on mech. interpretation of electricity. Died Vienna, Dec. 14, 1893.

ADLER, John Hans, economist; born Tachov, Czechoslovakia, Nov. 16, 1912; s. August and Lilly (Beck) A.; J.D., German U., Prague, Czechoslovakia, 1937; M.A., Yale, 1940, Ph.D., 1946; m. Vilma Joan Rabl, Sept. 12, 1939; children—Catherine M., Marcia V. Came to U. S., 1938, naturalized, 1945. Instr. Oberlin (O.) Coll., 1942-44; economist Fed. Res. Bd. Washington, 1944-45; chief fgn. exchange sect. U. S. Element Allied Commn. for Austria, 1945-46, dept. chief finance div., 1946-47; economist Fed. Res. Bank, N.Y.C., 1947-50; economist Internat. Bank for Reconstrn. and Devel., Washington, 1950-57, econ. advisers, 1957-61, dir. Econ. Devel. Inst., 1962-67, senior adviser, since 1967——. Member of the American Econ. Assn., Royal Econ. Soc., Soc. for Internat. Devel. Author: (with H. C. Wallich) Public Finance in a Developing Country, 1950; (with E. R. Schlesinger, E. C. Olsen) Public Finance and Economic Development in Guatemala, 1952; The Determinants of U. S. Import Trade, 1923-38, 1952; also articles. Home: 5620 Western Av., Chevy Chase, Md. 20015. Office: 1818 H St. N.W., Washington 20433.*

ADLER, Max, Austrian sociologist; b. Vienna, Austria, Jan. 15, 1873; m. Jenny Herzmark; 1 son, 1 dau.; became prof. U. Vienna. Author: Kausalität und Teleologie im Streite um den Wissenschaft, 1904; Marx als Denker, 1908; Wegweiser, Studieren zur Geistesgeschichte der Sozialismus, 1914; Engels als Denker, 1919; Der Marxismus als proletarische Lebenslehre, 1922; Das Soziologische in Kants Erkenntniskritik, 1924; Kant und das Marxismus, 1925; Lehrbuch der materialistischen Geschichte auffassung, 1930; Die Rätsel der Geschichte, 1926. Tried to create (on basis on Kant's and Marx's work) philos. founds. of modern critical socialism (transcendental socialism). Died Vienna, June 28, 1937.

ADLER, Robert, physicist; b. Vienna, Austria, Dec. 4, 1913; s. Max and Jenny (Herzmark) A.; Ph.D., U. Vienna, 1937; m. Mary Buehl, July 5, 1946. Came to U. S., 1940, naturalized, 1945. In charge Lab. Sci. Acoustics, Ltd., London, Eng., 1939-40; in charge lab. Asso. Research, Inc., Chgo., 1940-41; mem. research group Zenith Radio Corp., Chgo., 1941-52, asso. dir. research, 1952-63, dir. research, 1963——. Fellow I.E.E.E. (mem. adminstrv. com. electron devices group 1964——), A.A.A.S.; mem. Nat. Acad. Engring. Contbr. articles to sci. jours. Research on acousto-optical interaction; research on phasitron modulator used in early FM transmitters, receiving tubes for FM detection and color demodulation, transverse-field traveling wave tubes, electron beam parametric amplifier; electromech. I.F. filter, ultrasonic remote control for tv receivers. Home: 327 Latrobe Av., Northfield, Ill. 60093. Office: 6001 W. Dickens Av., Chgo. 60639.*

ADO, Andrey Dmitrievich, Russian pathophysiologist; b. 1909; grad. Kazan Med. Inst., 1931; D.Med.Sci., 1935; Specimen technician Kazan Med. Inst., 1926-31, asst. head chair path. physiology, 1931-38, head, 1938-52, also head Lab. Exptl. Immunology, Kazan Inst. Vaccines and Sera; prof. head chair path. physiology Pirogov 2d Moscow Med. Inst., 1952——, also head Allergological Research Lab., head Pathophysiol. Lab., Ivanovsky Inst. Virology. Del., 4th European Congress on Allergy, London, Eng. Mem. USSR Acad. Med. Sci. (corr.). Author over 80 works including Anaphylactic Shock and Allergic

Alteration of the Tissues, 1944, The Present State of Allergy Theory, 1951, Antigens as Extraordinary Stimuli of the Nervous System, 1952, Manual of Pathological Physiology, 1953, The Significance of Reflexes in the Development of Disease, 1955, The Pathophysiology of Phagcoytes, 1961. Co-editor Pathology and Morphology sects. Large Med. Ency., 2d edit.; mem. editorial bd. Jour. Soviet Medicine; dep. editor Pathological Physiology and Experimental Therapy. Research on path. physiology, allergy, immunity, effects of typhoid, dysentery and other antigens on acetylcholine metabolism, Address: 2d Moscow Med. Inst., Malaya Pirogovskaya 1, Moscow, USSR.

ADOLPH, Edward Frederick, Am. physiologist; b. Phila., July 5, 1895; s. William and Wilhelmina (Fleischmann) A.; A.B., Harvard, 1916, Ph.D., 1920; postgrad. Yale; m. Mary Grace Bagg, Apr. 4, 1921; children—Jean (Mrs. John H. Macpherson Jr.), Ruth (Mrs. Charles C. Hutton), Carl Edward. Faculty, Inst. U. Pitts., 1921-24, Marine Biol. Lab., Woods Hole, Mass., 1921-24; NRC fellow Johns Hopkins, 1924; faculty U. Rochester (N.Y.), 1925——, prof. physiology, 1948-60, prof. emeritus, 1960——; responsible investigator OSRD, 1942-45. Recipient Presdl. Certificate of Merit, 1948, Alumni Gold medal Sch. Medicine and Dentistry, U. Rochester, 1964. Mem. Am. Physiol. Soc. (past pres.), Am. Soc. Zoologists, Soc. Gen. Physiologists, Soc. for Study Growth, Am. Soc. Naturalists, Am. Acad. Arts and Scis., Sigma Xi, Alpha Omega Alpha. Author: Regulation of Size in Unicellular Organisms, 1931; Physiological Regulations, 1943; (with others) Physiology of Man in the Desert, 1947; also numerous articles. Research on role water in living organisms, physiol. regulation body fluids, body volumes, body temperatures; devel. in individuals self-regulation of fluids, heartbeats, and other characteristics. Home: 210 Castlebar Rd., Rochester, N.Y. 14610.*

ADRAIN, Robert, mathematician; b. Carrickfergus, Ireland, Sept. 30, 1775; m. Ann Pollock, 1 child. Came to Am., 1789; prof. natural philosophy and math. Queens Coll. (now Rutgers U.), New Brunswick, N.J., 1809-12, 26; prof. Columbia, N.Y.C., 1813-25; prof. math. U. Pa., 1827-34, vice provost, 1828-34; considered one of most brilliant mathematicians of his time in Am.; founder journals Analyst, or Mathematical Companion, 1808, The Mathematical Diary, 1825-33. Editor: Mathematics (of Hutton); Algebra (of Ryan), 1839, Contbd. 2 papers to Trans. Am. Philos. Soc. Suggested a curve which he named catenaria volvens; gave 2 proofs of exponential law of error (before Gauss), 1808. Died New Brunswick, Aug. 10, 1843.

ADRIAANSZOON, see Metius, Adriaen.

ADRIAENSZOON, Jacob Metius, Dutch optician; b. Alkmaar, Netherlands; s. Adriaen Authoniszoon; invented (independently of Z. Janszoon and J. Lippersheim) telescope, circa 1608.

ADRIAN OF CAMBRIDGE, Baron (Edgar Douglas Adrain), English physiologist; b. London, Eng., Nov. 30, 1889; s. Alfred Douglas and Flora Lavinia (Barton) A.; ed. Westminster Sch., Trinity Coll.; M.A., M.D., Cambridge; hon. degrees various universities; m. Hester Agnes Pinsent, June 15, 1923; 2 daus., 1 son. Lectr. in physiology Cambridge U., 1920-29, prof. physiology, 1937-51, master Trinity Coll., 1951-65; chancellor U. Leicester, 1957——. Fellow Royal Soc., 1923 (Foulerton research prof. 1929, Copley medal 1946, pres. 1950-55, fgn. sec. 1946); recipient (with Sherrington) Nobel prize for physiology and medicine for work on functions of neurons, 1932. Chevalier Legion of Honour (France); created baron, 1955; recipient Order of Merit, 1942. Pres. Royal Soc. Medicine, 1960-62; fellow Academia del Lincei; hon. mem. Am. Physiol. Soc., Nat. Acad. Medicine (Argentina), Acad. Nacional de Medicina (Mexico); fgn. asso. Nat. Acad. Scis. (USA); hon. fgn. mem. Am. Acad. Arts and Scis., Royal Acad. Medicine (Belgium), Royal Acad. Sci. (Amsterdam), Royal Flemish Acad. Sci.; corr. mem. Acad. Sci. (Paris); fgn. asso. Acad. Medicine (Paris), Kungl. Vetenshaps. Soc. (Upsala); mem. Am. Philos Soc. Author: The Basis of Sensation, 1928; The Mechanism of Nervous Action, 1932; The Physical Basis of Perception, 1947; also many papers. A foremost investigator of nervous system; one of 1st to realize the importance of brain wave research; carried out research on brains and ganglia of fish, insects, and mammals; (with Mathews) showed that change of electric potential in electroencephalograph is due to elec. activity of cortex. Home: Trinity College, Cambridge, Eng.*

ADRIANI, John, Am. physician; b. Bridgeport, Conn., Dec. 2, 1907; s. Nicola and Lucia (Caseria) A.; A.B., Columbia, 1930, M.D., 1934; m. Irene Ann Miller, Sept. 7, 1953; 1 son, John N. Fellow, N.Y. U., 1936-38, instr., 1938-41; faculty La. State U., New Orleans, 1941——, prof. surgery and pharmacology, 1941——; faculty Tulane U., New Orleans, 1944——, prof. surgery, 1950——; prof. gen. anesthesia Loyola Sch. Dentistry, New Orleans, 1947——. Recipient Distinguished Service award Am. Soc. Anesthesiologists, 1949, Internat. Anesthesia Research Soc., 1956; Guedel medal, Sigma Xi, 1959; certificate merit Mayoralty New Orleans, 1953. Fellow Am. Surg. Assn., Soc. Clin. Pharmacology and Chemotherapy; mem. So. Soc. for Clin. Re-

search, Soc. for Exptl. Biology and Medicine. Author: (with Thomas) Pharmacology of Anesthetic Drugs, 1941; Chemistry and Physics of Anesthesiologist, 1946; Techniques and Procedures of Anesthesiology, 1946; Nerve Blocks, 1955; Appraisal of Current Concepts of Anesthesiology, vol. I, 1961, vol. II, 1964, vol. III, 1966; Labat Regional Anesthesiology; also numerous articles. Research on mode of action of drugs used for anesthesia; devel. apparatus for adminstrn. anesthetics, local anesthetics, prevention serious reactions. Home: 67 N. Park Pl., New Orleans 70124. Office: Charity Hosp., New Orleans 70140.*

ADSON, Alfred Washington, Am. surgeon; b. Terril, Ia., Mar. 13, 1887; s. Martin and Anna (Bergeson) A.; B.Sc., U. Neb., 1912, A.M., 1918, D.Sc., 1948; M.D., U. Pa., 1914; M.S. in surgery, U. Minn., 1918; hon. D.Sc., St. Olaf Coll.; m. Lora G. Smith, Aug. 3, 1911; children—William Walter, Mary Louise, Martin Alfred. Successively fellow in surgery Mayo Clinic, 1st asst., jr. surgeon, neurol. surgeon, chief neurol. surgeon; later became neuro-surgeon, prof. neuro-surgery Mayo Foundation Grad. Sch. of U. Minn. Mem. Minn. Bd. Med. Examiners from 1929; mem. med. council VA; chmn. Am. Bd. Neurol. Surgery. Fellow A.C.S.; mem. A.M.A. (council on med. service and pub. relations), Am., Western surg. assns., Am. Neurol. Assn., Am. Neuro. surg. Assn. (pres. 1932), Internat. Neurol. Assn., Internat. Congress Surgeons, Sigma Xi, Alpha Omega Alpha, Research and publs. in devel. and improvement of surg. technique in removal of brain and spinal cord tumors; devel. of operations for treatment glossopharyngeal neuralgia, cervical ribs, Reynaud's disease, Hirschsprung's disease and essential hypertension. Died Nov. 12, 1951.

AEBY, Christopher Theodore, anatomist; born Gutenbrunnen, Switzerland, Feb. 25, 1835; student U. Göttingen (Germany), 1856-58; M.D., U. Basel (Switzerland), 1858. Faculty U. Basel, 1858-63, asso. prof. anatomy, 1863; apptd. prof. human and comparative anatomy U. Bern (Switzerland), 1863; apptd. prof. U. Prague (Czechoslovakia), 1884. Author: Die Symphysis Ossium pubis der Menschen, 1858; Untersuchungen über die Fortpflanzungsgeschwindigkeit der Reizung . . . , 1862; Die Schädelformen der Menschen und der Affen, 1867; Lehrbuch der Anatomie, 1871; Lehre von den Eingeweiden; Der Bronchialbaum der Saügethiere, 1880; Faserverlauf im Gehvin und Rückenmark, 1882; Das Hochgebirge von Grindelwald. Research on comparative anatomy of joints, also osteology; described Aeby's plane in nasion and basion perpendicular to median plane of cranium, 1867. Died Bilina, Czechoslovakia, July 7, 1885.

AEGIDIUS CORBOLIENSIS, see Giles of Corbeil.

AEPINUS, Franz Maria Ulrich Theodor Hoch, physicist; b. Rostock, Germany, Dec. 13, 1724; s. Franz Albert Aepinus; ed. in Rostock, also Jena, Germany; degree in math. Became pvt. docent U. Rostock; apptd. prof. astronomy Acad. Scis. Berlin, 1755, Acad. St. Petersburg, 1757; remained in Russia until 1798; mem. Imperial Acad. Scis. Russia. Author: Reflections on the Distribution of Heat on the Surface of the Earth, 1762; Electricity of Tourmaline, 1757; Tentamen theoriae electricitatis et magnetismi, 1759; essays on math., physics, astronomy. Attempted to develop fluid theory of Franklin; discovered electric properties of tourmaline; improved microscope; experimented with electricity and mag netism; devised double touch method of magnetizing; investigated pyroelectricity; explained virtually all elec. induction exclusively in terms of attraction, repulsion, flow of electricity in conductors; studied relation between conductors and non-conductors; designed parallel-plate electric condenser; investigated energy of Spitzen, dielec. properties of glasses, thin air layers; used differential calculus; discussed effects of parallax in transit of planet over sun's disk. Died Dorpat, Estonia, Aug. 10, 1802.

AETHELARD, see Adelard of Bath.

AETIOS (or AETIUS) OF AMIDA, physician; b. Amida, Mesopotamia, circa 502; ed. in Alexandria, Egypt; became physician to Emperor Justinian in Byzantium. Author med. ency. in 16 books, including best and most complete work of ancient times on ophthalmology, also books on internal medicine, surgery, obstetrics, gynecology, and description of diptheria, med. use of doves, use of ligatures to tie off blood vessels. Died circa 550.

AFANASEV (or AFANASIEV), Georgii Dmitrievich, Russian geologist, petrographer; born Mar. 17, 1906; graduated Leningrad Univ., 1930. With Petrographic Institute, USSR Academy of Sciences, 1930——, then with Institute of Geological Science., learned sec. dept. geology and geog. sci., 1948-53, dep. chief editor jour. News of USSR Acad. Sci.-Geology Series, 1954——, bur. mem. dept. geology and geog. sci., 1960——, chmn. Commn. for Exptl. Research on Deep Processes and Phys. Properties of Ores, 1962. Mem. USSR Acad. Sci. (corr.). Author: The Ground Deposits of Lake Sevan, 1933; Granitoids of Ancient Intrusive Formations in the Northwestern Caucasus, 1950; The Value of Granitization as a Process Leading to the Evolution of Granitoid Blocks, 1952; Basic Results from Studies of the Magmatic Geology of the

15

North Caucasus Folded Area, 1956; co-author: Date on Proving the Age of Boundaries between Certain Geological Systems and Epochs, 1963. Dep. chief editor Geology, 1956—. Research on magma formations in the Caucasus; study of petrogenesis, lake sedimentation, granitization. Address: USSR Acad. of Sci., Leninsky prosp. 14, Moscow, USSR.

AFONSO, Skoda, physiologist; b. Goa, India, Nov. 27, 1923; s. Bossuet and Berta (Dias) A.; came to U. S., 1959; grad. Med. Sch. Goa, 1946; M.S., in Physiology, U. Wis., 1961, Ph.D., 1965. Practice medicine specializing in heart diseases, Goa, 1948-59; research asso. medicine Cardiovascular Research Lab., U. Wis., Madison, 1961—. Mem. A.A.A.S. Research, publs. on cardiovascular physiology, measurement coronary blood flow; developed method for measurement cardiac heat prodn., thermodilution flowmeter for measurement coronary blood flow. Home: 200 S. Mills St., Madison, Wis. 53715.*

AFRICANUS, Sextus Julius, explorer, historian; b. Libya, circa 150; author world history from 5499 B.C. to 221 A.D.; credited with book which treats semaphore signaling and how to find width of a river from one side, circa 200.

AFSMANN, Paul Gustav Bernhard, German geologist; b. Dresde, Jan. 3, 1881; s. Gustav and Sophie (Pechstein) A.; ed. Techische Hogeschool Dresde, U. Berlin; doctorate degree; m. Herta Wunderlich, Jan. 26, 1933; children—Ilse, Mathilde, Charlotte. Asst. geology Acad. Mines, Berlin, 1906-08; geologist Prussian Inst. Geology, Berlin, from 1908; govt. geologist, from 1933. Mem. German Geol. Soc. Author: Fauna der oberschlesischen Frias; Das Berliner Diluvium; also other publs. on fish and crayfish. Address: Detmolder Str. 4, 1 Berlin 31, West Germany.

AFZELIUS, Adam, Swedish botanist; b. Larf, Sweden, Oct. 8, 1750; s. Frederick George Afzelius, became tchr. Oriental languages U. Uppsala (Sweden), 1777, demonstrator botany, 1785, prof. materia medica, 1812, founder Linnaean Inst., 1802; traveled to coast W. Africa, 1792; sec. Swedish Embassy, London, 1797-98. Fellow Royal Soc., 1798. Author several books on plants of Guinea coast; editor autobiography of Linnaeus, 1823. Died Uppsala, 1837.

AFZELIUS, Bjorn Arvid, Swedish zoologist; b. Stockholm, Sweden, June 30, 1925; s. Nils Arvid and Gretl (Thirring) A.; Fil.lic., Stockholm U., 1954, Fil.dr., 1957; m. Ulla Elisabeth Fogelberg, May 24, 1957; children—Gŏrel Tove Elisabeth, Boel Astrid Margareta. Research fellow in biophysics Johns Hopkins, 1957-58; research fellow Stockholm U., 1958—. Mem. Scandinavian Soc. for Electron Microscopists (vice chmn. 1966). Author: Anatomy of the Cell, 1966. Studies, publs. on fine structure of the cell and its organelles under the electron microscope; changes in egg and sperm during fertilization. Home: 21 Multragatan, Vallingby, Sweden.*

AFZELIUS, Johann, Swedish chemist; b. Larss Församling, Sweden, June 13, 1753; grad. U. Upsala (Sweden), 1776; prof. chemistry, metallurgy, pharmacy, U. Upsala, 1784-1820; mem. Acad Scis. Stockholm. Discovered lithia, 1817; taught Berzelius phlogistic principles; obtained (with Peter Öhrn) formic acid from distillation of ants and proved it distinct from acetic acid. Died Upsala, May 20, 1837.

AFZELIUS, Pierre, Swedish physician; b. Larf, Sweden, 1760; s. Frederick George Afzelius; ed. Paris, France and Edinburg, Scotland. Health service dir. Swedish army expdn. to Finland, 1789; prof. U. Uppsala (Sweden); physician to royal prince Charles Jean; insp. gen. mil. health service. Honored by Swedish govt. Mem. Acad. Scis. Stockholm (pres.). Died Uppsala, 1843.

AGABALYANTS, Georgii Gerasimovich, Russian agriculturist; b. 1904; grad. Don Inst. Agr. and Melioration; D.Agr. Sci., 1939. Prof., Krasnodar Inst. Wine Growing and Viticulture, 1937—; mem. staff All-Russian Research Inst. Wine Growing and Viticulture. Recipient Lenin prize, 1961. Research in wine chemistry and tech.; developed and introduced biochem. and physico-chem. methods of producing champagne by automated constant-flow process. Address: Inst. of Wine Growing and Viticulture, Krasnodar, RSFSR, USSR.

AGAFONOV, Fedor Dmitrievich, Russian pediatrician; b. Village of Lazarevka, Vyatka Guberniya, 1886; grad. Med. Faculty, Kazan U., 1911; D.Med. Sci. Vaccinator, asst. epidemiologist, 1909-11; intern Child Diseases Clinic, Kazan U., 1911-14; mil. physician, 1914-21; asso., then sr. asso. Gorky Pediatric Research Inst., asst. and lectr. dept. child diseases Gorky Med. Inst., head chair child diseases, 1931—. Cons. to various pediatric establishments, Gorky. Decorated Order of Lenin. Mem. Gorky Soc. Pediatricians (chmn.), All-Russian Soc. Pediatricians (bd. mem.). Author numerous works on pediatrics. Mem. editorial bd. Pediatriya. Address: Gorky Med. Inst., pl. Minina 10-11, Gorky, RSFSR, USSR.*

AGAPKIN, Ivan Nikitich, Russian dermatologist; b. Astrakhan, 1903; grad. Astrakhan Med. Inst., 1927; Cand. of Med. Sci. Intern, Dermatol. Clinic, Astrakhan Med. Inst., 1927-29; asst. Moscow Central Dermatol. and Venerological Inst., 1929-31; with Mongolian Peoples Republic, 1932-36; intern dermatologist Korolenko Hosp., Moscow, 1937-42; former head venerology sect. RSFSR Ministry Health 2 years; dir. Inst. Tb of Skin, 1934-38, also head dermatol. dept.; head skin dept. Tb Research Inst., RSFSR Ministry Health, 1958—. Co-author: Tuberculosis of the Skin, Tuberculosis of the Skin and the Subcutaneous Tissue. Author numerous works on pathogenesis, clin. apsects, treatment and measures against Tb of skin, osteo-articular disorders in persons with Tb of skin. Address: Inst. of TB, Ulitsa Dostoevskogo 4, Moscow, USSR.

AGAR, Wilfred Eade, zoologist; b. Apr. 27, 1882; s. Edward Larpent Agar; ed. King's Coll., Cambridge, fellow, 1907-13; m. Elizabeth MacDonald, 1908; 2 sons, 3 daus. Lectr. zoology U. Glasgow (Scotland); capt. 5th bn. Highland Light Inf., 1914-18; prof. zoology U. Melbourne (Australia). Fellow Royal Soc., 1921. Author: Cytology, 1920; The Theory of the Living Organism, 1943; also biol. articles. Died July 14, 1951.

AGARDH, Charles Adolphe, Swedish botanist; b. Baastad, Sweden, Jan. 23, 1785; ed. U. Lund (Sweden). Lectr. math., later prof. botany and rural economy U. Lund; named bishop, 1834 (gave up sci. for religion). Mem. Swedish Acad., Acad. Scis. Stockholm. Author: Systema Algarum (described 49 new algae species; gave 1st description of diatoms), 1824; Classes plantarum, 1825; an encyclopaedic work describing all known algae, 2 vols., 1828; Species, genera et ordines algarum, 8 vols., 1848-1901. Made (with Harvey) taxonomic systems; his and Lindley's classifications became known as parade of systems. Died Karlstad, Sweden, Jan. 28, 1859.

AGARDH, Jacob George, Swedish botanist; b. Lund, Sweden, Dec. 8, 1813; s. Charles Adolph Agardh; studied natural sci. Prof. botany U. Lund. Mem. Stockholm (Sweden) Acad. Sci. Studies on algae and their systematics, also cyonophyceae; prin. work is species, genus and orders Algarum, 1848-76. Died Lund, Apr. 27, 1885.

AGARWAL, Lalit Prakash, Indian ophthalmologist; b. Lucknow, India, Jan. 23, 1922; s. Madho Ram and (Kalavati) A.; M.B.B.S., King George's Med. Coll., Lucknow, 1946; D.O.M.S., Royal London Ophthalmic Hosp., London, Eng., 1947; D.O., Nuffield Lab. Ophthalmology, Oxford, Eng., 1947; M.S., King George Med. Coll., Lucknow, 1949; m. Savitri Agarwal, July 5, 1949; 1 dau., Kavita. Lectr. ophthalmology S.N. Med. Coll., Agra, India, 1950-57; reader, head dept. ophthalmology G.S.V.M. Med. Coll., Kanpur, India, 1957-59; prof. ophthalmology All India Inst. Med. Scis., New Delhi, 1959—; chief organizer, prof. ophthalmology Dr. Rajendra Prasad Centre for Ophthalmic Scis., New Delhi, 1966—. Barraquer Inst., fellow, Barcelona, Spain, 1959. Fellow Royal Soc. Medicine London, Indian Acad. Med. Scis.; mem. All India, U.K. ophthal socs. Author: Eye Diseases, 1962; Principles of Optics and Refraction, 1962; also numerous articles. Editor: Oriental Archives of Ophthalmology, 1963—. Research on theories of retinal detachment, aetiopathogenesis of glaucoma, cytological changes in trachoma, lytic cocktail for cataract surgery, mycotic keratits, Agarwal's operation for retinities pigmentosa, heterogenous keratoplasty, lens regeneration in primates and rodents, retinal vascular patterns, aetiopathogenesis of iridocyctitis, aetiology and treatment of Eales' disease, optic nerve degenerations, retinal glycogen, others. Home: C1/15, A.I.I.M.S. Campus, Ansari Nagar, New Delhi-16, India.*

AGARWALA, Rajendra Prasad, Indian metal physicist; b. Agra, U.P., India, Jan. 20, 1929; s. G. P. and Rajdulari Agarwala; B.S., Agra. U., 1948, M.S., 1950; Ph.D., Imperial Coll. Sci. and Tech., London (Eng.) U., 1953, Diploma Imperial Coll., 1953; m. Prabha Agarwala, May 5, 1950; 1 son, Sanjay. Wakefield Research scholar Imperial Coll., London, 1952-53; staff Nat. Chem. Lab., Poona, India, 1954-55; staff Atomic Energy Establishment, Trombay, Bombay, India, 1955-67, prin. sci. officer, 1966-67; prin. sci. officer Bhabha Atomic Research Centre, Trombay, Bombay, 1967—; faculty Bombay U., 1963—. 1st research grantee I.A.E.A., 1960-61. Mem. Electron Microscopy Soc. India, Sigma Xi. Fellow Indian Chem. Soc. Research, publs. on deformation and phase transformation using electron diffraction and electron microscopy, defect structures, solid state diffusion studies, gas solid diffusion studies; designed and fabricated high voltage electron diffraction unit with original modifications. Home: 19, Kenilworth, Peddar Rd., Bombay-26. Office: Chemistry Div., Bhabha Atomic Research Centre, Modular Labs., Trombay, Bombay-74, India.*

AGASSE, Edouard, French physician, inventor; b. Albi, France, July 2, 1877; s. Charles and F. Larrieu A.; ed. Lycée Henri-IV, Paris; M.D., licencié in lit.; m. Blance Bignon, Jan., 1928; 1 child, Germaine. Head clinic and lab. Faculté de médecine, Paris; med. cons. French Ministry of Labor; active mem. Internat. Bur. Labor Geneva. Recipient Legion of Honor. Author: Traité de laboratoire, 5 edits., 1911-53. Address: 44 quai Louis-Blériot, Paris 16, France.

AGASSIZ, Alexander (Emmanuel Rodolphe), naturalist; b. Neuchatel, Switzerland, Dec. 17, 1835; s. Jean Louis Rodolphe and Cecile (Braun) A.; came to U. S., 1849; A.B., Harvard, 1855; B.S., Lawrence Sci. Sch., 1857; LL.D., Harvard, 1885, St. Andrew's, 1901; Sc.D., U. Cambridge, 1887, Bologna, 1888; m. Anna Russell, Nov. 13, 1860; children—George, Maximillian, Rodolphe. Asst. on U. S. Coast Survey in Cal., 1859; asst. in zoology, Mus. Comparative Zoology, Harvard, 1860-65; developer and supt. Calumet & Hecla copper mines, Lake Superior 1865-69; dir. Anderson Sch. of Natural History on Penikese Island, 1874; mem. expdn. to S. America, 1875, where inspected copper mines of Peru and Chili and made surveys of Lake Titicaca; curator Mus. Comparative Zoology, 1874-85, dir., from 1902, contbd. valuable W. Indian, Central, S. America and Pacific zool. collection to mus., 1898. Assisted Sir Charles Wyville Thomson in classifying collections of expdn. of the Challenger in her voyage of 68,900 miles of deep-sea exploration, 1872-76; spent winters 1876-81 in deep-sea dredging in W. Indies on board U. S. Coast Survey steamer Blake; in charge of expdns. to the Sandwich Islands, the West Indies, Fiji Islands, Great Barrier Reef of Australia; in charge of expdn. of U.S.S. Albatross to Panamic regions and Galapagos, to central Pacific and eastern Pacific. Overseer, Harvard, 1873-78, 85, fellow, 1878-84, 86-90. Decorated knight Order of Merit (Prussia), 1902; officer Legion of Honor (France), 1896. Fellow Royal Soc., 1891. Pres. Nat. Acad. of Sciences; pres. Am. Acad. Arts and Sciences, 1898. Author: Seaside Studies in Natural History (with Mrs. Elizabeth Cabot Agassiz), 1865; Marine Animals of Massachusetts Bay, 1871; Explorations of Lake Titicaca; Three Cruises of the Blake, 1888; Revision of the Echini; Coral Reefs of Florida, Bahamas, Bermudas, W. Indies, of the Pacific, of the Maldives; Panamic Deep Sea Echini; Hawaiian Echini, Embryological Memoirs on Fishes, Worms, Echinoderms; North American Acalephae, 1865; Embryology of the Starfish, 1865. North American Starfishes, 1877; The Islands and Coral Reefs of Fiji, 1899. Conducted important investigations of invertebrate life; made careful oceanographic explorations of Pacific and Caribbean; hypothesized (on basis of similarity of deep sea animals of Caribbean and Pacific) that Caribbean Sea was bay of Pacific Ocean until the rise of Panama Isthmus in Cretaceous period; formulated theory of coral atolls, which differed from Darwin's. Died at sea, Mar. 27, 1910.

AGASSIZ, Jean Louis Rodolphe, zoologist, geologist; b. Motier-en-Vuly, Fribourg, Switzerland, May 28, 1807; attended Coll. Lausanne (France), 1822-24, U. Zurich (Germany), 1824-26, U. Heidelberg (Germany), 1826-27; M.D., U. Munich (Germany); 1827; Ph.D., U. Erlangen (Germany), 1829; m. Cecile Braun, Oct. 1833; children—Alexander, Ida, Pauline; m. 2d, Elizabeth Cabot Cary, 1850. Conducted zool. research under French naturalist Cuvier, Paris, 1831-32; prof. natural history U. Neuchatel (Switzerland), 1832-46; established Hôtel des Neuchatelois for glacier study; came to Am., 1846, toured East Coast lecturing on comparative embryology; went on sci. cruise along Mass. coast, 1847; prof. zoology and geology Lawrence Scientific Sch., Harvard, 1848-73; made exploration cruise of Fla. coral reefs, 1851; curator mus. of comparative zoology, Cambridge, Mass., 1859-73; Am. citizen, 1861; collected specimens for Cambridge Mus. in Brazil, 1863; nonresident prof. natural history Cornell U., Ithaca, N.Y., 1868; made scientific voyage around Cape Horn aboard Hassler, 1871-72; established Anderson Sch. Natural History, Penikese Island, Buzzard's Bay, 1873. Recipient Wollaston medal, 1836. Fellow Royal Soc., 1838; mem. French Acad. Scis., 1839 (phys. sci. prize, 1859); co-founder Nat. Acad. Scis., 1863; named to Hall of Fame for Great Americans, 1915. Author: The Fishes of Brazil, 1829; Recherches sur les Poissons (foundation of our present knowledge of fish), 1833-44; History of the Fresh Water Fishes of Central Europe, 1839-42; Etudes sur les Glaciers, (early exposition of glacial movements and deposits), 1840; Etudes Critiques sur les Mollusque fossilin, 1840-45; Nomenclator Zoologicus, 1842-46; Monograph on the Fossil Fishes of the Old Red or Devonian of the British Isles, 1844-45; Système Glaciare, 1846; Nouvelles études et expériences sur les Glaciers actuels, 1847; Contributions to the Natural History of the United States (contained Essay on Classification), 1857. Staunch antievolutionist, he opposed Darwinism and asserted new species arise only through divine intervention; still his zoological studies confirmed evolution; studied echinoderms and mollusks; worked on developing better zoological classification systems; in his glacier studies posited the existence of an "ice age" long eons ago. Died Cambridge, Mass., Dec. 14, 1873.

AGATHARCHIDES OF KNIDOS, naturalist, historian, geographer; b. Greece; flourished Alexandria, Egypt, circa 146 B.C. Author: On the Erythraean Sea (geography and ethnography of Arabia and Ethiopia); geography and history of Asia in 10 books, of Europe in 49 books. Gave 1st description of rhinoceros.

AGATHINOS OF SPARTA, physician; flourished 90; disciple of Athendeos; founder of eclectic (or episynthetic) sch. medicine, developed pneumatic sch.; wrote treatise on pulse, another on action of hellebore (partially based on experiments); recommended use of cold baths.

AGDE, Georg, German chemist; b. Halle, Germany, Aug. 25, 1889; s. Karl and Wilhelmine (Heine) A.; student chemistry U. Halle, 1904-14; m. Margret Schürenberg, Mar. 29, 1924; 2 sons, 1 dau. Chemist in mil. service; tchr. Indsl. U. Köthen (Germany), 1921-22; pvt. lectr. Darmstadt, Germany. Contbr. numerous articles in field. Achieved important results in coal research and utilizing and refining of coal products. Died Darmstadt, Sept. 12, 1944.

AGEEV, Nikolai Vladimirovich, Russian chemist, metallurgist; b. June 30, 1903; grad. Leningrad Poly. Inst., 1926; D.Chem. With Leningrad Poly. Inst., 1926-38, Inst. Gen. and Inorganic Chemistry, USSR Acad. Sci., 1938-40, 42-50; asso. Baykov Inst. Metallurgy, 1951—; chief editor jour. Metallurgy, 1956—. Mem. USSR Acad. Sci. (corr.). Author: Radiography of Metals and Alloys, 1932; Thermal Analysis of Metals and Alloys, 1936; The Chemistry of Metal Alloys, 1941; The Nature of the Chemical Bond in Metal Alloys, 1947; Diagrams of the State of Metal Systems, 1956; Research on Heat-Resistant Alloys, 1960; Production of Chromium Chloride, 1963. Research in chemistry of metal alloys. Address: Inst. of Metallurgy, Leninsky prosp. 49, Moscow, USSR.

AGELL, José, Spanish chemist; b. Masnou, Barcelona, Spain, Sept. 30, 1882; s. Eudaldo and Asunción A.; ed. Barcelona, Madrid (Spain) univs., U. Paris (France), Pasteur Inst., Paris; B.Ch., D.Sc., Ph.D. Prof. engring., indsl. schs., until 1952; dir. gen. Fibras Artificiales S.A.; mgr. Inst. Applied Chemistry; prof. chem. analysis Tech. Sch. Barcelona. Recipient Ultramar medal, medal Centenario de Pasteur, Gold medal Merito en el Trabajo, 1961. Mem. Royal Acad. Pharmacy, Assn. Textile Chemists and Colorists, Acad. Med. Scis. and Hygiene, Société de Chimie Industrielle (hon., Gold medal 1960, pres. Spanish sect.), numerous other sci. socs., acads. Author: Tratado de Analisis Quimico; Quimica General; El presente y el porvenir de las fibras artificiales; Estudio de las aguas potables; Estudio de los colorantes en la alimentacion. Work in chem. analysis, indsl. organics, especially polymers, fibers. Address: Avenida Republica Argentina, 19 Barcelona, Spain.

AGENJO, Ramón, Spanish entomologist; b. Santander, Spain, Jan. 21, 1908; s. Dionisio and Casilda (Cecilia) A.; Atty.-at-law, U. Madrid (Spain), 1931; m. Maria del Carmen Bullón, Oct. 2, 1946; son, Xavier Agenjo Bullon. Entomologist, Ministerio de Educacion y Ciencia, Madrid, Spain, 1945—; director of the Instituto Espanol de Entomologia, Madrid, 1967—; collabourer Servicio Especial de Plagas Forestales, Madrid, 1958—; sci. investigator Consejo Superior de Investigaciones Cientificas, Madrid, 1960—. Mem. Lepidopterists Soc. (past v.p.), Real Sociedad Española de Historia Natural (past gen. sec.), Author: Faunula Lepidopterologica Almeriense, 1952; Catálogo Ordenador de los Lepidopteros de España, 1946-66; also numerous articles, monographs. Research on reprodn., identification of Lepidoptera of Spain; author common names of Spanish butterflies. Home: 17, Marqués de Urquijo, Madrid 8. Office: 2 Gutiérrex Abascal, Madrid 6, Spain.*

AGER, Derek Victor, English geologist, paleontologist; b. Harrow, Eng., Apr. 21, 1923; s. Richard George and Violet Victoria (Green) A.; B.Sc. with 1st class honors, Chelsea Coll. Sci. and Tech., 1951; Ph.D., Diploma Imperial Coll., Imperial Coll. Sci. and Tech., 1954; m. Renee Eugenie Coleman, June 21, 1952; children—Kathleen Frances, Martin Charles. Faculty, Imperial Coll. Sci. and Tech., 1951—, lectr., 1954-58; reader, 1960—; vis. prof. U. Ill., 1958-59. Fellow Geol. Soc. London Murchison award, 1960; mem. Geologists' Assn., Paleontographical Soc., Paleontol. Soc. (U. S.), Paleontol. Assn. Author: Introducing Geology, 1961; Principles of Paleoecology, 1963; also articles. Research on Mesozoic braciopoda, palaeoecology, Mesozoic stratigraphy, and regional geology, morphology, evolution and ecology of fossil braciopods of Mesozoic era especially in relation to their geog. distbn. and adaptation to different environments. Home: 81 Denham Green Lane, Denham, Bucks., Eng. Office: Dept. Geology, Imperial Coll., London, S.W. 7., Eng.*

AGER, (or AGERIUS), Nicolas, physician, botanist; b. Itenheim, Alsace, 1568; s. Johann Heinrich Ager; ed. Basel, Switzerland, also Strasbourg, France; Ph.D., M.D.; prof. medicine and botany, Strasbourg, from 1618. Author: Theses physico-medicae de homine sano, 1593; Disputatio de zoophytis, 1625; Disputatio de anima vegetativa, 1629; De vita et morte; De nutritione. Editor: Reformierte deutsche Apotheke (Walther Hermann Ryff), 1602. Described several new plants and their medicinal value; studied nutrition, insomnia, psychologically disturbed persons. Died Strasbourg, June 20, 1634.

AGGELER, Paul Michael, Am. physician; b. Ferndale, Cal., Sept. 26, 1911; s. Siverenus Gregory and Marguerite (Griffith) A.; A.B., U. Cal., Berkeley, 1933; M.D., U. Cal., San Francisco, 1937; m. Dorothy Rowlands Meeker, Dec. 14, 1941; children—Judith (Mrs. Timothy Power), Candace. Research fellow in medicine U. Cal. Hosp., 1939-40, instr. in preventive medicine, 1941-44, asst. clin. prof., 1944-52, asso. clin. prof., 1952-59, clin. prof. medicine, 1959-64, prof. medicine, 1964—, acting chief hematology dept. internal medicine, 1961-63; chief hematology San Francisco Gen. Hosp., 1963—; attending physician Childrens Hosp., 1953-55, chief dept. medicine, 1955-62, cons. hematology, 1962—; asso. attending physician St. Mary's Hosp., 1953-62; cons. hematology Martinez VA Hosp., 1949—; attending physician San Francisco VA Hosp., 1953-55, cons. dept. medicine, 1955—; cons. hematology study sect. NIH, 1958-61, 1964-66, mem. tng. com. Nat. Heart Inst., 1962-64, member med. adv. council Nat. Hemophilia Found., 1962—; mem. Internat. Com. for Hemostasis and Thrombosis, 1964—, Recipient USPHS Career award, 1964—. Diplomate Am. Bd. Internal Medicine. Mem. Am. Soc. Hematology (mem. exec. com. 1967—), A.C.P., Am. Fedn. Clin. Research (pres. 1947), A.M.A., Am Soc. for Clin. Investigation, Cal. Acad. Medicine, Cal. Med. Assn., Internat. Soc. Hematology, San Francisco Med. Soc., Soc. for Exptl. Biology and Medicine, Western Assn. Physicians, Sigma Xi, Alpha Omega Alpha, Nu Sigma Nu. Author: (with S. P. Lucia) Hemorrhagic Disorders, 1949. Contbr. numerous articles in field to med., sci. jours. Authority on hemorrhagic diseases; discoverer important Plasma Thromboplastin Component (variant of hemophilia which helped initiate understanding of multiplicity of factors involved in coagulation of blood); analysis of pharmacodynamics of coumarin anticoagulant drugs and demonstration of genetic variation in response to their administration. Home: 319 Castenada Av., San Francisco 94116. Office: San Francisco Gen. Hosp., San Francisco 94110.*

AGGIUNTI, Niccolo, physicist; b. Borgo di S. Sepulcro, 1600; 1st scientist to record observations of capillary action. Died 1635.

AGILINUS, Gaulterus, physician; probably French; flourished middle 13th century; influenced by Giles of Corbeil. Author: Summa medicinalis (a complete spl. path. and therapeutical work based on uroscopy); Compendium urinarum; Liber pulsuum; Glossulae super versus Aegidii (lost); De febribus; Summa or Practica; De dosi medicinarum; said to have written treatises Perah ha-refuah (Flower of medicine), Ma'amar ba-eresim (treatise on poisons).

AGLADZE, Rafael Ilyich, Russian electrometallurgist; born Dec. 29, 1911; graduated from Georgian Industrial Institute, 1934. Instructor, Moscow Chemico-Technol. Institute, 1937-43; dir. Institute Metals and Mining, Georgian Acad Sci., 1945-51, v.p., 1947-51, chmn. dept. technol. sci, dir. Inst. Applied Chemistry and Electrochemistry, 1955—; instr. Georgian Poly. Inst., 1943—. Recipient Stalin prize, 1943. Mem. Georgian Acad. Sci. Author: Obtaining Metallic Manganese by Electrolysis of its Salts, 1939; The Technology of Obtaining Metallic Manganese by Electrolysis 1942; Electrochemistry of Manganese, Vol. 1, 1957, Vol. 2, 1963, Vol. 3, 1967; co-author Alloys of Manganese with Copper, Nickel and Zinc, 1954, Hydroelectrometallurgy of Chrome, 1959, Obtaining Pyrophoric and Nitrate Manganese from a Manganese Amalgam, 1963. Research on electrochemistry and electrometallurgy; developer new electrochem. method of obtaining metallic manganese and new method of obtaining potassium permanganate. Address: Politekhnichesky Institut, ulitsa Lenina 69, Tbilisi; also Georgian Acad. Scis., ulitza Dserzinskova 8, Tbilissi, Gruz, SSR, USSR.*

AGNESE, Carlos Alberto, Argentinian physician; b. Buenos Aires, Argentina, May 10, 1936; s. Lazaro and Hilda (Flores) A.; B., Gral Roca Coll., 1952; physician Buenos Aires U., 1957; D.Medicine, 1961; m. Lucile Rosa Rodriguez, Mar. 1, 1962; 1 dau., Andrea Lucila. Physician, Pirovano Hosp., Buenos Aires, 1957—, Met. Pvt. Hosp., Buenos Aires, 1964—; chief med. practice Buenos Aires U., 1963—, docent autorizado medicine, 1966—. Recipient Wilde award Pub. Health Dept., Republic of Argentina, 1958. Mem. Argentine Med. Assn., Buenos Aires Internal Medicina Soc., Argentine History of Medicine Soc., Argentina Transmissible Illness Soc. Research, numerous articles on histoplasmosis, dissecting aneurysm, rheumatism, leprosy, biliary cirrhosis, hypertrophic pyloric stenosis. Home: 2718 Roque Perez, Buenos Aires, Office: 2184 Gral. Martinez, Buenos Aires, Argentina.*

AGNESI, Maria Gaetana, Italian mathematician; b. Italy, 1718; ed. privately; substituted for father as prof., chmn. dept. math. U. Bologna (Italy), 1750. Author: Instituzioni analitiche, 1748. Worked on witch of Agnesi or versiera cubic curve. Died Milan, Italy, 1799.

AGNEW, Cornelius Rea, Am. ophthalmologist; b. N.Y.C., Aug. 8, 1830; s. William and Elizabeth (Thompson) A.; grad. Columbia, 1849, M.D., 1852; m. Mary Nash, 1856. Began practice of medicine, 1854; surgeon Eye and Ear Infirmary, N.Y.C., 1855-64; surgeon gen. N.Y. Militia, 1858; med. dir. N.Y. State Hosp. for Volunteers during Civil War; an organizer U. S. Sanitary Comm.; organizer Sch. of Mines, Columbia, 1864, an ophthalmic clinic at Coll. Phys. and Surg., (of Columbia), Bklyn. Eye and Ear Hosp., Manhattan Eye and Ear Hosp.; prof. diseases of eye and ear Coll. Phys. & Surg., 1869-88. Devised operative procedures, including cantholplasty, for correction of divergent squint, for chalazion, 1866. Died N.Y.C., Apr. 18, 1888.

AGNEW, David Hayes, Am. surgeon; b. Lancaster County, Pa., Nov. 24, 1818; s. Robert and Mrs. (Henderson) A.; grad. med. dept. U. Pa., 1838; m. Margaret Irwin, 1841. A founder Irwin & Agnew Iron Foundry, 1846; bought, revived Phila. Sch. Anatomy, 1852-62; surgeon Phila. Hosp., 1854, Wills Eye Hosp., 1863, Pa. Hosp., 1865, Orthopedic Hosp., 1867; became asst. prof. surgery U. Pa., 1854, prof. clin. surgery, 1870; prof. surgery, 1871-89, emeritus, 1889; pres. Phila. Coll. of Physicians, 1889; attending physician to Pres. Garfield when he was assassinated, 1881; one of his classes painted by Thomas Eakins. Author: Clinical Reports, 1859-71; Treatise on the Principles and Practice of Surgery, 3 vols., 1878-83. Invented spl. splint for treatment fractured patella; devised splint for fracture of metacarpus, 1878, method for operative correction of webbed fingers, 1883. Died Phila., Mar. 22, 1892.

AGNIHOTHRUDU, Vangala, Indian mycologist; b. Attili, India, May 15, 1930; s. Subrahmanyam and Subbamma Vangala; B.Sc. with honors, Presidency Coll., Madras, 1950; M.A., Madras, 1951, D.Sc., 1964; Ph.D., Madras U., 1955; m. Vittala Indira, May 20, 1959; 1 son. Research fellow Univ. Botany Lab., Madras, 1950-54; sr. research fellow U. Madras, 1954-56; mycologist Tocklai Exptl. Sta., Indian Tea Assn., Cinnamara, Assam, 1956-64; tech. adviser Spl. Coffee Research Assn., also tech. adviser in fertilizer and pesticides Rallis India Ltd., Bangalore, India, 1964—. Recipient Pulney Andy's Gold medal Madras U. Fellow Indian Bot. Soc., Linnean Soc. Contbr. to tech. publs. Pioneer in rhizosphere microbiol. work in India; stimulated reinvestigations of mycetozoa; research on world tea fungi. Home: 22 Bensoncross, Bangalore 6. Office: Rallis India Ltd. (Fanp), 6-A Cunningham Rd., Bangalore 1, India.*

AGOSIN, M. Kankolsky, biochemist; b. Marseille, France, Dec. 1, 1922; s. Abraham Wolf and Raquel (Kankolsky) A.; M.D., U. Chile, 1948; m. Frida Halpern, June 18, 1948; children—Cynthia, Marjory, Mario. Faculty, U. Chile Sch. Medicine, 1948-52, 56-66, prof. chemistry, 1961-66; Rockefeller Found. fellow NIH, Bethesda, Md., 1952-53, research asso., 1953-55; prof., head dept. parasitology Hebrew U., Haddassah Med. Sch., Jerusalem, Israel, 1966—; vis. prof. dept. entomology London (Eng.) U., 1965. Cons. on insecticide chemistry WHO, 1958, 62. Recipient Prize Corvalan Melgarejo, Med. Soc. Santiago, 1947. NIH grantee, 1958—; Wellcome Trust Grantee, Eng., 1966. Mem. Biology Soc., Am. Chem. Soc., Soc. Tropical Medicine and Hygiene, N.Y. Acad. Scis., Am. Acad. Microbiology. Research, numerous publs. on biochemistry of parasites and their vectors, resistance to DDT and cross resistance to other insecticides. Home: 4926 Simon Bolivar, Santiago, Chile. Office: Hebrew U., Haddassah Med. Sch., Jerusalem, Israel.*

AGOSTINELLI, Cataldo Luigi, Italian engr., mathematician; b. Ceglie Messapico (Briendisi), Italy, Dec. 16, 1894; s. Rocco and Addolorata (Chirico) A.; studied engring, pure math.; D. Eng. and Math. Scis.; m. Angola Gili, May 28, 1939. Prof. mechanics U. Turin; dir. math. physics dept., U. Turin. Mem. Acad. Sci. Turin (nat.), Acad. dei Lincei (corr.), Acad. Sci. Modène (hon.). Research, publs. on rational and analytic mechanics, math. physics. Home: corso Duca degli Abruzzi 34/B, Turin. Office: Istituto Mathematico, Universita, via Carlo Alberte 10, Turin, Italy.

AGRAMONTE, Aristides, bacteriologist; b. Camagüey, Cuba, June 3, 1869; s. Dr. Edward and Mathilde (Simoni) A.; brought to U. S., in infancy; student Coll. City of New York, 1885-87; M.D., Coll. Physicians and Surgeons (Columbia), 1892 (Harsen prize); M.B., Havana U., 1899, M.D., 1900; (Sc.D., Columbia U., New York, 1914); m. Frances Pierra, Apr. 17, 1895. Practiced, New York; asst. surgeon U. S. Army, May, 1898-Oct., 1902; chmn. Bd. of Infectious Disease and mem. Nat. Bd. of Health, Republic of Cuba; prof. bacteriology and exptl. pathology, U. of Havana, 1900—. Mem. Am. Acad. Sciences. Mem. U. S. Army bd. that discovered transmission of yellow fever by mosquitoes, 1901. Died Aug. 19, 1931.

AGRANOFF, Bernard William, biochemist; b. Detroit, June 26, 1926; s. William and Phyllis (Pelavin) A.; B.S., U. Mich., 1954; M.D., Wayne State U., 1950; m. Raquel Betty Schwartz, Sept. 1, 1957; 2 sons, William, Adam. Biochemist, Nat. Inst. Neurol. Disease and Blindness, Bethesda, Md., 1954-60; vis. scientist Max Planck Institut Fur Zellchemie, Munich, 1958-59; surgeon USPHS, 1954-60; asso. prof. biol. chemistry U. Mich., Ann Arbor, 1960-65, prof., 1965—; research biochemist Mental Health Research Inst., 1960—; cons. pharm. industry, govt. Mem. Am. Soc. Biol. Chemists, Am. Chem. Soc., A.A.A.S., Internat. Neurochem. Soc. Contbr. numerous articles to profl. jours. Research, publs. on identification of intermediate compounds and pathways in the biosynthesis of complex lipids; understanding the chem. basis of memory. Home: 2960 Overridge Dr., Ann Arbor, Mich. 48104.*

AGRELL, Ivar Per Sigurd, Swedish physiologist; b. Lund, Sweden, July 2, 1912; s. Per Sigurd and Anna (Osterman) A.; M.S., U. Lund, 1936, Lic., 1941, doctor, 1941; m. Barbro Wallér, May 15, 1953; children—Alexander, Jep. Faculty, U. Lund, 1941—, prof. zoophysiology, 1959—, head Zoophysiol. Inst., 1949—. Recipient Linnean prize, 1959. Mem. Royal

Physiographical Soc., Royal Danish Acad. Sci. Research, publs. on gen. ecology, biochem. characterization of insect metamorphosis, cellular div. and differentiation. Home: 1 V.Martensgatan, Lund, Sweden.*

AGRICOLA, Georg, German physician; b. Mimbach, 1530; magister, U. Wittenberg (Germany), 1553, M.D., 1570; m. Margaretha Volg, 1554; m. 2d, Veronica Steinhauser; 8 children. Tchr., Martin Sch., Amberg, Germany, 1554-55, rector, 1555-69; apptd. city physician Amberg (saved town from epidemics, 1571, 74), 1570. Died Amberg, Jan. 12, 1575.

AGRICOLA, Georg Andreas, physician, philosopher; b. Ratisbone, 1672; student Württemberg, Halle (both Germany). Fellow Royal Soc., 1699. Author: Neuer und nie erhörter, dock in der natur und Vernunft wohl begründeter Versuch der Universalvermehrung aller Baume, Stauden und Blumengewachse; Agriculture Parfaite ou Nouvelle Decouverte, 1720; writings concerned with nature and reason as related to multiplication of trees, bushes and plants; pretended to have discovered method by which plant growth could be greatly accelerated; pub. useful work on culture and propagation of plants, 1717. Died Ratisbone, 1738.

AGRICOLA, Georgius, (Georg Bauer), German mineralogist; b. Glauchau, Saxony, Mar. 24, 1494; s. Gregor Bauer; B.A., U. Leipzig (Germany), 1517; student medicine, chemistry, Bologna, Venice, Padua (all Italy); probably M.D.; m. Anna Meyner, 1527; m. 2d, Anna; 2 sons, 3 daus. Tchr. classics Municipal Sch., Zwickau, Saxony, 1518-22, prin., from 1520; lectr. U. Leipzig, 1522-23; practiced medicine, Joachimsthal (now Czechoslovakia), from 1527; began work in mineralogy, 1527; apptd. ofcl. historian principality Saxony (ann. allowance from Prince Maurice); physician, Chemnitz, Saxony, from 1533; mayor Chemnitz, 1546, 47, 51, 53. Author: Bermannus sive de re metallica, 1530; De mensuris et ponderibus Romanorum atque Graecorum libri V, 1533; De ortu et causis subterraneorum libri V, 1546; De natura fossilium libri X, 1546; De veteribus et novis metallis libri II, 1546; De re metallica libri XII, 1556; others. Known as father of mineralogy; made 1st attempt to impose sci. order on unsystemized knowledge of miners; described, illustrated mining and metall. processes; rejected theory that rock crystals are formed from water by intense frost; asserted that minerals and certain rocks originated from petrifying juice; divided rocks into 5 classes; attempted to classify minerals and ores; discovered several minerals; worked on analysis of ores, fluxes and cement powders; described preparation, separation, purification of metals, manufacture of tin-plate, steel, distillation of nitric acid, reduction of Aqua Regia, amalgamation of gold; gave 1st description of vicinal form on a crystal; devised system of classifying geol. specimens; expert on indsl. diseases; undertook joint revision of Greek text of Galen, 1525. Died Chemnitz, Nov. 21, 1555.

AGRICOLA, Johann, German chemist, surgeon, physician; b. Palatinate, 1589. Practiced medicine, Leipzig, Germany. Author: Commentariorium, Notarum, Observationum et Animadversionum in Johannis Poppii Chymische Medicin, 1638; Chirurgia Parva, 1674; others. Strongly supported work of Paracelsus and chem. remedies; gave long descriptions of metals and their compounds in relation to their medicinal value; authority for reputed cases of transmutation. Death date unknown.

AGRICOLA, Johann Beverle (or Peurle), physician; b. Gunzenhausen, 1496; M.D., U. Ingolstadt (Germany), 1528, became prof. Greek lit. apptd. prof. medicine, 1531. Author 1st pharm. synomyn lexicon; Herbariae Medicinae libri duo, 1539. Promoted use of herbs; rep. classical sch. medicine. Died Ingolstadt, Mar. 6, 1570.

AGRICOLA, Johann Georg, German physician; b. Amberg, Germany, 1558; s. Georg A. and Veronica (Steinhauser) A.; B.A., U. Heidelberg (Germany), 1585; postgrad. U. Wittenberg (Germany), 1588-93; m. Barbara Hartlieb; m. 2d, Sabine; 1 son, Johann Georg. Apptd. city physician, Amberg, 1594; practiced medicine, Regensburg, Germany, from 1629. Author books on deer, 1603, 17, hotsprings (med. uses), 1619. Died Regensburg, Nov. 20, 1633.

AGRIPPA, Heinrich Cornelius (or Agrippa von Nettesheim), German physician; b. Cologne, Germany, Sept. 14, 1486; ed. U. Cologne; m. 3 times; many children; supported by royal patrons in Germany, France, Italy, and the Netherlands; sent by Maximilian I on secret mission to Paris; invited to teach theology, Dôle, France, 1509-10; denounced as heretic, compelled to leave; attended Council of Pisa, 1511; lectured at U. Pavia, 1515; town advocate and orator, Metz, 1518-20; became physician to queen mother France at Lyons, 1524-28; archivist and historiographer to Emporer Charles V, 1529; persecuted for occult beliefs. Author: De occulta philosophia (a defense of magic), 1531; De incertitudine et vanitate scientiarum, 1531; De nobilitate et praecellentia feminei sexus. Attacked pretensions of contemporary schoolmen, urged return to beliefs of early Christian Church; held that man may know God and nature through natural magic; his mystical philosophy drew on neoplatonic, cabalistic, and Christian traditions; developed doctrine of 3 spheres (elements, stars, spirits) corresponding to physical, heavenly, and mental worlds. Died Grenoble, France, Feb. 18, 1535.

AGRIPPA, Marcus Vipsanius, Roman general, statesman; b. Rome, 63 B.C.; m. 3d, Julia (dau. of Augustus, widow of Marcellus), 3 sons, 2 daus. Naval comdr.; suppressed Aquitanian revolt in Gaul, 38 B.C.; defeated Sextus Pompeiius Magnus at Mylae and Naulochus, 36 B.C.; became consul, 37, 27 B.C., aedile, 33 B.C., gov. Syria, 23 B.C., recalled to become Augustus' chief minister, 23 B.C. Built Pantheon, 27-25 B.C., aqueduct of Nimes, 18 B.C.; completed survey of Roman Empire undertaken by Caesar. Died Campania, Italy, 12 B.C.

AGUILLAR, Peris José, Spanish physicist; b. Valencia, Spain, Mar. 27, 1924; s. Antonio and Amparo A.; ed. univs. Valencia, Madrid (both Spain), B.S. in Chemistry, Sc.D. in Physics; m. Concepción Civera, 1950. Prof. thermodynamics U. Madrid, 1960—. Mem. Royal Soc. Physics and Chemistry. Research, publs. on thermodynamics, nuclear physics, nuclear emulsions. Address: Faculty Sciences, U. Madrid, Madrid, Spain.

AGUILLON (or AGUILONIUS), François, Belgian mathematician, optician; b. Brussels, Belgium; circa 1566; mem. Soc. of Jesus; prof. philosophy, Douai, France; prof. theology, Antwerp, Belgium; director of the Jesuit College at Antwerp. Author: Opticorum libri VI, (treatise on optics, term stereographic projection used), 1613. Introduced study of math. in low countries; gave earliest description of stereographic principles; discovered and defined horopter (line in field of vision). Died 1617.

AGUIRRE-BATRES, Francisco Javier, Guatemalan chemist; b. Guatemala City, Guatemala, June 26, 1924; s. Guillermo and Julia (Batres) Aguirre; Chemist, U. San Carlos, 1947; M.A., U. Tex., 1949; postgrad. Vanderbilt, Oak Ridge Inst. Nuclear Studies, Harvard; m. Rosemary Matos, Sept. 15, 1951; children—Roberto, Francisco, Alejandro. Chief microbiol. sect. Inst. Nutrition of C.Am. and Panama, Guatemala City, 1949-55; dep. dir. C.Am. Research Inst. Indsl. Tech., Guatemala City, 1956—; prof. genetics U. San Carlos, Guatemala City, 1949-54, prof. indsl. microbiology, 1950—. Founder, mem. bd. Guatemala Nuclear Energy Commn., 1955-60; Guatemalan del. 2d Internat. Conf. Nuclear Energy, Geneva, 1958. Mem. Guatemala Acad. Scis. (dir.), Guatemala Coll. Chemists, A.A.A.S., C.Am. Inst. Bus. Admin. Contbr. articles to profl. jours. Research on devel. tech. for indsl. utilization of C.Am. raw materials; patentee new methods for utilizing corn, coffee packaging, developer techniques for Gamma irradiation of tropical fruits. Home: 1 Avenida 9-23, Guatemala City 10. Office: Av. Reforma y 4 Calle, Guatemala City 10, Guatemala.*

AHLFELD, Frederick, German physician; b. Alsleben, Oct. 16, 1843; s. Joseph Frederick Ahlfeld; m. Gabriele Wilhelmine Grunow (dec. 1890); m. 2d, Elisabeth Vollmer, Dec. 29, 1891; 1 dau., 1 son. Author: De la formation du front et de la figure, 1873; Nutrition de l'enfant par le lait de la mère, 1878; Difformites humaines, 1880-83. Studied obstetrics; engaged in physiol. research that improved postnatal care for mother and child; investigated disinfectant capabilities of alcohol. Died Marburg, Germany, May 24, 1929.

AHLFORS, Lars Valerian, mathematician; b. Helsinki, Finland, Apr. 18, 1907; s. K. Axel and Sievä (Helander) A.; Ph.D., U. Helsinki, 1932; LL.D., Boston Coll., 1951; m. Erna Lehnert, July 22, 1933; children—Cynthia (Mrs. Patrick Bertrand), Vanessa (Mrs. Alex Keneas), and Caroline (Mrs. Frank Mouris). Came to the United States, 1946, naturalized, 1952. Prof., U. Helsinki, 1938-44; U. Zurich, 1945-46; prof. Harvard, 1946—, W. C. Granstein prof., 1964—. Recipient Field's medal for math., 1936. Mem. Societas Scientiarum Fennica, Academia Scientarum Fennica, Nat., Royal Swedish acads. sci., Am. Math. Soc., Am. Math Assn. Author: Complex Analysis, 1953; (with L. Sario) Riemann Surfaces, 1960. Home: 236 Beacon St., Boston. Office: Harvard U., Cambridge, Mass.*

AHLQUIST, Raymond Perry, Am. pharmacologist; b. Missoula, Mont., July 26, 1914; s. Perry Karl and Elsa (Ekroth) A.; B.S., U. Wash., 1935, M.S., 1937, Ph.D., 1940; m. Dorotha M. Duff, Sept. 9, 1939. Faculty, Med. Coll. Ga., Augusta, 1944—, prof., chmn. pharmacology, 1948-63, asso. dean Sch. Medicine, 1963—. Mem. Nat. Bd. Med. Examiners, 1960-65; chmn. Council Med. Television, 1965-66. Contbr. numerous articles to profl. jours. Developed presently accepted ideas on basic mechanism of action of adrenaline and related compounds and adrenergic portion of the autonomic nervous system, new class of drugs, beta adrenergic receptor blocking agts. that have potential value in treatment of angina pectoris and related disorders. Home: 2707 Downing St., Augusta, Ga. 30904.*

AHMAD, Fakhruddin, Indian geologist; b. Barabanki, India, Jan. 4, 1916; s. Razi Uddin and Shamim (Quidwai) A.; student Govt. Inter. Coll. Fyzabad, 1932-34; B.Sc., Aligarh (India) Muslim U., 1936, M.Sc. in Geography, 1938, Ph.D., 1949; M.Sc. in Geology, U. Tasmania, Australia, 1954; m. Salma Ansari, Oct. 27, 1943; children—Najma, Khalid, Zurar, Farooque. Lectr. geography Aligarh Muslim U. 1938-39, prof., head dept. geology, 1964—; with Geol. Survey India, 1941-60, Exploratory Tubewell Orgn., 1960, Geol. Survey India, 1960-63. Mem. Geol. Soc. India, Geochem. Soc. India, Geol. Mining and Metall. Inst. India. Author: Palaeogeography of the Gondwana Period in Gondwanaland with Special Reference to India and Australia and its Bearing on the Theory of Continental Drift., 1961; also articles. Research on palaeogeography and palaeclimatology of India, theory of continental drift. Home: Doodhpur, Aligarh, U.P., India.

AHMAD IBN YUSUF (Abu Gafar) (Ahmad ben Yusuf ben Ibraihim ibn al-Daya al-Misri), Egyptian mathematician; flourished circa 912; sec. of Muslims who ruled in Egypt, 868-905. Author books pub. in Latin: Liber Hameti de proportione et proportionalite (influenced medieval thought through Leonardo da Pisa and Jordanus Nemoraius); Liber de Arcubus Similibus; Liber centum verborum Ptolomei cum Commento Haly.

AHMAD, Nafis, geographer; b. Gulaothi, India, Dec. 2, 1911; s. M. Abdul and Shaharbano (Begum) A.; B.A. with Honors, Muslim U. Aligar, India, 1934, M.A. in Geography, 1935; Ph.D., London Sch. Econs., U. London, 1953; m. Akhtar Jahan Begum, Apr. 25, 1943; children—Zulfia (Mrs. Asafuddowlah), Nadim Ahmad, Nawaid Ahmad. Prof. geography Islamia Coll., Calcutta, Sr. Bengal Ednl. Service; lectr. geography Calcutta U., 1941-47; reader, head dept. geography Decca U., 1947—, prof. geography, 1956—, dean Faculty Sci., 1964—. Mem. Nat. Com. Geodesy and Geophysics, Pakistan; adviser Gazetteer Rewriting Com., Govt. of E. Pakistan. Recipient Tamgha-i-Imtiaz for edn. service Pres. of Pakistan, 1961. Fellow Royal Geog. Soc.; mem. Am. Geog. Soc., Pakistan Geog. Assn. (v.p.), E. Pakistan Geog. Soc. (pres.), Internat. Geog. Union (mem. commn. on teaching geog. gen. assembly 1964). Author: Muslim Contributions to Geography, 1947; The Basis of Pakistan, 1947; Oxford School Atlas for Pakistan, 1958; An Economic Geography of East Pakistan, 1959; also articles. Research in land use in Rampal Union and Fydabad Area, population of Pakistan, geography of E. Pakistan. Office: Dacca U., Dacca-2, E. Pakistan.*

AHMAD, Nazeer, soil scientist; b. Dundee, Guyana, Jan. 27, 1932; s. Mohamed Yussuf and Saheedan (Khan) A.; D.I.C.T.A., Imperial Coll. Tropical Agr., Trinidad, W.I., 1952, A.I.C.T.A., 1953; postgrad. McGill U.; M.Sc., U. B.C., 1955; Ph.D., U. Nottingham (Eng.), 1957; postgrad., U. Ill.; m. Shafferun Aysha Hussain, July 15, 1954; children—Shireen Salima, David Nazeer, Fazia Nazreen. Head div. chemistry and soil sci. Guyana Ministry Agr., also co-mgr. UN Spl. Fund Soil Survey project, 1957-62; lectr. soil sci. Faculty Agr., U. W.I., Trinidad, 1962-65, sr. lectr., 1965—. Chmn. sugar com., Guyana, pres. staff assn. Ministry Agr., 1959-61; univ. rep. Com. for Crown Lands Devel., Trinidad, 1966—. U. Nottingham Postgrad. fellow, 1955-57; U. S. AID fellow, 1961; Carnegie Corp. travel grantee, 1964; Rockefeller Found travel grantee, 1967. Mem. Am. Soc. Agronomy, Brit., Internat. socs. soil sci., Assn. Tropical Biology. Sec. editorial bd. Jour. Tropical Agr., 1964—. Research on Caribbean and West Indian soils including Guyana. Home: Canada Hall, U. W.I., St. Augustine, Trinidad, W.I.*

AHMAD, S., physician; b. Wazirabad, West Pakistan, Feb. 19, 1905; s. N. and Rehmat (Bibi) A.; F.Sc. (Med.), Govt. Coll., Lahore, 1923; M.B.B.S.K.E., Med. Coll., Lahore, 1928; m. Agra Nazim, June 16, 1946; 4 daus., 2 sons. Apptd. in Indian Med. Service, 1932, employed in mil. service, also as civil surgeon, lectr. surgery U.P., India; prof., head dept. surgery Agra Med. Coll., 1946-47; on parbion aphad for Pakistan, 1947-53; surg. specialist Jinnah Central Hosp., also med. supt., prof. surgery Dow Med. Coll., K'chi, 1953-61; prof., head dept. surgery Dow Med. Coll., 1961—; supt. Civil Hosp., K'chi, 1956-58, 65—; hon. surgeon J.P.M. Centre, Naval Hosp., Liaquat Nat. Hosp.; surgeon Navud Clinic. Licentiate Royal Coll. Physicians, London. Fellow Royal Coll. Surgeons, Eng.; mem. Brit. Med. Assn., Med. Assn. Pakistan, Assn. Surgeons Pakistan; fellow and mem. other profl. orgns. Research and publs. on surg. treatment of cancer, also on grafting and repair of lower genito-urinary defects. Home: 3/160-A, P.E.C.H.S., Karachi. Office: Navud Clinic, Saddar, Frere Rd., Karachi, Pakistan.*

AHMANN, Donald Henry, Am. chemist; b. Struble, Ia., Jan. 9, 1920; s. Henry F. and Philomene (Wictor) A.; student Trinity Coll., 1937-39; B.S., Ia. State U., 1941, Ph.D., 1948; m. Anne Harvey, Sept. 12, 1945; children—Richard Stanley, J. Carol, Rebecca A., Sarah W., Kathryn D., Elizabeth H. Jr. chemist, Manhattan Project, Ia. State Coll., Ames, 1942-46, AEC at Ia. State, 1946-48; research asso. Gen Electric Co., Knolls Atomic Power Lab., Schenectady, 1948-50; mgr. phys. chemistry, 1950-55, mgr. chemistry, chem. engring., 1955-57, at Vallecitos Atomic Lab., Pleasanton, Cal., 1957-67, mgr. nuclear materials and propulsion operation, Cin., 1967—. Mem. Am. Chem. Soc., Am. Soc. Metals, Am. Nuclear Soc., Phi Lambda Upsilon, Phi Kappa Phi. Important work includes research in radiochemistry, chemistry related to nuclear reactors, metallurgy and ceramics of nuclear materials, reprocessing of nuclear fuels. Home: Cin. Office: P.O. Box 15132, Gen. Electric Co., Cin. 45215.*

AHMAVAARA, Yrjö Arvi, Finnish mathematician; b. Oulu, Finland, Aug. 7, 1929; s. Arvi E. and Jenni (Päkkilä) A.; fil.kand., U. Helsinki (Finland), 1950, Ph.D., 1954; m. Anita Vanhala, May 17, 1952; children—Riitta, Outi, Ilkka. Research scholar Nord. Inst. Theoretical Atom Physics, Copenhagen, Denmark, 1957-59; scientist Com. for Atomic Energy, Helsinki, Finland, 1959-60; lectr. math. physics U. Helsinki, 1961-63, lectr. math. behavioral sci., 1967—; prof. math. physics Finnish U. Turku, 1963-66; vis. prof. math. psychology Ohio State U., 1966-67. Research, publs. on formulation of relativistic quantum theory in terms of geometrical symmetry (group) properties; contbr. to founds. of behavioral sci., semantic information logic; developed transformation analysis (a statis. multivariate research method). Home: Kulmakatu 1 A 1, Helsinki 17, Finland.*

AHMED, physicist, mathematician; b. 9th century; s. Musa Ibn Sakir; one of three brothers whose works cannot be differentiated; made astron. observations from obs. in their home in Bagdad. Author: Book of Geometry by the Three Brothers; Kitba fi al-hival. Worked with problems in geometry, including insertion of 2 mean proportionals, balance and trisections of angles, constrn. of ellipses; experimented with automatic mus. instruments.

AHMED BEN UTMAN, Egyptian physician; flourished circa 1250; s. Qadi Gamal ad-din; held title of med. chief of Egypt. Author: Arab Ophthalmology by a Practitioner in Cairo in the 13th Century.

AHMED, Quazi Akhter, Pakistani mycologist, plant pathologist; b. Pabna, East Pakistan, Jan. 1, 1926; s. Quazi Afazuddin and Rahimun (Nessa) A.; I.Sc., Rajshahi Coll., 1943; B.Sc., Calcutta U., 1945, M.Sc., 1947; Ph.D., Queen's U., U.K., 1954; m. Nilufar Rahman, Mar. 10, 1957; children—Anis Ahmed, Amin Ahmed, Bonna A. Curator herbarium, mus. and garden, botany dept. Dacca (East Pakistan) U., 1948-49, lectr., 1949-57, plant pathologist Jute Research Inst., 1957—; hon. fellow dept. plant pathology and physiology U. Minn., 1963-64. Postdoctoral fellow Conf. Bd. of Asso. Research Councils, Washington, 1963-64. Mem. Brit. Mycol. Soc., Am. Phytopath. Soc., A.A.A.S., Pakistan Assn. for Advancement Sci. Translator: (to Bengali from English) A Guide to Plant Diseases and Their Control, by A. Hafiz. Studies, publs. on fungal flora of East Pakistan; control of plant pathogens with antagonistic soil microbes; resistance of plants to disease. Home: 47/2, Bhajahari Shaha St., Dacca 1, East Pakistan.*

AHMOSE (or AHMES, AAHMESU), Egyptian scribe; flourished circa 1700-1550 B.C.; scribe in court King Aauserre; copied Ahmose (or Rhind) Papyrus (earliest complete work showing Egyptian math. devel., including unit fractions, problems involving geometric and arithmetic series, methods for measuring areas and making constrns., geometrical progression, solutions to equations of 1 unknown, knowledge of proportion, 3.1604 . . . as value of pi, treatment of each problem as spl. case). Author: Directions for Obtaining the Knowledge of All Dark Things.

AHORNRAIN, see von Ahornrain.

AHRENDT, Myrl Howard, Am. mathematician; born Grand Forks, North Dakota, Mar. 17, 1907; son of Albert George and Mathilda (Kohlhase) A.; B.A., Friends Univ., 1929; M.A., Univ. Wichita, 1932; postgraduate work, Ball State Teachers College, U. Mich.; m. Dolly Morrison, July 27, 1930; 1 dau., Margery (Mrs. David Bowen). Tchr. high schs., Ind., 1939-47; chmn. math and physics depts. Anderson (Ind.) Coll., 1947-51; exec. sec. Nat. Council Tchrs. Math., Washington, 1951-63; asst. to head Inst. sect. Nat. Sci. Found., Washington, 1963-64; ednl. specialist Office Ednl. Programs and Services, NASA, Washington, 1964-66; instructional resources officer, 1966—; lectr. U. Md., Washington, 1953-54, asso. prof., 1963—. Del. White House Conf. on Edn., 1955; lectr. John Hopkins, Balt., 1954-58, U. Va. Richmond, 1955. Fellow A.A.A.S., Brit. Interplanetary Soc.; mem. Nat., Ind. (pres. 1947-48) councils tchrs. math., Math. Assn. Am., N.E.A., Central Assn. Sci and Math. Tchrs., Fed. Schoolmen's Club, Sigma Zeta (nat. pres. 1950-51). Author: The Mathematics of Space Exploration, 1965. Contbr. articles to tech. jours. Research in math. linkages, application of math. and sci. in space exploration. Home: 804 Gregorio Dr., Silver Spring, Md. 20901. Office: Universal Bldg. N., Florida Av., Washington 20546.*

AHRENS, Edward Hamblin, Jr., Am. clinical investigator; b. Chicago, Illinois, May 21, 1915; s. Edward Hamblin and Pauline (Forsyth) A.; B.S., magna cum laude, Harvard, 1937, M.D. cum laude, 1941; m. Gertrude Fobes, Sept. 12, 1940; children—Sandra Huntington (Mrs. Peter Sammis), Peter Forsyth, Burgess Keith. Asst., Rockefeller Inst., N.Y.C., 1946-49, guest investigator, 1949-51, asso., 1952-55, asso. prof., 1955-60, professor, since 1960—; sr. fellow NRC, 1949-50, 58-59, Nat Found. for Infantile Paralysis, 1950-52. Mem. metabolism study sect. USPHS, 1956-61, chmn., 1959-61; mem. bd. sci. counselors Nat. Heart Inst., 1963—; chmn.; 1965—; sci. adv. council Blythedale Children's Hosp., Valhalla, N.Y., 1965—. Mem. Am. Soc. Biol. Chemists, Am. Soc. Clin. Investigation, Assn. Am. Physicians, Atherosclerosis Discussion Group (Gt. Britain), Harvey Soc., Internat. Conf. on Lipid Biochemistry (v.p. 1965—), Practitioners' Soc. N.Y., Soc. for Exptl. Biology and Medicine, Soc. Pediatric Research, Century Assn., Phi Beta Kappa, Sigma Xi. Founder of the Journal of Lipid Research, 1959, editor, since 1963—; Concepts in Biochemistry, 1964. Clin. research on lipid metabolism, especially as related to coronary heart disease. Home: 3 Leonard Rd., Bronxville, N.Y. 10708. Office: Rockefeller U., 66th St. and York Av., N.Y.C. 10021.*

AHRENS, Felix Benjamin, chemist; b. Danzig, Poland, Nov. 22, 1863; ed. univs. Berlin (Germany), Breslau (now Wroclaw, Poland), Kiel (Germany), 1882-86; apptd. asst. to Prof. A. Ladenburg, Breslau, 1887; private lectr., 1889; became asso. prof., dir. agrl.-tech. inst. Breslau U.; founder Tech. U. Breslau, 1910. Author: Handbuch der Elektrochemie, 1896; Anleitung zur chemisch-technischen Analyse, 1901; Einführung in die praktische Chemie, 2 vols., 1901-02; Lehrbuch der chemischen Technologie der landwirtschafflichen Gewerbe, 1905. Pub. Chemische Zeitschrift, 1901-06. Research on mineral coal tar, acetylene, sulfitlye. Died Berlin, Nov. 14, 1910.

AHRENS, Wilhelm Ernst Martin Georg, German mathematician; b. Lübz, Germany, Mar. 3, 1872; student in Rostock, also Berlin, Freiburg, Leipzig (with Sophus Lie), Germany; recipient degree, 1895. Became tchr. German Sch., Antwerp, Belgium, 1895, later at Leipzig U., at constrn. and machine design schs., Magdeburg, Germany, 1897-1904. Author: Scherz und Ernst in der Mathematik, 1904; Mathematische Unterhaltungen und Spiele; Mathematischen Spiele (in Enzyklopädie der Mathematischen Wissenschaften), 2 vols., 1910-18; Mathematikeranekdoten, 1916; pub. correspondence of Jacobi bros. Studied group theory, sources of math. knowledge. Died Rostock, Apr. 23, 1927.

AHRLAND, Sten Harald, Swedish chemist; b. Mariestad, Sweden, Aug. 4, 1921; s. Harald and Märta (Ljunggren) A.; Fil.Dr., U. Lund (Sweden), 1952; m. Ingrid Leveaux, June 23, 1946; children—Gudrun, Bo, Gertrud, Asa. Asst., inst. chemistry U. Lund, 1944-52, docent, 1952-58, asst. research prof. (forskardocent), 1960-65; research chemist I.C.I. Akers Research Labs., Welwyn, Eng., 1954; head analytical and fundamental processes sects. Eurochemic Co., Mol, Belgium, 1958-60; research asso. Swedish Natural Sci. Research Council, Lund, 1965—. Cons. Swedish Atomic Energy Co. (Ab Atomenergi), 1957—. Recipient Norblad-Ekstrano Gold medal Swedish Chem. Soc., 1965. Contbr. papers to sci. jours. Research on ionic equilibria in solution in order to establish scope and reliability of such measurements, factors governing formation of complexes between metal ions and ligands of various types; investigations of inorganic ion exchangers, for use in radioactive solutions. Home: 10, Kavallerigränden, Lund. Office: Institute of Chemistry, Helgonavägen, Lund, Sweden.*

AICHEL, Otto, anatomist, anthropologist; b. Concepcion, Chile, Oct. 31, 1871; s. Oswald and Clementine (Geisner) A.; M.D., Ph.D.; m. Dora Timmermann; 1 son, 3 daus. Apptd. Prof. gynecology, Santiago, Chile, 1900; became prosector Anatomy Faculty, Munich (Germany) U., 1909, Halle (Germany) U., 1911, U. Kiel (Germany), 1913; apptd. prof. anatomy and anthropology U. Kiel, 1921, founder Anthrop. Inst. Author: Der deutsche Mensch, ein Vergleich deutschen mit chilenischbolivianischen Funden, 1933. Studied comparative anatomy; undertook expdn. to Chile and Bolivia for research in racial anthropology. Died Kiel, Jan. 31, 1935.

AICHOLZ, Johann Emerich, physician; b. Vienna, Austria, 1520; student Vienna U., from 1536, also after 1550; Wittenberg, Germany, from 1543; became magister, 1547; student in France, also Italy, after 1550; M.D., Ph.D., U. Padua (Italy), after 1555. Prof. anatomy, also practiced medicine, Vienna, after 1557; founder (probably with Paul Fabricius) bot. garden of rare plants. Author: Consilium in hydrope monstroso. Died Vienna, May 6, 1588.

AIDA, Yasuaki, Japanese mathematician; b. Uzen, Japan, 1747; engaged in riparian work Tone River, Kanto; founded Mogami Sch. abacus calculation; evolved Tenshei-ho (method of algebraical calculation). Author: Jizaimonogatari; Sampo Kokon Tsuran. Died 1817.

AIELLO, Tommaso Nicolo, Italian chemist; b. Bagheris, Italy, Jan. 2, 1903; s. Salvatore and Teresa (Mannara) A.; D. Chem. Scis.; m. Concetta Carmi, Oct. 10, 1931; children—Salvatore, Teresa, Francia, Enrico. Prof. pharm. chemistry, rector U. Palermo. Mem. Italian, Amer. Chem. socs., Soc. Pharm. Chemistry, Reale Accademia Scienze Palermo. Home: via Archirafi 32, Palermo. Office: Università di Palermo, Palermo, Italy.

AIGNER, Franz Johann, Austrian physicist; born in St. Pölten, Austria, May 13, 1882; the son of Franz Ignaz and Franziska (Haberl) A.; Ph.D., Vienna (Austria) Univ., 1906; married Karoline Marie Habigen, July 12, 1941. Physics laboratory asst. Vienna Tech. U., apptd. asst. prof., 1918, asso. prof., 1925, prof. tech. physics,, 1930; became dir. research inst. German Reichs Post, 1938. Recipient Haitinger prize Acad. Scis., Vienna, 1924. Author: Unterwasserschalltechnik, 1922. Contbd. articles to math. and natural sci. jours. Research in electro-optics, electro-acoustics (used for mil. devel.). Died Vienna, July 19, 1945.

AIGRAIN, Pierre Raoul, French physicist; b. Poitiers, France, Sept. 28, 1924; s. Marmi and Germaine (Ligault) student École Normale; M.S., Carnegie Inst. Tech., 1947, D.Sc., 1948; Doctorat ès-Scis., U. Paris, 1950; m. Francine Bogard, Feb. 12, 1947; children—Philippe, Yves, Jacques. Asso. prof. U. Lille (France), 1952-54; asso. prof. U. Paris, 1954-58, prof., 1958-65, 68—; sci. dir. Def. Research Directorate (DRME), French Ministry Def., 1961-65; dir. higher edn. Ministry Edn., 1965-68. Del. gen. Recherche Scientifique et Technique France, 1968. Fellow I.E.E.E.; mem. French Phys. Soc. (sec. gen. 1960-67). Author: (with F. Bupleat) Les Semiconducteurs, 1958; (with Coelbo and Ascorelli) Electronic Processes in Solids, 1962; also articles. Semiconductor research on photomagneto-electron phenomena, recombustion radiation, semiconductor lasers, wave propagation in solid state plasmas (Helicon), thermoelectricity; patentee solid state devices, electronic telephone switching. Home: 8, Spucoeheory Pate, Paris 16, France.*

AIKAWA, Jerry Kazuo, Am. physician; b. Stockton, Cal., Aug. 24, 1921; s. Genmatsu and Shizuko (Yamamoto) A.; A.B., U. Cal. at Berkeley, 1942; M.D., Bowman Gray Sch. Medicine, 1945; m. Chitose Aihara, Sept. 21, 1944; 1 son, Ronald K. Practice medicine, specializing in internal medicine, Winston Salem, N.C., 1950-53, Denver, 1953—; faculty Bowman Gray Sch. Medicine, 1950-53; investigator Am. Heart Assn., Denver, 1952-57; asst. prof. medicine U. Colo. Sch. Medicine, 1953-60, asso. prof., dir. lab. services, 1960—. Mem. A.M.A., Soc. Exptl. Biology and Medicine, Soc. Nuclear Medicine, So., Western socs. clin. research. Phi Beta Kappa, Sigma Xi, Alpha Omega Alpha. Author: Myxedema, 1961; The Role of Magnesium in Biologic Processes, 1963; Rocky Mountain Spotted Fever, 1966; also numerous articles. Studies, publs. on pathophysiology of immune reactions, Rocky Mountain spotted fever, trichinosis, and rheumatic fever; changes in body fluids and electrolytes in various disease states; clin. pharmacology of digitalis and diuretic compounds. Home: 619 S. Poplar Way, Denver 80222. Office: 4200 E. 9th Av., Denver 80220.*

AIKEN, Howard Hathaway, Am. mathematician; b. Hoboken, N.J., Mar. 9, 1900; s. Daniel H. and Margaret Emily (Mierisch) A.; S.B., U. Wis., 1923; postgrad. physics U. Chgo.; A.M., Harvard, 1937, Ph.D., 1939; Dr. Ing. E.h., Technische Hochschule, Darmstadt, Germany, 1952; LL.D., Wayne State U., 1954; m. Agnes Montgomery, Jan. 7, 1943. With Madison Gas & Electric Co. (Wis.), 1923-27, Westinghouse Electric Mfg. Co., Chgo., 1927-31; instr. physics Harvard, 1939-41, asso. prof. applied math., 1941-46, prof., dir. computation lab.; now faculty mem. U. Miami, Coral Gables, Fla. Recipient Palmes de l'Academie Francais, 1949; Ingeniors Vetenskaps Akademien, 1950; National Fonds voor Wetenschap Pelijk Onderzoek, 1951; chevalier Legion of Honor, 1952; Raymond E. Hackett award, 1954; officers cross Order of Crown (Belgium), 1955; Distinguished Pub. Service award USN, 1956; decoration for civilian service USAF, 1957; U. Wis. award, 1958, La Medaille de Vermeil (Paris, France), 1959. Fellow Am. Acad. Arts and Scis.; mem. A.A.A.S., Econometric Soc., Gesellschaft fur Angewandte Mathematik und Mechanik, Harvard Engring. Soc., Phi Beta Kappa, Sigma Xi. Gen. editor Annals of Computation Lab., 1946—. Contbr. articles to sci. jours. Research, publs. on switching theory and automatic data processing; inventor world's largest digital calculator; completed his 1st automatic sequence-controlled calculator (Mark I), 1944, then all electrical calculator (Mark II), 1947. Home: 1511 S. Ocean Dr., Ft. Lauderdale, Fla. Office: P.O. Box 8245, U. Miami, Coral Gables, Fla.

AIKIN, Arthur, English chemist; b. Warrington, Eng., May 19, 1773; s. John Aikin; ed. Mr. Barbauld's Sch., Palgrave, Eng.; studied chemistry under Priestley; lectr. on chemistry and chem. manufactures, London, 1799. Fellow Linnean Soc.; mem. Geol. Soc. (founder 1807), Chem. Soc. (1st treas. 1841), Acad. Dijon (France). Author: With Observations in Mineralogy and other branches of Natural History (in N. Wales and Shropshire, Eng.) 1797; Natural History of the Year, 1798; Manual of Mineralogy, 1814; An Account of the most recent Discoveries in Chemistry and Mineralogy, 1814; (with C. R. Aikin) Dictionary of Chemistry and Mineralogy, 1807-14. Translator: Travels in Egypt (Denon), 1801. Died London, Apr. 15, 1854.

AIKIN, Charles Rochemont, English physician, chemist; b. Warrington, Eng., 1775; s. John Aikin; ed. Mr. Barbauld's Sch., Palgrave, Eng.; m. Anne Wakefield; mem. Royal Coll. Surgeons; sec. London Med. and Chirurg. Soc. Author: Concise View of all the most important Facts that have hitherto appeared respecting the Cow Pox, 1800; (with Arthur Aikin) Dictionary of Chemistry and Mineralogy, 1807-14. Died Mar. 20, 1847.

AILHAUD, John, French surgeon; b. Lourmarin, 1674; M.D., Aix. Credited with a purgative medicine that bears his name. Author: Traité de l'origine des maladies et des effects de la poudre purgative (attempt to prove all illness has same cause and cure), 1740. Died Vitrolles, 1756.

AILLY, Pierre d', see d'Ailly.

AINGER, Lorin, Am. physician; b. Arcata, Cal., Oct. 1, 1925; s. Earl H. and Virginia (Bonta) A.; A.A., U. Cal. at Berkeley, 1946; A.B., McGill U., 1947; M.S., M.D., Tulane U., 1951; m. Marion Claire MacArthur, Jan. 9, 1954; children—Paul Jesse MacArthur, Marc Christopher, Joel Michael. Teaching asst. in physiology U. Cal. at Berkeley, 1946-47, Tulane U., New Orleans, 1947; instr. pediatrics U. Utah, Salt Lake City, 1954-55, asst. prof., 1955-58; asst. prof. pediatrics U. Tenn., Memphis, 1958-62, asso. prof. pediatrics and surgery, 1962—, chmn. sect. pediatric cardiology, 1958——. Mem. A.M.A., Am. Acad. Pediatrics, Western, So. socs. pediatric research. Research in steroids in children, rheumatology, electrophysiology. Home: 5593 Normandy St., Memphis 38117.*

AINSLIE, Sir Whitelaw, surgeon; b. Dunse, Berwickshire, Eng., Feb. 17, 1767; s. Robert Ainslie; m. dau. James Cuninghame; 1 dau., Jane Catherine (Mrs. James C. Grantduff). Apptd. asst. surgeon E. India Co., 1788; became garrison surgeon, Chingleput, India; became surgeon, 1794, superintending surgeon, 1810; superintending surgeon so. div. Army, Madras, India, 1814-15. Author: Treatise Upon Edible Vegetables and the Materia Media of Hindostan, 1813; Materia Indica, 1826; Observations of on the Cholera Morbus of India, 1825; (with A. Smith and M. Christy) Report on the Causes of Epidemical Fever which prevailed in the Provinces of Coimbatore, Madeira, Dinigal, and Tinivelly in 1809-10-11, 1816. Died Apr. 29, 1837.

AINSWORTH, William Francis, English geographer; b. Exeter, Eng., Nov. 9, 1807; s. John Ainsworth; student geology in London, England, also Paris, France, Brussels, Belgium; mem. Col. Chesney's expdn. to Euphrates, 1835; as agt. Bible Soc. and Geog. Soc. explored River Halys (now Kizil Irmak), Turkey, visited Christians of Kurdistan. Author: Researches in Assyria, Babylonia . . . , 1838; Travels and Researches in Asia Minor, Mesopotamia . . . , 1842; Travels in the Track of the 10,000 Greeks, 1844; A Personal Narrative of the Euphrates Expedition, 1888. Died London, Nov. 27, 1896.

AINSWORTH-DAVIS, James Richard, zoologist; geologist; b. Bristol, Eng., Apr. 1861; s. James Davis; M.A., Cambridge (Eng.) U; M.Sc., Bristol (Eng.) U.; ed. also Royal Sch. Mines; m. Jessie Coutts, 1885; 1 son. Army instr., 1915-19; chmn. Central Civilian Adv. Bd., G.H.Q.; asst. sec. Service Students' Bur., Bd. Edn., 1919-20; lectr. biology Middlesex (Eng.) Hosp. Med. Sch., 1920-22; examiner rural economy Oxford (Eng.) U.; examiner zoology Edinburgh U., Aberdeen U. (both Scotland), U. Wales; prin. Royal Agrl. Coll., Cirencester, Eng.; prof. U. Bristol, U. Wales; life gov., dean sci. faculty, prof. zoology and geology Univ. Coll. Wales, Aberystwyth. Lectr. Empire Marketing Bd., Brit. Empire Cancer Campaign. Fellow Coll. Preceptors, Zool. Soc., Inst. Certified Grocers (hon.). Author: Notes on Habits of Patella; (with others) Memoir on Patella; Bionomics of Gastropod Evolution; The Natural History of Animals; Life of T. H. Huxley; The Pursuit of Natural Knowledge; Principles of Agriculture; Food Resources of the Empire; Book of the Empire; Cooking Through the Centuries. Editor and co-author: Science in Modern Life; Fream's Elements of Agriculture. Died Apr. 7, 1934.

AIRD, John, Brit. engr.; b. London, Eng., Dec. 3, 1833; s. John and Agnes (Bennett) A.; ed. privately; m. Sarah Smith, Sept. 6, 1855; 2 sons, 7 daus. Became mem. father's firm, chief partner, 1870; mem. Parliament, 1887-1905; became 1st mayor, Paddington, Eng., 1900. Created baronet, 1901. Asso., Instn. Civil Engrs.; mem. Iron and Steel Inst. Constructed Aswan and Asyut dams, 1898-1902; built gas and water reservoirs, rys., docks. Died Beaconsfield, Eng., Jan. 6, 1911.

AIRD, Robert Burns, Am. neurologist; b. Provo, Utah, Nov. 5, 1903; s. John W. and Emily (McAuslan) A.; B.A., Cornell U., 1926; M.D., Harvard, 1930; m. Ellinor H. Collins, Oct. 4, 1935; children—Katharine (Mrs. Lynn Miller), Mary (Mrs. Goode P. Davis, Jr.), John, Robert Bruce. Faculty. U. Cal., San Francisco, 1935—, prof. neurology 1949—, chmn. dept., 1947-66, dir. electroencephalographic lab., 1941—, dir. electromyographic lab., 1943—. Cons. numerous hosps., including Langley Porter Neuropsychiat. Inst., San Francisco Gen., Laguna Honda Home, San Francisco, VA Hosp., Ft. Miley; mem. med. adv. bds. Myasthenia Gravis Found., Cal. Epilepsy Soc., San Francisco chpt. Muscular Dystrophy Assn., 1959——. Recipient 10 Year Service award Nat. Multiple Sclerosis Soc., 1963; Honors Achievement award Angiology Research Found. and Purdue Frederick, 1964-65; comdr. order of Hipolito Unanue, Peru, 1963. Fulbright grantee, 1957. Diplomate Pan Am. Med. Assn., Nat. Bd. Medicine, Am.

Bd. Psychiatry and Neurology; certified sub-specialty clin. electroencephalography. Mem. A.M.A., Am. Acad. Neurology, Am. Neurol. Assn. (v.p. 1955), Am. Electroencephalographic Soc. (pres. 1953), Am. Epilepsy Soc. (pres. 1959), Harvey Cushing Soc., Internat. League Against Epilepsy, Western Soc. Clin. Electroencephalography (pres. 1947), Western Soc. Clin. Research, Am. Med. Writers Assn, Alpha Kappa Kappa, numerous others. Cons. editorial bds. Electroencephalography and Clin. Neurophysiology, 1951-65, Jour. Nervous and Mental Disease 1961——. Research, publs. on permeability of blood-brain barrier, demyelinating and degenerative diseases; diagnostic techniques of pneumoencephalography and electroencephalography. Home: 80 Summit Av., Mill Valley, Cal. 94941. Office: U. Cal. Med. Center, San Francisco.*

AIRY, George Biddell, English astronomer; b. Alnwick, Eng., July 27, 1801; s. William and Ann (Biddell) A.; grad. Cambridge U., 1823, LL.D., 1862; D.C.L., Oxford U., 1844; LL.D., Edinburgh U.; m. Richarda Smith, Mar. 24, 1830; 9 children, including Osmond, Hilda (Mrs. Routh). Apptd. fellow Cambridge U., 1824, Lucasian prof. math., 1826, Plumian prof. astronomy, also dir. Cambridge Obs., 1828; headed Brit. expdn. to observe transit of Venus across Sun, 1874. Decor. Legion of Honor, 1856; named royal astronomer. Fellow Royal Soc., 1836, Geol. Soc.; mem. Royal Astron. Soc. (Gold medal 1846), French Acad. Scis. (asso.). Author: Report on the Progress of Astronomy, 1832; 11 vols. including treatises on Trigonometry, circa 1825; Gravitation, 1834, Partial differential Equations, 1866, Sound and Atmospheric Vibrations, 1868; Magnetism, 1870; 377 papers; 141 ofcl. reports and addresses. Contbr. The Figure of the Earth, also Tides and Waves to Ency. Metropolitana. Worked primarily in celestial mechanics, especially lunar and planetary motion; discovered Airy's spirals in connection with light polarization through quartz crystals; discussed requirements of good telescope ocular, 1824; introduced cylindrical-spherical eyeglasses for astigmatic eyes, 1825; gave 1st theory of diffraction rings of objectglasses, 1834; stated modern theory of rainbow, 1836; invented compensation of ship's compass, 1839; perfected theory of tides given by Laplace (1774), 1847; pub. principle of hydraulic brake, 1851; determined mass and density of earth, 1855; developed method of dealing with problem of sun's translation, 1855; constructed orbit sweeper, 1861; headed Brit. expdn. to observe transit of Venus across Sun, 1874; equipped Royal Obs. with instruments he designed; created magnetic and meteorol. depts. at Greenwich; reduced planetary and lunar observations made at Greenwich, 1750-1830; opposed Faraday's lines of force theory (mathematically justified by Maxwell), temporarily kept John Couch Adams from investigations which would have led to discovery planet Neptune. Died Greenwich (now part of London), Eng., Jan. 2, 1892.

AISENBERG, Alan Clifford, Am. physician; b. N.Y.C., Dec. 7, 1926; s. Jacob and Celia (Able) A.; S.B., Harvard, 1945, M.D., 1950; Ph.D., U. Wis. 1956; m. Nadya Margulies, Oct. 2, 1952; children—James, Margaret. Am. Cancer Soc. fellow U. Wis, 1954-56; research fellow medicine Mass. Gen. Hosp., 1956-57, staff, 1957—, asst. physician, head med. tumor group, 1961——; research fellow medicine Harvard Med. Sch., 1956-57, faculty, 1957—, asst. prof. medicine, 1962——. Guggenheim fellow Nat. Inst. for Med. Research, London, Eng., 1964-65. Mem. Am. Fedn. for Clin. Research, Am. Assn. for Cancer Research, Royal Soc. Medicine, Transplantation Soc., N.Y. Acad. Scis. Author: The Glycolysis and Respiration of Tumors, 1961; also articles. Research in tumor metabolism, Hodgkin's disease, immunological responsiveness. Home: 124 Chestnut St., Boston 02108. Office: Huntington Labs., Mass. Gen. Hosp., Boston 02114.*

AITA, John Andrew, Am. physician; b. Council Bluffs, Ia., June 20, 1914; s. John B. and Marie Louise (Faber) A.; M.D., U. Ia., 1937; Ph.D., U. Minn., 1944; m. Nellie Marie Fordyce, June 15, 1937; children—John Fordyce, Anne Marie. Practice medicine, specializing in neurology and psychiatry, Omaha, 1948—; mem. staffs Luth. Med. Center, Clarkson, Meth., Immanuel, U. Neb. hosps. (all Omaha); asso. prof. neurology and psychiatry U. Neb. Coll. Medicine, 1948——. Fellow Am. Psychiat. Assn., Am. Acad. Neurology; mem. A.M.A., Alpha Omega Alpha. Author: Neurologic Manifestations of General Diseases, 1964; Neurocutaneous Diseases, 1966; Congenital Facial Anomalies with Neurological Defects, 1967. Home: 2302 N. 55th St., Omaha 68104. Office: 105 S. 49th St., Omaha 68132.*

AITCHISON, James, Scottish orthodontist; b. Glasgow, Scotland, Mar. 24, 1899; s. Charles and Helen (Leed) A.; ed. Anderson's Med. Sch., Glasgow; m. 1929. With U. Glasgow, began as demonstrator anatomy, then asst. lectr. orthodontics, lectr. dental histology, clinic prof., prof. dental surgery, dir. dental sch. Mem. Gen. Dental Council of Brit. Dental Assn., Council Dental Edn. Author: Dental Histology for Students; Dental Anatomy and Physiology for Students; Racial Differences in Teeth; Sex Differences in Jaws, Skulls and Teeth; Dental Interest in Palaeontology. Editor: The Dental Magazine. Home: 10, Park Terrace, Glasgow C. 3. Office: U. Glasgow.

AITCHISON, James Edward Tierney, surgeon; b. Neemuch, India, Oct. 28, 1835; s. James and Mary (Turner) A.; M.D., Edinburgh (Scotland) U.; LL.D.; m. Eleanor Carmichael Craig, 1862. Became asst. surgeon Honourable E. India Co.'s Service; brigade-surgeon Her Magesty's Bengal Med. Service, 1858-1888; Brit. commr., Ladak, India, 1872; served with Kurrum Field Force, 1878, botanist to force, 1879-80; naturalist Afghan Delimitation Commn., 1884-85. Fellow Royal Socs. London, 1883, Edinburgh. Author: A Catalogue of the Plants of the Punjab and Sindh, 1869; Handbook of the Trade Products of Leh, 1874; The Flora of the Kurrum Valley; The Botany of the Afghan Delimitation Commission; The Zoology of the Afghan Delimitation Commission; Notes of the Products of Western Afghanistan and North-East Persia. Died Sept. 30, 1898.

AITKEN, John, Scottish physicist; born Falkirk, Scotland, 1839; the son of Henry Aitken; educated Glasgow (Scotland) Univ.; apprentice marine engineer at R. Napier and Sons, Lancefield, Glasgow. Fellow Royal Soc., 1889; Royal Soc. Edinburgh. Contributed papers to publications royal socs. Investigated atmosphere; inventor instruments for measuring dust content of atmosphere, chronometers, the koniscope; explained some instances of fog formation; showed that dew on plants is usually exuded; investigated cyclones, color and color sensation, rigidity produced by centrifugal force. Died Falkirk, Nov. 13, 1919.

AITKEN, John A., Scottish surgeon; b. Scotland, 18th century; student medicine, Edinburgh, Scotland, became surgeon, also lectr. in surgery, anatomy, midwifery, chemistry, 1779; mem. Royal Coll. Surgeons Edinburgh, 1770. Author: Essays on several important subjects in Surgery, chiefly with regards to the nature and cure of fractures, 1771; Essays and Cases in Surgery, 1775; Conspectus rei chirurgical, 1777; Medical Improvement; an Address to the Medical Society of Edinburgh, 1777; Elements of the Theory and Practice of Surgery, 1779; Outlines of the Theory and Cure of Fever, 1781; Principles of Midwifery or Puerperal Medicine, 1784; Osteology, 1785; Principles of Anatomy and Physiology, 1786; Essays on Fractures and Luxations. 1790. Devised bilateral pelviotomy for dystocia in cases of narrow pelvis, 1784; altered mode of locking midwifery forceps; invented flexible lever blade, also pair of forceps for dividing and diminishing bladder stones too large for entire removal by lithotomy. Died Sept. 22, 1790.

AITKEN, Thomas Henry Gardiner, Am. entomologist; b. Porterville, Cal., Aug. 31, 1912; s. Thomas Gardiner and Edith A. (McHarg) A.; student Ore. State Coll., 1931-32; B.S., U. Cal. at Berkeley, 1935, Ph.D., 1940; m. Virginia Gale, Nov. 13, 1948; children—Bruce Gardiner, Brian Tammas. Chief malariologist for Allied Command, Corsica, 1944; theatre malariologist U. S. Army, Italy, 1944-45; mem. field staff Rockefeller Found., N.Y.C., 1946—, with Trinidad Regional Virus Lab., 1954-66, Belém Virus Lab., Belém, Pará, Brazil, 1967——. Recipient U. S. Army Typhus Commn. medal, 1945; Gold medal Sardinian Govt., 1951. Fellow Royal Soc. Tropical Medicine and Hygiene; mem. A.A.A.S., Entomol. Soc. Am., Am. Soc. Tropical Medicine and Hygiene, Am. Mosquito Control Assn. Author: (with others) The Sardinian Project, 1953. Publs. on application of measures for control of malaria; the biology of mosquitoes and other arthropods; studies of virus transmission by arthropods. Address: Rockefeller Found., 111 W. 50th St., N.Y.C. 10020.*

AITKEN, William, physician; b. Dundee, Scotland, Apr. 23, 1825; s. William Aitken; M.D., Edinburgh, 1848; LL.D., univs. Edinburgh, Glasgow (Scotland); m. Emily Clara Allen, 1884. Asst. pathologist Med. Commn., Crimea, Russia, 1855; prof. pathology Army Med. Sch., Fort Pitt, Chatham, Eng. (later at Metley), 1860-92. Knighted, 1887. Fellow Royal Soc., 1873. Author: Handbook of the Science and Practice of Medicine, 1857; An Essay on the Growth of the Recruit and young Soldier. Compounded pill (Aitken's pill) of reduced iron, quinine sulfate, strychnine, and arsenic trioxide to stimulate blood cell formation. Died June 25, 1892.

AITKIN, John, surgeon; student medicine, Edinburgh, Scotland; became surgeon, lectr. surgery, Edinburgh, 1779; practiced physics, anatomy, midwifery, chemistry; mem. Royal Coll. Surgeons. Author: Essays on Several important Subjects in Surgery, chiefly with regard to Nature and Cure of Fractures, 1771; Essays and Cases in Surgery, 1775; Conspectus rei chirurgicae, 1777; Medical Improvement, an Address to the Medical Society of Edinburgh, 1777; Elements of the Theory and Practice of Surgery, 1779, repub. as Elements of the Theory and Practice of Phisic and Surgery, 1783; Outline of the Theory and Cure of Fever, 1731; Principles of Midwifery or Puerperal Medicine, 1784; Osteology, or a Treatise on the Bones of the Human Skeleton, 1785; Principles of Anatomy and Physiology, 2 vols., 1786; Essays on Fractures and Luxations, 1790. Introduced variation on method of locking midwifery forceps; invented pair of forceps for dividing and diminishing stone in bladder when too large to use lithotomy. Died Edinburgh, Sept. 22, 1790.

AITON, William, English botanist; b. Hamilton, Scotland, 1731; 1st curator Kew Gardens, England, from 1759. Author: (with Solander and Dryander) Hortus Kewensis (catalog of plants in Kew Gardens, 5600 species enumerated), 3 vols., 1789. Responsible for elevating Kew Gardens to important position; instrumental in sending Francis Masson on bot. expdn., 1772. Died Kew, Feb. 1, 1793.

AJIMA, Naonobu, Japanese mathematician; b. Japan, flourished 18th century; studied math. of Nakanishi Sch., also Seki Sch. Retainer resident, Yedo, Yamagata Prefecture, Japan. Works collected by pupils in book Fukyu Sampo. Simplified complex calculating methods of time (his method comparable to Western formula of definite integral); formulated original theory on circle; developed infinite series dealing with problems of logarithmic theory, cylindrical intersection, investigation of equations, others; 1 of 2 most important men in history of Japanese math.; developed yenri (type of integral calculus); solved problem of inscribing spheres within larger sphere. Died 1798.

AJL, Samuel Jacob, microbiologist; born Poland, Nov. 15, 1923; s. Joseph and Celia (Hertz) A.; came to U. S., 1939, naturalized, 1943; B.S., Bklyn. Coll., 1945; Ph.D., Ia. State Coll., 1949; m. Adele Davis, Sept. 15, 1946; children—Stephen Ira, Diane Frances, Leslie Judith. Faculty, Sch. Medicine Washington U., St. Louis, 1949-52; chief microbiology chemistry Walter Reed Army Inst. Research, 1952-56, asst. chief dept. bacteriology, 1956-58, profl. dir. metabolic biology NSF, 1959-60; dir. research Albert Einstein Med. Center, Phila., 1960—; prof. dept. biology Temple U., 1960, vis. prof. bacteriology, 1960—; mem. metabolic biology panel NSF, 1959, Council U. S. Govt., 1963—. NSF Sr. Postdoctoral fellow, Israel and Oxford, 1958; Alumnus Honors award Bklyn. Coll., 1964. Fellow N.Y. Acad. Scis., A.A.A.S.; mem. Am. Soc. Microbiology, Soc. Biol. Chemists, Soc. Exptl. Biology and Medicine, Am. Acad. Microbiology, Sigma Xi. Studies, numerous publs. on bacterial growth and respiration; discovery of several new enzymes; pathogenesis of infectious diseases and the molecular mechanisms of action of microbiol. toxins. Home: 2296 Bryn Mawr Av., Phila. 19131. Office: York and Tabor Rds., Phila. 19141.*

AJZENBERG-SELOVE, Fay, physicist; born Berlin, Germany, Feb. 13, 1926; d. Mojzesz A. and Olga (Naiditch) Ajzenberg; came to U. S., 1940, naturalized, 1946; B.S. in Engring., U. Mich., 1946; M.S. in Physics U. Wis., 1949, Ph.D. in Physics, 1952; m. Walter Selove, Dec. 18, 1955. Research fellow Cal. Inst. Tech., 1952, 54; lectr. Smith Coll., 1952-53; cons., fellow Mass. Inst. Tech., 1952-53; from asst. prof. to asso. prof. Boston U., 1953-57; asso. prof. Haverford (Pa.) Coll., 1957-62, prof., 1962—; vis. asst. prof. Columbia, vis. prof. Nat. U. Mexico, summer 1955; lectr. U. Pa., 1957, cons. 1962-63. Exec. sec. com. on physics faculties in colls., mem. adv. com. on manpower, Am. Inst. Physics, 1962—. Smith-Mundt fellow U. S. Dept. State, 1955; Guggenheim fellow, 1965-66. Fellow Am. Phys. Soc.; mem. Am. Assn. U. Profs., Phi Beta Kappa (hon.), Sigma Xi. Home: 520 Brookview Lane, Havertown, Pa. 19083.*

AKABANE, Jiro, Japanese pharmacologist; b. Matsumoto, Japan, May 5, 1909; s. Moichiro and Masu (Imai) A.; M.D., Tokyo (Japan) Imperial U., 1940; m. Masako Tokunaga, Oct. 28, 1936, 1 dau., Mari. Research fellow Tokyo Imperial U. Hosp., 1934-44; prof. pharmacology Matsumoto Med. Sch., 1944-50; prof. pharmacology Faculty Medicine, Shinshu U., Matsumoto, 1950—, dean students, 1964—. Mem. Japanese Pharmacological Soc., Japanese Med. Soc. Alcohol Studies, Japanese Assn. Phys. Medicine and Balneology. Contbg. author: Aldehyde and Related Compounds, 1967; also numerous articles. Research in pathogenesis of alcohol intoxication and methylalcohol poisoning, treatment of alcohol intoxication, cyanamide therapy of alcoholism, research in pharmacology of acetaldehyde and related compounds. Home: Metoba, Matsumoto, Nagano, Japan.*

AKAISHI, Jun, Japanese chemist; b. Tofukuji, Shinonoi, Nagano, Japan, July 7, 1932; s. Heitaro and Chiyoko (Hirata) A.; B.S., Tokyo Met. U., 1957; m. Michiko Saito, Dec. 3, 1966. Tech. asst. Shinshu U., Nagano City, Japan, 1951-52; research chemist Japan Atomic Energy Research Inst., Tokai, Japan, 1957—, group leader, 1965—. Mem. Chem. Soc. Japan, Atomic Energy Soc. Japan, Japan Soc. Analytical Chemistry, Japan Radioisotope Assn., Health Physics Soc., Japan Radiation Research Soc., Japan Health Physics Soc. Research, publs. on devel. of analytical methods for artificial and natural radioelements in biol. substances; use of these methods for radiation protection. Home: Genken Nagabori-jutaku D 7-1, Tokai. Office: Japan Atomic Energy Research Inst., Tokai, Ibaraki, Japan.*

AKASOFU, Syun-Ichi, geophysicist; b. Naganoken, Japan, Dec. 4, 1930; s. Shigenori and Kumiko (Koike) A.; B.S., Tohoku U., Sendai, Japan, 1953, M.S., 1957; Ph.D., U. Alaska, 1961; m. Emiko Endo, Sept. 24, 1961; 1 son, Ken-Ichi. Came to U. S., 1958. Sr. research asst. Nagasaki U., 1953-55; asso. prof. geophysics U. Alaska, College, 1962-64, prof., 1964—. Mem. Am. Geophys. Union. Contbr. numerous articles to profl. jours. Studies on aurora polaris. Address: U. Alaska, College, Alaska 99701.*

AKCASU, Hüseyin Alaeddin, Turkish physician; b. Izmir, Turkey, Aug. 21, 1921; s. Osman and Faika A.; B.S., Turkish Lycee, Izmir, 1938; M.D., U. Istanbul, 1944; m. Hanriet Umit, Sept. 7, 1964; 1 child, Ershet. Prof. pharmacology Sch. Medicine, U. Istanbul, 1954—. Mem. Turkish Acad. Medicine, Turkish Soc. Medicine. Research and publs. on mode of action of narcotic analgesics and local anesthetics. Home: Florya Senlik Mahallesi Blok, Istanbul, Turkey.*

AKELEY, Carl Ethan, Am. naturalist; b. Clarendon, N.Y., May 19, 1864; s. Daniel Webster and Julia M. (Glidden) A.; ed. State Normal Sch. Brockport, N.Y.; m. 2d, Mary L. Jobe, Oct. 18, 1924. With Mus. of Milw., 1895-1909; with Field Mus., Chgo., 1895-1909, headed expdns. to Africa for museum, 1896, 1905; with Am. Mus. Natural History, N.Y.C., 1909-26, headed their expdns., 1909, 20-22, 26. Served as cons. engr., div. of investigation, research and development, Engr. Dept., U. S. Army, also as spl. asst. concrete dept. Emergency Fleet Corp. Recipient medal Franklin Inst. Developer technique of mounting animals in habitat groups; inventor cement gun used in mounting animals, Akeley camera used by naturalists; executed bronze sculptures including The Wounded Comrade, Cape Buffalo, The Nandi Spearman and Lions. Author: In Brightest Africa, 1923. Died Belgian Congo, Nov. 17, 1926.

AKELEY, Robert Vinton, Am. horticulturist; b. Presque Isle, Me., Nov. 13, 1910; s. George W. and Beulah (Barton) A.; B.S., U. Me., 1937, M.S., 1942; m. Hope Greenlaw, June 11, 1937; children—Andrea, Mary, Sarah. With U. S. Dept. Agr., Plant Industry Sta., Beltsville, Md., 1939—, sr. horticulturist, 1956-62, prin. horticulturist, 1962-64, leader potato investigations, 1960—. Fellow A.A.A.S.; mem. Am. Potato Assn. (past pres., past mem. exec. com.), Genetic Assn. Contbr. numerous articles to tech. jours. Research on new varieties potatoes. Home: 7 Melbourne Av., Silver Spring, Md. 20901. Office: Plant Industry Sta., Beltsville, Md. 10705.*

AKENSIDE, Mark, English physician; b. Newcastle, Eng., Nov. 9, 1721; s. Mark and Mary (Lumsden) A.; Student U. Edinburgh (Scotland); M.D, Leiden (Netherlands) U., 1744. Practiced medicine, Hampstead, Eng., 1745-47; under patronage of Jeremiah Dyson, from 1747; asst. physician, then physician Christ's, St. Thomas' hosps., Lond., 1759-61; apptd. physician to Queen, 1761. Fellow Royal Soc., 1753, Royal Coll. Physicians; mem. Med. Soc. Edinburgh. Author: Pleasures of the Imagination, 1744; Hymn to Naiads (poem), 1746; On Dysentery, 1764; thesis on origin and growth of human fetus (attacked untenable theories of Leeuwenhoek and others). Died London, June 23, 1770.

AKERBERG, Erik Per Hjalmar, Swedish plant breeder; b. Barnarp, Sweden, Jan. 26, 1906; s. Per Olof Hjalmar and Anna Hansson; Agronom., U. Reading (Eng.), 1931, Fil.Lic., 1937, Fil. Doctor, 1941, D.Sc., 1960; m. Inga Frideborg, Palm, Oct. 25, 1933; children—Per Olof, Lars, Ingrid Thelin (Mrs. Clas Thelin), Gunnar. Asst., Seed Contr. Sta., 1931; plant breeder Weibullshom, 1932-39; head br. stas. Swedish Seed Assn., 1939-54; prof. Nat. Agr. Expt. Inst., Ultuna, Sweden, 1954-56; dir. Swedish Seed Assn., Svalof, 1956—. Vice pres. Swedish Agr. Research Council, 1963. Named knight Swedish Order of N. Star, Danish Dannebrog Order. Mem. Eucarpia (pres. 1966—), Scandinavian Agr. Research Workers (hon. mem.), Finnish, Swedish agr. acads., Royal Physiogr. Soc., Danish Tech. Sci. Acad., Swedish Seed Growers Assn. (v.p. 1960—). Author: (with A. Hagberg) Mutations and Polyploids in Plant Breeding, 1961; also numerous articles. Research on seed formation of Poa pratensis, plant ecology of crops in No. climates, plant breeding problems in developing countries. Home: Svalöf, Malmöhus, Sweden. Office: Swedish Seed Assn., Svalöf, Sweden.*

AKERMAN, Karol, Polish chemist; b. Krakow, Poland, Feb. 24, 1913; s. Henryk and Sophia (Selinger) A.; master degree in chemistry, U. Jagielonski, Kraków, 1936, Ph.D. in Chemistry, 1960; m. Celina Milska, Apr. 5, 1955; 1 dau., Sophia. Chemist cement factory, Górka, Poland, 1936-39; chief engr. gypsum factory Szczerzec, USSR, 1939-45; chief Polish nonferrous metals industry, Katowice, 1945-49; chief Polish chem. industry, Gliwice, 1949-52; dir. research Inst. Sulfuric Acid, Warsaw, Poland, 1952-54, Inst. Light and Rare Metals, Skawina, Poland, 1954-58; dir. dept. radioisotopes application in industry Nuclear Research Inst., Warsaw, 1958—; chmn. Inst. Chem. Tech., Maria Curie-Sklodowska U., Lublin. Decorated Sztandar Pracy 1st degree, Officer Cross Polonia Restituta; recipient govt. prizes 1st degree, 1951, 2d degree, 1953. Mem. Polish, German chem. socs., Lublin Sci. Soc. Author: Gypsum and Anhydrite, 1964; Germanium, 1967; also numerous articles. Devel. prodn. of sulfuric acid from anhydrite, prodn. of aluminum oxide from poor raw materials by fusing in shaft furnace; synthesis of specific sorbent for isolation of germanium from sulfuric solutions; applications of tracers in chem. industry; isolation of rare elements and high purity substances. Home: 11 Mazowiecka. Office: 16 Dorodna, Warsaw, Poland.*

AKERS, Sir Wallace Alan, English chemist; b. London, Sept. 9, 1888; s. Charles Akers; ed. Christ Ch., Oxford; hon D.C.L., Oxford; m. Bernadette La Marre, 1953. With Brunner Mond & Co., 1911-24; also Borneo Co. Ltd., 1924-28; research dir. Imperial Chem. Industries Ltd., 1928-53. Dir. atomic energy research Dept. Indsl. and Sci. Research, 1941-46. Fellow Royal Soc., 1952, Royal Inst. Chemistry, Soc. Chem. Industry. Mem. Faraday Soc., Biochem. Soc., Advisory Council, Sci. Research Council. Died Nov. 1, 1954.

AKERT, Konrad, Swiss neurophysiologist; b. Zurich, Switzerland, May 21, 1919; s. Conrad and Seline (Eichmann) A.; M.D., U. Zurich, 1949; m. Ruth Giger, July 18, 1947; children—Rudolf, Maria, Fritz, Christoph. Asst. physiology, Zurich, also asst. pathology, Saint-Gall, 1946-51; fellowship Swiss Acad. Med. Scis., 1951-53; asst. prof. physiology and neurosurgery Johns Hopkins, 1953; asst. prof. physiology U. Wis., 1955, full prof. anatomy, 1960; prof., dir. inst. for brain research U. Zurich, since 1961. Recipient Robert Bing prize in neurology. Author (with R. Emmers) Stereotaxic Atlas of the Brain of the Squirrel Monkey, 1963; (with P. Hummel) Anatomie und Physiologie des limbischen Systems, 1963; also contbr. articles, chpts. Research on brain, physiology, anatomy of diencephalon. Home: Bachtoldstrasse 1, Zurich. Office: August-Forelstrasse 1, Zurich, Switzerland.*

AKHIEZER, Aleksandr Ilich, Russian physicist; b. Cherikov (now Mogilev Oblast), 1911; grad. Kiev Poly. Inst., 1934. With Kharkov Physicotech. Inst., 1934-40; prof. Kharkov U., 1941—. Mem. Ukranian Acad. Sci. Co-author: Some Problems of the Theory of the Nucleus, 1950, Quantum Electrodynamics, 1953; author: Simple Waves in Magnetic Hydrodynamics, 1959; The Theory of Interaction between Charged Particles and Plasma in a Magnetic Field, 1961; Stability Conditions for the Function of Electron Distribution in Plasma, 1961; Theory of Reflection of Electromagnetic Waves from Plasma, 1963. Research on atomic nucleus, charged-particle accelerators, plasma, quantum theory of kinetic processes, quantum field theory. Address: Kharkov University, Universitetskaya 16, Kharkov, Ukrainian SSR, USSR.

AKHIEZER, Naum Ilich, Russian mathematician; b. Cherikov (now Mogilev Oblast), 1901; grad. Kiev Inst. Higher Edn., 1924; D.Physico-Math. Sci., 1936. With Kiev U., 1928-33, Alma-Ata Mining Inst., 1941-43, Moscow Power Engring. Inst., 1943-47; with Kharkov U., 1941-43, prof., 1947—. Recipient Chebyshev prize USSR Acad. Sci. Mem. Ukrainian Acad. Sci. (corr.). Author: Elements of the Theory of Elliptical Functions, 1948; A Lecture on the Theory of Approximations, 1947; The Theory of Linear Operators in Hilbert Space, 1950; Lectures on the Calculus of Variations, 1955; The Theory of Paired Integral Equations, 1957; Effective Boundary Condition on the Interface of Multiplying and Retarding Media, 1957; A Sturm-Liouville Equation on a Semiaxis, 1963. Research on theory of best approximations. Address: Kharkov University, Universitetskaya 16, Kharkov, USSR.

AKHMEDSAFIN, Ufa Mendybaevich, Russian hydrogeologist; born July 10, 1912; graduated from Central Asian Industrial Institute, Tashkent, 1935. Head department hydrogeology Institute Geological Sci., Kazakhstan br. USSR Acad. Sci., 1941-46; head dept. hydrogeology Inst. Geol. Sci., Kazakhstan Acad. Sci., 1946—. Mem. Kazakhstan Acad. Sci. Author: The Subterranean Waters of Kazakhstan, 1952; Pressure Waters in Some Areas of Kazakhstan, 1952; Subterranean Water Resources in the Arid Regions of Kazakhstan and Means of Using them for Flooding Pastures, Water Supply and Oasis Irrigation, 1957; Forecast Maps of Artesian and Subsurface Waters in Kazakhstan, 1963. Research on origin and nature of subterranean waters in arid regions of Kazakhstan and Central Asia, supply of water to virgin land devel. areas in No. Kazakhstan. Address: Kazakhstan Acad. Sci., Alma-Ata, Kazakhstan, SSR, USSR.

AKHUNBAEV, Isa Konoevich, Russian surgeon; b. Sept. 24, 1908; graduated from Central Asian Medical Institute, Tashkent, 1935. Deputy commissioner of health of Kirghiz, 1935-37; director Kirghiz Medical Institute; head of dept. general surgery and director Inst. Regional Medicine, Kirghiz Ministry Health, 1946—. Mem. Kirghiz Acad. Sci. (pres. 1954-60), USSR Acad. Med. Sci. (corr.), USSR Acad. Sci. (chmn. presidium Kirghiz br. 1952-54), All-Union Sci. Soc. of Surgeons (bd. mem.), All-Union Soc. Endocrinologists, Internat. Assn. Surgeons. Author: (monograph) Child Appendicitis, 1949. Co-editor surgery sect. Large Med. Ency., 2d edit.; mem. editorial council Problems of Endocrinology and Hormone Therapy; editor symposia of med. articles. Contbr. numerous articles to profl. publs. Research on endemic goiter, appendicitis in children, echinococcosis of the lung. Address: Kirghiz Med. Inst., ulitsa Voroshilova 1, Frunze, Kirghiz SSR, USSR.

AKIBA, Tomoichiro, Japanese physician; b. Oyama City, Japan, Mar. 30, 1903; s. Toyokichi and Miyo (Ariba) A.; M.D., U. Tokyo, 1926, D.Med.Sci., 1929; m. Ritsu Eda, Mar. 15, 1929; children—Toyohiko, Miwa (Mrs. Hiroshi Ookawara). Tech. expert Nat.

Hygienic Inst. Japan, 1929-44; prof. Tokyo U., 1944-63, chmn. dept. microbiology, 1946-63; prof. Nippon U., Tokyo, 1963—; dir. research labs. Chugai Pharm. Co. Ltd., 1963—. Recipient Asakawa award Japan Bacteriological Soc., 1962. Mem. Japan Mycol. Soc., Japan Soc. Chemotherapy, Japan Food-hygienic Soc., Japan Bacteriological Soc. Author books including: Medical Microbiology, 1963; also numerous articles. Discovered venerupin, a poison in baby clams and oysters, that drug resistance is transferred in Enterobacteriaceae by cell conjugation irrespective of sex factor. Home: 16-5-4 Betsusho, Urawa-shi Saitama-ken, Japan. Office: 8-41-3 Takada-minami-cho, Toshimaku, Tokyo.*

AKIN, Wallace Elmus, Am. geographer; b. Murphysboro, Ill., May 18, 1923; s. Samuel Elmus and Sarah Elizabeth (Lindsay) A.; B.A., So. Ill. U., 1948; student U. Mich., 1943-44; M.A., Ind. U., 1949; Ph.D., Northwestern U., 1952; m. Peggy Jean Holt, June 11, 1948; children—Dianna Jean, David Wallace. Field investigations in Mexico, 1948; instr. phys. geography Northwestern U., 1950-52; field team chief Rural Land Classification Program for P.R., 1950-51; instr. geography Austin Peay State Coll., summer 1952; instr. U. Ill., Navy Pier Chgo., 1952-53; mem. faculty Drake U., 1953—, prof. geography and geology, 1962—, head dept., 1953—; cons., geologist Ia. Natural Resources Council, summers 1954-61. Fulbright research scholar Inst. Geography, U. Copenhagen (Denmark), 1961-62. Mem. Association of Am. Geographers (chairman of West Lakes div. 1967), Arctic Inst. N.Am., Royal Danish Geog. Soc., Soil Conservation Soc. Am. (chmn. internat. relations com. 1963; ofcl. rep. XIX Internat. Geog. Congress, Stockholm, 1960). Editor, contbr. bulls. Participated in devel. and execution of survey which served as basis for agrarian portion of Puerto Rican devel. program; conducted spl. study Puerto Rican dairy industry under tropical conditions; helped analyze water resources problems of Ia. in light of future needs, also water law in humid states; studied role of land reclamation in devel. Danish agr. Home: 5800 Pleasant Dr., Des Moines 50312.*

AKOPYAN, Aleksandr Arkadevich, Russian physicist; b. Dec. 25, 1890; grad. Petrograd Poly. Inst., 1917. Instr., Yerevan U., 1921-29, prof., 1929—. Mem. Armenian Acad. Scis. Named Hon. Scientist and Tech. Workers Armenian USSR, 1940—. Author: The Laws of Reversible Adsorption from Gas and Vapor Mixtures, 1933; The Laws of Thermodynamic Equilibrium Shift, 1946; A Possible Substantiation of the Second Principle of Thermodynamics, 1946; General Thermodynamics, 1955. Research in gen. and chem. thermodynamics. Home: USSR, Yerevan Armenian SSR, Ul. Teryana 105, Politekhnicheskу Institut.

AKPINAR, Sait, Turkish physicist; b. Istanbul, Turkey, Mar. 28, 1913; s. Yahya and Ümmihan (Zekeriya) A.; student Frankfurt U., 1934-37; Ph.D., Göttingen U., 1940; postgrad. Mass. Inst. Tech.; m. Remziye Daci, Jan. 3, 1948; 1 dau. Faculty, Istanbul U., 1940—, prof. physics, 1957—. Research dir. Cekmece Nuclear Research and Tng. Center, 1961-62, dir., 1962—. Mem. Turkish Phys. Soc. Translator 4 books. Research, publs. on exptl. identification of a photo-sensitizing agt. in alkali-halogenid monocrystals, determining energy resolution of nuclear proportional counters for alpha radiation, discovery of new type of discharge in such counters and elucidation of its mechanism, radioactive fallout, use of isotopes in medicine, electronic isomer simulator, new kind of electro-luminesence in alkali-halogenid crystals. Home: 54, Ürgüplü Sokak, Yesilyurt, Istanbul. Office: P.O. Box 1, Hava Alani, Istanbul, Turkey.*

AKSNES, Gunnar, Norwegian chemist; b. Norway, Aug. 8, 1926; s. Olav and Magnhild (Westreno) A.; Magister Sci., U. Oslo (Norway), 1951, Dr. Philosophy, 1962; m. Milly, June 25, 1950; children—Jack, Cyro, Svein. Research chemist Norwegian Research Def. Inst., 1952-60; visitor U. Reading (Eng.), 1957-58; asst. prof. U. Bergen (Norway), 1960-66, prof. organic chemistry, 1966—. Mem. Norwegian Chem. Soc. Research, publs. on reaction kinetic studies of organic compounds, mostly organic phosphorus compounds. Home: 24, Stabbursvegen Fjösanger, Bergen, Norway.*

AKSOY, Muzaffer Baki, Turkish physician; b. Antalya, Turkey, Mar. 18, 1915; s. Mehmet Numan and Nadire (Sunay) A.; M.D., Istanbul (Turkey) Med. Sch., 1940; m. Fatma Nefise Nedime Yolalan, Sept. 28, 1942; children—Necla, Leyla, Atilla. Attending physician Mersin State Hosp., 1947-57; blood research fellow Blood Research Lab., Boston, 1952-53; attending physician Beyoglu Emergency Hosp., Istanbul, 1957-60; faculty Istanbul U., 1959—, prof. medicine, 1965—, mem. med. staff, 1961—. Mem. Am. Eugenics Soc. Research, numerous publs. on immunologic properties of hemoglobin in Cooley's anemia, types of hemoglobin in various blood disorders, epidemiological survey of anemia. Home: 139/2 Tesvikye Cad. Nisantas, Istanbul. Office: Vakif Guraba Hosp., Capa, Istanbul, Turkey.*

AKULOV, Nikolai Sergeievich, Russian physicist; born on Dec. 11, 1900; graduated from Moscow Univ., 1916. Then professor at Moscow Univ., 1921-54, Moscow Institute of Chem. Machine-Building., 1955-57, Moscow Geological Survey Institute, 1957—. Recipient Stalin prize, 1941. Member of Belorussian Academy Sciences. Author: Ferromagnetism, 1939; The Theory of Chain Processes, 1951; Dislocations and Plasticity, 1961; Metal Fatigue Theory, 1963. Research in physics of metals, particularly ferromagnetism; formulated law of anisotropy, 1928; developer system of differential equations in partial derivaties for calculating chain diffusion in physics, chemistry and biology, 1947; developer theory of phys. properties of ferromagnetics under dynamic conditions, 1953, theory of atomic magnetic moments of alloys above and below Curie point, magnetic detectoscopes, automatic magnetic micrometer and anisometer for recording structural changes in metal, 1956. Address: Moscow Geol. Survey Inst., pr. Marksa 18, Moscow, USSR.

AL-ABDARI (Muhammad ben Muhammad Abdari), geographer; b. Valencia, Spain; flourished 1289 and afterward; travelled to Mogador, 1289, and Mecca on pilgrimage. Author: Trip to the West (topographical data, information on Muslim life and scholarship of the time).

AL-ABHARI (Atir al-din al-Abhari) (al-Muffadal ibn 'Umar al-Abhari Athir al-din), astronomer, mathematician, philosopher; b. Iran, circa 1200; disciple of encyclopedist, Kamal al-din; a son Al-Abhari. Author: Guide to Wisdom; Commentary on the Isagoge of Porphyrios; several astron. treatises including Resume of Secrets; On the Astrolabe.

AL-AHWAZI (Abu al-Hasan), sci. translator; flourished late 8th and early 9th century; translated Siddhanta (Indian work on astronomy which influenced Islamic sci.); also work with tables of planetary motion taken from Aryabhata treatise.

ALAIN DE LILLE (or Alan de Insulis); French philosopher, theologian, natural scientist; b. Lille, France, 1114-28; received clerical edn.; prof., rector U. Paris; tchr., Montpelier, France; joined Cistercian order; retired in Citeaux, France. Author: De planctu naturae, 1160-70; De fide catholica contra haereticos; Anticlaudianus (didactic poem); Opera maralia, paraemetica, et polemica (edited by Carlo de Vischi 1653); writings in theology, philosophy, poetry, natural science. Defended Catholicism against Albigensian heresy. Known as Universal Doctor because of wide investigations; primarily a Neoplatonist, but aware of Aristotle and neo-Pythagoreans; held that reason as well as revelation is necessary for complete understanding of God and universe. Died Abbey of Citeaux, 1202 or 03.

ALAMPIEV, Petr Martynovich, Russian economist; b. 1900; grad. Central Inst. Pub. Edn. Organizers; D.Econ. Sci. Instr., Leningrad Communist U., Communist U. of Workers of East, Geog. Faculty, Moscow U., until 1941; asso. Inst. Geography, USSR Acad. Sci., 1950-55, asso. Inst. Econs., 1955-56; head econ. areas sect. Econ. Research Inst., USSR Gosplan 1956-61; with Inst. Economics of World Socialistic System, 1962—. Author: The Abolition of the Economic Inequality of the Peoples of the Soviet East; Economic Areas of the USSR; Economic Regionalization, 1959-63; Soviet Kazakhstan, 1960; Problems of Division into General Economic Areas at the Present Stage, 1960; Economic Regions of Our Country, 1961 International Social Division of Labor, 1967. Editor: Methodology of Economic Geography, 1962. Inst. of Economics of World Socialistic System, Yaroslavskaya ul. 13, Moscow, USSR.*

ALAN DE INSULIS, see Alain de Lille.

AL-ANSARI (Abu abd Allah), Sams al-din, al-Sufi al-Dimasqui (Abu Zayd al-Ansari), geographer; b. Damascus, Syria, 1254; wrote Choices of Weather, Cosmography; geographer and traveler whose writings were considered valuable by later scientists and historians. Died 1327.

ALANSON, Edward, Brit. surgeon; b. Lancashire, Eng., 1747; practiced medicine, Liverpool. Author: Practical Observations Upon Amputations, 1779. Developed amputation method in which skin and muscles are cut in circular manner, leaving stump shaped like hollow cone, 1779. Died 1823.

ALARCON-SEGOVIA, Donato, Mexican physician; b. Mexico City, Mexico, May 6, 1935; s. Donato G. and Maria Teresa (Segovia) A.; B.S., Centro Universitario Mexico, 1952; M.D., Nat. U. Mexico, 1959; m. Marta Riquelme, May 29, 1961; children—Marta Eugenia, Maria Claudia. Fellow internal medicine and rheumatology Mayo Grad. Sch. Medicine, Rochester, Minn., 1961-65; head immunology lab., asst. head research, cons. in rheumatology Hosp. de Enfermedades de la Nutricion, Mexico City, 1965—. Recipient John Edward Noble Found. award, 1964, Philip Hench award for excellence in rheumatology, 1966. Mem. Am. Rheumatism Assn., Asociacion Mexicana de Reumatologia, Am. Coll. Chest Physicians, N.Y. Acad. Scis., Asociacion Medica del Hosp. de Enfermedades de la Nutricion, Sigma Xi. Research, publs. on role immunological factors in devel. or persistence of diseases of small bowel, pancreas and liver, systemic lupus erythematosus. Home: 449 Rebsamen St. Office: 2-803 Plaza de Miravalle, Mexico City, Mexico.*

AL-ASFUZART, (Muzaffar al-Asfuzart), mathematician, physicist; wrote summary of Euclid's Elements; studied specific gravity. Died before 1122.

AL-ASMAI (Abd al-Malik ben Qurayb al-Asmai), Arabic zoologist; b. Basra, 739; flourished at Bagdad, also Basra. Author: Book About Savage Animals; Treatise on the Making of Man (demonstrated Arabic knowledge of human anatomy); also treatises on various animals such as horses, camels, sheep. Died Basra, 831.

AL-ASTURLABI (al-Badt' al-Asturlabi), astronomer; lived in Ispahan, circa 1116, later in Bagdad; dir. astron. observations in palace of Saljug Sultan of Iraq, Mughith al-din Mahmud; compiled Mahmudic Tables; expert on constrn. of astrolabes. Died Bagdad, 1139-40.

ALAYMO, Marco Antonio, physician; b. Sicily, 1590; practiced medicine, Palermo, regarded as 1st physician of his time in Sicily; founded med. acad., Palermo. Author: Diadecticon (gives account of different med. substances), 1637. Died 1662.

AL-AYYUBI (Abu al-Fada Ismail ben Imad), geographer; b. Damascus, Syria, 1273. Participated in capture of fortress Markab at the age of 12, in recapture of Tripoli in siege of St. Jean-Acre; awarded principality of Hama for services to Sultan of Egypt Malik Nasir, 1310; named vassal of Sultan, and assisted him in other campaigns. Wrote book on geography, 1329; also 2 histories. Revised and corrected works of Abu al-Hasan. Died 1331.

ALBAHARY, Claude, French physician; b. France, Sept. 3, 1914; s. Jacques and Boisseau A.; M.D., Paris, 1940; m. Micheline Gadenne, Apr. 19, 1949; children—Jacques, Philipp, Emmanuelle. Became head clinic Faculty Medicine, Paris, 1945, prof. agrégé, 1955; named head internal medicine services Hopital St. Denis, 1958. Med. expert for tribunals; cons. expert med. testing Minstre de la Santé; mem. nat. commn. on indsl. hygiene Minstre du Travail. Mem. Labor Medicine Soc., Hematology Soc., Therapeutics Soc., Author: Maldies médicamenteuses, 1953; also numerous articles. Research on hyperthyroid pathogeny, 1940, lead intoxication and mechanism of its effect, 1944—, profl. pathology, especially hematology; accidental poisoning, 1953—. Address: 97 Prony, Paris 17e, France.*

AL-BAKRI, Arabic geographer; b. Huelva, Spain before 1040; lived in Cordoba, Spain, also at Ct. of Almeria; student of poetry and philology. Author: The Book of Roads and Provinces (contained hist. and ethnographical information); Dictionary of Ancient Geography (especially Arabic geography); Book on the Plants and Trees of Andalusia. His geog. works are oldest still in existance. Died 1094.

AL-BA'LABAKKI (Qusta ben Luga al-Ba'labakki), Greek translator, mathematician; Christian of Greek origin; translator: Mechanics of Heron; also works by Theodosios, Diophantos, Aristanokos, Hypsikles, Autolykos; also attributed to him, De sphera solida (on use of astrolabe, translated into Latin by Stephanus Arnaldus). Died circa 912.

ALBAN, Ernst, German inventor; b. Neubrandenburg, Germany, Feb. 7, 1791; s. Samuel Friedrich August and Elizabeth Sophia J. (Spengler) A.; student theology, medicine, physics, mechanics. Oculist, Rostock, Germany; Practiced medicine (known for cataract operations); founded machine shop (for constrn. steam engines, agrl. machinery), Klein-Wehnendorf, 1830, Plau, Germany, 1840. Author: Die Hochdruckdampfmaschine, eine Richtigstellung ihren Westes, 1843. Pioneer in devel. high pressure steam engine; constructed boiler with vertical water pipes in housings filled with low melting alloys, 1825, also high compression and condensation engine, pump with direct control without rotating parts; used escaping steam for heating factories; developed (with Vallet) hand-driven, windmill-like propeller for aerial nav. Died Plau, June 13, 1856.

ALBANESE, Anthony August, Am. biochemist; b. N.Y.C., Feb. 12, 1908; s. August and Lillian (Imburgia) A.; B.S., N.Y. U., 1930; Am.-Hungarian Exchange fellow, U. Budapest, 1930-32; Ph.D., Columbia, 1940; m. Julette O'Rorke, Nov. 29, 1933. Dir research St. Luke's Convalescent Hosp., 1949-60; dir. nutrition, metabolic research div. Burke Found., White Plains, N.Y., 1959—; dir. research Miriam Osborn Meml. Home, 1959—; investigator Office Naval Research, 1945—, Air Research and Devel. Command, 1960—; guest lectr. U. Brazil, 1959, U. Tokyo, U. Osaka, 1960. Cons. govt. agys., pharm., chem. orgns. Corn Industries fellow, 1940-41. Fellow Am. Inst. Chemists, Am. Coll. Clin. Pharmacology and Chemothera peutics, Gerontol. Soc.; mem. Am. Chem. Soc., Harvey Soc., Am. Soc. Biol. Chemists, Soc. Exptl. Biology and Medicine, Am. Inst. Nutrition, Am. Soc. Clin. Nutrition, Am. Med. Writers Assn., A.M.A., Sigma Xi. Asso. editor N.Y. State Jour. Medicine, 1959—; editor Newer Methods of Nutritional Biochemistry, vols. 1-3, 1963, 64, 67. Research, numerous publs. on nutritional needs and problems of infants, pre-school children; food habits of teenagers, coll. students, crippled children; metabolic mgmt. of wounded mil. personnel and post-surg. patients; nutritional requirements of elderly men and women, stroke patients; effect of fats, cholesterol in heart disease: nutritional problems of prolonged submarine and space travel. Home: 90 Harrison Av., Harrison, N.Y. Office: Burke Rehab. Center, White Plains, N.Y. 10605.*

AL-BARQAMANI (Japhet ibn Abu al-Hasan al-Barqamani al-Isra'ili al-Iskanari), physician; flourished in Alexandria after Maimonides; wrote treatise in Arabic on hygiene called Beneficient Discourse on the Preservation of Bodily Health.

ALBARRAN Y DOMINGUEZ, Joaquín, surgeon, urologist; b. Sagua la Grande, Cuba, 1860; student in Havana, Cuba; also Barcelona, Spain; M.D., U. Barcelona; also Paris, France. Asso. hosps. Paris, 1883-89; chief of clinic of urinary diseases, from 1890; agrégé, 1892; prof. Faculty Medicine, Paris, from 1894; surgeon of hosps., from 1894; prof., from 1906. Recipient Gold medal, 1888; Laureate, Acad. Scis. (twice), Acad. Medicine (3 times), Faculty Medicine. Mem. Anat. Soc., Surgery Soc., French Soc. Urology (v.p.). Author: Les reins des urinaires, 1889; Les tumeurs de la vessie, 1892; Titres et travaux scientifiques, 1892; (with B. Motz) Etude sur le traitement de l'hypertrophie de la prostate, 1898; Maladies de la prostate et rétrécissement de l'urètre, 1899; Les tumeurs du rein, 1903; Exploration des fonctions rénales, 1905; Clinique des maladies des voies urinaires, 1906. Studies on diseases of urinary tract, kidneys, and prostate gland; research on kidney tumors; did basic work in bladder surgery; described subtrigonal glands (Albarran's glands) of urinary bladder, 1892; invented cystoscope with movable lens, 1897; credited with performing 1st perineal prostatectomy in France, pub. 1909. Died Jan. 17, 1912.

AL-BASSAM, Mohammed Ali, mathematician, educator; b. Baquba, Iraq, Dec. 7, 1923; s. Abdul-Amir Ali and Sadika (Kadum) Al-B.; license in sci. U. Baghdad, 1944; M.A. in Pure Math., U. Tex., 1948, Ph.D., 1951; m. Sabiha Ismail Hakki, June 21, 1944; children—Wayil, Miss Nudhar, Nayil, Ali Rafil. Came to U. S., 1957. Asst. prof. Huston-Tillotson Coll. (Tex.), 1951-53; asst. prof. Tex Luth. Coll., 1951-53; asst. prof. U. Baghdad, 1953-59; asso. prof. Tex. Tech., 1959-61, prof., 1961-64, vis. prof. Am. U. Beirut, Lebanon, 1965—. Mem. Am., Indian, London math. socs., Math. Assn. Am., Math. Soc. France. Contbr. articles and papers of original research to numerous internat. publs. Home: P.O. Box 4346, Tech. Sta., Lubbock, Tex. 79409. Office: Dept. Math., Am. U. Beirut, Lebanon.*

ALBATEGNIUS, see Al-Battani.

AL-BATTANI (abu'Abdullah Muhammad ibn-Jabir al-Battani; Albategnius; Albatenius), Arabic astronomer; b. at or nr. Haran, upper Mesapotamia, circa 858; devoted most of life to making astron. observations at Raqqa on Euphrates River, 877-918. Author: Zij (astron. tables), Latin trans., 12th century, Spanish trans., 13th century, 1st printed edit., 1537 (under title De motu stellarum). One of most important astronomers of medieval Islam; tables greatly influenced devel. astronomy in medieval Europe; demonstrated motion solar apogee since Ptolemy's time; proved apparent angular distance of sun varies (thus annular eclipses possible); determined obliquity of ecliptic, length of tropical year and seasons; corrected Ptolemaic value for precession of equinoxes; used sines (instead of chords) in math. calculations. Died nr. Damascus, Syria, 929.

ALBAUM, Harry Gregory, biochemist; b. Odessa, Russia, Feb. 9, 1910; s. Morris and Sonia (Rimberg) A.; brought to U. S., 1912, naturalized, 1934; A.B., Bklyn. Coll., 1932; M.Sc., N.Y. U., 1934; Ph.D., Columbia U., 1938; m. Frieda Goldberg, Nov. 22, 1936; 1 dau., Judith (Mrs. Robert Gorman). Fellow Bklyn. Coll., 1932-34, tutor, 1934-38, instr., 1938-46, asst. prof., 1946-50, asso. prof., 1950-53, prof., 1953-63, exec. officer, 1963-65, dean div. grad. studies, 1965—; asso. dean div. grad. studies City U. N.Y., 1965—. Biochem. cons. trustee Cold Spring Harbor Lab., 1963—. Fellow N.Y. Acad. Scis.; mem. Am. Soc. Biol. Chemists, Soc. Gen. Physiologists, Soc. of Biology and Medicine, Phi Beta Kappa, Sigma Xi. Contbr. numerous articles in field to sci. jours. Research on metabolism and energy utilization in plant tissues; changes in tissues brought about by acclimatization to high altitude and tissue trauma; enzyme changes in blood produced as a result of a variety of pathologies including cancer. Home: 108 Seaman Av., Rockville Center, L.I. 11570. Office: Div. Grad. Studies, Bklyn. Coll., Bedford Av. and Av. H., Bklyn. 11210.*

ALBEAUX-FERNET, Michel Charles Etienne, French endocrinologist; b. Paris, Feb. 16, 1905; ed. Ecole Massillon, U. Paris; m. Christiane de Meyer, Apr. 18, 1947. Apptd. head of clinic, 1935, hosp. physician, 1939; physician, chief of service Hôpital Laënnec, from 1947; charge of missions to U. S., 1948, to Canada, 1952, to Greece, 1960, to U. S., 1966. Recipient Legion of Honor, Pub. Health silver medal for service of mil. health, laureate Faculté de médecin, L'Académie des sciences. Author: Les hormones en thérapeutique; L'Année Endocrinologique, 20 vols.; Comment traiter l'obésite; L'Endocrinologie du médecin praticien, 1955; (with J. Romani) Les oedèmes, 1961; contbr. to Encyclopédie médico-chirurgicale. Study of metabolic and endocrine problems. Address: 91, av. Henri-Martin, Paris 16, France.*

ALBEE, Fred Houdlett, Am. surgeon; b. Alna, Me., Apr. 13, 1876; s. F. Huysen and Charlotte Mary (Houdlett) A.; A.B., Bowdoin, 1899; M.D., Harvard, 1903; traveled and studied in Europe; Sc.D., U. Vt., 1916, Bowdoin Coll., 1917; LL.D., Colby, 1930; Sc.D., Rutgers U., 1940; m. Louella May Berry, Feb. 2, 1907; 1 son, Fred H. Prof., dir. dept. orthopedic surgery N.Y. Post-Grad. Med. Sch.; prof. orthopedic surgery U. Vt. Coll. of Medicine; cons. surgeon to 24 hosps., also Pa. R.R. System and Seaboard Air Line; cons. in orthopedics, Byrd Antarctic Expdn.; founder and med. dir. Fla. Med. Center, Venice, editor in chief Rehab. Rev.; chmn. N.J. Rehab. Commn.; dir. U. S. Army Gen. Hosp. No. 3; mem. adv. orthopedic council to surgeon general U. S. Army. Decorated by govts. of Rumania, Cuba, Spain, Hungary, France, Italy, Venezuela, Brazil. Founding fellow, gov. A.C.S.; hon. fellow Royal Soc. Medicine (Great Britain); founder Internat. Soc. Orthopedic Surgery; pres. Am. Orthopedic Assn., Am. Acad. Orthopedic Surgeons, Pan-Am. Med. Assn.; hon. mem. Leningrad Soc. of Orthopedic Surgeons, Brazilian Coll. Surgeons; mem. A.M.A., Kappa Sigma, Phi Chi. Author: Bone Graft Surgery, 1915; Orthopedic and Reconstructional Surgery, 1919; Injuries and Diseases of the Hip, 1937; Bone Graft Surgery in Disease, Injury and Deformity, 1940; A Surgeon's Fight to Rebuild Men, 1943; also numerous pamphlets on surg. subjects. Demonstrated original surg. methods of bone grafting in Germany, England and France, 1914, also in mil. hosps. of France, 1916; ofcl. rep. of Med. Corps U. S. Army, to Inter-Allied Congress, Rome, Paris and Bologna, 1919; rep. U. S. Army at Netherland Orthopedic Congress, Amsterdam, 1923. Credited with 1st utilization of living bone grafts as internal splints, 1911, 15; devised Albee's saw for cutting bone grafts (electrically operated, double rotary, adjustable blades), 1919. Died N.Y.C., Feb. 15, 1945.

ALBEE, George Wilson, Am. psychologist; b. St. Marys, Pa., Dec. 20, 1921; s. George W. and Maude (Allen) A.; A.B., Bethany Coll., 1943; M.S., U. Pitts., 1947, Ph.D., 1949; m. Constance Impallaria, August 6, 1955; children—Alexander, Luke, Maud, Sarah. Research psychologist Western Psychiat. Institute, Pittsburgh, 1949-51; assistant executive secretary American Psychol. Assn., Washington, 1951-53; Fulbright prof. Helsinki U., Finland, 1953-54; asso. prof. psychology Western Res. U., 1954-56, professor, 1957—, chairman of the department of psychology, 1957-60, 63-66, Ladd prof. psychology, 1959—; cons. psychology VA, Surgeon Gen. Army, State Ohio; dir. task forces on manpower Joint Commn. Mental Illness and Health, Cambridge, Mass. and Cleve., 1957-59; on sabbatical leave in Rome, Italy, 1960-61; cons. on research Peace Corps, 1962-65; profl. adv. com. Cleve. Welfare Fedn., 1963—. Bd. dirs., mem. exec. com., v.p. Cleve. Mental Health Assn. Fellow Am. Psychol. Assn. (bd. profl. affairs, council reps.; bd. dirs. 1965—; pres. div. clinical psychology); mem. Eastern, Midwestern, Ohio (pres. 1963-64), Cleve. psychol. assns., A.A.A.S., Am. Assn. U. Profs., Sigma Xi. Author: Mental Health Manpower, 1959. Studied factors asso. with supply and demand for professional manpower in field of mental health, relations of concepts of mental disorder to kinds of professionals required. Home: 2541 Arlington Rd., Cleve. 44118.*

ALBERGAMO, Francesco, Italian sci. philosopher; b. Favara (Agrigente), Italy, Aug. 18, 1896; s. Giacoma and Giuseppina (Buttice) A.; Ph.D.; m. Maria Rizzo, June 23, 1932; children—Augusto, Massimo, Vitorio, Maria. Prof. theoretical philosophy U. Naples. Mem. Pontaniana Acad., Acad. Sci. and Letters Palermo. Author: La critica della scienza nel Novecento, 1950; Storia della logica delle scienza empirico, 1952; La critica della scienza oggi in Italia, 1953; Introduzione alla logica della scienza, 1956; also others, including translations of Kant, Berkeley, Poincaré. Address: via Simone Martini, parco Risorgimento, Naples, Italy.

ALBERIGI-QUARANTA, Alessandro, Italian physicist; b. Reggio, Italy, June 15, 1927; s. Massimiliano and Maria (Peviani) A.-Q.; student U. Bologna (Italy), 1944-47; Laurea in Fisica, U. Rome, 1958. Researcher physics dept. U. Rome, 1948-56; staff Italian Electrosynchrotron, Frascati, 1956-60; asst. prof. physics dept. U. Bologna (Italy), 1960-66; prof. electronics U. Modena (Italy), 1966—. Mem. Società Italiana di Fisica (mem. council), I.E.E.E. (sr.), Am. Phys. Soc. Research, publs. on nuclear physics, electronic instrumentation and components, econs. of electronic industry. Home: 10 Via Cilea, Bologna, Italy. Office: 4 Via Università, Modena, Italy.*

ALBERS, H. C., cartographer; b. 1773; modified Lambert's equal-areaconic projection; reduced errors of scale in N. and S. by use of cone intersecting globe at two standard parallels. Died 1833.

ALBERS, Johann Abraham, German physician; b. Bremen, Germany, Mar. 20, 1772; s. Johann Christoph and Marie Catharina (Retberg) A.; M.D., Jena (Germany) U.; m. Marie Wilhelmine Retberg, 1799; 1 son, Wilhelm. Practiced medicine, Bremen, from 1798, apptd. city physician, 1807. Contbd. articles to med. jours., including Annals of Medicine, Edinburgh. Tried to popularize Am. med. literature in Germany; research on eye, obstetrics, pediatrics, deliberate limping of children. Died Bremen, Mar. 24, 1821.

ALBERS, Johann Friedrich Hermann (or Johann Franz Hermann), German pharmacologist; b. Dorsten, Germany, Nov. 14, 1805; s. Hermann and Clara (Zimmermann) A.; ed. Bonn, Germany; apptd. prof. pharmacology U. Bonn, 1862. Author: Atlas der pathologischen Anatomie, 1832-62; Beobachtungen auf dem Gebiete der Pathologie und pathologischen Anatomie, 3 vols., 1836-40; Handbuch der allgemeinen Pathologie, 2 vols., 1842-44; Handbuch der allgemeinen Arzneimittellehre, 1853. Represented sch. of exptl. pharmacologists. Died Bonn, May 11, 1867.

ALBERS-SCHÖNBERG, Heinrich Ernst, German roentgenologist; b. Hamburg, Germany, Jan. 21, 1865; s. Augustus Heinrich and Amalia (Des Aris) A.-S. Practiced medicine, Hamburg, from 1894; studied roentgen rays at pvt. instn., 1897, at St. George Hosp., Hamburg, 1903; apptd. mem. faculty U. Hamburg, 1907, 1st prof. roentgenology, 1919; founder German Roentgen Soc., 1897. Discovered injury caused by roentgen rays to reproductive organs; generalized diffuse osteosclerosis named in his honor; invented compression diaphragm which cuts out secondary rays, thus intensifying action of roentgen rays, reported 1903; described condition (known as Albers-Schönberg disease, also marble bone, ivory bone, marmorknochen) in which there are bandlike areas of condensed bone nr. epiphyseal lines of long bones and areas of condensation nr. edges of small bones, 1903. Died Hamburg, June 4, 1921.

ALBERT I (or Albert-Honoré-Charles Grimaldi), French oceanographer; b. Paris, France; Nov. 13, 1848; prince of Monaco; founded Inst. Oceanography, also Inst. Human Paleontology, Paris, Mus. Oceanography, Monaco. Mem. French Acad. Scis., 1891 (asso.), French Acad. Medicine, Acad. Agr. Investigated Atlantic Ocean, also Mediterranean Sea; set up bathometric chart of oceans. Died Paris, June 26, 1922.

ALBERT, Abraham Adrian, Am. mathematician, univ. dean; b. Chgo., Nov. 9, 1905; s. Elias and Fannie (Fradkin) A.; B.S., U. Chgo., 1926, M.S., 1927, Ph.D., 1928; m. Frieda Regina Davis, Dec. 18, 1927; children—Alan Davis, Roy Davis (dec.), Nancy Elizabeth (Mrs. Perry Goldberg). Fellow, NRC, Princeton, 1928-29; instr. Columbia U., 1929-31; asst. prof. math. U. Chgo., 1931-37, asso. prof., 1937-41, prof., 1941—, chmn. dept. math., 1958-62, Eliakim Hastings Moore distinguished service prof., 1960—, dean div. phys. scis., 1962—. Vis. prof. U. Brazil, 1947, U. Buenos Aires, 1947, U. So. Cal., 1950, Yale, 1956-57, U. Cal. at Los Angeles, 1958; chmn. div. math. NRC. 1952-55; dir. communications research div. Inst. for Def. Analysis, 1961-62; cons. Nat. Security Agy., Mitre Corp., Office Edn; Fulbright Distinguished lectr., Argentina, Brazil, Chile, Peru, 1965. Hon. mem. Sociedad Matematica Mexicana, Union Matematica Argentina; mem. Nat. Acad. Scis. (chmn. sect. math. 1958-61), Brazilian Acad. Scis., Acad. Scis. Buenos Aires, Am. Math. Soc. (Frank Nelson Cole prize 1939, pres. 1965-66), Phi Beta Kappa, Sigma Xi. Author: Modern Higher Algebra, 1936; Structure of Algebras, 1939; Introduction to Algebraic Theories, 1940; College Algebra, 1941; Solid Analytic Geometry, 1947; Fundamental Concepts of Algebra, 1957. Editor, Colloquium Publs., 1951-57. Research and publications on theory of linear associative albegras (gen. number systems); theory of finite division algebras with applications to theory of finite projective planes; created theory of Jordan algebras. Home: 1359 East Park Pl., Chgo. 60637. Office: Eckhart Hall, 1118 E. 58th St., Chgo. 60637.*

ALBERT, Adrien, Australian med. chemist; b. Sydney, Australia, Nov. 19, 1907; s. Jacques and Blanche (Allan) A.; B.S., U. Sydney, 1932; D.Philosophy in Medicine, U. London (Eng.), 1937, D.Sc., 1947. Faculty, research staff Sydney U., 1938-47; prof., head dept. John Curtin Sch. Med. Research, Australian Nat. U., Canberra, 1948—. Cons. on drugs Australian Army Med. Directorate, 1942-46. Fellow Australian Acad. Sci.; mem. Royal Australian Chem. Inst. (chmn. heterocyclic sect. 1965—), Chem. Soc., Biochem-Soc., Royal Australian Chem. Inst., Am. Chem. Soc., Med. Research Club. Author: The Acridines, 1951; Heterocyclic Chemistry, 1959; (with E. Serjeant) Ionization Constants, 1962; Selective Toxicity, 1965; also numerous articles. Research on correlating chem., phys. and biol. properties of drugs and natural substances, chemistry of nitrogenous heterocyclic substances, chelators. Home: University House, Canberra, Australia.*

ALBERT, Eduard, surgeon; b. Senftenberg, Bohemia, Jan. 20, 1841; s. Katharine (Zipobuicka) A.; studied medicine, Vienna, Austria; m. Pietsch; became surg. asst. at Dumreicheis Clinic, 1869; prof., Innsbruck, Austria, from 1873, Vienna, 1881-1900. Mem. Austrian Herrenhause. Author: Diagnostik der chirurgischen Krankheiten, 1875; Lehrbuch der speziellen Chirugie, 1877. Introduced Lister's antiseptic methods into Austria; among first modern surgeons to perform a nerve transplantation; research on ligaments and skeletal structure; described condition of painful heel (Achillodynia, also Albert's disease, or Swediaur's disease), 1893. Died Sept. 26, 1900.

ALBERT, Joseph, German gynecologist; b. Munich, Germany, July 1, 1866; s. Joseph Albert and Rosa (Huber) A.; ed. Munich, Berlin univs.; lecturing prof., gynecol. clinic Munich U., 1892-98, prof. gynecology, from 1905. Author: Lehrbuch der mikroskopisch-gynukologischen Diagnostik, 1897. Outstanding surgeon, promoter of operative gynecology; maintained contact with other brs. medicine. Died Oct. 17, 1919.

ALBERT, Philippe, French chemist, metallurgist; b. Paris, France, Dec. 14, 1922; s. Edouard and Germaine (Cadot) A.; Licence ès Scis., Faculte des Scis. Paris, 1945, Docteur ès Scis., 1955; m. Yvonne Brion, Oct. 7, 1947; children—Chantal, Jean Luc, Bernard, Xavier, Odile, Agnès, Pierre, Irène. Research jr. and sr. doctor ès scis. Centre d'Etudes de chimie metallucrique du CNRS, Vitry, France, 1946-57, master of research, 1957-62, dir. research, 1962—. Prof. activation analysis Ecole Nationale Supérieure de Chimie de Paris, Faculté des Scis., Institut Nat. des Scis. et Techniques Nucléaires de Saclay. Laureat French Acad. Scis. Mem. French Soc. Metallurgy, Chem. Soc. France, Am. Nuclear Soc., Instn. Metallurgists, Inst. Metals, Gt. Britain. Author: L'analyse par activation, 1964; also papers. Research on purification of metals by zone melting, devel. of systematic analysis by radio-activation by thermal neutrons, photons and charged particles of highest purity metals. Home: 1 Gros Caillon, Paris 7, France. Office: 15 Georges, Urbain 94, Vitry, France.*

ALBERT, Richard David, Am. physicist; b. Elmira, N.Y., Aug. 9, 1922; s. Samuel and Irene (Mitchell) A.; B.S., U. Mich., 1943; Ph.D., Columbia, 1951; m. Marjorie Fine, Apr. 14, 1946; children—John Mitchell, Charles Frederick, Thomas Milton, Carolyn Jane. Sr. scientist Westinghouse Atomic Power Div., Pitts., 1949-51; research asso. Gen. Electric Knolls Atomic Lab., Schenectady, 1951-55; sr. physicist Lawrence Radiation Lab., U. Cal., Livermore, 1955-65; research physicist, space scis. lab. U. Cal. at Berkeley, 1965—. First exptl. proof of Fermi theory of beta decay for low energies; 1st measurement of fission cross-sections in space using a high altitude nuclear explosion; 1st direct detection of alpha particles and monoenergetic electrons in the earth's auroral zone. Home: 317 Hartford Rd., Danville, Cal. 94526. Office: Space Scis. Lab., U. Cal., Berkeley, Cal.*

ALBERT OF BRUDZEWO (or Brudzewski), mathematician, astronomer; b. Poland, circa 1440; prof. math. (pure geometry and algebra) U. Cracow (Poland), from 1476; tchr. of Copernicus. Author: Commentary on Peurbach's Theoricae novae planetarum, 1482.

ALBERT OF COLOGNE, see Albertus Magnus.

ALBERT OF SAXONY (or Albert of Helmstädt, Albertus Novus, Albertutius, Albertus Parvus, Albertilla), philosopher; b. probably Helmstädt, Saxony, Germany, possibly 1316; possibly s. Bernhard-Rike; may have studied at U. Prague and received masters degree in Paris, 1351. Became rector English nation U. Paris, 1353, procurator, 1358, rep. in concordat with Picard nation, 1358, also examinator, receptor; rector U. Vienna, 1365; bishop, Halberstadt, Germany, 1366-90. Author: Commentary on Aristotle's De caelo et mundo; Tractatus (liber) proportionum (on arithmetic and geometric proportions, speeds, other mech. problems); Tractus de latitudinibus formarum; Secreta de herbis, Lapidibus et mineralibus; Super octo libros physicorum. Aristotelian commentator and expositor of new (Occamist) logic; credited with influencing Leonardo da Vinci's math. theories; among 1st to deduce erosion is chief cause of shape of earth's crust. Died Halberstadt, July 8, 1390.

ALBERT THE GREAT, see Albertus Magnus.

ALBERTI, Friedrich von, see von Alberti.

ALBERTI, Leone Battista, Italian mathematician, architect; b. Genoa, Italy, Feb. 18, 1404; s. Lorenzo di Benedetto and Margherita di Messer Piero (Benini) A.; student U. Padua (Italy); Doctorate in Canon Law, U. Bologna; travelled in France, Belgium, Germany, 1428-32. Author: Della pittura, 1436; De re aedificatoria, written ca. 1452, pub. 1485; had a strong influence on architecture for several centuries. Prominent as architect (designed churches in Mantua and Rimini, Italy), also in painting, sculpture, music; his architectural principles derive from his study of the ancient mathematics of proportion. Beauty he thought was a function of harmonious proportions. In developing laws of perspective, he was a forerunner of projective geometry (developed 4 centuries later by Poncelet). Died Rome, Apr. 25, 1472.

ALBERTI, Solomon, German physician; b. Naumburg, Germany, Sept. 30, 1540; s. Johannes Alberti; ed. medicine U. Wittenberg (Germany); m. Ursula Burenius; m. 2d, Magdalena Kentman. Prof. physics U. Wittenberg, from 1575, prof. medicine, from 1577. Author: Oratio de studio doctrinae physicae, 1575; Disputatio de lacrimis Wittenberg, 1581; Historia pierarumque partium humani corporis Wittenberg, 1585. Discovered structure of bladder, ureter, and papilla of kidneys, valvula of veins. Died Dresden, Germany, Mar. 29, 1600.

ALBERTILLA, see Albert of Saxony.

ALBERTUS, Jacobus, see Aubert, Jacques.

ALBERTUS MAGNUS, Saint (Albert the Great, Albert of Cologne, Count von Bollstädt), natural scientist, scholastic; born Lauingen, Swabia, Germany, 1193; the son of Count von Bollstädt; educated at Cologne, Germany, also Paris, France; received doctorate, Padua, Italy. Joined Dominican Order, 1223; teacher in Dominican schools, Hildesheim, also Freiburg, Breisgau, Ratisbon, Strasbourg, from 1228, Paris, 1245-48; regent in Cologne, 1248-54; became provincial of Germany, 1245; bishop Ratisbon, 1260-62; lived in Cologne from 1262; attended Council Lyons, 1274. Beatified by Roman Catholic Ch., 1622, canonized, 1931. Author: De mineralibus et rebus metallicus libri quinque; Summa de Creaturis; Aristotelian commentaries; commentaries on the sentences of Peter the Lombard; De virtutibus herbarum; De vegetabilibus; De alchemia (probably spurious); Summa naturalium (descriptions of healing properties of plants influential in 16th century); Liber aggregationis; De secretis mulierum (probably spurious); Opera omnia, pub. 1651; 26 books on botany, zoology (many new descriptions of animals); treatise on climatology. Pioneer in scholastic method (developed further by student Thomas Aquinas); formulated scholastic syllabus dividing natural philosophy into metaphysics, math., physics; leading commentator of Aristotle after Averroës (whom he opposed); student of Arabic learning; work regarded as beginning of Exptl. Sch.; observed comet, 1240; recognized embryo in seed studies; described lunar rainbow; commented on sea movements, also generation of mountains; studied alchemy and chemistry; described 95 precious stones or minerals; denied possibility of transmutation; probably 1st to isolate metallic arsenic. Died Cologne, Nov. 15, 1280.

ALBERTY, Robert Arnold, Am. phys. chemist, univ. dean; b. Winfield, Kan., June 21, 1921; s. Luman H. and Mattie (Arnold) A.; B.S., U. Neb., 1943, M.S., 1944; Ph.D., U. Wis., 1947; m. Lillian Jane Wind, May 22, 1944; children—Nancy Lou, Steven Charles, Catherine Ann. From instr. to prof. U. Wis., Madison, 1947-67, dean Grad. Sch., 1963-67; dean Sch. Sci., Mass. Inst. Tech., Cambridge, 1967—. Recipient Eli Lilly award biol. chemistry, 1956. Guggenheim fellow, 1951. Mem. Phi Beta Kappa, Sigma Xi, Alpha Chi Sigma. Author: (with F. Daniels), Physical Chemistry, 1966; (with F. Daniels, J. W. Williams, P. Bender, C. D. Cornwell) Experimental Physical Chemistry, 1962. Home: 7 Old Dee Rd., Cambridge, Mass. 02138.*

ALBINI, Giuseppe, Italian physiologist; b. Milan, Italy, 1827; student Pavia, 1845-48; M.D., U. Vienna (Austria), 1852; postgrad. Berlin, Germany. Tchr., Cravonia, Italy, from 1857; prof. U. Naples (Italy), 1860-1905; described remains of fetal structures (in form of gray, sagograin nodules) sometimes seen on free edges of atrioventricular valves of infants, 1867; wrote on glandular secretions, physiology of nerves, embryology. Died Turin, Italy, 1911.

ALBINO DA MONTECALIERO, Italian physician; b. Montecaliero, Italy; physician to princes of Acaia, circa 1328-1348; author: De Sanitatis Custodia, circa 1341 (good specimen of med. lit. of 14th, 15th centuries). Died 1348.

ALBINUS, philosopher; flourished circa 151; pupil of Gaius; tchr. of Galen; author prologue to Plato (discussed classification, nature and order of dialogues), also textbook of Plato's philosophy; works combine Platonism with Peripatetic and Stoic doctrines, Neoplatonic doctrines touched upon.

ALBINUS, Bernard Siegfried, anatomist; b. Frankfort/Oder, Germany, Feb. 22, 1697; s. Bernard Albinus; M.D., U. Leiden (Netherlands), 1719; postgrad. in Paris. Apptd. lectr. in anatomy and surgery U. Leiden, 1719, titular prof., 1721, prof. medicine, 1745; became rector Acad. Leiden, 1726, also 1738; sec. Academic Senate, 1731, 59; pres. Coll. Surgeons, Leiden, 1738. Fellow Royal Soc., 1764. Author: Historia musculorum hominis, 1734; Tabulae sceieti et musculorum humani, 1747; Tabulae ossium humanorum, 1753. Pioneer in a new epoch of human anatomy; classified knowledge handed down from Renaissance on anatomy of bones. Died Leiden, Sept. 9, 1770.

AL-BIRUNI, Abu Raihan Muhammed ibn Ahmed, Persian astronomer, mathematician, philosopher, geographer, encyclopedist; b. Khwarizm (now Khiva, U.S.S.R.), 973. Lived in Jarjan, then traveled in India where he taught Greek sciences and observed life and customs of the people; learned Indian language; eventually conversant with Turkish, Persian, Sanskrit, Hebrew and Syriac; went with Mahmud of Ghazni to Afghanistan, 1017; taught in Ghazni, 1917-48. Author: al-Athar al-Baquiya, or Chronology of ancient nations (dealt with calendars and eras of various peoples), 1000; al Qanun al-Mas'udi (an astronomical encyclopedia), the Mas 'udic canon, 1030; a summary of mathematics, astronomy and astrology; Tarikh al-Hind, or History of India (Eng. translation, 2 vols., 1888); Kital al-Saydala (a materia medica); translated several works from Sanskrit into Arabic. One of greatest scientists of Islam; corresponded with Ibn Sina (Avicenna); possessed critical spirit, love of truth, and intellectual courage; presented best medieval account of Hindu numerals; investigated geometric progressions; studied albirunic problems (trisection of angle and other problems which are unsolvable with ruler and compass alone); simplified stereographic projection; made geodetic measurements; accurately determined latitudes; determined longitudes; discussed whether earth rotates on its axis (without reaching definite conclusion); investigated specific gravity; determined specific density of 18 precious stones and metals; asserted speed of light is immense as compared with speed of sound, explained workings of natural springs and "artesian" wells by hydrostatic principle of communicating vessels; described various abnormalities, including Siamese twins; studied flower petals; considered Indus Valley to be ancient sea basin filled with alluvions; transmitted Muslim knowledge to the Hindus. Died Ghazni, Afghanistan, Dec. 13, 1048.

ALBRECHT, Carl Eduard, German physician; b. Bremen, Germany, Mar. 28, 1902; s. Friedrich Carl and Mary Ladson (Robertson) A.; ed. Leipzig, Heidelberg, Marburg, Munich, Hamburg; M.D.; m. Ada Berg, Aug. 29, 1927; children—Friedrich Carl, Ernst Carl Julius, Adelheid Johanna, George Alexander. Became med. asst. Red Cross, 1928; physician House of Health, Bremen, Med. Clinic, Heidelberg, and Neurological Clinic, Hamburg-Eppendorf (all before 1939); practiced medicine specializing in internal illnesses, Bremen, 1932-62, Leuchtenburg, from 1963. Author: Psychologie des mystischen Bewussteins, 1951; Das mystische Erkennen, 1958. Died July 19, 1965.

ALBRECHT, Carl Theodor, German geodesist; b. Dresden, Germany, Aug. 30, 1843; s. Friedrich Wilhelm and Christiana Juliana (Pohle) A.; student math., natural scis. Polytechnikum, Dresden, astronomy Berlin, Leipzig (Germany) univs.; m. Marie Stiemer, 1875. Asst. to Geodesic Inst., Potsdam, Germany, from 1869, head astron. dept., from 1873; faculty U. Leipzig, from 1869, prof., from 1875. Author: Astronom.-Geodät Arbb. f.d. Gradmessung im Kgr. Sachsen; Die astronom Arbb, 1883-85; Formeln u Helfstafeln f. Geograph Ortsbestimmungen, 1908; contbr. many articles to Dresdner Zeitung. Introduced wireless telegraphy to high precision longitude measurements. Died Potsdam, Aug. 31, 1915.

ALBRECHT, Heinrich, Austrian bacteriologist; b. Vienna, Austria, July 24, 1866; ed. pathology, bacteriology (probably in Vienna); mem. Acad. Scis. commn. investigating plague, Bombay, India, 1897; asst. prof. Graz, Austria, 1899-1902, asso. prof., 1902-13, prof. in ordinary of path. anatomy, from 1913; prof. in ordinary of path. anatomy U. Vienna, from 1920; research, publs. on plague, meningitis, Tb, whooping cough, oncology. Died Vienna, June 28, 1922.

ALBRECHT, Heinrich, German physician; b. Berlin, Germany, Sept. 2, 1921; s. Otto and Elisabeth (Judis) A.; student medicine Göttingen, Hamburg, Berlin (all Germany) univs.; m. Gisela Albrecht, 1954. From asst. to chief physician Psychiatry Clinic, Hamburg-Eppendorf; prof. psychiat. and nerve clinic U. Hamburg, 1946—. Chmn. Deutschen Gesellschaft für Jugendpsychiatrie, Deutsche Vereinigung für Jugendpsychiatrie, Deutsche Vereinigung für Jugendgericht und Jugendgerichtshilfen. Research, publs. in areas of child, youth psychiatry, problems of adolescence. Address: Psychiatrische und Nervenklinik, U. Hamburg, Martinstrasse 52, 2 Hamburg 20, Germany.

ALBRECHT, Hermann Ulrich Emil Ferdinand Johann, German physician; b. Verden/Aller, Germany, Apr. 19, 1897; s. August and Margarete (Saathoff) A.; ed. Frankfort/Main, Friburg/Brg. univs.; m. Elisabeth Fastenrath, May 21, 1935; children—Karin, Gösta, Dagman, Kerl. Agrégation, 1924; prof. agrégé, 1930; prof. U. Frankfort, 1937; dir., owner Röntgen Inst., Osnabrück, from 1938; became prof. Acad. Medicine, Danzig, 1950. Mem. several med. socs. Author: Das Ulcusproblem im Lichte der modernen Röntgenforschung, 1930; Die Rö-Diagnostik des Verdauungskanals einschl. d. Leber- und Gallenwege, 1931. Address: Osnabrück, Lümannstrasse 36/40, West Germany.

ALBRECHT, Jean-Sebastien, biologist; b. Cologne, Germany, 1695; prof. natural philosophy, Coburg, Germany. Author: Spicilegium ad historiam naturalem scarabei platyceri; Commercium litterarium, 1731; studies on harmful effects of solanum furiosum, effects of belladonna; contbd. descriptions of natural anomalies to Annales de l'Académie des Curieux de la Nature. Died Cologne, 1774.

ALBRECHT, Max, German petroleum engr.; b. Liegnitz, Germany, Oct. 24, 1851; s. Samuel and Emilie (Czarnikow) A.; Ph.D., U. Halle (Germany), 1871; partner firm A. Oehlrich and Co., Riga, Germany, from 1877, founded new plants in Baku, Russia, 1883, Hamburg, Germany; founder Mineralölwerken Albrecht and Co., 1891. Contbr. articles sci. publs. Produced paraffin, mineral oils of brown coal on sci. basis; expanded prodn., transportation oil products (mainly machine lubricants) all over world; founder 1st infant hosp. in Hamburg. Died Hamburg, Dec. 12, 1925.

ALBRIGHT, Fuller, Am. endocrinologist; b. Buffalo, Jan., 1900; A.B., Harvard, 1921, M.D., 1924. House officer Mass. Gen. Hosp., 1924-26; asst. med. resident Hopkins Hosp., 1927-28; Moseley travelling fellow, Harvard, 1928-29; instr. Harvard Med. Sch.,

1931-39, asst. prof., 1939-42, asso. prof., from 1942; asst. physician Mass. Gen. Hosp., 1929-37, asso. physician, 1937, physician, from 1939; cons. physician Mass. Eye and Ear Infirmary, from 1931; Henry Pickering Walcott fellow, Harvard, 1931-34. Mem. Soc. Clin. Investigation, Assn. Physicians, Endocrine Soc., Am. Acad. Parathyroid Disease. Described Albright syndrome (or Albright-McCune-Sternberg syndrome) characterized by asymmetric bone disease, abnormal skin pigmentation, and precocious puberty in females; Showed that parathyroid hormone acts on kidney to promote excretion of phosphates leading to increase of calcium in blood.

ALBRIGHT, Joseph Finley, Am. biologist; b. New Tazewell, Tenn., Mar. 9, 1927; s. Philip Ninde and Louise (Harris) A.; B.S., Southwestern U., 1949; postgrad. Vanderbilt U.; Ph.D., Ind. U., 1956; m. Marcia Lorraine Geckler, Sept. 9, 1952; children—Emily Christine, Kendra Suzanne. NIH fellow Oak Ridge Nat. Lab., 1956-58, biologist biology div., 1960—; asst. research prof. dept. surgery Med. Coll. Va., 1958-60. Mem. Am. Soc. Zoologists, Sci. Research Soc. Am., Soc. Developmental Biology, Am. Assn. Immunologists. Research, publs. on kinetics antibody response, differentiation and potentialities cells which initiate antibody response, differentiation and growth immune response during embryonic, neonatal life and senescence. Home: 120 Miramar Circle, Oak Ridge, 37830.*

ALBRIGHT, William Foxwell, Am. archaeologist, orientalist; b. Coquimbo, Chile, May 24, 1891; s. Wilbur Finley and Zephine Viola (Foxwell) A.; A.B., Upper Ia. U., 1912, Ph.D., Johns Hopkins, 1916, LL.D., 1964; Litt.D. (hon.) Upper Ia., 1922, Yale, 1951, Georgetown U., 1952, U. Dublin, 1953, Loyola Coll., Balt., 1958; Loyola U., Chgo., 1960; D.H.L., Jewish Theol. Sem. in Am., Jewish Inst. Religion, 1936, Hebrew Union Coll., 1948, Coll. Jewish Studies, Chgo., 1950; Th.D., U. Utrecht (Netherlands), 1936, U. Uppsala (Sweden), 1952; D. hon. caus., U. Oslo (Norway) 1946; LL.D., Boston Coll., 1947, U. St. Andrews, Scotland, 1949, Franklin and Marshall Coll., 1953; D.Phil., Hebrew U., Jerusalem, 1957; D.C.L., Pace Coll., 1957; Ph.D., La Salle Coll., 1958; Litt.D., Harvard, 1962, Lake Erie Coll., 1966; L.H.D., Manhattan Coll., 1961, Colby Coll., 1966; H.H.D., Wayne State U., 1961, Brigham Young U., 1962; m. Ruth Norton, Aug. 31, 1921; children—Paul N., Hugh N., Stephen Foxwell, David Foxwell. Acting dir. Am. Sch. Oriental Research, Jerusalem, 1920-21, dir. 1921-29, 33-36; W. W. Spence prof. Semitic langs. Johns Hopkins, 1929-58, now emeritus; research prof. Jewish Theol. Sem. Am., 1957-59; dir. archeol. expdn., Palestine, 1922-34, mem. U. Cal. African Expdn., Sinai, 1947-48; chief archeologist S. Arabian Expdn., Am. Found. Study Man, 1950-51; pres. Palestine Oriental Soc., 1921-22, 34-35; Am. mem. consultative com., Internat. Congress Orientalists, 1931-48; pres. Am. Oriental Soc., 1935-36; 1st v.p. Am. Schs. of Oriental Research, 1937—. Mem. Am. Council of Learned Societies (vice chmn. council 1939, recipient $10,000 prize 1961), Am. Philos. Soc. (v.p. 1956-59), Archeol. Inst. Am. (v.p., 1949) Soc. Bibl. Lit. (pres. 1939), Linguistic Soc. Am. (v.p. 1941), Internat. Orgn. Old Testament Scholars, (pres. 1956-59), Nat. Acad. Scis., Royal Danish, Flemish & Irish Acads.; corr. mem. Institut de France & Austrian Acad. Scis.; fellow Am. Acad. Arts & Scis., German Archeol. Inst.; hon. or corr. mem. many other tech., profl. socs. Author over 800 publs. on archeol., Bibl. and oriental subjects, including: The Excavation of Tell Beit Mirsim, 1932-43; From the Stone Age to Christianity, 1940; Archaeology and the Religion of Israel, 1942; The Archaeology of Palestine, 1949; History, Archaeology and Christian Humanism, 1964. Sr. editor: Anchor Bible, 1956—. Authority on Bibl. langs.; studied and described Dead Sea Scrolls; established Ceramic chronology of Palestine from 3500 B.C. on; many other contributions to Near Eastern history and civilization. Home: 3401 Greenway, Balt. 21228.

ALBRINK, Margaret Joralemon (Mrs. Wilhelm S. Albrink), Am. physician, educator; b. Warren, Ariz., Jan. 6, 1920; d. Ira Beaman and Dorothy (Rieber) Joralemon; B.A., Radcliffe Coll., 1941; M.S., Yale, 1943, M.D., 1946, M.P.H., 1951; m. Wilhelm S. Albrink, Sept. 16, 1944; children—Frederick Henry, Jonathan Wilhelm, Peter Varick. Faculty, Yale Med. Sch., 1952-61; asso. prof. medicine W.Va. U. Sch. Medicine, Morgantown, 1961-66, prof., 1966—. Mem. Am. Soc. for Clin. Investigation, Am. Fedn. for Clin. Research, Am. Pub. Health Assn., Am. Heart Assn., Am. Assn. for Clin. Nutrition. Research, publs. on lipid metabolism, role of serum triglycerides in coronary artery disease, metabolic and nutritional factors influencing serum lipids. Home: 817 Augusta Av., Morgantown, W.Va. 26505.*

ALBRITTON, Claude Carroll Jr., Am. geologist; b. Corsicana, Tex., Apr. 7, 1913; s. Claude C. and Iris (Stapleton) A.; A.B., So. Meth. U., 1933, B.S., 1933; M.A., Harvard, 1934, Ph.D., 1936; m. Jane Christman, Aug. 5, 1944; children—Jane DeHart, Claude Carroll III, Elizabeth Ann. Faculty, So. Meth. U., Dallas, 1936—, prof. geology, 1947—, chmn. dept. geology, 1947-51, dean faculty arts and scis., 1952-57, dean grad. sch., 1957—, chmn. bd. publ., 1964—, dir., Graduate Research Center, Inc.; geologist U. S. Geol. Survey, Washington, 1942-59. Mem. at large Dallas County Bd. Edn., 1958—. Recipient De Witt medal, 1933, Faculty Achievement award So. Meth. U. Alumni Assn., 1963; J. B. Woodworth fellow, 1935-36. Mem. A.A.A.S. (dir. 1962—), Geol. Soc. Am. (councilor 1957-60), Paleontol. Soc. (councilor, treas. 1955-62), Phi Beta Kappa, Sigma Xi, Kappa Alpha. Author: Lead and Zinc Deposits of the Goodsprings District, 1954; (with others) The Midland Discovery, 1955. (with others) Geology of the Sierra Blanca Area, Hudspeth County, Texas. Editor, co-author The Fabric of Geology, 1963. Home: 3436 University Blvd., Dallas 75205.*

ALBUMASAR (Abu Ma'shar), Arab astronomer; b. Balkh, Khurasan, circa 786; flourished in Bagdad. Author: Flores Astrologici, pub. 1488; De Magnis Conjunctionibus, pub. 1489; Introductorium in Astronomiam, pub. 1506. Was most quoted Arabic astrologer in West; made valuable contbns. on theory of tides; held theory that world was created when 7 planets were in conjunction in 1st degree of Aries; and would be dissolved at similar conjunction in last degree of Pisces. Died Wasid, Central Asia, 886.

ALBURGER, David Elmer, Am. physicist; b. Phila., Oct. 6, 1920; s. Elmer R. and Josephine (Reid) A.; B.A., Swarthmore Coll., 1942; M.S., Yale, 1946, Ph.D., 1948; m. Mary V. Mickle, Oct. 6, 1945; children—David R., Mary Jo, Evelyn A., Andrew R. Radio, radar engr. Naval Research Lab., Washington, 1942-45; sr. physicist Brookhaven Nat. Lab., Upton, N.Y., 1948—. Fellow Am. Phys. Soc. Research, numerous publs. on determination of properties of energy levels in nucleus by measuring radiations emitted. Home: Beaver Dam Rd., Brookhaven, N.Y. 11719. Office: Brookhaven Nat. Lab., Upton, N.Y. 11973.*

ALBUZJANI, see Abul Wafa.

ALCABITIUS (al-Qabisi), Arabic astrologer; flourished circa 960; s. 'Utman b. 'Ali; studied under al-'Imrani; wrote introduction to art of astrology; also work on conjunctions of planets (both translated to Latin by Johannes Hispalensis).

ALCHAZIN (al-Khazin) (Abu Ja'far Alchazin), mathematician, astronomer; b. Khurasan; al-Khazin means the treas., or the librarian; wrote commentary on 10th book of Euclid, other math. and astron. writings; was 1st Arab (according to Omar Khayyam) to solve the cubic equation by conic sections. Died circa 961-71.

ALCHINDUS (or Alkindus, also Abu Yusuf Yaqub ben Ishaq al-Kindi), Arabian physician; b. Kufa (now in Iraq), circa 813; ed. Basra, Baghdad; flourished in Baghdad under Al-Mamun and Al-Mutasim, 813-42, persecuted during Al-Mutawakkic's orthodox reaction, 847-61. Author works on philosophy, arithmetic, medicine, technology, astronomy, chemistry, including: De medicinarum compositarum gradibus investigandis libellus, pub. 1531; De pluviis, imbribus et ventris ac aeris mutatione, pub. 1507; Liber Jacob Alkindi de causis diversitatum aspectus et dandis demonstrationibus geometricis super eas (translated by Gerard of Cremona); Kitab fi'l-san a 'a al-kubra. Called the philosopher of the Arabs; made a neoplatonic study of Aristotle; developed notation for musical pitch. Died circa 870.

ALCMAEON, physician, anatomist; b. Croton, Italy; flourished circa 500 B.C.; pupil of Pythagoras; influenced by Heraclitos. Author: On Nature (basic text of pre-Hippocratic medicine, now lost). Made 1st known dissections of human body; recorded optic nerve, Eustachian tubes, trachea; distinguished arteries from veins; gave explanations of sleep, origin of sperm, sense impression; believed brain the center of human activity; possibly 1st Greek to operate on eye; made 1st studies of devel. of chicken in incubated egg (thus known as father of embryology); observed that planets move from W. to E.; believed everything in nature is conflict between opposites and extremes; compiled list of contrary principles (recorded in Aristotle's Metaphysics).

ALCOCK, Alfred William, surgeon, naturalist; b. June 23, 1859; ed. Aberdeen (Scotland) U.; M.B., LL.D.; m. Margaret Cornwall, 1897. Asst. prof. zoology U. Aberdeen, 1883-85; with Indian Med. Service, 1885-1907, with Punjab Frontier Force, 1886-88, surgeon-naturalist to marine survey India, on board Investigator, 1888-92, to Pamir Boundary Commn., 1895; supt. Indian Mus., prof. zoology Med. Coll. Bengal (India), 1893-1907; later prof. med. zoology U. Lond., Lond. Sch. Tropical Medicine. Fellow Royal Soc., 1901; mem. Zool. Soc. London, Zool. Soc. Netherlands, Cal. Acad. Sci. (hon.). Author: A Naturalist in Indian Seas, 1902; (with P. Manson-Bahr) Life and Work of Sir Patrick Manson; numerous monographs, papers on zoology and zoo-geography. Died Mar. 24, 1935.

ALCOCK, Nathaniel Henry, physiologist; b. Feb. 12, 1871; s. D. R. Alcock; B.A., M.D., U. Dublin (Ireland); postgrad. U. Marburg (Germany); D.Sc., U. London; m. Nora Lilian Lepard Scott; 1 son, 3 daus. Demonstrator anatomy Victoria U., Manchester, Eng., 1896-97; asst. to prof. Inst. Medicine, Trinity Coll., Dublin, 1898-1902; demonstrator physiology U. London, 1903; lectr. physiology St. Mary's Hosp. Med. Sch., 1904-11, vice dean, 1906-10; Joseph Morley Drake prof. physiology McGill U., Montreal, Que., Can., from 1911. Author: (with Dr. Ellison) Textbook of Experimental Physiology, 1909; numerous papers on physiol. subjects. Died June 12, 1913.

ALCOCK, Norman Zinkan, Canadian physicist; b. Edmonton, Alta., Can., May 29, 1918; s. Joseph and Edith Alma (Zinkan) B.; B.Sc., Queen's U., 1940; M.S. Cal. Inst. Tech., 1941; Ph.D., McGill U., 1949; m. Patricia Christian Hunter, June 29, 1948; children—Stephen Christopher, David, Nancy. Jr. research engr. NRC, Ottawa, Ont., Can., 1941-43, Telecommunication Research Establishment, Great Malvern, Eng., 1943-45; research physicist. McGill U., Montreal, Que., Can., 1945-47; research physicist Atomic Energy Can., Chal. River, Ont., 1947-50; v.p. Isotope Products, Ltd., Oakville, Ont., 1950-57; dir. engring. Canadian Curtiss-Wright, Montreal, 1957-59; pres. Canadian Peace Research Inst., Oakville, 1959—, dir., 1961—. Mem. Internat. Peace Research Assn., Sigma Xi. Author: Bridge of Reason, 1961; also articles. Patentee in field. Research on antenna designs for ground and air-born micro-wave radar, neutron diffraction in gas molecules, radioactive measuring gauges. Home: 244 Lakewood Dr., Oakville, Ont. Office: Canadian Peace Research Inst., Clarkson, Ont., Can.*

ALCORN, Gordon Dee, Am. biologist; b. Olympia, Wash., Apr. 6, 1907; s. John H. and Rachel (Austin) A.; B.S., Coll. Puget Sound, 1930; M.S., U. Wash., 1933, Ph.D., 1935; m. Rowena Lung, Aug. 8, 1935; 1 dau., Patricia (Mrs. Jack Peterson). Faculty, Coll. Puget Sound, 1930-35, U. Ida., 1935-37, Grays Harbor Coll., 1937-46; faculty U. Puget Sound, Tacoma, summers 1930-45, asso. prof. biology, 1946, prof., 1947—, chmn. dept., 1951—; dir., curator Puget Sound Mus. Natural History, 1951—. With div. edn. War Manpower Commn., 1943-45. Mem. Cooper Ornithol. Soc., Am. Ornithol. Union, Pacific N.W. Bird and Mammal Soc. (pres., past trustee), Am. Assn. U. Profs., Izaak Walton League Am. (bd. dirs.), Sigma Xi, Pi Gamma Mu, Phi Sigma. Editor-in-chief The Murrelet, 1951—. Contbr. numerous articles to profl. jours. Research, publs. on tighter bonds that weld sci. to arts, social scis. and related humanities. Home: 3806 N. 24th St., Tacoma 98406.*

ALCUIN (or ALKUIN, ALKVIN), scholar; b. York, Eng., 735; ed. in York; became master York Sch., 778; travelled to Rome, 780; became head edn. system for empire of Charlemagne; established sch. and library at imperial palace, Aachen, Germany; taught Charlemagne to read and inspired him to publish De litteris colendis; apptd. abbot, Tours, France, 796, founded sch. (instrumental in raising Franks to cultural level of English monasteries and bringing about Carolingian Renaissance). Author: Compendia (textbook of rhetoric); Propositiones ad acuendos iuvenes; a life of Willibrord; manuals of grammar, rhetoric, dialectic; theol., philos. works. Encouraged study and preservation of ancient texts; instituted multiple reproduction by copyists; his copyists developed Carolingian miniscule. Died Tours, May 19, 804.

AL-DAMIRI (Kamal al-din Mohammed ibn Musa al-Damiri), zoologist, natural historian; b. Damira, Egypt, 1344; ed. Al-Azhar, Cairo; prof. of tradition Al-Azhar Mosque, Cairo, Egypt; lived in Mecca, 1379-1399. Author various treatises in prose and verse including: On the Unity of God, circa 1360-79; Life of the Animals (listed 931 animals mentioned in Koran and Arab lit.; includes use of animals in medicine, food and folklore). Died Cairo, Oct. 1405.

ALDER, Henry L(udwig), mathematician; b. Duisburg, Germany, Mar. 26, 1922; s. Ludwig and Otti (Gottschalk) A.; came to U S., 1941, naturalized, 1944; student Eidgenössische Technische Hochschule, Zürich, Switzerland, 1940-41; A.B., U. Cal. at Berkeley, 1942, Ph.D., 1947; m. Benne B. Daniel, Apr. 8, 1963; 1 son, Lawrence J. Faculty, U. Cal. at Berkeley, 1947-48; faculty U. Cal. at Davis, 1948—, prof. math., 1965—. Vice chmn. bd. govs. Pacific Jour. Math. Mem. Math. Assn. Am. (chmn. No. Cal. sect. 1956-57, nat. sec. 1960—), Am. Math. Soc., Inst. Math. Statistics, Sigma Xi, Mu Alpha Theta (Internat. Distinguished Service award 1965), Pi Mu Epsilon. Author: (with E. B. Roessler) Introduction to Probability and Statistics, 1960; also articles. Research on theory of partitions and compositions; generalizations of Rogers-Ramanujan identities. Home: 724 Elmwood Dr., Davis, Cal. 95616.*

ALDER, Joshua, English zoologist; b. Newcastle/Tyne, Eng., Apr. 7, 1792; student natudal history; worked on Brit. conchology, zoophytology; mem. Lit. and Philos. Soc. Newcastle. Author: (with Albany Hancock) On the British Nudibranchiate Mollusca, 1845-55; contbr. papers to zool. publs.; his collection of Brit. shells and zoophytes and his library presented to Natural History Soc. of Northumberland, Durham and Newcastle/Tyne. Died 1867.

ALDER, Kurt, German chemist; b. Königshütte, Germany, July 10, 1902; s. Joseph and Maria Alder; student in Königshütte, also U. Berlin; Ph.D., U. Kiel (Germany), 1926; M.D. (hon.), U. Cologne (Germany), 1950; doctorate (hon.), Salamanca, Spain, 1954. Became docent U. Kiel, 1930, prof., 1934; apptd. dir. Chem. Inst., also prof. U. Cologne, 1940. Dir. research I. G. Fabren Ind.; became hon. advisor Consejo Superior de Investigaciones Cientificas, Madrid,

1953. Recipient (with Otto Diels) Nobel prize for chemistry, 1950. Mem. Real Sociedad Espanola de Física y Química (hon.), Mathematisch-naturwissenschaftliche Klasse, Bayerische Akademie der Wissenschaften (corr.). Contbd. papers to chem. jours. Developed (with Otto Diels) practical method for making ring compounds from chain compounds by forcing them to combine with maleic anyhydride (Diels-Alder reaction); thus provided method for synthesis of complex organic compounds. Died Cologne, June 20, 1958.

ALDERFER, Russell Brunner, Am. soil physicist, educator; b. Lansdale, Pa., Sept. 27, 1913; s. Ulysses F. and Sallie (Brunner) A.; B.S., Pa. State U., 1936, M.S., 1940, Ph.D., 1947; m. Dorothy Mease Pearce, Aug. 16, 1941; children—Catharine Pearce, Margaret Vivian. Soil technologist Soil Conservation Service, U. S. Dept. Agr., Safford, Ariz., 1936-39, State College, Pa., 1939-43; instr. soil tech. Pa. State U., University Park, 1943-47, asst. prof., 1947-48, asso. prof., 1948-52, prof., 1952-54, chmn. dept. soils, 1954-63; prof. soils Rutgers U., New Brunswick, N.J., 1963—. Lectr., Inst. Nutrition Scis., Columbia, 1959—. Fellow A.A.A.S.; mem. Am. Soc. Agronomy (pres. N.E. br. 1955), Soil Conservation Soc. Am. (pres. N.J. chpt. 1956), Plant Food Edn. Soc. N.J. (pres. 1964), Soil Sci. Soc. Am., Internat. Soil Sci Soc., Sigma Xi, Alpha Gamma Rho, Phi Kappa Phi, Gamma Sigma Delta. Contbr. articles to sci. jours., bulls. Research on phys. properties soils, soil and water conservation. Home: 390 Independence Blvd., North Brunswick, N.J. 08902.*

ALDEROTTI, Taddeo, (or Thaddaeus Florentinus); Italian physician; b. Florence, Italy, circa 1223; student medicine, philosophy, Bologna, Italy; founded (with others) med. sch., Bologna. Author: Consilia medicinalia (collection of clin. records representing new form med. literature); De conservanda sanitate; commentaries on Hippocrates, Galen, Isogoge of Hunain Ibn Ishaq. Died 1303.

AL-DIMASHQI (Shams al-din Abu 'Adallah Muhammad ibn Ibrahim ibn Abu Talib al-Ansari al-Sufi al-Dimashqi), Syrian cosmographer; b. possibly in Damascus, Syria, circa 1256-57. Imam in Village of Al-Rabwa. Author: On Physiognomy in Relation to Government (1st Arabic treatise mixing physiognomy with astrology); On the Signs of Death According to Hippocrates. Died 1326-27.

ALDINI, Giovanni, Italian physicist; b. Bologna, Italy, Apr. 10, 1762; nephew of Luigi Galvani. Prof. U. Bologna; founder Nat. Inst. Italy; mem. Council of State, Milan, Italy; publs. on galvanism, hydraulic lever, steam power; researched application of galvanism to medicine. Invented fire-fighting apparatus named after him. Died Milan, Jan. 17, 1834.

ALDOBRANDINO OF SIENA, Italian physician; flourished at court of Frederick II Hohenstaufen, Sicily; Author: The Regimen of the Body (1st work on health perservation in vernacular; dominated by Arab and Salernian medicine), 1234; also wrote on Galen, Avicenna and Hippocrates. Died circa 1287.

ALDRICH, Charles Anderson, Am. pediatrician; b. Plymouth, Mass., Mar. 4, 1888; s. David E. and Laura Linwood (Perkins) A.; B.S., Northwestern U., 1914, M.D., 1915; postgrad. Harvard, 1921; m. Mary McCague, Oct. 3, 1916; children—Robert A., Cynthia, Stephen. Attending roentgenologist and jr. attending physician in medicine Evanston (Ill.) Hosp., 1916-21, attending pediatrist newly born service, 1930-36, asso. physician and chmn. pediatric dept., 1936-42 asst. attending physician Children's Meml. Hosp., Chgo., 1921-22, asso. attending physician, 1922-41; chief of staff, 1941-44; cons. pediatrist Municipal Tb Sanitarium, 1925-27; cons. physician on staff Chgo. Nursery and Half-Orphan Asylum, 1941-44; asso. in pediatrics, Northwestern U. Med. Sch., 1934-35, asst. prof. 1935-36, prof. pediatrics, 1941-44; prof. pediatrics, Mayo Found. Grad. Sch., U. Minn., from 1944. Dir. Rochester Child Health Project, 1944. Mem. White House Conf. on Children in a Democracy. Mem. Northwest Pediatric Soc., NRC (com. on child devel.); mem. A.M.A. (adv. bd. med. splty.), Am. Acad. Pediatrics (com. on mental health), Am. Board Pediatrics, (sec.-treas. 1934-44, pres. 1945-47), Nat. Assn. for Nursery Edn., Soc. Research in Child Devel.; mem. (hon.) Nat. Com. on Mental Hygiene; sec. Am. com. Internat. Pediatric Congress in London (Eng.), 1933. Author: Cultivating the Child's Appetite, 1927; Babies Are Human Beings (with Mary M. Aldrich), 1938 English reprint. Understand Your Baby, 1939); Feeding Our Old Fashioned Children (with Mary M. Aldrich), 1941; also numerous articles and book revs. Died Oct. 5, 1949.

ALDRICH, Clarence Knight, Am. physician; b. Chgo., Apr. 12, 1914; s. L. Sherman and Bessie (Knight) A.; B.A., Wesleyan U., Conn., 1935; M.D., Northwestern U., 1940; m. Julie Honore Murphy, Feb. 4, 1942; children—Carol K., Michael S., Thomas K., Robert F. Staff, USPHS, 1943-46; asst. prof. psychiatry Students' Health Service, U. Wis., 1946-47; asst. prof. psychiatry, Sch. Medicine, U. Minn., Mpls., 1947-51, asso. prof., 1951-55; prof. dept. psychiatry U. Chgo. Sch. Medicine, 1955—, chmn. dept., 1955-64; vis. prof. dept. psychol. medicine, U. Edinburgh, 1963-64. Cons. VA, 1949-55; cons. psychiatrist Family and Children's Service of Mpls., 1949-54; mem. Ill. Psychiat. Adv. Council, 1955-63. Author: Psychiatry for the Family Physician, 1955; An Introduction to Dynamic Psychiatry, 1966. Publs. on relocation of the aged; nature of grief. Home: 5747 S. Kimbark Av., Chgo. 60637.*

ALDRICH, John Warren, Am. biologist; b. Providence, Feb. 23, 1906; s. Wilfred Westcott and Effie (Dwinel) A.; Ph.B., Brown U., 1928; A.B., Western Res. U., 1933, Ph.D., 1937; m. Elizabeth Louise Kendall, June 15, 1933; children—Elizabeth Louise (Mrs. Charles Littleton Hanson), Jane Warren. With Cleve. Mus. Natural History, 1930-56, chief sect. birds and mammals, 1948-56; research staff specialist wildlife classification and distbn. U. S. Fish and Wildlife Service, Washington, 1957—; hon. research asso. in birds Smithsonian Instn., Washington, 1965—. Fellow of American Ornithologists Union (vice president since 1966—); member of the Wilson, Cooper ornithol. socs., Ecol. Soc. Am., Washington Acad. Sci., A.A.A.S. Author: (with Stanley G. Jewett, Walter P. Taylor, William T. Shaw) Birds of Washington State, 1953; also numerous articles. Research on distbn., migration, classification and ecology N. and Central Am. birds; described 10 new subspecies birds. Home: 6324 Lakeview Dr., Falls Church, Va. 22041. Office: U. S. Fish and Wildlife Service, Washington 20240.*

ALDRICH, L. Thomas, Am. physicist; b. Hopkins, Minn., June 28, 1917; s. Geroge Malcolm and Ruth (Alden) A.; B.A., U. Minn., 1939, M.A., 1946, Ph.D., 1948; m. Margaret E. Glockler, June 24, 1941; children—Carol L., Peggy L. Teaching asst. U. Minn., 1939-40, 45, 46, research asst., 1946-48; physicist USN, Washington, 1940-45; asst. prof. physics U. Mo., Columbia; mem. staff dept. terrestrial magnetism Carnegie Instn., Washington, 1950—. Vis. prof. Kyoto (Japan) U. Fellow Am. Phys. Soc., A.A.A.S., Am. Geophys. Union. Research on application radioactive age measurements to geologic problems. Home: 9414 St. Andrews Way, Silver Spring, Md. 20901. Office: 5241 Broad Branch Rd. N.W., Washington 20015.*

ALDROVANDI (or ALDROVANDUS), Ulisse, Italian naturalist; b. Bologna, Italy, Sept. 11, 1522; s. Teseo Aldrovandi; ed. Bologna; med. degree from Padua; m. Veronica D'Antonio Marescalchi. Prof. natural history, philosophy, Bologna, from 1560; founder, dir. bot. gardens, Bologna, from 1568. Author: Ornithologiae libri XII, 1599-1643; De animalibus insectis, 1602; De reliquis animalibus, 1606; De Piscibus et cetis, 1613; Quadrupedum omnium bisculorum historia, 1613; De quadrupedibus solidipedibus, 1616; De quadrupedibus digitatis viviparis et . . . oviparis, 1637; Serpentum et draconum historiae, 1640; Monstrorum historia, 1642; Musaeum metallicum (word geology 1st used in modern sense), 1648; Dendrologia naturalis, 1668. Specialized in zoology and medicine; studied mollusks; collected animal, mineral, and vegetable specimens; under patronage Popes Gregory XIII, Sixtus V, made collections which became basis for Bologna mus. and led to a system of animal classification. Died Bologna, Nov. 10, 1605.

AL-DUHAWI (Ayyub al-Duhawi), Syrian translator; b. Odessa, Russia, 760; lived at Ct. of Caliphs, Bagdad. Author: Book of Treasures (ency. of philos. scis. taught in Bagdad, circa 817). Translator numerous works from Greek and Syrian, including Hunayn ben Ishaq, 35 treatises of Galen. Died 835.

ALEEM, Anwar Abdel, Egyptian biol. oceanographer; b. Alexandria, Egypt, Dec. 13, 1918; s. Mohamad Abdel and Aisha (Gawad) A.; B.Sc., U. Cairo, 1941; M.Sc., U. Alexandria, 1945; Ph.D., U. London, 1949; m. Hoda El-Odwy, Oct. 19, 1961; children—Mohamad, Hosam. Asst., faculty of sci. Alexandria U., 1942-46; research fellowships in Sweden, France, Germany, 1949-52; asso. prof. U. Alexandria, 1953-59, prof. biol. oceanography, head dept., 1959—. Fulbright scholar, asso. prof. U. So. Cal., research fellow Hancock Found., 1954; research fellow Inst. Marine Resources and Scripps Instn. Oceanography, 1955. Mem. Internat. com. Assn. Theoretical and Applied Limnology, 1953. Recipient U.A.R. State award, 1954. Mem. Internat. Assn. Oceanography. Author: (in Arabic) Marine and Freshwater Resources of Egypt, 1961; Oceans and Seas and Their Natural Resources, 1964; also numerous articles. Mem. Internat. Indian Ocean Expdn., 1964; developed apparatus and methods for simulating tides in lab., quantitative methods for assessing plankton and diatoms; research on ecology and taxonomy of marine phytoplankton and relation to nutrients, productivity studies in inland lakes, world-wide distbn. of parasitic maritime fungi, taxonomic studies of diatoms, algae, marine fungi and grasses. Home: 136 Rue Galal-Desouky, Alexandria. Office: U. Alexandria, Egypt, U.A.R.*

ALEFELD, Friedrich Christoph Wilhelm, German botanist; b. Gräfenhausen, Germany, Oct. 21, 1820; s. Georg Wilhelm Karl and Katharina (Graff) A.; student medicine, natural scis. U. Heidelberg (Germany); M.D., U. Giessen (Germany), 1843; m. Luise Vogel, Nov. 11, 1847; 2 children. Practiced medicine Niedermoldau, Ober-Ramstadt (both Germany), for 25 years. Author: Bienenflora, 1856; Phytobalneologie, 1863; Landwirtschafliche Flora Mittel-Europas, 1886. Noted for his study of taxonomy of garden plants. Died Ober-Ramstadt, Apr. 28, 1872.

ALEKIN, Oleg Aleksandrovich, Russian hydrochemist; born Aug. 23, 1908; graduated from Leningrad Univ., 1938. Asso., Leningrad Hydrological Institute, 1929-51; director Hydrochemical Institute, USSR Academy Science, 1951-54, 60-61; rector Rostof Univ., 1954-60; Recipient, Stalin prize 1951; member editorial staff of Hydrochem. Materials, 1960—. Mem. USSR Acad. Sci. (corr.). Author: General Hydrochemistry, 1948; Hydrochemistry of the USSR's Rivers, 1948-49; Principles of Hydrochemistry, 1953; Analysis of Land Waters, 1954; Relation between Ion Discharge and the Discharge of Suspended Substances, 1962; Chemistry of the Ocean, 1966. Research on the chemical analysis of natural waters, Home: Mamdokhtenski prospect 98, Leningrad, USSR. Office: USSR Acad. Sci., Leninsky prosp. 14, Moscow, USSR.*

ALEKSANDROV, Aleksandr Danilovich, Russian mathematician; born Aug. 5, 1912; Doctor of Physico-Mathematical Science. Corresponding member of USSR Academy of Science, 1946-64, full member, 1964—; rector of Leningrad Univ., 1950—. Recipient of Stalin prize, 1942. Head Leningrad delegation of Harvard, 1959. Author: The Internal Geometry of an Arbitrary Convex Surface, 1941; The Existence of a Convex Polyhedron and a Convex Surface with Given Dimensions, 1942; The Internal Geometry of Convex Surfaces, 1948; Convex Polyhedrons, 1950. Founder Soviet school of geometry; constructed from most general hypotheses internal geometry of convex surfaces; investigated internal multiform geometry governed by natural geometric requirements. Address: Leningrad University, Leningrad, USSR.

ALEKSANDROV, Anatolii Petrovich, Russian physicist; born Feb. 13, 1903; graduated from Kiev Univ., 1930; Doctor of Physics and Mathematical Science. Teacher of physics and chemistry secondary sch., 1922-30; asso. Leningrad Physicotech. Inst., USSR Acad. Sci., 1930-46, dir. Inst. Phys. Problems, 1946-54, dep. dir. (sci.) Kurchatov Inst. Atomic Energy, 1955-60; dir. Kurchatov Inst. Atomic Energy, USSR State Com. for Use Atomic Energy, 1960—. Mem. Com. for Lenin Prizes for Sci. and Tech., USSR Council of Ministers, 1960—; Mem. Soviet sci. delegation Internat. Conf. on Peaceful Use of Atomic Energy, Geneva, 1958. Recipient Order of Lenin, 1959, 63, Stalin prize, 1942, Hero of Socialist Labor, 1963, also State prize. Mem. USSR Acad. Sci. (Presidium mem.). Research on physics of dielectrics, mech. and electric properties of high-polymer compounds; discovered and studied characteristics of polymerized styrene, made polystyrene condensers. Address: Inst. of Atomic Energy, ul. Kurchatova 46, Moscow, USSR.

ALEKSANDROV, Boris Kapitonovich, Russian hydraulic engineer; born Aug. 18, 1889; graduated from Petrograd Polytechnical Institute, 1917. Instructor in secondary and higher educational establishments, 1918-39; chief engineer, head Greater Volga Bd., All-Union Trust for Planning Hydropower Stas., 1939-46; instr. Moscow Power Engring. Inst., 1946-48, prof., 1948—. Mem. USSR Acad. Sci. (corr.). Author; Automatic Recording of Lines of Liquid Flow, 1963. Research on harnessing of major lowland rivers of European USSR; studied means of transferring flow of No. rivers of Pechora and Onega into Volga and Kama rivers; developer new types of bldgs. for hydroelectric plants and shipping locks; designer Kama Hydroelectric Plant and Kama multichamber shipping locks; worked on design of Moscow Canal and Rybinsky and Uglich hydropower networks. Address: Moscow Power Engring. Inst., Krasnokazarmennaya ul. 17, Moscow, USSR.

ALEKSANDROV, Leonid Naumovitsh, Russian physicist; b. Dnepropetzovsk, Russia, Sept. 27, 1923; s. Naum L. and Veza (Mazkorna) A.; kandidats degrees, Dnepropetzovsk U., 1950, doctors degree, 1954; m. Julia Aleksandrova Melnik, Aug. 19, 1953; children—Svetlana, Andrej. Tchr., Pedagogical Inst., Slavjansk, 1950-52, mgr. dept. physics, 1953-58; mgr. dept. phys. status solid, also prof. Mozdovian U., 1958-65; mgr. phys. lab. Inst. Light Sources, 1959-65; mgr. lab. film physics Inst. Phys. Semiconductors, also prof. Novosibizsk U., 1965—; cons. in sci. Mem. United Sci. Council Light Engring., United Sci. Council Gen. Applied Physics, United Sci. Council Elec. Engring., Soc. Energetics, Soc. Science; service mem. Tsheznov; lectr. on theory of metals. Decorated 3 medals. Author: Internat. Fraction of Refractory Metals, 1965, also numerous articles. Research on spectrum of metals; ascertained the existence of critical size of new phase crystals in condensed systems with grating transition or limited diffusional mechanisms. Home: 25 Academitzezkaj, Novosibizsk 72. Office: Inst. Physics Semiconductors, Novosibizsk 90, USSR.*

ALEKSANDROV, Pavel Sergeievich, Russian mathematician; b. Bogorodsk (now Noginsk, Moscow Oblast), May 7, 1896; graduated from Moscow Univ., 1917; Doctor of Physico-Mathematical Science, 1934. With Smolensk, then Chernigov Department Public Education, 1917-20; university reader in literature; lecturer at Moscow U., 1921-29, prof., 1929—. Dean, Soviet Sch. Topology. Recipient Stalin prize, 1943, Order of Lenin. Mem. USSR, Polish, Berlin (corr.) acads. sci., Göttingen Acad., U. S. Nat. Acad. Sci. (fgn.), Am. Philos. Soc., Moscow Math. Soc. (hon., pres.). Author over 300 sci. works including Basic Theorems of Duality for Unclosed Sets of n-Dimensional Space, 1947; The Topological Concept of Space, 1947; The Homeomorphism of Point Sets, 1955; co-author An International Symposium on General Topology, 1961; Completely Regular Spaces and Bicompact Expansions Thereof, 1962. Research on point set theory, theory of functions of real variables, homological dimension, duality. Address: Gosudarstvenny University, Leninskie gory, Moscow, USSR.

ALEKSANDROV, Vasilii Geógievich, Russian botanist; b. June 5, 1887; studied under V. I. Palladin. Staff, Bot. Gardens, also U. Tiflis, 1915-27; prof. Tiflis U., 1920—; prof. Tomsk U., 1927-29; staff All-Union Plant Inst., Leningrad, 1929-42; staff Bot. Inst., USSR Acad. Sci., 1942—. Author: Anatomy of Plants, 1954. Research and numerous publs. on plants, especially physiology, anatomy and morphology.

ALEKSANDROW, Dymitr, Polish physician; b. Piotrków, Poland, Aug. 24, 1909; s. Constantine and Natalie (Bazylewska) A.; grad. U. Warsaw (Poland), Med. Faculty, 1935, M.D., 1945, Habilit., 1950; m. Wanda Wysznacka, 1948; 1 son, Peter. Faculty, U. Warsaw, 1950—, prof. medicine, 1964—; dir. 2d Med. Clinic, Acad. Medicine, Warsaw, 1954-64; dir. 2d Med. Clinic, Warsaw Acad. Medicine, 1964—. Mem. com. experts on heart diseases WHO, 1960—. Decorated Polonia Restituta. Fellow Internat. Soc. Cardiology, Am. Coll. Cardiology, Am. Coll. Chest Physicians; mem. Polish Acad. Scis., Internat., Polish socs. internal medicine. Author: (with W. Wysznacka) Diagnosis of Heart Diseases, 1963, Emergency Care in Internal Diseases, 1967; Clinical Electrocardiography, 1963; also articles. Research on pathogenesis of atherosclerosis, nutritional factors, disturbances in lipid metabolism; clin. studies on differential diagnosis of arterial hypertension, epidemiology of coronary heart disease and hypertension in Poland. Home: Wilcza 35/41 m 18, Warsaw, Poland.*

ALEKSANDROWICZ, Julian, Polish physician; b. Krakow, Poland, Aug. 20, 1908; s. Józef and Zofia (Schneid) A.; M.D., Jagiellonian U., Kraków, 1934; m. Maria Tislowitz, Nov. 4, 1935; 1 son, Georg W. Faculty, Jagiellonian U., 1947—, prof., dir. III Clinic Med. Acad., 1950—. Recipient Sci. State prize, 1948, Virtuti, Mil. Cross, 1944, Golden Badge Red Cross, 1960. Mem. Polish Soc. Hematology (an organizer, past v.p.), Cracow Mental Health Soc. (organizer, past pres.), Polish Acad. Scis. (organizer, chmn. sect. comparative leukemia research 1965—), Internal Soc. Hematology, Folia Humanistica, Folia Clinica Internat. Author: Diseases of Blood Forming Organs in the Light of Bioptical Investigations, 1946; Handbook of Hematology, 1950; Electron Microscopy of Blood Cells, 1955; The Leukemias, 1963; also numerous articles. Research on etiopathogenesis of leukemias. Discovered phenomenon of disturbed RNA, metabolism in leukemia, method of treatment and rehab. of resistant diseases of nervous system, including neuritis retrobulbaris, nervi optici, with nitrogen mustard derivatives; orgn. of social and preventive hematology dept. Home: 4 Curie-Skodowskiej, Kraków, Poland.*

ALEKSANYAN, Arto Bogdanovich, Russian hygienist, epidemiologist; b. 1892; grad. Med. Faculty, Novorossiysk U., 1916; D.Med. Sci. Head, 1st Sanitation and Hygiene Lab., Armenian Research Inst. Microbiology and Epidemiology, Yerevan, 1927-35; head chair epidemiology Yerevan Med. Inst., 1935—; chief state san. insp. Armenian Peoples Commissariat of Health, 1935-40; chmn. learned med. council Armenian Ministry Health; hon. sci. worker Armenian SSR, 1960—. Mem. USSR Acad. Med. Sci., All-Union (Presidium mem.), Armenian (chmn.) socs. microbiologists, epidemiologists and infectionists. Author numerous works including The Problem of Diphtheria, 1944; Criticism of the Present Method of Immunization against Diphtheria in the Light of Immunobiological and Epidemiological Observations, 1948; The Organization and Development of Epidemiology in Armenia in the Past Thirty Years 1920-50, 1953. Co-editor: Epidemiology and Infectious Diseases sect. Large Med. Ency., 2d edit; mem. editorial council Jour. Microbiology, Epidemiology and Immunobiology. Research on Armenian regional pathology, epidemiology, immunity and prophylaxis of diphtheria and dysentery. Address: Yerevan Med. Inst., ul. Kirova 2, Yerevan, Armenian SSR, USSR.

ALEKSEEV, Aleksandr Yemelyanovich, Russian electrical engineer; born Nov. 27, 1891; graduated from Leningrad Electrical Engineering Institute, 1925. With St. Petersburg Elektrik Plant, 1908-19, Leningrad Elektrosila Plant, 1924-33; professor Leningrad Institute of Rail Transportation Engineering, 1936—; asso. Inst. Electromechanics, USSR Acad. Sci., 1953—. Recipient Order of Lenin, Order Red Banner of Labor, Stalin prize, 1949. Mem. USSR Acad. Sci. (corr.). Co-author: Turbogenerators, 1939; Author: The Design of Electrical Machines, 1949; Electric Traction Motors, 1951; The Locomotive of the Seven-Year Plan, 1959; Requirements and Ways of Coupling the Traction Motors of Single-Phase AC Electric Locomotives and Diesel Locomotives, 1963; Electric Traction Machines and Converters, 1967. Research on theory and methods of ventilation, heat calculation of electric machines. Address: Leningrad Inst. Rail Transp. Engring., Moskovsky prosp. 9, Leningrad, USSR.*

ALEKSEEV, Vladimir Feodorovich, Russian chemist; b. Oct. 12, 1852; s. Feodor Alekseev; developed rule of straight line diameter which established graphically critical temperature of 2 dissolving liquids, 1885. Died Sept. 12, 1919.

ALEKSEIEVSKII, Nikolai Evgenievich, Russian physicist; b. 1912; grad. Leningrad Poly. Inst., 1936. With Ukraine Phys.-Tech. Inst., Kharkov, USSR, 1936-41; X-ray technician Hosp. Evacuation Centers, 1941-42; later lectr. physics dept. Med. Inst., Stalingrad, USSR Acad. Scis.; staff Inst. Phys. Problems, USSR Acad. Scis., 1942—; with dept. physics low temperatures Physics Faculty, Moscow State U., 1947-60, prof., 1950-60; head dept. exptl. physics Moscow Physico-Technol. Inst., 1960—. Corr. mem. USSR Acad. Scis. Research and publs. on physics of low temperatures. Home: Vorob'evskoye Shosse, 2. Office: Exptl. Physics Dept., Moscow Physico-Technol. Inst., Moscow, USSR.

ALENIO, Giulio, mathematician; b. Brescia, Italy, 1582; mem. Soc. of Jesus; missionary in China, 1610-44. Author: (in Chinese) Life of Jesus Christ, Of the Sacrifice of Mass, The Origin of the World, The Dialogue of St. Bernard, The Theater of the World; account of 1612 eclipse observed at Macao. Died Fou-Tcheou, China, 1644.

ALERS, Perry Baldwin, Am. physicist; b. Bisbee, Ariz., Mar. 24, 1926; s. Perry Albert and Helen (Baldwin) A.; B.A., Rice U., 1948, M.A., 1950; Ph.D., U. Md., 1955; m. Flora Ann Campbell, June 11, 1955; children—Perry C., Ellen V., Paul E. With U. S. Naval Research Lab., Washington, 1950—, research physicist 1955—, head crystal br., solid state div., 1966—; vis. scientist Brigham Young U., Provo, Utah, 1964-65. Fellow Am. Phys. Soc.; mem. Research Soc. Am. Research, publs. in low temperature physics, metal physics, magneto-optics, paramagnetism, high pressure physics, nuclear radiation, detector work at Hardtack weapons test. Home: 2914 W St. S.E., Washington 20020. Office: Code 6430, U. S. Naval Research Lab., Washington 20390.*

ALERS-MONTALVO, Manuel, sociologist; b. San Sebastian, P.R., Dec. 8, 1920; s. Jose Bernabe and Antonia (Montalvo) A.; B.S., Coll. City N.Y., 1941; M.A., Columbia, 1951; Ph.D., Mich. State U., 1953; m. Carmen M. Mora Cheves, Dec. 21, 1954; children—Carlos, Sonia, Jorge, Roberto. Sociologist, Inter-Am. Inst. Agrl. Scis., Lima, Peru, 1953-55, prof., head dept. econs. and social scis., Turrialba, Costa Rica, 1955-62; prof. sociology Colo. State U., Ft. Collins, 1963—, chmn. Center for Latin Am. Studies, 1966—; vis. prof. U. P.R., 1959, 61, 63, 66; cons. sociologist. John Hay Whitney Found. fellow, Peru, 1954-55. Mem. Am., Rural sociol. socs., Soc. Applied Anthropology, Latin Am. Studies Assn. Author: Sociologia: Introduccion a su uso en programas agricolas rurales, 1960; Pucara: un estudio de cambio, 1967; also articles. Research on human factors in tech. devel. in Latin Am. countries. Home: 1504 Pitkin St., Ft. Collins, Colo. 80521.*

ALESSIO (Alexis of Piedmont), med. writer. Author most famous Renaissance book of secrets, de Secretis, 1st part of which printed in Italian, 1556, 69 edits., up to 1691, translated into French, Latin, English, Dutch, German, Polish, Danish, Spanish (most receipts med., pharm. in content); nothing known of author although attempts made to identify him with alchemist, Girolamo Ruscelli of Viterbo.

ALEXANDER, Boyd, zoologist, explorer; b. Cranbrook, Eng., Jan. 16, 1873; s. Boyd Francis and Mary Wilson (Boyd) A.; student Radley Coll., 1887-91. Entered 7th Bn. Rifle Brigade, 1893, became capt., 1898, retired as lt., 1907; leader sci. expdn. to Cape de Verde Islands, 1897; explored Zambesi and Kafuc rivers, 1898-99; apptd. to Gold Coast Constabulary, also participated in relief of Kumasi, 1900; leader sci. expdn. to Fernando Po (successfully ascended Mt. St. Isabel, discovered many new birds), 1904; leader Alexander-Gosling Expdn. across Africa from Niger to Nile, 1904-07. Recipient Gold medal Royal Geog. Soc. (Antwerp, Belgium), 1907. Fellow Royal Geog. Soc. (Gold medal 1908), Royal Scottish Geog. Soc. (hon.), Zool. Soc. Author: From the Niger to the Nile, 2 vols. 1907; contbd. Birds of Kent to Victoria History of England. Made various zool. studies. Died Nyeri, Wadau, May 3, 1910.

ALEXANDER, Charles Paul, Am. entomologist; b. Gloversville, N.Y., Sept. 25, 1889; s. Emil and Jane (Parker) A.; B.S., Cornell U., 1913, Ph.D., 1918; D.Sc., U. Mass., 1959; m. Mabel Marguerite Miller, Nov. 10, 1917. Faculty, U. Mass., 1922—, prof., dept. head, 1930-59, dean Sch. Sci., 1945-52; prof. U. Cal., Berkeley, 1964. Decorated Comdr. Order Al Merito Bernardo O'Higgins, Chile, 1952. Fellow Royal Entomol. Soc. (London); mem. Entomol. Soc. Am. (past pres.), Am. Entomol. Soc., Nat. Pest Control Assn. Contbr. numerous articles to profl. jours. Research, publs. on crane flies of the world, described more than 9500 new species of these flies. Home: 39 Old Town Rd., Amherst, Mass. 01002.*

ALEXANDER, Edward Gordon, Am. biologist, b. Rich Hill, Mo., Aug. 18, 1901; s. James Edward and Mary (Gordon) A.; A.B., Central Methodist Coll., 1923; M.A., Princeton, 1925, Ph.D., 1931; m. Marion Frances Isely, Sept. 2, 1926; children—Mary Anne (Mrs. Richard L. Bingham), Douglas Gordon. Faculty, U. Colo., Boulder, 1931—, prof. biology, 1939-66, head dept. biology, 1939-58, now prof. emeritus. Fellow A.A.A.S.; mem. Am. Soc. Zoologists, Ecol. Soc. Am., numerous others. Author: Biology, 1935; General Zoology 1940; General Biology, 1956. Contbr. numerous articles to profl. jours. Research in role of altitude in distbn. of insects in mountains. Home: 765 14th St., Boulder, Colo. 80302.*

ALEXANDER, Edward Lawson, Am. chemist; b. Lewiston, Me., Oct. 22, 1925; s. Irving Edward and Sarah (Drew) A.; B.S., U. Me., 1950, M.S., 1951; Ph.D. in Phys. Chemistry, Vanderbilt U., 1955; m. Phyllis Mary Maxwell, Aug. 20, 1949; children—Bruce E., Steven M., Jeffrey D., Beth L. Research asso. in nuclear and radiation chemistry Knolls Atomic Power Lab., Gen. Electric Co., Brookhaven Nat. Lab., 1955-57; asso. prof., asst. dir. reactor project, radiation safety officer Ga. Inst. Tech., Atlanta, 1957-58; mgr. radiol. scis., indsl. reactor labs. Columbia, Plainsboro, N.J., 1958-62; prof., dir. radiation sci. center Rutgers U., New Brunswick, N.J., 1962—. Chmn., Conf. on Radiol. Health, 1964-65. Mem. Health Physics Soc. (pres. Greater N.Y. chpt. 1962), Am. Chem. Soc., Am. Nuclear Soc., Am. Assn. U. Profs. Home: 152 Clover Lane, Princeton, N.J. 08540.*

ALEXANDER, Franz, psychoanalyst, psychiatrist; b. Budapest, Hungary, Jan. 22, 1891; s. Bernard and Regina (Brossler) A.; B.A., Humanistic Gymnasium, Budapest, 1908; M.D., U. Budapest, 1913; m. Anita Venier, Mar. 7, 1921; children—Sylvia, Kiki (Francesca). Came to U. S., 1930, naturalized 1938. Research asso. in physiology Inst. for Exptl. Pathology, U. Budapest, 1910-13, research asso. in bacteriology Inst. for Hygiene, 1913-14; head bacteriol. field lab. and malaria sta. Austro-Hungarian Army 1914-18; research and clin. asso. in psychiatry and neurology Neuropsychiatric Clinic, U. Budapest, 1918-19; postgrad. work Psychiatric Hosp. (Charité), U. Berlin, 1920-21; clin. asso., lectr. in psychoanalysis, Inst. for Psychoanalysis, Berlin, 1921-30; vis. prof. psychoanalysis U. Chgo., 1930-31; research asso. in criminology, Judge Baker Found., Boston, 1931-32; teacher and research work and dir. Chgo. Inst. for Psychoanalysis from 1932; asso. prof. psychiatry U. Ill., 1938-43, prof. psychiatry, 1943-56; attending physician Cook County Psychopathic Hosp., 1938-56; now clin. prof. psychology U. So. Cal., also chief of staff, psychiat. dept. Mt. Sinai Hosp., Los Angeles, also dir. psychiat. and psychosomatic research inst. Cons., Nat. Adv. Mental Health Council, 1947-49; chmn. sect. on psychotherapy, psychoanalysis and psychosomatic medicine Congrés Internat. de Psychiatrie, Paris, 1950; mem. com. on rheumatic diseases NRC, 1948—. Ford Found. fellow Center for Advanced Study in Behavioral Scis., Stanford, Cal., 1955. Recipient Samuel Ruben award for outstanding achievement in mental health, 1958; Semmelweis medal for basic and lasting contbns. in field exptl. and clin. psychiatry Am.-Hungarian Med. Assn., 1959. Author numerous books including: Psychosomatic Medicine, Fundamentals of Psychoanalysis, 1950. Co-editor: Dynamic Psychiatry, 1952; also Psychosomatic Medicine. Pub: Psychoanalysis and Psychotherapy, 1956. Contbr. articles med. jours. known for psychoanalytic research on personality devel., psychoneuroses and psychoses, criminology, social psychology; work in psychosomatic medicine includes studies of psychogenic theory of peptic ulcers, influence of emotional factors on gastrointestinal disorders, essential hypertension, bronchial asthma, thyrotoxicosis and rheumatoid arthritis; advanced theory of specificity of emotional factors in different organic diseases. Home: 1011 Cielo Dr., Palm Springs, Cal. Office: Mt. Sinai Hosp., 8720 Beverly Blvd., Los Angeles 48.

ALEXANDER, Frederick William, Brit. physician; b. Feb. 16, 1859; ed. Univ. Coll., London, also St. Bartholomew's Hosp., London; m. Diana Constance Pyke, 1889; 1 dau. Practiced medicine, specializing in elec. and light treatment; dep. med. officer Croydon Infirmary and Workhouse and Infectious Wards, 1881-83; surgeon Royal Mail Steamer, Castle Line: resident asst. med. officer Mile End Old Town Infirmary and Workhouse and Schs.; western dist. med. officer and pub. vaccinatory, Mile End Old Town; med. officer health Met. Borough Poplar. Diplomate in Pub. Health, Royal Coll. Physicians and Surgeons, Eng.; licentiate Royal Coll. Physicians Edinburgh; licentiate in midwifery. Mem. Royal Coll. Surgeons. Author: Hints regarding the Management of Infants; Reports upon Condition of Swimming Baths and Water used therein of Electrolytic Disinfecting Fluid; Crusade against Consumption; A Possible Cause and Prevention of Malignant Disease; Vitamins accessory Food Factors: What are they?; Salt and Bread; Lack of Iodine; Two Reports upon Ultraviolet Rays; Twenty Years' Working of the Electrolytic Fluid and its use in Swimming Baths; papers in various med. jours. Died Mar. 14, 1937.

ALEXANDER, Gustav, Austrian physician, otologist; b. Vienna, Austria, Dec. 18, 1873; M.D., U. Vienna, 1898. Chief dept. otology Vienna Polyclinic, prof. otology U. Vienna. Contbr. 200 publs. on anatomy, embryology, physiology of hearing organs, path. anatomy, clin. otology. Described new tech. method for study of descriptive and topog. anatomy of the labyrinth, preparation of celloidin sects. and wax-plate reconstrn. (facilitating research on anatomy of ear). First to state that primary atrophy of neuro-epithelium of nerve-ganglion apparatus is factor in producing profound deafness; suggested operation for thrombosis of lateral sinus in making a jugular skin fistula; described sign for hereditary syphilis; pioneer orgn. kindergarten classes for early tng. congenitally deaf children. Killed by insane patient, Apr. 12, 1932.

ALEXANDER, Harry Louis, Am. physician; b. N.Y.C., Feb. 16, 1887; s. Henry and Minnie (Alexander) A.; B.A., Williams Coll., 1910, D.Sc., 1957; M.D., Columbia, 1914; m. Janet Holmes, Dec. 23, 1929; children—Beverly, Lewis, Janet (Mrs. Richard P. Wood), Andrew. Instr., Cornell U. Med. Sch., 1919-24; faculty Washington U. Med. Sch., St. Louis, 1924—, prof. clin. medicine, 1942-52, prof. emeritus, 1952—, Harry Alexander Vis. prof. Mem. central com. consultants VA, 1951-54; mem. adv. bd. Nat. Inst. Allergy and Infectious Disease, NIH; 1957-61. Author: Bronchial Asthma, 1928; Reactions with Drug Therapy, 1955; also numerous articles. Asso. editor Immunological Diseases, 1965; editor Jour. Allergy, 1929-49. Research in allergy and internal medicine. Home: 7316 Pershing Av., St. Louis 63130. Office: 52 Maryland Plaza, St. Louis 63112.*

ALEXANDER, Howard Wright, mathematician; b. Toronto, Can., June 19, 1911; s. John Arthur and Alice (Wright) A.; B.A., U. Toronto, 1933, M.A., 1934; Ph.D., Princeton, 1939; m. Mary Alice Nace, Jan. 25, 1942; children—John Edwin, Margaret Alice, George Arthur, Rachel Ann, Jane Elizabeth. Instr. math. Lehigh U., 1937-40, Fenn. Coll., Cleve., 1940;41; staff Soil Conservation Research Sta., Coshocton, O., 1942-44, Lab. Physiol. Hygiene, 1944-46; prof. Math. Adrian (Mich.) Coll., 1946-52; Earlham Coll., Richmond, Ind., 1952—. Pres., Mary E. Hill Nursing Home, Richmond, 1958—. Mem. Math. Assn. Am., Am. Statis. Assn., Soc. for Social Responsibility in Sci., Fellowship of Reconciliation. Author: Elements of Mathematical Statistics, 1962. Home: 111 S.W. G St. Office: Earlham Coll., Richmond, Ind.*

ALEXANDER, Irving E., Am. psychologist; b. N.Y.C., May 16, 1922; s. Alex and May (Nisenson) A.; B.A., U. Ala., 1946, M.A., 1947; A.M., Princeton, 1948, Ph.D., 1949; m. Miriam Pearl Fisher, May 2, 1944; children—Stephen F., David J., Kay Robin. Asst. prof. Princeton, 1949-58; psychologist, chief psychology sect., tng. br. Nat. Inst. Mental Health, Bethesda, Md., 1958-63; prof., dir. clin. tng., chmn. dept. psychology Duke, Durham, N.C., 1963—; vis. prof. clin. psychology dept. social relations Harvard, 1966-67. Cons. VA, Washington, Atlanta, Durham, Halifax County (N.C.) Health Center. Recipient Grover Cleveland Hall award U. Ala., 1943. Gordon McDonald fellow Princeton, 1947-48, James Theodore Walker fellow, 1948-49; Bollingen fellow Jung Inst., Zurich, Switzerland, 1954-55. Mem. Am., Southeastern psychol. assns., A.A.A.S., Am. Assn. U. Profs., Sigma Xi. Author: Psychopathology: A Source Book, 1958. Research on effects of loud sound on structure and function of the ear, mortality as a factor in human functioning, relationship of Jung's theory of types to human behavior. Home: 1111 Watts St., Durham, N.C. 27701.*

ALEXANDER, James Waddell, Am. mathematician; b. Sea Bright, N.J., Sept. 19, 1888; s. John White and Elizabeth (Alexander) A.; B.S., Princeton, 1910, A.M., 1911, Ph.D., 1915, D.Sc. (hon.), 1947; post-grad. U. Paris, U. Bologna; m. Natalia Levitskaya, Jan. 11, 1918; children—Irina (Mrs. Walter W. Reed), John. Mem. faculty Princeton U., 1911-33, prof. math., 1928-33, prof. Inst. for Advanced Study, 1933-51; Rouse Ball lectr. Cambridge (Eng.) U., 1936. Mem. Nat. Acad. Sci., Am. Philos. Soc., Am. Math. Soc. (Bôcher prize 1929), Math. Assn. Am. Research in topology. Home: 29 Cleveland Lane, Princeton, N.J. 08540.*

ALEXANDER, Jerome, Am. chemist; b. N.Y.C., Dec. 21, 1876; s. Isaac and Annie Josephine Lewis (Jackson) A.; B.S., Coll. City N.Y., 1896, M.Sc., 1899; m. Gertrude Eleanor Hammerslough, Apr. 9, 1903 (dec. Dec. 1945); children—Eleanor Gertrude Jackson, Dorothy Alexander Livingston. Cons. chemist and chem. engr.; chmn. spl. com. on colloids NRC. Fellow A.A.A.S., Am. Inst. Chemists (hon. mem.); mem. Am. Inst. Chem. Engrs. (hon.), Am. Chem. Soc., Assn. Cons. Chemists and Chem. Engrs., Soc. de Chimie Industrielle (hon. sec. Am. sect.; hon. mem. parent société), Am. Genetics Assn., Phi Beta Kappa Assos., (dir.) Phi Beta Kappa (pres. N.Y. Alumni), Pi Gamma Mu, Pi Lambda Phi, Tau Beta Pi (hon.). Decorated Officer de l'Instruction Publique (France), 1931, also Chevalier de la Légion d'Honneur, 1936; Townsend Harris Medal, 1943. Author: Colloid Chemistry, 1919, and subsequent edits., Glue and Gelatin (Am. Chem. Soc. monograph), 1923; chpts. in Rogers' Industrial Chemistry, 1942; Allen's Commercial Organic Analysis, 1913 and 1931, Liddell's Handbook for Chemical Engineers, 1920; Colloid Chemistry, Theoretical and Applied (in collaboration with about 450 men of all nations). Volume I, 1926, II, 1928, III, 1931, IV, 1932, V, 1944, VI, 1946, VII, 1950; (verses) Essences from Life's Alembic and Retorts from a Chemist's Laboratory, 1941; Tribute to Gertrude, 1946; Life—Its Nature and Origin, 1948 Translator: Colloids and the Ultramicroscope (Richard Zsigmondy), 1909. Pioneer worker with the ultramicroscope in America; specialist in colloid chem. and its scientific and indsl. applications. Died Jan. 18, 1959.

ALEXANDER, John, Am. surgeon; b. Phila., Feb. 24, 1891; s. Lucien Hugh and Mazie (Just) A.; B.S., U. Pa., 1912, M.A., 1913, M.D., 1916 (hon. Sc.D., 1940); m. Emma Ward Woolfolk, July 11, 1936. Mem. surg. and teaching staff U. Pa.; prof. surgery, U. Mich.; surgeon in charge sect. thoracic surg., U. Mich. Hosp.; chief surg., Mich. State Sanatorium; cons. thoracic surgery, various sanatoria and hosps. Recipient Samuel D. Gross prize Phila. Acad. Surgery, 1925; Henry Russel award, U. Mich., 1930; Trudeau medal, Nat. Tb Assn., 1941. Henry Russel lectureship, U. Mich., 1944. Fellow A.C.S.; mem. Am. Surg. Assn., Soc. of Thoracic Surgery Gt. Britain and Ireland (hon.), Sociedad de Argentina de Cirujanos (hon.); Société Belge de Chirurgie (hon.); Sociedad Paraguaya de Tisiologia (hon.); Sociedad de Tisiologia de Cordoba, Argentina (corr.); Tb Assn. of India (corr.); Am. Trudeau Soc. (pres.), Am. Bd. Surgery (founders' group), Soc. Clin. Surg., Internat. Soc. Surgery, Am. Assn. for Thoracic Surg. (pres. 1935), Central Surg. Assn. (founders' group), A.M.A., Nat. Tb Assn. (dir. 1941-44, v.p. 1945), Alpha Omega Alpha, Alpha Mu Pi Omega, Nu Sigma Nu (hon.), Sigma Xi, Phi Kappa Phi, Delta Tau Delta. Author: The Surgery of Pulmonary Tuberculosis, 1925; The Collapse Therapy of Pulmonary Tuberculosis, 1937; also articles on thoracic surgery. Mem. Adv. editorial bd. Jour. Thoracic Surgery. Died July 16, 1954.

ALEXANDER, Leo, psychiatrist; b. Vienna, Austria, Oct. 11, 1905; s. Gustav and Giesela (Rubel-Schaefer) A.; A.B., Piaristengymnasium, Austria, 1923; M.D., Vienna, 1929; m. Phyllis Harrington, Apr. 22, 1936; 4 children. Instr. neurology Harvard, 1934-41; asso. prof. neuro-psychiatry, Duke U., 1941-46; asst. clin. prof. psychiatry Med. Sch. Tufts U., from 1946; co-dir. Chandler Hovey Unit Research and Treatment of Multiple Sclerosis, from 1961; asso. Boston State Hosp., 1935-41, 46-48, dir. neurobiol. unit, 1948; cons. U. S. sec. war, Nuremburg War Crimes Trial, Germany, 1946-47. Mem. Psychiat. Assn., Neurol. Assn., Neuropath. Assn., Harvey Cushing Soc., Soc. Clin. and Exptl. Hypnosis, also others. Author: Treatment of Mental Disorders, 1935; Objective Approaches to Treatment in Psychiatry, 1958; also numerous articles. Research on psychophysiology, conditioned reflexes, endocrine effects on central nervous system, prognosis and treatment of multiple sclerosis.

ALEXANDER, Lloyd George, Am. chem. engr.; b. Lynn, Ind., Feb. 2, 1919, s. Pierre Logan and Minnie (Humphrey) A.; B.Ch.E., Purdue U., 1941, Ph.D., 1948; m. Marion Howard, June 2, 1942; children—P. Martin, J. Kimble. Engr., Dixie Ordnance, Monroe, La., 1942-43, Pine Bluff Arsenal, Pine Bluff, Ark., 1943-44; fellow dept. chemistry Purdue U., 1944-48; asst. prof. chem. engring. U. Ill., 1948-50; asst. prof. chem. engring. U. Okla., 1950-52; sr. devel. engr. Oak Ridge Nat. Lab., 1952—. Mem. Am. Nuclear Soc., Am. Inst. Chem. Engrs., Sci. Research Soc. Am. Research, publs. on transport in free turbulent flow; molten-salt nuclear power reactor design; nuclear desalination of sea water. Patentee in field oxygen of induced nitration of paraffins. Home: 110 Oklahoma Av. Office: Oak Ridge Nat. Lab., Oak Ridge 37380.*

ALEXANDER, Lyle Thomas, Am. soil scientist; b. Athens, Tex., Dec. 3, 1903; s. James C. and Maude (Dalton) A.; B.S., U. Ark., 1928, LL.D., 1958; Ph.D., U. Md., 1935; m. Helen Goodwin, June 7, 1927; children—Thomas Goodwin, Jennie Lyle (Mrs. Joseph F. Hodgson), Alice Maude (Mrs. John W. Huffman), Marth Sue (Mrs. David H. Bowman). Soil scientist U. S. Dept. Agr., 1928—, chief soil survey labs. Soil Conservation Service, Beltsville, Md., 1954—; lectr. on tropical soils Cornell U., Ithaca, N.Y., 1957. Recipient Distinguished Service award U. S. Dept. Agr., 1956; Career Service award Nat. Civil Service League, 1959. Fellow A.A.A.S., Am. Soc. Agronomy; mem. Am. Chem. Soc., Soil Sci. Soc. Am., Sigma Xi. Research, publs. in soil chemistry, soil mineralogy, soil classification, distbn. of fission products of weapons testing. Home: 5308 42d Av., Hyattsville, Md. 20781. Office: Soil Survey Labs., Soil Conservation Service, Plant Industry Sta., Beltsville, Md. 20705.*

ALEXANDER, Martin, Am. microbiologist; b. Newark, Feb. 4, 1930; s. Meyer and Sarah (Rubinstein) A.; B.S., Rutgers U., 1951; M.S., U. Wis., 1953, Ph.D., 1955; m. Renee R. Wulf, Aug. 26, 1951; children—Miriam, Helen, Stanley William. Faculty, Cornell U., Ithaca, N.Y., 1955—, prof., 1966—; vis. prof. Hebrew U., Israel, 1961-62. Pres. sci. adv. com. Environmental Pollution Panel; mem. UNESCO-ICRO Microbiology Panel; cons. Gen. Electric Co., United Fruit Co. Recipient Soil Sci. award, 1966. Mem. Am. Soc. Microbiology, Am. Acad. Microbiology, Soil Sci. Soc. Am., Phi Beta Kappa, Sigma Xi. Author: Introduction to Soil Microbiology, 1961. Research, numerous publs. primarily in field of soil microbiology, biochem. ecology. Home: 301 Winthrop Dr., Ithaca, N.Y. 14850.*

ALEXANDER, Mary Louise, Am. geneticist; b. Ennis, Tex., Jan. 15, 1926; d. Emmett Franklin and Florence (Hill) Alexander; B.A., U. Tex., 1947, M.A., 1949, Ph.D., 1951. Instr., research asst. Genetics Found., U. Tex., 1944-51; post-doctoral fellow AEC biology div. Oak Ridge Nat. Lab., 1951-52; post-doctoral research fellow U. Tex., Austin, 1952-55; research asso. M. D. Anderson Hosp. and Tumor Inst., U. Tex., 1956-58, asst. biologist, 1959-62, research scientist Genetics Found., U. Tex., 1962—; research participant Oak Ridge Inst. Nuclear Studies, 1951—. Mem. Radiation Research Soc., Genetics Soc. Am., Am. Soc. Human Genetics, Sigma Xi, Gamma Phi Beta, Phi Sigma. Research, publs. on genetic injuries from radiations and chems. Home: 3711 A Bridle Path, Austin 78703. Office: Genetics Found. U. Tex., Austin 78712.*

ALEXANDER, Peter, radiobiologist; b. Munich, Germany, Jan. 27, 1922; s. Elias and Kate (Lommer) A.; B.Sc., London (Eng.) U., 1941, Ph.D., 1943, D.Sc., 1956; m. June Pickwoad, Apr. 3, 1950; children—Adam, Maxine, Leone. Research asst. in war work, 1941-45; mgr. research dept. Wolsey Ltd., 1945-50; staff Chester Beatty Research Inst., Inst. Cancer Research, London, 1950—, head radiobiology dept., 1954—, since professor of radiobiology, since 1966—. Author: Fundamentals of Radiobiology, 1956-60; Atomic Radiation and Life, 1957-64; also numerous articles. Research on mechanism of biol. action of ionizing radiation. Home: Old Denshott, Leigh, nr. Reigate Sy., Eng. Office: Inst. Cancer Research, Belmont, Sutton Sy., Eng.*

ALEXANDER, Richard Charles, Scottish surgeon; b. Edinburgh, Scotland, Sept. 18, 1884; s. Richard and Martha (Wallace) A.; ed. George Watsons Coll., U. Edinburgh, U. Paris; M.A., M.B., Ch.B.; LL.D., St. Andrews; m. Marjorie Linda Morgan; children—Ian, Joan, David. House surgeon Royal Infirmary, 1909; with Chalmers Hosp., 1910-11, tutor surgery clinic, 1912-19 (both Edinburgh); asst. dept. surgery U. Edinburgh, 1912-19; surgery cons. Royal Infirmary, Dundee, 1921-51; prof. surgery U. St. Andrews, Dundee, 1936-51; dir. surgery Eastern Region Scotland, 1939-45. Pres. Forfarshire Med. Assn., 1958. Fellow Royal Coll. Surgeons (v.p. 1954-57). Author: Anuria, 1935; Renal Tuberculosis, 1936; Prostatic Enlargement, 1942. Address: 261 A Perth Road, Dundee, Scotland.

ALEXANDER, Richard Dale, Am. zoologist; b. White Heath, Ill., Nov. 18, 1929; s. Archie Dale and Elizabeth (Heath) A.; A.A., Blackburn Coll., 1948; B.S., Ill. State Normal U., 1950; M.S., Ohio State U., 1951, Ph.D., 1956; m. Lorraine Kearnes, Aug. 19, 1950; children—Susan Dale, Nancy Lorraine. Grad. research fellow Ohio State U., Columbus, 1954-56, research asso. Rockefeller Found., 1956-57; faculty U. Mich., Ann Arbor, 1957—, prof. zoology, 1966—. Fellow Ohio Acad. Sci., A.A.A.S. (Newcomb Cleveland prize 1962); mem. Soc. for Study Evolution (council 1965-67), Soc. Systematic Zoologists, Animal Behavior Soc., Sigma Xi. Contbr. articles to profl. jours. Bd. editors Animal Behaviour, 1962—, Evolution, 1961-63. Research in reproductive isolating mechanisms and acoustical communication in insects; evolution of mating behavior in arthropods; discovered numerous new species of crickets, katydids, cicadas, co-proposed new hypothesis for speciation by accidental temporal (instead of geographic) separation of populations and its application to certain insects. Home: 5530 Warren Rd., Ann Arbor, Mich. 48105.*

ALEXANDER, Robert Jackson, Am. economist; b. Canton, O., Nov. 26, 1918; s. Ralph S. and Ruth (Jackson) A.; B.A., Columbia U., 1940, M.A., 1941, Ph.D., 1950; m. Joan Powell, Mar. 26, 1949; children—Anthony Robert, Margaret Frances, Asst. economist Bd. Econ. Warfare, 1942, USAAF, 1942-45, Office Inter-Am. Affairs, 1945-46; faculty Rutgers U., New Brunswick, N.J., 1947—, prof. econs., 1961—. Vis. prof. various univs., 1949-64; cons. Latin Am. Affairs to govt. agys., AFL, 1948-59. Mem. Am. Econ. Assn., Am. Assn. U. Profs., Inter-Am. Assn. Democracy and Freedom, League for Indsl. Democracy. Author: The Peron Era, 1951; Communism in Latin America, 1957; Prophets of the Revolution, 1962; The Bolivian National Revolution, 1958; (with Charles O. Porter) The Struggle for Democracy in Latin America, 1961; Today's Latin America, 1962; Labor Relations in Argentina, Brazil and Chile, 1962; Primer of Economic Developement, 1962; The Venezuelan Democratic Revolution, 1964; Organized Labor in Latin America, 1965; Latin America, 1964; Latin American Politics and Government, 1965. Research and numerous publs. on impact of indsl. and modern polit. revolutions in Latin Am. Home: 944 River Rd., Piscataway, N.J. 08854. Office: Rutgers Univ., New Brunswick, N.J. 08903.*

ALEXANDER, Robert Spence, Am. physiologist; b. Melrose, Mass., June 14, 1917; s. William B. and Ethel (Sargent) A.; A.B., Amherst Coll., 1938, M.A., 1940; Ph.D., Princeton, 1942; m. Eleanor B. Paddock, Apr. 4, 1942; children—Joan P. (Mrs. David C. Fordham), Helen S., Carolyn L., Beverly R., Bruce E. Faculty, Med. Coll. Ga., 1953-55; prof., chmn. dept. physiology Albany (N.Y.) Med. Coll., Union U., 1955—; cons. physiologist Albany Med. Center Hosp., 1955—. Fellow A.A.A.S.; mem. Am. Physiol. Soc., Am. Heart Assn., Harvey Soc., Am. Inst. Biol. Scis., Assn. Am. Med. Colls., Sigma Xi. Asst. editor Circulation Research, 1952-54. Contbr. numerous articles to profl. jours. Analysis of brain centers controlling blood pressure, arterial pulse dynamics, hemodynamic factors in shock; analysis of role of venous musculature in regulation of circulation, mech. properties and contractile mechanisms of blood vessels. Home: 20 Forest Rd., Delmar, N.Y. 12405. Office: 47 New Scotland Av., Albany, N.Y. 12208.*

ALEXANDER, Stephen, Am. astronomer; b. Schenectady, N.Y., Sept. 1, 1806; s. Alexander and Maria A.; grad. Union Coll., 1824; attended Princeton Theol. Sem., 1832; LL.D., Columbia, 1852; m. Louisa Meads, Oct. 3, 1826, 3 children; m. 2d, Caroline Foreman, Jan. 2, 1850, 2 children. Tutor in math. Coll. of N.J. (now Princeton), 1832, prof. astronomy, 1840, prof. math., 1845-54, prof. astronomy, mechanics, 1855-78; head expdn. to observe solar eclipse, Labrador, 1860, chmn. com. to organize observation of solar eclipse at Ottumwa, Ia., 1869. An original mem. Nat. Acad. Scis., 1862; mem. Am. Philos. Soc., Am. Acad. Arts and Scis., A.A.A.S. (pres. 1859). Hypothesized that Milky Way has spiral structure, 1852; determined difference of longitude by fall of meteors, 1839. Died Princeton, N.J., June 25, 1883.

ALEXANDER OF ABONITEICHOS, thaumaturgist; flourished circa 3d century B.C.; cultist who had considerable influence during his time; irrationality and fear of divinities reflected in his work rather than an empirical science.

ALEXANDER OF APHRODISIAS, peripatetic philosopher; b. Aphrodisias, Caria; flourished 193-217 under Septimius Severa and Caracalla; head Lyceum, Athens, sometime between 198-211; tchr. of Simplicius; Author: On Fate (opposed the Stoic doctrine of necessity), On the Soul, and commentaries on Aristotle's Prior Analytics, Topics, De Sensu, Metaphysics, and Meteorologica. Noted commentator on Aristotle, tried to free Aristotelian doctrine from Neoplatonic and Stoic influence; wrote on fevers; acquainted with distillation of sea water.

ALEXANDER OF MYNDOS, biologist; flourished 1st century; founder literature on zool. marvels. Author: History of Birds; History of Animals (Menagé, others believe this is the Alexon Myndius to whom Diogenes attributes work of mythology).

ALEXANDER OF TRALLES, physician; b. Tralles, Lydia, 525; studied at Alexandria. After extensive travel, practiced and taught medicine, Rome. Author: The Art of Medicine, 12 books; De lumbricis; works on intestinal worms and their eradication; also on fevers. Possibly 1st parasitologist; one of best Greek physicians after Hippocrates; writings translated into Syriac, Arabic, Hebrew, Latin. Died 605.

ALEXANDRINOS, see Asclepiodotos.

ALEZZANDRINI, Arturo Alberto, Argentinian physician; b. Buenos Aires, Argentina, Nov. 24, 1932; s. Arturo and Blanca (Brauer) A.; M.D., Buenos Aires, U., 1958; m. Erminda Fernandez, Nov. 11, 1958; children—Arturo A., Mónica Patricia. Practice medicine specializing in ophthalmology, Buenos Aires; dir. Primer Hosp. Privado de Ojos; adscript in ophthalmology U. Buenos Aires; ophthalmic surgeon, aggregated ophthalmic physician Lagleyze Inst. Mem. Consejo Argentino de Oftalmología, Argentina (sec. 1965-67), French, Paragua (corr.) ophthalmol. socs., Argentina Plastic Surgery Soc. Research, publs. on ocular tumors, ocular plastic surgery, fluorescence angiography of retina. Home: José M. Estrada 2987, Sarandí, Buenos Aires, Argentina. Office: Rivadavia 1902, Capital Federal, Argentina.*

AL-FARISI (Muhammad ibn Abu Bakr al-Farisi), Persian mathematician, astronomer; flourished 2d half of 13th century, Yaman. Author: The Highest Understanding on the Secrets of the Science of the Spheres; Stairs of Burning Thought, probably after 1266; possibly author of Signs of the Universe from the Most Appropriate Properties (on magic squares).

ALFONSO X EL SABIO, see Alfonso X of Castile.

ALFONSO X OF CASTILE (or Alfonso X el Sabio, Alphonso X the Wise), Spanish monarch; b. Burgos, Spain, Nov. 23, 1221; s. Ferdinand III and Beatriz. King of Castile and Leon, 1252-84; established sch. for Muhammad al-Riquti; completed incorporation U. Salamanca, 1254; established Latin and Arabic Coll., Seville, 1254; continued campaign of his father against Moors, captured Cadiz, 1262; put forward as candidate for Holy Roman Emperor, renounced those claims, 1275. Author: Siete partidas, 1265; Libros del saber de astronomía, 5 vols.; Tabulae alphonsinae (best planetary tables of Middle Ages), 1252; sponsored transl. Old Testament by Jews of Toledo and 1st history of Spain. A founder Spanish sci.; a chief intermediary between Arabic and European knowledge; moon crater Alphonsus named in his honor. Died Seville, Arp. 4, 1284.

ALFORD, Leon Pratt, Am. engr., management scientist; b. Simsbury, Conn., Jan. 3, 1877; s. Emerson and Sarah Merriam (Pratt) A.; B.S. in E.E., Worcester Poly. Inst., 1896, M.E., 1905, Dr. of Engring., 1932; m. Grace A. Hutchins, Jan. 1, 1900; 1 son, Ralph I. Shop foreman McKay Metallic Fastening Assn., Boston, 1896-97, McKay-Bigelow Heeling Assn., 1897-99; prodn. supt., McKay Shoe Machinery Co., 1899-1902; mech. engr. United Shoe Machinery Co., Boston, 1902-07; engring. editor Am. Machinist, 1907-11, editor in chief, 1911-17; editor Industrial Management, 1917-20, Management Engineering, New York, 1921-23, Mfg. Industries, 1923-28; v.p. Ronald Press Co., 1928-34; asst. engr. in charge mfg. costs, Fed. Communications Commn., 1935-37; prof. administrative engring., N.Y. Univ., 1937—. Past v.p. and mem. research coms. Am. Engring. Council. Melville Gold Medalist, 1927; Gantt Gold Medalist, 1931. Author: Bearings and Their Lubrication, 1912; Laws of Management, 1928; Life of Henry Laurence Gantt, 1934; The Principles of Industrial Management for Engineers, 1940. Editor: Artillery and Artillery Ammunition, 1917; Management's Handbook, 1924; Cost and Production Handbook, 1934. Pioneer of handbooks for management; interpreted and transmitted work of 1st management scientists. Died New York, N.Y., Jan. 2, 1942.

ALFRED OF SARASHEL, translator, sci. writer; flourished 1185-1217; translated: Kitab al-Sifa (Ibn Sina) into Latin, also a pseudo-Aristotelian work De plantis (attributed to Nikolas Damaskenos), other Arabic works on alchemy and sci.; wrote commentaries on Meteorology, Parva naturalia (both by Aristotle), author original treatises influenced by Arabic sci., also De motu cordis, 1210.

AL-GAFIQI (al-Ghafiqi) (Abu Ja'fer Ahmad ibn Muhammad al-Ghagiqi), Spanish physician, botanist; b. Ghafiq, Spain; flourished 12th century. Author: On Simples; also treatise on eye ailments. Gathered and described plants from Spain and Africa; described yellow amber, sal ammoniac. Died 1165.

AL-GARDIZI, geographer; flourished 1050; supposedly studied under al-Biruni. Author: Beauty of Information; wrote about Greek sci., ethnographic differences.

AL-GASSANI, biologist; b. Spain; flourished 16th century; physician practicing in Fas, Morocco. Author: Garden of the Flowers as an Explanation of the Nature of Herbs and Drugs; Essay on Botanical Classification.

AL-GAWZI (Abu al-Farg) ('Abd al-Rahman ben 'Ali Muhammad ibn al-Gawzi) (Gamal al-din ibn al-Gawzi), Arabic sci. writer; b. Bagdad, circa 1115; s. Nast Allah. Author: Book of the Quintessence of Discourse on the Interpretation of Dreams; also treatises on gen. medicine, history, spiritual medicine, autobiography. Died Bagdad, 1201.

AL-GAZARI, Arabic physicist; flourished 1205. Author: Book about the Knowledge of Geometric Shapes; also numerous other works, many with collaborators. Credited as Arab with most knowledge of Greek mechanics.

ALGER, Chadwick Fairfax, Am. polit. scientist; b. Chambersburg, Pa., Oct. 9, 1924; s. Herbert and Thelma (Drawbaugh) A.; B.A., Ursinus Coll., 1949; M.A., Johns Hopkins, 1950; Ph.D., Princeton, 1958; m. Elinor Reynolds, Aug. 28, 1948; children—Mark, Scott, Laura, Craig. Internat. relations specialist Dept. Navy, 1950-54; instr. Swarthmore Coll., 1957; faculty Northwestern U., Evanston, Ill., 1958—, prof. polit. sci., 1966—; vis. prof. UN affairs N.Y. U., 1962-63. Mem. Am. Polit. Sci. Assn., A.A.A.S., Internat. Studies Assn., Midwest Conf. Polit. Scis. (recipient prize 1966). Author: (with others) Simulation in International Relations, 1963; also articles. Asso. editor Jour. Conflict Resolution, 1967. Research on growth, devel., impact, consensus devel. in internat. orgns.; devel. field research techniques for collecting data on internat. orgns. Home: 2510 Simpson St., Evanston, Ill. 60201.*

ALGER, Cyrus, Am. iron-master, inventor; b. West Bridgewater, Mass., Nov. 11, 1781; s. Abiezer and Hepsibah (Keith) A.; m. Lucy Willis, 1804, 7 children; m. 2d, Mary Pillsbury, 1833. Established iron foundry, 1809; designed 1st cylinder stoves 1822; contbr. rapid devel. of South Boston; alderman South Boston, 1824-27; formed So. Boston Iron Co., 1827; his shop turned out 1st gun ever rifled, 1834; Cyrus Alger Primary Sch. named after him, 1881. Considered one of outstanding metallurgists of his time. Died Boston, Feb. 4, 1856.

ALGER, Philip Langdon, Am. elec. engr.; b. Washington, Jan. 27, 1894; s. Philip Rounseville and Louisa (Taylor) A.; B.S., St. Johns Coll., 1912, M.A., 1915; B.S., Mass. Inst. Tech., 1915; M.S., Union Coll., Schenectady, 1920; m. Catharine Emma Jackson, June 30, 1918 (dec. Sept. 1945); children—Augusta Jordan (Mrs. David C. Prince, Jr.), Philip Langdon (dec. 1929), Catherine Jackson (dec. 1935), John Rodgers Meigs, Andrew Dugald Langdon, Anne Vogdes (Mrs. Gert Ehrlich) (adopted); m. 2d, Helen Jackson Hubbell, Nov. 9, 1946. With Gen. Electic Co., Schenectady, 1919-50, cons. engr., 1950-59; adj. prof. elec. engring. Renssalaer Poly Inst., Schenectady, 1959—. Dir. Mohawk Devel. Service, Inc., Schenectady, Schenectady Bur. Municipal Research, Fellow Am. Soc. M.E., I.E.E.E. (past dir., Lamme medal 1959), A.A.A.S., Am. Soc. for Quality Control; mem. Inst. Elec. Engrs. (Brit.) Nat. Soc. Profl. Engrs., Am. Math. Soc., Newcomen Soc., Aztec Soc., Soc. Cin., Société Française des Electriciens, United World Federalists, Nat. Municipal League. Author: Mathematics for Science and Engineering, 1957; Nature of Induction Machines, 1965; (with N. A. Christensen, S. P. Olmsted) Ethical Problems in Engineering, 1965; also numerous articles. Home: 1758 Wendell Av., Schenectady 12308.*

ALGOEWER, David, German mathematician; b. Ulm, Germany, Dec. 30, 1678; s. Georg and Evan Christina (Eberking) A.; student theology, math. Altdorf, Germany, 1697-1701, also Helmstedt, Halle (both Germany); m. Anna Maria Merck, Oct. 28, 1710. Prof. math. Ulm, 1705-14, 29-37, prof. catechesis, 1714-29; mem. ministerium of clergymen. Author: Specimen Meteorologiae Parallelae, 1714. One of 1st to recognize necessity of simultaneous weather observations; recorded observations based on meteorol. instruments, 1710-37. Died Ulm, May 24, 1737.

ALGULINUS, Walter, see Agilinus, Gaulterus.

AL-HA'IM (Abu al-'Abbas ibn al-Ha'im) (Sihab al-din ibn al-Ha'im) (Ahmad ben Muhammad ben 'Imad al-Faradi ibn al-Ha'im), mathematician; b. Cairo, Egypt, circa 1352. Prof., Madrasa Salahiya, Jerusalem; Wrote books on arithmetic, inheritance problems, algebra. Died Jerusalem, 1412.

AL-HAJJAI IBN YUSUF, mathematician; flourished late 9th century in Baghdad, Iraq; made 1st Arabic translation of Euclid, also an early translation of Ptolemy.

AL-HAKAM, II; flourished 961-76; 9th Umayyad caliph of Cordova, Spain, 961-76; encouraged study of math., astronomy and medicine; his library contained 400,000 catalogued volumes.

AL-HAKIM (Halid al-Hakim) (Halid ben Yazid ben Mu'awiya al-Hakim), Arab alchemist; s. Caliph Yazid; a prince of Umayyad family, Egypt; encouraged translation of Greek works into Arabic. Died circa 704.

AL-HAMDANI (Abu Muhammad al-Hasan ibn Ahmad ibn Ya'qub al-Hamdani ibn al-Ha'ik), Arabic geographer; b. of Yemenite family. Author: A Geography of Arabia; The Crown, a History of Yeman; Astronomical Tables of Yeman; Description of the Arab Country; Book of the Countries. Died in prison, San'a, circa 945-46.

AL-HANAFI (Qaisar ibn Abu al-Qasim ibn Abd al-Ghani ibn Musafir Alam al-din al-Hanafi) (Alam al-din al-Hanafi), mathematician, astronomer, engr.; b. Asfun, Upper Egypt, circa 1178-79, or 1168-69; student, Egypt, Syria; studied music and scis. under Ibn Yunus. Returned to Syria; entered service of Muzaffar II Taqi al-din Mahmud (ruler of Hamah from 1229-44). Wrote treatise on Euclid's postulates. Built water mills on the Orontes, also fortifications; built 2d oldest existing Arabic celestial globe. Died 1251.

AL-HARIZI (Judah ben Solomon al-Harizi), Arabic-Hebrew translator; b. Granada or Toledo, Spain, circa 1170; student theology, philosophy, and medicine, either Toledo or Granada; travelled to Marseilles, France, 1190; returned to France, 1216, then went to Egypt, Palestine, Syria; in Jerusalem, 1218, Damascus, 1220; returned from Basra to Spain, circa 1230. Author: The Wise One (contained information on Jewish culture of the period); Healing of the Body (poem on diet). Translator many works from Arabic to Hebrew, including: pseudo-Galenic treatise on the soul, circa 1200; Sayings of the Philosophers (Hunain ibn Ishaq); Ethical Treatise (attributed to Aristotle); Maimonides commentary on Mishnah; Essay on Resurrection, a gynecol. treatise. Died before 1235.

AL-HASAN (or al-Marrakusi, also Abu 'Ali al-Hasan ibn 'Ali ibn 'Umar al-Marrakushi), mathematician, astronomer; b. Morocco, flourished there until 1262; wrote extensively on math. and astronomy; acquainted with works of al-Khwarizmi, al-Farghani, al-Battani, Abu-l-Wafa, al-Biruni, Ibn Sina, al-Zarqali, Jabir ibn Aflah. Author: The Uniter of the Beginnings and Ends (most elaborate treatise on astron. instruments and methods, trigonometry, gnomonics in Muslim West), probably completed 1229-30; also lost treatises on occupations at the apparition of the new moon, on the calendar, probably one on influence of planetary conjunctions and eclipses.

AL-HASAN (Musa ben Sakir), geometer; b. Bagdad; flourished 1st half 9th century; s. Musa ben Sakir. One of 3 brothers who were scientists, patrons of the scis. and astronomers (with obs. in their home) in Bagdad. Co-author: Book on Geometry by the Three Brothers; Book on the Balance; Book on the Measurement of the Sphere (transl. by Gherardo da Cremona as Liber trium fratrum de geometria).

AL-HASAN AL-A'SAM, (Hubays ben al-Hasan al-Asam), Arabic translator; b. 2d half of 9th century; nephew of Abu ya'qub Ishaq b. Hunayn b. Ishaq al-'Ibad; mem. sch. of Hanayn; translated books IX-XV of De anatomicus administrationibus (by Galen) into Arabic.

AL-HASIB, Habas (or Habas Ahmad ben Abd al-lah al-Marwazi al-Hasib, also Ahmed ibn Abd Aliah), Persian astronomer; b. Iran, 770; participated in geodetic and astron. measurements ordered by Abbasid calips al-Ma'mun at his court Bayt al-hikma (house of wisdom); made astron. observations, 1825-35; influenced by al-Farghani (Alfraganus); developed some useful astron. tables; introduced concept of umbra versa (corresponds to modern concept of tangent) and of cotangent, circa 860. Author: Of Nativities. Died 864.

AL-HASSAN (al-Rammah), mil. scientist; b. Syria; wrote 2 treatises on mil. strategy which have prompted comparison with Marcus Graecus; his 1st treatise contains various pyrotechnic recipes; considered saltpeter the fundamental substance of pyrotechnics, explained methods of preparing and purifying it by means of potash and repeated crystallizations. Author: Book on Cavalry; Deceptions of War. Died circa 1294-95.

AL-HASSAR (Muhammad al-Hassar), Arab Spanish mathematician; b. Spain; flourished 12th or 13th century; wrote treatise on arithmetic and algebra (translated into Hebrew by Moses ibn Tibbon, Montpellier, France, 1271); also wrote on geometry.

AL-HAYYAT, astrologer; Author: On the Laws of Birth, Book One (translated by Ioannes Hispalensis and Dominicus Gundisalvus). Died 815.

ALHAZEN (or Abu Ali al-Hasan Ibn al-Haitham, also Haytham, or al-Asam), Arabian mathematician, physicist; flourished 10th-11th century; lived in Egypt under Fatamid Caliph al-Hakim; in attempt to gain patronage claimed to be able to stop flooding of Nile, had to spend years feigning madness to avoid execution by the caliph. Author: Kitab al-manazir (or Opticae thesaurus Alhazeni, libri vii, standard work during Middle Ages, influenced Roger Bacon, Kepler, Witelo); commentaries on Aristotle and Galen; a book based on Ptolemaic astronomy. Gave sci. explanation of refraction, reflection, binocular vision, focusing with lenses, the pinhole camera, the rainbow; published theories on catoptrics, spherical and parabolic mirrors, spherical aberration, dioptrics, atmospheric refraction, phenomena of apparent increase in size of sun and moon nr. the horizon; gave accurate account of the eye and vision by opposing Ptolemaic theory that vision originates in eye, correctly suggested that light comes from object to eye; gave original solution to cubic equation by means of conic intersections; gave summation formulas for 1st powers of natural numbers; solved Alhazen's problem, al-Mahani's equation; formulated correct conception of air pressure, 1030; 1st to calculate height of atmosphere.

AL-HIMYARI, geographer; b. Spain; came from family of geographers; numerous writings preserved only in fragments, including geography and polit. chronicle of al-Andalus, Muslim Spain. Died 1494.

ALI, M(ohamed) A(ther), zoologist; b. Bangalore, India, July 16, 1932; s. Mohamed Anwar and Fatima (Hussain) A.; Intermediate in Sci., Govt. Arts Coll., Madras, India, 1949; student Aligarh (India) Muslim U., 1949-50; B.S., Presidency Coll., Madras, 1950-52, M.S., 1954; Dr.Phil., U. B.C. (Can.), 1958; m. Marie-Antoinette Martineau, June 12, 1959. Instr., U. B.C., Vancouver, 1954-58; asst. prof. zoology McGill U., Montreal, Que., Can., 1958-59; research asso. Queen's U., Kingston, Ont., Can., 1959; asst. prof. zoology Meml. U., St. John's, N.B., Can., 1959-61; sr. fellow Max-Planck-Inst., Seewiesen, Germany, 1964-65; faculty Université de Montréal, 1961—, asso. prof. zoology, 1963—. Mem. Comité Internat. de Photobiologie, 1960—. Alexander von Humboldt Sr. fellow, 1964-65. Mem. A.A.A.S., Am. Soc. Ichthyologists and Herpetologists, Am., Canadian socs. zoologists, Assn. Canadienne Française pour L'Avancement des Scis., German Physiol. Soc., Sigma Xi. Research, numerous publs. on embryology, ecology and taxonomy sponges; described complete devel. one species sponge; showed effect environment on distbn. and form sponges; described 2 new species; research on vision in vertebrates; effect of various environmental factors on retinal adaptation, relationship response eye and behavior Pacific salmon; described retinas large number vertebrates; research on activity and metabolic rates fish and invertebrates, influence environment on invertebrate vision. Home: 5795 Av. Decelles, Montreal 26, Que., Can.*

ALIBEKOV, Serazhutdin Yusupovich, Russian dermatologist, venereologist; b. 1903; grad. Med. Faculty, Azerbaijan U., 1928; Cand. Med. Sci., 1937. Dist. physician, Village of Nizhniy Dzhengutay, Daghestan, physician Tb hosp., Buynaksk, Daghestan, instr. med. sch., physician out-patients skin and venereal diseases clinic, Makhachkala, 1928-33; postgrad. chair skin and venereal diseases 1st Moscow Med. Inst., 1933-37; asst. lectr. dept. skin and venereal diseases Daghestan Med. Inst., 1937-52, head chair skin and venereal diseases, 1952—. Hon. physician, hon. sci. worker Daghestan ASSR. Mem. Daghestan (chmn.), All-Union (bd. mem.) socs. dermatologists and venereologists, Daghestan Znanie Soc. (Presidium mem.). Editorial adviser Talgi and Its Medicinal Properties. Research on dermatology, venereal diseases, regional pathology, spa treatment, Tb of skin, cutaneous leishmaniasis and fungoid diseases. Address: Daghestan Med. Inst., pl. Lenina, Makhachkala, RSFSR, USSR.

ALIBERT, Jean-Louis-Marie, French dermatologist; b. Villefranche, France, May 2, 1768; s. Pierre and Claudine (Alric) A.; ed. l'Ecole Normale de Paris, also l'Ecole de Santé. Became sec.-gen. Soc. médicale d'émulation de Paris, 1796; apptd. adjunct prof. St. Louis Hospital, 1801, prof. medicine, 1802, chief physician to Louis XVIII, 1814, also to Charles X. Author: Précis théorique et pratique sur les maladies de la peau, 1818, Physiologie des passions, 1825; Nosologie naturelle. First physician to teach dermatology systematically; may be considered founder modern sch. dermatology in France; described mycosis fungoides or pian fungoide (Alibert's disease), 1806; gave 1st clear description of keloid in Note sur la keloide, 1816; described sycosis barbae (Alibert's mentagra), 1825, oriental sore, 1829, cutaneous lesion of syphilis (which he named syphilide), 1832; coined term dermatolysis, probably term dermatoses, 1832; classified skin diseases in Monographie des dermatoses, 1832. Died Paris, Nov. 4, 1837.

ALICATA, Joseph E(verett), parasitologist; born Carlentini, Italy, Nov. 5, 1904; s. Antonio and Concetta (Vaccaro) A.; came to U. S., 1919, naturalized, 1926; A.B., Grand Island Coll., 1927; M.A., Northwestern U., 1928; Ph.D., George Washington U., 1934; m. Hannah L. Davis, Jan. 23, 1929; children—Betty Mae, William D.; m. 2d, Earleen E. Moyer, June 30, 1958. Jr. zoologist Bur. Animal Industry, U. S. Dept. Agr., Washington, 1928-35; parasitologist Hawaii Agr. Expt. Sta., U. Hawaii, Honolulu, 1935—; parasitologist Bd. Health, Honolulu, 1936-37, health, sanitation com. Hawaiian Sugar Planters Assn., 1943; sr. scientist USPHS, Amman, Jordan, 1953-54. Fulbright scholar, 1950-51. Fellow A.A.A.S.; mem. Washington Acad. Sci., Am. Soc. Parasitologists, Am. Soc. Tropical Medicine and Hygiene, Sigma Xi, Phi Kappa Phi. Author: Parasitic Infections of Man and Animals in Hawaii, 1964. Contbr. sect. to Advances in Parasitology, 1965. Research, numerous publs. primarily in field of biology and control of parasites of man and animals. Home: 2130 Hunnewell St. Office: 2525 Varney Circle, U. Hawaii, Honolulu 96822.*

ALIEV, Gasan Ali Rzaogly, Russian agrobotanist; b. Salyany (now Azerbaijan), 1902; grad. Biol. Faculty, Azerbaijan U., Baku, 1930; postgrad. Sukhumi Subtropical Inst. Humoid Subtropics, 1937. With Botany Inst., Azerbaijan br. USSR Acad. Sci. (now Azerbaijan Acad. Sci.), 1932-44, dir. Inst. Botany, 1949-52, Acad. sec., 1952-57, head lab. Inst. Pedology and Agrochemistry, 1957—, chmn. Commn. for Preservation of Nature, 1960—; sr. asso., learned sec., then dep. dir. Inst. Pedology and Agrochemistry, 1944-49. Mem Azerbaijan Acad. Sci. Author: The Introduction of Eucalyptus in Azerbaijan, 1945; The Soils of the Lower Reaches of Rivers in the Southwestern Declivity of the Greater Caucasus, 1948; The Winter Crop of Fodder Grasses—A Major Source for Fodder Increase and the Improvement of Soil Fertility, 1955; The Use of Manganese as a Trace Fertilizer on Agricultural Crops in the Azerbaijan, 1956; An Assessment of Fissure Collectors in the Upper Chalk and Paleogene Formations of the Eastern Caucasian Foothills, 1963. Co-author, co-editor: The Soils of the Azerbaijan SSR, 1953. Research on Azerbaijan soils, use of trace elements as fertilizers. Address: Azerbaijan Acad. Sci., Baku, Azerbaijan SSR, USSR.

ALIEV, Musa Mirzoevich, Russian geologist; b. Shemakha, Azerbaijan, April 11, 1908; graduated from Mining Faculty, Azerbaijan Polytechnical Institute, 1931, postgraduate student, 1931-37; Lecturer Azerbaijan Industrial Institute; head Geological Bureau Transcaucasian Geol. Trust, then dep. head and chief engr. Azerbaijan Geol. Bd., 1937-39; dir. Azer. Poly. Inst., 1939-41; dir. Main Bd. Tng. Establishments, USSR Peoples Commissariat of Petroleum Industry, then head Azerbaijan Geol. Bd., also instr. Moscow Petroleum Inst., Azerbaijan Indsl. Inst. Azerbaijan U., 1941-44. Del., 20th Internat. Congress Geologists, Mexico, 1957. Recipient Order of Lenin. Mem. Azerbaijan Acad. Sci. (pres. 1950-58). Author: Lagodekhi-Akhalsopeli. A Geological and Petrographic Outline of Part of the Southern Slope of the Main Caucasian Range, 1940; Inocerams of the Chalk Deposits of the USSR, 1957. Exec. editor: The Geology of Azerbaijan, 1957; editor Data on the Regional Stratigraphy of the USSR, 1963. Research on paleontology and stratigraphy of chalk deposits. Address: Azerbaijan Acad. Sci., Baku, Azerbaijan SSR, USSR.

ALI IBN AL-IMAN (Abu al-Hasan), Iranian physician, mathematician; b. 1106; practiced medicine and math; author: Ta rih hukama al-Islam (History of the Philosophers of Islam).

ALI IBN MUHAMMAD AL-KUCHDJI, Turkish astronomer; s. Mohammed al-kouchdji; student, Samarkand; after studies left for Kerman; later returned to Samarkand and completed astron. tables begun by Oulough-Beg; worked in Persia and Constantinople. Wrote several treatises on astronomy while in Constantinople. Died Constantinople, 1474.

ALI IBN VELI IBN HAMZA, Arabic mathematician; living in Mecca, circa 1590; author: Tuhfet alâclad li-davi al-roshd re'l sadad (The Gift of Numbers for the Possessors of Reason and Correct Insight, a work which dealt with elementary algebra).

ALIKHANIAN, Artemii Isaakovich, Russian physicist; born June 24, 1908; grad. Leningrad U., 1931. Director, Physics Inst., Armenian Acad. Sci. Stalin prize, 1941, 48. Mem. Armenian, USSR (corr.) acads. sci. Author: Research on Narrow Showers at an Altitude of 3,250 Meters above Sea Level, 1949; An Investigation on the End of the RaE Spectrum by Means of a Double Magnetic Spectrometer, 1940; The Soft and Hard Component of Cosmic Rays and Meson Spin, 1940; Man and Space, 1960; One Possibility of Detecting High-Energy Charged Particles, 1961; New Experimental Data on u-Mesons, 1962. Research on nuclear physics and cosmic rays with brother (A. I. Alikhanov); with brother and M. S. Kozodaev discovered emission of pairs (positron and electron) by energized nuclei, 1934; with brother and L. A. Artsimovich provided exptl. proof of laws of pulse conservation of positron and negative electron fusion, 1935; with brother and B. A. Nikitin began research resulting in discovery of varitrons, 1939; with brother set up cosmic radiation sta. on Mt. Aragats, Armenia, 1943-45; with Asatiani discovered showers with low particle count (narrow showers in cosmic rays). Address: Armenian SSR Acad. Sci.; Physics Inst., Yerevan, Armenia SSR, USSR.

ALIKHANOV, Abram Isaakovich, Russian atomic physicist; born Mar. 4, 1904; graduated from Leningrad Polytechnical Institute, 1931. With Physicotechnical Institute, USSR Academy Science, 1927—, bureau member department physicomathematical science, dir. Inst. Theoretical and Exptl. Physics (now State Com. for Use Atomic Energy, USSR Council Ministers), 1959—, chmn. Commn. for Study Cosmic Rays. Mem. USSR, Armenian acads. sci. Recipient Stalin prize, 1941, 48. Co-author: Research on Synthetic Radioactivity, 1936, New Data on the Nature of Cosmic Rays, 1945, Cosmic Rays, 1949; author Weak Interactions, 1960; Experiment, 1962, also others. Research on nuclear physics and cosmic radiation; 1st to study basic positron formation laws by means of gamma-rays, 1934-37; investigated problems of beta-decay study; started (with brother A. I. Alikhanian) systematic research on beta-spectra of synthetic radioactive elements; developer new method of studying composition and energy spectrum of cosmic rays; with brother made large spectrometer for studying mass spectrum of particles in cosmic radiation, 1946; put into operation 1st Soviet reactor with heavy water inhibitor, 1949; a designer of 1st Soviet atomic bomb; inst. he directs produced 7,000,000,000-volt proton accelerator, 1961. Address: USSR Acad. Sci., Inst. Theoretical and Exptl. Physics, B. Cheremushkinskaya ul, 89, Moscow, USSR.

ALIMARIN, Ivan Pavlovich, Russian chemist; b. Sept. 11, 1903. With All-Union Research Inst. Mineral Raw Materials, 1923-53; instr. Moscow Inst. Fine Chem. Tech., 1929-50, prof., 1950-53; with Inst. Geochemistry and Analytical Chemistry, USSR Acad. Sci., 1949—, bur. mem. chemico-tech. and biol. sci. sect. formerly dept. chem. sci., 1960—; prof. Moscow U., 1953—, Mem. Soviet delegation Symposium on Microchem. Analysis, Birmingham U., 1958. Mem. USSR Acad. Sci. Co-Author: Colorimetric Determination of Minute Quantities of Niobium in the Form of a Rhodanide Group, 1946; Quantitative Chemical Determination of Germanium in Mineral Ash, 1946, Qualitative Semi-Microanalysis, 1952; author The Use of Radioisotopes in Chemical Analysis, 1955; Spectrophotometric Study of the Reaction of Thorium with Datiscetin, 1963. Research in analysis of minerals and ores, analytical chemistry of rare elements, microchemistry and radiochem. analysis. Address: Faculty of Chemistry, Lomonosov Moscow University, Lenin Hills, Moscow, B-234, USSR.*

ALIMOV, Shakir Alimovich, Russian phthisiologist; b. 1912; grad. Samarkand Med. Inst., 1935; D.Med. Sci., 1951. Physician various instns. 1935-39; chief physician various phthisiatric hosps. and Khamza Khamin-Zade Tb sanatorium, 1939-51; founder work therapy sanatorium, 1958; dir. Uzbek Inst. Tb, 1951—; head chair pulmonary Tb, Tashkent Postgrad. Med. Inst., 1952—; prof., 1952—. Del., Internat. Tb Congress, New Delhi, 1956, Istanbul, 1959, Rome, 1962, Amsterdam, 1967; convenor 1st Congress Uzbek Phthisiologists, Tashkent, 1962; del. other congresses. Recipient monetary prize class 1, USSR Ministry Health; named meritorious sci. and tech. worker, 1962. Mem. All-Union (dep. chmn. 1957—), Uzbek (chmn.) socs. phthisiologists, Internat. Tb League. Author numerous works including The Khamza Khakim-Zade Sanatorium, 1946; Regional Epidemiology, Pathology and Clinical Aspects of Tuberculosis, 1951; Clinical Aspects and Treatment of Pulmonary Tuberculosis, 1960; Prophylaxis, Treatment and Rehabilitation of Tuberculosis Patients at the 15th International Tuberculosis Congress, 1960. Editor: Med. Jour. of Uzbekistan, 1953-64. Address: Majlisil, Tashkent 86, Uzbekistan SSR, USSR*.

AL-IMRANI, Muslim mathematician, astrologer; b. Mesal, Upper Mesopotamia; tchr. of Alcabitius. Author: A Commentary on the Algebra of Abu Kamil; On Elections; also astrological treatises, including one on choosing of auspicious days which was translated by Savasorda, 1133. Died 955.

ALINSKY, Saul David, Am. sociologist; b. Chgo., Jan. 30, 1909; s. Benjamin and Sarah A. (Tannenbaum) A.; Ph.B., U. Chgo., 1930, postgrad., 1930-32; LL.D., St. Procopius Coll., 1958; m. Helene Simon, June 9, 1932 (dec. Sept. 1947); children—David, Kathryn; m. 2d, Jean Graham, May 15, 1952. Sociologist, Inst. for Juvenile Research, Chgo., 1931-

33, 36-39; mem. state prison classification bd. div. criminology Ill. State Penitentiary System, Joliet, 1933-36; co-founder Back of the Yards Neighborhood Council, Chgo., 1939; exec. dir. Indsl. Areas Found., Chgo., 1940—. Lectr. on criminology, community orgn. and organized labor in various univs. Recipient award for Social Justice, Cath. Youth Orgn. Am., 1950. Mem. Authors League Am. Author: Reveille for Radicals, 1946; John L. Lewis, A Biography, 1949; also articles. Office: 8 S. Michigan Av., Chgo. 60603.*

ALI QUSGI, MULLA 'ALA' AL-DIN, astronomer; one of four collaborators of Ulaj Beg; succeeded Salah al-cin Musa as dir. of obs. of Ulug Beg, Samarkand; went to Azerbaijan; sent by Uzun Hasan as ambassador to Constantinople; apptd. as 1st prof. of astronomy and math. in Turkey by Muhammad II at Madrasa Sch. of Saint Sophia. Died 1474.

AL-IRAQI (Abu al-Qasim al-Iraqi) (Abu al-Qasim Muhammad ibn Ahmad al-Simawi al-Iraqi), physician; b. Cordova, Spain; flourished 961-1013; physician to Al-Hakan II (961-76). Author of med. ency. in 30 sects. Discussed arterial ligatures, preparation of drugs by sublimation and distillation, cauterization, hemostasis, bone and eye surgery.

ALISON, Archibald, Brit. philosopher, psychologist; b. Edinburgh, Scotland, Nov. 13, 1757; s. Patrick Alison; student Balliol Coll., Oxford U., 1784; LL.B., Kenley; m. Dorothea Gregory, June 14, 1784; children—Archibald, William. Clergyman, Shropshire, 1790-1800; apptd. prebend, Salisbury, 1791, vicar, High Ercal, 1794; minister Episcopal chapel, Cowgate, Edinburgh. Fellow Royal Soc., 1792. Author: Essays on the Nature and Principles of Taste (formed basis for Francis Jeffrey's assn. theory of esthetics), 1790. Adherent, Scottish common sense philosophy; maintained associationist tradition; tried to prove that beauty is a product of agreeable ideas aroused by objects asso. with simple emotions. Died Colinton, May 17, 1839.

ALISON, William Pulteney, Scottish physician; b. Boroughmuirhead nr. Edinburgh, Scotland, 1790; s. Archibald Alison; student fine arts U. Edinburgh, from 1803, M.D., 1811; D.C.L., Oxford (Eng.) U., 1850. Physician to New Town Dispensary, from 1815; prof. med. jurisprudence (apptd. by Crown), from 1820; insts. of medicine prof., from 1822; prof., practice of medicine, 1842-56; 1st physician to her majesty for Scotland. Author: Outlines of Physiology, 1831; Outlines of Physiology and Pathology, 1833; On the History of Medicine, 1834; Supplement to Outlines of Physiology, 1836; On Inflammation, 1840; Reply to Dr. Chalmer's Objections to the Improvement of the Legal Provisions for the Poor in Scotland, 1841; Outlines of Pathology and Practice of Medicine, 1844; Observations on the Epidemic Fever in 1843 in Scotland and its Connection with the Destitute Condition of the Poor, 1844; Remarks on a Report on the Poor Law for Scotland, 1844; Observations on the Famine of 1846-47 in Scotland and Ireland, 1847; On Vital Affinity; Defence of the Doctrine of Vital Affinity. Writings were important contbns. to knowledge of fevers, also supplied valuable materials for the history of epidemics; believed and proved connection between disease and destitution. Died Sept. 22, 1859.

ALIX, Edmond, biologist; b. 1823; M.D., 1848; prof. Catholic Inst. Paris (France); sec. gen., dir. memoires Soc. Med. Sci.; mem. Philomath. Soc., Soc. Anthropology. Author: Note sur le cancérule, 1855; Étude sur les effets des tractions exercées sur la main et sur l'avant-bras des enfants, 1862; Essai sur la forme, la structure et le développement de le plume, 1866; Essai sur l'appareil locomoteur des oiseaux, 1874; Sur le movement batrachoïde des tortues; Sur une classification myologique des mammifères, 1876; Sur un nouvel anthropoïde (Gorilla mayéna) provenant de la region du Congo, 1877; Sur le cerucau a l'état foetal, 1877; Ostéologie et myologie des manchots, 1878; De la classification en général, 1880. Investigated embryology, physiology, classification, also bird structures and devel., human foetal and infant devel. Died 1903.

AL-JAHIZ (Abu Uthman 'Ami ibn Bahr al-Jahiz), Arabic man of letters; flourished circa 850, lived to be more than 90 years old; wrote Book of Animals; had genuine interest in natural and anthrop. scis.; knew how to obtain ammonia from animal offals by dry distillation; his work contains germ of many later theories (evolution, adaptation, animal psychology). Died 869.

AL-JAWBARI (Abd al-Rahim ibn 'Umar al-Dimashqi al-Jawbari Zain al-din), Arabic writer; b. Jawbar, nr. Damascus; flourished 1222-31; traveled widely in lands of eastern caliphate, going as far as India; flourished at court of Urtuqid sultan of Amid and Hisn Kaifa, Rukn al-din Mudud (reigned 1222-31). Author: Revelations of Secrets and Tearing Off of Veils, book exposing frauds and deceptions of money-changers, quacks and alchemists (gives valuable background of Muslim alchemy and tech., as well as Muslim manners and superstitions).

AL-JAZARI (Abu al-Izz Isma'il ibn al-Razzaz), mechanician; b. Diyar Bakr, al-Jazirah; flourished circa 1181-1206, under Urtuqid rulers of Diyar Bakr (northernmost dist. of al-Jazirah); completed treatise on knowledge of geometrical (mech.) contrivances, dealing chiefly with hydraulic apparatus (best Arabic work for study of Muslim applications of Hellenistic mechanics), 1205-06.

AL-JILDAKI ('Izz al-din 'Ali ibn Aidamur ibn 'Ali al-Jildaki), Arabic alchemist; flourished 1339-42; last important Arabic writer on alchemy. Author: The Brilliant Moon on the Secrets of Elixir; The Wish of the Expert . . . ; The Scattered Pearls; Hidden Pearls; Limit of Joy; Removing of the Veils; The Lamp on the Secrets of Alchemy; End of the Search; Introduction to the Secrets of Alchemy. Died Cairo, Egypt, circa 1342-43, or 1360-61.

AL-JURJANI (Isma'il al-Jurjani), Persian physician; b. Jurjan east of Caspian; compiled med. treatise in Persian (1st med. ency. in Persian). Author: The Treasure of the King of Khwarizm, after 1110; The Aims of Medicine, completed circa 1127-1135; The Admonisher. Died circa 1135-36.

AL-KALSAKI, mathematician; fl. 15th century; author: Raising of the Veil of the Science of Gubar; Used algebraic symbolism profusely; gave approximation for square root, similar to method of continued fractions but without modern notation. Died 1486.

ALKARKI (al-Kharki), Muslim mathematician; flourished Bagdad, circa 1000. Author: al-kafi fi-l-hisab (Book on Arithmetic); also treatise on algebra. First to work with higher roots; solved quadratic equations; gave arithmetical and geometrical proofs; he based his algebra largely on Diophantos. Died 1025.

AL-KASHI (or al-Kasi, also Jemshid ibn Med'ud ibn Mahmud Giyat ed-din al-Kashi), astronomer, mathematician; asst. to Ulugh Beg; dir. Beg's obs., Samarkand. Author: Key of Arithmetic (short treatise in Persian on arithmetic and geometry), circa 1430; Treatise on the Circumference. First to use decimal fraction; worked on summation of series; gave rule for summing 4th powers; his works not widely known in the West. Died circa 1436 or circa 1424.

AL-KATIBI (al-Qazwini) ('Ali ibn 'Umar al-Katibi) (Najm al-din 'Ali ibn 'Umar al-Qazwini al-Katibi), Persian astronomer, philosopher; attached for some time to obs. at Maraja. Author: Source of the Principles of Logic and Philosophy (contains discussion of diurnal rotation of earth); also another logical treatise and a work dealing with logic, physics and metaphysics; prepared an edit. of Almagest. Died 1277.

ALKEMADE, Cornelis Theodorus Joseph, Dutch physicist, b. St. Oedenrode, Netherlands, Nov. 26, 1923; s. L. P. J. and Helena (de Jongh) A.; Master's degree U. Utrecht (Netherlands), 1949, doctor's degree with honours, 1954; m. Eugenie Marie Albertine Klinkenberg, Feb. 4, 1953; children—Kees, Paul, Els, Robert, Marian. Lectr. U. Utrecht, 1946-55, 56-60, prof. exptl. physics, 1960—; postdoctorate fellow NRC Ottawa (Ont., Can.), 1955; part-time prof. Cath. U., Nijmegen, Netherlands, 1960-63. Mem. Netherlands Phys. Soc. (mem. bd.), Combustion Inst. Author: (with R. Herrmann) Flammenphotometrie, 1960; also articles. Editorial bd. and Flame, Combustion, 1966—. Devel. chem. analysis by flame photometry; invention chem. analysis by atomic light absorption in flames; research on phys. and physico-chem. processes in flames with metal vapours; phys. noise sources in optical radiation measurements, solid state devices, condensor microphones; optical lasers; microwave absorption spectra of gases. Home: 76 Julianalaan Bilthoven, Netherlands.*

AL-KHALILI (Shams al-din Abu Abdallah Muhammad ibn Muhammad al-Khalili), Syrian astronomer; b. Damascus, Syria; compiled elaborate tables after pattern of tables of al-Hasan al-Marrakushi (XIII-I), serving for determination of time, circa 1378; another of his tables is calculated for latitude of Damascus, 1408; also wrote treatises on use of various kinds of quadrants. Died 1408.

AL-KHAZIN, see Alchazin.

AL-KHAZINI (Abu al-Fath 'Abd al-Rahman al-Mansur al-Khazini), astronomer; flourished circa 1115-21. Author: The Esteemed Sinjaric Tables (astron. tables giving positions of stars for year 1115-16); completed Book of the Balance of Wisdom (on mechanics, hydrostatics, physics, states theories of gravity and lever, observations on capillarity, also contains tables for specific gravities of liquids and solids), 1121-22.

AL-KHWARIZMI, Arabic mathematician; b. Khwarazm (now Khiva); lived under Caliph ad-Mamum (ruled 813-33), Bagdad. Compiled astron. tables; also work on arithmetic; wrote treatise on algebra based on Hindu work of Brahmagupta which introduced name, al-jabr, circa 825; work on algorism (term deriving from his name) introducing Arabic numerals, method of calculating by decimals. Improved text and maps of Ptolemy's geography; calculated more accurate value for pi; studied astrolabe.

AL-KUHI (Abu Sahl Wijan ibn Rustam al-Kuhi), astronomer, mathematician; b. Kuh, Tabaristan; flourished circa 988; leader of astronomers working at obs. built by Buwayhid Sharaf al-Dawla, in Bagdad, 988; believed to have written numerous math. and astron. works; devoted his attention to those Archimedian and Appollonian problems leading to math. equations of higher degree than the 2d; solved some of them, and discussed conditions of solvability; these investigations are among the best of Muslim geometry.

ALLAIRE, James Peter, Am. mechanic, engineer; b. 1785; m. Frances Roe; m. 2d., Calicia Tompkins. Began as brass founder, N.Y.C., 1813; founder Allaire Works (1st steam engine works in N.Y.), 1815, Howell Works, 1831; in his time most important steamboat engine producer in U. S.; built 1st compound-type marine engines; built 1st house designed as tenement, N.Y.C., 1833. Died May 20, 1858.

ALLAIS, Maurice Felix Charles, French engr., economist; b. Paris, May 31, 1911; s. Charles and Louise (Caubet) A.; ed. Ecole polytechnique, Ecole nationale supérieure des mines de Paris; D.Eng., U. Paris.; Dr. Honoris causa, U. Groningen, 1964; m. Jacqueline Bouteloup, Sept. 6, 1960; 1 dau., Christine. Mine engr. active service, 1936-48; dir. Bur. Documentation and Mining Statistics, 1943-48; then prof. sch. mining Inst. Statistics, U. Paris; dir. research Nat. Center Sci. Research; prof. Grad. Inst. Internat. Studies, Geneva. Recipient award Acad. Sci., 1933, Acad. Moral and Polit. Sci., 1954, 59, Lancaster prize Johns Hopkins and Am. Soc. for Operations Research, Galabert prize French Astronautical Soc., 1959, laureate Gravity Research Found., 1959. Mem. Internat. Statis. Inst., Internat. Econometric Soc., Am. Soc. Research Operations. Author: Traité d'Economie pure, 1943; Abondance ou misère, 1946; Economie et intéret, 1947; Fondements comptables de la macro-économique, 1954; L'Europe Unie, Route de la Prospérité, 1960; Le Tiers Monde au Carrefour, 1962; The Role of Capital in Economic Development, 1963; La Reformulation de la Théorie Quantitative de la Monnaie, 1965; L'Impôt sur le Capital, 1966. also others. Home: 15, rue des Gates-Ceps, Saint-Cloud. Office: 62, Bd Saint-Michel, Paris 6, France.*

ALLAMAND, Jean Nicolas Sébastien, naturalist, physicist; b. Lausanne, Switzerland, Sept. 18, 1713; s. Thomas Allamand; prof. philosophy and math. U. Leyden (Netherlands), 1742-87; published Hist. Dictionary (Prosper Marchano), 1758-59. Fellow Royal Soc., 1747; mem. French Acad. Scis., 1769. Gave 1st explanation for Leyden jar. Died Leyden, Mar. 2, 1787.

ALLAN, Francis John, Brit. physician; b. Kirkcudbrightshire, Scotland, Nov. 23, 1858; s. Francis Stallan Allan; student Edinburgh U.; M.D., D.I.H., Cambridge U.; m. Ellen Mary Browning. Med. officer health City of Westminster, Borough of Shoreditch, Strand dist.; physician Hosp. for Consumption; lectr. san. sci. Kings Coll.; lectr. pub. health U. London, Westminster Hosp. Med. Sch.; examiner med. jurisprudence, pub. health U. Aberdeen; dir. Med. Sickness and Accident Assn. Fellow Bot. Soc. Edinburgh, Royal Soc. Edinburgh, Royal San. Inst.; mem. French Soc. Hygiene, Med. Officers of Health Soc. (pres.), Instn. Civil Engrs. (asso.). Author: Aids to Sanitary Science; (with C. E. Allan) Housing of the Working Classes Acts; A Practical Guide to Disinfection; other publs. Editor: Public Health. Died July 28, 1932.

ALLAN, Frank Nathaniel, physician; born Proton, Ont., Can., Dec. 26, 1899; s. George and Margaret (Ewing) A.; M.B., U. Toronto, 1922, B.Sc. in Medicine, 1924, Hon. M.D., 1928; m. Lillian Christie, Sept. 22, 1934; dau., Christie Allan (Mrs. W. Stephen Piper). Came to U. S., 1925, naturalized, 1942. Instr., U. Minn., 1927-32; staff Mayo Clinic, 1927-32; co-dir. med. staff Lahey Clinic, Boston, 1932-47, exec. dir., chmn. med. dept., 1948-65; practice medicine specializing in internal medicine, Rochester, Minn. 1927-32, Boston, 1932-65; asso. clin. prof. medicine Boston U. Sch. Medicine, 1961—; clinical professor of Medicine Georgetown University, Washington, 1966—. Consultant to the FDA, 1965—. Recipient of the Starr Silver medal U. Toronto, 1928. Mem. Am. (founding editor Jour. 1951-55, adv. editor, past pres.), New Eng. (past pres.) diabetes assns., Greater Boston Diabetes Soc. (past chmn. exec. com.), Endocrine Soc., Am. Soc. Clin. Investigation, Central Soc. for Clin. Research, Am. Soc. Exptl. Pathology, A.A.A.S., Am., Internat., Mass. socs. internal medicine, Canadian Diabetic Assn. (hon.), Academia Nacional de Medcina (hon., Buenos Aires, Argentina), Sigma Xi, Alpha Omega Alpha. Contbg. author: The Cyclopedia of Medicine, 1934, 39, 41; Musser's Internal Medicine, 1951; Current Therapy, 1961, 62; also numerous articles. Editor: Am. Diabetes Assn. Diabetes Guide Book, 1950, 2d edit., 1956. Editorial bd. New Eng. Jour. Medicine, 1952—, dep. editor, 1965-66. Research on endocrinology and metabolism, vertigo, coma, convulsions and psychosomatic problems, diabetes. Home: 44 Barnstable Rd., West Newton, Boston, Mass. 02165.*

ALLAN, John Andrew, Canadian geologist; b. Aubrey, Que., Can., July 31, 1884; s. James and Margaret (Lang) A.; B.A., McGill U., 1907, M.Sc., 1908; Ph.D., Mass. Inst. Tech., 1912; m.

Gladys Parsons, June 4, 1913; children—J. Donald, Edward J. Mem. Canadian Geol. Survey, 1906-18; dept. head, prof. geology and mineralogy U. Alta., Edmonton, 1912-49. Fellow Royal Soc. Can. (pres. geol. sect.); mem. Canadian Inst. Mining and Metallurgy (pres. 1932-33), Engring. Inst. Can., Assn. Profl. Engrs. Alta. (pres. 1930). Author thesis on Ice River intrusive nr. Field, B.C.; other geol. research and publs. on coal, structural geology and physiography of Canadian Rocky Mountains and foothills. Died Edmonton, May 23, 1955.

ALLAN, John Hamilton, Am. physician; b. Stamford, Conn., Oct. 4, 1907; s. Charles Hamilton and Ann (Patterson) A.; A.B., Johns Hopkins, 1929, M.D., 1933; m. Dorothy Allen Boyd, June 24, 1942; children—David Laird, Susan Blair. Instr., U. Pa., 1940-42; orthopedic surgeon-in-chief Germantown, St. Christopher's, Chestnut Hill hosps., Phila., 1947-50; prof., chmn. orthopedics U. Va. Sch. Medicine, Charlottesville, 1950—. Area cons. VA.; mem. Gov. Va. Overall Adv. Com. on Needs of Handicapped Children, 1958-64. Mem. Nat. Acad. Sci. (mem. com. on prosthetics edn. and information), Alpha Omega Alpha, Omicron Delta Kappa. Study of ambulation potential of congenital paraplegic. Home: 24 E. Range St. Office: U. Va. Hosp., Charlottesville, Va.*

ALLANDI, Stephanus, translator; flourished 1st quarter 14th century; with sch. at Montpellier, France; vice-chancellor, 1319; translator: Kitab Al-amal Bil-Kurra Al-Fulkiya (Qusta Ibn Luga); De Sphaera Solida; Dietarium, or Traslatio dietarii A Costa ben Luca compositi; Regimem contra defectum coitus; a treatise on cataract; a treatise on bloodletting; Annotationes in Anatomicam Mundini; Isagoge in Hippocratis et Galeni physiologiae partem anatomicam.

ALLARD, Claude, Canadian chemist; b. Montreal, Que., Can., Aug. 15, 1922; s. Charles Amadee and Brochu Marie-Jeanne Allard; B.Sc., U. Montreal, 1944, Ph.D. in Chemistry, 1948; m. Therese Lefebvre, June 29, 1946; children—Francine, Helene, Diane. Prof. chemistry Sch. Vet. Sci., St-Hyacinthe, Que., 1947-50; chief biochemistry sect. Montreal Cancer Inst., Notre-Dame Hosp., 1950-59; research asso. Nat. Cancer Inst. Can., 1958-59; chief Metabolic Research Lab. Montreal Heart Inst., 1960—, chief dept. labs., 1965—; research prof. Faculty Medicine, U. Montreal, 1962—. Pres., Corp. Profl. Chemists Que., 1961-63; sci. adviser Expo-67, Montreal, 1965. Fellow Chem. Inst. Can.; mem. Canadian Physiol. Soc., Canadian Biochemists, Am. Heart Assn., C.P.C.Q. Research, publs. on enzymes and their behavior in cancer, arterial enzymes and thrombosis considered as a precursor to heart disease. Home: 5 McCulloch St., Montreal 8, Que., Can.*

ALLARD, Elisabeth M.A.A.J., social scientist; b. Geertruidenberg, Netherlands, May 26, 1904. d. H.J.M.C. and M.A.E.A. (Barge) Allard; ed. Nimegue, Leyde, Oxford univs.; social work in Netherlands and Gt. Britain, 1930-39; instr. sociology U. Djakarta, 1946-47, sec.-curator Inst. Linguistic and Cultural Research, 1948-52; prof. sociology non-Western socs. Catholic U., Nijmegen, Netherlands, 1958—. Mem. Royal Inst. Anthropology Gt. Britain and Ireland. Author: EEn grammaticaal onderzoek van het Proza van Hadewijck; Animistic Beliefs and Rites in the Malay Archipel; also others. Address: Burg. Van Schaeck Mathonsingel 83, Nijmegem, Netherlands.*

ALLARD, R(obert W(ayne), Am. geneticist; b. Los Angeles, Sept. 3, 1919; s. Glenn A. and Alma (Roose) A.; B.S., U. Cal., 1941; Ph.D., U. Wis., 1946; m. Ann Wilson, June 16, 1944; children—Susan, Thomas, Jane, Gillian, Stacie. With U. Cal., Davis, 1946—; prof., 1957—. Mem. Genetics Soc. Am., Bot. Soc. Am., Am. Soc. Agronomy (Crop Sci. award 1964), Am. Soc. Naturalists, Phi Beta Kappa. Author: Principles of Plant Breeding. Research, publs. on genetic structure of populations. Home: 16 Parkside Dr., Davis, Cal. 95616.*

AL-LATIF (Abd al-Latif) (Abu Muhammad 'Abd al-Latif ibn Yusuf ibn Muhammad ibn Ali Muwaffaq al-din al-Baghdadi), physician, philosopher; b. Bagdad, 1162; student, Bagdad. Worked under Kamal al-din Ibn Yunus, Musul, 1189-1190; prof. mosque of Damascus, in Jerusalem after its conquest by Salah al-din, (1187); went to Egypt, 1193; tchr. al-Azhar; tchr. al-Aziziya, Damascus, 1207-08. Wrote numerous works including an account of Egypt which discussed botany. Died 1231.

ALLAWAY, William Hubert, Am. soil scientist; b. Homer, Neb., Apr. 12, 1916; s. James and Mary (Adams) A.; B.S., U. Neb., 1938; M.S., Ia. State U., 1939, Ph.D., 1945; m. Mildred Holland, Nov. 12, 1939; children—Susan (Mrs. Lewis H. LaRue), William Hubert, Nancy (Mrs. Robert O. Lindsley). Prof. soils Ia. State U., 1945-50; soil scientist U. S. Dept. Agr., Beltsville, Md., 1950-61; dir. U. S. Plant Soil and Nutrition Lab., Ithaca, N.Y., 1961—; vis. prof. U. Ill., 1959; prof. soils Cornell U., Ithaca, 1961—; cons. Chilean Minister Agr., 1963. Fellow Am. Soc. Agronomy; mem. Am. Chem. Soc., Soil Sci. Soc. Am., Am. Soc. Animal Sci., A.A.A.S. Research, numerous publs. primarily in field of effect of soils upon nutritional quality plants. Home: R.D. 2. Office: U. S. Plant Soil and Nutrition Lab., Ithaca, N.Y. 14850.*

ALLBUTT, Sir Thomas Clifford, English physician; b. Dewsbury, Eng., July 20, 1836; s. Thomas and Marianne (Wooler) A.; student Gonville and Caius Coll., Cambridge, Eng., 1855-60; M.B., St. George's Hosp., London, Eng., 1861; postgrad., clinics of Armand Trousseau, G.B.A. Trousseau, A.P.E. Bazin, A. Hardy (all Paris, France); M. Susan England, 1869. Practiced medicine, Leeds, Eng., cons. physician, 1861-89; commissioner lunacy, London, 1889-92; Regius professor of medicine Cambridge Univ., 1892-1925. Fellow Royal Soc., 1880; decorated knight commander, Bath, 1907. Author: System of Medicine, 8 vols., 1896-99; Diseases of the Arteries and Angina Pectoris, 1915. Investigated tetanus, also hydrophobia; invented short clin. thermometer, 1866; called attention to endarteritis of cerebral vessels in a case of cerebral disease in a syphilitic patient, 1868; described joint symptoms asso. with locomotor ataxia, 1869; did research on the use of the ophthalmoscope and the pathology of the nervous system; demonstrated origin of angina pectoris in aorta, 1894; described hyperpiesia dissociated from kidney disease, 1895. Died Cambridge, Feb. 22, 1925.

ALLCHIN, William Henry, Brit. physician; b. Oct. 16, 1846; ed. Univ. Coll., London; m. Margaret Holland, 1880. Physician-extraordinary to king; cons. physician, v.p. Westminster Hosp.; cons. physician Victoria Hosp. for Children, Poplar Hosp. for Accidents, Western and Marylebone Gen. Dispensaries; examiner, mem. consultative bd. med. service Royal Navy; examiner in medicine U. London, U. Durham, Royal Coll. Physicians, Med. Dept. Army. Mem. senate U. London. Life gov. U. London. Fellow Royal Coll. Physicians (Harveian orator 1903), Royal Soc. Edinburgh; mem. Med. Soc. London (pres. 1901-02). Author: Duodenal Indigestion (Bradshaw lecture Royal Coll. Physicians), 1891; The Breaking Strain (oration Med. Soc. London), 1896; Some Relationships of Indigestion (Huntarian Soc. lecture), 1897; Nutrition and Malnutrition (Lumeian lectures Royal Coll. Physicians), 1905; contbr. to Quain's Dictionary Medicine, Keating's Cyclo. Diseases of Children, Allbutt's System Medicine, other publs. Editor: A Manual of Medicine, 5 vols. Died Feb. 8, 1911.

ALLCUT, Edgar Alfred, mech. engr.; b. Birmingham, Eng., June 30, 1888; s. Thomas and Mary Jane (Jones) A.; B.Sc. with honors, U. Birmingham, 1908, M.Sc. in Engring., 1909; M.E., U. Toronto (Ont., Can.), 1931; m. Annie Josephine Walker, June 7, 1919; children—Monica Mary (Mrs. Philip Fitz-James), Stella Margaret (Mrs. Robert Muir). Asst. engr. Humphrey Pump Co., London, Eng., 1910-13; mgr. engring. and testing dept. W. & T. Avery, Ltd., Birmingham, 1913-17; chief insp. materials Austin Motor Co. (Eng.), 1917-20; faculty U. Toronto, 1921-56, prof., 1931-56, head dept. mech. engring., 1944-56; cons. engr., Toronto, 1921—. Chmn. air pollution bd. City of Toronto, also cons. Met. Toronto, 1949—; mem. Labor Safety Council Ont., 1961-65; commr. Ont. Energy Bd., 1960-66; chmn. com. substitute fuels NRC Can., 1941-45, mem. asso. com. on nat. bldg. code for Can., 1960-66. Recipient Heslop medal U. Birmingham, 1909; Plummer medal Engring. Inst. Can., 1943, Gzowski medal, 1947, Duggan medal, 1953; Akroyd Stuart prize Inst. Mech. Engr., 1930. Fellow Royal Aero. Soc., Am. Soc. M.E. Author: Materials and Engineering Design, 1922; Engineering Inspection, 1922; Principles of Industrial Management, 1950; Heat Engines, 1943; also numerous articles. Research in fuels and combustion, gasoline and diesel engines, heat transmission and conservation, air pollution and its control. Address: 315 Lawrence Av. W., Toronto 12, Ont., Can.*

ALLEAUME, Jacques, French mathematician; b. France, 1562; s. Jean and Marguerite (Auaye) A.; pupil of Vieta; flourished in Paris; probably royal engr.; interested in astrology; responsible for mech. div. circle by compass; worked with burning glasses, also machine for shaping parabolic lenses; drew up table of longitudes. Author: Confutatio . . . problematis ab Henrico Monatholio, 1600; La perspective spéculative et pratique. Translator: Book of Rantzovius on Nativities. Died 1627.

ALLEE, Warder Clyde, Am. zoölogist; b. nr. Bloomingdale, Ind., June 5, 1885; s. John Wesley and Mary Emily (Newlin) A.; S.B., Earlham Coll., 1908, LL.D., 1940; S.M. U. Chgo., 1910, Ph.D., 1912; m. Marjorie Hill, Sept. 4, 1912 (dec. 1945); children—Warder, Barbara Elliott, Mary Newlin; m. 2d, Ann Silver, June 26, 1953. Asst. in zoölogy U. Chgo., 1910-12; instr. botany, U. Ill., 1912-13, in zoölogy, Williams Coll., 1913-14; asst. prof. zoölogy Okla. U., 1914-15; prof. biology Lake Forest Coll., 1915-21; asst. prof. zoölogy U. Chgo., 1921, asso. prof., 1923, prof., 1928-50, emeritus, from 1950; head prof. biology U. Fla., 1950-55; instr. Marine Biol. Lab., Woods Hole, Mass., summers 1914-21, invertebrate course, 1918-21, trustee, from 1932; lectr. zoölogy, U. Cal., winter 1923; prof. zoölogy Nat. Summer Sch., Logan, Utah, 1924-26. Fellow A.A.A.S. (v.p. 1942), Am. Soc. Entomologists, Am. Acad. Arts and Scis.; mem. Am. Soc. Zoologists (pres. 1936), Nat. Acad. Sci., Ecol. Soc. Am. (pres. 1929), Am. Soc. Naturalists, Gamma Alpha, Phi Beta Kappa, Sigma Xi. Author: Animal Aggregations, 1931; Animal Life and Social Growth, 1932; The Social Life of Animals, 1938; Cooperation among Animals, with Human Implications, 1951. Co-author: Jungle Island (with Marjorie Hill Allee), 1925; Nature of the World and Man, 1926; The World and Man, 1937; Ecological Animal Geography, 1937; A Laboratory Introduction to Animal Ecology and Taxonomy, 1939; Principles of Animal Ecology, 1949; also tech. articles. Mng. editor Physiological Zoology, 1930-55. Chmn. com. on revision zool. articles Ency. Brit., 1944-50. Research in ecology of marine, fresh water and tropical rain forest habitants, emphasizing individual animal reaction to environment and group behavior. Died Mar. 18, 1955.

ALLEGRI, Annibale, Italian physician; b. Codogno (Milan), Italy, Apr. 18, 1916; s. Emilio and Maria (Malvicini) A.; Dr. of medicine and surgery. Prof. pathology, clin. methodology, med. therapy and gen. clin. medicine; practice medicine specializing in diseases of digestive parts, heart and circulation; physician in chief Santa Clara civil hosp., Trento, Italy; pres. Surg. Medicine Soc. Trento. Mem. Pvt. Council Pub. Health, Soc. Internal Medicine, Italian Soc. Cardiology, Hematology, Rheumatology, Gerontology. Research, publs. on diseases of heart, liver and blood. Home: viale N. Bolognini 58, Trento. Office: Opsedale civile Santa Chiara, via S. Croce 65 Trento, Italy.

ALLEN, Arthur Augustus, Am. ornithologist; b. Buffalo, Dec. 28, 1885; s. Daniel Williams and Anna (Moore) A.; A.B., Cornell U., 1907, M.A., 1908, Ph.D., 1911; m. Elsa Guerdrum, Aug. 17, 1913; children—Constance, Glen Olaf, Phebe Laura, David Guerdrum, Prudence Lloyd. Asst., later instr. in zoology Cornell U., 1906-15, asst. prof. ornithology, 1915-25, prof., 1925-53, emeritus, 1953—, dir. lab. ornithology, 1955—; explorer and lectr. in ornithol. fields, 1912—. Ornithol. expdns. to Colombia, 1912, Labrador, 1918, 1945, 1953; Hudson Bay, 1934-1944, 1954; Europe, 1938. Panama, 1944-45; OSRD project on jungle acoustics, for War Dept. Mexico. 1946, 1955, Alaska, 1948. Fellow Am. Orinthol. Union; mem. Wilson Ornithol. Club, Cooper Ornithol. Club Am. Soc. Mammalogists, Am. Soc. Naturalists, Wildlife Soc. (pres. 1938), Sigma Xi, Gamma Alpha. In charge of ruffed grouse investigations of Am. Game Assn.; discoverer of method of rearing ruffed grouse in captivity. Recipient Outdoor Life medal, 1924, for research on ruffed grouse, Burr prize Nat. Geog. Soc., 1948. Author: American Bird Biographies, 1934; The Golden Plover and Other Birds, 1939; American Bird Songs and Voices of the Night, albums of phonograph records (with P. P. Kellogg), 1942, 48, 55; Stalking Birds with Color Camera, 1951; Book of Bird Life, 2d edit. 1961; Bird Songs in Your Garden, 1961; also articles on birds in encys., sci. jours., Nat. Geog. Mag., others. Made valuable studies of birds' hearing ability, roosting habits, feeding of water fowl, ruffed grouse in captivity, other aspects of bird life. Died Jan. 24, 1964.

ALLEN, Augustine Oliver, Am. chemist; b. San Rafael, Cal., July 16, 1910; s. Edward O. and Winifrede (Augustine) A.; B.S. U. Cal. at Berkeley, 1930; Ph.D., Harvard, 1938; m. Agnes Robinson, July 2, 1938; children—Edith (Mrs. Roy Schult), Barbara (Mrs. William Fuchsman), Winifred, Catherine, Edward. Harvard traveling fellow U. Manchester, Eng., 1935-37; staff Ethyl Corp., Research Lab., Detroit, 1937-43, Metall. Lab., Manhattan dist. Atomic Energy Project, Chgo., 1943-46; staff Oak Ridge Nat. Lab., 1946-48; sr. chemist Brookhaven Nat. Lab., Upton, N.Y., 1948—; vis. scientist Nuclear Research Center, Athens, Greece, 1965-66. Mem. Radiation Research Soc. (pres. 1963-64). Author: The Radiation Chemistry of Water and Aqueous Solutions, 1961. Editorial bd. Jour. Phys. Chemistry, 1960—. Research, numerous articles on radiation chemistry; established a rational basis for study of water decomposition in nuclear reactors. Home: Woodville Rd., Shoreham, L.I., N.Y. 11786. Office: Brookhaven Nat. Lab., Upton, L.I., N.Y. 11973.*

ALLEN, Bennet Mills, Am. zoologist, embryologist; b. Greencastle, Ind., July 4, 1877; s. Albert and Alice (Bennet) A.; Ph.B., DePauw, 1898; Ph.D., Chicago, 1903. Instr. anatomy U. Wis., Madison, 1903-08, asst. prof., 1908-13; prof. U. Kansas, 1913-22; asso. prof. biology U. Cal. at Los Angeles 1922-24, prof. zoology, 1924-47, emeritus, 1947; with Atomic Energy Project, from 1947. Mem. Soc. Zoology, Soc. Exptl. Biology, Anat. Assn. Research on embryonic devel. of ovary and testis of mammals, anatomy of Hirudo, eye of Bdellostoma, devel. of rete cords, sex cords and sex cells of Chrysemys, thyroid gland and influence of glands on growth; irradiation effects on cells. Died Dec. 12, 1963.

ALLEN, Charles Elmer, Am. botanist; b. Horicon, Wis., Oct. 4, 1872; s. Charles and Eliza (North) A.; B.S., U. Wis., 1899, Ph.D., 1904; Sc.D., U. Chgo., 1941; m. Genevieve Sylvester, June 20, 1902; children—Edith (Mrs. C. R. Slater), Harold Sylvester, Charles Rittenhouse. Ct. reporter, 1890-1901; instr. botany U. Wis., 1901-04, asst. prof., 1904-07, asso. prof., 1907-09, prof., 1909-43, emeritus, from 1943; research asst. Carnegie Inst. of Washington, at U. Bonn (Germany), 1904-05; vis. prof. Columbia U., 1924. Mem. NRC, 1925-31, chmn. div. biology and agr., 1929-30. Fellow A.A.A.S. (v.p.); mem. Nat. Acad. Scis., Am. Soc. Naturalists (pres. 1936),

Genetics Soc. Am., Am. Genetic Assn., Am. Microscopic Soc. (pres. 1944), Am. Philos. Soc., Bot. Soc. Am. (pres. 1921), Am. Assn. U., Profs., Phi Beta Kappa, Sigma Xi. Author: A Textbook of Botany (with Edward M. Gilbert), 1917; A textbook of General Botany (with Gilbert M. Smith and others), 1924. Editor-in-chief Am. Jour. Botany, 1918-26. Contbr. to scientific publs. Research in cytology (particularly cell division), meiosis (demonstrated that some algae divisions are meitoic), heredity (discovered number of mutations in vegetable characteristics). Died June 24, 1954.

ALLEN, Charles Francis Hitchcock, Am. chemist; b. Milford, N. H., Aug. 12, 1895; s. Frederick James and Maria (Hitchcock) A.; A.B., Boston U., 1919, A.M., 1920, D.Sc., 1944; A.M., Ph.D., Harvard, 1924; D.Sc., McGill U., Montreal, Que., Can., 1937; m. Alberta Currie, Apr. 3, 1920; children—Phyllis Allen (Mrs. James Hugh Richmond), Ruth Allen (Mrs. Gerald John DeBoer). Faculty, Tufts Coll., 1924-29, McGill U., 1929-37; asst. div. head in charge organic and polymer research Eastman Kodak Co., 1937-61; prof. chemistry Rochester (N.Y.) Inst. Tech., 1961—. Mem. Am. Chem. Soc. Contbg. author: Organic Chemistry An Advanced Treatise, 1942; Ency. Chem. Tech., 1952; also numerous articles. Sec., editor Organic Synthesis, 1927—. Patentee photographic chemistry devices. Research in organic chemistry. Home: 284 Maplewood Av., Rochester, N.Y. 14613.*

ALLEN, Clabon Walter, astronomer; b. Subiaco, Australia, Dec. 28, 1904; s. James B. Allen; ed. western Australia; m. Rose Smellie, May 25, 1937; 5 sons. Asst. Commonwealth Obs., 1926-51; prof. astronomy U. Coll., London, Eng., 1951—. Mem. solar eclipse expdns., 1936, 40, 54, 55; pres. Inter-Union Commn. on Solar-Terrestrial Relations. Mem. Royal Astron. Soc. (fgn. corr.), Internat. Astron. Union. Author: Astrophysical Quantities, 1955. Research, publs. on physics, spectroscopy. Home: 43 Lawrence Garden, Mill Hill. Office: U. London Obs., Mill Hill Park, London, N.W. 7, Eng.

ALLEN, Doris Twitchell (Mrs. Erastus Smith Allen), Am. psychologist; b. Old Town, Me., Oct. 8, 1901; d. Asa Howard and Cora (Snow) Twitchell; A.B., U. Me., 1923, M.A., 1926, Sc.D., 1965; Ph.D., U. Mich., 1930; m. Erastus Smith Allen, Oct. 26, 1935; 1 son, Erastus T. Dir. field lab. Child Edn. Found., N.Y.C., 1932-35; founder psychol. services Children's Hosp., Cin., 1936, attending psychologist, 1936-48; chief psychologist Longview State Hosp., Cin., 1944-57; founder Children's Internat. Summer Villages, Inc., Cin., 1946-51, internat. pres., 1951-56, U. S. pres., 1956-65, mem. internat. bd. dirs., 1951-65; hon. counselor to internat. bd., 1965—; faculty U. Cin., 1949—, adj. prof. psychology Grad. Sch. and Arts & Scis., 1962—; vis. prof. psychology U. Me., summers 1962—. Recipient Gold medal for Outstanding Work in Internat. Relations, City of Stockholm, 1953, Les Palmes Academiques French Govt., 1961. Diplomate Am. Bd. Examiners in Profl. Psychology. A.A.A.S., Am., Ohio, Cin. psychol. assns., Soc. for Projective Techniques, Am. Soc. Group Psychotherapy and Psychodrama, Internat. Soc. Applied Psychology, Interam. Soc. Psychology, Internat. Council Psychologists (award for contbrn. to internat. relations 1962, president-elect 1967); mem. Sigma Xi, Phi Beta Kappa, Phi Kappa Phi, Sigma Phi, Delta Kappa Gamma. Author: Twitchell-Allen Three Dimensional Personality Test, 1940, 41; (with W. P. Matthews) A Handbook for Procedure for Children's International Summer Villages, 1961. Research in internat. relations through Children's Internat. Summer Villages, Inc. Home: 2447 Clybourn Pl., Cin. 45219.*

ALLEN, Duff S(hederic), Am. surgeon; b. Lebanon, Mo., July 7, 1895; s. William Thomas and Mary Elizabeth (Casey) A.; student Washington U., 1913-15, M.D., 1919; m. Mildred Lucille Burns, 1926. House physician St. Luke's Hosp., 1919; asst. in surgery Washington U., 1921-24, instr. in surgery, 1924-26, asst. prof., from 1926; resident surgeon Barnes Hosp., 1923-24, later asst. surgeon; asst. surgeon St. Louis Maternity Hosp.; vis. surgeon, chief of unit St. Louis City Hosp. Editor Washington U. Med. Alumnus, 1924; asso. editor Jour. Thoracic Surgery, Diplomate Am. Bd. Surgery. Mem. Phi Beta Pi (nat. pres.). Worked out surg. procedure for reconstruction of the esophagus; devised needle which reduces time required for an operation 5 minutes; devised first method for doing operations inside a living heart under direct vision. Contbr. studies on goitre and on the etiology of emphysema and abscess of the lung, the mechanism of secretion of gastric juice, and method of absorption of egg white from gastro-intestinal tract. Died Dec. 7, 1958.

ALLEN, Durward Leon, Am. biologist; b. Uniondale, Ind., Oct. 11, 1910; s. Harley J. and Jennie M. (LaTurner) A.; A.B., U. Mich., 1932; Ph.D., Mich. State Coll., 1937; m. Dorothy Ellen Helling, Sept. 23, 1935; children—Stephen R., Harley W., Susan E. Game research biologist Mich. Dept. Conservation, 1935-46; wildlife research biologist U. S. Fish and Wildlife Service, Laurel, Md., 1946-50, asst. chief br. wildlife research, Washington, 1951-54; prof. wildlife mgmt. Purdue U., Lafayette, Ind., 1954—. Recipient medal of honor Anglers' Club of N.Y., 1956. Fellow A.A.A.S.; mem. Wildlife Soc. (pres. 1956-57, Ann. Tech. Publ. award, 1946, Ann. Conservation Edn. award, 1955), Washington Biologists' Field Club, Am. Soc. Mammalogists, Ecol. Soc. Am., Am. Fisheries Soc., Seminarium Botanicum, Sigma Xi, Phi Sigma, Xi Sigma Pi. Author: Michigan Fox Squirrel Management, 1943; Pheasants Afield, 1953; Our Wildlife Legacy, 1954; The Life of Prairies and Plains, 1967. Editor, Pheasants in North America, 1956. Home: Route 10, Box 24, West Lafayette, Ind. 47906.*

ALLEN, Edgar, Am. anatomist; b. Canon City, Colo., 1892; s. Asa and Edith (Day) A.; Ph.B., Brown U., 1915, A.M., 1916, Arnold fellow, 1916-17, Ph.D., 1921, Sc.D., 1936; m. Marion Robins Pfeiffer, June 26, 1918; children—Frances Isabelle, Marjorie Eleanor. Asst. in biology, Brown U., 1913-15, asst. in embryology and neurology, 1915-17; investigator, U. S. Bur. Fisheries, Woods Hole, Mass., summer 1919, Fairport, Ia., summer 1922; instr. and asso. in anatomy, Washington U. Sch. of Medicine, 1919-23; prof. anatomy, Univ. Mo. from 1923, also dean med. sch., 1929-33; prof. anatomy, Yale U. Sch. of Medicine from 1933. Mem. A.A.A.S., Assn. Anatomists, A.M.A., Soc. Study Internal Secretions, Soc. Zoologists, Phi Gamma Delta, Phi Beta Pi. Contbr. to profl. jours. Research (with E. A. Doisy) on extraction of active principle from ovarian hormones, 1923, on menstrual flow (demonstrating it is withdrawal expression of endometrium resulting from cessation of estrogen activity in body), 1927. Died Feb. 3, 1943.

ALLEN, Edwin Brown, Am. mathematician, educator; b. Westerly, R.I., Nov. 3, 1898; s. John and Mary F. (West) A.; E.E., Rensselaer Poly. Inst., 1920, M.S., 1930, Ph.D., 1934; A.M., Harvard, 1934; postgrad. Columbia, U. Chgo.; m. Helen Cornelia Mackay, Sept. 3, 1927. Faculty, Rensselaer Poly. Inst., Troy, N.Y., 1920—, prof. math. and astronomy, 1934-64, head dept. math., 1934-60, dean Grad. Sch., 1959-64, prof., dean Grad. Sch. emeritus, 1964—, coordinator U. S. Naval Flight Prep. Sch., 1942-43. Research worker Watervliet (N.Y.) Arsenal, 1943-45; mem. various coms. N.Y. State Dept. Edn., 1934—; dir. Gen. Electric Math. Fellowship Program, summers 1952-59, NSF Summer Inst., 1957-59. Mem. Am. Math. Soc., Math. Assn. Am. (bd. govs. 1949-52), Soc. for Symbolic Logic, History of Sci. Soc., Am. Oriental Soc., Am. Research Center in Egypt (trustee 1951-63), Am. Schs. Oriental Research, Sigma Xi, Tau Beta Pi, Pi Mu Epsilon, Eta Kappa Nu, Alpha Tau Omega. Author: (with D. Maly and S. H. Starkey) Vital Mathematics, 1944; various papers on history of sci. in Orient. Home: 6 Fairlawn Lane, Troy, N.Y. 12180.*

ALLEN, Fred Harold, Jr., Am. physician; b. Holyoke, Mass., Feb. 23, 1912; s. Fred H. and Harriet (Ives) A.; A.B., Amherst Coll., 1934; M.D., Harvard, 1938; m. Frances Williams Brown, July 16, 1938; children—Philip B., Mark H., Barbara J. (Mrs. Donald R. Brewster), Dwight B. Practice medicine specializing in pediatrics, Holyoke, 1946-47; asso. dir. Blood Grouping Lab. of Boston, 1947-63; asst. clin. prof. pediatrics Harvard, 1958-63; dir. labs. New York Blood Center, Community Blood Council Greater N.Y., 1963; clin. asso. prof. pediatrics Cornell U., 1963—. Recipient Landsteiner award Am. Assn. Blood Banks, 1963. Mem. Phi Kappa Psi. Author: (with Louis K. Diamond) Erythroblastosis Fatalis, 1958. Chief editor Vox sanguinis, 1963—. Contbr. articles to tech. jours. Discovered relationship of jaundice to brain damage in newborn infants and prevention by exchange transfusion; participated in discovery of several new blood-group factors. Home: 3 Merestone Terrace, Bronxville, N.Y. 10708. Office: 310 E. 67th. St. N.Y.C. 10021.*

ALLEN, Frederick Graham, Am. physicist; b. Boston, Feb. 2, 1923; s. Frederick W. and Agnes (Horner) A.; B. Mech. Engring., Cornell U., 1945; Ph.D. in Applied Physics, Harvard, 1956; m. Susan Kate Morse, Sept. 21, 1949; children—Warren Katherine, Peter. Instr., Robert Coll., Istanbul, Turkey, 1946-49; mem. tech. staff Bell Telephone Labs., Murray Hill, N.J., 1956-66, with Bellcomm, Inc., Washington, 1966—. Fellow Am. Phys. Soc. (sec.-treas. div. electron physics 1965-68); mem. Sigma Xi, Phi Kappa Phi, Tau Delta Phi. Editorial bd. Jour. Vacuum Sci. and Tech., 1964-67. Research in elec. properties of atomically clean semicondr. surfaces, field emission and photoelectric emission from silicon and germanium. Home: 3909 Aspen St., Chevy Chase, Md. Office: 1100 17th St., Washington.*

ALLEN, George Cyril, Brit. economist; b. Kenilworth, Eng., June 28, 1900; s. George Henry and Elizabeth (Sharman) A.; M.Com., U. Birmingham (Eng.), 1922, Ph.D., 1928; m. Eleanora Shanks, Dec. 21, 1929. Lectr. econs. Koto Shogyo Gakko, Nagoya, Japan, 1922-25; lectr. U. Birmingham, 1925-29; prof. econs. U. Coll., Hull, 1929-33; prof. econ. sci. U. Liverpool, 1933-47; prof. polit. economy U. London, 1947-67, prof. emeritus, 1967—. Mem. Monopolies Commn. U.K., 1950-62; mem. acad. planning bd. New U. Ulster, Eng., 1965—. Decorated commdr. Order Brit. Empire; Order Rising Sun 3d Class (Japan). Fellow Brit. Acad.; mem. Royal Econ. Soc. Author numerous books including: (with A. G. Donnithorne) Western Enterprise in Far Eastern Economic Development, 1954, Western Enterprise in Indonesia and Malaya, 1957; Japan's Economic Recovery, 1958; The Structure of Industry in Britain: A Study in Economic Change, 1961, 66; Japan's Economic Expansion, 1965; Japan as a Market and Source of Supply, 1967; Monopoly and Restrictive Practices, 1968. Research, publs. on indsl. orgn. and structure, especially Brit. industry and Brit. indsl. history, Japan's econ. devel., past and recent, Western enterprises in Far East; analysis of indsl. devel. including regional devel. Home: Quinces, Beech Close, Cobham, Surrey, Eng.*

ALLEN, (Charles) Grant, sci. writer; b. Kingston, Ont., Can., Feb. 24, 1848; s. Joseph and Charlotte (Grant) A.; ed. U. S., France and Eng.; A.B., Merton Coll., Oxford (Eng) U., 1871; m. 2d, Ellen Jerrard, 1873; 1 son. Prof. mental and moral philosophy in coll. for edn. Negroes, Spanish Town, Jamaica, 1873-76; engaged in writing in Eng., from 1876. Author: Physiological Aesthetics; Charles Darwin, 1885; The Evolutionist at Large, 1883; The Color Sense, 1879; Phillistia, 1884; Babylon, 1885; Colin Clout's Calendar, 1882; The Color of Flowers, 1882; Flowers and Their Pedigrees, 1883; White Man's Foot, 1888; Force and Energy, 1888; The Tents of Shem; Post-Prandial Philosophy, 1894; Anglo-Saxon Britain, 1881; The British Barbarians, 1895; Science in Arcady, 1892; Historical Guides to Paris, Florence and Belgium, 1897; The Evolution of the Idea of God, 1897; numerous others. An ardent propagator of Darwinian theories; cultural exponent of evolutionary idea in various aspects of biology and anthropology. Died Surrey, Eng., Oct. 24, 1899.

ALLEN, Sir Harry Brookes, Australian pathologist; b. June 13, 1854; s. Thomas Watts Allen; M.B., Melbourne (Australia) U., 1876, M.D., 1878, B.S., 1879; LL.D., Edinburgh (Scotland) U., 1912, U. Adelaide (Australia), 1914; m. Ada Mason, 1891; 3 daus. Demonstrator anatomy Melbourne U., from 1876, lectr., from 1881; prof. anatomy and pathology, from 1882, prof. pathology, 1906-24, dean Faculty Medicine; pathologist Melbourne Hosp., from 1876. Mem. Tb in Cattle Bd. (joint author report), Central Bd. Health, 1883-84; pres. Royal Commn. on Sanitary State of Melbourne, Intercolonial Rabbit Commn., 1888-89; gen. sec. Intercolonial Med. Congress, 1889; chmn. exec. and finance com. Australian Red Cross Soc. Mem. Med. Soc. Victoria (pres. from 1892), United Med. Socs. Victoria (1st pres. from 1908). Author: Typhoid Fever and Milk Supply, 1889; Tumors of Kidney, 1896; Tumors, 1903; Tubercle of Joints, 1904; Report to Government on Health Conditions at Panama, 1913; on Risks of Middle Age and Specific Diseases, 1916. Editor: Australian Med. Jour., 1879-83. Drafted model by-laws for local bds. health; advised revolutionizing constrn. freezing chambers for meat preserving cos., 1883; created mus. pathology Melbourne U. Died Mar. 28, 1926.

ALLEN, Harry Clay, Jr., Am. chemist; b. Saugus, Mass., Nov. 26, 1920; s. Harry Clay and Sarah E. (Thorburn) A.; B.S., Northeastern U., 1948; M.S., Brown U., 1949; Ph.D., U. Wash., 1951; postgrad. Cambridge U., Eng.; m. Carolyn Alderman Bliss, Feb. 1, 1948; children—Carol B., Paul T. AEC fellow Harvard, Cambridge, Mass., 1951-53; asst. prof. physics Mich. State U., East Lansing, 1953-54; physicist radiometry sect., atomic and radiation physics div. Nat. Bur. Standards, Dept. Commerce, Washington, 1954-61, chief analytical and inorganic chemistry div., 1961-63, chief inorganic materials div., 1963-65, dep. dir. Inst. for Materials Research, 1965-66; asst. dir. minerals research U. S. Bur. Mines, Washington, 1966—. Recipient Exceptional Service award U. S. Dept. Commerce, 1964; Samuel Wesley Stratton award Nat. Bur. Standard, 1965. Author: Molecular Vib-Rotors, 1963. Research on molecular structure, molecular dynamics, molecular spectroscopy, interaction of electromagnetic radiation with matter. Home: 13009 Carney St., Silver Spring, Md. Office: U. S. Bur. Mines, 18th and E streets N.W., Washington 20240.*

ALLEN, Harry Willis, Am. entomologist; b. Pelham, Mass., Aug. 8, 1892; s. Lyman Wallace and Mary (Towne) A.; B.S., U. Mass., 1913; M.S., Miss. State U., 1922; Ph.D., Ohio State U., 1926; m. Margaret Johnson Warren, Dec. 27, 1919; children—Dorothy Louise, Richard Warren. Sci. asst. Bur. Entomology, U. S. Dept. Agr., 1913-17, 19-20; faculty Miss. State U., 1920-26; entomologist. Agr. Research Service, U. S. Dept. Agr., Moorestown, N.J., 1926-58, collaborator, 1958—; research asso. Acad. Natural Scis. Phila., 1958-63, fellow, 1963—. Mem. Am. Entomol. Soc. (past pres.), Entomol. Soc. Am., Sigma Xi. Research, numerous publs. on biol. control insect pests, specialist in classification of wasps. Home: 411 S. Washington Av., Moorestown, N.J. 08057. Office: Dept. Entomology, Acad. Natural Scis., 19th St. and Parkway St., Phila. 19103.

ALLEN, Herbert Stanley, Brit. physicist; b. Bodmin, Eng., Dec. 29, 1873: s. R. Allen; M.A., U. Cambridge (Eng.); D.Sc., U. London; LL.D., U. St. Andrews; m. Jessie Macturk, 1907; 1 son, 1 dau. Became supt. Blythswood Lab., Renfrew, Eng., 1900; apptd. lectr. physics King's Coll., London, 1905; reader in physics U. London, later U. Edinburgh (Scotland); prof. natural philosophy U. St. Andrews (Scotland), 1923-44, dir. physics research lab. Fellow Royal Soc., 1930, Royal Soc. Edinburgh, Inst. Physics. Author: Photoelectricity: the Liberation of Electrons by Light; (with H. Moore) Text-book of Practical Physics 3d edit., 1948; The Quantum and its Interpretation; Electrons and Waves; (with R. S. Maxwell) Text-book of Heat. Contbd. papers to proc. royal socs. London, Edinburgh. Died April 27, 1945.

ALLEN, James Harrill, Am. ophthalmologist; b. Chattanooga, Jan. 31, 1906; s. George Henry and Mary (Harrill) A.; B.A., U. Tenn., 1926; M.D., U. Mich., 1930; M.S. in Bacteriology, U. Ia., 1938; m. Ruth Collin Sanford, Aug. 17, 1934; children—Mary Helen, George Sanford, John Robert. From intern to prof. U. Ia., 1930-50; prof. ophthalmology Tulane U., New Orleans, 1950-—, chmn. dept. ophthalmology, 1953-67, asso. dean, med. dir. Tulane Clinics, 1967-—; sr. surgeon, dir. tng. ophthalmology Eye, Ear, Nose and Throat Hosp., 1950-—, sr. surgeon, chief Tulane div. ophthalmology Charity Hosp., New Orleans, 1950-67. Cons. to hosps. USPHS; mem. med. adv. bd. Nat. Council to Combat Blindness, 1951-—, Am. Soc. Ocularists, 1960-—; Mem. Am. Bd. Ophthalmology, 1950-58. Recipient Beverly Meyers Nelson Achievement award, Am. Bd. Opticianry, 1958. Diplomate Am. Bd. Ophthalmology. Mem. N.Y. Acad. Scis., So. Ocular Pathology Club, World, Am., So. med. assns., Assn. Am. Physicians and Surgeons, Assn. Research Ophthalmology (chmn. 1960), Am. Ophthal. Soc., Pan. Am. Assn. Ophthalmology, Aerospace Med. Assn., Assn. Mil. Surgeons, Am. Soc. Microbiology, Internat. Council Ophthalmology, A.A.A.S., Verhoff Soc., Contact Lens Assn. Ophthalmologists, Ophthal. Soc. Chile (hon.), Sigma Xi, Gamma Rho, others. Author: (with Earl Fisher, Jr.) Ocular Microbiology, 1956. Editor: Stabismus, Ophthalmic Symposium I, 1950, II, 1958; May's Manual of Diseases of the Eye, 1963, also editor, mem. editorial bds. numerous profl. jours. Research, publs. on conjunctivitis, various eye infections, diabetic retinal changes of eye, surg. eye technique, devel. of diagnostic equipment. Home: 9104 Quince St., New Orleans 70118.*

ALLEN, Joel Asaph, Am. zoologist; b. Springfield, Mass., July 19, 1838; s. Joel and Harriet (Trumbull) A.; studied at Wilbraham Acad.; then studied zoology under Agassiz, Lawrence Scientific Sch., Harvard; hon. Ph.D., Ind. U., 1886; m. Mary Manning Cleveland, Oct. 6, 1874 (dec. Apr. 1879); m. 2d, Susan A. Taft, Apr. 27, 1886. Lectr., Harvard, 1871-73, asst. in ornithology, Mus. Comparative Zoology, 1871-85; curator mammalogy and ornithology Am. Mus. Natural History, N.Y.C., from 1885. Editor Bull. Nuttall Ornithol. Club, 1876-83, The Auk (ornithol. Quar.), 1884-1912, Bull. and Memoirs, Am. Mus. Natural History, 1889-1917. Recipient Walker Grand Prize, 1903. Mem. Nat. Acad. Scis., Am. Ornithologists Union (1st pres. 1883-90). Fellow Am. Acad. Arts and Scis. Author: The American Bisons, Living and Extinct, 1876; History of North American Pinnipeds; Monographs of North American Rodentia (with Elliott Coues); Mammals of Patagonia, Belgian Congo, others. Studied adaptation of birds of geog. variations and corroborated Lamarckian hypothesis; studied relation of fauna to climate and geography; theorized that seasonal bird migration is primarily due to climatic changes occasioned by glaciation. Died Aug. 29, 1921.

ALLEN, John, Am. dentist, inventor; b. Broome County, N.Y., Nov. 4, 1810; s. Nirum Allen; M.D., Cincinnati Med. Coll.; studied dentistry under Dr. James Harris, Chillicothe, O.; m. Charlotte Dana, 1835, 1 son, Charles; m. 2d, Mrs. Cornelia Reeder, 1 dau. Practiced dentistry primarily in N.Y.C.; devised new denture; recipient gold medal from Am. Soc. Dental Surgeons for devising method of restoring facial contours which fell in due to previous denture practices, 1845; granted patent for new denture method, 1851; made false teeth of porcelain with platinum base instead of gold (enabled him to devise continuous gum to prevent oral fluids from entering into crevices of teeth); involved in much litigation over invention with Dr. William Hunter, Mem. Am. Soc. Dental Surgeons; a founder Ohio Coll. Dental Surgery, 1845, N.Y. Coll. Dentistry, 1865. Died Plainfield, N.J., Mar. 8, 1892.

ALLEN, John F., inventor; b. Eng., 1829. Came to Am., 1841; engr. aboard ship Curlew in Long Island Sound, invented new form of high speed valve motion; became engr. for Henry A. Burr, felt-hat body mfr., N.Y.C., 1860; formed (with Charles Porter, inventor of engine governor) Porter-Allen Co., produced Porter-Allen engine (pioneer high-speed steam engine); invented inclined tube vertical water-tube boiler; opened his own shop, Mott Haven, N.Y.; inventor of 2 pneumatic riveting systems (1 by percussion, 1 by pressure). Died N.Y.C., Oct. 4, 1900.

ALLEN, John Frank, physicist; b. May 6, 1908; s. Frank Allen; B.A., U. Man. (Can.), 1928; M.A., U. Toronto (Can.), 1930, Ph.D., 1933; M.A., Cambridge U., 1936; m. Elfriede Hiebert, 1933 (dissolved 1951); 1 son. Bursar, student and fellow Nat. Research Council Can., 1930-33; fellow NRC, U. S., 1933-35; research asst. Royal Soc. Mond Lab., Cambridge, Eng., 1935-44; lectr. physics U. Cambridge, also fellow and lectr. St. John's Coll., 1944-47; prof. natural philosophy U. St. Andrews (Scotland), 1947-—. Chmn. very low temperature commn. Internat. Union Pure and Applied Physics. Fellow Royal Soc., 1949; mem. Royal Soc. Edinburgh, Phys. Soc. London. Numerous sci. papers on exptl. low temperature physics. Research in superfluid helium and superconductivity. Address: Sch. Phys. Scis., North Haugh, Univ. St. Andrews, St. Andrews, Scotland.

ALLEN, J(oseph) Garrott, Am. physician; b. Elkins, W.Va., June 5, 1912; s. James Edward and Susan H. (Garrott) A.; student Davis and Elkins Coll., 1930-32; A.B., Washington U., St. Louis, 1932-36; M.D., Harvard, 1938; m. Dorothy O. Travis, July 15, 1940; children—Barry Worth, Edward Henry, Nannette Susan, Lester Travis, Joseph Garrott. Faculty, U. Chgo., 1943-59, prof. surgery, 1951-59, research asso. Manhattan project, 1944-46; group leader Argonne (Ill.) Nat. Lab., 1946-59; prof. surgery Stanford, 1959-—, exec. dept. surgery, 159-61; staff Palo Alto-Stanford Hosp., 1959-—. Recipient John J. Abel prize for research in irradiation injury Am. Soc. for Pharmacology and Exptl. Therapeutics, 1948, Gold Medal for original research Ill. State Med. Soc., 1948, 54, Samuel D. Gross award Phila. Acad. Surgery, 1955, Diplomate Am. Bd. Surgery. Mem. A.M.A., Am. Surg. Assn., Am. Assn. Blood Banks (1st John Elliott award 1956), Soc. Exptl. Biology and Medicine, Am. Physiol. Soc., A.C.S. (Chmn. com. on blood and allied problems), Soc. Clin. Surgery. Author: (with Jonathan Rhoads, Henry Harkins, Carl Moyer) Surgery, Principles and Practice, 1957, 61, 65; also numerous articles. Editor: Peptic Ulcer, 1959; chief editor Archives of Surgery, 1962-—. Introduced protamine sulfate clinically to neutralize anticoagulant heparin; research on theoretical basis and proof that blood coagulation is dynamic and vital process, human plasma containing hepatitis virus rendered safe by 180-day or longer incubation at 89°F. constant temperature, protective value convalescent plasma, virus-inactivated given every 2 weeks between 8th to 60th day after infectious blood administered. Home: 300 Pasteur Dr., Palo Alto, Cal. 94304.*

ALLEN, John Stuart, Am. astronomer; b. Pendleton, Ind., May 13, 1907; s. Elwood D. and Stella (Anderson) A.; B.A., Earlham Coll., 1928; M.A., U. Minn., 1929; Ph.D., N.Y. U., 1936; Sc.D., U. Tampa, 1958; LL.D., Earlham U., 1960; m. Grace Carlton, Aug. 23, 1933. Faculty, Colgate U., 1930-42, asst. prof., 1935-41, dean, 1941-42; bd. regents, dir. div. higher edn. State of N.Y., 1942-48; v.p. U. Fla., Gainesville, 1948-55, acting pres., 1955-57; pres. U. South Fla., Tampa, 1957-—. Chmn., Fla. State Fulbright Com., 1950-57. Named Outstanding Alumnus, U. Minn., 1962. Fellow A.A.A.S.; mem. Am. Astron. Soc., Royal Astron. Soc. Can., Sigma Xi, Sigma Pi Sigma. Author: (with French, Henshaw, Trainer, Woodruff) Atoms, Rocks and Galaxies, 1938; Astronomy, 1945; also articles. Research on binary star orbits. Home: 10911 Carrollwood Dr., Tampa 33618.

ALLEN, Mary Belle, Am. microbiologist; b. Morristown, N.J., Nov. 11, 1922; s. Frederick Madison and Mary Belle (Wishart) A.; B.S., U. Cal. at Berkeley, 1941; Ph.D., Columbia, 1946. Du Pont fellow Columbia, 1945-46; NRC fellow Washington U., St. Louis, 1946-47; research asso. Mt. Sinai Hosp., N.Y.C., 1947-49, Hopkins Marine Sta., Stanford, 1949-53; asst. research biochemist dept. plant nutrition U. Cal. at Berkeley, 1953-58, asso. dir. Lab. Comparative Biology, 1957-59, dir., 1959-64, sr. scientist, 1964-65; research chemist NIH, Bethesda, Md., 1965-66; prof. microbiology U. Alaska, College, 1966-—. Recipient Darbaker prize phycology Bot. Soc. Am., 1962. Fellow N.Y.C. Acad. Scis. mem. Am. Soc. for Microbiology (past chmn. com. on aquatic microbiology), No. Cal. Photobiologists and Photochemists (past sec.), A.A.A.S., Am. Chem. Soc., Am. Soc. Biol. Chemists, Am. Soc. Plant Physiologists, Am. Soc. for Cell Biology, Phycological Soc. Am., Soc. Protozoologists, Sigma Xi. Author: Comparative Biochemistry of Photoreactive Systems, 1962; also numerous articles. Research on photosynthesis, high temperature bacteria, isolation, cultivation, metabolism and pigmentation of freshwater and marine algae; co-discoverer photosynthesis by isolated chloroplasts. Address: U. Alaska, College, Alaska 99701.*

ALLEN, Merlin W., Am. nematologist; b. Wellsville, Utah, Dec. 1, 1912; s. William J. and Margaret (Walters) A.; B.A., Utah State U., 1935, M.A., 1937; Ph.D., U. Cal., 1947; m. Lois M. Allen, Apr. 29, 1938. With U. S. Dept. Agr., 1937-42; research asso. U. Cal. at Berkeley, 1943-47, asst. nematologist, 1947-53, asso. prof. nematology, 1953-59, prof. U. Cal. at Davis, 1959-—, chmn. dept., 1962-—. Research in taxonomy of plant parasitic nematodes. Home: 1102 W. 8th St., Davis, Cal.*

ALLEN, Merrill James, Am. optometrist; b. San Antonio, Aug. 2, 1918; s. Millard Henry and Mildred (Windnagle) A.; B.Sc., Ohio State U., 1941, M.Sc., 1942, Ph.D., 1949; m. Mary Joan Gower, Dec. 25, 1942; children—Marjorie Jayne, Merrill James. Physicist, Frankford Arsenal, Phila., 1943-44; Am. Optical Co. fellow Ohio State U., 1946-49, asst. prof., 1949-53; faculty Ind. U., Bloomington, 1953-—, prof. optometry, 1959-—. Mem. Ind. Gov.'s Adv. Com. on Driving Licenses, 1965-—; Gen. Services Adminstrn. Adv. Panel on Automotive Safety, 1966-—. Mem. Am. (com. on standards), Ind. (chmn. vision of aged com.) optometric assns., Am. Acad. Optometry, A.A.A.S., Optical Soc. Am., Assn. for Research in Ophthalmology, Am. Assn. Automotive Physicians, Physicians for Automotive Safety. Research, numerous publs. on visions and reforms of automobile visual design to provide greater safety. Home: 1311 Valley Forge Rd., Bloomington, Ind. 47401.*

ALLEN, Oscar Nelson, Am. bacteriologist, educator; b. Corsicana, Tex., May 15, 1905; s. Oscar Andrew and Lillie Mae (Dugan) A.; B.A., M.A., U. Tex., 1927; Ph.D., U. Wis., 1930; postgrad. Rothamsted Exptl. Sta., Harpenden, Eng., London (Eng.) Sch. Tropical Hygiene, Brown Inst., London; m. Ethel Elia Kullmann, July 11, 1930. Research asst. agrl. bacteriology U. Wis., Madison, 1927-28, instr., 1929-30, prof. bacteriology, 1946-—; asst. prof. bacteriology and plant pathology U. Hawaii, Honolulu, 1930-35, bacteriologist Grad. Sch. Tropical Agr., 1931-45, Cooperating soil bacteriologist Pineapple Research Inst., 1931-41, collaborator bacteriology and plant pathology Hawaii Agrl. Expt. Sta., 1932-41, asso. prof., 1935-40, bacteriologist Queens Hosp. Nursing Sch., 1935-41, acting chmn. dept. botany, 1934-36, 39-40, prof. bacteriology, 1940-45, chmn. dept. botany, 1940-42, vice chmn. dept. biol. scis., 1942, chmn. dept. bacteriology, 1943-45; chmn., prof. dept. bacteriology U. Md., 1945-46, Vis. prof. bacteriology U. Tex., summers 1937-39; del., mem. various congresses, confs. George Ives Haight Travelling awardee U. Wis., Far East, South Pacific, 1962. Fellow Am. Soc. Agronomy; mem. Am. Soc. for Microbiology (ad hoc. com. taxonomy com. 1963-—), Soil Sci. Soc. Am. (sect. vice chmn. sect. 1947, chmn. 1948), Stain Tech., Am. Assn. U. Profs., Bot. Soc. Am., Am. Phytopath. Soc., Canadian Soc. Microbiology, Internat. Soc. Soil Sci., Soc. for Gen. Microbiology, Sigma Xi, Phi Sigma, Gamma Alpha, Phi Kappa Phi, Alpha Gamma Rho. Author: Experiments in Soil Bacteriology, rev. edit. 1949, 51, 57; also numerous articles. Asso. editor: Bergey's Manual of Determinative Bacteriology (D. H. Bergey), 1948; cons. soils editor Agronomy Jour., 1948-54; co-editor Microbiology and Soil Fertility, 1965. Home: 4142 Hiawatha Dr., Madison, Wis. 53711.*

ALLEN, Raymond Archie, Am. physician, educator; b. Lyman, Utah, Nov. 6, 1921; s. Samuel F. and Helen (Oldroyd) A.; B.S., U. Utah, 1946; M.D., U. Louisville, 1946; M.S., U. Minn., 1954; m. Ethel Currier, July 1, 1960; children—Susanna, Timothy. Asst. in pathology and microbiology Rockefeller Inst., 1954-56; faculty pathology U. Cal. Med. Sch. at Los Angeles, 1956-—, prof., 1961-—. Recipient various awards for papers, exhibits. Mem. Am. Assn. Pathologists and Bacteriologists, Soc. for Exptl. Pathology, Coll. Am. Pathologists, Am. Soc. Clin. Pathologists. Contbr. articles to sci. jours. Pathologic studies of eye diseases, electron microscopic studies of retinal anatomy. Home: 16837 Addison St., Encino, Cal. 91316. Office: U. Cal. Med. Center, Los Angeles 90024.*

ALLEN, Richard Aubrey, Am. physicist; b. Marlboro, Mass., Mar. 23, 1934; s. Aubrey Clifford and Edna (Cadieux) A.; grad. St. Mark's Sch., 1952; A.B., Harvard, 1956; m. Petronella Regina Maria Remkes, Sept. 11, 1965. Faculty, St. Mark's Sch., 1958-59; mem. staff Avco Everett Research Lab., Everett, Mass., 1961-—. Mem. Am. Inst. Aeros. and Astronautics. Studies on understanding non-equilibrium thermal radiation produced in air shock fronts at super-satellite re-entry velocities, dissociation rate of nitrogen molecule, non-equilibrium overshoot of atomic line radiation in air shock fronts, possible existence of N- ion from shock tube studies, oscillator strengths for several N2 band systems.

ALLEN, Richard Sweetnam, Am. physiologist; b. Pekin, Ill., Nov. 9, 1896; s. William Henry and Margaret Anne (Olt) A.; student U. Chgo., 1917-18, 26-27, 36-37; B.S., U. Rochester, 1922, M.S., 1925; m. Leone Mergaret Stuart McLoughlin, Aug. 23, 1924; 1 son, William Henry, II. Research asso. U. Rochester (N.Y.), 1922-23, research asst., 1923-25; instr. coll. medicine U. Tenn., Memphis, 1925-26; faculty physiology U. Ky., Lexington, 1927-—, prof., 1936-—, head dept. anatomy and physiology, 1931-—, prof. physiology Med. Sch., 1960-—, pre-med. and predental student adviser, 1927-—, chmn. pre-med. and pre-dental recommendation and adv. coms., 1939-—. Mem. Am. Legion, Assn. Am. Med. Colls., A.A.A.S., Am. Genetic Assn., Nat. Geog. Soc., 40 and 8, Sigma Xi. Research, publs. on insulin, altitude effects on heart size, various metabolic studies on animals. Home: 1836 McDonald Rd., Lexington, Ky. 40503.*

ALLEN, Robert Day, Am. biologist; b. Providence, Aug. 28, 1927; s. Richard Day and Mary (Cottrell) A.; A.B. in Biology, Brown U., 1949; Ph.D. (USPHS fellow) U. Pa., 1953; m. Margaret Dampman Allen, Dec. 23, 1950; children—Elizabeth Hughes, Wayne Edgar. USPHS fellow Wenner-Gren Inst., Stockholm, Sweden, also Stazione Zoologica, Naples, Italy, 1953-54; instr. U. Mich., Ann Arbor, 1954-56; faculty Princeton, 1955-66, asso. prof. biology, 1961-66; prof., chmn. dept. biol. scis. State U. N.Y., Albany, 1966-—. Editor, producer sci. and ednl. films; vis. mem. Queens Coll., U. Cambridge (Eng.), 1966. Guggenheim fellow Osaka Japan U., 1961. Fellow A.A.A.S.; mem. Fedn. Am. Scientists, Biophys. Soc., Soc. Gen. Physiologists (past treas.), Am. Soc. Zoologists, Internat. Inst. Embryology, Am. Soc. for Cell Biology, Am. Soc. for Sci. Film, Sigma Xi. Author: (with N. Kamiya) Primitive Motile Systems in Cell Biology, 1964; also numerous articles. Developed improved light microscopes for biol. research; research on cell movement and cytoplasmic streaming, spindle fibers as refractile structures in living dividing cells, movement in lower organisms and single cells. Office: Dept. Biol. Scis., State U. N.Y., Albany, N.Y.*

ALLEN, Robert Francis, Am. clin. psychologist; b. N.Y.C., Dec. 26, 1928; s. Edwin E. and Mary (Thomson) A.; B.S., N.Y. State U., 1949, M.A., 1951; Ph.D., N.Y. U., 1957; postgrad. Columbia, Inst. Group Psychotherapy; m. Elaine J. Bender, Nov. 19, 1956; children—Judd, Peter. Research project dir. Fund for Advancement of Edn., Ford Found., 1955; dir. psychol. services Oceanside (N.Y.) Pub. Schs, 1955-58; practice psychology in Long Beach, L.I., N.Y., also Union, N.J., 1955—; academic dean, prof. Monmouth Coll., N.J., 1957-58; prof. psychology, dir. grad. studies Newark State Coll., Union, N.J., 1958—; asso. dir. Lab. for Applied Behavioral Sci., Union, 1962—; pres. Sci. Resources Assos., Union, 1965—. Cons. to colls. and univs., indsl. groups, clinics and hosps. Recipient N.Y. U. Founders Day award, 1957; Outstanding Community Service award Oceanside Mental Health Assn., 1958. Mem. Nat. Tng. Reps. (asso.), Am. Psychol. Assn., Soc. Psychol. Study Social Issues, Am. Acad. Polit. and Social Scis., N.Y. Soc. Clin. Psychologists, Am. Assn. U. Profs. (pres. coll. chpt., rep. to state conf.), Am. Assn. for UN, Am. Civil Liberties Union, Phi Delta Kappa. Author: Social Integration of Puerto Rican Children 1956; Authoritarianism as Manifested in Formal Education, 1958. Devel. design for tng. of behavioral scientists and mgmt. tng. personnel. Home: 1383 Brookfell Av., Union, N.J. 07083.*

ALLEN, Robert Scott, Am. biochemist; b. Tabiona, Utah, Nov. 13, 1917; s. Robert Ernest and Genevieve (Michie) A.; B.S., Brigham Young U., 1939, M.S., 1940; Ph.D., Ia. State Coll., 1949; m. Louise Pierce, Sept. 4, 1940; children—Don Robert, Gary Wayne, Ross Michael. Research asso. Ia. State U., Ames, 1947-49, faculty, 1949-67, prof., 1957-67, chmn. dept. biochemistry and biophysics, 1960-63; prof., head department biochemistry Louisiana State University, Baton Rouge, Louisiana, 1967—. Recipient Am. Feed Mfrs. award in dairy cattle nutrition, 1955; named Distinguished Prof. Agr., Ia. State U., 1965. Mem. Am. Chem. Soc. (past chmn. Ames sect.), Am. Dairy Sci. Assn., Am. Inst. Nutrition, Sigma Xi (past pres. Ia. chpt.), Phi Kappa Phi, Phi Lambda Upsilon, Gamma Sigma Delta. Editor: (with others) Physiology of Digestion in the Ruminant, 1965. Research, numerous publs. on biochemistry of vitamin A and carotene metabolism, lipid absorption and metabolism, forage preservation and utilization, etiology of bloat in ruminants. Home: 256 Court St., Baton Rouge 70810.*

ALLEN, Sally Lyman, Am. geneticist; b. N.Y.C., Aug. 3, 1926; d. Alexander V. and Dorothy (Rogers) Lyman; A.B., Vassar Coll., 1946; Ph.D. (John M. Prather, USPHS fellows), U. Chgo., 1954; m. John M. Allen, Mar. 10, 1951; 1 dau., Susan L. Research asst. R. B. Jackson Lab., Bar Harbor, Me., 1946-48, summers 1949, 50, 51; research zoologist U. Mich., Ann Arbor, 1953-55, research asso. dept. zoology, 1955—, associate professor of botany, since 1967—. Fellow A.A.A.S.; mem. Genetics Soc. Am., A.A.A.S., Am. Inst. Biol. Scis., Soc. Protozoologists, Am. Soc. Zoologists, Am. Soc. Naturalists, Am. Soc. for Cell Biology, Soc. Developmental Biologists, Sigma Xi, Phi Beta Kappa. Research in genetics of tumor transplantation in mice; host range in bacteriophage; protozoan genetics; current contbns. in developmental genetics and isozymes. Office: Depts. Botany and Zoology, U. Mich., Ann Arbor, Mich. 48104.*

ALLEN (or ALLEYN), Thomas, Brit. mathematician; b. Uttoxeter, Eng., Dec. 21, 1542; B.A., Oxford U., 1563, M.A., 1567; studied sci. under patronage Henry, Earl of Northumberland, also Count of Leicester; compiled manuscripts on sci., philosophy, history, archaeology (some preserved in Bodleian library, Oxford); pub. copy Ptolemy's De astrorum judiciis with notes. Skills in math. and astrology led his contemporaries to consider him a magician. Died Gloucester Hall, Sept. 30, 1632.

ALLEN, Victor Thomas, Am. geologist, educator; b. Dubuque, Ia., Oct. 28, 1898; s. William F. and Albina (Hauser) A.; A.B., U. Minn., 1921, M.S., 1922; Ph.D., U. Cal. at Berkeley, 1928; m. Murl Lynch, Aug. 17, 1932; children—Rosella Mary (Mrs. Thomas Verdon, Jr.), Robert Thomas. Asst. in geology U. Minn., 1921-22; Western Res. U., 1922-25; faculty U. Cal. at Berkeley, 1925-28; asso. prof., summers 1938, 62, 64; faculty St. Louis U., 1928—, prof., 1936—, dir. geology dept., 1946—. Commodity geologist U. S. Geol. Survey, 1942-44, sr. geologist, 1946—; mem. com. on sedimentation NRC, 1935-36. Fellow Geol. Soc. Am., Mineral. Soc. Am., Geol. Soc. Washington, A.A.A.S., Internat. Inst. Arts and Letters; mem. Am. Soc. for Engring. Edn., Sigma Xi. Author: This Earth of Ours, 1939. Research, publs. on petrography and geology of Ione formation, bentonites, glacial deposits, clays, bauxite deposits. Home: 117 Cornelia St., St. Louis 63122.*

ALLEN, Willard M(yron), Am. physician; b. Macedon, N.Y., Nov. 5, 1904; s. Lewis F. and Marion E. (Hoag) A.; B.S., Hobart Coll., 1926, D.Sc., 1940; M.S., U. Rochester Sch. Medicine and Dentistry, 1929, M.D., 1932, D.Sc., 1957; m. Julia Bell Gardner, 1927 (dec. 1941); 1 dau., Lucille. m. 2d, Dorothy Dunn Esley, 1946. Fellow NRC, Strong Meml. Hosp., Rochester, N.Y., 1932-33; faculty U. Rochester, 1934-40, fellow Gen. Edn. Bd., 1934-36; prof. obstetrics and gynecology, head dept. Washington U. Sch. Medicine, St. Louis, 1940—; practice medicine specializing in obstetrics and gynecology, St. Louis, 1940—; obstetrician, gynecologist in chief Barnes, Allied hosps., St. Louis, 1940—. Cons. St. Louis City, Homer G. Phillips hosps., 1940—. Co-discoverer (with George Washington Corner) of progesterone, 1929; proved progesterone is necessary for preservation of early pregnancy; co-producer (with Oskar Wintersteiner) of crystalline progesterone, try for isolgation of progesterone in pure form, 1935; 1934; recipient Eli Lilly & Co. award in biol. chemiscited by U. Rochester,, 1950. Mem. Phi Beta Kappa, Sigma Xi., Alpha Omega Alpha. Home: 12538 Conway Rd., Creve Coeur, Mo. 63141. Office: 4911 Barnes Hosp. Plaza, St. Louis 63110.*

ALLEN, William, Brit. chemist; b. London, Aug. 29, 1770; ed. Rochester, Eng.; student of Joseph Gurney Bevan; m. Mary Hamilton, 1796 (dec. 1797); 1 dau., Mrs. Cornelius Hanbury; m. 2d, Charlotte Hanbury, 1806 (dec. 1816); m. 3d, Grizell Birkbeck, 1827 (dec. 1835). With father's silk mfg. bus., Rochester, 1802; lectr. chemistry Guy's Hosp., 1802-26; conducted pharmacy (later became firm Allen & Hanbury); at request H. Davy, lectr. Royal Instn.; founder 2 manual labor schs., also a founder agrl. community, Lindfield, Eng. Fellow Royal Soc., 1807; mem. African Instn. Friend of Wilberforce, opposed slave trade, investigated carbonic acid; established (with Pepys) oxygen is unchanged in vol. when carbon is burned in it to form carbon dioxide, 1807, that in respiration no water is produced by oxidation and no nitrogen absorbed, that more carbon dioxide is formed in respiration of oxygen than of air. Died Lindfield, Dec. 30, 1843.

ALLENDOERFER, Carl Barnett, Am. educator; mathematician; b. Kansas City, Mo., Apr. 4, 1911; s. Carl William and Winifred (Barnett) A.; B.S., Haverford Coll., 1932; B.A., Oxford U. (Eng.), 1932, M.A. (Rhodes Scholar), 1939; Ph.D., Princeton, 1937; m. Dorothy Holbrook, June 26, 1937; children—Robert Duff, James Holbrook, William Barnett. Instr., U. Wis., 1937-38; from instr. to prof. Haverford Coll., Pa., 1938-51; prof. U. Wash., Seattle, 1951—, chmn. dept. math., 1951-62. Vis. instr. N.Y. U., 1947, Inst. for Advanced Study 1948-49; Mass. Inst. Tech., 1950; Fulbright lectr. Cambridge, Eng., 1957-58, Australia, 1963. Mem. A.A.A.S., Am. Math. Soc., Math. Assn. Am. (pres. 1959-60), Soc. Indsl. and Applied Math, Inst. Math. Statistics, Nat. Council Tchrs. Math., Phi Beta Kappa, Sigma Xi. Author: Principles of Mathematics, 1955, 1962; Fundamentals of Freshman Mathematics, 1959, 1965; Mathematics for Parents, 1965; (with C. O. Oakley) Fundamentals of College Algebra, 1966. Home: 4300 53d St., N.E., Seattle 98105.*

ALLER, Lawrence Hugh, Am. astronomer; b. Tacoma, Wash., Sept. 24, 1913; s. Leslie E. and Lenabelle (Davis) A.; A.B., U. Cal., 1936; M.A., Harvard, 1938, Ph.D., 1942; m. Rosalind Duncan Hall, Apr. 24, 1941; children—Hugh Duncan, Raymond Donald, Gwendolyn Jean. Jr. fellow Soc. Fellows, Harvard, 1939-42, instr. physics, 1942-43; research physicist U. Cal. at Berkeley, 1943-45; asst. prof. astronomy Ind. U., 1945-48; asso. prof. U. Mich., 1948-54, prof. astronomy, 1954-62; vis. prof. Australian Nat. U., Canberra, 1960-61, U. Toronto, 1961-62; prof. astronomy U. Cal. at Los Angeles, 1962—. Guest investigator Dominion Astrophys. Obs., 1951, Mt. Wilson Obs. Mem. Internat. Astron. Union, Royal Astron. Soc., Am. Acad. Arts and Scis., Nat. Acad. Scis. Author: (with Leo Goldberg) Atoms, Stars and Nebulae, 1943; Astrophysics, 1954; Gaseous Nebulae, 1956; Abundances of the Elements, 1961; Atmospheres of the Sun and Stars, 1963. Theoretical and spectroscopic studies of stellar atmospheres and gaseous nebulae; cosmic abundances of elements; transition probabilities for spectral lines. Home: 18118 W. Kingsport Dr., Malibu, Cal. 90265.

ALLERS, Rudolf, psychiatrist; b. Vienna, Austria, Jan. 13, 1883; s. Marcus and Augusta (Grailich) A.; ed. Vienna and Sacred Heart of Milan univs.; M.D., Ph.D., prof. of law honoris causa; m. Carola Meitner, Nov. 26, 1908; 1 son, Ulrich Stephen. Agrégé in psychiatry Munich and Vienna univs.; prof. psychology and philosophy Cath. U. Am., Washington. Mem. Am. Assn. Philosophy, Am. Soc. Metaphysics, Acad. Scis. N.Y. Author: Das Werden der sittlichen Person; Leben und Werke Anselm Canterbury; Character Education in Adolescence; The Successful Error; Existentialism and Psychiatry; also others. Home: 4922 La Salle Rd., Hyattsville, Md. Office: Georgetown University, Washington.

ALLERSTEIN, August, mathematician, astronomer; b. Germany, circa 1700; mem. Soc. of Jesus; assigned to China mission; chief dept. math. Emperor Kiang-long, Peking; took census of China for 2 years. Died Peking, circa 1777.

ALLER ULLOA, Ramon Maria, Spanish astronomer; b. Lalin (Pontevedra), Spain, Nov. 3, 1878; s. Domingo Enrique and Camilla (Ulloa) A.; ed. La Gardia, Lugo, Saint-Jacques de Compostelle; doctor of theology and math. sci. Priest, prof. U. Saint-Jacques de Compostelle, dir. observatory; founder Algoritmia obs. Mem. Royal Acad. Scis. (cons., corr.). Author: Introduccion a la astronomia; also articles. Address: calle Astronomo Aller, 5, Santiago de Compostella, Spain.

ALLEWELDT, Gerhardt Erich, agriculturist; b. Brightview, Alta., Can., July 21, 1927; s. Erich and Nina (Schubert) A.; Dr.Agr., Justus Liebig U., Giessen, Germany, 1956; m. Agnes Wetzel, May 30, 1952; children—Christiane, Monika, Karin, Jurgen. Research fellow Research Inst. Grape Breeding Geilweilerhof, Siebeldingen, Landau, Pfalz, Germany, 1956-65; prof. U. Agr. Hohenheim, Stuttgart, Germany, 1965—; dir. Inst. Viticulture, Stuttgart, Hohenheim, 1965—. Mem. German Soc. Applied Botany, Viticultural Research Assn. Research, publs. on growth and devel. of grapes including influence of environmental factors. Home: Landau Pfalz, Mozartstr. 25. Office: Inst. fur Weinbau Landwirtschaftliche Hochschule, 7, Stuttgart-Hohenheim, Germany.*

ALLFREY, Vincent George, Am. biochemist; b. N.Y.C., June 28, 1921; s. Thomas R. and Margaret (Ryan) A.; B.S., Coll. City N.Y., 1943; M.S., Columbia, 1948, Ph.D., 1949; m. Joan L. C. Brice, July 9, 1943; children—Barbara Claire, Kevin Mark. Research asst., asso. Rockefeller Inst., 1949-57; asso. prof. Rockefeller U., N.Y.C., 1957-63, prof. cell biology, 1963—; vis. prof. biology Yale, 1964-65. Mem. research adv. com. Am. Cancer Soc., N.Y.C. Health Research Council. Mem. Am. Soc. Biol. Chemists, Am. Soc. Cell Biology, Phi Beta Kappa, Sigma Xi. Editor: Jour. Gen. Physiology, 1958; Jour. Exptl. Zoology, 1964; Recent Results in Cancer Research, 1965; Current Topics in Developmental Biology, 1965. Research, publs. on chemistry of cell nucleus and control of chromosomal activity; demonstrated protein synthesis in nuclei and its dependence on ribonucleic acid synthesis; described new reactions which influence histone structure and activity. Home: 24 Winthrop Ct., Tenafly, N.J. 07670. Office: Rockefeller U., N.Y.C. 10021.*

ALLIBONE, Thomas Edward, Brit. physicist, elec. engr.; b. Sheffield, Eng., Nov. 11, 1903; s. Henry and Eliza (Kidger) A.; B.Sc. in Physics, U. Sheffield, 1924, Ph.D. in Metallurgy, 1926; Ph.D. in Physics, Cambridge (Eng.) U., 1929, D.Sc., 1935; D.Sc. (hon.) Reading (Eng.) U., 1956; m. Dorothy Margery Boulden, Feb. 14, 1931; children—Daphne Mary (Mrs. Harris), Noreen Elizabeth (Mrs. Morrish). In charge high voltage lab. Met. Vickers Elec. Co., Manchester, Eng., 1930-46; mem. Brit. mission Atomic Bomb Manhattan Project, Berkeley, Cal., Oak Ridge, 1943-45; research dir. Asso. Elec. Industries A.E.I. Lab., Aldermaston Ct., Eng., 1946-63; dir. research and edn. A.E.I., Woolwich, Eng., 1950-63; chief scientist Central Elec. Generating Bd., London, 1963—. Fellow Royal Soc., 1948, Instn. Physics, Royal Inst. Elec. Engrs., Am. Inst. E.E. Author: Release and Use of Atomic Energy, 1961; also numerous articles. Research on high voltages, artificial disintegration expts., 1927-; x-ray research; metall. preparation of metals and alloys; thermonuclear research. Home: Round Hill, Newbury, Berks., Eng. Office: Central Elec. Generating Bd. Hdqrs., 25 Newgate St., London E.C. 1., Eng.*

ALLINSMITH, Wesley, psychologist; b. London, Eng., May 21, 1923 (parents Am. citizens); s. Harry Bryan and Corinne Elizabeth (Allin) A.; A.B., Princeton, 1947; M.A., U. Mich., 1949, Ph.D. in Psychology, 1954; m. Beverly Balch, June 30, 1947; children—Bryan Balch, Wendy, Craig Lewis. Research asso. Office Pub. Opinion Research, Princeton, 1947; teaching fellow psychology U. Mich., 1948-49, intern clin. and counselling psychology, 1949-52, staff psychologist counseling service, 1952-53, instr., research asso. psychology, 1953-55; asst. prof., research asso. Lab. Human Devel., Grad. Sch. Edn., Harvard, 1955-56, asst. prof., asst. dir. lab., 1956-57, lectr., asso. dir. lab., 1957-60; project dir. Joint Commn. Mental Illness and Health, 1956-59; asst. prof. psychology dept. psychiatry, coordinator research center Studies Human Devel., Harvard Med. Sch.-Mass. Gen. Hosp., 1960-61; prof. psychology, head dept., U. Cin., 1961—, prof. psychology dept. psychiatry, 1962—. Cons. research methodology Child Guidance Home Cin.; research com. Hamilton Co. Diagnostic Clinic Mentally Retarded; adv. com. div. mental health Cin. Bd. Health; tng. cons. VA, also mem. adv. council Columbus area; mem. Cin. Com. Spl. Reading Disability. Recipient Howard C. Warren sr. prize psychology Princeton, 1947. Fellow Am. Psychol. Assn., Am. Sociol. Assn., A.A.A.S.; mem. Soc. Research Child Devel., Am. Assn. U. Profs., Sigma Xi. Author: (with G. W. Goethals) The Role of Schools in Mental Health, 1962. Editor: (with Judy F. Rosenblith) The Causes of Behavior, 2d edit., 1966. Studies in influence of religious affiliation and socio-econ. status on polit. attitudes and voting behavior; devel. of conscience in children; effects on children's sch. achievement from compulsivity and anxiety. Home: 3850 Clifton Av., Cin. 45220.*

ALLIOT, Hector, archeologist; b. Chateau des Forestiers, Gironde, France, Nov. 20, 1862; s. Jehan Hector and Lelia (Boymier) A.; ed. Lycée Bordeaux; Acad. Medicine, Sch. of Tech., Montpelier (France); A.B., U. France; D.Sc., U. Lombardy; m. Laurena Moore, Aug. 20, 1893. Associated with Farah Pasha in explorations at Tyre, Asia Minor, 1891; dir. Cliff Dwellers exploration, Chgo. Expn., 1893; prof. art history, U. So. Cal. 1908-09; dir. South Archaeol. Inst. Museum, from 1909. Mem. So. Cal. Acad. Scis., Deviser of Metrical Color Standard, Marine

Colormeter, 1910; del. Internat. Congress Geog., Rome, 1911. Author: Bibliography of Arizona. Died Feb. 15, 1919.

ALLIS, William Phelps, Am. physicist, educator; b. Menton, France (parents Am. citizens), Nov. 15, 1901; s. Edward Phelps and Amedine (Sgrena) A.; B.S., Mass. Inst. Tech., 1923, M.S., 1924; D.Sc., U. Nancy (France), 1925; m. Nancy O. Morison, June 11, 1935; children—Amedine (Mrs. Igor Bella), Edward M., John C. Faculty; Mass. Inst. Tech.; Cambridge; 1925——; prof. physics; 1950——; asst. sec. gen. for sci. affairs NATO, Paris, France, 1962-64; vis. prof. Harvard, 1955, U. Tex., Austin; 1960. Cons. Los Alamos Sci. Lab., 1952——. Decorated Legion of Merit. Fellow A.A.A.S.; mem. Am. Phys. Soc., Phys. Soc. London, Am. Acad. Arts and Scis., Civil Liberties Union, Council for a Livable World. Author: Nuclear Fusion, 1960; (with S. J. Buchsbaum and A. Bers) Waves in Anistropic Plasmas, 1963. Contbg. author Handbuch der Physik, vol. XXI, 1956. Home: 33 Reservoir St., Cambridge, Mass. 02138.*

ALLISON, Franklin Elmer, Am. microbiologist; b. Oakland City, Ind., Nov. 10, 1892; s. James Alexander and Sarah (Hillyard) A.; B.S.A., Purdue U., 1914; M.S.A., Ia. State U., 1915; Ph.D., Rutgers. U., 1917; m. Bertha Edith Rheinbold, Dec. 10, 1931. Scientist, U.S. Dept. Agr., Washington, 1917-18, sr. chemist, 1926-43, chief soil scientist, sr. chemist Agr. Research Service, Beltsville, Md., 1943-62; bacteriologist Nat. Canners Assn., Washington, 1919-20; biochemist Fixed Nitrogen Research Lab., Washington, 1920-26. Cons. Internat. Minerals and Chem. Corp., Skokie, Ill., 1963; mem. numerous sci. coms. Recipient Superior Service award U. S. Dept. Agr., 1960. Fellow Am. Soc. Agronomy, A.A.A.S., Am. Acad. Microbiologists; mem. Am. Chem. Soc., Am. Soc. Microbiologists, Soc. Plant Physiologists, Soil Sci. Soc. Am. Internat. Soc. Soil Scientists, Soil Conservation Soc. Am., Washington Acad. Scis. Contbr. numerous articles to jours., chpts. to books. Research on nitrogen-fixing bacteria and algae., biol. and chem. denitrification in soils, role of soil organic matter. Home: 4930 Butterworth Pl. N.W., Washington 20016.*

ALLISON, Fred, Am. physicist, educator; b. Glade Spring, Va., July 4, 1882; s. Robert Clark and Rebecca Jane (Clark) A.; B.A., Emory and Henry Coll., 1904, LL.D., 1933; postgrad. Johns Hopkins; M.A., U. Va., 1921, Ph.D., 1922; D.Sc., Auburn U., 1931; m. Elizabeth Harriet Kelly, Aug. 24, 1915; children—Elizabeth Harriet (Mrs. R. T. Comer, Jr.), Fred. Prof. physics Emory and Henry Coll., 1908-20, vis. prof., 1953-55; instr. physics U. Va., 1920-22; prof., head dept. physics Auburn (Ala.) U.; 1922-53, dean Grad. Sch., 1949-53, dir. research found., 1949-51, prof. emeritus, 1953——; prof. physics Huntingdon Coll., Montgomery, Ala., 1956——. Vis. prof., cons. physics dept. Chulalongkorn U., Bangkok, Thailand, summer 1955; research cons. Holloman AFB, N.M., summers 1957, 59, 60. Recipient President and Visitors Research prize U. Va., 1925; Herty Research medal Ga. sect. Am. Chem. Soc., 1933; physics bldg. at Auburn U. named in his honor, 1964, also lab. at Henry and Emory Coll. Fellow Am. Phys. Soc. (chmn. S.E. sect. 1941), A.A.A.S.; mem. Phi Beta Kappa, Sigma Xi, Phi Kappa Phi. Author: College Physics Laboratory Instructions; also numerous publs. on optics. Co-developer magneto-optic method of analysis for detection rare elements and isotopes; co-discoverer element 87 (francium), 1930, element 85 (astatine), 1931. Home: 202 W. Magnolia Av., Auburn, Ala. 36830.*

ALLISON, Ira Shimmin, Am. geologist; b. Gardner, Ill., Mar. 16, 1895; s. John William and Eva Catherine (Shimmin) A.; A.B. magna cum laude, Hanover Coll., 1917; postgrad. U. Chgo.; Ph.D. in Geology, U. Minn., 1924; m. Sadie Crowe Gilchrist, Sept. 5, 1921; children—Margaret (Mrs. James Kilbourn Clauss), David Elmer, Frances (Mrs. John Charles Sunderland). Instr. geology U. Minn., Mpls., 1920-25, asst. prof., 1925-28; instr. geology U. Chgo., summers 1922, 23; prof. geology Ore. State U., Corvallis, 1928——, chmn. dept., 1950-60; with U. S. Engrs. Dept., summers, 1930-35; br. head U. S. Army U., Shrivenham, Eng., 1945; chief adviser Ore. State U.-Kasetsart U. contract Internat. Cooperation Administrn., Bangkok, Thailand, 1954-56; cons. geologist Pacific N.W. mineral resources, 1930——. Recipient Alumnus of Year award Hanover Coll., 1958. Fellow Geol. Soc. Am., A.A.A.S.; mem. Ore. Acad. Sci. (pres. 1953-54), Am. Inst. Mining and Metall. Engrs. (chmn. Ore. sect. 1938-39, Sigma Xi, Phi Kappa Phi, Sigma Gamma Epsilon, Phi Gamma Delta. Author: (with W. Emmons, C. Stauffer, G. Thiel) Geology, 5th edit., 1960. Contbr. articles to profl. jours. on glacial drainage, Columbia River basin, Willamette Valley, pluvial lakes, geomorphology. Home: 2310 Harrison St., Corvallis, Ore. 97330.*

ALLISON, Philip Rowland, English surgeon; b. June 2, 1907; s. Jesse Rhodes and Rhoda A.; ed. Hymers Coll., Hull, Eng., U. Leeds (Eng.); m. Kathleen Greaves, 1937; 2 sons, 1 dau. Nuffied prof. surgery Radcliffe Infirmary, Oxford, Eng., 1954——. Fellow Royal Coll. Surgeons; mem. Soc. Thoracic Surgeons Gt. Britain and Ireland, Thoracic Soc., Brit. Cardiac Soc., Internat. Soc. Surgeons. Contbr. articles, chpts. to med. jours., texts. Sci. research in area vascular thrombosis, particularly application of surgery to treatment coronary thrombosis. Address: Nuffied Dept. Surgery, Radcliffe Infirmary, Oxford, Eng.

ALLISON, Samuel King, Am. physicist; b. Chgo., Nov. 13, 1900; s. Samuel Buell and Carrie (King) A.; B.S., U. Chgo., 1921, Ph.D., 1923; m. Helen Catherine Campbell, May 28, 1928; children—Samuel Campbell, Catherine King (Mrs. David L. Marshall). NRC fellow Harvard, 1923-25; fellow Carnegie Instn., 1925-26; faculty physics U. Cal. at Berkeley, 1926-30; faculty U. Chgo., 1930——, prof., 1942-59, Frank P. Hixon distinguished service prof. physics, 1959——, dir. Enrico Fermi Inst. for Nuclear Studies, 1946-58, 63——. Cons. NDRC, Washington, 1940-41; chmn. tech. and scheduling com. Los Alamos Sci. Lab., 1944-46; chmn. com. on nuclear sci. NRC, 1961——. Recipient medal for merit, citation Pres. Truman, 1946; Encomienda de la Orden Civil de Alfonso El Sabio, Spanish Ministry Edn., 1959. Fellow Am. Phys. Soc.; mem. Nat. Acad. Scis. (past chmn. phys. sect.). Author: (with A. H. Compton) X-rays in Theory and Experiment, 1935. Contbr. articles on x-rays, nuclear physics, atomic physics and ionic beams to profl. jours. Died Sept. 15, 1965.*

ALLIX, André, French geographer; b. Gap, France, 1889; ed. univs. Grenoble, Paris, Lyons. Tchr., Lycées, Grenoble, 1920-28, Lyons, 1916-20; with Faculté des Lettres, Lyons, 1928-44, rector, 1944-60. Mem. French nat. com. UNESCO. Recipient Gold medal Royal Scottish Geog. Soc., 1960. Author: La France economique, 1919; La Géographie des foires, 1923; L'Oisans au Moyen Age, 1923; Les avalanches, 1925; Un Pays de haute géographique, 1929; Manuel de géographie général, 1935-50; L'Homme dans la géographie humaine, 1949; co-author: Les Principales puissances du monde, 1935-53; Géographie des textiles, 1957. Research on human geography, especially life in mountainous regions. Home: 16 rue de la Procession, Paris 15e, France.

ALLMAN, George James, Brit. biologist; b. Cork, Ireland, 1812; s. James Allman; B.A., Trinity Coll., Dublin 1839, M.D., 1843, M.D., 1847; M.D. ad eundem, Oxford U., 1847; m. Hannah Shaen. Prof. botany Dublin U., 1844-55; Regius prof. natural history Edinburgh U., 1855-70, also keeper natural history museum, from 1855. Brisbane medal Royal Soc. Edinburgh, 1877, Cunningham medal Irish Acad., 1878. Fellow Royal Soc., 1854 (council 1871-73, Royal Gold medal, 1878), Royal Coll. Surgeons (Ireland); mem. Brit. Assn. for Advancement Sci. (pres. 1879), Linnean Soc. (pres. 1874, Gold medal 1896). Author: Monograph of the Fresh Water Polyzoa, 1856; The Method and Aim of Natural History Studies, 1868; Monograph of the Gymnoblastic or Tubularian Hydroids, 1871-72; numerous papers on marine zoology. Investigated classification and morphology of coelenterata and polyzoa; described (with others) Rhabdopleura discovered by O. Sars in Lofoten Islands, 1866; studied hemicordata and urochorda, 1868; coined terms endoderm, ectoderm. Died Parkstone, Eng., Nov. 24, 1898.

ALLMAN, George Johnston, Brit. mathematician; b. Dublin, Ireland, Sept. 28, 1824; s. William Allman; B.A., Trinity Coll., Dublin, 1844, LL.B., 1853, LL.D., 1854; D.Sc. (hon.), Queen's U., 1882; m. Louisa Taylor, 1853 (dec. 1864). Prof. math. Queen's Coll., Galway, also Queen's U. in Ireland, 1853-93; senator Royal U. Ireland, 1880. Fellow Royal Soc., 1884. Author: On some Properties of the Paraboloids, 1874; The Collected Works of James MacCullagh, LL.D., 1880; History of the Greek Geometry from Thales to Euclid, 1889; articles in 9th edit. Ency. Brit. Traced rise and progress of geometry and arithmetic; shed new light on history of early math.; influenced by positivism; corresponded with A. Comte. Died Dublin, May 8, 1904.

ALLMAND, Arthur John, English chemist; b. Wrexham, Eng., 1885; s. Frank A.; ed. U. Liverpool, D.Sc., 1910, also tech. high schs. Karlsruhe and Dresden, postgrad. in Germany, 1910-12; m. Marguerite Malicorne, 3 children. Asst. lectr., demonstrator U. Liverpool, 1913; prof. chemistry King's Coll., 1919-38, Daniell prof. chemistry, 1938-50, asst. prin., 1937-43, also fellow; emeritus prof. chemistry U. London, 1950-51. Fellow Royal Soc., 1929, Royal Inst. Chemistry; mem. Faraday Soc. (pres. 1947-48). Author: Principles of Applied Electro-Chemistry, 1912; also papers in tech. jours. Noted for research in electrochemistry, photochemistry. Died Aug. 4, 1951.

ALLORGE, Pierre, French botanist; b. 1891; prof., dir. cryptogamic lab. Mus. Natural History, Paris; investigated plants, algae. Author: Les associations végétales du Vexin francais, 1922; works on coasts France, Spain. Contbd. papers on algae and embryology to jours. Died 1944.

ALLOU, Charles Nicolas, French engr.; b. Paris, France, 1787; ed. École Polytechnique; chief engr., head underground projects Dept. of Seine, Paris. Decorated chevalier Legion of Honor. Mem. Soc. Antiquarians (v.p.), Soc. History of France, Soc. Geography. Author: Description des monuments des divers ages observés dans la Haute-Vienne, 1821 (recipient gold medal French Acad.). Studied French hist. monuments. Died 1843.

ALLPORT, Gordon Willard, Am. psychologist; b Montezuma, Ind., Nov. 11, 1897; s. John Edwarc and Nellie Edith (Wise) A.; A.B., Harvard, 1919 A.M., 1921, Ph.D., 1922; grad. work U. of Berli and U. of Hamburg, 1922-23, Cambridge U., 1923-24; L.H.D. (hon.), Boston U., 1958; Ohio Wesleya U., 1962; D.Sc., Colby College, 1964; D.Litt. (hon.) Durham University (England), 1965; m. Ada Lufki Gould, June 30, 1925; 1 son, Robert Bradlee. Inst in English, Robert College, Istanbul, Turkey, 1919-20; instr. in social ethics, Harvard University, 1924-26; asst. prof. in psychology, Dartmouth College 1926-30; asst. prof. psychology, Harvard Univ., 1930-36, associate prof., 1937, prof. psychology, 1942-66 R. C. Cabot professor social ethics, 1966——. Mem S.A.T.C. Past mem. nat. com. for UNESCO. Pres. Edr Exchange Greater Boston; past dir. Nat. Opinion Research Center; past mem. Social Sci. Research an Nat. Research Councils. Recipient Gold Medal awar Am. Psychol. Found., 1963. Hon. fellow Brit. Psycho Soc.; hon. mem. Spanish, Italian psychological societies, Deutsche Gesellschaft fur Psychologie, Österreichische Arztegesellschaft fur Psychotherapie; member Am. (mem. council 1936-38; pres. 1939), Easter (pres. 1943) psychol. assns., Phi Beta Kappa. Author Studies in Expressive Movement (with P. E. Vernon) 1933; The Psychology of Radio (with H. Cantril) 1935; Trait-Names: A Psycho-lexical study (with H S. Odbert), 1936; Personality—a Psychological Interpretation, 1937; Psychology of Rumor (with L Postman), 1947; The Individual and His Religion 1950; The Nature of Personality, 1950; The Natur of Prejudice, 1954; Becoming: Basic Consideration for a Psychology of Personality, 1955; Personalit and Social Encounter, 1960; Pattern and Growth c Personality, 1961; Letters From Jenny, 1965. Edito Jour. Abnormal and Social Psychology, 1937-49. Research psychology of race relations, personality, expressive movement, prejudice. Home: 386 School St Watertown 72, Mass. Office: William James Hal Harvard U., Cambridge, Mass.

ALLRED, Dorald Mervin, Am. zoologist; b. Lehi Utah, July 11, 1923; s. Robert M. and Hazel (Beck A.; student U. Ill., 1942; B.A., Brigham Young U. 1950, M.A., 1951; grad. Duke, 1951; Ph.D., U. Utah 1954; m. Berna Brown, Mar. 4, 1952; children—Kevin, Anita, Kyle, Darin. Chief entomology an arachnology sect. ecol. research U. Utah, Dugway 1954-56; faculty Brigham Young U., Provo, 1956——asso. prof. zoology, 1963——. AEC grantee, 1959——NIH grantee, 1956——. Mem. Sigma Xi. Author: Laboratory Guide for Medical Entomology, 1963; als numerous articles. Research on natural history ecto parasites of mammals especially parasitic mites. Home 685 E. 2780 N., Provo, Utah 84601.*

ALLRED, John Caldwell, Am. physicist; b. Breckenridge, Tex., Apr. 24, 1926; s. Oran H. and Katherin (Miller) A.; B.A., Tex. Christian U., 1944; M.A., U Tex., 1948, Ph.D., 1950; m. Mary Elizabeth Bode June 4, 1950; children—Susan Elizabeth, Katherin Anne, John Renne. Asso. prof. physics U. Houston 1956-59, asso. dean Coll. Arts and Scis., 1959-61 prof., 1961——, asst. to pres., 1961-62, v.p., dea faculties, 1962——; vis. scientist Tex. Acad. Sci. NSF, 1960. Mem. Tex Gov.'s Com. on Atomic Energy 1959-60; cons. AVCO Mfg. Corp., Lawrence, Mass 1958-59, U. S. AEC, Geneva, Switzerland, 1958 Mem. Am. Phys. Soc., Am. Nuclear Soc. (charter) Acoustical Soc., Sigma Xi. Research on particle scattering, neutron interactions, nuclear reactor theory hydrodynamics, archtl. acoustics, solid state. Home 3418 Broadmead Dr., Houston 77025.*

ALLUARD, Émile, French meteorologist; b. Orléans, France, Oct. 5, 1815; ed. École Normale. Prof Rheims (France) Lycée, Clermont-Ferrand (France Faculté des Sciences. Studies atmospheric humidity invented condensation hygrometer, 1878; founde 1st mountaintop meteorol. obs. on Puy-de-Dôme, France 1876. Died Clermont-Ferrand, Aug. 20, 1908.

ALLUISI, Earl Arthur, Am. psychologist, educator; b. Richmond, Va., June 11, 1927; s. Humber Peter and Elizabeth (Dini) A.; student U. Minn 1944, Va. Poly. Inst., 1944-45, Yale, 1945; B.S. Coll. William and Mary, 1949; M.A., Ohio State U. 1950, Ph.D., 1954; m. Mary Jane Boyle, Dec. 16 1954; children—John Carroll, Jean Elizabeth, Pau David Julian, Janet Ann. With psychology dept. Research Found., Ohio State U., 1954-57, U. S. Arm Med. Research Lab., Ft. Knox, Ky., 1957-58, Lockhead Missile Systems div., Palo Alto, Cal., 1958 Stanford Research Inst., 1958-59; faculty Emory U. 1959-61; asso. scientist Human Factors Researc Lab., Lockheed-Ga. Co., Marietta, Ga., lectr. Ga. Inst Tech. Sch. Psychology, 1961-63; prof. psychology U Louisville, 1963——, asst. dean for research Grad Sch., 1966——. Cons. to surgeon gen. U. S. Army 1966——. Fellow A.A.A.S., Am. Psychol. Assn.; mem Psychonomic Soc., Psychometric Soc., Southeastern Ky. (pres.) psychol. assns., Soc. Engring. Psychologist (mem.-at-large exec. com.), So. Soc. for Philosoph and Psychology (pres.-elect), Ky. Acad. Sci., Am Assn. U. Profs., Phi Beta Kappa, Sigma Xi, Psi Chi Alpha Psi Delta. Research, numerous publs. on assessment of man's performance capabilities under norma and stressful conditions including work-rest scheduling and infectious diseases, perception of depth form perception, quantitative aspects of attentive an vigilance processes in man, research methodology an statis. applications to psychol. research. Home: 4093 Gilman Av., Louisville 40207.*

AL-MAGHRIBI (Muhyi al-din al-Maghribi) (Muhyi al-milla wal-din Yahya ibn Muhammad ibn Abu al-Shukr al-Maghribi al-Andalusi), Hispano-Muslim mathematician, astronomer; flourished 13th century, Syria, then Maraga. One of his works on chronology and calendar of Chinese and Uighurs, showed some knowledge of Chinese astronomy at Maraga caused by extension of Mongol empire; also wrote on geometry, trigonometry, astrology, chronology, astrolabe, Greek classics.

AL-MAHALLI (al-As'ad al-Mahalli) (Ya'qub ibn Ishaq As'ad al-din al-Mahalli al-Yahudi), physician; b. al-Mahallah; flourished 12th century. Went to Damascus, 1201; moved to Cairo and remained until his death. Wrote various works on medicine, including: Discourse on the Principles of Medicine; Book of Parity; A Book of Medical Questions.

AL-MAHANI (Abu Abdallah Muhammed ibn Isa al-Mahani), Persian astronomer, mathematician; b. Mahan, Kirman, Persia; made observations of eclipses and conjunctions, 853-66 (utilized by Ibn Yunus); translated Euclid and Archimedes with commentaries; was 1st to state the Archimedian problem in form of cubic equation. Died circa 874-84.

Al-MAJRITI (Maslama ibn Ahmad al-Majriti), Spanish astronomer, mathematician, occultist; b. Madrid; earliest Hispano-Muslim scientist of note; wrote a treatise on the astrolabe, commentary on Ptolemy's Planisphaerium, a comml. arithmetic, and possibly a book on the generation of animals; probably was scholar who added tangent functions to his revision of al-Khwarizmi's astron. tables. Died circa 1007.

AL-MA'MUN (Abu al-Abbas Abdallah), Muslim ruler; b. Bagdad, 786; s. Haroun al-Raschid; defeated his brother and ascended to throne, 813; his reign troubled by numerous rebellions; enforced by persecution the dogma that Koran was created, not eternal; considered a heretic by many Muslim scholars; died while prosecuting war against Theophilus in Cilicia. Celebrated patron of sci. and lit.; established an acad., obs. and library in Baghdad; founded colls., collected Greek and Hindu manuscripts, paid for their transl.; invited scholars and savants of all nations to Bagdad (his capital, center of all learning at that time). Died 833.

AL-MAQRIZI, Arabic geographer; b. Cairo, Egypt, 1364; author numerous works on various topics including: Exhortations and Considerations; works on geography and biography of Egypt. Died 1442.

AL-MARWARRUDI, Arabic scientist; flourished circa 835, Bagdad; one of scientists participating in project, commd. by Caliph al-Ma'mun, to prepare astron. and geodetic measurement and tables.

ALMASI, Lucretia Rotschild, Rumanian chemist; b. Viseul de sus, Rumania, Mar. 28, 1921; d. Filip and Ileana (Simon) Rotschild; chemist Chem. Faculty, U. Cluj (Rumania), 1949; D.Chem. Sci., Chem. Faculty, U. Leningrad (USSR), 1953; m. Nicolae Almasi, Sept. 1948 (div.). Chemist synthetical drug factory Terapia, Cluj, 1948-49; chief research sector organophosphorus compounds Chem. Inst. Acad. R.S. Rumania, Cluj, 1954-—; prof.'s asst. chem. chair Medico-Pharm. Inst., Cluj, 1949-50. Mem. Soc. Rumanian Physicist and Chemists, Chem. Soc. Mendelejeff (USSR). Research, publs. and inventions on new methods of synthesis, numerous new compounds; reaction-mechanism; thermodynamical functions determination of asso. monomer equilibriums in H-bondings; bond refractions values determinations; Hammett equation applications; tautomeric problems; Raman and IR spectroscopical studies. Home: 13 Umbroasa. Office: Chem. Inst. Acad., 65 Donat, Cluj, R.S., Rumania.*

AL-MASIH ('Abd al-Masih Na'ima de Hims), Arabic writer, translator; flourished circa 1st half of 9th century; translator: Theologia Aristotelis of pseudo-Aristotle to Arabic.

AL-MAS'UDI (al-Mas Udi), Arabic geographer; native of Bagdad, did much traveling; described the Dead Sea, earthquake of 955; earliest mention of wind mills is found in his work, also geol. discussions, views of evolution (from mineral to plant, from plant to animal, from animal to man). Author: Fields of Gold and Mines of Precious Stones (ency. of geography known in Europe under title Prairies d'Or), 947; The Book of Surveying and Elevation. One of most notable geographers of Arab world; distinguished for ample use of sources and keen sci. curiosity. Died Cairo, Egypt, 957.

ALMASY, Felix, Swiss phys. biochemist; b. Osijek, Yugoslavia, Apr. 4, 1901; s. Eugen and Sofie (Kraus) A.; Ing.chem., Tech. High Sch., Brünn, Czechoslovakia, 1925; postgrad. U. Zurich (Switzerland), 1925-29; Dr.es.sc.phys., U. Paris, 1933; m. Anny Sibylle Kreinz, Sept. 14, 1929. Asst., Inst. Phys. Chemistry, U. Zurich, 1927-32, expert for phys. biochemistry faculties of medicine and vet. medicine, 1932-48, head chem. research lab. Faculty Vet. Medicine, 1948-—, faculty, 1934-—, prof., 1946-—. Mem. Am. Chem. Soc., Schweizerische Chemische Gesellschaft, Schweizerische Academie der medizinischen Wissenschaften (mem. spl. commns.). Contbg. author: Lehrbuch der Veterinär-physiology, 1965; also numerous articles. Research on vapor spectrography; fluorescent, Raman spectrography; blood spectra; blood gases; mineral metabolism of cattle; toxic action of chinin and berberin; carcinogenic action and spectroscopic detection of tar fractions; spectrochromatography; liver metabolism of guinea pig; zinc metabolism; cobalt metabolism. Home: 74 Bederstrasse, Zurich, Switzerland.*

AL-MAZINI (Abu Hamid, also Abu 'Abdallah Muhammad ibn 'Abd al-Rahim ibn Salaiman al-Qaisi al-Mazini al-Andalusi al-Gharnati), geographer; b. Grenada, circa 1080; visited all the Islamic countries; gave important descriptions of then little-known Slavic countries (Bulgaria, Russia); witnessed the trade in fossil bones, exported as far as Khwarizm for combmaking, while staying among the Bulgars, 1136. Author: Gift of Spirits and Choice of Marvels, 1162; Singularities in Some of the Wonders of the Occident. Died 1169.

ALMELOUEEN, Theodorus Jannson van, see van Almeloueen.

ALMENAR, Juan, Spanish physician; b. Valencia, Spain. Flourished 1490. Author: De morbo gallico (one of earliest extant descriptions of syphilis; advocated mercury treatment), 1502.

AL-MISRI ('Ali ibn Ridwan ben 'Ali ben Ga'far al-Misri), Egyptian physician; b. circa 998; author: Book for Hunting What is Harmful to the Body in Egypt; The Book Useful in Apprehending the Art of Medicine; also commentary on Galen's Ars parva (translated to Latin by Gerard of Cremona); commentary on Opus quadripartitum of Ptolemy. Died 1061.

AL-MIZZI (Abu 'Abdallah Muhammad ibn Ahmad al-Mizzi al-Hanafi), Muslim Egyptian astronomer; b. 1291; student Cairo; meuzzin in gt. mosque of Damascus, Syria. Author numerous works including treatises on constrn. and use of astron. instruments (astrolabes and quadrants). Died 1349.

ALMQUIST, Herman James, Am. biochemist; b. Helena, Mont., Mar. 3, 1903; s. Harry and Mary (Ericson) A.; B.S., Mont. State Coll., 1925, D.Sc., 1952; Ph.D., U. Cal. at Berkeley, 1932; m. Viola D. Pimentel, Oct. 11, 1935; children—Alan James, Eric John. Dir. research, v.p. Grange Co., Modesto, Cal., 1948-67; dir. Ann. Revs., Inc. Recipient Borden award, 1937. Fellow A.A.A.S., Am. Inst. Nutrition, mem. Soc. Exptl. Biology and Medicine, Poultry Sci. Assn., Soc. Biol. Chemistry, Am. Chem. Soc., Inst. Food Technologists. Research, numerous publs. on discovery, isolation, structure and synthesis of Vitamin K, requirements and metabolism of amino acids in poultry, mechanism of auto oxidation of benzaldehyde, role of inhibitors, bile acids in metabolism of the chick, vitamin A requirements, nutritional values of proteins. Home: Route 1, Box 90, Kelseyville, Cal. 95451.*

AL-MUZAFFAR, Arabic astronomer, mathematician; s. Muhammad ibn al-Muzaffar; author: The Book of the Plane Astrolabe; invented linear astrolabe; refined astrolabe. Died circa 1213.

ALMY, Gerald M., Am. physicist; b. Ewing, Neb., April 24, 1904; s. Billings G. and Ada (Marks) A.; B.S., U. Neb., 1924, M.S., 1926; Ph.D., Harvard, 1930; m. Ruth Virtue, July 21, 1930; children—Cynthia (Mrs. Thomas K. Landauer), Catherine. From instr. to prof. physics U. Ill., Urbana, 1930-—, head physics dept., 1964-—. Fellow Am. Phys. Soc.; mem. Phi Beta Kappa, Sigma Xi. Important work includes research in molecular spectra and fluorescence, nuclear physics, betatron devel. Home: 509 S. Ridgeway, Champaign, Ill.*

AL-NABATI (Abu al-'Abbas al-Nabati) (al-Rumiya) (Abu al-'Abbas Ahmad ibn Muhammad ibn Mufarraj al-Nabati ibn al-Rumiya al-Hafiz), biologist; b. Seville, Spain, circa 1170; travelled to Orient and wrote on plants he observed; al-Malaqi, Islamic botanist, was his student. Author: Book of the Voyage; Explanation of the Names of Simples in Dioscorides; Treatise on the Composition of Drugs. Died circa 1239.

AL-NAKHSHABI (Shaykh Diya al-din al Nakhshabi), Persian translator; Author: Particularities and Generalities (describes various parts of body and diseases; believed their harmony proves God's greatness). Died 1350.

AL-NAZZAM (Ibrahim al-Nazzam), Arabic scientist, philosopher; Mu'tazilite opponent of atomistic doctrines of period; tchr. of al-Gahiz. Died 845.

AL-NUWAIRI, Egyptian historiographer; b. Upper Egypt, Apr. 5, 1279; govt. sec.; intendant of army, Tripolis, Syria; intendant adminstrn. in Egyptian provinces of Al-Dakhaliya and Al-Murjahiya. Author: Aim of the Intelligent in the Arts of Letters (or The Highest Degree of Perfection), 31 books divided into 5 parts which dealt with heavens and earth (cosmology, geography, man including ethics and politics), animals, plants, history. Died Cairo, Egypt, June 16, 1332.

ALONSO-MISOL MARTINEZ, Félix, Spanish gynecologist; b. Madrid, Spain, Dec. 7, 1910; s. Félix and Angeles (Martinez) A.; ed. U. Madrid; M.D. in endocrine medicine; m. Silvia Hernández Cadalso, Sept. 17, 1955; children—Felix, Maria Silva. Gynecologist, obstetrician for Red Cross, Press Assn., also social insurance, Madrid; developer pharm. products in various countries. Mem. Press Assn., Coll. Physicians Madrid, Gen. Soc. Spanish Authors. Author: Métodos para la correción quirurgica de la retroversion y del prolapso del útero; Bioquímica del puerperio; Bioquímica de la degeneración cancerosa. Research on biochemistry of cancerous degeneration of cell. Address: calle Arenal, 18, Madrid, Spain.*

ALPEN, Edward Lewis, Am. biophysicist; b. San Francisco, May 14, 1922; s. Edward Lawrence and Margarite (Shipley) A.; B.S., U. Cal., Berkeley, 1946, Ph.D., 1950; m. Wynella June Dosh, Jan. 6, 1945; children—Angela Marie, Jeannette Elise. Asst. prof. pharmacology U. George Washington Sch. Medicine, 1950-51; head biophysics br. U. S. Naval Radiol. Def. Lab., San Francisco, 1951-58, head biol. and med. scis. div., 1958-—. NSF fellow, 1958-59; Guggenheim Found. fellow, 1965-66; recipient Sustaining Members award for research Assn. Mil. Surgeons of U. S., 1961; Sec. of Navy's medal for Distinguished Achievement in Sci., 1962; Dept. Def. Distinguished Civilian Service award, 1963. Mem. Am. Physiol. Soc., Radiation Research Soc., Soc. Exptl. Biology and Medicine, Cal. Acad. Scis., Royal Soc. Medicine (Eng.), Sigma Xi. Contbr. numerous articles to sci. jours. Developed analytical models for the control of red-cell prodn.; research on the relationship between phys. characteristics and biol. action of various ionizing radiations; developed sensitive techniques for the measurement of intracellular components by single cell spectrophotometry. Home: 2933 Dolores Way, Burlingame, Cal. 94010. Office: U. S. Naval Radiol. Def. Lab., San Francisco 94135.*

ALPER, Thelma G., Am. psychologist; b. Chelsea, Mass., July 24, 1908; d. David and Mollie Herman) Gorfinkle; B.A., Wellesley Coll., M.A., 1933; Ph.D. Radcliffe Coll., 1943; m. Abraham T. Alper, Apr. 1, 1932 (dec. Dec. 1953). Asst. in psychology Wellesley (Mass.) Coll., 1929-38, dir. remedial reading program, 1938-42, asso. prof. psychology, 1952-54, prof., 1954-—, chmn. psychology dept., 1963-—; instr. Harvard, 1943-46, lectr. clin. psychology, 1946-48; asso. prof. Clark U., Worchester, Mass., 1948-52; psychologist Judge Baker Guidance Center, Boston, 1959-—. Cons. VA. 1946-52. Diplomate Bd. Clin. Psychology. Fellow Am., Mass. (pres. 1952-54) psychol. assns.; mem. Civil Liberties Union (sec. Mass., 1960-63), Phi Beta Kappa (pres. Radcliffe 1952-54, pres. Wellesly chpt., 1963-67), Sigma Xi. Co-author books for Armed Forces. Research in personality, selective memory, and learning. Home: 217 Kent St., Brookline, Mass. 01246. Office: Wellesley Coll., Wellesley, Mass. 02181.*

ALPER, Tikvah, radiobiologist; b. Cape Town, South Africa, Feb. 22, 1909; d. Abram and Tybil (Sugarman) Alper; B.A. with distinction, U. Cape Town, 1927, M.A. with distinction, 1929; postgrad. U. Berlin, Kaiser Wilhelm Inst.; M.S., Washington U., St. Louis, 1939; m. Max Sterne, Mar. 17, 1932; children—Jonathan Sterne, Michael Alper Sterne. Jr. lectr. U. Cape Town, 1930, tchr. hard of hearing class, 1941; lectr. physics U. Witwatersrand, 1942-47; head biophysics sect. South African Council for Sci. and Indsl. Research, 1948-52, sci. research worker Med. Research Council, Hammersmith Hosp., London, Eng., 1951-—, dir. exptl. radiopathology research unit, 1962-—. Hon. lectr. radiobiology Postgrad. Med. Sch., U. London, 1962-—. Recipient Jr. South Africa medal Brit. Assn., 1932. Fellow Inst. Physics; mem. Royal Assn. for Radiation Research (chmn. 1966-—), Hosp. Physicists Assn., Photobiology Group (internat. com. photobiology), Brit. Inst. Radiobiology, Radiation Research Soc., Biopyys. Soc. Publs. on research in measurement of delta rays; investigations of radiation effects on bacteriophage; mechanisms of lethal action of radiation on cells and oxygen effect in radiobiology. Home: 84 Philbeach Gardens, London, S.W. 5, Eng.*

ALPERN, Daniil Yevseevich, Russian path. physiologist; b. 1894; grad. Med. Faculty, Kharkov U., 1918; D.Med. Sci. In chair path. physiology Kharkov U., 1918-29; prof., head chair path. physiology Kharkov Med. Inst., 1929-—. Mem. Ukranian Acad. Sci. (corr.), All-Union (bd. mem.), Ukranian (dep. chmn.) socs. path. physiologists. Author over 120 works; best known monographs are The Theory of Fever, 1928; The Vegetative Nervous System and Metabolism, 1931; Chemical Factors of Nervous Excitation in the Human Body, 1944; (textbook) Pathological Physiology, 1954; Inflammation, Problems of its Pathogenesis, 1959. Editor various sci. symposia including The Vegetative Nervous System and the Metabolism of Tissues, 1935; Hypersensitivity, 1936; Hyper-sensitive Reactions, 1938; Mem. editorial bd. Archives of Pathology; co-editor Pathology and Morphology sects. Large Med. Ency., 2d edit.; mem. editorial council Pathological Physiology and Experimental Therapy. Address: Med. Inst., Sumskaya 41, Kharkov, Ukranian SSS, USSR.

ALPERT, Daniel, Am. physicist; b. Hartford, Conn., Apr. 10, 1917; s. Elias and Dora (Prechepa) A.; B.S., Trinity Coll., Conn., 1937; D.Sc. (hon.), 1957; Ph.D.,

Stanford, 1942; m. Natalie Boyle, Jan. 12, 1942; children—Amy Vincell, Laura Jane. Teaching asst. physics Stanford, 1939-41, research fellow, 1941-42; with Westinghouse Electric Corp., Pitts. 1941-57, research physicist, 1942-50, mgr. physics dept., 1950-55, asso. dir., 1955-57; prof. physics U. Ill., Urbana, 1957—, dir. coordinated sci. lab., 1959—, dean Grad. Coll., 1965—. Mem. Def. Sci. Bd., Washington, 1963—. Mem. A.A.A.S., Newcomb Cleveland award 1954), Am. Phys. Soc., Am. Vacuum Soc., Phi Beta Kappa, Sigma Xi. Important work includes basic contbns. to extension of vacuum tech. to very low pressures. Editor: Jour. Vacuum Sci. and Tech., 1964—. Home: 402 W. Pennsylvania Av., Urbana, Ill. 61801.*

ALPERT, Harry, Am. sociologist; b. N.Y.C., Oct. 12, 1912; s. Morris and Mary (Levine) A.; A.B., Coll. City N.Y., 1932; M.A., Columbia U., 1935, Ph.D., 1938; Certificat de Sociologie, U. Bordeaux, France, 1933; m. Anitra E. Fink, June 15, 1936; children—Spencer Ward, Geoffrey Philip. Instr. sociology, Coll. City N.Y., 1933-43, asst. prof., 1943-47; asso. prof. sociology, chmn. dept. anthropology-sociology Queens Coll., 1948-50; pub. opinion analyst Office War Information, Washington, 1943-44; rationing statistics analyst OPA, Washington, 1944-45; analytical statistician Bur. Budget, Washington, 1945-48, 1950-53; program dir. social sci. research NSF, Washington, 1953-58; prof. sociology, U. Ore., 1958—, dean grad. sch., 1958-63, dean faculties U. Ore., 1963—; dir. dept. social sci. UNESCO, 1967—; vis. prof. sociology Yale, 1947, U. Wash., 1955; Cons., USAF, 1948-50, Research and Devel. Bd., 1948-50. Mem. Am. Sociol. Assn. (v.p. 1959), Am. Assn. for Pub. Opinion Research (pres. 1955-56), Sociol. Research Assn. (pres. 1961), Pacific Sociol. Assn. (pres. 1962-63). Author: Emile Durkheim and His Sociology, 1939. Contbr. numerous articles to sci. jours. Editor Am. Sociol. Rev., 1960-62; Studies in sociology of sci. with special emphasis on interrelations of sci. and govt., analysis of devels. in French sociology. Home: 1766 Alder St., Eugene, Ore. 97401.*

ALPERT, Leo, Am. meteorologist; b. Boston, Oct. 31, 1915; B.S., Mass. State Coll., 1937; M.A., Clark U., 1939, Ph.D, 1946; m., 1943; children—Cheryl Jean, Stephen Edward. Environmental specialist Gen. Staff, Dept. Army, 1947-50; engr. intelligence specialist Engr. Intelligence Div., C.E., 1950-51; chief climatology lab., geophysics directorate USAF Cambridge Research Labs., 1952; geographer, engr. Strategic Intelligence Div., C.E., 1953-60; tropical environmental sci. Environmental Scis. Div., Army Research Office, Washington, 1961-63, 65—; chief sci. U. S. Army Tropic Test Center, Canal Zone, 1963-65. Cons. to cos. in U. S., S.Am. Recipient Meritorious Civilian Service award Dept. Army, 1965. Mem. Am. Meteorol. Soc., Am. Geophys. Union, Assn. Am. Geographers. Contbr. articles to profl. jours. First use of radar for operational weather forecasting and climatol. studies of clouds and precipitation; pioneer in studies of warm cloud rainfall, cloud seeding in tropics. Home: 4120 34th St. N., Arlington, Va. 22207. Office: 3045 Columbia Pike, Arlington, Va. 22204.*

ALPHANUS, sci. writer; flourished 11th century; bishop of Salerno, Italy; mem. Salernian sch. medicine (based on rediscovery of ancient Greek works). Author: De quattuor humoribus de quibus constat corpus humanum (On the four humors of which the human body is composed).

ALPHEN, Jan van, see Van Alphen.

ALPHONSO OF CORDOVA, Spanish physician; flourished 1340; one of 1st to distinguish between ordinary boils and abscesses and those caused by plague. Author: Epistola et regimen de pestilentia (on plague of 1348).

ALPHONSO X THE WISE, see Alfonso X of Castile.

ALPINO, Prospero (or Alpini, Alpinus), Italian botanist, physician; b. Marostica, Venice, Nov. 23, 1553; s. Francesco and Barlocomea (Tarsia) A.; med degree U. Padua, 1578. Practiced medicine nr. Padua; became physician to consul gen., Cairo, Egypt, 1580, physician to Prince Andrea Doria, Venice, 1586; apptd. prof. botany, dir. bot. garden U. Padua, 1593. Author: De medicina Aegyptorum libri IV, 1591; De balsamo dialogus, 1592; De plantis Aegypti liber, 1592; De praesagienda vita et morte aegrotantium, libri septem, 1601; De medicina methodica libri tredecim, 1611; De rhapontico disputatio, 1612; De plantis exoticis libri duo, 1627. First European to describe coffee plant; detected male and female sexes in plants and the fertilization of female flowers from the male; investigated causes and herbal remedies for epidemics in Egypt; described numerous Egyptian, Italian, Cretan, plants; Linneaus named the genus Alpinia of the order of Zingiberacae in honor of its discoverer. Died Padua, Feb. 6, 1617.

AL-QABAJAQI (Bailak al-Qabajaqi), Arabic mineralogist, astrologer; b. Upper Egypt; flourished in Cairo, 1242-82; wrote book Treasure of Merchants Concerning the Knowledge of (precious) Stones, for sultan, Al Mansur II; mentions in this book his observation of use of floating compass in 1242.

AL-QAISI (Uthman al-Qaisi) (Ahmad ibn 'Uthman al-Qaisi), (Qadi Fath al-din Abu al-'Abbas Ahmad ibn 'Uthman al-Qaisi), physician; b. Egypt; flourished 1240-49; s. Qadi Jamal al-din Abu 'Amr 'Uthman; flourished during rule of Ayyubid sultan al-Salih; achieved title of head doctor of Egypt; wrote treatise on eye diseases, Result of the Thinking on the Treatment of the Troubles of Vision.

AL-QALQASANDI; s. Ali; 1 son, Nagn al-din Muhammad; wrote on geography, history and agr. of Egypt (including work in 14 vols.); most important works continued by his son. Died 1418.

AL-QAZWINI (or Abu Yahya Zakariya ibn Muhammad ibn Mahmud al-Qazwini, also Zahariya ibn Muhammad al-Qazwini), Persian sci. writer; b. Quawin, Iraq Ajami, Persia, circa 1203; s. Muhammad b. Mahmud; pupil of al-Abhari; acquainted with Ibn Arabt; lived in Damascus. Author: The Wonders of Animals and the Singularities of Creatures, Part I, Traces of Countries and Histories of God's Servants, Part II (encyclopedic work on astronomy, natural history, geography, history). Known as the medieval or Muslim Pliny because of his considerable learning and lack of critical spirit. Died 1283.

AL-QAZWINI, see al-Katibi.

AL-QAZWINI (Hamd Allah ben Abi Bakr ben Hamd al-Mustawfi al-Qazwini), Iranian sci. writer, historian, b. circa 1281; author, Nuzhat al-qulub (ency. of scis.), 1349; also history of Persia in prose, 1330, in verse, 1335.

AL-QIFTI (ibn Yusuf al-Qifti), Arabic sci. historian; b. circa 1172-3; s. Yusuf; wazir to Halab; lived in Aleppo. Author: A Book Giving Information About the Condition of the Best in the History of Philosophers (one of main sources for history of Arabic sci.). Died 1248.

AL-QUARASHI (Annafis al-Quarashi) (ibn al-Nafis al-Qarshi al-Misri al-Shafi'i), physician; b. Egypt or Syria; studied under Ibn al Dakhwar, Damascus, Syria. Practiced in Damascus. Wrote commentaries on writings of Hippocrates, Hunain ibn Ishaq, Ibn Sina; also wrote treatises on eye diseases, diet; commentary on Qanun, Kitab Mujiz al ganun. Preceded Servetus, Columbus, and Harvey in stating that blood circulates from right to left ventricle via lungs (pulmonary transit). Died 1288.

ALQUIÉ, Alexis Jacques, French physician; b. Perpignan, France, 1812; prof. clin. surgery U. Montpellier (France), also chief surgeon l'Hôtel-Dieu, Montpellier. Author: Cours élementaire de pathologie chirurgicale d'apres la doctrine de l'école de Montpellier, 1845; Precis de la doctrine médicale de l'ecole de Montpellier, 1847; Chirurgie conservatoire et Moyen de restreindre l'utilite des operations, 1850; Clinique chirurgicale de l'Hôtel-Dieu de Montpellier 1852; Etude médicale et experimentale de l'homicide réel ou simulé par strangulation, relativement aux attentats dont Maurice Roux a été l'objet, 1864. Died Montpellier, 1864.

AL-RAHMAN, Abd al-Sufi (Abu al-Husayn); see Abd al-Rahman, Umar al-Sufi.

AL-RAMMAH, see al-Hassan (al-Rammah).

AL-RASULI (al-Mujahid 'Ali al-Rasuli), sultan of Yaman; b. Yaman, circa 1362-63; s. Fifth Sultan al-Mujahid-'Ali. Author: The Desired Book of Peasants on Useful Trees and Aromatic Plants (included list of medicinal plants in Yaman, also mentioned his father's discovery of iron ore in Yaman circa 133-34).

AL-RAZI, see Rhazes.

AL-RAZI (Fakhr al-din al-Razi) (Abu 'Abdallah Muhammad ibn 'Umar ibn al-Husain ibn al-Khatib Fakhr al-din al-Razi), Persian mathematician, astronomer, physician; b. Ray, Jibal, 1149; an important rep. of As'arite sch. of Islamic thought; engaged in geomancy, medicine and math. Author: treatise on Euclid's postulates, astrological treatise, treatise called The Hidden Secret, a history of Shafi'ite doctors, a very important commentary on Koran called Keys of Mystery, treatise on principles of physics and metaphysics called Oriental (or illuminating) Questions; also 2 encys. in Persian: Totality of the Sciences (treating of 40 branches of knowledge), Garden of Lights (treating of 60 Truths of the Secrets). Died Herat, 1210.

AL-SAGHANI (Abu Hamid Ahmad ibn Muhammed al-Saghani al-Asturlabi), mathematician, astronomer, inventor; flourished in Bagdad; worked in Sharaf al-Dawla's obs., perhaps made the instruments there; investigated trisection of the angle. Died 990.

AL-SAMARQANDI (Muhammad ibn Ashraf al-Samarqandi) (Sams al-din Muhammad ibn Ashraf al-Husain al-Samarqandi), Arabic mathematician, astronomer; flourished 1276; writings include: Figures or Difficulties of the Foundation (explains 35 propositions of 1st book of Euclid's Elements); also star calendar for year, 1276-77, works on logic and dogmatics.

AL-SAMARQUANDI (Najib al-din al-Samarqandi) (Abu Hamid Muhammad ibn 'Ali ibn 'Umar Najib al-din al-Samarqandi), Iranian physician; flourished 1st half of 13th century at Samarkand; Muslim med. practitioner and theoretician; killed by Tartars during sack of Herat, 1220-23. Natis b. Clwad al-Kirmani wrote commentary on his Causes and Symptoms, circa 1423 which was translated into Persian, circa 1700.

ALSARIO, Vincent (or Alzario Della Croce), Italian physician; b. Genoa, Italy, 1576; 1st physician to Pope Gregory XV; practiced at Bologna, Ravenna and Rome; prof. Collegium Romanum, 1612-32. Author: De invidia et fascino veterum, 1595; Epherimidum, id est diuturnarum observationum libri II, 1599-1600; De epilepsia . . . 1600; De sugillatione quam Graeci . . . , 1617; Providenza metodica per preservarsi dall' imminente peste, 1630; De haemoptysi seu sanguinis sputo, 1633. Died circa 1633.

ALSBERG, C(arl) Lucas, Am. biochemist; b. New York, N.Y., Apr. 2, 1877; s. Meinhard and Bertha (Baruch) A.; A.B., Columbia, 1896, A.M., 1900; M.D., Coll. of Phys. and Surg. (Columbia), 1900; univs. of Strassburg and Berlin, Germany, 1900-03; m. Emma B. Peebles. Asst. in physiol. chem., 1902-05, instr. in biol. chem., 1905-08, Harvard; chem. biologist Bur. Plant Industry, 1908-12, chief Bur. Chem., 1912-21, U. S. Dept. Agr.; dir. Stanford Food Research Inst., 1921-37, consultant, 1937—; dean of grad. study, Stanford, 1927-33; dir. Giannini Foundation of Agrl. Economics, U. of California. Investigator U. S. Bureau of Fisheries, 1906-08. Authority on biochemistry of foods; research on metabolism; enzyme and protein chemistry, mold toxicology, poisonous plants, food supply and population, commodity regulation. Died Nov. 1, 1940.

ALSCHINGER, Andreas, botanist; b. Angern/Budweis, Nov. 20, 1791; ed. Linz, Austria. Prof. history, classical philology, natural history Gymnasium, Zara, Yugoslavia; bot. expdns., Dalmatia, 1829, later mountain area Zara, Investigated fauna, mineralogy. Author: Hora jadrensis, 1832. Contbd. papers to bot. jours. Austria. Died Vienna, Jan. 10, 1864.

AL-SHADHILI (Sadaqa ibn Ibrahim al-Misri al-Hanafi al-Shandhili), Egyptian ophthalmologist; flourished 2d half 14th century; continued and introduced new aspects into Arabic ophthalmology (which dates to beginning 9th century); wrote an elaborate treatise, Ophthalmological Support for the Diseases of the Sight Organ, extending (in Munich manuscript) to almost 3,500 lines; in this work discussed the embryological origin of the eye, and was 1st to speak of prevalence of eye diseases among the Egyptians (which he ascribed to dust and sand, as well as to weakness of their brains); introduced each group of diseases with path. generalities; indicated seasons and ages of life when certain diseases are more common; regularly began the prognosis with the words: "This disease is curable," or "It is feared," or "It is incurable."

AL-SHAFRA (Muhammed al-Shafra) (Abu 'Abdallah Muhammad ibn 'Ali al-Qirbilyani al-Shafra), Hispano-Muslim surgeon; b. Crevillente, nr. Elche; flourished 1st quarter 14th century; a renowned herbalist. Author: Book of Deepening and Consolidation on the Treatment of Wounds and Inflammations or Tumors (complete treatise on surgery as understood at time, divided into 3 parts: causes, symptoms and treatment of inflammations and tumors; wounds, extraction of arrows and reduction of fractures and luxations; simple or composite drugs used in surgery).

AL-SHATIR (Ibn al-Shatir) (Ala al-din 'Ali ibn Ibrahim ibn al-Shatir al-Muwaqqit al-Shatir), Syrian astronomer; b. Mar. 1306; flourished Damascus. Author treatises describing constrn. and use of various instruments, astron. theories; The New Table (compiled by order of Murad-I, 'Uthmanli, sultan 1360-89). Determined obliquity of ecliptic, Damascus, Syria, 1363-64; invented 2 types of quadrant.

AL-SHIRAZI (Qutb al-din al-Shirazi), Iranian mathematician, astronomer, physician; b. 1236; s. Muslih; student of Al-Tusi. In service of Iranian rulers Ahmed and Argan. Studied philos., theol. and med. problems; proposed solar calendar corrected by 1° by fixing 17 intercalary days in every 70 years, leaving error of one day in every 1540 years. Died 1311.

AL-SIRAFI (Abu Zayd al-Hasan al-Sirafi), geographer; wrote numerous reports of Arab voyagers, including voyage of Sulayman; drew maps for various books including: Models of the Climates; Sights and Provinces (by al-Istahri). Died 940.

ALSTADT, Donald Martin, Am. physicist, chemist; born Erie, Pa., July 29, 1921; s. Reinhold Leonard and Jean (Martin) A.; B.S., U. Pitts., 1947. Devel. engr. Lord Mfg. Co., Erie, 1945-48, chief physicist, 1948-54, mgr. basic research, 1954-58, mgr. central research, 1958-62, v.p., 1960-65, mgr. Hughson Chem. Co. div., 1958—, v.p., gen. mgr. Lord Corp., 1965—, executive vice president, 1966—. Consultant, Transistor Products Company, Boston, 1952-56. Fellow Am. Inst. Chemists; mem. Am. Inst. Physics, Am. Chem. Soc., Electrochem. Soc., Faraday Soc. (Eng.), Inst. Mgmt. Scis., N.Y. Acad. Sci., Sco. Rheology. Research on thermodynamics metal surfaces, dynamics gas plasmas, surface properties semicondrs. Home: 228 Rosemont Av. Office: 1635 W. 12th St., Erie, Pa. 16512.*

ALSTED, Johann Heinrich (also Alstedius, Altstaedtius, Alstedt), philosopher, theologian; b. Herborn, Prussia (now Germany), 1588; prof. theology Protestant U. Heborn. Author: Physica harmonica, Book I, Physicam mosaicam, Book II, Physicam hebraeorum, Book III, Physicam peripateticam, Book IV, Physicam Chemicam; Cursus philosopia encyclopedia libri, 1620; Elementale mathematicum . . . , 1615. Considered Bible as source of med. and moral scis. Died Alba Iulia, Rumania, 1638.

ALSTEEN, Pierre, agronomical engr.; b. Hornu, Belgium, Mar. 5, 1898; s. Léon and Louise (Quenon-Namur) A.; ed. La Louviere, Gembloux; degree in agronomical engring.; m. Léone Lambot-Couty, Oct. 17, 1922; children—Antoine, Suzanne, Phillippe. Asst. phys. and chem. agrl. sta. Gembloux; dir. Ligue agrl. province, Namur; engr.-in-chief Comptoir salts and potassium manure. Mem. Confrérie Herdiers de Bastogne, Agronomical Engrs. Gembloux (pres.), Belgian Agrl. Soc. Luxembourg (provincial sec.). Author: l'Agriculture et la vie chère; Syllabus d'agriculture, 2 vols.; Cent ans d'agriculture dans la province de Namur. Address: 116, avenue de Mersch, Arlon, Belgium.

ALSTON, Charles, Brit. chemist, botanist; b. Eddlewood, Eng., Oct. 24, 1683; s. Thomas Alston; ed. Glasgow, Scotland, (under Boerhaave) Leiden, Netherlands; grad. in medicine, Edinburgh, Scotland, 1719. Lectr. in botany and materia medica U. Edinburgh, also supt. bot. gardens, 1716-60; Fellow Royal Coll. Physicians. Apocyneous genus Alstonia named in his honor by Robert Brown. Author index of plants in Edinburgh garden, 1740; A Dissertation on Quick-Lime and Lime Water, 1751; Tirocinium botanicum edinburgense (attack on Linnean system of classification), 1753; lectures on materia medica (edited by J. Hope), 2 vols., 1770. Studied lime water and crust formation. Died Nov. 22, 1760.

ALSTON, Edward Richard, Brit. zoologist; b. Stockbriggs, Dec. 1, 1845; mem. Linnean Soc. (sec. 1880); contbd. papers on birds, rodents, especially Am. squirrels, to Proceedings Zool. Soc.; contbd. sect. on mammalia to Biologia Centrali-Americana (Salvin and Godman); asst. to T. Bell in 2d edit. British Quadrupeds, 1874. Died Mar. 7, 1881.

AL-SUFI, Abd al-Rahman (Abu al-Husayn); see Abd al-Rahman, Umar al-Sufi.

AL-SUYUTI, b. Cairo, Egypt; 1445; s. Abu Bakr; wrote abridgments from works of other historians including an ency. of 14 scis.; A History of the Caliphs; History of Egypt; Book of Primary Knowledge (from al-'Askari). Died Cairo, 1505.

ALT, Eugen Johann, German climatologist; b. Augsburg, Germany, Aug. 4, 1878; s. Georg and Anna (Schilling) A.; ed. Munich, Germany. Became adjunct Weather Bur. Bavaria, Munich, 1902; apptd. dir. Dresden (Germany) Weather Bur., 1921; named hon. prof. meteorology Tech. U. Dresden, also U. Forestry, Tharandt; became adminstr. German Weather Service, 1935. Author: Das Klima, 1912; Die Wettervorhersage, 1919; Klimaatlas von Sachsen (Sächsioche Landeswetterwarte), 1924; numerous articles in German meteorol. jours. and handbooks. Investigated atmospheric radiation. Died Dresden, Sept. 25, 1939.

ALT, Fred, Am. engring. exec.; b. Vienna, Austria, Mar. 6, 1912; s. Joseph and Else (Shreier) A.; D. Law, U. Vienna, 1934; B.S. in Elec. Engring., Cooper Union Sch. Engring., 1949; m. Hansi Schwarz, May 18, 1941. Came to U. S., 1938, naturalized, 1944. Chief elec. engr. Holmes & Narver, Los Angeles, 1951-58; chief instrument engring. and devel. br. NIH, Bethesda, Md., 1959-64; dir. testing div. Naval Oceanographic Instrumentation Center, Washington, 1964—. Mem. Instrument Soc. Am. (dir. marine scis. div. 1966—), Engrs. Joint Council (chmn. instrumentation com. 1964—) Am. Soc. for Testing and Materials (chmn. task group on instrumentation for exposure sites 1966—), A.A.A.S., Nat. Soc. Profl. Engrs., Marine Tech. Soc., I.E.E.E. Editor: Biomedical Sciences Instrumentation, vol. I, 1963; Advances in Bioengineering and Instrumentation, 1966. Research, publs. on devel. and application of instrumentation in indsl. automation, cathodic protection of undersea structures, biol. and med. engring., oceanographic instrumentation. Home: 5200 Carlton St., Washington 20016. Office: Oceanographic Instrumentation Center, Washington 20390.*

ALT, Howard L., Am. physician; b. Chgo., July 28, 1900; s. Frank H. and Clara (Lang) A.; M.D., Northwestern U., 1924, Ph.D., 1934; m. Patricia Drew, June 25, 1935; children—Leslie (Mrs. John Mott), Abby (Mrs. L. Russell Cartwright), Robin (Mrs. Crocker Snow, Jr.), Howard Lang, Winston D., P. Brooke. With Northwestern U. Med. Sch., 1930—, prof. medicine, 1956—, dir. student health Chgo. Campus, 1931—, chief hematology sect., 1965—; practice medicine specializing in hematology, Chgo., 1930—; staff Passavant Meml. Hosp., Chgo., 1930—, chief medicine, 1958-63. Cons. to hosps., research founds. Diplomate Am. Bd. Internal Medicine. Fellow A.C.P.; mem. Central Soc. for Clin. Research (past councillor), Chgo. Soc. Internal Medicine (past pres.), Blood Club (past chmn.), Hematology Club sec. 1941—), Soc. for Exptl. Biology and Medicine, Inst. Medicine Chgo. (past pres.), Internat., Am. socs. hematology, Central Clin. Research Club, Chgo. Soc. Med. History, Assn. Am. Physicians. Research, numerous publs. on clin. hematology and physiologic aspects of erythropoiesis. Home: 1144 Michigan Av., Evanston, Ill. 60202. Office: 707 Fairbanks Ct., Chgo. 60611.*

ALTEMEIER, William Arthur, Am. physician; b. Cin., July 6, 1910; s. William Arthur and Carrie (Moore) A.; B.S., U. Cin., 1930, M.D., 1933; M.S. in Surgery U. Mich., 1938; m. Edna Ann Wyss, June 16, 1934; children—William Arthur III, Ann (Mrs. James Dearworth), George C. Asso. surgeon Henry Ford Hosp., Detroit, 1939-40; faculty surgery U. Cin., 1940—, Christian R. Holmes prof., chmn. dept., 1952—, dir. surg. research bacteriology lab., 1940—; dir. surg. services Cin. Gen. Hosp., 1952—; surgeon-in-chief Holmes and Children's Hosp., 1952—; responsible investigator research contracts and grants Surgeon Gen.'s Office U. S. Army, 1942—; USPHS, 1944—. Mem. Hist. and Philos. Soc., A.C.S., A.M.A., Soc. Clin. Surgery. (pres. 1962-64), Am. (sec. 1957-64), Central (treas. 1953-56, pres. 1958-59), Pan Pacific, Western surg. assns., Am. Soc. Clin. Investigation, Internat. Soc. Surg., Surg. Biology Club, Am. Assn. Microbiology. Contbg. author numerous books. Research, publs. on surg. bacteriology, surg. infections, chemotherapy, gastroenterology surgery, cancer. Home: 1368 Neeb Rd., Cin. 45238. Office: Cin. Gen. Hosp., Cin. 45229.*

ALTER, David, Am. physicist, inventor; b. Westmoreland County, Pa., Dec. 3, 1807; s. John and Eleanor (Sheetz) A.; grad. Reformed Med. Coll., N.Y.C., 1831; m. Laura Rowley, 1832; m. 2d, Elizabeth A. Rowley, 1844; 11 children. Invented an electric clock, model of electric locomotive; invented elec. telegraph, 1836; discovered method of purifying bromine, method of obtaining coal oil from coal, 2d law of spectrum analysis (various elemental gases have spectra peculiar to themselves). Died Sept. 18, 1881.

ALTERMAN, Zipora Stephania, Balaban, mathematician; b. Berlin, Germany, Aug. 6, 1925; d. Leon and Regina (Wischnitzer) Balaban; M.Sc., Hebrew U. Jerusalem, Israel, 1949, Ph.D., 1953; m. Israel Alterman, June 28, 1950; 1 son, Ilan M. Scientist, Israel Dept. Research and Devel., 1948-55; staff dept. applied math. Weizmann Inst., Rehovot, Israel, prof. applied math., 1961-67; vis. research asso. U. Chgo., 1960-61; vis. prof. Hebrew U., 1961-65; vis. prof. Tel-Aviv (Israel) University, 1965-66; dir. of the Institute of Space and Planetary Science, 1967—; visiting member of the Courant Institute Math. Sci. N.Y. U., 1966-67. Mem. Israel Math. Union (past hon. sec.), Israel Soc. Geodesy and Geophysics (chmn. sect. seismology 1964—), Am. Math. Soc., Am. Geophys. Union, Seismol. Soc. Am., Israel Math. Union, Israel Union for Information Processing. Contbr. articles to tech. jours., chpts. in books. First to compute frequencies of free oscillations of earth, research in terrestrial spectoscopy, hydrodynamic stability; obtained complete theoretical seismograms for explosive and other point sources in a sphere; found propagation constants for sound and heat in rarefied gases. Office: Inst. Space and Planetary Science, Tel Aviv U., Tel Aviv, Israel.*

ALTEVOGT, Rudolf, German zoologist; b. Ladbergen, Germany, Jan. 22, 1924; s. Rudolf and Wilhelmine (Kemper) A.; Dr. rer. nat., U. Munster, 1950; m. Rosamunde Brunne, Feb. 6, 1953; children—Dirke, Heike. Sci. asst. U. Munster Zool. Inst., 1950-56, dozent, 1956-61, prof. zoology, 1961—. Mem. Deutsche Zoologische Gesellschaft, Gesellschaft Deutscher Naturforscher und Arzte, Internationale Vereinigung für Biophonetik, Deutsche Ornithologengesellschaft, Bombay Natural History Soc. Research, publs. on brain size and learning abilities in chicks, elephants, behavior of tropical shore crabs, their ecology and physiology, problems of animal communications. Home: 9 Schulteweg. Office: 9 Badestrasse, 44 Munster, Germany.*

ALTHAUS, Julius, physician; b. Lippe-Detmold, Germany, Mar. 31, 1833; s. Friedrich and Julie (Draescke) A.; student U. Bonn (Germany); became student U. Gottingen, (Germany), 1851, then U. Heidelberg; M.D., U. Berlin (Germany), 1855; student Johannes Mueller; Sicily; m. Anna Wilhemina Pelzer, June 1859; 2 sons, 1 dau. Worked with Jean Martin Charcot, Paris; attached to King's Coll. Hosp., London, Eng.; founder, pres. Hosp. for Epilepsy and Paralysis, Regent's Park, 1866-94, cons. from 1894. Fellow N.Y. Acad. Medicine (corr.); mem. Royal Coll. Physicians. Decorated insignia Order Crown Italy. Author: A Treatise on Medical Electricity, 1859; The Spas of Europe, 1862; On Paralysis, Neuralgia and Other Affections of the Nervous System and Their Successful Treatment by Galvanism and Faradisation, 1864; On Sclerosis of the Spinal Cord, 1885; Influenza: its Pathology, Symptoms, Complications and Sequels, 1892; On Failure of Brain Power: its Nature and Treatment, 1894. Used electricity in therapy. Died June 11, 1900.

AL-TIFASHI (Abu al-'Abbas Ahmad ibn Yusuf Shihab al-din al-Tifashi), mineralogist; Muslim, flourishing in Egypt; influenced work of Al-Qabagogi. Author: Nuzhat al-albab fi ma la yujad fi kitab (one of best known Arabic books on erotics); also book on precious stones, circa 1242; other books. Died circa 1253-54.

ALTMAN, James W(illiam), Am. psychologist; b. Dravosbug, Pa., Apr. 14, 1928; s. Chauncey Fulton and Mary (Frew) A.; B.S., U. Pitts., 1950, M.S., 1952, Ph.D., 1954; m. Marilyn June Muchler, Sept. 21, 1951; children—Lynn Bess, Neal Wallace, Blake James, Leigh Anne. Research psychologist Am. Inst. for Research, Pitts, 1951—, v.p., dir. Pitts. office. Mem. Am. Psychol. Assn., Human Factors Soc., Am. Ednl. Research Assn., Am. Vocational Assn., Am. Ordnance Assn. Identification of desirable maintainability characteristics for elec. and mech. equipment, devel. quantitative methods for evaluating human factors in system design, research on psychol. adjustment to shelter confinement, study of vocational skills. Home: 715 Westwood Dr., Gibsonia, Pa. 15044. Office: 135 N. Bellefield Av., Pitts. 15213.*

ALTMAN, Joseph, physiol. psychologist; b. Budapest, Hungary, Oct. 7, 1925; s. Samuel and Honor (Teitelbaum) A.; student U. Heidelberg, 1947-49; Ph.D., N.Y. U., 1959; m. Elizabeth Conklin Scherer, Apr. 24, 1951; 1 dau., Magda. Fellow dept. anatomy Columbia, 1959-60; research asso. dept. neurology Bellevue Med. Center, N.Y.C., 1960-61; asso. prof. dept. psychology Mass. Inst. Tech., Cambridge, 1962—; prin. investigator research project U. S. AEC, 1962—. Mem. N.Y. Acad. Scis. Author: Organic Foundations of Animal Behavior, 1966; also articles. Research on electrophysiol., anat. and behavioral studies visual system, autoradiographic investigation DNA, RNA, protein metabolism brain during postnatal devel. and on behavioral manipulation exptl. animals. Home: 80 Grozier Rd., Cambridge, Mass. 02139.*

ALTMAN, Mieczyslaw, Polish mathematician; b. Kutno, Poland, Dec. 2, 1916; s. Stanislaw and Barbara (Skrzypek) A.; student math. Warsaw (Poland) U., 1937-39, Lvov (Poland) U., 1940-41; M.Sc., Tashkent U., 1944, Ph.D., 1948; m. Wanda Teresa Kusal, Oct. 20, 1953; children—Barbara, Thomasz Edward. Instr., Inst. Math., Polish Acad. Sci., Warsaw, 1949-54, asst. prof., 1954-57, asso. prof., 1957-65, prof. math., 1965—, chief numerical analysis dept., 1958—; faculty Warsaw U., 1955-56, 67-68; vis. prof. Cal. Inst. Tech., 1958-59, N.Y. U., 1959-60. Recipient Poland's Banach prize in functional analysis, 1958. Mem. Polish Math. Soc. (v.p. Warsaw div. 1962-63). Author: Approximation Methods in Functional Analysis, 1963; also numerous articles. Reviewer, Math. Revs., 1963—. Research on functional analysis, theory of functional equations, topological methods, approximation methods, summability, numerical analysis, numerical methods of linear algebra, scalar equations, finite difference methods for partial differential equations, linear and nonlinear programming, generalized gradient methods of optimization. Home: Wolnej Wszechnicy 4/26, Warsaw 22, Poland. Office: Institut. Mat. PAN, Sniadeckich 8, Warsaw 10, Poland.*

ALTMANN, Franz, physician; born Vienna, Austria, Apr. 10, 1901; s. Ignaz and Hedwig (Weiner) A.; M.D., U. Vienna, 1925; m. Wilhelmina Findeis, Oct. 19, 1939. Came to U. S., 1938, naturalized, 1944. Faculty dept. anatomy U. Vienna, 1925-26, dept. pathology, 1926-29, dept. otolaryngology, 1929-38; faculty dept. otolaryngology Columbia Coll. Physicians and Surgeons, N.Y.C., 1938—; clin. prof., 1952—; staff Presbyn. Hosp., N.Y.C.; cons. in otolaryngology Bronx VA Hosp., N.Y.C. Research in operative treatment of deafness, histopathology of the ear. Home: 863 Park Av., N.Y.C. 10021.*

ALTMANN, Richard, histologist; b. Germany, 1852; asso. prof. anatomy U. Leipzig (Germany); proposed theory (Altmann's theory) that fundamental units of protoplasm are granular particles, circa 1890; developed low-temperature dehydration fixing technique for microscpe slides (now known as freeze drying), 1894; isolated and named nucleic acid. Died 1900.

ALTMUTTER, Georg, Austrian technologist; b. Vienna, Austria, Oct. 6, 1787; s. Matthias Altmutter; ed. Vienna, also Prague, Czechoslovakia; asst. in physics Theresianum, Vienna; prof. mech. technology Poly. Inst., Vienna. Author: Beschreibung der Wertzeugsammlung am Polytechnischen Institut zu Wien, 1825. Numerous inventions in mfg. playing cards, pinheads, terrestrial and celestial globes. Died Vienna, Jan. 2, 1858.

ALTSCHUL, Aaron Mayer, Am. biochemist; b. Chgo., Mar. 13, 1914; s. Philip and Sophie (Fox) A.; B.S., U. Chgo., 1934, Ph.D., 1937; m. Ruth Braude, Oct. 24, 1937; children—Sandra (Mrs. Frank Norman), Judy. Instr., U. Chgo., 1937-41; with So. Utilization Research and Devel. div. U. S. Dept. Agr., New Orleans, 1941—, chief research chemist, seed protein pioneering research lab., 1958—; prof. chemistry Tulane U., New Orleans, 1963—; lectr.

Mass. Inst. Tech., Cambridge, 1964-—. Research cons. Nat. Cottonseed Products Assn.; mem. tech. adv. com. Inst. for Nutrition of Central Am. and Panama, 1964; mem. panel on nutrition U. S.—Japan Coop. Med. Sci. Program; spl. cons. to sec. agr. on protein foods for developing countries. Recipient Superior Service award U. S. Dept. Agr., 1955; Golden Peanut award Nat. Peanut Council, 1964. Mem. Am. Soc. Biol. Chemists, Am. Soc. Plant Physiology, Am. Soc. Cell Biology, Am. Chem. Soc. (Charles F. Spencer award Kansas City sect. 1965), Inst. Food Technologists, N.Y. Acad. Sci., Phi Beta Kappa, Sigma Xi, Phi Tau Sigma. Author: Proteins, Their Chemistry and Politics, 1965; also numerous articles. Editor: Processed Plant Protein Foodstuffs, 1958. Discovered enzyme Cytochrome C peroxidase; led research to make cottonseed protein concentrate suitable for feeding to nonruminant animals, children; developed concept that seed proteins are in subcellular packages, coined term aleurin to describe them. Home: 429 N. Street S.W., Washington 20024. Office: U. S. Dept. Agr., Washington 20250.*

ALTSHULLER, Aubrey Paul, Am. chemist; b. Hammond, Ind., Feb. 15, 1927; s. Nathan Dundee and Eleanor (Levinson) A.; B.S., U. Chgo., 1948; M.S., U. Cin., 1950, Ph.D., 1951; m. Lillis Martha Flatman, June 12, 1952; children—Sandra Ann, Beth Lillis. Aero. research scientist Lewis Research Center, NASA, Cleve., 1951-55; with commd. corps USPHS, 1955-—, chief chem. and phys. research and devel. program Nat. Center for Air Pollution Control. Cin., 1961-—. Mem. adv. bd. Internat. Jour. Air and Water Pollution, 1963-—, Environmental Sci. and Tech., 1967-—. Recipient Cin. Chemist award, 1967. Mem. Am. Chem. Soc. (chmn. com. on air pollution 1964-—), A.A.A.S., Air Pollution Control Assn. Research, numerous publs. in elec. properties of solutions, thermal properties of various classes of substances, methods of separation and analysis for small concentrations of gaseous substances, photochemistry. Home: 3534 Principio St., Cin. 45226. Office: 4676 Columbia Pkwy., Cin. 45226.*

ALTSZULER, Norman, Am. endocrinologist; b. Suwalki, Poland, Nov. 20, 1924; s. Meyer Ben and Fay (Kalet) A.; came to U. S., 1937, naturalized, 1943; B.S., George Washington U., 1950, M.S., 1951, Ph.D., 1954; m. Rita Mintz, June 24, 1956; children—Henry, Paula. Faculty dept. pharmacology N.Y. U. Sch. Medicine, 1953-—, asso. prof., 1963-—. Mem. Am. Physiol. Soc., Endocrine Soc., Am. Soc. Pharmacology and Exptl. Therapeutics, Soc. for Exptl. Biology and Medicine. Research, numerous publs. on influence of hormones of pancreas, adrenal gland, and pituitary gland on prodn. and utilization of carbohydrates and fats. Home: 18 Stuyvesant Oval, N.Y.C. 10009.*

AL-TUGHRA'I (Abu Isma'il al-Husain ibn 'Ali ibn Muhammad al-Tughra'i), Persian poet, alchemist; b. Ispahan. Lived in Bagdad, 1111-12; later became wazir to Saljug sultan, Mas'ud ibn Muhammad in Musul; Author various treatises on alchemy including Book of the Brilliant Stone for the Preparation of the Elixer; Collection of the Secrets and Composition of Lights; Keys of Mercy and Lamps of Wisdom; Truths of the Evicence submitted with Regard to Alchemy (discussed skeptical views on alchemy of Ibn Sina); Lamiyat al'Ajam, 1111-12. Put to death for atheism, circa 1121.

ALTUKHOV, Yurij Petrovich, Russian ichthyologist; b. Elan-Koleno, Vorone Schukaja, USSR, Oct. 11, 1936; s. Piotr Kornilovitch and Aleksandra M. (Andreeva) A.; grad. Ichthyology Sch., Moscow (USSR) Tech. Inst. Fishery Mgmt., 1959; B.S., Moscow U., 1964; m. Domrina Kimma, Oct. 6, 1956; 1 son, Michail; m. 2d, Volkova Elena, Sept. 20, 1966. Staff, Karadag Biol. Sta., Acad. Sci. Ukrainian USSR, Karadag, Crimea, 1959-61, Azov-Black Sea Research Inst. Marine Fishery Mgmt. and Oceanography, Kerch, 1961-62; chair cytology Biol. Faculty, Moscow U., 1962-—. Research, publs. on biol. role and physiol. mechanism of optic reflex of fish during their orientation in space and flight; cellular adaptation of fishes' temperature and other poikilothermic animals in connection with problem of sight. Home: House 203, V-427, Block 38a, Moscow, USSR.

ALTUM, Bernard, German naturalist; b. Münster, Germany, Jan. 31, 1824; s. Bernhard Theodor and Anna Gertrude Antoinette (Huder) A.; ed. Munster; studied zoology under Johannes Muller and M. Lichtenstein in Berlin, Ph.D., 1855. Tchr., Münster; became docent Münster U., 1859, prof. zoology, 1869. Author: Les mammifères du pays de Münster, 1867; Der Vogel und sein Leben, 1868; Forstzoologie, 4 vols., 1872, 76-82; Traité de zoologie, 1878. A founder of forest entomology; studied mammals and their comml. importance; held anti-Darwinian view that birds respond instinctively to external stimuli. Died Eberswalde, Feb. 1, 1900.

AL-TUSI (al-Muzaffar al-Tusi) (al-Muzaffar ibn Muhammad ibn al-Muzaffar Sharaf al-din al Tusi), mathematician, astronomer; b. Tus, Klurasan; tchr. of Masa ibn Yunus Kama; al-din; flourished in Bagdad or Musul. Wrote a treatise on astrolabe; also paper on sub-div. of a square into four parts for prince Shams al-dim., Hamadan, 1209-10. Invented linear astrolabe (Tusi's staff). Died circa 1213.

AL-TUSI (Nasir al-din al-Tusi) (Abu Ga'far) (Muhammad ben Muhammad ben al-Hasan Nasir al-din al-Tusi al-Muhaqqiq), Iranian astronomer, mathematician; b. 1201. Imprisoned by mongols, 1256; entered service of Mongol Halagu Han; wazir of prince Il-Han, Persia, 1256-65; assisted at capture of Bagdad, 1258; directed obs. of Maraga and set up library there. Author: Book of the Sciences Intermediary between Geography and Astronomy (on Greek writings of Euclid, Appolonios, Archimedes, Ptolemy, Autolykos, Aristarchos, Hypsikles, Theodosios, Menelaos); Memorandum on Astronomy (astron. theory in which he criticizes Ptolemaic theory); De postulatio quinto (discussion of 5th postulate of Euclid); Book on the Theorem of the Sector; also made 2 edits. of Elements of Euclid. Died 1274.

AL-URDI (Mu'ayyad al-din al-Urdi al-Dimishqi), astronomer; b. Damascus, Syria; flourished circa 1259. Began career in hydraulics, then built astron. instruments; joined Obs. at Maraga, 1259. Author: The Instruments of the Observatory at Maraga.

ALVAREZ, Luis W., Am. physicist; b. San Francisco, June 13, 1911; s. Walter C. and Harriet (Smyth) A.; B.S., U. Chgo., 1932, M.S., 1934, Ph.D., 1936, Sc.D., 1967; Sc.D., Carnegie-Mellon U., 1968; m. Janet L. Landis, Dec. 28, 1958; children—Walter S., Jean S., Donald L., Helen L. Research asso., instr., asst. prof., asso. prof. U. Cal. at Berkeley, 1936-45, prof. physics, 1945-—, asso. dir. Lawrence Radiation Lab., 1954-59; in radar research and devel. Mass. Inst. Tech., 1940-43, Los Alamos, 1944-45. Recipient Collier trophy, 1946, medal for Merit, 1948, John Scott medal, 1953, Einstein medal, 1961, Nat. medal of Sci., 1964; named Cal. Scientist of Year, 1960, Michelson award, 1965. Fellow Am. Phys. Soc. (v.p. 1968); mem. Nat. Acad. Scis., Am. Philos. Soc., Am. Acad. Arts and Scis., Phi Beta Kappa, Sigma Xi. Contbr. numerous articles to sci. jours. Investigations of properties of nuclei and of particle resonant states, co-discoverer of cosmic ray East-West effect, helium 3, tritium, nuclear electron capture, 1st strange baryon and meson resonant states, muon catalysis of nuclear fission; 1st measured neutron magnetic moment; designed 1st neutron time-of-flight spectrometer, 1st proton linear accelerator, 1st charge exchange (tandem) accelerator, several hydrogen bubble chambers, several bubble chamber film measuring systems; patentee in radar and optics. Home: 131 Southampton Av., Berkeley 94707. Office: Lawrence Radiation Lab., U. Cal., Berkeley, Cal. 94720.*

ALVAREZ, Walter Clement, Am. physician; b. San Francisco, July 22, 1884; s. Luis Fernandez and Clementina (Schuetze) A.; M.D., Cooper Med. Coll. (Stanford), 1905; postgrad. Harvard Med. Sch., 1913; D.Sc. (hon.), Hahnemann Med. Coll.; m. Harriet Skidmore Smyth, Feb. 22, 1907; children—Gladys (Mrs. Raymond Archibald), Luis Walter, Robert Smyth, Bernice (Mrs. Bradley Brownson). Asst. in clinic pathology, Cooper Med. Coll., Stanford, 1906-07; physician Cananea, Mexico, 1907-09; internist San Francisco, 1910-25; assistant in medicine Stanford, 1911-12; asst. in research medicine U. Cal., 1915-16, instr., 1916-20, asst. prof., 1920-24, asso. prof., 1924-26; successively sr. cons. Div. Medicine, Mayo Clinic, Rochester, Minn., 1926-—, emeritus, 1950-—; asso. prof. medicine, Mayo Found., U. Minn., 1926-34, prof., 1934-51, emeritus; professorial lectr. medicine U. Ill. Med. Sch., 1951-52, emeritus. Diplomate Am. Bd. Internal Medicine, Am. Bd. Gastroenterology. First Caldwell Lecturer, Am. Roentgen Ray Soc. Fellow A.C.P.; mem. Assn. Am. Physicians, Am. Physiol. Soc., Am. Soc. for Clin. Investigation, Soc. Exptl. Biology and Medicine, Am. Gastroenterol. Assn. (pres. 1928; Friedenwald medallist 1951), Am. Soc. Pharmacology Exptl. Therapeutics, A.M.A., Am. Anthropol. Assn., Am. Assn. Phys. Anthropologists, A.A.A.S., Alpha Kappa Kappa, Sigma Xi, Alpha Omega Alpha. Author: Mechanics of Digestive Tract, 1922; Nervous Indigestion, 1930; The Neuroses, 1951; Danger Signals, 1953; Nervousness, Indigestion and Pain, 1954, An Introduction to Gastroenterology, 1948; Danger Signals, 1957; Live at Peace with Your Nerves, 1958; Practical Leads to Puzzling Diagnosis, 1958; Minds that Came Back, 1961; Incurable Physician, An Autobiography, 1963; Little Strokes, 1966; also chpts. in Oxford Medicine, numerous articles and booklets. Formerly editor-in-chief Modern Medicine; former editor Geriatrics. Syndicated med. columnist, 1951-—. Introduced idea of gradients in intestinal wall; made 1st electrogastrograms of man; established normal standards of blood pressure, also standards for gastric juice; research on psychosomatic medicine and neuroses, migraine, little strokes, sensory epilepsy. Home: Hotel Pearson. Office: 700 N. Michigan Av., Chgo.*

ALVAREZ-DEL-VILLAR, José, Mexican ichthyologist; b. Zamora, Mich., Mexico, May 30, 1908; s. Jose and Guadalupe (del Villar) Alvarez; biologist Escuela Nacional de Ciencias Biológicas, 1948; M.S., U. Mich., 1945; Dr. C., Instituto Politécnico Nacional, 1963; m. Felisa Solórzano, May 18, 1933 (div. Oct. 1943); 1 son, Ticul Alvarez; m. 2d, Clemencia Tèllez-Czirón, Sept. 12, 1948. Prof. zoology and comparative anatomy Escuela Nacional de Ciencias Biológicas, Mexico City, 1945-—, head dept. zoology, 1961-—, sub-dir., 1959-61. Mem. Colegio de Biológos de Mexico, Sociedad Mexicana de Historia Natural, Am. Soc. Ichthyologists and Herpetologists. Author: Los Cordados, 1966; also articles. Research on fresh-water fishes of Mexico including discovery of many new species.*

ALVARINO DE LEIRA, Angeles, biologist, oceanographer; b. El Ferrol, Spain, Oct. 3, 1916; d. Antonio Alvariño Grimaldos and Carmen González Diaz-Saavedra; B. in Sci. and Letters summa cum laude, U. Santiago de Compostela, 1933; M.Natural Scis. summa cum laude, U. Madrid (Spain), 1941, Doctor in Scis. summa cum laude, 1951; m. Eugenio Leira, Mar. 16, 1940; 1 dau., Angeles. Came to U. S., 1958, naturalized, 1966. Prof. biology El Ferrol, 1941-47; biologist dept. sea fisheries Ministry Commerce, Madrid, 1948-52; histologist Superior Council Sci. Investigations, Madrid, 1949-51; biologist, oceanographer Inst. Oceanography, Madrid, 1950-57; biologist Marine Biol. Lab., Plymouth, Gt. Britain, 1953-54; biologist, planktonologist Woods Hole (Mass.) Oceanographic Inst., 1956-57; biologist Scripps Instn. Oceanographer, U. Cal. at San Diego, 1958-—. Brit. Council fellow, 1953-54; Fulbright fellow, 1956-57. Mem. A.A.A.S., Natural History Soc., Marine Biol. Assn. U.K., Ciencia, Sigma Xi. Author: Distributional Atlas of Chaetognatha in the California Current Regions During the Monthly Cruises of 1954 and 1958, Vol. 3, 1965; also articles, chpts. in books. Discovered 8 new species of Chaetognatha, 3 new species Siphonophorae; research on indicator species of water masses, zoogeography of Chaetognatha in Pacific, Atlantic and Indian Oceans, Chaetognatha, Siphonophorae, Medusae. Home: 7535 Cabrillo Av., La Jolla, Cal. 92037.*

ALVIK, Ivar, orthopedic surgeon; b. Borgund, Sept. 26, 1905; s. Elling and Petrine (Ytterland) A.; ed. Oslo, Stockholm and N.Y. univs.; M.D.; m. Hanna, June 12, 1934; children—Anne, Else, Gunhild, Elling, Gudleik, Astrid. Became surgeon, 1949, orthopedic surgeon, 1952; chief surgeon Orthopedic Inst. U. Oslo, since 1953, apptd. prof. orthopedic surgery, 1959. Mem. Norwegian and Scandinavian Assn. Surgeons and Orthopediatricians, Internat. Club Orthopediatricians. Author: Tuberculosis of the Spine Research, publs. on scoliosis and its surg. treatment, also treatment of arthritis of hip. Home: Vidarsvei, 16, Baerum. Office: Trondheimsveien, 132, Oslo, Norway.

ALVORD, Ellsworth Chapman, Jr., Am. pathologist; b. Washington, May 9, 1923; s. Ellsworth Chapman and Katharyn (Watson) A.; B.S., Haverford Coll., 1944; M.D., Cornell, 1946; m. Nancy Armstrong Delaney, Nov. 6, 1943; children—Ellsworth Chapman III, Katharyn Watson, Jean Armstrong, Richard Watson. Faculty, Cornell, 1947-48, George Washington U., 1950-55, Baylor U., 1955-60; faculty U. Washington, Seattle, 1960-—, prof. pathology Sch. Medicine 1962-—; mem. staff King County, Univ., VA hosps., Seattle; cons. in pathology to various hosps. Chmn. med. adv. bd. Puget Sound Area chpt. Nat. Multiple Sclerosis Soc., 1962-—. Diplomate Am Bd. Pathology. Fellow Pacific Soc. Neurology and Psychiatry; mem. A.A.A.S., Am. Assn. Neuropathologists (pres. 1963), A.M.A., Am. Soc. Exptl. Pathology, Soc. Exptl. Biology and Medicine, Assn. Research in Nervous and Mental Diseases, Am. Neurol. Assn. (asso.), Phi Beta Kappa, Alpha Omega Alpha, others. Mem. adv. bd. Jour. Neuropathology and Exptl. Neurology, 1964-—. Research, publs. on diseases of nervous system especially malformations of brain and allergic inflammations of brain. Home: 5547 N.E. Windermere Rd., Seattle 98105.*

AL-WAHSIYA (or Ibn al-Wahsiya al-Kaldani al-Nabati, also Abu Bakr Ahmad ben 'Ali al-Kaldani al-Nabati ibn al-Wahsiya), Arabian alchemist; flourished 10th century. Author: Kitab al-sumum; Kitab al-filaha arnebatiya (on nabatian agr. and superstitions); Kitab al-usulal-kabir; Asrar al-tabi 'iyat ti hawass al-nabat.

ALWENS, Walter, German physician; b. Stuttgart, Germany, June 24, 1880; s. Edmund and Johanna (Maret) A.; M.D.; agrégé, 1916; m. Auguste Kohl, Sept. 22, 1931; practiced medicine specializing in internal medicine; prof. med. clinic Tübingen, Breslau and Frankfort/Main univs.; dir. dept. internal medicine U. Frankfort/Main, until 1950. Author: Spätrichitis Osteomalacie, 1926; Röntgenuntersuchung d. Gefässe, 1934; Der Chromat-Lungenkrebs, 1938. Address: Frankfort/Main, Gartenstrasse 112, West Germany.

ALWOOD, William Bradford, Am. horticulturist; b. Delta, O., Aug. 11, 1859; s. David William and Ann Eliza (Bradley) A.; student Ohio State U., 1882-85, Columbian (now George Washington) U., 1886-88, Royal Pomol. Sch., Germany, 1900-01, Inst. Pasteur, Paris, 1907; m. Seffie S. Gantz, Mar. 6, 1884; children—Hubert Jackson, Helen Anna, Nellie Sarah, Mabel Seffie, William Bradford, Lewis Gantz, Esther, Richard Olney. Taught country sch., 1879-81; supt. Ohio Agrl. Expt. Sta., 1882-86; spl. agt. U. S. Dept. Agr., 1886-88; vice-dir. Va. Agrl. Expt. Sta., 1888-1904; conducted investigations in horticulture and mycology, 1888-1904; prof. horticulture and allied subjects, Va. Poly. Inst., 1891-1904; in charge investigations on fermentation of fruit products for Bur. Chemistry, U. S. Dept. Agr., 1900-06; enological chemist, 1907-14; fruit grower, 1914. Vice pres. Internat. Congress on Agrl. Edn., Paris, 1900; del.

U. S. Dept. Agr. to Internat. Congress on Viticulture, France, 1907, and v.p.; pres. Internat. Congress Viticulture, San Francisco, 1915. Awarded Gold and Commemorative medals, St. Louis Expn., 1904; decorated by French Govt, with Cross, Officer du Mérite Agricole, 1907; silver medal and diploma, Nat. Agrl. Soc., France; Meritorious Service to Agr. certificate Va. Poly. Inst., 1923. Fellow A.A.A.S.; Royal Hort. Soc., Gt. Britain; mem. Permanent Internat. Commn. on Viticulture, Nat. Council of Horticulture, 1907, Société des Chemistes Experts de France; mem. nat. council Nat. Econ. League. Author many pamphlets and bulls. on hort. subjects and the chem. composition of apples and grapes, and composition of wines and ciders fermented with pure yeasts. Died Apr. 13, 1946.

AL-YAB UDI, Arabic physician; Jacobite Christian; his statements on natural heat of chicken and small birds began medico-philos. controversy between Ibn Butlan of Bagdad and Ibn Ridwan of Cairo, Egypt.

AL-YANBUI, geographer; b. possibly in Iran; flourished 10th century; traveled to India and China, across Tibet to Indian Meridional with ambassador of an Indian prince; crossed Kashmir, Afghanistan, Sikhistan, then returned to court of his prince Nasr b. Ahmad b. Ismail, Bukara. Author: Miracles of the Countries.

AL-YA'QUBI (Ibn Wadih al-Ya'qubi), Arabic historian, geographer; author: Book of Lands (history of world, and geography of Islamic states). Died 897.

ALYMOV, Andrey Yakovlevich, Russian epidemiologist, microbiologist; b. 1893; grad. Kharkov Med. Inst., 1921; D.Med. Sci., 1939. Epidemiologist, USSR Commissariat of Health, also at mil. med. establishments, 1922-30; sr. asso., head Inst. Normal and Path. Physiology, Moscow, 1930-36; head chair epidemiology Leningrad Naval Med. Acad., 1942-45; chief epidemiologist Soviet Navy, 1942-45, Soviet Army, 1948-51; head chair epidemiology Mil. Faculty, Central Postgrad. Med. Inst., Moscow, 1948-54; head Lab. Infectious Pathology, Inst. Normal and Path. Physiology, USSR Acad. Sci., 1953—. Mem. USSR Acad. Med. Sci. (corr.), All-Union Soc. Microbiologists, Epidemiologists and Infectionists (dep. chmn.). Author more than 70 works including Persian Relapsing Fever, 1935; Spontaneous Complications of Experimental Tick Recurrence and Their Effect on the Course of Spirochetosis, 1937; Principles of Diagnosis and Prophylaxis of Parasitic Typhus and Fevers, 1939. Mem. editorial bd. Bull. Exptl. Biology and Medicine; co-editor Microciology sect. Large Med. Ency., 2d edit. Studied rickettsiosis in Iran, 1930-34; studied Marseilles fever and proved its existence in Crimea, 1936-40. Address: Inst. Normal and Path. Physiology, Baltiyskaya ul. 8/I, Moscow, USSR.

AL-ZAFFAN (Efraim ibn al-Zaffan), Jewish physician; flourished 2d half of 11th century; disciple of 'Ali ibn Ridwan; lived at court of Fatimides; known for work as copyist; Salama b. Rahmun was his disciple.

ALZAMORA, ALBENIZ, Antonio, Spanish physician; b. Palma de Majorca, Spain, June 12, 1912; s. Vincente and Enriqueta (Albeniz) A.; M.D.; ed. Barcelona and Madrid univs.; m. Gloria Figueras Dotti; children—Gloria, Maria Luisa, Christina, Angeles, Antonie. Surg. orthopedist Hosp. San Pablo, sec. sch. orthopedic surgery; consul Coll. Physicians. Mem. Internat., Spanish Orthopedic Surgery and Traumatology socs., Coll. Surgeons, Acad. Med. Sci. Author: Tratamiento fracturas de femur; Indicaci ones quirúrgicas de tumores malignos osteogénicos; Rehabilitación; Espalda inadecuada; Tratamiento fractures rotula. Home: calle Santa Catalina Sena (Pedrelbas), 30, Barcelona. Office: via Cayetana, 157, Barcelona, Spain.

ALZATE Y RAMIREZ (Azalte y Ramírez), José Antonio, Mexican astronomer, geographer; b. Ozumba, Mexico, 1738; priest; dir. sci. missions for govt. and Roman Catholic Ch.; founded Diario literario (newspaper), 1768. Mem. French Acad. Scis., 1771. Wrote on astronomy, physics, metallurgy, silk growing, use of ammonia against gases in mines; made extensive maps of Mexico. Died Mexico, Feb. 2, 1799.

ALZHEIMER, Alois, psychiatrist; b. Markbreit, Germany, June 14, 1864; s. Eduard and Therese (Busch) A.; ed Würzburg, also Berlin, Germany; m. Cecile Wallerstein, 1894; 1 dau. Asst., Nerve Hosp., Frankfort/Main, later Heidelberg, Germany; became head anat. lab. Psychiat. Clinic, Munich, Germany, 1904; apptd. prof. psychiatry and neurology, Breslau (now Wroclaw, Poland), 1912. Investigated pathology of senile and presenile dementia; described presenile condition (Alzheimer's disease) with symptoms of prog. mental weakness and asso. with cortical cerebral sclerosis, 1907; studied symptoms of schizophrenia and maniac depression. Modified Mallory's phosphomolybic acid hematoxylin stain for neuroglia changes, also Mann's eosin-methyl blue stain for neuroglia changes. Died Breslau, Dec. 19, 1915.

AMADOR, Elias, physician; b. Mexico City, Mexico, June 8, 1932; s. Elias O. and Edith (Weidenheim) A.; B.S., Centro Universitario Mexico, 1949; M.D., Escuela Nacional de Medicine Universidad Nacional Autonoma de Mexico, 1956; m. Betsey Baldwin Fisher, June 30, 1960; children—Franz Gustave, Eric. Asst. in medicine Peter Bent Brigham Hosp., Boston, 1960-64, asso. pathologist, 1964-66; instr. pathology Harvard Med. Sch., 1964-66; pathologist in charge clin. pathology, asso. prof. pathology Western Res. U., Cleve., 1966—. Diplomate Am. Bd. Pathology. Mem. A.A.A.S., Am. Chem. Soc., Am. Assn. Clin. Chemistry, Am. Coll. Cardiology, Am. Assn. Pathol. Bacteriology, Am. Soc. Exptl. Pathology. Contbr. articles to sci. jours. Devel. accurate and reproducible enzymatic assays; their application to detection and diagnosis of early disease. Home: 2346 S. Overlook Rd., Cleveland Heights, O. 44106.*

AMAGAT, Emile-Hilaire, physicist; b. Saint-Satur, France, Jan. 2, 1841; doctorate, 1872; prof. math., physics Central Gymnasium, Fribourg, Switzerland, 1867-72; prof. physics Catholic U. Lyons (France), 1877-92; asst. lectr. l'Ecole Polytechnique, 1892. Fellow Royal Soc., 1897; mem. French Acad. Scis. (La Caze prize). Author works on static of fluids; collection of results of studies on behavior of gases under various pressures and temperatures, 1893. Made precise studies on elastic extension and sideways contraction according to Wertheim's method; studied compressibility of liquids, 1877; observed volumetric ratios of gases under high pressure, 1880; investigated rarified gases, 1883; in expts. on change in melting point through pressure, kept carbon tetrachloride and benzol in solid states, 1888; followed compressibility of gases to higher pressures, also determined compressability curves of permanent gases to 3000 atmospheres and 200° Centigrade, found that the more gas is compressed, the more it deviates from Boyle-Mariotte law. Died Saint-Satur, circa Feb. 16, 1915.

AMALDI, Edoardo, Italian physicist; b. Carpaneto, Piacenza, Italy, Sept. 5, 1908; s. Ugo and Luisa (Basini) A.; Dr. Physics, U. Rome, 1929; m. Ginestra Giovene, Oct. 19, 1933; children—Ugo, Francesco, Daniela. Became prof. physics U. Rome, 1937; sec.-gen. European Orgn. for Nuclear Research, 1952-54; pres. Internat. Union Pure and Applied Physics, 1957-60. Fellow Lincei Nat. Acad.; mem. Royal Soc. for Scis. Uppsala, Acad. Scis. USSR, Royal Instn. Gt. Britain, Am. Philos. Soc., Am. Acad. Arts and Scis., Nat. Acad. Scis. U.S.A., Royal Acad. Netherlands, Acad. Leopoldino. Research and publs. on atomic, molecular and nuclear physics. Home: Viale Parioli 50, Rome, Italy.

AMANN, Joseph Albert, gynecologist; b. Helmprechting, Germany, Mar. 13, 1832; s. Johann Evangelist and Theresia (Neumair) A.; ed. Munich, also Würzburg, Germany; m. Rosa Höber; 1 son, Joseph Albert. Apptd. asso. prof. Gynecol. Polyclinic, Munich, 1874; founder, dir. Gynecol. Clinic, Munich, from 1884. Author: Über den Einfluss der weiblichen Geschlechtskrankheiten auf das Nervensystem mit besonderer Berücksichtigung der Weseus und der Erscheinungen der Hysterie, 1874; Klinik der Wochenbettkrankheiten, 1876. One of 1st gynecologists to study urology. Died Munich, Jan. 21, 1906.

AMAR, Arjan D., physician; born Shorkot, India, Jan. 4, 1929; s. S. Santokh Singh and Jaisi Bai (Aneja) A.; F.Sc., U. Panjab (India), 1946; M.B.B.S., Med. Coll. Amritsar, India, U. Panjab, 1951; M.S. in Surgery, U. Mich., 1958; m. Kamla Chabra, Dec. 10, 1956; children—Akhil Reed, Vikram David, Arun Paul. Came to U.S., 1953, naturalized, 1966. Instr., U. Rochester, 1959-60; chief dept. urology Kaiser Found. Hosp., Walnut Creek, Cal., 1961—; staff Presbyn. Med. Center, San Francisco, John Muir Hosp., Walnut Creek; chief dept. urology Permanente Med. Group, 1961. Recipient 1st pl., Silver medal Med. Coll. Amritsar, 1951; named Most Distinguished Med. Grad., 1951; recipient U. (Panjab) Silver medal 1951; Med. Coll. Council Gold medal 1951. Fellow A.C.S., Royal Coll. Surgery; mem. A.M.A., Cal., Alameda-Contra Costa County med. assns., Am. Urol. Assn. (Joseph M. McCarthy award Western sect. 1965). Contbg. author; Surgery of Prostate, 1958; Ency. of Urology, 1967. Research, publs. in reduction radiation exposure in diagnosis and treatment urologic disease, diagnosis and treatment urologic disease in gynecologic patients, use double dose contrast medium in urographic investigation to improve diagnostic accuracy, urinary tract infections in pyelonephritis devel. methods and tests in investigative urology including nonradiographic method for demonstration vesicoureteral reflux; improved methods in diagnosis and treatment urinary tract anomalies; calculus disease during pregnancy; improved surg. methods for treatment of vesicoureteral reflux. Home: 140 Cragmont Dr. Office: 1425 S. Main St., Walnut Creek, Cal. 94596.*

AMARO, A., Brasilian protozoologist; b. Rio de Janeiro, Brazil, May 29, 1922; s. José and Maria (Verissimo) Amaro Verissimo; ed. Faculty Philosophy, Scis., Letters, Rio de Janeiro. Prof. biology Colegio Estadual Rivadavia Correa, 1959—; protozoologist Instituto Oswaldo Cruz, 1960—. Mem. Soc. Protozoologists, Nat. Assn. Biology Tchrs., Soc. Biology Rio de Janeiro. Research and publs. on Brazilian opalinids; revised scheme of opalinid classification. Home: Caixa Postal 84, Agencia Lgo. Machado, Rio de Janeiro, GB, Brasil ZC-01. Office: Instituto Oswaldo Cruz, Caixa Postal 926, Rio de Janeiro, GB, Brasil.*

AMATUS LUSITANUS (Castello Bianco, Joao Rodriguez de), physician; b. Castelo Branco, Portugal, 1511; lectr., Venice, Italy; practiced medicine, Ancona, Italy; after persecution by Inquisition, retired to Salonika, Greece. Author: Curationum medicinalium centurae quatuor, Bas, 1556; account of 700 cases in medicine and surgery, 1551-66. Described valves in veins. Died 1568.

AMBARTSUMIAN, Viktor Amazaspovich, Russian astrophysicist; b. Tbilisi, USSR, Sept. 18, 1908; s. Amazasp Asaturovich and Ripsime A.; grad. Leningrad (USSR) U., 1928; postgrad., Pulkovo Obs., 1928-31. Faculty Leningrad U., 1931-46, prof., 1934-46; head, research br., Yelabuga, World War II; dir. Yerevan (Armenia) Obs., 1944-46; founder, supr. planning and constrn. Byurakan Obs., Armenia, 1946, also dir. prof. astrophysics Yerevan State U., 1947—. Participant numerous internat. congresses, confs. 1946—; dep. Supreme Soviet, 1950—. Order of Lenin, 1945-58, Order Red Banner of Labor, 1945-56, Recipient Stalin prize, 1946, 50; Gold medal British Royal Astron. Soc., 1960. Mem. Acad. Scis. USSR (academician 1953—, mem. presidium 1961—), Acad. Scis. Armenian Soviet Socialist Republic (pres. 1947—), Armenian Soc. for Propagation Sci. and Polit. Knowledge (pres.), Internat. Astron. Union (pres. 1961-64), numerous other academies of science and scientific societies in various countries. Author: Theoretical Astrophysics, 1939; Light Scattering by Planetary Atmospheres, 1942; Stellar Evolution and Astrophysics, 1947; Star Clusters, 1949; Scientific Works, 2 vols., 1960, Problems of Extragalactic Research, 1962. Research and numerous publs. on stellar physics, stellar astronomy, cosmogony, galaxies and their systems. Discoverer new type stellar systems, 1941-43, naming them stellar assns., 1947; devised method of computing mass ejected by nova stars and developed theory of radial equilibrium of planetary nebulae; made sci. trip to Australia and Indonesia, 1963. Address: Acad. Scis. Armenia Soviet Socialist Republic, Barekmutyan 24, Yerevan, Armenian Soviet Socialist Republic.*

AMBERSON, William Ruthrauff, Am. physiologist; b. Harrisburg, Pa., June 17, 1894; s. Presley Neville and Margaret (Ruthrauff) A.; Ph.B., Lafayette Coll., 1915; Ph.D. in Biology, Princeton, 1922; postgrad. U. Coll., London, Eng., U. Kiel (Germany), U. Uppsala (Sweden); m. Grace Dorsett, Aug. 18, 1923 (dec. Dec. 1957); children—Margaret Mary, Barbara Dorsett; m. 2d, Christiana Birckhead Bond, Aug. 23, 1958. Faculty, U. Tenn. Coll. Medicine, Memphis, 1922-23, prof. physiology, head dept., 1930-37; asst. prof. U. Pa. Sch. Medicine, Phila., 1923-30; prof., head dept. physiology U. Med. Sch. Medicine, Balt., 1937-59; investigator, NIH grantee Marine Biol. Lab., Woods Hole, Mass., 1959-66; vis. lectr. U. Vienna Sch. Medicine, Austria, 1952. Guggenheim fellow, 1928. Mem. Am. Physiol. Soc., Soc. Gen. Physiologists, A.A.A.S., Brit. Nutrition Soc., Phi Beta Kappa, Sigma Xi. Author: (with D. C. Smith) Outline of Physiology, 1939, 48; also numerous articles. Research on animal light, bioelectricity, cell respiration, membrane permeability, blood substitutes, fibrous proteins, enzymes of striated muscles. Home: 16 Katy Hatch Rd., Falmouth, Mass. 02540.*

AMBRONN, Leopold Friedrich Anton, German astronomer; b. Meiningen, Germany, Oct. 27, 1854; s. Richard and Wilhelmine (Amthor) A.; ed. Leipzig, Germany, also Vienna, Austria, Strasbourg, France; Dr. Phil., Göttingen, Germany, 1887; m. Johanna Nollenberger; m. 2d, Elise Diehler; 1 son, 1 dau. Sci. aid Marine Obs., Hamburg, Germany, 1880-89; observer Celestial Obs., Göttingen, 1889-1920; became pvt. docent U. Göttingen, 1892, asso. prof., 1902; founder Göttingen Mechanikerschule. Author: Handbuch der astronomischen Instrumentenkunde, 2 vols., 1899; Sternverzeichnis, 1900. Died Göttingen, June 8, 1930.

AMBROSE, Anthony Michael, Am. pharmacologist, toxicologist; b. Connellsville, Pa., Sept. 8, 1898; s. John D. and Mary (Fasson) A.; B.Sc., Phila. Coll. Pharmacy and Sci., 1924; M.S. Fordham U., 1927, Ph.D., 1929; m. Grace Elisabeth Carlstrom, Dec. 26, 1928. Instr. analytical chemistry Phila. Coll. Pharmacy and Sci., 1923-24; research biochemist H. K. Mulford Co., Phila., 1924-26; instr. Fordham U., 1926-29, asst. prof., 1929-33, cons. biochemist, 1933-34; staff Western utilization and devel. div. Agrl. Research Service, U. S. Dept. Agr., Albany, Cal., 1934-55, prin. pharmacologist, 1955-56; pharmacologist, toxicologist, chief toxicology div. U. S. Army Environmental Hygiene Agy., Army Med. Service, Dept. Army, Edgewood Arsenal, Md., 1956-61; prof. pharmacology Med. Coll. Va., Richmond, 1961—; research asso. pharmacology Med. Coll. Va., 1934-35, Stanford Sch. Medicine, 1935-38, 42-44; asst. prof. pharmacology and toxicology U. Louisville Sch. Medicine 1938-42. Mem. Am. Soc. for Pharmacology and Exptl. Therapeutics, Soc. for Exptl. Biology and Medicine, Soc. Toxicology, Sigma Xi. Research, numerous publs. on detoxication mechanism, metabolism of organic compounds, pharmacology and toxicology of chems. used in foods, drugs and cosmetics, positive pressure anesthesia, blood flow, capillary permeability, chronic toxicity, bioassay, cold injury, digestion and metabolism of synthetic fats, flavanoids, environmental health hazards, cholinesterase inhibitors, toxicology of insecticidal agts., repellents, fungicides and clothing impregnants. Home: 4111 Stratford Rd., Richmond, Va. 23225.*

AMBROSINI, Bartolomeo, Italian naturalist; b. Bologna, Italy, 1588; pupil of Aldrovandi; successively prof. philosophy, botany, medicine U. Bologna, dir. bot. garden; assisted during plague, Bologna, 1630; author med. works. Died 1657.

AMBROSINI, Giacinto, Italian botanist; b. 1605; apptd. dir. bot. garden, also prof. botany U. Bologna (Italy), 1657; author catalog of univ. garden plants, also bot. dictionary. Died 1672.

AMBRUS, Clara Maria Bayer, physician; born Rome, Italy, Dec. 28, 1924; d. Anthony and Charlotte (Schneider) Bayer; student U. Budapest, 1943-47; M.D., U. Zurich, 1949; postgrad. Sch. Scis., Sorbonne, 1949-50; Ph.D. in Med. Scis., Jefferson Med. Coll., 1955; m. Julian L. Ambrus, Feb. 17, 1945; children—Madeline, Peter, Julian, Linda, Steven, Katherine, Charles. Came to U. S., 1949, naturalized, 1955. Mem. faculty Phila. Coll. Pharmacy and Sci., 1950-55; sr. cancer research scientist Roswell Park Meml. Inst., Buffalo, 1955-64, asso. cancer research scientist, 1964—; asso. research prof. pharmacology Grad. Sch. State U. N.Y., Buffalo, 1955—, asst. research prof. pediatrics Med. Sch., 1965—. Diplomate Am. Bd. Clin. Chemistry. Fellow Internat. Soc. Hematology, Am. Coll. Clin. Pharmacology and Chemotherapy; mem. Am. Soc. Hematology, Am. Soc. Pharmacology and Exptl. Therapeutics, Am. Soc. Physiology, Physiol. Soc. Phila., Am. Assn. Cancer Research, Fedn. for Clin. Research, Soc. Exptl. Biology and Medicine, Am. Med. Women's Assn., Sigma Xi. Author: (with J. L. Ambrus, J. W. E. Harrisson, V. Rossi) Laboratory Manual and Notes of Pharmacology, vols. 1-3, 1950, also chpts. in other textbooks, numerous articles. Home: 143 Windsor Av., Buffalo 14209; also West Hill Farm, Emling Rd., Boston 14025. Office: 666 Elm St., Buffalo 14203.*

AMBRUS, Julian Lawrence, physician; b. Budapest, Hungary, Nov. 29, 1924; student U. Budapest, 1943-46; M.D., U. Zurich, Switzerland, 1949; postgrad. U. Paris; Ph.D. in Med. Sci., Jefferson Med. Coll., 1954; m. Clara M. Bayer, Feb. 11, 1945; children—Madeline S., Peter S., Julian Lawrence, Linda C., Steven G., Katherine A. Research asst. U. Budapest, 1943-46, demonstrator pharmacology, 1946-47; asst. dept. pharmacology U. Zurich, 1947-49; asst. dept. therapeutic chemistry and virology Inst. Pasteur, Paris, France, 1949-50; from asst. prof. to prof. pharmacology grad. sch. Phila. Coll. Pharm. Sci., 1950-55; prin. cancer research scientist Roswell Park Meml. Inst., N.Y. State Dept. Health, 1955-65, asst. dir. 1960—, dir. program USPHS, 1956—; tng. dir. Springville Labs. for Med. Research, Roswell Park, 1965—; asst. prof. internal medicine Med. Sch. also Grad. Sch., State U. N.Y., Buffalo, 1955, asso. prof. pharmacology, 1955-64, prof. biochem. pharmacology, 1964—. Mem. com. on thrombolytic agts. Nat. Heart Inst., NIH, USPHS, 1963—; cons. A.M.A. Council on Drugs, 1961—; mem. adv. com. on research grants and fellowships United Health Found. Western N.Y., Buffalo, 1961—. Fellow Internat. Soc. Hematology, Am. Coll. Clin. Pharmacology, Council on Clin. Cardiology, Am. Coll. Angiology; mem. Am. Soc. Hematology, Am. Heart Assn., Am. Soc. Pharmacology, Exptl. Therapeutics, Am. Phys. Soc., Soc. Exptl. Biology and Medicine, Fedn. Clin. Research, Am. Soc. Cancer Research. Contbg. author: Anticoagulants and Fibrimolysins, 1961; Surgical Hemorrhage, 1965. Editor: Medical and Biochemical Hematology. Publs., research in hematology; internal medicine; clin. and exptl. pharmacology; oncology. Home: 143 Windsor Av., Buffalo 14209. Office: 666 Elm St., Buffalo 14203.*

AMDUR, Isadore, Am. educator, chemist; b. Pitts., Jan. 24, 1910; s. Benjamin and Millie (Silberblatt) A.; B.S., U. Pitts., 1930, M.S., 1930, Ph.D., 1932; m. Alice Pauline Steiner, June 16, 1935; children—Stephen Benjamin, Nicholas John. NRC fellow Mass. Inst. Tech., 1932-34, faculty, 1934—, prof. phys. chemistry, 1951—. Spl. lectr. U. Pitts., summer 1962; vis. scientist U. S.-Japan coop. sci. program NSF, vis. prof. U. Kyoto (Japan), 1965-66. John Simon Guggenheim Meml. fellow, 1955-56. Fellow A.A.A.S., Am. Phys. Soc., Am. Acad. Arts and Scis. (rec. sec. 1946-47, council 1948-52); mem. Am. Chem. Soc., Sigma Xi, Phi Beta Kappa, Phi Lambda Upsilon, Pi Lambda Phi. Contbr. numerous articles and papers of original research on theoretical and exptl. aspects of rates of reaction, intermolecular forces and kinetic theory of gases. Home: 47 Bay State Rd., Belmont, Mass. 02178. Office: 77 Massachusetts Av., Cambridge, Mass. 02139.*

AMEGHINO, Carlos, Argentinian geologist, paleontologist; b. Argentina, flourished circa 1860; discovered Phororhacos, giant Miocene bird in S. Am., 1867; studied fossil mammals with bro., 1886; led 1st of series of expdns. to Patagonia, 1887; with bro. did 2d most important paleontol. work of last 2 decades of 19th century.

AMEGHINO, Fiorino (or Florentino), paleontologist, anthropologist; b. Italy, Sept. 18, 1854; dir. Mus. Natural History, Buenos Aires, Argentina, from 1902. Author: Obras completas y correspondencia cientifica, 24 vols., 1913-36. Investigated fossils and extinct animals S. Am.; discovered and organized many sequences of fossilized mammals; believed that all mammals, including man (homunculus patagonicus) originated in Argentina; his fossil finds rank 2d only to those discovered in Am. W. in late 19th century. Died La Plata, Argentina, Aug. 6, 1911.

AMELINCKX, Severin, Belgian physicist; b. Willebroek, Belgium, Oct. 30, 1922; s. Jaak and Eugenie (Genijn) A.; Licencié in Math., U. Ghent (Belgium), 1944, Licencié in Physics, 1951, Ph.D. in Physics, 1952, Agrégé de l'Enseignement Supérieur-Group Physics, 1955; m. Jeanne Alloo, Dec. 27, 1948; 1 dau., Christiane. Research asso. solid state physics lab. U. Ghent, 1948-58, prof., 1958-63; head solid state physics dept. S.C.K.-C.E.N., Mol, Belgium, 1959-63, dep. dir. gen. for sci. affairs, 1963—; prof. U. Brussels (Belgium), 1962—, U. Antwerp (Belgium), 1965—. Author: (with W. Dekeyser) Les Dislocations et la croissance des cristaux, 1955; The Direct Observation of Dislocations, 1964; also numerous articles. Research on crystal growth mechanism, use of etching, decoration and transmission electron microscopy to study dislocations, contrast theory in electron microscope, Mössbauer effect, elec. effects asso. with dislocations. Home: 34 Gouverneur Holvoetlaan, Deurne Belgium. Office: S.C.K.-C.E.N., Mol, Belgium.*

AMELL, Alexander Renton, Am. chemist; b. Clarksburg, Mass., Mar. 3, 1923; s. Louis and Agnes (Renton) A.; B.S., U. Mass., 1947; Ph.D., U. Wis., 1950; m. Allison Hamlin Moore, Sept. 5, 1945; children —A. Renton, Nancy Allison, Jane Anderson, Rebecca Susan. Instr. chemistry Hunter Coll., 1950-52; asst. prof. Lebanon Valley Coll., 1952-55; research asso. Brookhaven Nat. Lab., 1955; mem. faculty U. N.H., 1955—, prof. chemistry, 1962—, chmn. dept., 1961—. Study of kinetics; radiation chemistry. Home: 4 Chesley Dr., Durham, N.H. 03824.*

AMELUNG, Ludwig Franz, German psychiatrist; b. Bickenbach/Bergstrasse, Germany, May 28, 1798; ed. Berlin; chief physician Hesse-Darmstadt Country Hosp. (for elderly, incurable and mental illnesses), Hofheim, since 1821. Author: Beiträge zur Lehre von den Geistes-Krankheiten, 2 vols., 1832-36. Rep. of somatic sch. psychiatry; studied physiol. and path. aspects as basis for understanding psychol. problems. Died Hofheim, Apr. 19, 1849.

AMELUNG, Walther Oskar Ernst, German archeologist; born Stettin, Germany, Oct. 15, 1865; the son of Hermann Amelung; educated at Tubingen, Leipzig, Munich universities, 1884-88; married Antonia Lebrun. Prof., 1st sec. Archeology Inst. Rome; authority on classic epoch Greek art, also Florentine art. Author: Die Skulpturen des Vatikan Museum, 2 vols., 1903-08; translator Sophocles. Died Bad Nauheim, Germany, Sept. 12, 1927.

AMES, Bruce Nathan, biochemist, geneticist; b. N.Y.C., Dec. 16, 1928; s. Maurice U. and Dorothy (Andres) A.; B.A., Cornell U., 1950; Ph.D., Cal. Inst. Tech., 1953; m. Giovanna Ferro-Luzzi, Aug. 27, 1960; children—Sofia, Matteo. Chief sect. microbial genetics NIH, Bethesda, Md., 1953-68; prof. biochemistry U. Cal. at Berkeley, 1968—. Genetics Soc. rep. NRC. Recipient Eli Lilly award Am. Chem. Soc., 1964; Washington Acad. Sci. award, 1965, Flemming award, 1966. Mem. Am. Soc. Biol. Chemists, Genetics Soc. Research, publs. on intermediates and enzymes in pathway of histidine biosynthesis, control mechanisms governing this biosynthetic pathway; understanding operon, a cluster of genes turned on and off together. Home: 1324 Spruce St., Berkeley, Cal. 94709.*

AMES, Joseph Sweetman, Am. physicist; b. Manchester, Vt., July 3, 1864; s. George Lapham and Elizabeth (Bacon) A.; A.B., Johns Hopkins, 1886, Ph.D., 1890, LL.D., 1936; LL.D., Washington College, 1907, Univ. Pennsylvania, 1933; m. Mary B. (Williams) Harrison, Sept. 14, 1809 (dec. 1931). Asst. in physics, 1888-91, asso., 1891-93, asso. prof., 1893-99, prof. physics, 1899-1926, dir. Physical Lab., 1901-26, provost, 1926-29 pres. 1929-35, pres. emeritus from 1935. Chmn., Nat. Adv. Com. for Aeros., 1927-39; chmn. fgn. service com. NRC which visited France and Eng. to study origin and devel. sci. activities in connection with warfare, 1917. Recipient Langley Gold medal Smithsonian Instn., 1935. Hon. mem. Royal Instn. Great Britain; fellow Am. Acad. Arts and Scis.; mem. Nat. Acad. Scis., Am. Phys. Soc. Author: Manual of Experiments in Physics, 1896; Theory of Physics, 1897; The Discovery of Induced Electric Currents, 2 vols., 1900; Textbook of General Physics, 1904; The Constitution of Matter, 1913. Editor: Astro-Phys. Jour.; asso. editor Am. Jour. Sci.; editor-in-chief Scientific Memoir Series; spl. editor for physics and aeros., 2d edit. Webster's New Internat. Dictionary. Known for aerodynamics research done by NACA under his guidance. Died Balt., June 24, 1943.

AMES, Oakes, Am. botanist; b. North Easton, Mass., Sept. 26, 1874; s. Oliver and Anna Coffin Ray A.; A.B., Harvard, 1898, A.M., 1899; Sc.D., Washington U., 1938; m. Blanche Ames, May 15, 1900; children—Pauline, Oliver, Amyas, Evelyn. Asst. in botany, Harvard, 1898-1900, instr., 1900-10; asst. dir. Bot. Garden of Harvard, 1899-1909, dir., 1909-22; apptd. asst. prof. botany, Harvard, 1915, prof. botany, 1926-32, Arnold prof. botany, 1932-35, research prof. botany, 1935-41, research prof. botany, emeritus, from 1941, curator Bot. Mus., Harvard, 1923-27; supr. Bot. Mus., Arnold Arboretum, Atkins, Inst. Arnold Arboretum, Cuba (Harvard), 1927-35; dir. Bot. Mus., 1936-45, asso. dir., from 1945, also chmn. Council Bot. Collections, Harvard, 1926-35; dir. 1st Nat. Bank of Easton. Recipient gold medal for eminent service to orchidology, Am. Orchid Soc., 1924; Centennial medal for same, Mass. Hort. Soc., 1929; George Robert White medal of honor for eminent service in horticulture, 1935. Fellow A.A.A.S., Linnaean Soc. London, Am. Acad. Arts and Scis., Am. Orchid Soc. (v.p.); mem. Am. Soc. Naturalists, N.Y. Acad. Scis. Nat. Inst. Social Scis., N.E. Bot. Club Washington Acad. Scis., Assn. Internat. des Botanistes, Orchid Circle of Ceylon, C.Z. Orchid Soc. (hon. pres.), Sigma Xi. Author numerous papers mainly dealing with orchids, in botany periodicals, and contbrs. on orchid flora of Fla., serial work "Orchidaceae" (7 vols.), 1905-22; also Enumeration of the Orchids of the U. S., Can., 1924; Economic Annuals and Human Cultures, 1939. Died Apr. 28, 1950.

AMES, Stanley Richard, Am. biochemist; b. Madison, Wis., Dec. 18, 1918; s. Walter Ray and Mary Ellen (Fruit) A.; B.A., U. Mont., 1940; A.M. (Residence scholar), Columbia, 1942, Ph.D. (U. fellow), 1944; Rockefeller fellow, U. Wis., 1943-46; m. Edwina Gray Powers, Aug. 19, 1943; children—Edwina Marie, Stanley Richard, Suzanne Claire, Judith Ann. With Distillation Products Industries div. Eastman Kodak Co., Rochester, N.Y., 1946—, head biochemistry research labs., 1963—, sr. research asso., 1966—; chmn. Animal Nutrition Research Council, 1960-61. Fellow A.A.A.S.; mem. Am. Inst. Nutrition, Am. Chem. Soc. (councilor 1965-67), Am. Soc. Biol. Chemists, Am. Soc. Animal Sci., Am. Soc. Exptl. Biology and Medicine, Poultry Sci. Assn., Sigma Xi, Phi Sigma, Phi Mu Epsilon, Phi Lambda Upsilon. Author: Determination of Copper in Copper-Proteins Using the Dropping Mercury Electrode, 1944. Contbr. chpts. to Swine Nutrition, 1954; Nutrition of Turkeys, Ducks and Pheasants, 1955; Methods of Biochemical Analysis, 1957; Annual Reviews of Biochemistry, 1958; Cattle Nutrition, 1963, 65; The Vitamins, 1967; also numerous articles. Developed analytical methods in enzyme chemistry; publs. on biochemistry, analysis, metabolism, nutrition and bioassay of vitamin A, vitamin E and lipids; discovered biol. activities of isomers of vitamin A and vitamin E, stable aqueous isomeric vitamin A compositions and new type of injectable vitamin ADE. Home: 61 Biltmore Dr., Rochester 14617. Office: Distillation Products Industries, Rochester, N.Y. 14603.*

AMES, William Francis, Am. mathematician; b. Brandon, Man., Can., Dec. 8, 1926; s. Paul Main and Della (Hebel) A.; came to U. S., 1932, naturalized, 1946; B.S., U. Wis., 1949, M.S., 1950, postgrad.; m. Theresa Danielson, May 29, 1951; children—Karen, Susan, Pamela. Instr. math. U. Wis., Madison, 1954-56; sr. engr. E. I. duPont de Nemours & Co., Wilmington, Del., 1956-59; faculty U. Del., Newark, 1956-67, prof. mech. engring. and computer sci., 1965-67; prof. mechanics and hydraulics U. Ia., Iowa City, 1967—; vis. prof. Stanford, 1963-64. Cons. to chem. and rubber cos., USPHS, NSF. NSF Faculty fellow, 1963-64. Mem. Am. Math. Soc., Soc. for Indsl. and Applied Math., Am. Inst. Physics. Author: Nonlinear Partial Differential Equations in Engineering, 1965; Nonlinear Ordinary Differential Equations in Transport Processes, 1968. Editor: Nonlinear Problems of Engineering, 1964; Nonlinear Ordinary Differential Equations in Transport Processes, 1968. Research, publs. in nonlinear partial differential equations of engineering and applied sci. Home: 350 Hutchinson Av., Iowa City, Ia. 52240.*

AMI, Henry M., Canadian geologist; b. Belle Rivière, Que., Can., Nov. 23, 1858; s. Marc and Anne (Giramaire) A.; B.A., M.Gill U., 1882, M.A., 1885, D.Sc., 1907; D.Sc., Queen's, 1892; Doctor (hon.), U. Padua; LL.D., U. Man.; m. Clarissa J. Burland, 1892. In charge war metals and minerals; with Trade dept. Brit. Embassy, Washington, 1917-20; dir. Canadian Sch. Prehistory in France, Ottawa, Can., also hdqrs., Les Eyzies, Dordogne, France; geologist, paleontologist, mem. tech. staff Geol. Survey, Can., 1882-1912; cons. geologist. Fellow geol. socs. London (Bigsby Gold medal 1903), Switzerland, Washington, Am., Palaeontol. Soc. Am., Archaeol. Inst. Am., Royal Geog. Soc., Royal Soc. Can., A.A.A.S.; mem. Brit. Assn. for Advancement Sci.; Royal Anthrop. Inst. Great Britain, Brit. Sci. Guild, Am. Assn. Museums, Canadian Inst. Mining and Metallurgy. Author: Stanford's Compendium of Geography, Vol. I, North America: Canada and Newfoundland; Resources of the Country between Quebec and Winnipeg along the line of the National Transcontinental Railway; numerous ofcl. reports, also papers and publs. on graptolites and other paleozoic faunas in Eastern Can.; paleontology and chronological geology of Can. Editor, Ottawa Naturalist, 1895-1900. Died Jan. 4, 1931.

AMICI, Giovanni Battista, Italian astronomer, botanist; b. Modena, Italy, Mar. 25, 1786; s. Giuseppe and Maria (Dalloca) A.; grad. U. Bologna, 1808; m. Giuseppina; children—Cesare, Giuseppe A., Emilia Amici Massu. Named prof. geometry, algebra Modena Lyceum, 1810; prof. geometry, algebra, trigonometry U. Modena, 1815-25; engaged in research on devel. optical instruments, from 1825; apptd. prof. astronomy U. Pisa, 1831; became dir. Obs. Florence, 1835; named hon. prof., also worked in microscopy Florentine Mus., 1859. Recipient Gold medal Italian Inst. Sci. Contbd. papers to sci. lit. Studied intercellular spaces in plants and deduced role in conduction of gases, 1806; described movement in cell, 1823; discovered (with Brongniart) role of stigma and pollen tube in studies of plant fertilization, 1823-30; produced dioptric or achromatic microscope; probably originated water immersion method for improved achromatic lenses of compound microscope then in use; invented prismatic spectroscope; developed micrometer telescope, catadioptrical microscope, meridian telescope, other instruments; used telescope to observe double stars, also moons of Jupiter; made significant measurements of sun's diameter and distance of certain fixed stars. Died Florence, Apr. 10, 1863.

AMIEL, Jean François, French chemist; b. Toulouse, France, Aug. 21, 1905; s. Jean Toussaint and Marie (Suc) A.; Licence, Ecole Normale Supérieure, Paris, 1927, Agrégation, 1929; Doctorate es Sciences, Sorbonne, 1933; m. Marie Jeanne Boudouresque, Jan. 6, 1931; 1 dau., Agnès. Reader, Faculté des Sciences, Lille, France, 1939-42; prof. chemistry Faculté des Sciences, Poitiers, France, 1942-50; prof. Faculté des Sciences, Paris, 1950——, also dir. Institut de Préparation des Professeurs d'enseignements du second degré; staff Laboratoire de Chimie des Solides, Sorbonne. Mem. Am. Chem. Soc., Société chimique de France, Société de Chimie-Physique, Société francaise de physique. Author: Cours de Chimie, 3 vols., 1951-54; Cours de Chimie générale, 3 vols., 1961, 64, 67; also articles. Research on slow oxidation of gaseous benzene and chain reactions, magnetochemistry of salts, oxides and coordination compounds of copper, nickel, chromium, manganese and iron, contact catalysis, synthesis from acetylene, morphology of fine powders, chlorates, halogeno-compounds of tellurium, thermal decompositions of organic salts. Home: 24 rue Ferdinand Jamin, Bourgla-Reine (92) France. Office: 1, rue Victor Cousin, Paris 5e, France.*

AMIRASLANOV, Ali Agamaly Ogly, Russian geologist; b. Dec. 1900; grad. Moscow Mining Acad., 1930; mem. staff All-Union Inst. for Mineral Resources, also Sci. Geol. Gold Survey Inst., 1930-39; chief engr., chief geologic prospecting adminstrn. Non-Ferrous Ministry USSR, 1939-47, dir., 1948-53, chief geologist, 1954-62; faculty Moscow Geologic Intelligence (Geologic Prospecting) Inst., 1931-62, prof., 1950-62; mem. staff Inst. for Geology of Ore Deposits, Petrography, Mineralogy and Geochemistry, 1957-62. Corr. mem. USSR Acad. Scis. (dep. acad. sec. geologic and geog. sect. 1955-57). Author: Mineralogical Characteristics of Pyritic Deposits in the Urals and Secondary Processes in Them, 1937. Studied copper, lead, zinc, other nonferrous and rare metal deposits. Died Oct. 1962.

AMIRKHANOV, Khabibulla Ibragimovich, Russian physicist; born April 28, 1907; graduated from Azerbaijan Univ., 1930; Doctor of Physico-Math. Science. Instructor, Azerbaijan Indsl. Inst., 1930-42, prof., 1942-50; with physics sect., then Inst. Physics and Math., Azerbaijan Acad. Sci., 1932-44, dir. Inst. Physics Daghestan br. USSR Acad. Sci., 1950——. Mem. USSR (Presidium chmn.), dir. Inst. Physics and Math., 1944-50; Azerbaijan acads. sci. Co-author: The Possibility of Using the Thermal Method To Obtain Core Samples from Oil Wells, 1938, Determining the Absolute Age of Rocks from the Conversion of K40 into A40, 1956, The Hall Effect in Indium Antimonide in Strong Impulse Magnetic Fields, 1963; author Thermal Rectification, 1946; Research on the Thermal Conductivity of Copper Oxides, 1956. Research on physics of petroleum and semiconductors. Address: Inst. of Physics, Makhachkala, RSFSR, USSR.

AMIR-MOEZ, Ali Reza, mathematician, educator; b. Teheran, Iran, Apr. 7, 1919; s. Mohammad and Fatema (Moez) A.-M.; B.A. in Math., U. Teheran, 1943; M.A. in Math., U. Cal. at Los Angeles, 1951, Ph.D. in Math., 1955. Came to U. S., 1947, naturalized, 1961. Instr. Teheran Inst. Tech., 1943-47; asst. prof. U. Ida., 1955-56; instr., asst. prof. Queens Coll., Flushing, N.Y., 1956-60; asst. prof. Purdue U., Lafayette, Ind., 1960-61; asso. prof. U. Fla., Gainsville, 1961-63; prof. math. Clarkson Coll. Tech., Potsdam, N.Y., 1964-65; prof. Tex. Technol. Coll., Lubbock, 1965——. NSF grantee, 1966. Mem. Am. Math. Soc., Math. Assn. Am., Sigma Xi. Author: Elements of Linear Spaces, 1962; Matrix Techniques, Trigonometry, Analytic Geometry, 1964; Matrix Techniques; Math and String Figures; also articles on eigenvalues and singular values of matrices. Office: Math. Dept., Tex. Techol. Coll., Lubbock, Tex. 79409.*

AMIS, Edward Stephen, Am. chemist; b. Himyar, Ky., Nov. 9, 1905; s. Jack and Artie (Southard) A.; B.S., U. Ky., 1930, M.S., 1933; Ph.D., Columbia, 1939; m. Annie Velma Birdwhistle, Sept. 2, 1934; children—Edward Stephen, Velma Dianne. Faculty, La. State U., 1939-45, asso. prof., 1943-45; staff Carbide & Carbon Chems. Corp., Oak Ridge, 1945-47; prof. chemistry U. Ark., Fayetteville, 1947——. Fellow N.Y. Acad. Scis.; mem. Am. Chem. Soc. (So. Chemist award 1959, S.W. award 1960), Chem. Soc., Ark. Acad. Scis., Sigma Xi, Alpha Chi Sigma, Pi Mu Epsilon, Phi Lambda Upsilon, Sigma Pi Sigma. Author: Kinetics of Chemical Change in Solution, 1949;

A Book of Verse and Prose, 1965; Solvent Effects on Reaction Rates and Mechanisms, 1966; (with James F. Hinton) NMR Studies of Ions in Solution, 1968; also numerous articles. Research on electromotive chemistry, conductance and transference of electrolytes in solution, kinetics and mechanism of chem. reactions in pure and mixed solvents, solvation of ions in pure and mixed solvents using conductance, transference and nuclear magnetic resonance procedures, theories of ion-diplolar molecule reactions, temperature coefficients of reaction rates, electron exchange reactions. Home: 1655 Woolsey St., Fayetteville, Ark. 72701.*

AMMAN, Johann Conrad, Swiss physician; b. Schaffhausen, Switzerland, 1669; grad. U. Basel (Switzerland), 1687. Practiced medicine, Amsterdam, Netherlands from 1687; early writer on instrn. deafmutes. Author: Surdus loquens (methods of teaching deaf-mutes to speak), 1692. Died Warmoud near Leiden, Holland, 1724.

AMMAN, Paul, German physician, botanist; b. Breslau, Germany, 1634; D. Physic, U. Leipzig (Germany), 1662, apptd. asso. prof. medicine, then prof. botany, 1674, prof. physiology, 1682; mem. (under name Dryander) Soc. Naturae Curiosorum. Author: Medicina critica, 1670; Paraensis ad docentes occupata circa institutionum medicarum emendationem, 1673; Supellex botanica, 1675; Character naturalis plantarum, 1676; Irenicum numae pompilicum; 1689. Died Leipzig, 1691.

AMMANN, Karl, Swiss veterinarian; b. Frauenfeld, Switzerland, Oct. 6, 1905; s. Karl and Emma (Honegger) A.; ed. Zurich, Geneva and Vienna univs.; Dr. of vet. scis.; m. Helen Piraud, May 7, 1932; children—Heinz, Katharina. Practice vet. medicine, 1929-32; asst. vet. clinic U. Zurich, 1933-39, prof. agrégé, 1939, prof. vet. surgery, 1943, full prof., from 1948. Author: Die Chirurgischen Nähte; Handlexikon der tierärztlichen Praxis: Augenkrankheiten. Research in ophthalmology of domestic animals. Home: Rebwiesstrasse 36. Office: Winterthurerstrasse 260, Zurich, Switzerland.

AMMANN, Othmar Hermann, civ. engr.; b. Schaffhausen, Switzerland, Mar. 26, 1879; s. Emanuel and Emilie Rosa (Labhardt) A.; C.E., Swiss Federal Poly. Inst., Zurich, 1902, Dr. Tech. Sciences, 1930; Dr. Engring., N.Y.U., 1931, Bklyn. Poly. Inst., 1956; Sc.M., Yale, 1932; Dr. Engring., Pa. Mil. College, 1934; Sc.D., Columbia University, New York City, 1941, Fordham University, 1964; m. Lilly Selma Wehrli, July 24, 1905 (dec. 1933); children—Werner, George Andrew, Margaret; m. 2d, Klary Vogt Noetzli, 1935. Came to U. S., 1904, naturalized citizen, 1924. Investigation, design and bldg. of bridges, etc. in Europe and U. S., 1902-23; asst. chief engr. on design and constrn. Hell Gate Bridge, New York, 1912-18; cons. engr., N.Y. City, 1923-25; chief engr. of bridges, 1925-30, chief engr. Port of New York Authority, 1930-39, in gen. charge planning and constrn. Outer-bridge Crossing and Goethals Bridge across Arthur Kill, arch bridge across Kill van Kull at Bayonne, N.J., George Washington Bridge and Lincoln Tunnel across Hudson River at New York, and other projects; mem. bd. engrs. in charge Golden Gate Bridge, San Francisco, 1929-36; chief engr. Triborough Bridge. Authority in charge of planning and construction of Triborough Bridge and Bronx-Whitestone Bridge, both across East River, N.Y. City, 1934-39; cons. engr., N.Y.C., 1939-46; partner Ammann & Whitney, cons. engrs., 1946——; prin. projects include Verrazano-Narrows Bridge, Throgs Neck Bridge (N.Y.); Walt Whitman Bridge, Phila., N.J., Conn. turnpikes, Dulles Internat. Airport, Wash., 600 foot radio telescope for USN. Fellow N.Y. Acad. Scis.; mem. Am. Soc. C.E. (hon.), Swiss Soc. Architects and Engrs. (hon.), Am. Inst. Cons. Engrs., Am. Ry. Engring. Assn., Am. Soc. Testing Materials, Inst. Civil Engineers of Great Britain. Awarded Thomas Fitch Rowland prize, 1918, Ernest E. Howard award Am. Soc. C.E., Nat. Medal of Sci., numerous other awards. Designer of many bridges, highways, buildings; known for improvements in suspension bridges both structurally and esthetically. Died Rye, N.Y., Sept. 22, 1965.

AMMON, Otto Georg, German anthropologist; b. Karlsruhe, Germany, Dec. 7, 1842; s. Jacob and Emma (Wottlin) A.; m. Antonie Wönshoffer; 2 sons, 2 daus. Engr.; journalist, pub., then owner Koustanzer Zeitung until 1883; active as anthropologist from 1883; mem. anthrop. com. Karlsruhe Hist. Soc. (began 1st complete and continuous anthrop. survey, 1885-89). Author: Anthropologische Untersuchungen der Wehrpflichtigen in Baden, 1890; Die Natürliche Auslese beim Menschen, 1893; Die Bedeutung des Bauernstandes fur Staat und Gesellschaft, 1894; Die Gesellschaftordnung und ihre natürlichen Grundlagen, 3 vols., 1895-1900; Zur Anthropologie der Badener, 1899. Developed Ammon's law anthropology; wrote on Nordic race. Died Karlsruhe, Jan. 14, 1916.

AMMONIOS, Greek philosopher; flourished 5th century; s. Hermias; pupil of Proclos at Athens; leader sch. Alexandria; tchr. of Damascios, Philoponos, Simplicios; commented on Aristotle, also Porphyry; divided math. into arithmetic, geometry, astronomy, music.

AMMONS, Robert Bruce, Am. psychologist, educator; b. Denver, Feb. 27, 1920; s. Bruce and Margaret (Gates) A.; student U. Denver, 1937; B.A., San Diego State Coll., 1939; postgrad. U. Cal. at Los Angeles; M.A., State U. Ia., 1941, Ph.D., 1945; m. Carol Hamrick, Aug. 21, 1949; children—Carolyn, Carl, Bruce, Douglas, Elizabeth, Richard, Stephanie, Ailsa Glenyss Ellyn. Began career as assistant professor psychology U. Denver, 1946-48; asst. prof. psychology Tulane U., New Orleans, 1948-49; asst. prof. U. Louisville, 1949-54, research fellow, 1954-55; asso. prof. U. N.D., Grand Forks, 1956-57; prof. psychology Univ. of Mont., Missoula, 1957——. Dir. psychol. clinic, Denver, 1946-48, Louisville, 1949-53; dir. contract and grant research U. S. Air Force, U. Louisville, 1949-55; prin. investigator USPHS, 1962——. Mem. Am., Rocky Mountain, Mont. (pres. 1964-——, sec. bd. examiners 1961-65), Midwestern, Southeastern, Brit., Japanese, Indian, Canadian psych. assns., So. Soc. for Philosophy and Psychology, Psychometric Soc., Am. Statis. Assn., Soc. for Gen. Systems Research, Psychonomic Soc., Am. Speech and Hearing Assn., Artorga, Mensa. Author: (with James B. Stroud) Improving Reading Ability, 1949, 2d edit. (with Stroud, Henry A. Bamman), 1956; (with wife) Full-Range Picture Vocabulary Test; Quick Test. Contbr. numerous articles to profl. jours. Editor: Psychological Reports, Perceptual and Motor Skills, 1955——. Research in motor skills, perceptual learning, problem solving, measurement, behavior theory. Home: 411 Keith Av., Missoula, Mont. 59801.*

AMONTONS, Guillaume, French physicist; b. Paris, Aug. 31, 1663; apprentice astronomer to Le Fevre; employed as architect in pub. works. Mem. French Acad. Scis., 1699. Author: Remarques et expériences physiques sur la construction d'une nouvelle clepsydre, sur les baromètres, thermomètres et hygromètres, 1695; Nouvelle théorie du frottement. Investigated friction, 1699; found pressure of air at constant vol. increases 1/3 when heated from room temperature to that of boiling water; worked to improve instruments used in physical experiments; invented hygrometer, 1687, non-mercuric conical barometer (useful in sea travel), 1695, air pressure thermometer, 1702, also heat meter, long distance optical telegraph using series of stas., rotation pump, hot air machine; noted barometers are affected by heat as well as atmospheric pressure, 1704; concept of absolute zero as temperature at which gas pressure vanishes completely derives from his method of measuring temperature-variations in pressure of fixed mass of gas maintained at constant vol. Died Paris, Oct. 11, 1705.

AMORETTI, Carlo, Italian naturalist, geographer; b. nr. Genoa, Italy, 1741; mem. order St. Augustine; prof. law U. Parma (Italy), later curator Ambrosian Library, Milan, Italy; sec. Societa Patriotica Milan, 1783-98; mem. Inst. Italy. Author: A Journey from Milan to the Three Lakes, (detailed study of Como, Maggiore, and Lugano area), 1794; a life of Leonardo da Vinci, 1784; treatises on natural sci. Died Milan, 1816.

AMOREUX, Joseph-Pierre, French physician, naturalist; b. Beaucaire, France, 1741; librarian U. Montpellier (France); studied histories of medicine, vet. art, agr., natural sci. Author: Traité de l'olivier, 1784; Recherches et expériences sur les divers lichens dont on peut faire usage en medécine et dans les arts, 1787; Notice sur les insectes de France réputés venimeux, 1789; Etat de la vegetation sous le climat de Montpellier, 1809. Died Montpellier, 1824.

AMOS, (Dennis) Bernard, immunologist; b. Bromley, Eng., Apr. 16, 1923; s. Benjamin and Vera (Oliver) A.; M.B., B.S., Guy's Hosp., London, 1951, M.D., 1963; m. Solange M.-M. Labesse, Aug. 25, 1949; children—Susan, Martin, Christopher, Nigel, Irene. Welcome Physiological Research Labs., 1939-40; Radcliffe Infirmary, Oxford, England, 1940-45; research fellow Guy's Hosp., 1952-55; sr. research fellow Roswell Park Meml. Inst., Buffalo, 1955-56, asso. cancer research scientist, 1956-58, prin. cancer research scientist, 1958-62; prof. immunology, chief div. immunology Med. Center, Duke, Durham, N.C., 1962——. Cons. to orgns., govt. agys. Mem. Am. Assn. Immunologists, Transplantation Soc., Am. Soc. Human Genetics, Am. Assn. Cancer Research. Editor: (with H. Koprowski), Cell Bound Antibodies, 1965; also numerous articles. Research on transplantation immunology; detection and identification of tissue and leucocyte antigens in mouse and man: effects of antibodies and immune cells on transplanted tumors and normal tissues; matching of kidney donors and recipients for transplantation. Home: Route 3, Hillsborough, N.C. 27278. Office: Box 3010, Duke Med. Center, Durham, N.C. 27706.*

AMOS, Carrol E(ly), Am. mathematician; b. Fredericktown, O., Jan. 26, 1906; s. Monroe S. and Eva (Spry) A.; B.S., Denison U., 1928; Ph.D., Ohio State U., 1932; m. Mary Lucille Rodgers, Dec. 12, 1945. Prof. math. U. Toledo, 1946——. Cons. USAF Sci. Bd., Dept. Def.; cons. Research and Devel. Bd. Washington, 1949-50. Mem. Am. Math. Assn., Am. Assn. for Engring. Edn. Study of application of integral equations to diffraction and potential problems. Home: 4762 Carskaddon Av., Toledo 43615.

AMOSS, Harold L(indsay), Am. physician; b. Cobb, Ky., Sept. 8, 1886; s. David Alfred and Carolyn Waters (Lindsay) A.; B.S., State U. Ky., 1905, M.S., 1907; M.D., Harvard, 1911, Dr. P. H., 1912; hon. Sc.D., George Washington U., 1922; m. Marguerite Dupree Moore, May 17, 1917; children—Harold Lindsay, Dudley Moore, Chemist Ky. Agrl. Expt. Sta., 1905; asst. chemist Hygienic Lab., USPHS, 1905-07, Bur. Chemistry U. S. Dept. Agr., 1907-09; physiol. chemist Western Pa. Hosp., 1909; instr. preventive medicine and hygiene, Harvard Med. Sch., 1909-12; asst. in pathology and bacteriology, 1912-14, asso., 1914-19, asso. mem., 1919-22, Rockefeller Inst. Med. Research; asso. prof. medicine Johns Hopkins' U. and asso. physician Johns Hopkins Hosp., 1922-30; prof. medicine, Duke, 1930-33. Cons. in medicine Greenwich, Grasslands, White Plains, United and No. Westchester Hosps. editor "Medicine," 1922-25. Mem. A.M.A., A.C.P., A.A.A.S., Assn. Am. Physicians, Assn. Am. Chest Physicians, Am. Soc. Clin. Investigation, Am. Soc. Exptl. Pathology, Am. Climatol. Soc., Am. Assn. Immunologists Am. Heart Assn., Harvey Soc., Interurban Clin. Club, So. Interurban Clin. Club, Sigma Xi, Phi Beta Kappa, Omicron Delta Kappa, Pi Kappa Alpha, Phi Chi. Contbr. various articles on physiol. chemistry, infectious diseases, immunology, epidemiology and clin. medicine. Died Nov. 2, 1956.

AMPÈRE, André Marie, French physicist, mathematician; b. Lyons, France, Jan. 20, 1775; s. J. J. Ampère; ed. at home; m. Julie Carron, 1799; 1 son, Jean Jacques; m. 2d, Mlle. Potot, 1807; 1 dau., Albine. Mastered math. of day by age 12; pvt. tutor math., chemistry, langs., Lyons, 1796-1801; prof. chemistry, physics, astronomy, Bourg, France, 1801; apptd. prof. math. Lycée of Lyons, circa 1802; asst. lectr. math. analysis École Polytechnique, from 1805, prof., from 1809, insp. gen., from 1808; apptd. prof. exptl. physics Collège de France, 1824. Mem. Bureau Consultatif des arts et métiers, from 1806. Mem. French Acad. Scis., 1814, Fellow Royal Soc., 1827; mem. Swiss Soc., acads. Berlin (Germany), Stockholm (Sweden), Brussels (Belgium), Lisbon (Portugal), Philos. Soc. Cambridge (Eng.), Soc. Physics and Natural History Geneva (Switzerland). Author: Considérations sur la théorie mathématique du jeu, 1802; Recueil d'observations électrodynamiques, 1822; Precis de la théorie des phenomenes electro-dynamiques, 1826; Memoire sur la théorie mathématique des phenomènes electrodynamiques uniquement déduite de l' expérience, 1827; Essai sur la philosophie des sciences, 1834; others. Founded, named sci. electrodynamics; introduced term electrostatics; showed 2 parallel wires carrying currents attract each other if currents are in same direction, repel each other if in opposite directions (beginning of concept of lines of force); showed solenoid when carrying current behaved exactly like magnet; caused current carrying wire to behave like magnetic compass needle, 1822; (showing how Newton's 3d law applied to forces between currents) stated Ampère's law (fundamental law of electrodynamics) describing mathematically magnetic force between 2 elec. currents, 1827; determined law governing rotation of currents under magnetic influence, 1828; introduced measuring techniques for electricity; developed astatic needle, projected instrument (to be called galvanometer) to measure flow of electricity, 1821; constructed commutator, 1832; explained light and heat as homogeneous natural phenomena; proposed theory that currents of electricity circle earth, that magnetism is essentially phenomenon of electrified particles of matter; ampere (quantity of electric current passing given point in given time) named for him because he was 1st to differentiate rate of passage of current from driving force behind it; also studies in math., physics, chemistry, psychology, botany. Died Marseilles, France, June 10, 1836.

AMPFERER, Otto, Austrian geologist; b. Innsbruck, Austria, Dec. 1, 1875; s. Nikolaus and Gertraud (Zangerl) A.; ed. Innsbruck, 1895-1901; m. Olga Sander, Nov. 20, 1902. Became employee Austrian Geol. Dept. 1901, head geologist, 1919, vice dir., 1925, dir., 1935; mem. Austrian Acad. Scis., Halle Acad. Scientists. Editor: Jahrbuches der Geologie Bundesanstalt, 1925-37. Investigated alpine tectonics, diluvial geology, geomorphology; made geol. maps of western and no. Austrian Alps. Died Innsbruck, July 9, 1947.

AMPLATZ, Kurtz, Am. radiologist; b. Weistrach, Austria, Feb. 25, 1924; s. Anton and Mary (Rungg) A.; M.D., U. Innsbruch, Austria, 1951; m. Maxine Heinrich, July 3, 1955; children—Curtis, Maria, Grace, Caroline. Came to U. S., 1953, naturalized, 1956. With U. Minn., 1957—, asst. prof., 1963, asso. prof., 1963—; staff U. Minn. Variety Club Heart Hosp., 1957—. Contbg. author: The Science of Ionizing Radiation, 1964. Contbr. numerous articles to sci. jours. Constrn. of cardiovascular injectors; pneumotomographic chair; rapid film changers; cardiovascular shunt detectors; description of screening technique for reno vascular hypertension; coronary arteriography. Home: 10 Evergreen Rd., St. Paul, Minn. 55110. Office: 412 S.E. Union St., Mpls. 55455.*

AMSDEN, Charles Avery, Am. archaeologist; b. Forest City, Ia., Aug. 18, 1899; licence en droit, Toulouse, France, 1921; A.B., Harvard, 1922. Curator, S.W. Mus., Los Angeles, 1927-28, sec., treas., 1928-41, mem. explorations, Nev., Cal., from 1930, mem. adv. bd. Lab. Anthropology, 1936-41. Mem. Harvard expdns., Ariz., 1913-15, U. Cal. coastal exploration, 1928, Andover-Pedos Expdn., 1929. Mem. Anthrop. Assn. (councilor from 1931), Soc. Archeol., S.W. Archeol. Fedn. (sec. 1929-31). Investigated Am. archeology, aboriginal Am. textiles, pottery, stove implements, early man in Am. Died Mar. 3, 1941.

AMSEL, Abram, psychologist; b. Montreal, Que., Can., Dec. 4, 1922; s. Aaron H. and Annie (Levitt) A.; B.A., Queens U., Can., 1944; M.A., McGill U., 1946; Ph.D., State U. Ia., 1948; m. Tess Steinbach, June 11, 1947; children—Steven David, Andrew Jay, Geoffrey Neal. Came to U. S., 1948, naturalized 1957. Faculty, Tulane U.. New Orleans, 1948-60, prof., 1956-60; prof. U. Toronto (Ont., Can.), 1961—; vis. prof. U. Cal. at Berkeley, summer 1962. NSF postdoctoral fellow U. Coll., London, 1966-67. Mem. Soc. Exptl. Psychologists, Psychonomic Soc., Am., Midwestern, Eastern psychol. assns., A.A.A.S., Sigma Xi. Cons. editor: Jour. Exptl. Psychology, Research, numerous publs. on learning and motivation theory, particularly frustrative factors in behavior. Home: 14 Clarendon Av., Toronto, Ont., Can.*

AMSLER, Roger Louis, French physician; b. Limoges, France, Sept. 22, 1895; s. Ernest and Louise (Bordas) A.; M.D., Faculté de Médecine Paris; m. Anne Fourcher, July 19, 1944. Dr.-in-chief Antituberculin Dispensaries, Maine-et-Loire, also for pneumophthisiologic service Nat. Sch. Medicine, Angers. Recipient gold medal Acad. Medicine. Contbr. numerous articles to jours. Home: 8, rue de Bel-Air, Angers (M.-et-L.), France. Office: U. Nantes, Angers, France.

AMSLER-LAFFON, Jakob, mathematician; b. Stalden, Switzerland, Nov. 16, 1823; s. Jakob and Elisabeth (Amsler) A.; studied theology, Jena, Germany, 1843-44, math., physics, Königsberg, Germany, 1844-48; doctorate Schaffhausen, Switzerland, 1856; m. Elise Laffon; 3 sons, including Alfred, Albert. Worked at obs., Geneva, Switzerland, 1848-49; became pvt. docent in math. U. Zurich (Switzerland), 1849; engaged in precision mechanics, from 1851; established factory for measuring and testing instruments. Mem. French Acad. Scis. Author: Über die mechanische Bestimmung Flächeninhaltes, 1850; Zur Theorie der Verteilung des Magnetismus im wiechen Eisen; works on thermodynamics, also precision instruments. Investigated laws of heat conduction in solid bodies, 1852. Inventor: polar planimeter, 1854, integrator, instruments for measuring surface of spheres, instruments for determining Fourier coefficient of function. Died Schaffhausen, Jan. 3, 1912.

AMSTUTZ, Edward Delbert, chemist; b. North Athens, O., May 1, 1909; s. Platte T. and Louise (Allison) A.; B.S., Wooster Coll., 1930; M.S., Inst. Paper Chemistry, 1931; Ph.D., Cornell U., 1936; m. Frances VanSchaack, Aug. 30, 1935; children—Edward A., Jean L., William P. Instr., Chem. Union Coll., 1936-38; faculty Lehigh U., Bethlehem, Pa., 1938—, prof., 1947—, head dept., 1960—. Mem. Am. Chem. Soc. Contbr. numerous articles to sci. jours. Research on structure of sulfur compounds, mechanisms of reactions, synthesis and properties of organic medicinal agts. Home: 3465 Altonah Rd., Bethlehem, Pa. 18017.*

AMSTUTZ, Gerhard Christian, Am. geologist; b. Bern, Switzerland, Nov. 27, 1922; s. Jakob and Maria Amstutz; diploma Swiss Fed. Inst. Tech., 1947, Dr. Sc. Nat., 1952; postgrad. U. Wash., Mass. Inst. Tech. Harvard; m. A. Bachmann, June 27, 1959; children—Martin, George. Came to U. S., 1949, naturalized, 1961. Faculty, Swiss Fed. Inst. Tech., Zurich, 1947-49, 51-52; geologist, petrologist Cerro de Pasco Corp., Peru, 1952-56; asso. prof. geology U. Mo., Rolla, 1956-61, prof., 1961-64; dir. Inst. Mineralogy and Petrology, U. Heidelberg (Germany), 1964—; guest prof. U. Tubingen, (Germany), 1962-63. Fellow Geol. Soc. Am., Mineral. Soc. Am., A.A.A.S., Am. Geophys. Union, Soc. Econ. Paleontologists and Mineralogists, Geol. Soc. Can., Mineral. Soc. Can., Sigma Gamma Epsilon (life), Sigma Xi (life). Author: Sedimentology and Ore Genesis, 1964; Glossary of Mining Geology, 1968; also numerous articles. Founder, Mineralium Deposita, 1965, mng. editor; mem. adv. editorial bd. Main Currents in Modern Thought, 1967—. Research in ore and rock genesis. Office: Mineralog. Inst., U. Heidelberg, Berlinerstrasse 19, 69 Heidelberg, Germany.*

AMTHAUER, Rudolf, German psychologist; b. Iserlohn-W., Dec. 19, 1920; s. Herman and Elisabeth (Thiel) A.; Ph.D. in natural scis. and psychology; m. Ingeborg Traeuptmann, Aug. 2, 1952; children—Edgar and Renate. Specialist in applied psychology. Mem. Internat. Assn. Applied Psychology, Profl. Union German Psychologists, German Soc. Psychology, Soc. Pedagogical Research. Author: Intelligenz und Beruf; Zum Problem der Produktiven Begabung; Intelligenz-Struktur-Test; Leistungsunterschiede der Geschlechier. Home: Frankfort/Main, Merzigerweg 4. Office: Farbwerke Hoechst A. G., Frankfort/Main, West Germany.

AMUNDSEN, Roald Engelbregt Gravning, Norwegian explorer; b. Borge, Norway, July 16, 1872; s. Jens Amundsen; ed. Christiania, Norway; student medicine, 2 years. Mate on Belgica, 1897-99; ice navigator, 1901-02; made voyages on Gjoa (1st to sail N.W. Passage from E. to W. 1903, located site of N. Magnetic Pole 1904), 1903-06; leader Norwegian Antarctic Expdn. (1st to reach S. Pole 1911), 1910-12; with Maud expdn. to N.E. passage, 1918-21; flew over N. Pole by dirigible from Spitzbergen, Norway, to Teller, Alaska, 1926. Mem. French Acad. Scis. Author: Amundsen's North-West Passage, 1908; The South Pole, 1913; Our Polar Flight, 1925; My Life as an Explorer, 1927; The First Crossing of the Polar Sea, 1927. Conducted oceanographic research off northeast Greenland, 1901. Disappeared nr. Spitzbergen, June 18, 1928.

AMUNDSON, Neal Russell, Am. chem. engr.; b. St. Paul, Jan. 10, 1916; s. Oscar and Hazel (Cottrell) A.; B. Chem. Engring., U. Minn., 1937, M.S., 1941, Ph.D. in Math., 1945; m. Shirley Kathryn Dimond, Sept. 25, 1941; children—Gregg Russell, Beth Eva, Erik Neal. Process engr. Standard Oil Co. N.J., 1937-39; fellow Brown U., 1944-45; faculty U. Minn., Mpls., 1945—, head dept. chem. engring., 1949—, prof. 1950—. Mem. Am. Chem. Soc. (Indsl. and Engring. Chemistry award 1959), Am. Inst. Chem. Engrs. (William H. Walker award 1960), Am. Math. Assn. Author: Mathematical Methods in Chemical Engineering, 1966; also numerous articles. Developed math. techniques for solution transient and steady state problems related to chem. reaction systems coupled with transport phenomena. Home: 2297 Folwell St., St. Paul 55108. Office: Chem. Engring. Bldg., U. Minn., Mpls. 55455.*

AMUSSAT, Jean Zulema, French surgeon; b. Saint-Maixent, France, Nov. 21, 1796; entered army; asst. surgeon Salpétrière Hosp.; mem. Paris Faculty Medicine. Author: The Torsion of Arteries, 1829. First to adopt method of torsion of arteries in hemorrhage; developed new surg. method by successfully placing artificial anus, 1835; credited with performing 1st lumbar colostomy for relief of intestinal obstruction, reported 1839. Died 1856.

AMYCLAS OF HERACLEA, mathematician; flourished circa 420 B.C.; perfected geometry further than Plato and Eudoxus.

AMYOT, John Andrew, Canadian surgeon, pub. health ofcl.; b. Toronto, Ont., Can., July 25, 1867; s. John F. and Sophie (Fére) A.; student Assumption Coll., Ont.; M.B., U. Toronto, 1891; m. Mary Keller, 1895; 5 sons, 2 daus. House surgeon Toronto Gen. Hosp., 1891-92; demonstrator pathology U. Toronto, 1892-1900, asso. prof. pathology, 1900-09, prof. hygiene, 1909-18; asst. surgeon St. Michael's Hosp., Toronto, 1894-98, surgeon, 1898-1900; lectr. comparative physiology Ont. Vet. Coll., 1898-1908; dir. Provincial Bd. Health Lab., Ont., 1900-18; Dep. Minister Pensions and Nat. Health, Can. Cons. in sanitation Canadian Overseas Forces, Eng., 1916-19. Contbr. papers on pub. health questions in Can., U. S. Co-dir. Internat. Waterways' Commn., Can. and U. S.; investigations on pollution of Gt. Lakes. Died Feb. 17, 1940.

ANAGNOSTAKIS, Andreas, surgeon; b. Crete, 1826; ed. Athens, Greece, Berlin, Germany, Paris, France; apptd. dir. Ophthalmology Inst., Athens, 1854, prof. ophthalmology U. Athens, 1856; devised operation for treating entropion, involving resection of a strip of orbicularis muscle through an incision along length of tarsal plate. Died 1897.

ANANEV, Boris Gerasimovich, Russian psychologist; b. 1907; grad. Gorsky Pedagogical Inst., Ordzhonikdize, 1928; postgrad. course Bekhterev Brain Inst., Leningrad, 1930; D.Pedag. Sci., 1940, Prof., Philos. Faculty, Leningrad U., 1944—; dir. Leningrad Research Inst. Pedagogy, RSFSR Acad. Pedagogical Sci., 1951—. Mem. RSFSR Acad. Pedagogical Sci., Soviet Soc. Psychologists (v.p.). Author: The Theory of Inner Speech in Psychology, 1946; Outline of 18th-19th Century Russian Psychology, 1947; Material on the Psychological Theory of Sensations, 1948; Some Problems of the Theory of Perception, 1949; The Problem of Conception in Soviet Psychological Science, 1950; The Association of Sensations, 1955; Spatial Differentiation, 1955; Man as the General Problem of Modern Science, 1957; New Developments in the Theory of Spatial Perception, 1960; Complex Study of Man as a Routine Task of Modern Science, 1962; co-author The Sense of Touch in the Processes of Perception and Work, 1959. Mem. editorial bd. Problem of Psychology. Research on study of sensations, transitions from sensory perception to thought, pedagogical psychology, psychopathology, history of Russian psychology. Address: Leningrad Research Inst. of Pedagogy, n. Kutuzova 8, Leningrad, USSR.

ANARITIUS (al-Nayrizi), astronomer, mathematician; flourished 892; s. Hatim; wrote book on atmospheric phenomena, treatise on spherical astrolabe, commentary on Elements of Euclid; compiled astron. papers, commented on Ptolemy, used so-called umbra (equivalent of tangent) as a genuine trigonometric line; influenced Europe through transl. by Gerard of Cremona. Died 922.

ANASTASIJEVIC, Predrag, Yugoslavian nuclear engr.; b. Beograd, Yugoslavia, June 22, 1923; s. Milan and Branka (Dragicevic) A.; Electrotech, and nuclear engr., Beograd U., 1954; m. Kleopatra Harisijades, Apr. 24, 1954; 1 son, Dobrica. Research worker Boris Kidrich Inst. Nuclear Scis., Beograd, 1954—, head reactor physics dept., 1959-62, head heat transfer dept., 1966—; dir. tng. and fgn relations dept. Yugoslav Nuclear Energy Commn., 1962-64; asst. prof. thermodynamics Electrotech. Faculty, Beograd U., 1955-63. Recipient Medal for Good Merit to Nation, 1946, Medal Work with red flag 1st order, 1959. Mem. Yugoslav Soc. Heat Engrs. (sec. gen. 1965—). Author: Nuclear Powered Ships, 1957; also articles. Research on heat transfer from nuclear reactor fuel elements, methods for burn-out detection, devel. ceramic fuel elements; designed earliest nuclear research reactors in Yugoslavia. Home: 9 Takovska. Office: PoB. 522, Beograd, Yugoslavia.*

ANASTASSIADIS, John A., Greek mathematician; b. Smyrna, Asia Minor, Aug. 2, 1912; s. Anastasios and Sophia (Hatzikosta) A.; B.A., U. Athens, 1931; Dr. degree U. Salonica (Greece), 1935; postgrad. Sorbonne, U. Paris (France), 1937-38; m. Dimitra Tuya, Dec. 30, 1943; 1 son, Aghis. Faculty, U. Salonica, 1932—, prof. math., 1947—, head math. dept., 1955—. Mem. Greek, French math. assns. Author: Theory of Real Functions, 1951; Recherches algébriques sur le théorème de Picard-Montel, 1959; Definition des fonctions euleriennes par des équations fonctionnelles, 1964; also articles. Research on exceptional values of integral and meromorphic functions, functions defined by functional equations. Home: I, Anghelaki, Salonica, Greece.*

ANATOLI, Jacob, astronomer, philosopher; b. Marseilles, France, circa 1194; m. dau. of Samuel ibn Tibbon; lived in Languedoc, Narbonne, Beziers; invited by Frederick II to Naples, 1231, worked there with Michael Scot on translations from Arabic and Greek into Latin. Author: Malmad Ha-Talmidim (advocated study of langs. and secular scis. by rabbis). Translator: Ibn Rushd's intermediate commentaries on Porphyry's Isagoge, also on Aristotle's Categories, Interpretation, Prior and Posterior Analytics, 1232; Ibn Rushd's commentaries; Ptolemy's Alamest, circa 1233; Ibn Rushd's summary of Ptolemy's Almagest; Al-Farghani's Astronomy, circa 1235; works of Averroës (into Hebrew). Died 1256.

ANATOLIOS OF ALEXANDRIA, chronologist, mathematician; b. Alexandria, Egypt; prof. Aristotelian philosophy at Alexandria; became asst. bishop of Caesarea, Palestine; bishop of Laodicea, Syria, 269-282. Author: Canon Pascal; Recherches arithmetiques (in French transl.); also manual of arithmetic in 10 books; wrote on determination of Easter and Egyptian method of reckoning. Studies in philosophy, math., physics, astronomy, grammar and rhetoric. Died circa 283.

ANATOLIUS OF BERYTOS, Vindonios, agriculturalist; b. Beirut (then Syria), flourished 4th or 5th century; asso. of Ali al-Tabari; writings translated into Arabic (during Caliphate of Harun al-Rashid), influential in Abbasid period; compiled collection of Greek writings on agr. (12 books), which was one of main sources of Byzantine geoponica.

ANAXAGORAS, Greek philosopher; b. Clazomenae, Ionia, circa 500 B.C.; s. Hegesibulus; pupil of Anaximenes; 1st tchr. philosophy, Athens, from circa 465 B.C.-432 B.C.; later indicted for impiety (1st record of scientist tried by state religion) although acquitted, he retired to Lampascus, Traos, where he continued to teach until his death. Author book on nature (extant in fragments). Brought sci. rationalism of Thales and Ionian sch. to Greece; influenced Euripides, Pericles, possibly Socrates; rejected Empedocle's four element theory; held that existence resulted from ordering of seeds by infinite mind (similar to contemporary concept atoms); that intelligence (or nous) acts upon masses of particles to produce motion and change; his theory of nous distinguished between corporeal and incorporeal, mind and matter; it also introduced concept of teleology into philosophy; correctly explained phases of moon, solar and lunar eclipses; dissected animals, also brain; saw lateral ventricles; thought black or yellow bile caused acute diseases; attempted to square the circle; originated theory of perspective; correctly explained rise of Nile; Died Lampsacus (now Lapseki, Turkey), 428 B.C.

ANAXILAOS OF LARISSA, astronomer, chemist; b. Rome; flourished circa 20 B.C.; studied medicine and nature; accused of practicing magic. Author: Amusements. Invented infernal flame by burning sulphur.

ANAXIMANDER, Greek astronomer, philosopher; b. Miletos, Ionia, circa 610 B.C.; pupil of Thales. Author book on natural philosophy (1st Greek work of philosophy, also 1st prose treatise in Greek), postulated primary substance called apeiron containing four contrary elements (earth, air, fire, and water); these elements through periodic revolution, union, and disunion, created innumerable worlds; thus thought 1st to see universe as subject to 1 law. Forerunner of rationalistic mechanistic philosophy. Helped introduce sci. of Middle E. into Greece; credited with introduction of sundial, discovery of obliquity of ecliptic (in 58th olympiad or 548 to 544 B.C.), invention of geog. maps; thought earth was spherical; believed man achieved his physical state by adaptation, life evolved from moisture, and human life evolved from aquatic source; introduced gnomon and used it to determine equinoxes. Died circa 547 B.C.

ANAXIMENES OF MILETOS, Greek philosopher; b. circa 585 B.C.; possibly pupil of Anaximander; held that primary substance of universe is air, which when rarified becomes fire, when condensed, wind, then cloud, water, earth, stone (theory represents conception of phys. rather than moral law governing cosmos); knew that moonlight is reflected sunlight. Died circa 525 B.C.

ANBAR, Michael, chemist; b. Danzig, June 29, 1927; s. Joshua and Chava (Migdal) A.; M.Sc., Hebrew U., Jerusalem, 1950, Ph.D., 1953; m. Ada Komet, Aug. 11, 1953; children—Ran, Ariel. Research asst. Weizmann Inst. Sci., Rehovoth, Israel, 1950-53, sr. scientist, 1959—; postdoctoral fellow, instr. U. Chgo., 1953-55; dir. radioisotope Tng. Center, Rehovoth, 1956-59; head dept. radiation research Soreg Nuclear Research Center, 1961-66 head chemistry div., 1962-66. Research asso. Argonne (Ill.) Nat. Lab., 1963-64; vis. prof. Stanford, 1964-65, Tel Aviv (Israel) U., 1966-67; sr. research asso. U. S. Acad. Sci., NASA Ames Research Center, 1967-68. Recipient Meir award, 1960; Zondek award, 1963. Mem. Am. Chem. Soc., Chem. Soc. London, Faraday Soc., Israel Chem. Soc., A.A.A.S., Sigma Xi. Research, numerous publs. on redox reactions in inorganic chemistry, especially hydrogen peroxide, hypohalites, trivalent copper, chloramines; studies on reaction mechanisms of equated electrons, hydrogen atoms and hydroxyl radicals; molecular biochemistry; biol. inorganic chemistry, including behavior of iodine in biol. systems; mechanisms of calcification and decalcification, somolytic processes in geochemistry and chem. evolutes, somochemistry, mechanism of somolysis of water. Office: Weizmann Inst. Sci., Rehovoth, Israel.*

ANCEL, Paul-Albert, French biologist; b. Nancy, France, Sept. 21, 1873; prof. embryology Faculty Medicine, Strasbourg, France. Mem. French Acad. Scis. Discovered (with P. Bouin) functional connection between endocrine function of testicles and interstitial membranes, studied sterile male cells, uterus, produced pseudopregnancy which appeared not to interfere with other functions. Died 1961.

ANCHEL, Marjorie Wolff (Mrs. Herbert Rackow), Am. chemist; b. N.Y.C., May 6, 1910; d. Meyer and Lena (Shulman) Wolff; B.A., Barnard Coll., 1931, M.A., Columbia, 1933, Ph.D., 1939; m. Herbert Rackow, Dec. 30, 1942. Asst. chemist Queens Coll., Flushing, N.Y. 1939-41; research asst. biochemistry Coll. Phys. & Surg. Columbia, N.Y.C., 1941-43; research asso. Squibb Inst. for Med. Research, New Brunswick, N.J., 1943-46; research asso. N.Y. Bot. Garden, Bronx, 1946-61, sr. research asso., 1961—. Fellow A.A.A.S., N.Y. Acad. Sci., mem. Am. Chem. Soc., Am. Inst. Biol. Sci., Am. Soc. Biochemists; Mycol. Soc. Am., Torrey Bot. Club, Sigma Xi. Research, publs. on chemistry, biochemistry of natural products including antibiotics, other fungal metabolites. Home: 147-01 3d Av., Whitestone, N.Y. 11357. Office: N.Y. Bot. Garden, Bronx Park, Bronx, N.Y. 10458.*

ANCKER-JOHNSON, Betsy, Am. physicist; b. St. Louis, Apr. 27, 1929; d. Clinton J. and Fern (Lalan) Ancker; B.S. with high honors, Wellesley Coll., 1949; Ph.D. (Am. Assn. U. Women fellow, Horton-Hallowell fellow) Tuebingen (Germany) U., 1953; m. Harold H. Johnson, Mar. 15, 1958; children—Ruth, David, Paul, Martha. Jr. research physicist, lectr. physics U. Cal. at Berkeley, 1953-54; staff Inter-Varsity Christian Fellowship, Chgo., 1954-56; sr. research physicist Microwave Physics Lab., Sylvania Elec. Products, Inc., Palo Alto, Cal., 1956-58; tech. staff David Sarnoff Research Center, RCA, 1958-61; research specialist Plasma Physics Lab., Boeing Sci. Research Labs., Boeing Co., Seattle, 1961—; asso. prof. dept. elec. engring. U. Wash., Seattle, 1964—. Fellow American Physical Society; mem. I.E.E.E. (senior), Research Society, Phi Beta Kappa, Sigma Xi. Research, publs. in solid state physics, plasmas in solids, microwave and molecular electronics, ferrimagnetism and nonreciprocal effects, X-ray studies of imperfections in nearly perfect crystals. Office: P.O. Box 3981, Seattle 98124.*

AN DER LAN, Johannes Alfred Antonius Maria, meteorologist, zoologist; b. Rovereto, Italy, Dec. 19, 1909; s. Gotthardt and Helene (Buschman) A.; Ph.D., U. Innsbruck; m. Ilse Langsteiner, Sept. 12; children—Helge, Hannes, Wolfgang. Asst. inst. zoology U. Innsbruck since 1935; meteorologist-in-chief Southern Adriatic, Ionien, Aegean seas, during war. Mem. Asso. of Naturalist Physicians Innsbruck, German Soc. Zoology. Author: Meteorologische Besonderheiten der Aegaeis; Histopathologische Auswirkung von Insektiziden bei Wirbellosen und ihre cancerogene Beurteilung; Die Warmblüter-Toxizität der Insektizide; Die Fisch-Toxizität der Insektizide. Home: Innsbruck, Brandjochstr. 5, Austria. Office: Innsbruck, Universitatsstr. 4, Austria.*

ANDERS, Edward, Am. chemist; b. Libau, Latvia, June 21, 1926; s. Adolph and Erica (Leventals) Alperovitch; student U. Munich (Germany), 1946-49; A.M., Columbia, 1951, Ph.D., 1954; m. Joan Elizabeth Fleming, Nov. 12, 1955; children—George Charles, Nancy Elizabeth. Came to U. S.; 1949; naturalized 1955. Instr., U. Ill., Urbana, 1954-55; faculty U. Chgo., 1955—, prof. chemistry, 1962—; vis. prof. Cal. Inst. Tech., Pasadena 1960, U. Berne (Switzerland), 1963-64; sr. resident research asso. Goddard Space Flight Center, Greenbelt, Md., 1961. Cons. NASA, 1960—. Recipient U. medal for excellence Columbia, 1966. Fellow A.A.A.S. (Newcomb Cleveland prize 1959), Meteoritical Soc. (councilor 1966—); mem. Am. Astron. Soc., Am. Chem. Soc., Am. Geophys. Union, Geochem. Soc. (councilor 1967—), Internat. Astron. Union (com. on meteors and meteontes 1965—), Internat. Union Geodesy and Geophysics (mem. com. on problems of geochemistry 1964—), Sigma Xi. Asso. editor Geochemica et Cosmochimica Acta, 1966—. Research, publs. on origin, composition, chem. fractionation, and organic matter of meteorites. Home: 5415 Hyde Park Blvd., Chgo. 60615.*

ANDERSEN, Bjorn, chem. engr.; b. Frederikstad, Norway, June 29, 1897; s. Elling and Auguste Petrea (Lorenzen) A.; B.S., Norwegian Inst. Tech., 1918, Chem. E., 1920; m. Ingeborg Soelberg, Mar. 27, 1924; children—Bjorn Andreas, Thor Bjorn, Erik Bjorn, Nils-Olav, Lars-Rolf. Came to U. S., 1924, naturalized, 1930. Asst. prof. Norwegian Inst. Tech., 1921-24; chief chemist Norway Inst. Testing Materials, 1922-24; research chemist Guggenheim Bros, N.Y.C., 1924-26, mgr. research, 1926-28; tech. dir. Celluloid Corp., Newark, 1928-41; dir. research, tech. dir. Celanese Corp. Am., 1941-47; dir. central research and devel., 1947-51, tech. dir., v.p., 1951-55, v.p., gen. mgr. Plastic Co. div., 1955-59, v.p. Celanese Devel. Co., N.Y.C., 1959-62, ret., 1962. Mem. U. S. Dept. Commerce Devel. and Trade Missions to Australia, 1960, Burma, 1962, Finland, 1964, Bulgaria, Hungary, 1966. Mem. Chem. Club N.Y.C. Developed cycle leaching-electrolytic process for recovery of tin from Komplextin concentrates; pioneer in chemistry and processing of plastics from cellulose esters and development of such esters; patentee in field. Home: 38 Park Rd., Maplewood, N.J. 07040.*

ANDERSEN, Einar Anton, Danish geodesist; b. Copenhagen, Denmark, Sept. 16, 1905; s. Carl A. and Louise (Frandsen) A.; M.Sc., U. Copenhagen, 1928, D.Sc., 1932; m. Inger Remmer, Oct. 25, 1929; children—Ulla (Mrs. Hans C. Lohren), Toni (Mrs. Steffen Strobaek), Aino. Staff, Royal Danish Geodetic Inst., Copenhagen, 1926—, state geodesist, 1940-55, dir., 1955—; faculty U. Copenhagen, 1947—, prof. geodesy, 1956—. Pres. mng. com. Internat. Glaciological Expdn. to Greenland, 1963-66, Danish com. for Upper Mantle Project, 1965—; pres. various Danish sci. coms. Named commdr. Iceland Falcon, 1961; Order Dannebrog, 1963; recipient Medal of Def. Services, 1954. Mem. Internat. Union Geodesy and Geophysics (treas., mem. bur. 1960—), Internat. Council Sci. Unions (pres. nat. com. 1962—), Royal Danish. Acad. Scis. and Letters, Danish Acad. Tech. Scis. Author books, numerous articles. Research on Sway-correction by pendulum measurements, transfer geog. coordinates by vertical sects and Clarke's curve of alignment; adjustment of observations by matrices. Home: 8, Serridslevvej, Copenhagen. Office: 7 Rigsdagsgaarden, Copenhagen K, Denmark.*

ANDERSEN, Harold Torbjorn, Norwegian physiologist; b. Oslo, Norway, Jan. 3, 1933; s. Arne Johan and Ellen (Haug) A.; M.S., U. Oslo, 1957, Dr.Philos., 1963; Ph.D., U. Pa., 1961; m. Liv Bjorsvik, June 28, 1958; 1 dau., Ellen Charlotte. Research asst. U. Oslo, 1957, grad. research biologist Scripps Inst. Oceanography, 1958-59; USPHS fellow U. Pa., 1959-60; fellow Norwegian Research Council Sci. and Humanities, 1961-62; faculty U. Oslo, 1962—, acting chmn. dept. zoophysiology, 1963-64, reader physiology, 1965—, head physiology div. Inst. for Nutrition Research, 1965—; exchange visitor John B. Pierce Found. Lab., New Haven, 1964-65. Mem. Norwegian Soc. for Biochemistry and Physiology, Scandinavian Physiol. Assn., Soc. for Exptl. Biology. Research, numerous publs. on physiology for diving animals, central nervous control of temperature regulation, acclimatization to cold in man, electrophysiology, chemoreception, taste. Home: Holtvn. 13A, Bekkelagshogda, Oslo, Norway.*

ANDERSEN, Kristian Lange, Norwegian physiologist; b. Grytten, Norway, June 28, 1920; cand. med. U. Oslo (Norway), 1949, dr. med., 1959; m. Berit Hamre-Hansen, June 8, 1946; children—Dag Johan, Kristin. Research asst., inst. hygiene U. Oslo, 1950-52; physician Aker Sykehus, Oslo, 1952-53; research fellow Norwegian Research Council, 1953-56; research asst. Inst. Aviation Medicine, Oslo, 1957-58; head Inst. Work Physiology, Oslo, 1958-64; prof. physiology U. Bergen, 1964-67; dir. Sch. Physiotherapy, Oslo, 1967—. Mem. expdns. for study of human populations of arctic and tropic communities. Author: (with K. Evang) Physical Activity in Health and Disease, 1967; also articles. Research on physiology of work and sport performances, cold tolerance; established limits for human adaptation to phys. activity and to cold exposure. Home: 20 Monolitvn, Oslo. Office: 132 Trondhjemsvn, Oslo, Norway.*

ANDERSEN, Per Oskar, Norwegian neurophysiologist; b. Oslo, Norway, Jan. 12, 1930; s. David A. and Olga (Rodvei) A.; B.Sc., Oslo Cathedral Sch., 1947; M.D., Oslo U., 1954, Dr.med., 1960; m. Kari Sletten, July 5, 1955; children—Hege, Kristin,

Espen, Inger Line. Asst. research fellow Lab. Neurophysiology, Oslo, 1951-56, research fellow, 1956-58; asso. prof. Inst. Anatomy, Oslo, 1958-63; Rockefeller fellow Australian Nat. U., 1961-63; asso. prof. Lab. Neurophysiology, U. Oslo, 1964—. Publs. on studies of synaptic orgn. within the hippocampus of brain; mechanisms of central control of sensory information; brain rhythmic activity. Home: 48B Dragvegen, Blommenholm, Oslo, Norway.*

ANDERSON, Adam, Brit. physicist; rector Perth (Scotland) Acad.; prof. natural philosophy St. Andrew's U., Scotland; supr. constrn. Perth gasworks (introduced improvements to that led to econ. gas prodn.). Contbd. articles to Brewster's Edinburgh Ency., 1830, Ency. Brit., various jours. Wrote on measurement of mountain heights by barometer, hygrometric state of atmosphere, dew point, illuminating power of coal gas. Died 1846.

ANDERSON, Alan James, Brit. biochemist; b. London, Eng., Nov. 16, 1924; s. Alexander Fraser and Rose (Smith) A.; B.Sc., London U., 1949, M.Sc., 1950, Ph.D., 1952, D.Sc., 1964. Research asst. in chem. pathology Westminster Hosp. Med. Sch., London U., 1952-57; research asso. in preventive medicine and biol. chemistry U. Ill. Coll. Medicine, Chgo., 1957-59; research biochemist, lectr. pathology Inst. Orthopaedics, Brit. Postgrad. Med. Fedn., London, 1959-67; sr. research biochemist Lilly Research Center Ltd., Windlesham, Eng., 1967—. A.R.I.C., 1949; F.R.I.C. 1961. Member of the Biochem. Soc., Assn. Sci. Workers, Assn. Clin. Biochemists. Publs. on research in biochemistry of sugar-containing macromolecules in normal and abnormal tissues, serum and urine; use of electrostatic interactions between charged macromolecules in relation to protein fractionation and biol. activity. Home: Toft, Monk's Rd., Virginia Water, Surrey. Office: Lilly Research Centre Ltd., Erl Wood Manor, Windlesham, Surrey, Eng.*

ANDERSON, Alexander, mathematician; b. Aberdeen, Scotland, 1582; tchr. math., Paris, France. Author: Supplementum Apollonii Redivivi (amendment of Ghetaldi's restoration of Apollonius' lost book), 1612; Francisci Vietae de Equationum recognitione et emendatione tractatus duo (Vieta's improvements in transformation and reduction of algebraical equations, with appendix by Anderson showing solution of cubic equations can be dependent on trisection of angle), 1615; Ad angularium sectionum analyticen theoremata . . . (additions to Vieta's theorems on angular sects. demonstrations), 1615; Vindiciae Archimedis, 1616; Animadversiones in Franciscum Vietam a Clemente Cyriaco nuper editae brevis . . . , 1617; Exercitationum mathematicarum decas prima, 1619. Noted for careful philology and command of ancient Greek math. analysis. Died perhaps 1619.

ANDERSON, Alfred Ronald, English chemist; b. Gateshead, Eng., May 31, 1929; s. Alfred and Margaret (Yearsley) A.; B.Sc. with 1st class honors, Kings Coll., U. Durham (Eng.), 1950, Ph.D., 1954; m. Eileen Eades, Aug. 9 1952; children—Vicki Barbara, David Timothy. Staff, Atomic Energy Research Establishment, U.K. Energy Authority, Harwell, 1953-58, prin. sci. officer, 1961—; exchange fellow Argonne (Ill.) Nat. Lab., 1958-60. Mem. Royal Inst. Chemistry, Faraday Soc., Chem. Soc. Contbg. author: Fundamental Processes in Radiation Chemistry, 1967; Research, publs. on radiation chemistry of reactor coolants in particular carbon dioxide and water, calorimetric methods for dosimetry in nuclear reactors, studies of radiation chemistry of water and aqueous solutions, of gases, including carbon monoxide, carbon dioxide, water vapor. Home: 9 Galley Field, Abingdon, Berks. Office: Chemistry Div., Atomic Energy Research Establishment, Harwell, Berks., Eng.*

ANDERSON, Allan George, Am. mathematician; b. N.Y.C., Oct. 30, 1923; s. Richard J. and Marcella (Lennon) A.; B.S., U. Fla., 1946, M.A., 1947; Ph.D., U. Mich., 1951; m. Margaret Alice Elliot, Sept. 18, 1949 (div. April. 1955); children—William Peter, Barbara Jean, Donald George; m. 2d, Lucia Zylak Lewis, Apr. 30, 1955; 1 dau., Patricia Lynn. Instr. to asst. prof. Oberlin Coll., O., 1950-53; asst. prof. to asso. prof., chmn. math. dept. Duquesne U., Pitts., 1953-55; mathematician Jones & Laughlin Steel Corp., Pitts., 1955-56, Pitts. Plate Glass Co., Creighton, Pa., 1956-57; chief statistician Gen. Tire & Rubber Co., Akron, O., 1957-58; prof., head math. dept. Western Ky. State Coll., Bowling Green, 1958-65; prof., chmn. math. dept. Upsala Coll., East Orange, N.J., 1965-66; prof. math. Parsons Coll., Fairfield, Ia., 1966—. Vis. lectr. NSF summer inst. Memphis State U., U. Tenn., 1962, 64; mem. panel evaluators Nat. Council for Accreditation of Tchr. Edn., 1964-65, evaluating com. So. Assn. Colls. and Secondary Schs., 1964-65. Mem. Am. Math. Soc., Math. Assn. Am. (chmn. Ky. sect. 1960-61, vis. lectr. Ky. sect. 1961-62, secondary sch. lectr. 1961-62, 62-63), Inst. Math. Statistics, Am. Statis. Assn., Phi Beta Kappa, Sigma Xi, Phi Kappa Phi. Developed prediction formulas for quantitative characteristics in polygenic systems. Home: 1107 E. Broadway, Fairfield, Ia. 52556.*

ANDERSON, Arthur Wallace, American microbiologist, educator; born Lisbon, North Dakota, December 2, 1914; Paul and Alma (Anderson) A.; B.S., N.D. State U., 1943; M.S., U. Wis., 1947; Ph.D., Ore. State U., 1952; m. Jean M. Fuller, Nov. 26, 1948; children—Martin Paul, Sally Nell, Kathryn Jane. Microbiologist, Wis. Alumni Found., Madison, 1947-49; asst. prof. U. Cal., Berkeley, 1952-53; with Ore. State U., 1953—, prof., research asso. Mem. Am. Soc. Microbiology, Inst. Food Technologists, Sigma Xi. Discoverer method for lowering resistance of spore forming bacteria; working on methods for irradiation protection and mechanism of irradiation resistance. Home: Route 1, Box 354, Corvallis, Ore. 97330.*

ANDERSON, Carl David, Am. physicist; b. N.Y.C., Sept. 3, 1905; s. Carl David and Emma Adolfina (Ajaxson) A.; B.S., Cal. Inst. Tech., 1927, Ph.D., magna cum laude, 1930; hon. Sc.D., Colgate U., 1937, Gustavus Adolphus Coll. LL.D., Temple U., 1948; m. Lorraine Elvira Bergman; children—Marshall David, David Andrew. Coffin research fellow Cal. Inst. Tech., 1927-28, teaching fellow in physics, 1928-30, research fellow in physics, 1930-33, asst. prof. physics, 1933-37, asso. prof., 1937-39, prof., 1939—, chmn. div. physics, math. and astronomy, 1962—. Recipient gold medal Am. Inst. of City of N.Y., 1935; Nobel prize in physics (with Hess), 1936; Elliott Cresson medal Franklin Inst., 1937; John Ericsson medal Am. Soc. Swedish Engrs., 1960. Mem. Am. Phys. Soc., Am. Philos. Soc., Nat. Acad. Scis., Tau Beta Pi, Sigma Xi. Research on X-ray photoelectrons, 1927-30, on gamma rays and cosmic rays, 1930—; discovered positron (independent of Blackett's similar discovery), 1932; (with Neddermeyer) produced positrons by gamma irradiation, 1933, experimentally proved existence of meson (corroborating Yukawa's prediction), 1938. Home: 2915 Lorain Rd., San Marino, Cal. 91108. California Institute of Technology, Pasadena, Cal. 91109.*

ANDERSON, Charles, mineralogist; b. Stenness, Scotland, 1876; s. John and Margaret (Smith) A.; M.A., Edinburgh U., 1898, B.Sc., 1900, D.Sc., 1908; m. Elsie Robertson, 1902; 1 son, 2 daus. Observer Ben Nevis Obs., 1900-01; became mineralogist Australian Mus., Sydney, 1901, dir., 1921-40. Research and publs. on chemistry and crystallography of Australian minerals, also on Australian fossil vertebrates. Died Oct. 30, 1944.

ANDERSON, Charles Alfred, Am. geologist; b. Bloomington, Cal., June 6, 1902; s. Amel A. and Mary (Lyman) A.; A.B., Pomona Coll., 1924, D.Sc., 1960; Ph.D., U. Cal. at Berkeley, 1928; m. Helen Argall, July 25, 1927; 1 son, Robert A. Faculty, U. Cal., Berkeley, 1928-42, asso. prof., 1936-42; with U. S. Geol. Survey, Menlo Park, Cal., 1942—, chief mineral deposits br., 1953-58, chief geologist, 1959-64, research geologist, 1964—. Recipient Distinguished Service award, 1960. Mem. Geol. Soc. Am. (councilor 1947-49), Soc. Econ. Geologist (v.p. 1961), Nat. Acad. Scis., Am. Acad. Arts and Scis. Research on volcanic rock, copper deposits; study of Gulf of Cal., parts of Nev., Ariz. Home: 270 Waverley St. Office: 345 Middlefield Rd., Menlo Park, Cal. 94025.*

ANDERSON, Dean Albert, Am. microbiologist; b. Pleasant Grove, Utah, Sept. 27, 1905; s. G. Albert and Annie (Radmall) A.; B.S. with honors, Brigham Young U., 1929; M.S., Ia. State Coll., 1930,Ph.D., 1932, postgrad., 1932-33; m. Elva Brown, May 26, 1927 (dec. Aug. 1965); children—La Deane A. (Mrs. Edward D. Durham), Gordon K. Instr. bacteriology Weber Coll., Ogden, Utah., 1933-41; from asst. to asso. prof. Brigham Young U., Provo, Utah, 1941-50; prof. microbiology, chmn. div. natural scis. Los Angeles State Coll., 1950-56; head dept. biol. scis., 1956-59, head dept. microbiology and pub. health, 1959-64. Writer, biol. sci. curriculum study Am. Inst. Biol. Scis., 1960. Fellow A.A.A.S.; mem. Soc. Am. Bacteriologists (pres. Intermountain br. 1948-49, pres. So. Cal. br. 1959—), Am. Soc. Microbiology, Soc. for Gen. Microbiology, Soc. Indsl. Microbiology Royal Soc. Health, Am. Pub. Health Assn., Sigma Xi, Phi Kappa Phi, Phi Lambda Upsilon. Author: (with L. P. Gebhardt) Microbiology, 1954, 65, Laboratory Instruction in Microbiology, 1954, 65. Research on soils, soil and salt lake bacteria; effects of detergents on bacterial growth. Home: 8956 Ardendale Av., San Gabriel, Cal. 91775. Home: P.O. Box 17 Albany, Cal. 94716.*

ANDERSON, Don Lynn, Am. geophysicist; b. Frederick, Md., Mar. 5, 1933; s. Richard Andrew and Minola (Phares) A.; B.S., Rensselaer Poly. Inst., 1955; M.S., Cal. Inst. Tech., Ph.D., 1962; m. Nancy Lois Ruth, Sept. 15, 1956; children—Lynn Ellen, Lee Weston. With Chevron Oil Co., Mont., Wyo., Cal., 1955-56; with Air Force Cambridge Research Center, Boston, 1956-58; with Arctic Inst. N.Am., Boston, 1958; faculty Cal. Inst. Tech., Pasadena, 1962—, asso. prof. geophysics, 1964—, director seismological laboratory, 1967—. Sloan Found. fellow, 1965-67. Mem. Am. Geophys. Union (James B. Macelwane award 1966), A.A.A.S., Soc. Exploration Geophysics, Seismol. Soc. Am., Sigma Xi. Asso. editor Jour. Geophys. Research, 1965-67. Research on physics of earth's interior; theoretical seismology; physics of sea ice; planetary interiors. Home: 669 E. Alameda St., Altadena, Cal. 91001. Office: Seismol. Lab., Bin 2, Arroyo Annex, Pasadena, Cal. 91101.*

ANDERSON, Donald Rex, Am. radiobiologist; b. Ephraim, Utah, June 10, 1916; s. John R. and Camilla (Sorenson) A.; B.A. in Vertibrate Zoology, U. Utah, 1952, M.A. in Exptl. Biology, 1954, Ph.D., 1956; m. Mary Alice Branz, Sept. 1, 1960; children—Nancy, Scot Rex. Research officer radiobiology br. USAF Sch. Aerospace Medicine, Brooks AFB, Tex., 1955-56; Air U. Research awardee Austin Lab., Tex., 1956; head, biochemistry-physiology group Radiobiol. Lab., Austin, 1957-60; chief, exptl. radiobiology sect. radiobiology br., instr. resident physicians USAF Sch. Aerospace Medicine, Brooks AFB, Tex., 1960—. Mem. A.A.A.S., N.Y. Acad. Scis., Radiation Research Soc., Am. Inst. Biol. Scis., Transplantation Soc., Sigma Xi. Research on radiation effects on creatine-creatinine metabolism, use of bone-marrow cell transplantation to treat radiation injury; immunochemistry of irradiated primates; problems of protection against ionizing radiation. Home: 3715 Invicta St., San Antonio 78218. Office: Box 4229, USAF Sch. Aerospace Medicine, Brooks AFB, Tex. 78235.*

ANDERSON, Earl Jennings, Am. plant pathologist; b. Roy, Wash., June 8, 1908; s. Albert and Henrikka (Jensen) A.; B.S., Wash. State U., 1932, M.S., 1934; Ph.D., U. Md., 1937; m. Marion Louise Preston, Aug. 18, 1934; 1 son, David Preston. NRC fellow U. Minn., 1937; instr. N.D. State Coll., 1938; with Pineapple Research Inst., Honolulu, 1939-42, 45—, head dept. plant pathology, 1951-62, 64—, head sect. pest control, 1963; head dept. plant pathology, chmn. plant pathology Wash. State U., 1943-45; Fulbright scholar U. Sydney, 1957, vis. scientist, 1964; affiliate prof. U. Hawaii, 1955—; dir. agrl. research Dole Philippines, Inc., 1965—. Mem. A.A.A.S., Phytopath. Soc., Nematological Soc., Hawaiian Bot. Soc., Hawaiian Acad. Sci., Am. Inst. Biol. Sci., Sigma Xi, Phi Kappa Phi, Alpha Zeta. Address: P.O. Box 362, Commercial Center P.O., Makati, Rizal, Philippines.*

ANDERSON, Edgar Shannon, Am. botanist, b. Forestville, N.Y., Nov. 9, 1897; s. Anson Crosby and Inez (Shannon) A.; B.S., Mich. State U., 1918; M.S., Harvard, 1920, Sc.D., 1922; m. Dorothy Moore, June 21, 1923. Faculty, Washington U., St. Louis, 1922-31, 35—, Engelmann prof. botany, 1937—, emeritus, 1966—; with Mo. Bot. Garden, St. Louis, 1922-31, 35—, editor Bull., 1963—, botanist, 1966—. Decorated Order of Yugoslavian Crown, 1936; recipient Darwin Wallace Medal, 1958. Author: Introgressive Hybridization, 1949; Plants, Man and Life, 1952; also numerous articles. Pioneer in field of genetics-taxonomy now known as biosystematics; research on evolutionary importance of hybridization; demonstrated that maize is divided into about 200 races and sub-races; originated techniques for recognizing and classifying races of maize, surveying, describing them; located, collected, introduced drought and cold hardy strains of Boxwood, Hedera, Ligustrum form eastern Balkans. Home: 2201 Tower Grove Ave. Office: 2315 Tower Grove Av., St. Louis 63110.*

ANDERSON, Elizabeth Garrett, Brit. physician; b. Aldeburgh, Eng., 1836; M.D., Paris; licensed to practice medicine by Scottish Soc. of Apothecaries, 1865; lectr. in medicine London (Eng.) Sch. Medicine for Women; 1st woman mayor Aldeburgh; pioneer in opening med. profession to women. Died Aldeburgh, Dec. 17, 1917.

ANDERSON, Ernest Carl, Am. chemist; b. Rock Island, Ill., Aug. 23, 1920; s. Ernest Axel and Mary (McGaughey) A.; A.B., Augustana Coll., 1942; Ph.D., U. Chgo., 1949; m. Catherine D. Payne, June 21, 1942; children—Christopher F., Nicholas J., Catherine J. Analytical chemist Metall. Lab., U. Chgo., 1942-44; staff mem. biophysics Los Alamos N.M. Sci. Lab., 1944-46, 49—. Mem. Nat. Acad. Sci.-NRC com. on nuclear sci., 1962-66. Rask-Orsted fellow U. Copenhagen (Denmark), 1951-52; recipient E. O. Lawrence award U. S. AEC, 1966. Fellow A.A.A.S. Research, numerous publs. on devel. and application low level counting techniques to fallout studies, human body composition, radioactivity of meteorites and moon, studies of biochemistry and biophysics of mammalian life cycle; co-discovery natural radiocarbon and devel. radiocarbon dating. Home: 1610 S. Sage St. Office: Los Alamos Sci. Lab., Los Alamos, N.M. 87544.*

ANDERSON, Ernest Robert, Am. phys. oceanographer; b. Great Falls, Mont., Jan. 11, 1917; s. Ernest William and Eva (Lysall) A.; B.A., U. Mont., 1939; M.S., U. Cal. at Los Angeles, 1948, Ph.D., 1953; m. M. Josephine Buergey, Dec. 26, 1941; 1 dau., Janet. Instr., Wolf Point (Mont.) High Sch., 1939-41; meteorologist N.W. Airlines, Inc., Edmonton, Alta., Can., 1942; oceanographer cons. Pacific Missile Range, Point Mugu, Cal., 1959-60; oceanographer USN Electronics Lab., San Diego, 1948-59, supervisory physicist, 1960-67; supervisory physicist Naval Undersea Warfare Center, San Diego, 1967—; instructor of mathematics at San Diego City College, San Diego, 1957—. Cons. Dept. Pub. Utilities, Tacoma, 1950-51; research asso. Inst. Naval Studies, USN War Coll., Newport, R.I., 1960-61. Recipient Spl. Hon. award USN Electronics Lab., 1962. Research, publs. in exchange of energy between ocean and atmosphere, relation of interaction of propagation of acoustic energy in sea water and pertinent environmental factors, devel. of oceanometrics concept for summarizing hist. oceanographic data. Home: 3788 Elliott St., San

Diego 92106. Office: Naval Undersea Center, San Diego 92152.*

ANDERSON, Frank Marion, Am. geologist; b. nr. Phoenix, Ore., June 4, 1863; s. Jessie Marion Anderson; grad. Willamette U., 1889; bachelor's degree Stanford, 1895, Ph.D., 1930; M.S., U. Cal., 1897, Ph.D., 1899; m. Elinor Anglin, 1900 (dec. 1916); m. 2d., Theresa M. Barry, 1919; children—Mrs. Ralph W. MacIntyre, Francis Marion, Frank Barry. Tchr. pub. schs., S.W. Ore.; field asst. Cal. Mining Bur., 1899; curator, dept. paleontology Cal. Acad. Scis., 1904-14; geologist So. Pacific Co., 1902-11. Fellow Geol. Soc. Am.; mem. Cal. Acad. Scis. (life). Author: Cretaceous deposits of the Pacific Coast, 1902. Collected cretaceous fossils; introduced name Paskenta beds (stratigraphic unit of lower cretaceous), also formational names Domengine, Temblor, Krezenhagen, Etchegoin, Tulare, Kern River. Died Berkeley, Cal., Sept. 24, 1945.

ANDERSON, Gaylord West, Am. epidemiologist, educator; b. Mpls., Dec. 31, 1901; s. Frank Maloy and Mary G. (Steele) A.; A.B., Dartmouth, 1922; postgrad. Sorbonne (Paris), U. Zurich (Switzerland); M.D., Harvard, 1928, D.Pub. Health, 1942; m. Viola Dennis, Oct. 26, 1929; 1 dau., Gail Elizabeth (Mrs. Harvey Safeer). Epidemiologist, Mass. Dept. Pub. Health, 1929-30, asst. dir. div. communicable diseases, 1930-31, dir., 1931-37, dep. state health commr., 1931-37; prof., head dept. preventive medicine and pub. health U. Minn., Mpls., 1937-46, Mayo prof., dir. Sch. Pub. Health, 1946—. Cons. WHO, U. S. Army, USPHS, U. S. Dept. State. Recipient Harrington award Mpls. Jr. C. of C., 1959. Fellow Royal Soc. Health (hon.); mem. Am. Pub. Health Assn. (pres. 1952, Sedgwick Meml. medal for outstanding service in pub. health 1963), Am. Epidemiological Soc. (pres. 1952), Phi Beta Kappa, Sigma Xi, Alpha Chi Sigma, Alpha Omega Alpha. Author: (with M. G. Arnstein) Communicable Disease Control, 1941; (with others) Global Epidemiology, vol. 1, 1944, vol. 2, 1951, vol. 3, 1954. Home: 2261 Folwell St., St. Paul 55108. Office: Sch. Public Health, U. Minn., Mpls. 55455.*

ANDERSON, Henry Graeme, surgeon; b. Scotland, 1882; s. Nicol Anderson; ed. Glasgow (Scotland) U., King's Coll., London Hosp.; M.D.; Ch.B.; m. Gladys Hood, 1921; 1 dau. House surgeon St. Mark's Hosp., 1907-08; surg. registrar Metropolitan, Cancer, Royal Nat. Orthopaedic hosps., 1909-12; commd. surgeon-lt. Royal Navy, 1914, advanced through grades to maj.; with Air Service Expeditionary Force, France and Belgium, 1914, Polegate and Eastbourne, Eng., 1916; surgeon to Brit. Flying Sch., Vendome, France, 1917, Central RAF (transferred from Royal Navy), 1918-19; cons. surgeon RAF; surgeon St. Mark's Hosp. for Diseases of Rectum; sr. asst. surgeon Belgrave Hosp. for Children. Fellow Royal Soc. Medicine, Royal Coll. Surgeons. Author: The Medical and Surgical Aspects of Aviation; The Rectum—Encyclopaedia of Surgery; Combined operations for Excision of Rectum—System of Operative Surgery; Operative Treatment of Haemorrhoids (Brit. Med. Jour.); numerous articles on aviation and surgery. Died June 28, 1925.

ANDERSON, Henry James, Am. mathematician; b. N.Y.C., Feb. 6, 1799; grad. Columbia, 1818; grad. Coll. Physicians and Surgeons, N.Y.C., 1823. Prof. math. and astronomy Columbia, 1825-50, became trustee, 1851, prof. emeritus math. and astronomy, 1866; went to Europe, circa 1850, converted to Catholicism, geologist for Dead Sea expdn. while travelling in Holy Land; mem. Am. scientific expdn. to observe patterns of planet Venus; served as pres. St. Vincent de Paul Soc., N.Y.C. Died Lahore, Northern Hindustan, India, Oct. 19, 1875.

ANDERSON, Herbert Lawrence, Am. physicist; b. N.Y.C., May 24, 1914; s. Joseph and Sima (Goldberg) A.; A.B., Columbia, 1935, B.S. in Elec. Engring., 1936, Ph.D. in Physics, 1940; m. Jean Betty Clough, Jan. 11, 1947; children—Faith A., Clifton L., Kelley P., Dana Z. Physicist, Metall. Lab., U. Chgo., 1942-44, faculty, 1946—, prof. physics, 1950—, dir. Enrico Fermi Inst., 1958-63; group leader Los Alamos Sci. Lab., 1944-46. Cons. Stein, Roe & Farnham, 1957—. Mem. Am., Italian phys. socs., A.A.A.S., Nat. Acad. Sci. Research, patents, publs. on nuclear instruments, elementary particles and high energy physics. Home: 4923 S. Kimbark Av., Chgo. 60615.*

ANDERSON, Sir Hugh Kerr, Brit. physiologist; b. Hampstead, Eng., July 6, 1865; s. James and Eliza (Murray) A.; ed. Gonville and Caius Coll., Cambridge (Eng.) U., St. Bartholomew's Hosp.; m. Jessie Innes, 1894; 1 son, 1 dau. univ. lectr. physiology Cambridge U., elected fellow Gonville and Caius Coll., 1897, became mem. med. and financial bds., 1908. Fellow Royal Soc., 1907. Research and publs. (many with Prof. John Newport Langley) on physiology of nerves. Died London, Nov. 2, 1928.

ANDERSON, Ingram Fairfax, med. geneticist; b. London, Eng., Jan. 28, 1935; s. Digby John and Doreen (Michau) A.; M.B.B.Ch., U. Witwatersrand, Johannesburg, S. Africa, 1959. Tutorial registrar dept. medicine Johannesburg Gen. Hosp., 1961-64; USPHS scholar Johns Hopkins Hosp., Balt., 1964-65; sr. physician, sr. lectr. H. F. Verwoerd Hosp., Pretoria, S. Africa. Mem. S. African Genetic Soc. (mem. com. 1962-64), royal colls. physicians London, Edinburgh. Guest editor S. African Med. Jour. Publs. on application of genetics to clin. medicine, initial human chromosomal studies in S. Africa; surveys among mental defectives and small genetic isolates. Home: 9 Scholtz Av., Norwood, Johannesburg, S. Africa.*

ANDERSON, J. Wemyss, Brit. engr.; b. Devon, Eng., 1868; s. John Anderson; M.Eng., Liverpool (Eng.), U., 1907; m. Caroline Emily Taylor, 1897; 1 dau. Dean faculty engring., asso. prof. Liverpool U., 1903-20, prof. engring., refrigeration, until 1924. Sci. adviser to dir. cold storage Ministry of Food. Author: Refrigeration, and Elementary Text Book. Research and publs. on refrigerating efficiency, mech. refrigeration, insulation, cold storages, vapour-compression refrigerating machines. Died Apr. 27, 1930.

ANDERSON, James, Brit. agriculturist, economist; b. nr. Edinburgh, Scotland, 1739; engaged in farming; pub. monthly periodical Recreations in Agr., Natural History, Arts and Lit., nr. London, Eng., 1799-1802. Author: Essays relating to Agriculture and Rural Affairs, 1777; Inquiry into the Nature of the Corn Laws, 1777. Editor The Bee (weekly periodical), 1790-94. Inventor 2-horse plow without wheels (Scotch plow); developed new theory on rent, adopted later by Malthus and others. Died 1808.

ANDERSON, John, Brit. inventor; b. Roseneath, Scotland, 1726; ed. U. Glasgow (Scotland). Prof. oriental languages U. Glasgow, from 1756, prof. natural philosophy from 1760. Planned fortifications for defense of Greenock against Thurot, 1759; lectured to artisans on sci. subjects, circa 1760-96. Fellow Royal Soc., 1759. Author: Institutes of Physics, 1786. Contbd. papers to periodicals, including Observations upon Roman Antiquities lately discovered to Military Antiquities (Roy), 1793. Bequeathed land for founding Anderson's Coll. (earliest mech. inst., became Royal Tech. Coll. 1913); said to have directed Watt to Newcomen engine; helped obtain Roman collection for U. Glasgow; invented cannon in which condensation of air in carriage counteracted recoil; devised plan for smuggling French newspapers into Germany by means of small balloons; gave cannon and 2 essays on war and mil. instruments to Nat. Conv. French Revolution. Died Glasgow, June 13, 1796.

ANDERSON, John, Brit. naturalist; b. Edinburgh, Oct. 4, 1833; s. Thomas Anderson; M.D., U. Edinburgh, 1862, L.L.D., 1885; m. Grace Thoms. Apptd. prof. natural history Free Ch. Coll., Edinburgh; became curator Indian Mus., Calcutta, 1865; mem. sci. expdns. to Tunnan, 1867, Burma, 1875-76, Mergui Archipelago, 1881-82; prof. comparative anatomy Med. Sch., Calcutta; returned to London, 1886, retired, 1887. Recipient Gold medal in zoology U. Edinburgh. Fellow Royal Society, 1879, Linnean Society; mem. Royal Physical Society (a founder, early pres.). Author: Mandalay to Momein, 1876; Anatomical and Zoological Researches . . . (monograph on platanista and orcella), 1878-79; Catalogue of Mammalia in the Indian Museum, 1881; Catalogue of Archaeological Collections in the Indian Museum, 1883; Contributions to the Fauna of Mergui and its Archipelago, 1889; English Intercourse with Siam, 1889; A Contribution to the Herpetology of Arabia, 1898; Zoology of Egypt: Part I, Reptilia and Batrachia, 1898, Part II, Mammalia. Investigated vertibrates, especially mammals, dolphins of India, platanista; 1st described orcella; contbd. notes on apes, reptiles, birds. Died Buxton, Eng., Aug. 15, 1900.

ANDERSON, John F., Am. physician; born Fredericksburg, Va., Mar. 14, 1873; s. John Kerwin and Lucy Ella (Hundley) A.; ed. in pub. schs., Fredericksburg, Locustdale, Bowling Green, Va., and Washington; M.D., U. of Va., 1896; student Pathologische Institut, Vienna; Thompson Yates Labs., Liverpool, Eng., and Liverpool Sch. of Tropical Medicine, 1899-1901; m. Lucy Temple Hundley, Nov. 6, 1899; children—John Layton (dec.), Richard Hundley, Beverly Whiting (dec.). Asst. surgeon U.S.P.H. and Marine Hosp. Service, 1898; passed asst. surgeon, 1903; on epidemic duty in connection with yellow fever, 1898; quarantine officer Dry Tortugas, 1898-99; immigrant insp., Ellis Island, 1899; sanitary observer at Glasgow, Oporto and Liverpool, 1899-1900; sanitary attaché U. S. consulates-gen., Barcelona, Marseilles, Vienna, London and Liverpool 1899-1901; asst. dir. Hygienic Lab., Washington, 1902-09, dir. 1909-16; dir. research and biological labs., E. R. Squibb & Sons, 1916, also v.p.; vice-pres. chmn. bd. New Brunswick Savings Inst. Fellow A.A.A.S.; mem. Am. Assn. Pathologists and Bacteriologists (past pres.), Soc. Am. Bacteriology, Soc. Exptl. Medicine and Biology, A.M.A., Am. Pub. Health Assn. (past pres.), Assn. Am. Physicians, Phi Beta Kappa. Author tech. articles. Described Rocky Mountain Fever and tick responsible for it, 1903; described first mech. water filters, 1893; (with J. Gold) transmitted measles to monkeys, 1911; (with Rosenau) showed difference between anaphylaxie and antitoxin formation, 1906. Died Sept. 30, 1958.

ANDERSON, John M., Am. logician, engr.; b. Cedar Rapids, Ia., July 29, 1914; B.A. with highest honors, U. Ill., 1935, M.A., 1936; Ph.D., U. Cal. at Berkeley, 1939; m. 1936. With Sears Roebuck and Co., 1939-41; research engr. Elgin Nat. Watch Co., 1941-42; with Mpls.-Honeywell Regulator Co., 1942-45, adminstrv. engr., 1945; lectr. math. U. Minn., 1945-46; cons. engr. operations research H R B-Singer, Inc., 1950-51; faculty Pa. State U., 1946—, prof. philosophy, 1951—, head philosophy dept., 1948-49, 52-55, 58—, acting asst. dean for research, 1964; guest prof. U. New Zealand, Otago, 1955, Free U. Berlin, Germany, 1960-61. Mem. Math. Assn. Am., Am. Assn. U. Profs., Am. Philos. Assn., Assn. for Symbolic Logic, Western Pa. Philos. Soc. (past pres.), Phi Beta Kappa, Sigma Xi, Pi Gamma Mu, Alpha Tau Omega. Author: The Individual and the New World, 1955; Calhoun: Basic Documents, 1952; co-author: Industrial Management, 1942; Natural Deduction: The Logic of Axiom Systems, 1962; also articles, essay. Patentee polynomial equation computer, vector computer. Application of analogue computers to ballistics; devel. of math. models for social and tactical problems; theory of automatic controls; stress analysis of springs; theory of deductive systems. Home: 321 Corl St., State College, Pa. 16801. Office: Boucke Bldg., University Park, Pa. 16802.*

ANDERSON, Joseph Tomlinson, Am. biochemist; b. Hilton, N.Y., Dec. 10, 1909; s. Arthur R. and Grace (Tomlinson) A.; B.S., U. Rochester, 1930, M.S., 1932, Ph.D., 1947; m. Gwendolyn R. Peck, Nov. 27, 1935; children—Owen L., Jan Ellen. Instr., Med. Sch., U. Rochester, 1947-50; from asst. prof. to prof. Lab. Physiol. Hygiene, U. Minn., Mpls., 1950—, prof., 1959—. Mem. Am. Chem. Soc., Am. Inst. Nutrition, Am. Oil Chemists Soc., Fellowship of Reconciliation, Phi Beta Kappa, Sigma Xi. Research, publs. on relationships between diet and type and amount of lipids in blood serum and their correlation to coronary heart disease. Home: 1965 Autumn St., St. Paul 55113. Office: Lab. Physiol. Hygiene, U. Minn., Stadium Gate 27, Mpls. 55455.*

ANDERSON, Kenneth Ellsworth, Am. microbiologist; b. Ithaca, N.Y., Dec. 21, 1914; s. Grover M. and Bessie (Hunt) A.; B.S., Cornell U., 1937, Ph.D., 1943; M.S., U. N.H., 1940; m. Agnes E. Schieckel, Feb. 7, 1942; children—Kenneth, Vincent, Joan, Mary, Paul, Kathleen, Agnes, Patricia, William, John, Theresa. Chemist, bacteriologist N.Y. Water Service Corp., N.Y.C., 1937-38; bacteriologist Genesee Brewing Co., Rochester, N.Y., 1946; faculty St. Bonaventure (N.Y.) U., 1946—, prof., chmn. dept. biology, 1949—, mem. Bd. Instrn., 1952—, dean Sch. Arts and Scis., 1966—. Cons. microbiology problems relating to oil prodn. Mem. A.A.A.S., Soc. Am. Microbiologists, U. S. Pub. Health Assn., N.Y. Acad. Scis. Editor Sci. Studies 1947—. Research in bactericides and oil floods, physiology of Desulfovibrio desulfuricans, and others. Home: 96 Rock City Rd., Olean, N.Y. 14760. Office: Dept. Biology, St. Bonaventure U., St. Bonaventure, N.Y.*

ANDERSON, Kinsey A., Am. physicist; b. Preston, Minn., Sept. 18, 1926; s. Melvin R. and Allene (Michener) A.; B.A., Carleton Coll., Northfield, Minn., 1949; Ph.D., U. Minn., 1955; m. Lilica Athena Vassiliades, May 29, 1954; children—Danae, Sindri. Guggenheim fellow Royal Inst. Tech., Stockholm, Sweden, 1959-60; faculty U. Cal., Berkeley, 1960—, prof. physics, 1966—. Cons., NSF, NASA Mem. Am. Geophys. Union, Am. Phys. Soc., Phi Beta Kappa, Sigma Xi. Contbr. numerous articles to profl. jours. Research in direct measurement of low energy solar cosmic rays, auroral zone particles with balloons and rockets, satellite measurements of radiation zones and solar particles in interplanetary space. Home: 8321 Buckingham Dr., El Cerrito, Cal. 94530. Office: Physics Dept., U. Cal., Berkeley, Cal. 94720.*

ANDERSON, Laurens, Am. biochemist; b. Belle Fourche, S.D., May 19, 1920; s. Adolph and Mary E. (Slaughter) A.; B.S. in Chemistry, U. Wyo., 1942; M.S., U. Wis. in Biochemistry, 1947, Ph.D., 1950; m. Doris Elaine Young, Sept. 15, 1945; children—Eric E., Karl A., Kristin E. Merck fellow in natural scis. Eidg. Technische Hochschule, Zurich, Switzerland, 1950-51; faculty dept. biochemistry U. Wis., Madison, 1951—, prof., 1961—. Mem. Am. Chem. Soc. (past chmn. Wis. sect.), Am. Soc. Biol. Chemists, Sigma Xi. Research, publs. on organic chemistry and metabolism of cyclitols and other carbohydrates.*

ANDERSON, Leigh Charles, Am. chemist; b. Muskegon, Mich., Mar. 28, 1899; s. Pere Rudolph and Josephine (Torgeson) A.; B.S. in Chemistry, U. Mich, 1924, M.S. in Chemistry, 1922, Ph.D., 1924; m. E. Alloa Caviness, Feb. 2, 1925; children—Leighton Charles, Peter Caviness, Robert Keith. Faculty Chemistry U. Mich., Ann Arbor, 1924—, prof., 1944—, chmn. dept. 1948-66. Recipient Silver Beaver award Boy Scouts Am., 1953. Fellow A.A.A.S.; mem. Am. Chem. Soc., Sigma Xi, Phi Lamba Upsilon, Gamma Alpha, Alpha Chi Sigma. Author: (with Robert C. Elderfield, Peter A. S. S. Smith, Werner E. Bachman), A Manual for the Organic Chemistry Laboratory, 1953, rev., 1960. Contbr. articles to sci. jours. Home: 1120 Lincoln Av., Ann Arbor, Mich. 48104.*

ANDERSON, Lewis Edward, Am. botanist; b. Batesville, Miss., June 16, 1912; s. William Carl and Bessie May (Hawkins) A.; B.S., Miss. State U., 1931; A.M. Duke, 1933; Ph.D., U. Pa., 1936; m. Ruth Lucille Geckler, June 23, 1941; children—Sarah Jane (Mrs. Roger Miller), Philip Carlton, Nancy Ruth (Mrs. David Goodridge), Carol May (Mrs. Marshall Crosby), David Edward. Instr. in botany U.

Pa., 1935-36; faculty Duke U., 1936-—; prof., 1954-—. Contbr. numerous articles on the systematics and cytology of bryophytes to tech. jours. Home: 2020 Sunset Av., Durham, N.C.*

ANDERSON, Marlowe George, Am. zoologist, educator; b. Oakland, Neb., July 8, 1908; s. George E. and Anna (Anderson) A.; A.B., Neb. Wesleyan Coll., 1928; M.A., Northwestern U., 1930, Ph.D., 1934; m. Florence Mendenhall, Dec. 23, 1933; children—George T., Christopher M. Instr., Northwestern U., 1934-36; asst. prof. No. Mont. Coll., 1936-37; faculty zoology N.M. State U., University Park, 1937-—, prof., 1951-—, head dept. biology, 1955-—. Fellow A.A.A.S. (exec. sec. Southwestern and Rocky Mountain div. 1957-—), Am. Soc. Parasitology, Am. Soc. Zoologists, Soc. Systematic Zoology, Am. Microscopical Soc., Am. Inst. Biol. Scis., N.M. Acad. Sci., Sigma Xi, Phi Kappa Phi, Beta Beta Beta, Pi Gamma Mu, Sigma Pi. Contbr. articles to profl. jours. Research on germ cell cycles and life histories of Trematoda. Home: 2510 S. Solano Dr., Las Cruces, N.M. 88001. Office: P.O. Box 3 AF, University Park Br., Las Cruces, N.M. 88001.*

ANDERSON, Norman Carl, Am. physicist; b. Ludington, Mich., May 8, 1918; s. Carl and Bertha (Olsen) A.; B.S., Northwestern U., 1942, M.S., 1947; m. Lilli-Anne Luria, Sept. 27, 1941; children—Jere, Steven, James. Asso. physicist Northwestern U., Evanston, Ill., 1943-47; physicist Continental Electric Co., Chgo., 1947-48; v.p. Electronics Corp. Am., Cambridge, Mass., 1948-57, Infrared Industries, Inc., Santa Barbara, Cal., 1957-—; v.p., dir. Electro-Nuclear Labs., Inc., Mountain View, Cal. Mem. Am. Phys. Soc., Optical Soc. Am. Research on reproducible photoconductors; process and techniques for making thallium sulfide and lead salts efficiently. Home: 62 Greylock Rd., Wellesley, Mass. 02181. Office: 62 4th Av., Waltham, Mass. 02154.

ANDERSON, Oskar Nikolai, statistician; b. Minsk, Russia, Aug. 2, 1887; s. Nikolai A.; ed. Univ. of Kazan; Technical Univ., St. Petersburg; m. Margarethe von Hindenburg; 2 sons. Asst. in statistics, Technical Univ., St. Petersburg, 1911-15; lectr. mathematical statistics, Univ. Kiev, 1918-19; prof. economics and statistics, Commercial Univ. of Varna, 1924-33; Rockefeller Foundation Fellowship, 1934; dir., Statistical Inst. for Economic Research, Univ. Sofia, 1935-42; prof. statistics, Univ. Kiel, 1942-47; prof. statistics, Univ. Munich, from 1947. Coeditor Metrika. Hon. fellow, Royal Statistical Soc., London; Deutsche Statistische Gesellschaft; fellow, Econometric Soc.; Inst. of Mathematical Statistics; A.A.A.S.; Am. Statistical Assn.; International Statistical Inst. Author: Zur Problematik der empirisch-statistischen Konjunkturforschung, 1929; Die Korrelationsrechnung in der Konjunkturforschung, 1929; Einführung in die mathematische Statistik, 1935; Struktur und Konjunktur der bulgarischen Volkswirtschaft, 1938; Probleme der statistischen Methodenlehre in den Sozialwissenschaften, 1954. Died 1960.

ANDERSON, Paul Julius, Am. physician; b. Akron, O., Oct. 11, 1925; s. Julius E. and Helen (Lang) A.; B.S., Ohio U., 1949; M.D., U. Chgo., 1953; m. Joan E. Cross, Sept. 9, 1950; children—Jon Brian, Colin Evers, Kell Erland. Faculty, Columbia, N.Y.C., 1958-—, asst. prof. neuropathology, 1961-—; staff Mt. Sinai Hosp., N.Y.C., 1954-—, asso. attending pathologist, dir. div. neuropathology, 1961-—, asso. attending neurologist, 1962-—; prof. neuropathology, asso. prof. neurology Mt. Sinai Med. Sch., 1966-—. Cons. pathologist City Hosp., Elmhurst, N.Y., 1964-—. Diplomate Am. Bd. Psychiatry and Neurology, Am. Bd. Pathology. Fellow N.Y. Acad. Medicine; mem. Am. Acad. Neurology, Am. Assn. Neuropathologists, Am. Assn. Pathologists and Bacteriologists, Am. Neurol. Assn., Am. Soc. for Exptl. Pathology, Assn. for Research in Nervous and Mental Diseases, Council Biology Editors, Histochem. Soc., Soc. Med. Jurisprudence. Author: (with Dr. Tibor Barka) Histochemistry, Theory, Practice, and Bibliography, 1963. Asso. editor Jour. Neuropathology and Exptl. Neurology, 1961-—, Jour. Histochemistry and Cytochemistry, 1965-—. Research, publs. in field of neuropathology, with emphasis on neuromuscular diseases, using techniques of histology, histochemistry, and electron microscopy; devised histochem., electrophoretic and chromatographic methods for study of enzymes in neural and muscle tissue. Home: 470 West End Av., N.Y.C. 10024. Office: care Dept. Neuropathology, Mt. Sinai Hosp., 1 E. 100th St., N.Y.C. 10029.*

ANDERSON, Philip Warren, Am. physicist; b. Indpls., Dec. 13, 1923; s. Harry W. and Elsie (Osborne) A.; B.S., Harvard, 1943, M.A., 1947, Ph.D., 1949; m. Joyce Gothwaite, July 31, 1947; 1 dau., Susan Osborne. Staff, Naval Research Lab., 1943-45; mem. tech. staff Bell Telephone Labs., Murray Hill, N.J., 1949-—, chmn. theoretical physics dept., 1959-60. Loeb lectr. Harvard, 1964; prof. theoretical physics U. Cambridge (Eng.), 1967-—. Recipient Oliver E. Buckley prize, 1964; Fulbright lectr., U. Tokyo, 1953-54, Overseas fellow Churchill Coll., Cambridge, 1961-62. Fellow Am. Phys. Soc., Am. Acad. Arts and Scis.; mem. Nat. Acad. Scis., Phys. Soc. Japan. Author: Concepts in Solids, 1963. Research in quantum theory, especially theoretical physics of solids, spectral line broadening, magnetism, superconductivity. Office: Bell Telephone Lab., Murray Hill, N.J.*

ANDERSON, Richard Davis, Am. mathematician; b. Hamden, Conn., Feb. 17, 1922; s. John Edward and Dorothea (Lynde) A.; B.A., U. Minn., 1941; Ph.D., U. Tex., 1948; m. Jeanette Olliver, Apr. 12, 1943; children—Susan C., Virginia D., Richard D., Charlotte M., David M. Instr., U. Tex., 1941-42, 45-48; faculty U. Pa., Phila., 1948-56, asso. prof., 1955-56; prof. La. State U., Baton Rouge, 1956-59, Boyd prof. 1959-—; mem. Inst. for Advanced Study, Princeton, N.J., 1951-52; 1955-56; Colleague Mathematisch Centrum, Amsterdam, 1962-63. Writer, panelist Sch. Math. Study Group, 1959-—; panelist Com. on Undergrad. Program in Math., 1962-—, mem., 1965-—. Alfred P. Sloan Research fellow, 1960-63. Mem. A.A.A.S., Math. Assn. Am., Am. Math. Soc. Research in point set theory, dimension theory, structure of continua, homeomorphism theory. Home: 2954 Fritchie Dr. Baton Rouge, La. 70809.*

ANDERSON, Richard John, physician; b. Ballybot, No. Ireland, July 29, 1848; s. Robert and Elizabeth (Harcourt) A.; ed. Queen's Coll., Belfast (No. Ireland) hosps.; M.A., M.D.; m. Hannah Perry, 1889. Worked in Leipzig, London, Paris, Heidelberg, Naples; held med. and san. position, 1873-75; demonstrator, lectr. anatomy, 1875-83; prof. natural sci., including comparative anatomy, from 1883; prof. natural history, geology, mineralogy Queen's Univ. Coll., Galway, Eire; clin. lectr., attendant County Galway Infirmary, 1890-91; examiner in geology, biology, zoology, botany Nat. U. Ireland; examiner in geology Royal U. Ireland, 1900-09. Hon. pres. sect anatomy XV Internat. Congress Medicine, Lisbon, Portugal, 1906, sect. physiology XVI Congress, Budapest, Hungary, 1909. Fellow Linnean Soc.; mem. Royal Coll. Surgeons. Author booklets, pamphlets, over 300 papers. Coconductor Internat. Jour. Anatomy and Physiology, from 1884; original collaborator Anat. Anzeiger, 1887. Inventor revolving microscopic arrangement. Died July 24, 1914.

ANDERSON, Richard Orr, Am. fishery biologist; b. Evanston, Ill., Oct. 23, 1929; s. Anders A. and Grace (Orr) A.; student Beloit Coll., 1947-50; B.S., U. Wis., 1951; M.S., U. Mich., 1953; Ph.D., 1959; m. Joan L. Veh, June 13, 1951; children—Barbara Jo, Donna Lyn. Conservation aide Wis. Conservation Dept., Madison, 1951; research technician Inst. for Fisheries Research, Mich. Dept. Conservation, 1952-58, fishery biologist, Mattawan, 1959-63; asst. prof. U. Mich., Ann Arbor, 1958-59; leader Coop. Fishery Unit, Bur. Sport Fish and Wildlife, asst. prof. zoology, U. Mo., Columbia, 1963-—. Mem. Am. Fisheries Soc., Am. Soc. Limnology and Oceanography, Internat. Assn. Theoretical and Applied Limnology, Mich. Assn. Conservation Ecologists, Mo. Acad. Sci., Sigma Xi., Phi Sigma. Study of dynamics of benthic populations, 1953-55, exptl. studies on growth of fish, 1956-58, fish cultural research, 1959-63. Home: 605 Seymour Rd., Columbia, Mo. 65201.*

ANDERSON, Robert, mathematician; flourished in Eng., 2d half 17th century; experimented with cannon mounted at Wimbledon, Eng.; Author: Stereometrical Propositions variously applied, but Particularly intended for Gageing, 1668; The Genuine Use and Effects of the Gunne, 1674; also wrote on fall of heavy bodies.

ANDERSON, Robert van Vleck, Am. geologist; b. Galesburg, Ill., Apr. 18, 1884; s. Melville Best and Charlena (van Vleck) A.; B.A., Stanford, 1906; m. Gracella Rountree, Mar. 1923; children—Robert Playfair, Patricia Sage (Mrs. John Loveland Armstrong), Gracella Gurnee. Geol. and zool. work, Japan, 1905; assistant in investigations of Cal. earthquake for Carnegie Inst., 1906; geologic aid, asst. geologist and geologist U. S. Geol. Survey, 1906-13; made investigations of geol. and petroleum resources of Cal., also mem. Oil Land Classification Bd.; cons. geologist in various fgn. countries, from 1911; geologist for S. Pearson & son, Ltd., London, Eng., 1913-18; rep. U. S. War Trade Bd. in Sweden; Am. del. Inter-Allied Trade Com., Stockholm, 1918-19; dir. Whitehall Petroleum Corp., Ltd., London, 1919-23, chief geologist, 1923-26. Engaged in independent scientific work, 1927-34. Collaborateur, Service de la Carte Geologique de l'Algérie, 1930-32; research in Algeria under grant from Geol. Soc. America, 1933; with Socony Vacuum Oil Co., Inc., 1934-44; research asso. Stanford U., from 1945. Fellow Geol. Soc. Am. (chmn. Cordilleran sect. 1928), Cal. Acad. Scis.; mem. Paleontol. Soc. Am., Assn. Am. Geographers, Soc. Econ, Geologists, Archaeol. Inst. Am., Am. Assn. Petroleum Geologists, Société Geologique de France. Sigma Xi. Author: Geology in Coastal Atlas of Western Algeria; also various reports pub. by U. S. Govt., and other scientific papers. Died June 6, 1949.

ANDERSON, Rose Gustava, Am psychologist; b. Gothenburg, Neb., June 23, 1893; d. Mathew and Emily (Axling) A.; A.B., U. Neb., 1917, M.A., 1919; Ph.D., Columbia, 1925. Research asst. Minn. Dept. Instns., St. Paul, 1919-23; chief psychology Mpls. Bd. Edn., 1925-29; dir. edn. adjustment bur. Westchester (N.Y.) County Children's Assn., 1930-32; dir. children's div., cons. to women Psychol. Corp., N.Y.C., 1932-42, dir. Psychol. Service Center, 1942-57, v.p. Psychol. Corp., 1942-55; ind. research Personnel Press, Princeton, N.J., 1958-62. Diplomate Am. Bd. Examiners. Fellow Am. Psychol. Assn. (past council rep., past pres.), Am. Orthopsychiat. Assn.; mem. Am. Personnel and Guidance Assn. Author: A Critical Examination of Test-Scoring Methods, 1925; Kuhlmann-Anderson Tests of Academic Potential; also articles. Research on improvement clin. and counselling procedures and their more effective use in diagnosis and counselling individuals. Home: 15-01 U Meadow Lakes, Hightstown, N.J. 08520.*

ANDERSON, Rubert Sigfred, Am. chemist, physiologist; b. Spokane, Wash., June 12, 1898; s. Gustav A. and Lavinia (Tilderquist) A.; B.S., U. Wash., 1921; M.A., Columbia, 1923, Ph.D., 1925. Physiologist, Med. Labs. Army Chem. Center, Edgewood, Md., 1953-—. Fellow N.Y. Acad. Scis.; mem. Am. Physiol. Soc., Am. Chem. Soc., Soc. Gen. Physiologists, Radiation Research Soc., Sigma Xi. Research on competitive and noncompetitive inhibition of an enzyme by the products of the reaction. Home: Box 632, Edgewood, Md. 21040.*

ANDERSON, Sydney, Am. zoologist; b. Topeka, Jan. 11, 1927; s. Robert Grant and Evelyn Fern (Hunt) A.; student Baker U., 1944-49; B.A., U. Kan., 1950, M.A., 1952, Ph.D., 1959; m. Ratia Justine Klusmire, Aug. 5, 1951; children—Evelyn Lee, Charles Sydney, Laura Lynnette. Asst. curator Mus. Natural History, instr. dept. zoology U. Kan., 1955-59; asst. curator dept. mammalogy Am. Mus. Natural history, N.Y.C., 1960-64, asso. curator., 1964-—. Tech. cons. pvt. cos., 1962-—. NSF fellow, 1952-54. Fellow A.A.A.S.; mem. Am. Soc. Mammalogists, Deutsche Gesellschaft fur Saugertierkunde, Ecol. Soc. Am., Soc. for Study Evolution, Soc. Systematic Zoology, Wildlife Soc. Research, publs. on distbn., variation and relationships mammals. Home: 98 Wainwright Av., Closter, N.J. 07624. Office: Central Park W. at 79th St., N.Y.C. 10024.*

ANDERSON, Theodore Wilbur, Jr., Am. statistician; born Minneapolis, June 5, 1918; s. Theodore Wilbur and Evelynn (Johnson) A.; student North Park Coll., 1935-37; B.S. with highest distinction, Northwestern U., 1939; M.A., Princeton, 1942, Ph.D., 1945; m. Dorothy Fisher, July 8, 1950; children—Robert Lewis, Janet Lynn, Jeanne Elizabeth. Asst. dept. math. Northwestern U., 1939-40; instr. math. Princeton, 1941-43, research asso., 1943-45; research asso. Cowles Commn., U. Chgo., 1945-46; faculty Columbia, N.Y.C., 1946-—, prof., 1956-—, chmn. dept. math. statistics, 1956-60, 64-—, dir. project Office Naval Research, 1955. Cons. Rand Corp., 1949-—; chmn. com. on statistics NRC, 1961-63, mem. NRC Adv. Panel to applied math. div. Nat. Bur. Standards, 1964-—. Guggenheim fellow, 1947-48; fellow Center for Advanced Study in Behaviorial Scis., 1957-58. Fellow Am. Statis. Assn, Econometric Soc., Royal Statis. Soc., A.A.A.S., Inst. Math. Statistics (pres. 1963); mem. Am. Math. Soc., Biometric Soc., Conf. Bd. Math. Scis. (exec. com.), Am. Assn. U. Profs., Psychometric Soc. (dir.). Author: An Introduction to Multivariate Statistical Analysis, 1958. Editor, Annals Math. Statistics, 1950-52; editorial bd. Psychometrika, 1954-—. Contbr. to math. theory of statis. inference; devel. new statis. methods in areas of multivariate analysis, sequential tests, time series analysis, multiple decision procedures, factor analysis, goodness-of-fit tests. Home: 35 Hickory Hill Rd., Tappan, N.Y. 10983. Office: Dept. Math. Statistics, Columbia U., N.Y.C. 10027.*

ANDERSON, Thomas, botanist; b. Edinburgh, Scotland, Feb. 26, 1832; M.D., Edinburgh, 1853. Entered Bengal med. service, 1854; served in Calcutta, also Delhi, India; in Calcutta 1858; became acting dir., later dir. Calcutta Botanic Garden, 1860. Author: Florula Adenensis (on plants collected in Aden), 1860. Introduced cinchona, ipecacuanha, other med. plants in Calcutta; instituted expts. leading to successful cultivation of cinchona in India. Died Edinburgh, Oct. 26, 1870.

ANDERSON, Thomas, Brit. chemist; b. Leith, Scotland, July 2, 1819; M.D., U. Edinburgh, 1841; student of Berzelius in Stockholm, Sweden, 1842, of Liebig in Giessen, Germany, 1843, also studied in Bonn, Berlin, Vienna. Became extra-academical univ. tchr. chemistry, 1846; apptd. chemist Highland and Agrl. Soc. Scotland, 1848; Regius prof. chemistry U. Glasgow, from 1852. Fellow Royal Soc. Edinburgh (Keith medal 1855, Royal medal 1872); mem. Glasgow Philos. Soc. (pres. 1859). Brit. Assn. for Advancement Sci. (pres. chem. sect. 1867). Editor, Edinburgh New Philos. Jour. Contbd. papers to sci. lit. Discovered pyridine, other constituents of bone oil, and a new organic sulfide; investigated constitution of anthracene, also agrl. chemistry. Died Chiswick, Eng., Nov. 2, 1874.

ANDERSON, Thomas Foxen Am. biophysicist, educator; b. Manitowoc, Wis., Feb. 7, 1911; s. Anton Oliver and Mabel (Foxen) A.; B.S., Cal. Inst. Tech., 1932, Ph.D., 1936; m. Wilma Fay Ecton, Dec. 28, 1937; children—Thomas Foxen, Jessie Dale. Research in phys. chemistry U. Munich (Germany), 1932-33; instr. chemistry U. Chgo., 1936-37; investigator in botany U. Wis., Madison, 1937-39, instr. chemistry, 1939-40; RCA fellow NRC, Camden, N.J., 1940-42; faculty biophysics U. Pa., Phila., 1942-—, prof., 1957-—, sr. mem. Inst. Cancer Research, 1958-—. Chmn., U. S. Nat. Com. on Pure and Applied Biophysics, 1964-—; mem. commn. on molecular biophysics Internat. Orgn. for Pure and Applied Biophysics, 1964-—. Fulbright and Guggenheim fellow Inst. Pasteur, Paris, France, 1955-57, recipient Silver medal, 1957. Mem. Nat. Acad. Scis., Soc. Francaise de Milcroscopie Electronique (hon.), Internat. Fedn. Electron Microscope Socs. (pres. 1960-64), Biophys. Soc. (pres. 1965), Electron Microscope Soc. Am. (pres. 1955). Contbr. articles to profl. jours. Developed techniques for use of electron microscope in biology, determined structures of viruses and bacteria and their relation to function; research on genetics of viruses and bacteria. Home: 326 Zane Av., Phila. 19111.*

ANDERSON, Thomas Joel, Jr., Am. economist; b. Blue Springs, Mo., Sept. 22, 1898; s. Thomas Joel and Susan (Sammons) A.; B.S. in Bus. Adminstrn., U. Mo., 1922, A.M., 1923; Ph.D., N.Y.U., 1934; m. Jo Hutchinson, Aug. 27, 1922; children—Mary Maude (Mrs. James T. Downey, Jr.), Betty Jo (Mrs. Robert S. Bassett). From instr. to prof. econs. Kan. State Coll. A. and M., Manhattan, 1922-29; mem. faculty N.Y.U., 1929—, prof. econs., 1948—, chmn. dept. econs. Sch. Commerce, 1959-65. Mem. Am. Econ. Assn., Phi Kappa Phi, Beta Gamma Sigma. Author: Federal and State Control of Banking, 1934; Our Competitive System and Public Policy, 1958. Editor: (with others) General Economics: A Book of Readings, rev. edit., 1963. Research on constl. powers of Congress over currency, banking and securities distbn.; Am. policy affecting the dual banking system; competition, restraint of trade, market monopoly; devel. of pub. policy affecting these conditions. Died June 30, 1967.

ANDERSON, Sir Thomas M'Call, Scottish physician; b. Scotland; ed. Edinburgh, Glasgow, Dublin, Paris, Würzburg, Berlin, Vienna; M.D., Glasgow. Prof. clin. medicine, 1874-1900, Regius prof. medicine, from 1900, univ. rep. on Gen. Med. Council, 1903; examiner in medicine and pathology Brit. and Indian med. services, 1897-1901; hon. physician in Scotland to King, from 1907; sr. physician Western Infirmary Glasgow. Hon. mem. dermatol. socs. Vienna, Am.; fgn. corr. mem. Société Francaise de Dermatologie et de Syphilographie. Author: On Psoriasis and Lepra, 1865; The Parasitic Affections of the Skin, 2d edit., 1868; Treatment of Diseases of the Skin, with an Analysis of 11,000 Consecutive Cases, 1872; On Eczema, 3d edit., 1874; Curability of Attacks of Tubercular Peritonitis and Acute Phthisis (Galloping Consumption), 1877; On Syphilitic Affections of the Nervous System: their Diagnosis and Treatment, 1889; A Treatise on Diseases of the Skin, 2d edit., 1894; Contributions to Clinical Medicine, 1898. Died Jan. 25, 1908.

ANDERSON, Victor . Elving, Am. geneticist; b. Stromsburg, Neb., Sept. 9, 1921; s. Edwin L. and Olga (Elving) A.; A.A., Bethel Jr. Coll., 1941; student Bethel Theol. Sem., 1941-43; B.A., U. Minn., 1945, M.S., 1949, Ph.D., 1953; m. Carol Esther Rexion, Aug. 31, 1946; children—Catherine, Carl, Christine, Martha. Faculty dept. biology Bethel Coll., 1946-60; asst. dir. Dight Inst. for Human Genetics U. Minn., 1954—, asso. prof. zoology and genetics, 1961-66, prof. genetics, 1966—. Cons. Nat. Inst. Neurol. Disease and Blindness. Named Alumnus of Yr., Bethel Coll. and Sem., 1965. Mem. A.A.A.S. (pres. Acad. Conf. 1967), Am. Soc. Human Genetics (dir.), Minn. Acad. Sci. (pres. 1964-65), Am. Sci. Affiliation (pres. 1963-65), Phi Beta Kappa, Sigma Xi. Author (with H. O. Goodman and S. C. Reed) Variables Related to Human Breast Cancer, 1958, also articles. Research on genetics in human behavior and mental retardation. Home: 1775 N. Fairview Av., St. Paul 55113. Office: Dept. Zoology U. Minn., Mpls. 55455.*

ANDERSON, Walter Stirling, Canadian surgeon; b. Kenton, Man., Can., Sept. 16, 1910; s. Andrew Frankland and Mary English (Young) A.; B.Sc. Arts, U. Alta., 1930; M.D., U. Toronto, 1934, M.S., 1938; m. Lilian Ferris Hoar, May 28, 1941; children—Marjorie Carolyn, Mary Lynn, Jaqueline Green, Andrew Robert, William Walter. Chief dept. surgery Baker Clinic, Edmonton, Alta., Can., 1946—; head dept. surgery Royal Alexandra Hosp., Edmonton, 1950—, in charge grad. teaching, 1956—; prof. surgery, faculty of medicine U. Alta., Edmonton, 1955-65. Diplomate Am. Bd. Surgery. Fellow A.C.S., Royal Coll. Surgeons; mem. Alta. (pres. 1955-56), Canadian (chmn. cancer com. Alta. div. 1949-54, chmn. hosp. relations and profl. services com. 1956—, bd. dirs. 1954—), med. assns., Phi Gamma Delta, Alpha Omega Alpha. Home: 13723 Summit Point. Office: 10004-105th St., Edmonton, Alta., Can.*

ANDERSON, William, surgeon, naturalist; surgeon's mate in Capt. Cook's ship Resolution, 1772-75, naturalist on 3d voyage. Genus Andersonia named in his honor by Robert Brown; Anderson's Island named in his memory. Author manuscript lists of animals and plants noted during voyages (in Banksian Library, Brit. Mus.); 2 papers on poisonous fish and detached rock nr. Cape Town in Philos. Trans. Died Aug. 3, 1778.

ANDERSON, Sir William, engr.; b. St. Petersburg, Jan. 5, 1835; ed. King's Coll., London; D.C.L., Durham, Eng.; m. Emma Brown, 1856. Apprentice W. Fairbairn and Sons, Manchester, Eng.; mgr., then partner Courtney Stephens and Col., Engrs., Dublin, Ireland, built Erith Ironworks after joining Easton and Amos, Engrs., London, in 1864; dir.-gen. Royal Ordnance Factories, Woolwich, Eng., from 1889. Fellow Royal Soc., 1891. Author: Lectures on Hydraulics and on Hydropneumatic Gun-Carriages; Conversion of Heat into Work. Died Dec. 11, 1898.

ANDERSON, William Arnold Douglas, physician; b. Ontario, Can., Aug. 27, 1910; s. Thomas H. P. and Lottie H. (Cook) A.; B.A., U. Toronto, 1931, M.D., 1934, M.A., 1936; m. Harriott I. Gates, Sept. 3, 1934; children—Douglas Richard, Mary Carolyne (Mrs. George Durkin), Joan Christine, Judith Estelle. Practice medicine specializing in pathology; prof. pathology Marquette U. Sch. Medicine, Milw., 1945-53, U. Miami (Fla.) Sch. Medicine, 1953—; dir. path. labs. Jackson Meml. Hosp., Miami, 1953—; vis. prof., U. Capetown, South Africa, 1961. Recipient Sci. Products Found. award, 1959; Ward Burdick award Am. Soc. Clin. Pathologists, 1965. Mem. Coll. Am. Pathologist (pres. 1956-57). Author: Synopsis of Pathology, 1942, 46, 52, 57, 60, 64; Pathology, 1948, 53, 57, 61. Research on oncology, endocrine disorders, and diseases of kidney. Home: 2701 Columbus Blvd., Coral Gables, Fla. 33134. Office: Jackson Meml. Hosp., Miami, Fla. 33136.*

ANDERSSON, Göran Karl, Swedish physicist; b. Ludvika, Sweden, Aug. 24, 1920; s. Gösta and Signe (Lindholm) A.; Fil.Dr., U. Uppsala (Sweden), 1957; m. Karin Svenson, June 22, 1947; children—Hans, Ingrid. Head div. Gustaf Werner Inst., Uppsala, 1955-61; research asso. CERN, Geneva, Switzerland, 1961-62, cons., 1959-60, 63; asso. prof. Chalmers U. Tech., Gothenburg, Sweden, 1964—; profl. asso. Nat. Acad. Scis., 1957-58. Research, publs. on nuclear spectroscopic studies of short-lived, neutron deficient isotopes of lead, thallium, mercury, xenon, iodine, tellurium with application of electromagnetic isotope separation; design of fringing field, point focusing type of isotope separator. Home: 14 Ulvasvägen, Partille, Sweden. Office: Dept. Physics, Chalmers U. Tech., Göteborg 5, Sweden.*

ANDERSSON, Nils Johan, Swedish botanist; b. Feb. 20, 1821; student botany, Stockholm, Sweden; gathered sci. information on Swedish expdn. around world, 1851-53; prof. at Lund, also Stockholm; attendant bot. collections Acad. Scis.; author works on willows and flora, also on Scandinavia and Lapland botany. Died Stockholm, Mar. 27, 1880.

ANDING, Ernst, German astronomer; b. Seebergen nr. Gotha, Germany, Aug. 11, 1860; s. Wilhelm and Martha (Tenner) A.; ed. Jena, also Munich, Germany; degree in astronomy, 1888; m. Bertha Echsle, Jan. 28, 1901; m. 2d, Marie Schmidt, Dec. 18, 1924. Became private lectr., Munich, 1895; observer Bavarian Com. for Internat. Geodesics, 1896; apptd. asso. prof. Munich U., 1903; named dir. Obs. Gotha, 1906. Author: Kritische Untersuchung über die Bewegung der Sonne durch den Weltenraum, 2 vols., 1901-10; treatise on empirical (astron.) coordinate system and its relationship to ideal inertia system. Died Gotha, June 20, 1945.

ANDOYER, Marie Henri, French mathematician, astronomer; b. Paris, Oct. 1, 1862; student Ecole Normale Supérieure, 1881; D.Sc., 1886; master confs. U. Toulouse (France); master confs., then prof. astron. U. Paris, from 1903; mem. Bur. Longitudes, French Acad. Scis. Author: Nouvelles tables trigonométriques fondamentales (table improving and extending logarithms of trigonometric functions), 1911; works on celestial mechanics. Died Paris, June 12, 1929.

ANDRADE, Edward Neville da Costa, English physicist; b. London, Eng., Dec. 27, 1887; s. S. H. da C. Andrade; B.Sc. with 1st class honors U. London, 1907; Trouton scholar, Ellen Watson scholar, Jessel scholar Univ. Coll., London, 1907-10; 1851 Exhbn. scholar, 1910-13; Ph.D. summa cum laude Heidelberg U., 1911; studied Cavendish Lab., Cambridge, 1911-12, U. Coll., London, 1912-13; John Harling fellow U. Manchester, 1913-14; D.Sc., U. London; D.Sc. (hon.), Durham, Manchester; LL.D., U. Edinburgh; m. Katherine Barbara Evans, 1917, 2 children; m. 2d, Mona Wilkinson, 1938. Fellow Univ. Coll., London, 1916; prof. physics Arty. Coll., Woolwich, 1920-28; Quain prof. physics U. London, 1928-50, emeritus, 1950—; resident prof., dir. Davy-Faraday Research Lab., 1950-52. Sci. adviser to dir. sci. research Ministry of Supply, 1939-43; mem. Adv. Council Sci. Research and Tech. Devel. to Ministry Supply, until 1942. Recipient Holweck prize, 1947; Grande médaille Osmond, Société Francaise de Métallurgie, 1951; Hughes medal Royal Soc., 1958. Fellow Royal Soc., 1935, mem. council, 1942-44; fellow Inst. Physics; corr. French Acad. Scis., 1950. Author: The Structure of the Atom, 1923; Airs, 1924; The Atom, 1927; Engines, 1928; The Mechanism of Nature, 1930; (with Julian Huxley) Simple Science, 1934; The New Chemistry; The Atom and Its Energy; Isaac Newton; An Approach to Modern Physics; A Brief History of the Royal Society; Physics for the Modern World; Rutherford; also many papers on phys. and math. subjects. Editor for physics 14th edit. Ency. Brit. Research on viscosity of liquids; grain growth in metals; history of phys. sci. Home: 19, The Boltons, London S.W. 10. Office: Dept. of Metallurgy, Imperial College of Science, London S.W. 7, Eng.

ANDRADE, Jules Charles Frederick, French engr.; b. Paris, Sept. 4, 1857; ed. Ecole Polytechnique, l'Ecole d'application de Fontainebleu; D.Sc., 1890. Prof. engring. Rennes then Montpellier (France) faculties scis.; prof. engring Faculty Sci., Besancon, France, chmn. rational and applied mechanics, from 1902; founder sch. chronology, Besancon. Mem. French Acad. Scis., 1902. Author: Cours de mécanique physique, 1898; Chronométrie, 1908; Le mouvement, mesures et l'étendue et mesures du temps, 1911; Horlogerie et chronométrie, 1925. Well versed in analysis, geometry, mech. physics; investigated mechanics, clockmaking, chronometers; confirmed Reech's law. Died Brighton-Plage, France, Feb. 25, 1933.

ANDRADE, Zilton Araújo, Brazilian physician; b. Bahia, Brazil, May 14, 1924; s. Flavio Henrique and Guiomar (Araújo) A.; B.S., Bania Coll., Salvador, Brazil, 1944; M.D., U. Bahia, 1950; m. Sonia Lobao Gumes, June 27, 1953; children—Virginia, Flavio Henrique, Marúsia, Zilton, Carlos, Ivan. Pathologist, Fundaçao Gonçalo Moniz, Salvador, 1953-56, pathologist, hosp. prof. Edgard Santos, head dept. pathology U. Bahia, Salvador, 1953—, prof. pathology, 1959; asst. in pathology Faculty of Medicine, Ribeirao Preto, U. Sao Paulo, 1956; research fellow Mt. Sinai Hosp., N.Y.C., 1960. Rockefeller Found. fellow, 1951-53, NIH fellow, 1960-61. Mem. Brazilian Soc. Pathologists (pres. 1964-66), Internat. Acad. Pathology. Research, numerous publs. on aspects of immune pathology and pathogenesis of tropical diseases including schistosomiasis, Chagas' diseases, Kala-Azar. Home: 25 Rua Renato Medrado, Salvador, Bahia, Brazil.*

ANDRAL, Gabriel, French physician; b. Paris, Nov. 6, 1797; s. Guillaume Andral; Dr. Medicine, U. Paris, 1821. Appointed professor hygiene University of Paris, 1828, professor internal pathology, 1830, professor general pathology, also head department pathology, 1839-76. Mem. Acad. Medicine, French Acad. Scis., 1843. Author: Clinique médicale, 5 vols., 1823-27; Anatomie pathologique, 3 vols., 1829; Projet d'un essai sur la vitalité, 1835; Cours de pathologie interne, 1836-37; Sur la traitement de la fièvre typhoide par les purgatifs, 1837; Essais d'hemotologie pathologique, 1842; Traite élémentaire de pathologie et de therapeutique générale, 1843; Histoire de la medicine, 1852-54. Introduced term anemia into medicine; Andral's decubitus (position assumed in early stages pleurisy) named in his honor; credited with introducing term hyperemia, circa 1827; studied syphilitic changes of liver, 1834; treated heart diseases; used thermometry in hosps.; pub. (with Gabarret) figures for 6 cases intermittent fever, 1839; 1st to advocate examination blood in disease, possibly 1843; wrote probably 1st internal medicine textbook, 1st gen. comprehensive textbook of path. anatomy. Died Paris, Feb. 13, 1876.

ANDRÉ, Charles Louis François, French astronomer; b. Chauny, France, Mar. 14, 1842; D.Sc., 1876; became prof. astronomy U. Lyons (France), 1877; apptd. dir. obs., Lyons, 1880. Mem. French Acad. Scis., 1902. Author: L'astronomie pratique, 1874-78; Etude sur la diffraction dans les instruments d'optique, 1876; Observations du passage de Mercure sur le Soleil à Ogden (Utah), le 6 mai 1878, 1881; Traite d'astronomie stellaire, 1898-1901; Les planetes et leur origine, 1909. Made valuable observations on transit of Venus during visits to Rocky Mountains, also Australia. Died Lyons, June 6, 1912.

ANDRÉ, Emile, agronomist; b. Schnepfenthal, Mar. 1, 1790; s. Christian Charles André. Became forest conservator for Prince of Salm, 1807, forest ranger, Dietrichstein, 1812, chief forester Salmschen command, 1819, forest insp. Auerpergschen dominion, 1823; investigated agronomy in Prague, from 1825. Author: Essai d'organisation forestière selon les besoins de l'époque, 1823; Methode de culture forestière la plus simple, 1832. Studied forests in Austria-Hungary; developed methods for more econ. use of forests; investigated comml. use of beet sugar and byproducts. Died Kisberg, Hungary, Feb. 26, 1869.

ANDRÉ, Eugene, naturalist; b. Trinidad, Jan. 3, 1861; ed. St Mary's Coll., Port-of-Spain, Trinidad; collected orchids, Maturin, also Cumana, Venezuela, 1893; collected orchids for Tring Mus., Caura River, Venezuela, 1897, 1900-01; founded nursery Port-of-Spain; also collected insects, birds. Author: A Naturalist in the Guianas, 1904. Died Dec. 30, 1922.

ANDRÉ, Jean Marie Gustave, French chemist; b. Paris, Aug. 27, 1856; prof. agrl. chemistry Nat. Agronomic Inst.; head research sta. vegetable chemicals, Meudon, France, 1885-89; mem. French Acad. Scis., 1925, Acad. Agr. Contbd. to progress of agronomics through work (with Marcellin Berthelot) on chemistry of sun plants. Died Paris, May 11, 1927.

ANDRÉ, Nicolas, French physician; b. Dijon, France, 1658; student U. Paris. Author: L'orthopédie or l'art de prévenir et de corriger dans les enfants, les deformités du corps (1st book devoted entirely to orthopedics), 1741. Coined word orthopedics, or French equivalent, 1741. Died 1742.

ANDRÉ, Yves Marie, French mathematician; b. Chateaulin, France, 1675; student Coll. de Clermont; joined Soc. of Jesus, 1693; tchr. mathematics, Caen, France; canon regular, Caen; friend of Malebranche; adherent of Jansenism. Author: Essay on the Beautiful; Treatise on Man; The Art of Conversation (poem); many works on physics, optics, architecture. Died at Caen, 1764.

ANDREAE, Johann Gerhart Reinhart, chemist; b. Hanover, Germany, Dec. 11, 1724; s. Leopold and Catharina Elisabeth (Rosennagen) A.; ed. Berlin, Frankfort/Main, Germany, also Leiden, Netherlands; m. Ilse Sophie Müller, Nov. 23, 1751. Pharmacist, Hanover; active in chem. research and publishing. Author: Alchemishsche Briefe, 1767; Briefe aus der Schwiez, 1776; Dispensatorium Brunsvicense, 1777; other works on mineralogy. Noted for publs.; 1st to analyse soil specimens for agrl. purposes. Died Hanover, May 1, 1793.

ANDREAS, Alexander, German mathematician; b. Regensburg, Germany, circa 1475; ed. Cologne, Germany (probably magister). Prof. math. U. Leipzig (Germany), 1502-04; lectr. Euclid, arithmetic, perspective, music. Author: Mathemalogium . . . , 1504; Perspectiva Joannis Pisani, 1504. Important in teaching, popularizing, translating works on algebra; editor math. problems (emphasizing need of proof). Died after 1504.

ANDRE-BALISAUX, Georges Jean, Belgian physician; b. Ixelles, Bruxelles, Belgium, Feb. 8, 1924; s. George and Gelle (Emmanuelle) A.-B.; Docteur en medecine, U. Louvain (Belgium), 1950; m. Monique Marthe Balisaux, July 15, 1950; 1 dau., Emmanuelle. Asst., Neurogial Institut, U. Louvain, 1955-58; mgr. Institut Neurologique Belge, Brussels, 1958-—; founder la Sté Belge de Recherches cliniques, 1953. Decorated commandeur de l'Ordre de St. Sylvestre, l'Ordre de St. Gregoire, chevalier de l'Ordre de la Couronn. Mem. Union Professionnelle des Neuropsychiatres Belges (sec. 1955-—), Société Belge de Neurologie, Société d'electroencephalographie et des sciences connexes de langue francaise (Paris). Author: (with Richard Gonsette) Neuroradiologie Clinique, 1966; also articles. Research on application of electron microscope studies central nervous system, diagnosis, treatment neurol. disorders. Home: 56 Darvin St. Office: 152 Linthout, Brussels, Brabant, Belgium.*

ANDREE, John, Brit. surgeon; b. Eng., 18th century; s. John Andree; M.D., circa 1798; apprentice to Grindall, sr. surgeon London Hosp.; became surgeon Magdalen Hosp., also lectr. anatomy, London, 1766; apptd. surgeon Finsbury Dispensary, 1781, St. Clement Danes Workhouse, 1784. Author: Essay on Gonorrhoea, 1777; Observations on the Venereal Disease, 1779; Considerations on Bilious Diseases, 1788; Cases and Observations (in Surgery), 1799. Performed successful tracheotomy for relief of croup of larynx, 1782. Died circa 1819.

ANDREE, Karl Theodor, German geographer; b. Braunschweig, Germany, Aug. 20, 1808; s. Carl W. Ludwig and Catharina C. (Röhlicke) A.; ed. univs. Jena, Berlin, Göttingen, (all Germany), 1826-30; m. Adelheid Solbrig; 1 son, Richard. Journalist, various newspapers, 1830-55; engaged in geog. and ethnol. studies (under influence Karl Ritter, Alexander von Humboldt), from 1855; named consul gen. to Chile from Saxony, 1858; founder Globus (geog. jour.), 1862; founder (with S. Ruge) Dresden Geog. Soc. Author: Nordham. in geographischen und geschichtlicher Umrissen, 1850; Geographische Wanderungen, 2 vols., 1859; Geographie des Welthandels, 3 vols., 1862-77. Died Bad Wildungen, Germany, Aug. 10, 1875.

ANDREE, Richard, German geographer; b. Braunschweig, Germany, Feb. 26, 1835; s. Karl Theodor A.; studied natural scis. U. Leipzig (Germany); m. Marie Eysn. Foundryman in Czechoslovakia, 1859-63, became dir. of plant; named prof. U. Munich (Germany), 1902; co-founder, head pub. house, Leipzig, 1873. Author: Nationalitätsverhaltnesse und Sprachgrenze in Bohmen, 1870; Tschecnische Gange, 1872; Die Metalle bei den Naturvölkern, 1884; Die Masken in der Völkerkunde, 1886; Die Anthropophagie, 1886; Die Flutsagen, 1891; Braunschweiger Volkskunde, 1896; Votive und Weihgaben, 1904. Pub. geog. and ethnographical books; editor: atlases, 1881, Globus, 1891-1903. Died Braunschweig, Feb. 22, 1912.

ANDREE, Richard Vernon, Am. mathematician, educator; b. Mpls., Dec. 16, 1919; s. Richard A. and Marguerite (Eigner) A.; B.S., U. Chgo., 1942; Ph.M., U. Wis., 1945, Ph.D., 1949; m. Josephine Peet, Dec. 15, 1944; children—David, Peter, Suzanne, Jeanne. Grad. asst. U. Wis., 1942-49, acting instr., 1947-49; faculty U. Okla., Norman, 1949-—, research asso. computer sci., 1957-—, acting dir. computer lab., summer 1960, prof., 1961-—, chmn. dept. math. and astronomy, 1961-—, founder, editor Math. Letter, 1951-55. Lectr. numerous summer and year insts. high sch. and coll. tchrs. math., 1960-—; dir. several inst. and confs. coll. tchrs., 1960-—, summer research participation programs, 1963-65. Nat. Bur. Standards fellow for advanced tng. in numerical analysis, 1959. Fellow A.A.A.S.; mem. Math. Assn. Am. (sec.-treas. Okla. sect. 1951-—, vis. lectr. 1960-—, mem. high sch. traveling lecture bur. 1961-—, editor Math. Smorgasbord books 1960-—), Soc. for Indsl. and Applied Math. (vis. lectr. 1961), Pi Mu Epsilon (nat. sec.-treas. 1957-—), Mu Alpha Theta (founder). Author: (with John C. Brixey) Fundamentals of College Mathematics, 1955, rev. 1961; Selections from Modern Abstract Algebra, 1958; Programming the IBM 650 Computer, 1958; Introduction to Calculus with Analytic Geometry, 1962; Indices and Power Residues, 1962; Computer Programming and Related Mathematics, 1962, rev., 1964. Contbr. numerous articles to profl. publs. Asso. editor Am. Math. Monthly, 1958-62; editor Math. Log, 1960-—. Research on p-adic numbers, matrix theory, modern algebra, computer math., divergent sequences, group theory. Home: 627 E. Boyd St., Norman, Okla. 73069.*

ANDRÉE, Salomon August, aero. engr.; b. Grenna, Sweden, 1854; student engring. Stockholm Tech. Coll.; mem. Swedish meteorol. expdn., 1882-83; tchr. Stockholm Tech. Coll., 1886-89; made (with 2 others) 1st attempt to reach N. Pole by balloon, from Spitzbergen, Norway, 1895; undertook expedition of north polar region, 1897. Died White Island nr. Greenland, after Oct. 2, 1897.

ANDREEV, Nikolaí Nikolaevich, acoustical physicist; b. June 28, 1881; grad. U. Basel (Switzerland), 1909. Taught and directed research in univs. and research instns., 1917-40; with Physics Inst., USSR Acad. Scis., 1940-54, Acoustics Inst., 1945-—. Founder, Sch. Soviet Acoustical Engrs.; corr. mem. USSR Acad. Scis., 1933-53, academician, 1953-—, chmn. commn. on acoustics, 1961-—. Author: Electric Oscillations and their Spectra; A Theoretical Investigation, 1917; (with I. G. Rusakov) Acoustics of the Moving Medium, 1934. Research and publs. on phys. and tech. acoustics, theory of vibrations, theory of diffusion of sound in moving media, piezo-electricity, theory of telephone, mus. acoustics. Home: Leninskii Prospekt 13, Moscow. Office: USSR Academy of Sciences Institue of Acoustics, Ulitsa Televideniya, 4, Moscow, USSR.

ANDRÉOSSY, Count Antoine-Francois, French geographer; b. Castelnaudary, France, Mar. 7, 1761; general in Napoleon's Egyptian campaign; mem. French Acad. Scis., 1824. Author: Histoire du canal du Midi, 1800; Sur les dépressions de la surface du globe, 1826. Research on hydrostatics, Black Sea; developed theory on depressions in earth's surface. Died Montauban, France, Sept. 10, 1828.

ANDRÉOSSY, François, French mathematician, engr.; b. Paris, France, June 3, 1633; great-grandfather of Antoine-François. Employed by Riquet to work on canal of Languedoc; studied hydraulics, Italy, 1660. Author: Carte du Canal de Languedoc, 1669. Drew map of Midi Canal (sometimes credited with constrn. of canal); geodesic expert. Died Castelnaudary, France, 1688.

ANDREOTTI, Aldo, Italian mathematician; b. Florence, Italy, Mar. 15, 1924; s. Libero and Margherita (Carpi) A.; Degree in Math., U. Pisa (Italy), 1947; postgrad. Inst. Alta Mathematica, Rome, Italy, Paris (France); hon. degree Nice (France) U., 1966; m. Barbara Jenkins, Apr. 20, 1953; children—Margherita, Maria-Tecla, Libero, Felicita. Fellow, Princeton, 1950-51, Inst. for Advanced Study, Princeton, N.J., 1951; prof. U. Turin (Italy), 1951-56; asso. prof. U. Nancy, 1956; maitre de recherches U. Paris, 1956-57; Princeton, also Inst. for Advanced Study, 1957-59; prof. U. Pisa, 1959-—; Gauss prof. U. Göttingen (Germany), 1962. Mem. Unione Matematica Italiana, Am. Math. Soc., Soc. Math. de France. Research, publs. on algebraic geometry, several complex variables. Home: 50 Vis S. Andrea, Pisa, Italy.*

ANDRÉS, Juan, Spanish mathematician; b. Saragossa, Spain, 15th century; studied to become priest. Author: Contes vells baralles noves; Sumario breve de la práctica de la aritmética de todo el curso de arte mercantil bien declarado (1 of 1st books on subject in Castilian, contains summary on finger symbolism), 1515. Died early 16th century.

ANDRESEN, John William, Am. forester; b. N.Y.C., Feb. 26, 1925; s. John Bernard and Benita (Rieken) A.; B.S. cum laude, Syracuse U., 1953; Ph.D., Rutgers U., 1958; m. Lillian E. Larsen, Feb. 3, 1946; children—John William, Joyce Elizabeth, Gayle Denise, Bruce James. Field instr. in dendrology State U. N.Y., 1951-53; lab. instr. botany, 1952-53; instr. forestry Rutgers U., 1953-58; asso. prof. forestry Mich. State U., East Lansing, 1958-64; prof., chmn. dept. forestry So. Ill. U., Carbondale, 1964-—. Cons. in forestry to Am. Mus. Natural History, 1956-58; cons. dendrology U. S. Forest Service, 1959-61. Mem. Soc. Am. Foresters, Ecol. Soc. Am., Soc. Plant Taxonomists, Internat. Assn. Plant Taxonomists. Author: (with Raymond Clark) Trees of the Lake States, 1960; also articles. Research in taxonomy and ecology of pines of N.Am.; discovered, described 2 new species of pine. Home: 2005 Dale Dr., Carbondale, Ill. 62901.*

ANDRESEN, Momme, photochemist; b. Risum, Oct. 17, 1854; ed. Tech. U. Dresden (Germany); degree from Jena, Germany, Dr. phil. (hon.), 1940. Asst. to Schmidt, Dresden; researcher Casella firm, Germany, later in Buffalo, N.Y.; with Agfa Co., Berlin, Germany, from 1887, became head photog. dept., 1889. Author: Das latente Bild, 1913; Agfa Photo Buch, 1922; papers in various periodicals. Invented rodinal developer, 1891, photog. dry-plates, 1895, durable photo printing papers. Died Königsteinhof, Jan. 12, 1951.

ANDREW, Edward Raymond, English physicist; b. Boston, Eng., June 27, 1921; s. Edward Richard and Anne (Henderson) A.; B.A., U. Cambridge (Eng.), 1942, M.A., 1946, Ph.D., 1948, Sc.D., 1964; m. Mary Farnham, July 1948 (dec. 1965); children—Charmian Mary, Patricia Rosalind, Anne Margaret Elizabeth (dec. Jan. 1968). Sci. officer Royal Radar Establishment, Malvern, Eng., 1942-45; Stokes student Pembroke Coll., Cambridge U., 1947-49; Commonwealth Fund fellow Harvard, 1948-49; lectr. natural philosophy St. Andrews (Scotland) U., 1949-54; prof. physics U. Coll. N. Wales, Bangor, 1954-63; Lancashire-Spencer prof. physics. U. Nottingham (Eng.), 1964-—. Ct. govs. Loughborough U. Tech.; mem. acad. adv. com. Brunel U. Fellow Inst. Physics, Phys. Soc., Royal Soc. Edinburgh; mem. Am. Phys. Soc. Author: Nuclear Magnetic Resonance, 1955; also articles. Research on nuclear magnetic resonance as applied to solids, especially molecular motion in solids and fine structure revealed by rapid rotation of solid specimens; studies on superconductivity, size-dependence of resistivity at low temperatures, signal detection in radar systems, microwave attenuation in ionized gases. Office: Dept. Physics, Univ. University Park, Nottingham, Eng.*

ANDREW, Walter Jonathan, Brit. archaeologist; s. Ely Andrew; m. Sara Roebuck, 1883 (dec. 1923); m. 2d, Henrietta Tragett, 1927. Admitted solicitor, 1882. Mem. Com. Ancient Earthworks and Fortified Enclosures, to 1914. Recipient Gold medal Balloon Soc.; Saltus Gold medal. Fellow Soc. Antiquaries; hon. mem. Derbyshire Archaeol. Soc.; pres. Brit. Numismatic Soc., 1909, hon. sec., editor, 1903-26. Author: A Numismatic History of the Reign of Henry I.; A Numismatic History of the Reign of Stephen; Buried Treasure—some traditions, records, and facts; The Prehistoric Stone-Circles of Derbyshire; The Battle of Brunanburh; also numerous articles. Identified Anglo-Saxon ring in Ashmolean as installation ring of Eolla, 2d bishop of Selsey, 1926; discovered King John's house, Romsey, Eng., 1927; discovered Winchester Anglo-Saxon bowl (now in Brit. Mus.) during excavations of Oliver's Battery for Hampshire Archaeol. Soc. Died Jan. 14, 1934.

ANDREW, Warren, Am. anatomist; b. Portland, Oregon, July 19, 1910; s. John and Alice (Lucke) A.; B.A. summa cum laude, Carleton College, 1932; fellow Brown U., 1932-33, M.S., 1933; Ph.D., Yale, 1936, U. Ill., 1936; M.D., Baylor Med. Coll., 1943; m. Nancy Valerie Miellmier, Aug. 18, 1936; 1 dau., Linda Nancy. Asst. instr. zoology Yale, 1933-34, U. Ill., 1934-36; teaching fellow U. Ga., 1936-37, Instr., fellow anatomy, 1937-39; instr. anatomy, sch. medicine Baylor U., 1939-41, asst. prof. anatomy, 1941-43; asso. prof. histology Southwestern Med. Coll., Dallas, 1943-45, prof., 1946-47; vis. prof. U. Montevideo, Uruguay, 1945-46; prof. anatomy, chmn. dept. sch. medicine George Washington U., 1947-48, 1949-52, cons. anatomy George Washington U. Hosp., 1947-52; vis. prof. Washington U., St. Louis, 1948-49; cons. cytology V.A. Hosp., Martinsburg, W.Va., 1950-52; Am. Mus. Natural History vis. investigator, Lerner Marine Biol. Lab., Bimini, Bahamas, B.W.I., 1951; vis. scientist Barro Colorado Island, Canal Zone, 1952; advisor anatomy, mem. exhibits com. Second Internat. Gerontological Congress, 1951; prof. anatomy dir. dept. Bowman Gray Sch. Medicine, 1952-57; prof., chmn. dept. anatomy sch. medicine Ind. U., Indpls., 1957-—; lectr. Tokyo and Kypto med. schs. Japan, summer 1960, Karachi, Pakistan, also Bombay and Jamnagar, India, summer 1962. Ofcl. del. Pan-Am. Congress Gerontology, Mexico, 1956; editorial bd. Gerontologia, Basel, 1957-—; rep. biology, Internat. Research Com. for Gerontology; participant IX International Congress Cell Biology; IV Internat. Congress Gerontology, 1957. Research award, Gerontological Found., 1959. Mem. Gerontological Society, International Assn. Gerontology, Am. Assn. Anatomists, Am. Society Zoologists, A.A.A.S., Soc. Exptl. Biol. and Medicine, Philos. Soc. Washington, Tex. State Medical Assn., Tissue Culture Assn., Washington Soc. Pathologists, Washington Acad. Medicine, N.Y. Acad. Sci., Biol. Soc. Montevideo, Amigos de la Naturaleza de Montevideo, American Society Electron Microscopists, Internat. Soc. Cell Biology, Am. Med. Soc. Vienna (life), Sigma Xi, Phi Beta Kappa. Author: Comparative Histology, 1959; Comparative Hematology, 1965; Micro-fabric of Man; One World of Science, 1966. Mem. editorial bd. Gerontological Newsletter, 1959-—; Quar. Bull. Ind. U. Med. Center, 1959. Contbr. sci. articles, profl. publs. Research on cellular changes accompanying age. Home: 5275 N. Capitol Av., Indpls.*

ANDREWES, Christopher Howard, English pathologist; b. London, June 7, 1896; s. Frederick and Phyllis Mary (Hamer) A.; ed. St. Bartholomew's Hosp.; m. Kathleen Lamb, Mar. 26, 1927; 3 sons. House physician St. Bartholomew's Hosp., 1921-23, 25-26; asst. resident physician Rockefeller Inst. Hosp., N.Y.C., 1923-25; mem. staff Nat. Inst. Med. Research, London, from 1927, dep. dir. until 1961, now ret.; William Julius Mickle fellow U. London, 1931; Oliver-Sharpey lectr. Royal Coll. Physicians, 1934; dep. dir. Royal Coll. Surgeons, 1952-61; charge WHO world influenza centre, Mill Hill, London, 1949-61. Fellow Royal

Soc., 1939. Devised test named after him for discovering uremia in patient; (with Wilson Smith and P. P. Laidlaw) successfully infected lab. animals with matter from throat of influenza patients. Home: Overchalke, Coombe Bissett, Salisbury, Wiltshire, Eng.

ANDREWES, Sir Frederick William, English pathologist, bacteriologist; b. Mar. 31, 1859; ed. Christ Ch., Oxford (Eng.) U., St. Bartholomew's Hosp., London, Eng.; m. Phyllis Mary Hamer, 1895. Prof. pathology U. London, from 1912; pathologist St. Bartholomew's Hosp., London. Croonian lectr. Royal Coll. Physicians, 1910; Harveian orator, 1920. Mem. Med. Research Council. Fellow Royal Soc., 1915. Wrote extensively on bacteriology of streptococci, immunological subjects; described Shigella alkalescens, sometimes cause of diarrheal disease in man, 1918. Died Feb. 24, 1932.

ANDREWS, Charles William, Brit. geologist; b. Hampstead, Eng., 1866; B.A., London, 1887, B.Sc., 1890, D.Sc., 1900; Became mem. staff geol. dept. Brit. Mus., 1892, later asst. keeper; made expdns. to Christmas Island, Cocos-Keeling Islands, 1897-98, Christmas Island, 1908; collected fossil vertebrates, Fayum, Egypt, 1900-06. Recipient Lyell fund Geol. Soc., 1896, Lyell medal, 1916. Fellow Royal Soc., 1906. Author: A Descriptive Catalogue of the Tertiary Vertebrata of the Fayum, Egypt; (with others) A Monograph of Christmas Island (Indian Ocean); Catalogue of the Marine Reptiles of the Oxford Clay, vols. I and II; papers on fossil vertebrates. Died May 25, 1924.

ANDREWS, David Arthur, Am. geologist; b. Jacksonville, Mo., Apr. 6, 1906; s. David L. and Annie M. (Terrill) A.; A.B., U. Mo., 1927, M.A., 1928; postgrad. Yale; m. Claudia Maxine Henry, Nov. 17, 1931; children—David Henry, Jannane (Mrs. Arthur R. Kinkel), Richard Hall. With U. S. Geol. Survey, 1929-62, asst. chief fgn. geology br. for Asia; 1950-56, chief mineral div. AID, Indonesia, 1956-60, chief participant tng., Washington, 1960-62; project mgr. geol survey Iran for UN, Tehran, 1962—. Del. to Econ. Commn. Asia-Far East, Rangoon, Burma, 1952, Bangkok, Thailand, 1957, mem. subcom. on mineral resource devel., Tokyo, Japan, 1953, 56, 60, Bangkok, 1954, Calcutta, India, 1958, Manilla, 1963. Recipient Distinguished Service award U. S. Dept. Interior, 1962. Fellow A.A.A.S., Geol. Soc. Am., Am. Assn. Petroleum Geologists; mem. Sigma Xi, Gamma Alpha, Sigma Phi Epsilon. Contbr. papers to profl. jours. Home: 5112 Allan Rd., Washington 20016. Office: UN Box 1555, Tehran, Iran.*

ANDREWS, Donald Hatch, Am. chemist; b. Southington, Conn., June 11, 1898; s. Russell Gad and Mary (Hatch) A.; B.A., Yale, 1920, Ph.D., 1923; m. Elizabeth Howland, Sept. 23, 1950; 1 son, Donald H. Research asst. Yale, 1923-24; Nat. Research fellow U. Cal., Berkeley, 1924-25; Internat. Research fellow U. Leiden (Holland), 1925-26; Research fellow Franklin Inst., 1926-27; faculty Johns Hopkins, 1927—, prof., 1930-63, prof. emeritus, 1963—, chmn. chemistry dept., 1936-44; Distinguished prof. chemistry Fla. Atlantic U., Boca Raton, 1963—. Fellow Am. Phys. Soc., A.A.A.S., Am. Philos. Soc., Royal Chem. Soc. (London), N.Y. Acad. Scis.; mem. Am. Chem. Soc. (chmn. div. phys. and inorganic chemistry 1934-35), Calorimetry Conf. (chmn. 1956-57), L'Institut Internationale de Froid (mem. premier commn. 1932—, Am. Inst. Chemist (dir. 1935-39), Am. Math. Soc., Phi Beta Kappa, Sigma Xi, Gamma Alpha, Alpha Chi Sigma. Author: (with R. J. Kokes) Fundamental Chemistry, 1962, 2d edit., 1964; Quirnica Fundamental, 1963. First theory of Roman Spectra; studied calorimetry of organic compounds; inventor superconducting bolometer. Home: 750 N.E. 33d St., Boca Raton, Fla. 33432.*

ANDREWS, Frank Maxwell, Am. aero. pioneer; b. Nashville, Feb. 3, 1884; s. James David and Louise (Maxwell) A.; student Montgomery Bell Acad., Nashville, 1897-1901; grad. U. S. Mil. Acad., 1906; student Air Corps Tactical Sch., 1927-28, Command and Gen. Staff Sch., 1928-29; grad. Army War Coll., 1933; m. Jeannette Allen, Mar. 16, 1914; children—Josephine, Allen, Jean. Commd. 2d lt. cav., 1906; through grades to col., 1935; rank of lt. gen. (temp.) 1941. Served in Philippine Islands, 1906-07; in Hawaii, 1911-13; served with Signal Corps (aviation sec.), 1917-20; served with Am. Forces in Germany, 1920-23; returned to U. S. as exec. officer. Kelly Field, 1923-25; mem. War Dept. Gen. Staff, 1934-35; apptd. temp. brig. gen. Air Corps, 1935, temp. maj. gen., Dec. 1935; organizer, comdr. G.H.Q. Air Force, 1935-39; apptd. brig. gen. U. S. Army, 1939; mem. War Dept. Gen. Staff, 1939-40; comdr. Panama Dept., 1942, Middle East Command, 1942-43; apptd. comdr. U. S. forces ETO, 1943. Decorated Comdr. Order of the Crown (Italy). Contbd. to evolution of Am. bombardment aviation; advocated strategic air power; instrumental in devel. Boeing B-17 bomber. Killed in airplane accident over Iceland, May 3, 1943.

ANDREWS, Frederick Newcomb, Am. animal scientist; b. Boston, Feb. 5, 1914; s. Frederick Huntoon and Gertrude (Macomber) A.; B.S., U. Mass., 1935, M.S., 1936, D.Sc., 1962; Ph.D., U. Mo., 1939; m. Gertrude E. Martin, Sept. 3, 1938; children—Frederick Martin, Donna Elaine. Faculty, Purdue U., Lafayette, Ind., 1940—, prof. animal physiology, 1949—, asst. dean Grad. Sch., 1949-52, dean, 1963—, head dept. animal scis., 1962—, v.p. for research, 1963—, v.p. Purdue Research Found., 1964—. Cons. to fed. agys., pvt. cos., Rockefeller Found. Fellow A.A.A.S.; mem. Am. Soc. Zoologists, Am. Assn. Anatomists, Am. Soc. Animal Sci. (Morrison Research award 1961), Am. Dairy Sci. Assn., Sigma Xi (Purdue Sigma Xi Research award 1949). Author: (with Rice, Warwick, Legates) Breeding and Improvement of Farm Animals, 1957; also numerous articles. Physiol. research on reprodn., fertility, endocrinology, growth process, environment, nutritional requirements growth, nutritional and endocrine relationships. Home: 691 Sugar Hill Dr., West Lafayette, Ind. 47906. Office: Purdue U., Lafayette, Ind. 47906.*

ANDREWS, Gould A., Am. physician; b. Grand Rapids, Mich., May 8, 1918; s. Gould Arthur and Florence (Belford) A.; A.B., U. Mich., 1940, M.D., 1943; m. Eva Hodgens, July 25, 1955; 1 dau., Ellen. Hematologist, med. div. Oak Ridge Asso. Univs., 1949-50, chief clin. services, 1950-58, asso. chmn., 1959-61, chmn. div., 1961—. Diplomate Am. Bd. Internal Medicine. Fellow A.C.P. Editor: Radioisotopes in Medicine, 1953; Progress in Medical Radioisotope Scanning, 1963; Radioactive Pharmaceuticals, 1966. Mem. cons. editorial bd. Jour. Nuclear Medicine; editorial cons. in nuclear medicine The New Physician. Research in hematology, med. uses of radioisotopes, effects of totalbody irradiation in human being. Home: 163 Outer Dr., Oak Ridge 37830. Office: Oak Ridge Associated Universities, E. Vance Rd., Oak Ridge 37830.*

ANDREWS, Henry Nathaniel, Jr., paleobotanist, educator; b. Melrose, Mass., June 15, 1910; s. Henry Nathaniel and Florence (Hollings) A.; B.S., Mass. Inst. Tech., 1934; M.S., Washington U., St. Louis, 1937, Ph.D., 1939; postgrad. Cambridge (Eng.) U.; m. Elisabeth Claude Ham, Jan. 12, 1939; children—Hollings T., Henry Nathaniel III, Nancy R. Staff mem. botany dept. Washington U., St. Louis, 1939-64, prof., 1948-64, dept. chmn., 1950-64; chmn. botany dept. U. Conn., Storrs, 1965—, Paleobotanist, Mo. Bot. Garden, St. Louis, 1945-64; meml. staff U. S. Geol. Survey, 1958—. Guggenheim Meml. Found. fellow 1950-51, 57-58, Fulbright fellow Poona U., India, 1960-61. Mem Paleobot. Soc. India (hon.), Geol. Soc. Am., Paleontol. Soc., Bot. Soc. Am., Torrey Bot. Club, Phi Beta Kappa, Sigma Xi. Author: Ancient Plants, 1947, 64; Index of Generic Names of Fossil Plants, 1955; Studies in Paleobotany, 1961. Research, publs. on fossil ferns, early seed plants. Home: 33 Lynwood Rd., Storrs, Conn. 06268.*

ANDREWS, Howard L., Am. physicist; b. North Kingstown, R.I., Oct. 27, 1906; s. Lucius Elton and Maud (Matteson) A.; B.E.E., Brown U., 1927, M.S., 1928, Ph.D., 1931; m. Dorothy Jeanette Thayer, Aug. 13, 1937; children—Jean M., Howard R., Alice E. Instr. physics Brown U., 1930-34; biophys. research Bradley Hosp., East Providence, R.I., 1934-37; research in biophysics of drug addiction USPHS Hosp., Lexington, Ky., 1937-42; biophysicist NIH, Bethesda, Md., 1937-45, radiation safety officer, 1945-65, head dept. radiation safety, 1963-65; asst. dir. P.R. Nuclear Center, 1965-67; prof. radiation biology and biophysics Sch. Medicine and Dentistry, Rochester, N.Y., 1967—. Recipient USPHS Meritorious Service medal. Mem. Nat. Acad. Sci. (exec. sec. biol. effects of atomic radiation com. 1959-65), Health Physics Soc. (pres. 1964-65), Am. Phys. Soc., Radiation Research Soc. Author: (with R. E. Lapp) Nuclear Radiation Physics, 1949; Radiation Biophysics, 1961; (with A. L. Russell) Basic Boating, 1964; also numerous articles. Pioneer in electroencephalography in U. S., effects of drugs on electroencephalogram; components of pain relief by drugs, biol. effects of high doses of radiation. Home: 608 Suburban Ct., Rochester, N.Y. 14620.*

ANDREWS, J. Robert, Am. physician; b. Kent, O., June 10, 1906; s. William B. and Anna (Doyle) A.; Ph.B., Brown U., 1928; M.D., Western Res. U., 1932; D.Sc., U.Pa., 1935; m. Anne Cosgrove, 1935, (div. 1955); children—Catherine F. (Mrs. Albert Kapikian), William B., J. Robert II; m. 2d, Nicole Crozier, Dec. 31, 1962. Prof. radiology, dir. radiology Bowman Gray Sch. Medicine and N.C. Bapt. Hosp., Winston-Salem, N.C., 1950-55; chief radiation br. Nat. Cancer Inst., NIH, Bethesda, Md., 1955-64; prof. radiology, dir. radiotherapy, cancer coordinator, Georgetown U. Med. Center, 1964—; chief radiotherapy sect. Washington VA Hosp., 1965—; cons. clin. center NIH; cons. Gordon A. Friesen Internat. Mem. Am. Roentgen Ray Soc., Am. Radium Soc., A.M.A., Radium Research Soc., Am. Assn. Cancer Research, Radiol. Soc. Research, publs. in clin. radiobiology, radiotherapy of cancer; interrelationship of oxygen effect, linear energy transfer (LET) and fractionation in radiobiology and radiotherapy. Home: 4428 Volta Pl. N.W., Washington 20007. Office: 3800 Reservoir Rd. N.W., Washington 20007.*

ANDREWS, Jay Donald, Am. biologist; b. Bloom, Kan., Sept. 9, 1916; s. Jay Straney and Eva (Dilley) A.; B.A., Kan. State Coll., 1938; B.S., in zoology U. Wis., 1940, Ph.D., 1947; m. Mary Stuart Hornsby, Mar. 23, 1948; children—Donna Gay, Jay Stuart. With Va. Inst. Marine Sci., Gloucester Point, Va., 1946—, faculty Coll. William and Mary, 1946-67, prof., 1959-67; asso. prof. U. Va., 1961-67. Mem. Nat. Shellfisheries Assn. (pres. 1965-67), Ecol. Soc. Am., Am. Soc. Limnology and Oceanography, Soc. Systematic Zoology. Research and publs. on ecology of mollusks, particularly growth, reproduction, management of oyster culture; epidemiology of oyster diseases. Home: Cornwallis Rd., Yorktown, Va. 23490. Office: Va. Inst. Marine Sci., Gloucester Point, Va. 23062.*

ANDREWS, John Machunka, Am. physician; b. Schenectady, Aug. 6, 1907; s. John and Julia (Koscak) Machunka; student U. So. Cal., 1927-28; D.O., Col. Osteo, Physicians and Surgeons, 1933; M.D., Cal. Coll. Medicine, 1962; m. Grace Elvina Hagen, Dec. 31, 1934; children—John Morton, Nancy Jane. Pvt. practice osteo. medicine, Los Angeles, 1933-50; dir. clinic Coll. Osteo. Physicians and Surgeons, Los Angeles, 1950-53; prof., chmn. dept. phys. medicine and rehab. Cal. Coll. Medicine, Los Angeles, 1953—, Baylor U., Houston, 1964-65. Mem. med. adv. bd. Nat. Found. and Multiple Sclerosis Soc. So. Cal.; pres. So. Cal. chpt. Nat. Rehab. Assn., 1963-64. Mem. Alpha Phi Omega, Sigma Sigma Phi. Research in muscle re-education and musculo-skeletal functional problems. Home: 5314 N. Crown Av., La Canada, Cal. 91011.*

ANDREWS, Kenneth William, English phys. metallurgist; b. Spalding, Eng., Aug. 5, 1916; s. John Thomas and Lydia (Pine) A.; B.Sc., Manchester (Eng.) U., 1938, D.Sc., 1956; D.Phil., Oxford (Eng.) U., 1940; m. Gwyneth Owen, Apr. 8, 1944; 1 son, Richard Owen. Research asst. Oxford (Eng.) U., 1938-42; sr. research worker research and devel. dept. United Steel Cos., Ltd., Stocksbridge, Eng., 1942-47, sect. leader X-ray sect., Rotherham, Eng., 1947-61, theoretical and structural metallurgy, 1961-64, physics and theoretical metallurgy sect., 1964—. Vis. lectr. local tech. colls. and univs. Fellow Instn. Metallurgists, Inst. Physics (past pres. Yorkshire br.); mem. Inst. Metals (past chmn. Sheffield sect.), Sheffield U. Metall. Soc. (past pres.), Iron and Steel Inst. Author: (with D. J. Dyson and S. R. Keown) Interpretation of Electron Diffraction Patterns, 1966. Research and publs. on constitution and structure of alloy, refractory and other oxide systems; occurrence, characterization, crystallography and reactions of carbide and alloy phases in steels (some not previously reported); phys. methods applied to steels, phys. properties, theory of transformations; calculation methods. Home: Redstone, 182 Moorgate Rd., Rotherham. Office: Swinden Labs., Moorgate Rd., Rotherham, Yorkshire, Eng.*

ANDREWS, Octavius William, Brit. surgeon; b. 1865; s. H. C. and Louisa Augusta (Lord) A.; ed. M.B., B.S., Durham (Eng.) U.; postgrad. St. George's Hosp., London, later at Pasteur insts., Lille, Paris, France, 1893; D.P.H., Cambridge U.; m. Florence Ferrar, 1894; 1 son, 2 daus. Entered Royal Navy, 1889; promoted to staff surgeon, 1899; naval rep. 14th Internat. Congress on Hygiene and Demography; mem. Internat. San. Commn., Dardanelles. Recipient Chadwick Gold medal, Naval prize. Author: Seamarks and Landmarks; Public Health Laboratory Work; also many articles. Research on leprosy and depopulation in some Pacific islands. Died Mar. 13, 1936.

ANDREWS, Roy Chapman, Am. zoologist, explorer; b. Beloit, Wis., Jan. 26, 1884; s. Charles E. and Cora M. (Chapman) A.; B.A., Beloit Coll., 1906, hon. Sc.D., 1928; M.A., Columbia, 1913; hon. Sc.D., Brown U., 1926; m. Yvette Borup, 1914; children—George Borup, Roy Kevin; m. 2d, Wilhelmina Christmas, Feb. 21, 1935. Director American Museum of Natural History, until 1942; explored in Alaska, 1908; collected white whales Saquenay River, 1909; special naturalist U.S.S. Albatross, on voyage to Dutch East Indies, Borneo, Celebes, 1909-10; explored N. Korea, 1911-12; with Borden Alaska Expdn., 1913; specialized in study of whales and other water mammals until 1914; leader Asiatic expdns. of Am. Mus. Natural History, 1st expdn., Tibet, frontier, S.W. China and Burma, 1916, 17, 2d expdn., N. China and Outer Mongolia, 1919, 3d expdn., Central Asia, 1921-32. Awarded Elisha Kent Kane gold medal Phila. Geog. Soc., 1929; Hubbard gold medal Nat. Geog. Soc., 1931; Explorers' Club medal, 1932, Charles P. Daly gold medal Am. Geog. Soc., 1936; Vega gold medal Royal Swedish Anthropol. and Geog. Soc., 1937; Loczy medal Hungarian Geog. Soc., 1937; Silver Buffalo award Nat. council Boy Scouts Am., 1952. Fellow Nat. Geog. Soc., A.A.A.S., N.Y. Acad. Scis., Am. Geog. Soc.; mem. Am. Philos. Soc., numerous other sci. socs., Phi Beta Kappa. Author 22 books on exploration, adventure, popular science, including: On the Trail of Ancient Man, 1926; The New Conquest of Central Asia, 1932; Meet Your Ancestors, 1945; Heart of Asia, 1951; Beyond Adventure, 1954; The Natural History of Central Asia (sci. report of his Asian expeditions), 12 vols.; also many sci. papers, mag. articles. Opened Gobi Desert to use of motor cars for comml. purposes; mapped much new area in the Gobi Desert; made 1st accurate gen. map of Mongolia; discovered many geol. strata previously unknown; discovered some of richest fossil fields in world; also 1st dinosaur eggs, skulls and parts of the skeleton of the Baluchitherium, and many other fossil mammals and reptiles previously unknown to science; his researches proved Central Asia to be one of chief centers of origin and distbn. of world's reptilian and mammalian life. Died Mar. 12, 1960.

ANDREWS, Solomon, Am. physician, inventor; b. Herkimer County, N.Y., 1805; practiced medicine, also held various polit. offices, Perth Amboy, N.J.; organized Aerial Nav. Co. (to establish regular line between Phila. and N.Y.C.); constructed airships and made ascent, 1863. Patentee wickless oil burner, fumigators, gas lamps, combination lock. Died 1872.

ANDREWS, Thomas, Irish chemist; b. Belfast, Ireland, Dec. 19, 1813; s. Thomas John and Elizabeth (Stevenson) A.; studied chemistry, Glasgow, also Paris; student medicine, Trinity Coll., Dublin (LL.D. 1873), also Belfast; M.D., Edinburgh, 1835, LL.D., 1871; D.Sc., Queen's U. Ireland, 1879; m. Jane Hardie Walker, 1842; 4 daus., 2 sons. Prof. chemistry Med. Coll., Academical Instn., Belfast, until 1849; prof. chemistry, v.p. Queen's Coll., Belfast, 1849-79. Fellow Royal Soc., 1849; Royal Soc. Edinburgh (hon).; mem. Brit. Assn. (pres. Glasgow 1876, pres. chem. sect. Belfast 1852, Edinburgh 1871). Discovered and named critical temperature of gases; investigated ozone, heat of chem. combination, elec. conductivity of flames and hot air, 1826; deigned air pump. Died Belfast, Nov. 26, 1885.

ANDREWS, Thomas, English chemist, metall. engr.; b. Sheffield, Eng., Feb. 16, 1847; s. Thomas and Mary (Bolsover) A.; ed. Sheffield; m. Mary Hannah Stanley, 1870; 3 sons, 1 dau. Head, Wortley Iron Works, from 1871; cons. to Admiralty, also Bd. Trade; lectr. engring. Cambridge U.; helped establish Sheffield U. Fellow Royal Soc., 1888, Royal Soc. Edinburgh, Chem. Soc.; mem. Instn. Civil Engrs. (Telford medal 1884), Soc. Engrs. (Gold medal 1902). Followed up work of Henry C. Sorby in early studies of metals with microscope; determined resistance of metal to sudden concussion at varying temperatures down to 0° F. Died nr. Sheffield, June 19, 1907.

ANDREWS, William Horner, Brit. veterinarian; b. Eng., June 22, 1887; s. H. G. Andrews; ed. Math. Sch., Rochester, Eng., Royal Vet. Coll., London, Eng., Institut Pasteur, Paris, France; m. Doris Burls, 1916; 1 son, 1 dau. Asst. govt. vet. bacteriologist, Transvaal, 1909-12; vet. research officer Union South Africa, 1912-14; capt. S. African Vet. Corps., German South-West Africa, 1914-15; sr. vet. research officer, 1918; prof. physiology Transvaal Univ. Coll., Pretoria, Union South Africa; research asst. Research Inst. in Animal Pathology, London, 1924; dir. Vet. Lab., Weybridge, Eng., 1927-41, Imperial Bur. Animal Health, 1929-41. Mem. Royal Soc. Vet. Surgeons. Contbr. articles on plant poisoning to sci. jours. Died Mar. 19, 1953.

ANDRIANOV, Kuzma Andrianovich, Russian chemist; born December 28, 1904; graduated from Moscow Univ., 1930. With All-Union Electrotechnical Institute, 1930; instructor Moscow Chemico-Tech. Institute, 1930-41; instr. Moscow Power Engring. Inst., 1941-46, prof., 1946—; lab. head Inst. Elemental Organic Compounds, USSR Acad. Sci., asso., 1954—. Mem. delegation agrobiologists and chemists to Italy, 1954. Recipient Lenin prize, 1962, Stalin prizes, 1943, 46, 50. Mem. USSR Acad. Sci. Author: Silico-Organic Compounds, 1943; Heat-Resistant Silico-Organic Dielectrics, 1957; (with D. A. Kardashev) Practical Work on Artificial Resins and Plastics, 1946, (with M. V. Sobolevskii) High-Molecular Silico-Organic Compounds, 1949; (with S. A. Yamanov) Organic Dielectrics and their Use in the Communications Industry, 1949; Synthesis of Polyesters and Polyester Amides of Cycloreticular Structure, 1963. Editor: Electro-Technical Materials Manual, 1959. First to synthesize polyorganosiloxanes, 1937; developer method of obtaining new polyorganometallosiloxane polymers, 1947; dir. work on synthesis of heat-resistant insulating silico-organic resins and varnishes and composite materials based on them. Home: 102 Fersman. Office: 14 Wawilowa, Moscow, USSR.*

ANDRIEUX, Jean-Lucien, French chemist; b. Roybère, France, Oct. 18, 1887; s. Joseph and Amélie (Berger) A.; student U. Grenoble (France); Lic.Sc., U. Lyons (France), 1910, Sc.D., 1929; m. Jeanne Lenrouost, Apr. 12, 1924; 4 children. Asst. in chemistry Faculty Scis., Grenoble, lectr. electrochemistry, 1930-33, prof. chemistry, from 1933; asso. dir. Inst. Electro-Chemistry and Electro-Metallurgy, 1939-45, dir., from 1945. Mem. French Acad. Scis., 1953, recipient prix Cahours, 1930, prix Paul Marguerite de la Charlonie, 1946; mem. French Soc. Metallurgy, Am. Electro-Chem. Soc., French Chem. Soc. (prix Cincel 1915), French Soc. Electricians. Author: Travaux sur l'electroyles des sels.

ANDRISANO, Renato, Italian chemist; b. Verona, Italy, June 5, 1916; s. Angelo and Angela (Zonno) A.; Dr. in Chemistry and pharmacology; m. Silvana Cardeti, Apr. 8, 1946; children—Angela Maria, Angelo, Oreste, Enza, Roberto. Ex-dean Pharmacology Sch., U. Parma (Italy); dir. Inst. Tech. Chemistry U. Bologna (Italy). Mem. Giemia di Scienze, Italian, London, Am. chem. socs., Soc. Pharm. Scis. Research, publs. on organic and pharm. chemistry. Home: via Toscannini 10, Bologna, Italy. Office: via Risorgimento 4, Bologna, Italy.

ANDRONICOS OF CYRRHOS, Greek architect, astronomer; b. Cyrrhos, Macedonia, circa 100 B.C.; constructed wind tower (to indicate wind directions) nr. Agora Temple, Athens.

ANDRONIKASHVILI, Elevter Luarsabovich, Russian physicist; born December 25, 1910; graduated from Leningrad Polytechnical Institute, 1932; Doctor Physico-Math. Sci. Instr., Tbilisi U., 1934-45, prof., 1948—; asso. Inst. Physics, Georgian Acad. Sci., 1942-45, dir. Inst. Physics, 1951—. Recipient Stalin prize, 1952. Mem. Georgian Acad. Sci. Author: Direct Observation of Two Types of Movement in Helium, 1946; Research on the Viscosity of the Normal Components of Helium, 1948; Study of Cosmic Rays below Ground, 1955. Research on quantum hydrodynamics and physics of cosmic rays. Address: Gosudarstvenny University, prosp. Chavchavadze 45, Tbilisi, Gruz. SSR, USSR.

ANDRUS, Charles Frederick, Am. plant breeder; b. Mt. Carmel, Ill., Jan. 21, 1906; s. Charles S. and Clara (Seiler) A.; A.B., George Washington U., 1931, M.A., 1932; D.Sc. (honorary) Clemson University, 1967; m. Margaret Jane Grow, Nov. 26, 1931; children—Jan Frederick, Charles Andrew. Sci. aide U. S. Dept. Agr., Washington, 1928-31, plant pathologist, 1931-38, Beltsville, Md., 1935-38, Charleston, S.C., 1938-48, horticulturist, 1948-63; research horticulturist in charge U. S. Vegetable Breeding Lab., Charleston, S.C., 1963—. Recipient FFVA Research award, U. S. Dept. Agr. Superior Service award. Mem. A.A.A.S., S.C. Acad. Sci., Am. Genetic Assn., Am. Soc. for Hort. Sci., Am. Inst. Biol. Sci. Developed basic plant breeding stocks, new techniques, new varieties of vegetable crops.*

ANDRUS, E(dwin) Cowles, Am. physician; b. Kaatsbon, N.Y., Feb. 28, 1896; s. J. Cowles and Margaret (De Witt) A.; A.B., Oberlin Coll., 1916, A.M., 1917, D.Sc., 1941; M.D., Johns Hopkins, 1921; m. Miriam Jay Wurts, June 10, 1933. NRC fellow U. London (Eng.), also U. Vienna (Austria), 1923-25; faculty Johns Hopkins, Balt., 1926—, asso. prof. medicine, 1931-61, prof. emeritus., 1961—, asst. dean med. faculty, 1929-34; staff Johns Hopkins Hosp., Balt., 1921—, chief cardiovascular div., 1954-61, physician-in-charge adult cardiac clinic, 1946-61. Cons. to hosps., govt. agys.; chmn. cardiovascular study sect. Nat. Adv. Heart Council, 1946-51, mem. council, 1953-62; mem. Cultural Exchange Mission to Russia, 1961. Recipient Gold Heart award Am. Heart Assn., 1956—, Presdl. citation, 1950. Master A.C.P.; mem. Inter-Am. (council 1954—, v.p. 1960—), Internat. (council 1958-66) cardiological socs., Sociedad Peruana de Cardiologia (hon.), A.M.A., Am. Physiol. Soc., Am. Soc. for Study Arteriosclerosis, Am. Heart Assn. (past pres.), Am. Soc. for Clin. Investigation, Am. Clin. and Climatol. Assn., Assn. Am. Physicians, Med. and Chirurg. Faculty Md., A.A.A.S., Phi Beta Kappa, Alpha Omega Alpha, Sigma Xi. Research, publs. on heart disease due to overactive thyroid, mechanism producing irregularities of heart beat, renal effects on hypertension. Home: 209 E. Highfield Rd. Balt. 21218. Office: 550 N. Broadway, Balt. 21205.*

ANDRUSHKIW, Joseph Wasyl, mathematician, educator; b. nr. Lviw, Ukraine, Mar. 21, 1906; s. Wasyl and Apollonia (Jachowski) A.; Maturity diploma with excellence, State Classical Gymnasium, 1924; M.S., State U. Lviw, 1930, M.Ed., 1932; Ph.D. in Math., Ukrainian Free U., Munich, Germany, 1946, Dr. Habilitatus, 1947, Dr. Philosophia (hon.), 1962; m. Sophie Kovalchuk, July 16, 1932; children—Bohdan, Roman. came to U. S., 1949, naturalized, 1955. Prof. State Tchrs. Coll., State Gymnasium and Lyceum, Kenty, Poland, 1930-39; instr. math. U. Lviw, 1939-40; prof. math. Tchrs. Coll., Jaworiw, Ukraine, 1942-44; dir. Tchrs. Coll., Mittenwald, Germany, 1945-48; asst. prof. Ukrainian Free U., Munich, 1947-49; faculty Seton Hall U., South Orange, N.J., 1949—, prof. math., 1955—, chmn. dept., 1962—. Mem. Am., Indian, Australian, Edinburgh, London, Glasgow math. socs., Math. Assn. Am., N.Y. Acad. Sci., A.A.A.S., Shevchenko Sci. Soc. N.Y.C. (exec. v.p. 1961—), Société Mathematique de France, Unione Mathematica Italiana, Deutsche Mathematiker Vereinigung, Schweizerische, Österreichische math. Gesellschaften, Polskie Towarzystwo Mat., Svenska Matsf. Mem. staff reviewers Math. Revs., 1961—. Research, publs. on entire functions, infinite series, ordered rings. Home: 149 Milton Pl., South Orange, N.J. 07079.*

ANDRY, Charles Louis François, French physician; b. Paris, 1741; doctor-regent Faculty Paris; physician to Napoleon; mem. Royal Soc. Medicine. Propagator of vaccine; opposed Mesmer. Died Apr. 8, 1829.

ANDRY, Nicholas, French physician; b. Lyons, France, 1658; ed. Reims, also Paris, France; doctorate, 1697. Named prof. Coll. France, 1701; became mem. editorial com. Journal des Savants, 1701; apptd. dean Faculty Medicine, 1724. One of 1st to practice vertebral and orthopedic surgery. Died May 13, 1742.

ANDY, Orlando Joseph, Am. physician, neurosurgeon; b. New Britain, Conn., Jan. 21, 1920; s. Jack and Josephine (Berloni) A.; B.S., Ohio U., 1942; M.D., U. Rochester, 1945; m. Louise C. Murphy, June 11, 1960; children—Maria, Orlando Joseph, Patrick. Intern St. Francis Hosp., Hartford, Conn., 1945-46; surg. intern Baptist Meml. Hosp., Memphis, 1948-49; fellow neuropathology U. Ill. Neuropsychiat. Inst., Chgo., 1949-50; neurosurg. resident Bapt. Meml. Hosp. and John Gaston Hosp., Memphis, 1950-52, fellow, resident neurol. surgery John Hopkins, 1952-53, instr. neurol. surgery, also USPHS fellow div. neurol. surgery, 1953-55; prof., head div. neurosurgery U. Miss. Med. Center, Jackson, 1955-60, prof., head dept. neurosurgery, 1960—. Diplomate Am. Bd. Neurol. Surgery. Mem. Am. Acad. Cerebral Palsy, Am. Acad. Neurology, A.C.S., Am. Electroencephalagraphic Soc., Am. Epilepsy Soc., Am. Fedn. Clin. Research, Am. Heart Assn., Am., So. med. assns., Assn. Research Nervous and Mental Diseases, Carribean, So. neurol. socs., Congress Neurol. Surgeons, Eastern, So. electroencephalagraphic assns., Harvey Cushing Soc., Law-Sci. Acad., Midwestern Psychol. Assn., Neurol. Soc. Am., N.Y. Acad. Scis., Pavlovian Soc., Sigma Xi, Phi Beta Kappa. Author: (with Stephan Heinz) Septum of the Cat, 1964. Research on epilepsy, cerebral palsy. Home: 3815 Rebecca Ct., Jackson, Miss.

ANEL, Dominique, French surgeon; b. Toulouse, France, circa 1679; ed. Toulouse and Montpellier, France; surgeon Emperor of Austria's army, early 18th century. Developed Anel's operations for treatment of aneurysms by ligation, and for dilatation of lacrimal duct with a probe, followed by astringent injection. Died 1730.

ANFIMOV, N. V., Russian archaeologist; D.Hist. Sci. Sr. asso. Inst. Archeology, USSR Acad. Sci. Author: Map of Ancient Settlements and Tombs of the Kuban, 1937; New Findings on the History of the Asian Bosphorus, 1940; Earthern Tombs of the Sarmatian Era in the Underground Graveyards of the Kuban, 1947; Towns in the Eastern Borderland of the Bosphorine State, 1948; The Population of the Kuban in the Scythian Era, 1949; The Meotidian Population of the Eastern Azov Revion, 1950; Agriculture among the Meotidian-Sarmatian Tribes of the Kuban, 1951; New Findings on the Meotidian-Sarmatian Culture of the Kuban, 1952; The Ancient Populations of the Kuban in the Last Millenia B.C. and the First Centuries A.D., 1953; Principal Stages in the Development of the Meotidian-Sarmatian Tribes of the Kuban, 1954; Tribes of the Kuban in the First Centuries A.D., 1955. Research on archeol. relics of Sarmatian era and Kuban tombs, ancient settlements and population of Kuban in Scythian era. Address: Inst. Archeology, 1-ya Cheremushkinskaya ul. 19, Moscow, USSR.

ANFINSEN, Christian Boehmer, Am. biochemist; b. Monessen, Pa., Mar. 26, 1916; s. Christian Boehmer and Sophie (Rasmussen) A.; B.A., Swarthmore Coll., 1937; M.S., U. Pa., 1939; Ph.D., Harvard, 1943; D.Sc., Swarthmore College, 1965, Georgetown University, 1967; married to Florence Bernice Kenenger, November 29, 1941; children—Carol Bernice, Margot Sophie, Christian Boehmer. American-Scandinavian Found. fellow Carlsberg Lab., Copenhagen, 1939, Rockefeller fellow, 1954-55; sr. cancer research fellow Med. Nobel Inst., Stockholm, Sweden, 1947; asst. prof. biol. chemistry, Markle scholar Harvard Med. Sch., 1948-50, prof. biochemistry, 1962-63; Guggenheim fellow Weizmann Inst., Rehovot, Israel, 1958; chief lab. cellular physiology and metabolism Nat. Heart Inst., Bethesda, Md., 1950-62; chief lab. chem. biology Nat. Inst. Arthritis and Metabolic Diseases, Bethesda, 1963—. Bd. govs. Weizmann Inst. Sci., Rehosot, Israel. Mem. Am. Soc. Biol. Chemists, Am. Acad. Arts and Scis., Nat. Acad. Scis., Washington Acad. Scis., Fedn. Am. Scientists (treas. 1958-59, vice chmn. 1959-60). Author: The Molecular Basis of Evolution, 1959. Contbr. sci. publs. Research on structure-function relationship in proteins; protein isolation, proteolysis, and synthesis; genetic basis of protein structure. Home: 9202 Cedar Way, Bethesda 14. Office: National Inst. Arthritis and Metabolic Diseases, Bethesda 14, Md.

ANG, Dang Dinh, Vietnamese applied mathematician; b. Hadong, Vietnam, Mar. 16, 1928; s. Dang Dinh and Nguyen Thi (Chuc) Bat; B.S., U. Kan., 1955; Ph.D., Cal. Inst. Tech., 1958; m. Buithi Minh Thi, Dec. 4, 1950; 5 children. Research fellow aeros Cal. Inst. Tech., 1958-60, vis. sr. research fellow geophysics, 1963; staff U. Saigon, Viet Nam, 1960—, prof. math., 1967—; Asaihl itinerant prof., Thailand, Philippines, Malaysia, Hong Kong, 1966; vis. prof. math. U. Malaya, 1966. Mem. Am. Math. Soc., Engring. Sci. Soc., Am. Union Geophysics. Research, publs. on mechanics and applied functional analysis. Home: 57 Duy-Tan, Saigon, Viet Nam.*

ANGEL, J(ohn) Lawrence, anthropologist; born London, Eng., Mar. 21, 1915; s. John and Elizabeth Day (Seymour) A.; came to U. S., 1928, naturalized, 1937; grad. Choate Sch., 1932; A.B. magna cum laude, Harvard, 1936, Ph.D., 1942; m. Margaret Seymour Richardson, July 1, 1937; children—Elizabeth Richardson (Mrs. Joel H. Feigon), Stephen Bearne, Jonathan Seymour. Asst. anthropology Harvard, 1939-41; instr. U. Cal., Berkeley, 1941-42, U. Minn., Mpls., 1942-43; asso. anatomy and phys. anthropology Jefferson Med. Coll., Phila., 1943-50, asst. prof., 1950-57, asso. prof., 1951-62, prof., 1962, vis. prof., 1962—; curator phys. anthropology Smithsonian Instn., U. S. Nat. Mus., Washington, 1962—; research asso. U. Mus., Phila., 1946-62; professorial lectr. anthropology George Washington U., Washington, 1962-—; vis. prof. anatomy Med. Sch., Howard U. Washington, 1962-—. Staff mem. Agora Excavations, Athens, Greece, 1949-—; civilian cons. surg. anatomy U. S. Naval Hosp., Phila., 1957-62; adv. panel on anthropology and history and philosophy sci NSF, 1960-62. Guggenheim fellow, 1947. Mem., A.A.A.S., Am. Assn. Phys. Anthropologists (v.p. 1959-61), Am. Anthropol. Assn., Archaeol. Inst. Am., Am. Assn. Anatomists, Am. Soc. Human Genetics, Greek Anthropol. Soc., Phi Beta Kappa. Author: Troy, The Human Remains, 1951. Contbr. articles to tech. jours. Home: 5311 Wriley Rd. Washington 20016. Office Div. Phys. Anthropology, Smithsonian Instn. Washington 20560.*

ANGELAKOS, Evangelos Theodoru, physiologist, pharmacologist; b. Tripolis, Greece, July 15, 1929; s. Theodore and Aglaia (Tsiverioti) A.; came to U. S., 1948, naturalized, 1966; student Athens (Greece) U., 1947-48; Fordham U., 1948-50; Cornell U., 1950-51; M.A., Boston U., 1953, Ph.D., 1956; M.D., Harvard, 1959; m. Eleanor Pell, Aug. 28, 1954; 1 son, Theodore Angelakos. Faculty, Boston U. Sch. Medicine, 1955-—, prof. physiology, 1963-—; research asso. in biomath. Mass. Inst. Tech., 1959-61. Dir. Harvard Apparatus Co., Dover, Mass.; cons. to govt. agys. Recipient USPHS Research Career Devel. award, 1960-—. Mem. Am. Physiol. Soc., Am. Soc. Pharmacology and Exptl. Therapeutics, Biophys. Soc., Biometric Soc., A.A.A.S., Sigma Xi. Research, publs. on physiology and pharmacology heart and circulation, mode action drugs, evaluation effect nerves on action heart; devel. new techniques. Home: 21 College Rd., Wellesley, Mass. 02118. Office: 80 E. Concord St., Boston 02181.*

ANGELI, Stefana, Italian mathematician; b. Venice, Italy, 1623; mem. Soc. of Jesus; prof. math., Rome, later Padua, Italy. Author: De infinitis parabolis, de infinitisque solidis ex variis rotationibus ipsarum partiumque earum genitis, 1654; Problemata LX circa conos, 1658; Miscellaneum hyperbolicum, 1659; Miscellaneum geometricum. De infinitorum cochlearum mensuris, 1661; Accessio ad stereometriam et mechanicam, 1662; Considerazione sopra la forza del argomento del G.-B. Riccioli contra il moto diurno della terra, 1667; Seconda Considerazione, etc., 1668; Terza considerazione, etc., 1669; Della gravita dell'Aria e Fluidi, 1671, 72. Died Padua, 1697.

ANGELL, James Rowland, Am. psychologist; b. Burlington, Vt., May 8, 1869; s. James Burrill and Sarah Swope (Caswell) A.; A.B., U. Mich., 1890, A.M., 1891; A.M., Harvard, 1892; Litt.D., U. Vt., 1915; LL.D., Yale, Harvard, Princeton, Columbia, U. Chgo., Union, U. Cin., McGill, Wesleyan (Conn.), Brown, U. Mich., U. Cal., N.Y. U., Williams Coll., Dartmouth, Rutgers, U. Pa., others; hon. Ph.D., Rensselaer Polytech.; univs. of Berlin and Halle, 1893; traveled and studied at Vienna, Paris, Leipzig; m. Marion Isabel Watrous, Dec. 18, 1894 (dec. June 1931); children—James Waterhouse, Marion Waterhouse, Caswell (Mrs. William Rockefeller McAlpin); m. 2d, Mrs. Katharine Cramer Woodman, Aug. 2, 1932. Instr. philosophy U. Minn., 1893; asst. prof. psychology and dir. psychol. lab. U. Chgo., 1894-1901, asso. prof., 1901-05; prof. and head dept., 1905-19, sr. dean, 1908-11, dean univ. faculties, 1911-19, acting pres., 1918-19; pres. Carnegie Corp., 1920-21; Yale U., 1921-37; ednl. counselor NBC, from 1937. Exchange prof., The Sorbonne, Paris, 1914. Chmn. psychology com. NRC, 1919-20; v.p. Nat. Com. for Mental Hygiene, Internat. Com. for Mental Hygiene. Decorated officer Legion of Honor, 1930; grand officer Order Crown of Italy, 1935; Chinese Blue Grand Cordon Order of Jade, 1937; gold medal Nat. Inst. Social Science, 1937. Trustee Am. Museum Natural History; bd. dirs. Museum of Science and Industry. Fellow Am. Acad. Arts and Scis.; mem. Am. Psychol. Assn. (pres. 1906), Am. Philos. Soc., Nat. Acad. Sciences, English-Speaking Union (nat. pres.), Phi Beta Kappa, Sigma Xi, Kappa Delta Pi; hon. mem. British Psychol. Soc. Author: Psychology (4th edit.), 1908; Chapters from Modern Psychology, 1911; Introduction to Psychology, 1913; American Education, 1937; also many articles in scientific jours. Editor of Psychological Monographs. 1912-22. Did significant work (with A. W. Moore) on reaction times; a leading figure in rise of functional psychology in U. S. Died Hamden, Conn., Mar. 3, 1949.

ANGELL, Robert Cooley, Am. sociologist, educator; b. Detroit, Apr. 29, 1899; s. Alexis Caswell and Fannie (Cooley) A.; A.B., U. Mich., 1921, M.A., 1922, Ph.D., 1924; m. Esther Robbins Kennedy, Dec. 23, 1922; children—James Kennedy, Sarah Caswell (Mrs. Donald Holcomb Parsons). Faculty sociology U. Mich., Ann Arbor, 1922-—, prof., 1935-—, chmn. sociology dept., 1940-52, dir. honors program Lit. Coll., 1957-61, dir. Center for Research on Conflict Resolution, 1963-65. Research asso. Columbia Law Sch., 1928-29; acting dir. social sci. dept. UNESCO, Paris, France, 1949-50; Deiches lectr. Johns Hopkins, 1957. Decorated Bronze Star medal. Mem. Am. (pres. 1950, dir. project Sociol. Resources for Secondary Schs. 1966-—), Internat. (pres. 1953-56) sociol. assns. Author: The Campus, 1928; A Study in Undergraduate Adjustment, 1930; (with C. H. Cooley and L. J. Carr) Introductory Sociology, 1932; The Family Encounters the Depression, 1936; The Integration of American Society, 1941; The Moral Integration of American Cities, 1951; Free Society and Moral Crisis, 1958, 65. Editor, Am. Sociol. Rev., 1946-48. Research on social and family integration; cities; the national society; problems of world integration; the growth of transnational participation. Home: 1007 Berkshire Rd., Ann Arbor, Mich. 48104.*

ANGELUCCI, Theodore, Italian physician; b. Belforte, Italy, 1540; known for his disputes with Patrizi on Aristotle; Author: Sentencia quod metaphysica sit eadem quae physica, 1584; Exercitatonum cum patricio liber, 1585; other med. works. Translated Aeneid of Virgil into blank verse, 1649. Died Montagnana, Italy, 1600.

ANGELUS, Johannes, astronomer; b. Bavaria, 1463; ed. Ingolstadt, Germany; worked in Vienna; corrected Georg Purbach's planetary tables. Author: Astrolabium planum, an astrological work, 1488; edited other astronomical and astrological works published by Ratdolt at Augsburg. Died Vienna, Sept. 29, 1512.

ANGENHEISTER, Gustav Heinrich, meteorologist; b. Cleve, Germany, Feb. 26, 1878; s. Heinrich and Elisabeth Henrika (Jansen) A.; studied math. and natural scis. at Heidelberg, Munich, Berlin (all Germany); m. Edith Anna Mathilde Wilhelmine Tammann, May 20, 1914. Asst., Physics Inst., U. Heidelberg, 1902-03; asst. Geophys. Inst., U. Göttingen (Germany), 1905, apptd. prof. geophysics, 1911; observator Geophys. Observatorium of Samoa; dir. Samoa Obs., 1914-21; observator geophys. dept. Geodetical Inst. Potsdam (Germany), 1921-—, dir. geophys. dept., 1925-—; dir. Geophy. Inst., Göttingen, 1928-—; named prof. h.c. Tech. U. Berlin, 1926; mem. expdn. to Iceland. Editor in chief Zeitschrift für Geophysik. Research on earthquakes especially propagation of surface waves, earth magnetism. Died Göttingen, June 28, 1945.

ANGER, Hal Oscar, Am. elec. engr.; b. Denver, May 24, 1920; s. Oscar Christian and Rosa (Nichols) A.; B.S., U. Cal. at Berkeley, 1943. Research asso. Radio Research Lab. Harvard, 1943-46; engr., research asso. Donner Lab., U. Cal., Berkeley, 1946-—. Recipient John Scott award for invention Positron Scintillation Camera; Guggenheim fellow, 1966-67. Mem. Soc. Nuclear Medicine, I.E.E.E., Sigma Xi, Tau Beta Pi. Author chpt. Instrumentation in Nuclear Medicine, 1967. Research, numerous publs. in use of heavy charged particle beams from cyclotrons for radiation therapy of tumors; inventor of first clinically successful radioisotope camera used in hospitals for visualizing tumors and organs with radioactive isotopes and other instruments used in nuclear medicine. Office: 361 Donner Lab., U. Cal., Berkeley, Cal. 94720.*

ANGERER, Lorenz Ludwig Maximilian, German physicist; b. Würzburg, Germany, Feb. 2, 1881; s. Ottmar Ritter von and Anna (Fasolt) Ritter von A.; ed. univs. Munich, Würzburg, Göttingen; degree, 1905; m. Clara Ida Schrauth, Mar. 23, 1911, 2 sons. Asst. to W. Röntgen, 1907-11; became conservator physics inst. Tech. U., Munich, 1912, lectr., 1920, asso. prof., 1927. Author: Wissenschaftliche Photographie, 1931; Technische Kunstgriffe bei physikalischen Untersuchungen, 1924. Investigated cathode rays, spectrum of atoms and molecules, sound measurements, photography. Died Munich, Feb. 20, 1951.

ANGEVINE, Daniel Murray, pathologist; b. St. John, N.B., Can., Oct. 8, 1903; s. James Edwin and Edna (Irvine) A.; B.A., Mt. Allison U., 1924; M.D., McGill U., 1929; m. Dorothy Edna Shepherd, July 9, 1933; children—James Murray, Charles Douglas, Judith Melanie (Mrs. Eugene Flath). Came to U. S., 1932, naturalized, 1942. Faculty, Cornell U. Med. Sch., 1932-40; vis. asst. prof. pathology U. Pa. A.I. duPont Inst., Wilmington, Del., 1940-45; prof. pathology, chmn. dept. U. Wis. Med. Sch., Madison, 1945-—. Cons. to Surgeon Gen., U. S. Army, VA, NRC; chief research pathologist Atomic Bomb Casualty Commn., NRC-NIH, Hiroshima, Japan, 1962-63. Bd. dirs. Wis. Arthritis Found. Mem. Am. Soc. Exptl. Pathology (pres. 1954), Am. Assn. Pathologists and Bacteriologists (pres. 1962), Federated Soc. Exptl. Biologists (chmn. bd. 1955). Chief editor Archives Pathology, 1964-—. Contbns. in field of pathologic alterations of connective tissues including bone, joints, and muscle; investigations of infected joints, healing of fractures, and connective tissue injuries; studies on growth and regeneration of muscle. Home: 3406 Viburnum Dr., Madison, Wis. 53705.*

ANGHILERI, Leopoldo José, chemist; b. Luján, Buenos Aires, Argentina, Aug. 22, 1928; s. José and Maria (Orlando) A.; Licenciate in Chem. Scis., U. Buenos Aires, 1951, D.Chemistry, 1957; m. Gladys Cooper Montenegro, Dec. 12, 1952 (div. July 1961); 1 dau., María Luján. Researcher, Argentine AEC, 1956-60, 63-64; mem. ednl. staff, 1956-60; chief chemist Volk Radiochem. Co., Skokie, Ill., 1960-63; research staff Bourse Joliot-Curie, Institut du Radium-Laboratorie Curie, Paris, France, 1964-66; research staff Deutsches Krebsforschungszentrum-Institut für Nuklearmedizin, Heidelberg, Germany, 1966-—; vis. prof. Rio Grande Do Sul, U., Brasil, 1965. Mem. Am. Chem. Soc., Soc. Nuclear Medicine. Research, publs. on techniques of preparation of radiopharmaceuticals and studies concerning their physiol. behavior.*

ANGIER, Robert Bruce, Am. organic chemist; b. Litchfield, Minn., Mar. 24, 1917; s. Hugh I. and Doris M. Angier; B.S. summa cum laude, Hamline U., 1940; M.S., U. Neb., 1942, Ph.D., 1944; m. Wilma B. Lamb, June 1945; children—Bruce N., Karen A. Research chemist Lederle Labs. div. Am. Cyanamid Co., 1944-56; research asso., 1956-62, group leader, 1962-—. Mem. Am. Chem. Soc., N.Y. Acad. Scis. Research and publs. on folic acid analogues, chemo-therapy of cancer, antineo-plastic and antiviral compounds, pteridine, pyrimidine and purine chemistry; synthesized folic acid, 1946, also pteroylglutamic acid. Home: 215 Rockland Rd. Office: 2 Lederle Labs., Pearl River, N.Y.

ANGIULLI, André, Italian philosopher; b. Castellana, Italy, Feb. 12, 1837; prof. U. Bologna (Italy), 1872-76; founder la Rassegna critica (philos. and lit. review), 1880. Author: la Philosophie et la recherche positive, 1868; Rivista critica di scienze lettere ed arti, 1871; Questions de philosophie contemporaine, 1873; la Pédagogie, l'etate et la famille, 1876; Rassegna di opere filosofiche, scientifiche e letterarie, 1881; la Filosofia e la suola, 1888. Attempted to introduce sci. method to solve metaphys. problems; advocated state education of both sexes. Died Rome, Italy, Jan. 2, 1890.

ANGLE, Edward Hartley, Am. orthodontist; b. Herrick, Pa., June 1, 1855; s. Philip Casebeer and Isabel (Erskine) A.; grad. Pa. Coll. Dental Surgery, Phila., 1878; M.D., Marion Sims Coll. Medicine, St. Louis, 1897; Sc.D., U. Pa., 1915; m. Anna Hopkins, June 27, 1908; 1 dau., Florence Isabel. Began practice dentistry at Towanda, Pa., 1878; prof. orthodontia, U. Minn., 1887-92, Northwestern U., 1892-98, Marion Sims Coll. of Medicine, 1896-99, med. dept. Washington U., 1897-99; founder, 1900, and first pres. Angle Sch. of Orthodontia, St. Louis, also Angle Coll. of Orthodontia and Infirmary, Pasadena, Cal., 1917. Author: Malocclusion of the Teeth, 1887 (8 edits.). Established orthodontia as a specialty and was 1st orthodontic specialist; patented 40 orthodontic inventions; designed wire splint for use in securing lower to upper teeth in fracture of mandible; devised standard classification of various types of malocclusion. Died Pasadena, Cal., Aug. 11, 1930.

ANGLE, Edward John, Am. dermatologist; b. Cedarville, Ill., Apr. 1, 1864; s. John Bouslough and Jane (Bell) A.; B.Sc., U. Wis., 1886; M.D., Med. Coll. of Ohio (U. Cin.), 1887; A.M., U. Neb., 1898, postgrad. research in embryology, 1896-98; M.D., U. Pa., 1895; postgrad work, N.Y.; m. Agnes Lillian Wolf, June 6, 1889; children—Sarah Jane, Florence Bell (Mrs. Guy E. Reed), Edward Everett Dupuytren, Barbara Josephine, Agnes Evelyn (Mrs. Harry Stevens). Practiced at Lincoln, Neb., 1895-—; prof. skin and genito-urinary diseases, Neb. Coll. Medicine, 1905-10; dermatologist St. Elizabeth's and Lincoln Gen. hospitals; cons. dermatologist Bryan Meml. and Lincoln Vets. hosps. Editor dept. skin and genito-urinary diseases, Western Med. Rev., 1902-07. Author: Parents and Their Problems (series of vols.), 1914; American Medicine—Expert Testimony Out of Court (2 vols.); also articles on dermatology, venereal diseases. Died Apr. 24, 1940.

ANGLERIUS, Petrus Martyr, see d'Anghiera, Pietro Martire.

ANGLICE, William Shelley, see William of Conches.

ANGO, Father Pierre, physicist, mathematician; lived at Caen, France, between 1650-1700; proposed a wave theory of light, caused by vibrations of source (in his Optique, 1682).

ANGOT, André Marie, French mathematician, physicist; b. Le Puy, France, Feb. 15, 1905; s. Louis Gabriel and Louise (Giraud) A.; Ingenieur, École Polytechnique, Paris, 1929, École Supérieure d'électricité, Paris, 1933; m. Louba Gorodetsky, Sept. 22, 1942. Became dir. Service du Controle Technique des P. T. T. Lyon, 1942; dir. Sect. d'Etudes du Materiel de Transmission; became asst. dir. Centre Nat. d'Études des Telecommunications (CNET), Paris, 1946; apptd. dir. SEFT, Paris, 1952, ingeneiur gen., 1955; became tech. dir. Sociéte T.R.T., 1957, now sci. dir. Prof. applied math. Ecole Supérieure d'electricité, Paris, 1942-—; pres. French Nat. Com. of Sci. Radioelectricity (French URSI sect.), 1960-62. Fellow I.E.E.E.; mem. I.R.E. (v.p. 1962), Société Francaise des Electroniciens et Radioélectriciens (pres. 1960), other sci. assns. Author: Complements de mathématiques à l'usage des ingenieures de l'electrotechnique et des télécommunications, 1949, 5th edit., 1965, also fgn. lang. edits. Research in mil. telecommunications and engring. edn. Home: 2 rue Eugène Poubelle, Paris 16, France. Office: 5, Ave. Reaumur, Le Plessis-Robinson (92), France.*

ANGOT, Charles Alfred, French meteorologist, physician; b. Paris, July 4, 1848; ed. L'Ecole normale superieure; became mem. staff Central Meteorol. Bur. France, 1878, dir., 1908; prof. physics, meteorology Inst. national agronomique, Paris; prof. physics Collège Chaptal, Paris, 1879-1924. Corr. mem. Bur. Longitudes. Author: Traité de météorologie, des instructions meteorologiques . . . (meteorologic classic); other studies on French climate. Contbd. numerous articles to sci. jours. Died Paris, 1924.

ANGSTRöM, Anders Jonas, Swedish physicist, astronomer; b. Lögdö, Medelpad, Sweden, Aug. 13, 1814; ed. U. Uppsala (Sweden); m., 1 son, Knut Johan. Apptd. privat docent in physics U. Uppsala, 1839, prof. physics, from 1858; worked at Stockholm Obs., 1842; became observer Uppsala Obs., 1843. Fellow Royal Soc., 1870 (Rumford medal 1872). Corr. physics sect. French Acad. Scis., 1873. Author: Optiska Undersokninfar (most important work, presented to Stockholm Acad.), 1853; Recherches sur le spectre solaire, 1869. One of founders of spectroscopy; studied spectra of flames, electric arcs, the sun, planets; showed that common lines in all these spectra are due to the gas through which the spark is discharged, and that incandescent gas emits luminous rays of same refrangibility as those it can absorb, 1853 (later recognized as containing a fundamental principle of spectrum analysis); began spectral analysis of the sun, 1861; discovered hydrogen in solar atmosphere, 1862; pub. his great map of normal solar spectrum, 1868; made map of entire solar spectrum, 1869; was 1st to examine spectrum of aurora borealis, 1867; also studied heat and magnetism, investigated the double refraction of metals; demonstrated that thermal conductivity is proportional to elec. conductivity, discovered method of measuring thermal conductivity; Angström unit (measuring length of light waves) named in his honor, 1905. Died Uppsala, June 21, 1874.

ANGSTRöM, Knut Johan, Swedish physicist; b. Uppsala, Sweden, 1857; s. Anders Jonas Angström; prof. physics, Uppsala, from 1896. Invented electric compensation pyrheliometer, 1893, other apparatus for photography infra-red spectrum; investigated solar radiation. Died Uppsala, Mar. 4, 1910.

ANGUILLARA, Luigi, Italian botanist, physician; b. Anguillara, early 16th century; became supt. botanic gardens, Padua, Italy, circa 1546; prof. U. Padua. Author: Sur les simples (descriptions of 20 new plants). Tried to identify plants with those described by ancient Greek and Roman writers. Died Ferrara, Italy, circa 1570.

ANGYAL, Stephen John, organic chemist; b. Budapest, Hungary, Nov. 21, 1914; s. Charles and Maria (Szanto) A.; Ph.D., Pázmány Péter U., Budapest, 1937; D.Sc., U. New S. Wales (Australia), Sydney, 1964; m. Helga Ellen Steininger, Feb. 28, 1942; children—Annette Marie, Robert Stephen. Staff, Chinoin, Hungary, 1937-40, Nicholas Pty. Ltd., Melbourne, Australia, 1941-46; lectr. U. Sydney (Australia), 1946-52; Nuffield Brit. Dominion Travelling fellow, Cambridge, Eng., 1952; asso. prof. New S. Wales U. Tech., 1953; prof. U. New S. Wales, 1960—. Carnegie Vis. fellow, 1953. Fellow Royal Australian Chem. Inst. (H. G. Smith Meml. medal), Australian Acad. Sci. Author: (with E. L. Eliel, N. L. Allinger, G. A. Morrison) Conformational Analysis, 1965; also articles. Application conformational analysis (study shapes of molecules) to chemistry of sugars; developed chemistry of inositols, sugar-like natural products; expanded applications of stereochemistry. Home: 304 Sailors Bay Rd., Northbridge, N.S.W. 2063. Office: Sch. Chemistry, U. New S. Wales, Kensington, N.S.W. 2033, Australia.*

ANIANSSON, G., Swedish chemist; b. Sundby, Sweden, Jan. 31, 1927; s. Albert Sjunne and Karin (Johansson) A.; Techn. Dr., Royal Inst. Tech., Stockholm, Sweden, 1961; m. Ingrid AnnMarie Scharman, Nov. 8, 1949; children—Lars Jockum, Hans Gunnar. Asst., Royal Inst. Tech., 1948-60, research asst., 1960-63, docent in phys. chemistry, 1961; prof. phys. chemistry U. Göteborg (Sweden), 1963—. Research, publs. on methods in surface chemistry, effects chem. bonds on interaction of charged particles with molecular electrons. Home: 33 Aschebergsgatan, Göteborg C. Office: 5A Gibraltargatan, Göteborg S, Sweden.*

ANICH, Peter, cartographer; b. Oberperfuss, Germany, Feb. 22, 1723; s. Ingenuin and Gertrud (Hammer) A.; self-taught in math., astronomy, geodesy. Turner by trade; maker terrestrial and celestial globes; commd. by govt. of Tyrol (Austria) to survey area by triangulation and charting (maps completed by B. Bueber, in use until 1840), 1760. Author: Atlas Tyrollensis, 20 sheets, 1774. Died Sept. 1, 1766.

ANICHKOV, Nikolay Nikolaevich, Russian pathologist; born November 3, 1885; graduated from St. Petersburg Military Medical Academy, 1909; Doctor of Medical Science, 1912; D.Med. Fac. (hon.), Berlin Humboldt U., 1960. Specialized in path. anatomy, until 1912; on sci. mission in Germany, 1913-14; dr. in active Russian army, 1914; lectr. chair path. anatomy Leningrad Mil. Med. Acad., 1916-19, head chair path. physiology, 1920-39, head chair path. anatomy 1939-46; on staff Inst. Exptl. Medicine, USSR Acad. Med. Sci., 1920—, head dept. path. anatomy. Mem. learned council USSR Ministry Health; del. All-Union and Internat. congresses in Germany, Sweden, Japan, Holland, others. Recipient Stalin prize, 1942, Order of Lenin (2). Mem. USSR (Mechnikov prize 1952), GDR (corr.) acads. sci., USSR Acad. Med. Sci. (pres. 1946-53). Author over 150 works, including Handbook of Pathological Physiology, 4th edit., 1938; Pathology of the Blood Vessels, 1940; Experimental Autoinfection, 1947; The Present State of Experimental Atherosclerosis, 1956; co-author Lesions of the Arteries in Old Age and in Atherosclerosis, 1941, Morphology of the Healing of Wounds, 1951. Editor, co-editor various med. publs., including Soviet encys., Archives of Pathology, Soviet Medical Experience in the 1941-45 Great Patriotic War; mem. editorial council Path. Physiology and Exptl. Therapy; sci. editorial staff Living Conditions and Health. Research on pathology of blood vessels, correlation of medullary centers, course of functional disorders in connection with brain anemia, reticulo-endothelial system, atherosclerosis, atheromatosis, hypertonia, pathogenesis of infectious diseases, war injuries. Address: Inst. Exptl. Medicine, USSR Acad. of Med. Scis., Kirovsky pr. 69-71, Leningrad. USSR.

ANICHKOV, Sergey Viktorovich, Russian pharmacologist; b. 1882; grad. Petrograd Med. Inst., 1918; D.Med. Sci. asst. chair pharmacology Petrograd. Mil. Med. Ac., 1919-22; lectr. chair pharmacology Leningrad Med. Inst., 1922-24; prof., head chair pharmacology Leningrad Mil. Med. Acad., 1924-37; head chair pharmacology Leningrad San. and Hygiene Med. Inst., 1945—; head dept. pharmacology Inst. Exptl. Med., USSR Acad. Med. Sci., 1948—. Elected to Internat. Council Pharmacologists as USSR rep. Internat. Congress Physiologists, Brussels, 1956, v.p. pharmacology sect., Buenos Aires, 1959. Decorated Order of Lenin, also others. Mem. USSR Acad. Med. Sci. Author over 80 works, including The Vascular Activity of Isolated Fingers in Healthy and Sick Persons, 1922; Pharmacological Heart Therapy in the Light of Experimental Pharmacology, 1936; Some Results of Pharmacological Analysis of the Chemical Sensitivity of the Carotid Glomerule, 1951; Pharmacological Substances of Adrenolytic and Cholinolytic Action, 1952; co-author Testbook of Pharmacology, 1955. Co-editor Pharmacology sect. Large Med. Ency., 2d edit.; mem. editorial bd. various periodicals including Pharmacology and Toxicology. Research on pharmacology of cardiovascular system. Address. Inst. Exptl. Medicine, Kirovsky pr. 69-71, Leningrad, USSR.

ANIGSTEIN, Ludwik, microbiologist; born Warsaw, Poland, Feb. 2, 1891; s. Isadore and Helen (Steinkalk) A.; Ph.D. magna cum laude, U. Heidelberg, 1914; M.D., U. Dorpat, 1915; M.D., U. Poznana, 1923; postgrad. certificate London Sch. Trop. Medicine and Hygiene, 1923; m. Dorothy M. Whitney, May 2, 1958; children—Alice, Robert. Came to U. S., 1939, naturalized; 1946. Parasitologist, State Inst. Hygiene, Warsaw, 1919-39; lectr. microbiology U. Warsaw, 1926-39; mem. Malaria Commn., League of Nations, 1924-39; research asso., lectr. U. Cal. at San Francisco, 1940; faculty U. Tex. Med. Br., Galveston, 1940—, prof. preventive medicine and pub. health, 1946—. Cons. to govts. in Asia, Africa, Europe, UNRRA, AEC; Gabriel Kempner vis. prof. U. Hamburg, 1960. Fellow Royal Soc. Tropical Medicine and Hygiene (sec. for Tex.), Am. Acad. Microbiology, A.A.A.S., Tex., N.Y. acad. scis.; mem. Tex. Pub. Health Assn., Soc. Exptl. Biology and Medicine, Am. Soc. Exptl. Pathology, Sci. Research Soc. Am., Reticulo-endothelial Soc., Sigma Xi, Phi Delta Epsilon, Mu Delta. Research on cancer by immunology and chemotherapy. Home: 28 Manor Way, Galveston, Tex. 77550.*

ANKER, Mathias Josef, Austrian mineralogist; b. Graz, Austria, May 6, 1771; s. Mathias and Elisabeth (Reiser) A.; ed. in music, natural scis. and medicine; m. Theresia Goutta, May 10, 1795; surgeon in Stainz; became dist. surgeon, Graz, in 1807; became asso. with Prof. Friedrich Mohs of Johanneum, 1811; apptd. prof. mineralogy, curator mineral. collection, 1818. Author several books, including: Kurze Darstellung einer Mineralogie von Steiermark, 1809; Geognostische Karte von Steiermark, 1832; also numerous articles. Produced new edit. of the catalog of the collection of minerals and metals (following classification system of Mohs), 1828-29. Discovered and analysed ankerit, 1835. Died Graz, Apr. 3, 1843.

ANNAEV, Rukhi Guseynovich, Russian physicist; b. 1909; grad. Moscow U., 1936. Instr., Moscow U., 1936-41, Ashkhabad Pedagogical Inst., 1941-45; dir. Physicochem. Inst., Turkmen br. USSR Acad. Sci., 1941-51; prof. Turkmen Acad. Sci., 1950—. Mem. Turkmen Acad. Sci. (v.p. 1956-60). Author: The Anomaly of the Even Thermomagnetic Effect in Ni-Mn Alloys, 1950; Magneto-Electric Phenomena in Ferromagnetic Metals, 1952; The Ordering of Atoms in Certain Alloys (Superstructure), 1956; A Study of the Magneto-Electric Properties of a Germanium and Ferrite Monocrystal, 1957; The Galvano-Magnetic Effect of Iron-Molybednum Alloys. 1962. Research in physics of metals and magnetology, magneto-elec. phenomena in ferromagnetic metals, semicondrs. Address: Gosudarstvenny University, Ashkhabad, Turkmenia SSR, USSR.

ANNETT, Henry Edward, English pathologist; b. June 5, 1871; ed. Univ. Coll., Liverpool, Eng., Victoria U., Manchester, Eng.; grad. with honors, 1894; m. E. L. Bell, 1906. Lectr. comparative pathology U. Liverpool, 1903, prof. comparative pathology, 1906-11, lectr. in animal pathology, 1922-28, cancer researcher, 1923-30; Turner research fellow cancer, 1931-38; supt. Inc. Liverpool Inst. Comparative Pathology from 1902; supt. Research Labs., Higher Runcorn, 1911-22; med. officer health Runcorn Urban Dist., 1913-23; acting med. officer health Widnes Borough Dist., 1915-19; med. officer in command Runcorn Vicarage Mil. Hosp. Mem. 1st expdn. Liverpool Sch. Tropical medicine to W. Africa, 1891, dir. 2d expdn., 1900; dir. animal diseases expdn. to Uruguay, 1905, Colonial Office expdn. to W.I., 1906-07. Author (with others) Report of Malaria Expedition to Sierra Leone, 1899, Report of Expedition to Nigeria (malaria), 1900, Report of Expedition to Nigeria (filariasis), 1900. Died Apr. 10, 1945.

ANOKHIN, Petr Kuzmich, Russian physiologist; b. January 26, 1898; graduated from Leningrad Institute Medical Knowledge, 1926; Doctor of Medical Science. With Pavlov's Laboratory, 1926-30; professor chair physiology Gorky Med. Inst., 1930-34; head dept. gen. physiology of higher nervous system All-Union Inst. Exptl. Medicine, 1934-46; head chair physiology, pathology of higher nervous system Central Postgrad. Med. Inst., 1936-49, 53-55; dir. Inst. Physiology, USSR Acad. Med. Sci., 1946-49; head chair physiology Vishnevsky Inst. Surgery, USSR Acad. Med. Sci.; head Lab. Human Embryogenesis, Inst. Midwifery and Gynecology, RSFSR Ministry Health. Del., Internat. Congress Physiologists, Brussels, 1956, Buenos Aires, 1959, Paris, 1960. Mem. USSR Acad. Med. Sci., Internat. Orgn. for Brain Research (exec. com. Co-author: The Problem of the Central and Peripheral Nervous System in the Physiology of Nervous Activity, 1935; author: Problems of Higher Nervous Activity, 1940; Neuroplasty for Battle Injuries of the Peripheral Nervous System, 1944; Systemogenesis as a General Pattern of the Evolutionary Process, 1948; The Problem of Cortical Inhibition and Its Place in the Study of Higher Nervous Activity, 1948; General Principles of the Compensation of Functional Disturbances and their Physiological Basis, 1955; Electroencephalograph Analysis of the Conditioned Reflex, 1958; A New Conception of the Architecture of the Conditioned Reflex, 1959. Mem. editorial bd. Sechenov Physiol. Jour. of USSR; editor Physiology sect. Large Med. Ency., 2d edit. Developer secretory-motor method of conditioned reflexes, 1930; formulated idea of systemogenesis as gen. pattern of evolutionary process and proposed physiol. theory of neutral cicatrix, theory of pathogenesis of amputation pains and theory of pathogenesis of central paralyses. Address: 1-y Med. Inst., B. Pirogovskaya 2-6, Moscow, USSR.

ANOSOV, Pavel Petrovic, Russian metallurgist; b. 1797. Author: O bulatach (über Damaszenerstahl), 1541. Introduced microscopic study of metal grindings into Russia; developed Damascene steel for blades. Died 1851.

ANREP, Gleb Vassilievitch von, see von Anrep, Gleb Vassilievitch.

ANSAY, Joseph Fanny Marie, Belgian radiologist; b. Ranst, July 1, 1912; s. Leon and Anne (Verhaert) A.; ed. Free U., Brussels; M.D. in surg. medicine and obstetrics; m. Herminie Bauwens, Dec. 16, 1935; children—Evelyne, Jacques. Asst. hosps., Brussels; chief radiology service Med. Surg. Inst. Longchamps, Brussels. Mem. French Soc. Radiology, also several Belgian sci. socs. Author numerous publs. on radiology. Home: 16, avenue Ptolémée, Brussels 18. Office: 255, avenue Winston-Churchill, Brussels 18, Belgium.

ANSBACHER, Stefan, biologist; b. Frankfort/Main, Germany, Jan. 27, 1905; s. Benno and Lucy Diana Ansbacher; B.Sc., U. Frankfort/Main, 1923; M.S., U. Geneva (Switzerland), 1929, Sc.D., 1933; children—Eleonore (Mrs. S. Alan Rall), Benno, Rudi, M. Victoria; m. 2d, Maria Magdalena Barley, Mar. 19, 1951. Research instr. Med. Coll. Charleston (S.C.), 1930; med. researcher Borden Co., Bainbridge, N.Y., 1930-36; research asso. Squibb Inst. for Med. Research, New Brunswick, N.Y., 1936-42; research dir. Am. Home Products Corp., N.Y.C., 1942-45; Schenley Industries, N.Y.C., 1945-46; sci. and med. cons., 1946—. Mem. Ind. State Med. Assn. (hon.), Soc. for Exptl. Biology and Medicine, Am. Soc. Biol. Chemists, A.A.A.S., Am. Soc. Profl. Biologists, A.M.A. (Billings medal 1936), Grant County Med. Soc. Research, publs. on chemotherapy of Tb; nutrition; animal protein factor; animal disease; enzymology; pharmaceuticals; hormonology; vitamins B12 and K and analogues; precipitation of proteinaceous matter. Address: 115 Overlook Rd., Shady Hills, Marion, Ind. 46952.

ANSCHÜTZ, Ottomar, photographer; b. Poznan, Poland, May 16, 1846; took over father's photog. shop, 1868; pioneer in cinematography; perfected snapshot pictures; invented device which could produce a series of animated pictures. Died Berlin-Friedman, Germany, May 30, 1907.

ANSCHÜTZ, Richard, German chemist; b. Darmstadt, Germany, Mar. 10, 1852; s. Carl Ludwig and Henriette Friederike (Delp) A.; ed. in Darmstadt, also Heidelberg, Tübingen; student of Kekulé; LL.D., U. Aberdeen (Scotland), 1906; m. Anna Pfluger, Sept. 6, 1888; 1 son, Ludwig. Apptd. asst. lectr. in Bonn, 1875, pvt. lectr., 1878, asso. prof., 1884, prof., 1898, also head Inst. Chemistry, emeritus, 1922. Author: Die Destillation unter vermindertem Druck, 1887; Bearbeitung V. Richters Chemie der Kohleustoffverbindungen, 2 vols., 1894-96; Das chemische Institut der Universität Bonn, 1904; August Kerkulé, 2 vols., 1929; numerous articles. Showed that maleic and fumaric acids are unimolecular; synthesized hexaphenylethane, hydroxycoumarins, anthracene; studied properties of hydrocarbons, aluminum chloride, and various acids, including oxalic acid. Died Darmstadt, Jan. 8, 1937.

ANSELL, George Frederick, inventor; b. Carshalton, Eng., Mar. 4, 1826; apprentice to surgeon, 4 years; student Royal Coll. Chemistry, 1 year; apptd. to Royal Mint, 1856; retired to practice as analyst. Author: Royal Mint, 1870. Research on fire-damp (explosive mixture of methane and air) in coal mines; patentee fire-damp indicator. Died Dec. 21, 1880.

ANSELL, George Stephen, Am. metallurgist; b. Akron, O., Apr. 1, 1934; s. Frederick Jesse and Fanny (Soletsky) A.; B.Met.E., Rensselaer Poly. Inst., 1954, M.Met. E., 1955, Ph.D., 1960; m. Marjorie Boris, Dec. 18, 1960; children—Frederick Stuart, Laura Ruth, Benjamin Jesse. Phys. metallurgist U. S. Naval Research Lab., Washington, 1957-58; faculty Rensselaer Poly. Inst., Troy, N.Y., 1960—. Robert W. Hunt prof. metall. engring., 1965—. Dir. Ilikon Corp., Natick, Mass; cons. to pvt. cos. Mem. Am. Inst. Mining, Metall. and Petroleum Engrs. (Hardy Gold medal 1961), Am. Soc. for Metals (Alfred H. Geisler award Eastern N.Y. chpt. 1964), Electron Microscope Soc. Am., N.Y. Acad. Scis., Sci. Research Soc. Am., Sigma Xi, Tau Beta Pi, Phi Lambda Upsilon. Research, publs. on theoretical and exptl. analyses of relationship between defect structure and properties of crystalline solids. Home: 14 Beechwood Dr., Latham, N.Y. 12110. Office: Dept. Materials Engring., Rensselaer Poly. Inst., Troy, N.Y. 12181.*

ANSELONE, Philip Marshall, Am. mathematician; b. Tacoma, Feb. 8, 1926; s. Frank and Ada (Otis) A.; B.S., U. Puget Sound, 1949, M.A., 1950; Ph.D., Ore. State U., 1957; m. Joann Marie Smith, Sept. 20, 1951; 1 dau., Cheryl. Mathematician, Hanford Atomic, Richland, Wash., 1951-54; sr. scientist Johns Hopkins Radiation Lab., Balt., 1954-58; mem. Math. Research Center, U. S. Army, U. Wis., 1958-63, vis. prof., 1966-67; prof. math. Ore. State U., Corvallis, 1963—; vis. prof. Karlsruhe Technische Hochschule, Germany, 1966, Stanford, 1967. Mem. Am. Math. Soc., Math. Assn. Am., Soc. Indsl. and Applied Math. Editor: Nonlinear Integral Equations, 1964. Research in approximate solutions of integral and operator equations, numerical solutions of transport equations.*

ANSHÜTZ, Johann Matthäus, German mineralogist; b. Suhl, Germany, Apr. 12, 1745; s. George Daniel and Rosina Dorothea Andrea A.; m. Christiane Elisabeth Witthauer, Feb. 5, 1765; at least 1 son, Heinrich Daniel. Gun dealer; worked with G. A. Werner in Freiburg (Germany). Author several publs., including: Über die Gebirgs- und Steinarten des chursächsischen Hennebergs Leipzig, 1788. Worked on Werner's classification system; research on mineralogy. Died Suhl, June 5, 1802.

ANSIAUX, Nicolas-Gabriel-Antoine-Joseph, Belgian physician; b. Ciney, June 6, 1780; Ph.D., U. Liège (Belgium), 1803; authority on legal aspects of medicine. Author: Clinique chirurgicale, ou Recueil de mémoires et observations de chirurgie pratique, 1820; Questions de médicine légale, précis des mémoires du Dr. Pfeffer, 1821; Systema chirurgiae hodiernoe Henrici Callisen, edito quinta, 1821; Discours sur la médicine légale, 1825. Died Liège, Dec. 26, 1834.

ANSLOW, Gladys Amelia, Am. physicist; b. Springfield, Mass., May 22, 1892; d. John and Ella (Leonard) Anslow; A.B., Smith Coll., 1914, A.M., 1917, D.Sc., 1950; Ph.D., Yale, 1924. Faculty physics dept. Smith Coll., Northampton, Mass., 1914-65, prof., 1936-60, research prof., 1960-65, emeritus, 1965—; chief information and communications Office of Field Service, OSRD, 1944-45. Recipient Pres.'s certificate of Merit, 1948. Mem. Am. Acad. Arts and Sci., Am. Phys. Soc., Am. Soc. Physics Tchrs., A.A.A.S., Am. Optical Soc., Soc. for Applied Spectroscopists, Am. Assn. U. Profs., Phi Beta Kappa, Sigma Xi. Research, publs. on ionizaton of gases by electron collisions and neutron collisions, use of ultraviolet and infrared spectra in study of structure of biologically important molecules. Home: 28 Bancroft Rd., Northampton, Mass. 01060.*

ANSON, Barry Joseph, Am. physician; b. Muscatine, Ia., Mar. 21, 1894; s. Francis P. and Alice (Barry) A.; A.B., U. Wis., 1917; postgrad. U. Chgo.; M.A., Harvard, 1923, Ph.D. in Med. Sci., 1926; m. Gertrude R. McNutt, Feb. 12, 1918. Sci. asst. U. S. Bur. Fisheries, Washington, 1917-22; NRC fellow in medicine Harvard Med. Sch., 1922-26; faculty anatomy Northwestern U. Med. Sch., 1926—, prof. editor Quar. Bull., 1942-62; prof. emeritus, 1962—, chmn., Robert Laughlin Rea prof., 1956-62; research prof. dept. otolaryngology and maxillofacial surgery U. Ia. Coll. Medicine, 1962—. Mem. Am. Assn. Anatomists (pres. 1957), Am. Otol. Soc. (hon. medalist), Am. Laryngol., Rhinol. and Otol. Soc., Chgo. Laryngol. and Otol. Soc., Am. Acad. Ophthalmology and Otolaryngology, A.M.A., Am. Assn. History Medicine, Soc. History Med. Sci., Assn. Med Illustrators. Author (with Theodore H. Bast): The Temporal Bone and the Ear, 1949; An Atlas of Human Anatomy, 2d edit., 1963; (with Walter G. Maddock): Callander's Surgical Anatomy, 4th edit., 1958; (with Leo M. Zimmerman) The Anatomy and Surgery of Hernia, 1953, 2d edit., 1967; (with James A. Donaldson) Surgical Anatomy of the Temporal Bone and Ear, 1967; also numerous articles. Editor-in-chief Morris' Human Anatomy, 12th edit., 1966. Home: Woodlawn Apts., 20 Evans St., Iowa City, 52240.*

ANSTED, David Thomas, English geologist; b. London, Feb. 5, 1814; ed. Jesus Coll. (pupil of Sedgwick), U. Cambridge; degree, 1836. Prof. geology King's Coll., London, 1840-53; lectr., Addiscombe, also Civil Engring. Coll., Putney, Eng.; mining cons., from 1850. Fellow Royal Soc., 1844; mem. Geol. Soc. (asst. sec.). Author: Geology, Introductory, Descriptive and Practical with numerous illustrations, 2 vols., 1844; The Ancient World, or Picturesque Sketches of Great Britain, 1847; The Gold-Seeker's Manual; The Great Stone Book of Nature, 1863; Science of Physical Geography; The World We Live In, 1869; Physiography, 1877; Water, Physical, Descriptive and Practical, 1878; numerous memoires. Editor Jour. and Proc. Geol. Soc. Investigated practical applications of geology to mining engring. water supply. Died Woodbridge, Eng., May 1880.

ANSTIE, Francis Edmund, English physician; b. Denizes, Eng., Dec. 11, 1833; s. Paul Anstie; ed. medicine King's Coll., London, from 1853; M.B., U. London, 1857, M.D., 1859; m. Miss Wass, 1862; 1 son, 2 daus. Asst. physician Westminster Hosp., 1860-73, physician, from 1873, co-founder, 1st dean Med. Sch. for Women. Licentiate Soc. Apothecaries. Fellow Royal Coll. Physicians; mem. Royal Coll. Surgeons. Author: Stimulants and Narcotics (contained results of research, observations and experiments); Neuralgia and the Diseases which resemble it, 1871. Investigated action of alcohol on the body in health and disease; research nervous diseases, pub. health; promoted therapeutics. Died Sept. 12, 1874.

ANTELAVA, Nikolay Vardenovich, Russian surgeon; b. Senakhi, Georgia, 1893; grad. Med. Faculty, Rostov-on-Don U., 1920; D.Med. Sci. Surgeon hosps. and clinics, Georgia, 1921-38; founder surg. dept. hosp. in Sukhumi, 1921; practice medicine specializing in surg. therapy of Tb at hosp. in Sukhumi, 1922; founder surg. dept. Tb sanatorium in Gulripshi, Agudzery and Abastumani, 1930; head chair hosp. surgery, Daghestan Med. Inst., Makhachkala, 1939-41; chief surgeon Georgian Ministry Health, 1949-54; head 1st chair surgery Tbilisi Postgrad. Med. Inst., 1941—; hon. sci. worker Georgian SSR, 1957—. Mem. USSR Acad. Med. Sci. (corr.). Recipient Lenin prize, 1961. Author over 90 works on surgery, including 9 monographs; Operative Collapse Therapy of Pulmonary Tuberculosis, 1939; Surgery of Thoracic Viscera, 1952; Surgical Forms of Brucellosis, 1954. Co-editor Surgery sect. Large Med. Ency., 2d edit.; mem. editorial council Surgery Jour. Developer new variations of thoracoplasty. Address: Tbisili Postgrad Med. Inst., ulsita Lunacharskogo 12, Tbilisi, Gruz. SSR, USSR.

ANTEVS, Ernst Valdemar, research geologist; b. Vartofta, Sweden, Nov. 20, 1888; s. Clas F. and Ada (Johansson) Ericsson; Ph.Mag., U. Stockholm (Sweden), 1912, Ph. Lic., 1916, Ph.D., 1917; D.Sc., U. Ariz., 1965; m. Ada E. Bradford, Oct. 8, 1929. Came to U.S., 1920, naturalized, 1939. Docent, U. Stockholm, 1917-35; research asso. Am. Geog. Soc., 1921-22, Carnegie Instn. Washington, 1922-23, 28-29, 31-40, Geol. Survey Can., 1923-24, 29-30; Harvard, 1924-26, pvt. research, Globe, Ariz., 1940—. Recipient awards Research Corp., N.Y.C., 1930. Fellow Geol. Soc. Am.; mem. Phila. Acad. Natural Sci., Am. Geog. Soc., A.A.A.S., Soc. Am. Archaeol. Author: Last Ice Recession in New England, 1922; Retreat of the Last Ice-Sheet in Eastern Canada, 1925; The Last Glaciation, 1928; also numerous articles, other books. Research on climatic history of past 25,000 years, its chronology, variation and applications to spread of Paleo-Indian to Am. Address: The Corral, Globe, Ariz. 85501.*

ANTHEMIOS OF TRALLES, Greek architect, engr., mathematician; b. Tralles, Lydia; s. Stephanos. Commd. by Justinian to build Ch. of Hagia Sophia, Constantinople (masterpiece of Byzantine architecture, combining oriental dome with Roman basilica), 532; completed by Isadoros of Miletos, circa 537; later became mosque, now mus. Byzantine art, Author: Concerning Remarkable Mechanical Devices (on burning mirrors, includes 1st record of practical use of directrix to reflect light to concentration point). Died circa 534.

ANTHONY, Francis, English alchemist; b. 1550. Author: Medicinae Chymicae et veri potabilis Auri assertio, 1610; The Apologie, or Defence of a Verity heretofore published concerning a medicine called Aurum Potabiel, 1616. Prepared, sold potable gold in London, Eng. without license from Royal Coll. Physicians (prosecuted by them); attacked in print by Matthew Gwynne, 1611, Thomas Rawlin, 1611, John Cotta, 1623. Died 1623.

ANTHONY, John Gould, Am. zoologist; b. Providence, R.I., May 17, 1804; s. Joseph and Mary (Gould) A.; m. Anna W. Rhodes, Oct. 16, 1832. Moved with parents to Cincinnati, 1816, engaged in bus., until 1851; became interested in natural history, collected mollusks in Ohio River; corresponded with mollusk collectors and students in East and Europe, including Louis Agassiz and S. S. Holderman, from 1835; toured Ky., Tenn., Ga. for health reasons, also to collect mollusks, 1853; apptd. by Agassiz in charge of mollusk collection Mus. of Comparative Zoology, Cambridge, Mass., 1863; mem. scientific staff on expdn. to Brazil, 1865; spent later years in classifying and arranging collections at Cambridge, also gathering data on family history. Author papers: "Two Species of Fossil Asterias in the Blue Limestone of Cincinnati," 1846, "Descriptions of New Species of American Fluviate Gastropods," 1861, "Description of a New Species of Shells," 1865. Died Cambridge, Oct. 16, 1877.

ANTIGONE OF CARYSUS, Greek biologist; b. Eubeé, Greece; flourished circa 285 to 247 B.C.; author: Des Vies d'Ecrivains Celebres (lost work); Histoire des Animaux; Traite du Style; Antipater (poem); Metamorphoses.

ANTIPA, Grigore, Rumanian biologist; b. 1867; dir. Bucharest (Rumania) Mus. Natural History, 1892-1944; mem. Rumanian Acad. Author: Rumania's Ichthyological Fauna, 1909; Fishes and Fishing in Rumania, 1916; The Black Sea, 1940. Founder of the Rumanian school of hydrobiology and ichthyology; a creator of modern museology; introduced dioramas. Died 1944.

ANTIPHON, Greek philosopher, mathematician; b. Rhamnus, Attica, 480 B.C.; Sophist; epic poet; attempted to solve quadrature of circle by method of exhaustion (later criticized by Aristotle). Author work on truth. Executed for part in establishing govt. by Four Hundred, Athens, 411 B.C.

ANTISELL, Thomas, geologist; b. Ireland, 1817; came to U. S., 1848; practiced medicine until 1854, also lectr. chemistry at Woodstock, Vt., Pittsfield, Mass., Berkshire Med. Inst.; apptd. examiner Patent Office, 1856, 1877-91; surgeon Union Army, 1861-66; apptd. chief chemist Dept. Agr., 1866; expert in chem. tech., Tokyo, Japan, 1871-77; worked with Survey of Pacific R.R. Wrote on agrl. chemistry, Irish and Am. Geology; placed final uplift of Pacific Coast Range in post-Miocene epoch; theorized that San Francisco bay was formed by a rupture in superficial strata caused by an uplifting of consolidated crust to N. and S. Died 1893.

ANTOMMARCHI, C. Francesco, physician, surgeon, anatomist; b. Corsica, 1780; prof. anatomy, Florence, Italy, 1812; anat. dissector to a hosp., Florence; faculty U. Pisa (Italy); last physician to Napoleon Bonaparte, St. Helena, 1818-21; dir. mil. hosps., Warsaw, during Polish Revolution; practiced homeopathy, New Orleans, 1836. Author: Dernier moments de Napoleon, 1823; Anatomical Plates of the Human Body, 1823-26. Died San Antonio, Cuba, Apr. 3, 1838.

ANTON, Nicholas Guy, physicist, electronic engr.; b. Trieste, Austria, Dec. 14, 1906; s. Joseph and Anne (Mandl) A.; grad. Tech. Inst. Leonardo da Vinci, Trieste, 1924; postgrad. Columbia; m. Bernice Irene Skripsky, June 19, 1932; children—Joan Carol (Mrs. Bertram Pearlman), Linda Elaine, Nanci Helen. Came to U. S., 1925, naturalized, 1932. Exhaust dept. engr.; gen. supt. Duovac Corp., Bklyn, 1927-32; pres., dir. Amperex Electronics Products (formerly Electronic Labs.), Bklyn., 1932-48; pres., dir. Anton Electronic Labs., Bklyn, 1948-61; pres., dir research and devel. EON Corp., Bklyn., 1962—; chmn. vacuum tube com. N. Am. Philips, N.Y.C., 1945-48; chmn. bd. Anton-Imco Co., N.Y.C., 1960-61. Chmn. adv. com. radiol. div. OCDM, 1957-60; mem. centennial com. Poly. Inst. Bklyn., 1954. Fellow I.R.E. (rep. nuclear standard bd. 1957-61), Am. Phys. Soc., N.Y. Acad. Scis., A.A.A.S.; mem Am. Soc. M.E., Soc. Non-Destructive Testing, Electronic Ind. Assn., Am. Math. Soc., N.Y. Atomic Indsl. Forum. Important work includes designer, special vacuum tubes, nuclear detectors and systems, nuclear-med.-electronic equipment, radiation monitoring equipment for space. Home: 3172 Bedford Av., Bklyn 10. Office: 175 Pearl St., Bklyn. 1.*

ANTONIADES, Harry Nicholas, biochemist; born Thessaloniki, Greece, March 12, 1923; s. Nicholas H. and Eustratia (Manos) A.; B.S., Athens U., 1950, Ph.D., 1952; m. Maria Tomaras, Dec. 27, 1953; children—Harry Nicholas, Anna-Maria. Came to U. S., 1953, naturalized, 1957. Research fellow dept. biophys. chemistry Harvard, 1954-55, research asso. biochemistry dept. medicine Med. Sch., 1956-65, asst. prof. biochemistry dept. nutrition Sch. Pub. Health, 1966—; research asst. Protein Found., Inc., Jamaica Plain, Mass., 1954-56, research asso., 1956-57, asso. investigator, 1957-61, sr. investigator, 1961—; asso. staff medicine Peter Bent Brigham Hosp., 1961-65; vis. asso. prof. biochemistry Med. Sch. U. So. Cal., 1961-62; vis. prof. medicine U. Ala. Med. Center, 1963. Mem. Am. Soc. Biol. Chemists, Am. Chem. Soc., N.Y. Acad. Scis., A.A.A.S., Am. Diabetes Assn. (Eli Lilly award 1962). Editor: Hormones in Human Plasma, 1960. Contbr. numerous articles to profl. jours. Studies on biochem. mechanisms regulating transport and biologic activity of hormones in blood of human beings; introduced concept that diabetes mellitus may result not only from lack of insulin in pancreas, but also from malfunction of mechanism which regulate insulin activity in blood. Home: 21 Magnolia Av., Newton, Mass. 02158. Office: 281 South St., Jamaica Plain, Mass. 02130.*

ANTONIEWICZ, Wlodzimierz, Polish archeologist; b. Zamborst, Poland, July 15, 1893; ed. Cracow, Poznan, Poland, also Vienna (Austria) U.; keeper Wawel Castle Mus., 1916-19; prof. pre-history Polish Free U., 1919-39; prof. Warsaw (Poland) U., 1920, dean faculty humanities, 1934-36, rector, 1936-39, also prof. Polish archeology; dep. dir. State Mus. Archeology. Mem. Polish Acad. Scis. Author: Foundations of Prehistoric Archeology in Poland, 1926; Polish Archeology, 1928; also papers. Made research trips through Europe, Syria, Palestine, Egypt.

ANTONINI, Eraldo, Italian biochemist; b. S. Piero a Sieve, Florence, Italy, Apr. 20, 1931; s. Enea and Anita (Biferali) A.; M.D., U. Rome (Italy), 1954; m. Virginia Pettini, June 3, 1955; children—Giovanni, Paola, Andrea, Cristina. Faculty, U. Rome, 1955-64, prof. chemistry Med. Faculty, 1965—; prof. molecular biology U. Camerino, 1964—. Editorial bd. Biochimica et Biophysica Acta, 1964—. Research, numerous publs. on structure and functions of respiratory proteins, chem. modifications of proteins and consequent functional effects. Home: 37 Via Parione, Rome, Italy.*

ANTOPOL, William A., Am. physician; b. N.Y.C., Apr. 6, 1903; s. Israel and Mata (Elman) A.; B.S., Coll. City N.Y., 1923; M.D., L.I. Coll. Hosp., 1927; m. Bella Scholer, June 24, 1937; children—Michael Richard, Stephen Charles. George Blumenthal fellow pathology Mt. Sinai Hosp., N.Y.C., 1928-29, Theodore Escherich fellow pathology, 1929-30, asst. pathology, 1930-36; pathologist, dir. labs. Bayonne (N.J.) Hosp., 1930-35; pathologist, dir. labs. Beth Israel Hosp., Newark, 1935-49, cons. pathologist, 1955—; pathologist, dir. labs and research, Beth Israel Hosp., N.Y.C., 1949—; dir. research Joseph H. Yeamans Levy Found., 1949—; clin. prof. pathology N.Y. U. Sch. Medicine Center, 1949—; prof. pathology Mt. Sinai Sch. Medicine, City U. N.Y., 1966—. Cons. pathology Hudson County Tb Hosp., N.J., 1930-37, N.Y. Infirmary, 1956—, VA Hosp., Lyons, N.J., 1947—; mem. adv. bd. for germ free studies dept. biology Lobund Labs., U. Notre Dame, Ind., 1962—; vis. prof. U. Ky., 1967, U. Montreal; comparative pathologist Osborn Labs. Marine Scis. Herbert L. Celler fellowship grantee, 1930; Com. on Sci. Research A.M.A. grantee, 1934, 35. Fellow N.Y. Acad. Medicine, Am. Soc. Clin. Pathologists, Coll. Am. Pathologists; mem. Assn. Am. Pathologist and Bacteriologists, Soc. Exptl. Biology and Medicine, Am. Pub. Health Assn., Harvey Soc., Am. Soc. Zoologists, Am. Assn. Anatomists, Biochem. Soc. Gt. Britain, N.Y. State Soc. Pathologists, Reticuloendothelial Soc., Am. Assn. Cancer Research, Am. Soc. Exptl. Pathologists, Am. Astronautical Soc., Aerospace Med. Assn., Sigma Xi. Research in tumor pathogenesis, heart pathology; discovered selective effects of surg. talc, certain therapeutics. Home: 350 1st Av., N.Y.C. 10010. Office: Beth Israel Hosp., 10 Nathan D. Perlman Pl., N.Y.C. 10003.*

ANTYLLOS OF ALEXANDRIA, Greek physician; surgeon; flourished 1st half 2d century. Author of therapy manual, surgery text (fragments extant). Devised operation for double ligation and excision of aneurysm (only procedure for 16 centuries); considered effects of air and soil on health; gave precise directions (with full consideration of possible complications) for many operations, including cataract removal, tracheotomy, scar removal, fistulae of breast or abdomen removal, resection of bones and joints, surgery of blood vessels; suggested plastic surgery for nose, eyelid and ear defects.

ANVILLE, Jean-Baptiste, see d'Anville, Jean-Baptiste Bourguignon.

AOCHI, Rinso, Japanese scholar, geographer; b. Yedo, Japan, 1784; s. Kaian Aochi; studied medicine, then Dutch language under Sajuro Baba, physics; became ofcl. interpreter Shogunate, 1822; apptd. physician, lectr. Western learning to Lord of Mito; physician to Lord of Matsuyama, Shikoku; translated into Japanese book on world geography, 1827; author books on geography, including Kikaikanran. Died 1833.

AOKI, Kiyoshi, Japanese scientist; b. Tokyo, Japan, 1907; grad. Tohoku U., 1930. Became prof. Hokkaido U., also mem. staff Low Temperature Phenomena Research Inst., 1944. Specialist in cold resistance in animals and plants; research writings on cold resistance of farm crops, effects of low temperature processings, freezing of foodstuff.

AOKI, Konyo (1st name Atsubumi, called Bunzo), Japanese scholar; b. 1698; disciple Togai Ito; became student Dutch learning, Nagasaki, Japan, 1736. Custodian Shogunate archives; became magistrate Archives, 1767; appt. mem. Confucian sect. Supreme Council Shogunate. Author books on Dutch learning; book on sweet potato culture. Distributed sweet potato seedlings throughout Japan (called Kansho Sensei, or Sweet Potato Tchr.). Died 1769.

AOKI, Shuhitsu, Japanese physician; b. Choshu, Japan, circa 1803; pupil Toan Kumami, also Phillipp Franz Von Siebold, Nagasaki, Japan; persuaded Takachika Mori, lord of Choshu clan to establish Hogaku-kai (sch. for study Western sci.); vaccinated (with bro. Kenzo) Choshu clansmen. Died 1863.

AOUST, Louis Stanislas Xavier Bathélomew, French mathematician; b. Béziers, France, 1814; became prof. differential and integral calculus Sci. Faculty Marseille (France), 1854, then prof. math. Stanislas Coll., Paris, France. Author: Théorie des coordonées curvilignes guelconques; Recherches sur les surfaces du second ordre, 1864-68; L'homme et la science, 1867; Analyse infinitesimale des courbes tracées sur une surface quelconque, 1869; Etude sur là vie et les travaux de Saint-Jacques de Silvabella, astronome marseillais, 1871; Analyse infinitesimale des courbes dans l'espace, 1876; Le Verrier, sa vie, ses travaux, 1876; Du système astronomique produisant l'egalité des jours polaires, 1878, Des asymptotes paraboliques des courbes, 1884. Known for work on 2d order surfaces in calculus. Died Marseille, 1885.

APATHY, Stefan, Rumanian biologist; b. Cluj, Rumania, 1863; research histology of nervous system, introduced new staining techniques (as silver and gold impregnation) for study histology. Died 1922.

APELT, Ernst Friedrich, German philosopher; b. Reichenau, Mar. 3, 1812; prof. philosophy, Jena, Germany. Author: Epochen der Geschichte der Menschheit, 1845; Die Spharentheorie des Eudopus und Aristoteles, 1849; Keppler's astronomische Weltansicht, 1849; Die Reformation der Sternkunde, 1852; Theorie der Induktion, 1854; Parmenidis et Empedoclis doctrina de mundi structura, 1856; Metaphysik, 1857; Religionsphilosophie, 1860. Died Oppelsdorf, Germany, Oct. 27, 1859.

APÉRY, Roger, French mathematician; b. Rouen, France, Nov. 14, 1916; s. Georges and Louise (Van der Cruysan) A.; ed. École Normale Supérieure; m. Denise Bienaime, Aug. 6, 1947; children—Denys, François, Robert. Lectr., Rennes, France, 1947-49; with Faculty Scis., Caen, France, 1949—, prof., 1959—, became head dept. math., 1966. Mem. Société des agregés, Société Mathématique de France. Research on algebraic geometry, diophantine equations. Home: 3 place de la Résistance, Caen, France.*

APIAN, Philipp, German mathematician, geographer; b. Ingolstadt, Germany, Sept. 14, 1531; s. Peter Apian; m. Sabine Scheuchenstuel, 1564; 1 dau., Sabine. Apptd. prof. math. U. Ingolstadt, 1552, Tübingen, Germany, 1569-84. Author: Bairische Land-Tafeln, 1566. First modern topographer; surveyed Bavaria, 1554-61, and constructed 484 square foot map (master work of 16th century topography, formed basis of surveying in Bavaria into 18th century); built terrestrial globes showing papal demarcation line between Spanish and Portuguese areas, also celestial globes. Died Tübingen, Nov. 14, 1589.

APIANUS, Petrus (or Peter Bienewitz, Peter von Bennewitz), German geographer, astronomer; b. Leisnig, Germany, Apr. 16, 1495; s. Martin and Gertrud Bienewitz; ed. Leipzig. Germany, also Vienna; m. Katharina Mesner; children—Philipp A., Georg A. Prof. math. U. Ingolstadt (Germany), 1527-52, founder printing shop, 1526; tutor to Emperor Charles V.; pub. Jordanus Nemorarius' Mechanics. Author: Cosmographia, seu descriptio totius orbis, 1524; Eyn neue und wolgegründte underweisung aller Kauffmans Rechnung, 1527; Instrumentum sinuum sive primi mobilis (1st table of sines for all minutes and ray 1000 pub. in Europe), 1534; Astronomicum caesareum, 1540. Innovator in cartographic projections, astron. instruments; projected and pub. 1st oval map (globe) of earth; described, printed some of earliest maps of Am.; described instruments designed for reproducing movements of celestial bodies; 1st to propose observing moon movements to determine longitudes; believed solar eclipses provided best way to determine differences between meridians; noticed comet tails point away from sun. Died Ingolstadt, Apr. 21, 1552.

APKER, LeRoy, Am. physicist; b. Rochester, N.Y., June 11, 1915; s. Russell Woodard and Sarah (LeRoy) A.; A.B., U. Rochester, 1937, Ph.D., 1941; m. Jean E. Dickey, Dec. 27, 1950. Coffin fellow U. Rochester, 1940-41; research asso. Gen. Electric Research Lab., Schenectady, 1941-50, mgr. semiconductor studies gen. physics dept., 1950—. Recipient Buckley prize, Am. Phys. Soc., 1955. Fellow Am. Phys. Soc., A.A.A.S.; mem. Phi Beta Kappa, Sigma Xi. Discovered exciton-induced photoemission process in potassium iodide crystals; extended research with energy transfer in crystals.

APOLLODOROS OF DAMASCUS, architect; flourished 98-117; built bridge over Danube, 104-05, Trajan Forum, 107-13, Trajan Aqueduct, several roads. Author: Poliorketika (dedicated to Emperor Trajan), 120. Executed by Emperor Hadrian.

APOLLONIOS, Greek mathematician; b. Perga (now Murtana, Turkey), circa 262 B.C.; ed. Alexandria. Author: Thirteen known treatises including: Conic Sections (presented 400 theorems, mostly original, covered work of some predecessors, including Euclid); (attributed works) De Rationis Sectione, De Spatii Sectione, De Sectione Determinata, De Tactionibus, De inclinationibus, De Locis Planis. Established main properties of conics; showed that all conics may be considered sects. of same cone; introduced names parabola, ellipse, hyperbola; determined evolute of any conic; investigated geometric algebra; exponent, possibly inventor astron. hypotheses of eccentric circles and of epicycles; inventor system of tetrads for expressing large numbers. Died circa 190 B.C.

APOLLONIOS OF CITIUM, Greek physician; flourished middle of 1st century B.C.; disciple of Zopyros; author commentary on Hippocrates' book of articulations (treatise on collection of curationes, one of which deals with epilepsy), also treatise against Heraclides of Tarentum.

APPOLONIOS OF TYANA, Greek philosopher; flourished 1st century B.C.; exponent of Neo-Pythagoreanism; traveled within and outside Roman Empire as a teacher and thaumaturgist; author biography of Pythagoras (fragment extant), treatise on sacrifice; 97 of his letters preserved by Flavius Philostratus in an 8 book biography.

APOSHIAN, H(urair) Vasken, Am. biologist; b. Providence, Jan. 28, 1926; s. Leo and Manishag (Aghajanian) A.; B.S., Brown U., 1948; M.S., U. Rochester, 1950, Ph.D., 1953; m. Mary Mobareket Zaidan, Apr. 7, 1948; children—Christine Manishag, David Leo, Mary Ann. Faculty, Vanderbilt U. Sch. Medicine, 1954-59, asst. prof., 1956-59; USPHS Research fellow Sch. Medicine, Stanford, 1959-62; asso. prof. microbiology Tufts U. Sch. Medicine, Boston, 1962-67; prof., chmn. cell biology and pharmacology U. Md. Sch. Medicine, Balt., 1967—. Mem. Am. Soc. Biol. Chemists, Am. Soc. for Pharmacology and Exptl. Therapeutics, Am. Assn. for Cancer Research, Am. Chem. Soc., N.Y. Acad. Scis., A.A.A.S. Research, publs. on enzymology of virus infections, oral agts. for heavy metal poisoning. Office: 660 W. Redwood St., Balt. 21201.*

APOSTOL, Tom Mike, Am. mathematician; b. Helper, Utah, Aug. 20, 1923; s. Mike A. and Florence (Pappathanasopoulos) A.; B.S., U. Wash., 1944, M.S., 1946; Ph.D., U. Cal., Berkeley, 1948; m. Jane Clark Thornton, Sept. 16, 1959. Faculty, U. Cal., Berkeley, 1948-49, Mass. Inst. Tech., 1949-50; faculty Cal. Inst. Tech., Pasadena, 1950—, prof., 1962—, Mem. Am. Math. Soc., Math. Assn. Am. (past bd. govs., sect. chmn.), A.A.A.S. Author: Mathematical Analysis, 1957; Calculus, Vol. I, 1961, Vol. II, 1962; Calculus with Linear Algebra, 1967. Research in elementary and analytic number theory. Office: 1201 E. California St., Pasadena, Cal. 91109.*

APPEL, Otto, German botanist; b. Coburg, Germany, May 19, 1867; s. Georg Ludwig and Anna Christiane Mathilde (Kern) A.; m. Anna Schaufert, 1894; 3 sons. Became asst. Inst. for Hygiene and Bacteriology, Würzburg, Germany, also Agrl. Inst., Königsberg (now Kaliningrad, USSR), 1899; entered ofcl. state health service, became dir. Biol. Inst. for Agr. and Forestry, by 1920. Author: Taschenatlanten der Plfauzenkrankheiten, 1926; (with E. Riehm) Atlas der Krankheiten der landwirtschaftlichen Kulturpflausen, 1924-29. Organized modern plant protection policy; investigated diseases of potatoes. Died Berlin-Zehlendorf, Germany, Nov. 10, 1952.

APPELL, Paul-Emile, French mathematician; b. Strasbourg, France, Sept. 27, 1855; ed. l'Ecole Normale; D.Sc., 1876. Prof. rational mechanics Faculty Scis. Paris, from 1885; dir. Acad. Paris, 1920-25; prof. Sorbonne, also l'Ecole de Seures. Gold medal (2d prize) King Oscar II of Sweden. Mem. French Acad. Scis., 1892 (v.p. 1913, pres. 1914). Author: Cours de mécanique rationnelle, 1888; Lecons sur l'attraction et la fonction potentielle, 1892; Traité de mecanique rationnelle (cours de mécanique de la Faculte des Sciences), 4 vols., 1893-96; Eléments de la théorie des vecteurs, 1921; Fonctions hypergéometriques et hyperspheriques, polynomes d'Hermite, 1926; (with Goursat) Théorie des fonctions algébriques et de leurs intégrales, 1895; (with Lacour) Principes de la théorie des fonctions élliptiques, 1897. One of 1st to study hypergeometric and elliptical functions or automorphs. Died Paris, Oct. 24, 1930.

APPERT, Nicolas (François), French inventor, bacteriologist; b. Chalons-sur-Marne, France, Oct. 23, 1752; self-educated. Cook and confectioner in father's establishment, then for various noblemen until French Revolution, established 1st comml. cannery in world, House of Appert, Massy, France, 1812. Recipient prize of 12,000 francs from Napoleon for invention preservation method of heating food and sealing it from air to prevent spontaneous generation of microorganisms (basis for canning industry of today), 1809, award Soc. for Encouragement Nat. Industry, 1822. Author: Art de conserver les substances, animales et vegétales, 1810. The father of comml. canning; published method of heat and hermetic seal cooking, 1810; began using round white steel cans instead of glass, 1822; invented bouillon cube; devised method of extracting gelatin from bones without using acid. Died Massy, June 3, 1841.

APPLE, Andrew Thomas Geiger, Am. astronomer; b. Hamburg, Pa., Mar. 5, 1858; s. Rev. Dr. Joseph Henry and Elizabeth Ann (Geiger) A.; A.B., Franklin and Marshall Coll., 1878, A.M., 1881; m. Ada Krebs, Sept. 4, 1884 (died 1910). Prof. natural sci., Palatinate Coll., Myerstown, Pa., 1880-83; ordained ministry Ref. Ch. in U. S., 1883; pastor in Bedford County, Pa., Catawissa, Bedford, and Washington, D.C., until 1907; prof. math. and astronomy, also dir. Daniel Scholl Obs. of Franklin and Marshall Coll., from 1907. Made observations of total eclipse of sun, in connection with Franklin and Marshall expdn., Centerville, Va., May 1900. Frequent contbr. to sci. and popular mags. and to encys. Died Feb. 15, 1918.

APPLEBY, John Francis, Am. inventor; b. Westmorland, N.Y., May 23, 1840; s. James and Jane Appleby; ed. in Wis.; m. 1867; 3 children; farmer in Ia.; joined Parker & Stone, Beloit, Wis., 1872; organized Appleby Reaper Works, 1874. Constructed 1st model of Appleby knotter (a basic component of most grain binders), 1859; developed rifle cartridge, mag. and automatic feeder. Died Nov. 8, 1917.

APPLEBY, Ralph C., Am. dentist; b. Eldon, Ia.; Nov. 16, 1919; s. Clarence E. and Doris (Carson) A.; B.S., State U. Ia., 1941, M.S., 1953, D.D.S., 1951; m. Berly Kathleen Moss, July 10, 1944; children—Drew Carson, Gary Meredith. Mem. Mission of Econ. Affairs, U. S. embassy, London, 1944-45; from instr. to prof., head dept. denture prosthesis State U. Ia., Iowa City, 1951—. Cons. denture prosthesis VA Hosp., Iowa City. Decorated Bronze Star. Mem. Am. Assn. Dental Schs. (chmn. conf. sect. complete dentures), Am. Coll. Dentists, Delta Upsilon. Author: Outline for Construction of Partial Dentures, 1963; Outline for Construction of Complete Dentures, 1968. Research in vertical dimension in denture constrn. Home: 106 7th Av. N., Iowa City 52240.*

APPLEGARTH, Augustus, English inventor; b. London, Eng., June 17, 1788; s. Augustus J. Applegarth; printer, Nelson Sq., Blackfriars Rd.; built printing office, Duke St., Stamford St. Invented composition ball and roller, also steam printing press, machine for printing 6 colors at 1 time; constructed machines for printing bank notes in way that prevented forgery; invented (with Cowper) 4 cylinder machine; patented vertical machine. Died Dartford, Eng., Feb. 9, 1871.

APPLEMAN, Milo Don, Am. microbiologist; b. Wellston, Mo., Dec. 3, 1907; s. Milo D. and Emma (Faust) A.; B.S., U. Ill., 1931, M.S., 1936, Ph.D., 1940; m. Lucille Elizabeth Mieher, July 28, 1936; children—Milo Don, Research asso. U. Ill., 1940-43, asst. prof., 1943-47; faculty U. So. Cal., Los Angeles, 1947—, prof. microbiology, 1950—, head dept. bacteriology, 1949-62. Cons. in microbiology. Fellow Fund for Advancement Edn., 1954-55; hon. research fellow U. Aberdeen (Scotland), 1954-55. Mem. study sect. bacteriology and mycology NIH, 1964-67. Mem. Société des Diplomes U. Nantes (France), Royal Soc. Medicine (affiliate), Am. Soc. Microbiologists (past pres. So. Cal., past councilor), Assn. Food and Drug Ofcls. U. S., Sigma Xi (past pres.), Phi Sigma, Alpha Sigma Lambda, Gamma Sigma Delta. Research, numerous publs. on water pollution and methods purification, devel. animal foods, methods raising germ-free fish. Home: 4315 San Rafael Av., Los Angeles 90042.*

APPLETON, Arthur Beeny, Brit. anatomist; s. Arthur John and Emily Amelia (Beeny) A.; ed. Downing Coll., Cambridge, St. Bartholomew's Hosp.; M.A., M.D.; m. Eva Gertrude Plewman, 1918; 1 son, 1 dau. Univ. lectr., Cambridge, 1926, fellow Downing Coll., 1930; prof. anatomy Royal Acad. Arts; prof., examiner U. London; examiner univs. Liverpool, Manchester, Cardiff, Cambridge. Licentiate Royal Coll. Physicians. Fellow Zool. Soc.; mem. Royal Coll. Surgeons (Hunterian prof. 1944); pres. Anat. Soc. Gt. Britain and Ireland, 1947-49. Edit. bd. Acta Anatomica, Basel, Switzerland. Author: Guide to Vertebrate Dissection for Students of Anatomy, 1929; Surface and Radiological Anatomy, 1938; also papers on posture, lungs, locomotor apparatus. Contbg. author: Cunningham's Text-Book of Anatomy, 1949; British Surgical Practice. Died Apr. 22, 1950.

APPLETON, Sir Edward Victor, Brit. physicist; b. Bradford, Eng., Sept. 6, 1892; s. Peter and Mary (Wilcock) A.; M.A., St. Johns Coll., Cambridge; studied under Sir. J. J. Thompson, Lord Rutherford; hon. Sc.D., Oxford, Leeds, Brussels, Cambridge, Sydney; LL.D. (honorary), Aberdeen, Birmingham, London, Glasgow, St. Andrew's; m. Jessie Longson, 1915; children—Margery (Mrs. W. M. Lamont), Rosalind, Isabel. Wheatstone prof. physics King's Coll., London, 1924-36; Jacksonian prof., nat. philosophy U. Cambridge, 1936-39; sec., Dept. of Sci. and Indsl. Research, 1939-49; prin. and vice chancellor Edinburgh U. from 1949. Knight grand cross Order Brit. Empire, knight comdr. Order of the Bath. Fellow Royal Soc., 1927; mem. of Royal Inst. of Brit. Architects (hon.). Recipient Liebmann Meml. prize Am. I.R.E., Hughes medal Royal Soc.; Faraday Medal, Inst. Elec. Engrs., U. S. Medal of merit; Nobel prize for physics, 1947; Albert medal Royal Soc. Arts, 1950; Gunning Victoria Jubilee prize Royal Soc. Edinburgh, 1960; medal of Honor I.R.E. (U. S.), 1962. Mem. I.R.E. (U. S.), Newcomen Soc., Am. Acad. Arts and Sci., Instn. of Elec. Engrs., Internat. Sci. Radio Union (hon. pres.), Brit. Assn. for Advancement Sci. (pres. 1953), Pontificial Acad. Sci. Gave exptl. proof of ionosphere (with Barnett), 1925; research into radio reflection in atmosphere; established magnetoionic theory of ionosphere (with Hartree), 1927-32; discovered upper reflection layer of ionosphere (Appleton layer); developed prototype of radar; studied radio properties of sunspots; headed Brit. atomic bomb research during World War II, as tech. head Dept. Sci. and Indsl. Research; active in IGY. Died Edinburgh, Scotland, Apr. 21, 1965.

APPLETON, John Howard, Am. chemist; b. Portland, Me., Feb. 3, 1844; s. Elisha Williams and Martha Wylly (Hyde) A.; Ph.B., Brown, 1863, A.M., 1869, Sc.D., 1900; m. Louise Mumford Day, Feb. 24, 1875; children—Ruth (Mrs. George Albert Goulding), Everard, William Day, Alice, Paul, Marguerite. Asst. instr. and instr. analytical chemistry, 1863-68, prof. chemistry applied to arts, 1868-72, prof. chemistry, 1872-1914, emeritus, Brown. State sealer weights and measures many yrs.; chemist R.I. State Bd. Agr., and Providence Water Works; mem. U. S. Mint Commn., 1891. Author: The Young Chemist, 1878; Short Course in Qualitative Analysis, 1878; Quantitative Analysis, 1881; Laboratory Year-Book, 1883-92; Beginner's Hand-Book of Chemistry, 1884; Advanced Quantitative Analysis, 1889; Medical Chemistry, 1889; Lessons in Chemical Philosophy, 1890; Metals of the Chemist, 1891; Report Books for Chemical Work, 1891; Carbon Compounds, 1892; Chemistry of Non-metals, 1897; Easy Experiments of Organic Chemistry, 1898. Died Feb. 18, 1930.

APPLETON, Nathaniel Walker, Am. physician; b. Boston, June 14, 1755; s. Nathaniel and Mary (Walker) A.; A.B., Harvard, 1773, A.M., 1776; studied medicine under Edward Holyoke, Salem, Mass.; m. Sarah Greenleaf, May 24, 1780, 7 children. A founder, incorporator Mass. Med. Soc., 1781, recording sec., 1781-91; chmn. com. which produced 1st vol. of Medical Communications, 1790, contbr. papers "An Account of the Successful Treatment of Paralysis of the Lower Limbs, Occasioned by a Curvature of the Spine," "History of Haemorrhage from a Rupture of the Inside of the Left Labium Pudendi." Died Apr. 15, 1795.

APPLEYARD, Rollo, Brit. engr., physicist, inventor; b. Jan. 1, 1867; s. Septimus Appleyard; ed. Dulwich (Eng.) Coll.; m. Mabel Laming Evans, 1901. Mem. staff Royal Indian Engring. Coll., 1885-92; tech. adviser, writer Times, London, 1905-14; founder, dir., history sect. Admiralty, 1918. Author: The Conductometer and Electrical Conductivity; Measurement of Air-Speed; Height Measurements by Barometer and Thermometer; The Elements of Convoy Defence in Submarine Warfare; Pioneers of Electrical Communication (Hans Christian Oersted medal Copenhagen U. 1927); A Tribute to Michael Faraday; Charles Parsons—his Life and Work; The History of the Institution of Electrical Engineers, 1871-1931; also papers. Editor, Convoy Instructions. Mem. Instn. Civil Engrs. (Telford premium 1903, 20), Instn. Elec. Engrs. Inventor, Conductometer to measure elec. conductivity, also aero. and other instruments; investigated dielectrics, alloys, thermometry, surface-tension; discovered and tabulated length-function of solution for catenary problems. Died Mar. 1, 1943.

APRISON, Morris Herman, Am. biochemist, educator; b. Milw., Oct. 6, 1923; s. Henry and Ethel (Mollin) A.; B.S. in Chemistry, U. Wis., 1945, tchrs. certification, 1947, M.S. in Physics, 1949, Ph.D. in Biochemistry, 1952; m. Shirley Reder, Aug. 21, 1949; children—Barry, Robert. Grad. teaching asst. physics U. Wis., 1947-49, grad. research asst. pathology U. Wis. Sch. Medicine, 1950-51, grad. research asst. biochemistry, 1951-52; tech. asst. in physics Inst. Paper Chemistry, Appleton, Wis., 1949-50; biochemist, prin. investigator, head biophysics sect. Galesburg (Ill.) State Research Hosp., 1952-56; prin. research investigator biochemistry Inst. Psychiat. Research, asst. prof. depts. biochemistry and psychiatry Ind. U. Med. Sch., Indpls., 1956-60, asso. prof., 1960-64, prof., 1964—. Guest lectr. univs., socs. Mem. Am. Chem. Soc. (chmn. biannual symposium Ind. sect. 1962), Am. Physiol. Soc., Am. Inst. Physics, Biophys. Soc., Soc. Biol. Psychiatry, Soc. Exptl. Biology and Medicine, Sci. Research Soc. Am., Sigma Xi. Research, publs. in nitrogen fixation; pioneer research in correlation of cerebral biochem. changes and behavioral activity; research in neurotransmitter identification.*

APSTEIN, Carl Heinrich, zoologist; b. Stettin, Poland, Sept. 19, 1862; s. Adolph and Minna (Westphalen) A.; ed. at U. Leipzig (Germany), U. Freiburg (Germany), U. Kiel (Germany); Dr.phil., 1889; m. Anna Süverkrüp, June 24, 1905; 3 daus. Asst. Zool. Inst., Kiel; became lect. U. Kiel, 1898, asso. prof., 1906; mem. German deep-sea expdn. Valdivia, 1898-99; sci. officer Acad. Scis. from 1911. Mem. Internat. Nomenclature Com., German Zool. Assn. (sec. 1918-1946). Author: Dar Süsswasser Plankton, 1896; Tierleben der Hochsee, 1905; (with K. Wasikowski) Periodica Zoologica, 1938; also numerous articles on plankton, fish, algae. Editor in chief Tierreich, 1911-27; pub. Zoologischer Bericht, 1921-43. Research on plankton; numerous deep sea studies. Died Berlin, Germany, Nov. 14, 1950.

APSYRTOS, veterinarian; flourished circa 300 (has also been dated 150-250); chief vet. surgeon to Constantine the Great in campaign against Samaritans and Goths; later practiced in Bithynia; author 60 letters on gen. diseases and cures; gave best description of time of diseases and treatment by bloodletting, potion, ointment.

AQUILANO, Sebastiano, Italian physician; b. Aquila, Papal States; named prof. medicine in Florence (Italy), 1495. Author: De morbo Gallico, 1506; Quaestio de febre sanguinis ad mentem galeni. Recognized therapeutic value of mercury for syphilis. Died 1513.

AQUILINUS, Walter, see Agilinus, Gaulterus.

AQUINAS, St. Thomas, Italian theologian, natural philosopher; b. Roccasecca, terrs. of Naples, circa 1225; s. Landulf, Count of Aquino; ed. Monastery Monte Cassino, then U. Naples, circa 1235-41; studied philosophy and theology under Albertus Magnus at Dominican Sch., Cologne, 1244-45; B.Th., Paris, 1248. Became Benedictine oblate at Monte Cassino, 1230; entered Dominican order, 1244; apptd. 2d lectr., magister studentium at Cologne, 1248, created doctor theology, 1257; rendered many important services to church and state (particularly as adviser and councilor) through his courses and lectures in theology at U. Paris, 1252-59; at court of Pope Alexander IV at Anagni, 1259-61; at court of Pope Urban IV at Orvieto, 1261-65; at Dominican house of studies at Santa Sabina in Rome, 1265-67; at court of Pope Clement IV, Viterbo, 1268; at Paris, 1268-72; accepted chair at Naples at request of head of his order and of King Charles, 1272; was offered appropriate rewards for dedicated service to church, but refused all appointments to higher posts, including the archbishopric of Naples and abbacy of Monte Cassino; summoned by Pope Gregory X to attend Council of Lyons (convened in attempt to settle disputes between Greek and Roman Ch.), 1274, became seriously ill en route, died some weeks later. Canonized by Pope John XXII, 1323; festival of St. Thomas ranked on level with those of the 4 great Latin fathers by Pope Pius V, 1567; Roman Catholic clergy were directed to take his teachings as basis of their theol. position by Pope Leo XIII, 1879; declared patron of all Roman Cath. ednl. instns., 1880; his philos. system remains the basis of Cath. teaching to this day. Author: (major work) Summa Theologiae; Contra Gentiles; In Quatuor Sententiarum P. Lombardi libros; XII Quodlibeta Disputata; Quaestiones Disputae; Catena Aurea; In Omnes Epistolas Divi Apostoli Expositio; Super Isaiam et Jeremaiam; In Psalmos; Aristotelian commentaries including De Interpretatione, Posterior Analytics, Metaphysics, Physics, De Anima; Meditations; also many treatises. Held that there were 2 sources of knowledge: mysteries of Christian Ch. (revelation) and truths ascertained by human reason; stated that both are legitimate and respectable, but considered revelation the more important; brought to perfection tripartite method of medieval schoolmen; his philos. thinking reconciled Aristotelian philosophy and Cath. theology; his support of reason helped make natural philosophy (science) a respectable pursuit in Christian Europe again; opposed concept of innate ideas; believed all knowledge derives from sense experience; gave in comprehensive form ideas on origin of infinity; revitalized notion of a continuum; felt earth's roundness provable by mathematics and physics; suggested Ptolemy's astron. system could be supplanted. Died Cistercian Monastery of Fussa Nuova, diocese of Terracina, between Rome and Naples, Mar. 7, 1274.

AQUINAS (or AQUINUS), Suevus (Dacus), mathematician; lived in Nuremberg, Germany in or before 1471; at ct. of Prince Otto II of Pfalz-Moosbach, 1494; tchr. math.; pub. algebraic solutions to arithmetic problems. Author: De numerorum et sonorum proportionibus; Epistolae; Sermones de tempore.

ARAGO, Dominique-François-Jean, French astronomer, physicist; b. Estagel, France, Feb. 26, 1786. Sec., Bur. Longitudes; completed (with Biot) Delambre and Méchain's measurements of arc of meridian, Spain, 1806; apptd. prof. analytical geometry École Polytechnique, Paris, 1809; became dir. Paris obs., 1830; mem. Chamber Deps., 1830; minister war and marine, provisional govt., 1848. Fellow Royal Soc., 1818. Recipient Copley medal, 1825. Mem. French Acad. Scis. Author: Astronomie populaire, 4 vols., 1834-35. Discovered chromatic polarization; (with Fresnel) established fundamental laws of the polarization of light; invented polariscope; developed new photometer; studied compressibility of gases; made 1st precise measurement of density of gases; determined (with Biot) light refraction capabilities of different gases, 1806; determined (with Dulong) diffraction and dispersion properties of fluids and their vapors, 1816; continued work on Young's discovery of interference; clearly identified light with undulatory motion of transverse vibrations, thus promulgated wave theory of light; established (with Prony) speed of sound at 331.2 meters per second; investigated electricity and magnetism; discovered electromagnetic induction (failed to interpret it); made thermal studies; discovered rotation magnetism, 1824; differentiated 4 types of lightning in thorough descriptions of thunderstorm phenomena, 1838; precisely measured diameters of the planets; studied sun's corona and chromosphere; explained scintillation of stars as light interference, 1847; studied speed of light in different media; made double refraction micrometer, 1847. Died Paris, Oct. 2, 1853.

ARAKATSU, Bunsaku, Japanese physicist; b. Hyogo Prefecture, Japan, 1890; grad. Kyoto U., 1918;

studied in Europe and U. S., 1926. Asst. prof. Imperial U., Kyoto, then prof., 1936-50; prof. Taihoku U., Formosa; pres. Konan U., since 1951. Author: Genshikaku-no-Kenkyu (studies of atomic nucleus); also others on atomic matters and gamma rays. Central figure in establishment of a cyclotron at Kyoto U.

ARAKAWA, Edward Takashi, Am. physicist; b. Honolulu, Apr. 8, 1929; s. Wilfred S. and O. (Sasaki) A.; student Wayland Coll. (Tex.), 1947-49; B.S., Samford U., 1951; M.S., La. State U., 1953; Ph.D., U. Tenn., 1957; m. Harue Yamashiro, July 15, 1955; children—Sandra Kay, David Ken, Paul Edward, Carolyn Ann. Research asso. Oak Ridge Inst. Nuclear Studies, 1953-56; teaching asst. U. Tenn., 1956-57; physicist Oak Ridge Nat. Lab., 1957——. Mem. Optical Soc. Am., Am. Phys. Soc., Sigma Xi. Research on therapy of cancer patients with radioisotopes; radiation dosimetry of Japanese A-bomb survivors; discovery of plasma radiation in aluminum; devel. of polarizers for vacuum ultraviolet radiation, electronic properties of solids. Home: 111 Amherst Lane. Office: Oak Ridge Nat. Lab., Oak Ridge 37830.*

ARAKAWA, Hidetoshi, Japanese meteorologist; b. Fukushima, Japan, Aug. 4, 1907; s. Risaburo and Sei (Aoki) A.; student Tokyo U., 1928-31, D.Sc., 1938; m. Rikuko Motegi, Apr. 3, 1937; children—Katsutoshi, Kunitoshi, Michiko. Chief forecaster Central Meteorol. Obs., Tokyo, 1931-42, chief statis. sect., 1941-45, chief statis. div. 1945-49; chief forecasting research lab. Meteorol. Research Inst. Tokyo, 1949-64, dir. inst., 1966——; dir. Meteorol. Office, Fukuoka, 1964-66. Lectr., Tokyo U., 1941-64, Kyushu U., 1964-66. Recipient awards Minister of Transp., 1944, Army Weapons Comdr., 1945. Mem. Am., Royal meteorol. socs.; Meteorol. Soc. Japan. Author: Kisho Rikigaku (Dynamical Meteorology), 1941; Kisho Netsurikigaku (Thermodynamics in Meteorology), 1941; Typhoons, 1958; also articles. Editor: Climate of Asia, 1968. Research on air masses in Far East, proposal to use balloon bombs in World War II, analysis of structure of typhoons, climatic change as revealed by Japanese hist. documents. Home: Daita 3-5-27, Setagaya, Tokyo. Office: Koenji Kita 4-35-8, Suginami, Tokyo, Japan.*

ARAKI, Genataro, Japanese physicist; b. Kyoto, Japan, 1902; grad. in lit. and sci. Tokyo Bunrika U., 1918; Sc.D. Instr. Tokyo Bunrika U., for short time; then asst. prof. Kyoto U., later prof. Author: Atomic Theory; Molecular theory; Elemental particles. Research on mutual action between neutrons.

ARAKI, Huzihiro, Japanese math. physicist; b. Tokyo, Japan, July 28, 1932; s. Gentaro and Kiyoko (Kobayasi) A.; B.S., Kyoto U., 1955, M.S., 1957, D.Sc., 1961; Ph.D., Princeton, 1960; m. Chieko Kawaguchi, Mar. 26, 1961; 1 son, Isato. Faculty, Kyoto (Japan) U., 1957——, prof. physics 1966——; research asso. U. Ill., Urbana, 1960, asst. prof., 1962-64; guest lectr. Fed. Inst. Tech., Zurich, Switzerland, 1961-62. Mem. Phys. Soc. Japan, Japan Math. Soc. Research, publs. on quantum field theory, generalization of retarded functions, connection of spin and commutation relations between different fields, asymptotic behavior, developer idea of rings of local observables (R. Haag) to full math. theory.*

ARAKI, Shukuro, Japanese physician; b. Kumamoto-city, Japan, Jan. 1, 1927; s. Kanae and Sumiko (Namioka) A.; degree in med. sci. Kyushu U., 1960; m. Ryoko Okamoto, May 4, 1959; children—Eiichi, Lisa. Asst. in internal medicine Kyushu U., 1961-63, asso. prof. neurology Neurol. Inst., 1963——, vice chmn. dept. neurology, 1963——. Mem. Am. (asso.), Japanese (councilor) acads. neurology. Research, publs. on clin., biochem. subjects related to neurol. scis.; epidemiological studies on motor neuron disease, multiple sclerosis, homocystinuria, McArdle's disease, spinal cord metastasis. Office: Kyushu U., Fukuoka City, Japan.

ARAKI, Toshima, Japanese astronomer; b. Kumamoto Prefecture, Japan, 1897; grad. Kyoto U., 1923. Became asst. prof. Kyoto U., 1924, full prof., 1940-45. Mem. Japan Astronomers Assn. Author: Outline of Astronomy; Astronomy and the Cosmos; Kantian Theory of the Universe. Research in gen. theory of astronomy.

ARANTIUS (or ARANZI), Julius Caesar, Italian anatomist; b. Bologna, Italy, 1530; student of Andreas Vesalius; prof. surgery, anatomy, medicine U. Bologna; physician to Pope Gregory XIII. Author: De humano foetu (comprehensive text on embryology), 1564, 71. Said to have 1st recorded case of pelvic deformity, 1564; gave early description median cleft between 2 laminae of septum pellucidum (ventricle of Arantius, 5th ventricle, pseudocele), also described foramen ovale, ductus venosus (ductus venosus arantii), tubercles (bodies of Arantius) in semilunar valves of aorta and pulmonary arteries, arched ligament connecting diaphragm with 1st lumbar vertebra and lowest ribs, gave 1st clear description of inferior horns of lateral ventricles of brain, proved some external muscles of eye originate in margin of optic foramen (all 1571). Died Bologna, Apr. 6, 1589.

ARANZI, Julius Caesar, see Arantius, Julius Caesar.

ARAPOV, Dmitriy Alekseevich, Russian surgeon; b. 1897; grad. Med. Faculty, 2d Moscow U., 1925; D.Med. Sci. Intern, asst., dep. head surg. dept. Sklifosovsky First-Aid Research Inst., Moscow, 1929-41, head 1st Surg. Clinic, 1945——; chief surgeon No. Fleet, 1941-45; dep. chief surgeon Soviet Navy, 1945-50, chief surgeon, 1950——; lt. gen. Med. Service; dep. chief surgeon USSR Armed Forces, chief surgeon Soviet Navy, 1950——; hon. sci. worker or RSFSR, 1959——. Mem. USSR Acad. Med. Sci. (corr.), All-Union (bd. mem., bur. mem. sect. anesthesiologists), All-Russian (bd. mem.) socs. surgeons, Internat. Soc. Surgeons (hon.), also mem. learned council numerous sci. instns. in Moscow. Author of numerous works including Gas Gangrene, 1942; Inhalation Narcosis, 1949; Despeciated Serum, 1950; Anerobic Wound Infection, 1950; The Pathogenesis and Treatment of Burns, 1956; co-author The Gastointestinal Fistula and Its Surgical Treatment, 1963. Co-editor Surgery sect. Large Med. Ency., 2d edit. Research on wound infections, narcosis, acute appendicitis, ileus, burns, heart injuries, formation of artificial esophagus, blood substitutes. Address: Sklifosovsky First-Aid Research Inst., B. Kolkhoznaya pl. 3, Moscow, USSR.

ARATOS OF SOLI, astronomer, poet; b. probably Soli, Cilicia, 315 B.C.; pupil of Denys l'Heracleote; student Peripatetic, Stoic schs., Athens, Greece; lived at ct. of king of Macedonia (Antigonus Gonatas), also Antiochus I of Syria. Author: Phaenomena (poem on astronomy in 3 parts, the 1st two of which are based on the Phaenomena of Eudoxos, and the third on Theophrastos). Discussed common risings and settings of stars and weather signs, also course and influence of stars. Died Macedonia, 245 B.C.

ARAVANIS, Christ Ioannis, Greek physician; b. Lefkas, Greece, Nov. 30, 1923; s. Ioannis and Sofia (Stavraka) A.; M.D., U. Athens (Greece), 1948, Ph.D., 1958; m. Tula-Athena Theodos, June 12, 1958; children—John, Helen. Asso. div. cardiology Chgo. Med. Sch., 1954-58; faculty U. Athens, 1959——, asso. prof. cardiology, 1962——; dir. lab. electrocardiography and vectorcardiography, attending physician Hippocrates U. Hosp., Athens, 1962——; dir. program for study heart diseases in Greece, USPHS, 1960-—. Fellow Internat. Coll. Angiology (co-editor 1963-—, v.p. 1963——); mem. Greek Soc. Atherosclerosis (gen. sec. 1958——), Am. Greek heart assns., Am. Coll. Cardiology, N.Y. Acad. Scis., Hellenic Med. Soc., Chgo. Inst. Medicine. Author: Digitalis, 1957; Diseases of the Chest, 1960; Encyclopedia of Cardiology, 1960; also numerous articles. Research on phonocardiography and electrocardiography, epidemiology of cardiovascular diseases. Home: 21 Thesseos St. Philothei, Athens. Office: 47 Queen Sophia Av., Athens 140, Greece.*

ARAVIYSKY, Aleksandr Nikolaevich, Russian dermatologist, venerologist; b. Ishma, Tyumen Oblast, 1897; grad. Med. Faculty, Tomsk U., 1924; D.Med. Sci., 1946. Intern, asst., asst. lectr., lectr. Skin and Venereal Diseases Clinic, Tomsk Med. Inst., 1924-41; head chair skin and venereal diseases, dir. Novosibirsk Postgrad. Med. Inst., 1941-56; prof. 1941——; head chair skin diseases 1st Leningrad Med. Inst., 1956——. Mem. Leningrad Soc. Dermatologists and Venerologists (bd. mem.). Author over 60 works, including Variation of the Causative Organism and Clinical Polymorphism of Favus, 1946; co-author The Laboratory in the Dermatologist's and Venerologist's Practical Work, 1940. Address: 1-y Med. Inst., ulsita Lva Tolstogo 6-8, Leningrad, USSR.

ARBER, Agnes, English biologist; b. 1879; d. Henry Robert Robertson; B.Sc., U. Coll., London, Eng., 1899; M.A., Newnham Coll., Cambridge (Eng.) U.; D.Sc.; m. Edward Alexander Newell Arber, Aug. 5, 1909; 1 dau., Muriel. Instr. botany U. Coll., London, 1903-09; fellow Newnham Coll., Cambridge U., 1912-13, 18-20; Leverhulme fellow, 1936-38. Recipient Linnean medal, 1948. Fellow Royal Soc., 1946, Linnean Soc.; mem. Bot. Soc. Am. (corr.), Indian Bot. Soc. (hon.). Author: Herbals, 1912; Water Plants, 1920; Monocotyledons, 1925; The Gramineae: A Study of Cereal, Bamboo and Grass, 1934; The Natural Philosophy of Plant Form, 1950; The Mind and the Eye, 1954; The Manifold and the One, 1957. Contbr. articles on botany and its history. Contbd. to field of philosophy biol. study; distinguished between phylogenetic speculation reached by typological arguments and phylogenetic speculation reached by exam. of plant form in light of hist. evidence. Died Mar. 22, 1960.

ARBITER, Nathaniel, Am. metallurgist; b. Yonkers, N.Y., Jan. 2, 1911; s. David and Ida (Rockman), A.; A.B., Columbia, 1932; children by previous marriage—Jane Latiné, Jerome; m. 2d Carolyn S. Metz, March 21, 1961; children—Laurie, Robin Dorothy, Corinna. Began career as assistant at the Columbia Sch. Mines, N.Y.C., 1937-43, asso. prof., 1951-58, prof. 1958——; research engr. Batelle Meml. Inst., Columbus, O., 1943-44; Phelps Dodge Corp., N.Y.C., Morenci, Ariz., 1944-51. Cons. to numerous maj. metall. interests in N. and S.Am. Mem. Am. Inst. Mining, Metall. and Petroleum Engrs. (R. H. Richards award 1961, dir, treas. 1959-61). Editor: Milling Methods in the Americas, 1964; Transactions VII International Mineral Processing Congress, 1964. Patentee in field. Research in engring., engring.

sci. underlying processing of ores and minerals. Home: 331 Hudson Terrace, Piermont, N.Y. 10968.*

ARBOGAST, Louis-François-Antoine, French mathematician; b. Mutzig, Alsace, Oct. 4, 1759; prin. Coll., Strasbourg, France; prof. math. Sch. Artillery, Strasbourg, also Central Sch. Bas-Rhin; dep. Constituent Assembly, also Legislative Assembly; mem. Nat. Conv., circa 1793; asso. Nat. Inst., 1796. Recipient prize for essay on differential and integral calculus Acad. Petersburg, 1792. Author: Calcul des derivations, 1800. Developed Arbogast method and series in calculus; 1st to separate symbols of operation from those of quantity; redefined continuity. Died Strasbourg, April 8, 1803.

ARBUCKLE, Howard Bell, Am. chemist; b. nr. Lewisburg, W.Va., Oct. 5, 1870; s. John Davis and Elizabeth (Van Lear) A.; B.A., with first honor, Hampden-Sidney Coll., 1889, M.A., 1890; spl. student in chemistry, U. Va., 1894-96; Ph.D., in Chemistry, Johns Hopkins, 1898; m. Ida Meginniss, June 4, 1896; children—Howard Bell, Adele Taylor (Mrs. Donald Rohl). Prof. ancient langs., Seminary West of Suwanee (foundation for U. Fla.), 1891-94; teacher, and prof. Agnes Scott Inst. (later coll.), Decatur, Ga., 1898-1912; prof. chemistry Davidson (N.C.) Coll., from 1913. A founder of Continental Dorset Club for registry of pure bred Dorset sheep; contbg. editor Am. Sheep Breeder, 1900-20; founder Edgewood Stock Farm and imported selected sheep from Eng., 1904. Mem. Am. Red Poll Breeders Assn., Am. Aberdeen-Angus Breeders Assn., Am. Chem. Soc. (founder, pres. Ga. sect., pres. N.C. sect.), N.C. Acad. Sci. (pres. 1925), Pi Kappa Alpha (councilor princeps, 1900-05; grand councilor, 1913-33), Gamma Sigma Epsilon (grand chancellor 1920-28), Phi Beta Kappa, Omicron Delta Kappa, Scabbard and Blade. Author: Redetermination of the Atomic Weight of Zinc and Cadmium, 1898; Laboratory Manual in Household Chemistry, 1912; The Life and Habits of the Honey Bee, 1925. Contbr. numerous articles to chem. lit. and agrl. jours. Researches in corn proteins and cellulose products; discovered pyrolene. Died July 19, 1945.

ARBUTHNOT, John, Brit. physician; b. Scotland, 1667; possibly educated in Aberdeen, Scotland; University College, Oxford, 1692; M.D., St. Andrews University, 1696; children—Charles, George, 2 daus. Teacher of mathematics, London; attended Prince George of Denmark at Epsom; appointed physician in ordinary to Queen Anne, 1709; attendant in her last illness; became censor, 1723, Harveian orator, 1727. Elected Fellow of the Royal Society 1704; Royal College of Physicians. Author: An Examination of Dr. Woodward's Account of the Deluge, etc., 1697; Essay on the Usefulness of Mathematical Learning, 1700; Table of Grecian, Roman, and Jewish Measures, Weights and Coins reduced to the English Standard, 1707, 27; An Essay Concerning the Nature of Aliments and the Choice of Them, 1731; Practical Rules of Diet in Various Constitutions and Characters of Human Bodies, 1733; Essay Concerning the Effects of Air on Human Bodies, 1733. Leading physician of his time; friend of Swift, Pope and other lit. figures; writer poetry and satire, notably Law is a Bottomless Pit; or, the History of John Bull, 1712. Wrote articles on mathematics and numismatics. Died Hampstead, Eng., Feb. 27, 1735.

ARBUZOV, Aleksandr Erminingeldovich, Russian organic chemist; born September 11, 1877; graduated from Kazan U. 1900; D.Chem.Sci., 1914. Married Ekaterina Krotova; children—including Boris. Asst. Agrl. and Forestry Inst., Novaya Aleksandriya, 1901-—, asst. professor, 1906——; professor Kazan Univ. 1911-30, dean Physico-Math. Faculty, 1924——, dir. Butlerov Chem. Research Inst., 1929-30; prof. Kazan Inst. Chem. Tech., 1930——; Dir. Arbuzov Inst. of Chem. of Kazan br. USSR Acad. Sci., 1960——; honorary science worker of RSFSR. Mem. USSR Acad. Sci. (Presidium mem. 1960——, chmn. Presidium Kazan br. 1945——). Decorated Order of Lenin (2); recipient Stalin prize, 1943, 47. Author: The Structure of Phosphorus Acid and Its Derivatives, 1905; Woodworm and the Chemical Composition of the Resins of Some Conifers, 1925; Free Radicals, 1932; Tautomeric Conversions in Some Phosphoro-Organic Compounds, 1934; Obtaining Substituted Indoles by Catalytic Decomposition of Phenylhydrazines, 1936; The Reaction Mechanism of Binary Exchange of Metallic Derivatives of Tautomeric Organic Compounds, 1939; A Brief Outline of the Development of Organic Chemistry in Russia, 1948; M. V. Lomonosov, The Great Russian Scientist and Chemist, 1945; A. M. Butlerov, the Great Russian Chemist, 1949; Selected Works 1952; co-author Catalytic Decomposition of Phenylhydrazine by Monohaloid Copper Salts, 1913, The Effect of Haloid-Substituted Esters of Fatty Acids on Esters of Phosphorus Acid, 1914, The Structure of Boyd's Acid Chloride, 1929. Coordinating editor: Chemistry and the Use of Organic Phosphorus Compounds, 1962. Research on phosphorus-organic compounds; discovered Arbuzov Rearrangement reaction for intermediate esters of phosphoric acid, 1905; (with B. A. Arbuzov) discovered method of obtaining free radicals of triarimethyl series; developed important technique for collecting soft gums from conifers. Died Jan. 22, 1968.

ARBUZOV, Boris Alexandrovitch, Russian chemist; b. Nowaja, Alexandrija, Russia, Nov. 4, 1903; s. Alexandr Ermingeldovitch and Ekaterina (Krotova) A.; student Inst. Agr. and Forestry, Kazan, USSR, 1921; M.S., Kazan State U., 1926; m. Olga Michailova, Mar. 2, 1926; 1 dau., Marina. Docent, Kazan State U., 1929-30; faculty Kazan Inst. Chemistry and Tech., 1930-38, prof., head chair, 1938; head dept. Kazan br. USSR Acad. Scis., 1946-59, dir. Inst. Organic Chemistry, 1959-—; dir. Inst. Organic Chemistry, Kazan State U., 1960-—. Recipient A.M. Butlerov's prize Russian Soc. Physics and Chemistry, 1926, D. I. Mendeleev's prize, 1949, State prize, 1951. Mem. Chem. Soc., Société Chemique de France, USSR Acad. Scis. (academician.) Author: The Investigations in the Field of Isomeric Transformations of Bicyclic Terpenes and its Oxides, 1936; Co-author: Allyl Regroupings, 1949; also numerous articles. Research on phosphorus-organic chemistry and isomeric transformations of bicyclic terpenes and its oxides, numerous reactions of addition to some diene-containing systems, investigations of conformations of cyclic compounds by methods of dipole moments and nuclear magnetic resonance. Home: 3-b Schkolnaja, Kazan, USSR.*

ARBUZOV, Sergei Yakovlevich, Russian pharmacologist; b. 1903; grad. Mil. Med. Acad., 1933; D.Med. Sci.; postgrad. Mil. Chem. Acad., 1937. Asso., Chem. Research Inst., Moscow, 1937-40; jr. instr. dept. pharmacology Kuybyshev Mil. Med. Acad., 1940; asso. dept. physiology Mil. Med. Acad., 1933; sr. instr. dept. pharmacology and pharmacy Kirov Mil. Med. Acad., Leningrad, 1944-56, head chair pharmacology and pharmacy, 1960-—; sr. asso., head dept. radiobiology Inst. Exptl. Medicine, USSR Acad. Med. Sci., 1956-60, now head Lab. Radiobiology, Leningrad. Mem. Pharmacological Commn., USSR Acad. Med. Sci., USSR State Pharmacopeia Com.; mem. RSFSR Pharmacological Com. Mem. Soc. Pharmacologists (bd. mem., central council), Leningrad Soc. Physiologists, Biochemists and Pharmacologists (bd. mem.), All-Russian Soc. Pharmaceutists (chmn.). Author over 100 works including monographs. Mem. editorial bd. Fiziologichesky zhurnal SSSR im. I.M. Sechenova; mem. editorial council Pharmacology and Toxicology; co-editor Large Med. Ency., 2d edit. Research on pharmacology, toxicology, synthesis of drugs, radiobiology and radiation pharmacology; originator methods of synthesizing phenadin, pyridoxiphene. Address: Kirov Mil. Med. Acad., ulitsa Lebedeva 6, Leningrad 9, USSR.

ARCE, José, Argentine physician; b. Loboría, Buenos Aires, Argentina, 1881; s. Juvencio and Luisa Arce; ed. Brit. Acad., 1881-91, Colegio Salvador, 1892-95; M.D., Buenos Aires Faculty Medicine, 1903. Dep. dir. Nat. Univ. Inst. Normal Anatomy and Operative Medicine, 1900-03, dir. surg. clinic, 1904-05, substitute prof. clin. surgery, prof. ad interim, descriptive anatomy Faculty of Medicine, 1907-10; reorganized Faculty of Medicine, 1918, prof., from 1919; dean Buenos Aires Med. Sch., 1935-40; surgeon various hosps. Argentine rep. at many confs. and congresses. Mem. Buenos Aires province Chamber Deps., 1903-13; dep. to nat. legislature, 1924-29; ambassador to China; ambassador to UN, 1946. Mem. Argentine Med. Assn., Surg. Soc. Rio de Janeiro (hon.), acads. medicine Lima, Rio de Janeiro, Mexico, Paris, Madrid, Rome, Acad. Surgery, Paris, A.C.S., Soc. Obstetrics and Gynecology (founder), Surg. Soc., Buenos Aires. Author books and articles on pneumonectomies and inhibition of thoracic nerves; noted for work in chest surgery.

ARCELIN, Adrien, paleographer, archaeologist; b. Fuisse, 1838; ed. Sch. Paleography, 1860-64. Paleographer, Haute-Marne, France; in charge mission to Orient; collaborated on geol. map France. Author: Morimond et les milices cheualeresques d'Espagne et de Portugal, 1864; Les bulles pontificales des archives de la Haute-Marne, 1866; (under pseudonym Adrien Cranile) Indicator héraldique et genealogique du Macconais, 1865, Solutré ou les chasseurs de rennes de la France centrale; publs. on archeology, history, botany, philosophy. Died Saint-Sorlin, 1904.

ARCHAGATHOS OF PELOPONNESUS, Greek physician; b. circa 219 B.C.; s. Lysanias; probably 1st Greek surgeon to practice in Rome, became citizen, later banished as carnifex or butcher; worked to free slaves; invented type of plaster that bears his name.

ARCHAMBAULT, George Francis, Am. pharmacist; b. Springfield, Mass., Apr. 29, 1909; s. George Charles and Katherine V. (Mayette) A.; Ph.G., Mass. Coll. Pharmacy, 1931, Ph.C., 1933, D.Pharmacy, 1960; LL.B., Northeastern U., 1941; Sc.D., Phila. Coll. Pharmacy and Sci., 1951; LL.D., Temple U., 1961. Chief pharmacy br. div. hosps. Bur. Med. Services USPHS, 1947-65, pharmacy liaison officer to Office Surgeon Gen., pharmacy cons. div. Med. Care Adminstrn., Bur. State Services, 1965-—. Recipient Distinguished Service medal USPHS, 1965. Mem. Am. Pharm. Assn. (past pres.), A.A.A.S. (past v.p.), Assn. Mil. Surgeons. Editorial bd. Mil. Medicine, Am. Jour. Hosp. Pharmacy. Developed, implemented hosp. pharmacy service system for USPHS; research, publs. on investigatorial drug controls, hosp. pharm. planning techniques and practices. Home: 5916 Melvern Dr., Bethesda, Md. Office: USPHS, Washington.*

ARCHBOLD, Richard, Am. zoologist; b. N.Y.C., Apr. 9, 1907; s. John F. and May (Barron) A.; spl. studies Columbia U., 1931; Pres., Archbold Expeditions, 1933-—; resident dir. Archbold Biol. Sta., Lake Placid, Fla., 1941-—. Trustee Archbold Meml. Hosp. Mem. Am. Mus. Natural History. Leader zool. expeditions to Madagascar, 1929-31, New Guinea, 1933-34, 36-37, Netherlands Indies, 1938; research on animal behavior. Home and office: Route 2, Box 380, Lake Placid, Fla. 33852.*

ARCHDEACON, James William, Am. physiologist; b. Carlisle, Ky., Oct. 29, 1911; s. James M. and Carrie (Smith) A.; A.B., U. Ky., 1933, M.S., 1940; Ph.D., U. Rochester, 1943. Faculty U. Ky., Lexington, 1946-—, prof. arts and scis., 1956-—, prof. medicine, 1960-—. Fulbright lectr. U. Malaya, 1964-65. Mem. Am. Physiol. Soc., A.A.A.S., Sigma Xi. Contbr. articles to sci. jours. Research in physiology, food intake and bile secretion, membrane transport function. Home: 678 Bishop Dr., Lexington, Ky. 40505.*

ARCHELAUS, philosopher; flourished 5th century B.C.; b. Athens or Miletus; pupil of Anaxagoras; said to have taught Socrates; held view that air is source of water and fire, which form mud (basis for earth and heavenly bodies) and man and animals; avoided Anaxagoras' dualism by not distinguishing mind from air. Died 428 B.C.

ARCHENHOLD, Friedrich Simon, astronomer; b. Lichtenau, Oct. 2, 1861; s. Moses and Rosa (Blumenfeld) A.; ed. Strasbourg, France, also Berlin, Germany; m. Alice; 1 son, Günther, 3 daus. Asso. with Berlin Obs., from 1891; founder Weltall (astron. jour.), 1909. Suggested 68 centimeter objective diameter and 21 meter focal length telescope (now in Archenhold Obs., Berlin-Treptow). Died Berlin, Oct. 14, 1939.

ARCHER, Frederick Scott, Brit. inventor; b. Bishop Stortford, Eng., 1813; worked as sculptor; invented camera which contained processes for producing photograph; invented collodion process, 1850; said to have been 1st to use triplet lens. Died May, 1857.

ARCHER, Hugh M., Am. engr; b. Dover, N.J., June 22, 1916; s. Harvey G. and Helen (Morris) A.; B.E.E., Rensselaer Poly. Inst., 1937; postgrad. U. Mich., Bowdoin Coll.; m. Mary Jane Reed, May 11, 1940; children—June, Ruth, Lucy. Sr. Research engr. Detroit Edison Co., 1937-42,47-50; dir. Diffin Research Labs., Inc., Detroit, 1942-45; founder, gen. mgr. Archer-Reed Co., Dearborn, Mich., 1950-—; founder, pres. Spiratex Co., Dearborn, 1955-—; cons. in individual practice, Mich. area, 1943-—. Cons. to univs.; asst. tech. dir. Bikini Bomb Tests. Mem. Mich. Soc. Profl. Engrs., Soc. Plastic Engrs. (dir. 1964), Rensselaer Soc. Engrs., Sigma Xi, Tau Beta Pi. Patentee optic electronic, blood preservation devices; research in electo-optics for glass sheet manufacture and control, ultraviolet measurements. Home: 22210 Morley Av. Office: 23874 Kean Va., Dearborn, Mich. 48124.*

ARCHER, Walter E., Brit. ichthyologist; b. July 4, 1855; s. Clement R. Archer; m. Alice Lina Murray, 1878; 3 sons, 3 daus. Salmon researcher, 1884-92; insp. salmon fisheries for Scotland, 1892-98; chief insp. fisheries Bd. Trade, 1898-1903; asst. sec. Bd. Agr. and Fisheries, 1902-12. Brit. del. Internat. Conf. for Investigation of Sea, 1899; mem. Royal Commn. on salmon fisheries of Gt. Brit., 1900-02; mem. Tethyological Research Com., 1901; Brit. del. Inter-Nat. Council for study of sea, from 1904, pres., from 1908. Fellow Royal Soc. Edinburgh, Statis. Soc., London. Author salmon fishery reports Fishery Bd. for Scotland, 1892-98, sea and salmon fishery reports Eng. and Wales, from 1898. Studied salmon of Norway. Died Aug. 19, 1917.

ARCHER, William, naturalist; b. nr. Down, Ireland May 6, 1830; s. Richard and Jane Matilda (Verling) A.; Moved to Dublin, Ireland and began natural history studies, 1846; Librarian Royal Dublin Soc., 1876, Nat. Library Ireland (compiled catalogue), 1877-95. Fellow Royal Soc., 1875; Royal Irish Acad. (sec. for fgn. corr. 1875-80, Cunningham gold medal 1879), Dublin Micros. Club (sec.). Contbr. articles Quarterly Jour. of Micros. Sci., Proceedings of Dublin Micros. Club. Notable research on protozoa; discovered new species and genus of freshwater Sarcodic organism, 1875. Died Aug. 14, 1897.

ARCHER, William Harry, Am. oral surgeon; b. Ambridge, Pa., Mar. 6, 1905; s. William Harry and Marie (Morris) A.; D.D.S., U. Pitts., 1927, B.S., 1937, M.A., 1947; m. Louise R. Etzel, Apr. 26, 1933; 1 son, William Harry III. Faculty, U. Pitts., 1927-—, prof. oral surgery and anesthesia, chmn. dept. oral surgery, mem. grad. faculty Sch. Dentistry, 1946-—, lectr. Sch. Nursing and Sch. Pub. Health, 1955-—; mem. staff hosps.; also cons. Diplomate Am. Bd. Oral Surgery. Fellow Am. Coll. Dentists, Internat. Coll. Dentists, Internat. Coll. Anesthetists, Internat. Assn. Oral Surgeons.; mem. Am. Dental Assn., Internat. Dental Research Soc., Odontological Soc. Western Pa., Am. Soc. Oral Surgeons, Am. Dental Soc. Anesthesiology, Am. Acad. Cleft Palate Rehab., Am. Acad. Oral Pathology, Fedn. Dentaire Internat., Internat. Coll. Anesthetists, Pierre Fauchard Acad., Internat. Assn. Oral Surgeons, Omicron Kappa Upsilon, Sigma Xi, Phi Alpha Theta, others; hon. mem. Association des Licencies et Dentistes Universitaires de Belgique, Helenic Soc. Oral Surgeons, Athens, Greece, P.R. Dental Soc., others. Author numerous books including Oral Surgery, 1952, 4th edit., 1966; Cirugia Bucaldental, 1958; Dental Anesthesia, 1952, 2d edit., 1958; Anestesia En Odontologic, 1955; Oral Surgery Directory of the World, 1957, 3d edit., 1966. Research, publs. on anesthesia in oral surgery; complications of fractures, infections of face and mouth; improvement of dental edn. Home: 761 Osage Rd., Pitts. 15243.*

ARCHESILAOS, mathematician; flourished between the 4th and 3d century B.C.; taught by Autolycos; founded middle acad. which succeeded old acad. of Plato.

ARCHIBALD, Raymond Clare, mathematician; b. Colchester County, N.S., Can., Oct. 7, 1875; s. Abram Newcomb and Mary (Mellish) A.; B.A., U. of Mt. Allison Coll., N.B., 1894, LL.D., 1923; Harvard, 1895-98; B.A., 1896, M.A., 1897; U. Berlin, 1898-99; U. Strasbourg, 1899-1900, Ph.D., 1900; Sorbonne, Paris, France, 1909-10; U. Rome, 1922; hon. doctor, U. Padua, 1922. Prof. math., librarian, head violin dept. Mt. Allison Ladies' Coll., Sackville, N.B., 1900-07; prof. math. Acadia U., Wolfville, N.S., 1907-08; instr. math. Brown U., 1908-11, asst. prof., 1911-17, asso. prof. 1917-23, prof. 1923-43, emeritus from 1943; lectr. U. Cal., 1924, Harvard, 1931, Columbia, 1939-40. Rep. for U. S. and Can. of Euler Commn. of Swiss Soc. of Naturalists, 1922-39; mem. Am. sect. Internat. Math. Union (mem. Internat. Com. on Bibliography 1924-28; mem. div. phys. scis. NRC, 1928-31, 40-43, 44-47, chmn. Internat. Com. on Math. Tables and other Aids to Computation, 1939-50. Fellow Am. Acad. Arts and Sciences, A.A.A.S. (v.p., chmn. Sect. A, 1928, Sect. L, 1937); fgn. fellow Masarykova Akademie Prace (Czechoslovakia), Societatea de Stiinte of Cluj (Rumania); hon. mem. Polish Math. Soc., Math. Assn. (Eng.), Phi Beta Kappa; mem. Sigma Xi, London Math. Soc., Deutsche Mathematiker Vereinigung, Edinburgh Math. Soc., Unione Matematica Italiana, Am. Math. Soc. (asso. editor Bull. 1914-20, council, 1918-41; librarian, 1921-41; trustee 1923), Math. Assn. Am. (pres. 1922, trustee, 1923-30), History of Science Soc. (council 1924-40). Founder Mary Mellish and Archibald Mem. Library English and Am. Poetry and Drama, Mt. Allison U., Sackville, 1905-55. Author: The Cardioid and Some of Its Related Curves, 1900; Bibliography of Life and Works of Simon Newcomb, 1905, 24; Mathematical Instruction in France, 1910; Euclid's Book on Divisions of Figures with a Restoration, 1915; The Training of Teachers of Mathematics for the Secondary Schools of the Countries Represented in the International Commission on the Teaching of Mathematics, 1918; Benjamin Peirce (1809-1880), 1925; Bibliography of Babylonian and Egyptian Mathematics, 1927-29; Klein's Famous Problems of Elementary Geometry, rev. edit., 1930; Outline of the History of Mathematics, 1932, 6th edit., 1949, The Scientific Achievements of Nathaniel Bowditch, 1937; Semicentennial History American Mathematical Soc., 1888-1938, 1938; Fifty Mathematical Table-Makers, 1948; also articles in Ency. Brit., Dictionary Am. Biography. Editor: English transl. of J. Steiner's, Geometrical Constructions With a Ruler—Given a Fixed Circle, 1949; asso. editor Am. Math. Monthly, 1917-18, editor in chief, 1919-21; asso. editor Revue Semestrielle des Publications Mathématiques, 1921-34, Isis, from 1924, Scripta Mathematica, from 1932; founder, editor Math. Tables and Aids to Computation (quar.), 1943-49. Extensive contbr. to math. jours. and revs. Died July 26, 1955.

ARCHIBALD, Reginald MacGregor, Am. biochemist, physician; b. Syracuse, N.Y., Mar. 2, 1910; s. Eben Henry and Minnie (Archibald) A.; B.A., U. B.C., 1930, M.A., 1932; Ph.D., U. Toronto, 1934, M.D., 1939. Vis. investigator Rockefeller Inst., N.Y.C., 1940-43, spl. investigator, 1943-45, asso., 1946, mem., prof., 1948-—; prof., head dept. biochemistry Johns Hopkins Sch. Hygiene and Pub. Health, 1946-48; physician Rockefeller Inst. Hosp., 1948-—, sr. physician, 1955-—. Mem. adv. bd. Analytical Chemistry, 1956-58. Mem., A.A.A.S., Am. Chem. Soc., Am. Soc. Biol. Chemists, Endocrine Soc., Soc for Research in Child Devel., Am. Fedn. Clin. Research, Harvey Soc., Med. and Chirurg. Faculty Md., Coll. Physicians and Surgeons Ont., Biochem. Soc. (Great Britain), Soc. for Exptl. Biology and Medicine, Am. Bd. Clin. Chemists (v.p. 1957-63). Author: Chemical Characteristics and Physicological Roles of Glutamine, 1945. Editorial bd. Jour. Biol. Chemistry, 1948-58, Jour. Clin. Endocrinology, 1952-59. Research on methods of clin. chemistry, measurement of enzymes, influence of hormones on enzymes, problems of growth and devel. in children. Home: 266 Ancon Av., Pelham, N.Y. 10803 Office: Rockefeller Inst., 66th St. and York Av., N.Y.C. 10021.*

ARCHIBALD, Sir Robert George, physician; b. Secunderabad, India, July 4, 1880, s. W. F. and Edith (Cookson) A.; MB., Ch.B., Edinburgh (Scotland) U., 1902); M.D.; m. Olive Chapman Cant, 1919; 2 daus. Commd. officer Royal Army M.C., 1906; ambulance surgeon No. Hosp., Liverpool, Eng.; asst. med. officer Rainhill Asylum; house surgeon St. Mary's Hosp. for Women and Children, Plaistow; with Sleeping Sickness Commn., Uganda, 1907; attached Egyptian army, from 1908, with Blue Nile Operations, 1908, Mediterranean Expeditionary Force, Dardanelles, 1915, Darfur Expdn., Sudan, 1916; dir. Wellcome Research Labs., Khartum, from 1920, Stack Med. Research Labs., 1928-36; prof. bacteriology Farouk U., Alexandria, Egypt, from 1947, prof. parasitology, from 1949; pathologist County Lab., Poole, Eng.; med. supt. Leper Settlement, Trinidad, B.W.I. Recipient pathology prize Army Med. Coll., 1906. Created knight, 1934. Mem. Société Pathologie Exotique (corr.), Royal Soc. Medicine (Council Sect. Tropical Diseases and Parasitology), Brit. Med. Assn. (pres. Sudan br. 1935-36). Author: (with Andrew Balfour) Reviews of Recent Advances in Tropical Medicine, Supplements to Reports, Wellcome Research Labs. Editor: (with W. Byam) Practice of Medicine in the Tropics. Contbr. Oxford Index of Therapeutics, 1921. Died May 2, 1953.

ARCHIBALD, William James, Canadian physicist; b. Sydney, N.S., Can., Oct. 20, 1912; s. John Thomas and Maud (Burrows) A.; B.A., Dalhousie U., 1933, M.A., 1935; Ph.D., U. Va., 1938; D.Sc., U. N.B., 1961; m. Margaret Alva Armstrong, Aug. 14, 1936; children—Ian Donald, Malcolm Alexander Armstrong, Stephen Bronson. Faculty, Dalhousie U., Halifax, N.S., 1942-—, prof., 1955-—, head dept. physics, 1955-60, dean arts, sci., 1955-61, now Fales prof. theoretical physics. Postdoctoral Stirling fellow Yale, 1938, Postdoctoral fellow U. Va., 1953. Mem. Am., Can. assns. physicists, N.Y. Acad. Sci. Rearch in equation for solving problems connected with ultra centrifuge. Home: 1630 Walnut St., Halifax, N.S., Can.*

ARCHIGENES, physician; b. Apamea, Syria, circa 53; pupil of Agathinus; practiced in Rome during reigns of Domitian, Nerva, and Trajan, 81-117; mem. eclectic sch.; influenced by pneumatic theories; directed his therapeutics toward 8 bad temperaments; originated theory of pulse (later borrowed by Galen). Died circa 117.

ARCHIMEDES, Greek mathematician, physicist, engr.; b. Syracuse, Sicily, circa 287 B.C.; s. Phidias; probably studied in Alexandria with successors of Euclid; returned to Syracuse and devoted life to math. research and writing; an aristocrat of independent means, on intimate terms with King Hieron II of Syracuse. Author: On the Sphere and Cylinder; The Measurement of a Circle; On Conoids and Spheroids; The Sand Reckoner (Arenarius); On Spirals; On the Equilibrium of Planes; On Floating Bodies; On the Quadrature of the Parabola; On the Method. Often considered greatest math. genius of antiquity; calculated arithmetical approximation to the value for pi (more accurate than any previously obtained); devised an exponential-like system for expressing large numbers; found areas and volumes of special curved surfaces and solids, including surface and volume of sphere and segment of sphere, also area of segment of parabola and spiral, and volume of any segment of solids of revolution of 2d degree; determined ratio which cylinder bears to sphere it circumscribes; developed math. proofs using methods of exhaustion and reduction to absurdity; devised first math. exposition of principle of composite movements; worked on cubic equations; found means to calculate square roots by approximation; founded sci. of statics; developed elegant and mathematically sophisticated proof of law of lever on statical-geometrical principles; determined center of gravity in various bodies; invented sci. of hydrostatics; developed concept of specific gravity; investigated completely positions of rest and stability of right segment of paraboloid of revolution floating in fluid; demonstrated propositions concerning relative immersion and weight in fluid of solids less dense, as dense as, and more dense than fluid; relative to solids more dense than fluid, developed Archimedes' Principle: solid, when weighed in fluid, will be lighter than its weight in air by weight of fluid displaced; often obtained results by intuition or measurement, only later verifying them by rigorous math. proof; said to have invented device that easily drew fully-loaded ship from land to sea (of this, Plutarch reports Archimedes as saying: give me a place to stand on, and I will move the earth); invented a planetarium (mech. contrivance which could imitate motions of sun, moon, planets, fixed stars), also a hydraulic organ; contrived and used mil. engines in defense of Syracuse against Romans. Died during Roman sack of Syracuse, 212 B.C.

ARCHYTAS, Greek mathematician, philosopher, general; b. Tarentum, Italy, circa 428 B.C.; Pythagorean; mem. sch. Plato; gov. Tarentum 7 times. Credited with invention of analytic method math., also of pulley; developed theory of proportion and gave solution to doubling of cube; distinguished harmonic from arithmetical and geometrical progressions; applied geometry to mechanics and treated mechanics methodically; contbd. to theory of acoustics; constructed flying bird; possibly responsible for main contents of Book VIII, Euclid's Elements. Drowned in shipwreck, circa 47 B.C.

ARCO, Georg Wilhelm Alexander Hans Graf von, see von Arco, Georg Wilhelm Alexander Hans.

ARCOLANI, Giovanni, Italian surgeon; prof. medicine, surgery Bologna, then Padua (both Italy); treated many diseases neglected from Greek medicine almost to modern times, such as empyema, liver abscess, ascites, ileus, difficulties of urination. Author: Practica Medica; also commentaries on work of Rhazes and Avicenna. 1st to mention filling teeth with gold; gave excellent description of alcoholic insanity; wrote about conditions of nose and mouth, such as polyps. Died Padua, circa 1460-84.

ARCTOWSKI, Henryk, geophysicist; b. Warsaw, Poland, July 15, 1871; studied chemistry and geology, univs. of Paris, Liege and Zurich, 1888-96; British Mus., London, 1896-7; Ph.D., U. Lemberg, 1912; m. Arian Jane Addy, Mar. 28, 1900. In charge phys. observations Belgian Antarctic Expdn., 1897-9; afterwards assisted at Royal Ob. of Belgium until coming to Am., 1909; visited Spitzbergen, 1910; chief of sci. div. N.Y. Pub. Library, 1911-19; prepared reports on geography, mineral resources, ethnography, demography, agr. and industry of Poland for use of Am. delegation to Peace Conf., 1919; prof. geophysics and meteorology Lwow U., 1920-39; with Smithsonian Inst., Washington, from 1939. Sec. for meteorology Belgian Astron. Soc.; mem. Belgica Commn., Internat. Polar Conf. Fellow Royal Geog. Soc. (London), A.A.A.S., N.Y. Acad. Scis., Assn. Am. Geographers; mem. Nat. Inst. Social Scis.; corr. mem. Belgian Geol. Soc., Belgian Geog. Soc. Knight of Order of Leopold; medals and hon. distinctions from Belgian Royal Acad., City of Antwerp, Geog. Soc., London, etc. Author: Die Antarktischen Eisverhaltnisse, 1903; L'Enchainement des Variations Climatiques, 1909; Study of the Changes in the Distribution of Temperature, 1916; also reports pub. by the Belgian Govt., articles in Am. and European sci. jours. Discovered and studied geology of Antarctic Andes, established 1st complete record of meteorol. observations made in south polar regions. Died Feb. 21, 1958.

ARDEN-CLOSE (or CLOSE), Sir Charles (Frederick), geographer; b. Aug. 10, 1865; s. Frederick Close; Sc.D. (hon.), Cambridge (Eng.) U.; m. Gladys Violet Percival, 1913; 1 son, 1 dau. Entered army, 1884; Indian Survey, 1889-93; in charge Niger Protectorate, Kamerun Boundary Surgery, 1895; Brit. commr. Nyasa-Tanganyika Boundary Commn., 1898; served in S. African War, 1899-1900; head, geog. sect. Gen. Staff, 1905-11; dir.-gen. Ordnance Survey, 1911-22; Halley Lectr. Oxford (Eng.) U., 1914, ret., 1922. Del., Conf. on Frontiers between Uganda, German E. Africa, and Congo, Brussels, Belgium, 1910; boundary commr. under Representation of the People Bill, 1917. Created knight, 1918. Fellow Royal Soc., 1919; mem. Royal Geog. Soc. (Victoria Research medal 1927, pres. 1927-30), S.E. Union Sci. Socs. (pres. 1922), Geog. Assn. (pres. 1927), Internat. Population Union (pres. 1931-37), Internat. Geog. Union (pres. 1934-38), hon. mem. various fgn. geog. socs. Author: Text Book of Topographical Surveying; The Early Years of the Ordnance Survey, 1926; The Map of England, 1932; Geographical Byways, 1947; contbr. tech. papers. Died Dec. 19, 1952.

ARDERNE (or ARDEN), John, English surgeon; b. circa 1307; army surgeon Hundred Years' War; probably at Antwerp, Belgium, 1338, Algeciras, Spain, 1343, Bergerac, France, 1347; practiced medicine, Wiltshire, Eng., then Newark-upon-Trent, Eng., 1349-70, London, from 1370. Author: De arte medicinae, 1370; Practica de fistula, 1376. Considered father of English surgery; developed surg. practices similar to those of modern aseptic surgery; emphasized cutting boldly, keeping instruments and wounds clean, obtaining healing with minimum of suppuration, use of light dressings. Died 1377 or after.

ARDUINO, Giovanni, Italian geologist; b. Caprino, Italy, Oct. 16, 1713; ed. Verona, Italy; with govt. mining service; dir. mines, Tuscany; later prof. mineralogy and metallurgy, Venice, also Padua, Italy. Author: Osservzzione sulla fisica constituzione delle Alpi Venete, 1759; Saggio fisico-mineralogico de lythogomia e orognosia; Raccolta di memorie chimico-mineralogiche, metallurgiche e orittografiche, 1775. Known as father of modern Italian geology; 1st classified younger rocks as Primary, Secondary, Tertiary (used in part today, 1760); 1st to recognize volcanic origin of crystalline rocks; 1st to realize importance of applying paleontology and chemistry to chronology of formations. Died Venice, Mar. 21, 1795.

ARENA, Jay Morris, Am. physician; b. Clarksburg, W.Va., Mar. 3, 1909; s. Anthony and Rose (Sandy) A.; B.S., W.Va. U., 1930; M.D., Duke, 1932; m. Pauline Elizabeth Monteith, July 10, 1931; children—Rosanne, Jay Morris II, Carolyn Jean, Mary Margaret, Katherine, Pauline, Regina. Instr. pediatrics Vanderbilt U., 1936; asst. pediatrics Duke Hosp., 1936-40, asso. prof., 1940-56, prof., 1956-—; dir. Duke U. Med. Center Poison Control Center; editorial bd. Nutrition Today. Cons. for poison control centers FDA Nat. Clearing House. Mem. Am. Pediatric Soc., Am. Acad. Pediatrics (exec. bd.), N.C. Pediatric Soc. (pres. 1964), Am. Assn. Poison Control Centers (pres. 1967-68), Nat Assn. for Practical Nurse Edn. and Service, Phi Beta Kappa, Sigma Xi, Alpha Omega Alpha. Author: Poisoning: Chemistry, Symptoms, Treatment, 1963. Contbr. numerous articles in field pediatrics to med., sci., profl. jours. Editorial bd. Clin. Pediatrics, Pediatric News. An originator of poison control centers in U.S.; responsible for safety closures on drug bottles; research in caustic, petroleum distillate and other types of poisoning. Home: 1403 Woodburn Rd., Durham, 27705. Office: Box 3024, Duke U. Med. Center, Durham, N.C. 27706.*

ARENDS, Georg, German pharmacist; b. Chemnitz, Germany, Dec. 18, 1862; s. Emil Fridolin and Marie Helene (Gast) A.; ed. Leipzig, Germany; m. Elisabeth Marie Heyne, Oct. 8, 1891; 1 son, Johannes. Pharmacist, indsl. chemist, 1887-95; founder, owner Elisabeth-Apotheke (standardized mass prodn. of prescriptions), Chemnitz, 1908-37. Author numerous books and publs. on pharmacy and medicine. Sci. editor Pharmazeutische Zeitung, Berlin. Died Chemnitz, Dec. 23, 1946.

ARENDT, Rudolf Friedrich Eugen, German chemist; b. Frankfort, Germany, Apr. 1, 1828; s. August Ferdinand and Johanna Henriette (Schröder) A.; apprentice in pharmacy, then book bindery; studied natural scis. probably at Leipzig (Germany) U., from 1854; degree, 1859; m. Henriette Henschel; 1 son, 1 dau. Stenographer; asst. Agrl. Research Sta., Mockern, nr. Leipzig; tchr. natural scis. and tech. Pub. Sch. Commerce, Leipzig, from 1861, prof., from 1880. Author: Technik der Experimentalchemie, 1880; Lehrbuch der anorganischen Chemie, 1898. Chief editor Chemisches Zentralblatt, from 1862. Created systematic curriculum of elementary chemistry with textbooks and pedagogical writings. Died Leipzig, May 15, 1902.

ARENS, Josef Ferdinand, chemist; b. Malang (Java), Nov. 12, 1914; s. P. M. and A. S. (Stucky) A.; ed. U. Amsterdam; m. C. Peperzak, 1944. Asst. U. Amsterdam, 1939-41; lab. chemist N.V. Organon, Oss, 1941-48; prof. organic chemistry U. Bandung (Indonesia), 1948-53; prof. U. Groningen, 1953-60; prof. U. Utrecht, since 1960. Recipient Holleman prize in chemistry, 1959. Mem. Royal Acad. Scis. Research, publs. on synthesis of vitamin A, chemistry of acetylenic compositions, organic sulphides, peptides. Address: Soestdijkerstraatweg 94, Hilversum, Netherlands.*

ARENTSEN, Kaj, Danish psychiatrist; b. Copenhagen, Denmark, June 8, 1916; s. Helfred and Gerda (Nissen) A.; M.D., U. Copenhagen; m. Ellen Nielsen, Dec. 30, 1942; children—Vibeke, Jens. Became asst. in psychiatry Aarhus Psychiat. Hosp. 1954, med. dir. dept. legal psychiatry, 1958; med. dir. Glostrup Psychiatric Hosp., 1960-—. Author: Patients in Danish Psychiatric Hospitals, 1959; Alkoholister og deres behandling, 1960; also publs. on psychiatry from social view point. Home: Nordre Rignvej 69, Bolig 29, Glostrup. Office: Glostrup Psychiatric Hospital, Ringvej, Denmark.

ARENTZEN, Edward Sentman, Am. engineer; b. Stratford, N.J., Jan. 7, 1916; s. Edward and Mamie (Vogel) A.; B.S., U. S. Naval Acad., 1937; M.S. in Naval Arch., Marine Engring., Mass. Inst. Tech., 1943; m. Marcia Brooks Eddy, Mar. 29, 1941; children—Marcia (Mrs. Stephen Root), Karen Christina. Various engring. assignments U. S. N., 1943-58; prof. naval constrn. Mass. Inst. Tech., Cambridge, 1958-62, adminstrv. officer physics, 1962-—; cons. Gen. Elec., 1962-63. Recipient Joseph Linnard award, 19—. Mem. Soc. Naval Architects and Naval Engrs., Am. Phys. Soc. Devel. modern submersible submarine hull. Home: 47 Fletcher Rd., Belmont, Mass. 02178.*

ARETAEOS OF CAPPADOCIA, Greek physician; b. Cappadocia (now in Turkey), circa 81; follower of Hippocrates; lived in Rome, probably also Alexandria. Author: On the Causes and Signs of Acute and Chronic Diseases, 4 vols.; On the Treatment of Acute and Chronic Diseases, 4 vols. Mem. eclectic sch., also studied pneumatic theories; ranks after Hippocrates in observation and ethics; gave 1st description of aura in epilepsy, also diabetes mellitus, quinsy; recognized tetanus, nodular leprosy, pneumonia, asthma, diphtheria; distinguished cerebral from spinal paralysis; formulated X pattern of motor decussation. Died circa 138.

ARETINO, Guido, Italian ophthalmologist; b. Arezzo, Tuscany, Italy; flourished circa 1326 in Salerno; student of Roger of Salerno; may possibly be identical with author of the Liber Mitis Quem Edidit Guidus Aretinus (gen. med. treatise); Surgery of Master Roger (written by Arentino, approved by Roger).

AREY, Leslie Brainerd, Am. anatomist; b. Camden, Me., Feb. 15, 1891; s. Arthur Brainerd and Mary Josephine (Page) A.; A.B., Colby Coll., 1912, Sc.D. (hon.), 1937; Ph.D., Harvard, 1915; LL.D., Chgo. Med. Sch., 1934; m. Mary Edith Holt, July 1, 1926. Faculty, Northwestern U., Chgo., 1915-—, Robert L. Rea prof. anatomy, chmn. dept., 1924-56, prof. emeritus, 1956. Vis. prof. U. P.R., 1953; cons. NIH, 1956-—; chmn. Internat. Com. on Embryological Nomenclature, 1960-—. Recipient Alumni medal Northwestern U., 1959. Mem. Am. Assn. Anatomy (pres. 1952-54), Am. Soc. Zoologists (sec.-treas. 1925-30), Chgo. Acad. Scis. (pres. 1956-—), Phi Beta Kappa, Sigma Xi, Alpha Omega Alpha, Pi Kappa Epsilon. Author: Textbook of Embryology, 1917, 20; Developmental Anatomy, 1924, 30, 34, 40, 46, 54, 63; Human Histology, 1957, 63; Northwestern Medical School, 1859-1959, 1959; also numerous articles. Research in behavior and physiology of various invertebrates, physiology of retina, devel. of human embryo, histology of invertebrates and mammals. Home: 3440 Lake Shore Dr., Chgo. 60657.*

ARFKEN, George Brown, Jr., Am. physicist; b. Jersey City, Nov. 20, 1922; s. George Brown and Ann (Hill) A.; B.E. in Chem. Engring., Yale, 1943, M.S., 1948, Ph.D., 1950; m. Carolyn Irene Dines, June 18, 1949; children—Bruce, Peter, Cynthia. Physicist, Oak Ridge Nat. Lab., 1950-52; faculty Miami U., Oxford, O., 1952-—, prof., chmn. dept. physics, 1956-—. Cons. Los Alamos Sci. Lab., 1960-65. Mem. Am. Assn. Physics Tchrs. Author: Mathematical Methods for Physicists, 1966. Contbns. to understanding of behavior of nuclear particles and nuclear forces. Home: 5301 Coulter Lane, Oxford, O. 45056.*

ARFVEDSON (or ARFWEDSON), Johan August, Swedish chemist; b. Skagerholms-Bruk, Sweden, Jan. 12, 1792; ed. Uppsala, Sweden, later under Berzelius. Became sec. Royal Bur. Mines, Stockholm, Sweden; made geol. tour Fontainebleau Forest and Clermont, France area, 1819; set up own lab., Stockholm, 1819. Mem. of the Helvetian Scientific Society, Swedish (gold medal 1841) Academy of Science, French Academy of Sci., 1828. Investigated chem. composition of minerals; discovered lithium in some of its compounds (later isolated by Davy); worked on uranium compounds and alloys; prepared uranous oxide, 1823. Died Hedensoe, Sweden, Oct. 28, 1841.

ARGAND, Aimé, Swiss inventor, physicist, chemist; b. Geneva, Switzerland, 1755; inventor Argand lamp with circular wick, inside and outside of which air was induced, 1782-84; (used until invention of Bunsen burner). Died Oct. 24, 1803.

ARGAND, Emile, Swiss geologist; b. Geneva, Switzerland, 1879; ed. Paris, 1899, U. Lausanne (Switzerland); student of Maurice Lugeon, also Albert Heim, 1908-09. Architect; then prof. geology mineralogy, paleontology U. Neuchâtel (Switzerland), 1911-40; founder Geol. Inst. Neuchâtel. Fgn. corr. geol. socs. Am., London. Recipient prix William Huber, 1908, prix Spendiaroff, 1913, prix Marcel Benoist, 1927, prix Cuvier, Acad. Sci., Paris, 1927. Author: Les nappes de recouvrement des Alpes occidentales, 1911; Sur l'arc des Alpes occidentales, 1916; La tectonique de l'Asie (biography of earth in its entirety, his most important work), 1922. Mapped geology of Alps, Asia; studied tectonic problems of structural geology in Alps; advanced knowledge of genesis of mountain chains and continents; originated comparative tectonic anatomy and geo-embryology. Died Neuchâtel, 1940.

ARGAND, Jean Robert, mathematician; b. Geneva, Switzerland, July 18, 1768; a pioneer in use of complex numbers to show all algebraic equations have roots; wrote on existence of sq. root of negative number; introduced word modulus; Argand diagram for representation of complex number by point (probably based partly on simultaneous discoveries of Wessel); pioneer in use of complex numbers, treated functions of complex variables geometrically. Author: Essai sur une manière de representer les quantités imaginaires dans les constructions géométriques, 1806. Died Paris, France, Aug. 13, 1822.

ARGELANDER, Friedrich Wilhelm August, German astronomer; b. Memel, E. Prussia (now Klaipeda, USSR), Mar. 22, 1799; s. Johann Gottlieb and Dorothea Wilhelmine (Grünlingen) A.; Ph.D., Königsberg, Germany, 1822; m. Marie Charlotte Courtan; 2 daus. Asst. to Bessel at Königsberg, 1820, lectr.; became head. obs., Turku, Finland, 1823, Helsinki, Finland, 1832; apptd. prof., Helsinki, 1828; supr. constrn., dir. obs., Bonn, Germany, prof. U. Bonn, 1837-75, also rector. Fellow Royal Soc., 1846; mem. Societas Fennica Helsingfors, Royal Astron. Soc. London, Astronomische Gesellschaft (chmn., Germany), numerous fgn. acads. sci. Author: Untersuchungen über die Bahn des grossen Kometen von 1811, 1822; Über die eigene Bewegung des Sonnensystems, 1837; Uranometria nova, 1843; Aufforderung an Freunde der Astronomie, 1844; Bonner Durchmusterung (last atlas compiled without photography, lists position and brightness of over 324,000 stars in the northern heavens), 1859-63. Established location of sun in cosmos; continued F. W. Bessel's work on determining the positions of stars; introduced decimal div. of stellar magnitudes leading to accurate stellar photometry; developed step method for estimation of brightness of stars (still in gen. use); introduced modern system of naming variable stars; confirmed existence of apex. Died Bonn, Feb. 17, 1875.

ARGHITTU, Cristino, Italian physician; b. Nughedu S. Nicolo, Italy, July 24, 1910; s. Gavino and Antonica (Giua) A.; M.D., U. Sassari, 1935, Ph.D. in Hygiene, 1956; m. Fanny Cossu, Dec. 27, 1937; 1 dau., Maria Cristina. Chief, Radiopathology Lab., Camen, Italy; col. Italian Army Med. Corps, Italian Dept. Def., Rome; dir. Head Mil. Hosp., Rome. Decorated Three War's Cross, Knight of Italian Republic. Mem. Società Italiana d'Igniene, Società Italiana di Microbiologia, Società Italiana di Biologia e Medicina Nucleare, Soc. Nuclear Medicine, Acc. Lancisiana Roma. Author: Traumatologia della Esplosioni Nucleari, 1960. Research, publs. on hygiene and microbiology, radiobiology, radiopathology, nuclear medicine. Home: 15, Via Annia, Rome. Office: 50 P.za Celimontana, Rome, Italy.*

ARGO, Harold Virgil, Am. physicist; b. Walla Walla, Wash., Jan. 20, 1918; s. Emmett Gerdo and Bertha (Munns) A.; B.A., Whitman Coll., 1939; M.A., George Washington U., 1941; Ph.D., U. Chgo., 1948; m. Mary Frances Langs, Oct. 2, 1942; children—Leslie Ann, Paul Emmett, Peter Stanton, Theodore Robert. Jr. physicist Naval Ordnance Lab., Washington, 1942-44; staff mem. Los Alamos Sci. Lab., 1944-46, 1948—, alternate group leader in physics div., 1959—. Research on nuclear reactions in light elements, Vela nuclear detection satellites, solar x-ray measurements. Home: 2571 Walnut St., Office: P.O. Box 1663, Los Alamos.*

ARGYLL, Duke of (George John Douglas), aero. pioneer; b. England, 1823; ed. privately; m. Elizabeth Sutherland, 1844; m. 2d, Amelia Maria Anson, 1881; m. 3d, Ina-Erskin M'Neill, 1895. Lord privy seal, 1853-55, 59-66, 80-81; postmaster-gen., 1855-58; sec. state for India, 1868-74. Chancellor, U. St. Andrews; elder bro. Trinity House. Trustee, Brit. Mus. Fellow Royal Soc., 1851; mem. Aero. Soc. (1st pres. 1866-96). Author: Primeval Man, 1869; Unity of Nature, 1884; What Is Science? 1898; Organic Evolution Cross-examined, 1898. Studied flight of birds and possibility of imitation by man; advocated heavier-than-air craft. Died 1900.

ARGYRIS, John, physicist, engr.; b. Volos, Greece, Aug. 19, 1913; s. Nicolas and Lucie (Caratheodory) A.; ed. Athens, Berlin, and Zurich univs.; D.Sc., London; D.Eng., Munich; m. Inga Lisa, Sept. 10, 1953. Research (with J. Gollnow u. Son) on structures, Stettin, 1937-39; sr. officer charge structural research Royal Soc. Aeronautics, London, 1943-49; became reader dept. aeros. Imperial Coll. Sci. and Tech., U. London, 1949, prof., since 1955; dir. Institut für Statik und Dynamick der Luft- und Raumfahrtkonstruktionen, U. Stuttgart, 1959—. Fellow Am. Inst. Aeros. and Astronautics, Royal Aero. Soc., Wissenschaftliche Gesellschaft für Luftfahrt. Author: The General Theory of Cylindrical and Conical Tubes under Torsion and Bending Loads, 1947; Energy Theorems and Structural Analysis, 1960; Modern Fuselage and the Elastic Aircraft, 1963; Recent Advances in Matrix Methods of Structural Analysis, 1964. Research on application of high speed computers to engring. problems; finite element methods for analysis of engring. structures. Home: 53 Princess Gate, London S.W. 7., Eng.; also Solitude Haus II, Stuttgart, Germany. Office: Imperial College Science and Technology, Prince Consort Road, London S.W. 7, Eng.; also 225 Robert Leicht Strasse, Stuttgart-Vaihingen, Germany.

ARGYROS, Isaac, Byzantine mathematician; b. probably Constantinople; flourished mid-14th century; probably a monk and disciple of Nicephoros Gregoras; author scholia to Ptolemy, and to 1st 6 books Euclid, treatises on astrolabe and extraction square roots, astron. texts on geodesy, geometry, arithmetic of Nicomachus, trigonometry. Died 1372.

ARIES, Robert Sancier, chem. engr.; b. July 21, 1919; s. Robert Emile and Sophie (Presente) A.; grad. Ecole Superieure Chimie Industrielle, U. Lyon (France), 1940; M. Engring., Poly. Inst. Bklyn., 1943, D.Eng., 1947; M.A., U. Minn., 1946; M.Sc., Yale, 1948; m. Elsie Hassit, May 5, 1946 (div. July 1955); children—Vivian Sue, Lynn Oddette. Chem. engr. Conn. Rubber Co., New Haven, 1942, Sargent & Co., New Haven, 1943, Nat. Foam System, West Chester, Pa., 1955; pres. Aries Assos., cons. engrs., N.Y.C., Monte Carlo, 1945—; adj. prof. Poly. Inst. Bklyn., 1947-57; pvt. prof. U. Geneva, Switzerland, 1958—. Mem. Société de Chimie Industrielle (founder Am. br., 1948, v.p., 1949-61), Soc. Engring. Sci. (founding mem.), Chem. Market Research Assn., Assn. Cost Engrs., Institut International pou l'Etude et l' Expertise d'Oevres d'Art et Antiquités (pres. Paris 1964), Northeastern Wood Utilization Council (tech. dir. New Haven 1942-52), T.A.P.P.I. (recipient Chems. award 1956), Research Soc. Am., Sigma Xi. Author: (with R. D. Newton) Chemical Engineering Cost Estimation, 1950, rev., 1955; (with W. Copulsky) The Marketing of Chemical Products, 1948; La Rentabilité d'un Pruede Chimique, 1958; Les Faux dans la Peinture et l'Expertise Scientifique, 1965; also articles. Patentee chem. processes. Home: 76 Av. Paul Doumer, Paris 16, France. Office: Scala Palace, Monte Carlo, Monaco.

ARIETI, Silvano, psychiatrist; born Pisa, Italy, June 28, 1914; s. Elio and Ines (Bemporad) A.; B.A., Lycee Galileo, Pisa, 1932; M.D., Med. Sch. Pisa, 1938; diploma in psychoanalysis William A. White Inst., 1952; m. Marianne Thompson, Oct. 24, 1965; children—David, James. Came to U. S., 1939, naturalized 1943. Pvt. practice psychiatry, psychoanalysis, N.Y.C.; asso. prof. clin. psychiatry State U. N.Y. Coll. Medicine, 1953-61; prof. clin. psychiatry N.Y. Med. Coll., 1961—; faculty, tng. and supervising analyst William A. White Inst., 1962—. Recipient Milan Group Advancement of Psychoanalysis Gold medal award, 1964. Mem. William Allanson White Psychoanalytic Soc. (pres. 1964), Am. Acad. Neurology, Am. Assn. Neuropathologists, Am. Anthropol. Assn., Am. Psychiat. Assn., Acad. Psychoanalysis, A.M.A., Am. Psychopath. Assn., Assn. Advancement Psychotherapy Author: Interpretation of Schizophrenia, 1955; The Intrapsychic Self, 1967. Editor: Am. Handbook of Psychiatry, vols. I, II, 1959, vol. III, 1966. Research, numerous publs. on psychopathology, psychodynamics and psychotherapy of schizophrenia; relations between psychodynamics and thought disorders; pioneer work in office treatment of schizophrenic patient; research in creative process and in cultivation of creativity. Address: 103 E. 75th St., N.Y.C. 10021.*

ARIFOV, Ubai Arifovich, Russian physicist; b. June 15, 1909. Samarkand Pedogogical Academy, 1931; D.Physico-Math. Sci. Asst. chair physics Tashkent U., 1932—; dir. Physicotech. Inst., Uzbek Acad. Sci., 1945-56, dir. Tashkent Inst. Nuclear Physics, 1956—; hon. sci. worker Uzbek SSR, 1959—, Mem. Uzbek Acad. Sci. (pres. 1961—, chmn. presidium dept. physico-math. sci. 1956—). Author: Positive Surface Ionization of Atoms and Molecules, 1948; Methods of Secondary Emission during the Bombardment of Conductors by Ions, 1954; Study of Secondary Metal Emission under the Effect of Bombardment of Alkaline Elements of Positive Ions, 1956; Dependence of the Ion Dispersal Factor on the Mass Ratio of Colliding Particles, 1957; The Action of Gamma Rays on the Living Chrysalis of the Mulberry Silkworm, 1957; Electron Emission during the Bombardment of Potassium by Neon Ions, 1963. Research in applied and atomic physics. Address: Uzbek Acad. Scis., ulitsa Kuybysheva 15, Tashkent, Uzbek SSR, USSR.

ARIGA, Hisao, Japanese sericulturist; b. Nagano Prefecture, 1910; grad. Tokyo U., 1935; doctorate degree. Prof. Tokyo U.; councillor Japan Agr. Soc.; dir. Japan Sericulture Soc. Specialist in sericulture and heredity rules governing silkworms; research on x-ray effects on sex chromosome of silkworms.

ARIGA, Kizaemon, Japanese sociologist; b. Nagano-Ken, Japan, Jan. 20, 1897; s. Kizaemon VI and Sei (Niimura) A.; grad. Tokyo Imperial U., 1922; Ph.D., Tokyo U., 1956; m. Sada Ikegami, June 20, 1924; Instr., Tokyo U., 1946; prof. Tokyo U. Edn., 1946-56; prof. Keio U., 1956-65; pres. Japan Women's U., Tokyo, 1965—; chief dir. Nihon Jomin Bunka Kenkyujo. Mem. Japanese Sociol. Soc. (dir., past pres.). Author books including: On Japanese Family and Tenant Farmer; The Complete Books of Kizaemon Ariga, 10 vols.; also articles. Clarification of some characteristics of Japanese culture; studies of Japanese family and village life including comparative studies. Home: 12-3 Hisagi 3Chome, Zushi-Shi, Tokyo, Japan.*

ARIMA, Yoriyuki, Japanese mathematician; b. 1712; lord of Kurume, Fukuoka Prefecture, Japan; studied math. of Seki Sch.; author 20 books on math. Promoted study of math. in Japan. Died 1783.

ARING, Charles Dair, Am. neurologist; b. Dent, O., June 21, 1904; s. Fred and Alice (Dair) A.; B.S., U. Cin., 1929, M.D., 1929; m. Mary Shroder, Oct. 16, 1931; children—Dair (Mrs. David C. Rausch), Charles Shroder. Fellow in neurophysiology Yale, 1934-35; Rockefeller neurology Nat. Hosp., London, Eng., U. Madrid, Breslau U., 1935-36; instr. neurology U. Cin. Coll. Medicine, 1935-38, asst. prof., 1938-41, asso. prof., 1941-46; prof., 1947—; prof. neurology U. Cal. at San Francisco, 1946-47. Mem. editorial bds. Archives of Neurology, Psychosomatic Medicine, Jour. Nervous and Mental Disease, Jour. Asthma Research, Modern Medicine. Research, numerous publs. in clin. neurology, neurophysiology, neuropathology, psychiatry; med. evaluation and philosophy. Home: 2401 Ingleside Av., Cin. 45206.*

ARISTAEOS (the Elder); Greek philosopher; b. Croton, circa 320 B.C.; s. Damophon; mem. Platonic Sch.; m. Theano (widow of Pythagoras); succeeded Pythagoras as head of his sch. (according to Iamblichus). Author various works on math. (some used by Euclid), including 5 on solid loci connected with conics, regular solids. Skilled in analytic method.

ARISTARCHOS OF SAMOS, Greek astronomer; b. Samos, Greece, circa 320 B.C.; studied under Straton of Lampsacus, in Alexandria. Probably taught in Alexandria; accused of impiety for his astron. theories by the Stoic, Cleanthes. Author: (only extant work) On the Magnitudes and Distances of the Sun and Moon. Famous as author of heliocentric hypothesis in astronomy (though theory is not mentioned in only extant writing); according to quotations from Archimedes and Plutarch, he maintained that "the fixed stars and the sun remain unmoved and that the earth revolves about the sun" while at the same time rotating on its own axis, which is inclined to the plane of the ecliptic; seems to have thought that the sun is larger than the earth; from his own writing, it is evident that he regarded astronomy as mathematical rather than descriptive; made the 1st attempt to measure the sizes and distances of sun and moon (his methods were geometrically correct, but conclusions were erroneous because he lacked accurate measuring instruments); 1st to attempt to work out this distance problem by trigonometry; made observation of summer solstice, 281-80 B.C.; said to have added 1/1623d of a day to Callippus' estimate of 365 1/4 days as length of year; said to have invented improved sun-dial (scaphion), represented by pointer in hollow hemisphere. Died circa 250 B.C.

ARISTIPPUS, Henricius, Italian translator; flourished in 12th century; worked in Sicily; translated the 4th book of Aristotle's Meteorology into Latin.

ARISTOTLE, Greek philosopher; b. Stagira, Chaldice (nr. Macedon), Greece, 384 B.C.; s. Nicomachus (ct. physician to King Amyntas II of Macedon); student under Plato, Athens (Greece) Acad., 367-347 B.C.; m. Pythias (niece, adopted dau. Hermias of Atarneus), circa 347 B.C.; 1 dau., Pythias; m. 2d, Herpyllis of Stagira; 1 son, Nicomachus. Traveled to various parts Greek world, especially Asia Minor, 347-

342 B.C.; apptd. tutor by Phillip II of Macedon to son (later Alexander the Great), circa 342 B.C.; founder, tchr. Lyceum (or Peripatos), Athens, 335-323 B.C.; founded zoo, Athens (zool. specimens, animals supplied by Alexander); ret. to Chalcis, island of Euboea, upon death of Alexander, 323 B.C. Author: Categories; On Interpretation; Prior Analytics; Posterior Analytics; Topics; On Sophistical Refutations; Physics; On the Heavens; On Generation and Corruption; On the Soul; On Memory and Reminiscence; On Dreams; The History of Animals; On the Parts of Animals; On the Generation of Animals; Metaphysics; Nicomachean Ethics; Politics; Rhetoric; Poetics. Founder systematic study of logic and theory of syllogism; introduced variables into logic which enabled him to develop logic as sci. instead of collection of examples; held object of a sci. to be found in its principles, its elements or its causes; believed sci. is knowledge by causes; pioneer of classification in biology; collected facts on natural history from personal observation, dissections, works of others; distinguished cetaceans from fish; described placental dogfish, sexual reprodn. of cuttlefish, octopus, habits of bees, devel. of chick embryo, transformation of caterpillar to butterfly, testes of mammals, habits of electric fish, gall bladder of pelamid, stomach in rudiments; held heart to be center of heat and life (brain being merely cooling organ for blood); supported notion of epigenesis; distinguished between red-blooded, non-red-blooded animals; opted for teleology in biology (manifested teleological attitude throughout work). Accepted 4 element theory of Empedocles (earth, water, air, fire) for sublunar region, which was region of change; thought that each element had own natural place toward which it tended to move in straight lines; held things in translunar or celestial regions composed of 5th element (ether) whose natural motion was circular and in some sense divine (thus sublunar and celestial regions fundamentally different in character); expanded astron. notions of Callipus, Eudoxus; believed earth spherical and at rest in center of finite ordered world (cosmos); held earth surrounded by crystalline-like, transparent, indestructible, homocentric spheres carrying planets; held speed of object directly proportional to contact force propelling it and inversely proportional to resistance of medium through which it passes; rejected actual infinite and possibility of void or vacuum. These views and also his views on metaphysics, ethics, politics, literary criticism gained wide acceptance in learned world when introduced into Europe in 12th, 13th centuries A.D. Died Chalcis, 322 B.C.

ARISTOXENOS OF TARENTUM, Greek mathematician, philosopher; b. Tarentum, Greece; flourished 250 B.C.; s. Mnesias or Spintharus; ed. Mantinaea, Arcadia, under father and Lampyrus of Erythrae, later student of Xenophilus the Pythagorean, also Aristotle, Theophrastus (in Athens). Became Peripatetic; authority on mus. theory; held that notes of scale should be judged by ear rather than math. ratio; formulated laws of edn. and ideas on soul. Author: Elements of Harmony, 3 vols.; Elements of Rhythm; reputed author of approximately 450 other works on music, philosophy, history, literature, including biographies of Pythagoras, Archytas, Socrates, Plato.

ARISTYLLOS OF ALEXANDRIA, astronomer; b. Alexandria, Egypt; flourished 233 B.C.; author works on fixed stars (later used by Hipparchos and Ptolemy), commentaries on Aratus (now lost).

ARIYAMA, Kanetaka, Japanese theoretical physicist; b. Tokyo, Japan, 1904; grad. Tokyo U., 1928; postgrad. studies in Germany; doctorate 1943. Instr. Tokyo Higher Normal Sch., then prof. Peer's Sch.; became prof. Nagoya U., 1942; elected rep. theoretical physicists Sci. Council Japan, 1950. Author: Quantum Mechanics; On the Property of Matter; also others. Specialist in metallic electronics and property of matters; important contbr. to devel. of quantum mechanics and property of matters in Japan.

ARKEL, Anton Eduard van, see van Arkel.

ARKELL, William Joscelyn, Brit. paleontologist, geologist; b. Highworth, Eng., June 9, 1904; s. James and Laura Jane (Rixon) A.; B.A., Oxford U., 1925, B.Sc. (Burdett-Coutts scholar), 1927, M.A., D.Phil., 1929, D.Sc., 1934; m. Ruby Lillian Percival, 1929; children—Raymond, Julian, Mervyn. Asso., prehistoric survey U. Chgo. (investigations of geology and prehistoric archeology Nile Valley, Egypt), 1926-30; lectr. in geology, New College, Oxford, 1929-33, fellow, 1933-40; with Ministry War Transport, 1941-43; became fellow Trinity Coll., Cambridge, 1947; asso. with Sedgewick Mus., Cambridge. Recipient Mary Clark Thompson Gold medal Nat. Acad. Scis. (Washington), 1944; Lyell medal Geological Soc. London, 1949; von Buch medal German Geological Society, 1953. Fellow Royal Society, 1947; hon. or corresponding member of the Linnean Society. Normandy, geol. socs. France, Germany, Egypt, Paleontol. Soc. Am., Dorset Natural History and Archaeology Soc. Author: Jurassic System in Great Britain; The Geology of Oxford; Oxford Stone; The Geology of Weymouth, Swanage, Corfe and Lulworth; Jurrasic Geology of the World, 1956. Authority on Jurassic stratigraphy and paleontology, Middle and Upper Jurassic ammonites. Died Apr. 18, 1958.

ARKHANGELSKY, Vitalii Nikolaevich, Russian ophthalmologist; b. 1897; grad. Med. Faculty, 1st Moscow U. (now 1st Sechenov Med. Inst.), 1922; D.Med. Sci., 1938. Specialized in ophthalmology Moscow U. Clinic, later intern, asst., lectr.; head chair ophthalmology Kuybyshev Med. Inst. (now Kuybyshev Mil. Med. Acad.), 1938—; head chair ophthalmology 1st Sechenov Med. Inst., Moscow, 1953—; head chair ophthalmology Kiev Med. Inst., 1944; sr. ophthalmologist Ukrainian Ministry Health, 1944; dir. Gelmgolts Inst. Ophthalmology, Moscow, 1953. Decorated Order of Lenin; recipient Averbakh prize, 1962. Mem. USSR Acad. Med. Sci. (corr.), Kiev Oblast Soc. Ophthalmologists (hon.), All-Russian (chmn.), All-Union and Moscow (bd. mem.) sci. socs. ophthalmologists. Author over 90 works including Microscopic Techniques for Ophthalmologists; Eye Diseases; The District Physician's Manual, 1957; Morphological Principles of Ophthalmological Diagnosis, 1961. Editor: Herald of Ophthalmology, Med. Abstracts Jour., Eye Diseases (multi-vol. manual); dep. editor Ophthalmology Jour.; co-editor Ophthalmology sect. Large Med. Ency., 2d edit. Introducer blood transfusion into ophthal. practice; 1929; deviser numerous original operation methods, hermetic closure of operative incision after opening anterior eye chamber, 1943, hermetic closure in operations on lacrimal sac, 1952; developer new operations, diathermocoagulation oa iris neoplasm, 1955, diathermocoagulation of ciliary body in glaucoma, 1957; research on path. and anat. nature of clin. processes, syphilis, leukemia, vitamin deficiency, hypertonia, diseases of retina. Address: 1-y Med. Inst., B. Pirogovskaya 2-6, Moscow, USSR.

ARKIN, Aaron, physician; b. Lebau, Latvia, Sept. 6, 1888; s. Isaac C. and Lena (Gordon) A.; came to U.S., 1891; B.S., U. Chgo., 1909, M.D., 1912, Ch.D., 1913; M.A., U. Wis., 1910; m. Alice Kauffmann, Nov. 17, 1945. Fellow physiology U. Chgo., 1908; instr. pharmacology U. Wis., 1909; fellow pathology U. Chgo., Rush Med. Coll., 1910-13; prof. pathology, bacteriology W.Va. U. Med. Sch., 1914-22; research asso. U. Vienna, 1922-26; clin. prof. medicine Rush Med. Coll., 1927-58; clin. prof. medicine U. Ill., 1927-28, emeritus prof., 1958-66; attending physician Weiss Meml., Columbus, Michael Reese hosps. (all Chgo.); cons. physician Cook County, Mt. Sinai hosps., Chgo. Recipient L. C. P. Freer medal Rush Med. Coll. Mem. Israel Med. Soc., A.M.A., Am. Physiol. Soc., Am. Assn. Pathology and Bacteriology, Am. Pharmacol. Soc., Am. Heart Assn., Am. Diabetes Assn., A.C.P., Sigma Xi, others. Research, publs. in immunology, pathology, internal medicine; 1st demonstrated spirocheta pallida in acquired ulcers of bowel, 1909; 1st description of clin. and x-ray findings in double aortic arch in adults, 1926; 1st description of histologically healed end-stage of periarteritis nodosa, 1930; studies of carcinoma of lung with suggestion of its relationship to smoking, 1930, research in antibody prodn., chemotherapy, carcinoma of lung, pancreas, acute pancreatitis, periarteritis nodosa. Died Nov. 1, 1966.

ARKWRIGHT, Sir Joseph Arthur, English bacteriologist, physician; b. Thurlaston, Eng., 1864; s. Arthur William and Emma (Wooley) A.; ed. Trinity Coll., Cambridge (Eng.) U., St. Batholomew's Hosp., London; m. Ruth Wilson, 1893; 3 daus. Practiced medicine, Hales Owen, Eng., 1893-1904; Lister Inst. Preventive Medicine, London, 1906-27, asst. bacteriologist, 1908-27. Fellow Royal Coll. Physicians, Royal Soc., 1926. Author: (with J. C. G. Ledingham) The Carrier Problem in Infectious Diseases, 1912. Studied meningococci, diptheria and its effects; described stimulus-response variation of coli-typhoid-dysentery organisms; well versed in field botany and horticulture. Died London, Nov. 22, 1944.

ARKWRIGHT, Sir Richard, English inventor; b. Preston, Eng., Dec. 23, 1732; apprentice to Nicholson, Preston, a barber in Bolton, Edward Pollit, a wigmaker in Churchgate; m. Patience Holt, Mar. 31, 1755; one s., Richard; m. 2d, Margaret Biggins, Mar. 24, 1761; 1 daughter. In wig bus., then textile industry; built several cotton mills (mark origin of factory system), by 1782; by 1789 operated 5 mills in Derbyshire and 3 in other countries; knighted, 1786; patented cotton spinning machine (water frame powered by animals, falling water, then Watts's steam engine, 1790), 1769; invented carding machine, 1775; also other machinery for textile mfg.; originality of several inventions contested. Died Cromford, Eng., Aug 3, 1792.

ARLEY, Niels Henrik, Danish atomic physicist, mathematician; b. Copenhagen, Denmark, Feb. 7, 1911; s. Hjalmar Petersen and Ragnhild (Petersen); Mag. Scient., U. Copenhagen, 1935, Dr.Phil., 1943; m. Ellen Margrete Bruun, June 26, 1935; 1 dau., Else. Techr. math. Royal Danish Naval Acad., 1935-55; sci. pvt. sec. to Niels Bohr., 1938-40; research fellow Niels Bohr. Inst., for Theoretical Physics, U. Copenhagen, 1936-40, asst. prof., 1940-54, asso. prof., 1954—, dir. Geophys. Research Inst., 1947-53. Vis. scientist Norsk Hydros Inst. for Cancer Research, 1954—; guest lectr. various univs. in Norway, Sweden, Eng., Germany, U.S.A., 1937—. Mem. Danish Math. Soc., Danish Phys. Soc., Danish Cancer Soc., Danish Soc. for Protection Sci. Work (past pres.). Author: (with K. R. Buch) Introduction to the Theory of Probability and Statistics, 1940 (U.S. edit. 1950, USSR edit. 1951, German and Spanish edit. 1961). Theory of Stochastic Processes and Their Application to the Theory of Cosmic Radiation, 1943; Lectures on Radiobiology, 1954; (with H. Skov) Atomic Power: An Introduction to the Technical, Military, Medical and Biological Problems of the Atomic Age, 1959 (German edit. 1960). Asso. editor Jour. for Geophys. Research, 1952-58. Contbr. numerous articles to profl. jours. Research on theory of stochastic (random processes) and interpretation of interaction of radiation and matter in cosmic radiation, radiobiology, carcinogenesis. Home: 12 Polarvej, Hellerup, Denmark. Office: Niels Bohr Inst. U. Copenhagen, Denmark; also Norsk Hydros Institute for Cancer Research, Oslo 3, Norway.*

ARLOING, Saturnin, French bacteriologist; b. Cusset, France, Jan. 3, 1846; prof. physiology Faculty Medicine, Lyons, France, later prof. exptl. and comparative pathology; dir. Lyons Vet. Sch.; corr. French Acad. Scis.; mem. Soc. Agr.; wrote on bacteriology of symptomatic anthrax, pleuro-pneumonia, Tb, and other diseases; introduced sero-agglutination of tubercle bacilli in diagnosis. Died Lyons, Mar. 21, 1911.

ARLOW, Jacob, Am. physician; b. N.Y.C., Sept. 3, 1912; s. Adolph and Ida (Feldman) A.; B.S., N.Y. U., 1932, M.D., 1936; grad. Profl. Sch., N.Y. Psychoanalytic Inst., 1947; m. Alice Diamond, Oct. 31, 1936; children—Michael, Allan, Seth, Jonathan. Practice medicine specializing in psychiatry, N.Y.C., 1940; instr. Columbia Coll. Phys. and Surg., 1942-52; lectr. N.Y. Psychoanalytic Inst. Sch. Applied Psychoanalysis, 1948, 51, instr., 1956—; faculty State U. Coll. Medicine, N.Y.C., 1952—, clin. prof. psychiatry, 1962—; instr. Hillside Hosp., N.Y.C., 1954-55. Recipient A.A. Brill award N.Y. Psychiat. Inst., 1963; Clin. Essay prize Brit. Psychoanalytic Soc., 1957. Mem. A.M.A., Am. Psychiat. Assn., Am. (past pres.), Internat., N.Y. psychoanalytic socs., Am. Psychosomatic Soc. Author: The Legacy of Sigmund Freud, 1956; Psychoanalytic Concepts and the Structural Theory (with Charles Brenner), 1964; also articles. Research on origin of symptoms, character of psychosomatic disorders; use of psychology in group processes. Home: 94 Wildwood Rd., Great Neck, N.Y. 11023. Office: 120 W. 59th St., N.Y.C. 10019.*

ARLT, Carl Ferdinand Ritter von, see von Arlt, Carl Ferdinand Ritter.

ARMALY, M. F., ophthalmologist; b. Shefa Amer., Palestine, Mar. 1, 1927; s. Farid M. and Fadwa (Bahouth) A.; Matric. certificate, Govt. Arab Coll., Jerusalem, Palestine 1944; B.A., Am. U. Beirut, Lebanon, 1947, M.D., 1952; M.S., U. Ia., 1957; m. Aida A. Makdisi, July 2, 1950; children—Raya, Farid. Came to U.S., 1960, naturalized, 1966. Faculty, U. Ia., Iowa City, 1955—, prof. ophthalmology, 1966—; cons. Nat Inst. Neurol. Diseases and Blindness, 1963—. Mem. A.M.A., Assn. for Research in Ophthalmology, Am. Ia. acads. ophthalmology and otolaryngology, Am. Physiol. Soc., Soc. Exptl. Biology and Medicine, Jackson County Med. Soc., French Opthalmol. Soc., N.Y. Acad. Scis., Sigma Xi. Contbr. chpt. to International Ophthalmology Clinics, vol. 2, No. 1, 1962. Publs. on demonstration of parasympathetic controls of eye; devel. of methods in diagnosis of glaucoma; demonstration of heritable nature of steroid-induced ocular hypertension and its relationship to glaucoma and diabetes. Home: 147 Koser Av., Iowa City 52240.*

ARMATI, Salvino degli, see degli Armati.

ARMENTROUT, Steve, Am. mathematician; b. Eldorado, Tex., June 19, 1930; s. T. B. and Carrie (Mercer) A.; B.A., U. Tex., 1951, Ph.D., 1956; m. Ruth Elizabeth Evans, June 23, 1962; children—Mary Helen, John Steven, Thomas James. Faculty, U. Ia., Iowa City, 1956—, prof. math., 1964—; mem. Inst. for Advanced Study, 1964-65; vis. prof. U. Wis., Madison, 1966-67. Mem. Am. Math. Soc., Math. Assn. Am. Research and publs. on upper semicontinuous decompositions of topological spaces, especially monotone decompositions of 3-manifolds. Home: 1 Washington Pl., Iowa City 52240.*

ARMITAGE, Kenneth Barclay, zoologist; b. Steubenville, O., Apr. 18, 1925; s. Albert Kenneth and Virginia (Barclay) A.; B.S., Bethany Coll., 1949; M.S., U. Wis., 1951, Ph.D., 1954; m. Katie Lou Hart, June 5, 1953; children—Carol Virginia, Keith Barclay, Kevin Conner. Instr., U. Wis. Extension, 1954-56; faculty U. Kan., Lawrence, 1956—, prof. zoology, 1966—. Mem. A.A.A.S., Ecol. Soc. Am., Am. Soc. Zoologists, Am. Soc. Limnology and Oceanography, Soc. for Study Evolution, Am. Inst. Biol. Sci., Sigma Xi. Research, numerous publs. on behavioral and physiol. mechanism enabling animal populations to adapt to their environment, role social behavior in regulating population density animals; pioneered study ice-covered warm water lakes in Antarctica. Home: 2619 Belle Crest Dr., Lawrence, Kan. 66044.*

ARMITAGE, Peter, English statistician; b. June 15, 1924; s. Harry and Florence Armitage; ed. Huddersfield Coll.; M.A., Trinity Coll., Cambridge (Eng.) U., 1952; Ph.D., U. London (Eng.), 1951; m. Phyllis Enid Perry, 1947; 3 children. Mem. Ministry Supply, London, 1943-45, Nat. Phys. Lab., London, 1945-46; staff stat. research unit Med. Research Council, London Sch. Hygiene and Tropical Medicine, U. London, 1947-61, prof. med. statistics, 1961—. Mem. Royal Statis. Soc. (hon. sec. 1958-64), Internat. Statis. Inst., Biometric Soc. Author: Sequential Medical Trials, 1960. Research, publs. statis. methods biol. and med. fields. Home: 7 Plymouth Park, Sevenoaks,

Kent, Eng. Office: London Sch. of Hygiene and Tropical Medicine, U. London, Keppel St. W.C. 1, London, Eng.

ARMOUR, Donald John, surgeon; b. Cobourg, Ont., Can., 1869; s. John Douglas Armour; ed. Upper Can. Coll., also Univ. Coll., London, Eng.; B.A., M.B., U. Toronto; m. Louise Mitchel; 1 son, 2 daus. Arris and Gale lectr. Royal Coll. Surgeons, 1904-05, named Hunterian prof. surgery and Pathology, 1908; Lettsomian lectr. Med. Soc. London, 1927; sr. demonstrator anatomy Univ. Coll., London; surgeon Nat. Hosp. for Nervous Diseases, Belgrave Hosp. for Children, Italian Hosp.; surgeon specialist, mil. hosps., 1914-19; cons. surgeon W. London Hosp., Blackheath and Charlton Hosp., Acton Hosp. Fellow Royal Coll. Surgeons Jacksonian prize essayist 1906); mem. Royal Coll. Physicians, London; pres. Med. Soc. London, 1929-30, W. London Medico-Chirurg. Soc., 1927-28, neurol. sect. Royal Soc. Medicine, 1928-29, Assn. Brit. Neurol. Surgeons, 1930-32. Author: The Surgery of the Spinal Cord and its Membranes; Principles of Spinal Surgery; Injuries to the Brain and Spinal Cord; Surgical Diseases of the Skull, Brain and Spinal Cord; papers on head injuries, abdominal surgery, surgery of nervous system. Died Oct. 23, 1933.

ARMOUR, Philip Danforth, Am. indsl. pioneer; b. Stockbridge, N.Y.; May 16, 1832; ed. common schools; mined in Cal., 1852-56; in commission business, Milw., 1856-63; mem. Plankinton, Armour & Co., packers, 1863; later became interested in Chgo. grain commission firm of H. O. Armour & Co., which established a pork-packing plant, 1868, and became Armour & Co., 1870; head of firm, Chgo., from 1875; conducting the largest pork-packing, dressed-meat and provision business in world; founder of Armour Mission and Armour Inst. Tech. Introduced refrigeration and new principles of large-scale orgn. to meat packing industry; eliminated waste of previously discarded portions of pigs. Died 1901.

ARMSBY, Henry Prentiss, Am. agrl. chemist; b. Northbridge, Mass., Sept. 21, 1853; s. Lewis and Mary A. (Prentiss) A.; B.S., Worcester Poly. Inst., 1871; Ph.B., Yale, 1874, Ph.D., 1879; LL.D., U. Wis., 1904; m. Lucy A. Harding, Oct. 15, 1878. Instr. chemistry, Worcester Poly. Inst., 1871-72; tchr. Fitchburg (Mass.) High Sch., 1874-75; asst. in chemistry, Rutgers, 1876-77; chemist Conn. Agrl. Exptl. Sta., 1877-81; vice prin., prof. Storrs (Conn.) Agrl. Sch., 1881-83; prof. agrl. chemistry, U. Wis., 1883-87; organizer, dir. Pa. Agrl. Exptl. Sta., 1887-1907; dir. Inst. of Animal Nutrition, Pa. State Coll., from 1907. Chmn. com. on expt. sta. exhibit, Chicago Expn., 1893, Paris Expn., 1900; expert in animal nutrition, U. S. Dept. Agr., from 1898; mem. Commn. on Agrl. Research, 1906-08. Mem. agrl. com. NRC, 1917; del. Inter-Allied Scientific Food Commn., 1918. Author: Manual of Cattle Feeding, 1880; Principles of Animal Nutrition, 1903; The Nutrition of Farm Animals, 1917; Conservation of Food Energy, 1918; (finished by C. R. Moulton) The Animal as a Converter of Matter and Energy, 1925. Developed respiration calorimeter large enough for observation on farm animals; demonstrated that econ. loss is entailed in feeding to livestock agrl. products suitable for direct use by humans. Died Oct. 19, 1921.

ARMSTRONG, Sir Alexander, Brit. physician, naturalist; b. Croghan Lodge, Ireland, 1818; s. Alexander Armstrong; student Trinity Coll., Dublin; M.D., U. Edinburgh, 1841; m. Lady William King Hall, 1894. Entered navy as asst. surgeon, 1842; in med. charge party for exploration Xanthus, 1843; apptd. to royal yacht, 1846, surgeon, 1849; surgeon and naturalist in arctic expdn. under Robert John Le Mesurier Maclure; med. supt. Malta Hosp., 1859-64; dir.-gen. medical dept. navy, 1869-71. Gilbert Blane Gold medal. Fellow Royal Soc., 1873. Author: Personal Narrative of the Discovery of the North-West Passage, 1857; Observations on Naval Hygiene, 1858. Died Sutton-Bonnington, Eng., July 4, 1899.

ARMSTRONG, Edward Frankland, Brit. chemist; b. 1878; s. Henry Edward Armstrong; D.Sc.; Ph.D.; LL.D.; m. Ethel Turpin, 1878; 1 son, 1 dau. Mng. dir. Joseph Crosfield & Sons, Warrington, Eng., 1915-25; with William Gossage & Son, Widnes, Eng., 1915-20; with Brit. Dye Stuffs Corp., Manchester, Eng., 1925-28; sci. adviser Ministry Home Security; dir. S. Met., S. Suburban and Comml. gas cos. Chmn., Soap Mfrs. Employers' Fedn., Joint Indsl. Council for Soap Trade, 1920-25, organizing com. Internat. Congress Science Management, 1935. Fellow Royal Society, 1920 (vice-president 1942, 43), Royal Institute Chemistry, Imperial Coll. Science (hon.), City and Guilds of London Inst.; mem. Inst. Chem. Engrs., Civil Def. Research Com., Soc. Chem. Industry (pres. 1922-24), Assn. Brit. Chem. Mfrs. (chmn. 1930-33), Brit. Assn. Chemists (chmn. 1926), Brit. Standards Instn. (chmn. 1934-35, 37-38). Author: The Simple Carbohydrates, 1910; Chemistry in the Twentieth Century, 1924; The Glycosides, 1931; The Sea as a Storehouse, 1945; contbr. papers Royal Soc., Chem. Soc., others. Died Dec. 14, 1945.

ARMSTRONG, Edwin H(oward), Am. elec. engr.; b. N.Y.C., Dec. 18, 1890; s. John and Emily Gertrude (Smith) A.; E.E., Columbia, 1913, Sc.D., 1929; Sc.D., Muhlenberg Coll., 1941; m. Marian MacInnis, Dec. 1, 1923. Asst. in dept. elec. engring., Columbia, 1913-14; asso. with Michael I. Pupin in research, Marcellus Hartley Research Lab., at Columbia U., 1914-35; prof. elec. engring. Columbia, from 1934. Recipient Medal of Honor, I.R.E., 1917; Egleston medal, Columbia U., 1939; Holley medal, Am. Soc. Mech. Engrs., 1940; Franklin medal, Franklin Inst., 1941; John Scott medal, Bd. of City Trusts, City of Phila., 1941; Edison medal Am. Inst. E.E., 1943; Medal for Merit, 1947; Washington award Western Soc. Engrs., 1951; U. S. medal of Merit, end of World War II. One of 20 scientists honored in Pantheon of Union Internationale des Télécommunications, Geneva, Switzerland. Mem. I.R.E. Contbr. to tech. jours. Inventor regenerative circuit, 1912; superheterodyne receiver (basic AM circuit), 1918; super-regenerative circuit, 1920; wideband FM system, 1933; simultaneous broadcast of more than one FM program on same frequency, 1953; worked on application of FM to mil. communications, long-distance communication and long-range continuous-wave radar, during World War II. Died N.Y.C., Feb. 1, 1954.

ARMSTRONG, Frederick Edward, Am. petroleum engr.; b. Ludlow, Vt., June 25, 1920; s. Forrest C. and Mabyn (Ballard) A.; student U. Cal. at Los Angeles, 1941-44; E.E., Okla. State U., 1946; m. Normalyn Wilson, Oct. 4, 1944; children—Forrest C., Michel Lynn. Project leader nuclear research U. S. Bur. Mines Petroleum Research Center, Bartlesville, Okla., 1949—. Mem. Am. Nuclear Soc., Soc. Petroleum Engrs., Am. Inst. Mining and Metall. Engrs. Research, publs. on 1st use of radioisotopic tracers in petroleum prodn and reservoir engring. research. Home: P.O. Box 8, Copan, Okla. 74022. Office: P.O. Box 1398, Bartlesville, Okla. 74002.*

ARMSTRONG, George, pediatrician; b. Eng., 1719; 3 daus.; founded 1st European dispensary for poor children, London, Eng., 1769. Author: Essay on the Diseases Most Fatal to Infants, 1767; A General Account of the Dispensary for the Infant Poor, 1772. Died 1789.

ARMSTRONG, George Eli, Canadian surgeon; b. Leads, Que.; 1854; s. John and Harriet M. (Ives) A.; M.D., C.M., McGill U.; studied in Eng., Germany France; M.D., LL.D., D.Sc., Liverpool, Eng.; M.Ch , Dublin, Ireland; m. Mary Hadley, 1878; 1 son, 4 daus.; m. 2d, Jessie Reid, 1917. Prof. anatomy Bishops U., Lennoxville, Que., Can., 1877-83, prof. physiology, 1883-85; became surgeon Montreal Gen. Hosp., 1890; prof. surgery McGill U. from 1895; named chief surgeon Royal Victoria Hosp., 1911; cons. surgeon Western Gen. Hosp., Montreal Gen. Hosp., Protestant Hosp. for Insand, Verdun, Canadian Hosps. in Eng., 1916-19. Hon. fellow Royal Acad. Medicine in Ireland, Royal Coll. Surgeons in Ireland; pres. Canadian Med. Assn., 1910, Am. Surg. Assn., 1914, A.C.P., 1920. Contbg. author: Reference Handbook of the Medical Sciences (Buck); Surgery (Keen); American Practice of Surgery (Bryant and Buck). Wrote on surgery of pancreas, infectious diseases, tongue and salivary glands, surg. treatment of hemorrhage from stomach. Died May 25, 1923.

ARMSTRONG, Harry George, Am. physician; born De Smet, S.D., Feb. 17, 1899; s. Robert John and Lucy (Flantje) A.; B.S., U. S.D., 1923; M.D., U. Louisville, 1925; M.S., U. Cin., 1939; M.S., U. Toronto, 1941; m. Mary Evelyn Sutherland, May 4, 1925; children—Jeff, Robert. Pvt. practice medicine, 1926-30; commd. 1st lt. USAAF, 1930, advanced through grades to maj. gen. 1950; dir. aerospace research, 1934-49; surgeon gen. USAF, 1949-54; ret., 1958. Mem. A.M.A., Aerospace Med. Assn. Author: Aviation Medicine, 1939; Fit To Fly, 1941; Aerospace Medicine, 1961. Contbr. numerous articles to med. jours. Established and directed USAF aeromed. and aerospace research. Address: 103-A Tanglebriar Lane, San Antonio 78209.*

ARMSTRONG, Henry Edward, English chemist; b. London, Eng., May 6, 1848; s. Richard and Mary Ann (Biddle) A.; student Royal Coll. Chemistry, 1865, 67; Ph.D., Leipzig, Germany, 1870; LL.D., St. Andrews U., Scotland; D.Sc., Madrid, also Melbourne; m. Frances Louisa Lavers, 1877; 4 sons, 3 daus. Tchr., St. Bartholomew's Hosp., 1870-72; prof. chemistry London Instn., 1871-84, London Inst., 1884-1913; prof. Central Tech. Coll. City and Guilds London Inst., from 1879; asso. with Rothamstead Research Sta. Recipient Davy medal, 1911; Messel medal Soc. Chem. Industry, 1922; Albert medal Royal Soc. Arts, 1930; Horace Brown medal Inst. Brewing. Fellow Royal Society, 1876; member of the Chemical Society (president 1893-95), British Association. Author: An Introduction to Organic Chemistry, 1874, 80; Teaching of the Scientific Method, 1903; The Art and Principles of Chemistry, 1927; (with C. E. Groves) rev. edit. Organic Chemistry (Book III of W. A. Miller's Elements of Chemistry), 1880. Recognized importance of periodic law; called chem. change reversed electrolysis; proposed quinonoid theory of color, benzene substitution, also centric formula of benzene; adopted idea of residual affinities; main supporter (with Mendeléeff) Chem. theory of solution; developed hydroxyliation theory; proposed modern systematic organic names at Geneva Congress, 1892; suggested name catalyst instead of contact substance; investigated enzymes, naphthalene, camphor, crystallography. Died London, July 13, 1937.

ARMSTRONG, Marvin Douglas, Am. chemist; b. Wilmington, N.C., Apr. 15, 1918; s. Marvin D. and Marie (Curry) A.; B.S., U. S.C., 1938; M.S., U. Ill., 1939, Ph.D., 1941; m. M. Josephine Stahl, Feb. 28, 1946; children—Penelope, Joel S., Peter H., Douglas R., Alan C. Research asso. dept. biochemistry Cornell U., 1941-42, 46; asst. prof. U. Utah, Coll. Medicine, Salt Lake City, 1946-51, asso. research prof. biochemistry 1951-57; chmn. biochemistry dept. Fels Research Inst., Yellow Springs, O., 1957—. Decorated D.F.C., Air medal. Mem., Am. Chem. Soc., Am. Soc. Biol. Chemists, Biochem. Soc. (London), Soc. Exptl. Biol. Medicine, A.A.A.S., Phi Beta Kappa, Sigma Xi, Phi Kappa Phi, Alpha Chi Sigma. Research in phenylketonuria, urine acids; discovery of O-metylation pathway in metabolism of catechols; devel. diagnostic test for pheochromolytoma. Home: 139 E. Limestone St. Office: Fels Research Inst., Yellow Springs, O. 45387.*

ARMSTRONG, Sir William George, Brit. engr.; b. Newcastle/Tyne, Eng., Nov. 26, 1810; s. William Armstrong; LL.D., Cambridge (Eng.) U.; D.C.L., Oxford (Eng.) U.; M.E., Dublin (Ireland) U., m. Margaret Ramshaw, 1835 (died 1893). Solicitor, Donkin, Stables and Armstrong, Newcastle, Eng., 1832-47; founder, engr. Elswick Works, from 1847, later chmn.; engr. Rifled Ordnance, Woolwich, Eng., 1859-63. Fellow Royal Soc., 1846; mem. Brit. Assn. (pres. 1863). Author: A Visit to Egypt, 1873; Electric Movement in Air and Water, 1897. Contbr. papers on electricity of effluent steam, 1841-43. Invented improved rotary water motor, 1839, hydraulic crane, Armstrong gun (containing rifling in barrel, breech loading, reinforcement of barrel by spirals of wrought iron), 1855; built steam elec. machine, 1840; produced what is believed by some to be 1st modern armored cruiser, 1882; built steel coastal arty. and arty. towers used by Brit. navy. Died Rothbury, Eng., Dec. 27, 1900.

ARMY, Thomas James, Am. agronomist; b. Worcester, Mass., June 12, 1923; s. Thomas J. and Grace (Hanson) A.; B.S., U. Mass., 1948, M.S., 1950; Ph.D., N.C. State Coll., 1954; m. Greta Garland, Sept. 8, 1950; children—Kent James, Lori Gay. Instr. botany dept. U. Mass., 1948-50; soil scientist U. S. Dept. Agr., Amarillo, Tex., 1956-58, Bozeman, Mont., 1954-56; research investigation leader, Ft. Collins, Colo., 1958-62; sr. research asso. Internat. Minerals & Chem. Corp., Skokie, Ill., 1962—. Fellow A.A.A.S.; mem. Am. Soc. Agronomy, Am. Chem. Soc., Am. Soc. Plant Physiologists, Soil Conservation Soc., Phi Kappa Phi, Sigma Xi. Editor: with (F. A. Greer) Genes to Genus, 1965. Contbr. numerous articles on plant nutrition and crop prodn. to tech. jours. Derived math. relationships between crop yields and weather for semi-arid agr. leading to rational interpretation climate in terms crop growth; research on interactions between soil properties, crop mgmt., growth regulators and aerial invironment leading to improved fertilizer practices and higher prodn. major crops. Home: 2805 Weller Lane, Northbrook, Ill. 60062. Office: IMC Growth Scis. Center, Libertyville, Ill. 60048.*

ARNALD OF VILLANOVA, (Arnaldus Villanovanus), physician; born near Valencia, Spain, circa 1235; ed. Paris, also Montpellier, France, Naples, Italy. Lived in Paris, also Montpellier, Barcelona, Rome; served as physician to royalty and popes; prof. U. Montpellier; tchr. philosophy and medicine, Barcelona, also Paris. Author: Medicinalium introductionum speculum; numerous books on medicine, chemistry, astrology, theology. Authenticity of numerous books on alchemy (later of great influence) has been questioned. Both a mystic and exptl. scientist; accepted transmutation; constructed cosmology dominated by spiritus (vital force); introduced tinctures and medicated wines. Died at sea between Naples and Genoa, circa 1313.

ARNAUD, Paul Charles, French chemist; b. La Tronche, France, Jan. 28, 1930; s. Louis and Helen (Vellot) A.; Licence ès sciences, U. Grenoble (France), Doctorate ès Sciences, 1956; m. Anne F. Divisia, July 28, 1953; children—Paule-Catherine, Anne-Cécile, Véronique, Sophie, Geneviève. With Faculty Scis., U. Grenoble, 1947—, prof., 1964—. Dir., Ecole de Promotion Supérieure du Travail. Mem. Societé chimique de France, Am. Chem. Soc. Author: Cours de chimie Organique, 1964; also articles. Research on stereochemistry including absorption spectroscopy, nuclear magnetic resonance, chemistry of small ring compounds (cyclopropanes, epoxydes), carbonyl compounds. Home: 13, chemin des Résistants, 38 La Tronche, France. Office: Laboratoire de chimie organique, Domaine Universitaire de Grenoble, 38 St. Martin d'Hères, France.*

ARNAUD, Yves Noël, French biologist; b. Nimes, France, Dec. 24, 1929; s. Raoul Joseph and Lucie (Davin) A.; Diplôme d'Etat, Faculté de Pharmacie, Bordeaux, France, 1955; Licence ès science U. Paris, also U. d'Orsay, 1960; spécialiste de Radiobiologie, U. Paris, 1963; m. Arlette Marie Bousquet, July 1, 1954; children—Olivier, Lawrence, Pascale, Frédéric. Biologist asst. Centre d'Etudes de Recherches de Médecine Aéronautique, 1956-57; asst. Radiobiol. Lab., Commissariat à l'Energie Atomique, 1957-65; head molecular radiobiology sect. service de biophysique, 1965—; tchr. Certificate d'Etudes approfondies de radiobiologie, 1960—. Mem. Société

Francaise de Biophysique, Société Francaise de Chimie Biologique. Research, publs. on physico-chem. aspects of radiobiology, dosimetry of ionizing radiation and neutrons on biol. point of view, effects of irradiation on living organisms, mechanisms of radiation effects at molecular level, structure-function relationship of specific macromolecules. Home: 1, rue Buffon, Orsay, Essonne 91. Office: Dept. de Biologie B.P. N° 2 GIF s/Yvette 91, France.*

ARNAUDOV, Mihail Petrov, Bulgarian psychologist; b. Rousse, Bulgaria; s. Peter Simeonov and Droumka (Panova) Voinov; student U. Sofia (Bulgaria), U. Leipzig (Germany), U. Berlin (Germany), 1898-1900; Phil.Dr., U. Prague (Czechoslovakia), 1905; Dr.h.c., U. Heidelberg (Germany), 1939, U. Münster (Germany), 1939; m. Stefanka Assenova Simidova, Sept. 1915; children—Assen, Latka Hristoskova, Peter. High sch. tchr., Vidin, Sofia, Bulgaria, 1901-07; faculty Sofia U., 1907—, prof., since 1916—; dep. dir. Nat. Library; minister of edn. Decorated Order of Alexander 1st degree. Mem. Bulgarian Acad. (academician). Author: Studies on Bulgarian Rites and Fiction, 2 vols., 1920; An Introduction in Literary Science, 1920; Psychology of Literary Creative Work, 1931; Leaders of the Bulgarian National Revival, 9 vols.; Sketches on the Bulgarian Folklore, 1935; Personalities and Problems of European Literature, 1931; also monographs on Vaso, Yavorov, K. Hristov; numerous articles. Research on psychology of lit. creative work, theory of factors conditioning it, criteria of lit. devel. in gen., history of cults and fiction on comparative basis. Home: 7/b Slavianska, Sofia, Bulgaria.*

ARNAULD, Antoine, mathematician, theologian; b. 1612; 20th and youngest s. Antoine and Catherine Marie Arnauld; ed. Sorbonne, 1642; France's leading Jansenist controversialist; priest, mem. Sorbonne; expelled from Sorbonne and faculty theology, 1656; subsequently helped compile textbooks, Port-Royal-des-Champs; traveled to Belgium, 1679; controverted theology of grace with Malebranche. Author: De la fréquente communion (a pro-Jansenist attack on Jesuits), 1643; Eléments de géométrie, 1667; writings against Calvinists and free-thinkers; La logique, ou l'art de penser, 1662. Died, Aug. 8, 1694.

ARNDT, Helmut, German economist; b. Königsberg, Germany, May 11, 1911; s. Adolf and Louise (Zabeler) A.; ed. Berlin, Marburg and Munich univs.; Ph.D. in Law and Polit. Sci.; m. Elfriede, Jan. 25, 1947; children—Claudia, Rolf. Prof. agrégé U. Marburg, 1946; vis. prof. U. Syracuse (N.Y.), 1950-51; prof. U. Marburg, 1952, U. Istanbul, 1953, Technische Hogeschool, Darmstadt, 1954, Free U. Berlin, 1957; mem.-consul Commn. of Inquiry, 1961-62; vis. prof. U. Heidelberg, 1963, dir. Inst. Econs. Mem. Soc. Econ. and Social Sci. (com. of dir.), German Soc. Sociology, Am. Econ. Soc., Internat. Asso. Polit. Sci., Royal Econ. Soc. Author: Voraussetzungen des Martautomatisms, 1947; Schöpferischer Wettbew. u. Klassenlose Gesellschaft, 1952; Ausbeutung u. Marktform, 1959; Die Konzentration in der Wirtsch, 1960; Martmechanismus und heterogene Konkurrenz, 1964. Home: 1 Berlin 33, Ehrenbergstrasse 29. Office: 1 Berlin 33 Garystrasse 21, West Germany.

ARNDT, Rudolf Gottfried, psychiatrist; b. Bialken, Mar. 31, 1835; s. Heinrich Gottfried and Ernestine (Wendt) G.; ed. Greifswald, also Halle, Germany; M.D.; 1 son. Practiced medicine from 1861; army physician, 1864, 66, 70-71; head, asylum, Greifswald, from 1867; asso. prof. psychiatry, from 1873. Author: Lehrbuch der Psychiatrie für Ärzte und Studierende, 1883; Biologische Studien, Part I, Das biologische Grundgesetz, Part II, Artung und Entartung, 1892-95. Attempted to give psychiatry sci. basis; studied convulsions, nervous systems, progress of psychoses. Died Greifswald, Sept. 29, 1900.

ARNDT, Ulrich Wolfgang, physicist; b. Berlin, Germany, Apr. 23, 1924; s. Ernst J. and Clara M. (Juliusberg) A.; B.A., U. Cambridge (Eng.), 1945, M.A., Ph.D., 1949; m. Valerie Howard Hilton-Sergeant, July 29, 1958; children—Elizabeth, Caroline B., Annabel C. Research fellow metallurgy dept. Birmingham (Eng.) U., 1948-49; research worker Royal Instn., London, Eng., 1950-55, Dewar fellow, research asso. U. Wis., Madison, 1956-58; research scientist Med. Research Council Lab. Molecular Biology, Cambridge, 1962—. Cons. U.K. Atomic Energy Authority, 1960-66. Mem. Inst. Physics and Phys. Soc. Author: (with B. T. M. Willis) Single Crystal Diffractometry, 1966; also articles. Devel. new exptl. techniques in X-ray crystallography; inventor (with D. C. Phillips) automatic X-ray diffractometer. Home: 9, Courtyards, Little Shelford, Cambridge. Office: Med. Research Council Lab. Molecular Biology, Hills Rd., Cambridge, U.K.*

ARNDT, Walter, German zoologist; b. Landeshut, Germany, Jan. 8, 1891; s. Fedor and Klara (Conrad) A.; ed. U. Breslau (now Wroclaw, Poland); made ednl. expdns. to Norway, 1911, Corsica, 1913, Hohen Tauern Mountains, 1920. Voluntary asst. Zool. Inst. and Mus. Breslau, 1920-21; became asst. Zool. Mus. Berlin, 1921, curator, 1925, prof., 1931. Author: Tierwelt Nord und Ostsee, 1935; Spongiologica Unterssuchungen am Orchrida-See, 1938; also author, pub., editor numerous publs., many in cumulative or reference works. Contbd. to knowledge of flora and fauna of sea, anatomy, fungi, reptiles, systematics. Executed by Nazis, Brandenburg, Germany, July 26, 1944.

ARNEIL, Gavin Cranston, Scottish pediatrician; b. Glasgow, Scotland, Mar. 7, 1923; s. Loudon and Elizabeth M.S. (Hogg) A.; M.D. with honours, U. Glasgow, 1949, Ph.D., 1960, M.B., Ch.B., 1945, diploma in child health; m. June Barbara Lauder, Aug. 9, 1961. Practice internal medicine, Glasgow, 1948-50, pediatrics, Glasgow, 1950; lectr. pediatrics U. Glasgow, 1951-58, Leonard Gow lectr., 1962—; cons. pediatrician Royal Hosp. for Sick Children, Glasgow, 1959-61. Mem. panel to study radiation hazard Med. Research Council, 1962—. Recipient WHO Sr. Research prize, 1956, Watson Research prize, 1956, Hutchison Research prize, 1954. Fellow Royal Colls. Physicians Edinburgh, Glasgow, London; mem. Brit. Paediatric Assn. (mem. council 1964—), Brit. Med. Assn., Scottish Paediatric Soc., Nutrition Soc., Internat. Soc. Internal Medicine. Research, numerous publs. on diseases of kidneys in children, danger radio-active strontium-90 to children, malnutrition in infants and soldiers., Home: 31 Cleveden Rd., Glasgow, Scotland.*

ARNETH, Joseph, German physician; b. 1873; prof. at Munster; specialized in hematology. Devised Arneth's formula for classification of neutrophils of blood into groups based on nuclear lobulation (supposedly an indication of their phagocytic ability).

ARNETH, Joseph Calsanza Ritter, archaeologist; b. Leopoldschlag, Austria-Hungary, Aug. 12, 1791; ed. at Linz, also Vienna, Austria, with Cabinet of Medals and Antiques, Vienna, 1810-13, soldier 1813-1814; traveled in 1815; returned to Cabinet and became dir., 1840; prof. numismatics and archeology U. Vienna. Mem. French Acad. Scis. Author: Geschichte der osterreichischen Kaistertums, 1827; Synopsis numorum antiquorum, 2 vols., 1837-42; Die Monumente des k. k. Münz- und Antikenkabinetts, 3 vols., 1849-50. Died Karlsbad, Germany, Oct. 21, 1863.

ARNETT, Edward McCollin, Am. chemist, educator; b. Phila., Sept. 25, 1922; s. John Hancock and Katherine (McCollin) A.; A.B., U. Pa., 1943, M.S., 1946, Ph.D., 1949; m. Mary Hall Flounders, Aug. 18, 1951; children—Eric, Brian. Research dir. Max Levy & Co., Phila., 1949-53; asst. prof., asso. prof. Western Md. Coll., Westminster, 1953-55; research fellow Harvard, 1955-57; faculty chemistry U. Pitts., 1957—, prof., 1964—. Adj. sr. fellow Mellon Inst., Pitts., 1964—. Mem. Am. Chem. Soc., A.A.A.S., Chem. Soc. (London), Sigma Xi, Phi Lambda Upsilon. Research, publs. on physicochem. study of organic chemistry, role of solvents; inventor instrument to analyze for deuterium. Home: 211 Lytton Av., Pitts. 15213.*

ARNETT, Ross Harold, Jr., Am. entomologist; b. Medina, N.Y., Apr. 13, 1919; s. Ross Harold and Hazel Dell (Oderkirk) A.; B.S., Cornell U., 1942, M.S., 1946, Ph.D., 1948; m. Mary Catherine Ennis, Feb. 16, 1942; children—Ross Harold III, Michael J., Mary Anne, Barbara E., Frances X. C., Joseph A., Bernadette T. Matthew C. Instr., Cornell U., 1945-48; entomologist U. S. Nat. Mus., Washington, 1948-54; asso. prof. St. John Fisher Coll., Rochester, N.Y., 1954-58; prof. biology Cath. U. Am., Washington, 1958-66, head dept. biology, 1962-66, dir. Inst. Study Natural Species, 1961-66; professor of entomology at Purdue University, Lafayette, Indiana, 1966—. Member of the Entomological Society Am., Entomol. Soc. Washington (pres. 1964), Sigma Xi, Phi Kappa Phi. Author: Beetles of the United States, 1962; (with D. C. Braungart) Introduction to Plant Biology, 1965; also numerous articles. Standardized classification beetles U. S. and families beetles world; research on study speciation as it affects evolutionary process. Home: 550 Elston Rd., Lafayette, Ind. 47905.*

ARNHEIM, Rudolf, psychologist; b. Berlin, Germany, July 15, 1904; s. Georg and Betty (Gutherz) A.; Ph.D. U. Berlin, 1928; m. Mary E. Frame, Apr. 11, 1953; 1 dau., Margaret. Came to U. S., 1940, naturalized, 1946. Asso. editor publs. Internat. Inst. for Ednl. Films, League Nations, Rome, Italy, 1933-38; faculty Sarah Lawrence Coll., Bronxville, N.Y., 1943-68; vis. prof., grad. faculty New Sch. for Social Research, N.Y.C., 1943-68; prof. psychology Carpenter Center for Visual Arts, Harvard, 1968—. Visiting lecturer at Ochanomizu University, in Tokyo, 1959-60. Research grantee for study visual factors in concept formation U. S. Office Edn., 1966-67; John Simon Guggenheim Meml. Found. fellow, 1941-42. Mem. Soc. Aesthetics (past pres.), Am. Psychol. Assn. (past pres. div. psychology and arts), Coll. Art Assn. Author: Art and Visual Perception, 1954; Film as Art, 1957; Picasso's Guernica, 1962; Towards a Psychology of Art, 1966; also articles. Application of psychology especially visual perception to arts, creative process and visual thinking. Home: 49 Rockledge Rd., Bronxville, N.Y. 10708.*

ARNOLD, Aaron, Am. biochemist; b. Milw., Jan. 3, 1910; s. Charles and Anne (Spector) A.; B.S., U. Wis., 1933, M.S., 1936, Ph.D., 1937; m. Marion Asnes, Dec. 25, 1933; children—Stephen Arnold, Janet (Mrs. William Evans Schleicher). Head water-soluble vitamins lab. Nopco Chem. Co., Harrison, N.J., 1939-43; head spl. markets and nutritional research lab. Winthrop Chem. Co., Rensselear, N.Y., 1943-49; with Sterling-Winthrop Research Inst., Rensselaer, 1949—, sect. head biochemistry, 1963—. Fellow A.A.A.S.; mem. Am. Inst. Nutrition, Am. Bd. Nutrition, Am. Chem. Soc., Am. Assn. Cereal Chemists, N.Y. State Soc. Med. Research, Animal Nutrition Research Council, Harvey Soc. Studies, publs. on essential amino acids and vitamins; acetylcholinesterase inhibitors; iron sources; anabolic steroids, effect of estrogens on cholesterol biosynthesis; nitrogen balance. Home: 47 Terrace Av., Albany, N.Y. 12203. Office: Sterling-Winthrop Research Inst., Rensselaer, N.Y. 12144.*

ARNOLD, Aza, Am. inventor, patent atty.; b. Smithfield, R.I., Oct. 4, 1788; s. Benjamin and Isabel (Greene) A.; m. Abigail Dennis, July 28, 1815. As youth learned carpenter's and machinist's trades; worked in mfg. plant of Samuel Slater, Pawtucket, R.I., 1808; operated (with Larned Pritcher and P. Hovey) machine shop, Pawtucket until 1819; opened cotton mill, Great Falls, N.H., 1819; went back to R.I. few years later (North Providence), made machine for mfg. textile machinery; obtained patent for roving machine for spinning cotton, 1823, introduced into England, 1825, brought law suits because of infringement of patent rights (new code of patent laws passed largely because of his suits, 1836, however he received no compensation); operated Mulhausen Print Works, Phila., 1838-50; patent atty., Washington, D.C., circa 1850-65; invented self-raking and self-setting saw for sawing machines (his last invention, patented 1856). Died 1865.

ARNOLD, Bion Joseph, Am. elec. engr.; b. Casnovia, nr. Grand Rapids, Mich., Aug. 14, 1861; s. Joseph and Geraldine (Reynolds) A.; student U. Neb., 1879-80, E.E., 1897, D.Eng., 1911; B.S., Hillsdale (Mich.) Coll., 1884, M.S., 1887, MPh. (hon.), 1889, hon. diploma, 1903; post-grad. Cornell, 1888-89; D.Sc., Armour Inst., 1907; m. Carrie Estelle Berry, Jan. 14, 1886 (dec.); m. 2d, Margaret Latimer Fonda, Dec. 22, 1909. Chief designer, Ia. Iron Works, Dubuque; mech. engr., C.G.W.Ry.; later cons. engr. Chgo. office Gen. Electric Co.; became ind. cons. engr., 1893; pres. Arnold Engring. Co. Designed and built Intramural Ry., Chgo. Expn.; cons. elec. engr. Chgo. & Milw. Elec. Ry., Chicago Bd. of Trade, C.,B.&Q. R.R., Grand Trunk Ry. on electrification of St. Clair tunnel; cons. engr. Wis. Ry. Commn., 1905-07; devised plan for electrically operating trains of N.Y. Central R.R. in and out of N.Y.; mem. electric traction com. Erie R.R., 1900-04; cons. engr. to revise street ry. systems of city of Chgo., 1902; chief engr. rebuilding Chgo. traction system, from 1907; cons. engr. Pub. Service Commn., 1st Dist., N.Y., matters connected with subway and st. ry. properties; chief subway engr. city of Chgo., also cons. engr. on traction matters for cities of Pitts., 1910, Providence, Los Angeles, San Francisco, 1911, Toronto and Cin., 1912. Mem. of Chicago Ry. Terminal Commn. until 1921; mem. Traction and Subway Commn., 1916-17; adviser to Des Moines, Omaha, Winnipeg, Sacramento, New Orleans, Detroit, Harrisburg, Rochester, Syracuse, Jersey City, Toronto, others; mem. Naval Consulting Bd. Recipient Washington award Western Soc.; many other medals and prizes. Contbd. to devel. street ry. systems in many Am. cities; inventor magnetic clutch; innovator alternating current and single phase traction systems. Died Jan. 29, 1942.

ARNOLD, Bradford Henry, Am. mathematician; b. Chehalis, Wash., Oct. 14, 1916; s. Everett O. and Alice (Chase) A.; B.S., U. Wash., 1938, M.S., 1940; Ph.D., Princeton, 1942; m. Mary Ellen Forrest, June 8, 1941. Instr., Purdue U., 1942-44; aerodynamicist Boeing Airplane Co., 1944-46; asst. prof. Mont. State U., 1946-47; faculty Ore. State U., Corvallis, 1947—, prof., 1959—. Fulbright grantee, Iraq, 1957-58, China, 1962-63. Author: Intuitive Concepts in Elementary Topology, 1962; Logic and Boolean Algebra, 1963. Home: 125 N. 32d. St., Corvallis, Ore. 97330.*

ARNOLD, Chester Arthur, Am. botanist, educator; b. Leeton, Mo., June 25, 1901; s. Elmer and Edith (Funderburg) A.; B.S., Cornell U., 1924, Ph.D., 1929; m. Jean Davidson, Aug. 30, 1933; children—David G., Eric Bruce, Patricia Ann. Faculty botany U. Mich., Ann Arbor, 1928—, curator fossil plants, 1929—, prof., 1947—. Vis. scientist Birbal Sahni Inst. Paleobotany, Lucknow, India, 1958-59; fgn. del. Indian Sci. Congress, 1959. Mem. Internat. Orgn. Paleobotany (pres. 1959-64), Bot. Soc. Am., Geol. Soc. Am., A.A.A.S., Mich. Acad. Sci., Sigma Xi. Author: Introduction to Paleobotany, 1947. Research on Paleozoic, Mesozoic and Tertiary plants of N.Am., fossil flora of Mich. Home: 1911 Mershon Dr., Ann Arbor, Mich. 48103.*

ARNOLD, Christoph, German astronomer; b. Sommerfeld/Leipzig, Germany, Dec. 17, 1650; s. Hans and Sabine (Hainemann) A.; self-taught; m. Anna Straube, 1683. Worked in agr.; became interested in astronomy and installed obs. in home. Author: Göttliche Gnadenzeichen, in einem Sonnenwunder vor Augen gestellt, 1692. Discovered new comets, 1682 (8 days before J. Hevel), 1686; observed passage of Mercury calculations on moons of Jupiter. Died Sommerfeld/across sun, 1690; made numerous observations and Leipzig, Apr. 15, 1695.

ARNOLD, Godfrey Edward, otolaryngologist; born Olmuetz, Austria, Jan. 6, 1914; s. Anton and Maria (Schicht) A.; B.A., Theresian Acad., Vienna, Austria, 1932; M.A., State Acad. Music, Vienna, 1936; M.D.,

Vienna U., 1937; postgrad. U. Berlin (Germany); German diploma in otolaryngology, 1940, Austria, 1946; otolaryngology fellow N.Y. U. Postgrad. Med. Sch., 1949-51; m. Isolde Reuter, Feb. 14, 1942; 1 dau., Claudia Alexandra. Came to U. S., 1948, naturalized, 1956. Practice medicine specializing in otolaryngology, audiology and phoniatrics, Vienna, 1941-48; faculty U. Vienna, 1941-45, N.Y. U. Postgrad. Med. Sch., 1951-59; dir. research N.Y. Eye and Ear Infirmary, N.Y.C., 1956-63, asst. surgeon, 1951-63; prof. otolaryngology surgery, asst. prof. physiology U. Miss. Med. Sch., also dir. div. otolaryngology U. Hosp., Jackson, 1963-—. Spl. cons. NIH, USPHS, 1961-—; cons. VA Hosp., Jackson, 1964-—; Miss. chmn. Am. Deafness Research Found., 1965-—. Recipient Harris P. Mosher Meml. award, 1961. Diplomate Am. Bd. Otolaryngology. Mem. Am. Assn. U. Profs., A.C.S., A.M.A., Am. Acad. Ophthalmology and Otolaryngology, Am. Laryngol., Rhinol. and Otol. Soc., Am. Larngol. Assn., Pan-Am. (diplomate), So. med. assns., Internat. Assn. Logopedics and Phoniatrics, Austrian Otol. Soc. (corr.), Sigma Xi, numerous others. Author: (with R. Luchsinger) Lehrbuch der Stimm- und Sprachheilkunde, 1949, 2d edit, 1959, Voice-Speech-Language, 1965. Research, numerous publs. on integration of communications scis. with medi ine, developed injection procedure for laryngeal reconstrn. also discovered several types of hearing loss from brain injury. Home: Miranda Hall, Cedar Crest, Clinton, Miss. 39056. Office: U. Hosp., Jackson, Miss. 39216.*

ARNOLD, Harold DeForest, physicist; b. Woodstock, Conn., Sept. 3, 1883; s. Calvin and Audra Elizabeth (Allen) A.; Ph.B., Wesleyan U., 1906, M.S., 1907, D.Sc., 1930; fellow in physics U. Chgo., 1907-09, Ph.D., 1911; m. Leila Stone Beeman, Sept. 3, 1908; children—Audra Elizabeth, Dorothy Edith. Asst. in physics, Wesleyan U., 1906-07; prof. physics, Mt. Allison U., Sackville, N.B., Can., 1909-10; research engr., Western Electric Co., N.Y. City, 1911-24; dir. research, Bell Telephone Labs., N.Y.C., from 1925. Recipient John Scott medal and award, 1928, for development of 3-electrode high vacuum thermionic tube. Died July 10, 1933.

ARNOLD, Hubert Andrew, Am. mathematician, educator; b. Chgo., Nov. 15, 1912; s. Charles Harrison and Irma Carlotta (Sears) A.; A.B., U. Neb., 1933; postgrad. Sorbonne, Paris, France; Ph.D. in Math., Cal. Inst. Tech. Asst. instr. math. Cal. Inst. Tech., 1935-39; instr. U. Minn., 1939-40, Princeton, 1941-42; vis. research scholar U. Va., 1940-41; faculty math. U. Cal. at Davis, 1948-—, asso. prof., 1956-—. Mem. A.A.A.S., Am. Math. Soc., Math. Assn. Am. Contbr. articles on topology, complex variables, group theory, abstract differentials to profl. jours. Home: 533 E St., Davis, Cal. 95616.*

ARNOLD, James Sloan, Am. physicist; b. Ely, Nev., Sept. 27, 1911; s. Richard Peter and Alta (Campbell) A.; A.B., U. Cal. at Los Angeles, 1934; M.S., N.M. State U., 1951; postgrad. Stanford, 1953-55; m. Merceda Jewett, Feb. 20, 1942; children—Richard James, Elizabeth Gail. Engr., Acoustical Engring. Co., Los Angeles, 1936-38; elec engr. Trumbull Elec. Mfg. Co., Los Angeles, 1939-48; physicist Phys. Sci. Lab., N.M. State U., University Park, 1948-53, tchr., 1948-50; physicist Stanford Research Inst., Menlo Park, Cal., 1953-66; staff engr. McDonnell Aircraft Corp., St. Louis, 1966-—. Mem. Am. Rocket Soc. (past pres. No. Cal. sect.), Am. Inst. Aeros. and Astronautics, Sci. Research Soc., Combustion Inst. Contbr. articles to tech. jours. Patentee in field. Developed exptl. technique for observation ignition phenomena in hydrocarbon-air mixtures; developed, invented 1st successful non-destructive test method for structural adhesive bonds; developed techniques for describing vibration modes solids; invented modulated combustion loudspeaker illustrating pyroacoustic amplification. Home: 17 Chandler Ct., Florissant, Mo. 63031. Office: Dept. 315, McDonnell Aircraft Corp., St. Louis 63166.*

ARNOLD, John Oliver, Brit. metallurgist; b. Peterborough, Eng., 1858; s. D. Nelson Arnold; D.Met.; m. Miss England, 1883; 2 sons, 1 dau. With engring. dept. Brown, Bayley and Dixon's, Sheffield, Eng., Telford Premium Instn. C.E.'s, from 1887; spl. lectr. steel Brit. Assn. expdn. to S. Africa, 1905; acting prof. metallurgy Sheffield U., from 1889, dean metallurgy, 1917, retired, 1920; prof. emeritus. Telford premium and gold medal, 1895, Bessemer gold medal, 1905. Fellow Royal Soc., 1912; mem. Sheffield Soc. Engrs. and Metallurgists (pres. 1914-18), Iron and Steel Inst. (mem. council 1915, hon. v.p. 1925). Author: Steel Works Analysis, 4 editions; numerous papers including: The Influence of Elements on Iron; The Molecular Constitutions of High Speed Steels and their Correlations with Lathe Efficiencies, 1919; The Influence of Carbon on Iron; Factors of Safety in Marine Engineering. Research revealed remarkably powerful high speed steel. Died Mar. 27, 1930.

ARNOLD, Joseph, Brit. naturalist; b. Beccles, Eng., Dec. 28, 1782; apprentice to surgeon, Crowfoot; M.D., Edinburgh, 1807. Apptd. surgeon Brit. Navy, 1808; in med. charge Northampton (ship), 1815; made excursions in Java; naturalist with Stamford Raffles, Sumatra; became gov. Sumatra, 1816. Fellow Linnaean Soc. Discovered Rafflesia arnoldi (plant without stem or leaves); collected shells and fossils (bequeathed to Linnaean Soc.). Died Padang, Sumatra, July 1818.

ARNOLD, Kenneth James, Am. statistician; b. Pawtucket, R.I., Aug. 20, 1914; s. William Baxter and Eva (Mellor) A.; B.A., Mass. Inst. Tech., 1937, Ph.D., 1941; m. Pauline Frew, Aug. 22, 1939; children—William D., Robert J., Rebecca F. Instr. math. Mass. Inst. Tech., Cambridge, 1941-43; asst. prof. math. U. N.H., Durham, 1943-44; sr. math. statistician Stat. Research Group, Columbia, N.Y.C., 1944-45; asst. prof. math. U. Wis., Madison, 1945-52; asso. prof. math. Mich. State U., Lansing, 1952-55, dept., 1963-67. Fellow Inst. Math. Statistics, (sec.-treas. 1952-55); mem. Soc. for Indsl. and Applied Math. (mem. council 1958-63, editor Jour. 1958-—), Am. Statis. Assn., Am. Math. Soc., Math. Assn. Am., Sigma Xi. Home: 6049 Skyline Dr., East Lansing, Mich.*

ARNOLD, Magda Blondiau, psychologist; born Mt. Trübau, Austria, Dec. 22, 1903; d. Rudolf and Marie (Blondiau) Barta; B.A., U. Toronto (Ont., Can.), 1939, M.A., 1940, Ph.D., 1942; m. Robert Karl Arnold, Jan. 23, 1926 (div. 1950); children—Evelyn Joan, Margaret Anne (Mrs. E. A. Gaviller), Katherine Mary (Mrs. Robert Johnson). Came to the United States of Am., 1947, naturalized, 1954. Lectr., U. Toronto, 1942-47; dir. research and tng. Can. Vets. Affairs Dept., Toronto, 1946-47; vis. lectr. Wellesley (Mass.) Coll., 1947-48; asso. prof. psychology, head dept. Bryn Mawr (Pa.) Coll., 1948-50; lectr. Harvard, summers 1947, 48; prof. psychology, head dept. Barat Coll., Lake Forest, Ill., 1950-52; Helen Putnam fellow in advanced research Radcliffe Coll., Cambridge, Mass., 1952-54; prof. psychology Loyola University, Chicago, Illinois, 1954-—, dir. behavior lab., 1961-—. Guggenheim fellow, 1957-58; Fulbright research prof., 1962-63. Mem. Cath. Commn. for Intellectual and Cultural Affairs, Sigma Xi. Author: (with J. A. Gasson, S. J.) The Human Person, 1954; Emotion and Personality, 2 vols., 1960; Story Sequence Analysis, 1962. Research, publs. on theory of brain function including memory, imagination circuits; Story Sequence Analysis method. Home: 2223 Central St., Evanston, Ill. 60201.*

ARNOLD, Philipp Friedrich, anatomist; b. Edenkoben, Germany, Jan. 8, 1803; s. Zacharias and Susanne Margaretha (Brünings) A.; M.D., Heidelberg, Germany, 1825/26; m. Ida Eberhardine Gock, 1830; 5 children, including Julius, Ida. Became prosector, anat. inst. U. Heidelberg, 1826, asst. prof., 1833-34; apptd. prof. anatomy and physiology, Zurich, Switzerland, 1835, Freiburg, Germany, 1840, Tübingen, Germany, 1845, Heidelberg, 1852-73. Author: Über den Ohoknoten, 1828; Kopfteil des vegetativen Nervensystems . . . , 1831; Icones nervorum capitis, 1834; (with W. J. Arnold) Lehrbuch der Physiologie des Menschen, 4 vols., 1836-42; Bemerkungen über den Bau des Flirns und Rückenmark, 1838; Tabulae anatomicae, 1838-42; Handbuch der Anatomie des Menschen, 3 vols., 1845-57. First to describe Arnold's nerves and canal (part of vagus nerve complex), Arnold's ganglion (otic ganglion); described frontopontine tract of cerebral peduncle. Died Heidelberg, July 4, 1890.

ARNOLD, Wilhelm Karl, German psychologist; b. Nuremburg, Germany, Oct. 14, 1911; s. Simon and Johanna (Holhut) A.; Ph.D., U. Munich; certificate in psychology; m. Elisabeth Hergenroder, May 31, 1941; children—Irmingard, Hedwig. Army psychologist; chief service of psychology Fed. Bur. Placement and Unemployment; dir. Psychol. Inst. U. Würtzburg. Author: Psychologisches Praktikum, 1957; Begabungseandel und Erziehungsfragen, 1960; der Pauli-Test, 1961; Person, Charakter, Persönlichkeit, 1962; also articles. Research in social psychology; ability and personality testing. Home: Würzburg, Meyer-Olberslenstr. 7. Office: Würzburg, Domerschulstr. 13, West Germany.

ARNOLD, William Archibald, Am. biologist; b. Douglas, Wyo., Dec. 6, 1904; s. William Archibald and Nellie Agnes (O'Brien) A.; B.S., Cal. Inst. Tech., 1931; Ph.D., Harvard, 1935; m. Jean Irving Tompkins, Sept. 11, 1929; children—Elizabeth Irving (Mrs. G. Lew Choules), Helen Holbrook. Research asst., also asst. prof. Stanford, 1937-42; physicist Eastman Kodak Co., Rochester, N.Y., 1942-44, Oak Ridge, 1944-46; with Oak Ridge Nat. Lab., 1946-—, prin. biologist, 1946-—. Sheldon fellow, 1935-36; Gen. Edn. Bd. fellow, 1936-37; Rockefeller fellow, Copenhagen, Denmark, 1938-39; recipient Charles F. Kettering research award Kettering Found., 1963. Mem. Nat. Acad. Scis. Research on photosynthesis. Home: 115 Tabor Rd., Oak Ridge 37832. Office: Biology Div., Oak Ridge Nat. Lab., Oak Ridge 37830.*

ARNON, Daniel I., biochemist; born Poland, Nov. 14, 1910; s. Leon B. and Rachel (Chodes) A.; B.S., U. Cal., 1932, Ph.D., 1936; m. Lucile Jane Soulé, Feb. 24, 1940; children—Anne, Ruth, Stephen, Nancy, Dennis. Faculty, U. Cal., Berkeley, 1936-—, prof. plant physiology, 1950-60, prof. cell physiology, 1960-—, chmn. dept. cell physiology, 1961-—, biochemist Expt. Sta., 1958-—; Belgium-Am. Found. lectr., U. Liege (Belgium), 1948. Recipient Newcomb Cleveland prize Am., A.A.A.S., 1940, Gold medal U. Pisa, Italy, 1958, Charles F. Kettering award in photosynthesis, 1963; Guggenheim fellow U. Cambridge, 1947-48, Hopkins Marine Sta, Stanford, 1962-63; Fulbright Research scholar Max Planck Inst. für Zellphysiologie, Berlin-Dahlem, Germany, 1955-56. Mem. Nat. Acad. Scis., Am. Acad. Arts and Scis., Academie d'Agriculture de France (corr.), A.A.A.S., Am. Chem. Soc., Am. Soc. Biol. Chemists, Am. Soc. for Cell Biology, Am. Soc. Plant Physiologist (v.p. 1950, pres. 1952), Biochem. Soc. London, Scandanavian Soc. Plant Physiologists. Editor: Annual Rev. of Plant Physiology, 1948-55. Discovered molybdenum and vanadium as essential elements for green plants and algae, photosynthetic phosphorylation and complete photosynthesis by cell-free particles, role of ferredoxin in energy conversion reactions of photosynthesis; demonstrated that among bacteria there are mechanisms of noncyclical photophosphorylation comparable to those of chloroplasts; showed that energy necessary for oxidation of water in photosynthesis results from action of light. Home: 101 Southampton Av., Berkeley, Cal. 94707.*

ARNOTT, George, botanist; b. Edinburgh, Scotland, Feb. 6, 1799; A.M., Edinburgh U., 1818; admitted to faculty advocates, 1821; studied botany (became acquainted with Wight, Greville, later William Hooker); m. Mary Hay Barclay, 1831; 3 sons, 5 daus. Worked in Paris (France) herbaria, from 1821; assisted Hooker with account of plants collected in Capt. Beechey's voyage to Pacific and Bering Strait (pub. 1841), 1830-40; acting bot. lectr. (for Hooker), Glasgow, 1839, apptd. prof. botany, 1845. Contbd. papers to periodicals, including 2 on mosses, 1821, descriptions of new plants from S.Am., India, Senegambia; assisted Wight in Illustrations of Indian Botany, also Prodromus florae peninsulae indiae orientalis; asso. with William Hooker in 6th edit. British Flora, 1850. Made valuable collection of diatoms. Died Glasgow, Scotland, Apr. 17, 1868.

ARNOTT, Neil, Brit. physician, natural philosopher, inventor; b. Arbroath, Scotland, May 15, 1788; s. Ann (Maclean) A.; M.A., Marischal Coll., 1805; became student of Everard Home at St. George's Hosp., 1806; M.D., U. Aberdeen (Scotland), 1814; m. Mrs. Knight, 1856. Surgeon, E. India Co., China, 1807-09; made 2d voyage to China, 1810; practiced medicine, London, from 1811; apptd. physician to French Embassy, 1816, later to Spanish Embassy; apptd. physician extraordinary to Queen, 1837. A founder U. London, 1836, also original mem. senate. Decorated Gold medal Paris Exhbn., 1855. Fellow Royal Soc., 1838, (Rumford medal for Arnott's stove 1854). Author: Elements of Physics, 1827; Survey of Human Progress, 1861. Made observations on ocean currents, tides, waves, winds, other atmospheric phenomena during voyage to China, 1807-09); invented economical, smokeless grate (Arnott's stove), devised waterbed, 1832. Died Edinburgh, Mar. 22, 1874.

ARNOTT, William Melville, Brit. physician; b. Scotland, Jan. 14, 1909; s. Henry and Janetta Main (Melville) A.; M.B., Ch.B., B.Sc., M.D., Edinburgh (Scotland) U.; M.D., Birmingham (Eng.) U.; m. Doroth Eleanor Hill, Aug. 11, 1938; 1 son, Christopher. Lectr. therapeutics U. Edinburgh, 1936-46, dir. postgrad. med. studies, 1945-46; prof. medicine U. Birmingham (Eng.), 1946-—. Fellow Royal colls. physicians London (mem. council 1954-56), Edinburgh, Can., Coll. Pathology (founder), A.C.P. (hon.); mem. Assn. Physicians Gt. Britain, Physiol. Soc., Brit. Cardiac Soc., Med. Research Soc. Contbr. articles on renal hypertension, peripheral vascular affections, cardio-pulmonary physiology, med. edn. to Brit. med. jours. Home: 40 Carpenter Rd., Edgbaston. Birmingham 15. Office: Queen Elizabeth Hosp., Birmingham 15, Eng.

ARNOWITT, Richard Lewis, Am. physicist; b. N.Y.C., May 3, 1928; s. Leon and Belle (Feinberg) A.; B.S., Rensselaer Poly. Inst., 1948, M.S., 1948; Ph.D., Harvard, U. 1953; m. Young In Rhee, Apr. 21, 1961; children—Michael, Myron. Research asso. U. Cal. Radiation Lab., Berkeley, 1952-54; mem. Inst. for Advanced Study, Princeton, N.J., 1954-56; asst. prof. Syracuse U., 1956-59, asso. prof., 1959-62; prof. Northeastern U., 1962-—. Mem. Am. Phys. Soc. Studies and numerous publs. on high energy physics in the theories used to understand interactions between particles and resonances that arise in high energy scattering; relativity, on structure of Einstein's gravitational theory concerning the quantization of the theory and the nature of gravitational radiation. Home: 7 Wheeler Rd., Lexington, Mass. 02173. Office: Dept. Physics, Northeastern U., Boston 02115.*

ARNQUIST, Warren Nelson, Am. physicist; b. New Richmond, Wis., Aug. 12, 1905; s. C. William and Agnes (Libby) A.; B.S. magna cum laude in Physics-Math., Whitman Coll., 1927; Ph.D. cum laude, Cal. Inst. Tech., 1930; m. Grace Mabel Gardner, May 19, 1933; children—Clifford Warren, Ann Esther (Mrs. K. R. Mills), Carolyn Grace. Asso. prof. physics Ala. Poly. Inst., 1930-35; physicist Gulf Research & Devel. Co., 1935-41; research fellow Cal. Inst. Tech., 1941-45; asst. dept. head U. S. Naval Ordnance Test Sta., 1945-47; physicist USN Dept. Office Naval Research, Pasadena, Cal., Washington, London, Eng., 1947-55; physicist Army Office Research, Pasadena, Cal., 1955-58; asst. dir. research System Devel. Corp., Santa Monica, Cal., 1958-61; dir. sci. research Douglas Aircraft Co., Santa Monica, 1961-66, research scientist, Huntington Beach, Cal., 1966-—. Cons. to pvt. industry, govt. agys.; lectr. U. So. Cal., 1956-60. Recipient Bur. Ordnance Tech. award, 1945; Army-Navy Appreciation award, 1948. Fellow Am. Phys. Soc., A.A.A.S.; mem. Am. Geophy. Union, Am. Astron. Soc. Author: Range Testing of Rockets, 1947;

also articles. First reliable results on low energy electron scattering in Mercury; contbr. to dielectric phenomena; 1st re-entry radiation measurements; organizer, dir. 1st internat. airborne eclipse expdn., 1963; dir. Douglas Eclipse Expdn., 1966. Home: 8127 Delgany Av., Playa del Rey, Cal. 90291. Office: 5251 Bolsa Av., Huntington Beach, Cal. 92646.*

ARNSTEIN, Henry, Am. chem. and mech. engr.; b. N.Y.C., Nov. 10, 1886; s. Adolph and Rose (Markstein) A.; A.M., U. Budapest, 1906; Sc.D., U. Berlin, 1908; D.Eng., U. Heidelberg, 1910; m. Nettie Becker, June 3, 1917; children—Burnerd, Lawrence Hugo, J. Robert. Began with Krupp's Essen. Germany, 1908; chief chemist and plant mgr. Fleischmann Yeast Co. and Am. Diamalt Co., 1913-19; cons. engr., San Francisco, 1919-21; cons. chemist and indsl. engr., Phila., 1921—; pres. Farm Products Chem. Co. of Am. Spl. tech. adviser to govts. of Cuba, Argentine, Brazil, Colombia and Peru. Contbr. many papers to trade and tech. publs.; many of his works pub. by Cuban govt. Spl. research on indsl. fermentation and distillation, particularly of yeast and alcohol; developer of process for prodn. fuel alcohol from waste products. Recipient first prizes from Uruguayan govt. in world-wide competition for best and most practical engring. project dealing with utilization of surplus agrl. crops and establishment of industries. Died July 24, 1935.

ARNY, Henry Vinecome, Am. pharm. chemist; b. Phila., Feb. 28, 1868; s. Louis Christian and Sarah (Shinn) A.; Ph.G., Phila. Coll. Pharmacy, 1889; studied U. Berlin, 1893-94, U. Göttingen, 1892-93, 1894-96, Ph.D., 1896; m. Katharine Moody Smith, Apr. 22, 1903; children—Robert Allen, Sarah Elizabeth, Malcolm Moody, Francis Vinacomb. Prof. pharmacy and dean Coll. of Pharmacy, Western Res. U., Cleve., 1897-1911; prof. chemistry, 1911-37, dean, 1930-37, Coll. of Pharmacy, Columbia U. Editor of The Druggists' Circular, 1914-15, Year Book, Am. Pharm. Assn., 1916-22; tech. editor Am. Druggist, 1928-1936. Remington medalist, 1922; Ebert medalist, 1924. Mem. com. of revision U. S. Pharmacopaeia, com. of revision Nat. Formulary; pres. Am. Conf. of Pharm. Faculties, 1915-16; mem. exec. com. Am. Metric Assn., 1916-20; chmn. Nat. Conf. on Pharm. Research, 1922-29; fellow Chem. Soc. (Eng.); mem. Am. Pharm. Assn. (pres. 1923-24), Am. Chem. Soc., A.A.A.S. Franklin Inst.; hon. mem. German Pharm. Soc. (Berlin). Pharm. Soc. of Great Britain. Author: Principles of Pharmacy, 1909, 4th edit., 1936. Died Nov. 3, 1943.

ARON, Hermann, German physicist; b. Kempen, Prussia, Germany, Oct. 1, 1845; prof. Berlin U.; pioneer in radio communication; research on electric currents, clock pendulum. Died Homburg von der Höhe, Germany, Aug. 29, 1913.

ARONEY, Manuel James, Australian chemist; b. Sydney, Australia, Aug. 31, 1932; s. Dimitrios James and Stella (Aroney) A.; B.Sc. with 1st class honors, U. Sydney, 1954; M.Sc., 1956, Ph.D., 1960; m. Ann Paschalis, Feb. 7, 1960; children—Dimitrios James, Theodore. Faculty, U. Sydney, 1955—, sr. lectr. chemistry, 1965—; Commonwealth Office Edn. tutor chemistry Colombo Plan students, 1955-58; asst. examiner chemistry Colombo Plan students, 1955-58; asst. examiner chemistry New S. Wales Leaving Certificate, 1955-64. Fellow Chem. Soc. London; mem. Royal Australian Chem. Inst. (asso.). Research, publs. on molecular structure determination by phys. methods; studies of configurations and conformations of organic and organometallic molecules in solution state using dipole moments, electric birefringences, infrared and nuclear magnetic resonance spectroscopy. Home: 10 Molloy Av., South Coogee, Sydney, N.S.W., Australia.*

ARONHOLD, Siegfried Heinrich, mathematician; b. Angerburg, Germany, July 16, 1819; doctorate, Königsberg, Germany, 1851; tchr., Giessen, Germany, also Zurich, Switzerland, Heidelberg, Berlin, Königsberg. Author: Über eine fundamentale Begründung der Invariantentheorie, 1863; papers, especially in Crelle's jour. Contbd. to theory of invariants of algebraic forms; largely responsible (through Arnhold's notation) for modern symbolic methods of calculus in algebraic geometry; investigated plane curves of 3d and 4th orders. Died Königsberg, Mar. 13, 1884.

ARONOW, Saul, Am. biophysicist; b. N.Y.C., Oct. 4, 1917; s. Abraham and Minnie (Mirel) A.; B.E.E., Cooper Union Inst. Teach., 1939; Ph.D., Harvard, 1953; m. Alice Pearlman, Feb. 12, 1942; children—Victor, Frederick, David, Nathan, Louise, Jessie. Engr., U. S. Signal Corps., Phila. 1940-42; tech. officer U. S. Army Signal Corps, 1942-46; project engr. Harvey Radio Labs., Cambridge, Mass., 1946-50; teaching fellow Harvard, 1950-53; asso. applied physicist Mass. Gen. Hosp., Boston, 1953—; indsl. cons. med. engring., 1950—; dir. Med. Engring. Corp., Cambridge, Mass. NSF fellow, 1950. Mem. I.E.E.E., Internat. Fedn. for Med. Electronics, Soc. Nuclear Medicine, Am. Assn. Physicists in Medicine. Author: The Fallen Sky, 1960; Instrumentation in Nuclear Medicine, 1967. Contbr. numerous articles to sci. jours. Devel. instrumentation for med. applications, principally diagnosis with radioisotopes, in particular, a scanner for brain tumor localization, an image-intensifier camera, automated radioisotope counters. Home: 303 Franklin St., Newton, Mass. 02158. Office: Mass. Gen. Hosp., Boston 02115.*

ARONS, Arnold Boris, Am. physicist; b. Lincoln, Neb., Nov. 23, 1916; s. Solomon and Esther (Rosen) A.; M.E., Stevens Inst. Tech., 1937, M.S., 1940; Ph.D., Harvard, 1943; M.A., Amherst Coll., 1953; m. Jean Margaret Rendall, Aug. 17, 1942; children—Marion, Janet, Kenneth, Paul. Research group leader Woods Hole (Mass.) Oceanographic Instn., 1943-46, asso. phys. oceanography, 1950—, trustee, 1964—; faculty Stevens Inst. Tech., 1946-52, asso. prof., 1948-52; prof. physics Amherst (Mass.) Coll., 1952—; group leader in charge underwater blast measurements Operation Crossroads, Bikini, 1946. Cons. on explosion phenomena Naval Ordnance Lab., Silver Spring, Md., 1948—, Waterways Expt. Sta., Vicksburg, Miss., 1950—. Guggenheim fellow, 1957-58; NSF Faculty fellow, 1962-63. Fellow A.A.A.S.; mem. Am. Assn. Physics Tchrs. (pres. 1967-68, Distinguished Service citation 1964, editor Resource Letters in Physics 1962-67), Am. Phys. Soc., Am. Meterol. Soc., Am. Geophys. Union, Am. Geochem. Soc. Author: Development of Concepts of Physics, 1965; also articles. Editor: (with A. M. Bork) Science and Ideas, 1964. Research on characteristics explosion pressure waves, nature and effects explosion phenomena in air and water, hydrodynamics rotating systems, phys. oceanography. Home: 155 Woodside Av., Amherst, Mass. 01002.*

ARONS, Leon, German inventor; b. Berlin, Germany, 1860; prof. U. Berlin; (with Rubens) studied propagation of electricity in dielectrics, 1891; inventor mercury vapor lamp, 1892. Died 1919.

ARONSON, Lester R., Am. biologist, mus. ofcl.; b. N.Y.C., Apr. 9, 1911; s. Samuel and Hannah (Besthoff) A.; B.A., Cornell U., 1932, M.S., 1933; postgrad. Columbia; Ph.D., N.Y. U., 1945; m. Evelyn Rappaport, Sept. 26, 1936; children—Carl Henry, Frederick Richard. Staff, Am. Mus. Natural History, N.Y.C., 1938—, acting chmn., 1946-49, asso. curator, 1946-56, chmn., 1949—, curator, 1956—, dean council sci. staff, 1958-60. Lectr., Hunter Coll., 1946-53; adj. asso. prof. biology N.Y. U., 1951-58, adj. prof., 1958—. Fulbright research fellow, 1953-54. Fellow N.Y. Zool. Soc., N.Y. Acad. Sci., Am. Soc. Zoologists (chmn. sect. animal behavior 1959), Sigma Xi. Curator: Copeia, 1958—, editorial bd., 1962-63; asso. editor Animal Behaviour, 1962—. Contbr. numerous articles on neurology, endocrinology, physiology, sex behavior and learning in fish, amphibians and mammals, also species typical behavior patterns in relation to problems in genetics and evolution to sci. publs. Home: 47 Cedar St., Hillsdale, N.J. 07642. Office: Central Park W. at 79th St., N.Y.C. 10024.*

ARONSON, Seymour, Am. phys. chemist; b. New Britain, Conn., Jan. 23, 1929; s. Joseph Harry and Ida (Isaacson) A.; B.S. in Chemistry, Yeshiva U., 1950; M.S., U. Pitts., 1959; Ph.D., Poly. Inst. Bklyn., 1956; m. Judith Ralston, July 14, 1957; children—Debrah, Susan, Elaine. Sr. engr. chemistry Bettis Atomic Power div. Westinghouse Electric Corp., Pitts., 1954-60; chemist nuclear engring. dept. metallurgy div. Brookhaven Nat. Lab., Upton, N.Y., 1960—. NSF fellow, 1952-53. Mem. Am. Chem. Soc., A.A.A.S. Contbg. author Uranium Dioxide—Properties and Nuclear Applications, 1961. Research, publs. on phys. chemistry of solids of interest in nuclear reactor tech., thermodynamic and elec. properties of refractory compounds of uranium and thorium, elec. properties of graphite exposed to neutron irradiation, thermodynamic properties of layer compounds of graphite. Home: 34 Keswick Dr., East Islip, N.Y. 11730. Office: Brookhaven Nat. Lab., Upton, L.I., N.Y. 11973.*

ARONSON, Stanley Maynard, Am. physician; b. N.Y.C., May 28, 1922; s. Eliuh and Lena (Hassner) A.; B.S., Coll. City N.Y., 1943; M.D., N.Y. U. Coll. Medicine, 1947; m. Betty Ellis, June 3, 1947; children—Susan, Lisa, Sarah. Faculty, Columbia Coll. Phys. and Surg., 1951-54; Yale Sch. Medicine, 1964—; prof. pathology, asst. dean State U. N.Y., Bklyn., 1960—; dir. labs. Kings County Hosp. Center, Bklyn., 1965—. Cons. physician neuropathology Jewish Chronic Disease Hosp., Bklyn., 1951; NIH, 1962—, N.Y. regional office VA, 1965—. Mem. A.M.A., Am. Neurol. Assn., Am. Assn. Neuropathology, N.Y. Acad. Medicine, Am. Acad. Neurology, Am. Assn. Pathologists and Bacteriologists, N.Y. Neurol. Soc. Author: (with B. W. Volk) Cerebral Sphingolipidoses, 1962; Inborn Disorders of Sphingolipid Metabolism, 1966; also numerous articles. Research on genetics, pathology, and diagnostic features of infantile cerebral degenerative diseases, population dynamics, pathology and epidemiology of cerebral vascular disease and stroke. Home: 124 Neptune Av., New Rochelle, N.Y. 10805. Office: 451 Clarkson Av., Bklyn. 11203.*

ARP, Halton C., Am. astronomer; b. N.Y.C., Mar. 21, 1927; s. August C. and Anita (Cryst) A.; A.B. cum laude, Harvard, 1949; Ph.D. cum laude, Cal. Inst. Tech., 1953; m. Susanna Bixby Dakin, Sept. 13, 1962; children—Kristana Marta, Alissa Jan. Carnegie fellow Mt. Wilson and Palomar Obs., 1953-55; research asso. Ind. U., 1955-57; staff mem. Mt. Wilson and Palomar Obs., Carnegie Inst. Wash., Cal. Inst. Tech., 1957—; vis. prof. NSF, 1960—. Mem. Am. Astronom. Soc. (Helen Warner prize 1960), A.A.A.S. (Newcomb Cleveland prize 1961), Internat. Astronom. Union, Fedn. Am. Scientists (chmn. Los Angeles chpt. 1965), Sigma Xi. Author: Atlas of Peculiar Galaxies, 1965; also numerous articles. Established ratio between the absolute magnitude at maximum brightness of novae and the speed of decline of magnitude, 1956. Office: 813 Santa Barbara St., Pasadena, Cal. 91106.*

ARPPE, Adolf Edward, Finnish chemist; b. Rides, Finland, June 9, 1818; pupil of Wöhler; prof., rector U. Helsingfors (Sweden). Author: De ipdeto Bismutic; History of Chemistry in Finland. Discovered thymol; obtained nitraniline; studied bismuth compounds, morphine, other alkaloids. Died Apr. 14, 1894.

ARRHENIUS, Johan Peter, Swedish botanist; b. Jaereda, Sweden, Sept. 27, 1811; prof. U. Uppsala (Sweden); worked on Swedish agr.; author: Botanical Terminology; Course of Botany; Botany and Handbook of Farming. Died Jaereda, Sept. 5, 1889.

ARRHENIUS, Svante August, Swedish chemist, physicist; b. Schloss Wijk, nr. Uppsala, Sweden, Feb. 19, 1859; ed. U. Uppsala, 1876-81, Ph.D., 1884; ed. U. Stockholm, 1881-84. Began career as privat docent in phys. chemistry U. Uppsala; traveled to univs. Würzburg, Graz, Amsterdam and Leipzig, worked with Ostwald, Boltzmann, van't Hoff, 1886-88; lectr. physics U. Stockholm, 1891-95, full prof., 1895, rector, 1897-1902; lectr. U. Cal., 1904; dir. Nobel Inst. for Phys. Chemistry, 1905-27. Recipient Davy medal Royal Soc., 1902; Nobel prize in chemistry for theory of electrolytic dissociation, 1903. Fellow Royal Soc., 1910; mem. Swedish Acad. Scis.; corr. French Acad. Scis., 1911. Author: Recherches sur la Conductibilité galvanique des électrolytes: (1) conductibilité galvanique des solutions aqueuses extremement diluées, (2) Théorie chimique des électrolytes (presented to Stockholm Acad. Scis.), 1883; Uber die Dissociation der in Wasser gelösten Stoffe, 1887; Lärobok i teoretik elektrokemi, 1900; Lehrbuch der kosmischen Physik, 1903; Theorien der Chemie, 1906; Immunochemistry, 1907; Das Werden der Welten, 1907; Theories of Solutions, 1912; Quantitative Laws in Biological Chemistry, 1915; Die Schicksale der Sterne, 1915; Der Lebenslauf der Planeten, 1921; Erde und Weltall, 1926. Formulated theory of electrolytic dissociation (Arrhenius' theory), later extended theory to cover problems of rate of chem. reaction and phenomenon of atmospheric electricity; discovered expression for latent heat as function of the observed raising of boiling point through dissolution of a non-volatile compound; discovered speedup of chem. reactions with increasing temperature, 1889; interested in structure of universe, was one of first to appreciate light pressure and its part in cosmic physics (such as repulsion of comet's tail by radiation from sun); speculated on origin of life on earth as having come from other planets; opposed Clausius' theory of heat-death of universe; other research areas include: viscosity of solutions, osmosis, reaction velocities, immunochemistry, aurora borealis. Died Stockholm, Oct. 2, 1927.

ARRIGHETTI, Nicolo, Italian physicist; b. Florence, Mar. 17, 1709; joined Soc. of Jesus, 1724; tchr. philosophy, Spoleto, Prato, Sienna, Italy. Author: Ignis theoria, 1750; treatises on heat, light, electricity, barometric pressure. Died Jan. 31, 1767.

ARRILAGA (or ARRILLAGA), Francisco C., physician; b. Argentina, late 1800's; prof. clin. medicine U. Buenos Aires. Author: Arteritos Pulmonar; Cardiacos Negros; others. Introduced eponym of Ayerza's disease (form of blood disease marked by persistent oversupply of red blood cells).

ARROW, Gilbert John, English entomologist; b. Dec. 20, 1873; s. John Garner Arrow; m. Rachel Katharine Davis. Employee, architect's office, 5 years; entered Natural History Mus., 1896; became dep. keeper, dept. entomology Brit. Mus. (Natural History). Fellow Zool. Soc., Royal Entomol. Soc. London. Author 5 vol. work on Indian beetles (Fauna of India Series); numerous papers. Died Oct. 5, 1948.

ARROW, Kenneth Joseph, Am. economist; b. N.Y.C., Aug. 23, 1921; s. Harry I. and Lillian (Greenberg) A.; B.S., Coll. City N.Y., 1940; M.A., Columbia, 1941, Ph.D., 1951; LL.D., U. Chgo., 1967; m. Selma Schweitzer, Aug. 31, 1947; children—David Michael, Andrew Seth. Research asso. Cowles Commn. for Research in Econs., U. Chgo., 1947-49, asst. prof. econs., 1948-49; acting asst. prof. econs., statistics Stanford, 1949-50, asso. prof., 1950-53, prof., 1953-68, pres. Inst. Mgmt. Scis., 1962, exec. head dept. econs., 1953-56, 1962-63; professor economics Harvard Univ., 1968—; cons. RAND Corp., 1948—; mem. staff Council Econ. Advisers U. S. Govt., 1962. Center for Advanced Study in Behavioral Scis. fellow, 1956-57; overseas research fellow Churchill Coll., Cambridge U., 1963-64. Fellow Econometric Soc. (pres. 1956), Inst. Math. Statistics, Am. Statis. Assn.; mem. Am. Econ. Assn. (John Bates Clark medal 1956). Author: Social Choice and Individual Values, 1951, 2d edit., 1963; (with S. Karlin, H. Scarf) Studies in the Mathematical Theory of Inventory and Production, 1958; (with L. Hurwicz and H. Uzawa) Studies in Linear and Non-Linear Programming, 1958; (with M. Hoffenberg) A Times Series Analysis of Interindustry Demands, 1959; Aspects of the Theory of Risk-Bearing, 1965. Theoretical analysis of social choice, gen. impossibility of arriving at consensus by voting mechanisms; rigorous analysis of competitive equilibrium, its existence, stability and optimality; empirical analysis of substitutability between capital and labor in prodn.; possible implications of learning to econ. devel.; optimum behavior for bus. firms, particularly with regard to inventories; possibilities and limits of market mechanisms for handling uncertainty. Home: 895 Lathrop Dr., Stanford, Cal. 94305.*

ARROYAVE-BORJES, Guillermo, Guatemalan biochemist; b. Guatemala, Guatemala, June 29, 1922; s. Julio and Julia (Borjes) Arroyave; Chemist Pharmacist, U. San Carlos (Guatemala), 1947; Ph.D., U. Rochester, 1953; m. Cornelia Jane Barber, Sept. 24, 1949; children—María, Julia, John William, Elizabeth Ann, José Enrique, Nancy Jane. Chief labs. Inst. Nutrition Central Am. and Panama, 1949-50, chief div. physiol. chemistry, 1953—; vis. asso. prof. Mass. Inst. Tech., 1962. Cons. WHO, 1962—, mem. expert adv. panel on nutrition, 1962—; mem. com. on dietary allowances food and nutrition bd. Nat. Acad. Scis., 1965. Mem. Internat. Union Nutrition Scis. (mem. com. on procedures for appraisal protein-calorie Nutrition, Sociedad Latinoamericana de Nutrición, Colegio de Farmacéuticos y Químicos de Guatemala, Academia de Ciencias Físicas, Médicas y Naturales. Research, numerous publs. on biochemistry of nutrition especially metabolic alterations in children suffering from protein-calorie malnutrition; biochem. methods for evaluation of nutritional status in population groups. Home: 5a Av., no. 20-00, Guatemala City 14. Office: Inst. Nutrition of Central Am. and Panama, Carretera Roosevelt, Guatemala City 11, Guatemala.*

ARRUGA LIRO, Hermenegildo, Spanish ophthalmologist; b. Barcelona, Spain, Mar. 4, 1886; s. Edouard and Dolores L. (Forgas) A.; ed. U. Barcelona, Hôtel-Dieu, Paris, France, Charity Hosp., Berlin, Germany; D. medicine and surgery; m.; children—Dolorès, Maria, Thérèse, Alfred. Chief clinic of Dr. Menacho, Barcelona; consul Spanish commn. UNESCO. Recipient numerous awards in ophthalmology. Mem. Soc. Ophthalmology Barcelona (pres.), hon. Spanish-Am. Soc. Ophthalmology (pres.), Royal Acad. Medicine Barcelona, socs. ophthalmology Belgium, Venezuela, Switzerland, Brazil, Chile, Italy, Mexico, Cuba, Greece, Egypt, Colombia (all hon.), Am. Acad. Ophthalmology (hon), Royal Coll. Surgeons (hon.). Author: The Detached Retina; Ocular Sugery. Research, publs. on cataracts, glaucoma, corneal grafting. Address: Pasaje Méndez-Vigo, 3, Barcelona, Spain.

ARSHAVSKY, Ilya Arkadevich, Russian physiologist; b. 1903; grad. Med. Faculty, Rostov U., 1926; postgrad. dept. physiology 2d Moscow Med. Inst. and Physiology Lab., Obukh Inst., 1926-29, Leningrad U., 1929-32; D.Med. Sci., 1936. Lectr. dept. physiology Kazan U., 1932-35; founder Lab. Exptl. Developmental Physiology, Inst. for Care Children and Adolescents (now Lab. Developmental Physiology and Pathology, Inst. Normal and Path. Physiology, USSR Acad. Med. Sci.) 1935, head, 1935—. Research on exptl. basis of physiology and pathology at early ages, analysis of features of nervous regulation of cardiovascular activity, correlation of functioning of various systems, physiol. basis of antichem. protection at early ages, inhibition in organs and systems at various ages. Address; Inst. Normal and Path. Physiology, USSR Acad. Med. Sci., Baltiysky poselok 13, Moscow D-52, USSR.

ARSLAN, Michele, physician; b. Padua, Italy, Jan. 23, 1904; s. Yerwant and Antonia (De Besi) A.; Physician degree, U. Padua, 1927; Hon. Doctor, U. Uppsala (Sweden). 1963. m. Vittoria Marchiori, Apr. 5, 1936; children—Antonia (Mrs. Paolo Veronese), Paola, John, Edward, Charles. Asst. prof. U. Frankfurt (Germany), 1932-33, U. Berlin (Germany), 1935; faculty U. Padova, 1936—, prof. otorhinolaryngology Med. Faculty, 1936—, dir. Library, 1952—; head ear nose and throat dept. U. Hosp. Padova, 1936—. Decorated Chevalier de la Legion D'Honneur (France). Mem. Royal Soc. Medicine (London); hon. mem. otorhinolaryngol. socs. of Germany, Austria, Switzerland, Greece. Author: L'Evaluation de L'Invalidite Auditive, 1953; also numerous articles. Research on physiology of human vestibular apparatus, diseases of collagenous tissues of upper respiratory ways; discoveries and new applications of ultrasonics in surgery. Home: 37, Via Altinate, Padova, Italy.*

ARTAMONOV, Mikhail Illarionovich, Russian historian, archeologist; b. Vygolovo, Tver, Russia, Dec. 5, 1898; s. Illarion Artamonovich Balyabin and Natalya (Krosavtseva) A.; grad. Leningrad U., 1924, D.Hist. Sci., 1943; m. Anna Meinikove, Dec. 1922; 1 dau., Trina; m. 2d, Olga Poltavtseva, Dec. 11, 1936; 1 son, Oleg. Dir., Inst. History Material Culture (now Inst. Archeology), USSR Acad. Sci., 1938-43; prof. chair archeology Leningrad U., 1935—, head chair archeology, 1949—, pro-rector, 1948-51; dir. State Hermitage Mus. 1951-64. Mem. Soc. Antiquaries of Scotland (hon.), Polish Acad. Sci. Author: Medieval Settlements on the Lower Don, 1935; Essays on the Earliest History of the Khazars, 1936; The Origin of the Slavs, 1950; The History of the Khazars, 1962. Exec. editor: Slav Antiquities, 1962, Relics of the Bronze and Early Iron Ages in Eastern Europe, Southern Siberia and Central Asia, 1962. Specialist in history of Bronze and Iron Ages in south of present-day USSR; studied Bronze- and Iron Age relics along Don and North Caucasus in 1930's. condr. excavations at Khazar fortress of Sarkel, 1934-36, 49-51. Home: 27, Nalichnaya, Leningrad, USSR.*

ARTEDI, Peter, Swedish naturalist; b. Anund, Sweden, Feb. 22, 1705; ed. U. Uppsala (Sweden), from 1724. Partner (with Linné) in natural history work; used mus. Hans Sloane during visit to Eng., 1734; described fish collection of Seba (an apothecary), Amsterdam, Netherlands, 1735; materials for sci. study fishes edited by Linné, 1738; classified umbellifiers according to involucres. Drowned at Amsterdam or Leiden, Netherlands, Sept. 27, 1735.

ARTEMIADIS, Nicolas, mathematician; b. Constantinople, Turkey, May 17, 1917; s. Kyriacos and Despina (Docmejoglou) A.; M.Math., U. Solonica (Greece), 1939; diploma Institut Francais d'Athènes (Greece), 1951; Diplôme d'Etudes Superieures in Analyse Superieure, U. Paris (Sorbonne), 1954, in Algebre et Théorie de Nombres, 1955, Doctorat ès Sciences (mention très honorable), 1957; m. Zafiria Pyplidou, Aug. 14, 1963. Came to U. S., 1958, naturalized, 1963. Tchr. pvt. schs., Athens, 1949-51; asst. prof. math. U. Wis. at Milw., 1958-60, asso. prof., 1961—. Prof. math. U. Salonica, 1960-61; lectr. Math. Congress, Rome, Italy, 1961. Decorated Medal for Excellent Acts (Greece); recipient Stavropoulos award Greek Embassy, Paris, 1957; research grantee Army Research Center, Madison, Wis., 1960-61, U. Wis., 1959. Mem. Am., French, Greek math. socs. Reviewer Zentralblatt für Mathematik. Contbr. articles to tech. jours. New important results in real and complex analysis; fundamental theorems on Fourier transforms with application in holomorphic, typically real, univalent function; tauberians; generalization of Caratheodory's theorem concerning coefficients of holomorphic functions with nonnegative real part, to Dirichlet series; introduction and investigation of new class of functions much larger than class of typically real functions. Home: 1041 E. Knapp St., Milw. 53202.*

ARTEMIDOROS OF EPHESOS, Greek geographer; flourished circa 100 B.C.; Ephesian ambassador to Rome; author 11 vol. geog. work (purported to have dealt with entire world, attached importance to phys. geography and distances between places; work now lost); quoted by Strabo.

ARTEMOV, Vladimir Aleksandrovich, Russian psychologist, phonetician; b. 1897; D.Pedag. Sci. Prof. chair psychology, dir. Lab. Exptl. Phonetics and Psychology of Speech, Moscow Pedagogical Inst. Fgn. Langs. Mem. Internat. Soc. Phonetic Scis. (v.p.). Author: Practical Work in Experimental Psychology, 1927; Psychology of Perception, 1940; An Outline of Psychology, 1954; Experimental Phonetics, 1956; A Lecture Course on Psychology, 1958; The Interrelations of the Physical Properties, Perceptible Characteristics, Linguistic Meanings and Semantic Content of Speech, 1960; The Culture of Speech, 1960; Psychology of Foreign Language Teaching, 1966. Editor: A Symposium on Phonetics, 1959; Experimental Psychology, 2 vols., 1960. Research in exptl. phonetics, psychology of speech and fgn. lang. instrn. Home: G-146, 2-aya Frunsenskaya 10, Kv. 185. Office: 1-y Moscow Pedagogical Inst. Foreign Languages, Metrostroevskaya ulitisa 38, Moscow, USSR.*

ARTEPHIUS (or Artefius, Artesius), alchemist; possibly flourished 12th century, wrote Liber secretus clavis maioris sapientiae de transmutatione metallica; writings under his name are probably transls. from Arabic, perhaps from al-Tugra'i's work; cited Ibn Sina's work; Artephius was cited by Roger Bacon and William of Auvergoe.

ARTHABER, Gustav Adolph, Austrian paleontologist; b. Vienna, Austria, Oct. 21, 1864; s. Johann Joseph Rudolph and Elise Hermine (Clausz) A.; Ph.D., Vienna U., 1897; m. Bertha Rohn, 1903. Became lectr. Vienna U., 1892, lecturing prof. 1897, asso. prof., 1907, prof., 1921; apptd. adjunct Paleontol. Inst., 1899; made expdns. to Mediterranean area, also Armenia. Mem. Geol. Soc. Vienna (a founder). Contbd. articles including Die Cephalododenfauna der Reiflinger Kalke . . . , 1895-96, Die alpenen Trias der Mediterranean-Gebietes, 1903-08, Die Trias von Albaniens, 1911. Investigated paleontology and stratigraphy of Trias formations and significance as belemnitella quadrata; studied reptile fossils. Died Vienna, Apr. 29, 1943.

ARTHUR, Jett Clinton Jr., Am. phys. chemist; b. Hemphill, Tex., May 31, 1918; s. Jett Clinton and Anna Alice (Smith) A.; B.A., Stephen F. Austin State Coll., 1939; M.A., U. Tex., 1946; m. Oma LaVerne Pitts, June 2, 1941; children—Martha Ann (Mrs. Robert Lloyd Walters), James Clinton, Laura Jean. Research chemist So. Regional Research Lab., Agrl. Research Service, U. S. Dept. Agr., New Orleans, 1941-49, head protein products investigations, 1949-52, head biochem. investigations, 1952-56, head radiochemistry investigations, 1956-66, chief research chemist, 1966—. Recipient Citizenship award Mayor and City Council New Orleans for outstanding service to youth, 1958. Mem. Am. Chem. Soc., Fiber Soc., La. Congress P.T.A. (hon. life mem.), Sigma Xi, Alpha Chi. Contbg. author: Advances in Protein Chemistry, 1953; Technique of Organic Chemistry, 1959; Ency. Americana, Ency. Polymer Sci. and Tech. Editor phys. organic chem. sect. Chem. Abstracts, 1963—, abstractor, 1946—. Patentee in field. Research, publs. on radiation chemistry of natural products and the phys. organic chemistry of natural polymers, resulting in new food and textile products. Home: 3013 Ridgeway Dr., Metairie, La. 70002. Office: 1100 Robert E. Lee Blvd., New Orleans 70119.*

ARTHUR, Joseph Charles, Am. botanist; b. Lowville, N.Y., Jan. 11, 1850; s. Charles and Ann (Allen) A.; B.S., Ia. State Coll., 1872, M.S., 1877; postgrad. Johns Hopkins, 1879, Harvard, 1879, U. Bonn (Germany), 1896; Sc.D., Cornell, 1886; LL.D., State U. Ia., 1916; Sc.D., Ia. State Coll., 1920; Sc.D., Purdue U., 1931; m. Emily Stiles Potter, June 12, 1901. Instr. botany, Univ. Wis., 1879-81, U Minn., 1882; botanist, Expt. Sta., Geneva, N.Y. 1884-87; prof. botany, Purdue, U., Lafayette, Ind., 1887, prof. vegetable physiology and pathology, and botanist Ind. Expt. Sta., 1888-1915; prof. emeritus, Purdue U., from 1915. Speaker Internat. Congress Arts and Scis., St. Louis, 1904; del. Internat. Bot. Congress, 1905, 10, 30. Fellow A.A.A.S. (v.p., 1895); mem. Bot. Soc. Am. (twice pres.), Torrey Bot. Club, Deutsche Botanische Gesellschaft, Ind. Acad. Science (pres. 1903), Soc. Promotion Agrl. Science, Am. Acad. Arts and Scis., Ia. Acad. Science, Am. Philos. Soc., Am. Assn. Univ. Profs., Am. Soc. Naturalists, Mycol. Soc. Am., Am. Phytopathol. Soc. (pres.). Author: Handbook of Plant Dissection (with C. R. Barnes and J. M. Coulter), 1886; Living Plants and Their Properties (with Daniel Trembly MacDougal), 1898; Uredinales, in N. Am. Flora, 1907-29; The Plant Rusts (with others), 1929; Manual of the Rusts in United States and Canada, 1934. Research and publs. in physiology and pathology of vegetables, notably in reference to fungi and rusts. Died Apr. 30, 1942.

ARTHUS, André René, French physician; b. Lille, France, Jan. 27, 1903; s. Maurice and Marie-Thérèse (Weissenbach) A.; ed. sch. of medicine Lausanne, Lyon univs.; M.D.; m. Yvonne Sargnon, Dec. 8, 1928; children—Daniel, Manual, Florence, Dominique, Hélène. Practice medicine specializing in endocrinology and psychology; ex-privat docent Faculté de médecine Lausanne; ex-dir. Organotherapic Research Lab., Garches. Mem. Internat. Asso. Analytic Psychology, Inst. Psychology Geneva. Author: Comprendre pour mieux agir; Mariage; Un monde inconnu, nos enfants; Les mystères de la vie, expliqués aux enfants; Adolescence; also lab. manuals and articles. Home: L'Esplanade, 36, rue Louis-Thevenet, Lyon 4 (Rhone). Office: 3, rue Ney, Lyon 6, France.

ARTHUS, Nicolas-Maurice, French physiologist; b. Angers, France, 1862. Chief Physiol. Conf., U. Paris, 1890; became prof. physiology and microbiology Fribourg University, 1896, Pasteur Institute Laboratory, Lille, France, 1900; appointed professor physiology, Lausanne, Switzerland, 1907. Member French Academy of Sciences, 1943. Author: Coagulation des liquides de l'organisme, 1894; Précis de chimie physiologique, 1895; Natur des enzymes, 1898; injections répétées de serum du cheval chez Lapin, 1905; De l'anaphylaxie a l'immunitie, 1921; Precis de physiologie microbienne, 1921. Established role of calcium in blood coagulation, 1890; discovered Arthus phenomenon in allergy, 1903. Died Fribourg, Feb. 24, 1945.

ARTMAN, Michael, microbiologist; b. Lvov, Poland, July 23, 1924; s. Joshua and Deborah (Weinreb) A., M.Sc., Hebrew U., Jerusalem, Israel, 1952, Ph.D., 1955; m. Stephanie Lupak, Feb. 13, 1948; 1 son, Gregory. Staff dept. bacteriology Hebrew U. Hadassah Med. Sch., 1952-58, faculty, 1958—, prof. microbiology Inst. Microbiology, 1966—; vis. prof. U.C.H., Ibadan, Nigeria, 1966. Mem. Microbiol. Soc. Israel. Research, publs. on biochem. changes in animals with exptl. Tb, action streptomycin, ribosomal structure and function, bacteriophages. Home: 30 Hehalutz St., Jerusalem, Israel.*

ARTMAN, Robert Arthur, Am. physicist, educator; b. Freeport, Ill., Sept. 23, 1915; s. Frank Arthur and Geneva (Downing) A.; A.B., Ill. Coll., 1937; M.S., State U. Ia., 1939, Ph.D., 1951; m. Ruth Alice Chilton, Aug. 20, 1938; children—Judith (Mrs. Charles E. Mothersbaugh III), Bruce Chilton. Research asst. State U. Ia., 1937-39, 49; instr. Ala. Poly. Inst., 1939-41; head physics dept. USAF Pre-Flight Sch., Maxwell Field, Ala., 1941-46; asso. prof. physics Evansville (Ind.) Coll., 1946-50, prof., 1950-52, acting chmn. physics dept., 1951-52; supr. gen. physics sci. lab. Ford Motor Co., Dearborn, Mich., 1952-56, supr. applied physics, 1956-58; prof. physics Bucknell U., Lewisburg, Pa., 1958—, chmn. dept., 1959—. Vis. scientist Brookhaven Nat. Lab., summers 1959, 60; prin. investigator research project on ultrasonic waves, NSF, 1961-65. Mem. Am. Phys. Soc., Am. Assn. Physics Tchrs., Phi Beta Kappa, Sigma Xi, Sigma Pi Sigma, Pi Mu Epsilon. Research publs. in solid state physics. Home: R.D. 1, Lewisburg, Pa. 17837.*

ARTOBOLEVSKI, Ivan Ivanovich, Russian engineer; born Oct. 9, 1905; graduated Faculty Agricultural Machine-Building, Moscow Timiryazev Agricultural Academy, 1926; Doctor Technological Science, 1936. Instr., Moscow Electromech. Inst., 1927-29; instr. Moscow Inst. Chem. Tech., 1929-32, prof., 1932—; prof., head chair theory of mechanisms and machines Moscow Inst. Chem. Machine-Bldg., 1932; prof. Zhukovsky Air Force Engring. Acad., 1932-38, Moscow U., 1932-49; asso. Inst. Mech. Engrs., USSR Acad. Sci., 1937—, acting sec. Dept. Tech. Sci., 1942-54, head Lab. Machine Dynamics, Inst. Machine Sci., 1937—, bur. mem. Dept. Tech. Sci., 1960—; prof., head chair theory of mechanisms and machines Moscow Aviation Inst., 1941—; hon. sci. and tech. worker of RSFSR, 1945—. Mem. USSR Acad. Sci., All-Union (dep. chmn. 1947—), RSFSR (chmn.) socs. for dissemination polit. and sci. knowledge, Soc. Mech. Engrs. (pres. 1945-54), All-Union Council Sci. and Tech. Soc. Author: The Theory of Spatial Mechanisms, 1937; The Synthesis of Plane Mechanisms, Parts 1-2, 1939-42; The Synthesis of Mechanisms, 1944; Mechanisms, Vols. 1-4, 1947-51; Linear-Envelope Cycloid Curves, 1962; The Curvature of Linear-Envelope Cycloid Curves, 1962; co-author Analysis Methods for Automatic Machines, 1945. Specialist in machine theory and automation; research in theoretical and exptl. methods of studying dynamics of machines in operation. Address: Aviation Inst., Volokolamskoe shosse 8, Moscow, USSR.

ARTOM, Camillo, biochemist; b. Asti, Italy, June 5, 1893; s. Vittorio E. and Cemma (Pugliese) A.; M.D., U. Padua, Italy, 1917; Ph.D. in Physiology, U. Messina, Italy, 1923; Ph.D., in Biochemistry, U. Palermo, Italy, 1926; m. Bianca M. Ara, July 29, 1928; 1 son, George V. Came to U. S., 1939, naturalized, 1947. Prof., head dept. physiology U. Cagliari, 1930-35, U. Palermo, 1935-38; prof., head dept. biochemistry Sch. Med. Scis., Wake Forest Coll., Winston-Salem, N.C., 1939-41, prof., head dept. biochemistry Bowman Gray Sch. Medicine, 1941-61, prof. emeritus, 1963—. Rockefeller Found. fellow U. Naples, 1926. Fellow A.A.A.S.; mem. Am. Soc. Biol. Chemists, Soc. for Exptl. Biology and Medicine, Societe de Chimie Biologique (Paris). Contbr. numerous articles to profl. jours. Research in metabolism of fats and other lipids, especially phospholipids, problems of intermediary metabolism. Home: 1538 Overbrook Av., Winston-Salem, N.C. 27104.*

ARTSIKHOVSKY, Artemii Vladimirovich, Russian archeologist; born 1902; graduated Moscow Univ., 1925. Research worker Historical Mus., Moscow, 1925-38; Professor Moscow Univ., 1937—, head chair archeology History Faculty, 1939—; director. Novgorod Archeological Expedition, Institute Archeology, USSR Academy of Science, 1929-59; director archeological investigations during construction of Moscow Metro. Deputy chairman USSR National Committee Historians, 1956—. Mem. USSR Acad. Sci. (corr., Presidium's prize). Author: The Burial Mounds of the Vyatichi, 1930; Ancient Russian Miniatures as Historical Sources, 1944; Introduction to Archeology, 1947; Principles of Archeology, 2d edit., 1955; Excavations on the Slavna in Novgorod, 1949; New Finds in Novgorod, 1955; Excavations in Novgorod in 1952, 1953; The Novgorod Birch-Bark Documents, Vols. 1-5, 1953-62. Chief editor: Soviet Archeology, Moscow U. Herald, History Series. Specialist in ancient Russian archeology and history of Russian feudalism; leader expdn. which discovered ancient Russian birch-bark documents, 1951. Address: Soviet Archeology, 1-aya Cheremushkinskaya ulitsa 19, Moscow V-36, USSR.

ARTSIMOVICH, Lev Andreevich, Russian physicist; b. 1909; grad. Belorussian U., Minsk, 1928. Tchr., researcher at higher edul. and research insts. in Moscow and Leningrad, 1930—; with Physico-Tech. Inst., USSR Acad. Sci., 1930-48, acad. sec. dept. physico-math. sci., 1959—, asso. Inst. Atomic Energy. Mem. USSR Acad. Sci. Recipient Lenin prize, 1958. Author: Bremsstrahlung of High Energy Electrons, 1938; Angular Distribution of Fast Electrons Scattered by Aluminum Nuclei, 1946; The study of High-Current Pulse Discharges, 1956; Head Radiation of Pulse Discharges, 1956; co-author The Radiation of Fast Electrons in a Magnetic Field, 1946. Research in atomic and nuclear physics; (with others) confirmed laws of conservation in electron and positron annihilation, 1936. Research on interaction of fast electrons with matter, especially Bremsstrahlung and angular distbn. of scattered electrons, 1935-40; developer theory of chromatic aberrations in electronic optical systems, 1943-46; dir. work on devel. electromagnetic method of isotope separation and high power elec. discharges to find means of obtaining controlled thermonuclear reactions. Mem. editorial bd. Reports of USSR Acad. Scis., 1960—. Address: Inst. Atomic Energy, ulitsa Kurchatova 46, Moscow, USSR.

ARTUSIO, Joseph Francis, Jr., Am. physician, anesthesiologist; b. Jersey City, Nov. 26, 1917; s. Joseph Francis and Jennie A. (Cuneo) A.; B.S. in Chemistry, St. Peter's Coll., 1939; M.D., Cornell U., 1943; m. Mary Louise Ellis, Oct. 8, 1945; children—Marianne, Suzanne, Joseph Francis III, Evelyn Ellis, Mark Ellis, Douglas Ellis. Asst. attending anesthesiologist N.Y. Hosp., N.Y.C., 1947-48, attending anesthesiologist in charge, 1948-57, anesthesiologist in chief, 1957—; faculty Cornell U. Med. Coll., N.Y.C., 1946—, prof. anesthesiology in surgery, 1957—, prof. anesthesiology in obstetrics and gynecology, 1957—. Mem. Am. Soc. for Pharmacology and Exptl. Therapy, N.Y. Acad. Medicine, Soc. Exptl. Biology and Medicine, A.M.A., Med. Soc. County of N.Y., Am. Bd. Anesthiology, Am. Coll. Anesthesiologists, Am. Soc. Anesthesiologists, Assn. U. Anesthetists. Author: Practical Anesthesiology, 1963. Editor in chief Clin. Anesthesia series, 1963. Research in ether analgesia, halogenated ether, halogenated hydrocarbons. Home: 238 Corlies Av., Pelham, N.Y. Office: 525 E. 68th St., N.Y.C. 10021.*

ARTZ, Curtis Price, Am. surgeon; b. Jerome, O., Sept. 29, 1915; s. Curtis Emuel and Bertha (Price) A.; A.B., Ohio State U., 1936, M.D., 1939; M.S., Baylor U., 1952; m. Kathryn Lucille Risley, June 24, 1939; children—Susan Elizabeth, John Curtis, Joanne Risley. Commd. capt. M.C., U. S. Army, 1949, advanced through grades to lt. col., 1953; chief surg. sect. Research, Devel. Bd., Office Surgeon Gen., 1949-50; dir. surg. research team, Korea, 1951-52; comdg. officer surg. research unit Brooke Army Med. Center, Ft. Sam Houston, Tex., 1953-56; ret., 1956; practice medicine, specializing in surgery, Jackson, Miss., 1956-63, Galveston, Tex., 1963-65, Charleston, S.C., 1965—; faculty Sch. Medicine U. Miss., 1956-62; Shrine prof. surgery med. br. U. Tex., 1963-65; prof. surgery, chmn. dept. Med. Coll. S.C., 1965—; dir. Clin. Research Center U. Hosp., Jackson, 1962-63; dir. Shrine Inst. For Burns, Galveston, 1963-65; mem. surgery sect. NIH, 1966—. Coordinator 1st Internat. Congress on Research and Burns, 1960—; civilian cons. Surgeon Gen. U. S. Army, 1965—. Recipient Alumni Achievement award Ohio State U., 1964. Mem. Am., So. (Shipley award 1959) surg. assns., Soc. U. Surgeons, A.C.S., Southeastern Surg. Congress (pres. 1967), So. Soc. Clin. Research, Alpha Omega Alpha. Author: Battle Casualties in Korea, 1955; (with E. Reese) The Treatment of Burns, 1957; (with J. D. Hardy) Complications in Surgery and Their Management, 1960, 67; Research in Burns, 1962. Research, numerous publs. on surg. physiology; better understanding of shock and use of blood in shock, metabolic derangements that occur after burning injury; improvement of methods of local care and infections of burns. Home: 1615 Burning Tree Rd., Charleston, S.C. 29407.*

ARTZY, Rafael, mathematician, educator; b. Konigsberg, Germany, July 23, 1912; s. Eduard I. and Ida (Freudenheim) Deutschlander; student Konigsberg U., 1930-33; M.A., Hebrew U., 1934, Ph.D., 1945; m. Elly Iwiansky, Oct. 12, 1934; children—Ehud, Michal, Barak. Came to U. S., 1960. Tchr., principal Israel High Schs., 1934-51; instr., asst. prof. Israel Inst. Tech., 1951-60; research asso., lectr. U. Wis., Madison, 1956-58; asso. prof. U. N.C., Chapel Hill, 1960-61; prof. Rutgers U., New Brunswick, N.J., 1961-65, State U. N.Y., Buffalo, 1965-67, Temple Univ., Philadelphia, Pennsylvania, 1967—; with Inst. for Advanced Study, 1964. Mem. Am. Math. Soc., Math. Assn. Am., Israel Math. Union, Am. Assn. U. Profs. Author: Linear Geometry, 1965. Contbr. research papers in Am., Israeli, German jours., Hebrew Ency. Home: 415 Wischman Av., Oreland, Pa. 19075. Office: Dept. Math., Temple U., Phila. 19122.*

ARUTYUNOV, Vardan Yakovlevich, Russian dermatologist, venerologist; b. 1899; grad. Med. Faculty, Azerbaijan U., 1926; D.Med. Sci., 1939. Intern, Skin and Venereal Diseases Clinic, Azerbaijan Med. Inst., 1926-32; asst. Skin and Venereal Diseases Clinics, 1st and 2d Moscow Med. Inst., 1932-36; asst. Skin and Venereal Diseases Clinic, 4th Moscow Med. Inst., 1936-39, lectr., 1939-41; head Dermatology Clinic, Moscow Oblast Clin. Research Inst., 1945—; head venereal diseases dept. USSR Ministry Health, 1948-50; head dermatology and venerology clinic Moscow Dist. Clin. Research Inst., 1945-66; head dermatology clinic Moscow Research Inst. Cosmetology, 1966—. Mem. Moscow Oblast Soc. Dermatologists and Venerologists (chmn.), Purkinje Czech Med. Soc. (hon.), Czech Dermatol. Soc. (hon.). Author over 150 works including Skin and Venereal Diseases (manual), 1946; Modern Views on the Pathogenesis, Clinical Aspects and Treatment of Psoriasis, 1962; co-author Lesions of the Skin in Reticulosis, 1962, Hemodermia (manual), 1964. Research on pellagra, soft chancre, lupus erythematosus, allergic vasculitis, skin aging. Mem. editorial council jour. Vestnik dermatologii i venerologii. Home: Lenin Prospect 41. Office: Moscow Research Inst. of Cosmetology, Gorky Ulitsa 19A, Moscow, USSR.*

ARUTYUNYAN, Nagush Khachaturovich, Russian mech. engr.; b. 1912; grad. Yerevan Indsl. Technicum, 1930, Moscow Mil. Engring. Acad., 1936; postgrad. Leningrad Poly. Inst., 1936; D. Tech. Sci., 1948. Sr. engr. Bd. for Constrn. Hydropower Plants on Zangu River, 1936-41; asst. Leningrad Poly. Inst., 1936-41, instr., 1945-50; prof. chair mechanics Yerevan U., 1950, rector, 1961—; asso. sect. math. and mechanics Armenian Acad. Sci., 1945—, acad; sec. dept. tech. sci., 1952-56; prof., 1949—. Mem. Armenian Acad. Sci. (v.p. 1961-63). Author: Problems of the Theory of Creep, 1952. Research and publs. on theory of elasticity and gen. theory of creep; devised exact method for solving problems of twisting and bending of prismatic rods with polygonal cross section. Address: Verkhovny Soviet Arm. SSR, Yerevan, Armenia SSR, USSR.*

ARVIA, Alejandro Jorge, Argentinian chemist, phys. chemist; b. La Plata, Argentina, Jan. 14, 1928; s. Alejandro and Adela (Stupenengo) A.; Bachelor's degree, Nat. Coll., U. La Plata, 1951, chemist, 1950, D. degree, 1952; m. Angélica Ormaechea, July 10, 1952. Asst., Instituto Sup. de Investigaciones, U. La Plata, 1955-56, prof. electrochemistry, 1960-65, prof. phys. chemistry, 1963—, head electrochemistry sect. Research Inst., Faculty Chemistry, 1964—; asso. research fellow Northwestern U., Evanston, Ill., 1957; prof. U. Buenos Aires, 1958-59; mem. Researcher Career, NRC, 1962—, mem. adv. bd. chemistry, 1965—. Mem. adv. bd. Electrochimica Acta, 1962—; sec. for Argentina, Internat. Com. Electrochem. Thermodynamics and Kinetics, 1962—. Named one of Ten Outstanding Young Men of Argentina, Jr. C. of C. Buenos Aires, 1965. Mem. Faraday Soc. London, Electrochem. Soc. Japan, Asociación Química Argentina (mem. advisory board 1963—). Research, numerous publs. on electrochemistry, electrochem. kinetics, thermodynamics, galvanic cells with molten salts, gaseous kinetics, spectroscopy; interpretation of infrared spectra. Home: 48 No. 460, La Plata, Argentina.*

ARVIDSON, Johann, see Afzelius, Johann.

ARVY, Lucie, French comparative histoenzymologist; b. Eymoutiers, France, Feb. 11, 1907; s. Léonard and Amélie (Barbas) A.; Docteur in Medicine, Faculty Medicine, Paris, 1939; Docteur ès Scis., Faculty Scis., Paris, 1947. Preparator, Endocrinological Lab., Ecole des Hautes Etudes, 1939-42; researcher Centre National de La Recherche Scientifique Sorbonne, Paris, France, 1943-58, CNRZ, Jouyen-Josas, France, 1958—. Fellow N.Y. Acad. Scis., Histochem. Soc.; mem. Société zoologique de France, Société de Biologie de Paris, Assn. des anatomistes de langue francaise, Soc. d'Hematologie, Internat. Soc. Hematology, Soc. Parasitologie. Author: Les techniques actuelles d'histoenzymologie, 1958; Histoenzymologie des glandes endocrines, 1963; Splénologie, 1965; Les phosphatases du tissu nerveux, 1966; also numerous articles. Research on dimorphism sexuality of blood in mammals, blood in invertebrates, neurosecretion, endocrine glands and their enzymology; discovered leucopoietic organ in Lepidoptera, Diptera.*

ARYABHATA, Hindu mathematician, astronomer; b. Pataliputra, India, 476. Author: Aryabhatiya, 499 (important description of Indian numerical system); writings on rules of arithmetic, plane and spherical trigonometry, solutions of linear and quadratic equations. Worked on series, permutations, arithmetic progressions; gave gen. solution of diophantine equations of 1st degree; wrote on functions of angle, especially sine; gave accurate figure for pi; treated indeterminate equations by application of continued fractions (essentially method used today); one of 1st known to use algebra; maintained theory of diurnal rotation of earth; explained cause of eclipses of sun and moon. Died circa 550.

ARZACHAL (Al-Zarqali, Elzarakeel), Arabic astronomer, mathematician; b. circa 1029; worked in Toledo, Spain, 1080, also Cordova. Invented improved astrolabe; 1st to prove motion of solar apogee with reference to stars; edited Toledan tables; explained constrn. of trigonometrical tables. Died circa 1087.

ARZHANYKH, Ivan Semenovich, Russian mathematician; b. Village of Brusilov (now Kiev Oblast), 1914; grad. Leningrad U., 1935; D.Physico-Math. Sci., 1956. Prof. math. Central Asian U., Tashkent, 1957—. Mem. Uzbek Acad. Sci. (corr.). Author: The Application of Lie Groups to Integral and Differential Equations, 1947; Dynamic Systems of an Order Greater than Zero, 1949; The Integral Parentheses of a System of Common Differential Equations, 1951; Some New Problems in the Integration of Differential Hydrodynamics Equations, 1952; A Method of Integrating Differential Equations with Partial Derivatives of the First Order, 1954; Regular Integral Equations for the Dynamics of an Elastic Body, 1955; The Universal Value of Contact Transformations, 1956; Integral Representation of Solutions for a Field Theory System of Differential Equations, 1957; Equations for the Motion of a Meson in an Electromagnetic Field, 1960; New Stability Inequalities, 1961. Address: Gosudarstvenny University, ulitsa K. Marksa 32, Tashkent, Uzbekistan SSR, USSR.

ARZUMANYAN, Anushavan Agafonovich, Russian economist; b. 1904; grad. Inst. Red Prof. Rector, Yerevan U., 1937; reader, then asst. lectr. Azerbaijan State U., 1946-52; sect. head, then dep. dir. Inst. Econs., USSR Acad. Sci., 1952-56, acad. sec. dept. econ. sci., 1962—, dir. Inst. World Econs. and Internat. Relations, 1956—, mem. editorial bd. World Econs. and Internat. Relations jour. Mem. USSR Acad. Sci., All-Union Soc. for Dissemination Polit. and Sci. Knowledge (dep. chmn.). Author: The Problem of Class Nature and the Method of the Theory of Cost in the British Classical Political Economy, 1940; The Great October Revolution and the Crisis of Capitalism, 1957; The Rivalry of the Two World Systems and the Cairo Conference, 1958; The New Stage in the General Crisis of Capitalism, 1961; The Crisis of World Capitalism in Its Current Stage, 1962; Important Aspects of the Development of the World Economy, 1962; Modern Capitalism and the Class Struggle, 1963. Address: Inst. World Econs. and Internat. Relations, 2-ya Yaroslavskaya ulitsa 3, kerp. 8, Moscow, USSR.

ASADA, Goryu, Japanese astronomer; b. Kitsuki, Bungo Province, Japan, 1734; s. Keisai Ayabe; self-taught in astronomy; practiced medicine; tch. astronomy; formed original views on movements of heavenly bodies and calendar calculation. Died 1799.

ASADA, Tsunesaburo, Japanese scientist; b. Osaka, Japan, July 18, 1900; grad. Tokyo U., 1924; postgrad. studies in Germany; doctorate degree. Research worker Shiomi Phys. and Chem. Research Inst.; apptd. prof. Osaka U., 1933, then prof. Indsl. Sci. Research Inst., Konan U. Mem. Japan Physics Soc. Specialist in optical sci. and electricity; research, writings on microtic analysis of gold bullion and mercury discharge of electricity; expts. on artificial rainfall.

ASAI, Toshinobu, Japanese biochemist; b. Asahigawa, Japan, Mar. 15, 1902; s. Yūki and Taki (Suzuki) A.; grad. dept. agrl. chemistry Tokyo Imperial U., 1925, D.Agr., 1935; m. Toshiko Oikawa, Oct. 30, 1925; 1 dau., Masako (Mrs. Hiuga Saito). Faculty, dept. agrl. chemistry U. Tokyo, 1925-62, prof., 1944-62, dir. Inst. Applied Microbiology, 1957-62; dir. Brewing Sci. Research Inst., Sumidaku, Tokyo, 1962-67; prof. Coll. Agr., Nihon U., Tokyo., 1967—. Mem. adv. com. Internat. Assn. Microbiol. Socs., 1963-66. Recipient Japan Acad. prize, 1957. Mem. Agr. Chem. Soc. Japan (past v.p.). Author: Industry of Microorganisms, 1956; also numerous articles. Classification of acetic acid bacteria and study of its biochem. activities; research on amino acid fermentation. Home: 3-27-10 Miyasaka, Setagayaku, Tokyo, Japan.*

ASAKAWA, Norihiko, Japanese physician, bacteriologist; b. Kochi Prefecture, Japan, 1865; student Tokusuke Nakae, also Saisei Gakusha Sch., Tokyo; joined staff Infectious Disease Research Inst., Home Ministry, 1894. Evolved theory of natural immunity to tetanus through expts. in sera prodn. and serum treatment; studied toxicosis; made discoveries about treatment of erysipelas and typhoid fever. Died 1907.

ASANA, Jehangir Jamasji, Indian biologist, parapsychologist; b. Broach, India, July 21, 1890; B.A., Bombay (India) U., 1911, M.A., 1912; M.A., U. Cambridge (Eng.), 1925; m. Dina Nowroji Bonshah, 1926. Lectr. zoology Gujarat Coll., Ahmedabad, 1913-32, prof. biology, 1923-32, 1940-46; prof. biology Ismail Coll., Andheri, 1933-39. Founding mem. Family Planning Conf., India. Applied biology to study parapsychology; studied mysticism, Christian and Indian philosophy and religion. Died Dec. 16, 1954.

ASANO, Kiyoshi, Japanese geologist; b. Aichi Prefecture, Japan, 1910; grad. Tohoku U., 1935; doctorate degree. Prof. Tohoku U., since 1951; councillor Japan Geologists' Soc., Japan Paleontology Soc. Author: Paleontology; Outline of Geology; also others. Research on geologic evolution, stratification and fossilizations.

ASANO, Osuke, Japanese elec. engr.; b. Okayama Prefecture, Japan, 1858; grad. Engring. Coll.; apptd. 1st dir. Electric Test Sta., Communications Ministry, 1891; prof. Tokyo U., also Waseda (Japan) U. Authority on telegraphy, submarine cable communications; dir. laying of cable between Nagasaki, Japan and Formosa. Died 1940.

ASAPH HA-ROPHEH OF SYRIA, (or Asaf Judaeus), Syrian physician, biologist; flourished 6th or 7th century or possibly 9th; work influenced by Dioscorides, Galen and Hippocrates; author: Sefer Asaph (also known as Sefer Refuoth and Midrash Refuoth), earliest Hebrew med. manuscript extant (contains legendary history of medicine, explains composition of body from 4 elements, also discusses physiology of blood, organs, diseases, food and nutrition, antidotes and prescriptions, embryology, fever, herbs, other med. conditions); also astron. and cosmographic works.

ASBOE-HANSEN, Gustav, Danish physician; b. Haderslev, Denmark, Aug. 12, 1917; s. Jes and Frida (Dose) A.-H.; M.D., U. Copenhagen (Denmark), 1943; m. Thyra Hojberg Nielsen, Apr. 2, 1944; children—Sten, Per. Staff dermatologist Finsen Inst., Rudolph Bergh Hosp., 1949-52, Municipal Hosp., 1952-58, Univ. Hosp., 1958-59 (all Copenhagen); cons. dept. physiol. chemistry U. Minn. Med. Sch., 1958; vis. sr. cancer research internist dept. dermatology Roswell Park Meml. Hosp., Buffalo, 1958; asst. chief Finsen Inst., Copenhagen, 1959-60; prof. dermatology and venerology U. Copenhagen, 1960—, chmn. dept. skin and venereal diseases U. Hosp., 1960—, dir. Connective Tissue Research Labs., 1949—. Vis. prof. various univs., research centers; participant in numerous internat. symposia, confs. Recipient Honor prize Michael Hermann Nielsen Found., 1954, Danish Rheumatism Assn., 1955. Hon. mem. Pacific, Am., Australian, Brit., French, Indian, Iranian, Israeli, Polish dermatol. assns., rheumatologic associations Am., Finland, Turkey, Med. Assn. Sweden and Finland.; corr. mem. Finnish, Norwegian dermatol. assns.; mem. Internat. Soc. Cybernetic Medicine, Am. Soc. Exptl. Biology and Medicine, Am. Soc. Investigative Dermatology, Am. Soc. Geronotology, A.A.A.S., others. Author: On the Mucinous Substances of Connective Tissues, 1951; Connective Tissue in Health and Disease, 1954; Hormones and Connective Tissue, 1966; also numerous articles. Research on structure and functions of connective tissues, mesenchymal diseases, myxedema, various skin diseases including cancer; described Asboe-Hansen sign, 1950, Asboe-Hansen disease, 1953. Home: 18 Skovagervej, Copenhagen-Charlottenlund 2920, Denmark.*

ASCENZI, Antonio, pathologic anatomist; b. Boulogne-sur-Mer, France, May 4, 1915; s. Armando and Adreina (Allevy) A.; M.D.; m. Roberta Graziani, June 14, 1915; children—Paola, Maria-Grazia. Prof. pathologic anatomy U. Pisa (Italy). Mem. Acad. Medicine Rome, Italian Soc. of Pathology. Author: Cicero 111 6 (study of mandible of Neanderthal man), 1950. Research, publs. on relation of structure and ultrastructure of normal bones. Home: 21, via R. Pereira, Rome. Office: Istituto di Anatomia Patologica, via Roma 57, Pisa, Italy.

ASCH, Morris Joseph, Am. otolaryngologist; b. Phila., July 4, 1833; s. Joseph M. and Clara (Ulman) A.; grad. Jefferson Med. Coll., 1852, M.D., 1855. Clin. asst. to Dr. Samuel Gross, Jefferson Med. Coll.; asst. surgeon U. S. Army in Civil War, worked in surgeon gen.'s office, 1861-62; surgeon-in-chief to arty. reserve of Army of Potomac; med. insp. Army of Potomac, med. dir. 24th Army Corps, med. insp. Army of the James and staff surgeon to Gen. Philip H. Sheridan, 1865-73, served at battles of Chancellorsville, Mine Run, Gettysburg and Appomatox Ct. House; while on Sheridan's staff active in cholera epidemic of 1866 and yellow fever epidemic of 1867; with Sheridan in Chgo. until 1873; practiced medicine, N.Y.C., specialized in laryngology; a founder Am. Laryngol. Assn.; surgeon to throat dept. N.Y. Eye and Ear Infirmary, Manhattan Eye and Ear Hosp. Devised Asch operation for correction of deviated septum through cruciate incision over the deviation, 1890. Died Oct. 5, 1902.

ASCHAFFENBURG, Gustab, psychiatrist; b. Zweibrücken, Germany, May 23, 1866; s. Louis and Julie (Feibes) A.; m. Maja Nebel, 1901; 1 son, 3 daus. Worked in Vienna; became asso. prof., Heidelberg, Germany; prof. psychiatry, Cologne, Germany, 1903-34; dir. Asylum Lindenthal; pub. Monatsschrift für Kriminalpsychologie und Strafrechtreform, from 1904, Handbuch der Psychiatrie, 1911-23. Author: Das Verbrechen und seine Bekämpfung, 1903; Gefängnis oder Irrenanstalt, 1908; Die Sicherung der Gesellschaft gegen gemeingefährliche Geisteskranke, 1912; Lokalisierte und allegm. Ausfallserscheinungen nach Hirnverletzungen, 1917; Psychiatrie und Strafrecht, 1928. Worked in forensic psychiatry and sociol. aspects psychiatry; compiled library on criminal behavior. Died Balt., 1944.

ASCHAN, Adolf Ossian, Finnish chemist; b. Helsinki, Finland, May 16, 1860; Ph.D.; postgrad. studies in Finland, Germany, Sweden; s. Carl Achates; m. Elin Enquist, 1885. Became lectr., then asso. prof. chemistry U. Helsinki, prof., 1908-27; founder Apothecary Union in Finland. Hon. mem. Tech. Assn. Finland, Chem. Soc. Sci. (Stockholm, Sweden); mem. Swedish, Norwegian, Danish sci. acads., Soc. Sci. (Gottingen, Germany), Estonian Sci. Acad. in Finland. Editor (with others) books on chemistry, including Roscoe Schorlemmer's organic chemistry, 1895-1909. Studied stereochemistry, other aspects alicyclic compounds, including allyl formate, turpentine, camphor, naphthenes, napthenic acids. Died Feb. 25, 1939.

ASCHERSON, Paul Friedrich August, German botanist; b. Berlin, Germany, June 4, 1834; s. Ferdinand Moritz and Henriette Ferdinandine Auguste (Odenheimer) A.; M.D., Berlin, 1855; Ph.D. (hon.), Rostock, Germany. Asst., Bot. Garden, Berlin, 1860-76; prof. botany and plant geography U. Berlin, from 1863; apptd. custodian Bot. Mus., Berlin, 1884. Author: Flora der Provinz Brandenburg, 1864; Illustration de la flore d'Egypte, 1887; (with P. Graebner) Synopsis der mitteleuropaischen Flora, 7 vols. A leading 19th century German botanist; made bot. expdns. to Mediterranean; mem. Rohlfs' expdn. to Libyan desert, 1873-44, visited Tothmas desert, Egypt, 1887 (bot. discoveries recorded in Rohlfs' Reise von Tripolis nach der Oase Kufra, 1881). Died Berlin, Mar. 6, 1913.

ASCHHEIM, Selmar, German gynecologist, biochemist; b. Berlin, Germany, Oct. 4, 1878. Became privatdocent bio'l research in gynecology U. Berlin, prof., from 1931. Research (with B. Zondek) includes discovery of prolans A and B in human pregnancy urine, 1927, devel. of test for pregnancy (Aschheim-Zondek test), 1928, and demonstration that anterior lobe implants in immature rodents produced premature sexual devel.

ASCHKENASY, Alexandre, physician, biologist; b. Warsaw, Poland, Nov. 6, 1908; s. Maximilien and Marie (Jablonska) A.; student medicine Paris, Montpellier, 1927-34; M.D., U. Paris (France), 1939, Sc.D., 1954; m. Paule Lelu, Oct. 7, 1944. Asst., Hosp. Hôtel-Dieu, Paris, 1948-55; chief Lab., Hosp. Pitié, Paris, 1955-64; master research Nat. Center Sci. Research, 1954-61, dir. research, 1961—; prof. Coll. Medicine, Hosps. Paris, 1964—. Recipient Silver medal Nat. Center Sci. Research, 1964; Vermillion medal Nat. Acad. Medicine, 1964. Mem. French Soc. Biology, Internat. a. fr. Soc. Hematology, French Soc. Endocrinology, French Soc. Biol. Chemistry, French Assn. Physiologists, others. Author: Red Cell Diseases and Endocrine Disorders, 1952; also numerous articles, chpts. in books. Research on hormonal regulation of blood cell prodn. and influence of dietary proteins and amino acids on hemopoiesis and endocrine glands. Home: 10, rue Oudinot, Paris 7, France. Office: C.N.R.S., 74 rue Gabriel Péri, Montrouge 92, France.*

ASCHOFF, (Karl Albert) Ludwig, German pathologist; b. Berlin, Jan. 10, 1866; s. Ludwig and Blanka (Heinze) A.; ed. Bonn, Germany, also Strasbourg, France, Göttingen, Germany; degree, 1889; m. Clara Dieterichs, 1895; 2 sons, 2 daus. Asst. lectr. Gottingen; apptd. prof. path. anatomy, Marburg, Germany 1903, Freiburg, Germany, 1906-36; pub. Pathologische Anatomie, 1908. Author: Handbuch der ärztlichen Erfahrungen im Weltkriege, 1918; Über den Engpass der Magens, 1918; Vorträge über Pathologie (lectures in Japan 1924), 1925. Studied types and etiology of appendicitis; investigated cholecystitis, problems of icterus, thrombosis, Tb; described inflammatory nodule in heart muscle characteristic of rheumatic process (Aschoff's bodies) 1904; introduced term reticuloendothelial system, 1924. Died Freiburg, June 24, 1942.

ASCLEPIADES OF BITHYNIA, (or Asclepiades of Prusa), physician; b. ca. 130 B.C.; studied philosophy and medicine in Greece, perhaps in Alexandria; practiced medicine at Parion on the Hellespont, later at Athens, at Rome after 91 B.C. Author about 20 works (see fragments in C. G. Gumpert's Asclepiadis Bithyni fragmenta, 1798). Believed to be 1st to mention tracheotomy; introduced atomic theory of Democritos and Epicurus into medicine; opposed theory of humors and of healing power of nature; explained health and disease as unhindered or inhibited movement of bodily corpuscles; used dietetic therapy; early prophet of Methodists (med. sect); credited with inventing shower baths and swinging beds; divided diseases into acute and chronic; pioneer in humane treatment of mental patients, also in geriatrics. Died Rome, circa 40 B.C.

ASCLEPIODOTOS (or Alexandrinos, Askepiodotos), physician, mathematician; b. Alexandria, circa 470; disciple of Proclus; determined primary colors and nuances of secondary colors; studied properties of plants and animals; distinguished 500 species of trees; used white Hellebore as med. remedy; interested in magic; known as a thaumaturgist. Died circa 500.

ASCLEPIOS (Asklepios; Latin Aesculapius), Greek physician, Greek god of medicine; probably from Thessaly; ca. 1300 B.C. Mentioned by Homer in the Iliad as a skillful physician and as father of Podaleirius and Machaon, physicians to Greeks in Trojan War; later Asclepios became a hero and was considered a god; was described as a son of Apollo and the nymph Coronis; the legend was that the centaur Chiron taught him the art of healing; Asclepios angered Zeus by bringing Hippolytos, son of Theseos, King of Athens, back to life and was killed by a thunderbolt; Asclepios was worshipped in many parts of Greece as the god of healing; temples were built in his honor where patients slept and waited for the god to effect cures or prescribe remedies to them in their dreams; the staff and single serpent coiled around it were Asclepios' chief symbols (used today to signify med. profession); Asclepios was often accompanied by his daus. Hygieia, goddess of health, Panaceia and Isao (personifications of health); the cult of Asclepios was brought to Rome from Epidauros (the site of his most famous shrine); in 293 B.C. by order of the Sibylline books, or in hope of ridding the city of a pestilence, a temple was built for Asclepios on an island in the Tiber.

ASDELL, Sydney Arthur, physiologist; born Bramhall, Cheshire, Eng., Aug. 23, 1897; s. John and Amy (Love) A.; B.A., Cambridge U., 1922, M.A., 1925, Ph.D., 1926; m. Muriel Tregarthen Marrack, Aug. 2, 1923; children—Philip Tregarthen, Mary Kathleen. Came to U.S., 1930, naturalized, 1942. Faculty, Cornell U., Ithaca, N.Y., 1930—, emeritus prof. animal physiology, 1965—. Cons. U. S. Dept. Agr., 1940-55; Fulbright prof. Denmark, 1952-53. Fellow Am. Assn. for Advancement of Science; member of the Am. Physiol. Soc., Am. Assn. Anatomists, Am. Soc. Vet. Physiologists and Pathologists, Sigma Xi. Author: Patterns of Mammalian Reproduction, 1946; Cattle Fertility and Sterility, 1955; Dog Breeding, Reproduction and Genetics, 1966. Research, publs. on mammalian reprodn., growth, lactation and genetics. Home: 27 Renwick Heights Rd., Ithaca, N.Y. 14850.*

ASEEV, Dmitriy Dmitrievich, Russian physiologist; b. Ryazan, 1900; grad. Med. Faculty, 1st Moscow U., 1925; D.Med. Sci. Dist. physician rural areas of Ryazan Oblast, 1925-27; intern Ryazan Tb Out-Patients Clinic, also in sanatoria of Ryazan Health Dept., 1928-34; asso., head lung surgery dept., then learned sec Tb Research Inst., RSFSR Ministry Health, Moscow, 1935-50, chmn. Tb Commn., Learned Council, head diagnosis dept. and dep. dir. studies Tb Research Inst., 1951—; lectr. postgrad. courses Central Postgrad. Med. Inst. and Sechenov 1st Moscow Med. Inst., 1937—. Mem. Sci. Commn., Internat. Union for Combatting Tb, 1959—; del. 15th Internat. Conf. for Combatting Tb, Istanbul, 1959. Mem. All-Union (bd. mem.), All-Russian (chmn.) socs. phthisiologists. Author over 80 works including The Place of Therapeutic Pneumothorax in the Treatment of Persons Suffering from Destructive Forms of Pulmonary Tuberculosis, 1963. Dep. editor jour. Problemy tuberkuleza; editor Trudy Moskovskogo nauchno-issledovatelskogo instituta tuberkuleza Ministerstva zdravookhrasniya

RSFSR. One of 1st in USSR to use and approve Soviet gold preparation krizanol; described and developed original methods of investigation, cavernography, endoscopy of caverns, radiomorphological comparison, others. Address: RSFSR Ministry Health, Tb Research Inst., n. Bozhedomka 4, Moscow, USSR.

ASELLIUS, Gaspar (or Gasparo Aselli, Gaspari Asellio), Italian anatomist; b. Cremona, Italy, 1581; became prof. anatomy and surgery, Pavia, Italy; practiced medicine, Milan, Italy. Author: De lactibus seu lacteis veinis, 1627. Discovered lacteal vessels, also group of lymphatic nodes at root of mesentery nr. pancreas (Aselli's glands, Aselli's pancreas). Died Milan, Apr. 24, 1626.

ASENJO, Conrado Federico, Puerto Rican biochemist; b. San Juan, P.R., Dec. 6, 1908; s. Conrado and Mercedes (Diaz) A.; Chem. E., Rensselaer Poly. Inst., 1933, M.S., U. Wis., 1938, Ph.D., 1940; postgrad. U. Chgo.; m. Angeles Mayoral, Mar. 24, 1945; children—Miriam, Maria, Maureen, Jorge, Juan, Conrado. Faculty, Sch. Tropical Medicine, U. P.R., San Juan, 1933—, prof., head dept. biochemistry and nutrition, 1952—, asst. dean, 1959—, cons. agrl. Expt. Sta., 1941—. Cons. VA Hosp., San Juan, 1956—. Guggenheim fellow, 1937-39. Mem. Latin Am. Soc. Nutrition (pres. 1965-66), Sigma Xi, Phi Lambda Upsilon. Research on biologically active plant constituents, nutritional evaluation of proteins, nutritional liver necrosis, fat absorption in sprue, folic acid requirements, vitamins in tropical foods, nutrition surveys. Home: Duarte 120, Hato Rey, P.R. 00917. Office: Sch. Medicine, Univ. P.R., San Juan, P.R. 00905.*

ASERINSKY, Eugene, Am. physiologist; b. N.Y.C., May 6, 1921; s. Boris and Sonia (Hectin) A.; Ph.D., U. Chgo., 1953; m. Sylvia Simon, May 23, 1942; children—Armond, Jill; m. 2d Rita Goldman, Jan. 15, 1959. USPHS fellow U. Chgo., 1951-53; research asso. U. Wash., 1952-53; faculty Jefferson Med. Coll., Phila., 1953—, prof. physiology, 1966—; vis. research scientist Eastern Pa. Psychiat. Inst., 1962—. Mem. Am. Physiol. Soc., Assn. Psychophysiology Study of Sleep, N.Y. Acad. Scis., Am. Astronautical Soc., Am. Med. Writers Assn. Discovered occurrence of rapid eye movements in sleep of normal human subjects; provided first objective method of studying neural function and behavioral patterns asso. with dreaming. Home: 1846 Howe Lane, Maple Glen, Pa. 19002. Office: 1025 Walnut St., Phila. 19107.*

ASH, Edwin Lancelot Hopewell, Brit. neurologist; b. Jan. 16, 1881; M.D., B.S., Univ. Coll., London; m. Mabel Streeter, 1907; 3 sons; m. 2d, Christine (Chryssoula) Diamandis, 1952. Practiced medicine specializing in neurology, Harley St., for 35 years. Recipient Cheadle Gold medal St. Mary's Hosp. Mem. Royal Coll. Surgeons Eng., Brit. Med. Assn. (chmn. Westminster and Holborn div. 1938-39). Author: Mind and Health, 1910; Nerves and the Nervous, 1911; The Problem of Nervous Breakdown, 1919; Melancholia in Everyday Practice, 1934. Died Aug. 7, 1964.

ASH, John, English physician; b. Warwickshire, Eng., 1723; B.A., Trinity Coll., Oxford, 1743, M.A., 1746, M.B., 1750, M.D., 1754. Practiced medicine, Birmingham, Eng.; helped found Birmingham Hosp., became its 1st physician. Censor, U. London, 1789, 93; Harveian orator, 1790; Gulstonian lectr., 1791; Croonian lectr., 1793. Fellow Coll. Physicians, Royal Soc., 1787; mem. Eumelian Club (founder). Author: Experiments and Observations to investigate by Chemical Analysis the Properties of the Mineral Waters of Spa, Aix, etc, 1788. Noted for skill as physician. Died London, June 18, 1793.

ASH, John Sidney, English biologist; b. Gosforth, Eng., May 26, 1925; s. Sidney and Kathleen (Denley) A.; B.Sc., Durham (Eng.) U., 1945; D.I.C., Ph.D., Royal Coll. Sci., 1951; m. Jonquil Wilsford-Gudgeon, Nov. 21, 1951; 1 dau., Caroline Penelope-Jane. Game Biologist Imperial Chems. Industries Game Research Sta., 1951-60; head scientist Game Research Assn., Fordingbridge, Eng., 1960-65, research dir., 1965—. Mem. council and various coms. Brit. Trust for Ornithology and Brit. Ornithologists Union. Research, and numerous publs. on biology and ecology of bird ectoparasites, population dynamics of game birds, particularly partridges, pesticide/wildlife problems; population and bird-migration studies, especially in desert regions. Home: Godshill Wood, Fordingbridge. Office: Game Research Assn., Fordingbridge, Hampshire, Eng.*

ASHBURN, Percy Moreau, Am. mil. surgeon; b. Batavia, O., July 28, 1872; s. Allen W. (M.D.) and Julia M. (Kennedy) A.; M.D., Jefferson Med. Coll., 1893; m. Agnes Davis, July 6, 1896; children—Allen D. (dec.), Frank D., Ann Virginia. Apptd. contract surgeon U.S.A., 1898; commd. 1st lt. asst. surgeon, 1898, advanced through grades to col. M.C., 1917; served in Philippine Campaign, 1899; pres. army bd. for study of tropical diseases in P.I., 1906-07, 13; attached to U.S. Commn. to Republic of Liberia, 1909; gen. insp. Health Dept., Panama Canal, 1914-15; comdr. Med. Officers' Tng. Camp, Ft. Benjamin Harrison, Ind., 1917; on duty in A.E.F., 1918-19; comdt. Med. Field Service Sch., 1920-23; prof. mil. hygiene, U. S. Mil. Acad., 1923-27; librarian Army Med. Library, Washington, 1927-32, ret.; became supt. Columbia Hosp. for Women, 1934. Author: The Elements of Military Hygiene, 1909; History of the Medical Department of the United States Army, 1929. Demonstrated (with Charles Franklin Craig) that organisms responsible for dengue are a filtrable virus, 1907. Died Aug. 20, 1940.

ASHBY, Arthur Wilfred, agrl. economist; b. Aug. 19, 1886; s. Joseph Ashby; diploma with honors Oxford (Eng.) U., 1911, M.A. (hon.), 1923, M.A., 1946; postgrad. (hon. fellow polit. economy 1915) U. Wis.; m. Rhoda Dean; 1 son. Ministry Agr. research scholar in agrl. economics, 1912-15; with Food Prodn. Dept., 1917-18; sr. research asst. Agrl. Economics Research Inst., Oxford, 1920-24; apptd. adv. lectr. in agrl. economics Univ. Coll. Wales, Aberystwyth, 1924, prof., 1929-45; dir., inst. for research in agrl. economics Oxford U., 1946-52. Mem. Royal Commn. Agr., 1919; mem. Departmental Com. on Prices Farm Produce (Linlithgow), 1923-24; apptd. mem. Nat. Agrl. Wages Bd., 1924; mem. Council Agr. for Eng., 1920-46, chmn., 1939-40; mem. standing com., 1924-46; mem. adv. com. Ministry Agr., from 1924; mem. Council Agr. for Wales, 1927-46, chmn., 1944-45. Mem. Agl. Economics Soc. (pres. 1934-35, 52-53), Internat. Conf. Agrl. Economists (v.p. 1949-52), Royal Swedish Acad. Agr., Sci. Agrl. Soc. Finland. Author: One Hundred Years of Poor Law Administration; Oxford Studies in Social and Legal History (edited by P. Vinogradoff), 1912; Allotments and Small Holdings in Oxfordshire, 1917; (with P. G. Byles) Rural Education, 1923; (with I. L. Evans) The Agriculture of Wales, 1944. Contbd. papers to sci. jours. and reviews. Died Sept. 9, 1953.

ASHBY, Winifred Mayer, med. biologist; born London, Eng., Oct. 13, 1879; d. George Mayer and Mary-Ann (Brock) A.; B.Sc., U. Chgo., 1903; M.S., Washington U., St. Louis, 1905; Ph.D., U. Minn., 1921. Came to U. S., naturalized, 1892. Fellow immunology Mayo Found., 1917-24; med. bacteriologist, health officer St. Elizabeth Hosp., Washington, 1925-48, vol. researcher, 1948-58. Mem. Am. Assn. Immunologists, Am. Soc. Bacteriologists, Soc. Biol. Psychiatry, Am. Soc. Pub. Health, Soc. for Exptl. Biology and Medicine, Mayo Found. Contbg. author: George R. Minor Symposium of Hematology, 1949; Carbon Dioxide Therapy, 1958; also numerous articles. Pioneered technique for studying life tranfused blood cells; demonstrated that essential factor in pernicious anemia was not blood destruction; research on relationship between quantitative distbn. enzyme and functioning brain. Address: 5735 Mallow Trail, Lorton, Va. 22079.*

ASHCROFT, Michael Tansley, epidemiologist; b. Hoylake, Eng., Aug. 6, 1925; s. Charles Stuart and Georgina (Rose) A.; B.A. with 1st class honors, Oxford, Eng., 1946, M.A., 1950, B.M.B.CH., 1949, D.M., 1963; D.T.M.H., London (Eng.) U., 1955, D.P.H., 1959. House physician Radcliffe Infirmary, Oxford, 1950; med. research officer E. African Trypanosomiasis Research Orgn., Tanzania, Uganda, 1954-58; dir. Brit. Guiana Typhoid Research Unit, 1959-62; dep. dir. Med. Research Council Epidemiological Research Unit, U. Coll. W.I., Kingston, Jamaica, 1963—. Research, publs. on epidemiology of sleeping sickness in Africa, demonstration of efficacy of typhoid vaccines in Brit. Guiana, child growth and devel. in W.I. Address: Med. Research Council, Epidemiological Research Unit, U. W.I., Kingston 7, Jamaica.*

ASHE, William Francis, Am. physician; b. Braddock, Pa., Dec. 14, 1909; s. William F. and Catherine (Euwer) A.; A.B., Oberlin Coll., 1932; M.D., Western Res. U., 1936; m. Kathleen Terry Little, Dec. 24, 1945; children—Carl F. and Lynn. Practice medicine, specializing in internal medicine, Cin., 1940-42, 46-50; instr. internal medicine U. Cin., 1940-42, staff Kettering Lab., 1940-42, asso prof. indsl. medicine, 1946-50; chief med. service Holzer Hosp., Gallipolis, O., 1950-54; prof., chmn. dept. preventive medicine Ohio State U., Columbus, 1954—. Cons. on thermal environmental problems in industry Govt. India, 1954-56; cons. VA, Dayton, O., 1947—, Interdepartmental Com. on Nutrition Nat. Def.; Spain, 1958, Chile, 1960, Venezuela, 1963. Recipient Bernard O'Higgins award Chile, 1960. Fellow A.C.P., Am. Coll. Preventive Medicine (v.p. 1965—), A.A.A.S., Am. Acad. Occupational Medicine (pres. 1964), Aerospace Med. Assn.; mem. A.M.A., Am. Indsl. Hygiene Assn., Am. Pub. Health Assn. Research in internal medicine, indsl. toxicology, environmental physiology and nutrition. Died Feb. 26, 1966.

ASHERSON, Geoffrey Lister, English immunologist; b. London, Eng., Oct. 21, 1929; s. Rehemiah and Isabelle (Barnet) A.; ed. Univ. Coll. Hosp., London; M.A., D.M., U. Oxford; m. Miss Armstrong, Apr. 6, 1952; children—Claral, Richard, Philip. Commonwealth scholar Walker and Eliza Hall Inst. Med. Research, Melbourne, Australia; mem. sci. staff Inst. Med. Research, Mill Hill; now reader immunology U. London. Fellow or mem. Royal Soc. Medicine (hon. sec. sect. pathology, mem. council, library rep. sect. clin. immunology and allergy, mem. council sect. comparative medicine), Royal Instn. Gt. Britain (bd. visitors); mem. royal colls. physicians London, Edinburgh. Contbg. author: Systemic Lupus Erythematosis and Other Diseases. Contbr. articles to med. jours. Research on antibodies against nuclear and cytoplasmic cell constituents. Address: 21, Harley St., London W. 1, Eng.*

ASHFORD, Bailey Kelly, Am. physician; b. Washington, Sept. 18, 1873; s. Francis Asbury and Isabella Walker (Kelly) A.; student Columbian (now George Washington) U., 1 yr., M.D., Georgetown U., 1896, Sc.D., 1911; grad. Army Med. Sch., 1898; Sc.D., Columbia Univ., 1933, U. of P.R., 1933; hon. M.D., U. of Eyypt, 1932; m. Maria Asuncion Lopez, June 24, 1899; children—Mahlon, Gloria Maria, Margarita. Resident physician Children's Hosp., Washington, 1895-96; apptd. 1st lt. U. S. Army, 1897; advanced through grades to col., 1917; served with mil. expdn. to P.R., July 1898, and in Battle of Hormigueros, 1898; div. surgeon 1st Div., 1917; in charge battle tng. of med. officers. Zone of Armies, A.E.F., 1917-18. In 1899 determined cause of the anemia of agrl. class of Puerto Rico, later popularized as "hookworm" disease"; founded P.R. Anemia Commn., which began first campaign against disease in Western Hemisphere, 1904. Del. from U. S. to Internat Cong. Indsl. and Alimentary Hygiene, Brussels, 1910; mem. med commn. to Brazil, Rockefeller Found., 1916; U. S. del. to Internat. Cong. of Tropical Medicine and Hygiene, Cairo, 1928. Prof. tropical medicine and mycology Columbia U., collaborating with Sch. of Tropical Medicine (Puerto Rico). Hon. mem. and pres. Am. Soc. Tropical Medicine, Puerto Rico Med. Assn.; fellow A.C.P., A.C.S. Decorated D.S.M. Author: Anemia in Porto Rico, 1904; Uncinariasis in Porto Rico (with Gutierrez), 1911; The Organization and Administration of the Medical Department, in the Zone of the Armies (Keen's Surgery, Vol. VII); Sprue (Tice's Loose-leaf Medicine), 1931; A Soldier in Science, 1934. Died Nov. 1, 1934.

ASHFORD, Theodore Askounes, chemist; b. Greece, 1908; s. Nicholas Askounes and Catheryn (Togias) A.; B.S., U. Chgo., 1932, M.S., 1934, Ph.D., 1936; m. Venette Tomaras, 1933; children—Nicholas A., Theodore Askounes, Harold-James A. Came to U. S., 1922, naturalized, 1930. Asst. prof., U. Chgo., 1936-50; asso. prof., prof. U. St. Louis, 1950-60; prof. chemistry, dir. natural scis. and math. div., U. S. Fla., Tampa, 1960—, asso. dean liberal arts, 1965—. Sec., Fla. Bd. Examiners in Basic Scis. Fellow A.A.A.S.; mem. Am. Chem. Assn. (examinations com. 1940—, award in chem. edn.), Assn. for Higher Edn., Am. Assn. U. Profs., Fla. Acad. Scis., Sigma Xi, Sigma Pi Sigma. Author: From Atoms to Stars, 1960; The Physical Sciences, 1967. Editor: ACS-Examinations. Research on organometallic compounds. Home: 1832 Bearss Av., Tampa, Fla. 33612.*

ASHHURST, John, Jr., Am. surgeon; b. Phila., Aug. 23, 1839; B.A., U. Pa., 1857, A.M., M.D., 1860; LL.D., Lafayette, 1895; m. Sarah Stokes Wayne, Dec. 8, 1864, 7 children. Acting asst. surgeon U. S. Army, 1862-65; surgeon to numerous Phila. hosps.; prof. clin. surgery U. Pa., from 1877, Barton prof. surgery, from 1888; mgr. Episcopal Hosp., from 1880; v.p. surg. sect. Internat. Med. Cong., Phila., 1876. Mem. Am. Surg. Assn. (v.p. 1896). Fellow Coll. Phys., Phila. (pres. from 1898). Author: Injuries of the Spine, 1867; Principles and Practice of Surgery, 1871; De La Laparotomie, ou Section Abdominale, Comme Moyen de Traitement de L'Intussusception, translated by Dr. Lutaud, Paris, 1875. Editor: Internat. Ency. Surgery, 6 vols., 1881-86. Wrote nearly all of surg. revs. in Am. Jour. Med. Scis., 1867-77; known as a historian of surgery and surgeons. Died Phila., July 7, 1900.

ASHKIN, Julius, Am. physicist; b. Bklyn., Aug. 23, 1920; s. Isadore and Anna (Fishman) A.; A.B., Columbia, 1940, A.M., 1941, Ph.D., 1943; m. Claire Ruderman, Sept. 1, 1946; children—Beth, Laura. Research asso. Metall. Lab., Chgo., 1942-43; mem. staff Los Alamos (N.M.) Sci. Lab., 1943-46; asst. prof. physics U. Rochester (N.Y.), 1946-50; faculty physics Carnegie Inst. Tech., Pitts., 1950—, prof., 1958—, chmn. dept. physics, 1961—. Mem. Am. Phys. Soc., A.A.A.S., Sigma Xi. Research on physics of elementary particles. Home: 5136 Beeler St., Pitts. 15217. Office: Dept. Physics, Carnegie Inst. Tech., Pitts. 15213.*

ASHLEY, George Hall, Am. geologist; b. Rochester, N.Y., Aug. 9, 1866; s. Roscoe B. and Anna (Hall) A.; M.E., Cornell U., 1890, A.M., 1892; Ph.D., Stanford, 1894; Sc.D., Lehigh U., 1937; m. Mary E. Martin, July 11, 1895; children—Carlyle, Jr., Dorothy (Mrs. R. H. Ross), Paleontologist, Rochester, 1889-91; asst. geologist Geol. Survey of Ark., 1891-93; taught in Cal., 1894-96; asst. state geologist of Ind., 1896-1900; prof. biology and geology and curator of museum Coll. of Charleston (S.C.), 1900-03; prof. pharmacognosy, Med. Coll. State of S.C., 1901-03; asst. geologist U. S. Geol. Survey, 1901-05, geologist 1905-12, adminstrv. geologist, 1912-19; state geologist of Pa., 1919-46; consulting mining geologist from 1946. Acting prof. geology, Vanderbilt U., 1917. Fellow A.A.A.S., Geol. Soc. of Am., Pa. Acad. Scis., Am. Inst. Mining and Metall. Engrs.; mem. Coal Mining Inst. America, Soc. Econ. Geol. (pres. 1948). Author numerous geol. reports and articles in lit. and tech. jours. Died May 28, 1951.

ASHLEY, Laurence Marvin, Am. biologist; b. Battle Creek, Mich., June 14, 1908; s. Roy Vernon and Edith Leona (Marvin) A.; B.S., Mich. State. U., 1930; M.A., U. Mich., 1932; Ph.D., U. So. Cal., 1945; m. Ruth Louise Belknap, Dec. 21, 1931; children—Elizabeth Ann (Mrs. Bob Myers), Robert Burton. Fellow, Inst. for Fishery Research, U. Mich., 1930-33;

asso. prof. biology Columbia Union Coll., Takoma Park, Md., 1933-40; instr. anatomy Loma Linda (Cal.) U. Med. Sch., 1940-46, asst. prof., 1946-47; chairman sci. department, professor zoology Columbia Union College, 1947-49; prof. zoology Walla Walla (Wash.) College, 1949-60; fishery research biologist, histopathologist Western Fish Nutrition Lab., Cook, Wash., 1960—. Mem. Am. Fisheries Soc., Am. Inst. Biol. Soc., Wildlife Disease Soc., Am. Assn. Anatomists. Author: Laboratory Anatomy of the Shark, 1950; Laboratory Anatomy of the Turtle, 1954; also articles. Research on human heredity, microscopic anatomy, comparative anatomy, rainbow trout liver cancer, exptl. induction cancers with chem. carcinogens. Home: Star Route. Office: Western Fish Nutrition Lab., Star Route, Cook, Wash. 98605.*

ASHMEAD, William Harris, Am. entomologist; b. Phila., Sept. 19, 1855; s. Albert S. and Elizabeth (Graham) A.; A.M., Fla. Agrl. Coll., 1901; D.Sc., Western U. Pa., 1901; m. Harriet Holmes, 1878. Was with J. B. Lippincott Co., Phila.; with brother became pub. agrl. books. agrl. weekly, and daily, Jacksonville, Fla., 1876; edited scientific dept. of weekly, devoting self chiefly to investigation of injurious insects; spl. field entomologist U. S., Dept. Agr., 1887, asst. entomologist and investigator, from 1889; entomologist State Agrl. Coll., Lake City, Fla., 1888; spl. studies in Berlin, winter 1889-90; asst. curator, div. insects U. S. Nat. Mus., from 1897. Author: Orange Insects (treatise on beneficial and injurious insects in Fla.), 1880; Monograph of the North American Proctotryphidae, 1893; Classification of the Chalcid Flies or the Super-family Chalcidoidea, 1904; also numerous articles in jours. Described 607 new genera, 3100 new species; studied chiefly the hymenobtera and rhynchota, especially species affecting orange culture. 250 contributions to jours; Died 1908.

ASHMOLE, Elias, English alchemist, historian; b. Lichfield, Eng., May 23, 1617; M.D., Oxford U., 1690; m. Eleanor Mainwaring, 1638; m. 2d, Lady Mainwaring, Nov. 16, 1649; m. 3d, Miss Dugdale, 1668. Became solicitor under patronage Baron Pagitt, 1638; apptd. commr. excise, Lichfield, 1644; commd. (through George Wharton) in Ordinance; commr. of excise at Worcester, 1645-46. Fellow Royal Soc., 1663. Editor: Arthur Dee's Arcanum, or the great secret of the hermetical philosophy, 1650; Theatrum Chemicum Britannicum (collection medieval alchemical poetry in English), 1652; Author: Institutions, Laws and Ceremonies of the Order of the Garter, 1672. Bequeathed Ashmolean Mus. and valuable collection of manuscripts to Oxford U. Died London, May 18, 1692.

ASHTON, Floyd Milton, Am. plant physiologist; b. Indpls., Jan. 27, 1922; s. Floyd F. and Florence (Stark) A.; B.S., U. Ill., 1947; Ph.D., Ohio State U., 1955; m. Phyllis M. Randall, Jan. 24, 1942; 1 son, Mark M. Asst. biochemist Expt. Sta., Hawaiian Sugar Planters Assn., Honolulu, 1947-56; asso. prof. botany U. Cal. at Davis, 1956—. Chmn. research com. Western Weed Control Conf., 1959-61. USPHS Spl. fellow, 1962-63. Mem. Weed Soc. Am. (chmn. policy and site com. 1965—), Am. Soc. Plant Physiologists, Scandinavian Soc. Plant Physiologists, Sigma Xi, Phi Epsilon Phi. Research, numerous publs. on action herbicides, proteolytic enzymes in germinating seeds, effect moisture stress on photosynthesis, weed control. Home: 602 12th St., Davis, Cal. 95616.*

ASHTON, Geoffrey Cyril, geneticist; b. Croydon, Eng., July 5, 1925; s. Cyril Hanniss and Ethel (Pate) A.; B.Sc. with honors, Liverpool U., 1945, Ph.D., 1958, D.Sc., 1967; m. Kathleen J. Stanley, February 25, 1951; children—Carolyn Joy, Kathryn Alison, Melinda Jane, Jonathan Geoffrey. Research assistant Univ. of Toronto (Ontario), Canada, 1948-50; sect. leader Glaxo Labs., Eng., 1951-56; sr. sci. officer Farm Livestock Research Centre, Eng., 1956-58; prin. research officer Commonwealth Sci. and Indsl. Research Orgn., Australia, 1958-64; prof. genetics U. Hawaii, Honolulu, 1964—, chmn. dept., 1965—. Mem. blood group scientists panel FAO, 1963—, chmn. subcom. on protein polymorphism nomenclature, 1963—. Mem. Genetics Soc. Am., Am. Soc. for Human Genetics, Am. Soc. Naturalists. Initial discoveries of several protein and enzyme genetic polymorphisms in several species of domesticated animals. Home: 5414 Kirkwood Pl., Honolulu 96821.*

ASHTON, Norman Henry, English pathologist; b. London, Eng., Sept. 11, 1913; s. Henry James and Margaret Ann (Tuck) A.; grad. King's Coll., London, 1934, Westminster Hosp. Med. Sch., London, 1939; D.Sc., U. London, 1960. Pathologist, Westminster Hosp., 1947-48; dir. dept. pathology Inst. Ophthalmology, U. London, 1948-53, reader, 1953-57, prof. pathology, 1957—; vis. prof. ophthalmology Wilmer Inst., Johns Hopkins Hosp. Balt., 1959; cons. pathologist Moorfields Eye Hosp., 1948—. Recipient Nettleship Gold medal for sci. work in ophthalmology. 1955; Middlemore prize, 1955; Proctor Gold medal, 1957; Doyne medal, 1960; Bowman medal, 1965; Donders medal, Holland, 1967; Mackenzie medal, 1967; William Julius Mickle fellow U. London, 1961. Fellow Coll. Pathologists (founder, past council mem.); mem. American Academy of Ophthalmologists and Otolaryngologists (honorary), also European Ophthalmic Pathology Soc. (founding, life pres.), Internat. Acad. Pathology (past pres. Brit. div.), Diabetic Assn. Gt. Britain (hon. life), Assn. Clin. Pathologists (past mem. council), Path. Soc. Gt. Britain and Ireland, Ophthal. Soc. U.K., Order St. John (comdr.), Royal Coll. Surgeons Eng., Royal Coll. Physicians London. Research, numerous publs. on mechanism of injurious effect of oxygen on retinal vessels of premature baby thus elucidating cause of retrolental fibroplasia, blindness in diabetes and high blood pressure, eye disease. Home: 2, The Cloisters, Westminster Abbey, London, S.W.I., Eng.*

ASHTON, William Easterly, Am. gynecologist; b. Phila., June 5, 1859; s. Samuel Keen and Caroline M. (Smiley) A.; M.D., U. Pa., 1881; M.D., Jefferson Med. Coll., 1884; LL.D., Ursinus, 1904; m. Alice Elizabeth Rosengarten, Oct. 5, 1891; 1 dau., Dorothy. Mem. faculty of hosp. and Jefferson Med. Coll., 1884-92; gynecologist to hosp. and prof. gynecology Medico-Chirurg. Coll., 1892-1916; prof. gynecology Grad. Sch. Medicine, U. Pa., from 1916. Decorated D.S.C.; cited for award Congl. Medal Honor. Fellow A.C.S. Author: Essentials of Obstetrics, 1888 (trans. into Chinese); The Practice of Gynecology, 1905; also articles. Inventor surg. instruments and appliances; first who substituted pads of gauze for marine sponges. Died 1933.

ASHURBANIPAL; see Assurbanipal.

ASHWORTH, James Hartley, Brit. zoologist; b. 1874; s. James Ashworth; ed. U. Manchester (Eng.); B.Sc., London, 1895; D.Sc., 1899; m. Clara Hough, 1901. Jr. lectr., demonstrator in zoology U. Manchester, 1896-1900; lectr. in invertebrate zoology U. Edinburgh, from 1933; v.p. Zool. Soc. Scotland, from mology and protozoology, 1905-19, prof. zoology, 1919-27, prof. natural history, from 1927; vis. prof. U. Cal. at Berkeley, 1930. Recipient Keith medal Royal Soc. Edinburgh, 1916. Fellow Royal Soc., 1917 (council 1924-26); pres. Royal Phys. Soc. Edinburgh, 1918-21, sect. D Brit. Assn., 1923; gen. sec. Royal Soc. Edinburgh, from 1933; v.p. Zool. Soc. Scotland, from 1928. Author: Catalogue of Chaetopoda in British Museum, Part 1, Arenicolidae, 1912; papers on Rhinosporidium, also histology and anatomy of marine worms, distbn. of Anopheline mosquitoes in Scotland, nature and structure of giant nerve cells and fibres. Died Feb. 4, 1936.

ASHWORTH, Ural Stephen, Am. biochemist; b. Walla Walla, Wash., Sept. 11, 1905; s. William H. and Lillian (Varnd) A.; A.B., U. Mo., 1929, Ph.D., 1933; postgrad. Yale; m. Louise Main, Sept. 5, 1935; children—Lillian, William, Jack, Judy, Bobby. Asso. prof. U. Ark. Med. Sch., 1937-39; dairy chemist State Coll. Wash., 1939-55, prof. animal scis., 1955—. Recipient Borden (Co.) award in dairy mfg., 1962. Mem. Am. Dairy Sci. Assn. (mem. com. on milk composition 1961—), Am. Chem. Soc., Inst. Food Technologists, Sigma Xi. Asso. editor Jour. Dairy Sci., 1961—. Research, numerous publs. on milk proteins including devel. methods for total milk protein and undenatured whey protein., problems dry milk mfr., mechanism curd formation. Home: 1815 Creston Lane, Pullman, Wash. 99163.*

ASIMOV, Isaac, biochemist, sci. writer; b. Petrovichi, Russia, Jan. 2, 1920; s. Judah and Anna Rachel (Berman) A.; brought to U. S., 1923, naturalized, 1928; B.S., Columbia, 1939, M.A., 1941, Ph.D., 1948; m. Gertrude Blugerman, 1942; children—David, Robyn Joan. With Boston U. Sch. Medicine, 1949-, asso. prof. biochemistry, 1955—. Recipient James T. Grady award for science writing American Chemical Society, 1965. Author of: Pebble in the Sky, 1950; I, Robot, 1950; The Stars, Like the Dust, 1951; Foundation and Empire, 1952; Caves of Steel, 1954; End of Eternity, 1955; The Naked Sun, 1957; (textbook) Biochemistry and Human Metabolism, rev. edit., 1957; Nine Tomorrows, 1959; The Words of Science, 1959; The Living River, 1960; Kingdom of the Sun, 1960; Words from Myths, 1961; Realm of Algebra, 1961; Life and Energy, 1962; Words in Genesis, 1962; Fact and Fancy, 1962; Words on the Map, 1962; Words from the Exodus, 1963; The Human Body, 1963; The Genetic Code, 1963; Intelligent Man's Guide to Science, 1960; View from a Height, 1963; Kite that Won the Revolution, 1963; Asimov's Biographical Encyclopedia of Science and Technology, 1964; Human Brain, 1964; A Short History of Biology, 1964; Quick and Easy Math, 1964; Adding a Dimension, 1964; A Short History of Chemistry, 1965; The Greeks, 1965; Of Time and Space and Other Things, 1965, and many others. Well known internationally as one of world's foremost science writers and as craftsman of science fiction. Home: 45 Greenough St., West Newton 65, Mass. Office: 80 E. Concord St., Boston.*

ASINGER, Friedrich, chemist; b. Freiland, Austria, June 26, 1907; s. Rudolf and Elisabeth (Mitterer) A.; Dr.techn.Ac, Tech. U., Vienna, Austria, 1932; Dr. phil.habil, U. Graz (Austria), 1942; m. Ottilie Pollack, Dec. 18, 1938; 1 dau., Roswitha. Asst. prof. Tech. U., Vienna, 1932-35; dept. chief Vacuum Oil Co., Vienna, 1935-37; group leader J. G. Farbenin D. Leuna, 1937-45; leader research dept. USSR Ministry Chemistry, 1946-54; prof. organic chemistry U. Halle (Germany), 1954-57; prof. organic chemistry Tech. U. Dresden, Germany, 1957-58; prof. tech. chemistry Tech. U., Aachen, Germany, 1959—. Mem. Acad. Scis. Berlin, N.Y. Acad. Scis. Author: Chemistry and Technology of Paraffinic Hydrocarbon, 1956; Chemistry and Technology of Monoolefins, 1959; Introduction to Petrochemistry, 1960; Paraffins Chemistry and Technology; Monoolefins Chemistry and Technology, 1968; also numerous articles. Research on chemistry of paraffinic hydrocarbons, alkylation, chlorination,, concomitant reaction of sulphur and ammonia with ketones (Asinger reaction). Home: 50 Dr. Hahnstr., Aachen, 51, West Germany.*

ASKANAZY, Max, pathologist; b. Germany 1865; 1st to note assn. of decalcification and parathyroid tumors (in a case of osteitis fibrosa cystica 1904). Died 1940.

ASKARYAN, Gurgen Ashotovic, Russian physicist; b. Moscow, USSR, Dec. 14, 1928; s. Ashot and Asja (Asarjan) A.; grad. Moscow U., 1952; Ph.D., Acad. Sci., 1961. Staff, Lebedeb Phys. Inst., Moscow, 1961—, sr. sci. worker, 1964—. Research, publs. on self-focusing of intense electromagnetic and ultrasonic rays, Cezenkov radiation from electromagnetic waves, coherent radiation from cosmic showers, self acceleration of charged particle bunches. Home: Rojdestvenskii bulvar 19, Office: Leninskii prospekt, 53, Moscow, USSR.*

ASKENASY, Eugen, botanist; b. Tarnopol (now USSR), May 5, 1845; s. M. and Maria (Raffalovik) A.; degree from Heidelberg, Germany, 1866. Became lectr. in Heidelberg, 1872, asso. prof., 1881, hon. prof., 1897. Mem. Leopoldina, Halle and other sci. socs. Investigated physiology of growing, fluid transport in plant organism, algology; founder cohesions theory, 1905. Died Sölden, Austria, Aug. 24, 1903.

ASKEPIODOTOS, see Asclepiodotos.

ASK-UPMARK, Karl Erik Fritz, Swedish physician; b. Lund, Sweden, Oct. 4, 1901; s. August and Elisabeth (Ask) U.; M.D., U. Lund, 1928; m. Kathie Modigh, Aug. 16, 1946. Asst. prof. medicine U. Lund, Sweden, 1935-46; head physician Sahlgrenska Hosp., Gothenburg, Sweden, 1946—; prof. medicine U. Upsala, Sweden, 1946—. Guest lectr. Royal Soc. Medicine, London, Eng., U. Cal., 1947. Rockefeller Found. fellow, Boston, 1932-33. Mem. Royal Soc. Scis., Soc. Medicine Argentina, Soc. Angiology Brasil, Soc. Internal Medicine Norway, Soc. Medicine Finland. Author: Carotid Sinus, 1935; Heart Disease, 1945; Bed Side Medicine 1963; Acute Medicine, 1959, 64, 66; Nervous System and Internal Disease, 1966. Research, publs. on relationship of unilateral kidney lesion to hypertension, periodic fever, megacolon, viral etiology of neoplasms, prognosis of subarachnoid hemorrhage without surgery. Home: Svartbacksgatan 37 A, Upsala, Sweden.*

ASLING, Clarence Willet, Am. anatomist; b. Duluth, Minn., June 17, 1913; s. Clarence Daniel and Elizabeth (Woolverton) A.; A.B., U. Kan., 1934, M.A., 1937, M.D., 1939; Ph. D., U. Cal., Berkeley, 1947; m. Helen J. Alber, 1936 (div. 1963); children—Joseph H., Carol J.; m. 2d, Frances Goldfarb, Oct. 23, 1964. Member faculty Univ. California at Berkeley, 1947—, prof. anatomy, 1957—. Recipient Fulbright research award Gt. Britain, 1953-54; medal U. Liege, 1954; professeur invité Faculté de Medecine U. Geneva, 1963. Guggenheim fellow, 1962-63. Fellow A.A.A.S.; mem. Am. Assn. Anatomists, Anat. Soc. Gt. Britain and Ireland, Am. Inst. Nutrition, Soc. Exptl. Biology and Medicine, Teratology Soc., Congenital Anomalies Research Association Japan, Sigma Xi. Research, numerous publs. on endocrine and nutritional regulation of skeletal devel., exptl. induction of congenital defects in mammals, localization of radioisotopes, autoradiography and histochemistry. Home: 373 Gravatt Dr., Berkeley, Cal. 94705. Office: Dept. Anatomy, U. Cal. Med. Center, San Francisco 94122.*

ASMUNDSON, Vigfus Samundur, poultry scientist; b. Reykjavik, Iceland, Sept. 24, 1895; s. Asmundur Magnusson and Helga Vigfusdottir; B.S.A., U. Sask., Can., 1918; M.S. in Agr., Cornell U., 1920; Ph.D., U. Wis., 1930; LL.D., U. Cal. at Davis, 1963; m. Aline Mary McGrath, Aug. 24, 1935; children—Vigfus A., Craig M., J. Michael, Mary Irene, Roderick Vincent, Nancy A. Came to U. S., 1932, naturalized, 1940. Instr. poultry dept. U. B.C., 1920, asst. prof., 1921, asso. prof., 1927-32; asso. U. Cal. at Berkeley, 1932-33; faculty U. Cal. at Davis, 1933—, prof. poultry dept. Coll. Agr., 1946-63, prof. emeritus, 1963—, Faculty Research lectr., 1947. Fellow Poultry Sci. Assn.; mem. World's Poultry Assn., Genetics Soc. Am., Am. Genetic Soc., Soc. for Study Evolution, Soc. for Exptl. Biology and Medicine. Fellow A.A.A.S. Research, numerous publs. on genetic variability in leading food species of birds especially lethal and nonlethal mutation, effect of selection on composition of muscles of chicken with inherited muscular dystrophy and variation in interspecific hybrids of turkeys. Home: 517 Miller Dr., Davis, Cal. 95616.*

ASMUS, Erik, German chemist; b. Saint-Petersburg, Russia, Oct. 29, 1908; s. Wilhelm and Marie-Wilhelmine (Kluge) A.; ed. Technische Hochschule, Hanover, U. Munich; Ph.D. m. Herta Miersch, July 30, 1936; children—Klaus-Dieter, Ingebord. Ast. Breslau and Marburg; prof. Marburg, Münster, Berlin. Mem. Soc. German Chemists. Author: Einfuhrung in die höhere Mathematik; Physikalisch-chemische Rechenaufgaben; also numerous articles.

Home: Berlin-Dahlem, Wichernstr. 24. Office: Berlin 12, Hardenbergstr. 34, West Germany.*

ASMUSSEN, Erling, Danish physiologist; b. Naestved, Denmark, Aug. 6, 1907; s. Ejnar and Maren (Moeller) A.; cand.mag., U. Copenhagen (Denmark), 1932, Dr.phil., 1934; m. Edith Hansen, Dec. 23, 1932; children—Bodil (Mrs. Jörgen Nielsen), Lisbet (Mrs. Jono deCastro Lopo), Erik, Mette. Lectr., Ordrup Gymnasium, 1932-34; faculty Lab. for Theory Gymnastics, U. Copenhagen, 1934—, prof., 1961—; sci. dir. Danish Nat. Assn. Infantile Paralysis, 1956—. Mem. Royal Danish Acad. Scis. and Letters, Am. Acad. Phys. Edn. Author textbooks in physiology of phys. edn.; also numerous articles. Research on various aspects of phys. exercise. Home: 17 Kollegievej, Charlottenlund, Denmark. Office: 32 Juliane Maries Vej, Copenhagen, Denmark.*

ASMUSSEN, Robert Wirenfeldt, Danish chemist; b. Viborg, Mar. 26, 1903; s. Frederick and Petra (Sorensen) A.; ed. U. Copenhagen; doctor's degree; m. Grethe Abildgaard; Jan. 5, 1937; children—Nils, Jan, Peter. Prof. chemistry tech. univ., Copenhagen, from 1944, vice-chancellor, 1956-60. Mem. Royal Danish Acad. Sci. and Letters, Acad. Tech. Sci. Research, publs. on the magneto-chemistry of inorganic complex compounds. Home: Solvgade 83, Copenhagen K. Office: Chemical Lab. B., Technical University, Lyngby, Denmark.

ASPATURIAN, Vernon Varaztat, polit. scientist; b. Armavir, Russia, Feb. 16, 1922; s. Serop and Gayane (Ohanesian) A.; brought to U. S., 1923, naturalized, 1929; A.A., Los Angeles City Coll., 1943; B.A. with highest honors, U. Cal., Los Angeles, 1947, Ph.D., 1951; m. Suzanne Lee Dohan, Aug. 29, 1948; children—Heidi Jeanne, Nancy Lee. Lectr. in polit sci. U. Cal., Los Angeles, 1950-51; chief Russian Desk, Psychol. Warfare Sect., U. S. Command, Tokyo, Japan, 1952; asst. prof. polit. sci. Pa. State U., 1952-56, asso. prof, 1957-61, prof., 1961—, research prof., 1966—, dir. Slavic and Soviet Lang. and Area Center, 1965. Smith-Mundt vis. prof. internat. relations Grad. Inst. Internat. Studies, Geneva, Switzerland; vis. prof. Columbia U., 1963, Sch. Advanced Internat. Studies, John Hopkins, 1961-63, U. Cal. Los Angeles, 1964-65. Cons. RAND Corp.; cons. Army War Coll., 1963; research asso. Washington Center of Fgn. Policy Research; cons. Planning Research Corp., 1965. Rockefeller fellow, 1956-57. Mem. Am. Polit. Sci. Assn., Am. Assn. U. Profs., Phi Beta Kappa, Pi Sigma Alpha, Pi Gamma Mu. Author: The Union Republics in Soviet Diplomacy, 1959; Modern Political Systems: Europe, 1963; The Soviet Union in the World Communist System, 1966; Foreign Policy in World Politics, 1967. Contbr. numerous articles to sci. jours. Devel. of new conceptual frameworks for the analysis of Soviet polit. instns. and processes and Soviet fgn. policy and diplomacy. Home: 1154 William St., State College, Pa. 16801. Office: Pa. State U., University Park, Pa. 16802.*

ASPREY, Larned Brown, Am. chemist; b. Sioux City, Ia., Mar. 19, 1919; s. Peter and Gladys (Brown) A.; student Morningside Coll., 1936-38; B.S., Ia. State Coll., 1940; Ph.D., U. Cal. at Berkeley, 1949; m. Margaret Elizabeth Williams, May 3, 1944; children—Peter, Elizabeth, Barbara, Robert, Margaret, Thomas, William. Chemist, Campbell Soup Co., 1940-42, Metall. Lab., 1944-46, Radiation Lab., U. Cal. at Berkeley, 1946-49; chemist, staff mem. Los Alamos Sci. Lab., 1949—. Fellow A.A.A.S.; mem. Am. Chem. Soc. (past treas. div. inorganic chemistry), Sigma Xi. Contbg. author: Progress in Inorganic Chemistry, vol. 2, 1960. Research, numerous publs. on inorganic and phys. chemistry of actinide elements, fluorine chemistry, halogen fluorides, lanthanide chemistry; preparation and characterization of metals and compounds of actinium, thorium, protactinium, uranium, neptunium, plutonium, americium, curium. Home: P.O. Box 157, Fairview Sta., Espanola, N.M. 87532. Office: P.O. Box 1663, Los Alamos 87544.*

ASQUITH, Dean, Am. entomologist, educator; b. Lowell, Mass., Mar. 3, 1912; s. Sam and Ethel (Dean) A.; B.S., U. Mass., 1933, M.S., 1939; m. Doris Holway Redman, Oct. 29, 1932; children—Peter Dean, Kate Holway. Field entomologist Rohm & Haas Co., Phila., 1939-46; pvt. practice as cons. entomologist, Amherst, Mass., 1946-48; faculty entomology Fruit Research lab. Pa. State U., Arendtsville, 1948—, prof., 1957—. Mem. Entomol. Soc. Am., A.A.A.S., Entomol. Soc. Pa., Sigma Xi, Phi Kappa Phi, Theta Chi. Research, publs. on fruit insects, spl. emphasis on resistance of orchard mites to pesticides. Home: Box 176, Biglerville, Pa. 17307. Office: Pa. State University Fruit Research Lab., Arendtsville, Pa. 17303.*

ASRATIAN, Ezras Asratovich, Russian physiologist; born 1903; graduated Agricultural Faculty, Yerevan Univ., 1926, graduated Medical Faculty, 1930. With Pavlov's Laboratory, Physiological Institute, USSR Academy of Science, 1930-38, dir. Physiol. Lab., 1944-60, dir. Inst. Higher Nervous Activity and Neurophysiology, 1960—; with Bekhterov Brain Inst., 1935-41, Leningrad Pedagogical Inst., 1936-41; head dept. normal physiology Pirogov 2d Moscow Med. Inst., 1950—. Del. numerous Internat. Physiol. Congresses, 1935-65. Mem. USSR (corr., Pavlov gold medal 1961), Armenian acads. sci. Author: New Data on the Physiology of the Cerebrum, 1941; Outline Etiology, Pathology and Therapy of Traumatic Shock, 1945; Adaptive Phenomena in the Injured Body, 1948; I. P. Pavlov, His Life and Scientific Heritage, 1949; The Physiology of the Central Nervous System, 1953; Compensatory Adaptations, Reflex Activity and the Brain, 1965. Co-editor Physiology sect. Large Med. Ency., 2d edit., also other works on physiology; chief editor Pavlov Jour. Higher Nervous Activity. Research on devel. of recovery symptoms in injured body, demonstrating that cerebral cortex plays decisive role in restoring lost or impaired functions; developer new principle of switching in conditioned reflex activity, of sleep treatment for shock, paralysis, concussion and other organic disorders; developer anti-shock liquid used by Soviet Army in World War II, 1942. Address: Inst. of Higher Nervous Activity and Neurophysiology, Pyatnitskaya ulitsa 48, Moscow, USSR.*

ASSAGIOLI, Roberto, Italian psychiatrist; b. Venice, Italy, Feb. 27, 1888; s. Leone and Elena (Kaula) A.; studied medicine, psychology, philosophy; M.D.; m. Nella Ciapetti, Aug. 14, 1922. Practiced medicine specializing in nervous maladies; dir. Istituto di Psiosintesi, Florence; pres. Psychosynthesis Research Found. Author: Parapsychological Faculties and Psychological Disturbances, 1958; Dynamic Psychology and Psychosynthesis, 1959. Address: via San Domenico 16, Florence, Italy.

ASSHETON, Richard, Brit. biologist; b. Lancashire, Eng., Dec. 23, 1863; s. Ralph and Emily Augusta (Feilden) A.; B.A., Trinity Coll., Cambridge, 1886, M.A., 1890; Sc.D.; m. Frances Annette Ellen Bazley; 1 son, 2 daus. Lectr. in zoology Owens Coll., Manchester, Eng., 1889-93; researcher in biology, Cambridge, from 1893, univ. lectr. animal embryology, from 1911; lectr. biology Guy's Hosp. Med. Sch., U. London, 1901-14; lectr. vertebrate embryology Imperial Coll. Sci. and Tech., London, 1910; examiner in zoology U. Aberdeen. Fellow Royal Soc., 1914. Author: The Development of Gymnarchus Niloticus; The Geometrical Relation of the Nuclei in an Invaginating Gastrula; The Growth in Length of the Vertebrate Embryo; Variation and Mendel; Dolichoglossn Serpentinus. Died Oct. 23, 1915.

ASSMAN, Herbert, German physician; b. Danzig, Germany, Nov. 25, 1882; s. Edwin and Anne Emma Laura (Steimmig) A.; ed. Geneva, Switzerland, also Königsberg (now Kaliningrad, USSR); m. Eleonore Steimmig, Aug. 14, 1909; 3 sons, 2 daus. Became lecturing prof., Leipzig, Germany, 1913, extraordinary prof., 1919, asso. prof., 1922, prof., also dir. univ. polyclinic, 1927; apptd. dir. med. clinic, prof. U. Konigsberg, 1931. Author: Die Klinische Röntgendiagnostik der inneren Erkrankungen, 1922; (with L. Mohr, L. Staehelin) Handbuch der inneren Medizin; other med. handbooks. Used X-rays in diagnosis; studied devel. of Tb. Died Oldenburg, Germany, Feb. 27, 1950.

ASSMANN, Richard, meteorologist; b. Magdeburg, Prussia, Apr. 13, 1845; s. Adolf and Dorothea (Burckhardt) A.; M.D., Ph.D.; m. Johanna Andree; 1 dau. Became lectr. at Halle, Germany, 1885; dir. aero. obs. Prussian Meteorol. Inst., Berlin-Reinickendorf; founder, head Prussian Aero. Obs., Lindenberg, 1899-1914; founder, pub. Das Wetter (periodical), until 1918; hon. prof., Giessen, Germany, Mem. Royal Meteorol. Soc. London. Author: (with A. Berson) Wissenschaftliche Luftfahrt, 3 vols., 1899-1900. Discovered (with Leon Teisserenc de Bort) stratosphere, 1902; developed (with Hergesell) method of measuring wind velocity with balloons followed by theodolites; invented aspirations psychrometer, also method of studying air strata by sounding with self-expanding rubber balloons; founded jour. Das Wetter. Died Giessen, May 28, 1918.

ASSURBANIPAL (or Ashurbanipal), Assyrian king; flourished 7th century B.C.; s. Esarhaddon; king of Assyria, circa 669-circa 626 B.C.; defeated Tirhakah in Egypt and placed Necho in power, 664 B.C.; lost Egypt to Necho's son Psamtik I, 660 B.C.; put down revolt of brother in Babylon, 668-48 B.C.; defeated Elam, 642-39 B.C. Raised Assyria to height of power; made extensive collections cuneiform tablets in library at Nineveh which form basis of much of our knowledge of ancient Near Eastern thought. Died circa 626 B.C.

ASTALDI, Giovanni Ernesto, Italian physician; b. Dorno, Pavia, Italy, Sept. 22, 1914; s. Antonio Francesco and Francesca (Angeleri) A.; B.A., Lyceum Cairoli, Vigevano, Italy, 1933; M.D., U. Pavia (Italy), 1939; m. Stefana Gaggero, Feb. 12, 1942; children—Antonella (Mrs. Roberto De Nova), Alberto. Faculty, U. Pavia, 1939-52, prof. clin. pathology and internal medicine, 1948-52; prof. dept. internal medicine, dept. clin. pathology Municipal Hosp., Tortona, Italy, 1952—, dir. Blood Research Found. Center, 1956—. Mem. Internat., European socs. hematology, Internat. Soc. Cell Biology, European Tissue Culture Club, Tissue Culture Assn., Reticuloendothelial Soc., Internat. Soc. Chemotherapy, N.Y. Acad. Scis. Author books, numerous articles. Direct observation in vitro of maturation of megaloblasts to normoblasts; research on abnormal PAS-positive structures in erythorblasts from thalassemia, intestinal biopsy in different conditions, stimulation of normal and leukemic lymphocytes in tissue culture with phytohemaglutinin. Home: Via del Carmine 5, Pavia, Italy. Office: Blood Found. Center, Hosp. Tortona, Tortona, Italy.*

ASTAPOVICH, Igor Stanislavovich, Russian astronomer; b. 1908. Head, Ashkhabad Astrophys. Lab. Author: New Data on the Flight of the Large Meteorite of 30 June 1908 in Central Siberia, 1933; The Results of Studies on the Orbits of 66 Meteorites, 1939; The Counterglow Parallax, The Earth's Gas Tail, 1944; Meteor Stream Radiants from Observations in China during the First Millennium A.D., 1960; Successes of Meteor Astronomy 1958-61, 1963. Specialist in meteors and meteorites; made 37,000 observations of meteors, 1925-49; studied conditions of their entry into earth's atmosphere and crust; proved that earth possesses gas tail, 1944. Address: Ashkhabad Astrophysical Lab., Ashkhabad, Turkmenia SSR, USSR.

ASTARITA, Giovanni, Italian chem. engr.; b. Naples, Italy, Oct. 7, 1933; s. Tommaso and Carmen (Grimaldi) A.; D.Engring., U. Naples, 1957; M.Chem.Engring., U. Del., 1959; m. Nerina Giuliani, Sept. 8, 1958; children—Tom, Luca. Faculty, U. Naples, 1959, 61-65, prof. chem. engring., 1966—; asso. prof. U. Del., 1965-66. Mem. Brit. Instn. Chem. Engrs. (asso.), Associazone Italiana Ing. Chimici. Author: Principi di Ingegneria Chimica, 1962; Mass Transfer with Chemical Reaction, 1966; also articles. Research on phenomena where diffusion and chem. reaction take pl. simultaneously, flow of viscoelastic materials (materials having both liquidlike and solidlike characters). Home: 16, via Dei Mille, Naples, Italy.*

ASTAUROV, Boris Lvovich, Russian biologist; b. Kazan, Russia, Oct. 27, 1904; s. Lev Mikhailovich and Olga Andreyevna (Teeckenko) A.; grad. Moscow U., 1927, Cand. Biol. Sci., 1936, D.Biol. Sci., 1938; m. Tatiana Michilovna Yakovleva, Jan. 15, 1936; 1 dau., Olga; m. 2d, Natalia Sergaeevna Skadovskaya, Jan. 8, 1944; 1 dau., Natalia. Asst. genetics lab. Inst. Exptl. Biology, Moscow, 1924-26, sci. worker lab. developmental mechanics, 1935-47; sci. worker genetics dept. Commn. for Investigation Natural Resources, USSR Acad. Scis., Moscow, 1926-30, with Inst. Cytology, Histology and Embryology, 1939—, head Filatov Lab. Developmental Mechanics, 1947-48, sci. worker, 1948-55, head Filatov Lab. Exptl. Embryology, Severtzov Inst. Animal Morphology, 1955—, head lab. developmental cutogenetics Inst. Developmental Biology, 1967, dir. Inst., 1967—; sci. worker dept. genetics and breeding Middleasian Inst. Sericulture, Tashkent, 1930-36; Recipient Silver medal for invention heat shock method for thermic cure of silkworm nosema disease All-Union Exhbn. Achievements of Nation, 1963, G. Mendel 100th Anniversary Meml. medal Czechoslovak Acad. Sci., 1965. Fellow Internat. Inst. Embryology; mem. Internat. Soc. for Cell Biology, Moscow Soc. Naturalists (chmn. genetics sect.), Am. Soc. Zoologists (corr.), USSR Acad. Scis. (academician), Genetics and Breeding Soc. USSR (pres. 1966—. Mem. editorial bd. Cytology, 1959—, div. biology Bull. Moscow Soc. Naturalists, 1962—, Genetics, 1965—, Priroda, 1965—. Research and numerous publs. on cytogenetics and developmental biology of Drosphila and silkworms; developer method of thermal artificial parthenogenesis in unfertilized silkworm eggs, 1934, complete androgenesis for 1st time in animals especially for 1st time of interspecific androgenesis—devel. of adult exclusively male progeny from enucleated egg cytoplasms taken from one species with nucleus taken from another species; proved by means of androgenesis the nuclear versus cytoplasmic control of specific differences and nuclear versus cytoplasmic localization of injuries caused by ionizing irradiation, 1947, prodn. of exptl. tetraploid bisexual strain for 1st time in animal (silkworm), 1957-66. Home: 3 Demetr. Ulyanov ulitsa, Moscow B-333. Office: Inst. Developmental Biology, USSR Acad. Scis., Vavilov ulitsa 26, Moscow B-133, USSR.*

ASTBURY, William Thomas, English physicist; b. Longton, Stoke-on-Trent, Eng., Feb. 25, 1898; s. W. E. Astbury; ed. (scholar) Jesus Coll., Cambridge (Eng.) U., M.A., Sc.D., 1937; D.(h.c.), U. Strasbourg, 1946; m. Frances Gould, 1922; 2 children. Demonstrator in physics Univ. Coll., London, also asst. to Sir William Bragg, 1921-23; asst. to Bragg at Davy-Faraday Lab. of Royal Instn., 1923-28; became lectr. textile physics U. Leeds, 1928, reader in textile physics, 1937, dir. textile physics research lab., 1928, prof. biomolecular structure, hon. reader in textile physics, 1945—. Recipient Warner Meml. medal Textile Inst., 1935; Actonian prize Royal Instn. Gt. Britain, 1935; Silvanus Thompson medal Brit. Inst. Radiology, 1948; medal U. Brussels, 1949; Internat. Sci. Relations medal Am. Soc. European Chemists, 1953. Fellow Royal Soc., 1940, Croonian lectr., 1945, mem. council, 1946-47; fellow Inst. Physics; mem. Swedish Royal Acad. Scis. (fgn.), Royal Soc. Scis. Uppsala, N.Y. Acad. Scis. (hon. life), Instituto Lombardo di Scienze e Lettere (corr.). Author: Fundamentals of Fibre Structure, 1933; Textile Fibres under the X-rays, 1943. Research and publs. on X-ray analysis, crystal structure, fibres, proteins and molecular biology. Died June, 1961.

ASTEN, Friedrich Emil von, see von Asten.

ASTIER, André Lucien, French physicist; b. Villeneuve-les-Avignon, France, Apr. 7, 1922; s. Léon and Rose (Lacroix) A.; ed. Polytech. Sch., U. Paris; Dr. of phys. sci.; m. Jeanne Desplats, Nov. 11, 1945. Engr. of bridges and highways, Auch, 1947-49; Marseille, 1949-54; with lab. sci. research, Paris, also physics lab. Polytech. Sch., during 1954-60; under-dir. labor dept. of nuclear physics Coll. France, since 1962. Recipient cognacq-jay prize Acad. Sci., 1962. Mem. French, Italian Physics socs. Research, publs. on elementary physics. Home: 30, rue de la Saussaye, Massy (Seine-et-Oise). Office: Coll. de France, 11, place Marcelin-Berthelot, Paris 5, France.

ASTON, Bernard Cracroft, agrl. chemist; b. Beckenham, Eng., 1871; s. Murray Aston; ed. Otago (New Zealand) U., Victoria (Australia) Univ. Coll. Became chemist New Zealand Dept. Agr., 1899, retired, 1937. Dir. Empire Marketing Bd.'s Grant for Mineral Content New Zealand Pastures Research, 1928-33; pres. New Zealand Sci. Congress, 1929. Recipient Hector prize and medal for chem. research, 1925. Fellow Royal Inst. Chemistry, Royal Soc. New Zealand (v.p. 1935); mem. New Zealand Forest and Bird Protection Soc. (pres. 1934, 46, 47). Publs. on New Zealand botany and agr. Studied chem. aspects of agr.; proved iron starvation causes bush sickness in ruminants. Died May 31, 1951.

ASTON, Francis William, English chemist, physicist; b. Harborne, Eng., Sept. 1, 1877; s. William and Fanny Charlotte (Hollis) A.; student Malvern Coll., 1891-93; entered U. Birmingham (Eng.), 1893; B.A., Cambridge, 1912; several hon. degrees. Fermentation chemist with brewery firm, Wolverhampton, Eng., 1900-03; researcher for J. H. Poynting at U. Birmingham, 1903-08; aircraft engineer during World War I; became research asst. to J. J. Thomson, Trinity Coll., Cambridge, 1910, fellow, 1920 (worked under Lord Rutherford in the Cavendish Laboratory); researcher in Cavendish Laboratory, Cambridge, 1921-45. Named chmn. Internat. Com. on Atoms, 1935. Recipient Nobel prize for chemistry, 1922, Dudley medal Phys. Soc., 1944, John Scott medal, Phila., Italian Paterno medal. Fellow Royal Soc., 1921, (Bakerian lectr. 1927, Royal medal 1938), Royal Inst. Chemistry; mem. Brit. Assn. (pres. phys. and math. sect. 1935), Acad. Scis. USSR, Academia dei Lincei. Author: Isotopes, 1922; Mass-spectra and Isotopes, 1933. Introduced mass-spectrograph (based on J. J. Thompson's apparatus), 1919, and used it to discover nonradioactive isotopes, including those for neon and chlorine; established isotopy as basic phenomonon of nature; made mass analysis of metallic elements; predicted thermonuclear energy, 1922, synthesis of elements, 1936. Died Cambridge, Nov. 20, 1945.

ASTRAND, Per Olof, Swedish physiologist; b. Svenarum, Sweden, Oct. 21, 1922; s. Karl August and Martha (Thunander) A.; M.D., Karolinska Institutet, Sweden, 1952; D. Honoris Causa, U. Grenoble (France), 1966; m. Irma Ryhming, July 14, 1956; children—Elin Kristina, Per Gustaf. Faculty, dept. phys. edn. Coll. Phys. Edn., Stockholm, Sweden, 1946-52, dept. physiology, 1952—; asso. research physiologist U. Cal. at Berkeley, 1957; staff dept. physiology Middlesex Hosp. Med. Sch., Eng., 1952-53; Lankenau Hosp. dept. physiology, 1958. NSF scholar, 1957; recipient Ruhemann-Plakette Deutsche Sportärzteverbund, 1963, Am. Acad. Phys. Edn. Research award, 1953. Author: (with K. Rodahl) Textbook of Work Physiology, 1967; also numerous articles, books in Swedish. Research on phys. work capacity, tng., environment, respiration, circulation during work, sport physiology. Home: Orrspelsvagen 6, Näsbypark, Sweden. Office: Dept. Physiology, Coll. Phys. Edn., Stockholm, Sweden.*

ASTRE, Gaston Prosper, French geologist; b. Toulouse, France, Apr. 16, 1896; s. Pierre and Marguerite (Andrieu) A.; ed. U. Toulouse; Dr. Sci. in geology and paleontology; m. Jeanne Lafourcade, July 19, 1921; children—Christine, Jean. Lectr. geology Faculté des Sciences Toulouse; lectr. geology and hydrology Faculte de médicine et de pharmacie; head research Nat. Center of Research France; dir. Mus. Natural History Toulouse. Mem. Asso. Acad. Scis., Archeol. Soc. Southern France, Natural History Soc. France (v.p.). Author numerous publs., including: Documents de géologie luchonaise; Radiolitidés nord-pyrénéens. Special research on mammiferous fossils in Aquitaine Basin. Address: 47, rue Barrau, Toulouse, France.

ASTRUC, Jean, French anatomist, physiologist; b. Sauve/Languedoc, France, Mar. 19, 1684; prof. anatomy, Montpellier, France, then Toulouse, France, from 1710; became prof. medicine College Royale, circa 1731; cons. physician to King Louis XIV of France. Author: Tractatus de motus fermentativa causa . . . , 1702; Traité de la cause de la digestion . . . , 1714; De morbis venereis . . . , 1736; Conjectures sur les mémoires originaux, dont it parait que Moise s'est servi pour composer le livre de la Genèse, 1753; Traité des maladies des femmes, 6 vols., 1761-65; Mémoires pour servir d'histoire naturelle de la Faculté de Medecine de Montpellier, 1767. Placed reflex center (sensorium commune) in white matter of brain, where motor nerve receives impulse from nerve in medulla; held that digestion is caused by ferment rather than trituration (Iatro-math. theory). Died Paris, Mar. 5, 1766.

ASTRUP, Tage, biochemist; b. Esbjerg, Denmark, Nov. 5, 1908; s. Frede and Marie (Christensen) A.; M.Sc., Polytech. Inst. Copenhagen, Denmark, 1932; Ph.D., U. Copenhagen (Denmark), 1944; m. Henny Andersen, Oct. 17, 1934; children—Lise (Mrs. Brian Dalton), Nina, Marit. Came to U. S., 1961, Chemist Grindsted-vaeket, Grindsted, Denmark, 1932-34; research asso. Aarhus (Denmark) U., 1934; asso. Biol. Inst. Carlsberg Found., Copenhagen, Denmark, 1935; research chemist Leo Pharms., Copenhagen, 1935-36; with Biol. Inst. Carlsberg Found., Copenhagen, Denmark, 1937-60; dir. research James F. Mitchell Found., Inst. for Med. Research, Washington, 1961—. Recipient Esso award, Copenhagen, 1950; Thorvald Madsen Med. award, Copenhagen, 1959. Contbr. numerous articles in field to sci. jours. Studies of formation and resolution of fibric clots, fibrinolytic systems in the body, clot formation and resolution in the body, its role in thrombosis and the pathogenesis of arteriosclerosis. Home: 5304 Locust Av., Bethesda, Md. 20014. Office: 5401 Western Av. N.W., Washington 20015.*

ASTUNI, Enrico, Italian engr.; b. Avellino, Italy, Feb. 28, 1914; s. Enrico and Ida (Caretti) A.; Studied classics, indsl. and electro-tech. engring.; Dr. in engring.; m. Milena Liguori, May 25, 1935; children—Enrico, Domenico, Giulio, Ida. Dir. electrotech. inst. U. Genoa (Italy). Mem. Italian Electro-Tech. Asso. Genoa (pres.), Tech. Counsel of Telecommunications. Author 2 vols. on electro-tech. material and electronic machines, also study on type of commutator (switch for electronic machines). Research on theory of elec. circuits and linear transformations for elec. systems. Home: viale Gambaro 15, Genoa. Office: via Montallegro 1, Genoa, Italy.

ASTWOOD, E(dwin) B(ennett), physician; b. Bermuda, Dec. 29, 1909; s. Ernest Millard and Imogene (Clingman) A.; B.S., Washington Missionary Coll., 1929; student Coll. of Evangelists, 1929-31; M.D. and C.M., McGill U. Med. Sch., 1934; fellow Johns Hopkins Hosp., 1935-37; Rockefeller fellow Harvard, 1937-39, Univ. fellow, 1938-39, Ph.D. in Biology, Harvard, 1939; D.Sc., U. Chgo., 1967; m. Sara Merritt, June 3, 1937; children—Philip Merritt, Nancy Bennett. Asso. in obstetrics Johns Hopkins U., and asst. obstetrician Johns Hopkins Hosp., 1939-40; asst. prof. of pharmacotherapy Harvard Med. Sch., 1940-45; asso. in medicine Peter Bent Brigham Hosp., 1940-45; research prof. of medicine Tufts Med. Sch., 1945-52, prof. medicine, 1952—; endocrinologist J. H. Pratt Diagnostic Hosp., 1945-48; sr. physician New Eng. Med. Center Hosps. (Boston Dispensary); physician at Boston Dispensary, 1945—. Recipient Ciba award Assn. Study Internal Secretions, 1944; Cameron prize U. Edinburgh, 1948; John Phillips Meml. award A.C.P., 1949; Borden award Assn. Am. Med. Colls., 1951; Lasker award Am. Pub. Health Assn., 1954; Nat. Inst. Arthritis and Metabolic Disease research career awards. Mem. Am. Physiol. Soc., A.A.A.S., Am. Chem. Soc., Am. Chem. Soc., Endocrine Soc. (pres. 1961-62, Fred Konrad Koch award 1967), Assn. Am. Physicians, Am. Soc. Clin. Investigation, A.C.P., Am. Acad. Arts and Scis., Nat. Acad. Sci., Am. Coll. Clin. Pharmacology and Chemotherapy (regent 1964), Brit. Soc. Endocrinology. Research, publs. on mammary gland devel., sex hormones, corpus luteum function, radioactive iodine, pituitary hormones, metabolism of fat; introduced use of thiouracil in treatment exophthalmic goiter, 1945. Home: 30 Irving St., Brookline, Mass. 02146. Office: 171 Harrison Av., Boston 02111.*

ASUNDI, Rango Krishna, Indian physicist; b. Asundi, India, Aug. 14, 1895; s. Krishna Ramchandra and Kamaladevi Asundi; ed. Deccan, Fergusson, Wilson Colls.; B.A. with honors, King's Coll., London U., 1917; B.Sc., Bombay (India), 1919, M.Sc., 1924; Ph.D., London, 1929; m. Radha, 1918; 1 son, Krishna. Lectr. Bombay Univ. Colls., 1917-27; reader Aligarh (India) U., 1931-39; prof. Banaras (India) U., 1939-56; physicist Atomic Energy Establishment, Trombay, Bombay, 1956—. Chmn. physics adv. com. Dept. Atomic Energy, India, 1956-60; pres. physics Indian Sci. Congress, 1955; chmn. Conv. Spectroscopists, India, 1957—. Fellow Indian Acad. Scis., Nat. Inst. Scis. India. Editor Symposiums on Spectroscopy, 1957-61. Research and publs. on atomic and molecular spectroscopy, structure of molecules. Home: 60, Green St., Bombay 54 AS. Office: AEET, Trombay, Bombay, India.*

ASWATHANARAYANA, Uppugunduri, Indian earth scientist; b. Vallur, A. P., India, July 1, 1928; s. Sudarsanarao and Sitamma (Tanguturi) A.; B.Sc. with honors, Andhra U., Colls., Waltair, India, 1950, M.Sc., 1951, D.Sc., 1955; m. Sarojinidevi Aswathanarayana, Sept. 1949; children—Viswanath, Vasundhara, Srinivas. Research asst. Council Sci. and Indsl. Research, Scheme, India, 1951-54; faculty dept. geology Andhra, reader, 1959—. Sr. vis. research fellow Cal. Inst. Tech., Pasadena, Cal., 1957; vis. scientist Oxford (Eng.) U., 1963; former dir. research projects Council Sci. and Indsl. Research, Indian Dept. Atomic Energy, Internat. Upper Mantle Project. Fellow Geol. Soc. India, Geochem. Soc. Indian, Indian Geophys. Union; mem. Geochem. Soc. Research, publs. on dating ancient rocks of India by radioactivity decay methods, radioactivity of terrestrial materials, pattern of distbn. of some metals in rocks, soils, waters and plants, origin of heavy mineral beach sand deposits of W. and E. coasts of India. Home: 20 Sea Sands Quarters, Andhra U., Waltair, A.P., India.*

ATABEKOV, Grigorii Iosifovich, Russian elec. engr.; b. 1908; grad. Electromech. Faculty, Tbilisi Poly. Inst., 1930; D.Tech. Sci. Engr., Transcaucasian Regional Bd., Main Bd. Power Mgmt., sr. engr. Moscow Regional Bd. Power Mgmt., then sr. engr. All-Union Trust for Design and Planning Thermal Elec. Power Plants, Networks and Substas., 1930—; instr. elec. engring. Transcaucasian Indsl. Inst., Moscow Instr. for Mechanization and Electrification of Agr., Bonch-Bruevich Inst. Communications, 1945; head chair theoretical principles elec. engring. Moscow Aviation Inst., 1946—. Author 98 works including The Relay Protection of High-Voltage Networks; The Distance Principle of Protecting Long-Distance Transmission Lines; The Theoretical Principles of the Relay Protection of High-Voltage Networks; Harmonic Analysis and the Operator Method; Linear Electric Circuits. Mem. editorial bd. News of Higher Ednl. Instns., Power Engring., Invention in the USSR, 1957—. Has 40 inventions. Address: Moscow Aviation Inst., Volokolamskoe shosse 8, Moscow, USSR.

ATCHISON, William Franklin, Am. mathematician, computer scientist; b. Smithfield, Ky., Apr. 7, 1918; s. William Duncan and Mary Lou (Beatty) A.; A.B., Georgetown Coll., 1938; M.A. in Math., U. Ky., 1940; Ph.D. in Math., U. Ill., 1943; m. Lois Ethel Bruinkool, June 7, 1947; children—Allen Franklin, Glen Ray, Mary Beth, David Duncan. Instr., U. Ill., Urbana, 1943-44, asst. prof. math., 1946-55; head programming and coding group Ga. Inst. Tech., Atlanta, 1955-57, dir. Rich Electronic Computer Center, 1957-66, acting dir. Sch. Information Sci., 1963-64, research prof. math., 1963-66; dir. computer sci. center, prof. computer sci. U. Md., College Park, 1966—. Cons. on computer work NSF, 1964—; chmn. adv. com. on computers and computer sci. So. Regional Edn. Bd., Atlanta, 1960—, cons., 1964—. Mem. Assn. for Computing Machinery (chmn. com. on curriculum in computer sci. 1962—), Am. Math. Soc., Am. Math. Assn., Soc. for Indsl. and Applied Math. Author: (with C. H. Sisam) Analytic Geometry, 1955. Home: 10711 Gatewood Av., Silver Spring, Md. 20903. Office: Computer Science Center, U. Md., College Park, Md. 20742.*

ATEN, Adrian Hendrik Willem, physicist; b. Jan. 22, 1908; s. A. H. W. and A. H. J. (van Iterson) A.; ed. Hilversum Coll., U. Utrecht; m. M. R. Noeggerath, 1938. Researcher, Johns Hopkins, 1934-35; asst. Columbia, 1935-36; asst. Inst. Theoretical Physics, Copenhagen, 1937-38; researcher physics lab. Philips of Eindhoven, 1939-47; with Inst. Nuclear Physics, Amsterdam, since 1947; apptd. prof. radiochemistry U. Amsterdam, 1949. Mem. Royal Asso. Chemists, Netherlands, Am. Soc. Chemists, Physics Soc. Contbr. articles to prof. jours. Address: Pieter Lastmankade 31, Amsterdam, Netherlands.

ATHANASIADIS, Constantin, Greek statistician; b. Athens, Greece, 1898; s. Antoine and Diana Makri (Dimitriou) A.; Dr. ès sc., Athens Sch. Econ. Sci.; m. Dora Jovannou, 1962; children—Dora, Antoine, Alkis. Prof., Athens Sch. Econ. Sci.; ret. navy capt. Mem. Paris Statistics Soc., Greek Soc. Statistics (pres.). Author: Statistique, 4 vols.; Exercices de statistique, 3 vols. Research and publs. on math. statistics. Home: Odos Negri 9, Athens, Greece.

ATHANSIU, Jean, physiologist; b. Sascut, Moldavia, 1868; ed. Bucarest Vet. Sch., then Richet's Lab., Paris. Dir., Internat. Inst. Physiology, Paris, 1902-05; prof. physiology Bucarest Faculty Sci., 1906-26; rector U. Bucarest. Author: Bibliographia physiologica, 1893; Répertoire des travaux physiologie, 1894. Research in physiology, especially on nervous system. Died 1926.

ATHAVALE, Vishnu Traymbak, Indian chemist; b. Poona, India, Apr. 5, 1909; s. Tryambak Ramachandra and Parvati Athavale; student S.P. Coll., Fergusson Coll., Poona, 1927-31; B.Sc., Bombay (India) U., 1931, M.Sc., 1935, Ph.D., 1938; postgrad. Indian Inst. Sci, Bangalore; m. Kamala Vishnu, May 25, 1939; children—Devayani (Mrs. S. D. Chachad), Hemalata (Mrs. W. V. Ghat), Rajani. Asst. chemist Nat. Test House, Alipore, Calcutta, India, 1939-48; chemist rare minerals sect. Geol. Survey India, Calcutta, 1948-51, AEC, Bombay, 1951-56; head analytical div. Bhabha Atomic Research Centre, Bombay, 1957—. Recipient 1st prize for research paper All India competition Indsl. Research Bur., Govt. of India, 1938; R.R. Desai Gold medal Bombay U., 1939; Sudborough medal Indian Inst. Sci., Bangalore, 1938. Fellow Indian Chem. Soc.; mem. Royal Inst. Chemistry (U.K., asso.). Research, publs. on high temperature reactions, including hydrogenation of oils, refractive dispersion of oils and fats, hydrazine complexes of metals; studies and applications in electrochemistry, spectrophotometry, ion exchange, gravimetry and volumetric methods, activation analysis, geochemistry, low pressure methods, thermochemistry. Home: 427 Narayan Peth, Poona-2, India. Office: Analytical Div., Bhabha Atomic Research Centre, Modular Labs., Bombay-74AS, India.*

ATHAY, Russell Grant, Am. astrophysicist; b. Smith-field, Utah, Dec. 5, 1923; s. Henry Elzo and Mabel (Jacques) A.; student U. Wash., 1943, U. Cal. at Los Angeles, 1943-44; B.S. in Radio and Physics, Utah State U., 1947, postgrad.; Ph.D. in Astrophysics, U. Utah, 1953; m. Twila Jane Jensen, Apr. 20, 1945; children—Russell Jay, Carol Jane, Luann, Darrell Grant, Renee. Observer-in-charge High Altitude Obs., Harvard and U. Colo., 1950-52; sr. research staff High Altitude Obs., Nat. Center for Atmospheric Research, Boulder, Colo., 1953—, acting dir., 1961-62; research asso. Harvard Obs., 1955-56; chmn. dept. astrophysics, geophysics U. Colo., 1962-63, adj. prof., 1961—; head dept. physics U. Utah, 1963-64; vis. research scientist Max-Planck Institut, Munich, Germany, 1965-66. Mem. U. S. nat. com. Internat. Year of Quiet Sun, 1963—; mem. Nat. Acad. Sci.-NRC com. adv. to Nat. Bur. Standards, 1962—. Mem. Am. Astron. Soc., Am. Geophys. Union, Internat. Astron. Union. Author: (with R. N. Thomas) Physics of the Chromosphere, 1961; also numerous articles. Research on structure solar atmosphere, radiative energy transfer and theory spectral line formation. Home: 1029 Paragon Dr. Office: P.O. Box 1558, Boulder, Colo. 80302.*

ATHENAEOS OF ATTALEIA, physician; practiced medicine under Claudius in Rome, 41-54; founder sch. Pneumatists; considered med. knowledge part of gen. edn.; his system was important for speculative formulation rather than practical applications; physiology was based on Aristotle; basic elements included 4 qualities, with pneuma as 5th; explained health and disease in terms of good and bad temperaments.

ATHENAEOS OF CYZICOS, mathematician; flourished circa 350 B.C.; mem. of Academy; and conducted mathematical investigations with other mems.; famous in his time, for work in geometry.

ATKIN, Lawrence, Am. chemist; b. N.Y.C., Nov. 4, 1908; s. Joseph T. and Leona (Hausman) A.; B.S., Coll. City N.Y., 1932; Ph.D., N.Y. U., 1941; m. Hilda Zuckerman, Oct. 29, 1932; children—Susan Leona, Lucille Carol. With Standard Brands, Inc., 1928—, dir. research, Stamford, Conn., 1960—; research chemist Wallerstein Labs., N.Y.C., 1946-48, Chmn., Gordon Conf. on Food and Nutrition, 1947. Fellow Am. Inst. Chemists, N.Y. Acad. Sci.; mem. Am. Chem. Soc., A.A.A.S., Am. Assn. Cereal Chemists, Soc. Chem. Industries, Soc. Biol. Chemists, Am. Inst. Nutrition, Inst. Food Tech. Author: (with others) Enzymes and Their Role in Wheat Technology, 1946. Contbr. numerous articles to profl. jours. Identified growth factors for bakers and brewers yeasts; developed methods of analysis for B vitamins; contbr. to knowledge basic to enrichment of white bread. Home: 248 Westchester Ave., Tuckahoe, N.Y. 10707. Office: Betts Av., Stanford, Conn. 06904.*

ATKIN, Niels Bentzen, English biologist; b. London, Eng., July 20, 1913; s. Eric Edwin and Henriette (Bentzen) A.; B.A., Cambridge U., 1934, M.B.B.Ch., 1950; House surgeon Middlesex Hosp., London, 1939; with Med. Service, RAF, 1940-46; registrar in radiotherapy Middlesex Hosp., 1946-52; cytogeneticist dept. cancer research Mt. Vernon Hosp., Northwood, Middlesex, 1953—. Fellow Royal Soc. Medicine. Research, publs. on cytogenetics of human malignant disease, abnormalities of karyotypes, deoxyribonucleic acid content and sex chromatin content of malignant cells, chromosomal abnormalities in man other than those associated with malignant disease. Home: 37 Woodstock Av., Golders Green, London N.W. 11, Eng. Office: Mt. Vernon Hosp., Northwood, Middlesex, Eng.*

ATKINS, John, naval surgeon; b. Plaistow, Eng., 1685; entered navy as surgeon, circa 1703; served in various parts of world, including ships bound for Africa and W.I. Author: The Navy Surgeon, or a Practical System of Surgery, 1722; Voyage to Guiana, Brazil, and the West Indies, 1725. Gave 1st description African trypanosomiasis (sleeping sickness) in English, 1734; med. writing describes mineral springs, empirics, amulets, infirmaries, winds and currents around W.I. Died 1757.

ATKINS, Peter William, English theoretical chemist; b. Chalfont-St.-Giles, Eng., Aug. 10, 1940; s. William Henry and Ellen (Edwards) A.; B.Sc., U. Leicester (Eng.), 1961, Ph.D., 1964; postgrad. U. Cal. at Los Angeles; M.A., U. Oxford (Eng.), 1965; m. Judith Ann Kearton, Aug. 22, 1964. Harkness fellow Commonwealth Fund, U. Cal. at Los Angeles, 1964-65; fellow, tutor in phys. chemistry Lincoln Coll., Oxford, 1965—; univ. lectr. in chemistry Oxford U., 1965—. Fellow Chem. Soc., Faraday Soc.; mem. Am. Phys. Soc. Author: (with M.C.R. Symons) The Structure of Inorganic Radicals, 1966; also articles. Research on determination of structure of inorganic radicals using electron spin resonance, phenomena causing spin relaxation in molecules and relation of these to molecular interactions in solids, liquids and gases. Home: 7 Bridge St., Witney, Oxfordshire. Office: Lincoln Coll., Phys. Chemistry Lab., South Parks Rd., Oxford, Oxfordshire, Eng.*

ATKINS, William Ringrose Gelston, physiologist, biologist; b. Sept. 4, 1884; s. Thomas Gelston and Mary Eliza (Innes) A.; M.A., Sc.D., Trinity Coll., Dublin, Ireland; m. Ingaborg Jackson Miller, 1922; 1 son. Asst. to univ. prof. chemistry, Trinity Coll., 1906-11, univ. prof. botany, 1911-20; indigo research botanist Imperial Dept. Agr., India, from 1920; head, dept. gen. physiology Marine Biol. Assn. Lab., Plymouth, Devon, Eng., 1921-55; with Meteorol. Office, Air Ministry, 1943-45. Recipient Boyle medal Royal Dublin Soc., 1928. Fellow Royal Soc., 1925, Royal Inst. Chemistry, Inst. Physics. Author: Some Recent Researches in Plant Physiology, 1916; numerous reports, papers in sci. publs. Died Apr. 4, 1959.

ATKINSON, Daniel Edward Am. biochemist; b. Pawnee City, Neb., Apr. 8, 1921; s. Max and Amy (Neiswanger) A.; B.Sc., U. Neb., 1942; Ph.D., Ia. State U., 1949; m. Elsie Ann Hemmingson, Sept. 14, 1948; children—Kristine, Owen, Joyce, Ellen, David. Research fellow Cal. Inst. Tech., 1949-50; asso. scientist Argonne Nat. Lab., Chgo., 1950-52; asst. prof. chemistry U. Cal., Los Angeles, 1952-56, asso. prof., 1956-62, prof., 1962—. Mem. Am. Soc. Biol. Chemists, Am. Chem. Soc., Am. Soc. Microbiology, Am. Soc. Plant Physiologists. Contbr. articles to profl. jours. Study of metabolic regulation by modulation of enzyme activities; kinetics of regulatory enzymes; role of adenosine phosphates in control of energy metabolism; bacterial reduction of sulfite and nitrite. Home: 3123 Malcolm Av., Los Angeles 90034.*

ATKINSON, George Francis, Am. botanist; b. Raisinville, Mich., Jan. 26, 1854; s. Joseph and Josephine (Fish) A.; student Olivet Coll., 1878-83; Ph.B., Cornell, 1885; m. Lizzie S. Kerr, Aug., 1888. Asst. prof. entomology and gen. zoölogy, 1885-86. asso. prof., 1886-88, U. N.C.; prof. botany and zoölogy, U. S.C., and botanist of exptl. sta., 1888-89; prof. biology, Ala. Poly. Inst. and Agrl. and Mech. Coll. of Ala., and biologist of exptl. sta, 1889-92; asst. prof cryptogamic botany, Cornell U., Ithaca, N.Y., 1892-93, asso. prof., 1893-96, prof., head dept. botany, from 1896. Am. rep., bot. confs., 1905, 10. Mem. Am. Bot. Soc. (1st pres. 1905). Asso. editor The Botanical Gazette, 1896-98. Author: Heterodera radicicola; Biology of Ferns, 1894; Elementary Botany, 1898; Lessons in Botany, 1900; Studies of American Fungi, 1900; Mushrooms, Edible, Poisonous, Etc., 1903; First Studies on Plant Life, 1904; College Text-Book of Botany (enlargement of Elementary Botany), 1905. Contbr. 150 scientific papers to botan. jours. Died Nov. 14, 1918.

ATKINSON, Henry, Brit. mathematician; b. Great Bavington, Eng., June 28, 1781; s. Guthbert Atkinson; ed. by father. In charge Bavington Sch., circa 1794; opened sch. (with father), West Belsay, circa 1797, also superintended sch. at Woodburn; later kept sch. (with sister), Stamfordham; subsequently moved to Hawkwell, then Newcastle-on-Tyne, 1808. Contbd. papers on math. to jours. Discovered method for extracting roots of equations of higher orders (claim of priority supported by paper contbd. to Lit. and philos. Soc. Newcastle 1809; method pursued by Holched, also Nicholson, Horner), 1801. Died Jan. 31, 1829.

ATLEE, John Light, Am. surgeon; b. Lancaster, Pa., Nov 2, 1799; s. Col. William Pitt and Sarah (Light) A.; M.D., U. Pa., 1820; m. Sarah Franklin, Mar. 12, 1822, 3 children. First doctor to remove successfully both ovaries in 1 operation in a case of ovarian tumors, 1843; a founder Pa. Med. Soc., 1848, pres., 1857; an organizer A.M.A., Phila., v.p., 1865, pres. 1882; a promoter of Franklin and Marshall Coll., Lancaster, prof. anatomy until 1869; mem. Pa. Pub. Sch. Bd.; trustee Pa. State Lunatic Asylum; hon. fellow Am. Gynecol. Soc. Died Lancaster, Oct. 1, 1885.

ATLEE, Washington Lemuel, Am. surgeon; b. Lancaster, Pa., Feb. 12, 1808; s. William Pitt and Sarah (Light) A.; began study medicine with brother, John L. Atlee, 1824; grad. Jefferson Med. Coll., 1829; m. Ann Eliza Hoff, 10 children. Mem. staff Lancaster Hosp.; prof. med. chemistry Med. Coll. of Phila., 1844-52; pioneer of ovariotomy operation; pres. Pa. State Med. Soc., 1874, v.p. nat. soc., 1875. Author: Diagnosis of Ovarian Tumors, 1872. Died Sept. 6, 1878.

ATREYA, Hindu surgeon; flourished 6th century B.C.; tchr. at Taksasila on Jhelam River; one of 3 leading texts of Brahminical medicine based on work by Atreya (written by Caraka).

ATTENHOFER, Heinrich Ludwig, Swiss physician; b. Sursee, Switzerland, Apr. 2, 1783; ed. Vienna (Austria) U. hosp. physician, Petrograd, Russia, 1808-15; attained high polit. rank in Luzern canton; prescribed early successful treatment of syphilis with fever; also worked in demography, statistics. Author: Medizinische Topographie der Flaupt und Residenz Stadt St. Petersburg, 1817; Geschichtliche Denkwürdigkeiten der Stadt Sursee, Luzern, 1829. Died Sursee, June 25, 1856.

ATTFIELD, John, English pharmacologist; b. nr. Barnet, Eng.; 1835; s. John and Ann Attfield; entered Pharm. Soc., 1850; Ph.D., U. Tübingen (Germany); m. Martha Harvey, 1865; 2 daus. Demonstrator chemistry St. Bartholomew's Hosp., 1854-62; prof. practical chemistry Pharm. Soc., 1862-96. Cofounder Brit. Pharm. Conf., 1863, sr. sec., 1863-80, president, 1882-84, Fellow Royal Society, 1880; Instructor Chemistry (a founder council); member council chemistry society, 1874-78; honorary member 23 foreign societies and colls. of pharmacy. Author: Chemistry: General, Medical, and Pharmaceutical, 1st edit., 1867, 18th, 19th edits. (with Dobbin), 1903, 06; Water and Water-Supplies; also many papers, notes, lectures. Gen. Med. Council editor: (with others) Pharmacopoeia, 1885; Addendum to Pharmacopoeia, 1890; Pharmacopoeia, 1898; Indian and Colonial Addendum to Pharmacopoeia, 1900. Founder, reporter Gen. Med. Council Annual Reports on Progress of Pharmacy, adviser on pharm. chemistry, 1886-1900; founder, organizer imperialization of nat. Pharmacopoeia, 1886-93, 93-1900 (under Med. Council), also originator union of pharmacists with physicians in compilation, 1890. Died Mar. 18, 1911.

ATTHILL, Lombe, Brit. obstetrician, gynecologist; b. Ardess, Magheraculmoney, Ireland, Dec. 3, 1827; s. William and Henrietta Margaret (Eyre) A.; B.A., M.B., Dublin (Ireland) U., 1849, M.D., 1865; m. Elizabeth Dudgeon, Apr. 1850; m. 2nd, Mary Christie, June 1, 1872; 1 son, 9 daus. Began career as surgeon; head ward for treatment women's diseases Adelaide Hosp.; fellow Gonville and Caius Coll., Cambridge (Eng.) U.; pres. Dublin Obstet. Soc., 1874-76; pres. obstet. sect. Royal Acad. Medicine, 1884-85; pres. Irish Coll. Physicians, 1888. Author: Clinical Lectures on Diseases Peculiar to Women, 1883; Recollections of an Irish Doctor, 1911; Recollections of a Long Professional Life; also articles in profl. jours. Helped develop gynecology as a specialty; among first to perform successful ovariotomy in Ireland. Died Sept. 14, 1910.

ATTIX, Frank Herbert, Am. radiation physicist; b. Portland, Ore., Apr. 2, 1925; s. Ulysses Sheldon and Alma (Michelsen) A.; A.B. cum laude in Physics, U. Cal. at Berkeley, 1949; M.S. in Physics, U. Md., 1953; m. Shirley Adeline Lohr, Jan. 24, 1959; children—Shelley Anne, Richard Haven. Research physicist NIH, Bethesda, 1949-50, Nat. Bur. Standards, Washington, 1950-57; radiation shielding physicist ACF Industries, Washington, 1957-58; research physicist Naval Research Lab., Washington, 1958—, head dosimetry br., 1963—. Chmn. absorbed dose com. Internat. Commn. on Radiol. Units and Measurements, 1958-62; dosimetry cons. Nat. Commn. on Radiation Protection, 1957—. Mem. Am. Phys. Soc., Health Physics Soc., Research Soc. Am. Editor, coauthor: Luminescence Dosimetry, 1967; (with W. C. Roesch, E. Tochilin) Radiation Dosimetry, 1967; also numerous articles. Measurement radiation dose and related quantities; codeveloper of a theory of cavity ionization and exptl. verification of that theory; invented variable-length free-air ionization chamber; made (with V. H. Ritz) definitive measurement of gamma ray output of radium; measurement of dose by luminescent materials; clarified dosimetry concepts and units. Patentee in field. Home: 5125 27th Ave., Hillcrest Heights, Md. 20031. Office: Code 7620, Naval Research Lab., Washington 20390.*

ATTNEAVE, Fred, III, Am. psychologist, educator; b. Greenwood, Miss., Mar. 25, 1919; s. Fred II and Octavia (Rowell) A.; B.A., U. Miss., 1942; Ph.D., Stanford, 1950; children (1st. marriage)—Dorothy, Philip Henry; m. 2d, Alice Lynn Leeper, Dec. 29, 1959. Asst. prof. U. Miss., 1949-51; research psychologist Air Force Personnel and Tng. Research Center, San Antonio, 1951-56; fellow Center for Advanced Study in Behavioral Scis., Stanford, 1956-57; vis. asso. prof. U. Cal. at Berkeley, 1957-58; asso. prof. U. Ore., Eugene, 1958, prof., 1958—. Mem. Am. Psychol. Assn., A.A.A.S., Sigma Xi. Author: Applications of Information Theory to Psychology, 1959. Home: Route 4, Box 319-M, Eugene, Ore.*

ATTWOOD, Stephen Stanley, Am. elec. engr., educator; b. Cleve., May 29, 1897; s. William Rix and Mary (Hamilton) A.; B.S. Engring. in Mech. Engring., U. Mich., 1918, M.S. in Elec. Engring., 1923; m. Frieda Diekhoff, June 8, 1927; children—Stanley W., Julia C. Faculty elec. engring U. Mich., Ann Arbor, 1920—, prof., 1938—, chmn. elec. engring. dept., 1953-58, dean coll. engring, 1958—; staff mem. Mass. Inst. Tech., 1943-44. Mem. Ofcl. Govt. Mission to Eng. on Radiowave Propagation; 1943-44; dir. Wave Propagation Group, div. war research Columbia, 1944-45; cons. Detroit Edison Co., 1930, Fisher Body Corp., Detroit, 1939, Navy contract U. Tex., Austin, 1946. Recipient Pres.'s certificate of Merit for services in the field of radio propagation NDRC, 1948. Mem. Sigma Xi, Tau Beta Pi, Eta Kappa Nu. Author: Electric and Magnetic Fields, 1932; Radio Wave Propagation, 1949; also numerous articles. Home: 1520 Cambridge Rd., Ann Arbor, Mich. 48104.*

ATWATER, Wilbur Olin, Am. chemist; b. Johnsburgh, N.Y., May 3, 1844; grad. Wesleyan U., 1865; Ph.D., Sheffield Scientific Sch., Yale, 1869; student univs. of Leipzig and Berlin; m. Marcia Woodard, Aug. 26, 1874. Prof. chemistry East Tenn. U., 1871-72, Me. State Coll., 1873; prof. chemistry Wesleyan U., Middletown, Conn., from 1873; dir. Conn. Agrl. Expt. Sta., (1st state agrl. sta. in U. S.), 1875-77; dir. Storrs (Conn.) Expt. Sta., 1887-1902; founded, Office Expt. Stas., U. S. Dept. Agr., 1888, Dir., until 1891; spl. agt. Dept. Agr., from 1891, chief of nutrition investigations, from 1908. Writer over 150 papers on chem. and allied subjects, notably (with F. G. Benedict) An Experimental Inquiry Regarding the Nutritive Value of Alcohol. One of inventors of Atwater-Rosa-Benedict respiration calorimeter (linked phys. engry and human metabolism). Died 1907.

ATWELL, Wayne Jason, Am. embryologist; b. Fairfield, Neb., Oct. 19, 1889; grad. Neb. Wesleyan

U., 1911; A.M., U. Mich. 1915, Ph.D., 1917; M.D., U. Buffalo, 1934. Became head anatomy dept. U. Buffalo, 1918; grantee NRC, 1938; tchr. histology, anatomy, U. Mich. Contbr. articles to Contributions to Embryology, 1930; contbr. numerous biol. articles to sci. jours. Studied functions of the several parts of the hypophysis, also early human embryo; contbd. to knowledge of pituitary body. Died Mar. 27, 1941.

ATWOOD, George, English physicist, mathematician; b. London, Eng., 1746; ed. Trinity and Westminster colls., Cambridge, grad. 1769; fellow, tutor, prof. Trinity Coll., 1769-84; position in English customs service after 1784. Fellow Royal Soc., 1776. Author: Analysis of a Course of Lectures on the Principles of Natural Philosophy, 1784; Treatise on the Rectilinear motion and Rotating of Bodies, 1784; An Essay on the Arithmetic of Factors, 1786; Dissertation on the Construction of Arches, 1801. Inventor, Atwood's machine for verifying laws of falling bodies; rediscovered radix method for computing logarithms; investigated stability of bodies on surface water. Died Westminster, Eng., July 11, 1807.

ATWOOD, Wallace Walter, Am. geographer, geologist; b. Chicago, Ill., Oct. 1, 1872; s. Thomas Green and Adalaide Adelia (Richards) A.; B.S., U. of Chicago, 1897, Ph.D., 1903 D.Sc., Worcester Polytechnic Institute, 1943; LL.D., Clark University, Worcester, Mass., 1946; married Harriet Towle Bradley, Sept. 22, 1900; children—Rollin Salisbury, Wallace Walter, Jr., Mrs. Harriet Olmsted, Mrs. Mary Hedge. Asst. geologist, N.J. Geol. Survey, 1897, Wis. Natural History Survey, 1898-99; instr., Lewis Inst., Chicago, 1897-99, Chicago Inst., 1900-01; fellow asst. and asso., 1899-1903, instr. and asst. prof. physiography and gen. geology, 1903-10. asso. prof., 1910-13, U. of Chicago; prof. physiography, Harvard, 1913-20; president Clark Univ., Sept. 1920-1946, president emeritus since 1946; dir. Clark School of Geography, 1920-46. Asst. geologist, U. S. Geological Survey, 1901-09, geologist since 1909; geologist, Ill. Geol. Survey since 1906. Recipient, distinguished service award, U. of Chgo., 1945, distinguished service medal, Chgo. Geog. Soc., 1948. Pres. Nat. Parks Assn., 1929-33. Fellow Geol. Soc. Am., Am. Acad. Arts and Sciences, Am. Antiquarian Soc.; mem. Assn. Am. Geographers (ex-pres.), Chicago Geog. Soc., Ill. Acad. Sciences, Chicago Museum and Library Extension Council, Nat. Council Geography Teachers, pres. Pan-Am. Inst. of Geography and History, 1932-35, hon. pres.; mem. Commn. Internationale de l'Atlas des Formes du Relief Terrestre; hon. member National Academy of Science of Mexico, Mexican Society of Geography and Statistics. Capt., Harvard R.O.T.C., 1917. Author: Physical Geography of the Devil's Lake Region (with R. D. Salisbury), 1899; Physical Geography of the Evanston-Waukegan Region of Ill. (with J. W. Goldthwait), 1908; Interpretation of Topographic Maps, 1908; Glaciation of the Uinta and Wasatch Mountains, 1909; Mineral Resources of Southwestern Alaska, 1910; Geology and Mineral Resources of the Alaska Peninsula, 1911; New Geography, Book II, 1920; Home Life in Far Away Lands, 1928; The Americas, 1929; Nations Beyond the Seas, 1930; The United States Among the Nations, 1930; The World at Work, 1931; Physiography and Quaternary Geology of the San Juan Mountains, Colorado, 1932; The Growth of Nations, 1936; Physiographic Provinces of North America, 1940; The Protection of Nature in the Americas (publ. No. 50, Pan-Am. Institute of Geog. and History), 1941; The United States in the Western World, 1944; The Rocky Mountains, 1945; co author (with Ruth E. Pitt) Our Economic World, 1948; also numerous sci., and ednl. papers. Founder and editor of Economic Geography, 1925. Research in glacial and physiographic geology; studies on geography of U. S., other countries. Died Annisquam, Mass., July 24, 1949.

AUB, Joseph, Am. physician; b. Cin., Jan 16, 1846; M.D., Med. Coll. Ohio, 1866; apptd. oculist, aurist Cin. Hosp., 1871; prof. ophthalmology, otology Cin. Coll. Medicine and Surgery. Transplanted skin from arm to eyelid for cure of ectropion, 1877. Died Cin., May 14, 1888.

AUBERT, Hermann, German physiologist; b. Frankfurt, Germany, Nov. 1826; ed. Berlin, grad., 1850; prof. physiology, Breslau (now Wroclaw, Poland), later at Rostock, Germany, 1862-92. Author: Die cephalopoden des Aristoteles, 1862; Physiologie der Netzhaut, 1865; Handbuch der Physiologie; Shakespeare als Mediziner, 1873; Grunzüge der physiologischen Optik, 1876. Editor, Zeitschrift für Psychologie. Made 1st thorough exptl. measurement of zonal changes of color sensitivity; gave 1st description Aubert phenomenon; studied phenomena of dark and light adaption, also indirect vision, visual and cutaneous space perception, bodily orientation. Died Feb. 12, 1892.

AUBERT, Jacques Albert (or Jacobus Albertus), chemist, physician; b. Vendôme, France. Author: Des natures et complexions des hommes et d'une chacune partie d'iceux, et aussi des signes par lesquels on peut discerner la diversité d'iceles, 1571; De metallorum ortu et causis contra Chemistas brevis et dilucida explicatio, 1575; Duae apologeticae responsiones ad Josephum Quercetanum, 1576; Progymnasmata in Joannis Fernelii librum de abditis rerum naturalium causis, 1579; Institutiones physicae instar commentariorum in libros physicae Aristotelis, 1584; Semeiotice seu ratio dignoscendarum sedium male affectarum et affectuum praeter naturam, 1587. Died Lausanne, Switzerland, 1586.

AUBERTIN, Charles, French physician; b. 1876; M.D., 1916; became intern, 1900, certified physician, 1912; aggregate physician, 1923; dir. Le Mans Surg. Center; prof. therapeutics Paris Faculty Medicine, from 1939. Research in cardiology, also morphology of blood and blood-forming tissues. Died Paris, 1950.

AUBIN, André Léon, French physician; b. Meaux, France, July 16, 1887; s. Léon and Claire (Oudart) A.; studied medicine, otorhinolaryngology; M.D.; m. Antoinette Thierry, June 13, 1918; children—Jacqueline, Jean-Pierre, Philippe. Prof. med. clinic Faculté de Médecine, Paris. Mem. Acad. Medicine. Contbr. articles to French and Internat. Congress of Otorhinolaryngology. Address: 30, rue Guynemer, Paris 6, France.

AUCHTER, Eugene Curtis, Am. horticulturist; b. Elmgrove, N.Y., Sept. 14, 1889; s. William David and Florence Monroa (Curtis) A.; B.S., Cornell U., 1912, M.S., 1918, Ph.D., 1923; m. Catherine Elizabeth Beanmont, Aug. 25, 1914. Asst. in pomology, Cornell U., 1911-12; asst. and asso. prof. horticulture W.Va., U., 1912-17; head dept. of horticulture U. Md., 1918-28; prin. horticulturist in charge div. fruit and vegetable crops and diseases U.S. Dept. of Agr., 1928-38, asst. chief Bur. Plant Industry, 1935-38, chief Bur., 1938-42. adminstr. Agrl. Research, 1942-45; pres. and dir. Pineapple Research Inst. of Hawaii, also v.p. Pineapple Growers Assn. from 1945. Bd. mgrs. N.Y. Bot. Gardens. Hon. Fellow Royal Hort. Soc. London; fellow A.A.A.S.; mem. Am. Soc. Plant Physiologist, Wash. Acad. Sci., Am. Soc. Naturalists, Bot. Soc. Washington, Am. Phytopathology Soc., Am. Soc. Hort. Sci., Am. Genetics Assn., Sigma Xi, Phi Kappa Phi, Alpha Zeta. Author: (with H. B. Knapp) Orchard and Small Fruit Culture, 1929; Growing Tree and Small Fruit, 1929. Died July 8, 1952; buried Honolulu.

AUDEBERT, Jean Baptiste, French naturalist, painter, engraver; b. Rochefort, France, 1950/59; student natural history; miniature painter; produced plates illustrating animals, birds. Author: Natural History of Apes, Lemurs and Galeopitheci (with 62 plates printed by new oil color method Audebert invented), 1800; History of Humming-Birds, Fly-Catcher, Jacamars, and Promerops, 1802; unfinished works of birds later edited by Vieillot and Desray. Died 1800.

AUDIFFREDI, Giovanni Battista, Italian astronomer, mathematician, bibliographer; b. Saorgio, Italy, Feb. 2, 1734; joined Dominican Order; in charge Bibliotheca Casanetensis (began catalog of holdings); began bibliography of books pub. in major Italian cities; wrote on transits of Venus, also comets of 1769. Author: Celestial Phenomena Observed, 1753-56. Died Rome, Italy, July 3, 1794.

AUDOUARD, Matheiu François Maxence, French physician; b. Castres, France, July 29, 1776; ed. Montpellier, France. Served in French army; voted pension for research by French legislature; author treatises on intermittent fevers, yellow fever, 1807-24; tried to prove that Stegomyia fasciata mosquito (now known as Aedes aegypti) transmits yellow fever. Died 1856.

AUDOUIN, Jean-Victor, French entomologist, naturalist, anatomist; b. Paris, Apr. 27, 1797; ed. as physician; m. dau. Alexandre Brongniart, 1827. Apptd. prof. anthropology Natural History Mus., 1833; asst. librarian Inst. France. A founder Soc. Entomologique de France; mem. French Acad. Scis., Soc. Agr. Author: Histoire des insectes nuisibles à la vigne et particulièrement de la pyrale, 1842. Founder (with A. T. Buorgniart and J. B. A. Dumas) and editor, Annales des scis. naturelles, 1824. Contbd. to Règne animal (Cuvier). Investigated (with Milne-Edwards) Crustacea and Annelida; studied muscardine, plant parasites, fauna of coastal Atlantic. Died Paris, Nov. 9, 1841.

AUDRIETH, Ludwig Frederick, chemist, educator; born Vienna, Austria, Feb. 23, 1901; s. Ludwig Anton and Fredericka (Herrmann) A.; came to U.S., 1902, naturalized, 1912; B.S., Colgate U., 1922; Ph.D., Cornell U., 1926; postgrad. (Nat. Research fellow) U. Rostock (Germany); m. Maryon Laurice Trevett, Mar. 2, 1937; children—Kaaren (Mrs. James Robert Tague, Jr.), Elsa Craven, Anthony Ludwig. Research asst. Cornell U., 1926-28; faculty chemistry U. Ill., Urbana, 1928—, prof., 1946-61, prof. emeritus, 1961—. Indsl. cons., 1937—; cons. mil., govtl. depts., agys., 1946—; ofcl. investigator NDRC, 1940-42; mem. Chem. Corps Adv. Council, 1952-57; sci. attache Am. Embassy, Bonn, Germany, 1959-63; cons. on sci. affairs Dept. State, 1966—. Recipient Prechtel medal Tech. U. Vienna, 1965. Mem. Am. Chem. Soc., Gesellschaft Deutscher Chemiker, Ill. Acad. Scis., A.A.A.S., Phi Beta Kappa, Sigma Xi, Sigma Nu, Phi Lambda Upsilon (nat. pres. 1950-54, nat. editor 1938-42). Author: (with B.A. Ogg) The Chemistry of Hydrazine, 1951; (with J. A. Kleinberg) Non-aqueous Solvents, 1953. Founding mem. Inorganic Syntheses, 1934, editor-in-chief, 1945-50. Research, publs. on nitrogen products, rocket fuels, solvents; co-discoverer of sucaryl. Home: 1515 Waverly Dr., Champaign, Ill. 61822. Office: Dept. Chemistry, U. Ill., Urbana, Ill. 61803.

AUDUBON, John James, artist, ornithologist; b. Les Cayes, Santo Domingo (now Haiti), Apr. 26, 1785; illegitimate son of Jean Audubon and a Creole woman known as Mademoiselle Rabin, legalized by adoption (1794) as son of Jean and Anne (Moynet) A.; m. Lucy Bakewell, June 1808; children—Victor, John W. Went with father to live in France, 1789, baptized Jean Jacques Fougère Audubon, 1800; came to U. S., to avoid conscription during Napoleonic Wars, 1803; naturalized 1812; settled on father's estate "Mill Grove," nr. Phila., 1804; began studies of Am. birds, made 1st bird-banding expt. (thus initiating study of bird migrations), 1804; engaged in various bus. activities, N.Y.C and Louisville, Ky.; jailed for debt, 1819; taxidermist Western Museum, Cincinnati, 1819-20; began trip down Mississippi and Ohio rivers to observe birds, 1820, paid expenses by painting portraits; tutor, drawing tchr., street sign painter, New Orleans; made unsuccessful trip to Phila. to find a publisher for his bird drawings, 1824; taught music and drawing (to wife's pupils), St. Francisville, La., 1825; took drawings to Europe, 1826, met with favorable reception; elected to Royal Soc. Edinburgh (Scotland), 1827; Fellow Royal Soc., 1830; paid expenses by painting birds; engaged in preparing drawings and texts for his 1st book, Edinburgh, returned several times to continue work; returned to Am. with reputation as foremost American naturalist, 1831; had completed 435 natural history paintings by 1838; elected fellow Am. Acad., 1830; settled on estate "Minnie's Land" (now Audubon Park), N.Y.C., 1841; subject of much controversy during his life; responsible for drawings and literature for his books (most of scientific identification and nomenclature supplied by others); one of earliest Am. conservationists. Author: Birds of America (1st modern atlas of ornithology), 4 vols., 1827-38; Ornithological Biography, 5 vols., 1831-38; Synopsis of the Birds of North America, 1839; Viviparous Quadrupeds of North America (completed by his sons), colored plates pub. in 2 vols., 1842-45, text in 3 vols., 1846-54. Died N.Y.C., Jan. 27, 1851.

AUENBRUGGER, Leopold Edler von Auenbrugg, Austrian physician; b. Graz, Austria, Nov. 19, 1722; pupil of van Swieten, U. Vienna; physician Spanish Hosp., Vienna, 1751-62; became physician-in-chief Hosp. Holy Trinity, Vienna, 1751. Recipient patent of nobility Emperor Joseph II, 1784. Author: Inventum novum, ex percussione thoracis humani, ut signo, abstrusos interni pectoris morbus detegendi, 1761. A founder of anat. diagnosis; invented percussion method of phys. diagnosis. Died Vienna, May 18, 1809.

AUER, John, Am. pharmacologist, physiologist; b. Rochester, N.Y., Mar. 30, 1875; s. Henry and Luise (Hummel) A.; S.B., U. Mich., 1898; M.D., Johns Hopkins, 1902; m. Clara Meltzer, Oct. 1, 1903; children—James, Helen, John. Med. house officer, Johns Hopkins Hosp., 1902-03; fellow, asst., asso. and asso. mem. Rockefeller Inst. for Med. Research 1903-06 and 1907-21; instr. physiology, Harvard Med. Sch., 1906-07; prof. pharmacology and dir. dept., St. Louis U. Sch. of Medicine, from 1921; pharmacologist, St. Mary's group of hosps. from 1924. Mem. A.A.A.S., Am. Physiol. Soc., Am. Soc. for Pharmacology and Exptl. Therapeutics (sec. 1912-16; pres., 1924-27), Soc. for Exptl. Biology and Medicine (v.p. 1917-19), Assn. Am. Physicians, Harvey Soc., St. Louis Acad. Scis. Author: numerous publications on digestion, respiration, physiological action of various drugs; studies on war gases, tetanus, connective tissue, etc. (with Metzer) proved the anesthetic properties of magnesium, 1906. Died Aug. 30, 1948.

AUER, Karl (Baron von Welsbach), chemist; b. Vienna, Austria, Sept. 1, 1858; student of Bunsen in Heidelberg, Germany; created hereditary noble by Franz Josef I of Austria, 1901. Made extensive investigations rare metals; produced cerium from Welsbach mantle to improve gas lighting; separated supposed element didymium into prascodymium and neodymium; invented osmium incandescent lamp (longer lasting than carbon filament, prepared way for discovery of tungsten filament), 1898; discovered Mischmetal (commonly used as cigarette lighter flints). Died Welsbach Castle, Carinthia, Austria, Aug. 8, 1929.

AUER, Vaino, Finnish geologist and geographer; b. 1895; ed. Helsinki U.; Ph.D., Dr. H.C.; hon. degree U. Bonn (Germany). Mem. Forest Research Inst., 1918-29; geologist Geol. Survey Can., 1926; lectr. geography Helsinki U., 1922-29, prof., 1929-63; leader geog. expdns. to Patagonia and Tierra del Fuego, 1928-29, to Patagonia, 1937-38, 47-53, 57. Chmn. nat. com. Finland Internat. Geographer Union, 1955-57; mem. Commn. Natural Scis. of the State, 1956-61; hon. dir. Forestry Research of Argentina, 1947; mem. Finnish Sci. Acad. highest award of honour of Finnish Cultural Found. Recipient Fennia Gold medal, 1950; Kairamo Silver medal, 1950; Holmberg award Acad. de Ciencias Exactas, Argentina, 1950; Cajander medal; Martin Behaim Silver medal, 1959; Rueppell Silver medal, 1961; Helsinki U. medal; Rockefeller scholar, 1926. Mem. Geochem. Soc., Nat., Finnish (pres. 1959) acads. scis., Finnish

Geog. Soc., Finnish Forestry Assn. (pres. 1935-36). Author: Las Capas Volcánicas como Base de la Cronología Postglacial de Fuegopatagonia, Consideraciones científicos sobre la conservación de los recursos naturales de la Patagonia; Nuevos Aspectos de la Sequia en la Patagonia; The Pleistocene of Fuego-Patagonia: I The Ice and Inter-Glacial Ages, II History of the Flora and Vegetation, III Shoreline Displacements, IV Bog Profiles. Research in palynology, teprochronology, glacial geology and paleography. Address: Rakuunantie 4B 14, Helsinki, Finland.

AUERBACH, Friedrich, German chemist; b. Breslau, Germany (now Wroclaw, Poland), Aug. 23, 1870; s. Leopold and Arabella (Hess) A.; ed. Leipzig, also Breslau, from 1888; degree, 1893; m. Selma Sachs, 1897; 1 dau. Worked in chem. industry, also pvt. labs.; entered ofcl. state health service, 1904. Author: ((with Abegg) Handbuch für anorganische Chemie, 1908. Researcher in inorganic chemistry; expert in poisoning problems. Died Berlin, Aug. 4, 1925.

AUERBACH, Leopold, German anatomist; b. Breslau, Germany (now Wroclaw, Poland), April 28, 1828; student in Berlin, Leipzig, Breslau; M.D., 1849; m. Arabella Hess; children—Felix, Friedrich. Practiced medicine, from 1850; became lectr. U. Breslau, 1863, asso. prof., 1872. Contbd. papers to sci. lit. Described ganglion cells (Auerbach's ganglion) in plexus of autonomic nerve fibers between muscular coats of intestine, 1862, 63; gave 1st description nuclear substance (erythrophi) of cells with affinity for red dyes; credited with introducing prefix karyo- (in word Karyolis), 1874. Died Breslau, Sept. 30, 1897.

AUFFENBERG, Walter, Am. zoologist; b. Dearborn, Mich., Dec. 6, 1928; s. Walter and Ida (Lange) A.; B.S., Stetson U., Deland, Fla., 1951; M.S., U. Fla., 1953, Ph.D., 1956; m. Elinor Ann Wright, July 5, 1947; children—Walter, Kurt, Garth, Troy. Asst. prof. dept. biology, asso. curator Fla. State Mus., U. Fla., Gainesville, 1956-59, asso. curator, chmn. dept. natural scis. Fla. State Mus., asso. prof. dept. zoology, 1963-—; asso. dir. biol. scis. curriculum study U. Colo., 1959-63; USIS cons. Am. Sci. Film Forum, 1963; AID cons. Indian Edn. Project, 1964, NSF cons., 1967; chmn. film com. Biol. Scis. Curriculum Study, 1964, 67; cons. Center for Latin Am. Studies, 1967. Fellow A.A.A.S.; Mem. Am. Soc. Ichthyologists and Herpetologists (bd. govs.), also others. Inst. Biol. Scis. Contbr. numerous articles to profl. jours. Research in area of vertebrate paleontology and animal behavior. Home: Route 3, Box 191M, Gainesville, Fla.*

AUFHAMMER, Gustav Adolf Wilhelm, German agronomist, plant breeder; b. Larrieden, Germany, Feb. 22, 1899; s. Johann Georg and Babette (Herrmann) A.; Diplomlandwirt, Techn. Hochschule München, 1922, Assessor, 1925, Dr.techn.sciences, 1928; m. Else Niedermüller, Aug. 14, 1935; children—Walter, Dietlinde (Mrs. Bauhöfer), Armin, Wolfram, Renate. Profl. practice plant breeding farm, Mauern, Germany, 1922-23; referendar, assessor service plant breeding inspection, Ansbach, Germany, 1924-27; staff Technische Hochschule, München, Germany, 1927-31, prof., 1949-—, dir. Institute Plant Cultivation and Plant breeding, 1949-—, dir. research farm Roggenstein, 1960-—, rector magnificus, 1960-62, dean faculty agronomy, 1954-56, chief leader adminstrn. Weihenstephan, 1957-59; assessor Ackerbauschule Triesdorf, Germany, 1931-35; staff Regierungsrat Bayer. Landessaatzuchtanstalt Weihenstephan, 1936-49. Recipient Honor needle Ackerbauverband Mittelfranken Verdienstorden of Bavaria, 1965. Mem. Bundessortenamt, Deutsche Forschungsgemeinschaft, Forschungsrat für Ernährung, Landwirtschaft und Forsten, Sci. Council Brewing Barley, Bundesforschungsanstalt für Getreideverarbeitung, Gesellschaft für Pflanzenbauwissenschaften, Ernährungsbiologie, European Brewery Conv. (mem. barley com.), Landwirtschaftsausschuss des Wissenschaftstrates. Author: (with Kiessling), Bilderatlas zur Braugerstenkunde, 1931; (with P. Bergal, F. R. Horne), Barley Varieties, 1955; Auflage in Vorbereitung, 1964; Neuzeitlicher Getreidebau, 1964; (with G. Fischbeck, A. Haisch) Die Verwendung von Gammabestrahlungsanlagen zur Behandlung von Pflanzen, 1965; also numerous articles. Research on improving resistance against winter coldness of grain, for raising yields of wheat and barley, for improving baking quality of winter and summer wheat, brewery quality of summer barley; plant breeding of new varieties with resistance against diseases, higher yields and higher quality of wheats and barleys. Home: 9 Hohenbachernstrasse, 805 Freising, Bavaria, Westdeutschland. Office: Institut für Pflanzenbau und Pflanzenzüchtung der Technischen Universität München in 805 Freising-Weihenstephan.*

AUGE, Juan, Spanish mathematician; b. Barcelona, Spain, June 10, 1919; s. José and Rosa A.; licentiate in math. U. Barcelona, 1941; D. in Math., U. Madrid (Spain), 1943; m. Pilar Lidón Falcó, 1948. Prof. math. analysis U. Barcelona, 1948-—, sec. Faculty Sci., 1946-52, vice dean Faculty Sci., 1958-60. Mem. Royal Spanish, Am. math. socs. Author: Investigaciones sobre el método de Graeffe, 1943. Research, publs. on distbns., ordinary and partial differential equations. Address: Faculty Scis., Barcelona U., Barcelona 7, Spain.

AUGELLI, John P., geographer; b. Celenza, Italy, Jan. 30, 1921; s. P. John and Antoinette (Tacaruso) A.; B.A., Clark U., 1943; M.A., Harvard, 1949, Ph.D., 1951; m. Conchita Garcia de la Noceda, Aug. 24, 1948; children—John F., Robert W. Teaching fellow Harvard, 1947; asst. prof. U. P.R., 1948-52; asso. prof. U. Md., 1952-61, prof., 1961; prof. U. Kan., Lawrence, 1961-—. Cons. U. S. State Dept., Ford Found., Fgn. Area Fellowship Program, Fulbright Screening Com., Office Health, Edn. and Welfare; U. S. mem. Commn. on Geography Pan Am. Inst. Geography and History. Mem. Social Sci. Research Council (joint com. Latin Am. studies 1964-67), Assn. Am. Geographers, Am. Geog. Soc., Latin Am. Area Studies Assn. Author: (with Robert West) Middle America: Its Land and Peoples, 1966; also numerous articles. Research in human, econ., cultural geography of Latin Am., with spl. emphasis on Caribbean, So. Brazil, Central Am. Home: 1131 West Hills Pkwy., Lawrence, Kan. 66044.*

AUGENSTEIN, Leroy George, Am. biophysicist; b. Decatur, Ill., Mar. 6, 1928; s. Roy Henry and Minnie (Reifsteck) A.; student James Millikin University, 1944-46, Doctor of Science (honorary), 1967; B.S., U. Chgo., 1949; M.S., U. Ill., 1954, Ph.D., 1956; m. Elizabeth Schmalfuss, Sept. 23, 1950; 1 son, David Leroy. Research asso. U. Ill., 1951-56; biophysicist Brookhaven Nat. Lab., Upton, N.Y., 1956-58, 60-62; biophysicist AEC, 1958-60, lectr., 1962-—; sci. coordinator U.S. Sci. Exhibit, Seattle World's Fair, 1960-61; prof., chmn. biophysics dept. Mich. State U., East Lansing, 1962-—. Dir. U. Internat., East Lansing. Cons. McPherson Instrument Co., Acton, Mass., 1965-66, NIH, 1966-—, Am. Inst. Biol. Scis., 1962-—. Mem. Sigma Xi. Author: Bioenergetics, 1959; (with R. Mason, B. Rosenberg) Physical Mechanisms in Radiation Biology, 1964; also articles. Demonstrated that radiation produces highly specific bond rupture in proteins; research on adsorption of proteins at an interface, processing of information and data processing in humans. Home: 2708 Ramparte Path, Holt, Mich. 48842. Office: Mich. State U., East Lansing, Mich. 48823.*

AUGER, Pierre Victor, French physicist; b. Paris, France, May 14, 1899; s. Victor E. and Eugenie (Blanchet) A.; Docteur ès sciences, Ecole Normale Superieure de Paris, 1926; m. Suzanne Motteau, July 7, 1921; children—Mariette Berl, Catherine Mintz. Prof., U. Paris, 1937-—; dir. Higher Edn. for France, 1945-48. Dir. sci. dept. UNESCO, 1948-60; pres. French Space Commn., 1960-62; dir. gen. European Space Research Orgn., 1964-67. Named commdr. French Legion of Honour, 1961; recipient Foltrinelli Internat. Prize, 1961. Discoverer Auger-Effect in atomic physics, extensive Auger Showers (EAS) of cosmic rays. Home: 12 Emile Faguet, Paris 14, France.*

AUGHINBAUGH, William Edmund, American physician, explorer; b. Westmoreland County, Va., Oct. 12, 1871; s. William L. and Anna M. (O'Neill) A.; LL.B., Nat. Law Sch., Washington, 1892, LL.M., 1896; M.D., Columbian (now George Washington) U., 1897; m. Mary A. Douglas, Oct. 12, 1900. Admitted to U. S. Supreme Ct. bar; was asso. editor Leslies Mag. and fgn. and export editor N.Y. Comml.; then asso. editor N.Y. Daily Investment News; dir. Joint Securities Corp. Lectr. fgn. econs. and fgn. trade N.Y. U., also Columbia U. Author: Selling Latin America, 1915; A Port for Bolivia, 1916; Trademark Tragedies, 1916; Advertising for Trade in Latin America, 1922; Advertising for Trade Abroad, 1922; Volcanoes, Rats and Men, 1932; I Swear by Apollo, 1938. Traveled extensively; worked among lepers; studied bubonic plague in India; made health survey for Egypt. Died Dec. 18, 1940.

AUGURELLO, Giovanni Aurelio, Italian humanist, alchemist; b. Rimini, Italy, circa 1454. Author odes and epistles in style of Horace; best known for Chrysopoeia, Geronticon Liber, (poems of alchemy). Died circa 1537.

AUGUST, Ernst Ferdinand, inventor; b. Prenzlau, Germany, 1795; prof., Berlin, Germany; built psychrometer (humidity gauge), 1828, also differential barometer, heliostat, spiral hygroscope; investigated correction of thermometer; set up steam pressure formula of water steam. Died 1870.

AUGUSTIN, Mircea Octavian, Rumanian physican; b. Bucharest, Rumania, Apr. 30, 1921; s. Mircea S. and Elena (Angelescu) A.; M.D., U. Bucharest, 1947, D.Med. Scis., 1966. Med. researcher 1st degree Inst. Endocrinology, Acad. Rumania, Bucharest, 1948-—. Contbr. numerous articles to profl. jours. Research on endemic goiter, exophthalmos, pathogenesis of obesity and on metabolic disturbances in hypo- and hyperanabolic syndromes. Home: 45 Maria Rosetti, Bucharest 13. Office: 34 Bul. Aviatorilor, Bucharest, Rumania.*

AUGUSTINE, Saint (Aurelius Augustinus), philosopher, theologian; b. Tagaste, Numidia (now Souk-Ahras, Algeria), Nov. 13, 354; s. Patricius, a pagan, and Monnica, a Christian; ed. Tagaste, Madauros, Carthage (where he joined Manichean sect); took a mistress, who bore him a son, Adeodatus, 371; left Carthage for Rome, 383; held chair of rhetoric, Milan, 384, where he came under influence of Ambrose and neo-platonism; converted, 386; ordained priest, 391 at Hippo Regius, Numidia (now Bône, Algeria), auxiliary bishop, 395, and bishop shortly after. Author: Confessiones; De civitate dei; De doctrina christiana; De trinitate; De magistro; De vera religione; De libero arbitrio; Contra Faustum Manichaeum; De natura et gratia; Retractiones; many others. Successively a Manichaen, neo-platonist, and Christian; encouraged study of sciences, generally as an aid to accurate interpretation of scripture; held world created by God; that He implanted certain seeds or potencies (rationes seminales) in material things; thus new forms appear in course of time; this sometimes said to be a theory of evolution; held fetus receives soul during second month; in political theory, developed influential doctrine of two cities (heavenly and earthly), competing for allegiance in God's creation; in general, Augustine's ideas profoundly affected subsequent development of thought in Western world; one of most influential thinkers of all time. Died Hippo (during siege by the Vandals), Aug. 28, 430.

AUGUSTINE OF TRENT, (or Augustinus de Tridento), Italian med. astrologer; b. Brescia, Italy; flourished 1340; Augustinian monk; lectr. U. Perugia (Italy); author treatise on pestilence of 1340 (written before Black Death, covers astrological causes and conditions, regimen to avoid plague, diet, bloodletting and purgation, astrological lore of year).

AUINGER, Mathias, Austrian paleontologist; b. Lindach, Austria, Mar. 23, 1811; s. Matthias and Maria Anna (Pirschinger); with mil. service until 1849; became clerk Hof-Mineralien Kabinet (ofcl. geol. collection), Vienna, Austria, 1849, civil servant, 1871; corr. Geol. Inst. Author: (with R. Hoernes) Die Gasteropoden der Meeres-Ablagerungen (standard work on geology of Vienna basin), 1879-84; assisted B. Schwarz von Mohrenstein in publishing paleontol. monographs. Discovered numerous fossil shellfish species. Died Vienna, Oct. 11, 1890.

AULT, Wayne Urban, Am. geochemist; b. Monroe, Mich., Jan. 20, 1923; s. Percy Leroy and Grace (Love) A.; B.A., Wheaton Coll., 1950; M.A., Columbia U., 1956, Ph.D., 1957; m. Ruth Andrews, June 10, 1947; children—Leroy, Gary, Douglas, Gordon. Apprentice, Ford Motor Co., Dearborn, Mich., 1941-43; sci. tchr. Dundee (Mich.) Community Schs., 1947-48; instrument engr. Wheaton Engring., (Ill.), 1948-49; research asst. Lamont Geol. Obs., Columbia U., Palisades, N.Y., 1950-57; geochemist, volcanologist U. S. Geol. Survey, Hawaiian Volcano Obs., Hawaii Nat. Park, Honolulu, 1957-62; sr. scientist, dir. tech. relations Isotopes, Inc., Westwood, N.J., 1962-—. Recipient U. S. Dept. Interior Citation for documentary sci. film on volcanos. Mem. Am. Geophys. Union, Geol. Soc. Am., Geochem. Soc., Sigma Xi. Contbr. chpt to Researches in Geochemistry, 1959. Contbr. articles to sci. jours. Research on isotopic tracers in environmental pollution; instrumentation in application of radiation to indsl. gauging; research on application of radioisotope tracers for secondary recovery, primary pressure maintenance and underground gas storage for petroleum industry; application of geochronometry to geologic problems; natural radiation measurements to grading of potash ore; geochem. and geophys. research on volcanoes; geochemistry of sulfur isotopes and application to origin of ore deposits. Home: 41 Highview Av., Nanuet, N.Y. 10954. Office: 123 Woodland Av., Westwood, N.J. 07675.*

AURAND, Leonard William, Am. biochemist, educator; b. Shamokin Dam, Pa., Feb. 5, 1920; s. James Wilson and Esther (Weissinger) A.; B.S., Pa. State U., 1941, Ph.D., 1949; M.S., U. N.H., 1947; m. Eleanor May Nichols, Feb. 22, 1943; children—Rebecca Louise, Thomas James, Sarah Jane. Faculty dept. biochemistry N.C. State U., Raleigh, 1949-—, prof., 1960-—. Cons. nutrition sect. Dept. Health, Edn. and Welfare, 1960-—. Mem. Am. Chem. Soc., Am. Dairy Sci. Assn., Am. Inst. Nutrition, Inst. Food Technologists, Sigma Xi, Alpha Zeta, Alpha Chi Sigma, Phi Lambda Upsilon, Gamma Sigma Delta. Author: (with H. O. Triebold) Food Composition and Analysis, 1963. Research, publs. on oxidation of lipids, effects of heat on proteins of milk, off-flavors in foods. Home: 921 Trailwood Dr., Raleigh, N.C. 27606.*

AURELIANUS, Caelius, see Caelius Aurelianus.

AURIOL, Pierre, French natural philosopher; student U. Paris, France, 1304; D. and magister in theology, Paris, 1318; joined Franciscan order; lector Franciscan house, Bologna, Italy, 1312, Toulouse, France, 1314; provincial of Aquitanian, France, 1320; archbishop of Aix-en-Provence, France, 1321. Author: De principiis naturae, 1312; Quodlibeta, 1320. Discussed moot questions of metaphysics, psychology, ethics with Thomas Anglicus, Herve of Nedélés and the Scotists. Died Avignon, France, Jan. 1322.

AUSLASSER, Vitus, botanist; b. Vomp nr. Schwaz, Austria, 15th century; studied sci.; became Benedictine monk, Monastery St. Sebastian, Ebersberg, Upper Bavaria. Author: Herbarius (contains med. comments on plants, also 198 watercolor illustrations of plants, some showing phases of devel. of same specimen), 1479.

AUSMAN, Robert K., Am. physician; b. Milw., Jan. 31, 1933; s. Donald C. and Mildred (Shafrin) A.; student Kenyon Coll., 1950-52; student Marquette

U., 1952-53, M.D., 1957; postgrad. U. Minn. Damon Runyon Research fellow U. Minn., 1958-61; research fellow Roswell Park Meml. Inst., Buffalo, 1961; dir. Health Research, Inc., Buffalo, 1961—; asso. prof. U. Buffalo, 1963—. Recipient Minn. Medicine Writing award, 1962. Author: Gold and Blue-Fifty Years, 1963; also numerous articles. Developed disposable oxygenator for isolated cancer perfusion; developed 1st simplified automated med. records system. Home: 869 Delaware Av., Buffalo 14209. Office: 666 Elm St., Buffalo 14203.*

AUST, Joe Bradley, Am. physician; b. Buffalo, Sept. 8, 1926; s. Joe Bradley and Edith Catherine (Derby) A.; student U. Buffalo, 1943-44, M.D., 1949; student Union Coll., 1944-45; M.S., U. Minn., 1957, Ph.D., 1958; m. Constance Anne McMullen; children—Jay Bradley, Bonnie Jean, Barbara Ann, Linda Lee, Mary Louise, Tracy Roberta. Faculty, U. Minn., 1958-66, prof. dept. surgery, 1964-66; with U. Minn. Hosps., 1957-64, asso. prof., 1960-64; chief surgery Anoka State Hosp., 1962-65, dir. Tumor Clinic, 1958—, cancer coordinator, 1960—, prof. surgery, 1964—; prof., chmn. dept. surgery S. Tex. Med. Sch., San Antonio, 1966—. Fellow A.C.S.; mem. A.A.A.S., N.Y. Acad. Scis., Am. Assn. for Cancer Research, Soc. Head and Neck Surgeons, Soc. U. Surgeons, Soc. Exptl. Pathologists, Am. Assn. for History Medicine, Am. Surg. Assn., Am. Soc. Clin. Oncology, Soc. Exptl. Biology and Medicine (founding), Soc. for Surgery Alimentary Tract, Transplantation Soc. (founding). Research and publs. on diagnosis of intracranial lesions, blood volume measurements using RIHSA, surgery of intracardiac defects, regional perfusion, homografts. Office: 915 Stadium Dr., San Antonio.

AUSTEN, Ernest Edward, Brit. entomologist; b. London, Eng., Oct. 19, 1867; s. Ambrose Austen; ed. Heidelberg (Germany) U.; m. Cécile Mary Noël Buchanan, 1905 (died 1932); 2 daus. With Brit. Mus., from 1889, keeper of entomology, 1927-32; naturalist on bd. Siemens' Bros. Cable S. S. Farady, Amazon River, Brazil, 1895-96; 1st expdn. Liverpool Sch. Tropical Medicine to Sierra Leone, 1899. Mem. Colonial Office Departmental Com. on Sleeping Sickness, 1913-14, Tsetse Fly and Locust Coms., Econ. Adv. Council, Sci. Research and Entertainment Panels, Adv. Council Brit. Film Inst. Recipient Mary Kingsley medal Liverpool Sch. Tropical Medicine, 1920. Fellow Zool. Soc.; mem. Royal Soc. Tropical Medicine and Hygiene (v.p.) Author: A Monograph of the Tsetse Flies; British Blood-Sucking Flies; Illustrations of African Blood-Sucking Flies; A Handbook of the Tsetse Flies; (with E. Hegh) Tsetse Flies, their Characteristics, Distribution, and Bionomics, with some account of Possible Methods for their Control; the House-Fly, its Life-History, Importance as a Disease Carrier, and Practical Measures for its Suppression; Blood-sucking Flies, Ticks, etc., and How to Collect Them; Blood-Sucking and other Flies known or likely to be concerned in the spread of Disease; (with A. W. McK. Hughes) Clothes Moths and House Moths, their Life-History, Habits and Control; Bombyliidae of Palestine; contbr. articles Encyclopaedia Britannica, profl. jours. Prepared several tropical health exhibits Brit. and internat. exhbns. Died Jan. 16, 1938.

AUSTIN, Colin Russell, biologist; b. Sydney, Australia, Sept. 12, 1914; s. Ernest Russell and Linda (King) A.; B.V.Sc., U. Sydney, 1936, B.Sc., 1938, M.Sc., 1940, D.Sc., 1954; M.S., Cambridge (England) University, 1967; m. Patricia Constance Jack, Feb. 15, 1941; children—Richard, Mark. Research officer Commonwealth Sci. and Indsl. Research Orgn., Sydney, 1938-54; lectr. Sydney U., 1938-54; sci. staff Med. Research Council, London, Eng., 1954-64; head genetic and developmental disorders research program Delta Regional Primate Research Center, prof. embryology Tulane U., Covington, La., 1964-67; Charles Darwin prof. animal embryology Cambridge University, 1967—. Visiting professor Florida State University, 1963; corp. member Marine Biol. Lab., Woods Hole, Mass., 1962—. F. R. Lillie Meml. fellow, 1961. Author: The Mammalian Egg, 1961; Fertilization, 1965; also numerous articles. Editor, Jour. Reproduction and Fertility, 1959-64. Research on fertilization, devel. early embryo. Home: Manor Farm House Toft, Cambridge, Eng.*

AUSTIN, George Marion, Am. neurosurgeon; b. Phila., May 10, 1916; s. George and Gladys (Weisel) A.; A.B. in Math., Lafayette Coll., 1938; M.D., U. Pa., 1942; M.Sc. in Neurology and Neurosurgery, McGill U., 1951; grad. radioisotope course Inst. Nuclear Studies, Oak Ridge, 1953. Mem. faculty U. Pa. Med. Sch., 1952-57; prof., head div. neurosurgery U. Ore. Med. Sch., Portland, 1957-68; prof., chief sect. neurosurgery Loma Linda (Cal.) U. Med. Sch., 1968—. Diplomate Am. Bd. Neurol. Surgery. Mem. Research Soc. Neurol. Surgeons (chmn. 1963-64), Am. Physiol. Soc., Biophys. Soc, Soc. Indsl. and Applied Math., Soc. U. Surgeons, Am. Electroencephalographic Soc., Harvey Cushing Soc., A.C.S., A.M.A., Scandinavian, Western neurosurg. socs., Internat. Soc. Research in Stereoencephalotomy, Sigma Xi. Research and publs. in neurophysiology, neurology, neurosurgery. Address: Loma Linda University School of Medicine, Loma Linda, Cal.*

AUSTIN, James Bliss, Am. metallurgist; b. Washington Grove, Md., July 16, 1904; s. Harry A. and Pauline (Rearick) A.; Chem. E., Lehigh U., 1925, D.Sc. (hon.), 1962; Ph.D., Yale, 1928; m. Janet Evans, Oct. 7, 1930; children—Peter Allison, Winifred May. Phys. chemist research lab. U. S. Steel Corp., 1928-44, asst. dir. research lab., 1944-46, dir. research, 1946-54, asst. v.p., 1954-56, v.p., 1956-58, adminstv. v.p., 1958—. Fellow Metall. Soc. Am., Inst. Mining and Metall. Engrs. (dir. 1959-62), Am. Ceramic Soc. (trustee 1943-46), A.A.A.S., N.Y. Acad. Sci.; mem. Am. Soc. for Metals (pres. 1954), Iron and Steel Inst. London (hon. v.p.), Century Assn., Iron and Steel Institute Japan (Honorary), National Academy of Engineering, Phi Beta Kappa, Tau Beta Pi, Alpha Chi Sigma. Author: The Flow of Heat in Metals, 1942. Research on chem. thermodynamics, thermal expansion of metals and refractories, adsorption gases on metals. Home: 114 Buckingham Rd., Pitts. 15215. Office: 525 William Penn Pl., Pitts. 15230.*

AUSTIN, John, Brit. polit. theorist, jurist; b. Suffolk, Eng., Mar. 3, 1790; s. Jonathan A.; served with Brit. Army in Sicily, circa 1806-11; studied law, called to bar Inner Temple, 1818; read law with John Stuart Mill, 1820-21; studied in Heidelberg, Bonn (both Germany), 1826-28; m. Sarah Taylor, 1820; 1 dau., Lucie. Gave up law practice due to poor health, 1825; prof. jurisprudence U. London, 1826-32; apptd. mem. Criminal Law Commn., 1833; lectured on jurisprudence for soc. Inner Temple, 1834; apptd. (with Sir George Cornewall Lewis) royal commr. Island of Malta, 1836-38; returned to Eng., wrote for Edinburgh Rev.; lived in Paris, 1844-48, returned to Eng. at start of French Revolution. Author: The Province of Jurisprudence Determined, 1832; Lectures on Jurisprudence, 1869. Presented comprehensive analysis of principles underlying all legal systems; contbd. precision, order and logic to utilitarian doctrine in field of juristic thought; concisely re-defined and clarified such concepts as law, status, sovereignty; known as father of positive law and determinate sovereignty; an architect of modern jurisprudence, he held that law is based on power, that positive law is expression of will of sovereign authority (not to be confused with dictates of religion or ethics); 1st to draw distinction between legal and moral criteria in analysis of law. Died Weybridge, Surrey, Eng., Dec. 1859.

AUSTIN, Louis Winslow, Am. physicist; b. Orwell, Vt., Oct. 30, 1867; a. Prof. Lewis Augustine and Mary Louise (Taft) A.; A.B., Middlebury Coll., 1889, hon. D.Sc., 1920; student U. Strassburg, 1889-90, 91-93, Ph.D., 1893; fellow Clark U., 1890-91; m. Laura A. Osborne, Aug. 16, 1898. Instr., asst. prof. physics, U. Wis., 1893-1901; in German govt. service (Phys. Tech. Reichsanstalt), 1902-04; joined Bur. of Standards, Washington, 1904; head U. S. Naval Radiotelegraphic Lab., 1908-23; became chief Lab. for Spl. Radio Transmission Research, Bur. Standards, 1923. Fellow I.R.E. (pres. 1914, medal 1927); v.p. Internat. Union for Scientific Radiotelegraphy and chmn. Am. Sect.; mem. tech. adv. com. Conf. on Limitation of Armament, Washington, 1921. Author: Physical Measurement (with Prof. C. B. Thwing), 1896. Died June 27, 1932.

AUSTIN, Mary Lellah, Am. zoologist; b. Austinburg, O., Apr. 12, 1896; d. Henry Lewis and Harriet (Phillips) A.; B.A., Wellesley Coll., 1920, M.A., 1922; Ph.D., Columbia, 1928. Asst. custodian zoology dept Wellesley (Mass.) Coll., 1920-22, instr., 1928-31, asst. prof., 1931-40, asso. prof., 1940-52, prof., 1952-61; research scholar zoology dept. Ind. U., Bloomington, 1961—; asst. Barnard Coll., N.Y.C., 1925-26, lectr. 1926-28; exchange tchr. zoology dept. Isabella Thoburn Coll., Lucknow, India, 1932-33, lectr. 1935-36. Fellow A.A.A.S.; mem. Genetics Soc. Am., Am. Soc. Human Genetics, Soc. Protozologists, Am. Soc. Naturalists, Am. Soc. Zoology, N.Y. Acad. Sci., Phi Beta Kappa, Sigma Xi, Sigma Delta Epsilon. Research on physiology, genetics, serotypes, antigens, antibiotic response of paramecium aurelia. Home: 506 1/2 N. Indiana Av., Bloomington, Ind. 47403.*

AUSTIN, Oliver Luther Jr., Am. ornithologist; b. Tuckahoe, N.Y., May 24, 1903; s. Oliver L. and Elizabeth (Wise) A.; B.S., Wesleyan U., Conn., 1926; Ph.D., Harvard, 1931; m. Elizabeth Schling, Sept. 10, 1930; children—Anthony, Timothy Oliver. Mem. U. S. Biol. Survey, 1930-35; dir. Austin Ornithol. Research Sta., 1931-58; head wildlife br. Nat. Resources sect., SCAP, Tokyo, 1946-51; prof. zoology Air-War Coll., Maxwell AFB, 1952-57; curator birds Fla. State Mus., Gainesville, 1957—. Moderator, Town Wellfleet, Mass., 1935-42. John Simon Guggenheim fellow, 1952. Fellow Am. Ornithologists Union (v.p. 1963-65). Author: Birds of Newfoundland Labrador, 1932; Birds of Korea, 1948; Birds of Japan, 1953; Birds of the World, 1961; numerous articles. Editor The Auk, 1967—. Home: 205 S.E. 7th St., Gainesville, 32601. Office: Fla. State Mus., Seagle Bldg., Gainesville, Fla. 32601.*

AUSTRIAN, Robert, Am. physician; b. Balt., Apr. 12, 1916; s. Charles Robert and Florence (Hochschild) A.; A.B., Johns Hopkins, 1937, M.D., 1941; m. Babette Friedmann Bernstein, Dec. 29, 1963; children—Jill Bernstein, Toni Bernstein. Practice medicine, specializing in internal medicine, Bklyn., 1952-62, Phila., 1962—; vis. physician Kings County Hosp., 1952-62; attending physician Hosp. U. Pa., 1962—; asso. prof. medicine State U. N.Y. Downstate Med. Center, 1952-57, prof. medicine, 1957-62; vis. scientist dept. microbial genetics Pasteur Inst., Paris, France, 1960-61; John Herr Musser prof., chmn. dept. research medicine Sch. Medicine U. Pa., 1962—; Tyndale vis. lectr., prof. Coll. Medicine U. Utah, 1964; cons. Bklyn. Hosp., Maimonides Hosp., 1953-62, Phila. Gen. Hosp., 1962—. Mem. study sect. NIH, 1965—; com. mem. Armed Forces Epidemiological Bd., 1965—. Trustee Johns Hopkins University, Baltimore. Recipient U. S. A. Tyhus Commn. medal, 1947. Diplomate Am. Bd. Internal Medicine. Fellow A.C.P., N.Y. Acad. Scis.; mem. Assn. Am. Physicians, Am. Soc. Clin. Investigation, Am. Clin. and Climatol. Assn., Am. Soc. Microbiology, Soc. Exptl. Biology and Medicine, Harvey Soc., Am. Fedn. Clin. Research, Am. Assn. Immunologists, N.Y. Acad. Medicine, Phi Beta Kappa, Sigma Xi, Alpha Omega Alpha, Omicron Delta Kappa. Mem. editorial bd. Am. Rev. Respiratory Diseases, 1963-66, Jour. Bacteriology, 1964—, Bacteriological Revs., 1967—. Research, numerous publs. on pathogenesis, treatment and prevention of bacterial pneumonia and allied disorders, genetics and factors controlling virulence of pneumococci and streptococci. Home: 250 S. 17th St., Phila. 19103. Office: Hosp. U. Pa., Phila. 19104.*

AUTHIER, André, French crystallographer; b. La Fleche, France, June 16, 1932; s. Jean and Elizabeth (Hipper) A.; Ecole Normale Supérieure, 1951, Agrégation de Physique, 1955, Docteur ès sciences, 1961; m. Françoise Lefevre, 1959; children—Bertrand, Isabelle, Sophie. Under dir. lab. theoretical physics Collège de France, 1961-65; maitre de confs. lab. mineralogy crystallography Faculté des Sciences de Paris, 1965—, prof. sans chaire, 1968—. Recipient Médaille de Bronze, CNRS, 1961; prix Danton, 1963; prix Ancel, 1967. Mem. Société Française de Physique, Association Française de Cristallographie, Société Française de Minéralogie-Cristallographie. Research, publs. on crystal defects using X-ray topography-dynamical theory of X-ray diffraction. Home: 12 Degas, Paris 16, France.*

AUTOLYCOS OF PITANE, astronomer, mathematician; b. Pitane, Aeolia, circa 360 B.C.; older contemporary of Euclid; author treatise on motion of sphere, treatise on risings and settings of fixed stars (earliest extant complete texts on Greek math.). Showed skill in geometry; 1st to try to explain certain difficulties in theory of homocentric spheres. Died circa 300 B.C.

AUVERT, Louis-Rene, engr.; b. 1857; ed. École Polytechnique; inventor accumulator locomotive (proved unwieldy), 1899; asst. in installation direct current line from Fayet to Chamonix, France; builder (with Ferrand) locomotive which was supplied with single phase current, carried own transformer, 1911. Died 1930.

AUWERS, (Georg Friedrich Julius) Arthur von, see von Auwers, (Georg Friedrich Julius) Arthur.

AUWERS, Karl Friedrich von, see von Auwers, Karl Friedrich.

AUZOUT, Adrien, French physicist, astronomer; b. Rouen, France, Jan. 1622; mem. Royal Obs. Paris. Fellow Royal Soc., 1666; mem. French Acad. Scis., 1666. Author: Traité de micromètre, 1667. Discovered shadow of Saturn on its ring, 1662; improved thread network micrometer, 1667; measured (with Picard) meridian between Malvoisine and Amiens, France, 1669-70. Died Rome, Jan. 12, 1691.

AVDEEV, Mikhail Ivanovich, Russian physician; b. 1901; grad. Med. Faculty 1st Moscow U., 1924; D.Med. Sci., 1940. Postgrad., asst. dept. forensic medicine Med. Faculty, 1st Moscow U., 1925-31; asst. dept. path. anatomy Sechenov 1st Moscow Med. Inst., also forensic med. cons., 1931-40; head chair forensic medicine and criminology Mil. Legal Acad., 1940-42; chief forensic expert Soviet Army, 1942—; head Central Forensic Med. Lab., Mil. Med. Bd., USSR Ministry Def., 1942—; hon. sci. worker of RSFSR, 1962—; mem. learned council Inst. Criminology, Ministry Internal Affairs and Mil. Med. Bd. Mem. USSR Acad. Med. Sci. (corr.), All-Union Soc. Forensic Physicians and Criminologists (bd. mem.). Author over 70 works and 100 articles on forensic medicine and gen. pathology including Manual of Forensic Medicine for Law Schools; A Course in Forensic Medicine, 1959. Editor works on forensic medicine; co-editor Ency. Dictionary of Military Medicine; A Legal Dictionary; Large Med. Ency., 2d edit. Specialist in forensic medicine; founder Soviet mil. forensic med. service. Address: USSR Ministry of Defense, Arbatskaya pl., Moscow, USSR.

AVEBURY, see Lubbock, Sir John.

AVE-LALLEMENT, Robert Christian Barthold, physician, explorer; b. Lübeck, Germany, July 25, 1812; studied medicine, Berlin, also Heidelberg, Paris, Kiel; m. Meta Löwe, Feb. 28, 1841; 3 children; m. 2d, Ida Louise Löwe, Apr. 11, 1856; 2 children; m. 3d, Adamine Ulrike Wilhelmine Rosen, 1872. Practiced medicine, Rio de Janeiro, Brazil, from 1837, Lübeck, from 1859; mem. Brazilian Health Commn.; returned to Lübeck and joined A. Humboldt's Novara expdn. to Brazil, 1855; travelled down Nile, 1869.

Author: Reise durch Südbrasilien, 2 vols., 1859; Reise durch Nordbrasilien, 2 vols., 1860; Des Dr. Joachim Jungius aus Lübeck Briefwechsel mit seinem Schulern und Freunden, 1869; Wanderungen durch die Pflanzewelt der Tropen, 1880; Das Leben des Dr. med. J. Jungius 1587-1657, 1882. Died Lübeck, Oct. 10, 1884.

AVENARIUS, Richard Ernst Abund, German chemist; b. Koblenz, Germany, Feb. 9, 1840; s. Albert and Amalie (Maehler) A.; m. Angelika Maeckler, Mar. 25, 1867; 1 son. Served in army until 1871; discovered process of impregnating wood with creosote; formulated wood preserving fluid (carbolineum or Avenarius); founded plants in Germany, Austria, Hungary, Russia, from 1867. Died Gau-Algesheim/Rhine, Germany, Feb. 1, 1917.

AVENARIUS, Richard Heinrich Ludwig, philosopher, psychologist; b. Paris, France, Nov. 19, 1843; studied at Zurich (Switzerland), Berlin (Germany); grad. U. Leipzig (Germany), 1876; m. Maria T. Semper, Mar. 17, 1877; Prof. philosopher, Zurich, 1877-96; founder Akademischphilosophische Verein, Leipzig. Author: Über der beiden ersten Phasen der Spinozischen Pantheismus, 1866; Kritik der Reinen Erfahrung, 2 vols., 1888-80; Der menschliche Weltbegriff, 1891; Philosophie als Denken der Welt Gemäss den Princip des Kleinsten Kraftmasses, 1876. Tried to use principle of least action to relate pure experience to intellect and environment (empirioriticism); conducted (with K.F.W. Ludwig) physiol.-neurol. research; used introjection as connection between individual and his environment; introduced term independent in reference to experience underlying phys. fact. Died Zurich, Aug. 18, 1896.

AVENDS, Georg Adalbert, German horticulturist; b. Essen, Germany, Sept. 21, 1863; s. Karl and Sophie (Steckel) A.; student gardening and plant growing Inst. for Fruit, Wine and Garden Culture, Geisenheim, later in Breslau (Germany) Bot. Garden; m. Helen Pfeifer, May 14, 1891; 2 sons; established exptl. and ednl. instns. for garden culture. Recipient numerous awards; G.-A. Meml. medal founded in his honor. Mem. and officer German and fgn. assns. Author: Mein Leben als Gärtner and Züchter. Produced new breeds of flowers and improved existing ones. Died Wuppertal-Ronsdorf, Germany, Mar. 5, 1952.

AVENTINUS, Johannes (original surname Turmair), historian, cartographer; b. Abensberg, Germany, July 4, 1477; s. Peter Turmair; ed. Ingolstadt, also Vienna, Cracow, Paris; 2 children. Tchr., Paris, France, also Ingolstadt (Germany) U., from 1508; tchr. Greek and math., Cracow, Poland; tutor to bros. Duke William IV, Munich, Germany, from 1512; collected hist. information on Bavaria, Germany, for dukes Wilhelm and Ludwig, after 1510; produced map Bavaria (important regional map), 1523; helped found the Sodalitas Litteraria Angilostadensis, an organization for bringing old manuscripts to light. Author: Abacus atque vetustissima, veterum latinorum per digitos manusque numerandi . . . consuetudo (on finger symbolism), 1522; Annalium Boiorum libri VII, 1554; Baier Chronik, 1566. Called the "Bavarian Herodotus;" stressed humanistic approach to historical writing. Died Regensburg, Germany, Jan. 9, 1534.

AVENZOAR, see Ibn Zuhr.

AVERROËS (Abu al-Walîd Muhammad ibn-Ahmad ibn-Rushd), Muslim philosopher, physician; b. Cordova, Spain, 1126; student theology, jurisprudence, math., medicine, philosophy, Cordova; m.; several sons. Versed in Malekite system of law; lived in Marrakesh, Morocco, circa 1153-69; judge, Seville, Spain, from 1169, Cordova, from 1171; physician to Sultan Abu Ya' qub Yusuf, Marrakesh, 1182, Ya' qub al-Mansur, circa 1182-95; accused of heresy, exiled to Lucena, nr. Cordova, circa 1195; recalled to Marrakesh, restored to favor, 1198. Author: Destructio Destructionis; De Substantia Orbis; De Animae Beautitudine; Colliget; several commentaries on Aristotle (for which chiefly noted); various treatises on jurisprudence, astronomy, grammar, medicine, philosophy. One of most important medieval Muslim thinkers; philosophy (Averroism) in many ways similar to late Christian scholasticism (had impact on Christian thought which was coming to forefront as Islamic philosophy faded); in medicine, had clear understanding of function of retina; recognized immunity conferred by single episode of smallpox. Died Marrakesh, Dec. 10, 1198.

AVERY, Oswald Theodore, physician; b. Halifax, Nova Scotia, Oct. 21, 1877; A.B., Colgate; M.D., Coll. Phys. and Surg., Columbia, 1904; Sc.D., Colgate, 1921; LL.D., McGill U., 1935. Mem. Rockefeller Inst. for Med. Research, 1923-43, researcher emeritus, 1943-48. Recipient Paul Ehrlich Gold medal, Copley medal Royal Society; Kober Found. medal Georgetown U., 1946. Mem. Assn. Physicians, Am. Soc. for Clin. Investigation, Assn. Pathologists and Bacteriologists, Am. Soc. Exptl. Pathology, Soc. Am. Bacteriologists, N.Y. Acad. Medicine, Am. Acad. Arts and Sciences, Nat. Acad. Sciences, Der Norski Videnskaps Akademi (Olso), Académi Royale de Medécine de Belgique, Société Philomathique de Paris. Research on pneumococcus bacteria, their intercellular endo-enzymes and new serological types causing lobar pneumonia; (with Dochez) discovered specific soluable substances responsible for immunological action of pneumococci, 1947. Died Feb. 20, 1955.

AVERY, Samuel, Am. chemist; b. Lamoille, Ill., Apr. 19, 1865; s. Stephen B. and Mary T. Avery; A.B., Doane Coll., 1887; B.Sc., U. Neb., 1892, A.M., 1894; Ph.D., Heidelberg, 1896; LL.D., Doane and U. of Idaho, 1909; m. May B. Bennett, Aug. 4. 1897. Adj. prof. chemistry, U. Neb., 1896-99; prof. chemistry and chemist, Agrl. Expt. Sta., U. Ida., 1899-1901; prof. analytical and organic chemistry, U. of Neb., 1901; prof. agrl. chemistry and chemist, U. Neb. Expt. Sta., 1902-05; head prof. chemistry U. Neb., Lincoln, 1905-08, acting chancellor, 1908-09, chancellor, 1909, chancellor emeritus and prof. of research in chemistry, from 1927. Vice chmn. chemistry com. NRC, 1918. U. S. mem. Internat. Conciliation Commn. with Sweden, 1914-15. Died Jan. 25, 1936.

AVERY, William Hinckley, Am. physicist; b. Ft. Collins, Colo., July 25, 1912; s. Edgar Delano and Mabel (Gordon) A.; A.B., Pomona Coll., 1933; A.M., Harvard, 1935, Ph.D., 1937; m. Helen Wallace Palmer, July 18, 1937; children—Christopher, Patricia (Mrs. W. Randolph Bartlett, Jr.). Research chemist Shell Oil Co., St. Louis, Houston, 1939-43; head propulsion div. Allegany Ballistics Lab., NDRC, Cumberland, Md., 1943-46; cons. physics, chemistry Arthur D. Little & Co., Cambridge, Mass., 1946-47; with Applied Physics Lab., Johns Hopkins, Silver Springs, Md., 1947—, supr. aeros. div., 1961—; mem. Polaris Ad Hoc Group on Long Range Research and Devel., 1958-64; mem. coms. NASA, 1959—; mem. spl. adv. com. on propellants to dir. def. research and engring., 1962. Recipient Naval Ordnance Devel. award, 1947, C. N. Hickman award, 1951. Fellow Am. Inst. Aeros. and Astronautics; mem. Am. Chem. Soc., Am. Phys. Soc., Combustion Inst. (dir.), Phi Beta Kappa. Contbr. numerous articles to profl. jours. Research and patents in rocket and ramjet propulsion, chem. kinetics and combustion, molecular spectra; pioneer process of mfg. rocket propellants, rocket loading arrangement, liquid-solid rocket, spectrometer prism mountings, temperature measuring device. Home: 724 Guilford Ct. Office: 8621 Georgia Av., Silver Spring, Md. 20910.*

AVEZ, Andrew, French mathematician; b. Montrouge, France, Oct. 27, 1930; Doctorat de mathématiques, Sorbonne, Paris, 1963. Asso. prof. U. Minn., Mpls., 1966; prof. math. U. Paris (France), 1967—. Mem. Société Mathématique de France, Am. Math. Soc. Author: (with V. Arnold) Problèmes ergodiques de la mécanique classique, 1966; Ergodic Theory of Dynamical Systems, 1966; also articles. Research on gen. relativity (cosmological models, radiations), differential geometry (Lorentzian manifolds, harmonic spaces, conformal transformations, Einsteinian spaces), ergodic theory (Classical systems). Home: 3 Victor Basch, Montrouge 92, France. Office: I.H.P., 11 Rue Pierre Curie, Paris, France.*

AVIADO, Domingo Mariano, pharmacologist; b. Manila, Philippines, Aug. 28, 1924; s. Domingo and Severina (Mariano) A.; permanent resident U. S., 1952—; student U. Philippines Coll. Liberal Arts, 1940-42, Coll. Medicine, 1942-45; M.D., U. Pa., 1948; m. Asuncion P. Guevara, Aug. 15, 1953; children—Maria Cristina, Carlos G., Domingo G., Maria Asuncion. Faculty U. Pa., 1948—, prof. pharmacology, 1965—; mem. Phila. Gen. Hosp. staff, 1955—. Vis. lectr. anesthesiology Albert Einstein Med. Center, 1955—; vis. lectr. physiology Women's Med. Coll., 1961-62. Recipient Rockefeller Found. Travel award, 1961; Linnaeus medal 1st Internat. Pharmacological meeting, Stockholm, 1961, Purkinje medal 2d Internat. Pharmacological meeting, Prague, 1963. NIH fellow, 1948-50, Guggenheim Found. fellow, 1962-63. Mem. Am. Soc. Pharmacology and Exptl. Therapeutics, Am. Physiol. Soc., A.A.A.S., Am. Heart Assn., Internat. Union Physiol. Scis. (treas. sect. on pharmacology 1959-65), Internat. Union Pharmacology (treas. 1965-66), Physiol. Soc. Phila. (pres. 1959-60), Sigma Xi, Alpha Omega Alpha. Author: The Lung Circulation, 2 vols., 1965. Sect. editor Chem. Abstracts, 1952-58; asso. editor Circulation Research, 1958-62; editorial cons. Dorland Illustrated Med. Dictionary, 1963—. Research, publs. on action of drugs on cardiovascular and respiratory systems. Home: 1020 Indian Creek Rd., Phila. 19151.*

AVICENNA (Ibn Sina, or Abu Ali al Hussein ibn Abdallah), Persian physician, philosopher; b. Kharmaithen, nr. Bukhara, Persia, 980; early edn. under tutors and wandering scholars; self-taught in logic, geometry, and the Almagest. Apptd. physician to Amir of Bukhara, 997, saved his life and rewarded with access to royal library of Samanids; went to Urjensh (in modern Khiva) and received monthly stipend from vizier after end of Samanid dynasty, 1004; traveled through dists. of Nishapur and Merv, also to borders of Khorasan, circa 1012; moved to Rai (nr. modern Teheran), then to Hamadan; apptd. med. attendant by amir, Shams Addaula; became vizier, later banished, then restored at illness of amir, imprisoned by new amir until Hamadan was captured by Isfahan, 1024, escaped and fled to Isfahan; physician, gen. sci., lit. adviser to Prince Abu Ya'far'Als Addaula, to 1037. Author numerous works which had enormous influence on devel. medicine (standard med. texts through 17th century) including most famous work: The Canon of Medicine; also numerous philos. works. Greatest physician medieval times; systematically compiled entire theoretical and practical med. knowledge known to his day; based theories on work of Galen and Hippocrates, using methods of Aristotle; described in detail causes, symptoms, cures for innumerable diseases; studied drugs and pharmacological methods; stated that retina, not lens, was essential organ of vision; announced theory transmission certain infections by way of placenta; developed idea of tracheotomy; studied various nervous diseases and discussed therapeutic measures; philosophy (known as Avicennism) based on Aristotle but includes Neoplatonic ideas; made astron. observations; suggested speed of light must be finite quantity; wrote treatise on alchemy (rejected metallic transmutation). Died Hamadan, Persia, June, 1037.

AVOGADRO, Amedeo (Conte di Quarengna), Italian math. physicist; b. Turin, Italy, June 9, 1776; s. Felipe A.; ed. U. Turin (Italy); B.Jurisprudence, 1792, D.Eccles.-Law. 1796; studied math. and physics, 1800-05; m. Felicita Massèdi Biella; 6 children including Luigi, Felice. Practiced law for many years; became prof. math. and physics Royal Coll. Vercelli, 1809; apptd. to 1st Italian chair of math. physics at Turin, 1820, chair suppressed on polit. grounds, 1822, reoccupied chair, 1834-50, ret., 1850. Honored with monument, unveiled in Turin in 1911 (100 years after publ. of most famous work). Mem. Acad. Scis. Turin, Chief Ednl. Council, Turin. Author: Fisica di codpi ponderabili, 4 vols., 1837-41; Essai d'une manière de déterminer les masses relatives des molécules élémentaires des corps, et les proportions selon lesquelles elles entrent dans les combinaisons (best known work, pub. in Jour. de Phys.), 1811. Postulated that equal volumes of gases at the same temperature and pressure contain identical numbers of particles, or molecules which could be divided in chem. reactions (polyatomic molecules), 1811 (now known as Avogadro's law, largely ignored during his lifetime, but became basis for Cannizzaro's method for determining atomic weights, 1858); Avogadro's name also applied to number of molecules present in one mole (gram molecular weight) of any substance; also studied specific heat, electricity, expansion of liquids in heat, acid-alkali antagonisms; stated that all bodies have the properties of oxygenicity and oxydability in inverse proportions. Died Turin, July 9, 1856.

AVRON, Mordhay, Israeli biochemist; b. Tel-Aviv, Israel, Sept. 29, 1931; s. Isaiah and Hava (Honovsky) Abramsky; B.S., U. Cal. at Los Angeles, 1953, Ph.D., 1955; m. Nira Mitrany, July 10, 1955; children—Boaz, Dana. Postdoctoral fellow Johns Hopkins, 1955-58; faculty Weizmann Inst. Sci., Rehovot, Israel, 1958—, prof. biochemistry, 1966—; vis. asso. prof. Johnson Found., U. Pa., Phila., 1964-65. Mem. UNESCO Com. on Organized Metabolism, 1963—. Mem. Am. Soc. Plant Physiology, Israel Soc. Biochemistry, Internat. Cell Research Orgn. Research, publs. on oxidative phosphorylation in plant mitochondria; enzymes related to photosynthesis; mechanism and mode of action of photophosphorylation in chloroplast. Home: 9 Neve-Weizmann, Rehovot, Israel.*

AVSYUK, Grigorii Aleksandrovich, Russian glaciologist, geodesist, cartographer; b. 1906; completed advanced geodetic studies; D.Geog. Sci. With Main Bd. of No. Sea Route, 1932-37; with Inst. Geography, USSR Acad. Sci., 1937—, dep. acad. sec. dept. geology and geog. sci., 1957—. Author: Contemporary Studies of the Glaciers of the Soviet Arctic, 1959; Glaciological Research during the 1957-1958-1959 International Geophysical Year, 1960; Artificial Acceleration of the Melting of Glaciers To Increase the Flow of Rivers in Central Asia, 1962. Editor symposium Glaciological Research, 1960. Dir. systematic studies of glacial regions in various parts of Central Asia, expecially Tyan-Shan and Pamirs, 1939-47; observer no. slopes of Terskoy Alatau range from High-Altitude Tyan-Shan Physico-Geog. Sta., 1947—; developer photogrammetric methods of determining rate of movement of glacial ice, electrometric methods of measuring temperature of glaciers at extreme depths, methods of studying supply and expenditure of matter in glaciers, methods of artificially boosting glacial thaw for irrigation and hydropower requirements; dir. all geol. operations performed in USSR under IGY program. Address: Inst. Geography, 3-ya Meshchanskaya ulitsa 61-62, korp. 18, Moscow, USSR.

AVTSYN, Aleksandr Pavlovich, Russian pathoanatomist; b. 1908; grad. Sechenov 1st Moscow Med. Inst., 1933; D.Med. Sci. Asst., lectr., prof. chairs path. anatomy 1st and 3d moscow Med. Inst., head Vlasov Pathoanat. Inst., Moscow Infectious Disease Hosp., USSR Ministry Health, 1933-55, now dept. head Research Lab., Lenin Mausoleum, also mem. com. on antibiotics; now dir. Research Inst. Human Morphology, USSR Acad. Med. Sci., also head Pathoanat. Lab., Burdenko Inst. Neurosurgery. Del., 3d Internat. Congress on Clin. Pathology, Brussels, 1957. Decorated Order of Lenin, 1955. Mem. USSR Acad. Med. Sci., Moscow (dep. chmn.), All-Union (dep. chmn.) socs. pathoanatomists, Moscow Soc. Cytologists (chmn.). Author numerous works including An Outline of Military Pathology. Mem. editorial bd. Archives of Pathology; mem. editorial council Antibiotics, Korsakov Jour. Neuropathology and Psychiatry; co-editor Psychiatry sect. Large Med. Ency., 2d edit., multi-vol. manual on path. anatomy. Established angioparalytic effect of Rickettsia prowazeki toxin and proved its specificity; research on morphological lesions in typhus, especially conjunctival exanthema, spl. pathoanat. and clin. symptom of typhus known as Chiari-Avtsyn symptom; original method of impregnating nervous fibers, especially for myeloarchitectonic; research on path. anatomy, pathology of central nervous system, infectious and geog. pathology. Address: Research Inst. Human Morphology, USSR Acad. Med. Sci., 3-ya Meshchanskaya ulitsa 61-62, korp. 18, Moscow, USSR.

AWL, William Maclay, Am. physician; b. Harrisburg, Pa., May 24, 1799; s. Samuel and Mary (Maclay) A.; studied medicine, Harrisburg, circa 1817; attended U. Pa. Med. Sch., 1819; m. Rebecca Loughey, Jan. 28, 1830. Began practice medicine, Harrisburg; practiced in Lancaster, O., 1826, Columbus, O., 1833-76; mem. Ohio Legislature; advocated bill to place insane under state care (became law 1835); became supt. State Hosp. when it opened, 1838; pres. Assn. of Supts. and Asylums for Insane of U. S. and Can., 1838-51; drew up bill for founding ednl. schs. for blind and feeble-minded in Ohio. 1st surgeon West of Alleghanies to tie left common carotid artery; became interested in mental diseases, believed he could control any mentally unbalanced person by merely gazing into his eyes. Died Columbus, Nov. 19, 1876.

AXEL, Peter, Am. physicist; b. N.Y.C., May 12, 1923; s. Lewis and Anna (Light) A.; B.A., Bklyn. Coll., 1943; M.S., U. Ill., 1947, Ph.D., 1949; m. Shirley Thomas, June 12, 1954; 1 dau., Sarah Morris. Staff mem. Radiation Lab., Mass. Inst. Tech., 1943-46; faculty physics dept. U. Ill., Urbana, 1949—, prof., 1959—. Bd. dirs. Ednl. Found. for Nuclear Sci., Chgo. Guggenheim fellow, 1955; NSF Sr. postdoctoral fellow, 1962-63. Fellow Am. Phys. Soc. Research in nuclear physics. Home: 404 W. Delaware St., Urbana, Ill. 61801.*

AXELRAD, Arthur Aaron, Canadian biologist; b. Montreal, Que., Can., Dec. 30, 1923; s. Israel and Tema (Fox) A.; B.Sc., McGill U., 1945, M.D., C.M., 1949, Ph.D., 1954; postgrad. Karolinska Inst., Stockholm, Sweden; m. Barbara Pezim, Sept. 6, 1960; children—Robert A., David I., Elise T. Asst. prof. anatomy U. Toronto (Ont., Can.), 1957, asso. prof. med. biophysics, 1958-66, prof. anatomy (histology), 1966—; head subdiv. immunogenetics and cytology Ont. Cancer Inst., 1957—. Mem. Am. Assn. for Cancer Research, N.Y. Acad. Scis., Can. Assn. Anatomists, Biophys. Soc. Quantitative investigations on induction of mammalian leukemia (virology, genetics, immunology). Home: 3 Troon Ct., Willowdale, Ont. Office: Dept. Anatomy, U. Toronto; also 500 Sherbourne St., Toronto 5, Ont., Can.*

AXELROD, Abraham E., Am. chemist; b. Cleve., June 10, 1912; s. Max and Rose (Leiken) A.; B.A., Western Res. U., 1933, M.A., 1936; Ph.D., U. Wis., 1939; m. Velma Hellerstein, Sept. 12, 1939; children—Nancy, Philip. Research chemist Western Pa. Hosp., 1942-50; asso. prof. chemistry U. Pitts., 1945-50; asso. prof. Western Res. U., Cleve., 1950-54; prof. biochemistry Sch. Medicine, Univ. of Pitts., 1954—, associate dean School of Medicine, 1965—. Mem. Am. Chem. Soc., Am. Inst. Nutrition, Am. Soc. Biol. Chemists, Am. Assn. Immunologists, Biochem. Soc. (London, Eng.). Research on role of vitamins in immunological phenomena, tissue transplantation. Home: 5821 Walnut St., Pitts. 15232.*

AXELROD, Bernard, Am. biochemist; b. N.Y.C., Oct. 16, 1914; s. Alex and Rose Axelrod; B.S., Wayne State U., 1935; M.S., George Washington U., 1939; postgrad. U. Md.; Ph.D., Georgetown U., 1943; m. Sara Axelrod, 1934; children—Eugene, Judith Ann. Chemist, U. S. Dept. Agr., Washington, 1938-43, Cal., 1943-50, chief enzyme sect. Western regional research lab. Bur. Agr. and Indsl. Chemistry, 1952-54; asso. prof. biochemistry Purdue U., Lafayette, Ind., 1954-58, prof., 1958—, head dept. biochemistry, 1965—. Sr. research fellow Cal. Inst. Tech., 1950-52, NSF sr. fellow Carlsberg Laboratorium, Copenhagen. Fellow A.A.A.S.; mem. Am. Soc. Biol. Chemists, Am. Chem. Soc., Am. Inst. Biol. Sci., Am., Japanese socs. plant physiology. Research, publs. on mechanism of actron of phosphatase, carbohydrate metabolism in plants, lipoxidase, biosynthesis of lipo-amino acids, glycosidases. Home: 337 Hollowood Dr., West Lafayette, Ind. 47906.*

AXELROD, Daniel Issac, Am. paleobotanist; b. Bklyn., July 16, 1910; s. Morris and Augusta (Gallup) A.; A.B., U. Cal. at Berkeley, 1933, M.A., 1936, Ph.D., 1938; m. Nancy Robinson, June 1939 (div. Sept. 1965); 1 son, James Peter. Asst. prof. U. Cal., Los Angeles, 1946-52, prof. geology, 1952—, prof. geology and botany, 1962—. Decorated Bronze Star medal. Guggenheim fellow, 1952-53; NRC fellow U. S. Nat. Mus., 1939-41. Mem. Geol. Soc. Am., A.A.A.S., Paleontol. Soc., Soc. for Study Evolution. Research, publs. on evolution of Madro-Tertiary Geoflora; theory of angiosperm evolution; origin and age of desert vegetation; poleward migration of angiosperms in Cretaceous period; evolution of insular floras; determination of altitude of Tertiary floras; topographic history of Snake River Plain; evolution of subalpine Tertiary forests, role of equability in evolution and extinction. Home: 1805 Anderson Rd., Davis, Cal. 95616.*

AXELROD, Julius, Am. biochemist, pharmacologist; b. N.Y.C., May 30, 1912; s. Isadore and Molly (Leichtling) A.; B.S., Coll. City N.Y., 1933; M.A., N.Y. U., 1941; Ph.D., George Washington U., 1955; Doctor of Science (hon.), Univ. of Chicago, 1965; m. Sally Taub, Aug. 30, 1938; children—Paul Mark, Alfred Nathan. Chemist, Lab. Indsl. Hygiene, 1935-46; research asso. 3d N.Y. U., research div. Goldwater Meml. Hosp., 1946-49; asso. chemist sect. on chem. pharmacology Nat. Heart Inst., NIH, Bethesda, Md., 1949-50, chemist, 1950-53, sr. chemist, 1953-55, acting chief sect. on pharmacology Lab. Clin. Sci., Nat. Inst. Mental Health, 1955, chief sect. on pharmacology Lab. Clin. Sci., 1955—. Cons. George Washington U., 1959—; panelist Bd. U. S. Civil Service Examiners, 1958—. Recipient Meritorious Research award Assn. for Research in Nervous and Mental Diseases, 1965; Gairdner award for distinguished research, 1967. Fellow Am. Soc. for Neuropsychopharmacology; mem. German Pharmacological Soc. (corr.), Am. Chem. Soc., Am. Soc. Pharmacology and Exptl. Therapeutics, Am. Soc. Biol. Chemists, A.A.A.S., Sigma Xi. Editorial bd. Jour. Pharmacology and Exptl. Therapeutics, 1956—, Jour. Medicinal Chemistry, 1962—, Circulation Research, 1963—, Currents in Modern Biology, 1966; hon. cons. editor Life Scis., 1961—. Research, numerous publs. on biochem. mechanism action of drugs and their fate; discovery enzymes in drug and hormone metabolism, function pineal gland and its control. Home: 10401 Grosvenor Rd., Rockeville, Md. 20852. Office: NIH, Bethesda, Md. 20014.*

AXELROD, Leonard Richardson, Am. biochemist; b. Bklyn., Mar. 31, 1927; s. Philip and Rose (Richardson) A.; B.S., Coll. City of N.Y., 1948; Ph.D, U. Rochester, 1951; m. Phyllis Thelma Seiden, Aug. 18, 1951; children—Douglas Wayne, Judith Ellen, Mitchell Brian, Janet Ann. Mem. atomic energy project, dept. radiation biology U. Rochester (N.Y.), 1951-55; chmn. dept. biochemistry Southwest Found. for Research and Edn., San Antonio, 1955-65, dir. div. biol. growth and devel., 1965—; research prof. chemistry St. Mary's U., San Antonio, 1959—. Cons. AEC, 1953-55; mem. Nat. Council of Cystic Fibrosis, 1963—. Fellow Chem. Soc. London, Am. Inst. Chemists; mem. Am. Chem. Soc., N.Y. Acad. Sci., Am. Endocrine Soc., Soc. Nutrition and Endocrinology, Sigma Xi. Research, numerous publs. in steroid hormone methodology, steroid hormone metabolism, drug action in central nervous system, histogenesis and organogenesis in human and primate embryos. Home: 122 Rio Bravo, San Antonio 78213. Office: 10000 W. Commerce St., San Antonio 78206.*

AXENFELD, Theodor, ophthalmologist; b. Smyrna, Turkey, June 24, 1867; s. Julius Heinrich and Antonie (Link) A.; ed. Marburg, Germany, Breslau (now Wroclaw, Poland); m. Bertha Sturmer; 5 children. Became lectr., Marburg, 1895; apptd. prof. ophthalmology Rostock, Germany, 1897, Freiburg, Germany, 1901; founder ednl. and research inst. Author: Das Trachom, 1902; Blindsein and Blindenfürsorge, 1905; Die Bakteriologie in der Augenheilkunde, 1907; Lehrbuch und Atlas der Augenheikunde, 1907; Atiologie der Trachoms, 1915; Kriegsaugenheilkunde, 1923. Gave classical description of metastatic ophthalmia, 1894; isolated (independently of Morax) Hemophilus duplex, or Morax-Axenfeld bacillus, 1896; discovered (with Morax) organism (Morax-Axenfeld diplobacillus) which causes conjunctivitis. Died Freiburg, July 29, 1930.

AYALA-CASTANARES, Agustín, Mexican micropaleontologist; marine geologist; b. Mazatlan, Mexico, Aug. 28, 1925; s. Agustín Ayala and Maria Luisa Castañares; degree in biology U. Mexico, 1954, doctorate in biology, 1963; M.Sc. in Geology, Stanford, 1959; m. Alma Irma López, Dec. 14, 1957; children—Agustín, Alma Irma, Adriana Lilia. Micropaleontologist, Pemex, 1950-56; prof. paleontology Nat. Poly. U., 1955-60; researcher, head dept. micropaleontology and marine sci. Inst. Geology, U. Mexico, 1956—, prof. micropaleontology Faculty Scis., 1961—, head dept. biology, 1965-67, director of the institute of biology, 1967—. Councilor, Marine Geol. Com. Oceanographic coordination of Mexico, 1965-67. Mem. Acad. Sci. Investigation, Mexican Coll. Biology (pres. 1962-63), Mexican Geol. Soc. (sec. 1965-66), Soc. Econ. Paleontology, Geol. Soc. Am., Swiss Paleontol. Soc., Paleontol. Soc. Am., Mexican Soc. Natural History (pres. 1966, 67), Am. Soc. Limnology and Oceanography, Am. Soc. Ecology, Soc. Systematic Zoology. Research, publs. in Mexican and Caribbean fossil foraminifera, marine geology of Mexican Coastal lagoons. Office: Apartado Postal 70-157, Mexico 20, D.F., Mexico.*

AYANT, Yves, French physicist; b. Ollioules, France, Jan. 6, 1926; s. Lucien and Antonia (Décugis) A.; Licence es Sciences, Ecole Normale Supérieure, Paris, 1948, Diplome, 1950, aggregation, 1950; postgrad. C.N.R.S., 1950-54; m. Denise Ferrier, Mar. 26, 1958; children—Florence, Frédéric, Catherine. Staff lab. Collège de France, 1948-49; maitre de conférence U. Grenoble, 1959-60, prof., 1960—. Recipient Silver medal C.N.R.S., 1956, Langevin prize Société Française de Physique, 1957. Author: (with others) La Résonance Paramagnétique Nucléaire, 1955; also articles. Research on broadening of lives in nuclear magnetic resonance; discovered quantum correlation function; studied paramagnetism and paramagnetic resonance; obtained 1st values of crystalline field in rare earths gallium garnets; studied paramagnetic resonance of ions in cubic environments, broadening by charges compensatory in crystal. Home: 7 Montorge, Grenoble, Isère (38) France.*

AYER, James Bourne, Am. neurologist; b. Boston, Dec. 28, 1882; A.B., Harvard, 1903, M.D., 1907; Prof., later emeritus; James Jackson Putnam prof. neurology Harvard Med. Sch.; neurologist Mass. Gen. Hosp. Mem. A.M.A., Assn. Pathology and Bacteriology, Neurol. Assn., Royal Soc. Medicine. Extensive research on meningitis; introduced cisternal puncture for withdrawing cerebrospinal fluid, 1920; introduced term aseptic meningitis to describe form of disease having characteristic symptoms but showing no causative agent, 1920; devised method for determination of spinal subarachnoid block; developed Tobey-Ayer test (with George Loring Tobey Jr.) for lateral sinus thrombosis, 1925.

AYMEN, Jean Baptiste, French physician; b. Dordogne, France, 1729; student Bordeaux (France) Jesuit Coll.; docteur en medecine, U. Montpellier (France); faculty U. Montpellier; physician Castillon-sur-Dordogne, France. Mem. French Acad. Scis., Acad. Medicine, Acad. Dijon, France (prize 1751). Author: Dissertation dans laquelle on examine si les jours critiques sont les memes en nos climats qu'ils étaient dans ceux ou Hippocrate les a observés, 1752. Discovered use of turnip in preventing scurvy; plant species Aymenea named for him by Linnaeus. Wrote on medicine in time of Hippocrates. Died Castillon-sur-Dordogne, July 6, 1784.

AYMERICH, Giuseppe, Italian mathematician; b. Gagliari, Italy, Aug. 10, 1913; s. Carlo and Peppina (Asquer) A.; Ph.D. in mathematics; m. Giuseppina Asquer, Nov. 12, 1936; children—Carlo, Francesco, Maria-Camilla, Eugenio. Prof. rational mechanics U. Cagliari, since 1955. Mem. Italian Soc. Physics. Research, publs. on rational mechanics and phys. mathematics. Home: Piazza Martiri 10, Cagliari, Sardegna. Office: via Ospedale 72, Cagliari, Italy.

AYRES, John Clifton, Am. food technologist; b. Beckemeyer, Ill., Apr. 17, 1913; s. John and Barbara (Schmahl) A.; B.Ed., Ill. State Normal U., 1936; M.S., U. Ill., 1938, Ph.D., 1942; postgrad. U. Minn.; m. Helen L. Keil, July 26, 1935; children—John Edward, James Lee. Tchr., Bethel, Ill., 1931-33, Brookside, Ill., 1933-34, Normal (Ill.) Community High Sch., 1936-41; bacteriologist William S. Merrell Co., Cin., 1942-43; project leader bacteriological research Gen. Mills, Inc., Mpls., 1943-46; faculty Ia. State Coll., Ames, 1946-67, prof., 1952-67, prof. in charge food tech., 1953-67; chmn. food sci. div. U. Ga., Athens, 1967—. Mem. food protection com. Nat. Acad. Sci., chmn. food micro-biology subcom., 1964—, mem. com. on Salmonella, 1967—; guest scholar Kan. State University, 1964; member of the toxicology study section NIH, 1960-64, mem. pub. health service environmental health tng. study sect., 1966—. Recipient Poultry and Egg Nat. Bd. Achievement award, 1966. Mem. Inst. Food Technologists (exec. com. 1965—), Am. Soc. for Microbiology, Inst. Food Technologists, Soc. for Applied Bacteriology (Britain), Poultry Sci. Assn., Am. Meat Sci. Assn., Food and Container Inst. (asso.), Inst. Am. Poultry Industries (research council 1953—), Internat. Assn. Microbiol. Socs. (com. food microbiology and hygiene 1967—, Sigma Xi, Gamma Sigma Delta, Phi Sigma, Phi Tau Sigma (pres. 1962-63). Research, numerous publs. on devel. procedures for removing sugar from egg products and for improving sanitation of meats, poultry and eggs, demonstration that carbon dioxide will delay spoilage of meat, use of elevated storage temperatures to destroy Salmonella in dried egg products; established parameters for use of antibiotics in poultry and meats. Home: 1554 Lumpkin, Athens, Ga. 30601.*

AYRES, William Leake, Am. mathematician; b. Gatesville, Tex., June 26, 1905; s. Matthias Leake and Myrtie (Buckley) A.; A.B., Southwestern U., Georgetown, Tex., 1923; student U. of Texas, 1923-25; Ph.D., U. of Pa., 1927; student U. of Vienna, 1928-29; m. Juliette Pagenstecher, Sept. 3, 1926; children—Juliette (Mrs. D. G. Spears), Dorothy (Mrs. R. G. Rutishauser), William Leake III. National Research fellow, 1927-29; asst. prof. mathematics, U. of Mich., 1929-33, asso. prof., 1933-41; prof. mathematics, Purdue U., 1941-62, head dept. mathematics, 1941-48, asst. dean Grad. Sch. 1943-47, acting dean Sch. of Science, Education and Humanities, 1946-47, dean, 1947-62; v.p., provost So. Meth. U., Dallas, 1962-66, prof. math., 1966—; vis. prof. U. Va., summer 1938, U. of Calif. at Los Angeles, summer 1939. Mem. Ind. Commn. on Pub. Employee Retirement, 1953-54. Fellow Indiana Academy Science; member Am. Rose Soc. (pres. 1950-51), Am. Math. Soc. (asso. sec. 1938-46), Mathematical Assn. of America (vice president, 1946-47). Author: General College Mathematics (with C. G. Fry and H. F. S. Jonah), 1952, 60. Editor: Lectures in Topology (with R. L. Wilder), 1941. Research on properties of connected sets. Home: 5440 Del Roy Dr., Dallas 29.*

AYRTON, (Sarah) Hertha Marks, English physicist; b. Portsea, Eng., 1854; d. Levi and Alice Marks; ed. Girton Coll., Cambridge; m. William Edward Aryton, May 6, 1885; 1 dau., Barbara Bodichon. Recipient Hughes medal Royal Soc., 1906. Mem. Instn. Elec.

Engrs. Contbd. numerous papers to profl. jours.; collaborated with husband on admiralty reports on electric searchlights. Invented draftsman's line-divider, 1884; investigated electric arc; discovered connection between pressure in arc and current length; studied motion of water; discovered causes and process of formation of ripples in sand by seashore; invented anti-gas fan (used by Brit. army), 1915. Died Aug. 26, 1923.

AYRTON, Matilda Chaplin, physician; b. 1846; student London Med. Coll. for Women, 1867-69, Surgeon's Hall, Edinburgh, 1870-71; Bachelier ès scis., Bachelier ès lettres, U. Paris; M.D., Paris, 1879; m. William E. Ayrton, Dec. 21, 1871; 1 dau., Edith A. Founder, tchr. sch. for midwives, Japan, 1873-77; practiced medicine, also studied at Royal Free Hosp., London, from 1880, worked in Algiers and Montpellier, France. Helped establish club for women students, Paris, Somerville Club for women, London. Licentiate, King and Queen's Coll. Physicians in Ireland, 1879. Author: Recherches sur les dimensions générales et sur le développement du corps chez les Japonais, 1879; Child Life in Japan, 1879. Studied diseases of eye; tried to improve ednl. opportunities for women. Died London, July 19, 1883.

AYRTON, William Edward, English elec. engr., physicist; b. London, Sept. 14, 1847; s. Edward Nugent Ayrton; B.A., U. London, 1867; m. Matilda Chaplin, Dec. 21, 1871; 1 dau., Edith C.; m. 2d, Sarah Hertha Marks, May 6, 1885; 1 dau., Barbara Bodichon. Entered Indian Telegraph Service, studied electricity under William Thomson, Glasgow, 1867, asst. supt. Bombay, India, from 1868, sent to Alipur, 1871; in charge testing Great Western Ry. telegraph factory, Eng., 1872-73; prof. physics and telegraphy Imperial Engring. Coll., Tokyo, Japan, 1873-78; apptd. prof. City and Guilds London Inst. for Advancement Tech. Edn., 1879; became prof. applied physics, Finsbury, 1881; prof. physics and elec. engring. Central Tech. Coll., 1884-1908. Govt. cons. Admiralty (investigated electric searchlights, also heating of cables in warships); adviser Bd. Trade (work on elec. standards led to establishment testing lab.), 1889; apptd. to com. on elec. equipment Brit. ships, 1901. Fellow Royal Soc., 1881 (Gold medal 1901); mem. Instn. Elec. Engrs. (pres. 1892, chmn. editorial com. 1878-85). Author: Practical Electricity, 1887; numerous papers. Successfully demonstrated for 1st time that power can be distributed most safely and economically by high tension currents of small quantity, 1879; 1st to determine (with John Perry) dielectric constant of gases; invented (with Perry) ammeter, also electric power motor, various forms of voltmeter; invented transmission and absorption dynamometer, also dispersion photometer; showed (with W. E. Sumpner) theoretical law previously worked out for quadrant electrometers was not valid, 1891; studied terrestrial magnetism, electrolytic polarization, contact electricity, telegraphic tests, thermal conductivity of stone; investigated accumulators, Clark cells, galvanometer constrn., glow lamps, non-inductive resistances, electrostatic voltmeters, alternate current dynamos, ampere-balances, transformers, after 1891. Died Hyde Park, Eng., Nov. 8, 1908.

AYSCOUGH, Peter Brian, English chemist; b. Lincoln, Eng., May 5, 1927; s. Horace Nathan and Alice (Jackson) A.; student St. Catharine's Coll., Cambridge, Eng., 1944-46, 48-53. Research fellow Nat. Research Council Can., Ottawa, Ont., 1953-55; Imperial Chem. Industries Research fellow dept. Phys. chemistry U. Leeds (Eng.), 1955-58, lectr., 1958-65, sr., lectr., 1965—. Mem. Faraday Soc., Chem. Soc. (London). Author: Electron Spin Resonance in Chemistry, 1967; also articles. Research on reaction kinetics, especially fluorine compounds, applications of electron spin resonance in fields of photochemistry and radiation chemistry, reactions of free radicals. Home: 40, Bracken Edge, Leeds 8, Yorks., Eng.*

AYZERMAN, Mark Aronovich, Russian engr.; b. Daugavpils, Latvia, 1913; grad. Moscow Higher Tech. Sch., 1937; D.Tech. Sci., 1947. Asso., Inst. Automatics and Telemechanics, USSR Acad. Sci., 1939—; prof., 1955—. Author: The Effect of Non-Linear Characteristics on the Convergence of an Automatic Control Process and on Oscillation Generating Conditions, 1944; The Convergence of an Automatic Control Process after Large Initial Divergencies, 1946; Discount of Non-Linear Functions from Several Arguments in Studying the Stability of Automatic Control Systems, 1947; A Problem Relating to the Overall Stability of Dynamic Systems, 1949; An Introduction into the Dynamics of Automatic Engine Control, 1950; Adequate Stability Conditions for a Class of Dynamic Systems with Variable Parameters, 1951, also others; co-author: A Class of Dynamic Problems Agreeing with the Theory of Relay Systems, 1955; Determining Periodic Conditions in a Non-Linear Dynamic System with Piecewise Linear Characteristics, 1956; Stability in Linear Approximation of the Periodic Solution of a System of Differential Equations with Discontinuous Righthand Quantities, 1957; Finite Automatic Machines, 1960; Methods of Realizing Finite Automatic Machines whose Cycles Are Determined by Application of the Input State, 1960; Critical Cases in the Theory of Absolute Stability of Regulated Systems, 1963, also others. Address: Inst. Automatics and Telemechanics, USSR. Acad. Sci., Kalanchevskaya 15a, Moscow I-53, USSR.

AZARA, Félix de, see de Azara, Félix.

AZARNOFF, Daniel Lester, Am. physician; b. Bklyn., Aug. 4, 1926; s. Samuel Jacob and Kate (Asarnow) A.; B.S., Rutgers U., 1947, M.S., 1948; postgrad. U. Kan.; M.D., 1955; m. Joanne Stokes, Dec. 26, 1951; children—Rachel, Richard, Martin. Nat. Heart Inst. Research fellow U. Kan. Med. Center, Kansas City, 1956-58, faculty, 1962—, asso. prof. medicine, 1964—, asso. prof. pharmacology, 1965—, dir. clin. pharmacology study unit, 1964—; spl. trainee Nat. Inst. Neuro Diseases and Blindness, Washington U. Sch. Medicine, St. Louis, 1958-60; asst. prof. St. Louis U. Sch. Medicine, 1960-62. John and Mary R. Markle scholar in acad. medicine, 1962; Burrough Wellcome scholar in clin. pharmacology, 1964. Mem. Am. Soc. Pharmacology and Exptl. Therpaeutics, Central Soc. for Clin. Research, Am. Soc. Clin. Nutrition, N.Y. Acad. Scis., Am. Nutrition Inst., Sigma Xi. Research, publs. on drugs influencing lipid metabolism, drug metabolism and interrelations in therapeutics. Home: 2217 W. 50th St., Shawnee Mission, Kan. 66205. Office: U. Kan. Med. Center, Kansas City, Kan. 66103.*

AZAROFF, Leonid Vladimirovitch, physicist, educator; born Moscow, Russia, June 19, 1926; s. Vladimir I. and Maria (Odlen) A.; came to U. S., 1939, naturalized, 1945; B.S. cum laude, Tufts U., 1948; Ph.D., Mass. Inst. Tech., 1954; m. Carmen F. Wade, Mar. 9, 1946; Research physicist, sr. scientist Armour Research Found., Chgo., 1953-57; asso. prof., prof. metall. engring. Ill. Inst. Tech., Chgo., 1957-66; prof. physics, dir. Inst. Materials Sci., U. Conn., Storrs, 1966—. U. S. del. Internat. Union Crystallography Teaching Commn.; cons. Hilger-Watts, Inc., 1962-66, Philips Electronics Co., 1960—. Mem. Am. Phys. Soc., Am. Inst. Mining, Metall. and Petroleum Engrs., Mineral. Soc. Am., Am. Crystallographic Assn., Am. Soc. Engring. Edn., A.A.A.S. Author: (with M. J. Buerger) The Powder Method, 1958; Introduction to Solids, 1960; (with J. J. Brophy) Electronic Processes in Materials, 1963. Research, publs. on determination of atomic arrangements in crystals and their semicondr. properties, formation and properties of metal alloys. Home: 18 Mansfield Apts., Storrs, Conn. 06268.*

AZIZBEKOV, Shamil Abdulragimogly, Russian geologist; b. 1906; grad. Azerbaijan Indsl. Inst., 1930. With various geol. groups, 1930-37; instr. Azerbaijan Indsl. Inst., 1932-44, prof., 1944—; with Azerbaijan br. USSR Acad. Sci., 1936—; dep. chmn. presidium Azerbaijan Acad. Sci., 1941-44, v.p., chmn. dept. geol. and geog. sci., 1945-47, acad. sec. dept. geol. sci. and petroleum, 1959—; hon. sci. worker of Azerbaijan, 1960—. Mem. Azerbaijan Acad. Sci. Author: The Geology and Mineral Resources of Azerbaijan, 1944; The Geology and Petrography of Northeastern Caucasus Minor, 1947; The Structure and Genesis of the Gyumushlug Polymetallic Deposits, 1957; The Metallogeny of Azerbaijan, 1962. Chief editor: Geology of Azerbaijan. Petrography. Address: Azerbaijan Acad. Sci., Baku, Azerbaijan SSR, USSR.

AZZONE, Giovanni Felice, Italian biologist; b. Naples, Italy, Jan. 27, 1927; s. Domenico and Anna (Meola) A.; med. degree Med. Sch., Rome, 1950; m. Anna Rosa Zweifel, June 12, 1966; children—Fabrizio, Sara. Research asso. Inst. Biol. Chemistry, Rome, 1953-54; asst. prof. Inst. Gen. Pathology, Modena, Italy, 1954-59; sci. guest Wenner Grens Inst., Stockholm, Sweden, 1959-60; asst. prof. Inst. Gen. Pathology, Padua, Italy, 1961-64, prof. Med. Faculty, U. Padua, 1965—. Research, numerous publs. on isolation of a myofibrillar protein, gammamyosin which increases in contracted muscles, mechanism of actions of inhibitors on mitotchondrial and sarcoplasmic reticulum reactions, movements of ions and water in mitochondria, and mechanism of oxidative phosphorylation. Home: 69 Altinate, Padua, Italy.*

B

BAADE, Walter, astronomer; b. Schröttinghausen, Westphalia, Germany, Mar. 24, 1893; doctorate U. Göttingen (Germany), 1919; staff U. Hamburg (Germany), 11 years; came to U. S., 1931; staff Mt. Wilson Obs., 1931-48; astronomer Mt. Palomar Obs., 1948-58; returned to Göttingen, 1958. Recipient Gold medal Royal Astron. Soc. Gt. Britain, Bruce medal Astron. Soc. Pacific. Discovered Hidalgo (most distant known planetoid), 1920, innermost planetoid Icarus, 1948; research, (with R. L. B. Minkowski) on Andromeda galaxy, 1942; developed theory for Population I and Population II stars, 1944; confirmed distances in our own galaxy; showed cosmic distance scale between extra galactic bodies should be doubled; redetermined age of the universe; proved our galaxy is of average size; studied the great spiral in Andromeda; worked out new period-luminicity curve for outer galaxies, 1952; identified (with Minkowski) radio source in constellation Cygnus as two galaxies colliding, 1953. Died Göttingen, June 25, 1960.

BAADER, Franz Xavier von, see von Baader.

BAADSGAARD, Halfdan, geochemist; b. Mpls., Apr. 16, 1929; s. Paul and Inger (Petersen) B.; B.Sc., U. Minn., 1951; Ph.D., Swiss Fed. Inst. Tech., 1955; m. Anne Nora Peacock, June 9, 1957; children—Erik, Katherine, Garth, Kirsten, Karla. Sr. research asso. U. Minn., 1955-57; faculty U. Alta., Edmonton, Can., 1957—, prof. 1966—. NATO exchange prof. Can.-Denmark, 1964. Mem. Geochem. Soc., Am. Geophys. Union, Edmonton Geol. Soc. Research, publs. on isotopic variation in geol. materials, geol. age determination. Home: 5220 114 B St., Edmonton, Alta., Can.*

BAAR, Heinrich Siegfried, pathologist; b. Jagielnica, Austria, Feb. 15, 1892; s. Julius and Henriette (Ohrenstein) B.; Ph.D., U. Vienna (Austria), 1914, M.D., 1919, L.M.S.S.A., 1947; M.D., U. Birmingham (Eng.), 1948; m. Ida Turkl, Aug. 8, 1920; 1 dau., Stella. Chief physician Children's Out Patients Dept., Vienna, 1928-38; pathologist Children's Hosp., Birmingham, 1938-57; dir. pathology and clin. investigation Pineland Hosp., Maine, Eng., 1958-65; hon. lectr. pathology U. Birmingham, 1950-58. Fellow Coll. Pathologist; mem. Royal Coll. Physicians London, Assn. Clin. Pathologists, Path. Soc. Gt. Britain and Ireland, Brit. Paediatric Assn., Internat. Soc. Hematology, N.Y. Acad. Scis. Author: (with E. Stransky) Klinische Haematologie des Kindesalters, 1928; Disorders of Blood and Blood Forming Organs, 1963 (with Stella Baar, K. B. Rogers, E. Stransky); also numerous articles. Research on pathogenesis of alimentary anemias; discovered reactivated measles serum; described difference between homologous and heterologous sera; described pulmonary glomangioma and pulmonary ascaris granuloma. Died 1967.

BAARLI, Johan Vidar, physicist; b. Eidsvoll, Norway, June 23, 1921; s. Hans and Margit (Karlsen) B.; student Grimelands Tradesch., 1942-43; Cand. mag., U. Oslo (Norway), 1947, Cand. real, 1950, dr. Philos., 1962; m. Mildred Aadne, Apr. 30, 1950; children—Ase-Brit, John Vidar. Tchr., Grimelands High Sch., Eidsvoll, 1940-42; lectr. U. Oslo, 1948-50, asst. prof., 1950-52; research fellow U. Chgo., 1952-53; head dept. biophysics Norsk Hydros Inst. for Cancer Research, 1953-61; head health physics CERN, Geneva, Switzerland, 1961—. Sec. com. on nuclear research Norwegian Sci. and Tech. Research Council, 1954-62; mem. com. on sci. instrumentation Norwegian Research Council, 1959-61. Mem. Norwegian Physics Soc., Hosp. Physicists Assn., Health Physics Soc., A.A.A.S. Research, and publs. on plesiotherapy unit for cancer treatment, properties of back-scattered gamma rays in treatment of cancer, radiation dosimetry problems for biol. and health protection purposes. Home: 10, rue Boudine. Office: CERN, 1211 Geneva-23, Switzerland.*

BAAS, Johann Herman, German physician; b. Bechtheim, Germany, Oct. 24, 1838; s. Adam and Catharina (Cotheimer) B.; M.D., Giessen, Germany, 1860; practiced medicine, from 1860. Author: Grundriss der Geschichte der Medizin und des heilenden Standes, 1876; Zur Percussion, Auscultation und Phonometrie, 1887; William Harvey . . . , 1878; Medicinische Diagnostik, 1883; Grundriss der Hygiene, 1879. Developed phys. diagnosis; introduced use of phonometry for diagnosis. Died Worms, Germany, Nov. 10, 1909.

BABASINIAN, V(ahan) S(imon), chemist; b. Marsovan, Asia Minor, Nov. 28, 1876; s. Simon and Hripsimeh (Mallian) B.; A.B., Anatolia Coll., Asia Minor, 1895; A.M., Brown U. 1903, Ph.D., 1906. Came to U. S., 1897, naturalized, 1910. Instr. in chemistry, Brown U., 1903-06; instr. in chemistry, Lehigh, until 1909, asst. prof., 1909-11, asso. prof. 1911-22, prof. organic chemistry from 1922. With Chem. Warfare Service (research), Washington, 1918, du Pont Co., Wilmington, Del., 1919. Editor: Gattermann's Practical Methods of Organic Chemistry, 1914. Died May 24, 1939.

BABB, Albert Leslie, Am. nuclear engr., educator; b. Vancouver, B.C., Can., Nov. 7, 1925; s. Clarence S. and Mildred (Gutteridge) B.; B. Applied Sci. with 1st class honors, U. B.C., 1948; M.S., U. Ill., 1949, Ph.D., 1951; m. Marguerite Lois Henderson, Dec. 21, 1948; children—Eugene M., Philip L., Christine L. Came to U. S., 1948, naturalized, 1954. Research engr. Rayonier Inc., Shelton, Wash., 1951-52; asst. prof. chem. engring. U. Wash., Seattle, 1952-56, asso. prof., 1956-60, chmn. nuclear engring group, 1959-65, prof. chem. engring., 1960—, prof., chmn. nuclear engring. dept., 1965—, dir. Nuclear Reactor Labs., 1960—. Cons. bus. firms, govt. agys., hosp. Mem. Am. Soc. for Engring. Edn. (chmn. nuclear engring. div. 1965-66), Am. Nuclear Soc. (mem. edn. com 1962—, program com. 1965—), Am. Inst. Chem. Engrs., Am. Chem. Soc., Am. Soc. for Artificial Internal Organs, Sigma Xi, Tau Beta Pi. Research, publs. on molecular diffusion in liquids, separations and purification processes for nuclear fuels; nuclear reactor engring., nuclear fuel mgmt., hemodialysis. Home: 8218 Crest Dr., N.E., Seattle 98115.*

BABBAGE, Charles, English mathematician; b. nr. Treignmouth, Eng., Dec. 26, 1792; s. Benjamin B.; ed. Trinity Coll., Cambridge (Eng.) U., from 1811; grad. Perterhouse, 1814; M.A., 1817. Lucasian prof. math. Cambridge U., 1828-39. Fellow Royal Soc., 1816. Founder (with Herschel, Peacock and others) Analytic Soc. (led to intro. of Continental math. notation to Eng.), 1812, joint-founder Royal Astron. Soc. (1st Gold medal, 1823, sec. to 1824, later v.p., fgn. sec., mem. council), 1820, founder Statis. Soc. London, 1834; also mem. sci. socs. throughout world. Author: The Comparative View of the Various Institutions for the Assurance of Lives, 1826; The Table of Logarithms of the Natural Numbers from 1 to 108000, 1827; Decline of Science in England, 1830; Economy of Machinery and Manufactures, 1832; The Ninth Bridgewater Treatise, 1837; Exposition of 1851; Passages from the Life of a Philosopher (autobiography), 1864; also contbr. essays on calculus of functions to Philosophical Transactions, 1815-17. Repeated and extended (with Herschel) Arago's experiments on magnetisation of rotating plates, 1825; worked with and tried to perfect calculating machines; devised basic principles of modern computers (limited by mech. devices to implement them); received govt. subsidy to build computer (later lost subsidy and died before computer could be finished); contbr. to establishment of modern postal system in Eng., 1840; devised 1st reliable actuarial tables; inventor 1st speedometer, locomotive (cow catcher), opthalmoscope (1847). Died London, Oct. 18, 1871.

BABBAR, Om Prakash, virologist; b. Malkhanwala, West Pakistan, Sept. 1, 1919; s. Kirpa Ram and Kukmani (Devi) B.; F.Sc., Punjab (India) U., Lahore, 1937, L.V.P. with honors, 1941; B.V.Sc., Punjab U., Chandigarh; Ph.D., Agra (India) U., 1966; m. Prakash Luther, Oct. 8, 1940; children—Indra (Mrs. Krishna Kumar Sahdeu), Suman, Veena. Vet. surgeon Punjab Vet. Services, N. Western India, 1942-50; research asst. biol. product sect. Punjab Coll. Vet. Scis. and Animal Husbandry, Hissar, India; scientist virology div. microbiology Central Drug Research Inst., Lucknow, India, 1952—. Research, publs. on growth trachoma virus in chick embryo and its use in diagnosis, molecular chemistry of host-virus interaction, viral nucleic acids. Home: 17'B, Singar Nagar, Lucknow, Uttar, Pradesh, India. Office: Div. Microbiology, Central Drug Research Inst., Chattar Manzil Palace, Lucknow, India.

BABBITT, Benjamin Talbot, Am. inventor; b. Westmoreland, N.Y., 1809; s. Nathaniel and Betsey (Holman) B.; m. Rebecca McDuffie, 2 daus. Established machine shop, Little Falls, N.Y., 1831, manufactured pumps, engines, farm machinery; developed new, cheaper way to make baking soda, 1843; manufactured various brands of soap; obtained 1st patent for pump and fire engine, 1842; patented brush trimming machine, 1846; patented over 100 devices, including car ventilator, automatic boiler feeder, steam generator cleaning apparatus, rotary engine, balance valve, air pump, air compressor, wind motors, pneumatic propulsion, air blasts for forges; 1st to use free samples for advt. Died Oct. 20, 1889.

BABBITT, Isaac, Am. inventor; b. Taunton, Mass., July 26, 1799; s. Zeba and Bathsheba (Luscombe) B.; m. Sally Leonard; m. 2d, Eliza Barney; 9 children. Made 1st Brittania-ware manufactured in U. S., 1831, 1st brass cannon cast in U. S., 1834; patented journal-box lined with alloy known as Babbitt metal (an anti-friction bearing metal used in all railroad car axle-boxes), 1839; recipient Gold medal Mass. Mechanics Assn., 1841, also granted $20,000 by U. S. Congress for this invention, 1842. Died Somerville, Mass., May 26, 1862.

BABCOCK, Dale F(riend), Am. chem. engr.; b. Minneapolis, Kan., Oct. 5, 1906; s. Seth and Minnie (Reno) B.; B.S., Kan. State Coll., 1924; Ph.D., U. Ill., 1929; m. Marion Addison Gregg, June 17, 1930; children—Byron D., Ardis Lee (Mrs. Arthur J. Crull). Staff central research E.I. du Pont de Nemours Co. & Co., 1929-39, research supr. nylon div., 1939-42, tech. specialist atomic energy div., 1943-45, research mgr. Grasselli chems. dept., 1945-50, dir. nuclear engrng. sect. atomic energy div., 1950— (all Wilmington, Del.); tech. aide Nat. Def. Research com., 1942. Mem. Sigma Xi. Research, devels. on gas separations, nylon processing, atomic reactor engrng. Home: 711 River Rd. Wilmington 19898. Office: Nemours Bldg., DuPont Co., Wilmington, Del. 19898.*

BABCOCK, Ernest Brown, Am. botanist; b. Edgerton, Wis., July 10, 1877; s. Emilus Welcome and Mary Eliza (Brown) B.; student Lawrence Coll., 1895-96; grad. State Normal Sch., Los Angeles, 1898; B.S., U. Cal., 1906, M.S., 1911, LL.D., 1950; m. Georgia Bowen, June 24, 1908. First instr. in agrl. nature study State Normal Sch., Los Angeles, 1906-07; instr. U. Cal. at Berkeley, asst. prof. plant pathology, 1907-10, asst. prof. agrl. edn., 1910-13, prof. genetics 1913-47, emeritus, 1947; exec. v.p. Forest Genetics Research Found., 1952-54, pres., 1954. Research asso. Carnegie Inst., 1925-37. Pres., sect. exptl. taxonomy, VII Internat. Botany Congress, Stockholm, 1950. Mem. A.A.A.S., Am. Soc. Naturalists (v.p. 1934), Genetics Soc. Am., Western Soc. Naturalists, Am. Genetics Assn., Bot. Soc. Am. Cal. (president 1954), Nat. academy scis., Phi Beta Kappa, Sigma Xi, Alpha Zeta, Phi Sigma; hon. mem. Royal Bot. Soc. (Belgium), Japanese Bot. Soc. Author: Genetics in Relation to Agriculture, (with Roy E. Clausen), 1918, 27; Genetics Laboratory Manual (with J. L. Collins), 1918; The Genus Crepis (in Bibliog. Genetica), (with M. Navashin), 1930; The Genus Youngia (with G. Ledyard Stebbins, Jr.), 1937; The Am. Species of Crepis, 1938; The Genus Crepis, Univ. Calif. Publ. Botany, vols. 21 and 22, 1947. Died Dec. 8, 1954.

BABCOCK, George Herman, Am. inventor, mfr.; b. Unadilla Forks, N.Y., June 17, 1832; s. Asher M. and Mary (Stillman) B.; married 4 times. Chief draftsman Hope Iron Works, Providence; inventor Babcock & Wilcox high pressure boiler; co-inventor 1st polychromatic printing press; pres. Babcock, Wilcox & Co., boiler mfrs., 1881-93; pres. Am. Soc. M.E., 1887. Died Plainfield, Dec. 16, 1893.

BABCOCK, Harold Delos, Am. astrophysicist; b. Edgerton, Wis., Jan. 24, 1882; B.S., U. Cal., 1907; m., 1907; 1 son, Horace Welcome. Asst. in physics U. Cal., 1905-06, asst. instr. physics, 1908; lab. asst. Bur. Standards, 1906-08; physicist Mt. Wilson Obs., Carnegie Instn., 1909-48, cons., 1948, ret. Civilian with AEC; with Office Sci. Research and Devel.; solar eclipse expdns., 1918, 23, 30, 32. Fellow A.A.A.S. (Pacific div. prize 1929); mem. Nat. Acad. Scis., Astron. Soc., Astron. Soc. Pacific (pres. 1937, Bruce medal 1953); asso. Royal Astron. Soc. Research on standards of wave length in arc and solar spectra, analysis of sunspot and solar spectra in photog. infrared, ruling diffraction gratings, solar magnetism; discovered (with son) stellar magnetism; measured magnetic field of star 78 Virginis, thus providing 1st link between electro-magnetic and relativity theories. Address: 1820 Atchison St., Pasadena 7, Cal.

BABCOCK, Harriet, Am. psychologist; b. Westerly, R.I.; B.S., Columbia, 1922, A.M., 1923, Ph.D., 1930; m. H. Hobart Babcock. Psychologist, Manhattan State Hosp., 1924-25; chief psychologist Bellevue Hosp., 1926-28; began research in measurement of mental deterioration, 1924; engaged in clin. work in efficiency phase of mental functioning from 1931. Fellow Am. Assn. Applied Psychology. N.Y. Acad. Sci., Am. Psychol. Assn., A.A.A.S.; mem. eastern and local psychol. assns. Author: An Experiment in the Measurement of Mental Deterioration, 1930; Dementia Praecox; A Psychological Study, 1933; Revised Examination for the Measurement of Efficiency of Mental Functioning, 1941; Time and the Mind, 1941. Contbr. to tech. and sci. jours. Originator method of measuring and evaluating efficiency of mental functioning in normal and abnormal mental conditions by controlling abstract-verbal development; determined place of psychogenic psychoses between the normal and definitely abnormal; stated level-efficiency theory of intelligence; and showed relation of personality to basic mental functioning. Died Dec. 12, 1952.

BABCOCK, Horace Welcome, Am. astronomer; b. Pasadena, Cal., Sept. 13, 1912; s. Harold Delos and Mary G. (Henderson) B.; B.S., Cal. Inst. Tech., 1934; Ph.D., U. Cal., 1938; D.Sc., U. Newcastle-upon-Tyne, Eng., 1965; m. July 1, 1940 (div. 1957); children—Ann L., Bruce H.; m. 2d, Elizabeth M. Jackson, Aug. 30, 1958; 1 son, Kenneth L. Asst., Lick Obs., Mt. Hamilton, Cal., 1938-39; instr. Yerkes and McDonald Obs., 1939-41; staff Radiation Lab., Mass. Inst. Tech., 1941-42; with rocket project Cal. Inst. Tech., 1942-45; staff mem. Mt. Wilson and Palomar Obs., Pasadena, 1946-64, dir., 1964—. Mem. American Philosophical Society, National Academy of Sciences (recipient Acad. Draper medal 1957), Am. Acad. Arts and Scis., Am. Astron. Soc., Astron. Soc. Pacific, Royal Astron. Soc. London (Edington medal 1957), Internat. Astron. Union. Research on rotation Andromeda Galaxy, light night sky; discovered and investigated magnetic fields stars; (with H. D. Babcock) invented solar magnetograph; observed weak solar magnetic fields; proposed theory solar magnetic cycle; diffraction gratings. Home: 2189 N. Altadena Dr., Altadena, Cal. 91001. Office: 813 Santa Barbara St., Pasadena, Cal. 91106.*

BABCOCK, James Francis, Am. chemist; b. Boston, Feb. 23, 1844; s. Archibald D. and Fannie F. (Richards) B.; attended Lawrence Sci. Sch. Harvard, 1862; m. Mary Crosby, Mar. 28, 1869; m. 2d, Marion Alden, Aug. 24, 1892; 5 children. Prof. chemistry Mass. Coll. Pharmacy, 1869-74, Boston U., 1874-80; state assayer Mass., 1875-85, introduced 3 percent limit as defining intoxicating liquor; insp. milk Boston, 1885; inventor Babcock fire extinguisher. Died Dorchester, Mass., July 19, 1897.

BABCOCK, Stephen Moulton, Am. agrl. chemist; b. Bridgewater, N.Y., Oct. 22, 1843; s. Peleg B. and Cornelia B.; A.B. Tufts, 1866, LL.D., 1901; student chemistry, Cornell, 1872-75; Ph.D., U. Göttingen, 1879; m. May Crandall, Oct. 27, 1896. Instr. chemistry Cornell U., 1875-76; chemist. N.Y. Agrl. Expt. Sta., Geneva, 1882-87; prof. agrl. chemistry. U. Wis., 1887-1913 (emeritus); chief chemist Wis. Agrl. Expt. Sta. 1887-1913, asst. dir., 1901-13. Awarded bronze, medal by Wis. legislature, 1899; grand prize, Paris Expn., 1900; devised Babcock test for determining butterfat content of milk by centrifuging mixture of equal quantities of sulfuric acid and milk to be tested, 1890; invented an apparatus for determining the viscosity of liquids, research on fat solvents, milk sugar, metabolic water. Died July 2, 1931.

BABDZHANYAN, Gurgen Amayakovich, Russian geneticist; b. 1907; grad. Central Asian Cotton Inst., Tashkent, 1932. Head genetics sect. Armenian br. USSR Acad. Sci., 1938-39; dir. Inst. Genetics, Armenian Acad. Sci., (now Inst. Agr., Armenian Ministry Agr.), 1946-53, asso., 1946—, presidium mem., acad. sec. dept. agrl. sci., 1948-54. Mem. Armenian Acad. Sci. Author: The Selective Fertilization Capacity of Agricultural Plants, 1947; Notes on Sexual Mentor Phenomena in Plants, 1949; Diversity of Vitality and Heredity in Plants, 1950; The Florescence, Pollination and Fertilization of Wheat, 1955; Thrilling Prospects, 1962. Address: Armenian Acad. Sci., Yerevan, Armenia SSR, USSR.

BABEL, Jean, Swiss physician; b. Geneva, Switzerland, Feb. 17, 1910; s. Arthur and Louis (Fezio) B.; M.D., U. Geneva, 1934; m. Denise Sillig, Jan. 6, 1945; children—Jean-Francois, Nicole, Laurent. Practice medicine specializing in ophthalmology, Geneva, 1941—; faculty U. Geneva, 1948—, prof. ophthalmology, 1961—, dir. Univ. Eye Clinic, 1966—. Author: (with A. Franceschetti, J. Francois) Les Heredodegenérescences chiorioretiniennes, 1963; also numerous articles. Research on ophthalmology, eye pathology. Home: 16 Rue des Granges, Geneva, Switzerland.*

BABERO, Bert Bell, Am. parasitologist; b. St. Louis, Oct. 9, 1918; s. Andras and Bertha (Bell) B.; B.S., U. Ill., 1949, M.S., 1950, Ph.D., 1957; m. Harriett King, Feb. 19, 1950; children—Bert Bell, Andras. Med. parasitologist USPHS, Anchorage, 1950-53; head dept. biology Ft. Valley State Coll. (Ga.), 1957-59; prof. zoology So. U., Baton Rouge, 1959-60; lectr. Fed. Emergency Sci. Skeme, Lagos, Nigeria, 1960-62, cons., 1967-68; asso. prof. parasitology Baghdad (Iraq) U. Med. Sch., 1962-64; prof. zoology U. Nev., Las Vegas, 1965—. Mem. Am. Soc. Parasitologists, Helminological Soc. Washington, N.Y. Acad. Scis., Soc. Systematic Zoologists, Wildlife Disease Soc., Soc. Protozoology, Brazilian Soc. Biologists, Sigma Xi, Phi Sigma, Beta Beta Beta. Research, publs. on helminth parasites especially those of zoonotic importance, including taxonomy and morphology. Home: 3381 Rome St., Las Vegas 89109.*

BABES, Victor, physician, bacteriologist; b. Vienna, Austria, July 28, 1854; became researcher on rabies Pasteur Inst., Paris, 1886; demonstrated protective value of serum from an immunized animal, 1888; described (with P. Ernst) bodies (metachromatic granules in protoplasm of bacteria); devised mallein test used for early diagnosis of glanders, 1891; discovered Babesia genus of protozoans in blood of animals; described penetration of certain types of bacteria through unbroken skin or mucosa; developed antitubercular serum. Died Bucharest, Rumania, Oct. 19, 1926.

BABEUF, François Noël (known as Gracchus Babeuf), French polit. theorist; b. St. Quentin, Nov. 23, 1760; s. Claude Babeuf; ed. by his father; became commissaire à terrier, 1785; employed by a land surveyor at Roye; writer and pamphleteer against feudalism (for which he was twice arrested), from 1787; superintended publ. of his 1st work (Cadastre perpetuel), Paris, circa 1789; founder pamphlet Correspondant picard, imprisoned because of its violent nature; elected mem. municipality of Roye, but expelled, 1789; apptd. commr. to report on nat. property in Roye, 1791; elected mem. of council-gen. of dept. of Somme, 1792; transferred to post of adminstr. of dist. of Montdidier; fled to Paris after being accused of fraud, 1793, sentenced to 20 years in prison; at same time had been apptd. sec. to relief com. of Paris. Author: Cadastre perpetuel; L'an 1798 et le premier de la liberté française, 1789. Supporter of Reign of Terror; conspired to overthrow the Directory and establish a communistic society; believed that nature has given every man an equal right in enjoyment of all goods; advocated nat. ownership of large bus. enterprises, abolishment of inheritance and eventual nationalization of pvt. property; under his system prodn. and distbn. would be directed by an elected gov., food and clothing would be the same for all, and only those who did useful work would have polit. rights; these doctrines kept alive after his death by secret revolutionary groups; considered a forerunner of Karl Marx. Condemned for his socialistic propagandizing, executed Vendôme, France, Apr. 27, 1797.

BABICKY, Arnost, chemist; b. Vienna, Austria, Dec. 18, 1923; s. Antonin and Emilie (Ticha) B.; Ph.D., Charles U., Prague, Czechoslovakia, 1953; m. Miluse Rychlíková, Mar. 14, 1952; children—Katerina, Arnost. With Isotope Lab., Insts. for Biol. Research, Prague, 1952—, head dept. measuring methods and dosimetry, 1960—; with Charles U., Prague, 1964—. Mem. Czechoslovak Med. Soc., Biochem. Soc. Author: Radioisotopes in Biology and Medicine, 1960; (with J. Kolár, R. Vrabec), Physical Agents and Bone, 1965; also numerous articles. Research on influence of local injury to the whole mammalian organisms, changes in bone metabolism of whole skeleton due to such local injuries. Home: 230 Smetanova, Zbraslav I, Czechoslovakia. Office: 1083 Budejovická, Praha 4-Krc., Czechoslovakia.*

BABIKOV, Vladimir Vasil'evich, Russian physicist; b. Kirov, Russia, Jan. 25, 1931; s. V. G. and E.

(Kov'asina) B.; student Moscow (USSR) State U., 1949-51; Moscow Physics and Engring. Inst., 1951-55; Ph.D., U. Dubna (USSR), 1965; postgrad. Lebedev Phys. Inst., Acad. Sci. USSR. Sci. collaborator Joint Inst. for Nuclear Research, Labs. Nuclear Reactions and Theoretical Physics, Moscow, 1958—. Research, publs. on theory of Bremsstrahlung in hot hydrogen plasma, and problem of controlled thermonuclear reaction, theory of compound nucleus formation and decay in reactions induced by heavy ions, role of pion resonances in nuclear force problem, new methods in quantum theory of potential scattering. Home: Vavilov II-I, Dubna, USSR. Office: Head P.O. Box 79, Moscow, USSR.*

BABINET, Jacques, French physicist; astronomer; born Luisgnan, France, March 5, 1794; studied under Binet; teacher of mathematics, physics, and meteorology at various schools; became member of faculty of College de France, in 1838; professor of physics St. Louis Coll.; mem. French Acad. Scis. Author: Traité de Géométrie descriptive, 1850; Etudes et lectures sur les sciences d'observation et leurs applications, 1855-68; Télégraphie électrique, 1861; Paragenie, 1864. Invented a hygrometer, 1824, also three way stopcock of steam engine, a polariscope, an improved air pump, goniometer for measuring angles of very small crystals; 1st to suggest that wavelength of some particular ray of light should be a standard of length; research on Fraunhofer lines, 1829, also on double refraction, parhelion, magnetism, planet Mercury. Died Paris, Oct. 21, 1872.

BABINGTON, Benjamin Guy, English physician; b. London, Eng. 1794; s. William Babington; student Haileybury Coll., Guy's Hosp., Cambridge U.; M.D., 1830. Served with Navy, Walcheren, Netherlands, also Copenhagen, Denmark; apptd. to Madras presidency Indian Civil Service; became asst. physician Guy's Hosp., London, 1837, physician, 1840. Fellow Coll. Physicians, Royal Soc., 1828; pres. Royal Med. and Chirurg. Soc., 1861; founder, pres. Epidemiological Soc. Contbd. 2 memoirs on blood to Medico-Chirurgical Trans. First to employ expression liquor sanguinis for fluid portion of blood; devised the glottiscope, a crude form of laryngoscope, 1829. Died Apr. 18, 1866.

BABINGTON, Charles Cardale, Brit. botanist; b. Ludlow, Eng., Nov. 23, 1808; s. Joseph and Catherine (Whitter) B.; B.A., St. John's Coll., 1830, M.A., 1833; attended lectures J. S. Henslow; m. Anna Maria Walker, Apr. 3, 1866; apptd. prof. botany U. Cambridge (Eng.), 1861; named fellow St. John's Coll., 1882. Fellow Botanical Soc. Edinburgh, Linnean Soc., Geol. Soc., Royal Soc., 1851; member Ray Club (founder 1836, sec.), Entomol. Soc. (founder with others 1833). Author: Manual of British Botany, 1843; Flora of Cambridgeshire, 1860; The British Rubi, 1869. Editor: Annals and Magazine of Natural History, from 1842. Contbd. numerous papers, including 1st paper on Cambridge entomology in Mag. of Natural History, 1829. Bequeathed herbarium and bot. library to Cambridge U. Died July 22, 1895.

BABINGTON, William, physician, mineralogist; b. Portglenone, nr. Coleraine, Ireland, 1756; apprentice to practitioner at Londonderry; completed med. edn. at Guy's Hosp., London, Eng.; M.D., Aberdeen, Scotland, 1795; M.D. (hon.) Dublin, Ireland, 1831; children—Benjamin Guy, Mrs. Richard Bright. Asst. surgeon to Haslar (Naval) Hosp. for 4 years, beginning 1774; lectr. chemistry at med. sch., apothecary Guy's Hosp., named physician to hosp., 1795. Mineral Babingtonite named for him. Licentiate, fellow Coll. Physicians. Fellow Royal Soc., 1805; mem. Geol. Soc. (a founder; became pres. 1822), Hunterian Soc. (a founder). Author: A Systematic Arrangement of Minerals reduced to the Form of Tables, founded on the joint Consideration of their Chemical, Physical and External Characters, 1795; A New System of Mineralogy . . . , 1799; A Catalogue of the genuine and valuable Collection of Minerals of a Gentleman Deceased, 1805; Syllabus of the Course of Chemical Lectures at Guy's Hosp., 1789. Died London, Apr. 29, 1833.

BABINSKI, Joseph François Felix, French neurologist; b. Paris, France, Nov. 17, 1857; M.D., U. Paris, 1885. Intern, Legarand du Saulle, under Vulpian; employed labs.; asst. to Charcot, Salpetrière; chief, neurol. clinic, Hôpital de la Pitié, 1890-1927. Founder Société de Neurologie de Paris, 1907. Author: L'exposé des Travaux Scientifiques du Dr. J. Babinski, 1913. Demonstrated that there is no lessening of Achilles tendon reflex in hysterical paralysis, 1893; noted that in organic disease of pyramidal tract, stimulation of sole causes great toe to extend while others flex (sign distinguishes organic from hysterical hemiplegia, known as Babinski's sign), 1896; (with Alfred Fröhlich) investigated endocrinal disorder, adiposogenital dystrophy, known as Babinski-Fröhlich disease; suggested that manifestations of hysteria can be produced by suggestion, abolished by counter-suggestion, 1901. Died Paris, Oct. 29, 1932.

BABKO, Anatoliy Kirillovich, Russian chemist; b. 1905; grad. Kiev Poly. Inst., 1927; D.Chem. Sci. Instr., Kiev Poly. Inst., 1927-30, Kiev Inst. Food Industry, 1930-34; prof. Kiev U., 1934—; sr. asst. Inst. Gen. and Inorganic Chemistry, Ukranian Acad. Sci., 1941—. Mem. Ukranian Acad. Sci. Author: Colorimetric Analysis, 1951; Physicochemical Analysis of Complex Compounds, 1955; Quantitative Analysis, 1956; The Use of Ionic Chromatography To Determine the Polymerization Factor of Zirconium in Solutions, 1960; Modern Problems of Analytical Chemistry 1959; Methods of Obtaining Analytical Uranium Concentrates, 1963. Exec. editor chem. terminology dictionary pub. by Ukranian Acad. Sci. Research on analytical chemistry, colorimetry, chemistry of complex compounds. Address: Gosudarstvenny University, Vladimirskaya ulitsa 64, Kiev, Ukraine SSR, USSR.

BABO, Lambert Heinrich Clemens Carl, German chemist; b. Ladenburg/Neckar, Germany, Nov. 25, 1818; s. Lambert Josef and Karoline (Ehrmann) B.; ed. Heidelberg, Germany, Munich, Germany; M.D., 1842; student chemistry, Giessen, Germany; m. Elise Baumgärtner, Sept. 6, 1847; 2 sons, 1 dau. Became asst. prof. Freiburg, Germany, 1845, asso. prof., 1854, prof., 1859. Author: (with R. Fresenius) Arsen in Vergiftungsfällen, 1844; Zentrifugalkraft, 1852; Furfurol, 1853; Photographische Versuche, 1856. Research on alkaloids, vapor pressure of salt solutions, ozone; designed new lab. utensils; 1st to use centrifuge in the lab. Died Karlsruhe, Germany, Apr. 15, 1899.

BABSKII, Eugenii Borisovich, Russian physiologist; b. Goris, Armenia, Jan. 28, 1902; grad. Med. Faculty, Moscow U., 1924; D.Med. Sci., 1939; D.Biol. Sci. Asso. dept. physiology Timiryazev Biol. Inst., 1924-31; prof. Moscow Lenin Pedagogical Inst., 1932-49; head dept. gen. physiology Inst. Physiology, Ukrainian Acad. Sci., 1949-50; head Lab. Physiology, Inst. Thoracic Surgery, 1950-56; asso. USSR Acad. Med. Sci., 1952—, dir. Lab. Clin. Physiology, Inst. Normal and Path. Physiology. Mem. Ukrainian Acad. Sci. Author over 200 works including Manual on the Physiology of Man, 1937; The Formation of Physiologically Active Substances in Nerve-trunks, 1938; (co-author) Course of Normal Physiology, 1947; I. P. Pavlov 1849-1936, 1949; The Functional Role of Adenosine Triphosphate in the Activity of the Skeletal Musculature, 1950; Methods and Some Results of the Study of Mechanical Phenomena of Human Cardiac Action in Normal and Pathological Conditions, 1957. Research on formation, functional significance of chem. agts. in central nervous system, peripheral nerves and muscles; developer methods of dynamocardiography, also other methods of clin. physiology. Address: USSR. Acad. Med. Sci., Solyanka 14, Moscow, USSR.

BABUDIERI, Brenno, Italian microbiologist; b. Trieste, Italy, May 15, 1907; s. Antonio and Mary (Gangadi) B.; M.D., U. Pavia (Italy), 1931; diploma in respiratory diseases U. Rome, 1935; m. Raffaella Zoppi, Aug. 28, 1945; 1 son, Sergio. Docent in med. parasitology U. Rome, 1934, in gen. pathology, 1938; staff Istituto Superiore di Sanità, Rome, 1933—, now sr. research officer; prof. in charge microbiology Faculty Scis., U. Trieste, 1965—. Cons., WHO, Jordan, 1952, Spain, 1953-54; mem. experts com. on zoonoses WHO/FAO, 1959—, also dir. Internat. Reference Center Leptospirosis; mem. subcom. for leptospirosis Internat. Com. Microbiol. Nomenclature; mem. Italian Governmental Com. for Med. Specialities. Recipient G. Grassi award for parasitology, 1936. Mem. Acad. Medicine, Rome, Austrian Soc. Hygiene and Microbiology (hon.), Med. Soc. Buenos Aires, other med. assns. Contbg. author: Trattato italiano di malattie infettive; Trattato italiano di medicina interna. Contbr. articles to jours. Isolated and described some new types of pathogenic leptospirae; prepared widely used vaccine against leptospirosis; studied epidemiology of leptospirosis in rice fields; other research on Q fever and its epidemiology, trachoma, some new serological tests, morphology of many microorganisms, rickettsial and bedsonia infections, electron microscopy. Home: 22 L. Bodio, Rome. Office: 299 Viale Regina Elena, Rome, Italy.*

BACCAREDDA, Mario, Italian chemist; b. Rome, Italy, Aug. 15, 1907; s. Efisio and Vittoria (Boy) B.; Ph.D. in chemistry and engring.; m. Giulia Serafini, May 22, 1957; children—Maria-Vittoria, Carlo. Dir., Inst. Indsl. and Applied Chemistry, U. Pisa (Italy) Sch. Engring. mem. Società Chimica Italiana, Am. Chem. Soc. Contbr. numerous articles on indsl. chemistry to profl. publs. Research in nuclear technology and macromolecular chemistry. Home: via Nisi 1. Office: via Diotisalvi 2, Pisa, Italy.*

BACCELLI, Guido, Italian physician; b. Rome, Italy, Nov. 25, 1832; ed. U. Rome. Asst. prof. med. jurisprudence U. Rome, from 1856, later prof. clin. medicine; dep., div. Rome, 1874-1916; min. pub. instrn., 1880-1900; senator, 1890; minister agr., 1901-03. Corr. Académie des Scis. Author: Patologia del cuore e dell'aorta, 1864-67; La Malaria, 1878; Sur une Nouvelle Méthode du Traitement de Aneurismes del' Aorte. Research on malaria, heart disease, tumors; began use of mercuric chloride by injection as treatment for syphilis, 1893, treatment of tetanus by injection of carbolic acid, 1906. Died Rome, Dec. 10, 1916.

BACCETTI, B., Italian biologist; b. Florence, Italy, May 7, 1931; s. Bruno and Thea (Niccolini) B.; D. Agrl. Scis., U. Florence, 1954; m. Paola Barile, July 29, 1964; children—Nicola, Cosimo. Research staff Sta. Agrl. Entomology, Florence, 1954-65; prof. agrl. entomology U. Florence, 1959—; prof. zoology, biology and gen. zoology, dir. Inst. Zoology, U. Siena (Italy), 1966—. Mem. Nat. Acad. Entomology, Acad. Georgyofil, Acad. Fisiocritics in Italy. Research, numerous publs. on faunistics and zoogeography of Orthoptera; electron microscopy of several tissues and organs of invertebrate animals, physiology of insects, destruction of insects by ionizing radiations. Home: 216 Via Masaccio, Florence, Italy. Office: 4 Via P.A. Mattioli, Siena, Italy.*

BACCI, Andrea, Italian physician; b. St. Elpidio, Italy, 1550; prof. botany, Rome, Italy, 1567-1600; then physician to Pope Sixtus V. Author: De thermis, lacubus, fontibus, fluminibus et balneis totius orbis, 1588; De convivis antiquorum; De Venenis et antidotis Prolegomena, 1586; De Gemmis ac lapidibus pretiosis in Sacra Scriptura relatis, 1577; Tabula simplicium Medicamentorum, 1577; De Thermis libri VII, 1571; De naturali vinorum historia, 1596. Lectured on poisons; described curative properties of spas and mineral springs, baths. Died 1598.

BACH, Julius Carl von, see von Bach, Julius Carl.

BACH, L. M. N., Am. physiologist; b. San Francisco, Dec. 30, 1919; A.B., U. Cal. at Berkeley, 1940, M.A., 1943, Ph.D., 1945; m. Margaret Kandra, Apr. 1946; children—Stephen Matthew, Suzanne Margaret. Faculty, Tulane U., New Orleans, 1944—, prof. psychiatry and neurology, 1959—, lectr. Sch. Engring., 1953—, Sch. Social Work, 1956—, Sch. Architecture, 1958—; lectr. New Orleans Psychoanalytic Tng. Inst., 1958—, Law Sci. Inst., U. Tex., 1952-53, 61-65; exec. dir. Survey Physiol. Scis., Am. Physiol. Soc., Washington, 1953-56. Mem. Am. Physiol. Soc., Soc. for Exptl. Biology and Medicine (past chmn. So. sect., 1963-65), Am. Assn. U. Profs. (past pres. Tulane chpt.), Assn. Am. Med. Colls., Am. Acad. Neurologists, Assn. for Research in Nervous and Mental Diseases, Animal Behavior Soc., A.A.A.S., Sigma Xi. Research, publs. on hypothalamic regulation of growth hormone prodn. biotelemetry of brain activity, extravasation as a cause of coma, adrenalin interaction in prodn. CNS inhibition, orgn. brainstem respiratory, vasomotor and reflex regulatory patterns. Home: 7109 6th St., Harahan, La. 70123. Office: 1430 Tulane Av., New Orleans 70112.*

BACH, Max Hugo, German physician; b. Crösssuln, Germany, Sept. 22, 1859; s. Adolf and Henriette Eleonore (Beyer) B.; 2 sons, 1 dau. from 1st marriage; m. 2d, Marie Kaack, Jan. 15, 1913; 1 dau. Practiced medicine in resort of Hohenstein-Ernsttahl, later Elster; contbr. article on ultraviolet radiation to German med. jours.; developed mercury vapor quartz lamp (Kromeyer's invention) into what he called artificial sunlight lamp; expanded ultraviolet therapy of the time. Died Dresden-Bad Weisser Hirsch, Germany, Aug. 12, 1940.

BACHE, Alexander Dallas, Am. physicist; b. Phila., July 19, 1806; s. Richard and Sophia (Dallas) B.; grad. 1st in his class U. S. Mil. Acad., 1825; LL.D., N.Y. U., 1836; U. Pa., 1837; Harvard U., 1851; m. Nancy Fowler, 1829. Prof. natural philosophy and chemistry U. Pa., 1828-36, 42-43; 1st pres. Girard Coll., Phila., 1836-40; reorganized Phila. public schools, 1839-42; supt. U. S. Coast Survey, 1843-67; a founder A.A.A.S.; an incorporator, regent Smithsonian Instn., 1846; a founder, 1st pres. Nat. Acad. Scis.; pres. Am. Philos. Soc., 1855; adviser to the Pres. U. S., also v.p. Sanitary Commn. during Civil War. Hon. mem. Royal Acad. Turin, Imperial Geog. Soc. Vienna, French Acad. Scis.; Fellow Royal Soc., 1860. Author: Report on Education in Europe, 1839; Observations at the Magnetic and Meteorological Observatory of Girard College, 3 vols, 1840-47. Investigated terrestrial magnetism; made detailed magnetic survey of Pa.; supervised mapping of entire coastline; established 1st magnetic obs. in U. S. at Girard Coll., 1839. Died Providence, R.I., Feb. 17, 1867.

BACHE, Franklin, Am. physician, chemist; b. Phila., Oct. 25, 1797; s. Benjamin Franklin and Margaret (Markoe) B.; grad. U. Pa., 1810, M.D., 1814; m. Aglae Dabadie, 1818. Entered U. S. Army as asst. surgeon, 1813, promoted to full surgeon, 1814, served until 1816; practiced medicine 1816-24; physician to Walnut Street Prison, Phila., 1824-36; prof. chemistry Franklin Inst., Phila., 1826-32, Phila. Coll. Pharmacy, 1831-41, Jefferson Med. Coll., Phila., 1841-64. Fellow Coll. Physicians and Surgeons, Phila.; pres. Am. Philos. Soc., 1853-55. Author: System of Chemistry for the Use of Students in Medicine, 1819; Dispensatory of the United States of America, 1833. Co-editor N.Am. Med. and Surg. Jour., 1826-31. Died Phila., Mar. 19, 1864.

BACHELARD, Gaston, French philosopher of science; b. Bar-sur-Aube, Champagne, France, June 1884; s. Louis and Louise (Laurey) B.; Litt.D., U. Paris (France); m.; 1 dau. With Post & Telegraph Office, until 1914; prof. physics, chemistry Collège Bar-sur-Aube, 1919-30; prof. philosophy U. Dijon (France), 1930-40; prof. Sorbonne, Paris, 1940—, also dir. Inst. History of Sci. Author: Etude sur la propagation thermique dans les solides; La Valeur inductrice de la Relativité; Le pluralisme cohérent de la chimie moderne; La formation de l'esprit scientifique; Les institutions atomistiques; La dialectique de la durée; L'intuition de l'instant; Le nouvel esprit scientifique; L'experience de l'espace; others. Died, October, 1962.

BACHELET, Maurice Pierre, French chemist; b. Paris, France, July 6, 1906; s. A. and N. (Verlet) B.; ed. Collège Chaptal, École de physique et de chemie de Paris; Ph.D. in Phys. Scis.; m. N. Risacher, July 11, 1935; children—Daniel, Annie, Yves. Asst. to Mme. Pierre Curie at Radium Inst., Paris; prof. chemistry Faculté des sciences, Caen; dir. Collège science universitaire, Rouen. Research on chemistry of radioactive elements, particularly uranium. Patentee on treatment of poor minerals with uranium. Address: 18, rue Saint-Vincent, Fontenay-sous-Bois (Seine), France.

BACHELOT DE LA PYLAIE, Auguste Jean Marie, French botanist, archeologist; b. Fougères, France, May 25, 1786; ed. at École centrale de Laval, also Paris Natural History Mus.; studied under Cuvier and DeBlainville; visited Am.; worked in Brittany, 30 years. Author: Manuel de conchyliologie, 1828; Flore de Terre-Neuve, 1829; Alignements de Carnac, 1834. Numerous archaeol. expdns. in Brittany; pioneer in study primitive cults of prehistoric Brittany; research on marine algae, conchology. Died Oct. 12, 1856.

BACHEM, Albert, biophysicist; b. Bonn, Germany, Feb. 26, 1888; s. Joseph and Gertrude (Tonger) B.; Ph.D., U. Bonn, 1910; m. Erica Pietsch; children—Erica, Wolfgang Albert. Came to U. S., 1921, naturalized 1931. Served as asst. prof. physics, univs. Bonn and Frankfurt; prof. biophysics U. of Ill. Coll. Medicine, since 1924. Mem. Am. Physiol. Soc., Soc. Exptl. Biology and Medicine, Am. Congress Phys. Therapy, Sigma Xi. Author: Principles of X-Ray and Radium Dosage; also 6 physical therapy charts. Died April 3, 1957.

BACHER, Robert Fox, Am. physicist; b. Loudonville, O., Aug. 31, 1905; s. Harry and Byrl (Fox) B.; B.S., U. Mich., 1926; Ph.D., 1930, Sc.D., 1948; m. Jean Dow, May 30, 1930; children—Martha, Andrew Dow. Nat. Research fellow Cal. Inst. Tech., 1930-31, Mass. Inst. Tech., 1931-32; Alfred Lloyd fellow U. Mich., 1932-33; instr. physics, Columbia, 1934-35; instr. to prof. physics Cornell, 1935-49; radiation lab., Mass. Inst. Tech., 1940-45; head exptl. physics div. Los Alamos Lab. atomic bomb project, 1943-44, bomb physics div., 1944-45; mem. AEC, 1946-49; mem. President's Sci. Adv. Com., 1957-60, Naval Research Adv. Com., 1957-62; prof. physics Cal. Inst. Tech., 1949—, dir. Norman Bridge lab. physics, chmn. div. physics, math., astronomy, 1949-62, provost coll., 1962—; dir. Detroit Edison Co., Bell & Howell Co., Thompson Ramo Wooldridge, Inc. Trustee Asso. Univs., Inc., Atoms for Peace Awards, Carnegie Corp. of N.Y., Inst. Def. Analysis, Rand Corp., 1950-60. Recipient Medal for Merit, 1946. Mem. Am. Acad. Arts and Scis., Am. Phys. Soc. (pres. 1964), Nat. Acad. Sci., Am. Assn. U. Profs., Am. Philos. Soc., A.A.A.S., Sigma Xi. Compiled (with S. Goudsmit) study of atomic energy states as derived from analyses optical spectra. Home: 345 S. Michigan Av., Pasadena, Cal.

BACHET DE MEZIRIAC, Claude-Gaspar, French mathematician; b. Bourg-en-Bresse, France, Nov. 9, 1581; author: Problèmes plaisants et délectables, qui se font par les nombres, 1612; Chansons dévotés et saintes, sur toutes les principales fetes de l'année et sur autres divers sujets, 1615; pub. edit. with notes of Arithmetic of Diophantus, also 1st Graeco-Latin edit. Pappus' Arithmetic with commentary; translated several of Ovid's Epistles into French verse; modernized analytical pastime of Magic Squares. Died Feb. 18, 1638.

BACHMAN, G. Bryant, Am. chemist, educator; b. Kansas City, Mo., Aug. 18, 1905; s. Gustave William and Madge (Bryant) B.; B.A., U. Colo., 1926; Ph.D., Yale, 1930; m. Nancy Powell, Mar. 12, 1929; children—William P., Madge (Mrs. Edwin H. Harrison, Jr.), John B. Faculty, Ohio State U., 1930-36; chemist Eastman Kodak Co., Rochester, N.Y., 1936-39; faculty Purdue U., Lafayette, Ind., 1939—, prof. chemistry, 1941—. Cons. to chem. industries. Fellow Ind. Acad. Sci.; mem. Am., Ind. chem. socs., A.A.A.S. Author: Organic Chemistry, 1949. Bd. editors Jour. Organic Chemistry, 1936-56. Contbr. articles to chem. jours. Research in organic synthesis, drugs, plastics, rubbers, nitro compounds, combustion. Home: 923 N. Chauncey Av., W. Lafayette, Ind. 47906.*

BACHMAN, John, Am. naturalist; b. Rhinebeck, N.Y., Feb. 4, 1790; s. Jacob Bachman; ed. Williams Coll.; Ph.D. (hon.), U. Berlin (Germany), 1838; m. Harriet Martin, 1816. Sec. to Johannes Knickerbocker in an exploring expdn. and embassy to Oneida Indians; taught sch., Ellwood, Pa., later in Phila.; ordained to ministry, Lutheran Ch., 1814, served at St. John's Ch., Charleston, S.C.; founder S.C.'s Luth. Theol. Sem.; made collection of Southern animals and studied their habits and habitats, led to assn. with John Audubon (who used Bachman's work in his book on ornithology, 1830-35); a founder S.C. Hort. Soc., 1833; travelled in Europe with Audubon, 1838; tried to reconcile scripture with science during evolution controversy of 1850's; collaborated with Audubon on The Viviparous Quadrupeds of North America, 3 vols., 1845-59; The Unity of the Human Race, 1850. Died Columbia, S.C., Feb. 24, 1874.

BACHMANN, Friedrich, German mathematician; b. Wernigerode/Harz, Germany, Feb. 11, 1909; s. Hans and Elisabeth (Büchsel) B.; ed. U. Münster, U. Berlin; Ph.D.; m. Alexandra von Bredow, July 20, 1949; 1 son, Sebastian. Instr. U. Marburg, 1939—; prof. U. Kiel, 1949—. Mem. German, Austrian assn. mathematicians. Author: Aufbau der Geometrie aus dem Spiegelungsbegriff; Grundzüge der Mathematik. Home: Roonstrasse 7, Kiel. Office: University of Kiel, Olshausenstrasse 40/60, Kiel, Germany.

BACHMANN, Werner Emmanuel, Am. chemist; b. Detroit, Nov. 13, 1901; s. Arnold William and Bertha (Wurster) W.; student Detroit Jr. Coll., 1919-21; B.S.E., U. Mich., 1923, M.S., 1924, Ph.D., 1926; m. Marie Knaphurst, Sept. 14, 1927; children—Joan Marie, Roger Werner. Faculty chemistry U. Mich., 1925-51, prof. 1939-47, Moses Gomberg Univ. prof., 1947-51; asst. prof. U. Ill., summer, 1931, U. Zurich, 1928-29, Royal Cancer Hosp., London, 1935, U. Munich, 1935-36; research for Nat. Def. Com., OSRD, 1940-46. Recipient Henry Russel award, Naval Ordnance award, Presdl. Certificate of Merit (U. S.); King's medal (Eng.), 1948. Mem. Am. Chem. Soc., Nat. Acad. Scis., Sigma Xi, Phi Lambda Upsilon, Gamma Alpha, Tau Beta Pi, Alpha Chi Sigma. Author: (with L. C. Anderson) Laboratory Manual of Organic Chemistry, 1930. Bd. editors Jour. of Organic Chemistry, Organic Reactions. Syntheses in fields of sex hormones, cancer-producing compounds, penicillin and explosives; studies on free radicals and molecular arrangements. Died Mar. 22, 1951.

BACHMETIEV, Porfirii Ivanovich, physicist; b. Feb. 25, 1860; prof. Sofia (Bulgaria) U.; became prof. U. Moscow, 1913; applied his discovery of anabiosis to animals. Died Nov. 11, 1913.

BACHRACH, Peter, Am. polit. scientist; b. Chgo., June 19, 1918; s. Walter and Alice (Loeb) B.; A.B., Reed Coll., 1942; Ph.D., Harvard, 1952; m. Florence Helena Rice, Mar. 15, 1946; children—Lorein (Mrs. James JeDon), Catherine, Sarah, Ruth, Molly, David. Faculty, Bryn Mawr (Pa.) Coll., 1946—, prof. polit. sci., 1963—, chmn. dept. 1965—; vis. prof. Swarthmore Coll., 1954, U. P.R., 1956-57, Haverford Coll., 1962-63. Ford Found. fellow, 1952-53; Rockefeller Found. fellow, 1957-58, 64-65. Mem. Am. Polit. Sci. Assn., Am. Soc. Polit. and Legal Philosophy, Am. Assn. U. Profs., Am. Civil Liberties Union. Author: Problems in Freedom, 1954; The Theory of Democratic Elitism, 1967; also articles. Analysis of contemporary democratic theory especially its relationship to elite theory. Home: 673 Ardmore Av., Ardmore, Pa. 19003. Office: Library, Bryn Mawr Coll., Bryn Mawr, Pa. 19010.*

BACHYNSKI, Morrel Paul, Canadian physicist; b. Bienfait, Sask., Can., July 19, 1930; s. Nick and Caroline (Bachynski); B.Engring., U. Sask., 1952, M.Sc., 1953; Ph.D., McGill U., 1955; m. Slava Krkovic, May 30, 1959; children—Caroline Dawn, Jane Diane. Research asst. Eaton Electronics Lab., McGill U., Montreal, Que., Can., 1955; staff RCA Victor Co. Ltd., Montreal, 1955—, dir. micro and plasma physics lab., 1958-65, dir. research, 1965—; faculty McGill U., 1958—. Recipient David Sarnoff Gold medal RCA, 1963. Fellow Canadian Aeronautics and Space Institute; fellow Royal Society of Canada; asso. fellow Am. Inst. Aeros. and Astronautics; mem. I.E.E.E., Canadian Assn. Physics, Am. Phys. Soc., Am. Geophys. Union, Internat. Sci. Radio Union. Author: (with I. P. Shkarofsky, T. W. Johnston) Particle Kinetics of Plasmas, 1966; also articles. Research on wave propagation relating to communications over the earth, microwave physics and plasma physics especially electromagnetic properties of plasma, scaled geophys. plasma phenomena. Home: 78 Thurlow Rd. Office: 1001 Lenoir St., Montreal, Que., Can.*

BACIU, Clement Clement, Rumanian physician, orthopedic surgeon; b. Bucharest, Rumania, Sept. 2, 1922; s. Clement Stan and Maria (Seorsescu) B.; ed. Med. U., Bucharest; M.D.; m. Sultanica Niculescu, Oct. 30, 1946; children—Radu Valeriu, Mihnea Clement. With clinic for orthopedics and traumatology Brincovenesc Hosp., Bucharest, 1946—, preparator, 1949-52, secundarius, 1952-56, specialist, 1956-59, primarius, 1959—; asst. Faculty Med. Specialization, Bucharest. Mem. Rumanian, Belgian, French socs. orthopedics and traumatology. Author: (with A. Radulescu) Heine-Medin Disease, 1955, The Knee, 1964; The Callus Formation, 1958; Sport Traumatology, 1959; also papers. Research on paradoxical hypertrophy of suprarenals in fractures, fibroblastic metaplasia of adventitia vessels in fractures, surg. techniques for 1st metacarpal fracture, luxation, lesion of Achilles tendon; developed index for radiol. evaluation of collus. Home: 84 Icoanei. Office: Hospital Brincovenesc, 14 Sh. Cosbuc, Bucharest, Rumania.*

BACK, Ernest Adna, Am. entomologist; b. Northampton, Mass., Oct. 7, 1880; s. Adna and Mary Elizabeth (Young) B.; B.Sc., Mass. Agrl. Coll. (now Mass. State Coll.), 1904, Ph.D., 1907; B.Sc., Boston U., 1904; m. Clara Winifred Newcomb, Sept. 29, 1919; children—David Newcomb, Richard Chapell. Instr. entomology, Mass. Agrl. Coll., 1904-06, instr. botany, 1906-07; state nursery insp., 1902-06; agt. expert citrus insect investigations, U. S. Dept. Agr., 1907-10; entomologist and plant pathologist, Va. State Crop Pest Commn., and entomologist Va. Agrl. Expt. Sta., 1910-12; fruit fly investigations, H.I. and Spain, for U. S. Dept. Agr., 1912-16; entomologist in charge stored products and household insect investigations Bur. Entomology, U. S. Dept. Agr., 1916-34; prin. entomologist, research, div. of insects affecting man and animals Bur. Entomology and Plant Quarantine, U. S. Dept. Agr., from 1934. Mem. Am. Assn. Econ. Entomologists, Am. Entomol. Soc., Biol. Soc. Washington, Entomol. Regional Soc. Washington, Phi Kappa Phi, Alpha Sigma Phi; fellow A.A.A.S., Nat. Acad. Sciences, Am. Inst. Geneology. Wrote many tech. govt. publs. Investigated fruit fly in Hawaii, Bermuda, Spain; gave attention to fumigation methods for insects of household and stored products. Died 1959.

BACK, Frank Gerard, mech. engr.; b. Vienna, Austria; M.E., D.Sc., U. Vienna. Pres., Zoomar, Inc., Glen Cove, L.I., N.Y., 1945—. Recipient Gold Medal Ann. award Television Broadcasters Assn., 1947; Friedrich Voigtlander Gold Medal award Vienna, 1960. Fellow Royal Photog. Soc., Photog. Soc. Am., Soc. Photog. Scientists and Engrs., Soc. Motion Picture and Television Engrs. (Progress medal 1962); mem. Soc. Am. Mil. Engrs. Author books, articles. Research on design and engring. photog. optical equipment. Home: 1 Frost Creek Dr., Locust Valley, L.I. Office: 55 Sea Cliff Av., Glen Cove, L.I., N.Y.*

BACK, Kurt Wolfgang, social psychologist, educator; born Vienna, Austria, Oct. 17, 1919; s. Paul L. and Thekla (Fuchs) Baeck; came to U. S., 1938, naturalized, 1943; B.S., N.Y. U., 1940; M.A., U. Cal. at Los Angeles, 1941; Ph.D., Massachusetts Institute of Technology, 1949; m. Edith Bierhorst, May 15, 1949 (divorced 1965); 1 son, Allan. Asst. study dir. Research Center for Group Dynamics, 1946-49; social sci. analyst U. S. Bur. Census, 1949-51; research asso. Bur. Applied Social Sci., Columbia, N.Y.C., 1951-53, U. P.R., Rio Piedras, 1951-53; research asso. prof. U. N.C., Chapel Hill, 1956-59; asso. prof. sociology Duke, Durham, N.C., 1959-62, prof., 1962—. Research asso. Conservation Found., 1955-57. Recipient Burgess award Nat. Council for Family Relations, 1960. Mem. Am. Sociol. Assn., Am. Psychol. Assn., Am. Statis. Assn., Am. Assn. for Pub. Opinion Research, Soc. for Psychol. Study Social Issues (Helen L. Deroy award 1956), So. Sociol. Soc., Am. Gerontological Soc., Population Soc. Am. Author: (with L. Festinger and S. Schachter) Social Pressures in Informal Groups, 1950; (with R. Hill and J. M. Stycos) The Family and Population Control, 1959, The Survey Under Unusual Conditions, 1959; Slums, Projects and People, 1962; (with J. M. Stycos) The Control of Human Fertility in Jamaica, 1964; (with A. C. Kerckhoff) The June Bug, 1968. Publications on application of social psychology, methodology to such problems as group interaction, socio-biology, and theoretical investigation attempting math., empirical, and phenomenal unifications. Home: 2735 McDowell St., Durham, N.C. 27705.*

BACK, Nathan, Am. pharmacologist; b. Phila., Nov. 30, 1925; s. Joseph and Freda (Goldhirsh) B.; B.Sc., Pa. State U., 1948; M.Sc., Phila. Coll. Pharmacy, 1953, D.Sc., 1955; m. D. Toby Ticktin, June 17, 1951; children—Efrem Eli, Aaron Issar, Adina, Rachel Sivia, Sara Deborah. Neuropsychiat. asso. USN, Washington, 1944-46; dir. clin. labs. Israel Govt., 1948-49; research asso. Wyeth Inst. Phila., 1949-50; instr. pharmacology Phila. Coll. Pharmacy and Sci., 1953-55; sr. cancer research scientist Roswell Park Meml. Inst., Buffalo, 1955-61; prof., acting chmn. chmn. dept. biochem. pharmacology State U. N.Y. at Buffalo, 1961—, asso. pharmacology, 1961—; chairman department, 1968—. Fellow Royal Society Medicine, International Hematology Society; fellow N.Y. Acad. Scis., A.A.A.S.; mem. Am. Assn. Cancer Research, Soc. Pharmacology and Exptl. Therapeutics, Hematology Soc., Soc. Exptl. Biology and Medicine, Research Soc. Am., Internat. Hematology, Internat. Biochem. Pharmacology Soc., Sigma Xi, Rho Chi. Author: Laboratory Experiment in Pharmacology, 1961; (with Erdos, Sicuteri) Hypotensive Peptides, 1966; Chemistry, Pharmacology and Clinical Applications of Proteinase Inhibitions, 1967; also numerous articles. Editor: (with Pauletti and Martini) Pharmacology of Peptide Harmones and Proteins. Investigations on the mechanisms of thromboembolic phenomena and therapeutic approach to blood clot lysis using enzymes and synthetic agts., biochem. and physiologic aspects of shock states, role of histamine, serotonin, and hypotensive agts, pharmacology of cancer, chemotherapeutic and radioprotective agts. Home: 172 Sterling Av., Buffalo, N.Y. 14216.*

BACKEBERG, Curt, German botanist; b. Aug. 2, 1894; s. Heinrich and Hedwig (Bartel) B.; m. Emmy Marks, Nov. 11, 1919. Curator of old garden of King Leopold II, Saint-Jean-Cap-Ferrat (A.-M.), France. Mem. Am. Soc. Plants (past hon. v.p.), French, Austrian, German assns. plants (hon.). Author: Stachlige Wildnis; Wunderwelt Kakteen; Die Cactacaae; Das Kakteenlexikon; also manuals in German, Danish, Dutch. Editor numerous publs. Died Jan. 14, 1966.

BACKER, Hilmar Johannes, Dutch chemist; b. Dordrecht, Netherlands, Jan. 13, 1882; s. J. P. and E. P. (Mengel van Koetsveld van Ankeren) B.; doctorate in chemistry U. Leiden (Netherlands), 1911. Became chemist Treasury Dept., Amsterdam, 1914; prof. organic chemistry U. Groningen (Netherlands), from 1916. Mem. French Acad. Scis., 1945, also Am.,

French, Netherlands chem. socs. Contbr. to Recueil des travaux chimiques des Pays-Bas. Address: 10 Bloemsingel, Groningen, Netherlands.

BACKHAUSZ, Richard, Hungarian physician, immunologist; b. Budapest, Hungary, May 3, 1920; s. Oscar and Helen (Juszák) B.; M.D., Med. U., Budapest, 1944; Specialist, Clin. Lab., Budapest, 1953; m. Emily Hartmann, Jan. 28, 1950; children—Beata, Cecilia, Marian, Ladislaus. Med. bacteriologist, immunologist Phylaxia State Inst. for Sera and Vacines, Budapest, 1946-54; sci. dir. Human Inst. for Serobacteriological Prodn. and Research, Budapest, 1954—. Mem. Assn. Hungarian Microbiologists (sec.). Author: Immunodiffusion and Immunoelectrophoresis, 1966; also numerous articles. Editorial staff Immunochemistry, 1964—. Devel. new methods for immunochem. analysis, including quantitative immunoelectrophoresis, 1958, determination of diffuse coefficient of antigens and antibodies using angular immunodiffusion, 1964, elaboration of immunogram to determine immunological state of organism, 1966. Home: 5 Muzeum, Budapest. Office: 5 Szállás, Budapest, Hungary.*

BACKLUND, Johan Oskar, astronomer; b. Lenghem, Sweden, Apr. 28, 1846; ed. Uppsala, also Stockholm, Sweden; apptd. lectr. U. Uppsala, 1875; became asst. Stockholm Obs., 1876; named observator Dorpat (Estonia) Obs.; later became adj. astronomer Pulkova (Russia) Obs., apptd. dir., 1895. Fellow Royal Soc., 1911. Mem. French Acad. Scis., 1895, St. Petersburg Acad. Sci. Research on progressive decrease in period of Encke's comet; formulated theory of disturbances. Died Pulkova, Aug. 30, 1916.

BACKMAN, Jules, Am. economist; b. N.Y.C., May 3, 1910; s. Nathan and Gertrude (Schall) B.; B.S. cum laude, N.Y. U., 1931, A.M., 1932, M.B.A., 1933, D.C.S., 1935; m. Grace G. Straim, Oct. 18, 1935; children—Susan Patricia (Mrs. Charles R. Frank Jr.), John Randolph. Statistician, Sydeman Bros., N.Y.C., 1932-33; v.p. Economic Statistics, Inc., N.Y.C., 1933-35; asso. economist SEC, 1935; research asso. Madden & Dorau, N.Y.C., 1936-37; faculty N.Y. U. Sch. Commerce, 1938—, research prof., 1960—. Cons. adviser to various govtl. agys. and to industry; editorial writer N.Y. Times, 1943-48; editor Trusts and Estates mag., 1938-46. Recipient N.Y. U. Meritorious Service award, 1943, Madden award, 1960, Presdl. citation, 1964. Hon fellow Am. Statis. Assn.; mem. Am. Econ. Assn., Indsl. Labor Relations Assn., Soc. Bus. Adv. Professions, Author: Adventures in Price Fixing, 1936; Government Price Fixing, 1938; Investment Dynamics, 1939; Rationing and Price Control in Great Britain, 1943; Price Control and Subsidy Program in Canada, 1943; Economics of the Potash Industry, 1946; Surety Rate-Making, 1949; Bituminous Coal Wages, Profits and Productivity, 1950; Economics of Armament Inflation, 1951; War and Defense Economics, 1952; Price Practices and Price Policies, 1953; Rate Policies and Rate Practices of the Post Office, 1954; Administered Prices, 1957; Wage Determination, 1959; The Economics of the Electrical Machinery Industry, 1962; Studies in Chemical Economics, 1965; (with M. R. Gainsbrugh) Economics of the Cotton Textile Industry, 1946, Inflation and the Price Indexes, 1966. Home: 59 Crane Rd., Scarsdale, N.Y. 10583.*

BACKSTRöM, Johan Willem von, see von Backström.

BACKUS, George Edward, Am. geophysicist; b. Chgo., May 24, 1930; s. Milo Morlan and Dora (Dare) B.; Ph.B., U. Chgo., 1947, S.B., 1948, S.M. in Math., 1950, S.M. in Physics, 1954, Ph.D. in Physics, 1956; m. Elizabeth Evelyn Allen, Dec. 11, 1961; children—Benjamin Tom, Brian Nathaniel, Emily Anne. Junior mathematician Institute for Air Weapons Research, University of Chicago, Chicago, Illinois, 1951-53; physicist project Matterhorn, Princeton, 1957-58; asst. prof. math. Mass. Inst. Tech., 1958-60, asso. prof., 1960; asso. prof. geophysics, U. Cal., San Diego, 1960-62, professor of geophysics, 1962—. Member American Academy Arts and Scis., Am. Geophys. Union, Am. Math. Soc., Am. Phy. Soc., Soc. Indsl. and Applied Math., Am. Seismol. Soc. Soc. Exploration Geophysicists, N.Y. Acad. Scis. Proof of existence of self-sustaining, dissipative homogeneous dynamos, analysis of effect of earth's rotation on long seismic waves and short seismic and ocean waves, seismic effects of anistropy, analytic solution of inverse seismic problem for travel times on Great Circles, nonlinear inverse problems. Home: 7687 Hillside Dr., La Jolla Cal. 92037.*

BACON, Francis (Baron Verulam, Viscount St. Albans), Brit. philosopher, statesman, essayist; b. London, Jan. 22, 1561; s. Sir Nicholas and Ann (Cooke) B.; ed. Trinity Coll., Cambridge, 1573-75; studied law Gray's Inn., London, 1576, 79-82; m. Alice Barnham, May 10, 1606. Attached to Brit. embassy, France, 1576; admitted as outer barrister, 1582; mem. Parliament, 1584-1621; became confidential adviser to Earl of Essex, 1591, later was influential in having him convicted of treason, 1601; unofcl. mem. Queen Elizabeth's Learned Council, circa 1596; paid court to James I, nominated to King's Council, 1603; apptd. commr. for arranging England's union with Scotland, 1604; became solicitor gen., 1607, atty. gen., 1613, privy councilor, 1616, Lord-keeper, 1617, Lord chancellor, 1618; convicted for bribery and corrupt dealing in chancery suits, but pardoned, 1621; spent remainder of life writing in retirement. Knighted, 1603, created Baron Verulam, 1618, Viscount St. Albans, 1621. Author: Essays, 1597-1625; Advancement of Learning, 1605; De Sapientia Veterum, 1609; Novum Organum, 1620; Historia Ventorum, 1622; History of Henry VII, 1622; De Augumentis Scientiarum, 1623; Apophthegms New and Old, 1625; The New Atlantis, pub. 1627; Maxims of the Law, pub. 1630; Reading on the Statute of Uses, pub. 1642; also others. Best remembered as essayist and as philosopher; his insistence on expt. and induction was very influential; opposed the a priori methods of medieval scholasticism and reliance on authority in determination of truth; believed that science should concern itself with the actual world which is apparent to the senses. and that laws of science should be established as generalizations drawn out of vast mass of specific observations; held that heat is motion; conceived the phys. universe as an aggregate of matter, form and motion; identified magnetic force and gravitation; held that light travels with finite velocity; recognized idea of conservation of mass; made many measurements of specific gravities; showed that salts lower melting point of ice; noted that given quantity of acid dissolves only a certain weight of metal and no more; studied plant fertilizers; noted that putrefaction may be retarded by such processes as exclusion of air, drying and smoking; suggested that southern tips of Africa and S. Am. were homologous formations, 1620; suggested creation of universal European lang., 1605; his ideas were influential in creation of Royal Soc., 1660. Died London, Apr. 9, 1626.

BACON, George Edward, English physicist; b. Dec. 5, 1917; s. George H. and Lilian A. Bacon; ed. Derby Sch.; Emanuel Coll., Cambridge (Eng.) U.; m. Enid Trigg, 1945; 1 son, 1 dau. With Air Ministry, Telecommunications Research Establishment, 1939-46; dept. chief sci. officer, Harwell, 1946-63; prof. physics U. Sheffield, from 1963; chmn. neutron diffraction commn. Internat. Union Crystallography. Mem. Inst. Physics, Am. Crystallographic Assn. Author: Neutron Diffraction, 1955; Applications of Neutron Diffraction in Chemistry, 1963; X-Ray and Neutron Diffraction, 1966. Research, publs. on X-ray and neutron crystallography, carbon optics. Home: 28 Whirlow Park Rd., Sheffield 11. Office: dept. physics Sheffield University, Sheffield 19, United Kingdom.

BACON, Harold Maile, Am. mathematician, educator; b. Los Angeles, Jan. 13, 1907; s. Robert H. and Lura (Maile) B.; A.B., Stanford, 1928, A.M., 1929, Ph.D., 1933; m. Rosamond Clarke, Feb. 23, 1946; 1 son, Charles Robert. Clk. actuary's dept. Pacific Mut. Life Ins. Co., Los Angeles, 1929-30; faculty math. Stanford, 1930—, prof., 1950—. Instr. math. San Jose State Coll., winter 1934-35. Fellow A.A.A.S.; mem. Math. Assn. Am. (bd. govs. 1941-43, 57-60, sec. No. Cal. sect. 1939-45, chmn. 1949-50), Am. Math. Soc., Nat. Council Tchrs. Math., Cal. Math. Council, Am. Assn. U. Profs., Phi Beta Kappa, Sigma Xi, Phi Delta Kappa, Theta Delta Chi. Author: Differential and Integral Calculus, 1942, 2d edit., 1955; (with C. G. Jaeger) Introductory College Mathematics, 1954, 2d edit., 1962. Home: P.O. Box 4144, Stanford, Cal. 94305.*

BACON, John Mackenzie, English meteorologist, astronomer; b. Lambourn Woodlands, Eng., June 19, 1846; s. John and Mary (Lousada) B.; ed. Trinity Coll., Cambridge; m. Gertrude Myers, Apr. 11, 1871; m. 2d, Stella Valentine, Oct. 7, 1903; 2 sons, 2 daus. Began a cleric, later became lectr. on astronomy, meteorology. Fellow Royal Astron. Soc., Brit. Astron. Assn. Author: By Land and Sky, 1900; The Dominion of the Air, 1902. Made numerous balloon ascents to carry out acoustical, meteorol., geog. studies; observed solar eclipse, 1898, took 1st animated photographs. Died Dec. 26, 1904.

BACON, Roger, Brit. scholar; b. Ilchester, Somerset, Eng., circa 1214; studied at Oxford U., circa 1230; possibly took holy orders, 1233; possibly M.A. from U. Paris, circa 1244. Lectured as regent master U. Paris, 1241-47, then devoted himself to research; returned to Oxford, circa 1251, assumed Franciscan habit; sent back to Paris, circa 1257; requested by Pope Clement IV to write gen. treatise on the sciences, 1266; his works were condemned by Franciscan council (either because he was suspected of dealing in black magic, or because of his intemperate attacks on some of his contemporaries), 1277; imprisoned, 1277/8-92; returned to Oxford, remained there until death. Author: (principal work) Opus Majus, 1268; Opus Minus, 1268; Opus Tertium, 1268; Compendium Studii Philosophiae, 1271; Compendium Studii theologiae, 1292. De mirabile potestate artis et naturae, (possibly spurious), pub. 1542; Often considered the forerunner of modern sci. method because of his firm and outspoken belief in experimentation and math., and his rejection of principle of truth by authority; attempted to write a universal ency. of knowledge; wrote treatises on grammar, logic, math., physics, alchemy and moral philosophy; experimented in optics; seems to have stated the laws of reflection and an approximation of the laws of refraction; gave rough explanation of the causes of rainbow; pointed out inaccuracy of Julian calendar; interested in alchemy, believed that gold could be made from base metals; one of 1st to suggest that medicine should make use of remedies provided by chemistry; one of 1st to describe gun-powder (though he was not inventor of it). Died circa 1294.

BACON, Selden Daskam, Am. sociologist, educator; b. Pleasantville, N.Y., Sept. 10, 1909; s. Selden and Josephine (Daskam) B.; B.A. in History, Yale, 1931, M.A. in Govt., 1935, Ph.D. in Sociology, 1939; m. Cornelia Howard, Dec. 20, 1935; 1 dau., Cornelia Anne (Mrs. Lewis Jamieson); m. 2d, Margaret Keller, May 25, 1946; children—Selden Daskam, Michael McAlpine. Faculty sociology dept. Pa. State U., 1937-39; prof. sociology, dir. sect. on alcohol studies, dir. summer sch. alcohol studies Yale, 1939-62; prof. sociology, dir. Center Alcohol Studies, Rutgers, New Brunswick, N.J., 1962—. Mem. Conn. Commn. on Alcoholism, chmn. bd., 1945-59; mem. A.M.A. Commn. on Alcoholism, 1956-61, Nat. Safety Council Commn. on Alcohol and Drugs, 1960—; sec., treas. Nat. Council on Alcoholism, 1945-50; bd. dirs. N.Am. Assn. Alcoholism Programs, 1964—, Coop. Commn. on Study Alcoholism, 1962—; mem. National Adv. Com. on Alcoholism, 1966—. Mem. Am. Sociol. Assn., Eastern Sociol. Soc., Am. Criminological Soc., A.A.A.S. Author: (with R. Straus) Drinking in College, 1953; Sociology and Problems of Alcohol: Foundations for a Sociologic Study of Drinking Behavior, 1944; also articles. Editorial bd. Quar. Jour. Studies on Alcohol, 1948—. Studies, publs. on drinking customs, problems related to use of alcohol and social attempts to meet or avoid them; proposals, demonstrations, information processing. Home: Penn-Lyle Rd., Princeton Junction, N.J. Office: Center of Alcohol Studies, Rutgers U., New Brunswick, N.J. 08903.*

BADANOIU, Alexandru, Rumanian dermatologist; b. Tr. Magurele, Rumania, June 3, 1927; s. George and Olimpia (Paunescu) B.; physician, U. Bucharest (Rumania), 1951; d.med.scis., Rumanian Acad. Scis., Bucharest, 1960; m. Maria Penisoara, Feb. 5, 1951; Staff, Inst. Therapeutics, Rumanian Acad. Scis., Bucharest, 1951-60, lectr., 1958-60; lectr., head lab. for biol., functional and exptl. dermatology Inst. Dermato-Venerology, Bucharest, 1960—; asst. prof. Faculty Medicine, U. Bucharest, 1955-64. Recipient Victor Babes prize for med. research Rumanian Acad. Scis., 1965; medal for med. merit, 1963. Fellow Rumanian Dermatol. Soc., Internat. Assn. Dermatologists. Author: (with St. G. Nicolau) Elements of Pathophysiologic Dermatology, 1967; also numerous articles. Research in microbic allergy of skin, vascular allergy, immunopathology of skin, antitumoral immunity, pathophysiology of skin, exptl. dermatology, immunodiagnosis and therapeutics of syphilis. Home: 35, Bd Magheru. Office: Centrul Dermato-Venerologic, str. Dr. Grozovici nr. 1, Bucharest, Rumania.*

BADARAU, Eugen, Rumanian physicist; b. 1887; univ. prof.; dir. Inst. Physics, Bucharest, Rumania; mem. Acad. Socialist Republic Rumania. Author: Physics of Discharges in Gases, 1957. Co-author: Ionizing Gases, 2 vols., 1963, 65. Helped explain mechanism of luminescent and arc discharges.

BADDILEY, James, English chemist; b. Manchester, Eng., May 15, 1918; son of James and Ivy Logan (Cato) B.; B.Sc. with 1st class honors, U. Manchester, 1941, M.Sc., 1942, Ph.D., 1944, D.Sc., 1954; m. Hazel Mary Townsend, Sept. 20, 1944; 1 son, Christopher James. Sir Clement Royds Meml. scholar U. Manchester, 1942, Beyer fellow, 1943-44; fellow Swedish Med. Research Council, Stockholm, 1947-49; staff Lister Inst. Preventive Medicine, London, Eng., 1949-54; Rockefeller travelling fellow, 1954; prof. organic chemistry U. Newcastle upon Tyne (formerly U. Durham), 1955—. Mem. adv. bd. British National Com. for Biochemistry, 1961-66; mem. Govt. Grant Bd. Royal Soc., 1962-66; also mem. coms.; Karl Folkers lectr., 1962. Recipient Meldola medal Royal Institute of Chemistry, 1947. Fellow Royal Soc., 1961 (Leeuwenhoek lecturer 1967); fellow Royal Soc. Edinburgh, Chem. Soc. (past mem. council, Corday-Morgan medal and prize 1952, Tilden lectr. 1959); mem. Biochem. Soc. (mem. com. 1964), Soc. for Gen. Microbiology. Research, numerous publs. on nucleosides, penicillin, pyridoxal phosphate, active methionine, biosynthesis, nucleotide coenzymes, carbohydrates, bacterial cell walls, microbiol. chemistry. Home: 26 Wolsington Park S., Woolsington, Newcastle upon Tyne, Eng.*

BADEN, Martin William, Am. geologist, petroleum engr.; b. Independence, Kan., Nov. 25, 1878; s. John, Peter and Adelaide Elizabeth (Ballein) B.; B.A., Southwestern Coll., Winfield, Kan., 1929; Sc.D., (hon.), U. Pitt ., 1930; m. Grace Curnett, Jan. 15, 1900; 1 dau., Adelaide Elizabeth (Mrs. Frederick J. Barnard). Cons. geologist, owner, operator Martin W. Baden Research & Testing Lab., Winfield, 1917-59; v.p., geologist Trees Oil Co., 1921-31; processing engr. Boeing Airplane Co., Wichita, Kan., 1942-44; geologist Kanetex Refining Co., Arkansas City, Kan., 1945-46, Consol. Gas Utilities Corp., Oklahoma City, 1946-47; cons. petroleum geologist, engr., Winfield, Kan., 1947-57. Recipient Meritorious Contbn. award Wichita sect. Am. Chem. Soc., 1954. Research on problems in colorimetric chemistry, finding petroleum through use of scintillation radio counter; inventor or patentee ozone producing equipment for bleaching wheat flour, 1st x-ray equipment in Kan., preventive against corrosion in iron pipe; duplicating phonograph turntable; elec. and gas evaporator; photoelec. analyser equipment; radio antenna for autos; projector for projecting fossils on screen. Address: 1219 E. 9th Av., Winfield, Kan. 67156.*

BADER, Henri, glaciologist; b. Brugg, Switzerland, Jan. 15, 1907; s. Walter and Leonie (Bel) B.; Ph.D. in Mineralogy, U. Zürich (Switzerland), 1934; m. Adele Christen, Sept. 24, 1938. Came to U. S., 1945, naturalized, 1951. Engaged in avalanche research for Swiss Govt., 1935-38; mineralogist Lab. Quinico Nacional, Bogota, Columbia, 1940-41; quarry supt. Curacao Mining Co., Netherlands W.I., 1941-45; asso. research specialist Bur. Mineral Research, Rutgers U., 1945-49; chief scientist snow, ice and permafrost research establishment, U. S. Army, Wilmette, Ill., 1952-60; research prof. Sch. Engring., U. Miami (Fla.), 1960-63; sci. attaché Am. embassy, Bonn, Germany, 1963-64, Bern, Switzerland, 1965-67. Mem. Geol. Soc. Am., Geophys. Union, Glaciological Soc., Sigma Xi. Designed lab. and field instruments and methods for study snow and ice structure and mechanics; developed theories of plastic behavior of snow and ice under stress; introduced new concepts and practices of glaciological engring. Home: care Univ. Miami, Coral Gables, Fla. Cffice: American Embassy, Bern, Switzerland.*

BADER, Jean Pierre, French physician; b. Paris, July 17, 1926; s. René and Germaine (Launoy) B.; Bacalaureat, U. Paris; m. Nicole Leclerc, Dec. 11, 1951; children—Jean-Michel, Marie-Laurence, Jean-René. Asst. in hosps., 1959-64; prof. agrégé U. Paris, 1964—; sci. sec. INSERM (Nat. Inst. Health and Med. Research), 1965—. Mem. French Soc. Gastroenterology, French Soc. Pediatrics. Author: (with Pean) Médic. Pathology, 1955; also articles. Research on esophagus motility with electromanometrics and radiocinematographics methods, hiatal hernias and cardiospasm, Zollinger-Ellison syndrome using bioassays of urine of patients and biologic diagnostic tests; biol. studies on portacaval shunt; electron microscopic studies on pancreatic islets. Home: 7 bis, bld. Anatole France, Boulogne s/Seine (Seine), France. Office: Institut National de la Santé et de la Recherce Medicale 3, rue Léon Bonnat, Paris 16ème, France.*

BADER, Otto Nikolaevich, Russian archeologist; b. 1903; grad. 1st Moscow U., 1926; Cand. Hist. Sci., 1937—. Instr., asso. Anthropology Research Inst., 1st Moscow U., 1926-41; sr. asso. State Acad. History of Material Culture (later Inst. History of Material Culture), USSR Acad. Sci., 1932-41, sr. asso. Inst. Archeology, 1955—, also mem. Quaternary Commn.; lectr., head dept. archeology Perm U., 1946-55. Mem. Commn. for Study Geology and Geography of Karst. Mem. Internat. Assn. for Quaternary Research (Soviet sect.). Author: Archeological Monuments of Moscow and its Environs, 1947; Fortified Settlements of the Vetluga Basin, 1947; Archeological Monuments of Kama Basin, 1950; Monuments of the Turbino Period in the Middle Kama Basin, 1962; The Tomb of Balanovo, 1963; co-author: The Dawn of History in the Kama Basin, 1958. Research on history of primitive society; studied Mousterian sites of Volchiy Grotto and Chagaran-Koba in the Crimea, Upper Paleolithic sites of Tashtsky in Western Urals, Medved Stone in Eastern Urals, Paleolithic drawings in Kapovaya cavern in Urals, Neolithic monuments in Volga-Oka watershed, Bronze Age monuments in forest belt of European USSR; head expdn. to Moscow-Volga Canal site, Azov-Black Sea archeol. expdn., Kama and Votkinsk archeol. expdns. Address: USSR Acad. Sci., Archeol. Inst., 1-ya Cheremushkinskaya ulitsa 19, Moscow, USSR.

BADER, Otton Paulus Anthony, surgeon; b. Bonn Rhine, Germany, June 4, 1920; s. Edward and Michalina (Majewska) B.; physician diploma Med. Akademy, Lwów, Poland, 1944; doktor medicin Med. Akademy, Wroclaw, Poland, 1951; m. Elisabeth Lipczak, July 8, 1950; children—Barbara, Anna, Dorothea. Asst. surg. dept. Ins. Hosp., Katowice, Poland, 1945-49; asso. prof., docent Med. Akademy, Wroclaw, 1962—, teaching and cons. surgeon, 1956—. Recipient award sci. counsel Minister of Health, 1963, award of I degree, 1964. Mem. Polish Surgeons Soc., Société des Scis. et des Lettres de Wroclaw. Author: Metabolism, Acid-base Equilibrium and Hemodynamics in Deep Hypothermia and Extracorporeal Circulation, 1964; also numerous articles. Research on metabolism, acid-base equilibrium and hemodynamics in deep hypothermia and extracorporeal circulation, modification of extracorporeal circulation with use of own lungs as oxygenator; improvements in performance of extracorporeal circulation with use heart-lung machine. Home: 11/4 Pasteur. Office: 66 Curie-Sklodowska, Wroclaw, Poland.*

BADGER, Geoffrey Malcolm, Australian organic chemist; b. Port Augusta, South Australia, Oct. 10, 1916; s. John McDougal and Laura (Brooker) B.; student Geelong Coll., 1927-32, Gordon Inst., 1932-34; M.Sc., Melbourne U., 1938; Ph.D., London U., 1941; D.Sc., (I.C.I. research fellow), Glasgow U., 1949; m. Edith Maud Chevis, Apr. 27, 1941. Sr. lectr., reader, then prof. organic chemistry U. Adelaide (South Australia), 1949-64, dep. vice-chancellor, 1965—; mem. exec. C.S.I.R.O., 1964-65. Fellow Royal Inst. Chemistry. Author: The Structures and Reaction of Aromatic Compounds, 1954; The Chemistry of Heterocyclic Compounds, 1960; The Chemical Basis of Carcinogenic Activity, 1962. Research, publs. on aromaticity and aromatic character, carcinogenic activity. Home: 7 Rectory Walk, Springfield, South Australia. Office: U. Adelaide, Adelaide, South Australia.*

BADGER, George Franklin, Am. biostatistician; b. Everett, Mass., May 14, 1907; s. Benjamin Franklin and Elizabeth (Benson) B.; B.S., Mass. Inst. Tech., 1929; M.P.H., Johns Hopkins, 1932; M.D., U. Mich., 1938; m. Florence Loyola Collins, Aug. 30, 1934; children—George Franklin, Robert Lawrence, James Martin. Asst. epidemiologist Detroit Dept. Health, 1929-34; asst. prof. biostatistics Johns Hopkins, 1938-46; cons. Sec. of War, Office of Surgeon Gen., 1942-44; asso. prof. biostatistics Western Res. U. Med. Sch., Cleve., 1946-49, prof., 1949—, dir. div. biometry, 1963-67, dir. dept. biometry, 1967—. Home: 3323 E. Scarborough Rd., Cleveland Heights, O. 44118. Office: Univ. Hosps., Wearn Research Bldg., Cleve. 44106.*

BADGER, Richard McLean, Am. phys. chemist; b. Elgin, Ill., May 4, 1896; s. Joseph Stillman and Carrie Mabel (Hewitt) B.; Student Northwestern U., 1916-17, 19; B.S., Cal. Inst. Tech., 1921, Ph.D., 1924; M. Virginia Alice Sherman, July 8, 1933; children—Anthony Sherman, Jennifer Hewitt. Internat. Research fellow Univs. Gottingen and Bonn, Germany, 1928-29; faculty chemistry Cal. Inst. Tech., Pasadena, 1929-66, professor, 1945-66, emeritus professor, 1966—, also chairman of the chmn. faculty, 1961-63; lectr. chemistry. U. Cal. at Berkeley, 1931. Guggenheim fellow, 1960. Recipient award for Excellence in Coll. Chemistry Teaching, Mfg. Chemists Assn., 1961. Fellow Am. Phys. Soc., Am. Acad. Arts and Scis.; mem. Nat. Acad. Scis., Athenaeum (Cal. Inst. Tech). Contbr. articles to sci. jours. on spectra of molecules, molecular structures, hydrogen bonding. Home: 1963 New York Dr., Altadena, Cal. 91001.*

BADGER, Walter Lucius, Am. chem. engr.; b. Mpls., Feb. 18, 1886; s. Minor Campbell and Mary Helen (Albro) B.; B.A., U. Minn., 1907, B.S. in Chemistry, 1908, M.S., 1909; m. Helen Elizabeth Franklin, Apr. 8, 1913; 1 dau., Elizabeth Helen. Instr. chemistry U. Minn., 1908-09, spring 1910; chemist Gt. Western Sugar Co., fall 1909; asst. in chem. div. U. S. Bur. Standards, 1910-12; with U. Mich., 1912-37, prof. chem. engring., 1917-37; in charge research on water purification Detroit Edison Co., 1914-16; dir. research and cons. engr Swenson Evaporator Co., 1917—; mgr. cons. engring. div. Dow Chem. Co., 1936-44; cons. chem. engr., 1944—; pres. W. L. Badger and Assos., Inc., 1957—. Recipient Wm. H. Walker award, 1940. Mem. Am. Chem. Soc., Am. Inst. Chem. Engrs., Sigma Xi. Author: Heat Transfer and Evaporation, 1925; Inorganic Chemical Technology (with E. M. Baker), 1928; Elements of Chemical Engineering (with W. I. McCabe), 1931; Introduction to Chemical Engineering (with J. T. Bancherd), 1955. Died Nov. 19, 1958.

BADHAM, Charles, English physician; b. London, Eng., Apr. 17, 1780; M.D., Edinburgh, Scotland, 1802; B.A., Oxford (Eng.) U., 1811, M.A., 1812, M.B., M.D., 1817; m. Margaret Campbell; m. 2d, Caroline Foote, 1833. Entered Pembroke Coll., Oxford, as gentleman commoner, circa 1803, became censor, 1821; began practice medicine in London, 1803; physician to Duke of Sussex; physician Westminster Gen. Dispensary; lectr. physic, chemistry, medicine, London. Licentiate Royal Coll. Physicians Lond. Fellow Royal Soc., 1818, Royal Coll. Physicians. Author: Observations on the Inflammatory Affections of the Mucous Membrane of the Bronchiae, 1808; An Essay on Bronchitis with a Supplement Containing Remarks on Simple Pulmonary Abscess, 1814. Introduced term bronchitis, 1808, and established its history, diagnosis, treatment; also distinguished it from peripneumony, pleurisy, other conditions. Died London, Nov. 10, 1845.

BADIANO, Juan, botanist; b. Mexico; translated into Latin Martin de la Cruz bot. work on flora of New Spain.

BADIOU, Raymond, French mathematician; b. Bellerive-sur-Allier, Aug. 14, 1905; s. Léon and Jeanne (Taillepied) B.; ed. Collège d'Autun, Lycée de Lyon, Ecole normale supérieure; m. Marguerite Rouxbedat, Dec. 22, 1930; children—Jean-Paul, Alain. Agrégé in math.; prof. math.; mayor of Toulouse (France), 1944-58; dep. of Haute-Garonne, 1946-51; mem. municipal council of Toulouse. Address: 3, rue de Montségur, Toulouse, France.

BADOUX, Heli, Swiss geologist, paleontologist; b. May 7, 1911; Dr. sci., U. Lausanne (Switzerland), geology degree, 1934; m. Madeline Peitrequin. Geologist, Iraq Petroleum Ltd., 1935-41, Bur. Mines, Switzerland, 1941-45; prof., chmn. dept. geology U. Lausanne, 1950—. Mem. Société géologique suisse, Société géologique de France, Société géologique de Belgique. Research, publs. on geology of Swiss Alps, oil geology and underground water. Address: University of Lausanne, Palais de Rumine, Lausanne, Switzerland.

BAEDER, David H., Am. pharmacologist; b. Phila., June 23, 1925; s. Max and Ethel Baeder; A.B., Temple U., 1948, M.A., 1949, Ph.D., 1952; m. Evelyn Hornstein, Sept. 7, 1947; children—Jerome E., Larry C. Instr. physiology Temple U., 1947-48; lab. pharmacologist Wyeth Inst. for Med. Research, div. Wyeth Labs., Inc. div. Am. Home Products Corp., 1948-52, sr. research pharmacologist, 1952-59, supr. research and devel. Ives-Cameron Co. div. Am. Home Products Corp., 1959-62; dir. biol. scis. Mallinckrodt Chem. Works, 1962-66; prof. pharmacology, spl. asst. dean Faculties for Sci. and Tech., research asso. Space Scis. Center, U. Mo., Kansas City, 1966—; prof. biochem. pharmacology Sch. Pharmacy State U. N.Y., Buffalo, 1962-66; research prof. pharmacology Grad. Sch. Arts and Scis., U. Buffalo, 1959-60; vis. prof. Cité U., Marseilles, France, 1960, Hebrew U. Sch. Medicine, Jerusalem, Israel, 1962, Charles U., Prague, Czechoslovakia, 1963. Chief cons. Nat. Council for Research and Devel., Office of Prime Minister, Jerusalem, 1960—. Fellow N.Y. Acad. Scis., A.A.A.S.; mem. A.C.P. (hon.), Soc. for Exptl. Biology and Medicine, Am. Soc. for Pharmacology and Exptl. Therapeutics, Am. Heart Assn. (mem. council on arteriosclerosis), Pharm. Mfrs. Assn. (mem. drug toxicity commn.), Internat. Union against Tb, Internat. Union for Biochem. Problems of Lipid Metabolism, Internat. Union Physiol. Scis., Phi Beta Kappa, Sigma Xi, Rho Chi. Research, numerous publs. on devel. of antibiotics, drugs for cardiovascular diseases and Tb; participated in establishment research inst. for study drugs in Israel. Home: 61 Villa Coublag Dr., St. Louis 63131. Office: 5524 Trout Av., Kansas City, Mo.*

BAEHR, George, Am. physician; b. N.Y.C., Apr. 16, 1887; s. Herman and Sara (Gusky) B.; M.D., Columbia U., 1908; postgrad. in exptl. pathology, U. Freiburg (Germany), 1911-12, exptl. pharmacology, U. Vienna (Austria), 1912-13; m. Francine Gordon, Oct. 6, 1917; 1 dau. Barbara. Mem. A.R.C. San. Commn. in Russia and the Balkans, 1915-16; chief, 1st med. service, Mt. Sinai Hosp., 1927-50, dir. clin. research, 1944-50, distinguished service prof. Mt. Sinai Sch. Medidine, 1966; clin. prof. medicine Columbia U., 1929-50; cons. physician St. Joseph hosp. Monmouth Meml. Hosp. of Long Branch, N.J.; mem. Pub. Health Council, State of N.Y., 1935—, chmn., 1955—; pres., med. dir. Health Ins. Plan of Greater N.Y., 1950-57; spl. med. cons., 1957—; mem. Bd. Hosps., N.Y.C., 1950—; adminstrv. cons. Dept. Hosps., N.Y.C., 1933-45; apptd. chmn. Fed. Commn. on Phys. Rehab., 1940; med. dir. USPHS and chief med. officer U. S. Office Civilian Defense, 1941-44; del. from U. S. State Dept. to 4th, 5th and 6th Decennial Internat. Confs. to revise internat. list of causes of death, Paris, 1938-48. Fellow A.C.P., mem. N.Y. Acad. Medicine (pres. 1945-49, trustee 1950-60), A.M.A., Assn. Am. Physicians, Am. Public Health Assn. (v.p. 1952-53), Soc. for Clin. Investigation, Am. Assn. Pathologists and Bacteriologists, A.A.A.S., Sigma Xi. Co-author: Cecil's Textbook of Medicine, 8th edit., 1932; co-editor: Convalescent Care, 1940; Preventive Medicine, 1942; Medical Uses of Cortisone, 1954; Oxford Loose-Leaf Medicine, vol. IV, 1955. Chairman advisory editorial bd. Standard Nomenclature of Diseases and Operations, 1932-60. Contbr. pub. health and med. jours. Demonstrated renal lesions asso. with subacute bacterial endocarditis; described (with N. E. Brill, N. Rosenthal) giant follicular lymphadenopathy involving spleen and liver, characterized by multiple follicle-like nodules (Brill-Symmers disease). Home: 45 Sutton Pl. S., N.Y.C. 22. Office: 110 E. 80th St., N.Y.C.

BAEHR, Baron Waclaw (Venceslas), Polish cytologist; b. Poland, Oct. 6, 1873; s. Othon Baehr; ed. U. St. Petersburg (now Leningrad, Russia), U. Würzburg (Germany). Prof. cytology, dir. Cytology Inst., U. Warsaw (Poland). Mem. Polish Acad. Sci. and Letters, Sci. Soc. Warsaw. Author: Über die Zahl der Richtungskörper in parthenogenetisch sich entwickelnden Eirn von Bacillus rossii, 1907; Über die Bildung der Sexualzellen bei Aphididae, 1908; Ovogenese bei einigen viviparen Aphididen und die Spermatogenese von Aphis saliceti mit besonderer Berücksichtigung der Chromatinverhältnisse, 1909; Récherches sur la maturation des oeufs parthénogénétiques dans l'Aphis palmae, 1920; La spermatogenese et l'ovogenese chez le Saccocirrus major suivies d'une discussion générale sur le mécanisme de la réduction chromatique, 1920; Sur les bases cytologiques de l'hérédité, 1925; Bases cytologiques de phénomenes d'hérédité, 1930.

BAEKELAND, Leo Hendrik, chemist; b. Ghent, Belgium, Nov. 14, 1863; s. Karel L. and Rosalia (Merchie) B.; B.S., U. Ghent, 1882, D. Nat. Sc. maxima cum laude, 1884; laureate 4 Belgian univs., 1887; D.Ch., U. Pitts., 1916; D.Sc., Columbia, 1929; D.A.Sc., U. Brussels (Belgium), 1934; LL.D., U. Edinburgh, 1937; m. Celine Swarts, Aug. 8, 1889; children—Jenny, George W., Mrs. Nina Baekeland Wyman. Asst., later asso. prof. chemistry U. Ghent, 1882-89; prof. chemistry and physics Govt. Higher Normal Sch. Sci., Bruges, Belgium, 1885-87; came to U. S., 1889; founder, head Neperd Chem. Co. (mfr. Velox, other photog. papers of his invention), 1893-99; continued research chem. work. Cons. chemist and helped develop Townsend electrolytic cell for Hooker Electrochem. Co., Niagara Falls, 1905; pres. Bakelite Corp., 1910-39, mfg. Bakelite (a chem. synthesis from phenol and formaldehyde, replacing hard rubber and amber for uses in electricity and indsl. arts where former plastics are unsuited). Mem. U. S. Naval Cons. Bd., 1915-44; mem. U. S. Nitrate Supply Com., 1917, and chmn. com. on patents of NRC, 1917; Awarded Nichols Medal, Am. Chem. Soc., 1909; John Scott Medal, Franklin Inst., 1910; Willard Gibbs Medal, Am. Chem. Soc., Chgo. sect., 1913; Chandler Medal (first award), Columbia 1914; Perkin Medal for indsl. chem. research, 1916; grand prize, Panama-Pacific Expn., 1915; Pioneer trophy, Chem. Found., 1936; Messel medal, Soc. of Chem. Industry, London, 1938; Franklin medal, Franklin Inst., 1940. Chandler lectr., Columbia 1914; hon. prof. chem. engring., 1917-44. U. S. del. Internat. Congress Chemistry, 1909. Pres. Inventors' Guild, 1914; pres. sect. plastics, Internat. Congress Chemistry, 1912. Hon. mem. Electro-chem. Soc. (pres. 1909); mem. Am. Chem. Soc. (pres. 1924), Am. Inst. Chem. Engrs. (pres. 1912), Deutsche Chem. Ges., Soc. Chem. Industry of London (v.p. 1905), Nat. Acad. Scis., Am. Inst. City N.Y., Sigma Xi, Phi Lambda Epsilon (hon.), Tau Beta Pi; life mem. Am. Philos. Soc., A.A.A.S., Franklin Inst., Royal Soc. Arts (London), Société Chimique de France, Société de Chimie Industrielle of Paris (hon.); hon. mem. Royal Soc. Edinburgh, Soc. Belge des Electriciens, Am. Inst. Chemists. Author: Some Aspects of Industrial Chemistry, 1914. Invented bakelite; many patents U. S. and abroad, on the subjects of organic chemistry, elec. insulation, synthetic resins, plastics, etc. Contbr. numerous publs. on photochemistry, electro-chemistry, organic chemistry, chem. industries, patent reform, social and philos. subjects, etc. Died Beacon, N.Y., Feb. 23, 1944.

BAER, Erich Eugen Ferdinand, chemist; b. Berlin, Germany, Mar. 8, 1901; s. Eduard and Sofie (Lutz) B.; Ph.D. Friederich Wilhelms U., Berlin, 1927; m. Dorothy M. Hodder, Dec. 23, 1944. Research asso. U. Berlin, 1927-32; research asso. U. Basel, 1932-37, privat dozent, 1936; faculty U. Toronto (Ont., Can.), 1937—, prof., 1951—. Recipient Glycerine Research award Glycerin Producers Assn., 1953; Neuberg medal, 1961, award in lipid chemistry Am. Oil Chemists Soc., 1964. Fellow Royal Soc. Can., Chem. Inst. Can. (recipient medal 1962); mem. Am. Chem. Soc., Canadian Biochem. Soc., Am. Soc. Biol. Chemists, N.Y. Acad. Sci., A.A.A.S. Corr. editor: Jour. Lipid Chemistry, 1959—; asso. editor: Canadian Jour. Chemistry. Research, publs. on carbohydrates, lipids and their metabolic intermediaries. Home: 19 Glenaden Av. E., Toronto 18, Ontario, Can.*

BAER, Hans Helmut, chemist; b. Karlsruhe, Germany, July 3, 1926; s. Paul and Elsa (Menges) B.; cand. chem. Karlsruhe Tech. U., 1948; Dipl. Chem., U. Heidelberg (Germany) 1950, Dr. rer. nat., 1952; m. Gertrud H. Mackprang, May 1956; children—Thomas, Nicole. Research asso. Max Planck Inst. for Med. Research, Heidelberg, 1952-57; vis. asst. prof. dept. biochemistry U. Cal. at Berkeley, 1957-59; vis. scientist NIH, Bethesda, Md., 1959-61; faculty U. Ottawa (Ont., Can.), 1961—, prof. chemistry, 1965—. Fellow Chem. Inst. Can.; mem. Am. Chem. Soc., Gesellschaft Deutscher Chemiker. Research, publs. in carbohydrate chemistry, especially structure and synthesis of nitrogen-containing sugars, chemistry of antibiotics, chemistry of aliphatic nitro compounds. Home: 2887 Highfield Crescent, Ottawa 14, Ont., Can.*

BAER, John Elson, Am. pharmacologist; b. Cleve., Apr. 25, 1917; s. Joseph A. and Elsie (Royan) B.; A.B., Swarthmore, Coll., 1938; M.S., U. Pa., 1940, Ph.D., 1948; m. Dorothy Way, Aug. 30, 1947; children—Robert, John, Martha, Barbara. Chemist, N.Y. U. Sch. Medicine, 1943-46; instructor Haverford College, 1946-47; assistant prof. Carleton Coll., 1948-51; research asso. Sharp & Dohme div. Merck & Co., West Point, Pa., 1951-58, dir. pharmacological chemistry Merck Sharp & Dohme Research Labs., 1958—. Recipient N.A.M. Pioneers in Creative Industry Scroll, 1965. Mem. A.A.A.S., Am. Soc. Pharmacology and Exptl. Therapeutics, Am. Assn. Clin. Chemists. Research, publs. on drug metabolism, kidney physiology, diuretic agts. Home: 517 Tennis Av., Ambler, Pa. 19002. Office: Merck & Co., West Point, Pa. 19486.*

BAER, Joseph Louis, Am. surgeon; b. Chgo., Apr. 29, 1880; B.S., U. Chgo., 1902, M.S., 1903, M.D., Rush Med. Coll., 1904; postgrad. Berlin and Vienna, 1908; m. Gretchen Winslow Shattuck, July 28, 1913 (dec. Mar. 1926); m. 2d, Janet Bachrach, Jan. 22, 1931. Intern Michael Reese Hosp., Chgo., 1904-07, anaestetist, 1907-13, attending obstetrician and gynecologist, 1913-36, sr. attending gynecologist and obstetrician, 1936—; faculty dept. gynecology and obstetrics Rush Med. Coll., 1917—, prof., 1935-46, emeritus, 1946—. Bd. dirs. Blue Cross Plan for Hosp. Care, 1937—. Vice pres., dir. Am. Bd. Obstetrics and Gynecology, 1927—. Fellow A.C.S., Am. Gynecol. Soc.; mem. A.M.A., Chicago Gynecol. Soc., Chicago Inst. Medicine. Contbr. articles to med. press. Died Dec. 8, 1954.

BAER, Karl Ernst von, embryologist, biologist; b. Landgut-Piep, Estonia, Feb. 17/29, 1792; s. Magnus Johann and Juliane Louise B.; student medicine, Dorpat (now Tartu, Estonia), 1810-14, also Würzburg, Germany; m. Auguste von Medem, 1820; 5 sons, 1 dau. From prosector to prof. anatomy, zoology, dir. zool. mus., Königsberg, Germany, from 1817; prof. zoology, anatomy St. Petersburg (now Leningrad, USSR) Acad. Surg. Medicine, from 1834; apptd. librarian Acad. Scis. St. Petersburg, 1834. Recipient Copley medal Royal Soc., 1867. Mem. French Acad. Scis., 1858. Author: De ovi mammalium et hominis genesi, 1827; Über Entwicklungsgeschichte der Tiere, 1828-37; Untersuchungen über die Entwicklungsgeschichte der Fische, 1835; Selbstbiographie, 1867; Reden und kleinere Aufsätze, 1867-77. Considered founder modern embryology; discovered mammalian ovum within Graafian follicle, 1827, also notochord in early vertebrate embryo; investigated embryology of chick; demonstrated various vertebrate organs derived from germ layers by differentiation; favored epigenesis; nature philosopher (opposed Darwin's views); studied fishes of Baltic, Caspian seas; noted that in no. hemisphere, erosion is stronger on right banks of rivers, stronger on left banks in so. hemisphere (Baer's law). Died Dorpat, Nov. 16/28, 1876.

BAER, Reinhold, German mathematician; b. Berlin, Germany, July 22, 1902; s. Emil and Bianka (Timendorfer) Baer; student Hanover Inst. Tech., univs. Freiburg, Kiel; Dr.phil., U. Göttingen, 1925; m. Marianne Kirstein, Feb. 2, 1929; 1 son, Klaus. Faculty, U. Halle (Germany), 1929-35; hon. research fellow Victoria U., Manchester (Eng.), 1933-35; mem. Inst. Advanced Study, Princeton, N.J., 1935-37; asst. prof., U. Ill., Urbana, 1937-38, prof. math., 1944-57, emeritus, 1957—; prof. math. U. Frankfort/Main (Germany), 1956—. Mem. Am., German, London math. socs. Author: Linear Algebra and Projective Geometry, 1952, 66; (with H. Steinitz) Algebraische Theorie der Körper; also numerous articles. Research in foundations of geometry, theory of groups, rings. Home: 11 Gartenstrasse, 6243 Falkenstein/Taunus. Office: Robert-Mayer-Strasse 6-8, 6 Frankfort/Main, West Germany.*

BAER, Rudolf Lewis, physician; born Strasbourg, France, July 22, 1910; s. Ludwig and Clara (Mainzer) B.; abitur Musterschule Frankfurt A.M., 1928; student U. Heidelberg, U. Frankfurt, U. Berlin, U. Vienna; M.D., U. Basel, 1934; m. Louise Jeanne Grumbach, Nov. 6, 1941; children—John Reckford, Andrew Rudolf. Came to U. S., 1934, naturalized, 1940. Practice medicine, specializing in dermatology, N.Y.C., 1939—; prof., chmn. dept. dermatology N.Y. U. Sch. Medicine, 1961—; dir. dermatology U. Hosp.; dir. dermatology, syphilology Bellevue Hosp.; cons. dermatologist Manhattan VA Hosp., Goldwater Meml. Hosp., others. Recipient Dohi medal Japan Soc. Dermatology, 1965. Diplomate Am. Bd. Dermatology. Mem. A.M.A., Am. Acad. Dermatology (past dir.), Soc. Investigative Dermatology (past pres., pres. dir.), N.Y. Acad. Medicine, Am. Dermatol. Assn., Am. Acad. Allergy, Am. Coll. Allergy, A.A.A.S., Israel, Iranian, Brazilian, Polish, Venezuelan, Austrian, Danish, Finnish, French, Italian, Swedish dermatol. socs., French Allergy Soc. Author: Office Immunology, 1947; Atopic Dermatitis, 1955; Allergic Dermatoses Due to Physical Agents, 1956. Editor: Yearbook of Dermatology, 1956-65. Research, publs. on immunology with particular reference to skin: contact sensitivity, cross-sensitization, drug reactions, atopy; photobiology with particular reference to skin: phototoxic and photoallergic reactions mechanisms and clin. manifestations of reactions due to drugs, contactants, endogenous substances; biology of fungous infections of skin. Home: 1185 Park Av., N.Y.C. 10028. Office: 550 First Av., N.Y.C. 10016.*

BAER, William S., Am. orthopedist; b. Balt., Nov. 25, 1872; s. Robert and Mary (Corner) B.; A.B., Johns Hopkins, U., 1894, M.D., 1898; m. Ruth Adams, Oct. 15, 1901. Resident house officer Johns Hopkins Hosp., Balt., 1898-1900; head Orthopedic clinic, instr. Johns Hopkins Med. Sch., 1900-31, apptd. clin. prof. orthopedic surgery, 1926; founder, also dir. Children's Hosp., 1909; founder Md. League for Crippled Children, 1927. Mem. A.C.P., A.M.A., Am. Orthopedic Assn. Contbd. to orthopedic surgery; developed operations to restore motion to fused and stiffened joints; described treatment of chronic osteomyelitis with bluebottle fly larvae, 1931. Died Balt., Apr. 7, 1931.

BAERMANN, Georg Friedrich, German mathematician; b. Leipzig, Saxony, 1717; s. Georg Adam Baermann; student theology and math. U. Leipzig, from 1730, M.A.; postgrad U. Marburg; Master philosophy, U. Wittenberg (Saxony), 1737. Prof. math., Wittenberg, 1745; mem. German Soc., Leipzig. Author: De vectibus curvilineis, 1737; Analysis problematis geometrici (in Acta Eruditorum), 1748; De solutione cubicarum aliarumque aequationum ope sinuum, 1751; Theorematis algebraici demonstratio, 1751. Gave 1st gen. proof Newton's formulae of powers and equation roots; pub. edit. Euclid's complete works, 1744. Died Wittenberg, Feb. 6, 1769.

BAERTSCHI, Peter, Swiss chemist; b. Zofingen, Switzerland, July 15, 1919; s. Friedrich and Maria (Scheibler) B.; student Swiss Fed. Inst. Tech., Zürich, 1938-42, Dr.sc.nat., 1945; Ph.D., U. Basel (Switzerland), 1956; postgrad. U. Chgo., 1949; m. Maria Bruun, Oct. 20, 1948. Research asso. dept. phys. chemistry U. Basel, 1945-56; research fellow Inst. for Nuclear Studies, U. Chgo., 1949; head chemistry dept. Fed. Inst. for Reactor Research, Wureulingen, Switzerland, 1956—. Faculty, Swiss Fed. Inst. Tech., U. Basel. 1956—. Mem. Swiss Phys. Soc., Gesellschaft für Fachlentider Kerntechuik, Sigma Xi. Research, publs. on isotope effects and isotope separation (distillation), isotope geology, natural abundances of oxygen and carbon isotopes and their geochem. interpretation. Home: 1 Blanenstrasse, Birsfelden, Switzerland. Office: E.I.R., Wurculingen (AG), Switzerland.*

BAERWALD, Friedrich, economist; born Frankfort/Main, Germany, Oct. 14, 1900; s. Arnold and Charlotte (Lewino) B.; Baccalaureate, Musterschule Frankfurt/Main, 1919; LL.B., U. Frankfurt, 1922, LL.D., 1923; m. Franziska Schwarte, Aug. 23, 1948. Came to U. S., 1933, naturalized, 1940. With Fed. Ministry of Labor and Labor Adminstrn., Germany, 1926-33; faculty Fordham U., N.Y.C., 1935—, prof. econs. 1954—. Recipient Rockefeller award, 1953; Research award Germany, 1962. Mem. Am. Cath. (pres. 1953-54) econ. assns., Am. Sociol. Assn., Am. Arbitration Assn. (indsl. panel). Author: Fundamentals of Labor Economis, 1953; Economic System Analysis, 1960; Economics and Progress and Problems of Labor, 1967; also articles. Research on relation between productivity and wages, models of econ. devel. on a macrodynamic scale; concepts and applications in existential sociology.*

BAETJER, Anna Medora, Am. physiologist; b. Balt., July 7, 1899; d. J. Frank and Katherine (Cook) Baetjer; B.A., Wellesley Coll., 1920; Sc.D., Johns Hopkins, 1924; Dr. P. H., Woman's Med. Coll. Pa., 1953; D.Sc., Wheaton Coll., 1966. Faculty, Sch. Hygiene and Pub. Health Johns Hopkins, Balt., 1923—, prof. environmental medicine, 1962—; cons. Office Surgeon Gen. Dept. Army; trustee Indsl. Hygiene Found. Mellon Inst.; mem. Permanent Com. and Internat. Assn. Occupational Health. Mem. radiation control adv. bd. Md. Health Dept.; mem. com. on environmental physiology NRC, 1965—; mem. Commn. Environmental Hygiene Armed Forces Epidemiological Bd.; mem. adv. com. on safety of pesticide residues in foods Food and Drug Adminstrn., 1966—. Diplomate Am. Bd. Indsl. Hygiene. Mem. Md. Acad. Scis. (trustee 1965—), Am. Physiol. Soc., Am. Pub. Health Assn., Am. Indsl. Hygiene Assn. (Cummings Meml. award 1964, past pres.), Am. Conf. Govtl. Indsl. Hygienists, Am. Acad. Occupational Medicine (hon.), Phi Beta Kappa, Sigma Xi. Contbr. chpts. to Rosenau's Preventive Medicine and Pub. Health, 1965; Women in Industry-Their Health and Efficiency, 1946. Mem. editorial bd. Am. Jour. Epidemiology, 1952—, Archives of Environmental Health, 1960—. Research, publs. on effects of environmental temperature and humidity on susceptibility to agents of disease; self-protection of lungs against pneumoconioses; effects of potassium. Home: 4900 Roland Av., Balt. 21210. Office: 615 N. Wolfe St., Balt. 21205.*

BAEYER, Adolf von, see von Baeyer, Adolf (Johann Friedrich Wilhelm).

BAEYER, Johann Jakob, German geodesist; b. Müggelheim, Germany, Nov. 5, 1794; s. Jakob and Elisabeth (Tisch) B.; student of Bessel; m. Eugenie Hitzig, Jan. 7, 1826; 7 children. Joined army, 1813; joined trigonometric div. of Prussian gen. staff, 1821, named head of dept., 1835; named lt.-gen., 1858; founder, chmn. Geodetic Inst., Berlin, Germany, 1869-1885. Author of numerous publs. on geodetics including: Gradmessung in Ostpreussen und ihre Verbindung mit preussische und russische Dreiecksketten, 1838; Die Verbindungen der preussische und russische Dreiecksketten bei Thorn und Tarnowitz, 1857; Über der Grösse und Figur der Erde, 1861; Das Messen der Sphäroidische Erdoberfläche, 1862. Founder Europäische Gradmessung (1st internat. sci. assn.), 1864; promoted internat. and intercontinental geodesic relations. Died Berlin, Sept. 10, 1885.

BAEYER, Otto von, see von Baeyer, Otto.

BAEZ, Silvio, physician; b. Guarambare, Paraguay, July 6, 1915; s. Ricardo G. and Regina (Acosta) B.; B.A., Nat. Coll. Paraguay, 1935, B.S., 1936; M.D., Asuncion Med. Sch., 1942; m. Mildred L. Hotcaveg, May 14, 1950; children—Ricardo S., Ana Cleta, Pedro L. Faculty, Asuncion Med. Sch., 1943-44; fellow Inter-Am. Tng. Adminstrn, 1944-45; research asso. Med. Coll., Cornell U., N.Y.C., 1945-52, asst. prof., 1952-57; asst. prof. anesthesiology N.Y. U., N.Y.C., 1957-61; asso. prof. physiology Albert Einstein Coll. Medicine, N.Y.C., 1961—, dir. anesthesiology research labs. Mem. N.Y. Acad. Scis., A.A.A.S., Am. Physiol. Soc., Harvey Soc., Microcirculatory Soc., Assn. Gnotobiotics, Acad. Medicine and Surgery of Asuncion, Sigma Xi. Research and numerous publs. on mechanism concerning behavior of small living blood vessels; reaction of minute vessels in germ-free animals, laser irradiation, hypertension and shock; inventor method for controlled subcooling of cells under microscopic observation, method for measurement and electrographic recording of microscopic objects.*

BAFFIN, William, English explorer; b. circa 1584; pilot, ship Patience, voyage to Greenland, 1612; chief pilot, 7 ship fleet Muscovy Co. to Spitzbergen, 1613, 14; pilot, ship Discovery, expdn. to discover N.W. Passage (sighted Baffin Island, named for him in 1821), 1615, 16; master's mate, ship Anne Royal, E. India Co., voyage to India, 1617-19; master, ship London, E. India Co., voyage to India, 1620. Noted for explorations of Arctic regions; discovered, named Jones Sound, Lancaster Sound, and Smith Sound; determined latitudes, observed tides; recordings of orientation of compass needle led to 1st magnetic chart; 1st to try to determine longitude at sea by observation of moon. Died Qishm, Persia, Jan. 23, 1622.

BAGEHOT, Walter, English economist; b. Langport, Somersetshire, Eng., Feb. 3, 1826; s. Thomas Watson and Edith (Stuckey) B.; ed. Bristol, Eng.; B.A., U. Coll. London, 1846; M.A. 1848; later studied law; m.

Eliza Wilson, Apr. 21, 1858; became banker and ship owner in 1851; editor Nat. Rev., 1855-64, Economist, 1861-77; examiner in polit. economy U. London (Eng.). Author: The English Constitution, 1864; Lombard Street (study of the English banking system), 1873; Physics and Politics (a pioneer analysis of interrelations between social and natural scis.), 1875; Depreciation of Silver, 1877; Literary Studies, 1879; Economic Studies, 1880. Asserted the successful fusion of executive and legislative powers of the British cabinet guaranteed freedoms of the English Constitution; rejected Blackstone's tripartite division (separation of powers) analysis of the Constitution and substituted his dichotomy of the "dignified" (House of Lords, which preserves) and "efficient" (House of Commons and cabinet, which work and rule) parts. Died Langport, Mar. 24, 1877.

BAGELLARDI, Paolo, Italian physician; b. Fiume, Italy; author of Libellus de aegritudinibus infantium (1st text on infant diseases), contains original observations, although probably influenced by Arab treatises), 1472. Died 1494.

BAGGALEY, Andrew Robert, Am. psychologist; b. Cleve., Dec. 1, 1923; s. Walter and Jean (Brown) B.; A., Harvard, 1947, Ed.M., 1949; Ph.D., U. Chgo., 1952; m. Arlene D. Aberle, Sept. 3, 1949; children—Philip, Paula. Research asso. U. Ill., 1952-54; asst. prof., U. Wis., Milw., 1954-58, asso. prof., 1958-61; asso. prof. Temple U., Phila., 1961-64, prof., 1964-66; prof. University of Pennsylvania, 1966—. Mem. Am., Eastern, Pa. psychol. assns., Psychometric Soc., Sigma Xi, Phi Delta Kappa. Author: Intermediate Correlational Methods, 1964. Research on psychometrics, concept learning, and voting behavior. Home: 212 Hewett Rd., Wyncote, Pa., 19095.*

BAGGE, Erich Rudolf, German physicist; b. Neustadt/Coburg, May 30, 1912; s. Rudolf Leonhard and Elsa (Eckardt) B.; Dipl.Phys., Munich Inst. Tech., 1935; Dr.rer.nat., U. Leipzig, 1938; m. Herta Neulinger, Jan. 24, 1942; children—Elsa, Erika (Mrs. Heiner Weise), Eva. Faculty, U. Leipzig, 1941-48; asso. prof. U. Hamburg, 1948-57; prof., dir. Inst. Pure and Applied Nuclear Physics, U. Kiel, 1957—; became sci. dir. Soc. for Application Nuclear Energy to Shipbldg. and Navigation, Hamburg, 1956; named dir. Inst. for Reactor Physics, Research Reactor Station, Geesthacht, 1960. Mem. Soc. Promotion Application Nuclear Energy to Shipbldg. and Navigation (v.p.). Author: (with Diebner, Jay) Von der Uranspaltung bis Calder Hall; Die Entstehung der kosmischen Strahlung, 1966; also numerous articles. Developed (with Allkofer, Henning, Trümper), isotope separation apparatus, electronic triggered spark chamber; research on theory of acceleration of solar and galaxial cosmic ray particles. Address: 9 Roonstrasse, 23 Kiel, West Germany.*

BAGGENSTOSS, Archie H., Am. physician; b. Richardton, N.D., Apr. 13, 1908; s. Jacob R. and Louise (Kaiser) B.; B.A., U. N.D., 1930; B.S., 1931; M.D., U. Cin., 1933; M.S. in Pathology, U. Minn., 1938; m. Mildred Burkhart, Feb. 25, 1934; children—Roger D. Janet L., Ruth Ann. Practice medicine specializing in gen. practice, Cin. 1934-35; fellow in pathology Mayo Found., Rochester, Minn., 1935-38, first asst. pathol. anatomy, 1938, instr. pathology, 1939, cons. pathol. anatomy, Mayo Clinic, 1940, asst. prof. Mayo Found., 1943-47, asso. prof., 1947-52, prof. 1952—, head sect. path. anatomy, 1955—. Diplomate Nat. Pd. Med. Examiners. Fellow A.M.A. (sec. pathology and physiology sect.), Coll. Am. Pathologists; mem. Am. Assn. Pathologists and Bacteriologists, Internat. Acad. Pathology, Am. (mem. bd. censors 1953-55), Minn. (pres. 1947), soc. clin. pathologists, Am. Gastroent. Assn., Am. Assn. for Study of Liver Diseases Council, Phi Beta Kappa, Sigma Xi, Phi Chi. Editorial bd. Gastroenterology, 1956-66; Archives of Pathology, 1964. Histopathologic research, publs. on liver, pancreas; gastrointestinal tract. Home: 1166 Plummer Circle, Rochester, Minn. 55901.*

BAGGETT, Billy, Am. biochemist, educator; b. Oxford, Miss., Oct. 23, 1928; s. Lee and Estelle (Brown) B.; B.A., U. Miss., 1947; Ph.D., St. Louis U., 1952; m. Harriette Brady Lane, Nov. 23, 1949; children—Sallie Bodley, William Brown, Mary Lane, Teresa Lee, Harriet Lane. Faculty, Harvard Med. Sch., 1952-57; asst. in biochem. research Mass. Gen. Hosp., Boston, 1952-56, asst. biochemist, 1956-57; faculty U. N.C. Med. Sch., Chapel Hill, 1957—, asso. prof. pharmacology and biochemistry, 1959—. Cons. Research Triangle Inst., Durham, 1965—. Mem. Am. Soc. Biol. Chemists, Endocrine Soc., Am. Chem. Soc., Am. Assn. for Cancer Research, A.A.A.S. Research publs. on metabolism of sex hormones and adrenal cortical hormones, biosynthesis of sex hormones, biochemistry of abnormal gonads and adrenals, metabolism of cardiac glycosides. Home: Route 4, Box 483, Poplar Hills, Chapel Hill, N.C. 27514.*

BAGIOTTI, Tullio, Italian economist; b. Castione (Sondrio), Italy, Aug. 16, 1921; s. Giovanni and Stella (Negri) B.; Dr. Econs., U. Commerciale L. Bocconi, Milan, 1950; m. Anna Craveri, Oct. 20, 1954; children—Luisa, Giovanni. Asst. prof. econs. Bocconi U., Milan, 1950-54; Rockefeller fellow Johns Hopkins, U. Chgo., U. Cal., Columbia, 1954-55; prof. history econ. thought U. Pavia, 1957-60, Bocconi U., 1956-—; prof. polit. economy U. Padua, 1962—, also dir. Inst. Sci. Econs. Mng. editor Rivista Internazionale di Scienze Economiche e Commerciali, 1954-—. Mem. Italian Soc. Econs., Inst. Sci., Arts and Letters. Author: Storia della Valtellina, 1959; Storia delle dottrine economiche, 1961; Il Profitto, 1965; Economia, 1967; others; also articles. Research on past and present econ. theories, positive theory of profit. Home: 1 Teulie, Milan, Italy.*

BAGLIVI, Giorgio, Italian physician; b. Ragusa (now Dubrounik, Yugoslavia), Sept. 26, 1668; ed. U. Naples (Italy); physician to Pope Innocent XII; prof. anatomy Sacred Coll. in Rome, Italy. Fellow Royal Soc., 1698. Author: De praxi medica, 1696; Opera omnia medica practica et anatomica, 1704. May have been first to distinguish between smooth and striped muscles, probably 1700; discovered that when saliva evaporates it leaves a white saline residue; mem. of Iatrophys. Sch. which compared parts of the body to small machines; proposed theory that solid parts of body are seat of disease; rejected theory that motion of body was caused by continuous explosions in muscle. Died Rome, June 17, 1707.

BAGNARA, Joseph Thomas, Am. zoologist; b. Rochester, N.Y., July 26, 1929; s. Dominick Anthony and Fortunata (Zumbo) B.; B.A., U. Rochester, 1952; Ph.D., U. Ia., 1956; m. Mary Louise Schulze, Sept. 10, 1955. Faculty, U. Ariz., Tucson, 1956—, prof. dept. zoology, 1964—. Fulbright Research scholar U. Paris, 1963-64. Mem. Am. Soc. Zoologists, Am. Assn. Anatomists, Growth Soc., Ariz. Acad. Sci., Am. Inst. Biol. Sci. Research publs. on mechanisms controlling devel. and actions pigment cells, nature bright-colored pigment cells amphibians. Home: 3200 ViaCeleste St., Tucson 85718.*

BAGNERA, Giuseppe, Italian mathematician; b. Bagheria, Palermo, Italy, Nov. 14, 1865; degree in civil engring, 1890, in math., 1895; prof. algebra and analytic geometry U. Messina (Italy), 1901-09; prof. infinitesimal analysis U. Palermo, 1909-22; prof. Rome (Italy), 1922-27. Recipient (with M. De Franchis) Bordin prize French Acad. Scis., 1909. Studied by analysis Picard surfaces relating to Abelian functions. Died Rome, Mar. 12, 1927.

BAGNOLD, Ralph Alger, Brit. petroleum engr; b. Devonport, Apr. 3, 1896; s. Arthur Henry and Ethel (Alger) B.; ed. Royal Mil. Acad., Woolwich; M.A., U. Cambridge; m. Dorothy Alice Plank, May 8, 1946; children—Stephen Chester, Susanne Jane. Desert explorer, 1929-38; research dir. Shell Oil, 1946-48; researcher, adviser, 1948—. Fellow Royal Soc., 1944; Mem. Royal Geog. Soc. (Founders medal). Author: Libyan Sands, 1935; Physics of Blown Sand, 1941; Movement of Desert Sand, 1936; Flow of Cohesionless Grains in Fluids, 1956, also articles on hydraulics and desert exploration. Address: Rickwoods, Mark Beech, Edenbridge (Kent), Eng.

BAHA AL-DIN, see Muhammad ben Ahmed ben Abu Bisr Baha al-din.

BAHADUR, Krishna, Indian chemist; b. Allahabad, India, Jan. 20, 1926; s. Mahadeo and Bhagawati (Devi) Prasad; B.Sc., Allahabad U., 1944, M.Sc., 1946, D.Phil., 1949, D.Sc., 1956; D.I.C. (Nuffield Travelling fellow), Imperial Coll. Sci. and Tech., London, Eng., 1962; m. Ranganayaki, Sept. 28, 1952; children—Ranjana, Chandran, Mridula, Ila. Asst. prof. chemistry dept. U. Allahabad, 1950—, also hon. sec. delegacy. Life mem. Nat. Acad. Scis., India. Author: Synthesis of Jeewanu, the Proto-cell, 1966. Research, publs. on discovery of photochem. formation of amino acids and peptides in aqueous mixtures; synthesised Jeewanu, particles of life, capable of growth, reprodn. by budding and metabolic activity. Home: 68 Dilkusha, Allahabad. Office: Chemistry Dept., U. Allahabad, Allahabad, U.P., India.*

BAHCALL, John N., Am. physicist; b. Shreveport, La., Dec. 30, 1934; s. Malcolm and Mildred (Lazarus) B.; A.B., U. Cal. at Berkeley, 1956; M.A., U. Chgo., 1957; Ph.D., Harvard, 1961. Research fellow Ind. U., 1960-62; research fellow Cal. Inst. Tech., Pasadena, 1962-63, sr. research fellow, 1963-65, asst. prof., 1965-67; asso. professor, 1967—. Member American Physical Soc. Research and publications on weak interactions, nuclear physics, atomic physics, nuclear astrophysics, theory nuclear energy generation in stars by observing solar neutrinos. Office: Cal. Inst. Tech., Pasadena, Cal.*

BA HLI, Freddy, Burmese elec. engr.; b. Moulmein, Burma, June 28, 1922; s. Tan and Daw (Thein Hymyin) Ba Hli; B.Sc., U. Rangoon (Burma), 1942; M.Sc., Lehigh U., 1949; D.Sc., Mass. Inst. Tech., 1953; m. Ma Myint Thwe, Dec. 1, 1957; children—Ma Tin Tin Hlaing, Maung Tha Hlaing. Asst. dir. telecommunications, Burma, 1950-51; research asst. research lab. electronics, Mass. Inst. Tech., 1952-53; sr. research officer Union of Burma Applied Research Inst., 1953-55, dir. research, 1955-60, dir. gen., 1960—; lectr. elec. communications Faculty Engring., U. Rangoon, 1954-58; acting dir. Union of Burma Atomic Energy Center, 1955-57. Fellow Brit. Interplanetary Soc.; mem. Burma Sci. Assn. (pres.), Burma Research Soc. (v.p.), Sigma Xi. Research, publs. on discovery gen. method of time domain network synthesis for design elec. communications networks; presented elec. explanation of gravitational field. Patentee bamboo hardbd. manufacture. Home: 327 U Wisara Rd. Office: Union of Burma Applied Research Inst., Kanube, Rangoon, Burma.*

BAHN, Robert Carlton, Am. pathologist; b. Newark, July 24, 1925; s. Arlington Mott and Helen Esther (Houck) B.; student Albright Coll., 1942-43; M.D., U. Buffalo, 1947; Ph.D., U. Minn., 1953; m. Miriam Ruth Huer, July 30, 1949; children—David, Rebecca, Mark, Curtis. Fellow pathology Mayo Found. and Clinic, Rochester, Minn., 1950-53; cons. pathology, 1956—; with dept. physiol. chemistry U. Minn., Mpls., 1952-53; sr. asst. surgeon USPHS, NIH, Bethesda, Md., 1954-55; faculty Mayo Found., 1956-—, asso. prof. pathology, 1963—. Diplomate Am. Bd. Pathology. Mem. Endocrine Soc., A.M.A., Biol. Stain Commn., Am. Physiol. Soc., Histochem, Soc., Assn. for computing Machinery, Biomed. Information Retrieval Orgn., Am. Assn. Pathologists and Bacteriologists, Internat. Acad. Pathologists, Soc. for Exptl. Biology and Medicine, Am. Assn. Neuropathologists. Research, numerous publs. in endocrine pathology, neuro-endocrine physiology and applications of digital computers to medicine. Home: 1650 N.E. 11th Av., Rochester 55901. Office: Mayo Clinic, 200 S.W. 1st St., Rochester, Minn. 55902.*

BAHNER, Carl Tabb, Am. chemist; b. Conway, Ark., July 14, 1908; s. Gustavus Lunsford and Augusta (Moore) B.; B.S., Hendrix Coll., 1927; M.S., U. Chgo., 1928; Th.M., So. Bapt. Theol. Sem., 1931; postgrad. Yale, 1931-32; Ph.D., Columbia, 1936; m. Catharine Garrott, Sept. 17, 1931; children—Thomas Maxfield, Mary Catharine (Mrs. James Lewis Day), Frances Jane (Mrs. Ernest LeRoy Hendricks Jr.). Instr., Hendrix Coll., 1930-31; head physics dept. Union U., 1936-37; head chemistry dept. Carson-Newman Coll., 1937-67, coordinator research, 1967—; consultant Oak Ridge Nat. Labs., 1948—. Mem. American Chem. Soc. (chmn. East Tenn. sect. 1951), Tenn. Acad. Sci. (pres. 1951), A.A.A.S., Am. Assn. Cancer Research. Contbr. articles to sci. jours. Synthesis of compounds for use in study of cancer chemotherapy including several new compounds which prevent growth of certain types of tumors in animals; reactions of polyfluoroolefins; use of hydrotropic solutions in organic syntheses; invention of new processes for preparing derivatives of nitroparaffins; use of radioactive tracers in cancer research. Home: P. O. Box 549, Jefferson City, Tenn. 37760.*

BAHR, Gunter Friedrich, pathologist, biophysicist; born Hamburg, Germany, October 25, 1922; s. Carl Wilhelm and Elfriede (Wedekind) B.; M.D., U. Wurzburg, 1952, Karolinska Inst., Stockholm, Sweden, 1957; m. Karina Ingrid Edblad, Mar. 7, 1960; children—Josephine Karina, Nina Ingrid. Came to U. S., 1960, naturalized, 1968. Mem. staff Nobel Inst. for Cell Research, Stockholm, 1950-57; asst. prof. pathology Karolinska Inst., 1957-60; chief biophysics Armed Forces Inst. Pathology, Washington, 1960—. Vis. prof. Northwestern U., 1958; clin. asso. prof. pathology Georgetown U., Washington, 1962—. Recipient Meritorious Civil Service award U. S. Army, 1963, Research and Devel. awards, 1967; Maurice Goldblatt Cytology award, 1966. Member International Academy of Pathology; mem. International Acad. Cytology, Am. Soc. Exptl. Pathology, Am. Soc. Cell Biology, Histochem. Soc., Assn. Mil. Surgeons Am. (hon.), Am., Swedish electronmicroscopic socs. Author: (with Zeitler) Quantitative Electron Microscopy, 1965. Studies in contrast in electron microscopy; originator (with Zeitler) technique for weight determination. Home: 7207 Lenhart Dr., Chevy Chase, Md. 20015. Office: care Armed Forces Inst. Pathology, Washington 20305.*

BAIER, Johann Jakob, physician; b. Jena, Germany, June 14, 1677; s. Johann Wilhelm and Anna Catharina (Musäus) B.; ed. at Jena, Halle; twice married. Became prof. medicine Altdorf, Switzerland; 1704; transferred Imperial Acad. Nature Explorers to Altdorf, pres., from 1731; personal physician to Emperor; pharm. researcher Altdorf Bot. Garden. Author: Oryctographia Norica, 1730; Biographies of Physicians. Made copper plate prints of fossils. Died Altdorf, July 14, 1735.

BAIER, Othmar, German mathematician; b. Augsburg, Germany, Nov. 16, 1905; s. Richard and Paula (Krasmer) B.; ed. Technische Hochschule, Munich, U. Munich (Germany), Ph.D., 1935; m. Erna Zurl, 1938. Asst. in field, 1928-31, 39; asst. technische hochschulen Munich, Karlsruhe (Germany), 1931-37; prof. Technische Hochschule Stuttgart (Germany), 1952; dir. tech. scis. Technische Hochschule Munich from 1931, lectr. math., also substitute prof., from 1937, asst. prof., 1945-60, prof. geometry, also dir. Inst. Geometry, 1960—. Mem. Gesellschaft für Angewandte Mathematik und Mechanik, Deutsche Mathematik Vereinigung. Research, publs. on pure and applied math., geometry; specialist geometry, cinematics and its application; (with others) devel. math. theory of rotary piston engines, 1954—. Address: Hermelinweg 12, 8000 Munich 90, Germany.

BAIER, Vladimir N., Russian physicist; b. Kharkov, Russia, Sept. 27, 1930; s. Nicolai N. and Fanya S. (Ohercaskaja) B.; student Kiev State U., 1949-54; m. Lubov K. Roitburg, Apr. 27, 1957; children—Olga, Tatyana. Jr. scientist Lebedev Phys. Inst., Moscow, 1955-58; scientist Inst. Nuclear Physics, Moscow, 1959-61; sr. scientist Inst. Nuclear Physics, Novosibirsk, USSR, 1961—; prof. Novosibirsk State U., 1962—, chmn. physics dept., 1965—. Author articles. Research on quantum electrodynamics, high energy physics, colliding beams accelerator. Home: 34 Zolotodolinskaya. Office: Inst. Nuclear Physics, Novosibirsk 90, USSR.*

BAILAK AL-QABAKAQI, see al-Qabajaqi.

BAILAR, John Christian, Jr., Am. inorganic chemist; b. Golden, Colo., May 24, 1904; s. John Christian and Rachel Ella (Work) B.; B.A., U. Colo., 1924, M.A., 1925, Sc.D., 1959; Ph.D., U. Mich., 1928; Sc.D., U. Buffalo, 1959; m. Florence Leota Catherwood, Aug. 8, 1931; children—John Christian III, Benjamin Franklin. Asst. in chemistry U. Mich., Ann Arbor, 1926-28; faculty U. Ill., Urbana, 1928—, prof. chemistry, 1943—. Cons. to chem. industry, chem. book pubs. Recipient Frank Dwyer Meml. medal Chem. Soc. New South Wales, 1965; Alfred Werner medal Swiss Chem. Soc., 1966. Mem. Am. Chem. Soc. (pres. 1959, award in Chem. Edn. 1961, Priestley medal 1964), Internat. Union of Pure and Applied Chemistry (treas. 1963—), Phi Beta Kappa, Sigma Xi, Phi Lambda Upsilon, Alpha Chi Sigma (John R. Kuebler award 1962). Editor: Inorganic Syntheses, vol. 4, 1953; Chemistry of the Coordination Compounds, 1956. Author: (with B S. Hopkins) General Chemistry for Colleges, 4th edit., 1951, 5th edit., 1956; Essentials of Chemistry, 1946; (with T. Moeller, J. Kleinberg) University Chemistry, 1965; also articles. Discovered optical inversion and explained stereospecificity in complex inorganic reactions; developed method of double-bond hydrogenation. Home: 304 W. Pennsylvania Av., Urbana, Ill. 61801.*

BAILENGER, J., parasitologist, biologist; b. Rochefort, France, Nov. 2, 1924; s. F. and P. (Gaillard) B.; Pharmacist, Faculty Medicine and Pharmacy Bordeaux (France) 1948, Docteur en Médecine, 1956, Docteur ès Scis. Naturelles, Faculty Scis., 1951. With Faculty Medicine and Pharmacy Bordeaux, 1948—, became head tutorials, 1952, aggregate lectr., 1963, prof., 1965—. Recipient Bronze medal Bordeaux Hosps., 1949, Silver medal, 1950; laureate Acad. Medicine, 1962. Mem. Soc. Parasitology, Soc. Biology, Soc. Gastroenterology, Soc. Therapeutics, Soc. Phamacodynamics, World Assn. for Advancement of Vet. Parasitology. Author: Atlas de travaux pratiques de parasitologie humaine, 1952, 63; Coprologie parasitaire humaine, 1958; Coprologie parasitaire et fonctionelle, 1965. Research on coloration and cellular growth, parasitism of lower animals, human parasitism, hormonal influence on def. system of organisms, pathogenic power of protozoa and intestinal worms, diagnosis (coprology and immunology), therapeutics (anthelmintic, antiamebic, antifungi), epidemiology (intestinal parasites, hydatidosis, distomatosis, dermatophytes). Home: 52, rue d'Arcachon, Bordeaux 33, France.*

BAILEY, Sir Alfred John, Brit. engr.; b. 1867; s. William Bailey; ed. Victoria U., Manchester, Eng.; m. Ethel Ellis Johnson, 1902; 2 sons. Joined 3d V. B. Lancashire Fusiliers, 1896, advanced through grades to col., 1915, comdr. 3d line units E. Lancashire Div., 1915-16. Gov. Royal Tech. Coll., Salford, Eng.; governing dir. Sir W. H. Bailey & Co., Ltd., engrs., Manchester; chmn. bd. Grain Elevator Estate, Ltd., Manchester, Manchester Constn. Club Holdings Co., Ltd. Mem. E. Lancashire Territorial Army and Air Force Assn., Instn. Mech. Engrs., Inst. Metals. Inventor valves for high pressure steam pumps, oil testing machines, carburetors, silencers for gasoline engines, speed, water-level, pressure recording instruments. Died Aug. 18, 1940.

BAILEY, Alfred M(arshall), Am. ornithologist; b. Iowa City, Feb. 18, 1894; s. William H. and Mary (Jelly) B.; A.B., U. Ia., 1916; D.Sc. (hon.), Norwich U.; D. Pub. Service, U. Denver; m. Muriel Etta Eggenberg, June 16, 1917; children—Beth Elaine (Mrs. John Murphy), Patricia Jean (Mrs. James Witherspoon). Mem. staff La. State Mus., 1916-19, Bur. Biol. Survey, U. S. Dept. Agr., Alaska, 1919-21; staff Denver Mus. Natural History, 1921-26, dir., 1936—; staff Field Mus. Natural History, Chgo., 1926-27; dir. Chgo. Acad. Scis., 1927-36. Fellow A.A.A.S., Am. Ornith. Union; mem. Royal Australasian, Cooper ornith. socs., Wilson Ornith. Club, Soc. Mammalogists, Forestry Assn. Author: Birds of Arctic Alaska, 1948. Birds of New Zealand, 1955; Birds of Midway and Laysan, 1956; (with Robert J. Niedrach) Birds of Colorado, 2 vols., 1965. Research, publs. on extensive expdns. and collections for mus. displays; numerous natural history reports, primarily on birds. Home: 4340 Montview Blvd., Denver 80207. Office: Denver Mus. Natural History, City Park, Denver 80206.*

BAILEY, Benjamin Franklin, Am. elec. engr.; b. Sheridan, Mich., Aug. 7, 1875; s. William Martin and Lucy (Stead) B.; B.S. in Elec. Engring., U. Mich., 1898, A.M., 1900, Ph.D., 1907; m. Elsie Marion Egeman, Dec. 30, 1902; 1 son, Benjamin Franklin. Designer Edison Illuminating Co., Detroit, 1898; in testing dept., Gen. Electric Co., Schenectady, 1898-99; instr. electrotherapeutics U. Mich., 1900-01, instr. elec. engring., 1901-06, asst. prof., 1906-10, jr. prof., 1910-13, prof., from 1913, head dept. elec. engring., from 1925. Chief engr. Fairbanks-Morse Elec. Mfg. Co., 1908-09, cons. engr. Fairbanks-Morse Elec. Mfg. Co., Howell Elec. Motors Co. dir. Bailey Elec. Co., Howell Electric Motor Co.; v.p., dir. Fremont Motor Corp. Honored as Modern Pioneer by Nat. Assn. Mfrs. Fellow Am. Inst. E.E. (life); mem. Soc. Automotive Engrs., Sigma Xi, Tau Beta Pi, Eta Kappa Nu. Author: Induction Coils, 1903; Induction Motors, 1911; Elementary Electrical Engineering, 1913; Principles of Dynamo Electric Machinery, 1915; Alternating Current Machinery, 1934. Inventor Bailey elec. lighting, starting and ignition system, also single-phase condenser motor. Died Oct. 31, 1944.

BAILEY, Charles P., Am. heart surgeon; b. Wanamassa, N.J., Sept. 8, 1910; student Rutgers U., 1926-28; M.D., Hahnemann Med. Coll., 1932, LL.D., 1953; M.S., U. Pa., 1943, D.Sc., 1955; m. Lillian Dann; children—Donald, Robert, Patricia. Rotating intern Fitkin Meml. Hosp., Neptune, N.J., 1932-33; gen. practice medicine and surgery, 1933-37, thoracic surgery, 1940—; resident Sea View Hosp., S.I., N.Y., 1938-40; hon. prof. thoracic surgery U. Monterrey, Mexico, 1949; prof., head dept. thoracic surgery Hahnemann Med. coll. and Hosp., Phila.; guest lectr. Grad. Sch. U. Pa.; vis. surgeon, div. thoracic surgery Phila. Gen. Hosp.; cons. thoracic surgery VA; chief cardiovascular surgery West Jersey Hosp., Camden, N.J.; now chmn. prof. surgery Flower Fifth Av. Hosp., N.Y.C.; dir. cardiopulmonary sect. Albert Einstein Med. Center, Phila.; dir. Bailey Thoracic Clinic; research adminstr., dir. exptl. research Mary Bailey Found. for Heart and Great Vessel Research. Recipient man of the year interfaith award B'nai B'rith, 1950; gold medal A.M.A., 1951, Clarence E. Shaffrey award Med. Alumni St. Joseph's Coll., Phila., 1951, humanitarian award B'rith Sholom, 1955, Page One award Newspaper Guild Am., 1955, George B. Kunkel award Harrisburg Hosp., 1956. Diplomate Am. Bd. Thoracic Surgery (founder group). Fellow A.C.S., Internat. Coll. Surgeons, Am. Coll. Chest Physicians, A.M.A.; hon. fellow Surg. Soc. of Madrid; mem. Phila. County Med. Soc., Phila. Acad. Surgery, Am. Heart Assn., Heart Assn. Southeastern Pa., Laennec Soc., Hahnemann Alumin Soc., Am. Assn. Thoracic Surgery; hon. mem. Roman Med. Soc., Nat. Surg. Soc. Cuba, Nat. Med. Soc. Cuba. Author: Surgery of the Heart, 1955. Pioneered several types heart surgery; developed several new technics and instruments for heart surgery. Address: 34 E. 67th St., N.Y.C.

BAILEY, Donald Wayne, Am. biologist; b. Hutchinson, Kan., Feb. 28, 1926; s. George Elbert and Myra (Perkins) B.; student U. Mo., 1943-44; A.B., U. Cal. at Berkeley, 1949, Ph.D., 1953; m. Gertrude Elizabeth Cole, Sept. 13, 1948; children—Dennis Michael, Steven Duane. USPHS research fellow Jackson Lab., Bar Harbor, Me., 1953-55, research asso., 1955-57; asst. prof. U. Kan., 1957-59; chief genetics unit Lab Aids br. NIH, Bethesda, Md., 1959-61; asso. research geneticist Cancer Research Inst., U. Cal. Med. Center, San Francisco, 1961-67; staff scientist Jackson Lab., Bar Harbor, Me., 1967—; mem. subcom. genetic standards Inst. Lab. Animal Resources, NRC-Nat. Acad. Scis., 1959-61, 1965—. Mem. Genetics Soc. Am., Transplantation Soc., N.Y. Acad. Scis., Am. Genetic Assn. A.A.A.S., Am. Inst. Biol. Scis., Sigma Xi. Contbr. articles to sci. jours. Analyzed genetic and environmental interrelationships of various skeletal dimensions of mouse; determined spontaneous mutation rates of genes with quantitative effects in mouse; analyzed genetics of transplantation antigens in mouse.*

BAILEY, Edgar Henry Summerfield, Am. chemist; b. Middlefield, Conn., Sept. 17, 1848; s. Russell B. and Hannah (Miller) B.; Ph.B., Yale, 1873; Ph.D., Ill. Wesleyan, 1883; student Strassburg, 1881, Leipzig, 1895; m. Aravesta Trumbauer, July 13,1876; children—Kenneth Russell, Herbert Stevens, William Hotchkiss, Edgar Lawrence, Austin. Instr. chemistry Yale, 1873-74, Lehigh U., 1874-83; prof. chemistry and metallurgy U. Kan., 1883, dir. chem. lab., 1900. Chemist, Kan. Bd. Agr., from 1885, Kan. Bd. Health, 1899. Author: (with H. P. Cady) Laboratory Guide to Study of Qualitative Analysis, 1901; (with W. R. Crane) Gypsum (Vol. V); Mineral Waters (Vol. VII, Geol. Survey, Kan); Sanitary and Applied Chemistry, 1906; The Source, Chemistry and Use of Food Products, 1914; Laboratory Experiments on Food Products, 1915; Report on the Dietaries of some State Institutions under the care of the Board of Administration, 1921; (with H. S. Bailey) Foods from Afar, 1922. Died June 1, 1933.

BAILEY, Edgar Herbert, Am. geologist; b. Washington, Aug. 3, 1914; s. Herbert Stevens and Winifred (Everingham) B.; A.B., U. Redlands, 1934; Ph.D., Stanford, 1941; m. Gwendolen L. Cass, Dec. 17, 1946; children—Kathleen Marie, Mary Louise, Herbert George, Robert Allen, Arthur Edgar. With U. S. Geol. Survey, Menlo Park, Cal., 1938—, adminstrv. geologist, 1946-53, research geologist, 1953—; asst. Pomona Coll., 1939; instr. Stanford, 1946, guest lectr., 1947. Fellow Geol. Soc. Am., Mineral Soc. Am.; mem. Soc. Econ. Geology. Research, publs. on mercury deposits of U. S. and world, geology of Franciscan and related rocks of Cal., petrology and geology of glaucophane schists, ultramatic rocks, eugeosynclinal deposits. Home: 1835 Edgewood Rd., Redwood City, Cal. Office: 345 Middlefield Rd., Menlo Park, Cal.*

BAILEY, Edward Monroe, Am. chemist; b. New London, Conn., Aug. 27, 1879; s. Edward Monroe and Louise Maria (Hagan) B.; Ph.B., Yale, 1902, M.S., 1905, Ph.D., 1910; m. Myrtle Mix Studley, June 11, 1906; 1 son, Irving Monroe. Asst. chemist Conn. Agrl. Expt. Sta., 1902-17, chemist in charge from 1917; state chemist, Conn., from 1919. Mem. Joint Com. on Definitions and Standards for Food Products, U. S. Dept. Agr., 1922-38; cons. Council Pharmacy and Chemistry, 1920-30; mem. Council Pharmacy and Chemistry and Council on Foods, A.M.A., 1930-38. Mem. Am. Chem. Soc., Assn. Ofcl. Agrl. Chemists, Assn. Feed Control Officials, Assn. Dairy, Food, and Drug Control Ofcls., Sigma Xi. Contbr. numerous articles on chemistry and biochemistry of foods, drugs and agrl. products to profl. jours. Died Apr. 13, 1948.

BAILEY, Florence Augusta Merriam, Am. ornithologist; b. Locust Grove, N.Y., Aug. 8, 1863; d. Clinton L. and Caroline (Hart) Merriam; A.B., Smith Coll., 1886; LL.D., U. N.M., 1933; m. Vernon Bailey, Dec. 16, 1899 (dec. Apr. 1943). Fellow Am. Ornithologists' Union; mem. Cooper Ornith. Club (life), Nat. Audubon Soc. (emeritus, 1934), Biol. Soc. Washington. Recipient Brewster medal Am. Ornithologists' Union, 1931. Author: Birds Through an Opera Glass, 1889; My Summer in a Mormon Village, 1895; A-Birding on a Bronco, 1896; Birds of Village and Field, 1898; Handbook of Birds of Western United States, 1902; Birds of the Santa Rita Mountains in Southern Arizona, 1923; Birds of New Mexico, 1928; Birds, in Vernon Bailey's Wild Animals of Glacier National Park, 1918, and Cave Life in Kentucky, 1933; Among the Birds in the Grand Canyon Country, 1939; also numerous papers on bird life, especially of Western U. S. Died 1948.

BAILEY, Frederick Manson, botanist; b. Hackney, London, Eng., Mar. 8, 1827; s. John Bailey; m. Anna Maria Waite, 1856; 1 son, J. F. Went to S. Australia, 1839, Queensland, Australia, 1861; colonial botanist, Queensland, from 1881. Fellow Linnean Soc. Author: Handbook to Ferns of Queenland, 1874; The Fern-world of Australia, 1881; Synopsis of Flora of Queensland, 1883; Catalogue of Queensland Woods; Botany of Bellenden-Ker Expedition; A Half-century of Notes for Amateur Fruit-Growers, 1895; The Queensland Flora, 6 vols., 1899-1902; Weeds and Suspected Poisonous Plants of Queensland, 1906-07; Catalogue of Queensland, Flora, 1912. Died June 25, 1915.

BAILEY, Gilbert Ellis, Am. geologist; b. Pekin, Ill., Apr. 27, 1852; s. Gilbert Stephen and Sarah (Bunnell) B.; student old U. Chgo., 1868-72, U. Mich., 1872-73; Ph.D., Franklin (Ind.) Coll., 1881; m. Martha Cobb, 1876 (dec. 1879); m. 2d, Reba Boston, 1902. Prof. chemistry U. Neb., 1874-79; geologist Wyo. Ty., 1883-87; prof. metallurgy State Sch. Mines, S.D., 1888-89; asst., Cal. Mining Bur., 1900, 01; Death Valley explorations, 1901, 02, 03; prof. geology U. So. Cal., from 1909. Author: Saline Deposits of California, 1902; Mines and Minerals of San Bernardino County, California, 1902; California Soils, 1913; The Use of Explosives in Agriculture, 1914; Nitrating by Legumes, 1914; Vertical Farming, 1915; California, A Geologic Wonderland, 1924. Died Dec. 6, 1924.

BAILEY, Irving Widmer, Am. botanist; b. Tilton, N.H., Aug. 15, 1884; s. Solon Irving and Ruth Elaine (Poulter) B.; A.B., Harvard, 1907, M.F., 1909, D.Sc. (hon.), 1955; hon. D.Sc., U. Wis., 1931; m. Helen Diman Harwood, June 15, 1911; children—Harwood, Solon Irving, II. Asst. in botany, Harvard, 1909-10, instr., 1910, asst. prof., 1912, asso. prof., 1920, prof. plant anatomy, 1927-60, now prof. plant anatomy emeritus; chmn. Inst. Research in Gen. Plant Morphology, from 1946; research asso. Carnegie Instn. Mem. adv. bd. U. S. Forest Products Lab., 1914-16; mem. NRC, 1917-22 (exec. com. div. biology and agr. 1919-22, forestry com. 1917-20, biol. fellowship bd. 1928-32); mem. editorial bd. Am. Jour. Botany, 1915-18, Proc. Soc. Am. Foresters, 1914-16. Recipient Mary Soper Pope Award, 1954. Fellow A.A.A.S., Am. Acad. Arts Sci. (v.p. 1947-49); mem. emeritus Nat. Acad. Sci.; mem. Union Am. Biol. Socs. (council 1924-26), Soc. Am. Foresters, Bot. Soc. Am. (pres. 1945), Am. Soc. Naturalists, Bot. Soc. India (hon.), Royal Swedish Acad. Scis., Linnaeau Soc. London, Internat. Soc. Plant Microphologists (pres. 1960), Phi Beta Kappa. Contbr. articles in field. Demonstrated principles of evolutionary specialization (with W. W. Tupper) on basis of their investigations of living and fossil vascular plants. Home: 985 Memorial Dr., Cambridge 38, Mass.

BAILEY, Jacob Whitman, Am. botanist, chemist, geologist; b. Ward (now Auburn), Mass., Apr. 29, 1811; s. Isaac and Jane (Whitman) B.; grad. U. S. Mil. Acad., 1832; m. Maria Slaughter, Jan. 23, 1835; at least 1 son, 1 dau.; Served as 2d lt., arty., 1832-34; asst. prof. chemistry U. S. Mil. Acad., West Point, 1834-38, prof. chemistry, mineralogy, geology, 1838-57. Pioneer worker with microscope in U. S.; 1st Am. to detect diotomaseas in fossil state; investigated crystals found in tissues of plants; detected vegetable structures in anthracite ashes. Died Feb. 27, 1857.

BAILEY, John, Brit. agriculturist; b. Bowes, Eng., 1750; s. William Bailey; pvt. tutor; tchr. math., Witton-le-Wer, also land surveyor; land agt. to Lord Tankerville, Chillingham, Eng. Author: An Essay on the Construction of the Plough, 1795 (math. calculations used to show advantages of proposed alterations).

Promoted improvements in rural economy; studied mineralogy, chemistry, hydraulics, pneumatics. Died June 4, 1819.

BAILEY, John Hays, Am. bacteriologist; b. Chgo., May 3, 1900; s. George Troy and Clara (Koch) B.; B.S., U. Chgo., 1924, Ph.D., 1928; D.P.H., U. Mich., 1938; m. Gertrude Sarah Boyer, Apr. 20, 1939; children—Martha D. (Mrs. Peter Culver Brown), John Hoyne. Fellow Nelson Morris Research Inst., Chgo., 1928-29; Heusman fellow U. Ind. Sch. Medicine, Indpls., 1929-32; resident bacteriologist Municipal Contagious Disease Hosp., Chgo., 1932-35; sr. bacteriologist Ill. Dept. Health, Chgo., 1935-38; asst. prof. Loyola U. Sch. Medicine, 1938-41; bacteriologist Winthrop Chem. Co., Rensselaer, N.Y., 1942-43; chief bacteriologist Sterling-Winthrop Research Inst., Rensselaer, 1943-65; ret., 1965. Mem. Am. Soc. Microbiology, Am. Acad. for Microbiology (sec.-treas. 1957-59), Chgo. Inst. Medicine, A.A.A.S., Soc. Am. Bacteriologists (treas. 1957-59). Mem. editorial bd. Jour. Bacteriology, 1950-53, Research, publs. on occurrence of scarlet fever organisms in sch. children and relation of carriers to 2d cases in family, prodn. of penicillin, antibacterial agts. from plants, mechanisms of antibacterial action, surface active antibacterial agts. Home: 39 Huntington St., New Brunswick, N.J. 08901.*

BAILEY, Kenneth Claude, chemist; b. May 9, 1896; s. C. W. Bailey; M.A., U. Dublin (Ireland), 1922, Sc.D., 1925, Litt. D., 1929; Docteur, U. Toulouse (France), 1922; m. Dorothy Lavelle, 1923; 1 dau. Jr. dean U. Dublin, 1931-42, prof. phys. chemistry, 1935-47, registrar, from 1942, fellow Trinity Coll., from 1926. Mem. Royal Irish Acad. Author: Notes on the Natural History of Pliny in Hermathena, 1926, 31; (with Dorothy Bailey) Etymological Dictionary of Chemistry and Mineralogy, 1929; The Elder Pliny's Chapters on Chemical Subjects, Vol. I, 1929, Vol. II, 1932; The Retardation of Chemical Reactions, 1937; History of Trinity College, Dublin from 1892 to 1945, 1947. Contbd. papers on inhibition of chem. reactions, other chem. subjects to sci. jours. Died Sept. 18, 1951.

BAILEY, Liberty Hyde, Am. author, botanist, horticulturist; b. South Haven, Mich., Mar. 15, 1858; s. Liberty Hyde and Sarah (Harrison) Bailey; reared on farm; B.S., Mich. Agrl. Coll., 1882, M.S., 1886; LL.D., U. of Wis., 1907, Alfred U., 1908; Litt.D., U. of Vt., 1919; D.Sc., U. of Puerto Rico, 1932; m. Annette Smith, June 6, 1883 (deceased); children—Sara May (deceased), Ethel Zoe. Has given particular attention to botany, horticulture and other biological subjects and to rural problems and education; asst. to Asa Gray, Harvard, 1882-83; prof. horticulture and landscape gardening, Mich. Agrl. Coll., 1885-88; prof. horticulture, Cornell U., 1888-1903, dir. and dean Coll. of Agr., 1903-13. Awarded Veitchian silver medal, 1898, gold medal, 1927; George Robert White medal, 1927; gold medal Nat. Inst. Social Sciences, 1928; grande médaille Société Nationale d'Acclimatation de France, 1928; gold medal of honor, Garden Club of America, 1931; Arthur Hoyt Scott medal and award, Swarthmore Coll. and Hort. Socs., 1931; Distinguished Service Award, Am. Assn. of Nurserymen, 1931; Centennial Medal. Am. Pomol. Soc.; Green Thumb Medal, Nat. Victory Garden Inst.; Johnny Appleseed Medal, Men's Garden Club of Am.; Medal Award, Nat. Garden Inst. 1948. Chmn. Roosevelt Commn. on Country Life, 1898. Fellow Am. Acad. Arts and Scis., A.A.A.S. (pres. 1926); mem. Nat. Acad. Scis., Am. Philos. Soc., Bot. Soc. Am. (pres. 1926), Am. Soc. Naturalists; hon. mem. Royal Hort. Soc. (London), Hort. Soc. Norway, Japan Agrl. Soc., Hort. Soc. Japan, Chinese Soc. Hort. Science, hort. societies Mass., R.I. and Ind., New Zealand Inst. Horticulture, Am. Soc. Hort. Science (1st pres.); corr. mem. Phila. Acad. Natural Science, Royal Acad. Agr. (Turin), Société Lyonnaise d'Horticulture; honorary member Botanical Society of Edinburgh. Author numerous publications relating to field. Editor: Cyclopedia of American Horticulture, 4 vols.; Cyclopedia of American Agriculture, 4 vols.; Standard Cyclopedia of Horticulture, 6 vols. (reprinted in 3 vols.); Rural Science series; Rural Textbook series; Rural Manual series; Rural State and Province series. Influential in establishing horticulture as respected sci. Died Dec. 25, 1954.

BAILEY, Loring Woart, Am. geologist; b. West Point, N.Y., Sept. 28, 1839; s. Jacob Whitman and Maria (Slaughter) B.; A.B., Harvard, 1859, A.M., 1861; hon. Ph.D., U. of N.B. (Can.), 1873; LL.D., Dalhousie, N.S., 1896; m. Laurestine M. d'Avray, Aug. 19, 1863; children—Joseph Whitman, William d'Avray, Loring Woart, Margaret Marshall, Laurestine Marie. Prof. chemistry and natural sci. U. N.B., Fredericton, 1861-1901, biology and geology, 1901-06, ret. under Carnegie Found. Dir. Marine Biol. Sta. of Can., 1909. Author: Mines and Minerals of New Brunswick, 1864; Geology of Southern New Brunswick, 1870; Elementary Natural History, 1887. Died 1925.

BAILEY, Orville T., Am. pathologist; b. Jewett, N.Y., May 28, 1909; s. Milton O. and Ollie (Persons) B.; A.B., Syracuse U., 1928; M.D., Albany Med. Coll., 1932. House officer in pathology Peter Bent Brigham Hosp., Boston, 1933-34, asso. pathologist, 1940-43, asso. pathologist in neuropathology, 1946-49; resident pathologist Children's Hosp., Boston, 1934-35; vol. asst. Montreal (Can.) Neurol. Inst., 1935-37; instr. pathology Harvard Med. Sch., Cambridge, Mass., 1935-40, asso. in pathology, 1940-46, asst. prof., 1946-49; prof. neuropathology Ind. U. Sch. Medicine, Bloomington, 1947-51, prof., 1951-59; prof. neurology U. Ill. Coll. Medicine, Chgo., 1959—. Guggenhelm postwar fellow Cambridge (Eng.) U., 1946-47. Mem. A.M.A., Am. Bd. Pathology, A.A.A.S., Am. Assn. Neurol. Surgeons, Am. Assn. Neuropathology, Am. Acad. Neurology, Am. Neurol. Assn., Internat. Acad. Pathology, Chgo. Neurol. Soc., Chgo. Pathol. Soc. Research on dural sinus thrombosis, alloxan diabetes, myelin degeneration, brain tumors, radioactive materials in brain, physiology in relation myelin problems.*

BAILEY, Pearce, Am. physician; b. N.Y.C., July 22, 1902; s. Pearce and Edith L. (Black) B.; A.B., Princeton, 1924; M.A., Columbia, 1931; Ph.D., Sorbonne (Paris), 1933; honours course in chemistry U. London, 1934; M.D., Med. Coll. S.C., 1941; m. Georgette Dora Soudry, July 3, 1936. Asso. dir. psychol. center, Paris, 1933-36; chief neurol. resident Bellevue Hosp., 1942-44; chief, neurology sect. VA Central Office, Washington, 1946-48; attending neurologist D.C. Hosp., Washington, 1947—; attending neurologist Georgetown U. Hosp., Washington, 1947—; dir. Nat. Inst. Neurol. Diseases, NIH, 1951-59, Internat. Neurol. Research, Antwerp, Belgium, 1959-62; prof. clin. neurology, Georgetown U., 1950—; prof. clin. neurology U. P.R., 1963—. Named officer de la Sante Publique, France, 1949; comendador El Sol del Peru, Peru, 1963. Mem. Am. Acad. Neurology (pres. 1951-53), Nat. Epilepsy Soc. (pres. 1952-53), World Fedn. Neurology (sec., treas. gen. 1957—), Alpha Omega Alpha. Important work includes translations from French of important med. papers. Contbr. numerous papers to internat. jours. Home: Laguna Terr., 1-E, Calle Joffre 6, Santurce, P.R. 00907. Office: P.O. Box 3788, San Juan, P.R. 00904.*

BAILEY, Percival, Am. neurologist; b. Mt. Vernon, Ill., May 9, 1892; s. John Henry and Mattie Estella (Orr) B.; student So. Ill. State Tchrs. Coll., Carbondale, Ill., 1908-12; S.B., U. Chgo., 1914, Ph.D., 1918; M.D., Northwestern U., 1918; m. Yevnigé Bashbazirghanian, Oct. 25, 1923; children—Irene Anahid, Norman Alishan. Resident physician neurol. and psychopathic services, Cook County Hosp., Chicago, 1920-21; asst. à la Salpetrière, 1921-22, l'hospice St. Anne (both of Paris), 1925-26; asso. in surgery Peter Bent Brigham Hosp., 1922-25; attending neurologist New Eng. Deaconess Hosp. and Boston Dispensary, 1926-28; neurosurgeon Albert Merritt Billins Hosp., Chgo., 1928-39; neurologist U. Chgo. Clinics, 1933-39; teaching for different periods in various capacities, U. Chgo., Northwestern U., Harvard until 1928; asso. prof. surgery U. Chgo., 1928-29, prof. surgery, 1929-39, prof. neurology, 1933-39; prof. neurology and neurosurgery U. Ill.; dir. Ill. State Psychopathic Inst.; now dir. research Ill. State Psychiat. Inst., Chgo. Mem. A.M.A., Chgo. Med. Soc., Chgo. Neurol. Soc., Inst. Medicine Chgo., Central Neuropsychiatric Assn., Am. Assn. Anatomists, Am. Assn. Pathologists and Bacteriologists, Am. Neurol. Assn., Soc. Neurol. Surgeons, other Am. and fgn. socs. Research on heart anaesthetics. Home: 731 Lincoln St., Evanston, Ill. Office: 1601 W. Taylor St., Chgo.*

BAILEY, Richard William, Brit. engr.; b. Jan. 6, 1885; s. James William and Ann (Durley) B.; D.Sc.; m. Mary Florence Dorner Alderman, 1909, 2 dau. Cons. research engr., research dept. Metropolitan Vickers Elec. Co., Ltd., Trafford Park, Manchester, Eng. Fellow Royal Soc., 1949, Queen Mary Coll. (London U.); hon. asso. Manchester Coll. Tech.; mem. Instn. Mech. Engrs. (mem. council, pres. 1954). Died 1957.

BAILEY, Solon Irving, Am. astronomer; b. Lisbon, N.H., Dec. 29, 1854; s. Israel C. and Jane (Sutherland) B.; A.B., Boston U., 1881, A.M., 1884; A.M., Harvard, 1888; Sc.D. and hon. prof. astronomy U. of San Agustin, Peru, 1923; m. Ruth Poulter, 1883; 1 son, Irving Widmer. Sent to Peru, S.Am., to investigate conditions there in order to determine best location for a southern sta. for Harvard Coll. Obs., 1889; examined west coast from equator to Southern Chili, resulting in selection of Arequipa, Peru; in charge of work there, from 1892; established meteorol. sta. on summit of El Misti, at 19,000 feet elevation, where observations were carried on for 10 yrs., by far the highest scientific sta. in world, 1893; asst. prof. astronomy Harvard, 1893-98, asso. prof., 1898-1913, Phillips prof., 1913-25, acting dir. observatory, 1919-22. In 1908 visited S. Africa and carried on astron. observations on elevated plateau in northern part of Cape Colony. Discovered Cepheid variable stars of very short period (less than 20 hours) in a globular cluster (M.5), 1895; research and publs. on meteors, stellar photometry, variable stars, star clusters. Died June 5, 1931.

BAILEY, Thomas Pearce, Am. ethnologist; b. Georgetown, S.C., Aug. 18, 1867; s. Thomas Pearce and Maria Laval (Williams) B.; A.B., S.C. Coll., 1887; A.M., U. S.C., 1889, Ph.D., 1891; fellow in psychology, Clark U., 1892-93; m. Charlotte R. Burckmyer, Mar. 20, 1893 (dec. Sept. 1893); m. 2d, Minneola Davis, Aug. 1, 1895 (dec. Sept. 1931); children—Thomas Laval, James Preston, Minneola, Mary Belin; m. 3d, Carol Purse Oppenheimer, Sept. 12, 1935. Prin. graded sch., Georgetown, 1887-88; tutor English and history, 1888-89 U. S.C., sec., 1889-91, adj. prof. biology, 1891-92; prin. graded schs., Marion, S.C., 1893-94; asst. prof. edn. U. Cal., 1894-98, asso. prof. edn., 1898-1900; asst. prof. edn. in univ. extension div. U. Chgo., 1900-03; prof. psychology and applied psychology U. Miss., 1903-05, prof. psychology and edn., 1905-08, psychology and secondary edn., 1908-09, dean dept. of edn., 1905-09; supt. city schs., Memphis, Tenn., 1909-10; investigator for N.Y. Bur. of Municipal Research. 1910-11; dean All Saints' Episcopal Coll., Vicksburg, Miss., 1911-12; cooperative lectr. Miss. schs. and colls., 1912-14; prof. philosophy and psychology U. of South, 1914-26; cons. psychologist Miss. State Insane Hosp., summers, 1924-25; prof. philosophy, psychology, ethnology, and cons. psychologist Rollins Coll., 1926-44; emeritus since 1944. Mem. Phi Beta Kappa. Author: Love and Law, 1899; Race Orthodoxy in the South, 1914. Contbr. to ednl. and psychol. jours. Died Feb. 7, 1949.

BAILEY, Vernon, Am. biologist; b. Manchester, Mich., June 21, 1864; s. Hiram and Emily B.; student U. Mich., 1893, Columbian (now George Washington) U., 1894-95; m. Florence Augusta Merriam, Dec. 16, 1899. Served as chief field naturalist U. S. Biol. Survey, ret., 1933. Fellow A.A.A.S.; mem. Am. Ornithologists' Union, Cooper Ornithol. Club, Am. Forestry Assn., Washington Acad. Scis. Author: Spermophiles of the Mississippi Valley, 1893; Pocket Gophers of Mississippi Valley, 1895; Revision of Voles of the Genera Evotomys and Microtus, 1897; Mammals of District of Columbia, 1900, 1923; Biological Survey of Texas, 1905; Life Zones and Crop Zones of New Mexico, 1913; Revision of the Pocket Gophers of the Genus Thomomys, 1915; Wild Animals of Glacier National Park (mammals); Beaver Habits and Beaver Farming, 1923; Biological Survey of North Dakota, 1927; Animal Life of Carlsbad Cavern, 1928; Animal Life of Yellowstone National Park, 1930; Mammals of New Mexico, 1931; Mammals of Oregon, 1936; Cave Life of Kentucky. Died Feb. 14, 1942.

BAILEY, Wilford Sherrill, Am. vet. parasitologist; b. nr. Somerville, Ala., Mar. 2, 1921; s. Ollis Wilford and Bessie (Widener) B.; D.V.M., Auburn University, 1942, M.S., 1946; Sc.D., Johns Hopkins University, Baltimore, Maryland, 1950; m. Cratus Hester, May 30, 1942; children—Wilford Edward, Joe Sherrill, Margaret Ann, Sarah Jane. Faculty, Auburn (Ala.) U., 1942—, head prof. pathology, parasitology, 1951-62, asso. dean Grad. Sch., coordinator research, research prof. parasitology, 1962-66, v.p. for acad. affairs, 1966—. Mem. Am. Vet. Med. Assn. (chmn. com. on parasitology 1951-53), Am. Soc. Parasitologists (custodian 1954—, past v.p.), Am. Assn. Vet. Parasitologists, World Assn. for Advancement Vet. Parasitology (sec. organizing com. 1962), Sigma Xi, Phi Kappa Phi, Phi Zeta, Omicron Delta Kappa, Omicron Delta Epsilon, Alpha Psi (past pres. nat. council). Contbg. author: Canine Medicine, 1953, 59; Diseases of Cattle, 1956. Research, publs. on demonstration relationship esophageal worm in dog and cancer esophagus, parasitic infections of domestic animals particularly stomach and intestinal worms cattle, control parasitic gastroenteritis of cattle and sheep, studies on toxic hepatitis of dogs and hogs. Home: 778 Moores Mill Rd., Auburn, Ala. 36830.*

BAILEY, Wilfrid Charles, Am. anthropologist; b. Cicero, Ill., May 3, 1918; s. Frank Sidney and Mary (Walker) B.; B.S., U. Ariz., 1940, M.A., 1942; Ph.D., U. Chgo., 1955; m. Ethelene Scott, June 7, 1941; children—Mary Ann (Mrs. William Laney Littlejohn), Ruth Ellen, Charles Edward, James Arthur. Faculty, U. Tex., 1947-55, asst. prof., 1951-55; asst. prof. Miss. State U., 1955-62; prof. dept. sociology and anthropology U. Ga., Athens, 1962—. Fellow Am. Anthrop. Assn., Am. Sociol. Assn., Soc. for Applied Anthropology; mem. Soc. for Am. Archaeology, Rural, So. sociol. socs., Am. Ethnol. Soc., Am. Assn. U. Profs., Sigma Xi, Alpha Kappa Delta. Research, publs. on community orgn. and change, teaching materials for anthropology. Home: 225 Woodcrest Dr., Athens, Ga. 30601.*

BAILEY, Sir William Henry, Brit. inventor; b. Salford, Eng., May 10, 1838; s. John Bailey; ed. Manchester (Eng.) Grammar Sch.; m. Jane Dearden Dorning, 1866 (dec. 1904); 4 sons, 1 dau. Mayor of Salford; chmn., mng. dir. W. H. Bailey and Co., Ltd., Albion Works, Salford; dir. Manchester Ship Canal Co. Gov. John Rylands Library, Manchester. Fellow Royal Geog. Soc.; mem. Library Assn. U.K. (pres. 1906-07), Manchester Arts Club (pres.), Manchester Shakespeare Soc. (pres.). Author: Linnaeus and the Reign of Law. Contbr. lectures, pamphlets, articles, addresses. Early promoter Manchester Ship Canal; inventor perpetual motion actuated by tides for recording rise and fall and work on the sea-coast, also autographic instruments, pyrometers, other devices. Died Nov. 22, 1913.

BAILEY, William J(ohn), Am. chemist; b. East Grand Forks, Minn., Aug. 11, 1921; s. Admiral Ross and Erva (Stewart) B.; B.Chemistry, U. Minn., 1943; Ph.D., U. Ill., 1946; m. Mary Caroline Worsham, Aug. 27, 1949; children—Caroline Jane, John Robert, Barbara Ann. Arthur D. Little postdoctoral fellow Mass. Inst. Tech., 1946-47; asst. prof. chemistry Wayne

State U., Detroit, 1947-49, asso. prof., 1949-51; research prof. organic chemistry U. Md., College Park, 1951—. Chmn., Gordon Research Conf. on Organic Reactions, 1960; mem. NSF Postdoctoral Selection Com., 1963—, NRC Adv. Com. of Elastomers to U. S. Army Natick Laboratories, 1961—, chmn. Com. Macromolecular Chemistry, 1967—. Recipient Fatty Acid Producers Research award, 1955. Mem. Chem. Soc. Washington (pres. 1961), Am. Chem. Soc. (chm. elect div. polymer chemistry, 1967, mem. com. on nominations and elections Council), A.A.A.S., Am. Oil Chemists Soc., Phi Beta Kappa, Sigma Xi, Phi Kappa Phi, Phi Lambda Upsilon, Pi Mu Epsilon, Alpha Chi Sigma. Mem. editorial bds. Jour. Organic Chemistry, 1957-63, Macromolecular Synthesis, 1960—, Record Chem. Progress, 1950—, Jour. Macromolecular Science Chemistry, 1966—, Jour. Polymer Science, 1967—, Macromolecules, 1967—. Developed a new method for preparation of polymers; discovered several new polymers; produced a correlation between structure and properties in plastics and rubbers. Home: 6905 Pineway, University Park, Md.*

BAILEY, William Whitman, Am. botanist; b. West Point, N.Y., Feb. 22, 1843; s. Jacob Whitman and Maria (Slaughter) B.; grad. Brown U. 1864, Ph.B., 1873 A.M., 1893; LL.D., U N.B., 1900, studied botany, Columbia, 1872, Harvard Summer Sch., 1875, 76, 79; m. Eliza R. Simmons, Mar. 14, 1881. Asst. in chemistry Mass. Inst. Tech., 1866; asst. chemist Manchester (N.H.) Print Works, 1866; botanist U. S. Geol. Survey of 40th parallel, 1867-68; dep. sec. of State, R.I., 1868; asst. librarian Providence Athenaeum, 1869-71, dir., 1900-03; taught botany pvt. schs., Providence; instr. botany Brown U., Providence, 1877-81, prof., 1881-1906, emeritus, from 1906; spl. beneficiary of Carnegie Found. 1906. Author: Botanical Collector's Handbook, 1881; Among Rhode Island Wild Flowers, 1885; Botanical Note-Book, 1894; New England Wild Flowers, 1897; Botanizing, 1899; Poems, 1910 Died Feb. 20, 1914.

BAILLARGER, Jules Gabriel François, French neurologist; b. Montbazon, France, 1806; mem. staff Salpetrière Lunatic Asylum, Paris; mem. French Acad. Medicine. Author: De la Paralysie pellagreuse, 1848; Essai de Classification des Maladies mentales, 1854. Founded (with Longet and Cerise) Annales Medico-Psychologiques, 1843. Described white bands found in large pyramidal cell layer of cerebral cortex, 1840; described size of pupils of eyes in dementia paralytica. Died Dec. 31, 1891.

BAILLAT, Georges, French surgeon; b. Maury (P.-O.), France, June 15, 1896; s. Justin and Marie (Villa) B.; ed. Lycée et Faculté de médecine, Toulouse, France; M.D.; m. Geneviève Dubocq, Dec. 23, 1935; children—Raymond, Jean, Andre, Marie-Thérèse. Intern, Toulouse Hosps., 1921; chief surgery clinic Toulouse Med. Sch., 1925; surgeon hosps. of Perpignan, 1929—. Reporter, Congress of Gynecology, 1928. Mem. Order of Physicians of East Pyrenees (pres. of council). Home: 22, cours Palmarole, Perpignan (P.O.). Office: Clinique Saint-Jacques, Perpignan, France.

BAILLAUD, Édouard-Benjamin, French astronomer; b. Chalon-sur-Saône, France, Feb. 14, 1848; ed. Chalon, Lyons, Paris; D.Sc., 1876; children including Jules, René; prof. Faculty of Scis., also dir. Observatory, Toulouse, 1879, then at Paris, 1908-25; mem. Bur. Longitudes; corr. French Acad. Scis. Author: Cours d'astronomie, 1893-96. Made observations on satellites of Jupiter and Saturn, sunspots, occultations of stars by moon. Died Toulouse, France, July 8, 1934.

BAILLET, Marcel Jules Eugène, French pediatrician; b. Etampes, Dec. 1, 1892; s. Charles and Camille (Servant) B.; ed. U. Paris; M.D.; m. Louise Curriez, Sept. 8, 1921; 1 dau., Rose-Marie. Founder 1st med. clinic for children in France, Haute-Savoie, 1922; pediatrician at Saint-Gervais, 1922-55. Mem. Soc. Hydrology and Climatology, Nat. Syndicat of Med. Establishments for Children (hon. pres.). Contbr. articles to pediatric jours. Home: 106, rue de France, Fontainebleau. Office: 2, rue Pigalle, Paris, France.

BAILLIE, Matthew, anatomist; b. Shots, Lanarkshire, Scotland, Oct. 27, 1761; s. James and Dorothea (Hunter) B.; studied at U. Glasgow (Scotland), Balliol Coll. at Oxford (Eng.) U.; pvt. instruction Dr. William Hunter; M.D., 1789; m. Sophia Denman; 2 children. Elected physician St. George's Hospital, 1787; physician to George III. Fellow Royal Soc., 1790, College of Physicians. Author: Anatomy of the Gravid Uterus (left in manuscript by Dr. William Hunter), 1794; The Morbid Anatomy of Some of the Most Important Parts of Human Body (1st septematic textbook on morbid anatomy), 1795; Observations on paraplegia, 1822; Pulsation of the Aorta in the Epigastrium. Described transposition of viscera, 1788, dermoid cysts of ovary, 1789; first to observe floating kidney in human being, 1825; also first to define cirrhosis of liver, to distinguish common renal cysts from cysts of parastic hydatids of kidney; described gastric ulcer, also ulcers caused by typhoid fever; first to differentiate between nodular and infiltrating types of pulmonary Tb; described hepatization of lungs in pneumonia; 1st to notice miliary type of pneumonia; performed postmortem examinations; disproved theory that death was caused by a growth in heart; showed polypus was a mass of coagulated fibrin formed after death. Died Sept. 23, 1823.

BAILLIENCOURT-COURCOL, Albert de, French engr.; b. Neuilly-sur-Seine (Seine), France, Mar. 30, 1908; ed. Ecole polytechnique, Ecole supérieure d'electricité, Ecole supérieure des P.T.T.; m. N. Gueydon. Vice pres. commn. of indsl. prodn. and energy of Nat. Assembly; pres. Atomic Energy Commn.; mem. Superior Council of Sci. Research and Tech. Progress, 1956. Mem. Assn. for Expansion Tech. and Sci. Research (pres. 1957). Address: 38, av. Victor-Hugo, Paris 16, France.

BAILLON, Henri Ernest, French botanist, physician; b. Calais, France, Nov. 30, 1827; studied medicine in Paris, France; M.D.; qualified in natural scis.; prof. med. natural history Faculty Med., Paris, from 1863; later prof. hygiene, natural history École centrale des Arts et Manufactures; decorated Legion of Honor, 1867. Fellow Royal Soc., 1894. Author: Étude Générale du groupe des Euphorbiacées, 1858; Monographie des Buxacées et des Stylocérée, 1859; Recherches organogéniques sur la fleur femelle des Conifères, 1860; Recherches sur l'organisation, le developpement et l'anatomie des Caprifoliacées, 1864; Adansonia, recueil périodique d'observations botaniques, 10 vols., 1866-70; Histoire des Plants, 13 vols., 1867-95; Dictionaire de botanique, 4 vols., 1876-92; Histoire naturelle des plants de Madagascar, 3 vols., 1879; Traité de botanique médicale phanérogamique, 1883-84. Research particularly on phanerogames. Died Paris, July 19, 1895.

BAILLOU, Guillaume de, see de Baillou, Guillaume.

BAILLY, Jean Sylvain, French astronomer; b. Paris, France, Sept. 15, 1736; student La Caille; became dep. to Estates-Gen., 1789, also 1st pres.; mayor of Paris, 1789-91; executed by Jacobins for ordering nat. guard to fire into a mob at Champ-de-Mar, 1793. Mem. French Acad. Scis., 1763; Academie Francaise, 1784; Acad. Inscriptions, 1785. Recipient prize Berlin Acad., French Acad. Author: Essai sur la théorie des satellites de Jupiter, 1766; Sur les inégalités de la lumière des satellites de Jupiter, 1771; Histoire de l'astronomie ancienne, 1775; Histoire de l'astronomie moderne, 3 vols., 1779; Histoire de l'astronomie indienne et orientale, 1787; Essai sur les fables et sur leur histoire, 2 vols., 1799. Died Nov. 12, 1793.

BAILY, Francis, English astronomer; b. Newbury, Eng., Apr. 28, 1774; s. Richard Baily; apprenticed to London mercantile house, 1788-95; honorary D.C.L., Dublin (Ireland), 1835, Oxford (England) U., 1844. Member French Academy of Sciences (corresponding), 1836; Royal Astronomical Society, 1820 (a founder 1820, pres., v.p.), Brit. Assn. (became permanent trustee 1839), Geog. Soc. (became v.p. 1830). Fellow Royal Soc., 1821. Recipient Gold medal for preparation catalogue of 2881 stars Astron. Soc., 1827, for successfully repeating Cavendish expt., 1843. Author: Account of the Reverend John Flamsteed, 1835; Journal of a Tour in Unsettled Parts of North America in 1796-1797 (edited by De Morgan), 1856. Calculated mean density of earth using Cavendish method; revised several star catalogues; improved Nautical Almanac. 1st to describe bright spots (Baily's beads) along moon's disk during annular eclipse of sun, 1836. Died London, Aug. 30, 1844.

BAILY, Norman Arthur, Am. physicist. b. N.Y.C., July 2, 1915; s. Louis D. and Ida (Bolet) B.; B.S., St. Johns U., 1941; M.A., N.Y. U., 1943; Ph.D., Columbia, 1952; m. Rose Levine, Nov. 20, 1940; children—Phillip, Barbara. Instr., RCA Inst., N.Y.C., 1943-46; research scientist Radiol. Research Lab., Columbia, N.Y.C., 1946-52; radiation physicist Marine Biol. Lab., Woods Hole, Mass., 1946-52; operations analyst Hdqrs. Strategic Air Command, Omaha, 1952-54; prin. research scientist Roswell Park Meml. Inst., Buffalo, 1954-59; asst. prof. radiology U. Buffalo, 1954-59; asst. prof. biophysics, 1954-59; lectr. Canisius Coll., Buffalo, 1954-59; mgr. space scis. dept. Hughes Research Labs., Malibu, Cal., 1959—; clin. prof. radiology U. Cal. at Los Angeles, 1959-65, prof. in residence, 1965—. Mem. adv. bd. panel on bioscis. and aerospace medicine USAF, 1963—. Diplomate Am. Bd. Radiology, Am. Bd. Health Physics. Fellow Am. Coll. Radiology (asso.); mem. Am. Soc. for Testing and Materials (mem. E-10 com. on dosimetry 1959—), Am. Assn. Physicists in Medicine, Research Soc. Am., Biophys. Soc., I.E.E.E., Am. Phys. Soc., N.Y. Acad. Scis., A.A.A.S. Contbr. articles on research in nuclear physics, dosimetry and radioactivity Home: 1166 Tellem Dr., Pacific Palisades, Cal. 90272. Office: 3011 Malibu Canyon Rd., Malibu, Cal. 90265.*

BAILY, Walter Lewis, Jr., Am. mathematician; b. Waynesburg, Pa., July 5, 1930; s. Walter Lewis and Emily (Thompson) B.; S.B., Mass. Inst. Tech., 1952; M.A., Princeton, 1953; Ph.D., 1955; m. Yaeko Iseki, Jan. 7, 1963. Instr., Princeton, 1955-56, Mass. Inst. Tech., 1956-57; faculty U. Chgo., 1957—, prof. math., 1963—. Mem. Am. Math. Soc., Math. Soc. Japan. Research, publs. on compactification of orbit spaces of arithmetic groups acting on Hermitian symmetric spaces, moduli of Abelian varieties, moduli of curves. Office: Math. Dept., U. Chgo., Chgo. 60637.*

BAIN, Alexander, Brit. inventor; b. Watten, Caithness-Shire, Scotland 1810; apprentice to clockmaker, Wick, Scotland; went to London as journeyman in 1837; attended lectures at Adelaide Gallery. Claimed to have invented 1st printing telegraph; applied electricity to workings of clock and was one of 1st to work electrically several clocks from a standard time keeper; invented electric fire alarms, also sounding apparatus; became pioneer in modern high speed telegraphy by inventing chem. telegraph, 1843; independently discovered use of earth circuit. Died 1877.

BAIN, Alexander, Brit. psychologist, logician; b. Aberdeen, Scotland, June 11, 1818; s. George and Margaret (Paul) B.; grad. Marischal Coll., Aberdeen, 1840; LL.D., U. Edinburgh (Scotland), 1869; m. Frances A. Wilkinson, 1855; m. 2d, Barbara Forbes, 1893. Asst. to prof. moral philosophy Marischal Coll., from 1841; asst. sec. Met. San. Commn., 1848-50; lectr. Bedford (Eng.) Coll. Women; prof. logic, English Aberdeen U., 1860-80, lord rector, from 1890. Author: The Study of Character, including an Estimate of Phrenology, 1861; Logic, 1870; Mental and Moral Science, a Compendium of Psychology and Ethics, 1872; Mind and Body, 1872; The Senses and the Intellect, 1894; The Emotions and the Will, 1899. Editor: (with others) Aristotle (Grote), 1872; minor works by Grote, 1873; Autobiography (Grote), 1904. Contbd. to Internat. Sci. series, 1879. Founded periodical Mind, 1876; one of 1st to apply to psychology results of psychol. investigations; lucid exponent of posteriori sch. psychology; espoused materialistic system philosophy, utilitarian ethics. Died Sept. 18, 1903.

BAIN, Edgar Collins, Am. metallurgist; b. nr. La Rue, O., Sept. 14, 1891; s. Milton Henry and Alice Anne (Collins) B.; B.Sc., Ohio State U., 1912, M.Sc., 1916; D.Eng. (hon.), Lehigh U., 1936; D.Sc., Ohio State U., 1947; m. Helen Louise Cram, Feb. 18, 1927; children—Alice Anne (Mrs. Allan M. Mercer), David Erwin. Various sci. positions, 1913-28; metallurgist U. S. Steel Corp., 1928-35, asst. to v.p., 1935-43, v.p. research and tech., Pitts., 1950-55, asst. exec. v.p. operations, 1955-57; v.p. research and tech. Carnegie Ill. Steel Corp., 1943-50; cons. metallurgist, Sewickley, Pa., 1957—. Chmn. subcom. NDRC, 1939-40; bd. advisers chief of ordnance U. S. Army, 1942-43; chmn. engring. and indsl. research sect. Nat. Acad. Scis.-NRC, 1949-53. Recipient Benjamin Lamme medal Ohio State U., 1937, John Price Wetherell medal Franklin Inst., 1949, Grande Medaille de la Societe Francaise de Metallurgie, Paris, 1952. Fellow Am. Phys. Soc.; mem. Am. Soc. for Metals (past trustee, past pres., H. M. Howe medal 1930, Albert Sauveur Achievement award 1946, gold medal for research 1949), Nat. Acad. Scis., Am. Iron and Steel Inst. medal 1935. Mem. Am. Inst. Mining, Metall. and Petroleum Engrs. (Robert W. Hunt gold medal 1929), Am. Soc. for Testing and Materials, Iron and Steel Inst. Britain (hon. life), Iron and Steel Inst. Japan (hon. life), Japan Inst. Metals (hon. life, gold medal 1964), Sigma Xi, Alpha Sigma Mu, Phi Lambda Upsilon. Author: (with M. A. Grossman) High Speed Steel, 1931; Alloying Elements in Steel, 1939, rev. (with H. W. Paxton), 1961; rev. Principles of Heat Treatment (M. A. Grossman), 1964. Contbr. articles on steel metallurgy to profl. jours. Clarified atomic arrangement and mechanisms in metals, particularly alloy steel and its hardening; invented new steel-hardening process; discovered new steel constituent, bainite; identified basic roles of alloying elements in steel. Address: 434 Maple Lane, Edgeworth, Sewickley 15143.*

BAIN, H(arry) Foster, Am. mining engr.; b. Seymour, Ind., Nov. 2, 1872; s. William M. and Radie (Foster) B.; B.S., Moores Hill Coll., 1890; postgrad. Johns Hopkins, 1891-93; Ph.D., U. Chgo., 1897; m. Mary Wright, Dec. 1, 1902; 1 dau., Margaret. Asst. Ia. Geol. Survey, 1893-95; asst. state geologist, Iowa, 1895-1900; mgr. mines, Ida. Springs and Cripple Creek, Colo., 1901-03; geologist, U. S. Geol. Survey, 1903-06; dir. Ill. Geol. Survey, 1905-09; editor Mining and Scientific Press, San Francisco, 1909-15; editor Mining Mag., London, 1915-16; explorations in Far East, 1916-17, 19-20; asst. dir., U. S. Bur. of Mines, 1918-19, dir., 1921-24; cons. engr. Argentina, 1924-25, Colombia, 1929. Mem. Commn. for Relief in Belgium, 1915-16. Lectr. econ. geology, U. Ia., 1897, U. Chgo., 1903-04. Fellow Geol. Soc., Am.; mem. Am. Inst. Mining and Metall. Engrs. (sec. from 1925), Mining and Metall. Soc. Am., Canadian Mining Inst., Inst. of Mining and Metallurgy, Soc. Econ. Geologists (pres. 1926). Contbd. papers and reports on glacial and physiographic geology, coals of Ark., western interior coal field. Died Mar. 9, 1948.

BAIN, James Arthur, Am. pharmacologist, educator; b. Langdon, N.D., May 22, 1918; s. James Hamilton and Mabel (Aldritt) B.; A.A., Wayland Jr. Coll., 1938; B.S., U. Wis., 1940, Ph.D., 1944; m. Eleanor Theodora Hohaus, Dec. 5, 1947; children—Andrew J., Peter T. Research asst. McArdle Meml. Lab., U. Wis., 1940-44, Rockefeller fellow, 1946-47; research asso. U. Ill., 1947-50; faculty dept. pharmacology Emory U., Atlanta, 1950—, prof., 1954—, chmn. dept., 1957-62, dir. basic health scis., 1960—. Cons. to govt., nat. ays., industry, 1954—. Mem. Am. Chem. Soc., Soc. Exptl. Biology and Medicine, Am. Soc. Pharmacology and Exptl. Therapeutics, A.A.A.S., Am. Assn. Cancer Research, Sigma Xi. Research, numerous publs. on enzymatic processes applied to mode of action of drugs, those active in central nervous system and carcinogenesis, cause of epilepsy and neoplasia at cellular level.*

BAINBRIDGE, Francis Arthur, Brit. physiologist; b. Stockton-on-Tees, Eng., July 29, 1874; s. Robert Robinson and Mary (Sanderson) B.; ed. Trinity Coll.,

Cambridge (Eng.) U.; M.B., St. Bartholomew's Hosp., London, 1901, M.D., 1904; D.Sc.; m. Hilda Winifred Smith, 1905; 1 dau. Apptd. Gordon lectr. pathology Guy's Hosp., 1905; became asst. bacteriologist Lister Inst. Preventive Medicine, 1907; named prof. physiology Durham U., 1911, St. Bartholomew's Hosp., 1915. Delivered Milroy lectures, Royal Coll. Physicians, 1912. Fellow Royal Soc., 1919. Author: The Physiology of Muscular Exercise, 1919. Studied food poisoning bacilli, mechanism of lymph formation, urinary secretion, the effect of partial removal of kidneys; found that increase of pressure on the venous side of the heart accelerated the rate of beat. Died London, Eng., Oct. 27, 1921.

BAINBRIDGE, John, Brit. astronomer; b. Ashby-de-la-Zouch, Eng., 1582; s. Robert and Anne B.; B.A., Cambridge (Eng.) U., 1603, M.A., 1607, M.D., 1614. Practiced medicine; apptd. 1st Savilian prof. astronomy, Oxford (Eng.) U., 1619. Author: An Astronomical Description of the Comet of 1618, 1619; Procli sphaera et Ptolomaei de hypothesibus planetarum, 1620; Canicularia, 1648; Antiprognosticon. Made observations of latitude and longitude; emphasized need for lunar studies. Died Nov. 3, 1643.

BAINBRIDGE, Kenneth Tompkins, Am. physicist, educator; b. Cooperstown, N.Y., July 27, 1904; s. William Warin and Mae (Tompkins) B.; S.B., S.M., Mass. Inst. Tech., 1926; M.A., Princeton, 1927, Ph.D., 1929; M.A. (hon.), Harvard, 1940; m. Margaret Pitkin, Sept. 8, 1931; children—Martin Keeler, Joan (Mrs. Franklin Robinson Safford), Margaret Tompkins (Mrs. Donald Keith Robinson). NRC fellow Bartol Research Found., Swarthmore, Pa., 1929-31, Bartol Research Found. fellow, 1931-33; Guggenheim Meml. Found. fellow Cavendish Lab., Cambridge (Eng.) U., 1933-34; faculty physics Harvard, Cambridge, Mass., 1934—, prof., 1946-61, chmn. dept. physics, 1953-55, George Vasmer Leverett prof., 1961—. Tech. cons. NDRC, 1940-44, Mass. Inst. Tech. Radiation Lab., 1940-43, Los Alamos Lab., 1943-45; dir. Alamogordo Atomic Bomb Test, 1945. Recipient Louis Edward Levy medal Franklin Inst., 1933; Presdl. certificate merit for work on radar, 1948. Mem. Am. Phys. Soc., Nat. Acad. Scis., Am. Acad. Arts and Scis., Sigma Xi, Alpha Tau Omega, Tau Beta Pi. Contbr. articles to profl. publs. Patentee photoelectric cells, electronic multiplier, electromagnetic pumps. Office: Harvard U., Cambridge, Mass. 02138.*

BAINBRIDGE, William Seaman, Am. surgeon; b. Providence; son of William Folwell and Lucy E. (Seaman) B.; M.D., Columbia Coll. Phys & Surg., 1893, postgrad., 1896; 1893; grad. Presbyn. Hosp., 1895, Sloane Maternity Hosp., 1896; hon. A.M., Shurtleff Coll., Ill., 1899; M.S., Washington and Jefferson Coll., 1902; Sc.D., Western U. Pa., 1907; LL.D., Coe Coll.; Litt.D., Lincoln Meml. U., 1923; Dr. Honoris Causa, U. San Marcos (Peru), 1941; m. June Ellen Wheeler, Sept. 9, 1911; children—Elizabeth, William Wheeler, John Seaman, Barbara (Mrs. Angus McIntosh). Prof. operative gynecology, N.Y. Postgrad. Med. Sch., 1900-06; prof. surgery N.Y. Poly. Med. Sch. and Hosp., 1906-18; surgeon, N.Y. Skin and Cancer Hosp., 1903-18; surg. dir. N.Y. City Children's Hosps. and Schs., Manhattan State Hosp., Ward's Island; cons. surg. or gynecologist to 16 metropolitan and suburban hosps.; hon. prof. med. faculty, Univ. Santo Domingo, Dominican Republic Ofcl. rep. of U. S. Govt. at internat. congresses mil. medicine, surgery and sanitation, from 1921; pres. 8th session Internat. office Medico Military Documentation, Luxemburg, 1938, chmn. 9th session, Washington and New York, 1939; went on ofcl. mission to all republics of Central and South America for U. S. Navy Dept. and State Dept., 1941. Recipient numerous honors U. S. and other nations. Fellow Am. Assn. Obstet., Gynecol. and Abdominal Surgeons, Internat. Coll. Surgeons (internat. treas., 1935-46; chmn. bd. trustees U. S. chpt.), Am. Geriatrics Soc. (hon.), Internat. Coll. (hon.), Anesthetists, Royal Inst. Pub. Health (life), Royal Soc. Medicine (Eng.), A.M.A., N.Y. Acad. Medicine; mem. Assn. Mil Surgeons U. S. (pres. 1935), Internat. Med Club of New York (pres. 1934-38), Am. Acad. Phys. Medicine (pres. from 1941); hon. mem. profl. socs. in Belgium, Italy, Poland, France, Mexico, Hungary, Peru, Venezuela, Spain. Author: A Compend of Operative Gynecology, 1906; Life's Day Guide-Posts and Danger Signals in Health, 1909; The Cancer Problem, 1914 (French, Italian; Spanish, Polish, Arabic edits.); also brochures, med. papers and reports. Died Sept. 22, 1947.

BAIR, Joe Keagy, Am. nuclear physicist; b. Massillon, O., Mar. 10, 1918; s. D. Ray and Nell (Laughlin) Keagy; B.A., Rice Inst., 1940; postgrad. Columbia, 1940-41; m. Virginia Henderson, Aug. 2, 1941; 1 dau., Jeaninne (Mrs. R. C. Helmstetter). Asst. physicist Columbia, 1940-41; physicist Naval Ordnance Lab., 1941-47, Fairchild Engring. and Air Corps, Oak Ridge, 1947-51, Oak Ridge Nat. Lab., 1951—. Research, publs. on nuclear reactions induced by artificially accelerated particles. Home: 200 W. Fairview Rd. Office: P.O. Box X, Oak Ridge 37830.*

BAIR, William J., Am. radiation biologist; b. Jackson, Mich., July 14, 1924; s. William J. and Mona (Gamble) B.; B.A., Ohio Wesleyan U., 1949; Ph.D., U. Rochester, 1954; m. Barbara Sites, Feb. 16, 1952; children—William J., Michael Braden, Andrew Emil. NRC, AEC fellow U. Rochester, 1949-50, research asso. radiation biology, 1950-54; biol. scientist plant nutrition and microbiology Hanford Labs. (now Pacific N.W. Lab. Battelle Meml. Inst.), 1954-56, mgr. inhalation toxicology biology dept., Richland, Wash., 1956—, lectr. radiology Center for Grad. Study 1955—; U. S. participant IAEA, WHO Sci. Meeting on Diagnosis and Treatment of Radioactive Poisoning, Vienna, 1962, Symposium on Radiol. Health and Safety in Nuclear Materials Mining and Milling, Vienna, 1963; gen. chmn. Hanford Symposium on Inhaled Radioactive Particles and Gases, 1964. Mem. Radiation Research Soc., Health Physics Soc., N.Y. Acad. Scis., Soc. Exptl. Biology and Medicine, A.A.A.S., Reticuloendothelial Soc., Sigma Xi. Editor: Inhaled Radioactive Particles and Gases, 1965. Contbr. numerous articles to profl. jours. Demonstrated radiation effects on cell walls, and carcinogenicity of inhaled plutonium; radiation induced biochemical mutations in yeast, research on fate and biol. effects of inhaled radionuclides. Home: 102 Somerset St. Office: Pacific N.W. Lab., P.O. Box 999, Richland, Wash. 99352.*

BAIRD, David W., Am. physician; b. Baker, Ore., Oct. 21, 1898; s. David W. E. and Mamie (Berntson) B.; M.D., U. Ore., 1926; LL.D., U. Portland, 1946; m. Mary Alexander, Sept. 27, 1925; children —Mary (Mrs. Stanley Prouty), David Michael. Asst. in anatomy U. Ore. Med. Sch., Portland, 1923-28, faculty medicine, 1929—, med. dir. U. Ore. Hosps. and Clinics, 1935-43, asso. dean Med. Sch., 1937-42, acting dean, 1942-43, dean, prof. medicine, 1943—. Mem. Sigma Xi, Alpha Omega Alpha, Kappa Sigma, Nu Sigma Nu. Home: 2752 N.E. Thompson St., Portland, Ore. 97212.*

BAIRD, Derwood McVey, Am. animal nutritionist; b. Moorefield, Ky., July 13, 1922; s. John Lee and Mamie (McVey) B.; B.S., U. Ky., 1947, M.S., 1948; Ph.D., U. Ill., 1951; m. Allyne Higgason, Jan. 29, 1949; children—Kathy, Joan Lynn, Sue Ellen. Research asst. U. Ky., 1947-48, teaching assistant University of Illinois, 1948-51; with University of Georgia, 1951—, asso. animal sci., 1953—. Recipient Sears Roebuck Research award, 1960. Mem. Am. Soc. Animal Sci. (chmn. pasture sect. 1958), Am. Soc. Range Mgmt. Publs. on mgmt. of beef, sheep and hogs in relation to efficiency of prodn. and carcass characteristics. Home: 514 Brook Circle, Griffin, Ga. 30223. Office: Ga. Expt. Sta., Experiment, Ga. 30212.*

BAIRD, Donald, Am. vertebrate paleontologist; b. Pitts., May 12, 1926; s. George M.P. and Mary (Johnson) B.; B.S., U. Pitts., 1948; M.S., U. Colo., 1949; Ph.D., Harvard, 1955; m. Helen Lucille Bailey, Feb. 14, 1948 (dec. May 1963); children—Andrew B., Laurel J. Preparator, Carnegie Mus., Pitts., 1943; asst. geologist Pa. Geologic Survey, 1947-48; curator U. Cin. Mus., 1949-51; asst. curator vertebrate paleontology Mus. Comparative Zoology, Harvard, 1954-57; asst. curator vertebrate paleontology Princeton, 1957-63, asso. curator, 1963-67, curator Museum of Natural History, 1967—; research asso. Am. Mus. Natural History, N.Y.C., 1964—. Cons. N.J. State Mus., 1961—. Mem. Soc. Vertebrate Paleontology, Paleontol. Soc., Paleontol. Assn., Soc. for Study Evolution, Soc. Systematic Zoology, Am. Soc. Zoologists, Geol. Soc. N.J., Sigma Xi. Research publs. on Carboniferous fishes, amphibians and trackways, Triassic dinosaurs and other reptiles and trackways, Cretaceous marine reptiles, biostratigraphic geology eastern U. S. and Can.; fossil collecting. Home: 20 Edwards Pl., Princeton, N.J. 08540.*

BAIRD, James, biologist; b. Glasgow, Scotland, Feb. 10, 1925; s. William and Euphemia (Miller) B.; came to U. S., 1928, naturalized, 1943; B.S., U. Mass., 1951; m. Carol Whitford Shanklin, June 7, 1952; children—William Morgan, Margaret Miller, Robert Ferguson, Kathryn Graham. Naturalist, Nat. Capitol Parks, Washington, 1949-51; dir. Norman Bird Sanctuary, Middletown, R.I., 1955-60; dir. natural history services Mass. Audubon Soc., Lincoln, 1960—. Mem. Am. (elective), Brit. ornithologists unions, Wilson, Cooper ornithol. socs., Ecol. Soc. Am., Northeastern Bird-banding Assn. (pres. 1967—). Research on migration and bird biology. Home: 69 Hartwell Av., Littleton, Mass. 01460. Office: S. Great Rd., Lincoln, Mass. 01773.*

BAIRD, John Jeffers, Am. zoologist; b. North English, Ia., Jan. 1, 1921; s. William S. and Ruth (Jeffers) B.; B.A., State Coll. Ia., 1948; M.S., State U. Ia., 1953, Ph.D., 1957; m. Geraldine Garner, Oct. 13, 1945; 1 dau., Stephanie Lynn. Tchr. high sch., Muscatine, Ia., 1948-54; faculty Cal. State Coll., Long Beach, 1956—, prof. biology, 1964—, chmn. dept. biology, 1961-67, dir. NSF Inst. for High Sch. Tchrs., 1958-65; asso. dean, acad. planning office, chancellor California State Colleges, 1967—. Mem. A.A.A.S., Am. Soc. Zoologists, American Inst. Biol. Sci., Naturalists, So. Cal. Acad. Sci., Am. Assn. U. Profs., Sigma Xi. Research on devel. motor cells in amphibian spinal cord. Home: 3239 W. Ravenswood Dr., Anaheim, Cal. 92804. Office: 5670 Wilshire Blvd., Los Angeles 90036.*

BAIRD, John Logie, Brit. inventor; b. Helensburgh, Scotland, Aug. 13, 1888; son of John and Jesse M. (Inglis) B.; ed. Royal Tech. Coll.; Glasgow Univ.; m. Margaret Albu, 1931; 1 son, 1 dau.; Supt. engring. Clyde Valley Electric Power Co.; Mfr. boot-polish and jam; later interested in TV. Cons. tech. adviser, Cable and Wireless, Ltd., 1941—. Fellow TV Soc., Phys. Soc., Royal Soc. Edinburgh (hon.). Recipient Gold medal Internat. Faculty Scis., 1937. Inventor noctovisor for viewing objects in dark with infrared light, 1926, mech.-scanning TV, 1926; demonstrated color, daylight, stereoscopic TV, 1928, 1st transatlantic TV, 1928; began regular TV service and transmission of synchronizing impulses, 1929, simultaneous transmission of vision and sound, big-screen TV, 1930; televised public event during day, 1931; demonstrated ultra-short wave transmission, 1932; demonstrated color television, 1939; studied projection of images on cinema screen; his research on infrared rays led to devel. of modern directional devices. Died Sussex, England, June 14, 1946.

BAIRD, John Wallace, psychologist; b. St. Marys, Ont., Can., May 21, 1873; s. Charles and Agnes (Browning) B.; A.B., U. Toronto, 1897; postgrad. U. Leipzig, 1898; fellow in psychology U. Wis., 1899-1901, Cornell U., 1901-02, Ph.D., 1902; m. Barbara Morrison Sparks, 1914. Instr. in psychology Cornell U., 1902-1903; research asst. in psychology Carnegie Found., 1903-04; instr. psychology Johns Hopkins, 1904-06; asst. prof. psychology U. Ill., 1906-10; asst. prof. psychology Clark, 1910-13, prof., 1913-19. Exec. editor Am. Jour. Psychology; coöperating editor Psychol. Bull., Jour. Applied Psychology, Jour. Ednl. Psychology. Died Feb. 2, 1919.

BAIRD, Spencer Fullerton, Am. zoologist, naturalist; b. Reading, Pa., Feb. 3, 1823; s. Samuel and Lydia (Biddle) B.; A.B., Dickinson Coll., 1840, M.A., 1843, Ph.D. (hon.), 1856; M.D. (hon.), Phila. Med. Coll., 1848; LL.D., Columbia, 1875; m. Mary Helen Churchill, 1846; 1 dau., Lucy Hunter. Prin. founder Marine Lab., Wood's Hole, Mass.; prof. natural history Dickinson Coll., 1846-50; made explorations for U. S. Govt. in Wyo. Territory, 1850-60; sec. A.A.A.S., 1850-51; mem. Nat. Acad. Sci., 1864; 1st U. S. commr. Fish and Fisheries, 1871; elected asst. sec. Smithsonian Instn., 1850, sec., 1878. Author: (with C. Girard) Catalogue of North American Reptiles, 1853; Catalogue of North American Mammals, 1857; Catalogue of North American Birds, 1858; Review of American Birds, 1864-66; editor Iconographic Ency., 1849; The Annual Record of Science and Industry, 1871-77; (with T. M. Brewer, R. Ridgway) A History of North American Birds, 1874; The Annual Reports of Smithsonian Instn., 1878-87; prepared Smithsonian Instn.'s Instructions to Collectors. Established the Marine Biol. Lab., Wood's Hole, Mass.; organized expedition of research ship Albatross; introduced the method of field study of botany and zoology in Am.; his books on birds brought about the Baird Sch. of ornithol. description, emphasizing extreme accuracy. Died Wood's Hole, Aug. 19, 1887.

BAIRE, (Louis) René, French mathematician; b. Paris, Jan. 21, 1874; student l'École Normale Supérieure, 1892-95. Named prof. Lycée Nancy (France), 1895; faculty lycées Troyes, Bar-le-Duc, France; became lectr. U. Montpellier (France), 1902; successor to Meray at Faculty Dijon (France), 1905-14. Corr. mem. French Acad. Scis. Author: Leçons sur les fonctions discontinues, 1904; Théories générales de l'analyse, 1907; Théorie des nombres irrationnels, 1912. Investigated irrational numbers; divided notion of continuity into upper and lower semi-continuity; helped establish theory of functions of real variables. Died July 5, 1932.

BAISCH, Karl, German gynecologist; b. Gaildorf, Germany, Jan. 28, 1869; s. Friedrich and Berta (Schütt) B.; studied theology and philology, U. Tübingen (Germany), 1887-90; student, Munich, medicine, Freiburg, Germany; degree and state examination; m. Julie Mayer, 1910; 1 dau.; m. 2d, Irma Käppner, 1919; 1 dau. Asst., Path. Inst., Tübingen, also U. Women's Clinic, Tübingen, then became asst. prof. gynecology and obstetrics, 1904, asso. prof., 1910; dir. City Clinic Stuttgart (Germany), from 1913. Author: Reformen in der Therapie des engen Beckens, 1907; Ergebnisse der Geburtshilfe und Gynäkologie, 1909; Enges Becken, Praktische Ergebnisse, 1909; Leitfaden der geburtshilflichen und gynäkologischen Untersuchungen, 1911; Gesundheitslehre für Frauen, 1916; Lehrbuch der Geburtshilfe, 1926. Research on gynecological ray therapy, bacteriological problems of birth. Died Stuttgart, Jan. 8, 1943.

BAISSAS, Henri Pierre Lucien Roget, French physicist; b. Albi, Oct. 13, 1899; s. Pierre and Angèle (Groc) B.; ed. Lycée at Albi and Toulouse, Faculté des Sciènces at Toulouse and Rennes; agrégé in phys. sci.; m. Madeleine Pichon, Mar. 31, 1928; children —Jacques, Philippe, Didier. Prof. physics at Saint-Quentin, Albi, Hoch-Versailles; headmaster at Auch, Valenciennes; insp. acad. at Auch, Arras, Paris; insp.-gen. of pub. instrn.; mem. tech. council to cabinet of Georges Guille; dir. Center Nuclear Studies of Fontenay-aux-Roses, 1957-59; dir. cabinet of high-commr. of atomic energy, 1959-62; dir. physics of C.E.A., 1962—. Mem. French Physics Soc., Soc. of Electricians. Contbr. articles on electromagnetic oscillations and magnetohydrodynamics. Home: 11 bis, rue Villebois-Mareuil, Paris 16. Office: C.E.A., 29-33 rue de la Fédération, Paris 15, France.

BAIZE, Paule Achille Ariel, French physician, astronomer; b. Paris, France, Mar. 11, 1901; s. Charles and Jeanne (Garcelle) B.; ed. Lycée de Coutances, U. Caen, U. Paris; M.D.; m. Miss Roguet; m. 2d, Miss Pupaiz; children—Claire, François, Louis, Denis, Philippe. Intern, Paris hosps.; chief clinic at Med. Soc. of Paris; concentrated on pediatrics and astronomy; astronomer Paris Obs., 1932—. Mem. Nat. French Com. of Astronomy, Internat. Astron. Union, Astron. Soc. France. Contbr. numerous articles on astronomy and medicine to profl. publs. Address: 6, rue Daubigny, Paris 17, France.

BAIZER, Manuel Mannheim, Am. chemist; b. Phila., May 20, 1914; s. Joseph and Bessie (Baum) B.; B.S., U. Pa., 1934, M.S., 1937, Ph.D., 1940; m. Mary Martha Meshkov, Feb. 5, 1939; children—Joan, Eric, Carol. Control chemist J. Eavenson, Camden, N.J., 1934-35; asst. Phila. Inst. for Med. Research, 1937-38 instr. Bklyn. Coll., 1941-47; research asso. NDRC, U. Pa., 1942-44; project leader Gen. Chem. C., N.Y.C., 1944-46; head research dept. N.Y. Quinine and Chem. Works, 1946-58; sr. research chemist Monsanto Co., Dayton, O., St. Louis, 1958-62, scientist, St. Louis, 1962—. Recipient award Naval Ordnance Devel., 1944. Fellow Am. Inst. Chemists; mem. Am. Chem. Soc., A.A.A.S., Am. Inst. Chemists, Electrochem. Soc., Sigma Xi, Delta Phi Alpha. Contbr. articles to tech. jours. Patentee electro-organic syntheses, synthesis medicinals, opium alkaloids. Home: 856 Mission Hills Ct., St. Louis 63141. Office: 800 N. Lindbergh Blvd., St. Louis 63166.*

BAJAJ, Ishwar Dass, physician; b. Lahore, Pakistan, May 18, 1925; s. Tulsi Ram and Kesri (Devi) B.; M.B.B.S., Lucknow U., India, 1950, M. Surgery, 1956; m. Jogindra Gaind, Oct. 10, 1953; children—Sndeep, Ajay, Anil. House surgeon, demonstrator Med. Coll. Indore, 1950-53; sr. demonstrator, lectr. K.G. Med. Coll., Lucknow, India, 1953-56; lectr. Lady Hardinge Med. Coll., New Delhi, India, 1956-58; faculty Maulana Azad Med. Coll., New Delhi, 1958—, prof., head anatomy dept., 1959—, dir.-prin., 1963; mem. Faculty Medicine, Delhi U., 1959—. U. S. AID fellow various med. centres, U. S., 1964. Mem. Anat. Soc. India (exec. com. 1960—), Assn. for Advancement Med. Edn. India, Delhi Med. Assn. Author: Epiphyseal Union in Bones, 1956; also articles. Research on bone growth, appearance and fusion of epiphyseal centers in Delhi population, prostate gland, age changes in human vermiform appendix and relationship with appendicitis, comparative study of human temporomandibular joint. Home: Warden's Bungalow, Bahadur Shah Zafar Marg, New Delhi, India.*

BAJER, Andrews, cell physiologist; b. Czestochowa, Poland, Jan. 3, 1928; s. Jan and Maria (Morawska) B.; M.A., Jagellonian U., Cracow, Poland 1949, Ph.D., 1950, D.Sc., 1956; m. Jadwiga Alina Molè, June 21, 1951; children—Marieta Joanna, Anna Maria. Faculty, Jagellonian U., 1949-63, research asso. prof., 1956-63; research asso. U. Lund (Sweden), 1963-64; asso. prof. U. Ore., Eugene, 1964—; research Strangeways Research Lab., Cambridge, Eng., 1957-58, Chester Beatty Research Inst., London, Eng., 1958-59, King's Coll., London, 1959 dept. biophysics U. Chgo., 1959, U. Upsalla (Sweden), 1961, Inst. Genetics, U. Lund, 1963-64. Recipient 2d Individual prize Polish Acad. Scis., 1961, Ross G. Harrison prize XI Internat. Congress Cell Biology, 1964; Gold Eagle award Council on Internat. Nontheatrical Events, 1966; NIH Career Devel. award, 1967—. Mem. Lunds Bot. Soc., Internat. Soc. for Cell Biology, Internat. Assn. for Sci. Film, Am. Soc. for Cell Biology, A.A.A.S. Research, publs. on analysis of chromosome movements during cell div. in living plant and animal cells using microcinematography as research tool, devel. technique for keeping endosperm cells alive and normal in vitro, combining studies in vitro with those of fine structure. Home: 2830 Elinor St., Eugene, Ore. 97403.*

BAJON, Bertrand, physician; student medicine Toulouse, Paris (both France), 1751-56; asst. surgeon to army of Rhine, 1760-62; asst. to Dr. Tenon, Salpetrière Hosp., Paris, 1763-64; army surgeon, French Guiana, 1764-76. Recipient Gold medal Royal Acad. Surgery, 1773. Mem. French Acad. Scis., 1774. Author: Mémoires pour servir à l'histoire de Cayenne et de la Guiane, 1777-78. Discovered rubber tree, Pará, Brazil; 1st to explore some regions of French Guiana; studied Guiana's geography, animal life; extensive work on diseases of area. Died 1790.

BAJRAKTAREVIC, Mahmud, Yugoslavian mathematician; b. Sarajevo, Yugoslavia, Dec. 22, 1909; s. Muharem and Hasa (Porca) B.; student Philosophical Faculty, Belgrade, Yugoslavia, 1929-33; Doctorat d'université, Sorbonne, Paris, France, 1953; Doctorat Konjhodzic, Jan. 8, 1942; children—Hasa, Nedzad. Tchr. math. and physics secondary schs., Sarajevo, Yugoslavia, 1934-46, 46-49; prof. math. Superior Pedagogical Sch., Sarajevo, 1946, prof. physics, 1949-50; faculty dept. math. Philos. Faculty, Institute for Physics and Chemistry, Sarajevo, 1950—, prof., 1962—. Recipient prize for sci. work, 1966, Work decoration of 3d order, 1957. Mem. Acad. Scis. and Arts Socialist Republic Bosna and Hercegovina, Sarajevo Soc. Mathematicians, Physicists and Astronomers Yugoslavia. Research, publs. on resolution of certain functional and integro-functional equations, solutions of properties problems of generalized quasilinear means connected with entropy of information theory, summability proceedings connected with Stirling polynomials. Home: 20 Nadmlini, Sarajevo, Yugoslavia.*

BAJUSZ, Eors, physician; b. Kecel, Hungary, July 15, 1926; s. Mihaly and Maria (Gyugel) B.; M.D., Med. U. Budapest (Hungary), 1950; Ph.D., U. Montreal (Que., Can.), 1961; m. Tereza Maria Demjen, June 29, 1963. Instr., lectr. U. Budapest, 1950-57; research asso. Inst Exptl. Medicine and Surgery, U. Montreal, 1957-61, asst. prof. exptl. pathology dept., 1961-64; asst. research dir. Bio-Research Inst., Cambridge, Mass., 1964—. Cons. to cardiovascular research unit dept. medicine U. Vt., Burlington, 1962-64. Recipient Pfeiffer prize for exptl. medicine, 1960. Fellow Am. Coll. Chest Physicians (mem. adv. com. on pathology 1964—), Am. Coll. Cardiology, Internat. Coll. Surgeons, N.Y. Acad. Scis., Internat., Am. (Honors Achievement award 1965) colls. angiology; mem. Am., Canadian physiol. socs., Am. Soc. for Exptl. Pathology, Internat. Acad. Pathology, Endocrine Soc. Author: Conditioning Factors for Cardiac Necrosis, 1963; Nutritional Aspects of Cardiovascular Diseases, 1965; also numerous articles, chpts. in textbooks. Editor: Major Problems in Neuroendocrinology, 1964; Electrolytes and Cardiovascular Diseases, 1965; An Introduction to Clinical Neuroendocrinology, 1966; Physiology and Pathology of Adaptatation Mechanisms, 1966; editor-in-chief; Jour. Neuroendocrinology, Yearbook Methods and Achievements in Exptl. Pathology. Research on exptl. techniques cardiovascular disease studies, hereditary or congestive heart failure, nutritional factors in heart diseases, enzymatic changes in skeletal muscle disorders, environmental physiology, neuroendocrinology. Home: 8 Whittier Pl., Charles River Park, Boston 02114. Office: Bio-Research Inst., 9 Commercial Av., Cambridge, Mass. 02141.*

BAK, Borge, Danish physicist; b. Copenhagen, Denmark, Dec. 31, 1912; s. Thorvald and Hilmaria (Nielsen) B.; ed. U. Copenhagen; Ph.D. phil.; m. Sonja Jacobsen, Oct. 12, 1939; children—Ole, Kirsten, Niels, Ulla. Asst. prof. U. Copenhagen, 1944; master conf. in spectroscopy, 1950; prof. chemistry, 1957. Mem. Danish Acad. Sci., Danish Acad. Tech. Sci. Author: The Intramolecular Potential; Elementary Introduction to Molecular Spectra, 1963, also numerous publs. on molecular structure. Home: Soager 18, Gentofte. Office: Universiteitsparken 5, Copenhagen, Denmark.

BAK, Thor A., Danish chemist; b. Aarhus, Denmark, Apr. 28, 1929; s. A. K. and Karen (Moller) B.; M.Sc., Tech. U. Denmark, 1953; Ph.D., Columbia, 1956; postgrad., Brussels, Belgium, 1957-58; Dr. phil., U. Copenhagen (Denmark), 1959; m. Kate Faber, Aug. 6, 1953; children—Kristian, Frans. Asso. with Bell Telephone Labs., also Gen. Electric Co., Schenectady; with U. Copenhagen, 1956—, prof. thecretical chemistry, 1963—. Founder Severinus Found., 1966; dir. H. C. Orsled Inst., 1967—. Mem. Royal Danish Acad. Scis. and Letters, Danish Acad. Tech. Scis., Am. Inst. Physics, Am., Danish (chmn. 1966-67) chem. socs., Faraday Soc. Author: Mathematics for Scientists, 1966; also textbooks on chemistry and math., papers on chem. kinetics and statis. mechanics. Home: 65 Egernvej, Copenhagen 2000, Denmark.*

BAKER, Arthur Alan, Am. geologist; b. New Britain, Conn., Oct. 31, 1897; s. Frank and Caroline (Goodbred) B.; Ph. B., Yale, 1919, Ph.D., 1931; m. Clara Edith Graves, Sept. 29, 1925; 1 dau., Carolyn (Mrs. Yelverton Cowherd). With U. S. Geol. Survey, 1921—, asso. dir., Washington, 1956—. Recipient Distinguished Service award Dept. Interior, 1960—. Mem. Geol. Soc. Am., Am. Assn. Petroleum Geologist, Am. Geophys. Union, Washington Acad. Sci., Sigma Xi. Numerous publs. on research on geologic structure; stratigraphy; econ. geology especially relating to Utah. Home: 5201 Westwood Dr, Washington 20016. Office: U. S. Gel. Survey, Washington 20242.*

BAKER, Arthur Challen, entomologist; b. Belleville, Ont., Can., Feb. 5, 1885; Ph.D., George Washington U., 1918; became mem. staff Fed. Bur. Entomology, 1911; in charge lab., Mexico City, Mexico, from 1955; mem. Entomology Soc., Soc. Tropical Medicine and Hygiene, Washington Acad. Sociedad Mexicana de Historia Natural. Studied Aphidae, aleyrodids; experimented with fruitfly insecticides, repellants. Died 1959.

BAKER, Arthur Latham, Am. mathematician; b. Cin., May 7, 1853; s. John G. and Mary A. (Latham) B.; C.E., Rensselaer Poly. Inst., 1873; postgrad. U. Gottingen, 1896; Ph.D., (hon.), Lafayette Coll., 1889; m. Elizabeth Coit Hand, Sept. 26, 1878; 1 dau. Dorothy (Mrs. J. Roy Allen). Adj. prof. civil engring. Lafayette Coll., 1873-80; atty. at law, Scranton, Pa., 1880-89; prin. high sch., Scranton, 1882; editor Common Pleas Reporter and Weekly Digest, Scranton, 1885-87; prof. math. Stevens High Sch., Hoboken, N.J., 1889-91, U. Rochester, 1891-1901; head dept. math. Manual Tng. High Sch., Bklyn., 1901-17. Author: Annual Digest Pennsylvania Supreme Court Decisions, 1886-87; Graphic Algebra, 1892; Elliptic Functions, 1890; Solid Geometry, 1893; Conic Sections, 1893; The Art of Geometry, 1905; Quaternions as the Result of Algebraic Operations, 1910; Elementary Thick-lens Optics, 1911; Micrometry for the Amateur Microscopist. Died 1934.

BAKER, Benjamin, Brit. engr.; b. Keyford Frome, Eng., Mar. 31, 1840; s. Benjamin and Sarah (Hollis) B.; apprentice Neath Abbey Ironworks, 1856-60; asst. to W. Wilson, London, 1860; mem. staff John Fowler, 1861-75, partner, 1875-98; cons. engr. Philae, Assyut, Aswan dams (all Egypt), City and S. London Line (1st tube ry.). Decorated knight comdr. St. Michael and St. George, knight comdr. Bath; recipient George Stephenson medal Instn. Civil Engrs., 1881, Prix Poncelet (for Forth Bridge) Inst. France. Fellow Royal Soc., 1890; member Institution Civil Engineers, British Association, Royal Institution, Institution Mechanical Engineers, Iron and Steel Institute. Constructed (with John Fowler) Forth Bridge (using cantilever principle which he helped develop), 1890, London Metropolitan Ry., Victoria Sta. Died Pangbourne, May 19, 1907.

BAKER, Benjamin May, Am. physician; b. Norfolk, Va., Nov. 20, 1901; s. Benjamin May and Theodosia (Potts) B.; B.S., U. Va., 1922; M.A. with 1st class honours in Physiology (Rhodes scholar 1922-25), Oxford (Eng.) U., 1925; M.D., Johns Hopkins, 1927; m. Julia Scott Clayton, Feb. 20, 1939; children—Susan Vaughan (Mrs. J. B. Powell), Julia May, Benjamin May III, William Clayton. Resident physician Johns Hopkins Hosp., 1930-31; prof. medicine Johns Hopkins Med. Sch., 1965—; vis. prof. medicine Guys Hosp., London, Eng., 1966; cons. medicine U. S. Army; spl. research ballistiocardiography, diet and coronary heart disease. Mem. exec. com., chief investigator Nat. Diet-Heart Study, 1961—. Dir. Anderson Clayton Co. Served to col., M.C., AUS, 1942-46; PTO. Decorated Legion of Merit; recipient Research grants Nat. Heart Inst., 1948-62. Mem. A.M.A., Am. Heart Assn., Am. Clin. and Climatological Assn., Am. Soc. Clin. Investigation, Assn. Am. Physicians, Phi Beta Kappa, Delta Kappa Epsilon, Alpha Tau Omega. Systematic studies on malaria, coronary heart disease, and ballistocardiography; a number of isolated clinical observations. Home: Brightside Rd., Balt. 21212.

BAKER, Bernard Randall, Am. chemist; b. Los Angeles, Nov. 24, 1915; s. Jacob and Anna (Simon) Sacks; A.B., U. Cal., Los Angeles, 1937; Ph.D., U. Ill., 1940; m. Reba Ruth Brodsky, Aug. 1, 1937; children—Sharon (Mrs. Michael Stack, Jr.), Bonnie (Mrs. Willard Meyers, Jr.), Maureen. Group leader Lederle Labs. div. Am. Cyanamid Co., 1941-54; dir. organic chem. div. So. Research Inst., 1955-56; dir. cancer themotherapy Stanford Research Inst., 1956-61; prof. medicinal chemistry State U. N.Y., Buffalo, 1961-65; prof. chemistry U. Cal., Santa Barbara, 1966—; cons. Nat Cancer Inst., 1959—; mem. Nat. Adv. Cancer Council, 1960-62, Gordon Research Conf. Council, 1964-66. Recipient Research award, Am. Pharm. Assn., 1963, Ebert prize, 1964; First award in medicinal chemistry Am. Chem. Soc., 1966. Mem. Am. Chem. Soc., Am. Pharm. Assn., N.Y. Acad. Sci., Am. Assn. Cancer Research. Author: Design of Enzyme Inhibitors, 1967. Contbr. numerous articles to profl. jours. Research on prins. of drug design at molecular biology level, synthesis of natural products of medicinal interest, such as nucleosides, alkaloids, antibiotics, vitamins, structure and chemistry of enzymes. Home: 4542 Carriage Hill Dr., Santa Barbara, Cal. 93105.*

BAKER, Bruce Earle, Canadian chemist; b. Stanbridge East., Que., Can., Aug. 1, 1917; s. Harry Arnold and Blanche (Soule) B.; B.Sc. with honours, Bishops U., Que., 1940; D.Sc., Laval U., Que., 1944; m. Saxe Clare Cornell, May 22, 1948; children—Peter Cornell, Susan Jane, Philip Bruce, Robert Saxe, Jeffrey Arnold. Research chemist Mallinckrodt Chems., Montreal, Que., 1944-45, Monsanto Can., Montreal, 1945-46; agrl. faculty McGill U., Montreal, 1946—, prof. chemistry, 1964—. Mem. Nat. Dairy Council Can., 1962—. Mem. Am. Chem. Soc., Am. Dairy Sci. Assn., Canadian Food Tech. Assn., Sigma Xi. Research and numerous publs. on milk and milk products, protein hydrolysates, leguminous seeds. Address: P.O. Box 208, Macdonald Coll., Que., Can.*

BAKER, Burton Lowell, Am. anatomist; b. Fife Lake, Mich., Apr. 2, 1912; s. Albert Wesley and Elizabeth (Ries) B.; A.B., Kalamazoo Coll., 1933, Sc.D., 1958; M.S., Kan. State Coll., 1935; Ph.D., Columbia U., 1941; m. Hazel M. Hicks, Sept. 1940; children—Linda Sue (Mrs. Jeffrey Frank), Gary Jay. Faculty, U. Mich., Ann Arbor, 1941—, prof. anatomy, 1952—; Upjohn research asso., 1946-52; Mich. Meml.-Phoenix fellow, 1953-54; mem. cell biology B study sect. NIH, 1966—. Recipient Henry Russell award U. Mich., 1947, Jones, Stone and Oakley prizes, Alumni citation Kalamazoo Coll., 1951. Fellow A.A.A.S., N.Y. Acad. Scis.; mem. Am. Assn. Anatomists, Soc. Exptl. Biology and Medicine, Am. Physiol. Soc., Endocrine Soc. Author: (with D. J. Ingle) Physiological and Therapeutic Effects of Corticotropin and Cortisone, 1953. Contbr. numerous articles to profl. jours. Research on hormonal control over prodn. of digestive enzymes, response of tissues to adrenocortical hormones. Home: 1020 Belmont St., Ann Arbor, Mich. 48104.*

BAKER, Carleton Harold, Am. physiologist; b. Utica, N.Y., Aug. 2, 1930; s. Harold G. and Loretta (Darling) B.; B.A., Utica Coll., 1952; M.A. Princeton, 1954, Ph.D., 1955; m. Sara Frances Johnson, July 20, 1963; children—Elizabeth Ann, Janet Lee.

Faculty, Med. Coll. Ga., Augusta, 1955-67, asso. prof. physiology, 1961-67; prof. U. Louisville Sch. Medicine, 1967—. Mem. Am. Physiol. Soc., Am., Ga. heart assns., Am. Assn. U. Profs., Sigma Xi. Research publs. in adrenal gland function, regulating circulatory fluids, blood vol. and flow. Home: 214 Wenham Way, Anchorage, Ky. 40223. Office: Univ. Louisville Sch. Medicine, Louisville 40202.*

BAKER, Charles Fuller, Am. entomologist; b. Lansing, Mich., Mar. 22, 1872; s. Joseph Stannard and Alice (Potter) B.; B.S., Mich. Agrl. Coll., 1892; M.S., Stanford, 1903; m. Ninette Evans, Aug. 29, 1894. Asst. in zoology Mich. Agrl. Coll., 1891-92; asst. to zoologist and entomologist Colo. Agrl. Coll., 1892-97; zoologist Ala. Poly., and entomologist Expt. Sta., 1897-99; tchr. biology Central High Sch., St. Louis, 1899-1901; asst. prof. biology Pomona Coll., Claremont, Cal., 1903-04, asso. prof., 1908-09, prof. biology, 1909-12; chief dept. botany Estacion Agron. de Cuba, 1904-07; curator Bot. Garden and Herbarium, Museu Goeldi, Para, Brazil, 1907-08; dir.-elect Campo de Cultura Experimental Paraense, 1908; prof. agronomy, U. Philippines, from 1912. In charge Colo. zool. and forestry exhibit, Chgo. Expn., 1893; zoologist and asso. botanist Ala. Biol. Survey, 1897-98; botanist H. H. Smith exploring expdn. in Santa Maria Mountains, Colombia, S.Am., 1898-99; also conducted field explns. in So. Ill., Wis., Colo., N.M., Nev., Cal., Nicaragua, Cuba and Brazil. Mem. A.A.-A.S., Am. Assn. Econ. Entomologists, Entomol. Soc. Am., So. Cal. Acad. Scis. Author: (with C. P. Gillete) A Preliminary List of Hermiptera; Invertebrata Pacifica, 1903. Named at least 12 insects, including Cicadella circellata (Baker) and ignotus Baker; many insects taken from all over the world today bear his name. Died July 22, 1927.

BAKER, Charles Whiting, Am. civil engr.; b. Johnson, Vt., Jan. 17, 1865; s. Thomas Jefferson and Mattie (Whiting) B.; ed. Vt. State Normal Sch.; C.E. U. Vt., 1886; m. Rebekah Wheeler, June 4, 1890. Editor-in-chief The Engineering News, 1895-1917; in real estate brokerage business, 1920-26; mem. Baker, Simonds & Co., investment bankers, from 1926. Commr. Palisades Interstate Park, from 1913; cons. govt. engring. projects; adviser to Goethals on Panama Canal. Author: Monopolies and the People, 1889, 99; Pathways Back to Prosperity. Died June 5, 1941.

BAKER, Crosby Fred, Am. chemist; b. Hampden, Me., Feb. 15, 1887; s. Fred Crosby and Cora Ida (Cole) B.; B.S., Tufts Coll., Mass., 1910, M.S., 1911; m. Ruth Ellingwood, June 15, 1914; children—Crosby F., Barbara (Mrs. Stuart McKenzie), Betsy R. Instr. in chemistry Tufts Coll., 1911-19, asst. prof., 1919-24, prof., 1924-46, Henry Bromfield Pearson prof. natural sci., 1946-49, Robinson prof., from 1949, sec. dept. chemistry and chem. engring., 1939-46, chmn. dept. chemistry from 1946; cons. metallurgist, Eastern Smelting & Refining Co., 1915-20; metall. patent work, 1920-30, cons. and chem. engr. textile and finishing oils, from 1930; cons. Pepsodent Co. Sugar Found., from 1945. Mem. Am. Chem. Soc., Am. Inst. Mining & Metall. Engrs., Soc. Chem. Industry (London), Am. Soc. for Metals, Boston Microchem. Soc., Phi Beta Kappa, Sigma Xi, Alpha Kappa Pi. Died Dec. 9, 1954.

BAKER, Dale Burdette, Am. chem. engr., b. Bucyrus, O., Sept. 19, 1920; s. Omar Burdette and Bessie (Mollencopf) B.; B.Ch.E., Ohio State U., 1942, M.S., 1948; m. Rosemary Jean Johnston, Aug. 17, 1947; children—William Burdette, Daniel Dale, James Jay. Chemist-supr. E. I. du Pont de Nemours & Co., St. Paul, Pompton Lakes, N.J., Gibbstown, N.J., 1942-46; with Chem. Abstracts Service, Ohio State U., Columbus, 1946—, asso. editor, 1951-57, asso. dir., 1958, dir. 1958—. Mem. Am. Chem. Soc., A.A.A.S., Am. Documentation Inst., Ohio Acad. Scis. Home: 64 W. Dunedin Rd., Columbus, Ohio 43214. Office: Chem. Abstracts Service, Ohio State U., Columbus, O. 43210.

BAKER, E(dgar) G(ates) Stanley, Am. biologist; b. Peotone, Ill., June 7, 1909; s. Walter S. and Hallie E. (Gates) B.; A.B. (Rector scholar) DePauw U., 1931; postgrad. (Rector fellow) U. Chgo., 1931-33; Ph.D., Stanford U., 1943; m. Julia E. Chapman, Dec. 29 1935; children—Ann (Mrs. Michael H. Siegel), Edgar C., James S. Faculty, Wabash Coll., 1932-39; asst. prof. biology Cath. U. Am., 1946-50; faculty Drew U., Madison, N.J., 1950—, prof., head dept. zoology, 1951—. Chief reader biology advanced placement program Coll. Entrance Exam. Bd., 1965-68. NSF Sci. Faculty fellow, 1957-58. Fellow A.A.A.S.; mem. Am. Soc. Zoologists, Soc. Protozoologists, N.Y. Acad. Scis., Am. Assn. U. Profs. Sigma Xi, Gamma Alpha, Beta Beta Beta (pres. 1967—). Studies on physiology of protozoan populations; bacteria free cultures; and effects of nutrition on growth rate. Home: 165 Green Village Rd., Madison, N.J. 07940.*

BAKER, Frank, Am. anatomist; b. Pulaski, N.Y., Aug. 22, 1841; s. Thomas C. and Sybil S. (Weed) B.; M.D., Columbian (now George Washington) U., 1880; A.M., Georgetown U., 1888, Ph.D., 1890, LL.D., 1914; m. May E. Cole, Sept. 13, 1873. Prof. anatomy, Georgetown U., from 1883; supt. Nat. Zool. Park, 1890-1916; asst. supt. U. S. Life Saving Service, 1889-90. Editor Am. Anthropologist, 1891-98. Contbr. anat. articles to Wood's Reference Handbook of the Med. Scis., Standard Dictionary, Internat. Cyclo. Died Sept. 30, 1918.

BAKER, Frank Collins, Am. zoologist; b. Warren, R.I., Dec. 14, 1867; s. Francis Edwin and Anna Collins (Thurber) B.; ed. Brown U., 1888; Jessup scholar Acad. Natural Scis., Phila., 1889-90; B.S., Chgo. Sch. of Sci., 1896; m. Lillian May Hall, June 16, 1892 (dec. Aug. 1934). Mem. Mexican exploring expdn. sent out by Acad. Natural Scis., 1890; invertebrate zoologist Ward's Natural Scis. Establishment and sec. Rochester Acad. Scis., 1891-92; curator zoology Field Columbian Mus., Chgo., 1894; curator Chgo. Acad. Scis., 1894-1915 (life mem.); zool. investigator N.Y. State Coll. of Forestry, Syracuse U., 1915-17; curator Natural History Mus., U. Ill., 1917-39; cons. invertebrate Pleistocene paleontology Ill. Geol. Survey; field zoologist Wis. Geol. and Natural History Survey, 1920-22, Ill. Natural History Survey, 1931-32. Fellow A.A.-A.S., Geol. Soc. Am., Paleontol. Soc. Am.; mem. Am. Assn. Museums, Museums Assn. (British), Ecol. Soc. Am., Am. Malacological Union (pres. 1942), Limnological Soc. Am., Audubon Soc. (v.p. 1900-15), Sigma Xi; corr. mem. Zool. Soc., London. Author: A Naturalist in Mexico, 1895; Mollusca of the Chicago Area, 1898-1902; Shells of Land and Water, 1903; The Lynmoeidae of North and Middle America, 1911; Relation of Mollusks to Fish in Oneida Lake, 1916; Life of the Pleistocene, 1920; Mollusca of Big Vermilion River (in relation to sewage pollution), 1922; Fresh Water Mollusca of Wisconsin, 1928; The Mollusca of the Shell Heaps or Escargotieres of Northern Algeria, 1939; Fieldbook of Illinois Land Mollusca, 1939; Use of Animal Life by the Mound-Builders of Illinois. Contbr. to zoöl. and geol. jours., principally on mollusca. Died May 7, 1942.

BAKER, Sir George, Brit. physician; b. Devonshire, Eng., 1722; s. vicar of Modbury, Devonshire, Eng.; grad. King's Coll., Cambridge (Eng.) U., 1745; M.D., 1756; fellow King's Coll., Cambridge (Eng.) U.; practiced at Stamford, England; settled in London, 1761; physician to king and queen. Fellow Royal Soc., 1762, Coll. Physicians (pres.). Author: Harveian Oration, 1761; On the Epidemic Influenza and Dysentery of 1762, 1764; Enquiry into the Merits of a Method of Inoculating the Small-pox, 1766; Medical Tracts, 1818. Discovered endemic colic in Devonshire was due to lead poisoning from lead lining in cider presses. Died 1809.

BAKER, George, Brit. surgeon; b. Eng., 1540; mem. Barber Surgeon's Co., elected master in 1597; attached to household of Earl of Oxford; practiced in London. Author: The Composition or Making of the Most Excellent and Precious Oil called Oleum Magistrate and the Third Book of Galen, 1574; The Newe Jewell of Health Wherein is Contained the Most Excellent Secretes of Physicke and Philosophie divided into fower bookes, 1576; Antidotarie of Select Medicine, 1579; Translator: Questions (Guido), 1579; Chirurgical Works (Vigo) 1586. Early proponent of chemically prepared medicines. Died 1600.

BAKER, George, geologist; b. Coventry, Eng., Oct. 10, 1908; s. William and Edith (Duggan) B.; B.Sc., U. Melbourne (Australia), 1933, M.Sc., 1937, D.Sc., 1956; m. Margaret Kathleen Chisholm, June 3, 1950. Evening lectr. Melbourne Tech. Coll., 1935-38; faculty Melbourne U., 1939-48, evening lectr., 1945-48, staff Commonwealth Sci. and Indsl. Research Orgn., 1948—, prin. research officer, 1957-62, sr. prin. research scientist, 1962—, acting officer-in-charge, 1960-61. Hon. Asso. mineralogy Nat. Mus. Victoria, 1956—. Recipient David Syme Research medal, 1944, Royal Soc. Victoria Research medal, 1961. Fellow Mineral. Soc. Am., Meteoritical Soc.; mem. Mineral. Soc. London (life), Royal Soc. Victoria (life), past asst. sec.), Am. Geophys. Union, Geol. Soc. Australia, Australian and New Zealand Assn. for Advancement Sci., Marine Scis. Assn., Australasian Inst. Mining and Metallurgy, Internat. Commn. on Meteorites. Author: Tektites, 1959; Detrital Heavy Minerals in Natural Accumulates, 1962; also numerous articles. Research in mineralogy and petrology, mine agraphy, sand drift relative to harbours and detrital heavy minerals. Australian lektites and meteorites, phytoliths, human urinary calculi; 1st to recognize selenides in Kalgoorlie gold ores, microspherular pyrite in Australian sulphideores and rodingite in No. to explain australite shapes. Home: 145 Booran Rd., Tasmania; propounded aerodynamical control theory Glenhuntly, S.E. 9, Victoria, Australia. Office: Commonwealth Sci. and Indsl. Research Orgn., Mineragraphic Investigations sect., U. Melbourne, Parkville, N. 2, Victoria, Australia.*

BAKER, George Allen, Am. math. statistician; b. Robinson, Ill., Oct. 31, 1903; s. Edward Sheridan and Ida (Everingham) B.; B.S., U. Ill., 1926, Ph.D., 1929; postgrad (Milbank Meml Fund Research fellow) Columbia; m. Grace Elizabeth Cummins, June 12, 1930; children—George Allen, John Cummins. Asso. statistician USPHS, Washington, 1929; prof. math. Shurtleff Coll., Alton, Ill., 1931-34; prof. math. Miss. Woman's Coll., Hattiesburg, 1934-36; statistician Dept. Agr., Bur Home Econ., Birmingham, Ala., also Washington, 1937; faculty U. Cal., Davis, 1937—, prof. math., statistician, 1955—, Faculty Research lectr., 1955-56. Fellow Inst. Math. Statistics, A.A.A.S.; mem. Biometric Soc. (regional v.p. 1950), Am. Math. Soc., Math. Assn. Am., Statis. Assn., Econometric Soc., Sigma Xi, Pi Mu Epsilon, Gamma Sigma Delta. Author: Statistical Techniques Based on Probalistic Models, 1962; numerous research papers articles. Home: 606 C St., Davis, Cal. 95616.

BAKER, George Allen, Jr., Am. physicist; b. Alton, Ill., Nov. 25, 1932; s. George Allen and Elizabeth (Cummins) B.; B.S., Cal. Inst. Tech., 1954; Ph.D., U. Cal. at Berkeley, 1956; m. Elizabeth Ann Coles, Sept. 9, 1956; children—Constance Jean, Linda Ann, Deborah Jane. NSF fellow Columbia, 1956-57; staff mem. Los Alamos Sci. Lab., 1957-66; asso. research physicist U. Cal., San Diego, La Jolla, 1961-62; vis. prof. theoretical physics Kings Coll. U. London, Eng., 1964-65; physicist Brookhaven Nat. Lab., Upton, N.Y., 1966—; cons. Gen. Dynamics Corp., San Diego, 1962, Bell Telephone Labs., Murray Hill, N.J., 1964. Fellow Am. Phys. Soc. Contbr. numerous articles to profl. jours. Research on Pade approximant method of approximate analytic continuation applied to investigation of analytic functions near their singular points; statis. mechanics, numerical analysis, quantum theory and nuclear physics. Office: Applied Math. Dept., Brookhaven Nat. Lab., Upton, N.Y. 11973.*

BAKER, Henry, English naturalist; b. London, Eng., May 8, 1698; s. William Baker; m. Sophia Defoe, Apr. 1729; children—David Erskin, Henry. With father-in-law, Daniel Defoe, brought out Universal Spectator and Weekly Jour., 1728-31; Fellow Royal Soc., 1740; recipient Copley gold medal for microscopic work on saline particles, 1744. Author: The Microscope Made Easy, 1743; Employment for the Microscope, 1753; also verse including philos. poem on universe. Introduced Alpine strawberry and rhubarb in Eng. Died London, Nov. 25, 1774.

BAKER, Henry Frederick, English mathematician; b. Cambridge, Eng., July 3, 1866; s. Henry and Sarah Anne Baker; ed. St. John's Coll., Cambridge; Sc.D.; LL.D., Edinbrugh; m. Lily Isabella Homfield Klopp, 1893 (dec. 1903); 2 sons; m. 2d, Muriel Irene Woodyard, 1913; 1 dau. Lowndean prof. astronomy and geometry Cambridge U., 1914-36; fellow St. John's Coll. Fellow Royal Soc., 1898. Author: Abel's Theorem and Theta Functions, 1897; Multiple-periodic Functions, 1907; Principles of Geometry, 6 vols., 1922-33; Plane Geometry, 1943. Editor: Mathematical Papers (Sylvester), 4 vols., 1904-12. Contbd. to theory of functions, differential equations, continuous groups; founded a thriving sch. of geometry. Died Mar. 17, 1956.

BAKER, Herbert Brereton, English chemist; b. Livesey, nr. Blackburn, Eng., June 25, 1862; s. John and Caroline (Slater) B.; ed. Balliol Coll., Oxford; LL.D., Aberdeen, 1926; m. Muriel Powell, 1905; 1 son, 1 dau. Demonstrator chemistry Balliol Coll., also pvt. asst. to H. B. Dixon, 1883-85; named chemistry master, head sci. side Dulwich Coll., 1886; became headmaster Alleyn's Sch., Dulwich, 1902; apptd. Lee's reader in chemistry Christ Ch., Oxford, 1904; chief prof. chemistry Imperial Coll. Sci. and Tech., South Kensington, Eng., 1912-23. Adviser on gas warfare to War Office. Recipient Davy medal 1923. Fellow Royal Soc., 1902; mem. Chem. Soc. (Longstaff medal 1912, pres. 1926). Research on influence of water on chem. change; studied nitrogen trioxide, atomic weight of tellurium, 1907, also desiccating gases, poison gases. Died Gerrads Cross, Eng., Apr. 27, 1935.

BAKER, Herbert George, biologist; b. Brighton, Eng., Feb. 23, 1920; s. Herbert Reginald and Alice (Bambridge) B.; B.Sc., London U., 1941, Ph.D., 1945; m. Irene Williams, Apr. 4, 1945; 1 dau., Ruth Elaine. Research chemist, asst. plant physiologist Hosa Research Labs., Sunbury, Eng., 1940-45; lectr. Leeds U., Eng., 1945-54; sr. lectr. U. Coll. Gold Coast (Ghana), 1954-55, prof., 1956-57; asso. prof. botany, dir. Bot. Garden, U. Cal. at Berkeley, 1957-60, prof., dir. Bot. Garden, 1960—. Research fellow Carnegie Instn. Washington, 1948-49. Fellow A.A.-A.S.; mem. Ecol. Soc. Am. (chmn. Western sect. 1965-67), Soc. for Study Evolution (pres. elect 1968), Orgn. for Tropical Studies (member exec. com.), Am. Inst. Biol. Scis. (bd. govs.), Sigma Xi. Author: Plants and Civilization, 1965; also numerous articles. Editor: (with G. L. Stebbins) The Genetics of Colonizing Species, 1965. Research on reproductive biology of plants and effect of this on evolution and ecology of individuals, populations, communities. Home: 635 Creston Rd., Berkeley, Cal. 94708.*

BAKER, Herman, Am. microbiologist; b. N.Y.C., Jan. 22, 1926; s. Harry and Fannie (Becker) B.; B.S., Coll. City N.Y., 1946; M.S., Emory U., 1948; Ph.D., N.Y. U., 1956; m. Shirley Levitz, Nov. 15, 1952; children—Elliott, Joel. Research asso. Haskins Labs., N.Y.C., 1949—, Mt. Sinai Hosp., N.Y.C., 1951—; asso. prof. medicine N.J. Coll. Medicine, 1960—; dir. vitamin metabolism Roosevelt Hosp., N.Y.C., 1963-65; dir. vitamin labs. N.J. Coll. Medicine, 1965—. Mem. Soc. Exptl. Biology and Medicine, Am. Soc. Clin. Nutrition, Sigma Xi. Publs. on introduction of sensitive and specific methods for measurement of vitamins in biologic fluids and tissues. Home: 27 Wilk Rd., Fords., N.J. 08663. Office: N.J. Coll. Medicine, East Orange VA Hosp., 88 Ross St., East Orange, N.J.*

BAKER, Horace Burrington, Am. zoologist; b. Sioux City, Ia., Jan. 25, 1889; s. Robert Folen and Sophia J. (Burrington) B.; B.S., U. Mich., 1910, Ph.D., 1920; m. Bernadine C. Barker, Dec. 21, 1941; children—Elizabeth Coffin, Abigail Burrington (Mrs. Richard Woodhull Smith). Faculty, Colo. Coll., 1913-17; instr. U. Mich., 1919; faculty U. Pa., 1920—, prof. zoology, 1939-59, emeritus prof., 1959—, acting chmn. dept., 1955-58, grad. chmn., 1958-59; research asso. Bishop Mus., Honolulu, 1937—. Acad. Nat. Sci. Phila. research fellow, 1925—. Mem. Am. Soc. Naturalists, Am. Soc. Zoologists, Am. Ecol. Soc., Am. Malacological Union (past pres.), Am. Soc. Systematics, Pa. Acad. Sci., London Malacological Soc., Deutsche Malacological Gesellschaft, Sigma Xi. Author: Zonitid Snails from Pacific Islands, 1938; also numerous articles. Editor for mollusks Biol. Abstracta, 1925—, Nautilus, since 1932—. Research on morphology and systematics mollusks. Home: 11 Chelten Rd., Havertown, Pa. 19083.*

BAKER, Ira Osborn, Am. civil engr.; b. Linton, Ind., Sept. 23, 1853; s. Hiram Walker and Amanda (Osborn) B.; B.S., U. Ill., 1874, C.E., 1877, D. Eng., 1903; m. Emma Burr, Aug. 5, 1877 (dec. 1911); m. 2d, Angie Ewing Ritter, Aug. 7, 1913. Asst. in civil engring. and physics U. Ill., 1874-78, instr. civil engring. 1878-80, asst. prof., 1880-82. prof., from 1882. Author: Leveling, 1886; Treatise on Masonry Construction, 1889, 99, 1909; Engineer's Surveying Instruments, 1891; Treatise on Roads and Pavements, 1903, 13, 18. Died Nov. 8, 1925.

BAKER, Irvine Noel, Brit. mathematician; b. Adelaide, Australia, Aug. 10, 1932; s. Alfred Irvine and Rosa (Clifford) B.; B.Sc., U. Adelaide, 1953, M.Sc., 1955; Dr. Rer. Nat., U. Tübingen, 1957; m. Dorothy Gillian Hawkins, June 14, 1958; children—Stephen, Michael. Asst. prof. U. Alta., Can., 1957-59; lectr. Imperial Coll. London, Eng., 1959-64, sr. lectr., 1965—, asst. dir. math. dept., 1964—. Mem. London, Austrian math. socs., Math. Assn. Am. Research, publs. on various aspects of functions of a complex variable, expecially iteration of functions. Home: 40 Combemartin Rd., London S.W. 18. Office: Math. Dept., Imperial Coll., London S.W. 7, Eng.*

BAKER, James Andrew, Am. animal virologist; b. Garland, La., Dec. 16, 1910; s. William Benjamin and Mary (Baldridge) B.; B.S., La. State U., 1932, M.A., 1934; Ph.D., Cornell U., 1938, D.V.M., 1940; m. Hallie Dudley Dodson, Nov. 27, 1934; 1 son, Andrew Lindsay. Fellow Rockefeller Inst. Med. Research, Princeton, N.J., 1940, asst, 1941, asso., 1946; prof. virology Cornell U., Ithaca, N.Y., 1947—, dir. Vet. Virus Research Inst., 1950—; cons. virology to hosps. Mem. adv. com. Office of Dir. of Def. Research and Engring., Dept. Def., 1962—. Named Veterinarian of Year, 1951, Dogdom's Man of Year 1956, Gaines Poll. Mem. Am. Vet. Med. Assn., N.Y. Acad. Sci., Soc. Exptl. Pathology, Soc. Exptl. Biology and Medicine, Soc. Microbiology, Acad. Microbiology, U. S. Livestock San. Assn., Conf. Research Workers. Research and publs. on isolations and basic finding in studies of animal viruses through devel. and use of facilities in which the virus, the host species, and factors influencing their relationships were all carefully controlled, theory proven that some viruses may remain latent or persist in tissues of body; also theory of heterotypic vaccines, in which group related viruses can confer a sensitivity against each other to provide protection from a specific disease. Address: Vet. Virus Research Inst., Cornell U., Ithaca, N.Y. 14850.*

BAKER, John Fleetwood, Brit. civil engr.; b. Wallasey, Eng., Mar. 19, 1901; s. Joseph William and Emily (Fleetwood) B.; B.A., Cambridge (Eng.) U., 1923, Sc.D., 1937; D.Sc., U. Wales, 1927; LL.D., U. Glasgow (Scotland), 1962; D.Sc., U. Leeds (Eng.), 1953, U. Manchester (Eng.), 1958, U. Edingburgh (Scotland), 1963, U. Aston, 1966; D.Eng., U. Liverpool (Eng.), 1960; D.S., U. Ghent (Belgium), 1964; m. Fiona M. M. Walker, July 18, 1928; children—Joanna (Mrs. David Park), Dinah (Mrs. Nigel Recordon) B. Designer, Royal Airship Works, Cardington, Eng., 1924-26; asst. lectr. U. Coll. Cardiff, 1926-28; tech. officer Steel Structures Research Com., 1928-36; prof. civil engring. U. Bristol, 1933-43; head dept. engring., prof. mech. scis. U. Cambridge, 1943—; sci. adviser Ministry Home Security in charge design and devel. sect., 1939-43. Named officer Order Brit. Empire, 1941; created knight, 1961. Mem. Royal Inst. Brit. Architects (asso.), Instn. Civil Engrs. (mem. council 1947-66), Instn. Mech. Engrs., Instn. Structural Engrs., Brit. Welding Research Assn. (mem. council 1943—). Author: The Steel Skeleton, vol. I, 1954, vol. II, 1956; (with P. Field Foster) Differential Equations of Engineering Science, 1929; (with A. J. S. Pippard) Analysis of Engineering Structures, 1936; also numerous articles. Developed method of elastic design for steel framed structures; research on plastic behavior of steel structures; developed plastic or collapse method of design; invented devices for absorbing energy by plastic deformation. Home: 100 Long Rd., Cambridge. Eng.*

BAKER, John Gilbert, Brit. botanist; b. Guisborough, Eng., Jan. 13, 1834; student Soc. of Friends schs., Ackworth, also York; m. Hannah Unthank, 1860; children—1 son, 1 dau. Became 1st asst. botanist Herbarium of Royal Gardens, Kew, 1866, keeper, 1890-99. Recipient Victoria medal Royal Horticultural Soc., 1897, gold medal Linnean Soc., 1899, Veitch gold medal for hort., 1907. Fellow Royal Soc., 1878, Linnean Soc.; mem. Royal Irish Acad. Author: North Yorkshire; Flora of English Lake District; Flora of Mauritius; Handbooks of Fern Allies, Iridea, Amaryllideae, and Bromeliaceae; Monograph of British Roses; Ferns and Compositae of Brazil; (with Sir W. J. Hooker) Synopsis Filicum. Died. Aug. 1920.

BAKER, John Randal, English zoologist; b. Oct. 23, 1900; s. Julian A. Baker; ed. New Coll., Oxford; M.A., D.Phil., D.Sc., Oxford U.; m. Inezita Davis, 1923; 1 son, 1 dau.; m. 2d, Mrs. Helen Savage. Reader in cytology Oxford U., named professorial fellow New Coll., 1964. Recipient Oliver Bird medal, 1958. Fellow Royal Soc., 1958; pres. Royal Micros. Soc., 1964. Author: Sex in Man and Animals, 1926; Man and Animals in the New Hebrides, 1929; Cytological Technique, 1933; The Chemical Control of Conception, 1935; The Scientific Life, 1942; Science and the Planned State, 1945; Abraham Trembley of Geneva, 1952; Principles of Biological Microtechnique, 1958. Co-editor Quar. Jour. Micros. Science. Sci. expdns. to New Hebrides, 1922-23, 27, 33-34; research on chem. contraception. Address: The Mill, Kidlington, Oxford, Eng.

BAKER, Kenneth Frank, Am. plant pathologist; b. Ashton, S.D., June 3, 1908; s. Frank and May (Boyer) B.; B.S., Wash. State U., 1930, Ph.D., 1934; m. Katharine Cummings, June 17, 1944. Jr. pathologist U. S. Dept. Agr., Lincoln, Neb., 1935-36; asso. pathologist Pineapple Research Inst., Honolulu, 1936-39; faculty plant pathology U. Cal. at Los Angeles, 1939—, prof., 1948-60, at Berkeley, 1960—. Recipient Award of Merit, Cal. State Florists Assn., 1956, Cal. Assn. Nurserymen, 1966; Norman J. Colman award Am. Assn. Nurseryman, 1959. Mem. Am. (pres. Pacific div. 1959-60), Netherlands phytopathology socs., Bot. Soc. Am., Brit. Assn. Applied Biology, Brit. Mycol. Soc., Mycol. Soc. Am., A.A.A.S., Am. Inst. Biol. Scis., Am. Soc. Hort. Sci., Sigma Xi. Editor: Phytopathology, 1958-60, Ann. Rev., 1962—. Author: Ecology of Soil-Borne Plant Pathogens, 1965. Research on transmission and control of plant infections emphasizing thermotherapy. Home: 999 Middlefield Rd., Berkeley, Cal. 94708.*

BAKER, Lynn E., Am. psychologist; b. Tacoma, Jan. 25, 1909; s. Ernest Mason and Olive (Erland) B.; B.A., U. Wis., 1933, M.A., 1933, Ph.D., 1937; m. Marion Bernice Lowe, Aug. 3, 1935; children—Ann Mason (Mrs. William J. Furmage III), Deborah Wessel, John James, Nicholas Daniel. Instr., U. Wis., 1937-40; U. S. Bur. Census psychologist, 1940-42; U. S. Pub. Housing Adminstrn. area 3 personnel dir., Washington, 1942-44; mgmt. analyst UNRRA, 1944-47, OWI, 1947-48; mgmt. cons. Records Engring. Inc., Washington, 1948-49; research advisor, dir. manpower research USAF, 1949-53; chief psychologist Office Chief Research and Devel. U. S. Army, Washington, 1953—. Diplomate Bd. Indsl. Psychology. Fellow Am. Psychol. Assn. Home: 4324 Loyola Av., Alexandria, Va. 22304. Office: Highland Bldg., Columbia Pike, Arlington, Va.*

BAKER, Marcus, Am. geographer; b. Kalamazoo, Mich., Sept. 23, 1849; s. John and Chastina (Fobes) B.; grad. U. Mich., 1870; LL.B., Columbian U., 1896; m. Marian Strong, May 1899. Connected with U. S. Coast and Geodetic Survey, 1873-86; and with U. S. Geol. Survey, from 1886; spent several yrs. in explorations and surveys in Alaska and on Pacific coast; with William H. Dall, prepared the Alaska Coast Pilot. Sec. U. S. Bd. on Geographic Names; was cartographer Venezuelan Boundary Commn. Asst. sec. Carnegie Instn. of Washington. Author: Dictionary of Alaskan Geographic Names; Northwest Boundary of Texas; Survey of Northwestern Boundary of United States; and other bulletins and geog. and math papers. Died. 1903.

BAKER, Oliver Edwin, Am. econ. and sociol. geographer; b. Tiffin, O., Sept. 10, 1883; s. Edwin and Martha (Thomas) B.; B.Sc., Heidelberg Coll., Ohio, 1903, M.Sc., 1904, D.Sc., 1937; M.A., in Polit. Sci., Columbia, 1905; studied forestry, Yale U., 1907-08, agr., U. Wis., 1908-12, economics, 1919-21, Ph.D., 1921; Ph.D. (hon.), Göttingen (Germany), 1937; m. Alice H. Crew, 1925; children—Helen Thomas, Sabra Z., Edwin Crew, Mildred Coale. With Wis. Agrl. Expt. Sta., 1910-12; U. S. Dept. 1912-42; employed in research on farm population and on rural youth surveys, 1930-42; in charge of preparation, and editor of Atlas of Am. Agr., 1914-36; prof. geography, U. Md., from 1942, also in charge preparation Econ. Atlas of World. Mem. Assn. Am. Geographers (pres. 1931), Am. Meterol. Soc., Farm Econ. Assn., Am. Sociol. Soc. Author: (with A. R. Whitson) The Climate of Wisconsin and Its Relation to Agriculture, 1912; (with V. C. Finch) Geography of the World's Agriculture, 1917; (with M. L. Wilson and Ralph Borsodi) Agriculture and Modern Life, 1939; also edited Atlas Am. Agr. Contbr. to U. S. Dept. Agr. Year Books, 1915-38; and to geographic publs. Died Dec. 2, 1949.

BAKER, Paul Thornell, Am. anthropologist; b. Burlington, Ia., Feb. 28, 1927; s. Palmer Ward and Viola (Thornell) B.; B.A., U. N.M., 1951; Ph.D., Harvard, 1956; m. Thelma Marion Shoher, Feb. 21, 1949; children—Deborah Carol, Amy Laurel, Joshua Shoher, Felicia Beth. Research phys. anthropologist Q.M. Research and Devel. Center, 1952-57; research asso. biophysics Pa. State U., Univ. Park, 1957-58, faculty, 1958—, prof. anthropology, 1964—, acting head dept., 1964-65, dir. Andean Biocultural studies, 1964—. Cons. in phys. anthropology HRB-Singer, 1959—; mem. behavorial scis. fellowships rev. com. NIH, U. S. Internat. Biol. Program Human Adaptability Subcom.) Fulbright research scholar, Peru, 1962; Fulbright lectr., Brazil, 1962. Fellow Am. Anthrop. Assn.; mem. Am. Assn. Phys. Anthropologists (exec. com. 1962-65), A.A.A.S. Author: Evolution and Man, 1961; The Biology of Human Adaptability (with J. S. Weiner), 1966; also numerous articles. Research in human evolution biol. adaptation of man to heat, cold, altitude, nutrition, design of clothing and equipment to variation in human size. Home: Box 115E, R.D. 1, Bellefonte, Pa. 16823. Office: Pa. State U., University Park, Pa. 16802.*

BAKER, Perley Dustin, Am. chemist, educator; b. Bow, N.H., May 8, 1897; s. Rufus H. and Grace (Tuck) B.; B.S., Norwich U., 1920, D.Sc., 1956; A.M., Columbia, 1926, Ph.D., 1940; m. L. May Lloyd, Apr. 29, 1918. Faculty chemistry Norwich U., Northfield, Vt., 1920—, prof., 1939-50 head dept. 1946-50, 56-62, dean, 1950-56, prof. emeritus, 1962—. Cons. Office Civilian Def., Boston, 1941-42. Mem. Am. Chem. Soc. (chmn., councilor Vt. sect. 1953-60), Sigma Xi, Phi Lambda Upsilon, Epsilon Tau Sigma, Theta Chi. Research on sorption by colloids. Address: 18 Crescent Av., Northfield, Vt. 05663.*

BAKER, Philip Schaffner, Am. chemist; b. Mpls., Nov. 30, 1916; s. Ross Allen and Helen (Porter) B.; student Occidental Coll., 1933-34, Antioch Coll., 1934-36; A.B., DePauw U., 1938; M.A. U. Ark., 1939; Ph.D., U. Ill., 1943; m. Marilyn Francis Jarvis, Mar. 11, 1944; children—Ross Jarvis, Jean Carol, Susan Lynne. Research chemist PanAm. Refining Corp., Texas City, Tex., 1943-45, Inst. Paper Chemistry, Appleton, Wis., 1945-46; asst. prof. chemistry U. Vt., 1946-48; asso. prof. chemistry Bradley U., 1948-52; with Oak Ridge Nat. Lab., 1952—, asst. supt. Isotopes Devel. Center, dir. Isotopes Information Center, 1962—. Fellow Am. Inst. Chemists, A.A.A.S.; mem. Am. Chem. Soc., Am. Nuclear Soc., N.Y. Tenn. acads. sci., Sigma Xi. Author: (with G. M. Bradbury, J. W. Eichinger, E. A. Sigler) Chemistry and You, 1966; numerous articles. Editor: Isotopes and Radiation Technology. Research isotope prodn., applications; patentee preparation of lithium metal. Home: 104 Euclid Pl. Office: P.O. Box X, Bldg. 3047, Oak Ridge 37830.*

BAKER, Ralph, Am. biologist; b. Houston, Aug. 31, 1924; s. Reginald R. and Eleanora (Weiss) B.; B.Art and Sci., Colo. State U., 1948, M.S., 1940; Ph.D., U. Cal. at Berkeley, 1954; m. Eleanor Joanne Damer, June 19, 1965; children—Jack, Kit, Jennifer, Sean. Prof. botany, plant pathology Colo. State U., Ft. Collins, 1954—, asst. dir. Research Found., 1962-64; vis. prof. U. Cal. at Berkeley, 1963-64. Mem. com. on biol. control soil borne plant pathogens NRC, 1959-64. Recipient Florists's Mut. award for research on ornamental diseases, 1959, Pennock Distinguished Service award Colo. State U, 1965. Fellow A.A.A.S.; mem. Am. Phytopath. Soc., Am. Inst. Biol. Sci. (chmn. com. for MORL expts. 1965—), Sigma Xi. Author: (with W. D. Holley) Carnation Production, 1963; also articles. Research in biol. control plant pathogens in soil, diseases of ornamentals, math. biology, physiology of sexual reprodn. in fungi, control plant galls with animal anti-tumor agts., space biology. Home: 1216 Southridge Dr., Ft. Collins, Colo. 80521.*

BAKER, Richard Dean, Am. chemist; b. Hot Springs, S.D., June 9, 1913; s. Allen Henry and Emma B.; B.S., S.D. Sch. Mines and Tech., 1936; Ph.D., Ia. State U., 1941; m. Bonnie May Jourdan, Feb. 20, 1946; 1 son, Richard J. Research chemist U. S. Gypsum, Chgo., 1941-42; group leader Los Alamos Sci. Lab., 1943-56, div. leader, 1956—. Mem. Am. Chem. Soc. Research and publs. on plutonium and uranium chemistry. Home: 1999 Juniper St. Office: Los Alamos Sci. Lab., P.O. Box 1663, Los Alamos 87544.*

BAKER, Rollin Harold, Am. vertebrate zoologist; b. Cordova, Ill., Nov. 11, 1916; s. Charles Laurence and Minnie (Perkins) B.; B.A., U. Tex., 1937; M.S., Tex. A. and M.U., 1938; Ph.D., U. Kan., 1948; m. Mary Elizabeth Waddell, Mar. 21, 1939; children—Elizabeth Alice, Bruce Rollin, Byron Laurence. Field biologist Tex. Coop. Wildlife Research Unit, Coll. Sta., 1938-39; wildlife biologist Tex. Game and Fish Commn., Austin, 1939-43; asst. instr. zoology U. Kan., Lawrence, 1946-48, asst. prof., 1948-54, asso. prof., 1954-55, asst. curator Mus. Natural History, 1948-55, acting dir., 1950-51; dir. Mus., prof. zoology, fisheries and Wildlife Mich. State U., East Lansing, 1955—. Mem., Sigma Xi, Phi Sigma, Phi Kappa Phi, Alpha Epsilon Delta, Beta Beta Beta, Tau Kappa Epsilon. Research, publs. on classification, distbn., ecology of mammals, other vertebrates; discovered new animals; explorations in Latin Am., Mexico, Pacific Area. Home: 420 W. Grand River Av., East Lansing, Mich. 48823.*

BAKER, S. Josephine, Am. physician; b. Poughkeepsie, N.Y., Nov. 15, 1873; d. Orlando D. M. and Jennie Harwood (Brown) Baker; M.D., Woman's Med.

Coll., N.Y. Infirmary, 1898; Dr.P.H., Bellevue Med. Coll., N.Y. U., 1917. Asst. to commr. of health N.Y.C., 1907-08; dir. Bur. Child Hygiene, Dept. of Health, 1908-23; lectr. child hygiene Columbia U., N.Y. U.; cons. Children's Bur. of U. S. Dept. Labor, U. S. P.H.S., and various other orgns.; mem. bd. and cons. pediatrician, Clinton (N.J.) Reformatory for Women. mem. health com. League of Nations. Hon. pres. Children's Welfare Fedn. Trustee N.Y. Infirmary for Women and Children. Fellow A.M.A., Am. Pub. Health Assn., New York Acad. Medicine; mem. Am. Child Health Assn. (pres.) Am. Women's Med. Assn. (pres.); asso. mem. Am. Acad. Pediatrics. Author: Healthy Mothers, 1923; Healthy Babies, 1923; Healthy Children, 1923; Child Hygiene, 1925; Fighting for Life, 1939. Organized 1st bur. of child hygiene under govt. control in N.Y.C., leading to lowest baby death rate of any large city in America or Europe. Died Feb. 22, 1945.

BAKER, Samuel White, English engr., explorer; b. London, Eng., June 8, 1821; trained as engr.; M.A. (honorary), Cambridge (England) U., 1866; journeyed to Ceylon, 1845; founded agricultural settlement at Nuwara Eliya, Ceylon, 1848; supr. construction of railroad from Danube River to Black Sea, 1859-60; explored Nile tributary in Ethiopia, 1861-62; discovered Lake Albert, 1864; comm comdr. Egyptian expdn. in Central Africa, 1860-73; explored Cyprus, 1879; visited Syria, India, Japan, Am. Created knight, 1866; gold medal Royal Geog. Soc. Fellow Royal Soc., 1869. Author: The Rifle and the Hound in Ceylon, 1854; Eight Years Wandering in Ceylon, 1855; The Albert Nyanza, 1866; The Nile Tributaries of Abyssinia, 1867; Ismailia, 1874; Cyprus as I Saw It, 1879; Wild Beasts and Their Ways, 1890. Died nr. Newton Abbott, Devonshire, Eng., Dec. 30, 1893.

BAKER, Saul Phillip, Am. physician; b. Cleve., Dec. 7, 1924; s. Barnet and Florence (Kleinman) B.; B.S. in Physics, Case Inst. Tech., 1945; postgrad. Western Res. U.; M.S., in Physiology, Ohio State U., 1949, M.D., 1953, Ph.D., 1957. Sr. asst. surgeon gerontology br. Nat. Heart Inst., NIH, USPHS, Balt. City Hosps., also Johns Hopkins Hosp., 1954-56; asst. prof. medicine Chgo. Med. Sch., 1957-62; asso. prof. medicine Cook County Hosp. Grad. Sch. Medicine, 1958-62; practice medicine specializing in internal medicine, geriatrics, cardiology, Cleve., 1962—; head dept. geriatrics St. Vincent Charity Hosp., Cleve., 1964—. Cons. internal medicine disability determination sect. Old Age and Survivors Ins., Social Security Adminstrn., 1963—; cons. Bur. Workmen's Compensation, State of Ohio, 1964—. Fellow Am. Coll. Cardiology, Am. Geriatrics Soc., Gerontology Soc., A.A.A.S., Council on Arteriosclerosis, Am. Heart Assn.; mem. Central Soc. for clin. Research, Fedn. for Clin. Research, Soc. for Exptl. Biology and Medicine, Am. Phystol. Soc., N.Y. Acad. Sci. Research, publs. on test for atherosclerosis, lipid metabolism, thyroid function, basal oxygen consumption, body water compartments in aging, coronary heart disease, physiology of aging, geriatrics. Home: 2300 Overlook Rd., Cleveland Heights, O. 44106. Office: 14077 Cedar Rd., Cleve. (South Euclid, O.) 44118.*

BAKER, Thomas, Brit. mathematician; b. circa 1625; became battler Magdalen, Hall, Oxford, Eng., 1640; collated to vicarage of Bishop's Nymton, 1681. Fellow Royal Soc., 1684. Author: Geometrical Key, or Gate of Equations Unlocked, 1684. Solved biquadratic equations by geometrical constrn.

BAKER, Walter Wolf, Am. neuropharmacologist; b. Phila., Oct. 29, 1924; s. Jacob and Ethel R. (Brodsky) B.; B.S., Franklin and Marshall Coll., 1948; M.S., State U. Ia., 1950; Ph.D., Jefferson Med. Coll., 1953; m. Judith Marion Goldman, Aug. 3, 1952; children—Corinne Marcia, Andrew Wayne, Frederic Jay. Faculty, Jefferson Med. Coll., Phila., 1953—, asso. prof. pharmacology and psychiatry, 1959—; lectr. neuropharmacology Phila. Coll. Pharmacy and Sci., 1959—; dir. neuropharmacology Eastern Pa. Psychiat. Inst., 1963—; cons., lectr. Del. State Hosp.; cons. VA Hosp., Coatsville, Pa., Squibb Inst., New Brunswick, N.J. Mem. A.A.A.S., Soc. Biol. Psychiatrists, Assn. Research Nervous and Mental Diseases, Am. Soc. Pharmacology and Exptl. Therapeutics, Eastern Assn. Electroencephalography, Acad. Neurology, Soc. Exptl. Biology and Medicine, Phi Beta Kappa. Contbr. chpt. to Progress in Neurology and Psychiatry, 1959-62. Research, publs. on drugs, chems. used experimentally in animals to upset critically balanced neurotransmitter levels and functions in key brain areas; elec. activities of these malfunctioning brains when analyzed offer an electrochem. basis for understanding mechanisms underlying such clin. brain disorders as involuntary movements, epilepsy and abnormal behavior. Home: 905 Stratford Av., Melrose Park, Pa. 19126. Office: Henry Av. and Abbottsford Rd., Phila. 19129.*

BAKER, Will. C., Canadian physicist; b. ed. Queen's U., Kingston, Ont., also Cambridge (Eng.) U.; demonstrator in physics Queen's U., 1902-04, lectr., 1904-06, asst. prof., 1906-11, asso. prof., 1911-20, prof. physics, from 1920. Studied Hall effect in gold for low fields, also absorption of hydrogen by metals. Died 1940.

BAKER, William Hudson, Am. botanist, educator; b. Portland, Ore., Dec. 16, 1911; s. William T. and Helen T. (Hudson) B.; B.S., Ore. State U., 1935, M.S., 1942, Ph.D., 1949; m. Molly Ann Cochran, Nov. 17, 1934; 1 son, James William. Adminstr. pub. schs., Ore., 1935-42; asst. in botany Ore. State U., 1946-48; faculty U. Ida., Moscow, 1948—, prof. botany, 1958—, acting chmn. dept. botany, 1953-54, chmn., 1954—, acting head dept. biol. scis., 1955-56, head, 1956—. Ranger, naturalist Crater Lake Nat. Park, summers 1949-50. Fellow A.A.A.S. (council 1964); mem. Internat. Soc. Plant Taxonomists, Internat. Soc. Plant Biosystematists, Bot. Soc. Am., Am. Fern Soc., Am. Assn. U. Profs., Northwest Sci. Assn. (pres. 1963), Cal. Bot. Soc., Ida. Acad. Sci. (pres. 1959-60), New Eng. Bot. Club, Sigma Xi (pres. Ida. 1955-56), Phi Sigma Alpha Tau Omega. Author: (with others) Wildlife of the Northern Rocky Mountains, 1961. Research on floristics and plant distbn., flowering plants Northwestern Am., western grasses and ferns. Home: 403 N. Blaine St., Moscow, Ida. 83843.*

BAKER, William Kaufman, Am. geneticist; b. Portland, Ind., Dec. 2, 1919; s. Frank K. and Jennie (Shaeffer) B.; B.A., Coll. Wooster, 1941; Ph.D., U. Tex., 1948, M.S., 1943; m. Margaret Stewart, Mar. 4, 1944; children—Bruce S., Ann K., Brian D. Asst. prof. U. Tenn., 1948-51; sr. biologist Oak Ridge Nat. Lab., 1951-55; asso. prof. U. Chgo., 1955-58, prof. genetics, 1958—. Sr. fellow NSF, 1963-64. Mem. Genetics Soc. Am., A.A.A.S., Soc. for Study Evolution, Am. Soc. Naturalists (past sec.), Am. Soc. Zoologists, Phi Beta Kappa. Author: Genetic Analysis, 1965; also numerous articles. Co-editor: Am. Naturalist, 1965—. Research in genetics Drosophila, effects temperature and oxygen concentration on radiation-induced genetic changes, mechanism crossing over, basis heterochromatin and variegation, developmental genetics Drosophila. Home: 5505 S. Kenwood Av., Chgo. 60637.*

BAKER, William Morrant, English surgeon; b. London, Eng., Oct. 20, 1839; ed. St. Bartholomew's Hosp. Med. Sch.; m. Annie Mills, 1868; 2 sons, 4 daus. Apprentice to surgeon Payne; became resident in midwifery, also asst. to Dr. West, 1861; became asst. to Dr. James Paget, 1862; temporary head Martha Ward and maternity dept. St. Bartholomew's Hosp., apptd. demonstrator anatomy, physiology, 1865, elected warden, sec. to com. med. officers and lectrs., 1867, also gov.; lectr. physiology and gen. anatomy, 1869-1885; apptd. casualty surgeon, 1870; became asst. surgeon St. Bartholomew's Hosp., 1871, later in charge surg. out-patient room, charge skin dept., from 1875; examiner physiology bd. Royal Coll., 1878-87; full surgeon, 1882-92; examiner surgery, 1887-92; surgeon Bartholomew Close Dispensary, Evelina Hosp. for Children; examiner in surgery U. London, Durham (Eng.) U. Fellow Royal Med. and Chirurg. Soc.; mem. Path. Soc. London. Research and publs. on cysts and tracheotomy; described swelling behind knee known as popliteal bursitis or Baker's cyst, infectious dermatitis (erysipeloid). Died Pulbourgh, Sussex, Eng., Dec. 3, 1896.

BAKER, William Oliver, Am. research chemist; b. Chestertown, Md., July 15, 1915; s. Harold M. and Helen (Stokes) B.; B.S., Washington Coll., Chestertown, Md., 1935, Sc.D., 1957; Ph.D., Princeton, 1938; D.Eng., Stevens Inst. Tech., 1962; Sc.D., Georgetown U., 1962, U. Pitts., 1963, Seton Hall U., 1965; LL.D., U. Glasgow, 1965; m. Frances Burrill, Nov. 15, 1941; 1 son, Joseph Burrill. With Bell Telephone Labs., Murray Hill, N.J., 1939—, in charge polymer research and devel., 1948-51, asst. dir. chem. and metall. research, 1951-54, dir. research in phys. scis., 1954-55, v.p. research, 1955—. Trustee Aerospace Corp.; dir. Summit Trust Co., Babcock & Wilcox Corp. Bd. visitors USAF Systems Comd.; mem. Nat. Sci. Bd.; mem. Presidents Sci. Adv. Com., 1957-60, President's Fgn. Intelligence Adv. Bd., 1959—. Trustee Mellon Inst., Rockefeller Inst., Princeton U. Nat. Insts. Health lectr., 1958; recipient Am. Inst. Chemists Honor Scroll, 1962, Perkin medal Soc. Chem. Industry, 1963. Mem. Nat. Acad. Scis., Sci. Research Soc. Am. (bd. govs. 1961—), Sigma Xi, Omicron Delta Kappa. Contbr. to articles and books in field. Discoverer of new synthetic polymer, microgel, electronically active films and fibers; research in concept of organic ablative heat shields for missiles and space vehicles. Home: Spring Valley Rd., Morristown, N.J. 07960. Office: Bell Telephone Labs., Murray Hill, N.J. 07971.*

BAKER, Wilson, English organic chemist; b. Jan. 24, 1900; s. Harry and Mary B.; ed. Victoria U., Manchester, Eng.; B.Sc., M.Sc., Ph.D., D.Sc., U. Manchester; M.A., Oxofrd (Eng.) U.; m. Juliet Elizabeth Glaisyer, 1927; 3 children. Asst. lectr. chemistry U. Manchester, 1924-27; lectr., demonstrator chemistry Oxford U., 1927-44, fellow, praelector chemistry Queen's Coll., 1937-44; Alfred Capper Pass prof. Organic Chemistry U. Bristol (Eng.), from 1945, dean faculty sci., 1948-51, prof. emeritus, from 1965. Fellow Royal Soc., 1946, Royal Inst. Chemistry; mem. Chem. Soc. (v.p. 1957-60), Soc. Chem. Industry. Editor: (with T. W. J. Taylor) The Organic Chemistry of Nitrogen (N. V. Sidgwick), 2d edit., 1937. Research, publs. on organic chemistry including synthesis natural products, devel. synthetical processes, abnormal aromatic compounds, aromatic compounds such as non-benzenoid type, organic inclusion compounds, preparation large-ring compounds. Address: Sch. Chemistry, U. Bristol 2, Lane's End, Church Rd., Winscombe, Somerset, Eng.

BAKETEL, H. Sheridan, Am. physician; b. Hopedale, O., Nov. 15, 1872; s. Oliver Sherman and Rosie Lueretia (Mack) B.; student Phillips Exeter Acad., Boston U.; M.D., Dartmouth, 1895; postgrad. Harvard, also abroad; A.M., Holston Coll., 1908; m. Zada Call: m. 2d, Corinne Phillippi Sellers; children—Mary, H. Sheridan. Editor Gaillard's Med. Jour., 1905-08, Med. Times, 1908-15; began practice urology. N.Y., 1910; became lectr. med. econs. L. I. Coll. Medicine, Bklyn., 1915, prof. preventive medicine, 1915-31, emeritus prof., 1931— (trans. to med. coll. State U. Med. Center of N.Y.C.); co-founder editor-in-chief Med. Economics, from 1923; pres. physiol. labs. Reed & Carnrick, Jersey City, 1925-51; chmn. Reed & Carnrick Inst. for Med. Research, 1946-51; urologist to N.Y. hosps. Del. to U. S. Pharm. Conv., 1930-40. Fellow A.C.P., Am. Pub. Health Assn., N.Y. Acad. Medicine; pres. Am. Pharm. Mfg. Assn. 1929-31; mem. A.M.A., Am. Urol. Assn. (life), A.M.A., Am. Med. Editors Assn. (pres. 1920), N.H. Med. Soc., Newcomen Soc. Author: The Treatment of Syphilis, 1920; also monographs. Died July 7, 1955.

BAKEWELL, Robert, Brit. agriculturist, stockbreeder; b. Dishley, Eng., 1725; took charge of father's farm, 1755; pioneer in practice systematic inbreeding of stock; improved stock, chiefly long-horned cattle and sheep called by his name; produced Leicestershire sheep, Dishley, or New Leicestershire Longhorn cattle; 1st to establish large scale letting of rams for breeding. Died Oct. 1, 1795.

BAKEWELL, Robert, Brit. geologist; b. 1768; children—Robert, Frederick C.; lectured and exhibited sects. of rock formation and geol. map throughout country beginning in 1811; geol. instr.; visited Tarentaise, Graicin, and Pennine Alps in Switzerland and Auvergne, 1820-22. Author: Observations on Wool, 1808; Introduction to Geology, 1813; Introduction to Mineralogy, 1819; Observations on the Geology of Northumberland and Durham; Travels, 1823; Salt; Lava at Boulogne; Thermal Waters of the Alps; Mantell's Collection of Fossils; Pollen of Plants; Organic Life; Gold Mines in U.S., 1832; Fossil Elephants in Norfolk, 1835. Invented safety furnace for preventing explosions in coal mines; discovered fine scenite in large blocks in Charnwood Forest, 1812. Died Aug. 15, 1843.

BAKHMETEFF, Boris Alexander, civil engr.; b. Tiflis, Russia, May 14, 1880; s. Alexander Paul and Julia (Novitsky) B.; grad. Classical Gymnasium Tiflis, 1898; C.E., Inst. of Engrs. of Ways of Communication, St. Petersburg, 1903; postgrad. Poly. Inst., Zurich, 1903-04; D.Eng., Poly. Inst, St. Petersburg, 1911; m. Helen Speransky, July 15, 1905 (dec. 1921); m. 2d, Marie Helander Cole, June 7, 1938. Came to U. S., 1917, naturalized, 1935. Asst. docent, prof. of gen. and advanced hydraulics, hydraulic structures, water power engring., theoretical mechanics and applied mechanics Polytech. Inst. Emperor Peter the Great, St. Petersburg, 1905-17; cons. engr. specializing on water power, St. Petersburg, 1907-15; enlisted with Red Cross, beginning of World War; chief plenipotentiary Central War Indsl. Com. to U. S., 1915-16; mem. Anglo-Russian Purchasing Commn., 1915-16; apptd. under-sec. of state (vice minister) Ministry for Commerce and Industry of Provisional govt. of Russia under premiership of Prince Lvov, 1917; apptd. head Extraordinary Russian Commn. to U. S. and Russian ambassador representing Provisional (Kerensky) Govt., 1917; continued as ambassador of the State of Russia until 1922; cons. engr. N.Y.C., 1923—; chmn. bd. Lion Match Co.; pres., dir. No. Mercury Felt Corp.; dir. Potash Co. Am., Research Corp.; prof. civil engring., Columbia, 1931, hon. prof., 1951; mem. bd. cons. engrs. Panama Canal, 1946-47. Recipient Grand medal, Soc. Drs. Engring., 1946. Fellow Am. Geog. Soc., Fgn. Policy Council, N.Y. Acad. Scis., Inst. Aero. Scis. (asso.); mem. A.A.-A.S., Am. Soc. C.E. (hon. 1946; chmn. com. research; mem. bd. engring. found.), Am. Soc. M.E., Engring. Inst. (Can.), Conn. Acad. Arts and Scis., Sigma Xi, Tau Beta Pi. Author: Lectures on Hydraulics (Russian), 1912; Varied Flow of Liquid in Open Channels (Russian), 1912; Variable Flow of Liquids (Russian), 1914; Hydraulics of Open Channels, 1932; Mechanics of Turbulence, 1936. Contbr. to jours. Died July 21, 1951.

BAKKE, Arthur Lawrence, Am. botanist; b. Horace, N.D., June 26, 1886; s. Martin and Angnette (Sundt) B.; B.S., Ia. State U., 1904, M.S., 1909; Ph.D., U. Chgo., 1917; m. Jeanette D. Henderson, July 19, 1952. Faculty, Ia. State U., Ames, 1910—, prof. plant physiology, 1925—. Fellow A.A.A.S.; mem. N. Central Weed Control Conf. (hon.), Am. Soc. Plant Physiologists (life). Research, numerous publs. on control of weeds, water relations of plants. Home: 701 Ash St., Ames, Ia. 50010.*

BAKKE, Edward Wight, Am. economist; b. Onawa, Ia., Nov. 13, 1903; s. Oscar Christian and Harriet (Wight) B.; B.A., Northwestern U., 1926, LL.D., 1964; Ph.D., Yale, 1932; m. Mary Sterling, Sept. 1, 1926; children—Karl Edward, Carolyn Sterling (Mrs. Albert Bacdayan), William Wight. Instr. sci. soc. Yale, 1932-34, asst. prof., 1934-36, asst. prof. econs., 1936-38, prof., 1938—, Sterling prof., 1940—, dir. grad. studies in econs., 1940-50, dir. Yale Labor and Mgmt. Center, 1944—; asst. prof. sociology Harvard, 1936; Fulbright prof., Denmark, 1953. Cons., U.S. depts.

Navy and Labor, 1944—; mem. Nat. Manpower Policy Task Force, 1966—. Mem. Am. Econ. Assn., Indsl. Relations Research Assn. (past pres.), Conn. Acad. Arts and Scis., A.A.A.S., Delta Sigma Rho. Author numerous books including: The Unemployed Man, 1933; Citizens Without Work, 1940; Unions Management and the Public, 1947; Mutual Survival, 1947; Bonds of Organization, 1950; The Fusion Process, 1953; The Human Resources Function, 1958; A Norwegian Contribution to management Development, 1959; A Positive Labor Market Policy, 1963; also numerous articles. Research on unemployment, labor market, organizational structure and dynamics, indsl. relations, structure of orgn., fusion process, models for analysis of labor and student movements. Home: Rimmon Rd., Woodbridge, Conn. 06525. Office: 2 Hillhouse Av., New Haven 06520.*

BAKKER, Cornelis Jan, physicist; b. Amsterdam, Netherlands, Mar. 11, 1904; D. Physics, U. Amsterdam, 1931; hon. degree U. Geneva (Switzerland), 1957; m. Anna Margaretha Herwig, 1933; 1 son, 2 daus. Mem. sci. staff Phys. Lab. Philips, Eindhoven, Netherlands, 1933-46; faculty U. Amsterdam, from 1946, asso. prof. physics, from 1955; dir. Zeeman-lab., also dir. Inst. for Nuclear Research, Amsterdam, 1946-55; dir. gen. European Orgn. for Nuclear Research, Geneva, from 1955. Mem. Royal Netherlands Acad. Scis. Publs. on spectroscopy, phys. problems in radio, nuclear physics. Died Apr. 23, 1960.

BAKSHI, Trilochan Singh, botanist; b. Wardha, India, Apr. 1, 1925; s. Bishan Singh and Gurnam Kaur; B.Sc. with honors, St. Xavier's Coll., Bombay, 1947; M.Sc., U. Saugar, India, 1949; Ph.D., Wash. State U., 1958; m. Yeshwant Mahendra, Nov. 16, 1948; children—Jadgip, Kanwaldip, Ravi. Asst. prof. botany Birla Coll., Rajasthan, India, 1949-52; lectr. U. Delhi, India, 1952-53; prof. botany Nat. Coll., Quadian, India, 1953-54; postdoctoral fellow U. Sask., Can., 1958-59; botanist, ecologist Govt. Sierra Leone, West Africa, 1960-63; lectr. botany U. Ghana, 1963-64; asso. prof. botany Notre Dame U., Nelson, B.C., Can., 1964-65, prof. botany, head dept. biol. scis., 1965—. Fulbright fellow Wash. State U., 1954-58. Fellow Indian Bot. Soc.; mem. Internat. Soc. Plant Morphologists (life), Am. Inst. Biol. Scis., Sigma Xi. Author: (with R. A. Stobbs) The Soils and Geography of the Boliland Region of Sierra Leone, 1963; also articles. Explained evolution of inflorescence in family Amaranthaceae; showed that Pterosopora is not a saprophyte but is parasitic on fungi enveloping its roots; research on weeds in relation to their persistence, forest ecology of Kokanee Glacier Park, B.C. Home: 310 High St., Nelson, B.C., Can.*

BAKULEV, Aleksandr Nikolaevich, Russian surgeon; born December 7, 1890; graduated from Medical Faculty, Saratov Univ., 1915; D.Med. Sci. with Saratov Univ., 1922-26; with 2d Moscow Medical Institute, 1926-30, director of faculty of surgery Spasokukotsky Clinic, 1939—; prof., 1939—; founder, dir. Inst. Thoracic Surgery, USSR Acad. Med. Sci., 1956-59, sci. dir. Inst. Cardiovascular Surgery, 1959—, chmn. Sci. Coordination Center. Del., 3d Internat. Congress Surgeons, Rome, 1960. Recipient Stalin prize, 1949, Lenin prize, 1958. Mem. USSR Acad. Med. Sci. (pres. 1953-60), USSR, Serbian acads. sci.; hon. mem. Czechoslovakian Purkinje Med. Soc., Soc. Polish Surgeons, All-Union Soc. Surgeons. Author: Surgical Treatment of Spinal Cord Tumors, 1939; Conservative Treatment of Brain Abscesses by Puncture, 1940; Sunk Sutures in the Late Treatment of Craniocerebral Wounds, 1942; The Diagnosis and Treatment of Adhesive Pericarditis, 1948; Surgical Treatment of Diseases of the Heart and Major Vessels. Diagnosis, Practice and Prospects, 1952; Surgery of Acquired Diseases and the Aorta, 1954; coauthor: Pneumectomy and Lobectomy. Methods of Operation, 1949; Experience in the Use of Contrast Angiocardiography in Thoracic Surgery, 1951; Congenital Heart Defects, 1955; Surgical Treatment of Purulent Lung Diseases, 1961. Former editor Herald of USSR Acad. Med. Scis.; mem. editorial bd. Thoracic Surgery, Large Med. Ency., 2d edit., New Surg. Archives, Surgery. Research on cerebral, thoracic and cardiac surgery, operations for lung diseases; made detailed studies of kidney function, tumors, bone surgery; introduced encephalography and ventriculography into USSR; performed 1st operation in USSR on congenitally defective heart. Home: Kaluzhskaya 18. Office: 2d Moscow Med. Inst., Malaya Pirogovskaya 1, Moscow, USSR.

BAKWIN, Harry, Am. physician; b. Utica, N.Y., Nov. 19, 1894; s. Simon and Emma (Nodel) B.; B.S., Columbia, 1915, M.D., 1917; postgrad. univs. Vienna, Berlin; m. Ruth Morris, Feb. 2, 1925; children—Edward, Patricia (Mrs. Frederic R. Selch), Barbara (Mrs. William S. Rosenthal), Michael. Faculty, Cornell U. Med. Coll., 1919-24, Columbia Coll. Phys. & Surg., 1925-30; faculty N.Y. U. Med. Coll., N.Y.C., 1930—, prof. clin. pediatrics, 1945—; vis. physician Bellevue Hosp. Cons. N.Y. Infirmary, Horton Meml. Hosp., Middletown, N.Y., Mt. Vernon Hosp., N.Y., Phelps Meml. Hosp., Tarrytown, N.Y., Beth Israel Hosp., Newark, Bayonne (N.J.) Hosp. and Dispensary. Mem. Am. Acad. Pediatrics (pres. 1955-56), Am. Pediatric Soc., A.M.A., N.Y. Acad. Medicine, Sigma Xi, Alpha Omega Alpha. Author: (with Ruth Bakwin) Psychologic Care During Infancy and Childhood, 1942; Clinical Management of Behavior Disorders in Children, 1953, 3d edit., 1966; also articles. Home: 132 E. 71st St., N.Y.C. 10021.*

BAKWIN, Ruth Morris, Am. physician; b. Chgo., June 3, 1898; d. Edward and Helen (Swift) Morris; B.A., Wellesley Coll., 1919; M.D., Cornell Med. Coll., 1923; M.A., Columbia, 1929; m. Harry Bakwin, Feb. 2, 1925; children—Edward, Patricia (Mrs. Frederic R. Selch), Barbara (Mrs. William S. Rosenthal), Michael. Asst. pediatrician Fifth Av. Hosp., N.Y.C., 1925-35; asst. pediatrician Bellevue Hosp., N.Y.C., 1927-43; asst. vis. physician Children's Med. Service, 1943-48, asso. vis. physician, 1948-55, vis. physician, 1955—; with Child Guidance Clinic 1927-66; attending physician pediatrics Univ. Hosp., 1962—, prof. clin. pediatrics N.Y. U., 1961—, mem. Staff N.Y. Infirmary, 1929—, dir. dept. pediatrics, 1936-54, dir. emeritus, 1955-66, co-dir. dept. pediatrics 1966-67. Mem. finance Com., 1962—. Bd. dirs., mem. Adv. Council on Mental Illness, 1962—. Recipient Elizabeth Blackwell award, 1950; N.Y. Infirmary Merit award, 1960. Mem. Am. Acad. Pediatrics (state chmn. N.Y., 1965-67), Am. Med. Women's Assn. (councilor 1962-65), A.M.A., Women's Med. Soc. N.Y. Author: Psychologic Care During Infancy and Childhood, 1942, Clinical Management of Behavior Disorders in Children, 1953; 2d edit., 1960, 3d edit., 1966 (both with H. Bakwin). Contbr. numerous articles to profl. jours. Address: 132 E. 71st St., N.Y.C. 10021.*

BALAKRISHNA, Ramachandra, Indian economist; b. Bangalore, India, June 6, 1905; s. Ramachandra Mudaliar, Dharwar Balakrishna and Balambal B.; B.A., Maharaja's Coll., Mysore, India, 1926, M.A., 1928; Ph.D., London Sch. Econs., 1939; m. Rajeswari, May 3, 1934; children—B. Shantha Kumar, Leela (Mrs. A. G. Singaravelu). Faculty, U. Mysore, 1929-44, asst. prof. econs., 1939-44; faculty U. Madras (India), 1944-61, prof. econs., 1946-61; mem. India Tariff Commn., Bombay, 1961-65; dept. dir.-gen. Nat. Council Applied Econ. Research, New Delhi, India, 1965-66; prof. econs. Bangalore (India) U., Central Coll., 1966—; prin. Mahranis' Coll. for Women, Bangalore, 1940-41. Recipient Sir Hugh Daly gold medal, 1928. Mem. All India Econ. Assn. (past gen. sec.). Author: numerous books including: Regional Planning in India, 1948; International Economic Relations, 1957; Measurement of Productivity in Indian Industry, 1958; Review of Economic Growth in India, 1961; also articles. Research on econ. devel. and feasibility of further progress of a state, techno-econ. surveys and feasibility studies. Home: 51 Dacosta Sq., Cooke Town, Bangalore-5, India.*

BALAKRISHNA, S., Indian geophysicist; b. Guntur, India, Aug. 3, 1931; s. Suri Bhagavantam and Sita (Mahalakshmi) B.; B.Sc. with honors, Andhra U., 1952, M.Sc., 1953; Ph.D., Osmania U., 1955; m. S. Indira, Apr. 1954; children—Subba Laxmi, Prem Kumar, Ravi Kumar. Research fellow Nat. Inst. Sci. of India, 1954; postdoctoral research fellow Harvard, 1956-57; reader Osmania U., India, 1957-61, reader, head dept., 1961-62, hon. head dept., 1962-64; asst. dir. Nat. Geophys. Research Inst., Hyderabad, India, 1962-64, asst. dir., 1964—. Fellow Geol. Soc. Am., Am. Geophys. Union, Geol. Soc. Japan, Exploration Geophysicists of Japan, Indian Geophys. Union (sec.-treas. 1962). Research, numerous publs. on elasticity of rocks and their relations to certain geol. features, study of earth-tides in India, magnetic and gravity surveys, interpretation of geol. structures. Home: Sriniket, Tarnaka, Hyderbad-7, Office: Nat. Geophys. Research Inst., Hyderbad, India.*

BALAKRISHNAN, A. V., elec. engr.; b. Palghat, India, Dec. 4, 1922; s. A.S. and Ammukutty (Ammae) B.; M.A., U. Madras, India, 1947; A.M., M.S., U. So. Cal., 1950, Ph.D., 1954; m. Kay Moulton, Sept. 1952; children—David, Sally, Robby, Jerry, Ken. Came to U. S., 1947, naturalized, 1958. Instr, Yale, 1953-54; sr. engr. RCA Labs., Camden, N.J., 1954-56; prof. U. So. Cal., 1956-57; with Space Tech. Labs., 1959-61; faculty U. Cal. at Los Angeles, 1957-59, 61—, prof. engring. and math., 1963—. Cons. NASA. Fellow I.E.E.E.; mem. N.Y. Acad. Sci., Union Radio Sci. Internat. Author: Advances in Communication Systems, 1963; Space Communications, 1963; (with L. W. Newstadt) Computing Methods in Optimization, 1964; also articles. Research on design of signals for space communication; information theory; identification and control of complex systems and related computing techniques; description of systems from external measurements.

BALAN, Jozef, Czechoslovakian microbiologist; b. Bratislava, Czechoslovakia, May 9, 1929; s. Dezider and Renata (Donath) B.; Dipl.Ing., Slovak Poly. U., Bratislava, 1952; Dr., C.Sc., Slovak Acad. Scis., Bratislava, 1962; m. Jana Ebringerová, Nov. 27, 1965. Staff research and control lab. Slovakofarma, Hlohovec, Czechoslovakia, 1952-55; staff dept. microbiology Biol. Inst., Slovak Acad. Scis., Bratislava, 1956—, sci. worker, 1962—; faculty Kommenius U., Bratislava, 1964—, lectr. tech. microbiology, 1964—. Mem. Czechoslovak Microbiol. Soc., Slovak Biol. Soc. Research, publs. on discovery and codiscovery of several new antibiotics and new methods for search for antiprotozoal antibiotics. Home: 31 Obchodna, Bratislava. Office: 2 Jánska, Bratislava, Czechoslovakia.*

BALANDIN, Aleksei Aleksandrovich, Russian organic chemist; born December 20, 1898; graduated from Moscow Univ., 1923; Asso. with Moscow U., 1927—, professor, 1934—; head lab. Inst. Organic Chemistry, USSR Acad. Sci., 1935—; head Dept. Organic Catalysis, Moscow State U., 1959—. Decorated Order of Lenin; recipient Mendeleev prize, 1936, Lebedev prize, 1945, Stalin prize, 1946. Mem. USSR Acad. Sci. Author: Modern Problems of Catalysis and of Multiplet Theory, 1935; Catalytic Dehydrogenation of Hydrocarbons and its Use in the Synthesis of Rubber from Gases, 1942; Organic Catalysis Theory, 1947; Selective Catalysis Theory, 1956; The Kinetics of Alcohol Dehydrogenation, 1957; Thermodynamics of the Reaction of Demethylation of Toluene and Cresol with Steam, 1963. Research on organic catalysis; originator, developer multiplet theory of catalysis used in research on kinetics of catalytic hydrogenation and dehydrogenation. Address: Zelinskii Inst. Organic Chemistry, Leninskii prosp. 31, Moscow, USSR.

BALARD, Antoine Jérôme, French chemist; b. Montpellier, France, Sept. 30, 1802; ed. as apothecary at Montpellier; became demonstrator, chem. asst. Faculty Scis.; prof. chemistry Royal Coll. and Sch. Pharmacy, also Faculty Scis.; succeeded Thénard in chair of chemistry Sorbonne, Paris, France, 1842; apptd. prof. chemistry College de France, 1851; mem. French Academy Scis., 1844. Discovered bromine in sea water, 1826, hypochlorous acid, also chlorine monoxide, 1834; determined composition of Javelle water also bleaching powder; studied methods for obtaining soda and potash from sea water. Died Paris, Apr. 30, 1876.

BALAZS, Béla Arpád, Hungarian astronomer; b. Rakoshegy, Hungary, May 4, 1935; s. Lajos Zoltan and Ilona (Kovacs) B.; grad. in math. Roland Eötvös U., Budapest, Hungary, 1958, D., 1964; m. Kamilla Bock, Mar. 13, 1963. Sci. collaborator Konkoly Obs., Budapest, 1958—; lectr. Roland Eötvös U., 1962—; fellow Hamburg (Germany), 1961-62, 64. Mem. Internat. Astron. Union (mem. commn. 37, 1964—), Roland Eötvös Phys. Soc. Research, publs. on applied celestial mechanics, three-color photometry of open clusters, spectral classification of low dispersion spectra. Home: 2 Kecskeméti St., Budapest V, Hungary.

BALAZS, Endre Alexander, physician; b. Budapest, Hungary, Jan. 10, 1920; s. Endre Alexander and Vilma (Bonta) B.; M.D., U. Budapest, 1943; M.D. (hon.), U. Uppsala, 1967; m. Eva Tomes, Feb. 26, 1945; children—Marianne, Andre. Came to U. S., 1951, naturalized, 1956. Biomed. research specializing in biology and phys. biochemistry of connective tissues, Boston, 1951—; asso. dir. Retina Found, 1951-62, dir. dept. connective tissue research, 1961—; pres. Inst. Biol. and Med. Scis., 1962-63, 66—; instr. Med. Sch. Harvard, 1951-56, lectr., 1956—, biochemist Mass. Eye and Ear Infirmary, 1956-57. Fellow N.Y. Acad. Sci.; mem. Am. Rheumatism Assn., Biochem. Soc., Am. Chem. Soc., Assn. for Research in Ophthalmology (Friedenwald award 1963), Histochem. Soc., Soc. for Cell Biology, Tissue Culture Assn., Soc. Exptl. Biology and Medicine, Am. Soc. Biol. Chemists, A.A.A.S., Gerontol. Soc. Editor: (with R. Jeanloz) The Amino Sugars: The Chemistry and Biology of Compounds Containing Amino Sugars, 1965-67; co-editor-in-chief Experimental Eye Research, 1961—. Research, numerous publs. in molecular biology of intercellular space, chem. and phys. structure of vitreous body, synovial fluid and other connective tissues, biol. importance and phys. biochemistry of glycosaminoglycans. Home: 240 Pleasant St., Arlington, Mass. 02174. Office: 20 Staniford St., Boston 02114.*

BALBI, Adriano, Italian geographer, statistician; b. Venice, Italy, Apr. 25, 1782; prof. geography Coll. San Michele at Murano, 1808; prof. physics Lyceum of Fermo, 1811-13; later with customs office, Venice; visited Portugal, 1820, returned to Paris and remained there until 1832; Author: Essai statistique sur le royaume de Portugal et d'Algarve, 1822; Atlas ethnographique du globe ou classification des peuples anciens et modernes d'après leurs langues, 1826; Abrégé d Géographie, 1832. Died Padua, Italy, Mar. 14, 1848.

BALBIANI, Edouard Gérard, biologist; b. Haiti, Santo Domingo, July 31, 1823; chief histology physiol. lab French Nat. Mus. Natural History, 1867-73; apptd. to chair of comparative embryology Coll. de France, 1873; recipient grand prize in phys. scis. French Acad. Scis., 1873. Co-founder Archives d'anatomie microscopique. Author: Essai sur les fonctions de la peau, 1854; Recherches sur les phénomènes sexuels des infusoires, 1861; Sur la structure et la division du noyau chel le Spirochona semmipar, 1895. Research on sexual reprodn. in ciliated infusoria, vitelline body of young ovules of animals; discovered schizogenesis, studied devel. of plant louse. Died Meudon, France, July 25, 1899.

BALBOA, Vasco Nuñez de, Spanish explorer; b. Jerez de los Caballeros, Estremodura, Spain, circa 1475. Sailed with Rodrigo de Bastidas, 1501, to Hispaniola; failed as planter at Salvatierra and fled from creditors on Martin Fernandez de Encisco's vessel to San Sebastian, Panama, 1510; proposed they go to Darien, founded new town St. Maria de la Antigua del Darien, took over leadership from Encisco, sent him back to Spain as prisoner; heard that Encisco had complained to king and sentence was being issued condemning him and ordering him back to Spain; set out with expdn. to find Great Sea to West, 1513; 1st sighted Pacific Ocean, 1513, reached and claimed ocean and all shores it washed for Spain; visited Pearl Islands, returned with booty which he sent to Spanish crown; granted as reward title "adelantado of South Sea (or Admiral of Pacific) and gov. of Panama"; built 2 small brigantines and took possession of Pearl Islands for Spain; (meanwhile Pedro Arias de Avila had been sent to replace him); lured to Aela, Panama, by Pedro Arias (who had him tried and condemned for treason). Beheaded in public square, Aela, 1517.

BALBUS, Lucius Cornelius, Spanish mathematician; b. Cadiz, Spain, early 1st century B.C. Named Roman citizen by Pompeius for services against Sertorius in Spain; sided with Caesar against Pompeius in Rome; chief engr. to Caesar, traveled with Caesar to Spain, 61 B.C., Gaul, 58 B.C.; pvt. sec. to Caesar; became praetor, 43 or 42 B.C., consul 40 B.C. Author: Diary of Chief Events of Mine and Caesar's Times, 81.

BALCH, Charles Clive, Brit. animal nutritionist; b. Abergavenny, Mon., U.K., Dec. 8, 1924; s. Charles H. and Hilda (Day) B.; B.Sc., Reading U., 1945, Ph.D., 1950, D.Sc., 1964; m. Sina Martig, Mar. 12, 1960; children—Claire, John. With Nat. Inst. for Research in Dairying, Shinfield, Reading, Berks., Eng., 1945—, head feeding and metabolism dept., 1958—; vis. prof. Cornell U., 1952-53; vis. lectr. U. Minn., 1965. Cons. FAO, 1958, 63. Recipient Royal Agrl. Soc. Eng. Research medal, 1966. Mem. Nutrition Soc., Brit. Soc. Animal Prodn., Brit. Ornithologists Union. Chmn. editorial bd. Brit. Jour. Nutrition, 1965—. Research, numerous publs. on nutrition of dairy cattle and especially effects of phys. and biochem. processes of reticulo-rumen on productive processes. Home: Orchard House, Croft Rd., Shinfield nr. Reading. Office: Nat. Inst. for Research in Dairying, Shinfield nr. Reading, Berks., Eng.*

BALCH, Reginald Ernest, biologist; b. Sevenoaks, Eng., Dec. 29, 1894; s. Alfred E. and Sarah (Hawkes) B.; B.S., U. Toronto, 1923; M.S., Syracuse U., 1928, Ph.D., 1949; D.Sc., U. N.B., Can., 1963; m. Martha Agnes Bowman, June 4, 1929; children—Cynthia (Mrs. Sean Moore), Norval Edmund. Mem. Canadian Expeditionary Force, France, Belgium, Germany, 1914-18; agt. U. S. Dept. Agr., N.C., Ida., Wash., 1926-29; in charge Dominion Entomol. Lab., Fredericton, N.B., 1930-52, Forest Biology Lab., Fredericton, N.B., 1930-52, Forest Biology Lab., Fredericton, and research Atlantic Provinces, Can. Dept. Agr., 1952-60; research forest entomology, Fredericton, 1960—. Fellow Entomol. Soc. Am., A.A.A.S., Entomol. Soc. Can., Canadian Inst. Forestry (pres. 1949-51), Brit. Commonwealth Forestry Assn., Canadian Inst. Internat. Affairs. Author: The Ecological Viewpoint 1965. Studies, publs. on biology and control of numerous forest pests using ecol. approach and biol. methods; initiated use of virus diseases against forest insects. Home: 102 Alexandra St. Office: Forest Research Lab., Fredericton, N.B., Can.*

BALD, John Grieve, plant pathologist; born Melbourne, Vic., Australia, Sept. 1, 1905; s. S. George Robert and Rebecca (Grieve) B.; B.Agr. Sci., U. Melbourne, 1928, M. Agr. Sci., 1933; Ph.D., U. Cambridge, 1935; 1 dau. by previous marriage, Bridget (Mrs. Michael Perram); M. Suresht Renjen, Dec. 21, 1963; children—Suneel Robert, Vivek Arthur. Came to U. S., 1948, naturalized, 1957. Research officer Counsil Sci. Indsl. Research, Adelaide, S. Australia, Cambridge, Eng., Canberra, A.C.T., Australia, 1928-48; faculty U. Cal., Los Angeles, 1948—, prof. plant pathology 1952—. Guggenheim fellow, 1956-57, 63-64. Mem. Phytopath. Soc. Am., Am. Bot. Soc., N.Y. Acad. Scis., Assn. Applied Biologists, Eng. Helped establish nature, dispersion and control of some plant virus diseases, and soil borne fungus and bacterial diseases; studies in relationship between virus concentration and infectivity; synthesis of viruses in plant cells, and ecology of plant viruses. Home: 1023 Princeton St., Santa Monica. Cal. 90403.*

BALDACCI, Riccardo Francesco, Italian engr.; b. Pisa, Italy, Oct. 4, 1917; s. Torquato and Amalia (Pini) B.; ed. tech. sch.; Ph.D. in Engring.; prof. of sci. of constrn. at U. Genoa; m. Marta Lazzeri, Apr. 30, 1955; children—Marina, Nicoletta. Engr., creator projs. of structures in steel and reinforces concrete. Mem. Internat. Assn. Bridges and Scaffolding (Zurich), Internat. Assn. for Shell Structures (Madrid). Contbr. numerous articles on theory and constrn. of structures to sci. publs. Studies on physics and mechanics of solids and structures; exptl. stress analysis; steel and concrete engineering. Home: via Camila 22. Office: via Montallegro 1, Genoa, Italy.

BALDES, Edward James, Am. biophysicist; b. Fairfield, Neb., July 5, 1898; s. Joseph J. and Margaret (Lenzen) B.; A.B., U. Sask., 1918; A.M., Harvard, 1920, Ph.D., in Physics, 1924; Ph.D. in Physiology, U. Coll., U. London, 1936; LL.D. (hon.), U. Sask., 1955; m. Mary Cooney, June 27, 1934; children—Joseph James, Mary Margaret, Honora, Elizabeth Louise. Instr. Harvard, 1923-24; biophysics sect. Mayo Clinic, Mayo Found., U. Minn., Mpls., 1924-63, emeritus prof. biophysics, 1963—; sci. analysis branch Life Scis. div. OSRD, Dept. Army, Arlington, Va., 1963—. Recipient War Dept. Commendation for Exceptional Civilian Service, 1945; Presdl. Citation, 1948; chevalier de la Legion d'Honneur (France), 1951. Research, numerous publs. on biophysics, physiology and aviation medicine; studies in vapor pressure osmometry, blood flow electro-encephalography, effects of acceleration on man, capillary microscopy, use of hematoporphyrin in cancer detection. Home: 206 S. Jackson St., Arlington 22204. Office: Army Research Office, 3045 Columbia Pike, Arlington, Va.*

BALDI, Bernardino (Baldi d'Urbino), Italian mathematician; b. Urbino, Italy, 1553; studied math. and classical lit. at Padua, under tchrs. including Commandino. Became abbot of Guastalla; sec. to various prelates and to Duke of Urbino. Author: La nautica (didactic poem on seafaring), 1590; A Chronicle of Mathematicians (abridgement of work intended to contain biographies of over 200 mathematicians), pub. posthumously, 1707; In Mechanica Aristotelis Problemata Exercitationes (commentary, including assertion that motion is engendered by motion), 1582; Geography of Edrisi (transl. into Italian), also ecologues, prose dialogues, hist. works, valuable commentary on Vitruvius. Died Urbino, 1617.

BALDINGER, Ernst Gottfried, German physician; b. Gross-Vargula, Germany, May 13, 1738; s. Johann Baldinger; student medicine at Erfurt, Halle, Jena, (all Germany); m. Dorothea Friederike Gutbier, 1764; m. 2d, Caroline Lisette Drebing, 1787; children—Sophie Friederike Ernestine (Mrs. Goerg Theodor Christoph Handel), Friederike Wilhelmine Amalie (Mrs. Bernhard v. Gehren). Pvt. lectr. in Jena; mil. doctor in Prussian army during Seven Years War; practicing physician in Langensalza, 1763; became prof., Jena, 1768; named prof. medicine, dir. clinics, Göttingen, 1773; named by Landgraf Friedrich von Hessen-Kassel as dir. all med. facilities and instns. State of Hessen-Kassel, 1783; went to Marburg, 1785, and restored med. sch. Author: De militum morbis, 1763. Publisher and collector of med. writings; Darmstadt library purchased his library after his death; among his students were J. Arnemann, J. C. G. Ackermann, J. F. Blumenbach, J. F. Meckel, S. T. Sömmering. Died Marburg, Jan. 21, 1804.

BALDINI, James Thomas, Am. biologist; b. Paterson, N.J., Jan. 21, 1927; s. James Thomas and Emma (Ronca) B.; B.S., Rutgers U., 1947; M.S., Purdue U., 1949, Ph.D., 1951; m. Audrey Ina Stoner, Oct. 13, 1951; children—Pamela Allison, James Thomas III, Sandra Jean, John Brooks. Biologist, DuPont Co., New Brunswick, N.J. 1951-52, research biologist Stine Lab., Newark, Del., 1952-64, sr. research biologist in clin. research, Wilmington, Del., 1964—. Mem. N.Y. Acad. Sci., Am. Inst. Nutrition, Sigma Xi. Patentee; publs. on research to demonstrate that energy metabolism is affected by amino acid deficiency; discovered relation between energy content of diet and amino acid requirement of birds; showed what protein requirement may be met by free amino acids. Home: 904 Baylor Dr., Newark, Del. 19711. Office: 5062 DuPont Bldg., Wilmington, Del. 19898.*

BALDNER, Lienhardt, naturalist; b. Strasbourg, France, Jan. 9, 1612; s. Carl and Ursula (Mock) B.; m. Salome Friess, Jan. 27, 1636; 4 children; m. 2d, Anna Ursula Spengel, 1650; 4 children; m. 3d, Barbara Grosse, Apr. 13, 1665; 4 children. Animal and landscape painter; held various pub. offices; tchr. natural phenomena and history. Author: Das Vogel-Fisch- und Thierbuch, 1666, edited by R. Lauterborn, 1903. Research on living habits of birds, fish, animals. Died Strasbourg, Feb. 4, 1694.

BALDRIDGE, Robert Crary, Am. biochemist; b. Herington, Kan., Jan. 9, 1921; s. Ben F. and Lena (Bradshaw) B.; B.S., Kan. State Coll., 1943, M.S., 1948; Ph.D., U. Mich., 1951; m. Anne E. Dukelow, July 12, 1943; children—Patricia, Ben, Herbert, Thomas, Robert D. Faculty, Temple U., Phila., 1953—, prof., 1962— asso. dean Grad. Sch., 1965—. Cons. Med. div. Oak Ridge Inst. Nuclear Studies. Mem. Am. Soc. Biol. Chemists, Biochem. Soc. Gt. Britain, Soc. Exptl. Biology and Medicine, Sigma Xi, Phi Kappa Phi. Research, publs. on nature of enzymatic defect in human genetic disorder, histidinemia, studies on histidine metabolism. Home: 137 Stanley Av., Glenside, Pa. 19038. Office: 3420 N. Broad St., Phila. 19140.*

BALDUS, Richard, mathematician; b. Saloniki, Hungary, May 11, 1885; s. Wilhelm H. Phil. and Elizabeth (Schüsser) B.; ed. Munich, also Erlangen; degree, 1910; m. Bertha Elisabeth Dedreuk, Aug. 10, 1912; 2 sons, 2 daus. Lectr., 1911; became prof. geometry Tech. U. Karlsruhe (Germany), 1932; prof. math. Tech. U. Munich (Germany); mem. Acad. Scis. Heidelberg, Acad. Scis. Munich. Author: A Popular Work on Formalism and Intuitionism in Geometry, 1924. Basic research on non-Euclidean geometry, also research on topography, math. aspects of music. Died Munich, Jan. 28, 1945.

BALDWIN (Balduin, Balduinus), Christian Adolphus, German alchemist; b. Doebeln, Germany, 1632; magistrate Hayn or Grossenhayn, Saxony, Germany; mem. Academia Naturae Curiosorum; Fellow Royal Soc., 1677. Author: Aurum superius et inferius Aurae superioris et inferioris hermeticum, 1673; Phosphorus hermeticus seu Magnes Luminaris, 1675; Hermes curiosus sive experimenta physicochymica nova, 1680; other chem. treatises. Discovered phosphorescent chem. substance from a nitrate of lime (known as Baldwin's phosphorus) wrote on extraction of gold from air by universal magnetism; described copper meteorite that fell near Haina in 1677, performed 18 chem. expts. on it. Died Dec. 31, 1682.

BALDWIN, Ernest Hubert Francis, Brit. biochemist; b. Gloucester, Eng., Mar. 29, 1909; s. Hubert Charles and Nellie Victoria (Hailes) B.; B.A., Cambridge (England) University, 1931, Ph.D., 1934, Doctor of Science, 1967; m. Pauline Mary Edwards, Dec. 27, 1933; children—Nicola (Mrs. David Q. Milligan), Nigel St. John. Demonstrator in biochemistry Cambridge U., 1936-43, lectr., 1943-49; prof. biochemistry, head dept. Univ. Coll., London, Eng., 1950—. Vis. prof. biochemistry Scripps Instn. Oceanography, U. Cal. at La Jolla, 1956-57; Rose Morgan, prof. biochemistry U. Kan., 1965. Fellow N.Y. Acad. Scis., Inst. Biology; mem. Cambridge Philos. Soc., Biochem. Soc., Marine Biol. Assn., Soc. for Exptl. Biology, Renal Assn. (Found. mem.). Author: (with D. J. Bell) Cole's Practical Physiological Chemistry, 10th edit., 1949; (with J. Needham) Hopkins and Biochemistry, 1949; The Nature of Biochemistry, 1962; Introduction to Comparative Biochemistry, 1937; Dynamic Aspects of Biochemistry, 1947 (Cortina-Ulisse prize 1952). Research in biochem. evolution. Home: 8 Crofters Rd., Moor Park, Northwood, Middlesex, Eng. Office: Dept. of Biochemistry, University College London, Gower St., London, W.C.I., Eng.*

BALDWIN, Evelyn Briggs, Am. meteorologist, explorer; b. Springfield, Mo., July 22, 1862; s. Elias Briggs and Julia Cornelia (Crampton) B.; M.S., Northwestern Coll., 1885. Prin. high sch. and supt. city schs., Kan., 1887-91; observer U. S. Weather Bur., 1892-1900; insp.-at-large, signal corps U. S. Army; meteorologist Robert Edwin Peary's N. Greenland expdn., 1893-94; meteorologist, 2d in command Walter Wellman's polar expdn. to Franz-Josef Land, 1898-99; built and named Ft. McKinley; discovered and explored Graham Bell Land, 1899; organized head Baldwin-Ziegler polar expdn., 1901-02; established 4 depots of supplies from south to north coast of Franz-Josef Land as basis for proposed dash to North Pole, and 3 safety stas. on northeast coast of Greenland for use on return march. Author: Search for the North Pole; Franz-Josef Land, 1898-99; Auroral Observations Franz-Josef Land, 1898-99; also meteorol. reports. Died Oct. 25, 1933.

BALDWIN, Francis Marsh, Am. zoologist; b. West Upton, Mass., Jan. 16, 1885; s. Ellory Albee and Rosa Arbella (Wood) B.; A.B., Clark U., 1906, A.M., 1907; Ph.D., U. Ill., 1917; m. Bessie Mae Seay, July 15, 1912 (dec. Nov. 1949); children—Gwendolyn, Francis Marsh; m. 2d, Esther Pardee Harper, Aug. 5, 1950. Instr. nature and science, Ky. Normal Sch., 1908-11; prof. biology Western Maryland Coll., 1911-15; asst. zoology, U. Ill., 1915-17; research, Marine Biol. Lab., Woods Hole, summer 1915; asst. prof. zoology Ia. State Coll., 1917-19, asso. prof. 1919-20, prof. physiology, 1920-27; prof. zoology, U. So. Cal., Los Angeles, from 1927, chmn. dept., 1929-36; dir. Marine Biol. Sta., 1928-36; chmn. Biol. Div., 1936; Fellow Ia. Acad. Science; mem. A.A.A.S., Am. Assn. Univ. Profs., Am. Physiol. Soc., Soc. Exptl. Biol. and Medicine; mem. corp., Marine Biol. Lab., Woods Hole, Mass.; mem. So. Calif. Acad. Science, Sigma Xi, Phi Kappa Phi. Author and contbg. editor: Elementary Manual for Physiology, 1927; Practical Exercises in Human Anatomy, 1932; Manual for Advanced Physiology (metabolism), 1933; Manual for Neurology, 1935. Contbr. articles on marine biology to Ency. Brit. Book of Year, 1937-46; collaborator for biol. abstracts. Frequent contbr. scientific articles to jours. Died Feb. 2, 1951.

BALDWIN, George C., Am. physicist; b. Denver, May 5, 1917; s. Harry Lewis and Elizabeth (Watson) B.; B.A., Kalamazoo Coll., 1939; M.A., U. Ill., 1941, Ph.D., 1943; m. Winifred M. Gould, Apr. 27, 1952; children—George T., John E., Celia M. Instr. physics U. Ill., Urbana, 1943-44; research asso. Gen. Electric Co., Schenectady, 1944-55, nuclear engr., Cin., 1955-57, reactor mgr. Argonne Ill. Nat. Lab.; 1957-58, physicist, Schenectady, 1958-67; adj. prof. nuclear engineering and science. Rensselaer Polytechnic Institute, 1964-67, professor, 1967—. Mem. Phi Beta Kappa, Sigma Xi, Phi Kappa Phi Gamma Alpha. Discoverer nuclear dipole resonance; research publs., patents in photonuclear effect, particle accelerators, reactor kinetics, elec. propulsion, electron collison cross sects., stimulated emission devices. Home: 1046 Merlin Dr., Schenectady 12309. Office: Rennselaer Poly. Inst. Troy, N.Y. 12181.*

BALDWIN, Horace Strow, Am. physician; b. Englewood, N.J., Oct. 14, 1895; s. John Hall and Annie (Strow) B.; B.S., Wesleyan U., 1917; M.D., Cornell U., 1921; m. Florence Reed, Sept. 3, 1924; children—

Horace Reed (dec.), Judith (Mrs. James B. Hutcheson), Asso. clin. prof. Cornell U., 1942-61; asso. attending physician N.Y. Hosp., 1942-61, cons. medicine, 1961—, chief allergy clinic, 1935-61; mem. sci. and edn. council Allergy Found. Am. Mem. Am. Acad. Allergy (pres. 1952), N.Y. Acad. Pneumococcus Pneumonia. Research, publs. on antibody treatment of pneumococcus lobar pneumonia, physiology of asthma syndrome, cortical steroid mgmt., role of infection especially staphylococcus. Office: 136 E. 64th St., N.Y.C. 10021.*

BALDWIN, Ira Lawrence, Am. bacteriologist, college official; b. Oxford, Ind., Aug. 20, 1895; s. Thomas Atkinson and Eva (Mock) B.; B.S.A., Purdue U., 1919, M.S., 1921, D.Sc. (hon.), 1945; Ph.D., U. of Wis., 1926; m. Mary Eliza Lesh, Dec. 29, 1920 (dec.); children—Helen Lucile (Mrs. Maurice Guptill) Frances Mary (dec.), Robert Lesh; married 2d, Ineva R. Meyer, Apr. 17, 1954. Tchr. bacteriology, Purdue U., 1919-23, asst. prof., 1924-25, asso. physiology, exptl. station, 1926; asst. prof. agrl. bacteriology, Coll. of Agr., U. of Wis., 1927-29, asso. prof., 1929-32, prof. since 1932, head dept., 1941-44, asst. dean, 1932-42, dean Grad. Sch., 1944-45, dean Coll. of Agr. and dir. Agr. Expt. Station and Agr. Extension Service, 1945-48, v.p. academic affairs, 1948-58, special assistant to the president, 1958-66, emeritus v.p., 1966—. Mem. Natural Resources Com. of State Agencies Wis., 1952-66; cons. N. Central Assn. Colls. and Secondary Schs., bd. Directors, 1961-65; dir. rural devel. office Assn. State Univs. and Land-Grant Colls., 1963-64. Mem. NSF panel on regulatory biology. Served as 2d lt., F.A., U. S. Army, 1918; cons. U. S. Army, 1945, WPB, 1942-43, FSA, 1943-44, Research & Devel. Bd., 1946-53, USPHS, 1946-48, Dept. of Army; mem. Chem. Corps Adv. Bd., 1953-65, chairman, 1958-65; member Munitions Command Advisory Group, 1965—. Fellow Royal Society Arts; member American Academy of Microbiologists (member board 1956-61), 62—; chairman bd. 1957-60), Am. Soc. Bacteriologists (pres. 1944, hon. mem.), Am. Soc. Agronomy, Am. Podiatry Assn. (mem. spl. commn. on status of podiatry edn. 1961), Wis., Ind. acads. sci., A.A.-A.S., Am. Phytopathol. Soc., Am. Soc. Plant Physiologists, Sigma Xi, Alpha Zeta, Phi Lambda Upsilon, Phi Sigma. Studies on relationship between leguminous plants and root nodule bacteria, with particular reference to utilization of atmospheric nitrogen by higher plant and bacteria; also on industrial fermentations—production of bakers' yeast, butyl alcohol. Home: 1111 Dartmouth Rd., Madison 5, Wis.

BALDWIN, James Fairchild, Am. gynecologist; b. Orangeville, N.Y., Feb. 12, 1850; s. Cyrus H. and Mary P. (Fairchild) B.; A.B., Oberlin Coll., 1870, A.M., 1876; M.D., Jefferson Med. Coll., Phila., 1874; m. Fidelia Finch, 1874; children—Austin Guy, Fredriks Hull (Mrs. Fred. R. Hoover), Hugh A., Helen F. (Mrs. Helen F. Pease); m. 2d, Ida Strickler, 1889; children—Alice G. (Mrs. Harry B. Hall), Josephine F. (Mrs. Harry W. Yoxall). Prof. physiology and anatomy, Columbus Medical Coll., 1875-82; prof. surg. gynecology chancellor, Ohio Med. U., 1892-99; prof. clin. surgery, Ohio State U.; surgeon and chief of staff Grant Hosp. Author: Operative Gynecology, 1898; also articles in med. jours. Devised operation for formation of artificial vagina by interposing loop of intestine between rectum and urinary bladder, 1904. Died Jan. 20, 1936.

BALDWIN, James Mark, Am. psychologist; b. Columbia, S.C., Jan. 12, 1861; s. Cyrus H. and Lydia Eunice (Ford) B.; A.B., Princeton, 1884, A.M., 1887, Ph.D., 1889; hon. D.Sc., Oxford (Eng.) U., 1900 (first hon. degree in science ever given by Oxford), U. Geneva, 1909; LL.D., Glasgow U., 1901, S.C. Coll., 1905; studied Leipzig, Berlin and Tübingen; m. Helen Hayes Green, Nov. 22, 1888; children—Helen Green (Mrs. John A. Sterrett), Elizabeth Ford (Mrs. Philip M. Stimson). Instr. French and German, Princeton, 1886; prof. philosophy Lake Forest (Ill.) U., 1887-89, U. Toronto, 1889-93; prof. psychology Princeton, 1893-1903; prof. philosophy and psychology Johns Hopkins, 1903-09, Nat. U. Mexico, 1909-13. Hon. pres. Internat. Congress Criminal Anthropology, Geneva, 1896; pres.-elect Internat. Congress of Psychology, 1909-13; succeeded William James as mem. Acad. Moral and Polit. Sci. in Inst. France. Recipient gold medal, Royal Acad. of Denmark, 1897; Cross, Legion of Honor (France). Mem. Am. Psychol. Assn. (pres. 1897-98). Author: Handbook of Psychology, 2 vols., 1889-91; Elements of Psychology, 1893; Mental Development in the Child and the Race, 1896; Social and Ethical Interpretations in Mental Development, 1898; Story of the Mind, 1898; Fragments in Philosophy and Science, 1902; Development and Evolution, 1902; Thought and Things, or Genetic Logic, 3 vols., 1906-11; Darwin and the Humanities, 1909; The Individual and Society, 1910; History of Psychology, 2 vols., 1913; Genetic Theory of Reality, 1915; France and the War, 1915; American Neutrality, 1916; The Super-State, 1916; Paroles de Guerre, 1919; Between two Wars—Memories and Opinions, 2 vols., 1926. Joint Author: History of Psychology in Autobiography, 1931. Co-founder, editor Psychol. Rev., 1894-1909; editor Dictionary of Philosophy and Psychology, 1901-06. A pioneer of Am. psychology; expert on child psychology; 1st to thoroughly develop idea that individual differences work with inherited mutations to influence course of evolution; founded psychol. labs. at U. Toronto, Princeton and restored lab. at Johns Hopkins. Died Paris, France, Nov. 8, 1934.

BALDWIN, John Thomas, Jr., Am. botanist, educator; b. Chase City, Va., Sept. 5, 1910; s. John Thomas and Lona Earle (Price) B.; A.B., Coll. William and Mary, 1932; Ph.D., U. Va., 1937; postdoctoral fellow Cornell U., 1937-38. Asst. prof. biology Coll. William and Mary, Williamsburg, Va., 1937-39, prof., 1946—, head dept. biology, 1952-62; instr. botany U. Mich., 1939-42; agt., asso. cytogeneticist U. S. Dept. Agr., Amazon Valley, 1942-44; mgr. Blandy Exptl. Farm, asst. prof. biology U. Va., 1944-46. Horticulturist, Econ. Mission to Liberia, U. S. Dept. State, 1947-48; prin. botanist for cortisone studies U. S. Dept. Agr., West Africa and Mexico, 1949-50. Mem. A.A.A.S., Am. Soc. Naturalists; Am. Genetics Assn., Am. Soc. Plant Taxonomists, Internat. Assn. Plant Taxonomy, Am., New Eng., Cal. bot. socs.; Soc. Study Evolution, Phi Beta Kappa, Sigma Xi, Phi Kappa Phi, others. Editor: The Great Dismal Swamp of Virginia and North Carolina. Research, publs. on cytogenetics, cytotaxonomy, cytogeography; developer techniques for rapid handling of chromosomes. Home: Box 1588, Williamsburg, Va. 23185.*

BALDWIN, Loammi, Am. botanist, civil engr.; b. North Woburn, Mass., Jan. 21, 1745; s. James and Ruth (Richardson) B.; m. Mary Fowle, 1772; m. 2d, Margery Fowle, 1791; 2 sons, George, Loammi. Apprenticed in cabinet making; engaged in surveying and engring., Woburn, Mass., circa 1765; apptd. lt. col. 38th Infantry Regt., Continental Army, 1775; discharged, 1777; represented Woburn in Mass. Gen. Ct., 1778-79, 1800-04; sheriff Middlesex County (Mass.), 1780-circa 1785; chief engr. Middlesex Canal (connecting Charles and Merrimac rivers), 1793-1803. Mem. Am. Acad. Arts and Scis. Discovered strain of apple later known as Baldwin. Died Woburn, Oct. 20, 1807.

BALDWIN, Maitland, Am. neurosurgeon; b. N.Y.C., Sept. 29, 1918; s. Alvi Twing and Esther (MacLeod) B.; grad. Choate Sch., 1935; Harvard, 1935-38; M.D., C.M., Queen's U., 1943; M.Sc., diploma in neurosurgery McGill U., Montreal, Que., Can., 1952; m. Shirley Alexandria Lewis, Oct. 7, 1960; children—Joan, Francis, Raymond. Lectr., McGill U., 1948-52; asst. prof. U. Colo., 1952-53; chief br. surg. neurology NIH, Bethesda, Md., 1953—, clin. dir. Nat. Inst. Neurol. Diseases and Blindness, 1961—; faculty Georgetown U., Washington, 1953—, clin. prof. surgery, 1961—. Mem. space medicine adv. group NASA, 1963—. Fellow A.C.S., N.Y. Acad. Scis.; mem. Royal Soc. Medicine (affiliate), Soc. Neurol. Surgeons, Harvey Cushing Soc., Neurosurg. Soc. Am. Editor: Temporal Lobe Epilepsy. Contbr. numerous articles to tech. jours., chpts. to books. Research on physiology human temporal lobe, effects low temperatures on brain function, effects surg. lesions on brain stem. Home: 10413 Lloyd Rd., Potomac, Md. 20854. Office: 9000 Wisconsin Av., Bethesda, Md. 20014.*

BALDWIN, Matthias William, Am. inventor; b. Elizabethtown, N.J., Dec. 10, 1795; s. William Baldwin; m. Sarah Baldwin (cousin), 1827, 3 children. Began career as jeweller, nr. Phila.; devised and patented process for gold plating; a founder Franklin Inst. for Betterment of Labor, 1824; became toolmaker in Philadelphia, 1825; was 1st man in U. S. to manufacture calico printers' rolls and bookbinders' tools; manufactured stationary engines, 1827; locomotives, 1831; constructed Old Ironsides (one of 1st Am. locomotives used in transp.), 1832; founded sch. for Negro children, 1835; founder M. W. Baldwin (now Baldwin Locomotive Works); mem. Pa. Legislature, 1854. Mem. Am. Philos. Soc., Am. Hort. Soc. Died Phila., Sept. 7, 1866.

BALDWIN, Robert Lesh, Am. biochemist; b. Madison, Wis., Sept. 30, 1927; s. Ira Lawrence and Mary (Lesh) B.; B.A., U. Wis., 1950; Ph.D. (Rhodes scholar), Oxford U., 1954; m. Anne Theodora Norris, Aug. 28, 1965. Faculty dept. biochemistry U. Wis., 1955-59; faculty dept. biochemistry Stanford, 1959—, prof., 1964—. Guggenheim fellow, 1958-59. Mem. Am. Chem. Soc., Am. Soc. Biol. Chemists, Phi Beta Kappa. Asso. editor: Jour. Molecular Biology, 1964—. Developed methods for measuring molecular weight and homogeneity of macromolecules with analytical ultracentrifuge; study of DNA replication and related problems by phys. chemistry. Home: 1243 Los Trancos Rd., Portola Valley, Cal. 94025. Office: Dept. Biochemistry, Stanford, Palo Alto, Cal. 94304.*

BALDWIN, Robert William, Brit. biochemist; b. Huddersfield, Eng., Mar. 12. 1927; s. William and Doris (Mellor) B.; B.Sc. with 1st class honors, U. Birmingham (Eng.), 1948, Ph.D., 1951; m. Lilian Haston, July 19, 1952; 1 son, Neil. Dorothy Temple Cross Travelling fellow, 1951-52; sr. research fellow in cancer U. Nottingham (Eng.), 1952-61, dir. Brit. Empire cancer Campaign Labs., 1961—. Sec., Brit. and Irish Group for Co-ordinating Human Cancer Studies, 1964—. Mem. Brit. Assn. Cancer Research (chmn. 1966—), Biochem. Soc., Brit. Soc. for Immunology. Editor: (with J. H. Humphrey) Autoimmunity, 1965. Research, publs. in chem. carcinogenesis, significance of immunity in devel. cancer. Home: 29 Parkside Av., Long Eaton, Nottingham, Eng.*

BALDWIN, Samuel Prentiss, Am. naturalist; b. Cleve., Oct. 26, 1868; s. Charles Candee and Caroline Sophia (Prestiss) B.; A.B., Dartmouth, 1892, A.M., 1894, D.Sc., 1932; LL.B., Western Res. U., 1895; m. Lilian Converse Hanna, Feb. 15, 1898. Admitted to Ohio bar, 1894, began practice law in Cleve.; discontinued practice because of ill health, 1902, and devoted attention to science, principally ornithology; dir. Baldwin Bird Research Lab. for the study of live wild birds; originator of bird-banding method used in U. S. Biol. Survey; chmn. bd. The Williamson Co.; pres. The New Amsterdam Co. Research asso. in biology Western Res. U. Trustee Cleveland Mus. Natural History. Wrote: Bird Banding by Means of Systematic Trapping (Linnean Soc. of N.Y.), 1920; (with F. C. Lincoln) Manual for Bird Banding; Measurements of Birds; Physiology of the Temperature of Birds. Died Dec. 31, 1938.

BALDWIN, William, Am. physician, botanist; b. Newlin, Pa., Mar. 29, 1779; s. Thomas and Elizabeth (Garretson) B.; M.D., U. Pa., 1807; m. Hannah M. Webster, 1808. Practiced medicine, Wilmington, Del., 1808-12; naval surgeon, St. Mary's, Ga., 1812-16; made bot. surveys of Del. and Ga., lived, collecte[d] specimens among Creek Indians in Ga. for several months; travelled to S.Am., 1816-17; apptd. botanist on expdn. to Rocky Mountains under command o[f] Maj. Stephen H. Long, 1819. Died Franklin, Mo., Aug 31, 1819.

BALDWIN, William James (St. John), mech. engr. b. June 14, 1844, on shipboard; birth recorded a[t] Waterford, Ireland; s. John and Giovanna Caterina (San Giovanni) B.; ed. Boston, and Charlottetown (St Dunstan's), P.E.I., 2 yrs.' spl. tng. in naval architecture; studied navigation drawing, engring. and physics; m. Began in mech. engring. and naval architecture, 1863; spl. work in naval constrn. during Civil War; was with Donald McKay at East Boston, in the construction of 3 monitors and the conversion of several blockade runners into U. S. cruisers; in the Brazilian service, as asst. naval constr., 1866-67. Wa[s] 1st domestic engr. in the high bldgs. of N.Y.; cons engr. and designer for the U. S. War Coll., Washington U. S. Immigrant Station, N.Y. Harbor, U. S. Soldiers' Home, Tenn., others, for Dept. Health, N.Y.C. hosps. and power-plants; over 24 yrs. cons. engr. N.Y Telephone Co. and Empire City Subway Co. Asso. editor Engring. Record, 1880-89; lectr. and prof. thermal engring., Poly. Inst., Bklyn. Mem. Commn. of Am Soc. M.E. that formulated the standard pipe thread (known as "the Briggs formula") for U. S. and Can. 1886; mem. internat. commn. for the formulation o[f] an internat. standard for pipes and fittings; mem spl. com. Am. Soc. M.E. for elec. screw threa[d] standards. Life mem. A.I.A. Author: Steam Heatin[g] for Buildings, 1881; Hot Water Heating and Fitting 1887; Baldwin on Heating, 1890; Data for Heatin[g] and Ventilation, 1897; An Outline of Ventilating an[d] Warming, 1899; The Ventilation of the School-Room 1901. Died May 7, 1924.

BALDY, John Montgomery, Am. gynecologist; b Danville, Pa., June 16, 1860; s. Edward Hurley an[d] Henrietta Cooper (Montgomery) B.; M.D., U. of Pa. 1884; m. Edith Lyndsey, Aug. 5, 1896; m. 2d, Hele[n] M. Constien, May 1919. Practiced, Scranton, Pa. 1885-91, and subsequently at Phila. Credited wit[h] performing one of 1st gastrectomies in U. S., 1893 devised operation for suspension of the uterus by su[...] turing round ligaments to posterior surface of uteru[s] (Baldy-Webster operation), 1903. Died Dec. 13, 1934

BALE, William F., Am. biophysicist; b. Augusta N.J., Jan. 2, 1911; s. Robert Osborne and Cora (Bales B.; B.A., Cornell U., 1932; Ph.D., U. Rochester, 1936 m. Mary Ella Cardew, Apr. 28, 1939; children—Emi[ly] C., Karen L., Mary E. With U. Rochester (N.Y.) Sch Medicine and Dentistry, 1932—, chief, asso. di[v.] Manhattan Project, 1943-46, chief div. radiology an[d] biophysics, 1947-65, prof. dept. radiation biology biophysics, 1948—; radiobiologist U. S. AEC, Wash ington, 1949-51, mem. adv. com. biology and med cine, 1963—; mem. bd. sci. cons. Sloan-Ketterin[g] Inst. for Cancer Research, 1960—. Mem. Am. Phys iol. Soc., Radiation Research Soc., Soc. Exptl. Bio[l]ogy and Medicine, Am. Phys. Soc., Am. Indsl. Hygien[e] Assn., A.A.A.S., Sigma Xi. Research, publs. on us[e] antibodies as radioactive carriers in cancer therapy biolog. effects radiation, radioactive tracers. Home 132 Highland Pkwy., Rochester, N.Y. 14620.*

BALECH, Enrique, Argentinian planktologist; b Telén, Argentina, Aug. 17, 1912; s. Emilio Fernan[dez] and Juana (Capdeville) B.; Professor on Natur[al] Scis., Superior Inst. for Professorship, Buenos Aire[s] Argentina, 1937; m. Electra Isabel Megías, June 1[...] 1939. Staff, Lab. Protistology, Argentine Mus. Natu[]ral Sci., 1934-37, curator, 1937-47; faculty U. L[a] Plata, Buenos Aires, prof. planktologist, 1961-62 head div. marine biology and hydro-biol. sta. Arger[n]tine Mus. Natural Sci., 1962—; vis. scientist Frenc[h] Labs., 1952, Scripps Instn. Oceanography, 1957-5[8] Hopkins Marine Sta., 1959, Tex. A. and M. U 1964-65. Collaborator, Hydrographic Service, Arger[n]tina, 1954—; mem. Argentine Com. Oceanograph[y] 1965—, Latin Am. Council Oceanography 1961— Recipient Holmberg prize, Argentina, 1944; Frenc[h] fellow, 1952; Guggenheim fellow, 1957-58. Fello[w] Argentine Council for Sci. Research; mem. Argentir[ne] Assn. Natural Sci., Mediterranean Com. Planktolog[y]

Internat. Sci. Esperantist Assn., Internat. Physological Soc., Sigma Xi. Author: Fitoplankton Marino (with H. J. Ferrando), 1964; also numerous articles. Research on fresh-water and marine planktonic protists, systematics, morphology and ecology especially of dinoflagellate and tintinnids, plankton of Atlantic Pacific, Arctic, Antaractic oceans, Mediterranean, No. Seas, Gulf of Mexico and Caribbean Sea; described new family, genera, numerous new species. Home: Casilla de Correo 64, Necochea, Argentina. Office: Pto. Quequén, Argentina.*

BALES, Robert Freed, Am. social psychologist, educator; b. Ellington, Mo., Mar. 9, 1916; s. Columbus Lee and Ada Lois (Sloan) B.; B.S., U. Ore., 1938, S.M., 1941; A.M., Harvard, 1943; Ph.D., 1945; m. Dorothy Louise Johnson, Sept. 14, 1942. Research asso. Yale, 1944-45; faculty sociology Harvard, Cambridge, Mass., 1945——, prof., 1958——, dir. lab. social relations, 1960-67. Mem. Am. Sociol. Assn., Am. Psychol. Assn., Am. Acad. Arts and Scis., Boston Psychoanalytic Soc. and Inst. (affiliate). Author: Interaction Process Analysis, 1950; (with T. Parsons, E. Shils) Working Papers in the Theory of Action, 1953; (with T. Parsons) Family Socialization and Interaction Process, 1955; (with A. P. Hare, E. F. Borgatta) Small Groups, 1955. Devel. of, publs. on research method, interaction process analysis, 1949; co-developer computer method for content analysis, 1960. Home: 61 Scotch Pine Rd., Weston, Mass. 02193. Office: William James Hall, Harvard, Cambridge 38, Mass.*

BALEZIN, Stephan Afancisivich, Russian chemist; b. Volodinsk, Perm, 1908; s. Afanosy and Pelageya (Krotora) B.; D.Sc., Leningrad Hertsen State Pedagogical Inst., 1944; m. Tamara Kaplan; children—Mark S., Natalia, Alexander. Became asso. with Moscow Lomonosov State U., 1930; with Carper Physics-Chemistry Research Inst., 1932-36, Kuibyshev Med. Inst., 1936-38, Lenin State Pedagogical Inst., Moscow, 1938——. Recipient State prize; named hon. worker of scis., 1964. Mem. Mendeleyev Chemistry Research Soc. Co-author textbook on phys. and colloidal chemistry, 1948; author manual, other works, also articles. Research and inventions in mechanisms of metal corrosion inhibitors. Home: Novo-Peschanaya St., Moscow. Office: Moscow Lenin State Pedagogical Inst., Kibalchich stt. 6, Moscow, USSR.*

BALFOUR, Andrew, Brit. physician; b. Edinburgh, Scotland, Mar. 21, 1873; s. T.A.G. and Margaret (Christall) B.; M.B., C.M., Edinburgh, 1894, M.D. (Gold medal thesis), 1898, B.Sc. in Pub. Health, 1900; D.P.H., Cambridge, Eng., 1897; D.Sc., LL.D.; student Strasbourg, France; m. Grace Nutter, 1902; 2 sons. Civil surgeon S. African War, 1900-01; dir. Wellcome Trop. Research Labs., Khartoum, Sudan, 1902-13; med. officer health, Khartoum, 1904-13; dir.-in-chief Wellcome Bur. Sci. Research, London, until 1923; dir. London Sch. Hygiene and Tropical Medicine, from 1923. Sci. adviser to Inspecting Surg.-Gen. Brit. Expeditionary Force, E. Africa, 1917; pres. Egyptian Pub. Health Commn., 1918; health commr. Mauritius, 1921, Bermuda, 1923. Created knight comdr. St. Michael and St. George; 1930; recipient Cragg's Research prize in tropical medicine, 1905; Mary Kingsley medal of Liverpool (England) School of Tropical Medicine. Fellow Royal College Physicians, Edinburgh; President Royal Society Tropical Medicine and Hygiene, 1925-27. Co-author: Lewis and Balfour's Public Health and Preventive Medicine; Balfour and Scott's Health Problems of the Empire, 1924; Author: War against Tropical Disease, 1920; 1st, 2d, 3d, 4th reports Wellcome Tropical Research Labs, Khartoum; Memoranda on Medical Diseases in Tropical and Sub-Tropical War Areas, 1919. Died Jan. 30, 1931.

BALFOUR, Sir Andrew, Scottish botanist; b. Denmiln, Fifeshire, Scotland, Jan. 18, 1630; s. Michael and Joanna (Durham) B.; ed. St. Andrews, Oxford, Eng., also in Paris and Montpellier, France, and Padua, Italy; M.D., Caen, France, 1661. Traveled in France and Italy several years; practiced medicine, London, Eng.; traveled on continent again; became tutor to Earl of Ross; returned to St. Andrews to practice medicine; moved to Edinburgh, Scotland; founder Pub. Bot. Gardens, Edinburgh, circa 1680, also started (with Sir Robert Sibbald) hosp. which later became Edinburgh Royal Infirmary. 1st pres. Royal Coll. Physicians Edinburgh. Balfouria genus Australian plants named in his honor; cultivated over 1000 species in bot. garden. Died Edinburgh, Jan. 10, 1694.

BALFOUR, Francis Maitland, naturalist, biologist; born Edinburgh, Scotland, Nov. 10, 1851; s. James Maitland and Lady Blanche Balfour; B.A., Trinity Coll., Cambridge (England), 1873. Worked under Dr. Michael Foster; joined Stazione Zoologica, Naples, Italy, circa 1873; apptd. lectr. animal morphology, Cambridge, 1876, spl. professorship animal morphology founded for him, 1882. Fellow Royal Soc., 1878 (Royal medal 1881). Author: A Complete Treatise on Comparative Embryology, (basis of modern embryology), 2 vols., 1880-81. His studies on elasmobranch fishes added to knowledge of devel. of kidneys and allied organs, spinal nerves, changes in ovum and early stages of embryo; early exponent of recapitulation; studied relation of sharks to other vertebrates and invertebrates. Died Switzerland, July 19, 1882.

BALFOUR, Henry, English ethnologist, archeologist; born England, 1863; the son of Lewis Balfour; educated at Trinity College, Oxford, England; M.A.; married Edith Wilkins, 1887. Curator, Pitt Rivers Museum, Oxford, from 1891; elected research fellow Exeter Coll., Oxford, 1903. Fellow Royal Soc., 1924; Soc. Antiquaries, Royal Anthrop. Inst., Zool. Soc., Royal Geog. Soc. (pres.); pres. Anthrop. Inst. Great Britain and Ireland (also mem. council from 1891), 1903-04, anthrop. sect. Brit. Assn., 1904, 29, Oxford Ornithol. Soc., Folk-lore Soc., Ashmolean Natural History Soc., Museums Assn., Somersetshire Archeol. and Natural History Soc., Prehist. Soc. E. Anglia, Oxford Anthrop. Soc.; corr. mem. anthrop. socs. Paris, Rome, Florence, Washington, Upsala, Vienna; hon. fgn. corr. Zool. Survey India. Author: The Evolution of Decorative Art; The Composite Bow and its Affinities; The Natural History of the Musical Bow; also numerous papers on ethnol. and archeol. subjects. Died Feb. 9, 1939.

BALFOUR, Sir Isaac Bayley, Scottish botanist; b. Edinburgh, Scotland, Mar. 31, 1853; s. John Hutton Balfour; ed. univs. Edinburgh, Strasbourg (France), Würzburg (Germany); M.D., D.Sc., LL.D., M.A.; m. Agnes Balloch; 1 dau. Regius prof. botany U. Glasgow (Scotland), 1879-84; fellow Magdalen Coll., also Sherardian prof. botany U. Oxford (Eng.), 1884-88; prof. botany U. Edinburgh, also King's botanist in Scotland, Regius keeper Royal Botanic Garden, Edinburgh, 1888-1922. Fellow Royal Soc., 1884. Mem. transit of Venus expdn. to Rodriguez, 1874; explored Island of Socotra, 1880. Died Nov. 30, 1922.

BALFOUR, John Hutton, Scottish botanist; b. Edinburgh, Scotland, Sept. 15, 1808; ed. St. Andrews U., M.A., Edinburgh U., M.D., 1832, LL.D. (St. Andrews, Edinburgh, Glasgow). Practiced medicine Edinburgh from 1834. prof. botany Glasgow, Scotland, 1841, then Edinburgh from 1845; (Dean, Medical Faculty for 30 yrs.) Emeritus prof. botany, Edinburgh, 1879; curator Royal Botanic Garden, Edinburgh. Queen's botanist for Scotland. Fellow Roy. Coll. Surgeons, Roy. Soc., 1856. Author: A Manual of Botany, 1848; Botanist's Companion, 1860; Guide to the Royal Botanic Gardens, 1873; Elements of Botany, 1876. Editor, Annals of Natural History, Edinburgh New Philosophical Jour. Died Edinburgh, Feb. 11, 1884.

BALFOUR, Margaret Ida, physician; daughter of Robert Balfour; ed. Edinburgh, Scotland, also London, Eng.; med. officer in charge Zenana Hosp., Ludhiana, India, 1892-95; med. supt. Women's Hosp., Nahan, India, 1896-1902, Lady Dufferin Hosp., Patiala, 1903-13; asst. to insp.-gen. Civil Hosps., Punjab, 1914-16; joint sec. Countess Dufferin's Fund, Delhi, also Simla, 1916-24; chief med. officer Women's Med. Service, India, 1920-24. Decorated comdr. Order Brit. Empire, 1924. Fellow Royal Coll. Obstetricians and Gynaecologists. Author: First Lessons for Country Midwives, 1920; Hukm Dey and Her Baby, 1921; The Care of the Baby, 1922; Infant Mortality in India and England, 1923; Maternal Mortality in Childbirth in India, 1927; Anaemia of Pregnancy, 1927; Maternity Conditions of Women Millworkers, Bombay, 1930; Indian Mothercraft, 1932; A Study of the Effect on Mother and Child of Gainful Occupation during Pregnancy, 1938; (with others) The Work of Medical Women in India, 1929. Died Dec. 1, 1945.

BALIANI, Giovanni Battista, Italian mathematician, physicist; b. Genoa, Italy, 1582; Senator of the Republic. Author: De motu naturali gravium solidarum, 1638; De motu naturali gravium solidarum et liquorum, 1646; Opere diverse, 1666. Said to have distinguished clearly between mass and weight and to have stated law of falling bodies. Died 1666.

BALIK, Joseph, Czechoslovakian ophthalmologist; b. Prague, Czechoslovakia, Mar. 10, 1914; s. Joseph and Anezka (Rybinova) B.; M.D., Charles U., Prague, 1945, C.Sc., 1958, D.Sc., 1966; m. Zdenka Smiskova, Aug. 31, 1946; children—Georg, Eva. Faculty, 1st Ophthalmic Clinic, Charles U., 1945——, vice dir., 1959——, asso. prof., 1966——. Mem. Med. Soc. J. E. Purkyne. Research, numerous publs. on physiology and pathology of lacrymal fluid, especially in patients suffering from keratoconjunctivitis sicca. Home: 8 Jerevanska, Prague 10, Czechoslovakia.*

BALIKCI, Asen, anthropologist; b. Istanbul, Turkey, Dec. 30, 1929; s. Kosma and Nidela (Ashimaki) B.; licence ès sciences géographiques Université de Genève (Switzerland), 1952; Ph.D. in Anthropology, Columbia, 1961; m. Verena Jeanne Ossent, Oct. 1, 1955; children—Nicolas, Anna. Ethnologist, Nat. Mus. Can., Ottawa, Ont., 1954-61; asso. prof. anthropology U. Montreal, Que., Can., 1961-66, chmn. Arctic Studies Group, 1966——; co-chmn. Program in Ethnographic Film, Cambridge, Mass., 1966. Cons. Ednl. Services, Inc., Cambridge, 1963——. Mem. Am. Anthrop. Assn., Societe des Americanistes de Paris, Assn. Internationale des Sociologues de Langue Francaise. Author: Development of Basic Socio-Economic Units in Two Eskimo Communities, 1964; Vunta Kutchin Social Change, 1963; also articles. Research on survey culture change in No. N.Am., social conflict, infanticide, suicide, ecol. adaption hunter-gatherers. Home: 476 Av. Outremont, Montreal 8, Que., Can.*

BALINKIN, Isay, physicist; b. Odessa, Russia, Sept. 14, 1900; s. A. and Mary (Sloutsky) B.; B.S. in Mech. Engring., Robert Coll., Istanbul, Turkey, 1925; M.Sc. (Hanna fellow) U. Cin., 1926, Ph.D., 1929: m. Ausma Vulfs, May 19, 1960. Came to U. S., 1925, naturalized, 1936. Faculty, U. Cin., 1927——, prof. exptl. physics, 1952——; designer Central Sci. Co., Chgo., 1930-31; dir. research Cambridge Tile Mfg. Co., Cin., 1936-65. Mem. U. S. Nat Com. on Illumination, 1951——; cons. Color Center, Hall Sci., N.Y. World's Fair, Interchem. Corp., 1964-65. Named Engr. of Year, Tech. and Sci. Socs. Council Cin., 1964. Fellow Am. Ceramic Soc., Optical Soc. Am.; mem. Intersoc. Color Council (chmn. 1950-51, Godlove award 1965), Am. Assn. Physics Tchrs., British Ceramic Soc., Medievalists (charter mem.). Research on mercury and mercury amalgam vapor lamps, wave machines, penetrometer for ceramic tile, measurement of small color differences, control of color uniformity, filtergraphs. Home: 1337 North Bend Rd., Cin. 45224.*

BALINSKY, Boris Ivan, zoologist; b. Kiev, Russia, Sept. 10, 1905; s. Ivan Martin and Elizabeth (Radzimovsky) B.; student U. Kiev, 1922-26; D.Biol. Scis., Acad. Sci. Ukranian S.S.R., 1936; m. Elizabeth Stengel, Mar. 15, 1947; children—John, Helen. Sci. officer Acad. Scis. USSR, 1925-41; prof. U. Kiev, 1933-41; prof. U.N.R.R.A., U. Munich, Germany, 1946-47; staff Agr. Research Council Gt. Britain, Edinburgh, Scotland, 1947-49; faculty U. Witwatersrand, Johannesburg, South Africa, 1949——, prof., head dept. zoology, 1954——, dean faculty, 1965-67; dep. dir. Zool. Inst., Ukranian Acad. Sci., Kiev, 1936-37. Fellow Royal Soc. S. Africa; mem. Internat. Inst. Embryology, S. African Electron Microscope Soc. Author: Development of the Embryo, 1936; An Introduction to Embryology, 1960; also numerous articles. Research on exptl. embryology, devel. in animals with electron microscope, systematics of stoneflies and dragonflies with descriptions of new species; discovered limb induction in amphibians. Home: 19 Oban Av., Blairgowrie, Johannesburg, South Africa.*

BALINT, Michael, psychoanalyst, psychiatrist; b. Budapest, Hungary, Dec. 3, 1896; M.D., U. Budapest, 1920; Ph.D., U. Berlin (Germany), 1924; M.Sc., U. Manchester (Eng.), 1945; m. Enid Flora Albu, 1953; 2 daus., 1 son. Tng. analyst, lectr. Hungarian Psychoanalytical Inst., 1926-39; dep. dir., later dir. Budapest Inst. Psychoanalysis, 1931-39; med. dir., psychiatrist N.E. Lancashire and Preston (both Eng.) child guidance clinics, 1942-45; cons. psychiatrist Tavistock Clinic, London, Eng., 1950-61; asst. dept. psychiatric medicine U. Coll., London Hosp., 1961——. Vis. prof. psychiatry Coll. Medicine, U. Cin., 1957-58. Fellow Brit. Psychol. Soc. (chmn. med. sect. 1955); mem. Brit Psychoanalytical Soc. (sci. sec. 1950-53). Author: Primary Love and Psycho-Analytic Techniques, 1953; The Doctor, His Patient and the Illness, 1957; Thrills and Regressions, 1959; (with Enid Balint) Psychotherapeutic techniques in Medicine, 1962. Work with individual differences of behavior in early infancy and objectively recording them, 1948, psychoanalytic tng. system, 1948, early developmental states, primary object love, 1949, termination of analysis, changing therapeutical aims and techniques in psychoanalysis, new beginning, paranoid and depressive sydromes, 1952, dynamics in tng. groups for psychotherapy.

BALIS, M. Earl, Am. biochemist; b. Phila., June 19, 1921; s. Harry and Frances (Spector) B.; B.A., Temple U., 1943; M.S., U. Pa., 1947, Ph.D., 1949; m. Bernice M. Lamborg, Dec. 30, 1945; children—Frances Andrea, Ellen Joyce. With Sloan-Kettering Inst., Rye, N.Y., 1949——, head nucleoprotein metabolism sect., 1957——, asso. mem., 1960-65, mem., 1965——; asso. prof. Med. Coll. Cornell U., 1954-66; prof. 1966——; vis. lectr. Adelphi U., 1963-64. Recipient Research Career award USPHS, 1963——. Mem. Am. Chem. Soc. (past sect. chmn.), A.A.A.S., Am. Soc. Biol. Chemists, Harvey Soc., Am. Assn. Cancer Research, Sigma Xi. Research, numerous publs. on metabolism of purines in normal and malignant tissues; determined biochem. action of anticancer drugs. Home: 21 Stonewall Circle, White Plains, N.Y. 10607. Office: 145 Boston Post Rd., Rye, N.Y. 10580.*

BALK, Christina Lochman (Mrs. Robert Balk), Am. geologist; b. Springfield, Ill., Oct. 8, 1907; d. David Julius and Nellie (Stanton) Lochman; A.B., Smith Coll., 1929, M.A., 1931; Ph.D., Johns Hopkins, 1933; m. Robert Balk, Mar. 15, 1947. Faculty, Mt. Holyoke Coll., 1935-47, asso. prof., 1946-47; lectr. U. Chgo., 1947; lectr. N.M. Inst. Mining and Tech., Socorro, 1954, prof. geology, 1957——; stratigraphic geologist N.M. Bur. Mining and Mineral Research, 1955-57. NRC grantee, 1934; Geol. Soc. Am. grantee, 1936-47; NSF grantee, 1959. Fellow Geol. Soc. Am., A.A.A.S.; mem. Paleontology Soc., N.M. Geol. Soc., Nat. Assn. Geology Tchrs. Sigma Xi, Phi Beta Kappa. Contbr. articles on paleontology and stratigraphy to tech. jours. Research in Cambrian paleontology and stratigraphy. Home: 1304 Vista Dr., Socorro, N.M. 87801.*

BALK, Robert, geologist; b. Reval, Estonia, May 31, 1899; s. Hugo and Mary (Kock) B.; student U. Greifswald, U. Gottingen, U. Breslau (all Germany);

Ph.D., 1924. Came to U.S., 1924, Geologist N.Y. State Mus., 1925-26, Minn. Geol. Survey, 1930, U. S. Geol. Survey, 1938-45; instr. Hunter Coll., N.Y.C. to 1935; chmn. geol. dept. Mt. Holyoke Coll., South Hadley, Mass., 1935-47; prof. geology U. Chgo., 1947-52; prin. geologist N.M. Bur. Mines and Mineral Resources, 1952-55; instr. structural geology Cal. Inst. Tech., Pasadena, 1954. Contbd. more than 100 articles to profl. jours. Founded (with others) what was later called granitetectonics; studied structural behavior of igneous rocks, N.Y., geology of Newcomb quadrangel, Albany, N.Y. Died Sandia, N.M., Feb. 19, 1955.

BALKE, Clarence William, Am. chemist; b. Auburn, O., Mar. 29, 1880; s. William Frederick and Clara Jacobena (Class) B.; A.B., Oberlin Coll., 1902; Ph.D., U. Pa., 1905; m. Minnie Maude Coddington, Apr. 21, 1905; children—Claire Coddington, Roger Redfield, Barbara, Hildegarde, Abigail Strader. Acting prof. physics and chemistry Kenyon Coll., Gambier, O., 1903-04; instr. chemistry Oberlin Coll., summer 1903; instr. chemistry U. Pa., 1906-07; asso. in chemistry U. Ill., 1907-10, asst. prof. inorganic chemistry, 1910-13, prof., 1913-16; chem. dir. Fansteel Products Co., North Chicago, Ill., from 1916. Mem. Am. Chem. Soc., Sigma Xi. Contbr. papers on rare metals to profl. jours. Discoverer of new methods of dehydrating, amalgamating, welding and processing; also methods for manufacture of tantalum and columbium. Died July 8, 1948.

BALL, Sir Charles Bent, Irish surgeon; b. Dublin, Ireland, Feb. 21, 1851; s. Robert Ball; med. scholarship, sr. moderatorship and gold medal in natural sci., Trinity Coll., Dublin; surg. travelling prize U. Dublin. M.D., M.Ch.; m. Annie J. Kinahan; 3 sons, 4 daus. Regius prof. surgery U. Dublin, also surgeon Sir Patrick Duns Hosp.; cons. surgeon Steevens, Monkstown and Orthopaedic hosps.; hon. surgeon to the King in Ireland; mem. Gen. Med. Council; Lane lectr., San Francisco, 1902; mem. Advanced Bd. Army Med. Service; Erasmus Wilson lectr. Royal Coll. Surgeons Eng., 1903; lord chancellor's cons. visitor in lunacy; med. referee under Workmen's Compensation Act; commnr. for edn., Ireland. Fellow Royal Coll. Surgeons Ireland; hon. fellow Royal Coll. Surgeons Eng. Author: The Rectum and Anus; Their Diseases and Treatment, 2d edit., 1908. Devised an operation for the relief of pruritus ani by cutting the nerve fibres which supply the area involved, 1887. Died Mar. 17, 1916.

BALL, Elmer Darwin, Am. entomologist; b. Athens, Vt., Sept. 21, 1870; s. Leroy A. and Mary A. (Mansfield) B.; B.S., Ia. State Coll., 1895, M.Sc., 1898; Ph.D., Ohio State U., 1907; m. Mildred R. Norvell, June 14, 1899. Asst. in zoology and entomology Ia. State Coll., 1895-97; asso. prof. Colo. Agrl. Coll., 1898-1902; prof. zoology and entomology Utah Agrl. Coll., 1902-07; dir. Expt. Sta. and Sch. Agr., Utah Agrl. Coll., 1907-16; state entomologist of Wis., 1916-18; prof. zoology and entomology Ia. State Coll., also state entomologist Ia., 1918-21, on leave as asst. sec. agr., 1920-21 dir. sci. work U. S. Dept. Agr., 1921-25; in charge celery insect investigations, Fla. Plant Bd., Sanford, Fla., 1925-28; dean Coll. Agr., dir. Agrl. Expt. Sta., U. Ariz., 1928-31, later prof. zoölogy and entomology. Fellow A.A.A.S., Entomol. Soc. Am.; mem. Am. (pres. 1918), Pacific Slope (pres., 1915-16) assns. econ. entomologists, Ecol. Soc. Am., Sigma Xi, Phi Kappa Phi, Gamma Sigma Delta (nat. pres. 1921-22). Author of systematic and life-history studies of Membracidae, Cercopidae, Jassidae and Fulgoridae, econ. studies of codling moth, grasshoppers and leaf hoppers, causing curly leaf of sugar beets and hopper burn of potatoes, biol. studies of celery tyer, also studies of poultry breeding; pioneer in devel. drying spray method for codling moth control; developed and organized methods to eradicate Am. foul brood of bees in Wis. Died Pasadena, Cal., Oct. 5, 1943.

BALL, Ephraim, Am. inventor; b. Lake Twp., O., Aug. 12, 1812; m. Lavina Babbs, circa 1835. Carpenter, Stark County, O., circa 1832; built threshing machine (with brother), circa 1838, built factory for mfg. parts, Greentown, O., 1840, manufactured Blue Plough and Hussey Reaper during 1840's, reorganized firm with new partners, 1851; developed Ohio Mower (1st of 2-wheeled flexible mowers), patented, 1857, produced mower in his factory from 1859. Died Jan. 1, 1872.

BALL, Eric Glendinning, biochemist, educator; b. Coventry, Eng., July 12, 1904; s. Charles Sturges and Nellie (Glendinning) B.; came to U. S. 1905, naturalized, 1919; S.B., Haverford Coll., 1925, A.M., 1926, D.Sc., 1949; Ph.D., U. Pa., 1930; A.M. (hon.), Harvard, 1942; m. Grace Snavely, Sept. 10, 1927. Asso. in physiol. chemistry Johns Hopkins Med. Sch., 1933-40; asso. prof. biol. chemistry Harvard Med. Sch., Boston, 1941-46, prof. 1946-—, chmn. div. med. scis., 1952-—, Edward S. Wood prof., 1962. Trustee Marine Biol. Lab., Woods Hole, Mass., 1942-—, Woods Hole Oceanographic Inst., 1953-57. Recipient Eli Lilly award in biochemistry, 1940; Cruziero do Sol, Brazil, 1945; certificate of merit U.S.A., 1948. Guggenheim Meml. Found fellow, 1937-38; Commonwealth Fund fellow, 1959. Fellow A.A.A.S.; Am. Soc. Biol. Chemists, Am. Chem. Soc.; Am. Acad. Arts and Scis., Biochem. Soc. Gt. Britain, Nat. Acad. Scis., Soc. Gen. Physiologists, Endocrine Soc., Sigma Xi, Alpha Omega Alpha (hon.). Editorial bd. Jour. Biol. Chemistry, 1950-60; bd. asso. editors Biochemistry, 1962-—. Research in biol. oxidations, enzymatic reactions under hormonal influence, pigments, malaria, hibernation. Home: 5 Byron Rd., Weston, Mass. Office: 25 Shattuck St., Boston 02115.*

BALL, Gordon Harold, Am. zoologist; b. Warren, Pa., Sept. 20, 1899; s. Leon Gerald and Helene (Goodman) B.; B.S., M.S., U. Pitts., 1921; Ph.D., U. Cal. at Berkeley, 1924; m. Meridian Greene, Aug. 12, 1940. Mem. faculty U. Cal. at Los Angeles, 1924-—, prof. zoology, 1946-—, chmn. dept., 1961-63. Mem. research council San Diego Zool. Soc., 1946-—, chmn., 1959-60. Served with U. S. Army, 1918. Mem. Soc. Protozoologists (pres. 1958; mem. editorial com. jour. 1954-57), Am. Soc. Parasitologists (pres. 1963), Western Soc. Naturalists (pres. 1941), A.A.A.S., Am. Soc. Zoologists, Am. Soc. Tropical Medicine and Hygiene, Microscopical Soc., Soc. Exptl. Biology and Medicine, Soc. Pathologie Exotique. Contbr. profl. jours. Research on parasites of marine invertebrates; life histories of protozoan parasites; physiology of mosquitoes; in vitro culture of malaria; paramecium life history; variation in mussels; hydrogen ion concentration of living tissue. Home: 290 Bronwood Av., Los Angeles 90049.

BALL, Henry Price, Am. elec. engr.; b. Phila., Jan 8, 1868; s. Joseph and Sarah (Price) B.; B.S., U. Pa., 1887, M.E., 1888; m. Anna Crobsy Daily, May 30, 1891 (dec. Apr. 23, 1934); 1 dau. Mabel; m. 2d, Margaret A. Capeliss, Apr. 3, 1937. With United Edison Mfg. Co., 1888-93, designed and patented many devices for distbn. of electricity for lighting, ry. and marine work; with Ward-Leonard Electric Co., 1893-1900, designed and patented complete line of rheostat theater dimmers and circuit breakers for control of electric current; chief engr. Gen. Incandescent and Arc Light Co., designed and patented apparatus for distbn. of high tension currents and automatic safety devices for control of same, large central sta. equipment; remote control switches and switchboard apparatus used by Commonwealth Edison Co., Chgo., Bklyn. Edison Co., others; cons. engr. Gen. Electric Co., 1903-08, designed and patented automatic machinery for reproduction of music played on piano and for manufacture of music rolls for use in player pianos; engr. heating dept. Gen. Electric Co., Pittsfield, Mass., 1908-14, designed and patented complete line of electric heating devices for domestic and indsl. use; mem. firm, chief engr. and factory mgr. S. Sternan & Co., Bklyn., 1914-17; supt. enamel factory of Lalande & Grosjean Co., Woodhaven, L.I., 1917-20; cons. engr. from 1920. Mem. Am. Inst. E. E., Edison Pioneers. Died May 1, 1941.

BALL, John, Irish geologist, botanist; b. Dublin, Ireland, Aug. 20, 1818; s. Nicholas and Jane (Sherlock) B.; ed. Christ's Coll., Cambridge (Eng.) U.; attended lectures John Stevens Henslow; m. dau. Nobile Alberto Parolini, 1856; 2 sons; m. 2d Julia O'Beirne, 1869. Studied glaciers, Zermatt, Switzerland, 1845; called to Irish bar, 1845; apptd. asst. poor-law commr., 1846-47, 1849-51; mem. Parliament for County Carlow, 1852; under sec. for colonies, 1855-57; mem. bot. expdn., Morocco, 1871. Fellow Linnean Soc., Geog. Soc. (all London) Royal Irish Academy, Royal Soc., 1868, Christ's Coll. (hon.), Alpine Club (1st pres. 1857). Author: Alpine Guide, 1863-68; Spicilegium Florae Moroccanae, 1878; Distribution of Plants on the South Side of the Alps, 1896. Classified flora of Alps; thought existing modes of transport insufficient to create present distribution of flora. Died South Kensington, Eng., Oct. 21, 1889.

BALL, John Geoffrey, Brit. metallurgist; b. Oakengates, Shropshire, Eng., Sept. 27, 1916; s. Harold and Ethel (Tudor) B.; B.Sc. with 1st class honors, U. Birmingham, 1940; m. Joan C. M. Wiltshire, Apr. 10, 1941. With Brit. Welding Research Assn., 1941, 45-49, Atomic Energy Research Establishment, Harwell, Eng., 1949-56; prof. phys. metallurgy Imperial Coll., U. London (Eng.), 1956-—, head metallurgy dept., 1957-—. Mem. manpower resources com. Ministry Tech. and Dept. Edn. and Sci., 1965-—; chmn. engring. physics com. Aero. Research Council, 1964-—. Mem. Inst. Welding (council, pres. 1965-66), Iron and Steel Inst., Inst. Metals, Instn. Metallurgists (pres. 1966-67). Research publs. on alloy structural steels, metallurgy of plutonium, materials behavior in nuclear reactor devel. Home: 3 Sylvan Close, Limpsfield, Surrey, Eng. Office: Imperial College, London S.W.7, Eng.*

BALL, Leon Anton Carl de, see de Ball.

BALL, Louise Charlotte, Am. oral surgeon; b. N.Y.C., May 28, 1887; d. Robert Jemison and Louise S. M. (Hansen) B.; A.B., Hunter Coll.; D.D.S., Coll. Dental and Oral Surgery (now Columbia U. Sch. of Dentistry), 1915; m. John B. Bundren, June 20, 1917. Asso. with Prof. W. J. Gies as dental investigator Columbia U. Sch. of Medicine, 1914; dental clinician apptd. to Bellevue Hosp. and Neponsit Hosp. for Children, N.Y., 1915-16; founder, N.Y. Sch. Dental Hygiene, Hunter Coll., 1916, also dean; founder courses in oral hygiene Columbia U., 1916, dir. 1916-19, asst. to dir. of extension courses, 1916-18; expert examiner in dental hygiene and dentistry Municipal Civil Service Commn., City of N.Y., 1917-19; dir. courses in gen. and dental roentgenology War Service Tng. Sch., N.Y.C., 1917-18; dir. Yorkville Dist. Dispensary for Oral Hygien and Dental Diagnosis, 1918; mem. adv. council Dept of Health, N.Y.C., 1918; founder, chmn. Dental Council of N.Y. State, Nat. Women's Party 1923-29 founder Dental Research Club of Women Dentist 1923, chmn., 1926; founder, hon. pres. Internat. Dental Health Found. for Children, 1920-—; conducte free ednl. dental clinics in 7 South Am. countries 1923, and preventive dentistry campaign for sch. children in S. Africa, 1927; pvt. practice dentistry, N.Y C., from 1915; conducted ann. essay contests on dental health and nutrition. Mem. Am. Dental Assn., Am Assn. Women Dentists, Kimberly Dental Soc. of Unior of S. Africa (hon. founder mem.), other assns. Author (bull.) Denticuring—Home Care of the Teeth and Nutrition (2 million copies printed in several langs.) Dental Riddlegrams, and many articles on dental hygiene and diet in mags. Producer of Say It wit Pearls (motion picture on dental health and nutrition) Died June, 1946.

BALL, Max W(alte), Am. geologist; b. Henry County Ill., Sept. 9, 1885; s. Lewis Henry and Jennie Ann (Hoffstatter) B.; E.M., Colo. Sch. Mines, 1906; LL.M. Nat. U., Washington, 1914; m. Amalia Maeder, Aug 18, 1915; children—Douglas Schelling, Jean Katherine (Mrs. I. R. Kosloff). With U. S. Geol. Survey 1906-16, chmn. oil bd., 1910-16; mining engr., lav officer U. S. Bur. Mines, 1916-17; chief geologis Royal Dutch Shell oil interests, Rocky Mountain region, 1917-18, gen. mgr., 1918-21; pres. Wester Pipe Line Co., 1921-27, Marine Oil Co. and asso. cos. 1922-28, Argo Oil Co., 1925-28; cons. practice, 1928 46; pres. Abasand Oils Ltd., 1930-43, Royal Royaltie Ltd., Denver, 1931-44; spl. asst. to dep. petroleun adminstr. Petroleum Adminstrn. for War, 1944-46 dir. oil and gas div. Dept. Interior, 1946-48, oil an gas cons., 1948-—, cons. to govt. Israel, drafting petroleum laws, 1950-53, govt. Turkey, 1953-54. Recipient Medal of Merit for distinguished achievemen Colo. Sch. of Mines, 1947; Gold medal for distinguished service Dept. of Interior, 1948. Fellow Am Geog. Soc., Geol. Soc. Am.; mem. Acad. of Polit. Sci. Am. Inst. Mining and Metall. Engrs. (vice chmn. Colo chpt.), Am. Assn. Petroleum Geologists (v.p. 1922-23 pres. 1923-24), Am. Petroleum Inst., Canadian Inst Mining and Metallurgy (chmn. No. Alberta sect. 1942 43). Author: This Fascinating Oil Business; also bulls. articles on geology, econs., internat. relationship Died Aug. 28, 1954.

BALL, Sir Nigel Gresley, Brit. botanist; b. Dublin Aug. 8, 1892; s. Charles and Annie Julia (Kinahan B.; ed. St. Columba's Coll, Dublin and Trinity Coll. M.A.; Sc.D.; M.D.; m. Florine Isabel Irwin, Dec. 28 1922; children—Charles, Ronald, Valerie. Asst. t prof. botany Trinity Coll., 1920-24; prof. botany Univ Coll. of Colombo (Ceylon), 1924-43; lectr. King's Coll. London, Eng., 1944-45, corrector, 1955-57, spl. lect botany, 1957-59. Mem. Linnean Soc. London. Contbr articles on plant physiology to profl. jours. Address 4 Ennerdale Rd., Richmond, Surrey, Eng.

BALL, Oscar Melville, Am. biologist; b. Miami, Mo. Aug. 25, 1868; s. William Henry and Eliza Ann (Braden) B.; B.A., U. Va., 1897; fellow in botan U. of Va., 1898; student U. Bonn, 1900, Leipzig 1900-03; M.A., Ph.D., Leipzig, 1903; m. Mary B Moon, June 16, 1900; 1 dau. Julia B. (Mrs. Rober M. Lee). Instr. in biology U. Va., 1896-97; prof chemistry, Miller Sch., Va., 1897-1900; prof. botan and mycology Agrl. and Mech. Coll. Tex., 1903-09 botany and zoology, 1909-11, biology, from 1911 Fellow Tex. Acad. Science (v.p.); mem. A.A.A.S., Am Econ. Soc., Nat. Inst. Soc. Sciences, Paleontol. Soc Am., Deutsche Botanische Gesellschaft. Author c various papers on plant physiology, soil bacteriolog and palaeobotany. Research in fossil flora of th Eocene. Died Nov. 11, 1942.

BALL, Richard William, Am. mathematician; b Streator, Ill., Aug. 16, 1923; s. Donald Escol an Jessie (Arensman) B.; B.A., U. Ill., 1944, M.A 1945, Ph.D., 1948. Faculty, U. Wash., Seattle, 1948 54; faculty Auburn (Ala.) U., 1954-—, prof. math 1960-—. Recipient bronze tablet U. Ill. Mem. Am Math. Soc., Math. Assn. Am., Phi Beta Kappa, Sigm Xi, Phi Kappa Phi, Phi Eta Sigma, Pi Mu Epsilon. Author: (with R. A. Beaumont) Introduction to Moder Algebra and Matrix Theory, 1954; Principles c Abstract Algebra, 1963. Research in abstract algebr especially groups and semi-groups. Home: 422 V Magnolia, Auburn, Ala. 36830.*

BALL, Robert C., Am. biologist; b. Elyria, O., De 10, 1912; s. Fredrick L. and Blanche (Cragin) B.; student Otterbein Coll., 1931-32; B.S., Ohio State U 1936; M.S., 1937; Ph.D., U. Mich., 1942; m. Bet Hewett, June 14, 1964; children—Barbara Ann Susan Lynne. Biologist Ohio Div. Conservation, Columbus, 1937, Mich. Inst. for Fish Research, 1937-4 46; faculty Mich. State U., East Lansing, 1947-— Commd. officer USPHS, 1942-45. Contbr. numero publs. on limnology to profl. jours. Home: 4586 He ron Rd., Okemos, Mich. Office: Dept. Fisheries an Wildlife, Mich. State U., East Lansing, Mich.*

BALL, Sir Robert Stawell, Brit. astronomer, mathematician; b. Dublin, Ireland, July 1, 1840; s. Robe and Amelia (Hellicar) B.; student Trinity Coll., Dublin, 1857-65; m. Frances Elizabeth Steele, 1868; sons, 2 daus. Tutor to sons of 3d Earl of Rosse at B

Castle, 1865-67; became prof. applied math. and mechanics Royal Coll. Scis., Dublin, 1867; Andrews prof. astronomy U. Dublin, royal astronomer of Ireland, 1874-92; Lowndean prof. astronomy Cambridge (Eng.) U., 1892-1913. Sci adviser Irish Lights Bds., 1882-1913. Knighted, 1886; Fellow Roy. Soc., 1873; mem. Royal Irish Acad. (sec. 1877-80, v.p. 1885-92), Royal Astron. Soc. (pres. 1897-99), Math. Assn. (pres. 1899-1900), Royal Zool. Soc. Ireland (pres. 1890-92). Author: The Theory of Screws: A Study in the Dynamics of a Rigid Body, 1876; A Treatise on the Theory of Screws, 1900; Treatise on Spherical Astronomy, 1908; also popular works on astronomy, including The Story of the Heavens, 1885; The Story of the Sun, 1893. Research on math. (mainly from a geometrical rather than analytical approach), theory of screw motions and their relations. Died Cambridge, Nov. 25, 1913.

BALL, Sydney Hobart, Am. geologist, mining engr.; b. Chgo., Dec. 11, 1878; s. Farin Q. and Elizabeth (Hall) B.; A.B., U. Wis., 1901, Ph.D., 1910; m. Mary Ainslie, Dec. 8, 1913; 1 dau., Mary Virginia. Geologist Mo. Bur. Mines, 1901-02; instr. U. Wis., Madison, 1902-03; field geologist U. S. Geol. Survey, Nev., also Cal.; began mineral expl ration Ryan Guggenheim group, Belgium Congo, 1907; with Rogers, Meyer, and Ball (active mineral deposit field work), 1917-49. Mem. WPB, World War II. Decorated Chevalier de l'Ordre du Lion, Officer de l'Ordre Royal du Lion, Comdr. de l'Ordre de Leopold II (Belgium). Pres. Explorers Club, Mining and Metall. Soc. Am., Soc. Econ. Geologists. Contbd. sect. on gems Minerals Yearbook, U. S. Bur. Mines; pub. annual review diamond industry in Jewelers' Circular Keystone. Responsible for 1st topographic and geologic mapping Death Valley; established trace metals div. U. S. Geol. Survey; studied Tertiary dikes, pleistocene glacial deposits (1st recognition mountain glaciers of 2 distinct ages in Colo. Front Range); opened diamond fields, Belgian Congo, also Angola; studied geology Miller County, Mo. Died N.Y., Apr. 8, 1949.

BALL, Walter William Rouse, Brit. mathematician; b. 1850; ed. Univ. Coll., London, Trinity Coll., Cambridge, Eng.; M.A.; m. Alice Mary Ball, 1885. Dep. prof. for J. K. Clifford, London, 1877; apptd. lectr. Trinity Coll., Cambridge, 1878, dir. coll. math. studies, 1891, sr. tutor and chmn. coll. edn. com., 1898, univ. rep. on Borough Council, 1905. Mem. governing bodies Westminster Sch., Cambridge Perse Sch.; mem. various examining, numerous other bds. and syndicates. Author: A Short Account of the History of Mathematics, 1912; Mathematical Recreations and Essays; An Essay on Newton's Principia; History of Mathematics at Cambridge; Text-Book on Algebra; A Guide to the Bar; several histories of Trinity Coll.; also papers contbd. to math. jours. Died Apr. 4, 1925.

BALL (or BALLE), William, Brit. astronomer; b. Eng.; s. Sir Peter and Ann (Cooke) B.; m. Mary Posthuma Hussey, 1 son, William. Joined meetings of Oxonian Soc. at Gresham Coll., 1659; helped found Royal Soc., 1660, 1st treas., 1662-63; left London, 1665; resumed his astron. studies at father's residence, Mamhead House, Devonshire. Made observations and drawings of Saturn (agreed with those made by Huygens), 1656-59. Died 1690.

BALLABIO, Camillo Benso, Italian physician; b. Milano, Italy, Apr. 17, 1912; s. Arturo and Ines (Garavaglia) B.; M.C., U. Pavia (Italy), 1936, Docente Patologia Medica, 1951; m. Adriana Zanoli, July 6, 1940. Staff, U. Milan, 1952—, prof. rheumatology, 1962—, dir. rheumatological dept. Med. Clinic, 1956—. Recipient Premia Acqui, 1951; Premia Ganassini, 1949; Premia Marzotto, 1954. Mem. Italian Soc. Internal Medicine and Rheumatology (sec.), Internat. League Against Rheumatism (asst. pres.). Author: (with C. Cavallero, S. Sala) Rheumatism Cortisone; also numerous articles. Research on gout; demonstrated biochem. condition of uric acid in blood; studied rheumatoid arthritis using ultracentrifugal pattern of proteins, crotison, intrapleural and intrapericardial steroids. Home: 3 Maino, Milan, Italy.*

BALLAND, Joseph-Antoine-Felix, French rural economist; b. Saint-Jullen-sur-Reyssouze, France, Jan. 16, 1845; ed. mil. sch., Strasbourg, France, 1865, Val-de-Grace, France; prin. pharmacist French Army; corr. French Acad. Scis., 1912; asso. Acad. Medicine; mem. Acad. Agriculture. Author: Travaux scientifiques des pharmaciens militares français, 1882; Les aliments, 1907; other books. Expert on corn prodn.; developed new food products from corn; suggested method of grading corn for quality. Died Paris, France, Jan. 6, 1927.

BALLANTINE, Stuart, Am. radio engr.; b. Germantown, Pa., Sept. 22, 1897; s. Charles Mansfield and Mary Stuart (Beverland) B.; ed. Grad. Sch. Harvard U., 1920-21, 1923-24; m. Virginia Gregory Orbison, June 18, 1927. With Marconi Co., 1914-15; bacteriol. lab., H. K. Mulford Co., 1916; research engr., Radio Frequency Labs., 1922-23, dir. research, 1927-29; engaged in private research, 1924-27; pres. Boonton Research Corp., 1929-34; pres. Ballantine Labs., Inc., elec. communication apparatus, Boonton, N.J., from 1935. Fellow Am. Phys. Soc., Acoustical Soc. Am., I.R.E. (pres. 1935); mem. Radio Club of Am., Franklin Inst. (mem. com. on sci. and arts from 1935). Recipient award for devel. Navy radio compass, U. S. Navy, 1921; Morris Liebmann Meml. award I.R.E., 1931; Elliott Cresson medal Franklin Inst., 1934. John Tyndall fellow at Harvard, 1923-24. Author: Radio Telephony for Amateurs, 1922; also articles on elec. communication. Invented direction finding compensator for control of antenna effect, developed linear detection and delayed automatic volume control for radio receivers; new type electrostethoscope; new methods of manufacturing vaccines based on effect of high hydrostatic pressures on micro-organisms; formulated vertical radio aerial theory; devised remote cut-off principle (variable mu) of vacuum tube construction; also free field method of calibrating condensor microphones. Died May 4, 1944.

BALLANTYNE, Horatio, Brit. chemist; b. 1871; s. Thomas and Jane (Chalmers) B.; ed. W. of Scotland Tech. Coll., also Athenaeum, Glasgow; m. Katherine I. Russell, 1899; 3 daus. Asst. to city analyst Glasgow, 1886-96; cons. chemist (patents, inventions, chem. manufactures), London, 1896-1928; adv. dir. Unilever, Ltd. Mem. com. Tech. Instns. Conf. on Patent Law Amendment, 1918-19; mem. Interdepartmental Com. on methods of dealing with inventions made by Govt. servants . . . , 1922; mem. com. on patent law and practice Bd. Trade, 1929. Fellow Chem. Soc., Royal Inst. Chemistry. Pub. papers, lectures on chemistry. Died Jan. 25, 1956.

BALLANTYNE, John William, Scottish pathologist; b. Eskbank, Scotland, June 4, 1861; s. John and Helen P. (Mercer) B.; M.B., C.M. (Buchanan scholar in midwifery and gynecology), Edinburgh U. (Scotland) U., 1883, M.D., 1889; m. Emily Mathew, 1889. Sr. asst. to prof. midwifery U. Edinburgh, 1885-90, lectr. in antenatal pathology and teratology, 1899-1900, examiner in midwifery, 1901-05, lectr. on midwifery to women students, 1916; lectr. on midwifery and gynecology Sch. Medicine, Royal Colls. and Sch. Medicine for Women, Edinburgh, 1890-1916; examiner in midwifery U. Aberdeen, 1895-99, also univs. Glasgow (Scotland) and Liverpool (Eng.); extra physician (in charge antenatal dept.) Royal Maternity Hosp., Edinburgh. Fellow Royal Coll. Physicians (Cullen prize 1902, hon. librarian), Edinburgh, Royal Soc. Edinburgh, Glasgow Obstet. and Gynaecol. Soc. (hon.), Am. Assn. Obstetricians and Gynaecologists (hon.); hon. mem. Am. Child Hygiene Assn.; socio honorario Soc. Scientifica Protectora da Infancia, Brazil; pres. Edinburgh Obstet. Soc., 1906-07, Edinburgh Med. Missionary Soc., 1907-12. Author: Diseases of Infancy, 1891; Diseases of Foetus, 2 vols., 1892-95; Teratogenesis, 1897; Manual of Antenatal Pathology and Hygiene, 1902-04; Essentials of Obstetrics, 1904; Essentials of Gynaecology, 1905; Expectant Motherhood, 1914; also more than 400 articles. Editor Teratologia, 1894-95, 2 issues of Green's Ency. and Dictionary of Medicine, 2d edit. Ency. Medica, Quinquennium of Medicine and Surgery. Sub-editor for Scotland of Internat. Clinics. Established 1st clinic for care of expectant mother and unborn child, 1901. Died Jan. 23, 1923.

BALLER, Warren R(obert), Am. psychologist; b. Trenton, Neb., June 19, 1900; s. Albert Ernest and Mary Louise (Taylor) B.; A.B., York Coll., 1923; M.A., U. Neb., 1928, Ph.D., 1935; postgrad. Columbia, U. Minn., U. S. Internat. U., 1967—; LL.D., George Willims Coll., 1962; m. Dorothy Gwendolyn Jensen, Mar. 16, 1941; children—William Warren, John Timothy, Elizabeth Clair. High sch. prin., Calloway Neb., 1923-24; supt. schs., Cheney, Neb., 1925-28, tchr. psychology York Coll., 1928-34, dean, 1933-34; faculty U. Neb., Lincoln, 1935—, prof. ednl. psychology, 1943—, dir. Student Counseling Center, 1942-48, dean lower div., 1949-52, chmn. dept. ednl. psychology, 1960-66; vis. prof. U. Tex., 1949, George Peabody Coll. for Tchrs., 1950, Northwestern U., 1950, U. Cal. at Los Angeles, 1951, 55-57. Gen. Edn. Bd. Postdoctoral fellow U. Chgo., 1940-41. Fellow Am. Psychol. Assn.; mem. Midwestern Psychol. Assn., Nat. Soc. Coll. Tchrs. Edn. (pres. 1964—), Neb. Assn. Coll. Tchrs. Edn., N.E.A., Sigma Xi, Phi Delta Kappa. Author: (with D. C. Charles) The Psychology of Human Growth and Development, 1961; Reading in the Psychology of Human Growth and Develpement, 1962; also articles. Research on mentally retarded, enuresis. Home: 3622 Kemper Ct., San Diego 92110.*

BALLESTER, Manuel, Spanish chemist; b. Barcelona, Spain, June 27, 1919; s. Manuel and Isabel (Boix) B.; Licenciado, U. Barcelona, 1944; Dr.Sc., U. Madrid, 1948; m. Montserrat Rodés, Sept. 20, 1949; children—Manuel, Eugene, Montserrat, Albert, Luis, Richard. Asso. prof. U. Barcelona, 1946-49, prof. phys. organic chemistry, 1953—; research scientist Higher Council for Sci. Research, Barcelona, 1949—, head phys. organic chemistry sect., 1952—; research fellow Harvard, Cambridge, Mass., 1949-51; vis. asso. Ohio State U., Columbus, Aerospace Research Labs., Wright-Patterson AFB, 1961-62. Recipient Agell prize, 1946; City of Barcelona prize, 1964; Centennial medal French Chem. Soc., 1965; knight-comdr. Order Alfonso X The Wise, 1965. Mem. Real Spanish Soc. Physics and Chemistry (award 1964), Assn. Harvard Chemists, Am. Chem. Soc. Research, numerous publs. on mechanism of Darzens condensation; pioneer in perchlorocarbon chemistry; synthesis of heat-resistant and chemically inert polymers; discovery of new spectral law and effects in benzene derivatives, first Inert Free Radicals, first magnetic plastic. Address: 7 Monterolas, Barcelona, Spain.*

BALLESTER HOYS, Augustin, Spanish surgeon; b. Sanlucar de Barrameda (Cadix), June 20, 1918; s. Augustin and Dolores (Hoys) B.; ed. U. Seville; licencié in medicine; m. Rosario Angulo, June 24, 1944; children—Augustin, Filomena, Firmin, Rosario, Pedro, Maria Carmen, Maria Dolores, Luis Felipe. Physician at municipal surg. unit and hosp. clinic; traumatologist Seguro Oficial de Enfermedad and C.N.S.A.T.; founder, dir. sch. for invalid children, Seville, Castellar. Recipient Izquierdo prize Athenee of Seville. Mem. Seville Acad. Medicine and Orthopedic and Traumatological Surgery, Sevillian Soc. Traumatology, Nat. Soc. Rehab., Assn. Invalids. Address: calle Porvenir, 6, Seville, Spain.

BALLET, Louis Gilbert, French physician; b. Ambazac, France, Mar. 29, 1853; ed. Paris; prof. mental pathology, Faculty of Medicine Clinic, U. Paris, later prof. history med. Faculty Medicine, from 1907; became chief Charcot's clinic Salpetrière, 1882. Mem. French Acad. Medicine. Author: Recherches anatomiques et cliniques sur le faisceau sensitif et les troubles de la sensibilité dans les lésions du cerveau, 1881; Lecons de clinique médicale, 1897. Specialized in nervous and mental diseases; described myxoedema; studied movement of tongue, also aphasia; fought against alcoholism. Died Paris, Mar. 16, 1916.

BALLHAUSEN, Carl Johan, Danish chemist; b. Copenhagen, Denmark, Apr. 4, 1926; s. Carl and Vibeke (Stein) B.; mag.scient., U. Copenhagen, 1954, Dr.Phil., 1958; m. Ingrid Vesterdal, Aug. 2, 1961; 1 son, Carl Christian. Asst., Chem. Lab. A, Tech. U. Denmark, 1954, amanuensis, 1957; postdoctoral fellow Harvard, 1955-56; vis. scientist Bell Telephone Labs., N.J., 1957-58; asst. prof. U. Chgo., 1958; prof. head Dept. Phys. Chemistry, U. Copenhagen, 1959—. Vis. prof. Columbia, 1962; Arthur D. Little vis. prof. Mass. Inst. Tech., 1963; vis. prof. John van Geuns Found., Amsterdam, 1965, U. La., 1967. Mem. Acad. Tech. Scis., Royal Danish Acad. Scis. and Letters. Author: Introduction to Ligand Field Theory, 1962; (with H. Gray) Molecular Orbital Theory, 1964. Danish editor Acta Chemica Scandinavica. Research on electronic structure of molecules. Home: Johannevej 1D, Charlottenlund, Copenhagen, Denmark.*

BALLI, Antonio, biologist; Ph.D. in Agrarian Scis. Prof., U. Modène, also in charge biology, zoology, zooculture and entomology courses at sch. medicine, instr. agrarian zoology at sch. scis.; founder, pres. Exptl. and Didactic Center of Apiculture of Province of Modène; dir. Biology Inst., U. Costa Rica, 1957—. Mem. Assn. Apiculture (founder, pres.). Contbr. numerous articles to sci. publs. Address: University of Costa Rica, San José, Costa Rica.

BALLING, Karl Josef Napoleon, chemist; b. Gabrielahütte, Austria-Hungary, Apr. 21, 1805; s. Michael and Anna (Rössler) B.; ed. at Polytechnikum, Prague, Czechoslovakia; also studied mining law at Prague U.; a son, Carl Albert. Became adj. in chemistry Prague Polytechnikum, 1824; founder first pvt. lab. of chem. analysis for various industries in Bohemia and Austrian monarchy. Author: Die Eisenindustrie Böhmens, 1867; Die Gährungschemie Prag, 4 vols., 1844-47; Der Getreidestein (zeilithoid) und seine Anwendung zur Biererzeugung, 1855; also numerous other books, and articles. Research in metallurgy, blast-furnace tech., sugar prodn., fermentation. Died Prague, Mar. 17, 1868.

BALLINGALL, Sir George, Scottish surgeon; b. Forglen, Banffshire, Scotland, May 2, 1780; s. Robert Ballingall; ed. St. Andrews; M.D., U. Edinburgh, 1819. Asst. to Dr. Barclay, anatomy lectr.; asst. surgeon 2d battalion 1st Royals, 1806, served in India, surgeon 33rd Foot, 1815-18, served in Paris; practiced medicine, Edinburgh, from 1818; apptd. lectr. mil. surgery U. Edinburgh, 1823, prof., 1825-55. Fellow Royal Soc., London, Edinburgh; corr. mem. French Acad. Scis. Author: Observations on the Diseases of European Troops in India, 1823; Observations on the Site and Construction of Hospitals; Outlines of Military Surgery 1852; also articles to med. jours. Described maduromycosis or madura foot (Ballingall's Disease), circa, 1818; believed hosp. gangrene best treated by cautery; used closed plaster method to treat compound fracture. Died Blairgowrie, Perthshire, Scotland, Dec. 4, 1855.

BALLIS, William Belcher, Am. polit. scientist; b. Portland, Ore., June 8, 1908; s. William and Bertha (Belcher) B.; A.B., Stanford, 1929; Ph.D., U. Chgo., 1936; m. Eunice Minette Schuster, Nov. 20, 1933; children—Nancy Eunice, William Albert. Instr., U. Chgo., 1932-37; fellow in Russian, Rockefeller Found., 1937-39; asst. prof. polit. sci. Ohio State U., 1939-41; Russian analyst Nat. Def. and War Agys., Washington, 1941-42; with American Embassy, Moscow, USSR, 1945-46; chief polit. sect. Eastern European br. div. research for Europe U. S. Dept. State, 1946-47, chief br., 1947-48; prof. polit. sci. U. Wash., Seattle, 1948-57; dir. internat. studies Nat. War Coll., 1949; prof. polit. sci. U. Mich., Ann Arbor, 1957—, dir. Center for Russian Studies, 1961-65; prof. Naval War Coll., Newport, R.I., 1964-65. Adviser, Inst. for Study USSR, Munich, Germany, 1953-54;

cons. to Army War Coll., 1961-62. Hon. fellow Inst. for Study USSR; mem. Am. Polit. Sci. Assn., Midwest Polit. Sci. Conf., Am. Assn. for Advancement Slavic Studies. Author: Legal Position of War, 1937; also articles, numerous revs. Editor, author: (with R. A. Rupen) Mongolian People's Republic, 1956. Pioneered 1st analytical studies on Mongolia and outer Soviet politics and fgn. policy; juridical status of war from ancient Greece to modern times. Home: 3011 Geddes Av., Ann Arbor, Mich. 48104.*

BALLOCH, Edward Arthur, Am. surgeon; b. Somersworth, N.H., Jan. 2, 1857; s. George Williamson and Martha J. (Palmer) B.; A.B., Princeton, 1877; A.M., 1891; M.D., Howard U., Washington, 1879; m. Lillian F. McGrew, June 8, 1886; 1 dau., Agnes M. In practice of surgery at Washington, 1879; prof. surgery, 1904, dean med. dept., 1909. Howard U.; attending surgeon and chmn. adv. staff, Freedmen's Hosp. Fellow A.C.S., So. Surg. and Gynecol. Assn.; mem. A.M.A., Med. Soc. D.C. (pres.). Died March 2, 1948.

BALLONIUS, see de Baillou, Guillaume.

BALLOU, Nathan Elmer, Am. chemist; b. Rochester, Minn., Sept. 28, 1919; s. Sidney Vaughan and Josephine (Elmer) B.; B.S., Duluth State Tchrs. Coll., 1941; M.S., U. Ill., 1942; Ph.D. (NRC fellow), U. Chgo., 1947; m. Anna Scovel White, Jan. 25, 1945; children—Robert Kendall, Douglas Paul. Research asso. Metall. Lab., U. Chgo., 1942-43; staff Clinton Labs., Oak Ridge, 1943-44, 45-56, Hanford Engring Works, Richland, Wash., 1944-45, Radiation Lab., U. Cal., Berkeley, 1947-48; head nuclear and phys. chemistry br. U. S. Naval Radiol. Def. Lab., San Francisco, 1948-59, head nuclear chemistry br., 1961—; chief dept. chemistry Centre d'Etude de l'Energie Nucleaire, Mol, Belgium, 1959-61. Mem. com. on nuclear sci. Nat. Acad. Sci.-NRC, 1958—, chmn. sub-com. on radiochemistry, 1962—. Recipient Superior Service award USN, 1966. Mem. Am. Chem. Soc., A.A.A.S., Sigma Xi. Discovery and characterization of new radio-active species; devel. radiochem. methods. Home: 1531 Campus Dr., Berkeley, Cal. Office: U. S. Naval Radiol. Def. Lab., San Francisco 94135.*

BALLOU, William Hosea, Am. mycologist, ichthyologist; b. Hannibal, N.Y., Sept. 30, 1857; s. Ransome R. and Mary Abigail (Green) B.; student Northwestern, 1877-81, U. Pa., 1896; spl. studies in natural science; hon. Sc.D., Ft. Worth (Tex.) U., 1911; Litt. D., Chicago Law Sch., 1920; LL.D., Coll. of Oskaloosa, Ia., 1921. Recorder U. S. Lake Survey, 1875-77; U. S. sci. survey of Niagara Falls, 1876; asst. engr. U. S. Yellowstone River Survey, 1878; govt. naturalist and representative of Harper's Weekly, Greely Relief Expdn., 1884; conducted crusade making animals safe in transport at sea and other like crusades, 1892-95; editor N.Y. Despatch (weekly), 1895-98; sec. Greater N.Y. Pub. Co., 1895-96; founder, sec. Westchester Free Hosp., 1892-95; pres. Pocantico Water Works Co., 1893-96; v.p. N.Y. & Westchester Water Co., 1891-98; owner and editor Sci. News Service. Del., 7th Internat. Geol. Congress, St. Petersburg, 1897; conducted govt. war propaganda to catch and eat more fish, under Herbert Hoover, 1917-18. Fellow Geopractic Soc. Am. (adv. editor), Acad. Soc. Internat. History, France; mem. Nat. Inst. Social Scis., Am. Soc. Ichthyologists and Herpetologists, Am. Soc. Mammalogists, Soc. Am. Mil. Engrs. Advocate of Agassiz' theory of multiple origin of man and associated animals; discovered new fungus species; founder movement to conserve wild mushrooms, 1908; donor of large collections of natural history to Northwestern U., and of fungi to State Museum, Albany, N.Y., New York Bot. Garden, and Lloyd Inst., Cin.; made geol. survey of Central Kans., discovering fossils of pre-fish, pre-amphibia, pre-reptiles with associated bivalves, and 1st plants in Devonian and Carboniferous Rock strata, and Upper Cretaceous marsupial, 1923; discovered fossil brachiopods in Silurian rocks, West Virginia, by which Edward Drinker Cope and James Hall fixed geol. age of Appalachian System of mountains as Silurian, 1889; discovered cancer on tail of a boa constrictor in Honduras, 1890; 1st to define cancer as a fungus, originating in reptiles and fish and breeding by infinitely small spores, later investing entire reptile or fish and communicated to man by drinking of infected water; scheduled 300 snakes and iguanas killed by cancer. Author novels and poems. Adv. editor Northwestern U. Alumni News and Living Age. Contbr. on popular science to Hearst syndicates. Closter, Bergen County, N.J. Died Nov. 30, 1937.

BALLS, Arnold Kent, biochemist; born Toronto, Ontario, Canada, Apr. 2, 1891; s. Alfred Z. and Amelia (Arnold) B.; B.S., U. Pa., 1912; Ph.D., Columbia, 1915, U. Prague (Czechoslovakia), 1929; m. Elizabeth Charlotte Franke, Apr. 2, 1922; 1 son, Kent Franke. Faculty, U. Pa., 1920-28; sr. chemist to head chemist U. S. Dept. Agr., 1932-51, collaborator, Albany, Cal., 1962—; prof. biochemistry Purdue U., Lafayette, Ind., 1951-62. Recipient Patriotic Civilian Service award U. S. Army, 1945; Superior Service award U. S. Dept. Agr., 1949, Distinguished Service award, 1952; E. V. McCollum award Johns Hopkins, 1959, Charles F. Spencer award Am. Chem. Soc., 1962. Mem. Inst. Food Technologists (charter, Nicholas Appert award 1962), Nat. Acad. Scis., Am. Soc. Biol. Chemists (sec. 1941-46, councillor 1946-49, editorial bd. 1952-58). Initial crystallization of papain, 1937, chymopapain, 1940, Beta-amylase, 1946, various derivatives of chymotrypsin, 1948-57, Alpha-amylase, 1948; research on mode of action of pancreatic lipase, phosphorylation, acylation and inhibition of chymotrypsin. Office: 800 Buchanan St., Albany, Cal. 94710.*

BALLS, William Lawrence, botanist; b. Garboldisham, Norfolk, Sept. 3, 1882; s. William and Emma Mary (Lawrence) B.; ed. (fellow 1909-13) St. John's Coll., Cambridge U.; Sc.D.; m. Florence Edith Tyrrell, 1909; 1 son. Botanist, Khedivial Agrl. Soc., 1904-10, Egyptian Govt. Dept. Agr., 1911-13, 27-33; cotton technologist Egyptian Ministry Agr.; designer, head exptl. dept. Fine Cotton Spinner's Assn., Bollington, 1915-26. Recipient Walsingham medal, 1096. Fellow Royal Soc., 1923, Textile Inst. (hon.); mem. Egyptian Inst. Author: The Cotton Plant in Egypt; The Development of Raw Cotton; Egypt of the Egyptians; Spinning Tests for Cotton Growers; Studies of Quality in Cotton; Analyses of Agricultural Yield. Research and publs. on growth, heredity, irrigation, instruments related to cotton; designed Giza Cotton Exptl. Sta., also system of pure seed supply for Egyptian cotton. Died July 18, 1960.

BALLY, Dorel, Rumanian physicist; b. Bucharest, Rumania, July 29, 1923; s. Leon and Duduca (Adania) B; B.Sc., U. Bucharest (Rumania), 1947; D.Physics, State U. Moscow (USSR) and Rumanian Dept. Edn., 1953; m. Ioana Martianu, Jan. 10, 1955; 1 dau., Ruxandra. Scientist, U. Tech., Bucharest, 1947-49; staff U. Bucharest, 1949—, asst. prof. physics, 1953—; div. chief Inst. Atomic Physics, Bucharest, 1953—. Author book, numerous articles. X-ray diffraction and spectroscopical studies of copper-nickle-iron and iron-nickle alloys; research on total reflection of neutrons, scattering of neutrons in gases and liquids, critical scattering of neutrons in ferromagnetics. Home: 35 B-dul Magheru, Bucharest. Office: Inst. Atomic Physics, P.O. Box 35, Bucharest, Rumania.*

BALMER, Johann Jakob, Swiss mathematician, physicist; b. Lausanne, Switzerland, May 1, 1825; s. Johann Jakob and Elisabeth (Rolle) B.; ed. at Basel, Karlsruhe, Berlin; Dr. Degree, 1849; m. Christine Pauline Rinck, 1868; 6 children. Lectr. on descriptive geometry U. Basel (Switzerland), 1865-90. Author: Die Naturforschung und die moderne Weltanschauung, 1868; Die Wohnung des Arbeiters, 1883; Zur Projektion des Kreises, 1884, also numerous others. Pioneer in research on structure of atom and spectrum analysis; formulated rule for wave lengths of lines visible in hydrogen spectrum (Balmer Series), which played an important role in Niels Bohr's hydrogen atom theory, 1885; most famous work was done in atomic theory. Died Basel, Switzerland, Mar. 12, 1898.

BALMONT, Vladimir Aleksandrovich, Russian agronomist, zootechnician; b. 1901; Grad. Siberian Inst. Agr. and Forestry, Omsk, 1926. Instr. Kazakhstan Zool. and Vet. Inst. and Kazakhstan Sheep-Raising Inst., 1932-37; dep. dir. Cattle-Raising Inst., 1936-47, dir., 1950-52. Decorated Order of Lenin; recipient Stalin prize, 1946. Mem. Kazakhstan Acad. Sci. (corr.), All-Union Lenin Acad. Agrl. Sci., All-Union Acad. Agr. Sci. (v.p. 1957-60). Co-Author: Sheep-Raising in Kazakhstan and Means of Improving It, 1939; author: Kazakh Fine-Wool Sheep, 1948; Improvement Trends in Fat-Tail Sheep, 1949; The Breeding of New Sheep Strains in Kazakhstan, 1956; New Strains of Livestock Bred in Kazakhstan, 1960; Productivity of Kazakh Fine-Wool Sheep on the Main Pedigree Farms, 1962. Research on improvement of sheep-raising and theoretical cattle-raising problems in Kazakhstan. Address: Filial Vsesoyuznoy akademii selskokhozyaystvennykh nauk, Alma-Ata, Kazakhstan SSR, USSR.

BALO, Joseph Matthias, Hungarian physician; b. Budapest, Hungary, Nov. 10, 1895; s. Joseph and Irene (Demeter) B.; M.D., U. Budapest, 1919; Rockefeller fellow Johns Hopkins U., 1923-24; m. Ilona Banga, May 26, 1945; 1 son, Joseph-Matthias. Faculty, U. Budapest, 1919-28, prof. path. anatomy, 1945—; head physician dept. pathology St. Stephen's Hosp., Budapest, 1926—; prof. Francis Joseph U., Szeged (Hungary), 1928-39, rector, 1939-40, dean med. faculty. Recipient Kossuth prize, 1955. Mem. Internat. Acad. Pathology, Royal Soc. Medicine London, Leopoldina, Gerontological Soc.; corr. mem. Hungarian Acad. Sci., Assn. German Neuropathologists and Neuroanatomists. Author: Die Unsichtbaren Krankheitserreger, 1935; (with B. Korpassy) Warzen, Papillome und Krebs, 1936; Die Erkrankungen der Weissen Substanz des Gehirns und des Rückenmarks, 1940; Lungenkarzinom und Lungenadenom, 1957; also numerous articles in field. Research on diseases of nervous system: observed concentric sclerosis of brain (Baló's disease), atherosclerosis—discovered elastase; studied cancer, etiology of pemphigus. Home: 21 Németvölgyi ut, Budapest XII. Office: 26 Öllöi ut, Budapest VIII, Hungary.*

BALOGH, Thomas, economist; b. Budapest, Hungary, Nov. 2, 1905; s. Emil and Eva (Levy) B.; Dr.Rer.Pol., Budapest U., 1927; postgrad. U. Berlin, Harvard; M.A., Oxford (Eng.) U., 1927; m. Penelope Tower, Mar. 17, 1945; children—Stephen, Christopher, Tereza. Rockefeller fellow, 1928-30; staff League of Nations, 1931; economist City of London, 1931-38; staff Nat. Inst. Econ. Research, 1938-44; faculty Oxford U., 1939—, fellow Balliol Coll., 1945—, reader in econs., 1960—; cons. to prime minister, 1968—. Econ. adviser Brit. Cabinet, 1964—, Malta, 1955-58, Jamaica, 1956, 61-62, also agys. of UN. Author: Studies in Financial Organization, 1946; Dollar Crises, 1949; Unequal Partners, 2 vols., 1961; Economics of Power, Poverty, 1966; also articles. Established importance of unequal tech. advance on internat. trade and devel., need for reform of ednl. systems in less developed countries; developed new critique of classical econs. in their application to indsl. and poor agrl. countries. Home: Wellside Well Walk, N.W. 3., London. Office: 10 Downing St., London S.W.1., Gt. Britain; also Graduate Center, Balliol College, Oxford, Eng.*

BALON, Eugeniusz Kornel, Czechoslovakian biologist, zoologist; b. Orlová, Czechoslovakia, Aug. 1, 1930; s. Józef and Jaromila (Zavicaková) B.; RNDr, Charles U., Prague, Czechoslovakia, 1953, CSc., 1962; m. Tatiana Lisková, June 21, 1954; 1 son, Janusz. Demonstrator, Charles U., 1953; asst. Fishery Research Sta., Fishery Research Sta., Trnava, Czechoslovakia, 1954; 1st dir. Fishery Research Lab., Bratislava, Czechoslovakia, 1954—; fish biologist Research Lab. Slovakian Akad. Sci., 1956—; asst. prof. Fish Inst. in Agr., since 1962—. Mem. Slovakian Zool. Soc. (sec. 1965—), Sci. Group for Underwater Study, Am. Soc. Ichthyologists and Herpetologists. Author: Rust plotice, 1955; Ryby Slovenska, 1966; also numerous articles. Editorial bd. Jour. Potapec, 1964, Act.Rer.Nat.Mus. Slovenica, 1966—. Research on evolution, changes and dynamics of Danubian fishes, morphoecological adaptations, life history studies, periodization and its terminology from fertilization to death, critical periods without suitable food, abundance and ichtyomass studies of Danube water with poisoning, ecol. study Antill reefs, theory of evolution of trout. Address: 806 Zelezná Studienka, Bratislava, Czechoslovakia.*

BALSAMO, Giuseppe (Conte Alessandro Cagliostro), Italian alchemist, magician; b. Palermo, Italy, June 8, 1743; s. of a shopkeeper; ed. Monastery of Caltagirone, Sicily; expelled for misconduct; studied under the Greek, Althotas, Rhodes; m. Lorenza Feliciani. Fled from Sicily after trouble with the authorities; traveled in Greece, Egypt, Arabia, Persia, Rhodes, Malta; introduced to society in Rome and Naples, Italy; visited London, Eng., and Paris, France, 1771; posed as the Grand Copt of the order of Egyptian Masonry and organized many lodges; imprisoned in the Bastille, also Fleet prisons; went to Rome where he was tried and condemned to death for heresy; the sentence was commuted to life imprisonment; his wife was sent to a convent; with his wife was known as a forger, physician, alchemist, and mesmerist. Died in prison of San Leo nr. Rome, 1795.

BALSAMO-CRIVELLI, Giuseppe, Italian naturalist; b. Milan, Italy, Sept. 1, 1800; became prof. mineralogy and zoology U. Pavia (Italy), 1851, apptd. prof. zoology and comparative anatomy, 1863. Contbd. many memoirs in all areas natural sci.; research and publs. on heterogenesis; identified Botrytis Bassiana (fungus causing muscardine in silk worm), 1835. Died Pavia, Nov. 15, 1874.

BALTA ELIAS, Joseph, Spanish physicist; b. Vilafranc del Panadès, Aug. 3, 1893; s. Joseph and Josefa Elias; ed. U. Barcelona, U. Madrid; Ph.D. in Physics; m. Adela-Calleja, July 10, 1935; children—José, Maria, Ferdinand, Marie, François, Josefa. Asst. prof., meteorologist U. Barcelona, 1933-41; prof. physics Salamanca U., 1941-64; prof. electronics and electricity U. Madrid. Mem. Royal Spanish Soc. Physicists and Chemists, French Soc. Electronicians and Radioelectricians, Acad. Sci. Madrid. Author numerous works including: Apropos de l'antenne de Hertz, 1927; Enigmas actuales phanteodos por la radiacion cosmica; Ein genialer Voläfer unseres wissenschaftlichten Zeitalten; Leonardo da Vinci, 1955. Studies in electronics, radioastronomy, radioelectricity, solid state physics, nuclear physics, history of science. Home: calle Ministro Ibeñez Marten. Office: Faculty of Sciences, University of Madrid, Madrid, Spain.

BALTHAZARD, Victor, French physician; b. Paris, France, Jan. 1, 1872; studied at College Chaptal and l'École Polytechnique; became mem. faculty medicine U. Paris, 1919. Decorated officer Legion of Honor. Mem. Acad. Medicine. Author: Une plaie sociale; Les avortements criminels. Worked on radiology of alimentary canal; used bismuth to study movements of stomach and intestine. Died 1950.

BALTZER, Arnim Richard, geologist; b. Zwochau, Saxony, Germany, Jan. 16, 1842; s. Friedrich and Frederike Sophie (Rühlmann) B.; studied sci. Zurich, Switzerland, 1860-63; doctorate Bonn, Germany, 1864. Became prof., rector indsl. dept., canton sch. of Zürich, 1869; named prof. U. and Technische Hochschule Zurich, 1873; became prof., Bern, Switzerland, 1884. Author: Beiträge zur geologischen Karte der Schweiz, 1880. Publs. and research on central alpine granite massif, especially Aar massif; worked on deck theory of Alps, ice age sedimentation of Swiss central country and South side of Alps, volcanology, structural nature of contact of limestone to gneiss. Died Hilterfingen, Thunersee, Germany, Nov. 5, 1913.

BALTZER, Fritz, Swiss zoologist; b. Zurich, Switzerland, Mar. 12, 1884; s. Armin and Anna (Berthold) B.; Ph.D. in Zoology U. Wurzburg (Germany), 1908; Dr. honoris causa U. Strasbourg (France), U. Freiburg (Germany); m. Frieda Kocher-Steinhauslein. Prof. extraordinary zoology univs. Wurzburg, Freiburg; prof. in ordinary U. Berne (Switzerland) from 1921, rector, 1938-39, prof. emeritus (ret.), from 1954; vis. prof. embryology Ia. State U., 1948-49. Mem. Academia Nazionale dei Lincei, Rome, Italy, Zool. Soc. France, Zool. Soc. Belgium, Swiss Soc. Natural Scis. Author: Theodor Boveri, Leben und Werk. Research, publs. on genetics, biochemistry of devel. sea-urchins, sex determination (bonellia), devel. and morphology (echiurids), physiology of the sense (spiders), exptl. embryology, merogony, chimeras of amphibians; (with P. S. Chen) studies on chromosomes and bastards of sea-urchins. Address: Finkenhubelweg 6, Berne, Switzerland.

BALTZER, Heinrich Richard, German mathematician; b. Meissen, Germany, Jan. 27, 1818; s. Andreas Karl and Augusta Charlotte (Kenzelmann) B.; doctorate, U. Leipzig (Germany), 1841; m. Jenny Gottheiner. Faculty, Dresdener Kreuzschule, 1842-68; prof. U. Giessen (Germany), 1869; mem. Royal Sci. Assn. Saxony. Author 1st German textbook on determinants, 1857; Elemente der mathematik (introduced sci. rigor in elementary math.); Analytische Geometrie, 1882. Published 1st vol. of A. F. Möbius work; copublisher work of C. G. J. Jacobi, 1884. Died Giessen, Nov. 7, 1887.

BALWANI, Jethanand Hotchand, pharmacologist; b. Mirpukhas, Sind, Pakistan, June 10, 1920; s. Hotchand Lalchand Balwani; M.B.B.S., Bombay (India) U., 1948, M.Sc., 1957, M.D., 1957; F.C.P.S., Coll. Physician and Surgeon, Bombay, 1955; m. Vishini W. Alimchandani, Oct. 14, 1959; children—Priscilla, Shirley. Faculty, Govt. Med. Colls., Maharashtra, India, 1949-—, prof. pharmacology, 1960-—; ICMR fellow Central Drug Research Inst., Lucknow, India, 1955-56; TCM fellow Down State Med. Centre, Bklyn., 1960-61; dir. Hindustan Antibiotics, Pimpri, India. Mem. Assn. Physiologists and Pharmacologists India, Advancement Med. Edn. in India. Research, publs. on analgesics, liver antipruritics, antihistaminics. Home: 403 Sea Croft, 104 Wode House Rd., Bombay 5, India. Office: B.J. Med. Coll., Poona-1, India.*

BALY, Edward Charles Cyril, Brit. chemist; b. Feb. 9, 1871; s. Edward Ely Baly; ed. Univ. Coll., London, Eng.; m. Ellen Agnes Jago, 1902; 4 sons. Asst. prof. chemistry, lectr. in spectroscopy Univ. Coll., London, 1908-10; Grant professor of inorganic chemistry University of Liverpool (England), 1910-37; deputy inspector of high explosives Liverpool area, 1915-21. Fellow Royal Soc., 1909, Institute Chemistry. Author: Spectroscopy, 1929; Photosynthesis, 1940; also many papers. Showed by expt. that small quantities of sugars and other organic substances are generated from mixtures of simple substances such as water, ammonia, carbon dioxide, under the influence of light. Died Jan. 3, 1948.

BALY, William, English physician; b. Norfolk, Eng., 1814; student Univ. Coll., London, Eng., 1831, St. Bartholomew's Hosp., 1832, also in Paris, Heidelberg, Berlin, 1834; M.D., 1836; practiced in London, then Devonshire; apptd. physician Millbank Penitentiary, 1840; became lectr. St. Bartholomew's Hosp., 1841, asst. physician, 1854; apptd. physician attendant on Queen and Royal Family, 1859. Fellow Coll. Physicians, Royal Soc., 1847. Author: (with W. W. Aull) Report on Cholera, Reports on Epidemic Cholera, 2 parts, 1854. First to realize that dysenteric sloughs in large intestine are related to ulcers of enteric fever in small intestine. Died Jan. 28, 1861.

BÄLZ, Erwin Otto Eduard von, see von Bälz, Erwin Otto Eduard.

BALZALGETTE, Joseph William, Brit. engr., b. Enfield, Eng., Mar. 28, 1819; pupil Sir John Benjamin McNeill; engr., Westminster, 1842; became engr. Commn. on London Drainage, 1849; chief engr. Metropolitan Bd. Works, London, Eng., 1855-89; engr. London drainage system, 1858-75, Thames embankments, 1862-74. Created knight, 1874; decorated companion Bath, 1871. Mem. Instn. Civil Engrs. (pres. 1884). Brought about completion new drainage system, London, 1865; designed bridges across the Thames at Putney and Battersea. Died Mar. 15, 1891.

BALZER, Felix, French dermatologist, syphilologist; b. Chateaubriant, Loire-Inférieure, France, Apr. 4, 1849; M.D. 1878. Became physician, Rennes, France; went to Paris, 1873; dir. histological lab. amphitheatre of hosp., 1877-80, also at Hosp. St. Louis, 1880-85, chief of service, 1896-1916. Mem. Académie de Médecine, 1908; pres. Société Francais de Dermatologie et Syphilographie, 1909-14. Author: Thérapeutique des Maladies Vénériennes, 1894; Maladies Vénériennes, 1906; (with others) Maladies des bronches et des poumons, 1910. Initiator use arsenic compounds in treatment syphilis; 1st to use bismuth treatment syphilis, 1889; identified causes generalized infection gonorrhea; used lactic acid against pelada, med. baths and oil juniper against psoriasis. Died Paris, Mar. 16, 1929.

BAMANN, Eugen, German chemist; b. Gundelfingen-Danube, Jan. 14, 1900; s. Hermann and Anna (Diebold) B.; student chemistry, pharmacology, alimentary chemistry at Munich and Wurtzburg; Ph.D.; m. Hanna Bosch. Instr. at Munich, 1931; sect. dir. Technische Hochschule, Stuttgart, 1931; prof. U. Tübingen, 1935, U. Prague, 1941; faculty U. Munich, dir. Inst. Pharmacy and Alimentary Chemistry, 1948. Mem. com. of direction Inst. Control of Medications; mem. sci. council German Inst. Research of Alimentary Chemistry. Mem. German Pharm. Soc. (pres.), Real Acad. de Farmaci (Spain) (corr.), German Acad. Naturalists, Léopoldina (Halle), Acad. Pharmacy of Paris (fgn.). Author: Die Methoden der Fermentforschung, 1941; Chemische Untersuchung von Arzneigemischen, Arzneispezialitaten und Giftstoffen, 1951. Research in biochemistry and pharmaceutical chemistry. Home: Tizianstrasse 129, 8 Munich 19. Office: Sophienstrasse 10, 8 Munich 2, Germany.

BAMBERGER, Eugen, German chemist; b. Berlin, Germany, July 19, 1857; studied medicine, then chemistry (with R. Bunsen, W. A. Hofmann, and K. Liebermann) in Berlin and Heidelberg. Asst. to Rammelsberg in Berlin, 1882, to A. von Baeyer, U. Munich, 1884; became pvt. lectr., 1885, prof. organic chemistry, 1892; went to Eidgenössische Technische Hochschule, Zurich, Switzerland, 1893, resigned, 1905; afterwards did only pvt. experimenting. Numerous articles and pure research on guanidine derivatives, retene, pyrene; discovered benzazimide, isomeric nitroso-compounds, isodizo compounds; characterized alicyclic compounds; worked in diazo-chemistry; reduced naphthylamines; made photochem. investigations of benzaldehyde derivatives; discovered dimethylaniline. Died Ponte-Tresa/Tessin, Dec. 10, 1932.

BAMBERGER, Max Georg Matthias, Austrian chemist; b. Kirchbichl nr. Kufstein, Austria, Oct. 7, 1861; s. Matthias and Philamena (Sauter) B.; ed. Tech. U., Vienna, Austria; degree in Giessen, 1891; m. Minie Bauer, July 21, 1902; 1 dau. Became asst. to Weidel at Agrl. U., 1886, to A. Bauer at Tech. U., 1888; named lectr. for organic chemistry, 1892; prof. inorganic chemistry Tech. U., Vienna, 1905-—. Mem. Internat. Nobel Prize-Jury Commn. Mem. Leopoldina-Karolinischen Deutschen Akademie der Naturforscher (Halle). Research and numerous publs. on resins (Haitzinger prize Vienna Acad.); improved gas masks (with F. Böck); 26 patents. Died Vienna, Oct. 28, 1927.

BAME, Samuel Jarvis, Jr., Am. physicist; b. Lexington, N.C., Jan. 12, 1924; s. Samuel Jarvis and Stella (Davis) B.; student Catawba Coll., 1941-42, Pa. State U., 1943-44; B.Sc., U. N.C., 1947; Ph.D., Rice U., 1951; m. Joyce Carleton Fancher, June 21, 1956; children—Karen Joyce, Dorthe Ann, Barbara Joan. Mem. staff Los Alamos Sci. Lab., 1951-—. Fellow Am. Phys. Soc.; mem. Am. Geophys. Union, Sigma Xi, Phi Beta Kappa. Study of low energy nuclear interactions, neutron physics; exptl. studies of neutrons and protons in space, solar wind interactions with geomagnetic field of the earth, geomagnetic tail. Home: 164 Dos Brazos. Office: Los Alamos Sci. Lab., Los Alamos 87544.*

BAMFORD, Clement Henry, English chemist; b. Stafford, Eng., Oct. 10, 1912; s. Frederic Jesse and Catherine Mary (Shelley) B.; B.A., Trinity Coll., Cambridge (Eng.) U., 1934, M.A., Ph.D., 1938, Sc.D., 1953; m. Daphne Ailsa Stephan, Dec. 10, 1938; children—Stephanie Catherine, Alan Charles. Fellow, Trinity Coll., 1937; dir. studies in chemistry Emmanuel Coll., Cambridge U., 1937; with Inter-Services Research Bur., Ministry Econ. Warfare, 1941-45; with Fundamental Research Lab. Messrs. Courtaulds Ltd., Maidenhead, Eng., 1945-62, head lab., 1947-62; Campbell Brown prof. indsl. chemistry U. Liverpool (Eng.), 1962-—, dean faculty sci., 1965-—. Fellow Royal Soc., 1964, Royal Inst. Chemistry (Meldola medal 1941, past chmn. Thames Valley sect.), Chem. Soc., Soc. Chem. Industry; mem. Faraday Soc. (past mem. council, v.p. 1963-66). Author: (with A. Elliott, W. E. Hanby) Synthetic Polypeptides, 1956; (with W. G. Barb, A. D. Jenkins, P. F. Onyon) The Kinetics of Vinyl Polymerization by Radical Mechanisms, 1958; also numerous articles. Research in free radical polymerization, synthetic polypeptides. Home: Broom Bank, Tower Rd, Prenton, Birkenhead, Cheshire, Eng. Office: Dept. Inorganic, Phys. and Indsl. Chemistry, Donnan Labs., U. Liverpool, Liverpool 7, Eng.*

BAMMERT, Karl, German engr.; b. Neu-Ulm, Germany, Dec. 13, 1908; s. Karl and Sabina (Bucher) B.; diploma engring. U. Karlsruhe, 1939, Dr.Eng., 1943; m. Alwine Heithorn, Dec. 28, 1939; children—Karl, Werner, Ulrich. Power sta. engr., 1933-38; scientist U. Karlsruhe, 1939; head dept. aircraft establishment, Brunswick, Eng., 1940-45, head dept. turbomachinery RAF, 1945-47; cons. engr. Ministry of Supply, London, 1947-48; head devel. dept. Gutehoffnungshutte Sterkrade AG Oberhausen, 1948-55; prof., dir. Inst. Turbomachinery, Gas Dynamics and Nuclear Engring., U. Hanover, 1955-—. Recipient Ferdinand-Redtenbacher placque U. Karlsruhe, 1940. Mem. sci. adv. bd. German Atomic Commn., Bonn; v.p. turbine com. Internat. Fedn. Electricity, Paris, 1958-—. Research and numerous publs. on gas turbines, steam turbines, turbocompressors, gas dynamics, atomic power stations, aero- and hydrodynamics of reactors and turbomachinery. Home: 3 Hannover, Alleestrasse 3, Germany.*

BAN, Thomas Arthur, psychiatrist; b. Budapest, Hungary, Nov. 16, 1929; s. Geza and Elizabeth (Rona) B.; M.D., U. Budapest, 1954; Dipl. Psychiatry with distinction McGill U., 1960; m. Joan Evelyn Valley, Nov. 22, 1963. Sr. research psychiatrist Douglas Hosp., Verdun, Que., Can., 1961-66, asso. dir. research, 1966-—; asst. prof. psychiatry McGill U., 1965-—. Mem. Que. Psychopharmacological Research Assn. (exec. sec.), Collegium Internationale Activitatis Nervosae Superioris (sec.-treas.). Author: Conditioning and Psychiatry, 1964. Contbr. numerous articles to med., profl. jours. Contbns. application of conditioning in clin. psychiatry, several new psychotropic drugs in treatment. Home: 17 Stratford Rd., Montreal 29, Que. Office: 6875 LaSalle Blvd., Verdun, Que., Can.*

BAN, Yoshio, Japanese organic chemist; b. Tokyo, Japan, Apr. 15, 1921; s. Tsuneo and Nakako (Katsumata) B.; B.degree, U. Tokyo, 1945, Ph.D., 1955; m. Miwako Hara, Nov. 20, 1948; children—Yumiko, Takashi. Research asso. U. Cal. at Berkeley, 1955-56; faculty Hokkaido U., Sapporo, Japan, 1956-—, prof. organic chemistry, 1957-—, dean faculty pharm. scis., 1966-—. Mem. Pharm. Soc. Japan (award 1963), Chem. Soc. Japan, Chem. Soc. (London, Eng.), Soc. Chem. Industry (London). Research, publs. on determination of absolute configuration of yohimbine and reserpine, stereochemistry of rhynchophylline, determination of structure of rubremetine, total synthesis of aspidospermine; discovered synthetic method of Beta carbolines. Home: S. 11, W. 21, Sapporo, Hokkaido, Japan.*

BANACH, Stefan, mathematician; b. Mar. 30, 1892; Dr.Phil., 1920; prof. Lvov (Poland) U. Author: Theories of Linear Operations, 1931. Founder modern functional analysis; developed (with F. Riesz, Hahn and others) theory of topological vectoral space. Died 1945.

BANCKES, Richard, English physician; b. Eng., 16th century. Author: Blanckest Herbal facsimile edition An Herbal, 1525, one of 1st herbals printed in English, said to be translation bot. part Encyclopedia Bartholomaeus Angelicus, 13th century.

BANCROFT, Edward, inventor; b. Westfield, Mass., Jan. 9, 1744; went to Eng., circa 1770, became contbr. articles on Am. to Monthly Rev.; acquainted with Benjamin Franklin and served as his agt. in London, at outbreak of Am. Revolution; agt. for Silas Deane in France while under pay Brit. govt., until 1783; lived in Eng. after 1783; Fellow Royal Soc., 1773. Author: Essay on the Natural History of Guiana, 1769; Experimental Researches Concerning the Philosophy of Permanent Colors, 1794. Inventor textile dyes. Died Margate, Eng., Sept. 8, 1821.

BANCROFT, J(ohn) Sellers, Am. mech. engr.; b. Providence, Sept. 12, 1843; s. Edward and Mary (Sellers) B.; grad. Central High Sch., Phila., 1851; m. Beulah Morris Hacker, Oct. 17, 1907. Began as apprentice William Sellers & Co., Phila., 1861; admitted to firm, 1873, gen. mgr., 1886-1902; gen. mgr., v.p., treas., mech. engr., Lanston Monotype Machine Co., 1902-19. Held about 100 patents for mech. and elec. inventions. Died Jan. 29, 1919.

BANCROFT, Joseph, physician; b. 1836. Author: Pituri and Duboisia, 1877; Further Remarks on the Pituri Group of Plants, 1878; Lecture 1. On Diseases of Animals and Plants that Interfere with Colonial Progress, 1879. Discoverer threadlike, white worms causing elephantiasis and other diseases by blocking lymphatic circulation, 1877; discussed leprosy in Queensland, 1891; related scleroderma to filaria sanguinis hominis, also discussed enormous bony tumor of thigh, 1894. Died 1894.

BANCROFT, Thomas Lane, physician, zoologist; b. Australia, 1860; discovered female adult of Filaria (roundworm parasite), 1876, also discovered Wuchereria bancrofti (parasitic worm found in lymph channels and scrotum of man); related Pasteurism to rabbit pest in Australia, 1888; discussed materia medica in relation to pharmacology of Queensland plants, 1889, discussed whip-worm of rat's liver, 1893, physiol. action of snake venom and strychnine cure of snakebite, 1894, (with others) effects of increase of white corpuscles in filairial blood; credited with 1st demonstration that mosquito Aedes aegypti carries dengue, 1906. Died 1933.

BANCROFT, Wilder Dwight, Am. chemist; b. Middletown, R.I., Oct. 1, 1867; s. John Chandler and Louisa Mills (Denny) B.; A.B., Harvard, 1888; Ph.D., U. Leipzig, 1892; postgrad. Harvard, 1888-89, Strassburg, 1889-90, Berlin and Amsterdam, 1892-93; D.Sc., Lafayette, 1919, Cambridge, 1923; LL.D., U. So. Cal. 1930; m. Katharine Meech Bott, June 19, 1895; children—Mary Warner, Hester, John Chandler, George, Jean Gordon. Asst. prof. phys. chemistry Cornell U., 1895-1903, prof., 1903-37, emeritus. Chmn. div. chemistry NRC, 1919-20; adv. com. C.W.S. Tallman prof. Bowdoin Coll., 1937. Bd. visitors, Bur. Standards, 1922-25; v.p. Internat. Union Chemistry, 1922-25. Fellow Am. Acad. Arts and Scis. hon. mem. Am. Electro-chem. Soc. (pres. 1905, 1919), Am. Electroplaters Soc., English, French, Polish chem. socs.; mem. Am. Chem. Soc. (pres. 1910), Am. Phys. Soc., Nat. Acad. Sci., Am. Philos. Soc. Author: The Phase Rule, 1897; Applied Colloid Chemistry, 1932; also numerous articles sci. jours. Founder, editor Jour. Phys. Chemistry, 1896-1932; asso. editor Jour. Franklin Inst., from 1913. Demonstrated that 2 immiscible liquids become miscible if a third (soluble in both) is added; investigated ternary mixtures, equilibrium in 2-component systems; analysis of solid phases; described simple method of determining decomposition potentials; stated equations for electromotive forces in terms of chem. potentials. Died N.Y., Feb. 7, 1953.

BAND, William, physicist; b. Liverpool, Eng., Aug. 27, 1906; s. William David and Amy (Cooke) B.; B.Sc., Liverpool U., 1926, M.Sc., 1927, D.Sc., 1946; m. Claire May Edwards, Aug. 12, 1931. Came to U. S., 1946, naturalized, 1953. Lectr. physics Liverpool U., 1927-29; asso. prof. Yenching U., Peking, China, 1931-37, prof., chmn. dept. physics, 1937-44; mem. Brit. Council, Chungking, China, 1944-45; fellow Inst. Nuclear Studies, U. Chgo., 1946-—, research asso. Inst. Study Metals, 1947-49; prof. physics Wash. State U., Pullman, 1949-—, chmn. dept., 1960-67; sr. physicist Stanford Research Inst., Menlo Park, Cal., 1953-—. Recipient Oliver Lodge Physics prize Liverpool U., 1926. Fellow Am. Phys. Soc., Phys. Soc. and Inst. Physics (London). Author: Introduction to Quantum Statistics, 1955; Introduction to Mathematical Physics, 1959; Shock Waves in Solids, 1965. Research, publs. on thermoelectricity, relativity, statis. mechanics, quantum theory, shock waves in solids. Home: 1609 Fisk St., Pullman, Wash. 99163.*

BANDELIER, Adolph Francis Alphonse, archeologist; b. Berne, Switzerland, Aug. 6, 1840; m. Fanny Ritter, Dec. 30, 1893. Came to U. S. in youth; traveled under auspices of Archaeol. Inst. of Am. among native races of N.M., Ariz., Mexico and Central America, 1880-85; went to Peru and Bolivia, 1892, on scientific expdn. for Henry Villard, and later pursued exhaustive archaeol., ethnol. and hist. researches in those countries (for six years for Am. Mus. Natural History, for which he gathered its extensive collection of Peruvian and Bolivian antiquities). Resided in Santa Fe, N.M., 1885-92; in charge documentary studies for Hemenway Archaeol. Expdn., 1886-89; lectr. Columbia U. on Spanish-Am. lit. in its connection with ethnology and archeology, 1904-11. Author: The Art of War and Mode of Warfare, 1877; Tenure of Land and Inheritances of the Ancient Mexicans, 1878; On the Social Organization and Mode of Government of the Ancient Mexicans; Historical Introduction to Studies Among the Sedentary Indians of New Mexico; An Archaeological Reconnaissance into Mexico, 1884; A Report on the Ruins of the Pueblo of Pecos, 1881; Final Report of Investigations Among the Indians of the Southwestern U. S., 1880-85, part 1, 1890, part 2, 1892; The Delight Makers (novel of Pueblo Indian Life); The Gilded Man; An Outline of the Documentary History of the Zuñi Tribe, 1892; The Indians and Aboriginal Ruins of Chachapoyas, Peru, 1907; The Islands of Titicaca and Koati (pub. by Hispanic Soc. of America), 1910; The Ruins of Tiahuanaco in Bolivia, 1912. His thorough research work resulted in the discrediting of the romantic school of Am. Indian history. Died Spain, Mar. 19, 1914.

BANDIC, Ivan Milos, Yugoslavian mathematician; b. Kikinda, Yugoslavia, Mar. 10, 1903; s. Milos D. and Danica (Telecki) B.; ed. Math. Faculty, U. Belgrade (Yugoslavia), 1923-28; m. Vida Vujin, June 27, 1932; 1 son, Dusan. Prof. high sch., Sombor, Yugoslavia, 1929-33, Bihac, Yugoslavia, 1933-41, Belgrade, 1941-45; insp. ministry Edn., 1945-47; prof. math. Higher Pedagogic Sch., Belgrade, 1947-58; prof. math. faculty pharmacy U. Belgrade, 1947-—. Mem. adv. mem. council for Profs. Math. in Instns. Edn., 1945-—. Mem. Soc. Mathematicians. Author textbooks on math., books on math. methods; also numerous articles. Editor, Jour. Tchrs. Maths., 1952-63. Solution of ordinary differential equations in theoretical physics, introduction of new forms of integrable differential equations, geometric interpretation of theorems of algebra and theories of infinite products. Address: 35 Deligradska, Belgrade, Yugoslavia.*

BANDL, Ludwig, obstetrician; b. Himberg, Niederösterreich, Germany, Nov. 1, 1842; M.D., U. Vienna (Austria), 1867; privat-dozent U. Vienna, 1875, prof. obstetrics and gynecology, 1880. Described a ring-shaped thickening of uterus just above internal os observed during labor, 1876; the ring marks lower border of contractile portion of uterus; known as Bandl's ring. Died Vienna, Aug. 26, 1892.

BANDLER, Clarence G., Am. surgeon, b. Owego, N.Y., Nov. 6, 1880; s. William and Eva (Fox) B.; A.B., Columbia, 1901, M.D., 1904; m. Miriam R. Zack, Aug. 17, 1951, Intern Bellevue Hosp., N.Y.C., 1904, adj. attending urologist, chief of clinic, dept. urology, 1906-12; inst. asso. in urology, med. dept. Columbia, 1906-25; prof. urology N.Y. Post-Grad. Med. Sch. and Hosp. of Columbia U., 1909-—; attending urologist Post-Grad. Hosp., 1934-—, dir. dept. urology; cons. surgeon Home for Aged and Infirm, Yonkers, N.Y., 1908-—; cons. urologist St. Francis Hosp., Port Jervis, N.Y., St. Vincent's Hosp., S.I., N.Y., U. Hosp., N.Y. U. Bellevue Med. Center, Bd. dirs. Asso. Hosp. Service. Diplomate Am. Bd. Urology (v.p.). Fellow A.C.S.; mem. Associete Internationale d'Urologie, N.Y. State, N.Y. County med. socs. Author numerous med. articles, including Tumors of the Urogenital Tract in the Young, Nephroptosis and Nephropexy, Urinary Obstruction. Died Nov. 15, 1957.

BANDROWSKY, Ernst Titus von, see von Bandrowsky, Ernst Titus.

BANDY, Mark Chance, Am. mining engr.; b. Redfield, Ia., July 22, 1900; s. John Leroy and Hattie Elizabeth (Chance) B.; B.Sc., Drake U., 1922; M.A., Columbia, 1923, E.M., 1926; Ph.D., Harvard, 1938; m. Jean Arney, June 1, 1929. Engr., Bethlehem Steel, 1927-28, Chile Exploration Co., 1929-34, Patino Mines, 1936-47; engr., exploration, chief geologist, chief engr., gen. mgr. ECA-Mining Engr., 1949-51; cons., dir. So. Minerals, 1951-52; gen. mgr. Rhodesia Copper Ventures, 1952-54; lectr. St. Andrews (Scotland) U., 1954-55; cons. Utex Exploration, 1955-58; cons. 1956-58. Hydrous copper borate chloride named bandylite in his honor, 1935. Fellow Soc. Econ. Geologists; mem. Am. Mineral. Soc., Mineral Soc. Gt. Britain (hon.), Phi Beta Kappa. Translator: (with J. A. Bandy) De natura fossilium, 1955. Died June 3, 1963.

BANDY, Orville Lee, Am. geologist, educator; b. Linden, Ia., Mar. 31. 1917; s. Alfred Lee and Blanche (Meacham) B.; B.S., Ore. State U., 1940, M.S., 1941; Ph.D. (Shell Oil fellow), Ind. U., 1948; m. Alda Ann Umbras, June 10, 1943; children—Janet Lee, Donald Craig. Faculty geology U. So. Cal., Los Angeles, 1948-—, prof., 1954-67, chmn. department geological sciences, 1967-—. Bd. dirs. Cushman Foundation for Foraminiferal Research. Fellow Geol. Soc. Am.; mem. A.A.A.S., Soc. Econ. Paleontologists and Mineralogists (counselor 1963-64), Am. Assn. Petroleum Geologists (trustee research fund 1960-64, distinguished lectr. 1963-64), Am. Soc. Limnology and Oceanography, Ecol. Soc. Am., Schweiz. Geologische Gesellschaft, Sigma Xi (pres. U. So. Cal. chpt. 1960). Contbr. numerous articles on foraminiferal ecology, correlation of foraminiferal structure with environment, stratigraphic correlation with planktonic foraminifera. Home: 6536 Holt Av., Los Angeles 90056.*

BANERJEE, Dilip K., Indian chemist; b. Calcutta, India, Jan. 16, 1912; s. Sitala K. and Suniti (Chatterjee) B.; B.Sc., Calcutta U., 1930, M.Sc., 1932, D.Sc., 1941; postgrad. (research scholar) U. Coll. Sci. and Tech.; m. Kanak Mukerjee, May 19, 1945; 1 son, Dipankar. Sir. P. C. Ray postdoctorate fellow Calcutta U., 1941-46; hon. lectr. postgrad. dept., 1943-46; prof. organic chemistry Jadavpur U., 1946-54; Watumull Found. fellow, U. Wis., 1947-49, Wis. Alumni Research Found. fellow, 1948-49; prof. organic chemistry Indian Inst. Sci., Bangalore, 1954-—. Recipient Woodburn medal Indian Assn. for Cultivation Sci., 1950. Fellow Indian Acad. Scis., Nat. Inst. Scis.; mem. Soc. Biol. Chemists (past pres.). Research, publs. on devel. new methods for total synthesis of steroids and their analogues and other natural products and their degradation products, new methods for functionalization of isopropyl group in terpenes. Home: Bungalow 10, Indian Inst. Sci., Bangalore-12, India.*

BANERJI, Sudhansu Kumar, physicist; b. Dacca, East Pakistan, Apr. 27, 1893; s. Hari Har and Nirada (Chatterji) B.Sc. with Honors in Math. (Govt. scholar), Calcutta U., 1912; postgrad. Presidency Coll., Calcutta, 1913-14; M.Sc. in Applied Math. 1st class, Calcutta U., 1914; D.Sc., 1918; m. Amiya Bhattacharya, May 5, 1916; children—Satyabrata, Debabrata, Muktibrata. Ghosh scholar applied math. U. Coll. Sci., Calcutta, 1914-15, faculty 1915-21, Ghosh prof., 1918-21; dir. Colaba and Alibag observatories, Bombay, India, 1922-31; hon. prof. applied physics Royal Inst. Sci., Bombay, 1928-34; superintending meteorologist, officiating dir. gen. observatories India Meteorol. Dept., Poona, 1932-41, superintending meteorologist, 1942-44, dir. gen. observatories, 1944-50; prof. math. Coll. Engring. Tech., Jadavpur, Calcutta, India, 1950-55, prof. emeritus, 1955-—; Ripon prof. Indian Assn. for Cultivation Sci., Calcutta, 1953. Hon. Centenary fellow Royal Meteorol. Soc. (London) (life fellow); mem. World Meteorol. Orgn. (exec. mem. 1947-50, 1st pres. regional commn. for Asia), Calcutta Math. Soc. (past pres.), Royal Asiatic Soc. (past sec. phys. scis.), Indian Sci. Congress (past pres. physics sect.), Indian Statis. Inst. (past v.p.), Indian Assn. for Cultivation Sci. (past v.p.), Nat. Inst. Scis. India (past v.p.), Indian Phys. Soc. (past pres.), Am. Meteorol. Soc., Am. Geophys. Union. Author: Earthquakes in Himalayan Region; Meteorology for Aimen in India, 1949; Bhumikampa in Bengali, 1965; also numerous articles. Recipient Blue Ribbon, Calcutta U., Mouat Gold medal; Mahendra Lal Sircar gold medal Indian Assn. for Cultivation Sci., 1918. Founder, editor Indian Jour. Meteorology and Geophysics, 1949-50. Research on impulsive sound, depth of earthquake foci, microseisms, electricity of overhead thunderclouds and other aspects of geophysics such as discontinuous fluid motion, aeros., hydrology. Home: 3 Ramani Chatterjee Rd., Calcutta 29, India.*

BANG, Bernhard Laurits Frederik, Danish veterinarian; born Soro, Sjaelland, Denmark, June 7, 1848; educated at Univ. Copenhagen; graduated from Royal Vet. and Agrl. Coll. (Copenhagen), 1873; M.D., 1880; professor of pathology and therapy, later director of Royal Vet. and Agrl. Coll., 1880-—; chief of Vet. Service Denmark; mem. French Academy Scis., 1932. Author: Measures Against Animal Tuberculosis in Denmark, 1908. Discovered method for eradicating bovine Tb, 1892; discovered Bang's Bacillus (Brucella abortus) which caused infectious abortion in cattle and brucellosis in man, 1897; investigated smallpox vaccination, actinomycosis, bacillary necrosis, swine fever. Died Copenhagen, June 22, 1932.

BANG, Frederik Barry, Am. physician; b. Phila., Nov. 5, 1916; s. Axel F. and Carol (Klee) B.; A.B., Johns Hopkins, 1935; M.D., 1939; m. Betsy Garrett, June 1, 1940; children—Caroline Barry (Mrs. Frances Moyer), Molly Garrett, Axel Frederik II. NRC fellow Vanderbilt U. Med. Sch., 1940-41, Rockefeller Inst., 1941-46; faculty Johns Hopkins, Balt., 1946-—, prof. pathobiology, chmn. dept. Sch. Hygiene, 1953-—; dir. Johns Hopkins Center for Med. Research and Tng., Calcutta, India, 1961-—. Fulbright scholar Nat. Inst. for Med. Research, Eng., 1955-56; Guggenheim fellow Station Biologique, Roscoff, France, summers 1961, 64. Mem. Am. Soc. Microbiology, Soc. for Exptl. Medicine and Biology, Am. Soc. Immunologists, Interurban Clin. Soc., Am. Soc. Tropical Medicine. Research, numerous publs. on use of tissue culture in study viruses, pathogenesis of virus diseases, demonstration of antimonial effects on egg devel. in schistosomes, presence of antibody like substances in marine invertebrates, action of endotoxin on blood clotting in limulus. Home: 3956 Cloverhill Rd., Balt. 21218.*

BANG, Hans Olaf, Danish physician; b. Christian and Nina (Rieper) B.; ed U. Copenhagen; M.D.; m. Gertrud Prior, June 29, 1937; children—Marianne, Lone. With various hosp. depts., particularly internal medicine; dir. dept. clin. chemistry of municipal hosp. of Aalborg, 1952-—. Mem. Dansk Laboratorie-Laegers Orgn., Dansk Selskab for Klinisk Kemi og Klinisk Fysiologi, Dansk Selskab for Intern Medicin, Gastroenterologisk Selskab. Author: B1—Vitamin-studier ve Hjaelp af Phycomyces—Methoden, 1941; Laboratorieundersogelser, 1956. Home: Klostermarken 47. Office: Hôpital Municipal, Aalborg, Denmark.

BANGA, Ilona (Mrs. Joseph Baló), Hungarian biochemist; b. Hódmezovásárhely, Hungary, Feb. 3, 1906; d. Samuel and Maria Róza (Berényi) B.; Doctor's degree in chemistry summa cum laude, U. Szeged (Hungary), 1929; student U. Vienna (Austria), 1925-27; Candidate Med. Scis., Med. U. Budapest (Hungary), 1952, D.Biol. Scis., 1956; m. Joseph Baló, May 26, 1945; 1 son, Joseph Matthias. Staff, Med. Chem. Inst., U. Szeged, 1930-46, titular asst. prof., 1944-46; faculty Biol. Chem. Inst., U. Budapest, 1946-50, prof., 1946-50; asst. prof. 1st dept. path. anatomy and exptl. cancer research Med. U. Budapest, 1950-—. Recipient (with Joseph Baló) Kossuth prize II degree, 1956. Mem. Internat. Union Biochemistry (mem. nat. com. 1955-—), Deutsche Akademie der Naturforscher Leopoldina. Author: Structure and Function of Elastin and Collagen, 1966; also numerous articles. Research on large scale preparation of ascorbic acid from Hungarian pepper, catalytic role of C4-dicarboxylic acids in tissue respiration; discovered actomyosine, (with J. Baló) pancreatic elastase; demonstrated role of mucoproteins which are covalently bound to fibers in function of elastic and collagen fibers. Home: 21 Németvölgyi ut, Budapest, Hungary.*

BANGERTER, Alfred, Swiss physician; b. Bienne-Berne, Apr. 22, 1909; s. Arnold and Serena (Buser) B.; ed. at Bienne, Berne, also other countries; m. Esther Burkard, July 5, 1948; children—Fred, Serena, Dorothea, Walburga, Beatrice. Dr. in chief Ophthalmology Clinic, Berne, Hosp. of Bienne, Ophthalmology Clinic, Saint-Gall, 1946-—, OPOS of Saint-Gall, 1961-—. Vice pres. European Council of Strabismus; hon. prof. U. Berne, 1956-—. Mem. Swiss Acad. Sci., German Soc. Physics, Swiss Soc. Ophthalmology. Author: Création de la pléoptique (traitement de l'amblyopie) (principle work on ophthal. surgery and treatment of strabismus and amblyopia). Home: Peter- und Paulstrasse 37. Office: Augenklinik, St. Gallen, Switzerland.

BANGHART, Frank W., Am. biostatistician; b. Ill., Oct 30, 1923; s. Frank W. and Elizabeth (Schneider); B.A., Quincy Coll., 1949; M.A., Drake U., 1950; Ed. D., U. Va., 1957; m. Betty J. Manker, Oct. 5, 1946; children—Diane Marie, Deborah Lee. Faculty, U. Va., Charlottesville, 1956-64, asso. prof., 1961-64, asst. dir. div. ednl. research, 1956-57, dir., div. ednl. research, 1957-61, dir. Biomathematics Lab., 1961-64; prof., dir. Ednl. Systems Devel. Center, Fla. State U., Tallahassee, 1964-—. Prin. investigator Office Naval Research; bioastronautics div. USAF. Mem. A.A.A.S., N.Y., Va. acads. sci., Biometrics Soc., Am. Statis. Assn., Nat. Council Tchrs. Math. Contbr. articles to tech. jours. Home: 3374 Lakeshore Dr., Tallahassee.*

BANISTER, John, botanist; b. Twigwoth, Eng., 1650; s. John Bannister; grad. Magdalen Coll., Oxford (Eng.) U., 1671, M.A., 1674; m. 1688. Came to Charles City County, Va., 1678, owned land on Appomattox River and acted as minister to Bristol Parish; engaged in studies local flora and fauna from his arrival, corresponded about his studies with other scientists including Compton, Sloane and Ray; trustee Coll. William and Mary, from circa 1690; part of his herbarium now in collection Brit. Museum; many of his sci. articles published posthumously in Philos. Trans. of Royal Acad. Died nr. Roanoke River, Va., May 1692.

BANISTER, Richard, Brit. oculist; b. Stamford, Eng.; studied under surgeon John Banister; m. Anne; practiced at Stamford, Eng.; translated (from work of Jacques Guillemeau) Treatise of One Hundred and Thirteen Diseases of the Eyes and Eyelids, with Some

Profitable Additions of Certain Principles and Experiments by Richard Banister, Oculist and Practitioner in Physic, 1622; specialized in eye diseases; observed hardness of eyeball in glaucoma; performed many cataract operations. Died 1626.

BANK, Thöger S. V., Danish mathematician; b. Copenhagen, Denmark, June 27, 1917; s. Thomas B. and Bodil (Andersen) B.; cand.mag., Copenhagen U., 1939, dr.phil., 1946; m. Eva Möller, Sept. 28, 1957. With Copenhagen U., 1942—, prof., 1956—. Recipient R. of D. Mem. Danish Acad. Research, publs. on quasi-analytic functions, theory of numbers. Home: 8 Vagtelvej, Copenhagen F, 200, Denmark.*

BANKER, Howard James, Am. biologist; b. Schaghticoke, N.Y., Apr. 19, 1866; s. Amos Bryan and Frances Alcena (Welling) B.; A.B., Syracuse U., 1892; A.M. Columbia, 1900, Ph.D., 1906; m. Mary Eugenia Wright, Aug. 23, 1894. Teacher, Troy Conf. Acad., Poultney, Vt., 1892-95; minister Meth. Episcopal Ch., 1895-98; tchr. math. Dickinson Sem. Williamsport, Pa., 1900-01; tchr. biology Southwestern State Normal Sch., California, Pa., 1901-04; prof. biology, DePauw U., Greeencastle, Ind., 1904-14; investigator, Eugenics Record Office, Cold Spring Harbor, L.I., 1914-33; exec. com. and sec. sect. 2, 2d Internat. Congress of Eugenics. Fellow A.A.A.S.; mem. Phi Beta Kappa, Sigma Xi; Author: The Hydnacae of North America, 1906. Contbr. papers on mycology and eugenics. to sci. jours. Died Sept. 23, 1943.

BANKOFF, S(eymour) George, Am. chem. engr., educator; b. N.Y.C., Oct. 7, 1920; s. Jacob and Sarah (Rashkin) B.; B.S., Columbia, 1940, M.S., 1941; Ph.D. in Chem. Engring., Purdue U., 1952; m. Mary Jo Rendleman, Aug. 11, 1944 (div. Sept. 1958); children—Joseph R., Elizabeth A.; m. 2d, Esther Horowitz Holmes, Feb. 6, 1960; children—Laura L., Jay M. Chem. engr. Sinclair Refining Co., East Chicago, Ind., 1941-42; project leader duPont Co., Hanford, Wash., Arlington, N.J., 1942-48; asst. prof. chem. engring. Rose Poly. Inst., Terre Haute, Ind., 1948-52, asso. prof., 1952-53, prof., 1953-58, chmn. dept. chem. engring. 1954-58; research asso. Cal. Inst. Tech., Pasadena, 1958-59; prof. chem. engring. Northwestern U., Evanston, Ill., 1959—. Cons. TRW Space Tech. Labs., 1961-65, Argonne Nat. Lab., 1963-67. NSF Sci. Faculty fellow, 1958-59; Fulbright scholar, also Guggenheim fellow, 1966-67. Mem. Am. Inst. Chem. Engrs. (edn. and accreditation com. 1961—). Contbg. author Advances in Chemical Engineering, vols. V, VI. Research, articles on boiling heat transfer, 2-phase flow, theoretical heat transfer, optimal process control. Patentee in field. Home: 2141 Ridge St., Evanston, Ill. 60201.*

BANKS, Charles Vandiver, Am. chemist; b. Blandinsville, Ill., Mar. 22, 1919; s. Charles Clement and Cora (Webb) B.; B.Ed. with honors, Western Ill. U., 1941; M.S., Ia. State U., 1944, Ph.D., 1946; m. Margie Lober, Aug. 3, 1941; children—Beverly Anne, Barbara Anne, Charles Roger. Member faculty Iowa State University, Ames, Iowa, 1941-44, 46—, professor of chemistry, 1954—, acting head department of chemistry, 1965, sr. chemist Inst. for Atomic Research and Ames Lab. AEC, 1954—, acting div. chief, 1965, sect. chief, 1966—. Mem. adv. bd. Analytical Chemistry, 1966—; editorial adviser Analytica Chimica Acta, 1966—. Mem. Am. Chem. Soc. (sec.-treas. div. analytical chemistry 1965-67), Ia. Acad. Sci., A.A.A.S., Sigma Xi, Phi Lambda Upsilon, Phi Kappa Phi, Alpha Chi Sigma, Sigma Zeta, Kappa Delta Pi. Research, numerous publs. on factors responsible for selectivity of organic analytical reagents, synthesis and application of vic-dioximes, also bifunctional organophosphorus compounds, solution chemistry metal ions, separation techniques, determination gases and other trace impurities in metals and metal salts. Home: 2019 Ashmore Dr., Ames, Ia. 50010.*

BANKS, Edgar James, Am. archeologist; b. Sunderland, Mass., May 23, 1866; s. John Randolph and Julia Maria (Dunklee) B.; student Amherst, 1886-87; A.B., Harvard, 1893, A.M., 1895; Ph.D., U. Breslau, 1897; m. Emma L. Lyford, July 16, 1893; m. 2d, Minja Miksich de Also Lukavecz; children—Edgar de Miksich, Daphne. Am. consul. Bagdad, Turkey, 1897-98; organized expdn. to excavate Babylonian city of Ur, 1899, but Sultan refused permission after years waiting; acting prof. ancient history Robert College, Constantinople, 1902-03; pvt. sec. to Am. minister to Turkey, 1903; excavated Babylonian ruin, Bismya, 1903, for U. Chgo., discovering several thousand inscribed objects from 4500 B.C. to 2800 B.C. and the white statue of King David, a pre-Babylonian king of 4500 B.C.; also much earlier ruins; field dir. of Babylonian expdn. and instr. Turkish and Semitic langs., U. Chgo., 1903-06; prof. Oriental langs. and archaeology Toledo U., 1909; lectr. on Babylonia, Arabia, Turkey, from 1906. Dir. Sacred Films, Inc., 1921-22; pres. Seminole Films Co., Inc. First American to climb to summit of Mt. Ararat, 17,212 ft., 1912; crossed the Arabian desert by camel on an exploring expdn., 1912. Author: Babylonische Hymnen der Berliner Sammlung, 1897; Jonah in Fact and Fancy, 1899; Bismya, or The Lost City of Adab, 1912; Bible and the Spade, 1913; Armenian Princess, 1914; Seven Wonders of the Ancient World, 1917; also many articles on archaeol. and other subjects. Died May 5, 1945.

BANKS, Edwin Melvin, Am. zoologist; b. Chgo., Mar. 21, 1926; s. David Louis and Eleanor (Johnson) B.; Ph.B., U. Chgo., 1948, B.S., 1949, M.S., 1950; Ph.D., U. Fla., 1955; m. Hilda Markoff, June 20, 1950; children—Daniel Lewis, Ronald Alan. Faculty, U. Ill., Urbana, 1955-63, 65—, asso. prof. biol. sci., animal behavior, 1963, prof. zoology, 1968—; asso. prof. zoology U. Toronto (Ont., Can.), 1963-65. Mem. Animal Behavior Soc. (program officer 1964—), A.A.A.S., Am. Soc. Zoologists, Ecol. Soc. Am., Am. Inst. Biol. Sci., Psychonomic Soc., Sigma Xi. Contbg. author: Behaviour of Domestic Animals, 1962; also articles. Research on social behavior and social orgn. higher vertebrates. Home: 1904 Crescent Dr., Champaign, Ill. 61820. Office: Dept. Zoology, U. Ill., Urbana, Ill. 61801.*

BANKS, H. Stanley, Brit. physician; b. Blantyre, Scotland, Feb. 17, 1890; s. Joseph and Jean (Walker) B.; M.A., Glasgow U., 1910, M.B., Ch.B., 1913, M.D., 1929; D.P.H., Cambridge U., 1915; m. Ruby Miller, Aug. 4, 1916; 2 children. Med. officer health, Motherwell, Scotland, 1919-26; med. supt. City Isolation Hosp. and Sanatorium, Leicester, Eng., 1926-33; dir. health div. UNRRA, Yugoslav mission, 1944-45; dir., med. supt. Park Hosp. for Infectious Diseases, London County Council, from 1933; lectr. infectious diseases St. Bartholomew's Hosp., London; examiner Cambridge U., also Gen. Nursing Council, Fellowship of Post Grad. Medicine, London; Milroy lectr. Royal Coll. Physicians, London, 1945. Mem. Royal Coll. Physicians, London, Brit. Med. Assn., Royal Soc. Medicine, Soc. Med. Officers of Health (past pres. fevers group). Author: Intravenous and Intraperitoneal Antitoxin in Scarlet Fever, 1936; Sulphonamides in Cerebro-Spinal Fever, 1938, 39, 40, 41; Meningococcal Encephalitis, 1942; Meningococcal Adrenal Syndromes, 1943. Address: Park Hosp., London S.E. 13, Eng.

BANKS, Sir John, Brit. physician; b. London; s. Percival Banks; ed. Trinity Coll., Dublin; M.D., D.L.; LL.D., Glasgow, Scotland; D.Sc. (hon.) Royal U.; m. Alice Wright. Regius prof. physic U. Dublin, 1880-98; physician in ordinary in Ireland to Queen Victoria; hon. physician to King in Ireland; mem. Gen. Med. Council. Fellow Royal Coll. Physicians, (pres. Ireland); 1st pres. Royal Acad. Medicine, Ireland. Author: Clinical Reports of Medical Cases; Loss of Language in Cerebral Diseases; also article on typhus in Quain's Dictionary of Medicine. Died July 16, 1908.

BANKS, Sir Joseph, English naturalist; b. London, Feb. 13, 1743; s. William and Sarah (Bate) B.; ed. Oxford (Eng.) U., 1760-63, hon. D.C.L.; m. Dorothea Hugessen, Mar., 1799. Traveled in Newfoundland to collect plants, 1766; accompanied Cook on expdn. around the world, aboard the Endeavour, and collected valuable natural history specimens (Botany Bay named on this voyage), 1768-71; visited Ireland, 1772; assisted Benjamin Thompson in founding Royal Instn. Gt. Britain for Promotion of Sci. Fellow Royal Soc., 1766 (pres. 1778-1820); mem. Nat. Inst. France. Author: Short Account of the Disease called the Blight, Mildew and Rust, 1805; also articles; pub. Kaempfer's Icones Plantarium, 1791; dir. edit. Roxburgh's Coromandel Plants, 1795-1819. 1st to show almost all Australian mammals were marsupials and more primitive than placental mammals of other continents; interested in founding colonies in far regions of world (efforts led to founding 1st Australian colonies); called father of Australia; sought to transplant plants from native regions to other lands where they might be useful (efforts brought breadfruit plant from Tahiti to West Indies); helped make Kew Gardens an important botanical center; plant genus Banksia named for him; a great patron of sci., his house in Soho Sq. was a gathering place for men of sci. Died Isleworth, nr. London, June 19, 1820.

BANKS, Maxwell Robert, Australian geologist; b. Punchbowl, N.S.W., Australia, July 21, 1925; s. Robert Leslie and May (McAlpin) B.; B.Sc. with honors, U. Sydney (Australia), 1946; m. Doris Mary Ingram, July 30, 1955; children—Robert, Anne, Susan, Robin. Faculty, U. Tasmania, Hobart, Australia, 1947—, reader geology, 1966—; vis. lectr. geology Amherst (Mass.) Coll., 1955-56. Recipient Univ. medal U. Sydney, 1946; Deas Thomson scholar, 1946—. Mem. Tasmanian Caverneering Club (past pres.), Australian Inst. Cartographers (chmn. Tasmanian div.), Sci. Film Soc. Tasmania (past pres.), Geol. Soc. Australia (past chmn. Tasmania div.), Geol. Soc. Australia (pres. 1966—), Royal Soc. Tasmania, Royal Soc. N.S.W., Geol. Soc. Australia, Paleontol. Soc., Paleontol. Assn., Am. Paleontol. Inst., Australian Inst. Cartographers. Editor: (with A. H. Spry) Geology of Tasmania, 1962; Geological Maps, 1953. Research, publs. on geol. history of Tasmania especially stratigraphy and paleogeography of Permian rocks, structure and formation of coral skeletons. Office: U. Tasmania, G.P.O. Box 252C, Hobart, 7001, Tasmania, Australia.

BANKS, Nathan, Am. entomologist; b. Roslyn, N.Y., Apr. 13, 1868; s. Daniel Gerow and Maria (Hawxhurst) B.; B.S., Cornell, 1880, M.S., 1890; m. Mary A. Lu Gar. June 2, 1897 (dec. Feb. 1956); children—Ruth, Bessie, Harold Bryant, Nellie May, Gilbert Shelley, Waldo Hawthorne, Dorothy Alice, Elsie Lucile. Douglas Hartley (dec.). Asst. entomologist U. S. Dept. Agr., 1900-16; curator insects. Mus. Comparative Zoology, Harvard, 1916-36; also asso. prof. zoology Harvard, 1928-36, emeritus from 1936. Mem. various entomological socs., Sigma Xi; fellow Am. Acad. Arts and Scis. Author: Treatise on the Acarina, 1904; Catalogue of the Acarina, 1907; Catalogue Nearctic Neuroptera. 1909; How to Collect and Preserve Insects, 1909; Catalogue Nearctic Spiders, 1910; Catalogue Nearctic Heteroptera, 1911; Index Economic Entomology, 1917; also many scientific and tech. papers. Owned largest collection of Arachnida and Neuroptera in U. S. Died Jan. 24, 1953.

BANKS, Richard Charles, Am. zoologist; b. Steubenville, O., Apr. 19, 1931; s. Clinton S. and Elizabeth (Harter) B.; B.S., Ohio State U., 1953; M.A., U. Cal. at Berkeley, 1958, Ph.D., 1961. Research zoologist Cal. Acad. Scis., 1961-62; curator birds and mammals San Diego Natural History Mus., 1962-66, Bird and Mammal Labs., Div. Wildlife Research; U. S. Fish and Wildlife Service, Washington, 1966—. Mem. Am. Ornithologists Union, Cooper, Wilson ornithol. socs., Am. Soc. Mammalogists, Soc. Study Evolution, So. Cal. Acad. Sci. Contbr. articles to sci. jours. Research field distbn., taxonomy and ecology of birds and mammals of Southwestern U. S. and Northwestern Mexico; relationships and hybridization of hummingbirds. Office: U. S. Nat. Museum, Washington 20560.*

BANKS, Robert Blackburn, Am. civil engr.; b. Wichita, Kan., Oct. 12, 1922; s. Bernard Thomas and Georgia (Corley) B.; B.S. in Civil Engring., Northwestern U., 1947, M.S., 1948; Ph.D. (Hilp fellow), U. Cal. at Berkeley, 1951; D.I.C. (Fulbright fellow) Imperial Coll. Sci. and Tech., London, Eng., 1952; m. Gunta Matisons, Dec. 25, 1960; children—Steven Matisons, Erik Blackburn. Research engr. Inst. Engring. Research, Berkeley, 1949-51; research engr. Infilco Inc., Tucson, 1952-54; faculty Northwestern U., 1954-61, prof. engring. sci., 1958-61, chmn. dept. civil engring., 1956-58, asst. dean for research and grad. studies, 1960-61; dir. research, prof. engring. SEATO Grad. Sch. Engring., Bangkok, Thailand, 1961-63; dean engring., prof. fluid mechanics U. Ill., Chgo., 1963-67; program adviser for sci. and engring. Ford Found., Mexico and C.Am., also prof. engring. Nat. Autonomous U. Mexico, 1967—. Cons. to pvt. cos. Recipient Eshbach award Northwestern U., 1948, Alumni Merit award, 1965. Mem. Am. Inst. Aeros. and Astronautics, Am. Geophys. Union, Am. Soc. C.E., A.A.A.S., Am. Soc. for Engring. Edn., Western Soc. Engrs., Marine Tech. Soc. Research in fluid dynamics, especially boundary layers, open channel flow, porous media flow. Home: Fuego 560, Pedregal, Mexico 20, D.F. Office: Reforma 243, Mexico 5, D.F.*

BANKS, Sir William Mitchell, Brit. surgeon; b. Edinburgh, Scotland, Nov. 1, 1842; s. Peter S. Banks; M.D. with honors, U. Edinburgh, 1864, LL.D., 1899; m. Elizabeth Rathbone Elliott, 1874; 2 sons. Demonstrator anatomy U. Glasgow (Scotland); joined staff Infirmary Sch. Medicine, Liverpool, Eng., 1868; became prof. anatomy when Infirmary Sch. merged with Univ. Coll., prof. emeritus, 1894; asst. surgeon Royal Infirmary, Liverpool, 1875-77, full surgeon, 1877-1902. Created knight, 1899. Gave Lettsomian lectures Med. Soc. London, 1900. Recipient gold medal U. Edinburgh, 1864; Fellow Royal Coll. Surgeons; mem. Liverpool Biol. Assn. (founder, pres. 1885), Med. Instn. (pres. 1890). Work led to modern operation for breast cancer. Died Aix-la-Chapelle, France, Aug. 9, 1904.

BANNEKER, Benjamin, Am. mathematician; b. Ellicott's Mills, Md., Nov. 9, 1731. Inherited his father's farm; showed unusual mechanical ability, constructed a wooden clock with no previous training; began making astron. calculations for almanacs, circa 1773; accurately calculated an eclipse, 1789, soon after sold his farm, concentrated on study of mathematics and astronomy; assisted in survey of D.C., 1790; published an almanac, 1792. Died Balt., Oct. 1806.

BANNER, Albert Henry, Am. marine biologist; b. Bellingham, Wash., Aug. 23, 1914; s. Henry Chester and Edith (Weisgerber) B.; student Bellingham State Normal Sch., 1931-33; B.S., U. Wash., 1935, Ph.D., 1943; M.S., U. Hawaii, 1940; m. Dora May Conrad, June 20, 1938; children—Christopher H., Robert B., Alan C., Catherine. H. Marine biologist Wash. State Dept. Fisheries, 1942-43; jr. biologist U. S. Fish and Wildlife Service, 1941-42; faculty U. Hawaii, Honolulu, 1946—, prof. zoology, 1957—; Dir. Hawaii Marine Lab., 1955-63. Mem. coral atoll project Pacific Sci. Bd., 1951, Jaluit Typhoon Assessment Team, 1958. U. Hawaii Pacific Program fellow Bishop Mus., Yale, 1954; Fulbright scholar, Thailand, 1960-61. Fellow A.A.A.S.; mem. Internat. Toxinological Soc. (charter), Soc. Systematic Zoology, Ecol. Soc. Am., Am. Soc. Limnology and Oceanography. Research, numerous publs. on identification distbn. marine shrimp, shrimplike crustaceans, biology, chemistry, pharmacology of toxins found in tropical marine fish and invertebrates. Home: 46-099 Lilipuna Rd., Kaneohe, Hawaii 96744. Office: Hawaii Inst. Marine Biology, P.O. Box 1067, Kaneohe, Hawaii 96744.*

BANNER, Edward A., Am. physician, surgeon; b. Chgo., Mar. 15, 1912; s. Edward Benjamin and Bertha (Wittie) B.; B.S., U. Ill., and Northwestern U.; M.D., U. Loyola, Chgo.; M.S., U. Minn.; m. Alice Uhlein, June 28, 1941; children—Barbara, Susan, Edward A., Thomas. Chief women's surgery Ashford Gynecol. Hosp.; faculty Mayo Grad. Sch., U. Minn., Rochester, Diplomate Am. Bd. Obstetrics and Gynecology. now asso. prof. Fellow A.C.S., Am. Coll. Obstetrics and Gynecology, Am. Soc. Infertility; mem. A.M.A., Mayo Alumni Found., Pan Pacific Surg. Soc., Continental (pres.) Minn. obstet. and gynecol. socs. Research, numerous publs. on hemolytic disease uterus. Home: Sunny Slopes, Rochester. Office: Mayo Clinic, Rochester, Minn.*

BANNISTER, Bryant, Am. dendrochronologist, archeologist; b. Phoenix, Dec. 2, 1926; s. Kimball and Elizabeth (Davis) B.; student U. Wis., 1944; B.A., Yale, 1948; M.A., U. Ariz., 1953, Ph.D., 1960; m. Betty Carol Stanaway, Aug. 22, 1951; children—Nancy Beth, Frances Kimball. With Lab. of Tree-Ring Research, U. Ariz., Tucson, 1953—, research asst. archeol. collections, 1953, curator archeol. collections, 1954-58, instr. dendrochronology, 1959, asst. prof., 1960-63, asso. prof., 1964, prof., dir. Lab., 1965—. Fellow Ariz. State Mus., 1949, research asso. Mus. No. Ariz., 1960—, collaborator Nat. Park Service, 1960—. Fellow Am. Anthrop. Assn., A.A.A.S.; mem. Ariz. Acad. Sci.; Soc. Am. Archeol., Tree-Ring Soc., Sigma Xi, Sigma Alpha Espilon. Editor of Tree-Ring Bulletin, 1958—. Author: (with T. L. Smiley, S. A. Stubbs), A Foundation for the Dating of some Late Archaeological Sites in the Rio Grande area, 1953; Tree-Ring Dating of Archaeological Sites in the Chaco Canyon Region, New Mexico, 1965. Home: 2800 E. 3d St., Tucson 85716.*

BANTA, Arthur Mangun, Am. zoologist; b. near Greenwood, Ind., Dec. 31, 1877; s. James Henry and Mary (Mangun) B.; B.S., Central Normal Coll., Ind., 1898; A.B., Ind. U., 1903, A.M., 1904; Edward-Austin fellow, Harvard, 1905-06, Ph.D., 1907; m. Mary Charlotte Slack, July 26, 1906; children—James Jerry, Ruth, Leah Margaret. Instr. pub. schs., Ind., 1895-97; prin. high sch., 1899-1901, Johnson County Normal Sch., summer, 1901; asst. in zoology Ind. U., 1903-05, instr. biol. sta., summers, 1903, 04; asst. in zoology Harvard, 1905-06, teaching fellow, 1906-07; with U. S. Fish Commn., Woods Hole, Mass., summers, 1906, 09; prof. biology Marietta Coll., Ohio, 1907-09; resident investigator, sta. for exptl. evolution Carnegie Instn., Cold Spring Harbor, N.Y., 1909-30; professorial lectr. in genetics U. Minn., 1927; vis. prof. exptl. zoology Brown U., 1929-30, research prof. biology, 1930-45, prof. emeritus, from 1945; asso. Carnegie Instn. of Washington, 1930-32, 36-37. Mem. NRC 3d. Fellowships in Biol. Sciences, 1933-37. Fellow A.A.A.S.; mem. Am. Soc. Zoologists, Am. Soc. Naturalists (sec. 1935-37), Genetics Soc. of Am., Limnological Soc. Am., Soc. Exptl. Biology and Medicine, Ecol. Soc. Am., Sigma Xi. Author: The Fauna of Mayfield's Cave, 1907; Selection in Cladocera on the Basis of a Physiological Character, 1921; Studies on the Physiology, Genetics, and Evolution of some Cladocera, 1939. Contbr. to biol. jours. Research on effects of changed environment on cave animals, also on heredity, devel., longevity, sex determination in lower organisms, sex intergrades. Died Jan. 2, 1946.

BANTI, Guido, Italian physician, pathologist; b. Montebicchieri, Italy, June 18, 1852; student at Pisa and Florence (both Italy); M.D., Florence, 1877; asst. path. anatomy U. Florence, 1878-82; became 1st doctor in Hosp. Santa Maria Nuova, Florence, 1883; became tchr. gen. pathology, 1887 and in 1889 tchr. path. anatomy; dir. Inst. Path. Anatomy for 30 years. Author: Endocarditi e nefriti, 1895; Patologica del polmone, 1902; Anatomia patologia (incomplete), 1907. Improved sanitation in Florence, circa 1900; research on neurology, aphasia, path. anatomy; discovered several varieties of diplococci, area in brain which controls speech; 1st description of splenomegaly accompanied by anemia, cirrhosis of liver, leukopenia (Banti's disease), 1882; research on leukemia; developed methods for isolating microorganisms from path. material; a founder of modern hematology. Died Florence, Jan. 8, 1925.

BANTING, Sir Frederick Grant, Canadian physician; b. Alliston, Ont., Can., Nov. 14, 1891; med. degree U. Toronto, 1916; served as med. officer in WWI; practiced medicine, London, Ont., until 1921; prof. U. Toronto, from 1923, Banting-Best professorship established in his honor; recipient (with others) Nobel prize for physiology and medicine, 1923; knighted, 1934; Fellow Royal Soc., 1935; Banting Research Found. established for him. Began research on internal secretion of pancreas, 1921; discovered (with Charles H. Best, under direction J. J. R. Macleod) hormone insulin, 1922; studied cortex of adrenal glands, cancer, silicosis; stimulated research in aviation medicine. Killed in airplane accident, Newfoundland, Feb. 21, 1941.

BANTON, Michael Parker, English sociologist; b. Birmingham, Eng., Sept. 8, 1926; s. Francis Clive and Kathleen (Parkes) B.; B.Sc., London Sch. Econs., 1950; Ph.D., Edinburgh U., 1954, D.Sc., 1964; m. Rut Marianne Jacobson, July 13, 1952; children—Sven Christopher, Ragivhild Cecilia, Lars Nicholas, Dagmar Hulda. Lectr. social anthropology, reader U. Edinburgh, 1953-65; vis. prof. polit. sci. Mass. Inst. Tech., Cambridge, 1962-63; prof. sociology U. Bristol, Eng., 1965—. Malinowski Meml. lectr., London, 1964. Mem. Brit. Sociol. Assn., Royal Anthrop. Inst., Internat. African Inst., Assn. Social Anthropologists of Commonwealth. Author: The Coloured Quarter, 1955; West African City, 1957; White and Coloured, 1959; The Policeman in the Community, 1964; Roles, 1965; Race Relations, 1967; also articles. Editor: Sociology jour. Research on analysis of social life in terms of role-relationships. Home: 22 Falcondale Rd., Westbury-on-Trym, Bristol, Eng.*

BANYAI, Andrew Ladislaus, physician; b. Hungary, Jan. 19, 1893; grad. Med. Sch., Royal Hungarian U., Budapest, 1915. Clin. dir. Muirdale Sanitarium, 1928-58; clin. prof. medicine emeritus Marquette U., 1958—. Chmn. Council on Internat. Affairs; Howard Lilienthal lectr., 1952; Selman Waksman lectr., 1960. Recipient Carlo Forlanini Gold medal; medal for sci. achievements Am. Coll. Chest Physicians. Hon. fellow Brazil, Chile, Peru, Colombia, Valparaiso Tb assns.; mem. Am. Coll. Chest Physicians (master, past pres.), A.C.P., A.M.A. (chmn. sect. diseases of chest 1954-55), also hon. and corr. memberships. Contbr. to 8 books. Author: Pneumoperitoneum Treatment, 1946. Editor: Non-Tb Diseases of the Chest; Dyspnea. Advances in Cardiopulmonary Diseases, 4 vols. Editorial bd. Diseases of the Chest; Am. Jour. Occupational Therapy; GP; sect. on chest diseases Excerpta Medica. Introduced combined method of artificial pneumoperitoneum and phrenicotomy in treatment of pulmonary Tb, 1934 (no longer popular). Home: 330 Diversey Pkwy., Chgo.

BAPTISTA, Antonio Manuel, Portuguese physicist; b. Almeirim, Portugal, Mar. 26, 1924; s. Antonio M. and Maria Rose Baptista; grad. U. Lisbon (Portugal), 1948; postgrad. Internat. Sch. Nuclear Scis. and Engring., State Coll. N.C., and Argonne (Ill.) Nat. Lab., 1957; m. Jovita Ovidio Baptista, June 4, 1958; children—Ana Maria, Antonio Manuel, Teresa Maria, Cristina Maria. Asst. prof. physics U. Lisbon Sci. Faculty, 1949-55, cons., since 1959; asst. Nuclear Energy Studies Commn., Lisbon, 1955-61; prof. physics Mil. Acad., Lisbon, 1961—, dir. physics lab., 1966—; dir. Isotope Lab., Instituto Portuguese de Oncologia Francisco Gentil, 1961—, Centro de Estudios de Energia Nuclear of Nuclear Energy Studies Commn., Instituto de Alta Cultura, Lisbon, 1966—. Mem. Real Academia de Ciencias Spain (fgn.), Internat. Soc. Hematology, Sociedade Portuguesa de Física e Química, Sociedade de Ciencias Médicas de Lisboa. Research, publs. on electrochemistry (behavior of metals in solution), radioactivity (detection systems and measurements), dosimetry of ionizing radiations, especially med. and biol. applications; med. and biol. applications of radionuclides. Home: 27 50 Dt Aven. Sacadura, Cabral, Lisbon. Office: Laboratório de Isótopes, Instituto Portugues de Oncologia, Lisbon, 4, Portugal.*

BAR, Paul, French obstetrician; b. 1853; M.D., 1881; dir. maternity hosp., Tenon, 1885-89; dir. maternity ward St. Louis Hosp., 1889-97, St. Antoine, from 1897. Pres. Obstet. Soc. Paris. Author: Clinical and Experimental Research pour servir a l'histoire de l'embryotomie céphalique, 1889; Lessons in Obstetrical and Experimental Research pour servir a l'histoire couchements, 1907. Pioneer in use of antiseptics in obstetrics. Died 1945.

BARABASCHI, Sergio, Italian physicist; b. Parma, Italy, June 15, 1930; s. Arturo and Armida (Arduini) B.; Master, Politecnico, Milano, Italy, 1952; Ph.Dr., U. Rome, 1962; m. Maria Luisa Stangalini, Apr. 7, 1956; children—Silvia, Pietro. Scientist, Electronic Lab., CISE, Milan, 1952-57; dir. Servomechanisms Lab., CNEN, Ispra, Italy, Engring. Lab., Rome, 1960-63, dir. technol. divs., 1963-64, nuclear reactor div., 1964—. Mem. Società Italiana di Física, Associazione Elettrotecnica Italiana, Am. Soc. Mech. Engrs., Am. Soc. for Testing and Materials. Author: (with R. Tasselli) Elements of Servomechanism, 1965; also articles. Research on electronic currents theory, nuclear reactor dynamics, remote handling equipment, nuclear reactors control systems, servomechanisms. Home: 701 Camilluccia. Office: 15 Belisario, Rome, Italy.*

BARABASHOV, Nikolay Pavlovich, Russian astrophysicist; born March 29, 1894; graduated from Kharkov Univ., 1919. With Kharkov Univ., 1919-33; director of Kharkov Observatory, 1930—; professor, head chair astronomy Kharkov University, 1934—; chmn. Commn. for Study Phys. Conditions on Moon and Planets, Astron. Council, USSR Acad. Sci.; chmn. Council for Coordination Complex Sci. Achievements in Astrometry and Astrophysics, Ukrainian Acad. Sci. Decorated Order of Lenin (2). Mem. USSR, Ukrainian acads. sci. Author: Photographic Photometry of Jupiter's Disk, 1931; The Results of the Processing of Photograms of Jupiter, 1933; Changes on the Surface of Jupiter, 1941; The Atmosphere and Surface of Mars, 1946; Changes in the Color of the Martian Seas, 1947; Photometry of the Light and Dark Zones of Jupiter, 1948; Research on Physical Conditions on the Moon and Planets, 1952; Results of Photometric Studies of the Moon and Planets at the Kharkov University Astronomical Observatory, 1957; Soviet Science in the Service of the People, 1959; What Covers the Moon?, 1963; co-author: Photographic Photometry of Mars in Red and Blue Rays, 1940. Made numerous photometric observations of Mars, Saturn and its rings, through light filter; studies of the surface of so-called lunar seas; construction of spectroscope for sun study. Address: Kharkov University, Universitetskaya 16, Kharkov, Ukraine SSR, USSR.

BARACH, Alvan Leroy, Am. physician; b. Newcastle, Pa., Feb. 22, 1895; s. Nathan L. and Jennie C. (Silman) B.; student Coll. City N.Y. 1912-15; M.D., Coll. Phys. and Surg. (Columbia), 1919; m. Frederica P. Pisek, Apr. 24, 1933; children—Jeffrey Alvan, John Paul. Asst. in medicine, Harvard (research at Mass. Gen. Hosp.), 1920-21; asst. in medicine, Presbyn. Hosp., 1922-31, clin. prof., 1952-65; asst. in medicine Coll. Phys. and Surg., 1922-25, instr. in medicine, 1925-28, asso. in medicine, 1928-35; cons. medicine, Woman's Hosp., 1928-34, also asso. prof. clin. medicine, Columbia Coll. Phys. and Surg., 1936, cons. in medicine, 1960—. Mem. Nat. Inventors Council, Dept. of Commerce. Recipient bronze medal, Class I, for original investigation helium and oxygen in various types of dyspnea, by A.M.A., 1936; scroll of honour for meritorious research in clin. med. and gas therapy by Internat. Anesthesia Research Soc., 1936; Townsend-Harris medal, Coll. City of N.Y., 1940, award of merit Am. Assn. Inhalational Therapists, 1960; Redway medal N.Y. State Med. Soc., 1964. Fellow Am. Coll. Chest Physicians (Coll. gold medal 1961), A.M.A., N.Y. Acad. Medicine; mem. Am. Soc. Clin. Investigation, Assn. of Am. Physicians, Soc. Exptl. Biol. and Med., Am. Acad. Allergy, Trudeau Soc., Omega Alpha, Omega Pi Alpha. Author: Principles and Practices of Inhalational Therapy, 1944; Physiologic Therapy in Respiratory Diseases, 1948; (with Hylan A. Bickerman) Pulmonary Emphysema, 1956; also articles. Devised oxygen tents and oxygen chambers generally adopted for use in treatment of pneumonia, cardiac disease and other cardio-respiratory conditions; introduced helium as a new therapeutic gas in treatment of asthma and obstructive lesions in larynx and trachea, 1934. Home: 72 E. 91st St. Office: 1050 Fifth Av., N.Y.C. 10028.*

BARACH, Joseph H., physician; b. Calvary, Poland-Russia, 1883; s. Zorach and Deborah (Oppenheim) B.; came to U. S., 1888, citizenship derived from father; student Park Inst., 1895-99; M.D., U. Pitts., 1903; postgrad. Columbia, 1903; m. Edna S. Levy, Sept. 21, 1915; children—Joseph L., Richard L. Resident pathologist and intern West Pa. Hosp., Pitts., 1904; asso. prof. medicine U. Pitts., also med. dir. U. Clins. Sch. Medicine, from 1930; sr. staff med. center hosps. from 1910; cons. dept. health Carnegie Inst. Tech., Pitts., also med dir. U. Clins. Sch. Medicine, from 1930; sr. staff med. center hosps. from 1910; cons. dept. health Carnegie Inst. Tech., Pitts., 1910; cons. in medicine Sewickley (Pa.) Valley Hosp., from 1925. Chmn. metabolism and endocrinology sect., research grants div. USPHS, 1946-51, nat. council arthritis and metabolism sect., research grants div. since 1952. Fellow A.A.A.S., A.C.P., Am. Diabetes Assn.; mem. A.M.A., Sigma Xi. Author: Self Help for the Diabetic, 1934; Diabetes and Its Treatment, 1946; Diabetes, The Patients Book, 1948; Diabetes and Its Treatment, 1949; Diabetes, The Foods and Facts on Diabetes, 1949. Contributor many articles to med. lit. in U. S. and abroad. Died Mar. 7, 1954.

BARAFF, Gene Allen, Am. physicist; b. Washington, Dec. 27, 1930; s. Charles and Lillian (Robbin) B.; A.B., Columbia, 1952; M.S., N.C. State Coll., 1956; Ph.D., N.Y. U., 1961; m. Eugenia Luden, May 30, 1958; children—Leslie Robbin, Andrew Howard, David Emanuel. Reactor physicist Astra Inc., Milford, Conn., 1956-58; research physicist Bell Telephone Labs., Murray Hill, N.J., 1961—. Mem. Am. Phys. Soc. Research, publs. on pertubation methods for nuclear reactor design, many body methods in atomic systems, transport theory in semicondrs., avalanche breakdown, band structure of bismuth, solid state plasma theory. Home: 29 Sawmill Dr., Berkeley Heights, N.J. 07922. Office: Bell Telephone Labs., Murray Hill, N.J. 07971.*

BARAKAT, Mohamed Zaki, Egyptian chemist; b. Cairo, Egypt, June 7, 1914; s. Taha Ibrahim and Dawlat Hassan (Yousry) B.; B.Pharmacy, Cairo U., 1935, M.Sc. in Chemistry, 1940, Ph.D., 1949; m. Safia Saber el Hamalawi, Apr. 8, 1948; children—Amina, Amin. Pharmacist, Ministry of Pub. Health, Egypt, 1935-37; chemist Labs. Pub. Health, Cairo, 1939-42; demonstrator chemistry dept. Faculty Sci., 1942-47; lectr. chemistry High Tng. Coll., Cairo, 1947-49; lectr. B and A biochemistry dept. Faculty Medicine, Abbassia, 1949-52; vis. staff radiotherapeutic dept. Cambridge (Eng.) U., 1951-52; asst. prof. biochemistry Ain-Shams U., 1952-57; prof. biochemistry Faculty Vet. Medicine, Cairo U., 1957-65; prof., head biochemistry dept., Faculty Medicine, Azhar U., Madina Nasr, Cairo, 1965—. Fellow Brit. Chem. Soc.; mem. Egyptian Acad. Pharmacy, Sci. Soc. (vice prin. 1966—), Egyptian Pharm. Chem. and Vet. Soc. Author: (with R. Moubasher) Theoretical Organic Chemistry, 1947; also numerous articles. Editorial bd. Jour. Vet. Medicine, 1960—. Research on synthesis, reactions and constn. of organic sulphur, nitrogen- or phosphorus-compounds, food sci., radiobiology, stereo-chemistry, photochem. reactions, molecular weight determination of non-volatile substances, deficiency diseases, chem. and biol. studies on N-Bromosuccinimide, new methods for microdetermination of organic and inorganic compounds of physiol., med., tech. and indsl. interest, application of new micromethods in biochem. analysis and pharm. assays. Home: 33 El-Mahroussa St., Cubba Garden, Cairo, Egypt.

BARANGER, Jacques Marie Félix, French surgeon; b. Orléans, Nov. 7, 1894; s. Leonard and Marie-Adélaïde (Desbois) B.; ed. Collège Sainte-Croix-d'Orléans, M.D., U. Paris; m. Marie-Antoinette Mordret, Apr. 24, 1925; children—Françoise, Michel, Patrice, Monique, Pierre, Laurence, Martine, Marie-Bénédicte. Intern, Paris hosps.; surgeon Clinc Saint-Côme-du-Mans, hosp. center of Mans. Mem. Nat. Assn. of Acad. Surgery, Ambroise-Paré Club (founder, pres.). Author: (essay) L'Organisation de la chirurgie, also articles on amputation of rectum. Home: 27, rue des Chanoines. Office: 17, rue de l'Etoile, Le Mans (Sarthe), France.

BARANGER, Pierre, French chemist; b. Neuville du Poitou, France, Sept. 26, 1900; s. Joseph and Julienne (Robert) B.; Engr., Ecole Polytechnique, Paris, 1921; student Pasteur Inst., 1923-31; Doctorat-ès-Scis., Paris U., 1931. With gas warfare dept. War Office, 1931-37; prof. chemistry Ecole Polytechnique, Paris, 1937—. Mem. Am. Chem. Soc., chem. socs. London, France, N.Y. Acad. Sci., other sci. socs. Research and publs. in chemotherapy; discoverer new remedies for leprosy, Tb, cancer. Home: 6, rue de Seine, Paris 6. France.*

BARANKIN, Edward William, Am. statistician, educator; b. Phila., Dec. 18, 1920; s. Myer and Esther (Grossman) B.; A.B., Princeton, 1941; M.A., U. Cal. at Berkeley, 1942, Ph.D., 1946; m. Claire Chertcoff, June 22, 1941; children—Joseph Paul, Barry Alexander. Teaching asst. math. U. Cal. at Berkeley, 1941-42, physicist Manhattan Project, Berkeley and Oak Ridge, 1942-45, research asst. statis. lab., dept. math., 1945, teaching asso. math., 1945-46, instr., 1947-48, asst. prof., 1948-52, asso. prof. math., 1952-59, research mathematician Inst. for Numerical Analysis, Nat. Bur. Standards, U. Cal. at Los Angeles, 1952-53, prof. statistics, Berkeley, 1959—. Physicist, Palmer Phys. Lab., NDRC, Princeton, N.J., summer 1941; instr. math. Engring., Sci. and Mgmt. War Tng. Council, Berkeley, San Francisco, 1942-45; asst. Inst. for Advanced Study, Princeton, 1946-47; math. cons. USAF, Nat. Bur. Standards, Rand Corp., Logistics and Statistics brs. Office Naval Research, Bell Telephone Labs., Inc.; statis. cons. Arthur D. Little, Inc., 1956—; vis. prof., John Simon Guggenheim Meml. fellow Institut Henri Poincare, U. Paris (France), 1956-57; Fulbright research prof. Math. Inst., Kyoto (Japan) U., 1962-63; Fulbright-Hays vis. prof. Inst. Math., Nat. U. Mexico, Nat. Inst. Math. Statistics, Mexico City, summer 1964. Fellow Inst. Math. Statistics (program chmn. 1959-60); mem. Am. Math. Soc., Econometric Soc., Inst. Mgmt. Sci. (editorial bd. 1955—), Institut International pour la Statistique dans les Sciences Physiques, Phi Beta Kappa, Sigma Xi. Research, publs. on math. theory of behaviour, sufficiency, applications of linear space theory, philosophy of sci. Home: 20 Highland Blvd., Berkeley, Cal. 94707.*

BARANOV, Pavel Aleksandrovich, Russian botanist; b. July 28, 1892; grad. Moscow U., 1917; lectr. Central Asian U., Tashkent, 1921-28, prof., 1928-45; with central bot. gardens USSR Acad. Sci., 1945-54, dir. Komarov bot. inst., from 1945. Named hon. sci. worker Uzbek SSR, 1943. Corr. mem. USSR Acad. Sci. (chmn. presidium Moldavian br. 1949-54). Author: Osnovnye etapy razvitiya botaniki, 1933; Istoriya embriologii rasteniy, 1955; V tropicheskoy Afrike. Zapiski botanika, 1956; numerous other publs. Specialist on flora of Pamir and Central Asia; investigated biology and evolution of vine, anatomy of mountain plants, embryology of orchids. Deceased.

BARANOV, Vasiliy Gavrilovich, Russian endocrinologist; b. 1899; grad. Petrograd Mil. Med. Acad., 1923; D.Med. Sci., 1939. Head, Lab. Age Physiology and Human Pathology, Pavlov Inst. Physiology, USSR Acad. Sci., Leningrad; head endocrinology dept. Inst. Obstetrics and Gynecology, USSR Acad. Med. Sci., head endocrinology research group; dir. course endocrinology and metabolism 1st Leningrad Med. Inst. Mem. med. council Leningrad City Dept. Health. Mem. USSR Acad. Med. Sci., All-Union Soc. Endocrinologists (chmn. presidium mem. Leningrad dept.). Author over 50 works including Guiding Principles and Methods of Treatment for Diabetics, 1953; Diseases of the Endocrine System and Metabolism, 1955. Co-editor Internal Diseases sect. Large Med. Ency., 2d edit.; mem. editorial bd. Problems of Endocrinology and Hormone Therapy. Research on functional disturbances of endocrine glands, pathogenesis and therapy of diabetes mellitus and nervous regulation of blood pressure. Address: 1st Leningrad Med. Inst., ulitsa Lva Tolstogo 6-8, Leningrad, USSR.

BARANSKI, Stanislaw, Polish physician; b. Rubiezewicze, Aug. 28, 1925; s. Stanislaw and Janina (Wojcikiewicz) B.; grad. Med. Faculty, Warsaw (Poland) U., 1952, dr.med., 1961; m. Wanda Stodolnik, Aug. 25, 1960. Sci. worker dept. histology and embryology Med. Sch., Warsaw U., 1950—, sr. research worker, 1956—; staff Inst. Aviation Medicine, Warsaw, 1952—, sr. worker, dir. research, 1959—. State cons. in chief for aviation medicine, 1959—. Recipient state award for application of radioactive isotopes in biomed. researches IId degree, 1960. Mem. Internat. Acad. Aviation and Space Medicine, Polish Astronautical Soc. (pres. sect. for biology 1965—), Polish Anat. Soc. (mem. council 1957—) Author: (with others) Uklad Krwiotwórczy Zwierzat Laboratorynych, 1962; also articles. Research on influence of flight on organism especially microwave irradiation; application of radioisotopes in exptl. aviation and space medicine. Home: 5 Nowowiejska, Warsaw, Poland.*

BARANY, Michael, biochemist; born Budapest, Hungary, Oct. 29, 1921; s. Joseph and Angela (Schlichter) Freed; M.D., U. Budapest, 1951, Ph.D., 1956; m. Kate Foti, Oct. 20, 1949; children—George, Francis. Came to U. S., 1960, naturalized, 1965. Lectr., Inst. Biochemistry, U. Budapest, 1951-56; intermediate scientist Weizman Inst., Rehovoth, Israel, 1956-58; asso. mem. Max Planck Inst., Heidelberg, Germany, 1958-60; mem. Inst. for Muscle Disease, N.Y.C., 1961—. Mem. Am. Soc. Biol. Chemists, Am. Physiol. Soc., Am. Chem. Soc., Am. Biophys. Soc. Research, publs. on isolation and characterization of contractile proteins, actin and myosin from various muscles, studies on their interaction in vitro and in vivo.*

BARANY (or BARANYI), Robert, physician; b. Vienna, Austria, Apr. 22, 1876; s. Ignaz and Marie (Hock) B.; M.D., U. Vienna, 1900; m. Ida Berger, 1909; 2 sons, 1 dau. Asst., Politzer's Ear Clinic, Vienna, 1903-11; lectr. on ear diseases U. Vienna, 1909; army physician World War I; prof. otology, head Ear, Nose, Throat Clinic, U. Uppsala (Sweden). Recipient Nobel prize in medicine and physiology, 1914, Guyot prize U. Groningen, 1914; Swedish Med. Soc. medal, 1925. Author: Untersuchungen über den Vestibular-Apparat des Ohres . . . , 1906; Physiologie und Pathologie des Bogengangapparates bein Menschen, 1907; Primäre Excision und primäre Naht accidenteller Wundez, 1919; Die Radikaloperation des Ohres ohne Gehörgangsplastik, 1923. Research on physiology and pathology of balancing apparatus of internal ear; devised method for diagnosing disease of semicircular canals, also tests for disease of cerebellum; methods resulted in easier diagnosis of certain ailments in inner ear; pioneer in treatment of otosclerosis; founder contemporary ear medicine. Died Uppsala, Sweden, Apr. 8, 1936.

BARANZANO, Redento, Italian astrologer, astronomer; b. Serravalle Sesia, Italy, 1590; s. Pietro Francesco and Clara B.; Barnabite from Vercelli; tchr. physics and philosophy at Annecy, Savoy, 1615. Author: Uranoscopia seu de coelo, 1617; Nova de motu terrae Copernicano iuxta summi pontificis mentem disputatio, 1618; Novae opiniones physicae, 1619; Campus philosophicus in quo omnes dialecticae quaestiones agitantur, 1620. Refuted Aristotle and supported Copernican view of universe, but retracted his views under pressure from Church. Died 1622.

BARB, Albert L., nuclear engr., educator; b. Vancouver, B.C., Can., Nov. 5, 1925; s. Clarence S. and Mildred (Gutteridge) B.; B. Applied Sci. first class honors U. B.C., 1948; M.S., U. Ill., 1949, Ph.D., 1951; m. Marguerite Lois Henderson, Dec. 21, 1948; children—Eugene M., Philip L., Christine L. Came to U. S., 1948, naturalized, 1954. Research engr., Rayonier Inc., Shelton, Wash., 1951-52; asst. prof. chem. engring. U. Wash., Seattle, 1952-56, asso. prof., 1956-60, chmn. nuclear engring. group, 1960—, prof. chem. and nuclear engring., 1960—, dir. Nuclear Reactor Labs., 1960—. Cons., Puget Sound Bridge and Dry Dock Co., Seattle, R. W. Beck & Asso., Seattle. Asst. coach Sno-King Ice Hockey League for Boys. Registered profl. engr. Wash. Mem. Am. Soc. for Engring. Edn. (v.p. nuclear engring. viv. 1964—), Am. Nuclear Soc. (mem. edn. com. 1962—), Am. Inst. Chem. Engrs., Am. Chem. Soc. Sigma Xi, Tau Beta Pi. Presbyterian (elder). Contbr. articles to tech. jours. on molecular diffusion in liquids, separations and purification processes for nuclear fuels; nuclear reactor engring., hemodialysis. Home: 8218 Crest Dr., N.E., Seattle 98115.

BARBA, Alvaro Alonso, metallurgist; b. Lepe, Andalusia, Spain, probably 1569; Spanish priest, Tarabuco, S.Am., circa 1609; curate, Tiaguacano (or Tiguanoco), S.Am., 1615; resided at Lepas, Peru, 1617. Author: El Arte de los metales . . . , circa 1640. Described amalgamation process used in Potosi in South America, where he served as priest for many years; gave detailed account of ores and minerals of S.Am.; described generation of metals; discovered process for extraction of gold, silver and copper by boiling with salt solution and mercury in copper vessel, circa 1607; described extraction of above metals by fusion, also their refining and separation; illustrated a balance in glass case.

BARBA, Pedro, Spanish physician; b. Spain, 1608; taught at Valladolid; became archiatre to Philip IV, 1621. Author: Vera Praxis de Curatione Teritanae stabilutur (1st treatise on cinchona bark and its use in treatment of malaria), 1642; Resunta de la materia di peste, 1648. Died 1671.

BARBACKI, Stefan, Polish geneticist; b. Wieliczka, Poland, Sept. 1, 1903; s. Jan and Zofia (Warchalowska) B.; Magister, Faculty Biology and Agronomy, Jagiellonian U., Cracow, Poland, 1925, Doctor, 1929; m. Krystyna Jankowska, Sept. 12, 1931; 1 son, Andrzej. Staff, Govt. Inst. Agrl. Research, Putawy, Poland, 1925-45, chief genetics lab., 1944-45; faculty Agrl. U., Poznan, Poland, 1945—, prof., 1965—, prorector, 1951-53, dir. Inst. Applied Biology, 1964—; dir. Inst. Plant Genetics, Polish Acad. Scis., Poznan, 1954—. Rockefeller Found fellow Galton Lab., U. London, 1935-36. Decorated Golden Cross of Merit, 1951; officer Cross Polonia Restituta, 1954; Golden distinction of City of Poznan, 1958. Mem. Polish Acad. Scis. (mem. presidium 1966—), Soc. Scis. Poznan (pres. 1960—). Author: Experimental Methods in Agricultural Sciences, 1935; (with K. Miczynski, S. Lewicki, A. Slabonski) Polish Wheats, 1937; Analysis of Variance in Agricultural Research, 1939; Factorial Analysis in Field Experiments, 1950; Lupin, 1952; also numerous articles. Editor: Rev. Agrl. Research, 1938-39, 46-47; chief editor Genetica Polonica, 1960—. Genetical studies on morphological and physiol. characters of barley; genetical and varietal studies on wheat, 1928-37; analysis of variance applied to biol. and agrl. expts.; inheritance and variability in Lupinus; developmental and populations studies in Vicia villosa; physiol. properties of Lupinus; physiol. properties of tetraploid and diploid forms of clover and serradella; variability and inheritance of protein level in plants; ability to produce protein in crosses of different form in barley and lupin. Home: 53 Mazowiecka, Poznan, Poland.*

BARBAN, Stanley, Am. biochemist; b. N.Y.C., Mar. 16, 1921; s. Isidore and Pauline (Wagner) B.; B.S., Coll. City N.Y., 1943; M.S., U. Mich., 1949; Ph.D., Washington U., St. Louis, 1953; m. Barbara Rosenberg, June 18, 1950; children—Beth Ellen, Lisa Claire. Bacteriologist, Dept. Health, Syracuse, N.Y., 1950; instr. State U. N.Y. at Syracuse Med. Coll., 1952-53; USPHS fellow NIH, Bethesda, Md., 1953-54, biochemist, 1954—. Mem. Am. Soc. Biol. Chemists, Am. Soc. Microbiology, A.A.A.S., Sigma Xi, Phi Sigma. Research, publs. on mechanisms of drug resistance, metabolism of tissue cell cultures, cell-virus biochemistry, bacterial metabolism. Home: 6603 Pyle Rd., Bethesda 20034. Office: NIH, Bethesda, Md. 20014.*

BARBARO, Ermolao, Italian naturalist; b. Venice, Italy, May 21, 1454; ed. Rome, Italy, under Pomponius Laetus, U. Padua, Italy. Prof. philosophy U. Padua, 1477; patriarch of Aquileia, Italy, 1491, later banished. Early writer on natural history; translated Pliny's Natural History, 1491, and some of Aristotle's works. Died June 14, 1493.

BARBASHIN, Yevgeniy Alekseevich, Russian mathematician; b. Village of Uinsk, Perm Oblast, 1918; grad. Urals U., Sverdlovsk; D.Physico-Math. Sci. Prof., Urals Poly. Inst., 1951—. Author: The Local Characteristics of Ordinary Points for a System of Differential Equations, 1943; The Behavior of Points during Geomorphic Transformations of Space, 1946; Classification of the Integral Manifolds of an Equation System in Full Differentials, 1947; The Theory of Generalized Dynamic Systems, 1948; Homomorphisms of Dynamic Systems, 1950; The Existence of Plane Solutions for Certain Linear Equations in Partial Derivatives, 1950; The Cross-Section Method in the Theory of Dynamic Systems, 1951; The Stability of Motion in Toto, 1952; Conditions for Preserving the Stability of Solutions for Integral and Differential Equations, 1957; A Problem in the Theory of Dynamic Programming, 1960; Stability of Solutions of Integral Differential Equations, 1963; co-author: The Existence of Lyapunov's Functions in the Case of Asymptotic Stability in Toto, 1954; Stability According to the First Approximation, 1955; A Qualitative Investigation of Equations Describing the Movement of Interacting Points around a Circle, 1961. Address: Urals Poly. Inst., Vtuzgorodok, Sverdlovsk RSFSR, USSR.

BARBER, Bernard, Am. sociologist; b. Boston, Jan. 29, 1918; s. Albert and Jennie (Lieberman) B.; A.B., Harvard, 1939, A.M., 1942, Ph.D., 1948; m. Elinor Anne Gellert, Sept. 25, 1948; children—Leslie Marianne, Christine Ruth, Philip Gellert, John Robert. Tutor, teaching fellow, Harvard, 1946-48; instr. Smith Coll., Northampton, Mass., 1948-49, asst. prof., 1949-52; faculty Barnard Coll., Columbia, 1952—, prof. 1960—. Mem., Am. Sociol. Assn., Eastern Sociol. Soc. (v.p. 1962-63). Author: Science and Social Order, 1952; Social Stratification, 1957; Drugs and Society, 1967. Editor: Social Class in Europe, 1965; The Sociology of Science, 1962. Home: Braeside Lane, Dobbs Ferry, N.Y. Office: Barnard Coll., Columbia Univ., N.Y.C. 10027.*

BARBER, Norman Frederick, physicist; b. Castleford, Yorkshire, Eng., Dec. 31, 1909; s. Benjamin T. and Jane (Wilkinson) B.; B.Sc. with 1st honors in Physics, U. Leeds (Eng.), 1931, M.Sc., Dip.Ed., 1933, D.Sc., 1962; m. Ruth Enid Dowden, May 5, 1956; children—Peter Norman, Mary Ruth. Tchr., St. Helens Sch., Lancashire, 1934-35, Bradford Sch., Yorkshire, 1935-37; with Royal Naval Sci. Service (formerly Admiralty Sci. Pool), 1937-50, New Zealand DSIR, 1950-64; prof. theoretical physics Victoria U., Wellington, New Zealand, 1964—. Fellow Royal Soc. New Zealand. Author: Experimental Correlograms and Fourier Transforms, 1962; also articles. Research on demagnetisation of ships, def. against magnetic mines, origin, behavior and mode of travel of ocean waves. Home: 69A Duthie St., Karori, Wellington, New Zealand.*

BARBER, Raymond Jenness, Am. mining engr.; b. Epping, N.H., Aug. 12, 1884; s. Albert Gilman and Annie Estelle (Skerrye) B.; S.B., Mass. Inst. Tech., 1906; m. June 1906; children—Raymond Jenness, Cedric Leonard; m. 2d, Edith Hudson MacLeod, Apr. 19, 1922. Mine rodman, Bingham Canyon, Utah, 1906; mining in Mexico, 1907; invention and mfg. small telephones for secret service Boston, 1907-10; mining exploration, appraisal, development, Los Angeles, 1910-13; asso. with Timothy W. Sprague, cons. engr., Boston, 1913-16; gen. mgr. Roxbury mines, Siskiyou Country, Cal., 1916-18; cons. engr. to shipyards U. S. Shipping Bd., 1918-19; cons. engr. Gilboa Dam of Catskill Water Supply, N.Y., 1919; pres., mgr. Roxbury Gold Mines, Cal., 1920-21; cons. engr., San Francisco, 1921-37; research in nitrocellulose lacquers, Boston, 1924-25; chemist, dir. Hydro-Carbon Cos., San Francisco, 1926-42; cons. in geology, Div. of Hwys., State of Cal., 1934-35; lectr. in placer mining Stanford, 1927-29, 1935-36; prof. geology, mining, dean Sch. of Mines, U. Alaska, 1937-40; mining engr. mem. Bd. Engrs. and Architects' Examiners, Territory of Alaska, 1939-45; engring. personnel counselor Lockheed (Vega) Aircraft Corp., Burbank, Cal., 1941-46; lectr. engring., geology U. So. Cal., 1946-48, vis. asso. prof gen. engring., 1948-52; curator mineralogy, petrology Los Angeles County Mus., 1950—. Fellow Royal Soc. Arts London; mem. Am. Inst. Mining and Metall. Engrs., Soc. for Promotion Engring. Edn., Nat. Highways Assn. Contbr. to tech. jours. Died Oct. 28, 1955; buried Utica, N.Y.

BARBER, Sherburne F., Am. mathematician; b. Nunda, N.Y., Oct. 25, 1907; s. George F. and Ethelwyn (Clark) B.; B.A., U. Rochester, 1929, M.A., 1930; Ph.D., U. Ill., 1933; m. Virginia L. Roy, Apr. 8, 1944; children—John, David, Andrew. NRC fellow in math. Johns Hopkins, 1933-34, Princeton, 1934-35; instr. math. State U. Ia., Iowa City, 1935-37; faculty Coll. City N.Y., 1937—, prof. math., 1956—, asst. dean, 1953—. Mem. Phi Beta Kappa, Sigma Xi. Work on Cremona and birational transformations; algebraic geometry. Home: Setalcott Pl., Setauket, N.Y.*

BARBER, Theodore Xenophon, Am. psychologist; b. Martins Ferry, O., Jan. 29, 1927; s. Xenophon and Helen (Hanos) B.; B.A., Am. U., 1954, Ph.D., 1956; m. Catherine Spinos, Mar. 3, 1957; children—Xenophon, Rania, Elaine. Post-doctoral research fellow dept. social relations Harvard, 1956-59; research asso. Worcester Found. for Exptl. Biology, 1959-61, cons. statistics, exptl. design, research asso. Medfield (Mass.) Found., 1961; research asso. dept. psychiatry Boston U. Sch. Medicine, 1962—; cons. statistics, exptl. design Tufts U. Sch. Dental Medicine, 1963-65. Mem. Am. Psychol. Assn. Author: A Scientific Approach to Hypnosis; 1967; also numerous articles. Constrn. of Barber Suggestibility Scale; sci. analysis of hypnotic behavior, including hypnotic analgesia, hallucinations, age regression, amnesia. Home: 3 Solon St., Wellesley, Mass. 02181. Office: Medfield Found., Medfield, Mass. 02042.*

BARBER-RILEY, Geoffrey, Brit. physiologist; b. Leeds, Eng., May 24, 1920; student Manchester U., 1946-48; B.Sc., Liverpool U., 1955, M.B., Ch.B., 1958, M.D., 1960; m. Nell Putt Tyrer, Nov. 11, 1944; children—William Paul, Richard Mark, Annabel Jane, Gaynor Noel. Faculty, Liverpool (Eng.) U., 1955-60, asst. lectr., 1958-60; lectr. U. Cape Town (S. Africa), 1960-63; prof., head physiology dept. U. Malaya, 1963-67; prof. physiology WHO, Inst. Med. Sci., Haile Sellassie I U., Addis Ababa, Ethiopia, 1967—. Research, publs. on excretory mechanisms of liver, bile function especially changes resulting from biliary obstruction. Office: Inst. Med. Sci., Haile Sellassie I U., Addis Ababa, Ethiopia.*

BARBERI, Benedetto, Italian mathematician; b. Cittareale (Rieti), Nov. 7, 1901; s. Camillo and Mariana (Chieroni) B.; Ph.D. in Math. and Phys. Sci. Instr. statistics U. Siena; prof. statis. econs. U. Rome Sch. Polit. Sci.; with Central Inst. Statistics, 1930—, head Office of Studies, 1945—, later dir. gen. of Inst. Participant numerous nat. and internat. congresses; pres. Conf. European Statisticians, Geneva. Mem. Assn. Italian de Ricerca Operativa (Rome). Author: Rilevazioni statistiche, 1957; Elementi de statistica economica, 1958; Il metodo statistico, 1961; Nozioni di calculo statistico, 1962. Address: via dei Chiavari 6, Rome, Italy.

BARBIÉ DU BOCAGE, Jean Denis, French geographer; b. Paris, France, Apr. 28, 1760; ed. Coll. Mazarin; studied geography under d'Anville; m. Mlle. de la Haye, Feb. 16, 1792; a son, Jean-Guillaume. Arrested as suspected revolutionary at court; librarian, 1792; became curator of maps and charts Nat. Library, 1802; employed by Napoleon, circa 1810; became prof. geography Sorbonne, Paris, France, 1809. Mem. Geog. Soc. (founder), Acad. Inscriptions. Author: Recueil de cartes géographiques de l'ancienne Grèce, 1788; (with Barthélemy) Voyage du jeune Anacharsis, 1789; (with Sainte Croix) Mémoires historiques et géographiques sur les pays situés entre la mer Noire et la mer Caspienne, 1796; Atlas of Anacharsis. Applied classic studies to geography of lands of antiquity; mapped Constantinople and suburbs, Greece and her colonies, India, Dardanelles, also mapped conquests of Alexander I. Died Dec. 28, 1825.

BARBIER, Charles, French inventor; b. France, 1789; devised alphabet for the blind consisting of elevated dots instead of embossed lines (later modified by Braille), circa 1820. Died 1859.

BARBIER, François-Antoine-Philippe, French chemist; b. Luzy, Nièvre, France, Mar. 2, 1848; asst. to Berthelot at Collège de France, École de Pharmacie; prof. chemistry Lyons, France, from 1878; tchr. of Grignard; Mem. French Acad. Scis. Laid basis for Grignard's discovery of organo-magnesium compounds, also Grignard reaction for organic synthesis. Died Bandol, Var, France, Sept. 18, 1922.

BARBIER, Lucien Eloi Adolphe, French agronomical engr.; b. Dijon, Aug. 11, 1901; s. Maurice and Lucie (Trullard) B.; ed. Lycée at Dimon, Nat. Agronomical Inst., U. Paris; Ph.D.; degrees in agronomical engring. and engring.; m. Marie-Thérèse Decaudin, Oct. 15, 1929; children—Colette, Luce, Anne, Jacqueline, Pierre, Claude, Marie-France, Dominique, Yves. Dir. research Central Sta. Agronomy. Pres., Commn. of Soil Chemistry. Mem. German Assn. for Studies of Soil (fgn.), Internat. Assn. Sci. of Soil (pres.), Acad. Agr. France. Contbr. over 180 articles to sci. publs. Address: 18, rue de l'Orangerie, Versailles (S.-et-O.), France.

BARBIER, Michel, French chemist; b. Villabé, France, Sept. 12, 1928; s. Charles and Germaine (Guillon) B.; Dr., U. Paris, 1954; m. Renée Dagard, Aug. 7, 1951; 1 son, Alain. Staff, Institut de Biologie Physico-Chimique, Paris, 1950, Inst. Organic Chemistry, Basle, Switzerland, 1958-59; staff Institut de Chimie des Substances Naturelles, Gif-sur-Yvette, France, 1960—, lab. chief, 1962—; maitre de recherches, CNRS, 1962—. Research, publs on isolations, chem. structures, synthesis and biosynthesis of natural compounds, chemistry of micro-organisms, of insects, pollens and plant sterols. Home: 1 Avenue du Parc des Expositions, Paris 15, France. Office: Institut de Chimie des Substances Naturelles, Gif-Sur-Yvette, France.*

BARBONI, Elio, Italian veterinarian; b. Spello, Nov. 18, 1907; s. Agapito and Lorenza (Angelucci) B.; D.V.M. Prof. microbiology and immunization, 1938; prof. gen. pathology and pathologic anatomy, 1940; asst. gen. pathology and vet. pathologic anatomy U. Perugia (Italy), prof., 1942, full prof., 1945, now full prof. gen. pathology and pathologic anatomy, also dean Sch. Vet. Medicine. Author numerous publs. including: Elementi de parassitologia veterinaria, 1943. Address: viale Pellini 45, Perugia, Italy.

BARBOUR, Erwin Hinckley, Am. geologist; b. Springfield, Ind., Apr. 5, 1856; s. Samuel Williamson and Adeline (Hinckley) B.; B.A., Yale, 1882, Ph.D., 1887; m. Margaret Roxanna Lamson, Dec. 7, 1887. Prof. geology Ia. Coll., 1888; prof. U. Neb., 1891; asst., later dir. U. S. geol. surveys; curator Neb. State Mus., 1892; geologist Neb. Bd. Agr., 1893. Fellow Geol. Soc., Nat. Geog. Soc. Research and publs. on geology and paleontology; discovered new rhinoceros fossil, early man fossils; work on fossils and mineral resources of Neb. Died 1947.

BARBOUR, George Brown, geologist; b. Edinburgh, Scotland, Aug. 22, 1890; s. Hugh Freeland and Margaret (Brown) B.; M.A. with honors, Edinburgh U., 1911; B.A., Cambridge (Eng.) U., 1915, M.A., 1916; Ph.D., Columbia, 1928; m. Dorothy Dickinson, May 15, 1920; children—Hugh, Ian, Freeland (dec.). Faculty, Yenching U., Peking, China, 1921-34, U. London (Eng.), 1934-37; faculty U. Cin., 1937—, prof., 1938—, dean Coll. Arts and Scis., 1938-58, emeritus professor, 1960—. Vis. prof. geology Duke, 1961-62, U. Louisville, 1964-65. Fellow Geological Soc. Am., Royal Geog. Soc., Geol. Soc. London, Royal Soc. Edinburgh. Author: Geology of the Kalgan Area, 1929; Physiography of the Yangtze River, 1936; In the Field with Teilhard de Chardin, 1965. Geomorphology. Home: 3521 Cornell Pl. 45220.*

BARBOUR, Henry Gray, Am. pharmacologist; b. Hartford, Conn., Mar. 28, 1886; s. John Humphrey and Annie (Gray) B.; A.B., Trinity Coll., Hartford, 1906; M.D., Johns Hopkins, 1910, fellow, 1910-11; m. Lilla Millard Chittenden, Sept. 15, 1909; children—Henry Chittenden, Dorothy Gray (Mrs. John D. Hersey), Russell Chittenden. Researcher, Freiburg, Germany, 1911, Vienna, Austria, 1912, London, Eng., 1913; asst. prof. pharmacology Yale, 1912-21, asso. prof., 1931-37, research asso., chmn. pharmacology and toxicology, from 1937; prof. pharmacology McGill U. 1921-23; prof. physiology and pharmacology U. Louisville, 1923-31; U. S. gas investigator 1917-18. Fellow Internat. Coll. Anesthetists; mem. A.M.A., Am. Physiol. Soc., Soc. Pharmacology and Exptl. Therapeutics, Am. Soc. Biol. Chemists, Soc. Exptl. Biology and Medicine, Central Soc. for Clin. Research, Am. Soc. Anesthetists (hon.), Phi Beta Kappa, Sigma Xi, Alpha Omega Alpha, Nu Sigma Nu, Delta Phi. Author: Experimental Pharmacology and Toxicology, Contbr. article Heat Regulation and Fever, in Blumer's Practitioner's Library, also articles to jours. Asso. editor Archives Internat. de Pharmacodynamie et de Therapie. Authority on body temperature and the effect of heavy water on mammals. Died Sept. 23, 1943.

BARBOUR, Percy E., Am. mining engr.; b. Portland, Me., Aug. 1, 1875; s. Clifford S. and Clara A. (Ford) B.; B.S., Worcester Poly. Inst., 1896, C.E., 1908; m. Viola Grace Hackward, Mar. 21, 1909. Mgr. Mass. Fan Co., ventilating engrs., Boston, 1897-1900; engr. Bingham Consol. Copper Co., Utah and Boston, 1900-03; engr. at smelter Tenn. Copper Co., 1904; gen. mgr. Navaho Gold Mining Co., Bland, N.M., 1905-06; cons. mining engr., Goldfield, Nev., 1907-08; dep. sheriff Esmerelda County, Nev., 1907; engr. Am. Smelting & Refining Co., also U. S. Smelting Co., Salt Lake City, Utah, 1909-10; gen. supt. Salt Lake Copper Co. (Ore. and Ida.), also lessee copper mine at Tecoma, Nev., 1910-11; gen. mgr. Uwarra Mining Co., Candor, N.C., 1911-14; editorial staff and mng. editor Engineering and Mining Jour., N.Y., 1915-17; asst. sec. Am. Inst. Mining and Metall. Engrs., N.Y., founder and editor Mining and Metallurgy, 1919-25; asst. to mgr. exploration dept. St. Joseph Lead Co., 1925-27; cons. mining engr., since 1927. Mem. Am. Inst. Mining and Metall. Engrs., Mining and Metall. Soc. Am. (sec., treas.), Soc. Am. Mil. Engrs. (gold medalist charter mem., officer). Author: Secondary Copper and tech. articles and papers. Authority on economics of copper, gold and silver. Died May 4, 1943.

BARBOUR, Philip Foster, Am. pediatrician; b. Danville, Ky., Feb. 24, 1867; s. Lewis Green and Elizabeth Anne (Ford) B.; A.B., Central U. Ky., 1884, A.M., 1899; M.D., Hosp. Coll. Medicine, Louisville, 1890; m. Jessie Lemont, Oct. 29, 1891; children—Mrs. Ruth Lamb, Philip Lemont; m. 2d, Elizabeth Akin, Jan. 6, 1909; 1 dau., Mrs. Catherine Akin Maxson. Practice medicine specializing in pediatrics, Louisville, 1892-1940; prof. chemistry Hosp. Coll. Medicine, 1895-98, prof. diseases of children, 1898-1907; prof. diseases children Louisville Coll. Medicine, 1907-08; clin. prof. diseases children and head dept. pediatrics, Med. Dept., U. Louisville, 1908-40. Fellow A.C.P., Am. Acad. Pediatrics (state chmn.); mem. A.M.A., Assn. Am. Tchrs. of Diseases of Children, So. Med. Assn., Conf. of Social Workers (pres.). Died Nov. 1, 1944.

BARBOUR, Thomas, Am. naturalist; b. Martha's Vineyard, Mass., Aug. 19, 1884; s. William and Julia Adelaide (Sprague) B.; A.B., Harvard, 1906, A.M., 1908, Ph.D., 1910, hon. Sc.D., 1940; Sc.D., Havana U., 1930, Dartmouth, 1935; Sc.D., U. Fla., 1944; m. Rosamond Pierce, Oct. 1, 1906; children—Martha Higginson (dec.), Mary Bigelow, William, Julia Adelaide, Louisa Bowditch, Rosamond. Made zoöl. explorations in East and West Indies, India, Burma, China, Japan, South and Central America, for Museum Comparative Zoölogy, Cambridge; later dir. of Harvard U. Museum and Museum Comparative Zoölogy; prof. zoölogy; mem. faculty Peabody Museum, Harvard; custodian Harvard Biol. Station and Bot. Garden, Soledad, Cuba, from 1927; exec. officer in charge of Barro Colorado Island Lab., Gatun Lake, Panama, 1923-45. Del. Harvard U. to 1st Pan-Am. Scientific Congress, Santiago, Chile, 1907-08; del. at founding Nat. U. of Mexico, 1910. Fellow Royal Geog. Soc. (London), Royal Asiatic Soc., A.A.A.S., Am. Acad. Arts and Scis., Nat. Acad. Sci.; hon. fellow Acad. Scis., Havana, Cuba; mem. Am. Soc. Zoölogists, Am. Philos. Soc., Am. Antiquarian Soc., Washington Acad. Science (v.p.), Acad. Natural Scis. of Phila., Boston Soc. of Natural History (pres. 1924-27), Phi Beta Kappa, Sigma Xi; fgn. mem. Zoöl. Soc., London; Linnaean Soc., London; corr. mem. Nederlandesche Dierkundige Vereiningung, Amsterdam, Hispanic Soc. Am. Research and publs. on reptiles and fishes, their systematic classification and geog. distbn. Died Jan. 8, 1946.

BARCHUSEN, Johann Conrad, see Barckhausen, Johann Conrad.

BARCIA GOYANES, Juan José, Spanish physician; b. Saint-Jacques-de-Compostelle, Dec. 26, 1901; s. Juan and Angeles (Goyanes) Barcia; ed. U. Saint-Jacques-de-Compostelle, U. Madrid; M.D.; m. Maria Salorio, Dec. 18, 1927; children—Maria, Juan Luis, Demetrio, Maria Angeles. Prof. med. sch. of Salamanque, Valencia; dean sch. medicine at Valencia; dir. dept. neurology Inst. Cajal. Cons., Council Sci. Research. Mem. Spanish Soc. Neurology (past pres.), Spanish Soc. Neurosurgery (past pres.); hon. mem. French, English, Italian neurosurg. socs. Author: La vida, et sexo y la herencia, 1928; Los tumores cerebrales, 1941; La nomina anatomica de sena, 1948; La nomina anatomica de Paris, 1960. Address: Gran Via M. del Turia, 62, Valencia, Spain.

BARCKHAUSEN (or BARCHUSEN), Johann Conrad, chemist; b. Horn, Germany, May 16, 1666; became mem. faculty, Utrecht, Holland, 1694, where he lectured and gave practical demonstrations on chemistry; became M.D. and lector, 1698; extra-ordinary prof. chemistry, 1703. Plant Barkhausia named in his honor. Author: Pharmacopoeus synopticus . . . , 1698; Pyrosophia, Succincte atque breviter Iatro-Chemiam, . . . , 1698; Acromata, . . . , 1703; Historia Medicinae, 1710; Compendium Ratiocinii chemici more geometrarum concinnatum, 1712; Elementa chemiae, . . . , 1718. Discovered succinic acid; discussed difficulty of separating elements because of their affinities for one another. Died Leyden, Netherlands, Oct. 2, 1723.

BARCLAY, Bertram Donald, Am. botanist; b. Champaign, Ill., Nov. 9, 1898; Bachiller en Humanidades, Deutsche Schule, Santiago, Chile, 1916; S.B., Coll. of Wooster 1923; S.M., W. Va. U., 1926; Ph.D., U. Chgo., 1928; m. Harriet George, Sept. 4, 1928; children—Bertram Donald, Arthur Stewart. Instr. botany W. Va. U., 1924-26, prof. botany, head dept. U. of Tulsa (Okla.), since 1929; staff Rocky Mountain Biol. Lab., Crested Butte, Colo., from 1929, v.p., from 1940, acting dir., 1948. Fellow A.A.A.S., 1930; mem. Bot. Soc. of Am., Okla. Acad. Scis. (pres. 1937), Ecol. Soc. Am., Phi Beta Kappa, Sigma Xi. Author: Origin and Development of Adventitious Roots in Hedera Helix L. (unpub.), 1926; Organography of Elephantelia Groenlandica at Varying Altitudes (unpub.), 1936; Origin and Development of Tissues in Stem of Selaginella Wildenovi, Botan. Gazette, 1931; various articles. Died June 6, 1953.

BARCLAY, John, Scottish anatomist; b. Perthshire, Scotland, Dec. 10, 1758; obtained bursary and studied for ch. St. Andrew's (Scotland); M.D., U. Edinburgh (Scotland), 1796; m. Eleonara Campbell, 1881. Tutor to family of Mr. C. Campbell; became tutor to family of Sir James Campbell, 1789; became asst. to John Bell, anatomist, 1789; established himself as anat. lectr. Edinburgh, 1797; named lectr. on anatomy and surgery Edinburgh Coll. Surgeons, 1804. Fellow Edinburgh Coll. Physicians. Author: A New Anatomical Nomenclature, 1803; The Muscular Motions of the Human Body, 1808; Description of the Arteries of the Human Body, 1812; Lectures on Anatomy, 1827; also contributed to Ency. Britannica, 1797. Developed a nomenclature of human anatomy based on sci. principles; research on comparative anatomy; donated mus. of anatomy (Barcleian Mus.) to Edinburgh Coll. Surgeons. Died Aug. 21, 1826.

BARCLAY-SMITH, Edward, Brit. physician; s. W. E. and Louisa (Barclay) Smith; ed. Brighton Coll.; Downing Coll., Cambridge (Eng.) U.; London Hosp.; M.A., M.D., B.Ch., Cambridge U.; m. Ida Mary Rogers; 1 son, 3 daus. Lectr. advanced human anatomy, examiner, fellow U. Cambridge; prof. anatomy U. London (Eng.), emeritus, from 1927, also chmn. bd. intermediate med. studies, chmn. bd. human anatomy and morphology, examiner; dean med. faculty, fellow, mem. council King's Coll., London; examiner univs. Durham, Manchester, Birmingham (all Eng.), Royal Coll. Physicians, London. Mem. Assn. des Anatomistes; hon. sec. Anat. Soc. Gt. Britain, Coeditor Jour. Anatomy, Buchanan's Manual of Anatomy, Buchanan's Dissection Guide. Contbr. papers to med. and sci. jours. Died July 5, 1945.

BARCROFT, Joseph, Irish physiologist; b. July 26, 1872, Glen, Newry, Ireland; s. Henry and Anna (Richardson) B.; B.Sc., London, 1891, King's Coll., Cambridge (Eng.) U., M.D., 1896; M.D. (hon.), Sofia, Bulgaria, and Louvain, Belgium; D.Sc. (hon.), Queens U., Belfast, Ireland, Harvard; m. Mary Agnetta Ball, 1903; 2 sons. Reader, Cambridge U., 1919-25; Fullerian prof. physiology Royal Instn., 1923-26; prof. physiology Cambridge U., 1926-37; Dunham lectr. Harvard, 1929; Terry lectr. Yale; became dir. animal physiology Agrl. Research Council, 1941. Knighted, 1935; Fellow Royal Soc., 1910; mem. U. S. Acad. Scis., Nutrition Soc. (pres.). Recipient Royal medal, 1922, Baly medal, 1923, Copley medal, 1944. Author: Respiratory Function of the Blood; Features in the Architecture of Physiological Function; The Brain and Its Environment. Led expdn. to Andes to study acclimatization; invented differential blood-gas manometer, 1908; research on respiratory function of blood, the brain and its environment; pioneer work on hemoglobin, on oxygen physiology problems, and on metabolism of isolated organs of animal body; studied spleen and physiology of developing fetus. Died Mar. 21, 1947.

BARCUS, William Dickson, Am. mathematician; b. N.Y.C., Mar. 19, 1929; s. William D. and Kathleen (Doyle) B.; S.B., Mass. Inst. Tech., 1950; postgrad. Princeton; D. Phil. (Rhodes scholar) Oxford, 1955; m. Louise Anne Edmond, Aug. 3, 1955; children—Alexandra Louise, Victoria Anne. Instr., Princeton, 1955-56; instr. Brown U., Providence, 1956-57, asst. prof., 1957-61; asso. prof. State U. N.Y., Stony Brook, 1961—, acting chmn. dept. math., 1963—. Mem. Am. Math. Soc., Sigma Xi. Research in algebraic topology, homotopy theory, cohomology operations. Home: High Tide, Van Brunt Manor Rd., East Setauket, N.Y. Office: State U. N.Y., Stony Brook, L.I., N.Y.*

BARD, Allen Joseph, Am. chemist; b. N.Y.C., Dec. 18, 1933; s. John J. and Dora (Rosenberg) B.; B.S. summa cum laude, Coll. City N.Y., 1955; A.M., Harvard, 1956, Ph.D. (NSF fellow), 1958; m. Frances Joan Segal, June 15, 1957; children—Edward David, Sara Lynn. Research chemist Gen. Chem. Co., Morristown, N.J., 1955; faculty U. Tex., Austin, 1958—, professor, 1967—; cons. Phillips Petroleum Co., Bartlesville, Okla., E. I. DuPont de Nemours & Co., Wilmington, Del. Recipient Ward medal, 1955. Mem. Am. Chem. Soc., Electrochem. Soc., A.A.A.S., Sigma Xi. Author: Chemical Equilibrium, 1966. Editor: Electroanalytical Chemistry—A Series of Monographs on Recent Advances, 1966, 67. Contbr. numerous articles to profl. jours. Devised (with H. B. Herman) technique cyclic chronopotentiometry; originated use coulometric techniques to study rates of reactions. Home: 6202 Mountainclimb St., Austin, Tex. 78731.*

BARD, John, Am. physician; b. Burlington, N.J., Feb. 1, 1716; s. Peter Bard; m. Miss Valleau, 1 son, Samuel. Practiced in Phila. and N.Y.C.; leader in establishing 1st quarantine sta., Bedloe's Island; apptd. health officer N.Y.C., 1759. First pres. Med. Soc. State N.Y., 1795. One of 1st to conduct dissections for ednl. purposes; performed 1st recorded successful operation of ectopic pregnancy in U. S., 1759. Died Hyde Park, N.Y., Mar. 30, 1799.

BARD, Louis, physician; b. Mens, France, May 10, 1854; student medicine, Lyons, France; prof. clin. medicine, Geneva (Switzerland) U. Author: Précis d'anatomie pathologique, 1890. Classified clin. forms of pulmonary Tb, 1898; (with Sergre and Pic) showed rapid cachexia, great pain and jaundiced complexion go with cancer of pancreas, 1888; observed malignant cells always have histological structure of tissues they developed from, 1890. Died Paris, Feb. 20, 1930.

BARD, Philip, Am. physiologist; b. Hueneme, Cal., Oct. 25, 1898; s. Thomas Robert and Mary B. (Gerberding) B.; A.B., Princeton, 1923, D.Sc., 1947; A.M., Harvard, 1925, Ph.D., 1927; Sc.D., Washington and Lee U., 1949, Doctor honoris causa, U. de San Marcos de Lima, 1951, Universidad Catolica de Chile, 1951; m. Harriet Hunt, June 29, 1922 (dec. Apr. 1964); children—Virginia Hunt (Mrs. M. K. Johnson), Elizabeth Stanton (Mrs. J. P. Stephens); m. 2d, Janet Rioch, Jan. 25, 1965. Instr. physiology Harvard Med. Sch., 1926-28, asst. prof., 1931-33; asst. prof. biology Princeton, 1928-31; prof. physiology, dir. dept. physiology Johns Hopkins Sch. Medicine, 1933-64, dean med. faculty, 1953-57, prof. emeritus physiology, 1964—. Chmn. bd. dirs. Berylwood Investment Co., Somis, Cal. Mem. div. med. scis. NRC, 1935-46; mem. Nat. Bd. Med. Examiners, 1935-46; Harvey lectr., 1938, Hughlings Jackson lectr. Montreal Neurol. Inst., 1943; George Cyril Graves lectr. Ind. U., 1948; Mitchell lectr. Coll. Physicians, Phila., 1953. Recipient Mem. Am. Physiol. Soc. (pres. 1941-46, chmn. bd. publ. trustees 1960-62), Assn. for Research in Nervous and Mental Disease (pres. 1950), Soc. for Exptl. Biology and Medicine (pres. 1959-61), Nat. Acad. Scis., Am. Philos. Soc. (Lashley award in neurobiology 1962), Am. Acad. Arts and Scis., Assn. Am. Physicians, Am. Neurol. Assn. (hon., Jacoby award 1959), A.A.A.S., Harvey Soc., Sociedad Argentina de Biolgia (hon.), Phi Beta Kappa, Sigma Xi. Editor, contbg. author: Macleod's Physiology in Modern Medicine, 1938; 41, Med. Physiology, 1956, 61. Contbr. numerous articles to tech. jours. on neurophysiology. Home: 6 Meadow Rd., Balt. 21212.*

BARD, Samuel, Am. physician; b. Phila., Apr. 1, 1742; s. John and Suzanne (Valleau) B.; grad. King's Coll. (now Columbia), 1760; M.D., U. Edinburgh (Scotland), 1765; LL.D. (hon.), Princeton, 1816; m. Mary Bard, 1770, 10 children, including William. A founder med. sch. King's Coll., 1767, prof. medicine, 1769-89, dean faculty, trustee, prof. theory and practice of physics, 1792, pres. Coll. Phys. and Surg., N.Y.C., 1811-21; personal physician to George Washington in N.Y.C. after Am. Revolution; a founder City Library, N.Y. Hosp., 1st med. sch. in N.Y.C., 1768, N.Y. Dispensary; pres. Agrl. Soc. of Dutchess County, Hyde Park, N.Y.; founder Protestant Episcopal Ch., Hyde Park. Author: The Shepherd's Guide; De Viribus Opii, 1765; An Enquiry into the Nature, Cause, and Cure of the Angina Suffocativa or Sore Throat Distemper (classic monograph on diphtheria), 1771; The Use of Cold In Hemorrhage, 1807, also a standard manual on midwifery, 1807. Contbd. to knowledge of diphtheria, other diseases of humans; also breeding and diseases of sheep. Died Hyde Park, May 24, 1821.

BARDAWIL, Wadi Antonio, physician, pathologist; b. Linares, Mexico, May 13, 1921; son of Haical and Angela (Bardawil) B.; B.S., U. Nuevo Leon, Mexico, 1938; M.D., Nat. U. Mexico, 1946; m. Cosette Gannam, Dec. 6, 1941; children—Antonio, Ronald, Lawrence, Angela, Carol, Nancy. Came to U. S., 1950, naturalized, 1957. Instr., U. Vt., 1950-52; faculty Harvard Med. Sch., 1952-59, research asso. pathology, 1957-59; staff Robert Breck Brigham Hosp., Boston, 1954-56; founder, dir. dept. pathology and med. research St. Margaret's Hosp., Boston, 1957—; faculty Tufts U. Sch. Medicine, Boston, 1957—, asso. prof. pathology, 1961—, prof. obstetrics and gynecology (pathology), 1966—; lectr. pathology U. Boston, 1965—. Diplomate Am. Bd. Pathology. Fellow Coll. Am. Pathologists; mem. Am. Soc. for Exptl. Pathology, Mass., Norfolk med. socs., Boston Obstet. Soc., New Eng., Mass. socs. pathologists, New Eng. Gynecol. and Obstet. Soc., Am. Soc. Clin. Pathologists, Soc. Latin-Am. Pathologists. Research, numerous publs. on vascular disease, autoimmune diseases, biology and pathology of pregnancy particularly placenta. Home: 73 Pacella Dr., Dedham, 02026. Office: 90 Cushing Av., Dedham, Mass. 02125.*

BARDEEN, John, Am. physicist; b. Madison, Wis., May 23, 1908; s. Charles Russell and Althea (Harmer) B.; B.E.E., U. Wis., 1928, M.E.E., 1929; Ph.D., Princeton, 1936; m. Jane Maxwell, July 18, 1938; children—James M., William A., Elizabeth A. (Mrs. T. J. Greytak). Geophysicist, Gulf Research & Devel. Corp., Pitts., 1930-33; jr. fellow Harvard, 1935-38; asst. prof. U. Minn., 1938-41; physicist Naval Ordnance Lab., Washington, 1941-45; research physicist Bell Telephone Lab., Murray Hill, N.J., 1945-51; prof. elec. engring., physics U. Ill., 1951—. Mem. Pres. 's Sci. Adv. Com., 1959-62; cons., dir. Xerox Corp., Rochester, N.Y., 1960—. Recipient Stuart Ballantine medal, 1952, Buckley prize, 1954, John Scott medal, 1955, Nobel prize, 1956; Fritz London award, 1962, Vincent Bendix award, 1964, Nat. medal Sci., 1965. Fellow Am. Phys. Soc. (pres. 1967-68); mem. Nat. Acad. Scis., A.A.A.S., Am. Acad. Sci., Am. Philos. Soc. Contbr. numerous articles to sci. jours. Studies of theory of solid state and low temperature physics, particularly semiconductors and superconductivity; co-inventor (with W. H. Brattain and W. Shockley) of transistor. Home: 55 Greencroft, Champaign, Ill. 61820.*

BARDELEBEN, Heinrich Adolf von, see von Bardeleben, Heinrich Adolf.

BARDENHEUER, Franz Bernhard Hubert, German physician; b. Lamersdorf, Germany, July 12, 1839; s. Hubert Matthias Joseph and Anna Sibylla (Frings) B.; received degrees from U. Berlin (Germany), 1864; m. Hermine Henriette Thelen, July 19, 1870; 3 sons, 2 daus. Asst. to surgeon W. Busch, Bonn, Germany, to ophthalmologist O. Becker, to surgeon G. Simon in Heidelberg, Germany; named dir. mil. hosp. in Cologne, Germany, 1870; began practice as eye doctor, Cologne, 1872; chief physician surgery dept. Cologne Pub. Hosp., 1874-1913; became prof., 1884; 1st adminstrv. prof. Cologne Acad. for Practical Medicine, 1904-07. Author: Die Behandlund der Vorderarmfrakturen durch Federextension, 1890; (with R. Graessner) Die Technik der Extensionsverbände bei der Behandlung der Frakturen und Luxationene der Extremitäten, 1905; Die allgemeine Lehre von der Frakturen und Luxationen mit besonders Berücksichtigung der Extensionsverfahren, 1907. Pioneered surg. treatment of fractures; introduced antiseptics in Cologne; introduced and developed constant-tension treatment of fractures; devised operation for floating spleen by making pocket in parietal peritoneum, 1895. Died Lamersdorf, Aug. 13, 1913.

BARDIS, Panos Demetrios, sociologist; born Lefkohorion, Arcadia, Greece, Sept. 24, 1924; s. Demetrios George and Kali (Christopoulos) B.; came to U. S., 1948, naturalized, 1958; student Panteios Sch., Athens, Greece, 1945-47; B.A. magna cum laude, Bethany (W.Va.) Coll., 1950; M.A. Notre Dame U., 1953; Ph.D., Purdue U., 1955; m. Donna Jean Decker, Dec. 26, 1964; 1 son, Byron Galen. Faculty, Albion Coll., 1955-59; faculty U. Toledo, 1959—, prof. sociology, 1963—. Sec.-treas. World Student Relief, Athens, 1946-48; U. S. rep. Internat. Congress Social Scis., Spain, 1965, 66; participant World Congress Sociology, France, 1966. Recipient Couphos prize Anglo-Am.-Hellenic Bur. Edn., 1949. Fellow Am. Sociol. Assn. (nat. membership com.), A.A.A.S., Internat. Inst. Arts and Letters (life); mem. Am. Assn. U. Profs., Nat. Council Family Relations, Nat. Acad. Econ. and Polit. Sci. (dir.), Nat. Writers Club, Nat. Assn. Standard Med. Vocabulary (cons. 1963—), Alpha Kappa Delta, Pi Gamma Mu. Author: (novel) Ivan and Artemis, 1957; The Family in Changing Civilization, 1967; also articles, poems. Editor in chief Social Sci., 1959—, book rev. editor, 1963—; asso. editor Indian Sociol. Bull., Indian Psychol. Bull., Revista del Instituto de Ciencias Sociales (Spain), 1965—; book rev. editor Internat. Rev. History and Polit. Sci. (India), 1966—; asst. Am. editor Indian Jour. Social Research, 1965—; editorial bd. Darshana Internat. (India), Jour. Edn. (India), 1965—, Sociologia Religiosa (Italy), 1966—. Research, publs. on history of sci., measurement of attitudes, history of ideas and social instns., prejudice; discovered and named geometrical figure ditoxon. Home: 2833 Goddard Rd., Toledo 43606.*

BARDOS, Thomas Joseph, chemist; born Budapest, Hungary, July 20, 1915; s. Artur and Vilma (Brachfeld) B.; Diploma in Chem. Engring., Royal Hungarian Tech. U., Budapest, 1938; Ph.D., U. Notre Dame, 1949; m. Mary Jane Choate, Mar. 24, 1951. Came to U. S., 1946, naturalized, 1952. Chem. engr. Vacuum Oil Co., Budapest, 1938-46; research asso. U. Tex., 1948-51; sect. head Armour & Co., Chgo., 1951-60; prof. medicinal chemistry State U. N.Y., Buffalo, 1960—; cons. chemistry Roswell Park Meml. Inst., Buffalo, 1961—. Fellow A.A.A.S.; mem. Am. Chem. Soc., Chem. Soc. (London), Am. Assn. Cancer Research, Am. Pharmacological Assn., Internat. Soc. Chemotherapy, N.Y. Acad. Scis., Sigma Xi, Rho Chi. Research, publs. on synthesis and structure of folinic acid; DNA in cancer cells; antimetabolites, alkylating agts. and dual antagonists as potential anti-cancer drugs; relationships between chem. reaction mechanisms and biol. action of drugs. Home: 131 Burbank Dr., Buffalo 14226.*

BARDSLEY, James Lomax, Brit. physician; b. Nottingham, Eng., July 7, 1801; ed. Glasgow (Scotland) U.; M.D., Edinburgh (Scotland) U., 1823. Physician, Manchester (Eng.) Infirmary, 1823-43; (with Thomas Turner) mgr. Manchester Royal Sch.

Medicine. Created knight, 1853. Mem. Brit. Med. Assn., Manchester Med. Soc. (pres. 1834); Royal Med. Soc. (pres.). Author: Hospital Facts and Observations, 1830. Contbr. articles on diabetes and hydrophobia to Cyclopedia of Practical Medicine, 1833. Pioneer in recording use of emetine in treatment of amebiasis, 1829. Died Manchester, July 10, 1876.

BARENDSEN, Gerrit Willem, Dutch biophysicist; b. Groningen, Netherlands, Aug. 14, 1927; s. Willem Pieter and Cornelia (Bakker) B.; Ph.D. in Physics, Groningen State U., 1955; m. Klaziena Bos, Sept. 10, 1954; children—Maryke-Bouwina, Cornelia-Yvonne, Irene-Johanna, Helena-Edith, Paula-Anita. Research physicist Yale, 1955-56, Radiobiol. Inst., Health Research Orgn., Ryswyk, Netherlands, 1956—; tchr. Inst. Radiopathology and Radiation Protection, Leiden (Netherlands) State U., 1958—. Mem. Dutch Soc. Radiation Biology, Dutch Health Physics Soc., European Soc. for Radiation Biology. Publs. on devel. of a counter technique for measurement of age by natural carbon-14, measurement of fall-out in biosphere, measurement of cellular response to irradiation. Home: 134 P. Meyerslaan. Office: 151 Lange Kleiweg, Ryswyk (ZH), Netherlands.*

BARER, Robert, English anatomist, biophysicist; b. London, Eng., Oct. 16, 1916; s. Nathan and Rachel (Krovoi) B.; B.Sc. in Physics, London U., 1938, B.Sc. in Physiology, 1939; M.B.B.S. with honors, U. Coll. Hosp., 1942; M.A., Oxford (Eng.) U., 1948; m. Gwendoline Briggs, Mar. 4, 1943; children—Robin Geoffrey, David Howard, Michael Richard. Univ. demonstrator in anatomy Oxford U., 1946-63; prof. human biology and anatomy, chmn. dept., Sheffield (Eng.) U., 1963—. Johnstone Laurence and Moseley Research fellow Royal Soc., 1949-53. Hon. fellow Royal Micros. Soc. (pres. 1967—); mem. Physiol., Anat. (Council) Optical Group, Photobiology Group (mem. com. 1965—), Biophys. Soc. for Exptl. Biology. Author: Lecture Notes on the Use of the Microscope, 1953; Advances in Optical and Electron Microscopy (with V. E. Cosslett), 1966; also numerous articles. Devel. and application of new methods of microscopy, quantitative methods of determining mass and concentration in living cells by phase and interference microscopy. Home: 55 Ranmoor Rd., Sheffield, Eng.*

BARFURTH, Dietrich, German anatomist; b. Dinslaken, Germany, Jan. 25, 1849; s. Dietrich and Henriette (Nünninghoff) B.; student natural scis. and math., Göttingen (Germany), 1875; later studied medicine in Bonn (Germany); degree, 1882; m. Helene Lohmann, circa 1879. Tchr. math. and sci., schs., Cologne (Germany), 4 years; became lectr. prof. Bonn, 1883; prosecutor Anat. Inst., Göttingen; apptd. prof. comparative anatomy, histology, embryology, Dorpat (Estonia) 1889; dir. Ant. Inst., Rostock, Germany, 1896. Author: Regeneration und Transplantation, Rückblick über der Ergebnissen 25 Jährliche Festchrift, 1917; Methoden zu Erforschung der Regeneration bei Tieren, 1920. Research on mechanics of devel., regeneration of tissues in tails of amphibians, parthenogenesis of vertebrates, genetics. Died Rostock, Mar. 23, 1927.

BARGE, Jacques Henri, French architect; b. Chateauroux, Nov. 30, 1904; s. Alfred and Jeanne (Lorichon) B.; ed. Lycée at Chateauroux, Ecole national superieure des beaux arts de Paris; govt. diploma in architecture; m. Suzanne Nicoud, Feb. 18, 1943; children—Odile, Monique, Michel. Tech. cons. Nat. Ministry Edn., 1942; chief architect Ministry Reconstruction and Urbanism, 1946; architect in chief of civil bldgs. and nat. palaces; mem. Gen. Council of French Bldgs., 1953, Gen. Office of Bldgs., 1955. Recipient Gold medal 1937 Expn., Dejean prize, Duc prize, medal of honor for architecture Expn. French Artists, also others. Mem. Order of Architects (past sec.-gen. regional council 5), Provincial Assn. French Architects (v.p. 1949), Central Soc. Architects, Colegio official de Arquitectos de Cataluna y Baleares (hon.), Internat. Union Architects, Gen. Confedn. French Architects (past sec.-gen.). Prin. works include Centre social de Chateauroux, (ch.) Sainte-Odile à Paris, Météorologie nationale à Paris, lycées in Bourges, Chateauroux, Blois, Rambouillet. Address: 177, bd. Saint-Germain, Paris 7; also 15 bis, rue des Arts, Chateauroux (Indre), France.

BARGEN, Jay Arnold, Am. physician; b. Mountain Lake, Minn., Oct. 25, 1894; s. Jacob F. and Anna (Balzer) B.; B.S., Carleton Coll., 1918; M.D., U. Chgo., 1922; M.S. in Medicine, U. Minn., 1927; m. Catherine Ruth Burns, Aug. 27, 1921; children—Robert Burns, Terese Ann (Mrs. John R. Cagney), Loretta Mae (Mrs. Lynn Tracy), Cornelia Marie (Mrs. George R. Cunnington). Staff cons. medicine. Mayo Clinic, Rochester, Minn., 1927-60, chmn. dept. gastroenterology, 1949-60, prof. medicine, 1945-60, pres. staff, 1956; cons. medicine, head dept. gastroenterology Scott & White Clinic, Temple, Tex., 1960—, dir. med. edn., 1963—; dir. Digestive Disease Found. Am., 1964—. Caldwell lectr. Am. Roentgen Ray Soc., 1966. Recipient Minn. State Med. Assn. Distinguished Service award, 1959. Mem. A.M.A. (chmn. sect. gastroenterology 1947), Am. Gastroenterology Assn. (pres. 1949, Friedenwald medal 1967), Med. Assn. N.Am. (pres. interstate postgrad. 1964). Author (with others) The Colon, Rectum and Anus, 1932; The Management of Colitis, 1935; The Modern Management of Colitis, 1943; Chronic Ulcerative Colitis, 1951. Contbr. numerous articles field disorders of digestive tract to med. jours. Research on intestinal tract especially disease chronic ulcerative colitis. Home: 3204 W. Av. T. Office: S. 31st St., Temple, Tex. 76502.*

BARGER, A. Clifford, Am. physiologist; b. Greenfield, Mass., Feb. 1, 1917; s. Paul and Rose (Solomon) B.; A.B., Harvard, 1939, M.D., 1943; m. Claire Basch, June 6, 1943; children—Craig, Cheryl, Curtis. Faculty, Harvard Med. Sch., Boston, 1946—, Robert Henry Pfeiffer prof. physiology, 1963—. Mem. Am. Physiol. Soc. (chmn. publs. com. 1962-63, 66—), Am. Acad. Arts and Scis., Mass. Soc. Med. Research (pres. 1957—). Asso. editor Circulation Research, 1963-66. Research, publs. in cardiovascular-renal physiology, particularly control of renal circulation, salt water balance. Home: 14 Orchard Rd., Brookline, Mass. 02146.*

BARGER, George, chemist; b. Manchester, Eng., Apr. 4, 1878; s. Gerrit and Eleanor (Higginbotham) B.; student Univ. Coll., London, Eng., 1896-98, King's Coll., Cambridge, Eng., 1898-1901; hon. degrees from Liverpool, Eng., Padua, Italy, Heidelberg, Germany, Utrecht, Netherlands, Mich., Lausanne, Switzerland; m. Florence Emily Thomas, 1904; 2 sons, 1 dau. Demonstrator in botany under Leo Errera, Brussels, Belgium, 1901-03; became mem. staff Wellcome Physiol. Research Labs., 1903; apptd. head dept. chemistry Goldsmith's Coll., New Cross, Eng., 1909; named prof. chemistry Royal Holloway Coll., Englefield Green, Eng., 1913; joined staff Med. Research Com., 1914; became 1st prof. chemistry in relation to medicine, U. Edinburgh (Scotland), 1919; Regius prof. chemistry, Glasgow, Scotland, from 1937. Dohme lectr. Johns Hopkins, 1928. Harbury medal Pharm. Soc., 1934; Longstaff medal, 1936, Davy medal, 1938. Fellow Royal Soc., 1919; v.p. chem. Soc., 1939; pres. sect. B, Brit. Assn., 1929. Author: The Simpler Natural Bases, 1914; Some Applications of Organic Chemistry to Biology and Medicine, 1930; Ergot and Ergotism, 1931; Organic Chemistry for Medical Students, 1932. Made studies of alkaloids and simple nitrogenous compounds of biol. importance; (with F. H. Carr and H. H. Dale) isolated ergotoxine from ergot, 1906, also isolated histamine from ergot, 1910; synthesized thyroxine, 1927. Died Aeschi, Switzerland, Jan. 6, 1939.

BARGER, Gerald Lee, agrl. meteorologist; born Weldon, Ia., Feb. 21, 1916; s. Claude Lee and Hattie (Mitchell) B.; A.B., Simpson Coll., 1938; M.S., Ia. State U., 1942, Ph.D., 1948; certificate in meteorology U. Chgo., 1942; M.Pub. Adminstrn., Harvard, 1961; m. Margery Louise Hitzel, Sept. 4, 1946; children—Patricia Ann, Claudia Jean, Janet Elaine, Jeffrey Lee. Research asst. Soil Conservation Service, U. S. Dept. Agr., Ames, Ia., 1940-42; area climatologist U. S. Weather Bur., Ames, 1948-58; asso. prof. agrl. climatology Ia. State U., Ames, 1952-58, operations analyst, 1950-58; dep. dir. Nat. Weather Center, Asheville, N.C., 1958-61, dir., 1961-64; chief planning unit World Meteorol. Orgn., Geneva, Switzerland, 1964—. Cons. agrl. meteorologist, 1952—; operations analyst U. N.C., 1958—; lectr. Asheville-Biltmore Coll., 1959-64. Decorated Air Medal with oak leaf cluster. Fellow A.A.A.S.; mem. Am. Meteorol. Soc. (chmn. com. on agrl. meteorology 1958-, asso. editor 1958-61, councilor 1961-64), Internat. Soc. Biometeorology, Sigma Xi, Phi Kappa Phi, Gamma Sigma Delta, Epsilon Sigma. Author: (with J. Y. Wang) Bibliography of Agricultural Meteorology, 1962; also articles on drought, soil moisture, frequency distbn. precipitation amounts. Home: 24 Av. Ernest Pictet, Geneva. Office: World Meteorological Orgn., 41 Av. G. Motta, Geneva, Switzerland.*

BARGER, Vernon Duane, Am. theoretical physicist; b. Curllsville, Pa., June 5, 1938; s. Joseph F. and Olive (McCall) B.; B.S., Pa. State U., 1960, Ph.D., 1963. Research asso. U. Wis., Madison, 1963-65, asst. prof. physics, 1965—. Mem. Am. Phys. Soc., Phi Kappa Phi, Tau Beta Pi. Research in phenomenological description of high energy scattering processes; classification of fundamental particles as Regge recurrences.*

BARGHOORN, Elso Sterrenberg, Am. paleontologist, botanist; b. N.Y.C., June 30, 1915; s. Elso Sterrenberg and Elizabeth (Brust) B.; B.A., Miami U., Oxford, O., 1937; M.A., Harvard, 1938, Ph.D., 1941; m. Margaret Alden MacLeod, Aug. 16, 1941 (div. Jan. 1951); 1 son, Steven Frederick; m. 2d, Dorothy Dellmer Osgood, Oct. 31, 1964. Faculty, Amherst Coll., 1941-46, asst. prof., 1944-46; field service cons. OSRD, 1944-46; faculty Harvard, 1946—, prof. botany, curator botany, 1955—. Cons., U. S. Geol. Survey, 1954—. Recipient Civilian Service award USN, 1946. Member of the National Academy of Sciences, Geological Soc. Am., Bot. Soc. Am., Geochem. Soc., Am. Acad. Arts and Sci. Research, numerous publs. on biology and environmental relations marine fungi involved in degradation organic matter in sea, microscopic and chem. changes indegradation organic matter in formation peat and coal; description and interpretation oldest organisms known in geologic record life. Home: R.F.D. Cross St., Carlisle, Mass. 01741. Office: 22 Divinity Av., Cambridge, Mass. 02138.*

BARGMANN, Wolfgang Ludwig, German physician; b. Nuremberg, Jan. 27, 1906; ed U. Frankfurt-am-Main, U. Munich, U. Vienna, U. Berlin; M.D. Instr. at Zurich, 1935; asst. prof. anatomy at Leipzig, 1938; prof. at Königsberg, 1942, at Göttingen, 1945; full prof. at Kiel, 1945; rector U. Kiel, 1950. Vice pres. German Community for Research, 1955-61, rector, 1965-66. Mem. Royal Swedish, Norwegian acads. sci., Leopoldina, Acad. Mayence (corr.), Japanese Soc. Anatomy (hon.). Studies on nervous system and electron microscopy. Home: Niemannsweg 81. Office: New University, Haus 30, Kiel, Germany.

BAR-HEBRAEUS, see Abu al-Farag.

BARKA, Tibor, physician; b. Debrecen, Hungary, Mar. 31, 1926; s. Imre and Hajnal (Szekely) B.; M.D., Debrecen U., 1950; m. Katalin Szalay, Mar. 3, 1957. Came to U. S., 1958, naturalized, 1963. First research asso. dept. morphology Inst. Exptl. Medicine, Hungarian Acad. Sci., Budapest, 1954-56; research asso. Inst. Cell Research and Genetics, Karolinska Institutet, Stockholm, Sweden, 1956-58; research asso. Mt. Sinai Hosp., N.Y.C., 1958-62, asst. attending pathologist, 1962-64, asso. attending pathologist, 1964—; prof. dept. pathology Mt. Sinai Sch. Medicine, 1966—; chmn. dept. anatomy, 1967—. Mem. Histochem. Soc., Am. Soc. Exptl. Pathology, Council Biol. Editors, Am. Soc. Cell Biology. Author: (with G. Kiszely) Practical Microtechnique and Histochemistry, 1958; Histochemistry, Methods of the Experimental Medicine, 1959; (with P. J. Anderson) Histochemistry: Theory, Practice and Bibliography, 1963. Editor-in-chief, Jour. Histochemistry and Cytochemistry, 1965—. Contbr. numerous articles to profl. jours. Research in devel. and application of histochem. methods, studies on regulation of cell div.*

BARKALOW, Frederick Schenck, Jr., Am. zoologist; b. Marietta, Ga., Feb. 23, 1914; s. Frederick Schenck and Katherine Aurelia (White) B.; B.S. in Chemistry, Ga. Inst. Tech., 1936; student Auburn U., 1936-38; M.S. in Zoology, U. Mich., 1939, Ph.D. (Rosenwald fellow 1946-47), 1948; m. Joan Metzger, Nov. 23, 1937; 1 dau., Joanna. Instr. zoology Auburn U., 1936-39, instr. botany and plant pathology, 1946; chief biologist Ala. Dept. Conservation, 1939-41; mem. faculty N.C. State U. of N.C., Raleigh, 1947—, prof. zoology, 1950—, head dept., 1950-63; cons. wildlife, 1939—. Mem. Sec. Agr. Adv. Com. Multiple Use Nat. Forests, 1963-65; del. White House Conf. Conservation, 1962; panelist NSF, 1959—, mem. tropical biology panel U. Costa Rica, 1962; sr. vis. fellow to Great Britain for OEEC, 1960; cons. disease vector study in Alaska for Dept. Def., 1951. Chmn. edn. div. United Fund Raleigh, 1954, bd. dirs., 1955-58. Served to lt. col. AUS, 1941-46. Recipient Am. Motors Conservation award, 1967. Mem. Archaeol. Soc. N.C. (pres. 1959-60), N.C. Acad. Sci. (v.p. 1962), N.C. Wildlife Fedn. (sec. 1948, trustee 1960—, president 1965-66), Am. Soc. Mammalogists (dir. 1961—, chmn. land mammals com. 1961-66), Am. Assn. Advancement Sci., Am. Inst. Biol. Sci., Am. Ornith Union, Arctic Inst. N.A., Assn. Tropical Biology, Biol. Soc. Washington, Ecol. Soc. Am., Am. Soc. Systematic Zoology, Wildlife Soc. (chmn. nominations com. 1961-62), Phi Beta Kappa (pres. Wake County assn. 1954-55), Sigma Xi, Phi Kappa Phi (pres. N.C. State chpt. 1960-61), Phi Sigma. Author numerous articles in field. Research covering game populations; game inventories on a state-wide basis; taxonomy of mammals; management of wildlife resources. Home: 3439 Bradley Pl., Raleigh, N.C. 27607.

BARKAN, Otto, Am. ophthalmologist; b. San Francisco, Apr. 5, 1887; s. Adolph and Louise (Desept) B.; B.A., Oxford, 1909, M.D., Munich, 1914; m. Margit Park, 1921; children—Park Otto, Thomas Adolph. House physician St. Mary's Hosp., London, 1915-16; with eye clinics, univs. Munich and Vienna, 1916-17; asst. U. Munich, 1917-19, U. Zurich, 1919-20; practice of medicine, San Francisco, from 1921; mem. surg. staff San Francisco Hosp., 1921; asso. clin. prof. Stanford U. Med. Sch., from 1921; cons. opthalmologist Veterans' Hosp. Recipient Howe medal for devel. surgery to relieve glaucoma in infants A.M.A., 1954. Mem. Am. Acad. Ophthalmology, Western Ophthal. Soc., A.M.A., Assn. for Research in Ophthalmology, Pacific Coast Oto-Ophthal. Soc., Brit. Ophthal. Soc., Royal Coll. Surgeons (London), Alpha Kappa Kappa. Contbr. articles on ocular surgery to med. jours. Died Apr. 26, 1958.

BARKER, Benjamin Fordyce, Am. physician; b. Wilton, Me., May 2, 1818; s. John and Phoebe (Abbott) B.; grad. Bowdoin Coll., 1837, M.D., 1841; m. Elizabeth Dwight, Sept. 14, 1843. Prof. obstetrics Bowdoin Med. Coll., from 1844; incorporator, prof, obstetrics N.Y. Med. Coll., from 1849; prof. obstetrics, diseases of women Bellvue Hosp. Med. Sch., N.Y.C. from 1861. Mem. Conn. (pres. 1848), N.Y. State (pres. 1856) med. socs., Am. Gynecol. Soc. (founder, 1st pres.), N.Y. Acad. Medicine (pres. 1882). Author: Puerperal Diseases, 1874. 1st Am. physician to use hypodermic syringe. Died N.Y.C., May 30, 1891.

BARKER, Franklin Davis, Am. zoologist; b. Ottawa, Kan., Sept. 16, 1877; s. Albert Wentworth and Martha Ella (Luther) B.; A.B., Ottawa U., 1898, A.M., 1900; postgrad. U. Chgo., 1898-1900; Ph.D., U. Neb., 1910; fellow Harvard, 1912; m. Lena Lovett, 1905; 1 son, John Franklin. Prof. biology Ottawa U., 1898-1903; prof. med. zoölogy U. Neb., 1903-26;

prof. zoology Northwestern U., summer 1923, head prof., from 1926. Prof. zoölogy U. Ill., summer 1917; Investigator Harpswell Biol. Sta., 1907, 12, 14, Bermuda Biol. Sta., 1912, 15. Research grantee Nat. Acad. Sci. Author: Synopsis of the Parasites of Man, 1926; Unit System Laboratory Outlines in General Zoölogy, 1926; Unit System Laboratory Outlines in Parasitology, 1926. Asso. editor Jour. Parasitology, from 1915. Died July 10, 1936.

BARKER, George Frederick, Am. physicist, chemist; b. Charlestown, Mass., July 14, 1835; s. George and Lydia Prince (Pollard) B.; Ph.B., Yale, 1858; M.D., Albany Med. Coll., 1863; Sc.D., U. Pa., 1898; LL.D., Allegheny, 1898, McGill, 1900; m. Mary M. Treadway, Aug. 15. 1861. Prof. natural scis. Wheaton Coll., 1861; acting prof. chemistry Albany Med. Coll., 1863; prof. natural scis. Western U. Pa., 1864; asst. chemistry Yale, 1865-67, prof. physiol. chemistry and toxicology, 1867-73; prof. physics U. Pa., 1873-1900; prof. emeritus, from 1900. U. S. commr. Paris Elec. Exhbn., 1881; del. elec. congress and v.p. jury of award; U. S. commr. Elec. Exhbn., Phila., 1884; on jury of awards, Chicago Expn., 1893. Expert in poisons, criminal cases and in Edison, Berliner and other patent suits. Mem. A.A.A.S. (pres. 1879), Am. Chem. Soc. (pres. 1891). Author: The Forces of Nature, 1863; Textbook of Elementary Chemistry, 1870; Chemical Discoveries of the Spectroscope, 1873; Conversion of Mechanical Energy into Heat by Dynamo-Electric Machine, 1880; Physics, 1892. Asst. editor Am. Jour. Sci., 1868-1909; editor Jour. of Franklin Inst., 1874-75. First in America to demonstrate radium in radio-active bodies. Died 1910.

BARKER, Gordon Hitchcock, Am. sociologist; b. Chgo., 1905; s. Willis E. and Grace (Hitchcock) B.; B.S., Northwestern U., 1929, M.A., 1939, Ph.D., 1940; m. Victoria Siegfried, June 19, 1948. Faculty, U. Colo., Boulder, 1940—, U. Md., 1953-54, 56-57, U. Hawaii, summers 1961-64. Fulbright prof. Pierce Coll., Athens, Greece, 1964-65; cons. Pres.'s Commn. on Law Enforcement and Administration of Justice, Juvenile Delinquency Demonstration and Tng. Project, Federal Rehabilitation Project, Denver. Mem. Am. Sociol. Assn., Am. Assn. U. Profs. Research, publs. on juvenile delinquency, expecially significance of offenses of boys and girls, social structure of correctional instns. and racial factors in delinquency; established, now directing programs of assistance in rehab. of delinquents in instns. and through cts. Home: 835 8th St., Boulder, Colo. 80302.*

BARKER, Sir Herbert Atkinson, Brit. surgeon; b. Southport, Eng., Apr. 21, 1869; s. Thomas Wildman and Agnes (Atkinson) B.; student surgery under J. Atkinson; m. Jane Ethel, 1907. Practice manipulative surgery, from 1904; manipulative surgeon Noble's Hosp., Isle of Man, from 1941. Author: Leaves from my Life, an Autobiography, 1927. Contbr. articles on manipulative surgery to Ency. Britannica, 1926, other publs. Prin. operations include cure or alleviation of derangements of knee cartilages without surgery, correction of flat feet, shoulder dislocations, tennis elbow, hallux rigidus, metatarsalgia, sacroiliac displacements (all by manipulation). Died July 31, 1950.

BARKER, Horace A(lbert), Am. biochemist; b. Oakland, Cal., Nov. 29, 1907; s. Albert C. and Nettie (Hindry) B.; A.B., Stanford, 1929, Ph.D., 1933; m. Margaret McDowell, Aug. 29, 1933; children—Barbara, Elizabeth, Robert. Instr. soil microbiology U. Cal., Berkeley, 1936-40, asst. prof., 1940-45, asso. prof., 1945-46, prof., 1946-50, chmn. div. plant nutrition, 1949-50, prof. plant biochem. and microbiologist agrl. expt. sta., 1950-59, professor of biochemistry, 1959—, chmn. dept., 1962-64, chmn. dept. plant biochemistry, 1950-53; asso. editor Ann. Rev. Microbiology, 1946-53; editor Archives of Biochemistry, 1951-54; mem. editorial bd. Journal of Bacteriology, 1955-60, Journal of Biochemistry 1960-65. Recipient Sugar Research award, National Acad. Sci., 1945; Neuberg medal, Am. Soc. European Chemists, 1959, Borden award in nutrition, 1962; Cal. Scientist Yr. award Cal. Mus. Sci. and Industry, 1965. Mem. Am. Chem. Society, Society of American Bacteriology, Am. Soc. Biol. Chemists, Biochem. Soc. (Hopkins medal 1967), Nat. Acad. Sci. Author: Bacterial Fermentations, 1956; also tech. articles sci. jours. Research in biochemistry and physiology of microorganisms; soil microbiology. Home: 1045 Mariposa Av., Berkeley 7, Cal.

BARKER, John Charles, English physician, psychiatrist; b. Bickley, Kent, Eng., July 11, 1924; s. Frederick Charles and Norah (Hyne) B.; student Cambridge (Eng.) U., 1943-45, M.A., 1955, M.D., 1960; Diploma in Psychol. Medicine, 1956; m. Jane Rosamund Homfray, Nov. 12, 1948; children—Nigel, Josephine, Julian. Various med. posts, U.K., 1948-55; cons. psychiatrist Banstead Hosp., Sutton, Surrey, Eng., 1959-62; cons. psychiatrist Herrison Hosp., Dorchester, Dorset, Eng., also Shelton Hosp., Shrewsbury, Shropshire, Eng., 1962—. Fellow Royal Soc. Medicine; mem. Royal Coll. Physicians London, Royal Coll. Physicians Edinburgh, Royal Medico-Psychol. Assn., Brit. Med. Assn., Brit. Soc. for Psychical Research. Author: (with M. D. Enoch, W. H. Trethowan) Some Unusual Psychiatric Syndromes, 1967; Scared to Death, 1967; also articles. Research on hosp. addiction, behaviour therapy including first application to tranvestism and compulsive gambling, sedation threshold. Home: Barnfield, Athgarvan, Yockleton, nr. Shrewsbury. Office: Shelton Hosp., Shrewsbury, Shropshire, Eng.*

BARKER, Lewellys Franklin, physician; b. Norwich, Ont., Can., Sept. 16, 1867; s. James F. and Sarah Jane (Taylor) B.; ed. Pickering Coll., Ont., 1881-84; M.B., U. Toronto, 1890; postgrad. Leipzig, 1895, univs. of Munich and Berlin, 1904; hon. M.D., U. Toronto, 1905; LL.D., Queen's Univ., Kingston, Can., 1908, McGill Univ., Montreal, Can., 1911; Univ. Glasgow (Scotland), 1930; m. Lilian H. Halsey, Oct. 1903; children—John Hewetson, William Halsey, Margaret Taylor. Asso. in anatomy Johns Hopkins U., Balt., 1894-97; resident pathologist, Johns Hopkins Hosp., 1894-99; asso. prof. anatomy Johns Hopkins, 1897-99, pathology, 1899-1900, prof. medicine, chief physician of hosp., 1905-13; prof., head dept. anatomy, Rush Med. Coll. (U. Chgo.), 1900-05; Johns Hopkins med. commr. to P.I., 1899; mem. special commn. apptd. by sec. of treasury to investigate plague in San Francisco, 1901; chmn. bd. scientific dirs. Wistar Inst. of Anatomy; pres. Nat. Com. for Mental Hygiene, 1909-18; chmn. advisory bd. Federal Industrial Inst. for Women, Alderson, W.Va., also Med. Council U. S. Vets. Bur., Mem. Assn. Am. Physicians (pres. 1913), Am. Neurol. Assn. (pres. 1916), So. Med. Assn. (pres. 1919), Assn. for Study Internal Secretions (pres. 1919), A.M.A. (v.p. 1917), Am. Soc. for Control Cancer (v.p.), Swedish Med. Soc.; corr. mem. profl. socs. of Budapest, Edinburgh, Vienna; mem. Phi Beta Kappa. Author: The Nervous System and Its Constituent Neurones, 1899; Translation of Wernre Spalteholz's Hand Atlas of Human Anatomy, 1900; Laboratory Manual of Human Anatomy (with Dean De Witt Lewis and D. G. Revell), 1904; The Clinical Diagnosis of Internal Diseases, 1916; Tuesday Clinics at Johns Hopkins Hospital, 1922; Blood Pressure (with N. B. Cole), 1924; The Young Man and Medicine, 1927; Psychotherapy, 1940; also numerous med. papers and addresses. Co-Editor: Endocrinology and Metabolism, 1922. Died July 13, 1943.

BARKER, Nelson Waite, Am. physician; b. Evanston, Ill., Apr. 25, 1899; s. Earle Sherman and Ollive (Waite) B.; A.B., Dartmouth, 1921; M.D., U. Chgo., 1925; M.S. in Medicine, U. Minn., 1929; m. Florence Eleanor Buswell, Apr. 6, 1926; children—Sylvia (Mrs. Robert N. Thalman), David Nelson, Robert Earle. Practice medicine specializing in internal medicine and vascular disease, Rochester, Minn., 1926-64; staff, Mayo Clinic, 1930-57, head sect. in medicine, 1948-57, pres. clinic staff, 1955; cons. medicine Rochester State Hosp., 1957—; faculty U. Minn. Mayo Grad. Sch. Medicine, 1930—, prof. medicine, 1948-64. Diplomate Am. Bi. Internal Medicine. Fellow A.C.P.; mem. Am. Heart Assn., Council on Circulation Arteriosclerosis, Am., Minn., So. Minn. (past pres.) med. assns., Minn. Soc. Internal Medicine, Central Soc. for Clin. Research, Am. Soc. for Study Arteriosclerosis (past pres.), Sigma Xi, Gamma Alpha. Author: (with E. V. Allen, E. A. Hines, Jr.) Peripheral Vascular Diseases, 1946, 55, 62; (with Florence B. Barker) Bird Songs of Southeastern Minn., 1963; also numerous articles. Research on diseases blood vessels, hypertension, blood coagulation, anti-coagulant drugs, lipid metabolism; pioneered clin. use anticoagulant dicumarol. Home: 920 10th St. S.W., Rochester 55901. Office: 200 1st St. S.W., Rochester, Minn. 55902.*

BARKER, Roger Garlock, Am. psychologist; b. Macksburg, Ia., Mar. 31, 1903; s. Guy and Cora (Garlock) B.; A.B., Stanford, 1928, M.A., 1930, Ph.D., 1934; m. Louise Dawes Shedd, June 17, 1930; children—Celia Louise (Mrs. Stephen Lottridge), Jonathan Shedd, Lucy Garlock. Faculty, Harvard, 1937-38, U. Ill., 1938-42, Stanford, 1942-46, Clark U., 1946-47; prof. U. Kan., Lawrence, 1947—, chmn. psychology dept., 1947-51; co-founder Midwest Psychol. Field Sta., Oskaloosa, Kan., 1947. Mem. com. on child devel. NRC-Nat. Acad. Scis., 1956-58, mem. com. primary records in behavioral scis., 1957-61. Recipient Research Career award Nat. Inst. Mental Health, 1963. Fellow Center For Advanced Study in Behavioral Scis., 1957-58. Fellow Am. Psychol. Assn. (past div. pres., Distinguished Sci. Contbn. 1963); mem. Soc. For Research in Child Devel. (past pres.), Soc. Psychol. Study Social Issues (Kurt Lewin award 1963), A.A.A.S., Ecol. Soc. Am., Soc. Gen. Systems Research, Soc. Biol. Rhythm, Am. Assn. U. Profs., Sigma Xi. Author: (with others) Adjustment to Physical Handicap and Illness, A Survey of the Social Psychology of Physical Handicap and Illness, 1946, rev., 1953; (with H. F. Wright) One Boy's Day, 1951, Midwest and Its Children, 1955; (with others) Specimen Records of American and English Children, 1961; (with P. V. Gump) Big School, Small School, 1964. Editor: The Stream of Behavior, 1963. Publs. on devel. of methods and concepts. for rec. and analyzing behavior and its environing conditions as they occur in everyday life. Home: Oskaloosa, Kan. 66066.*

BARKER, Samuel Booth, Am. biologist; b. Montclair, New Jersey, March 3, 1912; son of Harry and Marion (Booth) B.; B.S., cum laude, U. Vt., 1932; postgrad. Yale; Ph.D., Cornell U., 1936; m. Justine Rogers, July 31, 1934. Asst. physiology Cornell U. Med. Coll., 1937-40, fellow medicine, 1938-41; faculty U. Tenn. Coll. Medicine, 1941-44, asst. prof. physiology, 1943-44; faculty State U. Ia. Coll. Medicine, 1944-52, asso. prof., 1946-52; prof. pharmacology Med. Coll. and Sch. Dentistry, U. Ala. Med. Center, Birmingham, 1952-62, dir. grad. studies, prof. physiology, asso. dean, 1965—; prof. pharmacology Coll. Medicine, U. Vt., Burlington, 1962-65. Cons. scientist NIH, NSF. Krichesky fellow endocrinology U. Cal. at Los Angeles, 1951-52. Mem. A.A.A.S., Am. Physiol. Soc., Soc. for Exptl. Biology and Medicine, Harvey Soc., Am. Assn. U. Profs., Ala. Acad. Sci. (past pres.), Am. Fedn. for Clin. Research, Endocrine Soc. Am. Chem. Soc., Biochem. Soc. (London), N.Y. Acad. Scis., Am. Thyroid Assn., Phi Beta Kappa, Sigma Xi. Author: (with J. H. U. Brown) Basic Endocrinology, 1962, 2d edit., 1966. also numerous articles. Research on relation of hormones to metabolism, micromethods in biochem. analysis. Home: 1812 Woodcrest Rd., Birmingham, Ala. 35209.*

BARKER, Sidney Alan, English biol. chemist; b. Birmingham, Eng., Apr. 13, 1926; s. Philip Henry and Gladys (Allen) B.; B.Sc. with 1st class honors, U. Birmingham, 1947, Ph.D., 1950; D.Sc., 1957; m. Ruth May, Mar. 8, 1952; children—Jane Alison, Helen Elizabeth. Research fellow U. Birmingham, 1950-51, Makinnson student Royal Soc., 1951-53, Brit. Rayon fellow, 1953-54, lectr. chemistry, 1954-62, sr. lectr., 1962-64, reader in carbohydrate chemistry, 1964—; Rockefeller fellow U. Cal. at Berkeley, 1955-56, Rutgers U., 1956. Invited lectr. USSR Acad. Scis., 1962, Am. Chem. Soc., 1964, 65. Fellow Royal Inst. Chemistry. Author: (with M. Stacey) Polysaccharides of Microorganisms, 1960; Carbohydrates of Living Tissues, 1962; also numerous articles. Research in biol. chemistry, devising of new methods of separation and structure determination of complex carbohydrates. Home: 1 Abdon Av., Selly Oak, Birmingham, Eng.*

BARKER, Thomas Vapond, mineralogist; b. 1881; continued work of E. Federov; undertook systematic study of structure of crystal substance, 1930. Died 1931.

BARKHAUSEN, Heinrich, German physicist; b. Dresden, Germany, Dec. 2, 1881; student, engring. sch., Munich, Germany, then studied physics, univs. Munich, Berlin, Göttingen (Germany); doctorate, 1907; mem. research staff Siemens Labs., 1907-11; became prof. communication engring. Dresden U., 1911, emeritus after World War II. Recipient Mauris Liebmann Meml. prize Inst. Radio Engrs., 1933. Author: Das Problem der Schwingungserzeugung, 1907; Elektronenröhren, 3 vols., 1923-29. Research on magnetism; conducted expts. with short wave radio transmissions; developed (with Kurz) Barkhausen-Kurz oscillator for very high frequencies; discovered Barkhausen effect (concerning sudden changes in magnetic properties of metal produced by slow and regular changes in magnetic field); formulated electron tube coefficients; studied subjective measurement of sound. Died Dresden, Feb. 20, 1956.

BARKLA, Charles Glover, Brit. physicist; b. Lancashire, Eng., June 7, 1877; s. John Martin Barkla; grad. in physics U. Coll., Liverpool, Eng., 1899; Student Trinity Coll., Cambridge (Eng.) U., then King's Coll., London; D.Sc., Liverpool U., 1904, LL.D., 1931; m. Mary Esther Cowell; 2 sons, 1 dau. Lectr. advanced electricity Liverpool U., 1907-09; Wheatstone prof. physics King's Coll., U. London, 1909-13; prof. natural philosophy U. Edinburgh, 1913-44; lectr., German Sci. and Med. Congress, 1913. Fellow Royal Soc., 1912. Recipient Nobel prize in physics, 1917, Hughes medal Royal Soc. London, 1917. Research and publs. on electric waves, X-rays, secondary waves; discovered characteristic X-rays of elements, polarization of X-rays; formulated laws of X-ray scattering, laws governing the transmission of X-rays through matter, and excitation of secondary rays. Died Edinburgh, Oct. 23, 1944.

BARKMAN, Jan Johannes, Dutch botanist; b. Medemblik, Mar. 15, 1922; s. Jan and Gezina (Wilhelmina) B.; ed. univs. of Leiden, Utrecht, Zurich, SIGMA Montpellier; Ph.D. in Sci.; m. Maria Hendrika van der Weel, Oct. 22, 1951; children—Francina Charlotte, Maartje Gezina, Hans Arthur. Asst., curator Herbier National of Leiden; prof. U. Leiden; head Biol. Sta. at Wijster. Mem. Royal Dutch Bot. Soc., Brit. Bryological Soc., others. Author: Phytosociology and Ecology of Cryptogamic Epiphytes. Address: Kampsweg 29, Wijster (Dr), Netherlands.

BARLET, Annibal, French chemist, physician; flourished 2d half 17th century; gave instrn. by demonstration in his lab., Paris; tchr. of Aberdeen physician Matthew Mackaile. Author: Le vray et méthodique cours de la physique résolutive ou chymie, 1653, other papers. His works are primarily devoted to displacing alchemy; spoke of indivisible parts called Atoms.

BARLETTA, Mariano Santo di, Italian physician; b. Italy 1490; improved and popularized operation for removal of urinary calculi (preceded lateral lithotomy), recorded 1535. Died 1550.

BARLOW, Charles Franklin, Am. physician; b. Mason City, Ia., Nov. 20, 1923; s. Frank Richard and Marie Gertrude Barlow; S.B., U. Chgo., 1945, M.D., 1947; M.A., Harvard, 1963; m. Patricia Keith, June 30, 1953; children—John Keith, Ellen, Margaret. Faculty, U. Chgo., 1954-63, asso. prof., 1960-63; neurologist-in-chief Children's Hosp. Med. Center, Boston, 1963—; Bronson Crothers prof. neurology Harvard Med. Sch., 1963—; Ford Found. Frankfurt-Chgo. exchange Pharmakologisches Institut, 1960.

Cons. neurology Peter Bent Brigham Hosp., 1963—. Recipient McClintock Teaching award U. Chgo., 1963. Diplomate Am. Bd. Neurology and Psychiatry, Mem. Am. Acad. Neurology, Am. Assn. Nueropathologists, Am. Neurol. Assn., New Eng. Pediatric Soc., Boston Soc. Psychiatry and Neurology. Research, publs. on blood-brain barrier with radioisotope labelled drugs and other drugs and other compounds, normal, pathophysiologic and pathologic changes. Home: 121 Dover Rd., Westwood, Mass. 02090. Office: 300 Longwood Av., Boston 02115.*

BARLOW, Harold Everard Monteagle, English physicist; b. London, Eng., Nov. 15, 1899; s. Leonard and Katherine (Monteagle) B.; B.Sc. in Engring., U. Coll., London, 1921, Ph.D., 1923; m. Janet Hastings Eastwood, Sept. 5, 1931; children—Colin Hastings, David Monteagle, Neil Wallace, Lindsay Margaret. With E. Surrey Ironworks & Barlow and Young Ltd., 1923-25; faculty U. Coll., London, 1925-68, Pender prof. elec. engring., 1950-68; dir. Marconi Instruments Ltd., St. Albans of Sanders Electronic Ltd., Stevenage. Cons. Barlow, Leslie & Partners, 1945—, Imperial Chem. Industries Ltd., 1956—, Marconi Instruments Ltd., 1963—. Fellow Institute of Electronic and Electrical Engrs. (Faraday medal), Royal Soc., 1961, Inst. Elec. Engrs., Inst. Mech. Engrs. Author: Microwaves and Waveguides, 1947; (with A. L. Cullen) Microwave Measurements, 1950; Radio Surface Waves (with J. Brown), 1962; also numerous articles. Research on devel. thermionic ammeter, new high frequency Hall effect and relation to radiation pressure and applications, long-distance waveguides for telcommunications and devel. bends in waveguides. Home: 12 Higher Dr., Banstead, Surrey, Eng. Office: Univ. Coll., Gower St., London, W.C.1, Eng.*

BARLOW, Horace Basil, physiologist; b. Chesham, Eng., Dec. 8, 1921; s. James Alan and Emma Nora (Darwin) B.; M.A., Trinity Coll. Cambridge, 1943, M.D., 1947, Sc.D., 1966; postgrad. Harvard, 1944-46; m. Ruth Chattie Salaman, Dec. 22, 1954; children—Rebecca Nora, Natasha Helen, Naomi Jane. Came to U. S., 1964. Research fellow Trinity Coll. Cambridge, 1950-54, asst. dir. research physiol. lab., 1954-64, lectr. physiology Kings Coll., 1954-64; instr. Wilmer Inst. Johns Hopkins, 1953; prof. physiol. optics U. Cal., Berkeley, 1964—. Mem. Physiol. Soc., Exptl. Psychology Soc. (both London). Research, publs. on mechanisms neural processing of visual information in vertebrate retina; discovered lateral inhibition in ganglion cells of frog retina, selectivity of response to direction of image motion in rabbit retina. Home: 825 Santa Barbara Rd., Berkeley, Cal. 94707.*

BARLOW, John Sutton, Am. neurol. biophysicist; b. Raleigh, N.C., June 10, 1925; s. David Henry and Anne (Sutton) B.; B.S., U. N.C., 1944, M.S., 1948; postgrad. Johns Hopkins; M.D., Harvard, 1953; m. Sibylle Ernestina Jahrreiss, Aug. 5, 1950; children—Thomas W., Robert S., Lisa K. Faculty, Harvard Med. Sch., Boston, 1953—, research asso. in neurology, 1961—; mem. staff Mass. Gen. Hosp., Boston, 1953—, neurophysiologist, neurology service, 1961—; faculty Mass. Inst. Tech., 1954—, research affiliate in elec. engring., 1961—. Research asso. in biophysics U. Cal. Sch. Medicine, Los Angeles, 1966; asst. examiner Am. Bd. Electroencephalography, 1965. Recipient USPHS Research Career Devel. award Nat. Inst. Neurol. Diseases and Blindness, 1962—. Diplomate Am. Bd. Electroencephalography. Mem. Internat. Brain Research Orgn., Am. Electroencephalographic Soc., Biophys. Soc., Am. Acad. Neurology, A.A.A.S., Am. Geophys. Union, Boston Soc. Psychiatry and Neurology, Friends Med. Soc., NIH (neurology study sect. B 1966—). Translator: Mathematical Analysis of the Electrical Activity of the Nervous System (from Russian); also articles from Bulgarian, Chinese, Czech. Researcher, lectr., contbr. articles on computer studies of brain potentials and related bioelec. potentials in man; devel. analog computer techniques for analysis of bioelec. signals; geophys. aspects of geog. pathology; neurophysiol. basis of animal navigation. Home: 38 Holden Wood Rd., Concord, Mass. 01742. Office: Mass. Gen. Hosp., Boston 02114.*

BARLOW, Peter, Brit. mathematician, physicist; b. Norwich, Eng., Oct. 13, 1776; became master math., Woolich, Eng., 1806, prof. until 1847; Copley medal for work on magnetism, 1825; Fellow Royal Soc., 1823; mem. French Acad. Scis., 1828. Author: Elementary Investigation of the Theory of Numbers, 1811; New Mathematical and Philosophical Dictionary, 1814; Essay on Magnetic Attractions, 1820; pioneered sci. work on strength of metals, wood, stone, cement; research on constrn. of achromatic object lenses, a negative lens (Barlow lens), and fluid concave lenses; sought to produce achromatic telescope; devised method of rectifying compass error on ships caused by metal; developed early electric motor (Barlow's wheel); studied stability of rails. Died Mar. 1, 1862.

BARLOW, Richard Eugene, Am. statistician; b. Galesburg, Ill., Jan. 12, 1931; s. Clarence E. and Annabell (Ramp) B.; B.A., Knox Coll., 1953; M.A., U. Ore., 1955; Ph.D., Stanford, 1960; m. Barbara June Ferguson, Sept. 15, 1956; children—Jeanne Ann, Elaine, Lawrence Thomas. Staff, Gen. Telephone Labs., Palo, Cal., 1962-63; asso. prof. dept. indsl. engring. and operations research U. Cal. at Berkeley, 1963—. Cons. Boeing Sci. Research Labs., 1962—, RAND, 1964—. Mem. Am. Math. Soc., Inst. Math. Statistics, Am. Statis. Assn., Phi Beta Kappa. Asso. editor Technometrics, 1966—. Author: Mathematical Theory of Reliability, 1965; also articles. Research in reliability theory and applied probability and statistics. Home: 93 Simpson Dr., Walnut Creek, Cal. 94598. Office: 4177 Etcheverry Hall, Berkeley, Cal. 94720.*

BARLOW, Sir Thomas, Brit. physician; b. Lancashire, Eng., Sept. 4, 1845; s. James and Alice (Barnes) B.; ed. Owen Coll., Manchester, Eng., U. Coll., London, Eng.; m. Ada Helen Dalmahoy, 1880 (dec. 1928; children—Sir (James) Alan (Noel) Barlow, Thomas, Pattrick, 2 daus. Physician, Hosp. for Sick Children, London, 1885-99, U. Coll. Hosp., London, 1885-1910; physician extraordinary to Queen Victoria, 1899-1901. Fellow Roy. Soc., 1909. Contbr. many articles to profl. jours. and textbooks. Described infantile scurvy (Barlow's disease) and distinguished it from rickets in 1883; distinguished between Tb and simple meningitis; research and discoveries in children's diseases. Died London, Jan. 12, 1945.

BARLOW (or BARLOWE), William, English physicist, mathematician; b. Pembrokeshire, Eng.; s. William and Agatha (Wellesbourne) B.; B.A., Balliol Coll., Oxford (Eng.) U., 1564; entered holy orders circa 1573; made prebendary of Winchester, 1581; prebendary and later treas. of Lichfield, 1588; rector of Easton; chaplain to Prince Henry (son of James I); archdeacon of Salisbury in 1614. Author: The Navigator's Supply, 1597. First Englishman to write on properties of magnet; improved hanging of compasses at sea; discovered magnetic differences between iron and steel; discussed proper way of touching magnetic needles, also piercing and cementing lodestones. Died May 25, 1625.

BARLOW, William H., Brit. engr.; b. May 10, 1812; s. Peter Barlow; ed. engring. dept. H. M. Dockyard, Woolwich; with Maudsley and Field, Constantinople, 1832, erected works and machinery for Turkish Ordnance; became resident engr. Midland Counties Ry. (Midland Ry.), 1842; cons. engr., 1857; constructed southern portion of London and Bedford Line, including St. Pancras Sta.; joint engr. Clifton Suspension Bridge, 1861; apptd. to investigate cause of fall of Old Tay Bridge, 1879; built New Tay Bridge, 1880-87; cons. for Forth Bridge, 1881. Fellow Royal Soc. (a v.p. 1881); pres. Inst. Civil Engrs., 1880. Research and publs. on lighthouse illumination, diurnal electric tides and storms, resistance of flexure in beams, the logograph. Died Nov. 12, 1902.

BARNARD, Chester Irving, Am. mgmt. scientist; b. Malden, Mass., Nov. 7, 1886; s. Charles H. and Mary E. (Putnam) B.; student Harvard, 1906-09; LL.D., Newark U., 1937, Brown U., 1943; Bloomfield Coll. and Sem., 1945; Princeton 1947, N.Y. U., 1952; D.Sc. Rutgers U., 1936, U. Pa., 1947; m. Grace F. Noera, Dec. 6, 1911; 1 dau., Frances (Mrs. C. Stuart Welch, dec.). Began in statis. dept. Am. Tel. & Tel. Co., 1909; comml. engr., 1913-22; asst. v.p. and gen. mgr. Bell Telephone Co. of Pa., and asso. cos., 1922-23, gen. mgr., 1923-25, v.p. and gen. mgr., 1925-26, v.p. in charge operations, 1926-27; 1st pres. N.J. Bell Telephone Co., 1927-48; pres. Rockefeller Found. and Gen. Edn. Bd., 1948-52; chmn. NSF, 1952-54. Vice chmn. President's Commission on Health Needs of Nation, 1952. dir. Nat. War Fund, 1943-46; mem. bd. consultants on atomic energy control Dept. of State, 1946, presdl. spl. com. on Integration of Med. Services in Govt., 1946. Mem. N.Y.C. Bd. Health, from 1957. Recipient Meritorious Civilian Service Award, U. S. Navy, 1944, President's Medal for Merit, 1946; Legion of Honor, France. Fellow Am. Acad. Arts and Sci., A.A.A.S.; mem. Am. Philos. Soc. Author: The Functions of the Executive (one of earliest recognitions of import of informal and formal organizational structure), 1938; Organization and Management. One of 1st mgmt. scientists to emphasize exec. communication responsibilities, analyze role of status and incentive systems in orgns. Died June 7, 1961.

BARNARD, Christiaan Neethling, South African surgeon; b. Beaufort, West, C.P., South Africa, Nov. 8, 1922; s. Adam Hendrik and Margrieta (de Swardt) B.; M.B., Ch.B., U. Cape Town (S. Africa), 1946, M.D., 1953, M.Med., 1953; M.S., Ph.D., U. Minn., 1958; m. Aletta Gertruida Louw, Nov. 6, 1948; children—Deirdre Jeanne, Andre Hendrik. Dir. surg. research U. Cape Town, 1958—, asso. prof., 1963—; sr. cardiothoracic surgeon Groote Schuur Hosp., Cape Town, 1958—, head cardiothoracic surg. unit, 1961—. Cecil Adams bursary and Dazian Found. scholar, 1956; Ernest Oppenheimer Meml. Trust grantee, 1960. Fellow A.C.S.; mem. S. African Med. Assn., Soc. Thoracic Surgeons, S. African Soc. Physicians, Surgeons, Gynaecologists (founder mem.). Research, numerous publs. on congenital intestinal atresia; introduction of open heart surgery to S. Africa; devel. new design artificial heart valves; performed world's 1st human heart transplant operation, Dec. 3, 1967; continues to pioneer in devel. of this surgical procedure. Home: The Moorings, Flamingo Cresc. Zeekoevlei, Cape Town, S. Africa.

BARNARD, Edward Chester, Am. topographer; commr.; b. N.Y.C., Nov. 13, 1863; s. Owen Howard and Anne E. B.; E.M., Columbia, 1884; m. Juliet Gill Rogers, Dec. 16, 1908. Topographer U. S. Geol. Survey, 1884-1907, geographer, 1907-15; mapped sections of Ky., Tenn., Va., N.Y., in the East, and Calif., Ida., Mont., Ore., Wash. in the West; had charge. of party sent to Alaska by U. S. Geol. Survey to map Forty-Mile Dist., 1898, Nome Dist., 1900; chief topographer U. S. and Can. Boundary Survey, 1903-15, surveying and relocating U.S. and Can. boundary line, along 49th parallel from Pacific Coast to Lake of the Woods and Lake of the Woods, Rainy River and Rainy Lake; apptd. commr. U. S. for defining and marking boundary between U. S. and Can., except on Great Lakes and St. Lawrence River, and for marking and surveying boundary between Alaska and Can., 1915. Died Feb. 6, 1921.

BARNARD, Edward Emerson, Am. astronomer; b. Nashville; Dec. 16, 1857; learned photography and began astron. studies in boyhood; grad. Vanderbilt U., 1887, Sc.D., 1893; A.M., U. of Pacific, San José, 1889; LL.D., Queen's U., 1909; m. Rhoda Calvert, Jan. 27, 1881. In charge Vanderbilt U. Obs., 1883-87; astronomer Lick Obs., Cal., 1887-95; prof. practical astronomy U. Chgo., and astronomer Yerkes Obs., from 1895. Mem. U. S. Naval Obs. Eclipse Expdn. to Sumatra, 1901. Recipient Lalande gold medal, Scis., 1892, Arago gold medal, 1893, Janssen Gold medal, 1900 (all French Acad. Scis.); Gold medal Royal Astron. Soc. of Great Britain, 1897; Janssen prize French Astron. Soc., 1906; Bruce gold medal, Astron. Soc. of Pacific, 1917. Mem. Nat. Acad. Scis.; asso. fellow Am. Acad. Arts and Scis. Discovered 5th satellite of Jupiter, 1892, also 16 comets, Barnard's Runaway Star (dim star with rapid motion) Dark nebulae (catalogued 182 of them), other discoveries; much work in celestial photography. Died Williams Bay, Wis., Feb. 6, 1923.

BARNARD, Frederick Augustus Porter, Am. mathematician; b. Sheffield, Mass., May 5, 1809; s. Robert and Augusta (Porter) B.; grad. Yale, 1828, LL.D. 1859; LL.D., Jefferson Coll., 1855; D.D., U. Miss., 1861; m. Margaret McMurray, Dec. 27, 1847. Prof. math. and natural history U. Ala., 1837-54; prof. math. and natural philosophy U. Miss., 1854-56, pres., 1856-58; chancellor, 1858-61; published Letter to the President of the United States by a Refugee, 1863; pres. Columbia, 1864-89, founder Law Sch., Sch. Polit. Scis., Sch. Mines, Barnard Coll.; U. S. commr. Universal Expn., Paris, 1867; asst. U. S. commr. gen. Paris Expn., 1878; mem. U. S. Coast Survey; originator system of teaching deaf and dumb; mem. bd. Am. Bur. Mines; pres. Am. Meteorol. Soc., 1874-80; a founder, pres. A.A.A.S.; a founder Nat. Acad. Scis. Author: (principal works) Treatise on Arithmetic, 1830; Analytical Grammar, 1836; A History of the United States Coast Survey, 1857; Recent Progress of Science, 1859; The Metric System, 1871. Editor: Johnson's Universal Cyclopedia. Advocated equal ednl. privileges for men and women; fostered devel. of scis. in Columbia curriculum. Died N.Y.C., Apr. 27, 1889.

BARNARD, Harry Everett, Am. chemist; b. Dunbarton, N.H., Nov. 14, 1874; s. Nelson H. and Celestia A. (Rider) B.; B.S., N.H. Coll. Agrl. and Mechanic Arts, 1899; Ph.D., Hanover (Ind.) Coll., 1913; D.Sc., U. N.H., 1928; m. Marion Harvie, June 20, 1901. Asst. chemist, N.H. Expt. Sta., 1899, U. S. Smokeless Powder Factory, Indian Head, Md., 1900-01; chemist, N.H. Bd. of Health, 1901-05; Ind. Bd. of Health, 1905-19; state food and drug commn. of Ind., 1907-19; state commr. of weights and measures of Ind., 1911-19; food and drug inspn. chemist, U. S. Dept. Agr., 1907-19; pres. H. E. Barnard, Inc. Pres. Ind. Sanitary and Water Supply Assn.; dir. Am. Chem. Soc. (founder and 1st pres. Ind. Sect.); mem. Am. Inst. Chem. Engrs., Soc. Official Agrl. Chemists, Nat. Assn. State Food Commrs., Federal Food Standards Com., Indianapolis Technical Soc.; president Lake Michigan Water Commission; secretary Indiana Branch National Conservation Assn.; mem. exec. com. Nat. Conservation Congress, 1912. Federal Food Administrator for Ind., 1917-19, Pres., Am. Inst. of Baking, 1919, 27; sec. Am. Bakers' Assn., 1921-25; dir. White House Conf. on Child Health and Protection, 1929-31, Corn Industries Research Foundation, 1931-34; dir. research Nat. Farm Chemurgic Council, 1935-40, professional specialist in chemistry, War Manpower Commn., since 1941; with Fed. Economics Assn., 1945. Trustee New Hampshire Coll., 1903-06. Mem. Sigma Alpha Epsilon, Alpha Chi Sigma, Phi Kappa Phi. Author of N.H. and Ind. bd. of health reports and papers and addresses on subjects of food, drugs, water, sanitation, nutrition, child welfare, chemurgy, etc. Died Dec. 31, 1946.

BARNARD, Jerry Laurens, Am. taxonomist; b. Pasadena, Cal., Feb. 27, 1928; s. John E. and Vera (Jones) B.; A.B., U. So. Cal., 1949, M.S., 1950, Ph.D., 1953; m. Charline A. Miller, July 15, 1949; children—Robert, Rodger, Gretchen. Research biologist, U. So. Cal., 1953-59; with Beaudette Found., Cal., 1960-64; asso. curator crustacea, Smithsonian Inst., Washington, 1964—. Mem. Soc. Systematic Zoology,

Brit. Ecol. Soc., Marine Biology Assn. U.K. Research, numerous publs. in field taxonomy, ecology marine amphipoda Pacific Ocean. Office: Div. Crustacea, Smithsonian Inst., Washington 20560.*

BARNARD, Joseph Edwin, Brit. physicist; b. London, Eng., 1870; 1 son, 1 dau.; hon. dir. dept. applied optics Nat. Inst. for Med. Research, Hampstead, Eng.; Fellow Royal Soc., 1924; pres. Micros. Soc.; developed ultra-violet light method of photographing organisms too small to be seen witn light microscope, 1925; demonstrated (with Dr. Gye) what was believed to be cancer virus. Died Oct. 25, 1949.

BARNER, Hendrick Boyer, Am. physician; b. Seattle, Feb. 23, 1933; s. Henry Adolf and Billie (Halvorsen) B.; B.S., U. Wash., 1954, M.D., 1957; m. Mechthild Brigitte Boehnke, Mar. 6, 1961; children—Boyer Hendrick, Bjorn Oluf. Research fellow Lahey Clinic, Boston, 1960-61; USPHS postdoctoral fellow, instr. surgery U. Rochester (N.Y.), practice medicine, specializing in surgery, St. Louis, 1965—; instr. surgery, fellow cardiovascular surgery St. Louis U., 1965-—, asst. prof. surgery, 1966-—. Mem. A.C.S., Soc. Cryobiology (editorial bd. 1966-—), Am. Coll. Angiology, Am. Fedn. Clin. Research. Research, publs. on preservation of tissues for transplantation by freezing; physiologic effects of gastric freezing, vascular surgery with emphasis on autologous and homologous vessel grafts and measurement of blood flow; function and innervation of transplanted heart. Home: 12516 Big Bend Blvd., Kirkwood, Mo. 63122. Office: 1325 S. Grand Blvd., St. Louis 63104.*

BÄRNER, Johannes Max Wilhelm, German botanist; b. Grossenhain, Jan. 6, 1900; s. Max and Ida (Ackermann B.; ed. in chemistry, botany, pharmacognosy; Ph.D., U. Berlin; m. Ida Schmidt, Apr. 17, 1946. Cons. to govt.; cons. on agr. and forestry Inst. Biology, 1927-45; dir. sect. study of viruses Central Inst. Agr. and Forestry, Kleinmachnow, 1950-51; sci. cons. Fed. Inst. for Agr. and Forestry, 1951-—; instr. applied botany U. Berlin, 1955-—. Mem. German Soc. Botany, Assn. for Applied Botany, Internat. Assn. Agrl. Librarians and Documentalists, Soc. for Orgn. Librarians and Documentalists in Agr. Author: Die Nutzhölzer der Welt; Bibliographie der Pflanzenschutzliteratur. Home: Markoraf-Albrechtstrasse 5, 1 Berlin 31. Office: Königin-Luisestrasse 18, 1 Berlin 33 (Dahlem), Germany.

BARNES, Albert Coombs, Am. physiol. chemist, b. Phila., 1872; M.D., U. Pa., 1892; postgrad. U. Berlin (Germany), 1894, U. Heidelberg (Germany), 1899; m. Laura Leggett, 1901; head Barnes Found., Phila. Invented argyrol and ovo ferrin. Died 1951.

BARNES, Arlie Ray, Am. physician; b. Jennings County, Ind., Apr. 24, 1892; A.B., Ind. U., 1915, M.A., 1916, M.D., 1919, D.Sc., 1949; fellow Mayo Clinic, 1920-22; M.S., U. Minn., 1929; m. 1916, 1 child. First asst. Mayo Clinic, 1922-25, asso. medicine, 1925-37, cons. and head sect., 1937-53; prof., asst. dir. Mayo Found., Rochester, Minn., 1936-57, emeritus, 1957-—, cons. grad. sch. Mem. Assn. Am. Physicians, Am. Heart Assn. (pres. 1946-47), Inter-Am. Soc. Cardiology. Research in cardiology; clin., pathol. and electrocardiographic investigations of heart disease; introduced (with Hench, Slocum, Smith Polley and Kendall) cortisone in treatment of rheumatic fever, 1949. Home: 207 5th Av. S.W. 200 1st Av. S.W., Rochester, Minn. 55901.*

BARNES, Charles A., Am. physicist; b. Toronto, Can., Dec. 12, 1921; s. Andrew A. and Adella M. (Davidson); B.A., McMaster U., 1943; Ph.D., Cambridge U., 1950; m. Phyllis J. Malcolm, Sept. 1950; children—Nancy, Steven. Came to U. S., 1953, naturalized, 1961. Physicist, Brit.-Can. Atomic Energy Project, Montreal, Chalk River, Can., 1944-46; prof. physics U. B.C., Vancouver, Can., 1950-53, 55-56, Cal. Inst. Tech., Pasadena, 1953-55, 56-—.

BARNES, Charles Reid, Am. botanist; b. Madison, Ind., Sept. 7, 1858; grad. Hanover (Ind.) Coll., 1877; A.M., 1880; Ph.D., 1886; postgrad. Harvard, 1877, 78, 85-86, 92; m. Mary King Ward, Dec. 25, 1882. Prof. natural history, Purdue U.; 1880-86; prof. botany, U. Wis., 1886-98; prof. plant physiology U. Chgo., from 1898. Mem. Bot. Soc. Am. (pres. 1903), A.A.A.S. (gen. sec. 1896, v.p. 1899). Author: (with J. C. Arthur and J. M. Coulter) Plant Dissection, 1886; Keys to the Genera and Species of North American Mosses, 1896; Plant Life, 1898; Outlines of Plant Life, 1900; numerous bot. papers. Co-editor Bot. Gazette, from 1883. Studied plant physiology and bryophytes. Died 1910.

BARNES, David Fitz, Am. geophysicist; b. Boston, May 23, 1921; s. David Donald and Margaret (Fitz) B.; B.S., Harvard, 1943, M.A., 1955; postgrad. U. Cal. at Berkeley; m. Ann Tyndale-Biscoe, Oct. 29, 1965. With Woods Hole (Mass.) Oceanographic Instn., 1943-49; geophysicist U. S. Geol. Survey, Washington, 1949-56, Menlo Park, Cal., 1958-—; with Air Force Cambridge Research Center, Bedford, Mass., 1956-58. Mem. Arctic Inst. N.Am., Am. European, Bay Area socs. exploration geophysicists, Am. Geophys. Union, Glaciological Soc., Geol. Soc. Am. Research in geophys. investigations in Arctic areas, Alaskan gravity surveys, geophys. exploration in permafrost, gravity changes in earthquakes, phys. environment of perennially frozen lakes, strength of melting ice, off-shore gravity surveys, infrared luminescence of minerals. Home: 107 Northam Av., San Carlos, Cal. 94070. Office: U. S. Geol. Survey, Menlo Park, Cal. 94025.*

BARNES, Fancourt, Brit. physician; s. Robert Barnes; ed. Coll. de Honfleur (France), Lincoln Coll., Oxford; M.D.; Ch.M., Aberdeen. Sr. physician Chelsea Hosp. for Women; physician Royal Maternity Charity of London; cons. obstetric physician St. George's Hosp.; cons. physician Prudential Assurance Co., Brit. Lying-in Hosp. Fellow Royal Soc. Edinburgh; mem. Royal Coll. Physicians, London, Gynecol. Soc. (hon. corr. Boston), Imperial Acad. Medicine (hon. Constantinople), Paris Soc. Gynecology and Obstetrics (corr.). Author: System of Obstetric Medicine; Manual of Midwifery for Midwives; Perinaeorrhaphy by Flap Splitting. Editor Brit. Gynecol. Jour.; sub-editor Brit. Med. Jour. Died Feb. 20, 1908.

BARNES, George Eric, Am. prof. civil engring.; b. Washington, Apr. 17, 1898; s. Raymond F. and Mattie (Van Slyck) B.; B.S., Mass. Inst. Tech., 1923; C.E. (hon.), Case Inst. Tech., 1935; M.A., Western Res. U., 1953; m. Mary Magdalene O'Hara, Sept. 23, 1920; children—Mary, Dorothy, Janet. Prof. engring., Univ. of Fla., 1923-29, designing engr. Metcalf & Eddy, Boston, Mass., 1927, Fuller & McClintock, New York, N.Y., 1929-32, dept. of sanitation, City of New York, 1932-33; prof. and head dept. civil engring. and engring. mechanics, Case Inst. Tech., Cleve., 1933-55, professor hydraulic and sanitary engring., 1955-63, prof. emeritus, 1963-—; prof. sanitary engring. Sch. Public Health, U. N.C., Chapel Hill, 1963-—; cons. engr. in sanitation and hydraulic engring. since 1933. Vis. prof. san. engring. Universidad Central de Venezuela, 1961. Mem. Am. Inst. Cons. Engrs., Am. Soc. C.E., Am. Soc. Mech. Engrs., Am. Soc. for Engring. Edn. Am. Water Works Assn., Water Pollution Control Assn., Am. Academy of Sanitary Engineers, Venezuelan Assn. San. Engrs. (hon.). Contbr. to Engineering for Dams (Creager, Justin and Hinds), 1944; Hydroelectric Handbook (Creager & Justin), 1949; Kent's Mechanical Engineers Handbook, 1950; Tratamiento de Aguas Negras y Despojos Industriales, 1966; articles in engring. publs. on hydraulics and sanitation. Home: 1303 Willow Dr., Chapel Hill, N.C.

BARNES, Howard Turner, physicist; b. Woburn, Mass., July 21, 1873; s. William S. Barnes; B.A. in Sci., McGill U., Montreal, Que., Can., 1893, M.A. in Sci., 1896, D.Sci., 1900; m. Ann Kershaw Cunliffe; 2 sons, 1 dau. Demonstrator physics McGill U., 1895-98, lectr., 1800-01, asst. prof., 1901-06, asso. prof., 1906-07, MacDonald prof., 1907-19, dir. MacDonald Physics Bldg., 1909-19, emeritus prof., 1919. Tyndall lectr. Royal Instn., 1912. Fellow Phys. Soc., Canadian Inst. Engrs., Royal Soc. Can. (pres. math. and physics sect. 1909), Royal Soc. London, 1911, Royal Meteorol. Soc., Am. Geog. Soc., Australian Acad. Sci. Asso. editor Jour. Phys. Chemistry. Contbns. in fields of electromotive force, continuous electric calorimetry, specific heats of water and mercury variation of specific heats of liquid with temperature, critical velocity of liquids, thermal and elec. measurements, ice formation, prevention and navigation, iceberg research. Died Oct. 4, 1950.

BARNES, Jasper Converse, Am. psychologist; b. Meigsville, O., Aug. 28, 1861; s. Abraham and Margaret (Welch) B.; A.B., Marietta Coll., 1890, A.M., 1893; student, summer session, Cornell; Ph.D., U. Chgo., 1911; LL.D. Maryville Coll., 1928; m. Alice Mary Hopkins, Aug. 13, 1890; 1 son, Mark Hopkins. Tchr. pub. schs., Morgan Co., O., 1880-84; supt. schs., Belpre, O., 1890-92; prin. prep. dept., prof. edn. Maryville (Tenn.) Coll., 1892-1901, acting prof. psychology and polit. sci., 1901-03, prof., 1903-31, dean 1914-31, head dept. psychology and edn., 1931. Faculty, Summer Sch. of South, 1915, 18, Asheville (N.C.) Normal Sch., 1921-22, Ohio State U., summer 1920, U. Tenn., summers, 1923-27, and 1930, U. Wyo., summer 1928. Author: (booklet) Development of Personality in the Afro-American, 1900; Voluntary Isolation of Control in a Natural Muscle Group, 1915. Died Sept. 13, 1931.

BARNES, Richard Henry, nutritionist; b. La Jolla, Cal., June 29, 1911; s. Willard Gray and Nell (Tribby) B.; A.B., San Diego State Coll., 1933; Ph.D., U. Minn., 1940; m. Marjorie Ironside, Mar. 24, 1942; children—Kyle, Marjorie Anne, Lisa. Research chemist Scripps Metabolic Clinic, La Jolla, 1933-37; faculty U. Minn., Mpls., 1937-44; dir. biochem. research, Sharp & Dohme, Inc., Glenolden, Pa., 1944-49, asst. dir. research, 1949-50, asso. dir. research, 1950-56; dir. biochemistry Merck Sharp & Dohme Research Labs., West Point, Pa., 1956; dean Grad. Sch. Nutrition, Cornell U., Ithaca, N.Y., 1956-—. Chmn. com. on nutrition, mem. gen. com. on foods, adv. bd. on mil. personnel Supplies Nat. Acad. Scis.—NRC, 1964-—; mem. sci. adv. council Masonic Med. Research Lab., Utica, N.Y., 1959-—. Fellow A.A.A.S., N.Y. Acad. Scis.; mem. Am. Chem. Soc., Am. Soc. Biol. Chemists, Am. Inst. Nutrition, Soc. for Exptl. Biology and Medicine, Nutrition Soc. (London). Contbr. articles to sci. jours. Research in general nutrition, fat, protein, carbohydrate metabolism, contributions of intestinal microflora. Home: 140. N. Sunset Dr., Ithaca, N.Y. 14850.*

BARNES, Robert, English physician; b. Norwich, Eng., Sept. 4, 1817; s. Philip and Harriet (Futter) B.; ed. Windmill St. Sch., Univ. Coll.; M.D., London, Eng., 1848; m. Eliza Fawkner; 1 son, R. S. Fancourt, 2 daus.; m. 2d, Alice Hughes; 1 son, 1 dau. Lectr. Hunterian Sch. Medicine, also Dermott's Sch.; asst. obstetric physician, then obstetric physician London Hosp.; apptd. obstetric physician St. Thomas's Hosp., 1865, also dean med. sch.; became obstetric physician St. George's Hosp., 1875, cons. obstetric physician, 1885; physician Seaman's hosp., E. London Hosp. for Children, Royal Maternity Hosp. Lumleian lectr. on convulsive diseases in women, Coll. Physicians, 1873. Licentiate, fellow Royal Coll. Physicians; fellow (hon.) Med. Soc. London, Royal Med. and Chirurg. Soc.; mem. Royal Coll. Surgeons, Obstet. Soc. London (a founder 1858, pres. 1865-66), Brit. Gynecol. Soc. (founder 1884). Author: Obstetrical Operations, 1870; Medical and Surgical Diseases of Women, 1873; Obstetric Medicine and Surgery, 1884; Causes of Puerperal Fever, 1887. Pioneer in operative gynecology and pathology of obstetrics; devised Barnes' Bags for inducing labor by dilating cervix, 1858; described cervical zone of uterus, 1880, a curve of pelvis (named after him), 1884. Died Eastbourne, Eng., May 12, 1907.

BARNES, R(obert) Bowling, Am. physicist; b. Montgomery, Ala., June 9, 1906; s. Elly Ruff and Ula (Bowling) B.; A.B., Birmingham-So. Coll., 1925, D.Sc., 1957; Ph.D., Johns Hopkins, 1929; m. Eva Hoffman, Aug. 8, 1933; children—Robert Bowling, George MacIlwaine. Dir. physics dept. Stamford (Conn.) Research Labs., Am. Cyanamid Co., 1936-48; v.p. Am. Optical Co., Stamford, 1948-51; pres. Olympic Devel. Co., Stamford, 1952-54; pres. Barnes Engring. Co., Stamford, 1954-—. 1st. recipient Beckman award in chem. instrumentation Am. Chem. Soc., 1955. Mem. Phi Beta Kappa, Sigma Xi, Omicron Delta Kappa, Alpha Tau Omega. Publs., patents on pioneer research in indsl. application of infrared spectroscopy and electron microscopy, application of physics in chem. industry and med. research. Home: 60 Westover Rd., Stamford 06902. Office: 30 Commerce Rd., Stamford, Conn. 06904.*

BARNES, Robert Lloyd, Am. plant biochemist; b. Royersford, Pa., Nov. 13, 1926; s. Arlington Wayne and Elsie (Krauss) B.; B.S. cum laude, Duke, 1950, M.F., 1951, Ph.D., 1958; m. Joanne Slocum, June 5, 1951; children—Robert Lloyd, Thomas E. Faculty, Sch. Forestry U. Fla., 1951-57; plant physiologist, project leader U. S. Forest Service, Durham, N.C., 1958-65, chief forestry scis. lab., 1963-65; prof. Forest biochemistry Sch. Forestry, Duke, Durham, 1965-—. Mem. A.A.A.S., Am. Soc. Plant Physiologists, Am. Inst. Biol. Scis., Soc. Am. Foresters, Assn. Southeastern Biologists, Phi Beta Kappa, Sigma Xi. Research in physiology of forest trees, including nitrogen nutrition, flowering and air pollution. Home: 5303 Revere Rd., Durham, N.C. 27707.*

BARNES, Thomas Cunliffe, Am. electroencephalographer; b. Montreal, Que., Can., June 4, 1904; s. Howard Turner and Ann Kershaw (Cunliffe) B.; A.B., Cornell U., 1926; D.Sc., Harvard, 1929. Asst. prof. Yale, 1931-41; asso. prof. Hahnemann Med. Coll., 1942-58, St. Francis Coll., Bklyn., 1959-61; pharmacologist Wells Labs., Jersey City, 1962-65; research asso. Phila. State Hosp., 1966-—. Research fellow Cambridge (Eng.) U., 1930. Mem. Am. Physiol. Soc., Eastern Assn. Electroencephalographers, Soc. for Pharmacology, Am. Psychol. Assn. Author: Textbook of Physiology, 1937, Laboratory Manual of Physiology, 1937; also numerous articles. Research on brain waves, wound healing, effects of tranquilizing drugs on animals. Home: 34 Gramercy Park, N.Y.C. 10003. Office: Phila. State Hosp., Phila. 19114.*

BARNES, Virgil E(verett), Am. geologist; b. Chehalis, Wash., June 11, 1903; s. Charles N. and Della (Matheny) B.; B.S., State Coll. Wash., 1925, M.S., 1927; Ph.D., U. Wis., 1930; m. Mildred Louise Adlof, Sept. 28, 1932; children—Virgil Everett II, Louise, Elizabeth (Mrs. Hugh Walter Thompson). Research fellow Am. Petroleum Inst., 1930-31; topographic engr. U. S. Geol. Survey, 1933-35; geologist Bur. Econ. Geology U. Tex., Austin, 1935-—, asso. dir., 1962-—, prof. geology, dir. tektite research, 1960-—. Mem. Geol. Soc. Am., Mineral. Soc. Am., Am. Assn. Petroleum Geologists, Soc. Exploration Geophysicists, Soc. Econ. Paleontologists and Mineralogists, Am. Geophys. Union, A.A.A.S., Geochem. Soc., Sigma Xi, Gamma Alpha. Contbg. author: Tektites, 1963; also numerous articles. Research on origin tektites, Cambrian, Ordovician and Devonian rocks, minerals, meteorites, size and shape rock masses and ore bodies; identified lechatelierite; found puddle-like deposits in S.E. Asia having composition soil indicating terrestrial origin. Home: 207 E. 33d St., Austin 78705.*

BARNES, William, Am. surgeon, entomologist; b. Decatur, Ill., Sept. 3, 1860; s. William A. and Eleanor (Sawyer) B.; student State Normal Sch., Normal, Ill., 1878, Ill. U., 1879; B.S., Harvard, 1883, M.D., 1886; postgrad. Boston City Hosp., also Heidelberg, Munich and Vienna; D.Sc., James Millikin U. Decatur, Ill., 1929; m. Charlotte L. Gillett, June 20, 1890; children—Joan Dean Gillett, William. Practiced at Decatur, 1890-—; one of builders, pres. Decatur and Macon County Hosp. Fellow A.C.S., mem. various entomol. socs. Owner of largest collection of N.Am.

Lepidoptera then in existence, consisting of several hundred thousand specimens—over 10,000 species and varieties and over 6,000 types of various kinds. Author: Contributions to the Natural History of Lepidoptera of North America, 23 parts, 1911-24; Check List of the Lepidoptera of Boreal North America, 1917; Illustrated Species of the Genus Catocala, 1918; also numerous articles in entomol. mags. Died May 1, 1930.

BARNETT, Alfred John, Australian physician; b. Dunolly, Victoria, Australia, Mar. 27, 1915; s. Alfred Thomas and Edith (Burke) B.; student Wesley Coll., 1930-33; M.B.B.S., Melbourne U., 1939, M.D., 1947; m. Hazel Virgie, July 24, 1942; children—Adrian, Mark, Derek. Asst. med. supt. Mackay Dist. Hosp., Queensland, 1941; asso. physician Royal Melbourne Hosp., 1946-47, acting out patient physician, 1947-48; Returned Servicemen's League scholar, asst. in med. unit St. Mary's Hosp., London, 1948-49; dep. dir. clin. research unit Alfred Hosp., Melbourne, 1950-55, asso. dir., 1955—; demonstrator Melbourne U., 1947, clin. tchr. Monash U., 1966—. Recipient Stawell Meml. prize, 1952. Fellow Royal Australian Coll. Physicians; mem. Royal Coll. Physicians London, Victorian Soc. Pathology and Exptl. Medicine, Australasian Cardiac Soc., Australasian Physiol. Soc., Australasian Soc. Nephrology. Author: (with J. Fraser) Peripheral Vascular Disease, 1955; also numerous articles. Research on effect of noradrenaline in man and relation to hypertension; effect of ganglion blocking drugs and use in hypertension; assessment of various modes of therapy in peripheral vascular disease; mechanism of orthostatic hypotension; clin. studies on scleroderma. Home: 55 Summerhill Rd., Glen Iris, Victoria, Australia. Office: Clin. Research Unit, Alfred Hosp., Commercial Rd., Prahran, Victoria, Australia.*

BARNETT, Cyril Harry, anatomist; b. London, Eng., Oct. 30, 1919; s. Barnett and Rose (Freedman) B.; M.B., B.Chir., Cambridge (Eng.) U., 1943; m. Sheila Catherine Arnold, Aug. 2, 1950; children—Phillip Anthony, Harriet Jane. Demonstrator, lectr. St. Thomas's Hosp. Med. Sch., London, 1947-55, reader anatomy, 1955-64; lectr. U. Melbourne (Australia), 1956-57; Found. prof. anatomy U. Tasmania, Hobart, Australia, 1964—. Fellow Royal Coll. Surgeons; mem. Anat. Soc., Heberden Soc. (asso.). Author: (with D. V. Davies, M. A. MacConaill) Synovial Joints, 1961; (with D. Taverner, H. G. Lumby) The Human Body, 1966; also articles. Analysis of normal joint movements in man and animals; research on normal gait, footwear, foot structure and function; analysis of synovial fluid viscosities and joint lubrication, electron microscopy of articular cartilage. Address: 21 Bessels Way, Bessels Green, Sevenoaks, Kent, Eng.*

BARNETT, Gordon James, Am. psychoanalyst; b. Upton, Me., Mar. 13, 1921; s. James and Grace (Bragg) B.; B.S., U. N.H., 1943; M.A., Columbia, 1945, Ph.D., 1950; postgrad. William Alanson White Inst. Psychiatry, Psychoanalysis and Psychology, 1954-61; children—Gordon James, James Bragdon, Jayson Wayne. Staff psychologist Guidance Bur., Salvation Army, 1945-46, sr. psychologist Social Welfare Dept., 1946-48; staff psychologist U. S. VA Hosp., North Port, N.Y., 1948-49; staff psychologist Mental Hygiene Clinic, N.Y.C., 1949-52, Supervisory psychologist, 1952-55, asst. chief clin. psychologist, 1955-57; individual practice psychoanalysis, N.Y.C., 1955—; clin. asso. prof. psychology, supr. psychotherapy, postdoctoral tng. program Adelphi U., Garden City, N.Y., 1963—. Mem. William Alanson White Psychoanalytic Soc., Kappa Delta Pi, Phi Delta Kappa, Pi Gamma Mu, Alpha Kappa Delta. Home: 69 Washington Av., Garden City, N.Y. 11530. Office: 11 E. 68th St., N.Y.C., 10021.*

BARNETT, Harold Joseph, Am. economist; b. Paterson, N.J., May 10, 1917; s. Abraham and Lena (Schiff) B.; B.S., U. Ark., 1939; M.S., U. Cal. at Berkeley, 1940; M.A. (Social Sci. Research Council fellow), Harvard, 1948, Ph.D., 1952; m. Mildred Denn, Aug. 4, 1940; children—Peter, Alexander, Katherine. Teaching asst. U. Cal. at Berkeley, 1939-40; economist Treasury Dept., 1941-42, Dept. State, also Dept. Interior, 1946-52; economist Rand Corp., Washington, 1952-55, cons., 1948—; economist, dir. econ. growth studies Resources For Future, Washington, 1955-59, cons.. 1959—; prof. econs., chmn. dept. Wayne State U., 1959-63; prof. econs., Washington U., St. Louis, 1963—, chmn. dept., 1963-66; cons. U. S. Office Emergency Planning, 1962—, NSF, 1964—, USPHS, 1966—, U. S. Office Edn., 1966—, Com. for Econ. Devel., 1966—. Decorated Legion of Merit. Mem. Am. Econ. Assn., Regional Sci. Assn., Soc. Internat. Devel., Am. Assn. U. Profs. Author: Energy Uses and Supplies, 1950; Malthusianism and Conservation, 1959; Economic Markets in Television, 1962; (with C. Morse) Scarcity and Growth, 1963; also articles. Studies on analysis relation natural resources to econ. growth, projections changing energy uses and supplies, econs. of nat. security. Home: 51 Crestwood Dr., Clayton, Mo. 63105.*

BARNETT, Henry Lewis, Am. physician; b. Detroit, June 25, 1914; s. Lewis and Florence (Marx) B.; student Dartmouth, 1931-32; M.D., Washington U., St. Louis, 1938; m. Shirley Blanchard, Oct. 19, 1940; children—Judith Florence, Martin David. Instr. Washington U. Sch. Medicine, 1941-43; asst. prof. dept. pediatrics Cornell U. Med. Coll., 1946-50, asso. prof., 1950-55; prof., chmn. dept. pediatrics Albert Einstein Coll. Medicine, 1955—. Sec. med. adv. bd. Nat. Kidney Found., Found. for Internat. Child Health. Bd. dirs. Assn. for the Aid of Crippled Children. Recipient E. Mead Johnson award for research in pediatrics, 1949. Mem. Soc. for Clin. Investigations, Am. Physiol. Soc., Am. Pediatric Soc., Soc. for Pediatric Research. Author (with Holt, McIntosh) Pediatrics, 13th edn., 1961. Contbr. numerous articles in field to sci. jours. Research on mechanisms which control the volume and composition of fluids of the body from time of birth through adulthood; study of cause, diagnosis and treatment of children with various types of kidney disease; study of medical education, medical care of infants and children. Home: 118 W. 79th St., N.Y.C. 10024. Office: Bronx Municipal Hosp. Center, Pelham Pkwy. and Eastchester Rd., Bronx, N.Y.C. 10461.*

BARNETT, Homer Garner, Am. anthropologist; b. Bisbee, Ariz., Apr. 25, 1906; s. Lee N. and Charlotte (McEuen) B.; A.B., Stanford, 1927; Ph.D., U. Cal. at Berkeley, 1938; m. Judith Hazel Skaggs, Apr. 12, 1941; children—Linda A., Susan M. Instr., U. N.M., 1939; faculty U. Ore., Eugene, 1939—, prof., 1950—, acting chmn. dept., 1940-41, 49-50; asst. curator Ore. State Mus. Anthropology, 1939, acting dir., 1940-41, 49-50; anthropologist Smithsonian Inst., Washington, 1944-46; staff anthropologist Trust Territory Pacific Islands, 1951-53. Adviser to Netherlands New Guinea Govt., 1955; mem. Research Council, South Pacific Commn., 1952-53. Sr. fellow Nat. Sci. Found., 1956-57; fellow Center for Advanced Study in Behavioral Scis., 1964-65. Mem. A.A.A.S. (committeeman sect. H. 1963—), Soc. for Applied Anthropology (pres. 1961), Am. Anthropol. Assn. Author: Palauan Society, 1949; Innovation: The Basis of Cultural Change, 1953; Coast Salish of British Columbia, 1955, Anthropology in Administration, 1956; Indian Shakers, 1957; Being a Palauan, 1960; The Potlatch, 1968. Home: 2327 Jefferson St., Eugene, Ore. 97405.*

BARNETT, Isaac Albert, mathematician; born London, England, May 7, 1894; son of Tobias and Sarah (Lichtenfeld) B.; came to U. S., 1904, naturalized, 1912; B.S., U. Chgo., 1915, M.A., 1916, Ph.D., 1918; m. Fannie Reisler, Mar. 10, 1920; children—Ethel (Mrs. W. W. Stead), Noami (Mrs. Arnold Buchheimer). Staff, Washington U., St. Louis, 1918-19, Franklin Pierce Inst., Harvard, 1919-20, U. Saskatchewan (Can.), 1920-23; with U. Cin., 1923-64; staff Am. U., France, 1945; vis. prof., U. N.C., Chapel Hill, 1964-65, Farleigh Dickinson U., 1965, Ohio U., Athens, 1966—. Specialist in higher edn. Mil. govt., Munich, Germany, 1946-47. Research in number theory and analysis.*

BARNETT, Sir Louis Edward, surgeon; b. Mar. 24, 1865; s. Alfred A. Barnett; ed. Wellington State Sch. and Coll., also Otago (New Zealand) U., Middlesex Hosp., M.B., C.M., Edinburgh (Scotland) U., 1888; m. Mabel Violet Fulton; 3 sons, 1 dau. Lectr., prof. surgery U. Otago, 1895-1925, emeritus, from 1925; house surgeon Middlesex Hosp., 1889-90; hon. surgeon Dunedin Hosp., 1895-1925. Chmn. Otago and Southland div. Brit. Empire Cancer Campaign, 1928-38; chmn. N.Z. Hydatid Research Com. Fellow Royal, Am. (hon.) colls. surgeons; pres. Australasian Med. Congress, 1927, Royal Australasian Coll. Surgeons (also a founder), 1937-39, New Zealand br. Brit. Med. Assn. Contbr. papers on hydatid disease to med. lit. Died Oct. 28, 1946.

BARNETT, S(amuel) J(ackson), Am. physicist; b. Woodson County, Kan., Dec. 14, 1873; student Coll. of Emporia, 1889-91, U. Chgo., 1892-93; A.B., Denver U., 1894; postgrad. U. Va.; Ph.D., Cornell U., 1898; m., 1904. Instr. physics and biology U. Denver, 1894-95; asst., astron. obs., U. Va., 1895-96; faculty Colo. Coll., 1898-1900, Stanford, 1900-05, Tulane U., 1905-11, Ohio State U., 1911-18; physicist, dept. terrestrial magnetism Carnegie Instn., 1918-24, research asso., 1924-26; prof. physics U. Cal. at Los Angeles, 1926-44, emeritus, 1944—. fellow, 1953—; research asso. Cal. Inst. Tech., 1924—. Aid and magnetic observer U. S. Coast and Geodetic Survey, 1902-04; mem. NRC. Recipient Comstock prize Nat. Acad. Sci., 1918. Fellow or mem. Phys. Soc., Geophys. Union, Am. Acad., Washington Acad., Philos. Soc. Washington. Research on magnetic and electromagnetic theory, geomagnetic and other electric and magnetic measurements; gave exptl. proof of existence of ionosphere, 1925; investigated gyromagnetic effects, electron-inertic effect in metals, polarization in insulators by motion in magnetic fields, torsional elasticity.

BARNHILL, John Finch, Am. surgeon; b. Flora, Ill., Jan. 2, 1865; s. Robert and Angeline (Shirts) B.; M.D., LL.D., Central Coll. Phys. and Surg., Indpls.; studied N.Y. Eye and Ear Infirmary, N.Y. Polyclinic, Central London Ear, Nose and Throat Hosp., ear dept. U. Vienna, ear and throat dept., U. Berlin; LL.D., Ind. U., 1929; m. Celeste Terrell, Feb. 13, 1889. Practiced in Indpls., from 1888; specialist in surgery of head and neck; formerly prof. otolaryngology, Ind. U. Sch. of Medicine, later prof. surgery head and neck. Fellow A.C.S.; mem. A.M.A. (chmn. ear and throat sect. 1904), Am. Ear, Throat and Nose Assn. (v.p. 1908), Am. Acad. Ophthalmology and Otolaryngology (pres. 1931), Am. Laryngol. Soc. (pres. 1938), Am. Otol. Soc., Am. Laryngol., Rhinol. and Otol. Soc. (pres. 1927-28), Am. Acad. Ophthalmology and Otolaryngology (pres. 1932), Sigma Xi, Nu Sigma Nu. Author: Text Book on Ear, Nose and Throat, 1928; (stories) Not Speaking of Operations. Co-Author: Barnhill and Wales Modern Otology, 1907; Diseases Ear, Nose and Throat, 1927; Surgical Anatomy of the Head and Neck, 1937; Hatching the American Eagle, A Narrative of the American Revolution, 1937. Died Mar. 10, 1943.

BARNICKEL, William Sidney, Am. chemist; b. Lagrange, Ky., May 18, 1878; s. John and Mary (Dawkins) B.; Ph.G., St. Louis Coll. Pharmacy, 1902, Ph.C., 1903; m. Olive Edgeworth, June 4, 1904. Began as analyt. chemist with Allen Pfeifer Chem. Co., St. Louis, 1903; chief chemist Judge & Dolph Drug Co., 1904-10; prof. chemistry Am. Med. Coll., St. Louis, 1905-10; pres. W. S. Barnickel & Co., from 1910. Specialized in chemistry of petroleum; inventor of gasoline process, also of chem. process for treating waste oil and petroleum emulsions. Died May 19, 1923.

BARNOLA, José, Venezuelan physician; b. Caracas, Venezuela, June 26, 1913; s. Pedro Barnola and Antonia Duxans (Güell) Puig; Bachelor's degree, San Ignacio Coll., 1931; M.D., Central U. Venezuela, 1938; m. Alejandrina Quintero Medina, Mar. 6, 1943; children—José Pedro, José Rosendo, Isaias. Served with med. service Venezuelan Army, 1939-43; asso. prof. bacteriology and parasitology, 1943-46; head lab. Children's Hosp., Caracas, 1946-49; dir. dept. tropical medicine Central U. Venezuela, 1949-50, Armed Forces Central Hosp. Lab., 1950-55; titular prof., head dept. parasitology Central U. Venezuela, Caracas, 1955—. Mem. Nat. Acad. Medicine, Fed. Dist. Med. Coll. Caracas, Venezuelan Soc. Child Care and Pediatrics, Venezuelan Soc. Path. Anatomy. Author: Guia de Trabajos Practicos de Parasitologia, 1964. Research, publs. in clin. bacteriology, mycology, hematology. Home: Avenida San Felipe 16, La Castellana Chacao, Caracas, Venezuela. Office: Avenida Andres Bello, Policlinica Mendez Gímón, Venezuela.*

BARNOTHY, Jeno Michael, physicist; born Kassa, Hungary, Oct. 28, 1904; came to U. S., 1948, naturalized, 1954; s. Julius Preinreich and Ida (Ruprecht) B.; Ph.D., Royal Hungarian U., Budapest, 1933; M. Madeleine Forro, Dec. 24, 1938. Prof. physics Royal Hungarian U., 1933-48, Barat Coll., Lake Forest, Ill., 1948-53; chief physicist Nuclear Instruments & Chem. Co., Chgo., 1953-55; owner, tech. dir. Forro Sci. Co., Evanston, Ill., 1955—; pres. Biomagnetic Research Found., 1961—. Recipient F. Weiss medal Royal Hungarian Acad. Sci., 1938, Eotvos medal, 1947. Mem. Am. Astron. Soc., Astron. Soc. Pacific, Biophys. Soc., Am. Phys. Soc. Author: Experimental Physics, 1945. Contbr. numerous articles to profl. jours. Constructed first cosmic ray telescope, 1928; developer Fib steady state cosmological theory. Patentee nuclear instrumentation, space survival devices. Address: 833 Lincoln St., Evanston, Ill. 60201.*

BARNOTHY, Madeleine Forro (Mrs. Jeno M. Barnothy), physicist; born Zsambok, Hungary, Aug. 21, 1904; d. Robert and Margit (Somlo) Forro; Ph.D., Royal Hungarian U., Budapest, 1927; m. Jeno M. Barnothy, Dec. 24, 1938. Came to U. S., 1948, naturalized, 1954. Faculty, Royal Hungarian U., 1929-48, asso. prof., 1940-48; prof. Barat Coll., 1948-53; research asso. Northwestern U., Evanston, Ill., 1953-59; faculty U. Ill. Coll. Pharmacy, Chgo., 1955—, prof. physics, 1964—; postdoctoral fellow U. Gottingen (Germany), 1928-29. Recipient medal Royal Hungarian Acad. Sci., 1937, Eovos medal, 1947; fellow Am. Assn. U. Woman, 1954-55. Mem. Am. Phys. Soc., Biophys. Soc., Am. Astron. Soc., A.A.A.S., Sigma Xi. Editor: Biological Effects of Magnetic Fields, 1964. Research, numerous publs. on absorption spectra alkali halides, first cosmic ray telescope measurements, neutrino component at great depth, lifetime mu-meson, meterol. factors influencing cosmic ray intensity, extragaletic origin cosmic radiation; in biophysics; hematological changes, counteraction radiation syndromes, inhibition bacterial growth, genetic changes produced by static magnetic fields. Home: 833 Lincoln St., Evanston, Ill. 60201. Office: 833 S. Wood St., Chgo. 60612.*

BARON, Christian Charles, French biochemist; b. Dijon, France, Oct. 10, 1930; s. Francois Paul and Yvonne (Gaudemet) B.; Agregation, Doctorat ès Sciences Physiques, Ecole Normale Superieure, Paris; m. Monique Beuchart, May 15, 1965; 1 son, Bruno. Faculty, Faculte des Sciences, Dijon, 1957—, titular prof. biochemistry, 1963—; dir. biochem. lab. Institut de Biologie Appliquee à la Nutrition et à l'Alimentation. Expert analyst Ministere de la Sante Publique; asst. sec. gen. I.C.S.U. Abstracting Bd. Decorated Palmes Academiques. Contbg. author: Traite de Biochimie Générale, 1962. Research and publs. on enzymology, catalase, vitamin D, good flavors, gas chromatography. Home: 44 du Chapitre. Office: 2 Boulevard Gabriel, Dijon 21, France.*

BARON, Denis Neville, Brit. chem. pathologist; b. London, Eng., Oct. 3, 1924; s. Edward and Lilian (Silman) B.; M.B., B.S., U. London, 1945, M.D., 1950, D.Sc., 1966; m. Yvonne Elsa Stern, Dec. 6, 1951; children—Leonora, Jessica, Olivia, Justin. House physician Hosp. Sts. John and Elizabeth, London, 1945-46; asst. pathology Middlesex Hosp., London, 1949-50, sr. lectr. clin. biochemistry, 1950-54; hon. cons. chem. pathologist, faculty Royal Free Hosp. and Sch. Medicine, London, 1954—, prof. chem. pathology, 1963—. Mem. standing med. adv. com. Ministry Health, 1966—; vis. prof., examiner U. Ibadan (Nigeria), 1966, U. Lagos (Nigeria) 1966-68. Rockefeller travelling fellow in medicine U. Chgo., 1960-61. Fellow Coll. Pathologists; mem. Royal Coll. Physicians, Biochem. Soc. (mem. com.), Assn. Clin. Pathologists, Assn. Clin. Biochemists, Soc. for Endocrinology, Brit. Assn. for Cancer Research, Med. Research Soc. Author: Essentials of Chemical Pathology, 1957; (with N. D. Compston, A. M. Dawson) Recent Advances in Medicine, 1965; also numerous articles. Editor, Clin. Sci., 1966—. Application of biochemistry to medicine, especially treatment of cancer by hypophysectomy, use of enzyme and isoenzyme assays, use of leucoytes for study intracellular electrolytes; relation to diagnosis of computer-aided taxonomy of diseases; biochem. properties of 3 alph-hydroxy-steroid dehydrogenases. Home: 47 Holne Chase, London N.2. Office: Royal Free Hosp., London W.C.1, U.K.*

BARON, Hyacinthe-Théodore, French physician, chemist; b. Paris, Aug. 12, 1707; s. Hyacinthe-Théodore and Marie (Pellemoine) B.; studied medicine, Montpellier, France; docteur en médecine, 1732. Became med. attaché to army, Corsica, 1741; named physician Hôtel-Dieu, Paris, 1748; dean Faculty Medicine, Paris. Mem. French Acad. Scis., 1752. Author: Formules de médicaments, 1747; Compendiaria medicoram, 1751; history of physicians of Paris Faculty, from 14th century. Died Mar. 27, 1787.

BARON, Jozef, Polish physician; b. Opole, Poland, July 22, 1923; s. Jozef and Lucja (Liguda) B.; physician diploma Acad. Medicine, Poznan, 1951; m. Irena Harmata, June 6, 1957; children—Andrzej, Hanna, Jerzy. Faculty, 1st Clinic of Obstetrics and Gynecology, Acad. Medicine, Poznan, 1951—, asst. prof., 1964—. Mem. Polish Endocrinol. Soc., Polish Gynecol. Soc. (sec. Poznan 1955-65, gen. sec. 1965—, chmn. endocrinol. sect. 1966). Research, publs. on gynecol. endocrinology especially hormonal and cytogenetic disorders in intersexuality; original lights of pathogenesis of polycystic ovarian disease, hirsute, infertile women. Home: 31 A, Grunwaldzka, Poznan, Poland.*

BARON, Louis Sol, Am. bacteriologist; b. N.Y.C., Jan 2, 1924; s. Benjamin and Genevieve (Shankin) B.; B.S. Coll City N.Y., 1947; M.S., U. Ill., 1948, Ph.D., 1951; m. Rhoda H. Cohen, Aug. 24, 1961; 1 dau., Susan Ellen. Teaching asst. U. Ill., Urbana, 1949-52; bacteriologist, dept. bacterial physiology Walter Reed Army Inst. Research, Washington, 1952-56, chief dept. bacterial immunology, 1956—. Mem. genectics study sect. USPHS, NIH, Bethesda, Md., 1956—. Recipient Ann. award for Sci. Achievement in Biol. Scis., Wash. Acad. Sci. Mem. Am. Soc. for Microbiology, Genetics So. Am., Am. Assn. Immunologists, Biophys. Soc., Sigma Xi. Mem. editorial Proc. Soc. Exptl. Biology and Medicine, bd. Jour. Bacteriology. Research on transfer, isolation of genetic material between generically different bacteria. Home: 713 Horton Dr., Silver Spring, Md., 29096. Office: Dept. Bacterial Immunology, Walter Reed Army Inst. Research Walter Reed Army Med. Center, Washington 20012.*

BARON, Samuel, Am. physician, virologist; b. N.Y.C., July 27, 1928; s. Harry and Gertrude (Lipnick) B.; A.B., N.Y. U., 1948; M.D., 1952; m. Phyllis Goodman, Feb. 1, 1951; children—Steven Baron, Clifton, Jeffrey, Jonathan, Jody Lynn. Research fellow U. Mich. Sch. Pub. Health, Ann Arbor, 1953-55; staff USPHS, NIH, Bethesda, Md., 1955—; vis. scientist Nat. Inst. Med. Research, London, Eng., 1960. Recipient Henry L. Moses award for Med. research Montefiore Hosp., 1956-62. Diplomate Am. Bd. Microbiology. Mem. A.A.A.S., Am. Assn. Immunologists, Am. Soc. Microbiology, Soc. for Exptl. Biology and Medicine. Contbg. author: Advanced Virus Research, vol. 10, 1963; Modern Trends in Medical Virology, 1966; Interferon, 1966; also numerous articles. Asso. editor Am. Jour. Epidemiology, 1965—. Research on pathogenesis viral infections with emphasis on host defs., recovery factors, interferon and its action. Home: 6422 Kenhowe Dr., Bethesda 20034. Office: NIH, Bethesda, Md. 20014.*

BARON, Seymour, Am. chem. engr.; b. N.Y.C., Apr. 5, 1923; s. Benjamin and Tillie (Schuster) B.; B.Engring., Johns Hopkins, 1944, M.Sc., 1947; Ph.D., Columbia, 1950; m. Florence Chill, Aug. 27, 1950; children—Richard Mark, Paul Lawrence. Chem. engr. U. S. Indsl. Chem. Co., 1944-47; research asst. chem. engring. dept. Columbia, 1947-50; with Burns & Roe, Inc., Oradell, N.J., 1950—, chief chem. and nuclear engring., 1959-62, v.p. engring., 1962—; dir. Sci. Devel., Inc., Caldwell, N.J., Saline Water Conversion Corp., Oradell, N.J.; adj. prof. Bklyn. Poly. Inst., 1962—. Cons. on nuclear studies Manhattan Coll., 1964—. Fellow Am. Nuclear Soc.; mem. Am. Soc. M.E., Am. Inst. Chem. Engrs., A.A.A.S. N.Y. Acad. Scis., Sigma Xi, Phi Lambda Upsilon. Contbg. author Advances in Nuclear Science and Engineering, 1962. Research publs. on thermodynamics of thermal power plants; design nuclear power plants, comml. heavy water prodn. plant; devel. and design seawater conversion plants. Home: 684 Iroquois St., Oradell, N.Y. 07549. Office: 700 Kinderkamack Rd., Oradell, N.J. 07649.*

BARONE, Robert, anatomist; b. Constantine, Algeria, May 3, 1918; s. Louis and Berthe (Karlin) B.; ed. Nat. Vet. Sch., Lyons, France, Faculty Scis. Lyons; m. Mlle. Bergerat, Dec. 23, 1955; 1 dau., Antoinette. Became head anat. research Lyons Vet. Sch., 1941; apptd. lectr. Toulouse Nat. Vet. Sch., 1957; prof., head anatomy lab. Lyons Nat. Vet. Sch., 1959—. Decorated officer agrl. merit. Mem. Lyons Soc. Vet. Scis. (pres.), Speleological Soc. France (life), World Assn. Anat. Veterinarians (gen. sec. 1957—), Vet. Acad. France (corr.). Author: Anatomie des equidés domestiques, 1950-64; Anatomie comparée des mammifères domestiques, 1966; also articles. Research on comparative and vet. anatomy, embryology, bone transplants, structure of nervous system. Address: Ecole Nationale Vétérinaire, 69 Lyon 9, France.*

BARONI, Vitold, Rumanian bacteriologist; b. Bacau, Rumania, May 23, 1883; s. Edvin and Maria B.; ed. U. Vienna, Pasteur Inst., Paris, Inst. for Med. Expts., Bucharest; M.D.; m. Elene Dumitrescu. Dir. hygienic lab., Galatz, 1910-20, Med. and Epidemiological Service of Internat. Danube Commn.; prof. bacteriology Cluj U., dir. Bacteriological Inst. of med. faculty in Cluj. Research and publs. on Tb, cancer, cholera, fever diseases, serology. Address: Cluj, Calea Victoriei 70, Rumania.

BARONIO, Guiseppe, Italian surgeon; b. 1759; performed successful transplantation of skin grafts on animals, 1804. Died 1811.

BARONOV, Vladimir Alekseevich, Russian neuropathologist; b. 1903; grad. Med. Faculty, 1st Moscow U., 1929; D.Med. Sci. Head neurol. dept. Chita Mil. Dist. Hosp., 1930-39; asst. instr., sr. instr., dep. head chair nervous diseases Kirov Mil. Med. Acad., Leningrad, 1930-39. Author numerous works on battle trauma of nerve trunks of limbs, reflex contractures and paralyses, causalgia, closed trauma of brain, neuropathology of traumatic shock, penetrating radiation. Address: Kirov Mil. Med. Acad., ulsita Lebedeva 6, Leningrad, USSR.

BAR OSEAS, see Berossus.

BAROZZI, Francesco, Italian mathematician, humanist; b. Venice, Italy, Aug. 9, 1537; studied Greek, Latin, math. and philosophy; brought before Inquisition for sorcery. Author: Il nobilissimo ed antichissimo giuco pitagorico chiamato rimomachia, 1572; De cosmographia, 1585; Geometricum problema, 1586; also wrote on perspective, math. problems, and chess. Translated commentary of Proclus on 1st book of Euclid's Elements. Died Venice, May 23, 1604.

BAROZZI, Francesco, Italian engr.; b. Budrio (Bologna), Sept. 20, 1913; s. Antonio and Ines (Martelli) B.; ed. in electrotechnics; Ph.D. in Engring.; m. Fernanda Guggi, Aug. 21, 1941; children—Anna-Giovanni. Prof. electronics, dir. Electrotechnic Inst., U. Trieste. Mem. Assn. Italian Electrotechnologists. Author various publs. on electrotechnology. Home: via Bonci 8, Bologna, Italy. Office: via F. Severo 158, Trieste, Italy.

BARR, Alexander, Brit. statistician; b. Belfast, North Ireland, Sept. 24, 1925; s. Alexander and Agnes (McCully) B.; B.Com.Sc., Queen's U., Belfast, 1945, M.Sc. Econ., 1953; Ph.D., U. Reading, 1965; m. Evelyn Johnston, June 24, 1954; children—Ian Alexander, Caroline Jayne. Chief statistician Oxford (Eng.) Hosp. Bd., 1955—; statistician population genetics unit Med. Research Council, 1957—. Mem. Royal Statis. Soc. (Frances Wood Meml. prize 1954), Genetics Soc. Research, publs. in morbidity and mortality patterns of disease, clin. trials of drugs and treatments, population genetics, study inherited diseases. Home: 203 Oxford Rd., Kidlington, Oxford. Office: Oxford Regional Hospital Board, Old Rd., Headington, Oxford, Eng.*

BARR, Bengt Olov, Swedish physician; b. Stockholm, Sweden, Sept. 29, 1916; s. Olov David and Stina (Lindstedt) B.; M.D., Karolinska Institutet, 1945; m. Ulla Bolinder, Dec. 20, 1941; children—Björn, Eva, Kerstin, Gunilla. Staff, Karolinska Hosp., Stockholm, 1945—, asso. prof. ear nose and throat dept., 1956—, head paedo-audiology dept., 1964—. Research, publs. on audiology. Home: 9 Föreningsvägen, Stocksund, Sweden. Office: Karolinska Hosp., Stockholm 60, Sweden.*

BARR, E(rnest) Scott, Am. physicist; b. Lincolnton, N.C., Nov. 27, 1905; s. Peyton Arthur and Ida Beatrice (Brindle) B.; A.B., Univ., of N.C., 1926, A.M., 1933; Ph.D., 1936; m. Phoebe E. S. Baughan, May 9, 1931. Public sch. teacher, Honolulu, T.H., 1926-28; fgn. rep. Vick Chem. Co., Newfoundland, Mexico, and countries around Mediterranean, 1928-32; instr., Tulane Univ., 1936-37, asst. prof. of physics, 1937-40, asso. prof., 1940-47; sr. physicist research work on naval ordnance, Applied Physics Lab., Johns Hopkins Univ., 1945-46; prof. of physics, U. Ala., 1957—; cons. to Redstone Arsenal, 1952-65. Fellow A.A.A.S., Am. Phys. Soc., Optical Soc. Am.; mem. Am. Assn. Physics Tchrs., Am. Assn. U. Profs., Am. Inst. Physics, Southeastern sect. Am. Phys. Soc. (sec. 1940-45; chmn. 1948-49), Sigma Xi, Sigma Pi Sigma, Delta Upsilon, Pi Mu Epsilon, Alpha Epsilon Delta. Episcopalian. Contbr. tech. articles chiefly in history of physics in Am. Jour. Physics, Applied Optics, Infrared Physics, Physics Teacher, others. Research involving spectroscopic studies in infrared region, particularly of liquids; instructional aspects of optics. Home: 926 25th Av. E., Tuscaloosa, Ala. Office: Box 714, University, Ala.

BARR, John Henry, Am. mech. engr.; b. Terre Haute, Ind., June 19, 1861; s. John Henry and Eliza T. B.; B.M.E., U. Minn., 1883, M.S., 1888; M.M.E., Cornell, 1889; m. Katherine L. Kennedy, June 4, 1884; 1 son, John H. Engaged in mech. dept. Calumet & Hecla Copper Mining Co. and Lake Superior Iron Works, 1883-85; instr., asst. prof. and prof. mech. engring., U. Minn., 1885-91; asst. prof. and asso. prof., 1891-98, prof. machine design, 1898-1903, Sibley Coll., Cornell U.; factory mgr. Smith-Premier Works, Syracuse, N.Y., Feb. 1903; cons. engr. Union Typewriter Co., N.Y., Sept. 1909-13, Remington Typewriter Co., May 1913-23; v.p. Barr-Morse Corp., 1923-37; trustee of Ithaca (N.Y.) Savs. Bank. Chmn. N.Y. State Voting Machine Commn., 1903-14, Syracuse Lighting Commn., 1907; mem. Syracuse Intercepting Sewer Commn., 1908-12. Maj. ordnance, U.S.R.C., 1917-19, in office of chief of ordnance, U. S. A. and A.E.F., Aircraft Armament Sect., Paris, France. Trustee Cornell U., 1905-15. Author: Kinematics of Machinery, 1899; Notes on Machine Design; (with D. S. Kimball) Elements of Machine Design, 1909. Home: Ithaca, N.Y. Died Mar. 29, 1937.

BARR, Mark, elec. engr.; b. Pa. (parents Brit. citizens), May 18, 1871; s. Charles and Ann (M'Ginnis) B.; studied with Nikola Tesla, U. S., also in lab. Dr. Pupin, Columbia, chemistry with Dr. Elliott, N.Y.; grad. in physics and math. Central Coll., London U., 1895; m. Mabel Mary Richie, 1897; 2 sons. Became asst. editor Elec. World, N.Y., 1891. Recipient Gold medal, Paris, France, 1900. Fellow City and Guilds of London Inst.; mem. Instn. Elec. Engrs. (Premium prize 1895-96), Brit. Assn. (com. on screw threads 1903, com. standards 1905). Contbr. papers on phys. and mech. subjects to jours. Developed theory of new calculating mechanisms, also new methods for computing equation roots, 1918-24. Died Dec. 15, 1950.

BARR, Martin, Am. pharmacologist; b. Phila., Nov. 11, 1925; s. Louis and Bella (Moskowitz) B.; B.Sc. in Pharmacy, Temple U., 1946; M.Sc. in Pharmacy, Phila. Coll. Pharmacy and Scis., 1947; Ph.D., Ohio State U., 1950; m. Nancy Lipschutz, July 15, 1951; children—Lawrence Allen, Richard Andrew, Debra Ann, Steven Bruce. Grad. asst., then instr. Ohio State U. Coll. Pharmacy, 1947-50; from asst. prof. pharmacy to prof. phys. pharmacy and pharm. research Phila. Coll. Pharmacy and Sci., 1950-61; prof. pharmaceutics Wayne State U. Coll. Pharmacy, 1961—, chmn. dept., 1961-63, dean, 1964—; cons. Dept. Health, Edn. and Welfare, 1964-66. Recipient Distinguished Service award Alumni Assn. Coll. Pharmacy, Temple U., 1957; named Distinguished Alumnus, Temple U., 1964. Fellow Am. Coll. Apothecaries, A.A.A.S.; mem. Am. Pharm. Assn. (pres. Phila. 1954-55, chmn. sci. sect. 1959-60; Ebert medal 1956), Am. Soc. Hosp. Pharmacists, N.Y. Acad. Scis., Soc. Cosmetic Chemists, Am. Assn. Colls. Pharmacy (chmn. conf. tchrs. pharmacy 1961-62), Sigma Xi, Rho Chi. Contbg. author: Pharmaceutical Compounding and Dispensing, 2d edit., 1956; Remington's Practice of Pharmacy, 11th edit., 1956, 12th edit., 1960, 13th edit., 1965. Profl. editor: Mid-Atlantic Apothecary, 1953-64, Apothecary, 1953-64, Central Pharm. Jour., 1961-64. Research in areas of product formulation, drug stability, emulsion systems, ophthalmic preparations, dermatol. medication, viscosity; patentee in use of attapulgite clays as drug ingredient. Home: 20285 Beechaven Drive, Southfield, Mich. 48075. Office: 5501 2d St., Detroit 48202.*

BARR, Murray Llewellyn, Canadian physician; b. Belmont, Ont., Can., June 20, 1908; s. William Llewellyn and Margaret (McLellan) B.; B.A., U. Western Ont., 1930, M.D., 1933, M.Sc., 1938; LL.D., Queen's U., 1963; LL.D., U. Toronto, 1964; m. Ruth Vivian King, July 5, 1934; children—Hugh, Robert, Carolyn (Mrs. Robert McMaster), David. Faculty dept. anatomy U. Western Ont., London, 1936—, prof., 1950—, head dept., 1953—. Hon. cons. Victoria, St. Joseph's hosps., London; cons. Children's Psychiat. Research Inst., London. Recipient Flavelle medal Royal Soc. Can., 1959; award of A.C.P., 1962; Joseph P. Kennedy Jr. Found. award, 1967; Award of Merit, Gairdner Found., 1963. Mem. Alpha Omega Alpha. Research on sex chromatin, sex chromosome abnormalities, mental retardation. Barr body, found in all female cells, named for him. Home: 452 Wonderland Rd., London, Ont., Can.*

BARR, Thomas, Scottish surgeon; b. Elderslie, Scotland, 1846; ed. Vienna, Austria, also London, Eng.; M.B., C.M. Glasgow (Scotland) U., 1868; M.D., Glasgow, 1870. Lectr., Anderson's Coll., 1879-95; lectr. on diseases of ear Glasgow U., from 1895; aural surgeon Glasgow Western Infirmary; hon. aurist Glasgow Sick Children's Hosp.; sr. surgeon Glasgow Hosp. for Diseases of the Ear, Nose, and Throat. Pres. otol. sect. Brit. Med. Assn., 1888, Glasgow Path. and Clin. Soc., 1899-1900, Otol. Soc. U.K., 1904-05.

Author: Manual of Diseases of the Nose and Throat in relation to the Ear; Effects upon the Hearing of those who Work amid Noisy Surroundings; Investigation into the Hearing of School Children; Giddiness and Staggering in Ear Diseases; Traumatic Affections of the Ear; The Operative Treatment of Intra-Cranial Conditions dependent upon Purulent Diseases of the Ear. Died Dec. 14, 1916.

BARR, Thomas Calhoun, Jr., Am. zoologist; b. Nashville, Sept. 6, 1931; s. Thomas C. and Gladys (Hutchison) B.; A.B., Harvard, 1953; M.A., Columbia, 1954, postgrad.; Ph.D., Vanderbilt U., 1958; m. JoAnn Barr, Apr. 5, 1962; children—Maridel, Melisa, Thomas III. Instr. Tex. Technol. Coll., Lubbock, 1957-58; asst. prof. biology Tenn. Poly. Inst., Cookeville, 1958-61; asst. prof. zoology U. Ky., Lexington, 1961-66, asso. prof., 1966—. Fellow A.A.A.S.; mem. Nat. Speleological Soc. (pres. 1965-67), Soc. for Study Evolution, Soc. Systematic Zoology, Ecol. Soc. Am., Am. Assn. U. Profs., Sigma Xi. Author: Caves of Tennessee, 1961; also numerous articles. Research on ecology and evolution cave animals, beetles So. Appalachian mountains. Office: Dept. Zoology, U. Ky., Lexington, Ky. 40506.*

BARR, William Frederick, Am. entomologist; b. Oakland, Cal., Oct. 20, 1920; s. Eugene Bryant and Christine (Hansen) B.; A.A., San Francisco Jr. Coll., 1941; B.S., U. Cal. at Berkeley, 1945; M.S., 1947, Ph.D., 1950; m. Audrey Karen Sorensen, Aug. 31, 1946; children—Michael Eugene, Karen Christine, Steven William. Teaching asst. entomology U. Cal., Berkeley, 1945-47; faculty U. Ida., Moscow, 1947—, prof., entomologist, 1958—, dir. Summer Inst. Entomology, NSF, 1962, 63 Sci. Faculty fellow NSF, 1959-60. Mem., Sigma Xi, Gamma Alpha, Alpha Zeta, Phi Sigma. Classification of Coleoptera; biological studies on range plant and desert shrub insects. Home: 1415 Borah Av., Moscow, Ida., 83843.*

BARRAL, Louis Marius, anthropologist; b. Monaco, May 18, 1910; s. Donatien and Adélaide (Lagostena) B.; ed. College Saint-Charles at Bordighera, U. Paris; licencié ès sciences; m. Claude Roussin, June 28, 1938. Curator, Mus. Prehistoric Anthropology. Mem. Assn. Prehistory and Speleology of Monaco (pres.), French Prehistoric Soc. Author: La grotte Barriera (article). Excavated in Barriera Grotto and Saint-Benoit Grotto, renovated grottoes of Exotic Garden. Home: 30, blvd. d'Italie Monte-Carlo. Office: Museum of Prehistoric Anthropology, Monaco, Principality of Monaco.

BARRANDE, Joachim, geologist; b. Sangues, France, Aug. 11, 1799; ed. at École polytechnique, Paris, France, circa 1819; tutor to Count de Chambord; exiled with Charles X, 1830; lived with royal family in Eng. and Bohemia; adminstr. for duke after king's death. Author: Système silurien du centre de la Bohème, 8 vols., 1852; other publs. on geology, paleontology. Classified fossils of Bohemia. Died Frohsdorff, Austria, Oct. 5, 1883.

BARRAQUER, Joaquin, Spanish ophthalmic surgeon; b. Barcelona, Spain, Jan. 26, 1927; s. Ignacio and Josefa (Moner) B.; D. in Medicine and Surgery, U. Barcelona, 1951; M.D., U. Guayaquil (Ecuador), 1957; Prof. L.D. in Ophthalmology, U. Rome (Italy), 1958; M.D. Ministry of Pub. Health, Colombia, S.Am., 1965; m. Mariana Compte, Mary 4, 1953; children—Ma Elena, Mariana, Rafael, Ignacio. Dir. postgrad. courses Instituto Barraquer, Barcelona, 1953-55, dir. sci. publs., 1955-57, v.p., 1957—, exec. dir., 1962—. Recipient Leonardo Torres Quevedo prize, 1955, Ramón y Cajal Gold medal, 1960; named officer Bernald O'Higgins Ordre of Merit, 964; recipient Encomienda con Placa of Orden de Alfonso X el Sabio, 1965. Fellow Am. Acad. Ophthalmology and Otolaryngology (hon.); hon. mem. A.M.A., Societá Oftalmologica Lombarda, Pan. Am. Assn. Opthalmology; mem. Asociación Oftalmológica de la Akademia de Scis. Médicas Barcelona (past pres.), Royal Soc. Medicine (hon.). Author: La Extracción Intracapsular del Cristalino, 1962; (with others) Cirugia del Segmento Anterior del Ojo, 1964; also monographs, numerous articles. Research on methods of cataract removal; devel. numerous instruments for ophthalmic surgery. Home: 314 Muntaner, Barcelona, Spain.*

BARRATT, Ernest Stoelting, physiol. psychologist; b. North Charleroi, Pa., Mar. 31, 1925; s. Robert Duff and Marie (Stoelting) B.; B.A., Tex. Christian U., 1947, M.A., 1949; Ph.D., U. Tex., 1952; m. Bobbye Lee Rheinlander, 1946 (div. Jan. 1967); children—Robin Rhein. Asst. prof. psychology U. Del., 1951-57; prof. Tex. Christian U., 1957-61; faculty U. Tex., Galveston, 1961—, research prof. dept. neurology and psychiatry, 1961—, dir. Behavioral Sci. Lab., 1961—. Spl. NIH fellow Brain Research Inst., U. Cal. at Los Angeles, 1960-61. Mem. Am. Psychol. Assn., Soc. for Psychophys. Research, Soc. Biol. Psychiatry, Soc. Multivariate Exptl. Psychologists, A.A.A.S. Research, publs. on relationship of brain processes to complex behavior patterns especially to impulse control. Home: Box 274, Galveston, Tex. 77550.*

BARRATT, Raymond William, Am. biologist; b. Holyoke, Mass., May 4, 1920; s. George A. and Elizabeth (Bretschneider) B.; B.S., Rutgers U., 1941; M.S., U. N.H., 1943; Ph.D., Yale, 1948; m. Helen Ruggles, July 4, 1943; children—Marguerite E., William R. Research asso. biology Stanford, 1948-53, research biologist, acting asst. prof., 1953-54; faculty Dartmouth, Hanover, N.H., 1954—, prof. botany, 1958-62, prof. biology, 1962—, lectr. microbiology Med. Sch., 1961—, chmn. dept. biol. scis., 1965—. USPHS Spl. fellow, 1961-62. Mem. A.A.A.S., Am. Inst. Biol. Scis., Genetics Soc. Am., Sigma Xi, Alpha Zeta, Phi Sigma. Studies on mutations in microorganisms, mutagenic chems., relation of mutagens to carcinogens, genetics and biochemistry of fungi, gene control and regulation over metabolism in fungi. Home: 7 Kingsford Rd., Hanover, N.H. 03755.*

BARRELIER, Jacques, French botanist; b. Paris, France, 1606; entered Order of St. Dominic, 1635; became priest; collected shells and plants during visits to convents in France, Spain, Italy; most of his writings were destroyed by fire after his death; however Antoine de Jussieu preserved and published the copper plates; author: Plants of France, Spain and Italy (with numerous figures of plants and shells); Plumier named genus Barrelia in his honor. Died 1673.

BARRELL, Joseph, Am. geologist; b. New Providence, N.J., Dec. 15, 1869; s. Henry Ferdinand and Elizabeth (Wisner) B.; B.S., Lehigh U., Pa., 1892, E.M., 1893, M.S., 1897 (Sc.D., 1916); Ph.D., Yale, 1900; m. Lena Hopper Bailey, Dec. 27, 1902. Pub. sch. teacher, 1886-87; instr. in mining and metallurgy, Lehigh U., 1893-97; asst. mining engr., Lehigh Valley Coal Co., 1894, Butte & Boston, and Boston & Montana mining cos., Butte, Mont., 1897-98; field asst., U. S. Geol. Survey, 1899-1901; asst. prof. geology, Lehigh U., and in charge of dept. of natural sciences, 1900-03; asst. prof. geology, 1903-08, prof. structural geology, 1908—, Yale U. Fellow Geol. Soc. Am., Am. Acad. Arts and Scis., Nat. Acad. Scis. Research on magmatic stoping; igneous intrusions; isostasy; evolution and genesis of earth; regional geology; geol. time and processes. Died May 4, 1919.

BARREME, François, French mathematician; b. Tarascon, France, July 7, 1638; 1 son, Gabriel; tchr. applied math. Author: Livre des comptes-faits, 1682. Notable for wide knowledge of arithmetic, comml. calculating, bookkeeping; wrote accounting manuals widely used in 17th and early 18th centuries. Died 1703.

BARRER, Richard Maling, chemist; b. Wellington, New Zealand, June 16, 1910; s. Thomas Robert and Nina (Greensill) B.; B.Sc., U. New Zealand, 1930, M.Sc., 1931; Ph.D., Cambridge (Eng.) U., 1935, Sc.D., 1949; D.Sc., U. New Zealand, 1937; m. Helen Frances Yule, Aug. 15, 1939; children—Peter Maling, Alison Margaret (Mrs. Michael Davies), Hilary Susan, Christine Helen. Research fellow Clare Coll., Cambridge U., 1935-39; head chemistry dept. Tech. Coll., Bradford, Eng., 1939-46; reader in chemistry London (Eng.) U., 1946-48; prof. chemistry Aberdeen (Scotland) U., 1948-54; prof. phys. chemistry Imperial Coll., London U., 1954—, head chemistry dept., 1955—, dean Royal Coll. Sci., 1963-66. Mem. govt. coms.; cons. in field. Fellow Royal Inst. Chemistry, Royal Soc.; hon. asso. Royal Coll. Sci.; mem. Chem. Soc., Soc. Chem. Industry (mem. council), Faraday Soc., Am. Chem. Soc. Author: Diffusion in and Through Solids, 1941; also numerous articles. Research on diffusion in microporous media, in polymers, in zeolites, molecular sieves, their synthesis, ion exchange properties, synthesis and applications for mixture separation, adsorption at gas-solid interface. Home: 1 London Lane, Bromley, Kent, Eng. Office: Chemistry Dept., Imperial Coll., London, S.W. 7, Eng.*

BARRERA, Alfredo, Mexican biologist; b. Merida, Yucatan, Mexico, May 11, 1926; s. Alfredo and Rita (Marin) Barrera-Vásquez; B.Biol. Scis., Luis Vives Inst., Mexico City, 1946; Biologist, Escuela Nacional de Ciencias Biologicas, 1951, D.Sc. in Biology, 1965; m. Isabel Bassols Batalla; children—Narciso, Dalia, Jacinto, Marco Barrera Bassols. Entomologist, Nat. Antimalaria Campaign, Mexician Ministry Health and Welfare, 1947-50; faculty biology and med. entomology Nat. Sch. Biol. Scis., Mexico City, 1951—; entomologist research div. Nat. Commn. for Malaria Eradication, 1957-59; chief planning dept. Poly. Inst., 1962-64; dir. Natural History Mus., City of Mexico, 1965—; head grad. div. Nat. Sch. Biol. Scis., 1965—. Mem. Mexican Soc. Natural History (past pres.), Sociedad Mexicana de Entomologia. Research, publs. on taxonomy and distbn. of Siphonaptera (especially fleas of sylvan rodents) and other ectoparasites; participated in discoveries of enzootic bubonic plague in Mexico City and of onchocerciasis in spider monkey in So. Mexico; studies in biology of brown dog tick (transmitter of spotted fever in No. Mexico). Home: Retorno 201 No. 7-B, Unidad Modelo, Mexico 13, D.F. Office: Museo de Historia Natural, Nuevo Bosque de Chapultepec APDO Postal 18 845, Mexico 18, D.F., Mexico.*

BARRÈRE, Pierre, French naturalist; b. Perpignan, France, circa 1690; studied medicine and botany faculty medicine U. Perpignan, 1722-25. Physician, prof. botany U. Perpignan; spent 3 years in Guiana and Cayenne. Author: Nouvelles relations de la France Équinoxiale, 1740. Research, writings on natural history, geography of Guiana and Cayenne, also compiled list of their natural products and described many new animals; genus of plants in Guiana named after him. Died Perpignan, Nov. 1, 1755.

BARRET, William Morris, Am. geophysicist; b. Shreveport, La., May 2, 1898; s. Thomas Charles and Lillian (Hollingsworth) B.; student U. South, 1916-17, Columbia, 1919-20; B.E., Tulane U., 1923, E.E., 1932; m. Lola Belle Holloway, May 4, 1938. Cons. geophysicist, Shreveport, 1927-30; pres. Engring. Research Corp., 1930—, William M. Barret, Inc., 1931— (both Shreveport). Mem. Soc. Exploration Geophysicists, European Assn. Exploration Geophysicists, Am. Inst. Mining, Metall. and Petroleum Engrs., Am. Assn. Petroleum Geologists, La. Acad. Scis., Ark.-La.-Tex. Geophys. Soc., Franklin Inst., Am. Geol. Inst. (a sustaining founder), Sigma Alpha Epsilon, others. Research, publs. on instrumentation and techniques of applied geophysics; pioneered in devel. of magnetic method of geophys. prospecting. Patentee on geophys. instruments and methods, and mech. apparatus. Home: 2524 Fairfield Av., Shreveport. 71104. Office: Linwood at Dalzell, Shreveport, La. 71103.*

BARRETO, Ruy Carlos Ramos, Brazilian biochemist; b. Rio de Janeiro, Brazil, Oct. 1, 1930; s. Eurico and Julia (Ramos) B.; grad. chemistry Nat. Sch. Chemistry, U. Brazil, 1952; m. Monique Colette Weingart, Dec. 30, 1954; children—Robert Weingart, Daniel W., Aline W. Biochemist Inst. Animal Biology, Rio de Janeiro, 1953-54, head dept. biochemistry, 1954; researcher Inst. Agrl. Chemistry, Rio de Janeiro, 1954-55; biochemist Central Lab. Tb., Rio de Janeiro, 1955-58, head research dept., 1958; head dept. biochemistry Inst. Phthisiology, Rio de Janeiro, 1958-66; pres. Orma, Inc., Rio de Janeiro, 1964—; dir. Assessa, Ltd., Promarino, Ltd. Mem. Biochem. Soc. (Eng.), Societé de Chemie Biologique (France), A.A.A.S., Brazilian Assn. Chemistry. Research, publs. on paper chromatographic studies on dopping detection, vitamins in maté tea, chemotherapy of Tb, drug resistance and biochemistry of tubercle bacilli, epilepsy, drug toxicity and blood coagulation, devel. spl. methods for biochem. research, chem. technology. Home: 7, Rua Duque Estrada, Rio de Janeiro GB-ZC-20, Office: N.590, Av. Pres. Vargas, Rio de Janeiro GB-ZC-00, Brazil.*

BARRETT, Alan Hildreth, Am. astronomer; b. Springfield, Mass., June 7, 1927; s. Raymond Lathrop and Sibyl (Jesseman) B.; B.E.E., Purdue U., 1950; M.S., Columbia, 1953, Ph.D., 1956; m. Virginia McCulloch, Sept. 3, 1949; children—Richard Alan, Bonnie Jean. Research asst. Columbia, 1953-56; NRC fellow U. S. Naval Research Lab., Washington, 1956-57; research asso., lectr. U. Mich., 1957-61; asso. prof. elec. engring. Mass. Inst. Tech., 1961-65 prof., 1965-67, prof. physics, 1967—; mem. adv. panel for astronomy NSF, 1967—; mem. planetary atmospheres adv. subcom. Space Sci. Steering Com. NASA, 1967—, co-chmn. passive microwave team Apollo Applications Feasibility Study, 1965—; exchange scientist to USSR, 1964. Mem. Am. Phys. Soc., Am. Astronom. Soc., Am. Geophys. Union, Internat. Astronom. Union, Internat. Sci. Radio Union. Research, numerous publs. on spectral line radio astronomy, centimeter-wave length observations of radio sources, planetary radio observations, interpretations; co-experimenter Microwave Radiometer Expt. Mariner-2 Venus Spacecraft, 1960-64; co-discoverer hydroxyl (OH) Lines in radio astronomy, 1963. Home: 3 Dane Rd., Lexington, Mass. 02173. Office: Dept. Physics, Mass. Inst. Tech., Cambridge, Mass. 02139.*

BARRETT, Channing Whitney, Am. gynecologist; b. Blissfield, Mich., Dec. 14, 1866; s. David Fowler and Martha C. (Dewey) B.; student Fayette (O.) Normal U., Hillsdale (Mich.) Coll.; M.D., Detroit Coll. Medicine, 1895; m. Luella May Alvord, July 22, 1896; children—Russell Alvord, Florence Louise, Helen Elizabeth, Ruth Esther. Intern St. Luke's Hosp., Detroit, 1893-95; house physician Harper Hosp., Detroit, 1895-96; asst. surgeon Marion Sims. Hosp., Chgo., from 1896; prof. gynecology Chgo. Clin. Sch., 1900-06; prof. gynecology Chgo. Policlinic Sch.; prof., chief dept. gynecology U. Ill. Med. Sch. to 1930; prof. gynecology Loyola U. Med. Sch.; chief dept. of gynecology Cook County Hosp. Fellow Assn. Obstetricians and Gynecologists, Am., Chgo. (pres.) gynecol. socs.; mem. A.M.A., kindred orgns. Contbr. to med. jours. Died Jan. 29, 1958.

BARRETT, Charles Sanborn, Am. metallurgist; b. Vermillion, S.D., Sept. 28, 1902; s. Charles H. and Laura (Dunham) B.; B.S., U. S.D., 1925; fellow U. Chgo., 1927-28, Ph.D., 1928; m. Dorothy A. Adams, Aug. 2, 1928; 1 dau., Marjorie A. Metallurgy dept. Naval Research Lab., 1928-32; metals research lab., dept. metall. engring. Carnegie Inst. Tech., 1932-46; prof. inst. for study metals U. Chgo. since 1946; exchange prof. U. Birmingham, Eng., 1951-52; visiting prof., U. Denver, 1961, Stanford U., 1963; Eastman prof. Oxford U., Eng., 1965-66. Member national com. on crystallography, 1950-54. Recipient Mathewson medal, Am. Institute Mining and Metall. Engrs., 1934, 44, 51; Howe medal, Am. Soc. Metals, 1939; Clamer medal, Franklin Institute, 1950; Heyn medal, Deutsches Gesellschaft für Metallkunde, 1966; Sauveur medal Am. Soc. Metals, 1966. Fellow Am. Phys. Soc., Am. Inst. Mining and Metall. Engrs. (chmn. Inst. Metals div. 1956); mem. Am. Crystallographic Assn., National Academy of Sciences, American Society of Metals (honorary member). Inst. Metals (London), Internat. Union Crystallography (editor metals sect. Structure Reports, 1949-51), Sigma Xi, Phi Beta Kappa. Author: Structure of Metals, 1943, (rev. edit.) 1952. Research on phys. metallurgy, crystallography at low temperatures; structure, deformation, and transformation of metals. Home: 5756 Blackstone Av., Chgo. 37.

BARRETT, James Thomas, Am. microbiologist; b. Centerville, Ia., May 20, 1927; s. Alfred W. and Mary Marjorie (Taylor) B.; B.A., State U. Ia., 1950, M.S., 1951, Ph.D., 1953; m. Nancy Ann Tabor, June 12, 1949; children—Sara Joann, Robert Wayne. Faculty, U. Ark. Sch. Medicine, 1953-57; faculty U. Mo. Sch. Medicine, Columbia, 1957-—, professor microbiology, 1967-—. USPHS, NIH spl. fellow, 1963-64. Mem. A.A.A.S., Am. Assn. Immunologists, Am. Soc. for Microbiology. Research, publs. on immunology of enzymes, drug allergies, fungi, bacterial metabolism. Home: 2809 Rollins Rd., Columbia, Mo. 65201.*

BARRETT, Louis Carl, Am. mathematician; b. Murray, Utah, Jan. 23, 1924; s. John T. and Louise D. (Dahl) B.; B.S., U. Utah, 1948, M.S., 1951, Ph.D., 1956; m. Betty J. Grist, June 13, 1947; children—Louis Lee Grande, Linda Jene, Lori Lynn, Louise Ann. Grad. and teaching asst. U. Utah, Salt Lake City, 1947-53, instr. math., 1953-56; asso. prof. math and physics Ariz. State U., Tempe, 1956-57, lectr. on applied math., engr. in-tng. program, 1957; asso prof. math. S.D. Sch. Mines and Tech., Rapid City, 1957-58, prof., 1958-59, prof., head math. dept., 1960-65; prof., chmn. math. dept. Clarkson Coll. Tech., Potsdam, N.Y., 1965-66; prof., head math dept. Mont. State U., Bozeman, Mont., 1967-—. Research mathematician, consultant Holloman Air Devel. Center, N.M., summers 1955-56; cons. U. S. Naval Ordance Test Sta., China Lake, Cal., 1957-—. Mem. Am. Math. Soc., Soc. for Indsl. and Applied Math, Math. Assn. Am. (chmn. Rocky Mountain sect. 1961-62), Sigma Xi, Sigma Pi Sigma. Contbr. articles to tech. jours., esp. on optimization of rocket trajectories. Home: 1721 S. Willson, Bozeman, Mont. 59715.*

BARRETT, Lucas, naturalist, geologist; b. London, Eng., Nov. 14, 1837; ed. Univ. Coll., also at Eberdsdorf; 1 son, Authur. Traveled to Shetland, also Norway, 1855; apptd. curator Woodwardian Mus., Cambridge, Eng., 1855; studied marine fauna Greenland, 1856, Vigo, Spain, 1857; became dir. geol. survey, Jamaica, 1859; Jamaican commr. Internat. Exhbn., Eng., 1862. Fellow Geol. Soc. London. Contbd. papers to Quarterly Jour. Geol. Soc., also Annals and Mag. Natural Hist. Discovered bird bones and pterodactyl remains in phosphate bed of upper Greens, also in Cambridge, 1858; found hippuretes shells including new genus (named Barrettia by Woodward) in cretaceous limestones, Jamaica; showed orbitoidal limestone forms base of miocene formation; suggested sea bed beyond hundred-fathom line forms nearly uniform area throughout world; collection of radiates, echinoderms, mollusks housed in Brit. Mus. and U. Cambridge (Eng.). Died Jamaica, circa 1862.

BARRETT, Michael Thomas, dentist; b. Huntingdon, Que., Can., July 27, 1881; s. Dennis and Catherine (Timlin) B.; came to U. S., 1900; D.D.S., U. Pa., 1903; hon. M.S., Villanova, 1915; m. Della MacDonald, June 29, 1921. Demonstrator prosthetic dentistry U. Pa., 1904-10, instr. normal histology, 1910-14, instr. oral pathology Grad. Sch. of Medicine. Discoverer of amoebae in pyorrhea, 1914; research and publs. on pyorrhea alevolaris and protozoa of mouth, internal anatomy of teeth, effects of thymus extract on teeth of rats, etiology of dental caries. Died Aug. 22, 1940.

BARRETT, Otis Warren, Am. agriculturist; b. Clarendon, Vt., Apr. 18, 1872; s. James and Alice W. (Kelley) B.; B.Sc., U. Vt., 1896; D.Sc. (hon.), 1934; m. Bessie Lou Stearns, Apr. 27, 1898. In Jamaica, 1894, in employ of West India Improvement Co.; aptd. traveling agt. of commn. for Mexican exhibit of Paris Expn. of 1900, 1898; hon. curator entom. collections Museo de la Comisión Geográfico-Exploradora at Tacubaya, Fed. Dist., Mexico, 1898-1900; entomologist and botanist P.R. Agrl. Expt. Sta., 1901-05; plant introducer Office Seed and Plant Intro. and Distbn., U. S. Dept. Agr., 1905-08; commd. by Agrl. Soc. of Trinidad and Tobago, B.W.I., to report upon cacao diseases in Trinidad, 1907; dir. of agr. for Mozambique, Portuguese E. Africa, 1908-10; chief of divs. expt. stas. and horticulture Bur. Agr., Manila, P.I., 1910-14; horticulturist C.Z., 1914-17; mgr. coconut plantations in Nicaragua, 1917; with U. S. Dept. Agr., 1917; carbon expert U. S. War Dept., 1918-19; agrl. adviser to Liberia, 1920-21; agrl. survey of Haiti, 1922; with Dept. of Agr., San Juan, P.R., 1923-29; horticulturist U. Hawaii, 1929-30. Fellow A.A.A.S.; mem. Bot. Club, Bot. Soc. Washington, Entomol. Soc. Washington, Soc. Am. Mammalogists, Philippine Acad., Porto Rico Ateneo. Author: The Changa, or Mole Cricket, in Porto Rico, 1902; The Yautias, or Taniers, of Porto Rico, 1905; Promising Root Crops for the South, 1910; Coconut Culture, 1911; The Philippine Coconut Industry, 1913; The Food Plants of Porto Rico, 1925; The Tropical Crops, 1928; The Animals on Postage Stamps, 1936. Died Oct. 6, 1950.

BARRETT, William Fletcher, Brit. physicist; b. Jamaica, B.W.I., Feb. 10, 1844; s. W. G. Barrett; received pvt. edn.; m. Florence Willey, 1916. Became asst. to Prof. Tyndall, 1863; named sci. master Internat. Coll., 1867; apptd. lectr. physics Royal Sch. Naval Architecture, 1869; prof. physics Royal Coll. Scis., Dublin, Ireland, 1873-1910. Chief founder Soc. Psychical Research, 1882 (also pres.). Knighted, 1912. Fellow Royal Socs. London (1899), Edinburgh, Dublin, Philos. Soc., Royal Soc. Lit.; mem. Royal Irish Acad., Instn. Elec. Engrs. Author: Thought Transference, 1882; Monograph on the so-called Divining Rod, Vol. I, 1897, Vol. II, 1900; On the Threshold of a New World of Thought, 1908; On Creative Thought, 1910; On Psychical Research, 1911; On Swedenborg, 1912; On the Threshold of the Unseen, 1917; Deathbed Visions, 1926. Investigated sensitive flames, also recalescence, shortening of nickel by magnetization; discovered elec. and magnetic properties of stalloy (used in constrn. transformers, dynamos), 1899, also studied other iron alloys; investigated entroptic vision (led to invention of entoptiscope, also new optometer), 1905-07; pioneer in study of thought transferences. Died May 26, 1925.

BARRIERE, Pierre Fernand, French archeologist; b. Confolens (Charente), France, May 18, 1892; s. Martial and Marie (Dousinet) B.; ed. U. Bordeaux; Dr. ès lettres; m. Marcelle Labeguerie, Sept. 28, 1921; children—Jean, Claude Geneviève, Françoise, Madeleine, Anne-Marie, Jacqueline. Agrégé of univ., prof. lycées of Rochefort-sur-Mer, Perigeux, Bordeaux; prof. French lit. U. Bordeaux. Mem. Nat. Acad. Bordeaux, also numerous archaeol. assns. in Bordeaux. Address: 74, rue Mazarin, Bordeaux (Gironde) et Piegut-Pluviers (Dordogne), France.

BARRINGER, Daniel Moreau, Am. mining engr., geologist; b. Raleigh, N.C., May 25, 1860; s. Daniel Moreau and Elizabeth (Wethered) B.; A.B., Princeton, 1879, A.M., 1882; LL.B., U. Pa., 1882; spl. course in geology, Harvard, 1889, in chemistry and mineralogy, U. Va., 1890; m. Margaret Bennett, Oct. 20, 1897; children—Brandon, Daniel Moreau, Sarah Drew, John Paul, Elizabeth Wethered, Lewin Bennitt, Richard Wethered, Philip Ellicott. Practiced law with brother, 1882-89; cons. mining engr. and geologist, 1890-—; pres. and dir. of several mining cos. Trustee Jefferson Med. Coll. and Hosp. Author: The Law of Mines and Mining in the United States, 1907; Minerals of Commercial Value, 1907. Demonstrated in 1905 that Meteor Crater in Ariz. originated from impact of meteoric mass, probably compact cluster of iron meteorites. Died 1929.

BARRINGTON, Ernest James William, English zoologist; b. London, Eng., Feb. 17, 1909; s. William B. and Harriet (Mitchell) B.; B.A., Oxford U., 1931, B.Sc., 1934, M.A., 1936, D.Sc., 1948; m. Catherine A. M. Clinton, Sept. 23, 1943; children—John William, Heather. Lectr., Univ. Coll., Nottingham, Eng., 1932-—, prof. zoology, head dept., 1949-—, dep. vice chancellor, 1956-59, pub. orator, 1964-—, Rockefeller Found. Fellow in comparative physiology, McGill U., 1939, Harvard, 1940. Mem. Soc. for Exptl. Biology, Zool. Soc. London, Linnean Soc., Marine Biol. Assn. U.K., Inst. Biology. Author: Introduction to General and Comparative Endocrinology, 1963; Hormones and Evolution, 1964; Biology of Hemichordata and Protochordata, 1965; Invertebrate structure and Function, 1967; The chemical Basis of Physiological Regulation, 1968. European editor: General and Comparative Endocrinology, 1961. Research, numerous publs. on alimentary and endocrine systems of lower vertebrates and related forms. Home: 5, Manor Ct., Bramcote, Nottingham, NG 9, 3DR, Eng.*

BARRIOL, Jean, French chemist; b. Saint-Martin, France, Feb. 1, 1909; s. Jules and Jeanne (Ducellier) B.; student École Normale Supérieure, 1928-32, agrégation de physique, 1932; Doctorat ès Sciences, U. Paris, 1946; m. Odette Tourdot, Dec. 22, 1936; children—Roger, Françoise. Prof. lycée, Chaumont, France, 1933-35, Nancy, France, 1935-36, Metz, 1936-38, Paris, 1938-47; researcher, Nancy, 1947-48; rector U. Sarre, 1948-50; prof. theoretical chemistry U. Nancy, 1950-—. Mem. Société française de Physique, Société française de Chimie. Author: Mécanique quantique, 1952; les Moments Diploires, 1957; Eléments de Mécanique Quantique, 1966; also articles. Research on theory of polarization of atoms and molecules, dielectric behaviour of matter, molecular interactions in solutions. Home: 34 Albert-Ier, Nancy (54), France.*

BARRNETT, Russell Joffree, Am. biologist; b. Boston, July 27, 1920; s. Thomas Warren and D. Jerolyn (Shopwick) B.; A.B., Ind. U., 1943; M.D., Yale, 1948; m. Elizabeth Gorgine Smith, May 1, 1948; children—Russell Joffree, William Thomas, Elissa Hilson. Faculty, Med. Sch. Harvard, 1949-59; asso. prof., prof. Sch. Medicine Yale, New Haven, 1959-—; adviser AEC, WHO; vis. investigator Rockefeller Inst.; Mem. Soc. Cell Biology, Am. Assn. Anatomy, Histochem. Soc. (past pres.), Biophys. Soc., Am. Soc. Electron Microscopy, Internat. Soc. Cell Biology, Sigma Xi. Author: (with H. W. Deane and A. M. Seligman) Enzyme Histochemistry, 1960. Asso. editor: Jour. Ultrastructive Research, 1966-—; Jour. Histochemistry and Cytochemistry, 1965-—; Anatomical Record, 1963-—. Research, numerous publs. on enzyme and protein cytochemistry, applications of cytochemistry to electron microscopy with spl. reference to problems in fields of lipid metabolism, transport, irritability and motility. Home: 55 Goodrich St., Hamden, Conn. 06514. Office: 333 Cedar St., New Haven 06511.*

BARROIS, Charles, French geologist; b. Lille, France, Apr. 21, 1851; Dr. Sci. in Mineralogy and Geology, Lille; in charge ofcl. geol. work in Brittany; mem. mineralogy sect. French Acad. Scis., 1904 (v.p. 1926, pres. 1927); Fellow Royal Soc., 1913; sec. Internat. Congress Geology, 1900. Author numerous writings on paleontology and geology, including Les terrains anciens des Asturies et de la Galice; Faune des Calcaire D'Erbray. Died Nov. 5, 1939.

BARRON, Donald Henry, Am. physiologist; b. Flandreau S.D., Apr. 9, 1905; s. George E. and Mae L. (Reed) B.; B.A., Carleton Coll., 1928; M.S., Ia. State Coll., 1929; Ph.D., Yale, 1932; M.A. (hon.), Cambridge (Eng.) U., 1936; m. Annette Marie LaCourciere, Oct. 22, 1932; children—Marie Annette (Mrs. Stephen G. McCarthy), Donna Marie (Mrs. Robert Gomez). Teaching asst. in plant physiology Ia. State Coll., 1928-29; asst. in zoology Yale, 1929-31, in anatomy, 1931-32; instr. anatomy Albany Med. Coll., 1932-33; fellow NRC, U. Berne (Switzerland), 1933-34; asst. prof. anatomy Albany Med. Coll., 1935-36; lectr. Cambridge U., 1936-37; fellow, dir. med. studies St. John's Coll., Eng., 1937-40; asst. prof. zoology U. Mo., Columbia, 1940-42; asso. prof., 1942-43; asso. prof. physiology Yale, New Haven, 1943-47, prof., 1947-—, asst. dean med. sch., 1945-48. Chmn. human embryology and devel. study sect. NIH, 1963-—. Rockefeller Found. Traveling fellow, 1937; Sterling fellow, 1940. Mem. Am. Assn. Anatomy, Physiol. Soc. Gt. Britain and Ireland, Anat. Soc. Gt. Britain and Ireland, Cambridge Philos. Soc., Am. Physiol. Soc., Phi Beta Kappa. Mem. editorial bd. Jour. Comparative Neurology, 1948-—, mng. editor, 1956-—. Research on physiology of pregnancy, electrophysiology of spinal cord. Home: 334 Yale Av., New Haven.*

BARRON, E. S. Guzman, biochemist; b. Huari, Peru, Sept. 18, 1898; s. Sebastian and Agripina (Barron) G.; M.D., U. San Marcos, Lima, 1924; D.Sc., U. Trujillo, Peru, 1947, U. Brazil, 1956; m. Cora Durkee, Aug. 8, 1930; 1 son, Richard. Came to U.S., 1926, naturalized, 1939. Fellow The Rockefeller Found., 1927-28; prof. biochemistry U. Chgo.; hon. prof. medicine, Faculty of Medicine, Lima, 1949; hon. prof. medicine and chemistry U. Uruguay, 1956; hon. prof. sci. U. Arequipa, 1956. Mem. Am. Soc. Biol. Chemists, Am. Assn. Physicians, Am. Chem. Soc., Soc. Exptl. Biology and Medicine. Died June 25, 1957.

BARRON, Milton Leon, Am. sociologist; b. Derby, Conn., Feb. 25, 1918; s. Harry Bernard and Anne (Tevlin) B.; B.A., Yale, 1939, M.A., 1942, Ph.D., 1945; m. Matilda Ann Cogan, June 1, 1947; 1 son, Benjamin Monte. Orgns. and propaganda analyst Dept. Justice, Washington, 1943; instr. in sociology St. Lawrence U., Canton, N.Y., 1943-44; asst. prof. Syracuse U., 1944-48; asst. prof. Cornell U., Ithaca, N.Y., 1948-54; asso. prof. Coll. City N.Y., 1954-60, prof., 1961-—; vis. prof. Wells Coll., Aurora, N.Y., 1949-50, N.Y. U., 1964-65, Columbia, summer 1960 Yeshiva U., spring 1964; Fulbright lectr. Bar-Ilan U., Israel, 1962-63. Author: People Who Intermarry, 1947; The Juvenile in Delinquent Society; 1954, The Aging American, 1961. Editor, American Minorities, 1957, Contemporary Sociology, 1964. Research on social problems of underprivileged and minority-status groups, especially ethnic minorities and aged. Home: 51 Appleton Pl., Dobbs Ferry, N.Y.*

BARRON, Moses, physician, pathologist; b. Kovno, Russia, Nov. 8, 1883; B.S., U. Minn., 1910, M.D., 1911; m. 1919; 4 children. Came to U. S., naturalized, 1896. Demonstrator pathology, bacteriology U. Minn., 1912-13, instr. pathology, 1913-16, asst. prof., 1916-25, asso. prof., 1925-33, prof., 1933-52, prof. emeritus, 1952-—, staff, Univ. Hosp., 1924-26; also pvt. practice medicine. Dir. pathology lab. City Hosp., Mpls., 1913-14, City and County Hosp., St. Paul, 1914-17; staff St. Mary's Hosp., 1921-—, Asbury Hosp., 1926-—, City Hosp., 1927-52. Fellow A.C.P.; mem. A.M.A., Central Soc. Clin. Research, Minn. Acad. Medicine, Diabetes Assn. Work on relation of islets of Langerhans to diabetes, 1920 (led to discovery of insulin), carcinoma of lung, diseases of pancreas, Hodgkins disease, hepatomegaly, splenomegaly.

BARROS, Fernando Brito, Portuguese physician; b. Libson, Portugal, June 1, 1913; s. Gaspar and Maria (Brito) B.; M.D., U. Lisbon, 1941; postgrad. Instituto de Medicina Tropical; m. Candida Barros, Apr. 1, 1931; 1 son, Fernando Neves. Biologist, Lisbon's U. Hosp., 1948-56, head lab. biochemistry, 1956-—. Recipient Pfizer award, 1960. Mem. Sociedade de Cincias Médicas, Sociedad Port. de Endocrinologia, Sociedad Port. de Medicina Laboratorial, Sociedad Port. de Bioquimica. Research, publs. on cardiology, prevention errors in hosp. labs., geog. pathology and biochem. anthropology, distbn. haptoglobins in Portuguese population, identification of abnormal hemoglobins in Continental Portugal. Home: 50 3° Esq., Campo dos Martires da Pátria, Lisbon. Office: 47, 1°D, Av., Fontes Pereira de Melo, Lisbon, Portugal.*

BARROW, Gordon Milne, chemist; b. Vancouver, B.C., Can., Nov. 13, 1923; s. Edward and Kathleen (Love) B.; B.A.Sc., U. B.C., 1946, M.A.Sc., 1947; Ph.D., U. Cal. at Berkeley, 1950; m. Harriet E. Heuser, Sept. 8, 1957; children—Andrew Gordon, Elizabeth Anne, Peter Jonathan. Faculty, Northwestern U., Evanston, Ill., 1951-59; prof., head, dept. chemistry Case Inst. Tech., Cleve., 1959—. Guggenheim fellow, 1957-58. Mem. Am. Chem. Soc., Am. Phys. Soc. Author: Physical Chemistry, 1961; Introduction to Molecular Spectroscopy, 1962; The Structure of Molecules, 1964; (with M. E. Kenney, J. D. Lassila, R. L. Litle, W. E. Thompson) Programmed Supplements for General Chemistry, 1963. Research, publs. on nature of hydrogen bonding and effect of hydrogen bonding and molecular assns. on absorption of radiation in infra-red spectral region. Home: 32445 Jackson Rd., Chagrin Falls, O. 44022. Office: Case Inst. Tech., Cleve. 44106.*

BARROW, Isaac, English mathematician; scholar; b. London, Eng., 1630; s. Thomas B.; B.A., 1648, M.A., Trinity Coll., Cambridge (Eng.) U., 1652; M.A., Oxford U., D.D. by royal mandate, 1670. Became fellow Trinity Coll., 1649; received holy orders, Church of Eng., 1659; joined faculty Cambridge U., 1654, toured the Continent, 1655-59, became lectr. Greek, 1659, prof. geometry Gresham Coll., London 1662, apptd. 1st Lucasian prof. math., Cambridge U., 1664, resigned in favor of Newton, 1669, named master Trinity Coll., Cambridge U., 1672, became vice chancellor, Cambridge U., 1675. Fellow Royal Soc., 1663. Author: Euclidis elementa, 1655; Euclidis data, 1657; Mathematicae lectiones, 1664-66; Lectiones opticorum phaenomenum, 1669; Lectiones opticae et geometricae, 1669, 74; Archimedes opera; Apollonii concicorum libri IV, 1675; Lectio in qua theoremata Archemedis de sphaera et cylindro per methodum indivisibilium investigata 1678. Considered by his contemporaries 2nd only to Newton as a mathematician; his math. achievements paved way for Newton's discoveries; introduced differential triangle; developed method of tangents that differs from differential calculus chiefly in notation; first to observe reciprocal relation between differentiation and integration; determined properties of refraction in glass lenses. Also influential in building the Trinity College Library. Died London, May 4, 1677.

BARROW, Sir John, Brit. geographer; b. Dragley Beck, Eng., June 19, 1764; LL.D., Edinburgh (Scotland) U., 1821; m. Anne Maria Trüter, circa 1800. Became timekeeper in Liverpool (Eng.) iron foundry, 1778; began 1st voyage in Greenland whaler 1781; teacher of mathematics; became private secretary to ambassador to China, 1792; sec. to Admiralty, 1804-06; 1807-48. Knighted, 1835; Fellow Royal Soc., 1805; mem. Geog. Soc. (founder, v.p. 1830). Barrow Straits, Cape Barrow and Point Barrow named in his honor. Author: Travels in South Africa, 2 vols., 1801-04; Travels in China, 1804; A Voyage to Cochin China, 1806; A Chronological History of Arctic Voyages, 1818; Voyages of Discovery and Research Within the Arctic Regions, 1846; An Autobiographical Memoir, 1847. Promoted arctic exploration. Died London, Nov. 23, 1848.

BARROWS, Harold Kilbrith, Am. civil engr.; b. Melrose, Mass., Nov. 9, 1873; s. Cyrus Moulton and Augusta (Kilbrith) B.; B.S. in Civil Engring., Mass. Inst. Tech., 1895; m. Mabel R. Jordan, Feb. 11, 1907; 1 son, Kilbrith Jordan. With city engr., Newton, Mass., later with Met. Water Bd., Boston; asso. prof. civil engring. U. Vt., 1901-04; dist. engr. U. S. Geol. Survey, New Eng. and N.Y., 1904-08; cons. hydraulic engr., Boston, from 1907; asso. prof. hydraulic engring. Mass. Inst. Tech., 1909-21, prof., 1921-41, in charge of hydro-electric option in civil engring., prof. emeritus, from 1941. Cons. on water power, water supply and flood control for numerous state public utility and power commns., atty. gen.'s offices, for many municipalities and pvt. corps.; mem. and cons. Adv. Com. Flood Control, State Vt., 1928-31; regional cons. Nat. Resources Com., 1934-41. Fellow Am. Acad. Arts and Scis.; mem. Am. Inst. Cons. Engrs., Am., Boston (pres.) socs. civil engrs. Author: Water Power Engineering, 1927, 3d edit., 1943; Floods—Their Hydrology and Control, 1948; also many state and national reports upon water power, storage, water supply, and flood control. Died Mar. 15, 1954; buried Winchester, Mass.

BARROWS, Howard Strong, Am. physician; b. Oak Park, Ill., Mar. 28, 1928; s. Raymond Hayes and Martha Francis (Church) B.; A.A., San Mateo (Cal.) Coll., 1947; A.B., U. Cal., Berkeley, 1949; M.D., U. So. Cal., 1953; m. Phyllis L. Lorange, June 14, 1953; children—Pamela, Kimberley, Allison, Rebecca. Faculty, Columbia U., 1959-60; practice medicine, specializing in neurology, Los Angeles, 1960—; faculty Sch. Medicine U. So. Cal., 1960—, now prof. neurology, dir. Neuromed. Service and Neurol. Residency Program Los Angeles County Gen. Hosp.; co-investigator Extra-cranial Arterial Occlusion; program dir. neurol. tng. grant NIH. Fellow Am. Acad. Neurology; mem. Fedn. Western Socs. Neurol. Sci. (dir.). Author: (with Theodore Kurze) The Value of Ultrasonic Encephalography An Adjunct to Neurology and Neurological Surgery, 1966. Editor, U. So. Cal. Medicine, 1961-67; editorial bd. Cal. Medicine, 1963—. Research in med. edn., particularly in field of neurology, ultrasonic diagnosis in neurology and muscle disease. Home: 2500 Monterey Rd., San Marino, Cal. 91108. Office: 1200 N. State St., Box 93, Los Angeles 90033.*

BARRUS, George Hale, Am. engr.; b. Goshen, Mass., July 11, 1854, s. Hiram and Augusta (Stone) B.; B.S., Mass. Inst. Tech., 1874; m. Louise C. Williams, Oct. 2, 1897; 1 dau., Bella D. (Mrs. Edwin L. Bowman). Asst. in design and constrn., steam engring. lab. Mass. Inst. Tech., 1874-75; cons. practice, Boston, 1875—. Judge of exhibits, Mass. Charitable Mech. Assn., Franklin Inst. Elec. Exhbn., Phila.; Mass. judge of power exhibits, World's Fair, Chgo., 1893; mem. Govt. Adv. Bd. on tests of fuels and structural materials. Inventor several forms of steam calorimeter, coal calorimeter, draft gauge, steam meter and drainage system. Died Apr. 1929.

BARRY, Guy Thomas, microbiological chemist; b. Montreal, Que., Can., Apr. 14, 1920; s. Thomas Eugene and Leonie (Prevost) B.; B.Sc., Sir George Williams Univ., Montreal, 1942; Ph.D., McGill U., 1946; m. Enid Mellquist, Aug. 9, 1948; children—Raymond, Janet. Came to U.S., 1946, naturalized, 1950. Vis. investigator Rockefeller Inst. for Med. Research, N.Y.C., 1946-50; asst., 1950-52, asso., 1952-58; prof. research Meml. Research Center and Hosp., U. Tenn., Knoxville, 1958-65; cons. E. Tenn. Bapt. Hosp. Clin. Labs., Knoxville, 1959-65; dir. biochemistry Squibb Inst. for Med. Research, New Brunswick, N.J., 1964—. NIH fellow, 1946-48, Am. Cancer Soc. fellow, 1948-50; recipient Travel award joint com. Am. Chem. Soc. and Am. Soc. Biol. Chemists to attend Vth Internat. Congress Biochemistry, Moscow, USSR. Mem. Am. Chem. Soc., Am. Soc. for Biol. Chemists, Am. Soc. for Microbiology, Am. Assn. Immunologists, Harvey Soc., Sigma Xi. Contbr. numerous articles to tech. jours., esp. on enterobacteriaceae. Home: Sycamore Lane, Skillman R.D. 1, N.J. 08558. Office: Squibb Inst. for Med. Research, New Brunswick, N.J. 08903.*

BARRY, Herbert, III, Am. psychologist; b. N.Y.C., June 2, 1930; s. Herbert and Lucy (Brown) B.; B.A., Harvard, 1952; M.S., Yale, 1953, Ph.D., 1957. USPHS fellow Yale, 1957-59, faculty, 1958-61, asst. prof., 1960-61; asst. prof. psychology U. Conn., Storrs, 1961-63; research asso. prof. pharmacology Sch. Pharmacy, U. Pitts., 1963—, asso. prof. cross-cultural research dept. anthropology, 1966—. Mem. A.A.A.S., Am. Assn. U. Profs., Am., Eastern, Pa., Pitts. psychol. assns., Psychonomic Soc., American Soc. Pharmacology and Exptl. Therapeutics, Phi Beta Kappa, Sigma Xi. Research, publs. on inhibition-reducing effects of alcohol and amobarbital, effects of various drugs on animal performance using new apparatus and exptl. designs; measured and interpreted intercorrelations of child tng. practices, use of alcoholic beverages in preliterate socs. Home: 240 Melwood Av., Pitts. 15213.*

BARRY, John Michael, English biochemist; b. London, Eng., Apr. 20, 1924; s. Thomas Ernest and Josephine (Furlong) B.; student Dulwich Coll., 1937-42; M.A., Queen's Coll., Oxford U., 1952, D.Phil., 1956; m. Elaine Morris, Sept. 8, 1956; children—Thomas C. M., Alexander M., Veronica M. A. Research fellow Nat. Inst. for Research in Dairying, U. Reading (Eng.), 1946-47, dept. biochemistry U. Cambridge (Eng.), 1947-48, U. Chgo., 1948-51; faculty dept. agr. U. Oxford, 1951—, U. lectr., 1951—; research fellow McCollum-Pratt Inst., Johns Hopkins, 1956-57; vis. lectr. U. Fla. Coll. Medicine, Gainesville, 1961-62. Author: Molecular Biology: Genes and the Chemical Control of Living Cells, 1964; also articles. Research on biochemistry of mammary gland, protein and nucleic acid biosynthesis. Home: Shilton House, Shilton, Oxford, Eng.*

BARRY, John Milner, Irish physician; b. nr. Bandon, Cork, Ireland, 1768; s. James Barry; M.D., Edinburgh (Scotland), U. 1792; m. Mary Phair, 1808; 2d son was John O'Brien Milner. Founder and 1st physician Cork Fever Hosp. and House of Recovery, from 1802; lectr. agr. Royal Cork Instn. until 1815. Monument in his honor erected on grounds of Fever Hosp., 1824. Contbr. articles on vaccination, fever to London Med. and Phys. Jour, 1800-01, also to History of the Contagious Fever Epidemics in Ireland (Harty), 1817, 18, 19, 21, Trans. Irish Coll. Physicians, Vol. II. 1st to use vaccination in Ireland, 1800. Died 1822.

BARRY, Kevin Gerard, Am. physician; b. Newton, Mass., May 12, 1923; s. Michael L. and Catherine (Coleman) B.; grad. The Citadel, 1943; student Johns Hopkins, 1944-45; M.D., Georgetown U., 1949; m. Frances Ellen Ryan, July 26, 1942; children—Kevin Gerard II, Paul, Janet, Joan. Bn. surgeon 418th Combat Engrs., Germany, 1950-51, med. officer 98th Gen. Hosp., Munich, Germany, 1951-52, bn. surgeon 43d Combat Engrs., Germany, 1952-53; resident in medicine Walter Reed Gen. Hosp., Washington, 1953-56; chief medicine U. S. Army Hosp., Ft. Jay, N.Y., 1956-58; research internist dept. metabolism Walter Reed Army Inst. Research, 1958-61, chief dept. metabolism, 1961-65, dir. div. medicine, chief dept. metabolism, chief enlisted male gen. medicine, metabolism sect., 1961-66; clin. asst. prof. medicine Georgetown U. Hosp., 1961—; chief renal metabolic service, dir. med. edn. Washington Hosp. Center, 1966—. Fellow A.C.P., Am. Coll. Clin. Pharmacology and Chemotherapy; mem. A.M.A., Internat. Assn. Internal Medicine, Am. Fedn. Clin. Research, Am. Soc. for Artificial Internal Organs, Am. Heart Assn., Third Internat. Congress Nephrology. Contbr. numerous articles to sci. jours. Pioneer in use of osmotic diuretic mannitol for prevention and treatment of acute functional renal failure; delineated value of mannitol diuresis for treatment of exogenous and endogenous intoxications; invented and applied an indwelling peritoneal cannula to permit repeated dialysis without repeated surgery; directed exploitation of peritoneal membrane for dialysis and other purposes. Home: 905 Hyde Rd., Silver Spring, Md. Office: Washington Hosp. Center, 110 Irving St. N.W., Washington 20010.*

BARRY, Martin, physician, physiologist; b. Fratton, Eng., 1802; studied medicine Erlangen, Heidelberg, Berlin (Germany), Edinburgh, Scotland, Paris, France, London (Eng.); M.D., Edinburgh, 1833. Lectr. physiology St. Thomas's Hosp., Edinburgh, 1843; visited Prague, Czechoslovakia, Guiessen; worked with Purkinje at Breslau, Germany; returned to Eng., 1853. Royal medal, 1839. Fellow Royal Society, 1840; mem. Royal Coll. Surgeons. Research and publs. on embryology; discovered presence of spermatozoa in ovum, 1843, segmentation of yolks in ovum of mammal, penetration of spermatozoa into zona pellucida. Died 1855.

BARSA, Joseph Albert, Am. psychiatrist; b. N.Y.C., May 28, 1916; s. Albert and Marie (Matouk) B.; A.B., Holy Cross Coll., 1937; M.A., Fordham U., 1940; M.D., Georgetown U., 1945; certificate in psychoanalysis N.Y. Med. Coll., 1954; m. Vivian Rosaria Scala, Dec. 24, 1944; children—Edward, John, Joanne, Patrice, Christine. Sr. psychiatrist Rockland State Hosp., Orangeburg, N.Y., 1951-53, supervising, psychiatrist, 1953—; pvt. practice psychiatry, N.Y.C. and Pearl River, N.Y., 1951—. Fellow Am. Psychiat. Assn., Acad. Psychoanalysis, A.A.A.S.; mem. Soc. Med. Psychoanalysts. Contbr. articles to profl. jours. Pioneer devel. drug therapy in psychiatry; pioneer defining dual action tranquilizers. Home: 133 Blaisdell Rd., Orangeburg, N.Y. 10962. Office: 174 W. Washington Av., Pearl River, N.Y. 10965.*

BARSCHALL, Henry H., physicist; b. Berlin, Germany, Apr. 29, 1915; A.M., Princeton, 1939, Ph.D., 1940; m. Eleanor A. Folsom; 2 children. Instr., Princeton, 1940-41, U. Kan., 1941-43; staff mem. Los Alamos Sci. Lab., 1943-46, cons.; faculty U. Wis., 1946—, prof., 1950—. Recipient Bonner prize Am. Phys. Soc., 1965. Research, publs. in nuclear physics. Home: 1110 Tumalo Trail, Madison, Wis. 53711.*

BARTECKI, Adam, Polish chemist; b. Stanistawów, Mar. 17, 1920; s. Maksymilian and Mina (Sekler) B.; M.Sc. in Chemistry, Tech. U., Wroclaw, Poland, 1950, D.Sc., 1960, Habilitated Dozent degree, 1965; m. Krystyna Radnicka, Dec. 28, 1950; children—Janusz, Malgorzata, Ewa. Asst. sci. worker Tech. U. Wroclaw Inst. Chemistry and Tech. Coal, 1949-54, dept. rare elements chemistry, 1954-67, head dept., 1967—; vis. scientist U. Szeged (Hungary), 1959, U. Rome Inst. Inorganic Chemistry, 1963. Recipient Poland's 1000 year Memory award, 1965, golden mark of distinction Techn. U., 1966. Mem. spectroscopic com. Polish Acad. Scis. (sec. 1965—), Polish Chem. Soc. (past sec. Wroclaw div.). Contbg. author: Experimental Methods in Coordination Chemistry, 1967; also articles. Editorial Bd. Absorption Spectra in the UV and Visible Region, 1959—. Research on properties and luminosity of coal diffusion flames, chemistry of rare elements complex compounds, electronic spectroscopy and structure of transition metal oxycations. Home: 24/4 Smoluchowskiego, Wroclaw, Poland.*

BARTELL, Lawrence Sims, Am. chemist; b. Ann Arbor, Mich., Feb. 23, 1923; s. Floyd Earl and Lawrence (Sims) B.; student U. N.C., 1940-41; B.S., U. Mich., 1944, M.S., 1947, Ph.D., 1951; m. Joy Hilda Keer, Aug. 16, 1952; 1 son, Michael Keer. Research asst. Manhattan project U. Chgo., 1944-45; faculty Ia. State U., 1953-65, prof. chemistry, 1959-65; prof. U. Mich., Ann Arbor, 1965—. Mem. Am. Chem. Soc., Am. Phys. Soc. Asso. editor Jour. Chem. Physics, 1963-66. Research, numerous publs. in molecular structure, surface chemistry, isotope effects. Home: 305 Sumac Lane, Ann Arbor, Mich. 48105.*

BARTELMEZ, George William, Am. anatomist; b. N.Y.C., Mar. 23, 1885; s. Theodore and Caroline (Osten) B.; B.S., N.Y. U., 1906; Ph.D., U. Chgo., 1910; D.Sc. (hon.), U. Mont.; m. Erminnie Eliza Hollis, Mar. 23, 1912 (dec. May 1919); children—Caroline Jane (Mrs. Randolph Moore), Erminnie Hollis, Theodore Lawrence. Faculty, U. Chgo., 1907-50, prof. anatomy, 1929-50; cons. dept. embryology Carnegie Instn. Washington, 1950-57; guest investigator dept. zoology U. Mont., Missoula, 1957—. Fellow Chgo. Gynecol. Soc. (hon.); mem. Nat. Acad. Scis., Am. Assn. Anatomists (past pres.), Am. Soc. Zoologists. Contbr. articles to tech. jours. Found evidence in pigeons that head and tail, right and left axes are predetermined before ovulation, cytology of synapse, early devel. human brain agrees in detail with that of other vertebratee; identified location retina mammals in earliest stages devel.; research on female cycle in primates. Home: 224 Agnes Av. Missoula, Mont. 59801.*

BARTELS, Aleksandr Vladimirovich, Russian obstetrician, gynecologist; b. Moscow, 1898; grad. Med. Faculty, Moscow U., 1922. Asso. dept. obstetrics and gynecology Sechenov 1st Moscow Med. Inst., 1922-25, asst., 1925-42, lectr., 1942-53, dean Med. Faculty, 1943-53; lectr., 1942—; head 2d obstet. dept. Inst. Obstetrics and Gynecology, USSR Ministry Health, 1952—. Mem. Moscow, All-Russian and All-Union Soc. Obstetricians and Gynecologists (bd. mem.). Del., Internat Symposium of Perinatal Mortality, Moscow, 1962. Author over 50 works on obstetrics. Dep. editor Obstetrics and Gynecology. Research on postnatal infections; drew up classification of postnatal diseases; one of 1st in USSR to use antibiotics for treatment of postnatal sepsis. Address: Inst. Obstetrics and Gynecology, B. Pirogovskaya ulitsa 2-6, Moscow, USSR.

BARTELS, Ernst Daniel August, German physician, botanist; b. Braunschweig, Germany, Dec. 26, 1778. Prof. anatomy Helmstadt, Merburg, Breslau, Erlangen (all Germany); chair clin. medicine Berlin, Germany; adviser King of Prussia. Author numerous works on physiology, pathology, biology, neurology. Died Berlin, June 26, 1838.

BARTELS, Heinz, German physiologist; b. Friedrichshafen, Germany, Oct. 21, 1920; s. Friedrich and Luise (Freudenberger) B.; student U. Munich (Germany), 1941-42, U. Strassbourg (Germany), 1942-44; M.D., U. Tubingen (Germany), 1947; m. Ruth Banhart, Nov. 7, 1942; children—Matthias, Barbara, Regine, Eva, Tim. Faculty physiology dept., U. Kiel (Germany), 1948-51, asso. prof., 1951-63; prof. applied physiology U. Tubingen, 1963-65; prof. physiology, dir. dept. Hannover (Germany) Med. Sch., 1965—. Mem. Deutsche Physiologische Gesellschaft, Royal Soc. Medicine (London, Eng.). Author: Lungenfunktionsprüfungen, Springer, 1959; Am. translation, 1963; also numerous articles. Research on devel. methods for measurement of blood oxygen tension, physiology of oxygen transport in mammals and birds, placental gas exchange. Home: 4 Weberstrasse, Misburg b. Hannover. Office: 5 Osterfeldstrasse, Hannover, Germany.*

BARTELS, Henry Arthur, Am. oral microbiologist; b. N.Y.C., Jan. 22, 1895; s. George H. and Christina (Geis) B.; B.S., R.I. State Coll., 1917; D.D.S., Columbia U., 1927; m. Helen C. Commander, Dec. 24, 1917; 1 son, Henry Arthur. Mem. faculty N.Y. U. Dental Coll., N.Y.C., 1927—, asso. clin. prof. microbiology, 1960—; part-time bacteriologist N.Y. State Dept. Health, 1919-51. Fellow N.Y. Acad. Scis.; mem. Internat. Assn. Dental Research, Am. Soc. Microbiologists, N.Y. State Dental Soc., Am. Dental Assn. Research and numerous publs. on oral micro-organisms, defensive factors of oral cavity, sterilization and disinfection of dental instruments, infectious diseases of mouth, antibiotics. Home: 9416 34th Rd., N.Y.C. 11372.*

BARTELS, Johann Martin Christian, mathematician; b. Braunschweig, Germany, Aug. 12, 1769; s. Heinrich Elias Friedrich and Johanna Christine Margarethe (Köhler) B.; student Collegium Carolinum, Braunschweig, 1788-91; studied law and math., 2 years in Helmstedt, then in Göttingen (Germany); m. Anna von Saluz, 1803; children—Eduard, Johanna (Mrs. Wilhelm von Struve). apptd. head math. inst. Carolinum, 1805; became prof. math. Russian U., Kazan, 1807; named prof. pure and applied math., Dorpat (Estonia); tchr. of Nikolai I. Lobachevski. Mem. St. Petersburg Acad. Author: Vorlesungen über mathematischen Analysis, 1833. Died Dorpat, Dec. 1836.

BARTELS, Julius, German geophysicist; b. Magdeburg, Germany, Aug. 17, 1899; Ph.D., U. Göttingen (Germany), 1923; worked with Adolph Schmidt, 1923-27; prof. meteorology and physics, head Meteorol. Inst., Eberswalde, 1927-41; lectr. U. Berlin (Germany), prof. geophysics, 1941-45; prof. geophysics, dir. Geophys. Inst., Göttingen, from 1945; dir. Max Planck Inst. for Aeronomy, Lindau, from 1956. Recipient Bowie medal Am. Geophys. Union, 1964. Author: (with Sydney Chapman) Geomagnetism, 2 vols., 1940. Developed statis. methods for geophysics, indices for study of solar wave and particle radiation and their effects on geomagnetic variation; demonstrated sun's surface is never totally active or quiet; described effects of moon's gravitational force on atmospheric tides. Died Mar. 6, 1964.

BARTELS, Maximilian Carl August, physician, anthropologist; b. Berlin, Germany, Sept. 9, 1843; s. Christian August and Veronica Juliane V. (Roebel) B.; studied medicine, Berlin, 1868; later study in Vienna; m. Anna Hertzog, Nov. 11, 1872; 1 son, Paul. Asst. (to his father, also to R. Wilms) Bethany Hosp., Berlin, 1869-72; then practiced medicine, Berlin; became prof., 1903. Author: Die Medizin der Naturvölker, 1893; pub. revised edit. of H. Ploss' Das Weib in der Natur-und Völkerkunde. Research, writings on rudimentary phenomena and atavisms; Darwinist, influenced by anthrop. soc. in Berlin (Anthropologische Gesellschaft). Died Berlin, Oct. 22, 1904.

BARTH, Carl G(eorge Lange), management scientist; mech. engr.; b. Christiania, Norway, Feb. 28, 1860; s. of Jacob Böckman and Adelaide Magdalene (Lange) B.; grad. High Sch., Lillehammer, Norway, 1875; grad. Tech. Sch., Horten, 1876; m. Hendrikke Jacobine Fredericksen, Mar. 4, 1882 (died Feb. 25, 1916); children—J. Christian, Carl G., I. Adelaide Elizabeth F.; m. 2d, Sophia E. Roever, Jan. 25, 1919. In machine shops, Norwegian Navy Yards; instr. in mathematics and mech. drawing, Tech. Sch., Horten, to 1880; came to America, 1881; mech. draftsman with Wm. Sellers & Co., Phila., 1881-90, and instr. in mech. drawing, evening schs. of Franklin Inst., 1882-88; engr. and chief draftsman with Arthur Falkenau, Phila., 1890-1901; designer, Wm. Sellers & Co., 1891-95; engr. and chief draftsman, Rankin & Fritch Foundry & Machine Co., St. Louis, 1895-97; designer St. Louis Water Dept., Feb.-June, 1897; with Internat. Corr. Schs., Scranton, Pa., 1897-98; instr. in manual work and mathematics, Ethical Culture Schs., New York, 1898-99; machine shop engr. Bethlehem Steel Co., 1899-1901; rep. of Tinius Olsen Testing Machine Co. in Japan, 1923-24; retired. Expert in shop management, Ordnance Dept., U. S. Army, 1909-18, and again during World War; lecturer on scientific management Harvard U., 1911-16, and 1919-23, U. of Chicago, 1914-16. Assoc. of Frederick W. Taylor, father of scientific management; introduced Taylor system of management in machine shops, 1901-23. Died Oct. 28, 1939.

BARTH, Gunther, German radiologist; b. Zwickau, Germany, Aug. 19, 1915; s. George and Antonie (Thiele) B.; student U. Frankfurt, U. Munich (Germany); dr. rerem naturalium summa cum laude, U. Leipzig (Germany), 1939; m. Aug. 25, 1945. Faculty, U. Erlangen, Hessen, Germany, 1952—, prof. radiology, 1960—, founder Sch. for Nursing Edn., Med. Tech. Sch. for Male Assts. Mem. Royal Soc. Medicine London. Author numerous books including Die Bewegungsbestrahlung, 1953; Movingfield X-ray Therapy, 1960; Radiotherapia Moderna; also numerous articles on radiobiology. Office: 25 Friedrichstrasse, 63 Giessen, Hessen, Germany.*

BARTH, Jean Baptiste, French physician; b. Sarreguemines, France, Sept. 24, 1806; studied medicine Paris, France; Staff, Hôtel Dieu, Paris; mem. Faculty Medicine, Paris. Author: Les Rétrécissements et les altérations de l'aorte, 1837; Traité pratique de l'auscultation, 1841; also articles. Described herniation of loops of small intestine (Barth's hernia), 1836; discovered (with Roger) stethoscope is often inaccurate, 1841; research on dilatation of bronchi, 1856. Died Paris, Nov. 30, 1877.

BARTH, Lester George, Am. biologist; b. Detroit, June 29, 1905; s. Charles and Mary (Schroeder) B.; B.A., Wayne U., 1926; M.A., U. Mich., 1928; Ph.D., U. Chgo., 1930; m. Lucena Jaeger, June 9, 1948. NCR fellow, Naples, Italy, Berlin, Germany, 1930-31; Nat. Edn. fellow U. Cal., 1941; prof. biology Columbia, 1950—; mem. staff, also trustee Marine Biol. Lab., Woods Hole, Mass. Mem. Internat. Inst. Embryology (Amsterdam), Soc. Growth and Devel., Am. Soc. Zoologists. Author: Embryology, 1950, 54; The Energetics of Development, 1954. Translator: (French) Chemical Embryology, 1950. Asso. editor Physiol. Zoology, 1945-55. Home: 26 Quissett Av. Office: Marine Biological Lab., Woods Hole, Mass.*

BARTH, Paul, German sociologist; b. Baruthe, Silesia, Aug. 1, 1858; prof. philosophy and edn., Leipzig, from 1897. Author (1st sociol. and philos. histories in German): Philosophy of History of Hegel and Hegelians, 1896; Philosophy of History of Sociology, 1897. Regarded society as an orgn. in which progress is determined by power of ideas. Died Leipzig, Germany, Sept. 30, 1922.

BARTH, Rudolf, biologist; b. Dortmund, Germany, Mar. 30, 1913; s. Christian Friedrich and Lina (Scheffen) B.; student U. Berlin, U. Bonn (Germany), 1932-37; Ph.D., U. Berlin, 1937; m. Lilli Meyer, Feb. 26, 1938; children—Ortrud Monika (Mrs. Hermann Schatzmayr), Ulrich. Asst. U. Bonn., 1936-38; regional chief vegetal san. control Ministry Agr., 1938-48; sci. cons. CELA, Boehringer de Ingelheim, Germany, 1948-50; scientist Inst. Oswaldo Cruz, 1950—, Research Inst. Brazilian Navy, Rio de Janeiro, 1960—; chief entomol. sect. Institute Oswaldo Cruz, 1964—; prof. embryology and histology U. Rio de Janeiro, 1956-58; chief scientist biol. groups Research Inst. of Brazil Navy and Research Vessel Almirante Aldanha, 1961—. Recipient medal of merit Oswaldo Cruz, Ministry of Health Brazil, 1959, Carlos Chagas, U. Belo Horizonte, 1959, Tamandare, Ministry of Navy, Brazil, 1966. Mem. Brazilian Acad. Sci. Brazilian Entomol. Assn., Brazilian Biol. Assn., Brazilian Cartographical Assn. Author: The Life of Animals, 1953; The Wildlife of the Itatiaia, 1960; General Entomology, 1967; also numerous articles. Research on composition and function of animal glands, application of optical methods in biology, biol. indicators of water masses in S. Atlantic, relation between density of life and phys.-chem. factors in ocean. Home: 285, Rua Uca, Rio de Janeiro, Ilha do Governador. Office: Instituto Oswaldo Cruz, Rio de Janeiro, Caixa postal 926, Brazil.*

BARTH, Tom Fredrik Weiby, Norwegian geochemist; b. Bolsoy, Norway, May 18, 1899; s. T. F. W. and Hanna (Kaurin) B.; Dr.Phil., U. Oslo (Norway), 1927; Dr.Phil. (hon.), U. Copenhagen (Denmark), 1950, U. Nancy (France), 1960, U. Kiel (Germany), 1965, U. Zurich, U. Liège (Belgium); 1967; m. Randi Thomassen, Dec. 16, 1922; children—Tone (Mrs. V. Hveding), T. Fredrik W. Asst. U. Oslo, 1924-27, prof. geochemistry, 1936-46, 49—, dir. Mineral.-Geol. Mus., 1949—; asst. prof. Technische Hochschule, Berlin, 1927-28, U. Leipzig, 1928-29; staff Geophys. Lab., Carnegie Inst. Washington, 1929-36; prof. U. Chgo., 1946-49. Mem. Internat. Commn. on Geochemistry, 1952—, pres., 1957-60. Decorated Royal Order St. Olav. Mem. Internat. Union Geol. Scis. (pres. 1964—). Author: Volcanic Manifestations, Hot Springs, and Geysers of Iceland, 1950; Theoretical Petrology, 1952; also numerous articles. Research on crystal structures, minerals and rocks. Home: Box 31, Voksenkollen, Norway. Office: 1 Sars, Oslo, Norway.*

BARTHELEMY, Antoine, see Clot Bey.

BARTHÉLEMY, Jean Jacques, French archaeologist; b. Cassis, France, Jan. 20, 1716; studied theology and ancient langs. under Jesuits at Marseilles, France; became keeper Royal Cabinet of Medals, 1753; studied archaeology in Italy, 1754-57. Mem. Acad. Inscriptions, French Acad. Fellow Royal Soc., 1755. Author: Voyage de jeune Anarchis en Grèce, 7 vols., 1788. Research and publs. on numismatics, ancient inscriptions, Greek life during Golden Age. Died Paris, Apr. 30, 1795.

BARTHÉLEMY, René, French electronic inventor; b. Nangis, France, Mar. 10, 1889; dir. Montrouge Center for TV Research; mem. French Acad. Scis. Gave 1st demonstration of TV in France, 1931; inventor isoscope in which light directly modulates high frequency. Died Antibes, France, Feb. 12, 1954.

BARTHÉLEMY, Toussaint, French physician; b. Nancy, France, Dec. 11, 1850; studied medicine, Nancy, then Paris; M.D., 1880; worked at Paris hosps. Lourcine, St. Louis, St. Lazare; mem. Soc. Legal Medicine; founded Soc. Dermatology. Author: Traité de Dermatologie, 1883, Syphilis et Santé Publique, 1890; Étude sur le Dermographisme, 1892. Expert on syphilis research; took X-ray pictures of hand and thorax. Died 1906.

BARTHEZ, Paul-Joseph, French physician; b. Montpellier, France, Dec. 11, 1734; student Narbonne, Toulouse; doctorate in medicine, Montpellier, 1753. Became physician to mil. hosp., Normandy, France, 1756; named cons. physician med. staff Army of Westphalia, 1757; apptd. prof. medicine, Montpellier, 1759, joint chancellor, 1774; named cons. physician to king, 1780. Mem. French Acad. Scis. Author: Nova doctrina de functionibus naturae humanae, 1774; Nouveaux eléments de la science de l'homme, 1778; Nouvelle méchanique des mouvements de l'homme et des animaux, 1798; Discours sur le génie d'Hippocrate, 1801; La théapeutique de fluxions, 1802; La colique iliaque chronique, 1802; Traité des maladies gotteuses, 1802; Traité du beau, 1807; Consultations de médecine, 1810. Joint editor Journal des savants, Encyclopédie méthodique. Introduced belief that nonmaterial life force is responsibile for all living things and is distinct from physico-chemical changes in the body and from the mind (vitalism principle) 1778. Died Oct. 15, 1806.

BARTHOLIN, Caspar Berthelsen, physician; b. Malmö, Sweden, Feb. 12, 1585; studied at U. Copenhagen (Denmark), also Rostock and Wittenberg, Germany; children—Thomas, Erasmus. Apptd. prof. medicine, U. Copenhagen, 1613, became rector, 1618, later prof. divinity; canon of Roskilde. Author: Institutiones anatomicae, 1611; Opuscula quatuor singularia, 1628; also treatises on green jasper, pygmies and medicine. Described single horned animals. Died Sorö, Zeeland, Netherlands, July 13, 1629.

BARTHOLIN, Caspar Thomèson, Danish anatomist, physician; b. Copenhagen, Denmark, Sept. 10, 1655; s. Thomas B.; studied at U. Copenhagen; prof. of anatomy there until 1701; physician to King of Denmark. Author of several med. and sci. publs., including De ovariis mulierum, 2 vols, 1675-78; De ductu salivati, 1685. Discovered pair of vaginal glands (Bartholin's glands), also a sublingual duct (Bartholin's duct). Died June 11, 1738.

BARTHOLIN, Erasmus, Danish physician, physicist; b. Roeskilde, Denmark, Aug. 13, 1625; s. Caspar Bartholin the Elder. Prof. geometry, medicine Copenhagen, Denmark. Author: De Cometis Annorum, 1664; Opusculum, ex Observationibus Hasniae Habitis Adornatum, 1665; Experimenta crystalli islandici disdiaclastici, 1669; De Naturae Mirabilibus Quaestiones Academicae, 1674. Discovered double refraction of Islandic feldspar (calcite), this discovery was in contradiction to Newtonian theory of light which was presented at same time.

BARTHOLINUS, Thomas, Danish anatomist, physician; b. Copenhagen, Denmark, Oct. 20, 1616; s. Caspar the Elder; studied at Copenhagen, Leiden (Netherlands), Paris (France), Montpellier (France); M.D., Basle (Switzerland), 1645; mem. Collegium Medicum, 1645, perpetual dean, 1654; children—Caspar, Thomas. Prof. math. U. Copenhagen, 1647, prof. anatomy, 1648-61; later librarian; apptd. King's physician, 1670. Author: Anatomia, 1640; De luce animalium, 1647; Historiarum anatomicarum et medicarum centuriae, 1654-1660; De medicana Danorum domestica, 1666; De morbis biblicis, 1672; Acta medica et philosophica Hafniensia, 1672-79. First to describe entire lymphatic system, 1653; confirmed existence of thoracic duct, 1652. Died Hagestedgaard, Denmark, Dec. 4, 1680.

BARTHOLOMEW, George Adelbert, Jr., zoologist; b. Independence, Mo., June 1, 1919; s. George A. and Esther (Carstensen) B.; A.B., U. Cal. at Berkeley, 1940, M.A., 1941; Ph.D., Harvard, 1947; m. Elizabeth Burnham, Nov. 7, 1942; children—Karen Elizabeth, Bruce Monroe. Faculty, U. Cal. at Los Angeles, 1947—, professor of zoology, 1959—; research on physiology and behavior desert vertebrates and marine birds and mammals. Mem. Am. Soc. Zoologists, Am. Soc. Mammalogy, Ecol. Soc. Am., Am. Ornithologists Union. Studies on water economy and temperature regulation of reptiles, birds, and mammals; on ecology of desert vertebrates; on social behavior of marine mammals and birds. Home: 551 W. Rustic Rd., Santa Monica, Cal. Office: University of California, Los Angeles.

BARTHOLOMEW, John George, Scottish geographer; b. Edinburgh, Scotland, Mar. 22, 1860; s. John B.; ed. Edinburgh Univ., LL.D.; m. Jennie Macdonald; 2 sons, 2 daus. Geographer and cartographer to the King; head, Edinburgh Geographical Institute. Fellow, Royal Soc., Edinburgh (on Council, 1909-12); Royal Geographical Soc. (Victoria Research Gold Medal, 1905); Royal Scottish Geographical Soc. (founder and hon. sec., 1884); sec., Section E, British Assn., 1892; hon. mem. Geographical Socs. Paris, Chicago, St. Petersburg. Author: Survey Atlas of Scotland, 1895-1912; Citizen's Atlas, 1898-1912; Atlas of Meteorology, 1899; Survey Atlas of England and Wales, 1903; Survey Gazetteer of British Isles, 1904; Atlas of World's Commerce, 1907; Imperial Indian Gazetteer Atlas, 1908; Atlas of Zoogeography, 1911. Introduced layer contour coloring in topography mapping; developed special maps and educational atlases. Died Apr. 16, 1920.

BARTHOLOMEW, Lloyd Gibson, Am. physician; b. Whitehall, N.Y., Sept. 15, 1921; s. Emerson F. and Minnie (Swinton) B.; Asso. B.A., Green Mountain Jr. Coll., 1939; B.A., Union Coll., 1941; M.D., U. Vt. 1944; M.S. in Medicine, U. Minn., 1952; m. Elisabeth Sheldon Thrall, Dec. 27, 1943; children—Suzanne (Mrs. Michael Garvey), Lynne, Lloyd Gibson, Deborah, Douglass Thrall. Asst. medicine Med. Sch., Dartmouth, 1948-49; cons. physician Mayo Clinic, Rochester, Minn., 1952—; faculty Mayo Grad. Sch. Medicine, U. Minn., Rochester, 1958—, asso. prof. medicine, 1963-67, professor of medicine, since 1967—; head section gastroenterology Mayo Clinic and Grad. Sch. Medicine, 1967—. Recipient Woodbury prize in medicine Carbee prize in obstetrics, 1944. Mem. A.M.A. (sec. gastroenterology 1962—, Billings award 1963), So. Minn. Med. Assn. (chmn. gifts and awards com. 1959—, pres. 1963-64), Am. Gastroent. Assn., Am. Assn. for Study Liver Diseases, N.Y. Acad. Sci., Sigma Xi. Contbg. Author: Current Concepts of Clinical Gastroenterology, 1965; Clinical Diagnosis by Laboratory Methods, 1963; Cancer of the Stomach, 1964. Research, numerous publs. on diseases of liver and pancreas. Home: 1201 6th St. S.W., Rochester, Minn. 55901.*

BARTHOLOMEW OF BRUGES, physician; flourished 14th century; M.A., Paris, France, 1307; M.D., U. Montpellier, France, before 1315; lectr. U. Paris, 1307-09; canon Andenne, Cambrai, 1331; physician to Guy I of Châtillon, circa 1330-42. Author: Notule in Ysagogas Johannitii; Glossule in libros Aphorismorum Ypocratis; Dicta super Prognostica; Scriptum super primum Canonis Avicennae; Remedium epydimie. Died circa 1354.

BARTHOLOMEW OF PARMA, Italian astrologer; b. Parma, Italy, flourished, 1286; wrote many astrological treatises, including Liber de occultis, 1280; Breviloquium astrological, 1286; Breviloquium or Ars geomantiae, 1288; Tractatus sphaerae, 1297. Died 1297.

BARTHOLOMEW OF SALERNO (Bartholomaeus Salernitanus); flourished 1st half 12th century. Author of a Practica, treatise on pathology and therapeutics popular in Western Europe; may have authored booklet on distilled waters of which a German transl. or adaptation is generally ascribed to Michael Puff (2d half 15th century).

BARTHOLOMEW THE ENGLISHMAN (Bartholomaeus Anglicus de Glanvilla), English physician; b. Suffolk, Eng.; flourished circa 1250; ed. at Oxford, Eng.; went to Paris, France, circa 1220, as lector; then to Magdeburg, circa 1230. Author: De proprietatibus rerum (ency. of natural history, including all known branches of med. sci., astronomy, geography).

BARTHOLOMEW, Tracy, Am. engr.; b. Austin, Tex., Nov. 14, 1884; s. George Wells and Hettie Julia (Cole) B.; student Ohio State U., 1902-03; E.M., Colo. Sch. Mines, 1906; m. Sarah Jane Anderson, Oct. 6, 1921; children—George Anderson, Jane Anderson. Constrn. engr. Fed. Lead Co., Flat River, Mo., 1906-07; designing and test engr. Nev. Consol. Copper Co., McGill, Nev., 1907-09; gen. mgr. Alkali-Proof Cement div. Colo. Portland (now Ideal) Cement Co., Denver, 1909-11; mgr. Rico Tropical Fruit Co., Garrochales, P.R., 1911-21, pres., from 1921; sr. fellow Mellon Inst. of Indsl. Research, Pitts., 1921-39; mgr. of research Duquesne Slag Products Co., 1929-40; cons. engr., from 1940. Mem. A.A.A.S., Am. Inst. Mining and Metall. Engrs., Am. Soc. C.E., Am. Chem. Soc., Am. Ceramic Soc., Am. Soc. Municipal Engrs., Am. Soc. Testing Materials, Am. Concrete Inst., Engrs. Soc. of Western Pa. Died Dec. 7, 1951.

BARTHOLOW, Roberts, Am. physician; b. Howard County, Md., Nov. 28, 1831; grad. in arts, Calvert Coll., Md.; M.D., U. Md., 1852; LL.D., St. Mary's Coll. Prof. theory and practice of medicine and clin. medicine, dean of faculty Ohio Med. Coll., Cin., 1864-79; emeritus prof. materia medica, gen. therapeutics and hygiene, Jefferson Med. Coll., Phila., from 1879. Royal Med. Soc. of Edinburgh (Scotland). Author: Qualifications for the Military Service; Hypodermic Medication; Treatise on Materia Medica and Therapeutics; Practice of Medicine; Medical Electricity; also many papers, essays. First to apply electrodes to human cortex and to demonstrate contralateral muscular contractions, 1874. Died 1904.

BARTISCH, Georg, German surgeon; b. Königsbrück, Germany, 1535; itinerant surgeon from age 13; apptd. court oculist to Duke August of Saxony. Author: Augendienst (1st German book on ophthalmology), 1583. Specialized in diseases of the eye, eye operations; credited with being 1st to excise the eye in cancer cases, 1583. Died probably Dresden, 1606.

BARTLE, Robert G(ardner), Am. mathematician; b. Kansas City, Mo., Nov. 20, 1927; s. Glenn G. and Wanda (Mittank) B.; B.A., Swarthmore Coll., 1947; S.M., U. Chgo., 1948, Ph.D., 1951; m. Doris Marie Sponenberg, Oct. 6, 1951; children—James, John. AEC postdoctoral fellow Yale, 1951-52, instr., 1952-55; faculty U. Ill., Urbana, 1955—, prof. 1964—; vis. scholar U. Cal., Berkeley, 1961-62. Mem. Phi Beta Kappa, Sigma Xi. Author: The Elements of Real Analysis, 1964; The Elements of Integration. Editor, Ill. Jour. Math, 1963—. Research, publs. in functional analysis. Home: 712 W. Park St., Champaign, Ill. 61820.*

BARTLETT, Albert Allen, Am. physicist; b. Shanghai, China, Mar. 21, 1923 (parents Am. citizens); s. Willard William and Marguerite (Allen) B.; student Otterbein Coll., 1940-42; B.A., Colgate U., 1944; M.A., Harvard, 1948, Ph.D., 1950; m. Eleanor Frances Roberts, Aug. 24, 1946; children—Carol Louise, Jane Elizabeth, Lois Jeanne, Nancy Marie. Research asst. Los Alamos Sci. Lab., 1944-46; faculty U. Colo., Boulder, 1950—, prof., physics, 1962—; faculty Harvard Summer Sch., 1952, 53, 55, 56; vis research worker Nobel Inst. Physics, Stockholm, Sweden, 1962-63. Mem. Am. Phys. Soc., Am. Assn. Physics Tchrs., Sigma Xi. Research, publs. on design and operation of high luminosity beta ray spectrometer and nuclear spectroscopic studies; design and devel. teaching apparatus for student labs. Home: 2935 19th St., Boulder, Colo. 80302.*

BARTLETT, Clarence, Am. physician; b. Bklyn., May 22, 1858; s. William F. and Margaret (Ritter) B.; M.D., Hahnemann Med. Coll. and Hosp., 1879; m. Anna C. Miller, Sept. 29, 1885 (dec. Nov. 1910); m. 2d, Mrs. Mary G. Wright, Apr. 4, 1912. Practiced medicine, Phila., from 1879; prof., head dept. medicine Hahnemann Med. Coll. and Hosp. Mem. Alumni Assn. Hahnemann Med. Coll. (pres.), Am. Inst. Homeopathy (hon. pres. 1923), Pa. Homeo. Soc., Phila. Acad. Medicine; hon. mem. N.Y., Va., Pa. (pres. 1922) homeo. med. socs., Brit. Homeo. Soc., others. Author: Clinical Medicine—Diagnosis, 1903; Clinical Medicine—Treatment, 1904; Practice of Medicine, 3 vols. 1923. Editor Hahnemannian Monthly. Deceased.

BARTLETT, Elisha, Am. physician; b. Smithfield, R.I., Oct. 6, 1804; s. Otis and Waite (Buffum) B.; grad. in medicine Brown U., 1826; m. Elizabeth Slater, 1829. Settled in Lowell, Mass., 1827, practiced medicine; prof. anatomy Berkshire Med. Instn., Pittsfield, Mass., 1832-40; an editor Med. Mag., Boston, 1832-35; mayor of Lowell, Mass., 1836; prof. medicine Transylvania U., Lexington, Ky., 1841, 46, U. Md., Balt., 1844-45, U. Louisville, 1849-50, N.Y.U., 1850-52, Coll. Phys. and Surg., N.Y.C., 1852-55. Author: Fevers in the United States 1842; History, Diagnosis, and Treatment of Edematous Laryngitis, 1850. Died Smithfield, July 19, 1855.

BARTLETT, Sir Frederic Charles, Brit. psychologist; b. Stow-on-the-Wold, Eng., Oct. 20, 1886; s. William Bartlett; B.A., St. John's Coll., Cambridge, 1909; M.A., London U., 1911; hon. doctorates Athens, 1937, Princeton, 1947, Louvain, 1949, Edinburgh, 1961, Oxford, 1962; m. Emily Mary Smith; 2 sons. Became asst. to dir. psychol. lab. Cambridge U., 1914, univ. reader in exptl. psychology, 1922, prof. exptl. psychology, 1931-52, dir. psychol. lab., 1922-52; editor Brit. Jour. Psychology, 1924-48. Huxley lectr., Birmingham, Eng., 1957-58. Recipient Huxley medal, 1943, Baly medal, 1943, Longacre award, 1952. Fellow Royal Soc., 1932, Brit. Psychol. Soc. (pres. 1950, hon. fellow); mem. Brit. Assn. (pres. sect. J 1929); hon. mem. or fgn. asso. Spanish, Swiss, Swedish, Exptl. psychol. socs., Am. Nat. Acad. Scis., Am. Acad. Arts and Scis., Internat. Assn. for Applied Psychology. Author: Exercises in Logic, 1913; Psychology and Primitive Culture, 1923; (with C. S. Myers) Text-Book of Experimental Psychology, Part II, 1925; Psychology and the Soldier, 1927; Remembering: An Experimental and Social Study, 1932; The Problem of Noise, 1934; Political Propaganda, 1941; The Mind at Work and Play, 1951; Thinking: An Experimental and Social Study, 1958; also papers. Editor: (with others) The Study of Society, 1939. Leader in exptl. psychology in Britain; devised numerous machines for testing servicemen during World War II. Home: 161 Huntington Rd., Cambridge, Eng.

BARTLETT, James Holly, Am. physicist; b. Bklyn., Nov. 2, 1904; s. James H. and Martha Maude (Walker) B.; B.C.E., Northeastern U., 1924; A.M., Harvard, 1926, Ph.D., 1930; m. Vera May Brothers, Dec. 26, 1932; children—Anne Holly, Jane Louise. Asst. prof. physics U. Ill., Urbana, 1930-36, asso. prof., 1936-45, prof., 1945. Cons. Lockheed Corp., 1958-59; sci. adviser USAAF, 1945-46; Exchange prof. U.S.S.R., 1961, 63. Rockefeller Found. fellow, 1940-41. Fellow Am. Phys. Soc., Electrochem. Soc. Research in nuclear shell model, 1932, exchange force, 1937, electrochem. potentiostat, 1941, restricted 3 body problem, 1964. Research, publs. on fast electrons, corrosion. Address: P.O. Box 1921, University, Ala. 35486.*

BARTLETT, Kenneth Alden, Am. entomologist; b. Dorchester, Mass., Mar. 26, 1907; s. William G. and Maude P. (Hutchinson) B.; B.S., Mass. Agrl. Coll., 1928; M.S., Harvard, 1929, Ph.D., 1931; m. Juanita F. Cedo, Dec. 14, 1954; children—Donald W., Kenneth Alden. Entomologist, U. S. Dept. Agr., Arlington, Mass., Mayaguez, P.R., 1931-41, acting dir. Fed. Expt. Sta., Mayaguez, 1941-43, dir., 1943-53; pres. V.I. Corp., Christiansted, St. Croix, V.I., 1953-62; exec. dir. Am. Freedom from Hunger Found., Washington, 1964-65. Mem. Lambda Chi Alpha. Publs. on insect parasites and fungus diseases of insects. Home: 1400 20th St., N.W., Washington.*

BARTLETT, Maurice Stevenson, Brit. statistician; b. Chiswick, June 18, 1910; s. William S. and Eva (White) B.; ed. Queen's Coll., Cambridge; M.A.; Ph.D. in Sci.; m. Sheila Rosemary Chapman, 1957. Statistician, Imperial Chem. Industries, 1934; lectr. math. Cambridge U., 1938; prof. math. statistics U. Manchester, 1947-60; prof. statistics London U., College, 1960-67; prof. bio-math. Oxford U., 1967—; Fellow Royal Soc., 1961, Royal Manchester Statis. Soc. (pres. 1966-67), Internat. Statis. Inst., Biometric Soc. Author: Introduction to Stochastic Processes; Stochastic Population Models in Ecology and Epidemiology; Essays in Probability and Statistics. Studies on statistical theory, methodology and biometrics. Home: 3 Southwood Hall, Highgate N.6. Office: Dept. Biomath., U. Oxford, Oxford, Eng.

BARTLETT, Neil Riley, Am. psychologist; b. Underhill, Vt., Aug. 10, 1917; s. Arthur N. and Elizabeth (Wall) B.; B.S., U. Vt., 1937; M.Sc., Brown U., 1939, Ph.D., 1941; m. Susan Carson, Aug. 22, 1942; children—David, Robert, William, Thomas. Research psychologist Brown U., 1941-43; asst. prof. Johns Hopkins, 1946-48; prof. Hobart Coll., Geneva, N.Y., 1948-58; prof., head dept. psychology, U. Arizona, Tucson, 1958—. Cons. USAF, 1951-58; physiol. psychologist USN Electronic Lab., San Diego, summers 1959—. Nat. Inst. Mental Health Fellow, Université de Paris, 1965. Mem. Phi Beta Kappa, Sigma Xi. Research on psychophysiological optics and on time factors in perceptual and motor responses. Office: Psychology Dept., U. Ariz., Tucson. 85721.*

BARTLETT, Paul Doughty, Am. chemist; b. Ann Arbor, Mich., Aug. 14, 1907; A.B., Amherst Coll., 1928, hon. Sc.D., 1953; A.M., Harvard, 1929, Ph.D. in chemistry, 1931; Sc.D., U. Chgo., 1954; m. 1931; 3 children. Nat. research fellow chemistry Rockefeller Inst., 1931-32; instr. U. Minn., 1932-34; faculty Harvard, from 1934, asst. prof., 1937-40, asso. prof. 1940-46; apptd. prof., 1946. Ofcl. investigator Nat. Defense Research Com., 1941-45. Recipient A.C.S. award, 1938, Roger Adams award, also Willard Gibb medal, 1963 (all Am. Chem. Soc.), A. W. von Hofmann medal German Chem. Soc. Mem. Nat. Acad. Scis., Am. Acad. Arts and Scis. Author: Nonclassical Ions, 1965. Demonstrated (with D. S. Tarbell) halogenation of double bond is 2-step process, 1936; synthesized (with A. Schneider) tri-tertiary-butyl carbinol; contbd. to understanding of indsl. reaction of paraffin alkylation as an ionic sequence; first to treat interconversion of molecular forms of elemental sulfer as problem in reaction mechanisms of ring compounds. Address: Harvard U., Cambridge, Mass.*

BARTLETT, Robert Abram, explorer; b. Brigus, Nfld., Aug. 15, 1875; s. William James and Mary J. (Leamon) B.; ed. Meth. Coll., St. Johns, Nfld.; passed exam. for Master of British Ships, Halifax, N.S., 1905; hon. A.M., Bowdoin Coll., 1920. Began explorations wintering with R. E. Peary, at Cape D'Urville, Kane Basin, 1897-98; comd. the Roosevelt, 1905-09, taking active part in Peary's expdn. to the pole, reaching 88th parallel; with Can. Govt. Arctic Expdn., 1913-14, as captain of the C.G.S. Karluk; sent by Nat. Geog. Soc. to locate bases for aircraft, N.W. Alaska, and shores Arctic Ocean, also recording times and currents and dredging for flora and fauna, 1925;

many other expeditions to Greenland, Baffin Land, Ellesmere Land, Siberia, and Labrador, 1926-41. Recipient Hubbard gold medal Nat. Geog. Soc., 1909; Hudson-Fulton silver medal, 1909; silver medal English Geog. Soc., 1910; Kane medal, Phila. Geog. Soc., 1910; silver medal, Italian Geog. Soc., 1910; awarded Back Grant, in recognition of splendid leadership after the 'Karluk' was lost, Royal Geog. Soc., 1918; gold medal Am. Geog. Soc. Hon. mem. Soc. of Dorset Men in London, Eng., Boy Scouts Am.; life mem. Am. Mus. Natural History; Am. Geophys. Union, corr. mem. Am. Geog. Soc. Author: Last Voyage of the Karluk, 1916; The Log of Bob Bartlett, 1928; Sails over Ice, 1934. Died Apr. 28, 1946.

BARTLETT, William H. C., Am. astronomer; b. Lancaster, Pa., Sept. 1804; student U. S. Mil. Acad., 1822-26; A.M., Princeton, 1837; LL.D., Hobart Coll., 1847; m. Harriet Whithorne, Feb. 4, 1829; 8 children. Second lt. C.E., U. S. Army, advanced through grades to col., ret., 1871; asst. engr. constrn. Ft. Monroe, 1828, Ft. Adams, 1829-32; asst. to chief engr., Washington, 1832-34; prof. natural and exptl. philosophy (set up obs.) U. S. Mil. Acad., 1836-71; actuary Mut. Life Ins. Co. N.Y. after 1871. Mem. Am. Acad. Arts and Scis., Am. Philos. Soc., Nat. Acad. Scis. (corporator 1863). Author textbooks on acoustics, optics, astronomy, mechanics, molecular physics. Made observations of comets; 1st Am. to use photography in astronomy. Died Feb. 11, 1893.

BARTLEY, Samuel Howard, Am. physiol. psychologist; b. Pitts., June 19, 1901; s. Edward G. and Mary (Byers) B.; B.S., Greenville Coll. 1923; M.A., U. Kan., 1928, Ph.D., 1931; m. Leola S. Bevis, June 25, 1938; children—Samuel Howard, Jeanne A., K. Joyce (Mrs. Dale J. Sherman). Instr., U. Kan., 1926-31; research asso. Washington U. Med. Sch., St. Louis, 1931-42; faculty Dartmouth Med. Sch.; prof. psychology Mich. State U., East Lansing, 1947—. Recipient Distinguished Prof. award Mich. State U., 1960, sr. award for meritorious research Sigma Xi, Mich. State U., 1962. Mem. Optical Soc. Am., Am. Psychol. Assn., Am. Physiol. Soc., Am. Acad. Optometry. Author: Vision, A Study of its Basis, 1941; Beginning Experimental Psychology, 1950; Fatigue and Impairment in Man, 1947; Principles of Perception, 1958; The Mechanism and Management of Fatigue, 1965. Described activity of brain portion of visual system; research on nature of human fatigue. Home: 348 Cowley Av., East Lansing, Mich. 48823.

BARTLING, Friedrich Georg (Gottlieb), German botanist; b. Hannover, Germany, Dec. 9, 1798; s. Heinrich Ludwig and Dorothea Juliana (Seeger) B.; ed. Göttingen, Germany; m. Horn (dec. 1845); 1 dau.; m. 2d, Eggeling; 4 daus., including Jenny (Mrs. Hall), Mary (Mrs. Macfadyan). Became prof., Göttingen, 1831; apptd. dir. bot. garden, 1837. Author (with H. Wendland) Beiträge zu Botanik, 1824-5; Flora der österreichischer Küstenländer, 1825; Ordines naturales plantarum, 1830. Set up new flora families and classified them; divided Cryptogamoea into Phaenogamous plants, cormophytes and thallophytes. Died Göttingen, Nov. 19, 1875.

BARTMANN, Karl, German physician; b. Hamm, Germany, June 18, 1920; s. Ferdinand and Maria (Boese) B.; student U. Leipzig (Germany), 1939, U. Rostock (Germany), 1940-42, U. Marburg (Germany), 1942-44; m. Gisela Zunkel, Jan. 8, 1947; children—Stefan, Dominik. Staff, II Medizinische Universitaetsklinik der Charité, Berlin, 1945-49, Pharmakologisches Institut, Freie Universität Berlin, 1949-52, Robert Koch Institut, Berlin, 1953-55; dir. central lab. Staedtische Klinik fuer Lungenkranke Heckeshorn, Berlin, 1955-67; head Inst. für Medizinische Mikrobiologie, Farbenfabriken Bayer, Wuppertal, 1967—; faculty medicine Freie U. Berlin, 1962—, apl. prof., 1967—. Recipient Franz Redeker Preis, 1961, Robert Koch Preis, 1966. Mem. Internat. Union Against Tb (chmn. com. on prophylaxis), Wissenschaftliche Arbeitsgemeinschaft fuer Therapie von Lungenkrankheiten, Deutsche Gesellschaft fuer Tuberkulose und Lungenkrankheisen, Paul-Ehrlich-Gesellschaft für Chemotherapy (sec.). Author: Isoniazid, Möglichkeiten und Grenzen, 1963; also numerous articles. Research on chemotherapy, especially Tb, prophylaxis for Tb by vaccination and drugs, bacteriology of Tb and nontuberculous chest diseases. Home: Julius-Lucas Weg. Office: Friedrich Ebert Strasse 217-319, 56 Wuppertal-Elberfeld, West Germany.*

BARTOLETTI, Fabritio, Italian physician; b. Bologna, Italy, Aug. 26, 1588; Ph.D., M.D., Bologna, 1613; became prof., Bologna, then at Pisa, Italy, 1620, Mantua, Italy, 1626. Author: Encyclopaedia hermetico-dogmatica . . . , 1619; Antidotarium chimico-dogmaticum; Methodus in dyspnoeam . . . , 1633. Credited with exposing hermitic impostures on temperaments, 3 principles, humors; discussed salt of antimony; extracted balsamic mercury from silver and gold; criticized Paracelsean dogmas of relations of parts of body to planets, salt, sulphur mercury theory; prescribed chem. remedies; described preparation of recrystallized milk sugar. Died Lendinara, Italy, Mar. 30, 1630.

BARTOLI, Adolfo, Italian physicist; b. Florence, Italy, Mar. 19, 1851; began studies at Pisa (Italy), 1874; degree U. Bologna (Italy); became prof. physics U. Sassari, 1878; named prof. Florence (Italy) Tech. Inst., 1879; apptd. prof., dir. obs., U. Catania, 1886; later became prof. at Pisa. Author: Sui movimenti provocati dalla luce e dal calore. Sulla costituzione degli elettroliti, 1882. Studied specific heat of water; by showing existence of radiation pressure, he confirmed Maxwell's theory that light exerts a certain pressure when reflected or absorbed, 1876; showed that a weak electric current using gold or platinum electrodes could dissassociate water, 1878; proposed Clausius-Bartoli hypothesis (an electrolyte contains some partly and entirely dissociated molecules). Died Pavia, Italy, July 18, 1896.

BARTOLI, Daniello, Italian physicist; b. Ferrara, Italy, Feb. 12, 1608; trained by Jesuits; entered Soc. Jesus, 1623; named Jesuit historiographer, 1650. Author: History of the Society of Jesus, 6 vols.; Del souno de tremori armonici e dell' udito, 1679; Della tensione e della pressione, 1677; Del ghiaccio e della coagulazione, 1681. Tried to reconcile methods of the speculative sch. with Galileo's expts.; studied effects of sound in air and water, also tension, pressure, ice, freezing. Died Rome, Italy, Jan. 12, 1685.

BARTOLOMMEO DA VARIGNANA, see da Varignana, Bartolommeo.

BARTON, Benjamin Smith, Am. naturalist, physician; b. Lancaster, Pa., Feb. 10, 1766; s. Thomas and Esther (Rittenhouse) B.; attended York Acad., Lancaster; M.D., U. Göttingen (Germany), 1789; studied medicine in England, 1786; m. Mary Pennington, 1797, 2 children including Thomas P. Became mem. Royal Med. Soc.; practiced medicine, Phila., 1789; prof. botany U. Pa., 1790-1813, prof. medicine, 1813-15. Author: Elements of Botany (1st elementary botany text by an American), 1803; Collections for an Essay Towards a Materia Medica of the United States (systematic treatise on medicinal plants), 1798-1804. Founder, editor Phila. Med. and Phys. Jour., 1805-08. Died Phila., Dec. 19, 1815.

BARTON, Derek Harold Richard, Brit. organic chemist; b. Gravesend, Eng., Sept. 8, 1918; s. William Thomas and Maude (Lukes) B.; B.Sc. with 1st Class Honors, Imperial Coll., Eng., 1940, Ph.D. in Organic Chemistry, 1942; D.Sc., U. London, 1949; D.Sc. (hon.), U. Montpellier (France), 1962, U. Dublin (Ireland), 1964; 1 son, William Godfrey Lukes. Research chemist on govt. project, 1942-44, Messrs. Albright and Wilson, Birmingham, Eng., 1944-45; asst. lectr. dept. chemistry Imperial Coll., 1945-46, Imperial Chem. Industries research fellow, 1946-49, prof., organic chemistry, 1957—; vis. lectr. chemistry natural products Harvard, 1949-50; faculty Birbeck Coll., 1950-55, prof. organic chemistry, 1953-55; Regius prof. chemistry U. Glasgow (Scotland), 1955-57; Arthur D. Little vis. prof. Mass. Inst. Tech., 1958; Karl Folkers vis. prof. U. Ill., also U. Wis., 1959. Recipient Hofmann prize Imperial Coll., 1940, Harrison Meml. prize Chem. Soc., 1948, Fritzsche medal Am. Chem. Soc., 1956, 1st Roger Adams medal, 1959. Fellow Royal Soc., 1954 (Davy medal 1961), Royal Soc. Edinburgh; mem. Am. Acad. Arts and Scis. (fgn. hon.). Contbr. numerous articles to Jour. Chem. Soc. Research in steroid and terpene fields led to opening of field of conformational analysis in organic chemistry, 1949; stated theory that structures of many phenols and alkaloids could be predicted which aided in understanding biosynthesis of many complex alkaloids, 1956; contbd. to pyrolysis of organic chlorides, 1945-52, devel. and applications of carbanion autoxidations, after 1960; pioneer in study of relationship of molecular rotation to structure in complex organic molecules. Home: 1A Grove Rd., Northwood, Middlesex, Eng. Office: Chemistry Dept. Imperial Coll., London S.W. 7, Eng.

BARTON, Donald Clinton, Am. geologist; b. Stow, Mass., June 29, 1889; s. George Hunt and Eva May (Beede) B.; A.B., Harvard, 1910, A.M., 1912, Ph.D., 1914; m. Margaret Dunbar Foules, June 26, 1923; 1 dau., Ann Foules. Instr. engring. geology Washington U., 1914-16; field geologist Empire Gas and Fuel Co., 1916-17; geologist Gulf Coast div. Amerada Petroleum Corp., 1919-23; chief geologist Rycade Oil Corp., 1923-27; chief Torsion Balance and Magnetometer div. Geophysical Research Corp., 1925-27; cons. geologist and geophysicist, 1927-34; research and cons. geologist and geophysicist Humble Oil and Refining Co., 1935—. Contbr. many papers on geology and geophysics; devel. torsion balance survey technique which led to discovery of Nash Dome in Texas. Died July 8, 1939.

BARTON, Edwin Henry, Brit. physicist; b. Oct. 23, 1858; s. John and Eliza (Lake) B.; ed. People's Coll. and Univ. Coll., Nottingham, Eng., also Imperial Coll. Sci., London, Eng., U. Bonn (Germany); D.Sc.; m. Mary Stafford, 1894; 2 sons. Various positions, including draftsman, engring. works, Nottingham; became mem. staff phys. dept. Univ. Coll., Nottingham, 1893, sr. lectr. and demonstrator, 1895, prof. exptl. physics, from 1905. Fellow Royal Soc., 1916, Royal Soc. Edinburgh, Inst. Physics. Author: A Text-Book of Sound, 1908; Analytical Mechanics, 1911; An Introduction to the Mechanics of Fluids, 1915; (with T. P. Black) Practical Physics for Colleges and Schools, 1912. Contbr. to Sci. Abstracts, Phys. Abstracts, from 1894; also contbr. papers on acoustics, magnetism, elec. waves to jours. Died Sept. 23, 1925.

BARTON, George Hunt, Am. geologist; b. Sudbury, Mass., July 8, 1852; s. George Washington and Mary S. (Hunt) B.; S.B., Mass. Inst. Tech., 1880; m. Eva May Beede, Sept. 18, 1884; children—Harold Beede, Donald Clinton, Helen Mary. Asst. in drawing, Mass. Inst. Tech., 1880-81, asst. in geology, 1883-84, asst. prof., till 1904; asst. on Hawaiian govt. survey, Honolulu, 1881-83; faculty Boston U., till 1904; lectr., 1915; later dir. Tchrs. Sch. Sci.; Lectr. geology Wellesley Coll., 1921-22; asst. geologist U. S. Geol. Survey; mem. 6th Peary expdn. to Greenland, 1896. Fellow Am. Acad. Arts and Scis. Author: Outline of Elementary Lithology, 1900; Outline of Dynamical and Structural Geology; also many geol. papers. Died Nov. 25, 1933.

BARTON, James Moore, Am. surgeon; b. Phila., Oct. 16, 1846; s. James M. and Esther (Rathvon) B.; M.D., Jefferson Med. Coll., 1868; m. Mary E. Craig, Feb. 4, 1908. Resident physician Episcopal Hosp., Phila., 1868-69; asst. to Prof. S. D. Gross, 1869-79; surgeon and prof. clin. surgery Jefferson Med. Coll., 1879-1900. Surgeon to Charity Hosp., 1869-79; lectr. operative surgery Phila. Sch. of Anatomy, 1877-82; surgeon to German Hosp., 1879-86, to Phila. Hosp., 1890-1900. In charge of surg. dept. Phila. Med. Times, 1884-89; asso. editor Sajous' Annual of Universal Med. Scis., 1889-92; author numerous surg. papers and transl. in various med. jours. Sr. fellow Am. Surg. Assn.; fellow Am. Acad. Medicine; mem. Vol. Med. Service Corps. Cons. surgeon to bd. for examining recruits for World War, 1917-18. Died June 28, 1926.

BARTON, John Rhea, Am. surgeon; b. Lancaster, Pa., Apr. 1794; s. William and Elizabeth (Rhea) B.; M.D., U. Pa., 1818; m. Susan Ridgeway. Practiced medicine, Phila., 1818-40; surgeon Phila. Hosp., 1920-22, Pa. Hosp., Phila., 1823; performed pioneer operation in case of anchylosis of hip joint, 1826; known for knowledge and treatment of bone fractures; author paper A New Treatment in a Case of Anchylosis, 1837; chair in his honor established by his widow at U. Pa. Med. Sch. Died Phila., Jan. 1, 1871.

BARTON, Robert Aitken, New Zealand animal scientist; b. Wanganui, New Zealand, Aug. 6, 1921; s. Frazer Burnett and Jeanie Cordiner (Reid) B.; Dip. Agr., Massey Agrl. Coll., 1940; Victoria U., 1944; m. Edna Joyce Eagle, June 18, 1949; children—James Philip, Pamela Joy, Steven Robert. Faculty, Massey U., Palmerston North, New Zealand, 1940—, sr. lectr., 1954-67, reader, 1967—. Collaborator, U. S. Dept. Agr., Beltsville, Md., 1960. Mem. New Zealand Soc. Animal Prodn. (past pres.), New Zealand Assn. Scientists, Inst. Meat (London), New Zealand Inst. Agrl. Sci., New Zealand Soc. Sci. and Tech. Pub., Am. Soc. Animal Sci., Am. Meat Sci. Assn., New Zealand Genetical Soc. Author: Quality Beef Prodn., 1959, Spanish edit., 1966. Editorial bd. World Rev. Animal Prodn., 1965—. Contbr. over 100 articles to profl. jours. Research on growth and devel. of meat animals, body and carcass composition, sheep fat characteristics, influence of environment and heredity on composition of sheep carcasses, muscle fiber dimensions, tenderness and flavor of meat, breeding of sheep and beef cattle for increased productivity, developer methods for rapid determination of body and carcass composition; demonstrator pasture species influences on composition of sheep carcasses, fat qualities and meat flavor; contbr. to concept of anat. constancy. Home: 16 Hardie St., Palmerston North, New Zealand.*

BARTON, Walter Earl, American psychiatrist; b. Oak Park, Illinois, July 29, 1906; s. Alfred J. and Bertha M. (Kalish) B.; A.B., U. Ill., 1928, M.D., 1931; m. Elsa Benson, July 2, 1932; children—John A., Gail M., Paul B. Various positions Worchester (Mass.) State Hosp., 1931-39, acting supt., 1939-40; clin. clk. Nat. Hosp., London, Eng., 1938; supt. Boston State Hosp., 1945-63; clin. prof. psychiatry Boston U. Med. Sch., 1954—; med. dir. Am. Psychiat. Assn., 1963—; lectr. George Washington Med. Sch., 1963—, Georgetown Med. Sch., 1963—. Mem. Baruch Com. Phys. Medicine, 1943-45. Diplomate Am. Bd. Psychiatry and Neurology (dir.). Recipient Nolan D. C. Lewis award Neuropsychiat. Inst. N.J., 1962. Fellow Am. Psychiat. Assn. (pres. 1961-62), A.M.A.; mem. Mass. Psychiat. Soc. (past pres.), Mass. Soc. for Research in Psychiatry (past pres.), Am. (past mem. bd. mgmt.), Mass. occupational therapy assns., Group for Advancement Psychiatry (past pres.), New Eng. Soc. Psychiatry, Royal Medico-Psychol. Assn. (hon.), Indian Psychiat. Soc. (corr.), Phi Beta Pi. Author: (with others) Observations on European Psychiatry, 1961; Administration in Psychiatry, 1962. Research, publs. on rehab. drug treatment of mental illness, ins., manpower devel., drug toxicity. Home: 3254 Arcadia Pl. N.W., Washington 20015. Office: 1700 18th St. N.W., Washington 20009.*

BARTON, William Henry, Jr., Am. engineer; born Balt., July 7, 1893; s. William Henry and Helen E. (Pritchett) B.; B.S. in C.E., U. Pa., 1917, C.E., 1921, M.S., 1923; m. Celia Mason, Aug. 19, 1920. Engring. work U. S. Bur. Public Roads, 1917-18, 19-20; tchr. civil engring., U. Pa., 1920-30, Pa. Mil. Coll., 1930-35, lectr. curator of astronomy Hayden Planetarium, from 1935. Mem. Brit. Astron. Assn., Am. Assn. of Museums, Am. Astron. Soc., Royal Astron. Soc. of Canada, Tau Beta Pi, Sigma Xi. Author: (with L. H. Doane) Sampling and Testing Highway Materials,

1925; (with S. G. Barton) Guide to the Constellations 1928; (with J. M. Joseph) Starcraft, 1935; An Introduction to Celestial Navigation, 1942; Stereopix, 1943. Contbr. articles to Sky mag. Died July 7, 1944.

BARTOW, Edward, Am. chemist; b. Glenham, N.Y., Jan. 12, 1870; s. Charles Edward and Sarah Jane (Scofield) B.; B.A., Williams Coll., 1892; Ph.D., U. Göttingen, 1895, Golden diploma, 1956; D.Sc., Williams Coll., 1923; m. Alice Abbott, Sept. 3, 1895 (dec. May 1951); 1 dau., Virginia. Asst. in chemistry Williams Coll., 1892-94, instr., 1895-97; instr. in chemistry, 1897-99, asso. prof., U. Kan.; 1899-1905; asso. prof. U. Ill., 1905-06, prof. san. chemistry, 1906-20; dir. State Water Survey, 1905-17; chief Water Survey Div., Dept. Registration and Edn., Ill., 1917-20; prof. and head dept. of chemistry and chem. engring. State Univ. of Ia., 1920-40, emeritus, since 1940; research cons. Johns-Manville Corp. 1940-41. Del. 9th Internat. Congress Chemistry, Madrid, 1934, 10th, Rome, 1938; mem. council Internat. chem. Union, 1922-25, 27-30, 33-38, v.p. for U. S. A., 1934-38; sec. Lake Mich. Water Commn., 1908; Commn. on Standards of Water for Interstate Carriers, 1913, 22; sec.-treas. Ill. Water Supply Assn., 1909-17. Recipient Medaille d'Honneur, des Epidémies, d'Argent (France). Mem. Am. Chem. Soc. (dir. 1933, pres. 1936), A.A.A.S., Societé Chim. Industrielle (France), Soc. Chem. Industry (Gt. Britain), Am. Water Works Assn. (trustee, 1913; v.p., 1921; pres., 1922), Am. Inst. Chem. Engrs. (dir. 1923-25, 1936-39), Am. Soc. C.E. (life mem. 1946), Franklin Inst., Am. Pub. Health Assn., Am. Soc. Testing Materials, Am. Pub. Works Assn., Nat. Inst. Social Sci., Acad. Polit. Sci., Am. Inst. Chemists, Spanish Acad. Sci. (corr.), other sci. socs., Sigma Xi. Author 14 vol. report on Ill. waters; also papers relating to field. Asst. editor Chem. Abstracts, 1911—. Died Apr. 12, 1958.

BARTRAM, John, Am. botanist; b. nr. Darby, Pa., Mar. 23, 1699; s. John and Elizabeth (Hunt) B.; m. Mary Morris, Jan. 1723; m. 2d, Ann Mendenhall, Sept. 1729; 11 children including John, William. First native Am. botanist; founded 1st bot. garden in U. S., Kingsessing, Pa., 1728; apptd. Am. botanist to King George III, 1765; made bot. and sci. journeys through the Alleghenies, Catskills, Carolinas, and Florida, adding new descriptions of plants and zool. specimens. Author: Observations on the Inhabitants, Climate, Soil, etc. made by John Bartram in His Travels from Pennsylvania to Lake Ontario, 1751; Descriptions of East Florida, 1769. Conducted 1st expts. in hybridization in N. Am., 1728; called "father of American botany"; name commemorated in Bartramia genus of mosses; introduced many American plants into Europe; designated by Linnaeus as greatest contemporary natural botanist. Died Kingsessing, Sept. 22, 1777.

BARTRAM, William, Am. naturalist; b. Phila., Feb. 9, 1739; s. John and Ann (Mendenhall) B.; Apprenticed to mcht., 1757-61; in business as trader, Cape Fear, N.C., 1761; explored St. John's River with father, 1765-66; engaged in extensive travels through Southern part of nation gathering specimens and seeds along with drawings for Dr. John Fothergill of London, 1773-78; partner with brother John in operating botanic garden founded by father on bank of Schuylkill River, nr. Phila., 1777-1812; elected mem. Am. Philos. Soc., 1786. Author: Travels (description of natural life in South of high literary quality), 1791; article "Account of the Species, Hybrids, and Other Varieties of the Vine in North America," 1804. Prepared drawings for Barton's Elements of Botany (1st elementary botany text by an American); his list of 215 native birds was most complete in existence until Wilson's American Ornithology; his name commemorated in genus Bartramia of bird known as upland plover, Bartram's plover, and Bartram's sandpiper. Died Kingsessing, Pa., July 2, 1823.

BARTRUM, John Arthur, New Zealand geologist; b. Geraldine, New Zealand, May 24, 1885; B.Sc., M.Sc., Otago (New Zealand) U. and Sch. Mines, 1904-08; m. Constance Loric, 1912; 1 son, 2 daus. With New Zealand geol. survey; tchr. Canterbury Agrl. Coll., Lincoln, New Zealand; lectr. Auckland (New Zealand) Univ. Coll., from 1914, later prof. until 1949. Fellow Geol. Soc. London, Geol. Soc. Am. (restricted), Royal Soc. New Zealand (Hector and Hutton research medals). Made classic contbn. to study of shore platforms. Died Rotorua, New Zealand, June 8, 1949.

BARTSCH, Alfred Frank, Am. biologist; b. Kaukauna, Wis., Nov. 30, 1913; s. Carl W. and Elizabeth (Miller) B.; B.A., U. Minn., 1936; Ph.D., U. Wis., 1939; m. Winnie V. Ireland, Sept. 1, 1937; children—Alfred Frank, Nancy Elizabeth. Biologist div. Water Supply and Pollution Control, USPHS, Portland, Ore., Washington, Cin., 1949-58, asst. chief research, Cin., 1958-61, dep. chief research, 1961-63, chief enforcement activities Pacific N.W., Portland, 1963-64; dir. research Pacific N.W. Water Lab. Fed. Water Pollution Control Adminstrn., Corvallis, Ore., 1964—; spl. cons. ICA, Brazilian Water Pollution Authority, 1956-57; chief of party investigating team Trust Terr. Pacific Islands, 1958; cons. Pan Am. Health Orgn., Brazil, 1965-66. Mem. Am. Fisheries Soc., Ecol. Soc. Am., Water Pollution Control Fedn., Am. Water Works Assn., A.A.A.S., Am. Soc. Limnology and Oceanography, Am. Inst. Biol. Sci. Research, publs. on control of induced eutrophication and related aquatic nuisances; stimulated amalgamation of biology with other profl. disciplines in effective detection and measurement of water pollution. Home: 3238 Gumwood Terrace. Office: 200 S. 35th St., Corvallis, Ore. 97330.*

BARTSCH, Jakob, astronomer; b. Luban, Germany (now Poland), 1600; student math., astronomy, Leipzig, Germany; medicine, Padua, Italy; Magister philosophiae; Dr. med., 1630; m. Susanna Kepler, 1630. Pub. numerous works including: Index, 1624; Planisphaerium, (mobile display showing position of planets, stars at any time), 1624; Planiglobium (4 astron. charts), 1628; introduced new constellations, divided stars into 11 magnitudes; with bro. Friedrich, made large sky globe, 1625; completed J. Schiller's Coelum stellatum christianum, 1627; assisted Kepler in calculations. Died Luban, Dec. 26, 1633.

BARTTER, Frederic Crosby, Am. physiologist; b. Manila, P.I., Sept. 10, 1914; s. George Charles and Frances (Buffington) B.; brought to U. S., 1927; B.A., Harvard, 1935, M.D., 1940; m. Jane Hazen Lillard, May 26, 1946; children—Frederic Crosby, Thaddeus C., Pamela. Dir. labs. USPHS Hosp., Sheepshead Bay, N.Y., 1942-44; med. officer in charge onchocerciasis investigation Pan-Am. San. Bur., Guatemala, Mexico, 1944-45; mem. staff Lab. Tropical Diseases, NIH, Bethesda, Md., 1945-46, chief clin. endocrinology br. Nat. Heart Inst., 1951—; research, clin. fellow in medicine Harvard Med. Sch., also Mass. Gen. Hosp., Boston, 1946-48, asst. in medicine, 1948-50, asso. in medicine, 1951, tutor in biochem. scis. Harvard Coll., 1946-51; asso. prof. pediatrics Howard U., 1958-64, prof. pediatrics, 1965—; asso. prof. medicine Georgetown U., 1960-64, clin. prof. medicine, 1965. Diplomate Am. Bd. Internal Medicine, Nat. Bd. Med. Examiners. Mem. Endocrine Soc., Am. Soc. for Clin. Investigation, Peripatetic Soc., Assn. Am. Physicians, Am. Physiol. Soc. Contbr. numerous articles in field to sci. jours. Research on physiology of adrenal and parathyroid glands, especially prodn. of aldosterone; defined biochem. basis of adrenogenital syndrome; described syndrome of juxtaglomerular hypoplasia with hypokalemia, alkalosis, normal blood pressure, syndrome of inappropriate ADH prodn. Home: 3332 36th St. N.W., Washington 20016. Office: Clinical Endocrinology Br., Nat. Heart Inst., Nat. Insts. Health, Bethesda, Md. 20014.*

BARUA, Arun Kumar, Indian chemist; b. Chandernagore, India, Nov. 1, 1929; s. Beni Madhab and Pankaja Barua; M.Sc., Calcutta U., 1950, D.Phil., 1955; m. Gita Ray, June 16, 1959; 1 son. Research fellow Bose Inst., Calcutta, India, 1955-57, lectr. chemistry, 1958-64 reader in chemistry, 1964—; postdoctoral research fellow U. Rochester (N.Y.), 1959-60, U. Buffalo, 1961. Fellow Indian Chem. Soc. Research, numerous publs. on chemistry of natural products, saponins, terpenoids, bitter principles from plants, antibiotics of microbial origin, synthesis of anti-cancer compounds. Home: 188/60 Prince Anwar Shah Rd. Office: Bose Institute, Calcutta 9, India.*

BARUA, Dhiman, Indian bacteriologist; b. India, Oct. 19, 1920; s. Dina Bandhu and Pramoda Barua; M.B.B.S. with honors, Calcutta (India) U., 1949; M.D. in Pathology and Bacteriology, Lucknow U., 1956; postgrad. Inst. Pasteur, Paris, 1958-59; m. Mridula Barua, July 8, 1953; children—Basab, Kausik. Demonstrator pathology and bacteriology Med. Coll., Calcutta, 1949-56; asst. prof. bacteriology Inst. Postgrad. Med. Edn. and Research, Calcutta, 1956-61; faculty Sch. Tropical Medicine, Calcutta, 1961-65. Cons. WHO, Geneva, Switzerland, specialist on cholera; mem. cholera expert com. Indian Council Med. Research, New Delhi. Recipient Gold medal Lucknow U., 1956. Mem. Indian Assn. Pathologists, Indian Med. Assn. Research, publs. on definition of type of diphtheria bacilli responsible for disease in Calcutta, type of E. coli causing urinary infection; 1st description in India of pathogenic E. coli; demonstrated drug resistance Staphylococci and M. tb; discovered cause of relapse of typhoid fever after chloramphenicol is due to its failure to sterilize gall bladder; found lepromin from Vole bacillus to differentiate two types of leprosy; described different methods for diagnosis of Cholera cases and carriers; developed lab. methods for differentiating different vibrios. Home: H/7 CIT Bldg., Calcutta-I, India. Office: Bacterial Diseases Unit, WHO, Geneva, Switzerland.*

BARUCH, Simon, physician; b. Schwersen, Germany, July 29, 1840; s. Bernard and Teresa (Green) B.; M.D., Med. Coll. Va., 1862; m. Isabel Wolfe, Nov. 27, 1867. Surgeon in Gen. R. E. Lee's army, 1862-65; practiced medicine Camden, S.C., 1865-81, then in N.Y.; cons. physician in chronic diseases; prof. hydrotherapy Coll. Phys. and Surg., Columbia. Chmn. S.C. Bd. Health, 1880. Hosp. erected in his honor, Camden, 1913; free municipal baths named in his honor, Chgo., 1910, N.Y., 1917. Author: Uses of Water in Modern Medicine, 1892; The Principles and Practice of Hydrotherapy, 3 edits. Diagnosed 1st case of perforating appendicitis successfully operated on, 1889; introduced free municipal bathhouses. Died June 3, 1921.

BARUCH, Sydney Norton, Am. research engr.; b. Mamaroneck, N.Y., Mar. 14, 1895; s. Joseph and Sophia (Van Kitzinger) B.; E.E., Cooper Union, 1911; D.Sc., Royal (Eng.), 1921; spl. courses in engring. Cooper Union; spl. study elec. phenomenon U. Cal.; unmarried. Chief engr. Fed. Telephone Co., radio div. Postal Telegraph Co., 1919-20, Gen. Petroleum Co. Am., 1921; pres. Pub. Service Corp. Cal., 1916-20; condr. pvt. research labs., N.Y.C., from 1930; dir., controller United Broadcasting chain of radio stations; chief research engr. Gen. Arc Lighting Co. Builder high power radio broadcast chain, 1925; designer broadcasting stations, CHCR, WBNY, WKBK, WKBQ, LIY (Bordeaux, France); cons. engr. spl. weapons div. U. S. Air Force, from 1943, U. S. Signal Corp, 1948. Recipient Gold medal Internat. Jury of Scientists, 1915. Fellow Royal Soc. of London; mem. I.R.E., Soc. Motion Picture Engrs. Inventor thermo relay and other devices, also thyraton and nortron type mercury rectifier tubes and sound recording on film, 1934; inventor depth bomb successfully used in destruction of submarines in World War I and World War II. Designer 300,000 volt direct current transmission system for Bonneville Project, U. S. Dept. Interior, 1941. Invented guided missile using jet propulson. Died Sept. 22, 1959.

BARUK, Henri Marc, French neuropsychiatrist; b. St. Avé, France, Aug. 15, 1897; s. Jacques and Marie (Brèchon) B.; ed. Faculty Medicine, Paris; m. Suzanne Sorano, Dec. 29, 1947. Physician in chief Maison Nationale de Charenton, Paris; prof. Faculté de medecine, Paris; dir. exptl. psychopathology lab. École Pratique des Hautes Etudes, Sorbonne, Paris. Author numerous books including: Troubles mentaux dans les tumeurs cérébrales, 1926; (with de Jong) Catatonie expérimentale par la bulbocapnine; Psychiatrie médicale expérimentale, 1938; Psychiatrie moral, 1945; Traité de psychiatrie, 1959. Research on cerebral localization, exptl. psychiatry in animals, moral consciousness and social psychiatry, Hebraic civilization and sci. of man. Home: 57 rue Maréchal Leclerc St. Maurice (Seine), France.*

BARUS, Carl, Am. physicist; b. Cin., Feb. 19, 1856; s. Prof. Carl and Sophia (Mollman) B.; attended Columbia, 1874-76; U. Wurzburg, Germany, 1876-80, Ph.D., 1879; LL.D., Brown, 1907, Clark U., 1909; m. Annie G. Howes, Jan. 20, 1887; children—Maxwell, Deborah Howes. Physicist, U. S. Geol. Survey, 1880-92; prof. meteorology U. S. Weather Bur., 1892-93; physicist Smithsonian Instn., 1893-95; prof. physics Brown U., Providence, 1895-1926, prof. emeritus, from 1926, dean grad. dept., 1903-26. Recipient Rumford medal Am. Acad. Arts and Scis. for various researches in heat. Mem. advisory com. Carnegie Instn., 1902; mem. hon. com. of Internat. Congress on Radiology, Brussels, 1905, 1910. Fellow Am. Acad. Arts and Sciences; mem. numerous socs. Author: The Electrical and Magnetic Properties of the Iron Carburets, 1885; Subsidence of Fine Solid Particles in Liquids, 1886; Physical Properties of the Iron Carburets, 1886; The Measurement of High Temperatures, 1889; Viscosity of Solids, 1891; Die Physikalische Behandlung Hoher Temperaturen, 1892; Compressibility of Liquids, 1892; Mechanism of Solid Viscosity, 1892; Volume Thermodynamics of Liquids, 1892; High Temperature Work in Igneous Fusion, 1893; Condensation of Atmospheric Moisture, 1895; Experiments with Ionized Air, 1901; The Structure of the Nucleus, 1902; Nucleation of the Atmosphere, 1905; Nucleation of the Uncontaminated Air, 1906; Condensation Induced by Nuclei and by Ions, 1907, part II, 1908, part III, 1909, part IV, 1910; Elliptic Interferences, 1911, 2d vol., 1913; 3d vol., 1915; Diffusion of Gases through Liquids, 1913; Interferences of Reversed and Non-reversed Spectra, vol. I, 1916, vol. II, 1917, vols. III and IV, 1919; Interferometer Experiments in Acoustics, vol. I, 1921, vol. II, 1923, vol. III, 1925; Acoustic Experiments with Pin-Hole Probe and the Interferometer, 1927. Edited The Laws of Gases, 1899. Emphasized (with others) that colloids are suspensions of extremely small particles in their ordinary states (2d phase in devel. colloid chemistry). Died Providence, Sept. 20, 1935.

BARUT, Asim Orhan, Am. physicist; b. Malatya, Turkey, June 24, 1926; s. Sabit and Serife (Evliya) B.; diploma Swiss Fed. Inst. Tech., 1949, Dr.Sc., 1952; m. Pierrette H. Gervaz, July 2, 1954; children—Turan, Susan, Sibel. Came to U. S., 1956, naturalized, 1962. Research asso. Swiss Fed. Inst. Tech., 1950-53; fellow U. Chgo., 1953-54; asst. prof. Reed Coll., 1954-55; staff NRC Can., also asst. prof. U. Montreal (Que., Can.), 1955-56; faculty Syracuse U., 1956-61, asso. prof., 1959-61; faculty U. Cal. at Berkeley, also staff Lawrence Radiation Lab., 1961-62; prof. U. Colo., Boulder, 1962—; staff Internat. Center for Theoretical Physics, Trieste, 1964-65. Fellow Am. Phys. Soc.; mem. Swiss Phys. Soc., N.Y. Acad. Scis., A.A.A.S. Author: Electrodynamics and Classical Theory of Fields and Particles, 1964; The Theory of Scattering Matrix, 1967; also numerous articles. Application of relativistic invariance in formulation theory of scattering matrix for interactions of fundamental particles, scattering matrix theory of quantum electrodynamics, dynamical groups, a new formulation of quantum theory with applications to elementary particles; research on dynamical symmetry and elementary particles interactions. Home: 760 12th St., Boulder, Colo. 80302.*

BARY, Heinrich Anton de, see de Bary, Heinrich Anton.

BASCH, (Samuel) Siegfried Carl von, see von Basch, (Samuel) Siegfried Carl.

BASCHIN, Adolf Karl Otto, German geographer; b. Berlin, Germany, Apr. 7, 1865; s. Karl and Minna (Steidel) B.; student pharmacy; student physics, chemistry, meteorology, geography U. Berlin, after 1885; m. Kathe Zimmermann. Made expdn. to Greenland, 1891, Lapland (for study of polar light), 1891-92; asst. Prussian Meteorol. Inst., Berlin, from 1892; custodian Geog. Inst., U. Berlin, 1899-1930; prof., from 1903. Editor: Bibliotheca Geographica, Geog. Soc. Berlin, 1893-1912. Important work on morphology of earth's surface, meteorology and polar research. Died Berlin, Sept. 4, 1933.

BASCOM, Florence, Am. geologist; b. Williamstown, Mass.; d. John and Emma (Curtiss) Bascom; A.B., B.L., U. Wis., 1882, B.S., 1884, A.M., 1887; Ph.D., Johns Hopkins, 1893. Instr. geology and petrography Ohio State U., 1893-95; lectr., asso. prof., Bryn Mawr Coll., 1895-1906, prof. geology, 1906-28, prof. emeritus from 1928; geol. asst. U. S. Geol. Survey, 1896-1901, asst. geologist, 1901-09, geologist, 1909-36. Mem. div. geology and geography NRC. Fellow Geol. Soc. Am. (councilor 1924-26; 2d v.p. 1930), A.A.A.S.; mem. Phila. Acad. Natural Scis., Geog. Soc. Phila., Washington Acad. Scis., Seismological Soc. Am., Soc. Women Geographers, Mineral. Soc. Am., Inst. Mineralogy and Meteorology, England, Am. Geog. Soc., Phi Beta Kappa, Sigma Xi. Joint author and author geologic folios; also bulletins and numerous papers in tech. jours. Asso. editor Am. Geologist, 1896-1905. Died June 18, 1945.

BASCOM, William Russel, Am. anthropologist; b. Princeton, Ill., May 23, 1912; s. George Rockwell and Litta Celia (Banschbach) B.; B.A., U. Wis., 1933, M.A., 1936; Ph.D., Northwestern U., 1939; m. Berta Montero-Sanchez, Nov. 26, 1948. Faculty, Northwestern U., Evanston, Ill., 1938-42, 46-57, prof. anthropology, chmn. dept., also acting dir. program of African studies, 1954-57; prof., dir. Robert H. Lowie Mus. of Anthropology, U. Cal. at Berkeley, 1957—. Spl. asst. OSS, 1942; sr. analyst Bd. Econ. Warfare, 1942-43; asst. spl. rep. Bd. Econ. Warfare, Africa, 1943-44, spl. rep., 1944-46; chief economist U. S. Comml. Co., Micronesia, 1946. Santa Fe Lab. of Anthropology fellow, summer 1935; Social Sci. Research Council fellow, Nigeria, 1937-38; Wenner-Gren Found. for Anthrop. Research grantee, 1948; Fulbright grantee, Nigeria, 1950-51; NSF sr. post-doctoral fellow, Eng., 1958. Mem. Am. (pres. 1952-54), Cal. (v.p. 1962—) folklore socs., Central States Anthrop. Soc. (pres. 1950-51), Am. Anthrop. Assn. (exec. council 1961-64), Am. Assn. Museums (council 1962-67), Oakland Museums Assn. (dir. 1963-64, 65—), Mus. African Art (adv. bd. 1964—), Cal. League for Am. Indians (dir. 1964-67), Phi Beta Kappa, Sigma Xi, Phi Eta Sigma, Alpha Kappa Delta, Alpha Zi Zeta. Author: (with Paul Gebauer) Handbook of West African Art, 1953, 64) The Sociological Role of the Yoruba Cult Group, 1944; Ponape, A Pacific Economy in Transition, 1965; African Arts, 1967; also articles. Editor: (with M. J. Herskovits) Continuity and Change in African Cultures, 1959, 62. Home: 624 Beloit Av., Berkeley, Cal. 94708.*

BASEDOW, Herbert, Australian physician, anthropologist; b. Kent Town, Adelaide, Australia, Oct. 27, 1881; s. M. P. F. Basedow; ed. Prince Alfred Coll., Adelaide, also Adelaide U., univs. Heidelberg (Germany), Göttingen (Germany), Breslau (now Wroclaw, Poland), Zurich (Switzerland); M.A., M.D., Ph.D., B.Sc.; m. Olive Nell Noyes. Asst. govt. geologist, S. Australia; chief med. officer, chief protector aborigines for Commonwealth, no. teretory; state and commonwealth med. commr. on aborigines; head several expdns. to explore Australia; with expdns. Geroge Le Hunte, Tom Bridges, Lord Stradbroke. Fellow Geol. Soc.; hon. mem. anthrop. socs. London, Göttingen, Adventurers of World (life), U. S.; mem. numerous sci. socs.; pres. Aborigines Protection League, 3 years. Author: The Australian Aboriginal, 1925; numerous works on Australian natural history and anthropology. Editor Bringa's Cooee Talks. Spl. corr. Australian and London mining jours. and other periodicals. Made numerous discoveries in Australian natural history and geography; authority on anthropology of Australian tribes. Died June 2, 1933.

BASEDOW, Karl Adolf von, see von Basedow, Karl Adolf.

BASEILHAC, Jean (Frère Cosme), French physician-surgeon; b. Pouy, France, Apr. 5, 1703; s. Thomas Baseilhac; student Hotel Dieu at Lyons, France, 1722-24. Physician to Bishop of Bayeaux, 1724-28; joined L'Ordre des Feuillants, 1729; began public practice, 1740; established hosp. for poor, 1753. Introduced lateral lithotomy to remove stones from bladder; invented numerous surg. instruments, including one used in lithotomy. Died July 8, 1781.

BASERGA, Renato Luigi, research pathologist; born Meda, Italy, April 11, 1925; s. Alessandro and Pina (Annoni) B.; M.D. summa cum laude (Maria Antonietta Della Casa scholar), U. Milan (Italy), 1949; m. Jane Conrad, Dec. 27, 1953; children—Susan, Janice. Came to U. S., 1952, naturalized, 1958. Asso. in oncology Chgo. Med. Sch., 1953-54; faculty Northwestern U. Med. Sch., 1958-65, asso. prof. pathology, 1964-65; research prof. Fels Research Inst., Temple U. Sch. Medicine, Phila., 1965—. Cons. Argonne Nat. Lab., Lemont, Ill., 1959-65, VA Research Hosp., Chgo., 1961-63. Recipient USPHS Research Career Devel. award, 1960-65. Mem. Am. Assn. for Cancer Research, A.A.A.S., Am. Assn. Pathologists and Bacteriologists, Am. Soc. for Exptl. Pathology, Radiation Research Soc., Am. Soc. for Cell Biology. Research, publs. on control of cell div. Home: 562 Manor Rd., Wynnewood, Pa. 19096. Office: Fels Research Inst., Temple U. Sch. Medicine, Phila. 19140.*

BASHANDY, Ekbal Khedr, Egyptian physicist; b. Minia, U.A.R., Mar. 9, 1934; daughter of Bashandy Khalafalla and Ehssan (Bashandy) Khedr; B.Sc. in Physics Cairo (Egypt) U., 1956, M.Sc. in Radiation Physics, 1958; Fil Lic., U. Uppsala (Sweden), 1959, Fil Doctor, 1960; m. Mohamed El-Nesr, Oct. 8, 1959; children—Osama Mohamed, Eman Mohamed. Demonstrator, researcher Atomic Energy Establishment, Cairo, 1956, chief nuclear spectroscopy div., 1962—; researcher Inst. Physics Uppsala, 1960-62; tchr. nuclear physics U. Cairo, 1963; supr. courses in nuclear physics Middle Eastern Regional Radio Isotope Centre, 1963—. Recipient State prize in nuclear physics, 1962. Mem. Math. and Phys. Soc. U.A.R. Research, publs. in nuclear spectroscopy; discovered new properties in nuclear structure. Home: 16 El-Sheikh Ali Mahmoud, Heleopolis, Cairo. Office: Atomic Energy Establishment, Cairo, U.A.R.

BASHAW, Elexis Cook, Am. geneticist; b. Mt. Juliet, Tenn., July 21, 1923; s. Lex C. and Mabel (Wright) B.; B.S., Purdue U., 1947; M.S., 1948; Ph.D., Tex. A. and M. U., 1954; m. Bettye Louise Wood, Feb. 19, 1945; children—Bettye Jane, Cheryl Jeannine. Asst. agronomist La. Agrl. Expt. Sta., 1948-50; asst. prof. agronomy Tex. Agrl. Expt. Sta., College Station, 1951-55; geneticist Agrl. Research Service, U. S. Dept. Agr., College Station, 1955—; prof. grad. faculty Tex. A. and M. U., 1952—. Mem. Crop Sci. Soc. Am., Am. Soc. Agronomy, Sigma Xi, Gamma Sigma Delta, Alpha Zeta. Research, publs. on cytogenetics of grasses including reproductive systems, radiation breeding and interspecific hybridization; reported genetic control of method of reprodn. in apomictic species; demonstrated potential value of apomixis in plant breeding. Home: 1208 Ashburn, College Station 77840. Office: Agronomy Dept., Tex. A. and M. U., College Station, Tex. 77843.*

BASHFORD, Ernest Francis, Brit. physician; b. Bowdon, Eng., 1873; s. William Taylor and Elizabeth (Booth) B.; student, univs. Edinburgh, Berlin, Frankfort/Main; M.B., Ch.B., U. Edinburgh, 1899; M.D., 1902; student of Ehrlich and Weigert-Grocers, Royal Prussian Inst. for Exptl. Therapeutics, Frankfort/Main; m. Elisabeth Alfermann, 1902; 1 dau. Asst. to dir. Pharmacological Inst., Berlin, 1901; asst. to prof. materia medica, pharmacology and clin. medicine U. Edinburgh, 1902; gen. supt. research Imperial Cancer Research Fund, 1903-14, dir. labs., 1903-14; Ingleby lectr. U. Birmingham, 1911; Middleton Goldschmidt lectr., N.Y., 1912; Von Leyden Meml. lectr., Berlin, 1912; hon. pres. 1st Internat. Cancer Congress, Heidelberg, 1906. Recipient Walker prize for quinquennial period, 1905-10, 1911. Editor, contbg. author Sci. Reports on Investigations of Imperial Cancer Research Fund, I-V; contbr. papers on pharmacology, biochemistry, immunity, cancer, pathology of wounds, to jours. Founder modern exptl. investigation of cancer in Gt. Britain; placed statis. and biol. investigation of cancer on comparative basis. Died Aug. 23, 1923.

BASHFORTH, Francis, Brit. ballistician; b. Thurnscoe nr. Doncaster, Eng., Jan. 8, 1819; s. John Bashforth; studied at St. John's Coll., Cambridge (Eng.) U.; m. Elizabeth Jane Pigott, 1869; 1 son. Learned math. of paths of projectiles during three years spent as practical civil engr.; adviser to War Office in 1873. Author: Capillary Action, 1883; The Bashforth Chronograph, 1890; Report on the Experiments Made with the Bashforth Chronograph, 1870; Mathematical Treatise on the Motion of Projectiles, 1873. Invented a precision chronograph; present knowledge of air resistance based on his expts. Died Feb. 13, 1912.

BASHKIN, Stanley, Am. physicist; b. Bklyn., June 20, 1923; s. Max and Bessie (Kovalik) B.; B.A. cum laude, Bklyn. Coll., 1944; Ph.D., U. Wis., 1950; m. Margaret Mary Turnbull, Aug. 22, 1957; children—James K., Margaret J., John S. Faculty, State U. Ia., 1953-62; asst. prof. La. State U., 1950-53; prof. physics U. Ariz., Tucson, 1962—; research fellow Cal. Inst. Tech., 1959. Cons. in optics. Fulbright Research scholar Australian Nat. U., 1959-60. Mem. Am. Phys. Soc., Am. Astron. Soc., A.A.A.S., Royal Soc. Arts, (London, Eng.). Contbg. author: Stars and Stellar Systems. Contbr. articles to tech. jours. Invented method of studying atomic systems using nuclear apparatus. Home: 4152 E. 6th St., Tucson 85711.*

BASHKIROV, Andrei Nikolaevich, Russian organic chemist; born December 22, 1903; graduated from Moscow Chemical and Technical Institute, 1929. With All-Union Research Institute of Gas and Artificial Liquid Fuel, and its Novosibirsk Siberian br., 1934-38; with Inst. Combustible Minerals, USSR Acad. Sci., 1939-47, asso. Petroleum Inst., 1947—; head chair Moscow Inst. Fine Chem. Tech., 1943—. Mem. USSR Acad. Sci. (corr.). Research and publs. on catalytic synthesis of hydrocarbons, spirits and amines on base of carbon monoxide and hydrogen monoxide; developer indsl. method of obtaining higher fatty alcohols by direct oxidation of hydrocarbons; desulfurization of gaseous and petroleum products; thermal processing of coal. Address: Moscow Inst. Fine Chem. Tech., M. Pirogovskaya ulitsa 1, Moscow, USSR.

BASILE, Attilio, Italian surgeon; b. Itala (Messine), Jan. 15, 1910; s. Francesco and Gaetana (Desalvo) B.; ed. in medicine, surgery; M.D.; m. Rosalia Lociero, June 28, 1953; children—Gaetanella, Francesco, Filadelfio, Guido. Prof. clin. surgery; dir. Centro Tumori, Catania, Italy; dir. sch. specialization in surgery, dir. sch. specialization in anesthesia. Author over 80 works on surgery, including treatise on surgery of internal secretory glands. Research in cardiovascular, thoracic, and abdominal surgery. Home: via Odorico da Pordenone 1. Office: Università degli Studi, Catania, Italy.

BASILEWSKY, Pierre, entomologist; b. Petrograd, Russia, Aug. 21, 1913; s. George and Helen (Stolnifoff) B.; ed. State Inst. Agronomic Studies, Gembloux, Belgium; Ingéneiur Agronome, 1936; Ingéneiur des Eaux et Forets, 1938; m. Emilie Linard, Oct. 16, 1939. Colonial agronomic engr.; forest conservation engr.; entomologist Royal Mus. Belgian Congo (now Royal Mus. Central Africa), Tervuren, Belgium, became head entomology sect., 1949, curator, 1958. Mem. Paris Natural History Mus. (corr.), Entomol. Soc. France (hon.), numerous sci. socs. Contbr. papers to sci. publs. Specialist in study of Coleoptera ground beetles of Africa; numerous research projects in systematic entomology of Africa; zool. exploration in Congo, 1952, Ruanda-Urundi, 1953, Kenya and Tanganika, 1957, St. Helena, 1965-66. Home: 17 Sq. Leopoldville, Bruxelles 4, Belgium. Office: Musée Royal de l'Afrique Central, Tervuren, Belgium.*

BASILIUS VALENTINUS, alchemist; supposed author of influential alchemical texts 1st published in 17th century which were probably written by their "editor", Johann Thölde. Basilius Valentinus was identified by 17th century authors as a Benedictine monk from Alsace, b. circa 1394, who traveled in Belgium, Holland, England, Egypt; lived at Benedictine monastery, Erfurt, circa 1413; his manuscripts allegedly discovered, 1600. Works ascribed to him include: Von dem grossen Stein der Uralten, 1602; Von dem natürlichen und ubernatürlichen Dingen, 1603; De Occulta Philosophia, 1603; Triumph Wagen Antimonii, 1604; Letztes Testament, 1626; his collected Chymische Schriften went through 6 ed., 1677-1775. Stressed macrocosm-microcosm analogies and 3 principles: salt, sulphur, mercury; described generation of metals in mystical terms, but showed excellent knowledge of mining and metallurgy; discussed purification of gold and its reduction to powder and amalgams of gold, also discussed transmutation and precipitation of 1 metal by another; gave detailed descriptions of antimony and its compounds, also made extensive use of antimony preparations in medicine; described arsenic and its compounds, mineral acids, alkalies. Works reveal influence of Paracelsus.

BASINSKI, Zigniew Stanislaw, metal physicist; b. Wolkowysk, Poland, Apr. 28, 1928; s. Antoni and Maria Z.A. (Hilferding) B.; B.A., Oxford (Eng.) U., 1950, B.Sc., 1952, D.Phil., 1954, D.Sc., 1966; m. Sylvia Joy Pugh, Apr. 1, 1952; children—Stefan Leon Hilferding, Antoni Stanislaw Hilferding. Research staff Mass. Inst. Tech., 1954-56; staff div. pure physics NRC, Ottawa, Ont., Can., 1956—, prin. research officer, 1966—; Ford Distinguished vis. prof. Carnegie Inst. Tech., 1965. Research, publs. on phase transformations, mech. properties of metals at very low temperatures theory of dislocations. Home: Box 589, 108 Delong Dr., Rural Route 1, Ottawa. Office: Sussex Dr., Ottawa 2, Ont., Can.*

BASKERVILLE, Charles, Am. chemist; b. Noxubee Co., Miss., June 18, 1870; s. Charles and Augusta Louisa (Johnston) B.; studied U. Miss., 1886-87; grad. U. Va., 1890; postgrad. Vanderbilt U., 1891, U. Berlin, 1893; Ph.D., U. N.C., 1894; m. Mary Boylan Snow, Apr. 24, 1895. Instr. U. N.C., 1891-94, asst. prof. chemistry, 1894-1900, prof. chemistry and dir. chem. lab., 1900-04; prof. chemistry and dir., Coll. City N.Y., from 1904. Involved in devel. Am. Chem. Soc. Author: School Chemistry, 1898; Key to School Chemistry, 1898; Radium and Its Applications in Medicine; General Inorganic Chemistry, 1909; Laboratory Exercises (with R. W. Curtis), 1909; Progressive Problems in Chemistry (with W. L. Estabrooke); Qualitative Analysis (with L. J. Curtman); Municipal Chemistry (and editor with other experts); Anesthesia (with J. T. Gwathmey); also numerous scientific, ednl. and technol. articles. Inventor processes for refining oils, hydrogenation of oils, plastic compositions, reinforced lead; discovered chem. elements, carolinium and berzelium; investigated chemistry of anaesthetics. Died Jan. 28, 1922.

BASMAJIAN, John Varoujan, anatomist; b. Constantinople, Turkey, June 21, 1921; s. Mihran and Mary (Evelian) B.; M.D. with honors, U. Toronto, 1945; m. Dora Lucas, Oct. 4, 1947; children—Haig, Nancy, Sally. From demonstrator to prof. U. Toronto, Ont., Can., 1947-57; research asso. St. Thomas Hosp., London, 1953, Hosp. for Sick Children, Toronto, 1953-57; prof., head dept. anatomy Queen's U., Kingston, Ont., 1957—. Sec., Banting Research Found., Toronto, 1955-57; exchange vis. scientist, USSR, 1963; chmn. Nat. Fitness Research Rev. Com. Can., 1967—; vice chmn. med. adv. com. Ont. div. Canadian Cancer Soc. Recipient Starr medal for med. research U. Toronto, 1956, Kazanjian Science award, 1967. Mem. Am. Assn. Anatomists, Anat. Soc. Gt. Britain and Ireland, Canadian Assn. Anatomy (Sec. 1965—), Pan Am. Assn. Anatomists (founding mem., councillor), Internat. Soc. Electromyographic Kinesiology (founder, chmn. exec. com.), Canadian Neurol. Soc., Am. Acad. Neurology, Mexican Soc. Anatomists (hon.). Author: Primary Anatomy, 1955; Muscles Alive, Their Functions Revealed by EMG, 1962. Editor: Grant's Method of Anatomy, 7th edit., 1965; editorial bd. Canadian Jour. Surgery, 1957—, Queen's Quar., 1961—, Electromyography, 1962—, Am. Jour. Phys. Medicine, 1966—, Electrodiagnostic-therapie, 1967—. Research, publs. in muscle anatomy and electronics; developed electronic devices; neuromuscular studies. Home: 55 Jane Av., Kingston, Ont., Can.*

BASOLO, Fred, Am. chemist; b. Coello, Ill., Feb. 11, 1920; s. John and Catherine (Marino) B.; B.E., So. Ill. U., 1940; M.S., U. Ill., 1942, Ph.D., 1943; m. Mary P. Nutley, June 14, 1947; children—Mary Catherine, Freddie, Margaret-Ann, Elizabeth Rose. Research chemist Rohm & Haas Chem. Co., Phila., 1943-46; faculty Northwestern U., Evanston, Ill., 1946—, prof. chemistry 1958—. Guggenheim fellow, 1954-55, NSF fellow, 1961-62. Mem. Am. Chem. Soc. (asst. editor jour. 1961-64), A.A.A.S., Chem. Soc., Sigma Xi, Phi Lambda Upsilon, Alpha Chi Sigma, Phi Kappa Phi. Author (with R. G. Pearson): Mechanisms of Inorganic Reactions, 1958. Contbr. to Chemistry of the Coordination Compounds, 1956. Asso. editor Chem. Revs., 1960—; editorial bd. Jour. Inorganic and Nuclear Chemistry, 1959—. Research, publs. in area of solution chemistry of metals. Home: 1125 Colfax St., Evanston, Ill. 60201.*

BASORE, Cleburne Ammem, Am. chem. engr.; b. Broadway, Va., May 13, 1893; s. Thomas Sidney Lee and Minnie (Ammen) B.; B.S., Auburn U., 1914, M.S., 1915; M.A., U. Mich., 1917; Ph.D., Columbia, 1929; m. Annie Elizabeth Terrell, June 12, 1920. Chemist, Tenn. Coal Iron & R.R. Co., Birmingham, Ala., 1915; research chemist Koppers Co., Pitts., 1917-20; faculty Auburn (Ala.) U., 1920-63, prof. chem. engring., 1929-63, prof. emeritus, 1963, head dept., 1939-63. Mem. Am. Chem. Soc. (Ala. chmn. 1940—), Am. Inst. Chem. Engring., Am. Soc. Engring. Edn., Royal Soc. Arts (London), Ala. Acad. Sci. (pres. 1949), S.A.R., Sigma Xi, Phi Kappa Phi, Phi Lambda Upsilon. Author: Introduction to Chemical Engineering, 1940; also articles. Research on utilization of cotton seed hulls, waste wood, coal by-products and wastes, low grade lignites, blast furnace slag, elimination or reduction of evaporation losses from petroleum storage tanks, iron and steel processing. Home: 121 Mitcham Av., Auburn, Ala. 36830.*

BASOV, Nikolai Ghennadievich, Russian physicist; b. Voronege, USSR, Dec. 14, 1922; s. Ghennadiy Fedorvich and Zinaida Andreevna (Molchanova) B.; grad. Moscow Mech. Inst., 1950, Cand. Phys. Math. Sci., 1953, D. Phys. Math. Sci., 1956; m. Qsenia Tihonovna, July 18, 1950; children—Ghennadiy, Dmitriy. With P. N. Lebedev Phys. Inst., USSR Acad. Sci., 1953—, vice dir. for sci. work, 1958—, head lab. quantum radio physics, 1963—; prof. solid state physics Moscow Inst. Phys. Engrs., 1963—. Mem. USSR Acad. Scis. Recipient Lenin prize, 1959; Nobel prize for fundamental research in quantum electronics resulting in creation of masers and lasers, 1964. Author over 150 works. Research on principle of molecular generator, 1952, realized molecular generator on molecular beam of ammonia, 1955, 3-level system for receiving states with inversal population suggested, 1955, proposed use of semicondrs. for creation lasers, 1958, realized various types of semicondr. lasers with excitement through p-n transitions, elec. and optical excitement, 1960-65, research on obtaining short powerful pulses of coherent light (lasers with double modulation). Address: P. N. Lebedev Phys. Inst., Leninsky Prospect 53, Moscow, USSR.*

BASOV, Vasilij Aleksandrovich, Russian surgeon, physiologist; b. 1812; first to bind gastric fistula, 1842; pioneered research on digestion physiology. Died 1879.

BASQUIN, Olin Hanson, Am. physicist; b. Dows, Ia., Jan. 30, 1869; s. Oliver William and Hannah (Valentine) B.; A.B., Ohio Wesleyan U., 1892; A.B., Harvard, 1894; A.M., Northwestern Univ., 1895, Ph.D., 1901, D.Sc., 1930; m. Jessie C. Guthrie (dec. 1907); m. 2d, Anna Stuart, Sept. 12, 1908; children—Harold G., Maurice H. Chief engr., Luxfer Prism companies, Chgo., 1897, London, 1898, Berlin, 1899; asst. prof. physics Northwestern U., 1901-09, prof. applied mechanics, 1909-26; with Haskelite Mfg. Corp., from 1926, became v.p. in charge engring. Asso. engr.-physicist in tests of steel columns U. S. Bur. Standards, 1916; in charge exptl. investigations on steel for Navy Dept., 1917; investigated engring. properties of plywood for use in mil. airplanes, 1918. Mem. Western Soc. Engrs. (Chanute medal 1915), Sigma Xi. Author: Tangent Modulus and the Strength of Steel Columns in Tests (Bur. of Standards), 1924; also papers on exptl. work in physics and in strength of materials. Died Mar. 30, 1946.

BASS, Allan Delmage, Am. pharmacologist; b. Marcus, Ia., Feb. 12, 1910; s. John Charles and Ethel Alice (Delmage) B.; B.S., Simpson Coll., 1931; M.S., Vanderbilt U., 1932, M.D., 1939; m. Sara Thompson, July 28, 1944; children—Allan Delmage, Sara J. Fellow, A.C.P., Yale, 1941-42, instr., 1942-43; instr. Vanderbilt Hosp., Nashville, 1943-44; prof. pharmacology State U. N.Y. at Syracuse, 1945-52; prof., chmn. dept. pharmacology Vanderbilt Med. Center, Nashville, 1953—. Chmn. pharmacol. test com. Nat. Bd. Med. Examiners, 1955-60; mem. study sect. NIH, 1960-64, project com. NIH Gen. Med. Scis. Program Com., 1964-67. Mem. A.A.A.S. (chmn. sect. med. sci. 1952-60), Am. Soc. Pharmacology and Exptl. Therapeutics (treas. 1958-60, pres. 1967-68), A.M.A. (council on drugs 1963—), Fedn. Am. Soc. Exptl. Biology (exec. com. 1966—), Am. Assn. Cancer Research, Sigma Xi, Alpha Tau Omega, Alpha Omega Alpha. Numerous publs. on research in endocrine pharmacology, autonomic pharmacology, cardiovascular pharmacology, psychopharmacology. Home: 4521 Price Circle Rd., Nashville 37205.*

BASS, Arnold M., Am. physicist; b. N.Y.C., Dec. 22, 1922; s. Nathan and Jean (Garfinkel) B.; B.S., Coll. City N.Y., 1942; M.A., Duke, 1943, Ph.D., 1949; m. Rosalyn Doren, Dec. 28, 1947; children—Jonathan, David, Daniel. Research asso. Mass. Inst. Tech., 1949-50; physicist Nat. Bur. Standards, Washington, 1950—, chief free radicals research sect. heat div., 1959-66, chief molecular energy levels sect., 1966—. Recipient Exceptional Service award U. S. Dept. Commerce, 1960. Fellow Optical Soc. Am., Washington Acad. Sci.; mem. Am. Phys. Soc., Philos. Soc. Washington, Royal Instn. Gt. Britain. Editor: (with H. P. Broida) Formation and Trapping of Free Radicals, 1960. Research, publs. on electronic spectra of small molecules in visible, ultraviolet and vacuum ultraviolet region, spectra of molecules and radicals in low temperature matrices, temperature measurement in hot gases by spectroscopic methods. Office: Nat. Bur. Standards, Washington 20234.*

BASS, Bernard Morris, Am. psychologist; b. N.Y.C., June 11, 1925; s. Alexander M. and Clara (Abrams) B.; student Coll. City N.Y., 1941-43; B.A., Ohio State U., 1946, M.A., 1947, Ph.D., 1949; m. Ruth Rothschild, Aug. 23, 1946; children—Robert, Jonathan. Faculty, La. State U., 1949-61, prof. psychology, 1957-61; vis. prof. U. Cal. at Berkeley, 1961-62; prof. U. Pitts., 1962-68; prof., dir. management research center University Rochester (N.Y.), 1968—; visiting professor Instituto Estudios Superiores de las Empresas, U. Navarre, Barcelona, Spain, 1966. Prin. investigator Aero-Space Lab., 1952-54, Office Naval Research, 1954—; cons. to pvt. cos., govt. Ford Faculty fellow, 1966-67. Diplomate Am. Bd. Examiners in Profl. Psychology. Mem. Am. Psychol. Assn., Internat. Assn. Applied Psychology. Author: Leadership, Psychology and Organizational Behavior, 1960; Organizational Psychology, 1965; Training in Industry: the Management of Learning, 1967; also numerous articles. Developed objective personality tests social interaction; verified numerous hypotheses about leadership and group behavior; developed psychol. tests for assessment and research, program standardized simulations for studying managerial behavior in different orgns. and different countries. Home: 5833 Aylesboro Av., Pitts. 15217. Management Research Center, Univ. Rochester, Rochester, N.Y.*

BASS, David Eli, Am. physiologist; b. Lowell, Mass., Aug. 15, 1912; s. Harry and Frieda (Rosengard) B.; A.B., Brown U., 1932; postgrad. Harvard; M.A., Boston U., 1951, Ph.D., 1953; m. Dorothy Kelly, May 12, 1945; children—Leslie, Jonathan, Nancy. Biochemist, Q.M. Climatic Research Lab., Lawrence, Mass., 1947-55; chief physiology br. Q.M. Research and Engring. Center, Natick, 1956-61; sci. dir. U.S.A. Research Inst. Environmental Medicine, Natick, 1961—; faculty Boston U. Sch. Medicine, 1952—, asso. prof. physiology, 1956—; adj. prof. zoology U. R.I., Kingston, 1965—, Mem. adv. bd. Inst. Environmental Biology, 1965—, Inst. Environomental Psychophysiology, U. Mass., 1961-65. Recipient Q.M. Research Dir.'s award, 1953; Sec. of Army Research and Study fellow., 1959-60. Fellow A.A.A.S., Am. Coll. Sports Medicine; mem. Am. Physiol. Soc., Am. Chem. Soc., Am. Soc. Clin. Chemists, Am. Fedn. for Clin. Research (sr.), Sigma Xi. Research, numerous publs. on physiology of heat and cold, mechanisms of acclimatization to heat, kidney function in cold, mechanisms of acclimatization to cold. Home: 24 Grey Birch Terrace, Newtonville, Mass. Office: U. S. Army Research Inst. Environmental Medicine, Natick, Mass. 02160.*

BASS, Frederic Herbert, Am. civil engr.; b. Hyde Park, Mass., June 19, 1875; s. George Walter and Elizabeth (Bellamy) B.; S.B., Mass. Inst. Tech., 1901; m. Lillian Leggett, June 27, 1903; children—Jason Parker, Elizabeth Bellamy. With engring. dept. Met. Water Works, of Mass., dam, aqueduct and filter constrn., 1896-99, 1901-02; with U. S. engr. corps, Boston Harbor Improvement, 1900; instr. in civil engring., U. Minn., 1901-04, asst. prof., 1904-10, prof. municipal and san. engring., 1910-19, head civil engring. dept., 1919-43; exec. dir. Am. Pub. Works Assn., 1943-45; v.p. Minn. Bd. Health, 1932-44, pres., 1936-39; mem. bd. engrs. Mpls.-St. Paul San. Dist.; designer numerous water works, sewerage systems and other pub. works; ret., 1947. Mem. Am. Soc. C.E., Am. Soc. Promotion Engring. Edn., Sigma Xi. Died May 13, 1954.

BASS, George Fletcher, Am. archaeologist; b. Columbia, S.C., Dec. 9, 1932; s. Robert Duncan and Virginia (Wauchope) B.; student U. Coll. S.W. of Eng., 1951-52; M.A., Johns Hopkins, 1955; postgrad. Am. Sch. Classical Studies, Athens, Greece; Ph.D., U. Pa., 1964; m. Ann Singletary, Mar. 19, 1960; 1 son, Gordon Wauchope. Research asst. U. Mus. U. Pa., Phila., 1961-64, asst. prof. classical archaeology, 1964—, asst. curator Mediterranean sect., 1964—, dir. Cape Gelidonya (Turkey) Excavation, 1960, Yassi Ada (Turkey) Excavation, 1961-64, Turkish Underwater Archaeol. Survey, 1965, Turkish Underwater Excavation and Survey, 1967. Named Outstanding Young Man Phila. Jr. C. of C., 1966. Mem. Archaeol. Inst. Am. Author: Archaeology Underwater, 1966. Directed first complete excavation of ancient shipwrecks under water, pioneer in underwater search, mapping and diving techniques, including underwater sterophotogrammetry and concept of research submarine for archaeology. Home: 4618 Larchwood Av., Phila. 19143. Office: U. Mus., 33d and Spruce Sts., Phila. 19104.*

BASS, Lawrence Wade, Am. chemical engineer; b. Streator, Ill., June 18, 1898; s. John Hiram and Sara (Leek) B.; Ph.B., Yale, 1919, Ph.D., 1922; studied Tulane Univ., Univ. of Lille, Sorbonne, Pasteur Institute, N.Y.U. Sch. of Law, 1923-27; married Edna Maria Becker, Nov. 23, 1935. Mem. science staff, Rockefeller Inst., N.Y. City, 1925-29; exec. asst. Mellon Inst. Industrial Research, Pittsburgh, 1929-31; dir. research Borden Co., New York, 1932-36; asst. dir. Mellon Inst., 1937-42; dir. New England Industrial Research Foundation, Boston, 1942-44; dir. chem. research, Air Reduction Co. 1944-48; dir. research and development, U. S. Indsl. Chemicals, Inc., N.Y.C., 1944-48, v.p., 1948-52; exec. staff Arthur D. Little, Inc., Cambridge, Mass., 1952-54, vice president, 1954-64, consultant, since 1964—; project director industrialization studies, Egypt, Iraq, 1953-55; mem. bd. Arthur D. Little Research Institute, Edinburgh, Scotland, 1957—. Chairman of the Com. on equipment and supplies Research and Development Bd., Dept. of Def., 1951-53. President's Certificate of Merit, 1947. Mem. A.A.A.S., Am. Inst. Chemists, Inst. Food Technologists, Am. Chem. Soc. (dir. 1946-49), Am. Inst. Chem. Engrs. (pres. 1945), Soc. Chem. Industry, Engrs. Joint Council (1946-49, chmn. 1948), Yale Engring. Assn. (pres. 1950-52), Sigma Xi. Author: Inorganic Complex Compounds, 1923; Nucleic Acids, 1931; Management of Technical Programs, 1965; numerous articles profl. jours. Investigated structure of nucleic acids and proteins, 1922-31; developed (with Andrew Fraser) methodology for quantitative study of economic and professional status of technical personnel, 1941; studied management of research and economic development of underdeveloped countries. Home: 220 Madison Av., N.Y.C. 10016. Office: 630 Fifth Av., N.Y.C. 10020.

BASS, Louis Nelson, Am. plant physiologist; b. Iola, Kan., Mar. 7, 1919; s. Herbert and Olive (Felker) B.; B.S., Upper Ia. U., 1940; M.S., State U. Ia., 1943; Ph.D., Ia. State U., 1949; m. Helen Jane Collins, Nov. 7, 1943; children—Colin David, Nelsa Louise. Plant breeder, Asgrow, Greeley, Colo., 1944; with Ia. State U., Ames, 1945-58, asst. prof. botany and plant pathology, 1949-58; plant physiologist Nat. Seed Storage Lab., U. S. Dept. Agr., Ft. Collins, Colo., 1958-63, research plant physiologist, 1963—. Mem. Assn. Ofcl. Seed Analysts (sci. ed. editor 1962—, exec, bd. 1966), Soc. Plant Physiologist, Ia. Acad. Sci., Sigma Xi, Gamma Sigma Delta. Research, publs. on germination requirements of seeds, effects storage condition on seed longevity; established use of 2, 3, 5-triphenyl tetrazolium chloride for rapid viability determination of Poa pratensis seeds. Home: 1117 Fairview Dr. Office: Colo. State U. Campus, Ft. Collins, Colo. 80521.*

BASS, Ludvik, math. physicist; b. Prague, Czechoslovakia, Mar. 9, 1931; s. Anthony and Agnes (Janda) B.; student Charles U., Prague, 1950-51; Dr. Phil., U. Vienna (Austria), 1954; M.A., Dublin (Ireland) U., 1959; m. Nina Schmahl, Mar. 11, 1961; 2 daus., Karen, Marianne. Research scholar Dublin Inst. Advanced Studies, 1954-56, research asso., 1958-63; lectr. in applied math. Trinity Coll., Dublin U., 1956-61; prin. mathematician B.S.A. Group Research, Birmingham, Eng., 1961-62; prin. lectr. math. Lanchester Coll. Tech., Coventry, Eng., 1962-65; professor of mathematics U. Queensland, Brisbane, Australia, 1967—; hon. vis. lectr. Sch. Chemistry, U. Newcastle on Tyne (Eng.), 1964-65. Fellow Inst. Physics (London); mem. Faraday Soc. Research, publs. on math. physics including relativity, classical and quantum electrodynamics, geophysics, stochastic processes, theoretical electrochemistry with applications to math, theories of nervous excitation and conduction. Home: 7 Tarbet St., Brisbane, Queensland, Australia.*

BASSETT (or Basset), Alfred Barnard, English physicist, mathematician; b. London, Eng., July 25, 1854; s. Alfred Bassett; ed. Trinity Coll., Cambridge (Eng.) U.; m. Edith Sarah Irwin de Chaundre, 1882 (dec. 1929); 1 dau. Thirteenth wrangler Cambridge U., 1877; barrister Lincoln's Inn, 1879. Mem. London Math. Soc. (v.p. 1892-94). Fellow Royal Soc., 1889. Author: A Treatise on Hydrodynamics, 1888; A Treatise on Physical Optics, 1892; A Treatise on Cubic and Quadric Curves, 1901; A Treatise on the

Geometry of Surfaces, 1910. Showed (with E. J. Routh) kinetic energy of a dynamical system in modified form of LaGrange's equations omitting certain velocities; studied spl. curves of fifth order; also research on theory of viscous fluids, elasticity. Died Dec. 5, 1930.

BASSETT, Homer Franklin, Am. entomologist; b. Florida, Mass., Sept. 2, 1826; ed. Berea U., Oberlin Coll.; tchr., Ohio, also Conn.; founded pvt. sch., Waterbury, Conn., 1859-67; in ins. and real estate bus.; librarian Bronson Library, Waterbury, 1872-1901. Mem. Entomol. Soc. Phila., Am. Entomol. Soc. Pioneer in study of Cynipidae; described many gallflies, including Cal. gallfly, live oak gallfly (collection housed in Phila. Acad. Scis.). Died Waterbury, June 28, 1902.

BASSETT, William Akers, Am. geologist; b. Bklyn., Aug. 3, 1931; s. Preston Rogers and Jeanne (Mordorf) B.; B.A., Amherst Coll., 1954; M.A., Columbia U., 1956, Ph.D., 1959; m. Jane Ann Kermes, Sept. 8, 1962; children—Kari Nicalo, Jeffrey Kermes, Penelope North. Research asso. Brookhaven Nat. Lab., 1960-61; asst. prof. geology U. Rochester, 1961-65, asso. prof., 1965—. Mem. Geol. Soc. Am., Mineral. Soc. Am., Am. Phys. Soc., Am. Geophys. Union, Geochem. Soc., Sigma Xi. Research and publs. on origin and crystal chemistry of sheet silicates, potassium-argon age determination of volcanic rocks, effects of high pressures and temperatures on phases proposed for the earth's interior. Home: 408 Dewey Rd., Churchville, N.Y. 14428.*

BASSHAM, James Alan, Am. chemist; b. Sacramento, Nov. 26, 1922; s. James Calvin and Helen (Baker) B.; B.S., U. Cal. at Berkeley, 1945, Ph.D., 1949; m. Leslie A. Groetzinger, Sept. 8, 1956; children—Eric, Glen, Helen, Frank. Research chemist, chem. biodynamics group Lawrence Radiation Lab., U. Cal. at Berkeley, 1949—. NSF Sr. Postdoctoral fellow Med. Research Council unit Oxford U., Eng., 1956-57; lectr. U. Cal. at Berkeley, 1957-59; cons. N.Am. Aviation Corp., 1965-66. Mem. Am. Chem. Soc., Am. Soc. Biol. Chemists, Am. Soc. Plant Physiologists. Author: (with Melvin Calvin) The Path of Carbon in Photosynthesis, 1955, The Photosynthesis of Carbon Compounds, 1962; also numerous articles. Mem. editorial bd. Analytical Biochemistry, 1964—. Participated in elucidation of basic photosynthetic carbon cycle; showed quantitative importance of cycle, studied interaction of photosynthetic and respiratory metabolism, measured direct contbrn. of photosynthesis, showed that not only sugars but amino acids and protein and other substances are direct products in chloroplasts, studied metabolic dynamics of photosynthesis. Home: 785 Balra Dr., El Cerrito, Cal. 94530. Office: Lawrence Radiation Lab., Berkeley, Cal. 94720.*

BASSI, Agostino, Italian physician; b. Lodi, Milan, Italy, Sept. 25, 1773; ed. U. Pavia (Italy); civil servant. Author: Del mal del segno calcinaccio o muscardino, 1835-36; also wrote on cultivation of potatoes, cheese, vinification, contagion, pellagra, cholera. In 1835 proved muscardine (disease of silkworms) was caused by fungus (named Botrytis bassiana in his honor); proposed that diseases were caused by animal or plant parasites (anticipated Pasteur by 10 years). Died Feb. 17, 1856.

BASSI, Laure Maria Catarina, Italian physicist; b. Bologna, Italy, Oct. 31, 1711; Ph.D., U. Bologna, 1733; m. Jean-Joseph Veratti, 1738, 12 children. Apptd. mem. faculty U. Bologna (1st woman to oc cupy chair of physics at a univ.), 1733. Author dissertations including: De Problemate quodam Mechanico; De Problemate quodam Hydrometrico. Died Bologna, Feb. 20, 1778.

BASSIE, V. Lewis, Am. economist; b. Chgo, Dec. 22, 1907; s. William Jones and Matilda (Rush) B.; Ph.B., U. Chgo., 1931, postgrad.; m. Janet Montgomery Hooks, Oct. 2, 1937; children—Carol Montgomery, William Chester. Economist consumer purchases study U.S. Dept. Agr., 1936-37; sr. economist Fed. Res. Bd., 1937-39; prin. economist U. S. Dept. Commerce, 1939-40, asst. to sec. commerce, 1945-48; chief civilian requirements div. Nat Def. Adv. Com., 1940-42; chief prodn. analyst WPB, 1942-44; adviser U. S. fgn. trade Fgn. Econ. Adminstrn., 1944-45; dir. Bur. Econ. and Bus. Research, prof. econs. U. Ill., Urbana, 1948—. Cons. to maj. indsl. and financial corps. Fellow Am. Statis. Assn.; mem. Am. Econ. Assn., Econometric Soc., Conf. on Income and Wealth, Nat. Bur. Econ. Research, Phi Beta Kappa. Author: Economic Forecasting, 1958; also numerous articles. Devel. methods for forecasting econ. change; analysis of stock-flow relationships for explaining cyclical behavior of investments; research on nat. econ. accounting. Home: 708 La Sell Dr., Champaign, Ill. 61820. Office: 408 David Kinley Hall, Urbana, Ill. 61801.

BASSINI, Edoardo, Italian surgeon; b. Pavia, Italy, Apr. 14, 1846; ed. U. Pavia; Joined Garibaldi's campaign, 1866; held chair clin. surgery at Pavia, and at Parma from 1878-80; dir. hosp. at Spezia (Italy), 1880-82; prof. pathology U. Padua (Italy), 1882-88, prof. clin. surgery, 1888-1921; elected senator, 1904. Author: Nuovo metodo operativo per la cura radicale dell'ernia inguinale, 1887; Un caso di rene mobile fissato col mezzo dell'operazione cruenta, 1882. Devised one of modern operations for inguinal hernia; also many operations for removal of humeroscapular sarcoma, ileo-caecal resection, nephropexy; introduced reconstructive operations by suturing layers. Died July 20, 1924.

BASSLEER, Roger Jose Renoit, Belgian cytologist; b. Liege, Belgium, July 1, 1931; s. Julien Prudent and Blanche (Bury) B.; M.D., U. Liege, 1956; m. Elyane Lascaris, July 19, 1958; children—Corinne, Nathalie. Asst., Inst. Histology, U. Liege, 1958—, chief asst. Mem. European Tissue Culture Club, Belgian Soc. Biology, Assn. French Lang. Anatomists. Research, numerous publs. on effects of antimitotic agents on mitosis, DNA synthesis, nuclear protein synthesis, nucleoli in tissue culture, DNA and nuclear protein metabolism in normal and cancer cells, organ culture in vitro-regeneration of skin, intoxication with ozone. Home: 37 Quai de l'Ourthe. Office: 20 Rue de Pitteurs, Liege, Belgium.*

BASSNAM, James Alan, chemist; b. Sacramento, Nov. 26, 1922; s. James Calvin and Helen (Baker) B.; B.S., U. Cal., Berkeley, 1945, Ph.D., 1949; m. Leslie Alberta Groetzinger, Sept. 8, 1956; children—Eric Alan, Glen James, Helen Sarah, Frank Lester. Research chemist Lawrence Radiation Lab. U. Cal., Berkeley, 1953—, lectr. dept. chemistry, 1957-59. Vis. fellow med. research unit U. Oxford (Eng.), 1956-57; cons. space and information div. N.Am. Aviation Co., Downey, Cal., 1964-65. NSF Sr. Postdoctoral fellow, 1956. Mem. Am. Chem. Soc., Am. Soc. Biol. Chemists, Am. Soc. Plant Physiology, A.A.A.S., Am. Inst. Biol. Scis., Phi Beta Kappa, Sigma Xi. Author: (with Melvin Calvin) The Path of Carbon in Photosynthesis, 1957, The Photosynthesis of Carbon Compounds, 1962. Research, numerous publs. on radioactive carbon as tracer to map pathway of photosynthetic carbon reduction in green plants; developed quantitative methods for kinetic studies biochem. pathways in living cells. Home: 785 Balra Dr., El Cerrito, Cal. 94530. Office: Lawrence Radiation Lab., Berkeley, Cal. 94720.*

BASSO, Sebastian, physician, chemist; flourished 1621; studied under Petrus Sinsonius at Acad. Pont, Mousson, Spanish Netherlands. Author: Philosophia naturalis adversus Aristotelem . . . , 1621; De diversitate partum compositarum . . . , 1621. Proposed 5 elements (spirit, oil, salt, earth, phlegm) instead of Aristotle's 4; held that all bodies are composed of small atoms of different kinds whose nature could not be changed; regarded ether as continuum producing material changes by causing motion and arrangement of atoms.

BASSON, Johan Kristof, South African physicist; b. Stellenbosch, South Africa, Feb. 21, 1928; s. Johannes and Christina (Hogewind) B.; B.Sc., U. Stellenbosch, 1945, M.Sc. in Physics, 1947; D.Sc. in Physics, U. Pretoria, 1953; m. Helga Lotz, Dec. 30, 1963; children—Kristof Johannes, Lise Rónel. Head biophysics and applied radioactivity div. Nat. Phys. Research Lab., Council for Sci. and Indsl. Research, Pretoria, South Africa, 1947-58; sr. lectr. physics U. Stellenbosch, 1958-59; dir. isotopes and radiation Atomic Energy Bd., Pretoria, 1960—. Vis. scientist Brookhaven Nat. Lab., 1955, Atomic Energy Research Establishment, Harwell, Eng., 1960-61. Mem. South African Inst. Physics, U. S. Health Physics Soc. Research, publs. on positron annihilation, standardization of isotopes and radiation (1st absolute counting of Alpha emitters in liquid scintillator), applications in industry and agr., med. physics in South Africa, research on radiobiol. effect, radiation dosimetry. Home: 485 Pienaar St., Brooklyn, Pretoria. Office: Atomic Energy Board, Private Bag 256, Pretoria, South Africa.*

BASSOT, (Jean-Antonin) Léon-Pierre, French geographer; b. Renève, Côte-d'Or, France, 1841; dir. geog. service French Army, 1898; staff Nice (France) Obs. Mem. Bur. Longitudes, French Acad. Scis., 1893. Author: (with Esteban) Détermination de la différence de longitude entre Paris et Madrid, 1899. Geog. and geodesic research especially relating to triangulation of France. Died Paris, France, 1917.

BASTIAN, Adolf, ethnologist; b. Bremen, Germany, June 26, 1826; M.D., Univ. Prague, 1850. Traveled around the world as ship's surgeon, 1851-59; expedition to Far East, 1861-66; prof. of ethnology, Univ. Berlin; keeper, Berlin Ethnological Museum. Mem., Berlin Anthropological Soc. (pres.; founder and editor, Journal of Ethnology); among founders of German Africa Society, 1878. Author: Der Mensch in der Geschichte, 1860; Die Völker des östlichen Asien, 1866-71; Ethnologische Forschungen, 1871-73; Die Kulturländer des alten Amerika, 1878; Der Buddhismus in seiner Psychologie, 1881; Indonesien, 1884; Der Fetisch an der Küste Guineas, 1885; Die mikronesischen Kolonien, 1899-1900; Die wechselnden Phasen im geschichtlichen Sehkreis und ihre Rückwirkung auf die Völkerkunde, 1900; others. Developed theory of psychic unity, stating all people have identical psychic disposition and ethnic cultural traits differ in form due to geography; important developer of ethnology. Died Port of Spain, Trinidad, Feb. 3, 1905.

BASTIAN, Henry Charlton, Brit. neurologist; b. Truro, Eng., Apr. 26, 1837; ed. U. London; m. Julia Orme; 3 sons, 1 dau. Became prof. path. anatomy, 1867; apptd. prof. clin. medicine U. Coll., London, 1878; became physician to U. Hosp., 1871. Fellow Royal Soc., 1868. Author: The Modes of Origin of the Lowest Organisms, 1871; The Beginnings of Life, 1872; Lectures on Paralysis, 1875; The Brain as an Organ of Mind, 1880; Memoirs on Nematoids; Studies in Heterogenesis, 1903; The Evidence for the Heterogenetic Origin of Bacteria, 1904; The Evolution of Life, 1907; The Nature and Origin of Living Matter, 1905; The Origins of Life, 1911. Research on diseases of brain, paralyses, aphasia, word blindness, word deafness; founder neurology in Eng.; promoted doctrine of abiogenesis (spontaneous generation); confirmed need for sterilization of med. instruments. Died Nov. 17, 1915.

BASTIANSEN, Otto Christian Astrup, Norwegian chemist; b. Balsfjord, Norway, Sept. 5, 1918; s. Alf Bjornskav and Hanna (Astrup) B.; Cand.Real, U. Oslo, 1945, Dr.Phil., 1949; m. Ragnhild Elisabeth Johnson, Dec. 15, 1945; children—Astrid Elisabeth, Tove Merete, Nerit Kristine. Research asso. U. Oslo, 1945-49, Cal. Inst. Tech., 1950; faculty U. Oslo, 1950-55, prof. phys. chemistry, 1962—; prof. theoretical chemistry Norwegian Inst. Tech., Trondheim, 1955-61; vis. prof. Ore. State U., 1961-62. Pres., Norwegian Research Council. Recipient Nat. Sci. prize Fridtjof Nansens Belonning. Mem. Norwegian, Royal Danish acads. sci. and letters, Royal Norwegian Acad. Author: Structural Chemistry Problems, 1948; also numerous articles. Research on structure of molecules, including conformational analysis and intra molecular motion. Home: 7 Vackerörn, Oslo 2, Norway.*

BASTIDE, Pierre Charles Mary Anthony, French biochemist, pharmacologist; b. Nancy, France, Sept. 11, 1927; s. Romain Camille and Rose (Delbos) B.; Chemist, U. Toulouse (France), 1952; B.S., U. Clermont-Ferrand (France), 1953, med. doctorate, 1961; Ph.D., U. Strasbourg (France), 1954; m. Cortial Janine, July 28, 1951; 1 son, Michel. Research staff dept. biol. chemistry and pharmacodynamy U. Clermont-Ferrand, 1951-54; staff U. Medicine and Chemistry, Clermont-Ferrand, 1954—, prof., 1958—; chemist Clermont-Ferrand Hosp., 1956—; expert toxicology and pharmacology Ministry of Health for Pharm. Spltys., 1961—. Mem. Soc. Biol. Chemistry, Soc. Pharmacodynamy, Soc. Chem. Biology, Soc. Therapeutical Chemistry, Ph.D. Socs., Physiologists' Assn., European Assn. for Toxicity Products, Internat. Round Table. Contbg. author: Pharmaceutical Actualities, 1958. Research and numerous publs. in gen. and comparative enzymology, pharmacodynamics, including permeability, effectors, gen. drug research. Home: 81 Bd. La Fayette-Clermont-Ferrand, Puy-de-Dome 63, France.*

BASTIDE, Roger Marius César, French sociologist, biologist; b. Nimes, Gard, France, Apr. 1, 1898; s. Marius and Delphine (Sevanier) B.; ed. Facultés des lettres, Bordeaux, Paris, Agrégé des lettres, 1924; Docteur ès lettres, 1958; Dr. Honoris causa, U. Sao Paulo, 1951; m. Jeanne Servan, Mar. 22, 1926; children—Suzanne, Christiane (Mme. Daniel Clesse). Prof. lycées of Cahors, Lorient, Valence, Versailles, 1924-38, U. Sao Paulo (Brazil), 1938-51; dir. studies l'Ecole Pratique des hautes études, 1951; prof. Faculté des lettres, Paris, 1959—, also dir. center of social psychiatry. Decorated chevalier Legion d'honneur, officer Cruzeiro do Sul. Mem. Assn. French-Speaking Sociologists. Author: Problèmes de la vie mystique, 1931; Eléments de sociologie religieuse, 1936; Sociologie et Psychanalyse, 1950; Brésil, terre des contrastes, 1957; le Condiarblé de Bahia, 1958; les Religions africaines au Brésil, 1960; Sociologie des Maladies Mentales, 1965. Did research for UNESCO on race relations in Brazil and France, 1951, on African students in France, 1953. Home: 48, rue du Général-Delestraint, Paris 16e, France.

BASTIEN, Paul, French metallurgist; b. Fontenay le Comte, France, May 26, 1907; s. Paul and Laure (Guillemet) B.; Ingénieur des Arts et Manufactures, 1929; Docteur ès sciences, 1933; Docteur en droit, 1936; Docteur honoris causa, U. Bruxelles, U. Gand; m. Fernande Collomé, Oct. 20, 1936; children—Bernard, Daniele (Mrs. Jean Triol), Catherine, Dominique. Became prof. École Centrale des Arts et Manufactures, 1942, also dir. research Sch. Engring.; sci. dir. Société des Foyes et Ateliers du Crensot. Mem. sci. council Compagnié d'Etudes es de Realisation de combustibles nucléaires, also Commissariat a l'Energie atomique. Mem. Sondure Inst. (pres.), French Civil Engrs. Soc. (past pres.), French Metallurgy Soc. (past pres.). Author: Réactifs d'attaque metallographiques; Le Magnesium; Techniques de l'Ingénieur; also numerous articles. Research on laws of conductivity of metals and alloys, mechanism of fragilization of steel by hydrogen, phenomena of differential ruptures, steel for natural gas wells, steels for parts of nuclear reactors, forgeability of steel. Home: 85 avenue Bosquet, Paris 7e. Office: 25 rue Pasquier, Paris 8e, France.*

BASTIN, Edson Sunderland, Am. geologist; b. Chgo., Dec. 10, 1878; s. Edson Sewell and Christiana (Boyd) B.; A.B., U. Mich., 1902, Sc.D. (hon.), 1941; M.S., U. Chgo., 1903, Ph.D., 1909; m. Elinor Norton, June 30, 1910. Mem. U. S. Geol. Survey, 1904-19, survey Me., Western mining dists., copper properties, Chile, 1916-17, chief, div. mineral resources, 1918-19; faculty U. Chgo., 1919-44, chmn. dept.

geology and paleontology, 1922-44. Mem. Ill. Bd. Nat. Resources and Conservation, 1922-44; chmn., div. geology and geography NRC, 1935-37. Fellow Geol. Soc. Am. (v.p. 1935), A.A.A.S. (v.p. 1930); mem. Soc. Econ. Geologists (pres. 1933), Am. Inst. Mining and Metall. Engrs., Phi Beta Kappa, Sigma Xi. Author: The Interpretation of Ore Textures, 1950. Contbr. field econ. geology, particularly investigation of mineral relations and origin of mineral deposits; interest in pegmatites; authority on superzene and hypozene silver enrichment. Died Oct. 9, 1953.

BASTOS ANSART, Manuel, Spanish surgeon; b. Saragossa, July 22, 1887; s. Atilano and Louise B.-A.; M.D., U. Saragossa; m. Consuelo Bastos Mora; July 22, 1918; children—Carmen, Victoria, Manuel, Elena, Aurora. Physician to Royal House of Spain; prof. sch. medicine at Madrid; surgeon Charity Hosp. of Madrid; dir. Inst. Re-education of Invalids at Madrid; pvt. practice medicine, Barcelona, Spain. Named physician honoris causa Am. Coll. Assn. Mem. Internat. Surg. Soc., Internat. Soc. Orthopedic Surgery and Traumatology, Spanish, German socs. surgery. Author 6 treatises on surgery, pathology, orthopedics and operational studies, also brochures and numerous articles. Address: Paseo Bonanova 22, Barcelona, Spain.

BASU, Rasanta Kumar, physician; b. Mymensingh, East Pakistan, Apr. 1, 1922; s. Praphulla Kumar and Niharkana (Guha) B.; B.Sc. with 1st class honours, U. Calcutta, 1941, M.B., 1946, D.O.M.S., 1951; m. Reba Ghosh, Nov. 18, 1948; 1 son, Pradip Kumar. Research asso., clin. affiliate in ophthalmology U. Cal. at San Francisco, 1957; research asso. in ophthalmology U. Toronto (Ont., Can.), 1957-59, asso. prof., dir. ophthalmological research, 1959—; senior ophthalmologist Toronto General Hosp., 1959—; associate Med. Research Council, 1965—; dir. Eye Bank Lab., Toronto, 1959-66. Mem. Canadian Ophthal. Soc., Assn. for Research in Ophthalmology, Toronto Acad. Medicine. Contbr. numerous articles to med. jours. Clin. and basic research in tissue culture and transplantation of eye tissues, cytology and immunology of eye, eye bank storage. Home: 19 Ringwood Crescent, Willowdale, Ont. Office: 1 Spadina Cres., Toronto 4, Ont., Can.*

BASU, Uma Prasanna, Indian chemist; b. Barisal, India, Jan. 1, 1904; s. B. B. and B. Roy (Choudhury) B.; B.Sc., Scottish Ch. Coll., 1924; M.Sc., Calcutta (India) U., 1926, D.Sc. in Chemistry, 1933; m. Sudharani Ray Choudhury, Aug. 11, 1924; children—Sabita Ghosh, Namita Ghosh, Basu N. K., Mamata Ghosh, Ajanta. Research scholar Calcutta U., 1927-31, lectr., 1931-35; chief chemist Bengal Immunity Co., Calcutta, 1935-47; dir. Bengal Immunity Research Inst., Calcutta, 1947—. Mem. Drugs Research Com., 1947—; mem. Bd. Industries, 1954-60; mem. Governing Body Med. Research, 1954-57. Recipient medal U. Nagarjuna Mouat Asutosh, Calcutta U., 1927-31; Barclay medal Asiatic Soc., 1962; Prem Chand Roy Chand scholar Calcutta U., 1935-36. Fellow Nat. Inst. Scis. India; mem. Instn. Chemists (past pres.), Indian Sci. Congress Assn. (past gen. sec.), Indian Pharm. Assn. (pres. 1966—), Indian Statis. Inst. (governing body 1966—). Research and numerous publs. on synthesis of organic chemicals, plant products, chemotherapy, oils and oily products, enzymes and ferments, methods of analysis and standardization, utilization of raw materials in industry; contributed to improvement of products in glass and glycerin industry. Home: 23/3 Gariahat Rd., Calcutta-19. Office: 39, Acharya Jagadish Bose Rd., Calcutta-16, W. Bengal, India.*

BATAILLON, Jean Eugène, French biologist, zoologist; b. Annoire, Jura, France, Oct. 22, 1864; s. Francois and Henriette (Dupuis) B.; Sc.B., Coll. Arbois, 1883; Lic.Sc., U. Lyons (France), 1886; Sc.D., U. Paris (France), 1891; Sc.D., U. Brussels (Belgium); m. Marie Wahl, Apr. 24, 1890; 5 sons. Lectr. zoology and physiology at Dijon (France), 1893; named prof. gen. biology, 1903; became prof. gen. biology U. Strasbourg (France), 1919; named rector U. Clermont-Ferrand, 1921; prof. zoology U. Montpellier (France), 1924-32; dean Faculty Scis., Dijon, 1907-18, U. Strasbourg., 1919-21; dir. zool. Inst. and Mus., U. Strasbourg. Mem. French Inst., Belgian Biol. Soc., Brussels Soc. Medicine and Natural Scis., Portuguese Soc. Natural Scis., Royal Belgian Acad. (asso.), French Acad. Scis., 1916. Author: Parthénogenèse traumatique des Batraciens, 1910; Analyse de la fécondation par le parthénogenèse . . . , 1929. Research on artificial parthenogenesis of frogs; discovered traumatic parthogenesis, 1910, also role of carbonic acid in maturation of egg, moving form in polyembryo of lamprey egg. Died Montpellier, 1953.

BATCHELDER, Esther L., Am. home economist; b. Hartford, Conn., May 19, 1897; d. Joseph Warren and Margaret (Odell) Batchelder; B.S., Conn. Coll. for Women, 1919; M.A., Columbia, 1925, Ph.D., 1929. Nutrition editor Butterick Co., N.Y.C., 1929-32; asst. prof. Wash. State U., 1932-34, U. Ariz., 1934-36; prof., dean home econs. R.I. State U., 1936-42; asst. dir. nutrition div. Agrl. Research Service, U. S. Dept. Agr., Beltsville, Md., 1942-56, dir. clothing, housing research div., 1956-65; cons. home econs., Bristol, Conn., 1965—. Recipient U. S. Dept. Agr. Distinguished Service award. Mem. Am. Inst. Nutrition, Am. Dietetic Assn., Am. Chem. Soc., Am. (chmn. research div. 1963-65), (pres. 1963-65) D.C. home econs. assns. Research, publs. on human requirements for Vitamins A, C, on quantities of these nutrients in common foods; directed research on human requirements for amino acids, minerals, vitamins and fatty acids and on household processing and use of foods to conserve palatability and nutritive value; functional requirements for clothing and housing and on use of textiles and equipment. Home: 53 Mark St., Bristol, Conn. 06010.*

BATCHELDER, John Putnam, Am. physician; b. Wilton, N.H., Aug. 6, 1784; s. Archelaus and Betty (Putnam) Batchelor; grad. Harvard. Settled in N.Y.C., lectr. anatomy and surgery, Castleton, Vt., also Berkshire (Mass.) Inst.; considered 1st Am. physician to perform operation to remove head of thigh bone; invented one-handed craniotome (instrument used in opening skull). Died N.Y.C., Apr. 8, 1868.

BATCHELOR, George Keith, applied mathematician; b. Melbourne, Australia, Mar. 8, 1920 ; s. George Conybere and Ivy (Berneye) B.; B.Sc., U. Melbourne, 1940, M.Sc., 1941; Ph.D., U. Cambridge (Eng.), 1948; Sc.D., U. Grenoble (France), 1959; m. Wilma Maud Ratz, Jan. 27, 1944; children—Adrienne, Clare, Bryony. Research officer Aero. Research Lab., Melbourne, 1940-44; faculty U. Cambridge, 1948—, prof. applied math., 1964—, head dept. applied math. and theoretical physics, 1959—. Recipient Adams prize U. Cambridge, 1951. Fellow Royal Soc.; fgn. hon. mem. Am. Acad. Arts and Scis. Author: The Theory of Homogeneous Turbulence, 1953; An Introduction to Fluid Dynamics, 1967; also numerous articles. Editor: Surveys in Mechanics (with R. M. Davies), 1956; Scientific Papers of G. I. Taylor, vol. 1, 1958, vol. 2, 1960, vol. 3, 1963; Jour. Fluid Mechanics, 1956—. Research on analysis of motion of fluids especially turbulent motion. Home: Cobbers, Conduit Head Rd., Cambridge, Eng.*

BATE, Charles Spence, English naturalist, dentist; born near Trure, England, March 16, 1819; son of Charles and Harriet (Spencer) B.; student surgery, dentistry; m. Emily Amelia Hele, June 17, 1847; 2 sons, 1 dau. Practiced dentistry. Licentiate Royal Coll. Surgeons; Fellow Linnean Soc., Royal Soc., 1861; member Odontological Soc. (v.p. 1860-62, pres. 1885), Brit. Dental Assn. (pres. 1883), Plymouth Instn. Author: Catalogue of the Specimens of the Amphipodous Crustacea, 1862; (with John Obadiah Westwood) History of the British Sessile-eyed Crustacea, 1863-68; Report on the Crustacea Macrura dredged by H.M.S. Challenger during the years 1873 and 1876, 1888. Contbr. articles on dentistry to profl. jours. Recognized as greatest authority of his time on crustacea. Died Devonshire, England, July 29, 1889.

BATEMAN, Alan Mara, Am. geologist, educator; b. Kingston, Ont., Jan. 6, 1889; s. George Arthur and Elizabeth (Mara) B.; B.S., Queens U., 1910, Ph.D., Yale 1913; m. Grace Hotchkiss Street, June 3, 1916. Came to U. S., 1910, naturalized, 1915. Secondary enrichment investigation Harvard, 1913-15; faculty Yale, 1913-59, Silliman prof. geology, 1925-59, chmn. dept., 1945-59, trustee Sheffield Sci. Sch. Cons. geology Kennecott Copper Co., N.Y.C., 1915—, Rhodesian Selection Trust, London and Rhodesia, 1929-32; head spl. U. S. Mission to Mexico, 1941-42; dir. metals and minerals Fgn. Economic Adminstrn., Washington, 1942-45; expert cons. SCAP, Tokyo, 1949; adviser ECA, ODM, Paley Com., Nat. Minerals Resource Com. Mem. Am. Inst. Mining Engrs., Mining and Metall. Soc. (pres. 1954), Mineral. Soc. Am., Geol. Soc. Am., Soc. Econ. Geologist (pres. 1941-42, Penrose Gold medal 1962), Am. Acad. Arts and Sci., Am. Assn. Petroleum Geologists, Soc. Geol. de Belge (hon.), Theta Xi. Author: Economic Mineral Deposits, 1950, Formation of Mineral Deposits, 1951. Editor, Jour. Econ. Geology, 1917—. Contbr. articles to sci. jours. Home: 450 St. Ronan St., New Haven.*

BATEMAN, Angus John, English geneticist; b. London, Eng., Sept. 2, 1919; s. John William and Lily (Bennet) B.; B.Sc. with honors in Botany, King's Coll., London, 1940, Ph.D., 1946, D.Sc., 1956; m. Eva Mary Dovey, June 8, 1945; children—Angela Elizabeth, Timothy John. With genetics dept. John Innes Hort. Instn., Merton Park, London, 1942-49, Bayfordbury, Herts, Eng., 1949-53; Intermediate BECC Research fellow cytogenetics lab. Paterson Labs., Christie Hosp. and Holt Radium Inst., Manchester, Eng., 1953-58, prin. research officer, 1958—. Mem. Genetical Soc. (mem. com. 1963-66), Assn. for Radiation Research. Research, numerous publs. in cross pollination of seed crops, genetics of self incompatibility in Criciferae, genetics of quantitative characters in barley and rye, induction of mutation by radio-isotopes in mouse and Drosophila; variation in X-ray sensitivity of germ cells of mouse and Drosophila during maturation, effects of X-rays on crossing-over in Drosophila, dominant lethal mutation in mammals. Home: 9 Finney Dr., Manchester 21. Office: Paterson Labs., Christie Hosp., Manchester 20, Eng.*

BATEMAN, Harry, mathematician; b. Manchester, Eng., May 29, 1882; s. Samuel and Marnie Elizabeth (Bond) B.; B.A., Trinity Coll., Cambridge, 1903, M.A., 1906; studied Göttingen and Paris, 1905-06; came to U. S., 1910; Ph.D., Johns Hopkins, 1913; m. Ethel Horner Dodd, July 11, 1912; children—Harry Graham (dec.), John Margaret. Prof. math., theoretical physics and aeros. Cal. Inst. Tech., from 1917. Mem. Am. Phys. Soc., Am. Math. Soc., London Math. Soc., Brit. Assn. Adv. Science, Am. Philos. Soc., Am. Acoustical Soc., Nat. Acad. Sci.; Fellow Royal Soc., 1928. Author: Electrical and Optical Wave Motion, 1915; Differential Equations, 1918; Partial Differential Equations of Mathematical Physics, 1931, 2d edit., 1944. Died Jan. 21, 1946.

BATEMAN, James, Brit. botanist; b. Redivals, Eng., July 18, 1811; s. John and Elizabeth (Holt) B.; B.A., Oxford (Eng.) U., 1834, M.A., 1845; m. Maria Sybilla Warburton, Apr. 24, 1838; children—John, Rowland, Robert, Katherine. Pub. illus. works on orchids. Fellow Linnean Soc., Royal Soc., 1838, Royal Hort. Soc. Author: Orchidaceae of Mexico and Guatemala, 1837-1843; Monograph of Odontoglossum, 1864-74; A Second Century of Orchidaceous Plants, 1867. 1st to bring carambola to maturity in Eng.; one of 1st to advocate cool orchid cultivation. Died Springbank, Eng., Nov. 27, 1897.

BATEMAN, Paul Trevier, Am. mathematician; b. Phila., June 6, 1919; s. Harold John and Anna (Yeager) B.; A.B., U. Pa., 1939; A.M., 1940, Ph.D., 1946; m. Felice Hilda Davidson, June 25, 1948; 1 dau., Sarah Elizabeth. Lectr., Bryn Mawr Coll., 1945-46; instr. U. Pa., Phila., 1946, vis. researcher, 1961-62; instr. Yale, 1946-48; mem. Inst. for Advanced Study, Princeton, N.J., 1948-50, 56-57; faculty U. Ill., Urbana, 1950—, prof., 1958—, acting head math. dept., 1963-64, head, 1965—; vis. prof. City U. N.Y., 1964-65. Mem. adv. panel for math. scis. NSF, 1963-66. NSF sr. postdoctoral fellow, 1956-57. Mem. Am. Math. Soc. (mem. council 1961-63, asso. sec. 1966—), Phi Beta Kappa, Sigma Xi, Pi Mu Epsilon. Editorial bd.: Ill. Jour. Math., 1959-64. Research and supervision of research in number theory and related parts of analysis and algebra. Home: 108 W. Meadows, Urbana, Ill. 61801.*

BATEMAN, Thomas, Brit. physician; b. Whitby, Eng., 1778; ed. pvt. schs.; student Windmill St. Sch. Anatomy, from 1797; M.D., U. Edinburgh, Scotland, 1801. Practice medicine London, Eng.; physician to pub. dispensary, London, Fever Hosp., London, from 1804; asso. Edinburg Med. and Surg. Jour. Licentiate Coll. Physicians; mem. Royal Med. and Chirurg. Soc. (1st librarian). Author: Synopsis of Cutaneous Diseases, 1813; Delineations of Cutaneous Diseases, 1817; A Succinct Account of the Contagious Fever of this Country in 1817 and 1818, 1818; Reports on the Diseases of London, 1819. Contbr. articles to Rees's Cyclopaedia. Important work on skin diseases; 1st to describe lichen urticatus, 1813; 1st to give accurate account of molluscum contagiosum, 1817. Died Apr. 9, 1821.

BATEN, William Dowell, Am. math. statistician; b. Ft. Worth, July 23, 1892; s. Anderson E. and Clara Kate (Williams) B.; B.A., Baylor U., 1914; M.A., U. Tex., 1918; Ph.D., U. Mich., 1929; m. Versie Jones Giles, Dec. 28, 1953; m. Allie Stephens Neel, Dec. 26, 1916; 1 son, James D. Tchr. high sch., 1914-19, N. Tex. A. and M. Coll., 1919-25, Bessie Tift Coll., 1925-27; asso. prof. U. Tenn., 1927-28, U. Mich., 1928-36; faculty Mich. State U., East Lansing, 1936-47, 49-61, agrl. statistician, 1949-61, ret., 1961. Fellow Am. Soc. Quality Control (mem. nat. com.), Inst. Math. Statistics (mem. nat. com.); mem. Math. Assn. Am. (pres. Mich. 1957), Sigma Xi. Research and numerous articles on applications of statistics to applied scis., probability theory. Home: 26 University Dr., East Lansing, Mich. 48823.*

BATES, Charles Carpenter, Am. earth scientist; b. Harrison, Ill., Nov. 4, 1918; s. Carl Albert and Vera (Carpenter) B.; B.A., DePauw U., 1939; M.A., U. Cal. at Los Angeles, 1943; Ph.D., Tex. A. and M. U., 1952; m. Pauline Barta, July 11, 1942; children—Nancy Ann, Priscilla Jane, Sally Jean. With Carter Oil Co., Tulsa, 1939-41; spl. asst. to pres. Am. Meteorol. Soc., 1946; oceanographic technician Marshall Islands, Woods Hole Oceanographic Inst., 1946; partner Bates & Glenn, Washington, 1946-47, asso. oceanographer to dep. dir. div. oceanography U. S. Naval Oceanographic Office, Washington, 1946-57; environmental surveillance coordinator U. S. Dept. Navy, 1957-60; chief Vela Uniform br. Advanced Research Projects Agy., U. S. Dept. Def., 1960-64; sci. and tech. dir. U. S. Naval Oceanographic Office, Washington, 1964—. U. S. Antarctic observer U. S. Dept. State, 1964—, Recipient U. S. Navy Meritorious Civilian Service award, 1952. Mem. Soc. Exploration Geophysicists (v.p. 1965-66, rep. to NRC 1963-66), Am. Geophys. Union (mem. council 1964—), Am. Assn. Petroleum Geologists (Pres.'s award 1953), Am. Meteorol. Soc., Seismol. Soc. Am. Research, publs. on delta formation, aerial ice observation, surf condition along enemy-held beaches, detection, location, and identification underground nuclear explosions for nuclear test ban purposes; pioneered oceanographic engring services to offshore oil operators. Home: 5807 Massachusetts Av. N.W., Washington 20016. Office: U. S. Naval Oceanographic Office, Washington 20390.*

BATES, David Robert, physicist; b. Omagh, Ireland, Nov. 18, 1916; s. Walter Vivian and Mary (Shera) B.; B.Sc., Queen's U., Belfast, 1937, M.Sc., 1938; D.Sc.,

U. Coll., London, 1951; m. Barbara Morris, Mar. 20, 1956; children—Katharine Mary, Adam David. Exptl. officer Admiralty Research Lab., Teddington, 1939-41, Mine Design Dept., Portsmouth, Eng., 1941-45; lectr. math. U. Coll., London, 1945-50, reader in physics, 1951; cons. U. S. Naval Ordnance Test Sta., Inyokern, Cal., 1950; prof. applied math. Queen's U., Belfast, Ireland, 1951—. Fellow Royal Soc., Inst. Math. and Its Applications; mem. Royal Irish Acad., Internat. Acad. Astronautics. Editor: The Planet Earth, 1957, 2d ed., 1964; Atomic and Molecular Processes, 1962; Editor-in-chief Planetary and Space Sci.; co-editor (with Immanuel Estermann) Atomic and Molecular Physics. Research in physics of upper atmosphere, in atomic physics. Home: 6 Deramore Park, Belfast 9. Office: Sch. Physics and Applied Math., Queen's U., Belfast 7, Ireland.*

BATES, David Vincent, physician; b. Kent, Eng., May 20, 1922; s. John Vincent and Alice (Dickins) B.; M.B., B.Ch., Cambridge U., 1945, M.D., 1954; m. Margaret Sutton, Mar. 24, 1948; children—Anne Elizabeth, Joanna Margaret, Andrew Vincent. Sr. lectr. medicine U. London, 1953-56; research fellow U. Pa., 1952; asso. physician Royal Victoria Hosp., Montreal, Que., Can., 1956—; asst. prof. medicine McGill U., 1956—, prof. exptl. medicine, 1965—, asso. dean grad. studies and research, 1964—, chairman department of physiology, 1967—; dir. respiratory div. joint cardiorespiratory service Royal Victoria, Montreal Childrens hosp., 1957—. Mem. Am. Soc. Clin. Investigation, Am., Canadian physiol. socs., Canadian Soc. Clin. Investigation, Physiol. Soc. London. Author: (with Christie) Respiratory Function in Disease, 1965. Contbr. numerous articles to sci. jours. Research on effect of lung disease on lung function, use of radioactive gases to study function different parts of lungs, lung function in athletes, old people, effect of air pollutants (ozone) on lung function. Home: 470 Portland Av., Montreal 16. Office: Dept. Physiology, McGill U., Montreal, Que., Can.*

BATES, Frederick (John), Am. phys. chemist; b. Marysville, Kan., Jan. 2, 1877; s. Charles A. and Harriett (Roberts) B.; B.S., U. Kan., 1900; A.M., U. Neb., 1902; m. Gertrude C. Coyle, Jan. 5, 1905. Chief magneto-optical and carbohydrate sect. Nat. Bur. Standards, 1903—; chief, optics div., 1941-47; cons. carbohydrate chemistry and optics, 1948; devised methods and prepared Treasury Dept. regulations for weighing, gauging, sampling, classifying and testing imported sugars; supr. govt. sugar labs. of the Customs Service, Treasury Dept.; pres. Internat. Commn. for Uniform Methods of Sugar Analysis. Mem. Am. Chem. Soc., Am. Phys. Soc., Washington Philos. Soc., Am. Optical Soc., Internat. Soc. Sugar Cane Technologists, Internat. Union of Pure and Applied Chemistry, Sigma Xi. Contbr. extensive researches in natural and magnetic rotary polarizations of light, and in transformations in silica, especially anomalous rotary dispersion; co-author of Bur. of Standards of Baumé scale. Co-author: (circular) Polarimetry, Saccharimetry, and the Sugars. Developed sensitive strip spectral polarizing system, Bates quartz compensating polariscope with adjustable sensibility, Bates cadmium-vapor arc lamp, Bates polariscope tubes, Bates sugar balance; research on inversion of quartz and of rotary polarization of magnetic elements at high temperatures. Died Nov. 1, 1958.

BATES, Grace Elizabeth, mathematician; b. Albany, N.Y., Aug. 13, 1914; d. Walter M. and Julia (Dexter) Bates; B.S., Middlebury Coll., 1935; Sc.M., Brown U., 1938; Ph.D., U. Ill., 1946. Tchr. math. George Sch., Pa., 1938-43; instr. Sweet Briar Coll., Va., 1943-44; faculty Mt. Holyoke Coll., South Hadley, Mass., 1946—, prof. math., 1958—. Mem. Am. Math. Soc., Math. Assn. Am., Inst. Math. Statistics. Author: (with F. Kiokemeister) The Real Number System, 1960; (with Johnson, Lindsey, Slesnick) Modern Algebra, 2d Course, 1963. Contbr. research articles in algebra and probability theory to tech. jours. Home: 16 College View Heights, South Hadley, Mass. 01075.*

BATES, Henry Walter, English naturalist; b. Leicester, Eng., Feb. 8, 1825; s. Henry B.; apprentice to Alderman Gregory, 1838; m. Sarah Ann Mason, Jan. 1861; 1 dau., 3 sons. Clk., Allsopp's Offices, Burton-on-Trent, Eng., 1845; travelled (with Alfred Russel Wallace) to Brazil, 1848; journeyed through Upper Amazons, 1851-59; returned to Eng., 1859; asst. sec. to Royal Geog. Soc., 1864-92. Fellow Linnean Soc., Royal Soc., 1881; mem. Entomol. Soc. (pres. 1869, 78). Species Callithea batessi named for him. Author: The Naturalist on the River Amazonas, 2 vols., 1863; Contributions to Insect Fauna of the Amazon Valley, 1867; Central America, West Indians, and South America, 1882; also numerous articles. Editor several works on natural history and topography. Discovered over 8000 new species in Amazons; proposed theory of mimicry; reinforced theory of natural selection. Died London, Feb. 16, 1892.

BATES, Leslie Fleetwood, English physicist; b. Mar. 7, 1897; s. W. F. Bates; B.Sc., U. Bristol (Eng.); Ph.D., Trinity Coll., Cambridge (Eng.) U.; D.Sc., U. London (Eng.); m. Winifred Frances Furze Ridler, 1925 (dec. 1965); 1 son, 1 dau. Lancashire-Spencer prof. physics Nottingham (Eng.) U., 1936-64, emeritus, 1964—. Sr. sci. adviser for civil defense N. midland region Eng.; Rippon lectr. U. Calcutta (India), 1960; Guthrie lectr., 1963. Recipient Holweck prize French, English phys. socs. Fellow Royal Soc., 1950, Inst. Physics (mem. bd. 1947-49); mem. Phys. Soc. (pres. 1950-52). Author: Modern Magnetism, 1939, 48, 51, 61, 63; Sir Alfred Ewing, 1946; Recent Advances in Physics, Science Progress, 1928-36. Research, publs. on electricity, magnetism. Address: Castlethorpe, Newcastle Circus, The Park, Nottingham, Eng.

BATES, Lindon Wallace, civil engr.; b. Marshfield, Vt., Nov. 19, 1858; s. William W. and Mary C. B.; ed. Yale; m. Josephine White; son—Lindon Wallace. Asst. engr. N.P. and Oregon Pacific rys.; contractor engr. or mgr. various ry., dock and terminal contracts in Ore., Wash., Mont., Kan., Mo., Ill., La., Cal., for transcontinental rys. or their subsidiary cos., on various projects, including Chgo. Drainage Canal,; built mammoth dredge Beta for U. S. Govt.; retained 1896-1902 by Belgian Govt. to prepare reports and projects for improvement of port of Antwerp; on Suez Canal on enlargement of canal; by Russian Govt. on rivers Volga, Dnieper and Bug, Azov Sea ports and channels, Black Sea ports,; by the Queensland Govt. designed 8 harbors and regulation of Brisbane, Mary, Fitzroy, Norman and Albert rivers; built large hydraulic dredge for Russian Govt.; with engrs. designated by govts. of Russia, Germany, Austria and Belgium prepared scheme for improvement of port of Shanghai; contracting engr. Galveston grade raising works; designed the Three Lake Panama Canal; dir. various works in Korea, Trinidad and Peru. Grand prix and decoration from French Govt., 1900, for distinguished services to science. Author: Retrieval at Panama; Colloidal Fuel Chmn. engring. com. Submarine Defense Assn., 1917. Inventor colloidal fuel. Died Apr. 22, 1924.

BATES, Marston, Am. naturalist; b. Grand Rapids, Mich., July 23, 1906; s. Glenn Freeman and Amy (Button) B.; B.S., U. Fla., 1927; A.M., Harvard, 1933, Ph.D., 1934; Sc.D. (hon.), Kalamazoo Coll., 1956; m. Nancy Bell Fairchild, Jan. 11, 1939; children—Marian Hubbard, Sally Norton, Barbara Fairchild, Glenn Peregrine. With Tecnico de Cooperacion Agricola, United Fruit Co., 1928-31, dir. servicio, 1930-31; research asst. Mus. Comparative Zoology, Harvard, 1935-37; staff internat. health div. Rockefeller Found., 1937-50, spl. asst. to pres., 1950-52; prof. zoology U. Mich., Ann Arbor, 1952—; dir. research U. P.R., 1956-57. Phi Beta Kappa Vis. scholar, 1962-63; fellow Center for Advanced Studies, Wesleyan U., Middletown, Conn., 1961; recipient Sci. Book award Phi Beta Kappa, 1960. Mem. Am. Acad. Arts and Scis., Am. Anthrop. Assn., A.A.A.S., Am. Geog. Soc., Am. Inst. Biol. Scis., Am. Soc. Mammalogists, Am. Soc. Naturalists (pres. 1961), Am. Sociol. Assn., Brit. Ecol. Soc., Ecol. Soc. Am., Nat. Assn. Biology Tchrs., Population Assn. Am., Soc. for Sci. Study of Sex, Soc. for Study Evolution, Soc. Systematic Zoology. Author: Natural History of Mosquitoes, 1949; The Nature of Natural History, 1950; The Forest and the Sea, 1960. Research, publs. on analysis of behavioral differences among mosquito species in So. Europe in relation to malaria transmission, devel. of explanation of occurrence of jungle yellow fever in S.Am., structure of tropical rain forest. Home: 630 Oxford St., Ann Arbor, Mich. 48104.*

BATES, Oric, Am. archaeologist; b. Boston, Mass., Dec. 5, 1883; s. Arlo and Harriet (Vose) B.; A.B., Harvard, 1905, A.M., 1914; postgrad. U. Berlin, 1906; m. Natica Y. Inches, June 5, 1913. In service Egyptian Govt. Archaeol. Expdn. in Nubia, 1906, 1908, of Palestinian Exploration Soc., in Syria, 1907; in charge Am. Archaeol. Soc. Expdn. to Cyrenaica, 1909, of Wellcome Archaeol. Expdn. at Gebel Moya, So. Sudan, 1910; pvt. exploration to Oasis of Siwah, 1911; pvt. archaeol. expdn. in Libyan Desert, 1913; Peabody Mus. Sudanese Expdn., 1914-15. Curator African dept., Peabody Mus., 1913—. Author: The Eastern Libyans, 1913. Editor, Harvard African Studies, 1917—. Died Oct. 10, 1918.

BATES, Robert Wesley, Am. endocrinologist, biochemist; b. Columbia, Ia., Jan. 31, 1904; s. Alfred Levi and Kate (Marshal) B.; B.A., Simpson Coll., 1925; Ph.D., U. Chgo., 1931; m. Mildred Brown, June 14, 1930; children—Mary Jane (Mrs. Robert M. Nichols, Jr.), Robert B., James M. Hormone biochemist NIH, Bethesda, Md., 1952—, endocrine study sect., 1958-61. Mem. Endocrine Soc. (mem. council), Am. Soc. Biol. Chemists, Soc. Exptl. Biology and Medicine, A.A.A.S., Am. Chem. Soc. Research, numerous publs. on isolation, bioassay and physiol. action of hormones of auterior pituitary gland, transplantable pituitary tumors which produce excess quantities of pituitary hormones causing gigantism or inducing permanent diabetes. Home: 5210 Danbury Rd. Office: NIH, Bethesda, Md. 20014.*

BATES, William Nickerson, Am. archaeologist; b. Cambridge, Mass., Dec. 8, 1867; s. Charles and Anna Pamela (Nickerson) B.; A.B., with honors, Harvard, 1890, A.M., 1891, Ph.D., 1893; postgrad. Am. Sch. Classical Studies, Athens, Greece, 1897-98; hon. L.H.D., U. Pa., 1940; m. Edith Newell Richardson, Dec. 28, 1901 (dec. Feb. 1926); children—William Nickerson, Robert Hicks. Instr. Greek, Harvard, 1893-95; with U. Pa., from 1895, as instr. Greek, asst. prof. and prof. Greek lang. and lit., also head dept., 1910-39, emeritus prof. from 1939. Incorporator Archaeol. Inst. Am., 1902; mem. mng. com. Am. Sch. Classical Studies in Athens from 1902, prof. Greek lang. and lit. and acting dir., 1905-06. Del. Internat. Congress for History of Religions, Leyden, 1912. Mem. governing bd. Am. Found. in France for Prehistoric Studies (now Am. Sch. Prehistoric Research), 1921-25. Fellow Am. Acad. Arts and Scis.; mem. Am. Philol. Assn., Hellenic Soc. (London), Phi Beta Kappa. Author: Date of Lycophron, 1895; Notes on the Theseum at Athens, 1901; The Old Athena Temple on the Acropolis, 1901; Etruscan Inscriptions, 1905; New Inscriptions from the Asclepieum at Athens, 1907; Five Red-Figured Cylices, 1908; Two Labors of Heracles on a Geometric Fibula, 1911; Euripides, a Student of Human Nature, 1930; Sophocles, Poet and Dramatist, 1940; also numerous articles in archeol. and philol. jours. Revised: Hertzberg's History of Greece, 1905. Editor: Iphigenia in Tauris (Euripides), 1904; Trans. of Univ. Museum, 1904-07; editor Am. Jour. Archaeology, 1908-20, editor-in-chief, 1920-24. Died June 10, 1949.

BATESON, Gregory, anthropologist; b. Grantchester, Eng., May 9, 1904; student Geneva U.; A.B., Cambridge, U., 1925, M.A., 1930; m. Margaret Mead; 1 dau., Margaret Catherine. Researcher among Baining, Gazelle Peninsula, 1927-28; among Sulka, Gazelle Peninsula, also Iatmul of New Guinea, 1928-30; tchr. Melanesian linguistics Sydney (Australia) U., 1929; research fellow St. John's Coll., Cambridge, 1931, 34; researcher, mountain community of Bali, 1936-38; William Wyse scholar Cambridge U., 1938; researcher among Iatmul of New Guinea, 1938; specialist Balinese material Am. Mus. Natural History, N.Y.C., 1940; anthrop. film analyst Mus. Modern Art, N.Y.C., 1942-43; with OSS, 1943-45; Guggenheim fellow, vis. prof. grad. faculty New Sch., N.Y.C.; vis. prof. Harvard, 1947-48; lectr. Langley Porter Clinic, 1948-50; ethnologist VA Hosp., Palo Alto, Cal., from 1950. Author: Naven, 1936; (with Margaret Mead) Balinese Character—A Photographic Analysis, 1942. Specialist in anthropology study of culture and personality, primitive social orgn.

BATESON, William, English biologist; b. Whitby, Eng., 1861; ed. Rugby Sch.; M.A., St. John's Coll., Cambridge (Eng.) U.; D.Sc. (hon.), Sheffield (Eng.) U.; m. Beatrice Durham, 1896; 1 son. Silliman lectr. Yale, New Haven, 1907; prof. biology Cambridge U., 1908-09; dir. John Innes Hort. Instn., Merton Park, Surrey, Eng., 1910-26; Fullerian prof. physiology Royal Instn., 1912-14. Trustee Brit. Mus., 1922-26. Fellow Royal Soc. (Darwin medal 1904, Royal medal 1920), 1894; mem. Brit. Assn. Advancement Sci. (pres., Australia 1914), N.Y., Nat. (Washington, fgn.) acads. sci., Royal Danish Acad., Royal Acad. Belgium, others. Author: Materials for the Study of Variation, 1894; Mendel's Principles of Heredity, 1902; Problems of Genetics, 1913; papers on biol. subjects. First to coin term genetics in reference to study of inheritance; through expts. discovered that not all characteristics are independently inherited, some inherited together; postulated that chordates were off-shoots of primitive echinoderm stock (theory now widely accepted). Died Feb. 8, 1926.

BATH, Markus, Swedish seismologist; b. Katrineholm, Sweden, July 29, 1916; s. Karl Ludvig and Sigrid (Friedner) B.; Ph.D., Uppsala (Sweden) U., 1949; m. Ingrid Viktoria Lundh, Mary 23, 1942; children—Kerstin, Hans, Eva, Anders. Asst., Meteorol. Inst., Uppsala, 1939-49; faculty, Uppsala U., 1949—, prof. seismology, 1967—, dir. Seismol. Inst., since 1961—. Mem. Am. Geophys. Union, Seismol. Soc. Am., Royal Meteorol. Soc. Research, numerous publs. in meteorology and seismology including microseisms, energy and magnitude of earthquakes, wave propagation. Home: Geijersgatan 43, Uppsala, Sweden.*

BATHER, Francis Arthur, Brit. geologist, zoologist; b. 1863; s. Arthur H. and Lucy (Blomfield) B.; ed. New Coll., Oxford, Eng.; M.A., 1890; D.Sc.; m. Stina Bergoo, 1896; 2 sons, 1 dau. With Brit. Mus. (natural history), from 1887, dep. keeper, 1902-24, keeper dept. geology, 1924-28. Fellow Royal Soc., 1909, Geol. Soc. (Wollaston Fund 1897, Lyell medal 1911); pres. Geol. Soc., 1927, Museums Assn., Brit. Assn. for Advancement Sci.; corr. Geol. Soc. Am. Author: Crinoidea of Gotland, Stockholm, 1893; Triassic Echinoderms of Bakony, 1909; Cystidea from Girvan, 1913; Studies in Edrioasteroidea, 1915; chpts. on echinoderms in Ency. Brit., Treatise on Zoology (Lankester), also contbd. tech. papers to sci. publs. Made detailed studies of fossils; expert on echinoderms and fossil crinoids. Died Mar. 20, 1934.

BATISTA, Augusto Chaves, Brazilian mycologist, microbiologist; b. Sto. Amaro, Brazil, June 15, 1916; s. José Otaviano and Teodora A. C. Batista; Agronomist Engring., Escola Agrícola da Bahia, 1937; m. Algezira A. Batista, Oct. 10, 1941. Dir. agr. State of Mato Grosso, Brazil, 1938-39; prof. phytopathology and microbiology Escola Agrícola da Bahia, 1939-46; prof. cathedratic of phytomathology Universidad Rural de Pernambuco, Brazil, 1946-54, chief plant pathology sect. Instituto Agronômico de Pernambuco; dir. Found., Instituto Demicologia, U. Federal de Pernambuco, 1954—. Recipient Medal of Merit of Agr., 1966. Mem. Academia Brasileira de Ciecias, Am. Phytopath. Soc., Brit. Mycol. Soc., Mycology Soc. Am. Author: (with R. Ciferri) The Chaetothyriales, 1957; Monografia dos Fungos Micropeltaceae, 1959; also numerous articles. Research on soil microbiology, improvement Latin Am. agrl. improvement, increase of food prodn. Home: 119 Teódulo Miranda, Recife, Pernambuco, Brazil.*

BATSAKIS, John George, Am. physician; b. Petoskey, Mich., Aug. 14, 1929; s. George John and Stella (Vlahakis) B.; student Va. Mil. Inst., 1947-48, Albion Coll., 1948-50; M.D., U. Mich., 1954; m. Mary Janet Savage, Dec. 28, 1957; children—Laura, Sharon, George. Asst. chief lab. service, chief clin. pathology Walter Reed Army Med. Center, Washington, 1959-61; faculty U. Mich. Med. Sch., Ann Arbor, 1962—, asso. prof. pathology, 1965—, asso. dir. clin. labs. Med. Center, 1965—; guest faculty U. S. Naval Med. Center, Bethesda, Md., 1966. Cons. Pontiac (Mich.) Gen. Hosp., 1966—, VA Hosp., Ann Arbor, 1962—. Recipient certificate of achievement Walter Reed Army Med. Center, 1961. Diplomate Am. Bd. Pathology. Mem. Coll. Am. Pathologists, Assn. Clin. Scientists, Assn. Mil. Surgeons U. S. A., A.A.A.S., Alpha Omega Alpha, Phi Kappa Phi. Author: (with S. E. Gould, P. R. Beamer, D. L. Hinerman) Microscopic Pathology, 1965; also numerous articles. Research in clinical enzymology, thyroid disorders, neoplasms of the head and neck, and gastro-intestinal pathology. Home: 2824 Colony Rd., Ann Arbor, Mich. Office: U. Mich. Medical Center, Ann Arbor, Mich. 48104.*

BATSCH, August Johann Georg Karl, German botanist; b. Jena, Germany, Oct. 29, 1761; s. Georg Laurentius and Johannetta Ernestina Margaretha (Francke) B.; student at Jena; dr. philosophy and medicine; m. Sophie Carolina Amalie Pfündel, Apr. 29, 1787; 3 children. Prof. natural history in Jena, later prof. medicine and philosophy. Author: Elenchus fungerum latine et germanice, 1783-89; Dispositio generum plantarum Jenensium secundum Linnaeum et familias naturales, quam speciminis inauguralis loco extulit, 1786; Botanik für Frauenzimmer, 1795; Tabulae affinitatum regni vegetabilis, 1802. Research on fungi; arranged domestic plants in 78 families and 9 classes; devel. gen. nature system. Died Jena, Sept. 29, 1802.

BATTAGLINI, Giuseppe, Italian mathematician; b. Naples, Italy, Jan. 11, 1826; prof. geometry U. Naples; apptd. prof. geometry U. Rome, Italy, 1872. Mem. Gionale di matematiche (a founder). Author: Trattato elementare di meceanica razionale, 1873; Sulla geometria immaginaria di Lobatschewsky, 1867. Translated many fgn. publs. into Italian. Research on principles of ruled geometry, algebraic forms, invariants, co-variants, applications to solution of certain differential equations; promoted Lobachevski's non-Euclidean geometry in Italy. Died Apr. 29, 1894.

BATTANDIER, Jules Aimé, French botanist, pharmacist; b. Annonay, France, Jan. 8, 1848; D.ès.S. in Natural Scis., Paris; became prof. Faculty Medicine and Pharmacy, Algiers, Algeria, 1879; named pharmacist, Douai, 1874; apptd. head pharmacist Mustapha Hosp., 1875. Mem. French Acad. Scis., 1918. Author: L'Algérie, 1898; Flore d'Alger et catalogue des plantes d'Algérie, 1903; Contribution à la flore atlantique, 1919. Studied flora of Africa. Died Algiers, Sept. 18, 1922.

BATTEN, Alexander William Chisholm, Brit. Naval officer; b. Sept. 28, 1851; s. E. Chisholm Batten; m. Brittie Ellen Wood; 1 son, 2 daus. Joined Royal Navy, 1865; mem. War Office Coms. Coast Def.; in Bangkok during French blockade; with ordnance dept. Admiralty; a.d.c. King, 1905; ret., 1907. Read paper to Royal Soc. (with Sir. N. Lockyer) on solar eclipse expdn. Died Nov. 2, 1925.

BATTERMANN, Hans Felix Heinrich, German astronomer; b. Bückeburg, Germany, June 20, 1860; s. Adolf Georg and Marie Johanne (v. Michalkowski) B.; student of W. Foerster and F. Tietjen, Berlin degree, 1881. Spent one year at marine obs., Hamburg, Germany, then lived in Berlin until 1904; named prof., dir. Königsberg (Germany) Obs., 1904; staff Göttingen (Germany) Obs., 1888. Author: Drei Sternkatalogae von mehrere Beobachtungsreihen von Sternbedeckungen in Beobachtungs-Ergebnisse der Königliche Sternwarte Berlin. Work included observation and compilation of star occultations, triangulation of bright Pleiades stars. Died Blankenburg, Germany, June 15, 1922.

BATTERSBY, Alan Rushton, English organic chemist; b. Lancashire, Eng., Mar. 4, 1925; s. William and Hilda (Rushton) B.; B.Sc., U. Manchester (Eng.), 1946, M.Sc., 1947; Ph.D., U. St. Andrews (Scotland), 1949, postgrad., 1949-53; m. Margaret Ruth Hart, June 18, 1949; children—Martin Keith, Stephen John. Lectr. chemistry U. St. Andrews, 1948-53, U. Bristol (Eng.), 1954-62; prof. organic chemistry U. Liverpool (Eng.), 1962—; Commonwealth Fund fellow Rockefeller Inst. for Med. Research, N.Y.C., 1950-51, U. Ill., Urbana, 1951-52. Fellow Royal Soc., Chem. Soc. (Corday-Morgan medal 1960); mem. Am. Chem. Soc. Editorial bd. Rodd's Chemistry of Carbon Compounds, 1966—. Research, numerous publs. on chemistry of natural products especially that of pharmacologically important alkaloids, biosynthetic pathways in plants using radioactive tracers. Home: 98 Osmaston Rd., Prenton, Birkenhead, Eng. Office: Robert Robinson Labs, U. Liverpool, Oxford Str., Liverpool 7, Eng.*

BATTERSBY, William S., Am. psychologist; b. N.Y.C., Aug. 18, 1918; s. William K. and Elcy (Post) B.; A.B., N.Y. U., 1946, M.A., 1948, Ph.D., 1951; m. Frances M. Canavan, Feb. 5, 1942; children—Karen, Kevin. Research asso. Mt. Sinai Hosp., N.Y., 1952-60, Ill. State Psychiat. Inst., 1960-64; faculty neurology Northwestern U. Med. Sch., 1960-64; prof. psychology Queens Coll., Flushing, N.Y., 1964—. Lectr., Bklyn. Coll., 1951-60, Coll. City N.Y., 1954-60; neurol. cons. Queens Gen., Mt. Sinai hosps., 1964—. Mem. Am. Psychol. Assn., Am. Physiol. Soc., Am. Neurol. Assn., A.A.A.S., Sigma Xi, Psi Chi. Author: (with H. L. Teuber, M. B. Bender) Visual Field Defects after Penetrating Missile Wounds of the Brain, 1960. Contbr. articles to profl. jours. Research in neuropsychology, neurophysiology, psychophysics. Home: 76 Sugar Maple Lane, Glen Cove, N.Y. 11542.*

BATTEY, Robert, Am. gynecologist; b. Augusta, Ga., Nov. 26, 1828; s. Cephas and Mary (Magruder) B.; student medicine, Rome, Italy; student Booth's Sch. Analytical Chemistry, Phila., 1855, Phila. Coll. Pharmacy, U. Pa.; M.D., Jefferson Med. Coll., 1857; m. Martha B. Smith, 1849; 14 children, including Henry H. Practiced medicine, Rome; established Martha Battey Hosp., Rome; in charge hosps., Ga., Miss., 1862-65; prof. obstetrics Atlanta Med. Coll., also editor Atlanta Med. and Surg. Jour., 1872-75. Performed early successful operation for vesicovaginal fistula, 1858; described simple treatment for congenital clubfoot, 1859; devised Battey's operation of ovariotomy for treatment of non-ovarian diseases (became important because of devel. of modern endocrinology), 1873. Died Nov. 8, 1895.

BATTIG, Karl Joseph, Swiss psychophysiologist; b. Hergiswil, Switzerland, May 4, 1926; s. Karl Joseph and Maria (Kneubühler); student U. Zurich, U. Göttingen, U. Paris; M.D., U. Zurich, 1953; m. Colette Burki, Dec. 1, 1956; children—Rainer, Basil, Patrick. Research fellow Fed. Sch. Tech., Zurich, 1955-57, research asst., 1959-62, lectr., 1962-65, asst. prof. hygiene and physiology of work, 1965—; vis. sci. NIH, Bethesda, Md., 1957-59. Recipient Barth award Paris Indsl. Medicine, 1964; award Swiss Inst. Phys. Edn., 1964. Mem. Swiss Physiol. and Pharmacol. Assn., Swiss Assn. Preventive Medicine, Internat. Brain Research Orgn. Editor: Die Toxikologie des Tabaks, Verlag Hand Huber, 1962. Editor, Swiss Jour. Preventive Medicine. Research, publs. on objective measurement of behavioral performance of rats effects of brain lesions, drugs and indsl. poisons on behavior. Home: 96 Oberdorfstrasse, Dubendorf 8600, Switzerland.*

BATTLE, Helen Irene, Canadian zoologist; b. London, Ont., Can., Aug. 31, 1903; d. Edward Barrow and Ida Elizabeth (Hodgins) Battle; B.A., U. Western Ont., 1923, M.A., 1924; Ph.D., U. Toronto, 1928. Faculty, U. Western Ont., London, 1924—, prof. zoology, 1949—, acting head dept. zoology, 1956-58. Fellow A.A.A.S.; mem. Canadian (pres. 1963-64); Am. socs. zoologists, Nat. Assn. Biology Tchrs., Society of Zoologists, Canadian Physiol. Society, Canadian Society of Cell Biologists, Canadian Assn. Anatomists, Canadian Fedn. Biol. Socs. Research, numerous publs. on normal and abnormal devel., action carcinogens purines, pyrimidines, antibiotics on growth and devel., early devel. salmon, zebrafish, goldeye. Home: 132 Mamelon St., London, Ont., Can.*

BATTLE, William Henry, Brit. surgeon; b. England, 1855; Hunt prof. surgery and pathology Royal Coll. Surgeons; lectr. St. Thomas Hosp. Med. Sch.; surgeon Royal Free Hosp.; cons. St. Thomas Hosp. Licentiate Soc. Apothecaries. Fellow Royal Coll. Surgeons, Royal Soc. Medicine; mem. Med. Soc. London, Assn. Surgeons. Author: (with E. M. Corner) Diseases of the Appendix Vermiformis and Their Surgical Complications; The Acute Abdomen; Traumatic Rupture of Intestine, 1907. Introduced vertical incision of rectus sheaths with inward retraction of rectus muscle, 1895; known as Battle-Jalaguier-Kammerer incision. Died 1936.

BAUCHOT, M., French ichthyologist; b. Haute Vienne, France, Jan. 19, 1928; s. Jean Marie and Jeanne Marie (Paillier) Boutin; ed. U. Paris, Faculty Scis.; Licencé d'Enseignement, 1950; Diplôme d'Études Supérieures; Doctorat d'État, 1959; m. Mlle. Bauchot, Aug. 4, 1953; children—Philippe, Jean Yves, Frédéric. Asst., Paris Faculty Scis., 1950-52, lectr. biol. oceanography, in charge lab. procedures, 1959—; asst. Nat. Mus. Natural History, Paris, 1952-59, asst. dir., 1959—. Mem. Soc. Zoology, Soc. Biogeography. Author: Les poissons; Les plus beaux poissons; also papers. Compiler descriptive catalogues of fish in collections Nat. Mus. Natural History. Research on Serrivomer larvae; taxonomic work on Lahridae, Blenniidae, Bathypteroidae, Scombridae. Home: 41 Fremicourt, Paris XV. Office: 57 Cuvier, Paris V, France.*

BAUDELOCQUE, Jean Louis, French physician; b. Heilly, France, Nov. 30, 1746; s. Jean-Baptiste Baudelocque; pupil of Solayrès de Renhac, Paris. Attached to Charité Hosp.; admitted to l'École Practique, Coll. Surgery; prof. obstetrics, l'École de santé, surgeon-in-chief, obstetrician, l'hospice de la Maternité; appointed accoucher to Empress Marie Louise by Napoleon I; prof. Health Inst. Author: Les principes sur l'art des accouchements, 1775; L'art des Accouchements, 1781. Contbr. articles Journal général de médecin. Invented pelvimeter, described external conjugate diameter of pelvis, 1781; established med. importance of pelvic measurements; studied positions of fetus in utero and best methods of delivery in each case. Died Paris, May 2, 1810.

BAUDENS, Jean-Baptiste Luciens, French surgeon; b. Aire-sur-la-Lys, France, Apr. 3, 1804; ed. Amiens Lycee, Sch. Mil. Medicine, Strassbourg; M.D., 1829. Attached to Val-de-Grace Mil. Hosp., Paris, 1825-30, became chief of sch. and hosp., 1841; surgeon-in-chief French Field Hosp. in Africa; surgeon-in-ordinary French army; 1st prof. Hosp. Med. Inst., Lille, France, surgeon-in-chief Mil. Hosp. Gras Crillon, France, 1837. Author: Nouvelle méthode des amputations, 1842; Des fractures des membre pelvien, 1854; Souvenirs d'une mission medicale a l'armeé d'Orient, 1857; La guerre de Crimée, 1858. Authority on gunshot wounds and methods of amputation; pioneer in methods of treating wounds with ice. Died Dec. 3, 1857.

BAUDERON, Brice, French physician; b. Parey, France, 1540; docteur en medecine, U. Montpellier (France); practiced at Macon. Author: Pharmacopée (became basic text for physicians of the time), 1588; Praxis medica in duos tractatus distincta, 4 vols., 1620, English transl., 1657. Research on pharmaceutics. Died 1623.

BAUDET, Pierre, Swiss chemist; b. Nyon, Switzerland, Dec. 25, 1921; s. Fernand Charles and Jeannine Baudet; student Florimont Institut, Geneva, Switzerland, 1936-42; U. Geneva, 1942-43; licencés ès scis. chimiques, Doctorat ès scis. chimiques; m. Mlle. Texier, Sept. 29, 1952; children—Jean-Philippe, Stéphane. Researcher, biochem. lab. Paris U., 1950; indsl. researcher, Roussel, France, 1951; became pvt. docent U. Geneva, 1952, chargé de cours, 1964, tchr. research, organic chemistry lab., 1952-66, also dir. research group in organic chemistry. Mem. Swiss, Geneva chem. socs. Research on purification of ocytocin, penicillinase, mechanism of transformation of casein into paracasein, nature of strepogenin, structure of fluromycin and saramycetin, minoinfrared spectrometry, isolation of new plant genins, new reactions of diayomethan; determined structure of saramycetic acid, saromycetoic acid. Home: 48 Malaendu, Geneva, Switzerland.*

BAUDOT, Emile, French engr.; b. Haute-Marne, France, 1845; mem. staff, post and telephone adminstrn.; telegraphic engr.; invented press telegraph (transmitted 3 or 4 times as many messages and replaced Hughes device on Paris-Rome line) 1877; installed telegraphic lines from Paris to Vienna, Austria, Bern, Switzerland, Berlin, Germany, London, Eng. Died 1903.

BAUDRIMONT, Alexandre Édouard, French chemist; b. Compiègne, France, Feb. 25, 1806; M.D., 1831; pharmacist in Paris, France; physician in Valenciennes; préparateur in chemistry Coll. de France; asst. prof. Med. Faculty; prof. chemistry in Bordeaux, France, 1848. Author: Introduction à l'étude de la chimie par la théorie atomique, 1833; Traité de chimie générale et expérimentale, 2 vols., 1844-46; Cours de la chimie Agricole, 1874; Histoire des Basques, 1867; Du sucre et de la fabrication, 1841; table analytique du bulletin et du Journal de Pharmacie (1809-30), 1831. Forshadowed unitary theory; claimed to have anticipated theory of types; proposed idea that atoms are set in motion by chem. changes, arrangement in product differing from initial reactants; formulated theory of geometrical forms of atoms and structures of bodies. Died Bordeaux, Mar. 1880.

BAUDRIMONT, Marie Victor Ernest, French chemist; b. Compiègne, France, Sept. 2, 1821; asst. in pharm. instns., Paris; became prof. Assn. Philotechnique, 1856. Mem. Acad. Physicians. Studied catalysts and their action in decomposing potassium chlorate (wrongly attributed action to contact). Died Paris, Sept. 14, 1885.

BAUER, Alexander Emil Anton, chemist; b. Altenburg, Hungary, Feb. 16, 1836; s. Alexander Josef and Josefine (v. Wittmann-Dengl'ár) B.; student math. and natural scis. Polytechnikum, also U. Vienna (Austria), 1853-56; m. an Englishwoman. Asst. to A. Schröther in Vienna, 1856; moved to Paris, 1859 where he worked with C. A. Wurtz in lab. École de médicine; lectr. for 8 years; named prof. chem. tech. (later gen. chemistry) Vienna Polytechnikum (became Technische Hochschule 1872), 1869; improved sewage system as mem. Vienna city council, 1871-73; gave up exptl. work in 1888 because of eye injury. Mem. acads. of Vienna and Halle, several domestic and fgn. natural sci. socs. including Soc. Biol. Chemistry (London). Author: Lehrbuch der Techn.-chemischen Untersuchungen, 1864; Chemie und Alchemie in Österrecih bis zu Beginn der 19 Jahrhunderts, 1885; also numerous monographs on hist. devel. chemistry. Research on derivatives and isomers of amylene oxide; (with J. Schuler) synthesized primelic acid, 1877; studied chemistry of artists' paints; obtained di-, tri-, and tetramylene; in later years became an expert in chem. history. Died Vienna, Apr. 12, 1921.

BAUER, Edmond Henri, French physicist; b. Paris, Oct. 26, 1880; Agrégé in physics, Dr. ès sciences,

Faculté des sciences, Paris; m. Renée Kahn, June 28, 1911; children—Anne-Marie, Etienne, Jean-Pierre. With lab. phys. chemistry of Jean Perrin, 1905-11; acting prof. Sch. Physics and Chemistry, Paris, 1913-14; lectr., then prof. theoretical physics Sch. Scis., Strasbourg, France, 1919-28; under-dir. physics lab. Collège de France, 1928-41; prof., then hon. prof. Sch. Scis., Paris, 1945-52. Mem. French Soc. Physics (hon. mem., pres.), Soc. Phys. Chemistry (hon. pres.). Author: L'électromagnétisme hier et aujourd'hui, 1949. Died 1963.

BAUER, Ferdinand, Austrian naturalist; b. Feldsperg, Austria, 1760; s. Lucas Bauer; flower painter on sci. expdn. (with J. Sibthorb) to Nr. East; contbd. descriptions new plants from Greece to Flora Graeca Sibthorpiana (Sibthorb); artist on expdn. (with R. Brown) to Australia and Tasmania, resulted in collection 4000 species dried plants, also book Illustrationes Novae Hollandia, 1806-13; made 1600 plant drawings during expdn. for Brit. govt. Died Vienna, Austria, Mar. 17, 1826.

BAUER, Franz, biologist; b. Feldsberg, Austria, 1758; s. Lucas Bauer; employed as illustrator of flowers; taken to Eng. by botanist Sir Joseph Banks; began work with anatomist, Sir. E. Home on sci. periodicals, 1816. Fellow Royal Soc., 1821. Research on structure of orchids, anat. structure of eye; studied blood to locate Hunter's globules; showed (with E. Home) that radial fibers are anterior to the circular. Died Kew, Eng., 1840.

BAUER, Gustav Conrad, German mathematician; b. Augsburg, Germany, Nov. 18, 1820; s. Conrad Michael and Luise E. (Graberg) B.; student Polytech. School in Augsburg, Vienna (Austria), (under P. G. Lejeune-Dirichlet) Berlin (Germany); degree Polytech. Sch. Erlangen (Germany), 1842; m. Amalie v. Schlichtegroll, 1862; children—Gustave, Elisabeth Charlotte (Mrs. Richard Schaupp), one other dau. Lived in Paris, France, 1842-43; tutor in house of Rumanian prince Ghika, 1845-55; after short stay in Eng., became lectr. U. Munich (Germany), 1857; prof., 1865-1905; mem. Bavarian Acad. Scis. Author: Vorlesungen über Algebra, 1903. Research on spherical functions, compound fractions, algebra and chiefly geometry; O. Perron used a Bauer compound fraction conversion to prove several formulas left by Indian Ch. V. Raman without proof, 1952. Died Munich, Apr. 3, 1906.

BAUER, Heinz, German mathematician; b. Nürnberg, Germany, Jan. 31, 1928; s. Hans and Elise (Gütler) B.; student U. Nancy (France); Dr.phil.nat., U. Erlangen, 1953, habilitation, 1956; m. Irene Pöllet, Oct. 4, 1957; children—Christian, Christina. Research fellow Centre Nat. Recherche Scientifique, Paris, France, 1956; prof., dir. Inst. für Versicherungsmathematik und Mathematische Statistik, U. Hamburg, 1962-65, dean Faculty of Sci., 1963-64; dozent U. Erlangen-Nürnberg, Germany, 1956-58, prof. math., co-dir. Math. Inst., 1965—; vis. asso. prof. U. Wash., Seattle, 1961-62; vis. prof. U. Paris, 1964, Cal. Inst. Tech., 1967. Mem. Deutsche Mathematiker-Vereinigung, GAMM, Am. Math. Soc., Société Mathematique of France. Author: Wahrscheinlichkeitstheorie und Grundzüge der Masstheorie, 1964; Harmonische Räume und ihre Potentialtheorie, 1966. Research, publs. on contbns. to integration theory, functional analysis, probability theory and abstract potential theory. Home: 17 Eschenweg, Erlangen. Office: 1½ Bismarckstrasse, Erlangen 852, Germany.*

BAUER, Julius, physician; born Nachod, Austria, Aug. 14, 1887; s. Ludwig and Clara (Schur) B.; M.D., U. Vienna, 1910; postgrad. U. Paris, 1914; m. Marianne Jokl, July 15, 1912; children—Franz K., Frederick K. Came to U. S., 1938, naturalized, 1944. Prof. medicine Vienna (Austria) U., 1919-38; prof. medicine La. State U., New Orleans, 1938-40; prof. medicine Loma Linda U., Los Angeles, 1941-61; lectr. medicine Univ. of Southern Cal., 1966—. Guest lectr., Germany, Holland, Belgium, France, Spain, U. S. Mem. Am., Cal. med. assns., A.A.A.S., Am. Heart Assn. Author: Konstitutionelle Disposition zu Inneren Krankherten, 1917, 21, 24; Innere Sekretion, 1927; Constitution and Disease, 1942, 45; the Person Behind the Disease, 1956; Differential Diagnosis of Internal Diseases, 1950, 55, 67; Errant Ways of Human Society, 1961; Medizin Kulturgeschichte des 20 Jahrhunderts in Rahmen einer Autobiographie, 1964; also numerous articles. Research on human constn., genetics, endocrinology, neurology, psychosomatic medicine, differential diagnosis internal diseases. Home: 615 N. Bedford Dr., Beverly Hills, Cal. 90210. Office: 1680 Vine St., Los Angeles 90028.*

BAUER, Karl Josef von, see von Bauer, Karl Josef.

BAUER, Karl Michael, German urologist; b. Fürth/Bavaria, Germany, Aug. 20, 1919; s. Anton and Marie (Kochenburger) B.; student medicine U. Erlangen, 1938-39, U. Kiel (Germany), 1939-40, U. Freiburg (Germany); M.D., certification U. Tübingen (Germany), 1943; m. Luise-Henriette Veith, July 29, 1950; 1 son, Michael. Specialist in urology, 1950; specialist in surgery, 1955; faculty U. Tübingen, 1959—, prof. urology, 1965—. Mem. Internat. Soc. Urology, Deutsche, Südwestdeutsche Gesellschaften für Urologie. Author: Die zystoskopische Diagnostik; co-author books. Research, numerous articles on functional bladder disturbances, organanastomosis of kidney and other exptl. studies on isotope-nephrography, cytostatic injuries of bladder, stone prophylaxis. Home: Stadt. Krankenhaus, 82 Rosenheim, Bavaria, Germany.*

BAUER, Louis Agricola, Am. astronomer; b. Cin., Jan. 26, 1865; s. Ludwig and Wilhelmina (Buehler) B.; C.E., U. Cin., 1888, M.S., 1894, D.Sc., 1913; Ph.D., A.M., U. Berlin, 1895; D.Sc., Brown U., 1914; m. Adelia Francis Doolittle, Apr. 15, 1891; 1 dau., Mrs. Dorothea Weeks. Astron. and magnetic computer U. S. Coast and Geod. Survey, 1887-92; docent in math. physics U. Chgo., 1895-96; instr. in geophysics U. Cin., 1896-97, asst. prof. math. and math. physics, 1897-99; insp. magnetic work and chief terrestrial magnetism div. U. S. Coast and Geod. Survey, 1899-1906; dir. dept. terrestrial magnetism Carnegie Instn., 1904-29, dir. emeritus and research asso., from 1930. Chief div. terrestrial magnetism Md. Geol. Survey, 1896-99; astronomer and magnetician, western boundary survey of Md.; lectr. in terrestrial magnetism, Johns Hopkins, from 1899. Mem. permanent com. on terrestrial magnetism and atmospheric electricity of Internat. Meteorol. Conf. and Internat. Assn. of Acads. Recipient Charles Lagrange prize (Physique du globe) of Académie Royale des Sciences, des Lettres et des Beaux-Arts de Belgique, 1905, Georg Neumayer Gold Medal, Berlin, 1913; Mem. NRC, from 1917; chmn. com. navigation and nautical instruments, Council Nat. Def., 1917-18; U. S. del. Brussels meetings, 1919, Internat. Research Council and Internat. Geodetic and Geophys. Union; U. S. del. Rome meeting of latter, 1922, Madrid, 24, Prague, 27; sec. and dir. central bur. sect. terrestrial magnetism and electricity of Internat. Geodetic and Geophys. Union, 1919-27, pres., 1927-30; chmn. Am. Geophys. Union, 1922-24 (chmn. sect. terrestrial magnetism and electricity, 1920-22 24-26). Fellow A.A.A.S. (v.p., chmn. sect. B, 1909), Am. Acad. Arts and Scis., Am. Geog. Soc. Founder, editor Terrestial Magnetism and Atmospheric Electricity, 1896-1928, co-editor, from 1928. Contbr. to sci. press on terrestrial magnetism, electricity physics. Influential in coordinating work on magnetism in various parts of world. Died Apr. 12, 1932.

BAUER, Louis Hopewell, Am. physician; b. Boston, July 18, 1888; s. Charles Theodore and Ada Marian (Shute) B.; A.B., Harvard, 1909, M.D., cum laude, 1912; honor grad. U. S. Army Med. School, 1914, U. S. Army School of Aviation Medicine, 1920; grad. Army War Coll., 1926; D.Sc., U. Sydney (Australia), 1955; m. Helena Meredith, Dec. 27, 1913; 1 son, Charles Theodore; m. 2d, Margaret Louise Macon, Aug. 9, 1930; 1 step-dau., Joan Macon (Mrs. William B. Lawrence, Jr.). Entered Med. Corps, U. S. Army, 1913, advanced through grades to lt. col. (emergency), resigned, 1926; med. dir. aeronautics branch U. S. Dept. of Commerce (now FAA), 1926-30; practice medicine specializing in cardiology, Hempstead, N.Y., 1930-53; chmn. bd. dirs. United Med. Service, Inc., 1954-59, now cons. Past mem. N.Y. State Public Health Council. Recipient John Jeffries award Inst. Aero. Scis., 1940; Theodore C. Lyster award Aero. Med. Assn. 1947; Bancroft medal, Queensland br. Brit. Med. Assn., Australia, 1955; hon. gold key, Med. Faculty Univ. Vienna, 1955; Paracelsus medal German Med. Assn., 1960. Diplomate Am. Bd. Internal Medicine, Am. Bd. Preventive Medicine (Aviation Medicine). Fellow A.C.P., Aero. Med. Assn. (pres. 1929-31); mem. World (sec. gen. 1948-61, now cons.), Am. (pres. 1952-53) med. assns., Med. Soc. State N.Y. (pres. 1947-48, Am. Heart Assn., Alpha Omega Alpha, Kappa Gamma Chi, Phi Beta Kappa. Editor in chief Journal of Aviation Medicine (bi-monthly), 1930-54, now editor-in-chief emeritus. Author: Aviation Medicine (1st Am. textbook on subject), 1926; Private Enterprise or Government in Medicine, 1947. Contbr. chpt. Medicine and Aeronautics in Tice's Practice of Medicine, 1942, Aviation Medicine chpt. in Cyclopedia of Medicine, Surgery and the Specialties, 1942, Aviation Medicine chpt. in Oxford Loose Leaf Medicine, 1943 (also pub. as monograph). Contbr. numerous articles on aviation medicine and cardiology to med. jours; The First Decade Report on the World Med. Assn., 1958. Home: 341 Harvard Av., Rockville Centre, N.Y. Office: 10 Columbus Circle, N.Y.C. 19.

BAUER, Max Hermann, German mineralogist; b. Gnadenthal, Germany, Sept. 13, 1844; s. Johann Hermann and Sophie Friederike (Faber) B.; student Stuttgart (Germany) Polytechnikum; degree from U. Tübingen (Germany); postgrad. mineralogy U. Berlin (Germany); m. Julie Schnurrer, 1874; 2 sons, 1 dau. Lectr. mineralogy and geology in Göttingen, 1871; became pvt. lectr. Berlin, 1872; apptd. prof. mineralogy and geology in Königsberg (Germany), 1875; named prof. mineralogy and petrography Marburg (Germany), 1884. Author: Lehrbuch der Mineralogie, 1886; Edelsteinkunde, 1896. Editor: Beitrage zur Mineralogie; Neues Jahrbuch für Mineralogie, Geologie, Paläontologie, 1880-91. Research and publs. on layer configurations and break-off lines in mica and other minerals, petrographical studies of Hessian basalt, studies in soil sci. on laterite; made geol. charts of Thuringia and Rhone. Died Marburg, Nov. 4, 1917.

BAUER, Oswald, metallurgist; b. Goldingen, Kurland, Latvia, Jan. 31, 1876; s. Theodor and Marie (Melville) B.; studied iron metallurgy Freiberger Adademie; diploma iron metall. engr., 1901; hon. dr. engring. Technische Hochschule Aachen; Asst. at Mining Acad. (Bergakademie), Berlin, Germany, 1902-03; became dept. head materials testing office in Berlin-Dahlem, 1918, dir., 1927, v.p., 1934; first dep. dir. Kaiser-Wilhelm Inst. for Metal Research, then became chief dept. head in 1924; temporary prof. iron metallurgy Technische Hochshule in Breslau, Germany; chmn. Deutsche Gesellschaft für Metallkunde from 1929. Author: (with E. Heyn) Metallographie, 1909; Unterssuchungen über Lagermetalle Antimon-Blei-Zinn-Legierungen, 1914; (with E. Deiss) Probenahme und Analyse von Eisen and Stahl, 1912; (with M. Hansen) Der Aufbau der Kupfer-Zinklegierungen, 1927; (with K. Memmler) Die Eigenschaften des Hartmessings, 1929; editor: (with O. Kröhnke, G. Masing) Die Korrosion metallischer Werkstoffe, 1936-40. Research and numerous articles on metallurgy, especially on composition and corrosion phenomena. Died Berlin-Dahlem, Aug. 2, 1936.

BAUER, Robert, German radiologist; b. Bruchsal, Dec. 11, 1898; s. Adolf and Camilla (Blersch) B.; ed. U. Freiburg, U. Munich; M.D.; m. Marie-Luise Gais; children—Christian, Camilla. Intern, 1925-31; radiologist U. Frankfort-am-Main, 1931-37; head dept. radiography U. Tübingen, 1937-41, now dir. Inst. Med. Radiology and dept. radiography of med. and surg. clinics; with Charity Clinic, Berlin, 1941-45. Mem. German Soc. Radiography (pres.), Swiss Soc. Radiology and Nuclear Medicine. Author: over 80 publs. on diagnostic radiography and biol. radiology. Co-editor: Fortschritte auf dem Gebiete der Röntgenstrahlen; Strahlentherapie; Nuklearmedizin. Home: Schwabstrasse 69. Office: Röntgenstrasse 11, Tübingen, Germany.

BAUER, Rudolf Wilhelm, physicist; b. Rothenburg, Germany, Nov. 28, 1928; s. Wilhelm Friedrich and Wilhelmina (Bach) B.; B.A., Amherst Coll., 1952; student U. Bonn, Germany, 1952-54; Ph.D., Mass. Inst. Tech., 1959; m. Margret Baer, July 12, 1958; children—Peter Walter, Andrew William, Barbara Ann. Came to U. S., 1954, naturalized, 1959. Instr. Mass. Inst. Tech., 1959-60, asst. prof. physics, 1960-64; vis. research physicist U. Cal. Lawrence Radiation Lab., Livermore, Cal., 1962-63, physicist, 1964—. Mem. Am. Phys. Soc., I.E.E.E., Sigma Xi. Research and publs. in radioactivity and nuclear spectroscopy; nuclear structure studies; nuclear reactions; neutron physics; plasma physics. Home: 1386 Kathy Court, Livermore, Cal. 94550. Office: L-24 Lawrence Radiation Lab., Livermore, Cal. 94551.*

BAUER, Sebastian Wilhelm Valentin, German inventor; b. Dillingen an der Donau, Germany, Dec. 23, 1822; s. Wilhelm and Wilhelmine (Hendinger) B.; apprenticed to a turner, 1835; non-commd. officer Denmark-German Bund war; built 1st German submarine (1850) which sank in 1851 in Kiel harbor; after several attempts built an improved model in Russia, 1855, which submerged 134 times before sinking in 1857; built vehicle to drive along sea floor, petroleum powered engine; designed semi-rigid steerable airship, 1856, flying machine, 1866. Died Munich, Germany, June 20, 1875.

BAUER, Siegfried Josef, space scientist; born Klagenfurt, Austria, Sept. 13, 1930; s. Rochus and Barbara (Plassnig) B.; Ph.D., U. Graz (Austria), 1953; m. Ingeborg Heiditsch, Dec. 20, 1954; 1 dau., Sonya. Came to U. S., 1953, naturalized, 1960. Physicist meteorol. br. U. S. Army Signal Research and Devel. Lab., Fort Monmouth, N.J., 1953-54, 55-57, sr. scientist Inst. for Exploratory Research, 1957-60; sect. head, sr. scientist Lab. for Space Scis., NASA-Goddard Space Flight Center, Greenbelt, Md., 1961-65; head planetary ionospheres br., 1965—. Lectr. dept. space sci. and applied physics Catholic U., Washington, 1964-65. Mem. A.A.A.S., Am. Geophys. Union, Internat. Sci. Radio Union (mem. commn. III 1959—, commn. IV 1962—), Internat. Assn. Geomagnetism and Aeronomy (mem. commn. VIII 1964—). Research, publs. on study upper ionosphere by radio reflections from moon, measurements distbn. and composition of charged particles in topside ionosphere by means high altitude sounding rockets and satellites. Home: 7 David Ct., Silver Spring, Md. 20904. Office: Goddard Space Flight Center, NASA, Greenbelt, Md. 20771.*

BAUER, Simon Harvey, chemist; born Kaunas, Lithuania, Oct. 12, 1911; s. Benzion and Goldie (Betten) B.; came to U. S., 1921, naturalized, 1927; B.S., U. Chgo., 1931, Ph.D., 1935; postgrad. Cal. Inst. Tech.; m. Miriam L. Rosoff, June 25, 1938; children—Frederick, Deborah, Ross. Instr., Pa. State U., 1937-39; faculty Cornell U., Ithaca, N.Y., 1939—, prof. chemistry, 1950—. Cons. to govt. agys., pvt. cos. Guggenheim fellow, 1949, Sr. NFS postdoctoral fellow, 1962. Fellow Am. phys. Soc., A.A.A.S.; mem. Am. Chem. Soc., Faraday Soc., N.Y. Acad. Scis., Am. Fed. Scientists. Asst. editor Jour. Am. Chem. Soc. 1964. Research, molecular structure by diffraction and spectroscopic techniques, thermochemistry boron hydrides, rates very fast reactions in gases, chem. kinetics in shock tubes. Home: 312 Comstock Rd., Ithaca, N.Y. 14850.

BAUER, Stanley, Am. physician; b. Newark, Nov. 18, 1928; s. David Lewis and Estelle (Sirkin) B.; B.S., Rutgers U. Coll. Pharmacy, 1950; M.D., Chgo. Med. Sch., 1954; m. Shirley Kaplan, Dec. 18, 1949; children—Barbara Ellen, Susan Gail, Donna Joy.

With Nat. Cancer Inst., NIH, Mt. Sinai Hosp., N.Y.C., 1954-57, 59-60; chief lab. service U. S. Naval Hosp., Corpus Christi, Tex., 1957-59; staff Beth Israel Hosp., N.Y.C., 1960-63, asso. pathologist, 1961-63; dir. labs. Princeton (N.J.) Hosp., 1963—; lectr. N.Y. U. Coll. Dentistry, 1960—; asso. bur. biol. research Rutgers U., 1964—, clin. asst. prof. Coll. Medicine, 1966—. cons. Princeton Labs., 1963—. Trustee, Chgo. Med. Sch., 1959—. Recipient Lehn and Fink medal, McKesson-O'Loughlin Meml. award, Merck prize Rutgers U., 1950; N.J. Pharm. Assn. prize, 1950; J. J. Sheinin Meritorious award Chgo. Med. Sch. 1954; Borden scholar, 1950. Mem. N.Y. Acad. Scis., Coll. Am. Pathologists, A.M.A., N.J. Pathologic Soc., Am. Soc. Clin. Pathologists, Assn. Clin. Scientists. Research, publs. on immunologic diagnosis streptococcal and mycoplasmal diseases; co-discovery Mono Test for infections mononucleosis. Home: Orchard Rd., Skillman, N.J. 08558. Office: 253 Witherspoon St., Princeton, N.J. 08540.*

BAUER, William Hans, pathologist; b. Prague, Czechoslovakia, s. Aloysius Bauer; M.D., St. Charles U. Prague, 1912; m. Mary Elizabeth Bauer, Nov. 4, 1913; children—John D., Inge Hynes, Anniliese Lamb. Research fellow in histology, med. sch. U. Innsbruck (Austria), 1912-14, asst., 1912-14, 18-24, pvt. docent, 1925-38, asso. prof., 1931-38, dir., 1934-38; prof. and dir. pathology, sch. dentistry St. Louis U., 1938—, affiliated with dept. pathology, med. schs., St. Louis U., also Marquette U.; coordinator Cancer Control Program (sch. dentistry). Asso. A.M.A.; mem. Am. Assn. Pathologist and Bacteriologist, Internat. Dental Research Assn., A.A.A.S., St. Louis Soc. Pathology, Sigma Xi. Collaborator: Pathology (W. A. D. Anderson), 1948. Died June 14, 1956.

BAUER, William Waldo, Am. physician; b. Milw., July 23, 1892; s. Robert W. and Anna (Bunteschu) B.; B.S., U. Wis., 1915; M.D., U. Pa., 1917; LL.D., George Williams Coll., 1960; m. Florence Anne Marvyne, Feb. 8, 1920; children—John Robert, Erminie Anne (Mrs. M. F. Wetzel), Charles Marvyne. Practice medicine specializing in pub. health, Chgo.; dir. health edn. A.M.A., 1932-62, dir. emeritus, 1962—; cons. Nat. Dairy Council, Mem. A.M.A., Am. Pub. Health Assn., Broadcast Pioneers, Am. Sch. Health Assn., Am. Assn. Health, Phys. Edn. and Recreation, Phi Beta Kappa, Alpha Omega Alpha, Phi Beta Pi. Author: Moving in to Manhood, 1963; (with Florence Marvyne Bauer) Way to Womanhood, 1965; (with Warren Schaller) Your Health Today, 1965; To Enjoy Marriage, 1967. Editor: Today's Health, 1950-58; Today's Health Guide, 1962-64. Publs. on communication and lang. understanding between drs. and pub. emphasizing mut. understanding, combatting of quackery; improvement of health practices and knowledge. Address: 400 E. Randolph St., Chgo. 60601.*

BAUEREISEN, Erich, German physiologist; b. Kiel, Germany, Feb. 18, 1913; s. Johann-Adam and H. (Siveke) B.; student univs. Heidelberg, Munich; M.D., U. Munich, 1937; m. H. Stahl, Apr. 22, 1953; 1 dau. Prof., dir. Physiol. Inst., U. Leipzig, 1952-58; prof., dir. Physiol. Inst., U. Würzburg (Germany), 1959—. Address: 9 Röntgenring, 87 Würzburg, West Germany.*

BAUERMAN, Hilary, metallurgist, mineralogist, geologist; b. London, Eng., Mar. 16, 1835; s. Hilary John and Anna Rosetta (Wychers) B.; became student Govt. Sch. Mines, 1851, later at Bergakademie, Freiburg, Germany. Apptd. asst. geologist Geol. Survey United Kingdom, 1855; apptd. geologist N.Am. Boundary Commn., Can., 1858; surveyed mineral deposits in several countries, including U. S.; became asso. Royal Sch. Mines, 1862; prof. metallurgy Ordnance Coll., Woolwich, Eng. Fellow Geol. Soc.; mem. Inst. Civil Engrs. (asso., Howard prize 1897), Iron and steel Inst. (hon.), Instn. Mining and Metallurgy (hon., Gold medal 1906). Author: Metallurgy of Iron, 1868; Systemic Mineralogy, 1881; Descriptive Mineralogy, 1884; (with J. A. Phillips) Elements of Metallurgy, 1891. Died Dec. 5, 1909.

BAUHIN, Gaspard, Swiss botanist, anatomist; b. Basel, Switzerland, Jan. 17, 1560; s. Jean the Elder and Jeanne (Fontaine) B.; ed. U. Padua (Italy), U. Montpellier (France), also in Germany; returned to Basel, 1580; M.D., 1581; student U. Bologna (Italy), U. Paris (France); m. Barbara Vogelmann, 1581; m. 2d, Maria Brüggler, 1596; m. 3d, Magdalena Burckhard, 1598; at least one son, Jean Gaspard. Apptd. prof. Greek, U. Basel, 1582; became prof. anatomy and botany, 1588 and prof. practical medicine, 1614; city physician; rector of univ., dean faculty. Eponym (with bro. Jean) of plant genus Bauhinia, named by Plumier. Author: Theatrum anatomicum, 1592; Pinax theatri botanici (1st classification of plants by genus and species), 1596; Prodromus theatri botanici, 1620. Described ileocecal valve (Bauhin's valve), 1588; gave potato its present name (Solanum tuberosum); collected plants in Italy, Germany, France; described many new species; anticipated Linnaeus' binomial arrangement. Died Dec. 5, 1624.

BAUHIN, Jean the Younger, botanist, physician; b. Basel, Switzerland, Feb. 12, 1541; s. Jean and Jeanne (Fontaine) B.; student U. Basel, to 1555, U. Tübingen (Germany) under Fuchs, 1560, also in France; doctorate, 1562; m. Denise Bornand, circa 1563; m. 2d, Anne Gregoire, 1598. Mem. sci. excursion (with Gesner), Switzerland; created pvt. garden and became tchr. plant study, Lyons, France, 1563; named prof. rhetoric U. Basel, 1566; also practiced medicine; apptd. physician to Duke of Württemberg, Montbeliard, France, 1570; dir. mus., Montbeliard. Honored (with bro. Gaspard) by Plumier with naming of plant genus Bauhinia. Author: Historia novi et admirabilis fontis balneique Bollensis, 1598; Historia plantarum nova et absolutissima (encompasses bot. writings of ancients known in his day), pub. 1650-51; Historia planatarum prodromus, pub. 1619. Considered father of botany by contemporaries. Died Montbeliard, Oct. 27, 1613.

BAULE, Bernhard, mathematician; b. Münden, Germany, May 4, 1891; s. Anton and Wilhelmine (Griesbert) B.; ed. in math. and physics at univs. of Kiel, Munich, Göttingen; m. Erna Kaulbers, 1920; m. 2d, Sigrun Kurschel, 1956. Asst. to Prof. Hilbert, U. Göttingen, 1913-19; asst. to W. Blaschke, U. Hamburg, 1919-21, in charge classes, 1920; prof. Tech. U. of Graz (Austria), 1921-38, prof. math., 1945—; sci. collaborator for physics and chemistry Kaiser Wilhelm Inst., Berlin-Dahlem, 1940-44; with Tech. U. of Munich, 1944-45. Author: Die Mathematik des Naturforschers und Ingenieurs, vols. 1-8, 1942-64, also numerous treatises on math., Physics, chemistry, agr. and biology, periodicals on differential geometry, theory of gas, heat condrs., analysis of structure and crossing plants. Address: Nibelungengasse 63, Graz, Austria.

BAULIG, Henri, French geographer; b. Paris, June 17, 1877; studied in France and U. S. Taught at Rennes, 1912-19; prof. geography U. Strasbourg, 1919-47. Author: Le Plateau central de la France et sa bordure Méditerranée, étude morphologique, 1923; l'Amérique septentrionale, 1935-36; Tome XIII de la Géographie universelle; Problèmes des terrasses, 1948; Essais de géomorphologie, 1950; Vocabulaire de géomorphologie, 1956. Mem. French Acad. Scis., 1949. Worked in area of relief forms.

BAUM, Frank George, Am. elec. engr.; b. Ste. Genevieve, Mo., July 18, 1870; s. Christian and Caroline (Kline) B.; A.B., Stanford U., 1898, E.E., 1899; m. Mary Dawson, July 18, 1900; children—Esther F., Helen E., Adah C. Elec. engr. Cal. Gas & Elec. Corp., 1902-07; cons. constrn. engr., 1907—; chief engr. hydro-electric work Pacific Gas & Electric Co., 1912—. Author: Alternating Currents, 1902; Alternating Current Transformer, 1903; Atlas of U. S. A. Electric Power Industry, 1923. Inventor of constant potential electric transmission system. Died 1932.

BAUM, Gerald Leonard, Am. physician; b. Milw., Dec. 19, 1924; s. Sanford Bert and Ruth (Baruch) B.; B.S., U. Wis., 1945, M.D., 1947; m. Charlotte Fishbain, Jan. 13, 1951; children—Susan, Michael, Stuart. Practice medicine specializing in internal medicine, Cin., 1952-54; research asso. mycology Jewish Hosp., Cin., 1956-58; chief pulmonary VA, asso. prof. medicine U. Cin. Coll. Medicine, 1958-65; chief pulmonary VA, Cleve., asso. prof. medicine Western Res. U., 1965—; research pulmonary disease Govt. Hosp., Tel Hashomer, Israel, 1963-64. Recipient Shared Histadrut prize, 1965. Mem. A.M.A., A.C.P., Am. Thoracic Soc., A.A.A.S., Central Soc. Clin. Research, Phi Beta Kappa. Editor: Textbook of Pulmonary Diseases, 1965. Contbr. numerous articles to med. jours. Home: 2164 Chatfield Dr., Cleveland Heights, O. 44106. Office: 10701 East Blvd., Cleve. 44106.*

BAUM, John D(aniel), Am. mathematician; b. Pitts., July 31, 1918; s. Hugo and Edyth (Solomon) B.; B.A., Yale, 1941, Ph.D., 1953; m. Marian Jewel Hooper, Aug. 22, 1948; children—Madeline Louise, Edyth Elizabeth, Ellen Gilbert. Asst. instrn. Yale, 1950-53; faculty Oberlin (O.) Coll., 1953—, prof., 1962—. Mem. Am. Indian, French, London math. socs., Math. Assn. Am. (cons. Com. on Undergrad. Program in Math. 1961—, mem. adv. group on communications 1963—; lectr. 1959-60), Phi Beta Kappa, Sigma Xi. Author: Elements of Point Set Topology, 1964; (with Richard J. Paul) Sets I, 1964; (with Richard J. Paul) Sets II, 1964; (with Roy A. Dobyus) The Structure of the Real Number System, 1967. Contbr. articles on topological dynamics to math. jours. Asso. editor Am. Math. Monthly, 1965—. Home: 97 Parkwood Lane, Oberlin, O. 44074.*

BAUM, Siegmund Jacob, physiologist; b. Vienna, Austria, Nov. 14, 1920; s. Joseph L. and Marie (Leiser) B.; came to U. S., 1939, naturalized, 1943; A.A., Los Angeles City Coll., 1947; B.A., U. Cal. at Los Angeles, 1949, M.A., 1950; Ph.D., U. Cal. at Berkeley, 1959; m. Arline R. Weber, Apr. 1, 1947; children—Jonathan W., Andrew M., Vicki M., Joseph L., Anthony P. Sr. project leader Recovery Project, U. S. Naval Radiol. Def. Lab., San Francisco, 1950-60; group leader space physiology and radiation biology Missle and Space div. Douglas Aircraft Co. Santa Monica, Cal., 1960-62; head cellular radiobiology div. Armed Forces Radiobiology Research Inst., Bethesda, Md., 1962-64, chmn. exptl. pathology dept., 1964—. Recipient 1st Ann. Sci. award U. S. Naval Radiol. Def. Lab., 1960. Mem. Radiation Research Soc., Am. Physiol. Soc., Transplantation Soc., Sigma Xi. Research, publs. on effects ionizing radiation upon endocrine and hematopoeitic system, radiation recovery and residual injury of radiosensitive organs, tissues and cells especially erythropoietic system, effect of radiation upon hematopoietic cellular kinetics. Home: 6600 Greyswood Rd., Bethesda 20034. Office: Armed Forces Radiobiology Research Inst., Def. Atomic Support Agy., Bethesda, Md. 20014.*

BAUM, William Alvin, Am. astronomer; b. Toledo, Jan. 18, 1924; s. Earle Fayette and Mable (Teachout) B.; B.A. summa cum laude, U. Rochester, 1943; M.S., Cal. Inst. Tech. 1945; Ph.D. magna cum laude, 1950; m. Ester Bru, June 27, 1961. Physicist Naval Research Lab., Washington, 1946-49; astronomer Mt. Wilson and Palomar Obs., Pasadena, Cal., 1950-65; research asso. Cal. Inst. Tech., 1950-65, dir. Planetary Research Center, Lowell Obs., Flagstaff, Ariz., 1965—. Guggenheim fellow Imperial Coll., London, 1960-61. Mem. Am. Astron. Soc., Royal Astron. Soc. (Britain), Astron. Soc. of Pacific, Internat. Astron. Union, Phi Beta Kappa, Sigma Xi, Theta Delta Chi. Research in observational cosmology, photometry of galaxies, stellar populations, photoelectric instruments, image tubes. Address: Lowell Obs., Flagstaff, Ariz. 86001.*

BAUMANN, Eugen, German chemist; b. Cannstadt, Germany, Dec. 12, 1846; s. Johannes and Caroline Elisabeth (Mayser) B.; ed. Polytechnikum Stuttgart (Germany), also at Tübingen, Germany, 1870-72; m. Theresia Kopp. Became asst. in Tübingen, 1870, later in Strasbourg, France; became pvt. lectr. Med. Faculty Strasbourg, 1876; apptd. head chemistry dept. Physiol. Inst. Berlin (Germany), 1877; became asso. prof. chemistry, Freiburg, Germany, 1882, prof., 1883. Editor, pub.: (with A. Kossel) Zeitschrift fur physiologische Chemie. Contbd. articles to Berichte der deutschen chemischen Gesellschaft, also to Liebig's Annalen. Discovered thryoxin (organic iodine compound in thyroid), 1895; introduced various hypnotic drugs, including sulfonal, tetronal, trional. Died Freiburg, Nov. 3, 1896.

BAUMANN, Germain, French physicist; b. Strasbourg, France, Apr. 21, 1931; s. Emile B.; student Lycée technique de Strasbourg, 1944-50; Université de Strasbourg, 1950-56; Doctorat d'état ès sciences physiques, 1963; m. Agnès Flesch, Mar. 29, 1959; children—Isabelle, Benedicte. Researcher, U. Strasbourg and Centre National de la Recherche Scientifique, from 1954, lectr. physics U. Strasbourg, 1963—. Research in slow energy nuclear physics, triton-proton elastic scattering at 3.8 MV; studies in high energy physics, eta-sigma interactions with study of hyperfragments, 1963; study hyperons and mesons with heavy liquid bubble chambers. Home: 22 Rue Gouraud, 67, Schiltingheim, France. Office: Rue du Loess, 67, Strasbourg 3, France.*

BAUMANN, Jean, French surgeon; b. Paris, France, May 7, 1906; s. Charles and F. (Blum) B.; ed. Coll. Rollin; M.D., Faculté de médecine, Paris; m. Jacqueline Bernard-Lévy, Apr. 26, 1937; children—Danielle, Jean-Pierre. Surgeon, prof. Faculté de médecine, Paris, 1957. Mem. Acad. Surgery, Soc. Thoracic Surgery, Tb Soc, French Soc. Anesthesia, Internat. Soc. Transfusion, others. Research on thoracic, war surgery, anesthesia, resuscitation. Address: 9 bis, rue Perignon, Paris 15, France.

BAUMANN, Jean-Aimé, Swiss physician; b. Lausanne, Switzerland, Sept. 17, 1910; s. Auguste and Marguerite (Georg) B.; M.D., U. Geneva; m. Jacqueline Frick-Cramer, Jan. 24, 1944; children—Anne-Lise, Claire-Pascale, Jeanne-Pauline. Head anatomy studies, reader, later prof.-sec. of senate U. Geneva; lt. col of health service. Sec.-gen. Internat. Bur. Differential Anthropology. Research on anatomy, histology, embryology, med. anthropology, mil. medicine. Home: Landecy/Geneva. Office: 20, rue école de médecine, Geneva, Switzerland.

BAUMANN, Paul, German physicist, chemist; b. Pforzheim, Dec. 13, 1897; s. Johann and Rosa (Strauss) B.; ed. Superior Tech. Sch., Karlsruhe, U. Heidelberg; Ph.D.; m. Gertrud Hering, Dec. 8, 1923; 1 dau., Rosemarie. With Badische Anilin & Soda-Fabrik, Ludwigshafen, from 1923; dir. Am. research firm, 1930-35; dir. prodn. Chemische Werke HULS, 1938, dir., 1941, titular dir., 1953, pres., 1958; pres. security council Gesellschaft mit beschränkter Haftung Faserwerke HULS. Mem. Union Lippe (Dortmund), Ind. Assn. Sci. (Essen), DECHEMA (Frankfurt), others. Author: Über die Erzeugung von Acetylen nach dem Lichtbogenverfahren; Erzeugung von Butadien und Synthesekautschuk; Die Produkte der modernen technischen Gas-Chemie mit spezialer Berucksichtigung der Erzeugung von Waschrohstoffen und Textilhilsmitteln. Died Mar. 29, 1967.

BAUMANN, Richard Wilhelm, German metallographer; b. Heilbronn, Germany, Oct. 24, 1879; s. Johann Adam F. Alexander and Luis (v. Gessler) B.; studied mech. engring. at Technische Hochschule Stuttgart (Germany); diploma, 1903; m. Helene Mathilde Süskind, Oct. 25, 1917. Asst., co-worker of C. von Bach at Materials Testing Inst., Stuttgart, became dir., 1924. Author: Die Festigkeitseigenschaften der Metalle in Wärme und Kälte, 1907; Kesselbleche mit Rissbildung, 1913; also numerous articles. Research on resistance of metals to heat and cold, bldg. materials of steam boilers, rivet joints of boiler plate; developed materials testing by use of metallography to discover causes of failure. Died Stuttgart, June 20, 1928.

BAUMÉ, Antoine, French chemist; b. Senlis, France Feb. 26, 1728; apprentice to chemist Claude Joseph

Geoffroy; owner chem. products bus., Paris, France, until 1780; returned to commerce during Revolution; became mem., then prof. chemistry École de Pharmacie, 1752. Mem. French Acad. Scis., 1773, French Soc. Med. Author: Eléments de pharmacie théorique et pratique, 1762; Manuel de chimie, 1765; Chimie experimentale et raisonnée, 1773. Developed processes for purifying saltpeter, bleaching silk, making sal ammoniac, dyeing cloth of two colors; improved manufacture of porcelain; invented hydrometer; devised Baumé scale for hydrometer readings; research on chem. affinity. Died Charenton le Pont, France, Oct. 15, 1804.

BAUME, Louis Joseph, Swiss dentist; b. Les Bois, Switzerland, Feb. 10, 1913; s. Louis Eugène and Marie-Elizabeth (Schill) B.; grad. U. Basle, 1937, Dr.med.dent., 1938; M.S., in Dentistry, U. Cal., 1950; m. Marguerite Arn, Oct. 18, 1944. Asst., Dental Sch. U. Basle, 1937-45; practice dentistry, Zurich, Switzerland, 1940-48; research asso. Hooper Found. Med. Research, 1948-54; asst. prof. U. Cal. at Berkeley Coll. Dentistry, 1951-54; prof. dental medicine U. Geneva, 1955-—; dir. Inst. Dental Medicine, 1961-65. Mem. council Swiss Soc. Research on Nutrition; chmn. Internat. Commn. on Standardization of Classification and Statistics on Oral Conditions; mem. expert panel WHO. Fellow Internat. Coll. Dentists (pres.), Am. Acad. A.A.A.S., Kaiserlich Deutsche Gesellschaft der Naturforscher, Leopoldina, Am. Coll. Dentists; mem. Fédération dentair internationale (v.p.), Société Suisse d'Odonto-stomatologie (past pres.), Sigma Xi, Delta Sigma Delta; hon. mem. numerous dental assns. in numerous countries. Research, numerous publs. on oral biology, dental edn. and oral statistics. Home: 2, Carrefour de Rive. Office: 30, rue Lombard, Geneva 1200, Switzerland.*

BAUMÉS, Jean Baptiste Timothée, French physician; b. Lunel, France, Jan. 20, 1756; M.D., U. Montpellier (France), 1777; became prof. Med. Sch. Montpellier, 1790. Author: l'Ictère ou jaunissé des enfants, 1785; la Maladie du Mésentère, 1787; la Phtisie pulmonaire, 1783; les Maladies qui résultent des emanations des eaux stagnantes, 1789; Essai d'un système chimique de la science de l'homme, 1798; Fondements de la science méthodique des maladies, 1801. Also articles in med. jours. Described retrosternal pain of angina pectoris (Baumès sign); developed theory which related all diseases to 5 primitive substances. Died Montpellier, July 19, 1828.

BAUMÈS, Prosper, physician; b. France, 1791; stated (with Abraham Colles) Colles-Baumes law that a child suffering from congenital syphilis will not necessarily cause ulceration in a woman nursing it; stated that Tb is an infectious disease which can be transmitted from mother to child. Died 1871.

BAUMGARTEN-TRAMER, Franziska (Mrs. M. Tramer), psychologist; b. Lodz, Poland; d. Raphael and Eleanor (Lubliner) Baumgarten; student U. Cracow (Poland), 1906, U. Paris (France), 1907; Dr.phil., U. Zürich (Switzerland), 1910; m. M. Tramer, May 17, 1924. Faculty, U. Bern (Switzerland), 1929-—, in charge of lectures, 1931-53, hon. prof., 1953-—; practice psychology, 1925-—. Mem. Internat. Soc. for Profl. Guidance (hon.), Internat. Assn. Applied Psychology (hon. gen. sec. 1951), Internat. Council Psychologists (dir.), World Fedn. for Mental Health (mem. profl. bd.), Acad. Human Rights, Inst. Internat. Human Problems of Labor (mem. adminstrv. bd.), Namur (Belgium). Author numerous books and articles. Research on problems of labor, profl. guidance, structure of psyche, moral problems in profl. life, gifts and talents. Address: 33 Thunstrasse, Bern, Switzerland.*

BAUMGARTNER, Andreas, b. Frieberg, Germany, Nov. 23, 1793; grad. U. Vienna, Austria, 1810; prof. physics U. Vienna, from 1823; minister commerce, trade, pub. works, 1851-55; tchr. weekly course mechanics, application of arts to industry, Vienna. Mem. Acad. Scis. Vienna (pres. 1851). Author: Die Mechanik in ihrer Anwendung auf Künste und Gewerbe, 1823; Die Naturlehre, 1823; Anleitung zum Heizen der Dampfkessel, 1841. Research on natural history; work in steam engines. Died July 30, 1865.

BAUMGARTNER, Eugen, Swiss physicist; b. Basel, Switzerland, June 21, 1926; s. Fritz and Anna (Hess) B.; Ph.D., U. Basel, 1953; m. Iris Bänteli, 1953; 4 children. Faculty, U. Basel, 1957-—, prof., 1960-—; nuclear physics researcher U. Rochester (N.Y.), 1956, Lawrence Radiation Lab., Berkeley, Cal., 1965. Dean, Faculty for Natural Scis., U. Basel, 1963. Mem. Schweiz. Natfd. zur förderung der Wissensch. Forschung. Research, publs. on structure of nuclear matter, polarization phenomena. Home: 22 Bammertackerweg, Benken Bl., Switzerland. Office: 82 Klingelbergstr., Basel, Switzerland.*

BAUMGARTNER, Leona (Mrs. N. M. Elias), Am. physician, educator; b. Chgo., Aug. 18, 1902; d. William J. and Olga (Leisy) Baumgartner; A.B., U. of Kan., 1923, M.S., 1925; grad. work (Rockefeller Research fellow), Kaiser Wilhelm Inst., Munich, 1928-29; Ph.D., Yale (univ. fellow, 1930-31, Sterling fellow, 1931-32), 1932, M.D., 1934; D.Sc. Women's Med. Coll., 1950, N.Y.U., 1954, Russell Sage Coll., 1955, Smith, 1956, Western Coll. for Women, 1960, U. Mass., 1963; L.H.D., Keuka Coll., 1963; LL.D. (honorary), Skidmore Coll., 1959, Oberlin College, 1965, Oberlin, O., m. Nathaniel M. Elias, 1942. Head biology dept. Colby (Kan.) Community High Sch., 1923-24; charge nursing edn. Kansas City Jr. Coll., 1925-26; charge div. bacteriology and hygiene U. of Mont., 1926-28; interne, asst. resident and asst. in pediatrics N.Y. Hosp. and Cornell Med. Coll., 1934-36; acting asst. surg. U.S.P.H.S., 1936-37; lectr. in nursing edn. Columbia, 1939-42; with N.Y. City Dept. Health, 1937-53, med. instr. child and sch. hygiene, 1937-38, dir. pub. health training, 1938-39, dist. health officer, 1939-40, dir. bur. child hygiene, 1941-48, asst. commr. charge maternal and child health, 1948-53; exec. director N.Y. Foundation, 1953-54; Commr. Health, N.Y.C. Health Dept., 1954-62; asso. chief U. S. Children's Bur., Fed. Security Agency 1949-50, cons., 1950-56; mem. faculty Med. Coll., Cornell, 1939-—, asst. prof. pub. health and preventive medicine, 1940-54, asso. prof. clin. pub. health and preventive medicine, 1954-58, prof. clin. pub. health and preventive medicine, 1958-—, asst. prof. pediatrics, 1944-57, asso. prof. pediatrics, 1957-—; lectr. sch. pub. health and adminstrv. Medicine, Columbia, 1957-67, pediatrician N.Y. Hosp., 1942-56, asso. attending pediatrician, 1956-—; visiting lecturer department of maternal and child health, Harvard Sch. Pub. Health, 1948-62, professor Harvard Medical School, Boston, 1966-—; asst. adminstr. Office of Technical Cooperation and Research, AID, Dept. of State, 1962-65. Adviser to Indian Minister of Health, 1955; member national advisory council to Peace Corps. Licentiate Am. Bd. Pediatrics, 1942, Am. Bd. Preventive Med. and Pub. Health, 1949. Recipient many awards, latest Albert Lasker Award, Am. Pub. Health Assn., 1954, Albert Einstein award 1964, numerous other awards. Mem. Harvey Soc., History Sci. Soc., Oxford Bibliographics Soc., American Public Health Assn. (pres. 1958-59, 2d woman in this position, American Academy Pediatrics, N.Y. Acad. Med., Am. Pub. Health Assn., Child Welfare League Am., World Health Orgn., Nat. Social Welfare Assembly (v.p.), Nat. Health Council (pres. 1956), Phi Beta Kappa, Sigma Xi, Mortarboard. Contbr. med. and sci. articles profl. jours. Research on cardiovascular disease; rehabilitation of the crippled child; polio vaccine; public health; syphilis; family planning; child care. Home: 56 Washington Mews, N.Y.C. 10003.

BAUMHAUER, Heinrich Adolf, German chemist; b. Bonn, Germany, Oct. 26, 1848; s. Mathias and Anna Margaretha (Käuffer) B.; studied at U. Bonn, 1866-70, later at U. Göttingen (Germany); m. Hedwig Mathilde Welter, Sept. 28, 1872; 3 sons, 3 daus. Prof. mineralogy U. Freiburg (Germany), 1895-1925; also prof. inorganic chemistry, 1906-07. Mem. Mineral. Soc. St. Petersburg (now Leningrad), Mineral. Soc. London. Author: Die Beziehungen zwischen dem Atomgewichte und der Natur der chemischen Elemente, 1870; Kurzes Lehrbuch der Mineralogie, 1884; Leitfaden der Chemie, 2 vols, 1884-85; Das Reich der Kristalle, 1889; Die Resultate der Ätzmethode, 1894, Darstellung der 32 möglichen Kristallographie Klassen . . . , 1899; Die neuere Entwicklung der Kristallographie, 1905. Designed a periodic system of elements arranged along a spiral (simultaneously with Mendeleev), 1869; research on mineralogy and symmetry of crystals; suggested index of chem. publs. (became Chemisches Zentralblatt) to be edited by German Chem. Soc. Died Freiburg, Aug. 1, 1926.

BÄUMLER, James, Swiss toxicologist, chemist; b. Basle, Switzerland, Mar. 10, 1925; s. Oskar and Elsa (Hunzinger) B.; Dr. phil. chemist, U. Basle, 1950, diploma federal nutrition chemist, 1956; m. Rose-Marguerite Piaget, Aug. 18, 1952; children—Markus, Esther. Mem. sci. staff U. Basle, 1950-—, chief toxicological dept. Inst. Forensic Medicine, 1960-—, forensic chemist dept. of police, 1960-—; cons. Swiss book for analytical examination of foodstuffs, 1960-—. Mem. European Orgn. for Research in Fluorine and Dental Caries Prevention (collaborater team III), Acad. Forensic and Social Medicine, German Soc. Forensic Medicine, Swiss Soc. Analytical and Applied Chemistry. Research, publs. on metabolism and detection of psychopharmaca; new procedures for microdetection of toxic agts. (mercury, fluorine). Home: 20 Sonnmattstr./4142 Munchenstein. Office: 22 Pestalozzistr. 4000, Basle, Switzerland.*

BAUMOL, William Jack, Am. economist; b. N.Y.C., Feb. 26, 1922; s. Solomon and Lillian (Itzkowitz) B.; B.Social Service, Coll. City N.Y., 1942; Ph.D., London (Eng.) U., 1949; LL.D., Rider Coll., 1965; m. Hilda Missel, Dec. 27, 1941; children—Ellen Frances, Daniel Aaron, Asst. economist U. S. Dept. Agr., 1942-43, 46; faculty London Sch. Econs., 1947-49; prof. econs. Princeton, 1949-—. Cons. Mathematica, Inc., Princeton, N.J., 1962-—; mem. econ. policy council State of N.J., 1966-—. Fellow Econometric Soc., mem. Am. Econ. Assn. (v.p. 1966-—). Inst. Mgmt. Sci. Author: Economic Dynamics, 1951; Welfare Economics and the Theory of the State, 1952; Economic Processes and Policies, 1954; (with L. V. Chandler) Economic Theory and Operations Analysis, 1961; Business Behavior, Value and Growth, 1959; The Stock Market and Economic Efficiency, 1966; (with W. G. Bowen) Performing Arts: The Economic Dilemma, 1966. Research on relationship between demand for money and volume of bus., effect of motivation of mgmt. on bus. decisions, causes and prospects for deficits in performing arts. Home: 214 Western Way, Princeton, N.J. 08540.*

BAUR, Erwin, German geneticist; b. Baden, Germany, Apr. 15, 1875; s. Wilhelm and Anna (Siefert) B.; student medicine and botany Heidelberg, Freiburg, Kiel (all Germany); doctorate degrees in medicine and botany; hon. dr. U. Uppsala (Sweden). Ship's doctor in S.Am.; became lectr. in Berlin (Germany), 1904; apptd. prof. botany Landwirtschaftliche Hochschule, Berlin, 1911; exchange prof., Madison, Wis., 1914; founder 1st genetics inst. in Friedrichshagen, 1922, an expanded inst. in Dahlem, Germany, 1922, Kaiser-Wilhelm Inst. for Plant Breeding and Genetic Research, 1929. Named hon. prof. U. Buenos Aires (Argentina). Author: Die wissenschaftliche Grundlagen der Pflanzenzüchtung, 1924; Einführung in die experimentelle Vererbungslehr, 1930; (with M. Hartmann) Handbuch der Vererbungwissenschaft, 1928. Editor several genetics jours. Research on hybridization of snapdragons; model analysis of genetic factors of snapdragon; bred lupine free of bitter principles; improved several cultured plants. Died Berlin, Dec. 2, 1933.

BAUR, Franz, German meteorologist; b. Munich, Germany, Feb. 14, 1887; s. Franz and Frida (Grandaur) B.; ed. U. Munich, U. Friburg; Dr. ès sciences naturales; hon. doctorate U. Giessen; m. Franziska Heider, Apr. 25, 1916; children—Franz, Walter, Herbert. Dir., Meteorol Obs. of Saint-Blaise; in charge research Prussian Meteorol. Inst., Berlin, Germany; dir. Inst. Research and Long-Term Meteorol. Forecasts; instr. U. Frankfurt, now hon. prof. Mem. Am., Hungarian (corr.) meteorol. socs. Author: Existenzeweis des Grosswetters; Meteorologische Taschenbuch; Physikalischstatische Regeln als Grundlagen für Wetterund Witterungsvohersagen. Address: Kaiser-Friedrich-promenade 115, Bad-Hamburg, Germany.

BAUSCH, Hans Albert, German chemist; b. Friburg /Br., Nov. 12, 1895; s. Otto and Ella (Bickel) B.; Ph.D. in Chemistry, U. Friburg/Br.; m. Elisabeth Herrmann, Apr. 1926; children—Manfred, Rolf. Asst. scientist Superior Sch. Agr., Berlin, Germany, 1921-25; chief dept. Inst. for Fermentation, Berlin, 1925-51; instr. comml. products Superior Sch. Economy, Berlin, 1926-45; hon. prof. U. Humboldt, Berlin, 1934, titular prof., 1950; instr. Tech. U., Berlin, 1951-52, titular of chair of tech. of fermentation, dir. Inst. Chemistry of Fermentation, dir. group alimentation and tech. of fermentation, 1952-—, prof. emeritus, 1962-—. Author: (with A. Binz) Chemotherapie des Gerstenbrandes, 1922; Arbeitsvorschriften zur chemisch-brautechnischen Betriebskontrolle, 1930; Chemische Technologie, 1950; Arbeiten der wiss. Metarbeiter der Versuchs- und Lehranstalt für Brauerei Berlin, 1953. Address: Pfeilstrasse 30, Berlin-Niederschonhausen, East Germany.

BAUSCH, Johann Lorenz, German botanist, physician; b. Schweinfurt, Germany, Sept. 30, 1605; s. Leonhard and Barbara (Büttner) B.; student Jena, Germany, 1623-26, Marburg, Germany, 1626-28, Padua, Italy, 1628-30; degree, Altdorf, Germany, 1630; m. Anna Margaretha Prückner, Nov. 9, 1630; 1 dau., Anna Maria. Practiced medicine, Schweinfurt, from 1630. Mem. Acad. Naturae Curiosum (1st such soc. in Europe, a founder, 1st pres. 1652). Author: Schediasmata bina curiosa de lapide haematite et aetite . . . , 1665; Schediasma curiosum de unicornu fossili, 1666; Schediasma posthumum de caeruleo et chrysocolla, 1668. Died Jan. 18, 1665.

BAUSCHINGER, Johann, German engr.; b. Nuremberg, Germany, June 11, 1834; s. Johann Michael and Anna Elisabeth (Busch) B.; ed. U. Munich (Germany); m. Katharine Strasser; 4 sons, including Julius B., 6 daus. Tchr. at trade sch. in Fürth, secondary sch. in Munich; became prof. tech. mechanics and graphic statics Munich Polytechnikum, 1868; dir. Inst. for Materials Testing. Author: Die Schule der Mechanik, 1861; Indicator Versuche an Locomotiven, 1869; Elemente der graphischen Statik, 1871; Instrumente zu Messen der Gestaltungsveränderung der Probekörper, 1882; also numerous articles. Research on durability of forged iron in modern constrn.; eponym for hardening of steel by straining it beyond elasticity limit (Bauschinger effect); founder Bauschinger confs. for establishment unified testing method of constrn. materials. Died Munich, Nov. 25, 1893.

BAUSCHINGER, Julius, German astronomer; b. Fürth, Germany, Jan. 28, 1860; s. Johann and Katherina (Strasser) B.; ed. U. Berlin (Germany); degree U. Munich (Germany), 1883; m. Katharina Schrauder; 2 sons, 3 daus. Mem. expdn. for investigation Venus transit, Hartford, Conn., 1882; became lectr. U. Munich, 1888; named prof. theoretical astronomy, dir. Inst. Astron. Calculation, Berlin, 1896; dir. univ. obs., prof. astronomy, U. Strasbourg (France), 1909; prof. astronomy, dir. univ. obs. U. Leipzig (Germany), 1920-30. Author: (with H. v. Seeliger) Münchner Sternverzeichnis, 1890; Münchner Sternverzuchungen, 1891; Untersuchungen über der astronomische Refraktion, 1896; Tafeln zur theoretische Astronomie, 1901; Bahnbestimmung der Himmelskörper, 1906; (with J. Peters), Logarithm-Trigonometriche Tafeln mit acht Dezimalstellen, 1910; also numerous articles. Research on determination of orbits of comets and planets. Died Leipzig, Jan. 21, 1934.

BAVENDAMM, Werner Hermann Theodor, German biologist; b. Berlin, Germany, Nov. 27, 1898; s. Emil and Louise (Biljes) B.; Ph.D., U. Berlin; m. Ingeborg Boden, Dec. 28, 1933; children—Jurgen, Dirk, Ute. With Natural Fibers Research Inst., Sorau, 1923-25,

Tech. U. Dresden, 1925-45; prof. agrégé U. Hamburg, 1950—. Cons. scientist, 1958—. Research and over 150 publs. on botany and microbiology, biology, pathology, conservation of forests and forest products. Home: Klosterbergenstrasse 18.*

BAWDEN, Sir Frederick Charles, Brit. plant pathologist; b. North Tawton, Eng., Aug. 18, 1908; s. George and Ellen (Balment) B.; M.A., Emmanuel Coll., Cambridge, 1930; m. Marjorie Elizabeth Cudmore, Sept. 6, 1934; children—Michael George, Peter Charles. Research asst. Potato Virus Research Sta., Cambridge, 1930-36; virus physiologist Rothamsted Exptl. Sta., 1936-40, head plant pathology dept., 1940-58, dep. dir. sta., 1950-58, dir., 1958—. Chmn., Agrl. Research Council Central Africa, 1964—; mem. Natural Environment Research Council, 1965; pres. Brit. Insecticide and Fungicide Council, 1965. Decorated knight bachelor Order Brit. Empire. Fellow Royal Soc., 1949; honorary life member Indian Botanical Soc., New York Academy Sciences; foreign member Royal Netherlands Academy Science; member Association Applied Biologists (pres. 1965). Author: Plant Viruses and Virus Diseases, 1964; Plant Diseases, 1950. Research, publs. on isolation of several plant viruses and characteristics, serology of plant viruses, factors that affect susceptibility of plants to infection with virus. Home: 1 West Common, Harpenden, U.K. Office: Rothamsted Exptl. Sta., Harpenden, U.K.*

BAXENDELL, Joseph, Brit. meteorologist, astronomer; b. Manchester, Eng., Apr. 19, 1815; s. Thomas and Mary (Shepley) B.; studied astronomy and meteorology in Robert Worthington's Obs.; m. Mary Anne Pogson, 1865; 1 son. Astronomer Manchester Corp., from 1859; meteorologist Southport Corp. Fellow Royal Astron. Soc., Royal Soc., 1884; mem. Manchester Lit. and Philos. Soc. (joint sec., editor). Research and numerous publs. on meteorology and terrestrial magnetism, variable stars; pioneer in detecting intimate connections between terrestrial and solar physics; advocated use storm signals. Died Birkdale, Eng., Oct. 7, 1887.

BAXTER, Alexander Duncan, Brit. mech. engr.; b. Liverpool, U.K., June 17, 1908; s. Robert Alexander and Violet (Austin) B.; B.Engring. with 1st class honors in Mech. Engring., 1930, M.Engring., 1933; m. Kathleen McClean, June 21, 1933; children—David Robert, Ann Elizabeth, Mary McClean. Sci. officer Royal Aircraft establishment, Farnborough, 1935-47; supt. Rocket Propulsion Establishment, Westcott, Eng., 1947-50; prof. aircraft propulsion Coll. Aeros., Cranfield, Eng., 1950-54, dep. prin., 1954-57; engring. dir. DeHavilland Engine Co., London, Eng., 1957-63; tech. exec. Bristol Siddeley Engines, Bristol, 1963—. Mem. various edn. adv. bodies, univ. external examiners. Fellow Royal Aero. Soc. (pres. 1966-67), Inst. Petroleum, Brit. Interplanetary Soc.; mem. Instn. Mech. Engrs., Council Engring. Instns. (mem. bd. 1962—). Author: Liquid Rocket Motor Design, 1955; also numerous articles, chpt. in book. Design and devel. of aero gas turbines, rockets, and ramjets for missiles and space research; spl. studies in high intensity combustion for these engines. Home: Court Farm, Pucklechurch, Gloucestershire, Eng. Office: Bristol Engine Div., Rolls Royce Ltd., P.O. Box 3, Filton, Bristol, Eng.*

BAXTER, Gregory Paul, Am. chemist; b. Somerville, Mass., Mar. 3, 1876; s. George Lewis and Ida Florence (Paul) B.; A.B., Harvard, 1896, A.M., 1897, Ph.D., 1899; Sc.D., U. Mich., 1929; m. Amy Bailey Sylvester, June 2, 1906; 1 dau., Elizabeth Paul. Asst. in chemistry Harvard, 1895-97, instr., 1897-99; instr. chemistry, Haverford Coll., Pa., 1899-1900; asst. prof. chemistry Swarthmore, 1900-02; instr. in chemistry Harvard, 1902-05, asst. prof., 1905-15, prof., 1915-25, Theodore William Richards prof., 1925-44, Theodore William Richards prof. emeritus, from 1944. Chmn. International Com. on Atomic Weights, 1930-47; Theodore W. Richards medalist, 1934. Fellow A.A.A.S., Am. Acad. Arts and Scis.; mem. Nat. Acad. Scis., Am. Chem. Soc., Phi Beta Kappa, Sigma Xi. Author: Researches upon the Atomic Weights, 1910, also papers in chemical periodicals. Studies in determination atomic weight of lead, establishing basis for estimation of age of minerals. Died, 1953.

BAY, Eberhard, German physician; b. Tübingen, Germany, Dec. 12, 1908; s. Eugen and Lydia (Otto) B.; student U. Tübingen, 1926-27, U. Kiel (Germany), 1927-28, U. Vienna (Austria), 1928-29; Dr.med. U. Berlin (Germany), 1932, Dr.med.habil., 1940; m. Elisabeth Uhlig, Oct. 12, 1940; children—Ellen, Eberhard, Herbert, Martin. Asst. neurologist U. Clinic, Berlin, 1934-39; lectr. U. Berlin, 1940-41; lectr. U. Heidelberg, Germany, 1941-48, asst. prof., 1948-55; prof. neurology U. Düsseldorf, Germany, 1955—, dir. Gesellschaft fur Neurologie, Société Francaise de Neurologie, Internat. League Against Epilepsy, Royal Soc. Med., Gerontological Soc., Gesellschaft Deutscher Naturforscher and Arzte. Author: Die Praxis der Erkennung und Beurteilung von Hirnverletzunge, 1941, Agnosie und Funktionswandel, 1950; also numerous articles. Research on brain injuries and their sequelae, disturbances of speech and other higher mental functions in brain diseases. Home: 84 Mooresstrasse. Office: 5 Moorenstrasse, Düsseldorf 4000, Germany.

BAY, Zoltan Lajos, Am. physicist, govt. ofcl.; b. Gyulavari, Hungary, July 24, 1900; s. Joseph and Julia (Boszormeny) B.; M.S., U. Budapest (Hungary), 1923, Ph.D., 1926; m. Julia Herczegh, July 7, 1947; children—Zoltan Kalman, Julia Rose. Came to U. S., 1948, naturalized, 1953. Prof. U. Szeged (Hungary), 1930-36, Budapest Inst. Tech., 1938-48; research prof. physics George Washington U., 1948-55; physicist Nat. Bur. Standards, Washington, 1955—. Recipient silver medal Dept. Commerce, 1962. Fellow Am. Phys. Soc.; mem. I.E.E.E. (sr.). Contbr. articles to profl. jours. Pioneer work in electroluminiscence, particle counting with electron multipliers; developed 1st moon radar, 1946. Home: 151 Quincy St., Chevy Chase, Md. 20015. Office: Nat. Bur. Standards, Washington 20234.*

BAYEN, Pierre, French chemist, pharmacist; b. Chalons-sur-Marne, France, Feb. 7, 1725; founder and organizer of French military pharmacy; inspector general of san. service of French army, named pharmacist-in-chief during the Seven Years War, member of French Academy Sciences, 1795. Author: Opuscules chimiques, 2 vols., 1798. Research on mineral waters of France; discovered fulminate of mercury; prepared oxygen by heating mercuric oxide, 1774, but did not recognize it as new substance; rejected phlogiston theory 3 years before Lavoisier. Died Paris, France, Feb. 15, 1798.

BAYER, Adolf von, see von Baeyer, Adolf (Johann Friedrich Wilhelm).

BAYER, Istvan, pharmacist; b. Budapest, Hungary, Oct. 18, 1923; s. Antal and Aranka (Farago) B.; Diploma in Pharmacy, U. Peter Pazmany, Budapest, 1945, Dr.Pharmacy, 1949; Candidate Pharm. Scis., Hungarian Acad. Scis., 1960; m. Eva Gathy, Sept. 30, 1950; children—Eva, Antal. Research, G. Richter Chem. Works, 1950-52, Inst. Drug Research, Budapest, 1952-54; chief sect. tech. devel. Hungarian Ministry Health, 1954-62; dir. Nat. Inst. Pharmacy, Budapest, 1962-67; social affairs officer Div. Narcotics, UN, Geneva, Switzerland, 1967—. Temporary lectr. Med. U., Budapest, 1957-65; mem. expert com. internat. pharmacopoeia and pharm. preparations WHO, 1966—; sec. Hungarian Pharmacopoeial Comm., 1954—. Mem. Hungarian Pharm. Soc., Hungarian Med. Assn., Hungarian Chem. Soc. Editor: Acta Pharmaceutica Hungarica, 1953—. Research, numerous publs. on drug analysis in fields of alkaloids (ergot, opium, others), Glycosides (digitale, rutin), Steroids and synthetic drugs, research on active principles of med. plants. Office: Palais des Nations, Geneva, Switzerland.*

BAYER, Johann, German astronomer; b. Rain, Bavaria, Germany, 1572; practiced law; author of Uranometria (first complete celestial atlas), 1603. Introduced system of cataloguing stars in order of brightness using Greek letters; charted 12 new constellations; unsuccessfully tried to introduce system of naming no. constellations from New Testament, so. from Old Testament. Died Augsburg, Germany, Mar. 7, 1625.

BAYER, Josef, Austrian anthropologist, paleontologist; b. Hollabrunn, nr. Vienna, Austria, June 10, 1882; s. Eduard and Elise (Schnotzinger) B.; ed. Gymnasium St. Polten; doctorate, U. Vienna; Mem. staff Natural History Mus. Vienna, from 1908; lectr. U. Vienna, from 1913, became dir. anthropol.-enthnol. dept., later dir. anthropol.-prehistoric dept. Author: Der Mensch in Elszeithalter, 1927; also numerous articles; pub. Die Eiszeit, from 1924. Made extended excavations in lower Austrian loess region; received internat. acclaim for ice age research; studied paleolithic (stone age) culture in Palestine and named uncovered culture Ascalonia. Died Vienna, July 23, 1931.

BAYES, Thomas, Brit. mathematician; b. 1702; s. Joshua and Ann Bayes; ed. privately. Ordained to Presbyn. ministry, London, Eng., pastor at Turnbridge Wells, Eng., 1720-52, ret., 1752. Fellow Royal Soc. 1742. Author: Divine Benevolence, or an Attempt to Prove that the Principle End of the Divine Providence and Government is the Happiness of His Creatures, 1731; An Introduction to the Doctrine of Fluxions, and a Defense of the Mathematicians against the Objections of the Author of the Analyst, 1736. Proposed theorem for inverse probability whereby probabilities of unknown causes are inferred from observed events; 1st to use probability inductively. Died Apr. 17, 1761.

BAYLE, Antoine Laurent Jessé, French physician; b. Vernet, France, Jan. 13, 1799; med. student of Laennec, 1815; M.D., U. Paris (France), 1822. Librarian, Faculty Medicine, Paris, from 1824; prof. Charenton, France, from 1827. Author: Recherches sur l'arachnitis chronique, 1822; Bibliothèque de thérapeutique, 4 vols., 1828-37; Petit manuel d'anatomie descriptive, 1823; Nouvelle doctrine de Maladies mentales, 1825; Traité des maladies du cerveau et de ses membranes, 1826; (with Holland) Manuel d'anatomie générale, 1827; Eléments de pathologie médicale, 2 vols., 1855-57. Editor: Encyclopédie des sciences medicales, 40 vols., 1835-46. Gave 1st description of progressive paralysis of insane (Bayle's disease), 1822. Died Paris, Mar. 1858.

BAYLE, François, French physician; b. St. Bertrand-de-Commines, France, 1622; Royal professor Toulouse Faculty Arts; physician, Toulouse. Mem. French Acad. Scis. 1699. Author: Problemata physicomedica, 1677; Histoire anatomique d'une quossesse de vingt-cing ans, 1679. Research on pregnancy; applied (following example of Boerhaave) physics and math. to medicine. Died Toulouse, Sept. 24, 1709.

BAYLE, Gaspard Laurent, French physician; b. Vernet, France, 1774; studied theology, law, medicine École de Medicine, Paris; degree, 1801; physician to Napoleon; staff Hôpital de la Charité; mem. Faculty Medicine Paris. Author: Remarques sur les tubercules, 1801; Recherches sur la phtisie pulmonaire, 1810. Introduced term miliary in describing acute miliary Tb; described tubercular laryngitis, lymphadenitis; advocated diagnosis of Tb by initial symptoms of emaciation and fever; correlated Tb of lungs with Tb of other organs; emphasized Tb as a specific disease rather than a degenerate condition following other diseases; 1st to use auscultation in diagnosis; pioneered study of cancer of liver; familiar with cancer of lungs. Died of Tb, 1816.

BAYLE, Pierre, French philosopher; b. Charlat, France, Nov. 18, 1647; ed. Toulouse, France. Went to Geneva, 1670; pvt. tutor, Geneva, Rouen, Paris; became prof., Sedan, 1675, Rotterdam, Netherlands, 1681, dismissed because of his beliefs, 1693. Author: Pensées diverses sur la comète, 1682; Critique générale de l'historie de Calvinisme de Maimbing, 1682. Compiler: Dictionnaire historique et critique, 1697. Founder critical jour. Novelles de la Republiques des lettres, 1684-87. Defended Calvinism; founder 18th century rationalism; his thought influenced both French and German Enlightenments; showed beliefs in saints and miracles are contradictory and without proof. Died Rotterdam, Dec. 28, 1706.

BAYLET, R., French virologist; b. Toulouse, France, June 9, 1923; s. Marcel Angèle (Daurie) B.; Docteur en Médecine, Lyons, France, 1947; postgrad. Institut Pasteur, 1952-55; m. Benazet, Nov. 1946; children—Michel, Jean François. Staff biol. labs. Centre recherche trypanosomiase africain, 1949-51, Centre transfu. Nord-Vietnam, 1952-54, biol. lab. Hôpital de Dakar, 1955-58, virology lab. Institut Pasteur, 1959-65; prof. agrégé in hygiene, hosp. biologist Laboratoire de'Hygiène et de Santé Publique, Faculty Medicine Dakar, Senegal, 1965—; prof. hygiene U. Dakar. Mem. Nat. Blood Transfusion Soc., Soc. Biology, Hygiene Soc., Psychopathology Soc. Author: (with Payet) Cliniques africaines; also numerous articles. Research on epidemiology of bacterial, viral and parasitic diseases of W. Africa; chromosomal studies of African monkeys. Address: Faculté de Médecine de Dakar-Fann, Sénégal.*

BAYLEY, Richard, Am. physician; b. Fairfield, Conn., 1745; studied medicine under John Charlton, N.Y.C.; studied anatomy under William Hunter, London, Eng., 1769-71; m. Miss Charlton. Went to London, 1769-71; during croup epidemic made study of disease's causes and treatment which cut mortality rate in half, 1774; in England, 1775-76; surgeon Brit. Army, Newport, R.I., 1776-77; practiced medicine, N.Y.C., 1777; prof. anatomy and surgery Columbia, 1792; health physician Port of New York, 1795. Author: Letter on the Croup, 1781; An Account of the Epidemic Fever which prevailed in the City of New York during Part of the Summer and Fall of 1795, pub. 1796; Essay on the Yellow Fever, 1797; Letters from the Health Office Submitted to the New York Common Council. 1st physician in Am. to amputate arm at shoulder-joint. Died Aug. 17, 1801.

BAYLEY, William Shirley, Am. geologist; b. Balt., Nov. 10, 1861; s. Robert P. and Emma (Downing) B.; A.B., Johns Hopkins, 1883, doctorate, 1886; m. Lucie Jacobs, 1894; 1 dau., Emily (Mrs. J. Howard Gellen). Prof. geology Colby Coll., 1888-1904; tchr. Lehigh U., 2 years; faculty U. Ill., from 1906, prof. geology, 1913-31, head dept., 1928-31, emeritus, from 1931. Asst. geologist, geologist U. S. Geol. Survey. Recipient prize Peabody Inst., 1879. Fellow Geol. Soc. Am. (v.p. 1936); mem. Mineral. Soc. Am. (pres. 1936), Soc. Econ. Geologists (v.p. 1932). Author: Elementary Crystallography, 1910; Minerals and Rocks, 1915; Descriptive Mineralogy, 1916; Guide to the Study of Non-metallic Mineral Products, 1930; also many papers. Asso. editor Am. Naturalist, 1886-87; reviewer Neues Jahrbuch fur Mineralogie, Berlin, Germany, 1890-1908; (business editor) Jour. Econ. Geology, 1905-42. Investigated igneous and metamorphic rocks, also underground waters in Me., geology of highlands in N.J. and S.W. N.Y., magnetic iron ores in eastern Tenn., magnetic and brown iron ores in western N.C., kaolins in N.C., geology of Tate Quadrangle in Ga., pre-Cambrian rocks and iron ores in Lake Superior area, Menominee iron-bearing dist. Mich. Died Feb. 13, 1943.

BAYLIS, Geoffrey Thomas Sandford, New Zealand botanist; b. Palmerston, North New Zealand, Nov. 24, 1913; s. Gerald T. de S. and Daisy (Aston) B.; M.Sc., Auckland (N.Z.) U., 1935; Ph.D., Imperial Coll., London (Eng.) U., 1938. Asst. mycologist Dept. Sci. and Indsl. Research, 1938-45; head botany dept. Otago (New Zealand) U., 1945—. Fellow Royal Soc. New Zealand; mem. New Zealand Ecol. Soc. (pres. 1965-66). Research, publs. on bot. exploration, plant succession, taxonomy of solanum, plant pathology especially ecol. significance of root inhabiting fungi.

Home: 367 High, Dunedin, New Zealand. Office Botany Dept., Otago U., Otago, New Zealand.*

BAYLISS, William Maddock, Brit. physiologist; b. Wolverhampton, Eng., May 2, 1860; s. Moses and Jane (Maddock) B.; ed. U. Coll., London, Wadham Coll., Oxford (Eng.) U.; m. Gertrude Ellen Starling, 1893; 3 sons, 1 dau. Tchr. physiology, Oxford, 1888; became mem. faculty U. Coll., London, 1888; spl. professorship gen. physiology created for him, 1912; named hon. fellow Wadham Coll., 1922; Croonian lectr., 1904, Oliver-Sharpey lectr., 1918, Sylvanus Thompson lectr., 1919, Herter lectr., 1922. Baly medal Royal Coll. Physicians, 1917, Royal medal, 1919. Fellow Royal Soc., 1903 (member council 1913-15; mem. Physiol. Soc. (sec. 1900-27, treas. 1922-24.) Research (with Rose Bradford) on electric currents in salivary glands; studied (with E. J. Starling) electric currents in mammalian heart, venous and capillary pressures, 1894, innervation of intestines, 1898-99, developed theory of hormonal control of internal secretion, discovered secretin (hormone responsible for secretions of pancreas) 1902; research on vascular system, brain circulation, vaso-motor reflexes, antidromic nerve fibers; introduced use of saline injections for treatment surg. shock. Author: The Nature of Enzyme Action, 1908; Principles of General Physiology, 1915; The Vaso-Motor System, 1923. Died Hamstead, Eng., Aug. 27, 1924.

BAYLOR, James Bowen, Am. geodesist; b. Mirador, Va., May 30, 1849; s. John Roy and Anne (Bowen) B.; hon. grad. Va. Mil. Inst., 1865; B.S., C.E., U. Va., 1872; LL.D., Baylor U., 1903; m. Ellen C. Bruce, Jan. 5, 1881 (dec.). Apptd. aid in U. S. Coast and Geod. Survey, 1874; made magnetic survey of N.C.; field officer U. S. Coast and Geod. Survey from 1874; determined the elements of earth's magnetism from Can. to Mexico in almost every state, and did geod., astron. and hydrographic work for survey in various sections of U. S.; oyster surveys in La.; Baylor survey of oyster grounds of Va., 1889-94; commr. Supreme Ct. of U. S. for boundary of Va. and Tenn., 1900-02; engr. Va.-Md. boundary, Pa.-N.Y. boundary and on U. S. and Can. Boundary Survey. Contbr. various reports of U. S. Coast and Geod. Survey. Died May 1924.

BAYLY, William, Brit. astronomer; b. Carions, Wiltshire, Eng., 1737. Asst. at Royal Obs.; sent by Royal Soc. to N. Cape to observe transit of Venus, 1769; astronomer Cook's 2d voyage to So. hemisphere, 1772, also 3d voyage, 1776; head-master Royal Acad., Portsmouth, Eng., 1785-1807. Author: Astronomical Observations made at the North Cape for the Royal Society; The Original Astronomical Observations made in the Course of a Voyage Towards the South Pole by W. Wales and W. Bayly, 1777; Original Astronomical Observations made in the Course of a Voyage to the Northern Pacific Ocean . . . in the Years 1776-1780, by Capt. J. Cooke, Lieut. J. King and W. Bayly . . . by order of the Board of Longitude, 1782. Contbd. astron. observations during voyages. Died Portsea, Eng., 1810.

BAYMA, Joseph, mathematician; b. Cirie, nr. Turin, Italy, Nov. 9, 1816; attended Royal Acad., Turin, Italy. Entered Jesuit novitiate, Chieti, 1832, ordained Jesuit priest, 1847; missionary to Algiers, 1847; asst. to astronomer Angelo Secchi, later dir. Osservatorio del Collegio Romano, Rome, Italy; rector Episcopal Sem., Bertinoro, Italy, 1852-58; prof. philosophy Stonyhurst Coll., Eng., 1858-69; pres. St. Ignatius Coll., San Francisco, 1869-72, prof. higher math., 1872-80; ret. to Santa Clara (Cal.) Coll. because of ill health, 1880. Author: (with Enrico Vasco) Il Ratio Studiorum adattato ai tempi presenti; De studio religiosae perfectionis excitando, 1852; Philosophia Realis, 1861; Elements of Molecular Mechanics, 1886. Asserted that phys. matter is reducible to unextended points, materially and mathematically non-continuous. Died Santa Clara Coll., Feb. 7, 1892.

BAYNES, John, inventor; b. Westmoreland, Eng., Aug. 24, 1842; s. Oswald and Agnes B.; took 5 years course at Ackworth (Soc. of Friends' Coll., Eng.); m. Helen A. Nowill, Apr. 24, 1867. Mem. Bengal Chamber of Commerce, 1864; engaged in gen. foreign shipping business until 1875; came to U. S., 1875; invented celluloid photographic films, 1884; gold etching photo process, 1885, and numerous inventions in arts, including photographic modeling; photographically modeled records of sound vibrations; invented process for producing musical and other sounds from graphic designs. Died 1903.

BAYNHAM, William, Am. surgeon; b. probably S.C., Dec. 7, 1749; s. John Baynham; studied surgery under Dr. Walker, probably S.C., 1764-69, St. Thomas Hosp., London, Eng., 1769-72. Prepared anatomical demonstrations and instructed in dissection for med. students at Cambridge (Eng.) U., 1772-81; practiced surgery, London, 1781-85; returned to U. S., 1785; practiced medicine specializing in surgery, Essex, Va., 1785-1814; gained reputation as surgeon in operations for stone, cataracts and extra-uterine pregnancy; descriptions of some of his operations in Vol. I of New York Med. and Surg. Jour. Died Essex, Dec. 8, 1814.

BAYRD, Edwin Dorrance, Am. physician; b. Chgo., Nov. 12, 1917; s. George Oliver and Helen (Dorrance) B.; A.B. magna cum laude, Dartmouth, 1939; M.D., Harvard, 1942; M.S. in Medicine, U. Minn., 1947; m. Muriel Helen Burns, Oct. 17, 1942; children—Edwin Dorrance, Garett Thomas, Deborah, George Oliver, Linda. Practice medicine specializing in internal medicine, Rochester, Minn.; faculty U. Minn., 1947—, prof. medicine, 1967—, chmn. hematology sect., 1967—; Norman Paul vis. prof. medicine Syndey (Australia) Hosp., 1963. Diplomate Nat. Bd. Med. Examiners, Am. Bd. Internal Medicine. Mem. A.M.A., Minn., So. Minn. med. assns., Am. Fedn. for Clin. Research, Internat., Am. (chmn. membership com. 1965) socs. hematology, Central Clin. Research Club, Minn. Soc. Internal Medicine, Central Soc. for Clin. Research, Hematology Club. Editor-in-chief Mayo Clinic Proc., 1963—; editorial bd. Minn. Medicine, 1964—, Year Book Cancer, 1964—. Research, numerous publs. on understanding and distinguishing plasmocytic diseases, their natural course and their treatment. Home: 1116 10th St. S.W. Office: Mayo Clinic, Rochester, Minn. 55901.*

BAYS, Severin, mathematician; b. LaJoux, France, June 2, 1885; s. Antonin and Marie (Vial) B.; student math., physics, chemistry U. Göttingen (Germany), Sorbonne U. Paris (France), 1906-11; Dr. math., 1911; Ph.D., 1919; m. Berthe Seydoux, 1911. High sch. tchr. Coll. St. Michel, Fribourg, Switzerland, 1911-19; extraordinary prof. math. U. Fribourg, 1921-25, ordinary prof., 1925—, rector, 1937-38. Municipal councillor City Fribourg, 1932—, pres., 1951. Mem. several Swiss and fgn. math. socs.; pres. Société helvétique des sciences naturelles, 1926, 45. Research, publs. on doctrine of combinations especially Steiner theory, cyclic systems of Steiner triples, group theory, and theory of numbers. Address: Route de Bertigny 41, Le Chalet, Fribourg.

BAZANOVA, Naylya Urazgulovna, Russian physiologist; b. Pishpek (now Frunze), 1911; grad. Alma-Ata Zool. and Vet. Inst., 1932, postgrad., 1932-35; D.Biol. Sci., 1945; Asst., Alma-Ata Zool. and Vet. Inst., 1936-38, head chair physiology farm animals, 1938-48, prof., 1946—; acad. sec. dept. biol. and med. sci. Kazakhstan Acad. Sci., 1951-62. Mem. Kazakhstan Acad. Sci. Decorated Order of Lenin. Author: The Development of the Regulation of Blood Circulation and Respiration in Various Farm Animals in Ontogeny, 1946; Changes Due to Age in the Type, Rhythm and Neuroregulation of Respiration in Camels, 1948; Physiological Principles for Increasing the Productivity of Farm Animals, 1954; Methods of Anastomosis of Steno's Duct in Sheep, 1958; co-author: Effect of Acidophilic and Lactic Corn Paste on the Zymotic Processes in the Rumen, 1961; Physiological Evaluation of New Silages, 1962. Address: Alma-Ata Zool. and Vet. Inst., prsosp. Abaya 8, Alma-Ata, Kazakhstan SSR, USSR.

BAZETT, Henry Cuthbert, physician, physiologist; b. June 25, 1885; s. Henry and Elisa Ann (Cruikshank) B.; M.D., M.A., Oxford U.; D.Sc., West Ont. Coll.; m. Dorothy Rufford Livesey, 1917; 1 son, 1 dau. Demonstrator in physiology St. Thomas's Hosp., 1910-11; Radcliffe travelling fellow Oxford U., 1912-15, fellow Magdalen Coll., 1912-20, demonstrator in pathology, 1913-16, Christopher Welch lectr. clin. physiology, 1919-21; prof. physiology U. Pa. Med. Sch., Phila., 1921-50. Vis. prof. med. research U. Toronto, 1941-43. Fellow Royal Coll. Surgeons; mem. Am. Physiol. Soc. (councillor 1946-50, pres. 1950-51). Research and numerous publs. on circulation and temperature control. Died July 12, 1950.

BAZHENOV, Vasilii Ivanovich, Russian architect; b. Moscow, Russia, Mar. 12, 1737; s. Ivan Bazhenov; ed. archtl. sch. of Prince Uktomsky, Moscow, 1751-55, Moscow U., 1755-58, Acad. Fine Arts, 1758-61, Acad. Architecture. Personal architect for Empress Catherine of Russia; employed by Paul I to erect bldgs. belonging to crown. Mem. St. Luke, Florence, Bologna acads.; Acad. Fine Arts St. Petersburg (v.p.). Designed model for reconstrn. Kremlin; considered 1st original Russian architect of modern times; made original design, began bldg. Palace of St. Michael; designed Kazan Ch., St. Petersburg, Russia. Died Aug. 13, 1799.

BAZIN, Antoine Pierre Ernest, French dermatologist; b. St. Brice-sous-Bois, France, Feb. 20, 1807; M.D., 1836; physician, St. Louis Hosp., 1847. Author: Lesions du poumon considerees dans les Fleùres dites essentielles, 1834; Acné varioliforme, 1851; Recherches sur la nature et traitement des teignes, 1853. Described psoriasis buccalis, 1861, also scabies and scrofulous ulcer of the leg (Bazin's disease); established dermatological classifications; showed difference between disease and infection; distinguished arthritic from lymphatic diatheses. Died Paris, Dec. 14, 1878.

BAZIN, Gilles Augustin, French botanist; b. Paris, France, 1681; practiced medicine Strasbourg, France; corr. Acad. Sci., 1737. Author: Observations sur les plantes et leur analogie avec les insects, 1741; Traité sur l'accroissement des plantes, 1743; Histoire naturelle des abeilles, 1744; others. Died Strasbourg, June 4, 1754.

BAZIN, Henri-Émile, French hydraulic engr.; b. Nancy, France, Jan. 10, 1829; student École polytechnique, from 1846; became Engr., 1848; engr. Bourgogne Canal, (enlarged it and made it profitable for commercial navigation), 1854, became head engineer of civil engineering, 1875, inspector general, 1886; member of French Academy Scis., 1900. Author: Recherches sur l'écoulement des eaux dans les canaux découverts, 1858; Recherches hydrauliques, 1865; Expériences nouvelles sur l'écoulement en déversoir, 1898. Discovered that current in canals can best be measured under water; explained sudden flux of water produced in estuaries and rivers. Died Feb. 7, 1917.

BAZIN, Suzanne, French biochemist; b. La Ferte-sous-Jouarre, France, July 4, 1912; d. Henri and Lucienne (Sallerin) B.; D.Pharmacy, U. Paris (France), 1941, D.Sc., 1944. Asst. hosps. Paris, 1935-38; staff Centre Nat. de la Recherche Scientifique, Paris, 1939—, research asst. Inst. Physico Chem. Biology, 1945-52, research asst. Pasteur Inst., 1952-60, master research dept. exptl. pathology, 1961—. Named Laureate, Faculty Pharmacy in Paris, 1936-37, Hosps. Paris, 1938, Acad. Scis., 1961. Mem. French Biochem. Soc., French Connective Tissue Club (founder). Research, numerous publs. on electrophoretic mobility of red and white blood cells, metabolic disturbances in leukocytes during phagocytosis and in hormones treated phagocytes, chem. components of connective tissue, mucopolysaccharides, glycoproteins, collagen, role of proteolytic enzymes in dynamics of inflammation, influence of anti-inflammatory drugs on chem. composition and on activity of proteolytic enzymes in inflamed tissues. Home: 49 rue du Val d'Or, Saint Cloud, Hauts de Seine. Office: Inst. Pasteur, Garches, Hauts de Seine, France.*

BAZY, Pierre Jean-Baptiste, French physician; b. Ariège, France, Mar. 23, 1853; Docteur en Médecine, Paris, France, 1886; physician Bicetre Hosp., 1886; later at Beaujon Hosp. (where he installed modern operating rooms), became dir. urinary surgery; member of Academy of Medicine, French Academy of Sciences, 1921, Association of Medicine, Seine (president). Author: Maladies des voies urinaires, 1892-1901. First to perform cystotomy (removal of tumors from bladder); invented probe for removing prostate gland; 1st to demonstrate opening of pleura as best way to examine lungs in case of surg. lesion. Died Jan. 22, 1934.

BEACH, Alfred Ely, Am. inventor; b. Springfield, Mass., Sept. 1, 1826; s. Moses Yale and Nancy (Day) B.; m. Harriet Eliza Holbrook, June 30, 1847. Founder (with Orson D. Munn) firm Munn & Co., publishers, 1846; editor Sci. Am., patentee typewriter, 1847, typewriter for blind, 1857, cable railways and tunneling shield, 1864; inventor pneumatic carrier system (now used in mail tubes); recipient Gold medal (for work on typewriter) Am. Inst., 1856. Died N.Y.C., Jan. 1, 1896.

BEACH, Frank Ambrose, Am. psychologist; b. Emporia, Kan., Apr. 13, 1911; s. Frank Ambrose and Bertha (Robinson) B.; B.S., Kan. State Tchrs. Coll., 1934, M.S., 1935; Ph.D., U. Chgo., 1940; D.Sc., McGill U., 1966; m. Anna Beth Odenweller, Mar. 5, 1935; children—Frank Ambrose III, Susan. Curator, chmn. dept. animal behavior Am. Mus. Natural History, N.Y.C., 1936-46; prof. psychology Yale, 1946-52, Sterling prof., 1952-58; prof. U. Cal. at Berkeley, 1958—. Recipient award for outstanding research in psychiatry Assn. for Research in Nervous and Mental Disease, 1958. Fellow Am. Psychol. Assn. (award for outstanding sci. contbns. 1958), N.Y. Acad. Scis.; mem. Soc. Exptl. Psychologists (Warren medal 1951), Am. Philos. Soc., Nat. Acad. Scis., Western Psychol. Assn. Author: Hormones and Behavior, 1948; (with C. S. Ford) Patterns of Sexual Behavior, 1952; also numerous articles. Editor: Sex and Behavior, 1965. Research on various vertebrate species to elucidate importance 3 groups factors affecting sexual activity, central nervous system, endocrine system and individual's previous experience and learning. Home: 8306 Terrace Dr., El Cerrito, Cal. Office: Dept. Psychology, U. Cal. at Berkeley, Cal. 94720.*

BEACH, John Youngs, Am. chemist; b. Washington, Nov. 19, 1912; s. S. H. and Annie (Youngs) B.; B.S., U. Cal., Berkeley, 1933; Ph.D., Cal Inst. Tech., 1936; m. Margaret Linforth, Jan. 4, 1938; children—Gordon L., Barbara Y. Nat. research fellow chemistry Princeton, 1936-38, instr. chemistry, 1938-41; with Chevron Research Co., Richmond, Cal., 1941—, supr. analytical and phys. measurements sect., 1953—. Fellow A.A.A.S.; mem. Am. Chem. Soc., Phi Beta Kappa, Sigma Xi, Pi Mu Epsilon. Research in electron diffraction by gas molecules, chem. quantum mechanics, mass spectrometry, infrared spectoscopy, x-ray diffraction. Home: 85 Tamalpais Rd., Berkeley, Cal. 94708. Office: Chevron Research Co., Richmond, Cal. 94802.*

BEACH, Moses Sperry, Am. publisher, inventor; b. Springfield, Mass., Oct. 5, 1822; s. Moses Yale and Nancy (Day) B.; Chloe Buckingham, 1845. Apprentice printer on N.Y. Sun, 1834, circa 1840; co-owner Boston Daily Times, 1845; operated N.Y. Sun (with brother and father), 1845-52, sole owner, 1852-60; devised new method for feeding paper to presses and pioneered printing both sides of sheet at once, 1862-68; retired to estate in Peekskill, N.Y., following sale of Sun to group represented by Charles A. Dana, 1868; travelled widely abroad. Died Peekskill, July 25, 1892.

BEACH, Moses Yale, Am. inventor; b. Wallingford, Conn., Jan. 15, 1800; s. Moses Sperry and Lucretia (Stanley) B.; m. Nancy Day, Nov. 19, 1819; children—Moses Sperry, Alfred Ely: Apprentice cabinet maker,

Hartford, Conn., 1814-18; partner in cabinet mfg. bus., 1819-circa 1828; part owner paper mill, Saugerties, N.Y., 1829-34; part owner N.Y. Sun (a leading "penny paper" in N.Y.C.), 1834-38, owner, publisher, 1838-48, increased circulation to 38,000 in 1843 by such devices as "Balloon Hoax" (1844), quick reporting of news through such methods as ship news service, spl. trains and horse expresses; established N.Y. Asso. Press (with other N.Y.C. newspaper publishers) to gather news in all major cities in nation during Mexican War; apptd. by Pres. Polk as spl. emissary to Mexico, 1846; publisher Weekly Sun (for farmers), Illustrated Sun and Monthly Lit. Jour.; lived in retirement, Wallingford, 1848-68. Developed engine using power of gunpowder explosions, 1819; invented rag cutting machine, circa 1826. Died Wallingford, July 19, 1868.

BEACH, S(ylvester) Judd, Am. ophthalmologist; b. Dedham, Mass., Apr. 7, 1879; s. Seth Curtis and Frances Hall (Judd) B.; grad. Phillips Exeter Acad., 1897; A.B., Harvard, 1901, M.D., 1905; m. Louise Harris, Oct. 7, 1909; children—Margaret Judd, Howell Williams, Edmund Beach. Surg. house officer Boston City Hosp., 1904-06; house physician Boston Lying-In Hosp., 1906, (acting) Mass. Eye and Ear Infirmary, 1907; practiced in Augusta, Me., spl. practice, 1909; mem. staff Augusta Gen. Hosp., Gardner Gen. Hosp.; spl. practice, Portland, since 1920; cons. ophthalmic surgeon Me. Eye and Ear Infirmary (pres. staff 1946-48); mem. staff local hosps.; guest lectr. grad. courses George Washington, Fla., Va., univs. Mem. council State Dept. of Health, 1916-24. Sec. Am. Bd. Ophthalmology; v.p. Found. for Vision; mem. exec. com. Ophthal. Study Council. Fellow A.C.S., A.M.A. (chmn. Ophthalmology sect.); mem. Am. Ophthalmol. Soc. (member council, pres. 1944), Am. Acad. Ophthalmology and Otolaryngology (mem. council 1930-35), Soc. for Research in Ophthalmology (mem. commn.). Mem. editorial bd. Quar. Rev. Ophthalmology. Author: Textbook of Refraction. Co-author: The Eye and its Diseases. Died Feb. 10, 1953.

BEACH, William Mulholland, Am. surgeon; b. Stoneboro, Pa., Sept. 15, 1859; s. Oliver and Ann Elizabeth (Mulholland) B.; A.B., Waynesburg (Pa.) Coll., 1882, A.M., 1885; M.D., Jefferson Med. Coll., 1889; m. Lucy Lazear Miller, 1882. Prof. Latin and Greek, Ozark (Mo.) Coll., 1882-85; pres. Odessa (Mo.) Coll., 1885-87; an organizer and surgeon Presbyn. Hosp., Pitts. 1895-1914; proctologist South Side Hosp., 1906-11; examining surgeon for pensions, 1893-97. Fellow A.C.S.; mem. A.M.A. (chmn. sect. gastroenterology, 1918-19), Am. Proctologic Soc. (pres., 1903-04), Am. Acad. Medicine. Contbr. 2 chpts. to Cook's Diseases of the Rectum and Colon. Inventor of proctoscope and colostomy supporter. Died Oct. 23, 1930.

BEACH, Wooster, Am. physician; b. Trumbull, Conn., 1794; attended Coll. Physicians and Surgeons, N.Y.C., 1825-29; m. Eliza de Grove, 1823, 2 sons. Author numerous articles and treatises on med. subjects often strongly critical of traditional med. opinion and practice, from 1825; established U. S. Infirmary (clinic where he treated patients many years), 1828; instrumental in orgn. of univ. med. dept., Worthington, O., 1830; established Electric Med. Jour., 1836; pres. Nat. Eclectic Med. Assn., 1855; also published Telescope and Ishmaelite (2 sheets expressing views on a variety of subjects). Author: The American Practice of Medicine (one of 1st med. textbooks dealing with relation between pathology and disease), 3 vols., 1833, later edits., 1846, 51; An Improved System of Midwifery (a treatise), 1851; Treatise on Pulmonary Consumption, 1840. Died N.Y.C., Jan. 28, 1868.

BEADLE, George Wells, Am. geneticist; b. Wahoo, Neb., Oct. 22, 1903; s. Chauncey Elmer and Hattie (Albro) B.; B.S., 1926, M.S., 1927, U. Neb.; Ph.D., Cornell U., 1931; D.Sc., Yale, 1947, U. Neb., 1949, Northwestern, 1952, Rutgers U., 1954, Kenyon Coll., 1955, Wesleyan U., 1956, Pomona Coll., 1961; M.A., Oxford, 1958, D.Sc., 1959; D.Sc., Birmingham U., 1959; m. Marion Cecile Hill, Aug. 22, 1928; (div. 1953); 1 son, David; m. 2d, Muriel Barnett, 1953; stepson, Redmond Barnett. Teaching asst. Cornell U., 1926-27, experimentalist, 1927-31; NRC fellow Cal. Inst. Tech., 1931-33, instr., 1933-35; guest investigator, Institut de Biologie, physico-chimique, Paris, 1935; asst. prof. genetics, Harvard, 1936-37; prof. biology (genetics), Stanford, 1937-46; prof. biology and chmn. div. of biology Cal. Inst. Tech., 1946-61; chancellor U. Chgo., 1961, pres., 1962-68. Eastman vis. prof. Oxford U., 1958-59. Recipient Lasker award, 1950, Dyer award, 1951, Emil C. Hansen Prize, 1953; Albert Einstein Commemorative award in sci., 1958; Nobel Prize in medicine and physiology (with E. L. Tatum, J. Lederberg), 1958; Am. Cancer Soc. award, 1959; Kimber Genetics award, 1960. Mem. Nat. Acad. Scis. (chmn. com. on genetic effects atomic radiation); Fellow Royal Soc., 1960; mem. Am. Philos. Soc., Genetics Soc. Am., A.A.A.S. (pres. 1955), Danish Royal Acad. Scis., Phi Beta Kappa, Sigma Xi. Author (with Alfred H. Sturtevant) An Introduction to Genetics, 1939; Genetics and Modern Biology, 1963; (with Muriel Beadle) The Language of Life, 1966. Research in genetics of Indian Corn and crossing-over in fruit fly; discovered (with E. L. Tatum) that genes act as biochem. regulators of even smallest details of an organism's biochem. reactions; that each gene is responsible for synthesis of a given enzyme (one gene—one enzyme); partly responsible for devel. of field of biochem. genetics; results of later work with Neurospora crassa (red bread mold) is credited with opening new areas in cancer research.

BEADNELL, Hugh John Llewellyn, geologist; b. Oct. 14, 1874; s. C. E. Beadnell; ed. Cheltenham Coll., King's Coll., London, Eng., Royal Sch. Mines; m. May Grace Thomson, 1904; 1 dau. With Geol. Survey, Egypt, 1896-1906; rancher, Can. 1912-15; surveyor Red Sea Coast, Central Sinai for Whitehall Petroleum Co., also Egyptian Govt., 1921-25; resident engr. Qattara Depression hydro-electric project Egyptian Govt., 1930-32. Fellow Geol. Soc., Royal Geog. Soc. (awards from both socs. for work in Egyptian deserts). Author: An Egyptian Oasis, 1910; Wilderness of Sinai, 1926; also sci. publs. on water-supply, sand dunes, geology of Egyptian deserts and oases. Mapped large sections of Nile Valley and Libyan Desert; discovered Arsinoitherium, also ancestors of modern elephants (Moeritherium, Palaeomastodon), in Fayum desert; sunk artesian bores to reclaim large tracts of desert in Kharga depression, 1906-10; mapped, sank deep wells Bir Sahra, Bir Messaha between Darb el Arbain and Oweinat, S. W. Libyan Deset, 1927-29. Died Jan. 2, 1944.

BEAHRS, Oliver Howard, Am. surgeon; b. Eufaula, Ala., Sept. 19, 1914; s. Elmer Charles and Elsa (Smith) B.; A.A., Chaffey Coll., 1934; B.A., U. Cal. at Berkeley, 1937; M.D., Northwestern U., 1942; M.S. in Surgery, U. Minn., 1949; m. Helen Edith Taylor, July 27, 1947; children—John Randolf, David Howard, Nancy Ann. Fellow, Mayo Found. Rochester, Minn., 1942, 46-49, staff, 1949—, head sect. div. surgery, 1950—; mem. surgical staff St. Mary's, Methodist hosps., Rochester, 1950—; prof. surgery Mayo Grad. Sch. Medicine, 1950—; vice chmn. bd. govs. Mayo Clinic. Trustee Mayo Foundation. Diplomate Am. Bd. Surgery. Fellow A.C.S.; mem. Am., Minn. surg. socs., A.M.A., American Thyroid Association, Central, Western surg. assns., Soc. Head and Neck Surgeons, Soc. for Surgery Alimentary Tract, Soc. Pelvic Surgeons, James Ewing Soc., Sigma Xi, Phi Kappa Epsilon, Phi Beta Pi, Theta Delta Chi. Research, numerous publs. on cancers of head and neck region, large intestine and rectum. Home: Route 1, Rochester, Minn. 55901.*

BEAL, Ernest Oscar, Am. biologist; b. Lancaster, Ill., Mar. 7, 1928; s. Oscar E. and Ruth E. (Smith) B.; B.A., N. Central Coll., 1949; M.S., State U. Ia., 1952, Ph.D., 1955; m. Sara Lou Fromm, Aug. 30, 1946; children—Ann Louise, Thomas Loren, Kenneth Ernest. Faculty, State U. Ia., 1952-54; faculty N.C. State U., Raleigh, 1954-68, prof., 1964-68, acting head dept. botany, 1963-64, dir. NSF Acad. Year Inst. Biology, 1965-68, prof. botany U. Minn., Lake Itasca, 1966-67; head dept. biology Western Ky. U., Bowling Green, 1968—. Recipient Research award for best paper submitted Assn. Southeastern Biologists Meeting, 1966, Outstanding Tchr. award N.C. State U., 1967. Fellow A.A.A.S.; mem. Ecol. Soc. Am., Bot. Soc. Am., Am. Assn. Plant Taxonomists, Am. Inst. Biol. Scis., Assn. Southeastern Biologists, Sigma Xi, Phi Sigma, Gamma Sigma Delta. Research in aquatic flowering plants in assn. with their environmental limitations in terms of pH, chloride concentration, organic matter, total soluble salts. Home: 60 Highland Dr., Bowling Green, Ky. 42101.*

BEAL, Jack Lewis, Am. pharmacognosist; b. Harper, Kan., July 7, 1923; s. Ellis Edward and Kathryn (Domnick) B.; B.S., Kan. U., 1948, M.S., 1950; Ph.D., Ohio State U., 1952; m. Earlene M. Maninger, Aug. 22, 1948; children—Linda Sue, Michael Lewis, Karen Ann. Instr., U. Kan., 1949-50; faculty Ohio State U., Columbus, 1952—, prof. pharmacognosy, 1963—, Mem. adv. panel Nat. Formulary, 1960—. Fellow Am. Found. Pharm. Edn., 1950-52; NSF Faculty fellow, 1958-59; Am. specialist Cultural exchange program Baghdad (Iraq) U., 1961. Mem. Am. Soc. Pharmacognosy (past pres.), Acad. Pharm. Scis. (sec. pharmacognosy and natural products 1966—), Am. Pharm. Assn., Am. Assn. Colls. Pharmacy (mem. conf. tchrs. 1962—). Research, publs. in isolation of constituents of possible therapeutic value from plants, biosynthesis of constituents of pharm. interest in plants, medicinal plant cultivation; devel. assays of constituents of biol. origin, chem. study poisonous plants. Editorial bd. Lloydia, 1966—. Home: 5544 Rockwood Rd., Columbus, O. 43224.*

BEAL, James Hartley, Am. pharm. chemist; b. New Phila., O., Sept. 23, 1861; s. Jesse and Mary B.; student Buchtel (now Akron) Coll., U. Mich.; B.Sc., Scio. (Ohio) Coll., 1884; A.B., 1888; LL.B., Cin. Law Sch., 1886; Ph.G. Ohio Med. Univ., 1894; Sc.D., Mt. Union Coll., 1895; Pharm. D., U. Pitts., 1902; Pharm.M., Phila. Coll. Pharmacy, 1913; m. Fannie Snyder Young, Sept. 29, 1886; children—George Denton, Nannie Esther. Dean dept. pharmacy and prof. chemistry and pharmacy Scio. Coll., 1887-1907, acting pres. 1902-04; dir. era course in pharmacy, New York, 1889-1909; prof. chemistry, metallurgy and microscopy Pitts. Dental Coll., 1896-1904; prof. theory and practice of pharmacy U. Pitts., 1903-11; gen. sec. and editor Jour. of Am. Pharm. Assn., 1911-14; dir. pharmacy research. U. Ill., 1914-17. Editor Midland Druggist and Pharm. Rev., 1908; chmn. com. on uniformity of legislation. methods of analysis and marking of food products, Nat. Pure Food and Drug Congress, 1898. Trustee U. S. Pharmacopoeial Conv., 1900 (pres. bd. trustees. 1910, 1940): pres. Am. Pharm. Assn., 1904-05. Am. Conf. Pharm. Faculties, 1907-08, Am. Druggists Fire Ins. Co., 1939-45; nat. councilor U. S. Chamber of Commerce, 1917-18; pharm. expert, War Industries Bd., 1918; pres. Nat. Drug Trade Conf., 1918-20. Remington medalist. 1919. Author: Notes on Equation Writing and Chemical and Pharmaceutical Arithmetic. 3d edition, 1903; Pharmaceutical Interrogations, 1896; Interrogations in Dental Metallurgy, 1900; Practical Pharmacy, 1907; Prescription Practice and General Dispensing, 1908; Principles of Theory and Practice of Pharmacy, 5 vols., 1910 Contbr. to pharm. jours. Died Sept. 20, 1945.

BEAL, John, Brit. sci. writer; b. Herefordshire, Eng., 1603; B.A., King's Coll., Cambridge (Eng.) U., 1632, M.A., 1636; D.D. Rector Yeovil, Somersetshire, Eng., after 1660; chief chaplain King Charles II, from 1665. Fellow Royal Soc., 1663. Author: Aphorisms concerning cider, 1664; Herefordshire Orchards, a Pattern for all England, 1656. Contbr. sci. papers to Philos. Transactions, letters to honorable Robert Boyle. Made observations on daily variation of barometer. Died 1683.

BEAL, Walter Henry, Am. agriculturalist, b. nr. Old Church, Va., Dec. 9, 1867; s. John and Charlotte Columbia (Ellett) B.; A.B., Va. Poly. Inst. 1886. M.E., 1886; m. Eleanor Gilliss Ashby, Apr. 27, 1910; children—Walter Henry, Mrs. Elizabeth B. Devlin, Anne Ashby, William Ashby. Asst. chemist Mass. Agrl. Exptl. Sta., 1887-91; specialist in agrl. meteorology, soils and fertilizers, editor and asso. in expt. sta. adminstrn. Office of Expt. Stas., U. S. Dept. Agriculture, 1891-1938. Fellow A.A.A.S.; mem. Agrl. Hist. Soc. Contbr. to Internat. Ency., Internat. Year Book, Webster's Internat. Dictionary, Ency. Americana. Died Jan. 1, 1946.

BEAL, William James, Am. botanist; b. Adrian, Mich., Mar. 11, 1833; s. William and Rachel (Comstock) B.; A.B., U. Mich., 1859, A.M., 1862, Ph.D. (hon.), 1880; S.B., Harvard, 1865; M.S., old U. Chgo., 1875; D.Sc., Mich. State Agrl. Coll., 1905; D.Agr., Syracuse U., 1916; studied under Agassiz, Asa Gray; m. Hannah Proud, Sept. 2, 1863. Tchr. natural sci., Friends' Acad., Howland Inst., Union Springs, N.Y., 1859-68; prof. botany, old U. Chgo., 1868-70; lectr. botany Mich. State Agrl. Coll., 1871, prof. botany and horticulture, 1871-81, prof. botany and forestry, also curator Bot. Mus., 1882-1903, prof. botany, 1903-10. Dir. State Forestry Commn. 1888-92. Mem. Soc. for Promotion Agrl. Sci. (1st pres.), Assn. Botanists of U. S. Expt. Stas., Mich. Acad. Sci. Author numerous publs. including: The New Botany, 1881; The Grasses of North America, 2 vols., 1887; Seed Dispersal, 1898. Pioneer in hybridization of grains. Died Amherst, Mass., May 12, 1924.

BEALE, Geoffrey Herbert, Brit. biologist; b. London, Eng., June 11, 1913; s. Herbert Walter and Elsie (Beaton) B.; B.Sc. in Botany, Imperial Coll. Sci. and Tech., London, 1935; Ph.D., U. London, 1938; m. Betty McCallum, Mar. 16, 1949; children—Andrew, Steven, Duncan. Research worker John Innes Horticulture Instn., London, 1935-40, Carnegie Inst., Cold Spring, Harbor, N.Y., 1947; Rockefeller fellow Ind. U., 1947-48; lectr. Edinburgh U. (Scotland), 1948-63, Royal Soc. Research prof., 1963—. Fellow Royal Soc., 1959; Author: The Genetics of Paramecium Aurelia, 1954. Research on genetics of flowering plants, inheritance of anthocyanin pigments, genetics protozoan Paramecium, inheritance of antigens, genetic basis of intracellular symbiots living in Paramecium. Home: 12 Middleby St., Edinburgh 9, Scotland.*

BEALE, Lionel Smith, Brit. physician, microscopist; b. London, Eng., Feb. 5, 1828; s. Lionel John and Frances (Smith) B.; M.D., King's Coll., London U., 1851; m. Frances Blakiston, 1859; 1 son, Peyton Todd Bowman. Anat. asst. Oxford U., from 1847; resident physician King's Coll. Hosp., 1850-51, prof. path. anatomy, hon. physician, from 1869, prof. emeritus, hon. cons. physician, from 1896; prof. physiology, anatomy King's Coll., from 1853, prof. anatomy, from 1876, mem. council, 1877-78, curator mus., 1876-88. Recipient Baly gold medal, 1871. Fellow Royal Soc., 1865, Royal Coll. Physicians. Author: The Microscope and its Application to Clinical Medicine, 1854; The Structure and Growth of the Tissues, 1865; Disease Germs, 1872; Bioplasm, 1872; The Liver, 1889; over 100 sci. articles. Introduced new methods micros. research; showed value of microscope to diagnosis in clin. medicine; 1st to practice method of fixing tissues by injecting them; anticipated by 5 years microbe theory of disease and Pasteur's doctrine of immunization; introduced new staining methods by which to distinguish component parts of tissues; discoverer Beale's cells. Died Mar. 28, 1906.

BEALL, Arthur Charles, Jr., Am. surgeon; b. Atlanta, Aug. 17, 1929; s. Arthur Charles and Clare (Scott) B.; B.S., Emory U., 1950, M.D., 1953; m. Cornelia Louise Williams, Aug. 22, 1949; children—Arthur Charles III, Vincent Lane. Research fellow Houston Heart Assn., 1960-61; with Baylor U. Coll. Medicine, 1954-56, 58—, asso. prof. surgery, 1966—; attending physician VA Hosp.; asso. attending physician Meth. Hosp.; asso. attending Jefferson Davis, Ben Taub Gen. hosps. (all Houston). Diplomate Am. Bd. Surgery, Bd. Thoracic Surgery. Mem. Harris County Med. Soc., Am., Pan. Am., Tex. med. assns., Am., Tex.

Houston heart assns., Am. Fedn. Clin. Research, So. Soc. Clin. Investigation, Houston, Tex. surg. socs., Southwestern Surg. Congress, Internat. Cardiovascular Soc., Soc. Vascular Surgery, Soc. Univ. Surgeons, Am. Gerontol. Soc., Am. Assn. Surgery of Trauma, Am. Assn. Thoracic Surgery, A.C.S., Am. Coll. Cardiology, Am. Coll. Chest Physicians, Western, So., Pan-Pacific surg. assns. Contbr. numerous articles to med jours. Devel., application techniques thoracic, cardiovascular surgery, particularly in area open-heart surgery. Home: 3218 Ella Lee Lane, Houston 77019.*

BEALL, George, Brit. inventor; b. Nov. 17, 1840; s. George Beall; m. Annie Robinson Grace, 1866; went to sea in service E. India Co., 1857; with John Bibby & Co.; given 1st command, 1865; examiner in nav. and nautical astronomy Her Majesty's tng. ship Conway; younger bro. Trinity House; prin. examiner master and mates Bd. Trade; ret., 1908. Recipient Gold medal and diploma Liverpool (Eng.) Internat. Exhbn., 1886. Author: Handbook to Deviascope, with short treatise on magnetism. Inventor deviascope for use in magnetic work. Died June 28, 1918.

BEALS, Carlyle S., Canadian astronomer; b. Canso, N.S., Can., June 29, 1899; s. Francis Harris and Annie (Smith) B.; B.A., Acadia U. (Can.), 1919, D.Sc. (hon.), 1948; M.A., U. Toronto (Can.), 1923; Ph.D., U. London (Eng.), 1926, D.Sc., 1934; D.Sc. (hon.) U. New Brunswick (Can.), 1956, Queens U. (Can.), 1960, U. Pitts., 1963; m. Miriam White Bancroft, Sept. 16, 1931. Asso. prof. physics Acadia U., 1926-27; astronomer Dominion Astrophys. Obs., Victoria, B.C., Can., 1927-46, dir. Dominion Observatories, Can., 1946-64; pvt. sci. cons., 1964—. Recipient gold medal Profl. Inst. Civil Service of Can. Fellow Royal Soc. Can. (H. M. Tory gold medal), Royal Soc., 1951; mem. Am. Astron. Soc. (pres. 1962-64). Research includes interpretation of spectra of Wolf Rayet and P Cygni stars, interpretation of interstellar absorption lines and the discovery and investigation of ancient meteorite craters. Address: Manotick, Ont., Can.*

BEALS, Ralph Leon, Am. anthropologist; b. Pasadena, Cal., July 19, 1901; s. Leon Edward and Elvina (Blickensderfer) B.; A.B., U. Cal. at Berkeley, 1926, Ph.D., 1930; m. Dorothy T. Manchester, June 13, 1923; children—Ralph C., Alan R., Marianna. With field div. edn. Nat. Park Service, Berkeley, Cal., 1933-35; faculty U. Cal. at Los Angeles, 1936—, Faculty Research lectr., 1953; dir. Latin Am. Ethnic studies Smithsonian Instn., 1942-43, U. Buenos Aires, 1962. Cons. schs., founds., govt. agys. Center for Advanced Studies in Behavioral Scis. fellow, 1955-56; Guggenheim Found. fellow, 1958-59. Mem. Am. Anthrop. Assn. (pres. 1950), A.A.A.S. (mem. council 1963-—), Soc. Am. Archeology, Soc. Applied Anthropology, Soc. Am. Sociologists, Soc. Am. Folklore, Soc. Mexicana de Anthropologia, Am. Ethnol. Soc., S. W. Anthrop. Assn. Author: (with H. Hoijer) An Introduction to Anthropology, 1953, 65; No Frontier to Learning, 1957; also numerous articles, monographs. Research in problems of stability and change in culture and soc. in Latin Am., chiefly in Indian peasant communities. Home: 16016 Anoka Dr., Pacific Palisades, Cal. 90272. Office: Dept. Anthropology, Univ. Cal., Los Angeles 90024.*

BEAM, Walter R(aleigh), Am. engineer; b. Richmond, Va., Aug. 27, 1928; s. Walter Raleigh and Rose Emma (Evans) B.; B.S. in Elec. Engring., U. Md., 1947, M.S. in Elec. Engring., 1950, Ph.D., 1953; m. Emma Victoria Reese, Mar. 22, 1951; children—David, James. Instr. elec. engring. U. Md., 1947-52; tech. staff RCA Labs., 1952-56; mgr. microwave advanced development RCA, 1956-59; head elec. engring. dept. 1959-62, prof. elec. engring. Rensselaer Polytech. Inst., 1959-64; IBM Research Center, Yorktown Heights, N.Y., 1964—, mgr. materials tech., and manager of exploratory memory, 1965-67, director general engring. tech., 1967—; consultant industry and govt. Sr. member Inst. Elec. and Electronic Engrs.; mem. Am. Phys. Soc., Sigma Xi, Tau Beta Pi, Eta Kappa Nu (hon. mention Outstanding Young Elec. Engr. award 1957), Phi Kappa Phi, Omicron Delta Kappa, Phi Kappa Sigma. Author: Electronics of Solids, 1965. Contbr. to handbook, author articles. Patentee microwave devices and systems. Contributions to understanding of wave phenomena, noise, and focusing of electron beams in microwave tubes and kinescope; development of new ways of depositing, measuring, and characterizing thin magnetic films. Home: 24 Hilltop Dr., Chappaque, N.Y. Office: IBM Research Center, 1000 Westchester Av., Harrison, N.Y.

BEAMENT, James William Longman, Brit. biophysicist; b. Crewkerne, Som., U.K., Nov. 17, 1921; s. Tom and Elizabeth May (Munden) B.; B.A. with 1st class honors, Queens' Coll., U. Cambridge (Eng.), 1943; Ph.D., London (Eng.) Sch. Tropical Medicine, 1945; m. Sara Juliet Barker, Aug. 18, 1962; children—Thomas Henry Horner, Christopher John Munden. Staff., Agrl. Research Council, 1946-60, prin. sci. officer, 1956-60; lectr. U. Cambridge, 1961-66, reader in insect physiology, 1966—, tutor Queens' Coll., 1962-66, adminstr. Zool. Lab., 1961-66, mem. gen. bd., bd. research, 1965—, chmn. faculty biology, 1966—. Fellow Royal Soc., 1964; mem. Soc. for Exptl. Biology (past sec.), Co. Biologists (past dir.), Internat. Congress Zoology (past recorder), Internat. Congress Entomology (past recorder), Zool. Soc. London (Sci. medal 1963), Cambridge Philos. Soc., Brit. Biophys. Soc., Royal Entomol. Soc. Author: (with others) Electronic Methods in Biology, 1956; also articles. Editor, Advances in Insect Physiology, 1963—. Research on metabolism, survival mechanisms of insects, animal salt and water pumps; inventor temperature-control, water loss measurement devices. Home: 19 Sedley Taylor Rd., Cambridge, U.K.*

BEAMS, Jesse Wakefield, Am. physicist; b. Belle Plaine, Kan., Dec. 25, 1898; s. Jesse Wakefield and Kathryn (Wylie) B.; A.B., Fairmount (Kan.) Coll. 1921; M.A., University of Wisconsin, 1922; Ph.D., University of Virginia, 1925; Sc.D. William and Mary College, 1941; Sc.D., Univ. of North Carolina, 1946; Sc.D., Washington and Lee University, 1949; married Maxine Sutherland, June 16, 1931. Instructor physics and mathematics, Ala. Poly. Inst., 1922-23; Nat. Research fellow U. of Va., 1925-26, Yale, 1926-27; instr. physics, Yale, 1927-28; asso. prof. physics, U. of Va., 1928-30, prof. since 1930. Francis H. Smith Prof. Physics; chmn. dept. of physics, 1948-63. Bd. dirs. Oak Ridge Inst. Nuclear Studies, Va. Inst. Sci. Research; mem. Nat. Research Fellowship com. NRC, 1949; mem. gen. adv. com. AEC, 1954-60. Recipient Potts medal, Franklin Inst., 1942; John Scott award, 1956; Lewis award Am. Philos. Soc., 1958. Fellow Am. Phys. Soc. (pres. 1958-59), A.A.A.S. (chmn. sect. B, 1943); mem. American Academy of Arts and Sciences, Am. Philos. Soc. (v.p. 1960-63), Am. Optical Soc., Va. Acad. Sci. (pres. 1947), Am. Physics Teachers Assn., Am. Assn. Univ. Profs., southeastern sect. Am. Physical Soc. (chmn. 1937-38), Nat. Research Council, Physics Division, 1933-36, 1937-40, Nat. Academy Sciences, Phi Beta Kappa, Sigma Xi. Contbr. numerous articles. Study of methods of acceleration of ions; ultra centrifuging; separation of isotopes and other substances by centrifuging; low temperature physics; electrical breakdown in gases; tensile strength of metals; magnetic balances; Kerr cells; constant speed rotation; ram jets; cavity oscillators. Home: 14 Monroe Hill, University of Virginia, Charlottesville, Va.

BEAN, John William, physiologist; b. Attercliff, Ont., Can., Sept. 8, 1901; s. Eusebius H. and Anna (Gieske) B.; student Stratford (Ont.) Collegiate Inst., 1920, N.Central Coll., 1921-23; A.B., U. Mich., 1924, M.S., 1925, Ph.D., 1930, M.D., 1936. Faculty dept. physiology Med. Sch. U. Mich., Ann Arbor, 1925—, prof., 1944—. Chmn. med. sect. on code revision N.Y. State Compressed Air. Mem. Am. Physiol. Soc., Soc. Exptl. Biology and Medicine, A.A.A.S., Sigma Xi, Phi Sigma, Phi Kappa Phi, Alpha Kappa Kappa. Editor: Biol. Abstracts U. Mich., 1954—. Studies, numerous publs. on respiration and its control, effects of O2 at high pressure and O2 toxicity, blood flow, brain and intestinal tract, hypertension, epileptiform convulsions, physiology of diving and hyperbaric medicine. Home: 810 W. Davis St., Ann Arbor, Mich. 48103.*

BEAN, R(obert) Bennett, Am. anatomist; b. Gala, Va., Mar. 24, 1874; s. William Bennett and Ariana Williamson (Carper) B.; B.S., Va. Poly. Inst., 1900; M.D., Johns Hopkins, 1904; m. Adelaide Leiper Martin, May 22, 1907; children—Mary Archer, William Bennett, Helen Holmes, George Martin. Asst. in anatomy John Hopkins, 1904-05; instr. anatomy U. Mich., 1905-07; dir. anat. lab., Philippine Med. Sch., Manila, 1907-10; asso. prof. and prof. anatomy Tulane U., 1910-16; prof. anatomy U. Va., from 1916. Pres. New Orleans Acad. Scis. Fellow A.A.A.S.; mem. Am. Anat. Assn., Am. Anthrop. Assn., Sigma Xi; corr. mem. Soc. Romana Antropologia. Author: The Racial Anatomy of the Philippine Islanders, 1910; The Races of Man, 1932; The Peopling of Virginia; also many papers pertaining to anatomy. Died Sept. 3, 1944.

BEAN, Tarleton Hoffman, Am. zoölogist; b. Bainbridge, Pa. Oct. 8, 1846; s. George and Mary (Smith) B.; M.E., State Normal Sch., Millersville, Pa., 1866; M.D., Columbian (now George Washington) U., 1876; M.S., Ind. U., 1883; m. Laurette H. Van Hook, Jan. 1, 1878. Curator, dept. fishes U. S. Nat. Mus., 1880-95; dir. N.Y. Acquarium, 1895-98; state fish culturist of N.Y., from 1906. Editor Proc. and Bulls., U. S. Nat. Mus., 1878-86, Report and Bull., U. S. Fish Commn., 1889-92; asst. in charge div. of fish culture, U. S Fish Commn., 1892-95; acting curator of fishes. Am. Mus. Natural History, N.Y.C., 1897. Rep. U. S. Fish Commn. at Chicago Expn., 1893, Atlanta Expn., 1895; dir. Forestry and Fisheries, U. S. Paris Expn., 1900; chief depts. fish and game, forestry St. Louis Expn., 1902-05. Author: The Fishes of Pennsylvania, 1893; The Salmon and Salmon Fisheries; Oceanic Ichthyology (with George Brown Goode), 1896; The Fishes of Long Island, 1902; The White World (part author), 1902; The Food and Game Fishes of New York, 1903; The Basses, Fresh-Water and Marine (part author), 1905; The Fishes of Bermuda, 1906. Known as most distinguished Am. fish culturist of his time; initiated nat. movement for preservation native Am. fish. Died Dec. 28, 1916.

BEAN, William Bennett, physician; b. Manila, P.I., Nov. 8, 1909; s. Robert Bennett and Adelaide Leiper (Martin) B.; B.A., U. Va., 1932, M.D., 1935; m. Abigail Shepard, June 17, 1939; children—Robert Bennett, Margaret Harvey, John Perrin. Teaching fellow Thorndike Meml. Lab., Boston, 1936-37, Harvard, 1936-37; sr. med. resident Cin. Gen. Hosp., 1937-38, asst. attending physician, 1941-46, clinician, attending physician, 1946-48; instr. medicine, fellow nutrition, Cin. Med. Coll., 1938-40; asst., then asso. prof. medicine U. Cin. Med. Coll., 1940-48; prof., head dept. internal medicine U. Ia. Coll. Medicine, Iowa City, 1948—, physician-in-chief univ. hosps., 1948—; sr. med. cons. VA, 1947—; cons. internal medicine Surgeon Gen., U. S. Army, 1954. Sci. bd. dirs. Nat. Vitamin Found., 1950. Recipient John Horsley Meml. prize U. Va., 1944. Diplomate Am. Bd. Internal Medicine, Am. Bd. Nutrition. Fellow Am. Coll. Chest Physicians (gov. Ia.), Am. Med. Writers Assn.; A.A.A.S.; member Am. College Cardiology (Groedel medal 1961; governor Ia.), A.M.A., Soc. for Experimental Biology and Medicine, Soc. Med. Consultant to Armed Forces, N.Y. Academy of Sciences, Archeology Inst. Am. (president Ia., 1956-58), Am. Assn. Med. History, Am. Soc. Tropical Medicine, Am. Heart Assn. (executive committee, science council), World Medical Assn., Assn. Am. Physicians, Sigma Xi, Alpha Omega Alpha. Author: Sir William Osler: Aphorisms, from His Bedside Teachings and Writings, 1950. Editor: Monographs in Medicine, 1951-52; Aphorisms from Latham, 1962; asst. editor Nutrition Revs., 1945-46; editorial bd., book rev. editor Cin. Jour. Medicine, 1946-48; asso. editor Jour. Clin. Investigation, 1947-52; editorial bd. Jour. Lab. and Clin. Medicine, 1948-54; editorial adv. bd. Am. Jour. Clin. Nutrition, Resident Physician; book rev. editor A.M.A. Archives Internal Medicine, 1954-61, editor-in-chief, 1961—, editorial cons. several publs. Research in cardiovascular disease, vitamin deficiencies, nutrition, rare diseases, moral responsibility in clin. research. Home: 723 Bayard St., Iowa City 52241.*

BEAR, Richard Scott, Am. biophysicist; b. Miamisburg, O., June 8, 1908; s. Harris V. and Georgianna (Scott) B.; S.B., Princeton, 1930; Ph.D., U. Cal. at Berkeley, 1933; m. Madelaine Borncamp, Aug. 4, 1931 (dec. 1940); children—Janet (Mrs. Donald McTavish), Andrea (Mrs. William A. Rugh); m. 2d, Josephine Hoyt, Sept. 4, 1942. Research asso. Washington U., St. Louis, 1934-38; asst. prof. Ia. State U.; Ames., 1938-41, dean Coll. Scis. and Humanities, 1957-61; faculty Mass. Inst. Tech., 1941-57, prof. biology, 1946-57; dean Boston U. Grad. Sch., 1961-66; prof. biology, 1961—. Mem. study sect. biophysics and biophys. chemistry NIH, 1961-64; mem. sci. adv. com. Helen Hay Whitney Found., N.Y.C., 1960-64. Fellow Am. Acad. Arts and Scis. (mem. membership com. 1964-66); mem. Biophys. Soc. (mem. council 1961-64), Am. Chem. Soc., A.A.A.S. Research, numerous publs. on molecular structure biol. tissues by optical means. Home: 75 Richardson Rd., Belmont, Mass. 02178. Office: Boston U., Boston 02215.*

BEARD, Charles Heady, Am. physician; b. Spencer County, Ky., Jan. 27, 1855; s. James P. and Emerin (Heady) B.; ed. Transylvania U., Lexington, Ky.; M.D., U. Louisville, 1877; postgrad. Post-Grad. Med. Sch., Polyclinic Knapp's Inst., N.Y., also London, Paris, Zurich, Vienna; m. Laura Clark, Sept. 24, 1888. Practiced medicine, Cannelton, Ind., 1877-83; house surgeon Manhattan Eye and Ear Hosp.; practiced medicine, specializing in eye and ear disorders, Chgo., from 1887; asst. surgeon Ill. Charitable Eye and Ear Infirmary 1887-90, surgeon, from 1890; oculist Cook Hosp., 1 yr.; attending phys. Central Free Dispensary, 1 yr.; oculist Passavant Meml. Hosp. Author: Ophthalmic Surgery, 1910-14; Ophthalmic Diagnosis, 1911-13; Collection of Pictures of the Fundus Oculi. Died June 3, 1916.

BEARD, David Breed, Am. physicist; b. Needham, Mass., Feb. 1, 1922; s. Daniel Breed and Anne (Curran) B.; B.S., Hamilton Coll., 1943; postgrad. Cal. Inst. Tech.; Ph.D., Cornell U., 1951; m. Eileen Mona Hersey, Mar. 5, 1945; children—Lawrence Bennett, Jonathan Breckenridge, Valerie Curran, Rosemary Diane. Instr., Cath. U. Am., Washington, 1950-51, U. Conn., Storrs, 1951-53; asst. prof. U. Cal. at Davis, 1953-56, 58-59, asso. prof., 1959-62, prof., 1962-64; staff scientist Lockheed Missiles & Space Co., Palo Alto, Cal., 1956-58; chmn. dept. physics and astronomy U. Kan., Lawrence, 1964—. Cons. to pvt. cos., govt. agys. Nat. Acad. Sci-NRC fellow, 1962, 63; NASA fellow, 1962-63; Fulbright scholar, Guggenheim fellow, U.K., 1965-66. Fellow Am. Phys. Soc., A.A.-A.S.; mem. Am. Assn. U. Profs., Am. Geophys. Union, Am. Assn. Physics Tchrs., Am. Assn. Engring. Scientists, Fedn. Am. Scientists, Sigma Xi, Sigma Pi Sigma. Author: Quantum Mechanics, 1963; also articles. Solved boundary of interaction between solar corona and geomagnetic field; developed physics and origin of comets and type 1 tails; calculated microwave emission from hot ionized gases; developed a meson theory of nuclear force; research on energetic nuclear reactions and nuclear force field. Home: 1200 Mississippi St., Lawrence, Kan. 66044.*

BEARD, David Franklin, Am. agronomist; b. Wood County, O., Aug. 1, 1912; s. James M. and Stella (Leimgruber) B.; B.S., Ohio State U., 1935, Ph.D., 1940; postgrad. Cornell U.; m. Kathryn B. Gillian, Mar. 15, 1936; children—Nancy C. (Mrs. Joseph Bell), Joan M. (Mrs. Michel Romig), Timothy V. Chief, Forage and Range Research br. U. S. Dept. Agr., Beltsville, Md., 1951-58; dir. research Waterman-Loomis Co., Adelphi, Md., 1958-64, v.p. research, 1964—. Fellow Am. Soc. Agronomy; mem. Internat. Crop Improvement Assn. (past dir.), Crop Sci. Soc. Am., Am.

Grassland and Forage Council, Am. Seed Trade Assn. Contbr. chpt. to Forages, 1962, sect. on grasslands to Ency. Brit., 1958. Research, numerous publs. on devel. of forage seed industry, seed certification and devel. of improved alfalfa varieties. Address: 10916 Bornedale Dr., Adelphi, Md. 20783.*

BEARD, George Miller, Am. physician; b. Montville, Conn., May 8, 1839; s. Spencer and Lucy (Leonard) B.; attended Phillips Acad., Andover, Mass., 1854-58; grad. Yale, 1862, Coll. Physicians and Surgeons, N.Y.C., 1866; m. Elizabeth Alden, Dec. 25, 1866. Began research in med. use of electricity, 1866, published 1st works in this field; lectr. diseases of nerves N.Y. U., 1868; mem. staff Demilt Dispensary, N.Y.C., from circa 1870; founded mag. Archives of Electrology and Neurology, 1874; del. Internat. Med. Congress, London, Eng., 1881. Author: Medical and Surgical Uses of Electricity, 1871; Hay Fever, 1876; The Scientific Basis of Delusions, 1877; Nervous Exhaustion, 1880; American Nervousness, 1881; numerous other publs. One of 1st neurologists in U. S.; 1st to formulate causes and treatment of seasickness; pioneer in reforms for care of insane; noted for researches in electro-therapeutics, neurasthenia or nervous exhaustion (Beard's disease). Died N.Y.C., Jan. 23, 1883.

BEARD, James Thom, Am. inventor; b. Bklyn., Oct. 19, 1855; s. Ira and Isabella O.B.; grad. Adelphia Acad., Bklyn., 1874; C.E., E.M., Columbia Sch. of Mines, 1877; m. Amelia E. Lawson, May 9, 1887; children—James Thom, Howard Iranaeus, Amelia Elizabeth. Asst. engr. Bklyn. Bridge, 1877-79; resident div. engr. C.,B.&Q. R.R., 1880-83; U. S. dep. mineral surveyor, Colo., 1883-85; mining engr. Ottumwa Fuel Co., 1885-91; propr. Iowa Coal Exchange, 1891-96; asso. editor Mines and Minerals and prin. Sch. of Mines, Internat. Corr. Sch., Scranton, Pa., 1896-1911; sr. asso. editor Coal Age, N.Y., from 1911. Sec. Ia. Mine Examining Bd., 1888-94. Founder, editor in chief Mine Inspectors' Inst. Author: The Ventilation of Mines, 1894; Design of Centrifugal Ventilators, 1899; Mine Gases and Explosions, 1908; Coal Age Pocket Book, 1916; Mine Gases and Ventilation, 1919. Compiler: Mine Examination Questions and Answers, 3 vols., 1923. Invented Beard-Mackie sight indicator for testing gas, Beard deputy safety lamp, Beard-Stine centrifugal mine fan. Died Dec. 26, 1941.

BEARD, John Stanley, ecologist; b. Gerrards Cross, Eng., Feb. 15, 1916; s. John Stanley and Amelia (Cheer) B.; B.A. in Forestry, Pembroke Coll., Oxford, Eng., 1937, B.Sc., 1941, D.Phil., 1945; m. Pamela Davey, July 11, 1940; children—Rowena, Elfrieda, Georgina. Asst. conservator forests Colonial Forest Service, Trinidad and Tobago, 1937-43; staff Colonial Devel. and Welfare Orgn., Windward and Leeward Islands, 1943-46; estates research officer Natal Tanning Extract Co., Ltd., Pietermaritzburg, S. Africa, 1947-61; dir. King's Park and Bot. Garden, Perth, Western Australia, 1961—; asso. Bot. Gardens Pietermaritzburg, 1954-61. Fellow Australian Inst. Park Adminstrn., Royal Soc. Western Australia, Brit. Ecol. Soc., Commonwealth Forestry Assn.; mem. Australian Ecol. Soc., Royal Hort. Soc., Bot. Soc. S. Africa. Author: Natural Vegetation of Trinidad, 1946; Natural Vegetation of Windward and Leeward Islands, 1949; Descriptive Catalogue of West Australian Plants, 1965; also numerous articles. Descriptive accounts of vegetation of W. Indian Islands and parts of tropical Am. and Western Australia, system of classification for vegetation types of tropical Am., taxonomy of genus Protea, Home: 2 Bellevue Terrace. Office: Kings Park, Perth, Western Australia, Australia.*

BEARD, Joseph Willis, Am. virologist; b. Athens, La., Nov. 5, 1901; s. Jasper Newton and Martha (Bridges) B.; B.S., U. Chgo., 1926; M.D., Vanderbilt U., 1929; m. Dorothy Lowe Waters, June 18, 1932. Asst. pathologist Rockefeller Inst., N.Y.C., 1932-35, asso., 1935-37; faculty Duke Sch. Medicine and U. Hosp., Durham, N.C., 1937—, James B. Duke prof. surgery, 1963—, prof. virology, 1965—. Mem. A.A.A.S., Soc. Exptl. Pathology, Soc. Exptl. Biology and Medicine, Soc. Bacteriology, Assn. Immunologists, Electron Microscope Soc., Assn. for Cancer Research, So. Soc. Clin. Research. Author: (with others) Zinsser's Textbook of Microbiology, 1964; also numerous articles. Research in surg. shock, chick embryo vaccine against equine encephalomyelitis, purification of animal viruses, influenza vaccines, virus-induced cancer. Home: Route 3, Hillsborough, N.C. 27278. Office: Duke U. Med. Center, Durham, N.C. 27706.*

BEARD, Rodney R., Am. physician; b. Guinda, Cal., Dec. 27, 1911; s. Aiton Holmes and Mathilda Anne (Rau) B.; A.B., Stanford, 1932, M.D., 1938; M.P.H., Harvard, 1940; m. Marion Louise Harper, July 3. 1938; children—Anne (Mrs. Frank Gerbode), Philip, Marian, Edin. Faculty Stanford Med. Sch., Palo Alto, Cal., 1940;—, prof., exec. dept. pub. health and preventive medicine, 1949—. Cons. state and govt. agys.; dir. Commn. Environmental Hygiene, Armed Forces Epidemiological Bd., 1955-66; vice chmn. gen. preventive medicine Am. Bd. Preventive Medicine, 1961-65. Diplomate Am. Bd. Preventive Medicine and Pub. Health, Am. Bd. Indsl. Hygiene. Fellow Am. Coll. Preventive Medicine, Am. Pub. Health Assn., A.A.A.S., Indsl. Med. Assn.; mem. Am. Indsl. Hygiene Assn., A.M.A., Western Indsl. Med. Assn. (past pres.), Assn. Tchrs. Preventive Medicine (past pres.), Sigma Xi, Delta Omega. Author: (with W. P. Shepard, C. E. Smith), Essentials of Public Health, 1948. Contbr. numerous articles to med. jours. Research in occupational health, rehabilitation, air pollution. Home: 542 Alvarado Row, Stanford, Cal. 94305.*

BEARDEN, Joyce Alvin, American physicist; born Greenville, South Carolina, October 19, 1903; the son of Joseph Sylvester and Annie (Haley) Bearden; A.B., Furman University, 1923; D.Sci., 1951; Ph.D. (fellow), U. Chicago 1926; m. Lillian Singleton, June 6, 1923; 1 son, Alan Joyce. With U. Chgo., 1926-29; faculty Johns Hopkins, Balt., 1929—, prof., 1939—; physicist Applied Physics Lab., Carnegie Inst., Washington, 1941-42; staff Applied Physics Lab., Silver Spring, Md., 1942-46, dir. Radiation Lab., 1943-55; cons. NDRC, 1940-42. Mem. Am. Phys. Soc., Phi Beta Kappa. Author: (with J. S. Thomsen) Atomic Constants, 1955; X-Ray Wavelengths, 1964. Research in absolute X-ray intensities and electron distributions; the Compton effect; X-ray wave lengths by ruled and crystal gratings; refraction of X-rays; determination of charge on electron and other constants. Home: 214 Lambeth Rd., Balt. 21218.

BEARDMORE, John Alec, Brit. geneticist; b. Burton-on-Trent, Eng., May 1, 1930; s. George Edward and Anne (Warrington) B.; B.Sc. with spl. honors, U. Sheffield, 1953, Ph.D., 1956; m. Anne Patricia Wallace, Dec. 26, 1953; children—Anne Virginia, James Wallace, Hugo John, Charles Edward. Research demonstrator in botany U. Sheffield, 1954-56, lectr. genetics, 1958-61; Commonwealth Fund fellow Columbia, N.Y.C., 1956-58; vis. asst. prof. plant breeding Cornell U., 1958; vis. research asso. in biology U. Rochester, N.Y., 1961; prof. genetics, dir. genetics lab. U. Groningen, Holland, 1961-66; NSF sr. fgn. fellow, vis. prof. genetics Pa. State U., 1966; prof. genetics, head dept. U. Coll. of Swansea, Wales, 1967—. Mem. Brit. Assn. for Advancement Sci. (sec. com. for cytology and genetics 1961-66), Soc. for Study Evolution, Genetics Soc. Am., Genetical Soc., Eugenics Soc., Sigma Xi (exec. com. internat. genetics congress 1963). Asso. editor, Evolution, 1961-65. Publs. on research on causes and nature of genetic variability in populations particularly effects of regularly varying environments. Home: 153 Derwen Fawr Rd., Swansea, Wales.*

BEARDSLEY, Heaekiah, physician; b. Am., 1748; credited with describing 1st case of congenital hypertrophic pyloric stenosis in Am., 1788. Died 1790.

BEARDSLEY, Richard King, Am. anthropologist; b. Cripple Creek, Colo., Dec. 16, 1918; s. Earl Pearson and Alice (Smith) B.; A.B., U. Cal. at Berkeley, 1939, Ph.D. 1947; m. Grace Cornog, Apr. 1, 1942; children—Elizabeth King, Kelcey Alison, Margaret Brangwyn. Instr., U. Minn., 1947; faculty U. Mich., Ann Arbor, 1947—, prof. anthropology, 1954—, field sta. dir. Center for Japanese Studies, 1953-54, dir., 1961-64. Mem. Am. Anthrop. Assn., Assn. for Asian Studies, A.A.A.S., Soc. for Am. Ethnology, Japanese Ethnol. Soc., Soc. for Am. Archaeology, Sigma Xi, Phi Beta Kappa. Author: (with John W. Hall, Robert E. Ward) Village Japan, 1959; (with R. J. Smith) Japanese Culture, 1962; (with J. W. Hall) Twelve Doors to Japan; 1965; also articles. Analysis of Japanese soc., prehistoric devel. Japanese culture also modern and prehistoric Soviet Asian cultures, gen. theory of community orgn., theory of sociopolit. modernization of complex socs. Home: 1121 Ferdon Rd., Ann Arbor, Mich. 48104.*

BEARMAN, Jacob Eleazer, Am. biostatistician; b. Mpls., June 6, 1915; s. Abraham and Etta (Zieve) B.; B.A. in Math. magna cum laude, U. Minn., 1936, M.A. in Math., 1938, Ph.D., 1947; m. Shirley Resnick, July 9, 1942; children—Kenneth D., Deborah Ann, Diane Louise, Abby Nehama. Asst. supt. computing project Minn. Dept. Edn., 1939-40; actuarial mathematician U. S. Bur. Census, 1940-41; teaching asst. U. Minn., Mpls., 1936-39, faculty, 1946—, head div. biostatistics, 1956—, prof., 1958—. Mem. adv. com. on epidemiology and biometry NIH, 1958-62, cons. surgeon gen.'s com. on smoking and health, 1963-64; staff Igbo-Ora project U. Ibadan, Nigeria, 1964—; cons. Rockefeller Found., N.Y.C., 1964-65; research adv. com. Community Health and Welfare Council, Mpls. also Hennepin County, 1954-60, 61-63. Fellow Am. Pub. Health Assn., Am. Statis. Assn.; mem. Biometric Soc., Am. Math. Assn., Inst. Math. Statistics, Am. Math. Soc., A.A.A.S., Phi Beta Kappa, Sigma Xi. Research in mathematical, industrial and medical statistics. Home: 1106 Washburn Av., N., Mpls. 55411.*

BEATER, Bernard Edwin, South African agrl. chemist; b. Durban, Natal, South Africa, June 8, 1908; s. George Eric and Margaret (Thompson) B.; B.A., U. South Africa, 1936; D.S., U. Victoria, 1952; m. Evelyn Finlay Mitchell, Nov. 11, 1938; children—Derk Finlay, Maureen Finlay. Soil scientist South African Sugar Assn., Mt. Edgecombe Expt. Sta., nr. Durban, 1946—. Fellow Royal Inst. Chemistry, Royal Soc. South Africa, Philos. Soc. England; mem. South Africa Chem. Inst., South Africa Geol. Soc. Author: Soils of the Sugar Belt: part I, Natal North Coast, 1957, part II, Natal South Coast, 1959, part III Zululand, 1962. Research, publs. on new and improved chem. methods of soil analysis with particular emphasis on soil fertility and rapid soil methods; inventor new and successful soil sampler. Home: 34 Monteith Pl., Durban N., Natal. Office: care S.A.-S.A. Expt. Sta., Mt. Edgecombe, Natal, South Africa.*

BEATON, George Hector, Canadian biologist; b. Oshawa, Ont., Can., Dec. 20, 1929; s. John Hector and Madelaine (Rogerson) B.; B.A., U. Toronto, 1952, M.A., 1953, Ph.D., 1955; m. Mary Patricia Clarke, Aug. 15, 1953; children—James Hector, Patricia Margo, Dorcas Eleanor. Faculty, U. Toronto, 1955—, prof., head dept. nutrition, 1963—; WHO fellow Inst. Nutrition Central Am. and Panama, 1961; mem. WHO expert panel on nutrition com. on nutrition in pregnancy, 1964; mem. FAO/WHO expert group on vitamin requirements, 1965; adviser to Dept. Nat. Health and Welfare, Canadian Council Nutrition. Mem. Nutrition Soc. Can. (pres.), Am. Inst. Nutrition. Author: (with E. W. McHenry) Basic Nutrition, 1963. Editor; (with E. W. McHenry) Nutrition: A Comprehensive Treatise, 1964, 66. Studies on nutrition in pregnancy, establishment of methods of estimating nutritional status for vitamin B6, interactions between protein and vitamin A. Office: Dept. Nutrition, Sch. Hygiene, U. Toronto, Toronto 5, Ont., Can.*

BEATON, John Rogerson, Canadian physiologist; b. Oshawa, Ont., Can., Sept. 7, 1925; s. John Hector and Madeline (Rogerson) B.; B.A., U. Toronto, 1949, M.A., 1950, Ph.D., 1952; m. Helen M. Douglas, Aug. 28, 1948; children—John Douglas, Catharine Jane, Eric Robert, Barbara Joy. Faculty, U. Toronto, 1952-55; sci. officer Def. Research Bd., Ottawa, Ont., 1955-59, cons., 1963—; sect. head Def. Research Med. Labs., Toronto, 1959-63; prof. physiology U. Western Ont., London, 1963-67; prof., head div. nutrition U. Hawaii, 1967—. Mem. Canadian Physiol. Soc., American Institute Nutrition, Nutrition Society Canada, Society of Biological Chemists, Soc. Exptl. Biology and Medicine, N.Y. Acad. Scis. Research, numerous publs. on biol. actions of vitamin B6 and of growth hormone, factors in regulation of food intake; fat-mobilizing substances in urine, metabolism in cold exposure and in hypothermia, metabolic aspects of obesity. Home: 46099 Lilipuna Rd., Kaneohe, Hawaii 96744. Office: Univ. of Hawaii, Honolulu 96822.*

BEATTIE, Edward James, Jr., Am. physician; b. Phila., June 30, 1918; s. Edward J. and Margaret (Stewart) B.; B.A. cum laude, Princeton, 1939; M.D. cum laude, Harvard, 1943; m. Joan Booth, May 22, 1948; 1 son, Bruce Stewart. Harvard Moseley Travelling fellow Hammersmith Hosp., London, Eng., 1946-47; surg. fellow, Markle scholar George Washington U., Washington, 1947-52; chief thoracic surgeon, chmn. dept. surgery Presbyn.-St. Luke's Hosp., Chgo., 1952-65; faculty U. Ill., 1952-65, prof. surgery, 1954-65; prof. surgery Cornell U., Ithaca, N.Y., 1965—, chief thoracic service Meml. Hosp., N.Y.C., 1965—, chief med. officer, chmn. dept. surgery, 1966—. Diplomate Am. Bd. Thoracic Surgery (vice chmn. 1965—, mem. adv. bd. for med. spltys. 1960—). Mem. A.C.S., A.M.A., James Ewing Soc., N.Y. Surg. Soc., N.Y. Med. Assn., Am. Western Central surg. assns., Soc. Clin. Surgery, Soc. Thoracic Surgeons, Am. Assn. Thoracic Surgery. Author: (with S. Economou) Atlas of Advanced Surgical Techniques, 1967; also numerous articles. Research on cardiovascular and respiratory physiology, exptl. tumors, transplanted and denervated lungs; developed techniques of thoracic surgery. Home: 430 E. 67th St., N.Y.C. 10021. Office: 444 E. 68th St., N.Y.C. 10021.*

BEATTIE, James Alexander, Am. chemist; b. Louisville, July 24, 1895; s. James William and Alice (Mount) B.; S.B., Mass. Inst. Tech., 1917, M.S., 1918, Ph.D., 1920; m. Doris A. Bunker, July 14, 1920; children—Pamela (Mrs. Charles T. Dotter), Barbara (Mrs. Preston S. Abbott), James William. Research asso. Mass. Inst. Tech., Cambridge, 1920-21, Nat. Research fellow, 1921-22, faculty phys. chemistry, asst. prof. 1923-29, asso. prof. 1929-38, prof. 1938-61, emer. prof. and lecturer 1961—; staff mem. Radiation Lab., 1941-46. Nat. Research fellow U. Leiden (Holland), 1922-23. Fellow Am. Acad. Arts and Scis., Am. Phys. Soc.; mem. Sigma Xi. Contbr. papers to sci. jours. on ionic theory, thermodynamics, gas constants, heats of vaporization. Home: 59 Stults Rd., Belmont, Mass. 02178.

BEATTIE, James Martin, physician; b. May 31, 1868; ed. U. Otago, N.Z., U. Coll., London, Eng.; M.A., U. New Zealand; M.D., U. Edinburgh, Scotland; D.Sc. (hon.), Nat. U. Ireland; m. 2d, Margaret; 1 son, 2 daus. Faculty U. Edinburgh, 1901-07; asst. pathologist Royal Infirmary, Edinburgh, 1902-05; prof. pathology and bacteriology, dean med. faculty U. Sheffield, Eng., 1907-12; prof. bacteriology U. Liverpool, Eng., 1912-34, dean med. faculty, 1914-17, prof. emeritus, 1935-55; city bacteriologist Liverpool, 1912-34. Examiner pathology numerous Brit. univs. Author: Textbook of General Pathology; Textbook of Special Pathology; Post Mortem Methods. Died Oct. 10, 1955.

BEATTY, Wallace, physician; b. Halifax, N.S., Can., Nov. 13, 1853; s. James and Sarah Hane (Burke) B.; B.A., Trinity Coll., Dublin, Ireland, 1876, M.B., B.Ch., 1879, M.D., 1886; m. Frances Eleanor Edge; 3 sons. Physician, Adelaide Hosp., Dublin; hon. prof. dermatology Trinity Coll. Dublin. Fellow Royal Coll.

Physicians in Ireland. Author: Lectures on Diseases of the Skin; also articles. Died Nov. 8, 1923.

BEAU, Joseph Honoré Simon, French physician; b. Collonges, Ain, France; b. 1806; attached successively St. Antoine's Hosp., Cochin Hosp. Mem. Acad. Medicine. Author: Recherches statistiques pour servir á l'histoire de epilepsie et le hysterie, 1836. One of 1st to apply physiol. ideas to pathology; maintained that functional disorders often precede anat. disorders; research on epilepsy, dyspepsia, anathesia, disorders of lungs and heart; 1st to describe transverse lines on fingernails apparent in cases of wasting diseases (known as Beau's lines); name also given to Beau's disease (cardiac insufficiency). Died Ferney, France, 1865.

BEAUCHAMP, Pierre-Joseph de, see de Beauchamp, Pierre-Joseph.

BEAU DE ROCHAS, Adolphe-Eugène, French inventor; b. Digne, France, 1815. Author: Théorie mécanique des télégraphes sous-marins, 1859; Recherches sur les conditions pratiques de la plus grande utilisation de la chaleur et, en general, de la force motrice, 1862. studied high pressure and adhesion locomotives; invented four-stroke combustion engine, 1862; worked on under-sea telegraph. Died 1893.

BEAUDETTE, F(red) R(obert), Am. veterinarian; b. Wichita, Kan., Apr. 15, 1897; s. Horace Fred and Cassie May (Leach) B.; D.V.M., Kan. State Coll., 1919, postgrad., 1921-23; D.Sc., Rutgers U., 1951; m. Velva Rader, Nov. 15, 1922; children—Robert Rader, John Horace, Thomas Rivers. Instr. bacteriology Kan. State Coll., 1919-21, asst. prof., 1921-23; asst. prof. poultry pathology Rutgers U., 1923-25, asso. prof. 1925-29, prof., 1929-—, chmn. dept. animal pathology, 1954-—; spl. cons. microbe & immunology study sect. USPHS. Recipient Borden award in poultry sci. 1944; citation N.J. State Grange, 1948; Tom Newman Meml. award N.Y. Acad. Sci., 1949; Centennial award Mich. State U., 1955; citation Rutgers Research Council Adv. Bd., 1955. Fellow N.Y. Acad. Sci., 1955; Poultry Sci. Assn.; mem. Am. Veterinary Med. Assn. (asso. editor jour.), Poultry Sci. Assn. (asso. editor Poultry Sci.), Am. Soc. Parasitologists, Am. Microscop. Soc., U. S. Livestock San. Assn., Research Workers of Animal Diseases in N.A., Acad. Med., Phys. and Natural Scis. of Havana (corr.), Sociedad Cubana de Historia Natural Felipe Poey (corr.). Sigma Xi. Died Jan. 17, 1957.

BEAUFORT, Sir Francis, Irish hydrographer; b. Flower Hill, Meathe, Ireland, May 7, 1774; s. Daniel Augustus and Mary (Waller) B.; m. Alicia Magdalena Wilson; 1 son, Francis Lestock. Entered navy, 1787, hydrographer, 1829-55. Fellow Royal Soc., 1814, Royal Astron. Soc., Royal Irish Acad.; mem. Académie des sciences (corr. geography sect.). Author: Karamania, or a Brief Description of the South Coast of Asia Minor, and of the Remains of Antiquity, 8 vols., 1817. Devised Beaufort's scale (for measuring wind velocity), 1805; Surveyed entrance to Rio de la Plata, 1807, S. coast of Asia Minor, 1811-12; formulated system of weather registration. Died Dec. 13, 1857.

BEAUFOY, Mark, Brit. astronomer, physicist; b. 1764; commd. col. Tower Hamlets militia, 1797. Fellow Royal Soc., 1815, Linnean Soc.; mem. Astron. Soc. (Silver medal for observations on Jupiter's satellites 1827), Soc. for Improvement of Naval Architecture (a founder 1791). Contbr. to Annals of Philosophy, 1813-26. Studied resistance of water; made best magnetic observations of his time to determine exact laws of diurnal variation, also to fix amount of maximum westerly declination in Eng.; 1st Englishman to ascend Mt. Blanc; 1st in Eng. to verify Euler's theorem on resistance of fluids, printed 1834. Died, circa 1830's.

BEAUJEU, Jacqueline Marthe Garnier, French geographer; b. Aiguilhe, France, May 1, 1917; d. Jacques and Marthe (Perrin) Garnier; ed. U. Paris; m. Jean Beaujeu, Dec. 28, 1942; 1 son, François. Asst., Sorbonne, Paris, France, 1941-42, prof., 1960-—; charge research Centre national de la recherche scientifique, 1946-47; master lecturers Faculté des lettres, Poitiers, 1947-48; prof. Faculté des lettres, Lille, 1948-60. Sec.-gen. Geographic Information, 1940-—. Author: Géographie de la population, 1956-58; Les iles Britanniques, 1963; Atlas du Nord de la France, 1961; Traité de géographie urbaine, 1964. Home: 6, rue Pierre-Curie, Paris 5. Office: Sorbonne, Paris, France.

BEAUMONT, see Élie de Beaumont.

BEAUMONT, George Ernest, English physician; b. Oxford, Eng., July 16, 1888; s. Edward and Emily Maria (Crow) B.; ed. Magdalen Coll. Sch., Univ. Coll., Oxford; M.D. with 1st class honors in Natural Sci., Middlesex Hosp., London, Eng; m. Norah Hamill, 1917; 1 dau., Rosalind Mary. Physician, Middlesex Hosp., Brompton Chest Hosp., London; med. examiner Oxford U. Recipient Mons medal, 1914. Fellow Royal Coll. Physicians (Edinburgh); mem. Royal Soc. Medicine. Author: Essentials of Medicine; Recent Advances in Medicine; A Pocket Medicine; Applied Medicine; Clinical Approach in Medical Practice. Address: 1 Hamilton Terrace, London, Eng.

BEAUMONT, William, Am. surgeon; b. Lebanon, Conn., Nov. 21, 1785; s. Samuel and Luctetia (Abel) B.; m. Mrs. Deborah Platt, 1821, 1 son, 2 daus. Apprenticed to physician, St. Albans, Vt., 1810-13; licensed to practice medicine by 3d Med. Soc. of Vt., 1812; surgeon 6th Inf., Plattsburg, N.Y., 1812-15; practiced medicine, Plattsburg, 1815-20; post surgeon Ft. Mackinac, Mich., 1820-25; treated patient (Alexis St. Martin) whose stomach was torn open by gun shot at close range (1822) which resulted in a large fistula, 2.5 inches in circumference, through which the process of digestion could be observed; conducted expts. with St. Martin until 1834 while stationed as post surgeon at Ft. Niagara, 1825-26, Ft. Howard, 1826-28, Ft. Crawford, 1828-34; corresponded with leading scientists about gastric fluids in digestion; served in St. Louis, 1834-39; resigned from Army Med. Corps, 1839; practiced medicine, St. Louis, until 1853. Author: Experiments and Observations on the Gastric Juice and the Physiology of Digestion (containing some 238 expts., greatest single contbn. ever made to study of gastric digestion), 1833. Died St. Louis, Apr. 25, 1853.

BEAUNE, Florimond de, see de Beaune.

BEAUNIS, Henri Etienne, French physiologist, psychologist; b. Amboise, France, 1830; prof. physiology Faculty Medicine Nancy (France); founder, dir. 1st French lab. physiol. psychology Sorbonne, U. Paris (France), 1889-92. Author: Anatomie générale et physiologie du système lymphatique, 1830; Impressions de campagne (notes on mil. medicine), 1871-72; Nouveaux elements de physiologie humain, comprenont les principles de la physiologie comparée et de la physiologie générale, 1876; Le somnabulisme provoque, 1886; L'Evocation du système nerveux, 1890. Studied hypnotism, also psychology of dreams; founder sch. of Nancy (opposed ideas of Charcot and sch. of Salpétrière). Died Le Cannet, France, 1921.

BEAUPHERTHUY, Louis Daniel, physician; b. Guadeloupe, W.I. 1803; med. student, Paris, France; studied yellow fever (in W.I.), blood, secretions in various types of fevers; demonstrated yellow fever is caused by a virus transmitted by mosquitoes, 1854; devised treatment of leprosy with bichloride of mercury. Died Brit. Guiana, Sept. 3, 1871.

BEAUTEMPS-BEAUPRÉ, Charles François, French hydrographer; b. Neuville-au-Pont, France, Aug. 6, 1766; engr., 1785; head nautical engr. for Navy; in charge of mapping Baltic Ocean, 1785; left with Entrecasteaux in ship La Pérouse to map Baltic coastline, 1791; returned to France in 1796; in charge of hydrographic work under Napoleon named keeper of Marine Depot in 1814; mem. French Acad. of Scis., 1810, Bur. Longitudes; named chevalier, also grand officer Legion of Honor. Author: Atlas of the voyage of Entrecasteaux, 1808; Neptune de la Baltique. Called Father of Hydrography; an island off coast of New Caledonia named in his honor; invented new method of map-making using bearing from stars and compass readings; undertook map work for invasion of Eng. under Napoleon. Died Mar. 16, 1854.

BEAUVALLET, Marcelle Jeanne, French physiologist; b. Fleury-les-Aubrais, France, Dec. 26, 1905; d. Aristide Armand and Alice (Bruneau) Beauvallet; CAPES degree École Normale Supérieure, Sèvres, France, 1928; Licence ès Sciences naturelles, Faculty Scis., Paris, 1930, Doctorate in Scis., 1935. Asst., Faculté de Médecine, 1941-44; asst. Sorbonne, 1946-55, head spl. projects, 1955-59, asst. lectr., since 1959-60; fellow Centre National de la Recherche Scientifique, 1934-44, in charge research, 1944-54, master researcher, 1954, head dir. research, 1961-—; research staff cardiovascular physiology Orsay (France) Faculty Sci., 1965-—. Decorated officier d'Académie, 1952; chevalier de l'Ordre du Mérite pour la Recherche et l'Industrie, 1963; named laureat l'Académie Nationale de Médecine, several times. Mem. l'Association des Physiologistes, Société francaise de Thérapeutique et de Pharmacodynamie, Société de Biologie. Author: L'excitabilité des cellules pigmentaires des écailles de poissons, 1935; also numerous articles. Researcher (under Louis lapicque) on electro-physiology of pigmentary cells of fish scales; discovered adrenergic fibers attached to cells which controlled their contraction; studied sympathetic nervous system and its mediators, variations in content of catécholamines in tissues under certain physiol. and pharmacological conditions. Home: 2, Rue Pierre Curie, Paris, France.*

BEAZLEY, Sir John Davidson, Brit. classical archaeologist; b. Glasgow, Scotland, Sept. 13, 1885; s. Mark John Murray and Mary Catherine (Davidson) B.; student Christ's Hosp.; M.A., Balliol Coll., Oxford (Eng.) U., 1910; Litt.D., Cambridge U., 1927, Ph.D., U. Marburg, 1927, LL.D., U. Glasgow, 1953, D. Litt., U. Durham, 1943, U. Reading, 1953, Oxford U., 1956, Dr.-ès-Lettres, U. Lyons, 1947; hon. fellow Balliol and Lincoln Colls., Oxford; hon. student Christ Church Oxford; Dr.-es Lettres, U. Paris, 1960; D.Phil., U. Thessalonike, 1963; m. Marie Henriette Bloomfield, Aug. 13, 1919. Student, tutor Christ Church, Oxford U., 1908-25, prof. classical archaeology, 1925-56; Sather prof. U. Cal., 1949. Knighted, 1949. Recipient Petrie medal U. London, 1937; Kenyon medal Brit. Acad., 1957; Companion of Honour, 1959; Premio Feltrinelli, 1965. Fellow Brit. Acad., Met. Mus. Art N.Y. (hon.); fgn. mem. Am. Philos. Soc. Am. Acad., Archaeol. Inst. Am., Accademia Pontificia Romana, Accademia dei Lincei, Royal Danish Acad., Austrian Acad.; associé étranger Academie des Inscriptions et des Belles-Lettres; fgn. mem. Athens Acad., Greek Archaeol. Soc. (hon. v.p.). Author: Attic Vases in American Museums, 1918; The Lewes House Collection of Ancient Gems, 1920; Attische Vasemalen des rotfigurigen Stils, 1925; Corpus Vasorum Antiquorum, Oxford, I-II, 1927-31; Greek Vases in Poland, 1928; Attic Black-figure; a sketch, 1929; Der Berliner Maler, 1930; Der Pan-Maler, 1931; (with Ashmole) Greek Sculpture and Painting, 1932; Der Kleophrades-Maler, 1933; Attic White Lekythoi, 1938; (with Magi) La raccolta Guglielmi, 1939; Attic Red-figure Vase-painters, 1942, 63; Potter and Painter in Ancient Athens, 1945; Etruscan Vase-painting, 1947; Development of Attic Black-figure, 1951: Attic Black-figure Vase-Painters, 1956. Home: 100 Holywell, Oxford, Eng.

BEBBER, Wilhelm Jacob van, see van Bebber, Wilhelm Jacob.

BEBERMAN, Max, Am. mathematician; b. N.Y.C., Aug. 20, 1925; s. Israel and Lillian (Miller) B.; B.S., Coll. City N.Y., 1944; A.M., Columbia U., 1950, Ed.D., 1953; m. Elizabeth Forrer Chapman, Jan. 18, 1947; children—Lynne, John, Alice, Martin, Ruth, Mary, James, Sarah. Faculty, U. Ill., Urbana 1950-54, 55-—, prof., 1957-—; asso. prof. Fla. State U., 1954-55. Inglis lectr. Harvard, 1958. Author textbooks in high sch. math. Home: 606 S. Highland St., Champaign, Ill. Office: 1210 W. Springfield, Urbana, Ill.*

BECCARI, Giacomo Bartolomeo, Italian physician, naturalist; b. Bologna, Italy, July 25, 1682; a founder Acad. Inquieti (later developed by Count Marsigli into Inst. Scis. and Arts); prof. physics Inst. Scis. and Arts, pres., 1723; named prof. logic U. Bologna, 1709, prof. medicine, 1712, 1st prof. chemistry in Italy, 1737. Mem. Royal Soc., 1728. Author: Consulti medici, 3 vols., 1777-81; De vi, quam ipsa per se lux habet . . . , 1757. Studied phosphorescence, sexual functions, chem. composition of wheat and milk, effect of light on silver salts. Died 1766.

BECCARI, Odoardo, Italian naturalist; b. Florence, Nov. 16, 1843; studied at Lucca, U. Pisa, U. Bologna (Italy); dir. bot. gardens, Florence, Italy; extensive explorations in New Guinea, Abyssinia, E. Indies Islands. Author: Malesia, raccolta d'osservazioni botaniche intorno alle piante dell'Arcipelago indomalese e papuano, 3 vols., 1877-89; Nelle fouste di Borneo, 1901; Asiatic Palms, 1908; Palme del Madagascar, 1912; Nuova Guinea, Selebes e Molucche, 1924; also articles. Founder, Nuovo Giornale Botanico Italiano, 1869. Discovered Amorphophallus titanum (largest flower in the world) in Sumatra. Died Florence, Oct. 25, 1920.

BECCARIA, Cesare Bonesana, marchese di, Italian economist, criminologist; b. Milan, Italy, Mar. 15, 1738; s. Marchese Beccaria Bonesana; ed. univ. in Pavia. Apptd. prof. law and economy (position created for him) Palatine Coll., Milan, 1768; became mem. supreme council econ. (for Austrian-ruled Milanese govt., 1771, mem. bd. for reform of jud. code, 1791. Author: Dei delitti e delle pene, 1764; Elementi di economia pubblica (posthumous publ. of lectures), 1804; Richerche intorno alla natura dello stile, 1770. Helped found and edit periodical Il Caffè. Applied math. to econs.; evolved wage and labor theories that anticipated Adam Smith, also production vs. population theories that anticipated Thomas Malthus; influenced local econ. reforms; advanced one of first arguments against capital punishment and inhuman treatment of criminals; advocated public rather than secret prosecution, also preventive rather than punitive justice; influenced penal codes throughout Europe. Died Milan, Nov. 28, 1794.

BECCARIA, Giovanni Battista, Italian physicist; b. Mondivi, Italy, Oct. 3, 1716; Tchr. Palermo, Sicily, Rome, Italy; prof. physics, Turin, Italy, from 1748; Fellow Royal Soc., 1755. Author: Dell'ellectricismo artificiale e naturale, 1753; Dell'elettricità terrestre atmosferia a cielo sereno, 1744. Extensive research on properties of electricity; experimentation with kites, and rockets in studies of atmospheric electricity; 1st to realize elec. charge on conductor is confined to surface. Died Turin, May 27, 1781.

BÉCHAMP, Pierre Jacques Antoine, French chemist; b. Bassing, nr. Dieuze, France, Oct. 16, 1816; Dr. Pharmacy and Sci., 1853; M.D., 1856; Apothecary in Strasbourg, France; prof. chemistry and pharmacy Med. Faculty U. Montpellier (France), 1857-75, later at U. Nancy (France); dean Catholic U. Lille (France). Mem. Acad Medicine. Author: Leçons sur la fermentation vineuse et sur la fabrication du vin, 1863; Lettres historiques sur la chimie addressee à M. le prof. Courty, 1876; Les Microzymas dans leurs rapports avec l'heterogenie, l'histogenie, la physiologie et la pathologie, 1883; also numerous articles. Research on enzymes, naphthalene; discovered method of preparing aniline from nitro-benzene; asserted fermentation was caused by molds in air. Died Paris, France, Apr. 15, 1908.

BECHER, Ernst Siegfried, German zoologist; b. Reinshagen, Germany, July 2, 1884; s. Ernst and Hulda (Kupper) B.; student U. Bonn (Germany), from 1902. Asst. to W. Spengel zool. inst. Giessen, Germany, 1908-14, dir., 1921-25; prof. Rostock, Germany, 1914-21; worked in Breslau, Germany, 1925-26. Contbr. numerous articles to sci. jours. Studies on form of calcium skeleton; improved polarization microscope; microtech. work on true coloring of cell nucleus; developed photographic process akin to pigment prints. Died Breslau, Apr. 1, 1926.

BECHER, Hellmut, German anatomist; b. Remscheid, Germany, Apr. 30, 1896; s. Ernst and Hulda (Küpper) B.; ed. Munster, Bonn, Munich, Giessen (all Germany) univs.; M.D., Ph.D.; m. Vera Padberg, 1951. Prof. anatomy, dir. Anat. Inst., U. Giessen, U. Marburg/Lahn (Germany), U. Munster (Germany), 1940—. Mem. Anatomische Gesellschaft, Zoologische Gesellschaft, Gesellschaft Deutscher Naturforscher und Arzte, Deutsche Gesellschaft für Elektronemikroskopie. Editor: Gegenbaurs Morphologisches Jahrbuch. Made parabiose expts. on rats; research on structure of sclera, kidney, comparative anatomy of placenta, deformed objects, result of hormones on activity of sperm, fine structure of retina, of hypothalamic tracts, chromatophores on skin of fish. Address: Hufferstrasse 59, Munster, Germany.

BECHER, Johann Joachim, physician, chemist, economist; b. Speyer, Germany, May 6, 1635; s. Joachim and Anna Margaretha (Gauss) B.; largely self-educated, but studied under Debus, Konrector of the Speyer Retscher-Gymnasium after 1644; traveled and studied in Sweden, Holland, Germany and Italy; M.D., U. Mainz (Germany), 1661, prof. medicine U. Mainz; ct. physician Mainz, 1663-64; ct. physician to elector of Bavaria, 1664-66; econ. adviser to Emperor Leopold I of Austria; founder Austrian Coll. Commerce; Author: Oedipus Chymicus, 1664; Physica Subterranea, 1669; Mineralogia, 1662; Tripus Hermeticus Fatidicus, 1689; other works in medicine, theology, politics, mineralogy, economics, and on the formulation of a universal language. Commd. by Dutch authorities to turn silver into gold by means of sand, 1678; favored strong government regulation of production, trade, finance and building of Rhine-Danube canal to open trade with Low Countries; advocated German colonization of S. Am.; proposed ednl. reforms to include technical edn; considered most important German mercantilist of the 17th cent.; rejected views of Aristotelians and Paracelsians on the elements; proposed 3 earth theory (vitrifiable, inflammable, and mercurial) of inorganic bodies; the further development of this theory by Georg Ernst Stahl laid basis for Phlogiston theory; showed sugar is necessary for fermentation, 1680; suggested distillation of coal to obtain tar; claimed to have invented thermoscope for regulating temperature of furnace automatically. Died London, Eng., Oct., 1682.

BECHER, Wolf, German physician; b. 1862; 1st to show outline of stomach by x-ray (in guinea pig) and lead contrast medium, 1896; showed possibility of radiol. diagnosis of gastric disease. Died 1906.

BECHERT, Karl Richard, German theoretical physicist; b. Nuremberg, Germany, Aug. 23, 1901; s. Karl and Hertel B.; Ph.D., U. Munich, 1925; m. Sibylle Lepsius, 1929. Prof. theoretical physics, dir. Inst. Theoretical Physics, U. Giessen, from 1933, rector, 1945-46; prof., dir. Inst. Theoretical Physics, U. Mainz (Germany), 1946—. Mem. Deutsche Physikalische Gesellschaft. Rockefeller fellow, Madrid, Spain, 1926-27. Work in biology, physics.

BECHHOFER, Robert Eric, Am. statistician; b. N.Y.C., Mar. 11, 1919; s. Julius and Lillian (Meyer) B.; A.B., Columbia, 1941, Ph.D. in Math. Statistics, 1951; m. Joan Edith Lebrecht, Oct. 4, 1952; children—Robin Ann, David Jay, Ellen, Laurie. Asst. prof. dept. indsl. engring. Columbia, 1951-52; faculty Cornell U., N.Y., 1953—, prof., 1957—, chairman of dept. obstetrics research, 1967—. Fellow Inst. Math. Statistics, Am. Statis. Assn. (dir. 1964-65, chmn. sect. phys. and engring. scis. 1962). Research in statis. multiple-decision ranking procedures; developed new procedures for ranking means, variances. Home: 36 Cornell St., Ithaca, N.Y. 14850.*

BECHHOLD, Heinrich Jakob, German chemist; b. Frankfort/Main, Germany, Nov. 13, 1866; s. Heinrich Hirsch and Fanny (Haymann) B.; student Freiburg and Berlin, Germany, also Strasbourg, France; m. Maria Johanna Neuburger, Mar. 8, 1896. With Paul Ehrlich's Inst. for Exptl. Therapy, Frankfort; founder Inst. Colloid Research, 1911. Author: Die Kolloide in Biologie und Medizin, 1911. Research on ultra filtration, virus, disinfection, pathology; pioneer in use of colloid chemistry in medicine. Died Frankfort/Main, Feb. 17, 1937.

BECHMANN, Rudolf Heinrich, physicist; born Nuremberg, Germany, July 22, 1902; s. Adolf and Anna (Hissinger) B.; Ph.D., U. Munich, 1927; m. Irmgard Bethke, Oct. 5, 1933 (dec.); m. 2d, Luella Hyers, Nov. 23, 1960. Came to U. S., 1953, naturalized, 1959. With Telefunken, Ltd., Berlin, Germany, 1927-46, ranking exec, 1942-46; dir. Oberspree Co., Berlin, 1946-48; prin. sci. officer Brit. Gen. Post Office, London, Eng., 1948-53; head dielectric phenomena sect. Brush Labs. Co., Cleve., 1953-56; cons. piezoelectric crystals and circuitry br. U. S. Army Electronics Lab., Fort Monmouth, N.J., 1956—. Recipient Civilian merit medal for outstanding performance German Govt., 1941. Fellow Am. Phys. soc., I.E.E.E. (mem. adminstrv. com. Profl. Group on Sonics and Ultrasonics 19——, asso. editor 1963, A.A.A.S.; mem. N.Y. Acad. Sci., German Com. for Standards Piezoelectric Crystal (corr. mem.). Author: The Piezo-optic and Electro-optic Constants of Piezoelectric Crystals, 1962. Contbg. author: Piezoelectricity, 1957; Landolt-Börnstein, Phys. Tables, Vol. 2, 1959. Research on crystal physics and piezoelectricity for telecommunication and ultrasonics. Discoverer AT-BT-,CT-,DT-cuts of quartz. Patentee in field. Home: 165 Park Av., Shrewsbury, N.J. 07704. Office: Solid State and Frequency Control Div., U. S. Army Electronics Labs., Ft. Monmouth, N.J. 07703.*

BECHSTEIN, Johann Matthaus, German ornithologist; b. Waltershausen, Germany, July 11, 1757; s. Johann Andreas and Catharina Elisabetha (Keysser) B.; student theology Jena, Germany. Tchr. Salzmannschon Erziehungsheim, Schnepfenthal, Germany, 1784-94; founder forestry sch. Waltershausen, 1794; dir., forestry acad., Dreissigacker, Germany, 1801-22. Author: Gemeinnützige Naturgeschichte Deutschlands, 1789-95; Naturgeschichte der Stubenvögel, 1795; Ornithologisches Taschenbuch, 1802-03; Vollständige Naturgeschichte aller schädlichen Forstinsekten, 1804; Forstbotanik, 1810; Die Forst-und Jagdwissenschaft nach allen ihren Teilen, 1818-27. Founder of ornithology in Germany. Died Dreissigacker, Feb. 23, 1822.

BECK, Adolf Franz, chemist; b. Chgo., Dec. 2, 1892; s. Otto and Ida (Pfeiffer) B.; student Firma Borsig, tech. sch., Berlin, Germany; m. Anna Gertrude Staub, June 19, 1919; 3 sons. With exptl. lab., chem. factory Griesheim-Elektron, Griesheim, Germany; owner Elektrochemische Werke, Bitterfield; dir. (name changed to Kaustik), 1945-49. Recipient Lilienthal Meml. medal, 1934. Author: Handbuch über Magnesium und seine Legierungen, 1937. Worked on making magnesium technically useful, finding applicable alloys, casting processes; aluminum alloy hydronalium derived from his research; discovered tech. use for light metal scrap. Died Bad Elster, Germany, Mar. 10, 1949.

BECK, Carl, surgeon; b. Neckargemuend, nr. Heidelberg, Germany, Apr. 4, 1856; s. Wilhelm and Sophia (Hoehler) B.; ed. Gymnasium of Heidelberg, 1869-74, also univs. Heidelberg and Berlin; M.D., U. Jena, 1879; m. Hedwig von Loeser, Feb. 16, 1889. Came to U. S., 1882; pres. St. Mark's Hosp., N.Y. from 1886; surgeon to St. Mark's, from 1886, German Poliklinik, from 1883; prof. surgery, N.Y. Post-Grad. Med. Sch., from 1890. Author: Manual on Surgical Asepsis, 1895; Text Book on Fractures, 1900; Die Röntgenstrahlen im Dienste der Chirurgie, 1902; Röntgen-Ray Diagnosis and Therapy, 1904; Röntgenchirurgie, Berlin, 1905; Amerikanische Streiflichter, 1905; Feuchtfroehliches and Feuchtunfroehliches, 1906; Der Schwedenkonrad (novel), 1906; Surgical Diseases of the Chest, 1907 (translated into German, 1909); Glimpses from Latin America, 1908; Röntgenuntersuchung der Leber und Gallenblase, 1909. Devised new method for extensive excision of ribs, new suture in hare lip operations; one of 1st to study applications of X-rays to medicine and surgery. Died June 9, 1911.

BECK, Claude Schaeffer, Am. surgeon; b. Shamokin, Pa., Nov. 8, 1894; s. Simon and Martha (Schaeffer) B.; A.B., Franklin and Marshall Coll., 1916, D.Sc., 1937; M.D., Johns Hopkins Univ., 1921; m. Ellen Manning, May 26, 1928; children—Mary Ellen, Kathryn Schaeffer, Martha Ann. Arthur Tracy Cabot fellow in research surgery, Harvard, and asso. surgeon Peter Bent Brigham Hosp., Boston, 1923-24; Crile fellow in surgery, Western Res. U., Cleve., 1924-25, various positions dept. surgery, 1925—, prof. neurosurgery, 1940-51, prof. cardiovascular surgery, 1951—; with Univ. Hosps., 1924—, asso. surgeon specializing in surgery of the heart, 1933—; chief cons. neurosurgery, Crile Vets. Hosp., Cleve., 1945—; vis. neurosurgeon Cleve. City Hosp.; cardiac cons. Mt. Sinai Hosp., Cleve. Decorated Legion of Merit, 1945. Spl. cons. Surgery Study Group, NIH, 1949—. Fellow A.C.S. Mem. Am. Surgery Assn., Assn. for Thoracic Surgeons, Soc. Clin. Surgery, Am. Soc. for Exptl. Pathology, A.M.A., Am. Heart Assn., Am. Bd. of Surgeons, Am. Bd. Thoracic Surgeons (founders group), Eastern, Central, Cleve. (pres.) surg. socs., Soc. for Exptl. Biology, Cleveland Heart Soc. (pres.). Contbr. chpts. and articles in med. books and jours. Devised several operations (with Vladimir L. Tichy) for increasing blood supply to heart, 1935; did extensive work on suture of wounds, lymph pressure, heart tumors. Home: 2272 Mount Vernon Blvd., East Cleveland 12, O. Office: 2065 Adelbert Rd., Cleve. 6.

BECK, Clifford Keith, Am. nuclear physicist, educator; b. nr. Salisbury, N.C., Apr. 12, 1913; s. Arthur Bradley and Zelma (Weant) B.; B.A., Catawba Coll., 1933, D.Sc. (hon.), 1952; M.S. in physics, Vanderbilt U., 1939; Ph.D. in Physics, U. N.D., 1943; m. Mary Beth Lassetter, May 28, 1943; children—Mary McLean, Clifford Keith, Barbara Gail, Jon Arthur. Instr. high sch. sci., math., coach, Salisbury, 1934-39; physicist wartime atomic energy Columbia, Oak Ridge, 1943-49; prof., head dept. physics, nuclear engring. N.C. State Coll., 1949-56, dir. constrn., operation 1st non-govtl. nuclear reactor 1949-56, trained 1st nuclear engrs. at doctorate level; dep. dir. regulatory staff U. S. AEC, 1956—. Vice-chmn. bd. dirs. Oak Ridge Inst. Nuclear Studies, 1950-56; U. S. del. Geneva confs. Peaceful Uses of Atomic Energy, 1955, 58, 64; tech. cons. indsl. govtl. orgns. Fellow Am. Phys. Soc.; charter mem. Am. Nuclear Soc. (dir. 1954-56). Initiated and directed laboratory for criticality study of fissionable materials, Oak Ridge, 1946-49; assisted in establishing early atomic energy regulatory programs and procedures. Home: Shiloh Church Rd., Boyds, Md. 20120.*

BECK, Edward Creer, Am. psychologist; b. Spanish Fork, Utah, Feb. 20, 1918; s. Spencer Robertson and Eiffel (Creer) B.; B.A., Brigham Young U., 1950; Ph.D., U. Utah, 1954; m. Leia Tendler, Jan. 22, 1944; children—James Spencer, Melanie, Jon Edward, Daniel David. Faculty, U. Utah Coll. Medicine, Salt Lake City, 1954—; 64—, asso. research prof. neurology, prof. psychology, 1967—; dir. neuropsychology labs. VA Hosp., Salt Lake City, 1958—, asst. research prof. psychiatry, neurology, pharmacoloy, 1959—. Sr. scientist NIH, Institut de Neurophysiologie, Faculte de Medecine, Marseilles, France, 1960-61. Cons. Utah prison, 1956-60, Utah Indsl. Sch., 1960-63, Wyo. State Hosp., 1963-66. Recipient Creative Talents award Am. Insts. for Research, 1963-64. Mem. Am. Psychol. Assn.; Utah Psychol. Assn. (pres. 1964-65); Inter-Am. Soc. Psychology, Am. Physiol. Soc., Assn. Research in Nervous and Mental Disorders, Am. Inst. Biol. Scientists, Sigma Xi. Research, publs. on elec. patterns of brain. Home: 3156 E. 4430 S., Salt Lake City 84117. Office: care VA Hosp., 500 Foothill Blvd., Salt Lake City 84113.*

BECK, Emil G., surgeon; b. Czechoslovakia, Mar. 27, 1866; s. Ignatz and Elizabeth (Pollack) B.; student U. Prague, 1878-83; M.D., U. Ill., 1896; m. Clara Hyde, June 28, 1897; 1 dau, Elizabeth (Mrs. E. Pardee). Came to U. S., 1886, naturalized, 1891. Dir. (founder-builder), North Chgo. Hosp., 1906-20. Mem. Am. Coll. Surgeons, A.M.A. (Gorgas medal), Western Surg. Assn., Chgo. Surg. Soc., Am. (Gold medal), North Am., German radiol. socs. Contbr. articles to jours. Developed Beck's paste containing bismuth subnitrate used to diagnose and treat Tb sinuses and cavities, also fistulous tracts, 1908. Died 1932.

BECK, Jay Vern, Am. microbiologist; b. American Fork, Utah, Jan. 15, 1912; s. James Vern and Gladys (Johnson) B.; A.B., Brigham Young U., 1933, M.A., 1936; Ph.D., U. Cal. at Berkeley, 1940; m. Faye Ellison, June 13, 1931; children—Dorthene (Mrs. Keith Richardson), Lynn (Mrs. Paul Ballard), Jacqueline (Mrs. Edward Foutz), David E., Bonnie J., John C. Tchr., N. Sevier High Sch., 1934; technician U. Cal., Berkeley, 1936-39; chemist FDA, San Francisco, 1939-44; asst. prof. U. Ida., Moscow, 1944-46; asso. prof. bacteriology Pa. State U., 1947-51; prof. bacteriology Brigham Young U., Provo, Utah, 1951—, acting research dir., 1961-62. Guggenheim fellow Sheffield U., Eng., 1957-58; NIH fellow, Stanford, 1965. Contbr. numerous articles to bacteriol. jours. on biochemistry and microbiology. Research on purine fermentation by bacteria; minor element deficiency in plants; microbial oxidation of sulfide minerals. Home: 1305 Elm Av., Provo, Utah 84601.*

BECK, John Christian, Am. physician; b. Audubon, Ia., Jan. 4, 1924; s. Vilhelm and Marie (Brandt) B.; B.S., McGill U., 1944, M.D., 1947, M.S., 1951, diploma in medicine, 1952; m. Frances Mary Mitchell, May 31, 1951; 1 son, Philip Norman. Practice medicine, specializing in internal medicine, Montreal, Que., Can., 1964—; physician-in-chief, of Royal Victoria Hosp.; dir. McGill U. Clinic; prof. medicine McGill U. Diplomate Am. Bd. Internal Medicine (dir.). Fellow A.C.P., Royal Coll. Physicians; mem. Endocrine Soc., Am., Canadian socs. clin. investigations, Royal Soc. Medicine, Assn. Am. Physicians, Am. Diabetic Assn., Canadian Physiol. Soc., Am. Fedn. Clin. Research, Sigma Xi. Research, publs. on adrenal cortical physiology and physiology and biochemistry of pituitary hormones, particularly growth hormone. Home: 592 Kenaston St., Town of Mount Royal, Que. Office: Royal Victoria Hosp., Montreal 2, Que., Can.*

BECK, Lewis Caleb, Am. chemist, botanist, mineralogist; born Schenectady, New York, October 4, 1798; the son of Caleb and Catherine (Romeyn) B.; graduated from Union College, Schenectady, 1817; student Coll. Phys. and Surg., 1816; grad. Union Coll., Schenectady, 1817; m. Hannah Smith, Oct. 17, 1825. Licensed to practice medicine in N.Y., 1818; lived in various parts of U. S., 1818-24, began bot. collection; became prof. botany Berkshire Med. Inst., 1824; named prof. chemistry and botany Vt. Med. Acad., 1826; named prof. chemistry Rutgers U., 1830, N.Y.U., 1836; prof. chemistry and pharmacy Albany (N.Y.) Med. Coll., 1840-53. Author: Botany of the Northern and Middle States, 1833; Mineralogy of New York, 1842. Made 1st chemical examinations of foodstuffs in America, 1848; studied adulterations in medicines, 1846; published on milk sickness, 1822. Died Albany, Apr. 20, 1853.

BECK, Paul A., metallurgist; born Budapest, Hungary, Feb. 5, 1908; s. Philip Odon and Laura (Bardos) B.; M.S. in Metallurgy, Mich. Coll. Mining and Tech., 1929, M.E., 1931; m. Lilian Heikkinen, Mar. 21, 1951; children—Paul John, Philip Odon. Came to U. S., 1936, naturalized, 1944. Research metallurgist Central Research Lab., Am. Smelting and

Refining Co., South Bend, N.J., 1937-41; chief metallurgist Beryllium Corp., Reading, Pa., 1941-42; supt. metall. lab. Cleve. Graphite Bronze Co., 1942-45; prof. metallurgy Notre Dame U., South Bend, Ind., 1945-51, U. Ill., Urbana, 1951—. Fellow Am. Inst. Mining, Metall. and Petroleum Engrs. (Mathewson Gold medal 1952); mem. Am. Phys. Soc., Am. Soc. Metals, Sigma Xi. Research on recrystallization, grain growth and textures in metals, electronic specific heat and alloy chemistry of transition elements. Editor: Theory of Alloy Phases, 1956; Electronic Structure and Alloy Chemistry of Transition Elements, 1963. Address: Metallurgy Bldg., Univ. Ill., Urbana, Ill. 61801.*

BECK, Samuel Jacob, psychologist; b. Tecuciu, Rumania, July 19, 1896; s. Abraham and Beatrice (Ciora) B.; A.B. cum laude, Harvard, 1926; M.A., Columbia, 1927, Ph.D., 1932; m. Anne Goldman, Sept. 14, 1926; children—James Ciora, Ruth Louise. Head psychology lab. dept. neuropsychiatry Michael Reese Hosp., Chgo., 1936-50, staff asso., 1950—; lectr. Northwestern U., 1943-60, U. Chgo., 1949—; v.p. Internat. Congress Mental Health, London, 1948. Fellow Am. Orthopsychiat. Assn. (past pres.), Am. Psychol. Assn. (Distinguished Contbn. award clin. div. 1961, past div. pres.); mem. Société Rorschach Internationale (v.p.), Soc. Projective Techniques and Personality Assessment (1st Ann. award 1965, past pres.), Sigma Xi. Author: Rorschach's Test, vol. III: Advances in Interpretation, 1952; The Six Schizophrenias, 1954; The Rorschach Experiment, 1960; (with others) Psychological Processes in the Schizophrenic Adaptation, 1965. Asso. editor: Am. Jour. Orthopsychiatry, 1929-56, 62—. Research, publs. on psychol. structure of the human personality, specifically, intelligence potential, emotions, imagination and their interactions; mental diseases. Home: 5236 Greenwood Av., Chgo. 60615.*

BECK, Stanley Dwight, Am. zoologist; b. Portland, Ore., Oct. 17, 1919; s. Dwight W. and Eunice (Dodd) B.; B.S., Wash. State U., 1942; M.S., U. Wis., 1947, Ph.D., 1950; m. Isabel H. Stalker, Aug. 29, 1943; children—Bruce Dwight, Diana Helene, Karen Christine, Marianne Elizabeth. Faculty, U. Wis., Madison, 1948—, prof. entomology, 1964—. Mem. Entomol. Soc. Am., Am. Soc. Zoologists, Wis. Acad. Scis., Arts and Letters, A.A.A.S. Author: The Simplicity of Science, 1959; Animal Photoperiodism, 1963; also numerous articles. Research on insect growth, nutrition, behavior, and host plant relations. Home: 6100 Gateway Green, Madison, Wis. 53716.*

BECK, Theodric Romeyn, Am. physician; b. Schenectady, N.Y., Aug. 11, 1791; s. Caleb and Catherine (Romeyn) B.; grad. Union Coll., Schenectady, 1807; M.D., Coll. Physicians and Surgeons, N.Y.C., 1811; m. Harriet Caldwell, 1814, 2 daus. Practiced medicine, Albany, N.Y., 1811-17; prin. Albany Acad., 1817-53; prof. medicine Western Coll. Physicians and Surgeons, Fairfield, N.Y., 1815-40, Albany Med. Coll., 1840-43; sec. N.Y. Bd. Regents, 1841-54; pres. N.Y. State Med. Soc., 3 terms; founder N.Y. State Library, N.Y. State Insane Asylum. Author: Elements of Medical Jurisprudence, 1823. Died Albany, Nov. 19, 1855.

BECK, William Carl, Am. surgeon; b. Chgo., Aug. 24, 1907; s. Carl and Eda (Stein) B.; attd. U. Wis., 1924-27; B.A., Northwestern U., 1928, M.B., M.D., 1932; m. Jane Murray, Dec. 13, 1945; children—Linda Jane (Mrs. Edward Stiger), Mary, Alice. Attending surgeon St. Joseph's Hosp., St. Vincent's Hosp., 1936-42 (both Chgo.); asso. surgeon Cook County, Ill. Research hosps., Chgo., 1936-42; chief surg. service, chmn. dept. surgery Guthrie Clinic, Sayre, Pa., 1946—; clin. prof. surgery Hahnemann Med. Coll., 1948-55; vis. lectr. surgery U. Pa. Sch. Grad. Medicine, 1952—; cons. VA, Bath, N.Y. Mem. Am. Surg. Assn., A.C.S., Soc. Internat. de Chir., Royal Soc. Medicine (Eng.) Central Surg. Assn., Am. Soc. Contamination Control, Sigma Xi. Editorial bd. Pa. Med. Jour., Group Practice, Contamination Control. Research, publs. in contamination control, cancer, edn. of ancillary surg. personnel; research in sterile and particle free airflow; devel. of equipment for controlling surg. infections. Home: 398 Pennsylvania Av., Waverly, N.Y. 14892. Office: Guthrie Clinic, Sayre, Pa. 18840.*

BECK, William Samson, Am. physician; b. Reading, Pa., Nov. 7, 1923; s. Myron Paul and Gertrude (Harris) B.; B.S. in chemistry, U. Mich., 1943, M.D., 1946; children—Thomas Russell, Peter Dean; m. 2d, Hanne Troedsson, July 20, 1964; 1 son, John Christopher. Med. researcher specializing in hematology and biochemistry, Los Angeles, 1948-57, Boston, 1957—; faculty U. Cal., Los Angeles, 1950-57, chief hematology sect. Atomic Energy Project, 1951-57; fellow biochemistry N.Y. U. Coll. Medicine, 1955-57; asst. prof. medicine, tutor biochem. scis. Harvard, 1957—; chief hematology unit Mass. Gen. Hosp., 1957—. Recipient Wenner-Gren Found. prize, 1954. Mem. Am. Soc. Biol. Chemists, Am. Chem. Soc., Am. Soc. Clin. Investigation, Am. Assn. Cancer Research. Author: Modern Science and the Nature of Life, 1957; (with G. G. Simpson) Life: An Introduction to Biology, 2d edit., 1965; The Human Body, 1968; also numerous articles. Mem. editorial bd. Blood, the Jour. of Hematology, 1962—. Research in biochemistry of leukocytes, enzymology of propionate metabolism, biochemical regulation of DNA synthesis, metabolic function of vitamin B12. Home: 60 Brattle St., Cambridge, Mass. 02138. Office: Mass. Gen. Hosp., Boston 02114.*

BECK, Wolfgang Maximilian, German chemist; b. Munich, Germany, May 5, 1932; s. Emil and Johanna (Geistbeck) B.; Ph.D., Munich Technische Hochschule, 1960; m. Gerda Lesch, Aug. 8, 1958; children—Markus, Gunter, Kathrin. Faculty, Technische Hochschule, München, Germany, 1963—, wissenschaftlicher Rat, 1966—. Mem. Gesellschaft Deutscher Chemiker. Research, publs. on infrared spectroscopic investigations of metal complexes, particularly of metal carbonyls and nitrosyls, spectroscopic and preparative investigations on fulminate, azide and other pseudohalide metal complexes. Home: 26 Melanchthon Str., Munich, Germany. Office: 21 Arcisstr., Anorganisch-Chem. Laboratorium, Technische Hochschule Munchen, Germany.*

BECKE, Friedrich Johann Karl, minerologist; b. Prague, Czechoslovakia, Dec. 31, 1855; s. Friedrich Becke; ed. U. Vienna (Austria); m. Minna Schuster; 1 son, 1 dau.; became asst. prof. petrography, 1880; apptd. asso. prof. mineralogy Czernowitz, USSR (formerly Rumania), 1882; became prof. Prague, 1890, Vienna, 1898; pub. Mineralogischen und Petrographischen Mitteilungen, from 1899. Wollaston medal Geol. Soc. London, 1929. Mem. French Acad. Scis. (corr. mineralogy sect.) 1911, Vienna Acad. Scis. (gen. sec. 1911). Author numerous books and articles. Research on symmetry of crystals using etched figues; invented method of identifying minerals by their light-refractive properties. Died Vienna, June 18, 1931.

BECKE-GOEHRING, Margot Lina Klara, German chemist; b. Allenstein, June 10, 1914; d. Albert and Martha (Schramm) Goehring; student univs. Munich, Halle; Diplom-Chem., 1936; Dr.sc.nat., 1939; m. Friedrich Becke, Mar. 9, 1957. Docent, U. Halle, 1944-46; docent U. Heidelberg, 1946-47, prof., 1947—, dean sci. faculty, 1961-62, rector 1966-67. Recipient Alfred Stock prize, 1961. Mem. Soc. German Chemists, German Bunsen Soc. Author: Ergebnisse und Probleme der Chemie der Schwefelstickstoffverbindungen, 1957; Medicus-Goehring, Qualitative Analyse, 1960; (with E. Fluck) Einführung in die Theorie der quantitativen Analyse, 1957; also numerous articles. Editor: Anorganische Chemie in Einzeldarstellungen. Research in chemistry of sulfur nitrogen and phosphorus nitrogen compounds, complex compounds with sulfur in nitrogen containing groups, inorganic ring systems, syntheses and new compounds in these fields. Address: 4 Scheffelstrasse, Heidelberg, West Germany.*

BECKENBACH, Edwin Ford, Am. mathematician; b. Dallas, July 18, 1906; s. Charlie Geiger and Lucy Emma (Richardson) B.; B.A., Rice U., 1928, M.A., 1929, Ph.D., 1931; m. Madelene Shelby Simons, Aug. 30, 1933 (div. June, 1960); children—Edwin Simons Beckenbach, Madelene Lenann, Sonya Suzann; m. Alice Judson Curtiss, June 24, 1960. Nat. research fellow Princeton, Ohio State U., U. Chgo., 1931-33; instr. Rice U., 1933-40; asst. prof. U. Mich., 1940-42; asso. prof. U. Tex., 1942-45; prof. U. Cal., Los Angeles, 1945— . Cons. Rand Corp., Santa Monica, Cal., 1949—; mem. Inst. for Numerical Analysis, Nat. Bur. Standards, 1948-50, Inst. for Advanced Study, Princeton, 1951-52; vis. scholar Swiss Fed. Inst. Tech., Zurich, Switzerland, 1958-59; mem. Sch. Math. Study Group, NSF, 1958-60. Guggenheim fellow, 1958-59. Fellow A.A.A.S., mem. Am. Math. Soc., Math. Assn. Am., Soc. for Indsl. and Applied Math., Indian Math. Soc., Société Mathematique de France, Circolo Mathematico di Palermo, Phi Beta Kappa, Sigma Xi, Pi Mu Epsilon. Author: Construction and Applications of Conformal Maps, 1952; Modern Math. for the Engineer, 1st series, 1956, 2d series, 1961; An Introduction to Inequalities, 1961; Inequalities, 1961; College Algebra, 1964; Modern Introductory Analysis, 1964; Applied Combinatorial Analysis, 1964; Essentials of College Algebra, 1965, others. Research, publs. on complex variables, minimal surfaces, convex and subharmonic functions, inequalities, also engineering and other aspects of applied math. Founder editor, Pacific Jour. Math. Home: 13478 Bayliss Rd., Los Angeles. 90049. Office: U. Cal., Los Angeles, Cal. 90024.*

BECKENKAMP, Jakob, German mineralogist; b. Horchheim, Germany, Feb. 20, 1855; s. Johann Cyrill and Katharina (Erben) B.; student math., natural scis. Bonn, Germany; grad. U. Strasbourg, France, 1881; m. Sophie Leikert, Sept. 4, 1883; 2 sons, 1 dau. Tchr., Mulhouse, Alsace, 1883-90, prof. physics, chemistry, 1890-97; faculty U. Freiburg, Germany, from 1885; prof. mineralogy, crystallography U. Wurzburg, Germany, from 1897. Mem. German Acad. Natural Scientists. Contbr. numerous articles to profl. jours. Research on relationship of phys. properties to crystal formation, application of equalization method to geometric crystallography, basis of crystal structure; worked on prodn. artificial silk, x-rays. Died Würzburg, Jan. 12, 1931.

BECKER, Arthur Dow, Am. physician; b. Austin, Minn., Aug. 20, 1878; s. Marcus and Sarah Growden (Blair) B.; D.O., S. S. Still Coll. Osteopathy and Surgery, 1903; D.O., Kirksville Sch. Osteopathy, 1910; B.S., Kirksville Coll. of Osteopathy and Surgery, 1925; D.Sc. in Osteopathy (hon.), 1938; D.Sc. (hon.), Coll. Osteo. Physicians and Surgeons, Los Angeles, 1944; m. Mabel Rollins, Oct. 17, 1906; children—Rollin Edward, Alan Robert. Practiced at Preston, Minn., 1903-08, 1911-15, Mpls., 1915-22, Seattle, 1926-28; mem. faculty and dean Kirksville (Mo.) Coll. Osteopathy and Surgery, 1922-26; mem. faculty and vice pres., 1928-35; pres. Des Moines Still Coll. Osteopathy, 1935-42; sr. cons. Detroit Osteo. Hosp., since 1944. Recipient Distinguished Service certificate Am. Osteo. Assn., 1941. Pres. Am. Coll. Osteo. Internists since 1944; pres. Am. Osteo. Assn., 1931-32, also trustee; pres. Asso. Colls. Osteopathy, 1938-39; twice pres. Minn. State Osteo. Assn. Member Minn. Bd. Osteo. Examiners 9 1/2 years. Contbr. to various osteopathic publs. Died May 16, 1947.

BECKER, Carl Johan, Danish archeologist; b. Copenhagen, Denmark, Sept. 3, 1915; s. Carl and Henny (Döcker) B.; M.A., Ph.D., U. Copenhagen; m. Birgit Hilbert, Oct. 1, 1949; children—Jytte, Karen, Anne-Lise. Asst., Nat. Mus., Copenhagen, 1934; asst. guardian, 1941; prof. prehistoric archeology U. Copenhagen, 1952. Mem. Royal Danish Acad. Sci. Author: Mosefundne Lerkar fra Yngre Stenalder, 1948; Die Mittelneolithischen Kulturen in Sudskandinavien, 1954; Forromersk jernalder i Syd-og Midjylland, 1961. Home: Egernvej, 23. Office: Frederiksholms Kanal 1220, Copenhagen, Denmark.

BECKER, Donald Eugene, Am. nutritionist; b. Delavan, Ill., Feb. 2, 1923; s. George Edwin and Esther (Peters) B.; B.S., U. Ill., 1945, M.S. in Nutrition, 1947; Ph.D., Cornell U., 1949; m. Elsie Jonas Hendrickson, Dec. 28, 1949; children—Esther A., Phyllis E., Donald Eugene, William E., Beth A. Faculty, U. Tenn., 1949-50; faculty U. Ill., Urbana, 1950—, prof. animal sci. 1958—, head of the department, since 1967—. Mem. subcom. on swine nutrition NRC, 1963—. Recipient award for Research in Nutrition Am. Feed Mfrs., 1957. Mem. Animal Nutrition Research Council (past chmn.), Am. Soc. Animal Sci., Am. Inst. Nutrition, Poultry Sci. Assn. Chief editor Jour. Animal Sci., 1967—. Publs. on definition of factors affecting protein nutrition, nutritive requirements for pregnancy and lactation, role of antibiotics in animal feeding. Home: 2209 Combes St., Urbana, Ill. 61801.*

BECKER, Fridolin, Swiss engr., cartographer, topographer; b. Linthal, Switzerland, Apr. 24, 1854; s. Bernhard and Elsbeth (Zweifel) B.; student engring. Eidgenössisches Polytechnikum, Zurich, Switzerland, 1872-76; m. Antonietta Pozzi, 1884; 2 sons, 3 daus. Mountain topographer, 1876-84; asst. Zurich Polytechnikum, 1884-90, prof. tech. drawing, cartography, 1890-1921, col. gen. staff, from 1901. Developer high mountain topography; contbr. treatises on cartography, maps and relief maps with color and hatching relief; improved rock and chart representations of topog. maps. Died Kusnacht-Zurich, Jan. 24, 1922.

BECKER, Friedrich Eberhard, German astronomer; b. Münster, June 12, 1900; ed. U. Münster, U. Berlin; Ph.D.; m. Maria Petz, Apr. 6, 1926. Dir. obs. U. Bonn, 1947, also prof. Mem. Deutsche Akademie d. Naturforscher, Akademie d. Wissenschaft u.d. Literatur, Arbeitsgem. f. Forschung (Dusseldorf). Contbr. articles on astronomy to profl. publs. Address: Poppelsdorferallee 49, Bonn, Germany.

BECKER, George Ferdinand, Am. geologist; b. N.Y.C., Jan. 5, 1847; s. Alexander Christian and Sarah Cary (Tuckerman) B.; B.A., Harvard, 1868; Ph.D., Heidelberg, 1869; passed final exam., Royal Sch. of Mines, Berlin, 1871; m. Florence Serpell Deakins, Feb. 11, 1902. Instr. mining and metallurgy U. Cal., 1875-79; U. S. geologist-in-charge, 1879-92 and from 1894; spl. agt. 10th Census, 1879-83. Examined gold and diamond mines of S. Africa, 1896. Detailed to serve as geologist with army in P.I., 1898-99; then in charge div. chem. and phys. research U. S. Geol. Survey; geophysicist Carnegie Instn. Mem. com. Nat. Acad. apptd. 1903, to prepare report at Pres. Roosevelt's request, on desirability of instituting scientific explorations of the P.I. Author: Atomic Weight Determinations, 1880; Geology of the Comstock Lode, 1882; Statistics and Technology of the Precious Metals (with S. F. Emmons), 1885; Geology of Quicksilver Deposits of the Pacific Slope, 1888; Gold Fields of Southern Appalachians, 1895; Gold Fields of Alaska, 1898; Gold Fields of South Africa, 1897; Geology of the Philippine Islands, 1901; Experiments on Slaty Cleavage, 1904; Tables of the Hyperbolic Functions (with C. E. Van Orstrand), 1908. Pioneer investigations of chemicophys. problems. Died April 20, 1919.

BECKER, Günther Hermann Rudolf, German biologist; b. Jutroschin, Germany, Nov. 25, 1912; s. Otto K. R. and Martha (Baade) B.; student univs. Munich, Göttingen, Rostock; Dr.phil., U. Rostock, 1936, Dr.-phil.habil., 1943; m. Elise Kutschwalski, Sept. 30, 1940; children—Raimund, Sabine, Christine. With State Materials Testing Inst., Berlin, 1937—; FAO expert Guatamala, 1951, India, 1956, Iran, 1956, Korea, 1967; OEEC expert Africa, 1962-63; dep. dir. bldg. materials div. Fed. Inst. Materials Testing, Berlin, 1963-67, dir. spl. fields div., 1967—; faculty Berlin Tech. U., 1952—, hon. prof., 1958—. Chmn. working groups FAO, OECD, Internat. Union Forestry Research Orgns., German Standards Orgn. Mem. German Soc. Applied Entomology (pres. 1956-63), German Soc. Wood Research (pres. 1966—), Internat. Acad. Wood Sci., numerous others. Editor: Wood and Organisms, 1966. Editor, Ann. Report on Wood Protection, 1951—. Materials and Organisms, 1965—. Research and numerous publs. on biology, physiology and ecology of wood-destroying insects and marine borers; influence of chemicals and magnetic fields on insects; testing and application of wood preservatives. Home: 14 Von-Laue-Strasse, 1000 Berlin 33, (Dahlem). Office: 87 Unter den Eichen, 1000 Berlin 45, West Germany.*

BECKER, Gustav Johann Eduard, watchmaker, inventor; b. Oels, Silesia, May 2, 1819; s. Johann Gottlieb and Henriette Caroline (Schwarz) B.; studied watchmaking, Germany, Austria, Switzerland; m. Louise Friedericke Henriette Seelig, Oct. 2, 1845; children—5 daus., 4 sons. Clockmaker, Frieburg, Germany, from 1847; trained weavers' sons to manufacture pendulum clocks; produced 1st regulators for clocks, 1849; commd. to make clocks for all postal adminstrn. offices, Prussia, 1854; gave 500,000th clock to Bismarck on 70th birthday, 1885. Died Berchtesgaden, Germany, Sept. 14, 1889.

BECKER, Heinrich Otto, ophthalmologist; b. Ratzberg, Germany, May 3, 1828; student philosophy Erlangen, Germany, math., natural history Berlin, Germany, medicine Vienna, Austria. Med. supernumerary Gen. Hosp., Vienna; prof. ophthalmology U. Heidelberg; founder Gräfe Mus., Heidelberg. Author: Beitrage zur Lehre vom Schender Dritten Dimension, 1861; Die Anomalien der Refraction und Accomodation, 1868; Atlas der pathologischen Topographie des Auges, 1874; Pathologie und Therapie der Linse, 1876; Zur Anatomie und Pathologie der gesunden und kranken Linsen, 1883. Inventor test chart for astigmatism using three parallel lines placed at various meridians, 1883. Died 1890.

BECKER, Howard (Paul), Am. sociologist; b. N.Y.C., Dec. 9, 1899; s. Paul John and Letitia Dickson (Stevenson) B.; B.S., Northwestern U., 1925, (A.M., 1926; exchange fellow, U. Cologne, 1926-27; Ph.D., U. Chgo., 1930; m. Frances Bennett, Mar. 16, 1927; children—Elizabeth Fairchild, Christopher Bennett, Ann Hemenway. Began work at 14 as unskilled laborer, later became indsl. engr. (Dort Motor Co., Internat. Harvester Co., etc.); admitted to coll. by exam., 1922; grad. instr. sociology, U. Chgo., 1928; instr. sociology U. Pa., 1928-31; asso. prof. sociology Smith Coll., 1931-37; prof. sociology U. Wis., Madison, from 1937, chmn. dept. sociology and anthropology, 1951-55; sociology lectr. Harvard, 1935, Birmingham, Eng., 1951, Toronto, Can., 1952, lectr. summer schools Harvard, Columbia and Stanford univs. Mem. Am. Sociol. Soc. (pres. 1959), German (hon.), Midwest (pres. 1946-47) sociol. socs., Sociol. Research Assn., Institut Internat. de Sociologie, Phi Beta Kappa. Research fellow abroad, Social Sci. Research Council, 1934-35. Author: Contemporary Social Theory (with others), 1940; German Youth: Bond or Free, 1946; Family, Marriage and Parenthood (with Reuben Hill, Marguerita Steffenson, editors) 2d ed., 1955; Through Values to Social Interpretation, 1950; Systematic Sociology (with Leopold von Wiese), rev. edit., Social Thought From Lore to Science, 2d edit., (with H. E. Barnes), 1952; Man in Reciprocity, 1956; Societies around the World (with Irwin Sanders, et al.), 1956; Modern Sociological Theory in Continuity and Change (with Alvin Boskoff, et al.), 1957; Man in Society, 1958; also articles. Book rev. editor Am. Sociol. Rev. Developed various theoretical and methodological devices, namely sacred-secular theory; research and publs. on sociology of Germany, problem of typology, tensions between supra-historically conceived forms and historically experienced content. Died 1960.

BECKER, Josef, radiologist; b. Billed/Banet, July 6, 1905; ed. Univs. Budapest, Vienna, Berlin, Heidelberg; m. Paula Becker, Dec. 21, 1937; children—Hildegard, Irene, Helmut, Hans. Agrege in med. radiology U. Heidelberg, 1940, prof., 1949, full prof., 1955. Mem. Deutsche Atomkommission und Strahlenschutz-kommission. Mem. Röntgen Soc. Germany, also hon. mem. numerous fgn. socs. Author: (with Sheer) Die radioaktiven Isotope in der Geburtshilfe und Gynäkolgie, 1956, Betatron und Telekobalttherapie, 1958; (with Shubert) Die Supervolt-Therapie, 1961; Lokalisierte Applikation künstlichradioaktiver Isotope, 1953, also numerous articles, revs. Research on therapeutics of rays, supervolt-therapy, nuclear medicine. Home: Albert Uebersestrasse 16-18. Office: Vossstrasse 3, Heidelberg, Germany.

BECKER, Karl Martin Ludwig, German statistician; b. Strohausen, Germany, Oct. 2, 1823; s. Johann Christian and Wilhelmine B.; student econs. U. Gottingen, 1851-53, U. Berlin, Germany; m. Johanne Schröder. Co-founder statis. bur. Oldenburg, Germany, 1853, dir., from 1855; co-founder Imperial Statis. Office, Berlin, dir., 1872-91. Editor: Statistische Nachrichten aus dem Grossherzogtum Oldenburg, 13 vols.; Statistik des Deutschen Reiches, 1873. Noted for orgn. statis. surveys; introduced household listing, Oldenburg, later control listing to improve dependability of census results; enlarged range of census questions; pub. death tables for German empire, Prussia. Died Berlin, June 20, 1896.

BECKER, Ralph Sherman, Am. chemist; b. Benton Harbor, Mich., Mar. 14, 1925; s. Ralph S. and Rose Mary (Koranda) B.; B.S., U. Vt., 1949; M.S., U. N.H., 1951; grad. student Mass. Inst. Tech., 1950-52; Ph.D., Fla. State U., 1955; m. JoAnn Cowles, June 7, 1952; children—Mark K., Sherryl D., Janet C., Scott M. Mem. faculty U. Houston, 1955—, prof. chemistry, 1963—, chmn. dept., 1961—; Fulbright vis. prof. U. Barcelona (Spain), 1962-63; vis. lectr. Weizmann Inst. Sci., Israel, 1963. Pres. Advance Research Inc., 1964—; v.p. C. E. Kaiser Co., 1962—; sec. Diverse Products Inc. Mem. Am. Chem. Soc., Am. Phys. Soc., Sigma Xi, Alpha Chi Sigma, Phi Kappa Phi, Phi Delta Theta. Contbr. articles, chpts. in books. Home: 518 Knipp Rd., Houston 77024.*

BECKER, Robert Adolph, Am. physicist; b. Tacoma, Feb. 10, 1913; s. Adolph Adam and Anna (Hurtienne) B.; B.S., Coll. Puget Sound, 1935; M.S., Cal. Inst. Tech., 1937, Ph.D., 1941; m. Dorothy May Wilkins, Sept. 10, 1944; children—Regina Ann, Doreen Maybell, Barbara Jean, Thomas Alan, Jennifer Kathleen. Physicist, Dept. Terrestrial Magnetism, Washington, 1941; asst. physicist Nat. Bur. Standards, Washington, 1941-42, asso. physicist, 1942-43; physicist Lawrence radiation lab. U. Cal. at Berkeley, 1943; prin. physicist applied physics lab. U. Wash., Seattle, 1943-44, sr. physicist, 1945-46; asst. prof. physics U. Ill., Urbana, 1946-51, asso. prof., 1951-55, prof., 1955-60; dir. space physics lab. Aerospace Corp., Los Angeles, 1960—. Guggenheim fellow, 1958-59. Fellow Am. Phys. Soc.; mem. Am. Astron. Soc., Am. Geophys. Union, Am. Optical Soc., Sigma Xi. Author: Introduction to Theoretical Mechanics, 1954. Research papers in nuclear physics, isotope separation, war research, astrophysics, space physics. Home: 5005 Oaklon Dr., Palos Verdes Peninsula, Cal. 90274. Office: P.O. Box 95085, Los Angeles 90045.*

BECKER, Roland Frederick, Am. anatomist, educator; b. Methuen, Mass., Aug. 12, 1912; s. William Curtis and Bertha (Schroeder) B.; B.S., U. Mass., 1935, M.S., 1937; Ph.D., Northwestern U., 1940; m. Florence Kingston Courtis, Sept. 8, 1936; children—Richard Frost, Constance Louise. Instr. to asst. prof. anatomy Northwestern U. Med. Sch., Chgo., 1940-46; chmn. anatomy U. Wash. Med. Sch., Seattle, 1946-49; dir. neuroanat. div. Daniel Baugh Inst. Anatomy, Jefferson Med. Coll., Phila., 1949-51; asso. prof. anatomy, dir. lab. perinatal sci. Duke Med. Center, Durham, N.C., 1952—. Med. educator Chiengmai Med. Sch., Thailand, AID, 1961-63. Mem. Am. Assn. Anatomists, Cajal Club, So. Soc. Anatomy, Am. Physiol. Soc., A.A.A.S., Am. Assn. Med. Colls., Phi Kappa Phi, Phi Beta Pi, Sigma Xi. Research in fetal physiology and conditions produced by maternal imbalances; olfaction and its mechanisms; learning and behavior in dog, rat, guinea pig. Home: 619 Hammond St., Durham, N.C. 27704.*

BECKERLEY, James Gwavas, Am. physicist; b. Chgo., Feb. 27, 1915; s. Gwavas Foster and Clara (Ungewitter) B.; A.B. cum laude, Stanford, 1935, Ph.D., 1945; m. Lucille Pool, Aug. 3, 1939; children—James Gwavas III, John D. Lectr. physics Rangoon (Burma) U., 1940-42, Columbia, 1942-45; professorial lectr. physics Stevens Inst., 1947-49, George Washington U., 1950-53; head div. tech. advisers U. S. AEC, N.Y.C., 1947-49, dir. classification, 1949-54; head engring. physics dept. Schlumberger Well Surveying Corp., Houston, 1954-59; physicist IAEA, Vienna, Austria, 1959-62; editor Siftor Project div. sponsored research Mass. Inst. Tech., 1962-65; pres. Radioptics, Inc., Plainview, N.Y., 1964—. Mem. planetology subcom. space sci. steering com. NASA, 1965—. Mem. Am. Nuclear Soc., A.A.A.S., Phi Beta Kappa, Sigma Xi. Research, publs. on devel. theory of radiation emitted by antenna arrays, theory of prodn. of X-radiation by heavy atoms; responsible for initial devel. new methods for prodn. nuclear graphite and heavy water; administered research in nuclear geophysics. Editor, co-author: Technology of Nuclear Reactor Safety, 1964; editor: Ann. Rev. Nuclear Sci., 1951-58, Nuclear Sci. and Engring., 1955-59, Nuclear Fusion, 1960-62. Home: 9 Jerusalem Lane, Cohasset, Mass. 02025. Office: 10 Dupont St., Plainview, N.Y. 11803.

BECKERS, Christian H., Belgian physician; b. Herstal, Belgium, Nov. 22, 1932; s. Henri and Marguerite (Jonlet) B.; M.D., U. Louvain (Belgium), 1956, Agrégé de l'Enseignement Superieur, 1960, postgrad.; m. Marie-Claire Chenu, Dec. 3, 1959; children—Vincent, Anne, Claire, Dominique, Francois. Research fellow Harvard Med. Sch., 1958-59; C.R.B. Grad. fellow B.A.E.F., 1958-59; maitre de confs. U. Louvain, 1959-63, charge de cours associé, 1963—. Radioisotope expert IAEA, 1963—; expert Pan Am. World Health Orgn., 1963, 66. Mem. N.Y. Acad. Scis., Am., European thyroid assns., Endocrinological Soc., Author: (with De Visscher) Le isotopes radioactifs en médecine, 1961; contbg. author: L'hormonogenèse dans les goitres endémiques et sporadiques, 1963. Research, numerous publs. on thyroid gland in physiol. and path. conditions, mechanisms underlying goiter diseases, especially endemic goiter. Home: 9E Oude Milse Baan, Haasrode, Belgium. Office: Lab. de Pathologie Générale, U. Louvain, 69 Brusselsestraat, Louvain, Belgium.*

BECKERS, William Gerard, chemist; b. Kempen, Germany, Feb. 12, 1874; s. Gerard and Maria Magdalena (Frantzen) B.; student Poly. Inst. Aix la Chapelle; Ph.D., U. Freiburg, 1897; m. Marie Antoinette Pothen; children—William Kurt, Elsa M. Came to U. S., 1902, naturalized, 1911. Prof. chemistry Royal Dye Sch., Crefeld, 1898-1900; became connected with Bayer Co., mfrs. dyestuffs and chemicals, Eberfeld, 1900, and came to U. S., 1902, in charge tech. depts. of Am. br.; founder, 1911, Beckers' Aniline & Chem. Works, also pres., chmn. bd.; co. consol. with other cos., 1917, as Nat. Aniline & Chem. Co., Inc., 1917, became dir., v.p.; dir. Allied Chem. & Dye Corp., other cos. Mem. Am. Chem. Soc., Soc. Chem. Industries (Eng.). Died Nov. 3, 1948.

BECKET, Frederick Mark, Am. chemist, metallurgist; b. Montreal, Que., Can., Jan. 11, 1875; s. Robert Anderson and Anne (Wilson) B.; B.A.Sc., McGill U., 1895, LL.D., 1934; A.M., Columbia U., 1899, postgrad., 1900-02, Sc.D., 1929; m. Geraldine McBride, Oct. 8, 1908; children—Ethelwynne (Mrs. Paul H. Folwell), Ruth Alene (Mrs. Ruth Becket Trauter). Came to U. S., 1895, naturalized, 1918. With Westinghouse Electric & Mfg. Co., East Pittsburgh, Pa., 1895-96, Acker Process Co., Jersey City, 1896-98, Niagara Falls, N.Y., 1899-1900, Ampere Electrochem. Co., 1902-03, Niagara Research Labs., 1903-06; with Electro Metall. Co. and Union Carbide Co., from 1906; cons. Union Carbide & Carbon Corp. Recipient Perkin medal, 1924, Acheson medal, 1937, Elliott Cresson medal, 1940. Howe Meml. lectr., 1938. Fellow A.A.A.S.; mem. N.Y. Acad. of Sci., Am. Inst. Mining and Metall. Engrs. (pres. 1933), Electrochem. Soc. (pres. 1925-26, hon. mem. 1936), Mining and Metall. Soc. Am., Am. Soc. for Metals, Iron and Steel Inst. of London. Contbr. to tech. publs. Died Dec. 1, 1942.

BECKETT, Arnold Heyworth, Brit. medicinal chemist; b. Blackpool, Eng., Feb. 12, 1920; s. Ernest Heyworth and Edith (Duckworth) B.; Ph.C., U. London, 1942, B.Sc. with 1st class honors, 1947, Ph.D., 1950; m. Miriam Eunice Webster, Apr. 1, 1942. Faculty, Chelsea Coll. of Sci. and Tech., London, Eng., 1951—, prof. pharm. chemistry, head Dept. of Pharmacy, 1959—. Recipient Stas medal Belgian Chem. Soc., 1962. Mem. Chem. Soc., Biochem. Soc., Brit. Pharmacol. Soc., Royal Inst. Chemistry, Pharm. Soc. Great Britain (mem. council, mem. MacGregor com.). Author: (with J. B. Stenlake) Practical Pharmaceutical Chemistry, 1963; (with H. S. Bean, J. E. Carless) Advances in Pharmaceutical Studies, vol. 1, 1964; also numerous articles. Research on design of drugs and importance of 3 dimension in biol. action, drug metabolism and distbn., kinetic studies in man. Home: 5 Blyth Rd., Bromley, Kent, Eng.*

BECKETT, Peter Gordon Stewart, physician; b. Dublin, Ireland, Oct. 11, 1922; s. Gerald Paul and Margaret (Collen) B.; B.S., Trinity Coll., Dublin, Ireland, 1943, M.B., B.Ch., B.A.O., 1945, M.D., 1960; m. Victoria L. Ling, Jan. 9, 1954; 1 son, Paul Timothy. Came to U. S., 1949, naturalized, 1954. Staff psychiatrist chief adult inpatient service Lafayette Clinic, Detroit, 1955-57, asst. dir. for research, 1957—; prof. psychiatry Wayne State U. Sch. Medicine, Detroit, 1965—. Mem. Brit., Irish, Am. med. assns., Am. Psychiat. Assn., Wayne County Med. Soc. Author: Adolescents out of Step: their Treatment in a Psychiatric Hospital, 1965; also articles. Research on clin. measurements in psychiatry, followup psychiat. illness, influence parent-child relationships, treatment adolescent psychiat. patients. Home: 1420 Anita St., Grosse Pointe Woods, Mich. 48236. Office: 951 E. Lafayette St., Detroit 48207.*

BECKETT, Philip Henry Trim, Brit. chemist, pedologist, geomorphologist; b. London, Eng., Mar. 25, 1928; s. William Eric and Katharine (Richards) B.; M.A., Oxford U., 1950, D.Phil., 1958; m. Elspeth McIntosh, July 4, 1954; children—Christopher, Victoria, Katharine, Philippa. Demonstrator, Oxford (Eng.), 1954-58, univ. lectr., soil scis. lab., dept. agr., 1958—; fellow St. Cross Coll., 1966—; field work on soils in numerous countries. Mem. Royal Geog. Soc. (Cuthbent Peck prize), Geologists Assn., Inst. Brit. Geographers, Soil Sci. Soc. Research, publs. on mapping soils by means of air photographs and on geomorphology as a guide to distbn. of soils, phys. chemistry of nutrient ions in soil, particularly Pandk, soil devel. Home: 16 Polstead Rd., Oxford, Eng. Office: Soil Sci. Lab., Dept. Agr., Oxford U., Oxford, Eng.*

BECKH, Harald J. von, see von Beckh, Harald J.

BECKMAN, Harry, Am. physician; b. Louisville, Aug. 14, 1892; s. Julius Victor and Florence (Stuber) B.; student U. Ky., 1911-12; M.D., U. Louisville, 1921; grad. study U. Vienna, 1930-31; m. Jane Smith, Mar. 6, 1943; children—Thomas Howell, John Ross. Intern N.Y. Skin & Cancer Hosp., N.Y. City, 1921-23; instr., dept. physiology and pharmacology, sch. medicine Marquette U., 1923, asst. prof., 1923-25, asso. prof., acting dir. dept. pharmacology, 1925-26, chmn., 1926-61, emeritus, 1962—; investigator malariology, Nat. Insts. Health, 1937—; cons. physician Milwaukee Co. and Columbia hosps., Milw.; honorary physician St. Mary's Hosp., Milw.; Houghton fellow clin. research Columbia Hosp., Milw. Recipient State Med. Soc. Wis. Council award, 1959, Marquette U. teaching excellence award, 1960. Fellow A.A.A.S., A.M.A.; mem. Internat. Pharmacopoeia of the World Health Orgn. (expert adv-panel), N.Y. Acad. Scis., Milw. Acad., Med. (pres. 1949), Wis. State and Milw. Co. med. socs., Soc. Exptl. Biology and Med., Am. Soc. Pharmacology and Exptl. Therapeutics (pres. 1956-57), Fedn. Am. Socs. Exptl. Biology, Central Soc. Clin. Research, American and Wis. heart assns., Milwaukee Society Internal Medicine, Am., Royal socs. tropical medicine and hygiene, Alpha Omega Alpha, Sigma Xi, Phi Chi, Alpha Tau Omega. Club: University (Milw.) Author: Treatment in General Practice, 1930; Pharmacology in Clinical Practice, July 1952; Drugs: Their Nature, Action and Use, 1958; Pharmacology: The Nature, Action and Use of Drugs, Dilemmas in Drug Therapy, 1967; writer of Drug Therapy, 1949—; contbg. editor Wis. Med. Jour., 1936—; mem. editorial adv. bd. The New Physician of Student A.M.A. Research in pharmacology and malariology; writing and editing in pharmacology and therapeutics; consulting in pharmacology and clinical medicine. Home: 2831 N. Shepard Av., Milw. 11. Office: Columbia Hosp., Milw.

BECKMAN, Herman F., Am. chemist; b. Ft. Worth, Nov. 18, 1925; s. Herman B. and Willie (Goforth) B.; B.A., Tex. Christian U., 1948; M.S., Tex. A. and M. U., 1953, Ph.D., 1959; m. Edwina De Arman, Dec. 21, 1946; children—Sharon, Chris, Alan, Glen. Chemist, U. S. Dept. Agr., Kerrville, Tex., 1948-55; asso. state chemist Tex. Agr. Expt. Sta., College Station, 1955-61; asso. chemist U. Cal., Davis, 1961—. Mem. Am. Chem. Soc., Entomol. Soc. Am., Sigma Xi. Research, numerous publs. in analytical chemistry relating to all types of pesticides including insecticides, fungicides, herbicides, nematocides, animal feed additives; devel. new instruments for analytical measurements. Home: 1114 Oeste Dr., Davis, Cal. 95616.*

BECKMANN, Ernst Otto, German chemist; b. Solingen, Germany, July 4, 1853; s. Johann Friedrich Wilhelm and Julie (Reusenhof) B.; student U. Leipzig, Germany, 1875-78; m. Bertha Oertel, Mar. 20, 1887; children—2 sons, 1 dau. Lectr. chemistry, pharmacy Tech. U. Braunschweig, Germany, from 1882; prof. phys. chemistry Giessen, Germany, 1891; prof. Erlangen, Germany, from 1892; dir., lab. for applied chemistry Leipzig, from 1897, Kaiser Wilhelm Institut, Berlin, 1912; noted for discovery of molecular rearrangement of oximes of ketones into acid amides or anilides (Beckmann molecular transformation), 1886; designed variable scale thermometer which he used to develop freezing and boiling point methods for determining molecular weights in solution; worked on camphor; contbd. to spectrum analysis. Died Berlin, July 12, 1923.

BECKMANN, Hermann, German physicist, engr.; b. Echem, Germany, July 29, 1873; s. Dietrich and Ulrike (Schecker) B.; student mech. engring. Technische Hochschule Charlottenburg, natural scis. U. Berlin (Germany); Ph.D., Tubingen, Germany, 1898. Employed by Elektrizitätswerk Linden, Accumulatoren-Werken E. Schulz, Witten, Germany; mem. staff prodn. dept., later head lit. bur. Accumulatoren-Fabrik Aktiengesellschaft, Berlin, 1905-33. Lectr. Technische Hochschule, Hannover, Germany, 1926-30, prof., from 1930. Contbr. sci. articles to English, French jours. Improved rapid forming process of accumulator plates; found microporous rubber particularly suitable substance for plate separators. Died Blankenburg, Germany, July 14, 1933.

BECKMANN, Johann, German technologist; b. Hoya an der Weser, Germany, June 4, 1739; s. Nicolaus and Dorothee Magdalena (Schüler) B.; ed. Göttingen, Germany; m. Sofie Louise Karoline Schlosser, 1767; 2 children. Tchr. St. Peter Gymansium, St. Petersburg, Russia, 1763; with Hannöverschen Magazin, 1764; became prof. philosophy, Göttingen, 1765, prof. econs., 1770. Author: De historia naturali veterum libellus primus, 1766; Anfangsgründe der Naturhistorie, 1767; Dan. Tilas Entwicklung einer schwedische Mineralhistorie, 1767; Gedanken von der Einrichtung ökonomischer Vorlesungen, 1767; Grundsätze der deutsche Landwirtschaft, 1769; Caroli a Linné Systema naturae . . . , 1771; Linnés Terminologia conchyliologiae, 1772; Anleitung zur Technologie, 1772; Einleitung in der Technologie (1st textbook on Technology), 1777; Grundsätze zur Vorlesungen über die Naturlehre, 1779; Beiträge zu Oekonomie, Technologie und Cameralwissenschaft, 12 vols., 1779-90; Beiträge zu Geschichte der Erfindungen, 5 vols., 1783-1805; Anleitung zu Handlungswissenschaft, 1789; Vorbereitung zur Waarenkunde, 2 vols., 1793; Vorrath kleiner Anmerkungen über mancherley gelehrte Gegenstände, 1795, 1803-06; Lexikon botanicum, 1801; Entwurf einer allgemeinen Technologie, 1806. Historian of inventions and tech.; most important exponent of cameralist sch. German agrl. sci. in latter half 18th century. Died Göttingen, Feb. 3, 1811.

BECKURTS, Karl Heinz Fritz Ferdinand, German physicist; b. Rheydt, Germany, May 16, 1930; s. Karl and Gisela (Brockdorff) B.; Dr.rer.nat., U. Göttingen, 1956; m. Regina Heuer, Aug. 3, 1956; children—Margarethe, Tobias, Johanna. Research asso. Max Planck Inst. Physics, Göttingen, 1956-58; research asso., group leader Kernforschungszentrum, Karlsruhe, 1958-62, dir. Inst. für Angewandte Kernphysik, 1963—; prof. reactor physics Tech. U., Karlsruhe, 1964—. Mem. Deutsche Physikalische Gesellschaft, European-Am. Nuclear Data Com. Author: (with K. Wirtz) Elementare Neutronenphysik, 1958, Neutron Physics, 1964; also articles. Devel. of method of pulsed sources in neutron and reactor physics; devel. methods for neutron cross section measurements. Home: 5 Osteroder Strasse, Karlsruhe, Germany.*

BÉCLARD, Auguste, French physician, physicist; b. Paris, France, Sept. 17, 1817; s. Pierre-Auguste Béclard; ed. Paris Faculty Medicine. Tchr. physiology, Paris Faculty Medicine; became dean Vulpiar Med. Sch., 1881. Mem. Acad. Medicine (perpetual sec., 1873). Author: Mémoire relatif à l'influence de la lumière sur le développement des animaux, 1858; Mémoire relatif à l'influence musculaire dans ses rapport avec la température animale, 1861; Mémoire sur l'influence de la température sur le développement comparé des systèmes organiques. Research on relationships between growth and various wave lengths of light, (1854), between muscular contraction and thermogenesis (1861), between spleen and portal vein. Died Feb. 9, 1887.

BÉCLARD, Pierre-Augustin, French anatomist; b. Angers, France, 1785; surgeon in chief Hosp. de la Charité, Paris; became prof. anatomy Sch. Medicine, 1818. Author: Eléments d'anatomie générale, 1823. Described femoral hernia; performed 1st removal of parotid glands, 1823. Died Paris, 1825.

BECQUEREL, Alexandre Edmond, French physicist; b. Paris, France, Mar. 24, 1820; s. Antoine César B.; studied under his father; 1 son, Antoine Henri; asst. to father musée d'Histoire Naturelle, later succeeded him as prof.; apptd. prof. Agronomic Inst., 1849; appointed to chair of physics Conservatoire des Arts and Métiers, 1853; member of French Academy of Sciences, 1863 (vice president 1879, president 1880); Fellow Royal Soc., 1888. Author: (with A. C. Becquerel) Eléments de physique terrestre et deméteorologie, 1847; La Lumière, ses causes et ses effets, 2 vols., 1869; Physico-Chemical Forces, 1875. Research on light, photochem. effects of solar radiation, phosphorescence of sulphides and uranium compounds, diamagnetism and paramagnetism; developed modification of Faraday's law; invented phosphoroscope, 1865. Died Paris, May 11, 1891.

BECQUEREL, Antoine César, French physicist; b. Châtillon-sur-Loing, Loiret, France, Mar. 8, 1788; studied at L'École Polytechnique; 1 son, Alexandre Edmond, Served with engrs. corps of army, 1810-15; ret. as major; professor physics Musée d'Histoire Naturelle, 1837—. Recipient Copley medal Royal Soc., 1837. Mem. Acad. Scis., 1829; Fellow Royal Soc., 1837. Author: Traité d'électricité et du magnétisme, 1834-40; Traité de physique dans ses rapports avec la chimie, 1842; Éléments de L'élecrochimie, 1843; Traité complet du magnétisme, 1845; (with Alexandre Edmond Becquerel) Éléments de physique terrestre et de météorologie, 1847; Des climats et de l'influence qu'exercent les sols boisés et déboisés, 1853. One of founders of electrochemistry; 1st to apply electrolysis on practical scale for recovery of medals from ores, 1836; refuted Volta's theory of contact; showed presence of chem. action is necessary for generation of electricity; constructed electric thermometer (used to measure temperatures of soil, atmosphere and internal body heat); established normal human mean temperature at 98.6 degrees F.; studied telegraphy and magnetism. Died Paris, Jan. 18, 1878.

BECQUEREL, Antoine Henri, French physicist; b. Paris, France, Dec. 15, 1852; s. Alexandre Edmond Becquerel; student École Polytechnique, Paris, 1872; 74; ingenieur Ecole des Poutes-et-Chaussées, 1877, ingenieur de première class, 1885, ingénieur en chef, 1894; D.-ès-Sc., 1888. children include Jean. Asst., Natural History Mus., Paris, 1878-92, prof. applied physics, 1892-95; lectr. École Polytechnique, 1876-95, prof. physics, 1895-1908. Decorated officer Legion of Honor; recipient (with Curies) Nobel prize, 1903; Rumford medal, Barnard medal, Helmholtz medal. Fellow Royal Soc., 1908; mem. Berlin Acad., French Academy Scis., 1889 (apptd. life secretary 1908); Academy of Science, Washington, Academia dei Lincei. Author: Déconverte des radiations invisibles émises par l'uranium, et des phénomènes produits par ces radiations; Recherches sur une propriété nouvelle de la matière, 1903. Discovered radioactivity in uranium and its salts (called Becquerel rays, later designated radioactivity by Curie), 1896; investigated magnetism, phosphorescence, light polarization and absorption in crystals; discovered Faraday effect in gases. Died Le Croisic, France, Aug. 25, 1908.

BECQUEREL, Jean Antoine Edmond Marie, French physicist; b. Paris, France, Feb. 5, 1878; s. Antoine Henri Becquerel; educated at Ecole polytechnique, University of Paris (France); Ingenieur Polytechnique; professor physics Natural History Museum, Paris, 1909-48; named honorary professor, 1948; examiner Ecole polytechnique. Mem. French Acad. Scis., 1946, Soc. Physics. Author: la Principe de la relativité et le théorie de gravitation; la radio-activité et la transformation des elements; l'Art musical dans ses rapports avec la physique. Research on phosphorescent light emission (at very low temperatures) of uranyl, rubies and emeralds, 1909; continued work of his father on optical and magnetic properties of crystals; discovered magnetic rotatory polarization. Died July 4, 1953.

BECQUEREL, (Alfred) Louis, French physician; b. Paris, June 3, 1814; s. Antoine César Becquerel; docteur en médecine, 1840; became head physician St. Perrine Hosp., 1852; named physician La Riboisière Hosp., 1855, Our Lady of Pity Hosp., 1856. Author: Séméiotique des urines, 1841; Recherches sur la composition du sang, 1844; Des applications du l'éléctricité à la pathologis, 1856. Described syphilitic changes of the liver, 1840; research on stuttering and its cure, composition of blood. Died Paris, Mar. 12, 1866.

BECQUEREL, Paul, French biologist; b. Paris, France, Apr. 14, 1879; s. Paul André and Clementine (Souchet) B.; B.A., Lic.Sc., Lyceum Henri IV, 1902; Sc.D., Faculty Scis., Paris, 1907; also Certificats d'études supérieures; m. Louise Dupont, 1904; 1 child. Prof., Faculty Scis., U. Poitiers (France), named asso. chief of studies, 1928; asst. in botany Faculty Scis., Paris; named dir. of the sta. of plant biology at Beau-Site, 1930; research staff Lab. Applied Physics and Natural Scis., Mus. Natural History, Paris. Decorated Knight of Legion of Honor, Officer Pub. Edn. recipient prize for plant physiology French Acad. Scis. Mem. Soc. Men of Letters, Bot. Soc. France, French Astron. Soc., Nat. Mus. Natural History (corr.), French Acad. Scis., 1945. Research and numerous publs. on physiology, anatomy, plant radioactivity; studied life at temperatures approaching absolute zero, discovered suspended animation could be produced by low temperatures, dehydration and high vacuum. Died Evian, France, June 22, 1955.

BEDDARD, Frank Evers, Brit. zoologist; b. Dudley, Worcestershire, June 19, 1858; s. John Beddard; ed. New Coll., Oxford; M.A., D.Sc. Naturalist, Challenger Expdn. Commn., 1882-84; prosector Zool. Soc., 1884-1915; lectr. biology Guy's Hosp.; examiner in zoology and comparative anatomy U. London; examiner in morphology Oxford U., U. New Zealand. Fellow Royal Soc., 1892. Author: Animal Coloration, 1892; Monograph on the Oligochaeta, 1895; A Textbook of Zoogeography, 1895; The Structure and Classification of Birds, 1898; Mamalia, 1902 (in Cambridge Natural History); Earthworms and their Allies, 1912. Investigated Oligochaeta order of worms (includes earthworms). Died West Hampstead, London, Eng., July 14, 1925.

BEDDOE, John, Brit. physician, anthropologist; b. Bewdley, Eng., Sept. 21, 1826; s. John and Emma (Child) B.; B.A., U. Coll., London, Eng., 1851; M.D., Edinburgh (Scotland) U., 1853, LL.D., 1891; m. Agnes Montgomerie Cameron Christison, 1858; 1 son, 1 dau. Med. staff in Crimea; physician to Bristol (Eng.) Royal Infirmary, 1862-73. Rhind lectr. 1891. Fellow Royal Soc., 1873, Royal Coll. Physicians; mem. Bristol and Gloucestershire Archeol. Soc. (co-founder 1875, pres. 1890), Anthrop. Soc. (pres. 1869-70), Royal Anthrop. Inst. (pres. 1889-91), Wiltshire Archeol. and Natural History Soc. (pres. 1909). Author: Contributions to Scottish Ethnology, 1853; Stature and Bulk of Man in the British Isles, 1870; The Races of Britain, 1885; Memories of Eighty Years, 1910. Contbr. to sci. publs., Brit. Assn. jour. Pioneer ethnol. research Europe; observed hair, eye colors W. Eng., 1846, Orkney, 1853; authority on phys. characteristics living European races; influenced devel. anthrop. sci. at home, abroad. Died July 19, 1911.

BEDDOES, Thomas, English physician; b. Shiffnal, Eng., Apr. 13, 1760; ed. Pembroke Coll., Oxford, Eng.; studied medicine, London, also Edinburgh; M.D., Oxford U.; m. Anna Edgeworth. Reader in chemistry, 1788-92; established Pneumatic Inst. (for treating disease by inhalation), Clifton, 1798; engaged Davy as asst., encouraged him. Author: Observations on the Nature of Demonstrative Evidence, 1793; Hygeia, 1801. Editor: Elements of Medicine (John Brown), 1795; Contributions to Physical and Medical Knowledge, principally from the West of England, 1799. Presented phenomena of disease vividly in writings; investigated medicinal properties of gases; observed adverse effects of carbon monoxide. Died Dec. 24, 1808.

BEDE (Venerable), English scholar; b. Jarrow, Eng., 673; given eccles. edn.; ordained priest, 703; most of his adult life spent in Benedictine monastery of Wearmouth and Jarrow; taught math., grammar, rhetoric, music, astronomy, Latin, Greek, and Hebrew. Author: Historia ecclesiastica gentis Anglorum, circa 734, 1st printed, 1474. Writings covered wide range of subjects; believed earth was sphere; used math. and astronomy for determining date of Easter; revived idea that phases of moon governed tides; 1st to realize that high tide occurs at different times in different places. Died Jarrow, May 26, 735.

BEDEL, Charles François Constantin, French chemist; b. Avranches, Mar. 14, 1889; s. François and Amélie (Manet) B.; ed. U. Rennes, U. Paris; Dr. ès sc., Dr. en droit, diplôme superièure in pharmacy; m. Madeleine Lemasle, Apr. 23, 1919; children—Simone, Jacques. Asst. of univ., prof. agrégé, master of lectures, titular prof. chair of legislation, titular prof. chair history of pharmacology; commr.-insp. dangerous, insalubrious or incommodious establishments of dept. of Seine. Mem. Acad. Pharmacy (pres.), Soc. Doctors in Pharmacy (pres.), Nat. Acad. Dental Surgery, Soc. History of Pharmacy of France (v.p.), others. Author: La participation de l'Etat et des autres personnes morales administratives à la production dans les industries physiques et chimiques. Research on products of polymerization of cyanhydric acid, phys. properties of sillicium and ferrosiliciums, intense magnetic fields and their applications. Home: 3 rue Gabriel-Péri, Montrouge (Seine). Office: Faculté de Pharmacie, Paris, France.

BEDELL, Frederick, Am. physicist; b. Bklyn., Apr. 12, 1868; s. Edwin Forrest and Caroline Louise (Cunningham) B.; A.B., Yale, 1890; M.S., Cornell, 1891, Ph.D., 1892; m. Mary L. Crehore, July 1, 1896 (dec. Mar. 1936); children—Eleanor (Mrs. Robert Cady (Burt), Caroline Cunningham (Mrs. Henry M. Thomas, Jr.); m. 2d, Grace Evelyn Wilson, July 19, 1938. Asst. prof. physics Cornell U., 1893-1904, prof. applied electricity, 1904-37, prof. emeritus from 1937; later cons. physicist in pvt. practice, Pasadena, Cal. Mem. jury of awards La. Purchase Expn., St. Louis, 1904; asso. editor, 1894-1913, mng. editor, 1913-22, The Physical Rev.; mem. Internat. Electrotech. Commn., 1913. Fellow A.A.A.S. (sec. of council 1898, gen. sec. 1899), Am. Phys. Soc., Am. Inst. E.E. Inst. (mgr. 1914-16, v.p. 1917-18), Phi Beta Kappa, Sigma Xi. Inventor systems of power transmission and communication and of scientific instruments, including aids for the deaf. Recipient award as Modern Pioneer on Frontier of Industry for invention of Cathode ray oscilloscope and bone conduction hearing aids, 1940; known for investigation in alternating currents of electricity. Author several works on electricity and aerodynamics; contbr. definitions in electricity to Webster's Internat. Dictionary. Died May 2, 1958.

BEDERSON, Benjamin, Am. physicist; b. N.Y.C., Nov. 15, 1921; s. Abraham M. and Lena (Waxlowsky) B.; B.S., Coll. City N.Y., 1946; M.A., Columbia U., 1948; Ph.D., N.Y. U., 1950; m. Betty E. Weintraub, Jan. 20, 1956; children—Joshua, Geoffrey, Aron, Benjamin. Staff Research Lab. Electronics, Mass. Inst. Tech., 1948-50; asst. prof. N.Y. U., 1952-57, asso. prof., 1957-59, prof. physics, 1959—; vis. scientist Lab. Gas Ionizzati, C.N.E.N., Frascati, Italy, 1963; Nat. Bur. Standards Atomic Physics div. adv. panel to Nat. Acad. Sci., 1962-64; sec., editor, mem. steering, program coms. Internat. Conf. on Physics of Electronic and Atomic Collisions, 1959—; dir., trustee Inst. for Med. Research and Studies, N.Y., 1960—. Fellow Am. Phys. Soc.; mem. Am. Assn. U. Profs., A.A.A.S., Sigma Xi. Contbr. articles to sci. jours. Research in atomic collision phenomena and atomic structure, using atomic beam techniques; plasma physics and lab. astro-physics; exptl. work in quiescent plasmas in thermodynamic equilibrium; measurements include determinations of atomic polarizabilities, electronatom, atom-atom and atom-ion collision cross sections; studies of atomic ineractions involving spin-selected and spin-analyzed atomic systems. Home: 8 Hall Av., Larchmont, N.Y. 10538. Office: Physics Dept., Gould Hall Tech., N.Y. U., University Heights, Bronx, N.Y. 10453.*

BEDFORD, Davis Evan, English cardiologist; b. Boston, Lincolnshire, Eng., Aug. 21, 1898; s. William and Lilian (Baxter) B.; ed. in London, Paris, Lyon, Vienna; M.D., London; m. Audrey Selina North, July 13, 1935; children—William, John Victor. Med. cons. Middlesex Hosp., Nat. Heart Hosp., London, Eng. Fellow Royal Coll. Physicians; mem. Acad. Medicine of Rome (corr.), Brit. Cardiac Soc. (pres.), European Soc. Cardiologists (hon. pres.), others. Former editor Brit. Heart Jour. Research and numerous articles on heart disease. Address: 33 Devonshire Pl., London, Eng.

BEDFORD, Gunning S., Am. physician; b. Balt., 1806; grad. St. Mary's Coll., Md., 1825; med. degree, Rutgers, 1829; postgrad. abroad. Apptd. prof. obstetrics at med. coll., Charleston, S.C., 1833; also taught in New College foundation, Albany, N.Y., and in N.Y.C. from 1836; founded Univ. Med. Coll., N.Y.C., 1840; founded 1st obstet. clinic for charity patients in N.Y.C., also Univ. Med. Coll., 1840. Author: Diseases of Women and Children, 1855; Principles and Practice of Obstetrics. Died Sept. 5, 1870.

BEDNAR, Blahoslav, Czechoslovakian physician, pathologist; b. Klobouky, Czechoslovakia, Dec. 18, 1916; s. Frantisek and Marie (Novotná) B.; MUDR, Charles U. Prague, Chechoslovakia, 1945, DrSc, 1962; m. Marie Sobotová, June 18, 1960; children—Magda, Marek, Pavel. With Charles U., 1947—, prof. pathology Med. Faculty, 1960—; dir. I. Hlava Institut Pathology, 1960—. Mem. Czechoslovak Med. Soc., J. E. Purkyne World Soc. Neuropathology. Author: (with R. Venecek) Základy anatomické patologie, 1962; (with others) Patologická anatomie, 1963; also numerous articles. Research on hepatitis, encephalitis, storiform neurofibroma, secondary uremic oxalosis, dystelectatic emphysema of lung, myeloproliferative syndrome. Home: 45 Nad Královskou oborou, Prague, CSSR.*

BEDWELL, William, mathematician; b. "Haslingburgensis A Saxo," ca. 1561; A.B., Cambridge (Eng.) U., 1585; A.M., 1588; scholar Trinity Coll., 1584; became rector St. Ethelburgh's, Bishopsgate St., 1601; named mem. Westminster co. of Bible translators, 1604. Author: Trigonicum architectonicum, 1612; Treatise on Geometrical Numbers, 1614. Translator: Arithmetic (Salignac); Way to Geometry (Ramus), 1636; Noted for his math. works and translations of Arabic math. works; explained use of carpenter's square and ruler in his writings. Died May 5, 1632.

BEEBE, Charles William, Am. naturalist; b. Bklyn., July 29, 1877; s. Charles and Henrietta Marie (Younglove) B.; B.S., Columbia U., 1898; postgrad., 1898-99; Sc.D. and LL.D., Tufts and Colgate U., 1928; m. Mary Blair; m. 2d, Elswyth Thane, 1927. Curator of ornithology N.Y. Zool. Soc., from 1899, also dir. of Dept. Tropical Research, N.Y. Zool. Soc., from 1919. Recipient Elliot medal, John Burroughs medal. Fellow N.Y. Acad. Scis., A.A.A.S., N.Y. Zool. Soc., Am. Ornithologists' Union; mem. Linnaean Soc., Soc. Mammalogists, Ecol. Soc., Audubon Soc., corr. mem. Zool. Soc., London, Société d'Acclimatation de France. Author books including: Pheasants—Their Lives and Homes, 1926; Pheasant Jungles, 1927; Beneath Tropic Seas, 1928; Nonsuch, Land of Water, 1932; Exploring with Beebe, 1932; Field Book of the Shore Fishes of Bermuda (with J. TeeVan), 1933; Half Mile Down, 1934; Zaca Venture, 1938; Book of Bays, 1942; Book of Naturalists, 1944; High Jungle, 1949; also many scientific papers and monographs relating to birds, fish and evolution. Led numerous sci. expdns.; made 1 of 1st deep sea descents (in specially constructed bathysphere, nr. Bermuda, to depth of 3028 feet), 1934. Died nr. Aruma, Trinidad, June 4, 1962.

BEECHER, Charles Emerson, Am. paleontologist; b. Dunkirk, N.Y., Oct. 9, 1856; s. Moses and Emily (Downer) B.; B.S., U. Mich., 1878; Ph.D., Yale, 1889; m. Mary Salome Galligan, Sept. 12, 1894; 2 daus. Asst. to paleontologist James Hall, Albany, N.Y., 1878-88; placed in charge invertebrate fossils Peabody Mus., 1888; asst. prof. hist. geology Sheffield Sch., Yale, 1892-97, prof., mem. governing bd. sch., 1897, curator geol. collections, trustee and sec. bd. trustees Peabody Mus., 1899, prof. paleontology, 1902. Fellow Geol. Soc. Am.; mem. Nat. Acad. Scis., Am. Assn. Conchologists, Malacol. Soc. London. Author: Studies in Evolution, 1901, also numerous papers in sci. periodicals. Made important contbns. to knowledge of devel. and structure of trilobites and brachiopods; discovered limbs and antennae on trilobites, 1893. Died New Haven, Conn., Feb. 14, 1904.

BEECHER, Henry Knowles, Am. physician, educator; b. Wichita, Kan., Feb. 4, 1904; s. Henry and Mary (Kerley) B.; A.B., U. Kan., 1926, A.M., 1927; M.D., Harvard, 1932; M.D. (hon.), U. Lund (Sweden), 1961; m. Margaret Swain, Nov 3, 1934; children—Jonathan French, Harriet (Mrs. Daniel Field), Mary Knowles (Mrs. Richard H. Price). Anaesthetist-in-chief, Mass. Gen. Hosp., Boston; Moseley Traveling fellow Harvard, Copenhagen, Denmark, 1935; Henry Isaiah Dorr prof. research in anaesthesia Harvard, 1941—. Mem. com. on shock NRC, 1940, subcom. on anesthesia, 1946; cons. surgeon gen. U. S. Army, 1946—, USAF, 1952-54, USN, 1953-56, USPHS, 1947-52, 57-58; mem. pharmacology tng. com. NIH, 1958-61; lectr., also participant numerous symposia. Decorated Legion of Merit (U. S.); Chevalier de la Legion d'Honneur (France). Recipient Warren Triennial prize Mass. Gen. Hosp., 1931, 37, Forum Anesthetists award, 1939, Distinguished Service to Humanity citation U. Kan., 1958, Distinguished Achievement award Modern Medicine, 1959. Fellow A.A.A.S., Am. Acad. Arts and Scis., mem. Internat. Soc. for Research in Anesthesia and Analgesia, A.M.A., Am. Soc. Pharmacology and Experimental Therapeutics, Soc. Clin. Investigation, Mass. Med. Soc., New York Acad. Sci. (life), Royal Soc. Medicine London (hon.), Argentina Soc. for Thoracic Surgery (hon.), Sociedad Nacional de Anestesiología (fgn. corr. mem.) Assn. Univ. Anesthetists, Am. Surg. Assn., Sigma Xi. Author: Resuscitation and Anesthesia for Wounded Men, 1949; Experimentation in Man, 1959; Measurements of Subjective Responses; Quantitative Effects of Drugs, 1959. Editor: Physiologic Effects of Wounds, 1951; Disease and the Advancement of Basic Science, 1959. Editorial bd. Jour. Pharmacology and Exptl. Therapeutics, Progress in Neuropharmacology; sci. council Jour. Neuropharmacology. Research, publs. on physiol., metabolic effects of anesthesia, quantitative effects of drugs, ethics of human experimentation. Home: 78 Green St., Canton, Mass. 02021. Office: Mass. Gen Hosp., Boston 20114.*

BEECHER, William John, Am. ornithologist; b. Chgo., May 23, 1914; s. Edward J. and Anne M. (Lawlor) B.; Ph.B., U. Chgo., 1946, B.S., 1947, M.S., 1949, Ph.D., 1954. Zool. asst. Chgo. Natural History Mus., 1937-54; sr. naturalist Cook County Forest Preserve Dist., 1954-57; dir. Chgo. Acad. Scis., 1958—. Cited for sci. edn. Adult Edn. Council Greater Chgo., 1963, Chgo. Area Tchrs. Sci., Assn., 1965. Mem. Conservation Council Chgo. (chmn. 1964—), Nature Conservancy (vice chmn. III. chpt. 1964—). Author: Nesting Birds and the Vegetation Substrate, 1942; Attracting Birds to Your Backyard, 1955. Research on phylogenetic relationship families songbirds, functional adaptations feeding mechanisms, bio-mechanics bird skull; inventor spectacle binocular with catadioptric system using mirrors. Home: 3048 N. Troy St., Chgo. 60618. Office: 2001 N. Clark St., Chgo. 60614.*

BEECKMAN, Isaac, Dutch physicist, mathematician; b. Holland, 1588; author: Mathematico-physica, 1644; atomist; friend of Descartes (whom he engaged to write Traité sur la musique, then attempted to attribute work to himself); studied impact of bodies; studied (with Descartes) and formulated a law of falling bodies; proposed that air had weight, also that vacuum existed; believed in principle of conservation of motion. Died 1677.

BEEDE, Joshua William, Am. geologist; b. Raymond, N.H., Sept. 14, 1871; s. Hiram Pratt and Lydia Maria (Brown) B.; B.S., Washburn Coll., Kan., 1896, A.M., 1897; Ph.D., U. Kan., 1899; m. Clara Frances McKee, Dec. 25, 1899; children—Genevieve (Mrs. G. G. Henderson), Lydia May (Mrs. T. O. Todd), Lucile Prosser, Clara Frances. Tchr. sci. Atchison County High Sch., Effingham, Kan., 1899-1901; instr. geology Ind. U., Bloomington, 1901-06, asst. prof., 1906-09, asso. prof., 1909-17; geologist, bur. econ. geology and technology, U. Tex., 1917-22; with Empire Gas and Fuel Co., 1922; with Dixie Oil Co., 1924-28; prof. geology and paleontology Ind. U., from 1928. Mem. Kan. Geol. Survey, 1896, 1898-99, 1903-10. Okla. Geol. Survey, 1911-17; aid, U. S. Geol. Survey, 1901-02. Research and publs. on carboniferous and permian formations and fossils from Neb. to Tex. and W.Va., and on origin of sediments and coloring matter of Red Beds, Kan.-Okla. Died Feb. 26, 1940.

BEEK, Leendert Klaus Hellinga van, see van Beek.

BEEKLY, Albert Leon, Am. geologist; b. Salina County, Neb., Oct. 29, 1883; s. William Henry and Alice (Halle) B.; ed. U. Neb.; m. Carmelita Quinn; children—Albert Lee, William, Betty. Asst., U. S. Geol. Survey, 1906-13; petroleum geologist, fgn. exploratory dept. S. Pearson and Sons, London, Eng., 1913-16; chief geologist Mid-Continent Petroleum Corp., 1916-49, cons., 1949-52. Mem. Geology Soc. Am., Am. Assn. Petroleum Geologists, Tulsa Geol. Soc. (pres.). Author: Structure of Typical American Oil Fields (paper). Made discoveries of petroleum resources in mid-continent region, including Ramsey Pool, Payne County, Okla.; prepared report on geology and coal resources of North Park, Colo. Died Tulsa, Sept. 29, 1952.

BEELER, Joe R., Jr., Am. physicist; b. Beloit, Kan., Aug. 13, 1924; s. Joe R. and Verneda (Dusenberry) B.; B.S., Kan. U., 1948, M.S., 1950, Ph.D., 1955; m. Mary Frances McKevitt, June 7, 1947; children—Joe R., III, Ann. Staff, Sandia Corp., Albuquerque, 1952-55; cons. physicist Gen. Electric Co., Materials and Propulsion Operation, Cin., 1955-67; prof. nuclear engring. and metallurgy N.C. State U., Raleigh, 1967—. Cons., Atomic Energy Research Establishment, Harwell, Eng., 1964, 66, Battelle Meml. Inst., Pacific N.W. Lab., 1967. Mem. Am. Phys. Soc., Am. Inst. Metall. Engrs., Sigma Xi, Tau Beta Pi. Research and publs. on devel. and use of gen. computer expt. systems for theoretical research in phys. metallurgy and solid state physics, prediction of and 1st treatment of energetic ion and atom channeling process in crystalline solids; radiation damage research. Office: Nuclear Engring. Dept., N.C. State U., Raleigh, N.C. 27607.*

BEEMAN, William Waldron, Am. physicist; b. Detroit, Oct. 21, 1911; s. Joseph John and Mary (Waldron) B.; student Wayne U., 1929-35; B.Sc. in Math., U. Mich., 1937; Ph.D. in Physics, Johns Hopkins, 1940; m. Eleanor Mildred Coswell, June 22, 1940; children—Ann Margaret, Richard William, John Michael, David Kevin. Research physicist Gen. Motors Corp., Detroit, 1940-41; faculty U. Wis., Madison, 1941—, prof. physics, 1952—, chmn. dept. physics, 1952-53, chmn. biophysics lab., 1964—. Cons. Argonne Nat. Lab., Los Alamos Sci. Lab., 1946-50; sci. adviser to Gov. John W. Reynolds, 1963-65. Fellow Am. Phys. Soc.; Bd. editors Rev. Sci. Instruments, 1958-60. Research, publs. in X-ray spectroscopy, small angle X-ray scattering, biophysics. Home: 5010 Tomahawk Trail, Madison, Wis. 53705*

BEER, August, German physicist; b. Trier, Germany, July 31, 1825; s. Joh. Georg and Maria Anna Antoinette Walburga Franzioka Josephine (Dupont) B., degree, U. Bonn (Germany), 1848; became lectr. U. Bonn, 1850; named asso. prof., 1855. Author: Einleitung in die höhere Optik, 1854. Research on optics; discovered law of absorption of light which is named after him. Died Bonn, Nov. 18, 1863.

BEER, Edwin, Am. urologist; b. N.Y.C., Mar. 28, 1867; s. Julius and Sophia (Walter) B.; A.B., Columbia U., 1896, M.D., 1899; postgrad. U. Berlin (Germany), U. Vienna (Austria), U. Prague (Czechoslovakia); m. Elsie Lilienthal, Aug. 20, 1902; children—Phyllis (Mrs. Edwin Koehler), Isabel S., Elizabeth. Became mem. staff Neurol. Inst., 1904; apptd. attending surgeon Mt. Sinai Hosp., N.Y.C., 1912, Bellevue Hosp., N.Y.C., 1916; attending physician to several N.Y. hosps. Recipient gold medal Internat. Urol. Congress at Brussels, 1917, gold key Am. Soc. Phys. Therapy, 1920. Mem. Am., N.Y. surg. socs., Internat. Surg. Assn., N.Y. Acad. Medicine (vice-pres.), Internat. Urol. Assn., Am., Italian urol. socs., N.Y. Path. Soc. (pres, Mt. Sinai Med. Bd.); hon. mem. Acad. Medicine Rome; corr. mem. German Urol. soc. Author: Collected Papers, 1931. Improved diagnostic and therapeutic methods in urology; research on diseases of urinary tract and asso. glands, discovered treatment of bladder tumors using high frequency waves, 1910. Died N.Y.C., Aug. 18, 1938.

BEER, Georg Joseph, Austrian ophthalmologist; b. Vienna, Austria, December 23, 1763; M.D., Univ. Vienna, 1786; professor Clinical Institute Vienna; author of numerous books on eye maladies, including Beobachtungen . . . über jene Augenkrankheiten welche aus allgemeinen Krankheiten des Körpers entspringen, 1791; Lehre der Augenkrankheiten, 2 vols., 1792; Methode den grauen Staar sammt der Kapsel auszuziechen, 1799; Geschichte der Augenkunde überhaupt und der Augenheilkunde insbesondere, 1813. Credited with establishing 1st eye hosp. in Europe, 1786; advocate of iridectomy, 1798; inventor surg. instruments, procedures, Beer's collyrium (eye lotion), Beer's knife (used in cataract operations), Beer's operation (flap method for cataract). Died Vienna, Apr. 11, 1821.

BEER, Hermann, Austrian engr.; b. Sept. 6, 1905; s. Josef and Kornelia (Valentinitsch) B.; ed. Superior Tech. Sch., Graz, Austria; Engr.; Ph.D. in tech. scis.; m. Anneliese Hartmann, Dec. 17, 1938; children—Armin, Gernot. Asst., Superior Tech. Sch.; engr., 1st engr., engr. in chief, prof. Technische Hochschule, Graz. Mem. Acad. Sci. Vienna (corr.), Acad. Sci. Tucuman (Argentina) (corr.), Internat. Assn. Bridges and Scaffolding (pres. Austrian group, pres. com. for stability of constrns.). Research and publs. on static, metal constrn., resistance of materials, projects for bridges, superstructures, barrages, spl. constrns. in steel. Home: Geidorfgürtel 46. Office: Rechbauerstrasse 12, Graz, Austria.

BEER, Wilhelm, German astronomer; b. Berlin, Germany, Jan. 4, 1797; banker; built obs.; recipient Lalande prize French Acad., 1836. Author: Mappa selenographica, 1834-36; Physische Beobachtungen des Mars in der Erdnähe, 1830; Der Mond nach seinen kosmischen und individuellen Verhältnissen, 2 vols., 1837. Spent 8 yrs. mapping prin. features of moon with great accuracy, measured heights of 1,000 mountains; mapped Mars, 1st to make definite picture of lighter and darker areas 1830. Died Berlin, Mar. 27, 1850.

BEERENS, Henri Pierre Edouard, French bacteriologist; b. Hellemmes-Lille, France, June 29, 1921; s. Henri Louis and Germaine (Dordin) B.; Docteur en Pharmacie, U. Lille, 1946; m. Briffaux, Dec. 26, 1944; children—Monique, Bernard. Became head lab. Institut Pasteur de Lille, 1950; instr. microbiology Institut de Hematologie; lectr. Centre de Études et de Recherche de bacteriologie alimentaire. Mem. French Microbiology Soc., Soc. Applied Bacteriology, Biol. Soc., European Soc. for Study Drug Toxicity. Author: Manuel de Techniques bacteriologiques, 1962; Infections humaines à bacteries anaerobes non toxiques, 1965; also numerous articles. Research in anaerobic bacteria, especially those with med. significance. Pres., Internat. Commn. on Bifide Nomenclature, Internat. Commn. on Nomenclature of Anaerobes. Home: 39 rue Faidherbe, Hellemmes-Lille, Nord 59, France. Office: 20 bd. Louis XIV, Lille Nord 59, France.*

BEERMAN, Herman, Am. physician, educator; b. Johnstown, Pa., Oct. 13, 1901; s. Morris and Fannie (Toby) B.; B.A., U. Pa., 1923, M.D., 1927, Sc.D. (Med), 1935; m. Emma N. Segal, May 13, 1924. Abbott fellow chemotherapeutic research U. Pa., 1932-46. Asst. chief dermatology clinic, 1938—, asso. in serology Pepper Lab., 1950—, head dept. dermatology and syphilology Grad. Hosp. U. Pa., 1953—, faculty U. Pa. Sch. Medicine, 1929—, prof. clin. dermatology and syphilology, 1951-56, prof. dermatology, 1956—, faculty Grad. Sch. Medicine, 1940—, prof., 1947—, chmn., 1949—; dermatologist out-patient dept. Pa. Hosp., 1929-36, dermatologist, chief out-patient service B, 1936-46, asso. dermatologist, 1946-47, dermatologist, head dept., 1947—; asst. dermatologist to radium clinic Phila. Gen. Hosp., 1938-40, dermatologist, 1940-53, cons., 1953—; dermatopathologist Skin and Cancer Hosp. Phila., 1947-54; surgeon USPHS Res., 1942-47, cons., 1937—; cons. dermatopathology VA Hosp., Phila., 1953—, U. S. Naval Hosp., Phila., 1954—; cons. lab. Children's Hosp. Phila., 1949—; cons. dermatologist Phila. Psychiat. Hosp., 1950-57; practice medicine specializing in dermatology, Phila., 1933—. Mem. panel venereal diseases subcom. infectious diseases, chemotherapy NRC, 1954—. Diplomate Am. Bd. Dermatology and Syphilology. Fellow A.C.P., A.A.A.S.; mem. Am. Acad. Dermatology and Syphilology (dir.), Am. Dermatol. Assn., Soc. for Investigative Dermatology (past pres., editorial bd., sec.-treas. 1950-65), A.M.A., Am. Med. Writers Assn., Pub. Health Soc. U. Pa., N.Y. Acad. Scis., Med. Soc. for Study Venereal Diseases (hon. life London), Sigma Xi. Editorial bd. Am. Jour. Med. Scis., 1939-67, Excerpta Medica. Research, publs. on therapy, manifestations of syphilis; various skin disorders, including drug eruptions. Home: 2422 Pine St., Phila. 19103. Office: 255 S. 17th St., Phila. 19103.*

BEERS, Charles Dale, Am. zoologist; b. Arcanum, O., Apr. 4, 1901; s. Charles and Alma (Bireley) B.; A.B., U. N.C., 1921, A.M., 1922; Ph.D., Johns Hopkins, 1925; m. Alma Holland, Sept. 16, 1941. Instr. biology Yale, 1925-26; research fellow Johns Hopkins, 1926-27; faculty U. N.C., Chapel Hill, 1927—, prof. zoology, 1937-60, prof. emeritus, 1960—. Fellow A.A.A.S.; mem. Am. Soc. Zoologists, Am. Soc. Parasitologists, Soc. Protozoologists, Am. Microscopical Soc., N.C. Acad. Sci. (Poteat award 1962). Contbr. numerous articles to profl. jours. on physiology of ciliated protozoa, host-parasite relations in sea-urchin ciliates, and cytochemistry of parasitic amoebae and ciliates. Home: 707 Gimghoul Rd., Chapel Hill, N.C. 27514.*

BEESON, W. Malcolm, Am. animal nutritionist; b. Meridian, Miss., Feb. 3, 1911; s. Malcolm Alfred and Effie (Harrison) B.; B.S., Okla. A. and M. Coll., 1931; M.S., U. Wis., 1932, Ph.D., 1935; m. Mary Louise Neff, Oct. 1, 1934; children—Joan (Mrs. Joseph W. Hemsky), Elizabeth (Mrs. Heber L. Short). Instr., A. and M. Coll. Tex., 1935-36; asst. prof. animal husbandry U. Ida., 1936-37, asso. prof., 1938-45; asso. prof. U. Ariz., 1937-38; asso. prof. animal sci., head animal nutrition research Purdue U., Lafayette, Ind., 1945-49, prof., 1949—, Lynn prof., 1959—. Chmn. com. on animal nutrition NRC, 1962—, chmn. subcom. on swine nutrition, 1953-64. Recipient Nutrition award Am. Feed Mfg. Assn., 1952; Outstanding Research award Okla. State U., 1953; CCC Fgn. Travel award, 1964; Morrison award, 1965. Mem. Am. Soc. Animal Sci. (past pres.), Am. Inst. Nutrition, Am. Dairy Sci. Assn., A.A.A.S., Internat. Union Nutritional Scis. (mem. U. S. nat. com.). Author: (with Julius E. Nordby, David L. Fourt) Livestock Judging Handbook, 1937; also numerous popular and sci. articles. Established amino acid and trace mineral requirements of swine; discovered growth stimulating effect diethylstilbestrol and antibiotics in cattle, factors contained in dehydrated alfalfa meal and distillers dried solubles for bacterial synthesis of protein from urea; devel. natural protein and high-urea supplements for beef cattle; isolated of unidentified growth factors from distillers dried solubles for swine and rat growth. Home: 1510 N. Grant St., West Lafayette, Ind. 47906. Office: Dept. Animal Sci., Purdue U., Lafayette, Ind. 47907.*

BEETON, Alfred Merle, Am. limnologist; b. Denver, Aug. 15, 1927; s. Charles Frederick and Edna (Smith) B.; B.S., U. Mich., 1952, M.S., 1954, Ph.D., 1958; m. Mary Eileen Wilcox, July 20, 1945; children—Maureen Ann, Heather Ann, Celeste Nadine; m. 2d Ruth Elizabeth Holland, June 4, 1966. Fishery biologist U. S. Bur. Comml. Fisheries, Ann Arbor, Mich., 1957-65, chief environmental research, 1960-65; prof. zoology, asst. dir. Center for Gt Lakes Studies, U. Wis., Milw., 1965—. Lectr. biology Wayne State U., 1957-61; lectr. civil engring. U. Mich., 1961-65; mem. research adv. com. Wis. Conservation Dept.; cons., U. S. Army C.E., 1967; Mem. Internat. Assn. Theoretical and Applied Limnology, Nat. Acad. Sci. (com. on eutrophication), Am. Assn. Limnology and Oceanography, Am. Soc. Zoologists, Internat. Assn. for Gt. Lakes Research (dir.), Sigma Xi. Contbr. chpt. to Limnology in North America, 1963. Research on oppossum shrimp Mysis relicta, Gt. Lakes eutrophication and limnology. Home: 2544 N. Frederick St., Milw. 53211.*

BEETZ, Friedrich Wilhelm Hubert von, see von Beetz, Friedrich Wilhelm Hubert.

BEEVERS, Harry, biologist; b. Durham, England, Jan. 10, 1924; s. Norman and Olive (Ayre) B.; B.Sc. with 1st class honors, Kings Coll., Durham U., Newcastle on Tyne, Eng., 1945, Ph.D., 1947; postgrad. Oxford (Eng.) U.; m. Jean Sykes, Nov. 19, 1949; 1 son, Michael Harry. Came to U. S., 1950, naturalized, 1958. Faculty, Purdue U., Lafayette, Ind., 1950—, prof. biology, 1958—; guest research collaborator Brookhaven Nat. Lab., 1953-55; vis. prof. Oxford U., 1956-57, Cambridge (Eng.) U., 1963-64. Cons. NSF, 1958—; Fulbright lectr., Australia, 1961. NSF Sr. fellow, 1963-64. Mem. Am. Soc. Plant Physiologists (past pres.), Soc. Biol. Chemists, Soc. for Exptl. Biology (Gt. Britain). Author: Respiratory Metabolism in Plants, 1961; also articles. Research on elucidation of plant respiration especially role of pentose phosphate pathway, established pathway of carbohydrate synthesis from fat in germinating seedlings, role glyoxylate cycle, biochem. aspects of germination, acid metabolism in plants. Home: 1710 Summit Dr., West Lafayette, Ind. 47906. Office: Dept. Biol. Scis., Purdue U., Lafayette, Ind. 47907.*

BEEVOR, Charles Edward, English neurologist; b. London, June 12, 1854; s. Charles and Elizabeth (Burrell) B.; ed. U. Coll., U. Coll. Hosp., London; M.B., 1879, M.D., 1882; postgrad. studies under Obersteiner, Weigert, Cohnheim and Erbat in Vienna, Leipzig, Berlin and Paris, 1882-83; m. Blanche Adine Leadam, Feb. 7, 1882; 1 son, 1 dau. Apptd. asst. physician Queen Square Hosp. and Gt. No. Hosp., 1885, became full physician and held position until his death; engaged in research (with Sir Victor Horsley) on localisation of cerebral functions, particullarly related to course and origin of motor tracts; lectr. to med. socs. Fellow Royal Coll. Physicians (Croonian lectr.); mem. Med. Soc. London (Lettsomian lectr.), Neurol. Soc. (pres. 1907-08), Assn. for Advancement of Medicine by Research (1st pres. corr. sect., hon. sec.). Author: Different Arteries Supplying the Brain, 1908; On Diseases of the Nervous System, 1898; also articles. One of greatest authorities on anatomy and diseases of nervous system; determined exactly the blood supply to different parts of brain; showed that distbn. of blood is purely anat. Died Dec. 5, 1908.

BEGAK, Maria Luiza, Brazilian cytogeneticist; b. Sao Paulo, Brazil, June 1, 1934; d. Lino Pires and Sylvia (Pires) de Camargo; B.Sc., M.Sc., U. Sao Paulo, 1956; m. Willy Begak, Sept. 6, 1959; children—Rubens, Rejane, Christina. Tchr. biology in high schs. and colls, Sao Paulo, 1956-58; research asst. in bacteriology Instituto Butantan, Sao Paulo, 1959-61, research asso. in cytogenetics, 1961—. Asst. fellow Inst. for Advanced Learning in Med. Scis., City of Hope Med. Center, Duarte, Cal., 1965—. Mem. Brazilian Soc. for Advancement Scis., Brazilian Soc. Genetics, Brazilian Soc. Biology. Editor, author (with husband) Biology, 4 vols., 1959. Contbr. numerous articles to profl. jours. Research on karyotypical evolution and sex determining mechanisms in ophidians and amphibians; discoverer natural bisexual species of tetraploid amphibians, cytological evidence of somatic segregation in sunfish, new chromosome trisomy in patients with congenital analgesia. Home: 2331 R. Alvarenga. Office: C.P. 65 Instituto Butantan, Sao Paulo, Brazil.*

BEGAK, Willy, geneticist; b. Mulhouse, France, Oct. 26, 1932; s. Szaja and Rosa (Marmelstein) B.; B.Sc., U. Sao Paulo, 1956, M.Sc., 1956, D.Sc., 1963; m. Maria Luiza Pires de Camargo, Sept. 6, 1959; children—Rubens, Rejane, Christina. Tchr. biology in high schs., colls. Sao Paulo, Brazil, 1954-58; research asso. in virology Instituto Butantan, Sao Paulo, 1956-60, head dept. genetics, 1960—; prof. med. genetics Faculty Med. Scis. of Santa Casa of Sao Paulo, 1964—; asso. mem. Inst. for Advanced Learning in Med. Scis., City of Hope Med. Center, Cal., 1963—. Mem. Brazilian Soc. Advancement Scis., Brazilian Soc. Genetics, Brazilian Soc. Biology, N.Y. Acad. Scis., A.A.-A.S. Author: (with Maria Luiza Becak) Biology, 4 vols., 1959; also articles. Discovered new chromosome trisomy in patients with congenital analgesia; discovery of heteromorphic sex chromosomes in ophidians; discovery of a natural bisexual species of tetraploid amphibians Odontophyrynus americanus; discovery of cytological evidences of somatic segregation in sunfish Lepomis cyanellus. Home: 2331 R. Alvarenga, Sao Paulo. Office: C.P. 65 Instituto Butantan, Sao Paulo, Brazil.*

BEGG, Alexander Swanson, Am. physician; b. Council Bluffs, Ia., May 23, 1881; s. Alexander Swanson and Lauretta (Slotterbeck) B.; B.S., Drake U., 1906; M.D., 1907; m. Grace Waers, 1908; children—John, Charles, Barbara. Instr. in pathology Drake U., 1907-09, asst. prof. pathology, histology and embryology, 1909-10, prof. histology and embryology, 1910-13; teaching fellow Harvard Med. Sch., 1911-12, instr. in comparative anatomy, 1913-18, dean Grad. Sch. Medicine, 1917-18, demonstrator in anatomy and instr. of histology, 1919-21; research asso., Carnegie Inst., 1915-16; prof. anatomy Boston U. Sch. of Medicine, from 1921, dean, from 1923. Served from lt. to col. Med. Res. Corps, 1917-40; active duty, office of surgeon gen. A.E.F., 1917-19. Died Sept. 26, 1940.

BEGG, Emil, meteorologist; b. Tartu, Estonia, Apr. 5, 1862; began studies in Tartu, 1881; candidate in polit. econs. and statistics, 1884; m. Anna Mottershead, 1888 (dec. 1890); m. 2d, Valerie Britzke, 1897; head dept. hydrometeorology Central Phys. Obs. St. Petersburg, Russia, 1885. Editor Ergebnisse der Niederschlagsbeobachtungen in Russland for 30 years. Organized Russian precipitation measurement network (2000 stas.) in 1914; these measurements were last dependable precipitation reports on Russia available to rest of world. Died Leningrad, Mar. 6, 1925.

BÉGHIN, Henri, French physicist, mathematician; b. Lille, France, 1876; ed. at École normale supérieure, 1921; became prof. at Montepellier (France) Faculty Scis.; in 1932 became prof. exptl. physics at Paris Faculty Scis., Ecole polytechnique; mem. French Acad. Scis., 1946; research on problems of friction; invented 1st French gyroscope compass usable in aircraft. Died 1946.

BEGOUEN, Henri (Comte), archeologist; b. Chateauroux, France, 1863; began career as journalist, later turned to study of prehistoric archeology; became prof. U. Toulouse (France), 1922. Discovered (with his 3 sons) the clay bisons of tuc d'Audoubert grotto, also engravings of Trois-Freres grotto (both in France). Died Ariège, France, 1956.

BEGUELIN, Nicolas de, see de Beguelin, Nicolas.

BÉGUILLET, Edme, French agronomist; b. Dijon, France, Oct. 13, 1729; counselor Dijon parliament; notary Bourgogne states; mem. French Acad. Scis., 1775. Author: Mémoire sur les avantages de la mouture économique, 1769; Dissertation sur liergot ou blé cornu, 1771; Traité général des subsistances des grains gui servent à la nourriture de l'homme, 1782. Studied culture of wheat, its uses, diseases and commerce; developed econ. method of milling wheat. Died May, 1786.

BEGUIN, Jean, French chemist; b. Lorraine, France, circa 1550; studied mining operations in Italy, Germany, Hungary; instr. chemistry, Germany; head sch. and lab. chemistry and pharmacy and metallurgy, Paris; almoner to Louis XIII. Editor of edition of Sendivogius' Novum lumen chymicum, 1608; Author: Tyrocinium chymicum e naturae fonte et mannali Experientia Depromptum (1st important chem. textbook; went through over 50 editions to 1669), 1610. Discovered calomel, 1608; probably 1st to mention acetone (obtained by distilling lead acetate; called burning spirit of Saturn) and other substances. Died circa 1620.

BÉGUYER DE CHANCOURTOIS, Alexandre-Émile, French geologist; b. Paris, France, Jan. 20, 1820; studied at École des mines; sent to Turkey; became prof. geology École des mines, 1875; apptd. insp. gen. mines, 1875. Author: Vis tellurique, 1864; l'Exploration géologique d'une partie de la turquie en Asie. Made geol. map of Haute-Marne, 1860; improved safety of anthracite mines; research on relationship of character of a people to their geol. habitat, demography; constructed a periodic table before Mendeleev (his method of presentation caused it to be ignored). Died Nov. 14, 1886.

BEHAIM, Martin, geographer, cartographer; b. Nuremberg, Bavaria; studied math. and astronomy under Regiomontanus. Mcht. in Antwerp, Belgium; travelled about Europe promoting Flemish trading interests; went to Lisbon, Portugal, circa 1481; apptd. by King John II to comm. to find better way of determining latitude; went with Diego Cam on exploratory voyage, W. coast of Africa, 1485-86; discovered mouth of Congo; prepared (probably with Hartmann Schedel) terrestrial globe, 1492; worked with astrolabe. Died Lisbon, Aug. 8, 1507.

BEHAL, Auguste, French chemist; b. Lens, France, Mar. 29, 1859; ed. Bethune, Paris; D.es.Sc. in Physics, 1888; also studied at Sorbonne, École Superieure de pharmacie; pharmacist-in-chief Maternity Hosp.; prof. organic chemistry Sch. Pharmacy, U. Paris, also Ecole de pharmacie. Mem. French Acad. Scis., 1921 (became pres. 1939), French Acad. Medicine, Soc. Pharmacy (pres.), Chem. Soc. France (pres.). Author: Modern Theories of Organic Chemistry; Nitrogen Derius and their Industrial Applications, 1889. Editor Dictionary of Pure and Applied Chemistry (Wurz). Research on acetylene carbide, acid anhydrides, malonic derius, phenolic ethers, chloral, camphors, creosote from wood. Died Mennecy, France, Feb. 2, 1941.

BÉHAR, Moises, Guatemalan physician; b. Huehuetenango, Guatemala, Aug. 28, 1922; s. Elías and Eugenia (Alcahé) B.; M.D., U. San Carlos, (Guatemala), 1949; postgrad. Sch. Medicine, U. Paris (France), Hopital des Enfants Malades, Paris France, Vanderbilt U.; M.P.H., Harvard, 1960; m. Beatriz Aldana, Aug. 14, 1954; children—Michelle, Jacqueline, Henri. Chief nutrition div. Pub. Health Service, Guatemala, 1951-53; asso. prof. hematology Sch. Medicine, U. San Carlos, 1951-55. Spl. cons. WHO in Kwashiorkor studies, 1951; cons. pediatrics Inst. Nutrition Central Am. and Panama, 1955-61, dir., 1961—; prof. nutrition in pub. health Columbia, N.Y.C., 1962—; vis. prof. nutrition Tulane U., 1962—; related dir. Gorgas Meml. Lab., Panama, C.Z., 1965—. Mem. Pediatric Assn. Guatemala (past pres.), Pediatric Assn. Costa Rica (corr.), Société Belge de Medicine Tropicale (corr.), Pediatric Soc. Panama (hon.), Am. Inst. Nutrition, Am. Assn. Pub. Health, Am. Soc. Pedatric Research, Fedn. Am. Socs. in Exptl. Biology, Am. Soc. Clin. Nutrition. Research, numerous publs. on characteristics, epidemiology, treatment and prevention protein calorie malnutrition in children. Home: Avenida Hincapié 25-10, Zona 13, Guatemala City, Guatemala. Office: Instituto de Nutrición de Centro América y Panamá, Carretera Roosevelt, Zona 11, Guatemala City, Guatemala, C.Am.*

BEHEM, Martin, see Behaim, Martin.

BÉHIER, Louis Jules, French physician; b. Paris, Aug. 26, 1813; intern under Guersant; Docteur en medecine, 1836. Named prof. pathology Paris Faculty Medicine, 1864; prof. clin. medicine Hôtel Dieu, Paris; physician Charité, Pitié, Hôtel Dieu hosps., Paris; physician to Louis Philippe. Decorated comdr. Legion Honor. Mem. Acad. Medicine. Author: De l'influence épidemique sur les maladies, 1844; Traité élémentaire de pathologie interne, 1844-58; Étude des mouvements et des bruits du coeur; Fissure congenitale du sternum, 1855; Origine de la gangrène des membres dans la fievre typhoide, 1857; Études sur la maladie dite fievre puerpérale, 1858; Conférences de clinique medicale, leçons faites à l'hopital de la Pitié, 1862; Pleuresies et épanchements modéris; Thoracentise avec trocarts capillaires et aspiration, 1873. Introduced treatment of typhoid with cold water in France; introduced English method of using alchohol to treat pneumonia; popularized new histological devels. in France; administered atropine and morphine as analgesics with J. F. B. Charrière's hypodermic syringe, 1859; performed successful blood transfusion at Hôtel Dieu, 1874. Died May 7, 1876.

BEHLE, William Harroun, Am. ornithologist, educator; b. Salt Lake City, May 13, 1909; s. Augustus Calvin and Daisy (Harroun) B.; A.B., U. Utah, 1932, M.A., 1933; Ph.D., U. Cal. at Berkeley, 1937; m. Dorothy M. Davis, July 11, 1934; children—Howard William, Raymond David. Faculty, U. Utah, Salt Lake City, 1937—, prof. zoology, 1951—. Fellow Am. Ornithologists Union, A.A.A.S.; mem. Cooper, Wilson ornithol. socs., Nat. Aubudon Soc., Utah Audubon Soc., Wildlife Soc., Am. Soc. Mammalogists, Soc. Study Evolution, Soc. Systematic Zoology, Utah Acad. Scis. Arts and Letters, Sigma Xi, Phi Sigma, Alpha Epsilon Delta, Phi Kappa Phi. Research, publs. on distbn., variation, ecology birds Western N. Am. Home: 1233 E. 8th South St., Salt Lake City 84102.*

BEHM, Alexander Karl, German physicist; b. Sternberg, Germany, Nov. 11, 1880; s. Anton and Paula (Prange) B.; m. Johanna Glamann, 1905; 1 son. Asst., Phys. Inst., Technische Hochschule, Karlsruhe, Germany; head, phys.-tech. research inst. Vienna, Austria; founder Behm-Echologt-Gesellschaft, Kiel, Germany, 1920. Author: numerous articles in tech. journals. Inventor echolot (sound plummet to determine ocean floor depth and track down obstacles), 1912. Died Tarp, Germany, Jan. 22, 1952.

BEHR, Lyell Christian, Am. chemist; b. Mpls., May 4, 1916; s. Christian and Elsie Alma (Schissler) B.; B.Chemistry, U. Minn., 1937; Ph.D., U. Ill., 1941; m. Patricia Ekander, June 5, 1954; children—Christopher, Barbara. Hormel postgrad. fellow U. Minn., 1941-42; chemist Chem. Warfare Service, Columbia, 1942-43, E. I. duPont de Nemours & Co., Inc., 1943-47; mem. faculty Miss. State U., 1947—, prof. chemistry, 1954—, head dept., 1963-64, dean Coll. Arts and Scis., 1964—. Mem. region VII selection com. Woodrow Wilson Fellowship Found., 1961—. Fellow A.A.A.S., Chem. Soc. (London, Eng.); mem. Am. Chem. Soc. (councilor), Sigma Xi, Blue Key, Alpha Chi Sigma, Phi Lambda Upsilon, Alpha Epsilon Delta, Omicron Delta Kappa, Phi Kappa Phi. Co-author: Brief Course in Organic Chemistry, 1959; Heterocylic Compounds, 1962, 66. Home: P.O. Box 644, State College, Miss. 39762.*

BEHRE, Ellinor H., Am. biologist; b. Atlanta, Sept. 28, 1886; d. Charles H. and Emilie (Schumann) Behre; A.B., Radcliffe, 1908; postgrad. Tulane U.; Ph.D., U. Chgo., 1918; children—(adopted) Emil, Charlotte (Mrs. Wendell B. Hatfield). Tchr. pvt. sch., Atlanta, 1908-12; faculty Newcomb Coll., New Orleans, 1912-15; staff Cold Spring Harbor Sta. for Exptl. Evolution, 1918-19; faculty Newcomb Coll., 1919-20; faculty La. State U., Baton Rouge, 1920-57, emeritus, 1957—, founder, dir. Marine Lab., Gulf Mexico, 1928-46; vis. prof. Mt. Holyoke Coll., 1936-37, Inter-Am. U., San Germán, P.R., 1964-65. Fellow A.A.A.S.; mem. Am. Soc. Zoologists, Am. Inst. Biol. Scis., La. State U., New Orleans, N.C. acads. sci., Sigma Xi, Phi Kappa Phi, Beta Beta Beta. Contbr. articles to tech. jours. Research on Gulf Mexico littoral fauna. Address: R.D., Black Mountain N.C. 28711.*

BEHREND, (Anton Friedrich) Robert, German chemist; b. Harburg (Elbe), Germany, Dec. 17, 1856; Ph.D., Leipzig, Germany, 1882; hon. dr. engring., Danzig, 1924. Became asst. Phys. Chem. Inst., 1881; joined 1st chem. lab. U. Leipzig, 1887; apptd. lectr. in chemistry, 1885, asso. prof. organic and phys. chemistry, 1889; became lectr. on organic chemistry Tech. U. Hanover (Germany), 1894, prof., 1895, permanent prof. organic chemistry, 1897, also of phys. chemistry until 1911, head organic chemistry lab., became emeritus, 1924. Recipient Gold medals, world's fairs of Chgo., 1893, St. Louis, 1904. Asso. mem. Saxon Soc. Scis. Leipzig. Research on synthesis of mucic and uric acids, derivatives of grape sugar (dextrose). Died Hanover, Sept. 15, 1926.

BEHREND, Bernard Arthur, elec. engr.; b. Villeneuve, Switzerland, May 9, 1875; s. Moritz and Rebecca (Wolf) B.; C.E., Univ. and Poly. Inst., Berlin, Germany, 1894; D. Eng. (hon.), Darmstadt, Germany, 1931; m. Margaret Plumer Chase, 1926. Chief engr. Bullock Electric Mfg. Co, Cin.; chief elec. engr., cons. engr. Allis-Chalmers Co., Allis-Chalmers-Bullock, Ltd., Montreal; adv. engr. Westinghouse Co. Nonresident lectr. McGill U., U. Wis., Leland Stanford Jr. U. Recipient Gold medal St. Louis Expn., for design for turbo-generators in modern form, 1904, John Scott medal, Franklin Inst., 1912. Mem. Am. Acad. Arts and Scis., Am. Inst. E.E. (sr. v.p.). Author: The Induction Motor—Its Theory and Design, 1900, in French, 1902, in German, 1903 (issued in new edit., 1922, under title, The Induction Motor and Other Alternating Current Motors); The Debt of Electrical Engineering to C. E. L. Brown, 1901; Engineering Education, 1907; The Work of Oliver Heaviside, 1928. Contbr. monographs to Am. and European jours. and trans. on theory of alternating current motors and generators. Inventor, patentee elec. devices and machinery; designer large machinery built by Bullock, Allis-Chalmers, Westinghouse cos. Died March 25, 1932.

BEHRENDT, John Charles, Am. geophysicist; b. Stevens Point, Wis., May 18, 1932; s. Allen Charles and Vivian (Frogner) B.; student Wis. State U., 1950-52; B.S. in Physics, U. Wis., 1954, M.S. in Geology, 1956, Ph.D. in Geophysics, 1961; m. Donna Miriam Ebben, Oct. 6, 1961; 2 sons, Kurt, Marc. Asst. seismologist Arctic Inst. N.Am., Ellsworth Sta., Antarctica, Filchner Ice Shelf Area, 1956-58; project asso. U. Wis., Antarctica, N.Am., S.Am., Europe, Asia, Australia, New Zealand, 1958-64; geophysicist Rocky Mountain area, Antarctica, U. S. Geol. Survey, Denver, 1964—. Recipient Antarctic Service medal for IGY, 1966. Fellow Royal Astron Soc.; mem. Am. Geophys. Union, Soc. Exploration Geophysicists, A.A.A.S., Glaciological Soc., Arctic Inst. N.Am., Am. Seismol. Soc., Colo. Sci. Soc. Contbr. articles to profl. jours. Geophys. studies of earth using gravity, seismology and earth magnetism, geologic and glaciologic structural investigations, standardization of world gravity base station network on all continents. Home: Box 171, Rural Route 2, Evergreen, Colo. 80439. Office: U. S. Geological Survey, Regional Geophysics Br., Federal Center, Denver 80225.*

BEHRENS, Harold, Radiochemist; b. Montevideo, Uruguay, July 28, 1925; s. Alfredo and María (LeBas) B.; student U. Uruguay, 1941-49, U. Exeter, Eng., 1950-51; Ph.D., U. Cambridge, Eng., 1954; m. Maria Rafaela Pellegrino, Oct. 13, 1951; children—Isabel, Ines, Margarita, Carmen, Veronica, Carolina. Asso. prof. physics Faculty Chemistry, U. Uruguay, 1946-54; prof. chemistry U. Concepcion, 1954-58; prof. gen. chemistry U. Chile, Santiago, 1958—, head dept. radiochemistry, 1960—. Mem. Chilean Chem. Soc. (dir. Santiago br.). Author: Review Guide to Users of Radioisotopes; Elementary Chemical Kinetics, 1965. Research, publs. on elucidation of mechanism of radioactive exchange in alkyl halides and soil fertilizer phosphorus uptake as well as electron exchange of thallous-thallic systems. Home: 9177 Las Condes. Office: Casilla 2777, Santiago, Chile.*

BEHRENS, Helmut, German inorganic chemist; b. Elsfleth, Germany, May 30, 1915; s. Wilhelm and Antonie (Roggemann) B.; ed. Freiburg (Germany) U., Munich (Germany) Inst. Tech.; D. Engring.; m. Liselotte Diess, 1940. Lectr. Munich Inst. Tech., from 1948, prof., 1956-60; prof. U. Erlangen-Nuremberg (Germany), 1962—. Mem. Gesellschaft Deutscher Chemiker, Royal Chem. Soc., Gesellschaft Deutscher Naturforscher und Aerzte. Research, publs. on metal carbonyls; investigations in liquid ammonia, co-ordination chemistry. Address: Institut für Anorganische Chemie II, Universitat Erlangen-Nuremberg, Fahrstrasse 17, 8520 Erlangen, Germany.

BEHRENS, Martin Gerhard, German biochemist; b. Giessen, Feb. 25, 1899; s. Dirk and Dorothea (Fahlberg) B.; Ph.D.; m. Elli Hartmann, Dec. 18, 1940; 1 son, Jürgen. Author: Isolierung von Zellkernen des Kalbsherzmuskels, 1932; Aufteilung der Leukozyten des Pferdeblutes nach der Grösse durch ein analog der Graigschen Gegenstromverteilung arbeitenden Trennverfahren, 1955; Gewinnung von Chloroplasten im nichtwässerigem Milieu, 1957; Gewinnung der eosinophilen Granulozyen des Pferdeblutes, 1962. Research on processes for isolation of organic elements in tissues of animal origin, 1929. Home: Aulweg 112. Office: Friedrichstrasse 24, Giessen, Germany.

BEHRENS, Rudolph August, physician; b. Germany, 17th century; Author: Examen aquarum mineralium Furstenau et Wechteldensium, 1724; De imaginario quodam miraculo in gravi oculorum morbo, ejusdemque spontanea atque fortuita sanatione, 1734. Described purpura hemorrhagica with the name morbus maculosus haemorrhagicus, 1735. Died. 1747.

BEHRING, Emil Adolf von, see von Behring, Emil Adolf.

BEHRING, Vitus, see Bering, Vitus.

BEIER, Ernst Gunter, psychologist; born Breslau, Germany, June 26, 1916; s. Paul and Hanna (Moses) B.; came to U. S., 1938, naturalized 1943; B.A., Amherst Coll., 1940; M.A., Columbia, 1947, Ph.D., 1949; m. Frances Redlich, Sept. 9, 1949; children—Paul, Lisa. Asst. prof. clin. psychology, head mental hygiene clinic Syracuse (N.Y.) U., 1948-53; prof. psychology, dir. clin. tng. program U. Utah, Salt Lake City, 1953—; cons. to VA Hosp., Salt Lake City, 1953—, Patton (Cal.) State Hosp., 1962—, Xerox Corp., 1965; practice psychology, Syracuse, 1949-53, Salt Lake City, 1953—. Diplomate Clin. Psychology, Am. Bd. Examiners in Profl. Psychology. Fellow Am. Psychol. Assn.; mem. InterAm. Soc. Psychology, Western Psychol. Assn. Author: The Silent Language of Psychotherapy, 1966; also articles. Research in communication processes among humans principally paralinguistic communication, gestures, as related to psychotherapy. Home: 1600 Michigan Av., Salt Lake City 84105.*

BEIER, Max Walter Peter, Austrian zoologist; b. Spittal, Drau, Austria, Apr. 6, 1903; s. Julius and Marie V. (Mitis) B.; Dr.phil., U. Vienna (Austria), 1927, Professor (hon.), 1961; m. Irmgard Zeitheim, May 7, 1931; children—Max, Irmgard, Gerhard-Peter. Staff. Mus. Natural History, Vienna, 1927—, dir. zool. dept., 1963—. Mem. Netherlands (hon.), Argentine (corr.), Finland entomol. soc. Research, numerous publs. on systematics of Pseudoscorpionidea, Mantidea and Pseudophyllinae; anat. and biol. studies on some insects. Home: 1 Proschkogasse, Vienna A-1060. Office: 7 Burgring, Vienna A-1014, Austria.*

BEIERWALTES, William Henry, Am. physician; b. Saginaw, Mich., Nov. 23, 1916; s. John Andrew and Fanny Elizabeth (Aris) B.; A.B., U. Mich., 1938, M.D., 1941; m. Mary Martha Nichols, Jan. 1, 1942; children—Andrew George, William Howard, Martha Louise. Faculty, U. Mich. Med. Center, Ann Arbor, 1944—, prof. internal medicine, 1959—, dir. Nuclear Medicine and Thyroid Research Lab., 1952—. Cons. to radioisotope service VA Hosp., Ann Arbor, 1953—. Mem. Soc. Nuclear Medicine (pres. 1965-66), Central Soc. for Clin. Research (councilor 1964—), Am. Thyroid Assn. (v.p. 1964-65), Assn. Am. Physicians, Am. Fedn. Clin. Research (pres. 1954-55).

Author: (with P. C. Johnson, A. J. Solari) Clinical Use of Radioisotopes, 1957; also numerous articles. Research on thyroid gland concerned with acid mucopolysaccharide accumulation, familial and hereditary biochem. abnormalities in cretinism and in thyroxine transport, sequestered nodular goiter, genesis of transplantable rat thyroid tumors; pioneer in photoscanning diseased kidney, myocardium, lung and parathyroid gland. Home: 1885 Fuller St. Office: 1405 E. Ann St., Ann Arbor, Mich. 48104.*

BEIGHTON, Henry, Brit. surveyor, engr.; b. Warwickshire, Eng., 1686; built steam engine with improved valve, Newcastle, 1718; prepared maps from his survey of Warwickshire, 1725-29; Fellow Royal Soc., 1720; editor Tipper's Ladies' Diary, 1713-43; contbd. papers on mech. design to Philos. Trans. Died Eng. Oct. 1743.

BEIGUELMAN, Bernardo, Brazilian geneticist; b. Santos, Brazil, May 15, 1932; s. Rael and Cecilia (Bilard) B.; student Faculdade Fil., Cienc. Letr., U. Sao Paulo. 1950-53, Ph.D., 1961; m. Sylvia Ehrenfreund, Aug. 8, 1961; children—Giselle, Evane, Lilian. Chmn. biology Faculty S is., Marília, Brazil, 1959-62; asso. research dept. gen. biology U. Sao Paulo, Brazil, 1962-63; chmn. dept. med. genetics Faculty Medicine, U. Campinas (Brazil), 1963—. Mem. Am. Eugenics Soc., Sociedade Brasileira de Genética, Sociedade Paulista de Leprologia, Sociedad Brasileira papa o Progresso da Ciência. Genetic and anthropol. studies and publs. on Japanese communities in Brazil, genetic mechanism of resistance to leprosy. Home: Rua Major Sertória, 557, 3°a. Office: Dept. Med. Genetics, U. Campinas, Caixa Postal 1170, Campinas, Estado Sáo Paulo, Brasil.*

BEIJERINCK, Martinus Willem, Dutch botanist; b. Amsterdam, Holland, 1851; student chemistry Delft Poly. Sch., Ph.D., 1877; Fellow Royal Soc., 1923; discovered tobacco mosaic disease was not caused by bacterial agt., 1898; used term filterable virus to describe infectious agt. Died 1931.

BEILBY, Sir George Thomas, Brit. indsl. chemist; b. Edinburgh, Scotland, Nov. 17, 1850; s. George Thomas and Rachel (Watson) B.; ed. Edinburgh U; hon. degrees, Glasgow, Birmingham; m. Emma Clarke Newman, 1877; 1 son, 1 dau.; became chemist Oakbank Oil Co., 1869; chemist Cassel Gold Extracting Co.; dir. Cassel Cyanide Co.. Castner-Kellner Alkali Co.; chem. governing body Royal Tech. Co., 1907-23; reported to Royal Commn. on Coal Supplies and Economy, 1902-04; mem. Royal Commn. on Fuel Oil Engines for Navy, 1912-13; dir., chmn. Fuel Research Bd., from 1917. Fellow Royal Soc., 1906; pres. Inst. Chemistry, 1909-12, Inst. Metals, 1916-18, chem. sect. Brit. Assn., S. Africa meeting, 1905; mem. Soc. Chem. Industry. Author: Aggregation and Flow of Solids, 1921. Inventor new process for retorting shale oil, 1881; devised synthetic process of manufacturing alkaline cyanides, also erected 1st factory for this purpose, Leith; experimented on crystalline and vitreous states of solids; made micros. studies of cell structure and properties of coke. Died Hampstead, nr. London, Aug. 1, 1924.

BEILSTEIN, Friedrich Konrad, Russian chemist; b. St. Petersburg (now Leningrad), Russia, Feb. 17, 1838; s. Friedrich B. and Katherine Margaret (Rutsch) B.; studied with Bunsen, Liebig, and Wöhler; ed. at Heidelberg (Germany), Göttingen (Germany) U. Asst. to Wöhler at Göttingen, 1860; prof. Inst. Tech., St. Petersburg, 1866-96; ret., 1896. Mem. Russian Imperial Acad. Scis., 1881. Author: Handbuch der organischen Chemie (describing thousands of organic compounds; in rev. and augmented form it remains standard reference book on organic chemistry), 2 vols., 1880-83. Extensive research in analytical and organic chemistry. Died Oct. 18, 1906.

BEISSEL, (Karl Christian) Stephen (Hubert), archaeologist; b. Apr. 21, 1841; s. Stephan and Elise (Jeghers) B.; student theology Bonn, Münster, Germany. Ordained priest Roman Catholic Ch., 1864. joined Soc. of Jesus, 1871. Author: Geschichte der Verehrung der Heiligen und ihrer Reliquien in Deutschland Wahrend des Mittelalters, 1885-92; Bilder aus der Geschichte der altchristlichen Kunst und Liturgie in Italien, 1900; Geschichte der Verehrung Marias in Mittelalter, 1909; Geschichte der Verehrung Marias im 16. und 17. Jahrhundert, 1910; Betrachtungs punkte für alle Tage des Kirchenjahres, 1907-11. Noted for works on archtl. history of ch. St. Victor, Xanten, and on worship of saints in medieval Germany. Died Valkenburg, Netherlands, July 30, 1915.

BEIT, Ferdinand, German chemist; b. Hamburg, Germany, July 27, 1817; s. Phillip Raphael and Philippine (Feidel) B.; student Polytechnikum Karlsruhe (Germany), later studied medicine in Munich, Germany; M.D., Berlin, Germany; m. Johanna Seligmann, 1850; children—Karl, Gustav, Eduard. Began as co-owner R. L. Beit & Co. founded (with brother Siegfried) the Chemie und Farbenfabriken, dealing in wholesale chems. Author: Über der Preisdifferenz des in der Hamburger Bank ein- und ausgehenden Silbers, 1845. Research on separation and refinery of gold and silver alloys especially safety against dangerous effects of sulphuric acid. Died Hamburg. Apr. 1, 1870.

BEKE, Charles Tilstone, Eng. geographer, explorer; born Stepney, Eng., October 10, 1800; studied law at Lincoln's Inn; Ph.D., University of Tübingen, Germany, 1834; m. grandniece of Sir J. Herschel; m. 2d, Emily Alston, 1856. Entered bus. career, 1820; Brit. acting consul, Leipzig, Germany, 1837-38; journey to Abyssinia, 1840; traveled to Syria and Palestine, 1861-62; went on mission to urge release of Brit. prisoners to King Theodore of Abyssinia, 1864; explored supposed location of Mt. Sinai, 1873-74. Fellow Soc. Antiquaries, Royal Geog. Soc., Statis. Soc. London, Syro-Egyptian Soc. London, Oriental Soc. Germany, Asiatic Soc. Recipient gold medals Royal Geog. Soc. London, Royal Geog. Soc. Paris. Author: Origines Biblicae or Researches in Primeval History, 1834; Discoveries of Sinai in Arabia and of Median; also articles on Oriental subjects. First to attempt the reconstruction of history using geology; investigated phys. structure of Abyssinia and Eastern Africa; demonstrated the Mountains of the Moon of Ptolemy are a portion of the meridional range; fixed latitude of numerous stas. using astron. observations; discovered river Gojeb, and depression of the Salt Lake Assal; constructed map of Gojam and Damot. Died Bromley, Eng., July 31, 1874.

BEKENCHONS, Egyptian engr.; b. 1293 B.C.; ed. mil. sch.; lived under Rameses II; priest Temple of Aman; erected granite obelisks. Died 1225 B.C.

BÉKÉSY, Georg von, see von Békésy.

BEKHTEREV, Vladimir Mikhailovich, Russian neuropathologist; b. Viatka Province, USSR, Jan. 22, 1857; Ph.D., Mil. Med. Acad., St. Petersburg (now Leningrad, USSR), 1881; postgrad. Leipzig, Berlin (both Germany) univs., U. Paris (France). Prof. mental diseases U. Kazan (USSR), from 1885; prof. mental and nervous diseases Mil. Med. Acad., St. Petersburg, from 1893; founder Psycho-Neurol. Inst., St. Petersburg, 1907, later dir.; founder Inst. for Study of Brain and Psychic Activity, 1918. Author: The Nerve Currents in Brain and Spinal Cord, 1882; The Functions of the Brain, 1903-07; General Diagnosis of the Diseases of the Nervous System, 1911, 15; Hypnosis, Suggestion and Psychotherapy, 1911; Collective Reflexology, 1921; Conduction Paths of Spinal Cords and Brain, 1926. Founder Nevrologui cheshiy vestnik (Jour. Neurology), 1885. Important contbns. to knowledge of anatomy, physiology of nervous system; research on localization of functions of brain; famous for work on nerve currents; identified layer of fibers in cerebral cortex (Bekhterev's fibers), 1891, ankylosing spondylitis (Bekhterev's disease), 1892, nucleus of gray matter which gives origin to fibers of roots of auditory nerve (Bekhterev's nucleus), 1892; proved that motor areas of cerebral cortex are base of individually acquired learned movements; established number of reflexes (both normal and path.); developed methods treatment for various mental disorders; later interested in parapsychology, studied thought suggestion. Died Moscow, USSR, Dec. 24, 1927.

BEKIERKUNST, Adam, microbiologist; b. Ozorkov, Poland, Mar. 19, 1914; s. Stanislav and Regina (Wisniewska) B.; M.Sc., U. Wroclav (Poland), 1948, Ph.D., 1950; m. Shulamith Szenfeld, July 22, 1941; 1 dau., Ruth Joanna. Sr. asst. dept. med. microbiology U. Wroclav Med. Sch., 1947-53; dir. dept. bacteriology and antibiotics Inst. Immunology and Exptl. Therapy, Wroclav, 1953-54; dozent Silesian Med. Acad. dept. med. microbiology, 1954-57; faculty Hebrew U., Hadassah Med. Sch., Jerusalem, Israel, 1958—, asso. prof., 1961—. mem. Am. Soc. for Microbiology. Research, numerous publs. on chemotherapy of bacterial infections, resistance to bacterial infections and biochem. aspects of host-parasite relationship. Home: 5, Rhov Shlain, B. Hakerem, Jerusalem, Israel.*

BEKKER, Johann Heinrich, Dutch bacteriologist; b. Amsterdam, Netherlands, Nov. 12, 1908; s. J. H. and C. (Best) B.; ed. U. Amsterdam; m. J. ter Heide, 1936. Bacteriologist, Nat. Inst. Pub. Health, Utrecht, Netherlands, dir. lab. bacteriology and serology, 1948—. Netherlands rep. various sci. congresses; mem. commn. of experts WHO. Mem. Am. Assn. Venereal Diseases. Co-author 2 manuals on bacteriology, also numerous articles in sci. jours. Address: Frans Halsstraat 63, Utrecht, Netherlands.*

BEKKI, Atsuhiko, Japanese geographer; b. Tokyo, Japan, 1908; grad. Kyoto U., 1932; prof. preparatory dept. Osaka Comml. U.; then dir. Inst. Study of Cultures of South Seas Area under Japanese Forces, Java; later apptd. prof. Fukui U., Kobe Comml. U., prof. geography Rikkyo U. Author: Civilization of the South; Civilization of the North. Research specialty topography of South-East Asia.

BEKKUM, Dirk Willem van, see van Bekkum.

BEKLEMISHEV, Nikolai Dmitrievich, Russian pathologist; b. 1915; grad. Vilnius U., 1939; D.Med. Sci., 1955. Dep. dir. Inst. Regional Pathology. Kazakhstan Acad. Sci., 1957—. Mem. Kazakhstan Acad. Sci. (corr.), All-Union Soc. Microbiologists, Epidemiologists and Infectionists (bd. mem., chmn. Consultative and Methods Commn. Kazakhstan br.). Author over 50 works including Chronic Brucellosis, 1958; Cortisone and Corticosteroids in Clinic, 1963; Chronic and Latent Brucellosis, 1965; Infectious Allergy, 1968. Research on diagnosis and treatment of brucellosis, allergology. Address: Inst. of Regional Pathology, Kazakhstan Acad. Sci., Alma-Ata 65, Kazakhstan SSR, USSR.

BEKRITSKY, Arkadii Arkadevich, Russian otorhinolaryngologist; b. Torzhok (now Kalinin Oblast), 1881; grad. Med. Faculty, Berlin U., 1910. Asst. various German otorhinolaryngol. clinics, 1909-11; intern dept. ear, nose and throat diseases Staroekaterina Hosp., Moscow, 1912-18; asst., later lectr. dept. otorhinolaryngology Central Postgrad. Med. Inst., 1922-38; head dept. ear, nose and throat diseases Mowcow Med. Stomatological Inst., 1938—; prof., 1939—. Mem. All-Union, Moscow socs. otorhinolaryngologists. Author over 50 works including Meninigitis of Aural Origin, 1938. Mem. editorial council Vestnik otorinolaringologii. Address: Moscow Med. Stomatological Inst., Kalyaevskaya ulitsa 18, Moscow, USSR.

BEKTUROV, Abiken Bekturovich, Russian chemist; b. 1901; grad. Siberian Inst. Agr. and Forestry, Omsk, 1931. Instr., Kazakhstan U., Alma-Ata, 1935-46, prof., 1946—; chmn. dept. mineral resources Kazakhstan Acad. Sci., 1946-54, dir. Inst. Chem. Sci., 1946—. Mem. Kazakhstan Acad. Sci. Author: Research on the Chemistry and Chemical Technology of Thermophosphates, 1947; co-author: Potash Thermophosphates, 1953; Physicochemical Study of the Saline Resources of the Chul-Adyr Deposits, 1955. Research on chem. tech. of minerals especially phosphates. Address: Kazakhstan University, Komsomolskays ulitsa 96, Alma-Ata, Kazakhstan SSR, USSR.

BELANGER, Leonard-Francis, Canadian physician; b. Montreal, Que., Can., Mar. 11, 1911; s. Joseph Amedee Deus and Frances Ann Kathleen (Leonard) B.; student Bourget Coll., Rigaud, Que., 1927-31; B.A., U. Montreal, 1931, M.D., 1937; M.A., Harvard, 1940; m. Marie Cecile Lefebvre, June 11, 1938; children—Nicole (Mrs. Michael Sakellaropoulo), Richard. Practice gen. medicine, Lacolle, Que., 1937; asst. dept. histology U. Montreal, 1938, asst. prof. histology and embryology, 1941; fellow Govt. Que. and Rougier-Armandie Found. Paris, France, 1939; research asso. dept. anatomy McGill U., 1945-46; organizer, head dept., 1st prof. dept. histology and embryology U. Ottawa Sch. Medicine, Ont., Can., 1946—, exec. council faculty medicine, 1957—. Mem. adv. bd. med. div. Nat. Research Council Can., 1950-54, asso. com. on dental research, 1959-65; chmn. sci. div. Nat. Film Inst., 1952-53; mem. clin. research com. Ont. Cancer Treatment and Research Found., 1955—; guest speaker various symposia, socs., confs., 1955—; in charge autoradiography course FAO Internat. Tng. in Isotope Techniques, Cornell U., Ithaca, N.Y., 1959, 62, instr. autoradiography Coll. Vet. Medicine, 1962-64; vis. prof. anatomy U. B.C., Vancouver, 1963-64. Fellow A.A.A.S., Royal Soc. Can.; mem. Am. Assn. Anatomists, Canadian Physiol. Soc. (treas. 1951-54, council 1954-57, chmn. symposium on cell ultra structure 1956), Histochem. Soc., Ottawa Biol. Soc. (1st pres.), Canadian Assn. Anatomists (sec. 1959-62). Asso. editor Canadian Jour. Biochemistry and Physiology, 1958, Jour. Histochemistry and Cytochemistry, 1960-64. Contbr. articles to profl. publs. Pioneer work in autoradiography, microradiography; research on sulphur and protein metabolism in bone, teeth, other tissues, life cycle bone cells, functional significance osteocytes. Home: 395 Templeton St., Ottawa 2, Ont., Can.*

BELARDINELLI, Giuseppe, Italian mathematician; b. Lesi, Ancôme, July 1, 1894; s. Domenico and Luisa (Montelli) B.; Ph.D. in Sci. and Math., U. Bologna. Prof. infinitesimal calculus and higher math. U. Milan (Italy). Mem. Istituto Marchigiano di Scienze, Lettere ed Arte, Istituto Lombardo di Scienze e Lettre (corr.). Contbr. numerous articles on math. analysis to profl. publs. Address: via Antonio Smarglia 9, Milan, Italy.

BELCHER, Donald Jenks, Am. civil engr.; b. Chgo., Feb. 11, 1911; s. O. Clifford and Helen (Jenks) B.; B.S. in Civil Engring., Purdue U., 1934, M.S., 1940, M.S.E., 1941, Ph.E., 1942; m. Nancy Foote, July 1, 1954; children—Marilyn Kay, Mary Candace, Mathew Brady, Mark Douglas, Neil Francis, Helen Stacey (dec.). Asst. prof., research engr. Purdue U., Lafayette, Ind., 1937-46; prof. civil engring., Cornell U., Ithaca, N.Y., 1947—; vis. lectr. Harvard, 1961—; pres. D. J. Belcher & Assos., Inc., Ithaca, 1947—; dir. Cornell Center for Aerial Photographic Studies, 1949—. Cons. Johns Hopkins Operations Research Office, 1950-54, govts. of Brazil, Burma, Columbia, Spain, Nigeria; chief UN Mission, Iran, 1950; cons. Calcutta Plan Commn., 1962-63. Recipient award Hwy. Research Bd., 1946; Civilian award medal UN, Korea, 1953. Author: Formation, Distribution and Engineering Characteristics of Soils, 1943; Formation, Distribution and Airphoto Identification of U. S. Soils, 1949; Analysis of Aerial Photographs, 1966; also articles. Research in uses of air photography for sci. purposes, soil identification, econ. geology, oil exploration, demographic surveys, transp. and land use evaluation; patentee instrument for measuring soil moisture and density using fast neutrons and gamma rays. Home: 1044 Cayuga Heights Rd., Ithaca, N.Y. 14850.*

BELCHIER, John, Brit. surgeon; b. Kingston, Eng., 1706; ed. Eton, St. Thomas's Hosp.; surgeon Guy's Hosp., from 1736, later gov.; gov. St. Thomas' Hosp.; Fellow Royal Soc., 1732; contbd. papers to Philos. Trans.; stained bones of living fowl by feeding madder-soaked bran which aided in study of osteogenesis, 1738. Died 1785.

BELDAMANDI, Prosdocimo de, see de Beldamandi, Prosdocimo.

BELDEN, Charles Dwight, Am. physician; b. Boonton, N.J., Feb. 16, 1845; s. Henry and Caroline (Wilcox) B.; ed. Williams Coll., 1861-62; served in Civil War, 1862-65; Coll. Phys. and Surg. (Columbia), 1865-66; M.D., N.Y. Homoe. Med. Coll., 1868; m. Mary E. Noble, Feb. 21, 1866; m. 2d, Katinka, Countess de Rudzinski, Apr. 26, 1907. In N.Y. banking firms, 1875-82. Discoverer of therapeutic use of venom of Gila monster for paralysis, locomotor ataxia and kindred disorders. Author: Orations and Addresses, 1902. Died July 27, 1919.

BELDING, Harwood Seymour, Am. physiologist; b. Simsbury, Conn., May 22, 1909; s. Anson Wood and Mary (Miller) B.; B.A., Wesleyan U., 1931; M.A., U. Conn., 1935; Ph.D., Stanford, 1949; m. Lola Dale Selley, Nov. 11, 1933; children—Marcia Dale (Mrs. David S. Watson), Jeffrey Harwood, William Anson. Instr. to asso. prof. U. Conn., Storrs, 1932-45; with fatigue lab., bus. sch. Harvard, Cambridge, Mass., 1942-46, asst. prof. indsl. physiology, 1945; research dir. Army Q.M. Climatic Research Lab., Lawrence, Mass., 1946-50; prof. environmental physiology, dept. occupational health, grad. sch. pub. health U. Pitts., 1951-—. Mem. Army Sci. Adv. Panel, Nat. Research Council bd. on Mil. Supplies and Equipment, subcom. on thermal factors in environment; adv. com. Army Surgeon Gen. on Environmental Medicine; U. S. adviser OEEC, 1955-59. Mem. Am. Physiol. Soc., Am. Indsl. Hygiene Assn., Ergonomics Soc. (Eng.). Editor, Ergonomics Jour., 1957-—, Jour. Applied Physiology, 1961-—. Research, publs. on thermal exchange and physiol. adjustments of man to environment. Home: 5818 Howe St., Pitts. 15232.*

BELEVTSEV, Yakov Nikolaevich, Russian geologist; b. Village of Orekhovo (now Kursk Oblast), 1912; grad. Dnepropetrovsk Mining Inst., 1937; D.Geol. and Mineral. Sci., 1952. With indsl. geol. instns., 1937-52; prov. Kiev U., 1953-—; dept. head Inst. Geol. Sci., Ukrainian Acad. Sci., 1953-—. Decorated Order of Lenin. Mem. Ukrainian Acad. Sci. (corr.). Author over 50 works (including monographs): Deep Oxidation Zones in the Rocks of the Krivoy Rog Basin, 1959. Research on geology and genesis of iron ores in Krivoy Rog Basin. Address: Kiev University, Vladimirskaya 64, Kiev, Ukraine SSR, USSR.

BELFRAGE, Gustaf Wilhelm, zoologist, entomologist; b. Sweden, 1834; came to Tex., 1867, lived there many years; collected groups of insects in Tex. area (many were described by Stal and other systemists). Died 1882.

BELGRADO, Giacopo, Italian mathematician; b. Udine, Italy, Nov. 16, 1704; ed. at Padua, Italy; became Jesuit, 1723; tchr., Venice, Italy, Parma, Italy; became confessor to Duke Philippo, also ct. mathematician, 1742; built obs., Parma, 1757; apptd. rector coll., Bologna (Italy), 1773. Mem. Acad. Scis. Paris (corr.), Inst. Bologna. Author: Ad disciplinam mechanicam, nauticam et geographicam acroasis critica et historica, 1741; De altitudine atmosperae aestimanda critica disquisitio, 1743; De gravitatis legibus acroasis physico-mathematica, 1744; De corporis elasticis disquisit, physicomathem, 1747; Della reflessione de corpi dall'acqua, 1753; Observatio defectus lunae, 1761; Delle senzazioni del calore, et del freddo, 1764; Il Trono di Nettuno illustrato, 1766; Theoria cochleae Archimedis, 1767; Dissertazione sopra i torrenti, 1768. Research on electricity, heat, geometry. Died Udine, Mar. 26, 1789.

BELGRAND, (Marie-François) Eugène, French hydrographic engr.; b. Ervy, France, Apr. 23, 1810; ed. at École polytechnique, École des ponts et chausées. Named insp. gen. bridges and hwys., 1874; became service dir. water and sewers City of Paris, France 1867; placed in charge of nav. of Seine between Paris and Rouen, 1852; dir. hydrometric system of the Paris Basin. Decorated cross Legion of Honor, 1871. Mem. French Acad. Scis., 1871, Acad. Library. Author: Études préliminaires, la Seine, régime de la pluie, des sources, des eaux courantes applications à l'agriculture, 1873; Les Travaux souterrains de Paris, les eaux, les aqueducs, 1875; Notice sur l'aqueduc romain de Sens; les Eauf anciennes de Paris, 1877; also numerous articles. Installed sewage system in Paris; built reservoirs of Montsouris; research on flow of Seine, also developed means of predicting its level for 3 days in advance; studied geology of Paris basin. Died Paris, Apr. 8, 1878.

BELIDOR, Bernard Forest de, see de Belidor, Bernard Forest.

BELIN, Jean Etienne Fernand, French sociologist; b. Nevers, Mar. 26, 1909; s. François and Valentine (Colin) B.; ed. schs. letters, sci. and law at Paris, sch. polit. sci., Collège de France, Sch. Advanced Studies, sch. sci. at Poitiers, Nat. Mus. History; Ph.D. in letters, sci. law; m. Félicité Godineau-Thareau, Dec. 28, 1939; children—Chantal, Etienne, Philippe, André, Solange, Marguerite. In charge of mission to cabinet of Ministry of Economy; prof. Lycée at Poitiers; prof. univs. of Bordeaux, Strasbourg, Dijon; in charge mission to Nat. Center Sci. Research; prof. Sch. Advanced Social Studies, Sch. for Advanced Internat. Studies, Sch. Anthropology. Pres. of Com. for Study of Man, Group for Study of Knowing for Understanding. Named laureat Acad. Moral and Polit. Scis.; recipient Sailliet prize. Mem. Louis Marin Inst. Ethical Research (pres.), French Inst. Sociology, Assn. French Speaking Sociologists, Social Mus., Soc. Ethnography, Soc. History of 3d Republic, Soc. Community Geography, French Group of Historians of Sci., French Soc. Cybernetics, Assn. for Devel. World Law, Syndicate of Sci. Press, Soc. History of French Revolution (past mem. com.), Sci. Soc. Clamecy, Soc. History of Yonne, Soc. Study of Toucy. Author: La conscience contemporaine et ses problèmes devant les faits; Les bases psychologiques de l'ordre social; La réforme de la connaissance; L'idée d'utilité sociale et la Révolution française; Terres et siècles; La Science nouvelle et les mécanismes politiques Promoter of reform of consciousness, notion of semantic cross-roads frequencies (with models). Address: 25, rue Gassendi, Paris 14, France.*

BELINFANTE, Frederik Jozef, physicist; born The Hague, Netherlands, January 6, 1913; the son of Johan Jacob and Louise E. (Ahn) B.; B.S., Rijksuniversiteit, Leiden, Netherlands, 1933, M.S., 1936, Ph.D. in Physics, 1939; m. Wilhelmina F. M. Beukers, Jan. 25, 1937; children—Hillegonda (Mrs. John C. Garrison), Johan Gijsbertus Frederik, Alexander Erik Ernst. Came to U. S., 1948, naturalized, 1955. Asst., Inst. for Theoretical Physics, U. Leiden, 1936-46, lectr., 1945-46; asso. prof. U. B.C., Vancouver, Can., 1946-48; vis. prof. Purdue U., Lafayette, Ind., 1947, asso. prof., 1948-51, prof. physics, 1951-—. Invited participant internat. sci. confs.; Am. rep. Internacia Scienca Asocio Esperantista, 1949-57. Recipient 1st award Gravity Research Found., 1956. Fellow Am. Phys. Soc., Ind. Acad. Sci.; mem. Soc. Engring. Sci. (charter), Am. Assn. Physics Tchrs., Nederlandse Natuurkundige Vereniging, Sigma Xi, Sigma Pi Sigma. Research publs. on relativistic quantum field theory, quantum electrodynamics, gravity theories. integrocausal quantum field theory, cosmology, Brownian motion in fluids; developed statis. reformulation 2d law thermodynamics, 1962. Home: 1809 Ravinia Rd., West Lafayette, Ind. 47906. Office: Dept. Physics, Purdue U., Lafayette, Ind. 47907.*

BELING, Willard A(dolph), Am. social scientist; b. Gt. Bend, N.D., Mar. 16, 1919; s. Adolph W. and Sadie (Worner) B.; B.A., U. Cal., Los Angeles, 1943; M.A., Ph.D., (teaching fellow), Princeton, 1947; m. Betty M. Melberg, Feb. 23, 1947; children—Janna R., Kristen L. Fellow Am. Council of Learned Societies Social Sci. Research Council, 1948; dir. socio-economic research div. Arabian Am. Oil Co., Saudi Arabia, 1949-58; research asso. Harvard, 1958-60; prof. internat. relations U. So. Cal., 1960-—, coordinator Middle East North African program, 1960-—, dir. Inst. World Affairs, 1965, 1967; cons. asso. Institut Internationale d'Etudes Sociales, Geneva, Switzerland, 1966-—. Fellow Social Sci. Research Council, Am. Council Learned Societies; mem. Am. Polit. Sci. Assn., Western Polit. Sci. Assn., Middle East Inst., African Studies Assn., Middle East Studies Assn., Phi Beta Kappa, Delta Phi Epsilon. Author: Pan-Arabism and Labor, 1960; Modernization and African Labor, 1965; editor, contbr. The Role of Labor in African Nation Building, 1967; editor (with G. O. Totten) Models and Developing Nations, 1967. Research and publs. on associational interest groups, internat. labor movements; devel. problems in the new nations of the world. Office: U. So. Cal., Los Angeles 90007.*

BELISARIO, John Colquhoun, Australian dermatologist; b. Sydney, Australia, Apr. 30, 1900; s. Guy Alexander Fernandez and Isobel (Colquhoun-Fraser) B.; M.B., Ch.M., U. Sydney, 1926, D.D.M., 1947, M.D., 1950; m. Freda Adele Sauber, Nov. 10, 1930; children—Mrs. Owen Grose, Dolores. Hon. med. officer, venereal dept. Royal Prince Alfred Hosp., Sydney, 1929-41, hon. physician diseases of skin, 1944-60, apptd. hon. cons. physician, 1960; hon. physician New S. Wales Masonic Hosp.; hon. dermatologist Renwick Hosp. for Infants, 1932-45; hon. cons. dermatologist Royal Hosp. for Women, Sydney, 1945-53; lectr. dermatology U. Sydney, 1945-61. Mem. research com. New S. Wales Cancer Council, 1957; mem. com. nomenclature and classification XI Internat. Congress Dermatology, Stockholm, 1957. Fellow Royal Soc. Medicine, Am. Acad. Dermatology and Syphilology (non-resident); mem. Brit. Assn. Dermatology (pres. New S. Wales br. 1945-46), Soc. Investigative Dermatology, U. S., N.Y. Acad. Scis.; hon., fgn. or corr. mem. German, New Zealand, Polish, Austrian, Israeli, Yugoslavian, Indian, Norwegian, Canadian, Brazilian, Netherlands, Mexican socs. or assns. dermatology, other med. socs. Author: Cancer of the Skin, 1959; also articles. Editorial bd. Exerpta Medica, 1957; hon. cons. editorial bd. Modern Medicine of Australia, 1958; bd. editors and collaborators Dermatologica, Basel, 1958; editorial bd. Acta Dermato-Venereologica; editorial adv. bd. Voice of Medicine. Research on electro-chemosurg. therapy of skin cancer with vitamin K5, topical cytotoxic therapy of skin cancer with combined cytotoxic agts. colcemid, methotrexate, thiocolciran. Home: 8 Rupertswood Av., Sydney. Office: 143 Macquarie St., Sydney, New S. Wales, Australia.*

BELITSER, Vladimir Aleksandrovich, Russian biochemist; born September 30, 1906; graduated from Moscow University, 1930; D.Biol. Sci. Assistant, 2d Moscow Medical Institute, 1930-35; with All-Union Inst. Exptl. Medicine, 1934-43, Lab. Physiol. Chemistry, USSR Acad. Sci., 1943-44; head Lab. Enzymes, Inst. Biochemistry, Ukrainian Acad. Sci., 1944-—. Mem. Ukrainian Acad. Sci. Author: Chemical Conversion in the Muscle, 1940; Action Mechanism of Enzymes, 1950; The Nature of the Conversion of Fibrinogen into Fibrin, 1952; Denaturization and its Concomitant Protein Changes, 1953; Denaturized Conversion of Proteins, 1955; Eliminating the Antigenic Properties of Serum Protein by a Combination of Partial Hydrolysis and Heat Processing, 1957; Structure of Proteins, 1960. Co-editor Chemistry sect. Large Med. Ency., 2d edit.; mem. editorial council Problems Med. Chemistry. Research on mechanism of cell respiration. Address: Ukrainian Acad. Sci., Vladimirskaya ulitsa 54, Kiev, Ukraine SSR, USSR.

BELKIN, John Nicholas, entomologist; born Petrograd, Russia, October 24, 1913; s. Nicholas Paul and Ina (Tardent) B.; came to U. S., 1928, naturalized, 1938; student Harvard, 1931-33; B.S., N.Y. State Coll. Agr. at Cornell U., 1938; Ph.D., Cornell U., 1946; m. Lorraine Lyla Marvin, Oct. 27, 1950; children—Nicholas J., Tanya, Natasha, Laura Therese, Michael J., Elisabeth J. Jr. entomologist TVA, 1942; asst. research specialist Rutgers U., 1946; asso. prof. biology Asso. Colls. Upper N.Y., 1946-49; head dept. biology Mohawk Coll., 1946-48; faculty U. Cal. at Los Angeles, 1949-—, prof., entomologist, lectr. pub. health, 1958-62, prof. zoology, 1962-—. Research grantee NSF, 1955-—, NIH, 1962-—, U. S. Army Med. Research and Devel. Command, 1963-—. Fellow A.A.A.S., Entomol. Soc. Am. (editorial bd. 1952-54, chmn. sect. A 1958); mem. Am. Soc. Zoologists, Am. Soc. Parasitologists, Soc. Systematic Zoologist (council 1962-—, sec.-treas. Pacific sect. 1952-55, pres. 1955-56), Sigma Xi, Phi Kappa Phi. Author: Mosquitoes of the South Pacific, 2 vols., 1962. Research, publs. on taxonomy and biology of mosquitoes of world, med. entomology, zoogeography. Home: 3631 Tilden Av., Los Angeles 90034.*

BELL, Agrippa Nelson, Am. sanitarian; b. Northampton County, Va., Aug. 3, 1820; s. George and Elizabeth (Scott) B.; student Tremont Med. Sch., Boston, Harvard Med. Sch.; M.D., Jefferson Med. Coll., Phila., 1842; A.M. (hon.), Trinity Coll., Hartford, 1860; m. Julia Ann Hamlin, Nov. 22, 1842. Asst. and past asst. surgeon USN, 1847-55; resigned; discovered steam as disinfectant in yellow fever, 1848, 1st used it nr. Vera Cruz, located in Bklyn., 1855, active in quarantine reform movement; mem. Quarantine Convs., 1857, 58, 59, 60, chmn. Com. on External Hygiene (reported system of quarantine regulations upon which subsequent regulations have been founded); supt. floating hosp. for yellow fever, N.Y. lower bay, 1861-62; drew substance N.Y. Quarantine Law, passed 1863; supervising commr. quarantine, N.Y., 1870-73; insp. quarantine Nat. Bd. Health, 1879; had charge of yellow fever extermination, New Orleans and Memphis, 1879. A founder Am. Pub. Health Assn. Author: Knowledge of Living Things, 1860; Records of Daily Practice—Scientific Visiting List, 1860; The Climatology and Mineral Waters of the United States, 1885. Founder, editor, pub. The Sanitarian, 52 vols., 1873-1904. Died 1911.

BELL, Alexander Graham, scientist, inventor; b. Edinburgh, Scotland, Mar. 3, 1847; s. Alexander Melville and Eliza Grace (Symonds) B.; ed. at Edinburgh and London U.; hon. Ph.D., Wurzburg, 1882; M.D., Heidelberg, 1886; LL.D., Harvard, 1896, Ill., Coll., 1896, Amherst, 1901, St. Andrews, 1902; Edinburgh, 1902; George Washington, 1913; Sc.D., Oxford, 1907; m. Mabel Gardiner Hubbard, 1877. Went to Can., 1870, to U. S., 1871, naturalized, 1882; became prof. vocal physiology, Boston U. 1873. Invented telephone (1st demonstration May 10, 1876; patent granted 1876); also invented photophone, audiometer, induction balance for painless detection of metal in the human body, for which he was awarded hon. M.D. by U. Heidelberg; with C. A. Bell and Sumner Taintor invented the graphophone, 1883. Recipient Volta prix French Govt., 1880; medal, London Soc. Fine Arts, 1902; Royal Albert medal; Elliott Cresson medal; John Fritz medal, 1907; Hughes medal Royal Soc. Arts, London, 1913; Edison medal, 1914; officer French Legion of Honor; elected to Hall of Fame for Gt. Americans, 1950. Founded and endowed Volta Bur. for increase of knowledge relating to deaf, 1887, Aerial Expt. Assn., 1907; founder, ex-pres. Am. Assn. to Promote Teaching of Speech to Deaf; regent Smithsonian Instn., 1898. Fellow Am. Acad. Arts and Scis., A.A.A.S.; mem. Nat. Geog. Soc. (pres. 1896-1904) Author of many scientific and ednl. monographs, including Memoir on the Formation of a Deaf Variety of the Human Race. Influential in founding mag. Science, 1880; produced 1st successful phonograph record; other inventions include flat and cylindrical wax recorders for phonographs, tetrahedral kite; studied deafness and investigated its heredity. Died Baddeck, Nova Scotia, Aug. 2, 1922.

BELL, A(udra) Earl, Am. geneticist; b. Providence, Ky., May 9, 1918; s. George Edgar and Ida (Cullen) B.; B.S., U. Ky., 1939; M.S., La. State U., 1941; Ph.D., Ia. State U., 1948; m. Floris Selina Thorning, June 12, 1941; children—Robert Earl, Diane Louise, Donald Wayne. Faculty, Purdue U., Lafayette, Ind., 1948—, prof. genetics, 1957—, chmn. Population Genetics Inst., 1957—. Genetical cons. Ind. Farm Bur. Coop., 1958—. Fulbright Research fellow, 1964; Guggenheim fellow, 1965; recipient Nat. Sci. Tchr. Star award 1960. Fellow A.A.A.S.; mem. Am. Inst. Biol. Scis., Genetics Soc. Am., Am. Soc. Naturalists, Am. Soc. Animal Sci., Am. Genetical Assn., Poultry Sci. Assn., Am. Assn. U. Profs. Research, numerous publs. on genetics and animal improvement. Home: 708 Sugar Hill Dr., West Lafayette, Ind. 47906. Office: Life Scis. Bldg., Purdue U., Lafayette, Ind. 47907.*

BELL, Barbara, Am. astronomer; b. Evanston, Ill., Apr. 1, 1922; d. George Irving and Hazel (Seerley) Bell; A.B., Radcliffe Coll., 1944, A.M., 1949, Ph.D., 1951. Research asst. Harvard Obs., Cambridge, Mass., 1951-57, sci. asst. to dir., 1957—. Mem. Am. Astron. Soc., Internat. Astron. Union. Research on assns. between solar phenomena and geomagnetic disturbaces. Office: 60 Garden St., Cambridge, Mass. 02138.*

BELL, Benjamin, Scottish surgeon; b. Dumfries, Scotland, Apr. 1749; s. George B.; apprentice to James S. Hill, surgeon; ed. Edinburgh (Scotland) Med. Sch., also Paris, France; m. Grizel Hamilton, 1775; 1 son, George. Surgeon Royal Infirmary, Edinburgh, 1772-circa 1801; named surgeon Watson's Hosp., 1778. Author: Theory and Management of Ulcers, 1779; System of Surgery, 6 vols., 1782-87. First to distinguish between syphilis and gonorrhea; one of 1st to emphasize pain prevention in surg. operations. Died Edinburgh, Apr. 5, 1806.

BELL, Charles, anatomist, surgeon; born Edinburgh, Scotland, November 1774, the son of William Bell; educated at U. of Edinburgh; married Marion Shaw, June 3, 1811. Surg. attendant Edinburgh Infirmary; began treating wounded from Corunna, Spain, at Haslar Hosp., 1809, from Waterloo at Brussels, Belgium, 1815; became prof. anatomy and surgery Royal Coll. Surgeons, 1824; prof. surgery, Edinburgh, 1836-42. Knighted; medal Royal Soc., 1829. Fellow Royal Soc., 1826, Coll. Surgeons Edinburgh. Author: A System of Dissections, 1798; Anatomy of Expression, 1806; System of Comparative Surgery, 1807; A New Idea of the Anatomy of the Brain, 1811; Letters on Diseases of the Urethra, 1820; Illustrations of Great Operations, 1821; The Nervous System of the Human Body, 1830; Institutes of Surgery, 1838; (with John Bell) Anatomy of the Human Body, 4 vols. 1793-1804; 1st to distinguish between sensory and motor functions of nerves (findings confirmed by Magendie 1822, now known as Bell-Magendie law), 1811; gave 1st description of facial paralysis from lesion of facial nerve (Bell's nerve), 1821. Died near Worcester, Eng., Apr. 28, 1842.

BELL, Eric Temple, mathematician; born Aberdeen, Scotland, February 7, 1883; the son of James Bell; ed. Stanford, Washington Univ.; Ph.D., Columbia, 1912; m. Jessie Brown, 1910; (dec. 1940); 1 son. From instr. to prof. math. U. Wash., 1921-26; prof. math. Cal. Inst. Tech., 1926-60. Recipient Bocher prize for math. research, 1921. Mem. Math. Assn. Am. (pres. 1931-33), Am. Math. Soc. (v.p. 1926), Nat. Acad. Scis., Am. Philos. Soc. Author: The Queen of the Sciences, 1931; Numerology, 1933; The Search for Truth, 1934; Handmaiden of the Sciences, 1937; Men of Mathematics, 1937; Man and His Lifebelts, 1938; The Development of Mathematics, 1940; The Magic of Numbers, 1946; Mathematics, Servant of Science, 1950; 15 sci. fiction novels (pseudonym John Taine). Contbr. articles math. jours. Died Watsonville, Calif., Dec. 21, 1960.

BELL, George Douglas Hutton, botanist, plant breeder; b. Oct. 18, 1905; s. George Henry and Lilian Mary Matilda B.; ed. Univ. Coll. N. Wales, Bangor, Cambridge (Eng.) U.; B.Sc., 1928, Ph.D., 1931; m. Eileen Gertrude Wright, 1934; 2 daus. Research officer Plant Breeding Inst., Cambridge U., from 1931, dir., 1947—, univ. demonstrator, from 1933, univ. lectr., from 1944, also fellow Selwyn Coll. Recipient Research medal Royal Agrl. Soc. Eng., 1956. Fellow Royal Soc., 1965; mem. Genetical Soc., Assn. Applied Biologists. Author: Cultivated Plants of the Farm, 1948; The Breeding of Barley Varieties in Barley and Malt, 1962. Research, publs. on genetics, physiology, crop improvement, methods of plant breeding, pathology and relevant scis. Home: 6 Worts Causeway, Cambridge. Office: Plant Breeding Inst., Trumpington, Cambridge, Eng.

BELL, George Howard, Scottish physiologist; b. Ayr, Scotland, Jan. 24, 1905; s. Thomas and Henrietta (Rodger) B.; M.D., Glasgow U.; m. Isabella M. Thomson, 1934; children—Howard, Alan. Instr. physiology U. Bristol, U. Glasgow; prof. physiology Queen's Coll., U. St. Andrews, Dundee, Scotland. Fellow Royal Coll. Physicians (Glasgow), Royal Soc. Edinburgh; mem. Physiol. Soc. Author: Textbook of Physiology and Biochemistry. Home: 80 Grove Rd., Broughty Ferry, Dundee. Office: Dept. of Physiology and Biochemistry, Queen's College, University of St. Andrews, Dundee, Scotland.*

BELL, George Irving, Am. physicist; b. Evanston, Ill., Aug. 4, 1926; s. George I. and Hazel (Seerley) B.; A.B. in Physics, Harvard, 1947; Ph.D., Cornell U., 1951; m. Laura Virginia Lotz, Jan. 13, 1956; children—Carolyn, George Irving. Staff, Los Alamos Sci. Lab., 1951—; prof. physics U. N.M., Los Alamos Grad. Center, 1960—; vis. lectr. Harvard, 1962-63. Mem. U. S. Fast Reactor Team to Britain, 1957; mem. air def. panel Pres.'s Sci. Adv. Com., 1961-62. Fellow Am. Nuclear Soc. (certificate merit 1966), Am. Phys. Soc., A.A.A.S. Research, publs. on theory neutron transport; developed theory for prodn. heavy nuclei by rapid neutron capture, math. models of cell growth and div. Home: 794 43d St. Office: T Div., Box 1663, Los Alamos 87544.*

BELL, Henry, Scottish engr.; b. Torphichen, Scotland, 1767; s. Patrick B.; trained in carpentry, engring. under Rennie; builder mills, ships; conceived idea of applying steam to navigation, 1786; built early Brit. steamship, the Comet, which, propelled by 3-power engine, travelled 7 miles per hour on Clyde River, 1812 (1st practical steamship in Europe); Bell's ideas led to Fulton's steamship; inventor: discharging machine, (improved calico printing). Died Helensburgh, Scotland, Nov. 14, 1830.

BELL, Isaac Lowthian, Brit. metall. chemist; b. Newcastle-on-Tyne, Eng., Feb. 15, 1816; studied in Denmark, Edinburgh (Scotland) U., U. Paris (France); studied new process for mfr. alkali, Marseilles, France; D.C.L., Durham, Eng., 1882; LL.D., Edinburgh, 1893; D.Sc., Leeds (Eng.) U., 1904; m. Margaret Pattison, 1842; children—Hugh Bell, Charles Lowthian, 3 daus. Entered office of Losh, Wilson & Bell, Newcastle, Eng., 1835, transferred to ironworks at Walker, 1836; founded chem. works, Washington, Eng., 1850; helped found amalgamation to build steel works at Clarence, 1899; Mayor Newcastle-on-Tyne, 1854-55, 62-63; dep. lt. sheriff Durham County, 1884. Created baronet, 1893; decorated Legion of Honor, France; recipient Bessemer Gold medal, 1874; George Stephenson medal Instn. Civil Engrs., 1900, Telford premium, 1900, Howard quinquennial prize, 1892. Fellow Royal Soc., 1874; mem. Iron and Steel Inst. (a founder 1869, pres. 1873-75), Instn. Mining Engrs. (a founder 1888, pres. 1904), Brit. Iron Trade Assn. (pres. 1886), Instn. Mech. Engrs. (pres. 1884), Soc. Chem. Industry (pres. 1889, Albert medal 1895), Author: The Development of Heat and its Appropriation in Blast Furnaces of Different Dimensions, 1869; Chemical Phenomena of Iron Smelting, 1871, 72; The Sum of Heat utilized in Smelting Cleveland Ironstone, 1875; The Separation of Carbon, Silicone, Sulphur and Phosphorus from Pig Iron, 1878; On the Value of Excessive Addition to the Temperature of the Air used in Smelting Iron, 1883; The Principles of the Manufacture of Iron and Steel, 1884; (with C. R. A. Wright) Chemical Phenomena of Iron Smelting, 1872. Determined basis for estimating heat balance of furnace, also for determining main sequence of chem. changes as descending fuel, ore and flux meet ascending furnace gases; determined capacity and height limits of furnaces. Died Dec. 20, 1904.

BELL, Jacob, Brit. pharmacist; b. London, Eng., Mar. 5, 1810; student Royal Inst., King's Coll.; founder Pharm. Soc. Great Britain, 1841; established, supervised Pharm. Jour. for 18 years; mem. parliament for St. Albans, from 1850. Fellow Chem. Soc., Linnean Soc., Zool. Soc., Soc. Arts. Author: Historical Sketch of the Progress of Pharmacy in Great Britain, 1843; Observations Addressed to the Chemists and Druggists of Great Britain, 1841; Chemical and Pharmaceutical Processes and Products, 1852. Formulated provisions for registration of pharmacists, system of edn. and exam., protection of pub. and other reforms in pharm. trade. Died June 12, 1859.

BELL, James, chemist; b. Armagh, Ireland, 1824; student U. Coll., London, Eng.; Ph.D., Erlangen, Germany, 1882; D.Sc. (hon.), Royal U. Ireland, 1886; m. Ellen Reece, 1858; 1 son, William James. Asst. in Inland Revenue Lab., 1846; dep.-prin. Somerset House Lab., 1868-74, prin., 1875-94; insp. lime and lemon juice Bd. Trade, also cons. chemist Indian Govt., 1869-94. Chem. referee under Sale of Food and Drug Acts, 1875-94. Fellow Royal Soc., 1884; mem. Inst. Chemistry (pres. 1888-91), Playfair Com. on Brit. and Fgn. Spirits. Author: Analysis and Adulteration of Food, 1881-83; Chemistry of Tobacco, 1887. Contbr. articles Jour. Chem. Soc., 1870. Research in grape and malt ferments, chemistry of tobacco. Died Hove, Eng. Mar. 31, 1908.

BELL, James Munsle, chemist; b. Chesley, Ont., Can., Apr. 19, 1880; s. John Charlton and Hannah (Munsie) B.; A.B., U. Toronto, 1902, A.M., 1905; Ph.D. (Sage fellow in chemistry), Cornell U., 1905; m. Mary E. Brawner, Apr. 2, 1909. Came to U. S., 1902, naturalized, 1908. Chemist, U. S. Bur. Soils, Washington, 1905-10; asso. prof. phys. chemistry U. N.C., 1910-13, prof., 1913-19, Smith prof., from 1919, head chemistry dept. from 1921, Kenan travelling prof., 1926-27, dean Sch. Applied Sci., from 1929. Author: (with Paul M. Gross) Elements of Physical Chemistry, 1929. Died Mar. 3, 1934.

BELL, Johann Adam Schall von, see von Bell, Johann Adam Schall.

BELL, John, Scottish surgeon; b. Edinburgh, Scotland, May 12, 1763; s. William Bell; student of Alexander Wood (surgeon in Edinburgh); m. Rosina Congleton, 1805. Became lectr. anatomy and surgery, Edinburgh, 1790; held appointment at Royal Infirmary until 1800; practiced surgery in Edinburgh, 20 years. Fellow Royal Coll. Surgeons Edinburgh. Author: (with Charles Bell) Anatomy of the Human Body, 4 vols., 1793-1804; Discourses on the Nature and Cure of Wounds, 2 vols., 1793-95; Engravings of the Bones, Muscles and Joints, 1794; Principles of Surgery, 3 vols., 1801-08. A founder of modern blood vessel surgery. Died Rome, Italy, Apr. 15, 1820.

BELL, John Milton, nutritionist; b. Islay, Alta, Can., Jan. 16, 1922; s. Milton Wilfred and Elsie (Larmour) B.; student Vermilion (Alta.) Sch. Agr., 1939-40; B.S.A., U. Alta., 1943; M.Sc., McGill U., Montreal, Que., Can., 1945; Ph.D., Cornell U., 1948; m. Edith Margaret Joan Smith, Sept. 21, 1945; children—Joyce, Donald, Marion, Douglas, Keith. Faculty, U. Sask., Saskatoon, Can., 1948—, head dept., prof. nutrition, 1954—. Chmn. asso. com. animal nutrition NRC Can., 1959; chmn. lab. animal nutrition com. of Nat. Acad. Sci.-NRC, 1965—. Fellow Agr. Inst. Can.; Mem. Canadian Soc. Animal Production (past pres.), Nutrition Soc. Can. (Borden award 1962, pres. 1966—), Am. Soc. Animal Sci., Am. Inst. Nutrition. Contbg. author: Introduction to Animal Science, 1962-66; also numerous articles. Editorial bd. Canadian Jour. Animal Sci., 1955-59. Research on nutrient requirements swine, effects bulk in diet, toxic factors in rapeseed meal, dairy calf nutrition, milk replacer formulas, protein quality evaluation. Home: 1530 Jackson Av., Saskatoon, Sask., Can.*

BELL, Joseph, Scottish surgeon; b. Edinburgh, Dec. 2, 1837; s. Benjamin and Cecilia (Craigie) B.; ed. U. Edinburgh; m. Edith Katherine Murray, 1865; 2 daus. Practiced medicine, specializing in surgery, Edinburgh; served from dresser to sr. surgeon, cons. surgeon Edinburgh Royal Infirmary; cons. Surgeon Royal Hosp. for Sick Children. Fellow Royal Coll. Surgeons Edinburgh. Author: Manual of Surgical Operations, 7th edit., 1894; Notes on Surgery for Nurses, 6th edit., 1906. Editor: Edinburgh Med. Jour., 1873-96. Died Oct. 4, 1911.

BELL, Louis, Am. elec. engr.; b. Chester, N.H., Dec. 5, 1864; s. Louis and Mary Anne Persis (Bouton) B.; A.B., Dartmouth Coll., 1884; Ph.D. (fellow in physics), Johns Hopkins, 1888; m. Sarah G. Hemenway, Dec. 3, 1893. Prof. applied electricity (organized elec. course) Purdue U., 1888-89; editor Elec. World, N.Y., 1890-92; chief engr. newly organized elec. power transmission dept. Gen. Electric Co.; cons. engr. mainly on work with elec. power transmission, 1895—. Lectr. on power transmission Mass. Inst. Tech., 1895-1905; lectr. on pub. lighting Harvard, also on illumination Harvard Med. Sch., 1914—. Fellow Am. Acad. Arts and Sciences; mem. Am. Inst. Elec. Engrs., Illuminating Engring. Soc. (past pres.). Author: (with Oscar T. Crosby) The Electric Railway, 1892; Power Distribution for Electric Railroads, 1896; Electric Power Transmission, 1897; The Art of Illumination, 1902; Boston Electrical Handbook (as chmn. Publ. Com.), 1904; The Telescope, 1922. Designed, installed 1st polyphase plants in U. S.; patentee power transmission and elec. apparatus. Died June 14, 1923.

BELL, Robert, Canadian geologist; b. Toronto, Ont., Can., June 3, 1841; s. Andrew Bell; studied practical chemistry and mineralogy under Dr. T. Hunt, 1861-63; B.A.Sc., M.D., C.M., D.Sc., McGill U.; D.Sc. (hon.), Cambridge, 1903; m. Agnes Smith; children—Edith, Alice. Joined Geol. Survey Can., 1857, later became asst. dir., chief geologist, then dir., 1901-06; prof. chemistry and natural scis. Queen's U., 1863-67. Mem. Commn. to Report on Mineral Resources of Ont., 1888-89; mem. expdns. to Hudson Bay, 1884, 85, 97; explored Baffin Island, 1897; del. from Can. to Internat. Geol. Congress, Vienna, 1903. Highest gold medals Royal Geog. Soc., Am. Geog. Soc.; King's gold medal. Fellow Royal Soc., 1897, Chem. Soc., Royal Soc. Can. Royal Astron. Soc. Can.; mem. Royal Scottish Geog. Soc., Geog. Soc. Paris. Writer numerous pamphlets on geography, geology, forestry, folklore. Pioneer in Canadian geography and geology; subdivided and mapped forests; made topog. and geol. surveys east of Rocky Mountains in Can., from 1857. Died Rathwells, Man., Can., June 17, 1917.

BELL, Ronald Percy, Brit. chemist; b. Maidenhead, Eng., Nov. 24, 1907; s. Edwin Alfred and Beatrice Annie (Ash) B.; B.Sc., M.A., Balliol Coll., Oxford, Eng., postgrad. U. Copenhagen (Denmark); LL.D., Ill. Inst. Tech., 1964; m. Margery Mary West, Apr. 16, 1931; 1 son, Michael John. Bedford lectr. in phys. chemistry Balliol. Coll., 1932-48, fellow, tutor, 1933-67, vicemaster, 1965-66; u. demonstrator, lectr. Oxford U., 1938-55, u. reader in phys. chemistry, 1955-67; prof. chemistry U. Stirling, Scotland, 1967—. George F. Baker lectr. Cornell U., Ithaca, N.Y., 1958. NSF fellow Brown U., Providence, 1964; recipient Meldola medal, 1936. Fellow Royal Inst. Chemistry, Royal Soc., 1944, Chem. Soc. (London); mem. Faraday Soc. (past pres.), Royal Danish Acad. Arts and Scis. (fgn. mem.). Author: Acid-Base Catalysis, 1941; Acids and Bases, 1952; The Proton in Chemistry, 1959; also numerous articles. Research on reaction kinetics in solution particularly acid-base catalysis. Home: 6 Victoria Sq., Stirling, Scotland.*

BELL, Thomas, English dental surgeon; naturalist; b. Poole, Eng., Oct. 11, 1792; s. Thomas B.; ed. Guy's St. Thomas' hosps., London; lectr. dental surgery, comparative anatomy Guy's Hosp. 1817-61; prof. zoology King's College, London, 1836. Fellow Royal Soc., 1828 (sec. 1848-53); mem. Royal Coll. Surgeons, Ray Soc. (pres. 1843-59), Linnean Soc. (pres. 1953-61). A founder of Zool. Jour. Author book on teeth, 1829; History of British Quadrupeds, 1837; History of British Reptiles, 1839; History of British Stalk-eyed Crustacea, 1853; Monograph of the Testudinata, 1833. Editor: Natural History of Selborne (Gilbert White), 1877. For a considerable period of time, the only capable surgeon who used sci. surgery for diseases of teeth. Died 1880.

BELL, Wendell, Am. sociologist; b. Chgo., Sept. 27, 1924; s. Wendell and Blanche (Leiferman) B.; B.A. with highest honors Fresno State Coll., 1946-48; M.A., U. Cal. at Los Angeles, 1951, Ph.D. (Social Sci. Research Council fellow), 1952; M.A., (hon.) Yale, 1963; m. Lora-Lee Edwards, June 15, 1947; children—Sharon Lee, David Howard. Asst. prof. sociology Stanford, 1952-54; asso. prof. Northwestern U., 1954-57; faculty U. Cal. at Los Angeles, 1957-63, prof., 1962-63; prof. Yale, 1963—, chmn. dept. sociology, 1965—. Mem. sociology selection panel com. on internat. exchange persons Fulbright-Hays awards, 1966—. Fellow Center for Advanced Study in Behavioral Scis., Stanford, Cal., 1963-64; Social Sci. Research Council Faculty Research fellow, 1956-59. Mem. postdoctoral research fellowship com. div. behavioral scis. Nat. Acad. Scis., 1966—. Mem. Am. Pacific (past v.p.), Eastern sociol. assns., Am. Polit. Sci. Assn., Acad. Polit. Sci., Sociol. Research Assn., Soc. for Study Social Programs. Author: Jamaican Leaders, 1964; (with E. Shevky), Social Area Analysis, 1955; (with R. J. Hill, C. R. Wright) Public Leadership, 1961; (with I. Oxaal) Decisions of Nationhood, 1964; also articles, chpts. in books. Editor, contbg. author: The Democratic Revolution in the West Indies, 1967. Series editor Internat. Studies in Polit. and Social Change, Schenkman Pub. Co., 1966—; asso. editor Am. Sociol. Rev., 1958-61. Research on urban analysis, devel. social area analysis, informal social relations in Am. cities, suburban shift in 1950's, transition to nationhood especially in English-speaking Caribbean; formulation theory of polit. and social change based on images of future. Home: Sperry Rd., Bethany, Conn. 06525. Office: Dept. Sociology, Yale, New Haven 06520.*

BELL, William Earl, physicist; b. Winnipeg, Man., Can., Apr. 2, 1921; s. Ross Earl and Ella (Sigurdson) B.; E.E., U. Alta, 1940; m. Frances Elizabeth Wright, Apr. 7, 1945; children—Ross Stuart, Robert Ian. Physicist Canadian Atomic Energy Project, Chalk River, Ont., 1945-49; research physicist Newmont Exploration, Ltd., Jerome, Ariz., 1950-54; atomic physicist Varian Assos., Palo Alto, Cal., 1955-61, cons., 1965—; co-founder, dir. research Spectra-Physics, Inc., Mountain View, Cal., 1961—; research asso. Stanford, 1956-65. Mem. Am. Phys. Soc. Publs. and patents on measurement of mumeson lifetime; discovered optically driven spin systems; invented gas ion laser. Home: P.O. Box 426. Office: P.O. Box M, Jerome, Ariz. 86331.*

BELLAIR, Pierre, French geologist; b. Mamers (Sarthe), June 12, 1910; s. Ernest and Camille (Gauthier) B.; ed. Ecole normale superièure, Sorbonne; Ph.D. in sci.; m. Renée Baudier, Jan. 15, 1940; children—Nicole, Alain, Mireille. Prof., Lycée of Algiers; dir. scis. Inst. Advanced Studies, Tunis; prof. geology Sorbonne. Contbr. numerous articles on Dauphiné, Sahara, Antarctic, French Austral Islands. Home: 25, rue de l'Yser, Sceaux (Seine). Office: 1, rue Guy de la Brosse, Paris 5, France.

BELLAMY, Edmund Henry, English physicist; b. Liverpool, Eng., Apr. 8, 1923; s. Herbert and Nellie (Ablett) B.; ed. King's Coll., Cambridge; M.A., Ph.D.; m. Joan Roberts, Oct. 31, 1946; children—Stephen Peter, Nigel Edmund, Richard Paul. Instr., U. Glasgow, 1951-58, senior lecturer, 1958-60; prof. physics, head dept. Westfield Coll., U. London, 1960—. Contbr. articles on physics, nuclear physics to sci. publs. Home: 7 Homefield Rd., Radlett, Hertfordshire, Eng. Office: Kidderpore Av., London N.W.3, Eng.*

BELLAMY, Edward, Am. polit. theorist; b. Chicopee Falls, Mass., Mar. 26, 1850; s. Rufus and Maria (Putnam) B.; attended Union Coll., Schenectady, N.Y., circa 1867-68; m. Emma Sanderson, 1882. Admitted to Mass. bar, circa 1877, began practice in Springfield; editorial writer N.Y. Evening Post, 1878; founder (with brother) Springfield Daily News, 1880; advocated nationalization of industry in his published works; devoted career to writing and propagation of socialist ideas, after 1885; founder New Nation, (weekly devoted to utopian socialism), Boston, 1891. Author: Dr. Heidenhoff's Process, 1880; Mrs. Ludington's Sister, 1884; Looking Backward (most famous novel, outlined his utopian econ. thinking), 1888; Equality, 1897; The Blind Man's World and Other Stories (collection short stories), 1898. Died Chicopee Falls, May 22, 1898.

BELLAMY, Frank Arthur, English astronomer; born 1864; ed. Magdalen College, Oxford (Eng.) U.; worker Radcliffe Obs., 1881-90; asst. Oxford U. Obs., 1890-1936; mem. Phenological Soc., Ashmolean Nat. Hist. Soc. (pres., author hist.) Prepared most photographs, measurements, final reductions of Oxford zone in Astrographic Catalogue; reduced measures on Vatican zone, also part of Potsdam zone. Died 1936.

BELLANCA, Angelo, Italian mineralogist; b. San Cataldo (Caltanissetta), Aug. 23, 1907. Prof., U. Palermo (Italy), 1948, prof. mineralogy Sch. Math., Phys. and Natural Scis., 1951—. Address: University of Palermo, Palermo, Italy.*

BELLANI, Angelo, Italian physicist; b. Monza, Italy, Oct. 31, 1776; canonicus in Milan, Italy. Research and numerous publs. on boiling of liquids; invented new types of aerometers, meterol. instruments; 1st in Italy to study thermometer; discovered zero point of glass thermometer shifts with time. Died Milan, Aug. 28, 1852.

BELLARDI, Luigi, Italian paleontologist, geologist; b. Genova, Italy, May 18, 1818. Author: Saggio di ditterologia messicana, 1859-60; Molluschi dei terreni Terziari del Piemonte e della Liguria, 1st 6 vols. (last 24 vols completed by F. Sacco), 1872-89; also articles. Research on fossil mollusks of Piedmont. Died Turin, Italy, Sept. 17, 1889.

BELLAVITA, Vito, Italian chemist; b. Perugia, Italy, Nov. 23, 1903; s. Antonio and Emilia (Biancucci) B.; Ph.D. in chemistry; m. Nera Cagnoli, Sept. 12, 1939; children—Claudio-Emilio, Maria-Letizia. Dir., Inst. Pharmacy, U. Perugia. Mem. Italian, French, German chem. socs., Chem. Soc. (London), Am. Pharm. Assn. Research on organic, pharm. and phys. chemistry. Home: via Degli Olivi' 24. Office: University of Perugia, Perugia, Italy.*

BELLAVITIS, Giusto, Italian mathematician; b. Bassano, Italy, Nov. 22, 1803; hon. degree in math.; became prof. math. Inst. Vincenza, 1843; named prof. geometry at Padua, Italy, 1845; nominated senator, 1866. Author: Saggio d'applicazioni del calcolo delle equipollenze, 1837; Metodo delle equipollenza, 1837. Research in projective geometry; 1st to use calculus of equipollence; introduced complex numbers into geometry; independently derived principle of inversion. Died Padua, Nov. 6, 1880.

BELLER, Fritz Karl, physician; b. Munich, West Germany, May 17, 1924; s. Carl Friedrich and Antonie (Woerz) B.; student Sch. Medicine, U. Berlin, 1942, Sch. Medicine, U. Prague, Czechoslovakia, 1944-45; Approbation U. Marburg, Germany, 1948, Dr.med., 1949; Dozent, U. Giessen, Germany, 1955; m. Marlis Duhl, Nov. 27, 1948; children—Verena, Christoph. Came to U. S., 1961, naturalized, 1966; asst. prof. U. Tübingen, Germany, 1956-58, asso. prof., head physician U. Womans Hosp., 1959-61, prof. obstetrics and gynecology, 1961; vis. asso. prof. N.Y. U. Sch. Medicine, 1961-63, asso. prof. obstetrics and gynecology, 1963—. Recipient Career Scientist award Health Research Council City of N.Y., 1961. Fellow Royal Soc. Medicine (London, Eng.), Am. Coll. Obstetrics and Gynecology; mem. N.Y. Acad. Scis., Am. Soc. Extpl. Pathology, Soc. for Gynecol. Investigation, Harvey Soc., European Soc. Hematology. Author: Das Blutgerinnungssytem bei der Schwangeren and beim Neugeboreen, 1957: Klinische Methoden der Blugerinnungsanalvses (with J. Juergens), 1959; also numerous articles. Research on physiology coagulation and fibronolytic enzyme system in pregnancy and newborn, oxytocinases in pregnancy, hemorrhagic diathesis, significance phenomenon disseminated intravascular coagulation and proteolysis, endotoxic shock in septic abortions, pathphysiol. mechanism endotoxins, diagnosis and treatment thromboembolic complications postoperatively and post partum. Home: 50-38 60th St., Woodside, N.Y. 11377. Office: 550 1st Av., N.Y.C. 10016.*

BELLEUAL, Pierre Richer de, see de Belleual.

BELLI, Lino, Italian physician; b. S. Vito Cadoze, Belluno, Italy, May 5, 1924; s. Albino and Domenica (De Vido) B.; M.D., U. Padua, Italy, 1960; m. Guiseppina Vinanti, July 5, 1954; children—Giovonna, Luca, Morine. Asst. in surgery U. Milan, Italy, 1948-53, 1st asst. in surgery, 1953-59, 1st asst. in thoracic surgery, 1960, asso. prof. surgery, 1960-65, asso. prof. thoracic surgery, 1965—; chief surgeon Ospeodale Maggiore Milan, 1967—. Mem. Soc. Surgery, Internat Coll. Surgery, Internat. Soc. Cardiovascular Surgery. Research, numerous publs. on physiopathology of liver, portal hypertension, vascular and gen. surgery, thoracic surgery. Home: 62 Sismondi St. Office: Ospedale Maggiore Niguarda, Milan, Italy.*

BELLINI, Lorenzo, Italian anatomist; b. Florence, Italy, Sept. 3, 1643; ed. U. Pisa (Italy); studied math. under Marchetti, medicine and anatomy under Redi, mechanics under Borelli. Pub. lectr. theoretical medicine, Pisa, 1663; prof. anatomy U. Pisa, 1664-94; physician to Grand Duke Cosimo III; cons. to Pope Clement XI. Author: Exercitatio anatomica de structura et usu renum, 1662; Gustus organum novissinum deprehensum, 1665; De urinis et pulsibus . . . , 1683. Discovered action of nerves on muscles; described excretory ducts of kidneys (Bellini's ducts), also uriniferous tubules (Bellini's tubules), 1662, papillae of tongue as organs of taste, 1665. Died Florence, Jan. 8, 1704.

BELLIS, Carroll Joseph, Am. surgeon; b. Shreveport, La., May 11, 1908; s. Joseph Edgar and Rose (Bloome) B.; B.S., U. Minn., 1930, M.S., 1932, Ph.D. in Physiology, 1934, M.D., 1936, Ph.D. in Surgery, 1941; m. Mildred Darmody, Dec. 26, 1951; children—Joseph C., David J. Practice medicine, specializing in surgery, Long Beach, Cal., 1946—; staff St. Mary's Community, Meml. hosps.; prof., chmn. dept. surgery Cal. Coll. Medicine, 1962-63; cons. surgery Long Beach Gen., San Pedro Community hosps. Cons. to Surgeon Gen., U. S. Army, 1941-45. Recipient Charles Lyman Green prize in physiology, 1934, Mpls. Surg. Soc. prize, 1938, Mississippi Valley Med. Soc. Ann. award, 1955. Diplomate Am. Bd. Surgery, Internat. Bd. Proctology. Fellow A.C.S., Internat. Acad. Proctology, Am. Coll. Gastroenterology, Am. Geriatrics Soc., Internat. Coll. Surgeons, Am. Soc. Abdominal Surgeons, Internat. Coll. Angiology, Phlebology Soc. Am.; mem. Am. Assn. Study Neoplastic Diseases, Am. Assn. History of Medicine, A.A.A.S., N.Y. Acad. Scis., Am. Med. Wristers Assn., Irish Med. Assn., Gerontol. Soc. Phi Beta Kappa, Sigma Xi, Alpha Omega Alpha. Author: Critique of Reason, 1940; Fundamentals of Human Physiology, 1945; Lectures in Medical Physiology, 1954. Research, publs. on hernia repair; infections of the extremities; pilonidal cystectomy; erythrocyte and plasma protein potentials; portal vein pressures; empyeme thoracis. Home: 3 S. Quail Ridge Rd., Rolling Hills, Cal. 90724. Office: 117 E. 8th St., Long Beach, Cal. 90813.*

BELLOCQ, Philippe Bertrand, French physician; b. Maspie (Basses-Pyrenees), France, June 18, 1888; s. Pierre and Marie (Poubet) B.; ed. U. Bordeaux, U. Toulouse; M.D.; m. Gabrielle Irague, Feb. 17, 1912. Intern in hosps., aide in anatomy, instr., later titular prof. med. sch. at Strasbourg (France); titular prof. med. sch. at Lille. Mem. Nat. Acad. Medicine (corr.). Home: 7, rue Salzmann. Office: Institut d'Anatomie Normale, Faculté de Médecine, Strasbourg, France.*

BELLOMO, Athos, Italian analytical electrochemist; b. Messina, Italy, June 14, 1927; s. Arturo and Maria (Buzzeghi) B.; grad. Lyceum Campailla Modica, 1945; degree in pure chemistry with maximum honors, U. Messina, 1951; m. Stefania Mannino, Sept. 1, 1965; 1 child. Asst. gen. chemistry inst., U. Messina, 1951—, docent analytical chemistry, 1963—, mem. Accademia Peloritana dei Pericolanti. Chief emergency radio service City of Messina. Recipient numerous awards for work as radio amateur. Mem. Amateur Radio Assn. Messina (pres.), Societa Peloritana Sci. Fisiche, Matematiche e Naturali, Chemistry Order and Nat. Assn. U. Assts. Research, publs. on high frequency conductometry; introduced and perfectioned system of frequenziometric measurements; realized new types of cells electrodeless and apparatus for galvanometric frequenziometric measurments; developed mechanism of a nomal curves, action of resistive factors, automatic high frequency apparatus; also has made analytical researches especially on analysis of alloys. Home: Via Fata Morgana n°1 - Is.459, Messina, Italy.*

BELLUIGI, Arnaldo, Italian geophysicist; b. Tolentino, Mar. 11, 1898; s. Umberto and Gamma (Romagnoli) B.; ed. in geoelectrics, electrogeocinetics, geophysics; Ph.D.; m. Maria Frassinelli, Jan. 11, 1919; children—Umberto, Bruno, Lity. Univ. prof. geophysics. Author 300 original memoirs in German, English, French, Russian, Spanish, Italian. Home: via Perugia (Prepo), Perugia. Office: via Valnerina 66, Rome, Italy.

BELON, Pierre, French naturalist, anatomist; b. Soulletière, Maine, France, 1517; s. Gilles and Anne (Baignol) B.; M.D., Paris; studied in Wittenberg, Germany, 1540. Travelled successively to Greece, Egypt and Palestine. Author: l'Histoire naturelle des Éstranges Poissons Marines, Paris, 1551; De aquatibilibus, 1553; De arboritum coniferis, 1553; La nature et diversité des poissons, 1555; l'Histoire de la Nature des Oyseaux. His writings foreshadow sci. of comparative anatomy; 1st (since Aristotle) to indicate comparative anatomy by portraying skeletons of man and bird with similar designations of corr. parts; introduced cedar to France. While collecting herbs, he was murdered by thieves, Paris, Apr., 1564.

BELOPOLSKY, Aristarch Apolonovich, Russian astronomer; b. Moscow, July 13, 1854; ed. U. Moscow. Became asst. astronomer Univ. Obs. Moscow, 1877; apptd. asst. to dir. Pulkovo (Russia) Obs., 1888, astrophysicist, 1890. Recipient Janssen Gold medal Inst. France, 1899. Mem. St. Petersburg Acad. Scis. (2 prizes). Discovered various spectroscopic binaries; investigated variable stars; made spectroscopic studies of planet rotations. Died 1934.

BELOUSOV, Vladimir Aleksandrovich, Russian pediatrician; b. 1895; grad. Med. Faculty, Kharkov U., 1917; D.Med. Sci., 1937. Asst., later lectr. Clinic Children's Diseases, Kharkov Med. Inst., 1921-38; head chair faculty and hosp. pediatrics, 1938—, prof., 1939—. Mem. learned med. council USSR Ministry Health, 1949—; chmn. Kharkov Oblast Council for Med. and Prophylactic Care Children, 1954—. Mem. USSR Acad. Med. Sci. (corr.), Ukrainian (dep. chmn.), Kharkov (chmn. 1953—) socs. pediatricians. Author over 50 works including Materials on the Problem of the Protein Regime in Health and Sick Schoolchildren Aged 8-11, 1937; Diagnostic Errors in Pediatric Practice, 1948; Clinical and Organizational Problems in the Treatment of Tuberculous Meningitisin Children, Mem. editorial bd.

Pediatrics, Obstetrics and Gynecology; mem. editorial council Pediatrics, also others. Research on treatment of Tb meningitis in children and other aspects of pediatrics. Address: Kharkov Med. Inst., Sumskaya ulitsa 41, Kharkov, Ukraine SSR, USSR.

BELOV, Nikolai Vasilevich, Russian crystallographer; born December 14, 1891; graduated from Leningrad Polytechnical Institute, 1921. With various tech. college and research institute labs., 1921-30; sr. asso. Leningrad Lomonosov Inst., USSR Acad. Sci., 1930-38, sr. asso. Inst. Crystallography, 1938-—; prof. Gorky U., 1946-—, head chair crystallography, 1953-—; prof. Moscow U., 1953-—. Decorated Order of Lenin, 1961. Mem. USSR Acad. Sci., Internat. Crystallography Union (v.p. 1957-—). Author: Some Elementary Characteristics of Minerals in the Light of their Fine Structure, 1945; The Structure of Ionic Crystals and Metallic Phases, 1947; The Crystalline Structure of Turmalin, 1949; Outline Structural Mineralogy, 1950; Some Applications of the Theory of Mineralizers, 1951; Structural Crystallography, 1951; (with R. P. Shabala) The Crystalline Structure of Wohlerite, 1962. Research and publs. on geometric crystallography, methods of x-ray analysis, and Fourier analysis; developer face-centered packing theory for atom components of crystals. Address: Moscow University, Leninskie gory, Moscow, USSR.

BELOZERSKII, Andrei Nikolaevich, Russian biochemist; born August 29, 1905; graduated from biology department of Physico-Mathematical Faculty, Central Asian University, Tashkent, 1927; postgrad. in plant physiology, 1927-30; D.Biol. Sci. Asst., later lectr. dept. plant biochemistry Moscow U., 1930-46, dir. plant biochemistry, 1946-—. Mem. USSR Acad. Sci. Author numerous works including The Metaphosphate-Nucleic Complexes of Yeast and the Chemical Nature of Volutine, 1955; co-author: Practical Manual on Plant Biochemistry, 1951. Coeditor Chemistry sect. Large Med. Ency., 2d edit.; mem. editorial council Problems of Med. Chemistry; Antibiotics. Research on presence of common proteins in nucleoproteids, nucleic acids and proteins, changes in nucleic acid content of bacteria in different phases and under influence of antibiotics. Address: Moscow University, Leninskie gory, Moscow, USSR.

BELT, Thomas, Brit. geologist; b. Newcastle-on-Tyne, Eng., 1832; geologist gold fields, Australia, 1852-62; charge of gold mines, N.S., Can., from 1862, Nicaragua, 1868-72; geol. expdn., Siberia and steppes of southern Russia, 1873-76. Fellow London Geol. Soc.; mem. Phila. Acad. Natural Scis. (corr.). Author: Mineral Veins: an Inquiry into their Origin, 1861; The Naturalist in Nicaragua, 1874; also zool. and geol. studies of various regions. Geol. research on glacial period; attributed formation of lower boulder clays, diluvium in Europe, also the destruction of great mammals and probably of paleolithic man, to glaciers flowing from Greenland. Died Denver, Sept. 21, 1878.

BELTRAMI, Eugenio, Italian mathematician; b. Cremona, Italy, Nov. 16, 1835; ed. Cremona, then under Brioschi at Pavia. Apptd. asso. prof. algebra and geometry U. Bologna (Italy), 1862, prof. rational mechanics, 1866; became prof. geodesy, Pisa, Italy, 1863, prof., Rome, 1873. Mem. Royal Acad. Lincei (pres., 1898); Senator, 1899. Author: Opere matematiche, 4 vols., 1902-20. Investigated non-Euclidian theory of hyperbolic space, also theory of surfaces, theory of space of constant curvature, various branches math. physics, including hydrodynamics, elasticity, magnetism, electricity, optics, theory of potential, conduction of heat, thermodynamics; elucidated Maxwell's work on electromagnetism. Died Rome, Feb. 18, 1900.

BELTRAN, Enrique, Mexican zoologist; b. Mexico, Distrito Federal, Mexico, Apr. 26, 1903; s. Enrique and Martina (Castillo) B.; Sc.D., U. Mexico, 1926; postgrad. Columbia U.; Sc.D. (hon.), U. Havana, Cuba, 1957, U. Michoacán, Mexico; m. Trini Gutierrez Games, Mar. 4, 1935; children—Enrique, Hector. Faculty, U. Mexico, Mexico City, 1921-64 prof. zoology, 1933-64; head dept. protozoology Inst. Tropical Diseases, 1939-52; prof. natural resources and history of biology Poly. Inst., 1940-52; dir. Marine Biol. Sta., Veracruz, Mexico, 1926-27, Biotech. Inst., 1934-35; tchr. secondary edn., 1937-39; dir. Inst. Natural Renewable Resources, Mexico City, 1952-—; undersec. forestry and game, 1958-64. Decorated Palmes Academique, Legion d'Honeur, Merito Florestal (Brazil); Merito Forestal 1st class (Mexico); Comandeur Merite Agricole (France), Aldo Starker Leopold medal (U. S. A.); others. Fellow Am. Acad. Microbiology; mem. Acad. Medicine, Mexican Soc. History Sci. and Tech. (pres. 1964-—), Soc. Natural History (sec. 1936-—), Soc. Zoologique (France), Soc. Zool., Soc. for History Sci. (Eng.), Assn. Forestal (hon. Argentina), A.A.A.S., Soc. Protozoologists, London Zool. Soc. (hon.), Cal. Acad. Sci. (hon.), Soc. Am. Foresters (hon.), Sigma Xi. Author: Temas Forestales 1961; La Batalla Forestal, 1964; others, also articles. Editor, Revista de la Soc. Mexicana de Histria Natural, 1938-—. Research on protozoa, especially parasitic protozoa of man, Mexican natural renewable resources. Home: Malaga 44, Mexico 19, D.F., Office: Dr. Vertiz 724, México 12, D.F.*

BELZONI, Giovanni Battista, archaeologist; b. Padua, Italy, Nov. 15, 1778. Earned living as strong man, London, (Eng.) 1803-12, and elsewhere; lived in Egypt, 1815-19; invented water wheel for use in Egypt, 1815; returned to Eng., 1820, and exhibited Egyptian collection in Piccadilly; died on expedition to Timbuktu. Author: Narrative of the Operations and Recent Discoveries in the Pyramids, Temples, Tombs, and Excavations in Egypt and Nubia, 1820. Pioneer in archaeology of Egypt; known as collector of antiquities; removed bust of Rameses II to British Museum, 1816; opened temple of Abu Simbel and tomb of Sethos I, 1817; opened 2d pyramid of Giza (tomb of Chephren), 1818; discovered tombs of Rameses I and Mentuherkhepesh; transferred his discoveries, including obelisks and sarcophagus of Sethos to British Museum. Died Gwato, Nigeria, Dec. 3, 1823.

BEMMELEN, see van Bemmelen.

BEN ANATOLI, Jacob, see Anatoli, Jacob.

BENASSI, Enrico, Italian physician; b. Parma, Italy, June 11, 1901; s. Umberto and Livia (Gocciadoro) B.; M.D., U. Parma, 1925. Prof. radiology U. Ferrara, 1947-54; prof. radiology, dir. radiol. dept. U. Turin (Italy), 1955-—; dir. Radiobiol. Centre Mario Ponzio, Turin Mem. Italian Soc. Med. Radiology and Nuclear Medicine (pres.), numerous others. Research, numerous publs. on med. radiology and nuclear medicine, diagnosis and therapy, immunology and exptl. oncology. Home: 32 Galliari. Office: 3 Genova, Torino, Italy.*

BENAZZI, Mario, Italian biologist; b. Cento, Ferrara, Italy, Aug. 29, 1902; s. Cesare and Benvenuta (Bergami) B.; D.Natural Scis., U. Bologna (Italy), 1925; m. Giuseppina Lentati, Sept. 8, 1931; Asst., Faculty Scis., U. Torino (Italy), 1926-34; asst. prof. U. Sassari (Italy), 1935-36; prof. gen. biology U. Siena (Italy), 1937-46; prof. zoology, dir. dept. zoology and comparative anatomy U. Pisa (Italy), 1946-—. Recipient Medaglia d'oro Merito Scientifico, 1965; Benemeriti della Scuola della cultura e dell'arte, 1965. Mem. Accademia Nazionale dei Lincei (corr.). Author: Problemi biologici della Sessualità, 1947; also numerous articles. Editor: Biologia e Zoologia generale-Genetica, 1966. Research on comparative endocrinology of thyroid and pituitary glands, hybridization in Urodela, cytogenetics in fresh-water planarians. Home: Via Dell' Ordine S. Stephano, 11-Marina di Pisa. Office: Istituto di Zoologia, Via A. Volta, 4 Pisa, Italy.*

BENCHIMOL, Alberto, physician; born Belem, Brazil, Apr. 26, 1932; s. Isaac and Nina (Siqueira) B.; M.D., U. Brazil, 1956; m. Helena L. Levy, Apr. 14, 1962; 1 son, Nelson. Came to U. S., 1957, naturalized, 1963. Instr. medicine U. Brazil, Rio de Janeiro, 1956; with U. Kan. Med. Sch., 1957-60, instr. medicine, 1959-60; asso. Scripps Clinic Research Found., La Jolla, Cal., 1960-66; dir. Inst. for Cardiovascular Diseases Good Samaritan Hosp., Phoenix, 1966-—. Mem. Am. Fedn. Clin. Research, Western Soc. Clin. Research, Am. Coll. Cardiology (gov. Ariz.), A.M.A., Am. Physiol. Soc., Am. Heart Assn. Cons. editorial bds. Am. Heart Jour., Am. Jour. Cardiology. Research, publs. on structure of conduction system of heart; effects of elec. currents on heart; basic control of cardiac functions; developed techniques to document early signs of coronary heart disease in man. Home: 7802 N. Central Av., Phoenix 85020. Office: 1033 E. McDowell Rd., Phoenix 85002.*

BENCINI, Alberto, Italian ophthalmologist; b. La Spezia, Italy, Aug. 15, 1897; s. Guido and Medea (Tardini) B.; Ph.D. in ophthalmology. Dir. opthalmology clinic U. Siena (Italy). Mem. Italian Ophthal. Soc., Siena Societá Internazionale della elettroretinagrafia clinica. Research and numerous publs. on herpes, toxoplasmosis; 1st in Italy to apply Gonin method for operations on detatched retina. Home: via Montanini 110 Office: via Monna Agnese 22, Siena, Italy.*

BENCOSME, Sergio A., pathologist; b. Monte Cristy, Dominican Republic, Apr. 27, 1920; s. Sergio Arturo and Floralba (Castillo) B.; B.Sc. in Biol. Sci., Escuela Normal Superior, Santo Domingo, Dominican Republic, 1938; M.D., U. Montreal (Que., Can.), 1947; M.Sc. in Pathology, McGill U., Montreal, 1948, Ph.D., 1950; m. Josefina Rojas, May 31, 1947; children—Rosanna, Yolanda, Violetta, Rolando, Humberto. Asst. prof. pathology U. Ottawa (Ont., Can.), asst. pathologist Ottawa Gen. Hosp., 1951-53; asso. prof. pathology Queen's U., Kingston, Ont., 1953-57, 59-65, prof., 1965-—; asst. research pathologist U. Cal. at Los Angeles, 1957-59. Mem. Am. Assn. Pathologists and Bacteriologists, Soc. for Exptl. Pathology, Soc. for Cell Biology, Internat. Acad. Pathology, Can. Assn. Pathologists, Canadian Physiology Soc. Research, publs. on devel. histological techniques especially pancreatic islets, characterization of RNA-containing structures at ultrastructural level, heart and structure of basement membrane; described histogenesis of islet cells, new cell of ventral origin in pancreas; demonstration of direct relationship between alpha cells and presence of glucagon in pancreas. Home: 37 Jorene Dr., Kingston, Ont., Can.*

BENCZE, George, Hungarian physician; b. Kiskunhalas, Hungary, Oct. 23, 1922; s. Joseph and Irma (Weiszberger) B.; med. dr. degree U. Med. Sch., Szeged, Hungary, 1950, sci. degree, 1960; m. Eva Fay, Oct. 13, 1951; children—Steve, Catharine. Staff, Path. Inst., U. Med. Sch., Szeged, 1948-50, 1st dept. medicine, 1950-64, 65-—; staff rheumatism research unit Canadian Red Cross Meml. Hosp., Taplow, Maidenhead, Eng., 1964. Mem. Heberden Soc., Royal Soc. Immunology, Internat. Soc. Rheumatology. Research, publs. on immunology and clin. relationship of rheumatic diseases especially collagen mesenchymal disease; discovered 2 types of lupus erythematosus cell factor and transferred disease to dogs. Home: 2/a Karolyi, Szeged, Hungary.*

BEN-DAVID, Joseph, sociologist; b. Gyor, Hungary, Aug. 19, 1920; s. David and Gisela (Mayer) Gross; to Palestine, 1941; student Hebrew U., Jerusalem, Israel, 1943-47, M.A. in History and Sociology of Culture, 1950, Ph.D. in Sociology, 1955; certificate in Social Adminstrn., London Sch. Econs., 1947-49; m. Miriam Sternberg, 1947; 3 children. With Nat. Service, 1939-51; with Hebrew U., 1951-—, asso. prof. sociology, 1965-—; vis. prof. U. Cal. at Berkeley, 1964-65; fellow Center for Advanced Study in Behavioral Scis., Stanford, Cal., 1957-58; vis. prof., U. Chicago, 1968-—. Mem. Nat. Council for Research and Devel., Prime Minister's Office, Jerusalem, 1964-—. Mem. Internat., Am. sociol. assns., Israel Hist. Soc., Israel Demographic Soc., Research Com. on Sociology Medicine, Research and publs. in social stratification, comparative sociology, sociology of sci. and professions in 19th century, social psychology, impact of edn. on career expectations and mobility. Home: 8 Berkhiahu, Jerusalem, Israel.*

BENDAVID (Ben David), Lazarus; German philosopher, mathematician; b. Berlin, Germany, Oct. 18, 1762; ed. U. Göttingen (Germany), U. Halle (Germany). Became prof., Vienna, Austria, 1793; returned to Berlin, 1797; dir. Berlin Freischule, 1806-25; accountant for comml. and financial instns. until death. Author: Über die Parallellinien, 1786; Vorlesungen über die Kritik der Praktischen Vernunft, 1796; Vorlesungen über die Kritik der Reihen Vernunft, 1796; Beiträge zur Kritik des Geschmacks, 1797; Versuch einer Geschmackslehre, 1798; Über der Ursprung unserer Kenntnisse (honored by Acad. Berlin), 1802; Versuch einer Rechtslehre, 1802; On the Jewish Calendar. Editor Spenerische Zeitung. Helped spread Kantian philosophy. Died Berlin, Mar. 28, 1832.

BENDER, Hans, German psychologist; b. Freiburg im Breisgau, Germany, Feb. 5, 1907; Ph.D., Bonn (Germany) U.; M.D., Strasbourg (France) U., 1941; m. Henriette Wichert, 1941; 2 daus., 1 son. Asst. Psychol. Inst., Bonn U., 1935-40; asst. prof. U. Strasbourg, 1941-42, prof. psychology, dir. Institut für Psychologie und Klinische Psychologie, 1942-45; prof. psychology U. Freiburg, 1946-54, prof. psychology and border areas in psychology, 1954-—; dir. Institut für Grenzgebiete der Psychologie und Psychohygiene, Freiburg, 1950-—. Mem. Deutsche Gesellschaft für Psychologie, Parapsychol. Assn., Societa Italiana di Parapsicologia (hon.), Society for Psychical Research, London, Eng. (corr.). Author: Parapsychologie—Ihre Ergebnisse und Probleme, 1935; Psychische Automatismsen: Zur Experimentalpsychologie des Unterbewusten und der Aussersinnlichen Wahrnehmung, 1936; Zum Problem der Aussersinnlichen Wahrnehmung, 1936. Editor: Zeitschrift für Parapsychologie und Grenzgebiete der Psychologie und Psychohygiene. Works on wide range of subjects in field of parapsychology including extrasensory perception, psychodinesis, poltergeists, mediumship, spontaneous phenomena, spiritual healing. Address: Institut für Grenzgebeite der Psychologie und Psychohygiene, Eichhalde 12, 78 Freiburg im Breisgau, Germany.

BENDER, Harry Albert, Am. mathematician, educator; b. Uhrichsville, O., Oct. 6, 1895; s. John and Minnie (Hite) B.; A.B., Ohio U., 1918; M.A., U. Ill., 1921, Ph.D., 1923; m. Agnes L. Harris, June 4, 1919; children—Roger H., Robert L., Richard A. Prof. math., physics Muskingum Coll., 1919-20; asst. in math. U. Ill., Urbana, 1920-23, instr. math., 1923-25, asso. in math., 1925-28; asst. prof. math. U. Akron (O.), 1928-36, asso. prof., 1936-44; faculty math. U. R.I., Kingston, 1944-—, prof. emeritus, 1964-—. Mem. Am. Math. Soc., Math. Assn. Am., Sigma Xi, Phi Kappa Phi, Pi Mu Epsilon. Author: Engineering Mathematics, 1941; College Algebra, 1950, 56; Bivariate Normal Distribution, 1957. Research, publs. on Abelian groups, especially isomorphisms and determination; also on constrn. of bivariate normal distbn. tables. Home: 32 Bayberry Rd., Kingston, R.I. 02881.*

BENDER, Israel Boris, dentist; born Russia, Nov. 4, 1905; s. Boris N. and Anna (Futeransky) B.; came to U. S., naturalized, 1914; D.D.S., U. Pa., 1930; m. Lillian Zebooker, Feb. 6, 1932; children—Andrew, Susan. Faculty, U. Pa., 1942-—, asso. prof. oral medicine, 1959-—, asst. prof. Grad. Sch. Medicine, 1954-—, chmn. dept. dentistry Albert Einstein Med. Center, 1963-—; vis. prof. U. So. Cal., 1960-61; vis. lectr. Dept. Stomatology Boston U., 1960; pvt. practice dentistry, 1932-—. Fellow A.A.A.S., Am. Coll. Dentists, Am. Bd. Endontists; mem. Am. Assn. Endodontists, Sigma Xi. Author: (with Samuel Seltzer) Dental Pulp, 1965. Research, publs. on oral medicine, post extraction bacteremia and its control with use of antibiotics; detection of systemic diseases by means of dental x-rays; effects of various dental manipulative procedures on dental pulp; detection of occult bone lesions in Mandible and Maxilla. Home: 7900 York Rd., Elkins Park, Pa. 19117. Office: 1551 Champlost Av., Phila. 19141.*

BENDER, Lauretta (Mrs. Henry B. Parkes), Am. child psychiatrist; b. Butte, Mont., Aug. 9, 1897; d.

John and Catherine (Irvine) Bender; B.S., U. Chgo., 1922, M.A., 1923; M.D., U. Ia., 1926; m. Paul Schilder, 1936; children—Michael, Peter, Jane; m. 2d, to Henry B. Parkes, May 1967. Senior psychiatrist Bellevue Hosp., N.Y.C., 1930-56, head children's psychiat. div., 1934-56; prof. clin. psychiatry N.Y. U., 1951-58; dir. Child Guidance Clinic, N.Y. Infirmary, 1952-60; prin. research scientist children's unit Creedmoor State Hosp., Queens Village, N.Y., 1956-60, dir. psychiat. research, 1960—; prof. psychiatry Adelphi Coll. Grad. Sch., Garden City, N.Y., 1957—; clin. prof. psychiatry Columbia Coll. Physicians and Surgeons, N.Y.C., 1960—. Recipient Elizabeth Blackwell award 1949, Adolf Meyer award 1955, N.Y. Infirmary award of merit, 1960; named Med. Woman of Year, N.Y. State, 1958. Mem. A.M.A., Am. Psychiat. Assn., Am. Neurol. Assn., Am. Orthopsychiat. Assn., Am. Acad. Child Psychiatry, Assn. for Research in Nervous and Mental Disease, Am. Assn. Mental Deficiency, Am. Soc. for Human Genetics, Internat. Congress Child Psychiatry, Am. Soc. Psycopathology of Expressian (v.p.), Soc. Biol. Psychiatry. Author: Visual Motor Gestalt Test and Its Clinical Use, 1937; Bellevue Studies vol. 1, 1952, vol. 2, 1953, vol. 3, 1954, vol. 4, 1956. Editor: Contributions to Developmental Neuropsychiatry (Paul Schilder). Research, publs. on application of Gestalt Test to children; childhood schizophrenia, brain damaged children; therapy, social psychiatry. Home: 44 Malone Av., Long Beach, N.Y. Office: Creedmoor State Hosp., Queens Village, N.Y. 11427.*

BENDER, Michael A., Am. biologist; b. N.Y.C., July 25, 1929; s. Clifford A. and Margaret (Rigg) B.; student Rutgers U., 1947-49; B.Sc., U. Wash., 1952; Ph.D., Johns Hopkins, 1956; m. Renee B. Parkyn, Sept. 16, 1950; children—Michele Alice, Leslie Claire. NIH-Nat. Cancer Inst. fellow Johns Hopkins, 1956-58; biologist, biology div. Oak Ridge Nat. Lab., 1958—. Fellow A.A.A.S.; mem. Genetics Soc. Am., Am. Genetics Assn., Radiation Research Soc., Am. Soc. Cell Biology. Research, numerous publs. on radiation-induced chromosome aberrations in mammals, devel. use chromosome aberration yields as method biol. dosimetry in cases accidental human radiation exposure, human chromosome aberration induction in connection with NASA manned spaceflight program. Home: 109 Wildwood Dr., Oak Ridge 37830. Office: Biol. Div., Oak Ridge Nat. Lab., Oak Ridge 37830.*

BENDER, Morris Boris, physician; born Uman, Russia, June 8, 1905; s. Boris and Anne (Nemirowsky) B.; came to U.S., 1914, naturalized, 1924; B.S., U. Pa., 1927, M.D., 1931; m. Sara Spirtes, June 28, 1936; children—Barbara (Mrs. Martin Steiner), Adam, Barnaby, Victor, Leila. Practice medicine, specializing in neurology, N.Y.C., 1938—; prof. clin. neurology N.Y. U., 1951-66; clin. prof. neurology Columbia U., 1953—; dir. Neurology Service, Bellevue Hosp., N.Y.C., 1951-61; dir. dept. neurology Mt. Sinai Hosp., N.Y.C., 1951-66; prof., chmn. dept. neurology Mt. Sinai Sch. Medicine, N.Y.C., 1966—. USPHS Spl. cons. to surg. gen., 1962-64. Recipient Jacobi medal Mt. Sinai Hosp., 1957. Mem. Am. Neurol. Assn., Internat. Congress Neurology, Am. Physiologic Soc., Am. Acad. Neurology, Am. Psychol. Assn., Am. Psychiat. Assn., Assn. Research Nervous and Mental Diseases, Soc. Exptl. Biology and Medicine. Author: Disorders in Perception, 1952; Visual Field Defects After Penetrating Missile Wounds of the Brain, 1960; The Oculomotor System, 1964; The Approach to Diagnosis in Modern Neurology, 1967. Editorial bd. Jour. of Mt. Sinai Hosp., 1953-57. Contbr. numerous articles to profl. jours. Research in vision, perception, oculomotor system and neurophysiology. Home: 400 E. Shore Rd., Great Neck, N.Y. 11024. Office: 1150 Park Av., N.Y.C. 10028.*

BENDER, Myron Lee, Am. chemist; b. St. Louis, Missouri, May 20, 1924; the son of Averam Burton and Fannie (Leventhal) Bender; B.S., Purdue U., 1944, Ph.D., 1948; postgrad. Harvard, U. Chgo.; m. Muriel Schulman, June 8, 1952; children—Alec Robert, Bruce Michael, Steven Pat. Instr., U. Conn., 1950-51; faculty Ill. Inst. Tech., 1951-60; asso. prof. Northwestern U., Evanston, Ill., 1960-62, prof., 1962—. Cons. Sinclair Research Labs., 1955-60, Melpar, Inc., 1960-65; mem. biochemistry study sect. NIH, 1963-67. Alfred P. Sloan Found. Research fellow, 1959-65; AEC fellow, 1949-50. Mem. Am. Chem. Soc., Nat. Acad. Scis., Chem. Soc. London, Am. Assn. U. Profs., A.A.A.S., Sigma Xi. Research, publs. on mechanism of ester hydrolysis, kinetic isotope effects, mechanistic studies on organic models of enzymes, mechanisms of action of proteolytic enzymes, theories of enzyme catalysis. Home: 2514 Sheridan Rd., Evanston, Ill. 60201.*

BENDER, Peter Leopold, Am. physicist; b. N.Y.C., Oct. 18, 1930; s. Clifford A. and Margaret (Rigg) B.; B.A., Rutgers U., 1951; Ph.D., Princeton, 1956, M.A., 1957; m. Bernice L. Koettgen, June 13, 1953; children—Carolyn Ann, Paul Rigg, Alan David. Nat. Acad. Sci.-NRC research asso. Nat. Bur. Standards, 1956-57, physicist Atomic Physics div., Washington, 1957-62, physicist Lab. Astrophysics div., Boulder, Colo., 1962—; mem. Joint Inst. For Lab. Astrophysics, 1962—; adj. prof. dept. physics and astrophysics U. Colo., 1964—. Fulbright grantee, Leiden U., 1951-52; Charles Coffin fellow, 1964-65; recipient Exceptional Service award Dept. Commerce, 1959, Samuel Wesley Stratton award Nat. Bur. Standards, 1962, Arthur S. Flemming award, 1967. Mem. Am. Phys. Soc., Am. Astron. Soc., Am. Geophys. Union, Optical Soc. Am., A.A.A.S. Research on fundamental constants, optical pumping, lasers and atomic collisions. Office: Joint Inst. for Lab. Astrophysics, Boulder, Colo. 80302.*

BENDINELLI, Goffredo, Italian archeologist; b. Città di Castello, Jan. 22, 1888; s. Torello and Clelia (Ravaioli) B.; ed. in letters, archeology, art history; m. Amelia Binotti, May 15, 1919; 1 son, Nello. Prof. archeology and art history U. Turin (Italy), 1925—. Mem. Nat. Inst. Archeology and Art History (Rome), Pontif. Accad. Rom. di Archeologia, German Archeol. Inst., Academy Arcadia (Rome). Author: Dottrina dell'archeologica e della storia dell'arte, 1938, also numerous publs. Specialist in classical archeology. Address: via Barbaroux 1, Turin, Italy.*

BENDITT, Earl Philip, Am. med. scientist, educator; b. Phila., Apr. 15, 1916; s. Milton and Sarah (Schoenfeld) B.; B.A., Swarthmore Coll., 1937; M.D., Harvard, 1941; m. Marcella Wexler, Feb. 18, 1945; children—John Milton, Alan Paul, Joshua Oliver. Instr. pathology U. Chgo., 1945-47, asst. prof., 1947-52, asso. prof., 1952-57; asst. dir. research LaRabida Childrens Sanitarium, Chgo., 1948-54; prof., chmn. dept. pathology, sch. medicine U. Wash., Seattle, 1957—. Mem. Phi Beta Kappa, Sigma Xi. Editorial bd. Lab. Investigation, 1962—, Jour. Exptl. and Molecular Pathology, 1962—; editor Electron Microscopic Atlas of Pathology, 1966. Home: 4528 W. Laurel Dr. N.E., Seattle 98105.*

BENDIX, Reinhard, Am. sociologist; b. Berlin, Germany, Feb. 25, 1916; s. Ludwig and Else (Henschel) B.; B.A., U. Chgo., 1941, M.A., 1943, Ph.D., 1947; m. Jane L. Walstrum, July 5, 1940; children—Karen, Erik, John. Faculty, U. Chgo., 1943-46, U. Colo., 1946-47; faculty U. Cal., Berkeley, 1947—, prof. sociology, 1956—, chmn. dept. sociology, 1958-61; vis. prof. Free U. Berlin, 1963, Theodore Heuss prof., 1964-65; vis. fellow St. Catherine's Coll., Oxford, 1965. Fulbright Research grantee, Germany, 1953-54; Carnegie Found. fellow, 1961-62. Mem. Am. (vice president 1964-65; recipient of the MacIver award 1958), Pacific sociol. assns., Internat. Sociol. Assn. (v.p.), Deutsche Gesellschaft für Soziologic, Phi Beta Kappa. Author: Higher Civil Servants in American Society, 1949; Social Science and the Distrust of Reason, 1951; Work and Authority in Industry, 1956; (with S. M. Lipset) Social Mobility in Industrial Society, 1959; Max Weber, An Intellectual Portrait, 1960; Nation Building and Citizenship, Studies in our Changing Social Order, 1964. Editor: (with S. M. Lipset) Class, Status and Power, 1953, 66. Contbr. numerous articles to profl. jours. Home: 3 Orchard Lane, Berkeley, Cal. 94704.*

BENDIX, Vincent, Am. engr., inventor, industrialist; b. Moline, Illinois, 1882; s. John Benedix; mech., engring. and gen. edn.; m. Elizabeth Channon, Apr. 6, 1922 (div.). Organized and chmn. bd. Bendix Aviation Corp., comprising allied group mfrs. here and abroad of automotive, aviation, marine and industrial apparatus until ret. in 1942; instrumental in devel. and formation corp. for manufacture Bendix Home Laundry; dir. Pioneer Instument Co., Bendix Eclipse of Can., Bendix Home Appliances, Jaeger Watch Co. Founder and sponsor Bendix Transcontinental Air Race and donor Bendix Trophy; sponsor Internat. Glider Meet, Elmira, N.Y., and donor Bendix Glider Trophy. Past pres. Soc. Automotive Engrs. Early pioneer in design and bldg. of automobiles; inventor Bendix drive which made automobile self starting practicable, has been widely used; introduced in U. S. first volume prodn. 4-wheel brakes for automobiles. Died Mar. 27, 1945.

BENECKE, Ernst Wilhelm, German geologist, paleontologist; b. Berlin, Germany, Mar. 16, 1838; s. Viktor and Emeline (Schunck) B.; doctorate Heidelberg, Germany; m. Emilie Fallenstein, 1866; 3 children, including Wilhelm, Marie (Mrs. Artur Benno Schmidt). Worked with A. Oppelin, Munich, Germany; became lectr. Heidelberg U., 1865, asso. prof., 1869; became prof., Strasbourg, France, 1872; commd. to establish and direct state inst. for geol. research and cartography of Alsace-Lorraine, 1892; publisher jours. Geognostische paläontologische Beiträge, 1865-76, Neue Jahrbuch für Mineralogie, Geologie, Paläontologie. Author: Über Trias und Jura in den Südalpen, 1866; Trias in Elsas-Lothringen und Luxemburg, 1875; Abriss der Geologie von Elsass-Lothringen, 1878; Geognostische Beschreibung der Umgebung von Heidelberg, 1879; Die Gliederung der Eisenerzformation in Deutsch-Lothringen und Luxemburg, 1892; Übersicht der geologischen Verhältnisse von Elgass-Lothringen, 1897; Geologische Führer der Elsass, 1900; Neue geologische Landesaufnahme von Elsass-Lothringen. Research on geology of Alsace-Lorraine, so. limestone Alps, triassic formations left of Rhine in so. Germany, jurassic formation in Lorraine and other areas. Died Strasbourg, Mar. 6, 1917.

BENEDEN, Edouard van, see van Beneden, Edouard.

BENEDETTI, Alessandro, Italian anatomist, physician; b. Legnano (Verona), Italy, 1460; degree U. Padua (Italy), 1475; physician with Venetian armies in Greece for several years; returned to Italy, circa 1490; founder anat. theater, tchr., Padua; named chief surgeon of Italian army against Carlo VIII, 1495; began teaching in Venice, Italy, 1497. Author: Anatomia sive historiae corporis humani, 1493; De abditis nonnullis mirandis morbis ac sanationis causis (prefigured work of Morgagni). Broke with tradition of Galen; one of 1st (with Antonio Benivieni) to use own observations as basis for med. practice. Died Venice, Oct. 31, 1512.

BENEDETTI, Gaetano, Swiss physician; b. Catane, July 7, 1920; s. Umberto and Vincenzina (Gaglio) B.; M.D., U. Catane; m. Hanni Straub, July 16, 1948; children—Christoph, Corradino, Iürg, Dorotea. Prof. agrégé psychiatry U. Zurich, 1953; prof. agrégé at Rome, 1956; prof. mental hygiene U. Basel (Switzerland), 1956. Mem. Internat. Soc. Psychoanalysis, Soc. Physicians of Basel, Soc. Med. Psychotherapy. Author: Forschungen über Schizophrenielehre, 1945; Alkoholhalluzinose, 1950; Klinische Psychotherapie, 1963. Home: Steinengraben 79. Office: Petersgraben 1, Basel, Switzerland.*

BENEDETTI, Giambattista, Italian physicist, mathematician; b. Venice, Italy, 1530; pupil of Tartaglia; became engr. for Duke of Savoy, 1567. Author: Resultio omnium Euclidis problematum aliorumque, 1553; De gnomonum umbrarumque solarium usu, 1574; Diversarum speculationum mathematicarum et physicarum liber, 1580; De coelo et elementis, 1585. Identified existence of centrifugal force; calculated balance of torque; worked on constrns. with ruler, also with ruler and compass; proposed theory of equilibrium of liquids in communicating vessels; stated water pressure on bottom of any vessel proportional to cross sect.; foreshadowed Pascal's hydraulic press; replaced Aristotle's notion of absolute heaviness and lightness with one of relative heaviness and lightness (velocity of body in free fall proportional to its absolute weight minus resistance of medium); his physics based on acceptance of doctrine of impetus; postulated concept of infinity in math. and physics. Died 1590.

BENEDICT, Albert Alfred, Am. microbiologist; b. Pasadena, Cal., Nov. 26, 1921; s. George G. and Irene (Jones) B.; B.A., U. Cal., Berkeley, 1948, M.A., 1950, Ph.D., 1952; m. Marion A. Felker, Dec. 28, 1947; 1 dau., Donna. Faculty, Med. br. U. Tex., Galveston, 1952-57, U. Kan., 1957-63; prof. U. Hawaii, Honolulu, 1963—, chmn., 1964—; guest investigator Rockefeller Inst., N.Y.C., 1960-61. Mem. Am. Soc. Microbiology, Am. Assn. Immunologists, Soc. Exptl. Biology and Medicine, A.A.A.S., Sigma Xi. Contbr. numerous articles to profl. jours. Studies on mechanism of antibody synthesis and on structure of antibodies. Office: Dept. of Microbiology, University of Hawaii, Honolulu 96822.*

BENEDICT, Francis Gano, Am. chemist, physiologist; b. Milw., Oct. 3, 1870; s. Washington Gano and Harriet Emily (Barrett) B.; A.B., Harvard, 1893; A.M., 1894; Ph.D., U. Heidelberg, 1895; Sc.D., Wesleyan U., Conn., 1911, U. Me., 1924; hon. M.D., U. Wurzburg, 1932; m. Cornelia Golay, July 28, 1897; 1 dau., Elizabeth Harriet. Instr. chemistry Mass. Coll. Pharmacy, 1892-94; instr. and asso. prof. chemistry Wesleyan U., 1896-1905, prof., 1905-07; dir. Nutrition Lab., Carnegie Instn. of Washington, 1907-37. Chemist Storrs Expt. Station, 1896-1900; physiol. chemist nutrition investigations U. S. Dept. Agr., 1895-1907. Fellow Am. Acad. Arts and Scis., A.A.A.S.; mem. Am. Chem. Soc., Am. Physiol. Soc., Nat. Acad. Scis., Am. Philos. Soc., Soc. Royale des Sciences Medicales et Naturelles de Bruxelles, Gesellschaft der Aerzte (Vienna), Kaiserlich Deutsche Akademie der Naturforscher (Halle); corr. mem. Royal Med. Soc. of Budapest, Hungary, 1934. Author: Elementary Organic Analysis, 1900; Chemical Lecture Experiments, 1901; also papers on organic and physiol. chemistry and numerous monographs pub. by Carnegie Instn. of Washington. Developed closed circuit respiration apparatus and calorimeter; studied oxygen content of atmosphere; examined conditions requisite for measurement and calculation basal metabolism; greatly expanded knowledge of comparative metabolism of vertebrate animals. Died May 14, 1957.

BENEDICT, Harris Miller, Am. botanist; b. Buda, Ill., Dec. 8, 1873; s. Miller Samuel and Anna Maria (Harris) B.; B.A., Doane Coll., 1894; B.S., U. Neb., 1896, A.M., 1897; Ph.D., Cornell, 1914; m. Florence Stevens McCrea, 1906; children—Harris, Jean, Ann, Martha, McCrea. Head of biol. dept. Lincoln (Neb.) High Sch., 1897-99, Omaha High Sch., 1899-1902; instr. biology U. Cin., 1902-03, asst. prof. 1904-08, asso. prof., 1908-11, prof. 1911-14, prof. botany, from 1914, organized dept. botany, 1914. Originator, 1908, and dir. Emery Bird Reserve (the first city bird reserve), Cin.; organized the first sch. garden courses, for tchrs. in Cin. schs., and a pre-agrl. course for univ. students. Research and publs. on senility in perennial woody plants, nature study and bird protection. Died Oct. 17, 1928.

BENEDICT, Manson, Am. chem. engr.; b. Lake Linden, Mich., Oct. 9, 1907; s. C. Harry and Lena I. (Manson) B.; B. Chemistry, Cornell, 1928; M.S., Mass. Inst. Tech., 1932, Ph.D., 1935; m. Marjorie Oliver Allen, July 6, 1935; children—Mary Hannah (Mrs. Myran C. Sauer, Jr.), Marjorie Alice (Mrs. Martin Cohn). NRC fellow chemistry, 1935-36; research asso. geophysics Harvard, 1936-37; research chemist M. W. Kellogg Co., 1938-43; in charge process design gaseous diffusion plant for uranium-235 Kellex Corp., 1943-46; dir. process devel. Hydrocarbon Research, Inc., 1946-

51; tech. asst. to gen. mgr. U. S. A.E.C., 1951-52; prof. nuclear engring. Mass. Inst. Tech., 1951—, head dept. nuclear engring., 1958—; sci. adviser Nat. Research Corp., 1951-58, dir., 1962-67; dir. Atomic Indsl. Forum. Mem. Mass. Adv. Com. on Radiation Protection; mem. gen. adv. com. U. S. AEC, 1958—, chmn., 1962-64. Recipient Indsl. and Engring. Chemistry award Am. Chem. Soc., 1962, Perkin medal Soc. Chem. Industry. Fellow Am. Acad. Arts and Sci.; mem. Am. Inst. Chem. Engrs. (William H. Walker award 1947), Am. Nuclear Soc. (pres. 1962-63, dir.), Nat. Acad. Scis. (Founders award 1965), Nat. Acad. Engring., Sigma Xi. Co-editor: Engineering Developments in the Gaseous Diffusion Process, 1949. Co-author: Nuclear Chemical Engineering, 1957. Devised practical procedure for separating U-235 from uranium hexafluoride; research on thermodynamics of gases, isotope separation. Home: 25 Byron Rd., Weston, Mass. 02193. Office: Dept. Nuclear Engring., Mass. Inst. Tech., Cambridge, Mass.*

BENEDICT, Ruth Fulton, Am. anthropologist; b. N.Y.C., June 5, 1887; d. Frederick S. and Beatrice J. (Shattuck) Fulton; A.B., Vassar Coll., 1909; Ph.D., Columbia U., 1923; m. Stanley R. Benedict, June 18, 1914 (dec. Dec. 1936). Lectr. in anthropology, Columbia U., 1924-30, asst. prof., 1930-36, asso. prof., 1936-48, prof., 1948; field trips to Am. Indian tribes, 1922-39; on leave with Bur. Overseas Intelligence, O.W.I., 1943-46. Author: Concept of the Guardian Spirit in North America, 1923; Patterns of Culture, 1934; Zuni Mythology (2 vols.), 1935; Race: Science and Politics, 1940; The Chrysanthemum and the Sword: Patterns of Japanese Culture, 1946. Contbd. toward enlarging scope of anthropology to include various culture motifs (or patterns) and the role of culture in devel. of individual personality; made important studies of the Zuni Indians (New Mexico), the Dobrians (New Guinea), and the Kwakiutl Indians (Brit. Columbia); was one of 1st to describe culture as an integrated whole; illustrated her concept of cultural integration by reference to certain Am. Indian tribes whose culture she believed to be integrated around an ideal personality type (Apollonian or Dionysian). Died N.Y., Sept. 17, 1948.

BENEDICT, Stanley R(ossiter), Am. chemist; b. Cin., Mar. 1884; s. Wayland R. and Anne (Kendrick) B.; B.A., U. Cin., 1906; Ph.D., Yale, 1908; m. Ruth Fulton, 1914. Instr. chemistry Syracuse U., 1908-09; asso. in biol. chemistry Columbia, 1909-10; asst. prof. chem. pathology, Cornell U. Med. Coll., N.Y., 1910-11, asst. prof. chemistry, 1911-12, prof. 1913—. Mem. Nat. Acad. Scis., A.A.A.S., Am. Soc. Biol. Chemists (pres. 1919-20), Am. Physiol. Soc., Harvey Soc. Mng. editor Jour. Biol. Chemistry. Discovered thiasine (ergothioneine) in blood; devised analytical methods for determining minute quantities of non-protein constituents of blood, thus introducing chem. analysis as useful tool in discovering chem. processes in normal body functioning; a pioneer in attempting to cause regression in tumors by disturbing metabolic processes. Died Dec. 22, 1936.

BENEDICTY, Mario Gustavo de, see de Benedicty.

BENEKE, Everett Smith, Am. mycologist; b. Greensboro, N.C., July 6, 1918; s. Herman H. and Grace (Smith) B.; B.S., Miami U., 1940; M.S., Ohio State U., 1941; Ph.D., U. Ill., 1948. Faculty, Mich. State U., East Lansing, 1948—, prof. botany and plant pathology, 1958—. Tech. acct. Escola Superior de Veterinaria, Belo Horizonte, Brazil, OAS, Washington, 1959; vis. prof. Instituto de Botanica, U. Sao Paulo, 1960. Mem. Mycol. Soc. Am. (pres. 1961), Am. Assn. Bioanalysts (sci. dir. sci. council 1963—), Am. Inst. Biol. Sci. (rep. on bd. 1964—), Internat. Soc. for Human and Animal Mycology, Am. Soc. for Microbiology, Sigma Xi (Jr. award for Distinguished Research Mich. State U. chpt. 1958), Phi Sigma. Author: Medical Mycology-Laboratory Manual, 1966; (with C. J. Alexopoulos) Laboratory Manual for Introductory Mycology, 1962. Research in antitumor substance in fungi; fungi found in aquatic habitats; diagnostic studies of pathogenic fungi and teaching mycology, both med. and indsl. Home: 1664 Forest Hill St., Okemos, Mich. 48864.*

BENEKE, Friedrich W., biologist; b. 1824; developed staining techniques in histology; used lilac aniline in acetic acid, 1862; recognized wide distbn. of cholesterol plants and animal tissues and fluids, 1862. Died 1882.

BENES, Václav Edvard, mathematician; born Brussels, Belgium, July 24, 1930; s. Bohus A. and Emilie (Zadna) B.; came to U. S., 1942, naturalized, 1956; A.B., Harvard, 1950; M.A., Ph.D., Princeton, 1953; m. Janet Franklin, June 7, 1951; children—Andrea, Nicholas. Research mathematician Bell Telephone Labs., Murray Hill, N.J., 1953—. Vis. lectr. Dartmouth, 1959-60. Mem. Am. Math. Soc., Soc. Indsl. and Applied Math., Assn. for Symbolic Logic, Inst. Math. Statistics, Mind Assn., Phi Beta Kappa. Author: General Stochastic Processes in the Theory of Queues, 1963; Mathematical Theory of Connecting Networks and Telephone Traffic, 1965; also articles. Research in analysis and probability with many applications to queues and telephone systems. Home: Mendham, N.J. 07945. Office: Bell Telephone Labs., Inc., Murray, Hill, N.J. 07971.*

BENETATO, Grigore Alexandru, Rumanian physiologist; b. 1905; prof. Medico-Pharm. Inst., Bucharest, Rumania; dir. Inst. Normal and Path. Physiology, Bucharest; mem. Acad. Socialist Republic Rumania. Author: Elementary Treatise on Physiology, 1948; Elements of Normal and Pathological Physiology, 2 vols., 1962. Contbd. to study of central nervous system in immuno-biol., physiol. processes of corticoadrenal gland.

BENFEY, Bruno George, pharmacologist; b. Dusseldorf, Germany, Oct. 9, 1917; s. Bruno and Adele (Sehlheim) B.; M.D., U. Hamburg, Germany, 1948; diploma in chemistry U. Gottingen, 1950; m. Jutta Nienstedt, Aug. 2, 1949; children—Matthias, Martin, Tillmann. Faculty, U. Gottingen, 1951; faculty McGill U., Montreal, Que., Can., 1952—, asso. prof. pharmacology, 1960—. USPHS fellow Yale, 1951-52. Mem. Am. Soc. for Pharmacology and Exptl. Therapeutics, Pharmacol. Soc. Can., Canadian Physiol. Soc. Research, publs. on isolation of antibiotics and hormones, biochemistry, pharmacology cardiovascular agts. Home: 116 Chestnut Dr., Baie d'Urfe, Que. Office: Dept. Pharmacology, McGill U., Montreal, Que., Can.*

BENGMARK, Stig Bertil Samuel, Swedish surgeon; b. Ostervala, Sweden, Apr. 10, 1929; s. Harry and Berta (Karlson) B.; M.D., U. Lund (Sweden), 1956, Ph.D. in Anatomy, 1958; m. Birgit Karlsson, Jan. 1, 1952; children—Camilla, Annette, Thomas, David, Cecilia, Samuel. Jr. instr. anatomy U. Lund, 1952-58, asst. prof. anatomy, 1958-59, resident in surgery, 1956-59; asst. prof. U. Goteborg (Sweden), 1960-62, resident surgery, 1959-62, asso. prof. surgery, dept. surgery, 1962—. Mem. Swedish Young Doctors Assn. (v.p. 1965). Author: Monograph on the Prostatic Urethra, 1958; also numerous articles. Research on surgery, pathology, anatomy, med. orgn., using tissue culture, histology and chem. methods, prostate and liver, regeneration after extensive liver resections in exptl. animals and man. Home: 24 Torild Wulffsgatan, Goteborg, SV, Sweden.*

BENHAMOU, Jean-Pierre, physician; b. Algiers, July 13, 1927; s. Edmond and Danan (Denise) b.; M.D., U. Paris, 1957; m. Francoise Teissier, July 3, 1952; children—Anne Francoise, Mathieu, Laurence. Practice medicine, Clichy, France, 1957—; prof. exptl. pathology Faculty Medicine, U. Paris, 1961—. Author: (with others) Semiologie Medicale, 1965; also numerous articles. Research on liver diseases, particularly hepatic hemodynamics. Home: 6 bis rue d'Auteuil, Paris 16, France. Office: Beaujon Hosp., 92-Clichy, France.*

BENIOFF, Hugo, Am. geophysicist; b. Los Angeles, Calif., Sept. 14, 1899; s. Simon and Alfrieda (Widerquist) Hamilton B.; B.A., Pomona Coll., 1921; Ph.D., Cal. Inst. Tech., 1935; m. Alice Silverman, Feb. 27, 1928; children—Paul, Dagmar (Mrs. E. Friedman), Elena (Mrs. E. G. Jackson, Jr.); m. 2d, Mildred Lent, Oct. 31, 1953; one daughter, Martha Gwen. Asst. Mt. Wilson Obs., 1917-21, seismol. research Carnegie Inst. Washington 1923-24; staff seismol. lab. Cal. Inst. Tech., 1934—, professor, 1950-64, seismology professor emeritus, 1964—; research engr. Submarine Signal Co., 1939-45, Baldwin Piano Co., 1946-62; mem. Ad Hoc Group on Detection Nuclear Explosions; mem. panel on seismic improvement Dept. State; mem. adv. com. for geophysics Air Force Office Sci. Research; chmn. cons. bd. for earthquake analysis Cal. Dept. Water Resources; consultant, Nat. Sci. Found., 1953; research on stress strain characteristics and structure of earth's crust. Recipient Arthur L. Day award Geol. Soc. America; William Bowie medal Am. Geophysical Union, 1965. Fellow Am. Academy of Arts Science, A.A.A.S., Geological Soc. Am.; mem. Nat. Acad. Sci., Am. Phys. Soc., Acoustical Soc. Am., Am. Geophys. Union, Seismol. Soc. Am., Royal Astron. Soc., Phi Beta Kappa. Patentee seismographs. Constructed variable reluctance seismograph, 1930's, which was basis of Geneva Conference nuclear test detection system; developed strain seismograph, Mercury tiltmeter, sensitive Microbarographs, magnetovariagraphs, underwater sound transducers, oscillographs, galvanometers; developed electronic pianos and string instruments; first to suggest that response spectrum was important parameter in antiseismic design; related distribution of aftershocks to dimensions of primary fault; proposed distribution of epicenters to be used as evidence for oceanic fault origin; showed global pattern of strain and release. Died Feb. 29, 1968.

BENIQUÉ, Pierre-Jules, French urologist; b. Paris, 1806; student L'école Polytechnique; M.D., Paris, 1835. Author: De la rétention d'urine et d'une nouvelle méthode pour introduire les bougies et les sondes dans la vessie, 1838; Reflexions et observations sur le traitement des retrecissements de l'urèthre, 1845. Invented surg. probe to dilate the urethra; named in his honor. Died Paris, 1851.

BENIRSCHKE, Kurt, physician; b. Glueckstadt, Germany, May 26, 1924; s. Fritz F. and Marie (Luebcke) B.; student U. Hamburg, 1942, U. Berlin, 1942, U. Wuerzburg, 1943; M.D., U. Hamburg, 1948; m. Marion E. Waldhausen, May 17, 1952; children—Stephen K., Rolf J., Ingrid M. Came to U. S., 1949, naturalized, 1955. Practice medicine, specializing in pathology, Boston, 1955-60, Hanover, N.H., 1960—; faculty Med. Sch. Harvard, 1953-60; pathologist Boston Lying-in Hosp., 1955-60; prof., chmn. dept. pathology Med. Sch. Dartmouth, 1960—; cons. Mary Hitchcock Meml. Hosp., 1960—, VA Hosp., 1960—, NIH, 1961—. Mem. Am. Soc. Pathologists and Bacteriologists, Am., New Eng. colls. pathologists, A.M.A., N.Y. Acad. Sci. Author: (with S. G. Driscoll) Pathology of Human Placenta, 1967. Research, numerous publs. on pathophysiology of human fetus, placental pathology, cytogenetics of mammals. Home: 16 Kingsford Rd., Hanover, N.H. 03755.*

BENIVIENI, Antonio, Italian physician, surgeon, anatomist; born Florence, Italy, circa 1440; ed. Florence, also Pisa, Siena, Italy. Practiced medicine, Florence; physician to Savonarola. Author: De abditis nonnulus ac mirancus morborum et sanationum causis (1st record in modern lit. of postmortem examinations made to find cause of death, also 1st book on path. anatomy, includes probably 1st record of transmission of syphilis from mother to fetus), 1507. One of 1st (with Alessandro Benedetti) to use own observations as basis for med. practice. Died 1502.

BENJAMIN, Asher, Am. architect; b. Greenfield, Mass., June 15, 1773; m. Achsah Hitchcock, Nov. 30, 1797; m. 2d, Nancy Bryant, July 24, 1805. Lived in Greenfield, Mass., 1790's, practiced as architect in area; began archtl. practice, Windsor, Vt., circa 1800, Boston, circa 1805; designed Old South Meeting House, Windsor, 1st Congl. Ch., Bennington, Vt., West Ch., Boston, Carew house, Springfield, Mass., Hollister House, Greenfield. Author: The Country Builder's Assistant, 1797; The American Builder's Companion, 1806; The Rudiments of Architecture, 1814; The Practical House Carpenter, 1830; numerous other works on bldg. style. Popularized Greek Revival architecture. Died Springfield, July 26, 1845.

BENJAMIN, Charles Henry, Am. engr.; b. Patten, Me., Aug. 29, 1856; s. Samuel E. and Ellen M. (Fairfield) B.: M.E. U. Me., 1881; D.E., Case Sch., 1908; m. Cora L. Benson, Aug. 17, 1879. Instr. and prof. mech. engring. U. Me., 1880-86; mech. engr. McKay Machine Co., Boston, 1886-89; prof. mech. engring., Case Sch. Applied Sci., 1889-1907; dean, engring. sch. Purdue U., Lafayette, Ind., 1917-21, dir. Engring. Expt. Sta. Author: Modern American Machine Tools, 1906; Machine Design, 1906; Steam Engine, 1909. Died 1937.

BENJAMIN, Fred Berthold, physiologist; b. Darmstadt, Germany, Oct. 24, 1912; s. Karl J. and Clara (Stern) B.; D.M.D., Bonn (Germany) U., 1935; M.S., U. Ill., 1950; Ph.D., Loyola U., Chgo., 1953; m. Rita Mullerheim, Apr. 22, 1942; children—Peter, Ronald. Came to U. S., 1946, naturalized, 1951. Asst. prof. Med. Sch. U. Pa., 1953-62; asst. to research dir. Naval Aviation Devel. Center, Johnsville, Pa., 1953-58; sr. research coordinator Republic Aviation Corp., N.Y.C., 1960-64; with Office Space Medicine NASA Hdqrs., Washington, 1964—, chief evaluation br., 1964-65, staff asst. Apollo Med. Support, 1965—. Professorial lectr. Georgetown U. Med. Sch., 1964—. Mem. Am. Physiol. Soc., A.A.A.S., Aerospace Med. Assn. Research, publs. on potassium release as the causative process in pain sensation, physiol. response of human body to environmental temperature, to space radiation and to exercise. Home: 11115 Easecrest Dr., Silver Spring, Md. 20902. Office: NASA Hdqrs., M-M, Washington 20546.*

BENJAMIN, Thomas Brooke, English mech. engr., mathematician; b. Wallasey, Eng., Apr. 15, 1929; s. Thomas Joseph and Ethel (Brooke) B.; B.Eng., U. Liverpool, 1950; M.Eng., Yale, 1952; Ph.D., Cambridge U., 1955; m. Helen Gilda Rakower Ginsburg, July 9, 1956; children—Lesley Anne Brooke, Joanna Helen Brooke, Peter Charles Brooke. Sr. asst. in research Cambridge (Eng.) U., 1955-58, asst. dir. research, 1958-66, lab. dir. dept. applied math. and theoretical physics, 1964—. Vis. scientist U. Mich., 1962; vis. prof. Inst. Geophysics and Planetary Physics, U. Cal. at San Diego, 1966-67; cons. English Electric Co., 1956—. Recipient L. F. Moody award Am. Soc. M.E., 1966. Fellow Royal Soc., Inst. Math. and Its Applications. Contbr. articles to profl. jours. Theoretical and exptl. research in fluid mechanics, stability of fluid motions, vortex flows, cavitation and water waves. Home: Inland Close, Grantchester, Cambridge, Eng.*

BENKESER, Robert Anthony, Am. chemist; b. Cin., Feb. 16, 1920; s. Carl A. and Teresa (Koller) B.; B.S., Xavier U., 1942; M.A., U. Detroit, 1944; Ph.D., Ia. State U., 1947; m. Abigail M. Stone, Oct. 19, 1946; children—Carol Ann, Robert G., Paul J., Donald E., Kenneth J. Faculty, Purdue U., 1946—, prof. chemistry, 1954—, asst. head chemistry dept., 1959—. Mem. Am., Brit. chem. socs., Sigma Xi, Alpha Chi Sigma, Phi Lambda Upsilon, Phi Kappa Phi. Research, numerous publs. on organic compounds containing metals chemically bounded to carbon portion especially how presence of metal influences chemistry of organic past of molecule. Home: 2113 Fairway Lane, West Lafayette, Ind. 47906.*

BENKOVSKY, Vasiliy Grigorevich, Russian organic chemist; b. 1912; grad. Chem. Faculty, Odessa U., 1940; D.Tech. Sci., 1952. Head, Lab. Physics and Chemistry of Oil, 1958; dep. dir. for sci. work Petroleum Inst., Kazakhstan Acad. Sci., 1958, now dir. Petroleum Inst. Mem. Kazakh Acad. Sci. (corr.). Author: Surface Tension of Hydrocarbons at Low Temperatures; co-author: Ion Exchange Tars from the Heavy Residues of Oil Refining, 1963. Address: Petroleum Inst., Kazakhstan Acad. Sci., Alma-Ata, Kazakhstan SSR, USSR.

BENMOSCHÉ, M(oses), surgeon; b. London, Eng., Dec. 5, 1883; s. Herman and Jane B.; student St. Mary's Male Acad., Norfolk, Va.; M.D., Med. Coll. Va., 1904; student Middlesex Hosp. and Coll., London, 1908-11; m. Simma Guttwoch, 1908; children—Elkanah, Jacob; m. 2d, Gwladys Goodman, Aug. 5, 1943. Instr. in histology, pathology, bacteriology Med. Coll. Va., 1904-06; asst. in cancer research Middlesex Hosp., London, 1908-09; dir. Pathol. Lab., Mobray Hosp., Capetown, S. Africa, 1911-12; dir. Pathol. Lab., Nashua, N.H., 1913-14; practice surgery, Detroit, 1915-19; gen. surgery practice and research in cancer, N.Y.C., from 1920; sr. in clin. surgery Mt. Sinai Hosp., N.Y.C. Del. to Internat. Congress on Tb, Washington, D.C., 1908. Author: Waifs and Orphans, A Book of Selected Poems; A Surgeon Explains to the Layman, 1940; also monographs on cancer research, surg. papers on method of reconstructing the internal ring in indirect inguinal hernia. First (with Frances I. Seymour) to describe magnification of a fertile human spermatzoon under electron miscroscope. Died Sept. 5, 1952.

BENNDORF, Friedrich August Otto, Austrian archeologist; b. Greiz, Sept. 13, 1838; s. Eduard and Lina (Gorsch) B.; doctorate, circa 1863; m. Sophie Wagner; 1 son. Travelled in Italy, Sicily, Greece, Asia Minor; became prof., Zurich, 1869, Prague, 1872, Vienna, 1877-98; dir. Austrian Archeol. Inst., from 1905. Author: (with R. Schone) Die antiken Bildwerke des Lateranischen Museums, 1867; Griechische und Sicilische Vasenbilder 1869; Die Metopen von Selinunt, 1873; (with G. Niemann) Reisen in Lykien und Karien, 1884, Das Heroon von Gjolbaschi-Trysa, 1889; also articles. Organized archeol. studies and research in Austrian empire; brought to Vienna many finds from Asia Minor, including artifacts from royal graves, Gjolbaschi, Ephesus. Died Vienna, Jan. 2, 1907.

BENNDORF, Hans, physicist; b. Zurich, Switzerland, Dec. 13, 1870; s. Otto and Sophie (Wagner) B.; studied in Vienna under Joseph Stefan, Felix Exner and Ludwig Boltzmann; m. Rosa Wagner, 1899; 3 sons, 1 dau. Asst. to Exner, 1899; prof. physics U. Graz (Austria), 1904-36. Contbr. numerous articles to German and Austrian sci. jours. Refuted Exner's theory on origin of atmospheric electricity by his own tests in Siberia, 1899; studied seismology from 1906; used measurements on earth's surface of volumetric velocity and direction of wave for every point along a ray to recognize larger underground abnormalities (Benndorf law). Died Graz, Austria, Feb. 11, 1953.

BENNE, Erwin John, Am. biochemist; b. Morrowville, Kan., May 21, 1902; s. Henry and Bertha (Thrun) B.; B.S., Kan. State U., 1928, M.S., 1931, Ph.D., 1937; m. Gladys Ethel Meyer, Aug. 22, 1928; children—Richard Gene, Max Erwin, Instr., Kan. State U., 1930-38; faculty Mich. State U., East Lansing, 1938—, research prof. dept. agrl. chemistry, 1946-61, prof. dept. biochemistry, 1961—. Fellow A.A.A.S., Assn. Ofcl. Agrl. Chemists (gen. referee 1960—); mem. Am. Chem. Soc. (chmn. Mich. State U. 1943-44, councillor 1943-44, 46-47), Mich. Acad. Sci. Arts and Letters, Sigma Xi, Phi Kappa Phi, Phi Delta Kappa. Research, publs. on plant and animal nutrition; devel. methods for investigation of composition of biol., agrl. materials. Home: 2353 Hulett Rd., Okomos, Mich. 48864.*

BENNET, Abraham, Brit. physicist; b. Eng., 1750; pastor, Worksworth, then Bentley, Eng. Fellow Royal Soc., 1789. Invented gold-leaf electroscope, 1786, also condensing electrometer; observed charges on evaporating liquids; 1st to observe contact electrification of metals; improved suspension of compass needle; experimented on magnetism of ferruginous bodies and brass. Died 1799.

BENNET, J., French radiologist; b. Pau, France, Apr. 24, 1924; s. Louis and Marie (Camps) B.; ed. Coll. Notre Dame de Sainte Croix, Neuilly, France; student medicine Faculté de Médecine de Paris; M.D., 1954. Became radiologist for Paris Hosps., 1960; asst. chief Hôpital des Enfant malades, Hôpital Saint Antoine; chief radiology dept. Hôpital E. Roux. Mem. French Electroradiology Soc., European Soc. Radiopediatry, French Gastroenterology Soc., Belgian Gastroenterology Soc., Royal Soc. Medicine London, Angeiology Soc. Research and publs. in gastroenterology, pediatric radiology, cardio-vascular and peripheric vascular diseases, abdominal diseases. Home: 38 bd. Marbeau, Paris 16e. Office: 14 rue Théodore de Banville, Paris 17e, France.*

BENNETT, Abram Elting, Am. physician; b. Alliance, Neb., Jan. 12, 1898; s. Charles E. and Bertha (Kinsey) B.; B.S., U. Neb., 1919, M.D., 1921; m. Mary Lou Lucas, Dec. 13, 1954; children—Foster Elting, Evelyn Ann, Jeanne (Mrs. Joseph Wetch). Practice medicine, specializing in neurology and psychiatry, Berkeley, Cal., 1948—; organizer, chief dept. psychiatry Herrick Meml. Hosp., 1948-63; asso. clin. prof. psychiatry U. Cal., 1950-66. Organizer, A. E. Bennett Neuropsychiat. Research Found. 1942; cons. VA; mem. psychiat. adv. com. Am. Bar Found. on mental illness and law. Recipient certificate of merit A.M.A., 1940, Hom. Mention awards, 1946, 56. Diplomate Am. Bd. Psychiatry and Neurology. Fellow Am. Psychiat. Assn. (life), A.A.A.S.; mem. A.M.A., Royal Medico-Psychol. Assn. (London), Assn. for Research in Nervous and Mental Diseases, Am. Neurol. Assn., Soc. Biol. Psychiatry (honorary member and past president), East Bay Psychiat. Assn., No Cal. Psychiat. Soc., Central Neuropsychiat. Assn. (past pres.), Alpha Omega Alpha, Phi Rho Sigma. Author: (with A. Purdy) Psychiatric Nursing Technic, 1940; (with others) The Practice of Psychiatry in General Hospitals, 1956. Contbr. numerous articles to profl. jours. Pioneered devel. psychiat. treatment in gen. hosps., convulsive shock therapy affective disorders; developed successful use of Curare preliminary to metrazol and electroshock in order to prevent traumatic complication of this therapy, Curare diagnostic test for Myasthenia Gravis, investigations in alcoholism, alcoholic brain disease, suicide prevention. Home: 3915 Happy Valley Rd., Lafayette, Cal. 94549. Office: 2000 Dwight Way, Berkeley, Cal. 94704.*

BENNETT, Alfred Rosling, Brit. engr.; b. London, Eng., 1850; ed. Bellevue Acad. Gen. mgr., chief engr. Nat. Telephone Co., Scotland and N. Eng., 1883-90, later Mut. Telephone Co., New Telephone Co.; engr. Municipal Telephone Exchange systems Glasgow, Scotland, Guernsey, Portsmouth, Hull, Brighton (all Eng.), Swansea, Wales. Vice chmn. exec. council Edinburgh (Scotland) Internat. Exhbn., 1890. Mem. Inst. Elec. Engrs., Instn. Locomotive Engrs. (v.p.). Author: A Cheap Form of Voltaic Battery, 1882; An Electrical Parcels Exchange, 1892; Telephone Systems of Continental Europe, 1895; A Convection-Scope and Calorimeter, 1897; Proposals for London Improvements, 1904; Historic Locomotives, 1906; The First Railway in London, 1912; A Saga of Guernsey, 1918; London and Londoners in the 1850's and 1860's, 1924. Contbd. pamphlets on telephonic subjects, papers, to sci. socs. Inventor telephonic translator, 1880, caustic alkali and iron battery, 1881, convection mill, 1896. Died May 24, 1928.

BENNETT, Alfred William, English botanist; b. Clapham, Eng., June 24, 1833; s. William and Elizabeth B.; B.A. with honors in chemistry and botany, U. London, 1853, M.A., 1855, B.Sc., 1868; m. Katherine Richardson, 1858. Bookseller, pub., London, 1858-68; became tutor for family of Gurney Barclay, 1868; apptd. lectr. botany Bedford Coll. also St. Thomas Hosp., 1868; biol. asst. to Dr. Norman Lockyer, 1870-74. Fellow Linnean Soc., Royal Micros. Soc. Author: A Narrative of a Journey in Ireland, 1847; (with G. Murray) Handbook of Cryptogamic Botany, 1889; Flora of Alps, 2 vols., 1896, 97. Translator, editor: (with (Thiselton-Dyer) Lehrbuch der Botanik (Julius Sachs), 3d. edit., 1875. Editor Jour. Royal Microsc. Soc., from 1897. Research on cryptogamic plants, especially fresh water algae, also on pollination and alpine plants. Died London, Jan. 23, 1902.

BENNETT, Clarence Edwin, Am. physicist, educator; b. Providence, May 23, 1902; s. George Wilfred and Clara (Wright) B.; Ph.B., Brown U., 1923, Sc.M., 1924, Ph.D., 1930; m. Ruth Nason Bennett, Sept. 8, 1928; children—Muriel Nason (Mrs. Arthur R. McAlister), Ronald Stokes. Instr. physics Brown U., Providence, 1924-31, Mass. Inst. Tech., Cambridge, 1931-34; faculty U. Me., Orono, 1934—, prof., 1940—, head physics dept., 1939—. With Office Naval Research, 1946-56; physics counselor Me., Am., insts. physics, 1962—. Fellow, Am. Phys. Soc. (past chmn. N.E. sect.), A.A.A.S., mem. Am. Assn. Physics Tchrs., Am. Soc. Engring. Edn. (mem. council 1949-51, physics editor 1955—), Am. Optical Soc., Phi Beta Kappa, Sigma Xi, Phi Kappa Phi, Tau Beta Pi, Sigma Pi Sigma (nat. councilor). Author: Physics, 1935; Physics Without Mathematics; 1949; Physics Problems, 1958; First Year College Physics, 1954. Research, publs. on optical properties of gases. Pioneered ednl. techniques in engring. physics. Home: 65 Forest Av., Orono, Me. 04473.*

BENNETT, Dorothea, Am. geneticist; b. Honolulu, Dec. 27, 1929; d. James William and Anna (Schorling) B.; A.B., Barnard Coll., 1951; Ph.D., Columbia, 1956. Research asso. dept. zoology Columbia, 1956-62; faculty Cornell U. Med. Coll., 1962—, asso. prof. anatomy, 1965—. Mem. Am. Soc. Human Genetics, Am. Soc. Naturalists, Am. Soc. Zoologists, Genetics Soc. Am., Soc. for Study Devel. and Growth, Sigma Xi. Research and publs. on effects of mutant genes on devel. and growth of mammalian embryo. Home: 1 Fairview Av., Tarrytown, N.Y. 10591. Office: Dept. Anatomy, Cornell U. Med. Coll., N.Y.C. 10021.*

BENNETT, Edward, Am. elec. engr.; b. Pitts., Oct. 26, 1876; s. Benjamin and Mary J. (Davis) B.; E.E., Western U. Pa., 1897; m. Ethel Moore, Aug. 16, 1911. Apprentice, Westinghouse Electric & Mfg. Co., 1897-98, research work, 1899; research engr. for George Westinghouse, in devel. Nernst lamp, later chief engr. Nernst Lamp Co., 1809-1904; mem. firm Beebe & Bennett, 1904-05; head electrician Nat. Electric Signaling Co., Washington, 1905-06; with Telluride Power Co., of Utah, 1906-09; with U. Wis., Madison, 1909-43, asso. prof. elec. engring. until 1913, prof. 1913-43, chmn. dept., 1918-40, prof. emeritus 1943; cons. engr., 1943-50. Del. Internat. Electrotech. Commn. as rep. Am. Inst. E.E., Paris, France, 1950. Fellow Am. Inst. E.E. (v.p. 1924-26), I.R.E., Am. Phys. Soc.; mem. A.A.A.S., Am. Assn. U. Profs., Soc. for Promotion Engring. Edn. (v.p. 1929-30), Wis. Acad. Sci., History of Philosophy Assn., Acad. Polit. Sci., Sigma Xi, Tau Beta Pi, Eta Kappa Nu, Phi Kappa Phi. Author: Introductory Electrodynamics for Engineers, 1926; also numerous bulls. and papers on edn. and engring. Died Jan. 10, 1951.

BENNETT, Edward Hallaran, Irish surgeon; b. Cork, Ireland, Apr. 9, 1837; s. Robert and Jane (Hallaran) B.; B.A., M.B., M.Ch., Trinity Coll., Dublin, Ireland, 1859; M.D., 1864; m. Frances Norman, Dec. 20, 1870; 2 daus. Anatomist, Dublin U., from 1864; prof. surgery Trinity Coll., Dublin, from 1873. Fellow Royal Coll. Surgeons Ireland (pres. 1884-86). Contbr. numerous articles to sci. publs. Formed collection fractures Path. Mus., Trinity Coll.; authority on fractures of bones; discovered Bennett's fracture, 1881; one of earliest to apply Listerian methods in surgery. Died June 21, 1907.

BENNETT, George Macdonald, English chemist; b. Lincoln, Eng., Oct. 25, 1892; s. J. E. and Hannah Martha (Grange) B.; B.A., Ph.D. (exhibitioner, scholar) Queen Mary Coll., London, Eng.; M.A., Sc.D. (exhibitioner, scholar) St. John's Coll., Cambridge (Eng.) U.; m. Doris Laycock, 1918 (dec. 1958). Fellow St. John's Coll., 1917-23; demonstrator in chemistry Guy's Hosp. Med. Sch., London; lectr. in organic chemistry U. Sheffield (Eng.), 1924-31, Firth prof. chemistry, 1931-38; fellow Queen Mary Coll., 1939; univ. prof. chemistry King's Coll., London, 1938-45; chemist Govt. Lab., London, from 1945. Decorated companion Bath. Fellow Royal Inst. Chemistry, Royal Soc.; mem. Chem. Soc. (hon. sec. 1939-46). Contbd. papers to Jour. Chem. Soc., other jours. Contbd. to classical stereochemistry, including work on pentamethylene sulphide and its derivatives; proposed now widely accepted order for electromeric and mesomeric effects of halogens; showed (with J. C. D. Brand and G. Williams) that nitric acid was converted into a cation which migrated toward the cathode during electrolysis. Died Feb. 9, 1959.

BENNETT, Granville Allison, Am. physician; b. Hiawatha, Kan., Sept. 23, 1901; s. Frank and Abby (Allison) B.; B.S., State U. Ia., 1923, M.D., 1925; A.M. (hon.) Harvard, 1942; m. Leonore Weber, May 15, 1926; 1 dau., Mary Allison (Mrs. Carlos Hudson). Faculty, Harvard, 1927-43, asso. prof., 1942-43; prof. pathology and bacteriology Tulane U., 1943-44; prof. pathology U. Ill., Chgo., 1944—, dean, 1954-68. Cons. Armed Forces Inst. Pathology, Washington, 1945. Mem. Coll. Am. Pathology, A.M.A., Soc. for Exptl. Pathology, Am. Soc. for Clin. Pathology, American Society for Clinical Investigation, Central Soc. Clin. Research, Assn. Am. Med. Colls., Sigma Xi, Phi Rho Sigma, Alpha Omega Alpha, Phi Kappa Phi. Author: Changes in the Knee Joint at Various Ages, 1942. Editorial bd. Archives Pathology. Contbr. articles to med. jours. Home: 3000 Sheridan Rd., Chgo. 60657. Office: 1853 W. Polk St., Chgo. 60612.*

BENNETT, Harold Earl, Am. physicist; b. Missoula, Mont., Feb. 25, 1929; s. Edward Earl and Linda (McCoy) B.; B.A., Mont. State U., 1951; M.S., Pa. State U., 1953, Ph.D., 1955; m. Jean Louise McPherson, Aug. 17, 1952. Physicist, Nat. Bur. Standards, Washington, 1953, Wright Air Devel. Center, Dayton, O., 1955-56; physicist Naval Weapons Center, China Lake, Cal., 1956-60, head phys. optics br., 1960—; tchr. extension program U. Cal. at Los Angeles, 1966—. Fellow Optical Soc. Am., (editor jour.); mem. Sci. Research Soc. Am. (past local pres., award for outstanding paper 1961), Sigma Xi. Contbg. author: Ency. Physics, 1966; Semiconductors and Semimetals III: Physics of III-V Compounds, 1967; Physics of Thin Films, vol. IV. 1967. Research, publs. on 1st precise interferometric wavelength determination at wavelengths longer than photog. infrared, precise determination velocity of light; derviation and exptl. verification of relation between roughness and normal incidence specular reflectance of a surface; exptl. proof that Drude theory holds for silver, gold and aluminum in infrared; design and devel. instrument for measuring roughness of very smooth surfaces; design and devel. instruments for accurately measuring specular reflectance. Home: 606 A Essex Circle. Office: Naval Weapons Center, China Lake, Cal. 93555.*

BENNETT, Henry Stanley, Am. anatomist, biologist; b. Tottori, Japan, Dec. 22, 1910 (parents Am. citizens); s. Henry James and Anna (Jones) B.; A.B., Oberlin Coll., 1932; M.D. cum laude in Anatomy, Harvard, 1936; D.Sc. (hon.), Monmouth Coll., 1962; m. Alice Roosa, July 28, 1935; children—Edith Roosa (Mrs. Richard Page), Anna Woodruff (Mrs. Matthew McNaught), Henry James, Patience St. John (Mrs. Richard Berkman). NRC fellow, instr. anatomy and pharmacology Harvard Med. Sch., Boston, 1937-42; asst. prof. Mass. Inst. Tech., 1945-48; prof. anatomy, chmn. dept. U. Wash., Seattle, 1948-60; dean div. biol. scis. including Sch. Medicine, prof. anatomy U. Chgo., 1961-65, Robert R. Bensley prof. biol. and med. scis., prof. anatomy, dir. Labs. for Cell Biology, 1966—. Lectr. various univs.; mem. numerous govt. coms. Trustee Home for Incurables, Salk Inst. Biol. Studies; bd. dirs. Ill. Soc. Med. Research, Infant Welfare Soc. Chgo., Nat. Soc. Med. Research. Mem.

Am. Acad. Arts and Scis., Am. Assn. Anatomists (pres. 1959-60), Anat. Soc. Japan, Coligio Anatomica Brasileiro, A.A.A.S., Am. Chem. Soc., Am. Physiol. Soc., Am. Soc. Cell Biology, Internat. Soc. Cell Biology, Biophys. Soc., Electron Microscopy Soc. Am., Histochem. Soc. (councilor 1954-58), Midwest Soc. Electron Microscopists, Pacific N.W. Bird and Mammal Soc., Assn. Am. Physicians, Am. Soc. Clin. Investigation, Western Assn. Physicians, A.M.A., Am. Assn. U. Profs. (pres. U. Wash. chpt. 1954-55), Sociedad Mexicana de Anatomia (hon.), Sigma Xi (pres. U. Wash. 1956-57), Alpha Omega Alpha. Asso. editor Jour. Histochemistry and Cytochemistry, 1956-59; adv. editor Internat. Rev. Cytology, 1957-61; editorial bd. Jour. Biophys. and Biochem. Cytology, 1954-60. Contbr. articles to sci. jours. Discoverer (with K. R. Porter) internal membrane system of striated muscle, (with DeRobertis) synaptic vesicles; classified capillaries; 1st synthesis of colored aryl mercurial for use as cytochem. reagent for sulfhydril groups. Home: 5827 S. Blackstone Av., Chgo. 60637.*

BENNETT, Hugh Hammond, Am. agrl. chemist; b. nr. Wadesboro, N.C., Apr. 15, 1881; s. William Osborne and Rosa May (Hammond) B.; B.S., U. N.C., 1903, LL.D., 1936; D.Sc., Clemson Coll., 1937; D.Sc., Columbia, 1952; m. Sarah Edna McCue, 1907 (died 1909); 1 dau., Sarah Edna (Mrs. Eugene Akers); m. 2d, Betty Virginia Brown, 1921; 1 son, Hugh Hammond. Soil scientist Soil Survey Div., Bur. of Soils, Dept. Agr., 1903-09, insp. in soil survey, 1909, 28; mem. agrl. expdn. Canal Zone, 1909; in charge explorative expdn., Alaska, 1914; mem. Chugach Nat. Forest Commn., 1916, Guatemala-Honduras Boundary Commn., 1919, Rubber Commn. to Central and S. America and W.I., 1923-24; in charge agrl. survey of Cuba, winters 1925-32; in charge, soil erosion and moisture conservation investigation, Bur. of Chemistry and Soils, U. S. Dept of Agr., 1928-33; dir. Soil Erosion Service, U. S. Dept. Interior, 1933-35; chief Soil Conservation Service, Dept. Agr., 1935-52, ret. Recipient Cullum medal Am. Geog. Soc. Fellow Soil Conservation Soc. of Am. (founder), Am. Soc. of Agronomy, Am. Geog. Soc., A.A.A.S.; mem. Canadian Conservation Assn. (hon.), Internat. Union for Protection of Nature (hon. pres.), Assn. Am. Geographers (pres. 1943), Washington Acad. Sci., Am. Soc. Agronomy, Friends of Land, Am. Forestry Assn., Phi Delta Theta. Author several books relating to field, 1913-—, including This Land We Defend, 1942. Contbr. to periodicals. A leader in soil conservation in U. S.; sometimes called the father of soil conservation. Died Burlington, N.C., July 7, 1960.

BENNETT, Ivan Loveridge, Jr., Am. physician; b. Washington, Mar. 4, 1922; s. Ivan Loveridge and Ruby (Jenrette) B.; A.B., Emory U., 1943, M.D., 1946; m. Martha Rhodes, June 24, 1944; children—Susan, Paul Bruce, Katherine, Jeffrey Ivan. Med. intern Grady Hosp., Atlanta, 1946-47, chief resident physician, 1951-52; asst. pathologist Johns Hopkins Hosp., Balt., 1949-50, physician, cons. bacteriology, 1954-—; asst. resident physician Duke Hosp., 1950-51; asst. prof. internal medicine Yale, 1952-54; asso. prof. medicine Johns Hopkins, 1954-57, prof. medicine, head biol. div., 1957-58, Baxley prof. pathology, also pathologist-in-chief Johns Hopkins Hosp., 1958-66; dep. dir. Office Sci. and Tech., Exec. Office of Pres., Washington, 1966-—; cons. VA Hosp., Balt., Clin. Center, Nat. Insts. Health, Bethesda; spl. cons. Surgeon Gen., U. S. Army. Mem. Pres.'s Sci. Adv. Com., 1966-—, Commn. on Epidemiological Survey, Armed Forces Epidemiology Bd.; research contract dir. Army Chem. Corps; mem. bd. sci. counselors Nat. Inst. for Dental Research; member executive committee of div. med. scis., mem. com. on pathology NRC; mem. bd. sci. advisers Armed Forces Institute Pathology. Diplomate Am. Bd. Internal Medicine, 1954. Fellow A.C.P., N.Y. Acad. Scis.; mem. Soc. Exptl. Biology and Medicine, Am. Psychosomatic Soc., Am. Fedn. Clin. Research (pres. 1957-58), Am. Soc. Clin. Investigation, Assn. Am. Physicians, A.M.A., Am. Assn. Pathologists and Bacteriologists, Am. Soc. Exptl. Pathology, Am. Clin. and Climatol. Assn., Am. Assn. Immunologists, Internat. Acad. of Pathology, So. Soc. for Clin. Investigation (pres. 1963-64), Johns Hopkins Med. Soc. (president 1963-64), also Phi Beta Kappa, Sigma Xi, Omicron Delta Kappa, Alpha Omega Alpha, Sigma Chi, Phi Chi. Author tech. articles sci. jours. Editor: Principles of Internal Medicine, rev. edit., 1958, 62, 66. Editorial bd. Ann. Rev. Medicine, Jour. Bio-chem. and Molecular Pathology. Investigation of pathogenesis of infection; infectious diseases; mechanism of fever. Home: 311 Broxton Rd., Balt. 12. Office: Office Sci. and Tech., Exec. Office of Pres., Washington.

BENNETT, James Edward, Am. physician, surgeon; b. Burlington, Wis., May 19, 1925; s. John Francis and Florence (Mauer) B.; student Notre Dame U., 1943-44; student Mass. Inst. Tech., 1944-45; M.D., Northwestern U., 1950; m. Ellen MacPherson, June 18, 1956; children—David, Martha, Thomas, Jonathan. Gen. practice medicine, Burlington, 1950-51; exchange fellow in plastic surgery, Wales, 1956-57; resident in plastic surgery U. Tex., Galveston, 1958-61; asst. prof. surgery, dir. plastic surgery Ohio State U., Columbus, 1961-64; prof. surgery, dir. plastic surgery Ind. U. Med. Center, Indpls., 1964-—. Mem. A.C.S., Plastic Surgery Research Council (sec. 1966), Frank A. Caller Surg. Soc., Am. Soc. Plastic and Reconstructive Surgeons, Am. Assn. for Surgery of Trauma, Am. Assn. Plastic Surgeons, Phi Rho Sigma. Research, publs. on detection of burn depth with radioactive isotopes; exptl. mgmt. of lymphedema. Home: 5865 Hunter Glen Rd., Indpls. 46226. Office: 1100 W. Michigan St., Indpls. 46207.*

BENNETT, John Hughes, Brit. physician; b. London, Eng., Aug. 31, 1812; apprentice surgeon at Maidstone; became student, Edinburgh, Scotland, 1833, LL.D., 1875; M.D., 1837; postgrad. Paris, 2 years, Germany, 2 years; 1 son; 4 daus. Became lectr. histology, Edinburgh, 1841; apptd. physician to Royal Dispensary, 1842; pathologist Royal Infirmary; prof. Insts. Medicine, Edinburgh, 1848-74; prof. U. Edinburgh, 1843-74. Mem. Royal Soc. Edinburgh, Edinburgh Coll. Physicians, Royal Med. Soc. (pres.), Parisian Med. Soc. (founder, 1st pres.). Author: Treatise on Codliver Oil as a Therapeutic Agent in Certain Forms of Gout, Rheumatism and Scrofula, 1841; also numerous articles. Described (simultaneously with Virchow) 1st recorded case of leucocythaemia (leukemia); advocated use of cod liver oil in Tb cases, prescribed fresh air treatment for pneumonia. Died Sept. 24, 1875.

BENNETT, Lawrence Herman, Am. physicist; b. Bklyn., Oct. 17, 1930; s. Harold and Irene (Kamel) B.; B.A., Bklyn. Coll., 1951; M.S., U. Md., 1955; Ph.D., Rutgers U., 1958; m. Devora Spintman, Mar. 22, 1953; children—Claire Ann, Charles L., Craig D. Physicist, U. S. Naval Ordnance Lab., 1951-53; tchr. U. S. Army Ordnance Sch., Aberdeen, Md., 1953-55; with Nat. Bur. Standards, Gaithersburg, Md., 1958-—, chief alloy physics sect., 1963-—; asso. prof. physics U. Md., 1961-—. Mem. Am. Inst. M.E., Am. Phys. Soc., Am. Soc. Metals. Research and publs. in nuclear magnetic resonance in metals, alloys, intermetallic compounds; determination of basic structures of metals, impurities under varied conditions. Home: 6524 E. Halbert Rd., Bethesda, Md. 20034. Office: Nat. Bur. Standards, Route 70 S and Quince Orchard Rd., Gaithersburg, Md. 20760.*

BENNETT, Leonard Lee, Jr., Am. biochemist; b. Savannah, Ga., Nov. 10, 1920; s. Leonard Lee and Edna (Reils) B.; student Armstrong Jr. Coll., 1938-40; B.A., Vanderbilt U., 1942, M.A., 1943; Ph.D., U. N.C., 1949; m. Martha Davis, June 8, 1949; children—Walter Lee, Deborah Ann, Brian. Instr., U. Ga., 1943-44; with So. Research Inst., Birmingham, Ala., 1945-—, head biochemistry div., 1956-64, dir. biochemistry research, 1964-—. Mem. adv. com. research therapy cancer Am. Cancer Soc., 1956-58, 60-63; mem. chemotherapy study sect. NIH, 1964-—. Mem. A.A.A.S., Am. Chem. Soc., Am. Assn. for Cancer Research, Soc. Biol. Chemists, N.Y. Acad. Scis. Research, publs. in biochemistry of cancer, mechanisms of drug action, biochemistry of purines and pyrimidines. Home: 608 Euclid Av., Birmingham 35213. Office: 2000 9th Av. S., Birmingham, Ala. 35205.*

BENNETT, Merrill Kelley, Am. econ. climatologist; b. Killingly, Conn., Feb. 13, 1897; s. Frank Wayland and Carrie (Williams) B.; Ph.B., Brown U., 1920, A.M., 1921, L.H.D., 1954; A.M., Harvard, 1926; Ph.D., Stanford, 1927; m. Dorcas Gallup, Aug. 24, 1922; children—John Francis, Helen Josephine (Mrs. Andrew D. Lucine), Stephen Williams. Instr. English, Brown U., 1921-22; faculty Stanford, 1923-62, dir. Food Research Inst., 1942-62, ret., 1962; vis. prof. Rice Inst., 1950-51; social sci. appraiser U. Pa., 1955-56; chief statistician Office Food Control, Hawaii, 1941-42; chief div. Food Allocations, Fgn. Econ. Adminstrn., 1943-45. Mem. Am. Econ. Assn., Am. Statis. Assn., Am. Geog. Assn., Assn. Am. Geographers. Author: Farm Cost Studies in the U. S., 1928; (with V. D. Wickiser) The Rice Economy of Monsoon Asia, 1941; (with E. S. Shaw) International Commodity Stockpiling, 1949; Food for Postwar Europe, 1944; The World's Food, 1954; also numerous articles. Research on trends in food consumption in U. S., relationship of food prodn. to climate, relationship of food consumption to incomes. Home: 15827 Poppy Lane, Monte Sereno, Cal.*

BENNETT, Miriam Frances, Am. zoologist; b. Milw., May 17, 1928; d. Stanley Edward and Dorothy (Wheeler) B.; A.B., Carleton Coll., 1950; M.A., Mt. Holyoke Coll., 1952; Ph.D., Northwestern U., 1954. Faculty, Sweet Briar (Va.) Coll., 1954-—, prof., chmn. dept. biology, 1964-—. NSF fellow, 1961. Fellow A.A.A.S.; mem. Am. Inst. Biol. Scis., N.Y. Acad. Scis., Ecol. Soc. Am., Am. Soc. Zoologists, Va. Acad. Scis., Assn. Southeastern Biologists, Marine Biol. Lab., Am. Micros. Soc., Sigma Xi. Research, publs. in biol. rhythmicity, timing mechanisms of clams, fiddler crabs, honey bees, earthworms; discovered overt tidal cycle running in crabs, 24-hour cycle food-collecting in bees maintained under constant conditions, 24 hour variation in rates of reactions in earthworms. Home: Sweet Briar, Va. 24595.*

BENNETT, Ralph Decker, Am. physicist; b. Williamson, N.Y., June 30, 1900; s. Edward A. and Sara Jane (Decker) B.; B.S. E.E. Union Coll., Schenectady, M.S.E.E., 1923; Ph.D. in Physics, U. Chgo., 1925; m. Mary Johnson, Sept. 1, 1934; children—Sarah Louise (Mrs. Richard Reichart), Ralph Decker. Faculty, Union Coll., Schenectady, 1921-23, 25-26; Nat. Research fellow Princeton, 1926-27, Cal. Inst. Tech., 1927-28; research asso. U. Chgo., 1928-31; prof. elec. measurements Mass. Inst. Tech., 1931-40; tech. dir. Naval Ordnance Lab., White Oak, Md., 1944-54; mgr. tech. dept. Knolls Atomic Power Lab., Schenectady, 1954-56; mgr. Vallecitos Atomic Lab., Gen. Elec. Co., Pleasanton, Cal., 1956-61; dir. research Martin Co., Balt., 1961, v.p., 1961-67, cons., 1967-—. Recipient Navy Distinguished Civilian Service award, 1950. Fellow Am. Phys. Soc., I.E.E.E.; mem. Atomic Indsl. Forum (dir. 1962-—). Research on x-rays confirming Compton effect, Geiger counters, prodn. of megavolt x-rays, variations of cosmic rays, mgmt. large scale research and devel. Home: 2818 Clay St., San Francisco 94115. Office: Martin Co., Friendship Internat. Airport, Md. 21240.*

BENNETT, Rawson, Am. electronics engr.; b. Chgo., June 16, 1905; s. Rawson and Cora (Jones) B.; B.S., U. S. Naval Acad., 1927; M.S. in Elec. Engring., U. Cal., Berkeley; m. Mary F. Wyman; m. 2d, A. Louise Holmes, Jan 2, 1948; children—Rawson, Sally Ann, Holmes, Gregory. Commd. ensign USN, 1927, advanced through grades to rear adm., 1961; service on various ships, 1927-34; with schs., 1934-37; exptl. engr., 1937-41; dir. electronics design Bur. Ships, 1942-46; dir. USN Electronics Lab., San Diego, 1946-50; chief Naval Research, 1955-61, ret., 1961; sr. v.p., dir. engring. Sangamo Electric Co., Springfield, Ill., 1961-63, corporate dir., research cons., 1963-—. Decorated Legion of Merit (2). Fellow, I.R.E., Acoustical Soc., Inst. Elec. Engrs., A.A.A.S., Inst. Aero. Scis. Developed sonic and supersonic underwater apparatus; anti-submarine attack teacher; supervised satellite project Vanguard, other Naval research programs. Home: 3200 N. Columbus St., Arlington, Va. 22207.*

BENNETT, Robert L(eo), Jr., Am. physician; b. Wilkinsburg, Pa., Dec. 18, 1911; s. Robert Leo and Nelle (McGuire) B.; B.S., U. Pitts., 1934, M.D., 1936, D.Sc., 1960; M.S., U. Minn., (fellowship Mayo clinic, 1937-40), 1940; m. Esther McDowell, July 10, 1937; children—Judith, Susan. Interne Mercy Hosp., Pittsburgh, 1936-37; asst. prof. physical medicine, Univ. Wis., 1940-41; dir. phys. medicine, Ga. Warm Springs Found., 1941-48, asst. med. dir., 1948-53, med. dir., 1953-58, exec. dir., 1958-—, prof. phys. medicine, Emory U. Med. Sch., 1945-—, chmn. dept. phys. medicine Emory U. Hosp., 1946-65; Horowitz vis. prof. N.Y. U. Inst. Phys. Medicine and Rehab., 1962. Mem. Meriwether County Hosp. Authority, Ga.; mem. com. prosthetic research and devel. Nat. Acad. Scis.-Nat. Research council; medical director Georgia Rehabilitation Center; consultant area 3, phys. medicine rehabilitation VA, 1945-—; cons. Crippled Children's div., State Ga. since 1947; cons. to Surgeon-Gen., U. S. A. in phys. med. since 1948. Mem. adv. coms. Dept. Health, Edn. and Welfare, Nat. Found., American Rehabilitation Foundation. Recipient Gold Key award, Am. Congress Phys. Medicine and Rehabilitation, 1955. Diplomate Am. Bd. Phys. Medicine and Rehabilitation (sec. treas. 1947-53, chmn. bd. 1953-63). Mem. Internat. (v.p. 1957, chmn. exec. com. 1960), Am. (pres. 1951-52), congresses phys. medicine, Internat. Soc. Rehab. of Disabled (Am. bd. dirs. 1965-—), Am. (chmn. sect. phys. medicine 1961), So. med. assns., Am. Rheumatism Assn., Am. Acad. Phys. Medicine and Rehab., Am. Acad. Cerebral Palsy, Am. Acad. Neurology, Sigma Xi. Author numerous papers in field. Electromyographic studies of denervated muscle tissue; apparatus for support of weakened body segments; methods of care of convalescent poliomyelitis patients. Home: Warm Springs Found., Warm Springs, Ga. Office: Emory Hospital, Atlanta, also Warm Springs Foundation, Warm Springs, Ga.

BENNETT, Wendell C(lark), Am. anthropologist; b. Marion, Ind., Aug. 17, 1905; s. William Rainey and Ethel (Clark) B.; Ph.B., U. Chgo., 1927, A.M., 1929, Ph.D., 1930; m. Hope Ranslow, Oct. 30, 1935; children—Lucy, Martha. Asst. in anthropology Am. Mus. Natural History, 1931-38; asso. prof. anthropology U. Wis., 1938-40; asso. prof. of anthropology Yale, 1940-45, prof. since 1945; specialized in Andean archeology. Exec. sec. Joint Com. Latin Am. Studies, 1942-44. Fellow Royal Anthrop. Inst. Gt. Britain and Ireland (hon.); mem. Am. Anthrop. Assn. (pres. 1952), Soc. Am. Archeology, Sigma Xi. Author: The Tarahumara (with R. M. Zingg), 1935; (monographs) Excavations at Tiahuanaco; Excavations in Bolivia; Excavations on the North Coast of Peru; The North Highlands of Peru; Archeological Regions of Colombia; Excavations in the Cuenca region, Ecuador; Northwest Argentine Archeology; Aneadn Culture History. Died Sept. 8, 1953.

BENNETT, Willard Harrison, Am. physicist, educator; b. Findlay, O., June 13, 1903; s. Harry and Elsie Mae (Ward) B.; student Carnegie Inst. Tech., 1921-22; A.B., Ohio State U., 1924; M.S., U. Wis., 1926; Ph.D., U. Mich., 1928; m. Mona D. Sheets, Sept. 8, 1928; div. 1949; children—Willard Harrison, Barbara, Bruce, Stephan; m. 2d, Helen Mae Sawyer, Oct. 24, 1949; children—Charles, Ward, Rebecca. N.R.C. fellow Cal. Inst. Tech., 1928-30; faculty Ohio State U., 1930-38; dir. research Electronics Research Corp., 1939-41; dir. applied research Inst. Textile Tech., 1945; physicist, sect. chief Nat. Bur. Standards, 1946-50; prof. physics U. Ark., 1950-51; br. head, div. cons. U. S. Naval Research Lab., 1951-61; Burlington prof. physics N.C. State U., Raleigh, 1961-—. Mem. Gov.'s Sci. Adv. Com., N.C., 1961-—; cons. Los Alamos Sci. Lab. Fellow Am. Phys. Soc., Wash. Acad. Sci.; mem. Sigma Xi, Sigma Pi Sigma. Coauthor text book. Contbr. articles to profl. jours. Discovered pinch effect; resolved infra-red spectra of symmetric top molecules; developed 1st negative ion source of any element, principle of tandem accelerator, non-magnetic mass spectrometer used in space research; modeled radiation belts in a lab. tube before their discovery in space. Patentee in field. Home: 5500 N. Hills Dr., Raleigh, N.C. 27609.*

BENNETT, William Ralph, Jr., Am. physicist; b. Jersey City, Jan. 30, 1930; s. William Ralph and Viola Mildred (Schreiber) B.; A.B. in Physics, Princeton, 1951; Ph.D., Columbia, 1959; M.A. (hon.) Yale, 1965; m. Frances Ellen Commins, Dec. 11, 1952; children—Jean, William Robert, Nancy. Faculty, Yale, 1957-59, 62—, prof. physics and applied sci., 1965—; tech. staff Bell Telephone Labs., Murray Hill, N.J., 1959-62. Cons. to govt. agys., pvt. cos. Alfred P. Sloan Found. fellow, 1963-65; Guggenheim fellow, 1967. Fellow Am. Optical Soc.; mem. Am. Phys. Soc., I.E.E.E. (Morris Liebmann award 1965, sr. mem.), N.Y. Acad. Scis., Sigma Xi. Research, publs. on inelastic processes in excited states of atoms and molecules, optical, microwave and radio frequency spectroscopy, basic research on gas lasers; co-patentee first several gas discharge lasers.

BENNINGHOFF, William Shiffer, Am. botanist; b. Ft. Wayne, Ind., Mar. 23, 1918; s. William N. and Edith Esther (Shiffer) B.; student Ind. U., 1936-37; B.S. magna cum laude, Harvard, 1940, M.Sc., 1942, Ph.D., 1948; m. Gladys Helen Kunst, Apr. 19, 1941; children—Valerie Anne, Jonathan. Botanist, U. S. Geol. Survey, Washington, 1948-57, acting in charge Alaska terrain and permafrost sect., 1949-50, chief sect. 1953-57; faculty dept. botany U. Mich., Ann Arbor, 1957—, prof., 1960—, palynologist Gt. Lakes Research div. Inst. Sci. and Tech., 1960-63; prof., asst. dir. U. Mich. Bot. Gardens, 1965-66; field expdns. in Alaska, 1948-51, 55-56, No. Greenland, 1953, Iceland, 1954. Sec. exec. com. Arctic Inst. N.Am., 1966—, bd. govs., 1958-63, 1966—; mem. com. on polar research panel on biol. and med. scis. NRC-Nat. Acad. Scis., 1963—, panel chmn., 1966—. Recipient Bausch & Lomb Sci. award, 1936; U. S. Dept. Interior Meritorious Service award, 1954; George B. Emerson, III fellow, Harvard, 1940-42. Fellow A.A.A.S., Am. Geog. Soc., Arctic Inst. N.Am., Geol. Soc. Am.; mem. Internationale Vereinigung für Vegetationskunde (v.p. 1965—), Ecol. Soc. Am. (asso. editor 1965—), Am. Soc. for Limnology and Oceanography, Soc. for Cryobiology (charter). Research, publs. on interrelationships of vegetation and frozen ground in evolution of boreal landscapes, phytocoenology, pollen analysis, phytogeog. significance of air spora. Home: 2705 Provincial Dr., Ann Arbor, Mich. 48104.*

BENNIS, Warren G., Am. psychologist, educator; B. N.Y.C., Mar. 8, 1925; s. Philip and Rachel (Landau) B.; A.B., Antioch Coll., 1951; Ph.D., Mass. Inst. Tech., 1955; m. Clurie Williams, Mar. 30, 1962; children—Katharine, John Leslie. Asst. prof. social psychology Mass. Inst. Tech., 1955-56, prof., 1959—, also chmn. orgnl. studies group; asst. prof. psychology, Boston U., 1956-59; vis. lectr., Harvard, 1958-59; provost State U. N.Y. at Buffalo, 1967—. Fellow Am. Psychol. Assn. Am. Social. Assn.; mem. A.A.A.S. Author: Changing Organizations, 1966; Personal and Organizational Change Through Group Methods, 1965; The Planning of Change, 1961; also numerous articles. Research on social psychology leadership, social and organizational change. Home: 184 Le Brun Circle, Eggertsville, N.Y. 14226. Office: Old Faculty Club, Buffalo.*

BENOIT, Henri, French physicist; b. Montpellier, France, July 11, 1921; s. Jean-Daniel and Henriette (Bois) B.; student École Normale Supérieure de Paris, 1941-45, Dr.ès-Sciences Physiques, 1950, Agrégé, 1945; m. M. T. Bigand, July 20, 1946; children—Nicole, Alain, Eric. Under dir. Centre de Recherches sur les Macromolécules, 1952-58, dir.-adj., 1958-67, dir., 1967—; prof. U. Strasbourg (France). Recipient Médaille d'Argent Centre National de la Recherche Scientifique, 1961; named officier l'Ordre National des Palmes Académiques, 1962, chevallier Ordre de la Légion d'Honneur, 1963. Mem. Société de Chimie Physique (mem. council 1966), Internat. Union Pure and Applied Chemistry (mem. div. macromolecules 1967—), Comité Scientifique Action Concertée de la Délégation Générale à la Recherche Scientitifique et Technique, Commission Reginale pour le Developpement de la Recherche Scientifique et Technique. Research, numerous publs. on phys. chemistry of macromolecules, especially solutions of high polymers, statistic of macromolecular chains copolymers structure determination. Home: 9, rue de Bruges, 67. Office: rue Boussingault, 67, Strasbourg, France.*

BENOIT, Henri, French physicist; b. Paris, Sept. 18, 1928; s. Jean and Marguerite (Vincent) B.; Agregation de Physique, Ecole Normale Superieure, 1954; m. Christiane Tchang, Mar. 7, 1964. Research staff Institut d'Electronique Fondamentale, Faculté des Sciences, Orsay, France, 1952—, Grivet dir.; staff Centre National de la Recherche Scientifique, 1959-63, maitre, 1962-63; with Faculté des Sciences, Paris, 1955—, prof. physics, 1964—. Mem. Comité National de la Recherche Scientifique, 1958—. Recipient Medaille de Bronze Centre National de la Recherche Scientifique, 1960, Prix Felix Esclangon, 1960; NATO grantee Jeffries Lab., physics dept. U. Cal. at Berkeley, 1960-61. Mem. Société Francaise de Physique. Author: Element de Physique, (with J. Bricard), 1966; also articles. Research on nuclear magnetic resonance in a flowing liquid, low frequency maser oscillator; high resolution, relaxation and multiphoton transition in earth magnetic field or below, nuclear gyroscope; high resolution of adsorbed liquids on coals, relaxations of protons in antiferromagnetic salts, relaxation at low temperature. Home: 14 rue Oudinot, 75 Paris 7, France. Office: Institut d'Electronique, Batiment 220, Faculté des Sciences, 91-Orsay, France; also Faculté des Sciences, Tour 23, 9 Quai St. Bernard, 75 Paris 5, France.*

BENOIT, Jacques, French physician; b. Nancy, France, Feb. 26, 1896; s. Auguste and Elisabeth (Geny) B.; M.D., med. sch. at Strasbourg; Ph.D. in sci.; m. Meryem Jacquemaire Dramard, Mar. 13, 1934; 1 dau., Danielle (by previous marriage). Agrégé histology and embryology at med. sch.; prof. agrege med. sch. of Strasbourg, 1930, prof. embryology, 1946; prof. histology and embryology sch. histology and embryology sch. medicine and pharmacy at Algiers, 1939; prof. histophysiology Collège de France, 1952. Mem. Nat. Acad. Medicine. Research and publs. on histophysiology of testicle and sexual features, exptl. inversion of sex, effect of light on sexual activity, neuroendocrinology, desoxyribonucleic acids and racial features. Home: 81, av. Niel, Paris 17. Office: Collège de France, place Marcelin-Berthelot, Paris 5, France.*

BENOIT, Rene (Justin-Miranda), French physicist; b. Montpellier, France, Nov. 29, 1844; s. Justin B.; licence ès scis., Montpellier, 1868; docteur en médecine, 1868; docteur ès scis., 1873. Dir. Internat. Bur. Weights, Measures, Sevres, France, 1889-1915. Mem. Bur. Longitudes, French Acad. Scis., 1903; French Soc. Physics (pres.), Nat. Bur. Weights and Measures. Author: Études experimentales sur la resistance electrique des metaux, 1873; Sur la determination de l'ohm, 1884. Evaluated (with Michelson, later with Perot and Fabry) length of meter and of light waves; research on pendulum, elec. resistance of metals and their variation under changing temperatures; perfected triangulation devices; studied internat. ohm under pressure of mercury column. Died Courbevoie, France, May 5, 1922.

BENOLIEL, Solomon D., Am. electrochemist; b. N.Y., June 1, 1874; s. David J. and Pauline (Wasserman) B.; B.S., Coll. City N.Y., 1893; E.E. A.M., Columbia, 1896; m. Therese Lindeman, June 1, 1897; children—D. Jacques, L. Osmond, Jean S. Tchr., Adelphi Coll., 1897-1901; electrochemist, and gen. mgr. Roberts Chem. Co. (now Niagara Alkali Co.), 1901-06; gen. mgr. Internat. Chem. Co., Camden, later Phila., from 1906. Lectr., Bds. of Edn. N.Y. and Bklyn., also Bklyn. Inst. Arts and Sci. 1899-1901. Developed process for production of caustic potash and chemically pure hydrochloric acid by means of electric current; perfected a number of scientific cleaners, lubricants and burnishing compounds for indsl. uses, particularly the metal mfg. trades. Writer on liquid air, photo-therapy and electrochemistry. Died Nov. 23, 1932.

BENSAUDE, Raoul, physician; b. France, 1866; introduced serodiagnosis in work (with E. C. Achard) describing paratyphoid fever, 1896, also isolated Bacillus paratyphosus B (now called Salmonella paratyphi B); introduced quinine urea anesthesia in treatment of anal fissure; popularized endoscopic methods in gastroenterology. Died 1938.

BENSLEY, Benjamin Arthur, Canadian zoologist; b. Hamilton, Ont., Can., Nov. 5, 1875; ed. U. Toronto, also Columbia U.; B.A., Ph.D.; m. Ruth Horton, 1904; 1 son. Prof. zoology, head dept. biology U. Toronto. Fellow Royal Soc. Can. Research and publs. on mammalian evolution. Died Jan. 20, 1934.

BENSON, Arthur H., Irish ophthalmologist; b. 1852; s. Charles and Maria (Andrews) B.; M.A., M.D., Dublin U.; postgrad., London, Eng., also Vienna, Austria; m. Ethel Martha Rawson. Lectr. Ledwich Sch. Medicine, Dublin; house surgeon St. Mark's Ophthalmic Hosp.; surgeon Dublin Throat and Ear Hosp., Royal Victoria Eye and Ear Hosp., Dublin; ophthalmic and aural surgeon Royal City of Dublin Hosp.; examiner in ophthalmology Trinity Coll., Dublin, Royal Colleges of Physicians and Surgeons. Mem. Royal Acad. Medicine, U.K., Heidelberg (Germany) ophthal. socs. Author monographs, papers on ophthalmic and aural subjects. Described an inflammation of the vitreous humor in which star-shaped bodies were present, circa 1890. Died 1912.

BENSON, Bruce Buzzell, Am. physicist; b. Choteau, Mont., Feb. 22, 1922; s. Harry Fort and Mary (Buzzell) B.; B.A., Amherst Coll., 1943; M.S., Yale, 1945, Ph.D., 1947; m. Lucy Peters Wilson, Mar. 30, 1950. Instr. physics Yale, 1944-46, research asst. in nuclear physics, 1946-47; faculty physics Amherst (Mass.) Coll., 1947—, prof., 1960—; asso. in physics Woods Hole Oceanographic Inst., 1957—. John Simon Guggenheim fellow in oceanography, 1958-59. Mem. Am. Phys. Soc., Am. Geophys. Union, Geochem. Soc., Am. Assn. Physics Tchrs., Am. Assn. U. Profs., A.A.A.S., Phi Beta Kappa, Sigma Xi. Research publs. in mass spectrometry and oceanography, exchange of gases between atmosphere and ocean, new sampling devices, processing equipment. Home: 46 Sunset Av., Amherst, Mass. 01002.*

BENSON, Francis Colgate, Jr., Am. physician; b. N.Y.C., 1872; M.D., Hahnemann Coll., 1894; m. Jennie McFetridge; children—William D., Frank C., Jean. Cons. surgeon, radiologist, Women's Homeopathic Hosp., St. Luke's Hosp., Children's Hosp. (all N.Y.C.), W. Jersey Homeopathic Hosp., Camden, N.J., McKinley Meml. Hosp., Trenton, N.J., Pa. Hosp. for the Insane, Allentown; asso. with Hahnemann Coll. and Hosp., Phila. Mem. A.C.P., Pan-Am. Med. Assn., A.M.A., Am. Inst. Homeopathy. Author books on surgery and cancer treatment. Devised (with Earl B. Craig) Craig-Benson operation for cancer; devised small radium bomb for postoperative destruction of remaining cancer cells. Died Feb. 1941.

BENSON, George Campbell, Canadian phys. chemist; b. Toronto, Ont., Can., July 25, 1919; s. George Challoner and Adele (Campbell) B.; B.A., U. Toronto, 1942, M.A., 1943, Ph.D., 1945; postgrad. U. Bristol (Eng.); m. Barbara Ann Brisbin, Dec. 6, 1946; children—Barbara Gayle, Frederick Challoner. With NRC, Ottawa, Ont., 1945—, prin. research officer div. pure chemistry, 1963—. Recipient Bronze medal in sci. Brit. Assn. for Advancement Sci., 1942. Mem. Am. Chem. Soc., Am. Phys. Soc., Royal Soc. Can., Chem. Inst. Can. Research, publs. on exptl. and theoretical investigations of surfaces of crystals, thermochem. studies solutions. Office: Div. Pure Chemistry, Nat. Research Council, Ottawa, Ont., Can.*

BENSON, Lyman David, Am. botanist, educator; b. Kelseyville, Cal., May 4, 1909; s. Charles A. and Cora (West) B.; A.B., Stanford, 1930, M.A., 1931, Ph.D., 1939; m. Evelyn Berniece Linderholm, Aug. 16, 1931; children—Lyman David, Robert Leland. Instr., Bakersfield Jr. Coll., 1931-38; instr., U. Ariz., Tucson, 1939-40, asst. prof., 1940-44; asso. prof. Pomona Coll., Claremont, Cal., 1944-49, prof., 1949—, head botany dept., 1944—, dir. herbarium. Recipient Greater Linnaeus medal Swedish Royal Soc. Scis. 1952; Wig Distinguished Prof. award, Pomona Coll., 1963. Mem. Am. Soc. Plant Toxonomists (pres. 1960), Cactus and Succulent Soc. Am. (pres. 1956-57), Western Soc. Naturalists (pres. 1955), So. Cal. Botanists (pres. 1949-50), Bot. Soc. Am. (pres. Pacific sect. 1948). Author: Plant Classification, 1957; The Cacti of Arizona, 1940, 2d. edit., 1950; The Trees and Shrubs of the Southwestern Deserts, 1945, 2d. edition, 1954; Plant Taxonomy, Methods and Principles, 1962. Research, publs. on organ.; interpretation and re-evaluation of plant taxonomy Sci. Home: 1430 Via Zurita, Claremont, Cal.*

BENSON, Margaret J., botanist; ed. Newnham Coll., Cambridge; grad. U. London, 1891, 94, D.Sc. Examiner in Honours Internal B.Sc., London; mem. Faculty Sci., 1903; prof. botany U. London, 1912-22. Fellow Linnean Soc. Author: Contributions to the Embryology of the Amentiferae, Part I, 1894, Part II, 1906. Research and publs. on paleobotany, root parasitism in Exocarpus, ovule in Juglans and allied genera, mazocarpon. Died June 20, 1936.

BENSON, Otis Otto, Jr., Am. physician; b. Sandstone, Minn., Sept. 14, 1902; s. Otis Otto and Minnie (Sprague) B.; B.A., U. Mont., 1924, Sc.D., 1955; M.S., U. Ia., 1925; M.D., U. Chgo., 1930; 1 son, Otis Otto III; m. 2d, Ela Dawn McMillan, June 4, 1962. Commd. 1st lt. M.C., USAF, 1930, advanced through grades to maj. gen., 1956; dir. med. staffing, edn. Hdqrs. USAF, 1953-56; comdr. Aerospace Med. Center, Tex., 1956-61; ret., 1961; cons. to aerospace industry, 1961-65; dir. biosci., bioengring. S.W. Research Inst., San Antonio, 1965—. Recipient John Jeffries award Inst. Aero. Scis., 1951; Lyster award Aeromed. Assn., 1955. Diplomats Am. Bd. Preventive Medicine. Fellow A.M.A., Aerospace Med. Assn. (Hubertus Strughold award 1964), A.C.P.; mem. Am. Physiol. Soc., Alpha Omega Alpha. Author: Book of Health: A Medical Encyclopedia for Everyone, 1952. Editor: Physics and Medicine of the Upper Atmosphere, 1952; Physics and Medicine of the Atmosphere and Space, 1960. Research, publs. in aviation and space medicine. Home: 7707 Broadway, San Antonio 78209. Office: 8500 Culebra St., San Antonio 78228.*

BENSON, Sidney William, Am. chemist, educator; b. N.Y.C., Sept. 26, 1918; s. Julius W. and Dora (Cohen) B.; A.B. with honors (Columbia scholar, N.Y. State scholar), Columbia, 1938; A.M., Ph.D. (Thayer scholar, duPont fellow), Harvard, 1941; m. Natasha Sorokine, May 17, 1955; 2 children. Research fellow Harvard, 1941-42; instr. chemistry Coll. City N.Y., 1942-43; group leader Manhattan Project, Kellex Corp., 1943; asst. prof. U. So. Cal., Los Angeles, 1943-48, asso. prof., 1948-51, vice chmn. faculty senate, 1949-50, prof. chemistry, 1951-64; chmn. dept. kinetics and thermochemistry Stanford Research Inst., Menlo Park, Cal., 1963—. Research asso. dept. chemistry and chem. engring. Cal. Inst. Tech., 1957-58; vis. prof. U. Cal. at Los Angeles, 1959, U. Ill., 1959; hon. Glidden lectr. Purdue U., 1961; lectr. Phillips Petroleum Co., 1964; cons. Goodyear Tire and Rubber Co., 1957-61, Douglas Aircraft Co., 1958—. Jet Propulsion Labs., 1961—. Aerospace Labs., 1961—. Guggenheim fellow, 1950-51; Fulbright Fellow, France, 1950-51; NSF sr. postdoctoral fellow, 1957-58. Fellow Am. Phys. Soc., A.A.A.S.; mem. Am. Chem. Soc., Am. Inst. Physics, Faraday Soc., Phi Beta Kappa, Sigma Xi, Pi Mu Epsilon, Phi Lambda Upsilon. Author: Outline of College Chemistry, 1939; Syllabus for General College Chemistry, 1948; Chemical Calculations, 1952, 2d edit., 1963; The Foundations of Chemical Kinetics, 1960. Editorial bd. Elsevier Pub. Co., 1965—; editor in chief Indsl. Jour. Chem. Kinetics, 1968—. Research publs. on relation between molecular structure and chem. reactivity and stability; also on theory of liquid state. Office: Stanford Research Inst., Menlo Park, Cal. 94025.*

BENSON, William Noël, geologist; b. London, Eng., Dec. 26, 1885; s. William Benson; B.Sc., Sydney (Australia) U., 1907, D.Sc., 1916; B.A., Cambridge (Eng.) Univ., 1913, M.A., 1954; m. Gertrude Helen Rawson, 1923. Acting lectr. Adelaide (Australia) U., 1908; demonstrator in geology Sydney U., 1909-10, Linnean Macleay research fellow, 1914-15, acting lectr. in geology, 1916; research scholar from Sydney U. at Cambridge U., 1911-13, at various European univs., 1913-14; prof. geology and mineralogy U. Otago, Dunedin, New Zealand, 1917-50, William Evans prof. geology, 1950, emeritus from 1951. Recipient Carnegie grant for research in Eng., 1933-34, Clarke medal Royal Soc. New S. Wales, 1945. Fellow Royal Soc., Geol. Soc. London (Lyell fund, 1923, Lyell medal 1939), Royal Geog. Soc., Royal Soc. New Zealand (Hector medal 1933, Hutton medal 1944, pres. 1945-47), Australian and New Zealand Assn. for Advancement Sci. (Mueller medal 1951); mem. Mineral. Soc. (hon.), Linnean Soc. New South Wales, Geology Soc. Am. (corr.) Studied Great Serpentine Belt, New South Wales, Cainozoic volcanic rocks, New Zealand; laid foundation for knowledge of Devonian rocks of western New Eng. and of Lower Carboniferous epoch, also added to knowledge of geomorphology, Tertiary basic flows and intrusions. Died Aug. 20, 1957.

BENT, Henry E., Am. chemist; born Oglesby, Ill., October 6, 1900; son of Henry Albert and Josephine (Roberts) B.; A.B., Oberlin Coll., 1922; M.S., Northwestern U., 1923; Ph.D., U. Cal. at Berkeley, 1926; m. Florence E. Domo, Aug, 4, 1924; children—Henry Albert, Robert Demo. Instr. chemistry Harvard, 1926-32, asst. prof., 1932-36; asso. prof. chemistry U. Mo., Columbia, 1936-38, prof. chemistry, dean, 1938-66, dean emeritus, 1966—; chief fellowship sect. Office Edn., Washington, 1959-60. Chmn., Council of Grad. Schs. in U. S. Govt., 1960-61. Mem. A.A.A.S., Am. Chem. Soc., Assn. Grad. Schs. (pres. 1952-53), Assn. Land Grant Colls. and State Univs. (chmn. council grad. work 1954-55), Phi Beta Kappa, Sigma Xi, Alpha Chi Sigma, Phi Delta Kappa, Phi Lambda Epsilon. Contbr. numerous articles to profl. jours. Research, publs. on organic free radicals; amalgams; solutions; gases at chem. warfare. Home: 210 Westwood St., Columbia, Mo. 65201.*

BENT, Robert Demo, Am. physicist, educator; b. Cambridge, Mass., Dec. 28, 1928; s. Henry Edward and Florence (Demo) B.; student U. Mo., 1945-46; B.A., Oberlin Coll., 1950; M.A. Rice U., 1952, Ph.D., 1954; m. Mary Alice Keating, June 9, 1956; children—Lisa Clare, Jason Robert, Alan Demo. Research asso. Rice U., Houston, 1954-55, Columbia, N.Y.C., 1955-58; vis. research asso. Brookhaven Nat. Lab., Upton, L.I., N.Y., summer 1955; asst. prof. physics Ind. U., Bloomington, 1958-62, asso. prof., 1962-66, prof., 1966—. Guggenheim fellow, Oxford, Harwell, 1962-63. Fellow Am. Phys. Soc. Contbr. articles in nuclear physics to profl. jours. Home: 1720 Devon Lane, Bloomington, Ind. 47401.*

BENTELI, (Emmanuel) Albrecht, Swiss mathematician; b. Schwarzenegg, Switzerland, Apr. 10, 1843; s. Gottlieb Abraham and Maria Julia (Lauterburg) B.; began studies at Zürich Polytechnium in 1860; diploma in civil engring., 1863; hon.dr. Bern, 1909; m. Anna Maria Elis, 1868; 2 sons, 2 daus.; tchr. at various canton schs.; became prof. geometry, Bern, Switzerland, 1874; became gymnasium rector, 1889. Mem. Bernische Naturforschenden Gesellschaft, Schweizerische Naturforschenden Gesellschaft. Developed 1st precipitation charts for Switzerland; did basic work on lighting constrn. and perspective, meterology. Died Bern, Nov. 10, 1917.

BENTHAM, George, English botanist; b. Stoke nr. Portsmouth, Sept. 22, 1810; s. Samuel, Bentham and dau. of Dr. George Fordyce; studied at Faculty Theology at Montcuban, nr. Toulouse, France, also Lincoln's Inn, Eng.; m. dau. of Sir Hartford Brydges, 1834. Lived in France, 1814-27; worked on descriptive botany at Kew, Eng., from 1861. Recipient Royal medal, 1859. Fellow Linnean Soc. (elected vice president 1858, president 1861-74); member Horticultural Soc. (hon. sec. 1829-40), Royal Soc., 1862, French Acad. Scis., 1875. Author: Catalogue des plantes indigènes des Pyrénées, 1826; Outlines of a New System of Logic (developed doctrine of qualification of predicate), 1827; Plantae Hartwegiana, 1839-57; Handbook of British Flora, 1858; (with Joseph Hooker) Genera plantarum, 3 vols., 1862-83; Flora Australiensis, 7 vols., 1863-78; also numerous articles on classification and description of flowering plants, flora of Hong Kong and Australia. Divided plants into Dicotyledons, Gymnosperms, and Monocotyledons. Died Sept. 10, 1884.

BENTHAM, Jeremy, English philosopher, jurist; b. London, Feb. 15, 1748; s. Alicia Grove Bentham and a lawyer; B.A., Queen's Coll., Oxford, 1763, master's degree, 1766; studied chemistry briefly; called to bar, Lincoln's Inn, circa 1767, became bencher, 1817; during visit to brother in Russia (1785) became interested in prison discipline, mgmt., criminal law; returned to Eng., 1788; his work on poor laws (circa 1790) led to their revision; became French citizen, 1792. Author: Fragment on Government or Comment on Commentaries, 1776; Introduction to the Principles of Morals and Legislation, 1789; Principles of Penal Law; Truth vs. Ashhurst, 1792; A Protest against Law Taxes and Supply without Burden, 1795; Rationale of Punishments and Rewards, 1811. Upheld concept of utilitarianism; believed that increase of happiness should be sole object of concern to legislators and moralists; identified happiness with pleasure; believed that under democracy greatest happiness for greatest number could be best achieved; believed that instns. are justified by their utility, that rational law should be based on considerations for what human affectation and good of society demands; devised a moral arithmetic for judging value of pleasure of pain; influenced major governmental reforms, including vote by ballot, establishment of system of pub. prosecutors, criminal law reforms, simplification of forms of statutes. Died London, June 6, 1832.

BENTIUS, Hugo, see Benzi, Ugo.

BENTLEY, Charles Albert, Brit. physician; b. Chipping Norton, Eng., Apr. 25, 1873; ed. Univ. Coll. Liverpool, Royal Colls., Edinburgh; M.B., C.M., Edinburgh U.; diploma in pub. health Cambridge U., 1905, in tropical medicine and hygiene, 1905; M.D. (hon.), Calcutta, 1931. Prof. hygiene Egyptian U.; jr. surgeon Royal So. Dispensary, Liverpool, 1898; dir. pub. health Govt. of Bengal (India); with Ceylon Tea Co., Assam, 1900-07; mem. Duar's Blackwater Fever and Malaria Commn., 1907-09; mem. Bombay Malaria Enquiry, 1909-11; dep. san. commr., Bengal, 1911-15; dir. London Cinchona Bur., from 1949. Recipient Kaiser-i-Hind Gold medal, 1916. Fellow Royal Soc. Medicine, Royal Soc. Tropical Medicine and Hygiene, Soc. Med. Officers of Health; mem. Brit. Med. Assn. Author: (with S. R. Christophers) Malaria in the Duars and Blackwater Fever in the Duars; Bombay Malaria, 1911; Malaria at Dinajpur, 1912; Report on Malaria in Bengal, 1916; Malaria and Agriculture, Bengal, 1925; also pub. health reports, from 1915. Discovered that ground-itch is caused by larval anchylostomes in infected soil, 1901; discovered 1st known leucocytozoan, 1903; discovered Leishman-Donovan parasites in cases of Kala-Azar in Assam, 1903. Died Nov. 23, 1949.

BENTLEY, Gordon Mansir, Am. zoologist, entomologist; b. Great Barrington, Mass., Sept. 23, 1875; s. Charles Harrison and Elvira E. (Mansir) B.; B.S.A., Cornell U., 1900, A.M., 1901; M.S., U. Tenn., 1928; m. Mary Catherine Elmore, June 12, 1912; children—Juanita Louise, Edna Elvira. Prof. botany and zoölogy Union Acad., Belleville, N.Y., 1901-04; instr. entomology N.C. State Coll., Raleigh, 1905; also asst. state entomologist N.C., 1905; instr. in zoölogy and entomology U. Tenn., 1905-08, asso. prof. entomology, 1908-23, prof., 1923-50, head dept. 1926-44, prof. entomology, 1945-50, ret., 1950; state entomologist and plant pathologist 1909-50; cons. entomologist and zoologist. Pres. So. Plant Quarantine Bd., 1919-30, 37-38, 39-44. Fellow A.A.A.S.; mem. Am. Assn. Econ. Entomologists (v.p. 1918-29, 36; pres. plant quarantine and inspection sect. 1925; pres. apiculture sect. 1929), Entomol. Soc. Am. (charter), Tenn. Acad. Sci., Cotton States Entomologists (sec. 1923-27; pres. 1927), So. Nurserymens Assn. (hon. life), other profl. socs., Sigma Xi. Author: Lectures and Laboratory Guide for Economic Entomology, 1929; Insect Taxonomy, 1929; also 120 bulls. on econ. entomology. Contbr. to Jour. Econ. Entomology. Editor of Proc. Tenn. State Hort. Soc., Tenn. Market Bull., State Dept. Agr. Reports. Died Oct. 8, 1954.

BENTLEY, Madison, Am. psychologist; b. Clinton, Ia., June 18, 1870; s. Charles Eugene and Persis Orilla (Freeman) B.; B.S., U. Neb., 1895, LL.D., 1935; Ph.D., Cornell U., 1898. Asst. in psychology, instr., asst. prof. Cornell, until 1912; prof. psychology and dir. psychol. labs. U. Ill., 1912-28; Sage prof. psychology Cornell, 1928-36; cons. for psychology, Library of Congress, 1938-40, lectr. in psychology Cornell U., 1942-44; chmn. div. anthropology and psychology NRC, 1930-31. Fellow A.A.A.S.; mem. Am. Psychol. Assn. (pres. 1925), Phi Beta Kappa, Sigma Xi. Author: Studies in Social and General Psychology, 1916; Critical and Experimental Studies in Psychology, 1921; The Field of Psychology, 1924; Studies in Psychology from the University of Illinois, 1925; The New Field of Psychology, Pt. 1, The Psychological Functions and Their Government, 1934; The Problems of Mental Disability in England, 1938; Cornell Studies in Dynasomatic Psychology, 1938; The Theater of Living in Animal Psychology, 1943; Sanity in the Life Course, 1946; Towards a Psychological History of the Hominids, 1947; Primary Factors in the Government of certain Biomechanical Systems, 1952; also sect. in Manual of American Literature, 1909. Editor Am. Jour. Psychology, 1903-51. Contbr. to encys. and mags. Died May 29, 1955.

BENTLEY, Ronald, biochemist; born Derby, Eng., Mar. 10, 1922; s. Douglas and Agnes (Webster) B.; B.Sc., Derby Tech. Coll., 1943; Ph.D., Imperial Coll. Sci. and Tech., 1945; D.Sc., U. London, 1965; m. Marian Louise Blanchard, June 19, 1948; children—Colin Christopher, Alison Louise, Peter Douglas. Came to U. S., 1951, naturalized, 1957. Researcher, Imperial Coll. Sci. and Tech., London, Eng., 1945-46; researcher Columbia Coll. Phys. and Surg., 1946-47, research asso., 1952-53; sci. staff Nat. Inst. for Med. Research, London, 1948-52; faculty U. Pitts., 1953-—, prof. biochemistry, 1960—. Neil Arnott student U. London, 1943-44; Commonwealth Fund fellow Columbia, 1946-47; Pub. Health Service Spl. Research fellow Inst. Biochemistry, U. Lund (Sweden), 1963-64; John Simon Guggenheim Meml. Found. fellow, 1964. Mem. Am. Chem. Soc., Am. Soc. for Microbiology, Am. Soc. Biol. Chemists, Biochem. Soc. (London), Chem. Soc. (London), A.A.A.S., Sigma Xi. Contbr. numerous articles to tech. jours. Research on elucidation mechanisms by which variety microbial products are formed from carbohydrates, reactivity carbohydrates in chem. and biochem. systems. Home: 37 Thornwood Dr., Pitts. 15228.*

BENTLEY, Wilson Alwyn, Am. meteorologist; born Jerichox, Vt., Feb. 9, 1865; s. Thomas E. and Fanny (Colton) B.; ed. pub. sch. Jericho. Tchr. music, 1885-86; student of snow crystals and other meteorol. subjects, from 1882; made 5,150 photomicrographs of snow crystals, 600 of frost crystals, 200 of ice and ice crystals, hail, many hundreds of dew, clouds, raindrops; author monographs on these studies, pub. by U. S. Weather Bur., 1902, 04, 05, 08, also articles on snow and frost in Ency. Americana, contbns. to mags. and lectures. Died Dec. 23, 1931.

BENTZON, Povl Georg, Danish surgeon; b. Copenhagen, Denmark, Feb. 22, 1891; s. Povl and Harriet (Drachmann) B.; M.D., U. Copenhagen; m. Johanne Bierrum, May 15, 1917; children—Lars, Clara, Margrete, Harriet. Physician pvt. orthopedic clinic at Copenhagen, 1922-36; chief surgeon orthopedic hosp. at Aarhus, 1935-52. Mem. French Orthopedic Soc. (hon.), Nordic Assn. Orthopedics. Author: The Pathogenesis of Duchenne Erb's; Obstetric Paralysis; Operation for Fracture of the Carpal Schaphoid. Address: Elverdalsvej, 147, Hoejbjerg, Denmark.*

BENZ, Carl Friedrich, German inventor, engr.; b. Karlsruhe, Germany, Nov. 25, 1844; s. Johann George and Jos. (Vaillant) B.; studied mech. engring. under Ferdinand Redtenbacher and Franz Grashof, 1860-64; m. Berta Ringer, 1872; 2 sons, Eugen, Richard; 3 daus. Worked with several mech. cos.; founded iron foundry and mech. workshop Mannheim, Germany, 1871, founded Benz & Cie. Gasmotorenfabrik (later merged with Daimler-Motren-Gesellschaft), Mannheim, 1883; withdrew from corp., 1908; founded Benz Söhne, Ladenburg, Germany. Recipient Gold medal Munich Machine Exhbn., 1888. Author: Lebensfahrt eines deutsche Erfineer, Erinnerungen eines Achtzigjahrigen, 1925. Often considered father of automobile; built 1 horsepower 2 cycle gas motor, 1877; designed and demonstrated 3 wheeled motor vehicle; patented gasoline vehicle with accumulator, plunger carburetor and water jacket cooling system, 1886; built 4 wheeled vehicle with axle journal steering, 1893, later made improvements in electric ignition, motor capacity, carburetor, cooling system. Died Ladenburg, nr. Mannheim, Apr. 4, 1929.

BENZÉCRI, Jean-Paul, mathematician; b. Oran, Algeria, Feb. 28, 1932; s. Emile Elie and Odette (Benyamine) B.; ed. École Normale Supérieure, Paris; Agrege de Mathematiques, Paris, 1953; Ph.D., Princeton Univ., 1955; Doctorat d'état' Paris, 1960; m. Françoise Le Roy, Oct. 18, 1960; children: Thérèse, Jean. Asst., École Normale Supérieure, Paris, 1955-58; operational research, French Navy, 1958-60; lectr., Faculty of Sciences, Rennes, 1960-63; prof. mathematics, Rennes, 1963-65; prof. statistics, Paris, 1965—. Mem. Math. Soc. of France; Am. Math. Soc.; ATALA. Research in geometry, statistics; in data analysis, developed structural features by automatic computations performed on huge data tables. Office: 9 Quai Saint Bernard, Paris (5^{e}), France.

BENZENBERG, Johann Friedrich, German physicist; b. Schöller bei Elberfeld, Germany, Apr. 5, 1777; s. Heinrich and Johanna Elisabeth (Fues(s) B.; ed. Herborn, Marburg, Göttingen (all Germany); doctorate, Duisburg, 1800; m. Charlotte Platzhoff, 1807. Tchr. math. Düsseldorf Lyceum; founder pvt. obs. in Bilk, nr. Düsseldorf, Germany, 1844; dir. survey of duchy of Berg, 1805-10; publicist, active in polit. and econ. affairs; fled from Napoleon to Switzerland, circa 1810. Author: (with H. W. Brandes), Versuche, die Entfernung, die Geschwindigkeit und die Bahnen der Sternschnuppen zu bestimmen, 1800; Über die Bestimmung der geographischen Länge durch Sternschnuppen, 1802; Versuche über das Gesetz des Falls, über den Widerstand der Luft und über die Umdrehung der Erde, 1804; Vollständiges Handbuch der angewandten Geometrie, 1813; also several polit. and econ. books. Performed (with H. W. Brandes) 1st meteorite tracking expts; gave 1st proof. of eastward deviation of free falling bodies (earth's rotation); showed speed of sound is dependent on temperature, 1811; wrote on astron., phys. and geodetic subjects. Died Bilk nr. Düsseldorf, June 7, 1846.

BENZER, Seymour, Am. molecular biologist, educator; b. N.Y.C., Oct. 15, 1921; s. Mayer and Eva (Naidorf) B.; B.A., Bklyn. Coll., 1942; M.S., Purdue U., 1943, Ph.D., 1947; m. Dorothy Vlosky, Jan. 10, 1942; children—Barbara Ann, Martha Jane. Faculty physics Purdue U., West Lafayette, Ind., 1945-67, prof., 1958-61, Stuart Distinguished prof. biology, 1961-67; prof. biology Cal. Inst. Tech., Pasadena,

1967—; biophysicist Oak Ridge Nat. Lab., 1948-49; research fellow Cal. Inst. Tech., Pasadena, 1949-51. Recipient award of honor Bklyn. Coll., 1956; Ricketts award U. Chgo., 1961; gold medal N.Y. City Coll. Chem. Alumni Assn., 1962; Gairdner Found. award of merit, 1964; Fulbright research fellow Pasteur Inst., Paris, France, 1951-52; NSF sr. postdoctoral fellow Cambridge (Eng.) U., 1957-58. Fellow A.A.A.S.; mem. Nat. Acad. Scis., Am. Acad. Arts and Scis., Am. Philos. Soc., Harvey Soc. (hon.), Sigma Xi (Purdue U. research award 1957). Contbr. numerous articles on molecular biology, especially fine structure of gene, to profl. publs. Patentee germanium semicondr. devices. Home: 195 S. Wilson St., Pasadena, Cal. 91106.*

BENZI, Ugo (or Hugo Bentius, Hugo de Siena, Hugo Senensis, Ugone de Benciis), Italian physician; b. Siena, Italy, circa 1360; M.D., U. Siena. Became prof. medicine U. Siena, circa 1395, again in 1417; prof., Bologna, also Parma, Italy, 1402-16, Pavia, Italy, 1422-27; physician to Charles VII of France. Author: Tractato utilissimo circa la conservazione della sanitade (1 of 1st med. texts in vernacular, contains observations on personal hygiene), 1481. Died Ferrara, Italy, circa 1439.

BENZINGER, Theodor Hannes, physician, born Stuttgart, Germany, August 23, 1905; the son of Theodor Johannes and Alma (Heincke) B.; student U. Munich (Germany); D.Sc., U. Tuebingen (Germany), 1929; postgrad. U. Berlin (Germany); M.D., U. Freiburg (Germany), 1933; m. Ilse Koss, July 26, 1934; children—Rolf, Angela (Mrs. Rolf Stuempel), Monica (Mrs. Eckhard Uetermann); m. 2d, Maria Henke Gerhartz, Mar. 3, 1960; children—Robert Henke, Fay Ann. Came to U.S., 1947, naturalized, 1955. Faculty, U. Tuebingen, 1929, U. Freiburg, 1932-33, U. Goettingen, 1934-44; dir. aeromed. dept. German Air Forces Testing Center, Rechlin, 1934-44; head calorimetry div., dir. bioenergetics labs. Naval Med. Research Inst., Bethesda, Md., 1947—. Recipient Golden Scheele medal Chem. Soc. Stockholm (Sweden), 1933. Mem. Am. Physiol. Soc., Aerospace Med. Assn., N.Y. Acad. Scis., Colloque Claude Bernard, Deutsche Physiol. Gesellschaft, Gesellschaft Deutscher Naturforscher und Aertze. Discovered quantitative mechanism of human thermoregulation, thermodynamic quantities of key biochem. reactions; demonstrated survival in explosive decompression, airembolism as cause of death from blast; invented gradient layer calorimetry, clin. ear thermometry, heat pulse-microcalorimetry, reaction calorimetry method determining free energy and entropy. Home: 6607 Broxburn Dr. Office: Bio-Energetics Labs., Naval Med. Research Inst., Nat. Naval Med. Center, Bethesda, Md. 20014.*

BERANBAUM, Samuel, physician; born Toronto, Ont., Can., Apr. 20, 1915; s. Mandel and Bessie (Markson) B.; B.A. (Daniel Wilson scholar, Edward Blake scholar), U. Toronto, 1937, M.D., 1940; m. Betty Adelaide Samson, Mar. 24, 1949; children—Joan Louise, Nancy Ann, John Arthur. Came to U. S., 1940, naturalized, 1944. Practice medicine specializing in radiology, N.Y.C., 1942—; dir. radiology St. Barnabas Hosp. for Chronic Diseases, N.Y.C., 1945—; staff N.Y. U. Hosp., 1948—; attending radiologist Bellevue Hosp., N.Y.C., 1966—; faculty N.Y. U. Sch. Medicine, 1947—, clin. prof. radiology, 1960—. Diplomate Am. Bd. Radiology. Fellow Am. Coll. Radiology, Am. Coll. Gastroenterology; mem. Radiologic Soc. N.Am., Am. Roentgen Ray Soc., N.Y. Roentgen Soc., N.Y. Gastroent. Assn., N.Y. Acad. Gastroenterology, Am. Thoracic Soc., N.Y. Acad. Medicine, N.Y. Acad. Scis., Am. Heart Assn., Soc. Med. Jurisprudence, World Med. Assn. Author: (with P. H. Meyers) Special Procedures in Roentgen Diagnosis, 1964; also articles. Research on gastro-intestinal radiology, 70 mm. cineflurography, early diagnosis of pancreatic carcinoma, biliary tract disease. Home: 275 Central Park W., N.Y.C. 10024. Office: 121 E. 60th St., N.Y.C. 10022.*

BÉRAND, Laurent, French astronomer, physicist; b. Lyons, France, May 5, 1702; s. Christophe and Marie (Mercier) B.; Jesuit priest; tchr. math. and philosophy, Vienna, Austria; became dir. Lyons Obs., 1740. Mem. Acad. Lyons. Author: Physique des corps animés, 1775; also memoirs on calcination and electricity. Studied passage of mercury past sun, May 6, 1753; made astron. observations similar to those of La Caille at Cape of Good Hope; recorded annular eclipse of sun, 1764 (1st eclipse of this type observed in France). Died June 26, 1777.

BERANEK, Leo Leroy, Am. acoustician; b. Solon, Ia., Sept. 15, 1914; s. Edward Fred and Beatrice (Stahle) B.; A.B., Cornell Coll., 1936, D.Sc., 1946; M.S., Harvard, 1937, D.Sc., 1940; m. Phyllis Knight, Sept. 6, 1941; children—James Knight, Thomas Haynes. Instr. physics Harvard, 1940-43, dir. research on sound, 1943-45, dir. electro-acoustics and systems research, 1945-46; asso. prof. communications engring. Mass. Inst. Tech., 1947-58, lectr., 1958—, tech. dir. Acoustics Lab., 1947-53; pres., dir. Bolt, Beranek, & Newman, research, cons., devel., Cambridge, 1953—; pres., dir. Boston Broadcasters, Inc., 1963—; chmn. bd. Mueller-BBN GmbH, Munich, Germany, 1963—. Guggenheim fellow, 1946-47; recipient Presdl. certificate of merit, 1948. Fellow Acoustical Soc. Am. (Biennial award 1944; mem. exec. council; v.p. 1949-50, pres.-elect, 1953-54, pres., 1954-55; asso. editor, 1946-60, Wallace Clement Sabine Archtl. Acoustics award 1961), Am. Acad. Arts and Scis., Am. Phys. Soc., A.A.A.S., Audio Engring., Soc. (pres. 1967-68), I.E.E.E. (chmn. profl. group on audio 1950-51); mem. Am. Standards Assn. (chmn. acoustical standards bd.), Nat. Acad. Engring., Am. Soc. Engring. Edn., Phi Beta Kappa, Sigma Xi. Author: Principles of Sound Control in Airplanes (with others), 1944; Acoustic Measurements, 1949; Acoustics, 1954; Noise Reduction (with others), 1960; Music, Acoustics and Architecture, 1962. Editor Magazine Noise Control, 1954-55; asso. editor Sound mag., 1961-63. Research and publs. on acoustics, especially design of anechoic chambers, calculation of intelligibility of transmitted speech, design of concert halls, opera houses, loudspeakers; criteria for allowable noise; quieting of aircraft, engine test cells, wind tunnels, ventilation systems; measurement of sound and audio equipment. Home: 7 Ledgewood Rd., Winchester, Mass. 01890. Office: 50 Moulton St., Cambridge, Mass. 02138.*

BÉRARD, Auguste, French surgeon; b. Verrains, France, Aug. 2, 1802; M.D., Paris, 1829; became prof. clin. surgery, Med. Faculty, Paris; became cons. to Louis-Philippe, 1845; mem. Anat. Soc. (founder with others), Acad. Medicine. Author: (with Denonvilliers) Compendium de chirurgie pratique, 1840. Innovator in staphylorrhaphy, wound irrigations, varices and erectile tumors, fracture reductions. Died Paris, Oct. 15, 1846.

BÉRARD, Auguste, French navigator; b. Montpellier, France, Feb. 24, 1796; ed. Ecole navale; circumnavigated globe, 1817-20; made hydrographic expdns. on Mediterranean Sea, 1831-33; dir. port, Toulon, France, from 1847; vice-adm., from 1848. Mem. French Acad. Scis., 1840. Died Oct. 6, 1852.

BÉRARD, Jacques Etienne, French physicist, chemist; b. Montpellier, France, Oct. 12, 1779; ed. under Berthollet in Arcueil's lab.; M.D., Montpellier, 1827. Named prof. chemistry Faculty Sci., also Sch. Pharmacy, Montpellier, 1827, became docent Faculty Sci., also corr. Inst., 1819, prof. gen. chemistry and toxicology Sch. Medicine, from 1832. Decorated comdr. Legion of Honor; (with Delaroche) grand prize chemistry, physics (for work on specific heat of gases), French Acad. Scis., 1813. Mem. French Acad. Scis., 1819, Acad. Medicine. Author: (with Delaroche) Sur la determination de la chaleur spécifique des différents gaz (memoire), 1813. Determined (with Delaroche) specific heat of gases with constant pressure, also pointed out constancy (law of Delaroche and Berard), 1813; made detailed study of radiation, 1814; distinguished calorific (infrared), colorific (visible) and chem. (violet and ultraviolet) rays; held that chem. effects of light were not caused by heat; observed intramolecular respiration (animal respiration in absence of oxygen), 1821. Died Montpellier, June 10, 1869.

BÉRARD, Joseph Frederic, French biologist, philosopher; b. Montpellier, France, 1789; student U. Montpellier, became prof. hygiene, 1825. Author: Essai sur les anomelies de la variole et de la varicelle, 1818; Doctrine médicale de l'école Montpellier, et comparaison de ses principes avec aux des autres écoles, 1819; Doctrine des rapports du physique et du moral pour servir de fondement à la physiologie dite "intellectuelle" et à la metaphysique, 1823. Defended doctrines of Montpellier sch., stressing spiritualism and vitalism; his philosophy lay between psychology and physiology. Died Montpellier, 1828.

BERBERIAN, Dicran Abraham, physician, educator; b. Diarbekir, Turkey, Aug. 9, 1903; s. Abraham Hartune and Shamiran (Nalbandian) B.; B.A., Am. U. Beirut (Lebanon), 1926, M.D., 1930; postgrad. (Rockefeller fellow in parasitology) London (Eng.) Sch. Hygiene and Tropical Medicine, (Rockefeller fellow in mycology) U. Paris (France); m. Armine Vartan Poladian, Dec. 24, 1932; children—Cynthia (Mrs. Thomas Hale III), Dicran Aram, Raffi Robert. Came to U. S., 1947, naturalized, 1953. Instr. bacteriology and parasitology Sch. Medicine, Am. U. Beirut, 1930-35, adj. prof., 1936-45, asso. prof., chmn. dept. bacteriology, 1945-47; dir. research in tropical medicine and parasitology, staff physician Sterling-Winthrop Research Inst., Rensselaer, N.Y., 1947—; faculty medicine Albany (N.Y.) Med. Coll., 1947—, prof. microbiology, 1964—. Adviser, Lebanon Ministry Health, 1943-47; cons in medicine VA Hosp., Albany, 1953—. Recipient Merite Libanaise, Lebanon, 1943. Fellow N.Y. Acad. Scis.; mem. Royal Soc. Hygiene and Tropical Medicine (London), Am. Soc. Tropical Medicine and Hygiene, Soc. Am. Bacteriologists, Am. Soc. Parasitology, A.A.A.S., Am., Albany County med. assns. Research, publs. in bacteriology, immunology, micology, parasitology. Home: 389 Loudonville Rd., Loudonville, N.Y. 12211. Office: Sterling-Winthrop Research Inst., Rensselaer, N.Y. 12144.*

BERBERICH, Adolf Joseph, German astronomer; b. Überlingen; Germany, Nov. 16, 1861; s. Michael and Katherine (Hirt) B.; studied astronomy in Strasbourg, France, 1880-84; m. Wilhelmine Auguste Plesse, June 11, 1916. Began as asst. Astronomishce Recheninstitut (inst. for astron. calculations), Berlin, Germany, 1884, received teaching position, 1897; named prof., 1903. Recipient Valz prize Acad. Scis. in Paris, 1894. Editor Astronomischer Jahresbericht, 1905-09. Contbr. articles to astron. jours. Developed method of astronomically determining paths of smaller planets and comets; interested in newly developing field of astrophysics. Died Berlin-Tempelhof, Apr. 27, 1920.

BERBLINGER, Walther Emil, physician; b. Karlsruhe, July 13, 1882; ed. Karlsruhe Gymnasium, univs. of Heidelberg, Munich, German U. of Strasbourg; M.D. Asst., Inst. Hygiene, Strasbourg; asst. clinician Hosp. of Karlsruhe; prosector Inst. Pathology, Zurich, Switzerland; prof. agrégé U. Marburg; prof. U. Kiel, 1916—, full prof., 1918; dir. Inst. Pathology, Marburg, 1922; prof. Research Inst. at Davos (Switzerland), 1922-38. Mem. Soc. Physicians of Vienna (corr.). Author: Innere Sekretion Form und Funktion, 1928; Pathologie der Hypophyse des Menschen, 1932; Der Schwund tuberkulöser Lungenkavernen, 1943; Herzhypertrophie bei Lungentuberkulose, 1947. Editor: Centralblatt für Pathologie, 1915-38; co-editor Medizinische Welt, 1924—, Endokrinologie, 1934—, Aerztliche Forschung, 1946—, Acta Davosiana, until 1954. Research and numerous publs. on internal secretion, Tb and various branches of pathology. Address: Muri-bei-Bern, Switzerland.

BERDJIS, Charles Choeib, physician; born Kashan, Iran, July 1, 1918; s. Yagoub and Hosni (Fakimi) B; student U. Paris, Sorbonne, 1933-37, M.D., 1949; M.D., U. Geneva, 1945; m. Odette R. Bezzola, Mar. 12, 1941; children—Mariam C., Mary F. Came to U. S., 1953, naturalized, 1958. Faculty, U. Cal. Sch. Medicine at San Francisco, 1954-57; commd. maj. U. S. Army, 1957, advanced through grades to lt. col. 1958; chief pathologist, comdg. officer 4th Army Med. Lab., 1957-59; sr. pathologist Walter Reed Army Med. Unit, Ft. Detrick, Md., 1959-63; chief exptl. medicine and trauma research dept. U. S. Army Edgewood Arsenal, Md., 1964—. Mem. Am. Soc. for Exptl. Pathology, Radiation Research Soc., A.M.A., Sci. Research Soc. Am., Assn. Mil. Surgeons U. S., Am. Ordnance Assn., Internat. Soc. Scis. and Arts, Internat. Congress Radiation Research, Washington Soc. Pathology, N.Y., Geneva acads, scis., others. Author: Hematology in Sulfanilamide Therapy, 1940; (with Wernley) Human Parathyroids, 1946; Behavioral Sciences, 1966. Research publs., on pathology, irradiation, cancer, infectious diseases, behavioral research, endocrinology and nutrition with hormones and vitamins. Address: Exptl. Medicine Dept., U. S. Army Edgewood Arsenal, Md. 21010.*

BERDYEV, Ata Abdurakhmanovich, Russian physicist; b. 1914. Dir. Physicotech. Inst., Turkmen Acad. Sci. Mem. Turkmen Acad. Sci. (corr.). Author: The Dependence of Ultrasonic Wave Absorption in Pure Liquids on Temperature; The Absorption of Ultrasound at the Critical Temperature of a Solution; The Coefficient of Volume, the Second Factor of the Viscosity of Liquids; The Absorption of Ultrasonic Waves in Certain Chloral-Spirit Mixtures, 1960, also others. Research in ultrasonics, absorption of ultrasonic waves in liquids, study of propagation and absorption of ultrasonic waves in bi-component liquid systems, propagation and absorption of ultrasonic waves at high frequencies, molecular and atomic spectroscopy. Address: Physicotech. Inst., Turkmen Acad. Sci., Ashkhabad, Turkmen SSR, USSR.

BERENBLUM, Isaac, biologist; b. Bialystock, Poland, Aug. 26, 1903; s. Paul and Michle (Slabodsky) B.; B.Sc. with honors, Leeds U., Eng., 1923, M.B., Ch.B., 1926; M.D. with distinction, 1930, M.Sc., 1936; m. Doris Lina Bernstein, Aug. 30, 1928; children—Teresa (Mrs. Cohen), Ann Cecily (Mrs. Szöke). Riley-Smith research fellow in cancer research Leeds U., 1927-36; Beit Meml. research fellow Dunn Sch. Pathology, Oxford U., 1936-40, univ. demonstrator, 1940-48; spl. research fellow Nat. Cancer Inst., NIH, Bethesda, Md., 1948-50; prof., head dept. exptl. biology Weizmann Inst. Sci., Rehovoth, Israel, 1950—; vis. prof. Hebrew U., Jerusalem, 1951-56. Mem. N.Y. Acad. Sci. (hon. life mem.), Israel Acad. Sci. and Humanities, World Acad. Arts and Scis. Author: Man Against Cancer, 1952; Cancer Research Today, 1967; also numerous articles. Research on mechanism of carcinogenesis; discovered anticarcinogenic action and cocarcinogenic action; co-discoverer 2-stage mechanism of carcinogenesis; research on metabolism of carcinogens, on radiation leukaemogenesis; discovered RLP factor in sheep spleen capable of inhibiting radiation leukaemogenesis. Home: 33 Ruppin St. Office: Weizmann Inst. Sci., Rehovoth, Israel.*

BERENDES, Julius, German otorhinolaryngologist; b. Elberfeld, Germany, Mar. 2, 1907; s. Rudolf and Helene (vom Hagen) B.; M.D., m. Anita Metzger, Feb. 22, 1934; children—Peter, Renate, Klaus, Ulrich. Asst., prin. physician at Heidelberg, 1933-44; physician in chief at Mannheim, 1957; full prof., dir. HNO Clinic at Marburg (Germany), 1957—. Mem. German Soc. Otorhinolaryngol. Physicians (pres. 1960-61), German Soc. Phoniatrics (pres. 1962—), French Assn. Exptl. Phonology (pres. 1963-65). Author: Bewegungsstörungen des Kehlkopfes, 1956; Anleitung zur Funktionsprufung des Ohres, 1957; Einführung in die Sprachheilkunde, 1963; HNO-Heilkunde, 1963. Address: Deutchhaus-Strasse 3, Marburg/Lahn, Germany.

BERENDT, Gottlieb Michael, German geologist; b. Berlin, Germany, Jan. 4, 1836; s. Michael and Veronica Rosalie (Huhn) B.; degree, 1863; m. Alwine Necker, 1866; 2 sons including Werner; 1 dau. Became asso. prof. U. Königsberg (Germany), 1873; state geologist, dir. lowlands dept. Geologische Landesanstalt, 1875-

1904; prof. Bergakademie, Berlin. Author: De formatione diluviana in Marchia provincia ac potissimum dren—Teresa (Mrs. Cohen), Ann Cecily (Mrs. Szöke). in vicinitate Postempiae, 1863; Geologie der Kur. Haffs, 1869; Die Umgebung von Berlin, 1877; Blätter der Geologischen Karte von Preussen 1:25,000 mit Erläuterungen; 12 Blätter der Geologischen Karte von Ostpreussen 1:000,000; also numerous geol. charts and maps. Pioneered research on lowlands geology and classification of formations; research on German diluvium; organized cartographic diluvian surveys in scale 1:25,000. Died Schreiberhau, Germany, Jan. 27, 1920.

BERENGARIO DA CARPI, Giacomo (or Jacopo), Italian surgeon; b. Carpi, Italy, 1470; s. Faustino Berengario da Carpi; studied medicine, Pavia, Italy; M.D., Bologna, Italy. Prof. surgery, Pavia, also Bologna, 1502-22; practiced medicine, Rome, also Ferrara, Italy, 1522-30. Author: Anathomia Mundini noviter impressa ac per Carpum castigata, 1514; Isagogae breves in anatomiam corporis humani, 1514; Tractatus perutilis et completus de fractura cranei, 1518; Commentaria cum amplissimis additionibus supra anatomiam mundini cum textu ejus in pristinum nitorum redacto, 1521. Pioneer in modern times in anatomy; described 2 cases of vaginal hysterectomy, also said to have been 1st to describe sphenoid sinuses and veriform appendix, 1521; described thymus accurately, also gave 1st description of operation for removal of uterus in a case of prolapse, 1522; 1st to discuss action of cardiac valves, to describe horse shoe kidney, to use mercury to treat syphilis; introduced term vas deferens; called restorer of anatomy by Fallopius. Died Ferrara, 1530.

BERENSON, Gerald Sanders, Am. physician; b. Bogalusa, La., Sept. 19, 1922; s. Meyer A. and Eva (Singerman) B.; B.S., Tulane U., 1943, M.D., 1945; m. Joan Seidenbach, Mar. 7, 1951; children—Leslie M., Ann M., Robert S., Laurie G. Faculty, Tulane U. Med. Sch., 1948-1952, U. Chgo., 1952-54; faculty La. State U. Sch. Medicine, New Orleans, 1954—, prof., 1963—; vis. staff Charity Hosp. La., New Orleans, 1948-52, 54—, Crippled Children's Hosp., 1955—. Fellow A.C.P.; mem. New Orleans Acad. Internal Medicine (pres. 1965), So. Soc. for Clinical Investigation (sec.-treas. 1965), La. Heart Assn., Am. Soc. Biol. Chemists, Soc. for Exptl. Biology and Medicine, A.A.A.S., A.C.S., N.Y. Acad. Scis., Am. Coll. Cardiology, Am. Heart Assn. Research, numerous publs. on biochemistry of connective tissue and role in cause of cardiovascular diseases; devel. techniques, methods and instruments for study chem. compounds from connective tissue. Home: 144 Audubon Blvd., New Orleans 70118. Office: 1542 Tulane Av., New Orleans 70112.*

BERÉNYI, Dénes, Hungarian physicist; b. Debrecen, Hungary, Dec. 26, 1928; s. D. and Vilma (Tóth) B.; Dipl.phys., Kossuth Lajos U., Debrecen, 1952, Ph.D., 1959; Candidate phys. sci., Hungarian Acad. Sci., 1963; m. Elvira Bódor, June 12, 1956; children—Rita, András. Asst., Kossuth Lajos U. Inst. Exptl. Physics, 1952-54; staff Inst. Nuclear Research, Hungarian Acad. Scis., 1954—, head nuclear spectroscopy div., 1963—; faculty Moscow (USSR) U., 1959-60. Research, publs. on beta and gamma ray spectroscopy especially electron capture to positron emission ratios and internal bremsstrahlung in forbidden transitions. Home: 12 Nemes. Office: 18/c Bem-tér, Debrecen, Hungary.*

BEREZHNOY, Anatoliy Semenovich, Russian chem. technologist; b. Village of Ostapovo (now Poltava Oblast), 1910, Mem. Ukrainian Acad. Sci. (corr.). Author: Silicon and its Binary Systems, 1958; Nikolay Semenovich Kurnakov: On the 100th Anniversary of his Birth, 1960; New Data on Binary Systems of Silicon, 1961; co-author: Solid State Reactions, 1949; (textbook) The Technology of Ceramics and Refractory Materials, 1962. Research on solid state reactions of silicate systems; developer new tech. processes for mfr. of chromite, heat-resistant magnesite and other refractory materials. Address: Ukrainian Acad. Sci., Vladimirskaya 54, Kiev, Ukraine SSR, USSR.

BERG, Albert Ashton, Am. surgeon; b. N.Y.C., Aug. 10, 1872; s. Moritz and Josephine (Schiff) B.; A.B., Coll. City of N.Y., 1894; M.D., Coll. Phys. and Surg., Columbia, 1891. House surgeon Mt. Sinai Hosp., N.Y.C., 1894-96, asso. surgeon, 1899-1912, surgeon, 1912-34, cons. surgeon, 1934—; cons. surgeon Hebrew Orphan Asylum, Hebrew Sheltering Orphan Asylum, Barnet Hosp., Paterson, N.J., Monmouth Meml. Hosp., Long Branch, N.J.; dir. surgery Beth Moses Hosp., Bklyn., Montefiore Hosp., N.Y.C. Mem. bd. visitors, Ray Brook, N.Y. State Hosp. Nat. Regent in surgery Internat. Coll. Surgeons (pres. 1946-48). Fellow A.M.A., surg. socs. Rome and Piedmont, Italy (hon.), N.Y. State Med. Assn., Am. Gastro-enterological Assn., N.Y. Acad. of Medicine; mem. Am. Bibliophile Soc., English Bibliog. Soc. Author: Surgical Diagnosis, 1905. Contbr. numerous articles and monographs to surg. jours. Collector of rare books, manuscripts and letters; presented great library of 50,000 rare volumes, manuscripts and letters of English and Am. lit. to N.Y. Pub. Library and established a trust fund for its maintenance and care; in memory of his brother, Dr. H. W. Berg, donated an Inst. for Research to Mt. Sinai Hosp., N.Y. Died July 1, 1950.

BERG, Benjamin Nathan, Am. physician; b. N.Y.C., Dec. 8, 1897; s. John and Rose (Lehr) B.; A.B., Coll. City N.Y., 1916; M.D., Columbia, 1920; m. Ethel Lichtenstein, Dec. 28, 1924; 1 son, Robert. Instr. surgery Columbia, 1927-30, Blumental fellow, 1928, asso. in pathology, 1927-64; asso. attending surgeon Harlem Hosp., N.Y.C., 1930-46, attending surgeon, 1946-63, cons. surgeon, 1963—; asso. attending surgeon Hosp. Joint Diseases, N.Y.C., 1938-48, attending surgeon, 1948-61, cons. surgeon, 1961—; research asso. Research Inst., N.Y.C., 1963-64; practicing physician and surgeon, N.Y.C., 1924—; med. dir. Ford Instrument Co., 1942-64; cons. Geigy Chem. Corp., 1964—. Dep. med. examiner Sanitation Dept., N.Y.C., 1963-65, chief med. examiner, 1966—. Diplomate Am. Bd. Surgery. Fellow A.C.S.; mem. A.M.A., Am. Soc. Exptl. Pathology, Soc. Exptl. Biology and Medicine, Harvey Soc., Gerontol. Soc., Teratology Soc., A.A.A.S., Alpha Omega Alpha. Research, numerous publs. in relation of nutrition to reprodn. and aging, endocrine factors in reprodn., teratology, nephrosis, muscular dystrophy, spontaneous diseases of lab. animals, pathol. physiology of stomach, gall bladder, liver, pancreas, mechanisms of fetal resorptions and malformations. Home: 40 E. 88th St. Office: 2 E. 88th St., N.Y.C. 10028.*

BERG, (Otto) Carl, pharmacologist; botanist, born Stettin, Poland, August 18, 1815; s. Johann Friedrich and Wilhelmine Friederike (Haase) B.; degree Berlin (Germany), 1848; m. Caroline Albertine Florentine Witthaus; 6 children. Joined faculty in botany and pharmacology Berlin, 1849, apptd. asso. prof., 1862. Author: Handbuch der Pharmazeutischen Botanik, 1845; Charakteristik der für die Arzneikunde und Technik wichtigsten Pflanzengenera in Illustrationen nebst erläuterndem Text, 1848; (with C. F. Schmidt) Darstellung und Beschreibung sämtlicher in den Pharmacopoea Borussica aufgeführten offizinellen Gewächse, 1853; Revisio Myrtacearum Americae hucusque cognitarum, 1855; Flora Brasiliensis Myrtographia . . . , 1855; Pharmazeutische Warenkunde, 1863; Anatomischer Atlas zur pharmazeutischen Warenkunde, 1865; Die Chinarinden der pharmakognostischen Sammlung, 1865. Developed pharmacology into independent sci.; established pharmaco-anatomy; research on S. Am. flora. Died Berlin, Nov. 20, 1866.

BERG, Clarence Peter, Am. biochemist; b. Mead, Neb., Aug. 30, 1900; s. Emil Gottfred and Rosella (Gibson) B.; A.B., Augustana Coll., 1924, LL.D., 1960; M.A., U. Ill., 1925, Ph.D., 1929; m. Esther Marie Carlson, June 7, 1927; children—James Peter, Howard Curtis, Rosella Marie. Instr., Luther Coll., 1925-26; asst. U. Ill., 1926-29; faculty U. Ia., Iowa City, 1929—, prof., 1945—; vis. prof. U. Tenn. Med. Sch., 1948. Fellow A.A.A.S., Soc. Biol. Chemists, Ia. Acad. Sci., mem. Am. Chem. Soc. (Ia. award for teaching and research 1963), Soc. Exptl. Biology and Medicine (editorial bd. Proc. 1946-51), Am. Soc. Biol. Chemists, Am. Inst. Nutrition, N.Y. Acad. Sci., A.A.A.S., Sigma Xi, Phi Sigma, Gamma Alpha. Contbg. author: Proteins and Amino Acids in Nutrition, 1948, 59. Editorial bd. Jour. Nutrition, 1956-60, Biol. Abstracts, 1934—. Contbr. articles to tech. jours, revs. Research in chem. changes involved in physiol. utilization for growth and metabolism of amino acids found in proteins but unsynthesizable by animal with spl. attention to separation and use of D isomers produced by chem. synthesis but not found as components of body tissues. Home: 528 N. Dubuque St., Iowa City 52240.*

BERG, Eduard, geophysicist; b. Trier, Germany, Nov. 9, 1928; s. Matthias and Maria (Gerner) B.; Diplom Physiker, Saarbrucken, 1953, Dr. Physics, 1955. With Inst. Sci. Research in Central Africa, Lwiro, Bukavu, Congo, 1955-63, head seismic and volcanic dept., Ex Belgian Congo, 1959-63, head Time Service, 1959-63; cons. Union Miniere du Haut Katanga, Seismology, 1959-61; asso. prof. geophysics Geophys. Inst., U. Alaska, College, 1963-67, prof., 1967—; mem. seismol. panel com. on Alaska earthquake Nat. Acad. Sci. Recipient Harry Oscar Wood award Carnegie Instn., Washington, 1962. Mem. A.A.A.S., Am. Geophys. Union, Seismol. Soc. Am., N.Y. Acad. Scis. Research and publs. on electromagnetic wave propagation in ducted troposphere, seismic determination of earth crustal structure in limited areas, gravimetric measurements, magna reservoir determination under volcanoes, triggering of earthquakes by oceanic tidal loads on earth crust in Gulf of Alaska. Office: Geophysical Inst., U. Alaska, College, Alaska 99701.*

BERG, Ernst Julius, elec. engr.; b. Ostersund, Sweden, Jan. 9, 1871; s. Ernst Victor Gabriel and Josefina (Hamren) B.; M.E., Royal Polytecknicum, Stockholm, 1892; Sc.D., Union U., 1910; m. Gwendoline O'Brien, June 15, 1904. Came to U. S., 1892. Engaged as elec. engr. Gen. Electric Co., Schenectady, 1892-1904, cons. engr., 1904-09; prof. elec. engring. U. Ill., 1910-13; cons. prof. elec. engring. Union U., 1907-09, prof. elec. engring., 1913—, dean of engring., 1932—. Author: Electrical Energy, 1908; (with W. L. Upson) First Course in Electrical Engineering, 1916; Advanced Course in Electrical Engineering, 1916; Heaviside's Operational Calculus, 1929. Research on rotary converters (for changing AC to DC) power transformation line disturbances, electromagnetic radiation. Died Sept. 9, 1941.

BERG, Fredrik Theodor, Swedish physician; b. Göteborg, Sweden, Sept. 5, 1806; docent anatomy U. Lund (Sweden); practiced medicine, Stockholm; mem. staff garrison gen. hosp.; mem. staff inf., gen. hosp. chief physician, 1842-49; tchr. legal medicine, then pediatrics Carolin Inst.; mem. hygiene commn., 1849, statistics commn., 1850; pres. Bur. Statistics, 1858-79. Author: l'Aphthe chez les enfants, 1846; Leçons cliniques sur les maladies de l'enfance, 1853; Matériaux pour la topographie médicale, 1853. Discovered and demonstrated that Candida albicans is the causative agent of thrush. Died Stockholm, May 7, 1887.

BERG, Hart O., Am. engr.; b. Phila., circa 1865; grad. in engring., Liège, Belgium; m. Lena Willets; asst. mgr. Colt Firearms Factory, Hartford, Conn.; head Belgium Nat. Gun Works; chief engr. Clément-Bayard, France; built automobile factories, Paris, France, also U. S.; built (with Simon Lake) submarines, Russia, 1904. Developed Browning gun in U. S.; helped introduce submarine to Europe; organized world's 1st aviation firm; laid out 1st flying field, Pau, France. Died Dec., 1941.

BERG, Hermann Johannes Gotthardt, German biophysioco-chemist; b. Greifswald, Germany, July 16, 1924; s. Heinrich and Johanna (Meissner) B.; Dr.rer.-nat., Tech. U., Dresden, Germany, 1953; m. Liebgard Peuker, July 17, 1954; children—Angelika, Dorothea, Albrecht. Mem. staff Inst. Microbiology and Exptl. Therapy, Jena, East Germany (of German Acad. Scis., Berlin, Germany), 1954—, habilitation, 1962, now head dept. biophysicochemistry; lectr. phys. chemistry Friedrich-Schiller U., Jena, 1965. Mem. Biophysikahische Gesselschaft in der DDR, Bunsengesellschaft, Gesselschaft Deutscher Naturforscher U. Arzte, Research, publs. on devel. of Photo-Polarography, 1960, disturbance of electrode-processes by light, and electrochem. measurement of free radical kinetics; polarographic techniques in molecularbiology, conformational variability and adsorption changes of nucleic acids and their complexes with pharmaca, mechanisms of photodynamic effects and their utilization against solid mice tumor. Home: 5 Am Gänseberg, Jena, East Germany.*

BERG, Irwin August, Am. psychologist; b. Chgo., Oct. 9, 1913; s. Bertil Sigfried and Clara (Anderson) B.; B.A. cum laude, Knox Coll., 1936; M.A., U. Mich., 1940, Ph.D., 1942; m. Sylvia Maria Taipale, Mar. 4, 1939; children—Karen Astrid (Mrs. Albert Charles Kirby). Asst. prof. psychology U. Ill., Urbana, 1942-47; asso. prof. Pomona Coll., Claremont, Cal., 1947-48, Northwestern U., Evanston, 1948-55; dean Coll. Arts and Scis., prof., chmn. dept. psychology La. State U., Baton Rouge, 1955—. Chmn. subcom. on testing and counseling U. S. Dept. Labor, 1964—; cons. psychology VA, 1947—. Fellow A.A.A.S., Am. Psychol. Assn. (pres. div. counseling psychology 1963-64), Am. Personnel and Guidance Assn.; mem. Am. Assn. Mental Deficiency, Southeastern (past pres.), Ill. (past pres.) psychol. assns. Author: Workbook in Psychology, 1955. Editor: (L. A. Pennington) An Introduction to Clin. Psychology, 1948; (with B. M. Bass) Objective Personality Assessment, 1959, Conformity and Deviation, 1961; also numerous articles. Research on theory and measurement deviant behavior. Home: 853 DuBois Dr., Baton Rouge 70808.*

BERG, Joseph Wilbur, Jr., Am. geophysicist; b. Essington, Pa., Oct. 6, 1920; s. Joseph Wilbur and Anne (Fullerton) B.; B.S., U. Ga., 1948; M.S., Pa. State U., 1952, Ph.D., 1954; m. Lillian Miriam Douglas, June 27, 1950; children—Anne Lillian, Joseph Wilbur III, Frederick Douglas. Instr., Armstrong Coll., Savannah, Ga., 1948-49; research asst. Pa. State U., 1949-55; asst. prof. dept. physics and geophysics U. Tulsa, 1954-55; asso. prof. dept. geophysics U. Utah, Salt Lake City, 1955-60; geophysicist Inst. for Def. Analyses, Washington, 1960-61; prof. dept. oceanography Ore. State U., Corvallis, 1961-66; exec. sec. div. earth scis. Nat. Acad. Sci., Washington, 1966—. Mem. Geophys. Soc. Am., Soc. Exploration Geologists, Seismol. Soc. Am. Research, publs. in generation and propagation of seismic waves, determination of earth structure from transit times of seismic waves, interpretation of earth gravity in terms geol. structure, conduction of electricity by rocks. Home: 1740 Proffit Rd., Vienna, Va. Office: 2101 Constitution Av., Washington 20418.*

BERG, Kaj, Danish zoologist; b. Aarhus, Denmark, May 13, 1899; s. Aage and Marie (Soerensen) B.; ed. Aarhus; Cand. mag., Copenhagen (Denmark) U., 1925; m. Agnete Odum, 1945; children—Torsten, Asger. Dir. freshwater biol. inst., prof. Copenhagen U., 1939—. Mem. Soc. for Conservation of Natural Scenery, Natural History Soc. Denmark. Editor: Folia Limnologica Scandinavia. Research and publs. on reprodn. and respiration. Home: 8 Chr. IV's vej, Hillerod. Office: 51 Helsingorsgade, Hillerod, Denmark.*

BERG, Paul, Am. biochemist; b. N.Y.C., June 30, 1926; s. Harry and Sarah (Brodsky) B.; B.S., Pa. State U., 1948; Ph.D., Western Res. U., 1952; m. Mildred Levy, Sept. 13, 1947; 1 son, John. Postdoctoral research fellow Am. Cancer Soc., Inst. Cytophysiology, Copenhagen, Denmark, 1952-53; faculty dept. microbiology Washington U. Sch. Medicine, St. Louis, 1954-59, asso. prof., 1957-59; asso. prof.

biochemistry dept. Stanford, Palo Alto, Cal., 1959-60, prof. 1960—. Recipient Eli Lilly award Am. Chem. Soc., 1959; named Cal. Scientist of Year, Cal. Mus. Sci. and Industry, 1964. Mem. Am. Soc. Bacteriologists, Nat. Acad. Scis., Am. Soc. Biol. Chemists. Editor: Biochem. and Biophys. Research Communications; adv. editor: Journal of Molecular Biology. Home: 838 Sante Fe St., Stanford, Cal.*

BERG, Rev Semjonovic, Russian naturalist; b. Benderakh, Russia, 1876; became tchr. ichthyology Moscow Faculty Sci., 1914; named prof. phys. geography U. Petrograd, 1916. Mem. Russian Acad. Scis., Soc. Geography (pres. 1940). Author: Les régions naturelles de l'U.R.S.S., French edit., 1941. Research on climate, vegetation, natural regions, geog. history of USSR; made expdns. to Aral Sea and Lake Ladoga. Died Leningrad, 1950.

BERG, Rolf Yngvar, botanist; b. Oslo, Norway, Dec. 2, 1925; s. Adolf Johan and Karen (Myhre) B.; student Sinsen Hoyere Skole, Oslo, 1940-45; Cand. mag., U. Oslo, 1950, Cand. real., 1952, Dr. philos., 1962; m. Tove Laura Kranoy, Aug. 4, 1951; 1 son, Rolf Audun. Sci. asst., bot. lab. U. Oslo, 1952-54, curator bot. mus., 1956-62, prof. botany, 1965—; research asso., depts. botany U. Cal. at Davis and Berkeley, 1954-56, asst. prof. at Davis, 1962-64, asso. prof., 1964-65 dir. univ. arboretum, 1964-65. Dir. Bot. Garden, also Bot. Mus., Oslo, 1965—. Mem. Norwegian Acad. Sci., Internat., Nordic assns. plant taxonomy, Internat. Soc. Plant Morphologists, Bot. Soc. Am., Am. Soc. Plant Taxonomists, Cal., Norwegian bot. socs., Norwegian Ecology Soc. Editor Nytt Magasin for Botanikk, 1959-62. Contbr. articles to jours. Research and publs. on ecol. and evolutionary relationships between ants and vascular plants, Norwegian flora, dispersal ecology as related to ants, plant embryology, evolutionary taxonomy of Liliaceae, Papaveraceae, Hydrophyllaceae. Home: 23 B, Trondheimsveien, Oslo 5. Office: Bot. Mus., Oslo 5, Norway.*

BERGE, Claude Jacques, French mathematician; b. Paris, June 5, 1926; s. Andre and Geneviève (Fourcade) B.; Ph.D. in math.; m. Jane Gentaz, Dec. 29, 1953; children—Jean-Michel, Vincent. Researcher, Nat. Center Sci. Research, 1954, dir. research, 1959; resident prof. Princeton, 1955. Author: Théorie générale des jeux à personnes, 1957; Théorie des graphs, 1958; Espaces topologiques, 1959. Home: 8, rue Claude-Matrat, Issy-les-Moulineaux (Seine). Office: 11, rue Pierre-Curie, I.H.P., Paris, France.*

BERGEIJK, Willem Andre van, see van Bergeijk, Willem Andre.

BERGEL, Franz, med. chemist; b. Vienna, Austria, Feb. 13, 1900; s. Moritz Martin and Betty (Spitz) B.; Dr.Phil. Nat., U. Freiburg Breisgau, 1924; Ph.D., London U., 1938, D.Sc., 1947; m. Phyllis Thomas, Apr. 6, 1939. Head dept. med. chemistry U. Chem. Inst., Freiburg, 1927-33, Venia Legendi, 1929; research worker U. Edinburgh and Lister Inst., London, 1933-38; dir. research Roche Products Ltd., Welwyn Garden City, Eng., 1938-52; prof. chemistry, head dept. Chester Beatty Research Inst., Inst. Cancer Research, Royal Cancer Hosp., London U., 1952-66, dean, 1963-66, prof. emeritus, 1966—. Hon. lectr. U. Coll., London, 1946—; vis. lectr. Harvard, Children's Cancer Research Found., Boston, 1959-60; chmn. com. for exptl. studies human cancers Commn. Exptl. Oncology, Internat. Union against Cancer. Fellow Royal Soc. London, Chem. Soc., Royal Inst. Chemistry, Inst. Biology, Royal Soc. Medicine; mem. A.A.A.S., Soc. Chem. Industry, Biochem. Soc., N.Y. Acad. Sci., Brit. Pharm. Soc. Author: Chemistry of Enzymes in Cancer, 1961; also numerous articles. Research on dehydrogenation of amino acids in vitro, constn. of hashish, synthesis of vitamins B and E, analgesics, synthesis of antitumorous agts., isolation of cystalline xanthine oxidase and studies of properties; discovered an enzymomimetic of cysteine-desulfhydrase; research on interaction of carbonyl and thiol and amino groups, human tumor studies. Address: Magnolia Cottage, Bel Royal, Jersey, C.I.

BERGELSON, Lev David, Russian chemist; b. Gaissin, Ukraine, Aug. 8, 1918; s. David and Cylia (Koucenog) B.; Ph.D., U. Moscow, 1941; candidate of sci., Inst. Organic Chemistry, Moscow, 1949, D.Sc., 1962; m. Noemi Ostrover, May 25, 1942; children—Marina, Ljuba. Research fellow Inst. Organic Chemistry, Moscow, 1954-59; sr. research worker Inst. Chemistry of Natural Compounds, Moscow, 1959-62. prof., head lipid lab., 1963—. Mem. Mendeleev Soc., Biochemistry Soc. USSR. Author: (with I. N. Nazarov) Chemistry of Steroid Hormones, 1955; (with A. S. Khokhlov and M. M. Shemyakin) Chemistry of Antibiotics, 1960; also articles. Co-editor Jour. Chemistry and Physics of Lipids, Jour. Chemistry Natural Compounds. Research in stereochemistry and mechanism of organic reactions; chemistry macrolide antibiotics and steroid hormones; lipid chemistry (discovery of new types of lipids). Home: 17 Lavroushensky per. Office: 18 ul. Vavilova, Moscow, USSR.*

BERGEN, Karl August von, see von Bergen, Karl August.

BERGER, Andrew John, Am. zoologist, educator; b. Warren, O., Aug. 30, 1915; s. Anton A. and Mary (Rodenberger) B.; A.B., Oberlin Coll., 1939; M.A., U. Mich., 1947, Ph.D.; 1950; m. Edith Grace Denniston, Aug. 13, 1942; children—John D., Diana M. From instr. to asso. prof. dept. anatomy U. Mich. Med. Sch., Ann Arbor, 1950-64; sr. Fulbright lectr. Maharaja Sayajirao U., Baroda, India, 1964-65; prof., chmn. dept. zoology U. Hawaii, Honolulu, 1965—. Hon. asso. in ornithology Bernice P. Bishop Mus., Honolulu, 1965—. Guggenheim fellow, 1963. Fellow A.A.A.S., Am. Ornithologists Union; mem. Am. Soc. Zoologists, Am. Assn. Anatomists, Wilson, Cooper ornithol. socs., Sigma Xi, Phi Sigma, Phi Kappa Phi, Psi Omega (hon.). Author: Bird Study, 1961; Elementary Human Anatomy, 1964; (with Josselyn Van Tyne) Fundamentals of Ornithology, 1959; (with J. C. George) Avian Myology, 1966. Research, numerous publs. on anatomy and life histories of birds. Home: 3262 Paty Dr., Honolulu 96822.*

BERGER, Christian Johann, physician; b. Vienna, Austria, 1724; 1st instr. Danish Sch. Midwifery; chief Lying-in-Hosp.; prof. U. Kiel (Germany), after 1772. Author: Sporsmaale over Menneskets Fodsel og Fodsels-Hielpen, 1766. Founder obstetrics in Denmark, 1st tchr. clin. obstetrics there; brought about normal vertex delivery. Died 1789.

BERGER, Edward William, Am. entomologist; b. Berea, O., Nov. 29, 1869; A.B., Baldwin-Wallace Coll., 1891, Ph.B., 1894; Ph.D., Johns Hopkins, 1899. Prof. sci. Baldwin-Wallace Coll., Berea, 1899-1901; tchr. Lincoln High Sch., Cleve., 1901-02; entomologist Exptl. Sta., Fla., 1906-11; state inspector nursery stock, 1911-15; entomologist State Plant Bd., 1915-43. Recipient Silver medal Royal Internat. Hort. Exhbn., London, 1912. Fellow Entomol. Soc. Am.; mem. Assn. Econ. Entomologists, Fla. Entomol. Soc. (pres. 1917), Fla. Hort. Soc. (hon.). Studied physiology and histology of cubo nedusae, Pseudocorpionida, biology and control of Aleyrodidae, insects of Fla., quantity prodn. entonogenous fungi and Vedalia, Radolia cardinalis. Died 1944.

BERGER, Emil, Austrian ophthalmologist; b. 1855. Author: Anatomie normale et pathologique de l'oeil; Les maladies des yeux dans leurs rapports avec la pathologie général. Described a pupil irregularity, usually elliptical, as a sign of early tabes dorsalis, 1889; invented refracting ophthalmoscope. Died 1926.

BERGER, Frank Milan, physician, pharmacologist; born Pilsen, Czechoslovakia, June 25, 1913; son of Otto and Martha (Weigner) B.; M.D., U. Prague (Czechoslovakia), 1937, State U. N.Y., 1948; m. Bozena Jahodova, Mar. 15, 1939; children—Franklin Milan, Thomas Jan. Came to U. S., 1947, naturalized, 1953. Research fellow physiology U. Prague, 1934-36, research asst. bacteriology, 1936-38; bacteriologist Czechoslovak State Inst. Health, 1938-39; sr. resident Monsall Hosp. Infectious Diseases, Manchester, Eng., 1941-43; chief pharmacologist Brit. Drug House, London, Eng., 1945-47; asst. prof. pediatrics U. Rochester (N.Y.), 1947-49; dir. research Carter Products, Inc., 1949-55, v.p., 1955-58; pres. Wallace Labs., Cranbury, N.J., 1958—. Mem. adv. council for biology Princeton, 1961—; mem. sci. adv. com. Inst. Microbiology, Rutgers U., 1960—. Fellow A.A.A.S., Am. Coll. Clin. Pharmacology and Chemotherapy, Am. Coll. Neuropsychopharmacology, N.Y. Acad. Scis., Royal Soc. Medicine; mem. Am. Assn. U. Profs., Am. Chem. Soc., A.M.A., Am. Soc. for Pharmacology and Exptl. Therapeutics, Am. Therapeutic Soc., Biometric Soc., Brit. Pharm. Soc., Eastern Psychiat. Research Assn., N.J., Mercer County med. socs., Soc. for Exptl. Biology and Medicine, Soc. for Gen. Microbiology, Soc. Am. Bacteriologists, Soc. Biol. Psychiatry, Theobold Smith Soc., Sigma Xi. Contbr. numerous articles to profl. jours. Developed purification of penicillin as a salt (World War II); introduced tranquilizer meprobamate; described muscle relaxant mephenesin (1946). Home: 227 Prospect Av., Princeton, N.J. 08540. Office: Wallace Labs., Half-Acre Rd., Cranbury, N.J. 08512.*

BERGER, Hans, German physician; b. Neuses, Germany, May 21, 1873; became asst. to Otto L. Binswanger, Jena, Germany, 1901, asso. prof., 1906, head physician, 1912, prof. psychiatry, 1919. Author: Über das Elektrenkephalogramm des Menschen. Devised system of electrodes to measure brain wave patterns (1st electroencephalogram for man), 1929. Died June 1, 1941.

BERGER, Katharina Bertha Charlotte, German zoologist; b. Breslau, Feb. 4, 1897; d. Georg and Emma (Schüller) Berger; ed. U. Breslau, U. Munich; Ph.D.; m. Osk Heinroth, Dec. 13, 1933. Asst. zoologist to profs. Koehler and von Frisch at Munich; sci. collaborator with husband; dir. zool. gardens, Berlin, German, 1945-56; instr. zoology Tech. U. Berlin, 1953—. Mem. Internat. Assn. Zoo Dirs., German Soc. Ornithologists, Fedn. for Protection of Nature, German Zool. Soc., Assn. for Protection of Birds, others. Address: Händel-Alee 7, Berlin 21, Germany.

BERGER, Marcel, French mathematician; b. Paris, France, Apr. 14, 1927; s. Jacques and Paule Lefebvre) B.; student École Normale Supérieure, 1948-51; Doctorat d'état ès Scis., Paris, 1954; m. Odile Moufle, Sept. 8, 1951; children—Isabelle, Anne, Benoit. With École Normale Supérieure, 1952-53; researcher C.N.R.S., 1953-55; master confs. Faculty Scis. Strasbourg (France), 1955-64; titulaire prof., Nice, France, 1964-66, Paris, 1966—. Research asso. Mass. Inst. Tech., 1956-57; vis. prof. U. Cal. at Berkeley, 1961-62; chargé du cours Peccot au Coll. de France, 1957. Recipient prix Maurice Audin, 1963. Mem. Math. Soc. France, Am. Math. Soc. Author: Lecture Notes on Geodesics in Riemannian Geometry, 1965. Studies include determination of possible holonomy groups for a Riemannian manifold, various results on Riemannian manifolds with strictly positive curvature (1/4-theorem, classification of homogeneous ones), uniqueness results for some Einsteinian manifolds, uniqueness results for spectrum of some Riemannian manifolds. Home: 11 bis ave. de Suffren, Paris 7, France. Office: I.H.P. 11 rue Pierre Curie, Paris 5, France.*

BERGER, Morroe, Am. sociologist; b. N.Y.C., June 25, 1917; s. Morris and Frieda (Trotiner) B.; B.Social Sci., Coll. City N.Y., 1940; A.M., Columbia, 1947, Ph.D., 1950; m. Paula Wainer, Mar. 7, 1943; children—Edward Morris, Kenneth Harry, Laurence Philip. Faculty, Princeton, 1952—, prof. sociology, 1958—, dir. Nr. E. program, 1962-68. Cons. to govt. agys., 1950-52, 65—; chmn. joint com. Nr. E. and Middle E. Social Sci. Research Council-Am. Council Learned Socs., 1963—. Mem. Am. Sociol. Assn., Middle E. Studies Assn. (pres. 1967). Author: Equality by Statute, 1952, revised 1967; Bureaucracy and Society in Modern Egypt, 1957; The Arab World Today, 1962; Madame DeStael on Politics, Literature and National Character, 1964; also numerous articles. Research on politics, social life Arab world, race relations in U. S. Home: 72 Clover Lane, Princeton, N.J. 08540.*

BERGER, Paul, French surgeon; b. Beaucort, France, 1845; prof. U. Paris (France); mem. Acad. Medicine. Author: De l'Infleunce des maldies constitutionelles sur la marche des lesions traumatiques. Introduced the gauze face mask in 1897; improved operative techniques for hernia and intestinal suturing; developed an operation for amputation of the arm at the shoulder girdle (Berger amputation) in 1883. Died Paris, 1908.

BERGER, William Victor, physician; b. Innsbruck, Austria, May 9, 1889; s. Thomas and Wilhelmine (Eyth) B.; ed. U. Munich; M.D., U. Innsbruck, Columbia; m. Mathilde Boesch, Sept. 29, 1924; children—Rosmarie, Wilhelm Frederique. Asst., U. Innsbruck, U. Basel, 1913-28; prof. U. Innsbruck, 1923, full prof., 1928; prof. in chief Hosp. of Salzburg (Austria), 1929-30; full prof. medicine, pres. dept. U. Graz (Austria), 1931-39; research asso. Columbia, 1940-45; asst. prof. N.Y. U. Sch. Medicine, 1945-59; ret., 1960. Mem. A.A.A.S., Am. Geog. Soc., Gesellschaft d. Aerzte (Vienna) (corr.). Author: Allergy Manual, 1940; Pancreatic Diseases: Diseases of the Liver, Pancreas and Piliary Tract, 1955; Arthritis and Tuberculosis: Protein in Bloodplasma; also articles on allergy, infectious diseases, plasmaprotein, gastrointestinal diseases. Address: Woodstock, Vt.

BERGERET, Pierre Marie, French physician; b. Gray, Jan. 24, 1894; s. Joseph and Marie (Petitot); ed. med. schs. at Nancy, Lyon; M.D.; m. Madeleine Besançon, Sept. 10, 1924; children—Bernard, Philippe, Jean-Marie, Guy. Aux. physician, 1914-18, Morocco, 1921, Germany, 1923, Ecole superièur de guerre, 1930; head lab. medico-physiol. studies of mil. aerons., 1933; physician gen. insp., tech. insp. aero. medicine, master research Health Service of Air. Mem. Internat. Aeromed. Assn. (past pres. French sect.), Internat. Acad. Med. Aerons., Assn. French Speaking Physiologists, Internat. Acad. Astronautics. Research on apparatus for inhalation of oxygen at various altitudes, apparatus for parachuting, for decompression explosives in pressurized cabins. Address: 7, rue du Docteur-Blanche, Paris 16, France.

BERGERON, Jules, French geologist; b. Paris, France, May 5, 1853; s. Etienne-Jules and Leroy (Dufour) B.; certified engr. Paris École Centrale; docteur ès scis. Began work on Service de la carte géologique, 1884; sent to Andalusia to study earthquakes, 1884; sent to study Black Mountain in Tarn region of France, 1889; prof. geology and mineralogy École centrale; named asst. dir. Paris Faculty Geol. Lab., 1897. Mem. French Geol. Soc. (became pres. 1892), Soc. Civil Engr. (pres.). Recipient prize Franco-Brit. Expn., London, Eng., 1908, prize from Brussels (Belgium) expn., 1910. Maps and theoretical studies on Meuse for Service de la carte geologique; research on coal formations, pub. health, problems of alcoholism. Died May 27, 1919.

BERGÈS, Aristide, French engr.; b. Toulouse, France, Sept. 4, 1833; s. Pierre Bergès; ed. École Centrale, Paris. Engr. Crédit Mobilier, Paris; built paper mill, Salat; built factory to produce mech. wood pulp, nr. Laney, 1867, introduced turbine power, 1869; inventor mech. stamp mill; patentee de-fibering, refining and straining process for paper pulp, 1864, also use of propellor in pulp trough, 1873; pioneer in use of expression white coal; cut through to Lake Croset at 1968 meters altitude to store water. Died Feb. 28, 1904.

BERGEY, David Hendricks, Am. bacteriologist; b. Montgomery County, Pa., Dec. 27, 1860; s. G. R. and Susan (Hendricks) B.; M.D., B.S., U. Pa., 1884; A.M., Ill. Wesleyan, 1894; Dr. Pub. Hygiene, U. Pa., 1916; m. Annie S. Hallman, June 5, 1884. First asst. Lab. of Hygiene, U. Pa., Phila., from 1896, asst. prof. bacteriology, 1903-16, asst. prof. hygiene and bacteriology, 1916-26, prof. hygiene and bacteriology,

1926-31, acting prof. hygiene, 1931-32; also dir. Lab. of Hygiene, 1928-31. Dir. research in biology, Nat. Drug Co., from 1931. Author: Handbook of Practical Hygiene, 1899; The Principles of Hygiene, 1901, 7th edit., 1921; also chpt. on domestic hygiene in Pyle's Personal Hygiene, 1904. Authority on classification of bacteria; prepared systematic arrangement for identification of class of microorganisms known as Schizomycetes, circa 1923. Died Sept. 5, 1937.

BERGHAUS, Heinrich, German geographer, cartographer; b. Cleves, May 3, 1797; ed. Marburg, Münster univs. With Lippe Hwy. Constrn. Dept., 1811, Prussian Trigonometrical Survey, 1816; tchr. applied geometry, cartographic drawing and machine bldg. Berlin Bauakademie, 1821-55, prof., from 1824; joint-founder sch. cartography and geography, Potsdam, 1839. Mem. Acad. Architecture. Author: Allgemeine Länder-und Völkerkunde, 6 vols., 1837-44; Physikalischer Atlas, 1838-48; Grundriss der Geographie in fünf Büchern, 1842; Atlas von Asien, 1843; Die Völker des Erdballs, 1845-47; Was man von der Erde weiss, 1856-60; Briefwechsel mit Alexander von Humbolt, 1863. Known for cartographic contbns.; made maps Netherlands, 1816, France, 1824, Germany, 1829, Asia, 1843, also atlas dealing with meteorology, climatography, hydrology, terrestrial magnetism, geology, geography of plants and animals, anthropology, and ethnography. Died Stettin, Germany, Feb. 17, 1884.

BERGHAUS, Hermann, German geologist; b. Kieve, Germany, 1828. Author: Carte universelle d'après la projection de Mercator, 1859; Carte du glacier de l'Oetzthal, 1861; Cartes physiques (murals); Physikalischer atlas (contained geol. maps of 5 continents to a scale of 1:80,000,000, also 2d comprehensive map of S. Am. to be published), new edit., 1892; also prepared maps for Stieker's and Sydow's German atlases. Died Gotha, Germany, 1890.

BERGIUS, Friedrich Karl Rudolph, chemist; b. Goldschmieden, Germany, October 11, 1884; the son of Heinrich and Marie (Haase) Bergius; educated at Breslau (now Wroclaw, Poland); Ph.D., Leipzig, Germany, 1907; m. Ohilie Krazert. Asst. to M. Brodenstein, Berlin, Germany; worked with Nernst in Berlin, also with Haber in Karlsruhe, Germany; prof. Tech. U., Hanover, Germany, 1909-14; with Goldschmidt Orgn., 1914 to end of 2d World War; lived in Austria, Spain, Argentina; technical adviser to combustibles division of Argentine ministry of industries. Recipient (with Carl Bosch) Nobel prize in chemistry for work on chem. high pressure methods, 1931. Author: The Use of High Pressure in Chemical Reactions, 1913. Invented process (Berginization) for conversion of coal dust into oil by action of hydrogen under pressure, also process for prodn. of sugar and cattle feed from hydrolysis of wood, 1946. Died Buenos Aires, Argentina, Mar. 30, 1949.

BERGIUS, Peter-Jonas, Swedish botanist; b. Stockholm, Sweden, 1730; pupil of Linnaeus; prof. natural history, Stockholm; genus Bergia named in his honor by Linnaeus; fellow Royal Soc. London, 1770; mem. Stockholm Acad. Scis. Author: Descriptiones plantarum ex Capita Bonae Spei (on flora of Cape of Good Hope), 1767. Died 1790.

BERGMAN, Norman A., physician; b. Seattle, Oct. 14, 1926; s. Sam L. and Anna (Greenberg) B.; B.A., Reed Coll., 1948; M.D., U. Ore., 1951; m. Betty Freeberg, June 26, 1952; children—David, Carol. Asso. in anesthesiology Columbia, N.Y.C., 1954-58; asso. prof. anesthesiology U. Utah, Salt Lake City, 1958—; vis. research asso. Royal Coll. Surgeons Eng., 1963-64. Mem. Am. Soc. Anesthesiologists, A.M.A., Sigma Xi. Research, publs. on alterations in pulmonary physiology, pulmonary gas exchange and mechanics of breathing during anesthesia and artificial respiration. Home: 4464 Fortuna Way, Salt Lake City 84117. Office: VA Hosp., Salt Lake City 84113.*

BERGMAN, Rudolf Anthony Marie, physician; b. Ierzeke, Zealand, Feb. 2, 1899; s. Johannes and de Munck Bergman; ed. med. sch. univs. Utrecht and Amsterdam; M.D.; m. Petronella Bertholee, May 16, 1928; children—Nelly Clazine Cato, Marjolene Suzanne, Louisette Rudolfa. With histology lab. at Utrecht, lab. U. Amsterdam, med. sch. at Batavia, of Dutch Indies, med. service N.I.W.O.E., Melbourne, Australia, med. sch. at Amsterdam, Royal Tropical Inst., Amsterdam. Mem. Sci. Assn., Acad. Coïmbre (Portugal) (corr.). Author: Inleiding tot de physische Anthropologie, 1957. Research and publs. on exptl. cytology, tissue culture, human typology, anatomy of serpents, anthropology. Home: J. M. Coenenstrasse 26. Office: Linnaeusstrasse 2a, Amsterdam, Netherlands.*

BERGMAN, Torbern (Olof), Swedish chemist, mineralogist; b. Katherinberg, Sweden, Mar. 20, 1735; educated at Inst. Skara; Ph.D., U. Uppsala (Sweden), 1758; studied under Linnaeus at Uppsala; 2 children. Named prof. natural history U. Uppsala, 1758, prof. math., 1761, prof. chemistry and mineralogy, 1767, ret., 1780. Mem. Acad. Scis., Stockholm, Sweden. Fellow Royal Soc., 1765. Author: Déscription physique de la terre, 1766; On the Aerial Acid, 1774; Dissertation on Elective Attractions, 1775; Opuscula physica, chemica et mineralia, 1779-90; Historia chemiae medicum aevum, 1782. Laid founds. for mineral. chemistry by classifying minerals according to chem. characteristics rather than appearances alone; developed theory of chem. affinity, also analytical methods in chemical analysis, 1775; did research in crystallography; studied carbon dioxide (which he called aerial air); first to obtain nickel in pure state; discovered elec. properties of tourmalines; developed methods for quantitative determinations of mineral composition; studied properties of cobalt, alum, bismuth; discovered hydrogen sulphide in mineral springs; predicted discovery of tungsten and molybdenum; accepted phlogiston theory. Died Medevi, Sweden, July 8, 1784.

BERGMANN, Ernst David, chemist; b. Karlsruhe, Germany, Oct. 18, 1903; s. Julius Yehuda and Hedwig (Rosenzweig) B.; Ph.D. summa cum laude, Berlin (Germany) U., 1924; Ph.D., Hebrew Tech. Inst., Haifa, Israel, 1959, U. Montpellier (France), 1965; m. Anna Itin, Nov. 26, 1952; Asst., lectr. U. Berlin, 1928-33; sci. dir. Sieff Inst., 1934-49, Weizmann Inst. Sci., Rehovot, Israel, 1949-51; dir. sci. dept. Ministry of Def., 1948-66; prof. organic chemistry Hebrew U., Jerusalem, Israel, 1952—. Chmn., Israel Atomic Energy Commn., 1952-66. Mem. Israel Acad. Sci., Nat. Council for Research and Devel., N.Y. Acad. Scis. (life), Argentine Chem. Soc. (hon.). Author: (with W. Schlenk) Organic Chemistry, 1932; Isomerism and Isomerization, 1946; also numerous articles, chpts. in books. Research on organic compounds of alkali metals, determination of structures from dipole moments, synthesis and phys. properties of polycyclic aromatic compounds and fulvene derivatives, organic fluorine compounds, syntheses in chloramphenicol series, insect chemistry. Home: 8 Keren Kayemeth St., Jerusalem, Israel. Office: Dept. Organic Chemistry, Hebrew U., Jerusalem, Israel.*

BERGMANN, Ernst von, see von Bergmann, Ernst.

BERGMANN, Maximinian, Austrian physician; b. Spital, Semmering, Austria, Apr. 3, 1922; s. Maximilian and Maria (Fanninger) B.; student U. Berlin (Germany), 1940-45; M.D., U. Würzburg, 1945; m. Gerda Scholz, Oct. 3, 1963; children—Gert, Maximinian, Michael, Maximilian. Asst., Hosp. Mürzzuschlag (Austria), 1946-55; oberarzt U. Clinic, Graz, Austria, 1955—, surg. specialist, 1956-59, urol. specialist, since 1959—; dozent U. Graz, 1964—. Recipient Kardinal Innitzer prize, 1964. Mem. Surg. and Traumatological Soc. Austria, Austrian Urol. Soc., Societe Internat. d'Urologie. Research and publs. on treatment of bladder cancer with isotopes, problems of kidney injuries, influence of cholic acid on intestinal motility. Home: 25 Dr. Robert Grafstrasse, Graz 8010, Austria.*

BERGMANN, Peter G(abriel), physicist; b. Berlin, Germany, Mar. 24, 1915; s. Max and Emmy (Grunwald) B.; student Tech. U., Dresden, Germany, 1931-32, U. Freiburg, Germany, 1932-33; Dr.rer. nat. U. Prague (Czechoslovakia), 1936; m. Margot Eisenhardt, May 23, 1936; children—Ernest Eisenhardt, John Eisenhardt. Came to U. S., 1936, naturalized, 1942. Research asst. to Albert Einstein, Inst. for Advanced Study, Princeton, N.J., 1936-41; asst. prof. Black Mountain (N.C.) Coll., 1941-42, Lehigh U., Bethlehem, Pa., 1942-44; staff mem. to asst. dir. Sonar Analysis Group, N.Y.C., 1944-47; asso. prof. physics Syracuse (N.Y.) U., 1947-50, prof., 1950—; prof., chmn. dept. physics Yeshiva U., N.Y.C., 1963-64. Adj. prof. Poly. Inst. Bklyn., 1947-57; lectr. vis. prof. Am., fgn. univs. Fellow Am. Phys. Soc.; mem. Fedn. Am. Scientists (vice chmn. 1961, chmn. 1964), German Phys. Soc., Am. Math. Soc., A.A.A.S., Sigma Xi. Author: Introduction to the Theory of Relativity, 1942; Basic Theories of Physics: Mechanics and Electrodynamics, 1949; Basic Theories of Physics: Heat and Quanta, 1951. Contbr. sect. to Ency. Physics, research articles on theoretical physics to profl. publs. Office: Dept. Physics, Syracuse U., Syracuse, N.Y. 13210.*

BERGNER, Per-Erik Emil, math. biologist; b. Örebro, Sweden, Apr. 13, 1932; s. Sten Erik Torwald and Karin (Bovin) B.; M.A., Uppsala (Sweden) U., 1956; B.M., Karolinska Inst., Stockholm, 1960, M.D., 1962; m. Eva Elisabet Sjöstedt, June 16, 1956; children—Sten-Erik, Patrik, Jonas. Chief research engr. Swedish Rd. Inst., Stockholm, 1961-62; asst. prof. med. physics Karolinska Inst. 1961-64; vis. scientist Oak Ridge Inst. Nuclear studies, 1964-65, sr. scientist med. div., 1965—. Mem. A.A.A.S. Research, publs. on tracer statis. mechanics known as tracer dynamics, theory of life-span data, cell survival; discovered an ergodic-like theorem for physicochem. kinetics in open heterogeneous systems; research on exchangeable masses of substances in heterogeneous systems. Home: 106 Morningside Dr. Office: Oak Ridge Asso. Univs., Med. Div., P.O. Box 117, Oak Ridge 37830.*

BERGQUIST, Lois Marie, Am. microbiologist; b. Jamestown, N.Y., Sept. 10, 1925; d. Dewey Elmer and Teckla (Stoneberg) Bergquist; A.B., U. So. Cal., 1950, M.S., 1959. Microbiologist, Los Angeles County Gen. Hosp., 1958-60; research asso. Cal. Coll. Medicine, 1959-66; asso. prof. Los Angeles Valley Coll., 1960—. Cons. in clin. chemistry and micrology in field diagnostics for clin. lab. Mem. A.A.A.S., N.Y. Acad. Scis., Am. Assn. Clin. Chemists, Am. Soc. Microbiology, Delta Kappa Gamma. Author: (with Ronald L. Searcy) Lipoprotein Chemistry in Health and Disease; also numerous articles. Research on diagnostic methods for clin. lab. including methods for cholesteroal measurements, lipoprotein analyses, fibrinogen estimation, application Berthelot reaction, serum urea nitrogen measurements, potassium analyses, methods for a more rapid identification of disease causing micro-organisms. Home: 6210 Riverton Av., North Hollywood, Cal. 91606. Office: 5800 Fulton Av., Van Nuys, Cal. 91401.*

BERGQUIST, Stanard Gustaf, Am. geologist; b. Ironwood, Mich., Aug. 13, 1892; s. Charles John and Ellen Elizabeth (Walquist) B.; A.B., U. Mich., 1915, M.Sc., 1927, Ph.D., 1933; m. Ada Evelyon Whitman, Aug. 6, 1924; 1 dau., Donna Jeanne. Instr. geology. Mich. State Coll., 1916, asst. prof., 1924, asso. prof., 1930, head dept. geology and geography, 1930, prof. geol., 1933, acting head dept. phys. sci., Basic College, 1945-58; head dept. phys. sci., 1948-52; member of soil survey, Michigan, summers, 1921, 24, marl survey, Mich. Geol. Survey, summers, 1925-26; geol. Land Econ. Survey. Mich., summers, 1928-33; mem. gov.'s natural gas fact finding com., Mich., 1937; apptd. by gov. to investigate iron mines in Gogebic Co. for State Tax Commn., summer 1942; field studies in glacial geology for Mich. Geol. Survey, summers 1935-46; cons. geol. Engrs. Research Assn., 1936-37. Chmn. non-metallic sect. Mich. Minerals Industries Conf., 1942-43. Fellow A.A.A.S., Geol. Soc. Am.; mem. Mich. Geol. Soc. (bus. mgr. 1936-37, 40-41), Mich. Acad. Scis. (pres. 1940), Sigma Xi. Contbr. papers on geology to sci. jours. Died Mar. 31, 1956.

BERGSON, Abram, Am. economist; b. Balt., Apr. 21, 1914; A.B., Johns Hopkins, 1933; M.A., Harvard, 1935, Ph.D., 1940; m. Rita S. Macht, Nov. 5, 1939; children—Judith, Emily, Lucy. Asst. prof. econs. U. Tex., 1940-42; faculty Columbia, 1946-56; prof., exec. com. Russian Research Center, Harvard, 1956—, dir. regional studies program: Soviet Union, 1961-64, dir. Russian Research Center, 1964—. Staff OSS, Washington, 1940-45, chief Russian econ. subdiv., 1945; cons. RAND Corp., 1948—, various fed. agys. Sheldon travelling fellow, 1937, 39-40; Ford Faculty fellow, 1962; Behavioral Sci. Center fellow, 1963-64. Fellow Econometric Soc., Am. Acad. Arts and Scis.; mem. Am. Philos. Soc., Com. on Economy Communist China, Council on Fgn. Relations. Author: The Real National Income of Soviet Russia Since 1928, 1961; The Economics of Soviet Planning, 1964; also articles. Editor: Soviet Economic Growth, 1953. Co-editor, contbr.: Economic Trends in the Soviet Union, 1963. Research in normative founds. of econ. analysis, rate of econ. growth in Soviet Union and underlying causes, econ. efficiency of socialist planning. Home: 113 Walker St., Cambridge, Mass. 02138.

BERGSON, Henri (Louis), French philosopher; b. Paris, France, Oct. 18, 1859; s. Michael and Kate B.; ed. at École normale supérieure, U. Paris. Tchr. philosophy Lycée Condorcet, Paris, 1889-97; joined faculty École normale supérieure, 1897; became mem. faculty Collège de France, 1900, prof. until 1921; Grifford lectr. Edinburgh (Scotland), 1912; tchr. in U. S. Recipient Nobel prize for lit., 1927; decorated grand cross of the Legion of Honor. Mem. Acad. française, 1914. Author: Essai sur les donées immédiates de la conscience, 1889; Matière et mémoire, 1896; Le rire, 1900; Introduction à la metaphysique, 1903; L'evolution créatrice, 1907; L'énergie spirituelle, 1920; Durée et simultanéité, 1922; Les deux sources de la morale et de la religion, 1932; La pensée et le mouvant, 1934. Known for philos. theory (Bergsonism); ultimate reality resting at base of his fundamental "intuition" is vital impulse (élan vital); devised theory of "creative evolution" and called mechanistic theories of evolution inadequate; developed theory of opposing tendencies, introduced concept of dealing with sci., theory of knowledge and metaphysics on same basis. Died Jan. 5, 1941.

BERGSTAND, Carl Östen Emanuel, Swedish astronomer; b. Stockholm, Sweden, Sept. 1, 1873; studied at Stockholm; Ph.D., Uppsala, Sweden, 1899; became professor of astronomy at Uppsala University, 1911; director of Royal Observatory, Uppsala. Member French Academy of Sciences, 1938, Internat. Astron. Union (became v.p. 1935). Author: Sensibilities des plaques photographiques. Used photography to determine exact position of certain stars and their parallaxes; research on differential refraction, satellites, Venus, solar corona. Died Uppsala, Sept. 27, 1948.

BERGSTEN, Folke S. E., Swedish hydrographer; b. Karlskoga, Sweden, Sept. 15, 1889; s. Erik and Agnes (Soderholm) B.; Swedish Fil. Lic., 1918; Swedish Fil. Dr., 1927; m. Svea Ericsson, 1923. With Swedish State Meteorol. and Hydrographical Inst., from 1919, state hydrographer, from 1932. Investigations of connection between winds and water level in sea, also Gulf Stream's importance for N. European climate. Address: Bromma, Sweden.*

BERGSTEN, Karl Erik, Swedish geographer; b. Risinge, Sweden, July 27, 1909; s. Abel and Ester (Jansson) B.; filosofie licentiatexamen, Lund (Sweden) U. 1937, filosofie doktor, 1943; m. Inga Linnea Sörensson, June 1, 1952; children—Anna Kristina, Karin Cecilia, Hans Holsten, Per Johan. With U. Lund, 1931-52, prof. geography, 1956—; prof. geography U. Göteborg (Sweden), 1952-56. Mem. Kungl. Fysiografiska Sällskapet. Author several books, articles. Research on climatology (dispersion of precipitation continentality), morphology of quaternary deposits, mining distrs. in Central Sweden, birth place fields, aggregation tendency in settlement. Home: Fakirens väg, 38, Lund, Sweden.*

BERGSTRÖM, Stig Olof Wilhelm, Swedish nuclear engr.; b. Linköping, Sweden, Mar. 28, 1925; s. Ernst Wilhelm and Lilly (Lindholm) B.; M.Engring., Royal Inst. Tech., 1949; m. Inga-Britt Bergström, Nov. 2, 1946; children—Björn, Mats, Pär, Ingela, Marika. With torpedo Dept. Royal Navy Bd., 1951-55; Am.-Scandinavian Found. Research fellow Hydrodynamics Lab., Cal. Inst. Tech., 1952-53; asso. prof. Research Inst. Nat. Def., 1955-57; tech. sec. Reactor Safeguards Com., 1957-59; head sect. for health and safety AB Atomenergi, Nyköping, Sweden, 1960—. Author: Stralning och stralkontroll inom kärnteknikken, 1965; also articles. Research on hydrodynamics, cavitation and sub-marine jet propulsion, environmental health, hazards from peaceful nuclear energy. Home: 8 Malmvägen, Nyköping. Office: AB Atomenergi, Studsvik, Nyköping, Sweden.*

BERGSTRÖM, Sune, Swedish biochemist; b. Stockholm, Sweden, Jan. 10, 1916; s. Sverker B. and Wera (Wistrand) B.; M.D., Karolinska Inst., Stockholm, 1943, D.Med.Sci., 1943; Ph.D. (hon.), U. Basel (Switzerland), 1960; m. Maj Gernandt, July 30, 1943. Faculty, Columbia, 1940-41; staff Squibb Inst. N.J., 1941-42, Med. Nobel Inst., Stockholm, 1942-46; faculty Basel U., 1946-47; prof. biochemistry U. Lund (Sweden), 1947-58; prof. Karolinska Inst., 1958—, dean med. faculty, 1963-66. Mem. Am. Acad. Arts and Scis., Am. Soc. Biol. Chemists (hon.), Swedish Acad. Scis., Swedish Acad. Engring. Research, numerous publs. on heparin, autoxidation, formation and metabolism of bile acids and cholesterol, isolation, structure and action of prostaglandins. Home: 12 Danderydsgatan, Stockholm, Sweden.*

BERIGARD, Claude Guillermet de, see de Berigard.

BERING, Vitus Jonassen (or Behring), navigator, explorer; born Horsens, Denmark, 1680. Joined Russian navy under Peter the Great, 1704; served against Swedes; appointed commander of Russian scientific expedition of the Sea of Kamtchatka, 1725, and explored coast of Kamtchatka, 3 years; returned to Russia, 1730; apptd. capt.-comdr., 1732; made voyage to find No. boundary of N.Am., 1741; attempted to return to Kamtchatka, but wrecked on Bering Island. Bering Sea named in his honor. Fragments of his voyage pub. under title Voyages et découvertes faites par les Russes, 1766. Discovered strait between Siberia and Alaska (Bering strait), also Big and Little Diomede Islands in strait, and St. Lawrence Island in Bering Sea, 1728-32; ascertained Asia is not joined to Am. discovered part of N.Am. coast, 1741. Died Bering Island, Dec. 19, 1741.

BERITASHVILI (BERITOV), Ivan Solomonovich, Russian physiologist; b. Village of Veczhini, Georgia, USSR, Dec. 29, 1884; grad. natural sci. dept. Physics, Math. Faculty, St. Petersburg U., 1910. Asst. natural sci. dept., later chair physiology St. Petersburg U., 1910-14; on sci. mission to Holland, worked in lab. of R. Magnus, 1914-15; sr. asst. prosector, lectr. chair physiology Novorossiysk U., Odessa, 1915-19; founder Physiol. Lab. (now Physiology Research Inst. of Georgian Acad. Sci.), 1935—; dir. Physiol. Research Inst., Georgian Acad. Sci., 1941-51, also head dept. gen. physiology Inst. Physiology; prof. Tbilisi U., 1919—. Decorated Order of Lenin; recipient Pavlov prize, 1938, Stalin prize, 1941, Sechenov prize, 1962. Mem. USSR, Georgian, N.Y. (hon.) acads. sci., USSR Acad. Med. Sci. Author over 300 works including The Theory of the Labyrinths and Cervical Tonic Reflexes, 1915; Essential Features of the Skeletal Musculature, 1916; Individually Acquired Activity of the Central Nervous System, 1932; General Physiology of the Muscular and Nervous Systems, 1947-48; Basic Forms of Nervous and Neuro-Psychic Activity, 1947; The Morphological and Physiological Principles of Temporary Connections in the Cortex of the Cerebral Hemispheres, 1956; Physiological Mechanisms of Behavior in Higher Vertebrate Animals, 1957; Neural Mechanisms of Higher Vertebrate Behavior, 1961. Research on forms of nervous and psychic activity, muscular coordination, and physiological mechanism of behavior in higher vertebrates. Home: 63 Lenin St., Tbilisi. Office: Tbilisi University, prosp. Chavchavadze, Tbilisi, Gruz. SSR, USSR.

BERK, J(ack) Edward, Am. physician; b. Phila., Nov. 24, 1911; s. Samuel and Esther (Pill) B.; B.A., U. Pa., 1932, M.Sc., 1939, D.Sc., 1943; M.D., Jefferson Med. Coll., 1936; m. Adeline Elizabeth Alberts, June 26, 1937; children—Philip Howard, Richard Hanna. Instr., U. Pa., Phila., 1941-46, vis. lectr. Grad. Sch. Medicine, 1961—; asst. prof. medicine Temple U., also research asso., asst. dir. Fels Research Inst., 1946-54; asso. prof. clin. medicine Wayne State U. Coll. Medicine, 1954-62, prof., 1962-63; prof., chmn. dept. medicine U. Cal. Irvine College of Medicine, 1963—; sr. attending physician, chmn. dept. medicine Los Angeles County Gen. Hosp., 1963—; cons. surgeon-gen. U. S. Army, 1947—; cons. numerous hosps. Diplomate Am. Bd. Internal Medicine and Gastroenterology. Fellow American College of Physicians; mem. Am. Soc. Gastrointestinal Endoscopy (pres. 1958-59, Rudolf Schindler award 1966), A.M.A., Am. Assn. U. Profs., A.A.A.S., Soc. Exptl. Biology and Medicine, N.Y. Acad. Scis., Internat. Soc. Internal Medicine, Am. Gastroent. Assn., Am. Fedn. Clin. Research, Bockus Internat. Soc. Gastroenterology (pres. 1967—), International Endoscopic Soc., Alpha Omega Alpha, Sigma Xi, Pi Gamma Mu, others. Mem. editorial bd. Current Therapeutic Research, 1959—, Cal. Medicine, 1963—. Research, publs. on demonstration of carrier state in infectious hepatitis, description of acidity in duodenal bulb in man and dog, liver changes in sarcoidosis, technique for differentiation of jaundice, detection of gastric cancer, characterization of serum amylases in man and animals, description of new methods for serum amylase and lipase determination, description of new instruments for demonstration proctosigmoidoscopy and esophageal varices. Home: 894-C Ronda Sevilla, Laguna Hills, Cal. 92653. Office: Orange County Med. Center, Orange Cal. 92668.*

BERKELEY, George, philosopher; b. near Kilkenny, Ireland, Mar. 12, 1685; B.A., Trinity Coll., Dublin, Ireland, 1704; founder (with others) philos. soc. for study of Boyle, Newton, Locke, 1705; became fellow Dublin U., 1707; ordained deacon, 1709; lectr. in div., Greek, Hebrew, 1721-24; presented to English court by Swift, 1713; visited France, Italy, 1713-28; became dean of Derry, 1724; received charter for Am. coll. in Bermudas (govt. grant never paid), 1725; in Am., 1728-31; became bishop of Cloyne, County Cork, Ireland, 1734-53; ret., Oxford, Eng., 1752. Author: Essay towards a New Theory of Vision, 1709; A Treatise concerning the Principles of Human Knowledge, 1710; Three Dialogues between Hylas and Philonous, 1713; De motu, 1721; Alciphron, or the Minute Philosopher (on freethinking as antithesis to his theory that nature is God's language), 1733; Analyst (pamphlet attacking founds. of calculus, attempted to prove Newton's logic faulty, demonstrated that, contrary to belief of Newton's followers, fluxion is not infinitely small in quantity), 1734; Siris, 1744. Immediate successor to Locke in sch. Brit. empiricism; developed system of subjective idealism (Berkeleianism) which held that perceptibility is essence of all so-called material things, which exist only by being perceived. Died Oxford, Jan. 14, 1753.

BERKELEY, Miles Joseph, English botanist; born Biggin Hall, near Audle, England, April 1, 1803; the son of Charles Berkeley; B.A. (scholar 1822-26), Christ's Coll., Cambridge (Eng.) U., 1825, M.A., 1828, hon. fellow, 1883. Ordained deacon, 1826; priest, 1827; curate St. John's Ch., Margate, 1829-33; perpetual curate Apethorpe and Wood Newton, Northamptonshire, 1833-68; rural dean Rothwell, 1853-68; vicar Sibbertoft, Northamptonshire, 1868—; examiner botany U. London 1865-70, 73-78; bot. dir. Royal Hort. Soc. Fellow Royal Soc., 1879 (Royal medal 1863); mem. Linnean Soc., Royal Hort. Soc. (editor). Genus of algae (Berkeleya) named in his honor. Author: Gleanings of British Algae, 1833; Fungi, vol. VI in English Flora (Smith), 1836; Introduction to Cryptogamic Botany, 1857; Outlines of British Fungology, 1860; Handbook of British Mosses, 1863; also numerous sci. articles. Founder Brit. mycology; made morphological classification of over 6000 species of fungi; research on plant pathology and fungi; assembled herbarium of over 9,000 species, now at Kew botanical gardens. Died Sibbertoft, July 30, 1889.

BERKELEY, William Nathaniel, Am. physician; b. Chestertown, Md., Nov. 8, 1868; s. Robert Carter and Fanny Campbell (Minor) B.; A.B., W.Va. U., 1886; Ph.D., U. Va., 1888, M.D., 1891; postgrad. U. Vienna, 1891-92; M.D., Bellevue Hosp. Med. Coll. 1896; m. Clara Helene Barker, 1908; children—Edmund C. Ella K. Began practice at Great Falls, Mont., 1892; moved to N.Y., 1895; attending physician Good Samaritan Dispensary, 1896-1922; dir. lab. exptl. medicine Cornell U., 1917-18. Author: Principles and Practice of Endocrine Medicine, 1926. Contbr. to Boston Med. and Surg. Jour., N.Y. Med. Jour. and Record, Am. Medicine. Research on parathyroid gland, pineal gland and pancreas. Deceased.

BERKEY, Charles Peter, Am. geologist; b. Goshen, Ind., Mar. 25, 1867; s. Peter and Lydia (Stutsman) B.; B.S., U. of Minn., 1892, M.S., 1893, Ph.D., 1897, hon. D.Sc., 1940; hon. Sc.D., Columbia U., 1929; m. Minnie M. Best, Sept. 4, 1894; children—Paul Ainsworth, Virginia Dale. Instr. in geology U. Minn., to 1903; tutor, instr. and asst. prof., asso. prof. and prof. geology Columbia U., 1903-39. A specialist on geology applied to engineering. Employed on state geol. surveys of Minn., Wis. and New York; cons. geologist N.Y. Bd. Water Supply, from 1906; also cons. geologist Met. Dist. Water Supply Commn. of Mass., Dept. of Water and Power of Los Angeles; cons. engr. U. S. Reclamation Bur., TVA; petrographer and geologist on many engring. and mining problems; chief geologist Central Asiatic Expdns., Am. Museum Natural History, from 1922; geologist Port of N.Y. Authority; mem. U. S. Colo. River Bd. Fellow A.A.A.S.; mem. N.Y. Acad. Scis. (pres.), Geol. Soc. Am. (sec., pres.), Geol. Soc. China, Am. Inst. Mining and Metall. Engrs., Am. Soc. C.E. (hon.), Am. Philos. Soc., Nat. Acad. Scis., Phi Beta Kappa, Sigma Xi, Tau Beta Pi; corr. mem. Geol. Soc. of London. Author: Application of Geology to Engineering Practice (known as Berkey Vol., pub. by Geol. Soc. Am.). Pioneer in engring. geology, advised on many major projects including Catskill Project, Grand Coulee and Shasta dams, Delaware aqueduct, George Washington, Whitestone and Triboro bridges, TVA structures. Died Aug. 22, 1955.

BERKEY, Donald Keith, Am. physicist, educator; b. Mt. Ayr, Ia., June 23, 1904; s. James Clifford and Maude (Baker) B.; B.Sc., Antioch Coll., 1929; M.A. (Baldwin fellow, Hanna fellow), U. Cin., 1931, Ph.D. (Laws fellow), 1932; m. Jean Rebecca Teeguarden, June 29, 1929; children—Christine (Mrs. Roger Harrison Berger), James Clifford II, David Paul. With Carborundum Co., Niagara Falls, N.Y., 1922-24, Detroit Edison Co., 1926-28, Beech-Nut Packing Co., Canajoharie, N.Y., 1932-40; faculty physics Colgate U., Hamilton, N.Y., 1940—, prof., 1957—. Mem. expdn. to investigate protons in aurora Nat. Geog. Soc. Ont., Can., 1950; cons. USAF, 1954—; cons. tech. panel on aurora and airglow U. S. Nat. Com. for IGY, 1955-57; vis. prof. Rensselaer Poly. Inst., summer 1960; mem. com. on aurora and airglow Arctic Inst. N.Am., 1963—. Fulbright research scholar Inst. for Theoretical Astrophysics, U. Oslo (Norway), 1952-53. Fellow A.A.A.S.; mem. Am. Inst. Physics, Am. Phys. Soc., Am. Assn. Physics Tchrs., Am. Assn. U. Profs., Am. Geophys. Union, Sigma Xi. Contbr. articles on auroral research, ultraviolet radiation to profl. publs. Home: 16 University Av., Hamilton, N.Y. 13346.*

BERKHEY, Johann Lefrancq van, see van Berkhey, Johann Lefrancq.

BERKMAN, Anton Hilmer, Am. biologist; b. Round Rock, Tex., Apr. 10, 1897; s. Gustav A. and Hilda (Forsman) B.; B.A., U. Tex., 1924, M.A., 1926, Ph.D., 1936. Asst. prof. biology Tex. A. and M. U., 1926-27; instr. U. Tex., 1927; faculty Tex. Western Coll., El Paso, 1927-66, prof. biol. scis., 1935-66, dean grad. div., 1958-59, dean arts and sci., 1959-63. Dir. research projects USPHS, 1948-51, 59. Fellow Tex. Acad. Sci., A.A.A.S.; mem. Am. Bot. Soc., Internat. Soc. Plant Morphology, Am. Soc. Bacteriologists. Research on ecology; hydrogen ion concentration value of soils in relation to woody plants; morphology; seedling anatomy of Cannabis sativa; microbiology; bacteriology of bronchiectasis; polluted irrigation water and its relation to health problems. Home: Box 751, Round Rock, Tex. 78664.*

BERKNER, Lloyd Viel, Am. physicist; b. Milw., Feb. 1, 1905; s. Henry Frank and Alma Julia (Viel) B.; B.S. in Elec. Engring., U. Minn., 1927; postgrad. physics George Washington U., 1933-35; D.Sc. (hon.), Bklyn. Poly. Inst., 1955, U. Calcutta, 1957, Dartmouth, 1958, Columbia U, 1959, U. Rochester, 1960, Tulane U., 1961; Ph.D. (hon.), Uppsala U., 1956; LL.D., U. Edinburgh, 1959; D.Engring., Wayne State U., 1962. Engr. in charge radio sta. WLB-WGMS, Mpls., 1925-27; elec. engr. airways div. U. S. Bur. Lighthouses, 1927-28; engr. 1st Byrd Antarctic Expdn., 1928-30; staff Nat. Bur. Standards, 1930-33; physicist dept. terrestrial magnetism Carnegie Inst., Washington, 1933-41, Australia, 1938-39, Alaska, 1941; head radar sect. Bur. Aeros., U. S. Navy, 1941-43, dir. electronic material br., 1943-45; exec. sec. research and devel. bd. U. S. Dept. Def., 1946-47; head sect. exploratory geophysics atmosphere dept. terrestrial magnetism Carnegie Inst., Washington, 1947-51; pres., trustee, chmn. exec. com. Asso. Univs., Inc., 1951-60; pres. Grad. Research Center S.W., 1960—; spl. asst. to U. S. Sec. of State, also to Dir. Fgn. Mil. Assistance Program, U. S. State Dept., 1949, chmn. internat. sci. steering com., 1949-50; mem. Nat. Security Resources Bd., 1952-53. Mem. Pres.'s Sci. Adv. Com., 1956—, chmn., 1957-59, 62—, cons., 1959—. Recipient U. S. Spl. Congl. Gold medal; Silver medal Aero. Inst.; Gold medal City N.Y., 1930; Sci. award Washington Acad. Scis., 1941. Fellow Am. Acad. Arts and Scis., Am. Inst. E.E., Am. Phys. Soc., Arctic Inst. N.Am., I.E.E.E. (dir. 1963—), N.Y. Acad. Scis.; fgn. mem. Royal Swedish Acad. Scis., Swedish Acad. Arts and Scis., Phys. Soc. India, New Zealand Electronics Inst., Royal Soc. Arts; mem. Nat. Acad. Scis. (treas., mem. council 1960—), A.A.A.S., Am. Geophys. Union (past pres., John A. Fleming award 1962), Am. Meteorol. Soc. (Cleveland Abbe award 1963), Am. Philos. Soc., Washington Acad. Scis., Aerospace Med. Assn. (hon.), Eta Kappa Nu (eminent mem.), Theta Tau. Contbr. articles to tech. jours. Office: Grad. Research Center S.W., P.O. Box 30365, Dallas 75230.*

BERL, Ernst, chemist; b. Freudenthal (formerly Austria), July 7, 1877; s. Max and Agnes (Hein) B.; student chem. engring., Tech. U. Vienna (Austria), 1894-98; Ph.D., U. Zurich (Switzerland), 1901; m. Margaret Karplus, Mar. 28, 1912; children—Herbert, Walter George. Came to U. S., 1933, naturalized, 1938. Asst., U. and Tech. U. Zurich, 1901, asst. prof., 1904-10; chief chemist rayon factory, Tubize, Belgium, 1910-14; chief chemist Austrian War Ministry, 1914-19; prof. chem. technology and electro-chemistry, Tech. U. of Darmstadt, Germany, 1919-33; research prof. Carnegie Inst. Tech. 1933. Mem. Am. Chem. Soc., Am. Inst. Chem. Engrs., Faraday Soc., Soc. Am. Mil. Engrs., Army Ordnance Assn., N.Y. Acad. Sci., Am. Inst. Chemists, Sigma Xi. Pub., author: (with G. Lunge) Taschenbuch für die anorganisch-chemische Grossindustrie, 1910-35, Chemisch-technische Untersuchungsmethoden, 1931-34. Died Pitts., Feb. 16, 1946.

BERLAGE, Hendrik Petrus, Dutch geophysicist; b. Amsterdam, Netherlands, Oct. 24, 1896; s. Hendrik Petrus and Marie (Bienfait) B.; student Eidg. Techn. Hochschule, Zürich, Switzerland, 1915-19, D.Tech. Scis., 1924; student U. Leyden (Netherlands), 1920-22; m. Elisabeth Smits, Apr. 28, 1924; children—

Elisabeth (Mrs. Kamerbeek), Cornelia (Mrs. Van der Want), Francisca (Mrs. Seutter), Cecilia (Mrs. Leeuwe). Sci. asso. Royal Magnetic and Meteorol. Obs., Batavia, Netherlands, E. Indies, 1925-46, dir., 1946-51; dir. Meteorol. and Geophys. Service, Netherlands Indies (later Republic of Indonesia), 1946-51; 1st sci. asso., dir. research in climatol. dept. Royal Netherlands Meteorol. Inst., De Bilt, Netherlands, 1951-62; asso. prof. gen. geophysics U. Indonesia, 1948-51; asso. prof. meteorology, climatology, oceanography U. Utrecht (Netherlands), 1954-66. Recipient Rosscha medal, Indonesia; decorated Officer Orange Nassau, Knight Netherlands Lion. Mem. Royal Netherlands Acad. Scis. Author: Het Onstaan van het Zonnestelsel, 1956; also articles. Research on fluctuations of meteoerl. elements throughout world of a few years, especially So. Oscillation, origin of solar system. Home: 387 Eykmanlaan, Utrecht, Netherlands.*

BERLEPSCH, Count Hans Hermann Carl Ludwig von, see von Berlepsch, Count Hans Hermann Carl Ludwig.

BERLIN, Alfred Franklin, Am. archaeologist; b. Cherryville, Pa., Jan. 12, 1848; ed. Easton, Pa.; m. Mary Ella Reed, June 6, 1871. Engaged in archaeol. research; clerk Ct. of Common Pleas, Lehigh County, Pa., 3 yrs. Mem. Anthropol. Soc. Washington (corr.), Univ. Archeal. Assn. U. Pa. (corr.), Royal Italian Didactic Soc., Rome (hon.), Lehigh County. Hist. Soc. (charter), Wyo. Hist. and Geol. Soc. (corr.), Hist. Soc. Ala. (hon.). Contbg. author: Sect. VI, The East Allegheny Section in Prehistoric Implements; Chapter III, The Lenni-Lenapé and Their Implements, in History of Lehigh County, Pa. Author: The German Immigrant in Pennsylvania before and during the Revolution. Died Oct. 2, 1925.

BERLIN, Nathaniel Isaac, Am. physician; b. N.Y.C., July 4, 1920; s. Louis and Gertrude (Sugarman) B.; B.S., Western Res. U., 1942; M.D., L.I. Coll. Medicine, 1945; Ph.D., U. Cal. at Berkeley, 1949; m. Barbara Ruben, June 14, 1953; children—Deborah J., Marc D. Post-doctorate research fellow Nat. Cancer Inst., Donner Lab., U. Cal. at Berkeley, 1948-50; research fellow, 1949-50, research asso., 1951-52; lectr., asso. in research medicine, div. medicine, physics dept. U. Cal. at Berkeley, 1952-53; spl. research fellow Nat. Heart Inst., Nat. Inst. Medicine, London, Eng., 1953-54; head metabolism service Nat. Cancer Inst., NIH, Bethesda, Md., 1956—, chief gen. med. br., 1959-61, clin. dir., 1961—. Cons. U. S. Naval Hosp., Bethesda, 1957—, U. S. Armed Forces Radiobiology Research Inst., 1965. Mem. N.Y. Acad. Sci., A.A.A.S., Internat. Soc. Hematology, Am. Fedn. Clin. Research, Soc. Exptl. Biology and Medicine, Radiation Research Soc., Am. Assn. for Cancer Research, Am. Physiol. Soc., Biochem. Soc., Am. Soc. Clin. Investigation, Sigma Xi, Alpha Omega Alpha, Zeta Beta Tau, Phi Delta Epsilon. Research publs. on red blood cell—formation, life span, polycythemia, anemia, bilirubin metabolism, application of radioactive isotopes.*

BERLIN, Rudolf, ophthalmologist; b. Germany, 1833; founder clinic, Stuttgart, Germany, 1861; pub. description of traumatic edema of retina (commotio retinae), 1873; described word blindness, 1877, also suggested term dyslexia. Died 1897.

BERLINCOURT, Ted Gibbs, Am. physicist; b. Fremont, O., Oct. 29, 1925; s. Weldon B. and Gladys (Gibbs) B.; B.S., Case Inst. Tech., 1949; M.S., Yale, 1950, Ph.D., 1953; m. Marjorie Kathleen Alkins, Feb. 28, 1953, 1 dau., Leslie Ellen. Project physicist U. S. Naval Research Lab., Washington, 1952-55; group leader Electronic Properties-Atomics Internat. div. N. Am. Aviation, Inc., Canoga Park, Cal., 1955-63; asso. dir. N.Am. Aviation Sci. Center, Thousand Oaks, Cal., 1964—. Fellow Am. Phys. Soc. exec. com. div. solid state physics; mem. A.A.A.S., Sigma Xi, Tau Beta Pi. Research in electronic and magnetic properties of metals and alloys at very low temperatures and very high magnetic fields; devel. and elucidation of high magnetic field superconducting materials. Home: 1430 Camino Magenta. Office: 1049 Camino Dos Rios, Thousand Oaks, Cal. 91360.*

BERLINER, Arnold, German physicist; b. Mittelnevland b. Neisse, Germany, Dec. 26, 1862; s. Siegfried and Marie (Mannheimer) B.; studied physics at U. Breslau (now Wroclaw, Poland); worked for Allgemeine Elektrizitätsgesellschaft (AEC), directed incandescent lamp prodn., ret. after 25 years; made several trips to Gen. Electric Co., Schenectady; after retirement became editor mag. Die Naturwissenschaften until forced by Nazis to retire, 1935. Recipient silver Leibniz medal of Prussian Acad. Author: Lehrbuch der Physik in elementaren Darstellungen, 1903. Editor: (with K. Scheel) Handwörterbuch der Physik, 1934. His physics textbook was considered one of the best for many years; had a wide knowledge of all natural scis. Died (suicide) Berlin, Mar. 22, 1942.

BERLINER, Emile, inventor; b. Hanover, Germany, May 20, 1851; s. Samuel and Sally (Friedman) B.; grad. Samson Sch., Wolfenbuttel, 1865; apprentice in printing shop; came to U. S., 1870; m. Cora Adler, 1881; 4 sons, 3 daus. Invented a loom, 1881, loose contact telephone transmitter or microphone, 1877; discovered that a loose contact will act as a telephone receiver, 1877, and was first to use an induction coil in connection with transmitters, 1878; joined Bell Telephone Co. (later Am. Tel. & Tel. Co.), 1877; patentee of other valuable inventions in telephony; invented the Gramophone, 1887, the first talking machine which utilized a groove of even depth and varying direction, and in which the record groove not only vibrated, but also propelled the stylus across the record (known also as the Victor Talking Machine), for which he was awarded John Scott medal and Elliot Cresson Gold medal by Franklin Inst., Phila.; also invented and perfected the present method of duplicating disc records; invented, the acoustic tile and acoustic cells for insuring good acoustics in halls, etc., 1925; engaged in ednl. campaign against dangers of raw milk and other dairy products, from 1901, planned and was mem. Washington Milk Conf., 1907. First to have made and used in aeronautical experiments light weight revolving cylinder internal combustion motor, 1908; under his general directions his son, Henry A., designed first successful helicopter, rising and sustaining himself in it, 1919. Pres. D.C. Tuberculosis Assn., 1915-21. Author: Health Rhymes. Died Washington, Aug. 3, 1929.

BERLINER, Ernst, Am. chemist; b. Kattowitz, Germany, Feb. 18, 1915; s. Joseph and Lucie (Selinger) B.; came to U. S., 1940, naturalized, 1949; student U. Breslau (Germany), 1935-36, U. Freiburg (Germany), 1936-38; M.A., Harvard, 1941, Ph.D., 1943; m. Frances Jean Bondhus, Sept. 11, 1947; 1 dau., Susan Lucy. Faculty, Bryn Mawr (Pa.) Coll., 1944—, prof. chemistry, 1953—, chmn. chemistry dept., 1950—. John Simon Guggenheim Meml. fellow, 1961-62; recipient Mfg. Chemists Assn. Coll. Chemistry Tchr. award, 1963. Fellow A.A.A.S.; mem. Am. chem. Soc., Chem. Soc. (London, Eng.), Sigma Xi. Research, publs. on phys.-organic aspects aromatic chemistry with emphasis on mechanism aromatic halogenation, effect alkyl groups, isotope effects, relative reactivities polynuclear aromatic systems. Home: 219 N. Roberts Rd., Bryn Mawr, Pa. 19010.*

BERLYNE, Daniel Ellis, psychologist; b. Salford, Eng., Apr. 25, 1924; s. Mark and Cissie (Spurgin) B.; B.A., Cambridge U., 1947, M.A., 1949; Ph.D., Yale, 1953; m. Hilde Strauss, Sept. 7, 1953; children—Judith Bettina, Deborah Joyce, Naomi Rhoda. Lectr., U. St. Andrews, Scotland, 1948-52; English-Speaking Union fellow Yale, 1951-52; substitute in psychology Bklyn. Coll., 1952-53; lectr. U. Aberdeen, Scotland, 1953-57; fellow Center for Advanced Study in Behavioral Scis., 1956-57; vis. asso. prof. U. Cal. at Berkeley, 1957-58; membre-résident Centre Intrenat. d'Epistémologie Génétique, U. Geneva, Switzerland, 1958-59; vis. scientist NIH, 1959-60; asso. prof. Boston U., 1960-62; faculty U. Toronto (Ont., Can.), 1962—, prof. psychology, 1963—. Fellow Brit. Psychol. Soc., Am., Canadian (dir. 1965—) psychol. assns.; mem. Exptl. Psychology Soc., Psychonomic Soc., A.A.A.S., Sigma Xi. Author: Conflict, Arousal and Curiosity, 1960; (with J. Piaget) Théorie du comportement et opérations, 1960; Structure and Direction in Thinking, 1965; also numerous articles. Research on attention, exploratory behavior and curiosity in animals and humans, relation structural and motivational aspects thinking. Home: 25 Clarendon Av., Toronto 7, Ont., Can.*

BERLYNE, Geoffrey Merton, Brit. physician; b. Manchester, Eng., May 11, 1931; s. Charles S. and Miriam H. (Rosenthal) B.; M.B., U. Manchester, 1954, M.D., 1966; postgrad. Columbia; m. Ruth L. Selbourne, June 7, 1959; children—Jonathan Daniel, Benjamin Gideon. House officer professorial med. unit Manchester Royal Infirmary, 1954-55, dir. dialysis unit, 1964—; faculty U. Manchester, 1959—, sr. lectr. medicine, 1966—; Mem. working party on dialysis Ministry Health, U.K., 1965—. Rockefeller Travelling scholar, U. S., 1961-62; fellow medicine Columbia, Presbyn. Hosp., 1961-62. Mem. Assn. Physicians Gt. Britain, Renal Assn. Gt. Britain, Med. Research Soc., Internat. Soc. Nephrology. Author: A Course in Renal Diseases, 1966; also articles. Research on physiology of urine concentration in health and disease states; devel. dietary methods for treatment acute and chronic renal disease. Home: 281 Kingsway, Cheadle, Cheshire, Eng. Office: U. Dept. Medicine, Manchester Royal Infirmary. Manchester 13, Eng.*

BERMAN, Leonard B., Am. physician; b. Boston, Jan. 17, 1924; s. Albert Malah and Esther (Olshansky) B.; student Amherst Coll., 1941-43; M.D., N.Y. U., 1947; m. Stasha Urban, June 25, 1945; children—Christopher, Peter, Alexandra. Faculty, Georgetown U. Sch. Medicine, 1956-61, Loma Linda U., 1961-63; asso. prof. medicine U. Louisville, 1963-66, prof., 1966—, chief sect. on renal diseases, 1963—. Cons. to surgeon gen. U. S. Army, 1965—. Mem. Am. Physiol. Soc., Soc. for Exptl. Biology and Medicine, Am. Fedn. for Clin. Research. Research, numerous publs. on heart-bladder reflexes, glucocortcoid influence on protein catabolism, serum amylase in renal disease. Home: 2425 Ransdell Av., Louisville 40207. Office: Louisville Gen. Hospital, Louisville 40207.*

BERMAN, Louis, Am. physician, b. N.Y.C., Mar. 15, 1893; s. Nathan and Dora (Rothfeld) B.; student Coll. City N.Y., 1909-11; M.D., Coll. Phys. & Surg. (Columbia), 1915; studied in Paris, Berlin and Vienna, 1922-23; B.S., Columbia, 1924. Began practice in N.Y.C., 1915; asso. in biol. chemistry. Sch. of Medicine, Columbia, 1921-28, and engaged in research on glands of internal secretion; conducted spl. study of endocrine glands in relation to criminology, at Sing Sing Prison, N.Y., 1932; also in active medical practice, specializing in internal glandular conditions; vis. endocrinologist Central Neurol. Hosp., 1934, asso. visiting neurologist, 1937. Bd. dirs. Nat. Crime Prevention Inst., 1936-37; apptd. exam. physician of rehab. Selective Service System; chmn. All Nations Com. for World Unity, 1942-43. Fellow N.Y. Acad. Scis.; mem. Am. Ethnol. Assn., A.M.A., Assn. for Study Internal Secretions, A.A.A.S., Am. Genetic Assn., Am. Therapeutic Soc. Author: The Glands Regulating Personality, 1921, rev. edit., 1928; The Personal Equation, 1925; the Religion Called Behaviorism, 1927; Food and Character, 1932; Study of Relation of Ductless Glands to Homosexuality, 1933; New Creations in Human Beings, 1938; Behind the Universe of a Doctor's Religion, 1943; Lectures on Psycho-Endocrinology; The Relations of the Psyche and the Internal Secretions, 1940; Discovery of Mode of Action of Insulin in Treatment of Insanity, 1942; also many articles on nutrition, metabolism, diet and endocrine glands. Editor dept. of endocrinology Journal of Medical Practice, 1935, 36, His books, Glands Regulating Personality, and Food and Character selected for micro-filming, for the Crypt of Civilization, 1938. Discovered internal secretion of parathyroid glands, 1924, also relation of internal secretion of ovaries to cholesterol metabolism, 1927; discovery of curative value of ovarian residue for certain breast tumors, 1929; discovery of value of adrenal cortex in treatment of Paget's Disease, 1930; discovery of ameliorative action of benzedrine on oculogyric crises of Parkinsonism, 1935. Died May 16, 1946.

BERMEJO-MARTINEZ, Francisco, Spanish chemist; b. La Coruña, Spain, Nov. 17, 1919; s. Gerardo Bermejo Pena and Mercedes Martinez Alamo; M.Scis. and Pharmacy, U. Santiago de Compostela, 1944; Ph.D., Madrid U., 1949; m. Antonia Barrera Ramallo, Aug. 5, 1949; children—Mercedes, Javier, José, Gerardo, Adela, Pilar, Ana Maria, Jacobo, Maria Teresa. Prof. faculty scis.-pharm. Santiago U., Santiago de Compostela, Spain, 1944-50, prof. analytical chemistry, 1951—; prof. Oviedo U., 1950. Decorated Cruz de Guerra, Cruz Roja del Mérito Militar, Medalla de la Campana; recipient Extraordinary prize Scis. Faculty, Santiago U., 1947. Mem. Real Sociedad Espanola de Fisica y Quimica, Instituto del Hierro y del Acero, Am. Chem. Soc., Am. Micrchem. Soc., Am. Soc. for Testing and Materials, Soc. for Analytical Chemistry London, Chem. Soc. (London), Société Chimique de France, Verin Österreichischer Chemiker, Biochem. Soc. (London), Koninklijke Nederlandse Chemische Vereiniging, A.A.A.S., Sociedad Latinoamericana de Normalizachiones. Author: Tratado de Quimica Analitica Cuantitativa, 1963; (with A. Prieto) Aplicaciones analiticas del Aedt y similares, 1960; also numerous articles. Research in analytical chemistry, especially analytical application of chelons. Home: 17-2, Huérfanas, Santiago de Compostela, Spain.*

BERN, Howard Alan, biologist; born Montreal, Que., Can., Jan. 30, 1920 (parents Am. citizens); s. Simeon and Ethel (Hyman) B.; B.A., U. Cal. at Los Angeles, 1941, M.A., 1942, Ph.D., 1948; m. Estelle Claire Bruck, Mar. 30, 1946; children—Alan, Lauren. Faculty, U. Cal. at Berkeley, 1948—, prof. zoology, research endocrinologist Cancer Research Genetics Lab., 1960—, chmn. grad. group in endocrinology, 1961—. Biology cons. Sci. Curriculum Improvement Study, Berkeley, 1963—; vis. prof. pharmacology U. Bristol; Eng., 1965-66. Guggenheim fellow, Paris, France, Cambridge, Eng., 1951-52; NSF Sr. fellow, Hawaii, 1958-59, Naples, 1965-66; fellow Center for Advanced Study in Behavioral Scis., Stanford, 1960. Fellow Zool. Soc. Calcutta (hon.); mem. Am. Soc. Zoologists (pres. 1967—), Soc. for Exptl. Biology and Medicine, Endocrine Soc., Am. Assn. Anatomists, Am. Assn. for Cancer Research, Société Zoologique de France. Author: (with Aubrey Gorbman) A Textbook of Comparative Endocrinology, 1962; also numerous articles. Editorial bd. Gen. and Comparative Endocrinology, 1961—, Endocrinology, 1963—, Jour. Exptl. Zoology, 1964—. Research on physiology reprodn. male mammals, nature process keratinization and influence vitamin A on skin, elucidation precancerous state in mouse mammary cancer, endocrinology mammary gland, cellular and organismal aspects neurosecretion, functional delineation caudal neurosecretory system, adrenal glands fishes, comparative physiology prolactin. Home: 1010 Shattuck Av., Berkeley, Cal. 94707.*

BERNACCHI, Louis Charles, physicist, explorer; b. Tasmania, Australia, 1876; s. A. G. D. Bernacchi; ed. Melbourne (Australia) U.; m. Winifred Edith Harris, 1906; 2 sons, 2 daus. Physicist, So. Cross Antarctic Expdn., 1898; went to Eng., 1900; Cuthbert Peek grantee Royal Geog. Soc.; physicist Nat. Antarctic Expdn. under Capt. Scott, 1901-04; explored Brit. Namaqualand, German S.W. Africa, 1905, primeval forests of Peru, Upper Amazon Basin, 1906, central Borneo, other remote areas of world. Recipient King's Antarctic medal, Royal Geog. medal. Mem. council Royal Geog. Soc., 1929-31. Author: To the South Polar Regions, 1901; A Very Gallant Gentleman (No Surrender Oates), 1933; Saga of the Discovery, 1938; also various publs. on sci. results of So. Cross and Discovery, on terrestrial magnetism, meteorology, seismology, gravity, atmospheric electricity. Died Apr. 24, 1942.

BERNAL, John Desmond, crystallography physicist; b. Nenagh, Ireland, May 10, 1901; ed. Stonyhurst Coll.; M.A., Emmanuel Coll., Cambridge (Eng.) U.; m. 1922; 2 sons. Research, Davy Faraday Lab., 1923-

27; lectr., later asst. dir. research in crystallography Cambridge U., 1934-37; prof. physics Birkbeck Coll., U. London, 1937-63, prof. crystallography, 1963-—. Hon. prof. Moscow (USSR) U., 1956. Recipient Lenin peace prize, 1953, Grotius medal, 1959. Fellow Royal Soc. (Royal medal 1945), 1937; mem. numerous European sci. acads., Mineral. Soc., Faraday Soc., Royal Instn., Cambridge Philos. Soc., World Fedn. Sci. Workers (v.p.). Author: The World, the Flesh and the Devil, 1929; The Social Function of Science, 1939; The Freedom of Necessity, 1949; The Physical Basis of Life, 1951; Marx and Science, 1952; Science and Industry in the Nineteenth Century, 1953; Science in History, 1954, rev. edits., 1957, 65; World without War, 1958, rev. edit., 1960; The Origin of Life, 1967. Contbr. articles on crystallographical, phys., biochem. subjects. Contbns. on sci., philos., social questions; studies of structures of organic, inorganic substances, liquids, also carbonaceous meteorites in relation to problem of origin of life, structure of simple and complex substances by methods of X-ray crystallography. Address: Dept. Crystallography, Birkbeck Coll., Malet St. W.C. 1, London, Eng.

BERNAL, Julián Nievas, Spanish analytical chemist; b. Zaragoza, Spain, Sept. 9, 1904; s. Julián and Concepción B.; D.Sc. in Chemistry, U. Zaragoza; m. Enriqueta Castejón, 1940. Prof. analytical chemistry Faculty Scis., U. Zaragoza, also dean. Mem. Royal Acad. Sci. Zaragoza (sec.), Royal Econ. Soc., Soc. Analytical Chemistry Gt. Britain, Royal Spanish Soc. Physics and Chemistry, Spanish Assn. Progress Sci. Author: Descapado de Materiales Ferrosos, 1936; Contribución al estudio de los depósitos electrolíticos de Zinc, 1940; La Enseñanza y la Investigación en las Ciencias Experimentales, 1948; Work in analytical chemistry, electrochemistry. Address: Plaza de Aragon 10, Zaragoza, Spain.

BERNARD, Claude, French physiologist; b. Saint-Julien, Rhone, France, July 12, 1813; s. Pierre Jean Francois and Jeanne (Saulnier) B.; studied Jesuit Coll. Villefranche, also Coll. Thoissey, in Ain; M.D., Collège de France, 1843; m. Marie Francoise Martin, 1845 (legally separated Apr. 1870); 2 daus. Became extern Hôtel-Dieu, 1836, intern, 1839, asst. to Prof. Magendie at Collège de France and Hôtel-Dieu, 1841; dep. prof. Collège de France, 1847-55, prof., from 1855; 1st to occupy chair of exptl. physiology Sorbonne, 1854-68; accepted chair gen. physiology Mus. Natural History, 1868, was provided with modern lab. facilities there by Napoleon III. Recipient croix Legion d'Honneur, 1859; Grand prize in physiology French Acad. Scis., 1845, 49, 51, 53; named senator of empire by imperial decree, 1869; 1st scientist honored by state funeral in France. Mem. French Acad. Medicine, 1854; mem. French Acad. Scis., 1854, v.p., 1868, pres., 1869; 1st v.p. French Biology Soc., 1847. Author: Introduction à la medécine expérimentale (main work); Rapport sur les progrès de la physiologie générale, 1867; Leçons de physiologie expérimentale, 1855/56; Leçons sur la physiologie et la pathologie du système nerveux, 1858; Leçons de pathologie expérimentale, 1872; Leçons sur le diabete et la glycogenèse animale, 1877. Established physiology as exact sci., laid foundations of exptl. method in that field; investigated digestion; did research on physiology and digestive action of pancreatic juice, 1848; discovered glycogenic function of liver, circa 1855; proved that animals can build up complex chems., as well as break them down; studied diabetes; discovered the vasomotor system of nerves (which controls dilation and contraction of blood vessels); held that body mechanisms strive to maintain a constant inner environment through self regulation, and that to accomplish this the various physiol. and chem. processes of the body have to be under an integrated central control (milieu intérieur); showed that red blood cells transport oxygen from lungs to tissues; investigated physiol. effects of various poisons (carbon monoxide, curare), showed that these generally work by specific and local effects, and gave one of 1st successful explanations of specific manner in which poisons act upon the body (thus was influential in placing pharmacology on an exptl. basis); also studied paralysis of nerves and independent excitability of muscle. Died Paris, Feb. 10, 1878.

BERNARD, Edward, English astronomer, philologist; b. Perry St. Paul, Eng., May 2, 1638; s. Joseph and Elizabeth (Lenche) B.; student Merchant Taylors' Sch., London, Eng., 1648-55; B.A., Oxford (Eng.) U.; 1659, M.A., 1662, B.D., 1668, D.D., 1684; m. Eleanor Howell, Aug. 6, 1693. Proctor, St. John's Coll., Oxford U., 1668, Savilian prof. astronomy, from 1673; dep. to Surveyor-Gen. (Christopher Wren), 1669; tutor to sons of Charles II, France, 1676-77. Fellow Royal Soc., 1673. Subject of monument, chapel St. John's Coll. Author: Tables of the Longitudes, Latitudes, Right Ascensions and Declinations of the Chiefest Fixt Stars, according to the Best Observers, 1684; The Observations of the Ancients concerning the Obliquity of the Zodiac; A Latin Letter to Mr. John Flamsteed, containing Observations on the Eclipse of the Sun, 1684; De mensuris et ponderibus antiquis libri tres, 1688; Private Devotions, 1689; Etymologicon Britannicum, 1689; supr. publ.: Catalog: librorum manuscriptorum Angliae et Hiberniae, 1697. Research, Arabic texts of Conics of Apollonius, Leyden, Germany; expert weights and measures. Died Eng., Jan. 12, 1696.

BERNARD, Francis, oceanographer; b. Mathieu, Calvados, France, Apr. 30, 1908; s. Noël and Marie Louise (Martin) B.; grad. Ecole Normale Supérieure, Paris, France, 1928; D.es Scis., Agregé de l'Université, Paris, France, 1937; m. Michelle Jourdan, June 6, 1951; children—Noël, Jean Felix. Agrégé-preparator Ecole Normale Supérieure, Paris, 1933-35; asst. Inst. Oceanographíque, Paris, 1935-39; maitre de conférences biology U. Lyon (France), 1940-41; prof. U. Algiers (Algeria), 1942-—, dir. Océanographical Inst., 1962-—. Expert, UNESCO for arid zone and oceanography, 1951-—. Mem. Zool. Soc., Entomol. Soc. Author: Faune d'Europe Formicidae, 1966; also numerous articles. Research on elementary marine fertility in warm seas and Mediterranean, Atlantic current nr. Algeria, fertility in deep water, Hymenoptera, ants. Home: 3 Rue Luciani, Alger, Algeria.*

BERNARD, Joseph (Marquis de Chabert de Cogolin), French astronomer; b. Toulon, France, 1724; mem. numerous naval expdns., 1748-84; became asst. at Mahé under Comte de Grasse, 1781; named vice adm., 1792; asso. of many European acads. Author: Account of a Voyage made on the Coasts of North America in 1750; Voyage sur les Côtes de l'Afrique septentrionale, 1753; Memoire sur Pusage des horloges marines, 1793; also numerous works on astronomy, physics, hydrography. Drew charts of Mediterranean coasts. Died Paris, France, 1805.

BERNARD, Leon, French biologist, physician; b. Paris, France, May 19, 1872; M.D., Paris, 1900. Tchr. histology at Faculty of Medicine, Paris; dir. (with Rist) Léon-Burgeois Dispensary at Laënic Hosp. beginning in 1922, also tchr.; titular prof. physiology, Paris, 1928. Mem. Hygiene com. League of Nations; mem. various pub. health coms. especially those concerned with Tb. Mem. Acad. Medicine. Author: Le pneumothorax artificiel dans le traitement de la tuberculose pulmonaire, 1907; La tuberculose pulmonaire, 1921; La défense de la santé publique pendant la guerre, 1929; (with Debre) Cours d'hygiène. Described (with E. Sergent) acute adrenal insufficiency caused by a disease of the suprarenal capsules; research on Tb especially as related to pub. health. Died Aug. 19, 1934.

BERNARD, Merrill, Am. hydrologist; b. Burlington, Ia., July 25, 1892; ed. Mil. Coll., of S.C., 1908-10, A. and M. Coll. of Okla., 1911; m. Claudia Turner, Aug. 1, 1914. Engaged in municipal irrigation and railroad engring. to 1916, civil engring., 1918-20; cons. engr. designing and supervising municipal, drainage and irrigation projects, La., Tex. C.Am., 1929-36; hydrologic specialist Miss. Valley Com., 1934-36; hydraulic engr. Soil Conservation Service, 1936-37; with U. S. Weather Bur., since 1937, chief River and Flood Div., 1937-39, hydrologic dir., 1939-46; chief, climatological and hydrological services, since 1946. Del., 220th anniversary Soviet Acad. Scis.; mem. Am. Meteorologic Mission to USSR, 1945, Am. Soc. C.E. (Norman medalist 1945), Am. Geophys. Union NRC (v.p. hydrology sect.), Washington Acad. Scis., Internat. Union Geodesy and Geophysics (pres. internat. assn. hydrol. 1948), Internat. Meteorol. Orgn. (v.p. hydrol. commn.), Research and Devel. Bd. (mem. com. on geophys. geog.), Am. Meteorol. Soc. Contbr. papers to govt. bulls. and engring. jour. Died Apr. 13, 1951.

BERNARD, Noël Pierre Joseph Leon, French physician, biologist; b. Beziers, France, Oct. 4, 1875; M.D., Ecole principale du Service de santé de la marine à Bordeaux, France. Served in French colonial health service, 1900-25; gen. dir. Pasteur Insts. Indochina, 1924-33; gen. sec. Overseas Pasteur Insts., Paris, France, 1933; under-dir. Pasteur Inst. Paris, from 1940, emeritus, from 1958. Named Grand officer, Legion of Honor, comdr. pub. health; recipient Croix de guerre, other French, fgn. awards. Mem. Overseas Acad. Scis. Author: De l'Empire colonial a la Union française; Versin, Pionnier, Savant, Explorateur; La Vie et la Oeuvre d'Albert Calmette. Showed part played by endophytes in germination and growth of orchid seeds (showed in particular that infection is prerequisite of successful germination). Home: 207, Rue de Vaugirard, Paris, France.

BERNARD, Pierre Eugène, French geophysicist; b. Antony, France, Jan. 15, 1915; s. Maurice René and Eugénie (Ranchet) B.; D.Sc., Faculté des Sciences Paris, (France), 1940; m. Suzanne Daviou, Mar. 27, 1951; children—Noële, Pascal. With Inst. de Physique du Globe, Paris, 1937-—, research officer, 1944-52, research master, 1952-—; Chief Microseismic Obs., Parc Saint Maur, France, 1965-—. Recipient Acad. Sci. prizes, 1940, 49, 54, Institut Oceanographique prize, 1943. Mem. Comité Nat. Geodésie et Géophysique, Société Astronomique de France; pres. European Sub-comm. on Microseisms, 1966. Author: Etude sur l'agitation microséismique et ses variations, 1941; also numerous articles, chpts. in books. Editor section Geophysics Bull. signalétique CNRS, 1942-—. Research on microseisms, discovered their cause in standing sea-waves in very center cyclonic lows, antipodal propagation of swell; made microseisms contribute to solar-weather relationships; proposed mechanisms for solar flares, geomagnetic storms and undecennal variation of geomagnetic field. Office: 9 quai St. Bernard, Paris V°, France.*

BERNARD, Pons-Joseph, French engr., mathematician; b. Trans, France, Oct. 1, 1764; s. Joseph and Catherine (Vacon) B.; ed. Oratory of Aix (France), Coll. of Marseilles (France). Prof. of math. and philosophy Oratory, Marseille; asst. dir. Naval Obs., Marseilles; head engr., dept. Var. Mem. French Acad. Scis. Works honored by acads. of Marseilles and Lyons. Author: Nouveaux principes d'hydraulique, appliqués à tous les objets d'utilité aux rivières, 1787; also publs. on agr., canals, embanking of rivers. Observed the satellites of Saturn, 1787. Died Trans, July 29, 1816.

BERNARD OF GORDON, see de Gordon, Bernard.

BERNARD OF VERDUN, French astronomer; b. Verdun, France; flourished late 13th century. Author: Tractatus optimus super totam astrologiam (comparison of homocentric and Ptolemaic theories). Works mark beginning of Ptolemaic supremacy.

BERNARDI, Enrico, Italian engr.; b. Verona, Italy, 1841; prof. physics and chemistry U. Pavia (Italy); became tchr. math., Padua, Italy, 1867; prof. Tech. Inst., Venice, Italy; prof. Sch. Engring., Padua. Inventor gas motor for sewing machines, 1884; helped develop head steered motor and injection carburetor; built 3-wheeled car with rear wheel drive, 1894, also 1st motorcycle. Died Turin, Feb. 21, 1919.

BERNARDINI, Gilberto, Italian physicist; b. Florence, Italy, Aug. 20, 1906; s. Alfredo and Elvira (Nannucci) B.; D. in Physics Scuola Normale Superiore, U. Pisa (Italy), 1928; D.Sc. honoris causa, U. Rochester, 1964; s. Nella Magherini, July 30, 1928; children—Maria Ludovica (Mrs. Antonio Zichichi), Nicola. Prof. physics U. Bologna (Italy), 1939-45, U. Rome (Italy), 1946-64; dir. Syncro-Cyclotron div., CERN, 1958-60, dir. research, 1960-63; dir. Scuola Normale Superiore, Pisa, 1964-—. Vis. prof. Columbia, 1949-51; research prof. U. Ill., 1951-57. Recipient Internat. Righi prize Acad. Sci. Bologna, 1942, Somaini Found. prize Academia Nazionale Lincei and Società Italiana di Fisica, 1956. Author: (with L. Fermi) Galileo and the Scientific Revolution, 1965; (with E. Amaldi) Fisica Generale, 1965; also articles. Fellow Am. Phys. Soc., mem. Italian Institut Nuclear Physics (pres. 1953-59), Societa Italiana di Fisica (pres. 1961-67), Academia Nazionale dei Lincei (Rome), acads. sci. Bologna, Modena, Bari. Research on cosmic rays and elementary particles. Home: Corliano Lastra a Signa, Florence, Italy. Office: Palazzo dei Cavalieri, Pisa, Italy.*

BERNARDO OF SEMINARA (Barlaam), Italian mathematician; b. Seminara, Italy, circa 1240; monk in Basilian monastery S. Elia de Copressimo in Galatro; lectr. in Constantinople, Turkey; abbot Monastery of the Saviour, Constantinople. Author: Logistica (computations with integers, ordinary fractions, sexagesimal fractions), 6 books; also commentary on theory of solar eclipses in Almagest. Gave arithmetical demonstrations of propositions in book II of Eculid.

BERNATEK, Erling Reinholdt, Norwegian chemist; b. Oslo, Norway, May 17, 1926; s. Walter Reinholdt and Elise (v. Schönenberg) B.; Mag.scient., U. Oslo, 1950, Dr.Philos., 1961; m. Ann Sohlberg, Sept. 24, 1951; children—Erik B., Oyvind B. Brit. Council scholar U. Coll., London, 1951-52; mem. faculty of sci., chemistry dept. U. Oslo, 1954-65, prof. dept. pharm. chemistry, 1966-—, dir. Pharm. Inst., 1967-—. Author: Ozonolyses in the Naphthoquinone and Benzofuran Series, 1960; also numerous articles. Research on quinone chemistry, mechanism of ozone reactions, solvolysis of ozonides. Home: 11D Folke Bernadottes Vei, Oslo 8, Norway.*

BERNAUER, Ferdinand; German geologist, mineralogist; b. Menzingen, Baden, Germany, July 23, 1892; s. Hermann and Lina (Dörschuck) B.; ed. Heidelberg U.; m. Sophie Glock, 1922. Asso. prof. mineralogy and petrography Technische Hochschule, Berlin, Germany from 1928. Author: Kolloidchemie al Hilfswissenschaft der Mineraloge, 1924; Gedrillte Kristalle, 1929; Junge Tektonik auf Island, 1943. Research on colloid mineralogy and crystallography; after 1930 worked only on volcanology including important work on bored crystals and volcanic and structural studies of the islands Vulcano and Iceland; invented polarization filter of organic substances (Bernotare). Died Berlin, May 16, 1945.

BERNAYS, Augusius Charles, Am. surgeon; b. Highland, Ill., Oct. 13, 1854; grad. McKendree Coll., 1872; M.D., U. Heidelberg (Germany), 1876; mem. Royal Coll. of Surgeons, London, 1877; settled in practice at St. Louis, 1878; inventor of improved methods in operative surgery; noted as tchr. of anatomy and surg. pathology and did much in introducing antiseptic method of surgery in the U. S. Author: Chips from a Surgeon's Workshop, 1880; Development of Valves of Heart, 1876; Development of Joints in General and of the Human Knee Joint, 1877; Golden Rules of Surgery, 1906. Possibly performed 1st excision of stomach in U. S., 1887; performed 1st successful Caesarean section in Mo.; developed several new surg. techniques. Died St. Louis, May 22, 1907

BERND, Joseph L., Am. polit. scientist; b. Macon Ga., Dec. 8, 1923; s. Laurence J. and Eva (Bloom) B. B.A., Mercer U., 1945; M.A., Boston U., 1953; Ph.D. Duke, 1957; m. Ruth Audrey Brady, July 2, 1960; 1 dau., Alison Ruth. Asst. prof. High Point Coll., 1957-59, asso. prof., 1959; asst. prof. polit. sci. So. Methodist U., 1959-62, asso. prof., 1962-65; prof., chmn dept. polit. sci. Va. Poly. Inst., 1965—; research asst. to Gov. M. E. Thompson, Atlanta, 1949, 50, 54; cons. Duke U. Press, U. S.C. Press, U. S. Commn. on Civil Rights, NSF. Fellow Social Sci. Research Council, Council of the Humanities; mem. Am. So. (mem. exec. com. 1966—), polit. sci. assns., Am. Assn., U. Profs., Am. Civil Liberties Union, NAACP. Author: Grass Roots Politics in Georgia, 1960; Mathematical Applications in Political Science, Vol. 1, 1965, Vol. 2, 1966, Vol. 3, 1967. Research and publs. of integration of systematic polit. theory connecting logical deductive models with empirical inductive models; effects of Georgia county unit system and polit. process and behavior. Home: Tom's Creek Rd., Blacksburg, Va. 24060.*

BERNDT, Franz, German chemist; b. Lubeck, Apr. 3, 1904; s. August and Lucie (Kummerow) B.; M.A. in Chemistry, Ph.D., U. Marburg; m. Elisabeth Scharloh, Dec. 7, 1928; children—Hans Detlef, Klaus Dieter, Sigred. Chemist, Accumulatorenfabrik Hagen, Emailschmelze Seibel Mettmann, Vereinigte Glanzstoff-Fabriken, Teltow Seehof and Sydowsaue; dir. Spinnfaser Aktiengesellschaft, Kassel, Germany. Home: Memelweg 10, 35 Kassel-Wilh. Office: Wohnstrasse 1, Kassel-B., Germany.*

BERNDT, Gerald Darwin, Am. math. statistician; b. North Mankato, Minn., May 7, 1920; s. Albert August and Lillian (Pittelkow) B.; student Gustavus Adolphus Coll., 1940-41, Mankato State Coll., 1941-42, 46-47; B.A., U. Minn., 1948; M.Sc., N.Y. U., 1951; postgrad. Am. U., George Washington U.; m. Elaine Evelyn Nienow, May 1, 1943. Climatologist, Weather Bur., Washington, 1951-53, U. S. Dept. Def., Natick, Mass., 1953-54; statistician Pres.'s Adv. Com. on Weather Control, Washington, 1954-57; chief math. analysis control div. Hdqrs. Strategic Air Command. Offutt AFB, Neb., 1957-60, math. statistician, operations analyst, 1960—. Cons. Colo. State U., Ft. Collins, 1959-61. Recipient Sustained Superior Performance award Dept. Air Force, 1961; Air Force certificate competence and trust, 1965. Fellow Royal Statis. Soc.; mem. Inst. Math. Statistics, London Math. Soc. (asso.). Contbr. articles to profl. jours. Derivation and application math. models toward analysis evaluation, prediction strategic operational capability in air; devel. and application war games models toward optimizing operational plans and strategies. Home: 1403 Little John Rd., Bellevue, Neb. 68005. Office: Operations Analysis, Hdqrs. SAC, Offutt AFB, Neb. 68113.*

BERNDT, Sune Bertil, Swedish aerodynamicist; b. Malmoe, Sweden, Feb. 26, 1923; s. Ernst R. and Elisabet (Roos) B.; civilingenjör (aeronautics) Royal Inst. Tech., Stockholm, Sweden, 1947, tekn. lic. (aeronautics and fluid mechanices), 1952; Ph.D., Cal. Inst. Tech., 1955; m. Ida Louise Guy, June 9, 1956; children—Julie Louise, Tore Fredrik. Research scientist Royal Inst. Tech., Stockholm, 1947-51, faculty, 1956—, prof. gas dynamics, 1959—; research scientist Aero. Research Inst. Sweden (FFA), 1951-59. Mem. Internat. Acad. Astronautics (corr.), Swedish Nat. Com. for Mechanics, Swedish Council Aeros. Contbr. papers to tech. jours. Theoretical research on compressible flow in particularly transonic flow. Home: 18 Fridhemsgatan, Stockholm K, Sweden.*

BERNE, Eric Lennard, psychiatrist; m. Montreal, Que., Can., May 10, 1910; s. David Hillel and Sara (Gordon) B.; B.A., McGill U., 1931, M.D., 1935, C.M., 1935; children—Ellen, Peter, Ricky, Terry Berne, Janice (Mrs. Michael Farlinger), Robin Way. Practice medicine specializing in psychiatry; staff Yale Psychiat. Clinic, 1936-38, N.Y. Psychoanalytic Inst., 1941-42, Mt. Sinai Hosp. N.Y. Psychiat. Out-Patient Clinic, 1941-43; with San Francisco Psychoanalytic Inst., 1947-56. Lectr. in psychiatry U. Cal.; adj. psychiatrist Mt. Zion Hosp., San Francisco; cons. in group therapy McAvley Clinic, San Francisco. Mem. Internat. Transactional Analysis Assn. (chmn. board), Indian Psychiat. Soc. (corr. mem.). Author: The Mind in Action, 1947; Layman's Guide to Psychiatry and Psychoanalysis, 1957; Transactional Analysis in Psychotherapy, 1961; The Structure and Dynamics of Organizations and Groups, 1963; Games People Play, 1964; Principles of Group Treatment, 1966. Editor, Transactional Analysis Bull. Publs. on use of game analysis for psychotherapy; originator of transactional theory of personality structure and devel. Home: Box 2111, Carmel, Cal. 93921.*

BERNE, Robert Matthew, Am. physiologist; b. Yonkers, N.Y., Apr. 22, 1918; s. Nelson and Julia (Stahl) B.; A.B., U. N.C., 1939; M.D., Harvard, 1943; m. Beth Goldberg, Aug. 18, 1944; children—Julie, Amy, Gordon, Michael. Faculty, Western Res. U. Sch. Medicine, 1949-66; prof. physiology, chmn. dept. U. Va. Sch. Medicine, Charlottesville, 1966—. Cons. NIH. Recipient Research Career award USPHS, 1962-66. Mem. Am. Physiol. Soc., Am. Soc. Clin. Investigation, Am. Heart Assn., A.A.A.S., Am. Assn. U. Profs., Phi Beta Kappa. Sect. editor Am. Jour. Physiology, 1965-66; editorial bd. Circulation Research, 1962—, Proc. Soc. Exptl. Biology and Medicine, 1962-64. Research, publs. in cardiovascular physiology, especially metabolism of heart and in mechanisms involved in regulation of blood flow in heart and skeletal muscle. Home: 1851 Wayside Pl., Charlottesville, Va. 22903.*

BERNER, Endre Qvie, Norwegian chemist; b. Stavanger, Norway, Sept. 24, 1893; s. Endre and Anna Marie (Gjemre) B.; student Norwegian U. Tech., Trondheim, 1913-18; Dr.tech., 1926; postgrad., Munich, 1922-23, Birmingham, 1928-29, London, 1954-55; m. Nathalia Weidemann, Sept. 16, 1922; m. 2d, Erna Gay, July 2, 1935. With Norwegian U. Tech., 1918-33, reader, 1922-33; prof. chemistry, U. Oslo, 1934-61. Mem. Norwegian Chem. Soc. (hon.), Acad. Sci. Oslo, Royal Norwegian Soc. Scis., Royal Swedish Acad. Engring. Scis., Chem. Soc., Royal Soc. Arts London, Soc. Chem. Industry (hon. fgn.). Author: Textbook in Organic Chemistry, 6th edit., 1964; also articles. Research in organic chemistry, thermochemistry, alkaloids, carbohydrates, phys. properties, structural problems, Diels-Alder reactions, Claisen condensations. Home: Gyldenloves Gate 13, Oslo 2, Norway.*

BERNFELD, Peter Harry William, Am. biochemist; b. Leipzig, Germany, June 1, 1912; s. Isidor and Elsa (Gutfreund) B.; M.S. in Chemistry, U. Leipzig, 1935; Ph.D. in Organic Chemistry, U. Geneva (Switzerland), 1937; m. Helen Cecily Kroch, Nov. 21, 1940; children—Michele Marion, Mark Raymond. Came to U. S., 1949, naturalized, 1955. Faculty dept. organic and inorganic chemistry U. Geneva, 1937-49, research fellow, 1937-39, chief chemis. 1939-49, privat-docent, 1947-49; asst. prof. biochemistry, nutrition Tufts U. Sch. Medicine, Boston, Mass., 1949-51, asso. prof., 1952-57; v.p. Bio-Research Cons., Inc., Cambridge, Mass., 1957—, dir. research, 1957—; v.p. Bio-Research Inst., Inc., Cambridge, 1957—, dir. research, 1957—. Recipient Werner medal Swiss Chem. Soc., 1948. Mem. A.A.A.S., Am. Soc. Biol. Chemists, Am. Chem. Soc., Am. Assn. Cancer Research, Soc. Exptl. Biology and Medicine, N.Y. Acad. Scis., Sigma Xi. Editor: Biogenesis of Natural Compounds, 1966. Research, publs. in cancer, vascular diseases, biochemistry of enzymes and starches, membrane permeability. Home: 10 Fairlee Rd., Newton, Mass. 02168. Office: 9 Commercial Av., Cambridge, Mass. 02141.*

BERNHARD, Fritz Georg Hermann, German physicist; b. Görlitz, Dec. 14, 1913; s. Hans and Susanne (Greinke) B.; Dipl.Ing., Berlin Inst. Tech., 1939; Dr. rer. nat. habil., Berlin, 1960; m. Elisabeth Schrottke, Oct. 2, 1940; children—Christian, Evemarie, Stefan, Niklaus, Irene, Julia. Dir. Inst. Nuclear Physics, German Acad. Sci., Berlin, 1955-60, dir. Inst. Chem. Physics, 1960-62; dir. I. Inst. Physics, Humboldt U., Berlin, 1962—, dean sci. faculty, 1968—. Recipient Nat. prize, 1957. Author: (with W. Hartman) Photovervielfacher in ihre Anwendung in der Kernphysik, 1957; also numerous articles. Developed (with W. Schüetze) method of measuring low ion currents down to 10 to the minus 19th amperes by counting individual ions via emitted electrons. Home: 19 Ohm-Kruegerstrasse, Berlin-Karlshorst. Office: 42 Juvalidenstrasse, 104 Berlin, East Germany.*

BERNHARD, William Francis, Am. surgeon; b. Bklyn., Dec. 11, 1924; s. William and Helen (Conroy) B.; B.A., Williams Coll., 1946; M.D., Syracuse U., 1950; m. June Horne, Sept. 17, 1948; children—Susan, William, Christine, Margaret, Catherine, John, Ann, James, Robert, Peter. Harvey Cushing fellow Peter Bent Brigham Hosp., Boston, 1954-55; research fellow in surgery Harvard Med. Sch., Boston, 1954-55, asst. in surgery, 1959-61, instr. surgery, 1961-62, asst. clin. prof. surgery; attending surgeon in thoracic and cardiovascular surgery VA Hosp., West Roxbury, Mass., 1960—; dir. surg. research lab. Children's Hosp., Boston, 1960—; asso. in surgery Children's Hosp. Med. Center, Boston, 1966—; practice medicine specializing in cardio-vascular surgery, Boston, 1959—; Mem. Soc. U. Surgeons, Am. Assn. Thoracic Surgery, A.C.S., Mass. Med. Soc., Am. Soc. Artificial Internal Organs, Am. Heart Assn. Research, numerous publs. in field of hyperbaric oxygenation for cyanotic congenital heart disease in infants and children, artificial heart. Home: 60 Singletary Lane, Framingham, Mass. 01700. Office: 300 Longwood Av., Boston 02115.*

BERNHARDI, Johann Jakob, German botanist; b. Erfurt, Germany, Sept. 1, 1774; prof. Erfurt U.; dir. Bot. Garden, Erfurt. Author: Systematische Verzeichniss der Pflanzen, welche in der Gegend um Erfurt gefunden werden, 1800; Anleitung zur Kentnniss der Pflanzen, 1801; Beob. über Pflanzengetässe und eine neue Art derselben, 1805; Ueber den Begriff der Pflanzenartes, 1834. Revised Linnean system of classification; studied pteridology (study of ferns), reeds or canals in leaves, germination; originated use of perpendiculars to faces drawn from center of crystal to determine shape and structure of a crystal. Died May 13, 1850.

BERNHARDUS TREVISANUS (Count of Trevigo), alchemist; b. Padua, Italy, 1406. Spent his life and fortune in study of alchemy and in pursuit of Philosopher's Stone. Author: Von der Hermetischem Philosophia, 1582; De Chymico Miraculo, quod Lapidem Philosophiae appellant; Liber de Secretissimo Philosophorum opere Chemico; Processus Lapidis Philosophorum ex Mercurio corporis; Responsio ad Thonam de Bononia de Mineralibus, et Elixiris compositione; De Transmutatione Metallorum Liber; Verbum Dimissum; De Chemia, opus historicum et dogmaticum. After 60 years effort, claimed to have succeeded in making the Philosopher's Stone, with all its virtues, in its purest form, 1481. Died 1490.

BERNHART, Finn Westelius, Am. biochemist; b. Sioux Falls, S.D., July 5, 1907; s. Petter K. and Fanny (Westelius) B.; Ph.C., S.D. State Coll., 1929, B.S., 1932; M.S., U. Minn., 1934, Ph.D., 1938; m. Adeline S. Puhr, July 7, 1930; 1 dau., Marie (Mrs. Harold Kaplan). Asst. biochemist Tulane U., 1933-35, instr., 1938-41; research asso. Cleve. Clinic, 1941-42, head biochem. sect. research div., 1942-43; head biochemistry dept. Research Lab., Wyeth, Inc., Chagrin Falls, O., 1943-45, Wyeth Inst. Applied Biochemistry, Phila., 1945-47; tech. dir. Nutritional div. Wyeth Labs., Radnor, Pa., 1947-49, dir. research and devel., 1949-56, head nutrition-endocrinology dept. Wyeth Inst. for Med. Research, 1956—. Mem. Am. Inst. Nutrition, Am. Chem. Soc., Soc. Exptl. Biology and Medicine, Sigma Xi. Research on isolation of penicillin, protein chemistry, nutritional value of milk. Home: 585 Cricket Lane. Office: Wyeth Labs., Radnor, Pa.*

BERNHEIM, Hippolyte, French physician; b. Mulhouse, France, Apr. 17, 1840; became prof. Faculty Strasbourg (France), 1868; later aggregate prof. (with splty. in hypnosis and auto-suggestion) med. clinic Faculty Medicine, U. Nancy (France). Author: Contribution à l'étude des localisations cérébrales, 1877; De la suggestion dans l'état hypnotique et dans l'état de veille, 1884; De la suggestion et de ses applications à la thérapeutique, 1886; Hypnotisme, suggestion et psychothérapie, 1890; L'Hypnotisme et la suggestion dans leurs rapports avec la médecine lé gale, 1897. First to use hypnotism to treat neuroses, 1884; developed Braid's idea that hypnosis is suggestion. Died 1919.

BERNHEIMER, Alan Weyl, Am. microbiologist; b. Phila., Dec. 9, 1913; s. Eugene and Helen (Weyl) B.; B.S., Temple U., 1935, M.A., 1937; Ph.D., U. Pa., 1942; m. Harriet Poller, Mar. 29, 1942; 1 son, Alan Weyl. Faculty, Pa. State Coll. Optometry, 1937-39; faculty N.Y. U. Coll. Medicine, 1941—, prof. microbiology, 1958—. Mem. coms. OSRD, NIH. Recipient Eli Lilly award in bacteriology and immunology, 1948; Research Career award NIH, 1962; Distinguished Alumni award Temple U., 1964. William H. Park fellow, 1960-61. Fellow A.A.A.S., N.Y. Acad. Sci., Am. Acad. Microbiology; asso. fellow N.Y. Acad. Medicine (sec. adv. com.. sect. on microbiology 1962-64); mem. Harvey Soc., Am. Soc. for Microbiology, Am. Micros. Soc., Mineral. Soc. Am., Am. Assn. Immunologists. Research, publs. in biochem. basis pathogenicity. Home: 51 Fifth Av., N.Y.C. 10003. Office: 550 1st Av., N.Y.C. 10016.*

BERNICK, Sol, Am. anatomist; b. St. Paul, Sept. 29, 1915; s. Abraham B. and Celia (Harris) B.; B.S., U. Minn., 1940; M.S., U. So. Cal., 1947, Ph.D., 1955; m. Ellen Rhoda Levy, July 11, 1944; children—James, Michael, Charles. Faculty, U. So. Cal., Los Angeles, 1947—, prof. anatomy Sch. Medicine, 1965—. Mem. Internat. Assn. for Dental Research, Am. Assn. Anatomists, Recticulo-endothelial Soc., N.Y. Acad. Scis., Am. Acad. Oral Pathology. Contbr. numerous articles to profl. jours. Research in growth and devel. of face and jaws, cause of atherosclerosis, relation of hormones to etiology of vascular disease. Home: 8450 W. 4th St., Los Angeles 90048. Office: 2025 Zonal Av., Los Angeles 90033.*

BERNOULLI, Christoph, Swiss naturalist; b. Basel, Switzerland, May 15, 1782; s. Daniel and Maria Magdalena (Burkhardt) B.; studied at U. Göttingen (Germany); m. Katherine Salome Paravicini, 1808; 9 children. Founder, prin. pvt. sch. Philotechnisches Institut, 1806-17; became prof. natural history U. Basel, 1818; became prof. indsl. scis., 1835. Author: Über das Leuchten des Meers, 1802; Psychische Anthropologie, 2 vols., 1804; Vademecum des Mechanikers, 1829; Handbuch der Technologie, 1833; Handbuch der Dampfmaschinenlehre, 1833; Handbuch die industriellen Physik, Mechanik und Hydraulik, 1834; Über Populationsitik, 1840. Died Basel, Feb. 6, 1863.

BERNOULLI, Daniel, mathematician; b. Groningen, Netherlands, Jan. 29, 1700; s. Jean Daniel B.; studied medicine univs. Basel, 1716-17, Heidelberg, 1718-19, Strassburg, 1719-20; M.D., U. Basel (Switzerland), 1721. Went to Venice to practice medicine, but turned to math. instead; prof. math. Acad. St. Petersburg, 1725-33; prof. anatomy and botany U. Basel, 1733-50, prof. exptl. philosophy, 1750-77. Fgn. asso. French Acad. Scis., 1748, recipient 10 prizes from acad.; Fellow Royal Soc. London, 1750; mem. nearly all learned socs. and acads. of Europe. Author: Exercitationis Mathematicae, 1724; Hydrodynamica, 1738; Specimen Theoriae Novae de Mensura, 1738. Devised a kinetic theory of gases, 1738; 1st to treat quantitatively the idea that gaseous pressure is due to molecular bombardment; deduced Boyle's law as a necessary consequence of this hypothesis; showed that as velocity of fluid flow increases, its pressure decreases; enunciated the theorem on hydraulic pressure which carries his name; developed an early formulation of principle of conservation of energy; did pioneer work in partial differential equations; showed how differential calculus could be used in theory of probability; introduced calculus of probability into epidemiology and insurance; 1st to use suitable notation for inverse trigonometric functions; expressed the roots of a quadratic equation in form of a recurring series (Bernoulli's equation); 1st to point out the usefulness of resolving a compound motion into motions of translation and rotation; also investigated vibrations of strings, oscillations and sounds of elastic plates, sound and timbre of organ pipes, other areas of acoustics; recommended use of the ship screw; observed phenomenon of cavitation; investigated use of water clocks on the sea, 1725; studied the tides, 1740. Died Basel, Mar. 17, 1782.

BERNOULLI, Jakob (or Jacques) I, Swiss mathematician; born Basel, Switzerland; December 27, 1654; the son of Nikolaus and Margaretha (Schönauer) Bernoulli; educated at Univ. Basel; married Judith Stupanus, 1684; 1 son, Nikolaus, 1 dau. Travelled in Germany, France, Holland, from 1676; named prof. math. U. Basel, 1687-1705. Author: Conamen novi systematis cometarum, 1682; Dissertatio de gravitate aetheris, 1683; Ars conjectandi, 1713; Opera omnia, 2 volumes, 1744; (with brother Johann) abstract mathematical treatises. Improved Leibnitz's differential calculus; coined term "integral"; laid founds. for calculus of variation; 1st to solve Leibnitz's problem of isochronous curve; proposed problem of catenary and solved complicated problems relating to it; determined elastic curve; devised formula for radius of curvature in rectangular and polar coordinates; studied logarithmic spirals, also cycloid, transcendental curves, equiangular spiral; stated limit to compressibility of gases; developed law of large numbers; proved binomial theorem for positive integral exponents; pioneered in theory of probability, applying it to statistics; advanced Bernoulli's theorem; invented Bernoulli numbers; Died Basel, Aug. 16, 1705.

BERNOULLI, Jakob (or Jacques) II, Swiss mathematician, physician; b. Basel, Switzerland, 1759; s. Johann and Susanna (König) B.; studied under Daniel Bernoulli at U. Basel; went to Russia, 1782, prof. math., St. Petersburg, from 1788; m. a granddau. of mathematician, Euler. Mem. St. Petersburg Acad. of Scis. Contbr. to Nova Acta Academia Petropolitania. Died St. Petersburg, Russia, July 3, 1789.

BERNOULLI, Johann (or Jean) I, Swiss mathematician, physicist; b. Basel, Switzerland, July 27, 1667; s. Nikolaus and Margaretha (Schönauer) B.; M.D., 1694; studied math. with bro. Jakob; m. Dorthea Falkner, 1694; 2 daus., 5 sons, including Nikolaus, Daniel, Johann. Taught Marquis de l'Hôpital math. while in France, 1690-92; apptd. prof. math. U. Groningen (Netherlands), 1695; succeeded Jakob as prof. math., Basel, 1705. Mem. French Acad. Scis., 1699, Berlin Acad., St. Petersburg Acad., Royal Soc., 1712, Sci. Inst. Bologna. Author: Lectiones mathematicae de methodo integralium, 1742; Commercium philosophicum et mathematicum, 1745; Korrespondenz mit Leibniz, 2 vols., 1745; Johannis Bernoulli opera omnia. Shared in prin. discoveries of bro. Jakob; furthered development of differential and integral calculus; defended case for Leibnizian priority in discovery of calculus against members of Royal Soc.; discovered exponential calculus; supported Leibniz's concept of "vis viva" (kinetic energy); 1st to use symbol g to denote acceleration caused by gravity; introduced abstract concept of function; worked on calculus of variations. Died Basel, Jan. 1, 1748.

BERNOULLI, Johann (or Jean) II, Swiss mathematician, physicist; b. Basel, Switzerland, May 18, 1710; s. Johann and Dorthea (Falkner) B.; ed. U. Basel; m. Susanna König; children—Johann, Jakob, Daniel. Prof. rhetoric, Basel, 1743-48; became prof. math., 1748. Mem. Berlin Acad. Scis., French Acad. Scis., 1782 (3 prizes). Research on light, the capstan, the magnet, and heat. Died Basel, July 17, 1790.

BERNOULLI, Johann (or Jean) III, Swiss mathematician, astronomer, geographer; born Basel, Switzerland, November 4, 1744; the son of Johann and Susanna (König) Bernoulli; studied mathematics and astronomy at Basel, philosophy at Neuchâtel, Switzerland; m. Veronica Beck, 1769; m. 2d. Carol Tempelhoff; 5 sons, 6 daus. Apptd. astronomer to Berlin (Prussia) Acad. by Frederick the Great, 1763; made several sci. trips through Europe; became dir. Berlin Obs., 1767; settled in Berlin permanently, 1779; became dir. math. class Berlin Acad., 1792; royal astronomer. Mem. acads. St. Petersburg (Russia), Stockholm (Sweden). Author: Lettres sur differents sujets, 1774-75; Reisen durch Brandenburg . . . , Russland, Polen, 6 vols., 1779-80; Lettres astronomiques, 1781. Pub., J. H. Lambert's work in 7 vols., also French transl. Elements of Algebra (Euler), 1785. Editor, Magazin pour les scis. mathematiques, 3 years. Died Köpernick, Germany, July 18, 1807.

BERNOULLI, Nikolaus (or Nicolas) I, Swiss mathematician; b. Basel, Switzerland, Oct. 20, 1687; s. Nikolaus and Ursula (Straehelin) B.; ed. by his uncles Jakob and Johann; m. Anna Birr, 1720; m. 2d, Katherine Battier, 1747; 1 dau. Named prof. math. U. Padua (Italy), 1716; apptd. prof. math. U. Basel, 1719, prof. logic, 1722; prof. law, 1731. Fellow Royal Soc., 1714. Publs. on determining sum of reciprocal squares. Editor: Ars conjectandi (Jakob Bernoulli). Solved many problems posed by uncles; studied infinite series and theory of probability; corresponded with Leibniz; acquainted Newton and J. Sterling in London, with P. de Montmort in Paris. Died Basel, Nov. 29, 1759.

BERNOULLI, Nikolaus (or Nicolas) II, Swiss mathematician; born Basel, Switzerland, January 27, 1695; the son of Johann and Dorthea (Falkner) Bernoulli; Ph.D. University of Basel, 1711; highest degree in law, 1715. Visited Italy and France; received chair of jurisprudence at U. Bern (Switzerland), 1720-23; became prof. math. Acad. St. Petersburg (Russia), 1725. Wrote on differential, integral and exponential calculus; proposed the paradox of St. Petersburg. Died St. Petersburg, July 26, 1726.

BERNSDORF, Wilhelm Richard Heinrich, German sociologist; b. Bielefeld, Apr. 6, 1904; s. Paul and Frederike Bernsdorf; Ph.D., U. Berlin. Chief div. statistics; instr. sociology. Mem. German Soc. Sociology. Author: Die Einheit der Sozialwissenschaften, 1955; Wörtebuch der Soziologie, 1955; Soziologie der Prostitution (in Handbuch der medizinischen Sexualforschung), 1955; Internationales Soziologenlexikon, 1959; Soziologie (in Handbuch der Wissenschaft und Bildung), 1960. Home: Fredericiastrasse 27, Berlin-19 (1). Office: Freie Universitat Berlin, Berlin, Germany.

BERNSTEIN, Arthur, Am. physician; b. N.Y.C., Nov. 15, 1909; s. Isaac and Lena (Sandman) B.; A.B., U. Pa., 1930, M.S., 1931, M.D., 1935; m. Grace Ellen Hoffman, June 26, 1935; children—Lory (Mrs. Louis Alan Greenbaum), Lawrence Joseph, Mark Henry, Penny Lee. Research bacteriology U. Pa., 1930-36, asso. cardiology Sch. Medicine, 1950—, research clin, pharmacology, screening of new compounds, 1947—; research asso. cardiovascular disease Newark Beth Israel Hosp., 1940—, attending in medicine, 1955—, pres. med. staff, 1965—, clin. coordinator Inst. Hosp.; attending cardiologist Babies Hosp., Newark, 1954—; attending cardiologist Clara Maass Hosp., Belleville, N.J., 1958—; cons. cardiology East Orange (N.J.) Gen., Fitkin Meml. hosps., Presbyn. Unit United Hosp., Newark, U.O.T.S. Children's Cardiac Home, N.J. Dept. Health; cons. medicine Newark City Hosp., St. Barnabas Med. Center; chmn. exec. editors Vascular Diseases. Diplomate Am. Bd. Internal Medicine. Fellow A.C.P., Am. Coll. Chest Physicians (pres. N.J. chpt. 1959-60), Am. Coll. Cardiology (gov. N.J., 1958-61), Am. Med. Writers Assn., Am. Geriatrics Soc., Internat. Am. colls. angiology, Am. Coll. Clin. Pharmacology and Chemotherapy; mem. Am. Therapeutic Soc., N.J. Heart Assn. (bd. trustees, sec. exec. com. 1952-65, pres. 1965—), A.A.A.S., N.Y. Acad. Scis., Am. Fedn. Clin. Research, Pan-Am., World med. assns., Essex County Med. Soc., Med. Soc. N.J., A.M.A., Acad. Medicine of N.J., Assn. Mil. Surgeons, Essex County Heart Assn. Research, publs. on cardiovascular diseases; digitalis and its actions; study of many of new oral diuretics with comments on problems of their uses; monitoring patients with acute myocardial infarction as well as arrhythmias discovered by this technique; research field atherosclerosis with particular reference to etiology; studies of therapy of patients with Angina and coronary artery disease; pioneer in use Beta Adrenergic stimulators in treatment of cardiogenic shock. Home: 100 Great Hills Rd., Short Hills, N.J. 07078. Office: 2130 Millburn Av., Maplewood, N.J. 07040.*

BERNSTEIN, Benjamin Abram, mathematician; b. Posvol, Lithuania, 1881; student Balt. City Coll., 1897-1902; A.B., Johns Hopkins, 1905; Ph.D., U. Cal., 1913; m. Rose Davidson, 1920; 1 child. Instr. math. U. Cal., 1907-18, asst. prof., 1918-23, asso. prof., 1923-29, prof., from 1929. Mem. Philos. Assn., Math. Soc., Math. Assn. Am. Assn. for Symbolic Logic. Research and publs. on Boolean algebra, math. of logics, founds. of math. Home: 2785 Shasta Rd., Berkeley, Cal.

BERNSTEIN, Heinrich Agathon, naturalist; b. Breslau, Germany (now Wroclaw, Poland), Sept. 22, 1828; s. Georg Heinrich and Agathe (Brüchner) B.; completed med. studies; became dir. sanatorium nr. Buitenzorg, Java; as official naturalist of Netherlands government (recommended by Hermann Schlegel) travelled through North Moluccas (Spice Islands), Waigeu, Obi and Papuan Islands; member of Academy Leopoldina in Halle, Germany. Research on biology and anatomy of birds in Java with many important publs.; many valuable objects he obtained as ofcl. Dutch naturalist are in Natural History Mus. in Leyden, Netherlands. Died Salawati, Papuan Islands, Apr. 19, 1865.

BERNSTEIN, Jeremy, Am. physicist; b. Rochester, ter, N.Y., Dec. 31, 1929; s. Philip Sidney and Sophy (Rubin) B.; B.A., Harvard, 1951, M.A., 1953, Ph.D., 1955. Research asso. Cyclotron Lab., Harvard, 1955-57; mem. Inst. for Advanced Study, 1957-59; NSF fellow, 1959-60; research asso. Brookhaven Nat. Lab., Upton, L.I., N.Y., 1960-62; asso. prof. physics, N.Y. U., 1962—; cons. Gen. Atomic Co., La Jolla, Cal., 1960—. Recipient Westinghouse-A.A.A.S. sci. writing award, 1964. Fellow Am. Phys. Soc. Author: The Analytical Engine, 1965; Ascent, 1966; also articles. Writer, New Yorker mag., 1961—. Research on weak interactions, properties of neutrino, indeterminate vector meson, partial couract conservation, radiative muon capture and K meson decys. Office: Physics Dept., N.Y. U., Washington Sq., N.Y.C. 10001.*

BERNSTEIN, Julius, German physiologist; b. Berlin, Germany, Dec. 8, 1839; prof. physiology U. Halle (Germany); Corr. mem. French Acad. Scis. Author: Untersuchungen über den Erregungsvorgang im Nerven—und Muskelsystem, 1871; described nerve impulse as wave of negativity passing along nerve, 1866; propounded theory that a nerve can be exhausted by continued stimulation, 1877; expanded and established Ostwald's membrane theory of nerve conduction, 1902. Died Halle, Feb. 6, 1917.

BERNSTEIN, Leon, mathematician; b. Shoden, Germany; s. Tobias and Jenny (Lekus) B.; ed. U. Kaunas, Lithuania, 1936-40; diploma U. Vilnius, Lithuania, 1941, Ph.D., 1945; m. Posia Zlotnik, Feb. 5, 1947; 1 son, Johanan. Lectr., U. Vilnius, 1941-45; lectr. Sch. Law and Econs., Tel Aviv, Israel, 1949-59, Tel Aviv br. Hebrew U., 1950-65, Tel Aviv U., 1965—. Mem. Am. Math. Soc., Israel Soc. Mathematicians. Author textbooks in math., articles in profl. jours. Studies in explicit classes of algebraic number fields, algebra, number theory. Home: 58 Sokolov, Tel Aviv, Israel.*

BERNSTEIN, Richard Barry, Am. phys. chemist; b. N.Y.C., Oct. 31, 1923; s. Simon and Stella (Grossman) B.; A.B., Columbia U., 1943, M.A., 1944, Ph.D., 1948; m. Norma B. Olivier, Dec. 17, 1948; children—Neil D., Minda D., Beth A., Julie L. With S.A.M. Lab., Manhattan Project, Columbia U., 1942-44; with C.E., U. S. Army, Manhattan Project, N.Y.C., Oak Ridge, 1944-46; faculty Ill. Inst. Tech., 1948-53, U. Mich., 1953-63; prof. chemistry U. Wis., Madison, 1963—. Cons. to industry. A. P. Sloan fellow, 1956-60; NSF Sr. Postdoctoral fellow U. London, 1960-61. Fellow mem. Am. Chem. Soc. (chmn. div. phys. chemistry 1965-66), Nat. Acad. Scis., Sigma Xi. Asso. editor Jour. Chem. Physics, 1962-65. Contbr. articles to sci. jours. Exptl. and theoretical studies of molecular collisions; elastic, inelastic and reactive scattering of molecular beams. Home: 3410 Lake Mendota Dr., Madison, Wis. 53705.*

BERNSTEIN, Seymour, Am. physicist; b. Chgo., Feb. 20, 1909; s. Isadore Samuel and Etta (Sher) B.; B.S. in Engring., U. Ill., 1930; Ph.D. in Physics, U. Chgo., 1939; m. Adelaide Rubin, Sept. 18, 1938; children—Ruth Alice (Mrs. Salo Hillel Hyman), Irene Susan. Engr., City of Chgo. Bur. Engring., 1930-34; instr. Austin Coll., Chgo., 1939-42; research asso. U. Chgo., 1942-44; chief physicist Oak Ridge Nat. Lab., 1944-64; prof. physics U. Ill., Chgo., 1964—; vis. prof. Israel Inst. Tech., 1955-56, U. Miami, Coral Gables, 1961-62. Sci. del. First Internat. Conf. on Peaceful Uses Atomic Energy, Geneva, Switzerland, 1955. Fellow Am. Phys. Soc. Research, publs. primarily in fields of slow neutron spectroscopy, total crosssect. of Xenon-135 as a function of energy, polarization of neutrons and nuclei and their interaction, total reflection of nuclear resonant radiation (Mossbauer effect), elementary particle physics, atomic energy, reactor physics. Home: 423 Homestead Rd., La Grange Park, Ill. 60525. Office: P.O. Box 4348, Chgo. 60680.*

BERNT, (Johann) Josef, physician; b. Leitmeritz, Bohemia, Sept. 14, 1770; doctorate U. Prague (Czechoslovakia), 1797; 4 sons, 3 daus.; was active in introducing smallpox vaccination in Bohemia; prof. forensic medicine in Prague, 1808, in Vienna, Austria, as successor to F. B. Vietz, 1813; co-editor state (Austrian) med. yearbooks from 1832. Author many books including: Systematisches Handbuch der gerichlichen Arzneikunde, 1813; Systematisches Handbuch der Staatsarzneikunde, 2 vols., 1816, 17; others on plague, St. Vitus dance and reviving the apparently dead. Responsible for elimination of unnecessary isolation measures in epidemics; recommended hydrostatic lung test to determine whether newborn baby has breathed. Died Vienna, Apr. 27, 1842.

BERNTHAL, Theodore George, Am. physiologist; b. Grand Rapids, Mich., May 5, 1904; s. Otto Christian John and Ernestine Louise (Fiebig) B.; B.A., U. Mich., 1925, M.S., 1927, M.D., 1930; m. Reva Elizabeth Derby, Aug. 18, 1940; children—David F. L., Margaret L., Theodore George, Douglas D. Prof. physiology, chmn. dept. Med. Coll. S.C., 1946—; cons. VA Hosp., Columbia, S.C., 1952-64, S.C. State Hosp., 1960-64; Career devel. panelist NIH, 1967—. Mem. Am. Physiology Soc., A.A.A.S., Soc. Exptl. Biology and Medicine, So. Soc. Clin. Research, Sigma Xi, Alpha Omega Alpha, Phi Sigma. Contbr. numerous articles to profl. jours. Research in chem. control of blood flow, especially reflex control through carotid body chemoreceptors, with both chem. and non-chem. factors in control of breathing, control of pulmonary circulation of blood and with regulation of the heart rate. Home: 1421 S. Edgewater Dr. Office: 80 Barre St., Charleston, S.C. 29401.*

BERNTHSEN, Heinrich August, German chemist; b. Crefeld, Germany, Aug. 29, 1855; s. Heinrich Friedrich and Maria Sylbilla (Potgiesser) B.; studied math. and natural scis., later chemistry in Bonn and Heidelberg (both Germany); Docteur h. c. Tech. U. Berlin (Germany), 1925, U. Heidelberg, 1926; m. Maria Magdalena Haubenschmied, 1884; 3 children. Became pvt. docent organic chemistry, Bonn, 1879, Heidelberg, 1882; asso. prof. Heidelberg, 1884-87, hon. prof., 1920; dir. Badische Anilin und Sodafabrik Lab., Ludwigshafen, 1887-1918. Author: Kurzes Lehrbuch der organischen Chemie, 1887, 1924; Textbook of Organic Chemistry, 1889, 1941. Research on acid derivatives, methylene blue; discovered formula for sodium hyposulphite; formulated phenoxazine dyes as quinonoid ammonium salts. Died Heidelberg, Nov. 26, 1931.

BEROSSUS (or Berosus, Bar Oseas), philosopher, historian; flourished 281-262 B.C.; priest, temple of Belus, Babylon; trans. into Greek the standard Babylonian work on astrology and astronomy; author history of Bablyonia in three books (fragments preserved by Eusebius and other Greek writers).

BERRENS, L(ubertus), Dutch biochemist; b. Utrecht, Holland, June 28, 1933; s. Willem and Anne M. (Van Dijk) B.; degree in biochemistry and biophysics State U. Utrecht, 1957, Doctoral thesis, 1959; m. Ans Diderik, Aug. 16, 1958; children—Anne N., Frank M., Michiel A. Research chemist dept. dermatology and allergy Acad. Hosp., Utrecht, 1960—; head research lab., head sect. biochemistry of skin, reader biochemistry skin U. Utrecht, 1966—. Mem. Royal Dutch Soc. Chemists, Dutch Soc. Cosmetic Chemists (pres. 1965). Research, publs. on proteins in psoriasis, furocoumarins in plants, porphyrins, reticulin; devel. theory on genesis of atopic allergens (formation of allergens from protein and sugar by Maillard reaction); isolated and purified several atmospheric allergens; discovered structural relationship among allergens; identified N-glycosidic lysine-sugar structures as common biologically active antigenic determinant. Home: 13 Pr. Hendrikstr., Bunnik (U.), Utrecht. Office: 101 Catharijnesingel, Utrecht, Holland.*

BERRY, Edward Wilber, Am. paleontologist; b. Newark, N.J., Feb. 10, 1875; s. Abijah Conger and Anna (Wilber) B.; educated privately; D.Sc., Lehigh U., 1931; m. Mary Willard, Apr. 12, 1898; children—Edward Willard, Charles Thompson. Pres., treas. and mgr. Daily News, Passaic, N.J., 1897-1905; asst. in paleobotany Johns Hopkins U., 1907-08, instr., 1908-11, asso. 1911-13, asso. prof. of paleontology, 1913-17, prof. from 1917, dean, 1929-42, provost, 1935-42; sr. geologist U. S. Geol. Survey from 1910; asst. state geologist of Md., 1917-42. Fellow Paleontol. Soc. Am. (pres. 1924). Geol. Soc. Am. (pres. 1945); Am. Acad. Arts and Sciences, A.A.A.S., Am. Soc. Naturalists; mem. Am. Philos. Soc., Nat. Acad. Scis., Washington Acad. Scis., Torrey Bot. Club, Société Géologique de France, Academia Nacional de Ciencias en Cordoba, Argentina, Sociedad Geologica del Peru. Recipient Walker prize, Boston Soc. Natural History, 1901; Mary Clark Thompson medal Nat. Acad. Scis., 1944. Lower Cretaceous of Maryland (Md. Geol. Survey), 1911; Upper Cretaceous of Maryland, 1916; Eocene Floras of Southeastern North America (U. S. Geol. Survey), 1916; Tree Ancestors, 1923; Paleontology, 1929; also over 500 articles on paleontol., geol. and biol. subjects in Am. and fgn. sci. periodicals. Editor, Paleontology, Biol. Abstracts; asso. editor Am. Jour. Scis. Specialized in classification and evolution of plants, particularly in Southeastern N. America, equatorial America and South America. Died Stonington, Conn., Sept. 20, 1945.

BERRY, Edward Willard, Am. paleontologist; b. Passaic, N.J., Nov. 24, 1900; s. Edward Wilber and Mary (Willard) B.; A.B., Johns Hopkins U., 1924, Ph.D., 1929; m. Dorothy Everett Pidgeon, Oct. 12, 1925; children—Mary Susan (Mrs. E. P. Robare), Edward Lewis, Samuel Stedman. Micropaleontologist, Internat. Petroleum Co., Negritos, Peru, 1925-28; instr. geology Ohio State U., 1929-30; prof., chmn. dept. geology Duke, 1936-66. Cons. geologist coal, oil, diamonds; mem. council 21st Internat. Geol. Congress, 1960; prof. geology U. Malaya, Kuala Lumpur, 1961-62. Fellow Geol. Soc. Am., Geol. Soc. London, A.A.A.S.; mem. Geol. Soc. South Africa, Geograph. Soc. New Zealand, PanAm. Inst. Mining Engring. and Geology, Am. Geophys. Union, Am. Assn. Petroleum Geologists, geol. socs. France, Switzerland, Peru, Mexico, Am. Inst. Mining Metall. Engrs., N.C. Acad. Sci. (pres. 1957-58), Yorkshire Geol. Soc., Paleontol. Inst., Paleontol. Assn., others, Sigma Xi, Kappa Sigma. Research publs. on identification and description of micro fossils of Coastal Plain of N.W. Peru, S.A.; study of Coastal Plain of Atlantic and Gulf Coast of U. S., both fossils animals and plants, vis. and studying Permian Gondwana Coals of So. Hemisphere. Home: 654 McClendon, Corpus Christi, Tex. 78404.*

BERRY, John, Scottish biologist; b. Edinburgh, Scotland, Aug. 5, 1907; s. William and Wilhelmina (Barns-Graham) B.; Ph.D., Trinity Coll., Cambridge (Eng.) U.; m. Bride Fremantle, Aug. 20; children—Margaret Wilhelmina, William, Peter Fremantle. Research on salmon for adminstrn. of fisheries in Scotland, 1930-31, sta. biol. research U. Southhampton, 1932-36; dir. research, 1937-39; biologist, dir. information hydroelectric adminstrn. of north of Scotland, 1944-49; Mem. Royal Soc. of Edinburgh, Brit. Orgn. for Conservation of Nature (dir. 1949—), Internat. Union for Conservation of Nature and Natural Resources (pres. 1954-56, v.p. 1956-60 Ecology Commn.). Author: The Status and distribution of wild geese and wild duck in Scotland, 1939, also publs. on devel. of hydroelectricity, fisheries, ornithology. Home: Tayfield, Newport on Tay. Office: Nature Conservancy, 12, Hope Terrace, Edinburgh, Scotland.

BERRY, Richard George, Am. physician; b. Bethel, Conn., Jan. 29, 1916; s. Frank A. and Rose (Ohler) B.; B.A., Wesleyan U., 1937; M.D., Albany Med. Coll., 1942; m. Jane Singleton, May 2, 1942; children—Richard George II, David A., Daniel S., John F., Patricia A. Commd. lt. (j.g.) USN, 1942, advanced through grades to comdr., 1953; staff neurologist U. S. Naval Hosps., Phila., 1947-50, Bethesda, Md., 1950-53; resigned, 1953, civilian cons., 1953-54; dir. neuropathology lab. Jefferson Med. Coll., 1954—, prof. neurology, neuropathology, 1959—; cons. neuropathology VA Hosps., Coatesville, Lebanon, Pa., U. S. Naval Hosp., N.J. State Hosp. at Ancora, Eastern Pa. Psychiat. Hosp. Fellow Coll. Physicians Phila.; mem. Phila. (pres. 1965), Am. Neurol. Assn., Am. Acad. Neurology, Am. Assn. Neuropathologists, A.A.A.S., Sigma Xi. Publs. on cerebral vascular disease; studies of small vessels of brain, lesions in the aging nervous system. Home: 108 N. Rolling Rd., Springfield, Pa. 19064. Office: 1025 Walnut St., Phila. 19107.*

BERRY, Richard Stephen, Am. chemist; b. Denver, Apr. 9, 1931; s. Morris and Ethel (Alpert) B.; A.B., Harvard, 1952, A.M., 1954, Ph.D., 1956; m. Carla Lamport Friedman, Sept. 3, 1955; children—Andrea, Denise, Eric. Instr., Harvard, 1956-57, U. Mich., 1957-60; asst. prof. Yale, 1960-64; asso. prof. chemistry U. Chgo., 1964—. Cons. Avco-Everett Research Labs., 1964—, Am. Oil Co., 1965—; vis. prof. U. Copenhagen (Denmark), 1967. Alfred P. Sloan fellow, 1962-66. Mem. Am. Phys. Soc., Am. Chem. Soc., Sigma Xi. Research, publs. on molecular properties, reactions, structure and spectra, interactions of electrons with atoms and molecules, measurement of atomic electron affinities, mechanism of ionization. Home: 5317 S. University Av., Chgo. 60615.*

BERRY, Roger Julian, radiobiologist; born N.Y.C., Apr. 6, 1935; s. Sidney Norton and Beatrice (Mendelson) B.; A.B., N.Y. U., 1954; B.Sc. in Medicine, Duke, 1957, M.D., 1958; m. J. Valerie J. Butler, Sept. 25, 1960. Locum and asst. radiation physicist Hosp. for Joint Diseases, N.Y.C., 1953-54, 56; USPHS Med. Student Research fellow, 1956-58; med. officer radiotherapy Churchill Hosp., Oxford, Eng., 1958, head radiobiology lab. radiotherapy dept., 1962—; Am. Cancer Soc. fellow in radiobiology Oxford U., 1959-60, Helen Hay Whitney fellow, 1962-65; sr. investigator in radiation NIH, USPHS, 1960-62; hon. lectr. Oxford U, 1963—, Med. Coll. St. Barts Hosp., London, Eng., 1963—. Tech. sec. U.K. Panel on Gamma and Electron Irradiation, 1965—. Recipient Borden award for undergrad. med. research, 1958. Mem. Radiation Research Soc., Brit. Inst. Radiology, Royal Soc. Medicine, Assn. for Radiation Research, Hosp. Physicists Assn., Brit. Soc. for Cell Biology. Research, publs. on devel. exptl. cancer systems, effects ionizing radiation in combination with specific metabolic blocking agts. in therapy. Home: Gate House, Charlbury, Oxford. Office: Radiobiology Lab., Radiotherapy Dept., Churchill Hosp., Oxford, Eng.*

BERRYAT, Charles Jean Jacques, French physician; b. Clamecy, France, Mar. 10, 1718; s. Charles and Marie (Gabelin) B.; docteur en medecine, U. Montpellier (France). Practice medicine Auxerre, France; royal physician; inspector of mineral springs. Mem. French Acad. Scis., 1750. Author: Observations physiques et medicinales sur les eaux minérales, 1751. Studied mineral water, also epidemic at Auxerre from which he died Jan. 16, 1754.

BERS, Lipman, mathematician; b. Riga, Latvia, May 22, 1914; s. Abraham Isaac and Bertha (Weinberg) B.; student U. Zurich, U. Latvia, Dr. Rer. Nat., U. Prague, 1938; m. Mary Kagan, May 14, 1938; children—Ruth (Mrs. Leon Ehrenpreis), Victor. Came to U. S., 1940, naturalized 1949. Research asso. Brown U., Providence, 1941-45; asst. prof. Syracuse N.Y. U. 1945-48, asso. prof., 1948-51; mem. Inst. for Advanced Study, 1949-51; vis. prof., N.Y. U., 1951-53, prof., 1953-64, chmn. grad. sch. dept. math., 1959-63, Guggenheim fellow, 1959-60; prof. Columbia, 1964—; vis. Miller research prof. U. Cal., Berkeley, 1968. Chmn.-designate math. div. NRC. Recipient Fulbright award, 1959-60. Fellow A.A.A.S., Am. Acad. Arts and Scis; mem. Am. Math. Soc. (v.p. 1967-68), Math. Assn. Am., Nat. Acad. Scis. Author: Mathematical Aspects of Subsonic and Transonic Gas Dynamics, 1958; (with John, Schechter) Partial Differential Equations, 1964; also research, publs. on partial differential equations, gas dynamics, generalization of analytic functions, quasi conformal mappings, Riemann surfaces and discontinuous groups.*

BERSOHN, Richard, Am. chemist; b. N.Y.C., May 13, 1925; s. Abraham P. and Jessie (Schaff) B.; B.S., Mass. Inst. Tech., 1943; M.A., Harvard, 1947, Ph.D., 1949; m. Virginia Straus, Aug. 17, 1947; children—Malcolm Mark, David Louis. With Atomic Bomb Project Oak Ridge Nat. Lab., 1944-46; postdoctoral fellow Columbia U., N.Y.C., 1949-51, prof., 1959—; prof. Cornell U., 1951-59. Alfred P. Sloan fellow, 1961-65. Fellow Am. Phys. Soc.; mem. Am. Acad. Arts and Scis. Contbr. numerous articles to profl. jours. Research in theory of magnetic resonance spectroscopy, optical pumping and emission spectroscopy. Home: 5271 Independence Av., Bronx, N.Y. 10471. Office: Havemeyer Hall, Columbia U., N.Y.C. 10027.*

BERSON, Solomon Aaron, Am. physician; b. N.Y.C., Apr. 22, 1918; s. Jacob and Cecelia (Lieberman) B.; B.S., Coll. City N.Y., 1938; M.D., N.Y. U., 1945, M.Sc., 1939; m. Miriam Gittelson, Dec. 24, 1942; children—Wendy, Deborah. George Blumenthal fellow N.Y. U., 1941-42; lectr. Hunter Coll., 1942-45; staff VA Hosp., N.Y.C., 1950—, chief radioisotope service, 1954—; sr. med. investigator VA, 1963—. Mem. nat. adv. council Nat. Inst. for Arthritis and Metabolic Diseases, NIH, 1961-64, mem. bd. sci. counselors, 1966—; editorial adviser Diabetes Lit. Index, 1966. Recipient Banting medal Am. Diabetes Assn., 1965, Eli Lilly award, 1957, William S. Middleton Med. Research award, 1960; Distinguished Alumnus award N.Y. U. Coll. Medicine, 1966. Fellow N.Y. Acad. Scis., N.Y. Acad. Medicine; mem. Assn. Am. Physicians, Am. Soc. Clin. Investigation, Am. Physiol. Soc., Harvey Soc. N.Y., Endocrine Soc., Soc. for Exptl. Biology and Medicine, A.A.A.S.; hon. mem. Peruvian, Chilean socs. endocrinology, Chilean Soc. Diabetes and Metabolic Diseases, Argentinean Soc. for Endocrinology and Metabolism. Research, numerous publs. on blood volume measurement with radioisotopes, kinetics of iodine metabolism and thyroid function, metabolism of albumin-iodine 131; demonstrated insulin antibodies in man using isotopic methods; developed radioimmunoassay for insulin, other peptide hormones. Home: 159 Yale St., Roslyn Heights, L.I., N.Y. Office: VA Hosp., Radioisotope Service, 130 W. Kingsbridge Rd., Bronx, N.Y. 10468.*

BERT, Paul, French zoologist, physiologist; b. Auxerre, (Yonne), France, Oct. 17, 1833; M.D., Faculty of Paris, 1863; D.-ès S., 1866; studied under Claude Bernard. Became prof. Faculty Sci., Bordeaux, France, 1867; named prof. gen. physiology, Sorbonne, Paris, France, 1871; became sec.-gen. Dept. Yonne, 1870; rep. in Legislature from Yonne, 1874-1886. French minister of public instruction, 1881-82; apptd. governor general of Tonkin and Annam, 1886. Recipient of grand biennial prize of French Academy of Sciences, 1857. Member of French Academy of Sciences, 1882, Société de biologie (became pres. 1878). Author: Notes d'anatomie et de physiologie comparées, 2 vols., 1867-70; La pression barometrique, 1877; La pression atmosphérique, 1878. Pioneered research in gas pressure and respiration; discovered caisson disease (the bends) was caused by the rapid liberation of nitrogen bubbles from the blood, 1871; contbd. to aviation medicine; studied toxic effects of oxygen at high pressure; proved symptoms of high altitude sickness was due to anoxemia, 1878; studied thermal death points, 1876; research on anesthesia and the grafting of animal tissues. Died Hanoi, Indochina, Nov. 11, 1866.

BERTALANFFY, Felix Dionysius, anatomist; b. Vienna, Austria, Feb. 20, 1926; s. Ludwig and Maria M. (Bauer) von Bertalanffy; student U. Vienna, Med. Sch., 1945-48; M.Sc., McGill U., Montreal, Que., Can., 1951, Ph.D., 1954; m. Gisele Lavimodière, Jan. 20, 1954. Med. faculty dept. anatomy U. Man., Winnipeg, Can., 1955—, prof., 1965—. Fellow Royal Microscopical Soc. (London), Pan American Cancer Cytology Society; member of Am., Canadian assns. anatomists, Am. Assn. for Cancer Research, Canadian Cytology Council, N.Y. Acad. Scis., Internat. Soc. for Stereology, A.A.A.S., Sigma Xi. Research, numerous publs. on normal and malignant cell proliferation, pulmonary histology, histochemistry and fluorescence microscopy, cancer diagnosis, cell kinetics of normal and regenerating tissues. Home: 886 Lindsay St., Winnipeg 9, Man. Can.*

BERTALANFFY, Ludwig von, Canadian biologist; b. Atzgerdorf, Austria, Sept. 19, 1901; s. Gustav and Charlotte (Vogl) von B.; student U. Innsbruck Vienna, Austria; Ph.D., U. Vienna, 1926; m. Maria M. Bauer, 1925; 1 son, Felix D., Faculty, U. Vienna, 1934-48; prof., dir. biol. research dept. U. Ottawa (Ont., Can.), 1949-54; dir. biol. research Mt. Sinai Hosp., vis. prof. U. So. Cal., Los Angeles, 1955-58; Sloan vis. prof., mem. research dept. Menninger Found., Topeka, 1958-60; prof. theoretical biology U. Alta., Edmonton, Can., 1961—. Fellow, Notgemeinschaft der Deutschen Wissenschaft, 1930-32, Rockefeller Found., 1937-38, Lady David Found., 1949, Center for Advanced Study in Behavioral Scis., Stanford, Cal., 1954-55, others. Fellow A.A.A.S.; mem. Internat. Acad. Cytology (cons.), Soc. for Gen. Systems Research (past v.p.), N.Y. Acad. Scis., Am. Soc. Naturalists, Canadian Physiol. Soc. Author numerous books including: Modern Theories of Development, 1933, 62; Theoretische Biologie, 1932, 42, 51; Problems of Life, 1952, 60; also numerous monographs, articles. Editor: Handbuch der Biologie, 1942—; General Systems, 1956—. Research in organismic conception in biology, gen. system theory, fluorescence cytodiagnosis of cancer, cell and comparative physiology, philosophy of sci. Home: 10929 86th Av., Edmonton, Alta., Can.*

BERTAPAGLIA, Leonardo, Italian physician; prof. surgery, Padua, Italy, circa 1420; pioneer in practice of complete anat. dissections. Author commentary on Avicenna (includes numerous personal observations revealing wide knowledge of anatomy), printed 1499. Died Padua, circa 1460.

BERTELE (Pertele), Georg August, German physician; b. Ingolstadt, Germany, Sept. 27, 1767; s. Vitus Bertele; m. Mrs. Fassmann; 5 children; m. 2d,

Walburg Berthold, 1 son, 2 stepsons; prof., Ingolstadt from 1793; Landshut, Germany, from 1800. Mem. Mineral. Soc. Jena. Author: Versuch einer Lebenserhaltungskunde, 1803; Lehrbuch der Minerographie, 1804; Handbuch der dynamischen Arzneimittellehre (used as textbook in several German univs.), 1805. Had unusual knowledge in chemistry, mineralogy, botany, zoology, pharmacy and pharmacology. Died Landshut, July 19, 1818.

BERTELLI, Timoteo, Italian physicist, seismologist; b. Bologna, Italy, Oct. 26, 1826; s. Francesco and Terese (Pallotti) B.; admitted to Barnabita Order, 1844; ordained 1850; lectr. on math. and physics in Order's houses at Moncalieri, Naples, Italy; lectr. Coll. della Querce, Florence, Italy, 1868-1905; apptd. dir. Vatican Obs., Rome, 1895; returned to Florence, 1898. Mem. Pontifical Acad. Nuovi Lincei (became pres. 1904). Founder microseismology; made a series of observations on earthquakes in Italy; made studies on compass, barometer, magnetic declination. Died Florence, Feb. 6, 1905.

BERTHELOT, (Pierre Eugène) Marcellin, French chemist; b. Paris, Oct. 25, 1827; pupil of Balard, Coll. de France, doctorate, 1854. Became mem. staff Coll. de France, 1851, prof. organic chemistry, 1865; named prof. organic chemistry École Supérieure de Pharmacie, 1859. In charge sci. defence Paris, 1870; elected senator for life, 1881; minister public instruction, 1886-87; minister foreign affairs, 1895-96. Member of the Academy of Medicine, French Acad. Scis., 1873 (permanent sec. from 1889). Fellow Royal Soc., 1877. Author many books including: Chimie organique fondée sur la synthèse, 2 vols., 1860; Sur la force de la poudre et des matières explosives, 1871; La synthèse chimique, 1875; Essai de mécanique chimique fondée sur la thermochimie, 2 vols., 1879; Les origines de l'alchimie, 1885; Science et philosophie, 1886; Collection des anciens alchimistes grecs, 3 vols., 1887-88; Introduction à l'étude de la chimie des anciens et du moyen age, 1889; La revolution chimique, Lavoisier, 1890; Thermochimie, 2 vols., 1897; Science et morale, 1897; Chaleur animale, 1899; Les carbures d'hyrogène, 1901. Helped found, also contbd. numerous articles to La Grande Encyclopedia. Pioneer in modern organic chemistry; one of 1st to synthesize organic substances that do not occur in nature, thus weakening old theory that vital force was necessary for existence of organic compounds; his research strengthened arguments of proponents of mechanistic view of life; developed law of distbn. for solutions in 2 nonmiscible solvents; designated terms exothermic and endothermic reactions; researched explosives extensively; a founder of thermochemistry; believed chemical phenomena can be explained in terms of mechanistic principles; stated that every spontaneous chem. reaction evolves heat; constructed calorimeter; demonstrated fixation of atmospheric nitrogen by bacteria in clay soils; devised method for liquefying gases. Died Paris, Mar. 18, 1907.

BERTHELOT, Paul Alfred Daniel, French physicist; b. Paris, France, Nov. 8, 1865; s. Marcellin Berthelot; D. in Phys. Sci., 1891. Asst. Mus. Natural History, from 1892; prof. pharmacy L'Ecole supérieure de pharmacie. Prix Jecker, 1898, prix Hughes, 1906 —Mem. French Acad. Sci., 1919, Acad. Medicine. Engaged in pyrometric research; founded new optic method using phenomena of interferences to evaluate temperature exactly by examining a luminous ray going through a gaseous mass, 1895-1900; devised method of limited denseness of determine accurately atomic weights, 1898; studied physico-chem. effects of light, especially ultraviolet, from 1910; reproduced assimilation of carbon in plants by use of ultraviolet rays. Died Mar. 8, 1927.

BERTHELSEN, Asger, Danish geologist; b. Aarhus, Denmark, Apr. 30, 1928; s. Oluf Valdemar and Charlotte (Jensen) B.; mag.scient., Copenhagen U., 1953, Dr.phil., 1960; postgrad. Neuchatel, 1954, Delft, 1956; m. Suoma Irene von Pahlman, Feb. 12, 1954; 1 child, Aino Grit. Sci. asst. Mineral. Mus., Copenhagen, 1953-56; with Geol. Survey Greenland, Copenhagen, 1956-61, state geologist, 1959-61; prof. geology Aarhus U., 1961-66, head Geol. Inst., 1961-66; reader Copenhagen U., 1960-62, prof. geology, head Inst. Gen. Geology, 1966—, chmn. geology dept., 1967—. Recipient Gold medal Copenhagen U., 1956. Mem. Danish Nat. Com. for Geology, I.U.M.P. (mem. working group on structural geology). Author: Geology of Tovgussap Nuna, 1960; (with Arne Noe-Nygaard) The Precambrian of Greenland, 1965; also articles. Research on rocks and structures of ancient and young folded mountain chains, such as W. Greenland, N. Norway, Himalaya. Home: 60 Österbrogade, 2100, Copenhagen Ö, Denmark.*

BERTHIER, Pierre, French geologist; b. Nemours, France, July 3, 1782; s. Pierre-Jean-Baptiste Berthier; ed. École Polytechnic, 1798-1801; joined Corps des mines, 1801; joined Council of Mines Lab., 1806; named engr. in chief, 1814; became prof., head lab. Sch. Mines, 1816; apptd. insp. gen., 1836. Grand medal of gold Central Agrl. Soc. France, 1853. Mem. French Acad. Scis., 1827. Author: Traité des essais par la voie sèche, 2 vols., 1833. First demonstration of lime in the pyrites of Guilt and of Wissant, 1818; described chloritic lime nodules from La Hève, 1820, zinc deposits of Combecave; research on iron ores, vegetables ashes, some problems of agr.; identified hydrate of alumina; discovered bauxite, precursor in treatment of titanium iron ores; analyzed the waters of Vichy, 1820. Died Paris, France, Aug. 17, 1861.

BERTHOLD, Arnold Adolph, German anatomist, physician; B. Soest, Germany, Feb. 26, 1803; s. Dieterich and Sophie (Adams) B.; degree medicine, Göttingen, Germany, 1823; studied comparative anatomy and zoology in Paris, France; m. Caroline Röder, 1827; 5 daus., 3 sons. Practiced in Berlin, 1924-25; pvt. lectr. Göttingen, also began practice medicine, 1925; prof., 1835; dir. Göttingen zool. collection from 1840; mem. Göttingen Acad. Scis.; articles in German physiol. and med. jours.; founder of hormone research; his most significant expt. was grafting testicular tissue onto castrated roosters proving they then displayed the same characteristics as normal roosters (1849). Died Göttingen, Feb. 3, 1861.

BERTHOLD, Gottfried, German botanist; b. Gahmen, Westphalia, Germany, Sept. 16, 1854; s. Franz Diederich Heinrich and Johanna Hermine Wilhelmine B.; m. Anna Brons; 2 sons, 1 dau.; prof., Göttingen, Germany, 1887-1923. Author Studien über Protoplasmamechanik, 1886; Untersuchungen zur Physiologie der Pflanzlichen Organisation, 2 vols., 1898-1904. First studied sea algae particularly those of Naples; later work was on the function of the cell from a phys. viewpoint (protoplasmamechanics); attempted to explain vegetable orgn. through causal, evolutionary anatomy avoiding teleological thought processes. Died Göttingen, Germany, Jan. 7, 1937.

BERTHOLLET, Claude Louis, chemist; b. Talloires, France, December 9, 1748; M.D. University of Turin (Italy), 1768; Paris, 1779. Pvt. physician to Phillip, Duke of Orleans; became superintendent of dyeing processes in France, 1784; a founder Ecole polytechnique, professor of chemistry, from 1794; science adviser to Napoleon in Egypt, 1798. Named senator by Napoleon, 1804, also was created count; created peer by Louis XVIII. Mem. French Acad. Scis., 1780; Fellow Royal Soc., 1789. Author: Eléments de l'art de la teinture, 1791; Recherches sur les lois de l'affinité, 1801; Essai de statique chimique, 1803; (with others) Méthode de nomenclature chimique, 1787. Proponent of Lavoisier's new theories of combustion; helped Lavoisier revise system of chem. nomenclature (still in use); known for analysis of ammonia; discovered bleaching properties of chlorine, 1789; showed presence of nitrogen in animal matter, 1791; determined composition of prussic acid; disproved Lavoisier's theory that all acids contain oxygen by analyzing hydrocyanic acid; experimented with dyeing; 1st chemist to introduce concept of mass as factor in determining chem. affinities; discovered reversibility of chemical reactions; discovered enzyme saccharase. Died Arcueil, France, Nov. 6, 1822.

BERTHOLON, Abbé Pierre-Nicolas, French physician; b. Lyons, France, circa 1742; ed. at Lazarist Sch., Lyons, 1756. Became tchr. theology at sem. of Béziers, 1773; prof. exptl. physics at Montpellier, France; prof. physics Sch. of Hérault. Mem. Acad. Lyons, Acad. Montauban. Recipient prizes for his work. Author: De l'éléctricité des végetaux, 1783; Des avantages que la physique et les arts peuvent retirer des aérostats, 1784; Théorie des incendies, 1787; La nature considerée sous ses aspects divers, ou journal d'histoire naturelle, 1787-89. First discussion on effect of electricity on plant movements, 1783; classified diseases into elec. and non-elec.; had lightning rods installed in Lyons, 1780, also Paris; invented portable lightning rods; used electricity for treatment of diseases. Died Apr. 21, 1800.

BERTHON, Edward Lyon, Brit. inventor; b. London, Eng., Feb. 20, 1813; s. Peter Berthon; med. studies under James Dawson, Liverpool, Eng., 1828; student Coll. Surgeons, Liverpool, Dublin, Ireland; M.A., Magdalene Coll., Cambridge (Eng.) U., 1849; m. June 4, 1834. Curate Lymington, Eng., 1845; held living Holy Trinity, Fareham, Eng., 1847-55, Romsey, Eng. Author: A Retrospect of Eight Decades, 1899. Inventor screw propeller for ships, nautical log, both condemned by Admiralty, 1834, 35; designer collapsible boat, approved by Admiralty, 1873. Died Oct. 27, 1899.

BERTHOUD, Ferdinand, inventor; b. Plancemont, Neuchâtel, Switzerland, Mar. 19, 1727; moved to Paris, 1745; clockmaker; mem. Nat. Inst., Fellow Royal Soc., 1764. Author: Essais sur l'horlogerie, 1763; Histoire de la mesure du temps par les horloges, 1802; made chronometers of high quality and accuracy; constructed elaborate ship's clock. Died June 20, 1807.

BERTILLON, Alphonse, French anthropologist-criminologist; b. Paris, France, April 23, 1853; s. Louis Adolphe Bertillon. As an expert on handwriting testified to the guilt of Alfred Dreyfus at both treason trials. Author: of Ethnographie Moderne des Races Sauvages, 1883; Photographie Judiciare, 1890; Identification anthropométrique, 1893; created a system of identifying persons anthropometrically by compiling measurements of selected parts of the body (Bertillon system); now largely replaced by fingerprint method, 1886. Died Paris, Feb. 13, 1914.

BERTILLON, Louis Adolphe, French anthropologist; b. Paris, France, Apr. 1, 1821; children—Jacques, Alphonse. Began career as physician; apptd. insp. gen. of benevolent instns. after revolution of 1870; a founder Paris Sch. Anthropology, became prof. demography, 1876; dir. Statis. Office of Paris. Author: Demographie figurée de la France (a statis. study of France's population), 1874. Died Feb. 28, 1883.

BERTIN, Exupère-Joseph, French physician, anatomist; b. Tremblay, France, Sept. 21, 1712; s. François and Marie (Piètre) B.; M.D., Rheims, France, 1737; served as the 1st physician to Moldavia, to 1744. Member of the French Academy of Sciences, 1748. Author: Traité d'ostéologie, 1754; Consultation sur la légitimité des naissances tardives; treatises on circulation of blood in fetus, anatomy of lacrymal system. Died Feb. 25, 1781.

BERTIN, Henri Léonard-Jean-Baptiste, French agronomist; b. Perigueux, France, Mar. 24, 1720; s. Jean and Lucrèce (de Saint-Chamans) B.; lawyer Bordeaux, 1741. Lt.-gen. police Paris, France, 1757-59; sec. finance France, 1759-61; French minister in charge of mines, manufactures, agr., 1761-74; attached to Fgn. Ministry, from 1774. Comdr. Holy Spirit and St. Michel. Mem. French Acad. Sci., 1763, Acad. Belles-lettres, Soc. Agr. Responsible for improving porcelain manufacture at Sèvres, France; founded vet. sch. Lyons, France; created Cabinet des Chartes. Died Sept. 16, 1792.

BERTIN, Léon, French zoologist; b. France, Apr. 8, 1896; licence en sci., 1917; aggregation, 1920; with herpetology lab. Paris Natural History Mus., became asst. dir. mus., 1938, head prof. herpetology, 1954. Author: L'atlas des poissons marins; Regarde sur la nature et ses mystères; La systematique et la biologie des épinoches, 1921. Specialized in abyssal fish; studied systematically fresh water stickleback fish; made original contbns. to biogeography of fish. Died Feb. 6, 1954.

BERTIN, Louis-Emile, nautical engr.; b. Nancy, France, Mar. 23, 1840; s. Pierre Julien and Anne (Dermier) B.; grad. in naval constrn. Ecole Polytechnique, Paris, France, 1863. Worked on prodn. arty. Franco-Prussian War; sent by French govt. to work for Japanese navy, 1885-90; engr. Toulon, France, also prof. steam engines Ecole du Genie maritime, 1890-92; head director French naval constructions, from 1892. Member of the French Academy of Sciences, 1903 (president 1922), Académie de la Marine. Author: Chaudières marines; Les grandes guerres civiles du Japon, 1894; La marine de Etats-Unis, 1896. Inventor div. of vulnerable sects. of battleships into water-tight compartments, 1871; studies on launching torpedoes by use of compressed air. Died Oct. 22, 1924.

BERTIN, Pierre Augustine, French physicist; b. Besancon, France, 1818; prof. in Strasbourg, France; research on magnetic rotating polarization; worked out refraction indices of dark plates with the microscope in 1849. Died 1884.

BERTIN, Réné-Joseph-Hyacinthe, French physician; b. Gohard, France, Apr. 10, 1767; s. Exupère-Joseph Bertin; student Paris, France; M.D., U. Montpellier (France), 1791. Began med. service in the army, 1791; sent to Plymouth as insp.-gen. health service of French prisoners, 1798; mem. campaign in Prussia and Poland, 1807; became chief physician l'hôpital Cochin, and later of l'hôpital des vénériens; became prof. hygiene Faculty of Medicine, Paris, 1822. Author: Traité des maladies du coeur et des gros vaisseaux, 1824; Traité de la maladie vénérienne chez les nouveau-rés, les femmes enceintes et les nourrices, 1810. One of the 1st physicians of France to use auscultation with the stethoscope; studied different forms of hypertrophy; from his work the idea of eccentric and concentric hypertrophy of the heart arose. Died Gohard, Aug. 1828.

BERTINI, Anton Francesco, Italian physician; b. Castelfiorentino, Italy, Dec. 28, 1658; s. Bernardo and Verdiana (Barchetti) B.; doctorate in philosophy and medicine, U. Pisa (Italy), 1678; m. Teresa Ghini, circa 1693; 1 son, Giuseppe Maria Saverio. Began practice medicine, Florence, 1680; also lectr. practical medicine Santa Maria Nuova Hosp. Mem. Accademia della Crusca. Author: La medicina difesa della calumnie, 1699; Risposta apologetica, 1700; Lospecchio che noradula, 1707. Described current medicine and treatments for various diseases. Died Dec. 10, 1726.

BERTINI, Eugenio, Italian mathematician; b. Forli, Italy, 1846; student Bologna (Italy) Sch. Cremona, 1863-66; grad., Pisa, Italy, 1866-68. Served with Garibaldi in War of Independence; became prof. geometry Pisa, 1875, Pavia, Italy, 1881, Pisa, 1893; mem. (with Cremona, C. Segre, Castelnuovo, Enriques, Servi) Italian Sch. Author: Introduzione alla geometria proiettiva degl' iperspazii, 1906; i Complenti di geometria proiettiva, 1927; contbg. author: Ricerche di geometria sulle superficie algebriche (Enriques), 1893. Contributed to theory of algebraic surfaces, also to classification of birational transformations. Died 1933.

BERTINI, Giuseppe, Italian physician; b. Florence, Italy, Sept. 15, 1772; s. Anton Francesco and Maria (Maddalena) B.; grad. with doctorate in medicine U. Pisa (Italy), 1794. Began practice, Florence, 1796; named master in residence Hosp. Santa Maria Nuova, 1802; charge of fight against yellow fever epidemic,

Livorno, Italy, 1804; became prof. med. history Med. Sch., Santa Maria Nuova, 1805; joined faculty U. Pisa, also Pisan Accademia, 1810. Mem. Accademia of Georgotili, Soc. Columbaria Florence, Soc. Filaratrica (founding). His work in history of medicine led to establishing tradition as an integral part in training physicians. Died Mar. 19, 1845.

BERTINI, Giuseppe Maria Saverio, Italian physician; b. Mar. 10, 1695; s. Anton Francesco and Teresa (Ghini) B.; doctorate in medicine, U. Pisa (Italy), 1714; m. Anna Maria Pucciozzi, 1718; 1 son, Antonio. Began practice, Florence, Italy, 1715; mem. staff Hosp. Santa Maria Nuova. Mem. Bot. Soc. Florence, Florentine Coll. Physicians. Author: Delluso esterno e interno del Mercurio, 1744; also articles. Studied treatment of fevers and diseases, especially internal and external use of mercury as curative (treatment rejected in his time, but later proved beneficial). Died Apr. 12, 1756.

BERTINO, Joseph Rocco, Am. physician; b. Port Chester, N.Y., Aug. 16, 1930; s. Joseph and Madalaine (Posillipo) B.; student Cornell U., 1947-50; M.D., Downstate Med. Center N.Y., 1954; m. Mary Patricia Hagemeyer, Sept. 29, 1956; children—Frederick, Amy Marie, Thomas Allen, Paul Phillip. USPHS Research fellow U. Wash. Sch. Medicine, Seattle, 1958-61; faculty Yale Sch. Medicine, 1961—, asso. prof. pharmacology and medicine, 1964—. Cons. USPHS, 1966—. N.Y. State scholar for medicine, 1950-54. Mem. Am. Soc. for Clin. Investigation, Am. Soc. Hematology, Biol. Chemists, Pharmacology and Therapeutics. Research, numerous publs. on mechanisms of action of folic acid antagonists, in man especially in leukemia and in cancer, drug resistance, cancer chemotherapy, metabolism of normal and leukemic white cells. Home: 384 Hill St., Hamden, Conn. 06514. Office: 333 Cedar St., New Haven 06511.*

BERTRAM, Douglas Somerville, entomologist; b. Glasgow, Scotland, Dec. 21, 1913; s. William Robertson and Katherine (McCaskill) B.; B.Sc. with honors U. Glasgow, 1935, Ph.D., 1940, D.Sc., 1964; m. Louisa Menzies McKellar, Aug. 2, 1947 (dec. July 1956); children—Rachel Ann, Katherine Louise. Strang-Steel scholar U. Glasgow, 1936, demonstrator dept. zoology, 1937-38; faculty Liverpool (Eng.) Sch. Tropical Medicine, 1938-40, lectr., 1946-48; reader dept. entomology London (Eng.) Sch. Hygiene and Tropical Medicine, 1948-56, prof., dir. dept. entomology, 1956—. Fellow Inst. Biology, Royal Soc. Tropical Medicine and Hygiene (councillor 1953—, treas. 1962—), Royal Entomol. Soc. London (councillor 1961-62); mem. Soc. for Exptl. Biology, Research Def. Soc., Brit. Soc. for Parasitology. Research, publs. on biology of mosquitos, ticks and mites and their role as carriers of filarial worms, viruses, malaria, aspects of insecticides, repellents and chemotherapy in pest and disease control. Office: Dept. Entomology, London Sch. Hygiene and Tropical Medicine, Keppel St., London, W.C.1, Eng.*

BERTRAND, Alexander, French archeologist; b. Paris, 1820; ed. Ecole normale supérieur, French Sch. at Athens, Greece. A founder Mus. Saint-Germain, became curator, 1862; instr. nat. archeology Louvre Sch., from 1883. Mem. Acad. Recordings and Belles-Lettres. Author: Essai sur les dieux protecteurs des héros grecs et troyans dans l'Iliade, 1857; Etudes de mythologie et d'archéologie grecque, 1858; les Voies romaines en Gaul, 1863; Archéologie celtique et gauloise, 1876; Cours d'archeologie nationale; la Gaule avant les Gaulois, d'après les monuments et les textes, 1884-86; la Religion des Gaulois, 1897; (with S. Reinach) les Celtes dans les vallées du Po et de Daunbe, 1894. Died 1902.

BERTRAND, Charles-Eugene, French paleo-botanist; b. Paris, Jan. 2, 1851; Docteur ès Scis., Paris, 1874; prof. botany Faculty Sci., Lille, France, 1881; dir, Archives botaniques du Nord de la France, 1881-87; mem French Acad. Sci.; recipient priz Kuhlmann, 1883, prix Boidin, 1875. Author: Anatomie comparée des tiges et des feuilles chez les gnétacies et chez les conifères, 1874; Les charbons, humiques et les charbons de purins, 1898; Recherches sur les tméspterides, 1883; Recherches sur les téguments séminaux des phanérogames (prix Boidin). Research on coals and geology of rocks of organic origin; improved knowledge of fossil types and their filiation; discovered (with B. Renault) that certain types of coal (bogheads) consist of an agglomeration of algae; studied role of vegetal and animal microorganisms in formation of hard coal. Died Aug. 18, 1917.

BERTRAND, Élie, Swiss naturalist; b. Obbe, Switzerland, 1712; pastor ch. Berne, Switzerland, from 1744; mem. acads. Stockholm, Sweden, Berlin, Germany, Florence, Italy, Lyons, France; founder Econ. Soc. Verdun, France. Author: Memoirs on the Interior Structure of the Earth, 1752; A General Dictionary of Fossils, 1763; other sci. works; also moral and religious treatises. Observer earthquake, 1755.

BERTRAND, Gabriel, French biochemist, bacteriologist; b. Paris, France, May 17, 1867; D. ès Sc., Paris, 1894; also studied at Mus. Natural History, Faculty of Pharmacy and Faculty of Sci., Paris. Mem. staff Maquenne Lab. applied vegetal physiology, later at Desclaux Lab. chem. physiology; hon. prof. Faculty Sci., Paris; became head biochemistry sect. Pasteur Inst., 1900. Decorated officier Legion of Honor, Commandeur des Palmes Académiques; Mem. Acad. Medicine, French Acad. Scis., 1923; Royal Instn. Gt. Brit. (hon.), Denmark, Holland, Sweden royal acads. Author: le Xylose, sucre de bois, 1894. Discovered manganese is essential to plant growth but needed only in traces; elucidated role of co-enzymes as activators of aerobic and anaerobic muscle, 1897; research on antivenomous vaccination, oxidases, rain waters, bacteria of sorbose, milk preservation. Died June 20, 1962.

BERTRAND, Joseph Louis Francois, French mathematician; b. Paris, Mary. 11, 1822; s. Alexandre and Françoise B.; ed. École Polytechnique, 1839-41, École Des Mines, 1841-44; Dr. ès Sci., 1839; 1 child, Marcel Alexandre. Prof., Coll. Saint-Louis; admissions examiner l'École Polytechnique, later tutor in analysis; head lectr. l'École Normale; prof. gen. physics and math. Collège de France, titular prof., 1862. Mem. French Acad. Scis. (perpetual sec., from 1874, succeeded J. B. Dumas, 1884), 1856. Fellow Royal Soc., 1875. Author: Sur la convergence des séries, 1842; Sur la propogation du son, 1846; Sur la théorie des phenomènes capillaries, 1848; Traité de calcul différentiel et intégral, 1864-70; Thermodynamique, 1887; Calcul des probabilités, 1889; Leçons sur la théorie mathématique de l'electricité, 1890; Blaise Pascal, 1890. Research on infinitesimal analysis, theory of functions, thermodynamics and probability theory. Died Apr. 5, 1900.

BERTRAND, Léon-(Louise-Théophile), French geologist; b. Arville, France, July 30, 1869; prof. geology Faculty Scis., Paris, France, Ecole centrale des arts et manufactures. Mem. French Acad. Scis., 1945. Research on structure of Pyrénées, oil of Madagascar, refractory materials, tectonics of Provence. Died Paris, Feb. 24, 1947.

BERTRAND, Louis, Swiss mathematician, geologist; b. Geneva, Switzerland, Oct. 30, 1731; ed. under Euler in Berlin, Germany; children include Jean-Élie; prof., Geneva, 1761-92. Mem. Acad. Berlin. Author: De l'instruction publique, 1774; Renouvellements periodiques des continents terrestres, 1799; Éléments de geometrie, 1812. Worked on theory of parallels, also on algebra; gave 1st reliable definition of angular quantity, 1778. Died Geneva, May 15, 1812.

BERTRAND, Marcel-Alexandre, French geologist; b. Paris, France, July 2, 1847; s. Joseph-Louis-Francoise B.; ed. École polytechnic, Sch. Mines. Became chief engr. mines, Vesoul, 1872; joined Central Service of Geol. Mapping, 1878; prof. geology Sch. Mines. Mem. French Acad. Scis., 1896, Geology Soc. (became pres. 1891). Author: La Chaîne de Alpes et la formation du continent européen, 1887; la Grande nappe de recouvrement de la basse Provence, 1899; Étude sur le bassin Louiller du Gard, 1900; Lignes directures de la geologie de la France, 1894. Research on mountain formation; showed profile of Alps is due to compression and horizontal displacement. Died Paris, Feb. 13, 1907.

BERTRAND, Paul, French botanist; b. Loos-es-Lille, France, 1879; s. Charles-Eugene Bertrand; prof. botany, Lille, then at Paris Mus. Natural History. Author: Liste provisoire des sphenoptéris du bassin houiller du Nord de la France, 1944. Studied plant fossils; made detailed studies of boghead algae; classified fossil ferns; demonstrated that flora in different coal fields in Europe varied. Died 1944.

BERTRUCCIO, Niccolo, Italian physician, anatomist; b. Lombardy, Italy; s. Ronaldino Bertuccio; studied under Mendino de'Luzzi; m. Giacoma. Prof. logic and medicine, Bologna, Italy; tchr. of Guy de Chauliac; continued human dissections in classrooms begun by Mon Dino. Author: Compendium; Commentary on Hippocrates' Aphorisms and on Galen's Tegni. Divided human organs into 10 categories; described brain; on diseases, he lists rational treatment, empiric treatment, treatment according to Qanun, then principal symptoms. Died 1347.

BERWALD, Ludwig, Czechoslovakian mathematician; b. Prague, Czechoslovakia, Dec. 8, 1883; s. Max and Friederike (Fischel) B.; ed. Munich, Germany; m. Hedwig Adler, 1915. Became mem. faculty German U. Prague, 1919, asso. prof., 1922, prof., 1927. Author, Krümmungseigenschaften der Brennflächen eines geradlinigen Strahlensystems und der in ihm enthaltenen Regelflächen, 1909. Contbd. numerous articles to math. jours. Investigated differential geometry, calculus of variations, theory of affined planes, Riemann geometry, parallel transference in spaces with gen. mass determination (used in devel. Cartan and Finsler theories). Deported by Nazis to ghetto of Lodz, Poland, 1941. Died Apr. 20, 1942.

BERWICK, William Edward Hodgson, Brit. mathematician; b. Bradford, Eng., 1888; ed. Clare Coll., Cambridge, fellow. 1921-24; D.Sc.; m. Daisy May Thomas, 1923. Asst. lectr. U. Bristol, 1911-13; lectr. math. U. Coll., Bangor, Wales, 1913-20; lectr. Leeds (Eng.) U., 1920-26, reader math. analysis, 1921-26; Prof. math. U. Coll. N. Wales, Bangor, 1926-41, emeritus, 1942—; tech. staff anti-aircraft exptl. sect. Munitions Inventions Dept., Portsmouth, Eng. Research and publs. on reworking of arithmetic of binary cubics, 1912. Died May 13, 1944.

BERZELIUS, Jöns Jacob Baron, Swedish chemist; b. Väfversunda, Sorgard, Sweden, Aug. 20 or 29, 1779; M.D., U. Uppsala, 1802; largely self-educated in chemistry; m. Johanna Elizabeth Poppius, 1835. Asst. prof. botany and pharmacy U. Stockholm, 1802-07, prof., from 1807; prof. chemistry and pharmacy, sch. surgery U. Stockholm, 1815-32. Mem. commn. on gun-powder, 1811-18; became mgr. Acad. for Agr., 1811; visited Eng., 1812, France, 1819. Recipient Copley medal Royal Soc. London, 1836; knighted, 1818, became baron, 1835. Mem. Swedish Acad. Scis., 1808, pres., 1810, perpetual sec., 1818; fgn. mem. Royal Soc. London, 1813; corr. chemistry sect. French Acad. Scis., 1816, fgn. asso., 1822. Author works including: Lehrbuch der Chemie (chemistry textbook), 1803; Theory of Chemical Proportions and the Chemical Action of Electricity, 1814; about 250 original memoirs. Writer, editor Swedish Acad.'s Economiska Annaler, 1807-09; published yearly rev. of chem. progress in which he editorialized on work of others, 1821-49. Did exptl. work in almost all branches of chemistry; determined by analysis the exact elementary constn. of some 2000 compounds, thereby validating the law of definite proportions; used oxygen as his standard of atomic weight; composed generally accurate table of atomic weights; collaborated with Hisinger in experimenting on electrolysis; proposed a dualistic or electrochem. theory (founded on supposition that atoms and radicals are electrically polarized) which exerted profound influence on chem. thought, 1812; one of chief founders of radical theory; introduced modern system of chem. notation; pioneered gravimetric analysis; discovered ceria, the oxide of cerium, selenium, silicon and thorium, 1828; obtained as elements, calcium, barium, stronium, tantalum and zirconium; investigated compounds of fluoric acid, and introduced term halogen; advocated use of a purely chem. system of mineral classification; developed improved analytical methods, introduced various improvements in lab. equipment. Died Stockholm, Aug. 7, 1848.

BESANÇON, François, French physician; b. Paris, France, Sept. 16, 1927; s. Justin Louis and Madeleine (Delagrange) B.; Docteur en Médecine Faculty of Médicine Paris, 1955; Docteur Ès Sciences Naturelles, Faculty of Scis. Paris, 1957; m. Béatrice Hoppenot, Apr. 5, 1951; children—Odile, Paul, Jean, Pascale, Hélène. Physician, Hôpitaux de Paris, 1961-66; prof. agrégé Faculty of Medicine Paris, since 1961—; chief cons. medicine Hôpital Lariboisiere, Paris, 1966—. Mem. Medicale des Hôpitaux, Gastro-enterologie, Therapeutique, Hydrologie. Author: L'anemie pernicieuse, 1956; also numerous articles. Research on exptl. lithiasis induced by dehydrocholate, new theory of megaoesophagus, pH metry in motility studies of small bowel, cineradiometry, exptl. incomplete cholestasis, serotonin as a physiol. factor of bowel motility, orthochloroprocainamide, pernicious anemia. Home: 14 Bld Emile Augier, Paris 16. Office: hôpital Lariboisiere, Paris 10, France.*

BESANEZ, Baron Eugen Franz Gorup von, see von Besanez, Eugen Franz Gorup.

BESANT, William Henry, English mathematician; b. Portsmouth, Eng., Nov. 1, 1828; s. Walter Besant; B.A. (sr. wrangler, 1st Smith prizeman), St. John's Coll., Cambridge (Eng.) U., 1850, Sc.D., 1883; m. Margaret Elizabeth Willis, 1861 (dec. 1911); 2 sons, 1 dau. Lectr., St. John's Coll., Cambridge U., for 35 years, also fellow; pvt. tutor. Moderator, Math. Tripos, 1856, 85; examiner U. London (Eng.), 1859-64; lectr., examiner various univs. Fellow Royal Soc., 1871. Author: Treatises on hydro-mechanics, geometrical conics, dynamics, roulettes and glissettes. Contbd. papers to Quarterly Jour. Math., Messanger Math., Ednl. Times. Died June 2, 1917.

BESCH, Paige Keith, Am. pharmacologist; b. nr. San Antonio, June 23, 1931; s. Henry Roland and Monette Helen (Kasten) B.; B.S., Trinity U., 1956, postgrad.; Ph.D., Ohio State U., 1960; m. Norma C. Fredenberg, Sept. 20, 1957; children—Kirsten C., Konrad N. Sr. biochemist clin. lab. Robert B. Green Meml. Hosp., San Antonio, 1954-55; sr. research asst. to chmn. dept. endocrinology S.W. Found. Research, Edn., San Antonio, 1955-58; research asso. biochemistry Columbus (O.) Psychiat. Inst. and Hosp., 1958-59; faculty Ohio State U. Med. Sch., Columbus, 1958—, asso. prof. dept. obstetrics, gynecology, 1966—; dir. div. steroid research dept. obstetrics, gynecology U. Hosp., 1960—; dir. steroid research labs. Grant Hosp. Research, Edn. Found., 1961-64; co-dir. USPHS Grad. Endocrine Tng. Grant, 1963—; USPHS Career Devel. Award fellow Nat. Inst. Child Health and Human Devel., 1964—; dir. Med. Research Cons., Inc.; cons. to indsl. firms, govt. agys. Ayerst Travel grantee Endocrine Soc., 1965. Fellow Am. Coll. Clin. Pharmacology and Chemotherapy; mem. Am. Physiol. Soc., Am. Fertility Soc., Endocrine Soc., Am. Assn. Clin. Chemists, Sci. Research Soc. Am., Am. Chem. Soc., A.A.A.S., Soc. Exptl. Biology and Medicine, Sigma Xi. Author: (with R. D. Barry) Steroid Determinations, 1964; (with H.C. Damm) Handbook of Biochemistry and Biophysics, 1966; (with others) Methods and References in Biochemistry and Biophysics, 1966. Publs. on work on in vivo and in vitro metabolism of steroid hormones, oral contraceptives, fetal pharmacology. Home: 5375 Banbury Dr., Worthington, O. 43085. Office: 650-N. Univ. Hosp., Columbus, O. 43210.*

BESICOVITCH, Abram Samoilovitch, mathematician; b. Berdiansk, Russia, Jan. 24, 1891; s. Samuel Abramovitch and Eve Lorn (Sauskan) B.; grad. U. St. Petersburg, 1912; m. Valentina Alexandrovna Oenissova, 1928. Tchr., U. Perm, 1914-20; dozent U. St. Petersburg, 1920-25, also prof. math. Inst. Ways and Communications and at Pedagogical Inst. and Naval Engring. Sch.; lectr. math. U. Liverpool (Eng.), 1926-27; Cayley lectr. math. Cambridge (Eng.) U., 1927-50, Rouse Ball prof. math., 1950-58; vis. prof. U. Pa., 1958-62, Dartmouth, 1962-63, Ore. State U., 1964-65, U. Wis., 1965, Cornell U., 1965. Recipient Adams prize, 1932; De Morgan medal London Math. Soc., 1950; Sylvester medal, 1952. Fellow Royal Soc., 1934. Author: Almost Periodic Functions, 1932; also papers. Made important contbns. to study of probabilities, theory of almost periodic functions, complex variables, differentiation in gen. theory of functions of real variable; also studied surfaces and theory of numbers.

BESKOW, Gunnar, Swedish geologist; b. Stockholm, Sweden, July 10, 1901; s. Natanael and Elsa (Maartman) B.; ed. univs. Uppsala, Stockholm; Dr. phil., U. Stockholm, 1929; m. Inga Ericsson, June 4, 1927; children—Dag, Alf. Became research geologist Geol. Survey Sweden (SGU), 1927; with reorganized geol. dept. Swedish Hwy. Research Inst., from 1932; prof. geology Chalmers Tech. U., Gothenburg, Sweden, 1949-67, emeritus, 1967—. Author: Södra Storfjället, 1929; Soil Freezing and Frost Heaving, 1935, English transl., 1947; also papers. Research on petrological equipment and tectonic architecture of Scandinavial Alps, soil physics and tech., frost action, processes of heaving and flowage, mechanics of destruction of foundations and hwys., oil-spill pollution of water. Home: 37 Saltholmsgate, Gothenburg, Sweden.*

BESLER, Basilius, German botanist; b. Nuremberg, Germany, Feb. 13, 1561; s. Michael Besler; m. Rosine Flock, Jan. 31, 1586; m. 2d, Susanne Schmidt, Dec. 1, 1596; 16 children. Owner pharmacy, 1589-1629; became mem. city council, 1594; owner of pvt. bot. garden. The Genus, Beslera, was named in his honor; Wrote numerous descriptions of gardens and rare plants; promoted and popularized botany. Died Nuremburg, Mar. 13, 1629.

BESOZZI DE CASTELBESOZZO, Nob Alessandro, Italian chemist; b. Milan, Italy, Dec. 21, 1898; s. Celeste and Enrichetta (Belloni) B. di C.; Ph.D. in Chemistry, U. Pavia; m. Olga Bonelli, May 24, 1930. Owner lab. for analysis and research. Mem. Assn. Chemists of Lombardia. Research and publs. on petroleum and nuclear chemistry, pigments. Home: via Statuto, 23. Office: via Leopardi, 12, Milan, Italy.

BESREDKA, Alexander, Russian bacteriologist, immunologist; b. Odessa, Russia, 1870; dir. Pasteur Inst., Paris. Author: Immunisation locale, 1925; Anaphylaxie et Antianaphylaxie, 1917. Studied serological reactions. Died Paris, 1940.

BESSEL, Friedrich Wilhelm, German astronomer, mathematician; b. Minden, Prussia, July 22, 1784; s. Carl Friedrich and Ernestine (Schrader); m. Johanna Hagen, 1812; 2 sons, Fritz Karl, Erich, 3 daus. Began career as accountant; became insp. Obs., Lilienthal, 1806; named dir. Obs., Königsberg (now Kaliningrad, USSR), 1810, also prof. U. Königsberg. Mem. French Acad. Scis., 1816. Fellow Royal Soc., 1825. Lalande prize of French Acad. for calculations on comet of 1807; Berlin Acad. prize for work on precession of equinoxes. Author: Fundamenta astronomiae, 1818; Tabulae Regiomontanae, 1830; Determination de la longueur du pendule simple à secondes; Tableau des recherches faites de 1835 à 38 . . . ; Astronomische Untersuchungen, 1841-42; Mesure de la parallaxe de la première composante de 61 Cygne; Biographie d'Olbers; Lectures populairs sur des questions scientifiques, 1848; (with Bayer) Mesure d'un degré dans la Prusse Orientale. Pioneer in exact astron. measurements; recalculated orbit of Halley's comet, 1804; introduced personal equation (differences in observational reports due to personal errors and brightness of star), 1833; worked out theory of instrumental errors; developed method for calculating solar eclipses; in determining parallax of star 61 Cygni, 1838, he made 1st authenticated measurement of a star's distance, also confirmed Copernican theory by showing earth's motion through space; predicted existence of planet (Neptune) beyond Uranus, 1840; 1st to suggest existence of dark stars; worked on star catalogs; studied atmospheric refraction; invented math. functions (Bessel functions) used in math. physics and astronomy; increased no. of accurately determined stars to 50,000. Died Königsberg, Mar. 17, 1846.

BESSEMANS, Joseph Francois Antoine Albert, Belgian physician; b. Saint-Trond, Belgium, Feb. 16, 1888; M.D., U. Louvain (Belgium), 1912; hon. degrees Montpellier, Strasbourg, Nancy, Lille, Lyons (all France) univs., U. Jassy (Rumania); m. Marie Jeanne Louisa Wiot, 1942; 1 son. Physician, prin. health insp. Belgium, 1922—; prof. Sch. Medicine, U. Ghent (Belgium), 1924—, now emeritus; prof. Sch. Criminology and Sci. Police Procedures, Ministry of Justice, Brussels, Belgium, 1926; Founder Belgian Com. for Sci. Investigation of Reputedly Paranormal Phenomena, 1949. Recipient 3 awards French Acad. Medicine, Distinguished Service Key Am. Congress of Physiotherapy of Medicine. Hon. or corr. mem. med., pub. health, sci. orgns. U. S., Europe. Contbr. articles on bacteriology, parasitology, hygiene, criminology. Research parapsychology, 1913—, research phenomena in fields of clairvoyance, water-divining, psychometry and fakirism (reports never having been able to establish an authentic paranormal phenomenon).

BESSEMER, Sir Henry, British engineer, inventor; born Charlton, Hertfordshire, England, January 19, 1813; a son of Anthony Bessemer; married Anne Allen, 1833; 2 sons; 1 daughter. Manufacturer gold chains and type founder in father's bus.; began trading art work in white metal in London, 1830; began manufacture bronze powders, gold paint by an original process, 1840; established steel works specializing in gun making, Sheffield, 1859. Recipient Albert Gold medal Society Arts, 1872. Fellow Royal Society, 1879; knighted, 1879; member Iron and Steel Institute (a founder; president 1871-73), Instn. Civil Engrs. Author: On the Manufacture of Malleable Iron and Steel Without Fuel, 1856; Manufacture of Malleable Iron and Steel, 1859; Some Earlier Forms of Bessemer Converter, 1886; On the Manufacture of Continuous Sheets of Malleable Iron or Steel Direct from the Fluid Metal. Invented perforated die, 1833 type composing machine, swinging saloon for seagoing vessels, 1875; patentee combination of cast iron and steel, manufacture steel by blowing pressurized air or steam through melted pig iron (Bessemer process), devised system of rollers for printing and embossing papers; improved telephones. Died London, Mar. 15, 1898.

BESSEY, Charles Edwin, Am. botanist; b. Milton, O., May 21, 1845; s. Adnah and Margaret (Ellenberger) B.; B.Sc., Mich. Agrl. Coll., 1869, M.Sc., 1872; studied with Asa Gray at Harvard, 1872-73, 75-76; Ph.D., State U. Ia., 1879; LL.D., Ia. Coll., 1898; m. Lucy Athearn, Dec. 25, 1873; 1 son, Ernst Athearn. Prof. botany Ia. Agrl. Coll., 1870-84, acting pres., 1882; prof. botany, U. Neb., Lincoln, from 1884, acting chancellor, 1888-91, 1899-1900, and 1907, head dean, from 1909. Bot. editor Am. Naturalist, 1880-97, of Science, 1897—, Johnson's Cyclo., 1893—. Mem. Neb. Rural Life Commn., 1911-13. Pres. A.A.A.S., 1910-11, Bot. Soc. America, 1895-96, Soc. Promotion Agrl. Science, 1889-91, dept. natural science N.E.A., 1895-96, Am. Micros. Soc., 1902. Author: Geography of Iowa, 1876; Botany for High Schools and Colleges, 1880; The Essentials of Botany, 1884; Elementary Botanical Exercises, 1892; The Phylogeny and Taxonomy of Angiosperms, 1897; Elementary Botany, 1904; Plant Migration Studies, 1905; Synopsis of Plant Phyla, 1907; The Phyletic Idea in Taxonomy, 1908; Outlines of Plant Phyla, 1909, 11, 12, 13. Pioneer in use simple lab. methods in teaching botany. Died Feb. 25, 1915.

BESSIS, Marcel Claude, physician; b. Tunis, Tunisia, Nov. 15, 1917; M.D., Sch. Medicine, Paris, France, 1944; m. Claude Perrot, Oct 20, 1952; children—Isabelle, Sophie, Frederique. Dir. research labs. Nat. Center Blood Transfusion, Paris, 1946-66; dir. Inst. for Cell Pathology, Hopital Bicetre, Paris, 1966—; prof. Sch. Medicine, Paris. Mem. Nat. Com. Research, 1960—. Decorated Chevalier Légion d'Honneur. Hon. mem. Harvey Soc., A.C.P. Author: Cytology of the Blood, 1956. Research and publs. on hematological diseases and blood cells. Home: 2 rue Saint Simon, Paris (7). Office: Institut de Pathologie Cellulaire Hopital Bicêtre 94 Kremlin Bicêtre, France.*

BESSMAN, Samuel Paul, Am. biochemist; b. Newark, Feb. 3, 1921; s. Edward S. and Sara R. (Greenserg) B.; student Coll. William and Mary, 1938-41; M.D., Washington U., St. Louis, 1944; m. Alice Neuman, July 3, 1945; children—David, Ellen. Practice medicine, specializing in pediatrics, Balt., 1947—; faculty U. Md., 1954—, prof. pediatric research, 1960—, prof. biochemistry, 1965—; dir. research Rosewood State Hosp., Owings Mills, Md., 1962. Diplomate Am. Bd. Pediatrics. Fellow A.A.A.S., Am. Acad. Pediatrics; mem. Soc. Pediatric Research, Am. Chem. Soc., Am. Soc. Biol. Chemistry, Biochem. Soc., Sigma Xi, Alpha Omega Alpha. Editor: Biochemical Medicine, 1966. Editorial bd. Analytical Biochemistry, 1958—; editorial cons. Current Med. Digest, 1965. Contbr. numerous articles to profl. jours. Research in treatment of lead poisoning, new enzymes, new chem. abnormalities in mental retardation, theoretical basis for hepatic coma, theory of insulin action, nuclear synthesis of hemoglobin, oxygen control of hemoglobin synthesis in a cell free system. Home: 1410 Woodcliff Av., Catonsville, Md. 21228. Office: U. Md. Med. Sch., Balt. 21201.*

BESSY, Bernhard Frenicle de, see de Bessy.

BEST, Charles Herbert, physician, educator; born West Pembroke, Me., Feb. 27, 1899; s. Herbert Huestis and Luella (Fisher) B.; B.A., U. Toronto, 1921, M.A., 1922, M.D. (J.U. Mackenzie, Ellen Mickle fellow) 1925; D.Sc., U. London, 1928, U. Chgo., 1941, U. Cambridge, 1946, U. Oxford, 1947, Laval U., 1952, U. Me., 1955, Northwestern U., 1959; Docteur, honoris causa, U. Paris, 1945; M.D. honoris causa, U. Amsterdam, U. Louvain, U. Liege, 1947; LL.D., Dalhousie U., 1949, Queen's U., 1950, U. Melbourne, 1952, U. Edinburgh, 1959; hon. degrees, univs. Chile, Uruguay, San Marcos, Peru, 1951; Doctor Med. (L.C.), Central U. Venezuela, 1958; M.D., Aristotelian U. Thessaloniki, 1963; hon. doctorate Freie U. Berlin, 1966; m. Margaret Hooper Mahon, Sept. 3, 1924; children—Charles Alexander, Henry Bruce Macleod. Dir. insulin div. Connaught Labs., Toronto, Ont. Can., 1922-25, asst. dir. labs., 1925-31, asso. dir., 1932-41, hon. cons., 1941—; faculty U. Toronto, 1923—, prof., head, dept. physiology, 1929-65, prof., head Banting and Best Dept. Med. Research, 1941-67, emeritus director, 1967—. Mem. numerous Canadian government agencies; member advisory com. on med. research WHO, 1963—. Bd. sci. dirs. internat. health div. Rockefeller Found., Roscoe B. Jackson Meml. lab., Bar Harbor, Me. Recipient numerous awards and decorations including Dale medal Soc., Endocrinology, 1959, John Phillips Meml. award A.C.P., 1953, Banting medal Am. Diabetes Assn., 1949, Sir Joseph Flavelle medal for med. research Royal Society Canada, 1950. Fellow Royal Society, 1938, Royal College Physicians London, Royal Society Medicine London (honorary), Royal Coll. of Physicians Edinburgh (honorary), Royal Society Can., Royal Coll. Physicians Can.; mem. Pontifical Acad. Scis., Fgn. Asso., Nat. Acad. Scis., many diabetes assns. throughout world (hon. pres.); hon. mem. European Assn. for Study Diabetes, Acad. Medicine Toronto. Author: (with N. B. Taylor) Physiological Basis of Medical Practice, 1937; The Living Body, 1938; The Human Body, 1932; Selected Papers of Charles H. Best, 1963. Co-discoverer insulin, 1921; discovered histaminase; studies in insulin and diabetes, heparin and thrombosis, and choline and liver damage. Home: 105 Woodlawn Av. W., Toronto 7, Ont. Office: Charles H. Best Inst., 112 College St., Toronto, Ont., Can.*

BEST, Maurice McDonald, Jr., Am. physician; b. Louisville, Ky., Feb. 8, 1920; s. Morris M. and Harriet (Neat) B.; student Ind. U., 1938-41; M.D., U. Louisville, 1945; m. Martha Virginia Fox, May 8, 1948; children—Susan, Michael, Sally. Faculty, U.. Louisville, 1948—, now asso. prof. medicine. Cons. Eli Lilly & Co. Diplomate Am. Bd. Internal Medicine. Mem. A.C.P., Am. Coll. Cardiology, Central Soc. Clin. Research, So. Soc. Clin. Investigation, N.Y. Acad. Sci., A.A.A.S., Sigma Xi. Research, publs. on relationship between lipid metabolism and human atherosclerosis. Home: 1233 Vance Av., New Albany, Ind. 47150.*

BESTHORN, Emil, German chemist; b. Frankfurt, Germany, Dec. 10, 1858; s. Johann Karl and Margarete Mathilde (Andreae) B.; ed. Heidelberg, Germany, Munich, Chemist, Frankfort; chemist, state labs., Munich, from 1890; contbd. research on quinoline series, preparation of long unknown py-sulfonic acid of quinoline; discovered red dye made from quinoline-alpha-carbonic acid. Died Munich, Nov. 15, 1921.

BÉTANCOURT Y MOLINA, Augustin de, see de Bétancourt y Molina.

BETH, Richard Alexander, Am. physicist; b. N.Y.C., Jan. 14, 1906; s. Otto Carl Heinrich and Frieda (Unger) B.; B.S., Worcester Poly. Inst., 1927, M.S., 1929, D.Sc. (hon.), 1963; Dr.Phil.Nat. (Am. exchange fellow, Alexander von Humboldt fellow) U. Frankfort, Germany, 1932; m. Hettie Sprague, June 26, 1943; children—Richard Sprague, Hettie Viginia Beth. Instr, physics Worcester Poly. Inst., 1927-29, asst. prof., 1932-39; asso. prof. math Mich. State U., 1939-40; research asso. physics Princeton, 1934-35, com. on passive protection against bombing NRC, 1940-42, mem. div. 2 NDRC, 1942-44, dept. head sta., 1942-46; mem. ALSOS Mission, War Dept. G2, Europe, 1945; chmn., Perkins prof. physics Western Res. U., Cleve., 1946-55; Fulbright Guest prof. U. Innsbruck, Austria, 1954-55, U. Bonn, Germany, 1963-64; physicist, accelerator dept. Brookhaven Nat. Lab., Upton, N.Y., 1955—. Mem. council reps. Argonne Nat. Lab., Lemont, Ill., 1947-54; mem. fellowship panel NSF, 1954, 59, 60, com. on fortification design, 1943-45; Joint Army and Navy Exptl. and Test Bd., Fort Pierce, Fla., 1943-45. Recipient U. S. Certificate Merit, 1948. Fellow Am. Phys. Soc., Cleve. Physics Soc. (pres. 1950-51), Sigma Xi (pres. Worcester 1937-38, Western Res. 1949-50), Tau Beta Pi Theta Chi. Contbg. author: Handbook of Physics, 1958, Encyclopaedic Dictionary of Physics, 1961. Asso. editor, Am. Jour. Physics, 1953-56. Research on orthoganal functions, detection and measurement of angular momentum of light, terminal ballistics, observation of scale effect in projectile penetration, magnetic fields, particle accelerators. Home: 8 Leisurely Lane, Bellport, N.Y. 11713. Office: Brookhaven Nat. Lab., Upton, N.Y. 11973.*

BETHARD, William Frederick, Am. physician; b. Richmond, Ind., Sept. 17, 1916; s. Fred D. and Edith (Thompson) B.; student San Diego State Coll., 1934-37; B.S., U. Chgo., 1940, M.D., 1942; m. Margery M. Smith, June 6, 1942; children—William S., Barbara J. Gen. practice medicine, Coronado, Cal., 1946-47; asst. resident in medicine U. Chgo., 1947-50, Damon Runyon sr. clin. research fellow, 1950-52, asst. prof. medicine, 1952-54; mem. staff Scripps Clinic and Research Found., La Jolla, Cal., 1954-59, now vis. mem.; med. dir. Gen. Atomic Div., Gulf Oil Company, San Diego, Cal., 1959—; cons. dept. nuclear medicine and lab. for radiobiology U. Cal. at Los Angeles Sch. Medicine; cons. dept. radiology U. S. Naval Hosp., San Diego. Research, articles in jours. and books, on iron metabolism; effects of irradiation on tissues, role of trace elements in red blood cell metabolism. Home: 1505 Buckingham Dr., LaJolla, Cal. 92037. Office: 10955 John Jay Hopkins Dr., San Diego, Cal.*

BETHE, Hans Albrecht, physicist; b. Strasbourg, Alsace-Lorraine, July 2, 1906; s. Albrecht Theodore and Anna (Kuhn) B.; ed. Goethe Gymnasium, Frankfurt on Main, U. Frankfort; Ph.D., U. Munich, 1928; D.Sc., Bklyn. Poly. Inst., 1950, U. Denver, 1952, U. Chgo., 1953, U. Birmingham, 1956, Harvard, 1958; m. Rose Ewald, 1939; children—Henry, Monica. Came to U. S., 1935. Instr. theoretical physics, univs. of Frankfort, Stuttgart, Munich and Tubingen, 1928-33; lectr. univs. Manchester and Bristol (Eng.), 1933-35; faculty Cornell U., 1935-—, prof. 1937-—; dir. theoretical physics div. Los Alamos Sci. Lab., 1943-46. Head of Presdl. Study of Disarmament, 1958; mem. Pres.'s Sci. Adv. Com., 1956-64. Recipient A. Cressy Morrison prize, N.Y. Acad. Sci., 1938, 40; Presdl. Medal of Merit, 1946; Henry Draper medal Nat. Acad. Sci., 1948; Max Planck medal, 1955; Enrico Fermi award AEC, 1961; Nobel Prize for physics, 1967. Fgn. mem. Royal Soc. London; mem. Am. Philos. Soc., Nat. Acad. Sci. (subcom. on penetration charged particles in matter), Am. Phys. Soc. (pres. 1954), American Astron. Soc. Author: Elementary Nuclear Theory, 1947; Mesons and Fields, 1955; Quantum Mechanics of One- and Two-Electron Atoms, 1957; Intermediate Quantum Mechanics, 1964. Contbr. to (books) Handbuch der Physik, 1933; Reviews of Modern Physics, 1936-37; Phys. Review. Originator (with Heitler) theory of shower of electrons and protons in cosmic radiation, theory of origin of sun's energy which assumes formation of heavy atoms from lighter ones (Bethe-Weizäcker cycle); many contbns. to theory of nuclear reactors, especially concerning the energy prodn. of stars; developed 1st theory of electron-positron pair creation; improved theory of how charged particles interact. Address: Lab. Nuclear Studies, Cornell U., Ithaca, N.Y.

BETHEL, James Samuel, botanist; b. New Westminster, B.C., Can., Aug. 13, 1915; s. Joseph and Ruth (Wilkinson) B.; came to U. S., naturalized, 1916; B.S.F., U. Wash., 1937; M.F., Duke, 1939, D.Forestry, 1947; m. Marinelle Rives, June 8, 1941; children—Ruth Anne, James Samuel, John P. Faculty, N.C. State Coll., Raleigh, 1949-59, prof. wood tech., dir. wood products lab., 1950-58, acting dean Grad. Sch., head wood product dept., 1958-59; head spl. projects sci. edn. sect. NSF, Washington, 1959-62; asso. dean Grad. Sch. U. Wash., Seattle, 1962-64, dean, prof. Coll. Forestry, 1964-—. UN cons. Govt. of Yugoslavia, FAO, Rome, Italy, 1952; adviser Govt. P.R., Econ. Devel. Adminstrn., San Juan, 1956; cons. Gov.'s Research Triangle, N.C., 1957-59; chmn. Wash. state Bd. Natural Resources, 1966-—. Mem. Internat. Union Forestry Research Orgns., Assn. State Coll. and U. Forest Research Orgns. (v.p. 1967-—), Soc. Am. Foresters, A.A.A.S., Council Forestry Sch. Execs., Am. Inst. Wood Engrs. (pres. 1958-59), Forest Products Research Soc., Soc. Wood Sci. and Tech., Am. Inst. Biol. Scis., Assn. Tropical Biologists, T.A.P.P.I., Inst. Wood Sci., Sigma Xi. Author: (with Nelson C. Brown) Lumber, 1958; (with E. S. Harrar, A. J. Panshin) Forest Products 1962; also articles. Research on structure and property of wood and application of statis. methods and linear algebra to control of quality of wood processing operations. Home: 3816 E. Mercer Way, Mercer Island, Wash. 98040. Office: Coll. Forestry, U. Wash., Seattle, 98105.*

BETHENOD, Joseph, French elec. engr.; b. Lyons, France, Apr. 28, 1883. Cons. engr. Société Alsacienne de constructions mécaniques. Mem. French Acad. Scis., 1942. Proved that revolution diagram can be expressed as a function of the slip by a linear substitution or projective transformation. Died Feb. 21, 1944.

BETTELHEIM, Bruno, psychologist; b. Vienna, Austria, Aug. 28, 1903; Ph.D., Univ. Vienna, 1938; m. Trude Weinfeld, May 14, 1941; children—Ruth, Naomi, Eric. Came to U. S., 1939, naturalized, 1944. Research asso. Progressive Edn. Assn., U. of Chicago, 1939-41; asso. prof. psychology Rockford (Ill.) Coll., 1942-44; asst. prof. ednl. psychol. U. of Chicago, 1944-47, asso. prof., 1947-52, prof. since 1952, Rowley prof. edn., prof. psychology and psychiatry, 1963-—; prin. orthogenic sch. 1944-—. Fellow Am. Psychol. Assn., Am. Orthopsychiatric Assn.; mem. Am. Philos. Assn., Am. Assn. Univ. Profs., Am. Sociol. Assn., Chgo. Psychoanalytic Soc. Author: Dynamics of Prejudice (with Morris Janowitz), 1950; Love is Not Enough—The Treatment of Emotionally Disturbed Children, 1950; Symbolic Wounds, 1954; Truants From Life, 1955; The Informed Heart, 1960; Dialogues with Mothers, 1962; The Empty Fortress, 1967. Research on development, psychoanalysis of children; emotionally disturbed, mentally retarded children. Home: 5725 Kenwood Av., Chgo. 60637.

BETTELHEIM, Frederick Abraham, chemist; b. Györ, Hungary, June 3, 1923; s. Anton and Elizabeth (Gyarfas) B.; student U. Szeged (Hungary), 1943-44, 45-46; B.S., Cornell U., 1953; M.S., U. Cal. at Davis, 1954, Ph.D., 1956; m. Annabelle Ganz, June 10, 1947; 1 son, Adriel A.A. Came to U. S., 1951, naturalized, 1963. Chemist, Agr. Expt. Sta., Rehovoth, Israel, 1946-51; instr. U. Mass., Amherst, 1956-57; faculty Adelphi U., Garden City, N.Y., 1957-—, prof. chemistry, 1964-—. Lalor fellow, 1958. Mem. Am. Chem. Soc., Am. Phys. Soc., Fedn. Am. Socs. Exptl. Biology, N.Y. Acad. Scis. Contbg. author: Carbohydrates, 1967; Physical Chemistry of Biological Polyelectrolyes, 1967. Research, publs. on structure of polymers in solid state, transitions from solid state to gel and concentrated solutions by X-ray diffraction, dichroism, birefringence, light scattering, dielectric dispersion and high vacuum vapor sorption techniques. Home: 450 Garden Blvd., Garden City, N.Y. 11530.*

BETTENDORF, Anton Joseph Hubert Maria, chemist; b. Ensival nr. Verviers, Belgium, June 1, 1839; s. Peter and Hubertine (Rhenasteine) B.; m. Gabriele David, 1866; asst. to Hans Landolt in Bonn, Germany, 1861-63; then pvt. research; research on rare earths and arsenic; discovered regular yellow arsenic in 1867 and in 1869 a sensitive reagent on arsenic called Bettendorf's reagent (solution of tin chloride in smoking hydrochloric acid). Died Bonn, Oct. 12, 1902.

BETTERIDGE, Walter, English metallurgist; b. Birmingham, Eng., June 1, 1911; s. Thomas Betteridge and Alice M. (Johnson) B.; B.Sc. with 1st class honors in Physics, Birmingham U., 1931, Ph.D. in Physics, 1933, D.Sc. in Metallurgy, 1958; m. Margaret Allen, Oct. 10, 1936; 1 dau., Margaret Ann. Research physicist Joseph Lucas Ltd., 1933-36; asst. chief metallurgist engine div. Bristol Aeroplane Co., 1936-47; with Internat. Nickel Ltd., London, 1947-—, supt. platinum metals div., devel. and research dept., 1959-—. research mgr., 1965-—. Fellow Inst. Physics, Inst. Metallurgists. Author: The Nimonic Alloys, 1959. Research on high temperature alloy devel. Home: 114 Barnfield Wood Rd., Park Langley, Beckenham, Kent, Eng. Office: Internat. Nickel Ltd., D & R Dept., Thames House, Millbank, London S.W. 1, Eng.*

BETTEX, Marcel Charles, Swiss physician, pediatrician; born La Tour de Peilz, Switzerland, April 11, 1920; the son of Marius and Helene (Widmer) Bettex; Fed. diploma medicine University of Lausanne (Switzerland), 1944, M.D., 1947; m. Micheline Galland, Aug. 19, 1952; children—Jean-David, Odile. Head dept. pediatric surgery U. Children's Hosp., Berne, Switzerland, 1958-—; faculty U. Berne Med. Sch., 1964-—, prof. pediatric surgery, 1965-—. Mem. Société Suisse de Chirurgie, Société Internationale de Chirurgie, Société Suisse d'Urologie, Societe Suisse de Pediatrie, Brit. Assn. Paediatric Surgeons, Soc. Paediatric Urol. Surgeons, Société Francaise de Chirurgie Infantile, Deutsche Gesellschaft für Kinderchirurgie. Author: Lehrbuch der Kinderchirurgie (with M. Grob, M. Stockmann), 1957; Ueber den vesikoureteralen Reflux beim Saugling und Kind, 1965; also articles. Research on paediatric surgery especially urology and gastrology. Home: 329 H. Sandbühl, 3122 Kehrsatz/BE, Switzerland. Office: 23 Frieburgstrasse, 3000-Berne, Switzerland.*

BETTI, Enrico, Italian mathematician; b. Pistoia, Italy, Oct. 21, 1823; D. Sc., U. Pisa (Italy), 1846. Prof., Florence, also U. Pisa, from 1857; elected to Parliament from Pistoria, 1862; became sec.-gen. pub. instrn., 1874. Author: Teorica delle forze newtoniane e sue applicazioni all'electtrostatica e al magnetismo, 1879; Lehrbuch der Potentialtheorie, 1885; Opere matematiche, 2 vols. pub., 1903-13; contbd. 1st rigorous exposition of Galois' theory of equations to B. Tortolini's Annali. Laid basis for topology through work on multidimensional spaces and Betti numbers (invariants related to connectivity of topological space of any number of dimensions); studied elliptic functions, linear substitutions in Galois field, also modular equations; research on potential theory, capillarity, and elasticity (including Betti reciprocity theorem). Died Pisa, Aug. 11, 1892.

BETTLEY, Francis Ray, English physician, dermatologist; b. London, Eng., 1909; s. Francis James and Charlotte (Wood) B.; grad. Whitgift Sch., 1927; M.D., U. London, 1935; postgrad. Vienna, Austria, Strassbourg, France; m. Jean McIntyre, May 1951; children—Katherine Rachel, Francis James. Cons. dermatologist Cardiff Royal Infirmary 1937-46; physician in charge dermatology dept. Middlesex Hosp., London, 1946-—; cons. physician St. John's Hosp., for Diseases of Skin, 1947-—; dean Inst. Dermatology, 1951-53. Recipient Territorial decoration; named comdr. Order St. Lazarus of Jerusalem. Fellow Royal Coll. Physicians, London; mem. Royal Soc. Medicine London, Brit. Assn. Dermatology, Dermatol. assns. France, Denmark, Holland, Poland, Israel, Venezuela. Author: Skin Disease in General Practice, 1940, 65. Hon. editor: Brit. Jour. Dermatology, 1950-60. Research, publs. primarily on epidermis, its permeability to water and other substances, effect of soap and other detergents. Home: Newton Vallence, Hampshire, Eng. Office: 19 Harley St., London, W.I., Eng.*

BETTMAN, Jerome Wolf, physician, ophthalmologist; born San Francisco, June 22, 1909; s. Jerome S. J. and Ellie (Wolf) B.; A.B., Univ. of California, 1931, M.D., 1935; m. Hortense Herz, Sept. 14, 1935; children—Jerome Wolf, Jean Louise (Mrs. Edward Lawrence Rossiter). Practice medicine, specializing in ophthalmology, Stanford, Cal., San Francisco, 1937-—; faculty Med. Sch. Stanford, 1937-—, clin. prof. surgery, 1957-—; clin. prof. ophthalmology Med. Sch. U. Cal., San Francisco, 1964-—; chief dept. ophthalmology Presbyn. Med. Center, 1957-—; cons., mem. visual scis. study sect. div. Research Grants NIH, 1962-—; cons. to com. sci. assembly Sci. Bd. Cal. Med. Assn., 1963-—. Recipient Honor award Am. Acad. Ophthalmology, 1958. Fellow A.C.S.; mem. Am. Acad. Ophthalmology and Otolaryngology, Pacific Coast Oto Ophthalmol. Soc. (past pres.), Western Sect. Assn. Research (past chmn.). Research, publs. on cataracts, agts. affecting vascular disorders of eye, tumors of eye. Home: 171 Benito Way, San Francisco 94127. Office: 2400 Clay St., San Francisco 94115.*

BETZ, Eberhard, German physiologist; b. Holzhausen, Germany, June 10, 1926; s. Emil and Berta (Kempf) B.; Dr.med., U. Marburg (Germany), 1947, dozent, 1965; m. Margarete Gebhardt, May 7, 1955; children—Annegret, Susanne, Christoph. Leader balneological inst. Bad Salzschlirf, 1953-55; staff Clinic for Circulatory Diseases, Orb, Germany, 1955, Med. Clinic Marburg/L, 1956-59, Physiol. Inst. Marburg/L, 1959-65; faculty, dozent in physiology Inst. Marburg, 1965-68; prof., dir. Inst. Applied Physiology, Tübingen; Germany, 1968-—. Mem. German Physiol. Soc. Research, publs. on pathophysiology of peripheral circulation, brain and myocardial circulation and nutrition, temperature regulation. Office: 5 Gmelinstrasse 5, Tübingen, Germany.*

BETZ, Johann Albert, German engr.; b. Schweinfurt, Dec. 25, 1885; s. Karl and Franziska (Plankl) B.; ed. Technische Hochschule and U.; Ph.D. honoris causa in engring., Ph.D. in tech. sci.; m. Ferdinand Loibl, July 14, 1919; children—Regina, Wiltrud. Asst., chief div., dir. Inst. Aerodynamic Research; dir. Max-Planck Inst. (formerly Kaiser Wilhelm Inst.); prof. U. Göttingen. Recipient Anneau Ludwig-Prandtl award Sci. Soc. Aviation. Mem. Inst. Aerodynamic Research (past dir.), Max-Planck Inst. Research, Acad. Sci. of Göttingen. Author: Windenergie und ihre Ausnützung durch Windmühlen; Konforme Abbildung; Einführungen die Theorie der Strömungsmachinen; Mechanik unelastischer und elastischer Flüssigkeiten;. Aehnlichkeitsmechanik. Died Apr. 16, 1968.

BETZ, Vladimir A., anatomist; b. Russia, 1834; discovered and described giant pyramidal cells of cerebral cortex (cells of Betz) in report pub. 1874. Died 1894.

BETZENDAHL, Walter Hans Wilhelm, German psychiatrist; b. Barmen-Wuppertal, July 4, 1896; s. Rudolf and Auguste (Rahlenbeck) B.; ed. univs. Tübingen, Friburg/Br., Berlin; M.D.; Ph.D.; m. Herta Gohr, May 18, 1949; 1 dau., Hedwig. Med. Asst. univ. clinic of neurology Charity, Berlin, Germany, 1939; prof., 1941; instr. psychology, 1949. Mem. Greek Soc. Otorhinolaryngology (corr.), Forschritte der Medizin, Psychiatrie, Neurologie und Medizinisch Psychologie. Author: Persönlichkeitsentwicklung und Wahnbildung, 1932; Die Ausdrucksformen des Wahnsinns, 1935; Das Bild des Hirnverletzten nach der ersten Auseinandersetzung mit dem Schaden, 1949; Der Wundstarrkrampf in chirurgischer und neurologischer Beureilung, 1953; Der menschliche Charakter in Wertung und Forschung, 1956. Address: Waitzstrasse 6, Kiel, Germany.

BEU, Karl Emil, physicist; b. Vienna, Austria, Mar. 23, 1921; s. Erich W. K. F. and Laura (Sommer) B.; B.S. in Chem. Engring., U. Mich., 1944, M. S. in Physics, 1953; children—Harold W., Nancy R., Walter E., Beverly A. Asst. chem. engr. Phillips Petroleum Research div. Bartlesville, Okla., 1944-48; sr. physicist Gen. Motors Research div. Gen. Motors Corp., Detroit, 1949-53; supervisor phys. measurement dept. Goodyear Atomic Corp., Piketon, O., 1955-—. Faculty, Ohio U., Athens, 1950-—. Mem. Am. Crystallographic Assn. (chmn. app. and standards com. 1962-65), Am. Soc. for Testing and Materials (chmn. task group on electron diffraction 1961-—, exec. council joint com. on powder analysis 1964-—), Am. Phys. Soc., Electron Microscopy Soc. Am. Contbg. author: Ency. of X-Rays and Gamma Rays, 1963; Handbook of X-rays, 1966. Patentee X-ray diffraction cameras. Research, publs. on polymer structure, phase composition and residual stresses in hardened steels, corrosion films, precision and accuracy in lattice parameter determination. Home: 191 Vine St., Chillicothe, O. 45601. Office: Box 628, Piketon, O. 45661.*

BEUDANT, Francois-Sulpice, French mineralogist, geologist; b. Paris, Sept. 5, 1787; ed. École Polytechnic, also École Normale. Named prof. math. Lycée d'Avignon, 1811; became prof. physics Lycée de Marseilles (France), 1813; made mineral. expdn. to Hungary, 1818; named prof. mineralogy Faculty Scis. Paris, 1820, later university inspector-general. Member of the French Academy of Sciences, 1824. Author: Voyage minéralogique et géologique en Hongrie, pendant l'anée 1818, 3 vols., 1822; Traité élémentaire de physique, 1824; Traité élémentaire de minéralogie, 1824; Cours élémentaire de minéralogie et de geologie, 1841. Studied crystallization; noted that a mineral substance could be variegated in form; proposed system based on chem. analogies for defining mineral species, 1830; experimented on fresh and salt water mollusks. Died Dec. 10, 1850.

BEULÉ, Charles Ernest, French archaeologist; b. Saumur, France, June 29, 1826; ed. Ecole Normale; prof. French Sch., Athens, Greece, from 1852; prof. archaeology Bibliothèque Imperiale. Author: L'Acropole d'Athens, 1854; Etudes sur le pecoponnèse, 1855; Les Monnaies D'Athenes, 1858; Histoire de l'Art Grec, 1870. Important archeol. discoveries in Greece, Carthage. Died Paris, France, Apr. 4, 1874.

BEUREN, Alois Joseph, German physician, cardiologist; born Düsseldorf, Germany, August 8, 1919; a son of Joseph and Maria (Raueiser) B.; student College Düsseldorf, University Bonn; M.D. U. Munich, 1945; m. Irmgard Kottmeier, Dec. 29, 1944; children—Stephan, Martin. Fellow cardiology Washington U., St. Louis, 1956, Johns Hopkins Hosp., Balt., 1957-59; prof. cardiology, chief dept. U. Göttingen, Germany, 1960—. Fellow Am. Coll. Cardiology; mem. German Soc. Internal Medicine. Author: The Angiocardiographic Demonstration of Congenital Heart Disease, 1966. Publs., research on cardial metabolism; left heart catheterization and congenital heart disease particularly supravalvular aortic stenosis and its assn. with vitamin D; hypercalcemic cardiovascular disease. Home: 4 Birkenweg Bovenden, Göttingen 34, Germany.*

BEURLING, Arne Karl-August, mathematician; b. Gothenburg, Sweden, Feb. 3, 1905; s. Conrad Verner and Elsa (Raab) B.; Philos-Cand., Uppsala (Sweden) U., 1926, Philos. lic., 1928, Philos. Dr. 1933; m. Brita Oestberg, July 7, 1936; 1 dau., Christina (Mrs. Albert Harbury); m. 2d Karin Vanja Lindblad, Dec. 31, 1950. Faculty, Uppsala U., 1930-52; vis. prof. Harvard, 1948-49; prof., permanent mem. Inst. Advanced Study, Princeton, N.J., 1954—; cons. Swedish State Dept., Dept. Def., 1939-45. Recipient Celsius medal Royal Soc. Scis., 1961, Distinguished Service award in sci. Yeshiva U., 1963. Mem. Royal Swedish Acad. Scis., Royal Soc. Scis. Uppsala, Royal Physiographical Soc. Lund, Danish, Finnish socs. sci. Contbr. numerous articles to profl. jours. Research in function theory, harmonic analysis, potential theory, Dirichlet series. Home: 102 Battle Rd. Circle. Office: Inst. for Advanced Study, Princeton, N.J. 08540.*

BEUTLER, Ernest, physician; b. Berlin, Germany, Sept. 30, 1928; s. Alfred D. and Kaethe (Italiener) B.; Ph.B., U. Chgo., 1946; B.S., 1948, M.D., 1950; m. Brondelle May Fleisher, June 15, 1950; children—Steven, Earl, Bruce, Deborah. Faculty, U. Chgo., 1955-59, U. So. Cal., 1960-64; clin. prof. medicine U. So. Cal. Sch. Medicine, 1967—; chmn. div. medicine City of Hope Med. Center, 1959—. Mem. Am. Fedn. Clin. Research, Internat., Am. socs. hematology, Central Soc. Clin. Research, Am. Soc. Clin. Investigation, N.Y. Acad. Sci., Am. Soc. Human Genetics, Western Assn. Physicians, A.M.A. Author: (with V. F. Fairbanks, J. L. Fahey) Clinical Disorders of Iron Metabolism, 1963. Contbr. numerous articles to profl. jours. Discovered abnormality of red blood cells leading to sensitivity to hemolytic anemia induced by drugs; demonstrated that portion of one X-chromosome in human females is genetically inactive, that iron enzymes are decreased in iron deficiency; investigator of iron metabolism in iron deficiency. Home: 1501 N. Highland Oaks Dr., Arcadia, Cal. 91006. Office: 1500 E. Duarte Rd., Duarte, Cal. 91010.*

BEVAN, Arthur Dean, Am. surgeon; b. Chgo., Aug. 9, 1861; s. Thomas and Sarah (Ramsey) B.; ed. Sheffield Sci. Sch., Yale, 1878-79; M.D., Rush Med. Coll., 1883; hon. A.M., Yale, 1916; m. Anna L. Barber, Feb. 1896. Prof. anatomy Ore. State U., 1886-87; prof anatomy, Rush Med. Coll., Chgo., 1887-99, asso. prof. surgery, 1899-1902, apptd. prof. surgery, 1902, head surg. dept., 1907; professorial lectr. surgery U. Chgo., from 1901; surgeon Presbyn. Hosp. Fellow Am. Surg. Assn. (pres. 1932); mem. A.M.A. (pres. 1917-18). Named officer Legion of Honor (France) in recognition of services as pres. A.M.A., 1918. Editor: American Edition of Lexer's General Surgery. Compiler: Text-book of Anatomy by American Authors: Text-book of Surgery by American Authors. Died June 10, 1943.

BEVAN, Edward John, English chemist; b. 1856, Birkenhead, Eng., ed. at Manchester, Eng.; asso. with Cross; public analyst to Middlesex County (Eng.) Council, invented present method of producing artificial silk by changing cellulose to viscose and ejecting it into sulfuric acid solution. Died 1921.

BEVAN, Lynne J(ohn), Am. hydraulic engr.; b. Atlanta, Ill., Dec. 27, 1881; s. John Luther and Armada Sarah (Thomas) B.; B.S., U. Chgo., 1903; B.S. in Mining, U. Cal., 1905; m. Elizabeth Alexandra Young, Dec. 29, 1914; children—John Alexander, Barbara Louise (Mrs. Jas. A. McGowan). Asst. engr. to John R. Freeman, cons. engr., 1905-06; asst. engr., later prin. engr. Viele, Blackwell & Buck, cons. hydraulic engrs., 1906-27; cons. engr., specializing in water supply and water power, N.Y.C., since 1928; spl. lectr. grad. course on water power engring., Polytechnic Inst. Bklyn., 1934; engring. counsel U. S. Bur. Internal Revenue and to corps. in litigation since 1921. Mem. Town Planning Bd., Montclair, N.J., 1932-38. Mem. Am. Soc. C.E. (chmn. power div. 1929-34), Mountclair Soc. Engrs. (pres.. 1931-32), Essex County Engring. Soc. Phi Beta Kappa, Sigma Xi. Compiler of Catalogue of Delta Upsilon, 1917; editor Cal. Jour. Tech. 1904-05. Contbr. articles to tech. jours. Died Jan. 29, 1952.

BEVELANDER, Gerrit, Am. histologist; born West Sayville, N.Y., Apr. 6, 1905; s. John and Alice (Lieuwen) B.; A.B., Hope Coll., 1926; M.A., U. Mich., 1928; Ph.D., Johns Hopkins, 1932; m. Alice McLaren, Feb. 2, 1949; 1 dau., Karen. Asst. scientist U. S. Bur. Fisheries, 1928-29; instr. Union Coll., N.Y.C., 1932-33; faculty N.Y. U., 1933-62, prof., chemn. dept. microanatomy, 1935-62; prof. micro-anatomy U. Tex., Houston, 1962—. Cons. USPHS, 1957-61, 63—. Fellow A.A.A.S., N.Y. Acad. Sci., Indian Soc. Zoologists; mem. Am. Soc. Anatomists, Histochem. Soc., Internat. Assn. for Dental Research, Am. Soc. Zoologists, Soc. for Exptl. Biology and Medicine, Tex. Electron Microscope Soc., Mt. Desert Island Bil. Lab, Bermuda Biol. Sta., Marine Biol. Lab. Author: Outline of Histology, 1963; Essentials of Histology, 1965; contbg. author Handbook Microscopical Technique; also numerous articles. Research on relation between fine structure and function epithelial cells, identification chems. and enzymes in tissues associated with mineralization, adverse effects antibiotics on formation skeletal structures crystal growth. Home: 689 Post Oak Lane, Houston. Office: 6516 John Freeman Av., Tex. Med. Center, Houston 77025.*

BEVER, Michael Berliner, metallurgist; b. Schmargendorf, Germany, Aug. 7, 1911; s. Rudolf and Maria (Bever) B.; came to U. S., 1934, naturalized, 1938; Dr.iur., U. Heidelberg, 1934; M.B.A., Harvard, 1937; S.M., Mass. Inst. Tech., 1942, Sc.D., 1944; m. Marion B. Gordon, Aug. 25, 1936; children—James G., Thomas G., Mary-Ivers. Faculty, Mass. Inst. Tech., Cambridge, 1940—, prof. metallurgy, 1956—; hon. research asso. metallurgy Harvard, 1966-67. Fellow Am. Acad. Arts and Scis.; mem. Am. Inst. Mining, Metall. and Petroleum Engrs. (Mathewson Gold medal 1965), Am. Soc. for Metals, Gordon conf. on Phys. Metallurgy (past chmn.), Inst. Metals (Gt. Britain), Sigma Xi. Co-editor: Basic Open Hearth Steelmaking, 1951. Editorial bd. McGraw Hill Materials Sci. and Engring. Series, 1960—; co-editor McGraw-Hill Metallurgy and Metall. Engring. Series, 1956—. Research, publs. on phenomena in metals and related materials, sci. aspects of technol. processes. Home: 23 Highland St., Cambridge, Mass. 02138.*

BEVERIDGE, William Ian Beardmore, veterinarian; b. Junee, Australia, Apr. 23, 1908; s. James and Ada (Beardmore) B.; ed. U. Sydney, Cantaberra; D.V. (Sydney), M.A., Cantaberra; D.V.M. honoris causa, Hanover; m. Patricia Thomson, May 31, 1935; 1 son, John Caldwell. Research bacteriologist McMaster Animal Health Lab., Sydney, 1930-37; Commonwealth Fund Service fellow Rockefeller Inst. for Med. research, Melbourne, 1941-46; in charge research Pasteur Inst., Paris, 1946-47; vis. prof. Ohio State U., 1953; prof. animal pathology Cambridge (Eng.) U., 1947. Fellow Australian Vet. Assn. (hon.); mem. Royal Coll. Vet. Surgeons (hon. asso.), Path. Soc. Gt. Britain and Ireland, Soc. for Gen. Microbiology. Author: The Art of Scientific Investigation, 1950; Foot-Rot in Sheep: A transmissible disease due to infection with F. nodosus, 1941, also articles. Home: 299 Milton Rd. Office: School of Veterinary Medicine, Madingley Rd., Cambridge, Eng.

BEVIS (or BEVANS), John, English physician and astronomer; b. Tenby, Eng., Oct. 31, 1693; B.A., Christ Ch., Oxford (Eng.) U., 1715, M.A., 1718. Physician London, Eng., before 1730; set up observatory, Stoke Newington, England, 1738. Fellow Royal Society, 1765; member of the Berlin (Germany) Academy of Sciences, French Academy of Scis., 1768. Author: Uranographia Britannica; Cymbalum Mundia; Tabulae Astronomicae, 1749. Independently discovered comet of 1744; observed transit of venus; strengthened charge of Leyden jar by applying a coating of tinfoil. Died Nov. 6, 1771.

BEWLEY, Loyal, Am. elec. engr.; b. Republic, Wash., Dec. 19, 1898; s. Frank W. and Sadie (Harvey) B.; B.S., U. Wash., 1923; M.S., Union U., 1928; m. Katheryne McAuley, Oct. 17, 1923; children—Robert Lynn, Donald Thomas. Design and research engr. Gen. Electric Co., 1923-40, cons. engring. edn. Gen. Electric Co., 1962-65; prof. elec. engring., dean engring. Lehigh U., Bethlehem, Pa., 1940-62. Recipient Coffin award Gen. Electric Co., 1934; Nat. Best Paper prize Am. Inst. Elec. Engrs., 1932; Hillman award Lehigh U., 1954. Fellow I.E.E.E. (Lamme award 1965). Author: Traveling Waves on Transmission Systems, 1933; The Dimensional Fields in Electrical Engineering, 1948; Alternating Current Machinery, 1949; Flux Linkages and Electromagnetic Induction, 1952; Tensor Analysis of Electric Circuits and Machines, 1961; also articles. Research on electric machine theory, traveling wave phenomena, electric transients, electromagnetic theory, electric and magnetic fields. Address: 2145 Orchard Park Dr., Schenectady 12309.*

BEWS, John William, botanist, ecologist; b. Kirkwall, Scotland, Dec. 16, 1884; s. James Bews; M.A., Edinburgh U., 1906, B.Sc., 1907, D.Sc., 1912; m. Williamina Elizabeth Cameron Mackay. Lectr. econ. botany Victoria U., Manchester, Eng., 1907-08; lectr. on plant physiology, asst. prof. botany U. Edinburgh, 1908-10; prof. botany Natal Univ. Coll., Pietermaritzburg, S. Africa, 1910-25, also from 1927, prin. coll., from 1930; dean faculty sci. U. S. Africa, 1921; prof. botany Armstrong Coll., U. Durham (Eng.), 1926-27. Mem. Union Govt. Research Grant Bd. Fellow Royal Soc. S. Africa, Linnean Soc.; pres. S. African Assn. for Advancement of Sci., 1931. Author: The Grasses and Grasslands of South Africa, 1918; The Flora of Natal and Zululand, 1921; Plant Forms and their Evolution in South Africa, 1924; The Ecological Evolution of the Angiosperms, 1927; The World's Grasses, 1929; Human Ecology, 1935; Life as a Whole, 1937; also papers on botany and ecology. Died Nov. 10, 1938.

BEXON, Gabriel, French naturalist; b. Remiremont, France, Mar. 10, 1747; s. Amé and Barbe (Pillement) B.; ed. sem. at Tout, also Besancon, France; D. Theology; became priest, 1771; tutor to Buffon's son. Author: Système de la fertilisation; Histoire des oiseaux. Collaborator (with Buffon) on animal descriptions for Histoire naturelle. Died Paris, Feb. 15, 1784.

BEY-BIENKO, Grigorii Yakovlevich, Russian entomologist; b. 1903; grad. Siberian Agr. Acad., Omsk, 1925. With All-Union Research Inst. Plant Protection, Leningrad, 1929-38 prof. Leningrad Agr. Inst., 1939—; with Inst. Applied Zoology of Phytopathology, 1946-48; head lab. Zool. Inst., USSR Acad. Sci., 1947—. Recipient Kholodkovsky prize, 1951, Stalin prize, 1952. Mem. USSR Acad. Sci. (corr.). Author: Dermaptera, 1936; Blattoidea, 1950; Tettigonidae. Subfamily. Leaf Grasshoppers, 1954; Certain Patterns of Change in Invertebrates during Reclamation of the Virgin Steppe, 1961; General Classification of Insects, 1962; Biological Systematics in the Service of Theory and Practice, 1963; co-author: Acridoidea Fauna of the USSR and Adjoining Countries, Parts 1-2, 1951; Agricultural Entomology, 1955. Address: Leningrad Agrl. Inst., Pushkino, ulitsa Komsomolskaya 10, Leningrad, USSR.

BEYER, Johann Hartmann, German mathematician; b. Frankfort/Main, Germany, Apr. 15, 1563; s. Hartmann and Katherine (Ligarius) B.; studied in Strassburg, France; Dr.med., U. Tübingen (Germany); m. 2d, Ursula Botzheim. Physician in Frankfurt/Main; mem. council, then mayor; corresponded with many famous mathematicians including Kepler; author of Logistica decimus, 1603; willed his sci. library and fund for pub. welfare and med. edn., 1624; although preceded by Stevin is supposed to have independently invented decimal fractions in Germany, 1596. Died Frankfort/Main, Aug. 1, 1625.

BEYER, Karl Henry, Jr., Am. pharmacologist, physiologist; b. Henderson, Ky., June 19, 1914; s. Karl Henry and Lennie (Beadles) B.; B.S., Western Ky. State Coll., 1935; Ph.M., U. Wis., 1937, Ph.D., 1940, M.D., 1943; m. Annette Weiss, Aug. 9, 1940; children—Annette Matilda (Mrs. Richard A. Ellison), Katherine Louise. Instr., Western Ky. State Coll., 1935-36, U. Wis. Med. Sch., 1939-43; with Sharp & Dohme, Glenolden, Pa., 1943-50, asst. dir. research, 1950-56; dir. Merck Inst. for Therapeutic Research, West Point, Pa., 1956-58, pres., 1961-65, v.p. for life scis. Merck Sharp & Dohme Research Labs., West Point, 1958-65, sr. v.p. for research, 1965—. Lectr. physiology Jefferson Med. Coll., 1959—; lectr. pharmacology Temple U. Med. Sch., 1956—, Grad. Med. Sch. U. Pa., 1955—; spl. lectr. pharmacology Woman's Med. Coll., Phila., 1958—. Recipient Merck Sci. award, 1959; Gairdner Found. award, 1964. Fellow A.C.P., A.A.A.S., N.Y. Acad. Scis., Royal Soc. Medicine; mem. Am. Chem. Soc., Am. Physiol. Soc., Soc. for Exptl. Biology and Medicine, Phila. Med. Soc., Phila. Physiol. Soc., Am. Soc. for Pharmacology and Exptl. Therapeutics (past pres.), Fedn. Am. Socs. for Exptl. Biology (pres. 1965-66), Canadian Pharmacological Soc., Assn. Am. Med. Colls., Am. Therapeutic Soc., Soc. Toxicology, Nat. Acad. Scis. (drug research bd.), Council on Circulation and Renal Sect., Am. Heart Assn., Editorial com. Ann. Rev. Pharmacology; editorial bd. Jour. Clin. Pharmacology and Exptl. Therapeutics; treas. Biol. Abstracts; 1965—. Research, numerous publs. on renal pharmacology, metabolism of drugs in body, enzymologic studies on secretory mechanisms of cells. Home: Gwynedd-Plymouth Rd., Gwynedd Valley, Pa. 19437. Office: Merck Sharp & Dohme Research Labs., West Point, Pa.*

BEYER, Kurt Friedrich August, German civil engr.; b. Dresden, Germany, Dec. 27, 1881; s. Alfred and Anna (Rossberg) B.; student Dresden; received his degree in 1908; m. Käthe Meissner, 1938; 1 son, 2 daus. Engr. for Siamese state r.r.'s until 1914; adviser to Siamese Ministry of Interior; built harbor installations of Bangkok, northern mountain ry., Bandara bridge, Petschaburi castle; became prof. in Dresden, 1919; designed plants for AG Sächsische Werke Böhlen and Kraftwerk Niederwartha. Mem. Deutschen Akademie der Wissenschaften, Sächsiche Akademie der Wissenschaften Leipzig, 1949; Bauakademie Berlin. Author: Statik im Eisenbetonbau, 1927; Technische Mechanik für Bauingenieure, 1954. Important static and constrn. work for open working of brown coal; until 1945 constructed almost all earth movers and rubble clearing bridges used; contributed to rebldg. of Elbe River bridges after war. Died Dresden, May 9, 1952.

BEYER, Robert Edward, Am. biochemist, educator; b. Englewood, N.J., Feb. 20, 1928; s. Edward I. and Rebecca (Lewis) B.; A.B., U. Conn., 1950, M.S., 1952; Ph.D., Brown U., 1954; m. Ernestine M. Schultz, Aug. 31, 1954; children—Timothy, Gwen, Amy. Research fellow U. Stockholm (Sweden), 1954-56; faculty Tufts U. Sch. Medicine, Boston, 1956-62, U. Wis., Madison, 1962-65; associate professor of zoology University of Mich., Ann Arbor, 1965—. Mem. Am. Soc. Biol. Chemists, Am., Canadian physiol. socs., Am. Soc. for Cell Biology, A.A.A.S. Research, numerous publs. on mechanism and regulation of oxidative phosphorylation, mode of action of hormones, mechanisms for adaptation of animals to cold environments. Home: 1606 Shadford Rd., Ann Arbor, Mich. 48104.*

BEYER, Robert Thomas, Am. physicist, educator; b. Harrisburg, Pa., Jan. 27, 1920; s. James Matthew and Mary (Gibney) B.; A.B., Hofstra U., 1943; Ph.D.,

Cornell U., 1945; M.A. (hon.), Brown U., 1957; m. Ellen Fletcher, Feb. 14, 1944; children—Catherine Elizabeth, Margaret Anne, Richard James, Mary Louise. Teaching asst. Cornell U., 1942-45; faculty physics Brown U., Providence, 1945-—, prof., 1958-—, executive officer, since 1966-—. Visiting professor Technische Hochschule, Stuttgart, Germany, 1961-62; cons. Raytheon, 1962-—, Am. Inst. Physics, 1957-—. Fellow Fund for Advanced Edn., U. Cal. at Los Angeles, 1953-54. Fellow Acoustical Soc. Am. (mem. exec. council 1956-59, pres. 1968-—), Am. Phys. Soc. Author: (with A. O. Williams, Jr.) College Physics, 1957; translator: Practical Analysis (F. A. Willers), 1948; Foundations of Quantum Mechanics (John Von Neumann), 1955. Research on structure of liquids and solids from ultrasonic measurements, nonlinear acoustics, underwater sound. Home: 132 Cushman Ave., East Providence, R.I. 02914. Office: Physics Dept., Brown U., Providence 02912.*

BEYRICH, (Heinrich) Ernst von, see von Beyrich.

BEYSCHLAG, Franz, German geologist; b. Karlsruhe, Germany, Oct. 5, 1856; s. Willibald and Maria (Clemen) B.; ed. U. Halle (Germany), Bergakademie Berlin (Germany); m. Käthe Studelmann, 1884; 1 son, Rudolf. Began work in mining, later became geologist; apptd. prof. geognosy and sci. of ore deposits Bergakademie Berlin, 1895; named sci. dir. Royal Prussian Geol. Inst. (under his direction became best geol. inst. in world), 1901, became pres., 1922; commd. to calculate world's iron ore and reserves, also to pub. world geol. map by XI Internat. Congress Geologists, 1910. Mem. German Geol. Soc. (chmn.). Author: Geognostische Skizze der Umgebung von Crock im Thüringer Wald, 1882; 17 Blätter der Geologischen Karte von Preussen mit Erläuterungen, 1890-1907; Carte Géologique internationale de l'Europe, 1893-1913; Die Lagerstätten der nutzbauen Mineralien und Gesteine, 2 vols., 1910-13; Mitteilung über die Eisenvorräte der Welt, 1910; Die Versorgung Deutschlands mit Stahlveredelungsmitteln, 1918; Geologische Karte der Erde, 1929-32. Pub., Zeitschrift für Praktische Geologie. Developed stratigraphic classification of lower new red sandstone of Thuringia; drew maps of usable ore beds in Germany; calculated coal, potash and iron ore in Germany. Died Berlin, July 23, 1935.

BEYTHIEN, Adolf Carl Heinrich, German chemist; b. Quakenbrück, Germany, Jan. 29, 1867; s. Adolf and Karoline (Wellinghoff) B.; ed. U. Göttingen, Germany, 1885-89; m. Margarete Rose, 1896; 1 son. With various factories, exptl. stas., chem. testing office Breslau, Germany; dir., municipal chem. testing office Dresden, Germany, 1899-1919; prof., from 1910; adviser to govt. on food laws. Recipient Joseph König Meml. medal, 1934, Brussels Gold medal, 1938. Author: Die Nahrungsmittelfälschungen und ihre Erkennung, 1914; Volksernährung und Ersatzmittel, 1922; Laboratoriumsbuch für den Lebensmittelchemiker, (remains an important text for food chemists today), 1931; Warenkunde der Süsswarenindustrie, 1942; Die Geschmackstoffe der menschlichen Nahrung, 1947; Einführung in die Lebensmittelchemie, 1950. Died Dresden, June 6, 1949.

BEZNAK, Margaret, physiologist; b. Budapest, Hungary, May 10, 1914; M.D., U. Budapest, 1938; m. A.B.L. Beznak, June 23, 1936 (dec.). Demonstrator dept. physiology, lectr., asst. prof. U. Budapest, 1934-45; research asso. Biol. Research Inst., Tihany, Hungary, 1945-48; with Nobel Inst. Dept. Biochemistry, Stockholm, Sweden, 1949, Med. Research Council Gt. Britain, 1949-53; sr. lectr. physiology, asst. prof. U. Ottawa, Ont. Can., 1953-58, asso. prof., 1959, prof., head dept. physiology, 1960-—. Mem. Phys. Soc. Eng., Am., Canadian phys. socs., Canadian Biochem. Soc. Research, numerous publs. on processes of hypertrophy and atrophy of organs, mainly enlargement of heart; endocrine influences on size and function of heart. Home: 61 Reid Av., Ottawa, Ont., Can.*

BEZOLD, Albert von, see von Bezold, Albert.

BEZOLD, Friedrich, otologist; b. Rothenburg, Germany, Feb. 9, 1842; ed. Vienna, Austria; grad., Berlin, Germany, 1866; became private docent U. Munich (Germany), 1877. Author: Lehrbuch der Ohrenheilk, 1906. Studied functions of tympanum, also of ossicles; provided early description of mastoiditis; presented new tests for audition in deaf-mutism, 1896, in unilateral deafness, 1897. Died Munich, Oct. 6, 1908.

BEZOLD, (Johann Friedrich) Wilhelm von, see von Bezold, (Johann Friedrich) Wilhelm.

BÉZOUT, Etienne, French mathematician; b. Nemours, France, Mar. 31, 1730. Became royal censor, 1758; named prof. and examiner, guards of pavillion of navy, 1763; examiner of students of the royal artillery corps. Member of the Marine Academy, French Academy of Scis. 1758. Author: Rectification des courbes, 1758; Cours de mathematiques, 4 vols., 1764-67; Théorie générale des équations algébriques, 1779. Worked on theory of equations; one of 1st to recognize value of determinants; devised method of elimination by symmetric functions, 1764; gave proof that 2 algebraic curves of degrees m plus n have common points, 1771. Died Basses-Loges, Avon, France, Sept. 27, 1783.

BEZREDKA, Alksander Michailovich, Russian microbiologist; b. Odesa, Russia, 1870; grad. U. Novorosiisk; student Pasteur Inst., Paris; worked with T. T. Mechnikcr. Author: Le choc anaphylactique et le principe de la desensibilization, 1930. Studied immunity. Died 1940.

BHABHA, Homi Jehangir, Indian physicist; b. Bombay, India, Oct. 30, 1909; student Elphinstone Coll., Royal Inst. Sci., Bombay; B.A., Cambridge U., 1930, Ph.D. (Rouse Ball traveling student math.), 1932; Isaac Newton studentship, 1934, sr. studentship of Exhbn. of 1851, 1936; D.Sc. (hon.), U. Patna, 1944, Lucknow U., 1949, Banaras U., 1950, Agra U., 1952; Porth, 1954, Allahabad, 1958, Cambridge, 1959, U. London, 1960, U. Padova, 1961, Andhra U., 1964, Aligarh U., 1964. Spl. reader theoretical physics Indian Inst. Sci., Bangalore, 1940, prof. cosmic ray research unit, 1942-45; dir., prof. theoretical physics Tata Inst. Fundamental Research, Bombay also sec. Govt. India Dept. Atomic Energy and chmn. AEC, India; dir. Atomic Energy Establishment Trombay, Bombay, Pres. Internat. Conf. on Peaceful Uses of Atomic Energy, Geneva, 1955; mem. sci. adv. coms. IAEA, UN. Recipient Adams prize, 1942, Hopkins prize, 1948. Padma Bhushan, 1954; hon. fellow Royal Soc. Edinburgh, Gonville and Caius Coll., Am. Acad. Arts and Scis.; fellow Royal Soc., World Acad. Arts and Scis.; fgn. assos. Nat. Acad. Scis. (U. S.); mem. Internat. Union Pure and Applied Physics (pres. 1960-63), Nat. Inst. Scis. India (pres. 1963, 64). Research and publs. on cosmic rays and quantum theory; introduced Cascade Theory (with Heitler) at same time as Carlson and Oppenheimer. Died Jan. 24, 1966.

BHANOT, Vidya Bhushan, Indian physicist; b. Lahore, Pakistan, Mar. 16, 1927; s. Balak Ram and Vidya Rani (Sharma) B.; M.Sc., Panjab U., 1948; Ph.D., U. Minn., 1959; m. Uma Sharma, Jan. 30, 1963; children—Sanjay, Rajeev. Lectr. physics Dayan and Anglo Vedic Coll., Jullundur, India, 1949-51, Aligarh Muslim U., India, 1951-54; teaching and research asst. U. Minn., Mpls., 1954-59; faculty Panjab U., Chandigarh, India, 1959-—, prof. physics, 1966-—. Fellow Indian Phys. Soc.; mem. Nat. Acad. Scis. India, Am. Phys. Soc., No. India Sci. Assn. Editorial bd. Everyday Sci. Studies on mass spectrometric determination of atomic masses in region from gadolinium to gold, calculations of nuclear binding energies. Home: G-5, Sector 14, Chandigarh 14, India.*

BHARGAVA, Pushpa Mittra, Indian biochemist; b. Ajmer, India, Feb. 22, 1928; s. Ram Chandra and Gayatridevi B.; B.Sc., Lucknow U., 1944, M.Sc., 1946, Ph.D., 1950; m. Edith Manorama Patrick, Aug. 11, 1958; 1 dau., Vaneeta. Lectr. chemistry Lucknow (India) U., 1950; research fellow Nat. Inst. Scis. India at Regional Research Lab., Ayderbad, India, 1950-52; lectr. chemistry Osmania U., Hyderabad, 1952-53; project asso. McArdle Lab. for Cancer Research U. Wis., Madison, 1953-56; Spl. Wellcome Research fellow Nat. Inst. for Med. Research, London, 1956-57; head biochemistry div., asst. dir. Regional Research Lab., Council Sci. and Indsl. Research, Govt. of India, Hyderabad, 1958-—. Mem. chem. research com. Council Sci. and Indsl. Research, New Delhi; sci. adv. com. Central Food Technol. Research Inst., Mysore, India; gen. council Central Family Planning Inst., India Ministry Health, New Delhi. Recipient Watumull prize for biochemistry, 1962. Mem. Internat. Cell Research Orgn. (panel on control processes and molecular biology), Andhra Padesh Akademi Scis. (exec. council), Societa Italiana per il Progresso della Zootecnica (corr.), Biochem. Soc. (London), Soc. Biol. Chemists India (v.p.), Indian Soc. for Study Reprodn. (past joint sec.). Co-editor: Nucleic Acids, 1965. Editorial bd. Indian Jour. Biochemistry. Discovered (with Heidelberg) nature of chem. linkage between carcinogen and tissue protein; studies in protein and nucleic acid synthesis in mammalian spermatozoa; developed method for preparation liver cells in suspension; research on aging process using cellular-chem. composition of mammalian liver; demonstrated permeability of cell is probably due to intercellular orgn., especially of non-cellular material, discovered new type of RNA in extracellular part of normal tissues. Home: C-3 Regional Research Lab. Housing Colony. Office: Regional Research Lab., Hyderabad-9, A.P., India.*

BHARGAVA, Rajeshwar Dayal, Indian mathematician; b. Kota, India, July 5, 1924; s. Raghubar Dayal and Chandra (Wati) B.; B.A. with honors, U. Delhi (India), 1945, M.A., 1947; Ph.D., London (Eng.) U., 1959, diploma Imperial Coll., 1960; m. Sneh Lata, May 8, 1947; children—Raj Rani, Raj Kumar. Lectr. math., asst. prof. in various colls. India, 1947-57; faculty Indian Inst. Tech., Kanpur, 1960-—, prof., 1963-—; sr. vis. fellow Imperial Coll., London, Eng. Cons. Nat. Phys. Lab., Teddington, U.K. Recipient Gold medal U. Delhi. Fellow Indian Phys. Soc. Editor: Labdev. Research, publs. in mechanics of solids, inclusion problems in theory of elasticity, problems in finite elasticity, second-order effects in elasticity theory plasticity, numerical analysis, solution of Laplace's eauation and biharmonic equation, ballistics. Home: B-Block, Civil Lines, Nayapura, Kota, India. Office: Indian Inst. Tech., Kanpur, India.*

BHARUCHA-REID, Albert Turner, mathematician, educator; b. Hampton, Va., Nov. 13, 1927; s. William T. and Mae (Beamon) Reid; B.S., Ia. State U., 1949, postgrad.; postgrad. U. Chgo., Columbia, (Polish Ministry Higher Edn. fellow) U. Wroclaw (Poland); m. Rodabe Phiroze Bharucha, June 5, 1954; 1 son, Kurush Firoze. Research asst. math. biology U. Chgo., 1950-53; research asso. math. statistics Columbia, 1953-55; asst. research statistician U. Cal. at Berkeley, 1955-56; instr. math. U. Ore., 1956-58, asst. prof., 1959-61; asso. prof. math. Wayne State U., Detroit, 1961-—, prof., 1965-—. Mem. Am., Polish, Indian math. socs., Inst. Math. Statistics, Indian Statis. Inst., Soc. for Indsl. and Applied Math., Royal Statis. Soc., Am. Phys. Soc., Sigma Xi. Author: Elements of the Theory of Markov Processes and their Applications, 1960. Contbr. articles on probability theory and applied math. to profl. jours. Reviewing staff Math. Revs. Research on theory stochastic processes USAF, 1953-54, prin. investigator project on stochastic theory epidemics, 1954-55; lectr. stochastic processes Indian Statis. Inst., Calcutta, summer 1956; prin. investigator stochastic processes project U. S. Army Research Office, 1956-62; co-prin. investigator project on stochastic processes and math. physics NSF, 1962-64. Research in mathematical biology, Markov processes and applied functional analysis. Home: 1812 Lafayette Towers W., Detroit 48207.*

BHASKARA, Atscharja, Hindu mathematician, astronomer; born 1114; sixth successor of Brahmagupta as head of the Ujjain, India, College of Astronomy. Author: Siddhānta S'iromani, completed circa 1150. Anticipated modern theory on convention of signs; solved indeterminate equations of 1st and 2nd degrees; used combinatorial analysis; anticipated Kepler's method for finding vol. and surface of sphere by crude integration process; suggested unknown quantities be represented by letters; reduced quadratic equations to single type and completely solved them; approached concept of quadratic and cubic residues; used cyclic method to solve Pellian equations; investigated right triangles and regular polygons to 384 sides; compiled elaborate sine tables; realized true meaning of dividing by zero; explained false position method. Died 1185.

BHATNAGAR, Prabhu Lal, Indian applied mathematician, astrophysicist; b. Kotah, Rajasthan, India, Aug. 8, 1912; s. Laxmi Lal and Ananda (Bai) B.; B.Sc. (Umang Lakshmi Kantilal Pandya Gold medal, Krishna Kumari Gold medal), Agra U., 1934, M.Sc. (Lord North Brook medal), 1936; D.Phil., Allahabad U., 1940, D.Sc., 1947; m. Anand Kumari, May 10, 1936; children—R. K., Kalpana (Mrs. Shri G. P. Bhatnager), Vinaya, Kamal, Brijendra. Lectr., St. Stephen's Coll., Delhi, India, 1939-47, head dept. math., 1947-56; reader U. Delhi, 1947-56; prof., head applied math. dept. Indian Inst. Sci., Bangalore, India, 1956-—; vis. lectr. Harvard Obs., 1952, vis. prof., 1960; distinguished vis. scientist Smithsonian Astrophys. Obs., 1964, 65. Recipient E. G. Hill Meml. prize Allahabad U., 1938-40. Fellow Nat. Inst. Sci. India, Indian Acad. Sci.; mem. Commn. on Magnetohydrodynamics and Variable Stars, Internat. Astron. Union, Indian Math. Soc. (pres. 1964-66). Author: (with D. H. Menzel, H. K. Sen) Stellar Interiors, 1964; (with C. N. Srinivasiengar) Theory of Infinite Series; Magnetofluid Dynamics; Determinants, Matrices and Dimensional Analysis, 1966. Editor, dir: Proceeds of Summer Seminar in Magnetohydrodynamics, 1963. Research, publs. on Bhatnagar-Gross-Krook model which gives an elegant, through approximate method for taking into account close collisions in gaseous assembly; work on secondary flow of highly viscous fluids called non-Newtonian fluids gives method for estimating non-Newtonian viscosity and visco-elasticity. Home: Varna Para, Kotah, Rajasthan, India. Office: Indian Inst. Sci., Bangalore 12, India.*

BHATNAGAR, Shanti Swarupa, Indian chemist; b. Feb. 1895; s. Lala Parmeshwari Sahai and Shrimati Parvati Devi; m. Shrimati Lajwanti, 1915; 2 sons, 2 daus.; ed. Lahore, India, London, Eng., Berlin, Germany; D. Sc. (hon.), Oxford, Patna, Benares, Allahabad, Agra, Lucknow, Delhi, Saugor. Prof. chemistry Benares (India) Hindu U., 1921-24; prof. chemistry, director of the chemical laboratories University of Punjab, Lahore, India (now West Pakistan), 1924-40; dir. sci. and indsl. research Govt. of India, 1940-55; sec. to govt. India Ministry Natural Resources and Sci. Research, 1951-55, Ministry Edn., 1952-55. Founder Lahore Research Scheme under Steel Bros. and Co. Ltd., London; sect. pres. Indian Sci. Congress, 1928, 38, gen. pres., 1944; pres. Nat. Inst. Scis. India, 1947-48; del. 10th Internat. Chem. Congress, Rome, Italy, 1938; mem. numerous other internat. sci. missions. Recipient Reddy prize in chemistry, 1947, Sir P. C. Ray Meml. medal, 1953, Durga Prasad Khaitan Meml. Gold medal, 1953; created knight, 1941; decorated Order Brit. Empire, 1936. Fellow Royal Soc., Inst. Physics, Chem. Soc., Punjab, Benares Hindu univs.; mem. Indian Chem. Soc. (past pres.). Author: Principles and Application of Magnetochemistry; Ilum-ul-Barq. Contbr. numerous papers on colloids, magnetism and photochemistry. Died Jan. 1, 1955.

BHOWMIK, Bijanendu, Indian physicist; b. Bengal, India, Feb. 1, 1918; s. Sashi Kumar and Krishnavamini (Majumdar) B.; B.Sc. with honors, Calcutta U., 1938, M.Sc., 1940; Ph.D., Delhi U., 1955; m. Jyotirmoyee Nath, Aug. 7, 1941; children—Anup, Arup. Lectr. physics Delhi (India) U., 1943-59, reader, 1959-

—. Mem. cosmic ray com. Govt. of India Dept. Atomic Energy, 1958-65. Contbr. articles to profl. jours. Research on primary cosmic rays and elementary particles produced by disintegration of atomic nuclei by high energy projectiles from accelerators, interaction and decay properties of strange particles, K-mesons and hyperons; discovered rare decay modes, leptonic decay of hyperons, hypernuclear physics. Home: 1-C Maurice Nagar, Delhi 7, India.*

BHUSSRY, Baldev R., dentist, anatomist; born Sialkot, India (now Pakistan), February 8, 1928; the son of Hans R. and Vidya (Hora) B.; B.D.S., Sir C.E. Dental College and Hospital, India, 1949; M.S., Univ. Rochester, 1953, Ph.D., 1956; m. Rose Mary Francesca, Nov. 22, 1956. Came to U. S., 1950, naturalized, 1961. Instr., U. Rochester, 1956-57; faculty Georgetown U., Washington, 1957-—, dir. research tng., 1957-65, chmn. dept. anatomy, 1963-—, prof. anatomy, 1964-—, Cons., Walter Reed Inst. Research—Dentistry, 1964-—, Naval Med. Center, 1964-—, Health Facilities Bd., USPHS, 1965-—. Fulbright scholar, 1950-53. Recipient Research Career Devel. award USPHS, 1958. Mem. A.A.A.S., Am. Assn. Anatomists, N.Y. Acad. Scis., Histochem. Soc., Soc. for Exptl. Biology and Medicine, Am. Acad. Oral Pathology, Am. Assn. Med. Colls., Internat. Assn. Dental Research, Sigma Xi. Research, publs. on evidence of changes in tooth enamel with age; demonstrated effect of toxic doses of sodium fluoride on developing teeth; research on role of mucopolysaccharides in collagen formation in healing wounds. Home: 9607 McAlpine Rd., Silver Spring, Md. 20901. Office: 3900 Reservoir Rd. N.W., Washington 20007.*

BIAGINI, Carissimo, Italian radiologist; b. Serre, Siena, Italy, Aug. 3, 1923; s. Giovanni and Irma (Gori-Martini) B.; M.D., U. Florence (Italy), 1948, Specialist Radiology, 1952; m. Bona Grilli, June 2, 1954; 1 dau., Bonizella. Asst., Inst. Gen. Pathology, U. Perugia, 1948-52; asst. Inst. Radiology, U. Florence, 1952-56, tchr. radiobiology Sch. Specialization in Radiology, U. Florence, 1952-56; asst. U. Rome (Italy), 1956-64, tchr. nuclear medicine, 1962-64; prof. radiology U. Sassari (Italy), 1964-—. Cons. for radiobiology, 1958-62; mem. commn. for radiol. protection Ministry Labor, 1959-63; mem. EURATOM Commn. Certificates of Health Physics, 1961-63; con. Research Center, Med. Corps., 1964-—. Mem. Italian Soc. Radiol. Nuclear Medicine (mng. com. 1960-63), Italian Assn. Medicine and Radiol. Biology (mng. com. 1960-—), Biometric Soc. Author: Introduction to the Radiobiology, 1955; Higher Energy Radiations (with C. Bompiani, P. G. Paleani Vettori), 1959; also numerous articles. Research on biol. effects ionizing radiations, radiol. protection problems, med. applications of radioisotopes, radiol. signs of cancer, diagnostic constrast mediums. Home: 10 Viale San Pietro, Sassari, Italy.*

BIALYNICKI-BIRULA, Iwo, Polish physicist; b. Warsaw, Poland, June 14, 1933; s. Stanislaus and Jadwiga (Bujalska) B.-B.; student Warsaw U., 1952-56; m. Zofia Wiatr, Jan. 31, 1956; 1 son, Piotr. Asst., Warsaw U., 1956-59; staff U. Rochester (N.Y.), 1959-61; prof. physics Warsaw U., 1961-—, chair quantum mechanics and field theory, 1962-—. Mem. Polish Phys. Soc. (award 1962). Author: (with Z. Bialynicka-Birula), Quantum Electrodynamics, 1967; also articles. Developed new method of describing internal degrees of freedom of elementary particles, 1958; research in quantum electrodynamics, elementary particles, field theory and statis. physics. Home: 9 Lipowa, Warsaw, Poland.*

BIANCANI, Giuseppe, Italian mathematician; b. Bologna, Italy, 1566; became Jesuit, 1592; prof. math., Parma, Italy. Author: Aristotelis loca mathematica ex universis ipsius operibus coccecta et explicata; Accessere de natura mathematicarum scientiarum tractatio, atque clarorum mathematicorum chronologia; Sphaera mundi seu cosmographia demonstrativa ac facili methodo tradita in qua totius mundi fabrica una cum novis Tychonis Kepleri Galilaei . . . adinventis continetur. Died Parma, June 17, 1624.

BIANCHI, Angelo, Italian geneticist; b. Pavia, Italy, July 14, 1926; s. Guerino and Agnese (Giorgi) B.; dr.Nat. Sci., U. Pavia, 1948, Dr. Biol. Sci., 1950, Libera Docenza in Genetics, 1954; m. Ida Bonomi, Sept. 12, 1953; children—Raffaele, Walter, Michele, Nicola. Postdoctoral fellow U. Pavia, 1948-51, asst., 1952-58, faculty, 1957-58; faculty U. Milan, Italy, 1959-64; tchr. Cath. U. Agr. Faculty, Piacenza, Italy, 1956-—; dir. Istituto di Allevamento vegetale, Bologna, Italy, 1964-—. Recipient Spallanzani award U. Pavia, 1951, 53. Mem. Italian Soc. for Agrl. Genetics (v.p. 1964-—), Italian Genetic Assn. (sec. 1966-—), Eucarpia, European Soc. Radiobiologists. Author: (with Manera) L'anemia macrocitica ereditaria, 1953; L'ABC della genetica delle piante, 1956; also numerous articles. Research on plant genetics, emphasis on mutagenesis and cytology in plant breeding, analysis of mutagenetic events following interspecific hybridization and treatment with chem. and phys. agts. Home: 37 Orfeo, Bologna. Office: 133 Corticella, Bologna, Italy.*

BIANCHI, Giovanni, Italian geologist; b. Rimini, Italy, Jan. 3, 1693; ed. Bologna, Italy, Padua, Italy; joined faculty in anatomy U. Sienna (Italy), 1741; revived Acad. Lincei, Rimini, 1774, also founder cabinet natural history. Author: De monstris ac rebus monstruosis, 1749. Died Rimini, Dec. 3, 1775.

BIANCHI, Giovanni Battista, Italian anatomist; b. Turin, Italy, Dec. 12, 1681; M.D., 1698; became dir. hosps. of Turin, circa 1698; founder anat. amphitheater, 1715; tchr. anatomy, chemistry, pharmacy, practical medicine; prof. anatomy U. Turin; 1st physician to King of Sardinia. Mem. French Acad. Scis. Author: Historia hepatica, . . . 1710; Ductus lacrymales novi . . . , 1715; De lacteorum vasorum positionibus et fabrica, 1743; De naturali in humano corpore vitiosa morbosaque generatione historia, 1761. Gave 1st description of valve at lower end of nasolacrimal duct; studied genital organs, anatomy of liver, lachrymal ducts. Died Turin, Jan. 20, 1761.

BIANCHI, Giuseppe, Italian chemist; b. Bergamo, Italy, Nov. 13, 1919; s. Egidio and Anna (Fumagalli) B.; D.Chem. Engring., Poly. Inst., 1943, D.Elec. Engring., 1945; m. Alessandra Marengoni, July 26, 1948; children—Margherita, Luigi, Mariarosa, Paola, Lucia, Marco. Asst. electrochemistry Polytechnic Inst., Milan, 1945; prof. phys. chemistry U. Camerino, 1955-56, U. Modena, 1956-59; prof. electrochemistry U. Milan, 1959-—, dir. Lab. Electrochemistry and metallurgy, 1960-—, prof. in charge corrosion and metals protection, 1963-—. Recipient Cavallaro Gold medal European Fedn. Corrosion, 1965. Mem. Electrochem. Soc., Nat. Assn. Corrosion Engrs., Inst. Metals. Research and numerous publs. on mechanism of corrosion by differential aeration of zinc, corrosion of interstitial compounds, anodic protection of stainless steel, hydrogen reference electrodes, cathodic reduction of oxygen, electrochem. oxidation of hydrocarbons. Home: 1 Piazza Libia, Milan, Italy.*

BIANCHI, Leonardo, Italian psychiatrist, neurologist; b. Bartolomeo, Galdo, Italy, Apr. 15, 1848; organizer psychiat. inst., Naples, 1882; minister pub. edn. Author 12 books on psychiatry, semeiology of nervous system, hemiplegia, diseases of brain and cerebral localization; also numerous articles. Showed that bilateral destruction of frontal lobes in monkeys causes character changes, 1920. Died Naples, Feb. 13, 1927.

BIANCHI, Luigi, Italian mathematician; b. Parma, Italy, Jan. 18, 1856; s. Francesco Saverio Bianchi; grad. U. Pisa (Italy), 1877; studied in Monaco, also Göttingen, Germany. Became docent Normal Sch., Pisa, 1881, succeeded Dini as director; became docent at the University of Pisa, 1886. Corr. member French Academy Sciences, 1920. Author: Lezioni di geometria differenziale, 1886; Lezioni sulla teoria dei gruppi finiti . . . , 1899; Lezioni sulla teoria delle funzioni di variabile complessa . . . , 1899; Geometria differenziale, 1902-03; Lezioni di geometria analitica, 1908; Lezioni sulla teoria delle forme quadratiche binarie e ternarie, 1911; Lezioni sulla teoria dei gruppi continui finiti di trasformazioni, 1918; Lezioni sulla teoria dei numeri algebrici, 1923. Studied problem of surfaces of constant curvature and application and deformation of surfaces emphasizing partial differential equations. Died Pisa, June 6, 1928.

BIANCHI, Umberto Guido, Italian chemist; b. Genoa, Italy, Aug. 20, 1931; s. Luigi and Italia (Malaspina) B.; D.Chemistry, Genoa U., 1955; Libero Docente in Macromolecular Chemistry, U. Rome, 1963; m. Nicla Gianati Ratto, May 11, 1957; 1 son, Eugenio Maria. Research, faculty Inst. Indsl. Chemistry, Genoa U., 1955-60, 62-—; NATO fellow dept. chemistry Manchester (Eng.) U., 1961-62. Mem. Societa Chimica Italiana. Research, numerous publs. on macromolecules in solution and solid state. Home: 45/1 Via Puggia. Office: 3 Via Pastore, Genoa, Italy.*

BIANCHINI, Francesco, (or Franciscus Blanchinus), Italian astronomer; b. Verona, Italy, s. Gaspare Bianchini; ed. Bologna, also Padova, Rome, Italy. Became librarian to Cardinal Ottoboni (later Pope Alexander VIII), 1684; named papal chamberlain, also canon Santa Maria Maggiore, 1689; employed by Pope Clement XI to form mus. Christian antiquities, also apptd. sec. commn. for calendar reform. Fellow Royal Soc., 1713. Author: De calendario et cyclo caesaris, 1703; Hesperi et Phosphori nova phaenomena, 1729; Astronomicae et geographicae observationes selectae, 1737; Opuscula varia, 1754; Storia universale, provata co'monumenti, e figurata co'simboli degli antichi, 1697; De vitis romanorum pontificum, 4 vols., from 1718. Observed spots on Venus; stated that Venus rotates in 24 1/3 days, 1729. Died Rome, Mar. 2, 1729.

BIANCONI, Giovanni Lodovico, Italian physicist, physician; b. Bologna, Italy, 1717; ed. U. Bologna; became physician to King of Poland, circa 1750. Author: Two Letters on Physics, 1746; Treatise on Electricity, 1748. Translated and published Collected Works on the Anatomy of the Human Body (Winslow), 6 vols., 1743-44. Originated idea that speed of sound varies with temperature of air, and studied it by observing flash and sound of cannon shot, 1740. Died Perugia, Italy, 1781.

BIASUTTI, Renato, Italian geographer, anthropologist; b. Spaniele de Friuli, Italy, Mar. 22, 1878; s. Luigi Fillippo and Teresa (Savio) B.; student Istututo Superiore di Studi Praxtici di Pertezionamento, 1897-99, also Florence, Italy; m. Dorothy Dimmick Du Point, 1906; m. 2d, Teresa Fontana, 1938; 4 children. Prof. geography U. Naples (Italy), 1913-27; prof. geography U. Florence (Italy), 1927-—; dir. Istitutto di geografia, Florence; pres. Centro di Studi di geografica etnografica, 1946. Mem. Società di Studi geografici Florence (pres. 1943-44, 46-—), Societé d'Anthropologie Paris, Istitulo di Studi etruschi. Author: Le Salse dell' Appennino settentrionale, 1907; Distribuzione dei caratteri e tipi antropalagici, 1912; Insehamenti rurale in Italia, 1931; Le razze e i popoli della terra, 3 vols., 1941; Geografia della popolazioni e delle sedi, 1945; Il volto della terra, 1946. Analysed osteologic and anthropometric material from various regions, including Tasmania, geol. and palont. chronology of Pleistocene Europe, forms of rural habitation in Italy; morphology of Italy.

BIBIKOV, Sergei Nikolaevich, Russian archeologist; b. Sebastopol, 1908; D.Hist. Sci., 1953. Dir., Inst. Archeology, Ukrainian Acad. Sci., 1955-—. Mem. Ukrainian Acad. Sci. (corr.). Author: Upper Paleolithic Finds in the Central Dniester Region, 1949; Luka Vrubevetskaya an Early Tripole Settlement on the Dniester, 1953. Mem. editorial bd. Soviet Archeology; exec. editor: Archeological Monuments in the Ukrainian SSR, 1955-60 (symposium) Archeology, 1957, 61; An Outline of the Ancient History of the Ukrainian SSR, 1959; (symposium) Archeological Monuments of the Ukrainian SSR, 1962. Address: Inst. Archeology, Ukrainian Acad. Sci. b. Tarasa Shevchenko 14, Kiev, Ukraine SSR, USSR.

BIBRA, Baron Ernst von, see von Bibra, Baron Ernst.

BIBRING, Grete Lehner, psychiatrist; b. Vienna, Austria, Jan. 11, 1899; d. Moritz and Victoria (Stengel) Lehner; M.D., U. Vienna, 1924; m. Edward Bibring, Dec. 18, 1921; (dec. Jan. 1959); children—George L., Thomas. Came to U. S., 1941, naturalized, 1946. Asst. dir. Vienna Psychoanalytic Clinic, 1926-38, tng. analyst, inst., 1933-38; mem., tng. analyst Brit. Psycho-analytical Soc. and Inst., 1938-41; tng. analyst Boston Psychoanalytical Soc., 1941-—, pres., 1955-58; faculty Harvard, 1946-—, clin. prof. psychiatry, 1961-65, emerita, 1965-—; staff Beth Israel Hosp., Boston, 1946-—, psychiatrist-in-chief, 1955-65, emerita, 1965-—, dir. psychiat. research, 1958-—; spl. lectr. analytic psychology Simmons Coll., Sch. Social Work, Boston, 1942-65; also vis. prof., lectr., cons. to various instns. Mem. adv. com. psychol. counseling center Brandeis U., 1954-—; cons. Radcliffe Inst. Ind. Study, 1961-.— Mem. Am. Psychoanalytic Assn. (pres. 1962-63), Internat. Psychoanalytical Assn. (v.p., 1959-63), N.Y. Acad. Sci., Group for Advancement Psychiatry, Am. Psychiat. Assn., A.A.A.S., Alpha Omega Alpha, others. Author: Old Age: Its Assets and Its Liabilities, 1966. Editorial adviser Children's Hosp.'s Ency. Child Care, Boston, 1965-—; mem. editorial bd. Mind and Medicine Monographs, 1960-—. Research and publs. currently on the psychol. course of pregnancy; psychoanalytic investigation of pregnant women; psychol. reactions to heart attacks. Home: 47 Garden St., Cambridge, Mass. 02138.*

BIBUS, Bertrand, Austrian physician; b. Vienna, Austria, Nov. 26, 1906; s. Bertrand and Rosa (Haas) B.; M.D., U. Vienna; m. Irene Gross, May 23, 1936; children—Brigitte, Johanna, Christine. Prof. urol. sect. Kaiser Franz-Josef Hosp. Mem. Order Physicians (Vienna), Altschotten Soc. (Vienna). Author: Die beiderseitigen Nierensteinkrankheit, 1947; Urologie in der Allgemeinpraxis, 1948; Die Zitronentherapie der Uratsteine des Hartraktes, 1959. Home: Währingerstrasse 134, Vienna XVIII. Office: Mariahilferstrasse 37, Vienna VI, Austria.

BICAIRE, H., French physician, endocrinologist; b. Paris, France, May 17, 1914; s. Pierre and Marie-Louise (Duvernoy) B.; Docteur en Médecine, Faculty Medicine Paris; m. Elisabeth Scheitel, Sept. 11, 1946; children—François, Brigitte. Head clinic, physician hosps., Paris; prof., Faculty Medicine Paris. Research on cortico-suprarenal gland, gradual incapacity of suprarenal gland, Cushing's syndrome, gen. endocrinology. Address: 40, rue Scheffer, Paris 16, France.*

BICHAT, Ernest Adolphe, French physician; b. Lunéville, France, Sept. 17, 1845; grad. physics Ecole normale, 1869; docteur ès sciences, 1873. Prof. physics, faculty sci. U. Nancy, France, from 1877. Mem. Acad. Sci., 1893, Chemical Inst., Electrotechnique Institute Nancy (founder both). Author: Introduction à l'étude de l'électricité statique, 1885. Contbr. numerous articles on electricity to sci. mags. Noted for studies on difference of potential between liquids and actino-electric phenomena. Died Nancy, July 26, 1905.

BICHAT, Marie-Francois-Xavier, French anatomist, physiologist; b. Thoriette, France, Nov. 11, 1771; s. Jean-Baptiste and Jeanne-Rose B.; ed. Coll. Nantua, also Jesuit Seminary Saint Irénée, Lyons, France; in Paris he studied surgery under Pierre Desault, 1793. Became professor of anatomy at Paris Medical School, 1795; physician Hôtel-Dieu, from 1800. Founder (with Alibert, Jean Burdin, Husson) Soc. Medicine, 1796. Author: Oeuvres chirurgicales de Desault, 1798; Traité des membranes, 1800; Recherches sur la vie et la mort, 1800; Anatomie générale, 2 vols., 1801; Traité de d'anatomie descriptive, 5 vols., 1801-03. Became editor Jour. Surgery, Paris, 1793. Founder of histology and path. anatomy; worked without microscope; 1st to show organs of body are composed of tissues; introduced word tissue; used dessication, putrefaction, maceration, other spl. techniques in study of body tissues; distinguished 21 tissues; believed all disease caused by tissue changes. Died Paris, July 22, 1802.

BICHEL, Jorgen, Danish physician; b. Odense, Denmark, Aug. 20, 1909; s. Peter Christian and Grethe (Jorgensen) B.; M.D., U. Copenhagen, 1935; m. Jo Helweg Andersen, Feb. 2, 1935; children—Peter, Claus, Svend, Grethe, Hanne-Lotte. Prof. gen. pathology Aarhus (Denmark) U., 1952—, head Cancer Research Inst., 1963—. Research and publs. on cancer and disorders of the blood. Home: 10, Hans Brogesvej, Brabrand, Denmark. Office: Cancer Research Institute, Radiumstationen, 8000 Aarhus C., Denmark.*

BICKEL, William Harold, American orthopedic surgeon; born Shamokin, Pennsylvania, July 4, 1909; son of Erwin Forrest and Florence (Simon) B., B.A. Lawrence College, 1931; M.S., Northwestern U., 1935, M.D., 1936; M.S., U. Minn., 1941; m. Annette Ray, Jan. 1, 1937; children—Barbara Bickel (Mrs. David Walvoord), Ruthann, (Mrs. John Seeley), Patricia, Priscilla. Practice medicine, Milw., 1937-38; fellow orthopaedic surgery Mayo Grad. Sch., Rochester, Minn., 1938-42, faculty, 1942—, prof. orthopaedic surgery, 1959—; staff Mayo Clinic, 1942-65, chief sect. orthopaedics, 1955-65. Exchange fellow Am. Orthopaedic Assn., Eng., 1949; recipient Distinguished Alumni award Lawrence Coll., 1963. Diplomate Am. Bd. Orthopaedic Surgery (past pres., examiner 1950—). Fellow A.C.S. (mem. adv. council 1960—), Am. Acad. Orthopaedic Surgeons (pres. 1964); Minn.-Dakota-Man. Orthopaedic Soc., Am.-Brit.-Canadian, Contemporary, Interurban orthopaedic clubs, Clin. Orthopaedic Soc., Internat. Soc. Orthopaedic Surgery and Traumatology, Blue Key, Mace, Sigma Xi. Asso. editor Jour. Bone and Joint Surgery, since 1958-61. Research and publs. on orthopedics. Home: MR72, Rochester, Minn.*

BICKERTON, Alexander William, astronomer; b. Alton, Eng., Jan. 7, 1842; s. Richard and Sophia Bickerton; ed. Royal Coll. Chemistry, also Royal Sch. Mines, London; m. Annie Phoebe Edward; m. 2d, Mary M. Wilkinson; 4 sons, 2 daus. Originator, London Popular Tech. Classes; organizer sci. Hartley Instn.; pub. analyst, Southampton, also Main Div., Hants, Eng.; sci. lectr. Winchester Coll.; prof. chemistry New Zealand U., 1874-1902; colonial analyst, New Zealand; lectr. Royal Colonial Inst., London, also Royal Instn. Recipient 2 internat. gold medals for new sci. equipment. Mem. Brit. Astron. Assn., astron. socs. France, New Zealand, New Zealand Inst. (astron. sect.), London Instn., Philos. Soc. Eng. (v.p.), Evolution Soc. (v.p.), Itinerant Research Soc. (v.p.), London, Internat. astron. socs. (pres.), Correlative Sci. Soc. (organizing pres.). Author: New Story of the Stars; Romance of the Earth; Romance of the Heavens; Atoms, Suns, and Systems; Birth of Worlds and Systems; Perils of a Pioneer; Materials for Lessons in Elementary Science; A Natural System of Map Drawing; papers contbd. to sci. jours. Commd. by New Zealand govt. to communicate new astronomy of cosmic impact; invented several pieces of sci. apparatus. Died Jan. 22, 1929.

BICKFORD, Reginald George, neurophysiologist; b. Brewood, Eng., Jan. 20, 1913; s. George and Jessie (Jones) B.; M.B., U. Cambridge, 1933, B.Chir., 1936; m. Joyce Audrey Davies, June 8, 1945; children—Michael, Christopher. Served with RAF as neuropsychiatric specialist, 1941-45; research asso. Mayo Found., Rochester, Minn., 1946-48; cons. in electro-encephalography Mayo Clinic, Rochester, 1948—; prof. physiology U. Minn., Rochester, 1954—. Mem. study group in computer research USPHS, 1964—, com. on research in epilepsy, 1965—. Mem. Am. Neurol. Assn., Am. Physiol. Soc., Am. EEG Soc. (pres. 1956). Author: EEG in Anesthesiology, 1960; KWIC Index to EEG Literature, 19—; also numerous articles. Inventor system for automatic adminstrn. of anesthetics; developed method of depth probing of human brain; discovered microreflexes to sensory stimulation; computer application to neurophysiology. Home: 1141 6th St. S.W., Rochester 55901.*

BICKHAM, Warren Stone, Am. surgeon; b. Shreveport, La., Aug. 23, 1861; s. Charles Jasper and Annie Augusta (Gray) B.; student U. of South, 1873-78, Yale, 1881, U. La., 1880-82; Pharm. M., M.D., Tulane U. 1883-86, LL.D., 1925; M.D., Coll. Phys. and Surgeons, 1887; postgrad. N.Y. Polyclinic, 1887; m. Flora Sabina Brandon, 1895 (dec. July 1930); m. 2d, Alice Martin, June 30, 1931. Intern, later surgeon Charity Hosp.; demonstrator operative surgery Tulane U.; jr. surgeon Touro Infirmary, New Orleans; instr. in surgery N.Y. Polyclinic; surgeon Manhattan State Hosp., N.Y.; instr. surgery N.Y. Post-Grad. Med. Sch. and Hosp.; asst. instr. operative surgery Columbia, 1900-06. Fellow N.Y. Acad. Medicine, A.C.S. Author: Textbook of Operative Surgery, 1903; Operative Surgery, 7th Vol., 1933; also sects. on amputations and ligations in Keen's System of Surgery, 1906. Died Dec. 1, 1936.

BICKLEY, William Elbert, Am. entomologist; b. Knoxville, Tenn., Jan. 20, 1914; s. William Elbert and Lucretia (Jordan) B.; B.S., U. Tenn., 1934, M.S. 1936; Ph.D., U. Md., 1940; m. Elizabeth Macgill, Apr. 5, 1941; children—Lucretia, James, David, Edith. Instr., U. Md., 1940-41; asst. prof. U. Richmond (Va.), 1946-49; asso. prof. entomology U. Md., 1949-57, prof., head dept. entomology, 1957—, asso. prof. Mountain Lake Biol. Sta., 1953. Cons., Va. State Health Dept., 1946-49. Mem. Entomol. Soc. Am., Am. Mosquito Control Assn., Am. Soc. Tropical Medicine and Hygiene, Washington Acad. Scis. Research and publs. on insect nervous system, classification lacewing-flies, mosquitoes, control Japanese beetle, vegetable insects. Home: 6516 40th Av., University Park, Hyattsville, Md. 20782. Office: Dept. Entomology, U. Md., College Park, Md. 20742.*

BIDDER, George Parker, Brit. biologist; b. May 21, 1863; s. George Parker and Anna (McClean) B.; ed. Trinity Coll., Cambridge; D.Sc. Sc.D., 1916; m. Marion Greenwood, 1899; 2 daus. With Univ. Table, Naples (Italy) Zool. Sta., 1886-88; lectr. on sponges U. Cambridge, 1894, 1920-27; researcher H.M.S. Vernon, 1914; chmn. Meeting of Brit. Zoologists, 1927-28. Contbr. article Sponges to Ency. Brit., 1929. Editor: Bibliography of Sponges (Vosmaer), 1938. Owner, Quar. Jour. Micros. Sci. Research and publs. on sponges, ocean-currents, death, senescence; researcher on sponges, Naples, Plymouth, Cambridge, 1886-1939; inventor bottom trailers for determining bottom currents of N. Sea, 1904. Died Dec. 31, 1953.

BIDDER, Heinrich Friedrich, physician; b. Gut Treppenhof, (Livland), Nov. 9, 1810; s. Ernst Christian Bidder; studied medicine Dorpat (Estonia); degree, 1834; m. Marie Johanna Rapp (dec. 1857); children—Ernst, Alfred. Faculty, U. Dorpat, 1836-69 beginning as prosector, prof. of anatomy until 1843, then became prof. physiology and pathology. Hon. mem. St. Petersburg Acad. Scis. Author: (with A. W. Volkmann) Die Selbständigkeit des sympathischen Nervensystems durch anatomische Untersuchungen nagewiesen, 1842; (with Carl Schmidt) Die Verdauungssäfte und der Stoffwechsel, 1852; Vergleichend-anatomische und histologische Untersuchung über die männlichen Geschlechts und Harnwerkzeuge der nackten Amphibien, 1846; Untersuchungen über die Textur des Rückenmarks, 1857; also articles in med. and sci. jours. Research on biochemistry including role of bile in digestion of fats, composition and effect of intestinal juices, protein metabolism; before Pavlov described psychic influence on the secretion of gastric juices; demonstrated the presence of free hydrochloric acid in gastric juices; discovered origin of sympathetic nervous system in the ganglia; ganglia found in auricular septum of frog's heart bear his name. Died Dorpat, Aug. 22, 1894.

BIDDULPH, Orlin, Am. plant physiologist; b. Hooper, Utah, Jan. 27, 1908; s. Samuel and Nellie (Quibell) B.; B.S., Brigham Young U., 1929, M.S., 1933; Ph.D., U. Chgo., 1934; m. Susann Fry, Aug. 26, 1942; children—Stuart. Ann (Mrs. James Brauer). Asst. prof., head dept. botany U. S.D., 1934-37; faculty Wash. State U., 1937—, prof. dept. botany, 1946—, dir. molecular biophysics lab., 1963—. Hutchinson fellow, 1933-34, Nat. Research Council fellow 1941, Guggenheim fellow, 1959. Fellow A.A.A.S.; mem. Am. Soc. Plant Physiology, Bot. Soc. Am., Am. Inst. Biol. Scis., Sigma Xi. Devel. of methods for use of radioisotopes in plant studies; demonstration of rapid circulation of elements in plants via the xylem and phloem systems; research on translocation of minerals and metabolites in the phloem of plants. Home: Rt. 2, Box 89, Pullman, Wash. 99163.*

BIDIE, George, surgeon; b. Backies, Banffshire, Apr. 3, 1830; M.B., U. Aberdeen, 1853; licentiate Royal Coll. Surgeons, licentiate in midwifery, Edinburgh, 1853; m. Isabella Wiseman, 1854; 2 sons, 5 daus. Prof. botany and materia medica Madras (India) Med. Coll.; supt. Lunatic Asylum, Madras, 1866-70; sec., statis. officer in head office Med. Dept., 1870-83; dep. surgeon gen., in charge Brit. Burma Div., 1884; san. commr. Madras Presidency, 1885-86; became surgeon-gen., 1886; apptd. hon. surgeon to Viceroy of India, 1887, Queen Victoria, 1898, King Edward VII, 1901, George V, 1910; fellow, pres. med. faculty Madras U. Recipient Coronation medals Edward VII, George V; decorated for discovery (1867) of preventive for insect pest of coffee plantations in southern India. Fellow Zool. Soc. Author: Report on the Ravages of Borer Insect on Coffee Estates, 1869; Handbook of Practical Pharmacy, 2d edit., 1883; Catalogue of Gold Coins in Government Central Museum, Madras, 1874; Neilgherry Parasitical Plants destructive to Forest and Fruit-trees, 1874; Descriptive Catalogue of Raw Products of S. India sent to Paris Exhibition (Gold medal), 1878; Native Dyes of Madras, 1879; Pagoda or Varaha Coins of S. India, 1883; Sand-binding Plants of Southern India, 1883. Died Feb. 19, 1913.

BIDLINGMAIER, Friedrich, German geophysicist; b. Lauffen, Germany, Oct. 5, 1875; s. Christof and Christine Carol (Schmid) B.; doctorate, Göttingen, Germany, 1900; m. Edith Ideler, 1906; m. 2d, Toni Schleissing, 1910. Asst., phys. inst. Technische Hochschule, Dresden, Germany, 1898-1900; mem. German expdn. to S. Pole under E. v. Drygalski, 1901-03; at Technische Hochschule Aachen (Germany), 1908; with Wilhelmshaven Obs., 1909-11; apptd. prof. U. Munich (Germany), 1912. also curator earth magnetic obs. Research and numerous articles, chpts. in handbooks on earth magnetism; invented double compass for measuring earth magnetic horizontal intensity. Died Argonne Forest, France, Sept. 23, 1914.

BIDLOO, Gottfried, Dutch naturalist, anatomist and surgeon; b. Amsterdam, Holland, Mar. 12, 1649. Prof. anatomy The Hague, Holland; prof. anatomy and chemistry Leyden, Holland; physician to William III of Eng. Mem. French Acad. Scis., 1699. Fellow Royal Soc., 1696. Author: Anatomia corporis humani, 1685; Vindicae quarundam delineationum anatomicarum, 1697. Chiefly noted as anat. scientist; excellent plates in his large work. Died Leyden, Apr. 30, 1713.

BIDWELL, Shelford, Brit. physicist; b. Thetford, Eng., Mar. 6, 1848; s. Shelford Clarke and Georgina (Bidwell) B.; B.A., Caius Coll., Cambridge (Eng.) U., 1870, LL.B., M.A., 1873, Sc.D., 1900; m. Wilhelmina Evelyn Firmstone, 1874; 1 son, 2 daus. Began law practice, 1873; began sci. study of electricity, magnetism, physiol. optics, circa 1880. Fellow Royal Soc., 1886 (council 1904-06), Royal Meteorol. Soc.; mem. Inst. Elec. Engrs., Phys. Soc. London (pres. 1897-99). Author: On Telegraphic Photography, 1881; The Influence of Friction on a Voltaic Current, On the Electrical Resistance of Carbon Contacts; The Electrical Resistance of Selenium Cells; On a Method of Measuring Electrical Resistances with a Constant Current; On the Sensitiveness of Selenium to Light and the Development of a Similar Property in Sulphur; On an Effect of Light upon Magnetism; On the Changes produced by Magnetisation in the Dimension of Rings and Rods of Iron and of some other Metals; On the Formation of Multiple Images in the Normal Eye; Some Curiosities of Vision. Research into photoelectric properties of selenium, 1880; devised instrument for electrically transmitting pictures of natural objects to a distance along a wire, 1881. Died Dec. 18, 1909.

BIEBER, Samuel, Am. biologist; b. N.Y.C., Feb. 5, 1926; s. Hyman and Pauline (Sussman) B.; B.A., N.Y. U., 1944, M.S., 1948, Ph.D., 1952; m. Roslyn Lilah Hewitt, Dec. 18, 1949; children—Susan Ellen, Scott Hewitt. Sr. research biologist Wellcome Research Labs., 1952-62; sci. collaborator N.Y. Aquarium, 1952-62; lectr., adj. asso. prof. L.I. U., Bklyn., 1957-62, asso. dean Grad. Sch., 1962-64, prof. biology, 1962-66, asso. dean for sci., 1963-64, dean Richard L. Conolly Coll., 1966—. Fellow N.Y. Acad. Sci.; mem. Am. Assn. for Cancer Research, Soc. for Exptl. Biology and Medicine, Soc. for Study Growth and Devel., Am. Chem. Soc., Am. Soc. Zoologists, A.A.A.S., Sigma Xi, Phi Sigma. Research, publs. on nucleic acid biochemistry and embryonic devel. and regeneration, use of antimetabolites in cancer chemotherapy and immunosuppression. Patentee cancer chemotherapy field. Home: 11 South Gate Dr., Spring Valley, N.Y. 10977. Office: L.I. U. Zeckendorf Campus, Bklyn, 11201.*

BIEBERBACH, Ludwig, German mathematician; b. Goddelau, Germany, Dec. 4, 1886; prof. Basel, Switzerland, Frankfurt/Main, Berlin (both Germany). Author: Einführung in der Konforme Abbildung, 1915; Lehrbuch der Funktionen theorie, 2 vols., 1921-24; Theorie der Differentialgleichungen, 1923; Vorlesungen über Algebra, 1928; Analytische Geometrie, 1930; Differentialgeometrie, 1932; Einleitung in den hohere Geometrie, 1933; Carl Friedrich Gauss, 1938. Research on functions of complex variables.

BIEBL, Richard, botanist; b. Salzburg, Austria, Feb. 12, 1908; s. Josef and Johanna (Scharfetter) B.; Ph.D., U. Vienna, 1930; m. Marianne Nemec, Aug. 9, 1941; children—Waltraud (Mrs. Erich Kussbach), Manfred, Wilfried. Tchr. several high schs., Vienna, Austria, 1931-39; asst. prof. U. Vienna, 1939-45, asso. prof., 1945-59, prof., 1959—; dir. Plant Physiology Inst., 1964—. Recipient Ehrenkreuz. Mem. Austrian Acad. Scis., Deutsche Botanische Gesellschaft, Zool. Botan. Gesellschaft in Wien. Author: (with H. Germ) Praktikum der Pflanzenanatomie, 1950, 2d edit., 1967; Protoplasmatische Okologie, vol. 12, 1962. Numerous publs. on plant anatomy, cell physiology, exptl. ecology, radiation botany. Home: 3 Mommsengasse, 1040. Office: U. Vienna, 1 Dr. Karl Luegerring, Vienna, Austria.*

BIEDENHARN, Lawrence Christian, Jr., Am. physicist; b. Vicksburg, Miss., Nov. 18, 1922; s. Lawrence Christian and Willetta (Lyons) B.; B.S., Mass. Inst. Tech., 1944, Ph.D., 1949; m. Sarah Jeffress Willingham, Mar. 25, 1950; children—John David, Sarah Willetta. Research asso. Mass. Inst. Tech., 1949-50; physicist Oak Ridge Nat. Lab., 1950-52, also cons.; asst. prof. Yale, 1952-54; asso. prof. Rice Inst., Houston, 1954-61; prof. physics Duke U., 1961—. Cons. Nat. Bur. Standards, Los Alamos Sci. Laboratory. Fulbright fellow, 1957, Guggenheim fellow, 1957, NSF post-doctoral fellow, 1964-65. Fellow Am. Phys. Soc., Inst. Physics (Gt. Britain); mem. Sigma Xi, Fedn. Am. Scientists. Author: (with P. J.

Brussaard) Coulomb Excitation, 1965; (with H. van Dam) Quantum Theory of Angular Momentum, 1965. Research in nuclear reactions and theoretical physics. Home: 2716 Sevier St., Durham, N.C. 27705.*

BIEGERT, Josef Otto Adolph, anthropologist; b. Rottweil, Germany, Sept. 21, 1921; s. Adolf and Maria (Voigt) B.; ed. Berlin, Würzburg, Frankfort (all Germany) univs. M.D., 1947; Ph.D., 1959; m. Marie-Christa Feustel, 1950; Ordinary prof. anthropology, dir. Anthrop. Inst., U. Zurich, from 1962. Mem. Swiss Soc. Anthropology, Ethnology, Natural Scis., Schweizerische Gesellschaft für Anthropologie und Ethnologie, Schweizerische Naturforschende Gesellschaft, Zoologische Gesellschaft Zurich, Schweizerishce Gesellschaft für Verebungsforschung. Research, publs. on primatology, human evolution, phys. anthropology. Address: Weidstrasse 9, Pfaffhausen/Zurich, Switzerland.

BIEL, Erwin Reinhold, climatologist; b. Vienna, Austria, Jan. 7, 1899; s. Hugo and Helene (Wittner) B.; Ph.D., U. Vienna, 1926; m. Margaret Horel, July 8, 1929; 1 dau., Renate Maria. Came to U. S., 1938, naturalized, 1944. Asst. instr. geography U. Vienna, 1927-29; head dept. climatology Breslau Obs., 1929-33, docent climatology, Breslau 1932-33; docent Volkshochschule, Vienna, 1935-38; vis. prof. climatology Rutgers U., New Brunswick, N.J., 1938-42, prof., 1942-63, chmn. dept., 1946-63, ret., 1963; vis. prof. U. Chgo., 1942-51, Fla. State U., 1965-67; Nat. Sigma Xi lectr. 1960, NSF lectr., 1960. Recipient Lindback Found. award, 1962; Rutgers U. award, 1963, citation Coll. of Agr., 1963. Asso. editor Jour. Meteorology, 1948-54. Research and publs. on gen. and regional climatology for Central Europe, Mediterranean area, and Asia; variability of precipitation and applied climatology.

BIELA, Wilhelm von, see von Biela, Wilhelm.

BIELIAUSKAS, Vytautas Joseph, psychologist; b. Plackojai, Lithuania, Nov. 1, 1920; s. Antanas and Anele (Kasparaite) B.; A.B., U. Vilkaviskis, Lithuania, 1940; Ph.D. in Psychology, U. Tuebingen, Germany, 1943; m. Danute G. Sirvydaite, Mar. 12, 1948; children—Linas A., Diana B., Aldona O., Cornelius V. Came to U. S., 1949, naturalized, 1955. Asst. prof. U. Munich, Germany, 1944-48; instr. King's Coll., Wilkes-Barre, Pa., 1949-50; faculty Sch. Clin. and Applied Psychology, Coll. William and Mary, 1950-58, prof., 1953-58, head dept. psychology, 1951-57; faculty Xavier U., Cin., 1958—, prof., 1960—, chmn. dept. psychology, 1959—; administrv. asst. to supt. Longview State Hosp., Cin., 1965—. Fellow Am., Ohio psychol. assns., A.A.A.S., Soc. for Projective Techniques; mem. Am. Soc. Applied Psychology, Am. Assn. U. Profs., Am. Acad. Psychotherapists, Am. Soc. Adlerian Psychology, Cath., Cin. (past pres.). psychol. assns., Internat. Cath. Assn. Med. Psychology. Author: Zmogus siu dienu problematikoje, 1945; House-Tree-Person Research Rev., 1963, 2d edit., 1965; also numerous articles. Research on House-Tree-Person projective technique, masculinity and feminity, borderline problems between psychology and philosophy. Home: 2968 Compton Rd., Cin. 45239.*

BIELING, Richard, physician, microbiologist; b. Gau-Algesheim, Sept. 3, 1888; s. Franz and Mäckler Bieling; ed. univs. Friburg, Munich, Berlin; M.D. Instr., U. Frankfurt/Main, 1923, full prof., 1927; full prof. Marburg/Lahn, 1935; head dept. serums vaccines I. G. Farben Industrie, Hoechst., Marburg; prof., dir. Inst. Hygiene, U. Vienna, 1952; hon. prof. microbiology and hygiene U. Friedrich-Wilhelm. Mem. German Soc. for Microbiology and Hygiene. Author: Viruskrankheiten des Menschen, 1938; Enstehung und biologische Bekämpfung typischer Infektionskrankheiten, 1937; Virusbuch, 1942; Die biologische Infektionsabwer, 1944; Die Grippe, 1949, also others. Died August, 1967.

BIELSCHOWSKY, Alfred, ophthalmologist; b. Namslau, Germany, Dec. 11, 1871; s. Hermann Bielschowsky; ed. Breslau, Heidelberg, Leipzig (Germany) univs.; Ph.D., U. Berlin, Germany, 1893. Came to U. S., 1934; lectr. ophthalmology Leipzig, from 1900, asso. prof., from 1906; prof. Marburg, Germany, from 1912, Breslau, from 1923; prof. ophthalmology, dir. Dartmouth Eye Inst., Hanover, N.H., from 1937. Author: (with A. von Graefe, Th. Samisch) Handbuch der gesamten Augenheilkunde. Work with physiology and pathology of eye movements and visual perception of space; aided persons blind from eye injuries during and after World War I. Died N.Y.C., Jan. 5, 1940.

BIELSCHOWSKY, Max, neurologist; b. Breslau, Germany (now Wroclaw, Poland), Feb. 19, 1869; studied in Breslau, Munich (Germany), received degree summa cum laude, 1892; postgrad. Senckenberg Inst., Frankfurt, Germany; 1 son, Franz. Asst. to E. Mendel; later worked in O. Vogt's Neurobiol. Inst.; became mem. brain research inst. Kaiser Wilhelm-Gesellschaft, 1925; emigrated to Eng. shortly before World War II. Author: Myelitis und Sehnervenentzündung, 1901; also articles in neurologic handbooks, jours. Discovered silver dyeing of neuro-fibrils; research in neurology and neuro-biology. Died Eng., Aug. 1940.

BIENAYMÉ, Irénée Jules, French mathematician; b. Paris, Aug. 28, 1796; student Ecole Polytechnique, 1815-16. With financial adminstrn., from 1820, became inspector soon after, inspector-general, from 1834; professor Sorbonne. Decorated Legion of Honor, 1844. Member of the French Academy of Sciences, 1852. Author: De la durée de la vie . . . , 1835; Probabilité des erreurs dans la méthode des moindres carrés, 1852; Remarques sur les differences qui distinguent l'interpolation de Cauchy . . . , 1853; Considérations à l'appui de la découverte de Laplace, 1853; Sur les fractions continues de M. Tchebychef, 1858; Sur un principe de M. Poisson . . . , 1869. Studied calculus of probabilities and its application to financial sci.; originated concept of mean convergence which implies probability convergence. Died Paris, Oct. 20, 1878.

BIENEWITZ, Peter, see Apianus, Petrus.

BIER, August, German surgeon; b. Helsen/Waldeck, Germany, Nov. 24, 1861; prof. U. Berlin (Germany), from 1907. Author Über zirkuläre Darmnaht, 1889; Hyperämie als Heilmittel, 1903; Die Seele, 1939. Introduced artificial hyperemia to treat surg. and other wounds, 1892; introduced cocaine in spinal anesthesia, 1899; forwarded use of hyperemia in treating surgical and other wounds. Died Sauen/Beeshow, Mar. 17, 1949.

BIER, Milan, biochemist; b. Vukovar, Yugoslavia, Dec. 7, 1920; s. Edmond E. and Ada (Diamant) B.; Lic. es Sci. Chim., U. Geneva (Switzerland), 1946; Ph.D., Fordham U., 1950; m. Rhoda M. Solvay, Mar. 27, 1952; children—Vicki M., James J., Robert E. Came to U. S., 1946, naturalized, 1952. Asst. prof. Fordham U., 1950-61; head dept. protein and enzyme chemistry Inst. Applied Biology, N.Y.C., 1953-62; adj. prof. Poly. Inst. Bklyn., 1957-59; vis. research prof. U. Ariz., Tucson, 1962—; research biophysicist VA, Tucson, 1962—; cons. Philips Roxane, Inc., St. Joseph, Mo., 1959—, USPHS, 1960-62. Pres. Electrofore Corp. Mem. Am. Chem. Soc., Am. Soc. Biol. Chemists, Am. Soc. Artificial Internal Organs. Editor: Electrophoresis, Theory, Methods and Applications, 1959, 67. Research and publs. on biochemistry of enzymes, stability of proteins, physico-chem. methods, chem. modifications of enzymes, large scale preparative electrophoresis, blood fractionation and preparation of gamma globulins, selective plasmapheresis; devel. of new artificial kidney, water purification by elec. means. Home: 5341 E. 7th St., Tucson 85711. Office: VA Hosp., Tucson 85713.*

BIERBAUM, Christopher Henry, American engineer, inventor; born Garnavillo, Iowa, February 14, 1864; B.S. Northern Illinois Normal School, 1886; M.E., Cornell University 1891. Instructor exptl. engring. Cornell U., 1892-96; cons. engr. Am. Stoker Co., Dayton, O., 1896-98; mem. Bierbaum & Merrick, engrs., Cin., 1898, Buffalo, 1901; founder, v.p. and cons. engr. in charge of research Lumen Bearing Co., from 1901; pvt. cons., Buffalo, 1903-28. Fellow A.A.A.S., Am. Soc. Mech. Engrs. (rep. on adv. com. for metall. research in U. S. Bur. Standards); mem. Am. Inst. Mining and Metall. Engring., Am. Soc. for Metals, Am. Micros. Soc., Sigma Xi. Contbr. many articles to tech. jours. and to engring, handbooks. Inventor of microcharacter, a device for studying physical properties of the microscopic constituents of metals, also determining the relative hardness of rolled metal sheets of less than two thousandths of an inch in thickness. Originator, patentee phosphor nickel bearing bronzes. Pioneer in studies of corrosive effect of oxidized mineral lubricating oils. Died June 15, 1947.

BIERER, Bert Worman, Am. poultry pathologist; b. Phila., June 2, 1911; s. Walter Steck and Bessie (Beckwith) B.; V.M.D., U. Pa., 1934; postgrad. arts and scis. U. S.C., 1950-51; m. Virginia Earl Murphy, Jan. 2, 1932; children—Ronald, Patricia (Mrs. David Bledsoe), Robert, Jody (Mrs. Don Shealy), Charles, Walter, Virginia. With Balt. City Health Dept., 1935-40, U. S. Dept. Agr., 1940-47; staff Clemson (S.C.) Livestock-Poultry Health Dept., 1947-—, lab. dir., 1956—; staff poultry sci. dept. Clemson U., 1960—, prof. poultry sci., 1964—. Recipient Distinguished Service award turkey com. Dixie Poultry Expn., 1964. Mem. Am. Poultry Sci. Assn., Am. Vet. Med. assn., Am. Assn. Avian Pathologists. Author: A Short History of Veterinary Medicine in America, 1955; also articles. Developed an egg wash germicide; evaluated nitrofuran drugs in Salmonella infections in poultry and use of injectable medicine, autogenous bacterin and water vaccine in poultry for fowl cholera; research on effect feed and water deprivation in poultry. Home: 5552 Sylvan Dr., Columbia 29206. Office: P.O. Box 1771, Columbia, S.C. 29202.*

BIERI, John Genther, Am. biochemist; b. Norfolk, Va., May 24, 1920; s. Bernhard H. and Elsie (Genther) B.; B.A., Antioch Coll., 1943; M.S., Pa. State U., 1944; Ph.D., U. Minn., 1949; m. Shirley J. Bloch, Sept. 19, 1943; children—Roger A., Barbara E., Nancy J. Faculty, U. Minn., 1948-49, Med. br. U. Tex., Galveston, 1949-55; biochemist NIH, Bethesda, Md., 1955—, chief sect. nutritional biochemistry Nat. Inst. Arthritis and Metabolic Diseases, 1964—. Recipient Mead Johnson award Am. Inst. Nutrition, 1965. Mem. A.A.A.S., Am. Chem. Soc., Am. Soc. Biol. Chemists, Am. Inst. Nutrition, Soc. Exptl. Biology and Medicine, Sigma Xi. Contbr. numerous articles to profl. jours. Research in fat soluble vitamins and essential fatty acids; biochem. studies of nutritional deficiencies. Home: 4616 Woodfield Rd. Office: NIH, Bethesda, Md. 20014.*

BIERMAN, Howard Richard, Am. physician; b. Newark, Jan. 27, 1915; s. Philip and Cecile (Cohen) B.; B.Sc., Washington U., 1935, M.D., 1939; m. Doris Rita Simmons, May 18, 1946; children—Barry, Tracy, Dana. Fellow in hematology sch. medicine, Washington U., 1939; house officer, dept. internal medicine Barnes Hosp., St. Louis, 1939-41; chief resident St. Louis Isolation Hosp. for Contagious Diseases, 1940; clin. physiologist Nat. Cancer Inst., 1946, chief clin. sect., lab. exptl. oncology, sch. medicine U. Cal., 1947-53, also asso. clin. prof. oncology; med. and sci. dir. Hosp. for Tumors and Allied Diseases, City of Hope Med. Center, Duarte Cal., 1953-56, med. and sci. dir. City of Hope Med. Centre, 1956-59, chmn. dept. medicine, 1954-59; clin. prof. medicine Loma Linda Univ., 1959—; dir. Inst. for Cancer and Blood Research, 1959—; sr. attending physician Los Angeles County Gen. Hosp. cons., Nat. Cancer Inst., 1953—; mem. of the biomechanics committee Protein Found., Harvard University. Served as comdr. USN, 1941-46. Diplomate Am. Bd. Internal Medicine; diplomate of Pan American Medical Association. Fellow A.C.P., A.A.A.S., Internat. Soc. Hematology, N.Y. Acad. Scis., Am. Soc. Hematology, Am. Coll. Angiology, Los Angeles Soc. Internal Medicine; mem. A.M.A., Cal., Los Angeles County medical societies, American, California socs. for internal medicine, Am. Soc. Clin. Investigation, Western Soc. Clin. Research, American Society of Clinical Oncology, American Federation Clin. Research, Soc. Exptl. Biology and Medicine, Am. Soc. for Pharmacology and Exptl. Therapeutics, Western Pharmacology Soc. (charter, pres. 1960-61, Sigma Xi. Author: Cancer Learning in Medical Schools, 1952, Cancer Learning in Dental Schools, 1954; Selective Arterial Catheterization, 1962, 67. Contbr. over 200 articles to profl. jours. Patentee. Developed concept of spreading accelerative forces equally to large areas for restraining belts harnesses and surfaces in aircraft; invented quick release device for oxygen masks and other safety devices in aviation; originated selective arterial catherization for diagnosis, therapeutic and investigative procedures in medicine; originated and 1st employed arterial therapy in cancer; first human cross-transfusion and leukocyte depletion (leukapherisis) procedures in man for leukemia and other diseases. Home: 300 Hilgard Av., Los Angeles 24. Office: 152 N. Robertson Blvd., Beverly Hills, Cal.*

BIERMANN, Ludwig Franz Benedikt, German astrophysicist; b. Hamm, Westfalen, Germany, Mar. 13, 1907; s. Franz and Thea (Schulte) B.; student U. München, 1925-27, U. Freiburg, 1927-28; Ph.D., U. Goettingen, 1932; m. Ilse Wandel, Jan. 3, 1942; children—Peter B., Christiane, Sabine. Exchange scholar U. Edinburgh, 1933-34, U. Jena, 1934-37, Univs. Sternwarte Berlin and Babelsberg, 1937-45; dozent U. Berlin, 1938-45, U. Hamburg, 1945-47; apl. prof. U. Hamburg, 1947, U. Goettingen, 1948; head astrophysics sect. Max Planck Inst. Physics, U. Goettingen, 1947-58; dir. Inst. for Astrophysics of Max-Planck-Inst. for Physics and Astrophysics, Munich, 1958—; hon. prof. U. Munich, 1959—; vis. prof. Cal. Inst. Tech., Haverford Coll. and Princeton, 1955, 61, U. Cal. at Berkeley, 1959-60, Sydney and Canberra (both Australia), 1960. Recipient Copernicus prize, 1943. Mem. Wissenschaftliche Leitung of Inst. for Plasma Physics, Munich-Garching, Max-Planck-Gesellschaft, Astronomische Gesellschaft, Physikalische Gesellschaft, Internat. Astron. Union, Gesellschaft für Angewandte Mathematik und Mechanik, Gesellschaft Deutscher Naturforscher und Arzte, Internat. Acad. Astronautics, Bayerische Akademie der Wissenschaften, Royal Astron. Soc., London, Société Royale des Scis. de Liége (Belgium) (corr.). Research on stellar evolution. Home: Munich 23, Rohmederstrasse 12. Office: Munich 23, Fohringer Ring 6 Germany.*

BIERMER, (Michael) Anton, physician; b. Bamberg, Germany, Oct. 18, 1827; s. Magnus and Karolina (Dinkel) B.; studied in Würzburg, Germany; m. Sophie Josepha Barbara Wahl, 1851; 1 son. Lectr., asst. Juliusspital, Würzburg; apptd. prof. specialized pathology, dir. med. clinic, Bern, Switzerland, 1861, Zürich, Switzerland, 1865; went to Breslau, Germany, 1874. Research and publs. on diseases of breathing organs, hematology, discovered pernicious anemia, 1872. Died Berlin, June 24, 1892.

BIERSTEDT, Robert, Am. sociologist; b. Burlington, Ia., Mar. 20, 1913; s. Henry F. and Bertha (Strauss) B.; A.B., U. Ia., 1934; M.A., Columbia, 1935, Ph.D., 1950; postgrad. Harvard, 1936-37; m. Betty MacIver, Dec. 26, 1939; children—Peter, Karen, Robin. Lectr. philosophy Columbia, 1937-39; lectr. social studies Bennington Coll., 1939-40; instr. Bard Coll., 1940-43; asst. prof. U. Wash., 1946, Wellesley (Mass.) Coll., 1946-47; faculty U. Ill., Urbana, 1947-53; prof., chmn. dept. sociology and anthropology City Coll. N.Y., 1953-59; Fulbright lectr. U. Edinburgh, Scotland, 1959-60; prof. N.Y. U., N.Y.C., 1960—, head dept. sociology, 1960-66; Barnett lectr. Oxford U., 1960; Fulbright lectr., London. Sch. Econs., 1966-67. Fellow Am. Sociol. Assn. (v.p. 1964-65); mem. Am. Assn. U. Profs., Am. Civil Liberties Union, Eastern Sociol. Soc. (pres. 1958-59), Brit. Sociol. Assn., Sociol. Research Assn. Author: The Social Order, 1957, 63; (with E. J. Meehan and P. A. Samuelson) Modern Social Science, 1964; Emile Durkheim, 1966; also articles. Editor: The Making of Society, 1959; adv. editor Dodd, Mead

& Co., 1957—. Research in clarification of sociol. concepts such as power, authority, etc. Home: 110 Bleecker St., N.Y.C. 10012.*

BIESBROECK, George van, see van Biesbroeck, George.

BIESELE, John Julius, Am. biologist, educator; b. Waco, Tex., Mar. 24, 1918; s. Rudolph L. and Anna (Jahn) B.; B.A. with highest honors, U. Tex., 1939, Ph.D., 1942; m. Marguerite Calfee McAfee, July 29, 1943; children—Marguerite Anne, Diana Terry, Elizabeth Jane. Fellow Internat. Cancer Research Found., U. Tex., Austin, 1942-43, prof. zoology, mem. grad. faculty, 1958—, cons. in biology M.D. Anderson Hosp. and Tumor Inst., Houston, 1958—, dir. Genetics Found., Austin, 1959—; fellow Internat. Cancer Research Found., Barnard Skin and Cancer Hosp., St. Louis, 1943, U. Pa., Phila., 1943-44; instr. U. Pa., 1943-44; temporary research asso. dept. genetics Carnegie Instn. Washington, Cold Spring Harbor, 1944-46; asst. Sloan-Kettering Inst. for Cancer Research, N.Y.C., 1946-47, research fellow, 1947, asso., 1947-55, head cell growth sect. div. exptl. chemotherapy, 1947-58, mem. 1955-58, asso. scientist, 1959—. Research asso. biology dept. Mass. Inst. Tech., 1946-47; asst. prof. anatomy Cornell U. Med Sch., N.Y.C., 1950-52, asso. prof. biology Sloan-Kettering div. Cornell U. Grad. Sch. Med. Scis., N.Y.C., 1952-55, prof. biology, 1955-58; mem. cell biology study sect. NIH, 1958-63; mem. adv. com. on research on etiology cancer Am. Cancer Soc., 1961-64; counsellor Cancer Internat. Research Coop., Inc., 1962—. Recipient Public Health Service Research Career award Nat. Cancer Inst., NIH, 1962-67. Fellow N.Y. Acad. Scis., A.A.A.S., Tex. Acad. Sci.; mem. Am. Assn. for Cancer Research (dir. 1960-63), Am. Assn. U. Profs. (v.p. U. Tex. chpt. 1963-64, pres. 1964-65), Am. Inst. Biol. Scis., Am. Soc. for Cell Biology, Am. Soc. Human Genetics, Am. Soc. Naturalists, Am. Soc. Zoologists, Electron Microscopy Soc. Am., Harvey Soc., Histochem. Soc., Internat. Soc. for Cell Biology, Tissue Culture Assn., Nat. Audubon Soc., Nature Conservancy, Phi Beta Kappa (exec. com. U. Tex. chpt. 1961-62, 64-65), Sigma Xi (pres. U. Tex. chpt. 1963-64), Phi Eta Sigma. Author: Mitotic Poisons and the Cancer Problem, 1958. Mem. editorial bd. Year Book of Cancer, 1959—. editorial adv. bd. Cancer Research, 1960-64. Studies, publs. on early chromosomal changes in cancer; effect of antimetabolite chems. on cancer cells; chromosomal abnormalities in certain retarded persons. Home: 2500 Great Oaks Pkwy., Austin, Tex. 78756.

BIETTI, Giambattista, Italian occulist; b. Padua, Italy, Apr. 29, 1907; s. Amilcare and Clelia (Engel) B.; M.D.; M.D. in surgery; m. Maria Titomanlio, June 24, 1934; children—Clelia, Amilcare. Dir. ophthal. clinic U. Rome; superior cons. mem. in instrn.; superior cons. mem. of health. Mem. Leopoldina Acad., Acad. Medicine of Rome, Italian Ophthal. Soc. (v.p.), Assn. Aero. and Spatial Medicine, Internat. Orgn. against Trachoma (pres.). Research and over 280 articles on ophthalmology. Address: via Cesare Beccaria 18, Rome, Italy.*

BIEZENO, Cornelis Benjamin, Dutch mech. engr.; b. Delft, Netherlands, Mar. 2, 1888; ed. Technol. U. Delft; Dr. honoris causa of applied sci. U. Amsterdam, Free U. Brussels, U. Gand. Prof. applied mechanics, 1914-58; vis. prof. tech. U. Bandung, 1929-30; rector Technol. U., 1937-38, 49-51; vis. prof. Mass. Inst. Tech., 1951-52. Mem. Royal Acad. Sci., Acad. Sci. Berlin (corr.), Acad. Sci. Madrid (corr.), Congress Physics and Medicine (pres.), Nat. Com. for Reorgn. Advanced Studies (pres. subdiv. tech. scis.), Internat. Congress Applied Mechanics (pres.), Conservatory of Center of Math. Author: (with Grammel) Technische Dynamik (translated into Spanish, Russian, English). Address: Maerten Trompstraat 27, Delft, Netherlands.

BIFFEN, Rowland, English botanist; b. Cheltenham, Eng., May 28, 1874; s. Henry John and Mary Biffen; ed. Emmanuel Coll., Cambridge (Eng.) U., Frank Smart student in botany, Gonville and Caius Coll., 1896; D.Sc. (hon.), Reading U., 1935; m. Mary Hemus, 1899. Studied rubber prodn., Brazil, Mexico, W.I.; became demonstrator in botany Cambridge U., also lectr. Sch. Agr., 1899, 1st prof. agrl. botany, 1908-31; dir. Plant Breeding Inst., 1912-36. Created knight, 1925; Darwin medal, 1920. Fellow Royal Soc., 1914. Pioneer in breeding rust-resistant wheat strains; found that Mendelian principles apply to inheritance of constitutional character in wheat, including resistance to specific plant diseases, 1903-05. Died Cambridge, July 12, 1949.

BIGELEISEN, Jacob, Am. chemist; b. Paterson, N.J., May 2, 1919; s. Harry and Ida (Slomowitz) B.; A.B., N.Y. U., 1939; M.S., Wash. State U., 1941; Ph.D., U. Cal. at Berkeley, 1943; m. Grace Alice Simon, Oct. 21, 1945; children—David M., Ira S., Paul E. Research scientist Manhattan dist. Columbia, 1943-45; research asso. Ohio State U., Columbus, 1945-46; fellow Enrico Fermi Inst., U. Chgo., 1946-48; sr. chemist Brookhaven Nat. Lab., Upton, N.Y., 1948—; vis. prof. Cornell U., 1953; NSF sr. fellow, vis. prof. Eidgen Techn. Hochschule, Switzerland, 1962-63. Trustee, Sayville Jewish Center, 1954—. Recipient Nuclear award Am. Chem. Soc., 1958, Gilbert N. Lewis lectr., 1963, E. O. Lawrence award, 1964. Fellow Am. Phys. Soc., Am. Chem. Soc., A.A.A.S.; mem. National Academy of Sciences; member of Sigma Xi, Phi Beta Kappa, Phi Lambda Upsilon. Mem. editorial bd. Jour. Chem. Physics, Ann. Rev. Phys. Chemistry, Jour. Phys. Chemistry. Research in photochemistry in rigid media, semiquinones, cryogenics, chemistry of isotopes, quantum statistics of gases, liquids and solids. Home: 47 Fairview Av., Bayport, L.I., N.Y. 11705. Office: Brookhaven Nat. Lab., Upton, N.Y. 11973.*

BIGELOW, Frank Hagar, meteorologist; b. Concord, Mass., Aug. 28, 1851; s. Francis Edwin and Ann (Hagar) B.; A.B., Harvard, 1873. A.M., 1880; B.D., Episcopal Theol. Sch., Cambridge, 1880; L.H.D., Columbian (now George Washington) U., 1899; m. Mary E. Spalding. Oct. 6, 1881. Astronomer, Cordoba Obs., Argentine Republic, 1873-76, 81-83; 1881-83; took part in Dr. B. A. Gould's exploration of So. heavens; prof. mathematics Racine Coll., 1884-89; asst. Nautical Almanac Office, 1889-91, mem. U. S. Eclipse Expdn. to W. Africa, 1889, Newberry, S.C., 1900, Spain, 1905; prof. meteorology U. S. Weather Bur., 1891-1910, chief, climatol. div.; prof. solar physics George Washington U., 1894-1910; prof. meteorology Oficina Meteorologica, Cordoba, Argentina, 1910-21; dir. Pilar Solar and Magnetic Obs., 1915-21. Rector, Natick, Mass., 1880-81; chaplain, Racine, Wis., 1885-89; asst. minister St. John's Ch., Washington, 1891-1910. In charge of researches into law of evaporation at Salton Sea and in U. S. generally; research and publs. on meteorology, solar physics, circulation and radiation of atmospheres of earth and sun, sun's radiation, other solar phenomena. Deceased.

BIGELOW, Henry Bryant, Am. zoologist; b. Boston, Oct. 3, 1879; A.B., Harvard, 1901, A.M., 1904, Ph.D., 1906, Sc.D. (hon.), 1946; Sc.D., Yale, 1941; Ph.D., U. Oslo (Norway), 1946; m. 1906; 2 children. Asst. in invertebrate zoology Mus. Comparative Zoology, Harvard, 1906-13, curator coelenterates, 1913-25, research curator zoology, 1925-27, curator oceanography, 1927—, asso. prof. zoology, 1927-31, prof., 1931-50, prof. emeritus, 1950—. Dir. Oceanographic Inst., Woods Hole, Mass., 1930-39; chmn. N.Am. Council Fisheries Investigations; mem. Agassiz expdn. Maldive Islands, 1901-02, W.I., 1907; Brown-Harvard expdn. Labrador, 1900; Albatross E. Pacific expdn., 1904-05; Recipient Agassiz medal, 1931, Bowie medal, Schmidt medal. Fellow Am. Acad. Arts and Scis., Royal Geog. Soc. (London); mem. Nat. Acad. Scis., Soc. Naturalists, Philos. Soc., Geog. Soc., Soc. Zool., Geophys. Union; London Zool. Soc., Marine Biol. Assn., others. Spl. work in cytology of Coelenterata, plankton, elasmobranchs, marine biology, oceanography, Siphonphorae.

BIGELOW, Henry Jacob, Am. surgeon; b. Boston, Mar. 11, 1818; s. Jacob and Mary (Scollay) B.; grad. Harvard, 1837, M.D., 1841; m. Susan Sturgis, May 8, 1847, 1 son, William Sturgis. Tchr. surgery Tremont Street Medical School, Boston, 1845-49; surgeon Massachusetts General Hospital, 1846-86; published first account of W. Morton's use of ether in surgical operation; asso. with discovery of surg. anesthesia, 1846; prof. surgery Harvard Med. Sch., 1849-84; did research and experiments on anatomy of hip-joint circa 1852; developed spl. type of lithotrite (Bigelow's lithotrite) for crushing bladder stones; 1st Am. surgeon to excise hip-joint, 1852; inventor operating chair. Author: Manual of Orthopedic Surgery (Boylston prize essay), 1844; Medical Education in America, 1871; writings collected in Works of Henry Jacob Bigelow (edited by William Sturgis Bigelow), 3 vols., 1900. Died Newtown Creek, Mass., Oct. 30, 1890.

BIGELOW, Jacob, Am. physician, botanist; b. Sudbury, Mass., Feb. 27, 1786; s. Jacob and Elizabeth (Wells) B.; grad. Harvard, 1806, LL.D., 1857; M.D., U. Pa., 1810; m. Mary Scollay; children include Henry Jacob. Began practice medicine, Boston, 1811; gave 1st lectures on botany at Harvard, 1812, active next decade collecting bot. specimens in New Eng.; prof. medicine Harvard Med. Sch., 1815-35; Rumford prof. applied sci. Harvard, 1816-27; founded Mt. Auburn Cemetery, Cambridge, Mass., as public health project, 1831; asso. with Mass. Gen. Hosp.; head of com. that formed Am. Pharmacopeia, 1820. Mem. Am. Philos. Soc., Am. Acad. Arts and Scis. (pres. 1846-63). Author: Florula Bostoniensis (standard manual of New Eng. botany until Gray's Manual of 1848), 1814; American Medical Botany, 3 vols., 1817, 18, 20; Treatise on Materia Medica, 1822; Discourse on Self Limited Diseases, 1835; Elements of Technology, 1829; Brief Expositions of Rational Medicine, 1858. Made important contbns. to med. theory and practice; several plants named Biglovia in his honor. Died Boston Jan. 10, 1879.

BIGELOW, Wilfred Gordon, Canadian surgeon; b. Brandon, Man., Can., June 18, 1913; s. Wilfred Abram and Grace (Gordon) B.; B.A., U. Toronto (Ont., Can.), 1935, M.D., M.S., 1938; m. Margaret Ruth Jennings, July 9, 1941; children—Mary Elizabeth (Mrs. Ian B. C. Currie), John Jennings, Dan David, William Gordon. Mem. faculty U. Toronto, 1948—, asso. prof. surgery, 1956—; head div. cardiovascular surgery Toronto Gen. Hosp., 1956—. Mem. Def. Research Bd. Can., 1966—; dir. Ont. Heart Found. Recipient Peters prize, also Lister prize, 1949; 1st medal Internat. Soc. Surgery, 1956; Gairdner award, 1959. Mem. Soc. Thoracic Surgeons Gt. Britain and Ireland (hon.), Assn. Surgeons of India (hon.), Am. Surg. Assn., Am. Assn. Thoracic Surgery, Soc. U. Surgeons, Royal Coll. Surgeons and Physicians Can., Internat. Cardiovascular Soc. Research and numerous publs. on hypothermia of low body temperature for heart surgery; developed (with J. C. Callaghan and J. A. Hopps) 1st elec. pacemaker to supply stimulus for slow or stopped heart; microscopic circulation and human valve transplants. Home: 137 Douglas Dr. Toronto 7, Ont., Can.*

BIGELOW, Willard Dell, Am. chemist; b. Gardner, Kan., May 31, 1866; s. William I. and Jennie (Lytle) B.; A.B., Amherst 1889; m. Nancy M. Nesbit, Apr. 9, 1896. Asst. prof. chemistry Ore. State Coll., 1889-90; instr. chemistry Washington High Sch., 1891-92; chemist, bur. chemistry U. S. Dept. Agr. 1892-1913, asst. bur. chief. 1903-13, chief, div. foods, 1901-13, mem. bd. food and drug inspection, 1913; chief chemist Nat. Canners' Assn., 1913-18, dir. research labs., 1918. Prof. chemistry Nat. U., Washington, 1893-98. Author bulls. on composition and adulteration of food, also on tech. of canning. Died Mar. 6, 1939.

BIGG, Edward Keith, Australian physicist; b. Armidale, Australia, Dec. 19, 1925; s. Edward Lionel and Hazel (Lee) B.; B.Sc., U. Sydney (Australia), 1945, B.Sc. with honors, 1946, M.Sc., 1948; Ph.D., U. London (Eng.), 1953; m. Robin Pursell, Mar. 21, 1953; children—Ian, Judith, Penelope, Alison. Commonwealth research asst. New Eng. U. Coll., 1947-48; radiophysicist Commonwealth Sci. and Indsl. Research Orgn., 1948—; sr. research scientist London U., 1951-53; vis. scientist Nat. Center Atmospheric Research, Boulder, Colo., 1964-65. Mem. Am. Geophys. Union, Royal Meterorol. Soc. Research and publs. on elec. discharges in low pressure gases, longwave radio propagation, supercooling of water, ice nuclei in atmosphere, study atmosphere by twilight scattering, radio emission from Jupiter. Home: 12 Wills Av., Castle Hill 2154, Australia. Office: Radiophysics Div., CSIRO, Pvt. Box, Epping 2121, Australia.*

BIGGART, J. Henry, Brit. neuropathologist; b. Nov. 17, 1905; s. John Henry and Mary (Gault) B.; ed. Queen's U., Belfast, Johns Hopkins Med. Sch.; M.B., 1928, M.D., 1931, D.Sc., 1937; M.D. (hon.), Dublin, 1957; m. Mary Isobel Gibson, 1934; 1 son, 1 dau. Pathologist Scottish Asylum's Bd., 1933-37; lectr. neuropathology Edinburgh U., 1933-37; regional dir. Blood Transfusion Service, 1939-46; prof. pathology Queen's U., Belfast, 1937—, dean faculty medicine, 1943—, dir. inst. pathology, 1948—. Chmn. lab. services com. Hosps. Authority, 1948, chmn. med. edn. and research com., 1950; mem. Gen. Med. Council, 1941; mem. Gen. Dental Council, 1959. Fellow Royal Coll. Physicians, Coll. Pathologists. Author: Text book of Neuropathology, 1936; also papers on gen. and nervous pathology. Produced (with George Lionel Alexander) diabetes insipidus in dogs by exptl. injury to hypothalamus, circa 1939. Address: 64 King's Rd., Belfast, Ireland.

BIGGER, Isaac Alexander, Am. physician; b. Bethel, S.C., June 25, 1893; s. Isaac Alexander and Mary Neel (Johnston) B.; student Erskine Coll., 1909-10, Davidson Coll., 1910-11, 12-13, U. Va., 1919, M.D., 1919; m. Beatrice Haslam, Sept. 9, 1922; children—Dorothy Neel (Mrs. Jack Grober), Barbara Norvelle (Mrs. George W. Reahm), Edith Millicent (Mrs. Claude S. Coleman). Intern U. Va., 1919-21, resident surgery, 1921-22, instr. surgery, 1922-23, asst. prof. surgery, 1923-27; asso. prof. surgery Vanderbilt U., 1927-30; prof. surgery and surgeon-in-chief Med. Coll. of Va. since 1930; mem. deans com. and cons. surgery, McGuire V.A. Hosp., Richmond, from 1946; cons. Portsmouth Naval Hosp. from 1950. Mem. subcom. thoracic surgery of coms. on mil. medicine, div. med. sciences NRC, 1941-42. Fellow A.C.S.; mem. Am. Surg. Assn., Am. Cancer Soc. (pres. Va. div. 1945, 47-48), Am. Assn. Thoracic Surgery (treas. 1937-44, v.p. 1944-45, pres. 1946), A.M.A., So. Med. Assn. (chmn. surg. sect. 1935-36), So. Surg. Assn. (pres. 1952-53; rep. on Am. Bd. Surgery 1945-51), Eastern Surg. Soc., So. Soc. Clin. Surgeons, Med. Soc. Va., Richmond Acad. Va. (pres., 1947). Author: Operative Surgery, 2 vols., 1937, new ed., 1940 (with Dr. J. Shelton Horsley). Contbr. articles to med. jours. Died Jan. 27, 1955; buried Richmond, Va.

BIGGER, Joseph Warwick, Irish physician; b. Belfast, Ireland, Sept. 11, 1891; s. Edward Coey and Maude Coulter (Warwick) B.; ed. St. Andrew's Coll., Dublin; M.B., Trinity Coll., Dublin, 1916; M.A.; M.D.; Sc.D.; D.P.H.; m. Patricia Mai Curtin, 1916; 1 son, 1 dau. Demonstrator in pathology, bacteriology U. Sheffield, Eng., 1916-19; pathologist, med. insp. Local Govt. Bd., Ireland, 1919-22; prof. preventive medicine, forensic medicine Royal Coll. Surgeons, Ireland, 1920-22; prof. bacteriology, preventive medicine Dublin U., 1924-50; dean sch. physic Trinity Coll., 1936-50. Fellow Royal Coll. Physicians in Ireland, Royal Coll. Physicians, Trinity Coll., Dublin (hon.); mem. Royal Irish Acad. Author: Man Against Microbe, 1939; Handbook of Hygiene, 1941; Handbook of Bacteriology, 1949. Contbr. Jour. Pathology and Bacteriology, Jour. Hygiene, Lancet, Irish Jour. Med. Sci., others. Died Aug. 17, 1951.

BIGGERS, John Dennis, physiologist; b. Reading, Eng., Aug. 18, 1923; s. Wilfrid Norman and Winifrid (Gardner) B.; B.Sc. in Vet. Sci., U. London, (Eng.), 1946, B.Sc. in Physiology, 1946, Ph.D., 1952, D.Sc., 1965; m. Barbara Joan Nevill Cobbold, July 24, 1948; children—David John, Philippa Jeanne, Jennifer Anne. Came to U. S., 1959. Lectr. univs. Sheffield, Eng., 1947-48, U. Sydney, Australia, 1948-55, Royal Vet. Coll., London, 1955-59; vis. scientist Strangeways Research Lab., also Commonwealth fellow Cambridge (Eng.) U., 1954-55; asso. mem. Wister Inst., Phila., 1959-61; King Ranch research prof. reproductive physiology U. Pa., 1961-66; prof. reproductive physiology Johns Hopkins Sch. Hygiene and Pub. Health, Balt., 1966—. Mem. corp. Marine Biol. Lab., Woods Hole, Mass. Fellow Royal Coll. Vet. Surgeons; mem. Anat. Soc. Gt. Britain, Soc. Study Reprodn. (pres. elect), Am. Inst. Biol. Scis. (gov.); Soc. Endocrinology U.K., Physiol. Soc. U.K., Internat. Biometrics Soc., Internat. Soc. Cell Biology, Soc. Fertility U.K., Am. Statis. Assn., A.A.A.S., Am. Soc. Cell Biology, Tissue Culture Association, Sigma Xi. Research on animal reproduction, action of hormones, uniformity of exptl. animals for bioassay, early mammalian devel., organ culture of cartilage and bone. Home: 313 Tuscany Rd., Balt. 21210.*

BIGGS, Burnard Storey, Am. chemist; b. San Marcos, Tex., Sept. 10, 1907; s. Kenner Dee and Lelia Lou (Storey) B.; B.A., S.W. Tex. Tchrs. Coll., San Marcos, 1927; M.A., U. Tex., 1931, Ph.D., 1933; m. Ruth McRell, Aug. 19, 1930; children—John McRell, Jacqueline Storey. Mem. tech. staff, coal research lab. Carnegie Inst. Tech., 1933-36; mem. tech. staff Bell Telephone Labs., 1936-58, asst. chem. dir., 1954-58; dir. materials and standards devel. Sandia Corp., Albuquerque, 1958-60, v.p. devel., 1960-61; v.p. Livermore Lab., Sandia Corp. (Cal.), 1961—. Chmn. Gordon Research Conf. Elastomers, 1957; chmn. sci. adv. com., prevention of deterioration center Nat. Acad. Scis.-NRC. Mem. Am. Chem. Soc. (chmn. N.J. sect. 1950-51), Am. Ordnance Assn. Research in nitrogen compounds; chem. structure of coal components; deterioration processes in organic materials. Home: 33 Golf Rd., Pleasanton, Cal. 94566. Office: P.O. Box 969, Livermore, Cal. 94550.*

BIGGS, Hermann Michael, Am. physician, bacteriologist; b. Trumansburg, N.Y., Sept. 29, 1859; s. Joseph Hunt and Melissa T. (Pratt) B.; A.B., Cornell, 1882; M.D., Bellevue Hosp. Med. Coll. (N.Y. U.), 1883; LL.D., N.Y. U., 1910, Rochester U., 1917; D.Sc., Harvard, 1920; m. Frances M. Richardson, Aug. 18, 1898. Vis. phys. Workhouse and Almshouse hosps., 1885-92; asst. pathologist Bellevue Hosp., 1886-92, pathologist, 1892-99; pathologist, City Hosp., 1886-92; lectr., prof. pathol. anatomy, Bellevue Hosp. Med. Coll., 1885-94; cons. phys. Hosp. for Contagious Diseases, from 1889; prof. therapeutics and clin. medicine, 1897-1907, asso. prof. medicine Univ. and Bellevue Hosp. Med. Coll., 1907-14, prof. medicine, from 1914; pathologist and dir. Bacteriol. Labs., 1892-1901, gen. med. officer N.Y. Dept. Health, 1901-14; state pub. health commr., N.Y., from 1914. Dir. Rockefeller Inst. for Med. Research from orgn. in 1901. Pres. Tuberculosis Preventorium for Children. Wrote works dealing with outbreak of cholera in N.Y., 1893, treatment of Tb, other med. subjects. Organized bacteriol. labs. of N.Y.C. Health Dept., 1892; introduced use of bacteriol. methods in san. surveillance of infectious disease; introduced use of diphtheria antitoxin in U. S., 1894; introduced division of infant and maternity welfare (as pub. health commr.). Died June 28, 1923.

BIGGS, Robert Mitchell, Am. economist; b. Mio, Mich., Aug. 29, 1915; s. Morgan S. and Myrtle A. (Mitchell) B.; B.A., Wayne State U., 1939; M.A., U. Mich., 1940, Ph.D., 1950; m. Jane E. Williams, June 6, 1953; 1 son, Richard Oliver. Teaching fellow U. Mich., 1941-45; instr. Wayne State U. 1946-49; from instr. to prof. U. Detroit, 1949-61; prof. econs. chmn. dept., also chmn. div. social sci. Jamestown (N.D.) Coll., 1961-63; prof. econs. U. Toledo, 1963—. Chmn. econs. sect. Mich. Acad. Sci., Arts and Letters, 1959. Mem. Am. Econ. Assn., Am. Finance Assn., Phi Beta Kappa, Phi Kappa Phi. Author: National Income Analysis and Forecasting, 1956; also articles. Empirical research in econ. behavior of consumer expenditures; bus. purchases of plant, equipment and additions to inventories, corporate profits. Home: 3145 Scarsborough St., Toledo 43615.*

BIGOT, Alexandre Pierre Désiré, French geologist; b. Cherbourg, France, May 15, 1863; grand nephew of Eudes-Deslonchamps; docteur ès scis.; prof. geology and paleontology Faculty Scis., Caen, France. Mem. French Acad. Scis., 1919. Published (with Deslongchamps) theoretical geol. study of Normandy. Died Mathieu, Calvados, France, Apr. 20, 1953.

BIGOURDAN, Guillaume, astronomer; b. Sistels, France, Apr. 6, 1851; ed. Toulouse (France) Obs.; docteur ès scis., Paris, 1886. With Paris Obs., from 1879; mission to Martinique, 1882, to Germany, Austria, Russia, 1883; became professor of astronomy Montsouris Observatory, 1890. Mem. of the French Academy of Sciences, 1904, Bur. Longitudes, Acad. Agr. Author: Observation de nebuleuses et d'amas d'étoiles, 1917. Undertook cataloging of some 7000 nebulae; discovered several feeble nebulae; studied deviation of positions of new nebulae, also star visilility in full daylight, star movements; contbd. to orgn. of transmission of wireless signals. Died Paris, Feb. 28, 1932.

BIGSBY, John Jeremiah, Brit. geologist; b. Nottingham, Eng., Aug. 14, 1792; s. John Bigsby; M.D., Edinburgh, Scotland, 1814. Commd. med. officer, 1817; reported on geology of upper Can., 1819; practiced medicine Newark, 1827-46, London, Eng., 1846-81; Brit. sec., med. officer Canadian Boundary Commn., 1881. Fellow Royal Soc., 1869, Geol. Soc. (Murchison medal, 1874). Author: Seaside Manual of Invalids and Bathers, 1841; Thesaurus Siluricus . . . , 1868; Thesaurus Devonico-Carboniferous . . . , 1878. Contbr. sci. and other articles to Transactions of Geol. Soc. and others. Died Gloucester, Eng., Feb. 10, 1881.

BIJL, Jacob Pieter, Dutch physician; b. Stavoren, June 3, 1880; s. Pieter Kornelis and Christina (Jäger) B.; M.D., U. Utrecht; m. Henriette Constance Momma, Oct. 10, 1911; 1 son, Henri Reginald. Asst. in path. anatomy U. Groningen; prof. Superior Sch. War; head bacteriology sect., underdir. central lab. of pub. health at Utrecht; dir. Inst. Preventive Medicine, Leyde; full prof. U. Leyde. Mem. council Hosp. of Psychiatry of Utrecht-den-Dolder; sec. San. Commn. of Utrecht. Mem. Dutch Soc. Sci., Soc. Scis. and Arts of Utrecht, also mem., sec., pres. numerous hygiene and preventive medicine orgns. Research and numerous publs. on hygiene, epidemiology, bacteriology, serology. Address: Rijksstraatweg 538, Wassenaar, Netherlands.

BIKFALVI, Andreas, physician; b. Borband, Transylvania, Romania, Jan. 1, 1916; s. Steph. and Anna (Krisan) B.; M.D.; U. Kolozsvar (Romania), 1933-40; postgrad. Carlo Forlanini Inst., Rome, Italy, 1942-43; m. D.M. Ildiko Kurjatko, July 13, 1954; children—Barbara, Andras, Ildiko. Asst., Carlo Forlanini Inst., Rome, 1945-47; asst. prof. surg. clinic U. Budapest Hungary, 1947-56; asst. prof. Clinic Chest Diseases Naples (Italy), 1946-57; asst. prof. surg. clinic U. Giessen (Germany), 1957—; head surg. dept. Lich, U. Hosp., 1966—. Mem. Deutsche Gesellschaft für Chirurgie, Hungarian Coll. Surgeons, N.Y. Acad. Scis. Research and publs. on devel. stapling device used in vascular surgery, plastic surgery of tracheo-bronchial tree. Home: 6302 Lich, Brunnerstrasse 20, West Germany.*

BILFINGER, Georg Bernhard, German philosopher, mathematician, physicist; b. Bad Cannstatt, Württemberg, Jan. 23, 1693; ed. in theology at Tübingen; studied under Christian Wolff at Halle, 1719. Prof. of philosophy, Halle, 1721-24; prof. of mathematics, 1724; prof. of physics, St. Petersburg, 1725-31; prof. theology, Tübingen, from 1731; privy councillor to Charles Alexander, Duke of Württemberg, 1735-37; member of regency council from 1737. Mem. St. Petersburg Acad. of Scis. Received prize of French Acad. of Scis., 1728. Author: Dilucidationes philosophicae de Deo, anima humana, mundo et generalibus rerum affectionibus, 1725. Defended Leibnizian doctrine of conservation of living forces against English Newtonians, French Cartesians, and Daniel Bernoulli; opposed Newtonian view of capillarity; was Wolffian leader of St. Petersburg Acad. of Scis., 1725-31; opposed Leibniz's view that monads, units of force, are both spiritual and physical at same time; asserted monads might be of different types, some spiritual and some physical; in opposition to Wolff and Leibniz, believed term "divine harmony" refers not to universe but only to relationship between soul and body. Died Stuttgart, Feb. 18, 1750.

BILHARZ, Theodor Maximillian, German physician, zoologist; b. Sigmaringen, Germany, Mar. 23, 1825; s. Anton and Elisabeth (Fehr) B.; student, Freiburg, Germany, from 1843, Tubingen, Germany, from 1845; M.D., 1849. Worked (with von Siebold) on comparative anatomy, Freiburg, Germany; became prosector Anat. Inst., Freiburg; accompanied Wilhelm Griessinger to Egypt, 1850, and entered govt. service; became hosp. dept. chief, 1853; named prof. Kar-el-bin Med. Sch., Cairo, Egypt, 1855, prof. descriptive anatomy, 1856, also official forensic physician. Contbd. numerous papers to med. and zool. publs. Discovered semitropical disease bilharziasis, 1851. Died Cairo, May 9, 1862.

BILIBIN, Aleksandr Fedorovich, Russian therapeutist, epidemiologist; b. Surazh (now Chernigov Oblast), 1897; grad. Kiev Med. Inst., 1922; D.Med. Sci., 1943. Dist. dr., head outpatient dept., later intern City Isolation Hosp., Bryansk, 1924-31; head isolation dept. Moscow Botkin Hosp., 1932-43; asst. Central Postgrad. Med. Inst., Moscow, 1933-35, lectr. chair infectious diseases, 1935-43; head chair infectious diseases 3d Moscow Med. Inst., 1944-49, Pirogov 2d Moscow Med. Inst., 1950—; sci. dir. Isolation Hosp., 4th Main Bd., USSR Ministry Health, 1950—; sci. dir. clinic Ivanovsky Inst. Virology, USSR Acad. Med. Sci.; prof., 1944—. Del., 2d Internat. Congress on Infectious Pathology, Milan, Italy. Mem. USSR Acad. Med. Sci., All-Union (bd. mem.), Moscow (bd. mem.) socs. epidemiologists, microbiologists and infectionists. Author over 100 works including The Semiotics and Diagnosis of Infectious Disease, 1950; A Differential Diagnostic Key to the Main Acute Infectious Diseases, 1955; The Treatment of Infectious Diseases: Basic Principles, 1956; Basic Tasks of Research and Practical Work for Reducing the Incidence of Sysentery, 1960; A Textbook of Infectious Diseases, 1962; coauthor: Typhoid and Paratyphoid, 1949; Course on Infectious Diseases, 1956. Co-editor Epidemiology and Infectious Diseases sects. Large Med. Ency., 2d edit.; mem. editorial bd. Therapeutic Archives; mem. editorial council Clin. Medicine, Med. Abstracts Jour., Jour. Microbiology, Epidemiology and Immunobiology, also others. Developer vaccine for treating tularemia, 1942, original classification of chronic forms of dysentery; devised method of using colloidally dispersed silver sulfathiazole for treating dysentery; 1st in USSR to describe torulosis, listerellosis and Omsk hemorrhagic fever. Address: Pirogov 2d Moscow Med. Inst., Malaya Pirogovskaya 1, Moscow, USSR.

BILLARD, Charles Michel, French physician; b. Pelouailles, France, June 16, 1800; M.D., Angers, France, 1828; practiced at Angers, from 1828. Author: Traité des maladies des enfants nouveau-nés et à la mamelle (on childhood diseases from birth to puberty), 1828. Translator: The Disease of the Eye (Lawrence), with "Precis on the Pathological Anatomy of the Eye," 1830; Elements of Chemistry (Thomson). Contbd. to study of pediatrics. Died Paris, Jan. 31, 1831.

BILLEN, Daniel, Am. biologist; b. N.Y.C., Nov. 27, 1924; s. Morris and Gertrude (Feinman) B.; B.S., Cornell U., 1948; M.S., U. Tenn., 1949, Ph.D., 1951; m. Gertrude Eleanor Berlin, Mar. 11, 1951; children—Rhonda Ann, Jerome, Robin Madoline. Instr. U. Tenn., 1950-51; biologist Oak Ridge (Tenn.) Nat. Labs., 1951-57; prof. biology U. Tex., also biologist, chief sect. radiation biology, M.D. Anderson Hosp. and Tumor Inst., Houston, 1957-66, prof. radiation biology U. Fla., Gainesville, 1966—; program dir. metabolic biology NSF, 1960-61; cons. NIH, 1964—. Recipient Assn. Southeastern Biologists Research award, 1953. Mem. Am. Soc. Microbiologists, Am. Acad. Microbiology, Am. Soc. Biol. Chemists, Radiation Research Soc. Studies, publs. showing that the ultimate fate of living systems to ultraviolet light and x-ray exposure is remarkably influenced by pre and postirradiation environment; the bacterial genome has been shown to be a primary target and its replication affected by radiation challenge.*

BILLET, Felix, French physicist; b. Fismes, France, Sept. 15, 1808; ed. Ecole normale Dijon, France, 1830, doctorat ès science, 1845. Lectr. physics Faculty Sci. Dijon, prof., dean, from 1873. Author: Traité d'optique physique, 1858-59. Chief works are on optics; inventor Billet's bi-lentil (an interferential apparatus). Died Dijon, Jan. 26, 1882.

BILLIG, Ernst, physicist, elec. mech. engr.; b. Vienna, Austria; B.Sc. in Mech. Engring., U. Vienna, B.Sc. in Elec. Engring., Dr.tech.; Dipl.Ing., U. Berlin (Germany); m. Gertrud Flor; children—Leslie, Linda, Ruth, David. Research engr. AEG Berlin High Voltage Lab.; in charge high voltage, cons. engr. Moscow Electric Supply Co.; in charge power generation and distbn. P.P.L. Dead Sea Works; in charge developmental lab. Mackbridge & Hewittic Electric Co., 1935-40; sr. research engr. Elec. Research Assn., 1940-45; head solid state physics sect. Research Lab., Asso. Elec. Industries, Aldermaston, Eng., 1946-61; asso. dir. physics research King's Coll., U. London, 1965—. Cons. to elec. and electronics industry, 1961—. Fellow Inst. Physics, Am. Inst. Physics; mem. Instn. Elec. Engrs., Royal Instn. Research, publs. on mech. stresses in large power transformers with fault conditions, elec. surges in power transmission; pioneered studies in purification of semiconductor materials and devices; research on crystal growth; discovered dendritic growth from melt, 1953. Home: 25 Woodside Av., Esher, Surrey, Eng. Office: Physics Dept., King's Coll., Strand W.C.2., London, Eng.*

BILLINGHAM, Rupert Everett, zoologist; b. Warminster, Eng., Oct. 15, 1921; s. Albert E. and Helen (Green) B.; B.A., Oriel Coll., U. Oxford (Eng.), 1943, M.A., 1947, Ph.D., 1950, D.Sc., 1957; m. Jean M. Morpeth, Mar. 29, 1951; children—John David, Peter Jeremy, Elizabeth Anne. Came to U. S., 1957. Lectr. dept. zoology U. Birmingham (Eng.), 1947-51; research fellow Brit. Empire Cancer Campaign, hon. research asso. dept. zoology University Coll., London, Eng., 1951-57; mem. Wistar Inst., Wistar prof. zoology U. Pa., Phila., 1957-65, prof., chmn. dept. med. genetics, 1965—. Mem. Allergy and Immunology Study sect. NIH, 1959-62. Recipient Alvarenga prize Coll. Physicians Phila., 1963, hon. award Am. Assn. Plastic Surgeons, 1964. Fellow Royal Soc. London, N.Y. Acad. Sci. Editorial bd. Transplantation Bull., 1958-60; adv. editor Jour. Exptl. Medicine, 1963—; asso. editor Jour. Immunology, 1964—, Jour. Expt. Zoology, 1964—. Research on pigment cell biology, wound healing, transplantation immunology, immunological tolerance, preservation of graft tissue; discovered (with L. Brent, P. B. Medawar) principle of immunologic tolerance. Home: 102 Anton Rd., Wynnewood, Pa. 19096. Office: Dept. of Medical Genetics, School of Medicine, University of Pa., Phila. 19104.*

BILLINGS, Frank, Am. physician; b. Highland, Wis., Apr. 2, 1854; s. Henry M. and Ann (Bray) B.; M.D., Northwestern U., 1881, M.S., 1890; studied in Vienna, London and Paris, 1885-86; Sc.D., Harvard, 1915; m. Dane Ford Brawley, May 26, 1887; 1 dau., Margaret (Mrs. Geo. R. Nichols, II). Demonstrator anatomy Northwestern U., 1882-86, prof. phys. diagnosis, 1886-91, prof. medicine, 1891-98; prof. medicine Rush Med. Coll. (affiliated U. Chgo.), from 1898, dean faculty, from 1900; professorial lectr. U. Chgo., 1901-05, prof. medicine, 1905-24. Chmn. A.R.C. Mission to Russia, 1917, Editor: Therapeusis of Internal Diseases,

2d and later edits., from 1914. Contbr. numerous articles to med. jours. Introduced concept of focal infection, 1912. Died Sept. 20, 1932.

BILLINGS, George Herrick, Am. metallurgist; b. Taunton, Mass., Feb. 8, 1845; s. Warren and Mary Frances (Caswell) B.; ed. Pittsburg and Mass. Inst. Tech.; m. Harriet Ann Goodwin, Apr. 24, 1879. Was employed in steel mills as roll turner, heater, roller, and later as chemist, mech. engr. and gen. mgr.; mfr. cold drawn steel, from 1889. Inventor of machines for drawing iron and steel bars for shafting and finishing rods. Contbr. to periodicals in field. Died Dec. 8, 1913.

BILLINGS, John Shaw, Am. surgeon, librarian; b. Switzerland County, Ind., Apr. 12, 1839; s. James and Abbie (Shaw) B.; A.B., Miami U., 1857, A.M., 1860; M.D., Med. Coll. Ohio, 1860; LL.D., U. Edinburgh, 1884, Harvard, 1886, Budapest, 1896, Yale, 1901, Johns Hopkins, 1902; M.D., Munich, 1889, Dublin, 1892; D.C.L., Oxford, 1889; m. Kate M. Stevens, Sept. 3, 1862. Demonstrator anatomy Med. Coll. Ohio, 1860-61; served in U. S. A., as asst. surgeon, Apr. 16, 1862; maj. surgeon, Dec. 2, 1876; lt. col. deputy surgeon gen., June 6, 1894; bvtd. capt., maj. and lt. col., 1865; in hosp. service during Civil War; later med. insp. Army of Potomac, in charge of library of surgeon-general's office until 1883; curator Med. Mus. and Library, 1883-95. In charge vital statistics 10th Census, vital and social statistics, 11th Census prof. hygiene U. Pa., 1891, and dir., 1893-96; chmn. bd. Carnegie Instn., from 1905. Fellow Am. Acad. Arts and Scis.; pres. A.L.A., 1901-02. Author: Principles of Ventilation and Heating, 1886; Index Catalogue of the Library of the Surgeon-General's Office U. S. A. (16 vols.), 1880-1894; National Medical Dictionary (2 vols.), 1889. Founded (with R. Fletcher) Index Medicus (monthy guide to current med. lit.), 1879; designed Johns Hopkins Hosp.; created N.Y. Public Library in its present form. Died March 11, 1913.

BILLINGS, Marland Pratt, Am. geologist; b. Boston, Mar. 11, 1902; s. George Bartlett and Helen Agnes (McDouough) B.; A.B., Harvard, 1923, A.M., 1925, Ph.D., 1927; D.Sc., Washington U., St. Louis, 1960; m. Katharine Stevens Fowler, Apr. 23, 1938; children—George Bartlett, Betty Jean. Faculty, Harvard, 1922-28, 30-—, prof., 1946-—, chmn. div. geol. scis., 1946-51; faculty Bryn Mawr Coll., 1928-30; staff U. S. Geol. Survey, part-time 1929-44, geologist, 1943-44. Civilian tech. obs. U. S. Army, 1944. Mem. A.A.A.S. (v.p. 1949), Nat. Acad. Scis., Geol. Soc. Am. (pres. 1959). Author: Structural Geology, 1942; Bedrock Geology of New Hampshire, 1956; Chemical Analyses of Rocks and Rock-Minerals from New Hampshire. Research on geology and engring.; geology of bedrock tunnels around Greater Boston. Home: 115 Dover Rd., Wellesley, Mass. 02181.

BILLINGS, William Dwight, Am. ecologist; b. Washington, Dec. 29, 1910; s. William Pence and Mabel (Burke) B.; A.B., Butler U., 1933, D.Sc., 1955; M.A., Duke, 1935, Ph.D., 1936; m. Shirley Ann Miller, July 29, 1958. Instr. botany U. Tenn., 1936-37; from asst. prof. botany to prof. U. Nev., 1938-52; asso. prof. botany Duke, Durham, N.C., 1952-58, prof., 1958-67, James B. Duke prof. botany, 1967-—. Ad. panel on environmental biology NSF, 1954-57. Recipient Merit award Bo. Soc. Am., 1960; Mercer award Ecol. Soc. Am.; Fulbright Research scholar, New Zealand, 1959. Mem. Ecol. Soc. Am., Brit. Ecol. Soc., Bo. Soc. Am. Author: Plants and the Ecosystem, 1964. Contbr. numerous articles to sci. jours. Editor, Ecology, 1952-57; Research in ecology of deserts and alpine regions, effects of severe environments on plant growth, physiol. ecology, metabolic rates of desert, alpine, arctic plants as controlled by temperature and drought stress; physiol. ecology of ecol. races. Home: 1628 Marion Av., Durham, N.C. 27705.*

BILLINGSLEY, Sir Henry, Brit. mathematician; b. probably Canterbury, Eng.; s. Roger Billingsley; scholar St. John's Coll., Cambridge (Eng.) U., 1551, also ed. Oxford (Eng.) U.; m. Elizabeth Boorne, 1572; children include Henry; m. 2d, Briget Draper. Became sheriff, London, Eng., 1584; became pres. St. Thomas's Hosp., 1594; became Lord Mayor, London, 1596; elected mem. Parliament, 1604. Made 1st English transl. Elements of Geometry (Euclid), 1570. Died Nov. 22, 1606.

BILLMAN, John Henry, Am. chemist; b. Bklyn., Feb. 8, 1912; s. Frederick and Marie (Van Schoonhoven) B.; B.S., U. Va., 1934; A.M., Princeton, 1935, Ph.D., 1938; m. Nannie Belle Humphries, Aug. 16, 1937; children—Betty V., William F. Faculty, U. Ill., 1937-39; prof. chemistry Ind. U., Bloomington, 1939-—; vis. prof. U. Wis., 1946, Yale, 1946, U. Del., 1958. Cons. numerous indsl. firms; responsible investigator OSRD; mem. pharmacology endocrinology fellowship panel Dept. Health. Sr. fellow USPHS, 1959-60. Fellow Ind. Acad. Sci.; mem. Am. Chem. Soc., Sigma Xi, Alpha Chi Sigma, Phi Lambda Upsilon, Alpha Epsilon Delta. Author: Methods of Synthesis in Organic Chemistry, 1954. Contbr. numerous articles to profl. jours. Research in areas of catalysis, spl. reducing agts., synthesis of amino acids, metal chelating agts., field of medicinal chemistry. Home: 910 S. Dunn St., Bloomington, Ind. 47401.*

BILLROTH, Christian Albert Theodore, surgeon, pathologist; b. Bergen, Rügen, Germany, Apr. 26, 1829; s. Karl Theodor and Christina (Nagel) B.; ed. univs. Griefswald, Göttingen (Germany), M.D., U. Berlin, 1852; visited med. schs., Vienna, Prague, Paris, other cities; m. Christel Michealis, 1858; 1 son, 3 daus. Asst. to Langenbeck, Berlin, 1852-60; prof. surgery, also dir. surg. clinic, Zurich, Switzerland, 1860-67, Vienna, Austria, from 1867; volunteer German ambulence service, 1870. Became mem. Austrian Herrnhaus, 1887. Author: Die allgemeine chirurgische Pathologie und Therapie, 1863; many other works. Founder, Vienna sch. surgery; studied histology and pathology; helped found modern intestinal surgery; devised many operations for gastro-intestinal surgery; performed 1st resection of esophagus, 1872, 1st resection of pylorus for cancer, 1881, 1st excision of larynx; introduced Lister's antiseptic methods into European operating rooms. Died Abbazia (now in Yugoslavia), Feb. 6, 1894.

BILLS, Charles Everett, Am. biochemist; b. Natick, Mass., Feb. 9, 1900; s. Charles Albert and Charlotte (Hunter) B.; B.S., Johns Hopkins, 1920, A.M., 1923, Ph.D., 1924; m. Gladys Colton Tebbs, Oct. 10, 1931; children—Charlotte Hunter (Mrs. Kenneth Grainger Edmunds), Richard Albert. Asst. in chemistry Johns Hopkins, Balt., 1919-21, vol. biochemist dept. medicine, 1959-—; lectr. biochemistry U. Alta., 1921-22; dir. research Mead Johnson & Co., Evansville, Ind., 1925-49; indsl. cons. govt. and industry, 1949-—. Mem. A.A.A.S., Am. Soc. Biol. Chemists, Am. Chem. Soc., Am. Inst. Nutrition, Soc. Exptl. Biology and Medicine, Biometric Soc. Contbr. to Biological Effects of Radiation (Duggar), 1936; Symposium on Nutrition (Herriott), 1953; The Vitamins (Sebreil and Harris), 1954. Patentee, publs. in field. Discovered multiple nature of vitamin D.; directed 1st comml. prodn. of synthetic vitamin D in U. S.; designed methods and apparatus for vitamin and mineral assays; studies on alcohols, alkaloids, protozoa, yeasts, cereals, vitamins, and sterols. Home: 6403 Murray Hill Rd., Balt. 21212.*

BILLWILLER, Robert August, Swiss meteorologist; b. St. Gallen, Switzerland, Aug. 2, 1849; s. Karl Ulrich and Johanna (Rade) B.; ed. Zurich, Switzerland, Göttingen, Leipzig, (both Germany); Ph.D. (hon.), Basel, Switzerland, 1901; m. Emma Schardt, 1874; 1 dau.; m. 2d, Elisabeth Küng, 1877; children—1 dau., 1 son. With Central Meteorol. Inst., Zurich, from 1872, dir., from 1874. Mem. Permanent internat. meteorol. com., from 1891. Contbr. numerous sci. articles to profl. jours. on precipitation and wind relationships. Improved observation network; established meteorol. sta. on Santis, 1882, obs., 1887. Died Zurich, Aug. 14, 1905.

BILTZ, Heinrich Johann, German chemist; b. Berlin, Germany, May 26, 1865; s. Carl and Auguste (Schlobach) B.; m. Freya de la Motte, 1901. asst. to Viktor Meyer, Heidelberg, Germany; to H. Limpricht in Greifswald; pvt. lectr. beginning in 1891; dept. head, asso. prof., Kiel, Germany, 1897-1911; in 1911 went to Breslau (now Wroclaw, Poland), to head inst. Biltz endowment set up in his honor to aid talented students. Author: Experimentelle Einführung in die unorganishche Chemie, 1898; Qualitative Analyse unorganischer Substanzen, 1900; (with W. Biltz), Übungsbeispiele aus der unorganischen Experimentalchemie, 1907, Die Ausführung quantitativer Analysen, 1930; Die neuere Harnsäurechemie, Tatsachen und Erklärungen, 1936; also 240 articles from 1888-1939. Determined vapor densities for several compounds and elements; developed uric acid chemistry. Died Breslau, Nov. 2, 1943.

BILTZ, Wilhelm Eugen, German chemist; b. Berlin, Germany, Mar. 8, 1877; s. Carl and Auguste (Schlobach) B.; Ph.D., Greifswald, Germany, 1898; hon. Dr. engring. Technische Hochschule Stuttgart, Germany; hon. Dr. techn. Deutsche Technische Hochschule, Prague, Czechoslovakia; asst. to W. Semmler in Greifswald, until 1900; asst. to O. Wallach in Göttingen, Germany, 1900-05; became prof., dir. chem. lab. Bergakademie, Clausthal, Germany, 1905; at Technische Hochschule in Hanover, Germany, from 1921; prof., Jena, Germany, 1926, Leipzig, Germany, 1928, hon. prof. Göttingen, 1929, emeritus, 1941. Mem. acads. of Göttingen, Halle, Leopoldina. Author: (with H. Biltz) Übungsbeispiele aus der unorganischen Experimental alchemie, 1907; Ausführung qualitativer Analysen, 1913; (with H. Biltz) Die Ausführung quantitativer Analysen, 1930; Raumchemie der festen Stoffe, 1934; Intermetallische Verbindungen; also numerous articles in sci. jours. Leader in the then new field of inorganic chemistry; research on nature of cohesive forces in solid bodies, especially crystals, colloids, salts, compounds of rare metals, molecular volumes of solid compounds. Died Heidelberg, Nov. 13, 1943.

BINAZZI, Maurizio, Italian dermatologist; b. Assisi, Italy, Nov. 12, 1916; s. Mario and Maria (Ceccarani) B.; Doctor Medecine, U. Perugia (Italy), 1940; Specialization Dermatology, U. Rome (Italy), 1945; m. Maria Grazia Carlani, June 15, 1949; children—Gianfranco, Roberto. Staff, dermatol. clinic Perugia Hosp., 1945-—, vice dir., 1950-66, dir., 1966-—; prof. dermatology U. Perugia, 1951-—. Mem. Italian Soc. Dermatology and Syphilography, Italian Soc. Exptl. Biology, Internat. Soc. Tropical Dermatology. Author: (with others); Il Cancro Cutaneo, 1956; also numerous articles. Outside editor Excerpta Medica-Dermatology and Venereology, 1953-—. Research on skin cancer, meteorol. influences on skin diseases, reactivity of human skin, genetics of granulosis rubra nasi and lipidoses; antibody formation in the skin; cutaneous mast-cells; keratoacanthoma; female alopeciae, phlogogenic effect of human serum; pathogenesis of burns; cutaneous effects of cortisone. Home: 29 Bontempi. Office: Policlinico Monteluce, Perugia, Italy.*

BINEAU D'ALIGNY, Armand, French chemist; b. Dové-la-Fontaine, France, 1812; student École Centrale, from 1832; then studied chemistry under J.-B. Dumas at Coll. de France; became prof. chemistry Faculty Scis. Lyons (France), 1839. Author: Observations sur l'absorption de l'ammoniaque et des azotates par les végétations cryptogramiques, 1851; Note sur l'onze atmosphérique, 1855. Determined (with Cahours) steam density under normal pressure, also showed it approaches nearer to law for ideal gas with increasing temperature; determined amount of ammonia in fog and rain water; studied chemistry in relation to agr. Died Feb. 10, 1861.

BINET, Alfred, French psychologist; b. Nice, France, July 8, 1857; studied medicine and law, Paris, from 1871. Founder first French psychol. lab. Sorbonne, Paris, 1889, dir., from 1894, also a founder first French psychol. jour. L'année psychologique, 1895. Author: La psychologie du raisonnement, 1886; Les alterations de la personalite, 1891; L'étude expérimentale de l'intelligence, 1903; (with Simon) Les enfants anormaux, 1907. Most prominent French psychologist of period; promoted interest in human intellectual capacities; stressed individual differences; developed (with Theodore Simon) tests to measure degrees of intelligence. Died, Paris, Oct. 18, 1911.

BINET, Jacques Philippe Marie, French mathematician, astronomer; b. Rennes, France, Feb. 2, 1786; ed. École Polytechnique. With dept. bridges and roads French Government; teacher of mathematics, also inspector general Ecole polytechnique; professor of astronomy College de France, from 1823. Member French Academy Sciences, 1843 (pres.). Author: Mémoire sur les intégrales définies eulériennes et sur leur application a la theorie des suites ainsi qu'à l'evaluation des fonctions des grands nombres, 1840; Memoire sur la formation d'une classe très étendue d'équations reciproques . . . , 1843. Investigated founds. of matrix theory; discovered multiplication rule of matrices, 1812; wrote on mathematics, physics, and astronomy. Died Paris, May 12, 1856.

BINET, Léon René, French physician; b. near Provins, France, Oct. 11, 1891; prof. physiology Faculty Medicine Paris, France; mem. Acad. Scis., Acad. Medicine, Acad. Surgery; worked with artificial serum and blood transfusions for wounded; studied effects of asphyxiation.

BING, R. H., Am. mathematician; b. Oakwood, Tex., Oct. 20, 1914; s. Rupert H. and Lula (Thompson) B.; B.S., S.W. Tex. State Tchrs. Coll., 1935; M.Ed., U. Tex., 1938, Ph.D., 1945; m. Mary Blanche Hobbs, Aug. 26, 1938; children—Robert H., Susan Elizabeth, Virginia Gay, Mary Patricia. Tchr. high sch. sci., Tex., 1935-42; faculty U. Tex., Austin, 1942-47; faculty U. Wis., Madison, 1947-—, prof. 1952-64, research prof., 1964-—, chmn. dept., 1958-60; vis. prof. V. Va., 1949-50; mem. Inst. Advanced Study, Princeton, 1957-58, 61-62, 67. Mem. Math. Assn. Am. (pres. 1963-64), Nat. Acad. Scis., Am. Math. Soc. (dir. Inst. Point Set Topology 1955, v.p. 1967-68), Conf. Bd. Math. Scis. (chmn. 1965-66), A.A.A.S. (v.p. 1959), Pi Mu Epsilon. Research in topology with spl. attention to metric spaces, Euclidean spaces. Home: 3509 Blackhawk Dr., Madison, Wis. 53705.*

BING, Robert, neurologist; b. Strasbourg, France, May 8, 1878; s. Berthold and Valerie (Guggenheim) B.; ed. U. Basle (Switzerland), Frankfort/Maine, Germany, Paris, London, Berlin. Became pvt. lect. Basle U., 1907, asst. prof., 1918, prof., from 1932; head neurol. ambulatory, from 1907; head neurol. lab., from 1926. Guest lectr. Liège U., 1926, Milan U., 1932; v.p. Internat. Neurol. Congress, 1931. Corr. mem., hon. pres. Swiss Neurologic Soc.; mem. acads. and sci. assns. in France, Germany, Belgium, Italy, Poland, Czechoslovakia, Switzerland. Author: Lehrbuch der Nervenkrankheiten, 1913; Kompendium der topischen Hirn- und Rückenmarksdiagnostik, 1909. Research in congenital cardiopathy. Died 1956.

BINGER, Carl Alfred Lanning, Am. psychiatrist; b. Long Branch, N.J., Aug. 26, 1889; s. Gustav and Frances (Newgass) B.; A.B., Harvard, 1910, M.D. cum laude, 1914; m. Clarinda Kirkham Garrison, June 3, 1926; children—David Garrison, Beatrice (Mrs. Norman Pettit), Katherine (Mrs. David Torrey Eames). Sheldon Travelling fellow Harvard, 1915-16; staff Johns Hopkins, 1915-17, Rockefeller Inst. Hosp., 1919-28; Rockefeller Inst. grantee, Germany, Switzerland, Eng., 1928-30; practice medicine specializing in psychiatry and psychoanalysis, N.Y.C., 1930-54; faculty Cornell U. Med. Sch., 1933-54; asso. prof. psychiatry Harvard Med. Sch. 1954-—; also lectr.; hon. physician Mass. Gen. Hosp., Boston, 1958-—. Decorated Cross Order St. George (Greece) Purple Heart; recipient Norton Med. award W. W. Norton & Co., Inc., 1945. Fellow Am. Acad. Arts and Sci.; mem. Am. Soc. Clin. Investigation, Am. Psychiat. Assn., Am. Psychoanalytic Assn. Am.

Psychosomatic Assn. (pres. 1963-64, editor-in-chief Am. Psychosomatic Jour. 1947-62). Author: The Doctor's Job, 1945; More About Psychiatry, 1948; Personality in Arterial Hypertension, 1948; Revolutionary Doctor, 1966; The Two Faces of Medicine, 1967. Research and publs. on physiology adrenals and calcium, oxygen therapy and pneumonia, psychosomatic studies. Home: 21 Lowell St., Cambridge, Mass. 02138. Office: Warren Bldg., Mass. Gen. Hosp., 275 Charles St., Boston 02114.*

BINGER, Louis Gustave, French explorer; b. Strasbourg, France, Oct. 14, 1856; s. Louis Gustave and Marie (Hummel) B. Served as 2d lt. Franco-Prussian War, 1874; topog. mission to Senegal, 1880. Mem. Acad. Scis., 1933. Author: Essai sur la langue bambara, 1886; Esclauage, islamisme et christianisme. Explorer French W. Africa; perfected 1st map of Senegal. Died Nov. 10, 1936.

BINGHAM, Hiram, Am. explorer; b. Honolulu, Nov. 19, 1875; s. Hiram and Minerva Clarissa (Brewster), B.; A.B., Yale, 1898; M.A., U. Cal., 1900; M.A., Harvard, 1901, Ph.D., 1905; Litt.D., U. Cuzeo, 1912; m. Alfreda Mitchell, Nov. 20, 1900; children—Woodbridge, Hiram, Alfred Mitchell, Charles Tiffany, Brewster, Mitchell, Jonathan Brewster; m. 2d, Suzanne Carroll Hill, June 28, 1937. Austin teaching fellow in history, Harvard, 1901-02, 04-05; preceptor in history and politics Princeton U., 1905-06; explored Bolivar's route across Venezuela and Colombia, 1906-07; lectr. on South Am. geography and history, Yale, 1907-09, asst. prof. Latin Am. history, 1909-15, prof., 1915-24; Albert Shaw lectr. on diplomatic history, Johns Hopkins, 1910. Del. U. S. Govt. to 1st Pan-Am. Scientific Congress, Santiago de Chile, 1908; explored Spanish trade route, Buenos Aires to Lima, 1908-09; dir. Yale Peruvian Expdn., 1911, discovered ruins of Machu Picchu; located Vitcos, last Inca capital, and made the first ascent of Mt. Coropuna, 21,703 ft.; dir. Peruvian expdns., 1912, 14-15, auspices of Yale U. and Nat. Geog. Soc.; adviser on South Am. collections in the Yale U. Library; lectr. on South Sea Islands, Naval Tng. Schs., 1942-43. Lt. gov., Conn., 1923, gov., 1925; Mem. U. S. Senate, 1925-33. Fellow Royal Geog. Soc.; hon. life mem. Nat. Geog. Soc.; mem. Hispanic Soc. Am. (hon. pres.), Am. Antiquarian Soc.; hon. mem. Nat. Acad. Hist. (Bogota); corr. member Lima Geog. Soc., Nat. Acad. Hist. (Caracas, Venezuela). Author: Journal of an Expedition across Venezuela and Colombia, 1909; Across South America, 1911; Vitcos, the Last Inca Capital, 1912; In the Wonderland of Peru, 1913; The Monroe Doctrine, An Obsolete Shibboleth, 1913; The Future of the Monroe Doctrine, 1920; An Explorer in the Air Service, 1920; Inca Land, 1922; Freedom under the Constitution, 1924; Machu Picchu, 1930; Elihu Yale, Governor, Collector and Benefactor, 1938; Elihu Yale—The American Nabob of Queen Square, 1939; Lost City of the Incas, 1948. Died June 6, 1956.

BINGHAM, Walter Van Dyke, Am. psychologist, mgmt. scientist; Swan Lake, Ia., Oct. 20, 1880; s. Lemuel Rothwell and Martha Evarts (Tracy) B.; student U. Kan., 1897-98; B.A., Beloit (Wis.) Coll., 1901 Sc.D., 1929; postgrad. U. Chgo., 1905-06, Ph.D., 1908; postgrad. U. Berlin, 1907; M.A., Harvard, 1907; Sc.D., Ill. Wesleyan U., 1950; m. Millicent Todd, Dec. 4, 1920. Instr. ednl. psychology Teachers Coll. (Columbia), 1908-10; asst. prof. psychology Dartmouth, 1910-15; prof. psychology and head div. of applied psychology, Carnegie Inst. Tech., Pitts., 1915-24, dir. Cooperative Research, 1921-24; dir. Personnel Research Fedn., Inc., 1924-34; mem. bd. Psychol. Corp. since 1920 (pres. 1926-28); professorial lectr. in psychology, Stevens Inst. Tech., 1930-40; cons. Occupational Information Service, U. S. Office of Edn., 1938-39; chief psychologist Adj. Gen's. Office, War Dept., 1940-47; exec. sec. com. on classification of personnel in the army, 1917-18; lt. col., Personnel Br. Gen. Staff, U. S. Army, 1918-19. First chmn. div. anthropology and psychology NRC, 1919-20. Chmn. com. on classification of military personnel, 1940-1946; chmn. bd. on clin. psychol. advisory to Surgeon Gen., U. S. Army, 1944-47; chmn. council, adviser to dir. personnel and adminstrn. Army Gen. Staff, 1946-48; cons. on personnel policies to Sec. Defense from 1949. Recipient sec. of war's Emblem for Exceptional Civilian Service, 1944. Hon. corr. Brit. Nat. Inst. Indsl. Psychol.; fellow A.A.A.S. (sec. council 1917); mem. Am. Psychol. Assn. (sec. 1911-14), Am. Assn. Applied Psychol. (pres. 1941); N.Y. State Assn. for Applied Psychology (pres. 1939); Psychometric Soc., Sigma Xi, Phi Beta Kappa. Author: Studies in Melody, 1910; Aptitudes and Aptitude Testing (still a classic in field), 1937, revised 1951. Joint author: Procedures in Employment Psychology, 1926; How to Interview, 1931; Psychology Today, 1932. Cons. editor Jour. Applied Psychology, Personnel Psychology. Often regarded as dean of Am. personnel psychologists. Died July 7, 1952.

BINI, Lucio, Italian neurologist, psychiatrist; b. Rome, Italy, Sept. 18, 1908; s. Antonio and Alessandra (Innocenzi) B.; ed. in neuropsychiatry U. Rome; m. Maria E. Galimberti, 1935; children—Alessandro, Armando, Andrea. Head neurology service Pio 1st. San Spirito e Ospedali Riuniti di Roma. Mem. numerous med. socs. Author: Traite de psychiatrie. Research and numerous publs., monographs on neuropsychiatry, electric shock treatment (a creator, 1937-38). Address: via dei Villini 18, Rome, Italy.

BINIECKI, Stanislaw, Polish scientific pharmacist; b. Slawno, Poland, July 29, 1907; s. Piotr and Maria (Czerniak) B.; grad. U. Poznan (Poland), 1933, dr. degree, 1937, medicina grad., 1946; M.A., U. Lublin (Poland), 1949; m. Alina Medynska, Apr. 13, 1946; children—Malgorzata, Krzysztof. Prof. medicinal drugs and chem. tech. Acad. Medicine, Warsaw, 1947—. Decorated Polonia Restituta Cross. Mem. Polish Acad. Sci. (pharm. sci. com.), Polish Pharm. Assn. (v.p.). Author: Preparation of Drugs; also articles. Research on synthesis of spasmolytics, blood pressure depressors, sympaticomimetics. Home: 38, Platnicza, Warsaw. Office: 25 Przemyslowa, Warsaw, Poland.*

BINKLEY, Otto Francis, Am. biochemist; b. Scottland, Ill., July 11, 1915; s. George E. and Minta (Kealey) B.; B.S. in Chemistry, U. Ill., 1938; M.S., U. Mich., 1939; Ph.D., Cornell U., 1942; m. Emily Katherine Shepard, Aug. 10, 1940 (dec. Aug. 1965); children—Taylor, Clark, Franklin, Kathryn. Asst., Rockefeller Inst., 1942-46; asso. prof. U. Utah, 1946-51; prof. Emory U., Atlanta, 1951—. Cons., VA, 1948—. Mem. Am. Chem. Soc., A.A.A.S., Am. Soc. Biol. Chemists, Harvey Soc., Sigma Xi. Research and publs. on metabolism of sulfur containing compounds, mechanism of action of enzymes, enzymatic basis of renal function. Home: 1960 Mason Mill, Decatur, Ga. 30033. Office: Emory U., Atlanta, 30322.*

BINNEY, Edward William, Brit. geologist; b. Morton, Eng., 1812; solicitor Manchester, Eng., from 1836; Fellow Royal Soc., 1856; mem. Manchester Geol. Soc. (founder, pres. 1857-59, 1865-67), Lond. Geol. Soc. Author: Sketch of the Geology of Manchester and its Vicinity; On Sigillaria and its Roots, 1858; Remarks on Sigillaria and Some Spores Found Imbedded in the Inside of its Roots. Contbr. numerous articles Manchester Geol. Soc. Furnished proof that all coal seams rest on old soils made entirely of vegetable matter, formed by decomposition of sigilloria; discovery important to mfr. industries. Died Dec. 19, 1881.

BINNIE, Alfred Maurice, Brit. engr.; b. Maldon, Eng., Feb. 6, 1901; s. David Carr and Ada G. (Ward) B.; M.A., Queens' Coll., Cambridge (Eng.) U., 1926. Demonstrator, lectr. engring. lab. Oxford U., 1925-44; Rhodes Travelling fellow, 1932-33; lectr. Cambridge U., 1944-54, reader engring., 1954—, fellow Trinity Coll., 1944—; sr. research fellow Cal. Inst. Tech., 1951-52. Scott lectr. Ormond Coll., Melbourne U., 1967; vis. scholar U. Cal. at Berkeley, 1967-68. Fellow Royal Soc.; mem. Instn. Civil Engrs., Instn. Mech. Engrs. Research, publs. on fluid mechanics. Home: Trinity Coll., Cambridge, Eng.*

BINSON, Boonrod, Thai. elec. engr.; b. Bangkok, Thailand, Sept. 18, 1915; s. Pad and Somporn (Kim) B.; B.Sc., Chulalongkorn U., Bangkok, 1934; M.S., Mass. Inst. Tech., 1938; D.Sc., Harvard, 1949; m. Payom Shulz, Nov. 14, 1945. Instr. elec. engring. Chulalongkorn U., 1934-36; head dept., 1952-63, prof., 1963—; sec. gen Nat. Energy Authority, Thailand Ministry Nat. Devel., Bangkok, 1954—. Mng. dir. N.E. Electricity Authority, Bangkok, 1963-66; Thailand dir. Mekong Devel. Com., 1957—. Mem. Engring. Inst. Thailand (pres. 1962-65), I.E.E.E., Registered Engrs. Thailand. Contbr. numerous articles to profl. jours. Research on filter circuit theory in elec. communication, transient technique in electric power network, rural electric-power devel. in developing countries. Home: 106 Klang, Bangkok. Office: Pibultham Villa, Krasutsuk, Bangkok, Thailand.*

BINSWANGER, Ludwig, Swiss psychiatrist; b. Kreuzlingen, Switz., Apr. 13, 1881; s. Robert and Berta (Hasenclever) B.; ed. College of Schaffhausen; Us. Lausanne, Heidelberg, Zurich; Univ. of Basle, hon. Ph.D., 1941; m. Hertha Buchenberger, Sept. 15, 1908; six children. Physician, Psychiatric U. Clinic Burghölzli, Zurich, 1906; asst., Psychiatric U. Clinic, Jena, 1907-08; asst., Sanatorium Bellevue, Kreuzlingen, 1908-10, chief, 1910—. Pres., Swiss Psychiatric Soc., 1926-29; mem., Comité de la Fondation "Lucerna"; Nat. Acad. Medicine, Madrid. Author: Einführung in die Probleme der allgemeinen Psychologie, 1922; Wandlungen in der Auffassung und Deutung des Traums, 1928; Grundformen und Erkenntnis menschlichen Daseins, 1942; Melancholie und Manie, 1960. Office: Bellevue Kreuzlingen, Switzerland.

BINSWANGER, Otto, German psychiatrist; b. Münsterlingen, Germany, Oct. 14, 1852; s. Ludwig Binswanger; ed. univs. Heidelberg, Strassburg, Zurich; m. Emilie Buedeker. Studied brain anatomy, Vienna; became asst. physician, Göttingen, 1877; research on path. anatomy, Breslau; dir. Psychiatric Sanitorium, Jena, 1879-1919; prof., Jena, from 1882. Author: Die pathologische Histologie der Grosshirnrindenerkrankung bei der allgemeinen progressiven Paralyse, 1893; Die Pathologie und Therapie der Neurasthenie, 1896; Die Epilepsie, 1899; (with Ernst Siemerling) Lehrbuch der Psychiatrie, 1904; also numerous others. Developed path.-anat. methods of examining organic brain diseases; research on progressive paralysis, epilepsy, hysteria, neurasthenia. Died Kreuzlingen, Germany, July 14, 1929.

BINZ, Arthur Heinrich, German chemist; b. Bonn, Germany, Nov. 12, 1868; s. Carl and Harriet Emily (Schwabe) B.; ed. Owens Coll., Manchester, Eng., London (Eng.) Inst.; studied chemistry in Bonn, Göttingen, Leipzig (all Germany); doctorate, 1893; hon. dr. engring., Karlsruhe, Germany, 1929; m. Junnita Reutlinger, 1901; 2 daus. Chemist in cotton factory Rolfts and Co., also at S. Schwabe & Co. in Manchester, 1894-97; became pvt. lectr. tech. chemistry, Bonn, 1899; named prof. Handelshochschule Berlin (Germany), 1906; dir. chem. dept. Speyer-Haus, Frankfort, Germany, 1918-21; prof. Landwirtschaftlichen Hochschule Berlin, until 1935. Mem. German Chem. Soc. (v.p. 1931-33), Am. Urol. Assn. (hon.). Author: Edelmetalle, Ihr Fluch und ihr Segen, 1943; also articles. Editor-in-chief Zeitschrift fürangewandte Chemie, 1922-23. Applied knowledge of organic and inorganic chemistry, especially sulfoxyl compounds, indigo, iodine, arsenic, antimony and selenium compounds, to med. sci.; developed new ways of diagnosing internal diseases, including those of urinary tract. Died Berlin, Jan. 25, 1943.

BINZ, Karl, German pharmacologist, chemist; b. Bernkastel, Germany, July 1, 1832; s. Peter Franz and Agathe (Day) B.; grad. in med. scis. Bonn (Germany) U., 1862; m. Harriet Emily Schwabe; children include Arthur. Named asso. prof. Bonn U., 1868, prof. pharmacology, 1873; founder Pharmacological Inst., Bonn, 1869. Author: Grundzüge der Arzneimittellehre, 1866; Vorlesungen über Pharmakologie, 1883-86; Experimentelle Untersuchungen über das Wesen der Chiniwirkung, 1868; Der Aether gegen den Schmerz, 1896. Investigated action of quinine, also alcohol, anesthetics, ethereal oils; described test for quinine in urine. Died Bonn, Jan. 12, 1913.

BION, Nicholas, French engr.; b. 1653; engr. to Louis XIV and Louis XV. Author: Usage des giobes célestes terrestres et des spheres suivant les différénts systemes du monde, 1699; Usage des astrolabes, 1702; Traité de la construction et des principaux usages des instruments de mathematiques, 1725; Description et usage d'un nouveau planisphère, 1727. Constructed terrestrial and celestial globes; developed math. and astron. instruments. Died Paris, 1733.

BIONDO, Michelangelo, Italian physician; b. Venice, Italy, 1497; practiced in Rome, Naples, Italy. Author: De Partibus-ic-tu-sectis citissime Sanadis. One of 1st to oppose use of heat in treatment of wounds, advocating plain water instead. Died 1565.

BIORCK, Gunnar Carl Wilhelm, Swedish physician; b. Gothenburg, Sweden, Apr. 4, 1916; s. Wilhelm K. A. and Louise (Petterson) B.; med. licentiate Karolinska Inst., Stockholm, 1942, M.D., 1949; m. Margareta Lundberg, May 26, 1944; children—Eva, Anders, Lena, Hans, Marie. Asst. prof. cardiology Karolinska Inst., 1949, prof. medicine, 1958—; asst. prof. internal medicine U. Lund, Sweden, 1950-58; head dept. medicine Serafirmer Hosp., Stockholm, 1958—, physician in chief, 1961—; physician to Royal Family, 1965—. Sci. mem. Royal Med. Bd., 1960—, Mil. Med. Bd., 1960—; mem. WHO Expert Adv. Panel, 1960—. Decorated knight Royal Order of Polar Star, 1963; Iranian Houmayon Order, 1965. Fellow Royal Coll. Physicians, London; corr. fellow A.C.P. Author: Our People and Its Future, 1940; Myoglobin in Man, 1949; If Your Heart Troubles You, 1953; Medicine for Politicans, 1953; Man's Possibilities, 1956; Conditions of Medical Care, 1966. Research, numerous publs. on biochemistry, physiology, clin. manifestations, epidemiology and rehab. of heart diesease. Home: 14 Bravallavägen, Djursholm, Sweden. Office: Serafimer Hosp., Stockholm, Sweden.*

BIOT, Jean Baptiste, French physicist; b. Paris, Apr. 21, 1774; s. Joseph Biot; ed. Ecole Polytechnique; m. Mlle. Brisson; 1 son, Edouard-Constant. Prof., l'Ecole Centrale de Beauvais (France); became prof. math. physics Coll. de France, 1800; named prof. astronomy Faculty Scis., U. Paris, 1809. Member French Acad. Scis., 1800, French Acad. comdr. Legion of Honor; Fellow Royal Soc., 1815 (Rumford medal 1840). Author: Essai de géometrie analytique, 1802; Traité élémentaire d'astronomie physique, 3 vols., 1805; Traité de physique expérimentale et mathématique, 4 vols., 1816; Mélanges scientifiques et litteraires, 3 vols., 1858. Made (with Gay-Lussac) 1st balloon ascent for sci. purposes; showed that earth's magnetism is not appreciably reduced in regions above earth's surface; worked (with Arago) on measurement of quadrant of meridian to standardize meter, 1806, also measured refraction indices of gases; discovered laws of rotary polarization of light; invented a polariscope; showed how polarimeter readings could be used to determine sugar concentrations, 1835; discovered (with Brewster) biaxial crystals; discovered (with Savart) law of force in magnetic field around straight current Biot-Savart law); helped develop electrostatics; studied saturation vapor pressure of water; from work on thermal conduction derived fundamental laws of heat flow; mineral biotite named for him. Died Paris, Feb. 3, 1862.

BIRCH, Arthur John, chemist; b. Sydney, Australia, Aug. 3, 1915; s. Arthur Spencer and Lily (Bailey) B.; B.Sc., Sydney U., 1937, M.Sc., 1938; D.Phil., Oxford U., 1940; m. Jessie Williams, Oct. 21, 1948; children—Susan Margaret, Michael John, Francis David, Rosemary Jane, Christopher Paul. Scholar, Royal Commn. for Exhbn. of 1851, Oxford, 1938-41, research fellow, 1941-45; Smithson fellow Royal Soc., Cambridge U., 1949-52; prof. organic chemistry U. Sydney, 1952-55; prof. organic chemistry Manchester U., 1955—; dean-elect Chemistry Sch., Inst. Advanced Studies, Australian Nat. U., 1966. Sci. adv. bd. Syntex Corp., 1958—. Recipient H. G. Smith medal Australian Chem. Inst., 1956; Fritzsche award Am. Chem. Soc., 1962. Fellow Australian Acad. Sci., Royal Soc., Royal Inst. Chemistry, Chem. Soc. (council 1958-59), Biochem. Soc. Author: New Chemistry Works, 1948. Numerous publs. on research of hydrogen addition to aromatic rings; total synthesis of nortestosterone; biosyntheses of plant and mold products from acetic acid. Home: 68 Parsonage Rd., Stockport, Cheshire, Eng. Office: Manchester U., Manchester, Eng.*

BIRCH, Bryan John, Brit. mathematician; b. Burton-on-Trent, Eng., Sept. 25, 1931; s. Arthur Jack and Mary (Buxton) B.; B.A., Trinity Coll., Cambridge, Eng., 1954, Ph.D., 1958; m. Gina Margaret Christ, July 1, 1961; children—Colin, Michael. Research fellow Trinity Coll., Cambridge, 1956-60; sr. research fellow Churchill Coll., Cambridge, 1960-62; faculty U. Manchester (Eng.), 1962-65; reader math. U. Oxford, fellow Brasenose Coll., 1966—. Mem. Cambridge Philos. Soc., London Math. Soc., Am. Math. Soc., Inst. Math. and its Applications. Editorial bd. Topology, 1966—, Inventiones Mathematicae, 1966—. Research, publs. on theory of numbers, especially Diophantine equations. Home: Green Cottage, Boars Hill, Oxford, Eng.*

BIRCH, Carroll LaFleur, Am. physician; b. Balt., Dec. 29, 1896; d. Charles Edward and Eva Virginia (Robinson) LaFleur; B.S. (Phila. City scholar 1917), U. Pa., 1920; M.D., U. Ill., 1925, M.S., 1927; certificate Sch. Tropical Medicine, Tulane U.; m. Richard B. Birch, June 17, 1921. Intern, Research Hosp., Chgo., 1925-26, resident in medicine, 1926-28; instr. medicine U. Ill., 1928-31, asst. prof. medicine, 1932—, asso. prof. internal medicine, 1942-50, prof., 1950—; dean Lady Hardinge Med. Coll., India, 1951-55; dir. USPHS, India, 1954-56; studied trop. diseases in Africa, 1949. Res. officer USPHS, with rank med. dir. Deaver scholar. Fellow A.C.P.; mem. Inst. Medicine, Am., Ill. med. assns., Chgo. Med. Soc., Chgo. Soc. Internal Medicine (v.p. 1932), Am. Soc. Trop. Medicine, English Assn. Geneticists, Am. Med. Womens Assn. (pres. Chgo. br. 1941-42; v.p. nat. assn., 1944-45), Central Soc. Clin. Research, A.A.A.S., Ill. Pub. Health Assn., Indian Assn. Pathology, Sigma Xi, Alpha Omega Alpha. Author: Hemophilia Clinical and Genetic Aspects, 1937. Research and publs. on hematology, especially hemophilia and trop. medicine. Home: 2045 Sedgwick St. (14). Office: 1853 W. Polk St., Chicago 12, Ill.

BIRCH, (Albert) Francis, Am. physicist; b. Washington, D.C., Aug. 22, 1903; s. George Albert and Mary Clayton (Hemmick) B.; B.S., in E.E., Harvard Univ., 1924; student Univ. of Strasbourg, France (Am. Field Service fellowship), 1926-28; M.A., Harvard, 1929, Ph.D., 1932 (John Tyndall scholarship, 1928-31); m. Barbara Channing, July 15, 1933; children—Anne Campaspe, Francis Sylvanus, Mary Narcissa. Engr., New York Telephone Co., 1924-26; instr. and tutor in physics, Harvard, 1931-32, research asso. in geophysics, 1932-37, asst. prof., 1937-43, associate prof. 1943-46, prof. since 1946; on leave Mass. Inst. Tech. 1941-42; Sturgis-Hooper professor of Geology, 1948. Served as lt. comdr. to comdr. U.S.N.R., 1942-45. Awarded Legion of Merit; Arthur L. Day medal Geol. Soc.; William Bowie medal, Am. Geophys. Union, 1960. Fellow Royal Astron. Soc., Am. Phys. Soc., Am. Acad. Arts and Scis., Geol. Soc. Am. (pres. 1964); mem. Am. Geophys. Union, Seismol. Soc. Am., Nat. Acad. Sci., Am. Philos. Soc., Sigma Xi. Editor: Handbook of Physical Constants. Geothermal studies; investigation of elasticity; properties of materials at high temperatures and high pressures.

BIRCH, John, Brit. surgeon; b. 1745 or 46; surgeon in army; settled in London; surgeon St. Thomas's Hosp., 1774-1815, also founder elec. dept.; surgeon extraordinary to prince regent. Author: Considerations of Efficacy of Electricity in removing Female Obstructions, 1779; A letter on Medical Electricity, 1792; An Essay on the Medical Applications of Electricity, 1802; other works on objections to small pox vaccinations. Introduced electricity as curing agt. Died Feb. 3, 1815.

BIRCH, Louis Charles, Australian biologist; b. Melbourne, Australia, Feb. 8, 1918; s. H. M. and E. (Hogan) B.; B.Agr. Sci., U. Melbourne, 1939; D.Sc., U. Adelaide (Australia), 1948. Vis. fellow U. Chgo., 1946, Oxford (Eng.) U., 1947; faculty U. Sydney (Australia), 1948—, Challis prof. biology, 1961—; vis. prof. U. Minn., 1959, Columbia U., 1954, U. Sao Paulo (Brazil), 1955, U. Cal., 1967. Fellow Australian Acad. Sci.; mem. Ecol. Soc. Am., Brit. Ecol. Soc., Soc. for Study Evolution (v.p. 1966—). Author: The Distribution and Abundance of Animals (with H. G. Andrewartha), 1954; Nature and God, 1966; Senior Science for High School Students; Biology, 1965. Research and publs. on population biology with devel. gen. theory of population ecology of insects; exptl. studies on evolution of natural populations of insects demonstrating natural selection; relevance of process philosophy to modern biology. Home: 35A Sutherland Crescent, Darling Point, N.S.W. Office: Zoology Bldg., U. Sydney, Sydney, N.S.W., Australia.*

BIRCH-HIRSCHFELD, Felix Victor, German pathologist; b. Kluvensiek, nr. Bovenau, Germany, May 2, 1842; s. Gustav and Mary Jane (Birch) B.-H.; m. Clara Baron; 1 son, Arthur; asst. Leipzig (Germany) path.-anat. inst.; prosector municipal hosp., Dresden, Germany, 1870; prof. gen. pathology and path. anatomy, Leipzig, 1885. Author Lehrbuch der allgemeinen und speziellen pathologischen Anatomie, 1877. Research on tumors and infectious diseases, infection entrances and tubercle structure in Tb; invented dyeing method for determination of amyloid. Died Leipzig, Nov. 19, 1899.

BIRCHER, Louis J(acob), Am. phys. chemist; b. St. Louis, Feb. 1, 1892; s. Jacob and Anna Helen (Wehrli) B.; A.B., B.S., U. Mo., 1915, M.A., 1917; Ph.D., U. Chgo., 1924; m. Mathilde G. Mann, Dec. 29, 1925; children—Helen Ann (Mrs. James E. Guillet), Marjorie M. (Mrs. Thomas E. Page). Faculty, Ia. State Coll., 1916-18, Wash. U., 1919-21; faculty Vanderbilt U., Nashville, 1921—, prof. phys. chemistry, 1927-62, chmn. dept., 1956-62, prof. emeritus, lectr. colloid chemistry, 1962—. Vis. prof. Mass. Inst. Tech., 1928-32. Fellow A.A.A.S.; mem. Am. Chem. Soc. (councillor 1948-50), Tenn. Acad. Sci. (pres. 1937), Sigma Xi (pres. Vanderbilt chpt. 1944-45). Author: Physical Chemistry—a Brief Course, 1940, also articles. Research in chemistry, particularly dealing with effect of temperature and pressure on overvoltage of metals. Home: 6140 Jocelyn Hollow Rd., Nashville. 37205.*

BIRCHER-BENNER, Maximilian Oskar, Swiss physician; b. Aarau, Switzerland, Aug. 22, 1867; s. Heinrich and Bertha (Krüsi) B.-B.; ed. in Zurich, Switzerland, Berlin, Germany; doctorate 1891; m. Elisabeth Benner-Schlumberger, 1893; 4 sons; 3 daus., including Ruth (Mrs. Alfred Kunz). Founder pvt. clinic on Zürichberg, 1897, head physician for 42 years; opened pub. sanatorium, Zürich, 1939. Author: Grundzüge der Ernährungstherapie auf Grund der Energetic, 1903; Früchtespeisen und Rohgemüse, 1924; Eine neue Ernährungslehre, 1924; Vom Werden des neuen Arztes, 1938. Pub. of mag. Der Wendepunkt im Leben und Leiden. Introduced a nutritional therapy; recognized curative powers of raw vegetables before discovery of vitamins. Died Zurich, Jan. 24, 1939.

BIRD, Frederic Dougan, surgeon; b. Richmond-on-Thames, May 27, 1858; s. Samuel Dougan and Catharine Emma (Tate) B.; ed. Scotch Coll., M.B., M.S., U. Melbourne (Australia); m. Lucy Clare Hopkins, 1882; 1 son, 1 dau. Demonstrator anatomy, examiner, lectr. surgery Melbourne U.; sr. surgeon Melbourne Hosp.; cons. surgeon Queen Victoria Hosp., Melbourne, also to Army in Egypt, Mediterranean Expeditionary Force, BEF, Salonica. Fellow Royal Coll. Surgeons; mem. Adelaide Med. Congress (pres. sect. surgery), Brit. Med. Assn. (v.p. sect. surgery London), Med. Soc. Victoria (pres.). Contbr. articles to surg. publs. Died May 29, 1929.

BIRD, Golding, English physician; b. Norfolk, Eng., Dec. 9, 1814; M.D., St. Andrews, 1838, M.A., 1840; m. 1842; 5 children. Lectr. natural philosophy Guy's Hospital, London, England, from 1836; lecturer materia medica College of Physicians, from 1847. Fellow Royal Society, 1846; mem. Linnean Society, Geological Society, licentiate Coll. Physicians, London. Author: Elements of Natural Philosophy, 1839; Urinary Deposits, . . . 1844; Lectures on Electricity and Galvanism in their Physiological and Therapeutical Relations, 1849. Med. treatises. Described Bird's disease, condition marked by presence of abnormal quantities of oxalic acid and oxalates in the urine, 1842; devised flexible stethoscope. Died Turnbridge Wells, Eng., Oct. 27, 1854.

BIRD, Herbert Roderick, Am. nutritionist; b. Madison, Wis., Jan. 7, 1912; s. Herbert R. and Minerva (Farwell) B.; B.S., U. Wis., 1933, M.S., 1935, Ph.D., 1938; m. Elenore Elkinton, Aug. 22, 1937; children—Herbert R., William H., Ellen, Mary. Asso. prof. U. Md., College Park, 1938-44; biochemist U. S. Dept. Agr., Beltsville, Md., 1944-48, head poultry investigations, 1948-53; prof. poultry nutrition U. Wis., Madison, 1953—, chief expdn. AID, Brazil, 1964-66. Recipient Am. Feed Mfrs. Award, 1948, Tom Newman award, Superior Service award U. S. Dept. Agr., 1949, Mem. Poultry Sci. Assn. (past pres., Borden award 1953), World's Poultry Sci. Assn. (v.p. 1962-66), Am. Inst. Nutrition, Am. Chem. Soc., Soc. for Exptl. Biology and Medicine, A.A.A.S., Soc. for Animal Sci. Research, numerous publs. on demonstration synthesis in digestive tract of B12, devel. practical poultry diets containing synthetic amino acids, demonstration in chickens of protein synthesis from dietary urea. Home: 5006 Hammersley Rd., Madison, Wis. 53711.*

BIRD, John, English instrument maker; b. Durham, Eng., 1709; cloth weaver by trade; engraver dial plates; came to London, circa 1740, became cutter instrument divs. for Sisson; after instrn. by Graham, founded own bus.; mech. coadjutor of Bradley. Author: The Method of dividing Astronomical Instruments, 1767; The Method of constructing Mural Quadrants exemplified by a Description of the Brass Mural Quadrant in the Royal Observatory at Greenwich, 1768. Built brass mural quadrant for Greenwich Obs., 1750, also for European observatories. Died Mar. 31, 1776.

BIRD, John Brian, geomorphologist; b. Birmingham, Eng., Aug. 28, 1923; s. George Harold and Edna (Attwood) B.; B.A. with honors, Cambridge (Eng.) U., 1947, M.A., 1949; m. Marjorie Beryl Briggs, Dec. 31, 1947; children—Joanne M., D. Neil, Colin R. Lectr., U. Toronto (Ont., Can.), 1947-50; faculty McGill U., Montreal, Que., Can., 1950—, prof. geography, 1962—, chmn. dept., 1967—. Fellow Arctic Inst. N.Am.; mem. Canadian Assn. Geographers (pres. 1959-60), Hakluyt Soc. Can. (hon. sec.), Am. Geog. Soc., Assn. Am. Geographers. Author: (with S. N. Namowitz and D. B. Stone) Earth Science, vol. 1, 1956, vol. 2, 1957; The Physiography of Arctic Canada, 1967. Research, publs. on description, analysis of spl. terrains, landforms of polar regions; numerous expdns. to arctic Can. Home: 27 rue de Lombardie, Preville, Que. Office: Geography Dept., McGill U., Montreal, Que., Can.*

BIRD, R. Byron, Am. chem. engr.; b. Bryan, Tex., Feb. 5, 1924; s. Byron and Ethel (Antrim) B.; B.S., U. Ill., 1947; postgrad. U. Amsterdam (Netherlands), 1950-51; Ph.D., U. Wis., 1950. Project asso. U. Wis., Madison, 1951-52, 53-55, faculty, 1955—, prof. chem. engring., 1957—, chmn. dept. chem. engring. U. Wis., 1964—; asst. prof. chemistry Cornell U., Ithaca, N.Y., 1952-53. Fulbright fellow Instituut voor Theoretische Physica, Amsterdam, Holland, 1950-51; Guggenheim fellow Technische Hogeschool, Delft, Holland, 1958. Petroleum Research fund research grantee, 1963. Mem. Am. Soc. for Engring. Edn. (Curtis McGraw award 1959, Westinghouse award 1960), Am. Inst. Chem. Engrs. (William H. Walker award 1962, Profl. Progress award 1965), Am. Chem. Soc., Am. Phys. Soc., Am. Soc. Rheology, Nederlandse Natuurkundige Vereniging, Koninklijk Instituut van Ingenieurs, Soc. Chem. Engrs. Japan. Author: (with J. O. Hirschfelder, C. F. Curtiss) Molecular Theory of Gases and Liquids, 1954, rev., 1964; (with W. E. Stewart, E. N. Lightfoot) Transport Phenomena, 1960, rev., 1966; (with W. Z. Shetter, Martinus-Nijhoff) Een Goed Begin . . . A Contemporary Dutch Reader, 1963. Research and publs. on molecular theory of transport phenomena, fluid dynamics of polymeric fluids, rheology of polymers, equation of state. Office: Chem. Engring. Dept., U. Wis., Madison, Wis. 53706.*

BIRD, Robert Montgomery, Am. chemist; b. Petersburg, Va., June 13, 1867; s. Henry van Leuvenigh and Margaret (Randolph) B.; B.A., B.S., Hampden-Sidney Coll., 1897; Ph.D., Johns Hopkins, 1901; m. Caroline Reid, June 11, 1902; children—Caroline Page, Robert Montgomery. Clk merc., r.r. and mfg. office 10 yrs.; prof. sci. and math. Frederick (Md.) Coll., 1898-99; acting prof. chemistry Miss. Agrl. and Mech. Coll., 1901-02; acting prof. chemistry U. Mo., 1902-04, instr., later prof. chemistry, 1904-07; prof. chemistry U. Va., 1907—. Author: Chemical Science Reader, 1911; Laboratory Course in General Chemistry, 1911; Typical Reactions of General Chemistry, 1912; Notes on Organic Chemistry, 1923; Typical Reactions of Organic Compounds. Died June 4, 1938.

BIRDSEYE, Clarence, Am. inventor; b. Bklyn., Dec. 9, 1886; s. Clarence Frank and Ada (Underwood) B.; student Amherst, 1910, M.A., 1941; m. Eleanor Gannett, Aug. 21, 1915; children—Kellogg G., Ruth, Eleanor, Henry S. Field naturalist, biol. survey, U. S. Dept. Agr., 1910-12; fur trader, Labrador, 1912-17; U. S. purchasing agt. U. S. Housing Corp., 1917-19; asst. to pres. U. S. Fisheries Assn., 1920-22; v.p. and pres. of companies pioneering in quick freezing dressed seafoods, 1923-29; with Birdseye-Frosted Foods, Inc., and Birdseye Lamps, 1930-34; pres. Birdseye Electric Co., 1935-38; engaged in devel. specialized food freezing and dehydrating processes and equipment, from 1939; pres. Process Evaluation & Devel. Corp. 1955-56. Mem. Am. Chem. Soc., Am. Fisheries Assn., Am. Soc. Refrigerating Engrs., Inst. Food Technologists (chmn. northeast sect. 1945-46), Internat. Assn. Milk Sanitarians Contbr. numerous papers and talks on food preservation. Granted numerous patents in fields of food preservation by refrigeration and incandescent light; studied dehydration of foods; developed process for making paper pulp from sugar cane "bagasse," straw, and other farm residues. Died Oct. 7, 1956.

BIRGE, Raymond Thayer, Am. physicist; b. Brooklyn, N.Y., Mar. 13, 1887; s. John Thaddeus and Caroline S. (Raymond) B.; A.B., U. of Wis., 1909, A.M., 1910, Ph.D., 1914; LL.D., University Cal., 1955; m. Irene Adelaide Walsh, Aug. 12, 1913; children—Carolyn Elizabeth, Robert Walsh. Instr. in physics Syracuse U., 1913-15, asst. prof., 1915-18; instr. in physics U. Cal., 1918, successively asst. and asso. prof., prof., 1926-55, prof. emeritus, 1955—, chmn. dept., 1933-55. Co-operating expert on Internat. Critical Tables; member advisory committee Office of Critical Tables. Physics Bldg., U. Cal. at Berkeley named in his honor. Fellow Am. Phys. Soc. (vice pres. 1954, pres. 1955), Optical Society America, A.A.A.S.; member National Acad. of Sciences, Am. Philos. Soc., Am. Assn. Physics Tchrs., Am. Inst. Physics (gov. 1955-58), Phi Beta Kappa, Sigma Xi. Contbr. to Physical Rev., etc. Original experimental verifications of quantum theory of molecular spectra;

discovered carbon 13 isotope; predicted existence of hydrogen 2 isotope, which led to its almost immediate discovery; made 1st comprehensive, critical analysis and calculation of values of all general physical constants—values which were adopted universally. Home: 1639 La Vereda St., Berkeley, Cal. 94720.*

BIRINGUCCIO, Vannoccio, Italian chemist, metallurgist; b. Siena, Italy, Oct. 20, 1480; s. Paolo and Lucrezia (di Bartolomeo) B. Served Pandolfo Petrucci, Tyrant of Siena, as engineer; improved knowledge of metallurgy and technology through visit to Germany; later served Duke of Parma, Duke Alphonso I of Ferrara, Republic of Venice, and was dir. of Arsenal in service of Pope Paul III at time of his death. Author: De la Pirotechnia (1st printed work to cover whole field of metallurgy, also deals with applied chemistry, fireworks, gunpowder and military arts), 1540. Here described liquation of silver from copper by means of lead; 1st description of amalgamation process for extraction of silver from its ores; described manufacture of steel and practical techniques of bell casting and cannon founding. One of 1st to note that lead increases 8 to 10% in weight when subjected to fire. Condemned claims of alchemists. Died, 1538 or 39.

BIRKBECK, George, Brit. physician; b. Seattle, Yorkshire, Eng., Jan. 10, 1776; M.D., U. Edinburgh, Scotland, 1799. Chair natural philosophy Andersonian Inst., Glasgow, Scotland, from 1799; free lectr., Glasgow, 1800-04; founder Mechanic's Inst., Glasgow, 1823, dir.; founder, president, London Mechanics Institution, 1824, (later named Birkbeck College); founder, councilor U. Coll., London, 1827. Believed in applying sci. to practical matters whenever useful; pioneered in making sci. information available to working classes. Died London, Dec. 1, 1841.

BIRKELAND, Jorgen Maurice, Am. microbiologist; b. Fergus Falls, Minn., Nov. 15, 1898; s. Mads and Helen (Notland) B.; B.S., M.S., N.D. Agrl. Coll., 1928; Ph.D., U. Chgo., 1933; m. Anna Marguerite Wyatt, Dec. 26, 1938; children—Jorgen Wyatt, Eric Wyatt. Bacteriologist, chemist Chgo. Bd. Edn., 1928-30; NRC fellow Rothamsted Ecpt. Sta., Harpenden, Eng., 1933-34; faculty Ohio State U., Columbus, 1935—, prof. microbiology, 1945—, chmn. dept., 1948-64; vis. lectr. Western Res. U. Med. Sch, 1944. Sci. attache U. S. Dept. State, Stockholm, Sweden, 1953-54; cons. AID, India, 1964. Recipient Alumni Achievement award N.D. State Agr. Coll., 1957. Mem. A.A.A.S., Am. Pub. Health Assn., Am. Soc. Microbiology, Soc. for Exptl. Medicine and Biology, Phi Beta Kappa, Sigma Xi, Phi Kappa Psi. Author: Microbiology and Man, 1942, rev., 1949; (with F. S. Crofts, William Wilkins), Workbook for Microbiology. Serological research and publs. on plant viruses, Tb. immunomology, virology, chemotherapy, bacterial variation. Home: 299 Piedmont Rd., Columbus, O. 43214.*

BIRKELAND, Olaf Kristian, Norwegian physicist; b. Oslo, Norway, Dec. 13, 1867; studied under H. Poincaré, Paris, France, also under H. Hertz, Bonn, Germany; became prof. physics, Oslo. Obtained luminous elec. effects suggesting cause of aurora borealis by exposing to cathode rays a magnetized model of earth in vacuum tube; gave (with Eyde) 1st solution for problem of fixation of nitrogen on indsl. scale, 1903. Died Tokyo, June 18, 1917.

BIRKHAUG, Konrad Elias, bacteriologist, physician; b. Bergen, Norway, Oct. 12, 1892; s. Karl Anderssen and Elisa Marie (Olsen) B.; came to U. S., 1911, naturalized, 1917; B.A., Jamestown Coll., 1917; M.D., Johns Hopkins; M.Sc., U. Rochester, 1927; m. Marie Mustad Berner, Mar. 8, 1938. Hosp. sec. Internat. YMCA, Russia, 1917-19, Verdun, France, 1919-20; asst., Charlton fellow Johns Hopkins Hosp., 1924-25; asso. prof. bacteriology U. Rochester, 1925-32; sous-chef Institut Pasteur Lab. Tb., Paris, France, 1932-35; mem. Christian Michelson Inst., Bergen, 1935-45, emeritus mem., 1956—; dir. Norwegian Nat. Tb. BCG Vaccine Lab., Bergen, 1937-45; prin. med. bacteriologist, dir. BCG Vaccine Lab., N.Y. State Dept. Health, Albany, 1946-56; asso. prof. pathology U. Albany, 1954-56; cons. USPHS, 1946-56. Recipient Gold medal Reconnaissance Francaise, 1920, Civic award Rochester, N.Y. for erysipelas antitoxin work, 1925, Gold medal Norwegian Nat. Tb Assn., 1945, Norwegian Red Cross for war service, 1946, Diploma of Honor, Am. Acad. Tb. Physicians, 1949, Gold medal N.Y. Med. Assn., 1954. Diplomate Am. Bd. Preventive Medicine and Pub. Health. Fellow A.A.-A.S., A.C.P., A.M.A.; mem. Norwegian Acad. Sci., Norwegian Med. Assn., Soc. Philomathique (Paris). Author: Telavaag—Battle in Western Norway, 1946; A Physicians's Cavalcade—Memoirs, 1965; also numerous articles. Discovered erysipelas antitoxin, Birkhaug test for rheumatism, iathergic immunity in exptl. Tb; research on preparation of BCG vaccine. Address: 51 Kalvedalsvei, Bergen, Norway.

BIRKHOFF, Garrett, Am. mathematician, educator; b. Princeton, N.J., Jan. 10, 1911; s. George David and Margaret (Grafius) B.; A.B., Harvard, 1932, Soc. of Fellows, 1933-36; postgrad. Cambridge (Eng.) U., 1932-33; Dr. Hon., U. Nacional Mexico, 1951, U. Lille (France), 1960, Case Inst. Tech., 1964; m. Ruth Wills Collins, June 21, 1938; children—Ruth Wills, John David, Nancy Collins. Instr., math. Harvard, Cambridge, Mass., 1936-38, asst. prof., 1938-41, asso. prof., 1941-46, prof., 1946—. Cons. govt. and pvt. industry, 1942—; Walker-Ames lectr. U. Wash., 1947; Taft lectr. U. Cin., 1947; chmn. organizing com. Internat. Congress Mathematicians, 1950. Guggenheim fellow, 1948. Mem. Am. Math. Soc. (v.p. 1958), Nat. Acad. Scis., Math. Assn. Am., Am. Acad. Arts and Scis. (v.p., 1966-68), Soc. Indsl. and Applied Math. (pres. 1967-68), Association for Computing Machinery, American Nuclear Society, Sociedad Mathematics Mexico (hon.), Acad. Ciencias Lima (hon.). Mem. Soc. of Friends. Author: (with S. MacLane) Survey of Modern Algebra, 1941, rev. edit., 1951; Lattice Theory, 1940, rev. edit., 1949; Hydrodynamics, 1950, rev. edit., 1960; (with E. Zarantonello) Jets, Wakes and Cavities, 1957; (with G. C. Rota) Ordinary Differential Equations, 1962. Contbd. to lattice theory, fluid mechanics, numerical analysis. Home: 45 Fayerweather St., Cambridge, Mass. 02138. Office: 2 Divinity Av., Cambridge, Mass. 02138.*

BIRKHOFF, George David, Am. mathematician; b. Overisel, Mich., Mar. 21, 1884; s. David and Jane Gertrude (Droppers) B.; student Lewis Inst., 1896-1902, U. Chgo., 1902-03; A.B., Harvard, 1905, A.M., 1906; Ph.D., U. Chgo., 1907; Sc.D., Brown, 1923, U. Wis., 1927, Harvard, 1933, U. Pa., 1938, Sofia, 1939; LL.D., St. Andrews, 1938; hon. Dr., Poitiers, 1933, Paris, 1936, Athens, 1937, U. Buenos Aires, 1942; member French Academy of Sciences, 1929; m. Margaret Elizabeth Graflus, September 2, 1908; children—Barbara (Mrs. Robert Treat Paine, Jr.), Garrett, Rodney. Instr. math. U. Wis., 1907-09; asst. prof. math. Princeton, 1909-11, prof., 1911-12; asst. prof. math. Harvard, 1912-19, prof., 1919-33, Perkins prof., from 1933, dean faculty of arts and scis., 1935-39; lectr. Collège de France, 1930. Decorated officier French Legion of Honor; named hon. mem. faculty San Marcos, Lima, 1942, U. Chile, 1942; recipient Querini-Stampali prize Royal Inst. Sci., Letters and Arts of Venice, 1918; Bocher prize for research in dynamics Am. Math. Soc., 1923; A.A.A.S. prize, 1926; Biennial prize (for research on systems of differential equations), Pontifical Acad. Scis., Vatican, 1933. Mem. Nat. Acad. (pres. 1936-37), Nat. Acad. Scis. Argentina, Royal Danish Acad. Scis. and Letters, Inst. of France, Göttingen, Lima acads. scis., Royal Acad. of the Lincei, Royal Inst. of Bologna, Pontifical Acad. Scis., Royal Irish Acad., Royal Soc., Edinburgh; hon. mem. Edinburgh, London math. socs., Peruvian Philosophic Soc., Sci. Soc. Argentina; mem. Phi Beta Kappa, Sigma Xi. Author: Relativity and Modern Physics, 1923; The Origin, Nature and Influence of Relativity, 1925; Dynamical Systems, 1928; Aesthetic Measure, 1933; Basic Geometry (with Ralph Beatley), 1941. Contbr. to math. jours. Editor, Annals of Mathematics, 1911-13, Trans. of Am. math. Soc., 1920-25, Am. Jour. Math., from 1943. Investigated linear differential equations and systems of difference equations; developed theory of graviton; investigated periodic orbits, three-body problems within general theory of dynamic systems and ergodic theory; studied mechanics of fluids in connection with math. treatment of viscosity. Died Nov. 12, 1944.

BIRKHOFF, Robert D., Am. physicist; b. Chgo., Jan. 29, 1925; s. Robert D and Ellen (Gleason) B.; B.S., Mass. Inst. Tech., 1945; Ph.D., Northwestern U., 1949; m. Ariel Frances Jewett, Nov. 4, 1945; Asso. prof. U. Tenn., Knoxville, 1949-55; physicist Oak Ridge Nat. Lab., 1955—, cons., 1950-55; head radiol. def. State of Tenn., 1952-55. Mem. Am. Phys. Soc., Health Phys. Soc. Author: Handbuch der Physik, Vol. 34, 1958; Health Physics, 1967; also articles. Measured cross sects. for plasmon excitation, electron flux in irradiated media, optical properties of metals, liquids, scintillators, electron diffusion in metals, plasmon and brems-strahlung light from irradiated metals. Home: 1433 Whitower Dr., Knoxville, Tenn. 37919. Office: Oak Ridge Nat. Lab., Oak Ridge, 37830.*

BIRKOFER, Leonhard, German chemist; b. Fürth, Germany, July 5, 1911; s. Karl and Marie (Benz) B.; Dr.rer.nat., U. Erlangen, 1935; m. Anneliese Leising, Nov. 15, 1941; 1 dau., Birgit. Asst., Max Planck Inst. Med. Research, Heidelberg, 1937-54; faculty U. Erlangen, 1944-49; docent Stuttgart Inst. Tech., 1949-54; asso. prof. U. Cologne, 1954-64, prof., 1964-65, dean sci. faculty, 1964-65; prof., dir. inst. organic chemistry U. Düsseldorf, 1965—. Mem. Soc. German Chemists, Gesellschaft Deutscher Naturforscher und Aerzte. Research and numerous publs. on synthesis of phenazines, B-amino acids, peptides; constitution of acylated anthocyanins from petunia, salvia splendens, raphanus sativus; organosilicon compounds and application in organic synthesis. Home: 21 Hardtstrasse, 5 Cologne-Klettenberg. Office: 127 ulenbergstrasse, 4 Düsseldorf, West Germany.*

BIRKS, John Betteley, English physicist; b. Eccles, Eng., Mar. 6, 1920; s. Walter Harry and Amy (Betteley) B.; scholar Queen's Coll., Oxford, 1938-40; B.A., Oxford (Eng.) U., 1946; Ph.D., Glasgow (Scotland) U., 1949, D.Sc., 1954; M.A., D.Sc., Oxford (Eng.) U., 1966; m. Margaret McNaughton, Aug. 30, 1941; children—Eleanor Irene, Harry John Betteley. Sci. officer Telecommunications Research Establishment, Swanage, Malvern, Eng., 1940-45; Imperial Chem. Industries fellow Glasgow U., 1946-48, lectr. natural philosophy 1948-51; prof. physics Rhodes U., S. Africa, 1951-54; research mgr. Brit. Dielectric Research, London, 1954-57; reader physics Manchester (Eng.) U., 1957—; vis. prof., NSF sir. fellow La. State U., 1965-66. Recipient Kelvin Gold medal Glasgow U., 1954. Fellow Inst. Physics, Instn. Elec. Engrs. (asso.), Brit. Biophys. Soc. Author: Scintillation Counters, 1953; Modern Dielectric Materials, 1960; Theory and Practice of Scintillation Counting, 1964; also numerous articles. Editor, Progress in Dielectrics, 1958—. Research on scintillation and luminescence of organic compounds. Home: 39 Spath Rd., Manchester 20, Eng.*

BIRKS, Thomas Rawson, English philosopher; b. Derbyshire, Eng., Sept. 28, 1810; student Dissenting Coll. at Mill Hill, Eng., Chesterfield (Eng.) Coll., Trinity Coll., Cambridge (Eng.) U.; m. Miss Bickersteth, 1844; 8 children; m. 2d. Vicar Trinity Ch., Cambridge, 1866-77; hon. canon Ely, Eng., 1871; prof. moral philosophy, Cambridge U., from 1872. Author: The Bible and Modern Thought, 1861; Scripture Doctrine of Creation, 1872; Modern Utilitarianism, 1874; Modern Physical Fatalism and the Doctrine of Evolution, 1876. Opponent of theory of evolution. Died Cambridge, July 19, 1883.

BIRMINGHAM, John, Irish astronomer; b. Milbrook, Ireland, 1816; discovered new star in Corona Borealis, 1866; began revision Catalogue of Red Stars (Schjellerup) to include Schmidt's list from Astronomische Nachrichten, 1872; discovered in Cygnus deep red star subsequently known by his name, 1881; discovered (with Webb and others) some 90 ruddy stars. Recipient Cunningham medal, 1884. Contbd. papers. to sci. jours., including Catalogue of Red Stars to Trans. Royal Irish Acad., 1879. Died Millbrook, Sept. 7, 1884.

BIRNBAUM, Milton, Am. physicist; b. Bklyn., Nov. 27, 1920; s. Louis and Dora (Stein) B.; A.B., Bklyn. Coll., 1942; M.S., U. Md., 1948, Ph.D., 1953; m. Mildred C. Delott, Nov. 24, 1957; children—Robin Sue, David Michael. Physicist, U. S. Naval Research Lab., Washington, 1946-53; head physics, elec. engring. dept. Bulova Research & Devel. Co., Woodside, N.Y., 1953-55; staff Microwave Research Inst. Bklyn. Poly. Inst., 1955-56, asso. prof. physics, 1956-61; tech. staff Aerospace Corp., El Segundo, Cal., 1961, asso. dept. head lasers and optics, 1968—; cons. Brookhaven Nat. Lab., Upton, L.I., N.Y., 1957-58, U. Cal. Radiation Lab., 1958. Mem. Am. Phys. Soc., Am. Assn. Physics Tchrs., Inst. Physics, Phys. Soc. London, Sigma Xi. Research, publs. on nuclear reactions and disintegrations, ruby lasers, semiconductor reflectivity modulation, semiconductor mirror Q switches, measurement of neutron binding energy. Patentee in field. Home: 4904 Elkridge Dr., Palos Verdes Peninsula, Cal. 90274. Office: P. O. Box 95085, Los Angeles 90045.*

BIRNSTEIL, Max Luciano, molecular biologist; b. Lichtensteig, Switzerland, July 12, 1933; s. Max and Dalila (Varella) B.; Diploma in Phys. Chemistry, Eidgenossische Technische Hochschule, Zurich, Switzerland, 1958, D.Natural Sci., 1959; m. Margaret Chipchase, Nov. 10, 1963; 1 son, Marcus Robert. Postdoctoral fellow div. biology Cal. Inst. Tech., Pasadena, 1960-63; faculty Inst. Animal Genetics, Kings Bldgs., Edinburgh, Scotland, 1963—, sr. lectr., 1966—. Mem. Biochem. Soc., Brit. Soc. for Cell Biology, Biophys. Soc., Schw. Botanische Gesellschaft, A.A.A.S. Research and publs. on functions of subcellular structures, plant ribosomes, the nucleolus and the cell nucleus and their relationship with protein synthesizing machinery of living cell, molecular aspects of differentiation and morphogenesis. Home: 17 Millar Crescent, Edinburgh, 10, Scotland.*

BIRREN, James E(mmett), Am. psychologist, gerontologist; b. Chgo., Apr. 4, 1918; s. August F. and Elsie (Kolkmann) B.; B.E., Chgo. Tchrs. Coll., 1941; M.A., Northwestern U., 1952, Ph.D., 1947; m. Elizabeth Solomon, Dec. 12, 1942; children—Barbara Ann, Jeffrey Emmett, Bruce William. Research psychologist, gerontology br. Nat. Heart Inst., Balt., 1946-51; chief sect. on aging Nat. Inst. Mental Health, Bethesda, Md., 1951-63; dir. aging program Nat. Inst. Child Health and Human Devel., Bethesda, 1963-65; dir. Rossmorr-Cortese Inst. Study Retirement and Aging, U. So. Cal., also prof. psychology, Los Angeles, 1965—, vis. prof. U. Chgo., 1959-60. Recipient Ciba Found. award 1956; meritorious service medal USPHS, 1965. Mem. Psychopath. Assn. (Stratton award 1960), Am. Physiol. Soc., Am. Psychol. Assn., Gerontol. Soc., Am. Geriatrics Soc., A.A.A.S., Am. Ednl. Research Assn., Psychometric Soc. Author: Psychology of Aging, 1964; also numerous articles. Editor: Handbook of Aging and the Individual, 1959; (with Alan T. Welford) Behavior, Aging and the Nervous System, 1965; Relations of Development and Aging, 1964. Demonstrated nature of functional changes in nervous system that occur with advancing age. Address: U. So. Cal., Los Angeles.*

BIRT, William Radcliffe, Brit. astronomer; b. 1804; fellow Royal Acad. Sci.; 1st pres. Selenographical Soc., circa 1877-81. Author of Hurricane and Sailor's Guide, 1850; Handbook of the Law of Storms, 1854, new edit., 1878. Reduced and arranged barometric observations of Sir John Herschel; research on atmopheric waves for Brit. Assn.; reduced and discussed elec. observations of 1848 at Kew. Died Essex, Eng., 1881.

BIRYUKOV, Dmitrii Andreevich, Russian physiologist; b. 1904; grad. Med. Faculty, Rostov-on-Don U., 1927; D.Med. Sci.; Hon. Dr., Prague U., 1958. Specialized in physiology Prof. N. A. Rozhansky Lab., Rostov-on-Don, Pavlov Lab., Leningrad, 1927-34; head chair physiology Rostov-on-Don, Voronezh, Moscow and Leningrad med. inst., 1935-43; dir. Voronezh Med. Inst., 1944-49; prof., 1935—; dir. Inst. Exptl. Medicine, USSR Acad. Med. Sci., 1950—; head chair normal physiology 1st Leningrad Med. Inst. Mem. USSR Acad. Med. Sci. Author: Unconditioned Salivary Reflexes in Man, 1945; Materials on Reflex Control of the Cardiovascular System, 1946; Comparative Physiology of Conditioned Reflexes, 1948; Further Results of Research on Comparative Physiology and Pathology of Higher Nervous Activity, 1955; Ecological Physiology of Nervous Activity, 1960; co-author: Physiological Methods in Clinical Practice, 1960. Chief editor Sechenov Physiol. Jour. USSR, 1950—; co-editor Physiology sect. Large Med. Ency., 2d edit., numerous others. Established features on unconditioned salivation in man, 1935; detected specific baroreceptors in dura mater, 1943. Address: Inst. Exptl. Medicine, USSR Acad. Med. Sci., Kirovsky prosp. 67-71, Leningrad, USSR.*

BISCHOF, Karl Gustav Christoph, German chemist, geologist; b. nr. Nürnberg, Bavaria, Jan. 18, 1792; s. Karl August Leberecht and Johanna (Oeder) B.; student Erlanger U.; m. Sophie Hesse; children—Georg Gustav, Karl Wilhelm. Prof. chemistry and tech. Erlanger U., from 1819, then in Bonn, from 1822. Author: (with G. A. Goldfuss) Geophysik-Statiskische Beschreibung des Fichtelgebirges, 2 vols., 1817; Lehrbuch der Stöchiometrie, 1819; Lehrbuch der Reinen Chemie, 1824; Die vulkanischen Mineralquellen Deutschlands und Frankreiches, 1826; also many others, including articles. Contbd. to changes in phys.-chem. consideration of geol. procedures; instrumental in exploiting geol. resources, including thermal springs for indsl. use; studied inflammable gases in coal mines; improved Davy safety lamp; research on internal terrestrial heat. Died Bonn, Nov. 30, 1870.

BISCHOFF, Carl Adam, German chemist; b. Würzburg, Germany, Apr. 8, 1855; s. Carl Adam and Katherine (Weis) B.; began studies of medicine in Würzburg, 1873; later studied chemistry under J. Wislicenus; under R. Fresenius, Wiesbaden Germany, 1876, under R. Bunsen, Heidelberg, Germany, 1877; dr. from Würzburg, 1879; m. Johanna Bitthäuser, 1882; 1 son, 1 dau. Asst. to Wislenus (accompanied him to Leipzig in 1885); became prof. in Riga, Latvia, 1887. Editor: (with P. v. Walden) Handbuch der Stereochemie, 1893-94; Materialien der Stereochemie, 1904. Published articles in German chem. jours. Research on organic synthesis, asymetric carbon atom; pioneer in stereochemistry. Died Munich, Germany, Oct. 18, 1908.

BISCHOFF, Christoph Heinrich Ernest, German surgeon; b. Hanover, Germany, 1780; prof. therapeutics, Bonn, Germany. Author (in German): Gall's Theory of Craniology, 1805; Relationship between Medicine and Surgery, 1842. Enlarged Gall's theories of brain and cranium. Died Bonn, 1861.

BISCHOFF, Fritz Emil, Am. chemist; b. Milw., May 21, 1899; s. Emil A. and Elsie (Meyer) B.; B.S., U. Wis., 1920, M.S., 1922, Ph.D., 1924. Research chemist Santa Barbara (Cal.) Cottage Hosp., 1925-28, dir. clin. chemistry labs., 1928-48, dir. research, 1939—; adv. trustee Internat. Cancer Research Found. DuPont fellow U. Wis., 1924. Diplomate Am. Bd. Clin. Chemistry. Fellow Am. Assn. Clin. Chemists (past mem. nat. exec. com.); mem. Am. Chem. Soc., Am. Inst. Nutrition, Am. Assn. Cancer Research, Am. Soc. Biol. Chemist, Phi Beta Kappa, Sigma Xi, Phi Lambda Upsilon. Bd. editors; Clinical Chemistry, 1966—. Research, publs. on alkyl titanates; principle of delayed resorption, red. cell estrone, solid state carcinogenesis, environmental factor in cancer, aliphatic sesqui-terpenes. Home: 1 El Vedado Lane. Office: 320 W. Pueblo St., Santa Barbara, Cal. 93105.*

BISCHOFF, Theodor Ludwig Wilhelm, German biologist, anatomist; b. Hannover, Germany, Oct. 28, 1807; s. Ernst and Juliane (Amelung) B.; dr. phil., U. Bonn (Germany), 1829, M.D., 1832; m. Kunigunde Tiedeman, 1839; 3 sons, 3 daus. Named lecturer, University of Bonn, 1833; assistant professor comparative and path. anatomy, Heidelberg, Germany, 1835; apptd. prof. physiology Geissen (Germany) U., 1843, also prof. descriptive and comparative anatomy, from 1844; professor at the University of Munich, 1855-78. Recipient numerous awards and decorations. Fellow Royal Soc., 1868. Mem. many sci. acads. Author: Entwicklungsgechichte der Säugetiere und des Menschen, 1842. Founder anat. mus. and physiol. inst., Giessen; studied embryology; described periodic ripening of ova in humans and other mammals; investigated differences between man and anthropoid apes; 1st to demonstrate presence of carbon dioxide and free oxygen in blood. Died Munich, Dec. 5, 1882.

BISEY, Sunker Abaji, inventor; b. Bombay, India, Apr. 29, 1867; s. Abaji Balvant and Nanibai (Durvey) B.; ed. high sch., Bombay; passed matriculation exam., 1888; m. Sushila Karnik, Feb. 22, 1893; children—Sonubai (Mrs. Narayan Laxuman Pradhan), Madhu, Reginald, Pramila. Came to U. S., 1916. In govt. service, India, 1889-98; as the "pioneer Hindu inventor," worked upon inventions in England, 1899-1915, later gen. mgr. Tata-Bisey Invention Syndicate, London; organized Am. Beslin Co., N.Y., 1918; founder, dir., tech. expert Bisey Ideal Typecaster Corp., 1920; dir. Am. Beslin Corp. Recipient gold medal Earls Ct. Exhbn., London, 1901. Hon fellow Soc. Sci., Letters and Arts (London), Inst. Inventor (N.Y.). Known as "The Edison of India," for invention of single and multiple typecasting machines, and in chemistry for water soluble non-irritating and non-poisonous iodine, known as Atomidine and Beslin. Died Apr. 7, 1935.

BISHOP, Donald Gerst, Am. polit. scientist; b. Altoona, Pa., May 16, 1907; s. Walter M. and Maona V. (Mason) B.; B.A., U. Akron, 1928; M.A., Princeton, 1929; Ph.D., Ohio State U., 1939; m. Iona Fay Maxwell, July 3, 1937. Faculty, Syracuse (N.Y.) U., 1938—, prof., 1952—, dir. citizenship program, 1946-51, chmn. internat. relations, 1950-65, chmn. dept. polit. sci., 1965-66; vis. prof. polit. sci. Tunghai U., Taichung, Taiwan, 1966-67. Cons. U. S. Govt., 1954—. Mem. Am., N.Y. (past pres.) polit. sci. assns., UN Assn. Author: Soviet Foreign Relations, 1952; Administration of British Foreign Policy, 1959; The Roosevelt-Litvinov Agreements, 1965; The Administration of U. S. Foreign Policy through the United Nations, 1967; also articles. Studied formulation and adminstrn. of fgn. policy by Am., Brit. and Soviet govts., Am.-Soviet diplomatic relations, 1933-43, problems of Am.-Chinese diplomatic relations, 1941—. Home: 833 Livingston Av., Syracuse, N.Y. 13210.*

BISHOP, Ernest Simons, Am. physician; b. Pawtucket, R.I., Nov. 29, 1876; s. Phanuel Euclid and Louise (Simons) B.; A.B., Brown, 1899; M.D., Cornell Med. Sch., 1908; m. Helen Earle, Jan. 20, 1912; children—Helen Kingsley, Amy. Intern, resident Bellevue Hosp., 1908-12; clin. prof. medicine N.Y. Polyclinic Med. Sch. Fellow A.C.P., Acad. Medicine, N.Y. Author: The Narcotic Drug Problem; also Chronic Drug Intoxications and Addictions in George Blumer edit. of Billings-Forchheimer's Therapeusis of Internal Diseases, 1925. Frequently called as expert witness in courts; influenced narcotic laws and other med. legislation; originator modern concept of narcotic drug addiction as curable phys. disease. Died Nov. 16, 1927.

BISHOP, George, English astronomer; b. Leicester, Eng., Aug. 21, 1785; with wine making bus. London, Eng., from 1803; builder obs. S. Villa, Regent's Park, England, 1836. Fellow Royal Society, 1848, Society of Arts; member Royal Astronomical Society (sec. 1833-39, treas. 1840-57, pres. 1857-58). Author: Astronomical Observations taken at the Observatory, South Villa, Regent's Park during the years 1839-51, 1852. Numerous astron. discoveries made at obs., including investigations of double stars by Rev. William Dawes, 1839-44. Died June 14, 1861.

BISHOP, George Robert, Brit. physicist; b. London, Eng., Jan. 16, 1927; s. George William and Lilian (Garrod) B.; B.A., Oxford U., 1948, D.Phil., 1950; m. Adriana Caberlotto, Aug. 14, 1952; children—James William, Georgina Rebecca. I.C.I. research fellow Oxford U., research fellow in physics St. Antony's Coll., Oxford, 1950-55; staff École Normale Supérieure, Paris, France, 1955-59; sci. adj. to dir. Lab. Linear Accelerator, Orsay, Paris, 1959-62; asso. prof. Faculty Scis., Orsay, 1962-64; Kelvin prof. natural philosophy U. Glasgow (Scotland). Author: Beta and Gamma Ray Spectroscopy, 1955; (with R. Hiban) Handbook of Physics, Vol. XLII, 1957; Electromagnetic Interaction with Nuclei, 1966; also articles. Early work on deuteron photodisintegration, oriented nuclear systems, polarization in nuclear reactions; research on nuclear structure via electromagnetic interaction, interaction of high energy electrons and photons with nuclei. Home: 17 Mirrlees Dr., Glasgow N. 2, Scotland.*

BISHOP, John Michael, English physician; b. Birmingham, Eng., Oct. 16, 1925; s. Arthur Claud and Dorothy (Jones) B.; M.B., Ch.B., U. Birmingham, 1948, M.D. with honors, 1956, D.Sc., 1966; m. Hilary Wishart, Aug. 15, 1952; children—Paul, Lucy. Faculty, U. Birmingham, 1958—, prof. medicine, 1966—; cons. physician United Birmingham Hosps., 1958—. Fellow Royal Coll. Physicians London; mem. Assn. Physicians Gt. Britain and Ireland, Physl. Soc., Cardiac Soc., Thoracic Soc., Med. Research Soc. Author: (with O. L. Wade) Cardiac Output and Regional Blood Flow, 1962; also articles. Research on physiol. measurement of various aspects of circulatory and respiratory function in health and diseases of heart and lungs especially effects of exercise. Home: 12 Birch Tree Grove, Solihull, Warwickshire, Eng. Office: Queen Elizabeth Hosp., Birmingham 15, Eng.*

BISHOP, Louis Faugères, Am. physician; b. New Brunswick, N.J., Mar. 14, 1864; s. James and Mary Faugères (Ellis) B.; A.B., Rutgers Coll., 1885, A.M., 1889, Sc.D., 1919; M.D., Coll. Phys. and Surg., Columbia, 1889; m. Charlotte D. Gruner, Nov. 14, 1899; 1 son, Louis Faugères. Resident physician St. Luke's Hosp., N.Y., 1889-92; prof. diseases of heart and circulation Fordham U.; heart specialist to Lincoln Hosp.; cons. in cardiovascular diseases Sea View Hosp. Author: Heart Troubles, Their Prevention and Relief; Heart Disease, Blood Pressure and the Nauheim Treatment; Arteriosclerosis: A Key to the Electrocardiograms; History of Cardiology. Died Oct. 6, 1941.

BISHOP, Peter Orlebar, Australian neurophysiologist; b. Tamworth, Australia, June 14, 1917; s. Ernest John Hunter and Mildred (Vidal) B.; student Barker Coll., Hornsby, Australia, 1932-34; M.B., B.S., Sydney (Australia) U., 1940, D.Sc., 1967; m. Hilare Louise Holmes, Feb. 20, 1942; children—Phillippa Leslie, Elisabeth Clare, Roderick Owen. Research fellow U. Coll., London, Eng., 1947-50, U. Sydney, 1950-51; mem. staff dept. physiology U. Sydney, 1951-67, prof. physiology, head dept., 1955-67; prof., head dept. physiology Australian Nat. U., 1967—. Mem. research adv. com. Nat. Health and Med. Research Council Australia, 1959-66. Fellow Australian Acad. Sci.; mem. Internat. Brain Research Orgn. (neurocommunications panel), Australian Physiol. Soc. (treas.), Postgrad. Med. Fedn. Australia (exec.), Physiol. Soc. (Gt. Britain). Editorial staff Physiology and Behavior, 1966—, Exptl. Brain Research, 1966—. Research, publs. in physiol. optics, neurology of vision using quantitative analysis of single nerve cell firing patterns to provide neural basis for visual perception; studies on formation of retinal image, image coding in optic nerve, transmission and data processing along visual pathway and data analysis by visual neurones in cerebral cortex. Home: 25 Lawson Crescent, Acton, A.C.T., Australia.*

BISQUE, Ramon E., Am. geochemist; b. Stambaugh, Mich., Sept. 1, 1931; s. Edward and Camilla (Zyskowski) B.; B.S., St. Norbert College, 1953; M.A., chemistry, Iowa State U., 1956; M.A., geology, Iowa State, 1957; Ph.D., Iowa State, 1959; m. Marie L. Young, July 31, 1953; children—Camile Luise, Stephen Michael, Laura L., Thomas Matthew, Daniel Ramon, Matthew Livingston. Asst. prof. chemistry, Colo. School of Mines, 1959-62; assoc. prof., 1962-68; dir., Earth Science Curriculum Project, 1965-67; prof. geochemistry, asst. graduate dean, Colorado School of Mines, 1967—. Board of Directors, co-founder, Earth Sciences, Inc.; consultant, Institute for Defense Analyses. Mem., Geochemical Soc.; Am. Chem. Soc.; Geological Soc. of Am.; Am. Institute of Prof. Geologists; Sigma Xi. Distinguished lectr., Am. Assoc. Petroleum Geologists, 1968. Research on global tectonics and earth's core; developed G. E. Rouse's theory that stresses at interface in earth's interior (globally) control distribution of surface features, such as mountains, deep oceanic trenches, mineral belts, island arcs, etc.; theory asserts that quakes, volcanoes, rifts fall into ordered pattern and are not chance happenings; provides possibility for prediction of earthquakes; with Rouse, asserted that intergalactic and interplanetary magnetic fields may react sufficiently with earth's magnetic field to cause core to rotate a little differently than earth does on regular turning around axis. Office: Colorado School of Mines, Golden, Colorado, 80401.*

BISSONNETTE, T. Hume, biologist; b. Dundas, Ont., Can., June 27, 1885; s. Julien Donald and Annie Isabel (Hume) B.; M.A., Queen's U., Kingston, Can., 1913; Ph.D. (Brit. Bd. Edn. scholar), U. Chgo., 1923; m. Julia Irene Powers, Mar. 1, 1924; children—Julien Hume, Donald King. Came to U. S., 1920, naturalized, 1931. Jr. master, Galt Coll. Inst., 1906-09; prin. Victoria Pub. Sch., Saskatoon, Can., 1913; science master, biology and chemistry Regina Coll. Inst., 1914-16; lectr. in biology, Queen's Univ., Can., 1919-20; asst. in zoology U. Chgo., 1921-23; instr. in zoology Y.M.C.A. Coll. of Liberal Arts, Chgo., 1922-23; prof. biology Coe Coll., Cedar Rapids, Ia., 1923-25; J. Pierpont Morgan prof. biology and head of dept. Trinity Coll., Hartford, Conn., since 1925; instr. in marine invertebrate zoology Marine Biol. Lab., Woods Hole, Mass., 1926-36, in charge of course, 1936-41; research vis. prof. Cambridge (Eng.) U., 1931-32. Recipient Walker Grand prize 1945, for investigations in Photo periodism. Fellow A.A.A.S., Ia. Acad. Sci.; mem. Assn. for Research in Internal Secretions, Am. Soc. Zoologists, Genetics Assn. Genetics Soc., Northeastern Birdbanding Assn., Am. Naturalists, Corp. of Marine Biol. Lab., Nat. Geog. Soc., Sigma Xi. Contbr. sci. articles to learned publs. Died Nov. 30, 1951.

BISTENI, A., Mexican physician; b. Tuxpan, Mexico, July 9, 1922; s. Angel and María (Adem) B.; M.D., Nat. U. Mexico, 1948; m. Faride Bustani, Jan. 6, 1952; children—Jorge, Luis Manuel, Fernando, Alejandro. With Inst. Nal. de Cardiología, Mexico City, Mexico, 1948-51, investigator, 1953—; with La. State U., 1952-53; asso. prof. medicine U. Mexico. Mem. Sociedad Mexicana de Cardiología, Sociedad Jaliscience de Cardiología, Am. Coll. Chest Physicians, Academia Nacional de Medicina Mexico. Author: Electrocardiografía y Vectocardiografía deductivas, 1964; also numerous articles. Research on electrocardiographic diagnosis of cardiovascular diseases, treatment of coronary heart diseases. Home: 145 Piacacho. Office: 300 Cuauhtemoc, Mexico D.F. 20, Mexico.*

BISTOLFI, Franco, Italian physician, radiologist; b. Genoa, Italy, May 18, 1926; s. Stefano and Marica (Papasaul) B.; Degree in Medicine and Surgery, U. Genoa, 1949, postgrad. specializing in radiology Inst. Radiology, 1951; m. Viviana Astolfi, May 30, 1960; 1 dau., Marica. With Inst. Radiology, U. Genoa, Italy, 1952—, ordinary asst., 1965—, libero docente in radiology, 1959-65, libero docente in radiobiology, 1965—, chief tomographic dept., 1957-59, chief roentgen and radiumtherapy dept., 1960—. Recipient V. Maragliano prize, 1952, Palmieri prize, 1965, prize Congress of Latin Culture Radiologists, 1957, 61, 64. Mem. Italian League Against Cancer, Accademia Medica. Author: (with others) La Stratigrafia assiale del cranio, 1959; (with N. Macarini, L. Oliva) La distribuzione cronologica della dose, 1963; also numerous articles. Research on tomography of skull, X-ray therapy of cancer; devised, developed new method for clin. dosage: chrono-biodosimetric method. Home: 63/2 L. Montaldo, Genoa, Italy.

BISWAS, Biswamoy, Indian zoologist; b. Calcutta, India, June 2, 1923; s. Saratlal and Malati (Neogy) B.; B.Sc., U. Calcutta, 1943, M.Sc., 1945, D.Phil., 1952. Zoologists in charge of bird and mammal sects Zool. Survey of India, Calcutta, 1950-62, superintending zoologist, 1963-—. Mem. Internat. Council Bird Preservation (sec. Indian Nat. Sect. 1952-—), Zool. Soc. (editor 1964-—), Internat. Ornithol. Congress (internat. com. 1958-—), Am. Ornithologists' Union (corr.). Research and numerous publs. on comparative morphology, taxonomy, ecology and zoogeography of bird and mammals of Indian region; discovered new sub-species and species of birds and mammals. Home: 4 Duff Lane, Calcutta 6. Office: Zool. Survey of India, Indian Mu., Calcutta 13, India.*

BISWAS, Nripen Nath, physicist; b. Calcutta, India, Aug. 1, 1930; s. Gaya Nath and Dhaneswari (Goswami) B.; B.Sc., Calcutta U., 1949, M.S., 1951, Ph.D., 1955; m. Lieselotte Ella Thiel, May 12, 1959; children—Indira Susan, Arun Kumar. Research asst. cosmic rays Bose Inst., Calcutta, India, 1952-55; research scholar physics Max Planck Inst. Physics, Göttingen, 1955-57; research physicist U. Bologna, 1957-58; vis. scientist Lawrence Radiation Lab., Berkeley, Cal., 1958-60; research scientist Max Planck Inst., Munich, Germany, 1961-65; asso. prof. physics U. Notre Dame, 1965-—. Research and publs. on elementary particles; with the discovery of K mesons and hyperons studies were made on the properties of these particles; studies of unstable particles (resonances) to find their properties, such as longevity, interaction with matter. Home: 52742 W. Cypress Ct., South Bend, Ind. 46637.*

BISWAS, Sukumar, Indian physicist; b. Jalpaiguri, West Bengal, India, July 1, 1924; s. Satyendra Prasad and Suhasini (Roy) B.; B.Sc., Presidency Coll., Calcutta, India, 1944; M.Sc., U. Coll. Sci., Calcutta, 1946; D.Phil., U. Calcutta, 1950; Ph.D. (UNESCO fellow), U. Melbourne (Australia), 1952; m. Reba (Bose) Dec. 11, 1953; children: Rana, Noopur. Research fellow Tata Inst. Fundamental Research, Bombay, India 1952-56, fellow, 1956-63; research asso. physics dept. U. Minn., Mpls., 1959-61; sr. postdoctoral research asso. NASA, Goddard Space Flight Center, Greenbelt, Md., 1961-63; reader Tata Inst. Fundamental Research, 1963-65, asso. prof., 1965-—. Mem. Am. Phys. Soc., Am. Geophys. Union. Research and publs. in nuclear physics, cosmic ray physics using nuclear emulsions in balloons and rockets; discovered method of measurement of mass of K-mesons and hyperons, spurious scattering in nuclear emulsions; developed knock-on electron method for energy measurements of heavy primaries in cosmic rays; research on helium and heavy nuclei in solar cosmic rays, composition and energy spectra of low energy cosmic rays. Home: Stardust, 80 Nepean Sea Rd., Bombay-6. Office: Tata Inst. of Fundamental Research, Colaba, Bombay-5, India.*

BITO, John Francis, Hungarian physicist; b. Szeged, Hungary, July 21, 1936; s. John Joseph and Irene (Katona) B.; grad. U. Szeged, 1958; D.Natural Scis. summa cum laude, State U., Szeged, 1960; Candidate Tech. Scis., 1967, Dr. Tech. Scis., 1968. Asst., U. Szeged, 1956-58; sci. worker Research Inst. for Electronics, 1958-64, sci. head discharge div., 1964-65; lectr. phys. faculty E.R. U., 1965-—; invited lectr. Tech. U., Budapest, 1966-—; specialist electronics Ministry of Machines, 1965-—, sci. head elec. discharge dept., 1964-—; mem. plasma physics group Roland Eötvös U., 1960-—. Recipient awards for eminent learning, 1950, 54, 58; Brody prize Soc. Hungarian Physicists, 1967. Mem. Assn. Hungarian Physicist, Assn. Hungarian Electrotechniscists, Commn. Internat. Eclarage, Assn. Vacuum Technologists, Central Astronautical Com. Author: Discharge Physics: Technico-physical problems of Astronautics (with J. Sinka), 1966; (with G. Szabó, J. Sinka) Introduction of Experimental Plasma Physics, 1967; (with I. Abouyi) Plasma Waves, 1967; Theoretical Plasma Diagnostics, 1968; others; also articles. Research on low pressure discharges, cold-plasma microparameters and relations between them, time resolution in plasma diagnostics at higher frequency discharges, boundary phenomena at cathode-plasma; microwave and laser beam diagnostics; diagnostic of space of oxide cathodes; energy balance of fluorescent lamp cathodes; built improved FL cathode; studied electronoptics in cathode space of discharges, physics of fluorescent lamps, theoretical investigations of hot, fusion plasmas, shock waves. Home: 22/A Korompai, Budapest, XII, Hungary.*

BITSADZE, Andrei Vasilevich, Russian mathematician; b. Georgia, USSR, May 22, 1916; s. Vasili and Mary (Bregwadze) B; grad. Tbilisi State U., 1940, M.Sc., 1945, D.Math., 1951; m. Nina Baderko, Oct. 18, 1958; 1 son, Andrei. With Inst. Math., Georgian Acad. Sci., 1941-48; sr. sci. worker Steklov Math. Inst., Moscow, 1948-59; head dept. Novosibirsk Math. Inst., 1959-—, Novosibirsk State U., 1958-—. Mem. USSR Acad. Sci. (corr.), also math. socs. of Moscow, Novosibirsk and Tbilisi. Author: Equations of the Mixed Type, 1959; Boundary Value Problems for Elliptic Second Order Equations, 1966. Research and numerous publs. on theory of elliptic boundary value problems, theory of equations of mixed type, theory of singular integral equations. Home: 73 Zolotodolinskaya. Office: Inst. Math., Novosibirsk 90, USSR.*

BITTER, Francis, Am. physicist, b. Weehawken, N.J., July 22, 1902; s. Karl and Marie (Schevill) B.; A.B., Columbia, 1924, Ph.D., 1928; postgrad. U. Berlin (Germany), 1925-26, (Nat. Research fellow), Princeton, 1928-30, Cal. Inst. Tech., 1928-30; m. Alice Coomara, May 31, 1928 (dec. July, 1958); m. 2d, Katharine Hodgson Welchman, Feb. 7, 1959. Research engr. Westinghouse Electric and Mfg. Co., East Pittsburgh, Pa., 1930-34, cons., 1934-42; asso. prof. physics Mass. Inst. Tech., Cambridge, 1934-51, prof. physics, 1951-—, asso. dean sci., 1956-60, master grad. house, 1962-65. Cons. Magnion, Burlington, Mass., 1958-—. Guggenheim Meml. fellow, 1933-34. Mem. Am. Phys. Soc., Am. Acad. Arts and Scis., Am. Assn. Physics Tchrs., Phi Beta Kappa, Sigma Xi. Author: Introduction to Ferromagnetism, 1937; Nuclear Physics, 1950; Currents, Fields, and Particles, 1956; Magnets: The Education of a Physicist, 1959; Mathematical Aspects of Physics, 1963. Research on magnetic properties of matter, high field magnet design. Died July 26, 1967.

BITTERMAN, Morton Edward, Am. psychologist; b. N.Y.C., Jan. 19, 1921; s. Harry Michael and Stella (Weiss) B.; B.A., N.Y. U., 1941; M.A., Columbia, 1942; Ph.D., Cornell U., 1945; m. Shirley Subke, Oct. 1, 1952 (div. Apr. 1964), children—Joan, Ann; m. 2d, Mary Gayle Foley, June 26, 1967. Asst. prof. Cornell U., 1945-50; asso. prof. U. Tex., 1950-54; vis. asso. prof. U. Cal. at Berkeley, 1954-55; mem. Inst. for Advanced Study, also vis. lectr. Princeton, U. Pa., 1955-57; faculty Bryn Mawr (Pa.), 1957-—, prof. psychology, 1960-—, chmn. dept. psychology, 1957-—. Mem. A.A.A.S., Am. Inst. Biol. Scis., Am. Psychol. Assn., Psychonomic Soc. Co-editor Am. Jour. Psychology, 1955-—. Research and publs. on comparative studies perception and learning in worms, mollusks, arthropods, fishes, reptiles, birds, and mammals. Home: 451 Gypsy Rd., King of Prussia, Pa. 19406. Office: Dalton Hall, Bryn Mawr Coll., Bryn Mawr, Pa. 19010.*

BITTMAN, Emil, Rumanian physician; b. Bucharest, Rumania, July 16, 1927; s. Isidor and Sofia (Victor) B.; M.D., Faculty Medicine, Bucharest, 1951, D.Medicine, 1956; m. Silvia Kahane, Sept. 20, 1951; 1 dau., Robina. Research fellow electrophysiol. dept. Inst. Normal and Path. Physiology D. Danielopolu, Acad. R.S. Romania, Bucharest, 1951-56, prin. research fellow clin. and neurophysiol. depts., 1956-—, sr. med. officer specializing in internal medicine and functional explorations, 1960-—; prof.'s asst. chair of physiology Faculty Medicine, Bucharest, 1954-57, 1960-61. Named Distinguished in medico-san. field, 1958. Mem. Rumanian Union Socs. Med. Scis., Internat. Electroencephalography Soc. Author: Electricitatea corpului omenesc, 1965; also numerous articles. Research on influence of cardiac nerves on elec. phenomena of heart in mammals, tonigen action of posterior hypothalamus on cerebral activity, existence of summation phenomena in autonomic nervous system, autonomic reciprocal innervation-type phenomena induced by mesencephalic reticular formation on some vegetative reflexes; designed stereotaxic instrument. Home: 27, Str. Bis. Amzei, Bucharest. Office: 11, Bd. 1 Mai, Bucharest, Romania.*

BITTNER, Alexander, paleontologist; b. Friedland, Bohemia, Mar. 16, 1850; s. Josef Bittner; began studies in Vienna, Austria, 1869; degree, 1873; asst. to Eduard Suess, 1873; geol. expdns. to Italy and Greece, 1874-75; asst. to state geol. inst., Vienna, 1877; became head geologist, 1897. Author: Die geologische Verhältnisse von Hernstein in Niederösterreich, 1882; Brachiopoden der alpinen Trias, 1890; Nachtrag, 1892; Lamellibranchiaten der alpinen Trias, 1895; Arbeiten zur Gliederung und Nomenklatur der alpinen Trias, 1896-99. Introduced problem of east alpine triassic stratigraphy into paleontologic-stratigraphic period of east alpine geology which led to explanation of marine triassic deposits; research on paleo-zoology. Died Vienna, Mar. 31, 1902.

BITTNER, John Joseph, Am. biologist; b. Meadville, Pa., Feb. 25, 1904; s. Martin and Minnie (Shults) B.; A.B., St. Stephen's Coll., Annandale-on-Hudson, N.Y., 1925; M.S., U. Mich., 1929, Ph.D., 1930; Sc.D., Bard Coll., 1950; Dr. Medicine and Surgery, U. Perugia (Italy), 1957; m. Mary Esther Mahaffy, June 23, 1930; children—Mary Margaret, Elizabeth Ann. Assistant in biology St. Stephen's College, 1923-25; master Donaldson School, Ilchester, Maryland, 1925-26; assistant in cancer research University Mich., 1927-30, incorporator Jackson Meml. Lab., Bar Harbor, Me., 1930, dir., 1935-42, treas., 1936-38, v.p., 1940-42, research asso., 1930-40; George Chase Christian prof. cancer research and dir. cancer biology, med. sch. U. Minn., from 1942; professor experimental pathology, from 1957; consulting associate scientist Sloan-Kettering Institute, from 1956. Mem. of Unitarian Service Com. (WHO-sponsored med. teaching mission to Austria), 1947. Recipient Alvaranga prize Coll. Physicians Phila., 1941; Comfort Crookshank award and lecture, London, 1951; Bertner award and lecture U. Tex. M.D. Anderson Hosp. and Tumor Inst., Tex. Med. Center, 1957. Mem. Am. Assn. Cancer Research (pres. 1947-48), NRC (chmn. milk factor panel, com. on growth, 1945-46, mem. exptl. genetics panel 1947-48), A.A.A.S., (editorial com. Gibson Island Symposium 1945), Am. Assn. U. Profs., Soc. Exptl. Biology and Medicine, Am. Cancer Soc., Harvey Soc., Soc. Physicians Vienna (corr. mem.), Royal Soc. Physicians London, N.Y. Acad. Scis., Sigma Xi. Mem. editorial bd. Cancer Research, 1941-50. Began research on cancer transmission in mice, 1930's; isolated milk factor (Bittner milk factor) in cancer transmission, 1949. Died Mpls., Dec. 14, 1961.

BIXBY, William Herbert, Am. elec. engr.; b. Indpls., Dec. 28, 1906; s. George Linder and Carrie (Tilton) B.; B.S.E. in Elec. Engring. and in Math., U. Mich., 1930, M.S. in Elec. Engring., 1931, Ph.D. in Elec. Engring., 1933; M.M.E., Chrysler Inst. Engring. 1935; m. Dorothy Bancroft, Jan. 17, 1963. Research engr. Detroit Edison Co. Research Lab., 1928-32; spl. problems engr. Chrysler Corp., Detroit, 1933-36; faculty Wayne U., Detroit, 1936-56, prof. elec. engring. 1950-56; cons. devel. engr. Power Equipment Co., Detroit, 1937-56, v.p. for applied research, 1956-62; v.p. for applied research North Electric Co., Galion, O., 1962-—. Fellow A.A.A.S., I.E.E.E.; mem. Engring. Soc. Detroit, Acacia, Sigma Xi, Phi Kappa Phi, Iota Alpha, Sigma Pi Sigma, Tau Beta Pi. Patentee electronically controlled rectifiers, line voltage regulators, generator excitor regulators. Home: 5274 Riverside Dr., Columbus 43221. Office: 1090 W. Henderson Rd., Columbus, O. 43214.*

BIXLER, Ray Herbert, Am. psychologist; b. Emerson, Neb., May 15, 1917; s. Ray Andrew and Josephine (Gress) B.; B.E., Ill. State U., 1939; M.A., Ohio State U., 1942, Ph.D., 1951; m. Marjorie Marie Martin, Jan. 28, 1956; children—Deveney, Danae, Dike, Minda. Psychologist, Akron (O.) Guidance Center, 1943-44; counselor U. Minn., Mpls., 1944-46; psychologist Minn. Psychiat. Inst., Mpls., 1946-48; faculty U. Louisville, 1948-—, head, dept. psychology, 1958-67. Dozent, U. Hamburg, Germany, 1952-53; cons. Ct. Domestic Relations, Louisville, 1950-52, housing com. Human Relations Commn., Louisville, 1962-67; Fulbright prof., Colombia, 1964. Bivins scholar, 1942. Mem. Am., Midwestern, Southeastern psychol. assns., Soc. Research in Child Devel., Soc. Philosophy and Psychology, Sociedad Interamericana de Psicologia, Phi Kappa Phi, Phi Delta Kappa, Delta Phi Alpha. Developed methods of test selection and interpretation in ednl. and vocational counseling; studied attitudes regarding desegregation and open housing; publs. on ethics of psychotherapy and behavioral research. Home: 3912 Chenoweth Run, Jeffersontown, Ky. 40299.*

BJERKEDAL, Tor, Norwegian physician; b. Oslo, Norway, Aug. 26, 1926; s. Sigurd and Herta (Toussaint) B.; M.D., U. Oslo, 1952, Dr.med., 1964; M.P.H., Harvard, 1958; m. Gudrun Kjerulff, Dec. 28, 1949; children—Anne, Nina, Tor Johan. Dist. health officer Norwegian Health Services, 1953-54; instr. Inst. Hygiene, U. Oslo, 1954-62, research fellow, 1962-64; vis. scientist Tb. Research Office, USPHS, 1958, 60-62; teaching fellow in biostatistics U. Cal. at Berkeley, 1964-65; head Inst. Hygiene and Social Medicine, U. Bergen (Norway), 1966-—. Physician in sch. and indsl. health service programs, 1955-64. Cons. Tb program USPHS, Washington, 1962-—. Mem. Norwegian Med. Assn., Internat. Union against Tb. (Paris, corr. mem. tech. com. on epidemiology, 1964-—). Author: Studies on Host-Agent Interaction in Experimental Tuberculosis, 1964; also articles. Research on immunity in Tb., mechanisms of BCG vaccination, and chemoprophylaxis of Tb with isoniazid, atypical mycobacterial infections, infectious hepatitis, blood pressure and body weight. Address: U. Bergen, Inst. Hygiene and Social Medicine, Bergen, Norway.*

BJERKNES, Carl Anton, Norwegian physicist; b. Oslo, Norway, 1825; m. Aletta V. D. Koren; 1 son, Vilhelm Friman Koren. Engr. in mines, Konsberg; sent on mission to Paris, France, 1855; became prof. U. Oslo, Norway, 1825; m. Aletta V. D. Koren; 1 son, Vilhonor Exposition Electricity, Paris, 1881. Author: Phénomènes hydrodynamiques, inversement analogues à ceux de l'électricité et du magnétisme, 8 vols., 1882; Nields Henrik Abel, sa vie et son action scientifique, 8 vols., 1885. Research on hydrodynamics. Died 1903.

BJERKNES, Jacob Aall Bonnevie, meteorologist; born Stockholm, Sweden, November 2, 1897; son of Vilhelm F. K. and Honoria (Bonnevie) B.; student at the University of Oslo (Norway), 1914-17, Ph.D., 1924; LL.D., Univ. California, Los Angeles; m. Hedvig Borthen, July 11, 1928; children—Vilhelm, Kirsten (Mrs. Per Kamsvaag). Meteorologist, Meteorol. Obs., Bergen, Norway; supt. Weathercast Center, Bergen, 1920-31; prof. meteorology U. Bergen, 1931-40; prof. U. Cal. at Los Angeles, 1940-—. Past cons. to meteorol. offices of Switzerland, Can., London, U. S. Weather Bur.; vis. lectr. Mass. Inst. Tech. Decorated Royal Order of St. Olav, 1947; recipient Nat. Medal Sci., 1967. Mem. Nat. Acad. Scis. Author (with V. Bjerknes) Physikalische Hydrodynamik, 1933; (with C. L. Godske) Dynamic Meteorology and Weather Forecasting, 1957. Contbr. numerous articles to sci. jours. Published theory of moving atmospheric fronts separating air masses of different geog. origin; many verifications followed by use of unmanned balloon and satellite observations; demonstrated the response of the global atmospheric wind system to the varying heat supply from the equatorial belt of the oceans, which explains to some extent the variability of climates from year to year. Home: 620 Adelaide Dr., Santa Monica, Cal. 90402.*

BJERKNES, Vilhelm Friman Koren, Norwegian physicist; b. Christiana, Norway, Mar. 14, 1862; s. Carl Anton and Aletta V.D. (Koren) B.; studied at U. Oslo (Norway), U. Paris (France), U. Bonn (Germany); Ph.D., U. Oslo, 1892; 1 son, Jacob Aall. Instr. math. and math. physics U. High Sch., Stockholm, Sweden, 1893-95; prof., 1895-1907; prof. mech. and math. physics U. Oslo, 1907-13, prof., 1926-32; prof. physics Leipzig (Germany) U., 1913-17; prof. Bergen (Norway) Geophys. Inst., 1917-26; organized network of meteorol. stas. throughout Norway; founder Bergen Sch. Meteorology. Mem. Royal Soc. London (fgn.), Nat. Acad. Scis. (fgn.). Author: Fields of Force, 1906; co-author: Dynamic Meteorology, and Hydrography, 2 vols., 1910, 11. Developed theory of electric resonance (important in devel. wireless telegraphy); (with Jakob Aall Bjerkener) originated polar front theory of cyclones (formed basis of modern weather forcasting). Died Oslo, Apr. 10, 1951.

BJERRUM, Jannik, Danish chemist; b. Copenhagen, Denmark, Apr. 5, 1909; s. Niels Janniksen and Ellen Emilie (Dreyer) B.; M.S., U. Copenhagen, 1932, Dr.phil., 1941; m. Grethe Vera Ehlers, Nov. 24, 1937; children—Niels Janniksen, Ole Jannik, Elsebet, Poul Jannik, Kirstine, Hans Jannik, Morten Jannik. Faculty, U. Copenhagen, 1936—, prof. chemistry, 1948—, pro-vice chancellor, 1965-67. Mem. Royal Danish, Royal Norwegian acads. scis., Danish Acad. Tech. Scis. Research, numerous publs. on coordination chemistry, especially on complex formation in solution. Author: Metal Ammine Formation in Aqueous Solution, 1941, 57. Home: Regensen, St. Kannikestraede 2, Kobenhavn K., Denmark.*

BJERRUM, Laurits, civil engr.; b. Farsö (Denmark), Aug. 6, 1918; s. C. A. and Henriette (Kragh-Hansen) B.; C.E., Tech. U. Denmark, 1941; Dr. techn., Fed. Inst. Tech., Zurich (Switzerland), 1952; Dr.sc.h.c., Loyola Coll., Balt., 1965; m. Gudrun Knudsen, Oct. 9, 1951; children—Chresten, Annette, Susanne. With lab. for harbor work and founds. Tech. U. Denmark, Copenhagen, 1941-42; head sect. soil mechanics and found. engrng. Chr. Ostenfeld, cons. engr., Copenhagen, 1941-42; mem. staff., founds. sect. Exptl. Inst. Harbor Work and Founds., Fed. Inst. Tech., Zurich, 1947-51; dir. Norwegian Geotech. Inst., Oslo, Norway, 1951—; vis. prof. Mass. Inst. Tech., Cambridge, 1957; vis. lectr. U. Ill., Urbana, 1956. Mem. Internat. Soc. Soil Mechanics and Found. Engrng. (pres. 1965-69), acads. tech. scis. Denmark, Norway, Norwegian Acad. Scis. and Letters, Brit. Instn. Civil Engrs. Author: (with A. Casagrande, Ralph B. Peck, A. W. Skempton) From Theory to Practice in Soil Mechanics; also numerous articles. Research on shear strength of soils and stability of clays, design and constrn. of founds. and excavation in soft clay. Home: 44 Trosterudveien. Office: 1 Forskningsveien, Oslo 3, Norway.*

BJERRUM, Niels, Danish chemist; b. Copenhagen, Denmark, Mar. 11, 1879; student of Julius Thomsen and Nernst; became prof. Royal Vet. and Agrl. Coll., Copenhagen, 1914; research and numerous publs. on electrochemistry, theory of acids and bases, measurement of hydrogen ion concentrations, amphoteric electrolytes, behavior of indicators, specific heats of steam; did significant work on infra-red spectra of polyatomic molecules, 1914. Died Sept. 30, 1958.

BJÖRK, Viking Olov, Swedish heart surgeon; b. Darlecarlia, Sweden, Dec. 3, 1918; s. Karl and Erika (Wikström) B.; grad. U. Lund (Sweden), 1944, doctorate, 1944; m. Ingegerd Ebba Christina Laurell, June 7, 1947; children—Agneta Birgitta, Ebba Christina, Anne Cecilia, Carolin, Anders Viking. Asso. prof. thoracic surgery Karolinska Instituet, Stockholm, Sweden, 1951-60; thoracic surgeon in chief U. Hosp., Uppsala, Sweden, 1960; prof. U. Uppsala, 1962-66; prof. thoracic surgery Karolinska Sjukhuset, Stockholm, 1966—. Guest prof., Brazil, Argentina, Chile, Peru, U.S.A., 1954, Russia, 1956, U. S., Japan, 1963, Germany, 1964. Scholar, Rome, 1942, London, 1944, U. S., 1950. Author: The Artificial Oxygenation of Blood for Brain Perfusion, 1948; also numerous articles. Description of disc oxygenator in heart-lung machine, 1947; introduced left heart catheterization, 1953, osteoplastic thoracoplasty in treatment of pulmonary Tb, 1954, artificial ventilation with respirator in postoperative treatment after thoracic surgery, 1955. Home: Radjursvägen, 4, Stocksund, Sweden. Office: Karolinska Sjukhuset, Stockholm, Sweden.*

BJORKMAN, Nils Halvorsson, Swedish histologist; b. Kristberg, Sweden, Mar. 29, 1919; s. Halvor N. and Astrid (Simonsson) B.; D.M.V., Royal Vet. Coll., Stockholm, 1954; m. Elisabeth N. Bengtson, June 4, 1949; children—Ylva K., N. Peder, B. Karsten. Faculty, Royal Vet. Coll., 1952—, research docent, 1965—. Postdoctoral fellow U. Wash., Seattle, 1960-61; vis. asso. prof. N.Y. State Vet. Coll., Cornell U., 1966-67. Fulbright scholar, 1966-67. Mem. Swedish Vet. Soc., Scandinavian Soc. for Electron Microscopy. Research, numerous publs. on fine structure and function of placenta of domestic and lab. animals, light and electron microscopic and histochem. studies female reproductive organs, pancreatic islets, other tissues, reproductive and absorptive organs in liver fluke. Home: Riddargatan 38, Stockholm, Sweden.*

BJORKSTEN, Johan Augustus, chemist; b. Tammerfors, Finland, May 27, 1907; s. Walter and Gerda (Ramsay) B.; M.S., U. Helsingfors, Finland, 1927, Ph.D., 1931; m. Christel E. Svedlin, Nov. 14, 1961; children—Sybil Joan (Mrs. Jurgen von Rennenkampff), Oliver J. W., Dargar William, Nils Johan. Came to U. S., 1931, naturalized, 1938. Guest research worker, U. Stockholm, 1927-28; Rockefeller post-doctorate fellow Edn. Bd., U. Minn., 1931-32; chief chemist Felton Chem. Co., Bklyn., 1933-35; head devel. Pepsodent Co., Chgo., 1935-36; chief chemist Ditto, Inc., Chgo., 1936-41, chem. dir. Quaker Chem. Products Corp., Conshohocken, Pa., 1941-44; v.p. A-B-C Packaging Machine Corp., Largo, Fla., 1940—; pres. Bee Chem. Co., Lansing, Ill., 1945-57; pres. Bjorksten Research Labs., Madison, Wis., 1944—; pres. Bjorksten Research Found., Madison, 1953—. Chmn. bd. dirs. Griffolyn Co., Houston. Chmn., organizer Gordon Research Conf. on Basic Chemistry Aging, Tilton, N.H., 1962. Recipient Merit award, Chgo. Tech. Socs. Council, 1964, CIBS award, 1965. Mem. Am. Inst. Chemists (pres. 1962-63, recipient Honor Scroll, 1959), Am. Chem. Soc., N.Y. Acad. Scis., A.A.S., Am. Geriatrics Soc., Gerontological Soc., Finnish Chem. Soc., Societas pro Fauna et Flora Fennica, Geriatrics Soc., Soc. Plastics Industry, Soc. Paint Tech., Am. Soc. Materials, Soc. Plastics Engrs., Sigma Xi, Gamma Alpha, Alpha Chi Sigma (hon. mem.). Author: (with Henry Torey, Betty Harker, James Henning) Polyester and the Applications, 1956. Research, publs., patents on polymers and plastic, composite structures, foaming of metals; originated theory that aging of humans, like plastics, is due mainly to crosslinkage of large molecules, 1941-42. Home: Route 2, Fish Hatchery Rd., Madison 53711. Office: P.O. Box 265, Madison, Wis. 53701.*

BLACET, Francis Edward, American chemist; b. Greenville, Ill., Feb. 25, 1899; s. Stephen and Rachel (Blacet) B.; A.B., Pomona Coll., 1922, A.M., 1924, D.Sc., 1965; Ph.D., Stanford, 1931; m. Kate Merrell, Aug. 7, 1924; children—Ann (Mrs. Don P. Mullally), Philip Merrell. Faculty, Pomona (Cal.) Coll., 1922-27, Stanford, 1931-32; faculty U. Cal., Los Angeles, 1932—, prof. chemistry, 1945—, chmn. dept., 1948-56, dean, div. phys. scis., 1957—. Ofcl. investigator NDRC, Northwestern U., 1942-45; cons. Los Angeles Co. Air Pollution Control Dist., 1948—. Recipient Presdl. Certificate of merit, 1948. Mem. Am. Chem. Soc., A.A.A.S., Phi Beta Kappa, Sigma Xi, Phi Lambda Upsilon, Alpha Chi Sigma. Contbns. in microgas analysis, photochemistry and mechanisms of gaseous reactions; chem. protection from toxic gases during World War II. Home: 849 Harvard St., Santa Monica, Cal. 90403. Office: 405 Hildegard Av., Los Angeles 90024.*

BLACHSTEIN, Artur Georg Israel, German physician; b. Dresden, Germany, Mar. 2, 1863; s. Viktor Blachstein; B.A., Cornell U.; studied medicine in Leipzig, Munich, Berlin and Göttingen (all Germany); state exam and doctorate from Leipzig. Worked first in Leipzig, later with L. Pasteur in Paris, France; physician for Red Cross in Bulgarian-Serbian campaign; physician at inst. for exptl. medicine, St. Petersburg, Russia; dir. Russian cholera expdn. to Baku, 1891; studied law and langs. in Leipzig; returned to medicine during World War I; took over newly created lectorate for stenography at U. Leipzig, 1908-12; writer articles, and monographs; successfully treated seemingly hopeless gastrointestinal catarrh in children; in stenography compared Gabelsberger and Pitman applying Gabelsberger to Slavic langs. and brought about its recognition as a discipline of gen. sci. of writing. Died Leipzig, Mar. 10, 1940.

BLACK, Alvin P(ercy), Am. chemist; b. Blossom, Tex., Aug. 30, 1895; s. Alexander Pere and Rena (Eades) B.; student Harvard, 1923-24; Ph.D., U. Ia., 1933; m. Lillian Virginia Russell, Sept. 6, 1917; children—Charles A., Virginia. Faculty, U. Fla., Gainesville, 1919-66, prof., 1923-57, research prof., 1957-66, research prof. emeritus, 1966—; cons. chem. engr. municipal and indsl., Gainesville, 1942—; pres. Black Crow and Eidsness, Cons. Engrs., Gainesville, 1947—. Mem. Coll. Electors Hall of Fame of Gt. Ams., N.Y. U., 1955—; mem. nat. adv. dental research council, USPHS, 1951-55. Recipient Fuller award Am. Water Works Assn., 1939, Goodell prize 1949, Diven medal, 1954, Water Purification div. award, 1959. Ralph Lloyd Jones award Nat. Assn. Soft Water Service Operators, 1956. Mem. Nat. Soc. Profl. Engrs., Cons. Engrs. Council, Am. Chem. Soc., Am. Water Works Assn., (past nat. pres.), So. Assn. Sci. and Industry (past pres.). Contbr. numerous papers to tech. lit. Introduced microelectrophoresis as an analytical tool in water research; pioneered recalcination of lime-soda softening sludges; new methods for determination of free chlorine and iodine residuals in water; research on iodine for disinfection of pub. water supplies and phys. effects on human population. Home: 544 N.E. 10th Av. Office: 700 S.E. 3d St., Gainesville, Fla. 32601.*

BLACK, Benjamin Marden, Am. surgeon; b. Salt Lake City, July 4, 1910; s. Benjamin W. and Jean (Blackburn) B.; A.B., Stanford, 1931, M.A. in Anatomy, 1933, M.D., 1936; M.S in Surgery, U. Minn., 1941; m. Eleanor Weber, Nov. 12, 1942. Practice surgery. Head sect. surgery Mayo Clinic, Rochester, Minn., 1942—; faculty Mayo Grad. Sch. Medicine, U. Minn., Rochester, 1942—, prof. surgery, 1957—. Mem. Am., Western, Central surg. assns., Internat. Soc. Surgery, A.C.S. Author: Hyperparathyroidism, 1953. Research and publs. on surg. aspects diseases thyroid, hyperparathyroidism, colon and rectum. Home: 432 10th Av. S.W., Rochester, Minn. 55901.*

BLACK, Davidson, Canadian anatomist, anthropologist; b. Toronto, Ont., Can., July 25, 1884; s. Davidson and Margaret (Delamere) B.; M.D., U. Toronto, 1906, D.Sc. (hon.); m. Adena Nevitt, 1913; 1 son, 1 dau. Instr. anatomy Western Reserve Coll. Medicine, 1909-11, asst. prof., 1913-16; prof. anatomy Peking (China) Union Med. Coll., 1918-21, prof. neurology, 1921-34; hon. dir. cenozoic research lab. Nat. Geol. Survey China. Recipient Grabau Gold medal Geol. Soc. China, 1929, King Gold medal Peking Soc. Natural History, 1932. Fellow Royal Soc., 1932, Nat. Inst. History and Philology (hon.), China; hon. corr. mem. geol. div. Academia Sinica, Galton Soc. N.Y., Field Mus. Natural History. Contbr. papers on neuroanatomy, phys. anthropology, human paleontology to sci. jours. Made anthropological studies of protohuman remains in Pleistocene cave deposits at Choukou, China. Discoverer Pithecanthropus pekinensis (Peking Man), 1929. Died Peking, China, Mar. 15, 1934.

BLACK, Edgar Clark, biologist; b. Davidson, Sask., Can., Aug. 14, 1908; s. John Marr and Annie (McDonald) B.; B.A., McMaster U., Brandon, Man., 1931; M.A., U. B.C., Vancouver, Can., 1935; Ph.D., U. Pa., 1940; m. Virginia Safford, June 22, 1940; 1 dau., Reta Katharine. Research asst. Banting Inst. U. Toronto, Onto., Can., 1940-44; asst. prof. physiology Dalhousie U., Halifax, N.S., Can., 1944-47; faculty U. B.C., 1947—, later became prof. physiology. Decorated Order Brit. Empire. Fellow Royal Soc. Can.; Mem. Am. Physiol. Soc. Research and numerous publs. on description of function of blood in freshwater fishes, aviation medicine, description of relation of severe muscular fatigue in fishes to carbohydrate metabolism; interpretation of result of fatigue to problems in fisheries. Died Mar. 11, 1967.

BLACK, Gordon, English physicist; b. Whitehaven, Eng., July 30, 1923; s. Martin and Gladys (Lee) B.; B.Sc., Hatfield Coll., Durham, 1945; Ph.D., D.I.C., Imperial Coll., London, Eng., 1953; m. Brenda Janette Balsom, Oct. 17, 1953; children—Janette Claire, Susan Catherine, Jonathan David Gordon, Roger Duncan Martin. Research physicist Brit. Sci. Instrument Research Assn., 1946-56; prin. sci. officer U.K. Atomic Energy Authority, 1956-58, sr. prin. sci. officer, 1958-60, dep. chief sci. officer, 1960-64; professor of computation Univ. of Manchester (Eng.) Inst. Sci. and Tech., 1964—. Dir., Nat. Computing Centre of U.K., 1965—. Fellow Inst. Physics; mem. U.K. Computer Bd. for Univs. Research and publs. on use of electronic computers in automatic numerical optimization, applied methods to optical systems and nuclear power stations, use of computers in bus. and mgmt. Home: High Lawn, Alan Dr., Hale, Cheshire, Eng. Office: Nat. Computing Centre of U.K., Quay House, Quay St., Manchester, Eng.*

BLACK, Greene Vardiman, Am. dentist; b. Scott County, Ill., Aug. 3, 1836; D.D.S., Mo. Dental Coll., St. Louis, 1877; M.D., Chicago Medical Coll., 1884; Sc.D., Ill. Coll., 1892; LL.D., Northwestern, 1898; m. Jane L. Coughennower, 1860 (dec. 1863); m. 2d, Elizabeth Akers Davenport, 1865. Began practice, Winchester, Ill., 1857, Jacksonville, Ill., 1864; lectr. pathology, histology, operative dentistry Mo. Dental Coll., 1870-80; prof. dental pathology Chgo. Coll. Dental Surgery, 1883-89; prof. dental pathology and bacteriology U. Ia., 1890-99; prof. operative dentistry, pathology and bacteriology Northwestern U. Dental Sch., 1891-1915, dean, 1897-1915. Recipient Internat. Miller prize, 1910. Mem. Nat. Dental Assn. (pres. 1901). Author: Formation of Poisons by Micro-Organisms, 1884; Periosteum and Peridental Membrane, 1887; Anatomy of the Human Teeth, 1891; Operative Dentistry, 2 vols., 1908; Diseases and Treatment of the Investing Tissues of the Teeth and the Dental Pulp, 1915; also many soc. papers and journal articles. Described path. histology of mottled enamel caused by fluorosis; formulated doctrine of extension of cavity for prevention of further decay; invented dental instruments. Died Chgo., Ill., Aug. 31, 1915.

BLACK, Harold Stephen, Am. research engr.; b. Leominster, Mass., Apr. 14, 1898; s. Stephen A. and Julia S. (Bushnell) B.; B.S., Worcester (Mass.) Poly. Inst., 1921, D.Eng., 1955; m. Meta C. Spreen, July 1, 1934. With engrng. dept. Western Electric Co., 1921-25; mem. tech. staff Bell Telephone Labs., Inc., Murray Hill, N.J., 1925-63; cons. prin. research scientist aerospace group Gen. Precision, Inc., Little Falls, N.J., 1963-66, communications cons., 1966—. Recipient Nat. Best Paper prize Am. Inst. E.E., 1934, Lamme gold medal, 1957; Modern Pioneer award N.A.M., 1940; John Price Wetherill medal Franklin Inst., 1941; certificate of appreciation U. S. War Dept., 1946; Research Corp. sci. award, 1952, John H. Potts Meml. Award, 1959. Fellow I.E.E.E., A.A.-A.S.; mem. Audio Engrng. Soc., N.Y. Acad. Scis., N.J. Acad. Sci., Franklin Inst., Telephone Pioneers Am., Sigma Xi, Tau Beta Pi. Author: Modulation Theory, 1953; also tech. articles. Research on amplifiers, multichannel carrier telephone and telegraph systems, pulse microwave radio relay systems, laminated conductors, advanced communication and guidance feedback techniques in aerospace field; inventor negative feedback amplifier (made possible the devel. of broadband transcontinental and transoceanic communications systems, also radar systems, servomechanisms theory, industrial control mechanisms). Home: 120 Winchip Rd., Summit, N.J. 07901. Office: General Precision, Inc., Aerospace Group, Little Falls, N.J.*

BLACK, James, Scottish biologist; b. Scotland, circa 1788; M.D., Glasgow, Scotland, 1820. Became asst. surgeon Royal Navy, 1809; practiced medicine, Bolton, Eng., until 1839, also 1848-56, Manchester, Eng., 1839-48, Edinburgh, Scotland, 1856-67. Licentiate Royal Coll. Surgeons, Eng., Royal Edinburgh Coll. Surgeons. Fellow Royal Coll. Physicians; pres. Brit. Med. Assn., 1842, Provincial Med. Assn., 1853, Manchester Geol. Soc. Author: An Inquiry into the Capillary Circulation of the Blood, 8 vols., 1825; A Comparative View of the More Intimate Nature of Fever, 1826; A Manual of the Bowels, 1840; Retrospective Address in Medicine, 1842; Observations and Instructions on Cold and Warm Bathing, 8 vols., 1846. Died Edinburgh, Apr. 30, 1867.

BLACK, James Francis, Am. chemist; b. Butte, Mont., Jan. 15, 1919; s. William Joseph and Agnes (Tillman) B.; B.S., U. Cal. at Berkeley, 1940; M.A., Princeton, 1943, Ph.D., 1943; m. Edna Frances Zekevitch, Oct. 15, 1955; children—Claudia Edna, Gregory James. With Esso Research & Engring. Co., Linden, N.J., 1943-—, group head, 1955-57, research asso., 1957-60, sr. research asso., 1960-67, asso. scientific adviser, 1967-—. Mem. adv. com. on isotopes and radiation devel. U. S. AEC, 1958-65, vice chmn., 1960-64; vice chmn. N.J. State Commn. on Radiation Protection, 1958-62. Mem. Am. Chem. Soc., Am. Nuclear Soc., Am. Meteorol. Soc., Am. Inst. Chem. Engrs. Research and publs. on measurement equilibrium distbn. of H3 between water and hydrogen; discovered 1st synthesis of molecular sieves, process of making biodegradable sodium alkane sulfonate detergents using gamma radiation; devel. new concept of weather modification utilizing large asphalt coatings to promote convective rainfall. Home: Canfield Rd., Convent, N.J. 07961. Office: P.O. Box 51, Linden, N.J. 07036.*

BLACK, James Harvey, Am. physician; b. Huntington, W.Va., Mar. 27, 1884; s. John Adam and Mary Nancy (Murphy) B.; student Southwestern U., 1900-02, Coll. of Physicians and Surgeons, Atlanta, 1903-05; M.D., So. Meth. U., Dallas, 1907; m. Alleen Patton, Sept. 4, 1913; children—Emily Anne (Mrs. Armand Garguilo), Lois (Mrs. Jack W. Crosland, Jr.). Intern St. Paul Hosp., Dallas, 1906-07; practiced medicine, Dallas, 1907-50, specializing in allergic diseases, from 1937; prof. bacteriology and pathology Soc. Meth. U. Med. Sch., 1907-15, dean, 1914-15; prof. bacteriology and preventive medicine Baylor U. Med. Sch., 1915-21, prof. preventive medicine, 1921-36, prof. clin. medicine, 1936-43; prof. clin. medicine Southwestern Med. Coll., Dallas, 1943. Diplomate Am. Bd. Internal Medicine. Fellow Am. Acad. Allergy, A.A.A.S., Am. Pub. Health Assn., A.C.P.; mem. Newcomen Soc., Tex. Philos. Soc., A.M.A., Am. Soc. Clin. Pathologists (pres. 1929-30), Assn. for Study of Allergy (pres. 1934-35), N.Y. Acad. Sci. Author: Practice of Allergy (with W. T. Vaughan), 1954. Editor sect. on allergy Biological Abstracts, from 1942; editorial bd. Jour. Allergy. Died Nov. 30, 1959; buried Dallas.

BLACK, John Angus, Brit. pediatrician; b. London, Jan. 11, 1918; s. Andrew Hogg and Mabel (Holder) B.; B.A., Caius Coll, Cambridge U., 1939; M.B., U. Coll. Hosp., U. London, 1942, M.D., 1951; m. Dorothy Florence Mary Burnett, Dec. 19, 1953; children—Rosemary Susan, Gabriel Ann, Patrick Jonathan, Emma Felicity. Registrar gen. medicine U. Coll. Hosp., 1947; staff Hosp. for Sick Children, 1948-54; cons. pediatrician, Glasgow, also Renfrewshire, Scotland, 1954-59; sr. lectr. child health Inst. Child Health, London, 1959-63; cons. pediatrician Sheffield (Eng.) Children's Hosp., 1963-—. Clin. tchr. child health U. Sheffield, 1963-—, examiner child health, 1965-—; examiner Gen. Nursing Council, 1962-—. Fellow Royal Coll. Physicians London; mem. Royal Soc. Medicine, Brit. Pediatric Assn., Renal Assn., Pediatric Research Soc., Neonatal Soc., Soc. for Study Inborn Errors of Metabolism. Contbg. author: Compendium of Emergencies, 1965; Recent Advances in Pediatrics, 1965. Research, publs. on kidney diseases in childhood, various metabolic (biochem. abnormalities) in children. Address: 118 Osborne Rd., Sheffield, Yorkshire, Eng.*

BLACK, Joseph, chemist; b. Bordeaux, France, April 16, 1728; the son of John Black and Robert Gordon's daughter; educated at Glasgow, also Edinburgh, Scotland; studied chemistry under William Cullen at University of Glasgow. Practiced medicine; prof. medicine, Glasgow, 1756-66; prof. medicine and chemistry, U. Edinburgh, 1766-97; 1st physician to George III in Scotland. Mem. Royal Soc., Edinburgh, Royal Coll. Physicians; fgn. asso. French Acad. Scis. Author: De humore acido acibis orto, et magnesia alba (doctoral thesis, important for pneumatic chemistry and quantitative analysis), 1754; An Analysis on the Waters of some Hot Springs in Iceland; Lectures on the Elements of Chemistry, delivered in the University of Edinburgh. First to show carbon dioxide (which he called fixed· air) could be formed by mineral decomposition as well as by combustion and fermentation; measured weight loss on heating calcium carbonate; helped lay foundations for modern quantitative analysis; invented ice calorimeter; originated concept of specific heat (capacities of bodies for heat), 1760; recognizing that heat quantity is different from heat intensity, he developed concept of latent heat (basis for modern thermal sci.), 1762, same theory may have led pupil James Watt to invention of steam engine. Died Edinburgh, Dec. 6, 1799.

BLACK, Paul Joseph, physicist; b. Cali, Colombia, Sept. 10, 1930; s. Walter and Susan (Burns) B.; B.Sc., U. Manchester (Eng.), 1950; Ph.D., U. Cambridge (Eng.), 1954; m. Mary Elaine Weston, Aug. 3, 1957; children—Simon Nicholas, John Benedict, Jeremy Paul, Michael Peter. Royal Soc. Jaffe Research student Cavendish Lab., Cambridge U., 1953-56; faculty physics dept. U. Birmingham (Eng.), 1956-—, reader crystal physics, 1966-67, joint organizer Nuffield A-level Physics Teaching Project, 1967-—. Hon. sec. edn. group Inst. Physics and Phys. Soc., 1965-—. Fellow Inst. Physics. Research, publs. on X-ray crystallography, structures of intermetallic compounds, structure of liquid metals and liquid alloys, Mössbauer effect, diffraction effects in scattering by nuclear resonance, use of electronic and nuclear resonances in crystallography. Home: 86 Bibury Rd., Birmingham 28, Eng.*

BLACK, Robert Foster, Am. geologist; b. Dayton, O., Feb. 1, 1918; s. Stanley C. and Margaret (Martin) B.; B.A., Coll. Wooster, 1940; M.A., Syracuse U., 1942; postgrad. Cal. Inst. Tech.; Ph.D., Johns Hopkins U., 1953; m. Hernelda R. Lone, Feb. 12, 1944; children—John R., Dean S. Geologist, Roosevelt Wildlife Conservation Dept., N.Y., 1942-43, U. S. Geol. Survey, Washington and Alaska, 1943-56; asso. prof. geology U. Wis., Madison, 1956-59, prof., 1959-—. Mem. A.A.A.S., Wis. Archeol. Soc., Wis. Acad. Sci., Arts and Letters, Brit. Glaciological Soc., Geol. Soc. Am., Soc. Econ. Geologists, Arctic Inst. N.Am., Am. Geophys. Union, Am. Inst. Profl. Geologists, Am. Soc. Photogrammetry. Research, publs. on permafrost, patterned ground, related cold climate geomorphology, Pleistocene geology, engring. geology. Home: 6112 Winnequah Rd., Madison, Wis. 53716.*

BLACK, Simon, Am. biochemist; b. Deerfield, Wis., Aug. 9, 1917; s. Bernard and Fanny (Holland) B.; B.S., U. Wis., 1938, M.S., 1941, Ph.D., 1942; m. Dorothy Gottlieb, Sept. 23, 1945; children—Bert, Roy Alvin, Franklin J. With U. Chgo., 1942-51, asst. prof. dept. medicine, 1949-51; fellow Soc. for Infantile Paralysis, Mass. Gen. Hosp., Boston, 1951-52; chemist NIH, Bethesda, Md., 1952-58, chief sect. biochemistry amino acids, 1958-—. Mem. Am. Chem. Soc., Am. Soc. for Biol. Chemists, Sigma Xi. Editorial bd. Jour. Biol. Chemistry. Research and publs. on discovery of enzymes involved in metabolic conversion and biosynthesis of several amino acids, intermediate substances and enzyme mechanisms. Home: 9209 Wadsworth Dr., Bethesda 20034. Office: Nat. Insts. Health, Bethesda, Md. 20014.*

BLACKALL, John, English physician; b. Exeter, Eng., Dec. 24, 1771; s. Theophilus and Elizabeth (Ley) B.; B.A., Balliol Coll., Oxford (Eng.) U., 1793; M.A., 1796; M.B., 1797; M.D., 1801; studied at St. Bartholomew's Hosp.; physician to Devon and Exeter Hosp., 1797, resigned, 1801, reapptd., 1807; became physician St. Thomas Lunatic Asylum, 1812; practiced medicine until circa 1851. Fellow Coll. Physicians. Author: Observations on the Nature and Cure of Dropsies, 1813; also observations on angina pectoris. Discovered dropsy is often asso. with albumin in urine and suspected this indicated diseased kidneys. Died Jan. 10, 1860.

BLACKBURN, Henry Webster, Jr., Am. physician; b. Miami, Fla., Mar. 22, 1925; s. Henry Webster and Francis (Smith) B.; B.S., U. Miami, Coral Gables, Fla., 1947; M.D., Tulane U., 1948; M.S., U. Minn., 1957; m. Nelly Paula Trocmé, Jan. 10, 1951; children—John Keith, Katherine Ann, Heidi Elizabeth. Staff, Methodist Clinics, Oriente Province, Cuba, 1949; med. officer in charge USPHS Fgn. Quarantine, Salzburg, Vienna, Austria, Munich, Germany, 1950-53; med. fellow U. Minn., 1953-56, faculty 1956-—, asso. prof. physiol. hygiene, 1961-—; med. dir. Mut. Service Ins. Cos., St. Paul, 1958-—. Chmn. tech. group on exercise electrocardiogram USPHS Heart Disease Control Program, 1965-—. Mem. Internat. Soc. Cardiology (chmn. tech. group 1963-—), American Epidemiological Society, American Heart Assn. Council on Epidemiology (vice chairman 1967-—), A.M.A., Am. Pub. Health Assn., A.A.-A.S. Author: (with Geoffrey Rose) Cardiovascular Survey Methods, 1966. Research and publs. on cardiovascular epidemiology, frequency and determinants of heart and blood vessel disease in different cultures and occupations, adaptation of clin. med. techniques to studies of diseases in large populations, electrocardiography. Home: 2108 Oliver Av., Mpls. 55405.*

BLACKENHORN, Max Ludwig Paul, German geologist; b. Siegen, Germany, Apr. 16, 1861; s. Carl and Sophie (Budach) B.; ed. in Göttingen, Bonn (both Germany); m. Maria Hermine Margarete Hattenbach, 1894; 2 sons, 2 daus. Became asst. to Oebbecke in Erlangen, 1888; paleontologist, geologist, Geol. Survey Egypt, 1897-99; mem. staff Prussian State Geol. Inst.; became prof., 1905. Recipient Turkish Liakat medal, 1908, Leibniz medal, 1923. Author 1st basic geology of Egypt. Laid founds. for modern stratigraphy of Syria and Palestine; charted numerous lower Hessian measurement tables. Died Marburg, Germany, Jan. 13, 1947.

BLACKETT, Patrick Maynard Stuart, English physicist; b. London, Eng., Nov. 18, 1897; s. Arthur Stuart and Caroline Francis (Maynard) B.; M.A., Magdalene Coll., Cambridge (Eng.) U.; m. Constanza Bayon, 1924; children—Giovanna, Nicholas. Prof. physics Birkbeck Coll., London, 1933-37, U. Manchester (Eng.), 1937-53; prof. physics Imperial Coll. Sci. and Tech., London, 1953-65, emeritus prof., sr. research fellow, 1965-—; Fellow Royal Soc., 1933 (pres. 1967). Fgn. Asso. Nat. Acad. Scis. Named Companion of Honour, 1965; recipient Am. Medal for merit, 1946, Nobel prize for physics, 1948. Author: Rayons Cosmique, 1934; Military and Political Consequences of Atomic Energy, 1948; Lectures on Rock Magnetism, 1956; Atomic Energy and East-West Relations, 1956; Studies of War, 1962. Research and publs. on nuclear and atomic physics with Rutherford, study cosmic rays by cloud chamber method, origin of magnetic fields of earth, sun, and stars, and magnetism of rocks. Home: 48 Paultons Sq., Chelsea, London S.W.3. Office: Royal Soc., 6 Carlton House Terrace, London S.W. 1, Eng.*

BLACKFORD, Staige D(avis), Am. physician; b. Alexandria, Va., Dec. 28, 1898; s. Launcelot Minor and Eliza Chew (Ambler) B.; B.S., U. Va., 1923, M.D., 1925; m. Lydia H. Fishburne, Aug. 20, 1927; children—Staige D., Linda H. Intern, Mass. Gen. Hosp., Boston, 1925-27, student physician and instr. in medicine, 1927-28; asst. prof. practice of medicine, U. Va., 1938-40, asso. prof., 1940-46, prof. from 1946; practiced medicine, specializing in internal medicine, Charlottesville, Va., from 1927; connected with U. Va. Hosp. Diplomate Am. Bd. Internal Medicine, Nat. Bd. Med. Examiners. Fellow A.C.P.; mem. Am. Clin. and Climatologic Assn., A.M.A., Sigma Xi. Died July 17, 1949.

BLACKMAN, Geoffrey Emett, English plant physiologist; b. London, Eng., Apr. 17, 1903; s. Vernon Herbert and Edith (Emett) B.; B.A., St. John's Coll., Cambridge (Eng.) U., 1926; m. Audrey Babette Seligman, Oct. 10, 1931. Head Botany sect. Jealott's Hill Agrl. Research Sta., Warfield, Berks., Eng., 1927-35; lectr. ecology Imperial Coll. Sci. and Tech., London, 1935-45; Sibthorpian prof. rural economy U. Oxford (Eng.) dept. agr., 1945-—, dir. Agrl. Research Council Unit Exptl. Agronomy since 1950-—. Chmn. chem. adv. com. Natural Rubber Industry, 1956; del. Clarendon Press, 1950-—. Fellow Royal Soc., 1959; mem. Inst. Biology (past pres.). Numerous publs. on agrl., ecol., physiol. and statis. investigations. Home: Woodcroft, Foxcombe Lane, Boarrs Hill, Oxford, Eng.*

BLACKMAN, Vernon Herbert, English plant physiologist; b. London, Eng., Jan. 8, 1872; s. Frederick and Katherine (Frost) B.; Sc.D., St. John's Coll., Cambridge (Eng.) U.; D.Sc. (hon.) U. Benares (India), U. Allahabad, m. Edith D. Emett, 1901 (deceased) children—Geoffrey Emett, Joan Delta; m. 2d, Thérèse E. Panisset, Dec. 29, 1941. Asst. botany dept. Brit. Mus., 1896-1906; lectr. botany Birkbeck, E. London colls., also lectr. vegetable cytology U. Coll. London, 1906-07; prof. botany Leeds U., 1907-11; prof. plant physiology Imperial Coll. Sci. and Tech., London, 1911-37, dir. research inst. plant physiology, 1913-43. Fellow, St. John's Coll., Cambridge, 1898-04. Fellow Royal Soc., 1913; mem. Royal Hort. Soc., Linnean Soc., Brit. Mycol. Soc. Editor, Annals Botany, 1922-47. Research and numerous publs. on plant cytology and plant physiology. Died Oct. 1, 1967.

BLACKWELDER, Richard E(liot), Am. zoologist; b. Madison, Wis., Jan. 29, 1909; s. Eliot and Jean Otis (Bowersock) B.; B.S., Stanford, 1931, Ph.D., 1934; m. Ruth Louise MacCoy, Jan. 3, 1935. Asso. curator U. S. Nat. Mus., Washington, 1941-54; asso. prof. St. John Fisher Coll., Rochester, N.Y., 1956-58; asso. prof. So. Ill. U., Carbondale, 1958-65, prof., 1965-—. Mem. Soc. Systematic Zoology (past pres.). Author: Directory of Zoological Taxonomists of the World, 1961; Classification of the Animal Kingdom, 1963; Taxonomy, 1967. Research and publs. in classification of animals, systematic zoology, particularly on insects. Address: Dept. Zoology, So. Ill. U., Carbondale.*

BLACKWELL, David, Am. mathematician, statistician; b. Centralia, Ill., Apr. 24, 1919; s. Grover and Mabel (Johnson) B.; A.B., U. Ill., 1938, A.M., 1939, Ph.D., 1941, D.Sc., 1966; m. Ann Madison, Dec. 27, 1944; children—Ann, Julia, David, Ruth, Grover, Vera, Hugo, Sara. Instr. math. Southern U., 1942-43, Clark Coll., 1943-44; asst. prof., then prof. math. Howard U., 1944-54; prof. statistics U. Cal. at Berkeley, 1954-—. Fellow Inst. Math. Statistics (pres. 1955); mem. Nat. Acad. Scis., Am. Math. Soc., Am. Statis. Assn. Co-author: Theory of Games and Statistical Decisions, 1954. Study of sequential analysis; Markoff chains. Home: 3021 Wheeler St., Berkeley 5, Cal.

BLACKWELL, Elizabeth, physician; b. Bristol, Eng., Feb. 3, 1821; came to U. S., 1832; ed. pvt. schs., Bristol and N.Y.; taught sch. in Ky., and the Carolinas; sought admission to several med. colls., but was refused until she entered the med. school at Geneva, N.Y., 1847, M.D. (1st med. degree awarded to woman in Am.), 1849; later studied in La Maternité and Hôtel Dieu, Paris; and St. Bartholomew's, London. Established practice in N.Y., 1851; founded a hosp.; founded (with sister, Dr. Emily Blackwell and Marie Zackrzewska) Woman's Med. Coll. of N.Y. Infirmary, 1857; lectured in Eng., 1858-59; registered as physician in England, 1859, practiced in London and Hastings, from 1869; prof. of gynecology London School Medicine for Women, from 1875; founded Nat. Health Soc. of London; aided in founding London Sch. Medicine for Women. Author: Physical Education of Girls, 1852; Religion of Health; Counsel to Parents on Moral Education; Pioneer Work in Opening the Medical Profession to Women, 1895; The Human Element in Sex, 1884; Decay of Municipal Representative Institutions. Died Hastings, Eng., May 31, 1910.

BLACKWOOD, Allister Clark, Canadian bacteriologist; b. Calgary, Alta, Can., Nov. 22, 1915; s. Allister Chester and Bessie (Saunders) B.; B.Sc., U. Alta., 1942, M.Sc., 1944; Ph.D., U. Wis., 1949; m. Mildred Marsh, May 1, 1943; children—Alan, Marsha, Susan. With NRC Can., 1944-46, 48-57, sr. research officer, Saskatoon, Saskatchewan, 1956-57; prof. chmn. dept. microbiology Macdonald Coll., McGill U., Montreal, Que., 1957—. Mem. Canadian Soc. Microbiologists (pres. 1964-65), Que. Société de Microbiologie, Am. Soc. Microbiology, Soc. for Gen. Microbiology, Soc. for Applied Bacteriology, Sigma Psi, Kappa Sigma. Research on microbial physiology and fermentations. Home: Windcrest Rd., Hudson, Que., Can.

BLAES, Gérard, see Blasius, Gerhard.

BLAEU, Jan, Dutch cartographer; b. 1596; s. Willem Janszoon Blaeu; Author: Novum et magnum theatrum urbium Belgicae regiae, 2 vols., 1650; Theatrum civitatum Italiae, 2 vols., 1663; Theatrum statuum Sabaudiae ducis, 2 vols., 1682; pub. several edits. of classics; (with his father) Atlas Magnus, 11 vols., 1650-62. Worked on map making. Died Amsterdam, Netherlands, Dec. 28, 1679.

BLAEU, Willem Janszoon, Dutch mathematician, cartographer, astronomer; b. Alkmaar, Holland, 1571; studied under Tycho Brahe; children—Jan, Cornelius. Founder firm known for globes, math. instruments, charts, Amsterdam, Netherlands, 1596. Author: Theatrum urbium et monunumentorum, 1619; Novus atlas, 1634-62; Institutio astronomica (sci. summary of Copernican doctrine), 1688. Began series of atlases in 1631 (finished by his sons). Died Amsterdam, 1638.

BLAGDEN, Charles, English physician; b. London, England, April 17, 1748; M.D., Univ. Edinburgh (Scotland), 1768. Medical officer British Army, 1768-1814. Fellow Royal Soc., 1772 (became secretary, 1784), corr. mem. French Academy Sciences, 1804. Author: Experiments and Observations in a Heated Room, 1775; On the Heat of Water in the Gulf Stream, 1781; History of the Congelation of Quicksilver, 1783; An Account of Some Late Fiery Meteors, 1784; On the Effect of Various Substances in Lowering the Point of Congelation of Water, 1788; Report on the Best Method of Proportioning the Excise on Spiritous Liquors, 1790; On the Tides of Naples, 1793; On Vision, 1813. First to demonstrate role of perspiration in maintaining constant body temperature, 1775. Died Mar. 26, 1820.

BLAGONRAROV, A(natolii) A(rkadyevich), Russian mech. engr.; b. Ankovo, Russia, June 1, 1894; student Mikhailovskaya Sch. Arty., then Higher Sch. Arty.; grad. Mil.-Tech. Acad., 1929; Dr. Tech. Scis., 1938. Charge dept. small arms Acad. Arty., later founder and dir. Inf. Weapons Research Center, also developed ednl. program for ordnance engrs.; charge Dzerzhinski Mil. Acad., World War II, specializing improvement arty. weapons; head Soviet Acad. Arty. Sci., 1946—. Head Soviet delegation Internat. Congress on Missiles, Paris, 1956; chmn. Soviet delegation Internat. Rocket and Satellite Conf., U. S., 1957; dir. Inst. Study of Machines, 1959. Served as comdr. units Red Army, Bolshevik Revolution; lt. gen. arty., World War II. Decorated Order of Lenin, Order of Red Banner; named meritorious sci. and tech. worker Soviet Union, 1940; recipient Stalin prize for contbn. to arty., 1941. Mem. USSR Acad. Scis. (presidum, dir. machine sci. sect. 1953, academic sec. tech. sci. dept. 1957—), Soviet Acad. Arty. Sci. (pres. 1946-50). Author: Basic Principles of Automatic Weapons, 1931; Material Part of a Shooting Weapon, I and II, 1945-46; also numerous sci. monographs on inf., aviation armament, ballistics, kinematics, rockets and automatic weapons. Editor-in-chief: Izvestiya Akad. Nauk S.S.S.R., Otdel. Tekh. Nauk. 1962—. Developed ground and aerial weapons for Soviet military forces; largely responsible for Soviet research program in astronautics. Address: Academic Secretary, Dept. Technical Sciences, Malyy Khariton'yevskii pereulok 4, Moscow, Russia.

BLAGRAVE, John, English mathematician; b. circa 1558; s. John and Anne (Hungerford) B.; self-educated in math.; m. a widow. Designed astron. and navigational instruments, designed and constructed sundials; lived in Reading, Eng. Author: The Art of Dialling; The Mathematical Jewel; Baculum familliare catholicon sive generale; Astrolabium uranicum generale, 1596. Published an astron. almanac. Died 1611.

BLAHA, Karel, Czech. chemist; b. Vejvanov, Czechoslovakia, July 29, 1926; s. Karel and Ruzena (Opatrnà) B.; Ing.chem., Inst. Chem. Tech., 1949; Dr. techn., Czech Tech. U., 1950; Ph.D., Czechoslovakian Acad. Sci., 1963; m. Vilma Ernstovà Sept. 13, 1952; children—Karel, Markéta. Predoctoral fellow dept. organic chemistry Inst. Chem. Tech., 1949-50; Czechoslovakian Acad. Sci. research fellow lab. Heterocyclic Compounds, 1960; leading research officer Inst. Organic Chemistry, Czechoslovakian Acad. Sci., 1960— (all Prague); reader organic chemistry Med. Faculty, U. Olomouc, 1967—. Mem. Czechoslovak Chem. Soc. (council), Czechoslovak Soc. for History Sci. Author: Reactions of Organometallic Compounds, 1960; Principles of Stereochemistry and Conformational Analysis, 1966; also articles. Editor-in-chief Collection Czechoslovak Chem. Communications, 1963—. Research on stereochem. properties of plant alkaloids, arrangement of polypeptide chains and properties of peptide bond. Home: 1 Pospisilova, Prague 3, Czechoslovakia. Office: 2 Flemingovo, Prague 6, Czechoslovakia.*

BLAHD, William Henry, Am. internist; b. Cleve., May 11, 1921; s. Mose E. and Rae (Lichtenstader) B.; student Western Res. U., 1939-40, U. Ariz., 1940-42; M.D., Tulane U., 1945; m. Ruth Netzorg, Feb. 14, 1945; children—Andrea, William Henry, Karen. With U. S. VA, 1948—, chief radioisotope service VA Center, Los Angeles, 1956—; prof. medicine U. Cal. Med. Sch., Los Angeles. Diplomate Am. Bd. Internal Medicine. Fellow A.C.P.; mem. A.M.A., Am. Fedn. Clin. Research, Western Soc. Clin. Research, Soc. Nuclear Medicine (v.p. 1965-66), Health Physics Soc., Alpha Omega Alpha. Author: (with F. K. Bauer, B. Cassen, C. C. Thomas) The Practice of Nuclear Medicine, 1958; also numerous articles. Editor: Nuclear Medicine, 1965. Developed techniques for application radioisotope traces to clin. medicine; research on cause primary muscle disease. Office: VA Center, Los Angeles 90073.*

BLAINVILLE, Henri Marie Ducrotay de, see de Blainville.

BLAIR, Edgar A., Am. physiologist; b. Nashville, Sept. 17, 1902; s. Edgar P. and Sarah A. (Williamson) B.; student U. Tenn., 1921-22; B.A., Vanderbilt U., 1925, M.S., 1927; Ph.D., Washington U., St. Louis, 1938; m. Virginia Helen Smith, Sept. 17, 1930; children—Margaret Malone (Mrs. Frank Dupras), Edgar Allan. Faculty, Washington U., St. Louis, 1930-46; commd. capt. U. S. Army, 1940, advanced through grades to Col. 1958; chief physiologist med. dept. Research Lab., Ft. Knox, Ky., 1946-48; physiologist U. S. Army chief basic sci. sect. Army Med. Service Sch., Ft. Sam Houston, Tex., 1949-58; ret., 1958; faculty U. Tex. Med. Br., Galveston, 1958—, prof., 1963—; vis. prof. Postgrad. Med. Center, Karachi, Pakistan, 1961-62. Mem. A.A.A.S., Am. Physiol. Soc., Royal Soc. Tropical Medicine and Hygiene, Sigma Xi. Research and publs. in electrophysiology of peripheral nerve. Home: 3503 Av. P, Galveston, Tex. 55770.*

BLAIR, Henry Alexander, physiologist, b. Winnipeg, Man., Can., Jan. 6, 1900; s. Edward and Isabella (McFarlane) B.; B.S., U. Man., 1925, M.S., 1927; Ph.D., Princeton, 1930; m. Eva Andrews, Aug. 4, 1926; children—Shirley Isabelle (Mrs. William Dodenhoff), Barbara Elizabeth (Mrs. John Murphy), Henry Alexander. Instructor in biophysics Western Reserve Medical School, Cleveland, Ohio, 1930-32; faculty physiology U. Rochester (N.Y.) Med. Sch., 1932—, prof., 1948—; dir. dept. radiation biology and atomic energy project, 1948—. Cons. biol. effects radiation Nat. Acad. Scis., Washington, 1955—, NRC, Washington, 1955—, USPHS, Cin., 1960—, Nat. Com. on Radiation Protection, Washington, 1955—, Armed Forces Radiobiology Research Lab., Bethesda, Md., 1964—, AEC, Washington, 1950—, sec. com. on indsl. med. fellowships. Mem. Am. Phys. Soc., Am. Physiol. Soc., Soc. Exptl. Biology and Medicine, Radiation Research Soc., Health Physics Soc., A.A.A.S., Sigma Xi. Editor: Biological Effects of External Radiation, 1954; editorial bd. Jour. Neurophysiology, 1945-55. Extended analysis of line spectra of silver and palladium; formulated kinetics of electrical excitation of nerve and muscle, dose-survival relations for mammals and insects exposed to ionizing radiation; developed method for measurement of gas in digestive tract; determined effects of common cardiac drugs on mechanism of cold blooded heart. Home: 1392 Clover St. Rochester 14610. Office: 260 Crittenden Blvd., Rochester, N.Y. 14620.*

BLAIR, John Edward, Am. microbiologist; b. Monroe, Me., May 30, 1899; s. John Stott and Alice E. (Staples) B.; A.B., Clark U., 1920; Sc.M., Brown U., 1921, Ph.D., 1923; m. Lorraine Hunter Ferguson, Sept. 6, 1923; children—Donald Ferguson, Malcolm John. Instr., bacteriology Stanford, 1923-26; head dept. bacteriology Hosp. for Joint Diseases, N.Y.C., 1927-64; head dept. microbiology Roosevelt Hosp., N.Y.C., 1964-68. Chmn., Internat. Subcom. on Phage-Typing of Staphylococci, 1958-66. Recipient Kimble Methodology Research award conf. state and Provincial Pub. Health Dirs. 1957. Mem. Am. Soc. for Microbiology (past pres.), Am. Acad. Microbiology, Am. Pub. Health Assn., N.Y. Acad. Medicine, A.A.A.S. Research and numerous publs. on staphylococci and staphylococcal infections especially indentification of staphylococci by phage typing. Home: 26 Paulin Blvd., Leonia, N.J. 07605. Office: 428 W. 59th St., N.Y.C. 10019.*

BLAIR, Patrick, surgeon, botanist; b. Dundee, Scotland, 1665; practiced medicine in Dundee, later in London, Eng., and finally in Boston, Eng. Fellow Royal Soc., 1712. Author: Miscellaneous Observations on the Practice of Physick, Anatomy, and Surgery, 1718; Botanick Essays, 8 vols., 1720; Osteographia elephantina, 4 vols., 1718; Pharmacobotanologia, 4 vols., 1723-28. Gave 1st description of congenital hypertrophic pyloric stenosis, 1717; confirmed new views on sexual character of flowering plants. Died Boston, 1728.

BLAIR, Robert, Scotish inventor, astronomer; b. Murchiston, near Edinburgh, Scotland; a son, Archibald Blair. Professor practical astronomy at U. Edinburgh, 1785-1828. Fellow Royal Soc. Edinburgh. Author: A Description of an Accurate and Simple Method of Adjusting Hadley's Quadrant for the Back Observation, 1783; Experiments and Observations on the Unequal Refrangibility of Light, 1794; Scientific Aphorisms, 1827. Invented aplanatic telescope by using fluids to produce colorless refraction; introduced lime juice as a scurvy preventative in navy. Died Dec. 22, 1828.

BLAIR, Vilray Papin, Am. surgeon; b. St. Louis, June 15, 1871; s. Edmund Harrison and Minnie (Papin) B.; A.B., Christian Brothers Coll., 1890, A.M., 1894; M.D., Washington U., 1803; m. Kathryn Lyman Johnson, 1907; children—Kathryn Lyman, Nancy Lucas, Mary Papin (dec.), Vilray Papin, John Bates Johnson. Practiced in St. Louis, 1893-48; prof. emeritus clin. surgery Washington U. Sch. Medicine, oral surgery, School Dentistry; vis. surgeon, Maternity and De Paul hosps.; asso. surgeon Barnes and St. Louis Children's hosps.; in charge plastic and oral surgery, section of the head, U. S. Army, 1917-18, chief cons. maxillo-facial surgery, AEF, 1918-19. Fellow A.C.S., Am. Laryngol. Assn., Am. Surg. Assn.; mem. Internat. Surg. Soc., Nat. Inst. Social Sciences, A.M.A., Assn. Am. Anatomists, So., Western surg. assns. Died Nov. 24, 1955.

BLAIR, W(illiam) Reid, Am. zoölogist; b. Phila., Jan. 27, 1875; s. William Reid and Jeannette (Houston) B.; D.V.S., McGill U., 1902, LL.D., 1928; m. Mildred Myrtle Kelly, Oct. 29, 1896. Veterinarian and pathologist N.Y. Zoöl. Park, 1902-22, asst. dir., 1922-26, dir. 1926-40; prof. comparative pathology, vet. dept. N.Y. U., 1905-17; cons. veterinarian, N.Y. State Dept. Agr., exec. sec. Am. Com. Internat. Wild Life Protection from 1938. Pres. Vet. Med. Soc., 1922-23; v.p. and trustee Bronx Soc. of Arts and Scis.; life mem. fellow N.Y. Zoöl. Soc.; fellow A.A.-A.S., Am. Geog. Soc.; mem. Quebec Zoöl. Soc. (Life), N.Y. Acad. Scis. (council), (corr.), Royal Zoöl. Soc. Ireland (corr.), Zoöl. Soc. of London (corr.), Internat. Soc. for Preservation of European Bison (corr.), Soc. for Preservation of Fauna of Empire (life), Am. Soc. of Mammalogists (trustee), Nat. Inst. of Social Sciences, Am. Vet. Med. Assn., Author: Diseases of Wild Animals in Confinement (pub. N.Y. Zoöl. Soc.). 1911; In the Zoo, 1929. Contbr. to sci. publs. on comparative medicine. Died Mar. 1, 1949.

BLAIR-BELL, William, Brit. physician; b. 1871; s. William and Helen Barbara (Butcher) B.; ed. King's Coll. Hosp., London; B.S., M.D.; LL.D., Glasgow, Scotland, also Liverpool, Eng.; m. Florence Bell. With Royal Infirmary, Liverpool, from 1905, pres., 1935, also cons. gynecol. and obstet. surgeon; named Arris and Gale lectr. Royal Coll. Surgeons, 1913, Hunterian prof., 1916; prof. obstetrics and gynecology U. Liverpool, 1921-31, also chmn. med. faculty; Ingleby lectr. Birmingham (Eng.) U., also Lloyd Roberts lectr. Manchester U., 1931. Fellow King's Coll., London; examiner in gynecology and obstetrics Manchester U., Royal Colls. Surgeons and Physicians, London, Durham U., U. Wales, Belfast U.; hon. cons. dir. Liverpool Med. (cancer) Research Orgn.; founder Gynecol. Vis. Soc. Great Britain, 1911. Chmn. exec. com. Brit. Congresses Obstetrics and Gynaecology. Bd. dirs. Jour. Obstetrics and Gynaecology Brit. Empire. Recipient John Hunter medal, also triennial prize for original work on pituitary body, council Royal Coll. Surgeons Eng. Fellow Royal Coll. Surgeons Eng.; hon. fellow A.C.S., Edinburgh Obstet. Soc., Am. Gynecol. Soc., Am. Gynecol. Club, A.M.A.; chmn. found. com., 1st pres. Brit. Coll. Obstetricians and Gynaecologists; pres. N. of Eng. Obstet. and Gynecol. Soc., Liverpool Med. Instn., sect. obstetrics and gynaecology Royal Soc. Medicine; hon. fgn. mem. Belgium Soc. Gynecology and Obstetrics. Author: The Principles of Gynaecology, 1910; The Sex Complex, 1916; The Pituitary, 1919. Contbg. author: The System of Treatment; The Practitioner's Encyclopedia; Watson Cheyne's and Burghard's system of Operative Surgery; New System of Gynaecology; also sci. papers in med. jours. Editor: Some Aspects of the Cancer Problem, 1930. Died Jan. 25, 1936.

BLAISDELL, B(aalis) Edwin, Am. mathematician; b. Lynn, Mass., May 11, 1911; s. Baalis Bullard and Hazel (Phillips) B.; B.S., Mass. Inst. Tech., 1932, Ph.D., 1935; m. Alice Grace Sarnessian, Mar. 29, 1941; children—Lucy Hazel (Mrs. Martin Russ), Thomas Franklin. Prof. chemistry Juniata Coll., Huntingdon, Pa., 1954-60, chmn. math., 1960—. Mem. Math. Assn. Am., Am. Math. Soc., Soc. Indsl. and Applied Math., Assn. Computing Machinery. Research on thermodynamic temperature scale, photochemistry of dyes, statis. theory of round-off errors. Home: 1431 Moore St., Huntingdon, Pa. 16652.*

BLAISE, Jean Edmond, French physicist; b. Raon-l'Étape, France, June 16, 1921; s. Edmond-Lucien and Alice (Kuehn) B.; Licence-ès-Sciences, U. Nancy (France), 1943; Doctorate-ès-Sciences, U. Paris, 1957; m. Yvonne Condé, Sept. 29, 1948; children—Véronique, François-Xavier, Cyrille. Staff, Centre National de la Recherche Scientifique, 1946—, dir. research, 1964—; resident research asso. Argonne (Ill.) Nat. Lab., 1959-60, cons., 1964. Mem. French Physics Soc., Optical Soc. Am. Research and publs. on determination of spins and nuclear moments, isotope shift in atomic spectra, classification of atomic spectra (rare earths, actinides). Home: 13 rue Fliane, Meudon (Hauts de Seine), 92, France. Office: C.N.R.S. 1 Place A. Briand, 92-Bellevue, France.*

BLAKE, Archie, Am. mathematician; b. Washington, Nov. 24, 1906; s. John Charles and Jean (Archibald) B.; B.S., U. Chgo., 1929, M.S., 1931, Ph.D., 1937; m. Bernice M. Lawrence, Jan. 11, 1947; 1 dau., Beverly Jean. With U. S. Dept. Commerce, 1931-44, Army Dept., 1944-50; applied math. cons. Mech. Research Corp., 1950-51, also dir.; with Cornell Aero. Lab., 1952-53; adv. engr. Westinghouse Electric Corp., Balt., 1954-56; systems staff mathematician Bendix Co., Ann Arbor, Mich., 1956-60; mgr. analytical sect. Raytheon, Bedford, Mass., 1960-62; sr. research asso. Martin Co., Denver, 1962-65; sr. research asso. Battelle Meml. Inst. Columbus, O., 1965—. Fellow Royal Statis. Soc.; mem. Am. Math. Soc., Inst. Mgmt. Scis., Inst. Math. Statistics, Math Assn. Am., N.Y. Acad. Scis. Phi Beta Kappa. Author: Canonical Expressions in Boolean Algebra, 1937. Research on new methods in Boolean algebra and statistics. Home: 135 Wilson Dr., Worthington, O. 43085. Office: N.Am. Rockwell Co., Columbus, O.*

BLAKE, Eli Whitney, Am. inventor; born Westborough, Mass., Jan. 27, 1795; s. Elihu and Elizabeth (Whitney) B.; grad. Yale, 1816; m. Eliza O'Brien, July 8, 1822, 6 sons, 6 daus. Partner firearms business with uncle Eli Whitney, New Haven, 1817-25, continued bus. with brother after uncle's death, 1825-36; propr. hardware factory, Westville, Conn., 1836-71; received patents for door lock, bedstead castors, stone and ore-crusher (Blake crusher); a founder Conn. Acad. Scis.; contbr. to Am. Jour. Sci.; Author: Original Solution of Several Problems in Aerodynamics, 1882. Wrote many scientific papers. Died New Haven, Conn., Aug. 18, 1886.

BLAKE, Emmet Reid, Am. zoologist; b. Abbeville, S.C., Nov. 29, 1908; s. John Rennie and Blanche (Ammen) B.; A.B., Presbyn. Coll. S.C., 1928, D.Sc., 1966; M.S., U. Pitts., 1933; m. Margaret Newcomb Bird, Oct. 18, 1947; children—Margaret Newcomb, Elizabeth Wier. Staff, Field Mus. Natural History, Chgo., 1935—, curator of birds, 1955—; explorations and research in Mexico, Guatemala, Brit. Honduras, Colombia, Venezuela, Guyana, Surinam, Brazil, Peru, W.I., U.S. Fellow Am. Ornithologists Union; mem. Wilson Ornithol. Club, Cooper Ornithol. Soc., Ill. Audubon Soc., Asociacion Ornitologica del Plata (hon.). Author books including: Birds of Mexico, 1953; Birds of Volcán de Chiriquí, 1958; Basic Science Dictionary, 1961; also numerous articles. Research on tropical Am. (neotropical) birds, especially origins, evolution, classification, distbn. and ecology, descriptions of new forms. Home: 1139 Judson Av., Evanston, Ill. 60202. Office: Roosevelt Rd. and Lake Shore Dr., Chgo. 60605.

BLAKE, Francis, Am. inventor; b. Needham, Mass., Dec. 25, 1850; s. Francis and Caroline (Trumbull) B.; high sch. edn., Brookline, Mass.; hon. A.M., Harvard, 1902; m. Elizabeth L. Hubbard, June 24, 1873. Served on U. S. Coat Survey 13 yrs., resigned; during last 2 or 3 yrs. was engaged in field work and its reduction to determine differences of longitude between observatories at Greenwich, Paris, Cambridge and Washington; devoted leisure to experimental physics, and in 1878 invented "Blake_ Transmitter," which played important part in devel. of telephony; patented many other elec. devices. Fellow A.A.A.S., Am. Acad. Arts and Scis. Died Jan. 1913.

BLAKE, Francis Gilman, Am. physician; b. Mansfield Valley, Pa., Feb. 22, 1887; s. Francis Clark and Winifred Pamelia (Ballard) B.; A.B., Dartmouth, 1808, Sc.D., 1936; M.D., Harvard, 1913; M.A. (hon.), Yale, 1921; Sc.D., Dartmouth College, 1936; m. Dorothy Dewey, June 1, 1916; children—Francis Gilman, William Dewey, John Ballard. Successively med. interne, asst. resident physician, resident physician Peter Bent Brigham Hosp., Boston, 1913-16; asst. at Hosp. of Rockefeller Inst., N.Y., 1916-17; asst. prof. medicine U. Minn., 1917-19; asso. in medicine Rockefeller Hosp., N.Y., 1919-20; asso. mem. Rockefeller Inst., 1920-21; John Slade Ely prof. medicine, Yale Sch. of Medicine, 1921-27, Sterling prof. medicine since 1927, dean 1940-1947. Physician-in-chief, New Haven Hosp., from 1921; sci. dir., med. research and devel. bd., surg. gen's. office Dept. of Army, 1952; mem. bd. sci. dirs. Rockefeller Inst., 1924-35; mem. NRC, Div. Med. Scis., 1925-36, chmn. of Div., 1933-36; Mem. bd. sci. dirs. Yerkes Lab. for Primate Biology, 1940-52. Recipient Charles V. Chapin Meml. award, Providence, 1945, U. S. Typhus medal, 1945, Medal for Merit, 1945. Fellow A.C.P. (regent 1939-47, v.p., 1948), mem. Assn. Am. Physicians (pres., 1949), Nat. Acad. Sci., Am. Philos. Soc., Am. Acad. Arts and Scis., Am. Soc. for Clin. Investigation (pres. 1931), Am. Soc. for Exptl. Biology and Medicine, Am. Soc. for Exptl. Pathology, Soc. Am. Bacteriologists, Am. Assn. Immunologists (pres. 1935), Am. Assn. Pathologists and Bacteriologists, Harvey Soc., A.A.A.S. (v.p. 1946), A.M.A. (chmn. sect. on medicine, 1938). Sigma Xi. Author: (with others) Epidemic Respiratory Disease, 1921; Studies on Exptl. Pneumonia, 1920; Studies on Measles, 1921; Treatment of Scarlet Fever with Anti-Toxin, 1924; Artificial Pneumothorax in Lobar Pneumonia, 1935; Chemo Therapy of Pneumonia, 1939; Penicillin Therapy, 1943; Tsutsugamushi Disease in New Guinea, 1945. Died Feb. 1, 1952.

BLAKE, John Frederick, Brit. geologist; b. Stoke next Guildfore, Eng., Apr. 3, 1839; s. Robert P. Blake; ed. Christ's Hosp., Caius Coll., Cambridge; M.A.; m. Miss Haslewood, 1866; 3 sons, 1 dau. Curate, Lenton, Eng., St. Mary's, Bryanston Sq.; math. master St. Peter's Sch., York, Eng., 1865-74; lectr. comparative anatomy Charing Cross Hosp., 1876-80; prof. natural sci. Univ. Coll., Nottingham, Eng., 1880-88; arranged Baroda Mus., India, 1895. Recipient Lyell medal. Fellow Geol. Soc. (pres. 1891-92). Author: Catechism of Zoology; British Fossil Cephalopoda; Astronomical Myths; Yorkshire Lias; Annals of British Geology, 4 vols.; Geology of Nottinghamshire (Victoria History of Counties); also article on cuttle-fish in Ency. Brit., papers in jours. Died July 7, 1906.

BLAKE, William Phipps, Am. mineralogist, geologist; b. N.Y.C., June 1, 1826; s. Elihu and Adeline (Mix) B.; Ph.B., Yale, 1852; LL.D., U. Cal. at Berkeley, 1910; m. Charlotte Haven Lord Hayes, Dec. 25, 1855. Chemist N.J. Zinc Co., Balt., 1852-54; geologist Pacific R.R. Survey, 1854-56; investigator mineral resources N.C., 1856-59; editor, proprietor Mining Mag. from 1859; mining engr. for govt. Japan, 1861-63; organized 1st sch. sci. Japan; explorer Stickeen River region, Alaska, discovered Stickeen glacier, 1863; minerologist, state bd. agr. Cal., prof. mineralogy and geology Coll. Cal., from 1864; prof. geology and mining, dir. Sch. Mines, U. Ariz., 1895-1905; territorial mineralogist and geologist Ariz., 1898-1910. Commr. sci. exhibits various expns.; chief sci. corps. of U. S. to Santo Domingo, 1871. Editor various sci. reports for U. S. govt. Died May 22, 1910.

BLAKELEY, George Henry, Am. engr.; b. Livingston, N.J., Apr. 19, 1865; s. Joseph and Mary A. (Gibson) B.; B.S., Rutgers Coll., 1884, C.E., 1894, ScD., 1924; m. Grace Delia Bogart, Apr. 12, 1893; 1 son, George Bogart. Pvt. engring. practice, 1884-1888; bridge eng. Erie R.R., 1888-90; chief engr, Passaic Steel Co., 1890-1902, mgr. of sales, 1902-05; structural engr. Bethlehem Steel Co., 1906-08, mgr. structural steel dept., 1908-27, v.p. from 1927, also dir.; pres. Bethlehem Steel Bridge Corp., 1916-23; pres. McClintic-Marshall Corp., 1931-35; dir. Bethlehem Steel Corp., Bethlehem Shipbuilding Corp., Bethlehem Steel Export Corp. Actively identified with constrn. of many important bridges and bldgs., including Delaware River Bridge (Phila.), Peace Bridge, Niagara River (Buffalo), Golden Gate Bridge (San Francisco), Merchandise Mart (Chgo.), Chase Nat. Bldg. (N.Y.C.), Field Museum (Chgo.). Fellow Am. Geog. Soc.; mem. Am. Soc. C.E., Am. Soc. Mech. Engrs., Am. Iron and Steel Inst., Tau Beta Pi, Phi Beta Kappa. Devised and developed improvements in manufacture and uses of structural steel, including broad flange structural steel sections, introduced 1908. Died Dec. 25, 1942.

BLAKEMORE, William Stephen, Am. surgeon; b. Stockdale, Pa., June 22, 1920; s. Isaac Thompson and Mary Jane (Crockett) B.; B.S., Washington and Jefferson Coll., 1942; M.D., U. Pa., 1945; postgrad. pharmacology George Washington U., 1946-47; m. Elainee Claire Hooven, Apr. 2, 1949; children—William Stephen, Holly Hooven, Karin Jane, Stephenie Elaine, Mary Jane Crockett, Laurel Claire. Lectr. pharmacology U. S. Navy Med. Sch., Bethesda, Md., 1947; faculty U. Pa., Phila., 1948—, J. William White asso. prof., 1960—, prof. surgery, 1962—, prof., chmn. dept. surgery Grad. Sch. Medicine, 1962—, surgeon-in-chief Grad. Hosp., 1962—, asst. to dir. Harrison Dept. Surg. Research, 1956-60, asst. dir., 1956-61, asso. dir., 1961—. Mem. dean's com. VA Hosp., Wilmington, Del., 1962-66. Trustee Washington and Jefferson Coll.; bd. Govs. Middle East Inst., Washington; exec. bd. Care-Medico, Phila. Recipient Nat. Humanitarian award Order Ahepa, 1962. Am. Cancer Soc. scholar, 1952-57; I. S. Ravdin traveling fellow, 1953-54. Diplomate Am. Bd. Surgery, Am. Bd. Thoracic Surgery. Mem. A.A.A.S., Am. Assn. for Cancer Research, Am. Assn. for Thoracic Surgery, Am. Assn. U. Profs., A.C.S., A.M.A., Am., Phila. Physiol. Soc., Am. Surg. Assn., Soc. Med. Consultants to Armed Forces. Internat. Cardiovascular Soc., N.Y. Acad. Scis., Internat. Soc. Surgery, Soc. Nuclear Medicine, Soc. U. Surgeons, Soc. for Vascular Surgery, Sigma Xi, Alpha Kappa Kappa, others. Asso. editor Surgery, 1965—. Surg. research and publs. on the heart and lung hypertension, cancer, and the mgmt. of disease. Home: 563 Heath Rd., Merion Station, Pa. 19066. Office: Grad. Hosp., U. Pa., 19th and Lombard Sts., Phila. 19146.*

BLAKESLEE, Albert Francis, Am. botanist, geneticist; b. Geneseo, N.Y., Nov. 9, 1874; s. Francis Durbin and Augusta Mirenda (Hubbard) B.; A.B., Wesleyan U., Conn., 1896, D.Sc., 1931; A.M., Harvard, 1900, Ph.D., 1904 (Bowdoin medal, 1905); D.Sc., U. San Marcos (Peru), 1925, U. Delhi (India), 1947, Yale, 1947; Wesleyan U., 1931, Sorbonne, France, 1951, Smith Coll., 1952, LL.D., U. Ark., 1947; m. Margaret Dickson Bridges, June 26, 1919 (dec.). Asst. botany Harvard, 1899-1900; instr. botany Radcliffe Coll., 1900-02; teaching fellow, Harvard, 1901-03; asst. in botany Summer Sch. of Cold Spring Harbor, L.I., 1901-02; collector in Venezuela for the Cryptogamic Herbarium of Harvard, summer 1903; investigator in Europe for Carnegie Instn., 1904-06; instr. botany Harvard, 1906-07; dir. Summer Sch. and prof. botany Conn. Agrl. Coll., 1907-14, prof. botany and genetics, 1914-15; resident investigator Carnegie Sta. for Exptl. Evolution, Cold Spring Harbor, 1912-13, in plant genetics 1915-41, asst. dir. dept., 1923-34, acting dir., 1934-35, dir., 1936-41, asso. in genetics Columbia U., 1940-42; William Allan Neilson research prof. botany Smith Coll., 1942-43, vis. prof., dir. Smith Coll. Genetics Expt. Sta., from 1943; vis. lectr. Harvard, 1948-49. Del. Carnegie Instn. to 3d Pan-Am. Scientific Congress, Lima, 1924-25; del. A.A.A.S. to Indian Sci. Congress, 1946-48; mem. div. biology and agr. NRC, 1931-33; bd. mgrs. N.Y. Bot. Garden, 1933-34, dir. Nat. Sci. Fund; trustee Biol. Abstracts, 1931-46, pres., 1942-46. Recipient A. Cressy Morrison prize, N.Y. Acad. Scis., 1926, 36, Henry deJouvenal prize, Palais de la Découverte, 1938. Hon. fellow, Nat. Inst. Scis. of India; fellow A.A.A.S. (pres. 1940), Am. Acad. Arts and Scis., mem. Am. Philos. Soc., Bot. Soc. Am. (pres. 1950), Nat. Acad. Scis., Am. Soc. Naturalists (pres. 1930), N.E. Bot. Club, Torrey Bot. Club (pres. 1933), Am. Genetics Assn., Genetic Soc. Am., Am. Eugenics Soc., Assn. for Research in Human Heredity, Human Genetics Soc., Soc. for Study Devel. and Growth (pres. 1945-46), Phi Beta Kappa. Sigma Xi; asso., fgn. or hon. mem. numerous other profl. socs. Author: Sexual Reproduction in the Mucorineae, 1904; New England Trees in Winter (with C. D. Jarvis), 1911, Trees in Winter (with C. D. Jarvis), 1913; Methods of Inducing Doubling of Chromosomes in Plants (with A. G. Avery), 1938. Contbr. to scientific jours. Discovered sexual reprodn. in bread molds; studied inheritance and geographical distribution of jimson weed; drew important conclusions about chromosome behavior, genic balance, and species evolution; discovered that chemical, colchicine, caused mutation in plants (1st chem. interference with heredity), 1937; did research (with East and Clausen) on response of plant to relative length of day and night. Died Northampton, Mass., Nov. 16, 1954.

BLAKESLEY, Thomas H., English physicist; b. 1847; s. J. W. Blakesley; ed. King's Coll., Cambridge, Eng.; M.A.; civil engr.; mem. Instn. Civil Engrs. Author: Alternating Currents of Electricity, 1899; Geometrical Optics, 1903; also papers on math. physics, geometrical optics, electricity. Discovered large mass of meteoric iron by local effect on magnetic declination; improved methods of measuring and defining properties of optical instruments; devised lenses and spectroscopes; investigated alternating currents, also related problems in telegraph cables, elec. power transmission; drew up tables of hyperbolic functions, invented portable barometer (amphisbaena) with mercury column of a few inches; studied properties of logarithmic spiral and devised linkage to develop families of such curves; invented mech. methods to solve cubic equations; pointed out relation of logarithmic curves to those of Spanish-Arabian architecture. Died Feb. 13, 1929.

BLALOCK, Alfred, Am. surgeon; b. Culloden, Ga., Apr. 5, 1899; s. George Zadock and Martha (Davis) B.; A.B., U. Ga., 1918; M.D., Johns Hopkins, 1922; M.D. honoris causa, U. Turin, 1951; Sc.M., Yale, 1946; Sc.D. (hon.), U. Rochester, and U. Chgo., 1951, Lehigh U., 1953, Emory U., 1954, Georgetown U., 1959; LL.D., Hampden-Sydney Coll., 1954; m. Mary Chambers O'Bryan, Oct. 27, 1930 (dec.); children—William Rice, Mary Elizabeth, Alfred Dandy; m. 2d, Alice Seney Waters, Nov. 12, 1959. Instr. surgery Vanderbilt Med. Sch., 1925-27, asst. prof., 1928-30, asso. prof., 1930-38, prof. 1938-41; prof. surgery and dir. dept. surgery Johns Hopkins U. from 1941; surgeon-in-chief John Hopkins Hosp. from 1941; vis. prof. surgery U. Rochester, 1959, also various other colls. and univs., from 1961; Agnew lectr., 1959. Mem. med. fellowship board NRC, 1939-51. Mem. com. John J. Carty Fund, Nat. Acad. Scis. Recipient Research medal So. Med. Assn., 1940, Gordon Wilson medal, 1941, Charles Mickle

Fellowship, 1947; Passano award, 1948; Rene Leriche award, 1949, Matas award, 1950, Distinguished Service award A.M.A., 1953, Internat. Feltrinelli prize for medicine, 1954, Lasker award, 1954, Roswell Park medal, 1955, Gairdner award, 1959; Modern Medicine award for distinguished achievement, 1960. Diplomate Am. Bd. Surg. (founder). Fellow A.C.S. (pres. 1954-55); mem. Nat. Acad. Sci., Nat. Soc. Med. Research (v.p.), Am. (pres. 1955-56), So. (pres. 1949) surg. assns., Am. Assn. Thoracic Surgery (pres. 1950), Internat. Soc. Surgery, A.M.A., Soc. Clin. Surgery (pres. 1950-52), Soc. Vascular Surgery (pres. 1951-52), Academie des Sciences, Institut de France, Royal Coll. Surgeons of Eng. and Edinburgh, Phi Beta Kappa, Alpha Omega Alpha, Nu Sigma Nu, Am. Coll. Cardiology (hon.), also various fgn. societies. Author: Principles of Surgical Care: Shock and Other Considerations. Asso. editor, Surgery, from 1936; mem. editorial bd. Archives of Surgery, from 1939; mem. adv. editorial bd. Jour. Thoracic Surgery, from 1946. Contbr. profl. jours. Invented and perfected operation for blue babies (with Helen Taussig), 1945; also did work in vascular surgery and treatment of shock. Died Sept. 15, 1964.

BLALOCK, Hubert Morse, Jr., Am. sociologist; b. Balt., Aug. 23, 1926; s. Hubert Morse and Dorothy (Welsh) B.; A.B., Dartmouth, 1949; M.A., Brown U., 1953; Ph.D., U. N.C., 1954; m. M. Ann Bonar, Aug. 13, 1951; children—Susan Lynn, Kathleen Ann, James Welsh. Instr. U. Mich., 1954-57, asst. prof., 1957-61; asso. prof. Yale, 1961-64; prof. U. N.C., Chapel Hill, 1964—. Mem. Am. Sociol. Assn., Am. Statist. Assn. Author: Social Statistics, 1960; Causal Inferences in Nonexperimental Research, 1964; Toward a Theory of Minority-Group Relations, 1967. Asso. editor Am. Sociol. Rev. 1962-65, Social Problems, 1964-66, Social Forces, 1964—. Contbr. articles to sci. jours. Methodological work on application of statis. theory to causal analysis, involving large numbers of variables, where experimental methods cannot be applied. Home: Laurel Hill Rd., Chapel Hill, N.C. 27514.*

BLANC, Daniel Louis, French physicist; b. Caen, France, Dec. 10, 1927; s. Auguste and Thérèse (Guille-Menet) B.; Licence es Sciences, U. Caen, 1948, Diplome Etude Supérieures, 1949, Agregation Physics, 1950; Doctorate es Sciences, U. Paris, 1956; m. Francoise Cartegnie, Oct. 18, 1954; children—Chatherine, Monique. Preparator, Faculty Sci., Caen, 1947-49, asst., Lille, France, 1949-52; asst. College de France, Paris, 1951-56; staff Faculty Scis., Toulouse, France, 1957—, prof. nuclear physics, 1960—, dir. Center Atomic and Nuclear Physics, 1957—. Decorated Chevalier des palmes académiques, chevalier de l'ordre du mérite agricole, chevalier de l'ordre du mérite pour la recherche et l invention. Mem. French, Italian phys. socs., French Soc. Health Physics, French Meteorol. Soc. Author: Détecteurs de particule, 1959; Eléments de Physique Nucleaire (with Ambrosino), 1960; Les radioéléments, 1966; also numerous articles. Research on dosimetry and spectrometry of nuclear particles (electronic detectors), ionic and atomic physics, interactions of fast neutrons with light nuclei, indsl. uses of radioisotopes. Home: 16 Alleé Fréderic-Mistral, 31-Toulouse-01, France.*

BLANC, (John Joseph Charles) Louis, Social and polit. theorist; b. Madrid, Spain, Oct. 29, 1811; studied law, Paris; m. Christina Groh, 1865; worked at attorney's clerk and math. tchr.; founder Revue du progrès, 1839; became mem. provisional govt. during revolution of 1848; apptd. to preside over govt. labor commn. established at Luxembourg to inquire into labor question; urged formation of social workshops (ateliers sociaux) in France, but plan was sabotaged; fled from Paris to Belgium, then to London after implication in workers' rebellion, 1848; returned to Paris after overthrow of Napoleon III in 1870; served as pvt. in Nat. Guard; elected mem. Nat. Assembly, 1871; advocated abolition of senate and presidency, 1878; introduced proposal for amnesty of communists, 1879, proposal carried. Author: Histoire de dix ans 1830-40, 1841; Histoire de la Revolution Française, 1847-62; Appel aux honnetes gens, 1849; Pages de l'histoire de la revolution de 1848, 1850; Discours politiques, 1847-81. Attributed evils of society to pressures of competition; demanded equalization of wages and merger of personal interests in the common good, to be accomplished by social workshops where men would unite their efforts for the common benefit; formulated basic idea that from each according to his abilities, to each according to his needs; his polit. and social ideas represent a link between utopian and sci. socialism and significantly influenced devel. of French socialism. Died Cannes, France, Dec. 6, 1882.

BLANCHARD, Arthur Alphonzo Am. chemist; b. Boston, May 4, 1876; s. Adolphus J. and Louise B. (Rand) B.; S.B., Mass. Inst. Tech., 1898; Ph.D., U. Leipzig, 1902; m. Eugenia M. Lord, Aug. 8, 1905; children—Shirley Louise (Mrs. William Hammond), Helen Lord (Mrs. Frederick D. Cowles), Malcolm Adolphus, Joseph. Asst. phys. chemistry Mass. Inst. Tech., 1898-1900; fellow, Leipzig, 1900-02; instr. chemistry, physics, N.H. Coll., 1902-03; instr. chemistry Mass. Inst. Tech., 1903-08; asst. prof., 1908-14, asso. prof., 1914-31, prof. from 1931, prof. emeritus from 1941; lectr. Harvard, 1929-30. Mem. Am. Chem. Soc., A.A.A.S., Am. Acad. Arts and Scis., N.E. Assn. Chemistry Tchrs., Am. Assn. Author: The Electrolytic Dissociation Theory (with H. P. Talbot), 1907; Synthetic Inorganic Chemistry, 1910; Foundations of Chemistry (with Frank B. Wade), 1914, Laboratory Manual for same, 1917. Translator: (with W. T. Hall) Laboratory Methods of Inorganic Chemistry, 1909. Research and publs. on metal carbonyls, viscosity of solutions, atomic structure and valence. Died Mar. 25, 1956. Home: 25 Evans Rd., Brookline 46, Mass.

BLANCHARD, Charles Émile, French naturalist, entomologist; b. Paris, France, Mar. 7, 1819; s. Émile Théophile Blanchard; ed. Mus. Natural History, 1834; prof. zoology, entomology Natural History Mus., 1862-95; prof. Nat. Agronomic Inst., 1876-88; (with Milne Edwards) expdn. to Sicily, 1844; mem. French Acad. Scis., 1862 (pres. 1883), Soc. Agr.; Author: Histoire naturelle de insectes, 1840, Organisation du règne animal, 1851-64; la Vie des etres animés, 1888; research on internal structure of vertebrates, comparative anatomy; explored Sicilian coastal ocean floors in 1894. Died Feb. 11, 1900.

BLANCHARD, Duncan Cromwell, Am. atmospheric physicist; b. Winterhaven, Fla., Oct. 8, 1924; s. Norman Harris and Edna (Perkins) B.; B.N.S., Tufts U., 1945, B.S. cum laude, 1947; M.S., Pa. State U., 1951; Ph.D., Mass. Inst. Tech., 1961; m. Roberta Claire Eike, June 1, 1957 (div.); children—Duncan Bruce, Rebecca Claire, Jonathan Stuart; m. 2d, Mary Lowe, Mar. 16, 1968. With Gen. Electric Research Lab., 1947-49; with Woods Hole (Mass.) Oceanographic Instn., 1951—, asso. scientist, 1964—. Mem. Am. Meteorol. Soc. (cons. ednl. movies, books), Sigma Xi, Sigma Pi Sigma. Research, publs. on raindrop size, shape, distbn., formation of rain; exchange of electricity between sea and air, lightning formation in oceanic volcanoes.

BLANCHARD, J. Ewart, Canadian physicist; b. Truro, N.S., Can., Mar. 22, 1921; s. Aubry and Agnes (Blair) B.; B.Sc., Dalhousie U., 1940; M.A., U. Toronto, 1947, Ph.D., 1952; m. Mary Helena, July 5, 1958; children—Jonathan Sandilands, Megan Blair. Geophysicist, N.S. Research Found., Halifax, 1949-66, v.p., 1966—; faculty Dalhousie U., 1949—, prof. geophysics, 1964-66, acting dir. Dalhousie Inst. Oceanography, 1966—. Mem. coms. NRC Can., 1951—; mem. Nat. Adv. Com. on Research in Geophys. Scis., 1965—. Mem. Canadian Assn. Physicists (com. on rock mechanics), Soc. Exploration Geophysicists, Am. Geophys. Union, Canadian Inst. Mining and Metallurgy, N.S. Inst. Sci., Seismol. Soc. Am., European Assn. Exploration Geophysicists. Research, publs. on gravity measurement in N.S., theory of elec. and electromagnetic methods of geophys. exploration, absolute stress measurements in crust of earth, deep seismic crystal expts. Home: 6470 Coburg Rd. Office: P.O. Box 1027, Halifax, N.S., Can.*

BLANCHARD, Jean Pierre François, French inventor, aeronaut; b. Les Andelys, France, July 4, 1750; m. Victoire Lebrun; m. 2d, Marie Madeleine Sophie Armand, 1796 or 97. Tried to build flying machine, 1770's; invented hydraulic machine, 1772; invented parachute; made 1st parachute jump, 1784; made 1st balloon crossing of English Channel, 1784; came to U. S., 1793; made balloon flights at Philadelphia and New York. Died Paris, Mar. 7, 1809.

BLANCHARD, Raoul, French geographer; b. Orléans, France, Sept. 4, 1877; s. Léon and Emilienne (Badinier) B.; ed. Ecole normale superièure; agrégé, Ph.D. in letters; Dr. honoris causa univs. Montreal, Gand, Que.; m. Jane de Lauwereyns, Apr. 10, 1901; children—Henriette, Guillaume, Antionette, Colette. Prof., Lycées of Doual, Lille, U. Grenoble, 1906-48, U. Chgo., 1927, Harvard, 1928-36, U. Cal. at Berkeley, U. Montreal, 1948-50, Que., 1952; dir. Alpine Inst. Geography. Recipient Gold medal of sci. research, Gold medal Am. Geog. Soc. Mem. numerous socs. in France and fgn. countries, Inst. de France. Author: Les Alpes occidentales; Le Canada français, 5 vols., 1935-52; La Flandre; L'Asie occidentale, 1929; Le Comté de Nice. Founder Revue de geographie alpine. Established study plan for urban geography, 1911. Address: 3, rue Marechal-Galliéni, Sèvres, France.

BLANCHARD, Raphael, French naturalist; b. St.-Christophe, France, Feb. 28, 1857; M.D., Paris, France 1880; prof Faculty Medicine, Paris; became prof. natural history, 1897, prof. parasitology, 1907; founder Inst. Colonial Medicine, Paris. Mem. Acad. Medicine (ann. sec. 1912-19), Zool. Soc. France (sec. gen. for 20 years). Author: Histoire naturelle medicale des moustiques, 1905; Traité de zoologie médicale, 1885. Founder, Archives des parasitologie, 1898. Founded parasitological medicine in France. Died 1924.

BLANCHARD, Thomas, Am. inventor; b. Sutton, Mass., June 24, 1788; s. Samuel and Susanna (Tenney) B. Invented machine that could produce 500 tacks per minute, 1806; devised a lathe which performed 2 different operations (turned rifle barrels externally and when breech was reached. cut both flat and oval portions); with Springfield (Mass.) Arsenal, 5 years; patented machine which could produce various irregular forms from a single pattern, 1820; invented steam-wagon, 1825; unsuccessfully promoted co. to build railroads, 1826; patented steamboat which was able to ascend rivers against strong currents and rapids, 1831; developed heavy timber-bending process, 1851. Died Boston, Apr. 16, 1864.

BLANCHET, Robert, French agronomist; b. Montlucon, France, Feb. 14, 1927; s. Jules and Marie Blanchet; Ingénieur Horticole, École Nationale d'Horticulture, 1952; Licencié ès-Scis., Faculty Scis. Paris, 1954, Docteur ès Scis., 1958; m. Huguette Chaminade, Apr. 6, 1956; children—Anne, Pierre, Claire. Entered French Nat. Inst. Agronomical Research, 1952; asst., then head of research Versailles Central Agronomical Center, 1952-59; dir. Chateauroux Agronomical Sta., 1959-64; dir. Toulouse Sta. Agronomy and Enology, 1964—. Mem. French Soc. Vegetal Physiology, various French and internat. assns. of soil science. Contbr. articles to sci. publs. Studies of prin. factors of mineral nutrition of plants in soil; research on methods of retention and movements of mineral ions in soil, their relationship with plant nourishment, application of this relationship to soil fertilization. Home: 95, ave de Lespinet, Toulouse 31. Office: 67 blvd. Deltour, Toulouse 31, France.*

BLANCHINUS, Franciscus, see Bianchini, Francesco.

BLAND, John Hardesty, Am. physician, educator; b. Globe, Ariz., Nov. 7, 1917; s. Walter Perry and Clara (Hardesty) B.; A.B., Earlham Coll., 1940; M.D., Jefferson Med. Coll., 1943; m. Elizabeth Lowry Carnman, Sept. 23, 1944; children—John Hardesty, Walter Perry III, Elizabeth Lowry, Linda Mendenhall. Faculty, U. Vt. Coll. Medicine, Burlington, 1949—, asso. prof., 1955—, dir. rheumatism research unit dept. medicine, 1950—; attending in medicine Mary Fletcher, DeGoesbriand Meml. hosps. Research fellow dept. rheumatology U. Manchester (Eng.), 1958-59. Fellow A.C.P.; mem. A.M.A., A.A.A.S., Am. Fedn. Clin. Research, Am., New Eng. rheumatism assns., N.Y. Acad. Scis., Sigma Xi. Author: The Clinical Use of Fluid and Electrolyte, 1952; Metabolism of Body Fluids, 1956; Arthritis, Medical Treatment and Home Care, 1960. Editor, contbr. Clinical Metabolism of Body Water and Electrolytes, 1963. Research and publs. on inter-relationships between water, salt and connective tissue chemistry, arthritis and rheumatism. Home: 343 S. Prospect St., Burlington, Vt. 05401.*

BLAND, Thomas, naturalist; born in Newark, England, October 4, 1809; the son Thomas and Mrs. (Shepard) B. Studied law in England. Practiced law, London; moved to Barbados, 1842; moved to Jamaica, began study of fauna of W.I., 1849; supt. of gold mine, Marmato, New Granada (now Columbia), 1850; moved to N.Y.C., 1852; wrote 72 papers on molluscs of U. S. and Antilles; Fellow Royal Geol. Soc. London; mem. Am. Philos. Soc. Author: (with William G. Binney) Land and Fresh Water Shells of North America, 1869. Died Aug. 20, 1885.

BLAND-SUTTON, John, English surgeon; b. Enfield Highway, Eng., Apr. 21, 1855; s. Charles William and Elizabeth (Wadsworth) S.; LL.D., Aberdeen, St. Andrews, Birmingham, Glasgow, Leeds; D.Sc., Toronto; M.Ch. (hon.), Dublin; m. Agnes Hobbs, 1882; m. 2d, Edith Heather-Bigg, 1899. Apptd. prosector, also pathologist Zool. Soc., 1881; asst. surgeon Middlesex Hosp., 1886-1905, apptd. surgeon, 1905, later cons. surgeon; surgeon Chelsea Hosp. for Women, 1896-1911; Erasmus Wilson lectr., also Hunterian prof. Royal Coll. Surgeons, 1886-91; Cavendish lectr., 1916; Bradshaw lectr., 1917; Hunterian Orator, 1923; Murphy orator, 1927. Created knight, 1912, baronet, 1925; recipient Jacksonian prize, Royal Coll. Surgeons, 1892. Hon. fellow A.C.S.; pres. Royal Coll. Surgeons, 1923-26, Royal Soc. Medicine, 1921-22; hon. mem. Odontological Soc., Great Britain. Author: Evolution and Disease, 1890; Ligaments, their Nature and Morphology; Surgical Diseases of the Ovaries, 1891; Tumours, Innocent and Malignant, 1893; Diseases of Women, 1897; Gall-stones and Diseases of the Bile Ducts, 1907; Man and Beast in Eastern Ethiopia, 1911; Fibroids of the Uterus, 1913; Selected Lectures and Essays, 1920; Orations and Addresses, 1924; The Story of a Surgeon, 1930; On Faith and Science in Surgery, 1930; Men and Creatures in Uganda, 1933. Worked in comparative anatomy, also comparative embryology, comparative pathology, gynecol. surgery. Died, London, Dec. 20, 1936.

BLANDAMER, Michael Jesse, Brit. chemist; b. Dorchester, Dorset, Eng., Dec. 19, 1934; s. Douglas Percy and Hylda (Bentley) B.; B.Sc., U. Southampton, 1956, Ph.D., 1960; m. Anne Elizabeth Taylor, Aug. 27, 1960; children—Peter Douglas, Ian Michael Jesie. Post-doctoral fellow NRC, Ottawa, Ont., Can., 1960-61; lectr. U. Leicester (Eng.), 1961—. Fellow Chem. Soc. London, Eng. Research, publs. on investigation of structures of salt solutions using spectroscopic and ultrasonic relaxation techniques; has clarified certain concepts concerning how ions interact with solvent and other ions in solution. Home: 74 Warerrey Rd., Leicester, Eng.*

BLANDAU, Richard Julius, Am. physician; b. Erie, Pa., Aug. 5, 1911; s. Richard Albert and Katie (Lubbers) B.; A.B., Linfield Coll., 1935; Ph.D., Brown U., 1939; M.D. with honor, U. Rochester, 1948; m.

Olive Lewellen, Oct. 9, 1937; 1 son, Richard Lewellen. Fellow depts. anatomy and psychobiology NRC, Yale, 1937-38; instr. Brown U., 1939-42, Harvard Sch. Medicine, 1942-43, faculty U. Rochester Sch. Medicine and Dentistry, 1943-48, asst. prof., 1948, Buswell fellow dept. urology and surgery, 1948-49; asso. prof. anatomy U. Wash., Seattle, 1949-51, asst. dean Sch. Medicine, 1955-60, asso. dean, 1960-64, prof. dept. biol. structure, 1951—. Cons, histopathology Manhattan Dist., 1946-48; Solomon Theron DeLee lectr. U. Chgo., 1954; spl. cons. Nat. Inst. Child Health and Human Devel., 1963, Cons. WHO, 1964—. Recipient Vienna Film Festival award, 1959. Mem. Am. Soc. Anatomists, Soc. for Problems in Growth, Soc. Exptl. Biology and Medicine, Am. Soc. for Study Sterility (Isidor Rubin award 1952, Ortho Research award 1956), Am. Soc. for Cell Biologists, Am. Soc. Teratologists, Am. Soc. Med. Illustrators (hon.), Sigma Xi, Alpha Omega Alpha (Borden Research award 1948), others. Asso. editor Am. Jour. Anatomy, 1957—, Am. Jour. Fertility and Sterility, 1954—. Home: 540 Edmonds Way, Edmonds, Wash. 98020. Office: U. Wash. Sch. Medicine, Seattle 98105.*

BLANDER, Milton, Am. chemist; b. Bklyn., Nov. 1, 1927; s. Benjamin and Yetta (Schwartzman) B.; B.S. in Chemistry, Bklyn. Coll., 1950; Ph.D. in Phys. Chemistry, Yale, 1953; m. Barbara Alice Swomley, Dec. 11, 1960; children—Julia, Kathryn, Daniel. Research asso. Cornell U., Ithaca, N.Y., 1953-55; group leader Oak Ridge Nat. Lab., 1955-62; mem. tech. staff N.Am. Aviation Sci. Center, Thousand Oaks, Cal., 1962—. Mem. Am. Chem. Soc., A.A.A.S., Meteoritical Soc.. Phi Beta Kappa, Sigma Xi. Editor: Molten Salt Chemistry, 1964; Research and publs. on theory of thermodynamic properties of molten salts and verified aspects of theory; calculated properties to be expected of primordial dust which is presumed precursor of our solar system; investigation of meteorites. Home: 872 Tamlei Av., Office: 1049 Camino Dos Rios, Thousand Oaks, Cal. 91360.*

BLANDIN, Philippe Frédéric, French surgeon; b. Aubigny, France, Dec. 8, 1798; began study of medicine at Paris, 1816; practiced medicine in Paris; named surgeon Hôtel Dieu, Paris, 1836; became prof. surgery, Faculty of Medicine, Paris, 1840. Mem. Acad. Medicine. Author: Traité d'anatomie topographique, 1826. Described mixed glands near tip of tongue (Blandin's, or Nuhn's glands), 1826; perfected operation for rhinoplasty; research on phlebitis and angioleucitis; formed theory of traumatic erysipelas; advocated ether as anesthetic. Died Paris, Apr. 16, 1849.

BLANE, Sir Gilbert, physician; b. Banefield, Ayrshire, Scotland, Aug. 29, 1749; s. Gilbert B.; began studies at U. Edinburgh (Scotland), at age 14; M.D. U. Glasgow (Scotland), 1778; m. dau. of Abraham Gardiner, July 11, 1786; 6 sons including Hugh Seymour, 3 daus. Became pvt. physician to Adm. Rodney, W.I., 1779; physician to fleet, 1779-83; came to Eng. with Rodney, 1781; physician St. Thomas's Hosp., Eng., 1783-95; became physician extraordinary to Prince of Wales, 1785, later physician in ordinary, also to George IV; commr. for sick and wounded seamen, 1795-1802; assisted in formation rules of quarantine, 1799; Licentiate College Physicians. Fellow Royal Society, 1784; member of the French Academy of Sciences. Author: On the Most Effectual Means for Preserving the Health of Seamen, particularly in the Royal Navy, 1780; Medical Logic, 1819; Observations on the Diseases of Seamen, 1795. Improved san. conditions of Navy; recommended use of lime juice to prevent scurvy in the navy, 1793. Died London, June 26, 1834.

BLANFORD, Henry Francis, English meteorologist, geologist; b. Whitefriars, Eng., June 3, 1834; s. William and Harriet (Simpson) B.; studied at schs. in Brighton, Eng., Brussels, Belgium; student Royal Sch. Mines; studied for 1 year at Freiberg (Germany); m. Charlotte Mackenzie, June 20, 1867; 2 sons, also daus. Apptd. to Geol. Survey India, 1855; prof. Presidency Coll., Calcutta, India, 1862-72; named meteorol. reporter to Govt. of Bengal, 1872, later to Govt. of India; ret. and returned to Eng., 1888. Fellow Geological Society, Royal Society, 1880; mem. Asiatic Society Bengal (president 1884-1885); honorary member several fgn. meteorol. socs. Author: (with Carl Johann August Theodor Scheere) An Introduction to the Use of the Blowpipe Together with a Description of the Blowpipe Characters of the Most Important Minerals, 1856; (with John William Salter) Palaeontology of Niti in the Northern Himalaya, 1865; (with J. E. Castrell) Report of the Calcutta Cyclone of 5 October 1864, 1866; The Indian Meteorologist's Vade Mecum, 1868; Rudiments of Physical Geography for the Use of Indian Schools, 1873; The Winds of Northern India, 1873; A Practical Guide to the Climates and Weather of India, 1889; An Elementary Geography of India, Burma, and Ceylon, 1889; also 50 articles on meteorology, geology. Classified cretaceous strata nr. Trichinopoly; discoveries on the origin of cyclones. Died Jan. 23, 1893.

BLANK, Albert Abraham, Am. mathematician; b. N.Y.C., Nov. 29, 1924; s. Harry Edward and Rose (Rosenstein) B.; B.A., Bklyn. Coll., 1944; M.S., Brown U., 1946; Ph.D., N.Y.U., 1951; m. Elizabeth Turner Clark, May 27, 1950; children—Sharon, Tamara, Deborah, Irina. Math. analyst Ophthal. Inst., Columbia, N.Y.C., 1950-52; asst. prof. math. U. Tenn., Knoxville, 1954-59; asso. prof. math. N.Y.U., N.Y.C., 1959-65, prof., 1965—. Cons., Sch. Math. Study Group, 1959—; mem. adv. bd. con. on sch. math. U. Ill., 1963-65. Fellow A.A.A.S., Am. Math. Soc., Math. Assn. Am.; Am. Phys. Soc., A.A.A.S., Sigma Xi. Author: Problems in Calculus and Analysis, 1966; Calculus, 1968. Asso. editor Am. Math. Monthly, to 1966. Research on math. theory of electromagnetism and optics, geometry of binocular space perception, magnetofluid dynamic theory. Home: 219 Pelhamdale Av., Pelham, N.Y. 10803. Office: Courant Institute, N.Y. University, N.Y.C. 10003.*

BLANK, Yakov Pavlovich, Russian mathematician; b. Liepaja, Latvia, 1903; grad. Kharkov Inst., Pub. Edn., 1926; postgrad. Kharkov Research Inst. Math. and Mechanics, 1929; D. Physico-Math. Sci., 1951. Prof., Kharkov U., 1952—. Author: Minimal Surfaces as Translation Surfaces, 1939; The Linear Surface Element in Spherical Geometry, 1940; Solution of an Engel Translation Surface Problem, 1948; Conjugated Conical Line Grids, 1949; Surfaces of Elliptical Space Shift, 1950; Conical Grids, 1952; W Congruencies, 1956; The Analogue of Translation Surfaces in Lobachevsky's Geometry, 1957, also others. Address: Kharkov University, Universitetskaya ulitsa 16, Kharkov, Ukraine SSR, USSR.

BLANKAART, Stephan, Dutch physician, chemist; b. Middleburg, Netherlands, Oct. 24, 1650; apprenticed to apothecary to learn botany and chemistry; Dr.Phil. and Medicine, 1674. Practiced medicine, Amsterdam, Netherlands. Author: Lexicon Medicum Graeco-Latinum, 1679; Die neue Heutiges Tages gebräuchliche Scheide-Kunst, oder Chimia nach den Gründen des Fürtreflichen Cartesii und des Alcali und Acidi eingerichtet, 1697; Neuscheinende Praxis der Medicinae, 1700; Blancards Theatrum Chimicum, 1700; other works on medicine, surgery, pharmacy. Collected works pub. 1701. Pioneered preparation of pills of ferrous iodide. Died Amsterdam, Feb. 23, 1702.

BLANKENHORN, David Henry, Am. cardiologist; b. Cleve., Nov. 16, 1924; s. Marion A. and Martha (Taggart) B.; student Dartmouth, 1942-43; M.D., U. Cin., 1947; m. Anne Wood Ramsey, June 15, 1948; children—David A., Mary, Susan, John C. Research asst. Rockefeller Inst. for Med. Research, 1952-54; instr. U. Cin., 1955-57; faculty U. Southern Cal., Los Angeles, 1957—, head cardiology sect., 1963—, prof. medicine, 1966—. Mem. Am. Fedn. Clin. Research, Am. Soc. for Clin. Investigation, Western Assn. Physicians, Am. Heart Assn., Council on Clin. Cardiology, Am. Coll. Cardiology. Research and publs. on calcification all major blood vessels, lipid chemistry. Home: 1165 Afton St., Pasadena, Cal. 91103. Office: 2025 Zonal Av., Los Angeles 90033.*

BLANKENHORN, Marion Arthur, Am. physician; b. Orrville, O., Nov. 13, 1885; s. Henry and Emma C. (Amstutz) B.; Ph.B., Wooster Coll., 1909; M.D., Western Res. U., 1914, M.A. in medicine, Grad. Sch., 1920; hon. D.Sc., Wooster Coll., 1939; m. Martha Finley Taggart; children—Martha, Mary Margaret, David Henry. Prof. biology Buena Vista (Ia.) Coll., 1909-10; resident physician in Lakeside Hosp., Cleve., 1914-17; instr. in medicine, 1919; voluntary asst. Rockefeller Inst. for Med. Research, 1925-26; prof. medicine Western Res. U., 1929-35; prof. medicine U. Cin., 1935-56, prof. emeritus, from 1956; responsible investigator aviation medicine NRC.; dir. edn. dept. internal medicine Jewish Hosp., Cin., from 1957. Fellow A.C.P. (regent v.p.; dir. survey of hosp. standards in internal medicine 1956-57); mem. A.M.A., Assn. Am. Physicians, Am. Soc. Clin. Investigation, Central Soc. Clin. Research, Am. Soc. Exptl. Pathology, Alpha Tau Omega, Nu Sigma Nu, Phi Beta Kappa. Alpha Omega Alpha. Sigma Xi. Died Sept. 3, 1957; buried Wooster, O.

BLANKENSHIP, Forrest Farley, Am. chemist; b. Gatesville, Tex., Oct. 14, 1913; s. Albert Sylvanus and Alva (Carruth) B.; student E. Tex. State Coll., 1929-31; B.A., U. Tex., 1932, M.A., 1933, Ph.D., 1943; m. Margaret Berry Burke, Oct. 18, 1934; children—Jane (Mrs. Carl Hunter Gibson) and Betty (Mrs. Arthur Tilo Alt) (twins). Prof. chemistry and physics Paris (Tex.) Jr. Coll., 1934-42; prof. chemistry U. Okla., Norman, 1943-51; chemist Oak Ridge Nat. Lab., 1951-60, sr. sci. adviser reactor chemistry div., 1960—. Fellow A.A.A.S.; mem. Am. Chem. Soc., Am. Soc. Metals, Research Soc. Am. Research and publs. on gas hydrates, use molten salts nuclear fuel, high temperature chemistry. Home: 627 Lakeshore St., Kingston, Ten. 37763. Office: Oak Ridge Nat. Lab., X-10, Oak Ridge 37830.*

BLANKS, Robert Franklin, Am. civil engr.; b. Maplehill, Kan., Sept. 4, 1900; s. Thomas Franklin and Vida Florence (Fairbanks) B.; B.S. in Civil Engring., Kan. State Coll., 1924. C.E., 1936, D.Sc., 1949; Laura Viola Denman, May 26, 1922; 1 son, Robert Franklin. Municipal engr., Burley, Ida., 1921; field engring. on location and constrn. Carey (Ida.) Irrigation Project, 1921; sci. and math. instr. Burley High Sch., 1921-22; asst. supt. Burley public schs., 1922-23; instr. in physics, chemistry and math., Hollister (Ida.) High Sch., 1924-25 17-28, athletic coach, 1924-25; asst. supt. and engr. irrigation project. Salmon River Canal Co., Hollister, 1925-27; made hydraulic investigations Am. Falls Reservoir and Basin Water Dist. No. 36, State of Ida., Idaho Falls, 1928; with office engring. Ft. Hall Irrigation project, U. S. Indian Irrigation Service, Blackfoot, Ida., 1928-29; office and field engring. Columbia Basin Investigations, Washington U. S. Engr. Office, Seattle, 1929-30; chief, research and geology div. U. S. Bur. of Reclamation, Denver, 1930-51. Vice pres., gen. mgr. Great Western Aggregates, Inc.; research cons. Ideal Cement Company. Chmn. Reinforced Concrete Research Council Engring. Found. Recipient Thomas Fitch Rowland prize for paper Deterioration of Concrete Dams Due to Alkali-Aggregate Reaction. Mem. Am. Soc. C.E. (chmn. com. on masonry and reinforced concrete), Am. Soc. for Testing Materials, Am. Concrete Inst. (pres. 1948)), Hwy. Research Bd., Colo. Soc. Engrs. (dir. 1951-53, pres. 1955), Sigma Xi. Contbr. to engring. publs. Patentee materials testing equipment for controlling rate of load application in testing materials. Died July 14, 1958.

BLASCHKE, Wilhelm, mathematician; b. Graz, Austria, Sept. 13, 1885. Prof., Prague, Czechoslovakia, Leipzig, Königsberg, Tübingen, Hamburg (all Germany), Balt., Chgo. Author: Kreis und Kugel, 1916; Vorlesungen über Differentialgeometrie und Geometrie; Grundlagen von Einsteins Relativitätstheorie, 1921-23; Vorlesungen über Integralgeometrie, 1937; Analytische Geometrie, 1948. Studied complex functions with several variables.

BLASCHKO, Alfred, German dermatologist; b. Freienwalde/Oder, Germany, Mar. 3, 1858; s. Hermann and Babette (Manuheimer) B.; grad. U. Berlin (Germany), 1881; m. Johanna Litthauer; 3 children. First worked in Stettin, Germany (now Poland); then practiced medicine, Berlin; later worked at Vienna skin clinic; worked with Köbner and Lewin, Berlin; specialized in dermatology after 1888; faculty dermatology U. Berlin; specialist at large Berlin health ins. clinic, 1883. Mem. Deutsche Gesellschaft zur Bekämpfung der Geschlechtskrankheiten (1st gen. sec. 1902, pres. 1916). Author: Das Sehzentrum bei Fröschen, 1880; Beiträge zur Architektonik der Oberhaut, 1887; Behandlung der Geschlechtskrankheiten in Krankenkassen und Krankenhäusern, 1890; Syphilis und Prostitution vom Standpunkt der öffentlichen Gesundheitspflege, 1893; Hygiene der Prositution und vener. Krankheiten, 1900; Die Lepra im Kr. Memel, 1909; Therapeutisches Taschenbuch der Hautund Geschlechtskrankheiten, 1911. Editor jour. Bekämpfung der Geschlechtskrankheiten. Research on anatomy, physiology and evolution of skin, serodiagnosis of syphilis, prophylaxis for venereal diseases; 1st pointed out paraffin embolism of lung with mercury injections. Died Berlin, Mar. 26, 1922.

BLASCHKO, Hermann Karl Felix, biochemist; b. Berlin, Germany, Jan. 4, 1900; s. Alfred and Johanna (Litthauer) B.; M.D., Freiburg U., 1924; Ph.D., Cambridge U., 1937; M.A., Oxford U., 1947; M.D. (hon.), Free U., Berlin, 1966; m. Mary Douglas Black, Dec. 29, 1944. Research asst. Kaiser-Wilhelm Inst. for Biology, Berlin, 1925-28, Kaiser-Wilhelm Inst. for Med. Research, 1929-32; demonstrator in physiology Cambridge U., 1934-44; with dept. pharmacology Oxford U., 1944-67, reader in biochem. pharmacology, 1964-67, fellow Linacre Coll.; hon. prof. Heidelberg (Germany) U., 1966—. Mem. neuropharmacology panel Internat. Brain Research Orgn., 1960. Fellow Royal Society, 1962; member British Pharmacological Society, Biochemical Society, Physiological Soc., Biophysical Society, Marine Biological Assn. U.K. Research on formation and biol. inactivation of amines of physiol. and pharmacol. interest. Home: 24 Park Town, Oxford. Office: Dept. Pharmacology, South Parks Rd., Oxford, Eng.*

BLASER, Roy Emil, Am. agronomist; b. Duncan, Neb., Mar. 3, 1912; s. Henry and Katie (Rupp) B.; B.S., U. Neb., 1934; M.S., Rutgers U., 1937; Ph.D., U. N.C., 1946; m. E. Catherine Agnew, June 30, 1935; 1 son, Dale Leroy. Prof. agronomy, sect. leader research div. Va. Poly. Inst., Blacksburg, 1949—. Cons. internat. agrl. problems. Recipient certificate of merit for outstanding service Am. Grassland Council, So. Pasture and Forage Crop Improvement Assn., 1963, plaque for 25 years distinguished service; Man of Year award So. Seedsmen Assn., 1956. Fellow Am. Soc. Agronomy; mem. Crop Sci. Soc. Am., Soil Sci. Soc. Am., Dairy Sci. Assn., Am. Grassland Council, Internat. Grassland Congress, Sigma Xi. Contbr. chpt. to Forages, 1966, numerous articles to profl. jours. Research in physiology and ecology of pasture, hay, silage and turf prodn. and utilization. Home: 104 York Dr., Blacksburg, Va. 24061.*

BLASIUS, Ewald, German analytical chemist; b. Berlin, Germany, Apr. 6, 1921; s. Otto and Erna (Mühlan) B.; D. Engring., Tech. U. Berlin; m. Ilse Welker, 1949. From asst. to sci. adviser Tech. U. Berlin, 1947-64; prof., dir. Institutes für Analytische Chemie und Radiochemie, Universitat des Saarlandes Saarbrucken, 1964—. Mem. Gesellschaft Deutscher Chemiker. Author: Chromatographische Methoden in der analytischen und präparativen anorganischen Chemie, 1958; (with others) Lehrbuch der analytischen und präparativen anorganischen Chemie, 1962; (with others) Einführung in der anorganisch-chemische Praktikum, 1963. Work in chromatographic, radiochem. methods of analytical chemistry. Address:

Institüt für Analytische Chemie und Radiochemie, Universitat des Saarlandes, 66 Saarbrucken 15, Germany.

BLASIUS, Gerhard, Dutch anatomist; b. circa 1625; prof. medicine U. Amsterdam (Netherlands), physician of hosp., librarian; pub. many edits. of best med. works of his day. Author: Anatome medullae spinalis, 1660; Miscellanea anatomica, 1673; Observationes anatomicae, 1674; Anatome animalium, 1681. Died Amsterdam, 1692.

BLASIUS, Johann Heinrich, German zoologist; b. Eckenbach, nr. Cologne, Germany, Oct. 7, 1809; s. Johann Heinrich and Louise Catherine (Eckenbach) B.; studied natural scis. in Berlin, Germany; m. Louise Thiele, 1841; 1 son, Rudolf. Prof. natural history, Collegium Carolinum, Braunschweig, Germany, 1836; built natural sci. collection; dir. newly founded Natural History Mus., Braunschweig, 1859. Author: (with A.v. Keyserling) Die Wirbelthiere Europas I, 1840; Reise in europäischen Russland in den Jahren 1840 und 1841, 1844; Fauna der Wirbelthiere Deutschlands I, 1857. Research on European mammals and birds; considered one of the leading German ornithologists after J. F. Naumann's death. Died Braunschweig, May 26, 1810.

BLASIUS, Wilhelm, meteorologist; b. Eckenbach nr. Cologne, Germany, July 24, 1818; s. Johann Heinrich and Louise Catherine (Eckenbach) B.; studied natural scis. in Bonn, Germany, 1842-47; m. Cäcilie Uhde, 1860; 1 son, 1 dau. Prof. natural history Lyzeum in Hanover, Germany, came to Cambridge, Mass. for polit. reasons; was in contact with Jean Louis Agassiz. Author: New Theory of Storms, 1852; Storms, Their Nature, Classification and Laws, 1875. Before V. Bjerknes, he formulated theory for formation of storms which follows basic ideas of modern polar front theoreticians, expanded H. W. Dove's theory; however since he set up his own terminology and failed to offer sufficient evidence for his ideas he has been generally forgotten. Died Braunschweig, Germany, Mar. 24, 1899.

BLASKOVIC, Dionyz, Czech. virologist; b. Jablonica, Czechoslovakia, Aug. 2, 1913; s. Koloman and Gabriela (Blaskovicová) B.; M.D., Charles U., Prague, Czechoslovakia, 1937; Sc.D., Czechoslovak Acad. Scis., 1953; m. Milada Janatová, May 7, 1941; children—Peter, Milada, Hana. Univ. asst. Charles U., Inst. Bacteriology Prague, 1937-39; univ. asst. Komensky U. Inst. Hygiene, Bratislava, Czechoslovakia, 1939-46, dir. dept. microbiology and epidemiology, 1946-53, asso. prof., 1946-53, prof. microbiology, 1954—; dir. inst. virology Czechoslovak Acad. Scis., Bratislava, 1953—. Recipient State prize, 1951; Purkinje medal, 1956; Order of Labor, 1963. Fellow N.Y. Acad. Sci.; mem. Slovak Acad. Scis. (v.p. 1953-59, 1965—, pres. 1961-65), Czechoslovak Acad. Scis. (v.p. 1961-65); fgn. mem. USSR Acad. Sci.; hon. mem. Austrian Soc. Microbiology and Hygiene, Polish Microbiol. Soc., All-Union Soc. Microbiologists, Epidemiologists and Hygienists. Author: (with others) Methods in Virology, 1954, The Circulation of Viruses, 1963; The Epidemy of Tick-Borne Encephalitis in Roznava, 1954; Natural Foci of Infections, 1956; Deliberative Vaccination of Domestic Animals with Tick-borne encephalitis virus, 1964; (with others) The Properties of Influenze A virus, type FF, 1958. Research, publs. on importance of Proteus Morgani, its serological classification, biology of influenza viruses, variability, immunogenity, structure of influenza viruses-subunits, chemistry and serology; ecology of tick-borne encephalitis virus, its epidemiology and control. Office: 1 Mlynska dolina, Bratislava, Czechoslovakia.*

BLATT, Albert Harold, Am. chemist; b. Cin., Jan. 9, 1903; s. Joseph and Fannie (Krebs) B.; B.S., Harvard, 1923, M.A., 1926, Ph.D., 1926; student Coll. of France, 1923-24; m. Therese Herman, Jan. 30, 1935; 1 son, Joel Herman. Research asso., Harvard, 1926-28, faculty, 1932-39; asst. to gen. mgr. Columbus-McKinnon Chain Co., 1928-30; research asso. U. Buffalo, 1930-32; faculty Queens Coll., City U. N.Y., 1939—, prof. chemistry, 1948—, chmn. dept., 1961—. Investigator, sci. liaison officer OSRD 1941-46. Recipient Naval Ordnance Devel. award, 1946; Presdl. Certificate Merit, 1948. Mem. Am. Chem. Soc., A.A.A.S. Author: (with James B. Conant) The Chemistry of Organic Compounds, 3rd, 4th, 5th editions, 1947-59. Editor: Organic Syntheses, vol. 1, 1941, vol. II, 1943; mem. editorial bd. Organic Reactions, vols. 4-15, 1948—. Research and publs. on establishing the reducing, enolizing and condensing action of organomagnesium reagents on carbonyl compounds; clarification of stereochemistry and some mechanisms of Beckmann rearrangement of several classes of ketoximes; recognition of chem. effects of hydrogen bonding. Home: 415 E. 52nd St., N.Y.C. 10022.*

BLATT, Frank Joachim, Am. physicist; b. Vienna, Austria, May 1, 1924; s. Paul and Gretta (Tschiassny) B.; came to U. S., 1939, naturalized, 1945; B.S., Mass. Inst. Tech., 1946, M.S., 1948; Ph.D., U. Wash., 1953; m. Gloria Freeman, Sept. 1, 1946; children—Michael Robert, Ann Margaret. Faculty, U. Wash., 1948-50; research asso. U. Ill., 1953-56; faculty Mich. State U., East Lansing, 1956—, prof. physics, 1961—; vis. prof. Eidgenössische Technische Hochschule, Zurich, Switzerland, 1963-64. Cons. U. S. Naval Research Lab., 1956-59, 65—; cons. Argonne Nat. Lab., 1961—, program rev. com. solid state scis. div., 1965—. NSF Postdoctoral Research fellow Oxford U., 1959-60. Mem. Am. Phys. Soc., N.Y. Acad. Scis., Schweitzer Physik Gesell. Author: Electronic Conduction in Solids, 1968; also numerous articles. Research in theory of transport properties in metals and semiconductors, magneto-optical effects in semiconductors, size effects in metals, thermoelectric power of metals and alloys, magnetocaloric effects. Home: 252 Lexington Av., East Lansing, Mich. 48823.*

BLATT, John Markus, physicist; b. Vienna, Austria, Nov. 23, 1921; s. Paul and Greta (Tschiassny) B.; B.Sc. with honours, U. Cin., 1942; Ph.D., Cornell U., 1945; Ph.D., Princeton, 1945; m. Sylvia Ray Epstein, Aug. 12, 1945; children—Ruth Ellen, David William Eli, Daniel Joseph, Miriam Greta. Research asst. Princeton, U., 1944-45; scientist RCA Labs., Princeton, N.J., 1945-46; research asso. Mass. Inst. Tech., 1946-49; faculty U. Ill., 1950-53, asso. prof. physics, 1951-53; reader Syndey (Australia), 1953-58; prof. applied maths. U. New S. Wales, Kensington, Australia, 1959—. Fellow Am. Phys. Soc., Australian Acad. Sci.; mem. Australian Math. Soc. (past mem. council), Am. Phys. Soc. Author: (with Victor Weisskop) Theoretical Nuclear Physics, 1952; Theory of Superconductivity, 1964; Research and publs. on nuclear 2-body and 3-body problems, nuclear forces, mechanism of superconductivity, various other fields including cosmic rays, statis. mechanics, quantum field theory, enzyme recognition length. Address: U. New South Wales Kensington, N.S.W., Australia.*

BLATTNER, Russell John, Am. physician; b. St. Louis, July 3, 1908; s. Rudolph Frederick and Lydia (Bergmann) B.; A.B., Washington U., St. Louis, 1929, M.D., 1933; m. Marian Koeneke, June 24, 1939 (dec. 1966); children—Frederick Russell, William Albert. Faculty, Washington U., 1937-47; prof., chmn. dept. pediatrics, Baylor U. Coll. Medicine, Houston, 1947—; physician in chief Tex. Children's Hosp., Houston, 1954—. Recipient Alumni citation Washington U. Sch. Medicine, 1956; certificate appreciation Tex. Assn. for Retarded Children, 1959, NIH, U. S. Dept. Health, Edn. and Welfare, 1962. Fellow Am. Acad. Pediatrics; mem. Pan Am. Med. Assn., Am. Pediatric Soc., Soc. for Pediatric Research, A.A.A.S., N.Y. Acad. Scis., Am. Acad. Soc. Human Genetics, Royal Soc. Medicine (affiliate). Author: (with W. A. Spencer), Treatment of Acute Poliomyelitis 1954; Research and publs. on disease of children with particular reference to cause, diagnosis and treatment. Home: 2227 Bellefontaine St., Houston 77025.*

BLATZ, Hanson, Am. physicist; b. Bklyn., Aug. 22, 1907; s. Edward B. and Henrietta (Hanson) B.; E.E., Poly. Inst. Bklyn., 1929; postgrad. Columbia, 1929-30, U. Rochester, 1945-48; m. Elizabeth J. Holland, Oct. 7, 1933; children—Anne (Mrs. Robert J. Trapani), Nils, Elizabeth (Mrs. James Craiglow). Asst. physicist N.Y. Cancer Inst., 1931-35; radiation physicist Bklyn. Cancer Inst., 1935-42; x-ray engr. Bulova Watch Co., 1943-44; x-ray designer Ritter Co., 1944-49; chief radiation br. AEC, N.Y.C., 1949-59; dir. N.Y.C. Office Radiation Control, 1959—; asso. prof. environmental medicine N.Y.U. Med. Center, 1952—; cons. AEC, USPHS, WHO, govts. Thailand, India. Recipient Superior Service award AEC, 1957. Fellow Am. Coll. Radiology, Am. Pub. Health Assn.; mem. Radiol. Soc. N.Am., Health Physics Soc. (dir.), Am. Indsl. Hygiene Assn.; A.A.A.S. Author: Radiation Hygiene Handbook, 1959; Introduction to Radiological Health, 1964. Research and publs. primarily in field of radiation from atomic energy; developed laws, regulations to protect against radiation and their adminlstrn. Home: 100 Bleeker St., N.Y.C. 10012. Office: 325 Broadway, N.Y.C. 10007.*

BLAU, Fritz, chemist; b. Vienna, Austria, Apr. 9, 1865; s. Josef and Johanna B.; doctorate U. Vienna, 1886; later studied for a year in Adolf v. Baeyer's chem. lab., Munich, Germany; hon. dr. engring. from Karlsruhe, Germany. Asst. to Lieben and Bahr; later joined faculty U. Vienna; adviser to Wiener Glühlampenfabrik Watt; joined Auergesellschaft Berlin (Germany) in 1902, dir. factory during World War I, dir. all sci. writing and patent dept. when incandescent lamp factories of AEG, the Auergesellschaft and Siemens & Halske AG merged into Osram GmbH K.G., Berlin, 1919. Held 185 patents including organic chemistry objects, mfg. tungsten metal and wire, discharging gases, radiation techniques, wireless telegraphy, electric stoves, X-ray technique. Died Berlin, Dec. 5, 1929.

BLAU, Henry Hess, Am. phys. chemist; b. Dayton, O., May 16, 1897; s. Samuel and Lena (Bowman) B.; B.S., Carnegie Inst. Tech., 1919; S.M., Mass. Inst. Tech., 1920; Chem. Engr., Carnegie Inst. Tech., 1928; Ph.D., U. Pitts., 1935; m. Edith Piersol, Sept. 9, 1924; children—Elizabeth (Mrs. John T. Stickney), Carolyn (Mrs. Thomas Morris Perot IV), Henry Hess. With Macbeth Evans Glass Co. (later merged with Corning Glassworks), 1920-41; with Fed. Glass Co., Columbus, O., 1941-63, v.p., dir., 1945-63; v.p. Fed. Paper Bd. Co., Bogota, N.J., 1958-63; prof. glass tech. Ohio State U., Columbus, 1945—. Sci. cons. Joint Chiefs of Staff, 1943-46. Recipient Alumni Merit award Carnegie Inst. Tech., 1965. Fellow Brit. Soc. Glass Tech., Am. Ceramic Soc. (S. B. Meyer Jr. award 1956, Toledo award 1963, A. V. Bleininger Meml. award 1965), A.A.A.S.; mem. Am. Chem. Soc., Am. Inst. Chem. Engrs., Am. Mgmt. Assn., Inst. Ceramic Engrs., Deutsche Glastechnische Gesellschaft. Author: The European Glass Industry, 1946. Research, publs. and U. S. and fgn. patents in glasses for illuminating, optical, signal, infra-red and structural purposes, refractories and processes for melting and fabricating glass products, high temperature phys. chemistry and tech., controlled nucleation and crystalline growths in glass systems and devel. of glass-crystal systems. Home: 55 N. Drexel Av., Columbus, O. 43209. Office: Ohio State U., 2041 N. College Rd., Columbus. 43210.*

BLAU, Ludwig Wilhelm, geophysicist; b. Graben, Germany, Aug. 9, 1894; s. Lewis and Wilhelmina (Suess) B.; came to U. S., 1909, naturalized, 1929; A.B., W. Tex. State Coll., 1925; M.A., U. Tex., 1926, Ph.D., 1929; m. Mollie Jane Herber, June 4, 1922; children—Esther Mollie (Mrs. Victor Grace, Jr.), Margaret Elisabeth (Mrs. Joe D. Clegg). Instr. physics U. Tex., 1926-29; research geophysicist Humble Oil & Refining Co., 1929-30, in charge geophysics research, 1930-37, in charge geophysics, prodn. and petroleum engring. research, 1937-42, cons. patent div., 1941-59. Mem. European Assn. Exploration Geophysicists, Tex. Acad. Sci., Math. Assn. Am., A.A.A.S., Am. Assn. Petroleum Geologists, Am. Geophys. Union. Research, publs., patentee in geophys. and geochem. exploration for petroleum, well-logging; discovered cholesterol-reducing effect of inositol, cherry control of gouty arthritis. Home: 2027 Colquitt Av., Houston 77006.*

BLAU, Monte, Am. chemist; b. Bklyn., June 17, 1926; s. Samuel and Rose (Cohen) B.; B.S., Bklyn. Poly. Inst., 1948; Ph.D., U. Wis., 1952; m. Guitta D. Drimer, June 30, 1946; children—Saul, Hannah, Chemist Geochronometric Lab. Yale U., 1953; chemist div. neoplastic diseases Montefiore Hosp., N.Y.C., 1953-54; with Roswell Park Meml. Inst., 1954—, prin. cancer research scientist dept. nuclear medicine, 1957—. Mem. Soc. Nuclear Medicine, Am. Chem. Soc., Am. Assn. Cancer Research. Research and publs. on application of radioactive materials to cancer diagnosis; developed new radioactive pharmaceuticals for the visualization of brain, pancreas, kidney and bone tumors; developed radioisotope scanning instruments for visualization of tumors. Home: 356 Maynard Dr., Buffalo, N.Y. 14226. Office: 666 Elm St., Buffalo, N.Y. 14203.*

BLAU, Peter Michael, sociologist, educator; b. Vienna, Austria, Feb. 7, 1918; s. Theodor I. and Bertha (Selka) B.; A.B., Elmhurst Coll., 1942; Ph.D., Columbia, 1952; m. Zena Smith, Aug. 7, 1948; 1 dau., Pamela L. Came to U. S., 1939, naturalized, 1943. With Wayne State U., 1949-51, Cornell U., 1951-53; asst. prof. U. Chgo., 1953-58, asso. prof., 1958-63, prof. sociology, 1963—. Predoctoral fellow Social Sci. Research Council, 1948-49; fellow Center Advanced Studies Behavioral Scis., 1962-63; sr. postdoctoral fellow National Science Foundation, 1962; Pitt professor Am. history and instns. Cambridge U., 1966-67. Served with AUS, 1943-45. Decorated Bronze star. Fellow Am. Sociol. Assn.; mem. Internat. Sociol. Assn., Am. Assn. U. Profs., Soc. Study Social Problems. Author: The Dynamics of Bureaucracy, 1955; Bureaucracy in Modern Society, 1956; (with W. R. Scott) Formal Organization, 1962; Exchange and Power in Social Life, 1964; (with Otis Dudley Duncan) The American Occupational Structure, 1967. Editor Am. Jour. Sociology, 1961-66. Research in theory of social exchange as a general process through which group structures develop; analysis of the characteristics of various kinds of formal organizations and bureaucracies; study of social mobility and the American occupational structure. Home: 5737 S. Dorchester Av., Chgo. 60637.

BLAXTER, John Harry Savage, biologist; b. London, Eng., Jan. 6, 1929; s. Kenneth William and Janet (Hollis) B.; 1st class honors degree in zoology Oxford (Eng.) U., 1952; m. Valerie Ann McElligott, Dec. 20, 1952; children—Julia Jane, Timothy John. Sci. officer Marine Lab., Aberdeen, Scotland, 1952-55, sr. sci. officer, 1955-64; lectr. natural history dept. U. Aberdeen, 1964—. Mem. Inst. Biology, Assn. U. Tchrs., Soc. Exptl. Biology, Ecol. Soc., Assn. for Study Animal Behavior, Marine Biol. Assn., Challenger Soc. Research and publs. on rearing of young marine fish including their behavior and physiology, behavior of adult marine fish especially different type of fishing gear, senses in marine fish. Home: 4 Queen's Rd., Aberdeen, Scotland.*

BLAXTER, Kenneth Lyon, Brit. physiologist, agriculturist; b. Norwich, Eng., June 19, 1919; s. Gaspard Culling and Charlotte Ellen (Lyon) B.; B.Sc., U. Reading, 1939, Ph.D., 1945, D.Sc., 1952, Nat. Diploma in Agriculture, 1939; m. M. Lillington Hall, Oct. 10, 1957; children—Alison, Mark, Piers. Sci. officer Nat. Inst. Research Dairying, 1939, 41-44, vet. lab. Ministry Agr., 1945-47; head nutrition dept. Hannah Dairy Research Inst. Scotland, 1949-65; dir. Rowett Research Inst., Aberdeen, Scotland, 1965—; hon. lectr. nutritional physiology U. Aberdeen, 1965—. Recipient Gold medal Royal Agrl. Soc. Eng., 1964, Baxter prize, 1960. Commonwealth Fund fellow, U. Ill., 1947. Fellow Royal Soc. Eng., Inst. Biology; mem. Étranger de'l Académie d'Agriculture de France, Physiol. Soc., Nutrition Soc., Royal Soc. Edingburgh, Soc. Endocrinology, Brit. Soc. Animal Prodn. Author: The Energy Metabolism of Ruminants, 1962; Energy Metabolism, 1965. Research and publs on the discovery of enzootic muscular dystrophy in cattle and its relation to tocopherol metabolism and unsaturated fatty acid metabolism; new methods of adjusting diets for ruminants to accord with their needs on basis of experiments in energy exchanges of animals; quantitative assessments of the effects of climatic components. Home: Wardenhill, Bucksburn, Aberdeen. Office: Rowett Research Inst., Bucksburn, Aberdeen, Scotland.*

BLEAKNEY, Walker, Am. physicist; b. Elderton, Pa., Feb. 8, 1901; s. Robert Wilson and Wilda (Hall) B.; B.S., Whitman Coll., 1924, D.Sc. (hon.), 1955; student Harvard, 1924-25; Ph.D., U. Minn., 1930; grad. study Princeton, 1930-32; m. Dorothy Clyde Thomas, July 16, 1931. Mem. faculty Princeton, 1932-—, prof. physics, 1945-—, chmn. dept., 1960-—. Fellow Am. Phys. Soc., Am. Acad. Arts and Scis.; mem. Nat. Acad. Scis., Phi Beta Kappa, Sigma Xi, Phi Delta Theta. Research on ionization in gases; istotopic constitution of the elements; nuclear physics; mass spectroscopy; fluid mechanics; shock waves; shock tubes. Home: 633 Prospect Av., Princeton, N.J.

BLEANEY, Brebis, English physicist; b. London, Eng., June 6, 1915; s. Frederick and Eva Johanne (Petersen) B.; B.A., Oxford U., 1937, M.A., D.Phil., 1939; m. Betty Isabelle Plumpton, Mar. 15, 1949; children—Michael Francis, Carol Heather. Faculty, Oxford (Eng.) U., 1945-—, fellow St. John's Coll., 1947-57, D'Lee's prof. exptl. philosophy, 1957-—, fellow Wadham Coll., 1957; vis. prof. Columbia U., N.Y.C., 1956-57; mellon vis. prof. U. Pitts., 1962-63. Named commander British Empire, 1965. Fellow Royal Society (Hughes medal 1962), 1950, Physical Soc. London; member of the American Physical Society. Author: (with B. I. Bleaney) Electricity and Magnetism, 1957. Research and publs. on magnetic properties using electron magnetic resonance. Home: Garford House, Garford Rd., Oxford, Eng.*

BLECHSCHMIDT, Erich, German physician; b. Karlsruch, Nov. 13, 1904; s. Eugen and Frieda (Vetter) B.; M.D.; m. Trautemaria Saenger, July 15, 1944; children—Ute, Meinulg, Martin, Cuon. Asst. in anatomy at Friburg, 1930; prosector at Giessen, 1940, Wurtzburg, 1941; full prof., dir. Inst. Anatomy of Göttingen, 1942; prof. U. Göttingen, 1947. Mem. profl. orgns. Author: Die Vorgeburtl. Entwicklungsstadien des Menschen, 1961; Der menschl. Embryo, 1963. Co-editor: Morphol Jahrbuch, also numerous articles. Home: Bruder Grimmallee 36. Office: Kreuzbergring 36, Göttingen, Germany.

BLEEKER, Wouter, Dutch meteorologist; b. Oudshoorn, Netherlands, Nov. 14, 1904; s. Pieter and Antje (Happe) B.; Ph.D. in Natural Scis., U. Utrecht; hon. doctorate U. Fla.; m. B. E. P. Eeuwens, Dec. 19, 1962; children—Frederike, Annie, Pieter. Sci. collaborator Royal Meteorol. Inst., 1930; dir. service of sci. research, 1942; full prof. U. Utrecht, 1946-—. Mem. Royal Dutch Acad. Sci., Am. Soc. Meteorology, Royal Meteorol. Soc. Author: Leerboek der Meteorologie, numerous other publs. Research in synoptic meteorology. Home: Soestoyhse weg 95. Office: Royal Netherlands Meteorological Inst., Ulrechtseweg 297, De Bilt, Netherlands.*

BLEGEN, Carl William, Am. archeologist; b. Minneapolis, Minn., Jan. 27, 1887; s. John H. and Anna B. (Olsen) B.; B.A., Augsburg Sem., Minneapolis, 1904, U. of Minn., 1907; B.A., Yale, 1908, Ph.D., 1920, hon. M.A., 1927; student Am. Sch. of Classical Studies, Athens, Greece, 1910-13; honorary doctorate, University of Oslo (Norway), 1951, Thessalonike (Greece), 1951, U. Athens, 1963; D. Litt., Oxford, 1957; LL.D., U. Cin., 1958; L.H.D., Hebrew Union College, Jewish Inst. Religion, 1963; Litt. D., Cambridge U., 1963; m. Elizabeth Denny Pierce, July 11, 1924 (dec. 1966). Sec. Am. Sch. Classical Studies, 1913-20, asst. dir., 1920-26, actg. dir., 1926-27; professor classical archeology Grad. Sch. Arts and Sciences, University of Cincinnati, 1927-57, prof. emeritus, 1957-—, became fellow, 1927; head department of classics, 1950-57; field dir. University of Cincinnati Archaeol. Expdn., Turkey and Greece; on leave of absence, with Office of Strategic Services, Washington, 1942-45; cultural relations attaché, American Embassy, Athens, Greece, 1945-46; dir. Am. Sch. Classical Studies, Athens, 1948-49. With the Am. Red Cross in Greece, 1918-19. Recipient gold medal Archaeological Inst. Am., 1965, gold medal Soc. Antiquaries of London, 1966. Corresponding fellow of British Academy. Fellow American Academy of Arts and Sciences; mem. Am. Philos. Soc., Am. Philol. Assn., Archeol. Inst. Am., Am. Assn. Univ. Profs., German Archeol. Inst., Archaeol. Soc. Athens (hon. v.p.), Soc. Promotion of Hellenic Studies, London, England (honorary), Royal Soc. Letters of Lund (Sweden), Swedish Royal Acad. Letters, History and Antiquities, Norwegian Academy of Science and Letters, also Phi Beta Kappa (honorary), Sigma Xi. Author: Korakou, A Prehistoric Settlement near Corinth, 1921; Zygouries, A Prehistoric Settlement in the Valley of Cleonae, 1928; Acrocorinth (with R. Stillwell, O. Broneer and A. Bellinger), 1930; Prosymna, the Helladic Settlement Preceding the Argive Heraeum (with Elizabeth Blegen), 1937; Troy, Vol. I (with J. L. Caskey, M. Rawson, J. Sperling), 1950, Troy, Vol. II (with J. L. Caskey and M. Rawson), 1951, Vol. III, 1953, Vol. IV (with C. Boulter, J. L. Caskey, M. Rawson), 1958; Troy and the Trojans, 1963; (with M. Rawson) The Palace of Nestor at Pylos, Vol. I, 1966. Contbr. archaeal. publs. Proposed Bronze Age chronology based on sherds excavated at Korakou; directed excavations of Troy, 1932-38, determining levels and setting VIIa as city of Trojan War; uncovered Linear B tablets in excavation of Pylos, 1939, 1952; developed scientific methods of archeology; investigations altered and added to knowledge of Greek Bronze Age. Home: 9 Plutarch St., Athens 139, Greece. Office: U. Cin., Cin. 21.

BLEGNY, Nicholas de, see de Blegny.

BLEICHER, Maurice, French physician; b. Aix-les-Bains, France, Sept. 9, 1903; s. Camille and Laure (Caffarel) B.; M.D., Med. Sch. at Nancy; m. Ginette Viscat, Feb. 21, 1941; children—Gérard, Bernard, Philippe, Jean-Pierre. Otorhinolaryngologist, former head anatomy work med. sch. at Nancy; former asst. otorhinolaryngology clinic at Nancy hosps.; pres. group med. specialists of Meurthe-et-Moselle. Named laureat of med. sch. of Nancy, Acad. Medicine of Paris. Mem. Order of Physicians of Meurthe-et-Moselle (v.p. council), Assn. Physicians of Meurthe-et-Moselle (v.p. council), Med. Soc. Nancy (v.p.), Biology Soc. (hon.). Research and publs. on med.-surg. anatomy of surrenal glands, anatomy and otorhinolaryngology. Address: 36, rue de la Ravinelle, Nancy (Meurthe-et-Moselle), France.

BLEICHER, Michael Nathaniel, Am. mathematician; b. Cleve., Oct. 2, 1935; s. David B. and Rachel (Faigin) B.; B.S., Cal. Inst. Tech., 1957; M.S., Tulane U., 1959, Ph.D., 1961; Ph.D., U. Warsaw (Poland), 1961; m. Betty M. Isach, June 4, 1957; children—Helene C., Laurence A., Benjamin D. Teaching asst. Tulane U., 1957-60; research asst. U. Warsaw, 1960-61; postdoctoral fellow U. Cal. at Berkeley, 1961-62; asst. prof. U. Wis., Madison, 1962-65, asso. prof., 1965-—. Mem. Am., Polish math. socs., Math. Assn. Am., Sigma Xi. Research and publs. on covering and packing convex sets in plane and higher dimensional space, analytic number theory, abstract algebra, logical found. of set theory. Home: 5133 Flad Ave., Madison, Wis. 53711.*

BLEIL, David Franklin, Am. physicist; b. Detroit, Dec. 4, 1908; s. David Winfred and Ella (Palm) B.; B.S. in Engring., U. Mich., 1934, M.S. in Math., 1937; Ph.D., Mich. State U., 1948; m. Katherine Ackerman, July 4, 1939; children—David Frederick, Richard Elliott, Katherine Susanne. Asst. physicist Mich. State U., 1936, physics instr., 1937-43; research physicist Naval Ordnance Lab., 1943-51, chief solid state div., 1951-52, chief physics dept., 1951-58, asso. tech. dir. for research, 1958-—. Recipient Superior Civilian Service award U. S. Navy Dept., 1963; Distinguished Civilian Service award Sec. Navy, 1966. Fellow Am. Phys. Soc., Washington Philos. Soc.; mem. Am. Inst. Physics (editor electricity and magnetism sect. Handbook 1957). Editor: Natural Electromagnetic Phenomena Below 30 kc/s, 1964. Research, publs. on geophys. prospecting by induced polarization, electricity and magnetisim and solid state physics. Home: 720 Springloch Rd., Silver Spring, Md. 20901. Office: U. S. Naval Ordnance Lab., White Oak, Silver Spring, Md. 20910.*

BLENK, Hermann, German engr.; b. Dec. 9, 1901; s. Gustav and Emilie (Nold) B.; Ph.D., U. Göttingen; m. Martha Schläwicke, Sept. 14, 1946; children—Luise, Marianne, Gertrud. Sci. collaboration, later dir. German Inst. Aero. Tests, Berlin-Adlershof, 1924-36; dir. dept. German Inst. Aero. Research, Brunswick, 1955. Mem. Soc. Aero. and Spatial Sci. (pres. 1952-59, Ludwig-Prandtl prize 1962), Soc. for Applied Math. and Mechanics, Sci. Soc. of Brunswick. Editor: ann. of Sci. Soc. Aeros., 1952-—; Zeitschrift für Flugwissenschaften, 1953-—. Home: Margaretenhöhe 32. Office: DFL, Flughafen, Braunschweig, Germany.

BLEULER, Ernst, physicist; b. Küsnacht, Switzerland, Jan. 4, 1916; s. Ernst and Berta (Zuppinger) B.; Diploma in Physics and Math, Swiss Fed. Inst. Tech., Zurich, 1938; Dr. Sc. naturalium, 1942, privat dozent, 1946; m. Ruth Keller, Mar. 10, 1945; children—Regula, Christina Cornelia, Elizabeth Susan. Came to U. S., 1947, naturalized, 1955. Assistant, Swiss Fed. Inst., Tech., Zurich, 1938-46, privat dozent, 1946-47; vis. prof. Purdue U., Lafayette, Ind., 1947-48, asso. prof., 1948-50, prof., 1950-64, J. S. Guggenheim fellow, 1961-62; prof. physics Pa. State U., University Park, 1964-—. Fellow, Am. Phys. Soc., mem. Am. Assn. Physics Tchrs., Am. Assn. U. Profs., Schweizerische Physikalische, Gesellschaft, Sigma Xi, Pi Sigma. Author: (with G. J. Goldsmith) Expermental Nucleonics, 1952. Editor: (with R. O. Haxby) Methods of Experimental Physics, Vol. II; Electronic Methods, 1964. Research in radioactive decay and nuclear reactions; study of nucleonics and electronics. Home: 115 Outer Dr., State Coll., Penn. 16801. Office: Accelerator Bldg., Pa. State U., University Park, Pa. 16802.*

BLEULER, Manfred Eugen, Swiss physician; b. Zurich, Switzerland, Jan. 4, 1903; s. Eugen Paul and Hedwig (Waser) B.; student univs. Geneva, Zurich; M.D., U. Zurich, 1928; m. Monica Bisaz, Mar. 30, 1946; 1 dau., Tinetta Hedwig. Sekundararzt psychiat. clinic, St. Priminsberg, 1933-38, univ. clinic, Friedmatt/Basel, 1938-42; prof. psychiatry U. Zurich, dir. univ. psychiat. clinic, Burghölzli/Zurich, 1942-—. Mem. Kaiserlich deutsch Acad. der Naturforscher; corr. mem. Royal Medico-Psychol. Assn. (London), Acad. Psychoanalysis (N.Y.), Swiss Soc. Physicians, Swiss Soc. Psychiatry, Am. Psychiat. Assn., German Acad. Physics, Leopoldina. Author: Krankheitsverlauf, Persönlichkeit und Familienbild Schizophrener, 1941; Lehrbuch der Psychiatrie, 1960; Entwicklungen d. Schizophrenielehre seit 1941, 1960; Entwicklungslinien psychiatrischer Praxis und Forschung, 1961. Research, publs. on psychopathology of endocrine patients; course, nature, and therapy of schizophrenia; acute cerebral diseases. Address: Lenggstrasse 31, Zurich 8, Switzerland.*

BLEWETT, John Pauls, physicist; b. Toronto, Ont., Can., Apr. 12, 1910; s. George J. and Clara (Woodsworth) B.; B.A., U. Toronto, 1932, M.A., 1933; Ph.D., Princeton, 1936; m. M. Mildred Hunt, June 9, 1936. Staff, Gen. Electric Research Lab., 1937-46; staff Brookhaven Nat. Lab., Upton, N.Y., 1947-—, sr. physicist, 1947-—. Fellow Am. Phys. Soc., I.E.E.E., N.Y. Acad. Scis. Author: (with M. S. Livingston) Particle Accelerators, 1962. Co-designer of Brookhaven particle accelerators. Home: Point Rd., Bellport, L.I., N.Y. 11713. Office: Brookhaven Nat. Lab., Upton, L.I., N.Y. 11973.*

BLEYER, Benno, German food chemist; b. Lindau, Germany, Feb. 16, 1885; s. Josef and Anna (Mulzer) B.; pharmacy apprentice in Munich, Germany; studied pharmacy and food chemistry U. Munich; student of Th. Paul and W. Prantl; m. Eleonora Kiessling, 1914; m. 2d, Emilie Greiner, 1920; 1 dau. Prof. chemistry Hocschule Weihenstephan, 1923; prof. pharmacy and food chemistry U. Munich (Germany), 1929; dir. Inst. Pharmacy and Food Chemistry, Deutsche Forschungsanstalt für Lebensmittelchemie, also Staatliche Chemische Untersuchungsanstalt. Recipient Josef-König meml. medal, 1945. Editor: Handbuch der Lebensmittelchemie; Handbuch der Milchwissenschaft. Research and numerous publs. on chemistry of beryllium, fermentation process, protein system in milk, vitamin B2; purification and determination of atomic weight of vanadium; developed micromethods of determining smallest amount of iodine, toxicologically important traces of copper, lead, and zinc in medicines and foods; promoted food laws in Germany; developed process for zymurgical extraction of citric acid. Died Munich, Dec. 24, 1945.

BLEZINGER, (Gustav) Adolf, German engr.; b. Ernsbach, Germany, Nov. 14, 1846; s. Christian August and Elise (Weyler) B.; studied iron metallurgy in Clausthal and Berlin, Germany; trained in shop work in 1869; m. Thusnelde Anna Pickhardt, 1881; 2 sons, 3 daus. Engr. in various iron works in Rhineland and upper Silesia; began work as independent engr. in fuel and furnace field, Duisburg, Germany, 1886; developed process of gasifying German low calorie rough lignites which was important to chem. industry in World War I; attempted to increase efficiency of smelting ovens while decreasing coal; developed gas generator with removable and revolving grate which simplified slag removal; Died Ludwigsburg, Germany, Apr. 5, 1927.

BLIEDEN, Harry Richard, Am. physicist; b. Chgo., Feb. 24, 1936; s. Alvin Davis and Ella Mae (Landwirth) B.; B.S., in Physics, Mass. Inst. Tech., 1957; M.S., U. Wash., 1960; Ph.D., Fla. State U., 1962; m. Nancy Baskin, Mar. 29, 1962; 1 dau., Molly Ann. Vis. scientist Niels Bohr Inst., Copenhagen, Denmark, 1962-63; vis. scientist European Orgn. for Nuclear Research, Geneva, Switzerland, 1963-64, staff scientist, 1964-65; staff scientist Brookhaven Nat. Lab., Upton, N.Y., 1965-—. Recipient Karl Taylor Compton award Mass. Inst. Tech., Cambridge, Mass., 1957; NSF fellow, 1962-64; Am. Swiss Found. fellow, 1964-65. Mem. Am. Phys. Soc. Research and publs. on levels in light nuclei and direct nuclear reaction mechanisms, devel. on-line computer digital readout, spectrometer systems with high date rate capabilities; discovery (with others) new charged non-strange bosons.*

BLINC, Robert, Yugoslavian physicist; b. Ljubljana, Yugoslavia, Oct. 31, 1933; s. Leo and Julia (Belak) B.; degree in physics U. Ljubljana, 1958, Ph.D., 1959; postgrad. (Sloan fellow), Mass. Inst. Tech.; m. Majda Stanovnik, Aug. 2, 1958; children—Ales, Marjeta. Research asso. Nuclear Inst. J. Stefan, Ljubljana, 1957-61, head solid state physics lab., 1965-—; docent U. Ljubljana, 1961-65, prof., 1965-—. Vis. prof. U. Wash., Seattle, 1965-66; mem. Yugoslav Fed. Council for Coordination of Research; mem. phys. sci. adv. bd. Yugoslav AEC. Recipient B. Kidric prize in physics for contbn. to quantum theory of ferroelectricity, 1961. Mem. Yugoslav Math. and Phys. Soc., Yugoslav, Am. chem. socs. Editor: Lectures on Solid State Physics, 1963. Editor Proc. 14th Colloque Ampère on Magnetic Resonance and Relaxation, 1966; editor Jour. Math. and Physics, Ljubljana; mem. adv. editorial bd. Physics of Condensed Matter, Solid State Electronics. Research and publs. on quantum theory of hydrogen bonded ferroelectrics, application of magnetic resonance to study of criticial phenomena in ferroelectrics; contbns. to understanding of hydrogen bonding, molecular motion in solids and crystal structure. Home: 1 Kumanovska, Ljubljana. Office: 39, Jamova, Ljubljana, Yugoslavia.*

BLINKS, Lawrence Rogers, Am. biologist; b. Michigan City, Ind., Apr. 22, 1900; s. Walter Moulton and Ella (Rogers) B.; student Kalamazoo Coll., 1918-19, Stanford, 1919-21; B.S., Harvard, 1923, M.A., 1925, Ph.D., 1926; m. Anne Hof, July 27, 1928; 1 son, John Rogers. Mem. faculty Rockefeller Inst., N.Y.C., 1926-33; asso. prof. plant physiology Stanford (Cal.) U., 1933-36, prof. biology, 1936-65, prof. emeritus, 1965, dir. Hopkins Marine Sta., 1943-65; prof. biology U. Cal. at Santa Cruz, 1965-—. Asst. dir. NSF, 1954-55, cons., 1955-66; mem. div. com. for biology and medicine, 1962-66; mem. com. on sci. and pub. policy U. S. Nat. Acad. Sci., 1962-66. Fellow Am. Acad. Arts and Scis., A.A.A.S. (v.p. 1955), Cal. Acad., Zoologisch-Botanische (hon. mem.) Gesellschaft of Vienna; mem. Bot. Soc. Am., Soc. Gen. Physiology (pres. 1952), Nat. Acad. Scis., Am. Soc. Plant Physiologists (Stephen Hales award 1952). Research and publs. on physiology of aquatic plants, accumulation of salts in cells, their elec. properties; demonstrated that elec. resistance of a stimulated cell fell greatly at peak of an action potential, that red pigments of red algae participated in photosynthesis. Home: Route 3, Box 522, Carmel, Cal. 93952.*

BLINOVA, Ekaterina Nikitichna, Russian meteorologist; b. Russia, Dec. 7, 1906; ed. N. Caucasus U. Sr. sci. research asso. Main Geophys. Obs., 1935-45; with Central Inst. Weather Forecasting, Moscow, USSR, 1943-—. Mem. USSR Acad. Scis. (corr.). Author many sci. works on long-range weather forecasting. Studies of conditions of atmospheric front stability, circulation of atmosphere, radiative equilibrium in atmosphere and quantitative explanation of centers of atmospheric action, origin of cyclones and anti-cyclones. Office: Central Inst. Weather Forecasting, Moscow, USSR.

BLIN-STOYLE, Roger John, English physicist; b. Leicester, Eng., Dec. 24, 1924; s. Cuthbert Basil and Ada (Nash) B.-S.; B.A. with honors in Physics, Wadham Coll., Oxford, Eng., 1949, D.Phil., Oxford U., 1951; m. Audrey Elizabeth Balmford, Aug. 30, 1949; children—Helena A., Anthony R. Pressed Steel Co. Research fellow U. Oxford, 1951-53; lectr. math. physics U. Birmingham (Eng.), 1953-54; sr. research officer in theoretical physics U. Oxford, 1954-62, fellow, lectr. physics Wadham Coll., 1956-62; prof. theoretical physics, dean Sch. Math. and Phys. Scis., U. Sussex, Brighton, Eng., 1962-—; vis. asso. prof. physics, Mass. Inst. Tech., 1960; vis. prof. physics U. Cal. at La Jolla, 1960. Fellow Inst. Physics and Phys. Soc.; mem. Am. Phys. Soc. Author: Theories of Nuclear Moments, 1957; also articles. Research on theory of nuclear structure especially electromagnetic phenomena and beta-decay, fundamental theory of beta-decay and manifestation of other weak interactions in nucleus. Home: 14 Hill Rd., Lewes, Sussex. Office: Sch. Math. and Phys. Scis., U. Sussex, Brighton, Sussex, Eng.*

BLISH, Morris Joslin, Am. chemist; b. Lincoln, Neb., Apr. 21, 1889; s. Frank May and Louise (Joslin) B.; B.S., U. Neb., 1912, M.A., 1913; Ph.D., U. Minn., 1915; m. Vera Buell, Apr. 21, 1921; 1 dau., Mary Louise (Mrs. Robert S. Black). With U. S. Dept. Agr., 1915-16, prin. chemist in charge protein dir. Western Regional Research Lab., Albany, Cal., 1939-43; asst. chemist Mont. Agr. Expt. Sta., Bozeman, 1916-22; chmn. dept. agrl. chemistry Agr. Expt. Sta., U. Neb., Lincoln, 1922-39; staff Internat. Minerals and Chem. Corp., Central Research Lab., Skokie, Ill., 1943-55, mgr. research in organic scis. until 1955. Recipient Thomas Burr Osborne medal Am. Assn. Cereal Chemists, 1936, Outstanding Achievement award U. Minn., 1953. Fellow A.A.A.S.; mem. Am. Chem. Soc., Am. Soc. Biol. Chemists, Am. Assn. Cereal Chemists (past pres.), Inst. Food Technologists, Am. Inst. chemists, N.Y. Acad. Scis., Sigma Xi, Alpha Chi Sigma, Gamma Sigma Delta, Phi Gamma Delta, Contbr. numerous articles to profl. jours. Research in field of protein chemistry with emphasis on cereals and cereal products, utilization of surplus agrl. products, poultry nutrition. Home: 330 Maryland Av., Phoenix 85013.*

BLISS, Chester Ittner, Am. biometrician; b. Springfield, O., Feb. 1, 1899; s. Chester Bradley and Henrietta (Ittner) B.; B.A., Ohio State U., 1921; M.A., Columbia, 1922, Ph.D., 1926; postgrad. U. Coll. London, Eng., 1933-35. asso. entomologist U. S. Dept. Agr., New Orleans, 1926-29, entomologist, Mexico City, Mexico, 1928, Whittier, Cal., 1930-33; mem. Inst. for Plant Protection, Leningrad, USSR, 1935-37; cons. biometrician Conn. Agr. Exptl. Sta., New Haven, 1938-—, biometrician, 1940-—; biometrician Storrs (Conn.) Expt. Sta., 1940-53; lectr. Yale, 1942-67, sr. research asso., 1963; vis. prof. Gonville and Caius Coll., Cambridge (Eng.) U., 1953-54, N.C. State Coll., 1946; vis. staff Commonwealth Sci. and Indsl. Research Orgn., Australia, 1961-67, Japanese Union Scientists and Engrs., Japan, 1961, 67, Indian Statis. Inst., 1961-62, UAR Dept. Statistics and Census, Cairo, Egypt, 1962. Fellow Am. Statis. Assn. (past vice president), Royal Statistical Association, A.A.-A.S., Institute Math. Statistics; mem. Biometric Soc. (hon. life, past pres., editorial bd. 1947-65), Entomol. Soc., Ecol. Soc., Soc. Pharmacology, Soc. for Quality Control, Internat. Statis. Inst. Author: The Statistics of Bioassay, 1952; (with D. W. Calhoun) An Outline of Biometry, 1954; Statistics in Biology, Vol. 1 (in press); numerous articles. Research on statis. methods in biology, biol. assay, exptl. design in biology, agr. and med. research. Home: 597 D Prospect St., New Haven 06511. Office: 123 Huntington St., New Haven 06511.*

BLISS, Eleanor Albert, American bacteriologist; born Jamestown, R.I., August 16, 1899; d. William J. A. and Edith G. (West) Bliss; A.B., Bryn Mawr Coll., 1921; Sc.D., Johns Hopkins, 1925; Sc.D. (hon.), Drexel Inst. Tech., 1956. Fellow in medicine Johns Hopkins, Balt., 1925-35, faculty, 1936-52; prof. biology, dean Grad. Sch., Bryn Mawr (Pa.) Coll., 1952-66, dir. 1945-52. adv. Cons. Med. Research div. U. S. Army Chem. Corps, 1945-50; bd. U. Pa., Phila., 1954-59; Fellow Am. Acad. Microbiologists, A.A.A.S.; mem. Bryn Mawr Civic Assn. (exec. bd.), Am. Soc. Bacteriologists, Am. Assn. Immunologists. Author: (with Perrin H. Long) Clinical and Experimental Use of Sulfanilamide, Sulfapyridine and Allied Compounds, 1939. Research and publs. on the hemolytic streptococcus; discovered group F, a minute hemolytic streptococcus of serological group; authority on use of sulfa drugs. Home: 310 Millbank Rd., Bryn Mawr, Pa. 19010.*

BLISS, Gilbert Ames, Am. mathematician; b. Chgo. May 9, 1876; s. George Harrison and Mary Maria (Gilbert) B.; B.S., U. Chgo., 1897, M.S., 1898, fellow, 1899-1900, Ph.D., 1900; hon. Sc.D., U. Wis., 1935; postgrad. U. Göttingen, 1902-03; m. Helen Hurd, June 15, 1912 (dec. Dec. 1918); children—Elizabeth, Ames; m. 2d, Olive Hunter, Oct. 12, 1920. Instr. Math. U. Minn., 1900-02; asso. in math. U. Chgo., 1903-04; asst. prof. math. U. Mo., 1904-05, Princeton, 1905-08; asso. prof. math. U. Chgo., 1908-13, prof., 1913-41, chmn. dept. math. from 1927, Martin A. Ryerson distinguished service prof., 1933-41, prof. emeritus 1941. Mem. Am. Math. Soc. (pres. 1921 22), Nat. Acad. Scis., Am. Philos. Soc., Am. Acad. Arts and Scis., Sigma Xi, Phi Beta Kappa. Author: Princeton Colloquium Lectures, 1913; Calculus of Variations, 1925; Algebraic Functions, 1933; Mathematics for Exterior Ballistics, 1944; Lectures on the Calculus of Variations, 1946. Asso. editor Annals of Mathematics, 1906-08, Trans. Am. Math. Soc., 1908-16. Died May 8, 1951.

BLISS, Nathaniel, Brit. astronomer; b. Bisley, Eng., Nov. 28, 1700; s. Nathaniel Bliss; B.A., Pembroke Coll., Oxford (Eng.) U., 1720, M.A., 1723; 1 son, John. Rector, St. Ebbe's, Oxford, from 1736, Savilian prof. geometry, from 1742; royal astronomer, 1762-64. Fellow Royal Soc., 1743. Observer transit of Venus, 1761, annular eclipse, 1764. Died Sept. 2, 1764.

BLITZER, Leon, Am. physicist; b. N.Y.C., Dec. 13, 1915; s. Jacob and Rebecca (Tropp) B.; B.S., U. Ariz., 1938, M.S., 1939; Ph.D., Cal. Inst. Tech., 1943; m. Paulin Meyer, June 21, 1942; children—Charles R., Miriam G. Instr., Cal. Inst. Tech., 1943-45; faculty U. Ariz., Tucson, 1946-—, prof. physics, 1950-—. Cons. TRW Systems, 1955-—. Mem. Am. Phys. Soc., Am. Assn. Physics Tchrs., Am. Geophys. Union, A.A.A.S. Author: (with L. Davis, Jr., J. W. Follin, Jr.) Exterior Ballistics of Rockets, 1958. Research and publs. on spectroscopy, exterior ballistics rockets, satellite orbit pertubation theory. Home: 2902 Calle Glorietta, Tucson 85716.*

BLIVAISS, Ben Burton, Am. physiologist; b. Chgo., July 4, 1917; s. Max and Esther (Zoot) B.; B.S., U. Chgo., 1938, M.S., 1940, Ph.D., 1946; m. Helen A. Friedman, Dec. 22, 1946; children—David, Mitchell, Howard. Research asst. U. Chgo., 1946, guest investigator, 1947 48; instr. physiology U. Ill., Urbana, 1946-47; faculty dept. physiology Chgo. Med. Sch., 1948-—, professor, 1966-—. Mem. Aerospace Med. Assn., Am. Fedn. for Clin. Research, A.A.A.S., Am. Physiol. Soc., Am. Soc. Zoologists, Endocrine Soc., N.Y. Acad. Sci., Soc. for Exptl. Biology and Medicine, Sigma Xi. Research and publs. on interrelations thyroid and gonads to devel. secondary sex characteristics, exptl. induction testicular tumors, relation adrenal cortex secretion to reproduction, effect simulated aerospace flight on adrenal cortex secretions, exptl. arthritis anti-inflammatory drugs, exptl. diabetes. Home: 7546 S. Clyde Av., Chgo. 60649.*

BLIXENKRONE-MöLLER, Niels, Danish Surgeon; b. Agerso, Dec. 2, 1907; s. Herman and Marie (Hansen) B.-M.; M.D., U. Copenhagen; m. Edith Larsen, June 28, 1941; children—Birgit, Helene, Niels. Chief surgeon communal hosp. of Aarhus, 1950-—; prof. surgery U. Aarhus. Named Jutlandia physician. Mem. nat. and internat. assns. of surgery, Collegium Regium Chirurgorum Universitatum Daniae. Research and numerous publs. on surgery, especially gastroenterology. Home: Lerbaekvej 16, Risskov. Office: Kommunehospital, Aarhus, Denmark.

BLIZARD, Everitt Pinnell, Am. physicist; b. Ottawa, Ont., Can., Sept. 30, 1916; s. John and Ethel (Risque) B.; brought to U. S., 1920, naturalized, 1938; B.S., Wesleyan U., 1937; M.A., Columbia U., 1938; D.Sc., Ottawa U., 1965; m. Barbara Rogers, Jan. 1, 1943; 1 son, Steven. Engring. physicist Foster Wheeler Ltd., London, Eng., 1938; research asst. Columbia U., 1939-41; physicist U. S. Navy, 1941-46; physicist group leader shielding research physics div. Oak Ridge Nat. Lab., 1946-54, asso. dir. physics div., 1954-55, dir. neutron physics div., 1955-—. Participant, Project Harbor, Nat. Acad. Scis., Woods Hole, Mass., 1963, Project Nobska, U. S. Navy, Nat. Acad. Scis., Woods Hole, Mass., 1956; sci. mem. U. S. Atoms for Peace Mission to Far East, 1957; ofcl. del. 2d, 3d Internat. Confs. on the Peaceful Uses of Atomic Energy, Geneva, Switzerland, 1958, 64; chmn. panel on reactor shielding IAEA, Vienna, Austria, 1964. Fellow Am. Phys. Soc., Am. Nuclear Soc., Nat. Acad. Scis. (com. underseas warfare, 1956-65, subcom. nat. com. on radiation protection and measurements, 1954-65, subcom. on techniques for distbn. of sci. information, 1964-65, subcom. on radiation shielding, adv. com. civil def., 1962-65, subcom. on symbols, units, nomenclature of com. on nuclear sci. div. phys. scis., 1960-65), Am. Standards Assn. (tech. subcom. 1964-65), Research and Engring. Soc. Am. Contbr. articles to sci. jours. Editor, Nuclear Sci. and Engring. Jour. Am. Nuclear Soc., 1959-65; editor Reactor Handbook, Vol. III, Part B., Sheilding, 2d edit., Interscience, 1962; editorial bd. Engineering Compendium of Radiation Shielding, IAEA, 1962-65. Pioneer in reactor shielding; shielding studies for first nuclear submarine; shielding studies for proposed nuclear propulsion aircraft, space vehicles; pioneer in organizing studies of shielding against nuclear weapons radiation. Died Feb. 22, 1966.

BLOCH, Bruno, Swiss dermatologist; b. Oberendingen, Switzerland, Jan. 19, 1878; s. Lemann and Mathilde (Guggenheim) B.; student medicine, Basel, Switzerland; m. Marguerite Bollag; 3 daus. Worked in skin clinic of Josef Jadassohn, Bern, Switzerland. chief physician Basel dermatol. dept., 1906; asso. prof., head new dermatol. clinic in 1913; went to Zurich, Switzerland, in 1916 where he established a clinic of his own design in 1924; a founder (with Jadassohn) of research in dermatology; one of the founders of biologically oriented dermatology; research on cancer, parasitic skin diseases, eczema, as an allergy, origin of melanin as main pigment. Died Zurich, Apr. 10, 1933.

BLOCH, David P., Am. biologist; b. Chgo., Feb. 10, 1926; s. Peter and Clare (Perskie) B.; B.S., Northwestern U., 1949; Ph.D., U. Wis., 1952; m. Jacqueline deGoumois, Oct. 31, 1952; children—Peter, Deirdre, Elizabeth. With Columbia Coll. Physicians and Surgeons, 1952-55, Am. Cancer Soc. fellow, 1952-54, Damon Runyon Found. fellow, 1954-55; asst. prof. zoology U. Cal., Los Angeles, 1955-61; prof. botany U. Tex., Austin, 1961-—. Guggenheim fellow U. Geneva, Switzerland, 1964-65. Recipient USPHS Research Career award, 1963-—. Mem. Internat. Soc. Cell Biology, Am. Soc. Cell Biologists, Soc. Study Devel. and Growth, A.A.A.S., Sigma Xi. Research on role of histones as possible regulators of gene activity during cell differentiation. Home: 3709 Gilbert St., Austin, Tex. 78703.*

BLOCH, Edward Henry, anatomist, physiologist; b. Berlin, Germany, Feb. 1, 1914; s. Ernst E. and Louise (Ehmer) B.; came to U. S., 1923, naturalized, 1936; B.S., U. Chgo., 1939, Ph.D., 1949, M.D., U. Tenn., 1946. Established investigator Am. Heart Assn., 1950-55; faculty Western Res. U., Cleve., 1953-—, associate professor of anatomy, 1956-—. Recipient Honors Achievement award Angiology Research Found., 1964-65. Mem. Am. Assn. Anatomists, A.A.A.S., Microcirculatory Soc., Am. Assn. Immunologists, Am. Heart Assn. Am. Phys. Soc., N.Y. Acad. Scis. Microcirculatory Soc. (exec. com. 1954-57, 64-—). Research and publs. on living structure of organs in body by direct observation, effect human diseases on flow properties of blood, study living events with high speed photography, devel. methods for study living structures at microscopic level. Home: 11428 Cedar Glen Pkwy., Cleve. 44106.*

BLOCH, Eugène, French physicist; b. Soultz, France, 1878; ed. Ecole normale; asst. Coll. de France; became prof. Lycée St. Louis, Paris, 1906; named lectr. Ecole normale, 1922; prof. theoretical and celestial physics Sorbonne. Author: La théorie cinétique; La théorie des quantas. Research on kinetic and quantum theories, ionization of air by phosphorus, spectography of ultraviolet rays, photo-electric effect of ultraviolet light on metal and dust. Died in concentration camp, 1944.

BLOCH, Felix, physicist; b. Zurich, Switzerland, Oct. 23, 1905; s. Gustav and Agnes (Mayer) B.; student Technische Hochschule, Zurich, 1924-27; Ph.D., U. Leipzig, 1928; hon. degree, Grenoble, 1959, Oxford, 1960, Jerusalem, 1962; m. Lore Clara Misch, Mar. 14, 1940; children—George J., Daniel A., Frank S., Ruth H. Came to U. S., 1934, naturalized, 1939. Asst. in physics Zurich, 1928-29; privatdozent, Leipzig, 1932-33; prof. physics Stanford U., (Cal.), 1934-—. Investigator Manhattan dist. Radio Research Inst. Harvard, 1942-45; dir. gen. Cern, Geneva, 1954-55. Recipient Nobel prize in physics, 1952. Horents fellow, Utrecht, Oersted fellow, Copenhagen, Rockefeller fellow Rome, 1928-34. Fellow Am. Phys. Soc. (pres. 1965); mem. Nat. Acad. Scis., Royal Dutch Acad. Exptl. and theoretical researcher in atomic and nuclear physics; developed nuclear induction method of measuring magnetic field of atomic nuclei. Home: 1551 Emerson St., Palo Alto, Cal. 94301.*

BLOCH, Hubert, Swiss microbiologist; b. Basle, Switzerland, May 26, 1913; s. Max and Susanne (Weil) B.; M.D., U. Basle, 1938; m. Friederike Barth, July 13, 1942; children—Judith S., Thomas P. Came to U. S., 1947, naturalized, 1955. Faculty, Basle U. Med. Sch., 1940-48, asst. prof. bacteriology and immunology, 1940-48; staff Pub. Health Research Inst., City of N.Y., Inc., 1948-56, chief div. Tb., 1953-56; vis. investigator Rockefeller Inst. Med. Research, N.Y.C., 1948-50; prof. microbiology, head dept. Sch. Medicine, U. Pitts., 1956-61; dir. research pharm. div. CIBA Ltd., Basle, Switzerland, 1961—. Spl. cons. USPHS, 1950-56; mem. various expert coms. WHO; mem. research council Fonds Nat. Suisse de la Recherche Scientifique, 1962—. Mem. Am. Acad. Microbiology, Deutsche Tuberkulosegesellschaft, Internat. Soc. Cell Physiology, Soc. Gen. Microbiology. Research and numerous publs. on Tb, immunology. Home: 25 St. Alban-Anlage, Basle. Office: CIBA Ltd., 141 Klybechstrasse, Basle, Switzerland.*

BLOCH, Iwan, German physician; b. Delmenhorst, Germany, Apr. 8, 1872; s. Louis and Rosette (Meyer) B.; ed. univs. Bonn, Heidelberg, Berlin; degree Würzburg, 1896; m. Rosa Heinemann; 1 son, Robert; m. 2d, Lisbeth Kuhn. Hosp. physician in dermatology, Berlin; lectr. evening sch. adult edn., Berlin-Lichtenberg. Author: Der Ursprung der Syphilis, 1901; Beitrage zur Aetiologie der Psychopathia sexualis, 1903; Das Sexualleben unserer Zeit, 1912; Die Praxis der Hautkrankheiten, 1908; Die Prostitution, 1912; (under pseudonym Eugen Dühren) Der Marquis de Sade und seine Zeit, 1899, Das Geschlechtsleben in England . . . , 1901-03. Editor: Handbuch der Sexualwissenschaft, from 1911. Founder of sexology; pioneer of sexual reform; exponenet Morbus Americanus theory that syphilis originated in Am. Died Berlin, Nov. 19, 1922.

BLOCH, Jack Herbert, Am. surgeon; b. N.Y.C., Nov. 25, 1930; s. Moses and Rosamond (Klein) B.; B.A. with high honors, U. Cal. at Los Angeles, 1952, M.D. with high honors, 1957; Ph.D. in Surgery U. Minn. Med. Schs., 1967; m. Gretchen R. Rokahr, July 14, 1959; children—Robert David, Paul Henry. Clin. asso. Nat. Cancer Inst., NIH, Bethesda, Md., 1958-60, postdoctoral fellow Nat. Heart Inst., 1963-66; fellow surgery U. Minn. Med. Sch., Mpls., 1960-67; established investigator, 1966—, asst. prof. surgery, 1967; asst. prof. surgery Cornell U. Med. Center, New York City, 1968—. asst. dir. Cancer Detection Center, U. Minn. Hosps., 1961-63. Recipient Hektoen Gold medal A.M.A., 1964. Mem. Phi Beta Kappa, Alpha Omega Alpha, Phi Delta Epsilon. Research and publs. on cancer biology especially tumor blood supply, preservation of transplantable tissues and organs by hypothermia or freezing and hyperbaric oxygenation, problems in surg. infection and wound healing, pathophysiology and treatment of shock syndrome. Home: 829 24th Av. S.E., Mpls. 55414. Office: N.Y. Hosp., Cornell U. Med. Center, 1300 York Av., N.Y.C. 10021.*

BLOCH, Konrad Emil, biochemist; b. Neisse, Germany, Jan. 21, 1912; s. Frederick D. and Hedwig (Streimer) B.; Chem.Eng., Technische Hochschule, Munich, Germany, 1934; Ph.D., Columbia, 1938; m. Lore Teutsch, Feb. 15, 1941; children—Peter, Susan. Came to U. S., 1936, naturalized, 1944. Instructor, Columbia U., 1939-46; asst. prof. biochemistry U. Chgo., 1946-50, prof., 1950-54; Higgins prof. biochemistry Harvard, 1954—. Recipient Nobel Prize in physiology and medicine (with F. Lynen), 1964. Fellow Am. Acad. Scis.; mem. Nat. Acad. Scis., Am. Chemical Soc. (Fritzsche Award, 1964), Harvey Soc. Author: Lipide Metabolism, 1961. With others, demonstrated that acetic acid furnishes building blocks for cholesterol in animal tissue, 1942; demonstrated acetate participation in synthesis of extended and branched chains and acetate's role in biosynthesis of steroids, 1950; showed sterols produced by ring closure from squalene, 1953; with others, illuminated sequence of cholesterol biosynthesis, 1946-58; early used radioactive labelling; important research on role of cholesterol in animal metabolism; investigated saturated and unsaturated acids illuminating biosynthesis of unsaturated fatty acids. Home: 16 Moon Hill Rd., Lexington, Mass. Office: 38 Oxford St., Cambridge, Mass.

BLOCH, Marcus Eliesar, German physician, ichthyologist; b. Ansbach, Germany, 1723; practiced medicine, Berlin. Author: Allegemeine Naturgeschichte der Fische, 12 vols., 432 plates in color, 1782-95; Systema ichthyologiae, 1801. Pioneer in research on intestinal worms and treatment for them; studied life cycle of fish, especially non-European fish. Died Karlsbad, Germany, Aug. 6, 1799.

BLOCH, Michel A., French physicist; b. Strasbourg, Mar. 22, 1930; s. Armand and Madeleine (Salomon) B.; Diplóme d'Ingenieur, École Polytechnique, Paris, 1952; Ph.D., Cal. Inst. Tech., 1958; m. Janine Levy, June 28, 1961; children—Thierry, Nathalie. Physicist physics labs. École Polytechnique; nuclear physics labs. Collège de France (both Paris), 1958—; lectr. École Polytechnique, 1964— with European Center Nuclear Research, Geneva, Switzerland, French AEC, Saclay. Research, publs. on high energy physics and elementary particles, photo-prodn. of pions, instrumentation in fields of bubble chambers and picture analysis. Home: 60 Rue Madame, Paris 6. Office: Nuclear Physics Lab., Collège de France, Paris 5, France.*

BLOCHMANN, Friedrich Johann Wilhelm, German zoologist; b. Karlsruhe, Germany, Jan. 21, 1858; s. Friedrich and Katherine Johanna (Zobel) B.; student in Karlsruhe; degree under O. Bütschli from U. Heidelberg (Germany), 1881; m. Anna Winkelmann, 1891; 3 sons, 2 daus.; lectr. at U. Heidelberg, Germany, 1885; asso. prof. zoology in Rostock, Germany, 1888; went to Tübingen, Germany, 1898; author of Untersuchungen über den Bau der Brachiopoden, 1892/1900; Die mikroskopische Tierwelt des Süsswassers, I: Protozoa, 1895; continued Butschli's Vorlesungen über vergleichende Anatomie (lectures on comparative anatomy); research on freshwater protozoa, evolution, parthenogenesis; classical representation of orgn. of brachipodes; symbiotic assn. of bacteria with a mollusk. Died Tübingen, Sept. 22, 1931.

BLOCK, George Edward, Am. surgeon; b. Joliet, Ill., Sept. 16, 1926; s. Edward and Florence (Hyland) B.; student Northwestern U., 1944-47; M.D., U. Mich., 1951, M.S., 1958; children—George Edward, John R. Mem. faculty U. Mich., 1958-60; faculty U. Chgo., 1960—, attending surgeon, 1960—, asso. prof. surgery, 1963-67, prof. surgery, 1967—; attending surgeon Cook County Hospital, Chicago, since 1967—. Recipient of McClintock award U. Chgo., 1964. Diplomate Am. Bd. Surgery. Fellow Am. Cancer Soc., A.C.S.; mem. Central, Western surg. assns., Soc. Surgery Alimentary Tract, Internat. Soc. Surgery. Research and publs. on hormonal control mammary carcinoma, treatment malignant melanoma, carcinoma of esophagus and stomach utilizing operation and irradiation, surg. therapy for various blood clotting abnormalities, anatomic approaches to dissection of lymph nodes of upper extremity. Home: Route 2, Yorkville, Ill. Office: 950 E. 59th St., Chgo. 60637.*

BLOCK, Henry David, Am. engr., mathematician; b. N.Y.C., Feb. 22, 1920; s. Isaac and Celia (Gottschall) B.; B.S. cum laude, Coll. City N.Y., 1940, B.C.E., 1943; M.S., Ia. State U., 1947, Ph.D., 1949; m. Phoebe T. Goggin, May 12, 1946; 1 son, David Lee. Exptl. flight test engr., stress analyst Goodyear Aircraft Corp., Akron, O., 1943-45; aerodynamicist Fairchild Engine & Aircraft Corp., Jamaica, L.I. N.Y., 1946; teaching fellow Ia. State U., 1946-47, faculty, 1947-53; asst. prof. U. Minn., 1953-55; faculty Cornell U., Ithaca, N.Y., 1955—, prof., 1961—. Mem. Am. Math. Soc., Math. Assn. Am., Soc. for Indsl. and Applied Math., N.Y. Acad. Scis., Assn. Computing Machinery, I.E.E.E., A.A.A.S., Phi Beta Kappa, Sigma Xi, Tau Beta Pi, Pi Mu Epsilon, Phi Kappa Phi. Author: An Introduction to tensor Analysis, 1962. Research and publs. on theoretical and applied math, statistics, math. econs.; research in theory of automata, control, learning systems, cybernetics, theoretical biology. Home: 23 Fairview Sq., Ithaca, N.Y. 14850.*

BLOCK, Martin M., Am. physicist; b. Newark, Nov. 29, 1925; s. George Perry and Sue (Ehrenkranz) B.; B.S., Columbia, 1947, M.A., 1948, Ph.D., 1951; m. Beate Sondhelm, Sept. 4, 1949; children—Steven Michael, Gail Frances. Instr., Columbia, 1947-48, research asst., 1947-51, scientist, 1951; research asso. Duke, 1951, asst. prof., 1952-57, asso. prof., 1957-61; prof. physics Northwestern U., Evanston, Ill., 1961—, chmn. dept., 1961-66. Guggenheim fellow, 1958-59. Fellow of the American Physical Society; member of the Italian Physical Society, N.Y. Acad. Sci., A.A.A.S., Sigma Xi, Sigma Pi Sigma, Pi Mu Epsilon. Contbr. numerous articles to sci. jours. Research elementary particle physics; inventor of liquid helium bubble chamber, measured parity of K meson; co-discoverer eta meson. Home: 624 Colfax St., Evanston, Ill. 60201.

BLOCK, Matthew Harold, Am. physician; b. Bklyn., Dec. 13, 1915; s. Leon and Ida (Frank) B.; B.S., Coll. City N.Y., 1937; M.S., U. Mich., 1938; Ph.D., U. Chgo., 1941, M.D., 1943; m. Ruth Lucas, June 2, 1942; children—William A., Leslie L., Richard A., Ilene B. Fellow, USPHS, 1947-49; asst. prof. medicine U. Chgo., 1949-53; faculty U. Colo. Med. Sch., Denver, 1953—, now prof. medicine. Cons. AEC, 1949-53, VA, Army, USAF hosps. Recipient Capps prize Chgo. Inst. Medicine, 1947; Med. Research award U. Chgo. Med. Alumni, 1964. Mem. Am., Central socs. clin. research, Am. Soc. Exptl. Pathology, Western Assn. Physicians, Am. Assn. Cancer Research. Research and numerous publs. on embryology and structure of hematopoietic tissues, clin. hematology, pathology of blood forming tissues. Home: 3921 S. Dexter St., Edgewood, Colo. 80110. Office: U. Colo. Med. Center, Denver 80220.*

BLOCK, Walter David, Am. chemist; b. Dayton, O., Oct. 16, 1911; s. Samuel and Minna (Larison) B.; B.S., Ohio State U., 1933; M.S., U. Mich., 1934, Ph.D., 1938; m. Thelma V. Levine, Sept. 7, 1941; children—Robert M., Margery E. Faculty, Med. Sch. U. Mich., Ann Arbor, 1939—, asso. prof. biochemistry, 1952—, also charge biochem. research labs. dept. dermatology; biochem. cons. Caylor Nickel Med. Research Found., 1958—, Ft. Wayne State Sch., 1960-63. Recipient Internat. Meml. award, 1959. Fellow A.A.A.S.; mem. Am. Soc. Biol. Chemists, Soc. Exptl. Biology and Medicine, Am. Inst. Nutrition, Am. Chem. Soc., Am. Heart Assn., Sigma Xi, Phi Lambda Epsilon. Author: (with K. VanGoor) Pharmacology and Therapeutic Uses of Gold Compounds, 1956; (with A. C. Curtis) Biochemical Changes in the Dermatitides, 1960. Research and publs. on amino acid metabolism in humans, metabolism of methionine and other sulfur amino acids in normal human adults, amino acid metabolism in diseases of the skin, protein nutrition in humans, pharmacology of gold compounds and epidermal enzymes, lipid and lipoprotein metabolism in various lipid abnormalities. Home: 1335 Glendaloch Circle, Ann Arbor, Mich. 48104.*

BLODGETT, Earle Comstock, Am. plant pathologist; b. Logan, Ia., Sept. 1, 1906; s. Charles E. and Irene Ramona (Osgood) B.; B.S., U. Ida., 1929, M.S., 1930; Ph.D., U. Wis., 1934; m. Ena Faye Colvin, June 16, 1931; children—James Comstock, John Emory, Ruth Jean. Plant pathologist Wash. State U., Wash. Dept. Agr., Prosser, Moxee City Quarantine Sta., 1946—; plant pathologist Wash. Dept. Agr., 1962—; FAO assignment to Israel, 1960-61, Turkey, 1965-67. Mem. Am. Phytopath. Soc., Am. Soc. Hort. Sci., Am. Pomol. Soc., Am. Inst. Biol. Sci., Sigma Xi, Alpha Xi, Alpha Zeta. Author: (with others) Handbook No. 10, U. S. Dept. Agr., 1951. Contbr. numerous articles to profl. jours. Research on tree fruit virus diseases; horticulture; agriculture. Home: 1309 Patterson Rd. Office: Research and Extension Center, Prosser, Wash. 99350.*

BLODGETT, Katharine Burr, Am. physicist and chemist; b. Schenectady, N.Y., Jan. 10, 1898; d. George Reddington and Katharine Buchanan (Burr) Blodgett; A.B., Bryn Mawr Coll., 1917; S.M., U. of Chicago, 1918; Ph.D., Cambridge (Eng.) U., 1926; D.Sc. (hon.), Elmira Coll., 1939, Brown U., 1942; Western Coll., 1942, Russel Sage Coll., 1944. Worked with Irving Langmuir in research lab., Gen. Elec. Co., Schenectady, 1918-24, 26—; Francis P. Garvan medal, 1951. Fellow Am. Phys. Soc.; mem. Optical Soc. Am. Inventor non-reflecting (invisible) glass; developed methods of constructing films of infinitesimal thickness; device for measuring thickness of films within one micro-inch; new kind of smoke-screen. Home: 18 N. Church St., Office: Research Laboratory, General Electric Co., Schenectady.

BLOEMBERGEN, Nicolaas, Am. physicist, educator; b. Dordrecht, Netherlands, Mar. 11, 1920; s. Auke and Sophia M. (Quint) B.; B.A., Utrecht U., 1941, M.A., 1943; Ph.D., Leiden U., 1948; M.A. (hon.), Harvard, 1951; m. Huberta D. Brink, June 26, 1950; children—Antonia, Brink, Juliana. Came to U. S., 1952, naturalized, 1958. Teaching asst. Utrecht U., 1942-45; research fellow Leiden U., 1948; mem. Soc. Fellows, Harvard, 1949-51, asso. prof., 1951-57, Gordon McKay prof. applied physics, 1957—. Recipient Buckley prize for solid state physics Am. Phys. Soc., 1958; Morris Liebmann award I.R.E., 1959; Stuart Ballantine medal Franklin Inst., 1961; Fellow Am. Phys. Soc., Nat. Acad. Scis., Am. Acad. Arts and Scis., I.E.E.E.; mem. Nat., Royal Dutch acads. scis., Dutch Physicists Soc. Author: Nuclear Magnetic Relaxation, 1948; Nonlinear Optics, 1965; also articles profl. jours. Developed method of three-level and multi-level pumping to energize masers; research in nuclear magnetic resonance, ferromagnetic resonance. Home: 3 Stonewall Rd., Lexington 73, Mass. Office: Pierce Hall, Harvard Univ., Cambridge, Mass. 02138.

BLOEMENDAL, Hans, biochemist; b. Fulda, Germany, Feb. 4, 1923; s. Michel and Minna (Hess) B.; B.Sc., U. Amsterdam, 1951, M.Sc., 1954, Ph.D., 1957; m. Rachel Braasem, Apr. 16, 1947; children—Betty, Michael, Asher. Asst., Lab. Embryology and Anatomy, U. Amsterdam, 1954-57; sr. research officer dept. biochemistry Netherlands Cancer Inst., 1957-65; prof. biochemistry U. Nymegen, The Netherlands, 1965—. Guest investigator Weizmann Inst., Israel, 1960, Rockefeller Inst., N.Y.C., 1964. Mem. Koninklijke Nederlandse Chemische Vereniging, Biochem. Soc., Council for Def. against Cancer. Author: Electrophoresis in Blocks and Columns, 1963. Research and publs. on isolation, structure and biosynthesis of proteins from eye lens, protein synthesis in rat liver, discoverer substructure of lens protein alpha crystallin. Home: 56 Willemsparkweg, Amsterdam, The Netherlands. Office: Dept. of Biochemistry, University of Nijmegen, Nijmegen, The Netherlands.*

BLOK, Harmen, Dutch mech. engr.; b. Amsterdam, Netherlands, Sept. 8, 1910; s. Pieter E. J. and Wikje H. (Poort) B.; ed. Superior Polytech. Sch. of Delft; degree in mech. engring.; m. Aleida Schut, Jan. 30, 1935; children—Pieter J. J., Hendrika A., Hendrik P., Johannes, Jacobus. Engr. in charge research on fundamental aspects of rubbing, usage and grease Delft lab. of Royal Dutch Shell, 1935-51; prof. mechanics Superior Poly. Sch. of Delft, charge lab., 1951—. Mem. Royal Inst. Engrs. (La Haye), Am. Soc. M.E. Author: Les températures de surface dans des conditions de graissage sous extrême pression, 1937; Fundamental Mechanical Aspects of Boundary Lubrication, 1940; Dissipation of frictional heat, 1955; Lubrication as a Gear Design Factor, 1958; Inverse Problems in Hydrodynamic Lubrication with Applications to Flexible Rubbing Surfaces. Home: Haagweg 192, Rijswijk. Office: Mekelweg 2, Delft, Netherlands.

BLOK, Jan, Dutch physicist; b. Leyde, 1918; ed. U. Amsterdam; m. J. E. Bisschop, 1951. Prof. secondary sch., 1943-48; prof. physics U. Amsterdam, 1948, also researcher Natuurkundig Lab. Research and publs. on nuclear and solid state physics, scattering of thermal neutrons by bound protons and deuterons. Address: Koninginneweg 172, Hs Amsterdam-Z., Netherlands.

BLOKHIN, Nikolai Nikolaevich, Russian surgeon; b. Lukoyanov, Nizhny Novgorod Guberniya, 1912; s. dr.; grad. Gorky Med. Inst., 1934; D.Med. Sci., 1946. Intern, postgrad. chair hosp. surgery, later asst. chair operative and hosp. surgery Gorky Med. Inst., 1934-41, founder Inst. Restorative Surgery, Orthopedics and Traumatology, head its clinic hosp. surgery, 1946, head chair gen. surgery, dir., 1950-52; prof., 1947-—; dir., head clin. dept. Moscow Inst. Exptl. and Clin. Oncology, USSR Acad. Med. Sci., 1952-—. Del., Internat. Congress on Plastic Surgery, Stockholm, 1955; head Soviet delegation Internat. Congress on Problems Phys. Tng., Brussels, 1958, Internat. Congress on Combatting Malignant Neoplasms, London, 1958; mem. exec. com. Internat. Anticancer Union and Internat. Union on Problems Plastic Surgery; pres. 8th Internat. Anticancer Congress, Moscow, 1962; del. Soviet-Am. Session on Cancer, N.Y.C., Internat. Conf. on Cancer, Tokyo. Decorated Order of Lenin; recipient Burdenko prize, 1956. Mem. USSR Acad. Med. Sci. (pres. 1960-—), Purkinje Czech Med. Soc. (hon.), Polish Acad. Sci. (fgn.), All-Union Soc. Oncologists (dep. chmn.). Author over 50 works including Skin Plastics in the Surgery of War Injuries, 1946; Skin Plastics, 1955. Editor: Vestnik AMN SSSR; co-editor Surgery sect. Large Med. Ency., 2d edit; mem. editorial bd. Problems of Oncology. Research on surg., chemotherapeutical and combined methods of treating tumors; developer numerous plastic operations. Address: Moscow Inst. Exptl. and Clin. Oncology, USSR Acad. Med. Sci., I-110, 3-ya Meshchanskaya 61/II, Korpus i, Moscow, USSR.

BLOKHINTSEV, Dmitrii Ivanovich, Russian physicist; b. Moscow, Russia, Jan. 11, 1908; s. Ivan Dmitrivich and Elisabeth (Zuikova) B.; grad. Moscow U., 1930; D.Theoretical Physics, 1937; D. H.C., Leipzig U., 1962, High Czechoslovakian Tec-. Sch., 1957; m. Seraphina Drapkina, 1930; children—Leonid, Tatiana, Igor. Research student Physics Faculty, Moscow U., 1930-33, docent, 1933-35, prof. dept. nuclear physics, 1937-—. Collaborator theoretical dept. Physics Inst., USSR Acad. Sci., 1935-50; dir. Physics-Energy Inst, USSR Atomic Energy Commn., 1950-56; dir. Joint Inst. Nuclear Research, Dubna, 1956-64, dir. theoretical dept., 1964-—. Recipient Stalin prize, 1952; Lenin prize for setting up 1st atomic power plant in world, 1957. Mem. USSR (corr.), Ukrainian (corr.), Hungarian, German acads. sci., Internat. Union Pure and Applied Physics (pres. 1966). Author: (with S. Drapkina) Theory of Relativity, 1940; Quantum Mechanics, 1944; Acoustics of Moving and Non-uniform Medium, 1946; Principal Problems of Quantitative Mechanics, 1966. Research, numerous publs. on theory of phosphoresence of solids on basis of quantum theory of semicondrs., 1934, devel. phenomenological theory of solid rectifiers, 1938, theoretical investigations in region of acoustics of moving medium, 1946, applied nuclear physics, constructed 1st periodic pulse reactor, 1957, problems of quantum mechanics, theory of elementary particles especially problems of causality and geometry in microworld. Home: Veksler 1, Dubna, USSR. Office: Head Post Office P.B. 79, Moscow, USSR.*

BLOM, Jakob, Danish chemist; b. Notmark, Nov. 22, 1898; s. Hans and Katrine (Mette) B.; ed. univs. Kiel, Copenhagen, Zurich, Rutgers U.; Ph.D.; m. Esther Kemp, Aug. 30, 1935; children—Jorgen, Sonja, Hans. In charge research in chemistry, physics, plant physiology, microbiology at univs. of Kiel, Copenhagen, Zurich, Rutgers U.; Rockefeller scholar; instr. superior sch. at Copenhagen; head Tuborg lab., 1929-—. Mem. com. of analysis European Conv. of Brewers. Mem. Danish Acad. Tech. Sci. Research and numerous articles on chemistry, biology, microbiology, plant physiology, analytical chemistry, steric structure of hydrates and carbon. Home: Gronnevej 26, Virum. Office: Brasserie de Tuborg, Hellerup, Denmark.

BLOMBÄCK, M. Wetter, Swedish physician, biochemist; b. Stockholm, Sweden, Oct. 1, 1925; d. Sten Erik and Maja (Carlander) Wetter; grad. Karolinska Institutet, Stockholm, 1958; m. Birger Blombäck, Aug. 11, 1951. Faculty, Karolinska Institutet. Cons. for diagnosis and treatment of coagulation disorders. Mem. World Fedn. Hemophilia (mem. medical com.). Research, numerous publs. on blood coagulation factors, assay blood coagulation factors, Willebrand's disease, hemophilia A; preparation factor VIII. Home: 118 Gröndalsvägen, 11746 Stockholm, Sweden.*

BLOMEFIELD, Leonard (formerly Leonard Jenyns), Brit. naturalist; b. Pall Mall, Eng., May 25, 1800; s. George Leonard Jenyns and the dau. of Dr. Heberden; B.A., St. John's Coll., Cambridge (Eng.) U., 1822; m. Jane Daubeny, 1844; m. 2d, Sarah Hawthorn, 1862. Ordained deacon, 1823, priest in Christ's Coll., 1824; vicar of Swaffham Bulbeck, Cambridgeshire, 1828-49; went to South Stoke, nr. Bath, Eng., 1850, to Swainswick, 1852, and to Bath, 1860; adopted name of Blomefield, 1871. Donor of Jenyns Library which included his herbarium of Brit. plants. Mem. Linnean Soc., Cambridge Philos. Soc., Zool. Soc. (charter mem.), Entomol. Soc. (charter), Ray Soc. (charter), Bath Natural History and Antiquarian Field Club (founder 1855, 1st pres.). Author: Manual of British Vertebrate Animals, 1835; Observations on Meteorology, 1858; an autobiography in 1889; also wrote a supplement to Natural History (White), 1846; numerous articles. Edited monograph on fishes for Zoology of the Voyage of H.M.S. Beagle, 1840; History of Selborne, 1843. At the end of his career known as the patriarch of natural history in Gt. Britain. Died Sept. 1, 1893.

BLOMQUIST, Harry Erik Eugen, Finnish physician; b. Jakobstad, Finland, Dec. 3, 1907; s. Anselm and Sofia (Lillqvist) B.; lic.med., Helsingfors (Finland) U., 1933, dr.med. et chir., 1940; m. Elvi J. Stenfors, Nov. 19, 1940; children—Anne-Sofie, Carl E., Anders. Asst. physician Municipal Hosp., Wasa, Finland, 1935-37, asst. chief physician Gen. Hosp. Nyslott, Finland, 1939-45; asst. surgeon Surg. dept. Maria Hosp., Helsingfors, 1945-48; Staff Helsingfors U., 1948-—, asst. surgeon in chief dept. II surgery, 1955-65, faculty, 1951-—, asst. prof. surgery, 1965-—. Recipient Medal del centenario de Academia Nacional de Medicina Mexico, 1964. Mem. Finnish Med. Assn., Finska Läkaresällskapet, Finnish Surg. Assn., Scandinavian Surg. Assn., Scandinavian Thoracic Surg. Assn., Internat. Soc. U. Colon and Rectal Surgeons, Internat. Coll. Surgeons. Research and publs. on anthorpometric and morphologic data of skulls of Finnish Lapps and clin. surgery. Home: 19 A 9 Mannerheimvagen, Helsingfors 25, Finland.*

BLONDEL, Andre, French engr.; b. Dijon, France, Aug. 28, 1863; ed. École Polytechnique, 1883, also Ecole des Ponts et Chaussées (doctor's degree); chief engr. central light house service until 1927; mem. French Acad. Scis., 1913. Founder of oscillographic methods; invented electro-magnetic oscillograph; theory of coupling of alternating current generators; theory of induction motors; research on wireless telegraphs, acoustics, and mechanics; developed electric tech.; one of 1st to build electric machines. Died Paris, France, Nov. 15, 1938.

BLONDEL, Charles, psychologist; b. Lyon, France, 1876; tchr. at Strasbourg, then Paris. Author: La conscience morbide, 1914; La psychanalyse, 1924; La mentalité primitive, 1926; Introduction à la psychologie collective, 1928; La psychologie de Marcel Proust, 1932; Le suicide, 1933. Research in abnormal psychology; developed concept of pure psychology; opposed mental submission to social pressure.

BLONDEL, François, French mathematician, architect; b. Ribemont, Picardy, France, June 10, 1618; son of Guillaume-François B.; professor mathematics Royal College; chosen by Louis XIV for math. tutor to dauphin; named dir. Acad. Architecture, 1672. Author: L'Art de jeter les bombes; Cours d'architectur, 1675-98; Nouvelle manière de fortifer, 1684. Built Porte St. Denis, Paris, 1672; advocated classical and rationalist doctrine of architecture; grasped nature of parabolic projections (with air resistance disregarded). Died Paris, Jan. 21, 1686.

BLONDLOT, René Prosper, French physicist; b. Nancy, France, July 3, 1849; prof. physics U. Nancy; mem. French Acad. Scis., 1894. Determined emission speed of x-rays; showed they are electromagnetic waves; determined speed of electric transmission in a metal wire periodically subjected to a sudden current; claimed to have discovered N-rays, 1903; claim debunked in 1904 by Robert W. Wood. Died Nancy, Nov. 24, 1930.

BLOOD, Frank Raymond, Am. biochemist; b. Denver, Sept. 13, 1910; s. Raymond F. and Rose (Bannick) B.; B.S. in Chem. Engring., U. Denver, 1934; Ph.D., U. Mich., 1940; m. Ethel S. Peterson, Dec. 28, 1935; 1 dau., Sally Sue. Biochemist, E. I. DuPont de Nemours & Co., Wilmington, Del., 1935-36, chief biochemist, 1940-42; prof. chemistry, asso. chmn. dept. U. Denver, 1942-50; prof. biochemistry, sch. medicine Vanderbilt U., Nashville, 1950-—; dir. clin. labs., 1959-—, asso. prof. pathology, 1960-—. Cons. Olin Chem. Corp., 1950-—; chmn. toxicology and safety evaluations sect. Gordon Research Conf., 1961; chmn. study sect. on toxicology USPHS, NIH, 1964-67; mem. subcom. Food Protection Com., 1953-—. Mem. Soc. Toxicology (charter mem.), Am. Physiol. Soc., Soc. Exptl. Biology and Medicine, Am. Soc. Pharmacology and Exptl. Therapeutics, A.A.A.S., Sigma Xi. Author: Laboratory Manual in Mammalian Physiology, 1954. Home: 3705 Woodmont Lane, Nashville 37215.*

BLOODGOOD, Joseph Colt, Am. surgeon; b. Milw., Nov. 1, 1867; s. Francis and Josephine (Colt) B.; B.S., U. Wis., 1888; M.D., U. Pa., 1891; m. Edith, Holt, Sept. 1, 1908. Resident physician Children's Hosp., Phila., 1891-92; asst. resident surgeon Johns Hopkins Hosp. Balt., 1892, resident surgeon, 1893-97; 1892; attended fgn. clinics and hosps., 1892-93; asso. in surgery Johns Hopkins U. and Hosp., 1897-1903; adj. prof. surgery. Johns Hopkins U.; chief surgeon. St. Agnes' Hosp. Mem. gen. med. com. A.R.C. adv. bd. Radiol. Research Inst. Fellow A.A.A.S. Editorial bd. (in charge surg. pathology) Am. Jour. Cancer. Devised inguinal hernia operation in which rectus muscle is transplanted and conjoined tendon is obliterated, 1918. Died Oct. 22, 1935.

BLOODWORTH, J. M. B., Jr., Am. pathologist; b. Atlanta, Feb. 21, 1925; s. J. M. B. and Elizabeth (Dimmock) B.; student Stanford, 1943-44; M.D., Emory U., 1948; m. Jean Stone, Nov. 26, 1947; children—Lowell Ann, Joyce Lynn, Elizabeth Carol. Asst. resident internal medicine U. Ia. Hosp., Iowa City, 1950-51; faculty Ohio State U., Columbus, 1951-62; prof. pathology U. Wis., Madison, 1962-—; chief labs. Madison VA Hosp., 1962-—. Mem. Gyro Internat., A.M.A., Am. Assn. Pathologists and Bacteriologists, Am. Soc. Exptl. Biology, Histochem. Soc., Endocrine Soc., Am. Diabetes Assn. (Research award 1963), Am. Heart Assn., Am. Fedn. Clin. Research, Soc. Exptl. Biology and Medicine, Internat. Acad. Pathology, Am. Soc. Clin. Pathology, Am. Assn. Neuropathologists, Nat. Soc. Med. Research, Am. Soc. Cell Biology, others. Research and publs. in degenerative vascular disease associated with diabetes mellitus. Home: 4401 Woods End, Madison, Wis. 53711.*

BLOOM, Charles James, Am. pediatrician; b. New Orleans, Oct. 23, 1886; s. Albert and Rose B.; B.Sc., Tulane U., 1908, M.D., 1912; postgrad. Harvard, 1914-16; m. Gladys Marie Reiss, Jan. 9, 1919; children—Charles James, Albert Reiss, Gladys Marie. Instr. zoology Tulane U., 1907, lectr. biology, 1908 instr., 1916-20, prof. pediatrics, in charge postgrad. med., dept., 1916-37; intern Touro Infirmary, New Orleans, 1912-14; pathol. intern Floating Hosp. for Children, also Infants' Hosp., Boston, 1914-16; prof. pediatrics and head dept., postgrad. med. dept. La. State U., 1937-39; sr. pediatrician Presbyn. Hosp., Lying-In Hosp., New Orleans Dispensary for Women and Children, also physician in charge at St. Vincent's Orphanage, 7th St. Protestant Orphans' Home, sr. pediatrician Hotel Dieu, vis. pediatrician Episcopal Home for Girls, examining pediatrician Jewish Fed. Camp; sr. vis. pediatrist Charity Hosp.; pediatrician in charge 7th St. Protestant Orphans' Home; sr. pediatrician, dept. of pediatrics Touro Infirmary, from 1914, mem. exec. com. and sec. of staff from 1943; co-chief dept. pediatrics Mercy Hosp.; cons. pediatrist Meml. Home, French Hosp., Flint Goodrich Hosp.; physician in charge Metairie Park Country Day School; mem. faculty So. Pediatric Seminar, Saluda, N.C.; founder and gen. chmn. Magnolia Sch. Mem. bd. Pure Milk Soc., New Orleans. Mem. A.M.A., La. State Med. Soc., So. Med. Assn., La. State Pediatric Soc. (pres. 1935), A.C.P., Am. Acad. Pediatrics, Mental Hygiene Soc. (dir.). Author: Care and Feeding of Babies in Warm Climates, 1922, rev. 1937. Contbr. articles to mags. Died Aug. 29, 1947.

BLOOM, Harris Julian Gaster, Brit. radiotherapist; b. Sheffield, Eng., June 30, 1923; s. Arthur and Olga (Levin-Epstein) B.; M.B., B.S., Middlesex Hosp. Med. Sch., U. London, 1947, M.D., 1950; m. Barbara Jean Snowman, Mar. 1, 1955; children—Pauline, Caroline, Michael. Cons. radiotherapist Royal Marsden Hosp. and Inst. Cancer Research, 1958-—, Royal Cancer Hosp., 1958-—, St. Peter, St. Pauls, St. Phillips hosps., 1959-—, Queen Mary's Hosp. for Children, 1960-—, West-End Hosp. for Neurology and Neurosurgery, 1960-—, St. Mary Abbott's Hosp., London, 1959-—. U.K. rep. Internat. Com. for Patient Care, Internat. Union Against Cancer, 1966-—. Recipient Laura de Saliceto award for cancer research U. London, 1953. Fellow Faculty Radiologists, Royal Soc. Medicine London; mem. Brit. Med. Assn., Brit. Inst. Radiology. Asst. editor Clin. Radiology, 1961-—. Research, publs. on pathology of breast cancer, treatment of genito-urinary tumors, pituitary and brain tumors; introduced concept of endocrine-dependency of cancer of kidney and hormone treatment of advanced cases; studies in radiobiology; tumor prodn., prevention and treatment. Home: 23 Raymond Rd., London S.W.19, Office: Royal Marsdan Hosp., London S.W.3., Eng.*

BLOOM, James Richard, Am. plant pathologist; b. Clearfield, Pa., Feb. 20, 1924; s. Raymond V. and Rozella (Dunlap) B.; B.S., Pa. State U., 1950; Ph.D., U. Wis., 1953; m. June Farwell, Sept. 11, 1947; children—James Richard, Heidi L., Gretchen E., Coralie A. Faculty, Pa. State U., University Park, 1953-—, prof. plant pathology, 1966-—. Mem. Am. Phytopath. Soc., Soc. Nematology. Research and publs. on role of nematodes in combination with fungi to produce plant diseases. Home: P.O. Box 317, Lemont, Pa. 16851. Office: Buckhout Lab., University Park, Pa. 16802.*

BLOOM, Myer, Canadian physicist; b. Montreal, Que., Can., Dec. 7, 1928; s. Israel and Leah (Ram) B.; B.S., McGill U., 1949, M.S., 1950; Ph.D., U. Ill., 1954; m. Margaret Holmes, May 30, 1954; children—David, Margot. Research fellow U. Leiden, 1954-56; research fellow U. B.C., Vancouver, 1956-57, faculty, 1957-—, prof., 1963-—. Fellow Am. Phys. Soc.; mem. Canadian Assn. Physicists. Studies and publs. on properties of solids, liquids and gases by measurement of rate at which nuclear spins come into thermal equilibrium with their surroundings. Home: 5669 Kings Rd., Vancouver 8, B.C., Can.*

BLOOM, Walter Lyon, physician; b. Ont., Can., Dec. 14, 1917; s. Jacob Isaac and Pauline (Breslav) B.; student Emory U., 1933-36; M.D., Yale, 1940; m. Suzanne Ferst, Aug. 2, 1942; children—Walter Lyon, Clement Alan. Physiol. investigator dept. medicine Columbia Coll. Phys. and Surg., 1943-44; James Hudson Brown research fellow dept. medicine and physiol. chemistry New Haven Hosp., Yale Sch. Medicine, 1946-47; asso. in biochemistry Emory U. Sch. Medicine, Atlanta, 1947-48, faculty medicine, 1947-55, lectr. biochemistry, 1952-57, asso. prof. medicine, 1955-57, dir. med. edn. and research Piedmont Hosp., 1957—. Contbr. numerous articles to professional publs. Home: Route 3, Bloomland Farm, Marietta, Ga. Office: 1968 Peachtree Rd. N.W., Atlanta 30309.*

BLOOM, William, Am. histologist; b. Balt., Sept. 15, 1899; s. Mayer Leon and Bertha (Singer) B.; A.B.; Johns Hopkins, 1919, M.D., 1923; Doctor honoris causa Jagiellonian U., Cracow, Poland, 1964; m. Margaret Abt, June 6, 1928. Asst. pathologist Michael Reese Hosp., Chgo., 1923; Douglas Smith fellow U. Chgo., 1926-28, prof. histology, 1929—, Charles H. Swift Distinguished Service prof., 1957—, prof. emeritus, 1965—. Mem. Nat. Acad. Sci., Am. Assn. Anatomists, Biophys. Soc., Soc. for Exptl. Pathology, Soc. for Cell Biology, Soc. for Growth and Devel., A.A.A.S. Author: (with A. A. Maximow) Textbook of Histology, 1930, (with D. W. Fawcett), 1962; Histopathology of Irradiation. Research on jaundice, inflammation, immunity, culture tissues outside body, malarial infections, blood cell formation and destruction, effects various radiations on cells, parts cells and entire animals, structure chromosomes and electron microscopy. Office: 5640 Ellis Av., Chgo. 60637.*

BLOOMFIELD, Lincoln Palmer, Am. polit. scientist; b. Boston, July 7, 1920; s. Meyer and Sylvia (Palmer) B.; B.S., Harvard, 1941, M.P.A., 1952, Ph.D., 1956; m. Irirangi Pamela Coates, July 17, 1948; children—Pamela, Lincoln, Diana. With U. S. Dept. State, Washington, 1946-57, various positions up to policy planner on UN affairs; faculty Mass. Inst. Tech., Cambridge, 1957—, dir. arms control project Center for Internat. Studies, 1960—, prof. polit. sci., 1963—. Cons. govt. agys., pvt. industry. Rockefeller fellow, 1954-55, Littauer fellow. Mem. Am. Polit. Sci. Assn., Council Fgn. Relations, Inst. Strategic Studies, Hudson Inst. Author: Evolution or Revolution?, 1957, The United Nations and U. S. Foreign Policy, 1960, (with others) Outer space, 1962, International Military Forces, 1964; Khruschev and the Arms Race, 1966. Devel. theoretical and analytical base for evaluating and planning U. S. policies toward the UN; developed and refined polit.-mil. gaming technique; investigated polit. and internat. security implications of disarmament and arms control, and of local conflict, particularly in developing regions. Home: 37 Beach St., Cohaaset, Mass. 02025.*

BLOOMFIELD, Maurice, Orientalist, philologist; b. Bielitz, Austria, Feb. 23, 1855; student U. Chgo., 1871-74, Furman U., 1876-77, M.A., 1877; fellow Johns Hopkins, 1878-79, Ph.D., 1879; postgrad., Berlin, Leipzig, 1879-81; several hon. degrees; m. Rosa Zeisler, June 20, 1855 (dec. 1920); children—Elinor Marie (Mrs. A. S. De Witt), Arthur Leonard; m. 2d, Helen Townsend Scott, July 9, 1921. Professor Sanskrit and comparative philology. Johns Hopkins U., 1881—. Recipient Hardy prize Royal Acad. Scis., Munich, 1908. Fellow Am. Acad. of Arts and Scis. Author: The Atharva-Veda, 1899; Cerberus, the Dog of Hades, 1905; A Concordance of the Vedas, 1907; The Religion of the Veda, 1908; Rig-Veda Repetitions, 2 vols., 1916; Life and Stories of the Jaina Savior Parçvanatha. Editor (1st time from original Sanskrit manuscript) Grihyasamgraha of Gobhilaputra, also Sultra of Kauçika; translator Atharva-Veda in Sacred Books of East (edited by Max Muller); editor (with Richard Garbe) Kashmirian, or Paippaláda-Veda, 1901. Died San Francisco, Cal., June 13, 1928.

BLOSS, James Ramsdell, Am. physician; b. Ceredo, W.Va., Sept. 5, 1881; s. Hiram Wesley and Carrie Lee (Ramsdell) B.; M.D., U. Va., 1905; m. Garnett Lucille Coleman, Oct. 17, 1906 (dec. Feb. 11, 1930); m. 2d. Muessette A. Hollobaugh, July 14, 1932. Asst. supt. Huntington State Hosp. for the Insane, 1906-13; practced at Huntington, W.Va., from 1913; obstetrician in charge of dept. C.&O. Hosp., Huntington; pres. staff St. Mary's Hosp., patron Smithsonian Instn.; founder and sustaining mem. U. S. com. World Med. Assn. Diplomate Am. Bd. Obstetrics, Gynecology. Fellow A.C.S., Acad. Internat. Medicine. A.M.A. (trustee 1935-45; chmn. sect. obstetrics and gynecology 1950). Am. Assn. Obstetricians. Gynecologists and Abdominal Surgeons (pres. 1948-49), Am. Psychiatric Assn.; mem. W. Va. Obstetric and Gynecol. Soc. (pres. 1943). W.Va. State Med. Assn. (pres. 1926), So. Med. Assn., Surgeons Assn., W.Va. Acad. Science. A.A.A.S. Editor W.Va. Med. Jour., 1916-26, 27-37. Died Apr. 21, 1951.

BLOSSER, Henry Gabriel, Am. physicist; b. Harrisonburg, Va., Mar. 16, 1928; s. Emanuel and Leona (Branum) B.; B.S., U. Va., 1951, M.S., 1952, Ph.D., 1954; m. Priscilla May Beard, June 30, 1951; children—William Henry, Stephan Emanuel, Gabe Fawley, Mary Margeret. Physicist 86" cyclotron group Oak Ridge Nat. Lab., 1954-56, group leader Cyclotron Analog I, 1956-58; asso. prof physics Mich. State U., East Lansing, 1958-61, prof. physics, dir. cyclotron project, 1961—. Cons. numerous instns. Mem. Am. Phys. Soc., Raven, Phi Beta Kappa, Sigma Xi, Kappa Alpha. Analysis of orbits in sectored cyclotrons, design of sectored cyclotrons, resonant extraction systems for cyclotrons. Home: 609 Beech St., East Lansing, Mich. 48823.*

BLOT, Claude P. H., French obstetrician; b. Paris, France, June 14, 1822; M.D., 1849; studied under Paul Dubois; began internship, 1845; named chief of obstetrics clinic, 1855; became prof. obstetrics, 1856; became head, vaccination service Paris Hosps., 1873. Mem. Acad. Medicine. Invented scissors for performing craniotomy in difficult labors (Blot's scissors). Died Mar. 13, 1888.

BLOUNT, Bertie Kennedy, Brit. chemist; b. Shoeburyness, Eng., Jan. 4, 1907; s. George Percy and Bridget (Bally) B.; B.A., Trinity Coll., Oxford U., 1928, B.Sc, 1929, M.A., 1932; Dr.Phil. Nat., U. Frankfurt (Germany), 1931. Faculty, lectr., dean St. Peter's Hall, Oxford U., 1931-38; dir. research, prin. tech. exec. Glaxo Labs. Ltd., 1938-40; asst. dir. research Wellcome Found., 1947-48; dir. research br. Control Commn. for Germany, 1948-49; dir. sci. intelligence Ministry of Def., 1939-51; dep. sec. Dept. Sci. and Indsl. Research, 1951-65; dep. sec. Ministry Tech., 1965-66. Decorated Comdr. of Bath. Mem. Internat. Inst. Refrigeration (pres. exec., mgmt. coms. 1963—). Research, publs. in organic chemistry, especially natural products, application of sci. and tech. in govt. Home: Tarrant Rushton House, Blandford Forum, Dorset, Eng.*

BLOUNT, Robert Estes, Am. physician; b. Bassfield, Miss., July 21, 1908; s. Estes Nathan and Mary Leola (Hathorn) B.; B.S., Millsaps Coll., 1928; M.D., Tulane U., 1932; m. Alice Boyd Ridgway, Oct. 17, 1930; children—Robert Estes, Richard B. R., Jane Elizabeth. Commd. 1st lt., M.C., U. S. Army 1933, advanced through grades to Brig. gen., 1960, chief of Medicine Brooke General Hospital, Ft. Sam Houston, Tex., 1951-55, cons. in medicine, Europe, 1955-58, chief medicine Brooke Gen. Hosp., 1959-60, dir. profl. service Office Surgeon Gen. Washington, 1960-62, spl. asst. for research and devel., 1962-65; comdr. William Beaumont Gen. Hosp., 1965-66; commander Fitzsimons General Hosp., 1966—. Decorated Legion of Merit. Fellow Am. Coll. Physicians; mem. A.M.A., Am. Clin. and Climatol. Assn., Am. Soc. Tropical Medicine and Hygiene, Am. Rheumatism Assn., Am. Diabetes Assn., Am. Heart Assn., Assn. Mil. Surgeons, Pi Kappa Alpha, Nu Sigma Nu, others. Research and publs. on infectious diseases and tropical medicine diseases, especially Malaria. Address: Fitzsimons Gen. Hosp., Denver 80240.*

BLOUNT, Samuel Gilbert, Jr., Am. physician; b. Providence, May 19, 1917; s. Samuel Gilbert and Hazel (Martin) B.; B.S., R.I. State Coll., 1939; M.D., Cornell U.; 1943; m. Jean H. Blount, Apr. 22, 1944; children—Lucy Miranda, Anne Gilbert, Donna Bradbury, Sarah Nelson, Lauren Havens. Pediatric cardiology fellow Cardiovascular Lab. Johns Hopkins Hosp., 1948-49, fellow Pediatric Cardiology Clinic, 1950; practice medicine, specializing in cardiology, Denver, 1950—; head div. cardiology, dir. Cardiovascular Lab. U. Colo. Med. Center, 1950—, prof. medicine, 1961—, mem. clin. research fellowship rev. panel Nat. Heart Inst. NIH, 1961-64, mem. com. cardiovascular tng. grant program dirs. heart tng. com., 1966—. Recipient Research Achievement award Am. Heart Assn., 1962, Cummings Humanitarian award Am. Coll. Cardiology, 1966. Fellow Am. Coll. Cardiology, Am. Heart Assn.; mem. Am. Clin. and Climatol. Assn., Am. Fedn. Clin. Research, Am. Soc. Clin. Investigation, Assn. Am. Physicians, Assn. U. Cardiologists (past pres.), Western Assn. Physicians, Western (emeritus), Central (emeritus) socs. for clin. research. Mem. editorial bd. Am. Heart Jour., 1957, Am. Jour. of Med. Scis., 1966, Cardiology Digest, 1966. Contbr. numerous articles to profl. jours. Research on pulmonary circulation and heart disease. Home: 190 High St., Denver 80218. Office: 4200 E. 9th Av., Denver 80220.*

BLUKET, Nina Aleksandrovna, Russian botanist; b. Oboyan (now Kursk Oblast), 1902; grad. Agronomical Faculty, Timiryazev Agrl. Acad., 1924, completed higher pedagogical courses, 1928; D.Biol. Sci., 1960. Instr., Moscow Polit. Ednl. Inst., 1926—; asst., later lectr. chair agro-botany All-Union Agrl. Pedagogical Inst., 1929—; dir. corr. tng. in botany various Moscow insts., 1930-31; instr. Acad. Socialist Agr., 1931-32; head botany dept. Timiryazev Biol. Mus.; lectr. Moscow Timiryazev Agrl. Acad., after 1943, now prof. Author: Methods of Teaching Carbon Assimilation, 1930; Practical Laboratory Work on Botany, 1934; The Plant and its Life, 1936; Starch in the Vegetative Organs of Angiospermae, 1960; The Study of the Elimination System in Plants, 1962. Address: Moscow Timiryazev Agrl. Acad., Novoe sh. 51, Moscow, USSR.

BLUM, Abraham Hyman, Am. psychologist; b. Erie, Pa., May 25, 1925; s. Charles and Gussie (Cohen) B.; B.S., U. Wis., 1949; M.A., Cornell U., 1951, Ph.D., 1953; m. Jeanne M. Moscowitz, Sept. 17, 1946; children—Kathe, Charles, Sarah. Dir., Ithaca Child Home, 1950; teaching asst. Cornell U., 1951-52; faculty Purdue U., 1952-59; asso. prof. psychology So. Ill. U., Carbondale, 1959-62; research asso. Nat. Inst. Mental Health, Heinz Werner Inst. Developmental Psychology, 1963-65; research asso., lectr. Clark U., Worchester, Mass., 1962-65; asso. prof. psychology Boston U., 1965-67; prof. U. Wisconsin-Milwaukee, 1967—. Dir. psychol. research sect. on mental retardation, Mass. Dept. Mental Health, 1964—; cons. Operation Headstart, 1965—. Fellow Soc. Research in Child Devel.; mem. Am. Psychol. Assn., A.A.-A.S., Sigma Xi. Publs. on research in cognitive-personality of presch. children; devised rigidity-flexibility scale and method for obtaining spatial localization of the self in very young children; modified abstract-concrete thinking and cognitive tasks for research with young children. Home: 2831 N. Prospect Av., Milw. 53211.*

BLUM, Baruch, neurobiologist; b. Zichron, Yaacov, Israel, Aug. 25, 1926; s. Nahaman and Shlomit (Frida Posener) B.; B.A., U. Minn., 1948, M.Sc., 1951; postgrad. (Distinguished scholar) U. Cal. at Berkeley, 1951-52; Ph.D., Georgetown U., 1955; m. Ruth Joseph, Jan. 24, 1957; children—Rachel, Barak Simon, Alice-Eenat. Head neurophysiology and pharmacology unit Weizmann Inst. Sci., Rehovoth, Israel, 1955—; med. research fellow Montreal (Que., Can.), Neurol. Inst., 1960-61; vis. asso. prof. dept. pharmacology McGill U., Montreal, 1961-62; vis. scientist NIH Bethesda, Md., 1962, dept. neurosurgery U. Wash. Sch. Medicine, Seattle, 1965-66. Mem. Israeli Soc. Physiology and Pharmacology, Israeli Soc. Electroencephalography and Clin. Neurophsiology, Israeli Soc. Microcirculation. Research and publs. on the discovery of gel of tungstic acid method for prodn exptl. epilepsy, the mountable stereotaxic instrument, and multiple microelectrode manipulator; research on convulsive phenomena and mechanisms of action of anti-convulsant drugs, temperal lobe epilepsies.*

BLUM, Ferdinand, biologist; b. Frankfurt/Main, Germany, 1865; 1st to use formaldehyde for tissue fixation, 1893; devised adrenaline glycosuria test for suprarenal disease. Died Zürich, Switzerland, 1959.

BLUM, Gerald S., Am. psychologist; b. Newark, N.J., Mar. 8, 1922; s. Benjamin Paul and Augusta (Cohen) B.; B.S., Rutgers U., 1941; M.A., Clark U., 1942; Ph.D., Stanford U., 1948; m. Myrtle Wolf, Mar. 3, 1946; children—Jeffrey, Nancy. Prof. psychology U. Mich., 1948—; cons. clin. psychology VA, 1949-59; cons. editor Bobbs-Merrill Reprint Series, 1961—. Fulbright fellow, 1954-55. Fellow Center for Advanced Study in Behavioral Scis.; mem. Soc. Projective Techniques (dir.), Am. Psychol. Assn., A.A.A.S., Soc. Projective Techniques, Sigma Xi. Author: Psychoanalytic Theories of Personality, 1953, A Model of the Mind, 1961, Psychodynamics, 1966. Research and publs. in exptl. psychodynamics; invention of the Blacky Pictures; use of hypnosis as a research tool. Home: 2641 Geddes Av., Ann Arbor, Mich. 48104.*

BLUM, Jacob Joseph, Am. physiologist; b. N.Y.C., Oct. 3, 1926; s. Paul and Anna (Brown) B.; B.S., N.Y. U., 1947; M.S., U. Chgo., 1950, Ph.D., 1952; m. Ruth Marsey, Mar. 19, 1950; children—Mark Douglas, Lisa, Laura. Staff phys. biochemistry di. Naval Med. Research Inst., 1952-56; asst. prof. biochemistry U. Mich., 1956-58; chief biophysics sect. gerontology br. Nat. Heart Inst., NIH, Balt., 1958-62; asso. prof. physiology Duke Med. Sch., Durham, N.C., 1962-66, professor physiology, 1966—. Merck fellow California Institute of Technology, 1952-53. Mem. Am. Physiol. Soc., Soc. Gen. Physiologists, Biophys. Soc., Soc. Protozoologists, Research and publs. on biochemistry of muscle proteins, role sulfhydryl groups in contractility, cell div. in protozoa, control cell div., diffusion in geometrically constrained enzyme systems. Home: 2525 Perkins Rd., Durham, N.C. 27706.*

BLUM, Jacques Marcel, French physicist, engr.; b. Paris, France, 1924; Licencie es Sciences, U. Paris, 1948, Dr. es Sciences, 1954; m. In charge research Nat. Center Sci. Research, Paris, 1948-56; research asso. U. Rochester, 1955-56; sci. adviser, dir. dept. indsl. applications of nuclear physics Intertechnique, Paris, 1956-58; cons. engr. Societe Ugine Kuhlmann, Paris, 1958—. Mem. Association Technique pour l'Energie Nucleaire (chmn. legislative group sect. radioisotopes users), Groupe Intersyndical pour l'Industries Nucleaire, Royal Photog. Soc. (asso.), Sigma Xi. Co-author: Atome et Industrie, 1958. Contbr. numerous articles to profl. jours. Research in uranium cycle (milling, purification, fuel element reprocessing, waste treatment and disposal), plants and reactors constrn., metallurgy of less common metals, applications of modern and nuclear physics. Patentee x-ray color photography control and regulation, and radioactive wastes treatment. Home: 5 rue Voltaire, Levallois-Perret, Hauts-de-Seine, France. Office: 25 Bd. Amiral Bruix, Paris XVI, France.*

BLUM, Leon Leib, physician; b. Telsiai, Lithuania, May 4, 1908; s. Karl C. and Henriette (Rabinowicz) B.; M.D., U. Berlin, 1933; m. Ernestine Berman, May 30, 1936; children—Leonore Sylvia (Mrs. Murray Hausner), Carolyn Ruth (Mrs. Robert Rodenfels). Came to U. S., 1934, naturalized, 1937. Pathologist, dir. labs. Asso. Physicians and Surgeons Clinic, Terre Haute, Ind., 1936—; founder, sr. partner Terre Haute Med. Lab., 1947—; mem. staff Union Hosp., Terre Haute, Putnam County Hosp., Greencastle, Clay County Hosp., Brazil, Freeman-Greene County Hosp., Vermillion County Hosp. (all Ind.); dir. med. tech., adj. prof. life scis. Ind. State U., Terre Haute. Trustee Am. Pathology Foundation, Private Practitioners Pathology Found. Recipient Achievement award Pathology Found., 1961. Diplomate Am. Bd. Pathology. Fellow Coll. Am. Pathologists (a founder), Am. Soc., Clin. Pathologists; mem. Internat. Acad. Pathology, Pan Am. Med. Assn., A.M.A., Jerusalem Acad. Medicine, Ind. Assn. Pathologists (pres. 1961). Introduced use of red cell transfusions, photoelectric method of red cell counting; research and publs. in clin. pathology and hematology. Home: 3200 Ohio Blvd., Terre Haute 47803. Office: 1505 N. 7th St., Terre Haute, Ind. 47808.*

BLUM, Lester, Am. surgeon; b. Bklyn., Nov. 5, 1906; s. Abraham and Marie (Gleitswan) B.; A.B., Columbia U., 1927, M.D., 1930; m. Jeanne Kaufman, July 1, 1951. Faculty, N.Y.U., N.Y.C., 1952—; asso. surgeon for peripheral vascular diseases Mt. prof. Mt. Sinai Med. Sch., N.Y.C., 1966—; asso. dir. surgery Becker Downtown Hosp., N.Y.C., 1962—; asso. surgeon for peripheral vascular diseases Mt. Sinai Hosp., 1960—. Mem. N.Y. Acad. Medicine, Phi Beta Kappa, Alpha Omega Alpha. Research and publs. on vascular surgery; developed first parabiotic blood pump. Home: 1185 Park Av., N.Y.C. 10028. Office: 101 E. 93d St., N.Y.C. 10028.*

BLUM, Louis, Luxemburger chemist; b. Ettelbrück, Luxemburg, Apr. 7, 1858; s. Johann Peter and Katherine (Diederich) B.; studied at agrl. trade sch., Ettelbrück; m. Margarete Brausch, 1885; 3 sons, 2 daus.; m. 2d, Veronika Schmitz, 1903; 1 dau. Began as office clk. Hochofenwerk, Metz & Co., Esch, Luxemburg 1873, transferred to Dommelding lab., 1882; became dir. lab. in Esch, 1888. Research and numerous articles on raw steel prodn., desulphurization in mixer and converter, occurence of vanadium in Luxemburg iron ores, pig iron prodn. especially blast furnace slag and alloying of ore, coal and limestone. Died Esch an der Alzette, Luxemburg, Feb. 30, 1920.

BLUM, Murray Sheldon, Am. entomologist; b. Phila., July 19, 1929; s. Jack B. and Betty (Fine) B.; B.S., U. Ill., 1952, M.S., 1953, Ph.D., 1955; m. Nancy Ann Hilpp, Sept. 5, 1953; children—Deborah Leighd, Darcelle Lynn, Dawn Lisette, Dana Lesley. Research entomologist U. S. Dept. Agr., Baton Rouge, 1957-58; prof. entomology La. State U., Baton Rouge, 1958-67; prof. University of Ga., Athens, U. S. NSF grantee, 1960-65; Sr. NSF fellow, 1965-66. Mem. Entomol. Soc. Am., A.A.A.S., Royal Entomol. Soc. London, Sigma Xi. Research and publs. on chemistry of insect defensive secretions, biochemistry insect spermatozoa, chemistry of insect communications chems. and their role in insect behavior, chemistry of ant venoms and anatomy of venom-producing glands. Home: 425 Ponderosa Dr., Athens, Ga. 30601.

BLUM, Victor Joseph, Am. geophysicist; b. Defiance, Ia., Mar. 30, 1907; s. John P. and Elizabeth (Ruschenberg) B.; A.B., St. Xavier U., 1931; M.S., St. Louis U., 1936, A.M., 1944; S.T.L., St. Mary's Coll., 1940. Faculty, St. Louis U., 1944—, prof. geophysics and geophys. engring., 1958—, dir. dept. engring., 1948-66, dean, 1956—, trustee 1955-67, trustee U. Corp. for Atmospheric Research, 1961-67. Registered profl. engr., Mo. Mem. A.A.A.S., Am. Geophys. Union, Am. Soc. for Engring. Edn., Nat. Soc. Profl. Engrs. Contbr. articles to tech. jours.; prepared detailed map geology and magnetic field distbn. Canon City area, Colo. Home: 221 N. Grand Blvd., St. Louis 63103.*

BLUMBERG, Joe Morris, Am. physician; b. Balt., June 27, 1909; s. Alexander Webster and Hortense (Morris) B.; B.S., Emory U., 1930, M.D., 1933; m. Margaret Catherine Weller, Aug. 29, 1935. Asst. resident Balt. City Hosp., 1934-36; asst. clin. pathologist U. Md. Hosp., Balt., 1936-41; commd. lt. U. S. Army, 1935, advanced through grades to maj. gen., 1966, chief lab. service hosp., Ft. Eustis, Va., 1941-44, pathologist Army Inst. Pathology, 1945-46, Oliver Gen. Hosp., Augusta, Ga., 1946-50, pathologist Walter Reed Gen. Hosp., 1950-54, comdg. officer 406 Med. Gen. Lab., Tokyo, Japan, 1954-57, cons. pathology and allied scis. Office of Surgeon Gen., 1957—, dep. dir. Armed Forces Inst. Pathology, Washington, 1957-63, dir., 1963-67, spl. asst. to surgeon gen. for research and devel. and comdg. gen. U. S. Army Med. Research and Devel. Command, 1967—. Member of the National Advisory Mem. Nat. Cancer Council, 1963-67; scientific dir. Am. Registry Pathology, 1960-63. Recipient Hektoen Bronze medal A.M.A. 1961, Stitt award Assn. Mil. Surgeons U. S., 1961, Seale Harris award So. Med. Assn., 1963. Mem. Coll. Am. Pathologists, A.C.P., Am. Soc. Clin. Pathology, Am. Assn. Pathology and Bacteriology, Internat. Acad. Pathology. Research and publs. on histochem. and electron microscopic studies in neuromuscular junction and muscle. Home: 5007 Jamestown Rd., Washington 20016. Office: U. S. Army Med. Research and Devel. Command, Washington 20315.*

BLUMBERG, Randolph, Am. oceanographer; b. San Antonio, Feb. 1, 1926; s. Randolph Julius and Toni (Wiederstein) B.; B.E.E., Tex. A. and M. U., 1948, M.E.E., 1950, Ph.D. in Phys. Oceanography, 1955; m. Ann Frances McClellan, Feb. 26, 1955; children—Randolph William, Richard James. Design engr. Fargo Engring. Co., Austin, Tex., 1949-51; research engr. S.W. Research Inst., San Antonio, 1951-52; research engr. Tex. A. and M. Research Found., 1953-55; petroleum engr. Humble Oil & Refining Co., Houston, 1955-57, sr. petroleum engr., 1957-60, sr. research engr., 1960-62; president Am. Sci. & Engring. Co., Houston, 1962—; asso. prof. Cullen Coll. Engring., U. Houston, 1964—; cons. Lower Colo. River Authority, 1949-51; cons. Project Mohole, 1963-64; lectr. oceanography, meteorology U. Tex. Ext. div. Sch. Offshore Drilling, Houston, 1966—. Fellow Tex. Acad. Sci.; mem. Am. Geophys. Union, Am. Meteorol. Soc, Nat. Soc. Profl. Engrs., I.E.-E.E., Am. Soc. Oceanography (v.p.), Sigma Xi, Tau Beta Pi. Contbr. articles to sci. jours. Pioneer in devel. of offshore installations; oceanographic and environmental engring.; hurricane, long range weather probability predictions; instrumentation design and devel. for measurement of winds, waves, tides, currents. Home: 5533 Redstart St., Houston 77035. Office: Humble Bldg., Houston 77002.*

BLUME, Carl Ludwig, botanist; b. Brunswick, Germany, June 9, 1796; 1st gen. practitioner at Java; prof. botany at Leyden, Netherlands; dir. bot. garden of Buitenzorg, 1822-26; dir. Leyden plant garden; mem. French Acad. Scis., 1851. Author: Aloude geschiedenis der Belgen of Nederduitschers, 1849. Research on flora of Philippines. Died Feb. 3, 1862.

BLUMENTHAL, Herman Theodore, Am. gerontologist; b. N.Y.C., Apr. 8, 1913; s. Samuel and Jennie (Applebaum) B.; B.S., Rutgers U., 1934; M.S., U. Pa., 1935; Ph.D., Washington U., St. Louis, 1938, M.D., 1942; m. Eleonore Gottlieb, Aug. 18, 1940; children—Daniel S., Frederick A. Practice medicine, specializing in pathology, St. Louis, 1943—; dir. labs. Jewish Hosp., 1950-58, dir. Inst. Exptl. Pathology, 1958-61; chief lab. service, dir. Clin. Research Program on Aging VA Hosp., Jefferson Barracks, St. Louis, 1961-66; gerontology research asso. Washington University, St. Louis, 1963-66, research professor gerontology, 1966—. Recipient Alexander Berg prize Washington U., 1941. Mem. Soc. Exptl. Biology and Medicine, Gerontol. Soc., Internat. Assn. Gerontology, Am. Heart Assn.; Am. Assn. Cancer Research, Am. Soc. Exptl. Pathology, Am. Soc. Pathologists and Bacteriologists, N.Y. Acad. Scis., Am. Diabetes Assn. Author: (with Probstein) Pancreatitis-A Clinico Pathological Correlation, 1959; Medical and Clinical Aspects of Aging, 1961; Cowdry's Arteriosclerosis A Survey of the Problem, 1966. Research and publs. on vascular disease, endocrinology, aging and cancer, influence of aging on diseases which are prevalent among old people, abnormal immune reactions in elderly people which are related to vascular disease, cancer, diabetes and others. Home: 1940 N. Geyer Rd., St. Louis 63131.*

BLUMENTHAL, Leonard Mascot, Am. mathematician, educator; b. Athens, Ga., Feb. 27, 1902; s. George and Henriette (Hirschfield) B.; B.S. in Civil Engring., Ga. Inst. Tech., 1923; M.S., U. Chgo., 1924; Ph.D., Johns Hopkins, 1927; m. Eleanor Berger, June 23, 1926. NRC fellow in math. Inst. for Advanced Study, Princeton, N.J., 1933-34, research asst., 1935-36; NRC fellow in math. U. Vienna (Austria), 1934-35; faculty U. Mo., Columbia, 1936—, prof. math., 1942—. Mem. fellowship bd. div. math. NRC, 1953-55; math. cons. Inst. Numerical Analysis, Nat. Bur. Standards, Los Angeles, 1951-52. Fulbright prof. U. Leiden (Netherlands), 1954-55, U. Madrid (Spain), 1962-63. Recipient Gold T, Ga. Inst. Tech., 1922; Distinguished Prof. award U. Mo., 1964. Mem. Royal Acad. Scis. Spain (corr. fgn. mem.), Math. Assn. Am. (bd. govs. 1942-46, vis. mathematician 1961-62, 63—), Am. Math. Soc. (symposium lectr. 1942), Am. Assn. U. Profs., Sigma Xi, Pi Mu Epsilon, Tau Epsilon Phi (Breitenbach medal 1922). Author: Distance Geometries, 1938; Theory and Applications of Distance Geometry, 1953; A Modern View of Geometry, 1961, Spanish edit. Geometria Axiomatica, 1965. Contbr. monographs to Am., German, Spanish jours. Collaborator Math. Revs., 1940—; asso. editor Am. Math. Monthly, 1942-51. Home: 205 E. Ridgeley Rd., Columbia, Mo. 65201.

BLUMENTHAL, (Ludwig) Otto von, see von Blumenthal, (Ludwig) Otto.

BLUMENTROST, Laurentius, physician; b. Bothenheiligen, Germany, Oct. 29, 1619; s. Wolfgang and Dorothea (Geisen) B.; student medicine Helmstedt, Leipzig, Jena, Germany; m. Anna Maria Donat, Aug. 29, 1648; m. 2d, Cäcilia Röver, Feb. 26, 1672; m. 3d, Anna Gosen, 1678; children—6 daus., 8 sons, including Christian, Johann Deodat, Laurentius. City physician Sangerhausen, Germany, 1649-51, Mühlhausen, Germany, from 1651; councilman Mühlhausen, from 1651; went to Moscow, Russia, 1668; physician to Czar Alexei, Russia, from 1672; accused (with other fgn. physicians) of poisoning Czar Fedor, 1682; exonorated, remained ct. physician. Author: De scorbuto, 1648; Pharmacopaea domestica et portalis, 1668. Died Moscow, Nov. 3, 1705.

BLUMER, Herbert, Am. sociologist; b. St. Louis, Mo., Mar. 7, 1900; s. Richard George and Margaret (Marshall) B.; A.B., U. of Mo., 1921, A.M., 1922; Ph.D., U. of Chicago, 1927; m. Marguerite Barnett, Aug. 16, 1922; 1 dau., Katharine; m. 2d, Marcia Jackson, Aug. 22, 1942; children—Linda, Leslie. Instr. sociology U. Missouri, 1922-25; instructor, sociology, University of Chicago, 1925-30, associate professor, 1931-47; professor sociology, U. of Michigan, 1936-37; prof. sociology, U. of Hawaii, 1939; professor sociology, U. Chgo., 1947-52; prof. sociology and chmn. dept. U. Cal., Berkeley 1952—; chairman social science council, 1956-58, director of Institute Social Sciences, 1958-66. Member of research staff Motion Picture Research Council, 1929-31; editor, Publs. of Am. Sociol. Soc., 1931-36; editor, Am. Jour. of Sociology, 1940-52; editor Sociology Series, pub. by Prentice-Hall, Inc., since 1934; principal liaison officer between Office of War Information and Bd. of Economic Warfare, Washington, D.C., 1943; pub. panel chmn., War Labor Board, 1943-44; permanent arbitrator, Armour & Company, 1944-45; chmn. bd. arbitration, U. S. Steel Corp., 1945-47. Mem. Institut Internat. de Sociologie, Soc. Study Social Problems (pres. 1955), Sociol. Research Assn., Internat. Soc. Sci. Study Race Relations, Am. (pres. 1956), Internat. (v.p. 1962-66) sociol. assns., Phi Beta Kappa, Delta Sigma Rho. Author sci. books and articles. Chmn. bd. editors Integrated Edn., 1965—. Developed theoretical outline of collective behavior, covering crowds and social movements; demonstrated that racial prejudice is an expression of social position and not of individual feelings; demonstrated that the process of industrialization is neutral in its social consequences; explained that social science operates with "sensitizing concepts" and not with "definitive concepts" of natural science. Home: 350 Pine Creek Rd., Walnut Creek, Cal. 94598. Office: U. Cal., Berkeley, Cal.*

BLUMER, Max, organic geochemist; b. Basel, Switzerland, Aug. 3, 1923; Ph.D., U. Basel, 1949. Came to U. S., 1950, naturalized, 1964. Faculty, U. Basel, 1943-50, U. Minn., 1950-51; staff Scripps Inst. Oceanography, 1951-52, Shell Oil, Amsterdam, Holland, 1952-53, Shell Devel. Co., Houston, 1953-58, CIBA, Ltd., Basel, 1958-59; staff Woods Hole (Mass.) Oceanographic Instn., 1959—, sr. scientist, 1964—. Mem. Am. Chem. Soc., A.A.A.S. Research and publs. on long term fate organic products organisms in nature, alteration these products in sea and in sediments, origin petroleum, devel. methods pertinent to this research. Address: Woods Hole Oceanographic Instn., Woods Hole, Mass. 02543.*

BLUMRICH, Josef, geologist; b. Raspenau, Bohemia, Jan. 13, 1865; s. Franz Blumrich; ed. Prague, Czechoslovakia; m. Helene Eugling, 1900; 2 sons. Tchr., Bregenz, Austria, from 1894. Author: Grundriss einer Geologie Vorarlbergs, 1921; publs. on geology, mineralogy, botany, muscology, zoology. Made 1st important compilation of geology of Vorarlberg (link between Eastern and Western Alps). Died Bregenz, Sept. 22, 1949.

BLUNT, George William, Am. hydrographer; b. Newburyport, Mass., Mar. 11, 1802; s. Edmund March and Sally (Ross) B.; m. Martha Garsett, 1821. Ran away from home to serve on mcht. ships, 1816-21; started publishing bus. with bro. E. & G. W. Blunt, N.Y.C., 1821, specialized in nautical works, made original surveys of many American harbors; helped organize present system of pilotage for New York, edited many publs., firm became large producer of charts; U. S. Hydrographic Office started mainly with purchases from his firm; served on Am. Bd. Pilot Commrs. for long periods; influential in causing U. S. government to organize Lighthouse Board and to adopt French System of lighthouses; U. S. commr. of immigration, 1852-54; pilot commissioner, 1855; harbor commissioner, 1867. Died N.Y.C., Apr. 19, 1878.

BLYDENSTEIN, John, botanist; b. Enschede, Holland, Dec. 29, 1929; s. Willem J. H. and Eleanore (Mysken) B.; came to U. S., 1955, naturalized, 1960; B.S., Wash. State U., 1956; M.S., U. Ariz., 1957, Ph.D., 1967; m. Rosa N. Abregó, Nov. 4, 1954; children—John Alfonso, Eleanore Rose, Mary Ann. Asst. mgr. Estancia Las Vertientes, Victoria, Argentina, 1952-54; botanist U. Cal. Bot. Garden, Berkeley, 1959; agrl. officer FAO assigned to Interam. Inst. Agrl. Scis., Venezuela, Colombia, Costa Rica, 1960—. Mem. Am. Soc. Range Mgmt., Ecol. Soc. Am., Internat. Soc. Tropical Ecology, Sigma Xi. Research and publs. on survey of tropical savanna vegetation in Latin Am., mgmt. tropical pastures in Latin Am. Address: Interam. Inst. Agrl. Scis., Zootecnia, Turrialba, Costa Rica.*

BLYTH, Colin R., mathematician; b. Guelph, Ont., Can., Oct. 24, 1922; s. Colin McDonald and Gladys (Martin) B.; B.A., Queen's U., 1944, M.A., U. Toronto, 1946; postgrad. U. N.C., 1946-48; Ph.D., U. Cal., 1950; m. Valerie Thompson, Aug. 27, 1955; children—Mary Alice, Georgina, Colin M., Heather, Alexander. Asst. prof., U. Ill., Urbana, 1950-54, asso. prof., 1954-59, prof. math., 1959—. Home: 701 W. Michigan Av., Urbana Ill., 61801.*

BLYTH, Edward, zoologist; b. London, England, Dec. 23, 1810; druggist at Footing; became curator Mus. of Asiatic Soc. of Bengal, Calcutta, India, 1841; returned to Eng. as pensioner, 1862. Author: Catalogue of the Mammals and Birds of Burma, 1875; also catalogue of mammals in the soc's. mus.; 1863; articles. Founder zoology in India. Died Dec. 27, 1873.

BOAK, Ruth A., Am. physician; b. Auburn, N.Y., May 25, 1906; d. Spencer J. and Jane (Clark) Boak; B.S., Cornell U., 1927, M.S., 1927, Ph.D., 1929; M.D., U. Rochester, 1940; m. Donald L. Ferris, May 30, 1942; children—Boak J., Don R. Asso., U. Rochester, 1930-47; asso. prof. infectious diseases, pediatrics U. Cal., Los Angeles, 1947-57, prof. infectious diseases, med. microbiology, pediatrics, 1957-67, prof. med. microbiology, immunology, pediatrics and pub. health, 1967—; vis. prof. Airlangga U. Sch. Medicine, Indonesia, 1963-66. Recipient Fulbright award, 1954, 1959, Los Angeles Times Woman of Year award in medicine, 1955. Fellow U. S. Pub. Health Mexican Border Assn.; mem. Johns Hopkins Med. and Surg. Soc., Am. Pub. Health Assn., Am. Venereal Disease Soc., Cal. Med. Research Assn. Contbr. numerous articles to sci. jours. Research in brucellosis syphilis, Coxsackie virus infections; infections of newborn. Home: 10639 Mason Av., Chatsworth, Cal. 91311.*

BOALT, Gunnar Hans Rudolf, Swedish sociologist; b. Ljustero, Sweden, Aug. 26, 1910; s. Anton and Ruda (Brodin) B.; Ph.D., Stockholm U., 1947; m. Carin Margareta Akerman, June 3, 1935; children—Siv, Birgitta (Mrs. Erik Alexius), Elsa Karin (Mrs. Tomas Fischer), Hans Ake, Sten Arne, Stella Margareta; m. 2d, Ulla Britta Margareta Halden, Aug. 8, 1957. Faculty, Stockholm (Sweden) U., 1947—, prof. sociology, 1954—, dean Faculty Social Work, 1954-60, Faculty Social Sci., 1965—. Mem. Nordic Council on Alcohol Problems, 1958-66; chmn. Central Union Social Work, 1956—; adj. prof. So. Ill. U., Carbondale, 1967. Author over 40 books, including Family and Marriage, 1965; (with T. Husén) Educational Research and Educational Change, 1968. Contbr. numerous articles to profl. journs. Research on mate selection and sociology of research work. Home: 64 Upplandsgatan. Office: 50 Regeringsgatan, Stockholm, Sweden.*

BOAS, Ernst Philip, Am. physician; b. Worcester, Mass., Feb. 4, 1891; s. Franz and Marie (Krackowizer) B.; B.S., Columbia, 1910, M.A., 1912, M.D., 1914; m. Helen T. Sisson, Dec. 25, 1917; children—Donald Philip, Norman Francis, Barbara Gertrud. Instr. in pathology Coll. Phys. and Surg. (Columbia), 1917, instr. in physiology, 1920-21; med. dir. Montefiore Hosp., 1921-29, attending physician, 1929-30; post-grad. tchr. diseases of heart Columbia, from 1926; asst. clin. prof. of medicine, 1938-51; asso. and cons. physician Mt. Sinai Hosp. Chmn. Physicians Forum for Study of Med. Care. Mem. A.M.A., Soc. for Clin. Investigation, Harvey Soc., N.Y. Acad. Med., Am., N.Y. heart assns., Phi Beta Kappa. Author: The Challenge of Chronic Diseases (with Nicholas Michelson), 1929; The Heart Rate, 1932; The Unseen Plague Chronic Diseases, 1940; Treatment of the Patient Past Fifty, 1941, 3d edit., 1947; Coronary Artery Disease, 1949. Died Mar. 9, 1955.

BOAS, Franz, anthropologist; b. Minden, Westphalia, July 9, 1858; s. M. and Sophie (Meyer) B.; student univs. Heidelberg, Bonn and Kiel, 1877-81; Ph.D., U. Kiel, Germany, 1881, M.D., honoris causa; LL.D., Sc.D., Oxford U., Clark U., Howard U. and Columbia; hon. citizen, honoris causa, U. Bonn, Graz; m. Marie A. E. Krackowizer, 1887; children—Mrs. Helene Marie Yampolski, Ernst P., Mrs. Marie Franziska Michelson. Explored Baffin Land, 1883-84; asst. Royal Ethnog. Museum, Berlin, and docent of geography U. Berlin, 1885-86; investigations in N.A., Mexico, P.R., 1886-1931; docent anthropology, Clark U., 1888-92; chief asst. dept. anthropology Chgo. Expn., 1892-95; lectr. phys. anthropology, Columbia U., 1896-99, prof. anthropology, 1899-1937, prof. emeritus from 1937; curator dept. anthropology Am. Mus. Natural History, 1901-05; hon. philologist Bur. Am. Ethnology, 1901-19. In Mexico, 1910-12; hon. prof. Nat. Mus. of Archeology, Mexico; corr. sec. Germanistic Soc. America, 1914; pres. Emergency Soc. German and Austrian Science, 1927; nat. chmn. Com. for Democracy and Intellectual Freedom, 1939-40, hon. chmn. from 1940; mem. Nat. Acad. Scis., Am. Philos. Soc., Am. Antiq. Soc., Am. Folklore Soc. (editor 1908-25, pres. 1931); fellow A.A.A.S. (v.p. 1895, 1907; pres. 1931), N.Y. Acad. Scis. (pres. 1910), Am. Anthrop. Soc. (pres. 1907, 1908), Am. Acad. Arts and Scis.; hon. or corr. mem. numerous European profl. socs.; hon. fellow Anthropol. Inst. of Great Britain and Ireland, Folk Lore Soc. of London; senator Deutsche Akademie, Munich. Author: The Growth of Children, 1896, 1904; Changes in Form of Body of Descendants of Immigrants, 1911; The Mind of Primitive Man, 1911, 1938; Kultur und Rasse, 1913; Primitive Art, 1927; Anthrop. and Modern Life, 1928-38; General Anthropology (with others), 1938; Race, Language and Culture, 1940; Dakota Grammar (with Ella Deloria), 1941; also publs. on anthropometry, linguistics and anthropology of North America. Editor Jesup N. Pacific Expdn., Internat. Jour. Am. Linguistics. Often regarded as father of Am. anthropology, his studies were internationally influential; re-examined premises of phys. anthropology; objectively analyzed and criticized the various schs. of ethnol. study; insisted on rigorous methodology in his work, to establish sci. value; one of 1st to apply statis. methods to biometric study; also one of 1st to do stratigraphic excavations in Mexico; stressed importance of linguistic analysis from the internal linguistic structure of Am. Indian langs.; collected vast anthrop. data in areas of culture, linguistics and archeology. Died Dec. 21, 1942.

BOAS, Ralph Philip, Jr., Am. mathematician; b. Walla Walla, Wash., Aug. 8, 1912; s. Ralph Philip and Louise (Schutz) B.; A.B., Harvard, 1933, Ph.D., 1937; m. Mary Elizabeth Layne, June 12, 1941; children—Ralph Layne, Anne Louise, Harold Philip. Instr. math. Harvard, 1936-37, lectr., 1943-45; NRC fellow Princeton U., 1937-38, Cambridge (Eng.) U., 1938-39; instr. Duke, 1939-42; exec. editor Math. Revs., Providence, R.I., 1945-50; lectr. in math. Mass. Inst. Tech., 1948-49; prof. math. Northwestern U., Evanston, Ill., 1950—. Chmn. Com. on Undergraduate Program in Math., 1968—. Guggenheim fellow, 1950. Mem. Am. Math. Soc. (v.p. 1959-60). Author: Entire Functions, 1954; (with R. C. Buck) Polynomial Expansions of Analytic Functions, 1958; A Primer of Real Functions, 1960. Integrability Theorems for Trigonometric Transforms, 1967. Contbns. to study of math. analysis. Home: 2440 Simpson St., Evanston, Ill. 60201.*

BOATO, Giovanni, Italian physicist; b. Rome, Italy, Sept. 3, 1924; s. Natale and Rita (Candusso) B.; Dottore in Chimica, U. Genoa (Italy), 1947; Dottore in Fisica, U. Rome (Italy), 1951; m. Maria Emilia Vallauri, Feb. 28, 1957; children—Alberta, Luisa, Anna. Research asso. Istituto di Fisica, U. Rome, 1951-53, Inst. for Nuclear Studies, U. Chgo., 1953-54; faculty 1st di Fisica, U. Genoa, 1954—, prof. physics, 1963—; vis. scientist physics dept. Rutgers U., 1965; mem. Capo Gruppo Ricerche in Struttura della Materia, Consiglio Naz. delle Ricerche, 1958—. Mem. Società Italiana di Fisica, Am. Phys. Soc. Research, publs. on mass spectrometry, measurement of isotope fractionation in natural environments, measurement of isotope separation in phase equilibria, solid argon, localized states in metals, superconductivity, detection of quantized flux lines in type II supercondrs. Home: 49 Via Orsini, Genoa, Italy.*

BOATRIGHT, Byron B(lackburn), Am. petroleum engr.; b. Colorado Springs, Feb. 10, 1900; s. William Louis and Minnie Ellen (Stump) B.; E.M., Colo. Sch. Mines, 1922; Ph.D., U. Colo., 1936; m. Sylva Dora Kerr, Mar. 10, 1922; children—Barbara Jeanne (Mrs. H. P. Oliver), William Gary. Flagman survey party U. S. Gen. Land Office, 1919; rouster, fireman, and tool-dresser Midwest Refining Co., Big Muddy and Salt Creek Fields, Wyo., 1922-25; jr. engr., asst. engr., later asso. engr. in charge oil and gas leasing div. U. S. Geol. Survey, state of Colo., 1925-26; dist. engr. Panhandle Dist., Marland Oil Co., 1926-28; head petroleum engring. dept. and prof. prodn. engring. Colo. Sch. Mines, 1928-37; cons. petroleum and natural gas engr., Golden, Colo., 1928-37; partner Parker Foran, Knode & Boatright (cons. engrs.), Houston, 1937-40, Foran, Knode, Boatright & Dixon, 1940-41, Foran, Boatright & Dixon, 1941-45, Boatright & Mitchell (cons. engrs. and geologists), 1945-46; vice pres. and chief engr. Republic Natural Gas Co., Dallas, 1946-48; 1st vice pres. Conroe Drilling Co., Austin, 1948-49, dir. 1949; gen. supt. Heep Oil Corp., Austin, 1948-49 cons. petroleum and natural gas engr., 1949-53; v.p., gen. mgr., dir. Houston Natural Gas Producing Co., from 1953. Registered engr., Tex. Mem. Am. Inst. Mining and Mech. Engrs., Am. Petroleum Inst., Sigma Xi. Contbr. articles to tech. jours. Died May 9, 1957.

BOBB, John Richard Ranvier, Am. physician; b. Herndon, Pa., Apr. 18, 1918; s. Clarence Curtin and Mertie Marie (Newman) B.; A.B. summa cum laude, Lafayette Coll., 1938; M.D., Temple U., 1942; m. Elma Faye Knorr, Jan. 18, 1946; 1 son, John Richard Ranvier. Faculty dept. physiology and pharmacology Bowman Gray Sch. Medicine, 1946-48, 56-62, instr. anesthesiology, 1962-63; asst. prof. pharmacology Temple U. Sch. Medicine, 1950-56; sr. staff anesthesiology Cleve. Clinic, 1963-65; staff VA Hosp., Cleve., 1965-66, chief staff VA Hosp., Long Beach, Cal., 1966—. Diplomate Am. Bd. Anesthesiology. Fellow Am. Coll. Anesthesiologists; mem. Am. Physiol. Soc., Soc. for Exptl. Biology and Medicine, Am. Assn. U. Profs. Contbr. articles to profl. jours. Research in cardiovascular physiology, cardiovascular responses to arterio-venous shunts. Home: 5901 E. 7th St. Office: VA Hospital, Long Beach, Cal. 90804.*

BOBBITT, James McCue, Am. chemist; b. Charleston, W.Va., Jan. 18, 1930; s. James Sterling and Grace (McCue) B.; B.S., W.Va. U., 1951; Ph.D., Ohio State U., 1955; m. Jane Ann Hickman, Mar. 15, 1952; children—John Sterling, Ann, Laura. Postdoctoral fellow Wayne State U., Detroit, 1955-56; instr. chemistry U. Conn., Storrs, 1956-59, asst. prof., 1959-63, asso. prof., 1963—. Vis. prof. U. East Anglia, Norwich, Eng., 1964-65. NSF fellow U. Zürich, 1959-60. Mem. Am. Chem. Soc., Chem. Soc. London, Am. Soc. Pharmacognosy, Phi Beta Kappa, Sigma Xi. Author: Thin-Layer Chromatography, 1963. Research in isolation, structure elucidation and synthesis of natural products; evolution of new synthesis of isoquinolins; devel. thin layer chromatography. Home: 6 Olsen Dr., Mansfield Center, Conn. 06250.*

BOBBITT, Oliver Beirne, Jr., Am. physician; b. Charleston, W.Va., Jan. 10, 1917; s. Oliver Beirne and Belle (Graves) B.; student Hampden Sydney Coll., 1934-36; B.S., U. Ga., 1939; M.D., U. Va., 1943; m. Betty Kearse, Nov. 13, 1943; children—Oliver B., Wesley R., Timothy G. Faculty, U. Va. Med. Sch., Charlottesville, 1947—, prof. pathology, 1957—, chmn. dept. clin. pathology, 1952—; dir. Clin. Labs., U. Va. Hosp., 1952—. Diplomate Am. Bd. Pathology. Fellow Am. Soc. Clin. Pathologists, Coll. Am. Pathologists. Contbr. articles to tech. jours. Research on human blood groups; co-discoverer new blood group antigen. Home: 1857 Westview Rd., Charlottesville, Va. 22902.*

BOBEK, Hans, Austrian geographer; b. Klagenfurt, Austria, May 17, 1903; s. Karl and Agnes Elisabeth (Petz) B.; Ph.D., U. Innsbruck; m. Helene Procopovici, July 27, 1929; 1 son, Hanspeter. Asst., U. Innsbruck, U. Berlin; prof. U. Friburg-B.; prof. superior sch. of commerce, Vienna, Austria, U. Vienna. Recipient Karl Ritter silver medal Soc. Geography of Berlin, Eduard Rüppel medal Geog. Soc. of Frankfurt. Mem. Austrian Acad. Sci.; hon. mem., corr. various geog. socs. in Munich, Belgrade, Zagreb, Edinburgh, Netherlands. Author over 100 works, including books on urban geography, alpine morphology, vegetation, desert formation in Iran, social geography of Orient, methodical geograph. Editor: Atlas der Republik Osterreich. Home: Mahlerstrasse 4, Vienna I. Office: Geogr. Institut, University of Vienna, Vienna, Austria.

BOBERG, Friedrich, German chemist; b. Bad Salzuflen, May 23, 1922; s. Gustav and Anna (Rasche) B.; Dipl.-Chem., Hanover Inst. Tech., 1949, Ph.D., 1951; m. Gisela Sievers, May 31, 1949; children—Michael, Doris. Sr. chemist dept. organic chemistry Inst. Petroleum Research, Hanover, Germany, 1952-63, sr. chemist, dept. labeled compounds and radiation chemistry, 1963—; docent Hanover Inst. Tech., 1965—. Mem. Soc. German Chemists, German Soc. Petroleum Research and Carbon Chemistry. Research, numerous publs. on nitro compounds, organic sulfur compounds, labeled compounds. Home: 14 Kiefkampstrasse, 3014 Misburg. Office: 30 Am Kleinen Felde, 3 Hanover, West Germany.*

BOBILLIER, Etienne, French mathematician; b. Chalons-sur-Marne, 1797. Dir., École de Arts et Métiers, Chalons. Author: Essai sur un nouveau mode de Recherche des Propriétés de l'Étendue, in Annales de mathématiques, 1827; Théorèmes sur les polaires successives, in Annales de Gregonne, 1828; Principes d'algèbre, 1825; Cours de géométrie, 1832; (with Plücker) Voresungen über synthetische Geometrie, 2 vols. (work in which authors devised synthetic methods of developing theory of polars and algebraic curves), 1867. Research on co-ordinate transformations, properties of curves, second-degree surfaces, gen. methods of analytical geometry. Died 1834.

BOBOCH, Samuel, biochemist, psychiatrist; b. Saskatoon, Sask., Can., Jan. 13, 1928; M.D., U. Toronto, 1951; Ph.D. in Biochemistry, Harvard, 1956; m. Elenore Ruth Sade, May 10, 1953; children—Samuel, Anne-Elenore. Teaching fellow, asst., instr. dept. psychiatry Harvard Med. Sch., Boston, 1956-61; dir. Neurochem. Research Lab., sr. psychiatrist Mass. Mental Health Center, Boston, 1956-61; faculty Boston U. Sch. Medicine, 1961—, asso. research prof. biochemistry and psychiatry, 1965—, dir. Found. for Research on Nervous System, 1961—. Mem. A.A.-A.S., Am. Psychiat. Assn., Am. Acad. Neurology. Author: (monograph) Biochemistry of Memory, 1968. Research, numerous publs. on brain, relationship of nervous system chemistry to behavioral states, psychopharmacology, demonstration of structure and membrane function of brain gangliosides in nervous system, demonstration that brain mucoids are involved in memory and learning. Home: 122 Carlton St., Brookline, Mass. 02215. Office: 36 The Fenway, Boston 02215.

BOBROV, Yergenii Grigorevich, Russian geog. botanist; b. Tver (now Kalinin), 1902; postgrad. Research Inst. Econ. Geography, Geog. Faculty, Leningrad, U., 1926-29; D.Biol. Sci., 1942. Mem. expdn. organized by Main Bot. Gardens (now Inst. Botany, USSR Acad. Sci.), 1921-22, herbarium asst., 1923-26, asso., 1926—, sec. dept. systematics, 1931—, dir., 1937-38; sr. asst. chair phytogeography Leningrad U., 1927-28. Author: The Vegetation of the Western Part of the Sterlitamak Canton of Bashkiria, 1927; Weeds of the Trans-Volga Area and the Irrigation Project, 1933; The Weeds of the Urals and the Campaign against Them, 1936; The Origin of Cultivated Red Clover, 1940; Modern Floristics and the Significance of V. L. Komarov's Studies for its Development, 1944; Some Forgotten Plants from the Journey of Gmelin the Younger in the Years 1768-73, 1958. Editorial sec. Flora of USSR, 1931-32; exec. editor Soviet Botany, 1937-38; mem. editorial bd. Flora SSSR. Research on vegetation of Urals, Bashkir, Central Asia, specialist in floristic systematics, explored upper and lower Volga tys. of Bashkir and Turkmen. Address: Inst. Botany, USSR Acad. Sci., ulitsa Popova 2, Leningrad, USSR.

BOCEK, Michael, Czechoslovakian physicist; b. Cheb, Czechoslovakia, May 9, 1928; s. Frantisek and Paula (Buchner) B.; RNDr., Charles U., Prague, Czechoslovakia, 1953, C.Sc., 1963; m. Eva Talacko, Aug. 1, 1953; children—Martin, Jaroslav. Faculty, Charles U., Prague, 1953—, sr. lectr., 1956—, head metal physics dept. Math. Phys. Faculty, 1965—; asst. prof. U. Mining, Freiberg/Sa, Eastern Germany, 1963-65. Mem. Union Czechoslovakian Math.

and Physics. Author: Lectures on Metal Physics, 1966; also articles. Research on plasicity of metal crystals especially hexagonal metals. Home: 261 Karlstejnská, Cernosice, Czechoslovakia. Office: 5, Ke Karlovu, Prague 2, Czechoslovakia.*

BOCHANTSEVA, Zinaida Petrovna, Russian cytologist, plant embryologist; b. Verny (now Alma-Ata), 1907; grad. Central Asian U., 1930; Cand. Biol. Sci., 1945. Asst., later lectr. Biol. Faculty, Central Asian U., 1926-41, asso. Bot. Gardens, Tashkent, 1930—. Author: Methods of Cytological Study of Plants, 1937. Research and publs. on intergeneric hybridization, caryosystematics, embryology, spermatogenesis, biology of florescence; mem. expn. to Central Asia to study wild decorative export plants from Trans-Ili Alatau, Chuili Mountains, Kirghiz Alatau, South Kazakhstan Olbast, Ferghana Mountains and Kermine area. Address: Central Asian University, ulitsa Karla Marksa 32, Tashkent, Uzbekistan SSR, USSR.

BOCHAROV, Arkadiy Alekseevich, Russian surgeon; b. Tutaev (now Yaroslavl Oblast), 1901; grad. Astrakhan Med. Inst., 1926; D.Med. Sci. With surg. dept, hosp., Rybinsk, 1926-32; with first aid sta., then intern Surg. Clinic, Sklifosovsky Inst. First Aid, Moscow, 1932-37, asst. Emergency Surgery Clinic, 1937-41; chief surgeon various mil. dists, 1946-48; dep. head chair hosp. surgery Naval Med. Acad., 1949-54; head chair naval surgery Kirov Mil. Med. Acad., 1955-61; maj. gen. med. service; dep. chief surgeon USSR Ministry Def., 1961—. Mem. Leningrad Soc. Naval Physicians (chmn.), Leningrad Pirogov Meml. Surg. Soc. Contbr. articles to Large Med. Ency. Research and numerous publs. on problem of blood transfusions from dead bodies, improved methods of treating diseases and injuries or organs in abdominal and thoracic cavities, transplantation of embryonic heterogenous tissues, pathogenesis and treatment of shock and craniocerebral injuries in acute stage. Address: USSR Ministry of Defense, Arbatskaya pl., Moscow, USSR.

BOCHEFONTAINE, L.-T., French physiologist; b. La Manche, France, 1840; demonstrated effect of brain gray matter on various organs of body; studied numerous toxic substances; injected himself with common bacilli to prove bacilli were not solely responsible for disease. Died Paris, France, 1886.

BOCHER, Maxime, Am. mathematician; b. Boston, Aug. 28, 1867; s. Ferdinand and Caroline (Little) B.; A.B., Harvard, 1888; Ph.D., Göttingen, 1891; m. Marie Niemann, July 21, 1891. Instr. math. Harvard, 1891-94, asst. prof., 1894-1904, prof., from 1904; exchange prof. U. Paris, 1913-14. Mem. Am. Math. Soc. (pres., 1909-10, an editor Trans. 1907-09, 1910-13). Author: Über die Reihenentwicklungen der Potential-theorie, 1894; Introduction to Higher Algebra, 1907; Introduction to the Study of Integral Equations, 1909; Leçons sur les methodes de Sturm (Sorbonne lectures of 1913-14). Contbr. to math. jours.; collaborator Encyclopäie der mathematischen Wissenschaften. An editor Annals of Mathematics, 1896-1900, 1901-07, 11-14. Observed that surfaces called cyclides afford unified geometric background for wide class of differential equations; generalized Lamé's differential equation to include as spl. cases the ordinary linear differential equations of 2d order, 1894. Died Cambridge, Sept. 12, 1918.

BÖCHER, Tyge Wittrock, Danish botanist; b. Copenhagen, Denmark, Oct. 25, 1909; s. Einar and Cathinca (Andersen) B.; Mag.Scient, U. Copenhagen, 1934, Dr.Phil., 1938; m. Kirsten Jensenius, Apr. 16, 1935; children—Jens, Ida (Mrs. Henning Urup), Henrik, Joakim. Asst., Royal Agrl. Coll., 1934-36; faculty U. Copenhagen, 1936—, prof. botany, 1954—. Exec. com. Arctic Sta., Disko, Greenland, 1950—; mem. Danish council Univ. Extension Work, 1946—. Mem. Internat. Orgn. Biosystematists (pres. 1964—), Bot. Soc. Brit. Isles (hon.), Danish Council for Nature Conservancy. Author books and numerous articles. Research on flora and vegetation of Greenland, evolutionary dynamics, biosystematics, cytotaxonomy, chromosome cytology, dwarf shrub heaths, dry grassland vegetation, and maritime communities. Home: 90 Fortunvej, Charlottenlund, 2920 Denmark. Office: Institute of Plant Anatomy and Cytology, Solvgade 83, 1307 Copenhagen, Denmark.*

BOCHNER, Salomon, mathematician, educator; b. Cracow, Austria-Hungary, Aug. 20, 1899; s. Joseph and Rude (Haber) B.; Ph.D., Berlin (Germany) U., 1921; m. Naomi Weinberg, Nov. 27, 1937; 1 dau., Deborah Susan (Mrs. Charles Frederick Kennel). Came to U. S., 1933, naturalized, 1938. Fellow Internat. Edn. Bd., 1925-26; lectr. U. Munich (Germany), 1927-33; faculty Princeton (N.J.), 1933—, prof., 1946—, Henry Burchard Fine prof. math., 1959—. Mem. Inst. for Advanced Study, Princeton, 1944-47; vis. prof. U. Cal. at Berkeley, 1953; cons. Los Alamos Project, Princeton, 1951, NSF, 1952. Fellow A.A.-A.S.; mem. Nat. Acad. Scis., Am. Math. Assn. Author: Fourier Integrals, 1932 (translated into English 1959); Several Complex Variables, 1948; Fourier Transforms, 1949; Curvature and Betti Numbers, 1953; Harmonic Analysis and the Theory of Probability, 1955; The Role of Mathematics in the Rise of Science, 1966. Developed Bochner theorem on positive-definite functions, Bochner integral, Bochner-Fejer summation of almost periodic functions. Home: 184 Springdale Rd., Princeton, N.J. 08540.*

BOCHVAR, Andre Anatolievich, Russian metallurgist; b. Aug. 8, 1902; s. A. M. Bochvar; grad. Moscow (USSR) Higher Tech. Sch. Faculty Moscow Higher Tech. Sch.; staff Moscow Inst. Non-ferrous Metals and Gold, 1930—, prof., 1934—. Recipient Stalin prize. Mem. USSR Acad. Scis. (corr.). Author: Study of the Mechanism and Kinetics of Crystallization of Eutectic Alloys, 1935; Basic Treatment of Alloys, 1940; The Various Mechanisms of Plasticity in Metal Alloys, 1948; Metallography, 1956. Gave theory for structural pecularities, anomalies of alloys; established temperature patterns of crystallization of metals and alloys; formulated structural theory heat resistance; developed methods of shaping castings by crystallization under pressure. Office: Moscow Inst. Non-ferrous Metals and Gold.

BOCK, August Carl, German anatomist; b. Magdeburg, Germany, Mar. 26, 1782; asso. prof., prosector Anatomisches Theater, Leipzig, Germany, from 1814. Author: Beschreibung des fünften Nervenpaares und seiner Verbindingen mit anderen Nerven vorzüglich mit dem Gangliensystem, 1817-21; Handbuch der praktischen Anatomie des menschlichen Körpers, 1819-22; numerous anat. and surg. publs. Discovered carotid ganglion (Bock's ganglion) in lower portion of cavernous sinus, also glosso-pharyngeal nerve. Died Leipzig, Jan. 30, 1833.

BOCK, Hieronymus, (Tragus), German botanist; b. Heidesbach, Germany, 1498; master of sch., Deuxponts; became physician; minister, St. Evangeline; supr. ducal gardens, personal physician to Duke of Nassau, Hornbach; traveled in mountains of central Europe. Wrote under Latin name Tragus. Author: Neu Kreutterbuch, 1539; also herbal of German plants, 1559. A founder of modern botany; attempted to construct a natural bot. arrangement according to phys. similarities. Genus Tragia named in his honor by Plumier. Died Hornbach, Feb. 21, 1554.

BÖCK, Josef, Austrian physician; b. St. Pölten, Austria, Oct. 13, 1901; s. Josef and Maria Anna (Artner) B.; student U. Graz; M.D., U. Vienna, 1926; m. Edith Zabrsch, Aug. 18, 1936; 1 dau., Susanne. Faculty eye clinic U. Vienna, 1927-44; became asst. dir. eye clinic U. Graz (Austria) 1944, prof., 1950-55, dean med. faculty, 1950-51, asst. dean, 1951-52, senator, 1953-55; prof., dir. II eye clinic, U. Vienna, 1955—, dean med. faculty, 1963-64, asst. dean, 1964-65. Recipient Silver medal for service to Republic Austria, 1965. Mem. Internat., European councils ophthalmology, Austrian, German, Swiss ophthal. socs., Ophthal. Soc. Vienna; hon. mem. ophthal. socs. Saloniki, Iran; Austrian delegate French Ophthal. Soc. Author: (with J. Meller), Augenärztliche Eingriffe, 5th and 6th edits.; Meller's Ophthalmic Surgery, 6th edition; also articles. Address: 2 Stadiongasse, 1010 Vienna, Austria.*

BOCK, Robert M., Am. molecular biologist; b. Preston, Minn., July 26, 1923; s. Glen E. and Hilda (Snyder) B.; B.S., U. Wis., 1949, Ph.D. in Chemistry, 1952; postgrad. Cal. Inst. Tech., 1955, U. Cambridge (Eng.), 1961; m. Ruth Golbien, Sept. 21, 1947; children—Karen, Susan. Faculty, U. Wis., Madison, 1952—, prof. molecular biology, 1965—, dean of the Graduate School, since 1967—. Mem. Am. Chem. Soc., Am. Soc. Cell Biologists, Soc. Exptl. Biologists. Contbr. numerous articles to sci. jours. Isolation and phys. chem. characterization several enzymes and determination their kinetic properties; developed methods for purifying and determining structure Ribonucleic acid molecules. Office: Molecular Biology Bldg., U. Wis., Madison, Wis. 53706.*

BOCKELMAN, Charles Kincaid, Am. physicist; b. San Francisco, Nov. 29, 1922; s. Bernhardt Jacob and Ruth (Kincaid) B.; Ph.B., U. Wis., 1947, Ph.D., 1951; m. Elizabeth Ann Button, June 18, 1950; 1 dau., Faith Lee. Research asso., physics dept. Mass. Inst. Tech., 1951-55; faculty physics Yale, New Haven, 1955—, prof., 1965—, asso. dir. Electron Accelerator Lab., 1961—, Nuclear Structure Lab., 1964—, director of physical sciences, since 1966—; fellow Inst. for Theoretical Physics, Copenhagen, 1958-59. Served with USAAF, 1942-45. Experimental investigation structure complex atomic nuclei. Fellow Am. Phys. Soc., N.Y. Acad. Sci. Home: 96 Killdeer Rd., Hamden, Conn. 06517.*

BOCKRIS, John O'Mara, electrochemist; b. Johannesburg, S. Africa, Jan. 5, 1923; s. Alfred and Emmeline (McNally) B.; student Xaverian Coll., 1934-38; B.Sc., Brighton Tech. Coll., 1943; postgrad. London U., 1943-45; Ph.D., Imperial Coll. Sci. & Tech.; 1945; m. Dorothy M. Sainty, 1946, (div. 1964); children—Anna Mary, Victor Francis; m. 2d, Halina S. Wroblowa, 1964. Came to U. S., 1953, naturalized, 1962. Lectr. Imperial Coll., London, 1945-54; faculty U. Pa., Phila., 1954—, prof. electrochemistry, 1962—. Cons. to bus. firms; dir. research in electrodics. Mem. Faraday Soc., Internat. Com. Electrochem. Thermodynamics and Kinetics, Electrochem. Soc. Author: Text Book of Electrochemistry, 1951, Modern Aspects of Electrochemistry, 4 vols. 1954-64, Technique of High Temperatures, 1959; Modern Electrochemistry, 1967. Research and publs. in electrode kinetics, electrochem. energy conversion, electrochemistry of ionic solutions, and other aspects of electrochemistry. Home: 37 University Mews, 45th and Spruce Sts., Phila. 19104.*

BOCKUS, Henry L(eroy), Am. physician; b. Newark, Del., Apr. 18, 1894; s. William Jones and Luella (Whiteman) B.; M.D., Jefferson Med. Coll., 1917, D.Sc. 1958; D.Sc., Dickinson College, 1946, University of Pennsylvania, 1961; Dr. Honoris Causa, Universidad Central Venezuela, 1965; m. Rosalynd Foss, January 13, 1935; 1 dau., Barbara Ann. Resident physician, Lenox Hill Hosp., N.Y.C., 1920-21; began practice as physician, Phila., 1921; then internist specializing in gastro-intestinal disorders and as gastroenterologist; organizer stomach clinic, Grad. Hosp., U. of Pa., 1921, and associated with the clinic since that date; prof. of gastroenterology, Grad. Sch. of Medicine, U. of Pa., since 1931, prof. and chmn. dept. medicine, 1949-60, prof. medicine emeritus, 1960—; consulting physician Graduate Hosp.; hon. civilian cons. surgeon gen., U.S.N.; chmn. bd. dirs. MEDICO; v.p. bd. directors CARE; cons. gastroenterologist Bryn Mawr Hosp., Abington Meml. Hosp., Phila. Naval Hosp.; hon. prof. medicine U. Antioquia (Colombia), 1964. Chmn. World Congress in Gastroenterology, Washington, 1958. Mem. bd. trustees Jefferson Medical College. Served as lt., M.C., USN, 1917-19; with 7th Regt., USMC in Cuba. Recipient Caldwell Medal, Am. Roentgen Ray Soc., 1950; Strittmater Award, Phila. County Med. Soc., 1951; Modern Med. award for achievement, 1962; decorated Comdr. Order Hipolito Unanue (Peru); Order al Merito de Chile. Fellow American College Physicians, Royal Society of Medicine London (honorary fellow), Royal Society of Arts, London, Nat. Acad. Medicine Mex. (hon.); honorary mem. Gastroenterol. Assns. of Central America, Chile, Cuba, Venezuela, Peru, Brazil, Argentina, Uruguay, Spain, India, Belgium, Germany, Colombia (hon.), Organization Mundialde Gastroenterologia (pres. 1958-62, hon. pres. 1962—), Am. Gastroenterol. Assn. (past pres., recipient Friedenwald medal 1962), A.M.A. (former chmn. sect. gastroenterology and proctology), Am. Bd. Internal Medicine (founder mem. subsplty. bd. gastroenterology), Phila. Coll. Physicians, Phila. Pathologic Soc.; hon. asso. Am. Proctologic Assn. Author: Gastroenterology, 3 vols., 1943-46; Postgraduate Gastroenterology, 1950; also contbr. sci. articles to med. jours. Mem. editorial bd. Jour. of Gastroenterology. Home: 1709 Rittenhouse Sq. Office: 250 S. 18th St., Phila. 3.

BODANSKY, Meyer, Am. biochemist, pathologist; b. Elizabetgrad, Russia, Aug. 30, 1896; s. Phineas and Eva (Geiro) B.; brought to U. S., 1907; A.B., Cornell U., 1918, Ph.D., 1923; M.A., U. Tex., 1922; M.D., U. Chgo., 1935; m. Eleanore Abbott, June 15, 1925; children—Samona, Eleanore Ruth. Instr. biol. chemistry U. Tex. Med. Sch., 1919-23, adj. prof., 1923-25; acting asst. prof. Stanford, 1925-26; asso. prof. U. Tex., 1926-30, prof. path. chemistry, 1930—; dir. labs. John Sealy Hosp., John Sealy Meml. Research Lab.; vis. prof. of physiol. chemistry Am. U. of Beirut, Syria, 1932-33. Certificate Am. Bd. of Pathology. Author: Introduction to Physiological Chemistry, 1927, 30, 34, 38; (with M. Fay) Laboratory Manual of Physiological Chemistry, 1928, 31, 35, 38; (with O. Bordansky) Biochemistry of Disease, 1940. Editorial bd., Am. Jour. Clin. Pathology. Contbr. to jours. Died June 14, 1941.

BODANSKY, Oscar, biochemist, physician; b. Elizabethgrad, Russia, Aug. 21, 1901; s. Phineas and Eva (Geiro) B.; came to U. S., 1907, naturalized, 1923, B.A., Columbia, 1921, M.A., 1922, Ph.D., 1925; M.D., U. Chgo., 1938; m. Barbara Biber, May 31, 1929; 1 dau., Margery (Mrs. Raymond S. Franklin). Research asso. in biochemistry U. Cal. at San Francisco, 1926-27; faculty U. Tex. Med. Sch., Galveston, 1927-30; instr. pediatrics N.Y. U. Coll. Medicine, N.Y.C., 1930-37; clin. asst. in pediatrics Beth Israel Hosp., N.Y.C., 1939-40, asst. adj., 1940-42, asso. in pediatrics, 1946-48; asso. prof. pharmacology Cornell U. Med. Coll., N.Y.C., 1946-48; attending clin. biochemist, chmn. dept. biochemistry, Meml. Hosp. for Cancer and Allied Diseases, N.Y.C., 1948—; chief div. biochemistry Sloan-Kettering Inst. for Cancer Research, N.Y.C., 1956—; prof. biochemistry Sloan-Kettering div. Cornell U. Grad. Sch. Med. Scis., 1951—. Decorated Legion of Merit; recipient Distinguished Service award Med. Alumni U. Chgo., 1952; Alfred P. Sloan Found. award in cancer research, 1962; Van Slyke award in clin. chemistry Am. Assn. Clin. Chemists, 1965. Mem. Coll. Am. Pathologists, N.Y. Acad. Medicine, Am. Soc. Biol. Chemists, Am. Assn. Cancer Research. Research, publs. in biochem. aspects of disease, mechanism of enzyme action, application of enzymology to medicine. Author: Biochemistry of Disease, 1952. Home: 535 E. 86th St., N.Y.C. 10028. Office: 444 E. 68th St., N.Y.C. 10021.*

BODANSZKY, Miklos, chemist; b. Budapest, Hungary, May 21, 1915; s. Lajos and Maria (Friedner) B.; diploma chem. Engring., Royal Tech. U. Budapest, 1939, D.Sc., 1949; m. Agnes A. Vadasz, Apr. 21, 1950; 1 dau., Eva D. Came to U. S., 1957, naturalized, 1964. Head dept. natural products Research Inst. Pharm. Chemistry, 1950-56; sr. lectr. pharm. chemistry Tech. U. Budapest, 1950-56; research asso. dept. biochemistry Med. Coll. Cornell U., N.Y.C., 1957-59; sr. research chemist, research supr., sr. research asso. Squibb Inst. Med. Research, New Brunswick, N.J., 1959-66; prof. chemistry Western Res. U., Cleve., 1966—. Mem. Am. Soc. Biol. Chemists, Am., Swiss chem. socs., Japan Antibiotics Research Assn. Author: (with M. A. Ondetti) Peptide Synthesis, 1966. Contbr. numerous articles to profl. jours. Research, publs. on advanced method for the synthesis of peptides, the nitrophenyl ester method, applied this method in stepwise synthesis of peptide hormones, such as oxytocin, vasopressins, secretin. Home: 18035 Fernway Rd., Shaker Heights, O. 44122. Office: Dept. Chemistry, Western Res. U., Cleve. 44106.*

BODE, Hans H., German inorganic, gen. chemist; b. Kiel, Germany, Jan. 18, 1905; s. Robert and Minna (Brammer) B.; Ph.D., U. Kiel; m. Ruth Bechmann, 1939. Prof. U. Kiel. Mem. Bd. Applied Electrochemistry. Mem. Gesellschaft Deutscher Chemiker, Deutsche Bunsengesellschaft, Electrochem. Soc. Author: (with Ludwig) Chemisches Praktikum für Mediziner, 1932, 9th edit., 1953. Research, publs. on inorganic, electrochem. chemistry, crystal structures, fluorine chemistry. Address: 624 Konigstein/Taunus, Sonnenhofstrasse 6, Germany.

BODE, Helmut Heinrich Ludwig, German analytical, inorganic chemist; b. Hannover, Germany, Jan. 27, 1917; s. Gustav B. and Marie (Reinecke) B.; ed. Technische Hochschule Hannover, U. Mainz (Germany); diploma in chemistry, 1947; dr. rer. nat., 1949; m. Gunhild Herschler, 1961. Asst., U. Mainz, 1950-54; lectr. Tech. U. Hannover, 1954, prof. 1957——. Work in photometric methods analysis, inorganic complex chemistry, trace analysis, flame photometry, extraction of chelate complexes. Address: Anorganisch-chemisches Institut, Callinstrasse 46, 3 Hannover, Germany.

BODE, Hendrik Wade, Am. engr.; b. Madison, Wis., Dec. 24, 1905; s. Boyd Henry and Bernice (Ballard) B.; B.A., Ohio State U., 1924, M.A., 1926; Ph.D., Columbia, 1935; m. Barbara Poore, Nov. 18, 1933; children—Katharine Anne, Beatrice Anne Hathaway. Mem. tech. staff Bell Telephone Labs., Murray Hill, N.J., 1926-44, in charge math. research group, 1944-52, dir. math. research, 1952-55, dir. research in phys. scis., 1955-58, v.p. for mil systems engring., Whippany, N.J., 1958-67; Gordon McKay prof. of systems engring. Harvard, 1967——. Trustee Research Analysis Corporation. Visiting fellow Princeton, 1948-49. Recipient Presdl. Certificate of Merit, 1948. Fellow I.E.E.E., Am. Acad. Arts and Scis., Am. Phys. Soc.; mem. Nat. Acad. Scis., Am. Math. Soc., Nat. Acad. Engring., Am. Inst. Aeros. and Astronautics, Soc. Indsl. and Applied Mechanics, Audio Engring. Soc. (hon.), Phi Beta Kappa, Eta Kappa Nu. Clubs: Beacon Hill (Summit, N.J.); Princeton (N.Y.C.). Author: Network Analysis and Feedback Amplifier Design, 1945. Patentee in electric curcuit theory, mil. devices. Home: 153 Beechwood Rd., Summit, N.J. Office: Harvard U., Cambridge, Mass. 02138.*

BODE, Johann Elert, German astronomer; b. Hamburg, Germany, Jan. 19, 1747; astronomer Berlin Acad. Sci., 1772-1825; dir. Berlin (Germany) Obs., 1786-1825. Fellow Royal Soc., 1789. Author: Anleitung zur Kenntnis des gestirnten Himmels, 1768; Représentation des Astres, 1782; Uranographia, 1801. Founder, Berliner astronomisches Jahrbuch, 1774, 54 vols., editor till 1829. Devised empirical arithmetical formula (Bode's law) for expressing distances of planets from sun (same discovery made by J. O. Titius), 1772; named planet Uranus; constructed atlas (Uranographía) of 17,240 stars and nebulae (12,000 more than in previous charts), 1801. Died Berlin, Nov. 23, 1826.

BODEA, Cornel, chemist; b. Vienna, Austria, Aug. 28, 1903; s. Isidor and Eugenia (Stanescu) B.; Dr. Ing.Chem., Technische Hochschule Berlin, Charlottenburg, Germany, 1928; m. Ileana Baltariu, Oct. 23, 1938; 1 dau., Ileana Bodea. Staff, Statiunea Chimica-Agronomica, Cluj, Rumania, 1928-31; chemist Institutul Agronomic, Cluj, 1931-41, prof. biochemistry, 1942——. Chief organic chemistry sect. Chem. Inst. Acad., Cluj, 1960——. Decorated Ordinul Muncii. Mem. Rumanian Acad., A.A.A.S. Editor: Tratat de Biochimie Vegetala, 4 vols., 1964-67. Research and numerous publs. on chemistry and biochemistry of Carotenoids, gen. plant biochemistry, chemistry of phenothiazine. Home: 52 Brasov, Cluj, Rumania.*

BODEMER, Charles William, Am. anatomist, med. historian; b. Denison, Ia., Jan. 4, 1927; s. Herman Charles and Blanche (Nicola) B.; student City Coll. San Francisco, 1947, U. Cal. at Berkeley, 1949; B.A. magna cum laude, Pomona Coll., 1951; M.A., Claremont Grad. Sch., 1952; grad. scholar Dartmouth, 1952; Ph.D., Cornell U., 1956; m. Sheila Campbell Hadley, June 20, 1948; children—Karen Hadley, Eric Charles, Brett William. Faculty, Dartmouth, 1952-53, Cornell U., 1953-56; faculty Sch. Medicine U. Wash., Seattle, 1956——, asso. prof. biol. structure, 1962——, head div. biomed. history, 1964——, asso. dean, 1965——. Vis. lectr. NSF div. history, philosophy Sci. Rev. Bd., 1965——, Am. Physiol. Soc., 1961-64; mem. Pacific Sci. Center's Vis. Scientist Program, 1964——; mem. manpower and tng. com. Nat. Library of Medicine. Mem. Am. Assn. Anatomists, Am. Soc. Zoologists, History of Sci. Soc., Am. Assn. History of Medicine, Soc. for Study Growth and Devel., Sigma Xi. Research, publs. on structure, devel. of components of nervous system; regeneration of lost body parts; 18th century med. scis. Home: 425 35th Av., Seattle 98122. Office: 1959 N.E. Pacific Av., Seattle 98105.*

BODEN, Brian, biologist; born Capetown, South Africa, Apr. 15, 1921; s. Thomas Henry and Iris Francis (Brimble) B.; student U. Stellenbosch, 1937-40; B.Sc. in Zoology, U. Capetown, 1945; Ph.D. in Oceanography, U. Cal. at San Diego, 1950; m. Elizabeth Maitland Kampa, June 3, 1949; children—Peter C., Karen E. Came to U. S., 1946, naturalized, 1959. Oceanographer, Bermuda Biol. Sta., 1951-54; research fellow Scripps Instn. Oceanography, La Jolla, Cal., 1954-56, asst. research biologist, 1956-67, asso. research biologist, 1967——. Mem. Am. Geophys. Union, A.A.A.S., Marine Biol. Assn., Soc. Systematic Zoology, Am. Micros. Soc., Challenger Soc. (London), Western Soc. Naturalists, Soc. Vis. Scientists (London), Sigma Xi. Research, publs. on taxonomy of diatoms and euphausiids, vertical distbn. of marine mesopelagic animals with emphasis on ones that migrate diurnally in response to changes in photoenvironment, bioluminescence of pelagic animals, transmitted sunlight at mesopelagic depths, spectral sensitivity of mesopelagic crustaceans. Home: 1339 Justin Rd., Cardiff, Cal. 92007. Office: P.O. Box 109, La Jolla, Cal. 92037.*

BODENSTEDT, Erwin, German physicist; b. Cologne, Germany, Jan. 25, 1926; s. Hermann and Lia (Bensberg) B.; Dr., U. Bonn, 1953; postgrad. Cornell U., 1954-55; Dr.habil., U. Hamburg, 1961; m. Ruth Susemihl, Dec. 27, 1956; children—Martin, Monika, Michael. With research lab. SSW, Erlangen, Germany, 1955-56; sci. adviser U. Hamburg (Germany), 1956-63; prof. U. Bonn (Germany), 1963——. Mem. German Phys. Soc. Research, numerous publs. on nuclear spectroscopy, gamma-gamma angular correlations, magnetic moments of excited states. Address: 2 Spreestrasse, Bonn, West Germany.*

BODENSTEIN, Adam von, see von Bodenstein, Adam.

BODENSTEIN, Dietrich H. F. A., biologist; b. Corwingen, East Prussia, Germany, Feb. 1, 1908; s. Hans and Charlotte (Lilienthal) B.; student U. Königsberg, Germany, 1926-28, U. Berlin, 1928-33; Ph.D., U. Freiburg, Germany, 1953; m. Jean Coon, July 22, 1947; 1 dau. by previous marriage, Evelina (Mrs. William C. Suhler). Came to U. S., 1934, naturalized, 1940. Research asst. Kaiser Wilhelm Inst. Biology, Berlin, 1928-33; research asso. German-Italian Inst. Marine Biology, Rovigno d'Istria, Italy, 1933-34, Stanford Sch. Biology, 1934-41; John Simon Guggenheim Meml. Found. fellow, dept. zoology, Columbia, 1941-43; asst. entomologist Conn. Agrl. Expt. Sta., New Haven, 1944; insect physiologist, med. div. Army Chem. Center, Md., 1945-57; embryologist Nat. Heart Inst., gerontology br. Balt. City Hospitals, 1958-60; Lewis and Clark prof. biology, chmn. dept. U. Va., 1960——. Mem. Am. Acad. Arts and Sciences, American Soc. Zoologists, Genetics Soc. Am., Am. Soc. Naturalists, Am. Assn. Anatomists, Soc. for Study Development and Growth, Nat. Acad. Sci., Soc. Biology Brazil (hon.), Sigma Xi. Author articles profl. and sci. publs. Research on experimental morphology of amphibians; developmental physiology of amphibians and insects; endocrinology of insects; developmental genetics of Drosphilia. Home: 536 Valley Rd., Charlottesville, Va.

BODENSTEIN, (Ernst August) Max, German chemist; b. Magdeburg, Germany, July 15, 1871; s. Franz and Elise (Meissner) B.; ed. at Heidelberg (Ph.D., 1893), Wiesbaden, Charlottenburg, Göttingen (all Germany); D.Sc. (hon.), Princeton; Dr.Ing. h.c., Germany; m. Maria Nebel, 1896; 2 daus. Asst. prof. in Heidelberg, 1899; later worked with Wilhelm Ostwald in Leipzig, Germany, 1900; asso. prof. Leipzig, 1904; dept. head Inst. for Phys. Chemistry, Berlin, 1906-08; prof. in Hanover, Germany, 1808-23; prof. Inst. for Phys. Chemistry, Berlin, 1923-36; emeritus, 1936. Fellow Chem. Soc. London; mem. Berlin Acad. Sci., Bulgarian Chem. Soc., Göttingen Soc. Scis. (corr.), Hungarian Acad. Sci. (fgn.). Author: Handbuch der Katalyse I, 1941; also numerous essays in German sci. jours. Research in phys. chemistry and electrochemistry, especially on equilibrium and velocity of gaseous reactions and photo chem. reactions; thought to have discovered chain reaction. Died Berlin, Sept. 3, 1942.

BODIAN, David, Am. anatomist; b. St. Louis, May 15, 1910; s. Harry and Tillie (Franzel) B.; S.B., U. Chgo., 1931, Ph.D., 1934, M.D., 1937; m. Elinor Widmont, June 26, 1944; children—Helen, Marion, Brenda, Alexander, Marc. NRC fellow U. Mich., 1938; research fellow Johns Hopkins, 1939-40, asso., 1942-45, faculty, 1946——, prof. anatomy, dir. dept. anatomy, 1957——; asst. prof. Western Res. U., Cleve., 1940-42. Cons. USPHS vaccine coms., 1955-65, vaccine adv. com. Nat. Found. for Infantile Paralysis, 1954-57. Recipient E. Mead Johnson award, 1941; Dist. Service award U. Chgo. Alumni, 1955; USPHS award, 1956; named to Polio Hall of Fame, Ga. Warm Springs Found., 1957. Mem. A.A.A.S., Am. Assn. Anatomists, Am. Assn. U. Profs., Am. Soc. Cell Biologists, Am. Physiol. Soc., Assn. for Research in Nervous and Mental Diseases, Nat. Acad. Scis., Soc. Francais de Neurologie Paris (hon.). Author: (with H. A. Howe) Neurol Mechanisms in Poliomyelitis, 1942; also mumerous articles. Mng. editor Am. Jour. Hygiene, 1948-57. Invented staining method for nerve fibers; research on structure of nervous tissue, behavior of poliovirus in body, embryonic devel. of nervous system; discovered several aspects of nerve cell interconnections, of maj. types of poliovirus, basic structure of neurohypopysis, mechanisms of polio virus spread to nervous system by way of blood. Home: Rolandvue Av., Towson, Md. 21204. Office: 709 N. Wolfe St., Balt. 21205.*

BODIN, Jean, French polit. scientist, economist; b. Angers, France, 1530; studied law, Toulouse; m. 1576. Lectr. jurisprudence, Toulouse; settled in Paris, as advocate; apptd. King's atty., Laon, 1576; elected by 3rd Estate of Vermandois as its rep. in States-Gen., Blois (distinguished himself for defense of religious freedom and peace); sec. to Duke of Alencon, 1581; spent final years in Laon. Author: Methodus ad facilem historiarum cognitionem, 1566; Six Livres de la Republique, 1576; Responsio ad Paradoxa Malestretti, 1588; Universale Naturae Theatrum, 1594. Developed first comprehensive system of polit. sci. since Aristotle; champion of authority of the sovereign, subject to divine and natural laws; noted distinction between form of state and form of govt.; justified individual property; stressed importance of climate; subscribed to many mercantilist ideas, but also called for free trade under certain conditions, and for tax adjustments; analyzed factors governing rises in prices. Died of plague, Laon, 1596.

BODINE, Joseph Hall, Am. zoologist; b. Lake Hopatcong, N.J., Sept. 19, 1895; s. Gilbert and Sarah Annie (Hall) B.; A.B., U. Pa., 1915, Ph.D., 1920; m. Sarah Olivia Heimach, Nov. 19, 1919 (dec. June 1950); 1 son, Joseph Hall; m. 2d, Eunice Willis Beardsley, June 1951. Instr. zoölogy U. Pa., 1915-16 and 1920-25, asst. prof., 1925-28, prof., 1928-29; prof. and head dept. zoölogy. U. Ia., from 1929. Chmn. scientific advisory com., Biol. Lab., Cold Spring Harbor, N.Y.; mem. exec. com. div. of agr. and biology, NRC, 1933-34, vice-chmn. of div., 1934-35, mem. fellowship bd., 1941; mem. Ia. Basic Sci. Bd., 1935-50; mem. Atomic Energy Commn. Fellowship Bd., 1948——; dir. Iowa Lakeside Laboratory, from 1932. Fellow A.A.A.S. (rep. Am. Physiol. Soc. 1948-49); mem. Am. Assn. Zoologists (v.p. 1933-34, pres. 1947, exec. com. 1948-52), Nat. Acad. Scis., Am. Physiol. Soc., Soc. Exptl. Biology and Medicine, Am. Soc. Naturalists (v.p. 1938, pres. 1940), Am. Micros. Soc., Sigma Xi. Mem. editorial bd. Physiol. Zoölogy; asso. editor. Jour. Morphology, Am. Naturalists. Died July 23, 1954.

BODLÄNDER, Guido, chemist; b. 1855. Prof. in Breslau. Research on electrolytical, optical properties of solutions; set up valence theory based on electrical affinities; studied optical properties of crystals. Died 1904.

BODMAN, Geoffrey Baldwin, soil physicist; b. Barking, Eng., Oct. 11, 1894; s. George and Ann Elizabeth (Perry) B.; B.S.A., U. Sask., 1919; M.S., U. Minn., 1924; Ph.D., 1927; m. Helen Mary English, Sept. 23, 1920; children—Elizabeth Helen (Mrs. W. A. Shuey), Margaret Jean (Mrs. Harding Rose), Anne (Mrs. G. G. Van Vlack). Came to U. S., 1922, naturalized, 1936. Asst. prof. U. Minn., 1927; mem. faculty U. Cal. at Berkeley, 1927——, prof. soil physics, 1939-62, chmn. dept. soils, 1948-55, emeritus, 1962——; soil scientist mil. geol. unit U. S. Geol. Survey, 1943-45. Vis. prof. Nat. Taiwan U., 1956, U. Alexandria, 1962-63; cons. editor profl. mags. Recipient Outstanding Achievement award U. Minn., 1952. Fellow Am. Soc. Agronomy, A.A.A.S.; mem. Soil Sci. Soc. (pres. 1958), Western, Brit. socs. soil sci., Am. Geophys. Union, Internat. Soc. Soil Sci., Sigma Xi, Phi Lambda Upsilon, Gamma Sigma Delta. Research and numerous publs. on phys. properties of soils; water retention and energy relationships of soil and water; conductivity of water by, entry into soils; soil and soil clay interaction with synthetic organic polymers; viscosity of clay suspensions; soil aggregation, compaction, soil profiles. Home: 1590 Le Roy Av., Berkeley, Cal. 94708.*

BODMER, Johann Georg, inventor; b. Zurich, Switzerland, 1786; lived in Eng., Austria. Patentee steam engines, ship engines; invented cylinder with opposite pistons, expansion slide-valve, rotary slide-valve, valveless air pump with piston plunging into condensed water; built cotton spinning machine that revolutionized the industry, introduced in Manchester, Eng., 1824. Died 1864.

BODOT, Hubert, chemist; b. N.Y.C., Jan. 25, 1933; s. Bernard and Marguerite (Deschamps) B.; Ph.D., Faculté des Sciences, Montpellier, France, 1957; m. Paulette Guirard, Nov. 24, 1955. Research trainee Centre National de la Recherche Scientifique, Montpellier, 1954-56; asst. Faculte des Sciences, Montpellier, 1956-58, supr., 1958-60; lectr. U. Abidjan, Ivory Coast, West Africa, 1961-63, prof., 1964; prof. Faculté des Sciences, Marseille, from 1964. Mem. French, Am. chem. socs. Research, publs., on phys. organic chemistry, especially kinetic study chlorohydrin solvolysis (mechanisms of neighboring group participation), also conformational analysis. Home: Lotissement des Cauvelles, Allauch 13, France. Office: Traverse de la Barasse, Marseille 13, France.*

BODROGI, György, Hungarian cardiologist; b. Nagyenyed, Hungary, June 8, 1906; s. Janos and Roza (Bartok) B.; ed. Bethlen Gabor Coll., 1926, Pazmany Peter U., Budapest, Hungary, 1926-30, Sch. U., Paris, France, 1931; Rockefeller research fellow Inst.

Hygiene, Budapest, 1931-32; m. Zelma Palfi, Apr. 29, 1939; children—Ildikó, György. Internist, St. István, St. Rókus Hosp., Budapest, 1933-45; dir. Lorántffy Zs. Hosp., Budapest, 1943-52; cons. cardiologist Juvenile Welfare Center for Heart Diseases, Budapest, 1952-58; research dir. dept. orgn. Nat. Inst. Cardiology, Budapest, 1958-61; dir. Juvenile Welfare Center for Heart Diseases, Budapest, 1961——. Mem. Sodalitas Cardiologorum, Hungaroum. Research, publs. on incidence in Hungary of diseases caused by Entamoeba hystolitica, value of external mechanograms in estimating intraventricular pressure, practical importance of auscultation and phonocardiography, diagnostic value of phlebogram and other mechanograms in cases of heart failure. Home: 2 Fehérhajó. Office: 27 Rosenberg, Budapest, V., Hungary.*

BOECKMANN, Johann Lorenz, German physicist, mathematician; b. Lübeck, Germany, May 8, 1741; s. Peter and Catherine Elisabeth (Häcker) B.; ed. Jena, Germany; m. Margarete Dorothea Eichrodt; 1766; 3 sons, including Karl Wilhelm. Apptd. prof. math. and physics Fürstenschule, Karlsruhe, Germany, 1764; became mem. Consistory, Baden, Germany, 1769; founder Baden meteorol. Inst., 1778; established sci. secondary schs., teaching seminars, Baden, also 16 weather stas. Author: Beyträge zur Geschichte der Mathematik und Naturlehre in den badischen Landen, 1787. Built working model of steam powered vehicle (after description of one built by Verbiest 1678), 1775. Died Karlsruhe, Dec. 15, 1802.

BOEHLER, Jorg, Austrian physician; b. Bozen, South Tyrol, Austria, Dec. 15, 1917; s. Lorenz and Leopoldine (Settari) B.; M.D., U. Vienna, 1941; m. Susi Foest-Monnshoff, June 18, 1947; children—Nikolaus, Elisabeth, Alexander, Peter. Asst., Accident Hosp., Vienna, Austria, 1945-51; chief service Accident Hosp., Linz, Austria, 1951——; prof. accident surgery U. Vienna, 1964——. Mem. numerous nat., internat. med. socs. Author: (with Lorenz Böhler) The Treatment of Fractures, 1965. Research, numerous publs. on treatment of fractures, wounds, brain, abdominal, vascular and nerve injuries. Address: 1 Blumauerplatz, A4020 Linz, Austria.*

BOEHM, Felix Hans, physicist; b. Basel, Switzerland, June 9, 1924; s. Hans G. and Marguerite (Philippi) B.; M.S., Inst. Tech., Zurich, 1948, Ph.D., 1951; m. Ruth Sommerhalder, Nov. 26, 1956; children—Marcus F., Claude N. Came to the U. S., 1952, naturalized 1964. Research asso. Inst. Tech., Zurich, Switzerland, 1949-52; Boese fellow Columbia, 1952-53; faculty Cal. Inst. Tech., Pasadena, 1953——, prof. physics, 1961——. Fellow, Am. Phys. Soc. Research on nuclear physics, nuclear beta decay. Home: 2510 N. Altadena Dr., Altadena, Cal. Office: Cal. Inst. Tech., Pasadena, Cal.*

BOEHM, Joseph Anton, Austrian botanist; b. Gross-Gerungs, Austria, Mar. 13, 1831; ed. under E. Fenzl, Franz von Unger, Vienna. Tchr., Commerce Sch.; prof. botany, Vienna, from 1869. Discovered water circulation in plant cells. Died Vienna, Dec. 2, 1893.

BOEHM, Martin, see Behaim, Martin.

BOEKE, Hendrik Enno, mineralogist; b. Wormerveer, Holland, Sept. 12, 1881; s. Isaak Hermann and Sara Maria (Van Gelder) B.; studied chemistry in Amsterdam from 1900, m. Leonore Mirandolle, 1906. Worked under G. Tamman in Göttingen, Germany 1905-06; asst. to mineralogist F. Rinne in Hanover, Germany, 1906; joined faculty U. Königsberg, 1908; became 1st German prof. of new dept. of physicochem. mineralogy and petrography in Leipzig, 1910; to Halle, Germany, 1911; to Frankfort/Main, Germany, 1914; part-time at U. Ghent (Belgium) during World War I. Mem. Leopoldina in Halle, Bataafsch Genotschap der Proefondervindelijke Wijsbegeerte in Rotterdam. Author: Die Anwendung der stereographischen Projektion bei kristallographischen Untersuchungen, 1911; Die gnomonische Projektion in ihrer Anwendung auf kristallographischen Aufgaben, 1913; Die Grundlagen der physikal-chemischen petrographie, 1915. Stressed math. and physico-chem. thinking and working methods in petrography as opposed to purely descriptive heretofore used. Died (suicide) Frankfort/Main, Germany, Dec. 6, 1918.

BOEKELHEIDE, Virgil, Am. organic chemist; b. Chelsea, S.D., July 28, 1919; s. Charles F. and Eleonor (Toennies) B.; A.B., U. Minn., 1939, Ph.D., 1943; m. Caroline A. Barrett, Sept. 1, 1945; children—Karl, Anne, Erich. Instr. U. Ill., 1943-46; from asst. to prof. organic chemistry U. Rochester, 1946-60; prof. organic chemistry U. Ore., 1960——; cons., 1962——. Mem. adv. panel NIH, 1962——. Guggenheim fellow, 1953-54; Swiss-Am. Found. lectr., 1960. Mem. Nat. Acad. Scis. Bd. editors Organic Synthesis, 1956——. Organic Reactions, 1956——, Jour. Am. Chem. Soc., 1962——. Synthesis of novel molecules; elucidation of structure of eurarl alkaloids; synthesis of unusual, novel heterocycles. Home: 2017 Elk Dr., Eugene, Ore. 97403.*

BOELL, Edgar John, Am. biologist; b. Rudd, Ia., Oct. 30, 1906; s. Albert Emil and Gertrude (Van Der Las) B.; B.A., U. Dubuque, 1929, D.Sc., 1952; Ph.D., State U. Ia., 1935; postgrad. (Rockefeller Found. fellow), Cambridge U., 1937-38; m. Mildred Cottingham, June 3, 1932; children—Carl David, Dorothy Eleanor. Faculty, Yale, New Haven, 1938——, prof. biology, 1946, Ross Granville Harrison prof. exptl. zoology, 1947——. Cons. editor in zoology McGraw-Hill Book Co., 1949-62. Fulbright fellow, Carlsberg Lab., Copenhagen, 1953-54; Guggenheim fellow Universita di Roma, 1963-64; recipient Distinguished Alumnus award U. Dubuque, 1965. Fellow Am. Acad. Arts and Scis.; mem. Am. Soc. Cell Biology, Am. Soc. Zoology (exec. com. 1960-63), Am. Physiol. Soc., Internat. Soc. Cell Biology, Soc. for Developmental Biology (pres. 1952-53), Internat. Inst. Embryology, Soc. Gen. Physiology, Conn. Acad. Arts and Scis., A.A.A.S., Am. Inst. Biol. Sci., Sigma Xi. Co-mng. editor Am. Jour. Exptl. Zoology, 1968——. Research in enzyme synthesis and functional differentiation of various tissues in insect, amphibian, avian and mammalian embryos. Home: 577 Skiff St., North Haven, Conn. 06473. Office: Kline Biology Tower, Yale U., New Haven 06520.*

BÖER, Karl Wolfgang, physicist; b. Berlin, Germany, Mar. 23, 1926; s. Karl Wilhelm and Charlotte (Gruhlke) B.; Dipl.Phys., Humboldt U., 1949, Dr. rer.nat., 1952, Dr. rer.nat.habil., 1955; m. Ingeborg Krause, Sept. 3, 1947 (div.); m. Margit Krause, Feb. 27, 1961 (div.); 1 son, Ralf-Reinhard; m. 2d, Renate Schröder, May 19, 1967. Faculty, Humboldt-U., Berlin, 1949-61, prof., 1961; research prof. N.Y. U., 1961-62; asso. prof. physics, U. Del., Newark, 1962-67, prof., 1967——; head lab. Diel. Breakdown, German Acad. Sci., Berlin, 1955-61. Cons. numerous pvt. cos., govt. agys. Fellow Am. Phys. Soc.; mem. German Phys. Soc. Author: (with (Grimsehl, D. Lehrb) Physik IV, 1961; Festkörper Physik; (with Simon-Suhrmann, Photozellen) Photoleitfahgkeit, 1958; also numerous articles. Founder, editor-in-chief Physica Status Solidi, 1961; editor Fortschritte der Physik, 1953-54. Discovered fieldinstabilities in homogeneous semicondrs.; 1st exptl. observation of Franz-Keldysh effect, 1st electroluminescence of single crystals, dielectric breakdown, photo-conductivity, defectstructure, electron noise, photochem. reactions, dielectric aftereffects, influence of ambient atmosphere on photocondrs. Home: 1205 Nottingham Rd., Newark, Del. 19711.*

BOERBOOM, Anne Johannes Hendrik, Dutch physicist; b. The Hague, Netherlands, Feb. 28, 1922; s. Jan Harman and Antje (Schuwer) B.; ed. Tech. U., Delft, Holland, 1947, Municipal U., Amsterdam, Holland, 1952; postgrad. State U., Leiden; m. Adriana VandenBerg, July 22, 1953; children—Hessel S. K., Simon K. Asst. prof. Municipal U., Amsterdam, 1947-52; sr. scientist Lab. Atomic and Molecular Physics, Amsterdam, 1952——. Bd. dirs. Found. Isotope Geol. Research. Author: (with C. A. McDowell) Mass Spectrometry, 1963; (with R. I. Reed) Mass Spectrometry, 1965. Contbr. articles to profl. jours. Research on exptl. and theoretical mass spectrometry including high temperature thermochemistry and electron-atom collisions, calculation in field of ion optics as ion orbits and electric and magnetic fields in spl. electrode and pole piece configurations. Home: 25 II Albrecht Dürerstraat. Office: 407 Kruislaan, Amsterdam, Holland.*

BOEREMA, I., Dutch surgeon; b. Uithuizen, Netherlands, Oct. 14, 1902; s. Waalko and Annechien (Kruize) B.; M.D., U. Groningen, 1928; Ph.D., U. Glasgow (Scotland); m. Agatha Johanna Tolk, Aug. 3, 1933; children—Waalko Joost, Eegje (Mrs. H. A. de Ruyter), Bram, Annechien (Mrs. J. W. Menso), Liesbeth. Staff surg. dept. U. Groningen, 1928-46; prof. surgery U. Amsterdam, Wilhelmina Gasthuis, Amsterdam. Decorated knight Order Dutch Lion; officer Santé Publ. de France; officer la Legion d'Honneur. Hon. fellow A.C.S.; hon. mem. Soc. Thoracic Surgeons Gt. Britain and Ireland, Am. Assn. Thoracic Surgeons, Internat. Coll. Surgeons. Contbr. numerous articles to tech. jours. Invented (at same time as another dr.) hypothermia; introduced steel artificial joints, slow ligature of vessels, total gastrectomy without suture, ligation of oesophageal varices, application of hyperbaric oxygen in surgery. Home: 90 Apollolaan. Office: Wilhelmina Gasthuis, le Helmersstraat, 104, Amsterdam, Netherlands.*

BOERHAAVE, Hermann, Dutch physician, scientist; b. Voorhout, Netherlands, Dec. 31, 1668; s. James and Hagar B.; Ph.D., U. Leyden, 1689; M.D., Harderwijk, Netherlands, 1693; m. Marie Drolenraux, 1710; 4 children. Practiced medicine, Leyden; pvt. tchr. math. and chemistry; became lectr. in theoretical medicine U. Leyden, 1701, prof. medicine, 1709, prof. botany, 1709-29, prof. practical medicine, 1714, rector, 1714, prof. chemistry 1718-29. Fellow Roy. Soc., 1730 (Philos. Trans. dedicated to him 1737); asso. French Academy Scis., 1728. Author: Institutiones medicae, 1708; Aphorismi de cognoscendi et curandis morbis in usam doctrinae, 1709; Index plantarum, 1710; Institutiones et Experimenta Chemiae, 1724; Elementa Chemiae, 1732. Editor anat. and surg. works of Vesalius. Founded modern method of clin. teaching; revived the Hippocratic method of clinical observation and instruction; gave 1st description of sweat glands (Boerhaave's glands), 1693; established that smallpox is spread by contact alone; made earliest expt. in chem. calorimetry; proved experimentally that heat is weightless; distinguished mixtures from compounds; formulated laws of solubility of air in liquids; improved Leyden botanical gardens; studied sexuality of plants; disseminated Newtonian science on continent; studied pathology, hygiene, therapeutics, and botany; texts on chemistry and physiology were standard works of his time. Died Leyden, Sept. 23, 1738.

BOERMAN, Willem Everhard, Dutch geographer; b. Arnhem, May 27, 1888; s. Jan Willem and Louisa (Hendriks) B.; ed. U. Utrecht; Dr. honoris causa U. Durham; m. Jacoba Voorthuis, 1917; children—Anna Jacoba Louise, Louise. Tchr., Rotterdam, 1908-10; asst. geography inst. U. Utrecht, 1910-12; prof. Netherlands Lyceum, La Haye, 1912-21; lectr. social and econ. geography Ned. Econ. Hogeschool, Rotterdam, 1920-23; full prof., 1923-49, now prof. emeritus. Mem. geog. socs. of Edinburgh, Louvain, Vienna, Netherlands (hon.), Netherlands Soc. Geography, Soc. Econ. and Social Geography, Bataafse Gen., Council Urbanization, Internat. Geographers Union (v.p. 1935-49). Author: Tijdschrift v. Economische en Sociale Geografie (autobiography), 1958. Initiator of Wetenschappelijke Atlas van Nederland, 1931. Address: Van Beuningenstr. 18C, Rotterdam-4, Netherlands.

BOERSCH, Hans Paul, German physicist; b. Berlin, Germany, June 1, 1909; s. Rudolf and Gertrud (Rössel) B.; student Technische Hochschule Berlin-Charlottenburg, Germany, 1930-33; Dr.phil., U. Vienna (Austria), 1935, Dr.phil.habil, 1942; m. Maria Theresia Haar, Feb. 22, 1963; children—Margot, Peter, Jutta, Sabine. Sci. co-operator AEG-Forschungsinstitut, Berlin-Reinickendorf, Germany, 1935-40; faculty U. Vienna, 1941-46, U. Innsbruck (Austria), 1946; leader physics lab. Institut de Recherches Scientifiques, Tettnang, Wurttemberg, Germany, 1946-48; hon. prof., also leader lab. physikalisch-Technische Hochschule Braunschweig, Germany, 1948-54; prof. (ordentlicher prof.), dir. I. Phys. Inst., Technische Universitat Berlin, 1954——. Recipient Silberne Leibnizmedaille, Preuss. Acad. d. Wissenschaften, 1941. Mem. Braunschweigischen Wissenschaftlichen Gesellschaft (corr.), Deutsche Physikal. Gesellschaft, Gesellschaft für angew. Optik, Gesellschaft für Elektronenmikroskopie. Research, publs. on light and electron-optics; investigation on interchanging of elementary particles with materie and electromagnetic fields. Home: 17 Glockenstrasse, West Berlin 37. Office: 34 Hardenbergstrasse, West Berlin 12, 1 Germany.*

BOESCH, Ernest E., psychologist; b. St.-Gall, Switzerland, Dec. 26, 1916; ed. Geneva, Zurich (both Switzerland) univs.; dir. Internat. Inst. Child Study (research on devel. Thai children), UNESCO, Bangkok, Thailand, 1955-58; research in Thailand, 1960; dir. Inst. Psychology, U. Saarbrucken, Germany. Mem. Swiss, German, French psychol. assns., Internat. Psychoanalytical Assn. Work in developmental, ednl., clin. psychology, psychotherapy, comparative social and cultural psychology. Address: Eichendorffstrasse 18, Scheidt/Saar, Germany.

BOESCH, Hans Heinrich, Swiss geographer; b. Zurich, Switzerland, Mar. 24, 1911; s. Paul B. and Elsa Maria (Carpentier) B.; ed. U. Zurich, Clark U.; Ph.D., D.Sc. honoris causa; m. Henny Wild, 1940. Geologist, Brit. Oil Devel. Co. Ltd., 1937-38; prof. extraordinary U. Zurich, from 1941, prof., dir. Geog. Inst., from 1942; v.p. Union International de Geographie, 1949-56, sec. gen., 1956——. Vice pres. Internat. Council Sci. Unions. Mem. Royal, Scottish Royal geog. socs. Author: Wasser oder Oel, 1943; Die Wirtschaftslandschaften der Erde, 1947, Zentralamerikaheute, 1952; USA-die Erschliessung eines Kontinents, 1956; Der Mittlere Osten, 1959; Flugbild der Schwietzer Stadt, 1963; Geography of World Economy, 1964; Weltwirstschaftsgeographie, 1966. Research in econ., regional geography. Address: Geographisches Institut der Universitat, Blumlisalpstrasse 10, 8006, Zurich, Switzerland.

BOESLER, Felix Carl August, German polit. scientist; b. Leipzig, Germany, Mar. 22, 1901; s. Arthur and Marie (Beyer) B.; Ph.D. in Polit. Sci., U. Leipzig; m. Elizabeth Schulz, Sept. 8, 1928; children—Klaus-Achim, Albrecht. Promotion at U. Leipzig, 1925, head budget archives, 1928; agregation, 1930; mem. commn. on teaching adminstrv. scis., 1930; full prof. U. Kalingrad (Königsberg), 1936; full prof. social sci. U. Jena, 1939, also dir. Inst. History and Practice of Finance; mem. commn. on teaching research of structure of urban and regional constrns. Technische Hochschule of Stuttgart: in charge Inst. Research of Structures, Stuttgart. Mem. Regional Sci. Assn., Akademie für Raumforschung und Landesplanung (corr.). Research and publs. on finance, adminstrv. sci., statistics, urbanization, structures, exam. of structures, infrastructure. Home: Kirchaldenstrasse 6, Stuttgart-Botn. Office: Silcherstrasse 9, Stuttgart, Germany.

BOETHIUS, Anicius Manlius Severinus, Roman philosopher, statesman; b. Rome, circa 480; s. Flavius Manlius B.; m. Rusticiana, dau. of Symmachus; children—Anicius Manlius Severinus, Q. Aurelius Memmius Symmachus. Became favorite of Theodoric (Ostrogoth ruler of Rome from 500); became consul, 510 (his 2 sons jointly held same honor 522); shortly thereafter was accused of conspiring against Theodoric, condemned and imprisoned in Pavia, finally decapitated without trial, 524. Author: On the Consolation of Philosophy; transls. and commentaries on logical works of Aristotle; on Porphyry's Isagoge and on Cicero's Topica; transl. and revision of arithmetical and mus. works of Nichomachus of Gerasa; also wrote on Euclid's Elements in a work now almost completely lost, on astronomy (works completely lost) and several theol. treatises. Often considered the last rep. of the great Greco-Roman tradition, was perhaps the last Roman with an understanding of Greek; his greatest work, On the Consolation of Philosophy (written in prison) is a philos. dialogue between himself and figure representing philosophy; though a Christian, he employed arguments of pagan philosophy, particularly in discussions on virtue and free will; famous also as transmitter of Aristotelian and Platonic thought to the early Middle Ages; main contbn. to science was his transl. of certain of Aristotle's works and his commentaries on them, thus making an important part of Aristotelian logic available to medieval scholars; wrote summaries on various sci. subjects and treatises on logic, arithmetic, music, theology, geometry and astronomy; introduced a considerable improvement on the old abacus, supposedly used numerals of Indian origin; developed a rule for finding the number of combinations of "n" things taken two at a time. Died Pavia, 524.

BOETTGER, Oskar, German zoologist, paleontologist; b. Frankfurt/Main, Germany, Mar. 31, 1844; s. Rudolph Christian and Sophie Elisabeth (Harpke) B.; studied at Bergakademie, Freiberg, Germany, 1863-66; Ph.D., Würzburg, 1869. Tchr. Offenbach/Main, 1873, Frankfurt Musterschule, 1873-76; hon. curator of the paleontol. and herpetological collections Senckenberg Mus.; prin. Frankfurt Wöhler-Realgymnasium; prof. Research and numerous publs. on taxonomy of modern and prehistoric molluscs, especially clausiliidae, modern amphibians and reptiles; described over 1000 new types, many new species. Editor mag. Der Zoologische Garten, later Zoologischer Beobachter, 1896-1910. Died Frankfurt/Main, Sept. 25, 1910.

BOETTGER, Otto Werner, German physicist; b. Halle, Germany, June 18, 1923; s. Otto Karl and Frieda (Zaulig) B.; Dipl.-phys., Martin-Luther-U. Halle, 1951, Dr.rer.nat., 1955, Dr.rer.nat.habil., 1958; m. Eva Maria Zempel, Dec. 28, 1948; children—Birgit, Erik and Thomas (twins). Tchr., lectr. II Inst. Physics, U. Halle, 1952-58; staff research Inst. Allgemeine-Electricitaets-Gesellschaft, 1958-60, group leader, 1960-62; lectr., apl. prof. U. Mainz, 1962-66, prof., 1966. Mem. German Phys. Soc. Author: (with T. Mohr, S. Methfessel) Thin Layers, 1953; (with G. C. Moench, U. Zorll) Interference of Light, 1954. Research and publs. on different works on semiconductor materials for thermoelectricity; on isotope separation; inventor of prodn. of single crystals of silicon. Home: 21 Adolf-Miersch-Str., 6000, Frankfurt-am-Main, Germany.*

BOFFI, Vinicio Claudio, Italian physicist; b. Montopoli, Sabina, Italy, Aug. 17, 1927; s. Galliano and Candida (Merennoni) B.; m. M. Vittoria Bortolani, Aug. 8, 1964; 1 dau., Francesca Romana. With C.N.E.N., atomic energy commn. Italian Govt., Bologna, 1957—, dir. nuclear data lab. C.N.E.N. Computation Center, 1965—; asst. prof. U. Bologna. Research, publs. on theoretical neutron physics and related topics on math. aspects and solutions of Boltzmann equation in study of slowing down and thermalization of neutrons. Home: 43 La Spezia, Rome, Italy. Office: 2 Mazzini, Bologna, Italy.*

BOGAARDT, Maarten, physicist, mech. engr.; b. Djakarta, Indonesia, Apr. 2, 1922; s. Henri and Maartje A. J. (Haallebos) B.; B.S., U. Groningen, 1944; postgrad. Coll. de France; Ph.D., U. Utrecht, 1953; Sci. officer, Found. for Fundamental Research of Matter, 1950-55; design engr. Royal Dutch Oil Co., The Hague, 1955-58; dir. research design and devel. Netherlands Reactor Centre, The Hague, 1958—; prof. mech. engring. Tech. U. Eindhoven (Netherlands), 1958—. Author various books, numerous articles. Design research reactors; devel. marine reactors; research fluid flow in 2-phase systems. Home: 43 Van Hoornbeekstraat, The Hague, Netherlands. Office: Technol. U. Eindhoven, P.B. 513, Eindhoven, Netherlands.*

BOGACZ, Jaime, neurophysiologist; b. Poland, Sept. 15, 1925; s. Abraham and Ena (Gerwer) B.; B.S., Pub. Edn. Ministry, Uruguay, 1945; M.D., Faculty Medicine, Uruguay, 1954; m. Blanca Fressola, June 14, 1944; children—Ana, Alicia, David, Daniel. Staff, Faculty Medicine, Montevideo, Uruguay, 1954—, research asst. Lab. Neurophysiology, Inst. Neurology, 1960-62, head Central EEG Lab., Hosp. de Clinicas, 1962—. Cons., EEG, Child Welfare Council, Uruguay. Internat. USPHS fellow, 1963. Mem. Sociedad de Neurologia, Sociedad Sudamericana de EEG y Neurofisiologia Clinica. Research, publs. on vegetative phenomena in subcortical epilepsies, EEG abnormalities in hepatic insufficiency, sensory evoked potentials in man, cerebral elec. correlates with motivation. Home: Augusto Turenne 3364, Montevideo. Office: Laboratorio de EEG, Hosp. de Clinicas, Montevideo, Uruguay.

BOGARDUS, James, Am. inventor; b. Catskill, N.Y., Mar. 14, 1800; s. James and Martha (Spencer) B.; m. Margaret Maclay, Feb. 12, 1831. Awarded gold medals Am. Inst. N.Y. for 8-day, 3-wheeled chronometer clock, 1828, for invention of dry gas meter, 1835; won award for engraving machine and new method of making postage stamps, 1839; other inventions include: ring flier (cotton spinning), dynamometer, pyrometer, deep sea sounding machine, drilling machines; introduced cast-iron usage for frames, floors and all bldg. supports; built Bogardus Bldg. (1st all cast iron bldg. in world), 1850; designer Balt. Sun Bldg., Birch Bldg. Chgo., Pub. Ledger Bldg., Phila.; made numerous improvements in manufacture of tools and machinery. Died N.Y.C., Apr. 13, 1874.

BOGDAN, Alexei Vasilyevich, botanist; b. Saratov, Russia, Mar. 1, 1905; s. Vasilii Semonovich and Zinaida (Stavrovskaya) B.; agronomist degree Kuban Agrl. Inst., 1928; degree biol. scis. Rostov U., 1936; m. Valentina Ivanova, Feb. 11, 1932; 1 dau., Natalie (Mrs. John Hopewell). Mem. faculty Kuban Agrl. Inst., Krasnodar, U.S.S.R., 1928-35, Rostov U., 1935-42; pasture research officer Ministry Agr., Kenya, Nairobi-Kitale, 1947-66; sci. information officer Commonwealth Bur. Pastures and Field Crops, Hurley, Eng., 1966—. Fellow Linnean Soc. London; mem. Am. Ecol. Soc., Brit. Ecol. Soc., Brit. Grassland Soc., Wild Life Soc. Author: (with D. C. Edwards) Important Grassland Plants of Kenya, 1951; A Revised List of Kenya Grasses, 1958; also numerous articles. Revised Kenya pasture grasses and clovers as resources for introduction into cultivation; introduced into cultivation Kenya grass varieties; established breeding and pollinating habits of some tropical grasses and clovers; devel. method of determining depth of emergence of grass seedlings. Home: 13 Tunstall Ct., Hatherley Rd., Kew, Surrey, Eng. Office: Commonwealth Bur. Pastures and Field Crops, Hurley, Berks, Eng *

BOGDANOV, Fedor Rodionovich, Russian orthopedist, traumatologist; b. 1900; grad. Med. Faculty, 1st Moscow U., 1925; D.Med. Sci., 1938. Asso., Skilfosovsky Inst. First Aid, Moscow, 1926-32; asst. Surg. Clinic, Sverdlovsk Med. Inst., 1933-37, head chair gen. surgery, 1938-58; head clin. records dept. Sverdlovsk Research Inst. Restorative Surgery, Traumatology and Orthopedics, RSFSR Ministry Health, 1933-43, dir., 1944-58; prof., 1938—; chief surgeon Urals Mil. Dist., 1941-48; head chair orthopedics and traumatology Kiev Postgrad. Med. Inst., 1959—; sci. dir. Ukrainian Research Inst. Orthopedics and Traumatology, Ukrainian Ministry Health, 1959—. Mem. USSR Acad. Med. Sci. (corr.), All-Union Soc. Surgeons (bd. mem.). Author over 90 works including Interarticular Fractures, 1949; Orthopedo-Traumatological Apparatus and Instruments, 1949; Surgical Treatment of Foot Injuries and Diseases, 1953. Co-editor Surgery sect. Large Med. Ency., 2d edit.; mem. editorial council Clin. Medicine; mem. editorial bd. Orthopedics, Traumatology and Protheses. Research on gen. mil. surgery, orthopedics, traumatology and neurosurgery. Address: Ukrainian Research Inst. Orthopedics and Traumatology, ulitsa Vorovskogo 27, Kiev, Ukrainian SSR, USSR.

BOGDANOV, Ivan Lukyanovich, Russian epidemiologist; born 1903; grad. Medical Faculty, Smolensk U., 1927; D.Med. Sci. Head chair infectious diseases Sverdlovsk Med. Inst., 1941-51; chief epidemiologist Urals Mil. Dist., 1941-47; dir., head poliomyelitis dept. Kiev Inst. Infectious Diseases, USSR Acad. Med. Sci., 1951—. Del., 2d Internat. Congress on Infectious Diseases, Milan, 1959. Mem. USSR Acad. Med. Sci. (corr.). Co-editor Epidemiology and Infectious Diseases sect. Large Med. Ency., 2d edit.; mem. editorial council Jour. Microbiology, Epidemiology and Immunobiology; mem. editorial bd. Med. Affairs, Problems of Virology. Research and numerous publs. on etiology, clin. aspects, pathology, epidemiology, prophylaxis and therapy of infectious diseases. Home: Pecherski Spusk, 18-87, Kiev 11. Kiev Inst. Infectious Diseases, Tsitadell 11, Kiev, Ukraine SSR, USSR.*

BOGDONOFF, Morton David, Am. physician; b. N.Y.C., Dec. 8, 1925; s. M. Myron and Minnie (Alpher) B.; B.S., Cornell U., 1944, M.D., 1948; m. Jano Segal, July 1, 1951; children—Reid, Ladd, Jesse, Drue. Eli Lilly Research fellow div. endocrinology and metabolism dept. medicine Duke, 1953-54; faculty U. Miami (Fla.) Med. Sch., 1956-57; faculty Duke Med. Sch., Durham, N.C., 1957—, prof. medicine, 1962—, asst. dean grad. med. edn., 1967—. Diplomate Nat. Bd. Medical Examiners, Am. Bd. Internal Medicine. Fellow A.C.P.; mem. Am. Fedn. for Clin. Research (past pres., editor Clin. Research 1959-64), Am., So. socs. for clin. investigation, Assn. Am. Physicians, A.A.A.S., Endocrine Soc., Psychosomatic Soc. (past nat. councillor), Soc. for Exptl. Biology and Medicine, Gerontological Soc., N.C. Heart Assn., N.Y. Acad. Scis., Am. Diabetic Assn., A.M.A. (chief editor Archives Internal Medicine 1967—), Alpha Omega Alpha. Research, publs. on diabetes, social and situational stress effects on heart and endocrine function. Home: 2425 Wrightwood Av., Durham, N.C. 27705.*

BOGDONOFF, Seymour Moses, Am. aero. engr.; b. N.Y.C., Jan. 10, 1921; s. Glenn and Katie (Cohen) B.; B.S. in Engring., Rensselaer Poly. Inst., 1942; M.S. in Aer. Engring., Princeton, 1948; m. Harriet Eisenberg, Oct. 1, 1944; children—Sandra Sue, Zelda Lynn, Alan Charles. Asst. sect. head fluid and gas dynamics NACA, Langley, Field, Va., 1942-46; research asso. Princeton, 1946-50, faculty, 1950—, prof. aero. engring., 1957—, Robert Paterson prof., 1963—, head gas dynamics lab. Mem. sci. adv. bd. USAF, 1960—, Northrup Norair, 1963—; mem. research adv. com. NASA, 1963—. Dir. Therm., Inc., Ithaca, N.Y. Fellow Am. Inst. Aeros. and Astronautics (dir.); mem. Sigma Xi. Contbr. numerous articles to tech. jours. Research on axial flow compressors and turbins, viscous effects in aerodynamics, hypersonic aerodynamics, hypersonic aerodynamics at very high altitudes; developed new research tools for high speed aerodynamics. Home: 39 Random Rd., Princeton, N.J. 08540.*

BöGEL, Konrad Winfried, German virologist, veterinarian; b. Cannstatt, Germany, Jan. 16, 1931; s. Erich and Renate (Lenz) B.; student Vet. Sch., U. Vienna (Austria), 1954, U. Stuttgart (Germany), 1950; Dr. med. vet., U. Giessen (Germany), 1956; diploma Immunology, Pasteur Inst., Paris, France, 1965; m. Eleonore Erb, Oct. 26, 1956; children—Uta, Susanne. Field veterinarian, 1956; staff State Vet. Inst., Stuttgart, 1957; staff Fed. Research Inst. for Virus Diseases of Animals, Tübingen, Germany, 1958-63, French Vet. Inst. for Tropical Countries, Ft. Lamy, Republic of Tchad, Africa, 1964; virologist Fed. Research Inst. for Virus Diseases of Animals, Tübingen, 1965-67; with vet. pub. health unit, div. communicable diseases WHO, 1967—. Mem. German Vet. Med. Assn., Orgn. for Grads. Pasteur Inst. Research and publs. on viral enteric and respiratory diseases of cattle, isolation, characterization and grouping of bovine enteroviruses and bovine rhinovirus, epizootological investigations of bovine muscosal disease and respiratory diseases, devel. living tissue culture vaccine against parainfluenza 3, diagnosis of rinderpest. Home: 6 Ave. de Budé, 1211 Geneva. Office: Vet. Pub. Health Unit, Div. Communicable Diseases, WHO, Geneva, Switzerland.*

BOGERT, Marston Taylor, Am. chemist; b. Flushing, N.Y., Apr. 18, 1868; s. Henry A. and Mary B. (Lawrence) B.; A.B., Columbia, 1890, Ph.B., 1894, Sc.D., 1929; LL.D., Clark U., 1909; R.N.D., Charles U. (Prague); m. Charlotte E. Hoogland, Sept. 12, 1893; children—Annette B. (Mrs. Frank B. Tallman), Elise B. (Mrs. F. K. Huber). Asst. organic chemistry Columbia U., N.Y.C., 1894-97, tutor, 1897, instr., 1897-1901, adj. prof., 1901-04, prof., 1904-39, emeritus prof. in residence from 1939; rep. Columbia U. on bd. trustees N.Y. Coll. of Pharmacy, 1930-36. Lecturer organic chemistry, N.Y. U., 1919-20. Mem. Am. Advisory Com. of Honor 7th Internat. Congress of Applied Chemistry, London, 1909; pres. Organic Sect., 8th Internat. Congress Applied Chemistry, Washington, and N.Y., 1912; mem. of White House Conf. on Conservation of Natural Resources, 1908; affiliated in both exec. and adv. capacities with many govt. agys. Fellow A.A.A.S., London Chem. Soc., Royal Soc. of Edinburgh (hon.); mem. Assn. chimica Italiana, Société Chimique de Paris, Nederland Chem. Ver., Swiss Chemists' Soc., Am. Inst. Chemists (medalist for 1935-36), Am. Chem. Soc. (pres., Nichols medal, 1905; Priestley medal, 1938); hon. mem. Soc. Chem. Industry Eng. (ex-pres.), Chemists Club (pres.), Nat. Acad. Scis. (chmn. chem. sect., 1926-29), Washington, N.Y. acads. scis., Am. Philos. Soc., Am. Acad. Arts and Scis., Am. Assn. Univ. Profs., Phi Beta Kappa, Delta Phi, Sigma Xi. Asso. editor Jour. Am. Chem. Soc., 1924-30; mem. bd. editors and editor trustee, Jour. Organic Chemistry from 1937; contbr. profl. jours. Did internationally recognized work in synthesis organic compounds, especially aromatic compounds. Died Mar. 21, 1954.

BOGGESS, William Randolph, Am. forester; b. Oakvale, W.Va., Apr. 9, 1913; s. Bernard F. and Maude (Boyd) B.; A.B., Concord Coll., 1933; M.F., Duke, 1940; m. Effie Cowan, Apr. 9, 1938; children—Randolph, Sam, Mary (Mrs. Carol Ray Daniel), Elizabeth. Biology tchr. Mercer County Schs., 1933-35; staff Ala. Poly. Inst., Auburn, 1939-43, asso. forester, 1946-47; faculty Dixon Spring Agr. Center, U. Ill., 1948-58, prof. forestry, Urbana, 1958—. Mem. Am. Water Resources Assn. (dir. 1965—, editor Water Resources Bull., 1965—), Soc. Am. Foresters, Ecol. Soc. Am., Soil Sci. Soc. Am., A.A.A.S., Am. Geophys. Union, Soil Conservation Soc., Am., Ill. Acad. Sci. Research and publs. on effect of environmental factors on tree growth with spl. emphasis on water relations. Home: 509 W. Washington St., Urbana, Ill. 61801.*

BOGGS, S(amuel) Whittemore, Am. geographer; b. Coolidge, Kans., Mar. 3, 1889; s. Charles F(airman) and Lillian Louise (Whittemore) B.; B.L., Berea Coll., 1909, D.Sc., 1949; student Yale U., 1912-13; M.A. Columbia U., 1924; m. Amy Burt Bridgman, Aug. 16, 1916; children—Mary Lillian, Barbara Bridgman. Pvt. sec. to Pres. Frost, Berea Coll., 1909-12; secretarial work U.P.R.R., Omaha, Neb., 1913-14; secretarial and exec. work Internat. Com. of Y.M.C.A. and other orgns., N.Y., 1914-19; geog research, map compilation, editor Am. Book Co., 1916-24, edited maps for World Missionary Atlas, 1921-24; geographer U. S. Dept. State, from 1924; lectr. on internat. boundaries, Columbia U., summers 1939, 40, 41, 42; lectr. Am. Univ., 1945-46. Tech. adviser U. S. delegation, Conf. for Codification of Internat. Law, The Hague, 1930; ofcl. del. Internat. Geog. Congresses, Cambridge, 1928, Paris, 1931, Warsaw, 1934, Lisbon, 1949, Pan Am. Inst. Geography and History, 3d gen. assembly, Lima, 1941, 4th gen. assembly Caracas, 1946, 1st consultation on geography Rio de Janeiro, 1949, 5th gen. assembly, Santiago, Chile, 1950, 3d consultation, Washington, 1952, 6th consultation on cartography, Ciudad Trujilio, 1952; mem. U. S. Geog. Bd., 1924-34, chmn. exec. com., 1927-34, U. S. Bd. on Geog. Names, 1947—, chmn., 1949-51. Fellow

A.A.A.S., Royal Geog. Soc.; mem. Assn. Am. Geographers (councillor 1941-42), Am. Soc. Profl. Geographers, Am. Council Learned Soc. (sec.-treas. 1942-48, treas., 1948-50), Acad. Polit. Sci., Am. Geophys. Union, Wash. Acad. Sci., Mexican Soc. Geography and Statistics (hon.), Lima Geog. Soc. (corr.). Author: International Boundaries—A Study of Boundary Functions and Problems, 1940; Classification and Cataloging of Maps and Atlases (with Dorothy C. Lewis) 1945. Contbr. to Dist. Am. History and various mags. Died Sept. 14, 1954.

BOGOLEPOV, Nikolay Kirillovich, Russian neuropathologist; b. 1900 grad. Med. Faculty, 1st Moscow U., 1926; D.Med. Sci., 1946. Asso., Central Inst. Work Capacity Expertise, 1930-47, head dept. neuropathology, sci. dir., 1948-52; head chair med. labor expertise Central Postgrad. Med. Inst., Moscow, 1952-58; prof., 1952—; head chair nervous diseases Pirogov 2d Moscow Med. Inst., 1958; chief neuropathologist 4th Main Bd., USSR Ministry Health. Recipient Stalin prize, 1951. Mem. All-Union (bd. mem.), Moscow (bd. mem.), All-Russian (dep. chmn.) socs. neuropathologists and psychiatrists. Author of 140 works including Problems of Neuropsychiatric Outpatients Practice, 1936; Comatose States, 1950; Disorders of the Motor Functions in Vascular Lesions of the Brain, 1953; co-author: Medical Labor Expertise in Cerebral Traumas, 1957. Dep. editor Korsakov Jour. Neuropathology and Psychiatry; co-editor Neuropathology sect. Large Med. Ency., 2d edit. Address: 2d Moscow Med. Inst., Malaya Pirogovskaya 1, Moscow, USSR.

BOGOLIUBOV, Nikolai Nikolaevich, Russian mathematician; b. Nizhnii Novgorod, USSR, Aug. 8, 1909; s. Nikolai Bogoliubov; Ph.D. honoris causa Ukrainian Acad. Scis. With Ukrainian Acad. Scis., 1928—; prof. Kiev, Moscow (both USSR) univs., 1936-50; dir. Lab. Theoretical Physics, Joint Inst. for Nuclear Research, 1951—. Cons. Kharkov (USSR) Aviation Inst., Inst. Indsl. Power. Recipient Merlani prize Acad. Bologna (Italy), 1928, Stalin prize, 1947, Lenin prize 1958. Mem. Ukrainian, USSR acads. scis., French Math. Soc., Inst. Math. Established dept. nonlinear mechanics, Kiev, Sch. Theoretical Physics, Moscow; proved importance and laws of dispersion of elementary particles; work on statis. physics (theory of non-linear oscillation 1946), plasma physics, function theory, differential equations, theory of vibrations, theory of stability; gave new synthesis of Bohr's theory of quasi-periodic functions; developed methods for asymptotic integration of nonlinear equations describing oscillatory processes; gave gen. theory dynamic systems, investigations in superfluidity, superconductivity; obtained 1st results in theory of degeneration (Bose-Einstein) on non-ideal gases; evolved systematic micros. theory of superconductivity; attempted to reject Hamiltonian formalism in quantum field theory and replace it by phys. conditions, notably causality. Home: Leninskiye gory, Korp. L, Moscow. Office: Dept. Theoretical Physics, A.B.V. Steklov Mathematical Inst. USSR, 1-y Academicheskii Proyezd, 28, Moscow, USSR.

BOGORAD, Lawrence, biologist; b. Tashkent, USSR, Aug. 29, 1921; s. Boris and Florence (Bernard) B.; came to U. S., 1922, naturalized, 1935; B.S., U. Chgo., 1942, Ph.D. in Botany, 1949; m. Rosalyn G. Sagen, June 29, 1943; children—Leonard Paul, Kiki Lee. Faculty U. Chgo., 1948-51, 53-67, prof. botany, 1961-67; prof. biology Harvard, 1967—; visiting investigator Rockefeller Institute, New York City, Inst., N.Y.C., 1951-53. Mem. NSF adv. panel for science facilities, 1962-67, NIH cell biology study sect., 1966—. Recipient NIH Research Career award, 1963. NRC-Merck fellow, 1951-53; Fulbright Research scholar, 1960; NSF fellow, 1960-61. Mem. A.A.A.S., Am. Soc. Biol. Chemists, Am. Soc. Cell Biologists, Am. Soc. Plant Physiologists (pres. elect 1967), Bot. Soc. Am., Sigma Xi. Asso. editor Bot. Gazette, 1959-67, Plant Physiology, 1965-66; editorial com. Annual Review of Plant Physiology, 1963-67. Study of mechanism of biol. formation of porphyrins, including heme and chlorophyll; investigations of hereditary material in chloroplasts and the structure and devel. of chloroplasts. Home: 2 White Pine Lane, Lexington, Mass. 02173. Office: Harvard U., Biol. Labs., 16 Divinity Av., Cambridge, Mass. 02138.*

BOGORODITSKY, Nikolai Petrovich, Russian elec. engr.; b. Tashkent, 1902; grad. Leningrad Poly. Inst., 1929; D.Tech. Sci., 1940. With Leningrad Electrophys. Inst., 1929-30; lectr. Leningrad Poly. Inst., 1930-37; with Budenny Mil. Electrotech. Acad., Leningrad, 1933-42; with Leningrad Electrotech. Inst., 1935-42, prof., head chair electrotech. materials, 1945-49, dep. rector, 1949, now rector; dir. works lab., 1941-45. Decorated Order of Lenin. Mem. All-Union Bur. Elec. Insulation (chmn.), Soc. Radio Engring. and Communications (chmn. radio equipment and materials reliability sect.). Author: High-Grade Dielectrics, 1938; High-Grade Absolute Dielectrics, 1948; Electrotechnical Materials, 1951; co-author: Electrophysical Principles of High-Frequency Ceramics, 1958; The Graph Analysis Method of Calculating the Thermal Sparkover Voltage of High-Frequency Insulators, 1961; Theory of the Thermal Sparkover of Dipole Dielectrics, 1962; The Effect of a Covalent Bond in a Titanium Dioxide Crystal on the Value of the Dielectric Penetration Coefficient, 1962. Developer and patentee radio materials known as tikond, mycalex, highgrade glass, radio porcelain and ultra porcelain, 1933-42; organizer new field of specialist studies in physics and tech. of dielectrics and semiconductors. Address: Leningrad Electrotech. Inst., ulitsa Prof. Povoda 5, Leningrad, USSR.

BOGOROV, Veniamin (Benjamin) Grigorievich, Russian oceanographer; b. Dec. 24, 1904; grad. Moscow U., 1926. On staff Inst. Fisheries and Oceanography, 1930-41; with Lab. Oceanography (now Inst. Oceanology), USSR Acad. Sci., 1941-61, head plankton lab., dep. dir. Recipient State prize, 1951. Mem. USSR Acad. Sci. (corr.). Author: Plankton of Polar Basin, 1940; Role of Biological Indicators for Knowledge of the Hydrological Regime of the Sea, 1945; Sea Life, 1954; The Principle of the Unity of Nature in Oceanic Research, 1959; The Unity of Oceanic Natural Life, 1959; Leap over the Ocean, 1960; The Ocean Deeps and the Problem of Dumping Radioactive Waste in Them, 1960; A Further Voyage on the Vityaz, 1961. Research on typology of seas, geog. zones of oceans, productivity of seas, daily migration of plankton, specialist in hydrobiology. Address: Inst. Oceanology, Acad. Sci., 1 Sadovaja, Moscow, J-387, USSR.

BOGREN, Hugo, Swedish radiologist; b. Jönkoping, Sweden, Sept. 1, 1933; s. Gunnar and Signe (Holmstrom) H.; Candidate of medicine U. Lund (Sweden), 1954; Licenciate of medicine U. Gothenburg (Sweden), 1958, D. Medicine, 1964; m. Elisabeth Faxén, Nov. 1, 1956; children—Cecilia, Annika (dec.), Niclas, Joakim. Asst. radiology roentgen diagnostic dept. I and II, Sahlgrenska Sjukhuset, Gothenburg, 1958-68; asst. prof. radiology Ekmanska Sjukhuset, Gothenburg, 1968—; researcher Inst. Med. Biochemistry and Histology, Gothenburg, 1958—. Author: The Composition and Structure of Human Gall Stones, 1964; also articles. Research on devel. photoradiographic methods for measurement exophthalmos, localization of eye splinters, chem. composition of human gall stones with discovery of their essential uniformity. Home: 12 Patrullgatan, V. Frölunda. Office: Ekmanska Sjukhuset, Gothenburg, Sweden.*

BOGUE, Donald J., Am. sociologist, demographer; b. Ogden, Utah, Feb. 2, 1918; s. Lloyd L. and Anna (Stringer) B.; B.A., State U. Ia., 1940; M.A., State U. Wash., 1941; Ph.D., U. Mich., 1949; m. Elizabeth Mullen, Dec. 23, 1944; children—Erna Lynne, Gretchen Elaine. Asst. dir. Scripps Found. for Research in Population Problems, Miami U., Oxford, O., 1947-54; faculty U. Chgo., 1954—, prof. sociology, 1956—, dir. Community and Family Study Center, 1961—; tech. asst. expert UN, Bombay, India, 1959-60, Santiago, Chile, 1964. Cons., Office Statis. Standards, 1958-65. Fellow Am. Statis. Assn.; mem. Population Assn. Am. (past pres.), Am. Sociol. Assn., Internat. Union for Sci. Study Population Problems. Author: Structure of the Metropolitan Community, 1949; The Population of the United States, 1959; Skid Row in American Cities, 1963; Contributions to Urban Sociology (with E. W. Burgess), 1963; Principles of Demography, 1968; also articles. Editor, Demography, 1963—. Research on problems city and metropolis, migration and adjustment migrants, alcoholism and Skid Rows, motivation and mass communication to accelerate adoption fertility control practices among low edn. populations to help solve world population crisis. Home: 5801 Dorchester Av., Chgo. 60637.*

BOGUSH, Lev Konstantinovich, Russian surgeon, phthisiologist; b. 1905; son of dr.; grad. Med. Faculty, Nizhniy Novgorod U., 1928; D.Med. Sci., 1943. Head surg. dept., chief physician Vacha Rayon Hosp., Gorky Oblast, 1930-33; chief physician Gorky Oblast Hosp., 1933-34; head supr. dept. for lung, bone and joint Tb Gorsky Oblast Tb outpatients dept., asst. chair faculty surgery Gorky Med. Inst., 1934-40; sr. asso. Tb Inst., USSR Acad. Med. Sci., 1946—, head surg. dept., 1951—; prof. chair Tb Central Postgrad. Med. Inst., Moscow, 1951—; sci. dir. surg. dept. Moscow City Central Gen. Hosp. Head orgn. for surg. aid to pulmonary Tb patients in outlying regions. Recipient Lenin prize, 1961. Mem. USSR Acad. Med. Sci., Moscow (chmn. thoracic surgery sect.), All-Union (bd. mem.) socs. phthisiologists. Author over 100 works including The Surgical Treatment of Lung Tuberculosis, 1952; Cavernotomy in Patients with Lung Tuberculosis, 1955; Segmentary Resection in Tuberculosis, 1956; co-author: The Surgical Treatment of Tubercular Empyemas, 1961. Mem. editorial bd. Problems of Tb. Developer and pioneer method of ligating pneumolobular veins in lung Tb, 1935, introduced method of surg. ablation of short pleural adhesions for patients with ineffective artificial pneumothorax, 1945, perfected cavernotomy operation, 1947, developer operation of isolated ligation of lobular and segmentary bronchi, 1954. Address: Tb Central Postgrad. Med. Inst., G-242, pl. Vosstaniya 1-2, Moscow, USSR.

BOGUSLAVSKI, Sergei Anatolevich, Russian physicist; b. Russia, Nov. 19, 1883; s. Anatolii Boguslavski; prof. Moscow U. Discovered pyroelectricity, theory of formation of dielectrical crystals. Died Sept. 3, 1923.

BOGUSLAWSKI, Palm Heinrich Ludwig von, see von Boguslawski, Palm Heinrich Ludwig.

BOHANNAN, Paul James, Am. anthropologist; b. Lincoln, Neb., Mar. 5, 1920; s. Hillory and Hazel (Truex) B.; B.A., U. Ariz., 1947; B.Sc. Oxford U., 1949, D.Phil., 1951, M.A., 1952; m. Laura Marie Smith, May 15, 1943; 1 son, Denis Michael. Lectr. social anthropology Oxford (Eng.) U., 1951-56; faculty Princeton U., 1956-59; prof. anthropology Northwestern U., Evanston, Ill., 1959—, Stanley G. Harris prof. social sci., 1967—; field work Nigeria, 1949-53, Kenya, 1955, No. Cal., 1963-64; exec. dir. Human Environments in Middle Africa Project, Nat. Acad. Scis., 1958-61. Recipient Legion of Merit U. S. Army, 1945; fellow Center for Advanced Study in the Behavioral Scis., 1963-64. Mem. Social Sci. Research Council (past dir.), African Studies Assn. (past pres.), Am. Anthropol. Assn., Royal Anthropol. Inst., Assn. Social Anthropologists Gt. Britain, Am. Ethnol. Soc., A.A.A.S., Internat. Africa Inst. Author: (with Laura Bohannan) The Tiv of Central Nigeria, 1954; Justice and Judgment among the Tiv, 1957; Homicide and Suicide in Africa, 1960; Social Anthropology, 1963; Africa and Africans, 1964; (with Laura Bohannan) Tiv Economy, 1968; (with George Dalton) Markets in Africa, 1963; also articles. Editor: Museum Sourcebooks in Anthropology. Research on social orgn. and culture African peoples, family orgn. and divorce among Am. middle classes, concept folk system analysis ethnographic reporting. Home: 405 Deerfield Rd., Deerfield, Ill. 60015. Office: Northwestern U., Evanston, Ill. 60201.*

BOHE, Harald August, Danish mathematician; b. Copenhagen, Denmark, Apr. 22, 1887; s. Christian and Ellen (Adler) B.; Ph.D., U. Copenhagen, 1910; hon. Dr., U. Grenoble (France); m. Ulla Borregaard, Dec. 17, 1919; 3 children. Prof., Poly. Inst., Copenhagen, 1915; prof. U. Copenhagen, from 1930, also dir. Math. Inst. Mem. many Scandinavian Socs.; hon. mem. Am. Philos. Soc., London Math. Soc. Author: Matematisk Analyse, 1917; Fast periodische Funktionem, 1933. Formulated theory of almost periodic functions; research on Dirichlet series and their application to analytic number theory through Riemann zeta function. Died Copenhagen, Jan. 22, 1951.

BÖHLER, Lorenz, Austrian surgeon; b. Wolfurt, Vorarlberg, Jan. 15, 1885; ed. U. Vienna, M.D., Mayo Clinic; m. Leopoldine Settari, 1912. Prof. traumatology, U. Vienna; dir., accident hospital, Vienna. Author: Die Technik der Knochenbruchbehandlung. Founder of modern accident surgery; developed technique for setting broken bones and treatment of wounds; advocated separation of accident surgery as separate branch of surgery; organized model accident hospital in Vienna. Office: Vienna IX, Severingasse 1, Austria.

BÖHM, August Josef Georg (Elder von Böhmersheim), Austrian geologist; b. Vienna, Austria, Apr. 27, 1858; s. Carl and Marie (Kautz) B.; studied under E. Suess, F. Simony, F. von Richthofen; m. Marie Hoffman, 1895; 1 son, 1 dau. Pvt. lectr., Vienna, from 1887; became asso. prof., 1902; prof., Czernowitz, 1908; hon. prof., Graz, 1920. Author: Fuhrer durch die Hochschwabgruppe, 1881; Einteilung der Ostalpen; 1887; Geschichte der Moranankunde, 1901. Pioneer research (with E. Brückner, R. Richter) on explanation of alpine glacial forms; gave classification (still valid today) of eastern Alps; known as great mountain climber. Died Graz, Austria, Nov. 19, 1930.

BOHM, David Joseph, physicist; b. Wilkes-Barre, Pa., Dec. 20, 1917; s. Samuel and Freda (Popky) B.; B.S., Pa. State Coll., 1939; postgrad. Cal. Inst. Tech., 1939-41; Ph.D., U. Cal. at Berkeley, 1943; m. Sarah Woolfson, Mar. 15, 1956. Staff, U. Cal. Radiation Lab., 1943-46; asst. prof. Princeton, 1946-51; prof. Universidade de Sao Paulo, Brazil, 1951-55, Technion Maipa, Israel, 1955-57; researcher Bristol (Eng.) U., 1957-61; prof. physics Birkbeck Coll., U. London (Eng.), 1961—. Author: Quantum Theory, 1951; Causality and Chance in Modern Physics, 1955; Special Theory of Relativity, 1965. Research, publs. on theory of plasmas extending concept to electrons in metals and liquid helium, fundamentals of quantum theory, topology and geometry of spacetime. Office: Birkbeck Coll., Malet 47, London W.C.1, Eng.*

BÖHM, Georg, geologist, paleontologist; b. Frankfort, Germany, Dec. 21, 1854; s. Simon and Rosalie (Willert) B.; ed. Strasbourg, France, also Berlin, Göttingen, Germany; degree, 1877; m. Ella Werthauer, 1890; 1 son, 1 dau. Asso. with K. A. von Zittel, Munich; apptd. asso. prof., Freiburg, Germany, 1888. Specialist in bivalves; worked on fossil fauna of Mesozoic layer in Dutch E. Indies and New Zealand, 1899-1902, also in Europe, Turkestan; refuted theory of Jurassic mollusk continent by discovery of richly developed marine Mesozoic in Sula Islands. Died Freiburg, Mar. 18, 1913.

BÖHM, Richard Johann Constantin, German zoologist; b. Berlin, Germany, Oct. 1, 1854; s. Ludwig and Franziska Louise (v. Meyerinck) B.; student zoology; doctorate (under Ernst Haeckel), U. Jena (Germany), 1877; With P. Reichard on zool. expdn. to Zanzibar, from there reached E. bank of Tanganyika Lake, then S.W. to Upämba Lake (discovered by him), 1880-84. Author: Von Sansibar zum Tanganjika (letters), 1888. Contbr. notes, sketches in Journal für Ornithologie, 1882-87. Notes and collections (brought back by Reichard) furthered ornithol. knowledge of E. Africa; 1st zoologist to reach border zone between E. African steppes and W. African jungles; discovered many new types birds, mammals. Died nr. Katapana, Katanga, Mar. 27, 1884.

BOHM-BAWERK, Eugen (or Bohm von Bawerk), Austrian economist; b. Brunn, Austria, Feb. 12, 1851; ed. univs. Vienna (Austria), Heidelberg, Leipzig, Jena (all Germany). Prof. polit. economy U. Innsbruck (Austria), 1880-89; entered Austrian fgn. ministry, 1889, fgn. minister, 1895, 97-98, 1900-04; prof. political economy U. Vienna, 1904-11. Pres. Vienna Acad. Scis., 1911. Author: Kapital und Kapitalzins, 1884, 89; Karl Marx at the Close of His System, English transl., 1898. Founder, leading exponent of Austrian sch. of econs.; noted critic of Karl Marx and classical sch. of econs.; developed theory of marginal utility based on comparison between present and future values; held that enjoyment of present goods is more valuable than prospective enjoyment of same type of goods at future time; defined interest as representative of that difference in value. Died Vienna, Aug. 28, 1914.

BOHMAN, Verle Rudolph, Am. nutritionist; b. Peterson, Utah, Dec. 29, 1924; s. Victor R. and Nancy (Fernelius) B.; student Cornell U., 1950-52; B.S., Utah State U., 1950, M.S., 1951, Ph.D., 1952; m. Renee Jorgensen, June 22, 1945; children—Margaret Louise, Verle Duane, Jolene Renee, Van Reid, Gregory Nathan. Faculty U. Nev., Reno, 1952-—, prof. nutrition, 1962-—, chmn. animal sci. div., 1960-—. Mem. Am. Inst. Nutrition, A.A.A.S., Am. Dairy Sci. Assn., Am. Soc. Animal Sci., Phi Kappa Phi, Sigma Xi, Alpha Zeta. Editor: Livestock Production, 1966. Research and numerous publs. on devel. techniques in range livestock nutrition, toxicity molybdenum, utilization of lipids by ruminant, measurement of fallout products using domestic animals. Home: 916 Sbragia Way, Sparks, Nev. 89531. Office: Animal Sci. Div., U. Nev., Reno 89507.*

BÖHME, Jacob (Boehme, Behmen), German mystic, philos. theologian; b. Altseidenberg, nr. Görlitz, Prussia, Apr. 24, 1575. Worked as shoemaker; had mystical experience under influence of Lutheran pastor Martin Moller, 1600; writings condemned by eccles. and civil authorities. Author: Aurora oder Morgenröthe im Aufgang, 1612; Von der Gradenwahl, 1623; Erklärung über das Erste Buch Mosis (or Mysterium Magnaum), 1623; Der Weg zu Christo (collected devotional works), 1623; also others. Early proponent of life philosophy; revived mystic approach to religion and life; greatly influenced Western philos. and religious thought, especially in Romantic Age, and on Schelling, Hegel, Schopenhaur, Nietzsche, Hartmann, Bergson, Heidegger; used alchemical phraseology and imagery to illustrate religious views; his religious ideas influenced George Fox and early Quakers, German Pietists. Died Görlitz, Nov. 17, 1624.

BÖHMER, Philipp Adolph, German anatomist; b. Halle, Germany, Aug. 25, 1717; s. Justus Henning and Eleonora Rosina (Stutzing) B.; doctorate, U. Halle, 1738; m. Johanna Dorothea Neumann; m. 2d, Marie Sophie v. Brandenstein. City physician Eisleben, Germany; Saxon-Weimar physician in ordinary; prof. anatomy U. Halle, from 1741; rector Medizinische Academie, until 1789. Translated into German, Artis obstetriciae compendium (R. Manningham), 1746, also added to it small treatise of own, De usu et praestantia forcipis anglicanae (in it 1st described G regoire's forceps); 1st to use German word Zange for forceps. Died Halle, Nov. 1, 1789.

BOHMONT, Dale Wendell, Am. agronomist; b. Wheatland, Wyo., June 7, 1922; s. John E. and Mary (Armann) B.; certificate, Wheatland Normal Sch., 1941; B.S., U. Wyo., 1948, M.S., 1950; Ph.D., U. Neb., 1952; M.P.A. (Conservation fellow), Harvard, 1959; m. June Rains, Dec. 25, 1941; children—Dennis E., Craig W. Faculty, U. Wyo., 1948-60; asso. dir. Colo. Agr. Expt. Sta., Colo. State U., 1960-63; dean, dir. Coll. Agr. U. Nev., Reno, 1963-—; chmn. Nat. Expt. Sta. Div., 1967. Fellow A.A.A.S., Am. Soc. Agronomy; mem. Agrl. Research Inst., Gamma Sigma Delta (past internat. pres.). Contbg. author: Research Methods Academy of Science, 1962. Contbr. numerous articles to profl. jours. Research in cereal investigations; developed, established range improvement methods for control of big sagebush with 24D, use of shale oil for agr. Home: 6995 Kimlick Lane, Reno.*

BOHN, Georges, French biologist; b. France, May 28, 1868; ed. under Giard at Sorbonne. Author: Mécanismes respiratoires chez les Crustacés decapodes, 1901. A precursor in behavioral analysis; investigated physico-chem. determinism of biol. processes. Died 1948.

BOHN, Johann, German chemist, physiologist; b. Leipzig, Germany, July 20, 1640; studied in various univs.; M.D., Leipzig, 1665; 17 children. Apptd. prof. anatomy Leipzig, 1668, prof. therapeutics, 1691; became dean Med. Faculty, 1700. Author numerous books, including De Aëris in sublunaria influxu, 1675; Dissertationes chymico-physicae . . . , 1685; Circulus anatomico-physiologicus . . . , 1686; Circulus anatomico physica . . . , 1710; De alkali et acidi . . . , 1681; Experimenta ac dubia . . . , 1683. Showed reflex phenomena in decapitated frog to be entirely mech. as opposed to then current view of vita (spirits in nerve fluid); attacked de la Boé's chem. system of physiology and laid foundation of a scientific treatment of forensic med. Died Leipzig, Dec. 19, 1718.

BOHN, René, German chemist; b. Dornach, nr. Mülhausen, Germany, Mar. 7, 1862; s. Charles and Caroline (Bourry) B.; studied chemistry at Tech. Sch., Zürich, Switzerland; doctorate U. Zürich, 1883; m. Hedwig Schoch, 1895; 1 son, 1 dau. Became chemist Badische Anilin- & Soda-Fabrik, Ludwigshafen, Germany, 1884, mem. bd. dirs., 1906 apptd. prof., 1914. Discovered indanthrene blue (1st non-indigo vat dye) 1901; made other discoveries in alizarin dyes. Died Mannheim, Germany, Mar. 6, 1922.

BOHNENBERGER, Johann Gottlieb Friedrich von, see von Bohnenberger, Johann Gottlieb Friedrich.

BOHR, Aage Niels, Danish physicist; b. June 19, 1922; Ph.D., U. Copenhagen (Denmark); postgrad. research work, London, Eng., 1943-45. Research asst. Inst. Theoretical Physics, U. Copenhagen, from 1946, prof. physics, 1956-—; dir. Niels Bohr Inst. Recipient Dannie Heineman prize, 1960, Pope Pius XI medal, 1963. Mem. Danish, Norwegian acads. scis., Royal Physiograph. Soc. Lund, Sweden, Am. Acad. Arts and Scis., Am. Philos. Soc. Author: Rotational States of Atomic Nuclei, 1954. Research quantum physics; specialist nuclear physics. Home: Granhogen 10, Hellerup, Copenhagen. Office: Inst. Theoretical Physics, U. Copenhagen, Blegdamsvej 15-17, Copenhagen O, Denmark.

BOHR, Christian, animal physiologist; b. Denmark, 1855; studied under Karl Ludwig; discovered effect of carbon dioxide on dissociation curve of hemoglobin; studied exchange of gases in respiration. Died 1911.

BOHR, David Francis, physiologist; b. Zurich, Switzerland, June 22, 1915 (parents Am. citizens); s. Frank and Mildred (Lombard) B.; M.D., U. Mich., M.D., 1942; m. Kathleen Frederica Schleede, Apr. 6, 1940; children—John Nicholas, Barbara, Louise. Lab. officer, detachment comdr. Dutch Hosp. Ship, 1943-46; research fellow U. Cal. at San Francisco, 1946-47, vis. prof. pharmacology, 1955-56; faculty U. Mich., Ann Arbor, 1948-55, prof. physiology, 1956-—; vis prof. Physiologische Institute, U. Heidelberg (Germany), 1961-62. Mem. study sect. adv. grouo USPHS, 1960-64. Mem. Am. Physiol. Soc., Soc. Exptl. Biology and Medicine, Am. Heart Assn., Microcirculatory Soc., Council High Blood Pressure Research, Contbr. numerous articles to jours., chpts. to books. Research in cardiovascular physiology with particular attention to behavior vascular smooth muscle. Home: 355 Sumac Lane, Ann Arbor, Mich. 48105.*

BOHR, Niels Henrik David, Danish physicist; b. Copenhagen, Denmark, Oct. 7, 1885; s. Christian and Ellen (Adler) B.; Master's degree U. Copenhagen, 1909, Doctorate, 1911; hon. doctorates from more than 30 univs., worldwide; m. Margrethe Norlund, Aug. 1, 1912; 6 children, including Hans Henrik, Erik, Aage, Ernest. Carlsberg Found. grantee for study abroad, in Eng.; worked under J. J. Thomson at Cavendish Lab., Cambridge, Eng., 1911-12, under Rutherford at Manchester, Eng., 1912-13; lectr. physics U. Copenhagen, 1913-14, Victoria U., Manchester, 1914-16; prof. theoretical physics U. Copenhagen, 1916-62; a founder Inst. for Theoretical Physics, 1920, dir., 1920-62; visited U. S., 1938, 39; fled from German occupation in World War II, went 1st to Sweden, then Eng., finally to U. S., 1943, and served in adv. capacity Los Alamos Atomic Labs.; returned to Copenhagen, 1945, concerned himself with developing peaceful uses of atomic energy; chmn. Danish Atomic Energy Commn.; organized 1st Atoms for Peace Conf., Geneva, 1955. Recipient Gold medal Acad. Scis. Copenhagen, 1908; Nobel prize in physics, 1922; 1st Atoms for Peace award, 1957. Pres., Royal Danish Acad. Scis.; fgn. mem. Royal Soc. London, 1926; mem. French Acad. Scis., 1937, fgn. asso., 1945; mem. nearly all learned socs. of Europe. Author books including: (in transl.) Theory of Spectra and Atomic Constitution, 1922; Atomic Theory and the Description of Nature, 1934; The Unity of Knowledge, 1955; also many articles. Adapted Planck's quantum theory to Rutherford's model of atomic structure, thus devising Bohr's theory of the atom, 1913, which represents the atom as a dynamic system of electrons rotating around a nucleus (atom emits or absorbs electromagnetic radiation only when an electron passes from 1 orbit to another of different energy level); his was 1st reasonably successful attempt to use spectroscopic data to describe the internal structure of atom; was among those who pointed out that electrons exist in shells and that the electron content of the outer-most shell determines the chem. properties of the atoms of a particular element; developed principle of complementarity according to which certain phenomena may be examined in each of 2 mutually exclusive ways, with each being valid in its own terms; developed the correspondence principle, 1916. Died Copenhagen, Nov. 18, 1962.

BOHREN, Bernard Benjamin, Am. geneticist; b. Omaha, Aug. 15, 1914; s. Benjamin F. and Jessie (Osterman) B.; B.S., U. Ill., 1937; M.S., Wash. State U., 1940; Ph.D., Kan. State U., 1942; m. Dorothy May Morgan, Apr. 15, 1938; children—James Frederick, Linda May, Lois Ann (Mrs. Stephen S. Bigley). Faculty, Purdue U., Lafayette, Ind., 1943-—, prof. genetics, 1950-—. Recipient Poultry Sci., Assn. Research award, 1944. Mem. Poultry Sci. Assn., Genetics Soc. Am., Am. Genetic Assn., Biometrics Soc., A.A.A.S. Research and publs. on population genetics as applied to selection theory, biology of quantitative traits of fowl. Home: Rural Route 13, Box 451, Lafayette, Ind. 47905.*

BOHRMANN, Alfred, German astronomer; b. Mannheim, Feb. 28, 1904; Ph.D. in natural scis. Prof. astronomy U. Heidelberg (Germany), 1939-—; chief observer Landessternwarte auf dem Königstuhl. Address: Landessternwarte, Heidelberg, Germany.

BOHROD, Milton George, Am. pathologist; b. Chgo., Sept. 2, 1904; s. George and Fanny (Feingold) B.; B.S., U. Ill., 1924, M.D., 1927; m. Ruth Horodesky, May 30, 1926; children—Roberta (Mrs. Warren B. Cheston), John R. Pathologist Methodist, St. Francis hosps., Peoria, Ill., 1931-40, St. Francis Hosp., Miami Beach, Fla., 1941-42; pathologist, dir. labs. Rochester Gen. Hosp. (N.Y.), 1942-—; clin. asso. prof. pathology U. Rochester Coll. Medicine and Dentistry, 1957-—; pathologist Park Av. Hosp., Rochester, 1949-61. Mem. N.Y. State Bd. Med. Examiners, 1958-63; cons. to hosps. Diplomate Am. Bd. Pathology. Mem. A.M.A., N.Y. State, Monroe County med. assns., N.Y. State Lab. Assn. (past prs.), Am. Soc. Clin. Pathologists. Fellow Am. College of Pathologists, also Alpha Omega Alpha, Sigma Xi. Research and publs. on pathology of allergice and collagen diseases, lathyrism, med. semantics including definitions, methods in med. photography. Home: 234 Westminster Rd., Rochester 14607. Office: 1425 Portland Av., Rochester, N.Y. 14621.*

BOHUN, Antonín, physicist; b. Dolní Bojanovice, Czechoslovakia, June 14, 1913; s. Vavrinec and Marie (Bohunová-Vesela) B.; M.Maths. and Physics, Charles U., Prague, Czechoslovakia, 1945, D.Rerum Naturalium, 1948; m. Růzena Filipovicová, Mar. 8, 1943. Lectr. High Tech. Sch., Prague, 1946; chief physicist State Radiol. Inst., Prague, 1947-51; chief physicist ionic crystals dept. Inst. Solid State Physics, Czechoslovak Acad. Scis., Prague, 1952-—; lectr. radiol. physics Inst. for Post-grad. Physicians Edn., Prague, 1950-—; lectr. solid state physics Charles U., Prague, intermittently 1966-—. Sec., Radiation Protection Com. for Czechoslovakia. Mem. Soc. Czechoslavak Mathematicians and Physicists. Author: (with F. Běhounek, J. Kumpar) Radiologic Physics, 1954; also numerous articles. Elaborated proposal on radiation protection; explanation of connections of exoelectron emission with coloring and luminescence in ionic crystals; elaboration of models of some color and luminescence centres, mechanism of exoelectron emission. Home: 8 Pod Sporilovem, Prague 4. Office: 10 Cukrovarnická, Prague 6, Czechoslovakia.*

BOILEAU, Pierre Prosper, French mechanician; b. Metz, France, Feb. 19, 1811; ed. École Polytechnique, 1831; prof. mechanics Metz Sch. Instrn.; organized New Sch. of Arty. Versailles, France, 1856; also prof. of arty. Mem. French Acad. Scis., 1875. Author: Instruction pratique sur les scieries, 1855; Applications de la mécanique aux machines, 1872; Traité de la mesure des eaux courantes, 1854; Notions nouvelles d'hydraulique, 1881. Research on use of vulcanized rubber, paddle wheels; invented apparatus for measuring surface and volume of running water. Died Versailles, France, Sept. 11, 1891.

BOIS, Pierre, Canadian anatomist; b. Oka, Que., Can., Mar. 22, 1924; s. Henri and Ethier (Germaine) B.; M.D., U. Montreal (Que.), 1953, Ph.D., 1957; m. Joyce Casey, Sept. 8, 1953; children—Monique, Marie, Louise. Research fellow pathology U. Montreal, 1957-58, faculty, 1958-—, asst. prof. dept. pharmacology, 1960-64, prof., 1962, head dept. anatomy, 1964-—; asst. prof. histology Ottawa (Ont., Can.), 1958-60. Mem. Am., Canadian assns. anatomists, N.Y. Acad. Scis. Research, numerous publs. on morphological effects hormones, histamine and mast cells in magnesium deficiency, muscular dystrophy, exptl. thymic tumors. Home: 25 Elmwood St., Outremont, Que. Office: 2900 Mt. Royal Blvd., Montreal, Que., Can.*

BOISBAUDRAN, Paul Émil Lecoq de, see de Boisbaudran.

BOISSEVAIN, Charles Hercules, physician; b. Amsterdam, Holland, Oct. 18, 1893; s. Charles Ernest Henri and Maria Barbera (Pijnappel) B.; grad. Gymnasium, Amsterdam, 1911; M.D., U. Amsterdam, 1919; m. Countess Marie Theresa Zwetana von Hartenan, 1925; m. 2d, Ruth Davis Dangler, 1928; children—Menso, Maria Barbera. Came to U. S. 1923, naturalized, 1930. Research asso. Institut Pasteur de Brabant, Brussels, 1921-23; vis. research prof. Colo. Coll., Colorado Springs, since 1924; lab. dir. Colo. Found. for Research in Tb, from 1924. Mem. A.M.A., Soc. for exptl. Biol. and Medicine. Died Oct. 18, 1946.

BOISSIER, Jacques R., French pharmacologist; b. Avrances, France, Dec. 2, 1921; s. Gabriel J. A. and Louise (Lemarquand) B.; Pharmacien, Faculté de Pharmacie, Paris, 1945, Docteur en Médecine, Faculté de Médecine, 1953, Docteur ès Scis., Faculté des Scis., 1958; m. Monique Gaume, June 18, 1965; 1 dau., Juliette. Pharmacist, Hôpitaux de Paris, 1949; became asst. pharmacology Faculté de Médecine de Paris, 1947, asst. prof., 1958, chmn. chair of pharmacology, 1966. Vis. tchr., Afghanistan, 1950-52, Lebanon, 1960, Cambodia, 1964; pharmacist, biologist Hopital R. Poincaré. Mem. commns. French

Pharmacopoea, Nat. Inst. for Medicine and Pub. Health. Mem. French Soc. Therapeutics and Pharmacology, Internat. Collegium Neuro-psycho, European Soc. for Study of Drug Toxicity. Author: (with R. Hazard, J. Cheymol, J. Lévy, P. Lechat) Manuel de pharmacologie; also articles. Research in psychopharmacology; studies on synthesis of new drugs, mechanism of drug action, prediction of drug activity in animal behaviour, toxicity of drugs and side-effects. Home: 17, Chaussée de la Muette, F-75, Paris 16, France.*

BOISSIER, Pierre-Édmond, Swiss botanist; b. Geneva, Switzerland, May 10, 1810; mem. French Acad. Scis., 1885. Author: Flora Orientalis, 5 vols., 1867-84; Voyage botanique dans le midi de l'Espagne, 1839-42. Research on Oriental flora. Died Valeyres, Switzerland, Sept. 25, 1885.

BOITEAU, Pierre Louis, French biologist, plant chemist; b. Cognac, France, Dec. 3, 1911; s. Jean Marius and Suzanne (Brun) B.; Ingénieur Horticole, Nat. Sch. Horticulture, Versailles, France; Ingénieur d'Agronomie Tropicale, Inst. Nat. d'Agronomie Tropical, Nogent-sur-Narne; m. Marthe Cauby, Aug. 4, 1934; 6 children. Dir., Bot. and Zool. Garden of Antanarivo (Madagascar), 1934-45; prof. cellular biology, Antanarivo, 1943-46; dir. Inst. Sci. Research Madagascar, 1945-47; formerly asst. Faculty Medicine Paris (France), then asst. dir. lab. physiologie nutritionneles, 1947—, also mem. staff Inst. Malgache de Recherches Appliquees, Tananarive. Cons. Union Francaise, 1949-58, sec. assembly, 1953-58. Corr. Mus. Nat. d'Histoire Naturelle Paris; mem. Académie Malgache, Société Chimie Biologique, Soc. Chimique de France, Soc. Nutrition et Diététique, Soc. Nationale d'Horticulture de France, others. Author: (with A. Rakoto Ratsimananga, Bozena Pasich) Les Triterpénoides en Physiologie animale et végétale, 1964, also others. Research, publs. on plants of Madagascar and their applications, drugs, role of hormones and hormone-like substances in nutrition; contributor to discovery of hormone-like activity in animals of plant constituents. Home: 77 rue de l'Abbe Carton, 75, Paris 14e. Office: 12 rue de l'Ecole de Medecine, Paris 6e, France; also Institut Malgache de Recherches Appliquées, B.P. 489, Tananarive, Madagascar.*

BOIVIN, Albéric, physicist; b. Baie St.-Paul, Que., Can., Feb. 11, 1919; s. Joseph-Francois and Marie-Lydie (Perron) B.; B.Applied Sci., U. Laval, 1944, M.S., 1947, D.Sci., 1960; m. Marie-Ange Boivin, June 25, 1945; children—Hélène, Lise, Richard, Brigitte. Spectroscopist, Aluminum Co. Can. Ltd., 1943; mem. faculty U. Laval, 1944—, prof., 1955—; cons. Canadian Armament and Research Establishment, Valcartier, Que., 1952, 57, 64, RCA, Moorestown, N.J., 1963; guest scientist Lab. d'Electronique et de Physique appliquée, Paris, France, 1952; sr. research asso. U. Rochester, 1962-63. Guggenheim fellow, 1962-63; Laureate of Prix David, Province Que., 1965; recipient Spl. award Corp. des Opticiens d'Ordonnance du Que., 1966. Fellow Optical Soc. Am.; mem. Assn. Canadienne francaise pour l'Avancement des Sciences (bd. dirs. 1965-67), Canadian Assn. Physicists (councillor 1966-67), Royal Astron. Soc. Can. Author: Théorie et calcul des figures de diffraction de révolution, 1964. Research in area of diffraction studies of optical imaging; full theory of imaging by zone plates, effects of spherical aberration on diffraction images, discovery and investigation of spl. functions Ap(z) of interest in diffraction; new formulation of campling theorem in communica tion theory, investigation of modes in laser resonators, work in microwave interferometry and electromagnetic diffraction in wide-angle systems and antennas. Home: 834 Eymard, Quebec 6. Office: Dept. Physics, Univ. Laval, Ste-Foy, Quebec 10, Que., Can.*

BOIVIN, André-Félix, French chemist; b. Auxerre, France, Apr. 18, 1895; studied medicine at Marseilles, France, Faculty of Medicine, Strasbourg, France; docteur en scis., 1931. Faculty med. chemistry Bucharest (Rumania), 1930-36; head research Pasteur Inst., Garches, France; professor of biological chemistry at the University of Strasbourg, from 1946. Mem. French Academy of Sciences, 1948, Acad. Medicine. Research on genetics of unicellular organisms; determined nature of typhus endotoxin; proved existence of cytoplasmic RNA in bacteria Giemsa; formulated (with Vendrely) hypothesis explaining difference between genes and viruses, 1947. Died July 7, 1949.

BOJANOVIC, Jelena, chemist; b. Istanbul, Turkey, Aug. 6, 1918; d. Jasa and Amalija (Starek) B.; grad. U. Belgrad, 1949; Ph.D., Serbian Acad. Sci., 1953. Asst. prof. Chem. Inst., Serbian Acad. Sci., 1949-55; asso. prof. Faculty Medicine of Belgrad, 1955-66, prof. chemistry, 1966—. Mem. Serbian Chem. Soc., Yugoslave Soc. Physiologists. Author: On Bisamides, 1956; also numerous articles. Research on synthesis and activity of bisamides, effect of insulin on metabolism of proteins, lipids and carbohydrates in schizophrenic persons and animals, role of proteins, lipids and glucides in ageing. Home: 18 Radoja Domanovica, Belgrad, Yugoslavia.*

BOK, Bart Jan, astronomer; b. Hoorn, Holland, Apr. 28, 1906; s. Jan and Gesina-Annetta (Van der Lee) B.; student U. Leiden, Holland, 1924-27; Ph.D., U. Groningen, Holland,, 1932; m. Priscilla Fairfield, Sept. 9, 1929; children—John Fairfield, Joyce Annetta (Mrs. John B. Ambruster). Came to U. S., 1929, naturalized, 1938. Aggaziz fellow, Harvard, 1929-30; R. W. Willson teaching fellow, 1930-33; asst. prof. astronomy, 1933-39; asso. prof., 1939-46; R. W. Willson prof. astronomy, 1947-57; prof. astronomy Australian Nat. U., also dir. Mt. Stromlo Obs., Canberra, 1957-66; prof. astronomy U. Ariz., also dir. Steward Obs., Tucson, 1966—. Recipient Oranje-Nassau medal, Netherlands, 1957. Fellow Royal Astro. Soc., Inst. Physics; mem. Nat. Acad. Scis., Am. Astro. Soc., A.A.A.S., Am. Acad. Arts. and Scis., Royal Astro. Soc. Can. (hon.), Royal Netherlands Acad. Arts and Scis. (corr.), Internat. Astro. Union (pres. commn. 33 structure and dynamics of Galactic System, 1961-67), Sigma Xi. Author: The Distribution of Stars in Space, 1937; (with Priscilla F. Bok) The Milky Way, 1941, rev. 3d edit., 1957; (with F. W. Wright) Basic Marine Navigation, 1944; The Astronomer's Universe, 1958. Research and publs. on the Milky Way, primarily So. Milky Way; cosmic evolution, star clouds of Magellan and radio astronomy. Research in interstellar matter and galactic structure and dynamics. Home: 200 Sierra Vista Dr., 85721.

BOK, Ivan Ivanovich, Russian geologist; b. 1898; grad. Urals Poly. Inst., Sverdkovsk, 1928. With Urals Geol. Bd., until 1936; with Kazakhstan Geol. Trust, Semipalatinsk, 1936-38; with Inst. Geol. Sci., Kazakhstan br. USSR Acad. Sci., 1938-46; instr. Sverdlovsk Mining Inst., later Kazakhstan Mining and Metall. Inst., 1938-46; prof., 1948—; asso. Inst. Geol. Sci., Kazakhstan Acad. Sci., 1946—. Mem. Kazakhstan Acad. Sci. Author various works including Agronomical Ores, 1955; Observations of Minerals in Geological Investigations, 1957. Discovered several mineral deposits of nickel, cobalt, asbestos, helped compile metallogenetic and prognostic map of Kazakhstan; on mineral. expdn. to No. and Central Kazakhstan. Address: Kazakhstan Acad. Sci., Alma-Ata, Kazakhstan SSR, USSR.

BÖKER, Hans, anatomist, zoologist; b. Mexico City, Mexico, Nov. 14, 1886; s. Henirich and Luise (v. der Nahmer) B.; student medicine Freiburg, Kiel, Berlin (all Germany); m. Gerda Hasenclever, 1913; 1 dau.; m. 2d, Marie Juliane vom Berg, 1917; 3 sons, 1 dau. From asst. to asso. prof. Freiburg Anatomisches Institut, 1912-32; prof., dir. Anatomisches Institut, Jena, Germany, 1932-38; in Cologne, Germany, 1938. Author: Einfuhrung in die vergleichende biologische Anatomie der Wirbeltiere, 2 vols., 1935-37. Founded idea of comparative biol. anatomy (no longer accepted) which was based on analogy rather than homology research; believed living organism underwent changes in constrn. to adapt to changed environment. Died Cologne, Apr. 23, 1939.

BOKHARI, Mohammed Sibtain, physicist; b. Batala, India, Dec. 2, 1928; s. Sayed Mohammed Hussain and Ume-Kalsoom B.; F.Sc., M.A.O. Coll., Panjab U., 1948, B.Sc., Govt. Coll., 1950, M.Sc. (Prev.), 1951, M.Sc. (Final), F.C. Coll., 1952; Ph.D., U. Liverpool, Eng., 1959; postgrad. Western Res. U., 1956; m. Khurshid Rizvi, May 21, 1962; children—Raza Hussain, Manazza Zehra, Raza Ali. Lectr. physics Islamia Coll. for Men, Lahore, 1952-54, Islamia Coll. for Women, 1954; lectr. physics Edn. Dept., West Pakistan, 1954-60; sr. sci. officer Pakistan Atomic Energy Commn., 1960—, head physics div. Lab. Karachi, 1960, dir. lab., 1960-61, dep. sec. tng. and tech. affairs, 1961-62, sec. electronic com., 1961-62; asst. sci. adviser Pres.'s Secretariat, Sci. and Tech. Research Div., Govt. of Pakistan, 1965-66; prin. research officer Def. Sci. Orgn., Chaklala, 1966—. Mem. Am. Phys. Soc., Am. Nuclear Soc. Research and publs. on neutron component in cosmic rays, role of spin-orbit interactions in stripping and pickup reactions and elastically scattered spin 1/2 particles in medium range energies, fast neutron transport in water shields, neutron source distbn. in pool-type reactor; devel. technique for measurement of differential reaction cross-section in light nuclei simultaneously with fast neutrons of all energies from research reactor; designed and developed fast neutron spectrometer for reactor environments, strain-gauge transducer for measurement light dynamic loads without hysteresis. Home: House 8, Lane 39, Krishan Nagar, Lahore, Pakistan. Office: D.S.O. Labs., Rawalpindi, Chaklala, Pakistan.*

BOKII, Georgii Borisovich, Russian phys. chemist; b. Oct. 9, 1909; grad. Leningrad Mining Inst., 1930. Instr., Moscow U., 1939-44, prof., 1944—; asso. Inst. Gen. and Inorganic Chemistry, USSR Acad. Sci., 1939—. Mem. USSR Acad. Sci. (corr., Presidium's prize 1954). Author: The Crystal Chemistry of Complex Compounds, 1948; The Theory of Daltonides and Bertollides, 1956; co-author: A New Method of Determining the Structure of Complex Compounds, 1952, also others. Research on crystallography and crystal chemistry, crystal chemistry of complex compounds, proposed method for quantitative determination of transinfluence values by measuring interatomic intervals in crystals of complex compounds, developer atomic structure theory for daltonides and bertollides. Home: 1-ya Meshchanskaya, Moscow. Office: Inst. Gen. and Inorganic Chemistry, Leninsky prosp. 31, Moscow V-71, USSR.

BOKSAN, Slavko, Yugoslav engr., b. Djurdjevo, nr. Novi-sad, Yugoslavia, June 18, 1889; s. Paja and Helene B.; ed. coll. in Novi-sad, Tech. U., Berlin, Germany; m. Gertrude Jarius. Mem. bd. dirs. Radio Corp.; partner firms Ing., S. Boksan & Co., Elekroindustrija Bitolj (all Belgrade). Built up Elect. Central Stas. in Bitolji, Negotin, Arandjelovac, Lozniza, and other places in Yugoslavia; research, publs. related to inventions of Nikola Tesla on magnetic field and polyphase elect. current. Address: Mirocka 4, T. 225-65, Belgrade, Yugoslavia.

BOLAND, Frank Kells, Am. surgeon; b. Indpls., May 3, 1875; s. Kells Hewitt and Louise (Bright) B.; A.B., U. Ga., 1897, Sc.D., 1926; M.D., Emory U., 1900; student Johns Hopkins; m. Molly Horsley, Apr. 25, 1905; children—Frank Kells, Joseph Horsley. Resident surgeon St. Joseph's Hosp., Balt., 1900-03; practiced medicine, Atlanta, since 1903; with Emory U. since 1903, prof. surgery, 1921-30, prof. clin. surgery since 1930, prof. surgery, 1942-45, prof. anatomy, Sch. Dentistry, 1907-19, prof. physiology, 1919-49; vis. surgeon Grady Municipal Hosp., Emory U. Hosp. Fellow A.C.S., Am. Surg. Assn.; mem. A.M.A. (vice chmn. surg. sect. 1930), So. Med. Assn. (chmn. and sec. surg. sect.; v.p. 1926-27; chmn. council 1934; pres. 1937), So. Surg. Assn. (v.p. 1926; pres. 1934), Am. Assn. for Thoracic Surgery, Amer. Assn. for Traumatic Surgery, Med. Assn. of Ga. (pres. 1925-26), Fulton County Med. Soc. (pres. 1921), Southeastern Surg. Congress (pres. 1932-33). Société Internationale de Chirurgie, Phi Beta Kappa. Author: The First Anesthetic, Story of Crawford Long, 1950. Died Nov. 11, 1953.

BOLD, Harold Charles, Am. botanist; b. N.Y.C., June 16, 1909; s. Edward and Louise (Krüsi) B.; A.B., Columbia, 1929, Ph.D. in Botany (Univ. fellow), 1933; M.S., U. Vt., 1931; m. Mary E. Douthit, June 8, 1943. Faculty Vanderbilt U., 1932-39, 45-57, Columbia, 1939-43; prof. botany U. Tex., Austin, 1957—, chmn. dept. 1960—. Mem. Phycology Soc., Brit. Phycology Soc., Bot. Soc. Am. (pres. 1966, editor Am. Jour. Botany, 1958-65), Torrey Bot. Club (corr. sec. 1941). Author: The Plant Kingdom, 1960; Morphology of Plants, 1967; (with C. J. Alexopoulos) Algae and Fungi, 1967. Research in plant morphology and phycology. Office: Dept. Botany, U. of Texas, Austin, 78712.

BOLDREY, Edwin Barkley, Am. physician, educator; b. Morgantown, Ind., July 17, 1906; s. Edwin Howe and Florence (Barkley) B.; A.B., DePauw U., 1927; M.A., Ind. U., 1930, M.D., 1932; M.Sc., McGill U., 1936; m. Helen Burns Eastland, June 16, 1932; children—Nancy Jeanne, Edwin Eastland, Susan Ellen. Faculty, U. Cal. at San Francisco Sch. Medicine, 1940—, chmn. Dept. neurol. surgery, 1951-56, prof. neurol. surgery, 1960—. Cons. neurol. surgery VA, mil. hosps.; surgeon gen. USPHS, 1960-64; 66—; lectr. Oakland Naval Hosp., 1966. Diplomate Am. Bd. Neurol. Surgery (past dir.). Mem. A.M.A., A.C.S., A.A.A.S., Am. Acad. Neurol. Surgeons (past pres.), Am. Neurol. Assn., Soc. Neurol. Surgeons (pres. 1965-66), Harvey Cushing Soc. (past v.p.), Canadian, Western (past pres.), San Francisco (past pres.) neurol. socs., Am. Assn. Neuropathologist. Contbr. numerous articles to profl. jours. Research in human cerebral localization, biology and therapy of brain tumors, intracranial hemorrhage, human hypophysectomy, surg. therapy in epilepsy. Home: 924 Hayne Rd., Hillsborough, Cal. 94010. Office: U. Cal. Med. Sch., San Francisco 94122.*

BOLDT, Hermann Johannes, gynecologist; b. nr. Berlin, Germany, June 24, 1856; s. Hermann and Amalie (Krüger) B.; came to U. S., 1865; ed. public grammar and high schs.; studied and practiced pharmacy; M.D., Univ. Med. Coll., N.Y.U., 1879; m. Hedwig Krüger, 1891. Began practice medicine, 1879, specializing in gynecology, from 1891; prof. gynecology, N.Y. Post-Grad. Med. Sch. and Hosp. (now post-grad. dept. Columbia U.), 1890-1923; was cons. gynecologist, Stuyvesant Polyclinic, Post Graduate, Beth Israel, St. Vincent's and Union hosps.; retired, 1929. A.C.S. (founder mem., mem. bd. govs.); hon. mem. Am. Gynecol. Soc., Am. Gynecol. Club; mem. Nat. Soc. Sciences, N.Y. Acad. Sciences, Internat. Gynecol. Soc. and Obstet. Soc. (pres), N.Y. Acad. Medicine (ex-chmn. gynecol. sect.), Southern Surg. Soc., Mil. Surgeons U. S. A., Royal Soc. of Medicine (London), Gynecol. Soc. of Germany. Research on physiol. action of cocaine, gynecol. pathology; inventor various gynecol. instruments, operating table for abdominal surgery (won medal at Paris Expn. 1900), examining table for office use. Died Jan. 12, 1943.

BOLDUAN, Charles Frederick, hygienist; b. Bielefeld, Germany, May 7, 1873; s. William and Juliane (Dreibholz) B.; Ph.G., Coll. Pharmacy, N.Y.C., 1893; M.D., Coll. Phys. and Surg., Columbia, 1901; postgrad. U. Berlin, 1903; m. Adele Jönsson, Sept. 15, 1906; 1 son, Nils W.; m. 2d, Herma Engelsdorff, Mar. 1, 1928. Prof. bacteriology and hygiene Fordham U., 1905-08; bacteriologist Dept. of Health, City N.Y., 1904-07; asst. to gen. med. officer Dept. of Health, 1907-13; dir. Bur. Pub. Health Edn., 1913-18; chief, sec. Public Health Edn., USPHS, 1918-21; surgeon USPHS, detailed U. S. consular service, Europe, 1921-28; dir. bur. health edn., Dept. of Health, N.Y.C., 1928-43; lectr. preventive medicine and hygiene Columbia, 1918-22, Sch. Sociology and

Social Service, Fordham, 1928-42. Sec. N.Y. Tb Preventorium for Children. Founder, hon. pres. N.Y., Diabetes Assn. Fellow Am. Pub. Health Assn., N.Y. Acad. Medicine; mem. N.Y. Soc. Med. History, Am. Diabetes Assn. (hon.). Author: Immune Sera, 5th edit. 1918; Applied Bacteriology for Nurses 8th edit. 1941; Public Health and Hygiene, 3d edit. 1941; Spanish edit. 1943. Translator: Suppression of Tuberculosis, 1905; Serum Diagnosis, 1905; Collected Studies on Immunity, 1906. Contbr. papers to med. jours. Editor Bull., N.Y.C. Dept. Health. Made numerous statistical studies on cancer, heart disease, infant mortality, typhoid fever, hist. studies on health conditions in N.Y.C. during 19th century. Died Apr. 4, 1950.

BOLDYREV, Tikhon Yefimovich, Russian epidemiologist, hygienist; b. 1900; grad. Leningrad Mil. Med. Acad., 1926; D.Med. Sci., 1939. In Red Army, 1919-53; prof., 1940—; chief Anti-Epidemic Bd., Main Mil. San. Bd., Soviet Army, 1941-45; head chair epidemiology Med. Faculty, Central Postgrad. Med. Inst., later head chair epidemiology, 3d Moscow Med. Inst., 1942-49; dep. minister of health of USSR, chief san. insp. of USSR, 1947-54; sr. counselor Ministry Health, Chinese Peoples Republic, 1954-56; head chair epidemiology Central Postgrad. Med. Instr., Moscow, 1956—; head epidemiology sect. Gamaleya Instr. Epidemiology and Microbiology, USSR Acad. Med. Sci., 1956—. Mem. USSR Acad. Med. Sci. (corr.), All-Russian Med. Soc. Epidemiologists, Mircobiologists and Infectionists (chmn. 1956—). Author over 100 works including Bathhouse and Laundry Problems in the Red Army, 1934; Bath, Laundry and Disinfection Services in the Army, 1940; Problems of Epidemiology, 1943; co-author: Practical Manual of Troop Disinfection, 1934. Co-editor Encyclopedia Dictionary of Military Medicine, Encyclopedic Dictionary and Handbook for Medical Orderlies; editor Hygiene and Sanitation; mem. editorial bd. Soviet Medical Experience in the 1941-45 Great Patriotic War, many vols.; mem. editorial bd., editor Epidemiology and Infectious Diseases sects. Large Med. Ency., 2d edit., mem. editorial council Jour. Microbiology, Epidemiology and Immunology. Research on epidemiology, protection of troops against chem. toxic agts., disinfection and pub. health orgns. Address: Inst. Epidemiology and Microbiology, USSR Acad. Med. Sci., Malaya Shchukinskaya ulitsa 13, Moscow, USSR.

BOLEY, Forrest Irving, Am. physicist; b. Fort Madison, Ia., Nov. 27, 1925; s. Ira Everett and Olive (Conlee) B.; B.S., Ia. State U., 1946, Ph.D., 1951; M.A. ad eundum gratium, Wesleyan U., 1959; m. Marjorie Lovell, Dec. 26, 1946; children—Kathleen, Sandra, Philip, John. Prof. physics Wesleyan U., Middletown, Conn., 1951-61; physicist Lawrence radiation lab. U. Cal. at Berkeley, 1961-64; prof. physics and astronomy Dartmouth Coll., Hanover, N.H., 1964—. Editor Am. Jour. Physics, 1966—. Research in plasma physics, cosmic rays, astrophysics. Home: 20 Occom Ridge, Hanover, N.H.*

BOLIN, Bert Richard, Swedish meteorologist; b. Nyköping, Sweden, May 15, 1925; s. Richard Johannes and Karin (Johansson) B.; B.S., U. Uppsala (Sweden), 1946; M.S., U. Stockholm (Sweden), 1950, Ph.D., 1956; m. Ulla Karin Irene Frykstrand, June 7, 1952; children—Dan, Karina, Göran. Faculty, U. Stockholm, 1951-61, asst. prof., 1956-61, prof., 1961—. Dir., Internat. Meteorol. Inst., Stockholm, 1957-65, 67—; sci. dir. European Space Research Orgn., 1965-67. Mem. Royal Swedish Acad. Scis. Author: Introduction to Meteorology, 1968; also articles. Contbr. to early devel. in field numerical weather forecasting; research on chemistry of atmosphere and problems of global air pollution, especially that of carbon-dioxide, circulation and transfer studies of oceans using chem. and radioactive tracers. Home: 44 Ekbacksvägen, Danderyd, Sweden. Office: 41 Tulegatan, Stockholm VA, Sweden.*

BOLL, Jacob, geologist, naturalist; b. Bremgarten, Canton Aargau, Switzerland, May 29, 1828; s. Henry and Magdalena (Peier) B.; attended U. Jena; m. Henriette Humbel, 1854. Owned pharmacy, Bremgarten, 1854-74; studied natural history of Canton Aargau; visited Tex., 1869-70, collected specimens for Louis Agassiz; studied geology and natural history of Tex., 1874-80; collected fossils and reptiles in North and N.W. Tex.; employed by U. S. Entomol. Commn. for study of Rocky Mountain locust, 1877-80; commd. by Canton Aargau to collect specimens of Colo. potato beetle, seeds of woody plants, fresh water and marine mollusks of Tex. Died Wilbarger County, Tex., Sept. 29, 1880.

BOLLAY, William, aero. engr.; b. Stuttgart, Germany, Jan. 14, 1911; s. Frederick Jacob and Dorothea Frieda (Kramer) B.; came to U. S., 1924, naturalized, 1929; B.S., Northwestern U., 1933, Sc.D. (hon.), 1959; M.S., Cal. Inst. Tech., 1934, Ph.D., 1936; m. Jeanne Marie Brinsley; Aug. 30, 1934; 1 dau., Melody (Mrs. George A. Kladnik). Mem. faculty Cal. Inst. Tech., 1933-37; Harvard, 1937-41; in charge exptl. projects jet propulsion and gast turbine engines Bur. Aeros., Navy Dept., 1941-45; tech. dir. aerophysics lab. N. Am. Aviation, 1945-51; pres. tech. dir. Aerophysics Devel. Corp., Santa Barbara, Cal., 1951-58, cons. engr., 1959—; vis prof. Mass. Inst. Tech., 1962-63, Stanford, 1963—. Inst. Def. Analyses, 1962—; mobility adv. group U. S. Army, 1960—. Decorated Legion of Merit; recipient Citàtion of Honor USAF Assn., 1950. Fellow Am. Inst. Aeros. and Astronautics (Wright Bros. Lecture award 1950); mem. Commn. Engring. Edn. (dir. 1962-65), Nat. Inventors Council, Am. Soc. Engring. Edn. Research and publs. in aerophysics, especially jet and rocket powered aircraft, ballistic and guided missiles, artificial satellites, engring. edn. Home: 4592 Via Vistosa, Santa Barbara, Cal. 93105. Office: Sch. Engring., Stanford U., Stanford, Cal. 94305.

BOLLE, August Franz Friedrich Albert, German botanist; b. Potsdam, Aug. 13, 1905; s. Friedrich and Helene (Albrecht) B.; Ph.D., U. Berlin; m. Irmgard Bolle, 1935. With Acad. Sci., Berlin; asst. scientist Pflanzenreichs; with Office of Protection of Plants, Kiel, Germany; asst. scientist, head labs. Mem. German Soc. Botany, Assn. Applied Botany, Assn. Botany of Province of Brandesburg. Author: Uebersicht über die Gattung Geum, 1933; Resedaceae, 1936; Theorie der Blattstellung, 1939; Theorie der Blütenbände, 1940. Home: Hohenbergstrasse 13. Office: Westring 383, Kiel, Germany.

BOLLE DE BAL, Marcel, Belgian sociologist; b. Brussels, Belgium, July 29, 1930; s. Jean and Paula (Coblyn) B. De B.; doctorate in Law, U. Brussels, 1953, M.A. in Econs., 1954; postgrad. indsl. relations U. Chgo.; m. Francoise Leclercq, Mar. 2, 1956; children—Marion, Pierre. Research staff, Inst. Sociology, U. Brussels, 1955—, research dir., 1964—, prof. sociology of work, 1962—. Cons. govtl. offices, banks, cos., trade unions. Mem. Société Belge de Sociologie, Association des Sociologues de Langue Francaise. Author: Relations humaines et relations industrielles, 1958; le Salaire à la production, 1966; la vie de l'enterprise, 1967; also articles. European critical analysis of human relations movements; devel. sociol. theory of payment by output; synthesis of theories concerning evolution of working class. Home: 82 drive des Fauvettes, Linkebeek, Belgium. Office: Institut de Sociologie, 44, av. Jeanne, Brussels 5, Belgium.*

BOLLÉE, Amédée, French inventor; b. Le Mans, France, 1844. Invented steam automobile which ran between Le Mans and Paris, 1873; patented front axle with 2 journals (made possible correct position of each front wheel in curves). Died 1917.

BOLLER, Reinhold, Austrian physician; b. Vienna, Austria, June 9, 1901; s. Victor and Maria-Anna (Scheimpflug) B.; ed. univs. Vienna, Graz, Kiel; M.D.; m. Edith Boller, Mar. 21, 1963; 1 dau., Brigitta. Substitute physician Clinic of Wenckebach, 1925; asst. Eppinger Clinic; agrégé internal medicine Inst. Clin. Medicine; dir. Inf. Spital, Vienna; maj. physician; physician in chief sect. destroyed hosps. of Vienna; instr. internal medicine. Mem. German and French Soc. Gastroenterology (council), Austrian Soc. Nutrition Research (v.p.); hon. mem., corr. Belgian, Spanish, Greek, Argentine socs. gastroenterology. Author 185 sci. works, including 6 books, collaborator 2 others. Died Jan. 7, 1968.

BOLLEY, Henry Luke, Am. botanist, pathologist; b. Manchester, Ind., Feb. 1, 1865; ed. Purdue U. Instr., Purdue U., 1888-90; asst. botanist Ind. Agrl. Exptl. Sta.; prof. botany and zoology N.D. Agrl. Coll., from 1890; seed commr. N.D., 1909-29. Author, N.D. pure seed laws, 1908-29. Carried out exptl. research in plant diseases; developed method for preventing diseases of cereal grains by using formaldehyde as disinfectant; discovered parasitic nature of potato scab and method of prevention; discovered cause of wheat-sick and flax-sick soil; developed wilt-resistant varieties of flax, also rust-resistant varieties of wheat. Died 1956.

BOLLIGER, Adolph, biologist; b. Zurich, Switzerland, Oct. 8, 1897; s. Adolf and Eugenie (Ackermann) B.; Ph.D., U. Basel, 1922; D.Sc., U. Sydney (Australia), 1956; m. Clara Coradi, 1922; m. 2d, Dorothy Dark, 1937; children—Walter, Harry, Peter, John, Stephen. Research asso. Henry Ford Hosp., Detroit, 1925-28; dir. Gordon Craig Research Lab., U. Sydney (Australia), 1930-62. Fellow Royal Australian Chem. Inst., Australian Acad. Sci. Research, publs. on Australian marsupials, organic chemistry, parasitology, exptl. pathology. Died Sydney, Oct. 1962.

BOLLINGER, Clyde John, Am. geographer, climatologist; b. Eaton County, Mich., Oct. 24, 1888; s. Charles and Emma Loretta (Perry) B.; A.B., U. Mich., 1915; M.A., U. Chgo., 1920; postgrad. Clark U., 1929-30; m. Hazel Jessie Guest, June 21, 1916; children—Ruth Florence, Ralph Perry, Helen Francis. Tchr., adminstr. pub. schs., 1909-17; faculty U. Okla., Norman, 1920—, asso. prof. geography, 1925-60, chmn. dept., 1920-45, prof. emeritus, 1960—. Fellow Okla. Acad. Sci., Am. Geog. Soc., Am. Meterol. Soc.; mem. Soc. Research Geophysicists. Author: Elementary Geography of Oklahoma, 1930; also articles. Research in field of solar energy, effects of solar energy on terrestial weather and climate, sun-spots, climatic cycles. Home: 725 Chautauqua St., Norman, Okla. 73069.*

BOLLINGER, Lowell Moyer, Am. physicist; b. Greene County, Va., Apr. 28, 1923; s. Amsey Floyd and Florence (Moyer) B.; A.B., Oberlin Coll., 1943; Ph.D., Cornell U., 1951; m. Margaret Jeffries, Nov. 5, 1944; children—Lesley, Jeffrey, Priscilla. Physicist aircraft engine research lab., NACA, Cleve., 1943-46; neutron physicist Argonne Nat. Lab., 1951—, dir. physics div., 1963—; guest physicist Atomic Energy Research Establishment, Harwell, Eng., 1961-62. Fellow Am. Phys. Soc. Research and publs. on interactions between slow neutrons and nuclei. Home: 1741 Prairie Av., Downers Grove, Ill. Office: Argonne Nat. Lab., Argonne, Ill.*

BOLLINGER, Otto, German pathologist; b. Altenkirchen, Germany, Apr. 3, 1843; s. Heinrich and Anna Maria (Conrad) B.; student medicine, natural scis. Berlin, Germany, Vienna, Austria, 1868-69; doctorate in comparative pathology, vet. medicine Munich, Germany, 1870; m. Hedwig Usener, 1874; 2 daus. Battalion physician in field hosp. Franco-Prussian war; faculty vet. sch. Zurich, Switzerland, lectr. U. Zurich and Polytechnikum, from 1871; prof. pathology Munich Sch. Vet. Medicine, asso. prof. comparative pathology U. Munich, from 1874; dir. Munich Path. Inst., from 1880, rector, from 1908. Author: Atlas und Grundriss der pathologischen Anatomie, 2 vols., 1901. One of creators of comparative pathology; diagnosed actinomyces, origin of cowpox from human vaccination, identity of murrain in cattle, human Tb, mycotic nature of meat poisoning; wrote study on idiopathic heart hypertrophy (Munich beer heart). Died Munich, Aug. 13, 1909.

BOLLMAN, Jesse Louis, Am. physician; b. Springfield, Ill., Feb. 19, 1896; s. George Ennison and Emma (Frichot) B.; A.B., U. Ill., 1917, B.S., M.D., 1923; M.S., U. Mich., 1918; m. Mildred Lee Montgomery, Nov. 29, 1922; children—Elizabeth (Mrs. J. Kummer), Jean (Mrs. E. T. Rulison), Jesse Louis. Instr. bacteriology U. Mich., 1917-18, U. Ill., 1919-23; faculty Mayo Found., U. Minn., Rochester, 1923—; prof. physiology, 1943-61. Mem. physiology study sect. NIH, 1954-58, gen. medicine study sect., 1960-64. Recipient John Phillips Meml. award A.C.P., 1964, Mem. Am. Gastroenterol. Assn. (Julius Friedenwald medal 1961), Am. Soc. Exptl. Pathology (pres. 1941), Am. Assn. Liver Disease (pres. 1954), A.M.A., A.A.A.S., Am. Chem. Soc. Contbr. numerous research papers on physiology and pathology of liver and gastrointestinal tract to profl. jours. Home: 410 6th Av. S.W., Rochester, Minn. 55901.*

BOLLOBAS, Béla, Hungarian physician; b. Karácsond, Hungary, Aug. 28, 1911; s. István and Anna (Herczeg) B.; diploma U. Pázmány Péter, Budapest, Hungary, 1936; m. Emma Varga, July 21, 1940; children—Emöke, Béla, Boglárka, Enikö. Surgeon, Mil. Hosp., 1936-45; asst. lectr. ear-nose-throat U. Pázmány, 1945-47; staff Hosp., László, 1947-55, Hosp. János, 1955-64; head physician Hosp. István, 1964-67, Hospital Arpád, 1967— (all located in Budapest, Hungary). Member of Korányi Sándor Soc. Author: Ear, Nose, and Throat Complications of Acute Infectious Diseases, 1957; also numerous articles. Devel. new instruments for ear, nose, and throat diseases; introduced medicinal and diagnostical methods for antrum perfusion; described new syndrome rhinitis vasomotorica climacterica; physiol. and anat. research on spatium parapharyngeum, ear, nose and throat operational agressions, ear complications of acute infectious diseases; submacroscopical anat. elements of hearing mechanism. Home: 4 Orlay. Office: IV Arpád U. 126, Budapest, Hungary.*

BOLOGNARI, Arturo, biologist; b. Copenhagen, Denmark, Jan. 25, 1917; s. Antonio and Sophie (Riis) B.; Degree in Natural Scis. with Honors, Catania (Italy) U., 1940; m. Lidia Scimone, May 18, 1940; children—Mirella, Velleda, Daniele. Staff, Messina (Italy), U., 1944—, prof. histology and embryology, 1963—, dir. Zoology Inst., 1962—. Recipient prize Lyon Internat. for Biology, 1960, prize Marzotto for Medicine, 1961. Mem. Unione Zoologica Italiana, Messina Accademia Peloritana. Author: A. Bolognari-Lezioni di Istologia ed Embriologia, 1964; also numerous articles. Research on migration of sperm-whales to Mediterranean Sea, white body of Cephalopoda, neurosecretion of Amphibia, glutathion in eggs of sea urchins, vitellogenesis in oocytes of Mollusca and Echinodermata, cytopathic action of polio and Coxsackie viruses, tumoral cell; discovered nucleolonema does not exist in nucleolus but nucleolini does; produced exptl. proteic synthesis in tumoral cell; developed hypothesis that nucleolus is first to feel effects of oncogen factors. Home: 85 Corso Italia, Messina, Italy.*

BOLOS DEMOCRITOS OF MENDES, see Democritos.

BOLSHAKOV, Kirill Andreievich, Russian chemist; born December 24, 1906; graduated Kazan University, 1930. With Moscow Institute Rare and Minor Metals, 1930-48; instructor Moscow Institute. Fine Chem. Tech., 1933-48, now rector; prof., 1948—. Decorated Order of Lenin; recipient Stalin prize, 1941. Mem. USSR Acad. Sci. (corr.). Research and publs. on physico-chem. principles of tech. processes for obtaining rare elements. Address: Moscow Inst. Fine Chem. Tech., M. Pirogovskaya ulitsa 1, Moscow, USSR.

BOLSI, Dino, Italian neurologist, psychiatrist; b. Rimini (Forli), Sept. 29, 1898; M.D. Prof. nervous and mental diseases clinic of U. Torino (Italy), 1938—, also dean med. sch. Inst. Medicine and Sch. Specialization in Psychiatry and Neuropathology.

Mem. Italian Soc. Psychiatry (com.), Acad. Physicians of Torino, Soc. Exptl. Biology, Italian Soc. Neurology (pres.). Co-dir. Rivista di patologia nervosa e mentale, Rivista di endocrinoneurochirurgia, also surg. sect. Minerva Medica. Research and publs. on degenerative diseases of cerebellum, Adie's syndrome, hystological studies of microglia, mental illness, mesenchymona. Address: via Cosseria 11, Torino, Italy.

BOLSTER, Calvin Mathews, Am. aero. engr.; b. Ravenna, O., Aug. 17, 1897; s. William Harvey and Lottie (Mathews) B.; B.S., U. S. Naval Acad., 1919; M.S., Mass. Inst. Tech., 1923; M.S. in Aero. Engring., Cal. Inst. Tech., 1936; m. Agnes Elizabeth Ryan, Jan. 5, 1925; children—Calvin Mathews (dec.), Carolyn E. (Mrs. William H. Dougherty), Dennis R., Mary A. Commd. ensign USN, 1919, advanced through grades to rear admiral, 1949, asst. chief bur. aeros. for research and devel. Navy Dept., 1948-50, chief naval research, 1951-53, ret., 1953; head research and devel. Gen. Tire Co., Akron, O., 1954-62; dir. ind. research and devel. adminstrn. Aerojet Gen., El Monte, Cal., 1962-64, cons., 1965 . . . , mem. tech. adv. bd.; mem. sci. bd. Gen. Tire and Rubber Co. Decorated Legion of Merit (U. S.); hon. comdr. Mil. Order of Brit. Empire. Recipient Robert Goddard Meml. award Am. Rocket Soc., 1949; Nat. Air Council award, 1950. Fellow Am. Inst. Aeros. and Astronautics. Patentee stern handling beam for mechanical handling of large rigid airships; in charge rocket jet-assisted take-off devel. in USN, World War II. Home: 2450 Roanoke Rd., San Marino, Cal. 91108. Office: 9100 E. Flair Av., El Monte, Cal. 91735.*

BOLSTERLI, Mark, Am. physicist; b. New Haven, Oct. 3, 1930; s. Arthur Alfred and Frances (Hipple) B.; student Swarthmore Coll., 1947-49; A.B., Washington U., St. Louis, 1951, Ph.D., 1955. Fulbright fellow U. Birmingham (Eng.), 1955-56; research asso. Washington U., 1956-57; faculty U. Minn., Mpls., 1957—, professor physics, 1967—; vis. prof. U. Copenhagen, 1961-62. Cons. Los Alamos Sci. Lab., 1963—. Guggenheim fellow U. Oxford (Eng.), 1964-65. Fellow Am. Phys. Soc. Research and publs. on problems in theoretical nuclear physics, theory of systems with large numbers of particles, quantum field theory. Home: 106 France Av. S., Mpls., 55416.*

BOLT, Bruce Alan, mathematician; b. Largs, Australia, Feb. 15, 1930; s. Donald Frederick and Arlene (Stitt) B.; B.Sc. with honors, New Eng. U. Coll., 1952; M.Sc., Sydney (Australia) U., 1954, Ph.D., 1959; m. Beverley Bentley, Feb. 11, 1956; children—Gillian, Robert. Math. master Sydney Boys' High Sch., 1953; lectr. U. Sydney, 1954-61, sr. lectr., 1961-62; research seismologist Columbia, N.Y.C., 1960; dir. seismographic stas. U. Cal. at Berkeley, 1963—, prof. seismology, 1963—. Mem. com. on seismology Nat. Acad. Scis., 1966—. Fulbright scholar, 1960. Fellow Am. Geophys. Union; mem. Seismol. Soc. Am. (editor Bull. 1965—, dir. 1965—), Assn. Seismology and Physics Earth's Interior (exec. com. 1964-67), Australian Math. Soc., Royal Astron. Soc., Sigma Xi. Research on dynamics, elastic waves, reduction geophys. observations; discovered more detail on structure transition region between liquid and solid portions earth's core. Home: 1508 Le Roy Av., Berkeley, Cal. 94708.*

BOLTON, Benjamin Meade, Am. bacteriologist; b. Richmond, Va., Apr. 7, 1857; s. James and Anna Maria (Harrison) B.; student Charlottesville Inst., 1871-75; M.D., U. Va., 1879; postgrad. S.C. Coll., 1882-83, U. Heidelberg, 1883-84, U. Göttingen, 1884-86, U. Berlin, 1886; m. Johanna Heriette Louise Liebau, 1886; children—Meade, Theodore; m. 2d, Laetitia Todd, 1898; 1 dau., Laetitia Todd. Asst., Johns Hopkins, 1886-88, asso., 1892-95; prof. hygiene and bacteriology S.C. Coll., 1888-89; dir. bacteriol. dept. Hoagland Lab., 1889-92; dir. of lab. Bd. Health, Phila., 1895-96; prof. pathology and bacteriology U. St. Louis, 1901-04; dir. lab. State Bd. of Health, N.J., 1907-08; expert in exptl. therapeutics Bur. of Animal Industry, Washington; biologist Md. Agrl. Expt. Sta., 1912-14; expert in diseases of animals Cuban Agrl. Expt. Sta., 1914-15; dir. Virchow Lab., St. Louis, 1917-19; pathologist New Samaritan Hosp., Sioux City, Ia., 1920-21; pathologist St. Joseph's Hosp., Paterson, N.J., 1921—; bacteriologist R.I. Bd. Health. Died Aug. 12, 1929.

BOLTON, Edward Richards, Brit. chemist; b. 1878; s. J. A. Bolton; ed. King's Coll., London; m. Norah Binning, 1902. Mng. dir. Tech. Research Works, Ltd.; analyst, tech. cons. chemist. Mem. delegacy, fellow U. London, King's Coll.; mem. adv. council on plant and animal products Imperial Inst.; rep. Great Britain on Commn. Internationale pour l'Etude des Matières Grasses. Fellow Royal Inst. Chemistry, Chem. Soc. (council 1927-29); mem. Instn. Chem. Engrs., Soc. Pub. Analysts (pres. 1926-28), Inst. Chemistry Great Britain and Ireland (v.p. 1925-28), Medico-Legal Soc. London (v.p.), Brit. Standards Instn. (chmn. com. on vegetable oils). Author: Oils, Fats, and Fatty Foods: Their Practical Examination; Co-author: Oils, Fats, Waxes, and Resins. Contbr. to Ency. Brit., also to Allen's Commercial Organic Analysis, papers pub. in sci. jours. Patentee various chem. processes. Died Feb. 10, 1939.

BOLTON, Ellis Truesdale, Am. biophysicist; b. Linden, N.J., May 4, 1922; s. Elliott L. and Elizabeth (Lindsay) B.; B.S., Rutgers U., 1943, Ph.D., 1950; m. V. Elaine Alber, Sept. 11, 1943; children—Roger T., Craig E. Instr. zoology Rutgers U., 1946-49; with Carnegie Instn., Washington, 1949—, dir. dept. terrestial magnetism, 1966—. Cons., NIH. Mem. Washington Acad. Sci. (biol. achievement award 1959), A.A.A.S., Biophys. Soc. (mem. council 1964—), Am. Geophys. Union, Am. Soc. Plant Physiologists, Sigma Xi. Contbg. author: Studies of Biosynthesis in Escherichia coli, 1963; also articles. Devel. agar technique to immobilize single strands of DNA for studies of hybridization reactions; pioneer in devel. techniques in molecular biology. Home: 1 Briggs Ct., Foxhall, Silver Spring, Md. 20906. Office: 5241 Broad Branch Rd. N.W., Washington 20015.*

BOLTON, Elmer Keiser, Am. chemist; b. Phila., June 23, 1886; s. George and Jane (Holt) B.; A.B., Bucknell U., 1908, hon. D.Sc., 1932; A.M., Harvard, 1910, Ph.D., 1913; student Kaiser Wilhelm Inst. für Chemie, Berlin, 1913-1915; D.Sci. (honorary), University of Delaware; married Marguerite L. Duncan, December 6, 1916; children—Duncan G., Marjorie L., Elmer K. With E. I. du Pont de Nemours and Co. 1915-51, successively asst. mgr. Lodi Works, mgr. organic div. of chem. dept., dir. chem. sect. of dyestuffs dept., asst. chem. dir. of chem. dept., 1929-30, chem. dir., 1930-51; retired, 1951; now member tech. advisory panel on materials Dept. of Def. Trustee Bucknell University. Member visiting committee, Harvard University; mem. advisory bd., Rutgers Univ. Research Council; mem. bd. mgrs., The Wilmington Institute Free Library. Mem. Nat. Acad. of Sciences, Am. Chemical Soc. (dir., 1940-43), Am. Inst. Chem. Engrs., Sigma Xi. Recipient Chemical Industry Medal, 1941, Perkin Medal, 1944; Willard Gibbs gold medal, Am. Chem. Soc., 1954. Isolated, determined chem. structure of pigments of geraniums, scarlet sage and red chrysanthemums; leader in establishing U. S. dyestuff industry; research leading to 1st synthetic rubber; basic research on polymerization. Home: 2310 W. 11th St., Wilmington, Del.

BOLTON, Francis John, Brit. army officer, electrician; b. Eng., 1831; s. Thomas Wilson Bolton; m. Julia Mathews, 1866. Commd. ensign Gold Coast Arty., Brit. Army, 1857, advanced through grades to col. (hon.), 1881. Created knight, 1884. Mem. Soc. Telegraph Engrs. and Electricians (founder 1871, v.p., editor Jour.). Author: Description of the Illuminated Fountain and of the Water Pavilion, 1884. Developed (with Philip Howard Colomb) system of visual signalling; inventor oxy-calcium light for night signalling; designer, controller colored fountains and electric lights S. Kensington Exbns., 1883-86. Died 1887.

BOLTON, Herbert, Brit. geologist; ed. Royal Coll. of Sci., London, Owens Coll., Manchester; D.Sc.; m.; 2 daus. Asst. keeper Manchester Mus., Victoria U., 1890-98; curator Bristol Mus., 1898-1911; dir. Bristol Mus. and Art Gallery, 1911-30; mem. ct. govs. Bristol U., to 1930, reader in paleontology, 1911-26; mem. adv. council Victoria and Albert Mus., 1920-33. Recipient Murchison Geol. fund, 1922. Fellow Royal Soc. Edinburgh, Geol. Soc.; pres. Museums Assn., 1923-24; mem. N.Y. Acad. Scis. (corr.), Manchester Geol. and Mining Soc. (hon.), Inst. Mining engrs. (hon.). Research and publs. on geology and paleontology of Brit. Coal Measures, fossil insects of Coal Measures, also on museums. Died Jan. 18, 1936.

BOLTON, James Robert, chemist; b. Swift Current, Sask., Can., June 24, 1937; s. James Linden and Margaret (McFadden) B.; B.A., U. Sask., 1958, M.A., 1960; Ph.D., Cambridge U., Eng., 1962; m. Wilma Burdette Hall, Dec. 26, 1959; children—Judith Louise, James Thomas. Research asso. Columbia, 1962-64; asst. prof. U. Minn., Mpls., 1964-66, asso. prof., 1966—. Mem. Am. Chem. Soc., Am. Phys. Soc., Am. Assn. U. Profs. Research, publs. on devel. electron spin resonance spectroscopy as tool to provide detailed information about electronic structure of molecules with unpaired electrons. Home: 1912 Carl St., St Paul 55113. Office: Dept. Chemistry, U. Minn., Mpls. 55455.*

BOLTON, Werner, chemist; b. Tiflis, USSR, Apr. 8, 1868; s. William and Katherine (Kölle) B.; student Berlin, Leipzig (both Germany), 1888-95; doctorate, Leipzig, 1895; m. Paula Goldschmidt. Electrochemist, firm Siemens and Halske. Used tantalum (particularly hard, tough metal) as lamp filament to find replacement for carbon filament, 1st manufactured, 1903 (eventually replaced by tungsten filament). Died Berlin-Westend, Germany, Oct. 28, 1912.

BOLTWOOD, Bertram Borden, Am. chemist, physicist; b. Amherst, Mass., July 27, 1870; s. Thomas Kast and Matilda (Van Hoesen) B.; Ph.B., Sheffield Sci. Sch., Yale, 1892, Ph.D., 1897; post-grad. studies Munich, Leipzig and Yale; John Harling fellow, U. Manchester, 1909-10. Asst. analytical chemistry, 1894-96; instr., Yale, 1896-1900, asst. prof. physics, 1906-10, prof. radio-chemistry, 1910-1927; acting prof. chemistry, dir. Kent Chem. Lab., 1918-22, supervised bldg. Sloan Physics, Sterling Chemistry labs. Fellow Am. Acad. Arts and Scis. Translator: Quantitative Analysis by Electrolysis (A. Classen), 1898; Physical Chemistry for Beginners (C. H. Van Deventer), 1899. Proved radium is a disintegration product of uranium; discovered element ionium, 1907; established basis for study of isotopes and used radioactivity for calculation of the age of geol. strata. Died Aug. 15, 1927.

BOLTZ, Hans, German geodesist; b. Elbing, Germany, Apr. 9, 1883; s. Hugo and Johanna (Reich) B.; student chemistry Königsberg, Germany, 1902-04; doctorate Berlin, Germany, 1908 (studied under M. Planck, H. v. Strure, F. R. Helmert); m. Helene Koreuber, 1921; 1 son, 1 dau. Probably observator with Geodätisches Institut, Potsdam, Germany, 1908-10, 12-22, dept. head, prof., from 1927, dir., from 1946; asst., Königsberg Obs., 1910-11; Contbr. articles to Preussisches Geodätisches Institut, other German geod. jours. Participated in Turkish land survey program, 1932-34; research in modern calculation methods for fixing of surveying foundations; famous for devel. process and substitution process named after him; 1st grand adjustment of triangulation networks with 673 conditional adjustments as a whole; worked on important tables for practical calculations in higher geodesics. Died Potsdam, Mar. 28, 1947.

BOLTZMANN, Ludwig Eduard, Austrian physicist; b. Vienna, Austria, Feb. 20, 1844; s. Ludwig and Katherine (Pauernfeind) B.; degree U. Vienna, 1866; Dr.h.c., U. Oxford (Eng.); became asst. Physikalisches Institut, Vienna, and prof. theoretical physics, 1867; joined U. Graz (Austria), 1869-73, became dir. Phys. Inst., 1876-90; apptd. prof. math. at U. Vienna, 1873-76, prof. theoretical physics, 1894-1900, and prof. theoretical physics and natural philosophy, 1902-06; apptd. prof. theoretical physics U. Munich (Germany), 1889, U. Leipzig (Germany), 1900-02. Fellow Royal Soc., 1899. Mem. acads. of Göttingen, Vienna, Berlin, Stockholm, Uppsala, Turin, Rome, Amsterdam, St. Petersburg, N.Y., London, Paris, Washington, St. Louis. Author: Vorlesungen über Maxwells Theorie der Elektrizität und des Lichtes, 2 vols., 1891-93; Vorlesungen über die Prinzipe der Mechanik, 2 vols., 1897-1904; Vorlesungen über kinetische Gastheorie, 2 vols., 1896-98; Populäre Schriften, 1905; Wissenschäfte Abhandlungen, 3 vols., 1909. With Maxwell founded equipartition theory; demonstrated Stefan-Boltzmann law on radiation from black body; research on electro-magnetic theory; thermodynamics kinetic theory of gases leading to theory of statis. dynamics. Died Duino, near Trieste, Sept. 5, 1906.

BOLYAI, Farkas, Hungarian mathematician; b. Szeklerland, Hungary, Feb. 9, 1775; student U. Jena (Germany); student U. Göttingen (Germany), 1796-99; 1 son, János. Staff, U. Göttingen; prof. math., physics, and chemistry Reformed Coll., Maros-Vásárhely, Hungary, 1804-51. Author: Tentamen juventutuem studiosam in elementa matheseos purae introducendi, 2 vols., 1832-33. Independently of Lobachevsky, he and his son developed non-Euclidean geometry. Died Maros-Vásárhely, Hungary (now Mures, Rumania), Nov. 20, 1856.

BOLYAI, János, Hungarian mathematician; b. Kolozsvar, Hungary (now Cluj, Rumania), Dec. 15, 1802; s. Farkas Bolyai; entered Royal Coll. Engring., Vienna, Austria, 1817. Became an officer of engineers, 1822. Author: Appendix scientiam spatii absolute veram exhibens (appendix to father's work Tentamen juventutem studiosam in elementa matheseos purae introducendi, 1832-33), 1832. Helped develop (with father, independently of Lobachevsky) non-Euclidean geometry. Died Maros-Vásárhely, Jan. 27, 1860.

BOLZA, Oskar, (pseudonyn F. H. Marneck), German mathematician; b. Bergzabern, Germany, May 12, 1857; s. Emil and Luise (König) B.; student U. Berlin (Germany), 1875-79, 85-86, univs. Strasbourg (Germany, now France), Berlin, Göttingen (Germany), 1879-82, U. Frieburg (Germany), 1883-85, Ph.D., Göttingen, 1886, postgrad., 1887-88; m. Anna Neckel, 1898; no children. Student tchr. Gymnasium, Freiburg, 1882-83; reader in math. Johns Hopkins, 1888-89; asso. in math. Clark U., 1889-93; asso. prof. math. U. Chgo., 1893-94, prof., 1894-1910, nonresident prof., 1910-42; hon. prof. math. U. Freiburg, 1910-26, 29-33; pvt. researcher in Sanskrit and religious psychology, 1926-29, in math. and psychology, Freiburg, 1933-42. Author: Lectures on the Calculus of Variations, 1904; Vorlesungen über Variationsrechnung, 1908; Lectures on Integral Equations, 1913; Gauss und die Variationsrechnung (Gauss' Werke, Vol. 10), 1922; Glaubenlose Religion, 1931; Aus meinen Leben, 1936. Reduced hyperelliptic integrals to elliptic integrals; studied elliptic and hyperelliptic functions, also calculus of variations, especially problem of Bolza. Died Freiburg, July 5, 1942.

BOLZANO, Bernhard, Czechoslovakian mathematician, logician; b. Prague, Czechoslovakia, Oct. 5, 1781; ed. U. Prague; ordained priest 1805; apptd. prof. philosophy of religion U. Prague, 1805, removed because of liberal teachings, 1819; retired after refusing to recant his beliefs on absurdity of militarism and war, and need for socioeconomic and educational reforms, 1824. Author: Athanasia, 1827; Functionenlehre, 1834; Lehrbuch der Religionswissenschaft, 4 vols., 1834; Versuch einer neuen Darstellung der Logik, 4 vols., 1837; Rein analytischer Beweis, 1817, Paradoxien des Unendlichen, 1851. In mathematics stressed need for rigorous definition and strict methodology, research on real functions and aggregates; gave proof binomial formula 1816; postulated means of distinguishing between finite and infinite classes; held advanced views on variables, limits, continuity; in philosophy studied logic and epistemology; pioneer of modern logic; in physics investigated force, space, and wave propagation; criticized Kant's theories of space and time. Died Prague, Dec. 18, 1848.

BOMBARD, Alain Louis, French biologist; b. Paris, France, Oct. 27, 1924; s. Gaston and Marie (Stodel) B.; ed. U. Paris; m. Ginette Brunn, June 10, 1952; children—Renaud, Anne, Nathalie, Christophe, Antoine. Intern, Hosp. of Boulogne-sur-Mer, 1950-52; researcher Oceanographic Mus. of Monaco, 1952; survival experience on board Hérétique, 1952; dir. marine lab. La Coryphène, 1958-61; dir. marine lab. Captain Cap, 1961—. Physician for cities of Paris and Ostende, Gand. Mem. Soc. French Explorers. Author: Naufragé volontaire, 1953; Rapport technique sur la survie en mer, 1955. Home: 16, rue Maurice-Thédié, Amiens. Office: Captain Cap, Saint-Pierre-des-Embiez, par le Brusc, France.

BOMBASTUS VON HOHENHEIM, see Paracelsus, Theophrastus Bombastus von Hohenheim.

BOMBELLI, Rafaello, Italian mathematician; b. Bologna, Italy, circa 1530; author treatise on algebra, 1572. Laid found. of better knowledge of imaginary quantities by pointing out reality of apparently imaginary expression assumed by a root; gave approximation processes to roots of affected equations; used continued fractions to find sq. root.

BOMER, Aloys Wilhelm Joseph Hubert, German food chemist; b. Haus Sobberinghoff, Germany May 16, 1868; s. Adam and Antonie (Becker) B.; student natural scis. Münster, Berlin (both Germany), 1887-91; m. Maria Beckmann; 1 son, 1 dau. Asst. at agrl. research sta. (Landwirtschaftliche Versuchsstation), Münster, dir., from 1911; prof. Münster, 1921-35. Recipient Joseph Konig meml. medal, 1934. Mem. Reichsgesundheitsrat, Assn. German Food Chemists (chmn.). Co-editor: Zeitschrift für Untersuchung d. Lebensmittel, 1911-36; (with A. Juckenack and J. Tillmans) Handbuch der Lebensmittel-chemie, 1932. Worked with chemistry of fats; developed new methods of identifying fats and fat mixtures including phytosterinacetate method and melting point difference method; instrumental in working out modern food laws, regulations in Germany. Died Germany, Oct. 9, 1936.

BOMKE, Hans Alexander, physicist; b. Berlin, Germany, May 26, 1910; s. Karl Christian and Charlotte (Teucher) B.; Dr. Habil., Tech. U. Berlin, 1935; Ph.D., U. Berlin, 1931; Venia Legendi, U. Munich, 1949; m. Ilse Cornelia Nagel, Jan. 22, 1932; 1 son, Thomas Roger. Came to U. S., 1952, naturalized, 1957. Faculty, U. Berlin, 1930-33, Tech. U. Berlin, 1933-39; research physicist Kaiser Wilhelm Inst., Berlin, 1939-43; sect. chief Aerial Nav. Research Inst., Oberpfaffenhofen, Bavaria, 1943-45; head biophysics lab. Sch. Medicine, U. Munich, 1946-52; with U. S. Army Electronic Labs., Ft. Monmouth, N.J., 1952—, prin. sci. Exptl. Research Inst., 1961—; adj. prof. physics Fordham U., 1961-62; project officer U. S. High Altitude Atomic Tests, 1962; mem. lab. astrophysics panel NSF, 1961-63. Recipient U. S. Army Research and Devel. Achievement award, 1964. Mem. Am., German phys. socs., Am. Geophys. Union, Optical Soc. Am., Am. Rocket Soc., Am. Physics Tchrs. Assn., A.A.A.S., N.Y. Acad. Sci. Author: Vacuum Spectroscopy, 1937; Atom and Ion Beams, 1939; Theory of Propagation in Wave Guides, 1950; also numerous articles. Research in uranium fission physics, 1939-42, radar, 1943-45, Farady effect inertia, 1952-54, hydromagnetic phenomena in inosphere and exosphere, 1958—; inventor micro dosimeter for use in radiation therapy, sensitive magnetometer. Home: 408 Central Av., Spring Lake, N.J. 07762. Office: Inst. for Exploratory Research, Ft. Monmouth, N.J.*

BöMMEL, Hans Eberhard, physicist, educator; b. Munich, Germany, Oct. 15, 1912; s. Otto and Valentina (Wiwodzoff) B.; Ph.D. in Physics, U. Zurich, 1943; m. Elisabeth A. Feitknecht, Apr. 27, 1946; children—Thomas, Valentina. Mem. sci. staff Inst. for Testing Materials, Zurich, Switzerland, 1944-46; research supr., lectr. U. Zurich, 1946-53; mem. tech. staff Bell Telephone Labs., 1953-61; prof. physics U. Cal. at Los Angeles, 1961—. Sci. staff mem. N.Am. Aviation Sci. Center, Thousand Oaks, Cal., 1963—. Fellow Acoustical Soc. Am.; mem. Am., Swiss (life) phys. socs. Contbr. articles to profl. jours. Developed use of ultrasonic waves as tool in solid state physics; discovered that ultrasonic absorption can increase enormously in metals at low temperature due to electron-lattice interaction, magneto-acoustic effect; developed methods to generate microwave frequency ultrasonic waves. Home: 874 Malcolm Av., Los Angeles 90024.

BOMPIANI, Enrico, Italian mathematician; b. Rome, Italy, Feb. 12, 1889; s. Arturo and Domenica (Gaifani) B.; Ph.D. in math.; Dr. honoris causa U. Gröningen. Prof., Poly. Sch. of Milan, U. Bologna; full prof. analytical and descriptive geometry; dean Inst. Math., U. Rome; dir. Centre internat. du mathém; instr., lectr. at univs. in Europe, U. S. and Indies. Recipient Gold medal Italian Soc. Science, plant physiologists, Soc. Study Devel. and Growth, 1926. Mem. numerous sci. acads. and coms. in Italy and fgn. countries. Research and numerous publs. on math, theories of geometry and space. Address: via Verona 22, Rome, Italy.

BONADUCE, Antonio, Italian veterinarian; b. Mosciano S. Angelo (Teramo), June 10, 1916; s. Domenico and Argia (Londrillo) B. Univ. prof., 1952; titular prof. infectious diseases, prophylaxy and vet. control, dir. inst. U. Messina (Italy), also instr. sch. vet. medicine. Mem. Italian Soc. Exptl. Biology, Soc. Vet. Scis. Collaborator: La Nuova Veterinaria; La Clinica Veterinaria; Zooprofilassi. Address: via S. Cecilia 30, Messina, Italy.

BONAFOUS, Mattieu, agronomist; b. Lyons, France, Mar. 7, 1793; docteur en medecine, Montpellier, France; practiced medicine Turin, Italy; helped popularize vaccination; raised silkworms. Mem. French Acad. Scis., 1835. Agronomical Inst. Turin. Author: l'Education de vers à soie, 1821; la Culture du mûrier, 1825; l'Histoire naturelle, agricole et économique du mäis, 1836. Worked with acclimatization of fgn. plants, also culture of sugar beet; translated fgn. works on silkworm raising by Morikounia and Marco Vida. Died Paris, France, Mar. 23, 1852.

BONAPARTE, Charles Lucien, ornithologist; b. Paris, France, May 24, 1803; s. Lucien Bonaparte; nephew of Napoleon Bonaparte; ed. in Italy; married Zenaide (his cousin). Came to U. S. (Phila.), 1822; completed Wilson's Ornithology, (listing at least 100 species of birds he discovered), 1825-33; contbr. articles on ornithology to sci. journals; returned to Italy to continue studies, 1828; became prince of Canino and Musignano (upon death of his father), 1840; entered Italian politics, joined anti-papal faction, served as v.p. of republican assembly; v.p. Constitutional Assembly, Rome; mem. Junto at Rome, 1848; fled Italy, 1848, settled in France, 1850; became dir. Jardin des Plantes, 1854. Author: Geographical and Comparative List of Birds of Europe and North America, 1838; several other studies, published 1827-58. Died Paris, July 29, 1857.

BONAPARTE, Napoleon, French emperor, patron of science; b. Ajaccio, Corsica, Aug. 15, 1769; s. Carlo Maria and Letizia (or Laetitia) (Ramolino) de Buonaparte; m. Josephine de Beauharnais, Dec. 1, 1804; m. 2d, Maria Louise, Apr. 2, 1810, 1 son, Napoleon Francis Joseph Charles (Napoleon II); 1 illegitimate son, Comte Alexandre Walewski. First consul of France, 1800-04; emperor of France, 1804-14; relinquished idea of building colonial empire in Am. after having been unsuccessful in squelching native revolt in Santo Domingo (which he had hoped to make center of overseas colonies), 1803; sold La. to U. S. for 60 million francs, 1803; exiled to Elba, 1815, escaped, returned to Paris; defeated by Continental forces at Waterloo, Belgium, Mar. 17, 1815, exiled to St. Helena, 1815-21. Founded the Inst. of France, composed of 5 acads., including Acad. of Scis., 1795. Died St. Helena, May 5, 1821.

BONAPARTE, Roland-Napoléon, French geographer; b. Auteuil, France, May 19, 1858; s. Pierre and Eléonore (Ruffin) B.; ed. St. Cyr Mil. Sch.; m. Marie-Félix Blanc, 1880; 1 dau., Marie (Princess George of Greece). Served in army; made anthropol.-geog. expdns. to Eng., Holland, Norway, Switzerland; installed stas. for study of glaciers in Alps; supported Burgeois' expdn. to Ecuador; sci. French Acad. Scis. Mem. Soc. Geography (pres.), Internat. Inst. Geography elected pres. 1922), French Soc. Glaciers (pres.), French Acad. Scis., 1907 (founder fund for encouragement of sci.). Author: Les habitants de Suriname, 1884; Anthropologie laponne, 1886; Le glacier d'Aletsch et le lac de Maerjelen, 1889; Documents de l'époque mongole des XIIe et XIVe siècles, 1895. Died Apr. 14, 1924.

BONATTI, Guido, Italian astrologer; b. Cascia, Italy; in Ravenna and Bologna, Italy, 1223, Forli, 1233; adviser to Frederich II; entered service of Ezzelino III da Romano, 1259; became astrologer to Guido di Montefeltio, 1260. Author: Tibes astronomicus (deals with signs of zodiac, planets, meteorology, and astrological forecasts of weather), circa 1261-77. Died 1297.

BONAVENTURA, Federigo, Italian physician; b. Ancône, Italy, 1555; m. Panthesile Capegna; 12 children; minister, later extraordinary ambassador Duke of Urbino. Author: De octomestris partus natura adversus vulgatam opinionem peripatetica disputatio, 1596; Della ragion di stato e della prudenza politica libri IV, 1623. Tried to prove 8 month fetus can live. Died Mar. 1602.

BONCOMPAGNI, Balthasar, Italian mathematician; b. Rome, May 10, 1821; mem., later librarian Accademia dei Lincei; founder, editor math. bull., 1868-87. Research and publs. on math. Editor works of Pietro Cossalo, Leonardo da Vinci; translator Arabic math. treatise. Died Apr. 13, 1894.

BOND, Charles Farrington, Am. zoologist; b. Chester, Pa., June 30, 1920; s. Charles Martin and Elizabeth (Stults) B.; B.A., Bucknell U., 1942; M.A., Cornell U., 1947, Ph.D., 1957; m. Amy Llewellyn Stevenson, Apr. 29, 1945; children—Anne Elizabeth, Charles Stewart. Instr., Wyoming Sem., Kingston, Pa., 1942-43, Bucknell U., 1943-45, U. Mass., 1948-50; asst. prof. s. U. Vt., Burlington, 1950-58, asso. prof., 1958-64, prof. zoology, 1964—. Recipient Danforth Found. Tchr. award, 1955. Fellow A.A.A.S.; mem. Green Mountain Audubon Soc. (v.p. 1964—, chmn. bd. 1964—, pres. 1966), Am. Soc. Zoologists, Soc. Study Evolution, Nat. Audubon Soc., Phi Beta Kappa, Sigma Xi (pres. Vt. chpt. 1966—). Contbr. articles to tech. jours. Research on influences various hormones on circulating blood volume in mammals, adaptions for underwater activity in birds, beaver populations, application computer math. to biol. problems. Died Feb. 1, 1967.

BOND, Douglas D(anford), Am. psychoanalyst; b. Waltham, Mass., July 2, 1911; s. Earl Danford and Grace (Newson) B.; A.B., Harvard, 1934; M.D., U. Pa., 1938; fellow, Inst. of Pa. Hosp., 1940-41; research fellow in physiology, Harvard Med. Sch., 1941-42; Sc.D. (hon.), Heidelberg Coll., 1953; m. Helen Cannon, July 3, 1937; children—Peter Danford, Thomas Cannon, Sharon, Barbara. Chief, lab. psychiatry, Sch. Aviation Medicine, Randolph Field, Tex., 1942-43; chief, div. of psychiatry 1st Central Med. Establishment, 8th Air Force, 1943-45; chief cons. in psychiatry Army Air Forces, Washington, 1945; prof., dir. dept. psychiatry Sch. Medicine, Western Res. U., Cleve., 1946—, dean Sch. Medicine, 1959—, dir. psychiat. services Univ. Hosps., 1946—; dir. psychiatry Rainbow Hosp., Cleve., 1950—. Cons. to various govt. agys.; med. adviser A.R.C., 1948-51; mem. project com. mental health program Nat. Inst. Mental Health, 1961-64. Chmn. bd. Am. Fund for Psychiatry, 1960-63; trustee Grant Found, 1962—; bd. dirs. Founds.' Fund for Research in Psychiatry, 1964—. Mem. Am. Psychoanalytic Assn. (sec. 1955-57), Ohio (pres. 1952-53), Am. psychiat. assns., A.A.A.S. A.M.A., Author: Love and Fear of Flying, 1952; also articles on physiology and psychiatry to publs. Home: 3017 Fairmount Blvd., Cleveland Heights, O. 44118. Office: 2065 Adelbert Rd., Cleve. 44106.*

BOND, George Phillips, Am. astronomer; b. Dorchester, Mass., May 20, 1825; s. William Cranch and Selina (Cranch) B.; grad. Harvard, 1845; m. Harriet Harris, Jan. 27, 1853, at least 3 children. Asst. observer Harvard Obs., 1845-59, dir. 1859-65, Phillips prof. astronomy, 1859-65; received gold medal from Royal Astron. Soc. of London; papers include Cometary Calculations, The Method of Mechanical Quadrature, Memoir on the Donati Comet of 1858. With father, credited with discovery of Hyperion (8th satellite of Saturn), 1848, and crape ring, 1850; investigated perturbations of cometary orbits and theory of constitution of Saturn's rings; discovered 11 new comets; founder photographic astronomy; performed 1st double-star photography, 1857; 1st to suggest measurement of star's magnitude using diameter of image on photog. plate, 1858. Died Cambridge, Mass., Feb. 17, 1865.

BOND, Henry, Brit. mathematician; b. circa 1600; tchr. applied math. Royal Dockyard, Chatham, Eng.; editor Tapp's Seaman's Kalendar for about 20 years; published works on determination of longitude; best known for prediction of ann. rate of variation of compass, method of using this to determine longitude. Died 1678.

BOND, Thomas, Am. physician; b. Calvert County, Md., 1712; s. Richard and Elizabeth (Chew) B.; studied medicine under Dr. Alexander Hamilton, Annapolis, Md.; studied medicine in Paris; m. Sarah Roberts, 7 children. Began practice of medicine, Phila., circa 1734; founder Pa. Hosp., Phila., oldest hosp. in U. S., 1752; delivered 1st course in clin. lectures in U. S., 1766; mem. original bd. trustees Coll. of Phila. (now U. Pa.); a founder Am. Philos. Soc., 1768; mem. Phila Com. of Safety, 1776; pres. Humane Soc. of Phila., 1780. Author: An Account of Worm Bred in the Liver; A Letter to Doctor Fothergill on the Use of the Peruvian Bark in Scrofula; Essay on the Utility of Clinical Lectures. Died Mar. 26, 1784.

BOND, Victor Potter, Am. physician, biophysicist; b. Santa Clara, Cal., Nov. 30, 1919; s. George Milton and Lois Edna (Potter) B.; A.B., U. Cal. at Berkeley, 1943; M.D., U. Cal. at San Francisco, 1945; Ph.D., U. Cal. at Berkeley, 1952; D.Sc., L.I. U., 1963. m. Mary Scherba, Feb. 23, 1946; children—Barbara Ann, Frederic William, Robert Michael, Beverly Jean. Intern, USN Hosp., Astoria, Ore.; head exptl. pathology br. USN Radiol. Def. Lab., San Francisco, 1948-54; scientist Med. Research Center, Brookhaven Nat. Lab., Upton, N.Y., 1955-57, head div. microbiology, 1957-62, chairman of the medical department, 1962-67, associate director, 1967—, Director commission on radiation and infection Armed Forces Epidemiol. Bd. mem. radiobiology com. Nat. Acad. Sci., radiation study sect. NIH; lectr. Am. Inst. Biol. Scis. Mem. Setauket (N.Y.) Sch. Bd., 1963-67. Served to lt. comdr. M.C., USN, 1945-54. Author: Radiation Injury in Man, 1960; Radiation Injury, Its Basis in Cellular Kinetics, 1965. Home: Cedar Lane, Strongs Neck, Setauket, N.Y. 11785. Office: Dir.'s Office, Brookhaven Nat. Lab., Upton, N.Y. 11973.*

BOND, William Cranch, Am. astronomer; b. Portland, Me., Sept. 9, 1789; s. William and Hannah (Cranch) B.; A.M. (hon.), Harvard, 1842; m. Selina Cranch, July 18, 1819; m. 2d, Mary Roope Cranch, 1831; 6 children including William Cranch, George P. Established pvt. observatory at Dorchester, Mass.; astronomical observer to Harvard College, 1840; director Harvard Observatory, 1843-44. Member American Academy Arts and Sciences, Am. Philosophical Soc., Royal Astronomical Soc. (England). Independent discoverer Comet of 1811; did pioneer work on rates of chronometers, meteorology and magnetism, 1831-39; 1st dir. Harvard Observatory, 1839-59, did intensive studies of planets, sun spots, also Orion and Andromeda nebulae; constructed 1st seagoing chronometer made in Am.; made 1st photograph of a star (Vega), 1850; worked with his son George P. (who succeeded him as dir. Harvard Obs.) in discovering 8th satellite of Saturn, 1848, the crape ring, 1850, and in developing chronograph for automatically recording positions of stars; used chronometer and telegraph for determining longitude. Suggested that size of star could be determined from size of its image on a photographic plate, 1858. Died Cambridge, Mass., Jan. 29, 1859.

BONDARCHUK, Vladimir Gavrilovich, Russian geologist; born August 11, 1905; graduated Volhynian Institute Public Education, Zhitormir, 1924. Associate, Ukrainian Geological Board, 1926-38; with Kiev Univ., 1930-41, professor, rector, 1944-51; dep. chmn. Ukrainian Council Ministers, 1951-53; dir. Inst. Geol. Sci., Ukrainian Acad. Sci., 1953-63. Mem. Ukrainian Acad. Sci. Author: Tectoorogeny, 1946; The Geological Structure of the Ukrainian SSR, 1947; The Principles of Geomorphology, 1949; The Geomorphology of the Ukrainian SSR—Geological Relief Development of the Ukrainian SSR, 1949; The Soviet Carpathians, 1957; The Structure of the Earth's Crust, 1962. Exec. editor Terminological Dictionary of Geology, 1959. Research on geol. structure of Quaternary deposits and geomorphology of Ukraine. Address: Ukrainian Acad. Sci., Vladimirskaya 54, Kiev, Ukraine SSR, USSR.

BONDARÉNKO, Igor Ilyích, Russian physicist; b. 1926; grad. coll., 1949. Staff Physics Inst. for Utilization Atomic Energy, State Com., Council Ministers USSR, Moscow, 1950—; prof., 1960—. Recipient Lenin prize (for participation in sci. investigations of physics of nuclear reactors on fast-moving neutrons), 1960. Author works on physics of fast-neutron reactors. Office: USSR Acad. Scis., B. Kaluzhskaya, 1L, Moscow V-71, USSR.

BONDELID, Rollon Oscar, Am. physicist; b. Grand Forks, N.D., Jan. 8, 1923; s. Oscar Anton and Stella (Gravelle) B.; B.S., U. N.D., 1945; M.S., Washington U., St. Louis, 1948, Ph.D., 1950; m. Phyllis Maxine Peterson, June 16, 1946; children—Claudia Joanne, Timothy Rollon, Janet Kim. Physicist, Naval Research Lab., Washington, 1952—, head Cyclotron br., 1963—. Mem. Am. Phys. Soc., Washington Acad. Scis., Sci. Research Soc. Am. Research on charged particles nuclear scattering, neutron cross sect. measurements, absolute determination of nuclear interaction energies, high resolution resonance and threshold measurements, distbn. of energy loss. Office: Code 7610, Naval Research Lab., Washington 20390.*

BONDI, Amedeo, Am. microbiologist; b. Springfield, Mass., Dec. 13, 1912; s. Amedeo and Amelia (Montevert) B.; B.S., U. Conn., 1935; M.S., U. Mass., 1937; Ph.D., U. Pa., 1942; m. Virginia Carstens, Apr. 4, 1940; children—Peter, John, Barbara Ann, Edward. Prof., head dept. microbiology Hahnemann Med. Coll., Phila., 1947—. Recipient Linback Teaching award, 1963. Mem. Am. Soc. Microbiology (past chpt. pres.), A.A.A.S., N.Y. Acad. Scis., Am. Assn. Immunologists, Am. Soc. Tropical Disease, Am. Pub. Health Assn., Soc. Exptl. Biology and Medicine. Author: (with others) Laboratory Manual of Medical Microbiology, 1962. Contbr. numerous articles to profl. jours. Studies, publs. on active immunization against Brucellosis, antigenic structure of Bordetella pertussis, methods of bacterial antibiotic sensitivity testing, mechanisms of resistance of microrganisms to antibiotic agts. Home: 2064 Horace Av., Abington, Pa. 19001. Office: 235 N. 15th St., Phila. 19102.*

BONDI, Hermann, mathematician; b. Vienna, Austria, Nov. 1, 1919; s. Samuel and Helene (Hirsch) B.; B.A., Trinity Coll., Cambridge (Eng.) U., 1940, M.A., 1944; m. Christine Mary Stockmann, Nov. 1, 1947; children—Alison, Jonathan, Elizabeth, David, Deborah. Lectr. maths. U. Cambridge, 1948-54; prof. applied maths. King's Coll., U. London, 1954—; vis. prof. Cornell U., Ithaca, N.Y., 1960; on leave as dir. gen. European Space Research Orgn., 1967—. Chmn. astronomy policy and grants com. Sci. Research Council, 1965-67; chairman of the Nat. Com. for Astronomy, 1964-67; pres. Internat. Com. on Gen. Relativity and Gravitation. Fellow Royal Soc., 1959, Royal Astron. Soc., Cambridge Philos. Soc. Author: Cosmology, 1960; The Universe at Large, 1961; Relativity and Common Sense, 1964; also numerous articles. Research in constn. of stars, structure and evolution of universe, gen. relativity, especially propagation of gravitational disturbances; known for steady state theory of expanding universe. Home: East House, Buckland Corner, Reigate Heath, Surrey, Eng. Office: Esro, 114 Av. des Neuilly, Neuilly s/s France.*

BONDUELLE, Michel Jean, French physician; b. Château-Gontier, France, Apr. 19, 1912; s. Henri Gustave and Adrienne (Festa) B.; M.D., Paris, 1944; m. Yvonne Collette, Dec. 20, 1944; children—Francois, Claude-Marie, Philippe. Became head clinic Faculté de Médecine, Paris, 1945; became asst. physician Hôpitaux de Paris, 1946; named physician, chief neurol. dept. Hôpital Saint-Joseph, 1950. Expert for tribunals, 1965; mem. Collège de Médecine des Hôpitaux de Paris. Mem. French Neurology Soc., Neurosurgery Soc. French Lang., Med. Soc. Paris Hosps. Author: Automatismes et Fugues épileptiques, 1963; also numerous articles. Research on description of postural tremor, 1952, myoclonias, multiple sclerosis, lateral amyotrophic sclerosis, neurol. disorders in porphyria, epilepsy, treatment facial neuralgia, polyneuropathies. Home: 15 Boulevard des Invalides, Paris 7, 75, France. Office: Hôpital Saint-Joseph I, Rue Pierre-Larousse, Paris 14, 75, France.*

BONDY, Philip Kramer, Am. physician; b. N.Y.C., Dec. 15, 1917; s. Eugene Lyons and Irene (Kramer) B.; B.A., Columbia, 1938; M.D., Harvard, 1942; M.S., Yale, 1962; m. Sarah B. Ernst, Mar. 18, 1949; children—Jonathan Lyons, Jessica, Steven Marks. Alexander Brown Cox fellow physiol. chemistry Yale, 1948-49, mem. faculty, 1948-49, 52—, prof., 1961-65, C. N. H. Long professor of medicine, chairman of the department, 1965; associate in medicine at Emory U., 1949-51, asst. prof., 1951-52. Fellow N.Y. Acad. Sci.; mem. Endocrine Soc. (mem. publs. com. 1960—, council 1965-68), Am. Soc. Clin. Investigations, Am. Fedn. Clin. Research, So. Soc. Clin. Research, Soc. Exptl. Biology and Medicine, Assn. Am. Physicians, Phi Beta Kappa, Sigma Xi, Alpha Omega Alpha. Asso. editor: Cecil-Loeb Textbook of Medicine, 1964. Editorial com. Jour. Clin. Investigation, 1956-57, editor in chief, 1957-61; editorial bd. Conn. Medicine, 1959-63, Yearbook Medicine, 1955—, Medicine. Contbr. numerous articles to jours. Research on factors controlling concentration glucose in blood, factors related to control plasm concentrations adrenal corticosteroids, biosynthesis adrenal corticosteroids, intermediary products corticosteroid breakdown, mode action adrenal corticosteroid cortisol on protein synthesis in liver. Home: Chestnut Lane, Woodbridge, Conn. 06525. Office: 333 Cedar St., New Haven 06525.*

BONE, William Arthur, Brit. chemist; b. Stockton-on-Tees, Mar. 19, 1871; s. Christopher Bone; grad. Victoria U., Manchester, 1891; postgrad. under Victor Meyer, U. Heidelberg (Germany); m. Kate Hind (dec. 1914); 1 son, 2 daus.; m. 2d, Mabel Isabel Liddiard. Elected Berkeley fellow Owens Coll., fellow Victoria U., 1892; apptd. head chem. dept. Battersea Poly., London, 1896; lectr. chemistry and metallurgy Owens Coll., Manchester, 1898-1905; prof. applied chemistry, 1st Livesey prof. coal gas and fuel industries Leeds U., 1905-12; chief prof., head dept. chem. tech. Imperial Coll. Sci. and Tech., London U., 1912-36; research dir. Bone Research Assn. Recipient Howard Potts Gold medal Franklin Inst., Phila. for sci. discoveries in and inventions in surface combustion, 1913; medal Royal Soc. Arts for work on brown coals and lignites, 1922; Melchett medal Inst. Fuel, 1931; medal Soc. Chem. Industry, 1933. Fellow Royal Soc., 1905, Bakerian lectr., 1932, Davy medallist, 1936; fellow Chem. Soc.; hon. mem. Instn. Gas Engrs., Am. Gas Inst., Inst. Fuel, Inst. Iron and Steel Inst. Author: Coal and its Scientific Uses, 1918; Flame and Combustion in Gases, 1927; Gaseous Combustion at High Pressures, 1930; Coal, its Constitution and Uses, 1936; also papers. Inventor Bonecourt system of surface combustion and radiophragmbeating; research on combustion and explosions, constitution of coal; developed (with Armstrong) hydroxylation theory of hydrocarbon combustion. Died London, June 11, 1938.

BONELLI, Franco Andrea, Italian zoologist; b. Cuneo, Italy, Nov. 11, 1784; s. Tommaso-Michelle Bonelli; studied mechanics, later zoology; m. Ferdinanda D'Ancona, 1818; 3 daus., 2 sons. Became prof. zoology Imperial U., Turin, Italy, 1811; named asst. dir. Mus. Zoology, 1815, dir., 1825. Mem. numerous acads. including Société Linneene di Parigi, Acad. Sci. Phila. Died Nov. 18, 1830.

BONERA, Gianni Franco, Italian physicist; b. Bergamo, Italy, Jan. 11, 1936; s. Mario and Laura (Pastore) B.; Laurea, U. Pavia (Italy), 1958, Libera docenza in Structure of Matter, 1963. Prof. advanced physics Inst. Physics, Pavia (Italy) U., 1960—. Mem. Italian, Lombarda phys. socs. Research, publs. on theoretical and exptl. works in nuclear magnetic resonance, particularly a method for measurement of nuclear relaxation times, exptl. determination of deuterium quadrupole constant in Benzene and Acetone, theoretical and exptl. study of quadupolar echoes in solids. Home: 96, Corso Carioli. Office: Instituto di Fisica, Università di Pavia, Pavia, Italy.*

BONET, Federico, zoologist, paleontologist; b. Madrid, Spain, June 18, 1906; s. Jose and Amparo (Marco) B.; B.S., Madrid U., 1923, M.S., 1928, Ph.D., 1932; m. Alicia Ceballos; children—Trinidad (Mrs. Angel Escorial), Alicia, Amparo, Carmen, Federico, Arturo. Curator, Museo Ciencias Naturales, Madrid, 1930-39; asst. prof. U. Madrid, 1930-33, prof., 1933-39; prof. Instituto Politecnico Nacional, Mexico City, Mexico, 1939—, head dept. zoology and paleontology, 1946—; research fellow Instituto Geologia Mexco U., Unite, 1950—; cons. paleontologist Petroleos Mexicanos, 1950—. Guggenheim Found. fellow, 1947; Research fellow Instituto Mexicano del Petroleo, 1966. Mem. Sociedad Mexicana de Historia Natural, Sociedad Mexicana de Geologos Petroleros. Author on geology and paleontology of Mexico, 1952, 52, 56; also numerous articles. Exploration of caves and discovery of cave animals in Spain, Morocco, Guinea, Mexico, U. S.; elaboration of geol. time scale in terms of microfossils for use in petroleum geology; study of fossil reefs and its bearing on petroleum geology. Home: Fuego 430, Mexico 20, D.F., Mexico. Office: Carpio y Plan de Ayala, Mexico D.F., Mexico.*

BONET, Juan Pablo, Spanish physician; b. Aragon, Spain, 1575; under patronage of Constable of Castile. Author: Reducción de las letras y artes para enseñar á hablar á los mutos, 1620 (contained alphabet for deaf and dumb). Devised method of teaching deaf to speak and communicate with finger-spelling. Abbé Charles Michel de l'Epée (1712-89) acknowledged indebtedness to Bonet's work. Died circa 1630.

BONET, Théophile, Swiss anatomist; b. Geneva, Switzerland, Mar. 6, 1620. Author of Sepulchretum sive anatomie practica, 1679; Labyrinthus medicus extricatus, 1679; Medicina septentrionalis collatitia, 1685, 86. Summarized most important med., surg. and pharmacological findings known, both historical and contemporary. Died Mar. 29, 1689.

BONET-MAURY, Paul, French biophysicist; b. Paris, France, July 2, 1910; s. Georges and Elizabeth (Minder) B.; D.Sc., Faculty Scis., 1930; Dr. Pharmacy, Faculty Pharmacy, 1933; div.; children—Daniel, Alain Marc. Mem. staff Centre Nat. de la Recherche Scientifique, 1938—, chief radioprotection service Inst. Radium, 1948—; prof. Nat. Inst. Sci. and Nuclear Tech., 1963—. Decorated chevalier Legion of Honor. Mem. Soc. Physics, Soc. Biology, Assn. Microbiologists, Internat Assn. Radioprotection (sec.-gen. 1961—). Research on radio vaccines, photocolorimetry, bio-photometry, irradiation of viruses, radiobiology of particles of high energy. Home: 3 Sq. Albin Cachot, Paris. Office: 11 Rue Pierre Curie, Paris, France.*

BONETTI, Alberto Mario, Italian physicist; b. Genova, Italy, Sept. 6, 1920; s. Mario and Eva (Porchietti) B.; D.Physics, U. Genova, 1946; 1 dau., Stella Serena. Asst., U. Genova, 1945-52, U. Milano (Italy), 1952-62, faculty U. Bari (Italy), 1954-67, prof. exptl. physics, 1962-67, dir. Computing Center, 1964-67; staff cosmic ray and space physics group Mass. Inst. Tech., 1961-62; prof. space physics U. Florence (Italy), 1967—. Mem. Italian Phys. Soc., Am. Geophys. Union. Research, publs. on elementary particle physics, observations on hyperons, K'mesons and hypernuclei; studies on nuclear emulsion technique, solar wind physics, celestial infrared. Home: 2 viale Galileo, Florence, Italy.*

BONFA, Jean, French astronomer; b. May 30, 1638; b. Nimes, France, May 30, 1638; s. Firmin and Jeanne (Anse) B.; ed. Jesuit Coll. Avignon (France), 1654; ordained priest; tchr. math. at Avignon for 12 years; became tchr. hydrography, Marseilles, 1681; mem. French Acad. Scis., 1699. Author: Observations du mouvement d'une tache qui a paru dans le soleil en 1681. Observed comet of 1680, 81, solar eclipse, 1683; made geog. map of Comtat region of France; invented instrument for taking trigonometric measurements. Died Dec. 5, 1724.

BONFILS, Immanuel Ben Jacob, astronomer, mathematician; flourished in Tarasçon, also Avignon, Orange (all France), 1340; many writings on Jewish calendar, calculation of position of stars, tables of Venus and Mercury, accurate determination of eclipses, computation of pi, extraction sq. roots, arithmetical rules, decimal fractions; anticipated exponential calculus (as found in Nicholas Chuquet more than a century later) and decimal system (developed by Elijah ben Abraham Mizrahi, and completed by Simon Stevin).

BONGIOVANNI, Alfred Marius, Am. physician, educator; b. Phila., Sept. 22, 1921; s. Joseph N. and Elisa (DiSilvestro) B.; B.S., Villanova Coll., 1940; M.D., U. Pa., 1944. Asst. physician Rockefeller Inst., N.Y.C., 1948-50; asso. in pediatrics U. Pa., Phila., 1950-52, asso. prof. pediatrics, 1954-64, prof., chmn. dept., 1964—; asst. prof. pediatrics Johns Hopkins U., Balt., 1952-54; physician in chief Children's Hosp., Phila., 1963—. Mem. study sect. NIH, 1963-66; U. S. del. to Soviet Union on Physiol. Devel. Child, 1965. Served with USNR, 1943-46. Recipient Ciba award Endocrine Soc., 1956; Mead Johnson award Am. Acad. Pediatrics, 1957; League of Children's Hosp. award, 1962; Schaffrey medal St. Joseph's Coll., Phila., 1965. Mem. Soc. Clin. Investigation. Editorial bd. Pediatrics, 1964-67, Jour. Clin. Endocrinology and Metabolism, 1964-67, Am. Jour. Med. Sci., 1960-66. Research in endocrine, steroid metabolism. Home: 2222 Locust St., Phila. 19103. Office: 1740 Bainbridge St., Phila. 19146.*

BONHAM, Russell A., Am. chemist; b. San Jose, Cal., Dec. 10, 1931; s. Aubrey Russell and Margaret (Wallace) B.; B.A. with honors, Whittier Coll., 1954;

Ph.D., Ia. State Univ., 1958; m. Miriam Anne Dye, Mar. 23, 1957; children—Frances, Margaret, Anne. Faculty Ind. U., Bloomington, 1958-60, 1960—, prof. chemistry, 1965—. Nat. Acad. Scis.-NRC post-doctoral research asso., 1960; asst. prof. U. Md., 1960. Alfred P. Sloan Found. fellow, 1964-66; Guggenheim Found. fellow, 1964-65; Fulbright scholar, Tokyo, 1964-65. Mem. Am. Crystallographic Assn., Am. Chem. Soc., A.A.A.S., Sigma Xi, Phi Lambda Upsilon. Research and numerous publs. on hindered rotation in simple organic molecules by use of gas electron diffraction, new techniques for evaluation of molecular integrals, use of electron scattering to investigate electron density distbns. around simple molecules. Office: Dept. of Chemistry, Indiana U., Bloomington, Ind. 47401.*

BONHOEFFER, Karl Friedrich, physical chemist; b. Breslau, Jan. 13, 1899. Prof. in Frankfurt, Leipzig, Berlin; dir., Max Planck Inst. for physical chemistry, Göttingen. Died Göttingen, Germany, May 15, 1957.

BÖNI, Albert, Swiss physician; b. Schänis, Jan. 10, 1912; s. Albert and Emma (von Kraaz) B.; ed. univs. Zurich, Friburg/Br., Berlin, Hamburg, Bonn; M.D.; m. Rosmarie Salzmann, Jan. 30, 1942; children—Lukas, Martin. Surgeon under Prof. Brun, Lucerne, Switzerland, phys. medicine under Prof. V. Neergaard, Zurich, Switzerland, Inst. Serums, Copenhagen, Clinic of Rhumatology, Lund; full prof., 1949. Mem. Fed. Commn. Rhumatology; expert WHO. Hon. mem. 8 rheumatology leagues. Collaborator: Lehrbuch der Therapie; Handbuch der praktischen Geriatrie; Der vozeitig verbrauchte Mensch, also over 100 publs. on rhumatology and phys. therapeutics. Home: Enzenbuhlstrasse 44. Office: Gloriastrasse 25, Zurich 6, Switzerland.

BONIFACIO, Giovanni, Italian physician; b. 1574; author: L'Arte de cenni, 1616; devised sign language for deaf-mutes, 1616.

BONIFAS, Valentin H., Swiss microbiologist; b. Geneva, Switzerland, June 19, 1918; s. Paul A. and Alice E. (Sordet) B.; student medicine U. Geneva; m. Gilberte A. Kunz, Oct. 19, 1946; children—Caroline, Elisabeth. With Inst. Hygiene, U. Geneva, head bacteriological, virus labs.; asso. prof. St. Louis U. Sch. Medicine; now head virus lab. U. Berne (Switzerland). Office: Virus Lab., U. Berne, Berne Switzerland.*

BONIN, Gerhardt von, see von Bonin, Gerhardt.

BONINE, Chesleigh Arthur, Am. educator, geologist; b. Sorrento, Fla., Mar. 25, 1888; s. Joel Carter and Lola Ida (Hemry) B.; E.M., Lehigh U., 1912; postgrad. Johns Hopkins; m. Beulah Howell Whiteman, Nov. 22, 1913; 1 dau., Ann Whiteman (Mrs. Walter Allard Snow). With U. S. Geol. Survey, Washington, 1912-17; instr. geology Lehigh U., Bethlehem, Pa., 1917-18; faculty Pa. State U., University Park, 1918-47, prof. earth sci., head dept., 1923-47, prof. emeritus, 1947. Dir., geologist Rocky Mountain Oil Producing Co., 1921-62. Cons. geologist; Pa. State U. del. Internat. Geol. Congress, Madrid, Spain, 1926; mem. Del. State Archeol. Bd., 1956-58; mem. com. on sedimentation, origin oil and gas NRC. cil. Mem. Soc. Econ. Geologists, Am. Assn. Petroleum Geologists, Am. Inst. Mining and Metall. Engrs., Pa. Acad. Scis., Geol. Soc. Washington, Am. Assn. U. Profs., Sussex Soc. Archeology and History, Lewes Hist. Soc., Dutch Colonial Soc. Del., Sigma Xi, Phi Delta Theta, Tau Beta Pi, Sigma Gamma Epsilon. Contbr. articles to geol. bulls., jours. Address: 63 Henlopen Av., Rehoboth Beach, Del. 19971.*

BONINI, William Emory, Am. geologist; born Washington, Aug. 23, 1926; s. John E. and Thelma (Scrivener) B.; B.S. in Engring., Princeton, 1948, M.S. in Engring., 1949; Ph.D., U. Wis., 1957; m. Rose Rozich, Dec. 4, 1954; children—John Allen, Nancy Mara, James Prior. Faculty Princeton, 1952—, prof. dept. civil and geol. engring., 1966—. Fellow Geol. Soc. Am., Royal Astron. Soc.; mem. Assn. Engring. Geologists, Am. Geophys. Union, Soc. Exploration Geophysicists, European Assn. Exploration Geophysicists. Editor: (with H.D. Hedberg, J. Kalliokoski) Role of National Governments in Exploration for Mineral Resources, 1964. Research, publs. in gravity anomalies and its relation to geology; geophys. methods in exploration for water and in foundation engring., rock magnetism and structural geology, crustal geophysics. Home: 74 Robert Rd., Princeton, N.J. 08540.*

BONNAFONT, Jean Pierre, French physician; b. Plaisance, France, 1805; M.D., Montpellier, France, 1834; medic in army on Algerian campaign, 1835-37; head physician Staff Hosp., Paris, 1842. Author of Reflexions sur l'Algerie, 1846; De la Surdi-Mutité, 1853; Traité théorique et pratique des maladies de l'oreille, 1860; Pérégrination en Algérie, 1884; also many memoirs. Used tuning fork to study deafness, also studied wounds from firearms, battlefield first aid, soldier hygiene.

BONNARDEL, Raymond Georges, French physiologist and psychologist; b. Limay, France, 1901; s. René and Lucie (Grosjean) B.; ed. Faculté des sciences de médicine des lettres, U. Paris; m. Gabrielle Chastel, Feb., 1922. Adj. asst. Gen. Physiology Lab., U. Paris, 1929; asst. lab. for physiol. orgn. l'Ecole pratique des hautes études, 1931; chief Nat. Conservatory Arts and Crafts, 1932; head research Nat. Center Sci. Research, 1935. Decorated chevalier Legion of Honor. Mem. French Soc. Psychology (pres. 1952), Internat. Assn. Applied Psychology (sec.-gen. 1951). Author numerous articles. Research in psychology and physiology.

BONNER, James Frederick, Am. biologist; b. Ansley, Neb., Sept. 1, 1910; s. Walter Daniel and Grace (Gaylord) B.; A.B., U. Utah, 1931; Ph.D., Cal. Inst. Tech., 1934; m. Harriet Rees, Jan. 1, 1939; children—Joey, James Jose. NRC Research fellow, Utrecht, Leiden, Netherlands, Zurich, Switzerland, 1934-35; with Cal. Inst. Tech., 1935—, asso. prof., 1943-46, prof., 1946; Eastman vis. prof. Balliol Coll., Oxford (Eng.) U., 1963-64. Mem. Nat. Acad. Scis., A.A.A.S., Am. Chem. Soc., Am. Soc. Plant Physiology. Author Plant Biochemistry, 1950, (with J. Varner) 2d edit., 1965; Principles of Plant Physiology, 1952; (with H. Brown, J. Weir) The Next Hundred Years, 1957; (with P. Ts'o) The Nucleohistones, 1964; The Molecular Biology of Development, 1965. Contbr. numerous articles to sci. jours. Developed methods for study of gene action in test tube; studied chem. agts. which control gene activity and growth of fertilized egg into embryo and adult. Home: 1201 E. California St., Pasadena, Cal. 91109.*

BONNER, John Tyler, Am. biologist, educator; b. N.Y.C., May 12, 1920; s. Paul Hyde and Lilly Marguerite (Stehli) B.; grad. Phillips Exeter Acad., 1937; B.Sc., Harvard, 1941, M.A., 1942, Ph.D., 1947; m. Ruth Anna Graham, July 11, 1942; children—Rebecca, Jonathan Graham, Jeremy Tyndall, Andrew Duncan. Asst. to asso. prof. Princeton (N.J.), 1947-58, prof. biology, 1958—. Lectr. embryology Marine Biol. Lab., Woods Hole, Mass., 1951-52; spl. lectr. U. London (Eng.), 1957; trustee Biol. Abstracts, 1958—. Served from pvt. to 1st lt. USAAF, 1942-46. Sheldon traveling fellow, Panama, 1941; Rockefeller traveling fellow, France, 1953; Guggenheim fellow, Scotland, 1958; NSF sr. postdoctoral fellow, Eng., 1963; recipient Selman A. Waksman award for contbns. to microbiology Theobold Smith Soc., 1955. Mem. Am. Soc. Naturalists, Soc. Gen. Physiologists, Soc. Growth and Devel., Mycol. Soc. Am., Phi Beta Kappa, Sigma Xi. Author: Morphogenesis, 1952; Cells and Societies, 1955; The Evolution of Development, 1958; The Cellular Slime Molds, 1959; The Ideas of Biology, 1962; Size and Cycle, 1965. Editor: Growth and Form, 1961. Editorial bd. Am. Scientist, 1961—, Jour. Gen. Physiology, 1962—, Growth, 1955—, Am. Naturalist, 1958 60, 66—, Princeton U. Press, 1964-68. Research on devel. of cellular slime molds; publs. on devel., evolution. Home: 148 Mercer St., Princeton, N.J. 08540.*

BONNER, Oscar Davis, Am. chemist; b. Jackson, Miss., May 9, 1917; s. Oscar Davis and Bertha (Basser) B.; B.S. Millsaps Coll., 1939; M.S., U. Miss., 1948; Ph.D., U. Kan., 1951; m. Vaudie Vee Ball, Aug. 3, 1940; children—Davis Roy, Richard Edward, Timothy George. Chemist, Mississippi Testing Laboratories, Jackson, 1940-42, Filtrol Corp., Jackson, 1946-47; faculty University of South Carolina, Columbia, 1951—, prof., head dept. chemistry, 1960—; research participant Oak Ridge Nat. Lab., 1951. Chmn., Gordon Research Conf. on Ion Exchange, 1965. Fulbright research prof. for Germany, 1957-58. Mem. Am. Chem. Soc., S.C. Acad. Sci., Sigma Xi, Phi Lambda Upsilon. Contbr. articles to sci. jours. Research on elucidation properties solutions. Home: 5012 Furman Av., Columbia, S.C. 29206.*

BONNER, Walter D., Am. physiologist; b. Salt Lake City, Oct. 22, 1919; s. Walter Daniel and Grace (Gaylor) B.; B.S., U. Utah, 1940; Ph.D., Cal. Inst. Tech., 1946; m. Josephine Silberberg, May 13, 1944; children—Andrew, Brian. Research asst. Cal. Inst. Tech., 1943-45; research asso. in biology Harvard, 1946-49; Am. Cancer Soc. fellow Cambridge U., Eng., 1949-50, USPHS fellow, 1949-50; biochemist Smithsonian Instn., Washington, 1952-53; faculty Cornell U., 1953-59; prof. phys. biochemistry, plant physiology Johnson Found. U. Pa., 1959—. Mem. Am. Chem. Soc., Am. Soc. Biol. Chemists, Biochem. Soc., Biophys. Soc., Am., Canadian, Japanese socs. plant physiologists, Soc. Study Devel. and Growth, Sigma Xi. Research, publs. in uncovering some mechanisms that lead to conservation of energy during cellular respiration and photosynthesis in plants. Home: 4047 Pine St., Phila. 19104.*

BONNER, William A., Am. chemist; b. Chgo., Dec. 21, 1919; s. Francis Augustis and Celestine (Horine) B.; A.B., Harvard, 1941; Ph.D., Northwestern U., 1944; m. Norma C. Ballentyne, Apr. 8, 1961; children—Randolph N., Joseph D., Jay F., Candace K., Constance C., Teresa A. Faculty, Northwestern U., 1944-46; faculty Stanford, 1946—, prof., 1958—. Sr. chemist Oak Ridge Nat. Lab., 1952-53; cons. NIH, 1962-66. Guggenheim fellow, 1952. Mem. Am. Chem. Soc. Author: (with A. J. Castro) Essentials of Modern Organic Chemistry, 1965. Research numerous publs. on reaction mechanisms, heterogeneous catalysis, stereochemistry, electrochemistry. Home: 1630 California Av., Palo Alto, Cal. 94304.*

BONNET, Amédée, French surgeon; b. Ambérieu, France, Mar. 19, 1809; studied medicine in Lyons (France) and Paris (France); docteur en médecine, 1832; became head surgeon Hôtel-Dieu, Lyons, 1833; appointed professor of clinical surgery School of Medicine, Lyons, 1839; corresponding member French Academy Scis., 1855. Author: Traité des sections tendineuses et musculaires, 1841; Traité des maladies des articulations, 2 vols., 1845; Traité de thérapeutique des maladies articulaires, 1853. Considered renovator of articular surgery; developed various techniques including a splint for fractures, apparatus for straightening knee, operation for clubfoot. Died Dec. 2, 1858.

BONNET, Charles, Swiss naturalist; b. Geneva, Switzerland, Mar. 13, 1720; studied law, doctorate, 1743; m. Mlle. De la Rive. Mem., Grand Council of Geneva, 1752-68. Fellow, Royal Soc., London, 1743; French Acad. Scis., 1740. Author: Traité d'insectologie, 1745; Recherches sur l'usage des feuilles dans les plantes, 1754; Essai de psychologie, 1754; Essai analytique sur les facultés de l'ame, 1760; Considerátions sur les corps organisés, 1762; Contemplation de la nature, 1764-65; Palingénésie philosophique, 1760-70; Oeuvres d'histoire naturelle et de philosophie (18 vols.), 1779-1788. Discovered parthenogenesis while studying aphids; advocated theory of preformation; 1st to use term evolution; held catastrophic theory of evolution; that all forms of life step up a notch after each catastrophe; discovered pores through which larvae and insects breathe (stigmata); studied regeneration of lost parts in various animals; precursor of physiological psychology. Died Genthod, Switzerland, May 20, 1793.

BONNET, Mireille, French ophthalmologist; b. Toulon, France, Feb. 23, 1933; s. Ernest and Berthet (Marcelle) B.; grad. U. Lyons Faculty Medicine, 1952. Practice medicine specializing in ophthalmology, 1962—; head U. Clinic, Lyons, France, 1963—; dir. Lyons Ophthal. Clinic, 1963—; aggregate prof. Lyons Faculty Medicine, 1966—. Mem. several ophthal. socs., including French, Belgian. Author: Corineal Fistulas, 1962; also numerous articles on medicine and surgery of eye. Clin. study and treatment of corneal fistulas after cataract extraction and after glaucoma operations; research on ophthal. manifestations of coagulative diseases. Home: 156 cours A. Thomas, Lycn 8e, Franre, Office: Hopital de Grenoble, Latronche (Isere), France.*

BONNET, (Pierre) Ossian, French mathematician; b. Montpellier, France, Dec. 22, 1819; studied at l'Ecole polytechnique 1838; became prof. math. astronomy Faculty Scis., Paris, France, 1878; Mem. Bur. Longitudes, French Acad. Scis., 1862. Corrected and expanded Cauchy's method of solving partial differential equations of 1st order having any number of variables; devised simpler forms of logarithmic criteria. Died Paris, June 22, 1892.

BONNET, Pierre Numa Louis, French zoologist; b. Villefranche-de Rouergue (Aveyron), Sept. 1, 1897; s. Eugène-Pierre and Clotilde (Jean-Baptiste) B.; ed. Sch. Sci., Montpellier and Toulouse: Ph.D. in sci.; m. Camille Cluzon, Sept. 15, 1923; children—Suzanne, Marthe. Prof., lectr. sch. sci. of Toulouse, 1950—; hon. prof., 1962. Mem. Internat. Commn. Nomenclature, Soc. Natural History (Toulouse), Zool. and Entomol. Soc. France. Author: Bibliographia Araneorum, also numerous works on entomology. Address: 43, route de Narbonne, Toulouse (Haute-Garonne), France.

BONNEY, Sherman Grant, Am. physician; b. Cornish, Me., July 15, 1864; s. Calvin Fairbanks and Harriott O. (Cheney) B; A.B., Bates Coll., 1886, A.M., 1889; M.D., Harvard, 1889; m. Nancy Brooks Little, Nov. 23, 1886; m. 2d, Mrs. Jessie Elwood Ray, Dec. 1, 1915. Began practice, Lewiston, Me., 1889; moved to Denver, 1891; ret. from practice, 1930; prof. medicine and dean Med. Dept. U. Denver; prof. medicine U. of Colo.; vis. phys. St. Luke's Hosp., physician St. Joseph's Hosp. Fellow A.C.P.; mem. Am. Climatol. Assn., A.M.A., Colo. State Med. Soc., Denver Clin. and Pathol. Soc. (pres.), Nat. Assn. Study and Prevention of Tb, Am. Soc. Tropical Medicine. Author: Pulmonary Tuberculosis and Its Complications, 1908, 2d edit, 1910. Donor of Bonney Meml. Library in Cornish, 1929, named in honor of his parents. Died Nov. 19, 1942.

BONNEY, Thomas George, English geologist; b. Rugeley, Eng., July 27, 1833; s. Thomas and Eliza Ellen (Smith) B.; B.A., St. John's Coll., Cambridge (Eng.) U., 1856, Sc.D., Cambridge U.; D.Sc., Dublin Ireland, Sheffield, Eng.; LL.D., Montreal, Que., Can. Ordained priest, 1858; became math. master Westminster Sch., 1857; became fellow St. John's Coll., 1859, jr. dean 1861, tutor, 1868, lectr. geology, 1869, life fellow, 1905; named Yates-Goldsmith prof. geology U. College, London, Eng., 1877; resigned professorship, 1901; Whitehall preacher, 1876-78; Hulsean lectr., 1884; Rede lectr., 1892. Named hon. canon Manchester Cathedral. Mem. Brit. Assn. (elected asst. gen. sec. 1881, pres. 1912), Geol. Soc. London (sec., pres.), Mineral. Soc. (pres.). Author: The Story of Our Planet, 1893; Charles Lyell and Modern Geology, 1895; Ice-Work, 1896; Volcanoes, 1898; The Building of the Alps, 1912; The Present Relations of Science and Religion, 1913; Memories of a Long Life, 1921; also numerous articles. Pioneer in research in petrology and glaciology; pub. 1st tech. descriptions of many Brit. rocks. Died Cambridge, Dec. 9, 1923.

BONNEY, Victor, Brit. gynecologist; s. W. A. Bonney; ed. St. Bartholomew's Hosp., Middlesex Hosp.; qualified, 1896; M.S., M.D., B.Sc., London U.; m. Annie Appleyard. Resident officer, Middlesex Hosp., obstetric tutor, 1903, cons. gynecologist in surgery, also surgeon, mil. br., Clacton-on-Sea, Eastern Command, 1914-18; mem. staff or cons. Chelsea Hosp. for Women, Brit. Post-Grad. Med. Sch., Royal Masonic Hosp., Miller Hosp., Queen Alexandra's Mil. Hosp., County of London; Emden research scholar Middlesex Hosp. Cancer Research Inst.; Hunterian orator, Hunterian prof., Bradshaw lectr. Royal Coll. Surgeons Eng.; examiner to Conjoint Bd. of Eng. Fellow Royal Soc., Coll. Surgeons Eng., Royal Australian Coll. Surgeons (hon.), Royal Coll. Obstetricians and Gynaecologists (hon.), Assn. Surgeons, Am. Gynecol. Soc. (hon.), Assn. Surgeons Gt. Britain and Ireland. Author: A Text-book of Gynecological Surgery; Difficulties and Emergencies of Obstetric Practice; The Technical Minutiae of Extended Myomectomy and Ovarian Cystectomy . . . Perfected radical hysterectomy for cancer of cervix (introduced by Wertheim 1900-01). Died July 4, 1953.

BONNIER, Gaston, French botanist; b. Paris, France, Jan. 2, 1853; ed. Ecole Normale. Master, Ecole Normale; prof. botany U. Paris, from 1887, founder, lab. for bot. research; founder, lab. vegetable biology, Fontainebleau, 1889, La Revue générale de botanique. Mem. Acad. Scis., 1896, Acad. Agr. Author: Les nectaires, 1879; Recherches sur la respiration et l'assimilation des végétaux, 1883; Flore du nord de la France et de la Belgique, Nouvelle Flore des environs de Paris, Nouvelle Flore pour la détermination facile des plante, Plantes des champs et des bois, all 1887; La synthèse des lichens, Petit Flore, Eléments de botanique, all 1889; Biologie végétale, La géographie botanique, 1894; La végétation de la France, 1894; Observations sur les modifications de végetaux suivant les conditions physique du milieu; Flore complete illustree. Studied constn. of lichens, gaseous exchanges among plants, respiration of vegetables, tissues deprived of chlorophyll, mushrooms, green tissues in darkness, comparative cultures at various altitudes, influence of electric light on plants. Died Paris Jan. 2, 1923.

BONNOT DE CONDILLAC, Étienne, French philosopher; b. Grenoble, France, Sept. 30, 1715; s. Gabriel B de C.; ed. Jesuit College, Lyon. Mem., French Academy, 1768. Author: Essai sur l'origine des connoissances humaines, 1746; Traité des systèmes, 1749; Traité des sensations, 1754; Oeuvres complètes (23 vols.), 1798. Important Physiocrat and assoc. of Rousseau, Diderot, Duclos; studied progress and influence of language; demonstrated workings of analytic empiricism; developed idea that value of commodity depends on its usefulness. Died Beaugency, France, Aug. 3, 1780.

BONOMINI, Bruno, Italian physician, radiologist; b. Verona, Jan. 31, 1905; s. Tullio and Adele (Lombardi) B.; M.D.; diploma in radiology; m. Maria Alessandra Bolognini, Mar. 29, 1941; children—Alessandro, Lorenzo, Georgio. Prin. dir. sect. radiotherapy of tumors Civil Hosp., Padua, Italy; prof. radiology U. Padua. Mem. Italian Soc. Med. Radiology, Italian Soc. Radiobiology, Italian Soc. Cancerology, others. Contbr. over 150 articles to profl. publs. Home: via dei Concariola 14. Office: Ospedale civile, Padua, Italy.

BONPLAND (Goujaud), Aimé Jacques Alexandre, French explorer, naturalist, botanist; born La Rochelle, France, Apr. 28, 1773; s. Jacques Simon and Marguerite (de La Coste) B.; studied medicine, Paris, France. Med. surgeon in navy; returned to Paris, 1798; left for S.Am. with Humboldt, 1799; returned to Paris and named intendant of gardens of Malmaison and Navarre, 1804; made voyage to Buenos Aires, Argentina, 1818; tchr. natural history in univ., Buenos Aires; journey to Bolivia and Brazil, 1821; when he entered Paraguay he was captured by dictator Francia and imprisoned; freed, 1830. Mem. French Acad. Scis., 1817. Author: Plantes equinoxiales recueillés au Mexique, 1808-09; Monographie des mélastromées, 1806; (with Humboldt) Voyage aux régions équinoxiales du nouveau continent, 1815; Description des plantes rares que l'one cultive à Navarre et à Malmaison, 1817; Mimoses et autres plantes légumineuses du nouveau continent. Brought back (with Humboldt) 6,000 S. Am. plants, which he presented to the Jardin des Plantes, Paris, 1804; explored Rio Negro, provinces of Sante-Fé and Gran Chaco, foot of Andes mountains. Died May 11, 1858.

BONSALL, Frank Featherstone, mathematician; b. 1920; ed. Bishop's Stortford Coll., Merton Coll., Oxford (Eng.) U.; M.A., D.Sc.; m. Gillian Patrick, 1947. Prof., U. Newcastle/Tyne, (Eng.), 1959-65, Edinburgh (Scotland) U., 1965—. Vis. prof. Tata Inst. Fundamental Research, Bombay, India, 1960-61. Mem. London Math. Soc. Work on functional analysis. Address: Math. Inst., U. Edinburgh, 20 Chambers St., Edinburgh 1, Scotland.

BONTE, Frederick James, Am. physician; b. Bethlehem, Pa., Jan. 18, 1922; s. Frederick R. and Harriett (Stoudt) B.; B.S., Western Res. U., 1942, M.D., 1945; m. Mary Helen Hawke, 1953 (div. 1964); 1 son, Frederick W. Asst. radiologist Univ. Hosp., Cleve., 1952-56; radiologist Parkland Mem. Hosp., Children's Med. Center, Presbyn. Hosp., Dallas, 1956—; prof., chmn. radiology U. Tex. Southwestern Med. Sch.; mem. radiology tng. con. Nat. Inst. Gen Med. Scis. NIH, 1966—. Fellow Am. Coll. Radiology; mem. A.M.A. (residency rev. com. radiology 1966—, Soc. Nuclear Medicine, Am. Roentgen Ray Soc. Radiol. Soc. N.Am., A.A.A.S., Assn. U. Radiologists. Author (with J. R. Williams) Roentgenological Aspect of Nonpenetrating Chest Injuries, 1961. Contbr. numerous articles to sci. jours. Research, publs. on mammalian cell radiobiology; exptl. nuclear medicine radionuclide scanning of internal organs; radioisotope therapy. Home: 3337 Whitehall Dr., Dallas 75229. Office: 5323 Harry Hines Blvd., Dallas 75235.*

BONTEMPS, Georges, glassmaker; b. France, 1799; ed. Ecole polytechnique, Paris; became dir. Choiseq-le-Roi Glassworks, 1823; visited Eng., 1848; became dir. Chance Glassworks, Eng. Author: Peinture sur verve au XIXe siecle, 1845; Guide du verrier, 1868. Revitalized glassmaking industry in France; studied mass prodn. of stained glass, lab. glass, flint glass, crown glass; provided glass for Foucault's telescope in Paris Obs. Died 1884.

BONTING, Sjoerd Lieuwe, Dutch biochemist; b. Amsterdam, The Netherlands, Oct. 6, 1924; s. Sjoerd Lieuwe and Johanna (Hagedoorn) B.; B.Sc., U. Amsterdam, 1944, M.Sc. cum laude, 1950, Ph.D., 1952; m. Susanna Maarsen, Jan. 10, 1951; children—Marion Suzanne, Paul Sjoerd, Elizabeth Johanna, Peter Jan. Instr., U. Amsterdam, 1947-52; research asso. State U. Ia., Iowa City, 1952-55; asst. prof. U. Minn., Mpls., 1955-56; asst. prof. U. Ill. Med. Sch., Chgo., 1956-60; head sect. on cell biology, ophthalmology br. NIH, Bethesda, Md., 1960-65; prof., chmn. dept. biochemistry U. Nijmegen, The Netherlands, 1965—. Recipient Fight for Sight award Nat. Council to Combat Blindness and Assn. Research Ophthalmology, 1961, 62; Merit award Internat. Rescue Com., 1963; Arthur S. Flemming award, 1964; Heinz Karger prize, 1964. Mem. Am. Soc. Biol. Chemists, Am. Soc. Cell Biology, Am. Histochem. Soc., A.A.A.S., Tissue Culture Assn., Assn. Research Ophthalmology, Netherlands Chem. Soc., Netherlands Biochem. Soc., Sigma Xi. Research, numerous publs. on relations between enzymes and cell function, between function and enzyme activity in various parts of renal nephron, relation between active cation transport and Na-K activated ATPase activity in cells and tissues and in formation of aqueous humor, cerebrospinal fluid in man and salt secretion in marine birds and Elasmobranch fishes and in visual process. Home: 290 St. Annastraat, Nijmegen, The Netherlands.*

BONTIUS, Jacobus, physician, naturalist; b. Denmark, 1592; s. Gerard Bontius; visited India and Persia; beginning in 1627 practiced medicine, Batavia. Wrote works on diseases, botany and natural history of the East. Am. plant genus named in his honor by Plumier. Said to have contbd. 1st modern sci. account of beriberi (observed in E. Indies), pub. 1642. Died 1631.

BONUS, Petrus, Italian physician, alchemist; fl. ca. 1494; Author: Introductio in artem chemiae integra, 1602; Epistola; De secreto omnium secretorum dei dono; Pretiosa margarita novella ex concordantiis sapientum collecta, de thesauro et lapide philosophorum; Rationes pro alchimia et contra.

BONUZZI, Silvio, Italian surgeon, biologist; b. Verona, Nov. 26, 1890; s. Giuseppe and Elisa (Zecchinato) B.; M.D.; M.D. in surgery. Hon. mem. Center Study and Research, U. Verona. Contbr. articles on medicine, biology to profl. publs. Address: Santa Maria di Zevio, Verona, Italy.

BONVICINI, Dante, Italian engr.; b. Chieti, Aug. 13, 1892; Ph.D. in engring. Univ. prof., 1941; now prof. sci. of constrn.; instr. constrn. in wood, metal and cement; dir. Inst. Applied Mechanics, U. Padua (Italy). Mem. Acad. Patavina Scis., Letters and Arts (corr.). Address: piazza Mazzini 12, Padua, Italy.

BONWILL, William Gibson Arlington, Am. surgeon, dentist inventor; b. Camden, Del., Oct. 4, 1833; hon. degree Dental Coll., 1865, Jefferson Med. Coll., 1865; m. Abigail Elizabeth Warren, June 13, 1891. Worked as carpenter, cabinet-maker and sch. tchr., 1847-53; practiced dentistry, Dover, Del., 1854-71, then in Philadelphia. Author: Geometry and Mechanics Deny Evolution; papers on dental and oral surgery. Invented first dental and surgical engine, 1869; first electrical mallet, 1869; mechanical mallet, 1878; the all-porcelain tooth-crown, 1879; first removable bridge, 1889; also numerous other dental appliances. Invented also first injector for boilers, 1856; first binder to a wheat reaper, 1856; the safety-pointed pin, 1863; the first machine to carve marble and rock by power, 1869, and many others. Died 1899.

BOODT, Anselmus Boëthius de, see de Boodt.

BOOK, Jan Arvid, Swedish geneticist; b. Malmo, Sweden, Nov. 4, 1915; s. Arvid and Wivi (Haag) B.; Ph.D., U. Lund (Sweden), 1943, M.D., 1949; D.M.S., U. Uppsala (Sweden), 1954; Dipl.Sci. h.c., Salerno (Italy) Med. Sch., 1961; m. Ruth Kerstin Alskog, May 31, 1947; children—Eva Christine, Kerstin Marianne, Marie Jeanette. Research asst. Inst. Genetics, U. Lund, 1940-46; dir. med. genetics research unit U. Lund, 1946-49; spl. research fellow USPHS Nat. Inst. Mental Health U. Minn., 1949-51; asst. dir. State Inst. for Human Genetics, Uppsala, 1951-55, dir. 1955-57; asso. prof. med. genetics U. Uppsala, 1954-57, now prof. genetics, dir. Inst. for Med. Genetics, 1957—; Mem. expert panel on human genetics WHO, 1961. Mem. Swedish Med. Soc., Royal Soc. Medicine London, Am. Soc. Human Genetics, Mendelian Soc. Lund, Am. Soc. Mental Deficiency. Author: A Genetic and Neuro-Psychiatric Investigation of a North Swedish Population, 1953; also numerous articles. Defined genetical disease entities by extensive epidemiological-genetical investigations of psychiat. and neurol. disorders; developed clin. cytogenetics; described several new chromosome syndromes in man such as triploid-diploid mosaicism and meiotic irregularities causing infertility. Home: 1 Granebergsvagen, Uppsala, Sweden.*

BOOK, William Frederick, Am. psychologist; b. Princeton, Ind., June 10, 1873; s. Christian Henry and Mary Elizabeth (Bussdicker) B.; A.B., Ind. U., 1900; Ph.D. (fellow) Clark U., 1906; postgrad. U. Chgo., Columbia; m. Mary Roach Cougle, Sept. 3, 1907 (dec.); m. 2d, Clara D. King, June 3, 1926. Prof. psychology, Univ. Mont., 1906-12; prof. ednl. psychology Ind. U., 1912-13 prof. psychology, dir. psychol. lab., 1917—; dir. vocational edn. Ind., 1913-17; Instr. various summer sessions Ind. U., Columbia, U. Wis., U. Hawaii. Fellow A.A.A.S. Author: The Psychology of Skill, 1908; The Intelligence of High School Seniors, 1921; The Psychology and Pedagogy of Skill, 1925; Learning to Typewrite—with a Discussion of the Psychology and Pedagogy of Skill, 1925; Learning How to Study and Work Effectively, 1926; How to Succeed in College, 1927; Economy and Technique of Learning, 1931 (with Merrill T. Eaton) Experimental Workbook on Psychology of Learning. Editor Jour. Applied Psychology. Died May 22, 1940.

BOOKER, Henry George, scientist; b. Barking, Essex, Eng., Dec. 14, 1910; s. Charles Henry and Gertrude Mary (Ratcliffe) B.; student Palmer's Sch., Grays, Essex, 1921-30; B.A., Christ's Coll., Cambridge, 1933, Ph.D., 1936; m. Adelaide Mary McNish, July 9, 1938; children—John Ratcliffe, Robert William, Mary Adelaide, Alice. Came to U. S., 1937, naturalized, 1952. Fellow Christ's Coll., 1935-48; sci. officer Ministry Aircraft Prodn., London, 1940-45; lectr. Cambridge U., 1945-48; prof. elec. engring. Cornell U., 1948-59, dir. sch. elec. engring., 1959-63, asso. dir. Center Radio Physics and Space Research, IBM prof. engring. and applied math., 1962-65; prof., chmn. dept. applied electrophysics U. Cal. San Diego, 1965—. Formerly internat. chmn. commn. on tropospheric radio propagation Internat. Sci. Radio Union, chmn. commn. on magnetosphere. Entrance scholarship, Christ's Coll., 1930, Allen scholarship, 1934-35, Smith's prize, 1935; Duddell, Kelvin and instn. premiums Instn. Elec. Engrs., London, 1948-50; Guggenheim fellow Cambridge U., 1954-55. Fellow Royal Meterol. Soc., I.E.E.E.; mem. Am. Astron. Soc., Nat. Acad. Scis., Am. Meteorol. Soc., Am. Phys. Soc., Am. Geophys. Union, Instn. Elec. Engrs. Soc. Engring. and Edn., Sigma Xi. Author: An Approach to Electrical Science, 1959. Research and publs. on antennas, on radio wave propagation. Home: 8696 Dunway Dr., La Jolla 92037. Office: Dept. Applied Electrophysics, U. Cal. at San Diego, La Jolla, Cal. 92038.*

BOOKHOUT, Cazlyn Green, Am. marine zoologist; b. Gilboa, N.Y., Jan. 28, 1907; s. Eugene Anson and Mae (Green) B.; A.B., St. Stephen's Coll., 1928; A.M., Syracuse U.; 1929; Ph.D., Duke, 1934; m. Elizabeth Circle, Aug. 21, 1936; children—Beverly Ann (Mrs. William Lovell), Glenn Allen. Instr. biology Woman's Coll. U. N.C., 1929-31; asso. prof. Elon Coll., 1934-35; faculty Duke, 1935—, prof., 1954—, chmn. Marine Lab. exec. com., 1949-63, dir. Marine Lab., 1950-63, 1964-68, acting program dir. oceanographic program, 1967—; chmn. biol. oceanography com., 1960-63. Recipient Pryn prize in biology St. Stephen's Coll., 1927. Mem. Am. Soc. Zoologists, Assn. Southeastern Biologists, Atlantic Estuarine Research Soc., N.C. Acad. Sci. (Poteat award 1950), Soc. Limnology and Oceanography, Sigma Xi (pres. Duke chpt. 1941-42). Research, publs. on polychaetes, barnacles, crabs in lab. from egg to adult; effects of various environmental factors on devel. and survival of crab larvae. Home: 1105 Front St. Office: Marine Lab., Duke U., Beaufort, N.C. 28516.*

BOOLE, George, mathematician, logician; b. Lincoln, Eng., Nov. 2, 1815; ed. Nat. and Comml. schs., Lincoln; studied classics on his own, later higher math., from 1832; LL.D., U. Dublin, hon. D.C.L., Oxford U.; m. Mary Everest, 1855, 5 daus. Tchr. pvt. schs., Doncaster, then Waddington, circa 1831-35; established own sch., Lincoln, circa 1835; prof. math. Queen's Coll., Cork, Ireland, 1849-64; also pub. examiner for degrees Queen's U. Fellow Royal Soc., 1857 (Royal medal 1844); Recipient Keith medal Royal Soc. Edinburgh, 1857. Author: An Investigation of the Laws of Thought (most important work), 1854; Treatise on Differential Equations, 1859; Treatise on the Calculus of Finite Differences, 1860. Best known for his math. analysis of logic and his adaptation of symbolic lang. and notation to logical processes; dealt with adaptation of these principles to theory of probability; his system (Boolean algebra, which forms basis of what is now called symbolic logic) provides means of solving problems of far greater degree of complexity than had previously been thought possible; did considerable original research on differential equations and calculus of finite differences; developed theory of invariance and covariance. Died Cork, Ireland, Dec. 8, 1864.

BOOLOOTIAN, Richard A., Am. physiologist; b. Fresno, Cal., Oct. 17, 1927; s. Vanig and Vivian (Ohannesian) B.; B.A., Fresno State Coll., 1951; M.A., Fresno State Coll., 1953; Ph.D., Stanford, 1957; m. Georgia Lee Johnston, Apr. 9, 1950, (div. 1966); children—Mark, Alan, Craig. Prof. dept. zoology U. Cal. at Los Angeles, 1957-66; vis. investigator Marine Biol. Lab., Woods Hole, Mass., 1964-65; cons. Biol. Scis. Curriculum Study, 1964—, Commn. on Undergrad. Edn. in Biol. Scis., 1965—. Recipient Lalor Found. Faculty Research award 1959, 61; fellow NIH, 1965. Mem. A.A.A.S., Am. Soc. Zoologists, Challenger Soc. London, Marine Biologists Assn. U.K., N.Y. Acad. Sci., Soc. for Exptl. Biology, Soc. Gen. Physiologists, Am. Sci. Film Assn. (v.p. 1961-—), Sigma Xi. Author: (with Donald Heyneman) An Illustrated Laboratory Text in General Zoology, 1962; (with Gideon Nelson and Gerald Robinson) Fundamental Concepts of Biology, 1967. Editor, contbg. author: Physiology of Echinodermata, 1966. Author numerous articles. Research on comparative physiology marine invertebrates to elucidate factor regulating, influencing, or controlling reproductive processes.*

BOONE, William Werner, Am. mathematician; b. Cin., Jan 16, 1920; s. William Edward and Mazie (Werner) B.; A.B., U. Cin., 1945; M.A., Princeton, 1948, Ph.D., 1952; m. Eileen Georgeanna Herweh, June 13, 1949; children—William John, Theodore Sebastian. Instr. Princeton, part-time 1945-47; instr. Rutgers U., 1947-50; asst. prof. Cath. U. Am., 1950-54; mem. Inst. for Advanced Study, Princeton, N.J., 1954-56, 1964-66; Fulbright research scholar, U. Oslo (Norway), 1956-57; John Simon Guggenheim Meml Found. fellow Oxford (Eng.) U., Manchester (Eng.), U., Münster U., Westfalen, Germany, 1957-58; asso. prof. U. Ill., Urbana, 1958-60, prof. 1960—, asso. mem. Center for Advanced Study, 1960-62. Cons., USMC, 1952-53, logic project Am. Car and Foundry Co., Alexandria, Va., 1954-56, System Devel. Corp., Santa Monica, Cal., 1963—, Argonne (Ill.) Nat. Lab., 1966—; guest prof. Göttingen (Germany) U., 1967. Mem. Assn. for Symbolic Logic (exec. com. 1963-65), Am. Math. Soc., Phi Beta Kappa, Sigma Xi. Asso. editor Proc. Am. Math. Soc., 1967—. Recursive unsolvability of the word problem for finitely presented groups. Home: 406 W. California Av., Urbana, Ill. 61801.*

BOORDE, Andrew (Borde), Brit. physician; b. nr. Cuckfield, Eng., circa 1490; ed. Oxford, Eng., U. Glasgow (Scotland); studied medicine abroad. Joined Carthusians; became suffragan bishop of Chichester, Eng., circa 1521; took oath of conformity, London Charterhouse, 1534; sent abroad by Cromwell to report on feelings about Henry VIII, 1535; traveled to Jerusalem, 1538; settled at Montpellier, France, circa 1538-42; imprisoned in Fleet, London, 1549. Author: Fyrst Boke of the Introduction of Knowledge, 1542; Handbook of Europe; Dyetary, 1542; Brewyary of Health, 1547; Boke of Berdes. Wrote 1st printed handbook of Europe, 1st printed specimen of Gypsy language, earliest modern work on hygiene. Died Apr. 25, 1549.

BOORSE, Henry Abraham, Am. physicist, educator; b. Norristown, Pa., Sept. 18, 1904; s. Henry Agustus and Martha (Godshall) B.; grad. with distinction U. S. Naval Acad., 1926; M.A., Columbia, 1933, Ph.D. (Barnard fellow), 1934; postgrad. (Lydig fellow) Cambridge (Eng.) U., 1934-35; m. Margaret V. Hazelton, Mar. 12, 1931; children—Ronald H., Michael H. (dec.), M. Suzanne. Instr. Coll. City N.Y., 1935-37; asst. prof. physics Barnard Coll., Columbia, N.Y.C., 1937-43, asso. prof., 1943-48, prof., 1948—, dean faculty, 1959—, acting pres., fall 1962. Cons. AEC, Brookhaven Nat. Lab., 1946-58; del. Fedn. Am. Scientists to Internat. Conf. Atomic Scientists, Oxford, Eng., 1946, also observer for Carnegie Endowment for Internat. Peace; chmn. Calorimetry Conf., 1956-57. Ernest Kempton Adams fellow, 1938-40. Fellow Am. Phys. Soc.; mem. Am. Assn. Physics Tchrs., Phi Beta Kappa, Sigma Xi. Contbr. articles on low temperature physics, properties liquid helium, superconductivity to profl. jours. Home: 338 Summit Av., Leonia, N.J. 07605. Office: Barnard Coll., N.Y.C. 10027.*

BOOT, Henry Albert Howard, English physicist; b. Birmingham, Eng., July 29, 1914; s. Henry and Ruby May Boot; B.Sc., Ph.D., U. Birmingham; m. Penelope M. Herrington, May 1, 1948; children—Nicholas J. A., Christopher J. Nuffield research fellow, Birmingham, 1945-48; now prin. sci. mem. Royal Naval Sci. Service, also dean sci. mems. Recipient Thomas Gray Meml. prize Royal Soc. Arts, prize of Royal Commn., John Scott prize. Mem. Inst. Physics, Instn. Elec. Engrs. (London). Research and publs. on cavity magnetron (inventor with J. T. Randall, 1939), contents of plasma and optical masers. Home: Old Mill Cottage, Rushden, Buntingford, Hertfordshire. Office: Services Electronic Research Lab., Baldock, Hertfordshire, Eng.

BOOTH, Charles, English sociologist; b. Liverpool, Eng., Mar. 30, 1840; s. Charles and Emily (Fletcher) B.; ed. Royal Institution School, Liverpool; hon. D.Sc., Cambridge; hon. D.C.L., Oxford; hon. LL.D., Liverpool; m. Mary Macaulay, 1871; 3 sons, 4 daus. Trained in Liverpool office of Lamport & Holt's Steamship Company; entered partnership with brother, 1862; chairman, Booth Steamship Co., until 1912; mem., tariff commission, 1903-04. Fellow, Royal Soc., 1899; pres., Royal Statistical Soc., 1892-94; privy councillor, 1904. Author: Occupations of the People, 1886; Life and Labour of the People in London (17 vol.), 1891-1903; Pauperism: a Picture; and the Endowment of Old Age, 1892; The Aged Poor, Condition, 1894; Old Age Pensions, 1899; Poor Law Reform, 1910. Used descriptive statistics to describe life in London during last decade of 19th century; forerunner of modern social surveys; showed relationship between casual employment and poverty; advocated old age pension. Died Whitwick, Eng., Nov. 23, 1916.

BOOTH, James, mathematician; b. Lava, Leitrim, Ireland, Aug. 25, 1806; s. John and B.; B.A., Trinity Coll., Dublin, Ireland, 1832, M.A., 1840, LL.D., 1842; m. dau. of Daniel Watney. Prin., Bristol (Eng.) Coll., 1840-43; ordained minister, 1842; vice prin. Liverpool (Eng.) Collegiate Instn., 1943-48; became vicar of Stone, nr. Aylesbury, 1859. Fellow Royal Society, 1846, Royal Academy of Science; member Liverpool Literature and Philosophical Society (president 1846-49), Soc. Arts (treas., chmn. council 1855-57). Author: A Treatise on Some New Geometrical Methods, 1st vol., 1873, 2d vol., 1877; On the Application of a New Analytic Method to the Theory of Curves and Curved Surfaces, 1840; also publs. on math. and edn. Invented (independently of Plücker) tangential co-ordinates (Boothian co-ordinates). Died Apr. 15, 1878.

BOOTH, James Curtis, Am. chemist; b. Phila., July 28, 1810; s. George and Ann (Bolton) B.; A.B., U. Pa., 1829; studied analytic chemistry with F. Wöhler and H. G. Magnus in Germany, 1833-36; LL.D., Lewisburg U. (now Bucknell U.), 1867; Ph.D. (hon.), Rensselaer Poly. Inst., 1884; m. Margaret Cardoza, Nov. 17, 1853. Taught chemistry Flushing, Long Island, 1830-32; opened lab. for instrn. in chemical analysis and applied chemistry, Phila., 1836; prof. chemistry Franklin Inst., 1838-45; taught chemistry at Phila. Central High School, 1842-45; melter and refiner Phila. Mint, 1849-88; prof. chemistry U. Pa., 1851-55; mem. 1st Geol. Survey of Pa.; state geologist Del. Mem. Am. Philos. Soc., Acad. Natural Scis., Md. Inst. for Promotion Mech. Arts, Phila. Soc. for Promotion of Agr., Hist. Soc. Pa., Am. Chem. Soc. (pres. 1883-85). Author: Encyclopedia of Chemistry, 1850; Our Recent Improvements in the Chemical Arts, 1851. Devised methods of refining gold-silver bullion; studied nickel ores in Pa. Died Haverford, Pa., Mar. 21, 1888.

BOOTH, Mary Ann Allard, Am. microscopist; b. Longmeadow, Mass., Sept. 8, 1843; d. Samuel Colton and Rhoda (Colton) B.; ed. Wilbraham (Mass.) Acad., and under pvt. teachers. Devoted to research with microscope; lectured before many sci. socs. in U. S. and Can.; made photomicrographs of germ-bearing fleas of rats, for steropticon slides, during campaign agains bubonic plague in San Francisco, 1907-09; made large collection of parasite specimens and photomicrographs. Recipient medals and diplomas New Orleans Expn., 1885, St. Louis Expn., 1904, San Francisco Expn., 1915. Editor Practical Microscopy, 1900-07. Died Sept. 15, 1922.

BOOTH, Vernon Hollis, English biochemist; b. Herne Bay, Eng., June 5, 1903; s. Abram and Mary (Paine) B.; B.A., U. Cambridge, 1933, Ph.D., 1937; m. Barbara Kathleen Church, 1941; children—Daphne (Mrs. G. T. Lawrence), Robin, Pauline, Peter. Mem. Trinity Coll., U. Cambridge, 1930—, Ramsay Meml. fellow biochemistry, 1938-39, Beit. Meml. fellow physiology, 1939-42, mem. sci. staff Agrl. Research Council, Dunn Nutritional Lab., 1942-63; mem. internat. staff UN Devel. Programme Beirut, 1966-67; lysine research dept. agr. U. Cambridge, 1967—. Free lance scientist, 1963—; tchr., cons., editor. Author: Preservation of Carotene in Dried Green Crops, 1955; Carotene, Its Determination in Biological Materials 1957. Research on enzymes, vitamins, mapping plant lipids by paper chromatography, leading to discovery of new compounds from leaves. Home: 10 Thornton Close, Girton, Cambridge. Office: Sch. Agr., Downing St., Cambridge, Eng.*

BOOTHBY, Walter Meredith, Am. med. researcher; b. Boston. July 28, 1880; s. Alonzo and Marie Adelaide (Stodder) B.; student Boston U. Sch. of Medicine. 1901-05; A.B., Harvard, 1902, M.D., 1906, M.A., 1907; m. Catharine Burns. Nov. 15, 1930; children (by previous marriage)—Gertrude (Mrs. Louis Schulze). Nancy (Mrs. Robert Reinhardt). Practiced surgery in Boston, 1909-16; in charge metabolism and respiration labs., Peter Bent Brigham Hosp., 1913-16; instr. in anatomy Harvard Med. Sch., 1910-16. also lectr. on anesthesia. 1914-16; head of sect. of metabolic research. Mayo Clinic. 1916—; asst. in medicine, Mayo Found., 1917-23, asso. prof. in medicine, 1923-26; prof. exptl. metabolism. 1936-48; chmn. Mayo Aero Med. Unit for Research in Aviation Med., 1942-48; guest prof. aviation medicine Inst. of Physiology. U. Lund (Sweden), 1948-50; advisor on research Sch. of Aviation Medicine. prof. physiology, Air U., Randolph Field. USAF, Tex., 1950-51; Lovelace Found. for Med. research. from 1951, head of dept of respiratory physiology. Recipient (with W. R. Lovelace) Collier Trophy for med. edn. and research, 1938; Fellow A.C.S., A.C.P.; mem. A.M.A., Am. Physiol. Soc., Am. Soc. Biol. Chemists, Am. Soc. for Clin. Investigation. Soc. for Exptl. Biology and Medicine. Assn. Am. Physicians, Am. Inst. Nutrition, Am. Soc. for Exptl. Pathology, Am. Soc. for Pharmacology and Exptl. Therapy, Am. Soc. Anesthetists, Aero Med. Assn. of U. S., Inst. Aero. Scis., Nat. Aeronautic Assn., Mass. Med Soc., Alumni Assn. of Mayo Found., Sigma Xi. Writer numerous papers on respiration, metabolism, thyroid diseases, aviation medicine. devised (with F. J. Cotton) apparatus for adminstrn. nitrous oxide-oxygen-ether anesthesia, 1912; introduced aminoacetic acid (also known as glycine and glycocoll) in treatment of myasthenia gravis, 1932. Died July 4, 1953.

BOOTHBY, William Munger, Am. mathematician; b. Detroit, Apr. 1, 1918; s. Thomas Franklin and Florence (Munger) B.; A.B., U. Mich., 1941, M.A., 1942, Ph.D., 1949; m. Ruth Robin, June 8, 1947; children—Daniel, Thomas, Mark. Instr., Northwestern U., 1948-50, asst. prof., 1951-58; asso. prof. Washington U., St. Louis, 1959-61, prof., 1962—; Am. Swiss Found. for Sci. Exchange fellow Swiss Fed. Inst. Tech., Zurich, 1950-51; mem. Inst. for Advanced Study, Princeton, N.J., 1961-62, NSF Sr. Postdoctoral fellow, 1961-62, U. Geneva (Switzerland), 1965-66. Mem. Am. Math. Soc., Math. Assn. Am., A.A.A.S., Sigma Xi. Research on homogeneous spaces of Lie groups. Address: Washington U., Dept. Mathematics, St. Louis 63130.*

BOPP, Franz, German philologist; b. Mainz, Germany, Sept. 14, 1791; ed. Aschaffenburg, Paris; prof. oriental languages and gen. philology, Berlin, 1821-67. Author: Über das Conjugations-System in Vergleichung . . . , 1816; Vergleichende Zergliederung des Sanskrits und der mit ihm verwandten Sprachen, 1824; Ausführliches Lehrgebäude der Sanskrit-Sprache, 1827; Über die Verwandtschaft der malayisch-polynesichen Sprachen mit den indisch-europaischen, 1841; Vergleichendes Assentuations-System des Sankriti und Greichischen, 1854; Comparative Grammar of the Sanskrit, Zend, Greek, Latin, Lithuanian, Gothic, German and Slavonic Languages (translator E. B. Eastwick, 1856). Founder sci. of comparative philology. Died Berlin, Oct. 23, 1867.

BORASS, Emil, German geodesist; b. Forsthaus, Büssen, Feb. 22, 1856. Asso. with Geodetic Inst., Potsdam, Germany, from 1881, dept. head and prof., 1901-21. Author: Bericht uber die relativen Uessingen der Schwerkraft mit Pendellapparaten in der Zeit von 1808-1909 und uber ihre Darstellung im Potsdamer Schweresystem. Rep., authority in matters of internat. gravity (gravitation) measurements; unit of terrestrial gravitation of Potsdam internationally accepted. Died Potsdam, Germany, Nov. 16, 1930.

BORCHARDT, Karl Wilhelm, German mathematician; b. Berlin, Germany, Feb. 22, 1817; s. Moritz and Emma (Heilborn) B.; student Berlin; Ph.D., Königsberg, 1843; m. Rosa Oppenheim. Faculty, Berlin U., from 1851; mem. Berlin, French acads. scis.; editor: Journal für die reine und angewandte Mathematik; noted for 1st publ. on a cubic equation (with aid of which secular disturbances of planets could be determined.) Died Berlin, June 27, 1880.

BORCHERS, (Johann Albert) Wilhelm, metallurgist; b. Goslar, Germany, Oct. 6, 1856; s. Albert and Pauline (Landahn) B.; studied in Greifswald, Erlangen and Munich, Germany, 1875-79; doctorate from Erlangen, 1878; studied at Bergakademie (mining acad.) Clausthal, 1891-92; hon. dr. engring. Technische Hochschule Breslau; m. Lucie Probst, 1886; 4 sons. Chemist, firm E. de Haën, Hannover, Germany, 1879, dir. metall., mech., electric and furnace works, 1887; with Colonial Chem. Works, West Medford, Mass., 1882; tchr. Maschinenbau- und Hüttenschule in Duisburg, 1892-97; prof. Technische Hochschule Aachen, (Germany), 1899; founded inst. for metallurgy and electrometallurgy; rector Aachen Technische Hochschule, 1904-09. Author: Elektrometallurgie, 1891; Die elektrischen Öfen, 1897; Hüttenwesen, 1908; Die Metallhüttenbetriebe, 4 vols., 1915-24; (with Wust) Eisen und Metallhüttenkunde, 1899; (with W. Nernst) Handbuch der Elektrochemie, Vols. I-VI, 1894-99. Editor: (with A. Wilke) Zeitschrift für Elektrotechnik und Elektrochemie from 1894; founder, editor (with Wust) Metallurgie, from 1906. First in Germany to prepare calcium carbide, wrote over 100 sci. papers from his inst. for metallurgy and electrometallurgy on various areas of foundry work, studies on rust-resistant alloys which paved way for devel. of stainless steel. Died Goslar, Germany, Jan. 6, 1925.

BORCHGREVINK, Carsten Egeberg, Norwegian naturalist, explorer; b. Christiana (now Oslo), Norway, Dec. 1, 1864. Emigrated from Norway to Australia, 1888. Surveyor Australia, climber Mt. Lindsay, 1888; instructor in languages in Cooerwell Coll., New South Wales; mem. 1st Antarctic exploration party, 1895; comdr. 1st expdn. wintering on Antarctic continent, Cape Adare, 1899; explorer Ross Sea, on land S. to

78° 50′ latitude, southernmost expdn. to that time; sailed to West Indies to investigate volcanic conditions, 1902. Fellow Royal Geog. Soc. (Patron's medal 1930); mem. Royal Scottish Geog. Soc. (medallist, hon.). Author: First on the Antarctic Continent, 1901; The Game of Norway, 1920-25. Discoverer 1st known antarctic vegetation, mosses, lichen. Died Christiana, Apr. 21, 1934.

BORCHGREVINK, Christian Fredrik, Norwegian physician; b. Oslo, Norway, Sept. 1, 1924; s. Henrik Borchgrevink, and Minnie (Schibbye) B.; student U. Oslo Med. Sch., 1945-51, doctorate, 1961; m. Louise Storm, June 15, 1951; 1 son, Axel. Sr. adviser BCG Campaign, Indonesia, WHO, 1953-56; research fellow Inst. for Thrombosis Research, U. Oslo., 1957-61, cons. in hematology Inst. for Pharmaco-therapy; sr. research fellow Cardiovascular Research Inst., San Francisco, 1962-63; staff med. dept. VII, Ulleval Hosp., Oslo, 1963—. Mem. Internat., European hematol. assns. Editoral bd. Jour. Atheroscl. Res. Research, publs. on mechanism of hemostasis showing importance of blood platelets in sealing small cut in vessels, blood coagulation in secondary hemostasis, anticoagulant therapy in ischemic heart disease. Home: Gullerasveien 17 b, Office: Dept. VII, Ulleval Hosp., Oslo, Norway.*

BORDA D'ORO, see de Borda, Jacques François.

BORDAGE, Edmond, French biologist; b. Pons, France, 1863; dir. St. Denis Natural History Mus., Reunion Island; head Giraud Lab., Paris; research on neglected aspects on insect metamorphosis; studied orthoptera, hymenoptera, some crustaceans; translator several of John Lubbock's books into French. Died Paris, 1924.

BORDE, Andrew, see Boorde, Andrew.

BORDEN, Gail, surveyor, inventor; b. Norwich, N.Y., Nov. 9, 1801; s. Gail and Philadelphia (Wheeler) B.; m. Penelope Mercer, 1828; m. 2d, Mrs. A. F. Stearns; m. 3d, Mrs. Emeline Eunice (Eno) Church, 1860; several children. Moved with family to Covington, Ky., 1815, to Ind. Territory, 1816; sch. tchr. Ind. Ty., 1820-22; moved to Amite County, Miss., 1822; county surveyor, U. S. dep. surveyor Amite County, 1822-circa 1829; moved to Tex., circa 1829; supt. ofcl. surveys of Tex. colonies; del. San Felipe (Tex.) Conv., 1833; made 1st topog. map and planned layout of Galveston, Tex.; land agt. Galveston City Corp., 1839-51; became interested in concentrated foods, invented meat biscuit exhibited at London Fair, 1851; joined Shaker Colony at New Lebanon, N.Y.; developed a form of condensed milk, Lebanon, N.Y., 1853, patented it, 1856; established New York Condensed Milk Company, Wassaic, N.Y., 1858; moved to Borden, Tex., after Civil War; continued research into concentration of foods, patented process of concentrated juices and fruit, 1862. Died Borden, Tex., Jan. 11, 1874.

BORDEN, Simeon, Am. civil engr.; b. Fall River, Mass., Jan. 29, 1798; s. Simeon and Amy (Briggs) B. Moved with family to Tiverton, R.I., 1806; began work in machine shop, 1826, supt., 1828; constructed base bar used for measuring base line in trigonometrical map survey of Boston (most accurate map in its time in U.S.), 1830; mem. Mass. Legislature, 1832-33, 44-45, 49; chief surveyor Boston Map Survey, 1834 41; chief surveyor railroads in Me., N.H., Mass., Conn., 1841-51; began railroad construction, 1848; surveyed r.r. line between R.I. and Mass. Author: A System of Useful Formulae Adapted to the Practical Operations of Locating and Constructing Railroads, 1851. Made 1st geodetic survey in U. S.; devised surveying instruments. Died Oct. 28, 1856.

BORDEN, William Cline, Am. surgeon; b. Watertown, N.Y., May 19, 1858; s. Daniel J. and Mary L. (Cline) B.; M.D., Columbian (now George Washington) U., 1883, Sc.D., 1931; m. Jennie E. Adams, Oct. 27, 1883; children—Daniel Le Ray, William Ayres. Apptd. 1st lt. asst. surgeon U. S. Army, 1883, advanced through grades to lt. col., ret., 1909. Comd. Gen. Hosp., Key West, Fla., during Spanish-Am. War; Gen. Hosp., Washington, 1898-1907; also prof. mil. surgery, Army Med. Sch., and prof. surg. pathology and mil. surgery, Georgetown U., 1898-1907; comd. div. hosp., Manila, 1908; prof. surgery, dean med. dept. George Washington U. and surgeon in chief George Washington U. Hosp., 1909-31, chief of surg. service Walter Reed Army Gen. Hosp., 1917-19. Fellow A.C.S. (a founder); mem. A.M.A. Author: Use of the Röntgen Ray by the Medical Department of the United States in the War with Spain, 1898, 1900; also several sects. in standard surg. works and many med. monographs and articles. Died Sept. 29, 1934.

BORDENAVE, Toussaint, French surgeon; b. Paris, France, Apr. 10, 1728; s. Pierre Bordenave; prof. Coll. Surgery of Paris; prof. surgery Sch. St. Côme, 1740—. Mem. Royal Acad. Surgery (dir.), French Acad. Scis. Author: Essai sur la physiologie ou physique du corps humain, 1756; Précis d'observations sur les maladies du sinus maxillaire (foreshadows sci. of otolaryngology); Essai sur le mécanisme de la nature dans la formation des os (disputes ideas of Duhamel), 1760. Research on properties of tissues, irritability and insensitivity of fibrous tendons. Died Mar. 12, 1782.

BORDERS, Huey Ingles, Am. plant pathologist; b. Bessemer, Ala., Dec. 18, 1905; s. Edwin Rice and Edith (Baldwin) B.; B.S. in Agr., U. Fla., 1933; M.S., U. Minn., 1938, Ph.D., 1947; m. Bonita Ruth Bailey, Feb. 24, 1934; children—Ann (Mrs. Peter Webb), John F. Plant pathologist Ga. Agrl. Extension Service, Tifton, 1939-40; asso. plant pathologist Fla. Agrl. Expt. Stas., 1944-46; asso. plant pathologist, Tifton, U. S. Dept. Agr., 1947-48, plant pathologist, 1948-55, sr. plant pathologist, 1955-59, asst. br. chief, vegetable and ornamental crops research br., Beltsville, Md., 1959; asso. plant pathologist U. Fla. Agrl. Expt. Sta., Plantation Field Lab., Fort Lauderdale, 1959—; vis. NSF lectr. Wash. State U., 1962. Mem. Am. Phytopath. Soc., A.A.A.S., N.Y. Acad. Scis., Mycol. Soc. Am., Fla. Hort. Soc., Sigma Xi, Alpha Zeta, Alpha Gamma Rho. Research on causes and control of fungal infections of plants, particularly wheat rust; on ornamental plants and turf grasses; plant nutritional deficiencies. Home: 3400 S.W. 26th St., Fort Lauderdale, Fla. 33312.*

BORDET, Jules Jean Baptiste Vincent, Belgian bacteriologist, physiologist; born Soignies, Belgium, June 13, 1870; M.D., University of Brussels, 1892; worked under Mechnikov, Pasteur Institute, Paris, France, 1894-1901; founder, director Pasteur Institute, Brussels, Belgium, from 1901; professor bacteriology University of Brussels, 1907-35. Recipient Nobel prize in medicine and physiology, 1919. Fellow Royal Soc., 1916. Mem. French Acad. Scis. Author: Traité de l'immunité, 1920. Worked in immunology and serology; gave 1st description of bacterial hemolysis, 1898; discovered (with Octave Gengou) complement-fixation reaction (basis for Wasserman syphilis test, also led to gen. method for fever diagnosis), 1900; discovered bacillus of whooping cough (Hemophilus pertussis) and method of immunization, 1906; studied microbe biology and serology. Died Brussels, Apr. 6, 1961.

BORDEU, Théophile de, see de Bordeu.

BORDIN, Edward S., Am. psychologist; b. Phila., Nov. 7, 1913; s. Morris and Jennie (Zarovsky) B.; B.S.C., Temple U., 1935, M.S., 1937; Ph.D., Ohio State U., 1942; m. Ruth Birgitta Anderson, June 20, 1941; children—Martha Christine, Charlotte Anna. With U. Minn., 1939-42, War Dept. 1942-45; acting dir. student counseling bur., asst. prof. U. Minn., Mpls., 1945-46; dir. student counseling center, asso. prof. psychology, Washington State Coll., Pullman, 1946-48; faculty U. Mich., Ann Arbor, 1948—, prof. psychology, 1955—, chief counseling div., 1948—. Cons., VA, 1948-57. Mem. Am. Psychol. Assn. (pres. div. counseling psychology 1956), A.A.A.S., Am. Coll. Personnel Assn. Author: Psychological Counseling, 1955; also numerous articles. Editor: Jour. Cons. Psychology, 1959-64. Research on role ambiguity therapist and therapeutic task in psychotherapy, influence personality on vocational choice, psychotherapy, evaluation counseling, personality devel. and vocational choice. Home: 210 Montgomery St., Ann Arbor, Mich. 48103.*

BORDLEY, James, III, Am. med. adminstr.; b. Centreville, Md., Dec. 7, 1900; s. James and Margaretta (Hollyday) B.; Ph.D., Yale, 1923; M.D., Johns Hopkins, 1927; Sc.D., Hartwick Coll., 1953, Union U., 1964; m. Julia Peabody Ross, July 4, 1936; children—Patricia (Mrs. Roderic D. Wiltse), James, Donald R. NRC fellow Johns Hopkins, 1930-32, asso. prof. medicine, 1937-47; dir. M. I. Bassett Hosp., Cooperstown, N.Y., 1947—; clin. prof. medicine Columbia, N.Y.C., 1948—. Cons. to Surgeon Gen., U. S. Army, 1948—. Recipient William H. Howell award Johns Hopkins, 1926. Mem. Century Assn., Phi Beta Kappa, Alpha Omega Alpha, Beta Theta Pi. Author: (with A. M. Harvey) Differential Diagnosis, 1955. Mng. editor Bull. Johns Hopkins Hosp., 1937-42. Home: 13 Main St. Office: M. I. Bassett Hosp., Cooperstown, N.Y. 13326.*

BORDLEY, John Beale, Am. agriculturist; b. Annapolis, Md., Feb. 11, 1727; s. Thomas and Ariana (Frisby) B.; m. Margaret Chew, 1750; m. 2d, Mrs. Sarah Mifflin, Oct. 8, 1776. Prothonotary, Baltimore County, Md., 1753-62; judge Md. Provincial Ct., 1766, Ct. of Admiralty, 1767; commr. for settlement of Md.-Del. boundary, 1768; engaged in agrl. experimentation with new machinery, seeds, crop rotation on extensive lands at mouth Wye River and Pool's Island, Md. from 1770; a founder Phila. Soc. for Promoting Agr., 1785, v.p., 1785-1804. Author essays: A Summary View of the Courses of Crops in the Husbandry of England and Maryland, 1784; Money, Coins, Weights, and Measures, 1789; Essays and Notes on Husbandry and Rural Affairs, 1801. Died on farm nr. Joppa, Md., Jan. 26, 1804.

BORDO, Louis, French inventor; b. Lyons, France, 1700; perfected windlass; worked on machine for perfecting prodn. of glass for mirrors; invented mech. divider usable with all math. instruments. Died Lyons, 1741.

BOREAS, Theophilos, psychologist; b. Athens, Greece, Dec. 16, 1876; DD.L., U. Athens; Ph.D., U. Leipzig (Germany), 1899; m. Despina Levantis, 1927. Dir. normal schs., Tripoli, N. Africa, from 1900; prof. normal schs., Athens, from 1905; prof. philosophy U. Athens, also dir. psychol. lab., from 1912, rector, from 1926. Mem. Greek Acad. Arts and Scis., Kant-Gesellschaft, Deutsche Philosophische Gesellschaft. Author: Das weltbildende Prinzip in der platonischen Philosophie, 1899; Heredity and Environment, their Influence upon Intelligence and Character, 1906; Entrails as Seat of the Soul, 1909; Experimental Psychology and Psychological Laboratories, 1913; Greek Philosophy after the Capture of Constantinople by the Turks, 1929; Experimental Studies of Memory; The Process of Development of Memory, 1930; The Rate of Forgetting, 1930; Training and Transfer in Memory, 1931; Methods of Memorisation, 1931; Reactiontime and Intelligence, 1932-34; Logic, 1932; Psychology, 1933; Introduction to Philosophy, 1935.

BOREI, Nans Georg, biologist; b. Stockholm, Sweden, Feb. 7, 1914; s. Johan Emil Pahlson and Elise Edlund; M.A., U. Stockholm, 1937, Ph.D., 1940, Fil. Dr., 1945; m. Maj Ellen Österlin, 1938; children—Karin (Mrs. Richard Hodges), Sven, Ragnar. Came to U. S., 1951, naturalized, 1960. Research asst. biophysics, exptl. zoology U. Stockholm, 1937-45, asst. prof. 1945-47, asso. prof., 1947-52, also head dept. devel. physiology and genetics Wenner-Gren Inst., 1947-50; head dept. biophysics, 1950-52, acting head Inst., 1948-50; vis. prof. Cal. Inst. Tech., 1951; vis. prof. U. Pa., Phil., 1951, prof. zoology, 1953—; research at Carlsberg Lab., Copenhagen, Denmark, marine biology stas., Europe, U. S., Can. Trustee Mt. Desert Island Biology Lab. Rockefeller fellow Molteno Inst., Cambridge (Eng.) U., 1948-49. Fellow A.A.A.S.; mem. Soc. Gen. Physiology, Brit. Biochemistry Soc., Brit. Soc. Exptl. Biology, Internat. Soc. Cell Biology, Sigma Xi. Contbr. articles on embryology and biochemistry to profl. jours. Research on baker's yeast, effect of fluoride on activity of respiratory enzymes, respiration in sea urchin eggs before and after fertilization, biochemistry of pigments in granules of jelly coat of sand dollar eggs, biochemistry cyanide-utilizing enzymes in marine lower animals, respiratory pigment changes during devel. in honey bee flight muscles, faunal analysis intertidal communities of No. Me. coast. Home: 3725 Hamilton St., Phila. 19104.*

BOREL, Félix Édouard Émile, French mathematician; b. St. Affrique, France, Jan. 7, 1871; D.Sc., École normale supérieure, 1894. Faculty, U. Lille (France), 1893-96; became sci. dir. École normale supérieure, 1911; dir. Inst. Henri-Poincaré, from 1927. Mem. Chamber Deputies, 1924-36; minister navy, 1925. Mem. Bur. Longitudes, French Acad. Scis. Author: La théorie des fonctions, 1897-1922; Traité du calcul des probabilités et ses applications, 1924-34; Probabilité et certitude, 1950. Worked in infinitesimal calculus and calculus of probabilities; developed 1st effective theory of measure of sets of points, thus becoming founder (with Baire and Lebesgue) of modern theory of functions of real variables; found elementary proof of Picard's theorem, 1896, from this created theory of integral functions and distbn. of their values; developed 1st systematic theory of divergent series. Died Paris, Feb. 4, 1956.

BOREL, Pierre, French chemist, optician; b. Castres, France, 1620; docteur en médecine, Montpellier, France, 1640; appointed physician to Louis XIV, 1654; m. Esther de Bonnafous, 1663. Mem. French Acad. Scis. Author: Bibliotheca chimica, 1654; De vero telescopii inventore (history of telescope), 1655; other works on natural history, chemistry, optics, astronomy, antiquities, philology, bibliography. Demonstrated that oil of vitriol dissolves marble best when diluted; described distillation of sal ammoniac with iron; pub. analysis of urine, 1688. Died Paris, June 1689.

BORELIUS, Carl Olof Gudmund, Swedish physicist; b. Falun, Sweden, Apr. 18, 1889; s. Carl-Aron and Gertrud (Frank) B.; Dr.Phil., U. Lund (Sweden), 1915; m. Kerstin Magnhild Tornberg, Apr. 22, 1916; children—Kerstin (Mrs. Nils Emmelin), Anna, Elisabet (Mrs. Horacio Pontis). Docent, U. Lund, 1915-22; prof. Royal Inst. Tech., Stockholm, Sweden, 1922-55. Author: Grundlagen des metallishen Instandes, 1935; also numerous articles. Research on physics of metals, low temperature physics, phase transitions. Home: 17 Föreningsvägen, Diursholm, Sweden. Office: Royal Inst. Tech., Stockholm 70, Sweden.*

BORELLI, Dante, med. mycologist; b. Parma, Italy, Feb. 19, 1920. s. Tersilio and Almerina (Bertani) B.; grad. summa cum laude, Università di Parma, 1944, degree in DermatoVenereology Dermatol. Clinic, Pavia, Italy, 1946; m. Maria Luisa Coretti, Sept. 24, 1953; 1 son, Fabius. Asst. prof. dermatology, Parma, 1946-48; rural physician Llanos, Venezuela, 1948-52; now titular prof. Facultad de Ciencias Medicas, Universidad Central de Venezuela, Caracas, chief med. mycology sect. Instituto de Medicine Tropical, Caracas, owner pvt. lab. of med. mycology, Caracas. Mem. Sociedad Venezuela Dermatologia, Colegio Ibero-Latino-Americano de Dermatologia, Internat. Soc. Human Veterinary Mycology, Royal Soc. Tropical Medicine, London (Eng.). Research, publs. on devel. of method of culturing pathogenic fungi that is more practical, cheaper and more effective than existing procedures. Home: Qta Po, Calle Guanaro, Caracas. Office: Instituto Médico del Este, Av. Casanova, Caracas, Venezuela.*

BORELLI, Giovanni Alfonso, Italian mathematician, physicist, astronomer, physiologist; born Naples,

Italy, June 28, 1608; educated Florence, also Rome, Italy. Became professor of mathematics, Messina, Italy, 1649-56, Pisa, Italy, 1656-67, returned to Messina, 1667; became polit. refugee, Rome, 1674; friend of Galileo, also Marcello Malpighi (who interested him in anatomy). Mem. Accademia del Cimento, Florence. Author: Theoria planetarum, 1666; De motionibus naturalibus a gravitite pendentibus, 1670; De motu animalium, 2 vols., 1680-81. Founder Iatro-math. sch. (which attempted to apply math. methods to medicine); explained muscular action mechanically in terms of system of levers, also tried to apply principles to heart, lungs, stomach, other organs; discovered comets travel in parabolic orbits, 1665; tried to explain motion of Jupiter's satellites by laws of attraction; inventor heliostat; gave more precise measure of speed of sound; investigated capillary phenomena; drew up plans for submarine; worked on problems of air pressure and surface tension of water. Died Rome, Dec. 31, 1679.

BOREN, Hollis Grady, Am. physiologist; b. Dallas, Mar. 24, 1923; s. Henry Grady and Lida Virginia (Higgins) B.; B.A., So. Meth. U., 1943; M.D., Baylor U., 1946; m. Lois Catherine Wriborg, Dec. 6, 1963; children—Douglas, Cynthia, Debra, Jay, Lisa. Staff VA Hosp., Houston, 1952-63, chief pulmonary function lab., 1953-63, asso. chief of staff for research and edn., 1961-62, chief med. service, 1962-63; faculty Baylor U. Coll. Medicine, 1952-63, asso. prof., 1962-63; dir., mem. Trudeau Found., Inc., Saranac Lake, N.Y., 1963-66; asso. prof. medicine U. Colo., Denver, 1966——. Recipient Billings Bronze medal, A.M.A., Miami, Fla., 1960. Mem. Am. Thoracic Soc., A.C.P., N.Y. Acad. Scis. Diplomate Am. Bd. Internal Medicine. Fellow A.C.P. Contbg. author Textbook of Pulmonary Diseases, 1965. Research and publs. on cause of pulmonary emphysema, utilizing new methods of preparing lung for study. Home: 6674 S. Franklin St., Littleton, Colo. 80120. Office: VA Hosp., Denver.*

BORESKOV, Georgii Konstantinovich, Russian phys. chemist; b. April 20, 1907; grad. Odessa Chem. Inst., 1928. With Ukrainian Chemico-Radiol. Inst. (now Ukrainian br. Inst. Rare Metals, Odessa), 1928-37; instr. Odessa Inst. Chem. Tech., 1930-37, Odessa U., 1934-37; dir. Catalysis Lab., Moscow Fertilizer, Insecticide and Fungicide Research Inst., 1937-49; asso. Karpov Physico-chem. Inst., 1946-60; prof. Moscow Inst. Chem. Tech., 1949-58; dir. Inst. Catalysis, Siberian dept. USSR Acad. Sci., prof. Novosibirsk U., 1958——. Recipient State prize 1942. Mem. USSR Acad. Sci. Research and publs. on sci. bases for selection and preparation catalysts and design of contact apparatus; proposed vanadium as catalyst for prodn. sulphuric acid. Address: Inst. of Catalysis, Siberian Dept., USSR Acad. Sci., Novosibirsk 90, RSFSR, USSR.

BORG, Göran Lars, Swedish mathematician; b. Kumla, Sweden, Nov. 19, 1913; s. Eric Johan and Elna (Petterson) B; Fil.mag., U. Uppsala (Sweden), 1937, Fil.lic., 1941, Fil.Dr., 1945; m. Gunborg M. Sjölinder, Oct. 1, 1939; children—Lars E., Erik S., Anna-Karin M. Docent, U. Uppsala, 1945-53; prof. U. Stockholm, 1953——. Mem. Swedish Math. Soc. (past chmn.), Swedish Nat. Research Council, Tekniska fak. beredning, Acad. Engring. (chmn. sec. VII). Author: Eine Umkehrung der Sturm-Liouvilleschen Eigenwertaufgabe, 1945. Home: 118 Alviksvägen, Stockholm, Sweden.*

BORG, Sidney Fred, Am. educator, engr.; b. N.Y.C., Oct. 3, 1916; s. Herman Leo and Pauline (Leibman) B.; B.S. in Civil Engring., Cooper Union Inst. Tech., 1937; M.C.E., Poly Inst. Bklyn., 1940; E.D., Johns Hopkins, 1956; M.Eng. (hon.), Stevens Institute Tech., 1958; m. Audrey Iva Elliott, Apr. 4, 1944; children—Nicholas Elliott, Andrew Douglas, Jill Debora, Kenneth. Engring. positions City of N.Y., U. S. War Dept., Turner Constrn. Co., Gen. Motors Co., 1937-43; with Grumman Aircraft Corp., 1951-52; asst. prof. civil engring., U. Md., 1943-45; asst. and asso. prof. aero. engring., U. S. Postgrad. Sch., 1945-51; head dept. civil engring., asso. prof. civil engring. Stevens Inst. Tech., 1952-56, prof., 1956——; Fulbright lectr. Royal Danish Tech. U., 1965-66; vis. prof. Technische Hochschule, Stuttgart, Germany, 1966; consulting engineer. Member board of directors Kreisler-Borg Constrn. Company, Scarsdale, New York Recipient distinguished alumnus award Poly. Inst. Bklyn., 1957. Licensed profl. engr. Fellow Am. Assn. for Advancement of Science, New York Acad. Sci., American Society C.E.; member Am. Soc. Engring. Edn., Sigma Xi, Tau Beta Pi. Author textbook, An Introduction to Matrix-Tensor Methods in Applied Mechanics, 1956; Advanced Structural Analysis, 1959; Fundamentals of Engineering Elasticity, 1962; Matrix Tensor Methods in Continuum Mechanics, 1963. Editor textbook series. Basic contribution to wedgewater entry problem; obtained similarity type solutions in various fields of applied mechanics; applied matrix-tensor theory in applied mechanics and dimensional theory. Home: 2 9th St., Hoboken, N.J.

BÖRGEN, Carl Nicolay Jensen, German astronomer; b. Schleswig, Germany, Oct. 1, 1843; s. Andreas Ulrich Christian and Nicoline Margarete (Jensen) B.; student Copenhagen, Denmark, Kiel, Germany, from 1863; Ph.D., Göttingen, Germany, 1869. Participant E. Greenland expdn., 1869-70; with obs. Leipzig,

Germany; dir. obs. Wilhelmshaven, Germany, from 1874; leader expdn. Kerguelen. Recipient Georg Neumayer medal, 1900. Concerned with geomagnetic observation, time service, tides; wrote treatise on tides which opened way for sci. work in this field in Germany. Died Wilhelmshaven, June 8, 1909.

BORGHESE, Elio, Italian biologist; b. Crevacuore, Italy, July 11, 1909; s. Aroldo and Maria (Chiesa) B.; M.D., U. Pavia, 1932, Dr.Sc. Biol., 1944; m. Luigia Provasi, Feb. 4, 1939; 1 son, Paolo. Asst., Inst. Human Anatomy, Pavia, Italy, 1932-56; prof., dir. Inst. Histology, prof., dir. Inst. Anatomy, U. Cagliari (Italy), 1956-62; prof., dir. Inst. Topographic Anatomy, prof., dir. Inst. Human Anatomy, U. Napoli (Italy), 1962-64; prof. human anatomy, 1965——; guest worker biology, Freiburg, Germany, 1939, Cambridge, Eng., 1948, Galveston, Tex., Balt., 1952. Organizer, dir. lab. embryology Nat. Com. for Nuclear Research, Frascati, Italy, 1958-62. Research, numerous publs. in human anatomy, cytology, embryology, including reciprocal influence of tissues in devel. mammalian organs in vitro, effects of radiation on embryos, microcinenatography. Home: 276 via Posillipo, Napoli, Italy.*

BORGI, Pietro (Borgo or Borghi) Italian mathematician; b. Venice, Italy, flourished 15th century. Author: Addiones in quibus etian sunt replicae Mathei Boringii, 1483; Libro de Abacho de arithmetica; Arithmetica, 1484; De arte mathematiche. Wrote best known Italian works on arithmetic of 15th century, including comml. arithmetic.

BORGNIS, Fritz Edward, physicist, educator; b. Mannheim, Germany, Dec. 24, 1906; s. Max F. and Adele (Dyckerhoff) B.; student Munich (Germany) U., 1924-29; Ph.D., Inst. Tech., Munich, 1936; m. Gerda M. Maschinda, Mar. 10, 1964. Asst. electrophysics lab. Munich Inst. Tech., 1932-38; centimeter wave research Telefunken Co. Labs., Berlin, Germany, 1938-40; lectr. physics U. Graz (Austria), 1940-47, head dept. applied physics, 1942-47; lectr., research physicist dept. indsl. research Fed. Inst. Tech., Zurich, Switzerland, 1947-50, prof., dir. Institut für Hochfrequenztechnik, 1960——; research asso. Wesleyan U., Middletown, Conn., 1950-51; research physicist Cal. Inst. Tech., Pasadena, 1951-54, dept. engring. and applied physics Harvard, Cambridge, Mass., 1954-57; dir. research Philips Research Labs., Hamburg and Aachen, Germany, 1957-60. Vis. prof. U. Innsbruck (Austria), 1948-49; vis. McKay prof. elec. engring. U. Cal. at Berkeley, 1955; hon. prof. physics U. Hamburg, 1959——. Fellow I.E.E.E., Am. Phys. Soc.; mem. Am. Acoustical Soc., Swiss, German phys. socs., Swiss Soc. for Natural Sci., Swiss Soc. Elec. Engrs. Author: (with C. H. Papas) Randwertprobleme der Mikrowellenphysik, 1956; Electromagnetic Waveguides and Resonators (Ency. Physics), 1957; also articles. Research on ultrasonics, waves in crystals, electromagnetic theory, plasma and microwave physics, phys. electronics. Home: 99 Bergstrasse, 8032, Zurich. Office: 7 Sternwartstrasse, 8006, Zurich, Switzerland.*

BORGO, Luca de, see de Borgo.

BORGOGNONI, Hugh (Ugo da Lucca, Hugh of Lucca), Italian physician, surgeon; b. Lucca, Italy; flourished circa 1214; 4 sons, possibly including Theodoric; accompanied Bolognese crusaders to Syria and Egypt; attended siege of Damietta; founder surg. sch. Bologna, circa 1214. Died circa 1252-58.

BORGOGNONI, Theodoric, (Teodrico Borgognoni, Theodoricus Cerviensis), Italian physician; b. Lucca, Italy, 1205; s. possibly of Hugh Borgognoni, who he studied under at Bologna, Italy. Became Dominican; penitentiary to Innocent IV (pope, 1243-54); bishop of Bitonto (Bari della Publie), Lucca, 1262-66; bishop of Cervia, Bologna, 1266-98. Author: De sublimatione arsenici; De aluminibus et salibus (both now lost); Practico equorum (Liber de medela equorum also Mulomedicina); Chirurgia, 1498. also treatises on facolnry and surgery. Used mercury salts for skin diseases and observed resultant salivation; developed father's method of using wine to prevent festering in wounds; suggested improvements in use of spongia soporifera to induce narcosis. Died 1298.

BORGSTROM, Bengt Ernst Gunnar, Swedish physician; b. Malmö, Sweden, Mar. 2, 1922; s. Ernst Lorenz and Ester (Wahlgren) B.; M.D., U. Lund, 1952; m. Brita-Lundh, Apr. 17, 1948; children—Anders, Per, Maria, Pia, Anna. Asst. prof. physiol. chemistry U. Lund, Sweden, 1952-56, prof., chmn. dept., 1960——; research fellow Swedish Med. Research Council, 1956-58; asso. prof. chemistry Caroline Inst., Stockholm, 1958-60; Rockefeller Found. fellow Johns Hopkins, Balt., also Enzyme Research Inst., Madison, Wis., 1954. Numerous publs. on research on mechanism of digestion and absorption of fats from intestinal tract of human and higher animals. Home: 1 Sjogatan. Office: 39 Solvegatan, Lund, Sweden.

BORGWARDT, Erich, German chemist; b. May 18, 1892; s. Wilhelm and Marie (Hertzig) B.; Ph.D.; m. Elisabeth Junge, Dec. 27, 1927; children—Wilhelm, Uwe. Asst. to Prof. Diels, Superior Sch., 1920-31; chemist S. A. Schering, 1921-33, dir., 1936-61; dir. S. A. Duco, 1949-61. Address: Waldseeweg 36, Berlin 28, Germany.

BORHEGYI, Stephan Francis de, see de Borhegyi, Stephan Francis.

BORICKY, Emanuel, Czech mineralogist; b. Milin, nr. Pribram, Czechoslovakia, Dec. 12, 1840; Ph.D., Prague, Czechoslovakia; became asst. Mineral. Mus. Prague, 1865, conservator, 1869, named asso. prof. mineralogy U. Prague, 1871, prof., 1880. Research and numerous publs. on minerals of Bohemia. Worked on specific crystalline forms produced when minerals are treated with acids to form salts, 1777; leader of petrographic sch. Died Prague, Jan. 26, 1881.

BORING, Edwin Garrigues, Am. psychologist; b. Phila., Oct. 23, 1886; s. Edwin McCurdy and Elizabeth Garrigues (Truman) B.; M.E., Cornell U., 1908, A.M., 1912, Ph.D., 1914; A.M., Harvard, 1942; Sc.D., U. Pa., 1946, Clark U., 1956; m. Lucy May Day, June 18, 1914; children—Edwin Garrigues II, Frank Henry, Mollie Day, Barbara (dec. 1950). Asst. in psychology, Cornell U., 1911-13, instr., 1913-18; prof. exptl. psychology dir. Psychol. Lab., Clark U., 1919-22; faculty Harvard, 1922——, prof. psychology, 1928-56, Edgar Pierce prof. psychology, 1956-57, prof. emeritus, 1957——, dir. Psychol. Lab., 1924-49; Lowell tv lectr., 1956-57; spl. univ. lectr. U. London (Eng.), 1959. Recipient Gold medal Am. Psychol. Found., 1959. Phi Beta Kappa vis. scholar, 1958-59; vis. scholar Univ. Center in Va., 1960. Fellow A.A.A.S., Soc. Exptl. Psychologists, Am. Acad. Arts and Scis.; mem. Am. Psychol. Assn. (past pres.), Nat. Acad. Scis., Am. Philos. Soc.; hon. mem. Brit. Psychol. Soc., Soc. Franc. de Psychol., Soc. Espan. de Psychol., Soc. Ital. di Psichol. Sci. Author: A History of Experimental Psychology, 1929; Physical Dimension of Consciousness, 1933; Sensation and Perception in History of Experimental Psychology, 1942; Psychologist at Large, 1961; History, Psychology and Science, 1963; (with R. J. Herrnstein) Source Book in the History of Psychology, 1965; also numerous articles. Cons. editor Basic Books, 1961——; editor Contemporary Psychology, 1956-61. Research on physiology cutaneous sensibility and qualities internal sensation, constancy apparent size receding visual objects, history exptl. psychology. Home: 21 Bowdoin St., Cambridge, Mass. 02138.*

BORISOVA (BEKRYASHEVA), Antonia Georgievna, Russian geog. botanist; b. St. Petersburg, 1903; grad. Geography Faculty, Leningrad U., 1927, finished postgrad. course, 1931; Cand. Biol. Sci., 1935. Asso. Bot. Inst., USSR Acad. Sci., 1927-41, sr. asso., 1941——; mem. expdns. to Rybinsk Uezd, Yaroslavl Guberniya, 1925, Kingisepp Rayon, Leningrad Oblast, 1927-29, So. Altay, Narym Mountains, Lake Zaysan, Saur Mountains and Monrak Mountains (Kazakhstan), 1930, Kopet-Dagh Mountains, 1931, 34, Leningrad Oblast, 1932-33, Magnitogorsk area in So. Urals, 1935, Alma-Ata, Dzharkent, Ili Valley and Dzhungar Alatau in Kazakhstan, 1938, Novosibirsk Oblast, 1943-44. Author: A New Species for the Flora of Central Asia, 1933; From Lake Zaysan to the Narym Mountains, 1935; New Leguminous Plants of the Pamirs and the Shugman, 1936; New Fibrous Plants of the Leguminosae Family, 1938; Tea, its Characteristics and Economic Significance, 1941; co-author; The Flora of Turkmenia, 1960. Specialist in floristic systematics. Address: Botanical Inst., USSR Acad. Sci., ulitsa Popova 2, Leningrad, USSR.

BORMAN, Aleck, Am. biochemist; b. Toledo, July 13, 1919; s. William and Gertrude (Glass) B.; B.S., Toledo, 1941; Ph.D., U. Ill., 1945; m. Roberta Axelrod, June 9, 1946; children—Elizabeth S., Judith L. With E. R. Squibb & Sons, New Brunswick, N.J., 1945——, dir. physiology research Squibb Inst. for Med. Research, 1962——. Fellow A.A.A.S., N.Y. Acad. Scis.; mem. Endocrine Soc., Soc. Exptl. Biology and Medicine, Am. Chem. Soc., Am. Assn. Cancer Research, Phi Beta Kappa, Sigma Xi. Contbr. numerous articles to sci. jours. Developed methods for isolation of Vitamin B12 from fermentation broths; developed method for purification of insulin from hog pancreas glands; discovered that chemically modified steroids can be much more active than naturally occurring hormone, hydrocortisone, many such modified steroids are now in gen. use in med. practice. Home: 42 North Dr., East Brunswick, N.J. 08816. Office: Georges Rd., New Brunswick, N.J.*

BORN, Axel, German geologist, paleontologist; b. Prenzlau, Germany, Feb. 5, 1887; s. Oskar and Emmy (Ponath) B.; studied in Berlin, Hannover, Göttingen (all Germany); doctorate, 1911; m. Grete Steinhaeussler, 1914; 2 daus. Prof. geology and paleontology Technische Hochschule and Bergakademie (mining sch.), Berlin-Charlottenburg, 1926. Contbr. to German jours., annuals. Geol. research work primarily asso. with practical concerns, industry; his geological interpretation of gravity field measurements led to design of isanomal charts of gravitational disturbances of Germany-Iberian peninsula; pioneer of applied geophysics with Kossmat. Died Berlin, Germany, Sept. 1, 1935.

BORN, Ignaz von, see von Born, Ignaz.

BORN, Kendall Eugene, Am. geologist; b. Chester, Ill., Feb. 15, 1908; s. Michael and Margaret M. (Dietrich) B.; B.S., McKendree Coll., 1930; M.S.,

Vanderbilt U., 1931, postgrad., 1931-32; postgrad. Washington U., 1932-35; m. Hazel Gentry, Feb. 15, 1936; children—Michael, Kendall Eugene. Asst. geologist Tenn. Geol. Survey, 1935-39, asso. geologist, 1939-45; asst. state geologist, Tenn., 1945-46; asst. prof. Mo. Sch. Mines, 1946. Mem. Am. Assn. Petroleum Geologists, Geol. Soc. Am., Paleontol. Soc., Soc. Econ. Paleontologists and Mineralogists, Mineral. Soc. Am., Am. Geophys. Union, Am. Inst. Mining Engrs. Contbr. numerous geol. articles to publs. Pioneer researcher in oil, gas, chem. limestone geology of Tenn. Died St. Louis, Sept. 21, 1947.

BORN, Max, physicist; b. Breslau, Germany (now Poland), Dec. 11, 1882; s. Gustav and Margaret (Kauffmann) B.; student Breslau, Heidelberg, Zurich univs.; Ph.D., Göttingen U., 1907; D.Sc. (hon.), Bristol U., Eng., 1928; M.A, Cambridge U, Eng, 1933; D.Sc. (hon.), Bordeaux U., 1948, Oxford U., 1954; Dr. rer. nat., Freiburg U., 1957, Berlin U., Frankfurt U.; LL.D., Edinburgh U., 1957: Dr. ing. (hon.), Stuttgart Tech. U., 1960; D.Sc. (hon.), Oslo U., 1961, Brussels University, 1961; Dr. rer. nat. (honorary), Frankfurt University, 1964; m. Hedwig Ehrenberg, August 2, 1913; children—Irene (Mrs. Newton-John), Margaret (Mrs. Pryce), Gustav V. R. Privatdocent, Göttingen U., 1909, prof., 1921; guest lectr. U. Chgo., 1912; prof. U. Berlin, 1915, Frankfurt University, 1919, Göttingen, 1921; guest lecturer at Massachusetts Inst. Tech., 1925; Stokes lectr. Cambridge U., Eng., 1933; guest prof. Indian Inst. Science, Bangalore, 1935-36; prof. natural philosophy, Edinburgh, Scotland, 1936-53, emeritus, 1953-—. Decorated Stokes medal, Cambridge, 1934; Macdougall-Brisbane and Gunning-Victoria Jubilee prize, Royal Soc. Edinburgh, 1945, 1950; Max Planck medaille, Germany, 1948; Hughes medal Royal Soc. London, 1950; Freedom City of Göttingen, 1953; Nobel prize, (with W. Bothe) 1954; Grotius Medal, Munich, 1956. Member of the Royal Society of Edinburgh, Royal Society of London, 1939; mem. acads. Berlin, Göttingen, Copenhagen, Stockholm, Moscow, Dublin, Am. Acad. Arts and Scis., Nat. Acad. Sci. Author: Dynamik der Kristallgitter, 1915; Die Relativätstheorie Einsteins, 1920; Atom Theorie des festen Zustandes, 1923; Probleme der Atomdynamik, 1926; Elementare Quantenmechanik (with Jordan), 1930; Optik, 1933; Atomic Physics, 1935; The Restless Universe, 1935; Experiment and Theory in Physics, 1943; Natural Philosophy of Cause and Chance, 1949; Kinetic Theory of Liquids (with Green), 1949; Dynamic Theory of Crystal Lattices (with Kun Huang), 1954; Physics in my Generation, 1956; Principles of Optics (with Wolf), 1959; Physics and Politics, 1962; Collected Papers (2 vol.), 1963; and about 300 articles. Helped develop quantum mechanics, which forms the basis of modern atomic and nuclear physics; in quantum mechanics interpreted wave function statistically and (with Jordan and Heisenberg) developed 3 dimensional particle motion; discovered Heisenberg's formalism to be identical with matrix calculus; investigated atomic structure, kinetic theory of fluids, and theory of relativity; gave rise to "uncertainty principle" by showing better experimental results were obtained from conceiving electrons as "statistical probabilities;" provided theoretical base of modern theory of solid state through his work on lattice dynamics; helped formulate Born-Oppenheimer theory of molecules; proposed new type of symmetry in physical systems in his principle of reciprocity; has discussed political responsibilities of scientists and condemned warlike uses of scientific knowledge. Home: Marcardstr. 4, Bad Pyrmont, Germany.

BORNEFF, Joachim Otto, physician, hygienist; b. Coburg, Bavaria, Germany, Oct. 2, 1920; s. Hugo and Frida (Berndsen) B.; student U. Erlangen (Germany), 1939-40, Dr. med., 1946; student U. Würzburg, 1942-45; m. Margarete Sieges, Oct. 19, 1946; children—Marianne, Christine. Research fellow U. Erlangen, 1947, 51, asst. Inst. Hygiene and Bacteriol., 1951-56, privat-dozent, 1956-61; prof., head dept. Inst. Hygiene, U. Mainz (Germany), 1961-—. Recipient Preis der Oberfrankenstiftung, Bayreuth, 1961. Mem. Deutsche Gesellschaft für Hygiene und Mikrobiologie, Deutsche Forschungsgemeinschaft (chmn. kommission für früh-invaliditaet). Contbg. author: Handbuch der Arbeitsmedizin, 1962. Research and publs. on detection cancerogenic compounds in surroundings of men, bacteriol. tests for milk and milk products, biol. effects of fluorescent lamps, infra red radiation effects on man in industry, feeding tests with detergents. Home: 33 am Eselsweg, 65 Mainz, Rheinl-Pfalz, Germany. Office: 65 Mainz, Hygiene-Institüt d. Univ., Augustusplatz, Germany.*

BORNEMISZA, Georg, physician; b. Felvinc, Apr. 14, 1916; s. Julius and Charlotte (Meszaros) B.; Dr. Med., U. Cluj (Rumania), 1939; m. Panna Bornemisza, June 24, 1948; 1 dau, Esther. Asst., 1st surg. clinic U. Budapest (Hungary), 1940-51; docent, med. sch. U. Debrecen (Hungary), 1951-—, also dir. inst. surg. anatomy and operative surgery. Founding mem. European Soc. Exptl. Surgery. Contbr. to med. publs. Research on auto-alloplastic method for tissue substitution, transplantation. Home: 4/b Thomas Mann, Debrecen, Hungary.*

BORNET, Edouard, French botanist; b. Guerigny, France, Sept. 2, 1828; mem. French Acad. Scis. Author: Notes algologiques, 1874-80; Etudes phycologiques, 1878. Discovered (with G. Thuret) impregnation of flowers; determined exact nature of lichens; described reproductive processes of red algae. Died Paris, Dec. 18, 1911.

BöRNMULLER, (Friedrich Nicolaus) Joseph, German botanist; b. Hildburghausen, Germany, Dec. 6, 1862; s. Franz and Meta (Meyer) B.; student hort. inst. Potsdam, Germany, 1880's; Ph.D. (hon.) Jena; m. Frida Amelung, 1895. Made many research trips; insp. bot. gardens, Belgrade, Yugoslavia, 1887-88; dir. Orient Herbarium, Weimar, Germany, 1903-38. Guest of Russian Fedschenko expdn. to E. Turkestan, 1913; worked in German occupied Macedonia, 1917-18. Mem. Turkistanische Wissenschaftliche Gesellschaft Taschkent (hon.). Author: Repertorium specierum novarum regni vegetabilis, 1938. Expert on oriental plants, his collection sold to Bot. Mus., Berlin-Dahlem. Died Weimar, Dec. 19, 1948.

BöRNSCHEIN, Hans, Austrian physiologist; b. St. Pölten, Austria, Feb. 26, 1920; s. Richard and Frieda (Pohl) B.; M.D., U. Vienna, 1945; m. Elsa v. Hlatky-Verbay, Oct. 18, 1944; 1 dau., Christine. Faculty dept. physiology U. Vienna, 1947-—, asso. prof. physiology, 1959-64, prof., head dept. gen. and comparative physiology, 1964-—; vis. scientist NIH, Bethesda, Md., 1955-56. Recipient award City of Vienna, 1954. Mem. Austrian Acad. Scis. (corr.), Austrian Soc. for Pure and Applied Biophysics (past pres.). Research, numerous publs. on neurophysiology of sense organs (vision, hearing, vestibular functions). Home: 3-5 Nikolsdorfergasse, Vienna 1050, Austria.*

BöRNSTEIN, Ernst, German chemist; b. Königsberg, Germany, June 19, 1854; s. Levin and Friederike (Meyer) B.; studied natural scis. in Heidelberg, Germany, from 1873, in Bonn, Germany, from 1874; doctorate, 1877; m. Berta Röhmann, 1886. Became asst. in Strasbourg, Germany (now France), 1879; with chemische inst. Landwirtschaftliche (Agrl.) Hochschule, Berlin, Germany, 1883-84; became mem. faculty Technische Hochschule, Berlin, 1903, prof., 1920. Author: Zersetzung der Steinkohle, 1906; Einführung in die Chemie und Technologie der Brennstoffe, 1926. Concerned with solid fuels; wrote definitive work on low temperature tar (importance not recognized until shortages in World War I); also wrote on fuel derivative by-products including dyes. Died Berlin, Germany, Feb. 21, 1932.

BöRNSTEIN, Richard Leopold, German meteorologist; b. Königsberg, Germany, Jan. 9, 1852; s. Levin and Friederike (Meyer) B.; studied in Heidelberg; in Göttingen from 1870, doctorate from Göttingen; m. Helene London, 1879; 3 sons, 1 dau. Asst. to G. H. Quinckes, Heidelberg, Germany; went to Landwirtschaftliche (agrl.) Akademie in Proskau, (transferred to Berlin as Landwirtschaftliche Hochschule 1881) 1878; built meteorol. sta., Berlin. Author: Leitfaden der Wetterkunde, 1913; (with H. Landolt) Physikalisch-chemische Tabellen, 1883. Invented wind pressure gauge, 1883, improved rainfall tabulation; pioneer of modern weather forecasting and was instrumental in devel. of public weather service; in R. Assmann and R. Suring's balloon expdns. in 1890's, conducted atmospheric electricity observations (then a new science). Died Berlin, Germany, May 13, 1913.

BORODIN, Alexander Porfiryevich, Russian chemist; b. St. Petersburg (now Leningrad), Russia, Nov. 11, 1833; student of Erlenmeyer, Sr.; M.D., 1858; became prof. chemistry Medico-Surgical Academy St. Petersburg, 1864; also well-known composer; promoted higher education of women; discovered way of obtaining a halide containing one less atom of carbon from an organic acid; research on derivatives on benzidine and valeraldehyde; prepared fluorobenzene. Died St. Petersburg, Feb. 14, 1887.

BOROVIK, Yevgenii Stanislavovich, Russian physicist; b. Petrograd, 1915; grad. Leningrad Poly. Inst., 1937; D.Physico-Math. Sci. Prof., Kharkov U.; lab. dir. Physicotech. Inst., Ukrainian Acad. Sci., Kharkov. Mem. Ukrainian Acad. Sci. (corr.). Research and publs. on thermal conductivity of rarefied gases in wide range of temperatures and pressures, behavior of electrons in metals under strong elec. and magnetic fields. Address: Kharkov University, Universitetskaya ulitsa 16, Kharkov, Ukraine SSR, USSR.

BOROWITZ, Sidney, Am. physicist; b. Bklyn., June 12, 1918; s. Morris and Rose (Cohen) B.; B.S., Coll. City N.Y., 1937; M.S., N.Y.U., 1941, Ph.D., 1948; m. Ruth Aaron Meyer, June 20, 1943; children—Michael Joseph, Elizabeth Ann. Jr. physicist Navy Dept., 1941-42; engr., sect. chief engring. Western Elec. Co., 1942-45; editor Public Domain mag., 1945-46; instr. N.Y.U., 1946-48, Harvard, 1948-50; mem. faculty N.Y.U., 1950-—, prof. physics, 1959-—, chmn. dept., 1961-—, asso. dir. electromagnetics div., Courant Inst. Math. Sci., 1957-59. Cons. panel lasers Nat. Acad. Sci., 1963; mem. radiation weapons analysis study group USAF, 1960-63. John F. Kennedy Meml. fellow, 1965. Fellow Am. Phys. Soc.; mem. Phi Beta Kappa, Sigma Xi. Author: (with Arthur Beiser) Essentials of Physics, 1966. Research on structure of atoms and nuclei, interaction of atoms with light and charged particles. Home: 900 W. 190th St., N.Y.C., 10040.*

BORREL, Amédée, French physician; b. Cazouls, France, 1867; M.D., Pasteur Inst. Asso. with Pasteur Inst.; prof. bacteriology and hygiene U. Strasbourg (France). Recipient Prix Albert 1st de Monaco for body of work Acad. Medicine Paris, 1929. Contbd. articles to publs. Pasteur Inst., French Acad. Scis. Investigated cancer, cellular div. of neoplasms, action of glycogen on tumors, exptl. Tb, bacteriophages, plague; developed (with Calmette) vaccine in which bacilli were killed by formaldehyde or heat for use against bubonic plague, 1895; diagnosed sheep-pox, 1898; believed that virus infected by parasite caused cancer. Died Cazouls, Sept. 14, 1936.

BORRELLY, Louis-Alphonse-Nicolas, French astronomer; b. Roquemaure, France, Dec. 8, 1842; studied at Marseille (France) Obs.; head astron. research at stas. of Valence, Orange, and Barcelonnette. Research on flying stars, 1869-71, intramercurial spaces (led to conclusion that no planets come together there); discovered a number of variable stars, nebulae, comets, 1873-1903. Died July 1926.

BORRER, William, Brit. botanist; b. Henfield, Eng., June 13, 1781; traveled throughout Britain; tried to cultivate all important Brit. species, also hardy exotic plants; justice peace, Sussex, Eng. Several plants, including genus Borreria (lichen) of Acharius named for him. Fellow Royal Soc., 1835, Linnean Soc., Wernerian Soc. Author: (with Dawson Turner) Lichenographia Britannica. Contbd. descriptions of species Myosotis, Rosa, most of Rubus to British Flora (W. Hooker), 1830, also some descriptions in supplement to English Botany. Died Jan. 10, 1862.

BORRI, Gioseppe Francesco, Italian surgeon, b. Milan, Italy, circa 1620; ed. Jesuit seminary. Went to Rome, later returned to Milan; condemned as heretic by Spanish Inquisition; studied chemistry and medicine, Strasbourg, France; traveled to Hamburg, Germany and Copenhagen, Denmark, then Austria where he was arrested, condemned to imprisonment for life. Author: La Chiare del Gabinetto del Cavagliere Gioseppe Francesco Borri Milanese, 1681. Studied in chemistry. Died in Castle St. Angelo, Aug. 20, 1695.

BORRICHIUS, Olaus (Borch, Oluf), Danish physician, chemist; b. Borch, Jutland, Denmark, 1626; studied at U. Copenhagen from 1644; travelled for six years; prof. chemistry, botany, Copenhagen, Denmark, 1666; Author: De ortu et progressu chemiae (most important 17th century history of chemistry), 1668; Metallische Probier-Kunst, 1680; Conspectus Scriptorium Chemicorum Illustriorum, 1697. Entered into controversies with Densing, Kircher, and Conringius over the antiquity of chemistry. Died Copenhagen, 1690.

BORRIE, John, New Zealand thoracic surgeon; b. Dunedin, New Zealand, Jan. 22, 1915; s. William Henry and Helen I. P. (Whyte) B.; M.B., Ch.B., U. Otago, 1938; Ch.M., F.R.C.S., F.R.A.C.S.; m. Mavis Helen Merrifield, Mar. 4, 1950; children—Michael John, Philip Merrifield, Louise Anne. Anatomy demonstrator U. London Middlesex Hosp., 1945; surg. registrar Southend-on-Sea, Eng., 1946; New Zealand Nuffield surg. fellow Gen. Infirmary, Leeds, 1947, U. Edinburgh, 1948; asst. thoracic surgeon Newcastle-upon-Tyne, Eng., 1948-51; Hunterian prof. Royal Coll. Surgeons, Eng., 1951; thoracic surgeon Green Lane Hosp., Auckland, New Zealand, 1952; 1st sr. regional thoracic surgeon Dunedin and Wakari Hosps., Dunedin, 1952-—; sr. lectr. U. Otago, 1952. Rockefeller fellow in heart surgery Houston and Mayo Clinic, 1960; sec. med. postgrad. com. U. Otago, 1954-—; sec. New Zealand Postgrad. Med. Fedn., 1954-62. Decorated mem. Order Brit. Empire, 1946; Jacksonion prize Royal Coll. Surgeons, Eng., 1952. Mem. Pan-Pacific Surg. Assn., Cardiac Soc. Australia and New Zealand, Thoracic Surg. Soc., Royal Australian Coll. Surgeons, Soc. Thoracic Surgeons of Gt. Britain and Ireland, Royal Soc. Medicine, London. Author: Emergencies in Thoracic Surgery, 1958; Lung Cancer, Surgery and Survival, 1965; also numerous articles. Research on disorders of heart, esophagus, lung, hydatid disease and surg. treatment; lung transplantation; tracheal stenosis in infancy. Home: 37 Newington Av. Office: Dunedin Hosp., Dunedin, New Zealand.*

BORRUS, Christopher, astronomer, mathematician; b. Milan, Italy, circa 1583. Became Jesuit, 1601; missionary, Macao and Cochin, China, 1618-22; tchr. math. U. Coimbra (Portugal); Cistercian, from 1632. Author: Rebatione della nuova missione delli pp. d.c.d.g. al regno della Cocincina, 1631; Collecta astronomica ex doctrina P. Christophori Borri mediolanensis ex Societate Jesus de tribus caelis aereo sydereo empyreo issu et studio domini Gregorii de Castelbranco. Studied magnetic variation of compass, determining longitude at sea, magnetic charts of Atlantic and Indian oceans; wrote on ancient astronomy, aerial, sidereal and empyrean heavens. Died Rome, Italy, May 24, 1632.

BORSCHE, Walther Georg Rudolf, German chemist; b. Stassfurt-Leopoldshall, Germany, May 31, 1877; s. Georg and Elisabeth (Dann) B.; m. Marianne Furbringer, 1913; 5 sons. Worked for short time with A. v. Baeyer, Munich, Germany; at organic dept. univ.

inst. in Göttingen under Otto Wallach, 1899-1926; taught tech. chemistry from 1912; to Frankfort, 1926; became dir. inst. for organic chemistry, Frankfort, 1935. Contbr. circa 230 publs. in Chemische Berichte and Liebig's Annalen der Chemie, 1898-1952. Explained constitution of meta- and isopurpuric acid; multi-nuclear condensed systems with heterocyclic rings. Died Frankfort/Main, Germany, Mar. 17, 1950.

BORSELLINO, Antonio, Italian physicist; b. Reggio Calabria, July 11, 1915; s. Giovanni and Lina (Barreca) B.; Ph.D. in phys. sci.; m. Jole De Rosa, Aug. 8, 1944; children—Giuliana, Giovanni. Mem. Italian Soc. Physics, Am. Phys. Soc., Am. Biophys. Soc. Research and publs. on cosmic rays, biophysics, statistics, electrodynamics. Home: via M. Corno, 68. Office: viale Benedetto XV 5, Genoa, Italy.

BORSOS, Tibor, immunochemist; b. Budapest, Hungary, Mar. 12, 1927; s. Edmund and Anna (Borsos) Borsos-Nachtnebel; came to U. S., 1949, naturalized, 1956; student Creighton U., 1951-52; B.A., Catholic U. Am., 1954; Sc.D. Sch. Hygiene and Pub. Health, Johns Hopkins, 1958; m. Ruth Moser, July 17, 1950; children—Michael Bela, David Julian. Nat. Cancer Inst. fellow Johns Hopkins Sch. Medicine, 1956-58, research asso., Nat. Heart Inst. fellow, 1958-60, asst. prof., NIH Research Career Devel. fellow, 1960-62; sr. investigator Nat. Cancer Inst., NIH, Bethesda, Md., 1962-66, head immunology sect., 1966—. Mem. Am. Assn. Immunologists. Research and publs. on immune reaction on molecular basis; showed that action of complement occurs at single sites on cell surfaces; pioneered purification of complement components and elucidation of their mode of action; invented new complement fixation test. Home: 4703 Edgefield Rd., Bethesda 20014. Office: Nat. Cancer Inst., NIH, Bethesda, Md. 20014.*

BORST, Lyle Benjamin, Am. physicist, educator; b. Chgo., Nov. 24, 1912; s. George William and Jean Carothers (Beveridge) B.; A.B., U. Ill., 1936, A.M., 1937; Ph.D., U. Chgo., 1941; m. Ruth Barbara Mayer, Aug. 19, 1939; children—John Benjamin, Stephen Lyle, Frances Elizabeth. Research asso. U. Chgo., 1940-43; sr. physicist Clinton Labs., Oak Ridge, 1943-46; chmn. dept. physics Brookhaven Nat. Lab., Upton, N.Y., 1946-51; prof. U. Utah, 1951-54; chmn. dept. physics, prof. N.Y. U., 1954-62; prof. physics State U. N.Y. Buffalo, 1962—. Cons. atomic power, 1946—; cons. AID, S.Am., 1962. Fellow Am. Phys. Soc. (chmn. vis. scientist program in physics 1960-63); mem. Am. Civil Liberties Union (nat. bd. 1959-63), Fedn. Am. Scientists (chmn. 1950-51), Sigma Xi (nat. lectr. 1964). Supervised constrn. and operation Brookhaven reactor; research on neutron diffraction, 1945-54, statis. mechanics liquid helium, 1962—. Home: 17 Twin Bridge Lane, Williamsville, N.Y. 14221. Office: State U. N.Y., Buffalo, N.Y. 14214.*

BORST, Maximilian, German pathologist; b. Würzburg, Nov. 19, 1869; s. Ferdinand and Christine (Oderaine) B.; M.D., Univ. Würzburg, 1892; m. Lucrezia Wimmer (d. 1911); m. 2nd. Maria Margarete Fronius, 1926; 1 son. Joined faculty Univ. Würzburg, 1897; prof., Medical Academy, Cologne, 1904; Univ. Göttingen, 1905; Univ. Würzburg, 1906; Univ. Munich, 1910—; dir. of rebuilding Pathological Institute, Univ. Munich, 1928-30. Mem., Bavarian Acad. Scis., numerous other German and foreign socs. Author: Pathologische Histologie, 1922; Die Lehre von den Geschwülsten (2 vol.), 1902; Allgemeine Pathologie der malignen Geschwülste, 1924. Research in pathology of growth and development, regeneration and transplantation; morphology of tumors; cell theory of cancer. Died Munich, Germany, Oct. 19, 1946.

BORSUK, Karol, Polish mathematician; b. Warsaw, Poland, May 8, 1905; s. Marian and Zofia (Maciejewska) B.; Master, U. Warsaw, 1927, Ph.D., 1930, Habilitation, 1934; m. Zofia Paczkowska, Apr. 26, 1936; children—Elisabeth Bieniewska, Maria Magdalena Bialynicka. Faculty, U. Warsaw, 1929—, prof. math., 1946—, dir. Math. Inst., 1952-64; mem. Inst. for Advanced Study, Princeton, N.J., 1946-47; vis. prof. U. Cal. at Berkeley, 1959-60, U. Wis., Madison, 1963-64; vice dir. Inst. Math., Polish Acad. Scis., 1956—. Corr. mem. Polish Acad. Scis., Bulgarian Acad. Scis. Author: Geometria Analityczna Wielowymiarowa, 1950; (with Wanda Szmielew) Foundations of Geometry, 1959; Theory of Retracts, 1966; also numerous articles. Editor, Rozprawy Matematyczne, 1952-66; vice editor Fundamenta Mathematicae, 1952—. Introduced idea of retract and devel. of theory of retracts; proved (with S. Ulam) antipodal theorem; introduced cohomotopy groups; discovered relation between decomposition of Euclidean n-dimensional space by compacta and mappings of compacta into (n-1)-dimensional sphere; studies on uniqueness of decomposition into Cartesian products. Home: 63 Filtrowa, Warsaw, Poland.*

BORTELS, Carl Louis Hermann, German microbiologist; b. Göttingen, Germany, Feb. 4, 1902; s. Heinrich and Mathilde (Hecke) B.; ed. U. Göttingen, U. Munich; Ph.D.; m. Ilse Catosie, 1929; children—Dietrich, Ingrid, Gisela, Jörg, Stefan. Asst. scientist, superior sci. cons., dir., prof. Mem. German Center Hygiene and Microbiology, Internat. Com. Biometrics, German Assn. Geology, Microbiology Center of Berlin. Author: Über die Beteutung von Eisen, Zink und Kupfer für Mikroorganismen; Molybdän als Katalysator der biologischen Stickstoffbindung; Öber Beziehungen zwische epidemiologischen und meteorologischen Geschehen unter besenderer e perimentaller Berücksichtigung der Inhiberwirkung; Das gefrieren unterkühleten Wassers in Beziehung zu interdiurnen Luftdruckänderingen und zur Solaraktivität. Home: Forbacherstrasse 17, 1 Berlin-West 37. Office: Königin Luisestrasse 19, 1-Berlin-West 33, Germany.

BORTHWICK, Harry Alfred, Am. botanist; b. nr. Mpls., Jan. 7, 1898; s. Alfred E. and F. Estella (Humphrey) B.; student U. Minn., 1917-19; A.B., Stanford, 1921, M.A., 1924, Ph.D., 1930; m. Myrtis V. Hall, June 4, 1923; 1 son, Howard Hall. Staff, U. Cal. at Davis, 1922-36; morphologist Plant Industry Sta., U. S. Dept. Agr., Beltsville, Md., 1936-44, botanist, 1944-48, plant physiologists, 1948—. Recipient Leonard H. Vaughan award, 1957; Distinguished Service award U. S. Dept. Agr., 1959, Hoblitzelle Nat. award in agrl. sci.; Norman Jay Colman award Am. Assn. Nurserymen, 1962; Golden Plate award Am. Acad. Achievement, 1964. Mem. Am. Soc. Plant Physiologists, Washington Bot. Soc., Washington Acad. Sci., Nat. Acad. Sci., Bot. Soc. Am. (Merit award 1963), Am. Soc. Hort. Sci., A.A.A.S., Biol. Soc. Washington. Contbr. numerous articles to tech. jours. Research on effect light on biol. response plants, isolation and purification plant pigment phytochrome; discovered that photoperiodic control arises from reversible photoreaction which not only regulates flowering many species but also germination light-sensitive seed, etiolation. Home: 4606 Beechwood Rd., College Park, Md. 20740. Office: Agr. Research Service Plant Industry Sta., U. S. Dept. Agr., Beltsville, Md. 20705.*

BORTKIEWICZ, Ladislaus von, see von Bortkiewicz.

BORUKAEV, Ramazan Aslanbekovich, Russian geologist; born January 23, 1899; graduated Leningrad Mining Institute, 1931. Leader of geological survey teams, chief engineer, head Kazakhstan Geol. Bd., 1931-38; with Kazakhstan br. USSR Acad. Sci., 1938—; dep. dir. Inst. Geol. Sci., Kazakhstan Acad. Sci., 1940, dir., 1964—, acad. sec. dept. mineral resources (now dept. universal and terrestrial sci.), 1956—. Mem. Kazakhstan Acad. Sci. (Presidium mem. 1956—). Author: The Prepaleozoic and Lower Paleozoic Eras in Northeastern Central Kazakhstan, 1955. Mem. editorial bd. Vestnik Kazakhstan Acad. Sci.; exec. editor Conference on the Unification of Stratigraphical Charts of the Prepaleonzoic and Paleozoic Eras in Eastern Kazakhstan, 1960. Research on regional geology, stratigraphy, tectonics, vulcanism and metallogeny, helped compile geol. maps of USSR and Kazakhstan. Address: Kazakhstan Acad. Sci. Alma-Ata, Kazakhstan SSR, USSR.

BORY DE SAINT VINCENT, Jean Baptiste George Marie, French naturalist, geographer; b. Agen, France, July 6, 1778. Naturalist with Baudin's voyage around world, 1800; visited Maritius, Réunion, St. Helena, and Canary Islands, 1801-02; employee Ile de France; served in campaigns of Empire; present at battles of Ulm, Austerlitz, and Waterloo; exiled from France until 1819; head sci. expdn. to Moree, Greece, 1829, later to Algeria, 1839. Mem. French Acad. Scis. Author: Essai sur les îles Fortunées . . . , 1802; Voyage dans les quatre principales îles des mers d'Afrique . . . , 1803; Voyage dans les quatre principales îles des mers d'Afrique, 1804; Voyage souterrain ou Description du plateau de Saint-Pierre de Maetricht, et de ses vastes cryptes . . . , 1821; Voyageur en Espagne, 1823; l'Homme; De la materière sous les rapports de l'Historie naturelle, 1824; Essai zoologique . . . , 1827; Histoire et description des îles de l'océan, 1839. Editor: Dictionnaire classique de l'histoire naturelle. Contbr. articles on natural history to Encyclopédie (Didot). Explored and mapped Réunion; developed doctines of Buffon and Lamarck; concluded that character of species is determined by heredity, also that there is less variation of species in older lands and continents than in newly formed ones. Died Paris, Dec. 22, 1846.

BORY, Louis Pierre, French physician; b. Sète (Hérault), June 20, 1880; s. Honoré and Marie (Eymond) B.; M.D., M.A. in Sci., U. Paris; m. Marcelle Berthe Bertrand; children—Madeleine, René, Philippe. Former intern Paris Hosps.; former instr. med. sch.; in charge clinic of med. sch. at Saint-Louis Hosp. Named laureat Acad. Medicine. Mem. Soc. Therapeutics (hon. pres.), Soc. Dermatology and Syphilology. Research and publs. on dermatology and syphilology, sulphur therapeutics and transpulmonary aerosols, cryotherapy of tumors and cutaneous cancer, comparative pathology. Address: 3, rue Gounod, Paris, France.

BORZENKOV, Jacob Andreevich, Russian anthropologist; b. Russia, 1825; prof. U. Moscow, 1870; author: Lectures on Comparative Anatomy, 1884. Died 1883.

BOSANQUET, Robert Carr, English archeologist; b. London, Eng., June 7, 1871; s. Charles Bertie Pulleine and Eliza (Carr) B.; ed. Trinity College, Cambridge; m. Ellen Sophia Hodgkin, 1902; 2 sons, 4 daus. Asst dir., British School of Archaeology, Athens, 1899; dir., 1900-05; prof. of classical archaeology, U. Liverpool, 1906-20. Dir., Cretan exploration fund; mem., Royal commission on ancient monuments in England; vice pres., Roman Soc. Author: Borcovicium, 1904; Phylakopi (with others), 1904. Excavated Phylakopi on Melos; Palaikastro and Praesos, Crete, 1901-04; Sparta, Laconia, 1905-6; Roman sites in Wales. Died Apr. 21, 1935.

BOSC D'ANTIC, Louis-Augustin-Guillaume, French naturalist, agronomist; b. Paris, France, Jan. 29, 1759; s. Paul and Marie (d'Hangest) B.D'A.; student Dijon, France, botany under Jussieu, Paris; m. Suzanne Bosc, Apr. 9, 1799. Worked in govt. office and other public offices, Paris, from 1777-93; fell from favor because of Girondist sympathies, took refuge in Montmorency forest, 1793-96; French consul to U. S., 1797-99; apptd. administrator of prisons, 1799; insp. nurseries, gardens of Versailles, France, from 1803; prof. botany Jardin des Plantes. Mem. French Acad. Scis., Acad. Medicine, Soc. Agr. Author: Histoire naturelles des coquilles, 1797; L'histoire naturelle des vers, 1800; Encyclopedie methodique d'agriculture, 1813; Histoire naturelles des crustacés, 1828; Dictionnaire raisonné d'agriculture; Dictionnaire d'histoire naturelle. Research in botany. Cons. in agr. to Tsar of Russia, Emperor of Austria, 1814. Died Paris, July 10, 1828.

BOSC D'ANTIC, Paul, French mfr., physician; b. Tarne, France, July 8, 1726; s. Pierre Bosc D'Antic; student medicine Montpellier, France, physics Paris, France; m. Marie d'Hangest; 1 son, Louis-Augustin-Guillaume; m. 2d, Mlle. Lallemand; 1 son, Joseph-Antoine. Physician, Netherlands, 1755-57; asso. of Reamur, Paris, 1757; mfr. mirrors, glass, chinaware, French provinces, 1758-76; royal physician to Louis XV, Paris, from 1776. Mem. French Acad. Scis. Author: Oeuvres, 1780. Perfected glass making technique, studied cause of bubbles, manufacture of white glass, bottle glass. Died Paris, June, 1784.

BOSCH, Carl, German chemist; b. Cologne, Germany, Aug. 27, 1874; s. Carl and Paula (Liebst) B.; ed. Charlottenburg, Germany, 1894-96; Ph.D., Leipzig, Germany, 1898; m. Else Schilbach, 1902; 1 son, 1 dau. Became chemist Badische Anilin und Soda Fabrik, 1899; general manager, 1919; creator exptl. sta. for prodn. ammonia, Oppau; succeeded Planck as dir. Kaiser Wilhelm Inst., 1935; chairman of I. G. Farben industrie, 1935-40. Recipient (with Friedrich Bergius) Nobel prize in chemistry, 1931. Studied catalysts; during World War I contributed to manufacture of synthetic gasoline; helped invent and develop chem. high pressure methods; adapted Haber's synthesis of ammonia for comml prodn. explosives and fertilizers. Died Heidelberg, Germany, Apr. 26, 1940.

BOSCH, Robert August, inventor, engr.; b. Sept. 23, 1861; ed. in U. S. with Siegmann Bergmann and Thomas Edison; 3 daus., 1 son. Employee, Siemens works, Woolwich, Eng.; mfr., inventor magnetos, also spark plugs, lamps, horns, oil pump devices, Stuttgart, Germany. Recipient Eagle plaque German Republic. Died Stuttgart, Mar. 12, 1942.

BOSCOVICH, Ruggiero Giuseppe, mathematician, physicist, astronomer; b. Ragusa, Dalmatia, May 18, 1711; ed. Collegium Romanum. Mem. Soc. of Jesus; apptd. prof. math. Collegium Romanum, 1740, at Pavia, Italy, 1764; with Vatican suppression of Jesuits became naturalized Frenchman, 1773; dir. marine optics French Navy; in Bassana to supervise printing part of his writings, 1783. Tchr. Milan, Italy, from 1783. Fellow Royal Soc., 1761. Author: De inequalitate gravitatis in diversis terrae locis, 1741; De determinanda orbita planetae, 1749; Elementa universae matheseos, 3 vols., 1754; Philosophiae naturalis theoria, 1758; Opera pertinentia ad opticam et astronomiam, 5 vols., 1785. Made valuable astron. observations; anticipated creation of achromatic telescope in expts. and observations of transit of Venus, 1761; popularized Newton's law of gravitation; helped remove hostility to Copernican system; worked on gen. problem of determining comet orbits; obtained data for defining spherical shape of earth on voyage to Brazil and Ecaudor; planned observatories at Collegium Romanum and Coll. of Brera, Milan; suggested and supervised draining of Pontine marshes; recommended use of iron bands to remedy cracking of St. Peter's dome; made archeol. studies in Poland, also Constantinople; invented ring micrometer, other instruments; evolved molecular theory of matter, proposed that ultimate particles of matter might be nothing more than centers of force, forces being constituent elements of matter. Died Milan, Feb. 13, 1787.

BOSE, Ajay Kumar, chemist; b. Silchar, India, Feb. 12, 1925; s. Abinash C. and Amita (Chanda) B.; B.S., U. Allahabad (India), 1944, M.S., 1946; Sc.D., Mass. Inst. Tech., 1950; M.Eng. (hon.), Stevens Inst. Tech., 1963; m. Margaret Logan, Sept. 13, 1950; children—Narayan, Ranjan, Indrani, Indira, Krishna, Rajendra. Research fellow Harvard University, Cambridge, Massachusetts, 1950-51; lecturer and assistant professor at Indian Inst. Tech., Kharagpur, 1950-56; research asso. U. Pa., 1956-57; research chemist Upjohn Co., Kalamazoo, 1957-59; asso. prof. Stevens Inst. Tech., Hoboken, N.J., 1959-61, prof. chemistry, 1961—; cons. numerous indsl. firms. Recipient Ward-Vidyant Gold medal, Jubilee Gold medal U. Allahabad, 1946, Meghnad Saha Meml. prize for popular sci. writing India, 1957. Mem. Am. Chem. Soc. (councillor). Contbr. chpt. to Interpretive Spectroscopy,

1965. Contbr. numerous articles to profl. jours. chemistry of natural products, biosynthesis, mass spectroscopy, synthesis of steroids and compounds related to penicillin. Home: 248 Morris Av., Mountain Lakes, N.J. 07046. Office: Stevens Inst. Tech., Castle Point Sta., Hoboken, N.J. 07030.*

BOSE, Akshayananda, Indian physicist; b. Dacca, India, June 14, 1911; s. Ramesh Charan and Hemnalini (Ray) B.; B.Sc. with honors in Physics 1st in Class I, Dacca U., 1932, M.Sc. 2d in Class I, 1933, D.Sc., 1945; m. Chameli Roy, Mar. 4, 1942; children—Malini, Ketaki, Kumardeb. Research fellow Indian Assn. for Cultivation Sci., Calcutta, 1933-38, research asst., 1942, research officer, 1947-54, reader physics, 1954-56, prof. physics, 1956—; prof. physics St. Edmund's Coll., Shillong, Assam, India, 1938-42; research asst. Allahabad U., 1942-45; prof. physics Jagannath Coll., Dacca, also part time lectr. Dacca U., 1945-47. Fellow Nat. Inst. Scis. India, 1956. Fellow Indian Phys. Soc., Royal Soc.; mem. Indian Assn. for Cultivation Sci. Sec. editorial bd. Indian Jour. Physics, 1964—. Research and numerous publs. on crystalline electric fields in paramagnetic crystals of salts of iron group elements; showed existence of effect of charged atoms outside ligand cluster upon magnetic behavior of paramagnetic ions in crystal lattice; changes in ligand fields from salt to salt of same paramagnetic ion and with temperature; developed and invented new types of magnetic balances, cryostats, and cylindrical metal dewars with reduced neck length. Home: Kamdahari, Post Office Garia, 24 Parganas, West Bengal. Office: Indian Assn. for Cultivation Sci., Jadapur, Calcutta-32, India.*

BOSE, Amarendra Nath, Indian pharmacologist; b. Patna, India, Nov. 8, 1910; s. J. C. and Hemlata (Ghosh) B.; student Calcutta (India) Med. Coll., 1928-35; M.B., Calcutta U., 1935, Biochem. Stand. Lab. (Cent. Drugs Lab.) 1938-39, D.Phil., 1958; postgrad. Sch. Tropical Medicine, Calcutta; m. Bela Dutta, May 10, 1938; 1 son, Pradip. Clin. pathologist S. Calcutta Clin. Lab., 1937-38; pharmacologist research dept. Bengal Immunity Co., Ltd., Calcutta, 1938-47; head dept. pharmacology Bengal Immunity Research Inst., Calcutta, 1947—, supt., 1952—; divisional surgeon 21st Satya Bhama div. St. Johns Ambulance Brigade, 1938—. Mem. Immunity Sci. Assn. (sec.), Assn. Physiologists and Pharmacologists India (past sec. Calcutta br.), Indian Med. Assn., Physiol. Soc. India, Indian Sci. Congress Assn. Research, numerous publs. on chemotherapeutics of sulphonamides, synthetic antimalarials, filaricidals, amoebicides, oral hypoglycaemic, agts., synthetic antispasmics, protein hydrolysates; studies on standardization of liver extracts and assay of vitamin B12 in liver; established new chemotherapeutic agts. in field of sulfa, antispasmodic and antimalarial therapy. Home: 39 Acharya Jagadish Bose Rd., Calcutta, West Bengal, India.*

BOSE, George Matthias, German physicist; b. Leipzig, Germany; Sept. 22, 1710; became assessor Faculty Arts Leipzig, 1727; apptd. prof. physics, Wittenberg, Germany, 1738. Fellow Royal Soc., 1757. Author: De obstetricum erronibus a medico forensi pervestigandis, 1733; Meteora heliaca, 1754; paper defending Tournefort's system of plant classification. Used isolated conductor to improve elec. machine; demonstrated human body can charge electrically on isolated base and can radiate light, circa 1743. Sent as hostage by Prussians to Magdeburg, Germany, 1760, died Sept. 17, 1761.

BOSE, Jagadis Chandra, Indian botanist, physicist; b. Mymensign, India, Nov. 30, 1858; s. Bhagaban G. Bose; B.A., D.Sc. (hon.), Calcutta, India; M.A., Cambridge (Eng.) U.; D.Sc. (hon), Lahore, Pakistan, also Allahabad, Benares, India; LL.D., Aberdeen, Scotland; m. Abala Dass. Prof. sci. Coll. Presidency, Calcutta, 1885-1915; founder, dir. Bose Research Inst., Calcutta, 1917-37. Delegate, International Science Congress, Paris, 1900; science member deputation to Europe, America, 1907, 14, 19; president Indian Science Congress, 1927; member internat. committee on intellectual cooperation League Nations. Created Knight, 1917. Fellow Royal Society, 1920; member of the Academy of Sciences Vienna (corr.), Societas Scientiarum Fennica (hon.). Author: Response in the Living and Non-Living, 1902; Plant Response; Comparative Electro-physiology, 1907; Irritability of Plants, 1913; The Physiology of the Ascent of Sap, 1923; Life Movements in Plants; The Physiology of Photosynthesis, 1924; The Nervous Mechanism of Plants, 1926; Plant Autographs and their Revelations; Collected Physical Papers, 1927; The Motor Mechanism of Plants, 1928; Growth and Tropic Movements of Plants, 1929. Studied plant physiology; water economy in plants; gave vitalistic explanations; inventor instrument (crescograph) for measuring plant growth; research on electrical radiation; demonstrated with electrical instruments fundamental similarity in responses of plant and animal tissues. Died Giridih, India, Nov. 23, 1937.

BOSE, Raj Chandra, statistician, mathematician; b. Hoshangabad, India, June 19, 1901; s. Protap Chandra and Ushangini (Mitra) B.; B.A. with honors, Punjab U., 1922; M.A. in Applied Math., Delhi U., 1924; M.A. in Pure Math., Calcutta U., 1927; D. Litt., 1947; m. Sandhya Lata Datta, Sept. 15, 1932; children—Purabi, Sipra (Mrs. Maurice Glen Johnson). Lectr. math. Asotosh Coll., India, 1930-34; statistician Indian Statis. Inst., Calcutta, 1934-40; lectr. Calcutta U., 1941-45, head statistics dept., 1945-49; prof. statistics U. N.C., Chapel Hill, 1949—; vis. prof. Columbia, 1947, U. Coll. London, 1955, Case Inst. Tech., 1959-60, U. Geneva, 1962; guest lectr. numerous fgn. univs. Pres. statistics sect. Indian Sci. Congress, 1947. Fellow A.A.A.S., Inst. Math. Statistics, Royal Statis. Soc., Nat. Inst. Scis. India; mem. Internat. Statis. Inst., Am. Math. Soc., Canadian Math. Soc., Math. Assn. Am. Important work includes co-discovery of Bose-Choudhury codes, 1959; numerous papers, publs. of original research in fields of statistics and math., especially combinatorial problems of exptl. design and information theory. Home: 108 Jones St., Chapel Hill, N.C. 27514.*

BOSHAMER, Kurt, German physician; b. Witten-Ruhr, May 11, 1900; s. Paul and Elizabeth (Gordes) B.; ed. univs. Münster, Göttingen, Munich, Greifswald; m. I. Koob, May 11, 1934. Prof., Lena, 1934; physician-in-chief Kwangsi Mil. Hosp., Nanking, China, 1934-36, Evangelical Hosp., Gelsenkirchen, 1936-45; physician-in-chief surg. and urol. clinic at hosp. establishment of Wüppertal, 1945-62; physician-in-chief urology clinic of Wüppertal, 1962. Mem. Soc. Urology, Soc. Fight against Cancer of North Westphalia (hon.), Soc. Surgery of Madrid, Soc. Urology of Italy (corr.), Internat. Coll. Surgeons (v.p.). Author: Lehrbuch der Urologie; Morphologie und Gense der Harnsteine in Kdb. der Urologie, 1962. Address: Heusnerstrasse 401, Wüppertal, Germany.

BOSKOFF, Alvin, Am. sociologist; b. Bklyn., Aug. 28, 1924; s. Benjamin and Pauline (Lazaroff) B.; B.Social Sci., City Coll. N.Y., 1945; M.A. (N.Y. State Regents scholar), Columbia, 1948; Ph.D., U. N.C., 1950; m. Priscilla W. Sutherland, July 19, 1950; children—Katharine Julia, Andrew Daniel, Alexander Julian. Faculty, U. Ill., 1950-51, Drake U., 1951-54, Old Dominion Coll., Norfolk, Va., 1955-58; prof. sociology Emory U., Atlanta, 1958—, acting dir. grad. studies in sociology, 1964—. Mem. bd., chmn. program policy com. Family Counseling Center Atlanta, 1960—. Mem. Am. Sociol. Assn. (council 1966—), So. Sociol. Soc. Author: Sociology of Urban Regions, 1962; (with Howard Becker) Modern Sociological Theory, 1957; (with H. Zeigler) Voting Patterns in a Local Election, 1964; (with W. Cahnman) Sociology and History, 1964. Asso. editor Am. Sociol. Rev., 1963-65. Research in understanding diversity in suburban families as part of urban devel., role of differences in civic responsibility in voting choices; explanation of major social changes in modern urban regions. Home: 1802 Ridgewood Dr. N.E., Atlanta 30307.*

BOSQUILLON, Édouard-Francois-Marie, French physician; b. Montdidier, France, 1744; prof. Greek, Coll. de France. Author: Aphorismes et prognostics d'Hippocrate, 1784. Advocated bleeding; enlarged med. knowledge in France through transls. of works by Hippocrates and Brit. physicians. Died Paris, 1816.

BOSS, Jeffrey Mark Newman, English cell biologist; b. London, Eng., Feb. 19, 1926; s. Philip Reginald and Mary (Newman) B.; M.B., B.S., U. London, 1948, B.Sc., 1949; Ph.D., U. Cambridge (Eng.), 1953; m. Elizabeth Parsons, Mar. 28, 1953; 1 dau., Sarah Jane. Med. Research council research student, U. Cambridge, 1949-52; histologist Middlesex Hosp. Med. Sch., London, 1952-54; Rockefeller fellow Tex. U., also U. Cal. at Berkeley, 1954-55; sci. staff Med. Research Council, Strangeways Lab., Cambridge, 1955-58; lectr. physiology U. Bristol (Eng.), 1958—. Mem. Internat., Brit. socs. for cell biology, Physiol. Soc., Soc. for Exptl. Biology. Contbr. articles to tech. jours. Research on how cells divide and secrete, their behavior when in contact with one another, their reaction to changes in concentration solids in their fluid. Office: Dept. physiology, Med. Sch., U. Bristol, University Walk, Bristol 8, Eng.*

BOSS, Lewis, Am. astronomer; b. Providence, Oct. 26, 1846; s. Samuel P. and Lucinda (Joslin) B.; A.B., Dartmouth Coll., 1870; LL.D., Union Univ., 1902; m. Helen M. Hutchinson, Dec. 30, 1871. Civilian asst. U. S. Northern Boundary Commn., 1872-76; div. Dudley Obs., from 1876, and prof. astronomy Union U. Mem. govt. expdn. to observe total eclipse in Colo., 1878; chief govt. expdn. to observe transit of Venus, Santiago, Chili, 1882; state supt. weights and measures, N.Y., 1883-1906; editor and mgr. Albany Express 1885; dir. Dept. of Meridian Astronomy, Carnegie Instn., Washington, 1906; editor of Astron. Journal, 1909. Recipient Gold medal, Royal Astron. Soc., London, 1905; Lalande prize Acad. Scis., Paris, 1911. Mem. Nat. Acad. Scis. Author: Declinations of Fixed Stars, 1878; Catalogue of 8,241 Stars, Leipzig, 1890 (Astronomische Gesellschaft). Monographs: The Solar Motion, and related papers, 1888; Prize Essay on the Physical Nature of Comets, 1881; Division Correction of the Olcott Meridian Circle, 1896; 179 Southern Stars, 1898; Solar Motion and Related Researches, 1901; Positions and Motions of 627 Standard Stars, 1903; Preliminary General Catalogue of 6,188 Stars, 1910; Catalogue of 1,059 Standard Stars, 1910, others. Died Oct. 5, 1912.

BOSS, Medard, Swiss psychiatrist, psychotherapist; b. St. Gall, Oct. 4, 1903; s. Medard and Klara (Schmid) B.; ed. univs. Zurich (Switzerland), Vienna (Austria); postgrad London, Berlin; M.D. in nervous diseases; m. Gertrude Wissler. Asst., Psychiat. Clinic Burgholzli, Zurich, 1928; 1st asst. Psychiat. Polyclinic, Zurich; chief physician Sanatorium Schloss Knonau; pvt. practice medicine specializing in nervous diseases, 1939—; lectr. Inst. Adult Edn., Zurich; charge lectures on psycho-analysis and psychomatic medicine U. Zurich, 1947—, training analyst psychiat. clinic, 1948—, prof. psychotherapy, 1953—. Pres. Interat. Fedn. Med. Psychotherapy; hon. mem. Indian Psychiat. Soc., corr. mem. Royal Medico-Psychol. Assn. Eng., mem. Swiss Soc. Psychoanlysis. Author: Körperliches Kranksein als Folge seelischer Gleichgewichtsstörungen, 4th edit., 1940; Die Bedeutung der Psychologie für die menschlichen Lebens- und Arbeitsgemeinschaften, 1943; Vom Weg und Ziel der tiefenpsychologischen Therapie, 1948; Sinn und Gehalt der sexellen Perversionen, 1947; Die Grundlagen einer psychosomatischen Medizin, 1949; Der Traum und seine Auslegung, 1953; Indienfahrt eines Psychiaters, 1959. Work on devel. of analytic approach psychopathology and psychotherapy. Address: Theaterstrasse 12, 8001 Zurich, Switzerland.

BOSSARD, James Herbert Siward, Am. sociologist; b. Danielsville, Pa., Sept. 29, 1888; s. John Henry and Augusta Minerva (Oplinger) B.; A.B., Muhlenberg Coll., 1909, D.H.L., 1948; M.A., U. Pa., 1911, Ph.D., 1917; Dorothy M. Lemie, May 14, 1929; children—Barbara, Constance. Prof. history and social sci., Muhlenberg Coll., 1911-17; asso. editor Allentown (Pa.) Morning Call, 1917-18; lectr. sociology Lafayette Coll., 1917-18, prof. sociology and economics, 1918, head dept. economics and govt., 1918-20; asst. prof. sociology Wharton Sch., U. Pa., 1920-25, prof. 1925—, also prof. of sociology, med. sch.; William T. Carter prof. child devel., and dir. William T. Carter Found. for Child Devel., U. Pa., 1938—; vis. prof. sociology, fellow Pierson Coll., Pierson Yale, 1947-48, Tchr. summers at U. Cal., 1929, 30, 33, 35. Pres. West Phila. Community Conf., 1922-1925; dir. social service dept. U. Pa. Hosp., Maternal Health Center, 1929-38; pres. Eastern Sociol. Conf., 1934-35; pres. Pa. Birth Control Fedn., 1936-39; mem. Philadelphia Commn. on Federal Housing Projects. Fellow Soc. Research in Child Devel.; mem. Am. Acad. Polit. and Social Sci., Am. Sociol. Soc. (mem. exec. com.; 1st v.p., 1941); Am. Council Learned Socs. Author: Ritual in Family Living, 1950; Toward Family Stability, 1950; Parent and Child: Studies in Family Behavior, 1953; The Large Family System, 1956; One Marriage, Two Faiths, 1957. Co-author: Successful Marriage, 1947; Marriage, The Family, and Parenthood, 1948; Courtship and Marriage, 1953. Part author: Introduction to Sociology, 1952. Spl. editor and co-author: Annals of Am. Acad. Polit. and Soc. Sci. Contbr. articles to mags. Died Jan. 29, 1960.

BOSSCHA, Johannes, Dutch physicist; b. Breda, Netherlands, Nov. 18, 1831; dir. Delft Poly. Sch., from 1878; mem. French Acad. Scis., 1910, Dutch Soc. Scis. (perpetual sec.); developed Bosscha corollaries, propositions which simplify applications of Ohm's law; investigated speed of sound over short distances; cause of galvanic polarization. Died Heemstede, Holland, Apr. 15, 1911.

BOSSE, Abraham, French mathematician, engraver; b. Tours, France, 1602; studied art, Paris; became tchr. perspective Royal Acad. Painting and Sculpture, 1648, named hon. engraver, 1651, counsillor, 1655, excluded from Acad., 1661; set up rival sch. (soon closed by govt.). Author numerous works on art and engraving, also theoretical writing on perspective inspired by Desargues, including: Traité des manières de graver en douce-taille, 1645; Leçons de géométrie et de perspective pratique, 1648; Traité des pratiques géométrales et perspectives, 1653. Developed original method of copper plate engraving; engraved over 1,400 pieces, represented Fr. civil life and costumes during the reign of Louis XIII. Died Paris, Feb. 14, 1676.

BOSSHARDT, David Kirn, Am. nutritionist; b. Rochester, Minn., Apr. 15, 1916; s. Elmer Harrison and Norma Delta (Kirn) B.; Asso. Sci., Va. Jr. Coll., 1934; B.S., U. Minn., 1938; M.S., Rutgers U., 1940, Ph.D., 1943; m. Dorothy Janet Clark, Feb. 7, 1943; 1 son, Bruce Clark. Research fellow Rutgers U., 1938-43; asst. N.J. Agr. Expt. Sta., 1938-43; research asso. Merck Sharp & Dohme Research Lab. div. Merck & Co., Inc., Rahway, N.J., 1943—. Mem. Am. Inst. Nutrition, Soc. for Exptl. Biology and Medicine. Research, publs. in silage preservation, carotinoid pigments in feed and butterfat, protein, amino acid, energy metabolism in small animals, vitamin requirements in small animals, unidentified growth factors, cholesterol and bile acid metabolism. Home: 41 Saw Mill Dr., Berkeley Heights, N.J. 07922. Office: Merck & Co., Rahway, N.J. 07065.*

BOSSOLASCO, Mario, Italian geophysicist; b. Torino, June 30, 1903; s. Giovanni and Guglielmina (Bertolina) B.; Ph.D. in math. and physics. Dir. Geophys. Sta., Mogadiscio, Italian Somaliland, Africa, 1932-33; prof. geophysics U. Messina (Italy), 1936-45, U. Genoa, 1946. Mem. Italian Soc. Geophysics and Meteorology (founder, pres. 1953—), Internat. Div. Sci. Commns. (pres.). Founder, dir. Geofisica pura e applicata, 1939. Research and over 150 articles on meteorology, geomagnetism, solar radiation, ionosphere, physics of ocean. Home: via Balbi 30. Office:

Institute of Geophysics and Geodetics, University of Genoa, P.O. Box 3145, Genoa, Italy.

BOSSUT, Charles, French mathematician, physicist; b. Tartaras, France, Aug. 11, 1730; ed. Lyon, Paris; prof. hydrodynamics and math. Sch. Engring., Mézières, France, 1752-68; examiner École Polytechnique, 1768; lost positions during French Revolution. Mem. French Acad. Scis. (pres. 1768). Author: Mécanique et dynamique, 1763; Mécanique statique, 1771; Traité d'hydrodynamique, 1772; Géometrie, 1773; Arthmétique, 1774; Cours complet de mathématiques, 7 vols., 1795-1801; Traité theorique et expérimental d'hydrodynamique, 2 vols., 1786-87; Essai sur l'historie des mathématiques, 1802. Editor collection Pascal's works, 15 vols., 1779. Contbd. to Encyclopédie. Investigated water pressure on agitated water wheels; investigated (with d'Alembert and Condorcet) resistance (rheostat) of lead. Died Jan. 14, 1841.

BOSTICK, Warren Lithgow, Am. physician; b. Dallas, July 28, 1914; s. John Benton and Caroline (Hackett) B.; A.B., U. Cal. at Berkeley, 1936; M.D., U. Cal. at San Francisco, 1940; m. Virginia Lee, Aug. 18, 1939; children—W. Alan, Bruce H. Faculty, U. Cal. Med. Sch., San Francisco, 1942-64, prof. pathology, 1954-64; prof. pathology, dean Cal. Coll. Medicine, U. Cal. at Irvine, Cal., 1964—. Recipient Gold Cane award U. Cal. San Francisco, 1940. Mem. Cal. Med. Assn. (past pres.), A.M.A. (delegate since 1959), Phi Beta Kappa, Alpha Omega Alpha. Author: (with Anderson, Johnson) Amebiasis, 1956; also numerous articles. Research on virus cause cancer, Hodgkin's Disease, amebiasis. Home: 1515 E. California Blvd., Pasadena, Cal. 91106. Office: 1700 Griffin St., Los Angeles 90031.*

BOSTICK, Winston Harper, Am. physicist; b. Free Port, Ill., Mar. 5, 1916; s. William Frederick and Alice (Johnson) B.; B.S. in Physics, U. Chgo., 1938, Ph.D., 1941; m. Virginia Halton Lord, June 16, 1942; children—Joel Lord, Verity Jo, Kent Anthony. Teaching fellow U. Chgo., 1938-41; staff mem. Radiation Lab., Mass. Inst. Tech., 1941-46, Research Lab of Electronics, 1946-48; asso. prof. physics Tufts U., 1948-54; staff mem. Lawrence Radiation Lab., Livermore, Cal., 1954-56; prof., head physics dept., Stevens Inst. Tech., Hoboken, N.J., 1956—. Recipient 1st prize Gravity Research Found. Essay Contest, 1961. NSF Postdoctoral fellow, 1961-62. Fellow, Am. Phys. Soc., Natural Food Asso. (pres. N.J. chpt. 1963—), Certified Natural Foods Found. (pres. 1964—), Phi Beta Kappa, Sigma Xi. Producer of plasmoids in lab., galactic-like configurations with plasma in a magnetic field. Demonstrated that some plasmoids are plasma eddies which rotate like rigid bodies. Home: Route 2, Box 987, Parker Rd., Chester, N.J. 07930. Office: Stevens Inst. Tech., Hoboken, N.J. 07030.*

BOSTOCK, John, Brit. physician; b. Liverpool, Eng. 1773; s. John Bostock; M.D., U. Edinburgh (Scotland), 1798; lectr. Guy's Coll. Fellow Royal Soc., 1818 (v.p. 1832); mem. Geol. Soc. (pres. 1826). Author: Remarks on the Nomenclature of the New London Pharmacopocia, 8 vols., 1810; An Elementary System of Physiology, 1824. Gave 1st complete description of hay fever (Bostock's catarrh), 1819. Died Aug., 1846.

BOSTON, L(eonard) Napoleon, Am. physician; b. Town Hill, Pa., Mar. 18, 1871; s. Alfred H. and Bethiah (Bacon) B.; grad. with highest honors Phila. Sch. Anatomy, 1895; M.D., Medico-Chirurg. Coll., Phila., 1896; A.M. (hon.), Ursinus Coll., Pa., 1902; m. Caroline Crandall, Oct. 28, 1905; children—Barbara C., La Barre. Bacteriologist Phila. Hosp., 1898-1901, Ayer Clin. Lab., Pa. Hosp., 1901; instr. in obstetrics Medico-Chirurg. Coll., 1897-99, medicine, 1899-1901, dir. clin. labs., 1901-05, asso. in medicine, 1904, adj. prof. medicine, 1905, prof. phys. diagnosis, 1912-16; phys. to Phila. Hosp. (Blockley), 1905; dir. clin. labs. of research Am. Hosp. for Diseases of Stomach, Phila., 1906; pathologist Frankford Hosp., 1909; prof. phys. diagnosis U. Pa., 1916-17, asso. prof. medicine Post-Grad. Sch., 1919-26, prof. phys. diagnosis Grad. Sch.. 1927; prof. principles and practice of medicine and clin. medicine Woman's Med. Coll. Pa., 1928—. Author: Text Book of Clinical Diagnosis by Laboratory Methods, 1904; (with James M. Anders) Text Book of Medical Diagnosis, 1911, 3d edit., 1925. Died July 4, 1931.

BOSWELL, Percy George Hammall, Brit. geologist; b. Woodbridge, Eng., 1886; s. G. J. Roswell; ed. Royal Coll. Sci., Royal Sch. Mines; D.Sc.; m. Hope Dobell, 1939. Demonstrator geology Royal Coll. Sci., 1914-17, also asso.; George Herdman prof. geology U. Liverpool, 1917-30; prof. geology, head dept. geology, oil tech., mining geology Imperial Coll. Sci. and Tech., U. London, 1930-38, emeritus prof., from 1939. Sci. adviser (geol.) to Ministry of Munitions of War, 1915-19; mem. E. African Archeol. Expdn., 1934-35; geol. adviser Met. Water Bd., 1934-54. Fellow Royal Soc, 1931; fellow or mem. Brit. assn. (sec. 1921-25, gen. treas. 1935-43), Prehistoric Soc. (pres. 1936), Geol. Soc. London (sec. 1932-34, v.p. 1934-36, pres. 1940-41), Geologists' Assn., geol. socs. India, Am., Belgian, Soc. Glass Tech. (founder mem.). Author: American Foundry Practice; Silurian Rocks of North Wales; Resources and Consumption of Water in the Greater London Area; The Muddy Sediments; books on sedimentary rocks, sands for glassmaking and refractory purposes; also papers. Died Dec. 22, 1960.

BOTALLUS, Leonard (Botalld, Botal, Botalli), Italian physician, b. Asti, Sardinian states, circa 1530; studied under Fallopius. Physician to Charles IX and Henry III. Gave 1 of earliest descriptions ductus arteriosus (ductus Botalli), also described foramen ovale of heart, circa 1575.

BOTELHO, Stella Yates, physiologist; b. Japan, Jan. 14, 1919; d. Francis M. and Emily (Evans) Botelho; came to U. S., 1920, naturalized, 1933; B.A., U. Pa., 1940; M.D., Woman's Med. Coll. Pa., 1949. Faculty U. Pa., 1949—, asso. prof. physiology, 1957—; vis. prof. Cambridge (Eng.), 1957-58; clin. neurophysiologist Phila. Gen. Hosp., 1954-63, sect. chief, 1963—; cons. neurophysiology Children's Hosp. Phila., 1963—; cons. basic scis. U. S. Naval Hosp. Phila., 1963—. Mem. Am. Physiol. Soc., Am. Acad. Neurology, Assn. for Research in Nervous and Mental Diseases, Assn. for Research in Ophthalmology, Am. Assn. U. Profs., A.A.A.S., Sigma Xi, Alpha Omega Alpha. Research and publs. in electrophysiology, basic nerve and muscle physiology, neurol. disorders, mechanisms of secretion and control of secretion of lacrimal gland. Home: Rittenhouse Claridge, Rittenhouse Sq., Phila. 19103.*

BOTERO, David, Colombian physician; b. Sonson, Colombia, Apr. 21, 1930; s. Leopoldo and Luisa (Ramos) B.; M.D., U. Antioquia, Medellin, Colombia, 1955; M.P.H., Colombia, N.Y.C., 1957; diploma in tropical medicine and hygiene U. London, Eng., 1960; m. Maria Cristina Gutierrez, Dec. 12, 1958; children—David Andres, Carlos Alejandro, Catalina. Prof. med. parasitology U. Antioquia Sch. Medicine, 1957—, dir. Sch. Lab. Technicians, 1957—, exec. dir. Sch. Medicine, 1966—, head dept. microbiology, parasitology, 1962—; asst. prof. tropical medicine Hosp. Universitario San Vicente de Paul, Medellin. Cons. Internat. Tropical Medicine Tng. Program, La. State U. Recipient U. Antioquia Gabriel Torra Villa prize. Mem. Am. Royal socs. tropical medicine and hygiene, Colombian Soc. Parasitology and Tropical Medicine (pres. 1965-68), Latin Am. Soc. Parasitologists. Research, publs. on tropical medicine, med. parasitology, especially antiparasitic drugs. Home: Carrera 27B, 8A-45. Office: Sch. Medicine, U. Antioquia, Medellin, Colombia.*

BOTHE, Walther, German physicist; b. Oranienburg, Germany, Jan. 8, 1891; s. Friedrich and Charlotte (Bothe) Bothe; Dr. phil. (student of Planck), U. Berlin; m. Barbara Below, 1920 (dec. 1951); 2 daus. Faculty, Physikalisch-Technische Reichsansstalt, Berlin, 1913-30; prof. physics, dir. inst. physics U. Giessen (Germany), 1930-32; dir. inst. physics U. Heidelberg (Germany), 1932-34, prof., 1945-57; dir. inst. physics Max Planck Inst. for Med. Research, Heidelburg, 1934-circa 1945, then supr. Mem. acads. scis. Heidelberg, Göttingen, Saxon Acad. Scis. (corr.). Decorated knight Order Pour le Mérite, peace class, 1952; Planck medal, 1953; Grosses Verdienstkreuz der Bundesrepublik Deutschland, 1954; Nobel prize for physics (with Max Born), 1954. Author: (with others) Geiger-Scheel's Handbook of Physics, 1933; (with W. Gentner and H. Maier-Liebnitz) Atlas of Typical Cloud Chamber Photographs, 2d edit., 1954; numerous sci. papers. Pioneer in the development of nuclear physics; discovered time-coincidence method for measuring particle direction, 1924, also (with H. Geiger) scatter processes and exptl. confirmation of Compton effect, 1923-26; investigated rates of fission reactions, 1927; showed existance and paths of penetrating charged particles in cosmic rays, 1929; discovered (with H. Becker) neutron radiation, 1930; worked on diffusion theory of neutrons and related measurements, 1939-45; responsible for construction of Germany's 1st cyclotron, 1944; discovered and developed method for measuring time with great precision, 1954. Died Heidelberg, Feb. 8, 1957.

BOTHNER-BY, Aksel Arnold, Am. chemist; b. Mpls., Apr. 29, 1921; s. Aksel Conrad and Merle Marie (von Hagen) Bothner-By; student U. Nanking (China), 1939; B.Chemistry, U. Minn., 1943; M.S, N.Y.U., 1947; Ph.D., Harvard, 1949; m. Christine Treuner, Oct. 15, 1949; children—Peter Ole, Anne Sigrun. Scientist, Brookhaven Nat. Lab., 1949-53; fellow Am. Cancer Soc., 1952-53; instr., lectr. Harvard, 1953-58; cons. Retina Found., 1957-58; staff fellow Mellon Inst., 1958—, dir., 1960-61, mem. adv. com., 1962— : Fulbright lectr. U. Munich (Germany), 1962-63; adjunct professor University of Pitts., 1964—. Served with AUS, 1943-45. Mem. Am. Chem. Soc., Schweizerische Chemische Gesellschaft, Sigma Xi. Author papers theoretical organic chemistry. Research on the mechanisms of organic reactions; research on the theory and application of nuclear magnetic resonance. Home: 6317 Darlington Rd., Pitts. 17. Office: Mellon Inst., 4400 5th Av., Pitts. 13.

BOTT, Raoul, mathematician; b. Budapest, Hungary, Sept. 24, 1923; s. Rudolph and Margit (Kovacs) B.; B.Engring., McGill U., 1945, M.Engring., 1946; D.Sc., Carnegie Inst. Tech., 1949; A.M. (hon.), Harvard, 1959; m. Phyllis Hazell Aikman, Aug. 30, 1947; children—Anthony, Jocelyn, Renee, Candace. Came to U. S., 1947, naturalized, 1959. Mem. Inst. Advanced Study, Princeton, 1949-51, 55-57; instr. math. U. Mich., 1951-52, asst. prof., 1952-55, prof., 1957-59; prof. math. Harvard, 1959—. Recipient Veblen prize in geometry, Am. Math. Soc., 1964. Member National Academy of Science, American Math. Soc. (council), American Acad. Arts and Sci. Editor: Topology. Asso. editor: Annals of Math., 1958—. Special research network theory, topology, and geometry. Home: 77 Kirkstall Rd., Newtonville 60, Mass. Office: 2 Divinity Av., Cambridge 38, Mass.

BOTTA, Paul Émile, archaeologist, physician; b. 1802; s. Carlo Giuseppe Guglielmo B.; ed. in medicine. Became physician to Mohammed Ali, 1830; apptd. French consul in Alexandria, 1833; French consular agent in Mosul (now in Iraq), 1840. Author: Monuments de Ninive découverts et décrits par Botta, mesurés et dessinés par E. Flandin, 4 vols., 1847-50. Early pioneer in Assyriology; excavated mound Khorsabad, 1843-46; discovered palace of Sargon; identified Khorsabad as site of Nineveh; found clay tablets at Kuyunjik giving 7th century B.C. dictionary to Class III cuneiform script. Died 1870.

BOTTAZZI, Filippo, biologist; b. 1867; his studies assisted in affirming that proptoplasm is a colloid. Died 1941.

BÖTTCHER, Arthur (or Boettcher), German physician, zoologist; b. Baucke, Germany, July 13, 1831; ed. Dorpat (now Tartu, Estonia), also Berlin (under Virchow), Vienna, Paris; became extraordinary prof. gen. pathology and path. anatomy, 1861, prof., 1862. Author: Über die Entwicklung und den Bau des Ohrlabyrinths, nach Untersuchungen an Säugethieren, 1868; Neue Beiträge zur Literatur des Gehörlabyrinths, 1872. Discovered spermin crystals (Böttcher crystals) in prostatic fluid; discovered and described dark polyhedral cells (Böttcher's cells) found in single layer on basilar membrane, 1868; introduced regressive staining method of overstaining followed by destaining or differentiation with alcohol, 1869. Died 1889.

BOTTEMA, Oene, Dutch mathematician; b. Groningen, Netherlands, Dec. 25, 1901; s. Rinze and Wytske (Hoekstra) B.; student U. Groningen, 1919-24; Ph.D., U. Leyden (Netherlands), 1927; LL.D., U. Leeds (Eng.), 1958; m. Femmina Berendsen, Aug. 21, 1930; children—Willemina (Mrs. Willem La Groz), Rinze. Tchr. secondary schs., 1924-33; dir. secondary schs., 1933-41; prof. math. and mechanics Technol. U. Delft (Netherlands), 1941—, rector magnificus, 1951-59. Author: Elementaire meethunde van het platte vlak, 1938; also numerous articles. Research in projective geometry, especially line geometry; elementary geometry; theoretical mechanics, especially kinematics. Home: 2 Ch. de Bourbonstr., Delft, Netherlands.*

BÖTTGER, Johann Friedrich, German chemist; b. Schleiz, Germany, Feb. 4, 1682; rediscovered method of making true hard porcelain; originator of Dresden China. Died 1719.

BÖTTGER, Rudolph Christian von, see von Böttger, Rudolph Christian.

BÖTTGER, (Carl) Wilhelm, German chemist; b. Leisnig, Saxony, Germany, Oct. 2, 1871; s. Heinrich Hermann and Johanna Josephine (Bernhardt) B.; practical edn. in Chemnitz, Berlin, Switzerland; studied pharmacy in Leipzig, Germany, doctorate, 1897; m. Anna Marie Bernhardt, 1910; 1 son, 1 dau. Asst. to O. Wallach, Göttingen, Germany; asst. until 1937, then dept. head at Physikalisch-Chemische Institut, Leipzig; faculty analytical and phys. chemistry U. Leipzig, 1903; research asso. Mass. Inst. Tech., Boston, 1904-05; became asso. prof. in Leipzig, 1910; hon. prof., 1922. Author: Grundriss der qualitativen Analyse vom Standpunkt der Lehre von den Ionen, 1902. Contbr. numerous articles to chem. jours. and periodicals. Sought to substitute empirical character of analytical chemistry with sci. basis by application of new knowledge of phys. chemistry. Died Hanover, Germany, Oct. 23, 1949.

BÖTTICHER, Gustav Robert Wilhelm Werner, German chemist; b. Zwickau, Mar. 25, 1900; s. Wilhelm and Marg. (Poehl) B.; state exam. for chemistry of foodstuffs; Tech. U. Dresden; Ph.D. in tech. sci.; m. Gisela Tanner, Aug. 22, 1924; children—Wolfgang, Ingrid, Gudrun. Dir., Establishment of Chem. Control of City of Dresden; dir. central sect. Research and Utilization of Mushrooms, Munich. Mem. Soc. German Chemists (group specializing in alimentary chemistry), Soc. Biol. Nutrition. Research and publs. on cider, sci. utilization of mushrooms and fruits of woods. Home: Theodorlindenstrasse 81. Office: Leopoldstrasse 175, Munich 23, Germany.

BOTTIN, Sébastien, French statistician; b. Le Meurthe, France, 1764; became sec. gen. for Bas-Rhin Dist., 1794. Author: L'almanach du commerce de Paris (basis for current Bottin bus. directory), 1801-59; L'annuaire statistique du département du Nord (1st annual statistics on comml. activities of Paris), 1811. Died 1853.

BOTTLINGER, Kurt Felix Ernst, German astrophysicist; b. Berlin, Germany, Sept. 12, 1888; s. Karl and

Frieda (Schmid) B.; student of Karl Schwarzschild and Hugo von Selliger; doctorate, 1912; m. Ilse Kelpe. Became asst. Babelsberg obs., 1919, observer, 1921; named prof., 1927; lectr. astrophysics U. Berlin from 1932. Contbr. articles to Babelsberg obs. publs., astrophysics handbooks, others. Research on giant and dwarf stars; works on photoelectric and photographic measurements of star colors—spectrometry, colorimetry, on interstellar absorption, rotation of Milky Way, on universe and theory of relativity. Died Berlin, Germany, Feb. 19, 1934.

BOTTOMLEY, James Thomson, Irish engr.; b. William Bottomley; ed. Queen's Coll., Belfast, Trinity Coll., Dublin; B.A., M.A., D.Sc.; LL.D. (hon.), U. Glasgow; m. Annie Elisabeth Heap; 1 son; m. 2d, Elisa Jennet (dec. 1913). Asst. to Prof. Thomas Andrews; demonstrator in chemistry, later physics King's Coll., London; Arnott and Thomson demonstrator U. Glasgow, 29 years. Fellow Royal Soc., 1888. Author elementary books on dynamics and hydrostatics, math. tables, also articles. Died May 18, 1926.

BOTTOMLEY, Warwick, Australian chemist; b. Sydney, Australia, Mar. 13, 1924; s. Percy and Grace (Armstrong) B.; B.Sc. with honors, U. Sydney, 1944; Ph.D., U. Western Australia, 1949; m. Edith Patricia Gaynor, Dec. 15, 1950; children—Matthew, Philip Mark, Stephen. Research asst. Western Australian Dept. Indsl. Devel., 1945-49; research scientist div. plant industry Commonwealth Sci. and Indsl. Research Orgn., Canberra, A.C.T., Australia, 1950—; univ. research fellow U. Nottingham, Eng., 1951-53; postdoctoral scholar dept. chemistry U. Cal. at Los Angeles, 1963; research asso. dept. biology Yale, 1964. Mem. Australian, Am. socs. plant physiologists, Australian Biochem. Soc., Royal Soc. Canberra, Am. Chem. Soc. Research, publs. on chemistry of substances isolated from plants including first structure of a leucoanthocyanidin, a new class of natural product; contbns. to biosynthesis of pyrolizidine alkaloids, the effects of light on flavonoid metabolism in plants and partial purification of plant cell div. hormones (cytokinins). Home: 26 Lynch, Hughes, A.C.T. Office: Commonwealth Sci. and Indsl. Research Orgn., Div. Plant Industry, Canberra, A.C.T., Australia.*

BOTTOMORE, Thomas Burton, sociologist; b. Nottingham, Eng., Apr. 8, 1920; s. Thomas Joseph and Margaret (Bacon) B.; B.Sc., U. London, 1943, M.Sc., 1949; m. Mary Kathleen Greasley, Jan. 30, 1953; children—Katherine Helen, Stephen, Eleanor Jane. Fellow Rockefeller Found., 1951-52; lectr., reader sociology London Sch. Economics and Polit. Sci. U. London, 1952-65; head dept. polit. sci., sociology and anthropology Simon Fraser U., Burnaby, B.C., Can., 1965-67, prof., 1968—; prof. sociology U. Sussex (Eng.), 1968—. Editor Current Sociology, 1953-62; English editor European Jour. Sociology, 1960—. Mem. Pacific Sociol. Assn., Assn. Asian Studies, Assn. British Orientalists, Aristotelian Soc., Mind Assn. Author: (with M. Rubel) Karl Marx: Selected Writings in Sociology and Social Philosophy, 1956, Sociology: A Guide to Problems and Literature, 1962, Karl Marx: Early Writings, 1963, Elites and Society, 1964, Classes in Modern Society, 1965, Social Criticism in North America, 1966. Studies, publs. in sociol. theory and the history of social thought; social stratification; the devel. of modern Indian soc. Home: 1419 W. Keith Rd., North Vancouver, B.C. Office: Simon Fraser U. Burnaby 2, B.C., Can.*

BOUCHARD, Charles Jacques, French physician, biologist; b. Montier-en-Der, France, Sept. 6, 1837; intern, Lyons, France, 1857; M.D., Paris, 1866. Head clinic Faculty Medicine Paris, from 1868, mem. faculty from 1869, physician, central bur., from 1870, titular prof. gen. pathology and therapeutics, from 1879. Decorated grand cross Legion Honor. Mem. Acad. Medicine, French Acad. Scis. Author: Les maladies par ralentissement de la nutrition, 1882; Leçons sur les auto-intoxications dans les maladies, 1887; Les microbes pathogènes, 1892; Traité de pathologie générale, 1895-97. Worked in urotoxics; advocated using lab. in conjunction with clinic; stressed usefulness of microbiology to med. studies; used X-rays to diagnose bone diseases, fractures, dislocations; discovered that inefficiency of gastric muscles may cause dilation of stomach (Bouchard's disease); discovered ratio between amount of urine and total solids of urine (Bouchard's coefficient). Died Sainte-Foy-les-Lyon, France, Oct. 28, 1915.

BOUCHARD, Francois, physician; b. Geneva, Switzerland, June 13, 1917; s. Henri and Suzanne (Schneller) B.; student Faculty Medicine, Paris, 1938-45, Inst. Cardiology Mexico, 1948-51; m. Mary Fontayne, Sept. 9, 1954; children—Isabelle, Hélène. Chief hemodynamic lab. clinic cardiology Hosp. Broussais, Paris, 1952—; chief hemodynamic lab. Hosp. Marie-Lannelongue, Paris, 1952—. Decorated chevalier Ordre de la Sante Publique. Mem. Société Francaise de Cardiologie, Société Chilienne de Cardiologie, Société Argentine de Cardiologie. Author: (with P. Soulie) Le Cathétérisme au Micromanometre, 1961; also numerous articles. Research on cardiac disease, intracardiac pressure and sounds. Address: 25 rue de l'Yvette, Paris 16°, France.*

BOUCHARDAT, Apollinaire, French pharmacist; b. Isle-sur-le-Serein, France, July 23, 1806; ed. Paris; intern in pharmacy, 1827, M.D., 1832. Admitted to staff, faculty medicine, U. Paris, 1833, apptd. prof. hygiene, 1852; chief pharmacist Hosp. St. Antoine, then at Hôtel-Dieu, 1834-55. Mem. Fr. Acad. Medicine (pres.). Author: Formulaire magistral, 1840; Formulaire vétérinaire; Traité de la glycosurie, 1875; Traité d'hygiène publique et privée basée sur l'étiologie, 1881; (with A. Delondre) Quinologie. Discovered, measured optic properties of phloridzine, also of amygdaline, salicine, many vegetable alkaloids; studied (with Sandras) ferments; made classic researches on digestion; observed histological changes of pancreas in diabetes; pioneer in prescribing sugar-free diets for diabetics; held contact theory of electrochemistry; established causes of scurvy and suggested remedies. Died Paris, Apr. 7, 1886.

BOUCHARDAT, Gustave, French chemist; b. Paris, June 4, 1842; s. Apollinaire Bouchardat; M.D., 1869; D. ès Scis. Physiques, 1872. Asst. to Berthelot, Coll. de France.; admitted to staff Paris Faculty Medicine, 1872; apptd. prof. Sch. Pharmacy, 1882. Mem. Fr. Acad. Medicine (pres. 1917). Author: Histoire des matières albuminoides, 1873. Investigated sugars, terpenes, quinine, cinchonine; synthesized rubber from isoprene and hydrogen chloride. Died Nov. 22, 1918.

BOUCHER, Carl Opdycke, Am. prosthodontist; b. Ft. Wayne, Ind., Oct. 14, 1904; s. Charles Foster and Winnifred (Opdycke) B.; student Ohio State U., 1922-23, D.D.S., 1927; m. Florence L. Griess, May 9, 1931; 1 son, James Bradford. Practice dentistry specializing in prosthodontics, Columbus, O., 1927—; faculty Ohio State U. Coll. Dentistry, Columbus, 1928—, chmn. div. prosthodontics, 1940—, prof. prosthodontics, 1942—, chmn. div. dental lab. technology, 1947-57. Cons. Walter Reed Army Med. Center, 1948—, Ft. Knox, U. S. Army, 1965—; mem. joint Commn. for Accreditation of Dental Labs., 1963—. Diplomate Am. Bd. Prosthodontics (past pres.). Fellow Acad. Denture Prosthetics (past pres.), Am. Coll. Dentists; mem. Columbus Dental Soc. (past pres.), Ohio State Dental Assn. (past pres.), Am. Dental Assn., Internat. Assn. for Dental Research, Assn. Dental Editors, Pierre Fauchard Acad., Fedn. Prosthodontic Orgns. (past pres.), Am. Cleft Palate Assn., A.A.A.S., Internat. Assn. Dental Research, Federation Dentaire Internationale, Am. Equilibration Soc., Sigma Xi, Omicron Kappa Upsilon, Psi Omega (past supreme grand master), others. Author: Prosthetic Dental Laboratory Manual, 1948; also articles. Editor: Jour. Prosthetic Dentistry, 1951—; Complete Dentures (Swenson), 1964; Current Clinical Dental Terminology, 1963, Ohio Dental Jour., 1958-60; cons. editor Dental Abstracts, 1961—. Research on anatomy of the mouth, impressions for dentures, jaw movements, displaceability of tissues under dentures. Home: 3107 Halesworth Rd., Columbus 43221. Office: 21 E. State St., Columbus, O. 43215.*

BOUCHER DE CRÈVECOEUR DE PERTHES, Jacques, French archeologist; b. Rethel, France, Sept. 10, 1788; s. Jules-Armand-Guillaume B.; traveled with Napoleon to Italy, Germany, Austria, and Hungary; Author: De la création, 5 vols., 1839-41; Antiquités celtiques et antédiluviennes, 1847-64; De l'homme antédiluvien et ses oeuvres, 1860; Nouvelles découvertes, 1865. Made extensive reports on archeology of southern France; discovered several prehistoric flint hatchets, nr. Abbeville, France, 1847, and described period of their origin; showed man had existed in Pleistocene epoch; discovered fossil human jawbone in quarries at Maulin-Quignon; supported theory of ante-diluvian man, and disputed theory of diluvial catastrophism. Died Abbeville, France, Aug. 5, 1868.

BOUCHEROT, Paul, French elec. engr.; b. Paris, France, 1869; ed. Paris Sch. Physics and Indsl. Chemistry. Attaché, Elec. Services, French Nat. R.R.'s; became lectr. Higher Sch. Electricity, Paris, 1898; apptd. prof. Paris Sch. Physics and Indsl. Chemistry, 1907. Vice pres. French Com. Radiotelegraphy; del. Internat. Elec. Commn. Recipient several awards from French Acad. Scis. and Soc. Civil Engrs. Mem. Union French Engrs. (founder). Research and publs. on vibrations of alternating current motors, influence of induction resistance on their attenuation, disadvantages of phase delay; created idea of wattless power. Died 1943.

BOUCHUT, Jean Antoine Eugène, French physician; b. Paris, May 18, 1818; M.D., Paris, 1842. Became mem. Faculty Medicine Paris, 1853; physician Enfants-Malades Hosp., from 1856; chief med. clinic Hôtel Dieu, from 1849; staff Sainte-Eugenie Hosp. Author: Traité de diagnostic des maladies du système nerveaux des enfants par l'ophthalmoscope, 1865; Atlas d'ophthalmoscopie medicale et de cerebroscopie, 1877; Traité pratique des maladies des nouveau-nés, 7th edit., 1879. Worked in fields of infantile diseases, gen. pathology, internal surgery; discovered choroid tubercules are diagnostic sign of tuberculous meningitis; invented tube for glottis to replace tracheotomy in treatment of croup (method not accepted), 1856; described neurasthenia, 1860. Died Nov. 1891.

BOUDET, Jean-Pierre, French pharmacist; b. Paris, 1778; pharmacist in Paris. Author: Mémoire sur l'éther phosphorique; Thèse sur les propriétés du phosphore, 1815. Research on embalming, also phosphoric ether, phosophorus. Died 1849.

BOUDET, Jean-Pierre, French pharmacist; b. Rheims, France, 1748; prof. chemistry, Rheims; placed in charge munitions manufacture, 1793; chief physician Charity Hosp.; participant in Napoleon's Egyptian campaign; mem. Egyptian Inst., Acad. Medicine. Author: Mémoire sur le phosphore, 1815. Died Paris, 1829.

BOUDIER, (Jean-Louis-) Émile, French mycologist; b. Garnay, France, Jan. 6, 1828. Mem. French Acad. Scis. Author: Icones mycologicae (600 original plates); Histoire et classification des discomycètes d'Europe, 1909. Recognized taxonomic value of characteristics linked to dehiscence of ascus by dividing mushroom subclass Discomycetes into Opercules and Inopercules. Died Blois, France, Feb. 4, 1920.

BOUÉ, Ami, geologist; b. Hamburg, Germany, Mar. 16, 1794; founder Geol. Soc. France, 1830. Author: La Turquie d'Europe, 4 vols., 1840; Essai de carte géologique de globe terreste, 1845; Recueil d'itinéraires dans la Turquie d'Europe, 2 vols., 1850. Pioneer in internat. sci. collaboration; found part of human skeleton with fossil mammals (Cuvier later concluded skeleton was recent), Lahr, Germany, 1823; pointed out age differences among mountain chains and asserted Alps had undergone repeated crustal movements, 1827. Died Vienna, Austria, Nov. 22, 1881.

BOUELLES, Charles de, see de Bouelles.

BOUGAINVILLE, Louis Antoine de, see de Bougainville.

BOUGEUR, Pierre, French mathematician, hydrographer; born Croisic, France, February 16, 1698; son of John and Françoise (Jousseau) B. Succeeded father as professor of hydrography, Croisic; became professor of hydrography, Le Havre, France, 1730; sent (with La Condamine and Godin) by French Acad. Scis. to measure length of degree of meridian, Peru, 1735; made (with Pingre, Camus, Cassini) new observations to verify meridian degree between Paris and Amiens, France, 1757. Mem. French Acad. Scis. (prizes 1727, 28, 29). Author: Essai d'optique sur la gradation de la lumière, 1729; Théorie de la figure de la terre, 1749; Nouveau traité de navigation et de pilotage, 1753. A founder photometry; made 1st quantitative measurements of comparative luminosities of sun and moon; invented photometer, also studied atmospheric absorption of light, 1729; invented heliometer (perfected by Fraunhofer); 1st to compare mass of earth with that of mountain, 1740. Died Paris, Aug. 15, 1758.

BOUILLAUD, Jean Baptiste, French physician; b. Garat, France, Sept. 16, 1796; M.D., Faculty Medicine, Paris, 1823. Asso. with St. Louis Hosp., Paris; intern. 1819; physician hosps., 1831; apptd. prof. med. clinic Charité, 1831; became dean Faculty Medicine, 1848. Decorated commander of the Legion of Honor. Member of the French Academy of Scis., Academy of Medicine (president 1861). Author: Anévrismes de l'aorte, 1823; (with Bertin) Traité des maladies du coeur, 1824; Traité de l'encé-phalite, 1825; Traité des fievres essentielles, 1826, Recherches cliniques et expérimentales, sur les fonctions du cervlet, 1827; Recherches expérimentales sur les fonctions du cerveau, 1830; Traité clinique des maladies du coeur, 1835; Nouvelles recherches sur le rhumatisme articulaire aigu, 1836; Essai sur la philosophie médicale, 1836; Clinique médicale de l'Hôpital de la Charité, 1837; Traité clinique du rhumatisme articulaire aigu, 1840; Traité dé nosographie médicale, 1846. Phrenologist; last gt. blood letter; 1st to localize speech center in middle of left cerebral hemisphere, 1825; used auscultation in diagnosis; studied cirrhosis; helped establish connection between rheumatic fever and heart disease, 1840; recognized value of digitalis. Died Paris, Oct. 29, 1881.

BOUILLON, Albert Jules, biologist; b. Lamorteau, Aug. 25, 1916; s. Jules and Marie (Cobert) B.; ed. U. Louvain, U. Paris; Ph.D. in sci. Dir., Med. Found., U. Louvain-Congo, Kalenda Center; prof. animal biology U. Lovanium of Leopoldville, also 1st dean sch. sci., founder, dir. Mus. Zoology. Mem. Congolese Soc. Natural Scis. (dir. cultural com.), Antennes (founder, dir.). Dir. (rev.) Zooléo. Research and publs. on spiders, termites. Home: 5, rue du Pont, Lamorteau, Belgium. Office: University of Lovanium, B.P. 220, Leopoldville XI, Congo.

BOUILLY, Georges, French surgeon; b. Orleans, France, 1848; aggregate physician; practiced surgery, specializing in abdominal and pelvic operations. Author: Manuel de pathologie externe; Quatre agrégés. Died 1903.

BOUIN, Jean Théodose, French astronomer; b. Paris, France, Feb. 26, 1715; studied astronomy with Pingré; astronomer Rouen, France; mem. Acad. Rouen, Frech Acad. Scis. Contbr. Memoires de l'Academie. Studies concerning comets of 1757, 1759; aurorae borealis, passage of Venus over sun. Died Boisguillaume, France, 1795.

BOUIN, Pol André, French physician; b. Vendresse, France, June 11, 1870; M.D.; prof. histology Faculty Medicine Strasbourg, Germany (now France); decorated comdr. Legion of Honor; mem. French Acad. Scis. (non-resident), Acad. Medicine. Investigated ergastoplasm; discovered (with P. Ancel) that endocrine function of testicle is localized at interstitial cell. Died 1962.

BOUIS, Jules, French chemist; b. Perpignan, France, 1822; prof. chemistry École supérieure de pharmacie; mem. Acad. Medicine. Author: Cours de chimie analytique, 1871; Traité élémentaire de chimie légale, 1873. Died Paris, 1886.

BOUISSON, Étienne-Frédéric, French physician; b. Mauquio, France, June 14, 1813; studied medicine, Montpellier, France; grad. in surgery, 1836. Became prof. physiology Faculty Strasbourg (France), 1837; apptd. prof. clin. surgery Faculty Montpellier, 1840; dean Faculty Medicine, 1868-78; founder Montpellier med. jour., 1864. Mem. French Acad. Scis. Author: Traité théoriques et pratique de la méthode anesthésique appliquée à la chirurgie, 1850; Tribut à la chirurgie, 2 vols., 1858-61. Investigated cauterization of wounds, longitudinal bone fractures, traumatisms of spinal column, also humors and secretions, including chyle, lymph, tears, bile. Died Montpellier, May 26, 1884.

BOUIX, Maurice, French mathematician, physicist; b. Caveirac, France, Oct. 20, 1913; s. Josué and Rose (Guerin) B.; licencé, agregation, École Normale Supérieure, 1937, Docteur ès Scis., 1952; m. Calvette Fonvieille, Oct. 8, 1937; 1 child, Michele. Prof., Lycée de Tunis, 1938, 40-43; chief engr. Centre Nat. d'Études des Telécommunications, Paris, 1946-52; chef de travaux Faculty Scis. Paris, 1952-58; prof. Faculty Scis. Alger, 1958-62, Faculty Scis. Montpellier, 1962-64, Faculty Scis. Rouen, 1964; dir. Centre d'Études Théoriques de la Detection et des Communications, Paris, 1963——. Mem. Math. Soc. France, French Soc. Radioelectricians, French Nat. Com. Sci. Radioelectricity. Author: Les fonctions généralisées ou distributions, 1964; Les discontinuités du rayonnement électromagnétique Dunod, 1966. Research on theory of radiowaves and antennas, new methods for boundary problems, theory of generalized functions in view of signal and information. Home: 5, rue Jules Simon, Paris 15. Office: CETHEDEC, 5 Bis, Av. Porte-De-Sèvres, Paris 15, 75 France.*

BOULANGER, Nicolas Antoine, French civil engr., geologist; b. Paris, France, Nov. 11, 1722; ed. Coll. Beauvais (France). Accompanied Baron of Thiers during campaigns of 1743, 44; constructed bridge at Vaucouleurs; began constrn. bridge at Foulain; ret. as insp., 1758. Author: Recherches sur l'origine du despotisme oriental, 1761; L'Antiquité devoilée, 1766. Collaborator, Encyclopédie. Believed age of earth does not follow chronology of Bible; held that superstition and religion originated in terror produced by scourges. Died Paris, Sept. 16, 1759.

BOULANGER, Paul, Canadian veterinarian; b. St.-Leon, Que., Can., May 10, 1918; s. Adelard and Marie (Brault) B.; D.M.V., U. Montreal, 1941; M.Sc., Cornell U., 1944; m. Ella Lefebvre, Aug. 26, 1946; children—Andre, Jacques, Pierre. Tech. officer Can. Dept. Nat. Health, 1946; with Can. Dept. Agr., 1942-46, 47—, research scientist, Hull, Que., 1966—, head research serology unit Animal Diseases Research Inst., 1960——. Mem. Can., Que. (pres. 1966—) vet. med. assns., Am. Immunol. Soc., Can. Soc. Microbiology, U. S. Conf. Research Workers in Animal Diseases. Research and publs. in complement fixation methods for swinesera, Rinderpest virus; Leptospirosis studies in cattle and swine; modified direct complement fixation text for cattle and swine antibodies. Home: 82 Bourque St. Office: 101 Gamelin St., Hull, Que., Can.*

BOULDING, Kenneth Ewart, economist; b. Liverpool, Eng., Jan. 18, 1910; s. William C. and Elizabeth A. (Rowe) B.; B.A. with 1st class honors, Oxford (Eng.) U., 1931, M.A., 1939; m. Elise Biorn-Hansen, Aug. 31, 1941; children—John Russell, Mark David, Christine Ann, Philip Daniel, William Frederic. Came to U. S., 1932, naturalized, 1948. Asst., U. Edinburgh (Scotland), 1934-37; instr. Colgate U., 1937-41; economist League of Nations, 1941-42; prof. Fisk U., 1942-43; asso. prof. Ia. State Coll., 1943-46; prof., chmn. dept. McGill U., Montreal, Que., Can., 1946-47; prof. Ia. State Coll., 1947-49; prof. U. Mich., 1949-67; prof. U. Colo., Boulder, 1967——. Recipient Am. Council Learned Socs. prize for distinguished scholarship in humanities, 1962. Mem. Soc. for Gen. Systems Research (past pres.), Am. Econ. Assn. (John Bates Clark medal 1949, pres. 1968—), A.A.A.S. (v.p., chmn. sect. K 1966—). Author numerous books including: Economic Analysis, 1941; also numerous articles. Research on capital theory, liquidity preference theory of market prices, theory of macro-econ. distbn., theory of orgn., gen. systems theory. Home: 890 Willowbrook Rd., Boulder, Colo. 80302.*

BOULDUC, Gilles François, French apothecary; b. Paris, France, Feb. 20, 1675; s. Simon Boulduc. Apothecary to Duchess of Orléans; physics demonstrator Jardin du Roi; 1st apothecary to king and queen, France. Mem. French Acad. Scis. Research on purgatives, Glauber's salt, Seignette salt, preparation of sublimates from aromatic cinchona; analysed mineral waters of Passy, France. Died Versailles, France, Jan. 17, 1742.

BOULE, (Pierre) Marcellin, French geologist, paleontologist; b. Montsalvy, France, Jan. 1, 1861; ed. Toulouse, Paris; doctorate, Museum Nat. Hist., Paris, 1892. Preparateur, 1892, asst., 1894, prof., Museum Natural History, 1902-36; dir., l'Anthropologie, 1893-40. Founded Institute of human paleontology, 1920; Annales de paléotologie; Archives de l'Institut de paléontologie humaine. Commander, Legion of Honor; pres., French Geological Soc. Author: La plateau de Lanemezan, 1896; Les volcans de la France centrale, 1900; Les Grottes de Grimaldi, 1906-19; Les hommes fossiles, éléments de paléontologie humaine, 1921. Research on mountain geology in Central France; investigated human fossils; reconstructed 1st complete Neanderthal skeleton, 1908; correlated geological and archaeological evidence in determining chronology of remote eras. Died Montsalvy, France, July 4, 1942.

BOULENGER, George Albert, zoologist, botanist; b. Brussels, Belgium, Oct. 19, 1858; LL.D., St. Andrews U.; D.Sc., Louvain, Belgium; m. Emilie (dec. 1936). In charge reptile collection, dept. zoology Brit. Mus.; later engaged in study wild roses Crepin Herbarium, Botanic Gardens, Brussels. Fellow Royal Soc., 1894; mem. French Acad. Scis., 1916. Author numerous works on zoology including 9 vols. on reptiles and batrachians, 4 vols. on African fresh-water fishes, 2 vols. on roses of Europe, 3-part revision of Asian roses. Died Saint Malo, France, Nov. 23, 1937.

BOULENGER, Julius Caesar, see Bulenger, Julius Caesar.

BOULEY, Henri-Marie, French veterinarian; b. Paris, May 17, 1814; s. Jean-Francois Bouley; ed. Alfort. Became head clin. services, Alfort, 1837; apptd. prof. comparative pathology Mus. Natural History, 1879, also Vet. Sch., Alfort; gen. insp. vet. schs. Mem. Acad. Medicine, Soc. Agr., French Acad. Scis. (v.p. 1884, pres. 1885). Author: Traité de l'organisation du pied du cheval, 2 vols., 1851; Leçons de pathologie comarée, 2 vols., 1882-83; (with Reynal) Nouveau Dictionnaire de médecine, de chirurgie et d'hygiène vétérinaires, 1856. Editor Bull. Central Soc. Vet. Medicine, 1847-85. Responsible for laws regulating sanitary conditions for animals; supported doctrines of Pasteur; introduced lithotrite in vet. surgery; studied glands, contagious pneumonia, typhoid of horned animals. Died Paris, Nov. 30, 1885.

BOULIGAND, Georges Louis, French mathematician; b. Lorient, Oct. 13, 1889; s. Louis and Adolphine (Bonard) B.; ed. Ecole normal superieure; Ph.D. in sci.; Dr. honoris causa, U. Louvain, 1956; m. Jeanne Glain, Nov. 11, 1920; children—Anne-Marie, Louis, Hélène, Paulette, Marie-Thérèse, Marie, Georges, Marcelle, Yves. Prof. lycées of Tours, 1914-16, Renes, 1916-20; lectr. Sch. Sci., Rennes, 1920-21; prof. rational mechanics Sch. Sci., Poitiers, 1921-32, prof. differential and integral calculus, 1932-38; full prof. Sch. Sci., Paris, 1938-48, titular prof., 1948-61. Mem. Acad. Sci. Letters (corr.), Royal Soc. Sci. at Liège, Acad. Sci. Paris (sect. for mechanics). Author: Fonctions harmoniques; Géométrie infinitésimale directe; Epistémologie et henrisitique. Home: 46, rue Saint-André-des-Arts, Paris 6. Office: Inst. H-Poincare, 11, rue Pierre-Curie, Paris, France.

BOULLAY, (Félix) Polydore, French pharmacist; b. Paris, 1806; s. Pierre François Guillaume Boullay; ed. under Dumas. Studied (with father) percolation; investigated changes in vol. caused by chem. combination, also action of ammonia gas on oxalic ester; suggested that cane sugar can be converted into grape sugar by vegetable acids; showed (with Dumas) esters are compounds of acid anhydrides and ether, also worked out their chem. formulae. Died as result of ether fire, May 24, 1935.

BOULLAY, Pierre Francois Guillaume, French pharmacist; b. Caen, France, Apr. 27, 1777; student pharmacy, Rouen, also Paris, France; D. ès Scis., 1815. Mem. Acad. Medicine (dean). Author: (with son Félix Polydore) Méthode de déplacement, 1833. Founder, editor Bull. Pharmacy. Discovered poisonous element (picrotoxine) in Indian berry, 1811; 1st to use phosphoric acid to transform alcohol into ether; investigated possibility of phosphorus in sugars, also action of some substances on muriatic and acetic ethers, muriates of mercury, phosphoric ether. Died Nov. 3, 1869.

BOULLIAU, Ismaël, French astronomer, inventor; b. Loudon, France, Sept. 28, 1605; s. Ismaël B.; ordained priest, 1630; friend, adviser Christian Huygens, Johannes Hevelius. Author: De natura lucis, 1638; Philolaus seu de vero systemate mundi, 1639; Astronomia philolaica, 1645; Ad astronomos monita duo, 1657; Opus novum ad aritheticam infinitorum, 1682. Work considered classic in history of astronomy; 1st to give plausible explanation for variation in stellar light; stellar variability; 1st used term evection for irregular motion of moon caused by solar attraction; constructed 1st known mercury thermometer, 1659. Died Paris, Nov. 25, 1694.

BOULTER, Donald, English plant physiologist; b. Cambridge, Eng., Aug. 25, 1926; s. George and Vera (Medland) B.; B.A. Christ Church, Oxford, 1951, M.A., 1953, Ph.D., 1953; m. Margaret Eileen Kennedy, Sept. 1, 1955; children—Susan Elizabeth, Sara Jane, Catherine Annabel, Vanessa Clare. Cecil fellow Yale, 1953-54; asst. lectr. King's Coll., U. London, 1955-57; lectr. U. Liverpool, Eng., 1957-64, sr. lectr., 1964-66; prof. botany U. Durham, Eng., 1966——. Mem. Soc. for Exptl. Biology, Biochem. Soc., Phytochem. Group, Soc. Scandinavian Physiologists. Research, numerous publs. on structure, role and biosynthesis of fungal and higher plant proteins. Home: 45 South St., Durham, Eng.*

BOULTON, Matthew, English engr., inventor; b. Birmingham, Eng., Sept. 3, 1728; s. Matthew Boulton; m. Anne Robinson, 1760. Conducted father's mfg. bus.; founder Soho works, 1762; partner with Watt, 1772-1809; minted copper coins for East India Company and several fgn. govts.; established new copper coinage for Gt. Britain, 1797. Fellow Royal Society London (1785), Edinburgh) member of the Lunar Soc. Discovered new method for inlaying steel; built (with Watt) 1st Watt steam engine, draw pump machine, Cornwall, water pump machine, London, also 1st hauling machine, 1st machine for corn grinding, machines for breweries, spinning and saw mills, improved coining machinery, 1790; patented various axial motions, governors, throttles, devices for expansive use of steam. Died Soho, nr. Birmingham, Aug. 17, 1809.

BOULTON, Matthew Robinson, English management scientist; b. Midlands, England, 1770; s. Matthew and Anne (Robinson) B.; ed. by private tutor and in Paris. With James Watt, Jr., in charge of business making letter-copying press, 1790; Boulton, Watt & Sons formed to manufacture steam engine, 1794; co-manager, Soho Foundry, 1796; formed Boulton, Watt and Company, 1800; formed banking company, Matthew Robinson Boulton & James Watt & Company, 1802; banking company dissolved, 1833. Became Squire of Manor, Great Tew, Oxfordshire, 1817. With Watt, Jr., developed and anticipated scientific management techniques; used market research, planned site location and machine layout; production planning and standards; excellent cost accounting procedures; employed work training and executive development; developed personnel welfare plan. Died 1842.

BOULTON, William Savage, Brit. geologist; b. Old Swinford, nr. Stourbridge, Eng., Aug. 8, 1867; ed. Mason Coll., Birmingham, Eng, Royal Coll. Sci. (scholar); M.Sc., D.Sc.; m. M.K. Munns, 1898; 1 son. Lectr. geology, geography U. Coll., Cardiff, Wales, from 1897; prof. geology, 1904; prof. geology, dean sci. faculty U. Birmingham, 1913-32, emeritus, 1932-54. Mem. Safety in Mines Research Bd., 1922-28, Govt. Inland Water Survey Com. Fellow Geol. Soc. (v.p. 1934-35); asso. Royal Coll. Sci.; mem. Brit. Assn. (pres. geol. sect. 1916), Inst. Mining Engrs. Editor, contbr. Practical Coal Mining, 6 vols., 1907. Contbr. numerous papers on geology, water supply, others. Died Sept. 14, 1954.

BOUMAN, Maarten Anne, Dutch ophthalmologist; b. Utrecht, Netherlands, Mar. 29, 1919; s. Elbertus and A. A. (Bongers) B.; Ph.D. in Natural Scis., U. Utrecht; m. Th. Verhaar, June 13, 1949; children—Bert, Pieter, Anne, Lydi, Monique, Maarten. Assn. prin. natural scis. U. Utrecht; dir. Inst. for Perception ot Toegepast Natuurwetenschappelijk Onderzoek, Soesterberg, 1949—; prof. med. physics and physiology at Utrecht, 1962—. Mem. adv. group on human factors NATO, Paris. Mem. Optical Soc. Am., Brit. Ergomoncical Soc., Brit. Illuminating Engring. Soc., Soc. Eclairagistes of France, Am. Acad. Optometry, Phys. Soc. of Netherlands, Ophthal. Soc. Netherlands, Physiol. Soc. Netherlands, Internat. Commn. for Optics, Internat. Soc. for Electroetinography. Address: Breitnerlaan 14, Utrecht, Netherlands.

BOUQUET, Jean-Claude, French mathematician; b. Morteau, France, Sept. 7, 1819; Docteur ès Scis., l'Ecole Normale, 1843. Tchr. lycee, Marseille, France, 1843; mem. Faculty Lyons (France), 1844-51; tchr. math. Bonapart Lycee, 1852-58, Louis-le-Grand Lycee, 1856-67; prof. phys. and exptl. mechanics Sorbonne, 1870, prof. differential and integral calculus, 1874-84; lectr. l'Ecole Normale, 1868-85. Mem. French Acad. Scis. Author: Leçons de géométrie analytique, 1847; Leçons de trigonométrie, 1862; (with Briot) Théorie des fonctions doublement périodiques (provided didactic form for Cauchy's views), 1859, Théorie des fonctions elliptiques, 1875. Introduced (with Briot) term holomorph to describe monodromic and monogenic function which does not become infinite; studied octagonal surfaces. Died Paris, Sept. 9, 1885.

BOUQUET DE LA GRYE, Jean-Jacques-Anatole, French hydrographer, astronomer; b. Thiers, France, May 29, 1827; s. Antoine and Marie (de la Brosse) B. de la G.; ed. Ecole Polytechnique; engr., 1849. Mapped coastline New Caledonia, 1853; mapped port, Alexandria, 1861; named dir. Hydrog. Service, 1866; reorganized Bur. Longitudes. Mem. French Acad. Scis., 1884 (observer 1st passage Venus, Puebla, 1882); commander French Legion of Honor. Author: Le pilote des côtes ouest de la France, 1869-73. Author: Guide des manoeuvres en cas de cyclone, 1881; Dynamique de la mer; port de la Rouchelle, 1882; Amelioration de la Seine; Paris port de mer, 1884; Rapport sur le régime de la Loire navigable, 1885. Investigated nav. Loire River; studied coast line France, 1863; observed passage of sun over Mercury, 1865; worked on port of New Rochelle; calculated value of solar parallax; invented and improved astronomical instruments. Died Paris, Dec. 22, 1909.

BOUR, Edmond, French mathematician; b. Gray, France, May 1832; student l'Ecole Polytechnique,

1850-52; Docteur-ès-Scis. Coll. France, 1855. Prof., Sch. Mines, St. Etienne, France, 1856-59; apptd. prof. l'Ecole Polytechnique, 1859. Recipient 1st prize math. scis. French Acad. Scis., 1861. Author: Déformation des surfaces, 1861; La dynamique et l'hydraulique; Traité de cinématique, 1865; Cours de méchanique et machines, 1865-74. Studied concept of geodesic curvature and minimal surfaces, triply orthogonal systems, applicable surfaces (popularized by O. Bonnet); rediscovered Jacobi's theorems, 1855; made known method of solving 3d degree equations numerically with 2 slide rules, 1857. Died Paris, Mar. 8, 1866.

BOURBAKI, Nicolaus, collective pseudonym of group of 10 or 20 mathematicians, mostly French (H. Cartan, C. Chaubauty, J. Dieudonné, C. Ehresmann, A. Weil, others), generally at one time students École Normale Supérieure, Paris; pseudonym, Bourbaki (chosen in jest), name of gen. in war of 1870 whose efforts marked by inglorious humiliation, disastrous failure. Author numerous math. texts, part of comprehensive treatise (Élements de Mathématique), purpose of which is to present math. in modern way (following Hilbert), also to display its axiomatic structure (project begun mid 1930's—).

BOURDEAU, Philippe François, biologist; b. Rabat, Morocco, Nov. 25, 1926; s. Michel Edgard and Lucienne (Imbrecht) B.; Ingenieur agronome State Inst. Agronomy, Belgium, 1949; M.A., Duke, 1951, Ph.D., 1954; m. Flora Gorirossi, Jan. 26, 1954; children—Jacqueline, Michele, Isabelle. Research asst. Duke, 1952-54; asst. prof. botany N.C. State Coll., Raleigh, 1954-56; asst. prof. forest ecology Yale, 1956-58, 60-62; prof. botany, dean agr. faculty U. Belgian Congo and Ruanda-Urundi, 1958-60; prin. research officer, biology div. European Atomic Energy Community, 1962-66, div. chief, 1966—, head biology service Euratom Joint Research Center, Ispra, Italy, 1962—. Mem. Ecol. Soc. Am., Biometric Soc., A.A.A.S. Publs. on research in physiol. ecology of plants, statis. analysis of vegetation, movement of radioactive isotopes in ecol. systems, effects of ionizing radiation on plants. Home: 9 via Piave, Bardello, Italy. Office: Euratom Joint Research Center, Ispra, Italy.*

BOURDEL, Léone, French psychologist, anthropologist; b. Paris, France, Dec. 20, 1907; d. Leon and Marie (Donnadieu) Bourdel; M.A. in Philosophy and Letters, Sorbonne; m. 2d, Jacques Genevay; 1 son, Richard-Olivier. Sec.-gen. Nat. Inst. Prof. Orientation, 1928-37; prof. Sch. Sci. Orgn. of Work, 1935-48, mem. council, 1937—; prof. Superior Sch. Anthrobiology, 1948—; dir. Lab. Applied Psychology-Léone Bourdel, 1937—. Named laureat of Inst., Internat. Found. Louis and Auguste Lumière; recipient Joseph Saillet prize Acad. Moral and Polit. Sci. Mem. Assn. Counselors of Dirs., Soc. Statistics (Paris), Soc. Men of Letters, Assn. Counselors of Pvt. Economy. Author: La Connaisance des hommes par la psychobiologie; La Mission de la France; Sang, Tempéraments, Travail et Races; Groupes sanguins et Tempéraments; Les Tempéraments psychobiologiques. Address: 32, rue de l'Assomption, Paris 16, France.

BOURDELIN, Claude, French chemist, pharmacist; b. Lyons, France, 1621; children—Claude, Francois. Apothecary, House of Orléans; mem. French Acad. Scis. Found that iron filings increase in weight on rusting, 1683; examined mineral waters; attempted dry distillation analysis of plants; analysed more than 2000 substances; opposed bleeding as therapeutic. Died Paris, Oct. 14, 1699.

BOURDELIN, Louis-Claude, French chemist; b. Paris, Oct. 18, 1696; s. Francois Bourdelin. Prof. chemistry Jardin du Roi, circa 1727-circa 1770; dean Faculty Medicine Paris, from 1736; physician to daus. Louis XV. Mem. French Acad. Scis. Attempted to show presence of alkalai in plants before combustion; burned succinic acid with saltpetre containing chloride in attempt to find hydrochloric acid; found copper salts (blue vitriol) color alcohol flame green. Died Paris, Sept. 13, 1777.

BOURDIEU, Pierre, French sociologist; b. Denguin, France, Aug. 8, 1930; s. Albert and Noemie (Duhau) B.; student École Normale Supérieure, 1961-64; agrégé in philosophy, Licence de Philosophie, Ecole Normale Supérieure; m. Marie-Claire Britard, Nov. 2, 1962; children—Jérome, Emmanuel. Asst., Faculté d'Alger, 1959-60, Faculté Paris, 1960-62; lectr. Faculté des Lettres, Lille, France, 1962-64; dir. studies École des Hautes Etudes, 1965—; asst. dir. Centre de Sociologie Européene. Author: Sociologie de l'Algérie, 1958; The Algerians, 1962; Travail et travailleurs en Algérie, 1964; (with Abdel Malek Sayad) le Déracinement, 1964; (with J. C. Passeron) Les étudiants et leurs études, 1964, Les Héritiers, 1964; Un Art Moyen, 1965; (with A. Doubel) L'Amour de l'Art, 1966; also articles. Research on Kalyl culture, culture of knowledge and teaching system in France, a theory of culture. Home: 24 Avenue Aristide Briand, Antony, France. Office: 10 rue M.-le-Prince, Paris 6e, France.*

BOURDIN, Pierre, French mathematician; b. 1595; mem. Soc. of Jesus; tchr. rhetoric, La Flèche, also Paris, France, 22 years. Author: Le cours de mathematique; Sol flamma . . . eiusque pabulo . . . aphorismi analogici parvi mundi ad magnum, magni ad parvum, 1646. Died 1653.

BOURDON, Eugène, French inventor; b. Paris, Apr. 8, 1808; employed by silk mfr., then optician, later in factory mech. constrns.; founder workshops for constrn. machine tools and steam machines, 1835. Invented steam condenser, 1848, metallic manometer, 1849, also pneumatic clock, speedometer, steam pump, 1857; conducted expts. on resistance of air to trains at high speeds. Died Paris, Sept. 29, 1884.

BOURDON, René, radiologist; b. Valparaiso, Chile, Feb. 29, 1916; s. H. and M. L. (Mococain) B.; ed. U. Paris, (France); m. M. Boyer, July 18, 1959. Apptd. head radiology dept. Paris Hosp., 1947; chief, dept. radiations Hosp. St. Louis, Paris, 1958—; chief roentgen and cobalt therapy dept., Am. Hosp. of Paris; prof. Coll. Medicine, Hosps. of Paris. Mem. Dermatol. Soc., Radiol. Soc., Gastrologic Soc., Neurol. Soc., Lymphologic Soc., Hematologic Soc. Author: Radiology of Small Intestine; also articles. Research in lymphography, hematology, treatment of cancer of blood and skin; inventor machine for radiology of knee. Home: 61, Bd. Victor-Hugo, Neuilly-sur-Seine, France. Office: Hopital St.-Louis, 2, pl. du Dr. Alfred-Fournier et 12 bis, rue G. aux Belles, Paris X, France.*

BOURGELAT, Claude, French veterinarian; b. Lyons, France, Mar. 27, 1712; dir. Vet. Sch. Paris; founder 1st Vet. Sch. (prototype for European vet. colls.), Lyons, 1761; researched anatomy and diseases of horse and domestic animals Royal Acad. Lyons. Author: Eléments d'hippiatrique, ou nouveaux principes sur la connaissance et sur la medicine des chevaux, 1753; Matière médicale raisonnée a l'usage de l'école vétérnaire, 8 vols., 1765; Reglement sur les écoles vétérinaires, de France, 1777; Traité de la conformation extérieure du cheval, 1779. Works on vet. medicine provided complete course of instruction. Died Paris, Jan. 3, 1779.

BOURGEOIS, Leon Zéphirin, French mineralogist; b. Paris, Aug. 26, 1856; preparator inorganic natural history Coll. de France, 1880-83; preparator organic chemistry Mus. Natural History, 1883, apptd. asst. in organic chemistry, 1890; apptd. asst. in chemistry Polytechnic, 1884. Made 1st attempts (with Fuchs, Fouque, Michel-Levy) to create minerals and igneous rocks artificially; used qualitative analysis to investigate specific crystalline forms produced by salt-forming acid attacks on minerals.

BOURGEOIS, Louis, French physician; studied under Ambroise Paré and famous midwives; physician to French Ct.; attended Marie de Medici in delivery of 6 princes. Author: Observations diverses sur la sterilite, perte de fruicts, recondite, accouchements, maladies des femmes et nouveaux naiz, 1609; Recit veritable de la naissances des messeigneurs et dames les enfants de France, 1625; Apologie contre les rapports des medecins, 1627. Observed detachment of placenta; pioneered sci. of midwifery. Died 1636.

BOURGEOIS, Paul Eugene Edouard, Belgian astronomer; b. Feb. 13, 1898; docteur ès sciences, physiques et mathematiques, Ph.D., U. Brussels (Belgium). Asst. Belgian Royal Obs., 1924, asso. astronomer, 1936, astronomer, from 1938, dir., 1947-63; prof. Université Libre de Bruxelles. Vice-pres. Inst. Sci. Research in Central Africa, Belgian Nat. Com. Internat. Co-operation Geophysics; pres. abstracting bd. Internat. Council Sci. Unions; mem. Bureau des Longitudes. Recipient Quinquennial prize statistics, 1929-33. Mem. (asso.) Academie Royal des Sciences d'Outre-Mer. Research, publs. on meridian astronomy, astrometry, astro-physics, stellar statistics, observational and statis. astronomy. Address: 31 Paul Hankar St., Brussels 18, Belgium.

BOURGEOIS, (Joseph-Emile) Robert, French geodesist; born Haut-Rhin, France, February 21, 1857; educated at Ecole polytechnique, Paris, France, 1876, military school at Fontainebleau, France. Member of military campaign to Tunisia, 1880; mem. several topographic expdns. to Algeria, 1887-93; head topog. expdn. to Madagascar, 1895; head geodetic expdn. to Ecuador to measure Quito meridian arc, 1901-02; prof. astronomy and geodesy Ecole polytechnique, Paris, 1908-28; became gen. during World War I; apptd. dir. Army Geog. Service, 1911. Mem. Bur. Longitudes, Soc. Geography (became v.p. 1909, Gold medal 1907), French Acad. Scis. Decorated Grand Cross Legion of Honor. Mem. Soc. Geography, Soc. Physics. Organized system of scouting and observation during World War I; determined (with Basset) new meridian of France; made observations on intensity of gravity in France and Algeria. Died Paris, July 8, 1945.

BOURGERY, Jean Baptiste Marc, French physician; b. Orleans, France, May 19, 1797; health officer, Romilly, France, 8 years; M.D., Paris, 1827. Author: Traité de petite chirurgie, 1829; Traité complet d'anatomie de l'homme, comprenant la médecine opératoire (significant work on human anatomy), 8 vols., 1830-45; Anatomie élémentaire, 1834-42; Des annexes du foetus et de leur developpement, 1846; articles in publ. French Acad. Scis., 1836-48. Died Paris, June 1849.

BOURGERY, Marc Jean, French anatomist, surgeon; b. France, 1797; described posterior ligament of knee (ligamentum popliteum obliquum or Bourgery's ligament). Died 1869.

BOURGET, Justin, French mathematician; b. Savas, France, 1822; Docteur ès mathématiques, École normale, 1852; became rector Aix Acad., 1882. Author: Théorie élémentaire des approximations numériques, 1860. Died 1887.

BOURIQUET, Gilbert, French phytopathologist; b. Paris, France, Oct. 17, 1903; s. Philippe and Annette (Pessiot) B.; Ph.D. in Sci. and Engring., Sch. Sci., Paris; Founder lab. vegetal pathology of Tananarive, dir. for 15 years; insp. gen. Labs. Agr. of France Overseas. Mem. Internat. Mus. (corr.). Author: Les Maladies des plantes cultivées a Madagascar; Le Vanillier et la Vanille dans le monde. Discovered 1st method to germinate vanilla grains. Address: 3, rue Régis, Paris 6, France.

BOURNE, Sir Alfred Gibbs, Brit. zoologist; b. Lowestoft, Eng., Aug. 8, 1859; s. Alfred Bourne; ed. Royal Sch. Mines; D.Sc., Univ. Coll., London; m. Emily Tree Glashier, 1888; 1 son, 1 dau. Asst. to E. Ray Lankester, 1879-85; researcher Zool. Sta., Naples, Italy, 1883-85; apptd. to Madras, India, 1885; registrar U. Madras, 1891-99; botanist to Govt. of Madras, 1897-98; warden Victoria Hostel, Madras, 1900-02; prof. biology Presidency Coll., Madras, Indian Educational Service, 1886-1903; dir. pub. instrn., Madras, also commr. for govt. examinations, additional mem. council Ft. St. George, 1903-14; dir. Indian Inst. Sci., Bangalore, 1915-21; mayor of Dartmouth, 1922-23. Fellow Royal Soc. Research and publs. on anatomy of Pelomyxa, Limnododium, Chaetobranchus, leeches, earthworms, Rotifera, pearly nautilus, scorpion poison; discovered adult male of species Filaria, 1888. Died July 14, 1940.

BOURNE, Geoffrey Howard, anatomist, primatologist; b. Perth, Western Australia, Nov. 17, 1909; s. Walter Howard and Mary Ann (Mellon) B.; B.Sc., U. Western Australia, 1930, B.Sc. with honors, 1931, M.Sc., 1932, D.Sc., 1935; D.Phil., U. Oxford, 1943; m. Maria Nelly Golarz, Oct. 31, 1964; children—(by previous marriage) Peter, Mervyn. Came to U. S., 1957, naturalized, 1962. Biologist, Australian Inst. Anatomy, Canberra, 1934-35; biochemist Commonwealth of Australia Adv. Council on Nutrition, 1936-37; demonstrator in physiology U. Oxford, 1941-47, reader in histology U. London, 1947-57; chmn. anatomy Emory U., Atlanta, 1957-62, dir. Yerkes Regional Primate Research Center, 1962—. Cons., Sch. of Aerospace Medicine, 1964—. (Beit Meml. fellow, Mackenzie Mackinnon fellow) Royal Coll. Surgeons, also Royal Coll. Physicians, 1941-43. Fellow Zool. Soc.; mem. anat. socs. U. S. and U.K., geronotological socs. U. S. and U.K., Am. Rocket Soc., Soc. Aerospace Medicine, Internat. Primatological Soc., Internat. Soc. Cell Biology, others. Author books including: Structure and Function of Muscle, 1962; Biochemistry and Physiology of Bone, 1956; Starvation in Europe, 1943; Division of Labour in Cells, 1963, others; also numerous articles. Editor: Cytology and Cell Physiology, 1941; (with J. F. Danielli) International review of Cytology, 20 vols., 1950-66; World Review of Nutrition and Dietetics 6 vols, 1900-1906. Pioneered in devel. of new sci. of histochemistry; showed (with T. R. Shanthaveeroappa) that nerves have protective covering continuous with the covering of the brain; 1st to demonstrate vitamin C in cells, 1933. Home: 2659 Varner Dr., Atlanta 30322.*

BOURNE, Gilbert Charles, English zoologist; b. Worcestershire, Eng., July 5, 1861; s. Robert and Anna (Baker) B.; ed. New Coll., Oxford, also under August Weismann at Freiburg-im-Breisgau, Germany; m. Margaret Graham Croft, 1887; 1 son, 1 dau. Became fellow New Coll., 1887; 1st dir. Marine Biol. Lab., Plymouth, Eng.; Linnacre prof. zoology and comparative anatomy Merton College, Oxford, 1906-21. Member of the advisory committee on fisheries Development Commission. Fellow Royal Society, 1910. Author: Introduction to the Study of the Comparative Anatomy of Animals, 1900-02. Investigated structure, devel., classification of corals and related animals; studied origins coral reefs at atoll Diego Garcia, Indian Ocean; opposed Darwin's theory of reef formulation by subsidence; made morphological studies of some mollusca; defended cell theory against Adam Sedgwick's attack. Died Abington, Eng., Mar. 9, 1933.

BOURNE, Lyle Eugene, Jr., Am. psychologist; b. Boston, Apr. 12, 1932; s. Lyle E. and Blanche (White) B.; A.B., Brown U., 1953; M.S., U. Wis., 1955, Ph.D., 1956; m. Vera Sakwa, Apr. 17, 1954; children—Barbara Ann, Elizabeth Jane, Andrew Lyle. Asst. prof. U. Utah, 1956-61; asso. prof. U. Cal. at Berkeley, 1961-62; faculty U. Colo., Boulder, 1963—, prof. psychology, 1965—. Research cons. VA Hosps., 1958—. Mem. Am., Midwestern psychol. assns., Psychonomic Soc., Psychometric Soc., Sigma Xi. Author: Human Conceptual Behavior, 1966; also articles. Editorial cons. Jour. Exptl. Psychology, 1962—, Psychol. Reports, 1960—, Psychol. Monographs, 1965—. Research on verbal and conceptual behaviors humans, quantitative theoretical analysis concept attainment, structural alalysis conceptual processes. Home: 1455 Alpine St., Boulder, Colo. 80302.*

BOURNE, William, Brit. mathematician; b. Gravesend, Eng., flourished 1565-83; s. William Bourne. Innkeeper, Gravesend; gunner Brit. army; tchr. mathematics. Author numerous books including: Rules of Navigation, 1571. Noted for giving 1st practical example of triangulation. Died Mar. 22, 1583.

BOURNEVILLE, Désiré Magloire, French physician; b. Garancières, France, Oct. 21, 1840; intern Paris, 1865, M.D., 1870. Asso. with Jour. Mental Medicine, also Contemporary Medicine; founder (with N. Pascal) Mouvement Medical, 1864-83; in med. service Prussian War, 1870; founder Med. Progress, 1873; municipal councellor, Saint-Victor, France, 1876-83; elected dep. from Paris (voted 1st funds for retarded children and epileptics), 1883, from Seine, 1885. Mem. Cremation Soc. (pres.); Med. Soc. Hosps.; Biology Soc.; Anat. Soc.; Soc. Pub. Medicine and Profl. Hygiene. Author: Leçons sur les localisations dans les maladies cerebrals; Lecons sur les maladies de foie et des voies biliaires; Etudes cliniques et thermométriques sur les maladies du système nerveux, 1862; (with P. Regnard) l'Iconographie photographique de la Salpêtrière, 1876-80, l'Hystéro épilepsie, 1878. Combined (with Brissaud) various kinds mental degeneration under name tuberous sclerosis of brain (Bourneville's disease), 1880; recognized cretinism and myxedema as the same condition. Died Paris, May 30, 1909.

BOURNS, Arthur Newcombe, Canadian chemist; b. Petitcodiac, N.B., Can., Dec. 8, 1919; s. Evans Clement and Kathleen (Jones) B.; B.Sc. with honors in Chemistry, Acadia U., N.S., 1941; Ph.D. in Chemistry, McGill U., Montreal, Que. Can., 1944; m. Marion Harriet Blakney, June 20, 1943; children—Barbara Ellen, Susan Kathleen (Mrs. William Milne), Robert Evans, Brian Hugh. Research chemist Dominion Rubber Co., Guelph, Ont., Can., 1943-44; lectr. Acadia U., 1944-45; asst. prof. U. Sask., 1945-46; faculty McMaster U., 1949—, prof. chemistry, 1952—, dean Faculty Grad. Studies, 1957-61, chmn. chemistry dept., 1965—. Mem. Com. on U. Affairs for Province of Ont., 1964—. Fellow Royal Soc. Can., Chem. Inst. Can.; mem. Am. Chem. Soc., Chem. Inst. Can., Royal Soc. Can. Asso. editor Canadian Jour. Chemistry, 1966—. Research, publs. on mechanism of various reactions in organic chemistry; pioneered research using difference in reaction rate of isotopically-labeled molecules to obtain information on highest energy state (transition state) through which a reacting system passes in going from reactants to products. Home: 122 Judith Crescent, Ancaster, Ont. Office: Dept. Chemistry, McMaster U., Hamilton, Ont., Can.*

BOURQUELOT, (Élie-) Émile, French biologist; b. Jandun, France, June 21, 1851; docteur ès scis., 1884. Intern, 1876; chief pharmacy Enfants-malades Hosp.; became prof. l'Ecole Supérieure, 1897. Mem. Acad. Medicine (v.p.), French Acad. Scis. Author: Les fermentations, 1893; Les ferments solubles, 1896. Studied soluble ferments, also methods of researching glucosides and sugars; showed (with Bridel) that enzymes usually act successively producing enzymatic chain, 1912. Died Paris, Jan. 26, 1921.

BOURRELLY, Pierre Paul Charles, French botanist; b. Miramas, France, July 12, 1910; s. Marius and Joséphine (Blanc) B.; Licencie ès Sciences, U. Paris, 1937, Docteur ès sciences, 1954; m. Berthe Bricard, Aug. 8, 1933. Asst., Museum National Histoire Naturelle, Paris, 1945, later became asst. dir., lectr. cryptogamy lab., 1954. Mem. Internat. Limnology Assn., French Phycology Soc., French Bot. Soc., Phycological Soc. U.S.A. Author: (with E. Manguin) Algues d'eau douce de la Guadeloupe, 1952; Recherches sur les Chrysophycées, 1957; Les Algues d'eau douce I. Les Algues vertes, 1966; also numerous articles. Research in fresh-water algae, including systematics, biology and culture, fresh-water flora, flagellate algae of France, Africa, Madagascar, Can. Home: 5 avenue Gambetta, St. Mandé (Val de Marke) 94, France. Office: 12 Rue de Buffon, Paris 5e (Seine) 75, France.*

BOURSEUL, Charles, inventor, engr.; b. Brussels, Belgium, 1829; became engr. French Telegraph System, 1851; perfected Morse telegraph for use in France; began research on elec. transmission of words. Died St.-Céré, France, 1912.

BOUSFIELD, Edward George Paul, Brit. physician; b. 1880; s. W. R. Bousefield; ed. Bristol (Eng.) U., Central Tech. Coll., St. Bartholomew's Hosp.; m. R. Wenger. Mgr., Exptl. Works of Henry Simon, Ltd., Manchester, Eng., 1902; mng. dir. Saxon Iron and Steel Works, Stokes-on-Trent, Eng., from 1904, Metal Finishers, Ltd., London, Eng., from 1910; surg. receiving officer St. Bartholomew's Hosp., casualty officer Queen's Hosp. for Children, 1916; res. med. officer Am. Women's Hosp. for Officers, 1917; demonstrator in morbid anatomy, asst. curator mus., St. George's Hosp., 1918; physician to London Neurol. Clinic, 1919. Author: Sex and Civilisation; The Omnipotent Self; Elements of Practical Psycho-Analysis; Pleasure and Pain; Functional Nervous Diseases; others. Editor: St. Bartholomew's Hosp. Jour., 1914-18. Specialized in treating nervous diseases. Died Nov. 21, 1957.

BOUSFIELD, Edward Lloyd, Canadian biologist; b. Penticton, B.C., Can., June 19, 1926; s. Reginald H. and Marjory F. (Armstrong) B.; B.A., U. Toronto, 1948, M.A., 1949; Ph.D., Harvard, 1954; m. Barbara Joyce Schwartz, June 20, 1953; children—Marjorie Anne, Jessie Katherine, Mary Elizabeth, Kenneth Lloyd. Invertebrate zoologist Nat. Mus. Can., Ottawa, Ont., 1950-64, chief zoologist, 1964—. Mem. Ottawa Field Naturalists Club (past pres.), Ecol. Soc. Am., Soc. Systematic Zoology, Am. Soc. Limnology and oceanography, A.A.A.S., Arctic Inst. N.Am., Sigma Xi. Author: Canadian Atlantic Sea Shells, 1960. Research, publs. on taxonomy, distbn., ecology, life history of shallow-water cirripede crustaceans Am. Atlantic region, intertidal ecology, taxonomy of shallow-water amphipod crustaceans of the families Gammaridae, Haustoriidae, Talitridae on world-wide basis, fresh-water amphipods of N.Am. Home: 48 Farlane Blvd., Ottawa 5. Office: Nat. Mus. Can., Ottawa, Ont., Can.*

BOUSFIELD, Guy William John, English physician; b. London, Eng., Oct. 20, 1893; s. Edward and Clara (Henman) B.; ed. St. Olave's Sch., St. Thomas's Hosp., London; M.D., B.S.; m. Phyllis Doyle, Dec. 15, 1923; 1 son, William Edward Doyle. House physician, house surgeon, asst. dept. cardiography St. Thomas Hosp.; dir. Camberwell Lab., Pub. Health Lab. at Denmark Hill. Mem. Brit. Med. Assn. Author: The Schick Test and Diphtheria and Scarlet Fever Immunisation; A Preliminary Course of Hygiene and Food Handling, also articles. Specialist in immunology. Address: Well Cottage, Goose Green, Warnham, Horsham (Sussex), Eng.

BOUSFIELD, Weston Ashmore, psychologist; b. Sao Ching, China, Apr. 22, 1904; s. Cyril Eustane and Lillie (Snowden) B.; came to U. S., 1904, naturalized, 1906; B.Mech. Engring., Northeastern U., 1927; A.M., Boston U., 1928; A.M., Harvard, 1932, Ph.D., 1933; m. Thelma Knight, July 4, 1935; children—Brenda Lee (Mrs. David Ambrose Remley;, Aldridge Knight. Mem. faculty Tufts U., 1929-39; faculty U. Conn., Storrs, 1939—, prof., 1946—, head, psychology dept., 1939-60. Prin. investigator research in human learning project Office Naval Research, 1951-62. Fellow Am. Psychol. Assn.; mem. Eastern (treas. 1947-50), Conn. (pres. 1947) psychol. assns., Sigma Xi, Phi Kappa Phi. Contbns. include theory and research methods; devel. math. formulas for analyzing verbal data; originator clustering, the tendency to group related items in recall. Home: S. Eagleville Rd., Storrs, Conn. 06268.*

BOUSSINESQ, (Valentin-) Joseph, French physicist; b. St.-André-de-Sangonis, France, Mar. 13, 1842; s. Jacques and Anne-Marie (Cavalier) B.; docteur ès scis., 1867. Tchr. math. Gap (France) Coll., 1866-72; became prof. differential and integral calculus Faculty Sci. Lille (France), 1872; apptd. prof. phys. and exptl. mechanics Faculty Scis. Paris, 1886. Recipient Prix Poncelet, 1872. Mem. French Acad. Scis. Author: Essai théoretique sur l'équilibre des massifs pulvérulents comparé a celui des massifs solides, et sur la poussée des terres sans cohésion, 1876; Essai sur la théorie des eaux courantes, 1877; Conciliation du véritable déterminisme méchanique avec l'existence de la vie et de la liberté morale, 1878; Étude sur divers points de la philosophie des Sciences, 1879; Application des potentiels à l'étude de l'équilibre et du mouvement des solides élastiques, 1885; Leçons d'analyse infinitésimale, en vue d'applications méchaniques et physiques, 2 vols., 1887. Adopted Poisson's theory of elasticity; developed Helmholtz's theory of beats; discoveries in electromagnetism (with those of Helmholtz) laid found. for Lorentz's rudimentary electron theory. Died Paris, Feb. 19, 1929.

BOUSSINGAULT, Jean-Baptiste-Joseph-Dieudonné, French agrl. chemist; b. Paris, France, Feb. 2, 1802; ed. St. Etienne Mining Sch.; D.-ès-S., 1832. Dir. mine explorations, S.Am.; mem. staff Simon Bolivar during insurrection Spanish colonies; prof. chemistry Faculty of Sciences, Lyons, France; professor of agricultural chemistry Conservatoire des Artes et Métiers, Paris, 1839-87. Member of French Assembly, from Bas-Rhin, 1848-52. Member of French Academy of Sciences, 1839, Society Agr. Author: Traité d'économie rurale, 1844, revised as Agronomie, chimie agricole, et physiologie, 8 vols., 1860-74; Études sur la transformation du fer en acier, 1875; (with J. B. A. Dumas) Essai de statique chimique des êtres organisés, 1841. Founder exptl. agrl. chemistry; studied plant physiology and nutritive value of fodders; studied (with Dumas) gases of air; showed nitrogen is essential to plants and animals; gave rough account of nitrogen cycle; showed plants absorb nitrogen from nitrates in soil rather than air; proved carbon is assimilated from carbon dioxide in atmosphere; recommended use of iodized salt for cure of goiters, 1833; research on atomic weights and properties of steel alloys. Died Paris, May 12, 1887.

BOUTIGNY, Pierre-Hippolyte, French chemist; b. Colleville, France, May 11, 1798; became pharmacist, Rouen, France, 1821, Evreux, France, 1823-42; conducted researches, La Villette, France; worked in indsl. chemistry, London, Eng. Mem. Soc. Pharmacy (pres.), Soc. Civil Engrs. Author: Études sur les corps à l'état sphéroidal. Investigated calefaction, also cures for drunkeness, legal medicine, adulteration of foodstuffs. Died Evreux, Mar. 17, 1884.

BOUTLEROV, Aleksandr M. (Butlerov), chemist; b. Kazan, Russia, Sept. 6, 1828; ed. U. Kazan; became prof. chemistry Kazan U., 1852; made trip to Paris, 1857-58; went to St. Petersburg, Russia, 1868. Author: Introduction à l'étude complète de la chimie organique, 1864; Sur les propriétés de l'acide triméthylacétique, 1874. Research and publs. on methylene compounds, tertiary alcohols, simple acids, polymerization ethylene compounds, atomic weights of chem. elements, phys. properites of ice under light pressure. Died Biarritz, France, 1886.

BOUTROUX, Emile, French philosopher; b. Montrouge, France, July 28, 1845; ed. École Normale Supérieure; U. Heidelberg (Germany), 1869-70; Docteur ès Lettres, 1874; at least 1 son, Pierre. Prof., Lycee de Caen; prof. philosophy univs. Montpellier, 1874, Nancy, 1876; Paris, from 1885; maitre de conférences École Normale Supérieure, 1877; dir. Thiers Found., from 1902. Mem. French Inst., 1898, Académie Française, 1912, St. Petersburg Acad. Scis., Accademia dei Lincei, Rome, Royal Acad. Denmark, Brit. Acad., Instituto Lombardo di Scienze e Lettere, Milan. Author: De la contingence des lois de la nature, 1874; Leibniz, La monadologie, 1880; Socrate fondateur de la science de la science morale, 1883; De l'idée de loi naturelle dans la science et la philosophie, 1895; Questions de morale et d'éducation, 1895; Études d'histoire et la philosophie, 1897; Pascal, 1900; La philosophie de Fichte, 1902; Psychologie du mysticisme, 1902; Science et religion dans la philosophie contemporaine, 1908; William James, 1911; The Beyond that is Within, 1912; Philosophy and War, 1916; Études d'histoire de la philosophie allemande and Novelles études d'histoire de la philosophie, 1927. Translator: Edward Zeller's Die Philosophie der Greichen (from German into French, with introduction), 3 vols. Strongly against prevailing scientism of day, argued that since laws of nature deal with necessary connections, they express only what is permanent and stable in being; but world also manifests genuinely irreducible contingency, dynamism, which is creative action of God; therefore no opposition should exist between God and religion on one hand and exptl. sci. on the other, and that metaphysics and sci. should complement each other (these views exerted considerable influence in late 19th to early 20th century France). Died Paris, France, Nov. 22, 1921.

BOUTROUX, Pierre Leon, French mathematician; b. Paris, Dec. 6, 1880; s. Emile and Aline C. (Eugenie) B.; ed. École normale superieure, Paris, also U. Paris, 1900; asso. with U. Montpellier (France); prof. integral calculus U. Poitier (France), 1908-1920; vis. prof. U. Nancy (France), 1909; guest lectr. Collège de France, 1911-12, prof. gen. history sci., 1920-22; prof. higher math., chmn. grad. math. dept. Princeton, 1913-14. L'imagination et les mathématiques selon Descartes, 1900; Lecons sur les fonctions définies par les equations différentielles du premier ordre, 1908; Oeuvres de Blaise Pascal, pub—suivant l'ordre chronologique avec documents complementaires, 1908; Les Principes de l'analyse mathématique. Exposé historique et critique, 2 vols., 1914-19; L'ideal scientifique des mathematiciens dans l'antiquité et dans les temps modernes, 1920; (with Vita Volterra, Jacques Hadamard, Paul Langevin) Henri Poincare: l'oeuvre scientifique, l'oeuvre philosophique, 1914. also articles. Studied multiform functions, singularities of differential equations; continued Paul Painlevé's work on possibility of infinite number of solutions for differential equations. Died Aug. 1922.

BOUTRY, George Albert, French physicist; b. Le Mans, France, May 27, 1904; s. Albert G. L. and Angela (Urel) B.; student U. Lille (France). 1924-27; Agrégé de Physique, Paris (France) U., 1927; Docteur ès Sciences, 1934; D.Sc., Laval U., Quebec, Que., Can., 1952; m. Louise Roque, June 20, 1925; children—Claire Francoise (Mrs. D. Luis Segura Lopez), Jean Luc Yves Patrick. Dir. Laboratoire d'Essais, Paris, 1936-48; prof. Inst. d'Optique Paris, 1940-48; prof. vacuum physics Conservatoire Nat. des Arts et Metiers, Paris, 1944—; founder, pres. Laboratoires d'Electronique et de Physique Appliquée, Paris, Limeil, Brevannes, France, 1950—; vis. prof., lectr. various univs. Decorated Commandeur des Palmes academiques, Officier de la Legion d'Honneur; recipient Television Inter. award, Montreux, 1962. Fellow I.E.E.E., Phys. Soc. (London); mem. Internat. Council Sci. Unions (pres. abstracting bd. 1963—), Internat. Union Pure and Applied Physics (chmn. publ. com. 1952-62). Société des Ingénieurs Civils de France (past pres.), Société Francaise de Physique, Société Francaise des Electricens et Radioélectriciens. Author: Les Phénomènese Photoélectriques, 1934; Optique Instrumentale, 1950; Physique Appliquée aux Industries du Vide et de l'Electronique, 2 vols., 1962, 64; also numerous articles. Research on photoelec. phenomena, electroptics, TV. and vacuum electronics. Home: 94 Villecresnes, France. Office: P.O. Box 15, 94 Limeil-Brevannes, France.*

BOUTY, Edmond, French physicist; b. Nant, France, Jan. 12, 1846; D. ès Scis., Ecole normale, Paris, 1874. Tchr. physics Lycée St. Louis, Paris, 1876-83; apptd. lectr. Sorbonne, 1883, prof. physics, 1885. Recipient Prix Lacaze, 1895. Mem. French Acad. Scis., Internat. Soc. Electricians (pres.), French Soc. Physics (pres.). Author: La vérité scientifique, sa poursuite, 1908. Editor Jour. de Physique. Studied electric conductivity of concentrated salt solutions, 1884-88, dielectric stability of mica, 1891, acoustical lens and sensitive flame, 1896, magnetism, passage of electricity through gas. Died Paris, Nov. 5, 1922.

BOUVARD, Alexis, French astronomer; b. Contamines, France, June 27, 1767; ed. under Laplace; attended free lectures College de France; student astronomer Paris Obs., 1793-94, apptd. adjunct astronomer, 1795, later dir. Recipient prize French Inst., 1800. Mem. French Acad. Scis., 1803, Bur. Longitudes, 1804. Fellow Royal Soc., 1826. Computed new tables for Uranus previously thought to be a star) in unsuccessful attempt to settle discrepancies; 1st to point out irregularities in Uranus' orbit, 1821, (Leverrier's and Adams' discovery of Neptune, 1846, confirmed Bouvard's 1821 hypothesis); discovered 8 comets, assisted Laplace in preparation of Traité de Mécanique céleste (1799-1825). Died Paris, June 7, 1843.

BOUVART, Michel Philippe, French physician; b. Chartres, France, Jan. 11, 1707; s. Claude Bouvart; student medicine Rheims, France. Dir., hosp. Chartres, from 1730; chair medicine Royal Coll., Paris, 1747-56; prof. physiology Sorbonne, U. Paris; regent doctor Fac. Med., Paris. Author works on vaccination, premature birth. Noted Parisian physician. Died Paris, Jan. 19, 1787.

BOUVEAULT, Louis, French chemist; b. Nevers, France, Feb. 11, 1864; D. ès Scis. Physiques, Ecole Polytechnique, 1890. Asst. prof. Paris Faculty Scis.; became lectr. chemistry Faculty Medicine, Lyons, France, 1892; apptd. asso. prof. Faculty Medicine, Nancy, France, 1899. Developed method of preparing alcohols from esters by reducing acid part of molecule; developed process for synthesizing aldehydes. Died Paris, Sept. 6, 1909.

BOUVERET, Léon, French physician; b. France, 1850 or 51; ed. Lyons, also Paris, France; M.D., 1878; prof., Lyons, 1878-1929. Author: Traité de l'empyème, 1888; Traité des maladies de l'estomac, 1893. Described paroxysmal tachycardia (Bouveret's syndrome), 1889; studied (with Tripier) treatment of typhoid fever by cold baths; used stomach pump to continue Leube's studies of digestive processes. Died Feb. 1929.

BOUVET, Joachim, mathematician; b. Le Mans, France, July 18, 1656; mem. Soc. of Jesus; sent by Louis XIV to collect sci. data, China, 1688, 99; interpreter for emperor's son during 2d stay, dir. survey of empire, also mapped several provinces, 1708-15. Mem. French Acad. Scis. Author: l'État présent de la Chine, 1697; Portrait historique de l'empéreur de la Chine, 1697; math. works in Tartar, later translated into Chinese. Contbd. to sci. devel. in China. Died Peking, China, June 28, 1732.

BOUVIER, Louis-Eugène, French naturalist, entomologist; b. Saint-Laurent-Grandvaux, France, Apr. 9, 1856; grad. École normale primaire de Lons-le-Saunier, 1872; docteur ès scis., 1887. Tchr. various schs.; became forest probationer Mus. Natural History, 1887, titular prof. entomology, 1895-1932; head practical work, lab. comparative zoology Hautes Etudes, dir. lab. anat. zoology, 1894; became prof. Higher Sch. Pharmacy, 1889. Pres., Zool. Soc. France, 1894, Entomol. Soc., 1897, French Acad. Scis., 1926. Author: Vie psychique des insectes, 1919; Habitudes et métamorphoses des insectes, 1919; le Communisme chez les insectes, 1926; Monographie des lépodoptères saturnides, 1934; Decapodes marcheurs de la faune de France, 1940. Investigated mollusks, articules, onychophores; determined origin of gastropods and evolution of their asymmetry; affirmed homogeneity of prosobranches and hermophroditic gastropods; studied adaptation of mollusks, geog. and bathymetric distbn. also anat. constrn. peripates and pagurides. Died Paris, Jan. 14, 1944.

BOUVIER, Pierre Bernard, Swiss physicist; b. Geneva, Switzerland, 1917; s. B. Rolfes; ed. Geneva, Zurich (Switzerland) univs., Sc.D.; m. Valentine Zullig, 1943. Tchr. physics Coll. Geneva; asst. prof. physics, astronomy; prof. astrophysics Geneva U.; astronomer Geneva Obs. Mem. Internat. Astron. Union, Swiss Soc. Physics, Soc. Geophysics, Meteorology, Astronomy. Research, publs. on astronomy, physics, meteorology, astrophysics; specialist stability problems stellar systems, dynamics of clusters, numerical integrations of n-body problems. Address: Observatoire Astronomique, 1920 Sauverny, Geneva, Switzerland.

BOUWKAMP, Christoffel Jacob, Dutch mathematician; b. Hoogkerk, June 26, 1915; s. J. and G. (Bakker) B.; ed. U. Groningen; m. Alida H. van der Veer, 1943. Curator lab. physics U. Groningen; mem. research lab. Philips N.V. at Eindhoven, 1941——; prof. applied math. U. Utrecht, 1954-58, Technol. U. Eindhoven, 1958. Mem. Royal Netherlands Acad. Sci., Am. Soc. Math., numerous others. Author: Theoretical and Numerical Treatment of the Diffraction by a Circular Aperture. Research and numerous publs. on physics, pure and applied math., wave propagation, electromagnetic theories. Address: Goorstraat 10, Eindhoven, Netherlands.

BOUYOUCOS, George John, soil physicist; b. Tripolis, Greece, May 23, 1890; s. John D. and Paris (Kapsimalis) B.; came to U. S., 1901, naturalized, 1925; B.S., U. Ill., 1908; Ph.D., Cornell U., 1911; postgrad. Göttingen, Germany, U. Paris, France, U. London, Eng.; D.Sc., U. Thessaloniki, 1951; m. Agnes Robb, Oct. 8, 1940; 1 son, John. Research soil physicist Mich. State U., East Lansing, 1911-20, research prof., 1920-58. Cons. cane sugar industry, coffee plantations, S.Am., E. Africa. Decorated Cross of Our Savior, Greece, 1919. Fellow Am. Soc. Agronomy, A.A.A.S.; mem. Am. Soc. Agronomy, Internat. Soil Sci., Am. Chem. Soc., Sigma Xi (award 1943). Developed dilatometer method to classify water in soils, hydrometer method for mech. analysis soils, plaster of Paris elec. resistance method measuring soil moisture, elec. hygrometer to measure relative humidity of air. Address: 706 N. Hagadom Rd., East Lansing, Mich.*

BOVEE, Eugene Cleveland, Am. protozoologist; b. Sioux City, Ia., Apr. 1, 1915; s. Earl Eugene and M. Nora (Johnson) B.; B.A., State Coll. Ia., 1939; M.S., State U. Ia.; 1948; Ph.D., U. Cal. at Los Angeles, 1950; m. Maezene Belle Wamsley, May 18, 1942; children—Frances Anne, Gregory Joe, Matthew Wamsley. Asst. bus. mgr. State Coll. Ia., 1939-40, instr., 1946-48; faculty State U. Ia., 1940-41, Greene High Sch., 1941-42; faculty U. Cal. at Los Angeles, 1948-50, asso. research zoologist, 1962-65, research zoologist, 1965——; asst. prof. biology Cal. State Poly. Coll., 1950-52; asso. prof. zoology, acting chmn. dept. N.D. State U.; 1952-53; asst. prof. U. Houston, 1953-55; instr. biology Sacred Heart Coll., Houston, 1953-55; asst. prof. zoology U. Fla., Gainesville, 1955-58, asso. prof., 1958-62. Fellow Ia. Acad. Sci.; mem. Am. Micros. Soc. (exec. com. 1960-62), Am. Inst. Biol. Sci., Am. Soc. Zoologists, Biophys. Soc., Soc. Indsl. Microbiology, Soc. Protozoologists, Soc. Taxonomic Zoology, West. Soc. Naturalists, Sigma Xi. Contbr. articles to tech. jours. Research on effects temperature on death rate, growth rates crustacea, physiology and biomechanics protoplasmic movement, locomotion microorganisms, ameboid protozoa. Home: 2347 Malcolm Av., Los Angeles 90064.*

BÖVENTER, Edwin von, see von Böventer, Edwin.

BOVERI, Theodor Heinrich, German zoologist; b. Bamberg, Germany, Oct. 12, 1862; s. Walter Boveri; M.D., Munich, Germany, 1885; m. Marcella Imelda O'Grady, 1897; 1 dau. Apptd. asst. prof. zoology and comparative anatomy Munich U., 1887, asst. zool. inst., 1891; became prof., Würzburg, Germany, 1893. Mem. acads. Munich, Berlin, Copenhagen, Petersburg, Phila., N.Y. Author: Zellen-Studien, 6 vols., 1887-1907; Problem der Befruchtung, 1902. Added to knowledge cellular structure; 1st to apply word centrosome to part of animal cell (rediscovered by Flemming and van Beneden), also 1st to point out chromosomes split as part of mechanism of reproduction, 1888; made diagrammatic representation of spermatogenesis still in use, 1892; investigated abnormal fertilization. Died Würzburg, Oct. 15, 1915.

BOVET, Daniel, pharmacologist; b. Neuchatel, Switzerland, Mar. 23, 1907; s. Pierre and Amy (Babut) B.; Doct.Sc., U. Geneva (Switzerland), 1929; Doct.hon.causa, U. Palermo (Italy), U. Geneva, 1958, U. Do Brazil, 1958, U. Montpellier (France), 1959, U. Paris (France), 1960, U. Nancy (France), 1962, U. Strasbourg (France), 1963, Georgetown U., 1964; m. Filomena Nitti, Mar. 19, 1938; 1 son, Daniel-Pierre. Research asst. Institut Pasteur, Paris, 1929-47; head dept. therapeutic chemistry Istituto Superiore di Sanita, Rome, 1947-64; prof. pharmacology Sch. Medicine, U. Sassari (Italy), 1964——; vis. prof. Sch. Medicine, U. Cal. at Los Angeles, 1965——. Named Chevalier Legion d'Honneur (France) Cav. gr. cr. (Repubblica Italiana); recipient Nobel Laureate prize in physiology and medicine, 1957. Mem. Acad. Naz. dei XL, Acad. Naz. Lincei, Royal Soc. London (fgn.), Acad. Sci. Paris (corr.), Am. Acad. Arts and Scis. (fgn. hon.). Author: (with F. Bovet-Nitti) Medicaments du Système Nerveux Vegetatif, 1966; (with F. Bovet-Nitti, G. B. Marini Bettolo) Curare; also numerous articles. Research on pharmacological structure-activity relationships; co-discoverer antibacterial properties of sulfanilimids; pioneered in antihistamines, curare-like anesthetics. Home: 30 Via G.B., de Rossi, Roma, Italy. Office: 1, Via Rolando, Sassari, Italy.*

BOVEY, Frank Alden, Am. chemist; b. Mpls., June 4, 1918; s. John Alden and Margaret (Jackson) B.; B.S., Harvard, 1940; Ph.D., U. Minn., 1948; m. Shirley Elfman, June 19, 1941; children—Margaret Alden (Mrs. Frederick Sandback), Frank Alden III, Victoria Adams. Research chemist central research dept. Minn. Mining & Mfg. Co., St. Paul, 1942, head, polymer sect., 1948-55, research asso., 1955-62; asst. chief chemist Nat. Synthetic Rubber Co., Louisville, 1942-45; mem. tech. staff Bell Telephone Labs., Murray Hill, N.J., 1962——, head polymer research department, since 1967——. Recipient award Union Carbide Chems., 1958. Mem. Am. Chem. Soc. (Minn. award 1962, Am. Phys. Soc., Am. Soc. Biol. Chemists, A.A.-A.S., Sigma Xi, Phi Lambda Upsilon. Author: (with I. M. Kolthoff, A. I. Medalia, E. J. Meehan) Emulsion Polymerization, 1955; The Effects of Ionizing Radiation on Natural and Synthetic High Polymers, 1958; NMR Data Tables, 1967. Engaged in research in the chemistry of preparation of synthetic rubbers and plastics; devel. new plastics for spl. uses; study of stereochemistry of polymers by use of nuclear magnetic resonance spectroscopy; kinetics of internal motions in molecules; fundamental study of optical properties of biol. polymers. Home: 19 Rolling Hill Dr., Morristown, N.J. Office: care Bell Telephone Labs., Murray Hill, N.J. 07971.*

BOVEY, Henry Taylor, civil engr.; b. Devonshire, Eng.; ed. Cambridge U.; LL.D., McGill U., Queen's U.; m. Miss Redpath; 2 sons, 3 daus. Elected fellow Queen's Coll., Cambridge, hon. fellow, 1906; rector Imperial Coll. Sci. and Tech.; asst. engr. Mersey Docks and Harbour Works; apptd. prof. civil engring. and applied mechanics McGill U., Can., 1887, dean dept. applied sci., 1888; rector Imperial Coll. Sci. and Tech.; Fellow Royal Soc., 1902; Royal Soc. Can. (pres. sect. III 1896); founder Canadian Soc. Civil Engrs., pres. 1900; mem. Instn. Civil Engrs., Eng., Liverpool Soc. Civil Engrs. (a founder), Nat. Electric Light Assn. U. S. (hon.), Brit. Assn. (v.p. mech. sect. 1897). Author: Applied Mechanics, 1882; Theory of Structures and Strength of Materials, 1893, 6th edit., 1905; Hydraulics, 1895, 6th edit., 1904; also papers. Died Feb. 2, 1912.

BOVIE, William T., Am. biophysicist; b. Augusta, Mich., Sept. 11, 1882; s. William and Henrietta (Barnes) B.; A.B., Mich., 1908; A.M., U. Mo., 1910; Ph.D., Harvard, 1914; Sc.D., Albion Coll., 1929; m. Martha Adams, Sept. 15, 1909; 1 son, William Adams. Prof. geology and biology Antioch Coll., 1906-07; research fellow, Cancer Comm., Harvard U., 1914-20; instr. bacteriology Harvard, 1920-21, asst. prof. biophysics, 1920-27; prof. biophysics Northwestern U., 1927-29; lectr. social technology. Colby Coll., 1939-48. Recipient John Scott medal Franklin Inst. 1928. Mem. Botanical Soc. Physiol. Soc., Physical Soc., Soc. for Cancer Research, Soc. Tropical Medicine Am. Acad. Arts and Scis., Am. Chem. Soc. (chmn. biol. sect. 1919-23). Developed pioneer methods for therapeutic use of radio active substances; perfected electric apparatus for bloodless surgery and for prevention of metastasis of cancer cells; inventor various biophysical instruments. Died Jan. 1, 1958.

BOWDEN, Frank Philip, physicist; b. Hobart, Tasmania, May 2, 1903; s. Frank Prosser and Grace (Hill) B.; M.Sc., U. Tasmania, 1926, D.Sc., 1932; Ph.D., Cambridge (Eng.) U., 1929, Sc.D., 1933; C.B.E., 1956, F.R.S., 1948; m. Margot Hutchinson, Dec. 12, 1931; children—Piers, Humphrey, Jonathan, Sophie (Mrs. Barnaby Milburn). Rockefeller Internat. Research fellow, sr. 1851 exhibitioner, 1930; fellow Gonville and Caius Coll., dir. studies Cambridge U., 1931-37, Humphrey Owen Jones lectr., 1937, reader, 1946-57; head tribophysics lab. Commonwealth Sci. and Indsl. Research Orgn., Melbourne, Australia, 1938-45; professorial fellow, dir. lab. physics and chemistry of solids Cavendish Lab., Cambridge, 1957-65, prof. surface physics, 1966——. Dir. English Electric Co. Ltd.; cons. Tube Investments Research Lab., 1953——. Recipient Redwood medal, 1953; Elliott Cresson medal Franklin Inst., 1955; Rumford medal Royal Soc., 1956. Mem. Faraday Soc. (past v.p.), Cambridge Philos. Soc. (past pres.). Fellow Inst. Physics. Author: (with D. Tabor) Friction and Lubrication of Solids, part I, 1954, part II, 1964; Initiation and Growth of Explosion in Solids, 1952; (with A. Yoffe) Fast Reactions in Solids, 1958. Research on electrochemistry, surface and strength properties of solids, friction and lubrication, initiation and growth of explosion. Home: Finella W., Queens' Rd., Cambridge, Eng.*

BOWDEN, Kenneth Frank, English oceanographer; b. Southampton, Eng., Dec. 23, 1916; s. Frank and Margaret (Walker) B.; B.Sc., U. Southampton, 1937, M.Sc., 1939, D.Sc., 1956; m. Lilias Tweed Miller Nicol, Apr. 10, 1946; 1 dau., Margaret. Sci. officer Admiralty Anti-submarine Establishment, 1939-45; lectr. oceanography U. Liverpool, 1945-52, prof., 1954——; prin. sci. officer Nat. Inst. Oceanography, Wormley, Surrey, 1952-54. Fellow Inst. Physics, Royal Astron. Soc. Publs. on studies of turbulent character of tidal currents, influence of wind on currents in coastal waters, dynamics of circulation in estuaries, processes which determine temperature and salinity structure of ocean. Home: 100 Meols Parade, Hoylake, Cheshire, Eng.*

BOWDITCH, Henry Ingersoll, Am. physician; b. Salem, Mass., Aug. 9, 1808; s. Nathaniel and Mary (Ingersoll) B.; attended Boston Latin Sch., 1823-25; grad. Harvard, 1828, M.D., 1832; studied medicine, Paris, France, under Pierre Louis, 1832-34; m. Olivia Yardley, 1838. Mem. staff Mass. Gen. Hosp., Boston, 1831-32, 38-92; practiced medicine, Boston, from 1834; abolitionist leader from 1830's, friend of William L. Garrison; active in case of runaway slave George Latimer, 1842-43; prof. clin. med. Harvard Med. Sch., 1859-67; an original mem. Mass. Bd. Health, 1869-79. Fellow Am. Acad. Arts and Scis.; mem. Nat. Acad. Scis., A.M.A. (pres. 1877). Author: The Young Stethoscope, 1846; Public Hygiene in America, 1877; also many med. papers. Expert on diseases of chest. especially Tb; pioneer in performing operations for pleural effusions with suction pump; noted for work in preventive medicine and public sanitation; introduced several new surg. techniques; influential in organizing ambulance service during Civil War. Died Boston, Jan. 14, 1892.

BOWDITCH, Henry Pickering, Am. physiologist; b. Boston, Apr. 4, 1840; s. Jonathan Ingersoll and Lucy Orne (Nichols) B.; A.B., Harvard, 1861, A.M., 1866, M.D., 1868; studied in Europe under Claude Bernard, Jean Charcot, K.F.W. Ludwig; D.Sc., Cambridge, U., 1898; LL.D., Edinburgh, 1898, U. Toronto, 1903, U. Pa., 1904; m. Selma Knauth, of Leipzig, Sept. 9, 1871. Asst. prof. physiology Med. Sch., Harvard, 1871-76, prof. physiology, 1876-1903, George Higginson prof. physiology, 1903-06, dean

Med. Sch., 1883-93, established 1st physiol. lab. in Am. at Harvard, 1871; Trustee Boston Pub. Library, 1895-1902. Served as lt., capt. and maj. U. S. Vol. Cav., 1861-65. Fellow Am. Acad. Arts and Scis., A.A.A.S. (v.p., 1886, 1900); mem. Nat. Acad. Scis., Am. Soc. Psychical Research (founding mem.), Soc. Psychical Research, London (corr.), A.M.A., Am. Physiol. Soc. (1st pres. 1887), Am. Soc. Naturalists (pres. 1898), Congress Am. Physicians and Surgeons (pres.-1900). Author: Growth of Children, 1877; Hints for Teachers of Physiology, 889; Is Harvard a University; 1890; Are Composite Photographs Typical Pictures?, 1894; Advancement of Medicine by Research, 1896. Editor: Jour. Physiology, 1877-98. Demonstrated that nerve fibers are indefatigable (Bowditch's law), 1890. Died 1911.

BOWDITCH, Nathaniel, Am. navigator, astronomer, mathematician; b. Salem, Mass., Mar. 26, 1773; s. Habbakkuk and Mary (Ingersoll) B.; no formal schooling after 1783; A.M. (hon.), Harvard, 1802; m. Mary Boardman, Mar. 25, 1798; m. 2d, Mary Ingersoll, Oct. 28, 1800; 8 children including Jonathan Ingersoll, Henry Ingersoll. Made 5 sea voyages between 1795-1803; prepared 1st Am. edit. The Practical Navigator (J. H. Moore), 1799, revised and enlarged under title The New American Practical Navigator, 1802, 9 edit. published during his lifetime, 56 reprints or edits, published since his death; portions of the work reprinted under title Bowditch's Useful Tables, 1844; his skill in math. led to positions as pres. Essex Fire & Marine Ins. Co., 1804, actuary Mass. Hosp. Life Ins. Co., 1823-38; made survey of Salem harbors, 1804-06; published translation with commentary of 1st 4 vols. of Méchanique Céleste (LaPlace) (most important sci. work); best known papers include one concerning meteor which exploded over Weston, Conn., 1807, another discussing the motion of a pendulum. Mem. Am. Acad. Arts and Scis., 1799, pres., 1829-38, pub. 23 papers on nautical and astron. subjects in acad.'s Memoirs, 1804-20; Fellow Royal Soc., 1818, Royal Soc. Edinburgh, Brit. Assn., other fgn. socs. Died Boston Mar. 17, 1838.

BOWDOIN, James, Am. colonial gov., amateur scientist; b. Boston, Aug. 7, 1726; s. James and Hannah (Pordage) B.; grad. Harvard, 1745; LL.D., Edinburgh U.; m. Elizabeth Erving, Sept. 15, 1748. Elected to Mass. Gen. Ct., 1753-56; mem. Mass. Council, 1757; pres. Mass. Provincial Congress, 1775, Mass. Constl. Conv., 1779; gov. Mass., 1785-87; suppressed Shay's Rebellion, 1786; mem. Mass. Conv. to ratify U. S. Constn., 1788; a founder, 1st pres. Am. Acad. Arts and Scis., 1780-90, presented various papers on natural philosophy before Acad.; fellow Harvard, 1779; fellow royal socs. of London, 1788, and Edinburgh; Bowdoin Coll. named in his honor. Died Boston, Nov. 6, 1790.

BOWEN, Charles Clark, Am. biologist; b. Detroit, Mar. 18, 1917; s. Charles Clark and Geraldine (Jarvis) B.; student U. Mich., 1935-39; B.A., Mich. State U., 1949, M.S., 1951, Ph.D., 1953; m. Vada Robinson, Aug. 28, 1947; children—Clark, Gail (Mrs. Marvin Lindmark), Jean (Mrs. Dean Smith), Lecia. Postdoctoral fellow NIH at Brookhaven Nat. Lab., 1953-55; faculty Ia. State U., Ames, 1955—, prof., 1962—, chmn. cell biology, 1966—; NSF vis. scientist; dir. Botany Electron Microscope Facility. Mem. Bot. Soc. Am., Am. Soc. Cell Biology, Radiation Research Soc., Electron Microscope Soc. Am., A.A.A.S., Sigma Xi. Research, publs. on cell nucleus and its changes during the course of cell division at fine structure level, higher plant cells and cells of fungi and algae, influence of radiation and certain drugs upon chromosomal structure. Home: 3218 West St., Ames, Ia. 50010.*

BOWEN, Edmund John, English phys. chemist; b. Worcester, England, April 29, 1898; the son of Edmund Riley and Lilias (Kamester) Bowen; M.S., Balliol College, Oxford (England) University, 1915, D.Sc., 1947; m. Edith Moule, July 8, 1924; children—Margaret Lillias (Mrs. John Pinsent), Humphry John Moule. Fellow, Univ. Coll., Oxford U., 1922-65, Aldrichian praelector in chemistry, 1952-65. Recipient Davy medal, 1965. Fellow Roy. Soc., 1935; member Chem. Soc. London, Faraday Soc., Phys. Soc. Author: The Chemical Aspects of Light, 1946; (with F. Wokes) The Fluorescence of Solutions, 1953; also numerous articles. Research in photochemistry and on fluorescence. Home: 10 Park Town, Oxford, Eng.*

BOWEN, Edward George, physicist; b. Swansea, Wales, Jan. 14, 1911; s. George and Ellen Ann (Owen) B.; B.Sc., U. Coll. Swansea (Eng.), 1930; M.Sc., U. Wales, 1931; Ph.D., U. London, 1934; D.Sc., U. Sydney (Australia), 1957; m. Enid Vesta Williams, Dec. 27, 1938; children—Edward, David, John. Sci. officer radar devel. team Brit. Air Ministry, 1935-36, Bawdsey Research Sta., Eng., 1936-40; rep. of Brit. Ministry of Aircraft Prodn. at Radiation Lab., Mass. Inst. Tech., 1940-43; dep. chief radiophysics div. Commonwealth Sci. and Indsl. Research Orgn., Sydney, 1944-46, chief, 1946—. Decorated comdr. Order Brit. Empire; medal for Freedom U. S. Govt.; recipient Thurlow award Am. Inst. Navigation. Fellow Am. Geophys. Union, Australian Acad. Scis., Royal Meteorol. Soc.; mem. Am. Acad. Arts and Scis. (fgn.). Editor: Textbook of Radar, 1954. Research, numerous publs. on radar, devel. airborne radar, atmospheric physics and radio astronomy, mem. team which built first air def. system; helped convert Australian research program to peacetime basis. Home: 174 Edinburgh Rd., Castlecrag, Sydney. Office: City Rd., Sydney, N.S.W., Australia.*

BOWEN, Howard Rothmann, Am. economist; b. Spokane, Wash., Oct. 27, 1908; s. Henry G. and Josephine (Menig) B.; B.A., Wash. State U., 1929, M.A., 1933; Ph.D., U. Ia., 1935; LL.D., Cornell Coll., Mt. Vernon, Ia., 1956, Knox Coll., 1964; L.H.D., Loras Coll., 1964; Litt.D., Grinnell Coll., 1964; H.H.D., Coe Coll., 1965; m. Lois B. Schilling, Aug. 24, 1935; children—Peter Geoffrey, Thomas Gerard. Faculty, U. Ia., 1935-42; with U. S. Govt., 1942-45; economist Irving Trust Co., 1945-47; dean Coll. Commerce and Bus. Adminstrn., U. Ill., Urbana, 1947-52; prof. econs. Williams Coll., 1952-55; pres. Grinnell (Ia.) Coll., 1955-64; U. Ia., Iowa City, 1964—. Chmn. Nat. Commn. on Tech., Automation and Econ. Progress, 1964-66; member board of commissioners National Commn. on Accrediting, 1965—. Mem. Phi Beta Kappa, Beta Gamma Sigma, Phi Kappa Phi. Author: English Grants in Aid, 1939; Toward Social Economy, 1948; Social Responsibility of the Businessman, 1953; Christian Values and Economic Life, 1954; Automation and Economic Progress (with Garth Mangum), 1966; also numerous articles. Organized regular statis. reporting on number, turnover, and structure of bus. enterprises in U. S.; pioneered work on non-market mechanisms for econ. decision-making and for measurement of social welfare. Home: 102 Church St., Iowa City 52240.*

BOWEN, Humphry John Moule, English chemist; b. Oxford, Eng., June 22, 1929; s. Edmund John and Edith (Moule) B.; M.A., D.Phil., Magdalen Coll., Oxford U., 1947-53; m. Ursula Hill Williams, Aug. 9, 1953; children—Jonathan P., William G., Benjamin A. Research fellow A.E.R.E., Harwell, Eng., 1953-55, sr. sci. officer, 1955-64; lectr. Reading (Eng.) U., 1964—. Mem. Chem. Soc., Soc. Analytical Chemistry. Author: (with D. Gibbons) Radioactivation Analysis, 1963; Introduction to Botany, 1965; Trace Elements in Biochemistry, 1966; also articles. Research on methods of ultramicroanalysis of various elements, effect of ionizing radiation on biol. material, devel. biol. standard material. Home: 20 Winchester Rd., Oxford, Eng. Office: Chemistry Dept., Reading U., Reading, Berkshire, Eng.*

BOWEN, Ira Sprague, Am. astronomer; b. Seneca Falls, N.Y., Dec. 21, 1898; s. James Henry and Philinda May (Sprague) B.; A.B., Oberlin Coll., 1919; postgrad. U. Chgo., 1919-21; Ph.D., Cal. Inst. Tech., 1926; D.Sc., Oberlin Coll., 1948; Ph.D., U. Lund, 1950; Sc.D., Princeton, 1953; m. Mary Jane Howard, July 12, 1929. Asst. in physics U. Chgo., 1919-21; lectr. in physics Cal. Inst. Tech., 1921-26, asst. prof., 1926-28, asso. prof., 1928-31, prof. 1931-45; dir. Wilson Obs., 1946-64, also Palomar Obs., 1948-64, distinguished service staff mem., 1964—; Morrison research asso. Lick Obs., 1938-39. Recipient Draper medal Nat. Acad. Scis., 1942; Potts Medal, Franklin Inst., 1946; Rumford Premium, Am. Acad. Arts and Scis., 1949; Ives medal Optical Soc. Am., 1952; Bruce medal Astron. Soc. of Pacific, 1957. Mem. Nat. Acad. Scis., Am. Philos. Soc., Am. Acad. Arts and Scis. Am. Astron. Soc., Astron. Soc. Pacific, (asso.) Royal Astron. Soc. (Gold medal 1966), Indian, Swedish Royal acads. scis. Contbr. articles on atomic structure, cosmic rays, spectra of gaseous nebulae to sci. jours. Presented final evidence needed to prove that entire universe is composed of same elements found on earth, 1927; demonstrated that nebulae are made up largely of hydrogen and helium, and that nebulium could be evidence of presence of oxygen and nitrogen under spl. conditions prevailing in nebulae, 1934; developed image slicer (Bowen's device) which permitted utilization of more light collected by spectrograph, 1938. Home: 2388 N. Altadena Dr., Altadena, Cal. 91001.*

BOWEN, Lucius Murray, Am. psychiatrist; b. Waverly, Tenn., Jan. 31, 1913; s. Jess Sewell and Maggie May (Luff) B.; B.S., U. Tenn., 1937, M.D., 1937; m. LeRoy Ellis, Jan. 3, 1944; children—Susan, Joanne, Kathleen, Charles. Psychiat. tng. Menninger Sch. Psychiatry, Topeka, 1946-49; psychoanalytic tng. Topeka Inst. Psychoanalysis, 1948-54, Wash. Psychoanalytic Inst., 1954-60; gen. practice medicine, 1938; mem. psychiat. staff Menninger Clinic, 1946-54; clin. research in psychiatry Nat. Insts. Mental Health, Bethesda, Md., 1954-59, acting chief adult psychiatry, 1954-55, chief family study sect., 1954-59; asso. prof. clin. psychiatry Georgetown U. Med. Center, Washington, 1959—, chief family psychiatry, 1960—; pvt. practice as family psychotherapist, Chevy Chase, Md., 1954—; cons. family psychotherapy Psychiat. Inst., U. Md., Balt.; 1959—. Charter mem. Shawnee Guidance Center, Topeka, 1948-49, clin. dir., 1949-52, sr. psychiat. cons., 1952-54. Fellow Am. Psychiat. Assn., Am. Orthopsychiat. Assn.; mem. Group for Advancement Psychiatry, Family Com., Sigma Chi, Phi Chi. Research, publs. in devel. of family concept of emotional illness and method of family psychotherapy. Home: 4903 DeRussey Pkwy., Chevy Chase, Md. 20015. Office: Georgetown U. Med. Center, 3800 Reservoir Rd., Washington 20007.*

BOWEN, Norman Levi, geologist; b. Kingston, Ont., Can., June 21, 1887; s. William Alfred and Elizabeth (McCormick) B.; A.M., Queen's U., 1907, B.S., 1909, LL.D., 1941; Ph.D., Mass. Inst. Tech., 1912; hon. Sc.D., Harvard Tercent; m. Mary Lamont, Oct. 3, 1911; 1 dau., Mrs. Jerold Orne. Came to U. S. 1909. Field investigator Ont. Bur. Mines, 1907-09, Geol. Survey Can., 1910-11; petrologist Carnegie Instn., Washington, 1912-18, and 1920-37; Charles L. Hutchinson distinguished service prof. petrology U. Chgo., 1937-47, chmn. dept. geology, 1945-47; prof. mineraology Queen's U., 1919-20; petrologist Carnegie Instn., from 1947. Supr. optical glass prodn. War Industries Bd., World War I. Mem. Geol. Soc. Am. (pres. 1946), Mineral Soc. Am. (pres. 1937), Mineral. Soc. London, Geophys. Union, Am. Acad. Arts and Scis., Am. Philos. Soc., Nat., Washington, Indian (hon.) acads. scis., Kaiserlich deutsch Akademie der Naturforscher (Halle), Soc. Geol. Belgique, Finnish Acad. Sciences, All-Russian Mineral. Soc. Recipient Bigsby medal Geol. Soc. London, Eng., 1941, Penrose medal Geol. Soc. Am., 1941. Miller medal Royal Soc. Canada, 1943. Yacht. Author: The Evolution of the Igneous Rocks, 1928; also articles in profl. jours. Joint discoverer of Mullite, the fundamental constitutent of fire clay refractories. Died Sept. 11, 1956.

BOWEN, William Gordon, Am. economist; b. Cin., Oct. 6, 1933; s. Albert A. and Bernice (Pomert) B.; A.B., Denison U., 1955; Ph.D., Princeton, 1958; m. Mary Ellen Maxwell, Aug. 25, 1956; children—David Alan, Karen Lee. Faculty, Princeton, 1958—, prof. econs., 1965, dir. Grad. Studies, Woodrow Wilson Sch. pub. and Internat. Affairs, 1964-66, now provost, 1967—. Cons. Council Econ. Advisers, Office Edn. Mem. Am. Econ. Assn., Indsl. Research Assn., Phi Beta Kappa. Author: The Wage-Price Issue: a Theoretical Analysis, 1960; Wage Behavior in the Postwar Period: An Empirical Analysis, 1960; The Federal Government and Princeton University: A Report of the Effects of Princeton's Involvements with the Federal Government on the Operation of the University, 1962; Economic Aspects of Education: Three Essays, 1964; (with W. J. Baumol) Performing Arts: The Economic Dilemma, 1966; also articles. Editor: Labor and the National Economy, 1965; (with F. H. Harbison) Unemployment in a Prosperous Economy, 1965. Developed comprehensive analysis of econ. problems of live performing arts; showed labor force participation rates are sensitive to econ. conditions; analyzed role of wage settlements in inflationary process; demonstrated that debt finance can cause a transfer of real income from one generation to another. Home: 10 Maclean Circle, Princeton, N.J. 08540.*

BOWER, Frederick Orpen, Brit. botanist; b. Ripon, Eng., Nov. 4, 1855; s. Abraham and Cornelia (Morris) B.; ed. Trinity Coll., Cambridge, grad., 1877; studied under Julius Sachs, Wurzburg, 1877-78, also under Heinrich Anton de Bary, Strasbourg, 1879 (both Germany). Lectr. (introduced instrn. in modern botany) Christ Coll., Cambridge, 1877; lectr. botany S. Kensington, 1882; research at Jordrell Lab., Kew; regius prof. botany U. Glasgow (Scotland), 1885-1925. Recipient Neill prize Royal Soc. Edinburgh (pres. 1919-24), 1926, Darwin medal, 1938. Fellow Royal Soc., 1891 (medal 1910), pres. Brit. Assn., 1930. Author: The Origin of a Land Flora, 1908; the Botany of a Living Plant, 1919; Sixty years of Botany in Britain, 1938; Primitive Land Plants (also known as Tharcheoniat de). Research on plant morphology, including evolutionary morphology of pteridophyta; formulated generally accepted explanation for gradual increase in importance of asexual generation. Died Ripon, Apr. 11, 1948.

BOWERBANK, James Scott, English geologist; b. London, Eng., 1797; partner in distillery, London; lectr. on botany, 1822-24, on human osteology, 1831; a founder London Clay Club, 1836, Paleontol. Soc., 1847. Fellow Royal Soc., 1842. Author: Fossil Fruits of the London Clay (standard work), 1840; numerous papers. Worked on classification of spongidae, also studied their vital powers and spiculate elements. Died London, Mar. 8, 1877.

BOWERS, Albert, chemist; b. Manchester, Eng., July 16, 1930; s. Albert and Mary (Munn) B.; B.Sc., London U., 1951; Ph.D., U. Manchester, 1954; Fulbright postdoctoral fellow Wayne State U., 1954-55; m. Eileen B. Easthope, Sept. 26, 1953; children—Anne Christine, Karen Elizabeth, Deborah. Sci. officer Brit. Atomic Energy Authority, Radiochem. Center, Amersham, Eng., 1955-56; with Syntex S.A., Mexico D.F., 1956—, dir. research, 1961-63, v.p., dir. research, 1963-65, v.p. Syntex Corp., 1965—. Fellow Chem. Soc., London, Eng.; mem. Academia de la Investigación Científica (Sci. prize 1964), Am. Chem. Soc. Research, publs. and patents on steroid hormones; developed basic new procedures for the selective introduction of fluorine into steroid hormones, and new syntheses of the biologically important 19-nor steroids; synthesized several new classes of androgenic, progestational, and corticoid hormones. Home: 16, Rio Escondico, Lomas Hipódromo, Mexico 10, D.F. Office: 1457 Insurgentes Sur, Mexico 19, D.F., Mexico.*

BOWERS, Alfred William, anthropologist; b. Sandford, N.S., Can., June 1, 1901; s. Dorris Chadbourne and Nellie (Killam-Harris) B.; brought to U. S., 1907, naturalized, 1915; certificate Dickinson Coll., 1923; B.S. magna cum laude Beloit Coll., 1928; Ph.D., U. Chgo., 1948; m. Gladys E. Monson, July 16, 1939; children—Lois May (Mrs. Gary Chesnut), Norman Arnold. Tchr., N.D. rural schs., 1918-21; tchr. Mandan, (N.D.) Jr. High Sch., 1921-23; supt. schs., Dunn Center, N.D., 1923-27; research asst. Beloit Coll., 1929-32; field agt. U. S. Dept. Agr., N.D., 1934-38; research asst. U. Chgo., 1932-34, 1938-39, 1946-49; prof. anthropology U. Ida., 1949-67; anthropologist for Coeur d'Alene Indians before Indian Claims Commn., 1952-53, for Arikara, Mandan, Hidatsa before Indian Claims Commn., 1964-66. Recipient Sigma Xi award for best published research, 1965. Mem. Am. Anthropol. Assn., Soc. Am. Archaeology, Plains Anthropologist, Ida. Acad. Sci., Ida. Hist. Soc., Sigma Xi, Phi Kappa Phi. Author: Mandan Social and Ceremonial Organization, 1950; Hidatsa Social and Ceremonial Organization, 1965; A History of the Mandan and Hidatsa, 1948. Contbr. articles in field to sci. jours. Pioneer research in archaeol. history of semi-sedentary agrl. communities of the Mandan, Hidatsa and Arikara situated along the Missouri River; research in and an analysis of their social, econ., polit., religious orgn.; conducted intensive excavations in archeol. sites near Mobridge, S.D.; active in archeol. investigations of untested areas in Ida.; active in assisting various Indian tribes and U. S. Govt. in determining tribal lands. Home: 618 N. Howard, Moscow, Ida. 83843.*

BOWERS, Alphonzo Benjamin, Am. inventor, engr.; b. West Baldwin, Me., Sept. 25, 1830; s. Wilder and Sarah Hay (Thompson) B.; ed. Me. Wesleyan Sem., Bridgewater State Normal Sch., Phillips Acad., and by pvt. study. Built 1st dam at 16; went to Cal., 1853; engaged in mining, teaching, writing and lecturing; studied law, became his own atty. in suits against infringers of his patents; chief clerk office surveyor gen., 1863, dep. surveyor gen., 1864; in charge of sales of state lands, 1863-67. Mem. Internat. Congress of Commerce and Nav. Brussels, 1898; mem. Permanent Internat. Assn. Nav. Congresses. Invented method of cheap transp. of earth by stream of water on down grade in open flume for bldg. dams and embankments, 1858; flexibly connected floating pipes and developed method of building levees from old style dredges; used rotary excavator in hydraulic dredging, transp., filling, and hydraulic dredge. Died Jan. 24, 1926.

BOWERS, John Zimmerman, Am. physician; b. Catonsville, Md., Aug. 27, 1913; s. John C. and Adelaide (Schuman) B.; B.S., Gettysburg Coll., 1933; Sc.D., 1958; M.D., U. Md., 1938, Sc.D., 1959; D.H.L. (hon.), Womans Medical College, 1967; m. Imogene Clapp, Oct. 21, 1943; children—John C., Mary Imogen, David Warren. Practice medicine specializing in edn., Balt., 1945-47; dep. dir. U. S. AEC div. biology and medicine, 1947-50; dean U. Utah Med. Sch., 1950-55, U. Wis. Med. Sch., 1955-61; vis. prof. Kyoto (Japan), U. Med. Sch., 1962-64; vis. prof., 1965-—; staff Rockefeller Found., N.Y.C., 1964-65; pres. Josiah Macy Jr. Found., N.Y.C., 1965-—. Mem. sci. adv. com. for TV, Bell Labs., 1955-62; mem. Council on Med. Edn. and Hosps., 1957-62; mem. adv. com. W. K. Kellogg Found., 1955-62; cons. Ford Found., 1952-59. Mem. A.A.A.S., A.M.A., A.C.P., N.Y. Acad. Medicine. Author: Medical Education in Japan, 1965; also numerous articles. Editor, Jour. Med. Edn., 1957-62. Studies on biol. effects radiation on electrolytes; med. students and med. education in Asia, Japan, Philippines. Home: 1 W. 54th St. Office: 277 Park Av., N.Y.C.*

BOWERS, Raymond, physicist; b. London, Eng., July 11, 1927; B.Sc., U. Coll., London U., 1948; D.Phil., Oxford (Eng.) U., 1951; Came to U. S., 1951, naturalized, 1964. Research fellow U. Chgo., 1951-53; research physicist Westinghouse Electric Corp., 1954-60; prof. physics Cornell U., Ithaca, N.Y., 1960-—. Cons. to various cos. Fellow Am. Phys. Soc. Research on physics of solids at low temperatures. Home: 109 Hanshaw Rd., Ithaca, N.Y. 14850.*

BOWERS, Roy Anderson, Am. pharm. chemist; b. Racine, Wis., May 11, 1913; s. Sidney E. and Dagmar (Anderson) B.; B.S., U. Wis., 1936; Ph.D., 1940; m. Harriett Teresa Byer, Aug. 19, 1940; children—Clarke G., Mary J. Asst. prof. U. Toledo, 1940-41; faculty U. Kan., 1941-43; asso. prof., 1943-45; dean, prof. pharmacy U. N.M., Albuquerque, 1945-51, Rutgers U., Newark, 1951-—. Bd. dirs. Am. Found. for Pharm. Edn., Am. Pharm. Assn. Found. Mem. Am. (past 1st v.p., past mem. council), N.J., N.M. pharm. assns., Am. Assn. Colls. Pharmacy (past pres.), N.J. Soc. Hosp. Pharmacists, N.Y. Acad. Scis., Am. Inst. History Pharmacy, Sigma Xi, Kappa Psi, Alpha Chi Sigma, Rho Pi Phi (hon.), Alpha Zeta Omega (hon.), Delta Sigma Theta (hon.), Phi Kappa Phi, Phi Lambda Upsilon. Author: (with David L. Cowen) The Rho Chi Society, 1961; contbg. Author: American Pharmacy, 1960; also articles. Isolated and identified constituents carnauba wax; research on fats, oils and waxes. Home: 76 Franklin St., Cedar Grove, N.J. 07009. Office: 1 Lincoln Av., Newark 07104.*

BOWIE, Clifford Pinckney, Am. engr.; b. Phillipsburg, Mont., June 28, 1879; s. Clifford Napoleon and Mary Ellen (Irvine) B.; B.S., U. Cal., 1904, C.E. (hon.), 1923; m. Alice Frances Jones, June 8, 1906 (dec. Aug., 1923); 1 dau., Barbara Alice; m. 2d, Nan Mountjoy, June 2, 1928. Asst. engr. U. S. Reclamation Service, 1903; asst. city engr., Hayward, Calif., 1904; asst. to city engr., San Francisco, 1904-06; city engr., Chico, Cal., 1906-07, asst. engr., later chief engr. Asso. Pipe Line Co., 1907-16; supervising engr. U. S. Bur. Mines Petroleum Sta., San Francisco, 1916-46; cons. engr. since 1946. Mem. Am. Inst. Mining and Metall. Engrs. Author bulls. U. S. Bureau Mines; Construction and Operation of a Single Tube Cracking Furnace for Making Gasoline, 1916; Oil Storage Tanks and Reservoirs, 1918; Extinguishing and Preventing Oil and Gas Fires, 1918; Oil Camp Sanitation, 1921; The Bowie-Gavin Process, 1926; Salvage of Materials in the Oil Industry, 1929; Transportation of Gasoline by Pipe Line, 1932; Hardening of Mud Sheaths in Contact with Oil, and a Suggested Method for Minimizing Their Sealing Effect in Oil Wells, 1937, Contbr. to jours. Inventor Bowie-Gavin process. 1920, mud lining oil and gas wells, 1936, oil well formation slotter, 1939, drilling wells through heaving formations, 1940. Died Apr. 1, 1952.

BOWIE, Edward Hall, Am. meteorologist; b. Annapolis Junction, Md., Mar. 29, 1874; s. Thomas John and Susanna Hall (Anderson) B.; M.S., St. John's Coll., Annapolis, Md., 1920; m. Florence C. Hatch, Dec. 12, 1895; children—Mrs. Helen McKinstry Prentiss, Mrs. Margaret Lowndes Wallace, Mrs. Susanna Anderson Lindquist. Entered service U. S. Weather Bur., 1891; asst. observer at Memphis, 1891-95, Montgomery, Ala., 1896-98; observer, Dubuque, Ia., 1898-1901; section dir., Galveston, Tex., 1901-03; local forecaster, St. Louis, 1903-09; chief forecast div. U. S. Weather Bur., 1910-12, nat. forecaster, 1909-24, prin. meteorologist and dist. forecaster, Pacific States, from 1924. Mem. cons. meteorology and sci. hydrology, Am. Geophys. Union. Fellow Am. Meteorol. Soc., Cal. Acad. Scis.; mem. Philos. Soc. Washington, Washington Acad. Scis. Died July 29, 1943.

BOWIE, Stanley Hay Umphray, Brit. geologist; b. Shetland, Scotland, Mar. 24, 1917; s. James Cameron and Mary (Nicholson) B.; B.Sc. with 1st class honors in Geology, U. Aberdeen (Scotland), 1941; m. Helen Elizabeth Pocock, Nov. 1, 1948; children—Roderick Cameron, Antony Roy. Geophysicist, Wilkins & Devereux Ltd., part-time 1940-41; sr., prin. geologist atomic energy div. Geol. Survey Gt. Britain, 1946-55; chief geologist atomic energy div. Geol. Survey Gt. Britain, Kensington, London, 1955-67; head geochem. div. Inst. Geol. Scis., London, 1967-—. Cons. geologist U.K. Atomic Energy Authority, 1955-—. Recipient Silver medal Royal Soc. Arts, 1958. Fellow Mineral Soc. Am.; mem. Instn. Mining and Metallurgy (editorial bd. Applied Earth Sci.), Mineral Soc. (chmn. com. on geochemistry), Internat. Mineral. Assn. (sec. commn. on ore microscopy), Soc. Econ. Geologists, Mineral. Soc. Gt. Britain, Geol. Soc. London. Contbg. author: Nuclear Geology, 1954; Physical Methods in Determinative Mineralogy, 1967. Research, publs. on econ. geology of atomic energy raw materials, techniques of discovering radioactive mineral deposits, including devel. several types of Geiger-Müller and scintillation counters, beryllium detectors and portable X-ray fluorescence analysers; developed quantitative system of determining opaque minerals and applied autoradiograph techniques to study of uranium and thorium ores; discovered large quantities of helium in Orange Free State. Office: 64/78 Gray's Inn Rd., London W.C.1, Eng.*

BOWIE, William, Am. geodesist; b. Annapolis Junction, Md., May 6, 1872; s. Thomas John and Susanna A(nderson) B.; St. John's Coll., Annapolis, Md.; B.Sc., Trinity Coll., Conn., 1893; C.E., Lehigh U., Pa., 1895, Sc.D., 1922; M.A., Trinity, 1907, Sc.D., 1919; LL.D., U. Edinburgh (Scotland), 1936; Sc.D., George Washington U., 1937; m. Elizabeth Taylor Wattles, June 28, 1899; children—William Bladen (dec.), Clagett. Mem. field force, U. S. Coast and Geod. Surveys, 1895-1937, retired; engaged in field on coast or geod. surveys in many states of U. S., also in P.R., P.I., and Alaska, 1895-1909; in charge Div. of Geodesy, 1909-36; commd. hydrographic and geodetic engr., 1917. In charge summer course in practical astronomy and geodesy, Columbia U., 1912-17. Del. from U. S. to numerous internat. scientific confs. Pres. Internat. Geod. Assn., 1919-33, Internat. Geod. and Geophys. Union, 1933-36; chmn. bd. Surveys and Maps of Federal Govt.; 1922-24. Spl. lectr. Lehigh U., 1922-36. Recipient Cresson medal Franklin Inst., 1937, William Bowie medal Am. Geophys. Union, 1939. Wrote number of publs. of the Coast and Geodetic Survey on the various branches of geodesy, including measurement of base lines, triangulation, gravity, isostasy; also geodetic and engring. articles in scientific and engring. mags. Author: Isostasy, 1927. Died Aug. 28, 1940.

BOWKER, Albert Hosmer, Am. math. statistician; b. Winchendon, Mass., Sept. 8, 1919; s. Roy C. and Kathleen (Hosmer) B.; B.S., Mass. Inst. Tech., 1941; Ph.D. (NRC fellow), Columbia, 1949; m. Rosedith Sitgreaves, Sept. 26, 1964; children—Paul, Caroline and Nancy (twins). Asst. statistician Mass. Inst. Tech., 1941-43; asst. dir. statis. research group Columbia, 1943-45; asst. prof. statistics Stanford, 1947-50, head dept. statistics, 1948-59, asso. prof., 1950-53, dir. applied math. and statistics labs., 1951-63, prof. math. and statistics, 1953-63, dean grad. div., 1958-63; chancellor U. City N.Y., 1963-—. Cons. statis. engring. sect. Nat. Bur. Standards, 1949-—; mem.-at-large div. math. NRC-Nat. Acad. Scis., 1962-65; mem. adv. com. Office Statis. Standards on Statis. Policy, 1963; chmn. fellowship com. Assn. Grad. Schs., Assn. Am. Univs.; chmn. mem. com. Council Grad. Schs. in U. S. Fellow Inst. Math. Statistics (council 1952-—, pres. 1961-62), Am. Statis. Assn. (dir. 1959-61, pres. 1964), Am. Soc. Quality Control (editorial bd., past chmn. sect. com. on fellows), A.A.A.S.; mem. Biometric Soc., Am. Assn. for Computing Machinery, Operations Research Soc. Am., Soc. for Indsl. and Applied Math., Sigma Xi (exec. com. 1963-66). Author: (with Henry P. Goode) Sampling Inspection by Variables, 1952; (with Gerald J. Lieberman) Handbook of Industrial Statistics, 1955; (with Gerald J. Lieberman) Engineering Statistics, 1958; also articles. Home: 128 E. 95th St., N.Y.C. 10028.*

BOWLAND, John Patterson, Canadian animal nutritionist; born Manitoba, Canada, February 10, 1924; the son of Herbert John and Gertrude (Patterson) B.; B.S.A., U. Manitoba, 1945; M.S., Wash. State U., 1947; Ph.D., U. Wis., 1949; m. Helen May Campbell, May 28, 1946; children—Anne, Lynne. Agrl. supr. Can. Dept. Agr., Brandon, Man., 1945-46; faculty U. Alta., Edmonton, Can., 1949-—, prof. animal nutrition, 1962-—; research Nat. Inst. for Research, Darrying, Shinfield, Reading, Eng., 1959-60. Recipient Borden award Nutrition Soc. Can., 1966. Fellow A.A.A.S.; mem. Agrl. Inst. Can., Alta. Inst. Agrology, Am. Soc. Animal Sci., Canadian Soc. Animal Prodn., Nutrition Soc. Can., Sigma Xi. Author (with D. R. Clandinin, L. R. Wetter) Rapeseed Meal for Livestock and Poultry, A Rev., 1965. Research, numerous publs. on energy protein and energy amino acid relationships in diets of swine and rats, copper metabolism in pigs, milk composition of swine. Home: 11243 79th Av., Edmonton, Alta., Can.*

BOWLER, John Pollard, Am. surgeon; b. Cambridge, Mass., Jan. 3, 1895; s. John William and Ellen (Pollard) B.; A.B., Dartmouth, 1915, D.Sc., 1952; M.D., Harvard, 1919; M.Sc., Mayo Found., Mayo Clinic, 1924; m. Madelaine Gile, May 28, 1928; children—Patricia (Mrs. Richard D. Leggat), Janet. Mem. surg. staff Hitchcock Hosp., Hanover, N.H., 1925-60, chief of staff, chmn. Hitchcock Clinic, 1927-60; faculty Dartmouth Med. Sch., 1927-60, dean, 1927-44, prof. surgery, 1937-60; cons. VA. Chmn., N.H. med. adv. com. to SSS, 1940-—. Mem. N.H. Hill-Burton Commn., 1948-54; N.H. State chmn. Nat. Found., 1964-—. Recipient Robins award N.H. Med. Soc., 1962. Mem. Theta Delta Chi, Alpha Kappa Kappa. Contbr. articles to med. publs. Home: 5 Hovey Lane, Hanover, N.H. 03755.*

BOWLES, Edward Lindley, Am. elec. engr.; b. Westphalia, Mo., Dec. 9, 1897; s. Samuel Addison and Julia (Johnson) B.; B.S., Washington U., St. Louis, 1920; M.S., Mass. Inst. Tech., 1922; D.Sc., Norwich U., 1945; m. Lois Wuerpel, June 17, 1922; children—Edmund Addison, Frederick Wuerpel. Faculty, Mass. Inst. Tech., 1920-—, prof. elec communications, 1937-63, prof. emeritus, 1963-—; cons. engr., 1923-—; pres. Whitin Machine Works, Whitinsville, Mass., 1965-66; dir. White Consol. Industries, Cleve., 1966-—, Jaffell Ash, Waltham, Mass.; asst. to pres. Aneley Corp., Boston, 1966. Cons. to govt. agys., pvt. cos.; chmn. ad hoc adv. com. on VHF-UHF, TV allocations Senate Com. On Interstate and Fgn. Commerce, 1956-58. Decorated D.S.M., Order Brit. Empire; recipient Presdl. medal of merit, 1948; Distinguished Alumni citation Washington U., 1955. Fellow Am. Inst. E.E., I.R.E., Am. Phys. Soc., Am. Acad. Arts and Scis. (past v.p.); mem. A.A.A.S., Soc. for Promotion Engring. Edn., Operations Research Soc., Sigma Xi. Patentee in field. Invention and devel. electronic devices especially in communications; research on air navigation. Home; 15 Greylock Rd., Wellesley Hill, Mass. 02181.*

BOWLES, Gordon Townsend, Am. anthropologist; b. Tokyo, Japan, June 25, 1904 (parents Am. citizens); s. Gilbert and Minne (Pickett) B.; B.A., Earlham Coll., 1925; Ph.D., Harvard, 1935; m. Elizabeth Jane Thomas, Feb. 6, 1932; children—Anne, Barbara (Mrs. Thomas E. Swann). Asst. in anthropology Harvard, 1932-35, research asso., 1935-38; asst. prof. U. Hawaii, 1938-42; mem. Govt. Mission to China, 1942; research asso. in phys. anthropology Bishop Mus., Honolulu, 1938-42; with Fgn. Econ. Adminstrn., also U. S. Dept. State, 1942-47; exec. sec. Conf. Bd. Assn. Research Couns., Com. Internat. Exchange of Persons, 1947-51; vis. prof. anthropology Tokyo U., 1951-58; faculty Columbia, 1959-60; faculty Syracuse (N.Y.) U., 1960-62, 1960-—, prof. anthropology, 1962-—; Asso. mng. dir. Internat. House Japan, 1952-58. Recipient Fulbright-Hays lectureship award for Japan, 1967-68. Fellow Anthrop. Assn.; mem. Japan Ethnol. Soc. (hon. life), Assn. Phys. Anthropology, Am. Inst. Biol. Scis., Assn. for Asian Studies, A.A.A.S., Assn. for Asian Studies. Author: New Types of Old Americans at Harvard and Eastern Women's Colleges, 1932; (with Earl W. Count) Fact and Theory in Social Science, 1964. Research on environmental influences and size increase by generations among eastern seaboard old white Americans; comparative anthropometric studies in Tibetan highlands, W. China, N. India; research on inbreeding and assortive mating patterns in Japanese mountain community, genetic isolates, environmental pressures and population control, such as econ. distress and famines in relation to infanticide, abortion. Home: 128 Dorset Rd., Syracuse, N.Y. 13210.*

BOWLES, Oliver, mining engr., geologist; b. nr. Lindsay, Ont., Can., Jan. 10, 1877; s. William Henry and Sarah A. (Glaspell) B.; B.A., U. Toronto, 1907, M.A., 1908; postgrad. U. Mich., U. Minn.; Ph.D., George Washington U., 1922; m. Eva H. Workman, 1908; children—William George, Edgar Oliver. Came to U. S., 1908, naturalized, 1914. Field work Bur. Mines, Ont., summers, 1908-10; tchr. geology and mineralogy U. Mich., 1908-09, U. Minn., 1909-14; with State Geol. Survey, Minn., summers 1911-13; temporary geologist U. S. Geol. Survey, parts of 1912-14; quarry technologist U. S. Bur. Mines, 1914-17, mineral technologist, 1917-23; supervising engr. nonmetallic minerals expt. sta., New Brunswick, N.J., 1923-28, supervising engr. bldg. materials sect., 1928-37, asst. chief non-metal economics div., 1937-42, chief, 1942-47, ret., 1947; parttime research prof. U. Md., 1948—. Recipient Gold medal Dept. Interior for distinguished service in Bur. Mines; Hardinge award Am. Inst. Mining and Metall. Engrs. Mem. Am. Inst. Mining and Metall. Engrs., Mineral Soc. Am., Soc. Econ. Geologists, N.Y. Acad. Sci., Sigma Xi; hon. mem. Inst. Quarrying of Gt. Britain. Author: The Stone Industries; also numerous bulletins, reports and articles in tech. press on stone, slate, lime, cement, asbestos. Died Aug. 1, 1958.

BOWLEY, Sir Arthur Lyon, Brit. mathematician, economist; b. Nov. 6, 1869; s. J. W. L. Bowley; ed. Christ's Hosp., Trinity Coll., Cambridge; D.Sc., Manchester; D.Litt., Oxford U.; m., 1904; 3 daus. Asst. master St. John's Sch., Leatherhead, 1893-99; Newmarch lectr. Univ. Coll., London, 1897-98, 1927-28; math. lectr. Univ. Coll., Reading, 1900-07; prof. math. and econs., 1907-13, lectr. econs., 1913-19; prof. statistics U. London, 1919-36; acting dir. inst. statistics U. Oxford, 1940-44; lectr. London Sch. Econs. and Polit. Sci., from 1895. Recipient Cobden prize, 1892; Adam Smith prize, 1894. Fellow Royal Statis. Soc. (Guy Gold medal 1935, pres. 1938-40), Royal Econs. Soc. (v.p.), Internat. Inst. Statisticians (hon. pres.), Econometric Soc. (pres. 1938), Brit. Assn. (pres. 1906). Author: England's Foreign Trade in the 19th Century, 1893; Wages in the United Kingdom in the 19th Century, 1900; Elements of Statistics, 1901; National Progress in Wealth and Trade, 1904; An Elementary Manual of Statistics, 1910; A General Course of Pure Mathematics, 1913; Measurement of Social Phenomena, 1914; War and External Trade, 1915; Change in Distribution of Income, 1920; The Mathematical Groundwork of Economics, 1924; Some Economic Consequences of the Great War, 1930; Studies in National Income, 1924-38, 42; also articles. Co-author: Livelihood and Poverty, 1915. Died Jan. 21, 1957.

BOWMAN, John Eddowes, Sr., English naturalist; b. Nantwich, Eng., Oct. 30, 1785; s. Eddowes Bowman; m. Elizabeth Eddowes, July 6, 1809; children—Mrs. George S. Kenrick, Eddowes, Henry, William, John Eddowes, Jr. Partner, banking bus., 1813-16; br. mgr. Beck & Co. bank, Welshpool, Wales; mng. partner bank, Wrexham, Wales, 1824-30; engaged in sci. studies, Manchester, Eng., from 1837. Fellow Linnean Soc., Geol. Soc.; a founder Manchester Geol. Soc. Contbd. papers to sci. jours. Investigated mosses, fungi, parasitical plants; Enctothyra Bowmanni (fossil) named in his honor. Died Dec. 4, 1841.

BOWMAN, John Eddowes, Jr., Welsh chemist; b. Welshpool, Wales, July 7, 1819; s. John Eddowes and Elizabeth (Eddowes) B.; ed. under Daniel at King's Coll., London. Became demonstrator chemistry King's Coll., 1845, 1st prof. practical chemistry, 1851; a founder Chem. Soc. London. Author: A Lecture on Steam Boiler Explosions, 1845; An Introduction to Practical Chemistry, 1848; A Practical Handbook of Medical Chemistry, 1850; papers in sci. jours. Died Feb. 10, 1845.

BOWMAN, Robert Irvin, biologist; born Saskatoon, Sask., Can., Nov. 19, 1925; s. Irvin and Phoebe (Herner) B.; B.A., Queen's U., Kingston, Ont., Can., 1948; Ph.D., U. Cal. at Berkeley, 1957; m. Margret Helene Seckel, Dec. 15, 1951; children—Paul, Peter, Carl. Came to U. S., 1948, naturalized, 1957. Faculty, San Francisco State Coll., 1955—, prof. biology, 1965—; mem. survey Galapagos Islands, Ecuador, UNESCO, 1957; co-dir. Galapagos Internat. Sci. Project, U. Cal. at Berkeley, 1963-64. Recipient Medal of Honor, Republic of Ecuador, 1964. Mem. Am. Ornithologists' Union, Ecol. Soc. Am., Cooper Ornithol. Soc. Author: Morphological Differentiation and Adaptation in Galapagos Finches, 1961; also articles. Editor: The Galapagos, 1966. Research in evolutionary biology, vertebrates Galapagos Islands. Home: 1069 Sterling Av., Berkeley, Cal. 94708. Office: 1600 Holloway Av., San Francisco 94132.*

BOWMAN, Thomas E., Am. zoologist; b. Bklyn., Oct. 21, 1918; s. Karl Murdock and Eliza (Stearns) B.; S.B., Harvard, 1941; M.A., U. Cal. at Berkeley, 1948; Ph.D., Scripps Inst. Oceanography, 1954; m. Mary Jo Coogan, Mar. 6, 1943; children—Judith Stearns, Kathleen Isabell, Susan Abbott. Grad. research biologist Scripps Inst. Oceanography, 1948-53; asst. prof. marine biology U. R.I., 1954; asso. curator, div. marine invertebrates Smithsonian Instn., Washington, 1954—. Home: 13210 Magellan Av., Rockville, Md. 20853. Office: Div. Marine Invertebrates, Smithsonian Instn., Washington 20560.*

BOWMAN, William, English ophthalmic surgeon, anatomist; b. Nantwich, England, July 20, 1816; s. John Eddowes and Elizabeth (Eddowes) B.; educated General Hospital, Birmingham, England; King's College, London; and on the continent; m. Harriet Paget, Dec. 24, 1842; 7 children, including Paget Bowman. Apprentice to Joseph Hodgson, Birmingham; jr. demonstrator anatomy, mus. curator King's Coll., apptd. prof. physiology and gen. and morbid anatomy, 1848, elected asst. surgeon Hosp., 1840, surgeon, 1856; surgeon Royal Ophthalmic Hosp., Moorfields, 1851-76. Established (with Robert Bentley Todd) St. John's House and Sisterhood. Created baronet, 1884. Fellow Royal Soc., 1841; Royal Coll. Surgeons; mem. Ophthal. Soc. U.K. (1st pres.). Author: Physiological Anatomy and Physiology of Man (with Robert B. Todd), 5 vols.; 1845-56; Lectures . . . on the Eye, 1849; Collected papers, 1892. Early expert in use ophthalmoscope; investigated minute structure of eye, kidney, striped muscles; Bowman's membrane, glands, capsule named for him; considered father gen. anatomy in Eng. Died Joldwynds, Eng., Mar. 29, 1892.

BOWMAN, William Cameron, Brit. pharmacologist; b. Carlisle, Eng., Apr. 26, 1930; s. John William and Esther (Cameron) B.; B.Pharm., University of London (England), 1952, Ph.D., 1955, D.Sc., 1967; m. Kathleen Margaret Sheerer, December 20, 1952; children—Alison, Ewen Cameron. Mem. faculty, U. London, 1952-55, 57-66, reader pharmacology, 1963-66; prof. exptl. pharmacology U. Strathclyde, Glasgow, Scotland, 1966—. Fellow Royal Soc. Medicine, Inst. Biology; mem. Pharm. Soc., Brit. Pharmacological Soc., Physiol. Soc. Author: (with G. B. West, M. J. Rand) Textbook of Pharmacology, 1967; also articles. Research into mechanism of action of drugs affecting striated muscle. Home: 17 Belmont Rd., Kilmacolm, Renfrew, Scotland. Office: U. Strathclyde, Glasgow, Scotland.*

BOX, George Edward Pelham, statistician; b. Gravesend, Eng., Oct. 18, 1919; s. Harry and Helen (Martin) B.; B.Sc., Univ. Coll., U. London (Eng.), 1947, Ph.D., 1952, D.Sc., 1961; m. Joan Gunnhild Fisher, Dec. 12, 1959; children—Helen Elizabeth, Harry Christopher. Came to U. S., 1956. Statistician, head statis. techniques research sect. Imperial Chems. Industry, Blackley, Manchester, Eng., 1948-56; vis. research prof. U. N.C., 1952-53; dir. statis. techniques research group Princeton, 1957-59; prof. statistics, chmn. dept. U. Wis., 1960—; cons. to industry, 1956—. Recipient Profl. Progress award Am. Inst. Chem. Engrs., 1963. Ford Found. vis. prof. Harvard Grad. Sch. Bus. Adminstrn., 1965-66. Fellow Royal Statis. Soc. (Guy medal 1964). Am. Statis. Assn. (past v.p.), Inst. Math. Statistics, Am. Soc. Quality Control, A.A.A.S.; mem. Internat. Statistics Inst., Biometrics Soc. Author: (with others) Statistical Methods in Research and Production, 1957; Design and Analysis of Industrial Experiments, 1959; also research publs. Originator response surface methodology and evolutionary operation for designing expts. to improve indsl. processes; research on robustness of statis. procedures, building and estimation of sci. models, forecasting of time series and control of indsl. processes. Home: 3437 Edgehill Pkwy., Madison, Wis. 53705.*

BOXER, George Ernst, biochemist; b. Vienna, Austria, Feb. 2, 1915; s. Arnold and Helene (Kreitner) B.; M.D., U. Vienna, 1938; Ph.D., Columbia, 1944; m. Lily Behar, Jan. 23, 1940; children—Steven G., Peter A. Demonstrator med. chemistry, Vienna, 1935-38; asst. in biochemistry, Cambridge, Eng., 1939; faculty Columbia, 1941-43; research asso. OSRD, 1943-45, research asso. Merck & Co., 1945-56, dir. enzymology Merck Inst., 1956-62, dir. biol. cancer research Merck, Sharp & Dohme Research Labs., 1962-66, exec. dir. Merck Inst. Therapeutic Research, Rahway, N.J., 1966—. Mem. Am. Chem. Soc., A.A.A.S., Am. Soc. Biol. Chemists, Soc. Exptl. Biology and Medicine, Royal Soc. Medicine, Am. Assn. Cancer Research. Research, numerous publs. in use isotopes in studies metabolism, antibiotics, disturbances in lipid metabolism, enzymology, intermediate metabolism of normal, intermediate metabolism normal and malignant tissues. Home: 513 Fairmont Av., Westfield, N.J. 07090. Office: Merck Inst. Therapeutic Research, Rahway, N.J. 07065.*

BOYARSKY, Louis Lester, Am. biophysicist; b. Jersey City, Sept. 5, 1919; s. Samuel and Emma (Newman) B.; B.S., Coll. City N.Y., 1941; M.S., Purdue U., 1945; Ph.D., U. Chgo., 1948; m. Lila H. Benjamin, Dec. 25, 1941; children—Amy, Gregory. Psychophysiologist, Inst. for Juvenile Research, Chgo., 1949-50; faculty U. Ky., Lexington, 1950—, prof. biophysics, 1958—. Recipient Alumni Research award U. Ky., 1959. Fulbright fellow U. Milan, 1958. Mem. Am. Physiol. Soc., Biophysics Soc., Am. Acad. Neurology, Phi Beta Kappa, Sigma Xi. Contbr. articles to sci. jours. Researcher in neurology. Home: 1729 Traveller Rd., Lexington, Ky. 40504.*

BOYARSKY, Saul, Am. urologist, physiologist; b. Burlington, Vt., July 22, 1923; s. Samuel and Ethel (Kaplan) B.; B.S., U. Vt., 1943, M.D., 1946; m. Rose Eiseman, June 17, 1945; children—Myer William, Terry Linda, Hannah Gail. Instr., Duke, 1954, prof. urology, asst. prof. physiology Sch. Medicine, 1963—; USPHS fellow physiology N.Y. U. Sch. Medicine, 1954-55, instr., 1955-56; faculty Albert Einstein Coll. Medicine, 1956-63, asso. prof., 1962-63; chief urology VA Hosp., Durham, 1963—. Cons. urology Watt, Lincoln hosps., 1965—, NIH Inst. Arthritis and Metabolic Diseases. Fellow N.Y. Acad. Scis.; mem. A.A.A.S., Am. Urology Assn., A.M.A., Am. Fedn. Clin. Research, Mt. Desert Island Biol. Lab., Am. Med. Writers Assn., Am. Physiol. Soc., Am. Fertility Assn., Am. Paraplegia Soc., Pan Pacific Surg. Assn., Soc. Exptl. Biology and Medicine, Pan Am. Med. Assn., Am. Assn. U. Profs., A.C.S., Am. Assn. Med. Colls., Internat. Soc. Nephrology, Sigma Xi. Author: Neurogenic Bladder, 1966; also numerous articles. Research on renal, ureteral, bladder physiology, cinefluorography urinary tract in urology, ureteral pharmacology, neurogenic bladder disease. Home: 2713 Sevier St., Durham, N.C. 27705.*

BOYCE, Earnest, Am. civil and san. engr., educator; b. Winterset, Ia., July 11, 1892; s. Marcus James and Grace (Smith) B.; B.S., Ia. State Coll., 1917, C.E., 1930; M.S. in Engring., Harvard, 1932; m. Elsie Jane Green, Sept. 21, 1919; 1 son, James Earnest. Faculty, U. Kan. Sch. Engring., 1920-41; dir. div. sanitation, chief engr. Kan. State Bd. Health, 1924-41; chief sanitation facilities sect. San. Engring div. USPHS, Washington, 1941-44; prof. municipal and san. engring. Coll. Engring., prof. pub. health engring. Sch. Pub. Health, U. Mich., Ann Arbor, 1944-61, chmn. dept. civil engring., 1947-61; cons. san. engr., 1961—. Mem. Pub. Health Planning Team, Germany, summer 1951; vis. team WHO, Indonesia, 1953, cons. san. engr., Western Pacific region, 1958-60; cons. san. engr. ICA, India and Pakistan, 1960; cons. Pan Am. Health Orgn., Latin Am., 1962—. Recipient Charles Alvin Emerson medal Water Pollution Control Fedn., 1957. Diplomate Am. Acad. San. Engrs. Mem. Water Pollution Control Fedn. (hon., past pres.), Am. Soc. C.E., Am. Pub. Health Assn. (v.p. 1961-62), Am. Water Works Assn., Am. Soc. Engring. Edn., Engring. Soc. Detroit (pres. 1956), Sigma Xi, Phi Kappa Phi, Delta Omega, Tau Beta Pi. Research, publs. on san. and pub. health engring. edn. Home: 1601 Granger Av., Ann Arbor, Mich. 48104.*

BOYCE, John Shaw, Jr., Am. forest pathologist; b. Portland, Ore., May 25, 1921; s. John Shaw and Lillian (Jameson) B.; B.S., Yale, 1942, M.F., 1948; Ph.D., Duke, 1951; m. Anne Stackhouse Tilghman, Dec. 20, 1946; children—Merrill, Rebecca, Jacqueline, Olivia. Pathologist, S. E. Forest Expt. Sta. U. S. Forest Service, Asheville, N.C., 1950-61; prof. forest pathology U. Ga., 1961-66; Soc. Am. Foresters vis. scientist, 1966. Mem. Soc. Am. Foresters, Am. Phytopath. Soc., Asso. editor Jour. Forestry, 1961-66. Contbr. numerous articles to profl. jours. Research in forest disease, pine needle diseases, oak wilt, root rots of pines.*

BOYCE, Joseph C(anon), Am. physicist; b. Pitts., Jan. 23, 1903; s. David C. and Mary L. (Wright) B.; A.B., Princeton, 1922, A.M., 1923, Ph.D., 1926; postgrad. U. London, Cambridge U.; m. Emily M. Hughes, June 15, 1934; children—Mary Hughes (Mrs. Nelson A. Gelfman), Emily Jane (Mrs. Thomas E. White), Elizabeth Rogers, Frances Julia (Mrs. Ronald R. Swann), Katharine Louise. Faculty, Princeton, 1926-32, Mass. Inst. Tech., 1932-41, N.Y. U., 1945-50; with OSRD, Cambridge, Mass., Washington, 1941-45; asso. lab. dir. Argonne Nat. Lab., Chgo., 1950-55; v.p., grad. dean Ill. Inst. Tech., Chgo., 1955-62; with Office Sci. Personnel, Nat. Acad. Scis.-NRC, Washington, 1962—, dep. dir., 1964—. Co-dir. Harvard-Mass. Inst. Tech. Solar Eclipse Expdn., 1936; research asso. Carnegie Instn. Washington, 1939-50; cons. Dept. Def., 1948-51. Recipient Presdl. certificate of merit, 1947 (U.S.); King's medal for service, 1948 (U.K.); medal Nat. Found. Sci. Research, 1952 (Belgium). Fellow Am. Phys. Soc., A.A.A.S., Royal Astron. Soc.; mem. Am. Astron. Soc. Editor: New Weapons for Air Warfare, 1947. Contbr. articles to sci. jours. Participant identification neon and argon in spectra gaseous nebulae, 1934-36; research in spectroscopy vacuum ultraviolet, wavelength standards, spectra rare gases with applications to astrophysics. Home: 2500 Que St., Washington 20007. Office: 2101 Constitution Av. N.W., Washington 20418.*

BOYCE, Sir Rubert William, English pathologist, hygienist; b. London, Apr. 22, 1863; s. Robert Henry and Louisa (Neligan) B.; ed. U. Coll., London; M.B., London U., 1889; m. Kate Ethel Johnston (dec. 1902); 1 dau. Asst. prof. pathology U. Coll., London, 1892; prof. pathology U. Coll., Liverpool, 1894-1911; bacteriologist Liverpool Corp., 1898; founder (with Alfred Jones), also dean Liverpool Sch. Tropical Medicine, after 1898; organized trips to investigate diseases in Tropics, 1901, to examine yellow fever epidemics in New Orleans and Brit. Honduras, 1905, then to West Indies, 1909, West Africa, 1910; public analyst City of Liverpool; mem. Royal Commn. on Tb. Fellow Royal Soc., 1902. Author: A Textbook of Morbid Histology, 1892; Papers on Tropical Sanitation, 1894-95; Mosquito or Man, 1909; Health Progress and Administration in the West Indies, 1910;

(Joint) Handbook of Pathological Anatomy. Research on comparative pathology, med. entomology, biochemistry, tropical diseases, sanitation. Died June 18, 1911.

BOYCE, Stephen Gaddy, Am. plant ecologist, forester; b. Ansonville, N.C., Feb. 5, 1924; s. William Henry and Louise (Gaddy) B.; student Davidson Coll., 1942-43; B.S., N.C. State U., 1949, M.S., 1951; Ph.D., 1953; m. Helen Ashley, Dec. 23, 1951; children—Stephen Ashley, Thomas Mark. Instr., Meredith Coll., Raleigh, N.C., 1951-52, N.C. State U., 1952-53; asst. prof. botany Ohio U., 1953-57; research forester U. S. Forest Service, Carbondale, Ill., 1957-64; asst. dir. Central States Forest Expt. Sta., Columbus, O., 1964-66; chief br. genetics research U. S. Forest Service, Washington, 1966-67, asst. to dep. chief for research, 1967—. Lectr., So. Ill. U., 1958-59; cons. A.E.C., 1961. Recipient Poteat award Elisha Mitchell Sci. Soc., 1951. Mem. Ecol. Soc., Soc. Am. Foresters, Am. Inst. Biol. Sci., Sigma Xi, Phi Kappa Phi. Contbr. numerous articles to tech. jours. Determined way oceanic salts get into atmosphere, effects oceanic salts on coastal plants, genetics dune plants, embryology and seed devel. in yellow poplar trees, adaption plants to strip-mined lands, structure wood in yellow poplar, eastern cotton wood, black walnut trees, techniques for growing high value woods. Home: 925 N. Madison St., Arlington, Va. 22205. Office: U. S. Dept. Agr. Bldg., Washington 20250.*

BOYCE, William Henry, Am. urologist; b. Ansonville, N.C., Sept. 22, 1918; s. William H. and Louise (Gaddy) B.; B.S., Davidson Coll., 1940; M.D., Vanderbilt U., 1944; m. Anna Doris Shore, June 5, 1948; children—Lockhart, Catharine, Barbaba, Frederick. Instr. biology Davidson (N.C.) Coll., 1940-41; faculty Bowman Gray Sch. Medicine, Winston-Salem, N.C., 1952—, prof., sect. urology, 1960—; urologist-in-chief N.C. Bapt. Hosp.; attending urologist Forsyth Meml., Kate Bitting Reynolds Meml. hosps. Mem. Nat. Acad. Scis.-NRC. Diplomate Am. Bd. Urology. Mem. A.M.A., Am. Urol. Assn., A.C.S., Am. Assn. Genitourinary Surgeons, A.A.A.S., Soc. Univ. Surgeons, So. Soc. Clin. Research, Soc. Exptl. Biology and Medicine, Soc. Univ. Urologists, N.Y. Acad. Scis., Société D'Urologie Internationale, Alpha Omega Alpha, Sigma Xi. Research on urinary calculi, mechanisms of formation and theory, chemotherapy in cancer of bladder and kidney, urinary bladder physiology as related to artificial electrically induced urination. Home: 1970 Georgia Av. N.W., Winston-Salem, N.C. 27104.*

BOYD, David Armitage, Jr., Am. psychiatrist; b. Detroit, Jan. 14, 1906; s. David Armitage and Laura May (Staffin) B.; A.B., U. Mich., 1926, M.S. in Neuropsychiatry, 1939; M.D., Jefferson Med. Coll., 1930; m. Cathleen Jane Singer, Aug. 29, 1932; 1 son, David Armitage IV. Asst. prof. psychiatry U. Mich., Ann Arbor, 1936-39; chmn. dept. neurology and psychiatry Ind. U. Sch. Medicine, Indpls., 1939-49; cons. Mayo Clinic, Rochester, Minn., 1949—, sr. cons. in psychiatry, 1963—, prof. psychiatry Mayo Found., U. Minn. Grad. Sch., 1949—. Exec. sec. Am. Bd. Psychiatry and Neurology, 1951—. Home: 1140 Plummer Circle, Rochester, Minn. 55901.*

BOYD, Eldon Mathews, Canadian physician; b. Kingston, Ont, Can., Sept. 1, 1907; s. Joseph Reuben and Annabella (Prittie) B.; B.A., Queen's U., Kingston, 1928, M.A., 1929, M.D., 1932; m. Eleanore Marion Rowland, Sept. 1, 1933; children—Carl Edmund, Marion Agnes (Mrs. Ronald Price-Jones). Practice medicine, specializing in pharmacology, Kingston, 1934—; faculty Queens U., 1934—, prof., head dept. pharmacology, 1938—; cons. Kingston Gen. Hosp. Mem. Soc. Pharmacologists, Pan-Am., Can. med. assns., Soc. Toxicologists, Canadian Fedn. Biol. Socs., Pharmacological Soc. Can. (past pres.), Canadian Soc. Chemotherapy, Queens U. Alumni Assn. (dir.). Contbr. chpt. to Drill's Pharmacology in Medicine, 1965. Studies, numerous publs. on fat metabolism, hormones, water metabolism, expectorant drugs, cancer, drug toxicity. Home: 211 Union St. W., Kingston, Ont., Can.*

BOYD, James, geophysicist; b. Kanowna, West Australia, Dec. 20, 1904; s. Julian and Mary Innes (Cane) B.; B.S., Cal. Inst. Tech., 1927; M.Sc., Colo. Sch. Mines, 1932, D.Sc., 1934; m. Ruth Ragland Brown, Aug. 17, 1932; children—James Brown, Harry Bruce, Douglas Cane, Hudson. Instr. in geology Colo. Sch. Mines, 1929-34, asst. prof. mineralogy, 1934-37, asso. prof. econ. geology, 1938-41, dean faculty, 1946-47; asst. to sec. of interior of mineral matters, 1947; dir. Bur. of Mines 1947-51 also Def. Minerals Adminstr., 1950-51; exploration mgr. Kennecott Copper Corp., 1951-55, v.p. exploration, 1955-60; pres. Copper Range Co., 1960—; chmn. bd. White Pine Mining Co., 1960—; dir. Detroit Edison Co. Geologist U. S. Geol. Survey, 1933-34; cons. geology, mining and geophysics, 1935-40; pres., gen. mgr. Goldcrest Mining Co., 1939-40. Chmn. com. on mineral research NSF, 1952-57. Decorated Legion of Merit with oak leaf cluster; recipient D.S.M., Colo. Sch. Mines, 1949, Rand gold medal, Am. Inst. Mining Engrs., 1963; Distinguished Alumni award Cal. Inst. Tech., 1966; award Soc. Mining Engrs. Mem. Nat. Acad. Engring., Mining and Metall. Soc. America (pres. 1960-63), Am. Inst. Mining Engrs., Am. Soc. Econ. Geologists, Geol. Soc. Am., Canadian Inst. Mining and Metallurgy, Soc. Exploration Geophysicists, Acad. Polit. Sci. Home: 122 Old Church Rd., Greenwich, Conn. Office: 630 Fifth Av., N.Y.C.*

BOYD, John Smith Knox, Brit. physician; b. Largs, Scotland, Sept. 18, 1891; s. John Knox and Margaret (Smith) B.; M.B., Ch.B. with Honors, Glasgow (Scotland) U., 1913, M.D. with Honors, 1948, LL.D., 1958; D.P.H., D.T.M.H., Cambridge (Eng.) U., 1924; m. Elizabeth Edgar, June 11, 1918 (dec. 1956); m. 2d, Ellen Mary Harvey Murphy, June 3, 1957. Commd. H. M.C., Royal Army, 1914, advanced through grades to brigadier, 1945, ret., 1946; pathologist, 1916-46; dir. Welcomme Labs. Tropical Medicine, London, 1946-55, trustee, 1956-66. Decorated Order Brit. Empire; created knight bachelor, 1958; recipient Parkers Meml. Gold medal, 1923, Leishman Meml. medal, 1935, Bellahouston Gold medal, 1948. Fellow Royal Soc. London, Royal Coll. Physicians London, Royal Coll. Physicians Edinburgh, Scotland, Inst. Biology, Royal Soc. Tropical Medicine and Hygiene (past pres.), Royal Soc. Medicine (hon.); mem. Path. Soc., Soc. for Gen. Microbiology, Soc. for Immunology. Research and publs. on classification of dysentery bacilli including discovery of several new types; initiated treatment of bacillary dysentary with sulphonamides (with N. Hamilton Fairley); studied treatment of relapsing malaria; introduced immunization against tetanus in Brit. Army; studies on bacterial viruses, lysogenization. Home: 6 The Covert, Northwood, Middlesex, Eng. Office: 52 Queen Anne St., London, Eng.*

BOYD, Linn John, Am. physician, educator; b. Detroit, Sept. 30, 1895; s. David A. and Laura (Staffin) B.; M.D., U. Mich., 1918; m. Madeline H. Young, June 8, 1918. Asso. prof. U. Mich., 1919-25; prof. medicine N.Y. Med. Coll., 1926-58, cons. 1958—. dir. Sch. Grad. Sci., 1958-62, cons., 1962—; pres. medical board Metropolitan Hospital, New York City, 1935-47, Bird S. Coler Hospital, New York City, 1937-47; cons. USPHS Hospital, Stapelton, Staten Island, 1932—. Mem. Sigma Xi, Alpha Omega Alpha, Alpha Kappa Kappa. Author: Textbook of Electrocardiography, 1935; Cardio-vascular Disease, 1938; X-Ray Diagnosis of the Heart and Great Vessels, 1955, 65. Research on diseases, radiology of heart and great vessels, electrocardiology; history of medicine. Home: 611 Palmer Rd, Yonkers, N.Y. 10701.*

BOYD, Louise Arner, Am. explorer, author; b. San Rafael, Cal., Sept. 16, 1887; d. John Franklin and Louise Cook (Arner) Boyd; ed. Miss Stewart's Sch., San Rafael, Miss Murison's Sch., San Francisco; LL.D., U. Cal., 1939; LL.D., Mills Colls., 1939. Explorer of East Greenland; explorer polar region, N.E. and W. Greenland (Spitzbergen and Fronz Josef Land); flew pvt. chartered plane over North Pole (1st woman to reach North Pole by flight), 1955. Cons. to U. S. War Dept., 1942-43. Decorated Chevalier Legion of Honor (France), St. Olaf of Norway (1st fgn. woman to receive award); awarded Andree plaque by Swedish Anthropol. and Geog. Soc., Cullum gold medal by Am. Geog. Soc., medal King Christian Xth of Denmark; certificate of Appreciation, U. S. Army. Royal Horticultural Soc. (London), Am. Polar Soc. (hon., dir.), Cal. Acad. Sci. (hon.), Am. Soc. Photogrammetry, Am. Geog. Soc. (1st woman councilor 1960), Assn. Pacific Coast Geographers, Brit. Glaciological Soc., Am. Hort. Soc., Soc. Woman Geographers, Cal. Bot. Soc., Nat. League Am. Pen Women, Geog. Soc. Phila. (hon.), Sigma Delta Epsilon. Author: Fiord Region of East Greenland, 1935; Polish Countrysides, 1937, Coast of Northeast Greenland. Contbr. to Geog. Rev. Known for her Polar expdns.; credited by Am. Polar Soc. with having contbd. more to knowledge of Greenland, Greenlansea, Spitzbergean and Franz Joseph Land than any other explorer; 1st to sail to inner ends of Ice Fjord, Greenland (for which achievement Danish govt. named terr. in vicinity of de Geer glacier Louise A. Boyd Land). Home: 1055 California St., San Francisco. Office: 210 Post St., San Francisco 8.

BOYD, William Clouser, Am. immunochemist; b. Dearborn, Mo., Mar. 4, 1903; s. William Oliver and Wilmuth (Clouser) B.; A.B., Harvard, 1925, A.M., 1926; Ph.D., Boston U., 1930; m. Lyle A. Gifford June 9, 1931; 1 dau., Sylvia Lyle. Teaching fellow Boston U. Med. Sch., 1926-30, asst. prof., 1935-38, asso. prof., 1938-48, prof. immunochemistry, 1948—. Guggenheim fellow, 1935-36, 37-38, 61; Fulbright fellow, Pakistan, 1952. Fellow Am. Acad. Arts and Scis.; mem. Am. Assn. Immunologists (pres. 1959-60), Am. Assn. Human Genetics (pres. 1957), Am. Assn. Phys. Anthropology, A.A.A.S. Boston Mycol. Club (pres. 1960). Author: (with Fritz Schiff) Blood Grouping Technic, 1942; Fundamentals of Immunology, 1943, 47, 56; Genetics and the Races of Man, 1950; (with B. S. Walker and I. Asimov) Biochemistry and Human Metabolism, 1952, 54, 57; (with Asimov) Races and People, 1955; Introduction to Immunochemical Specificity, 1962. Research, publs. on blood groups of mummies, in racial studies, in immunology, chemistry of Rh blood antigens; discoverer lectins, 1948; nomogram for chi-square, 1965. Home: 24 Edward St., Belmont, Mass.*

BOYD-ORR OF BRECHIN MEARNS, John, Baron, Scottish physiologist, nutritionist; born Ayrshire, Scotland, September 23, 1880; the son of Robert Clark and Annie (Boyd) B.-O.; B.Sc., U. Glasgow (Scotland), 1911, M.A., 1900, M.B., 1913, M.D., 1914, D.Sc., 1920; numerous hon. degrees Brit., Scottish univs.; m. Elizabeth Pearson Callum, Feb. 22, 1915; children—Elizabeth Joan (Mrs. Ken Barton), Noel Donald (dec.), Helen Annie (Mrs. David Lubbock). Dir., Rowett Research Inst., 1913-45, Imperial Bur. Nutrition, 1926-45; editor Nutrition Abstracts and Revs., 1932-45; dir. gen. FAO, 1945-48; rector Glasgow U., 1945-46, chancellor, 1946—; mem. Parliament, 1945—. Recipient Nobel Peace Laureate, 1949; Commdr. Legion of Honor, 1949; Grocers award of U. S. A.; Harben medal Royal Inst. Pub. Health, U.K. Fellow Royal Soc., 1932; Royal Soc. Edinburgh; hon. mem. N.Y. Acad. Medicine. Author numerous books on sci. of nutrition econs. and policies. Research, publs. on malnutrition due to dietary deficiencies and its extent in the U.K. Address: Brechin Angus, Scotland.*

BOYDEN, Edward Allen, Am. anatomist; b. Bridgewater, Mass., Mar. 20, 1886; s. Arthur Clarke and Katherine (Allen) B.; A.B., Harvard, 1909, A.M., 1911, Ph.D., 1916; postgrad. Anat. Inst. Freiburg, Breisgau; m. Margaret Lorinda Hilsinger, Nov. 12, 1916; children—Mary Scarborough, Arthur Clarke. Prof. anatomy U. Minn., 1931-54, chmn. dept., 1940-49, head, 1949-54, prof. emeritus, 1954—; vis. prof. U. Wash., Seattle, 1954-56, research prof., 1956—. Recipient citation Minn. Med. Assn., 1955; Achievement award Modern Medicine, 1960; Coll. Film award of chest physicians, 1962; Distinguished Service award Minn. chpt. Sigma Xi, 1963. Hon. fellow Am. Coll. Chest Physicians; mem. Am. Assn. Anatomists (past pres.), Am. Physiol. Soc., Am. Acad. Arts and Scis. (hon.), Alpha Omega Alpha. Author: A Laboratory Atlas of the Pig Embryo, 1933; (with S. I. Levin) The Kosher Code of the Orthodox Jew, 1940; Segmental Anatomy of the Lungs, 1955. Mng. editor Anat. Record, 1928-48; asso. editor: Disease of Chest, 1962—; redactor Acta Anatomica, 1960—. Contbr. numerous articles to profl. jours. Research on exptl. analysis of factors controlling formation of kidney and its anomalies; discovered that egg yolk empties cat gall bladder providing standard meal for demonstrating that evacuation of human gall bladder is retarded in pregnancy, speeded up in ulcer patients and returns to normal rate after vagotomy, proving it is under hormonal and not nervous control; demonstrated postnatal mode of growth of human lung. Home: 3426 Meridian Av. N., Seattle 98103.*

BOYDEN, Guy Lee, Am. otolaryngologist; b. Brooking, S.D., Feb. 17, 1885; s. Francis and Mary Emaline (Darling) B.; B.S., S.D. State Coll., 1905, D.Sc., 1955; M.D., Northwestern U., 1911; m. Zella Thompson, Sept. 26, 1917; children—Mary (Mrs. Wm. C. Lindsell, Jr.), Catherine (Mrs. O. T. Coffelt, Jr.), Thomas, Jean (Mrs. Francis L. Hales). Intern Mayo Clinic, 1911-12, resident, 1912-13; gen. practice medicine, Pendleton, Ore., 1913-21; practice limited to otolaryngology, Portland, Ore., from 1921; sr. cons. U. S. Vets. Hosp., Portland, from 1948; mem. faculty U. Ore. Med. Sch., from 1948, asst. prof. otolaryngology, 1932, asso. prof., 1946, prof. and head dept. otolaryngology, from 1946. Recipient award for distinguished services in ednl. programs Am. Acad. Ophthalmology and Otolaryngology, 1958. Mem. A.M.A., Am., Ore. acads. ophthalmology and otolaryngology, Am. Triologic Soc., Pacific Coast Soc. Ophthalmology and Otolaryngology (pres.), Ore., Multnomah County, Portland med. socs., Am. Laryngol. Assn. Contbr. to med. jours. Died Sept. 9, 1958.

BOYDEN, Seth, Am. inventor; b. Foxborough, Mass., Nov. 17, 1788; s. Seth and Susanna (Atherton) B.; m. Abigail Sherman; children—Susan Obadiah, Matilda, George, Seth. Inventor process for making patent leather, 1819, malleable iron, 1826; manufactured stationary steam engines; developed forerunner of "Am. Process" furnace grate bar, 1847, an inexpensive process for manufacturing sheet-iron, a hat-forming machine; made 1st Am. daguerreo-type; originated machines for manufacturing nails and cutting files; described by Thomas Edison as one of America's greatest inventors. Died Hilton, N.J., Mar. 31, 1870.

BOYDEN, Uriah Atherton, Am. engr., inventor; b. Foxborough, Mass., Feb. 17, 1804; s. Seth and Susanna (Atherton) B.; Worked in mgf. with bro. Seth, Newark, 1825; took part in 1st survey Boston and Providence R.R., worked on constrn. other railroads; opened engring. office, Boston, 1833; supr. constrn. Nashua and Lowell R. R., 1836-38; as engr. Amoskeag Mfg. Co. designed hydraulic works, Manchester, N.H.; designed turbine water-wheel 1844, also three 190 horsepower turbines (all for Appleton Cotton Mills, Lowell, Mass.); best-known for developing spiral approach (admits water to turbine at uniform velocity); left most of his money to Bd. Trustees, Foxborough, Mass. for constrn. of observatories. Died Oct. 17, 1879.

BOYE, Martin Hans, chemist, geologist; b. Copenhagen, Denmark, Dec. 6, 1812; ed. Borgerdyskolen, Copenhagen; grad. U. Copenhagen, 1832; Polytechnic Sch., Copenhagen, 1835; Univ. Pa., med. dept., 1844; spl. studies in analytical chemistry and physics, hon. A.M., U. Pa., 1844. Came to U. S., 1836; assisted Robert Hare in chem. investigations, 1837-38; asst. geologist and chemist 1st geol. survey of Pa., 1838-43; jointly discovered new compound, chloride of platinum with binoxide of nitrogen, 1839; jointly discovered perchloric ether, the most explosive of all substances, 1841; conducted lab. for analysis and instrn. in analytical and practical chemistry, 1842-45; discovered, and with others applied the first process of refining cotton seed oil, 1845; specimens exhibited at Centennial Expn., Phila., 1876, received 1st premium; prof. chemistry and natural philosophy, Central High Sch., Phila., 1845-59; delivered many lectures; ret., 1859. Author: A Treatise on Pneumatics, being the Physics of Gases, 1855; Chemistry, or the Physics of Atoms, 1857. Wrote also, Explosive Power of Perchloric Ether in Proc. Am. Philos. Soc., Perchloric Ether in Trans. Am. Philos. Soc.; "Analysis" in Booth's Chemical Ency.; Analysis of the Bittern of a Saline on the Kiskeminetas; of Magnetic Iron Pyrites Containing Nickel; of Schuykill, Croton and Rock Creek Waters, etc., in Silliman's and Franklin Inst. Journals, 1842-45. Home: Coopersburg, Pa. Died Mar. 5, 1909.

BOYER, Fernand Alexandre, French chemotherapist; b. Fesselines, Creuse, France, Aug. 18, 1905; s. Alexandre and Marguerite (Leger) B.; ed. Conservatoire National des Arts et Metiers, Faculté des Sciences, Paris; Docteur, U. Paris, 1951; m. Marie Roux, Jan. 22, 1927; children—Janine (Mrs. Tuygressier), Rolande. Staff, Institut Pasteur, 1934—. Mem. French Microbiology Soc. Research and numerous publs. on sulfonic action, reaction of sulfamides, vaccination of lab. animals with streptococci, antitoxin action of cortisone, terrain factor in exptl. infections, endotoxin reactions. Home: 207 rue de Vaugirard, Paris 15e, France.*

BOYER, Ralph L., Am. mech. engr.; b. Botkins, O., Aug. 4, 1901; s. Calvin O. and Ethel (Lucas) B.; B.S., Ohio State U., 1924, M.E., 1930; m. Doris Dormire, June 7, 1924; 1 dau., Jean (Mrs. Davis C. Marshall). Co-developer compound diesel engine Sperry Gyroscope Co., Bklyn., 1922-26; with Cooper Bessemer Corp., Mt. Vernon, O., 1926-65, v.p., chief engr., 1938-55, v.p., dir. engring., 1955-61, dir., 1950-65; ret., 1965. Recipient Lamme medal, 1951. Fellow Am. Soc. M.E. (citation Gas Turbine Power div. 1965, Diesel and Gas Power div. 1966); mem. Sigma Xi. Contbr. articles to profl. jours. Developer high compression gas engine, aircraft jet type indsl. gas turbine. Home: 1011 New Gambier Rd., Mt. Vernon, O. 43050.*

BOYER, Raymond Foster, Am. polymer physicist; b. Denver, Feb. 6, 1910; s. Charles Foster and Margaret (Daly) B.; B.S. in Physics, Case Inst. Tech., 1933, M.S., 1935, D.Sc. (hon.), 1955; m. Frances Klobucar, Aug. 17, 1936; children—Charles, James, Rima, Margaret. With Dow Chem. Co., Midland, Mich., 1935—, asst. dir. Research Lab., 1945-47, dir., 1947-52, dir. plastics research, 1952—. Mem. various govt. adv. panels. Mem. Am. Phys. Soc. (chmn. high polymer div. 1953), Am. Chem. Soc. (chmn. div. polymer chemistry 1956), N.Y. Acad. Scis., Soc. Plastics Engrs. Contbr. articles to tech. jours. Patentee in field. Research on ultraviolet light stability plastics, behaviour plasticizers, nature glass transition in plastics, molecular weight distbns. plastics. Home: 415 W. Main St., Midland, Mich. 48640.*

BOYES, John Wallace, Canadian biologist; b. Sundridge, Ont., Can., Jan. 27, 1907; s. Richard Edward and Orinda Parney Benton (Louden) B.; B.Sc. with high honours in Biology, U. Sask. (Can.), 1933, M.Sc. 1936; Ph.D., U. Wis., 1939; m. Beatrice Chamberlain Westcott, Aug. 28, 1937; children—Philip Westcott, Barbara Grace, Alan Ramsay, Margaret Ruth. Jr. research asst. applied biology NRC, Ottawa, Ont., 1939-40; asst. prof. genetics, plant breeding U. Alberta, Can., 1940-42; from asst. prof., acting chmn. genetics dept. to asso. prof., chmn. genetics dept. McGill U., Montreal, Que., Can., 1945-55, John and Anne Molson prof., chmn., 1955—. Pres. genetics sect. Internat. Union Biol. Scis., 1958-63; pres. Internat. Commn. for Genetic Nomanclature, 1963; v.p. XI Internat. Congress Genetics, 1963. Mem. editorial bd. Canadian Jour. Genetics and Cytology. Research, publs. in plant embryology, interspecific crosses in wheat, cytology of Diptera, and related fields. Home: 484 Strathcona Av., Montreal, Que., Can.*

BOYKO, Hugo Nathanael, botanist; b. Vienna, Austria, Oct. 6, 1892; s. Adolf and Caroline (Hirschberger) B.; Ph.D., U. Vienna, 1930; m. Elisabeth Spitzer, May 25, 1920; children—Eva (Mrs. Michael Avi-Yonah), Maya, Herbert Gabriel. Indsl. mgr., Austria, 1919-26; faculty plant sociology Vienna U., 1928-35; pvt. sci. research, Israel, 1935-44; govt. ecologist Israel Dept. Forests, 1944-47; ecol. adviser Ministry of Agr. and to Prime Minister, 1948-61; ecol. cons., 1961—. Lectr. numerous univs., congresses; hon. cons. UNESCO, 1952—; presiding mem. Council for World U., 1965—. Recipient Weizmann prize, 1953; Fleming medal, 1959. Fellow A.A.A.S.; mem. P.R. Acad. Scis. (corr. mem.), Internat. Union Biol. Scis. (chmn. internat. commn. on applied ecology 1947-61), Internat. Soc. for Bio-Meteorology (v.p.) 1956-61, (chmn. internat. commn. on ecol. climatography 1956-64), Internat. Union for Conservation of Nature (internat. ecol. commn. 1960—), World Acad. Art and Sci. (founder, hon. sec. gen., pres. 1965—). Editor, contbg. author: Science and the Future of Mankind, 1961; Salinity and Aridity—New Approaches to Old Problems, 1966; also monographs, articles. Devised new methods for plant ecology; new plant geog. subdiv. of Middle East; geo-ecol. law, biol. rules for climatic extremes, global salt circulation, basic law of universal balance; plant growing under direct irrigation with natural seawater; biol. soil desalination. Home: 1 Ruppin, Rehovot, Israel.*

BOYLAN, David Ray, Am. engr.; b. Belleville, Kan., July 22, 1922; s. David Ray and Mabel (Jones) B.; B.S. in Chem. Engring., U. Kan., 1943; Ph.D. Chem. Engring., Ia. State U., 1952; m. Juanita R. Sheridan, Mar. 24, 1944; children—Sharon Rae, Gerald Ray, Elizabeth Anne, Lisa Diana. Instr., U. Kan., 1942-43; project engr. Gen. Chem. Co., Camden, N.J., 1943-47; sr. engr. Am. Cyanamid Co., Elizabeth, N.J., 1947; plant mgr. Arlin Chem. Co., Elizabeth, 1947-48; faculty Ia. State U., Ames, 1948—, prof. 1956—, asso. dir. Engring Expt. Sta., 1959—. Mem. Am. Chem. Soc., Am. Soc. Engring Edn., Am. Inst. Chem. Engring., Sigma Xi, Phi Lambda Upsilon, Sigma Tau, Phi Kappa Phi, Tau Beta Pi. Research in transient behavior and flow of fluids through porous media, unsteady state, and fertilizer tech., devel. fused-phosphate fertilizer processes, theoretical and exptl. correlation of filtration; patents and paper in field. Home: 1516 Stafford St., Ames, Ia. 50010.*

BOYLAND, Eric, English biochemist; b. Manchester, Eng., Feb. 24, 1905; s. Alfred E. and Helen (Walton) B.; B.Sc.Tech., Manchester U., 1926, M.Sc., 1928; Ph.D., U. London (Eng.), 1930, D.Sc., 1936; m. Margaret Esther Maurice, Sept. 24, 1931; children—Michael Morris, Brian Robert, Helen Margaret. Research asst. physiology Manchester U., 1926-28; Grocers' Co. scholar, Beit Meml. fellow for med. research Lister Inst. for Preventive Medicine, London, 1928-30, Kaiser Wilhelm Inst. für Medizinische Forschung, Heidelberg, Germany, 1930-31; physiol. chemist Royal Cancer Hosp., London, 1931; reader biochemistry U. London, 1935-47 prof., 1948—; research officer Ministry of Supply, 1941-44, Ministry of Agr., 1944-45; prof. biochemistry Chester Beatty Research Inst., Inst. Cancer Research, 1948—; bd. govs. Royal Marsden Hosp. Mem. panel on food additives WHO, 1958—. Author: The Biochemistry of Bladder Cancer, 1962; also numerous articles. Research mechanisms by which substances cause cancer with a view to prevention of the disease. Home: 42 Bramerton St., London, S.W. 3. Office: Chester Beatty Research Inst., Fulham Rd., London, S.W. 3, Eng.*

BOYLE, Paul E., Am. dentist, educator; b. Somerville, Mass., Apr. 20, 1900; s. John Andrew and Olive (Berry) B.; student Dartmouth Coll., 1918-19; D.M.D., Harvard, 1923; m. Rosalie Dunlap, Nov. 11, 1937; children—Paul E., Margaret Ann (Mrs. Richards Newcomer), Alorie (Mrs. Craig W. Parkhill). Practice dentistry, Boston, 1923-41; faculty Harvard, 1926-44, U. Pa., 1944-56; dean Sch. Dentistry, prof. Western Res. U., Cleve., 1956—. Mem. bd. sci. counselors Nat. Inst. for Dental Research, 1960-64. Fellow Internat. Coll. Dentists; mem. Am. Dental Assn., Am. Coll. Dentists, Am. Acad. Dental Medicine, Am. Acad. Oral Pathology (pres. 1949), Am. Assn. Pathologists and Bacteriologists, Internat. Assn. for Dental Research (pres. 1955-56), A.A.A.S. (v.p. 1963), Am. Soc. for Exptl. Pathology, Sigma Xi, Omicron Kappa Upsilon. Editor: Histopathology of the Teeth and Their Surrounding Structures (Kronfeld), 1949. Research on vitamin deficiencies as a cause of dental abnormalities in tooth formation, role of dentist in detection and diagnosis of mouth cancer. Home: 15914 Chadbourne Rd., Shaker Heights, O. 44120. Office: 2165 Adelbert Rd., Cleve. 44106.*

BOYLE, Robert, Brit. natural philosopher, chemist; b. Lismore Castle, Munster, Ireland, Jan. 25, 1627; s. Richard Boyle, Earl of Cork, and Catherine Fenton; ed. Eton Coll., circa 1635-38, traveled on continent and studied with pvt. tutors, 1638-44; hon. D.Physics, Oxford U., 1665. Became active participant and mem. Invisible Coll., London, 1644, which orgn. became Royal Soc., 1662; established a lab., Oxford, 1654; became dir. East India Co., and gov. corp. for spread of Gospel to New Eng., 1661-89; declined provostship of Eton, 1665; financed much missionary work in Orient, assumed large part of financial responsibility for printing Bibles for the various Brit. colonies; moved to London, 1668, remained there until death; bequeathed his mineral. collection to Royal Soc.; in his will founded and endowed Boyle lectures on def. of Christianity against unbelievers. Repeatedly declined honor of being raised to peerage. Fellow Royal Soc., 1663, elected pres. Royal Soc., 1680, declined because of form of oath of office. Author numerous works including: New Experiments Physico-Mechanicall touching the Spring of the Air, 1660, new edit. with a Defense Against Linus, 1662; The Sceptical Chymist, 1661; Origin of Forms and Qualities (According to the Corpuscular Philosophy), 1660; Certain Physiological Essays, 1661; Some Considerations touching the Usefulness of Experimental Naturall Philosophy, 1663, Vol. 2, 1671; Experiments and Considerations Touching Colours, 1664; Memoirs for the Natural History of the Human Blood, 1684; numerous moral and religious essays. Best known for his discovery of Boyle's law (volume of a gas is inversely proportional to its pressure) which was suggested to him by Richard Townley; observed correlation of boiling points of liquids and pressure; investigated compressibility of water, showed that it expands when it freezes, and noted that this expansion begins prior to actual freezing; defined the term, element, and showed that current methods of analysis did not conclusively prove the existence of the Aristotelian four elements or the Paracelsian three principles; distinguished clearly between mixtures and compounds; believed in corpuscular theory of matter; one of 1st scientists to collect a gas (1st to collect hydrogen and nitric oxide); designed a vacuum pump, experimented in pneumatics; investigated specific gravities, refractive powers, crystals, colors, electricity, capillary action and combustibility of hydrogen; used color indicators to classify substances as acidic, basic or neutral; suggested that alcohol should be used as preservative for biol. specimens; demonstrated experimentally that air is the normal medium by which sound reaches the ear; was 1st chemist in Eng. to use hermetically sealed thermometer; invented compressed-air pump; made use of an evacuated cylinder to demonstrate Galileo's assertion that in a vacuum all bodies fall at same velocity; studied chemistry of combustion and respiration. Died London, Dec. 30, 1691.

BOYLE, Robert William, physicist; b. Oct. 2, 1883; s. Albert D. and Sophie Boyle; ed. Manchester U.; B.Sc., McGill U., 1905; M.A., 1906, Ph.D., 1909; M.A., LL.D.; 1851 Exhbn. Sci. research scholar, research on radioactivity U. Manchester, 1909-11; lectr. physics McGill U., 1911, asst. prof., 1912; prof. physics U. Alta., 1912, 19, dean Faculty Applied Sci., 1921. mem. Alta. Council Sci. and Indsl. Research, 1923-29; dir. div. physics and elec. engring. Nat. Research Labs., Ottawa, 1929-48, ret., 1948. Mem. Royal Soc. Canada (past pres. sect III, Flavelle medal for sci. 1940), Assn. Profl. Engrs. of Alta (past pres.). Research and publs. on properties of matter, radioactivity, ultrasonics. Died Apr. 18, 1955.

BOYLSTON, Zabdiel, Am. physician; b. Brookline, Mass., Mar. 9, 1679; s. Dr. Thomas and Mary (Gardner) B.; m. Jerusha Minot, Jan. 18, 1705, 8 children. First physician to introduce small pox inoculation into Am. during Boston epidemic, 1721, for which he was persecuted and his life frequently threatened; wrote books in defense of inoculation, including The Little Treatise on the Small pox, 1721; An Historical Account of the Small pox Inocculation in New England, 1726; Fellow Royal Soc., 1726. Died Brookline, Mar. 1, 1766.

BOYNTON, Damon, Am. biologist; b. Chgo., Sept. 27, 1908; s. Percy Holmes and Lois (Damon) B.; student Amherst Coll., 1927-28; B.S., Cornell U., 1931, Ph.D., 1937; m. Mary Fuertes, May 30, 1931; children—Louis Fuertis, Katharine (Mrs. Roger Payne), Maria. Farm mgr., Amherst, Mass., 1931-34; faculty Cornell U., Ithaca, 1934-64, prof. pomology, 1947-64, dean Grad. Sch., 1959-64; sr. adviser on research and edn. Inst. Agr. Sci., FAO, Turralba, Costa Rica, 1964-66; acad. adviser to Agrarian U. Peru, La Molina, also vis. prof. N.C. State U., La Molina, 1966—. Mem. adv. council on financial aid to students N.Y. State Bd. Regents, 1961-64; mem. N.Y. State Sci. Adv. Bd., 1963-64. Guggenheim fellow, 1945-46; Travelling fellow Rockefeller Found., 1959; NSF grantee, 1958-62; U. S. Office Edn. grantee, 1960-64. Fellow A.A.A.S.; mem. Assn. Grad. Schs. (mem. com. on testing, 1959-64, chmn., 1961-64), Assn. Land Grant Colls. and Univs. (grad. study dir. exec. com. 1960-64), Am. Inst. Biol. Scis. (governing bd. 1960-61), Am. Soc. Hort. Sci. (editor, bus. mgr. Proc. 1955-59), Phi Beta Kappa, Sigma Xi. Research, numerous publs. on effects soil and nutrient requirements, absorption, and metabolism fruit plants. Home: R.F.D. 2, Ovid, N.Y. Office: N.C. State U. Project, Agrarian U., La Molina, Peru.*

BOYS, Sir Charles Vernon, Brit. physicist; b. Wing, Rutland, Mar. 15, 1855; s. Charles Boys; ed Royal Sch. Mines, Cambridge; LL.D., Edinburgh; m. Marion Amelia Pollock, 1892 (div. 1910), 1 dau., 1 son. Recipient Duddell medal Phys. Soc., 1925; Elliott Cresson Gold medal Franklin Inst., 1939. Fellow Royal Soc., 1888 (Royal medal 1896, Rumford medal 1924), Royal Soc. (Edinburgh, hon.); hon. mem. N.Y., Acad. Scis., Phys. Soc. Moscow; mem. Phys. Soc. London (pres. 1916-17), Röntgen Soc. (pres. 1906-7). Knighted, 1935. Author. Soap Bubbles, Their Colours and the Forces which Mould Them; The Natural Logarithm, 1935; Weeds, Weeds, Weeds, 1937; also numerous papers on integrating mechanisms, quartz fibres, radiomicrometer, heat of moon and stars, constant of gravitation, solid dipleidoscope prisms, gas calorimetry, speed of lightning, other subjects. Invented moving lens lightning camera, 1900 (used by Schonland in S. Africa and MacEachron in U. S. in discovering gradual devel. of lightning flash); designed improved torsion balance, 1895, with which he determined value of Newton's constant of gravitation, arriving at value of 5.5270 for mean density of earth; invented radio-micrometer, with which he could measure heat radiation from moon and planets, proved that temperature of Jupiter's surface is less than 100° centigrade; recommended use of quartz fibres (rather than silk) for delicate suspension instruments; designed calorimeter to measure thermal power of coal gas. Died Andover, Eng., Mar. 30, 1944.

BOYTCHEV, Boytcho Boytchev, Bulgarian physician; b. Lom, Bulgaria, June 12, 1902; s. Boytcho Georgiev and Kichka (Draganova) B.; grad. Med. Faculty, Modena, Italy, 1928; m. Ann Antonova, Nov.

28, 1945; 1 dau., Mariana. Practice gen. medicine, 1929-30; asst. I. Surg. U. Clinic, Sofia, Bulgaria, 1930-35, chief orthopedic dept., 1937-45; fgn. asst. Istituto Ortopedico Rizzoli, Bologna, Italy, 1936-37; dir. U. Clinic for Orthopaedics and Traumatology, Sofia, 1946-65; dir. Clinic for Orthopaedics and Traumatology at Postgrad. Med. Tng. Inst., Sofia, 1951—; dir. Center for Bone Tumours, Bulgarian Acad. Scis., 1962—. Chief cons. specialist in orthopaedics and traumatology Bulgarian Ministry Pub. Health. Recipient 3 golden medals for sci. activity. Mem. Bulgarian Assn. Orthopedic Surgeons (pres.), Bulgarian Assn. History of Medicine (pres.); hon. mem. Italian, Soviet, Hungarian, Polish, Czechoslovak, Yugoslav orthopedic-trauma assns. Author: Orthopaedics and Traumatology, vol. I, 1943, vol. II, 1946; (with B. Conforty, C. Tchokanov) Operative Orthopedics and Traumatology, 1955; Bone Tumors, 1962; (monographs) Operative treatment of Congenital Dislocation of the Hip Joint, 1945, Contribution to the Operative Treatment of Congenital Dislocation of the Patella, 1943; also numerous articles. Research on habitual dislocation of shoulder, treatment of hallux valgus, tunnelizing arthrodesis, elongative arthrodesis of knee, wide resection of knee for tumors, onepole resection of joints for bone tumors and replacement with preserved homogenous bone grafts. Editorial bd. Operative Surgery, 6 vols.; several fgn. orthopedic jours.; editor in chief Orthopaedics and Traumatology, 1964—, Chirurgia. Home: 157-A Rakovsky St., Sofia, Bulgaria.*

BOZLER, Emil, physiologist; b. Steingebronn, Germany, Apr. 5, 1901; s. Carl and Ursula (Hilning) B.; Ph.D., U. Munich, 1923; m. Klara Hoppe, Aug. 25, 1933; children—Ruth E. (Mrs. Donald J. Moorhead), Carl O., Hans M. Came to U. S., 1932, naturalized 1939. Privat dozent U. Munich, Germany, 1924-32, fellow Rockefeller Found., 1928-29; research fellow Johnson Found., Phila., 1932-36; prof. physiology Ohio State U., Columbus, 1936—. Mem. German Physiol. Soc. (hon.). Research on physiology of muscle. Contbr. articles to tech. jour. Research in muscle physiology; demonstrating fiber to fiber conduction in smooth muscle, role of calcium and magnesium in contraction and relaxation; discovered electric potentials of vertebrate hearts. Home: 203 Acton Rd., Columbus 43214.

BOZOKI, György, Hungarian physicist; b. Kunhegyes, Hungary, June 21, 1930; s. Gyula and Erzsébet (Kronstein) B.; Ph.D., Roland Eötvös U., candidate phys. scis. Hungarian Acad. Scis., 1963; m. Éva Gombosi, Nov. 2, 1956; 1 dau., Andrea Katalin. Jr., sr. researcher, group leader, dep. leader cosmic ray dept. Central Research Inst. Physics, Budapest, Hungary, 1953-64; asst. prof. faculty atomic physics Roland Eötvös U., Budapest, 1959—; sr. researcher, group leader Joint Inst. for Nuclear Research, Dubna, USSR, 1964-65. Mem. Roland Eötvös Phys. Soc. (Schmid prize 1965). Research, publs. on physics of cosmic rays (penetrating showers, jets), high energy accelerated particles (nuclear interactions). Home: 15/b Nagyenyed, Budapest. Office: Konkoly-Thege, Budapest, Hungary.*

BOZOKY, László, Hungarian physicist; b. Nagyvárad, Hungary, May 4, 1911; s. Gyula and Gabriella (Mártonffy) B.; Ph.D., Pázmány Péter U., Budapest, Hungary, 1936, candidate phys. sci., 1952, D.phys.-sci., 1963; m. Mária Mészáros, May 10, 1947. Asst., Tech. U. Budapest, 1936-44; physicist Eötvös L. Rádium and Röntgen Inst., 1937-49; head phys. dept. Nat. Inst. for Oncology, Budapest, 1949—; head radiol. dept. Central Inst. for Phys. Research Budapest, 1952-59; lectr. Eötvös Loránd U., Budapest, 1952-65, prof., 1965—. Pres. radiol. protection com. Hungarian Standard Office, 1954—; mem. Atomic Energy Commn., pres. health and safety com., 1956-58; cons. Isotop Inst., 1960—. Mem. Eütvös L. Phys. Soc. (pres. health physics dept. 1962—), Hungarian Biophys. Soc. (mem. presidium 1962—), Internat. Radiation Protection Assn. (exec. council 1966—). Author: (with J. Tigyi) Investigation Methods of Exptl. Medicine, 1965; also articles, chpts. in books. Research on determination spectroscopical constants, health physics and radiotherapy especially with cobalt, constrn. whole body counting apparatus; introduction of radioisotopes in metal and oil industry. Home: Szabolczka M. u. 1, Budapest XI, Hungary.*

BOZORTH, Richard Milton, Am. physicist; b. Salem, Ore., Apr. 10, 1896; s. Scott and Elizabeth (Dearborn) B.; A.B., Reed Coll., 1917; Ph.D., Cal. Inst. Tech., 1922; m. Louise Huntley, Aug. 3, 1921; children—Katherine (Mrs. Eugene Edward Beyer, Jr.), Alison (Mrs. James Warren Fowle). Mem. tech. staff Bell Telephone Labs., Murray Hill, N.J., 1923-61, head magnetic research, 1926-46; Fulbright research prof. U. Tokyo, 1961-62; cons. IBM Research Center, Yorktown Heights, N.Y., 1961—; cons. Naval Ordnance Lab., Silver Spring, Md., 1961—. founder ann. Conf. Magnetism and Magnetic Materials. Author: Ferromagnetism, 1951. Contbr. to Ency. Brit., 1947. Research, publs. on basic research on magnetic phenomena and materials especially properties of crystals; Home: 20 Park Rd., Short Hills, N.J. 07078. Office: IBM Research Center, Yorktown Heights, N.Y. 10598; also Naval Ordnance Lab., Silver Spring, Md. 20910.*

BRAAMS, Reinier, Dutch physicist; b. 's-Hertogenbosch, Netherlands. Aug. 29, 1923; s. Reinier and Bastiana (Straver) B.; Candidate, U. Utrech, 1945, M.Sc., 1949; Ph.D., Cambridge U., 1952; m. Maria Tekla Beversluis, July 10, 1951; children—Antoinette Tekla, Reinier, Margaretha W., Anna Maria Elisabeth. Physicist, Utrecht U. Hosp., 1952-55; research asso. dept. biophysics Yale, 1955-57; sr. scientist Physics Lab., Utrecht U., 1957-63, prof. molecular biophysics, 1965—, head dept. radiation biophysics, 1965—; vis. scientist dept. radiation biology Karlsruhe Kernforschungszentrum, 1961; research asso. Argonne Nat. Lab., 1964. Mem. Radiation Research Soc., Assn. for Radiation Research (Gt. Britain), European Assn. for Radiobiology. Research, publs. on measuring properties of therapeutical beams of x-rays and fast electrons, effects of ionizing radiations on proteins, mechanism of inactivation of enzymes, nature of damage in irradiated proteins, mechanisms of protection against molecular damage especially of proteins irradiated in solid state, study of reactivity of hydrated electron with amino-acids, peptides and proteins.

BRAARUD, Trygve, Norwegian marine biologist; b. Verdal, Norway, Sept. 15, 1903; s. Carl Johan and Julie Hansen (Gjelvold) B.; B.A., 1927; Ph.D., 1936. Sci. asst. bot. lab. U. Oslo (Norway), 1926-33, other positions, 1936-47, prof. marine biology, 1947—. Mem. Sci. Acad. Author biology textbook. Research plant plankton and higher vegetation in Norwegian water; participated in internat. Passamaquoddy fisheries commn. investigations in Bay of Fundy, Gulf of Me., 1933-34.

BRACE, DeWitt Bristol, Am. physicist; b. Wilson, N.Y., Jan. 5, 1859; s. Lusk and Emily C. B.; A.M., Boston U., 1882; spl. student Mass. Inst. Tech., 2 yrs.; student physics, Johns Hopkins, 1881-83; U. of Berlin, 1883-85, under von Helmholz and Kirchhoff, Ph.D., 1885; m. Elizabeth Russell Wing, Oct. 16, 1901. Acting asst. prof. physics, U. Mich., 1887; prof. physics, U. of Neb., from 1888. Fellow and v.p. A.A.A.S. Author: Laws of Radiation and Absorption, 1901. Studied measure of contraction of light, 1904; invented Brace spectrophotometer, 1899, Brace half shade elliptic polarizer and compensator, 1904; proved that refractive index of clockwise and counterclockwise vibrations in a material medium differed enough to permit observable separation of a plane polarised incident beam into 2 beams circularly polarised in opposite directions, 1901. Died 1905.

BRACE, John Wells, Am. mathematician; b. Evanston, Ill., Jan. 19, 1926; s. George W. and Marcia (Campbell) B.; B.A., Swarthmore Coll., 1949; M.A., Cornell U., 1951, Ph.D., 1953; m. Patricia A. Demarest, June 16, 1950; children—James, George, Ann, Nancy, Catherine. Faculty math. U. Md., 1953—, prof., 1963—, asso. chmn. dept. math., 1964-66; research asso. U. Cal. at Berkeley, 1959-60. Mem. Woodrow Wilson Fellowship Regional Com., 1962—. Fellow A.A.A.S.; mem. Am. Math. Soc., Math. Assn. Am., Nat. Council Tchrs. Math., Société Mathematique de France, Indian Math. Soc., Sigma Xi, Phi Kappa Phi, Pi Mu Epsilon. Contbr. numerous articles on functional analysis to profl. publs. Office: Math. Dept., U. Md., College Park, Md. 20742.*

BRACELAND, Francis James, Am. physician; b. Phila., July 22, 1900; s. John Joseph and Margaret (L'Estrange) B.; A.B., LaSalle Coll., Phila., 1926, Sc.D., 1941; M.D., Jefferson Med. Coll., 1930, D.Litt., 1965; L.H.D., Canisius Coll., Buffalo, 1956; Sc.D., Coll. Holy Cross, Worcester, Mass., 1956; LL.D., Manhattan Coll., 1956; Sc.D., Cath. U. Am., 1957, Northwestern U., 1957, Trinity Coll., Hartford, Conn., 1958, Fairfield (Conn.) U., 1961; D.Litt., U. Hartford, 1964; m. Hope Van Gelder Jenkins, June 1, 1938; children—Mary Faith, John Michael. Practice medicine, specializing in psychiatry, 1932—; psychiatrist-in-chief Inst. Living, 1951-65, sr. cons., 1965—; clin. prof. psychiatry Yale, 1951—; lectr. on psychiatry Harvard, 1960—; cons. Hartford Hosp., St. Francis Hosp., Hartford, Reiss Mental Health Pavilion St. Vincents Hosp., N.Y.C. Recipient Signum Fidei medal La Salle Coll., 1955; Clarence E. Shaffrey, S.J. award St. Joseph's Coll., Phila., 1956; Distinguished Pub. Service award State Bar Assn. Conn., 1961; Laetare medal Notre Dame U., 1962; Cardinal Stritch award Loyola U., Chgo., 1965. Diplomate Am. Bd. Psychiatry and Neurology (past pres.). Fellow Am. Acad. Arts and Scis.; mem. Am. (past pres.), World v.p. 1961-66 psychiat. assns., A.M.A., Assn. For Research Nervous and Mental Disease (past pres.). Author: (with Michael Stock) Modern Psychiatry-Handbook For Believers, 1963. Editor: Faith, Reason and Modern Psychiatry, 1955. Research, publs. on interpretation aims and potentialities of psychiatry to Congl., other govt. coms. Home: 43 Ledyard Rd., West Hartford, Conn. 06117. Office: 200 Retreat Av., Hartford, Conn. 06102.*

BRACHET, Albert Toussaint Joseph, Belgian embryologist; b. Liège, Belgium, Jan. 1, 1869; prof. anatomy, embryology Brussels (Belgium) Faculty Medicine; mem. French Acad. Scis., 1918. Fellow Royal Soc., 1928. Author: Traité d'embryologie des vertébrés, 1921; La vie créatice des formes, 1927; L'oeuf et les facteurs de l'ontogégèse, 1915. Used frogs to study origin of nervous system, symmetry of egg, and devel. of liver and pancreas. Died Brussels, Dec. 27, 1930.

BRACHET, Jean Louis, Belgian biochemist; b. Brussels, Belgium, Mar. 19, 1909; s. Albert and Marguerite (Guchez) B.; M.D., U. Brussels, 1934; m. Francoise de Barsy, July 23, 1935; children—Etienne, Lise, Philippe. Asst., U. Brussels, 1934-38, charge de cours, 1938-43, prof. 1943—; vis. prof. U. Pa., 1946. Fgn. asso. Nat. Acad. Scis., Am. Acad. Arts and Scis. Author: Chemical Embryology, 1950; Biochemical Cytology, 1957; Biochemistry of Development, 1962. Spl. research nucleic acids in cell differentiation. Home: 26 av. des Cailles, Brussels, Belgium. Office: 67 rue des Chevaux, Rhode St. Genese, Belgium.

BRACKENRIDGE, William D., botanist; b. Ayr, Scotland, June 10, 1810; studied under Friedrick Otto, Berlin, Germany, Head gardener of Patrick Neill's Grounds, Edinburgh, Scotland; came to Phila. in service of Robert Buist, 1837; mem. U. S. Govt. expdn. to explore Pacific, 1838-42 (collected 10,000 species of plants representing 40,000 specimens; findings formed core of Nat. Herbarium); in charge of greenhouse, Washington, entrusted with care of living plants and preparation of report on ferns of expdn., 1842-55; ret. to Balt., 1855, purchased 30-acre farm; hort. editor Am. Farmer for some years. Author: Filices, Including Lycopodiaceae and Hydropterides, Vol. XVI, 1854. Died Balt., Feb. 3, 1893.

BRACONNOT, Henri, French natural historian; b. Commercy, France, May 29, 1781; ed. Strasbourg, also Paris, France; prof. natural history Lyceum, also dir. Botanic Garden, Nancy, France. Mem. French Acad. Scis., 1823. Author: L'agriculture élémentaire, 1838. Investigated vegetable and animal chemistry; studied saponification; discovered some acids from fruits and other plants. Died Nancy, Jan. 13, 1855.

BRADA, Zbynek, Czech. cancer biochemist; b. Roznov, Czechoslovakia, Jan. 16, 1926; s. Ladislav and Jarmila (Mensiková) B.; Ph.D., U. Brno (Czechoslovakia), 1949; C.Sc., Acad. Sci., Prague, Czechoslovakia, 1957; m. Vera Bradová Syrová, Oct. 19, 1949; 1 dau., Vera. Predoctoral fellow U. Bern (Switzerland), 1948, Cancer Assn. Brno, 1949-50; research worker research dept. Masaryk Radiotherapy Hosp., Brno, 1951-56; head research dept. Oncological Inst., Brno, 1957-60; head dept. biochemistry Cancer Research Inst., Brno, 1961—. Author: (with M. Rybák, I. Hais) Saulenchromatographie an Cellulose-Ionenaustauschern, 1966; also numerous articles. Research on tumour biology, analytical chemistry of proteins, biochem. aspects of tumour host relationship including research of tumour spread. Home: 29 Sady Osvobození, Brno. Office: 7 Zluty kopec, Brno, Czechoslovakia.*

BRADBURY, Norris Edwin, Am. physicist; b. Santa Barbara Cal., May 30, 1909; s. Edwin Perly and Elvira C. (Norris) B.; B.A., Pomona Coll., 1929, D.Sc., 1951; Ph.D., U. Cal., 1932; LL.D., U. N.M., 1953; D.Sc., Case Inst. Tech., 1956; m. Lois Platt, Aug. 5, 1933; children—James Norris, John Platt, David Edwin. NRC fellow in physics, Mass. Inst. Tech., 1932-34; asst. prof. physics, Stanford, 1934-37, asso. prof., 1937-42, prof. 1942-50; prof. physics U. Cal., 1951—; dir. Los Alamos Sci. Lab., 1945—. Mem. Sci. Adv. Bd. Nuclear Weapons, 1941-45. Decorated Legion of Merit. Fellow Am. Phys. Soc.; mem. Nat. Acad. Sci. Contbr. tech. articles to phys. revs. and jours. Expert on conduction electricity in gases, on properties of ions, on atmospheric electricity. Home: 1451 47th St. Office: Box 1663, Los Alamos.*

BRADBURY, Robert Hart, Am. chemist; b. Phila., Sept. 25, 1870; s. Robert and Margaret C. (Hart) B.; A.B., Central High Sch., Phila., 1887; Ph.D., U. Pa., 1893; m. Mabel Bradner, June 27, 1901; children—Robert Hart, Mabel Campbell. Asst. chemist Cambria Iron Co., 1889; chemist S. P. Wetherill Paint Co., 1891-92; prof. chemistry Central Manual Tng. Sch., Phila., 1893-1907; head dept. of sci. So. High Sch., Phila., 1907-30. Cons. chemist, Am. Chem. Paint Co.; lectr. on phys. chemistry, dept. of philosophy U. Pa., 1894-95. Mem. Am. Chem. Soc., Franklin Inst. Author: An Elementary Chemistry, 1903; A Laboratory Manual of Chemistry, 1903; An Inductive Chemistry, 1912; Laboratory Studies in Chemistry, 1912; A First Book in Chemistry, 1922, 3d edit. rev., 1938; New Laboratory Studies in Chemistry, 1923; Looseleaf Work-book for the Chemistry Laboratory, 1934. Died Mar. 27, 1949.

BRADBURY, Samuel, Am. physician; b. Germantown, Phila., Pa., Apr. 30, 1883; s. Samuel and Martha Washington (Chapman) B.; M.D., U. of Pa., 1905; m. Althea Norris Johnson. Sept. 26, 1914; children—Samuel, Emily Carey (Mrs. Alfred Vail), Althea Norris (Mrs. David Loshak), Wilmer Johnson. With Germantown and Pa. hosps., Phila., until 1911; asst. vis. physician City Hospital, N.Y., 1912-17, vis. physician, 1917-24; asst. vis. physician Belleville Hosp., 1924-27; chief in medicine Cornell Clinic, asst. prof. clin. medicine Cornell U. Med. Coll., N.Y., 1921-27; cons. phys. N.Y. Infirmary for Women and Children, 1924-27; dir. outpatient dept. Pa. Hosp., Phila., since 1927; dir. service and coordinator dept. med. Germantown Hosp. Diplomate Am. Bd. Int. Med. Mem. A.M.A. Med. Soc. State of Pa., Phila. County Med. Soc., Coll. of Physicians of Phila., Am. Climatol. and Clin. Assn., Acad. Medicine of N.Y., A.A.A.S., Harvey Soc. Author: Internal Medicine—Treatment, 1923;

(monograph) What Constitutes Adequate Medical Service? (pub. 1927 by a com. of United Hosp. Fund of N.Y.); also contbr. to Cyclopedia of Medicine and Text Book of Medicine by Am. Authors, Adequate Medical Care, 1937. Died Aug. 30, 1947.

BRADFORD, Donald Comnick, Am. geophysicist; b. Spokane, Wash., July 6, 1910; s. Donald R. and Lenore (Comnick) B.; B.S., U. Wash., 1932; M.S., St. Louis U., 1934; postgrad. U. Pitts.; m. Doris Alberta Bersche, June 6, 1933; children—Barbara Ann, William Donald. Geophysicist, Gulf Oil Corp., 1936-38; dir. Geophys. Obs., U. Pitts., 1939-43; chief radiosonde sect. U. S. Signal Corps Lab., Ft. Monmouth, N.J., 1943-44; chief engr. Serdex Corp., Boston, 1945-47; sect. chief Raytheon Mfg. Co., Waltham, Mass., 1947-52; mem. tech. staff Hughes Aircraft Co., Culver City, Cal., 1952-60; research scientist N.Am. Aviation, Inc., Downey, Cal., 1960-66, prin. scientist, aeronutronic div. Philco-Ford, Newport Beach, Cal., 1966—. Instr. U. Pitts., 1939-43. Mem. Research Soc. Am. (pres. br. 1965-66), Am. Geophys. Union, Soc. Exploration Geophysicists, A.A.A.S., Seis. Soc. Am., Sigma Xi; profl. mem. Am. Meteorol. Soc. Research, publs. in nature and origin of microseisms, measurement of microbarometric oscillations, geophys. edn., characteristics of earthquakes, meteoroid environment in solar space, origin of meteors, meteoroids in cis-lunar space, seismicity of Mars, probe instrumentation for atmospheres of Mars and Venus. Home: 1106 Nottingham Rd., Newport Beach 92660. Office: Philco-Ford Aeronutronic Div., Newport Beach, Cal. 92663.*

BRADFORD, Edward Hickling, Am. surgeon; b. Boston, June 9, 1848; s. Charles F. and Eliza E. (Hickling) B.; A.B., Harvard, 1869, A.M., 1872, M.D., 1873; m. Edith Fiske, June 20, 1900. Clin. instr. orthopedic surgery, 1881-82, asst. clin. surgery, 1881-86, instr. surgery, 1886-93, instr. orthopedics, 1889-93, 1st prof. orthopedic surgery, 1903-12; dean, from 1912. Harvard Med. Sch.; surgeon Children's Hosp., 1878-1909, Boston City Hosp., 1880-94. A founder Boston Indsl. Sch. for Crippled and Deformed Children. Fellow Am. Acad. Arts and Scis. Author: Treatise on Orthopedic Surgery, (with R. W. Lovett), 1895. Invented Bradford frame for handling children with tuberculosis of spine, later modified for use in fracture of the thigh, joint disease, etc. Died May 7, 1926.

BRADFORD, John Rose, English physician, physiologist; b. London, Eng., May 7, 1863; s. Abraham Rose and Ellen (Littleton) B.; B.Sc., Univ. Coll., London, 1883, M.D., 1889; numerous hon. degrees; m. Mary Roberts, 1899. Resident house physician, Univ. Coll. Hosp., from 1886, tchr., researcher anatomy and physiology, 1886-89, full physician, charge wards, 1900-23; asst. physician Nat. Hosp. Diseases of Nervous System, 1893-96; prof., supt. Brown Animal Instn., 1895-1903. With Trop. Diseases Bur., London Sch. Trop. Medicine, from 1908; governing bd. Lister Inst., 1899-1918, chmn., 1912-14; sr. med. adviser Colonial Office, Gt. Britain, 1912-24; cons. physician BEF, 1914-19. Fellow Royal Soc., 1894 (trop. diseases com. 1907; sec. biol. com. 1908-15), Royal Coll. Physicians (pres. 1926-31). Research (with William Bayliss) on elec. charges accompanying nervous stimulation of salivary glands; kidney research on men and dogs. Died London, Apr. 7, 1935.

BRADLEY, Charles Schenck, Am. inventor; b. Victor, N.Y., Apr. 12, 1853; s. Alonzo and Sarah (Schenck) B.; student U. Rochester, 1872; m. Emmaretta Orcutt Brackett, Feb. 16, 1876. Began with Edison Illuminating Co., N.Y., 1881; successively in employ various elec. cos., 1883-1914; with U. S. Reduction Co., from 1916. Patentee process for production of aluminum, three phase transmission of power, rotary convertor, fixation of atmospheric nitrogen, others. Died 1929.

BRADLEY, Dan Fordham, Am. chemist; b. Toledo, Apr. 26, 1929; s. Dan Theodore and Eloise (Smiley) B.; A.B., Oberlin Coll., 1951; Ph.D., U. Cal. at Berkeley, 1953; m. Raya Weinberg, Mar. 24, 1964; children—Debra Ann, Dan David, Jeremy David. Teaching asst. U. Cal., Berkeley, 1951-52, instr. chemistry, 1953-55; scientist USPHS, Nat. Insts. Health, Bethesda, Md., 1955-61, chief phys. chemistry sect., 1961—, lectr. Grad. Sch., 1958—; instr. Am. U., 1965—; Recipient Harry Holmes prize, 1951. Mem. Am. Chem. Soc., A.A.A.S., N.Y. Acad. Scis., Biophys. Soc., Soc. Gen. Physiologists, Sigma XI, Phi Beta Kappa. Editorial adv. bd. Biopolymers Jour., 1963-65, editorial bd. Biopolymers 1965—. Research, numerous publs. mechanism of converting light to chem. energy in plants and transient concentration changes in plant, chromatography and melting of nucleic acids, devel. dye-stacking theory of metachromasy, devel. theory of optical properties of biopolymers; computer simulation of biochem. reactions. Home: 5813 Cherrywood Lane, Greenbelt, Md. 20770. Office: 3N 321 Bldg. 10, NIH, Bethesda, Md. 20014.*

BRADLEY, Daniel Joseph, Irish physicist; b. Londonderry, Jan. 18, 1928; s. John C. and Margaret (Keating) B.; B.Sc. in Math., U. London (Eng.), 1953, B.Sc. in Physics, 1957, Ph.D., 1961; m. W. M. T. O'Connor, July 18, 1958; children—Sean, Mairead, Donal, Ronan. Primary sch. tchr., 1947-53; asst. lectr. Royal Holloway Coll., U. London, 1957-60, reader, 1964-66, lectr. Imperial Coll., 1960-64; prof. physics, dept. chmn. Queens U., Belfast, No. Ireland, 1966—. Cons. on optics, interferometry, lasers. Mem. Optical Soc. Am. Fellow Phys. Soc. (U.K.). Research, publs. on devel. interferometers for plasma diagnostics, laser spectroscopy and space research, 1st practical oscillating fabry perot interfero meter, giant pulse lasers, interaction of intense laser beams with matter. Home: 11, Sans Souci Park, Belfast 9. Office: Dept. Physics, Queens U. Belfast, Belfast 7, U.K.*

BRADLEY, Harold Cornelius, Am. physiol. chemist; b. Oakland, Cal., Nov. 28, 1878; s. Cornelius Beach and Mary S. (Comings) B.; B.A., U. Cal., 1900; Ph.D., Yale, 1905; m. Josephine Crane, July 8, 1908; children—Charles, Harold, David, Stephen, Joseph, Richard, William; m. 2d, Ruth Aiken, Nov. 27, 1957. Asst. prof. chemistry Yale Med. Sch., 1905-06; asst. prof. physiol. chemistry U. Wis. Med. Sch., Madison, 1906-10, prof., head dept., 1910-48, prof. emeritus, 1948—. Mem. Am. Soc. Biol. Chemists, Am. Chem. Soc., Am. Physiol. Soc., Sigma Xi. Research and publs. on human pancreatic juice, autolysis, cathepsin, atreophy; discovered zinc in invertebrate tissue and blood, 1904, manganese, 1907. Home: 2639 Durant Av., Berkeley, Cal. 94704.*

BRADLEY, James, English astronomer; b. Sherborne, Eng., Mar. 1693; s. William and Jane (Pound) B.; B.A., Balliol Coll., Oxford, 1714, M.A., 1717; D.D., Oxford U., 1742; m. Susannah Peach, June 25, 1744; 1 dau., Susannah (Mrs. Samuel Peach). Became vicar Bridstow, 1719; chaplain to bishop of Hereford; elected Savilian prof. astronomy Oxford U., 1721, lectr. exptl. philosophy, 1729-60; apptd. royal astronomer, 1742. Fellow Royal Soc., 1718 (Copley medal 1748, council 1752-62); mem. Berlin Acad. Sci.; mem. Fr. Acad. Scis. Author: Astronomical Observations made at the Royal Observatory at Greenwich from the year 1750 to the year 1762, Vol. I, 1798, Vol. II, 1805. His attempt to measure parallax of stars resulted in discovery of aberration of light, 1727, which enabled him to measure speed of light more accurately (thus confirmed Roemer's observations, also proved Copernican theory earth was in motion, supported Newton's particle theory of light); measured diameter Jupiter, 1733; discovered nutation of earth's axis; observed Jovian system; prepared astron. tables; stated laws of refraction and showed necessity for correcting for barometric pressure and temperature. Died Chalford, Eng., July 13, 1762.

BRADLEY, James Chester, Am. entomologist; b. West Chester, Pa., Feb. 11, 1884; s. Daniel Hiester and Virginia (Blanchard) B.; A.B., Cornell U., 1906, Ph.D., 1910; M.S., U. Cal. at Berkeley, 1907; m. Ruth Stephens Baker, June 27, 1940. Faculty, Cornell U., Ithaca, N.Y., 1905—, prof. entomology, 1918-52, prof. emeritus, 1952—; commr. Internat. Commn. Zool. Nomenclature, 1942-63, pres., 1952-63; v.p. Internat. Congress Zoology, Copenhagen, 1953. Hon. fellow Royal Entomol. Soc. London; fellow A.A.A.S.; hon. mem. Entomol. Soc. Netherlands, Entomol. Soc. Am., (past president), International congresses entomology; mem. Am. Entomol. Soc. Phila. (corr., past pres.), Am. Soc. Zoologists, Soc. Systematic Zoology, Soc. Study Evolution, Internat. Union Study Social Insects, Paleontol. Research Instn., Am. Malacological Union, Sigma Xi, Gamma Alpha, Pi Kappa Alpha, numerous others. Author: (with E. L. Palmer) Insect Life, 1915; A Manual for Genera of Beetles of America North of Mexico, 1930; Laboratory Guide to Study of Wings of Insects, 1935, rev., 1939. Research, numerous publs. on knowledge of taxonomy of Hymenoptera and Coleoptera, morphology of insects wings, rules of zool. nomenclature. Home: 604 Highland Rd., Ithaca, N.Y. 14850.*

BRADLEY, John Edmund, Am. physician; b. Balt., Oct. 31, 1906; s. Charles Edward and Mary (Henry) B.; B.S., Loyola Coll., M.D.; Georgetown U.; m. Kathryn Davis, Sept. 21, 1933; children—Mark Edmund, Mary Marcia. Faculty, U. Md. Med. Sch., Balt., 1934-66, prof., head pediatrics, 1948-66, emeritus prof., 1966—; pediatrician-in-chief Univ. Hosp., 1948—; head pediatrics Lutheran Hosp. Cons. to hosps., surgeon gen.; mem. Md. Bd. Health and Mental Hygiene. Mem. Am. Soc. Pediatrics, A.A.A.S., Am. Assn. Pediatrics, A.M.A., Alpha Omega Alpha. First to observe high blood pressure in children with Wilms (kidney) tumors; brought attention to high incidence of lead poisoning in lower socioecon. groups; instituted recommendations for treatment of epidemic viral vomiting. Home: 14 W. Cold Spring Lane, Balt. 21210. Office: U. Md. Hosp., Balt. 21201.*

BRADLEY, John Newton, English chemist; b. Marple, Eng., Nov. 29, 1931; s. George and Isobel (Dobson) B.; B.Sc., U. Birmingham (Eng.), 1952, Ph.D., 1955, D.Sc., 1965; m. Winifred Eleanor Pattinson, May 4, 1957; children—Karen Rosalind, Glenn Martin. Commonwealth Fund fellow Harvard, 1957-59; lectr. chemistry U. Liverpool (Eng.), 1959-64; prof. chemistry U. Essex (Eng.), 1964—. Recipient Meldola medal, 1960. Fellow Royal Inst. Chemistry; mem. Chem. Soc., Faraday Soc., Am. Phys. Soc., Combustion Inst. Author: Shock Waves in Chemistry and Physics, 1962; also articles. Research on rates and mechanisms of reactions in gas phase, especially reactions in shock waves at high temperatures, reactions in combustion systems and reactions of atoms and radicals; solid-state reactions, liquid-phase reactions of radicals. Home: 8 Elianore Rd., Colchester, Essex, Eng.*

BRADLEY, Philip Benjamin, English pharmacologist; b. Bristol, Eng., Aug. 17, 1919; s. John James and Edith Lilian (Smith) B.; B.Sc. with honors, Bristol U., 1949; Ph.D., U. Birmingham (Eng.), 1952, D.Sc., 1959; m. Joan Salter, June 10, 1945; children—Anthony Philip, Pamela Joan. Faculty, Birmingham U., 1949—, prof. neuropharmacology, 1964—, head dept. neuropharmacology, 1962-64. Exec. bd. C.I.N.E., 1957—, treas., 1964—. Mem. E.E.G. Soc. (fgn. sec. 1963—), Physiol. Soc., Pharmacological Soc. Contbg. author: Physiological Pharmacology, vol. I. Editor-in-chief for Europe and Middle E., Internat. Jour. Neuropharmacology. Research on functioning of brain in terms of neurophysiol. mechanism and behaviour, elucidation of mode of action of drugs, especially those used in psychiat. treatment, nature of chem. transmission in nervous system. Home: 72 Oakfield Rd., Selly Park, Birmingham 29, Eng.*

BRADLEY, Richard, English physician, botanist; b. 18th century; became prof. botany, Cambridge, 1724; became lectr. materia medica, 1729. Fellow Royal Soc., 1712. Genera named in his honor. Author: Treatise on Agriculture and Gardening, 1724. Studied generation of plants, plant anatomy. Died Cambridge, Nov. 5, 1732.

BRADLEY, Richard Crane, Am. physicist; b. Chgo., May 14, 1922; s. Harold Cornelius and Joseph (Crane) B.; A.B., Dartmouth, 1943; Ph.D., U. Cal. at Berkeley, 1953; m. Dorothy Alice Holden, June 7, 1947; children—Richard Crane, Helen Louise, Josephine Crane, David Holden. Teaching asst. U. Cal., Berkeley, 1947-51, research asst., 1951-53; research asso. Cornell U., Ithaca, N.Y., 1953-56, instr., 1956-57, asst. prof., 1957-61, asso. prof., 1961-62; asso. prof. Colo. Coll., Colorado Springs, 1962—. Cons. Kaman Nuclear Corp., Colorado Springs, 1963,—, Cancer Inst., Penrose Hosp., Colorado Springs, 1964—. Fellow, Am. Phys. Soc., mem. Am. Assn. Physics Tchrs., A.A.A.S., Am. Assn. U. Profs, Wilderness Soc., Nat. Parks Assn., Sigma Xi, Phi Beta Kappa. Research in physics of surfaces of solids, surface ionization, sputtering, ion emission phenomena, mass spectrometry, field emission microscopy. Home: 1730 N. Cascade Av., Colorado Springs, Colo. 80907.*

BRADLEY, Rupert Stevenson, English chemist; b. Devenport, Devon, Eng., Aug. 5, 1907; s. Robert and Ellen (Berry) B.; B.A., St. John's Coll., Cambridge, Eng., 1927, M.A., 1930; Sc.D., 1954; m. Phyllis Alice Lane, Aug. 4, 1930; children—Clare Elizabeth (Mrs. J. McFarlane), Jane Catherine (Mrs. P. D. Greene). With U. Leeds (Eng.) Sch. Chemistry 1927—, sr. lectr., 1946-57, reader inorganic and structural chemistry, 1957—. Mem. Chem. Soc., Faraday Soc., Royal Inst. Chemistry. Author: (with D. C. Munro) High Pressure Chemistry, 1966. Editor: Evaporation and Droplet Growth in Gaseous Media, 1963, Advances in High Pressure Research, vol. I, 1966; Author, editor High Pressure Physics and Chemistry, 1965. Research, numerous publs. on surface chemistry, solid reactions, nucleation, low vapour pressures, high pressure chemistry. Home: 25 Batcliffe Dr. Office: Sch. Chemistry, U. Leeds, Leeds, Eng.*

BRADLEY, Sterling Gaylen, Am. microbiologist; b. Springfield, Mo., Apr. 2, 1932; s. Benn and Lora (Brown) B.; B.A., B.S., S.W. Mo. State Coll., 1950; M.S. in Microbiology, Northwestern U., Ph.D. in Microbiology (NSF predoctoral fellow), 1954; certificate med. mycology Duke, 1957; m. Lois Evelyn Lee, May 13, 1951; children—Don, Evelyn, John, Phillip. Grad. teaching asst. Northwestern U., Evanston, Ill., 1950-51, Abbott research asst., 1951-52, instr. biology, 1954; instr. dept. bacteriology and immunology U. Minn., Mpls., 1956-57, asst. prof. dept. bacteriology, 1957-59, asso. prof. dept. microbiology, 1959-63, grad. faculty genetics, 1961—, prof., 1963—, chmn. genetics faculty group, 1964. Eli Lilly postdoctoral fellow U. Wis., 1954-55, NSF postdoctoral fellow dept. genetics, 1955-56. Mem. Am. Soc. for Cell Biology, A.A.A.S., Am. Assn. U. Profs., Am. Soc. for Microbiology (subcom. actinomycete nomenclature 1959-64), Soc. for Gen. Microbiology, Henrici Soc. Minn., Soc. for Indsl. Microbiology (pres. 1964-65), Am. Inst. Biol. Sci. (dir.), Mycol. Soc. Am., Soc. for Exptl. Biology and Medicine (nat. mem. com. 1961—, chmn. Minn. chpt. 1964-66), Genetics Soc. Am., Torrey Bot. Club (life), N.Y. Acad. Scis. (life), Minn. Acad. Sci., Sigma Xi. Editorial bd. Conf. on Antimicrobial Agts., 1960, Jour. Bacteriology, 1964—. Research, publs. on genetics of actinomycetes and actinophages, immunogenetics, antibiotics. Home: 4141 Reservoir Blvd., Mpls. 55421.*

BRADNER, Hugh, Am. physicist; born Tonopah, Nev., Nov. 5, 1915; s. Donald Byal and Agnes (Mead) B.; B.S. in Physics, Math., U. Miami, O., 1937, D.Sc., 1961; Ph.D., Cal. Inst. Tech., 1941; m. Marjorie Hall, Sept. 26, 1943; 1 dau., Barbara Claire. Research staff U. S. Naval Ordnance Lab., Washington, 1941-42, Los Alamos Lab., 1942-45, Radiation Lab., U. Cal., Berkeley, 1945-61; prof. aerospace sci. Inst. Geophysics, U. Cal., La Jolla,

1963—. Mem. mine adv. com. Nat. Acad. Sci., Washington, 1958—; cons. in accelerator design, exptl. nuclear physics, seismology, thermoelectric conversion, submarine operations Gen. Atomics, La Jolla, Wm. Brobeck Assos., Berkeley, 1960—. Fellow Am. Phys. Soc.; mem. Seismol. Soc. Research, publs. on sub-nuclear particles and solid-earth geophysics, accelerators, bubble chambers, explosives, mine warfare. Home: 827 Muirlands Vista Way, La Jolla, Cal. 92037.*

BRADWARDINE, Thomas, English mathematician; b. probably at Chichester, Eng., circa 1290; ed. Balliol Coll., Oxford U., 1321, Merton Coll., 1323. Proctor, Oxford U., 1325-26 26-27, left circa 1335; apptd. chaplain to Richard Bury, Bishop of Durham, London; became chancellor St. Paul's Cathedral, London, Eng., 1335; named chaplain-confessor to King Edward III, 1338; named archdeacon of Norwich, 1347; elected Archbishop of Canterbury, 1348; consecrated at Avignon, France, 1349. Author: Tractatus de proportionibus; Tractatus de continuo; De geometrica speculativa; De quadratura circuli, De arithmetica speculativa; De causa dei contra Pelagium; De praescientia et praedestinatione. Studies in geometry, math., moral philosophy, theology; application of exponential function to phys. theory in finding functional relationship between speed, force and resistance; developed math. formuale for phys. laws, especially in kinematics (leading to quantitative measurement of phys. processes). Died Lambeth, London, Aug. 26, 1349.

BRADY, Allan Jordan, Am. biophysicist; b. Fairview, Utah, May 23, 1927; s. J. Frank and Laura (Garlick) B.; student Utah State Coll., 1947-49; B.A., U. Utah, 1951, M.S., 1952; Ph.D., U. Wash., 1956; m. Dorothy Abbott, Sept. 5, 1952; children—Susan, David. Am. Heart Assn. research fellow U. Wash., 1956-57, Cambridge (Eng.) U., 1957-58, U. Cal. at Los Angeles, 1958-60, established investigator, 1960—. Recipient Career Devel. award NIH, 1966—. Mem. Biophysics Soc., Am. Physiol. Soc., A.A.A.S., Am. Heart Assn. (basic sci. council). Research on mechanism of heart contraction; developed electrode which can be inserted in single beating heart cell. Home: 3200 Federal Av., Los Angeles 90066. Office: U. Cal. at Los Angeles Medical Center, Los Angeles 90024.*

BRADY, Fred Charles, Am. physician; b. Charleroi, Pa, Dec. 19, 1912; s. Fred W. and Mary (Phillips) B.; B.S. in Chemistry, Va. Mil. Inst., 1934; M.D., Jefferson Med. Coll., 1939; m. Adelaide Aschmann, Feb. 14, 1942; children—Mary Lee, Fred C., Timothy P., Heidi A., Kathleen A. Dir. postgrad. course in applied anatomy U. Pitts., 1949—; fellow Lahey Clinic, Boston, 1947-48; mem. sr. surg. staff Mercy Hosp.; sr. cons. VA Hosp., Pitts.; chief surg. service, chief cons. gen. surgery Mayview State Hosp.; pvt. practice medicine specializing in surgery, Pitts. Mem. A.C.S. (past pres. S.W. chpt.), Pa. Med. Soc., Soc. Nuclear Medicine, Am. Geriatrics Soc., N.Y. Acad. Medicine, Allegheny County Med. Soc., Pitts. Surg. Soc., Pitts. Acad. Medicine. Contbr. articles to med. jours. Home: 209 Tennyson Av., Pitts. 15213. Office: M.D. Bldg., 1501 Locust St., Pitts. 15219.*

BRADY, Henry Bowman, Brit. naturalist, pharmacist; b. Gateshead, Eng., Feb. 23, 1835; s. Henry and Hannah (Bowman) B.; apprentice to Thomas Harvey, Leeds, Eng., 1850; student Newcastle (Eng.) Coll. Medicine; passed exam. Pharm. Soc., 1855; LL.D. Aberdeen (Scotland), 1888. Owner pharm. business, 1855-76; lectr. botany Durham (Eng.) Coll. Medicine. Genus Bradyina named in his honor, 1878. Fellow Linnean Soc., Royal Soc., 1874. Author: Report on the Foraminifera Collected by H. M. S. Challenger, 1884; Monograph of the Foraminifera of the Crag, (Part I with William Kitchin Parker, J. Rupert Jones) 1866; Monograph of Carboniferous and Permian Foraminifera, 1876; (with J. D. Siddall Chester), Catalogue of British Recent Foraminefera, 1879; also articles in jours. and books. Research on Foraminifera. Died Jan. 3, 1891.

BRADY, Roscoe O., Jr., Am. biochemist; b. Phila., Oct. 11, 1923; s. Roscoe O. and Martha (Roberts) B.; student Pa. State U., 1941-43; M.D., Harvard, 1947; postgrad. U. Pa., 1948-49. NRC fellow U. Pa., 1948-50, USPHS spl. fellow, 1950-52; sect. chief Nat. Inst. Neurol. Diseases and Blindness, NIH, 1954-67, acting chief lab. neurochemistry, Bethesda, Md., 1967—; professorial lectr. George Washington Sch. Medicine, 1963—; faculty Georgetown U. Sch. Medicine, 1967—. Recipient Superior Service award Dept. Health, Edn., Welfare, 1966. Mem. Am. Soc. Biol. Chemists, Am. Chem. Soc., Am. Acad. Neurology, Am. Acad. Mental Retardation, Soc. Exptl. Biology and Medicine. Author: (with Donald B. Tower) Neurochemistry of Nucleotides and Amino Acids, 1960; also numerous articles. First demonstration of enzyme system for fatty acid synthesis; research on role of malonyl coenzyme A in fatty acid synthesis; biosynthesis of myelin sheath lipids, nature of metabolic defects in Gaucher's disease, Niemann-Pick disease, and Fabry's disease; diagnostic tests for Gaucher's, Niemann-Pick, Fabry's diseases; metabolism of sphingolipids in neoplastic diseases. Home: 1600 S. Joyce St., Arlington, Va. 22202. Office: NIH, Bethesda, Md. 20014.*

BRADY, Wray Grayson, Am. mathematician; b. Benton Harbor, Mich., July 20, 1918; s. Wray G. and Mildred (Sauters) B.; B.S., Washington and Jefferson Coll., 1940, M.A., 1942; Ph.D., U. Pitts., 1953; m. Emilie Peterson, Apr. 30, 1943; children—Susan Annene, Wray Gordon. Instr. math. U. Wyo., Laramie, 1947-50; faculty Washington and Jefferson Coll., Washington, Pa., 1953-65, prof., 1958-65, chmn. dept. math., 1955-65; Bernhard prof. math., chmn. dept. U. Bridgeport, Conn., 1965—; prof. U. Ariz., Summer Inst., 1963-67. Cons. to Bettis plant Westinghouse Electric Co., 1955-60. Fellow A.A.A.S.; mem. Math. Assn. Am., Am. Math Soc., Am. Assn. U. Profs., Phi Delta Theta. Author: (With Maynard Mansfield) Calculus, 1960; Analytical Geometry, 1961. Research on infinite series, numerical analysis, transport theory, diffusion theory. Home: 421 Walnut Tree Hill Rd., Huntington, Conn. 06484. Office: Dept. Math., U. Bridgeport, Bridgeport, Conn. 06022.*

BRAEKKAN, Olaf Rikard, Norwegian biochemist; b. Narvik, Norway, Dec. 14, 1918; s. Kolbein and Berntine (Normann) B.; grad. Narvik, Coll., 1938; Mag.scient., U. Oslo (Norway), 1938, Dr. philos., 1964; m. Muriel Pearson, Feb. 9, 1949; children—Elsa-Mari, Anton, Martin. Biochemist, Nyegaard & Co. A/S, Oslo, 1940-45; with vitamin div. U. S. Dept. Agr., biochemistry dept. U. Wis., Madison, 1948-49; head govt. vitamin lab. Norwegian Fisheries Research Inst., Bergen, 1948—; asso. prof. U. Bergen, 1966—. Chmn., CODEX com. on fish and fishery products WHO/FAO World Food Standard Program, 1965—. Mem. Soc. for Advancement Sci. Bergen. Research, publs. on vitamins especially vitamin studies in fish, vitamin A utilization in chicks; presented hypothesis on function of red muscle in fish, 1956. Home: Natlandsvei 44. Office: Govt. Vitamin Lab., P. box 1874, Bergen, Norway.*

BRAESTRUP, Frits Wimpffen, Danish zoologist; b. Hornbaek, Denmark, Sept. 20, 1906; s. Tycho Cosmus W. and Inger (Bentzon) B.; cand mag Copenhagen (Denmark) U., 1931, Dr.Phil., 1941; m. twice; children—Mikael, Janne, Mie Merete. Staff, Zool. Mus., Copenhagen U., 1937—, chief curator dept. nonavian vertebrates, 1960—. Mem. Am. Soc. Mammalogists, Soc. for Study Evolution, Deutsche Gesellschaft fur Säugetierkunde, Dansk naturhist. Forening. Author: Hjortebogen, 1952; Fuglenes Verden, 1953; Dyrenes Udvikling, 1954; also articles. Research on vertebrate systematics and zoogeography, animal behavior regarding problems of ecology and evolution. Home: 42 Vester Paradisvej, Holte, Denmark. Office: 15 Universitetsparken, Copenhagen, Denmark.*

BRAGG, Arthur Norris, Am. zoologist; b. Pittsfield, Me., Dec. 18, 1897; s. Nathan M. and Emma (Cates) B.; B.S., Bates Coll., 1924; postgrad. Johns Hopkins; M.A., Boston U., 1934; Ph.D., U. Okla., 1937; m. Mary Georgiana Keirstead, Dec. 24, 1924; children—Betty Lee (Mrs. M. L. Amnesley), William Norris, Marita Jean (Mrs. Percival Marcus Lowell, Jr.). Grad. asst. in biology Johns Hopkins, 1924-25; asst. prof. zoology Marquette U., Milw., 1925-33; faculty U. Okla., Norman, 1934—, prof. zoology, 1953—. Vis. zoologist N.M. Highlands U., Las Vegas, 1940; curator amphibians Stovall Mus. Sci. and History, 1951—; field zoologist Okla. Biol. Survey, 1948-50, herpetologist, 1951—. Fellow Okla. Acad. Sci.; mem. Herpetologists League, A.A.A.S., Am. Soc. Zoologists, Am. Soc. Herpetologists and Ichthyologists, Am. Soc. Systematic Zoologists (charter), Southwestern Assn. Naturalists (charter), Ecol. Soc. Am. (asso.), Am. Assn. Biol. Soc., Animal Behavior Soc., Sigma Xi, Gamma Alpha, Phi Sigma. Author: Gnomes of the Night, 1965. Contbr. articles on protozoology, embryology, cytology, amphibians to profl. publs. Home: 108 W. Symmies St., Norman, Okla. 73069.*

BRAGG, Robert Henry, Am. physicist; b. Jacksonville, Fla., Aug. 11, 1919; s. Robert Henry and Camille (McFarland) B.; B.S., Ill. Inst. Tech., 1949, M.S., 1951, Ph.D., 1960; m. Violette McDonald, June 14, 1947; children—Robert Henry III, Pamela. Asst. physicist Research Lab. Portland Cement Assn., Skokie, Ill., 1951-54, asso. physicist, 1954-56; asso. physicist Research Inst., Ill. Inst. Tech., Chgo., 1956-57, physicist, 1957-59, sr. physicist, group leader, 1959-61; research physicist Lockheed Palo Alto (Cal.) Research Lab., 1961-63, sr. staff scientist, 1963—, sr. mem. Research Lab., 1964—. Cons. metallurgy dept. Argentina Atomic Energy Commn., 1966. Mem. Am. Phys. Soc., Am. Crystallographic Assn., Am. Soc. for Metals, A.A.A.S., Am. Inst. Mining and Metall. Engrs., Sigma Xi, Sigma Pi Sigma. Contbr. articles to sci. jours. Conducted x-ray, light scattering and small angle x-ray scattering investigations of colloidal silicate hydrates, surface energies of oxides and vapor pressures of hydrates; studies of electronic and thermal properties of thin films and single crystals of semiconductors, radiation effects in semiconductors, radiation effects on catalysis; x-ray studies of texture and defect structure of re-entry materials, beryllium and advanced carbons and graphites. Home: 4160 Coulombe Dr., Palo Alto 94306. Office: 3251 Hanover St., Palo Alto, Cal. 94304.*

BRAGG, Sir (William) Lawrence, physicist; b. Adelaide, Australia, Mar. 31, 1890; s. Sir William and Gwendoline (Todd) B.; student St. Peters Coll., 1900-05; M.A., Adelaide U., 1908, Trinity Coll., Cambridge U., 1911; m. Alice Grace Jenny Hopkinson, Dec. 10, 1921; children—Stephen, David, Margaret (Mrs. Mark Heath), Patience (Mrs. David Thomson). Fellow, lectr. in natural scis. Trinity Coll., Cambridge, 1914; Langworthy prof. physics Manchester U., 1919-37; dir. Nat. Phys. Lab., 1937-38; Cavendish prof. exptl. physics Cambridge U., 1938-53; Fulerian prof. chemistry, dir. Royal Instn., London, 1953-66. Recipient (with father) Nobel prize for physics, 1915; Barnard medal, 1914; Fellow Royal Soc., 1921 (council, 1931-33, Hughes medal, 1931, Royal medal, 1946, Copley medal, 1966); Roebling medal Min. Soc. Am., 1948; decorated comdr. Order of Leopold of Belgium, 1961; Companion of Honor, 1967. Fgn. assoc. Nat. Acad. Scis. Author: X Rays and Crystal Structure, 1915; The Crystalline State, 1934; Electricity, 1936; Atomic Structure of Minerals, 1937; Crystal Structure of Minerals, 1965; also numerous articles. With Sir William Henry Bragg originated science of X-ray crystallography 1912, continuous research in crystallography; elucidated structure of silicates, 1925, and demonstrated how this revelation changed chemistry of these salts; proposed classification system for silicates. Address: 6 The Boltons, London S.W.10, Eng.

BRAGG, William Henry, physicist; b. Wigton, Cumberland, Eng., July 2, 1862; studied at King's Coll., Isle of Man, Trinity Coll., Cambridge (Eng.); 1 son, William Lawrence. Prof. math. and physics Adelaide U., Australia, 1885-1908; Cavendish prof. physics U. Leeds, (Eng.), 1909-15; Quain prof. physics U. Coll., London, Eng., 1915-23; became prof. chemistry, dir. Royal Instn., 1923; named dir. Davy-Faraday Research Lab., 1923. Mem. bd. inventions and research Brit. Admiralty, 1916-19. Recipient (with son) Nobel prize for physics, 1915; created knight, 1920. Fellow Royal Soc., 1907 (Rumford medal, 1916, Copley Medal, 1930, pres. 1935-40); mem. French Acad. Scis. (corr.). Author: Studies in Radioactivity, 1912; (with W. L. Bragg); X Rays and Crystal Structure; also popular works which include: The World of Sound, 1920; Concerning the Nature of Things, 1925; The Universe of Light, 1933. Pioneered the study of crystal analysis by using X-rays; (with son) developed X-ray spectrometer for measuring the wave lengths of X-rays by crystal diffraction; determined the arrangement of atoms in crystals. Died London, Mar. 12, 1942.

BRAGONIER, Wendell Hughell, Am. botanist; b. Ia., Aug. 5, 1910; s. Robert Jacob and Cora M. (Hughell) B.; B.A., Ia. State Tchrs. Coll., 1933; fellow U. Chgo., 1939; M.S., Ia. State U., 1941, Ph.D., 1947; Rockefeller Found. fellow, Mexico, 1945-46; m. M. Alice Dowden, June 18, 1934; children—J. Robert, James W., Mary-Ellen. Tchr. Ia. high schs., 1933-34, 35-39, Tenn., 1934-35; faculty Ia. State Univ., 1939-63, beginning as fellow, successively instr., research asso., asst. to dir. Indsl. Sci. Research Inst., asso. prof. botany, 1947-50, prof., head botany, 1950-63; dean grad. sch. Colo. State U., Ft. Collins, 1963—, prof. botany and plant pathology; asso. dir. Camp Dodge br. Ia. State Univ., 1946-47. Chmn. sci. adv. com. Am. Seed Research Found. Mem. A.A.A.S., Botanical Soc. Am., Am. Phytopath. Soc., Ia. Acad. Sci., Sigma Xi, Phi Kappa Phi, Gamma Sigma Delta. Discovered cause of Umbrella disease of Sumac to be a fungus—Botrosphaeria ribis G. & D; directed research leading to discovery that beetles carry fungus causing wilt disease. Home: 1100 Ellis, Ft. Collins, Colo.

BRAHAM, Roscoe Riley, Jr., Am. meteorologist; b. Yates City, Ill., Jan. 3, 1921; s. Roscoe R. and Edith L. (Bowman) B.; B.S., Ohio U., 1942; S.M., U. Chgo., 1948, Ph.D., 1951; m. Mary Ann Moll, Mar. 12, 1943; children—Ruth Ann, Nancy Kay, Richard Riley, Jean Lou. Staff, U. S. Weather Bur., 1946-49; faculty U. Chgo., 1949-50, prof. meteorology, 1951—; faculty N.M. Inst. Tech., 1950-51; prof. U. Ariz., 1954-56. Recipient Losey award Inst. Aero. Scis., 1950; Silver medal U. S. Dept. Commerce, 1950. Mem. Am., Royal meteorol. socs., Am. Geophys. Union, Sigma Xi. Author: (with H. R. Byers) The Thunderstorm, 1949; also numerous articles. Discovered cell circulation in thunderstorms; research in weather modification. Home: 57 Longcommon Rd., Riverside, Ill. 60546. Office: 1100 E. 58th St., Chgo. 60637.*

BRAHANA, Henry Roy, Am. mathematician; b. Lowell, Vt., Aug. 16, 1895; s. Patrick H. and Mary (Richardson) B.; A.B., Dartmouth, 1916; A.M., Princeton, 1917, Ph.D., 1920; m. Myrtle Van Wart, July 3, 1918; children—John V., Thomas R., Frances M. m. 2d, Melinda Dunbar, Feb. 12, 1966. Faculty, U. Ill., Urbana, 1920-63, prof. math, 1940-63, prof. emeritus, 1963—. Research on classification of two dimensional surfaces, generators of simple groups. Metabelian prime-power groups. Home: Dennis, Mass. 02638.*

BRAHE, Tycho, Danish astronomer; b. Knudstrup, Scania (then part of Denmark), Dec. 14, 1546; s. of noble family, adopted and raised by his uncle, Jorgen Brahe; studied philosophy and rhetoric U. Copenhagen, 1559-62, law U. Leipzig, 1562-65, chemistry U. Augsburg, 1569-71 or 72; married peasant girl, 1573. Became interested in astronomy, 1560; set up lab. at Herritzvad Castle, Knudstrup (family estate), circa 1571; by royal command delivered lectures in Copenhagen, 1574-75; traveled through Germany to Venice, 1575; under patronage of King Frederick II of Denmark built an important astron. obs. on island of Hveen (now Ven), beginning 1576, equipped it with unprecedentedly large and accurate instruments which he used to survey entire sky; was visited there by scholars and rulers from all of Europe; after Frederick's death, 1588, had frequent arguments with new king, Christian IV, thus was deprived of his revenues, 1596; left Hveen, 1597, settled in Prague at invitation of Emperor Rudolf II, there engaged young Johann Kepler as his asst. Author works including: Astronomiae Instauratae Progymnasmata (principle work, edited by Kepler), 1602/3; De Nova Stella, 1573; De Mundi Aetherii Recentioribus Phaenomenis, 1588; Liber Secundus, 1588; Epistolae Astronomicae, 1596; Astronomiae Instauratae Mechanica, 1598. One of the greatest naked-eye astronomers; by his systematic survey of the sky, he improved almost every important astron. measurement, and frequently obtained results correct to within 2 minutes of arc (approximately the theoretical limit for naked-eye observations); never accepted Copernican system; instead, created Tychonic solar system, which postulated that planets other than earth rotate around sun, and that the sun, in turn, rotates about immobile earth; observed a nova or exploding star in constellation Cassiopeia, Nov. 1572; this, along with his study of comets, struck a blow against Aristotelian notion that heavens are perfect and unchanging; discovered the variation in the inclination of lunar orbit, found 4th inequality of the moon's motion; determined length of year correct to less than a second (this was instrumental in establishment of Gregorian calendar 1582). Died Benatky, Prague, Oct. 24, 1601.

BRAHMAGUPTA, Hindu mathematician; b. circa 598. Author: The Brāhma-Sphuta-Siddhānta (improved system of Brahma), circa 628; The Khandakhadyaka (on astron. system of Aryabhatta), English transl., 1934. Research on gen. solution of determinate and indeterminate equations of 1st and 2d degree. Died circa 660.

BRAIBANTI, Ralph John, Am. polit. scientist; b. Danbury, Conn., June 29, 1920; s. Daniel V. and Jane (Helena) B.; B.S., Conn. State Coll., 1941; A.M., Syracuse U., 1947, Ph.D., 1949; m. Lucy Kauffman, Feb. 19, 1943; children—Claire (Mrs. Lex Kingsbury Larson), Ralph Lynn. Asst. prof., asso. prof. Kenyon Coll., 1949-53; prof. polit. sci. Duke, Durham, N.C., 1953-68, James B. Duke prof., 1968—, also chmn. Center for Commonwealth Studies. Chief adviser Civil Service Acad. Pakistan, 1960-62; cons. AID, 1959—. Mem. Am. Polit. Sci. Assn. (asst. dir. 1950-51), Am. Soc. for Pub. Adminstrn., Am. Soc. Polit. and Legal Philosophy, Assn. Asian Studies. Author: Administrative and Economic Development in India, 1961; Tradition, Values and Socio-Economic Development, 1963; Asian Bureaucratic Systems, 1966; Research on Bureaucracy of Pakistan, 1966. Research, publs. on theory of polit. devel., role of legal instns., bureaucracy in developing states of Asia. Home: 2614 Stuart Dr., Durham, N.C. 27707.*

BRAID, James, Brit. surgeon, hypnotist; b. Rylaw House, Scotland, 1795; ed. U. Edinburgh; apprentice to Dr. Anderson and son Dr. Charles A., Leith; surgeon to miners Earl Hopetoun's Works, Lanarkshire; practiced medicine, Manchester, Eng. Author: Neurypnology, or the Rationale of Nervous Sleep, 1843; The Power of the Mind Over the Body, 1846; Observations on Trance, 1850; Magic, Witchcraft, Animal Magnetism and Electro-biology, 1852; Hypnotic Therapeutics, with an Appendix on Table-Moving and Spirit-Rapping, 1853; Observations on the Nature and Treatment of Certain Forms of Paralysis, 1855. Introduced term neuro-hypnotism, later shortened to hypnotism, to replace terms mesmerism, animal hypnosis, 1843; showed hypnotism is achieved by suggestion; pioneered research into unconscious mind. Died Manchester, Mar. 25, 1860.

BRAIDWOOD, Robert J., Am. archeologist; b. Detroit, Mich., July 29, 1907; s. Walter J. and Rhea (Nimmo) B.; A.B., U. Mich., 1932, A.M., 1933; Ph.D., U. Chgo., 1942; m. Linda Schreiber, 1937; children—Gretel, Douglas. Archeol. field work, Iraq, Syria, Iran, Turkey, also Ill., N.M., 1930—; faculty Oriental Inst., 1933—, prof. Old World prehistory, 1954—; faculty U. Chgo., 1940—, prof. dept. anthropology, 1954—. Fellow Am. Acad. Arts and Scis., Nat. Acad. Scis., Am. Philos. Soc.; mem. Am. Anthrop. Assn. (exec. bd. 1962-64), Internat. Union Pre-and-Protohistric Scis. (U. S. del. permanent council), Conf. Asian Archaeology of New Delhi (found. mem.); corr. mem. Deutsche Archaologische Institut, Instituto Italiano di Preistoria e Protostoria, Jysk Arkaeologist Selskab, Österreichische Akademie der Wissenschaft; mem. Am. Philos. Soc. Study of village society development in old world. Office: Oriental Inst., U. Chgo., Chgo.

BRAIKENRIDGE, William, Brit. mathematician; b. circa 1700; clergyman, Edinburgh, Scotland; obtained independently some of Maclaurin's geometrical results; Braikenridge-Maclaurin theorem states if the sides of a polygon are restricted to pass through fixed points and all the vertices except one lie on fixed straight lines, the free vertex describes a conic section or a straight line. Died after 1759.

BRAILLE, Louis, French inventor; b. Coupuray, France, 1809; became blind at 3; studied music and science at Paris Inst. for the Blind, became organist in Paris. Author: Procédé pour écrire les paroles, la musique, et le plain-chant au moyen de points. Developed Braille system of raised-dot writing (suggested by Charles Barbier 1820) as alphabet for blind, 1830. Died Paris, Jan. 6, 1852.

BRAINARD, Daniel, Am. surgeon; b. Oneida County, N.Y., May 15, 1812; s. Jeptha and Catherine (Comstock) B., Jr.; studied medicine under Dr. R. S. Sykes, Whitesboro, N.Y., Dr. Harold H. Hope, Rome, N.Y.; attended Fairfield Med. Coll., N.Y.; M.D., Jefferson Med. Coll., 1834; m. Evelyn Slight, Feb. 6, 1845. Practiced medicine, Chgo., 1836; went to Paris, France, 1839-41, influenced by French sch. of surgery; founder, organizer, and prof. anatomy and surgery Rush Med. Coll., Chgo., 1843-66; a founder Ill. Med. and Surg. Jour., 1844; also contbr.; corr. mem. Societé de Chirurgie, 1853; pres. Ill. Med. Soc., 1854. Author papers including The Venom of Rattlesnakes; The Effects of the Venom, and the Means of Neutralizing its Absorption, 1853. Invented a bone drill; studied deformities and fractures. Died Chgo., Oct. 10, 1866.

BRAINERD, Henry Dean, Am. physician; b. San Francisco, Dec. 3, 1914; s. Herbert K. and Myrtle (Healy) B.; A.B., U. Cal. at Berkeley, 1935, M.D., 1939; m. Harriet Hall, Jan. 30, 1955; children—Henry Dean, Alan, Karen, David, Eleanor. Faculty, U. Cal. at San Francisco, 1942—, William Watt Kerr asso. prof. medicine, 1951-55, William Watt Kerr prof. medicine, 1955—, chmn. dept. medicine, 1956-64; chief med. services San Francisco Gen. Hosp., 1951-56, 64—. Fellow A.C.P.; mem. Western Assn. Physicians, Western Soc. for Clin. Research., Pacific Interurban Clin. Club, Am. Bd. Internal Medicine (chmn. 1965-67). Author: (with M. Chatton, S. Margen), Medical Treatment, 1949, Current Diagnosis and Treatment, 1962; also numerous articles. Research on diagnosis and treatment infectious disease, clin. pharmacology antibiotic drugs. Home: 64 Rock Rd., Kentfield, Cal. 94904. Office: San Francisco Gen. Hosp., San Francisco 94110.*

BRAINERD, Henry Green, Am. neuropsychiatrist; b. Londonderry, N.H., May 23, 1852; s. Timothy G. and Lucinda R. (Dewey) B.; A.B., Dartmouth, 1874; postgrad. State U. Ia.; M.D., Rush Med. Coll., 1878; postgrad., N.Y., London, Eng.; m. Alma Loomis, 1879 (dec. 1882); m. 2d, Fannie Howard, 1887; children—Henry Howard, Fred Lindley. Asst. supt. Ia. Hosp. for Insane, 1878-86; prof. mental and nervous diseases Med. Dept. U. So. Cal., 1886-1911, dean faculty, 1889-1902; organizer, dean dental dept. 1896-99; supt. Los Angeles Gen. Hosp., 1888-92; chmn. Cal. commn. to locate and built hosp. for insane, at Norwalk, also mem. bd. mgrs.; v.p. Southern Calif. Sanitarium for Nervous Diseases. Died July 22, 1928.

BRAITHWAITE, John, English engr.; b. London, Eng., Mar. 19, 1797; s. John Braithwaite; student engring. Engr. in chief Eastern Counties R.R., 1836-43; co-founder, editor Railway Times, 1837; surveyed ry. lines in France, 1844, 46. Fellow Soc. of Antiquaries; mem. Inst. Civil Engrs. Author: Supplement to Captain Sir John Ross's Narrative of a Second Voyage in Search of a North-west Passage, . . . , 1835; Guideway Steam Agriculture, 1857. Builder 1st practical steam fire engine, 1829, caloric engine (with Ericsson), 1833; responsible for installing air pumps to ventilate Ho. of Lords, 1820; built donkey engine, 1822. Died Paddington Sta., London, Sept. 25, 1870.

BRAITHWAITE, John the Elder, Brit. engr.; children included Francis, John; maintained a factory; active in recovering valuable items from sunken ships; invented 1st successful forms of diving bell, also machine for cutting sunken ships. Died Westbourne Green, June 1818.

BRALEY, Alson Emmons, Am. ophthalmologist; b. Lake Mills, Ia., Jan. 9, 1906; s. Harry J. and Evelyn (Emmons) B.; M.D., State U. Ia., 1931; m. Hazel Deming, June 22, 1931; children—Alson Deming, Janet, William Harry. Instr. U. Hosps., Iowa City, 1937-39; faculty Wayne U., 1939-41, Columbia, 1941-49, N.Y. U., 1949-50; staff Receiving Hosp., Detroit, 1939-41, Columbia-Presbyn. Med. Center, N.Y.C., 1941-49; prof., head dept. ophthalmology Coll. Medicine, State U. Ia., Iowa City, 1950—. Recipient award for starting eye bank network, A.M.A. Diplomate Am. Bd. Ophthalmology. Mem. Am. Opthal. Assn., Am. Opthal. Soc., Am. Acad. Opthalmology and Otolaryngology, Assn. Research Ophthalmology, Ia. Acad. Ophthalmology and Otolaryngology, N.Y. Acad. Medicine, N.Y. Acad. Scis., World Med. Assn., Pan-Am. Assn. Ophthalmology, A.C.S., Sigma Xi. Research and publs. on tumors of eyelids, trachoma, inclusion blennorrhea, glaucoma, antibiotics in occular infection; started eye bank network. Home: 720 McLean St., Iowa City 52240.*

BRALLIAR, Floyd Burton, Am. naturalist; b. Richland, Ia., Sept. 11, 1875; s. Washington George and Martha A. (Hornbeak) B.; student Battle Creek (Mich.) Coll.; B.Sc., Walla Walla Coll., 1895; M.S., Emmanuel Missionary Coll., 1919; Ph.D., George Peabody Coll., 1921; m. Ada Conard, 1898 (dec. 1901); children—Ena Marie (Mrs. Thomas Abernathy), Ada Conard (Mrs. William Cheek); m. 2d, Mertie Boynton, 1904 (dec. Mar. 1951); children—Alice Isabelle (Mrs. Paul Rolen), Floyd Boynton, John Seward, Max Burton. Began teaching in rural schs., Ia., 1894; prof. of pedagogy Union Coll., Neb., 1903-04; ednl. supt. Central Union Conf., 7th Day Adventists Ch., 1904-06; prin., bus. mgr. Ia. Indsl. Acad., 1905-09; prin. Hillcrest Sch. Farm, Nashville, 1909-17; became prof. biology and related sci., Nashville Agrl. and Normal Inst., 1917; dean and head dept. of biology Madison Coll.; operator ornamental nursery and plant breeding sta.; lectr. for Sch. Assembly Assn.; first regional v.p. Am. Iris Soc. for South. Fellow Am. Assn. for Advancement of Science; mem. Ia. and Tenn. State hort. socs., Tenn. Bee-Keeper's Assn. (v.p.), Tenn. State Florists' Assn., Soc. of Am. Florists (lecture bur.). Author: Elo the Eagle, and Other Stories, 1908; Knowing Insects Through Stories, 1918; Knowing Birds Through Stories, 1922; Zip the Coon and Other Stories, 1931; Grape Culture in the South. Editor garden dept. in several daily newspapers. Sci. breeder of small fruits, also iris, cannas. Died Sept. 5, 1951.

BRAM, Joseph, anthropologist; b. Ekaterinburg, Russia, July 17, 1904; s. Nahum and Sophie (Rosenzweig) B.; grad. U. Paris, 1930; Ph.D., Columbia, 1941; m. Jean Rhys, May 1946; children—Susan, Elizabeth, Margaret. Came to U. S., 1935, naturalized, 1942. Mem. faculty Queens Coll., N.Y.C., 1940-48; faculty N.Y. U., N.Y.C., 1948—, now prof. sociology and anthropology. Mem. N.Y. Acad. Scis. (past chmn. div. anthropology), Am. Ethnol. Soc. (past v.p.) Am. Anthrop. Assn., Am. Sociol. Assn., Am. Soc. for Sci. Study Religion, Am. Mus. Natural History. Author: An Analysis of Inca Militarism, 1941; Language and Society, 1955; also articles, revs. Research in field of collective thought, collective imagination. Home: 4 Prospect St., Baldwin, N.Y. 11510.*

BRAMAH, Joseph, English engr.; b. Stainborough, Eng., Apr. 13, 1748; cabinet-maker in London; patentee safety lock, 1784, fountain pen; invented hydraulic press, 1795, patent lock, ship's propellor, moveable support for turning lathe, lavatories with water tanks and siphon, numerical printing machine for banknotes, planing machines, bottling machinery; suggested possibility of screw propulsion for ships, 1785, and hydraulic transmission. Died London, Eng., Dec. 9, 1814.

BRAMBELL, Francis William Rogers, zoologist; b. Dublin, Ireland, Feb. 25, 1901; s. Louis Alfred and Amelia Mary (Rogers) B.; C.B.E., B.A., U. Dublin, 1922, Ph.D., 1924, D.Sc. (hon.), 1966; D.Sc., U. London, (Eng.), 1927; m. Margaret Lilian Adgie, Dec. 27, 1927; children—Anne Elizabeth Rogers (Mrs. William Vincent Denard), Michael Rogers. Scholar, Royal Commn. for Exhbn. of 1851, 1924-26; Internat. Edn. Bd. fellow, 1926-27; lectr. zoology King's Coll., London, 1927-30; Lloyd Roberts prof. zoology U. Wales, U. Coll. N. Wales, Bangor, Caernarvonshire, 1930—; hon. dir. unit embryology, Argl. Research Council, 1953—. Fellow Royal Soc. (mem. council 1954-56, Royal medal 1964); mem. Zool. Soc. London, Soc. Exptl. Biology, Soc. Immunology, Soc. Developmental Biology, Internat Inst. Embryology, Soc. Endocrinology, Linnaen Soc., Marine Biol. Assn. U.K., Brit. Assn. (past pres. sect. D). Author: The Development of Sex in Vertebrates, 1930; (with W. A. Hemmings, M. Henderson) Antibodies and Embryos, 1951; also numerous articles. Research on estrus cycles, prenatal mortality, transmission of maternal immunity in animals. Home: y-Gwylain, Bangor, Caernarvonshire, Gt. Britain.*

BRAMBILA, Francesco, Italian economist; b. Milan, Italy, June 22, 1913; s. Carlo and Paschetta Maria Brambila; ed. in econs.; D.Sc.; m. Lida Longoni, Oct. 1, 1951. Prof. statistics U. L. Bocconi, Milan; mem. Internat. Inst. Statistics. Named laureat of Marzotto prize. Research and publs. on econs. and statistics. Home: via Guastalle 5. Office: via Sarfatti 23, Milan, Italy.

BRAMBILLA, Johann Alexander, see von Brambilla, Johann Alexander.

BRAME, Arden Howell, Jr., Am. herpetologist; b. Los Angeles, Mar. 19, 1934; s. Arden H. and Marguerite Lucile (Adams) B.; student U. Cal. at Los Angeles, 1956-57; B.A., U. So. Cal., 1957, M.S., 1966; m. Susan Diane Bronn, Aug. 23, 1964. Student profl. worker Los Angeles County Mus., 1959-65; supr. Eaton Canyon Nature Center, Pasadena, Cal., 1965—. Mem. Am.. Soc. Ichthyologists and Herpetologists, Herpetologist's League, Brit., Ohio (editorial bd.), Phila., N.Y. herpetological socs., Am. Soc. for Study Evolution, Soc. Systematic Zoologists, Ecol. Soc. Am., A.A.A.S., Am. Inst. Biol. Scis., Southwestern Soc. Naturalists, Sigma Xi, Phi Sigma. Co-author: The Salamanders of South America, 1963; also articles. Research on systematics. taxonomy, evolution in salamanders; described 18 new species salamanders. Home: 1690 N. Altadena Dr., Altadena, Cal. 91001. Office: 1750 N. Altadena Dr., Pasadena, Cal. 91107.*

BRAME, John Samuel Strafford, Brit. chemist; b. 1871; ed. Royal Coll. Sci., South Kensington, Eng.

Demonstrator Royal Naval Coll., Greenwich, Eng., 1897; from instr. to prof. chemistry, 1910-32. Mem. Instn. Petroleum Technologists (pres. 1921-23, chmn. standardization com.). Author: Treatise on Fuel-Solid, Liquid and Gaseous; Service Chemistry, 5th edit. Contbr. articles to profl. jours. Died Dec. 10, 1952.

BRAMER, Benjamin, mathematician; b. Hesse, Germany, 1580; Author: Apollonius Cattus, oder Geometrischer Wegweiser; Geometrisches Triangular-Insturment; Explication et Usus linealis proportionalis. Died 1650.

BRAMHALL, Ervin Hicks, Am. geophysicist; b. Palo Alto, Cal., Jan. 14, 1905; A.B., Stanford, 1926; M.Sc., Mass. Inst. Tech., 1928; Ph.D., Cambridge (Eng.) U., 1931. Research asso. Mass. Inst. Tech., 1932-33; physicist Byrd Antarctic Expdn., 1933-34; prof. physics Alaska U., 1935-41; physicist in charge coll. obs. dept. terrestrial magnetism Carnegie Inst. Tech., 1941-44; prof. physics U. Hawaii, 1945-48; physicist operations research office Johns Hopkins, 1948-50; dep. chief scientist research div. U. S. Dept. State, 1950-51; chief phys. sci. sect. Army Ordnance, Washington, 1952-56; analyst weapon systems evaluation group Inst. for Def. Analysis, Washington, 1956-59; chief weapons systems analysis Solar Aircraft Co., San Diego, 1959-60; staff scientist Cubic Corp., San Diego, 1960-61; mem. sr. planning staff Northrop Space Labs., Hawthorne, Cal., 1962-63; analyst Radio Corp., Am. Service Co., 1963—. Mem. Am. Geophys. Union, I.E.E.E., Am. Inst. Aeros. and Astronautics, Am. Ordnance Assn. Research and publs. on geophysics relating to radio propagation, oceanography, conditions in space. Home: 1188 Peebles Dr., Fairborn, O. 45324.

BRAMLETTE, Milton Nunn, Am. geologist; b. Bonham, Tex., Feb. 4, 1896; s. William A. and Eula L. (Nunn) B.; B.A., U. Wis., 1921; Ph.D., Yale, 1936; LL.D., U. Cal. at San Diego, 1965; m. Valerie Jourdan, Nov. 30, 1931 (dec.); 1 dau., Emily (Mrs. M. M. Assami). With U. S. Geol. Survey, 1921-24, 31-40, Gulf Oil Corp., 1924-28; faculty dept. geology U. Cal. Los Angeles and San Diego, 1941-61, prof., 1946-61, emeritus prof., 1961—. Recipient Distinguished Service medal Dept. Interior, 1963. Mem. A.A.A.S., Nat. Acad. Sci. (Thompson medal 1964). Research, publs. on petrology of sediments, stratigraphy, petroleum geology, econ. geology, submarine geology, micropaleontology.*

BRAMLEY, Arthur, physicist; b. Preston, Eng., Sept. 23, 1900; s. Thomas and Helen (Smith) B.; came to U. S., 1919, naturalized, 1931; A.B., U. Ore., 1922; Ph.D., Princeton, 1924; m. Jenny Eugenie Rosenthal, Jan. 13, 1943; children—Alan Keith, Timothy Bruce, Eleanor Gail. Fellow, asst. to dir. Bartol Found., Swarthmore, Pa., 1925-36; collaborator Fixed Nitrogen Lab., Washington, 1937-40; physicist War Dept., Washington, 1940-43; supr. Nat. Union Radio Corp., Orange, N.J., 1943-48; sect. head Du Mont Labs., Passaic, N.J., 1953-57; sci. Republic Aviation, Farmingdale, N.Y., 1960-62; cons. Gen. Dynamics Corp., San Diego, Cal., 1957-60; physicist Bramley Cons., Falls Church, Va., 1962—. Jacobus fellow Princeton, 1923-25. Fellow A.A.A.S., Am. Phys. Soc.; mem. Ore. Alumni Assn., Phi Beta Kappa. Patentee in field. Research in theory of relativity, electrodynamics, beta particles, 1922-36; optical characteristics of water, 1932; radioactive tracers, thermal diffusion, 1936-40; photoconductive response, electroluminescent displays, interaction of light with nonlinear media, 1950—. Home: 7124 Strathmore St., Falls Church, Va. 22042.*

BRAMWELL, Byrom, Brit. physician; b. N. Shields, Eng., Dec. 18, 1847; s. John Byrom and Mary (Young) B.; ed. Cheltenham; qualified in medicine Edinburgh, Scotland, 1869; m. Martha Crighton, 1872 (dec. 1919); 3 sons, 2 daus. Physician Newcastle Royal Infirmary, 1874-79; cons. Edinburgh, from 1879; pathologist Edinburgh Royal Infirmary, 1882-85, asst. physician, 1885-97, physician, 1897-1912. Created knight, 1924. Fellow Royal Soc. Edinburgh. Author: Diseases of the Spinal Cord, 1881; Diseases of the Heart and Thoracic Aorta, 1884; Intracranial Tumours, 1888; Atlas of Clinical Medicine, 3 vols., 1892-96; Clinical Studies, 1903-10. Noted and described functional part played by hypothalamus, 1888. Died Edinburgh, Apr. 27, 1931.

BRAMWELL, Edwin, Scottish physician; b. Jan. 11, 1873; s. Sir Byrom Bramwell; ed. univs. Edinburgh, Freiburg, Heidelberg, Paris; grad. U. Edinburgh, 1896, M.D., LL.D.; M.D. (hon.), Melbourne U. Cons. physician Royal Infirmary, Edinburgh, asst. physician, 1907, physician, 1919; cons. med. officer Scottish Union and Nat. Ins. Co.; Morison lectr. Royal Coll. Physicians, Edinburgh, 1917-18; Moncrieff-Arnott prof. clin. medicine U. Edinburgh, 1922-34; Bradshaw lectr. Royal Coll. Physicians, London, 1925, Croonian lectr., 1937. Fellow Royal Coll. Physicians (Edinburgh) (pres. 1934-35), Royal Coll. Physicians (London), Royal Soc. of Edinburgh, Royal Soc. Medicine (past pres. neurology sect.); mem. Assn. Brit. Neurologists (hon., past pres.), Brit. Med. Assn., Royal Med. Soc. (hon., pres. 1896-97), Phila. Neurol. Soc. (fgn. corr.), Neurol. Soc. of Paris (corr.). Contbr. chpts. on nervous system diseases to System of Medicine (Allbutt), Modern Medicine (Osler), Oxford Medicine, Ency. Medica, also numerous articles. Died Mar. 21, 1952.

BRAND, Ernst, German physician; b. Feuchtwangen, Germany, Jan. 2, 1827; M.D., Erlangen, 1851; became clin. asst. to Canstatt, Erlangen, 1849; practiced medicine, Stettin, Germany. Author: Die Hydrotherapie des Typhus, 1861; Zur Hydrotherapie des Typhus, Bericht über St. Petersbürg . . . , 1863; Die Heilung des Typhus, 1868; Die Wasserbehandlung der Typhösen Fieber, 1877. Introduced cold bath to treat typhoid fever, 1861. Died Mar. 7, 1897.

BRAND, Hennig, German chemist; b. Hamburg, Germany flourished 1670; m. Margarete; 1 son, 1 dau. Originally mil. officer, then physician in Hamburg, falsely (presumably) assuming title of Dr. med.; turned to alchemy and experimented with urine, hoping to find liquid to turn silver into gold; sold rights to his discovery and precipitated much argument; defended by Leibniz as true discoverer; Leibniz arranged job as mining expert; returned to Hamburg where all trace of him is lost. Discovered phosphorus (accidently in experimentation with urine) 1669.

BRAND, James John Cantley, inventor; b. Silverton, Eng., June 29, 1880; s. Alex Brand; ed. St. Joseph's Coll., Rockhampton, Australia; m., 1907; cons. engr., Sydney, Australia, 1907-11; with Royal Australian Navy, 1911-29. Mem. Inst. Marine Engrs. Contbr. to tech. publs. Inventor Brand system of powdered coal burning, also modern short flame burner. Died Sept. 12, 1952.

BRAND, K(arl Friedrich) Gerhard, biologist, physician; b. Luebeck, Germany, June 10, 1922; s. Johannes and Kaethe (Hoefer) B.; M.D., U. Hamburg, 1949, diploma tropical medicine, 1954; P.D., U. West Berlin, 1956; m. Inge Hoellein, Aug. 19, 1949; hon. diploma Facultad de Ciencias Medicas de Buenos Aires, 1960; children—Juliane, Bettina. Came to U. S., 1957, naturalized, 1964. Asso. prof. dept. microbiology Tropical Inst. Hamburg, 1952-55, Inst. Microbiology, Free U. West Berlin, 1955-57; practice medicine, specializing in microbiology, Mpls., 1957—; faculty Med. Sch. U. Minn., 1957—, now prof. dept. microbiology. Mem. Am. Assn. Immunologists, Am. Soc. Microbiologists, Soc. Exptl. Biology and Medicine, German Microbiology Soc. Research, numerous publs. on immunology in influenza, mumps, psittacosis, atypical pneumonia, antigenic structure of influenza virus, antigenic composition of human tissues regarding problems of transplantation, carcinogenesis, aging and evolution of species, carcinogenic events in tumors from plastic inserts. Home: 4710 DuPont Av. S., Mpls. 55409. Office: Dept. Microbiology, U. Minn. Med. Sch., Mpls. 55455.*

BRAND, Louis, Am. mathematician, educator; b. Cin., Sept. 27, 1885; s. Louis William and Josephine (Zingsheim) B.; Ch.E., U. Cin., 1907, E.E., 1908, A.M., 1909, Doctor of Science (honorary), 1956; Ph.D., Harvard University, 1917; m. Lulu Edith Shinkle, Aug. 14, 1915; children—Sara Josephine (Mrs. Oliver Marcy), Martha Louise (Mrs. Bruce Raymond). Head dept. mathematics and mechanics, coll. engring. U. Cin., 1919-56, head dept. mathematics, 1935-56, prof. emeritus, 1956—; Whitney vis. prof. Trinity Coll., 1956-57; M.D. Anderson professor of mathematics University of Houston, Tex., 1957—; vis. prof. University of Brazil, 1963. Fellow grad. sch. arts and scis. 1932—. Fellow A.A.A.S., Ohio, Tex. acads. scis.; mem. Am. Math. Soc., Am. Math. Assn., Sigma Xi, Tau Beta Pi, Phi Kappa Phi. Author math. textbooks, also articles math. jours. Study of infinite systems of linear integral equations; vector and tensor analysis and its applications to mechanics; vectorial mechanics; electrodynamics and geometry. Home: 4121 Taum, Houston 77004.

BRAND, Theodore Curt von, see von Brand, Theodore.

BRANDE, William Thomas, English chemist; b. London, Eng., Feb. 11, 1788; D.C.L. (hon.), Oxford, Eng.; m. dau. of Charles Hatchett; apprenticed to brother as apothecary, 1802; became lectr. physics, chemistry and medicine, London, 1808; apptd. prof. chemistry Apothecaries' Co., 1812; prof. chemistry Royal Instn., 1813-54; examiner in chemistry, 1846-58; named senate U. London. Fellow Royal Soc., 1809 (a sec. 1816-26), Royal Soc. Edinburgh. Author: Manual of Chemistry, 1819; Dictionary of Pharmacy and Materia Medica, 1839; Dictionary of Science and Art, 1842; also articles. (with Michael Faraday) joint editor Quar. Jour. Sci. and Art, 1816-36. Discovered naphthalene in coal tar; research on albumin, uranium oxides and salts, electro-chem, phenomena. Died Tunbridge Wells, Feb. 11, 1866.

BRANDEN, Carl Ivar, Swedish chemist; b. Lappland, Sweden, May 14, 1934; s. Henry M. and Greta (Nilsson) B.; D.Sc., Uppsala (Sweden) U., 1964; m. Lisbet Wikander, July 10, 1958; 1 son, Henrik Branden. Research asst. Uppsala U., 1958-62; research scientist Med. Research Council Lab. for Molecular Biology, Cambridge, Eng., 1962-63; asso. prof. Agr. Coll., Uppsala, 1963—; vis. prof. dept. biochemistry U. Cal. at Berkeley, 1966. Mem. Chem. Soc. Uppsala (chmn. -1965—), Chem. Soc., Soc. for Exptl. Biology. Research, publs. on allosteric transitions, X-ray crystallographic studies of enzyme horse liver alcohol dehydrogenase and refinement studies of myoglobin, small molecules of biol. interest and complexes between metal halides and oxygen or sulphur donor molecules. Home: 11A Marmorvagen, Uppsala, Sweden.*

BRANDER, Gustavus, Brit. naturalist, geologist; b. London, 1720; mcht., antiquary trader, London; curator, trustee Brit. Mus., 1761-87; dir. Bank of Eng. Fellow Royal Soc., 1754. Collected tertiary fossils; contributed collection of fossils discovered in cliffs around Christchurch to Brit. Mus. Died London, 1787.

BRANDES, Heinrich Wilhelm, German astronomer; b. Groden, Germany, July 27, 1777; student Göttingen, Germany; became prof., Breslau (now Wroclaw, Poland); named prof. physics, Leipzig, Germany, 1826. Author: Manual of Higher Geometry; also treatises on astronomy and hydraulics. Observed meteors; drew up 1st weather charts; studied shooting stars and demonstrated their earthly origin, 1798. Died Leipzig, May 17, 1834.

BRANDES, Rudolph, German pharmacist, chemist; b. Salzuflen, Germany, Oct. 18, 1795; s. Johann Gottlieb and Friederike (Nolte) B.; practical tng. in Osnabrück and Erfurt; doctorate, Halle, Germany, 1817; hon. dr. pharmacy, Marburg, Germany; m. Henriette Luise, 1824; 4 children; m. 2d, Johanne Luise Wessel, 1836; 3 children. Took over family pharmacy in Salzuflen (still in family ownership); founded jour. (with A. P. Du Menil and E. Witting) Archiv der Pharmazie, 1820; founded (with P. L. Geiger and J. v. Liebig) Annalen der Chemie, 1832; founded (with Du Menil, Witting and Beissenhirtz) Apothekerverein im nördlichen Deutschland, 1820, highest dir., until 1842. Author: Beiträge zur Witterungskunde. Contbr. articles to Archiv der Pharmazie. Discovered alkaloids delphinine, 1819, atropine, hyoscyamine; analyzed minerals and mineral waters; investigated various acids & coconut; set up 1st weather charts as meteorologist; corresponded with Goethe; his Archiv der Pharmazie was and still is influential. Died Salzuflen, Dec. 3, 1842.

BRANDHORST, Wilhelm, oceanographer; b. Bremen-Vegesack, Germany, July 28, 1929; s. Theodor and Erna (Illies) B.; Dr.rer. nat. summa cum laude, U. Kiel (Germany), 1955; m. Gertrud Neymanns, Aug. 24, 1960; children—Constanze, Claus-Peter, Denise. Sci. asst. Inst. Oceanography, U. Kiel, 1955-56; scientist inter-Am. tropical tuna commn. Scripps Inst. Oceanography, La Jolla, Cal., 1956-58; fisheries advisor tech. asst. program Fed. Republic of Germany, 1958-63; head Lab. Oceanography and Fisheries Biology, Valparaíso, Chile; chief oceanographer UN FAO Fisheries Devel. Inst., Santiago, Chile, 1963—. Research, publs. on relationships between phys. conditions and life cycle of Clupeid fishes, fertilizing mechanisms and organic decomposition, descriptive oceanographic studies. Office: Div. Fisheries Resources and Exploitation, Dept. Fisheries, FAO, Rome, Italy.*

BRANDT, Georg, Swedish chemist; b. Riddarhytta, Sweden, 1694; ed. under Boerhaave in Leiden, Netherlands; M.D., 1726. Became dir. chem. lab. Bur. Mines, Stockholm, Sweden, 1727; apptd. assay master mint, 1730. Discovered cobalt, circa 1730; made 1st accurate researches on arsenic; discovered zinc in calamine; distinguished between soda and potash, 1746. Died Stockholm, Apr. 29, 1768.

BRANDT, J. Leonard, physician; b. N.Y.C., Aug. 3, 1919; s. William W. and Anna (Wolff) B.; B.A., U. Mich., 1940; M.D., L.I. Coll. Medicine, 1943; m. Irma Silver, Aug. 15, 1950; children—Stephanie Ann, James David. Asso. prof. medicine State U. N.Y. Coll. Medicine, 1956-59; physician-in-chief Jewish Gen. Hosp., Montreal, Que., Can., 1959—; lectr. medicine McGill U. Faculty of Medicine, Montreal, 1959—; practice medicine specializing in internal medicine, nephrology, Montreal, 1959—. Diplomate Am. Bd. Internal Medicine. Fellow A.A.A.S., Canadian Soc. Clin. Investigation, A.C.P.; mem. Am. Physiol. Soc., Harvey Soc., Am. Fedn. Clin. Research, Sigma Xi. Research, numerous publs. on basic physiology of renal and hepatic blood flow and renal function in gen. Home: 4252 Isabella Av. Office: Jewish Gen. Hosp., Montreal, Que., Can.*

BRANDT, Johann Friedrich, naturalist; b. Jüterbog, Germany, May 25, 1802; insp. curricula, St. Petersburg, Russia; dir. Zool. Mus. Berlin (Germany). Mem. French Acad. Scis., 1870, Acad. Scis. St. Petersburg. Author: (all in German) Berlin Flora, 1825; Medical Zoology, 1827-34; Description of Poisonous Plants Growing Wild in Germany, 1838; Vertebrate Animals of Western Siberia, 1845; On the Classification of Fish, 1865; Natural History of the Mammoth, 1866. Responsible for orgn. library comparative anatomy, St. Petersburg. Died Merekul, Estonia, July 3, 1879.

BRANDT, William Henry, Am. botanist; b. Great Falls, Mont., May 25, 1927; s. Henry M. G. and Marion (Shoults) B.; B.A., Mont. State U., 1950; M.Sc., Ohio State U., 1951, Ph.D., 1953; m. Jane Ellen Gaethke, Feb. 3, 1953; children—Eric Franklin, Ellen Elizabeth. Asst., Ohio Agrl. Expt. Sta., Wooster, 1952-54; research biologist B.F. Goodrich Research Center, Brecksville, O., 1954-56; faculty Ore. State U., Corvallis, 1956—, asso. prof. botany, 1965—, acting dir. gen. botany teaching, 1966—. Mem. Bot. Soc. Am., Mycol. Soc. Am., Microbiol. Soc. Am. Author: (with others) Laboratory Problems in Botany I, 1966; also articles. Discovered 1st endogenous rhythm in fungus Neurospora crassa; showed that oak wilt does not decrease decay resistance in oak wood; analyzed respiration variation and morphogenesis in fungus Verticillium. Home: 1340 N. 14th St., Corvallis, Ore. 97330.*

BRANICAN, George Francis, Am. engr.; b. Shelby, Neb., July 21, 1903; s. Thomas and Jennie (Fyfe) B.; B.S. in Civil Engring., 1927; M.S., Kan. State U., 1933; m. Marion L. Eimers, June 16, 1928; children—George E., Thomas L., Susan M. (Mrs. Snowden Armstrong). Faculty, Kan. State U., 1927-42; faculty Bradley U., 1942-46, prof., dean engring., 1943-46; prof. mech. engring. Ia. State U., 1946-48; dean engring. U. Ark., Fayettesville, 1948—. Mem. coms. Ark. Bd. Registration, pres., 1956, 63. Mem. Nat. (chmn. edn. com.), Ark. (v.p. 1949-51) socs. profl. engrs., Am. Soc. E.E., (exec. com. 1963-65), Engring. Colls. Administrv. Council, Nat. Council Engring. Examiners (pres. 1968—), Am. Soc. C.E., Theta Tau, Tau Beta Pi, Theta Xi (pres. 1966-68). Contbr. articles on engring. to tech. jours., popular mags. Home: 1776 Applebury Pl., Fayetteville, Ark. 72701.*

BRANLEY, Edouard, French physicist; b. Amiens, France, Oct. 23, 1844; ed. St. Quentin, France; student Ecole Normale Supérieure, 1865; degrees in math. and natural sci.; doctorate Sorbonne; med. degree Catholic U. Paris, 1882; Tchr., Sorbonne, 3 years; prof. physics Cath. U. Paris, from 1875. Mem. French Acad. Scis. Author texts on physics. Investigated elec. waves; studied effects of ultraviolet rays, also elec. conductivity of gases; discovered principle of wireless telegraphy (developed by Marconi) by inventing coherer for reception wireless telegraphic waves, 1890; evolved forerunner of receiving antennae. Died Paris, Mar. 24, 1940.

BRANNER, John Casper, Am. geologist; b. New Market, Tenn., July 4, 1850; s. Michael T. and Elsie (Baker) B.; B.S., Cornell, 1874; Ph.D., Ind. U., 1885; LL.D., U. Ark., 1897, Maryville Coll., 1909. U. Cal., 1915; Sc.D., U. Chgo., 1916; m. Susan D. Kennedy, June 22, 1883. Geologist Imperial Geol. Commn., Brazil, 1875-77; asst. engr. and interpreter S. Cyriaco Mining Co., Minas Geraes, Brazil, 1878-79; spl. botanist in S. America, 1880-81; agt. U. S. Dept. Agr. in Brazil, 1882-83; topog. geologist Geol. Survey of Pa., 1883-85; prof. geology Ind. U., 1885-92; state geologist Ark., 1887-93; prof. geology Leland Stanford Jr. U., 1892-1915, acting pres., 1898-99, v.p., 1899-1913; pres., 1913-15, emeritus. Dir. Branner-Agassiz expdn. to Brazil, 1899; mem. Cal. Earthquake Commn., 1906-07; spl. asst. Geol. Survey of Brazil, 1907-08; dir. scientific expdns. to Brazil, 1899, 1911. Mem. Nat. Acad. Scis., Seismol. Soc. Am. (founder, pres. 1910-14). Author of reports and papers on phys. geography and geology. Asso. editor Journal of Geology. Investigated mineral deposits and suitability of soil for cotton culture in Brazil, 1874-84; studied earthquakes in No. and So. Am. Died Palo Alto, Cal., Mar. 1, 1922.

BRANSON, Branley Allan, Am. ichthyologist; b. San Angelo, Tex., Feb. 11, 1929; s. Branley Allan and Era Elizabeth (Rogers) B.; A.A., Northeastern Okla. A. and M. Coll., 1954; B.S., Okla. State U., 1956, M.S., 1957, Ph.D., 1960; m. Mary Louise Lewis, June 3, 1964; 1 son, Rogers McGowan. Asst. prof. biology Kan. State Coll., Pittsburgh, 1960-64; asso. prof. biology Eastern Ky. State U., Richmond, 1964—. Recipient Sci. award Okla. A. and M. Coll., 1953. Fellow Okla. Acad. Sci.; mem. Southwestern Assn. Naturalists (bd. govs. 1965—), Am. Malacological Union, Soc. for Study Evolution, Kan., Ky. acad. scis., Soc. Systemic Zoologists, Am. Soc. Zoologists, Am. Soc. Ichthyologists and Herpetologists, Sigma Xi, Phi Theta Kappa, Phi Kappa Phi. Research, numerous publs. on description several species unknown animals; described structural workings laterl-line system in various fishes; research on olfactory system, geog. distbn. fishes and mollusks. Home: 905 Vickers Dr., Richmond, Ky. 40475.*

BRANSON, Edwin Bayer, Am. geologist; b. Belleville, Kan., May 11, 1877; s. John McDowell and Harriet Melviney (Bullen) B.; A.B., A.M., U. Kan., 1903; Ph.D., U. Chgo., 1905; m. Grace Muriel Colton, Aug. 24, 1905; children—Carl Colton, Edwin Robert. Instr. in geology Oberlin Coll., 1905-07, asso. prof., 1907-09, prof., 1909-10; prof. geology U. Mo., 1910—. Served as geologist Mo. Bur. Geology and Mines intermittently, 1914-28; geologist Gypsy Oil Co., parts of 1920-28. Fellow Geol. Soc. Am., Paleontol. Soc. Am., A.A.A.S., Am. Assn. Petroleum Geologists; mem. Mo., St. Louis acads. sci., Phi Beta Kappa, Sigma Xi. Author: Geology of Missouri, 1918; Geology and Geography of Middle Eastern Costa Rica; Devonian of Missouri, 1923; Conodonts, 1933; Introduction to Geology, 1935, 2d edit., 1941; The Lower Mississippian of Missouri, 1938; Geology of Missouri, 1945. Contbr. to sci. jours. Died Mar. 12, 1950.

BRANT, Arthur Albert, geophysicist; b. Toronto, Ont., Can., Oct. 23, 1910; s. Stephen Barnett and Mary-Jane (Little) B.; B.A., U. Toronto, 1932; postgrad. Princeton; Ph.D., U. Berlin, 1936; m. Lilli Tekla Umbach, Oct. 4, 1940; children—Karin L., Roxanne H., Roderick L., Heidi L. A. Came to U. S., 1948, naturalized, 1954. Faculty, U. Toronto, 1932-33, 36-47; geophys. cons. to various mining cos., 1946-49; dir. geophys. dept. Newmont Exploration Ltd., Danbury, Conn., 1949—, also dir. Mem. geol. adv. com. Princeton, 1960—; mem. adv. com. Geophysics Inst., U. Cal. at Los Angeles, 1959—; Regents lectr. U. Cal. at Berkeley, 1966. Mem. Am. Inst. Mining Engrs. (Jackling award 1964), Soc. Exploration Geophysicists (hon.), European Assn. Exploration Geophysicists, Am. Geophys. Union, Profl. Engrs. Ont. Contbr. articles to profl. jours. Asso. editor Economic Geology, Exploration Geophysics. Patentee in field. Research on original devel. and application work on induced polarisation and electromagnetic methods of mining. Home: 1 Old Oscaleta Rd., Ridgefield, Conn. 06877. Office: 44 Briar Ridge Rd., Danbury, Conn.*

BRANT, John Weber Alexander, Am. research scientist; b. Electric, Mont., June 12, 1919; s. John W. and Elsie (Homer) B.; B.S. with distinction U. B.C., 1949; M.S., U. Conn., 1951; Ph.D., U. Ill., 1953; m. Rose Josephine Viczko, July 6, 1944; children—Edward P., Michael C. Research asst. U. Conn., 1949-50; research asst. U. Ill., 1950-53, cons. research scientist, 1953—; advisor, cons. Govts. Ceylon, Spain, Ecuador, 1954-56; expert, animal scientist UN Orgn., N.Y.C., 1953-54; prof. vet. medicine and agronomy U. Guayaquil, 1955-56; dir. metabolic lab., research asso. in physiology Cal. Coll. Medicine, 1956-57; prof. Howard U., 1958-62; research scientist, human factors systems mgr., head bioastronautics NASA Flight Research Center, Edwards, Cal., 1962-63; vis. scientist Ore. Med. Research Found., 1963-64; dir. sci. research, prof. applied sci. Portland State Coll., 1964-65; dir. environmental health, chief scientific div. Buchart-Horn Consulting Engrs., York, Pa., 1966—. Govt. of Can. scholar, 1949-51. Mem. Air Pollution Control Assn., Am. Chem. Soc., Am. Genetic Soc., Am. Indsl. Hygiene Assn., Am. Statis. Assn., Human Factors Soc., Internat. Oceanographic Found., N.Y. Acad. Sci., Water Pollution Control Fedn., Sigma Xi, Phi Sigma. Contbr. numerous articles to sci. jours. Pioneered studies of air pollution and human health in Los Angeles; developed sci. agr. in Ceylon, Spain, Ecuador; invented apparatus and developed method for determining instantaneous acceleration of bio-regulatory events; invented air water depollution system. Home: 2100 S.W. Ecole Av., Beaverton, Ore. 97005, also 107 Highland Rd., Southwood Hills, York, Pa. 17405. Office: 55 S. Richland Av., York, Pa. 17403.*

BRANTIGAN, Otto Charles, Am. physician; b. Chattanooga, Aug. 31, 1904; s. August and Olive (Von Allman) B.; B.S., Northwestern U., 1931, M.D., 1934; m. Edith May Reinhardt, Sept. 2, 1939; children—Charles, John, Thomas, Martha. With Balt. City Hosps.; 1933-55, chief surgery, 1948-56; faculty U. Md., 1938—, prof. clin. anatomy, prof. surgery, 1944—; chief thoracic surgery Ch. Home and Hosp., St. Joseph Hosp., Balt.; cons. thoracic surgery Ft. Howard VA Hosp., Balt. Mem. A.C.S. (gov. 1953-59), Am. Coll. Chest Physicians (gov., chmn. bd. govs.), Am, So. surg. assns., Am. Assn. Thoracic Surgery, many others. Author: Clinical Anatomy, 1963. Contbr. articles field clin. surgery to med. jours. Anatom. research human knee, on methods of reconstrn. repair of thoracic trachea; research on change of position of diaphragm in animals and man; research bronchial glands carried out on man and animals; research endeavors surg. treatment pulmonary emphysema. Home: 3 Paddington Ct., Balt. 21212. Office: 104 W. Madison St., Balt. 21201.*

BRASAVOLA, Antonius Musa, physician, botanist; b. Ferrara, Italy, Jan. 16, 1500; ed. Padua, also Bologna, Italy, Paris, France; doctorates in law, medicine, theology. Tchr. logic, natural philosophy, theory medicine, Ferrara; physician to Pope Paul III. Author: Examen omnium simplicium medicamentorum . . . (on physiology, plant remedies, opium, cinnabar, camphors, metals, stones, earths, salts, med. uses silver sublimate, mercury, other substances), 1536. Died July 6, 1555.

BRASDOR, Pierre, French surgeon; b. Maine, France, Dec. 19, 1721; M.A., Paris, 1752. Prof. anatomy and therapeutics, dir. Acad. Surgery; contbr. reports to jours. Acad. Surgery, Jour. Medicine. Pioneer in amputation at elbow and knee; invented method of binding carotid artery (named in his honor). Died Sept. 28, 1798.

BRASHEAR, John Alfred, Am. astronomer, engr.; b. Brownsville, Pa., 1840; s. B. B. and Julia B.; ed. pub. schs.; Sc.D., Western U. Pa., Princeton, 1911; LL.D., U. Wooster, 1896, Washington and Jefferson; D.Eng., Stevens Inst. Tech., 1912; m. Phoebe Stewart, Sept. 25, 1862. Learned machinist's trade; mech. engr., 1860-70; began constrn. of astron. and phys. instruments, 1870; actively engaged in mfg. same, from 1880. Acting dir. Allegheny Obs., 1898-1900; acting chancellor, Western U. Pa. Asso. with founding of Carnegie Inst. Tech.; planned orgn. of Frick Ednl. Commn. Known for his astron. lenses and precision instruments; manufactured speculum-metal plates for Rowland diffraction gratings; designed an astron. spectroscope. Died Pitts., Pa., Apr. 8, 1920.

BRASHER, Rex, Am. ornithologist; b. Bklyn., July 31, 1869; s. Philip Marston and Laura Alida (Bull) B.; grad. St. Francis Coll., Bklyn., 1884. Mem. Rex Brasher Associates, publs. Author: Secrets of Friendly Woods, 1926; Birds and Trees of North America (12 vols.), 1929. Contbr. to Compton's Ency., Nature Lovers Library, other publs. Made life-size, natural color paintings of all N.Am. bird species. Died Feb. 29, 1960.

BRATKOWSKA-SENIOW, Barbara (Mrs. Adam Wiktor), Polish physician; b. Warsaw, Poland, Nov. 8, 1923; d. Jan and Feliksa Bratkowska-Seniow; grad. Med. Faculty, Warsaw U., 1947; m. Adam Wiktor, Jan. 29, 1954; children—Jacek, Ewa. Asst. then adj. II Clinic, Internal Diseases, Med. Sch., Wroclaw, Poland, 1947-64; chief physician Municipal Hosp., Wroclaw, Poland, 1964—. Cons. internal diseases Hosp. for Obsterics. Decorated Golden Cross Merit; recipient Sci. award Soc. Polish Internists, 1959. Mem. Polish Med. Soc., Soc. Polish Internists, Polish Soc. History Medicine. Author: Erythematodes visceralis, 1964; also articles. Discovered correlation between dynamics and serum proteins in systemic hemopoietic diseases; research on diagnosis of lupus erythematosus and thyroid disorders. Home: 11/4 Dworcowa, Wroclaw 2. Office: 22/28 Rydygiera, Wroclaw 1, Poland.*

BRATT, Elmer Clark, Am. economist; b. Arapahoe, Neb., Nov. 12, 1901; s. Reuben Wilkinson and Daisy (Clark) B.; A.B., U. Neb., 1925, A.M., 1926, LL.D., 1955; postgrad. U. Cal., 1926-27; Ph.D., U. Wis., 1935; m. Bertha Margaret Brodfuehrer, June 16, 1928; 1 dau., Margaret (Mrs. Wayne Akers). Tchr., Furnas Co., Neb., 1918-20; prin. Stockville (Neb.) pub. sch., 1920-21; supt. Holstein (Neb.) pub. schs., 1921-22; supt. Danbury (Neb.) pub. sch., 1923-24; research asst. Nat. Bur. Econs., 1928-29; asst. prof. econs. Lehigh U., 1929-38, asso. prof., 1938-41, prof., 1941—, also head, dept. econs., 1958-65, dir. Bus.-Econs. Center, asso. dean bus. and econ. research, 1966—; chief econ. analyst. Bur. Fgn. and Domestic Commerce, Dept. Commerce, working on problems of econ. reconstruction after the war, 1942-45; prof. finance, Biarritz Am. U. in France and U. S. Army lectr. in Germany, 1945-46; statis. cons. Bethlehem Steel Co., 1930-31, 1949-53; econ. cons. U. S. Govt. Central Statis. Bd., 1934; indsl. economist, Nat. Bur. Econ. Research, 1936; dir. Consumers Mail Order Coöp, N.Y. 1938-42; cons. Fed. Res. Bd., 1954-55 Bur. Budget, 1956—; dir. data clearing house project Nat. Assn. Bus. Economists, 1962—; dir., v.p. Growth Found. Am., 1964—. Fulbright research scholar, evaluation rural devel. program Ceylon, 1955-56. Fellow Royal Econ. Soc., Nat. Assn. Bus. Economists; mem. Am. Econ. Assn., Am. Statis. Assn., Econometric Society Am. Finance Assn., Omicron Delta Kappa, Beta Gamma Sigma, Alpha Kappa Psi, Pi Gamma Mu. Author: Problems of Economic Change, 1936; Business Cycles and Forecasting, 1937, 5th rev. edit. 1961; This Unbalanced World, 1940; (with others) Economic Problems of War, 1942; Business Forecasting, 1958. Contbr. articles to Am. Econ. Jour. Am. Statis. Assn., Mgmt. Rev., Jour. Bus., also mags. and profl. jours. Originated statis. appraisals of information on constrn., inventories, several price areas for U. S. Govt.; Contbns. to long-term forecasting and turning point forecasting in short term. Home: 1521 W. Broad St., Bethlehem, Pa. 18018.*

BRATTAIN, Walter Houser, physicist; b. Amoy, China, Feb. 10, 1902 (parents Am. citizens); s. Ross R. and Ottilie (Houser) B.; B.S., Whitman Coll., 1924, D.Sc., 1955; M.A., U. Ore., 1926; Ph.D., U. Minn., 1929; D.Sc., U. Portland, 1952, Union Coll., 1955, U. Minn., 1957, Gustavus Adolphus Coll., 1963; L.H.D., Hartwick Coll., 1964; m. Keren Gilmore, July 5, 1935 (dec. Apr. 1957); 1 son, William G.; m. 2d, Emma Jane Miller, May 10, 1958. With radio sect. Bur. Standards, 1928-29; research physicist Bell Telephone Labs., Murray Hill, N.J., 1929-67; with div. war research Columbia, 1942-43; vis. lectr. Harvard, 1952-53; vis. lectr. Whitman Coll., 1962-67, adjunct-professor, since 1967—. Recipient John Scott award City of Phila., 1955; (with William Shockley, John Bardeen) Nobel Prize in Physics, 1956. Fellow Am. Phys. Soc., A.A.A.S., Am. Acad. Arts and Scis.; mem. Franklin Inst. (Stuart Ballantine medal 1952), Nat. Acad. Scis., Phi Beta Kappa, Sigma Xi. Co-inventor of transistor; research in semiconductors. Office: Whitman Coll., Walla Walla, Wash. 99362.*

BRATTLE, Thomas, Am. astronomer; b. Boston, June 20, 1658; s. Thomas and Elizabeth (Tyng) B., grad. Harvard, 1676; never married. Inherited father's estate and business, 1683; treas., Harvard, 1693-1713; involved in controversy with the Mathers over orgn. of Harvard corp., also in intense religious dispute with the Mathers, condemned Salem witchcraft proceedings, 1692. Wrote on astron. subjects; observed and sent accounts of several eclipses to Royal Soc., London. Died Boston, May 18, 1713.

BRATTSTROM, Bayard Holmes, Am. zoologist; b. Chgo., July 3, 1929; s. Wilber LeRoy and Violet (Holmes) B.; B.S., San Diego State Coll., 1951; M.A., U. Cal. at Los Angeles, 1953, Ph.D., 1959; m. Cecile D. Funk, June 15, 1952; children—Theodore Allen, David Arthur. Dir. edn. Natural History Mus., San Diego, 1949-51, asst. curator herpetology, 1951-56; teaching asst. zoology U. Cal., Los Angeles, 1951-56, asso. in zoology, 1954-56; research fellow in paleoecology Cal. Inst. Tech., Pasadena, 1955; instr. in biology Adelphi Coll., Garden City, N.Y., 1956-60; asst. prof. Cal. State Coll., Fullerton, 1960-61, asso. prof., 1961-66, professor, since 1966—; associate professor of zoology U. Cal., Los Angeles, summers 1962-63; hon. research asso. in herpetology, vertebrate paleontology Los Angeles County Mus., Los Angeles, 1961—; pres. Fullerton Youth Mus. and Natural Sci. Center, 1962-64, dir., 1962-66. Recipient research grants Am. Philos. Soc., Mexico, 1958, Panama,

1959, NSF, 1964-66; NSF sr. postdoctoral fellow Monash U.; Australia, 1966-67. Fellow, A.A.A.S., Herpetological League; mem., Am. Soc. Ichthyologists and Herpetologists (bd. govs. 1962—, v.p. western div. 1955), Orange County Zool. Soc. (mem. bd. 1962-65, pres. 1962-64), So. Cal. Acad. Sci. (dir. 1964-67), Ecol. Soc. Am., Soc. Study Evolution, Soc. Systematic Zoology, San Diego Soc. Natural History, Soc. Vertebrate Paleontology, Am. Soc. Mammalogists, Cooper, Am. ornithol. socs., Am. Soc. Zoologists, Am. Inst. Biol. Sci. Research, publs. in osteology, paleontology, systematics, behavior, ecology, physiology (especially temperature regulation), zoogeography of vertebrates, especially amphibians and reptiles, repopulation of volcanic islands, social behavior. Home: 2788 Verde Av., Anaheim, Cal. 98206. Office: Dept. Biology, Cal. State Coll., Fullerton, Cal. 92631.*

BRATTSTRÖM, Hans Olof, marine biologist; b. Skara, Sweden, July 10, 1908; s. Gustaf and Nanna (Beer) B.; Fil.dr., Lund U., 1941; m. Ingrid Göransson, Oct. 8, 1945; children—Göran, Kerstin. Docent zoology Lund U., 1941-49; prof. zoology U. Bergen (Norway), 1949-62, prof. marine biology, 1962—, dir. Zool. Lab., 1949-54, dir. Zool. Mus., 1949-63, dir. Biol. Sta., 1949—. A Leader Lund U. Chile Expdn., 1948-49; participant Internat. Indian Ocean Expdn., 1964. Mem. Royal Norwegian Acad. Sci., Soc. for Promotion Sci., Nordic Council for Marine Biology (sec.). Author: Bli med til Chile, 1952; also articles. Editor: Sarsia, 1961—. Research on echinoderms and ascothoracidans, marine ecology; marine biol. studies on Caribbean. Home: 10 Professorveien, Bergen-Minde, Norway. Office: Biologisk stasjon, Espegrend, Blomsterdalen, Norway.*

BRAUDE, Ernest Alexander Rudolph, Brit. chemist; b. June 8, 1922; s. R. and E. Braude; ed. Birkbeck Coll.; B.Sc. (Frank Hatton prizeman 1942), Imperial Coll. Sci. and Tech., Ph.D., D.I.C., 1945; D.Sc., U. London; m. Catherine Renate Zander, 1946. Asst. lectr. Imperial Coll. Sci. and Tech., 1946, lectr., 1947, prof. organic chemistry, 1955—. Reader in organic chemistry Imperial Coll., U. London (Eng.), 1952; Tilden lectr., 1956. Recipient Meldola medal. Fellow Royal Inst. Chemistry; mem. Chem. Soc. (council 1954), Royal Coll. Sci. (asso.).

BRAUDE, Raphael, animal nutritionist; b. Lodz, Poland, May 17, 1914; s. Marcus and Natalia (Buber) B.; Dipl.Agr., U. Danzig (Poland), 1936; Animal Husbandry Diploma, U. Edinburgh (Scotland), 1938; Ph.D., U. Reading (Eng.), 1945, D.Sc., 1956; m. Eva Tyfenbach, June 15, 1948; 1 son, Jeremy Mark. With Nat. Inst. for Research in Dairying, U. Reading (Eng.), 1939—, head pig husbandry dept., 1958—; tech. sec. pig research com. Agrl. Research Council, 1954—. Mem. panel animal nutrition experts FAO, 1958—. Mem. Pig Health Control Assn. (chmn.), European Assn. for Animal Prodn. (mem. pig commn.). Research and numerous publs. on pig nutrition, including use of feed additives to improve animal performance; discovered high doses of copper added to pig diet improves their performance. Home: 10 Cockney Hill, Reading. Office: Nat. Inst. for Research in Dairying, Shinfield, Reading, Eng.*

BRAUELL, Friedrich August, German bacteriologist; b. Weimar, Germany, Dec. 11, 1807; ed. Jena, Berlin, Germany; Copenhagen, Denmark; Ph.D., Erlangen, Germany, 1834. Worked in Wilna, Lithuania, 1838; asso. prof. Kasan, Russia, 1841-46; prof. Dorpat, Estonia, 1848-68, retired 1868; prof. (hon.) Leipzig, Germany, 1871, lectr. path. anatomy of domestic animals. Discoverer of Bacillus anthracis in man, 1857; did not believe it was peculiar to the disease. Died Dec. 10, 1882.

BRAUER, August, zoologist; b. 1863. Mem. German deep sea expdn., 1898-99; dir. zool. mus., Berlin, Germany from 1906; prof. from 1914. Important studies on devel. and anatomy of gymnophiones, also classification and comparative anatomy of deep sea fish. Died 1917.

BRAUER, Fred Günther, mathematician; b. Königsberg, Germany, Feb. 3, 1932; s. Richard and Ilse (Karger) B.; B.A., U. Toronto (Ont., Can.), 1952; S.M., Mass. Inst. Tech., 1953, Ph.D., 1956; m. Esther Luterman, June 22, 1958; children—David, Deborah, Michael. Instr., U. Chgo., 1956-58; faculty U. B.C., Vancouver, Can., 1958-60, asst. prof., 1959-60; faculty U. Wis., Madison, 1960—, prof. math., 1966—. Mem. Am. Math. Soc., Math. Assn. Am., Canadian Math. Congress. Author: (with John A. Nohel) Ordinary Differential Equations, 1966; also articles. Research on differential equations including boundary value problems, uniqueness theorems, stability, and asymptotic behavior of solutions. Home: 5113 Coney Weston Pl., Madison, Wis. 53711.*

BRAUER, Friedrich, German entomologist; b. 1832; originated modern classification of Insecta, 1885. Died 1904.

BRAUER, Ludolph, physician; b. Germany, 1865; 1st to suggest cardiolysis for relief of adhesive pericarditis (performed by Dr. Petersen), 1902; used nitrogen to produce an artificial pneumothorax in treatment of Tb, 1906; performed 1st radical thoracoplasty. Died 1951.

BRAUER, Richard Dagobert, mathematician; b. Berlin, Germany, Feb. 10, 1901; s. Max and Caroline (Jacob) B.; Ph.D., U. Berlin, 1925; M.A., Harvard, 1952; m. Ilse Karger, Sept. 17, 1925; children—George U., Fred G. Came to U. S., 1933, naturalized, 1954. Privatdozent U. Konigsberg, 1927-33; vis. prof. U. Ky., 1933-34; asst. Inst. for Advanced Study, Princeton, N.J., 1934-35; faculty U. Toronto (Ont. Can.), 1935-48, U. Mich., Ann Arbor, 1948-52; prof. math. Harvard, 1952—, Perkins professor of mathematics, 1966—; vis. prof. Nagoya U., 1959. John Simon Guggenheim fellow, 1941-42. Fellow Royal Soc. Can.; mem. Am. (pres. 1957-58, recipient Cole prize, 1949), London (hon.) math. socs., Nat. Acad. Scis., Am. Acad. Arts and Scis., Akademie der Wissenschaften Göttingen (Gauss prof. 1964) Home: 15 Franklin St., Belmont, Mass. 02178.*

BRAUN, Alexander Heinrich, German botanist; b. Regensburg, Germany, May 10, 1805; s. Alexander and Henriette (Mayer) B.; studied medicine under G. W. Bischoff, Dierbach and F. J. Schelver; m. Mathilde Zimmer, 1835; 2 sons, 3 daus. including Marie (Mrs. R. Caspary) and Cecilie (Mrs. Georg Mettenius); m. 2d, Adele Messmer, 1844; 1 son, 3 daus. Prof. in Freiburg i. Br., Giessen and Berlin, (both Germany); dir. Berlin bot. garden. Mem. French Acad. Scis., Leopoldina, Prussian Acad. Author: Vergleichende Untersuchung über die Ordnung der Schuppen an den Tannenzäpfshen, 1830; Betrachtungen über die Erscheinung der Verjüngung in der Natur, 1849/50; Individuum der Pflanzen, 1852; Algarum unicellularium genera nova et minus cognita, 1855; Parthenogenese von Coelebogyne, 1856. Contbr. articles to bot. jours. Research on algae (protoplasm in plant cells) and conifers (the cone as a blossom), Schimper-Braun theory of leaf arrangement, blossom morphology; attempted to relate plant morphology and phylogenesis; important contbns. in inductive research, phytography and classification; also natural philosopher. Died Berlin, Mar. 29, 1877.

BRAUN, Armin Charles John, Am. biologist; b. Milw., Sept. 5, 1911; s. Adolph and Ella (Schreiber) B.; B.S., U. Wis., 1934, Ph.D., 1938; predoctoral study European sci. labs., 1936-37. With Rockefeller U., 1938—, successively fellow, asst., asso., asso. mem., asso. prof., 1938-59, mem. and prof., 1959—, head dept. plant biology, 1955—; mem. sci. panel biology and medicine NSF, 1958-61; mem. sci. adv. bd. Inst. Cancer Research Phila., 1959—; sci. adv. panel Brookhaven Nat. Lab., 1960—. Recipient Newcomb Cleveland award, A.A.A.S., 1949. Mem. Nat. Acad. Scis., Soc. Study Devel. and Growth (exec. com.), Harvey Soc. (hon.), Am. Phytopathol. Soc., Am. Bot. Soc., Sigma Xi, Phi Sigma, Alpha Zeta, Delta Theta Sigma. Author numerous sci. papers, books. Research on physiology of tumor formation in crown gall disease of plants and nature of cellular changes that occur; physiology and biochemistry of disease and disease resistance in plants. Home: Ridgeview Rd., Princeton, N.J. Office: Rockefeller Univ., 66th St. and York Av., N.Y.C. 10021.

BRAUN, Charles Ernest, Am. organic chemist, educator; b. Bklyn., Mar. 10, 1900; s. Charles and Anna (Keilbach) B.; B.S. in Chemistry cum laude, Poly. Inst. Bklyn., 1922; A.M., Columbia, 1923, Ph.D. in Organic Chemistry, 1925; m. Isola Everton Oakes, Oct. 3, 1927; 1 son, Carl Everton. Research chemist DuPont Co., 1925-26, Bklyn. Edison Co., 1926-27, Barrett Co., 1927-28; faculty U. Vt., Burlington, 1928-63, prof. organic chemistry, 1942-63, chmn. dept. chemistry, 1942-60, coordinator research, 1950-53, dean Grad. Coll., 1952, 53, 60-63. Recipient Alumni Achievement award Poly. Inst. Bklyn., 1956. Mem. Am. Chem. Soc. (chmn., councilor Western Vt. sect. 1939-59), Sigma Xi, Phi Lambda Upsilon, Alpha Chi Rho. Author: Organic Syntheses, 1946-61. Contbr. articles to profl. jours. Research on organic synthesis, metallic nitration, mono and biguanidines, sugar derivatives of sulfonamides. Home: 173 De Forest Rd., Burlington, Vt. 05401.*

BRAUN, Dietrich, German chemist; b. Leipzig, Germany, Nov. 28, 1930; s. Wilhelm and Charlotte (Quehl) B.; student U. Leipzig, 1949-51, U. Mainz, 1952-55, Diplom-Chemiker, Dr.ner.nat., 1957; m. Margarete Jacobi, July 23, 1960; children—Bettina, Barbara. Asst., Organischchemisches Inst., U. Mainz, Germany, 1957-59, privatdozent, 1960; head chemistry dept. German Plastics Inst., Darmstadt, Germany, 1959—; privatdozent organic and macromolecular chemistry Tech. U. Darmstadt, 1961-66, prof. pure and applied polymer sci., 1966—. Mem. Gesellschaft Deutscher Chemikar, Gesellschaft Deutscher Naturforscher und Ärzte, Kolloid-Gesellschaft. Author: (with H. Cherdron, W. Kern), Praktikum der makromolekularen organischen Chemie, 1966. Editor: Angewandte Makromolekulare chemie; co-editor: Die Makromolekulare Chemie. Research, publs. on chem. reactions on polymers, organometallic polymers, polyradicals stereospezific polymerization and tacticity of polymers. Home: 56 Jakob-Jung-Str., Darmstadt-Arheilgen. Office: 6 R Schlossgartenstr., Darmstadt, Germany.*

BRAUN (Brunus, Bruin), Georg, German geographer; b. 1542; s. Melchior B. Archdeacon in Dortmund, Germany, then canon and dean at St. Maria ad Gradus, Cologne, Germany for 37 years. Author: religious writings, Civitates orbis terrarum, 6 vols., 1572-1618; portrayed important cities of world in topographically clear manner. Died Cologne, Mar. 10, 1622.

BRAUN, Gustav Oskar Max, geographer; b. Dorpat (now Tarta, Estonia), May 30, 1881; s. Maximilian and Toni (Leisterer) B.; studied in Königsberg (doctorate under Friedrich Halm, 1903) and Göttingen, Germany m. Ilse von Horn, 1910. Taught in Greifswald, Germany, 1907; turned down invitation to U. Peking, 1910; dept. head Inst. für Meereskunde (Oceanography), Berlin, Germany, 1911; prof. geography in Basel, Switzerland, 1912; in Greifswald, 1918-33. Author: Grundzüge der Physiogeographie, 1911; Deutschland, 1916; Ostpreussens Seen, 1903; Das Ostseegebiet, 1912; Die Nördlichen Staaten, eine soziologische Länderkunde, 1924; Nordeuropa, 1926. Influenced by geomorphological teachings of W. M. Davis; strived to improve terrain observations along with charts, best seen in Deutschland (a morphogenesis of relief and evolutionary analysis of cities). Died Oslo, Norway, Nov. 11, 1940.

BRAUN, Heinrich Friedrich Wilhelm, German surgeon; b. Rawitsch, Germany, Jan. 1, 1862; asst. to R. Volkman and V. Bramann, Halle, Germany; became private docent, 1894; prof. Leipzig (Germany) U., from 1905. Author: Uber Infiltration anaesthesia und regionare Anaesthesia, 1898; Local Anaesthesia, Its Scientific Basis and Practical Use, 1914. Synthesized novocaine ovprocaine and added adrenalin for clin. use, 1905; introduced operation to prevent complications in appendicitis by ligating ileocolic vein. Died Germany, Apr. 26, 1934.

BRAUN, Karl Ferdinand, physicist; b. Fulda, Germany, June 6, 1850; s. Konrad and Franziska (Gohring) B.; ed. U. Marburg (Germany); Ph.D., U. Berlin, 1872; m. Amelie Bühler, 1885; 2 sons, 2 daus. Asst., U. Würzburg (Germany), 1872-74; lectr. St. Thomas Gymnasium, Leipzig, Germany, 1874-76; extraordinary prof. U. Marburg, 1876-80; with U. Strasbourg (France), 1880-83, prof. physics, dir. phys. inst., from 1895; prof. physics Tech. High Sch., Karlsruhe, Germany, 1883-85; at Tübingen, Germany, 1885-95, helped found phys. Inst. Recipient (with Marconi) Nobel prize in physics, 1909. Inventor cathode-ray-tube oscilliscope (Braun tube), 1897; devised coupled transmitter and coupled receiver for improved wireless performance (Nobel Prize citation); investigated aspects of radiotelegraphy including directional transmission, crystal detectors, radio transmissions as beacons for nav. Died Bklyn., Apr. 20, 1918.

BRAUN, Maxmillan, German parasitologist; b. Myslowitz, Silesian Prussia, Sept. 30, 1850. Became asst. Zool. Inst., Wurzburg, Germany, 1876, private docent in zoology, 1877; apptd. prosector Inst. Comparative Anatomy, Dorpat (now Tartu, Estonia), 1879, prof. zoology, 1883; named prof. zoology, Rostock, Germany, 1886. Author: Parasiten des Menschen, 1883; Zur Entwickelungsgeschichte des breiten Bandwurms (Bothriocephalus latus. Brems.), 1866. Discovered intermediary host of bothriocephalus. Died 1930.

BRAUN, Otto, German dermatologist; b. Sarrebruck, Apr. 25, 1922; s. Andreas and Rosa (Falco) B.; ed. U. Münster/W., Mayence; M.D.; m. Franziska Golling, Sept. 1, 1951. Agrege, prof. dermatology, dir. univ. clinic of dermatology at Marburg (Germany). Mem. German, Uruguay (hon.), Polish (hon.) socs. dermatology, N.Y. Acad. Scis. Author numerous publs., including Lehrbuch für Dermatologie und Venerologie. Home: Goethestrasse 16, Cappel b. Marburg. Office: Deutschhausstrasse 9, Marburg, Germany.

BRAUN, Werner, microbiologist; b. Berlin, Germany, Nov. 16, 1914; s. Simon and Edith (Brach) B.; Ph.D., U. Göttingen (Germany), 1936; m. Barbara Melnikow, June 7, 1942; children—Renee, Stephanie, Robin. Came to U. S., 1936, naturalized, 1941. Research asso. U. Cal. at Berkeley, 1937-48; chief variation br. Chem. Corps. Biol. Labs., Frederick, Md., 1948-55; prof. microbiology Inst. Microbiology Rutgers The State U., New Brunswick, N.J., 1955—; vis. prof. U. P.R. Med. Sch., 1957-65, Hebrew U. Med. Sch., Jerusalem, Israel, 1962-63. Cons. U. S. Dept. Army, 1948—, U. S. Dept. Def., 1962. Recipient Barnett L. Cohen award Md. chpt. Soc. Am. Bacteriologists; Chem. Corps Superior Accomplishment award, 1954. Fellow Am. Acad. Microbiology, mem. Am. Assn. Immunologists, Am. Soc. for Microbiology, Genetics Soc. Am., A.A.A.S., Soc. for Exptl. Biology and Medicine, Sigma Xi, Phi Sigma. Author: Bacterial Genetics, 1953, 65; also numerous articles. Editor: (with M. Landy) Bacterial Endotoxins, 1964. Research on factors controlling bacterial population changes, factors influencing resistance to microbial pathogens, regulation immune responses. Home: 72 Mason Dr., Princeton, N.J. 08540. Office: Inst. Microbiology, Rutgers U., New Brunswick, N.J. 08903.*

BRAUN-FALCO, Otto, German physician; b. Saarbrücken, Germany, Apr. 25, 1922; s. Andreas and Rosa (Falco) B.-F.; civil service exam. and grad., 1948; M.D., U. Mainz (Germany), 1948; m. Franziska Golling, Sept. 3, 1951; Lectr. dermatology and venerology Mainz U., 1954-60; faculty Marburg (Germany) U., 1960—, prof., 1961—, dir. dermatology and outpatients dept., 1961—; dir. Clin. Hosp. for Dermatology, 1961-67; dir. Clin. Hosp. for Dermatology, U. Münich, 1967—. Mem. N.Y. Acad. Scis., German Soc.

Dermatology, German Soc. Histochemistry, Assn. S.W. Germany Dermatology, Leopoldina-Halle, Belgian Soc. Dermatology, Netherlands Soc. Dermatology, French Soc. Dermatology, Med. Soc. Germany; hon. mem. Polish Soc. Dermatology, Italian Soc. Dermatology, Soc. Dermatology in Hungary, Yugoslavian Soc. Dermatology. Author of textbooks in dermatology and venereology; contbg. author: The Histochemistry of the Hair Follicle, 1958; Die Histochemie der Haut, 1961; zur Histochemie d. Lupus erythematodes 1961; Allgemeine Pathologie des Bindegewebes, 1964; also numerous articles. Research on metabolic disorders of skin, disorders of keratinization, hair diseases, histochemistry and electron microscopy of skin diseases. Home: 11 Frauenlobstrasse. Office: 9 Frauenlobstrasse, 8 München 15, Germany.*

BRAUNE, Christian Wilhelm, German anatomist; b. Leipzig, Germany, July 17, 1831; prof., Leipzig, Author: Topogr.-Anatomische Atlas, 1872; Das Venensystem des menschlichen Körpers, 1884-89,, 2 vols.; Die Lage des Uterus und Foetus am Ende der Schwangerschaft, 1872. Introduced surg. techniques involving freezing of tissue. Died Leipzig, Apr. 29, 1892.

BRAUNER, Bohuslav, chemist; b. Prague, Czechoslovakia, May 8, 1855; ed. Prague, also under Bunsen, Heidelberg, Germany, then under Roscoe, Manchester, Eng., 1880-82. Became lectr. Charles U., 1882, asst. prof., 1890, prof., 1897-1925. Investigated chemistry of tellurium, rare earths, grouping of elements in periodic table, atomic weights; predicted 61st element (promethium), 1902. Died Feb. 15, 1935.

BRAUNS, Reinhard Anton, mineralogist, petrographer; b. Eiterfeld, nr. Kassel, Germany, Aug. 20, 1861; s. Carl and Mathilde (Meurer) B.; studied in Marburg, Germany, from 1881, doctorate, 1885; hon. dr. agr., Bonn, 1935; m. Karoline Wirth, 1895; 2 sons, 1 dau., Gertrud (Mrs. Prof. Dr. Schultz-Brauns). Lectr. in Marburg, 1887; prof. mineralogy and geology Technische Hochschule Karlsruhe, 1894; to Giessen, Germany, 1895; to Kiel, Germany, 1904; in Bonn, Germany, 1907-34. Mem. Deutsche Mineralogische Gesellschaft (a founder 1908); hon. mem. various sci. socs., also Halle Akademie. Author: Chemische Mineralogie, 1896; Anleitung zur Bestimmen von Mineralien, 1898; Das Mineralreich, 1903; Die optischen Anomalien der Krystalle, 1891; Mineralogie, 1893; Vulkane und Erdbeben, 1913; Flussige Kristalle und Lebewesen, 1931; Der kristallische Schiefer des Laacher Seegebietes . . . , 1911. Editor: Neues Jahrbuch fur Mineralogie und Geologie; Zeitblatt für Mineralogie. Published compendium of all known facts on optical anomalies, 1891; made use of phys. chemistry in mineralogy; wrote on diabase stones, underlined relationship between mineralogy and geology; wrote much on precious stones, particularly synthesis from 1883. Died Bonn, Germany, Jan. 28, 1937.

BRAUNSTEINER, Herbert, Austrian internist; b. Vienna, Austria, Mar. 10, 1923; s. Franz and Margaret (Pennenberg) B.; B.A., U. Paris, U. Vienna, 1941, M.D., 1948; m. Elisabeth Schmitz, July 1, 1948; children—Cathrine, Andrea, Teresa. Mem. staff dept. internal medicine U. Vienna, 1948, 51-52, 54-64; research fellow Institut de Recherches sur le Cancer, Paris, France, 1949-50, Sloan Kettering Inst., N.Y., 1953-54; prof. medicine, head dept. internal medicine U. Innsbruck (Austria), 1964—. Mem. Internat. Soc. Hematology (sec.-gen.). Author: Physiology and Pathophysiology of Leucocytes, 1962; Lipoproteinlipase, 1963; Thrombocytoasthenia, 1955. Research, publs. on hematology, discovery of an ergastoplasma in plasma cells, 1952, proof of their secretory mechanism; dynamics of blood platelets (thrombocytes), proof of a defective mechanism in thrombocytoasthenia, 1955; correlation of the function of basophil leucocytes with blood lipis, 1958; metabolism, determination of lipoproteinlipase and role in various diseases, turnover rates of lipids in various diseases, also studies on hyperlipemia. Home: 1 Falk. Office: 35 Anich, Innsbruck, Austria.*

BRAUNWALD, Eugene, physician; b. Vienna, Austria, Aug. 15, 1929; s. William and Clare (Wallach) B.; A.B., N.Y. U., 1949, M.D., 1952; m. Nina Helen Starr, May 25, 1952; children—Karen, Denise, Adrienne. Commd. asst. surgeon USPHS, 1954, advanced through ranks to med. dir., 1963; ret., 1963; chief cardiology br. Nat. Heart Inst., NIH, Bethesda, Md., 1960-66, clin. dir., 1966-68; clin. prof. medicine Georgetown U., 1966-68; prof., chmn. dept. medicine U. Cal., San Diego, 1968—; lectr. medicine Johns Hopkins Sch. Medicine, 1961—. Recipient Jacobs award for cardiovascular research U. Tex., 1961; Arthur Fleming award Fed. Govt., 1965; Abel award for research in pharmacology Am. Soc. Pharm. and Exptl. Therapy, 1965. Mem. Am. Heart Assn. (chmn. publs. com. 1965—, v.p.), Soc. for Clin. Investigations, Am. Physiol. Soc., Am. Coll. Cardiology (v.p.), Am. Pharm. Soc., A.C.P. Author: Idiopathic Hypertrophic Subaortic Stenosis, 1964; Mechanism of Contraction of the Normal and Failing Heart; also numerous articles. Editor: Yearbook of Cardiovascular Diseases, 1965. Editorial bd. Circulation, 1960—, Circulation Research, since 1966—, Jour. Clin. Investigation, 1964—, Am. Jour. Physiology, 1965—, Annals Internal Medicine, Am. Jour. Cardiology. Application physiol. methods of cardiovascular diagnosis, eludication mechanisms controlling function heart; clarification pharmacology digitalis; description idiopathic hypertrophic subaortic stenosis. Home: 7006 Longwood Dr., Bethesda 20014. Office: San Diego County-Univ. Hosp., 225 W. Dickinson St., San Diego.

BRAUS, Hermann, German anatomist; b. Aachen-Burtscheid, Germany, Aug. 15, 1868; s. Otto and Bertha (Ernest) B.; studied in Bonn, Jena, Berlin, Heidelberg, (all Germany), from 1888; m. Elisabeth Fürbringer, 1899; 3 daus. Asst. to M. Fürbringer, joined faculty in Jena, 1896; also influenced by Ernst Haeckel; prosector with A. Koelliker in Würzburg, Germany, 1899-1901; with Fürbringer in Heidelberg again, became his successor, 1912; to Würzburg, 1921. Author: Lehrbuch der Anatomie, 3 vols. (3d vol. compiled by C. Elze). Contbr. articles to med. jours. Exptl. work on formation and devel. of front extremities of embryos and young amphibian larvae; main contbn. to anatomy in 1st half of century was above-mentioned textbook with many original ideas; revolutionary influence of devel. of anat. teaching at univ. level. Died Würzburg, Germany, Nov. 28, 1924.

BRAVAIS, Auguste, French physicist; b. Annonay, France, Aug. 23, 1811; s. Francois-Victor Bravais; entered Polytechnic, 1829. Participant, hydrographic exploration on Algerian coasts; mem. expdn. to make astron., magnetic and meteorol. observations in bay Bellesmont; became prof. applied math. in astronomy Faculty Scis. Lyons (France), 1840; apptd. prof. physics Polytechnic, 1845. Mem. French Acad. Scis., 1854; a founder Meteorol. Socs., Hygrometric Soc. Author: Études cristallographiques, 1851; numerous articles. Introduced lattice theory (geometry used to explain phenomena of crystals); investigated optics and magnetism; conducted (with Martins) 1st sci. ascent Montblanc; introduced idea of molecular polyhedrons. Died Versailles, France, Mar. 30, 1863.

BRAVO, Francisco, physician; b. Osuna, Spain; flourished 16th century. Author: Opera medicinala (1st med. book pub. in Ams., included 1st description tabardillo, Mexican or Spanish typhus), 1570.

BRAY, John Leighton, Am. metallurgist; b. Millbridge, Me., Aug. 11, 1890; s. Charles Ambergh and Vinetta (Cook) B.; B.S. Mass. Inst. Tech., 1912. Ph.D., 1930; m. Jean Shaw, Aug. 23, 1925; children—Barbara Vilora, John Leighton. Metallurgist, Braden Copper Co., Rancagua, Chile, 1912-15, Consol. Mining & Smelting Co., Trail, B.C., 1915-16, Blackbutte (Ore.) Quicksilver Co., 1916-17, N.Y. & Honduras Mining Co., Honduras, C.A., 1918-20; prof. metallurgy, N.S. Tech. Coll., 1920-21; metallurgist for U. S. Tariff Commn., 1921-23; prof. metallurgy, metall. engring. Purdue U., W. Lafayette, Ind., from 1947, head Sch. Chem. and Metall. Engring., 1935-47. Mem. Am. Inst. Mining and Metall. Engrs., Soc. of Metals, Am. Chem. Soc., Am. Inst. Chem. Engrs., Electrochem. Soc., Inst. of Metals (Eng.), Soc. for Promotion Engring. Edn., Sigma Psi, Tau Beta Pi, Phi Lambda Upsilon, Omega Chi Epsilon. Author: Textbook of Ore Dressing (with R. H. Richards and C. E. Locke), 1925; Principles of Metallurgy, 1930; German Grammar for Chemists, 1937; Introductory Readings in Technical German, 1940; Non Ferrous Production Metallurgy, 1941; Ferrous Production Metallurgy, 1942; Patent Law and Procedure, 1948. Died Dec. 6, 1952.

BRAY, John Roger, ecologist; b. Belleville, Ill., June 20, 1929; s. Roger Hammond and Doris (Neuhaus) B.; B.A. high honors in Botany, U.Ill., 1950; Ph.D., U. Wis., 1955; m. Gwendolyn Jessica Struik, Dec. 23, 1961; children—Naomi, Hanna. Teaching and research asst. U. Wis., Madison, 1950-55; vis. lectr., research asso. U. Minn., Mpls., 1955-57; asst. prof. U. Toronto (Ont.), 1957-62; research asso. U. Sask., Saskatoon, Can., 1963; prin. sci. officer Dept. Sci. Indsl. Research, Palmerston North, New Zealand, 1963-66; consulting ecologist, 1966—. Member Ecological Society of America (associate editor Ecology, 1961-63), Brit., New Zealand ecol. socs. Research on quantitative techniques for classifying plant communities, philosophic bases of ecologic theory, prodn. utilization and efficiency of vegetation, chlorophyll content of vegetation, forest growth and glacier chronology in relation to climate and solar activity, climatology. Home: R.D. 2, Upper Moutere, Nelson. Office: P.O. Box 494, Nelson, New Zealand.*

BRAY, Philip James, Am. physicist; b. Kansas City, Mo., Aug. 26, 1925; s. Harry James and Ruth (Moerdyke) B.; Sc.B., Brown U., 1948; M.A., Harvard, 1949, Ph.D., 1953; m. Marion Rebecca Cooperman, Sept. 27, 1951; children—Carolyn Ruth, Philip James, Katherine Mary Elizabeth. Asst. prof. physics Rensselaer Poly. Inst., 1952-55; faculty Brown U., Providence, 1955—, prof., 1958—, chmn. dept. physics, 1963—; post-doctoral fellow, vis. prof. U. Sheffield, Eng., 1961-62. Fellow Am. Phys. Soc. (chmn. New Eng. sect. 1965-67), Am. Acad. Arts and Scis.; mem. Am. Ceramic Soc., Soc. Glass Tech., N.Y. Acad. Arts and Scis., Sigma Xi. Asso. editor Revs. Modern Physics, 1963-65. Research and numerous publs. on atomic arrangements, chem. bonding, defects in glasses and crystalline compounds, electron distbrns. in molecules of biol. significance. Home: 106 Highland Av., Barrington, R.I. 02806. Office: Brown U., Providence 02912.*

BRAY, William Crowell, chemist; b. Wingham, Ont., Can., Sept. 2, 1879; s. William Thomas and Sarah Jane (Willson) B.; B.A., U. Toronto, 1902; Ph.D., U. Leipzig (Germany), 1905; m. Nora Thomas, June 30, 1914. Research asso. Mass. Inst. Tech., 1905-10, asst. prof. physico-chem. research, 1910-12; asst. prof. chemistry U. Cal. at Berkeley, 1912-16, asso. prof., 1916-18, prof., from 1918, chmn. dept. chemistry, 1943-45; asso. dir. Fixed Nitrogen Research Lab., Washington, 1919. Fellow Am. Acad. Arts and Scis.; mem. Nat. Acad. Scis., Am. Chem. Soc., Am. Electrochem. Soc. Contbr. papers on inorganic and phys. chem. to Jour. Am. Chem. Soc.; studies in catalytic decomposition of hydrogen peroxide by halide-halogen couples; oxidation of hydrazine; hydrolysis of ferric ion and the standard potential of the ferrous-ferric electrode; preparation of hopcalite, a mixed-oxide catalyst for the low temperature oxidation of carbon monoxide. Died Feb. 24, 1946.

BRAYLEY (the younger), Edward William, Brit. sci. writer; b. London, 1802; s. Edward Wedlake Brayley; studied sci., London, Royal instns.; attended chemistry lectures of prof. Brandes. Joint-librarian London Instn.; joint-editor Annals of Philosophy, Zoological Journal, Philosophical Magazine, 1822-45. Fellow Royal Soc., 1854; mem. (original) Chem. Soc., Zool. Soc. Editor-contbr. English Cyclopaedia; contbr. Companion to the Almanac; prepared last authentic edit. of Parkes' Chemical Catechesm, 1834. Writings on mineral. chem., geology, zoology, meteors, seismology, phys. constitution and functions of sun, biographies of men of sci. Died Feb. 1, 1870.

BRAYTON, George B., Am. engr.; b. 19th Century. Engr. Exeter, N.H., Boston. Built 1st gas motor with constant pressure; patented 1st practical gasoline motor, 1872, converted it to light gasoline 2-stroke motor with mechanically controlled valves, 1874; this motor was widely used, built into an automobile and supposedly into an airship in 1876.

BRAYTSEV, Vasiliy Yakovlevich, Russian surgeon; b. Village of Zabelyshino, Mogilev Guberniya, 1902; grad. Med. Faculty, Moscow U., 1927; D.Med. Sci., 1946. Sr. asst. surg. dept. Bogdanov Inst. Blood Transfusion, Moscos, 1929-35; supernumerary asso. Faculty Surgery Clinic, Moscow U., 1929-35; chief physician Seredinsky Dist. Hosp., Moscow Oblast later asst. Surg. Clinic, Vladimirsky Moscow Oblast Clin. Inst., asst. Surg. Clinic, Central Postgrad. Med. Inst., 1935-41; chief surgeon med. health battalion, 1941; chief surgeon Orenburg Dept. Health, 1942; dep. head Kremlin Med. Health Bd., 1943-52; prof. chair gen. surgery Pirogov 2d Moscow Med. Inst., 1953; prof., 1953—; head surg. dept. 6th Clin. Hosp., Moscow Dept. Health, Mem. editorial bd. Meditsinskaya sestra, 1956—. Research and numerous publs. on blood transfusion, surgery of spleen, anerobic and purulent infections, diseases of organs of thoracic and abdominal cavities, aepsis; designer styptic fabric wadding adopted by Soviet Army, 1935, functional tire for transporting persons with fractures of hip and pelvis. Address: 6th Clin. Hosp., Moscow Dept. Health, Novaya Basmannaya ulitsa 26, Moscow, USSR.

BRAZHDYUNAS, Povilas Povilovich, Lithuanian physicist; b. 1897; grad. Kaunas U., 1925, Zurich U., 1930. Faculty, Kaunas U., 1926-40; faculty Vilnius U., 1940—, now prof., head chair exptl. physics; asso. Inst. Physics and Math., Lithuanian Acad. Sci. Mem. Lithuanian Acad. Sci. Research and publs. on physics of semicondrs., optical and photoelectric properties of sulphur, selenium and tellurium compounds, elec. properties of systems in semicondr. layers. Address: Vilnius University, ulitsa Universiteta 3, Vilnius, Lithuanian SSR, USSR.

BRAZZEL, James Roland, Am. entomologist; b. Hico, La., May 2, 1921; s. Lee and Benson (Campbell) B.; B.S., La. State U., 1951, M.S., 1953; Ph.D., Tex. A. and M.U., 1956; m. Allie B. Fowler, Jan. 28, 1941; 1 dau., Carolyn (Mrs. James Gholston). Asst. entomologist La. State U., Baton Rouge, 1955-57; faculty Tex. A. and M. U., College Station, 1957-62, prof., 1960-62; prof., head dept. entomology Miss. State U., State College, 1963—. Mem. Entomol. Soc. Am., Ecol. Soc. Am., Miss. Entomol. Soc., Sigma Xi. Research, numerous publs. on cotton insect control including boll weevil, pink bollworm, bollworm, evaluation of econ. levels of pest populations, diapause in boll weevil and diapause control, insecticide resistance studies of cotton pests, devel. diets and procedures for rearing cotton prests and biol. control. Home: Drawer M, State College, Miss. 39762.*

BREASTED, James Henry, Am. Egyptologist; b. Rockford, Ill., Aug. 27, 1865; s. Charles and Harriet N. (Garrison) B.; A.B., N. Central Coll., Naperville, Ill., 1888; postgrad. Chgo. Theol. Sem., 1888-90, hon. B.D., 1898; A.M., Yale, 1892; A.M., Ph.D., U. Berlin, 1894; LL.D., U. Cal., 1918, Princeton, 1929; Litt.D., Oxford, 1922; m. Frances Hart, Oct. 22, 1894; children—Charles, James Henry. Astrid. Became asso. with U. Chgo., 1894, asst. in Egyptology, 1894-96, asst. dir. Haskell Oriental Museum, 1895-1901, dir., 1901-31, instr. Egyptology and Semitic langs., 1896-98, asst. prof., 1898-1902, asso. prof., 1902-05, dir. Egyptian expdn., 1905-07, prof. Egyptology and Oriental history, 1905-33, collector for univ. in Egypt, 1894-95, chmn. dept. Oriental langs. and lits., 1915-33, first expdn. of Oriental Inst. in Egypt and Western Asia, 1919-20; relieved of all responsibility for instruction after 1925, in order to take full charge of work in Oriental Inst. in the Near East and related research projects; dir. Oriental Inst., from 1919. Collaborator on Egyptian Dictionary at Berlin, 1899-1900; apptd. 1900, on mission to museums of

Europe by commn. of Royal Acads. of Germany (Berlin, Leipzig, Munich and Göttingen), to copy and arrange Egyptian inscriptions in those museums for the Berlin Egyptian Dictionary, 1900-01. Recipient gold medal, Geog. Soc. Chgo., 1929; Rosenberger gold medal for contbn. to history of civilization, 1929; gold medal, Holland Soc. of N.Y., 1930. Pres. Am. Oriental Soc., 1918, History of Science Soc., 1926, Am. Hist. Assn., 1928; mem. Nat. Acad. Sci. Author: De Hymnis in Solem sub Rege Amenophide IV Comceptis, 1894; Erman's Egyptian Grammar (English edit.), 1894; A New Chapter in the Life of Thutmose III, 1900; The Battle of Kadesh, 1903; Egypt Through the Stereoscope, 1905; A History of Egypt, 1905, 2d edit., 1909; also fgn. edits. Ancient Records of Egypt, Historical Documents (5 vols.), 1906; The Temples of Lower Nubia, 1906; A History of the Ancient Egyptians, 1908; The Monuments of Sudanese Nubia, 1908; Development of Religion and Thought in Ancient Egypt, 1912; Outlines of European History, (with J. H. Robinson), 1914; Short Ancient History, 1914-15; Ancient Times—A History of the Early World, 1916, revised edit., 1934; Survey of the Ancient World, 1919; History of Europe, Ancient and Medieval (with J. H. Robinson), 1920; General History of Europe (with Robinson and Smith), 1921; The Oriental Institute—A Beginning and a Programme, 1922; Oriental Forerunners of Byzantine Painting, 1924; Conquest of Civilization, 1926; Edwin Smith Surgical Papyrus (2 vols.), 1930; The Oriental Institute (vol. XII of U. Chgo. Survey), 1933; The Dawn of Conscience, 1933. Asso. editor Am. Jour. Semitic Langs. Authority on history of Egypt; conducted archeological investigations in Egypt and Near East; established research lab to study and interpret archeological finds. Died Dec. 2, 1935.

BREATHNACH, Aodan Seosamh, anatomist; b. Roscrea, Ireland, Aug. 20, 1922; s. Michael and Nora (O'Halloran) B.; B.Sc., U. Coll., Dublin, Ireland, 1943, M.Sc., 1945, M.B., B.S., 1947, M.D., 1957; m. Rigmor Eyde, Aug. 9, 1949; children—Stephen Michael, Richard Stuart. Lectr., U. Coll., Dublin, 1945-46; sr. house physician Mater Hosp., Dublin, 1947-48; house physician St. Luke's Hosp., Guildford, Surrey, Eng., 1948-49; faculty St. Mary's Hosp. Med. Sch., U. London (Eng.), 1949-—, reader in anatomy, 1961-67, prof., 1967-—, mem. acad. bd., 1966-—, bd. studies in human anatomy U. London, 1956-—. Recipient Margaret Gordon Research prize Wright-Fleming Inst. Microbiology, 1961. Mem. Anat. Soc. Gt. Britain, Brit. Med. Assn., Assn. U. Tchrs. Author: Frazer's Anatomy of the Human Skeleton, 1958; also articles. Research on nervous system of aquatic mammals, histology and electron microscopy of human adult and fetal skin. Home: 17 Cranbook Dr., Esher, Surrey, Eng. Office: St. Mary's Hosp. Med. Sch., London W.2, Eng.*

BRÉAUTÉ, see de Bréauté.

BRECHER, Gerhard Adolf, physiologist; b. Goldap, Germany, June 14, 1909; s. Otto Ernst and Hedwig (Wulst) B.; M.A., Duke, 1930; Ph.D., U. Hamburg, 1932; M.D., U. Kiel, 1937; m. Eleanor Baker, Apr. 23, 1941; children—Armin G., M. Herbert, Elisabeth E. Practice medicine, specializing in physiology, Cleve., 1948-55, Columbus, O., 1955-57, Atlanta, 1957-67; faculty Sch. Medicine Western Res. U., 1948-55; prof. depts. physiology, ophthalmology, dir. Inst. For Research Vision, Coll. Medicine Ohio State U., 1955-57; prof., chmn. dept. physiology Emory U., Sch. Medicine, 1957-67; cons. Gen. Electric Co., Cleve., 1954-67; prof. dept. physiology U. Okla. Med. Center, Oklahoma City, 1967-—. Mem. Am. Physiol. Soc., Assn. Research in Ophthalmology, Am. Soc. Internal Artificial Organs, Am. Heart Assn., Am. Coll. Cardiology. Author: Venous Return, 1956; Heart Lung Bypass, 1962. Research, publs. on basic concepts of atrial hemodynamics, effect of ventricular diastolic suction on ventricular filling, hemodynamics of venous return as affected by normal and artificial respiration; devel. various blood flowmeters; contbr. to heart-lung bypass techniques, measurement of power of normal and abnormal binocular fusion, evaluation of aniseikonia, optokinetic nystagmus. Home: 7708 Rumsey Rd. Oklahoma City 73132. Office: U. Okla. Med. Center, Oklahoma City 73104.*

BRECHLING, Frank Paul Richard, economist; b. Wismer, Germany, Dec. 6, 1931; s. Heinrich and Eva (Block) B.; B.A., Trinity Coll. Dublin (Ireland), 1955; m. Brigid Gertrude Quinlan, Feb. 7, 1959; 1 dau., Vanessa Jane. Asst. lectr. U. Liverpool (Eng.), 1956-62; economist Nat. Econ. Devel. Office, London, 1962-64; economist Nat. Inst. Econs. and Social Research, London, 1964-65; vis. asso. prof. Mass. Inst. Tech., 1965-66; prof. econs. Northwestern U., Evanston, Ill., 1966-—, chairman of the department of econs., 1967-—. Mem. Royal Econ. Soc., Econometric Soc. Editor: (with F. H. Hahn) The Theory of Interest Rates, 1965. Research, publs. on monetary theory and policy, application modern statis. techniques to implementation of econ. policy. Office: Dept. Econs., Northwestern U., Evanston, Ill. 60201.*

BRECHT, Walter Hermann Paul, German paper technologist; b. Augsburg, Germany, June 29, 1900; s. Berthold and Sophie (Brezing) B.; Dipl.-Ing., Technische Hochschule Darmstadt, 1924, Dr.-Ing., 1925; m. Elizabeth Volz, Apr. 9, 1927; children—Ingeborg (Mrs. Fritz Luhde), Britta (Mrs. Niels Roeder). Staff, Hammermill Paper Co., Erie, Pa., 1925-26; mill mgr. G. Haindl'sche Papierfabriken Augsburg (Germany), 1926-31; prof. Technische Hochschule Darmstadt (Germany), 1931-—, dir. Institut für Papierfabrikation, 1931-—. Mem. Verein der Zellstoff und Papier-Chemiker und Ingeniure (Alexander Mitscherlich meml. medal 1936), Royal Swedish Acad. Engring. Scis. Stockholm (corr.). Author: (with Zippel) Wege zur Verbesserung des Wasserhaushalts von Papierfabriken, 1963; also numerous articles, chpt. in book. Research in paper tech.; invention and devel. measuring instruments for phys. testing of cellulose, wood pulp and paper, also for improvement waste water from paper factories. Home: 59 Ohlystrasse, 61 Darmstadt, Germany.*

BRECHTEL, Stephan, German calligrapher, mathematician; b. Bamberg, Germany 1523; s. Stephan Brechtel; pupil of calligrapher Johann Neudorfer, Nuremberg, Germany, 1540; later math. studies in Leipzig; m. Veronica Heyden, Aug. 29, 1548; 4 sons including Christoph Fabius, Franz Joachim, Stephan; 7 daus. including Katherine. Master gunner in siege of Leipzig, 1546-47; dir. sch. for writing and arithmetic, Nuremberg; became citizen, 1548; in Bamberg, 1561-63; councilman in Nuremberg, 1563. Attempted to improve German letters proportionally by using geometry; wrote on math. and field surveying; edited valuable edition of 15 books of Euclid. Died Nuremberg, June 27, 1574.

BRECKENRIDGE, Lester Paige, Am. mech. engr.; b. Meriden, Conn., May 17, 1858; s. Moses Paige and Lucretia L. (Wetherell) B.; Ph.B., Sheffield Sci. Sch., Yale, 1881; M.A., Yale 1909; Eng.D., U. Ill., 1910; m. May Brown, Dec. 19, 1883; children—Blanch F. (Mrs. Henry B. Dirks), Gladys S. (Mrs. Earl D. Finch), May H. (Mrs. D. B. Luckenbill); m. 2d, Susan W. Ford, July 26, 1911. Instr. mech. engring. Lehigh U., 1882-91, except 2 yrs. engaged in engring. work; prof. mech. engring. Mich. Agrl. Coll., 1891-93; prof. mech. engring. U. Ill., 1893-1909, dir. engring. expt. sta., 1905-09; prof. mech. engring. Sheffield Sci. Sch., Yale, 1909-23. Engr. in charge boilder div. U. S. Geol. Survey Fuel Testing Plant, St. Louis, from 1904. Mem. Am. Soc. M.E. (v.p. 1907-09), Soc. Promotion Engring. Edn., Western Soc. Engrs. (v.p. 1905-06), Am. Soc. Heating and Ventilating Engrs. (hon.). Writer articles in tech. jours., reports, bulls. Contrived and equipped dynamometer cars, 1897-99; invented automatic recording machine, 1901; also research on smokeless combustion and power prodn. Died N. Ferrisburg, Vt., Aug. 22, 1940.

BRECKENRIDGE, Robert George, Am. physicist; b. Jamestown, N.Y., Nov. 14, 1915; s. Robert and Catherine (Hilliard) B.; B.A., Cornell U., 1938, M.A., 1940; Ph.D., Mass. Inst. Tech., 1942; m. Alma Natalie Strout, Mar. 7, 1947. Research asso. Mass. Inst. Tech., Cambridge, 1942-46, asst. prof. elec. engring., 1946-49; chief solid state physics sect. U. S. Nat. Bur. Standards, Washington, 1949-55; head physics br. USN Office Naval Research, Washington, 1953-54; dir. research Nat. Carbon Co. div. Union Carbide Corp., Parma, O., 1955-59, dir. research Parma Research Lab., 1959-63, dir. research Union Carbide Corp. Research Inst., Tarrytown, N.Y., 1963-64; dir. physics dept. Atomics Internat. div. N.Am. Aviation, Canoga Park, Cal., 1964-—. Vice chmn. NRC Conf. on Elec. Insulation, 1954, chmn., 1955; mem. vis. com. for phys. sci. Western Res. U., 1960-67. Fellow Am. Phys. Soc.; mem. Washington Acad. Sci., N.Y. Acad. Scis., Am. Chem. Soc., Chemists Club of N.Y.C., Sigma Xi, Alpha Chi Sigma. Research on dielectric relaxation effects in ionic crystals, 1948, high dielectric constant ceramics, 1944, semicondrs., 1951, titanium oxide rectifiers, 1952, semiconducting compounds, 1953. Editor: Photoconductivity, 1956. Home: 19252 Kinzie St., Northridge, Cal. Office: 8900 DeSoto St., Canoga Park, Cal.*

BREDERECK, Hellmuth, German chemist; b. Frankfort/Main, Germany, May 29, 1904; s. Richard and Ottilie (Weber) B.; student chemistry U. Frankfort (Germany), 1922-25; doctorate U. Greifswald (Germany), 1927; m. Elisabeth Niedergerke, Nov. 14, 1933; children—Karl, Hans-Joachim, Peter-Michael. Asst., U. Greifswald, 1927-30, with U. Leipzig (Germany), 1930-41, prof., 1939-41; prof., dir. Institut für Organische Chemie und Biochemie, U. Jena (Germany), 1941-45; staff chem. lab., Heidenheim/Brenz, Germany, 1945-47; guest prof. U. Stuttgart (Germany), 1947; dir. Inst. für Organische Chemie, prof. Technische Hochschule Stuttgart, 1948-—, dean faculty arts and scis., 1956-58, rector, 1959-61. Mem. Sci. Council, 1964-67; chmn. sci. Commn., 1964-—. Recipient Emil-Fischer medal Gesellschaft Deutscher Chemiker, 1966. Mem. Am. Chem. Soc., Gesellschaft Deutscher Chemiker (pres.). Author: Vitamine und Hormone, 1936; also numerous articles. Research on nucleic acids, purines, pyrimidines, synthesis in the heterocyclic series, acid amides, new catalysts for polymerization and auto-oxidation; research and syntheses in the carbohydrate series. Home: 149 Zeppelinstrasse, 7000, Stuttgart-West, Germany. Office: 14 Azenbergstrasse, 7 Stuttgart-Nord, Germany.*

BREDICHIN, Fedor Aleksandrovich, Russian astronomer; b. Nicolaev, Russia, Nov. 26, 1831; s. Aleksander Bredichin; ed. Odessa, also Moscow, Russia. Prof. U. Moscow, from 1857; dir. obs., 1873-90; dir. obs., Pulkovo, Russia, 1890-95. Author: Untersuchungen über die Kometenformen, 1903; Études sur l'origine des météores cosmiques et la formation de leur courants, 1903. Investigated comet forms in connection with his theory of meteors; described structure of meteors; distinguished 3 kinds of comet tails according to substance, length, curvature. Died St. Petersburg, Russia, May 14, 1904.

BREDIG, Georg, chemist; b. Glogau, Poland, Oct. 1, 1868; asst. to Ostwald in Leipzig, Germany; became prof. phys. chemistry Zürich (Switzerland) Poly., 1910; prof. phys. chemistry and electrochemistry Karlsruhe (Germany) Hochschule, 1911. Pioneer in research on anomalous atomic weight of lead from different sources; studied catalytic action, especially that of colloidal platinum and poisoning of catalysts; developed method of preparing colloids by elec. disintegration (Bredig method); investigated relation of ionic mobilities to other phys. and chem. properties. Died N.Y., Apr. 24, 1944.

BREDON, Simon, English mathematician, astronomer, physician; b. Winchecombe, Eng.; M.D., Oxford (Eng.) U., 1330; fellow Balliol, later Merton. Author: Tabulae chordarum, Calculationes chordarum (both on trigonometry); Arithmetica theoretica; Exposition in quaedam captia Almagest; Tabula declinationis solis; Theoretica plaetarum; Oxford almanack for 1344. Followed Merton tradition in writings; made large collection books and instruments. Died after 1368.

BREDT, (Konrad) Julius, German chemist; b. Berlin, Germany, Mar. 29, 1855; s. August and Amalie v.d. (Leyen) B.; studied in Leipzig, Frankfurt (both Germany); doctorate, Strasbourg (Germany, now France), 1880; 1 adopted dau. Teaching asst. in Strasbourg until 1882; with dyestuffs and dyeing industry several years; joined faculty in organic chemistry, Bonn, Germany, 1889; named prof. Technische Hochschule Aachen (Germany), 1897. Co-editor: Journal für praktische Chemie. Contbr. numerous articles to Annalen der Chemie, Berichte der Deutschen Chemischen Gesellschaft, Journal für praktische Chemie. Interested in compounds of fat series, particularly fat acids; discovered and proved camphor formula, 1893; new important concepts in streochemistry, Bredt's Law. Died Aachen, Sept. 21, 1937.

BREED, Charles Blaney, Am. engr.; b. Lynn, Mass., Nov. 28, 1875; s. Charles Otis and Sarah (Guilford) B.; S.B., Mass. Inst. Tech., 1897; children—Charles Alfred (by 1st marriage), David Edson, Nancy Eleanor (by 2d marriage). Began with city engring. dept., Lynn, 1894; resident engr., Walden Pond Dam, Lynn; cons. engr. for state commns. on pub. utility projects, also cities and railroads on elimination of grade crossings and transp. econ. problems; prof. r.r. and hwy. transp. Mass. Inst. Tech., 1906-45, prof. emeritus, from 1945, head dept. civil engring., 1935-45; Mem. Am. Soc. C.E. (dir.), Am. Railway Engring. Assn., Am. Road Builders' Assn., N.E. Railroad Club (pres.), Chi Epsilon, Sigma Xi, Tau Beta Pi. Author: (with George L. Hosmer) Principles and Practice of Surveying (vols. I and II), 1906, 1908; Surveying, 1942; other publs. on hwy. transp. econs. Asso. editor American Civil Engineers' Pocketbook, 1911, American Mining Engineers' Handbook, 1913. Died Aug. 9, 1958.

BREED, Robert Stanley, Am. bacteriologist; b. Brooklyn, Pa., Oct. 17, 1877; s. Robert Fitch and Emma Marie (Beers) B.; B.S., Ph.D. Harvard, 1902; post-grad. U. Göttingen, 1910. U. Kiel, 1911; m. Louise Miller Helm, Aug. 23, 1899 (dec. Dec. 1905); 1 dau., Alice Fitch; m. 2d, Emma Margaret Edson, July 2, 1913. Instr. biology U. Colo., 1898-99; asst. in zoology Harvard, 1900-02; prof. biology Allegheny Coll., 1902-13, sec. faculty, 1907-10; bacteriologist N.Y. State Agrl. Expt. Sta., Cornell U., 1913-47, emeritus prof. bacteriology, from 1947; Vice pres. World Dairy Congress, Washington, 1923; London, 1928; del. Berne, 1914, Rome, 1934; permanent sec. Internat. Com. on Bact. Nomenclature, Internat. Assn. Microbiology and Internat. Bot. Congress, from 1930; del 2d Inter-Am. Conf. Agr., Mexico City, 1942. Fellow A.A.A.S. (council 1932-34), Am. Pub. Health Assn. (chmn. lab. sect. 1933-34); mem. Soc. Am. Bacteriologists (pres. 1927), Am. Dairy Sci. Assn., Am. Biol. Soc. Internat. Assn. Milk Sanitarians, Brit. Soc. Applied Bacteriology (corr.), Cuban Soc. Microbiologists, Geneva Hist. Soc. (pres. 1935-42), Phi Beta Kappa, Sigma Xi. Author bulls. and articles on biol. subjects, especially milk hygiene and systematic bacteriology. Editor in chief Bergey's Manual Determinative Bacteriology, from 1937; asso. editor Jour. of Bacteriology and of Biol. Abstracts. Died Geneva, N.Y., Feb. 10, 1956.

BREESE, Gerald William, Am. urban sociologist; b. Horseheads, N.Y., June 4, 1912; s. Bert M. and Leona B. (Goodrich) B.; A.B. cum laude, Ohio Wesleyan U., 1935; B.D., Yale, 1938; Ph.D. (Marshall Field fellow) U. Chgo., 1947; m. Alice Janette Bailey, July 4, 1937; children—Adele (Mrs. Robert Richards, Jr.), James B., Dana Sue Bailey, Brinda Sue Bailey. Mem. faculty Pacific U., Forest Grove, Ore., 1938-41, Shrivenham Am. U., Eng., 1945, U. Chgo., 1947; research planner Chgo. Plan Commn., 1942-43; sec. com. housing research Social Sci. Research Council, Washington, 1947-49; faculty Princeton (N.J.), 1949-—, prof. sociology, 1959-—, dir. Bur. Urban Research, 1950-66. Coordinator, sociologist cons. team for preparation master plan for Delhi, India, Ford Found., 1957-

58; vis. lectr. U. Natal, S. Africa, 1963; planning and research cons., 1950—. Recipient Demoblzn. award Social Sci. Research Council, 1946. Fulbright prof. Am. U. Cairo, 1954-55; Fulbright fellow Inst. Advanced Studies, Australian Nat. U., 1966. Mem. Am. Sociol. Assn., Am. Inst. Planners, Population Assn. Am., Am. Soc. Planning Ofcls., Eastern Sociol. Soc., Phi Beta Kappa. Author: Daytime Population of Central Business District of Chicago, 1949, Regional Analysis: Trenton-Camden Area, 1954; Industrial Site Selection, 1954; Urbanization in Newly Developing Countries, 1966; (with others) Accelerated Urban Growth in Metropolitan Fringe Area, 1954, Impact of Large Installations on Nearby Areas, 1966. Studies in socio-econ. regional analysis; pioneer in world-wide comparative analysis of structure and growth of large urban areas in newly developing countries. Home: 195 Russell Rd., Princeton, N.J. 08540.*

BREFELD, Oscar, German botanist, bacteriologist; b. Westphalia, Germany, Aug. 19, 1839; ed. univs. Halle, Munich, Würzburg (Germany); lectr. at Berlin, 1875-78; prof. univs. Eberswald, 1878-84, Munster, 1884-98 (both Germany), Breslau (now Wroclaw, Poland), 1898-1907. Noted for studies in mycology; discovered stigma of flowers infect wheat flowers with bunt spores; introduced new methods in mycological studies particularly use of gelatine cultures in bacteriology. Died Schlachtensee, Jan. 12, 1925.

BREGUET, Abraham Louis, inventor; b. Neuchatel Switzerland, Jan. 10, 1747; apprentice in Paris, also Versailles, France. Employed as mechanic and clockmaker; mem. Council Arts and Manufactures, during Restoration; mem. French Acad. Scis., 1816, Bur. Longitudes. Author: Essai sur les forces animales et sur le principe du mouvement voluntaire. 1811; Traité d'horlogerie (pub. under Moinet's name). Built Chappe's flügel telegraph, also precision stop watches, accurate chronometers, sympathetic pendulums, metal thermometers; improved watches by using ruby as bearing; perfected self-winding clocks. Died Paris, Sept. 17, 1823.

BREGUET, Antoine, French physicist; b. Paris, 1851; s. Louis Francois Clément; head Breguet workshops. Author: Telephone à ficelles machine gramme. Studied all phases of theoretical and applied electricity; research on string telephone; invented a mercury anephone, 1878. Died 1882.

BREGUET, Louis-Francois-Clement, French inventor; b. Paris, France, Dec. 22, 1804; s. Louis Breguet; clockmaker, mfr. elec. devices; dir. French navy's chronometer dept., from 1826; mem. Telegraphy Commn., 1845; mem. French Acad. Scis., 1874, Bur. Longitudes. Compared (with Fizeau) speed of light in air and water, 1830; built (with Masson) electric telegraph with 2 needles (tested in Rouen, France 1846-52), 1838-45; produced dial telegraph used by French ry.; invented system of electrical clocks for distant transmission of time; constructed device to control trains speeds, 1849, also device for determining speed of projectiles, 1850; determined (with Werthein) speed of sound in iron, 1851; built 1st telegraph line between Paris and Rouen. Died Paris, Oct. 27, 1883.

BREHM, Alfred Edmund, German biologist, naturalist; b. Renthendorf, Germany; Feb. 2, 1829; s. Christian Ludwig Brehm; ed. Jena, Germany, also Vienna, Austria; m. Math. Reiz, 1862; 2 sons, 3 daus. Made sci. voyage to N.E. Africa, 1847-52; in Leipzig, Germany, 1858, Norway, 1860; with Duke Ernest of Coburg-Gotha, in territory of Bogos, 1860-62; became dir. Zool. Gardens, Hamburg, Germany, 1862; helped found Berlin Aquarium, 1867, dir., 1867-74; traveled (with Finsch and Count Waldburg) to Occidental Siberia and Turkestan, 1877, also (with Crown Prince Rudolph of Austria) to Danubian area, 1878, Spain, 1879; later visited Am. Author: Reiseskizzen aus Nord-Ostafrica, 1855; Das Leben der Vögel, 1861; Ergesnisse einer Reise nach Habesch, 1863; Tierleben, 6 vols., 1864-69; Gefangene Vögel, 2 vols., 1872-76; (with Rossmassler) Die Tiere des Waldes, 2 vols., 1866-67. Research in ornithology and natural history. Died Renthendorf, Nov. 11, 1884.

BREHM, Christian Ludwig, German ornithologist; b. Schönau, nr. Gotha, Germany, Jan. 24, 1787; s. Carl August and Sophia Christiana (Heimberger) B.; ed. Jena (Germany) U., 1807-09, Dr. Med. (hon.), 1856; m. Amalie W. Wachter, 1813; 8 children, including Alfred Edmund. Ch. pastor, Renthendorf, Germany, from 1812. Mem. Leopoldina (Halle, Germany). Author: Beiträge zur Vögelkunde, 1820-22; Handbuch der Naturgeschichte aller Vögel Deutschlands, 1831; Der vollständige Vogelfand, 1855. Worked in classical ornithology; founder collection 15,000 birds (donated by Walter Rothschild to N.Y. mus., 1930). Died Renthendorf, June 24, 1864.

BREHM, John Joseph, Jr., Am. physicist; b. Memphis, Dec. 6, 1934; s. John Joseph and Mattie (Thornell) B.; B.S., U. Md., 1956, Ph.D., 1963; M.S., Cornell U., 1959; m. Mary Ellen Kempers, Nov. 26, 1959; children—John Joseph, Robert Dennis, Richard Matthew. Physicist, U. S. Naval Ordnance Lab., Silver Spring, Md., 1958-62; faculty Northwestern U., Evanston, Ill., 1963-67, asso. prof. physics, 1966-67, asso. prof. U. Mass., Amherst, since 1967—. NSF fellow, 1962-63. Mem. Am. Phys. Soc. Research in theory of strong interactions of elementary particles, particularly devel. application of methods of analytic function theory and symmetry theory to understanding of basic forces responsible for occurrence of known particles and resonant states. Home: 8 Applewood Lane, Amherst, Mass. 01002.*

BREHM, Vincent, Austrian zoologist; b. Duppau, Jan. 1, 1879; s. Alöis and Marie (Nonner) B.; Ph.D., U. Innsbruck; m. Grete Hübel, Nov. 13, 1915. Zoologist biology sta., Lunz, Austria. Mem. Internat. Soc. Limnology, Acad. Sci. of Prague (corr.). Research and numerous publs. on entomostraces, including Einführung in die Limnologie; Copepoda und Branchiura. Address: Biological Sta., Lunz am See, Austria.

BREHMER, Hermann, German physician; b. Kurtsch, Germany, Aug. 14, 1826; founder sanatorium (model establishment in Europe), Görbersdorf, Germany, 1859. Author: Die Therapie der Lungenschwindsucht, 1887. Introduced balneotherapy for treatment Tb, 1836. Died Gorbersdorf, Dec. 22, 1889.

BREININ, Goodwin Milton, Am. physician; b. N.Y.C., Dec. 10, 1918; s. Louis and Mary (Mirsky) B.; B.S., U. Fla., 1939; M.A., Emory U., 1940, M.D., 1943; m. Rose-Helen Kopelman, June 22, 1947; children—Bartley James, Constance. Practice medicine, specializing in ophthalmology, N.Y.C., 1947—; dir. eye service U. Hosp., Bellevue Hosp. (both N.Y.C.), 1959—; hon. research asso., dept. physiology U. Coll., London, Eng., 1966-67; cons. N.Y. Eye and Ear Infirmary, Manhattan Eye, Ear and Throat Hosp., St. Vincent's, St. Clare's, French hosps. (all N.Y.C.), USPHS Hosp., Stapleton, N.Y., Manhattan VA Hosp., N.Y.C. Daniel B. Kirby prof., chmn. dept. ophthalmology N.Y. U. Sch. Medicine, N.Y.C., 1959—; cons. Lenox Hill Hosp.; member various coms. NRC, USPHS, NIH.; mem. med. adv. com. Myastenia Gravis Found., Nat Council to Combat Blindness. Recipient Knapp medal sect. on ophthalmology A.M.A., 1957; Holmes citation and award Inst. Medicine Chgo., 1959. Fellow Am. Acad. Ophthalmology and Otolaryngology (instr. home study courses), A.C.S., N.Y. Acad. Medicine (chairman of the section on ophthalmology); mem. A.M.A. (sec. sect. on ophthalmology 1966—), Am., N.Y. ophthal. socs., Assn. for Research in Ophthalmology, Pan Am. Assn. Ophthalmology, Assn. for Research in Strabismus, Société Francaise D'Ophthalmology, Eye Study Club, Harvey Soc., A.A.A.S., Am. Orthoptic Council, N.Y. State, N.Y. County med. socs., Sigma Xi, Alpha Omega Alpha. Author: Electrophysiology of Extra Ocular Muscle, 1962. Mem. editorial bds. Archives Ophthalmology, Investigative Ophthalmology. Research, publs. on electromyography of eye muscles; methods for control of glaucoma; physiology of eye. Home: 912 Fifth Av., N.Y.C. 10021. Office: 560 1st Av., N.Y.C. 10016.*

BREISKY, August, gynecologist, obstetrician; b. Klattau, Bohemia, Mar. 25, 1832; s. Vincenz Anton and Elisabeth v. (Hasslinger) B.; studied medicine in Prague, Czechoslovakia, then in pathol.- anat. inst. of Wenzel Treitz, Prague, 1855; m. Pauline von Less; 1 son, Walter. Asst. in Prague obstet. clinic, 1858; prof. obstetrics and gynecology in Salzburg, Austria, 1866-67, Bern, Switzerland, 1867-74, Prague, 1874-86, Vienna, Austria, 1886. Contbr. articles to med. jours., handbooks. Developed method for exact measurement of pelvis; described malformations in female genitalia; one of most successful gynecol. surgeons of his time. Died Vienna, Austria, May 25, 1889.

BREISLAK, Scipio, Italian geologist; b. Rome, Italy, 1748; ed. Ragusa, Italy. Prof. Nazarene Coll., Rome; engaged in mining, Naples; tchr. Royal Mil. Coll., Naples, 1802-26; govt. insp. nat. manufacture saltpetre and gunpowder. Mem. Italian Inst., Milan. Author: Saggio d'Osservasioni; Topographia Fisica della Campagne; Voyages dans la Campanie; Introduzione alla Geologia, 1811. Engaged in improving manufacture of saltpetre. Died 1826.

BREIT, Franz Xaver, gynecologist; b. Mieders, Austria, July 1, 1817; s. Michael and Anna (Hartler) Preith; ed. Vienna, Prague, Padua, 1837-41; m. Helene Henriette Volz, 1848; 2 sons, 2 daus. Became asst. 1. Wiener Gebarklinik (Vienna birth clinic), 1844; apptd. asso. prof. obstetrics, Tübingen, Germany, 1847, prof., dir. obstet. clinic, 1848. Studied small pelvis and its dangers in birth process; pioneer in artificial induction premature birth (rather than caesarean sect.); probably 1st in Germany to use chloroform in childbirth. Died Tübingen, Aug. 17, 1868.

BREIT, Gregory, physicist; b. Russia, July 14, 1899; s. Alfred and Alexandra (Smirnova) B.; came to U. S., 1915, naturalized, 1918; A.B., Johns Hopkins, 1918, A.M., 1920, Ph.D., 1921; D.Sc., U. Wis., 1954; m. Marjorie MacDill, Dec. 30, 1927. NRC fellow U. Leyden (Netherlands), 1921-22, Harvard, 1922-23; asst. prof. physics U. Minn., 1923-24; math. physicist, dept. terrestrial magnetism Carnegie Inst. of Washington, 1924-29, research asso., 1929-44; prof. physics N.Y. U., N.Y.C., 1929-34, U. Wis., Madison, 1934-47; prof. physics Yale, 1947-58, Donner prof., 1958—; resident Technische Hochschule, Zurich, Switzerland, 1928; Vis. mem. Inst. for Advanced Study, Princeton, N.J., 1934-35; with Degaussing Naval Ordnance Lab., Washington, 1940-41; sect. mem. OSRD, NDRC, 1940-42; co-ordinator fast neutron-project, metall. lab. U. Chgo., 1942; with radiation lab. Johns Hopkins, 1942-43; with Aberdeen Proving Ground., 1943-45; mem. at large, div. phys. scis. NRC, 1932-35, rep. Am. Phys. Soc., 1938-41; chmn. reference com. on isotopes, 1940-44. Recipient Franklin medal Franklin Inst. Phila., 1964. Fellow I.E.E.E., Am. Phys. Soc., A.A.A.S., Geophys. Union, Phys. Soc. London, Am. Acad. Arts and Scis.; mem. Nat. Acad. Scis. (asso. editor Proc. 1958-60). Asso. editor Phys. Rev., 1927-29, 39-41, 54-56, 61-63, Il Nuovo Cimento, 1964—. Author: Handbuch der Physik, vol. 41, 1959. Research in nuclear physics; quantum theory; quantum electrodynamics; hyperfine structure; ionosphere studies. Office: Sloane Physics Lab., Yale, New Haven 06520.

BREITENBACH, Johann Wolfgang, Austrian phys. chemist; b. Vienna, Austria, June 22, 1908; s. Johann and Amalie (Dirnbacher) B.; Ph.D., U. Vienna; m. Poldi Mandl, Dec. 20, 1942; children—Renate, Ilse. Asst., instr., later dir. Mem. Austrian Assn. Chemists, Viennese Chemico-Physics Soc., German Chemistry Soc. Research and publs. on kinetics of formation of macromolecular substances, determination of molecular weights, distbn. of molecular weights of high polymers, initiation of polymerization by electrolysis, emulsion, proliferation. Home: Kreindlgasse 23, 1190 Vienna, Austria.*

BREITENBERGER, Ernst, physicist; b. Graz, Austria, June 11, 1924; s. Julius and Anna (Wiesinger) B.; student U. Graz, 1946-47; Dr.phil., U. Vienna (Austria), 1950; Ph.D., Cambridge (Eng.) U., 1956; m. Janine Dufaure, Jan. 28, 1954; children—Roland, Caroline, Gisela, Eric. Research asso. Institut für Radiumforschung, Vienna, 1950-51; Brit. Council research scholar Cavendish Lab., Cambridge, 1951-54; lectr. U. Malaya, Singapore, 1954-58; faculty U. S.C., 1958-63, prof. physics, 1960-63; prof. physics, Ohio U., Athens, 1963—. Mem. Math. Assn. (London), Am. Phys. Soc. Research, publs. on accelerators, radiation detectors, nuclear decay, statistics, stochastic processes in physics. Home: 47 Mulligan Rd., Rural Route 4, Athens, O. 45701.*

BREITENSTEIN, Niels Jakob, Danish archeologist; b. Skellerup, Apr. 22, 1904; s. Vilhelm and Maren (Nielsen) B.; ed. U. Copenhagen; m. Thora Favrholdt, Oct. 24, 1930; children—Thorkild, Jorgen. Curator, Nat. Mus. of Copenhagen, 1927, dir. antiquities collection, 1949, chief curator sect. Oriental and classical antiquities, 1952-61. Mem. Soc. for History Danish Civilization, German Archeol. Inst. (corr.). Author: Catalogue des terres cuites, 1941; Sylloge Nummorum Graecorum, (Royal Collections of Coins and Medals), 1942-48; Christian VIII's Vase-cabinet, 1951; Hans West, 1955. Address: Fjords Allé 16 N.J., Copenhagen, Denmark.

BREITHAUPT, Friedrich Wilhelm, German inventor; b. Kassel, Germany, July 23, 1780; s. Johann Christian and Susanne Margarete (Strack) B.; m. Christiane Frömbling, 1803; 4 sons including Georg August, Wilhelm Friedrich, 5 daus.; m. 2d, Jeanette Gissot, 1844. Apprenticeship in Freiberg, Germany; entered father's workshop (eventually Instrumenten-Fabrik F. W. Breithaupt und Sohn) for math., astron. and surg. instruments, 1798; named ofcl. Hessian mining mechanic (Bergmechanikus), 1806, Hessian ct. mechanic and mint master, 1814. Recipient golden medal of prince-elector of Hessia for circular dividing machine, 1824. Mem. Saxon Econ. Soc. (hon.). Improved Howgre level instrument, 1810; invented differential screw, 1826; invented vernier cover, 1827. Died Kassel, Germany, June 20, 1855.

BREITHAUPT, Johann August Friedrich, German mineralogist; b. Probstzella, Germany, May 18, 1791; s. Friedrich Gottlob and Anna Elisabeth Emanuela (Ehrlicher) B.; student math. and sci., Jena, Germany, 1809-11; studied geology under Werner, Freiburg, Germany; hon. doctor, Jena, Marburg; m. Agnes Ulricke Winkler, 1816; 3 daus. Insp. acad. collections, also adj. prof. Acad. Mines, Freiburg, 1813-26; prof. mineralogy Acad. Mines, 1826-66. Mem. acads. of Göttingen, Germany, Munich, Florence, Italy, Madrid, Halle, Germany, also numerous others. Antimonickel renamed breithauptite in his honor. Author: Ueber die Echtheit der Krystalle, 1816; Vollstandige Charakteristik des Mineralsystems, 1820; Die Bergstadt Freiberg, 1825; Ubersicht des Mineralsystems, 1830; Die Paragenesis der Mineralien, 1849; Mineralogischer Studien, 1866; Vollständiges Handbuch der Mineralogie, 3 vols., 1836-47. Described pseudomorphism; independently discovered 47 mineral species; originated law of paragenesis in minerals; made over 3,000 measurements of calcite; increased knowledge of crystallography; studied amorphous crystal structure, mineral. nomenclature. Died Freiburg, Sept. 22, 1873.

BREITKOPF, Johann Gottlob (Immanuel), German typographer, publisher; b. Leipzig, Germany, Nov. 23, 1719; s. Bernard Christoph B. and Maria Sophia (Hermann) B.; univ. studies under Johann Christoph Gottsched, Johann Friedrich Christ and Johann Jakob Mascow; m. Friderica Constantia Brix, Sept. 25, 1746; 5 sons including Bernhard Theodor, Christoph Gottlob, 4 daus. including Theodora Sophia Constantia. Entered father's printing shop, 1745, also father's pub. house, 1762. Mem. Gesellschaft der freien Künste, Ökonomische Societät (in Leipzig). Author: Nachricht von der Stempelschneiderei und Schriftgiesserei, 1777;

Über den Druck der geographischen Charten, 1777; Über die Geschichte der Erfindung der Buchdruckerkunst, 1779 (Fragment); Versuch, den Ursprung der Spielkarten . . . zu erforschen, Teil I, 1784; Über Buchdruckerei und Buchhandel, 1793; Uber Bibliographie und Bibliophilie, 1793 (all small works; magnum opus on history of art of printing never finished). Attempted to open new avenues of relief printing; created Britkopf-Fraktur (famous example of classical German books); made typog. improvements in music note printing known in music world for its clean, seamless image; experimented in printing maps but not commercially successful. Died Leipzig, Germany, Jan. 28, 1794.

BREITNER, Burghard, Austrian physician, surgeon; b. Mattsee, Salzburg, June 10, 1884; ed. Eiselsberg clinic, Vienna; Us. Graz, Kiel. Asst., Eiselsberg Clinic; lectr., prof., dir., Rudolfsspital, Vienna; prof., Innsbruck. Pres., Austrian Red Cross; mem., Soc. of Physicians, Vienna; Assoc. of Surgeons, Switz.; Assoc. Gastro-Enterolog., Belgium; Assoc. of Surgeons, Italy; Am. Goiter Assoc. Research on goiter problems, blood transfusion, bisexuality, sport medicine; published 22 handbooks and numerous articles. Died Innsbruck, Austria, Mar. 28, 1956.

BREITWIESER, Joseph Valentine, Am. psychologist; b. Jasper, Ind., Mar. 31, 1884; s. John Conrad and Katherine Elizabeth (Baitz) B.; A.B., Ind. U., 1907; A.M., 1908; Ph.D., Columbia, 1911; m. Ruth Fowler, Jan. 1910; children—Charles John, Katherine Rebecca, Janice, Joseph Valentine, Roland. Tchr. pub. schs., Tipton County, Ind., 1901-03; asst. at Ind. U., 1906-08; asst. in psychology Columbia, 1908-10; asst. prof. psychology and edn. Colo. Coll., 1910-11, prof., 1911-18; asso. prof. edn. U. Cal., 1918-27; prof. edn., dir. grad. div., and dean Sch. of Edn., U. N.D., Grand Forks, from 1927. Fellow A.A.A.S.; mem. Am. Psychol. Assn., Am. Genetic Assn., Am. Social Hygiene Assn., Nat. Inst. Social Sciences, Phi Delta Kappa, Sigma Xi. Author: Attention and Movement to Reactions, 1911; Psychological Experiments, 1914; Psychol. Advertising, 1915; Psychological Effects of Altitude, 1917; Psychological Education, 1925; The Education of the Emotions. Research and publs. in psychometry and edn. Died Mar. 7, 1950.

BREKHOVSKIKH, Leonid Maksimovich, Russian physicist; b. 1917; grad. Perm U., 1939; D.Physico-Math. Sci. Prof., Moscow U., 1953—; dir. Acoustics Inst., USSR Acad. Sci., 1954—, bur. mem. dept. physico-math. sci., 1960—. Del., Geneva Conf. Tech. Experts for Study Means Detecting Nuclear Explosions 1958. Recipient Stalin prize, 1951. Mem. USSR Acad. Sci. (corr.). Research and publs. on acoustics and theory wave propagation; scattering of X-rays in crystals and liquids. Address: Acoustics Inst., ulitsa Televideniya 4, Moscow, USSR.

BRELAND, Osmond Philip, Am. zoologist; b. Decatur, Miss., Sept. 17, 1910; s. Oscar Phillips and Lida Ruth (Adams) B.; B.S., Miss. State U., 1931; Ph.D., Ind. U., 1936; m. Virginia Nell Ellington, Aug. 4, 1931; children—Osmond Philip, William Michael. Instr., N.D. State U., 1936-38; faculty U. Tex., Austin, 1938—, prof. zoology, 1950—. Fellow Entomol. Soc. Am.; mem. Am. (regional dir. 1958-62), Tex. (pres. 1964) mosquito control assns., Am. Soc. Zoologists, Sigma Xi. Author: Animal Facts and Fallacies, 1948; Animal Friends and Foes, 1957; Animal Life and Lore, 1963; (with W. G. Whaley, others), Principles of Biology, 1964; (with A. Lee) Biology in the Laboratory, 1965; also numerous articles. Research on biology and ecology insects, spermatogensis insects, chromosomes mosquitoes. Home: 3604 Meredith St., Austin, Tex. 78703.*

BRELOT, Marcel Emile, French mathematician; b. Chateauneuf-sur-Loire, Dec. 29, 1903; s. Emile and Marguerite (Houy) B.; ed. Ecole Normale Supérieure; agrégé, 1927, Doctorat d'Etat, 1931; m. Alice Bautrant, 1935; children—Claude, Alain. Instr., lectr. at Algiers, 1933-38; titular prof. at Bordeaux, 1938-42, Grenoble, 1942-53, Paris, 1953—; tchr. Ecole Normale Supérieure, 1954-62. Mem. French, Italian, Am. math. socs. Author: Principes mathématiques de la mécanique classique, 1945; Éléments de la théorie classique du potentiel, 1959; Lectures on Potential Theory, 1960; Axiomatique des fonctions harmoniques, 1966; also numerous articles. Research in potential theory; notion of thinness, convergence theorem for superharmonic functions and potentials, resolutivity theorem for the Dirichlet problem, axiomatic theory of harmonic functions. Home: 3, rue Ernest-Cresson, Paris 14. Office: Institut H. Poincaré, 11, rue Pierre-Curie, Paris 5, France.*

BREM, Walter Vernon, Am. physician; b. Charlotte, N.C., Nov. 5, 1875; s. Walter Vernon and Hannie (Caldwell) B.; student Va. A. and M. Coll., 1899-90, Trinity Coll., Durham, N.C., 1890-92; B.S. U. N.C., 1896; postgrad. Union Theol. Sem., N.Y., 1897-98, U. N.C. Sch. Medicine, 1898-1900; M.D., Johns Hopkins, 1904; m. Marion Wolcott Winkler, Aug. 5, 1905; children—Phyllis, Lorna, Gwendolyn, Thomas Hamilton, Frederick Winkler, Walter Vernon. Med. house officer Johns Hopkins Hosp., 1904-05; physician and acting pathologist Ancon Hosp., C.Z. 1905-07; chief of med. clinic Colon Hosp., C.Z. 1907-11; prof. pathology and bacteriology. Los Angeles dept. U. Cal. Sch. Medicine, 1911-14; mem. Brem, Zeiler & Hammack, lab. of clin. pathology, Los Angeles, 1911—. Dir. so. div. Calif. Hygienic Lab., 1913-16; mem. Calif. Bd. Med. Examiners, 1923-25; cons. epidemiologist Calif. Bd. Health, 1924-1932. Died Nov. 18, 1937.

BREMER, Anders Hillegard, Norwegian horticulturist; b. Sarpsborg, Nov. 27, 1889; s. Kristen and Dordi (Hillegard) B.; Ph.D. in Agr. Scis., Norwegian Inst. Agronomy, 1930; m. Anna Björkum, July 8, 1919; children—Turid, Rannveig, Astrid, Jon. Genetics asst. at Darlington, Eng., 1908; prof. horticulture, 1914-18; dir. agr. and hort. research forum, 1920-45; prof. Norwegian Inst. Agronomy, 1946-60. Recipient Gold medal of merit, 1960. Mem. Union Linguists of Aas (hon.). Author: Grouping Vegetables; Temperature and Plant Growth; Genetische Untersuchungen mit Salat; Day Length and Plant Growth; Salat (in Handbuch der Pflanzensüchtung). Address: Kajav 26, Vollebekk, Norway.

BREMER, Frederic Gaston Nicolas, Belgian physician; b. Brussels Arlon, June 28, 1892; s. Gaston and Amélie (Leyder) B.; M.D., U. Brussels; Dr. honoris causa univs. Aix-en-Provence, Montellier, Strasbourg, Utrecht; m. Claire Baar, 1922; children—Antoine, Thérèse. Prof. gen. pathology U. Brussels, 1932-62. Mem. Royal Acad., Royal Acad. Medicine, Acad. Scis. of Inst. of France (corr.). Research and numerous publs. on physiology and neurology. Home: 9, rue Isidore-Verheyden, Brussels 5. Office: 115, bd. de Waterloo, Brussels, Belgium.

BREMER, John Lewis, Am. embryologist; b. N.Y.C., Nov. 3, 1874; s. John Lewis and Mary (Fransworth) B.; A.B., Harvard, 1896, M.D., 1901; post-grad. Oxford, 1896-97; m. Mary C. Bigelow, Sept. 29, 1906. Asst. in histology and embryology Harvard, 1902, instr., 1903-06, demonstrator in histology, 1906-12, asst. prof., 1912-15, asso. prof., 1915-31, Hersey prof. of anatomy, 1931-41, emeritus, from 1941. Fellow Am. Acad. Arts and Scis.; mem. Am. Assn. Anatomist. Died Dec. 25, 1959.

BREMERMANN, Hans Joachim, mathematician, biophysicist; b. Bremen, Germany, Sept. 14, 1926; s. Bernard and Berta (Wicke) B.; Ph.D., U. Muenster (Germany), 1951; Staatsexam. math., physics, 1951; m. Maria Isabel Lopez Perez-Ojeda, May 16, 1954. Came to U. S., 1952, naturalized, 1965. Asst., U. Muenster, 1951-52, 54-55; research asso. Stanford, 1952-53, vis. asst. prof., 1954; research fellow, Harvard, 1953; mem. Inst. for Advanced Study Princeton, N.J., 1955-57, 58-59; asst. prof. U. Wash., Seattle, 1957-58; asso. prof. math. U. Cal. at Berkeley, 1959-66, prof., 1966—, vice chmn. dept., 1961-62. Mem. Am., Austrian, German math. socs., German Soc. Applied Math. and Mechanics. Author: Distributions, Complex Variables and Fourier Transforms, 1965. Important work includes research in quantum field theory, biophysics, several complex variables, fundamental limit of data processing. Home: 1873 San Ramon Av., Berkeley, Cal. 94707.*

BREMIKER, Carl, German astronomer, geodesist; b. Hagen, Germany, Feb. 23, 1804; s. Johann Carl Bremiker; m. Ida Alwine Streuber, 1842; 1 son. Surveyor with Rhine-Westfalen land survey office; worked on a Berlin astron. annual; edited Nautisches Jahrbuch, 1850-77; insp. planning commn. in Prussian ministry commerce; dept. head Prussian Geodetic Inst., 1868. Author: Studien über höhere Geodäsie, 1869; Logarithmisch-trigonometrische Tafeln, 1872. tables, also own logarithmic-trigonometric tables of Discovered a comet, 1840, calculated comet orbits, revised and pub. new edition of Crelle calculation great accuracy (reprinted many times). Died Berlin, Germany, Mar. 26, 1877.

BREMS, Hans Julius, economist; b. Viborg, Denmark, Oct. 16, 1915; s. Holger and Andrea (Golditz) B.; cand.polit., U. Copenhagen, 1941, dr.polit., 1950; m. Ulla Constance Simoni, May 20, 1944; children—Lisa, Marianne, Karen Joyce. Came to U. S., 1951, naturalized, 1958. Asst. prof. U. Copenhagen (Denmark), 1943-51; asst. prof. U. Cal. at Berkeley, 1951-54; prof. econs. U. Ill., Urbana, 1954—. Recipient Gold medal U. Copenhagen, 1942. Mem Am. Econ. Assn., Royal Econ. Soc., Econometric Soc. Author: Product Equilibrium under Monopolistic Competition, 1951; Output, Employment, Capital and Growth, 1959; Quantitative Economic Theory, 1968; also articles. Contbr. to theory of monopolistic competition, theory of growth and internat. trade. Address: U. Ill., Commerce West, Urbana, Ill. 61801.*

BRENCHLEY, Winifred Elsie, Brit. botanist; b. Aug. 10, 1883; ed. Swanley Hort. Coll., U. Coll., London, Eng.; B.Sc., 1905; D.Sc., 1911. Head bot. dept. Rothamsted Exptl. Sta., 1907-48. Fellow U. Coll., Linnean Soc., Royal Entomol. Soc. Author: Weeds of Farm Land, 1921; Manuring of Grassland for Hay, 1924; Inorganic Plant Poisons and Stimulants, 2d edit., 1927; (with H. C. Long) Suppression of Weeds by Fertilisers and Chemicals, 3d edit., 1949. Contbr. numerous articles to profl. jours. Died Oct. 27, 1953.

BRENDEL, (Otto Rudolf) Martin, German astronomer; b. Niederschönhausen, nr. Berlin, Germany, Aug. 12, 1862; s. Otto and Minna (Immich) B.; studied math. and astronomy in Berlin, Munich, Stockholm, Paris, London; doctorate from Berlin, 1890; m. Emilie Marcelline Le Gargan; 1 son. Joined faculty of Greifswald, Germany, 1892; asso. prof. theoretical astronomy in Göttingen, 1898, lectr. actuarial math. and geodetics from 1902; taught math. and actuarial math. Akademie für Sozial-und Handelswissenschaften, Frankfort/Main, 1907; prof., dir. obs. at newly-founded Frankfort U., 1914. Recipient Prix Damoiseau (for Theorie der kleinen Planeten) Paris Acad. Scis., 1894. Contbr. numerous articles to periodicals. Concerned mainly with theoretical astronomy, wrote numerous famous studies on movement of planets and moon, perturbation calculations, methods of orbit determination; founder in Frankfort his own planetarium institute with aid of Paris and Stockholm acads., 1913; edited edition of works of Gauss (1898-1929) and materials for a Gauss bibliography (with F. Klein and L. Schlesinger). Died Freiburg im Breisgau, Germany, Sept. 6, 1939.

BRENEMAN, Abram Adam, Am. chemist; b. Lancaster, Pa., Apr. 28, 1847; s. Abraham and Anna B.; B.S., Pa. State Coll., 1866, M.Sc., 1871. Instr. chemistry Pa. State Coll., 1867-68, prof., 1869-72; asst. prof. indsl. chemistry Cornell U., 1875-79, prof. 1879-82; analyst, chem. expert, writer and lectr., from 1882; also in cons. practice. Chmn. internat. jury on mineral waters, Chicago Expn., 1893, St. Louis Expn., 1904; expert mem. municipal explosives commn., N.Y., 1906-09. Author: (with Prof. G. C. Caldwell) A Manual of Introductory Laboratory Practice, 1875; Report on the Fixation of Atmospheric Nitrogen, 1889; Report on Sewer and Conduit Explosions in New York, 1909, 1910. Editor, Jour. Am. Chem. Soc., 1884-93. Inventor Breneman process of rendering iron non-corrodible. Died May 10, 1928.

BRENNAN, Michael Joseph, Am. economist; b. Chgo., Aug. 29, 1928; s. Michael J. and Nora (McHugh) B.; B.S., DePaul U., 1952; M.A., U. Chgo., 1954, Ph.D., 1956; m. Isabel Thomas, Dec. 4, 1954; children—Mark Etienne, Moira Sioban, Keelin Marta. Faculty, Brown U., Providence, 1956—, prof. econs., 1964—, dean Grad. Sch., 1966—. Cons. industry, govt., 1962—. Mem. Am. Econ. Assn., Econometric Soc. Author: Preface to Econometrics, 1960; Patterns of Market Behavior, 1965; Theory of Economic Statics, 1965; (with Philip Taft and Mark B. Schupack) Economics of Age, 1967. Specialist in econometrics. Theoretical and empirical studies on commodity inventories, supply of agrl. commodities, age structure of the labor force, econs. of edn., geog. migrations of labor and capital among U. S. regions. Home: 44 Oriole Av., Providence 02906.*

BRENNEMANN, Joseph, Am. pediatrician; b. Peru, Ill., Sept. 25, 1872; s. Joseph and Mary (Schaefer) B.; Ph.B., U. Mich., 1895; M.D., Northwestern U., 1900; St. Luke's Hosp., 1900-02; m. Bessie D. Daniels, Jan. 2, 1905; children—Mary Elizabeth, Barbara, Deborah. Chief of staff Children's Meml. Hosp., prof. pediatrics U. Chgo., 1921-41; med. dir. Children's Hosp., Los Angeles, prof. Pediatrics, U. S. Cal. Med. Sch., 1941-43. Am. Chgo., Brit. (corr.) pediatric socs., Am. Acad. Pediatrics. Inst. of Medicine Chgo., Soc. for Pediatric Research (hon.). Chgo. Med. Soc. Contbr. to pediatric Lit. Editor Brennemann's Practice of Pediatrics. Died July 2, 1944.

BRENNER, Alexander, Austrian surgeon; b. Vienna, Austria, Feb. 22, 1859; pupil of Billroth. Chief surgeon in Linz. Improved surgical methods for inguinal hernia, appendectomy; ulcus ventriculi; advocated improvement in urological methods; improved cystoscope; developed new bladder stitch. Died Linz, Austria, 1936.

BRENNER, Marten Withmar, Finnish physicist; b. Helsingfors, Finland, May 28, 1926; s. Widar Magnus and Helmi (Granroth) B.; fil. dr., U. Helsinki (Finland), 1959; m. Andrea Barbara Stackelberg, Nov. 7, 1953; children—Mikaela, Markus, Helena, Richard. Asst., dept. physics U. Helsinki, 1953-62, docent, 1962—, chief radiotherapy clinic Univ. Central Hosp., Helsinki, 1962-66; prof. Abo Akademi, Turku, Finland, 1966—. Asso. prof. Inst. Tech., Helsinki, 1962. Mem. Fysikersamfundet, Finl. Fysiker Förening, Sv. Tekn. Vetensk. Akad., Finska Vet. Soc., Nord. Fören. Klin. Fysik., Nord. Fören. Stalskydd, Radiologfören. Research and publs. on nuclear physics, reactor physics, radiation physics, growth of cancer. Home: 2 Svartmundegränd, Abo, Finland.*

BRENNER, Paul Albert, German metallurgist; b. Stuttgart, Germany, Mar. 30, 1897; s. Paul and Elise (Kurz) B.; ed. Techs. Schs. Stuttgart, Berlin; Ph.D. in Engring.; m. Else Staehelin, Sept. 10, 1925; children—Rosemarie, Marion. With Inst. Aero. Proofs, Deutscher Versuchsanstalt für Luftfahrt, Berlin-Adlershof; dir. research V.L.W., Bonn, Germany; prof. Tech. Sch. at Hanover, Claustahl-Zellerfeld Acad. of factories. Recipient Guido Donegani gold medal Italian Assn. Metallurgy, 1961. Mem. Metall. Inst. (London), Am. Soc. Metals, Am. Inst., Mining, Metall. and Petroleum Engrs., German Assn. Metall. Scis. Research and numerous publs. on aviation and light metals. Home: Horionstrasse 6, Bad Godesberg, Germany. Office: Vereinigte Leichtmetall Werke G.m.b.H., Am Nordbahnhof, Bonn, Germany.

BRENTANO, Franz, German philosopher, psychologist; b. Marienberg, Jan. 16, 1838; ed. Us. Berlin,

Munich, Tübingen. Docent, U. Würzburg, 1866; prof. philosophy, U. Würzburg, resigned 1873; prof., U. Vienna, 1874-80. Author: Psychology of Aristotle, New Riddles, 1874; Psychologie vom empirischen standpunkte, 1874; Neue Rätsel, 1878; Vom Ursprung Sittlicher Erkenntnis, 1889; Untersuchungen zur Sinnespsychologie, 1907; Aristoteles und seine Weltanschauung, 1911; Aristoteles Lehre vom Ursprung des menschlichen Geistes, 1911; Von der Klassifikation der Psychischen Phänomene, 1911. Leader of school of "old" psychology; attempted to interrelate Aristotle and scholasticism with psychology; believed mental processes the material for psychological study. Died Zurich, Switzerland, Mar. 17, 1917.

BRENTANO, John Christian Michael, physicist; b. Vienna, Austria, June 27, 1888; s. Franz and Ida (Lieben) B.; student Lyceum Alpinum, Zuoz, Switzerland, 1902; licenza liceale honors Liceo Dante, Florence, Italy, 1908; Ph.D., U. Munich (Germany), 1914; Dr.Sc., U. Manchester (Eng.), 1935; m. Sophie Marie Leembruggen, June 29, 1925. Came to U. S., 1940, naturalized, 1946. Asst., lectr. Polytechic Inst., Zurich, Switzerland, 1917-21; faculty U. Manchester, 1921-40; faculty Northwestern U., Evanston, Ill., 1940-53, prof., 1946. Fellow Inst. Physics (Gt. Britain), Am. Phys. Soc.; mem. Société Sciences Naturelles (Switzerland), Am. Crystall. Assn., Sigma Xi. Research semicondr. properties of pressed powders and superposed thin layers non-semicondrs. on lattice space measurements and determination of F values in relation to x-ray wave length, solid state physics; devel. parafocusing powder layer method permitting use of counters; introduced scatter densitometry. Home: Les Scillas Blonay (Vaud) Switzerland.*

BREQUET, Louis Charles, French inventor and aero-pioneer; b. Paris, France, Jan. 2, 1880; mng. dir. Société Anonyme des Ateliers d'Aviation Louis Brequet, Maison Brequet (elec. equipment mgr. co.); mem. Union Aero. Constrn. (pres.); built and equipped 1st helicopter to go aloft perpendicularly with a passenger, 1909; later made model with pair superimposed coaxial rotors turning in opposite directions. Died 1955.

BRESCHET, Gilbert, French anatomist; b. Clermont-Ferrand, France, July 7, 1784. Intern, Paris; apptd. head anat. studies Paris Faculty Medicine, 1819; became chief surgeon Enfants-trouves Hosp., 1820; prof. anatomy Paris Sch. Medicine. Mem. Acad. Medicine, French Acad. Scis., 1835. Author: Nouvelles recherches sur la structure de la peau, 1835; Le système lymphatique, 1836; Traité d'anatomie humaine (lectures). Studied bones of skull and their diseases; investigated hydrocephalus; provided 1st accurate description of canals of diploe (Breschet's canals); 1819; named and described sphenoparietal sinus, circa 1827; 1st to use term helicotrema for structure of cochlea, 1834; studied hearing in fish and other animals. Died May 10, 1845.

BRESGEN, (Karl) Maximilian (Hubert), German physician; b. Ahrweiler, Rheinland, Mar. 1, 1850; s. Franz Joseph Hubert and Maria Josepha Sibilla (Knieps) B.; studied in Jena, Heidelberg, Berlin (all Germany), from 1868; doctorate from Heidelberg, 1872; state exam. in Berlin, 1873. Worked with Virchow at Pathologisches Institut, Berlin, Germany; asst. to Karl Störck, Vienna, Austria; settled in Frankfort/Main as specialist, 1877; popular specialist and med. writer in Wiesbaden, Germany from 1899. Author: Grundzüge einer Pathologie und Therapie der Nasen-, Mund-, Rachen-, und Kehlkopfkrankheiten, 1884. Developer of rhinology, particularly by numerous works on influence of hindered nasal breathing on mental devel. and capacity of school children. Died Wiesbaden, Germany, June 2, 1915.

BRESLOW, Ronald Charles, Am. chemist, b. Elizabeth, N.J., Mar. 14, 1931; s. Alexander E. and Gladys (Fellows) B.; A.B. summa cum laude, Harvard, 1952, M.A., 1953, Ph.D., 1955; m. Esther Greenberg, Sept. 7, 1955; children—Stephanie, Karen. NRC fellow Cambridge (Eng.) U., 1955-56; mem. faculty Columbia, 1956—, prof. chemistry, 1962—; cons. to industry 1958—; editor Benjamin, Inc., 1962—. Mem. medicinal chemistry panel NIH, 1964—. Fellow Am. Acad. Arts and Scis.; member of the Nat. Acad. Scis., Am. Chem. Soc. (Pure Chemistry award 1966), Phi Beta Kappa (first marshall 1952), Phi Lambda Upsilon (Fresenius award 1966). Author: Organic Reaction Mechanisms, 1965; also articles. Mem. editorial bd. Organic Syntheses, 1964—. Discovered chemical mechanism of thiamine action; synthesized first derivatives of cyclopropenyl cation and cyclopropenone and demonstrated that they represent simplest aromatic system; confirmed quantum mechanical prediction that some 4 pi electron compounds have ground triplet states. Home: 275 Broad Av., Englewood, N.J. 07631. Office: Dept. Chemistry, Columbia Univ., N.Y.C. 10027.

BRESNICK, Edward, Am. biochemist; b. Jersey City, N.J., Sept. 7, 1930; s. Frank and Tillie (Lobel) B.; B.S. magna cum laude, St. Peter's Coll., 1952; M.S., Fordham U., 1954, Ph.D., 1958; m. Eta Krupitsky, Jan. 20, 1957; children—Eric, Emory. Research asso. Med. br. U. Tex., Galveston, 1957-58; research biochemist Wellcome Research Labs., Tuckahoe, N.Y., 1958-59, sr. research biochemist, 1959-61; faculty Baylor U. Coll. Medicine, Houston, 1961—, asso. prof. pharmacology, 1965—. Recipient Lederle Med. Faculty award, 1966—. Mem. Am. Chem. Soc., A.A.A.S., Biochem. Soc., Am. Assn. Cancer Research, Am. Soc. Cell Biology, Am. Soc. Pharmacology and Exptl. Therapeutics, Am. Soc. Biol. Chemists. Research numerous publs. on mechanisms related to control of growth processes in normal and neoplastic tissues. Home: 6110 Paisley St., Houston 77035.*

BRESSE, Jacques Antoine Charles, French engr.; b. Vienne, France, Oct. 9, 1882; ed. École des Ponts et chaussées, Paris, France, 1841. Insp. 2d class civil engring.; prof. mechanics École polytechnique, École des ponts et chausées, 1859-65. Mem. French Acad. Scis., 1880 (Poncelet award). Author: Recherches analytiques sur le flexion et la résistance des pièces courbes, 1854; Calcul des moments de flexion dans une poutre à plusieurs travées solidaires; memoirs on applied mathematics. Died May 22, 1883.

BRESSLAU, Ernst Ludwig, zoologist; b. Berlin, Germany, July 10, 1877; s. Harry and Caroline (Isay) B.; studied medicine and natural scis. (especially zoology) in Munich, Germany, doctorate, Strasbourg, Germany (now France), 1902; m. Luise Hoff, 1907; 2 sons, 2 daus. Zool. expdn. to Central and Northeastern Brazil, 1913-14; head zool. dept. Georg-Speyer-Haus, Frankfort/Main, Germany, 1919; prof., dir. new zool. inst. U. Cologne, (Germany), 1925-33; emigrated to Brazil, 1934, became dir. zool. inst. U. Sao Paulo. Recipient Kaiser Nikolaus II prize for work in comparative anatomy and evolutionary history Internat. Congress Zoologists, Monaco, 1913. Works include: Die Entwicklung des Mammapparates der Monotremen, Marsupialier und einiger Placentalier, ein Beitrag zur Phylogenie der Säugertiere, 1907-13; The Mammary Apparatus of the Mammalia, 1920. Known for thorough studies on turbellaria (flatworms), devel. of mammary glands of mammals, way of life of bees and mosquitos, protozool. studies on significance of environment. Died Sao Paulo, Brazil, May 9, 1935.

BRETON, Jules Louis, French engr.; b. Pas-de-Calais, France, Apr. 1, 1872; s. Jules Breton; ed. Collège de France. Dir., Office Sci. and Indsl. Research, Paris, France, Revue scientifique et industrielle; senator from 1920. Mem. French Acad. Scis., 1920. Contbr. articles on animated photography, carbide of calcium, x-rays. Inventor automobile to destroy tank communication lines, World War I. Died Bellevue, France, Aug. 2, 1940.

BRETONNAYAU, René, physician; b. Loches, France, in late 15th century; 1 son, Théodore; author of a treatise on medicine in verse, La génération de l'homme, 1583; wrote on plague of 1597 in Loches; gave cosmetic advice to women.

BRETONNEAU, Pierre-Fidèle, French physician; b. Tours, France, Apr. 3, 1778; student Sch. Medicine, Paris, 1795; M.D., 1815; m. Marie-Thérèse Adam, 1801; m. 2d, Sophie Moreau, 1856. Chief physician, Hosp. Tours, from 1815. Corr. mem. French Acad. Scis., 1835. Named diphtheria, 1821; performed 1st successful tracheotomy for laryngeal diphtheria, 1825; distinguished scarlet fever from diphtheria, 1826; described typhoid fever and foresaw its differentiation from typhus; proponent of doctrine of morbid specificity, which foreshadowed germ theory of disease, 1855. Died Passy, nr. Paris, Feb. 18, 1862.

BRETSCHER, Egon, physicist; b. Zurich, Switzerland, May 23, 1901; s. Julius and Anna (Martin) B.; diploma in chem. engring. Fed. Inst. Tech., Zurich, 1925; Ph.D., U. Edinburgh (Scotland); 1928; m. Hanna Greminger, July 21, 1931; children—Scilla, Mark, Peter, Iris, Anthony. Research in organic chemistry and physics Eidgenössische Technische Hochshule, Zurich, 1928-34; Rockefeller Found. Internat. fellow in physics Cavendish Lab., Cambridge, Eng., 1934-36; Clerk Maxwell scholar, 1936-39; lectr. U. Cambridge, 1940-44; mem. Brit. mission to Los Alamos Lab., 1944-46; head chemistry div. Atomic Energy Research Establishment, Harwell, Berkshire, Eng., 1947, head nuclear physics div., 1948—. Vis. prof. U. Cal. at Los Angeles, 1953. Fellow Am. Phys. Soc. Home: Salix, Conduit Head Rd., Cambridge, Eng. Office: Atomic Energy Research Establishment, Harwell, Berkshire, Eng.*

BRETSCHNEIDER, Charles Leroy, Am. oceanographer; b. Red Owl, S.D., Nov. 9, 1920; s. Charles and Jennie (Seifert) B.; B.S. in Physics, Hillsdale Coll., 1947; M.S. in Civil Engring., U. Cal. at Berkeley, 1950, Ph.D., Tex. A. and M. Coll., 1959; m. Yveline Kerr, June 1, 1948; children—Eric Charles, Anne Denise. Mem. faculty Hillsdale (Mich.) Coll., 1946-48, U. Cal. at Berkeley, 1950-51, Tex. A. and M. Research Found., College Station, 1951-56; mem. beach erosion bd. U. S. Army C.E., Washington, 1956-61; dir. Washington office Nat. Engring Sci. Co., 1961-64, v.p. Eastern operations, McLean Va., 1964-66; prof., chmn. ocean engring. U. Hawaii, Honolulu, 1966—. Cons. engr. various offshore operations; lectr. to various instns. Registered profl. engr., Tex., D.C. Mem. Am. Soc. C. E., Am. Geophys. Union, Am. Meteorol. Soc., Permanent Internat. Assn. Navigation Congresses (life), Marine Tech. Soc. (v.p. 1964—), Sigma Xi. Research and publs. on physics of waves; field studies of waves. Office: 2565 The Mall, U. Hawaii, Honolulu 96821.*

BRETT, John Watkins, Brit. telegraphic engr.; b. Bristol, Eng., 1805; s. William Brett; established telegraphic communication between Eng. and France, 1850; laid cable from Am. to Eng. (failed after several messages), 1858. Author: On the Origin and Progress of the Oceanic Telegraph . . . , 1858; also several papers. Founder submarine telegraphy. Died Dec. 3, 1863.

BRETTE, R., French physician; b. Lyons, France, Apr. 26, 1920; s. Paul and Cecile (Gonnet) B.; ed. Faculty Medicine Lyons; m. Hélène Truchot, Oct. 28, 1954; children—Lawrence, Dominique, Jean-Paul. Became head clinic Paris, 1948; began practice medicine, 1955; aggregate prof., 1955—. Mem. French Soc. Gastroenterology, Med. Soc. of Hosps. of Lyons. Research, publs. on hepatic infections and infections of digestive organs. Home: 26, rue Vendôme, Lyons. Office: Hopital de l'Anquaille, Lyons, France.*

BRETTON, Henry L., polit. scientist; b. Berlin, Germany, May 18, 1916; came to U. S., 1938, naturalized, 1944; B.A. with honors Yale, 1947; M.A., U. Mich., 1948, Ph.D., 1951; m. Marian M. More, Sept. 8, 1951; children—Elizabeth, Alexander. Faculty, U. Mich., Ann Arbor, 1951—, prof. polit. sci., 1963—; vis. prof. U. Ghana, 1964-65; vis. prof., head dept., U. Coll., Nairobi, Kenya, 1965-66. Fulbright lectr., Innsbruck, Austria, 1956-57; mem. adv. council African affairs U. S. Dept. State, 1961—. Mem. Am. Polit. Sci. Assn., African Studies Assn. Author: Streseman and the Revision of Versailles, 1953; Power and Stability in Nigeria, 1962; The Rise and Fall of Kwame Nkrumah, 1967; also articles, chpts. in books. Research on conceptual problems in study polit. devel.; traced, analyzed and described aspects of power, influence and polit. stability in Africa. Home: 1925 Austin St., Ann Arbor, Mich. 48104.*

BREUDA, Botho Erich Curt, physician, chemist, biologist; b. Vienna, Austria, Dec. 18, 1892; s. Alsoi and Mathilde (Nedorost) B.; ed. U. Vienna, U. Graz, John Hopkins; M.D.; m. Franziska Musil, Oct. 18, 1920. Sci. collaborator Rockefeller Found., 1925-26; asst., instr. exptl. pathology U. Graz, 1927-33; med. specialist research scis., dir. Biopharma firm of pharm. products, Vienna, 1933; dir. Ortskrankenkasse labs., Vienna, 1938, Landesversicherungsanstalt, Vienna, from 1939, Dianostisches Zentralinstitut, Vienna, 1940-45; dir. firm of pharm. products of Dr. Kurt Breuda, Vienna, came to U. S. as sci. collaborator, 1951; dir. Chem. Hormone Corp., N.Y.C., 1953-63; prof. Am. Biochem. Corp., N.Y.C.; prof. research U. Fla., 1962—; former dir. Inst. Research, S. Gauning Childs Hosp., Vienna. Mem. German Soc. Pathology, Soc. Physicians of Vienna, Notgemeinschaft deutscher Wissenschaft. Author: Carcinom-Mesenchymin; Askerinad Atherosclerosis-Hypophyse. Address: 2054 Broadway, N.Y.C. 10023.

BREUER, Heinz, German biochemist; b. Bonn, Germany, May 2, 1926; s. Heinrich and Paula (Meder) B.; Dipl.-chem., Rheinische Friedrich-Wilhelms-U., Bonn., 1949, Dr.rer.nat., 1952, Physikum, 1954; m. Annlott v. Werder, Nov. 19, 1957; children—Dorothee Maria, Klaus Heinrich, Katharina Elisabeth. Asst. Physiologisch-chemisches Institut, Bonn, 1952-54; Brit. Council scholar dept. biochemistry U. Sheffield (Eng.), 1954-55; head dept. clin. chemistry and biochemistry surg. clinic U. Bonn, 1956—. Mem. Gesellschaft Deutscher Chemiker, Gesellschaft für biologische Chemie, Deutsche Gesellschaft für Endokrinologie, Biochem. Soc. (London), Internat. Study Group for Steroid Hormones, Deutsche Gesellschaft für klinische Chemie, Royal Soc. Medicine. Research and numerous publs. on biogenesis, metabolism, determination of steroid hormones, particularly estrogens; electrolyte metabolism, clin. biochemistry. Home: Am Zehnthof 2, 534 Bad Honnef, West Germany. Office: Chirurgische Universitätsklinik 53, Bonn-Venusberg, West Germany.*

BREUER, Josef (Robert), Austrian physician, psychologist; b. Vienna, Austria, Jan. 15, 1842. Practiced medicine, Vienna; was physician to many members of Viennese medical faculty. Mem., Viennese Acad. of Sci., 1894. Author: Studien über Hysterie (with S. Freud), 1895. Studied respiratory cycle and nervous control of respiration; discovered Hering-Breuer reflex, 1868; used hypnosis in treating hysterical patient, 1880; discovered that condition improved when patient recalled past experiences; work excited Freud and they worked together until circa 1900; Freud credited Breuer with making basic contributions to development of psychoanalysis. Died Vienna, June 20, 1925.

BREUIL, (Abbé) Henri Édouard Prosper, anthropologist, archeologist, paleontologist; b. Mortain, France, Feb. 28, 1877; s. Albert and Lucie Morio (de l'Isle) B.; ed. Sorbonne, Faculty of Scis., Paris, France; numerous hon. degree from fgn. univs. Privat dozent, U. Fribourg, 1905-10; prof. Institut de Paléontologie Humanine, Paris, from 1910; prof. College de France, 1929-47; mem. numerous acads. sci. Author: 400 centuries of Cave Art, 1952. Authority on paleolithic art; among 1st to interpret and record paleolithic art of rock carvings and cave paintings in Europe and Africa. Died 1961.

BREUSCH, Robert H., mathematician; b. Freiburg, Germany, Apr. 2, 1907; s. Friedrich and Luise (Stehle) B.; student U. Freiburg, Bonn, Berlin, 1925-30, Ph.D., 1932; M.A. (hon.) Amherst Coll., 1954; m. Kate Dreyfuss, July 25, 1936. Came to U. S., 1939, naturalized, 1945. Tchr., pvt. sch., Germany, 1932-36; prof. Universidad Santa Maria, Chile, 1936-39; tchr. Shady Hill Sch., Cambridge, Mass., 1940-43; prof. Amherst (Mass.) Coll., 1943-—. Contbr. articles to math. jours. Home: 19 Dana Place, Amherst, Mass. 01002.*

BREWER, A(ubrey) Keith, Am. physicist; b. Richland Center, Wis., Oct. 20, 1893; s. Edward and Hattie (Dove) B.; B.A., U. Wis., 1915, M.S., 1920, Ph.D., 1924. NRC fellow Cal. Inst. Tech., Pasadena, 1924-27; physicist fixed Nitrogen Research Lab., Washington, 1927-39, Nat. Bur. Standards, 1939-46, Navy Dept., 1946-—. Fellow Am. Phys. Soc.; mem. Washington Acad. Sci., Sigma Xi. Patents; Isotope Separator, 1946; Liquid-Liquid Extractor, 1953; Super Centrifuge, 1964; research in mass spectra, thermionic and photoelectric emission, electric discharge, etc. Home: 3247 38th St. N.W., Washington. Office: Navy Dept. Bldg. 52, U. S. Naval Obs., Washington.*

BREWER, Douglas Forbes, Brit. physicist; b. Cardiff, Wales, May 22, 1925; s. Stanley Leonard and Winifred Helen (Forbes) B.; M.A., Pembroke Coll., Oxford (Eng.) U., 1950, D.Phil., 1953; m. Jean M. Ross, July 26, 1952; children—Jonathan A., Charlotte D., Mark P., Sophia C., Lucy A. Staff Gen. Electric Co. Research Labs., Wembley, Middlesex, Eng., 1945-48; Nuffield Found. Research fellow Oxford U., 1953-57, 59-60, research asso., 1960-62; vis. faculty Ohio State U., 1957-59, asso. prof., 1958-59, vis. prof., 1961; faculty U. Sussex (Eng.), 1962-—, prof. physics, 1965-—. Fellow Phys. Soc. Research, publs. in hydrodynamic and thermal properties of condensed helium isotopes nr. absolute zero, and into properties of metallic alloys. Home: Broughton Astley, Kingston Rd., Lewes, Sussex. Office: U. Sussex, Falmer, Brighton, Sussex, Eng.*

BREWER, George Emerson, Am. surgeon; b. Westfield, N.Y., July 28, 1861; s. Francis B. and Susan H. (Rood) B.; A.B. Hamilton Coll., 1881, A.M., 1884, LL.D., 1916; M.D., Harvard, 1885; hon. D.Sc., Columbia, 1929; m. Effie Leighton Brown, June 29, 1893. Cons. surgeon to Muhlenberg Hosp., Plainfield, N.J., Perty Amboy (N.J.) Hosp., Presbyn., Roosevelt, Woman's St. Vincent's, St. Bartholomay's, City hosps., Ophthalmic and Aural Inst., House of the Holy Comforter (all N.Y.C.); asst. demonstrator anatomy, Coll. Phys. and Surg. (Columbia U.), 1892-1900, instr. surgery, 1900-03, clin. lectr., 1903-04, prof. clin. surgery, 1904-13, prof. surgery, 1913-17, emeritus. Research asso. Somatic anthropology, Am. Mus. Natural History, N.Y.C. Died Dec. 24, 1939.

BREWER, Leo, Am. chemist; b. St. Louis, June 13, 1919; s. Abraham and Hannah (Resnik) B.; B.S., Cal. Inst. Tech., 1940; Ph.D., U. Cal. at Berkeley, 1943; m. Rose Strugo, Aug. 22, 1945; children—Beth A., Roger M., Gail L. Mem. faculty U. Cal. at Berkeley, 1946-—, prof. phys. chemistry, 1955-—, faculty lectr., 1966, research asso. Lawrence Radiation Lab., 1943-61, head inorganic materials div., 1961-—; Robert W. Williams lectr., Mass. Inst. Tech., 1963; Henry Werner lectr. U. Kan., 1963; O. M. Smith lectr. Okla. State U., 1964; G. N. Lewis lectr. U. Cal., 1964, Hugh Huffman Meml. lectr. Calorimetry Conf., 1966. Mem. rev. com. reactor chem. div. Oak Ridge Nat. Lab.; research asso. Manhattan District, 1943-45; sec. gas subcom. high temperature commn. Internat. Union Pure and Applied Chemistry, 1957-60; chmn. materials adv. bd. Com. Investigation Application Plasma Phenomena, 1959-60. Great Western Dow fellow, 1942; Guggenheim fellow, 1950; recipient Ernest Orlando Lawrence Meml. award, 1961. Mem. Nat. Acad. Scis. (exec. com. Office Critical Tables 1964-66), Am. Assn. U. Profs., A.A.A.S., Am. Chem. Soc. (Leo H. Baekeland award 1953), Am. Electrochem. Soc., Am. Plant Life Soc., Coblentz Soc., Combustion Inst., Faraday Soc., Fedn. Am. Scientists, Internat. Plansee Soc. Powder Metallurgy, Am. Phys. Soc., Nat. Park Assn., Sigma Xi, Alpha Chi Sigma, Tau Beta Pi. Author: (with others) Thermodynamics, 1961. Asso. editor Jour. Chem. Physics, 1959-63; editorial adv. bd. Journal Physics and Chemistry Solids, Progress Inorganic Chemistry. Research in theory of solutions, reaction rates, high temperature chemistry and thermodynamics. Home: 15 Vista del Orinda, Orinda, Cal. Office: Dept. Chemistry, Univ. California, Berkeley, Cal. 94720.*

BREWER, Thomas Mayo, Am. ornithologist; b. Boston, Nov. 21, 1814; s. James Brewer; grad. Harvard, 1835, Harvard Med. Sch., 1838; m. Sally R. Coffin, 1849. Practiced medicine; gave up practice to write for Boston Atlas, later became editor; asso. with publishing firm Swan & Tileston (later Brewer & Tileston), until 1875; mem. Boston Soc. Natural History, 1835-80; published Supplement to Prof. Hitchcock's Catalogue of the Birds of Massachusetts, 1837, Wilson's Ornithology, 1840, North American Oology (describing eggs of all known bird species), 1857; mem. Boston Sch. Com., 1844-80; compiled (with Baird and Ridgeway), History of North American Birds, 1875. Died Jan. 23, 1880.

BREWER, William Henry, Am. botanist; b. Poughkeepsie, N.Y., Sept. 14, 1828; s. Henry B.; Ph.B., Yale, 1852, A.M., 1859; student Heidelberg, Germany; Munich, Germany; Paris, France. Ph.D., Washington and Jefferson, 1880; LL.D., Yale 1903; m. Angelina Jameson, Aug. 15, 1858 (died 1860); m. 2d, Georgiana Robinson, Sept. 1, 1868 (died 1889). Prof. geology and chemistry Washington Coll., Pa., 1858-60; asst. to J. D. Whitney on geol. survey of Cal., 1860-64; prof. chemistry, U. of Calif., 1863-64; prof. agriculture Yale, 1864-1903; pres. Conn. Bd. of Health, 1892-1909; made topog. survey Conn., sci. survey P.I., 1903; went to Arctic regions in the Miranda, 1904. Mem. Arctic Club (pres. 1894-1909), U. S. Forestry Commn., 1896, Nat. Acad. Scis. Author: (with others) Botany of California, 1875. Helped establish in Conn. 1st agrl. exptl. sta. in U. S., also Yale Forestry Sch., 1900; did survey work in the Sierra Nevadas; Mount Brewer named for him. Died New Haven, 1910.

BREWSTER, Sir David, Scottish physicist; b. Jedburgh, Scotland, Dec. 11, 1781; s. James and Margaret (Key) B.; ed. U. Edinburgh (Scotland); L.L.D., St. Andrews (Scotland) U., 1807; M.A., Cambridge (Eng.) U.; m. Juliet Macpherson, July 31, 1810; 1 dau.; m. 2d, Jane Kirk Purnell, 1857; 1 dau. Licensed preacher, 1804; became editor Edinburgh Mag., 1802, Edinburgh Ency., 1807-29; became 1st dir. Royal Scottish Soc. Arts, 1821; helped organize Brit. Assn. for Advancement of Sci., 1831; became prin., united colls. St. Salvador and St. Leonard, U. St. Andrews, 1837-59; asso. with Scottish Free Ch. movement, 1844; named vice-chancellor Edinburgh U., 1860. Fellow Royal Soc., 1815 (Copley medal 1815, Rumford medal 1818, Royal medal); corr. mem. French Acad. Scis., 1825; pres. Royal Soc. Edinburgh 1864. Knighted, 1882. Author: Some Properties of Light, 1813; Treatise on New Philosophical Instruments; Treatise on the Kaleidoscope, 1819; On the Periodical Colours produced by Grooved Surfaces; Elliptic Polarisation by Metals; The Optical Nature of the Crystalline Lens; The Optical Conditions of the Diamond; The Colours of Film Plates; Treatise on Optics, 1831; Letters on Natural Magic, 1831; Life of Sir Isaac Newton, 1832; Treatise on Magnetism; Martyrs to Science, 1841, 46; More Worlds than One, the Creed of the Philosopher and the Hope of the Christian, 1854. Contbd. numerous papers to sci. jours. Noted for researches into polarization of light; discovered that beam of light could be split into reflected and refracted portions (Brewster's law), 1811; discovered (with Biot) biaxial crystals; did research on absorption of light; formulated laws of metallic reflection; invented kaleidoscope, 1816; discovered method of producing interference fringes; improved stereoscope; invented polyzonal lenses for lighthouses, 1835. Died Alterby, nr. Melrose, Feb. 10, 1868.

BREWSTER, William, Am. ornithologist; b. Wakefield, Mass., July 5, 1851; grad. Cambridge High Sch., 1869; hon. A.M., Amherst, 1880, Harvard, 1899; m. Caroline F. Kettell, Feb. 9, 1878. Asst. in charge of collection birds and mammals, Boston Soc. Natural History, 1880-87; in charge dept. of mammals and birds, Cambridge Museum Comparative Zoölogy, 1885-1900, curator, dept. of birds, from 1900; most of his time devoted to managing private museum of ornithology at his place in Cambridge. Author: Bird Migration, 1886; Birds of Cambridge, 1906. Died July 11, 1919.

BREY, Wallace Siegfried, Jr., Am. chemist; b. Schwenksville, Pa., June 6, 1922; s. Wallace S. and Roxie (Lichty) B.; B.S., Ursinus Coll., 1942; M.S., U. Pa., 1946, Ph.D., 1948; m. Mary Louise Van Natta, Apr. 7, 1955; children—William Wallace, Paul David. Chemist, Warner Co., Paoli, Pa., 1942-44; faculty de Pauw U., Greencastle, Ind., 1948-49; faculty St. Joseph's Coll., Phila., 1949-52, asso. prof. chemistry, 1951-52; faculty U. Fla., Gainesville, 1952-—, prof., 1964-—. Cons. Peninsular Chem. Research, 1959-—, Air Products and Chems., 1966-—. Mem. Am. Chem. Soc. (chmn. Fla. sect. 1966), Am. Phys. Soc., Faraday Soc., Am. Assn. U. Profs., A.A.A.S., Sigma Xi. Author: Principles of Physical Chemistry, 1958; Physical Methods of Determining Molecular Geometry, 1965; also articles. Research on preparation of heterogeneous catalysts and reactions at their surfaces, reaction rates and molecular conformation by nuclear magnetic resonance, chem. shifts and coupling constants in nuclear magnetic resonance spectra of fluorine compounds, structure and behavior of water and clathrates of water. Home: 800 N.W. 37th Dr., Gainesville, Fla. 32601.*

BREYER, Maria Gerdina Brandwijk (Mrs. Jan Hendril Breyer), biochemist; b. Culemborg, Netherlands, Nov. 21, 1899; d. Nicolaas and Judith (Pijselman) Brandwijk; Phil. Doct., U. Utrecht (Netherlands), 1923, state exam. pharmacy, 1924; D.Sc., U. Witwatersrand (S. Africa), 1935; m. Jan Hendril Breyer, June 25, 1927; 1 dau., Judith (Mrs. Colin Denis Colclough). Lectr. pharmacology, med. faculty U. Witwatersrand, 1925-40; cons. chemist, research labs., Transvaal and Orange Free State Chamber Mines, S. Africa, 1943-65; pvt. cons. work, Johannesburg, S. Africa, 1965-—. Mem. govt. commn. tng. chemist and druggist in S. Africa, 1951-—. Mem. Royal Soc. S. Africa, S. African Assn. Advancement Sci., S. African Chem. Inst., Internat. Fedn. U. Women. Author: (with J. M. Watt) The Medicinal and Poisonous Plants of Southern Africa, 1932, The Medicinal and Poisonous Plants of Southern and Eastern Africa, 1962; also numerous articles in field. Research on phytochemistry, biochem. toxicology, anatomy and allergic problems of plants; folk, native medicines; protection of timber and textile materials in mines from fungal organisms. Address: 4 Petronelle, Johannesburg, South Africa.*

BRIAN, Luigi, Italian anthropologist; b. Genoa, Italy, Nov. 29, 1915; s. Alfonso and Anna (Reale) B.; ed. in natural scis. Prof. biology and gen. zoology, genetics and biology of races; titular of chair of anthropology U. Genoa; founder Provincial Commn. for San. Culture of Population, Genoa; co-founder, pres. Centro Studi Problemi Sociali; lectr. Mem. Italian Inst. Anthropology, Acad. Ligure of Scis. and Letters (Genoa), Italian Soc. Anthropology and Ethnology, Internat. Bur. Differential Anthropology (Geneva), Internat. Com. Standardization in Human Biology (Paris), Nat. Geog. Soc. (U. S.), others. Author over 40 works including Antropometrographia, 1960, also over 250 articles. Home: corso V. de Michiel 25/13b, Chiavari. Office: University of Genoa, via Balbi 5, Genoa, Italy.

BRIANCHON, Charles-Julien, French mathematician; b. Sèvres, France, Dec. 19, 1783; student Ecole Polytechnique, 1804-08. Commd. lt., arty., 1808; apptd. adj. to dir. gen. arms manufactures, France, 1815; became prof. arty. sch. Royal Guard, 1818; served under Napoleon in Spain and Portugal. Author: Memoire on Curved Surfaces of the Second Degree (paper), 1806; Mémoires sur les lignes du second ordre, 1817; Mémoires sur la poudre à Tirer, 1823; Essai chimique sur les reactions joudroyantes, 1825. Investigated gen. methods of analytical geometry, also co-ordinate transformations, properties of curves and 2d-degree surfaces; studied constrns. with ruler (led to projective geometry and theory of transversals); made notable contributions to projective geometry during its formative period; used polar theory to demonstrate duality of Pascal's hexagram, 1806, also to study problems connected with projective theory of conics, 1817; derived Brianchon's theorem from Pascal's theorem on hexagon; reintroduced concept of anharmonic relation. Died Versailles, France, Apr. 29, 1864.

BRIBEAUVAL, Jean-Baptiste Vaquette de, see de Bribeauval, Jean-Baptiste Vaquette.

BRICAUD, H., French physician; b. Cenon, France, Sept. 18, 1925; M.D., Faculty Medicine, Bordeaux, France, 1955. Research fellow CNRS, 1952-55; chief cardiology dept. CHR, Bordeaux, 1958-—; prof. asso. Faculty Medicine, Bordeaux, 1958-—, prof., 1963-—. Adviser, WHO, 1967-—. Mem. French Soc. Cardiology, French Soc. Radiology. Author: (with P. Broustet, G. Cabanieu) Le Coeur pulmonaire, 1964; also numerous articles. Research on myocardial functional capacity, especially work performance, pulmonary capacity of cardiac patients; studies on local alterations or regulation of coronary, cerebral, hepatic, lung circulation in health and diseases, diagnosis of congenital or acquired cardiovascular disease, immunological aspects of atherosclerosis and blood coagulation. Office: Hopital du Tondu, Bordeaux, France.*

BRICE, John Chadwick, English physicist; b. Minehead, Somerset, Eng., Feb. 7, 1934; s. Joseph Isaac and Grace (Davis) B.; B.A., Queens' Coll., Cambridge, Eng., 1965, M.A., 1959; m. Catherine Elaine Sykes, Apr. 20, 1957; children—Michael, Alison, Caroline. Physicist, Mullard Research Labs., Redhill, Surrey, Eng., 1956-—, leader crystal growth sect., 1960-—. Asso. Inst. Physics. Author: The Growth of Crystals from the Melt, 1965; also articles. Research on preparation and properties of materials used in solid state physics especially effects of preparation conditions on properties of crystals and evaporated layers. Home: 20 Kitsmead, Copthorne, Sussex, Eng. Office Mullard Research Labs. Redhill, Surrey, Eng.*

BRICE, Neil Mather, physicist; b. Brisbane, Australia, Feb. 27, 1934; s. Henry W. and Edith (Howie) B.; B.Sc., U. Queensland, 1954, Hons., 1958, M.Sc., 1959; Ph.D., Stanford, 1965; m. Marilyn Ruth Jordan, Mar. 25, 1961; children—Henry William, Amy Elizabeth Jordan, Betsy Laurel. Came to U. S., 1959. Radio-physicist, Macquarie Island Antarctic Expdn., 1957; leader Stanford Antarctic Very Low Frequency Traverse, 1960-62; asst. prof. Carleton U., Ottawa, Ont., Can., 1964-66; asso. prof. elec. engring. Cornell U., Ithaca, N.Y., 1966-—. Cons. Def. Research Bd., Can., 1964-—. Mem. A.A.A.S., Am. Geophys. Union, Sigma Xi. Research, publs. on theory radio wave generation by Van Allen radiation and developed new diagnostic techniques for upper atmosphere temperature and composition using radio wave propagation. Home: 121 Winston Dr., Ithaca, N.Y. 14851.*

BRICK, Robert Maynard, Am. metallurgist, cons.; b. Atlantic City, N.J., Nov. 18, 1908; s. Walter Clark and Anna (Kain) B.; Met. Engr., Lehigh U., 1929, Ph.D., Yale, 1933; m. Dorothy Smith, Dec. 12, 1936; children—Robert M., Randolph H., Dean C., Deborah V. Research fellow Yale, 1933-38; mem. faculty, 1938-45; fellow, Silliman Coll., 1941-45; dir. materials engring. N.H. Jr. Coll., 1945; prof. and dir. dept. metall. engring. U. Pa., 1946-55; dir. metall. dept.

central research div. Continental Can Co., 1955-61, gen. mgr. Central Research div., 1961-64, gen. mgr. corporate research and devel., 1964-66, cons. metal division, research and devel., 1966—; cons. Los Alamos Sci. Lab., 1951—; legal consultant in field of physical metallurgy since 1940. Recipient American Institute M.E. award for best research publication in Institute of Metals, 1935. Mem. sch. bd., Orange, Conn., 1942-45. Mem. Am. Inst. M.E. (chmn. Inst. of Metals, 1951). Nat. Research Council (mem. com. on ship steel 1950-54), Inst. Metals (London), Am. Soc. Metals, Am. Soc. Engring. Edn. (Am. Inst. M.E. rep. on engrs. joint council com. on nat. sci. leg. 1948-53). Author: Structure and Properties of Alloys, 1942, 3d edit., 1965. Contbr. jours. Research on deformation and recrystallization textures of metals, on segregation upon solidification, on X-ray diffraction analysis of solid solubility, on correlation of properties with microstructure; 21 U. S. patents on metals related to containers and container manufacture. Home: 326 E. 6th St., Hinsdale, Ill. Office: 1200 W. 76th St., Chgo.

BRICKER, Neal S., Am. physician; b. Denver, Apr. 18, 1927; s. Eli D. and Rose (Quiat) B.; B.A., U. Colo., 1946, M.D., 1949; m. Miriam Thalenberg, June 24, 1951; children—Dale, Cary, Susan. Practice medicine, specializing in internal medicine; faculty U. Colo., 1952-54, Harvard Med. Sch., 1954-56; faculty Washington U. Med. Sch., St. Louis, 1956—, prof. medicine, 1965—. Chmn. sci. adv. bd. Nat. Kidney Found., 1965—; vis. investigator Inst. Biol. Chemists, Copenhagen, Denmark, 1960-61; mem. gen. medicine study sect. NIH, 1965—, chmn., 1966—. Recipient Career Research award USPHS, 1964. Diplomate Am. Bd. Internal Medicine. Fellow Assn. Am. Physicians; mem. Am. Heart Assn. (med. adv. bd.), Am. Fedn., Central Soc. for clin. research, Am. Soc. Clin. Investigation, Am. Physiol. Soc., Soc. for Exptl. Biology and Medicine, A.C.P., Sigma Xi, Alpha Omega Alpha. Asso. editor Jour. Lab. and Clin. Medicine, 1961—; mem. editorial com. Jour. Clin. Investigation, 1964—. Studies on normal, path. physiology of kidney; electrolyte transport by isolated biologic membranes. Home: 6624 Waterman St., St. Louis.*

BRICKWEDDE, Ferdinand Graft, Am. physicist; b. Balt., Mar. 26, 1903; s. Ferdinand Henry and Virginia (Graft) B.; A.B., Johns Hopkins U., 1922, M.A., 1924, Ph.D., 1925; m. Marion Langhorne Howard, July 28, 1934; children—Marion Langhorne (dec.), Ruth Catherine (Mrs. Lance Eugene Cooper), Langhorne Virginia. Munsell research asso. in calorimetry Nat. Bur. Standards, Washington, 1925-26, chief low temperature lab., 1926-46, chief heat and power div., 1946-56, chief thermodynamics sect., 1946-52; staff U. Cal. Radiation Lab., Livermore, 1952-53; dean Coll. Chemistry and Physics, Pa. State U., 1956-63, Evan Pugh Research prof. physics, 1963—. Cons. cryogenic group Los Alamos Sci. Lab., 1948—; pres. Comité Consultatif de Thermométrie, Comité International des Poids et Mesures, Sèvres, France, 1965—. Recipient Chem. Soc. Washington Hillebrand prize, 1940; Washington Acad. Scis. award for Outstanding Achievement in Sci., 1941. Fellow Am. Phys. Soc. (bd. editors 1952-55), A.A.A.S.; mem. Am. Chem. Soc., Acoustical Soc. Am., Am. Assn. Physics Tchrs., Internat. Inst. Refrigeration (commn. I), Philos. Soc. Washington (pres. 1939), NRC, Internat. Union of Pure and Applied Physics, also member of Washington Acad. Scis., Sigma Xi. Editor: Part I Basic Concepts, Standards and Methods, Vol. III Temperature, Its Measurement and Control in Science and Industry, 1962. Research and numerous publs. on phys. phenomena at very low temperatures in particular on calorimetry and thermodynamic properties at low temperatures; one of team first to liquefy helium in U. S. at-269° C.; one of team which discovered deuterium, the heavy isotope of hydrogen; initiated, planned the Cryogenic Engring. Lab. of Nat. Bur. Standards at Boulder, Colo.; in charge design of hydrogen liquefiers. Home: 630 W. Fairmount Av., State College, Pa. 16801.* Office: Osmond Lab., Pa. State U., University Park, Pa. 16802.*

BRIDEL, Samuel Élisée, Swiss naturalist; b. Crassier, Switzerland, 1761; made various diplomatic missions to Berlin, Paris, Rome. Author: Muscologia recentiorum, 1797-1803; Methodus nova muscovrum da naturae norman, 1819; Bryologia universa, seu systematica, 1826-27. Research on mosses; laid founds. of modern bryology. Died Gotha, Switzerland, 1828.

BRIDFERTH (Byrhtferth), Anglo-Saxon monk; b. Eng., flourished 10th century; student under Abbot of Fleury; monk of Thorney; monk of Ramsey after 970; tchr. astronomy, the calendar, principles of mathematics. Author: De Temporum Ratione, printed 1612; De Natura Rerum; De Indignatione; De Ratione Unciarum; De Principiis Mathematicis; De Institutione Monachorum; Computus Latinorum ac Graecorum Hebraeorumque et Aegyptiorum necnon et Anglorum; also reputed to have written a life of St. Dustan.

BRIDGE, Josiah, Am. geologist; b. Norwood, O., July 17, 1890; s. Herbert Sage and Theresa (Hill) B.; B.A., U. Cin., 1913; M.S., U. Chgo., 1917; Ph.D., Princeton, 1929; m. Lucy Atwater Brown, June 5, 1918; children—Herbert Sage, Richard Benedict, James Andrew. Faculty Mo. Sch. Mines, 1914, 1920-30; field geologist Standard Oil Co., Mo. Bur. Mines and Geology, Tenn. Geol. Survey, 1929; specialist early Paleozoic stratigraphy U. S. Geol. Survey, 1930; field work Mascat-Jefferson City (Tenn.) zinc dist., 1937. Recipient posthumous award U. S. Dept. Interior, 1953. Fellow Geol. Soc. Am.; mem. Paleontol. Soc., Geol. Soc. Washington. Contbr. geol. bulls. and articles. Work on bauxite, Tenn. zinc, Micronesia projects. Died Takoma Park, Md., July 17, 1953.

BRIDGE, Norman, Am. physician; b. Windsor, Vt., Dec. 30, 1844; s. James Madison and Nancy Ann (Bagley) B.; M.D., Chgo. Med. Coll. (now med. dept. Northwestern U.), 1868, Rush Med. Coll., 1878; A.M. (hon.), Lake Forest U., 1889; LL.D., Occidental, 1920, U. Cal., 1922; Sc.D., Northwestern U., 1923; m. Mae Manford, May 21, 1874. Prof. clin. medicine Rush Med. Coll. (U. of Chgo.), 1887-98, prof. medicine, 1898-1901, emeritus prof., 1901—; v.p. Pan Am. Petroleum & Transport Co. Author: The Penalties of Taste and Other Essays, 1898; The Rewards of Taste and Other Essays, 1902; Lectures on Tuberculosis, 1903; House Health, 1907; Fragments and Addresses, 1915; The Marching Years (autobiographical), 1920; Mental Therapeutics and Other Papers, 1922; also papers on med. and cognate subjects in med. jours. and books. Died Jan. 10, 1925.

BRIDGE, Thomas William, Brit. zoologist; b. Birmingham, Eng., Nov. 5, 1848; s. Thomas Bridge and Lucy (Crosbee) B.; ed. Birmingham and Midland Inst., Eng.; B.A., Trinity Coll. (Found. scholar), Cambridge (Eng. U., 1876, M.A., 1880, Sc.D.; M.Sc., Birmingham. Demonstrator comparative anatomy U. Cambridge; prof. zoology Royal Coll. Sci., Dublin, Ireland; Mason prof. zoology and comparative anatomy U. Birmingham. Fellow Royal Soc., 1903, Linnean Soc., Zool. Soc. Contbr. papers, research on anatomy of fishes especially swim or air bladder. Died Birmingham, June 30, 1909.

BRIDGES, Calvin Blackman, Am. geneticist; b. Schuyler Falls, N.Y., Jan. 11, 1889; s. Leonard Victor and Amelia C. (Blackman) B.; B.S., Columbia, 1912, Ph.D., 1916; m. Gertrude Frances Ives, Sept. 7, 1912; children—Philip Newell, Norman Ives (dec.), Betsy Blackman, Nathan Ives. Research work in heredity under grants from Carnegie Instn., 1915-19; mem. staff Carnegie Instn., 1919—. Mem. Nat. Acad. Scis. Author: The Mechanism of Mendelian Heredity, 1915; (with T. H. Morgan) Sex-linked Inheritance in Drosophila 1916; Contributions to the Genetics of Drosophila Melanogaster (with T. H. Morgan), 1919, 23; Genetics of Drosophila, 1925. Formulated theory of genic balance; an originator of concepts of research in Drosophila leading to formulations of concepts of modern genetics, including proof of part played by chromosomes in conveying hereditary characteristics; studied giant salivary chromosomes in relation to the position of genes. Died Dec. 27, 1938.

BRIDGES, Robert, Am. physician; b. Phila., Mar. 5, 1806; s. Culpepper and Sarah (Clifton) B.; grad. Dickinson Coll., 1824; M.A., U. Pa., 1828. Became tchr. chemistry Phila. Coll. Pharmacy, 1831, trustee, 1839, prof. gen. and pharm. chemistry, 1842, prof. emeritus, 1879; vaccine physician, several years; apptd. dist. physician during cholera epidemic, 1832; prof. chemistry Franklin Med. Coll., 1846-48; became mem. Acad. Natural Scis., 1835, presented index of genera in its Herbarium (with Dr. Paul B. Goddard), 1835, revised index, 1843, pres. acad., 1864; became fellow Coll. Physicians of Phila., 1842, librarian, 1867-79, catalogued Urinary Calculi in Mutter Mus.; became mem. Franklin Inst. of Phila., 1836, Am. Philos. Soc., 1844. Asso. editor Am. Jour. Pharmacy, 1839-46, mem. com. for revision of pharmacopoeia, 1840, 70; editor Am. editions of Elementary Chemistry (George Fownes), Elements of Chemistry (Thomas Graham). Died Feb. 20, 1882.

BRIDGMAN, Percy Williams, Am. physicist; b. Cambridge, Mass., Apr. 21, 1882; s. Raymond Landon and Mary Ann Maria (Williams) B.; A.B., Harvard, 1904, A.M., 1905, Ph.D., 1908; Sc.D., 1934, 39, 41, Princeton, 1950; Yale, 1951; Dr. honoris causa, Paris, 1950; m. Olive Ware, July 16, 1912; children—Jane, Robert Ware. Fellow, Harvard, 1908-10, instr. physics, 1910-13, asst. prof. 1913-19, prof. 1919, Hollis prof. math. and natural philosophy, 1926-50, Higgins U. prof. Harvard, 1950-54, emeritus. Fellow Am. Acad. Arts and Scis.; mem. Am. Philos. Soc., Washington, Nat. acads. scis., Am. Phys. Soc.; corr. mem. Academia Nacional de Ciencias (Mexico); hon. fellow, Phys. Soc. (London). Foreign mem. Royal Soc. (London), Indian Acad. of Scis. Recipient Rumford medal Am. Acad. Arts and Scis., 1917; Cresson medal, Franklin Inst., 1932; Rozeboom medal, Royal Acad. Scis. Amsterdam, 1933; Comstock prize Nat. Acad. Scis., 1933; Research Corporation award, 1937; Nobel prize in physics for researches in high pressure, 1946. Mem. Phi Beta Kappa, Sigma Xi. Author: The Logic of Modern Physics, 1927; The Physics of High Pressure, 1931; Thermodynamics of Electrical Phenomena in Metals, 1934; The Nature of Physical Theory, 1936; The Intelligent Individual and Society, 1938; The Nature of Thermodynamics, 1941; Reflections of a Physicist, 1950; Studies in Large Plastic Flow and Fracture, 1952; The Nature of Some of Our Physical Concepts, 1952; The Way Things Are, 1959. Showed that viscosity increases tremendously with pressure (except for water); obtained new form of phosphorus by heating under pressure, 1921; his work led to prodn. of synthetic diamonds in 1955; also studied electrical conduction in metals and properties of crystals; vigorous exponent of operationalism as a philosophy of science. Died, a suicide, Randolph, N.H., Aug. 20, 1961.

BRIEGER, Ludwig, German internist; b. Glatz (Silesia), July 26, 1849; s. Salomon Brieger; studied in Breslau, Strasbourg, Vienna, Berlin; m. Adele Pacully. Asst. to H. I. Quincke at internal clinic in Bern; joined faculty in Berlin, 1881; founder pvt. polyclinic for internal diseases and pvt. lab., 1887; asso. prof. 1890; dept. head Königliche Institut für Infektions Krankheiten, 1891; taught physical-dietetic health methods U. Berlin, 1900; dir. U. Inst. for Hydrotherapy, 1901; hon. prof., 1919. Mem. Deutsche Gesellschaft für Volksbäder (pres.), Balneologische Gesellschaft (chmn.). Author: Über Ptomaine, 3 parts, 1885-86; (with M. Mayer) Licht als Heilmittel, 1904; (with A. Laqueur) Moderne Hydrotherapie, 1904, Physikalische Therapie der Erkrankungen der Muskeln und Gelenke, 1906; (with W. Krebs) Grundriss der Hydrotherapie, 1909. Pub. important work on metabolic and infectious diseases; known for work on ptomaines and toxins, proof of anti-enzyme content in blood serum in cancer known as Brieger reaction. Died Berlin, Germany, Oct. 18, 1919.

BRIESKORN, Carl Heinz, German chemist; b. Königsberg, Prussia, Germany, Nov. 10, 1913; s. Herbert and Katharina (Schuchardt) B.; pharmacist exam U. Munich (Germany), 1939, food chem. exam (certification), 1941, Dr.rer.nat., 1942; postgrad. U. Strasbourg (France); 19—; m. Endrich Elfriede, Apr. 9, 1942; children—Gerhard, Norbert, Renate. Asst., U. Strasbourg, 1941-44; staff State Research Inst. for Food Chemistry, Munich, 1945-47; lectr. U. Munich, 1947-52; dir. U. Instanbul (Turkey), 1952-55; sci. councillor U. Münster (Germany), 1955-59; asso. prof. U. Tübingen (Germany), 1959-60; prof. U. Würzburg (Germany), 1960—. Mem. Deutsche pharm. Gesellschaft, Gesellschaft deutscher Chemiker, Am. Chem. Soc., Deutsche Gesellschaft für Heilpflanzen forschung, Physico Medica Würzburg. Author: Grundlagen der Galenischen Pharmazie, 1955; also numerous articles. Research on volatile oils and their components especially terpenes; bitter principle (isolation of carnosol from sage, of picropoline from germander sorts); chemistry of color reactions; determination of connective tissue in meats on basis of tryptophane; composition of vegetable cuticula with respect of effect pesticides. Home: Trautenauer Strasse 45, 87 Würzburg, Germany.*

BRIFFAULT, Robert Stephen, surgeon, anthropologist; b. London, Eng., 1876; s. Frederic and Margaret M. (Stewart) B.; ed. privately Florence, Italy, London; M.B.; Ch.B.; m. Anna Clarke (dec. 1919); 1 son, 1 dau.; m. 2d, Herma Hoyt. Practice surgery New Zealand, from 1894, ret. circa 1919. Author: The Making of Humanity, 1919; Psyche's Lamp, 1921; The Mothers, 1927; Rational Evolution, 1930; Sin and Sex, 1931; Breakdown, 1932; Europa (novel), 1935; Reasons for Anger, 1936; Europa in Limbo (novel), 1937; The Decline and Fall of the British Empire, 1938; Fandango (novel), 1940; Les Troubadours et le Sentiment romanesque, 1945; New Life of Mr. Martin (novel), 1946. Analyzed matriarchy in primitive soc.; made controversial strongly pro-communist studies of modern polit., social conditions. Died Hastings, Eng., Dec. 11, 1948.

BRIGGS, Basil Hugh, physicist; b. Bradford, Eng., Aug. 30, 1923; s. Horace Franklin and Elsie (Smith) B.; B.A., Cambridge (Eng.), U., 1945, M.A., 1948, Ph.D., 1952; m. Gillian Awbery, Aug. 12, 1954; children—Keith, Jennifer, Alan, Margaret, Janet. Sci. officer Telecommunications Research Establishment, 1943-46; research student Cavendish Lab., 1947-51, Dept. Sci. and Indsl. Research appointee, 1955-60; demonstrator physics Cambridge U., 1951-55, asst. dir. research in physics, 1960-62; faculty U. Adelaide (South Australia), 1962—, reader in physics, 1963—. Fellow Inst. Physics, Australian Inst. Physics; mem. Am. Geophys. Union. Research, publs. on irregularities in ionosphere and their movements, upper atmosphere winds using radio methods. Home: 5 Euston Av., Highgate, South Australia. Office: Dept. Physics, U. Adelaide, South Australia.*

BRIGGS, George Edward, English plant physiologist; b. Grimsby, Eng., June 25, 1893; s. Walker Thomas and Susan (Townend) B.; M.A., St. John's Coll., Cambridge (Eng.) U.; m. Nora Burman, May 1, 1920; children—Peter George, Janet Mary (Mrs. Peter Berners Fellgett). Faculty, Cambridge U., 1919—, prof. botany, 1948-60, prof. emeritus, 1960—, fellow St. John's Coll., 1920—, pres., 1952-63. Fellow Royal Soc., 1935. Author: (with A. B. Hope, R. N. Robertson) Electrolytes and Plant Cells, 1962; Movement of Water in Plants, 1968; also numerous articles. Research on devel. photosynthetic activity in seedling leaves, analysis growth plants, velocity enzyme reactions in relation to concentration substrate, mechanism by which plant cells accumulate electrolytes in vacuoles. Home: 10 Luard Rd., Cambridge, Eng.*

BRIGGS, George McSpadden, Am. nutritionist; b. Grantsburg, Wis., Feb. 21, 1919; s. George McSpadden and Mary (McNelly) B.; B.S., U. Wis., 1940, M.S., 1941, Ph.D., 1944; m. Eleanor Reese, June 21, 1941; children—Cathy, Marilyn, Nancy. Chief nutrition unit lab. nutrition and endocrinology Nat. Inst. Arthritis and Metabolic Diseases NIH, U. S. Dept. Health, Edn. and Welfare, Bethesda, Md., 1951-58, exec. sec. biochemistry and pharmacology tng. coms. div. Gen. Med. Scis., 1958-60; prof. nutrition, biochemist Agrl. Expt. Sta., chmn. dept. nutritional scis. U. Cal., Berkeley, 1960—; scientist USPHS, 1957—. Mem. research com. Nutrition Found and Nat. Live Stock and Meat Bd., 1962—. Recipient Borden award Poultry Sci. Assn., 1958. Fellow Am. Pub. Health Assn., A.A.A.S.; mem. Am. Chem. Soc., Am. Dietetic Assn. (editorial bd. Jour. 1963-66), Am Inst. Nutrition (pres. 1967-68), Am. Soc. Animal Sci., Am. Soc. Biol. Chemists, Animal Nutrition Research Council, Inst. Food Technologists, Nutrition Soc. Gt. Britain, Poultry Sci. Assn., Soc. Exptl. Biology and Medicine, Sigma Xi, Phi Kappa Phi, Gamma Alpha, Phi Lambda Upsilon, Phi Eta Sigma, Alpha Zeta, Alpha Nu. Author: (with others) Nutrition and Physical Fitness, 1966. Asso. editor: Nutrition Revs., 1954-58. Mem. editorial bd. Fedn. Proc., 1957-60; Jour. of Nutrition, 1962—. Publs. on discovery, requirements of newer mems. of the vitamin B-complex, specifically vitamins B10, 11, 12; use of synthetic diets for chickens and guinea pigs. Home: 877 Revere Rd., Lafayette, Cal. 94549. Office: Dept. Nutritional Scis., U. Cal., Berkeley, Cal. 94720.*

BRIGGS, Henry, English mathematician; b. Warley Wood, Halifax, Eng., Feb. 1561; B.A., St. John's Coll., Cambridge (Eng.) U. 1581, M.A., 1585; M.A., Oxford U., 1619. Became fellow Cambridge, 1588; 1st prof. geometry Gresham Coll., London, 1596-1620; Savilian prof. astronomy, from 1619; fellow-commoner Merton Coll. Author: A Table to find the Height of the Pole, the Magnetical Declination being given, 1602; Tables for the Improvement of Navigation, 1610; Logarithmorum chilias prima, 1617; Lucubrationes et annotationes in opera posthuma J. Neperi, 1619; Euclidis Elementorum sex libri priores, 1620; A Tract on the Northwest Passage to the South Sea through the Continent of Virginia, 1622; Mathematica ab antiquis minus cognita; Arithmetica logarithmica, 1624; Trigonometria Britannica, 1633. Invented modern method of long division; calculated logarithmic tables to 14 decimal places; proposed system of common (or Briggsian) logarithms (tabulated to base 10) instead of Naperian system. Died Oxford, Jan. 26, 1630.

BRIGGS, John Carmon, Am. biologist; b. Portland, Ore., Apr. 9, 1920; s. Revoe C. and Jessie (Carmon) B.; B.S., Ore. State U., 1943; M.A. Stanford, 1947, Ph.D., 1952; m. Terry L. Beecher, Nov. 19, 1954; children—Linda, David, Daniel, Carleton, Douglas, Katherine, Samuel, Elizabeth. Aquatic biologist U. S. Fish and Wildlife Service, Corvallis, Ore., 1945-46; biologist Cal. Div. Fish and Game, Orick, 1950-51; research asso. Natural History Mus., Stanford, 1952-54; faculty U. Fla., Gainesville, 1954-58; asst. prof. U. B.C., Vancouver, Can., 1958-61; research scientist Inst. Marine Sci., U. Tex., 1961-64; prof. zoology, chmn. zoology and oceanography U. S. Fla., Tampa., 1964—. Recipient Phipps and Bird award Fla. Acad. Sci., 1959. Mem. Am. Soc. Ichthyologists and Herpetologists (past v.p., bd. govs. 1953-58, 59-64, 65—), Marine Biol. Assn. U.K., A.A.A.S., Am. Inst. Biol. Scis., Soc. for Study Evolution, Soc. Systematic Zoologists, Am. Fisheries Soc., Systematics Assn., Soc. for Study Bibliography Natural History, Biogeog. Soc. Japan, Marine Biol. Assn. Ind. Author: (with LF. DeBeaufort) Fishes of the Indo-Australian Archipelago, vol. XI, 1964; also articles, monograph. Research on evolution, distbn. and behavior fishes, gen. marine zoogeography, relation between marine paleotemperatures and endemism in marine animals around oceanic islands, effect geog. location on relative evolutionary success. Home: 11503 Lake Ridge Rd., Tampa, Fla. 33618.*

BRIGGS, Joseph Emmons, Am. surgeon; b. Dighton, Mass., Mar. 13, 1869. S. Albert and Sarah Jane (Simmons) B.; ed. Bristol Academy, Taunton, Mass., M.D., Boston U., 1890; post-grad. work U. of Vienna; m. Carrie A. Moore, Sept. 20, 1893 (died 1900); m. 2d., Flora A. Toulmin, Sept. 10, 1903. Practiced medicine, Boston, 1890; prof. surgery, Boston U. Sch of Trustee Mass. Mem. Hosps., Boston U. Contbr. med. Medicine, 1918; cons. surgeon Mass. Mem. Hosp.; articles. Invented a dilatable rubber bag used to control bleeding after suprapubic prostatectomy (Brigg's bag), 1906. Died Jan. 3, 1942.

BRIGGS, L(loyd) Vernon, Am. psychiatrist; b. Boston, Aug. 13, 1863; s. Lloyd and Sarah Elizabeth E. (Kent) B.; student Harvard Med. Sch., Tufts Med. Coll., Dartmouth Med. Coll.; M.D., Med. Coll. Va., 1899; m. Mary T. Cabot, June 1905; 1 son, Lloyd Cabot. Practiced in Boston, 1899—; visited hosps. for insane in Eng. and Scotland, 1905, Germany, Austria, Holland, Switzerland, 1907; formerly dir. N.E. Hosp. for Women and Children; commr. alien insane for Mass., 1912-16; mem., sec. Mass. State Bd. of Insanity, 1913-16; pres. of staff, physician to mental dept. Boston Dispensary; mem. Commn. for Reorganization Dept. of Mental Diseases of Mass., 1937-38; chief of neuropsychiatric service Base Hospital, Camp Devens, Ayer, Mass. Author: The Manner of Man that Kills; History of the Boston Psychopathic Hospital; Around Cape Horn in the Bark Amy Turner, 1926; Kent Genealogy; Genealogy of the Cabot Family in America; Experience of a Medical Student in Honolulu, 1926; California and the West, 1931; Arizona and Mexico, 1932; History and Genealogy of the Briggs Family, 1937. Author Briggs Law of Mass. requiring mental examination before trial of all persons indicted for felony in Mass. Died Feb. 28, 1941.

BRIGGS, Lyman James, Am. physicist; b. Assyria, Barry Co., Mich., May 7, 1874; s. Chauncey L. and Isabella (McKelvey) B.; B.S., Mich. State Coll., 1893, Sc.D., 1932; M.S., U. Mich., 1895, LL.D., 1936; Ph.D., Johns Hopkins, 1901; D.Eng., S.D. Sch. Mines, 1935; Sc.D., George Washington U., 1937, Georgetown U., 1939; Columbia, 1944; m. Katherine E. Cook, Dec. 23, 1896; children—Mrs. Isabel Myers, Albert Cook (dec.). In charge of Phys. Lab. Div. (now Bur. Soils), U. S. Dept. Agr., 1896-1906; physicist in charge Biophys. Lab., Bur. Plant Industry, 1906-12, in charge of biophys. investigations, 1912-20; detailed to Bur. Standards by exec. order, 1917-19, chief div. of mechanics and sound, 1920-33, asst. dir. research and testing, 1926-33, dir. Nat. Bureau of Standards, 1933-45, dir. emeritus, 1945—. Mem. Nat. Advisory Com. for Aeronautics, 1933-45; vice chmn. 1942-45, chmn. sub-com. on aircraft structures 1937-45, mem. aero-dynamics subcom., 1922-30; chmn. Fed. Specifications Bd., 1932-40, Fed. Fire Council, 1933-39; pres. Nat. Conf. on Weights and Measures, 1935-45; mem. Internat. Ice Patrol Bd., 1933-45; chmn. Washington Biophys. Inst. Council, 1933-39; bd. dirs. Am. Standards Assn., 1933-45; mem. U. S. Nat. Com. for Internat. Geophys. Year; exec. com. engring. div. NRC, 1945-50; dir. scientific program stratosphere balloon flights. Trustee George Washington U., 1945—. Recipient Magellan medal (with Paul R. Heyl), 1922. Medal for Merit, 1948; Gold medal, U. S. Dept. of Commerce for exceptional service. Fellow Am. Coll. Dentists (hon.), A.A.A.S., Am. Phys. Soc. (pres. 1938); mem. Nat. Acad. Scis., Am. Soc. M.E., Washington Acad. Sci. (pres. 1917); Philos. Soc. Washington (pres. 1916), Am. Philos. Soc., Am. Acad. Arts and Scis., Inst. Aero. Scis., Phys. Soc. Eng. (hon.), Newcomen Soc. (Eng.), Washington Acad. Medicine (pres. 1945-46), Tau Beta Pi, Sigma Xi, Sigma Pi Sigma. Research in aerodynamics and aircraft materials; collaborator (with Paul R. Heyl) on devel. earth inductor compass; studied behavior of objects in wind streams of high velocity, soil analysis, liquids under negative pressures. Contbr. to govt. and tech. publs. Died, 1963.

BRIGGS, Robert William, Am. biologist; b. Watertown, Mass., Dec. 10, 1911; s. Robin John and Bridget (McGonigle) B.; B.S., Boston U., 1934; Ph.D. (Austin Teaching fellow), Harvard, 1938; m. Janet Elizabeth Bloch, Sept. 27, 1940; children—Evan W., Alexander B., Meredith. Research fellow McGill U., Montreal, Que., Can., 1938-42; biologist Inst. for Cancer Research, Phila., 1942-56; prof. Ind. U., Bloomington, 1956-63, research prof., 1963—. Panelist, NSF, 1954-56. Mem. Nat. Acad. Scis., Am. Acad. Arts and Scis., Am. Soc. Zoologists. Research, publs. on function cell nucleus in early devel.; showed by nuclear transplantation that nuclei do not begin to exert specific effects in devel. until gastrulation. Home: 1900 Atwater Av., Bloomington, Ind. 47401.*

BRIGGS, William, Brit. physician, oculist; b. Norwich, Eng., 1642; s. Augustine Briggs; M.A., Cambridge (Eng.) U., 1670, M.D., 1677; postgrad. under Vieussens at Montpellier, France; 1 son, Henry Briggs. Became fellow Corpus Christi Coll., Cambridge U. 1668; physician St. Thomas's Hosp., 1682-89; physician in ordinary to William III, from 1696. Fellow London Coll. Physicians (censor 1685, 86, 92). Author: Ophthalmographia, 1676; Theory of Vision, Part I (Part II pub. in Philos. Trans. 1683), 1682. Recognized retina as expansion containing fibers of optic nerve; emphasized hypothesis of vibrations as explanation for nervous action; provided 1st recorded description of nyctalopia, 1684. Died Sept. 4, 1704.

BRIGGS, William Egbert, Am. mathematician; b. Sioux City, Ia., Mar. 26, 1925; s. Egbert Estabrook and Berenice (Reynolds) B.; B.A. magna cum laude, Morningside Coll., 1948; M.A., U. Colo., 1949, Ph.D., 1953; m. Muriel Mae Lambert, Aug. 29, 1947; children—William L., Roger P., Barbara Ellen, Lindsey Ann. Asst. instr. Morningside Coll., 1947-48; math. tchr. Elwood (Ia.) High Sch., 1948, part-time instr. U. Colo., Boulder, 1948-53, research asst., 1953-54, staff math. dept., 1955-63, acting dean, 1963-64, dean Coll. Arts and Scis., 1964—; math. tchr. Baseline Jr. High Sch., Boulder, 1953-54. Dir. Acad. Year Inst. for High School Tchrs. Sci. and Math., acting chmn. math. dept., 1959-60; hon. research asso. Univ. Coll., London, Eng., 1961-62. Mem. Am. Math. Soc., Math. Assn. Am. (gov. 1963-66), Am. Assn. U. Profs., Sigma Xi. Author: (with others) Analytic Geometry, 1964. Research and publs. in analytic number theory and sieve-generated sequences. Home: 1440 Sierra Dr., Boulder, Colo. 80302.*

BRIGHAM, Albert Perry, Am. geologist; b. Perry, N.Y., June 12, 1855; s. Horace A. and Julia (Perry) B.; A.B., Colgate U., 1879, A.M., 1882, LL.D., 1925; postgrad. Hamilton Theol. Sem., 1879-82; A.M., Harvard, 1892; Sc.D., Syracuse, 1918; L.H.D., Franklin, 1921; m. Flora Winegar, June 27, 1882; children—Charles Winegar (dec.), Elizabeth (Mrs. Lawrence V. Roth). Ordained Bapt. ministry, 1882; pastor Stillwater, N.Y., 1882-85, Utica, 1885-91; became prof. geology, Colgate, 1892, emeritus prof. Instr. Harvard Summer Sch. of Geology, 4 summers; prof. Cornell Summer Sch., 1901-04, U. Wis., 1906; lectr. Oxford U. Sch. of Geography, 4 summers and Hilary term, 1924, also at U. of London, 1924. Vice chmn. div. geol. geography NRC; mem. gen. geog. com. U. S. Commn. Washington Bicentenary. Author: A Text-Book of Geology, 1900; Introduction to Physical Geography (with G. K. Gilbert), 1902; Geographic Influences in American History, 1903; Students' Laboratory Manual of Physical Geography, 1904; From Trail to Railway Through the Appalachians, 1907; Commercial Geography, 1911; Essentials of Geography (with C. T. McFarlane), 1916; Cape Cod and the Old Colony, 1920; Manual for Teachers of Geography (with C. T. McFarlane), 1921; The United States of America (U. of London Press), 1927; Glacial Geology and Geographic Conditions of Lower Mohawk Valley, 1929. Collaborator, N.Y. State Museum; consultant in geography, Library of Congress. Contbg. editor Geog. Rev. Died Mar. 31, 1932.

BRIGHAM, Amariah, Am. physician; b. New Marlboro, Mass., Dec. 26, 1798; s. John Brigham; trained in offices of Drs. E. C. Peet (New Marlboro) and Plumb (Canaan, Conn.); m. Susan C. Root, Jan. 23, 1833. Practiced medicine, Enfield, Mass., 1820-22, Greenfield, Mass., 1822-29, Hartford, Conn., 1830-37; prof. anatomy and surgery Coll. Phys. and Surg., N.Y.C.; physician, supt. Retreat of Insane, Hartford, 1840, N.Y. State Lunatic Asylum, 1842-49; founder Am. Jour. Insanity. Author: Remarks on the Influence of Mental Cultivation on Health, 1832; A Treatise on Epidemic Cholera, 1832; Observations on the Influence of Religion on the Health of Mankind, 1835; An Inquiry Concerning the Diseases and Functions of the Brain, the Spinal Cord, and Nerves, 1840; Died Utica, N.Y., Sept. 8, 1849.

BRIGHAM, Carl Campbell, Am. psychologist; b. Marlboro, Mass., May 4, 1890; s. Charles Francis and Ida (Campbell) B.; Litt.B., Princeton, 1912, A.M., 1913, Ph.D., 1916; m. Elizabeth Duffield, Feb. 10, 1923; 1 dau., Elizabeth Hollister. Began as coll. Instr., 1916; asst. to chief Fed. Bd. for Vocational Edn., 1919-20; asst. prof. psychology Princeton, 1920-24, asso. prof., 1924-27, prof. since 1928. Mem. A.A.A.S., Am. Psychol. Assn. Author: Two Studies in Mental Tests, 1917; Study of American Intelligence, 1923; Study of Error, 1932. Contbr. to sci. publs. Died Jan. 24, 1943.

BRIGHT, Sir Charles Tilston, English telegraph engr.; b. London, Eng., June 8, 1832; ed. Merchant Taylor's Sch. Joined Electric Telegraph Co., 1847; became engr.-in-chief Atlantic Telegraph Co., 1856, which laid 1st Transatlantic cable between Ireland and Nfld., Can., 1858; cons. engr. Magnetic Co., 1860-70; cons. engr. on laying Atlantic cables of 1865 and 66; laid cables on land and sea between Port Patrick and Donaghadee, Ireland; laid cables in Mediterranean Sea, 1861-73, W.I., 1865-68; Liberal mem. Parliament, 1865-68; Mem. Inst. Elec. Engring. (pres. 1886-87). Knighted, 1858. Patentee acoustic telegraph (Brights Bells), 1855. Improved (with Josiah Latimer Clark) method for application asphalt covering to submarine cables; laid 1st cable between Ireland and Scotland, 1853; 1st cable between Am. and Europe, 1858. Died Kent, Eng., May 3, 1888.

BRIGHT, Harold Frederick, Am. statistician; b. Smethport, Pa., Aug. 6, 1913; s. Stanley and Florence (Dunn) B.; B.A. in Math., Lake Forest Coll., 1937; M.A. in Math., U. Rochester, 1944; Ph.D. in Psychology, U. Tex., 1952; m. Elizabeth Korhumel, Mar. 23, 1938; children—Stanley J., Beverly Ann. Instr., Tamalpais Sch., San Rafel, Cal., 1937-39; chmn. dept. math. San Angelo (Tex.) Coll., 1941-43, register, dir. guidance, 1945-49; asst. prof. math. Denison U., Granville, O. 1943-44; asst. prof. math. U. Rochester (N.Y.), 1944-45; asso. dir. research Am. Assn. Jr. Colls., Austin, Tex., 1949-52; chief tech. services Human Resources Research Office, Washington, 1952-54, dep. dir., 1954-56; specialist Operations Research and Synthesis, Gen. Electric Co., Plainville, Conn., 1957-58; prof. statistics George Washington U., Washington, 1958—, exec. officer, 1958-64, asso. dean faculties, 1964-66, vice president for academic affairs, 1966—, research staff mem. Patent, Trademark and Copyright Found., 1959—, cons. U. S. Army Logistics Research Project, 1960-62. Research asso. Tex. Legislative Council, summer 1950; mem. math. steering group U. S. Dept. Army, 1954-56; vis. lectr. grad. faculty Trinity Coll., Hartford, Conn., 1957-58; cons. statis. problems in research Human Resources Research Office, 1958—. Mem. Math. Assn. Am., Am. Psychol. Assn., Inst. Math. Statistics, Royal Statis. Soc., Am. Assn. U. Profs., Sigma Xi. Research on computer applications in statistics. Home: 314 Branch Circle S.E., Vienna, Va. 22180. Office: Dept. Statistics, George Washington U., Washington 20006.*

BRIGHT, Norman Francis Henry, chemist; b. Poole, Eng., May 9, 1913; s. Frank Wells and Ellen (Gradage) B.; B.Sc. with honors in Chemistry, U. Bristol (Eng.), 1934, Ph.D., 1937; m. Marion Ferguson Laidlaw, June 4, 1940; 1 son, Michael Wells. Research staff Imperial Chem. Industries Ltd., Ardeer, Scotland, 1937-39, 44-51, Cornwall, Eng., 1939-44; postdoctorate fellow NRC Can., Ottawa, Ont., 1951-53;

head phys. chemistry sect. mineral scis. div. Canadian Dept. Energy, Mines and Resources, Ottawa, 1953—. Fellow Royal Inst. Chemistry (London), Am. Ceramic Soc., Chem. Inst. Can. (mem. council 1965—); mem. Mineral. Assn. Can. (founding), Internat. Union Pure and Applied Chemistry (Canadian rep. on commn. 1960—). Research, publs. on inorganic chem. reactions occurring at elevated temperatures, phase equilibrium studies in oxide systems, chem. aspects of tech. of electronic and magnetic ceramics. Fgn. mem. editorial bd. Revue des Hautes Températures et des Réfractaires, 1965—. Home: 200 Rideau Terrace, Ottawa 2. Office: 555 Booth St., Ottawa 4, Ont., Can.*

BRIGHT, Richard, English physician; b. Queen Square, Bristol, Eng. Sept. 28, 1789; s. Richard Bright; M.D., Edinburgh (Scotland) U., 1812; Student Guy's Hosp., London, Eng., Peterhouse, Cambridge (Eng.) U.; M.D., m. Miss Babington; 1 son; m. 2d, Miss Follet; 3 sons, 2 daus. Traveled on continent, 1818-20; physician Guy's Hosp., 1824-43, became cons. physician, 1843; apptd. physician extraordinary to Queen Victoria, 1837; Gulstonian lectr., 1833; censor, 1836, 39; Lumleian lectr., 1837; mem. council, 1838, 43. Recipient Monthyon medal Inst. France. Fellow Coll. Physicians, Royal Soc., 1821. Author: Reports of Medical Cases, 1827; Address at the Commencement of a Course of Lectures on the Practice of Medicine, 1832; Gulstonian Lectures on the Functions of the Abdominal Viscera, 1833; Clinical Memoirs on Abdominal Tumours, 1860; also contributed to Guy's Hospital Reports, 1836. Explained that kidney disorders caused certain dropsy symptoms; these kidney disorders known collectively as Bright's disease, 1827; studied (with Bouillard) cirrhosis and described syphilitic changes of liver; discovered (with Chopart and Bouchardat) histological changes of pancreas in diabetics. Died London, Dec. 16, 1858.

BRIGHT, Timothy, Brit. physician, inventor; b. Cambridge, Eng., circa 1551; M.D., Trinity Coll., Cambridge, 1579; student medicine, Paris. Physician, St. Bartholomew's Hosp., 1586-90; granted exclusive privilege by Queen Elizabeth for teaching and printing shorthand, 1588; became rector Methley, 1591, Berwick-in-Elmet, 1594. Author: Treatise on Melancholy (thought to have inspired Robert Burton's Anatomy of Melancholy), 1586; Characterie, 1588. Inventor modern shorthand with alphabetical basis for initial letters, 1588. Died York, Eng., Nov., 1615.

BRIL, Mark Timofeevich, Russian dermatologist, venerologist; b. 1890; grad. Med. Faculty, Yurev U., 1917; D.Med. Sci., 1939. Physician, Glavche Skin and Venereal Diseases Polyclinic, Odessa, 1917-20; head skin and venereal diseases dept. Petrozavodsk Central Hosp., 1920-30; asso. Leningrad Skin and Venereal Diseases Inst., 1930-31; sr. asso. Bashkiria Skin and Venereal Diseases Inst., Ufa, 1930-31; head chair skin and venereal diseases Tomsk Med. Inst., 1940-50, Volgograd Med. Inst., 1950—; prof., 1940—. Mem. Volgograd Dermatol. and Venerological Soc. (chmn.). Author circa 70 works including Saltless Diet in Tuberculosis of the Skin, 1945; Chronic Pyococcus Ulcers of the Tibia, 1947; Child Eczema, 1961. Mem. editorial council Jour. Dermatology and Venerology. Address: Volgograd Med. Inst., ulitsa Pavshikh bortsov 1, Volgograd, RSFSR, USSR.

BRILL, Abraham Arden, psychiatrist; b. Austria, Oct. 12, 1874; s. Philip and Esther B.; Ph.D., N.Y. U., 1901; M.D., Coll. Phys. and Surg. (Columbia), 1903; m. K. Rose Owen, May 21, 1908; children—Gloia Bernheim, Edmund. Served as asst. phys. Central Islip State Hosp., and in clinic of psychiatry, Zurich, Switzerland; chief of clinic in psychiatry, Columbia U., and lectr. psychoanalysis and abnormal psychology N.Y. U.; was asst. prof. psychiatry, Postgrad. Med. Sch.; later lectr. on psychoanalysis and psychosexual scis. Columbia; clin. prof. psychiatry N.Y. U. mem. Am. Psychopathol. Assn., Am. Psychoanalytic Assn., A.M.A., N.Y. Acad. Medicine, Am. Psychiat. Assn., Am. Therapeutic Assn., A.A.A.S., Anthropol. Ethnol. Soc. Author: Psychoanalysis—Its Theories and Practical Application; Fundamental Conceptions of Psychoanalysis, 1921, 1922; Freud's Contribution to Psychiatry, 1944; Lectures on Psychoanalytic Psychiatry, 1946; also numerous pamphlets on Psychiatric subjects. Translator: Jung's Psychology of Dementia Praecox, 1909, 1936; Freud's Selected Papers on Hysteria, 1909; Three Contributions to the Théory of Sex (Freud), 1910, 30; Freud's Interpretation of Dreams, 1913. 1932: Psychopathology of Everyday Life. 1914; Wit and Its Relations to the Unconsicious, 1916; Leonardo da Vinci, 1916; The History of the Psychoanalytic Movement, 1917; Totem and Taboo, 1918; Reflections on War and Death, 1918; Psychoanalysis—Exploring the Hidden Recesses of the Mind, 1924. Editor: English edition of Bleuler's Text Book of Psychiatry, 1925; Breuer and Freud—Studies in Hysteria, 1936; The Basic Writings of Sigmund Freud. 1938. Known chiefly for his translations of Jung and Freud, and for his exposition of their theories. Died N.Y.C., Mar. 2, 1948.

BRILL, Alexander Wilhelm von, see von Brill, Alexander Wilhelm.

BRILL, Dieter Rudolf, physicist; b. Heidelberg, Germany, Aug. 9, 1933; s. Rudolf and Else (Rudloff) B.; A.B., Princeton, 1954, Ph.D., 1959. came to U. S 1950. Faculty, Princeton, 1958-60; Flick Exchange fellow U. Hamburg, Germany, 1960-61; faculty Yale, 1962—, asso. prof. physics, 1967—. Cons. Sci. Teaching Center, Mass. Inst. Tech., 1966—. Mem. Am. Phys. Soc., Phi Beta Kappa, Sigma Xi. Contbg. author: Gravitation—An Introduction to Current Research, 1962. Research, publs. on gravitation, rotating masses inertial frames, Mach's prin., founds. quantum mechanics; proof that gravitational waves carry positive energy. Home: 453 Calhoun Coll., Yale, New Haven. 06520.

BRILL, Nathan Edwin, Am. physician; b. N.Y.C., Jan. 13, 1860; s. Simon and Adelheit (Frankenthal) B.; A.B., Coll. City N.Y., 1877, A.M., 1883; M.D., N.Y. U., 1880; m. Elsa M. Josephthal, June 8, 1899; children—Elisabeth Joyce, John Lewis. Attending phys. 1st med. div. Mt. Sinai Hosp., from 1893; prof. clin. medicine, Coll. Phys. and Surg. (Columbia), from 1910; cons. physician Mt. Sinai Hosp. Trustee N.Y. Acad. Medicine; mem. Pub. Health Com.; fellow New York Acad. Medicine. Translator: Klemperer Clinical Diagnosis, 1898. Discovered previously unrecognized form of typhus fever (Brill's disease), 1910; active in study of diseases of blood-forming organs; among 1st to introduce operation of splenectomy for thrombocytopenic purpura into U. S. Died Dec. 13, 1925.

BRILL, Norman Quintus, Am. physician; b. N.Y.C., Aug. 2, 1911; s. Louis and Ella (Applebaum) B.; B.S., Coll. City N.Y., 1930; M.D., N.Y. U., 1934; m. Doris R. Corcoran, Jan. 21, 1937; children—James C., Peter L., Mary C. Private practice of neurology and psychiatry, N.Y.C., 1939-41; pvt. practice, cons. Walter Reed Gen. Hosp., 1946-53; prof. chmn. dept. neurology Georgetown U. Med. Sch., 1946-49; chief research br. neurology and psychiatry div. VA central office, 1946-49; prof., chmn. dept. psychiatry U. Cal. Med. Sch., Los Angeles, 1953—; med. supt. neuropsych. inst., U. Cal. Med. Center, 1953—; sr. cons. psychiatry Brentwood VA Hosp., Los Angeles, 1953—; cons. Met. State Hosp., Norwalk, Cal.; national consultant in psychiatry, USAF, 1962—. Bd. regents Nat. Library Medicine, 1961-65, dep. dir. neuropsychiatry div. Office Surgeon Gen., 1945-46. Fellow Am. Psychiatric Association, American, Psychoanalytic Assn., Am. Coll. Psychiatrists; mem. A.M.A. (former mem. mental health council), A.A.A.S., Sigma Xi. Author: Psychiatry in Medicine, 1962. Research on electroencephalography; 1st definitive follow-up study on war neuroses, relationship between age and resistance to mil. stress, stimulation of tissue by magnetic fields, adrenal gland activity in mental disorders. Home: 230 N. Cliffwood Av., Los Angeles 90040. Office: U. Cal. Med. Sch., Los Angeles 90024.*

BRILLOUIN, Louis Marcel, French physicist; b. Saint Martin de Melle, France, Dec. 19, 1854; s. Marcel and Mascart B.; entered École normale supérieure, 1874; received doctorate; m. Stefanie Prussak, Nov. 1913; 1 dau. Mem. faculties Nancy, Dijon, Toulouse (all France); named maitre de confs. École normale, 1887; apptd. prof. Coll. de France, 1900. Mem. French Acad. Scis., 1921. Author: Propagation de l'électricité, 1904; Lecons sur la viscosité des liquides et des gaz, 2 vols., 1904; Mémoire sur l'ellipticité du géoïde dans le tunnel du Simplon, 1908; Stabilité des aeroplanes, 1910; Actions héréditaires discontinués et équations différentielles qui en résultent, 1920; Recherches sur la structure des cristaux et l'anisotropie des molecules; Etude des conditions de similitude dans le champ électrique; Études théoriques dur l'élasticité des solides isotropes; Travaux sur la relativite: les points singuliers de l'univers d'Einstein. Developed math. theory of gas diffusion, 1900; research on viscosity of liquids and gases, 1907, stability of airplane, 1910, reproduction of waves in dielectric cables, 1939, also distribution of weight on globe, structure of crystals, radiotelegraphy, relativity; proved that the specific heat of a transmitting medium is proportional to cube of absolute temperature. Died Paris, June 21, 1948.

BRIN, Myron, Am. chemist; b. N.Y.C., July 1, 1923; s. Philip and Frances (Kraut) B.; B.S., Cornell U., 1947, M.S., 1948; Ph.D., Harvard, 1951; m. Phyllis Diana Bletcher, June 4, 1944; children—Kenneth Philip, Steven Charles, Mitchell Francis. Research asso. Thorndike Lab., Boston, 1953-56; instr. Med. Sch. Harvard, 1953-56; chief biologist Food & Drug Research Labs., Inc., N.Y.C., 1956-58; asso. prof. biochemistry, medicine Upstate Med. Center State U. N.Y., Syracuse, 1958—. Mem. Am. Soc. Biol. Chemists, Am. Inst. Nutrition, Am. Chem. Soc., Soc. Exptl. Biology, N.Y. Acad. Scis. (A. Cressy Morrison award in natural scis. 1962), A.A.A.S. Editor: (with Robert Dunlop) Chemistry and Metabolism of L(+) and D(−) Lactic Acids, 1965; Annals N.Y. Acad. Scis., vol. 119 1965. Research, publs. on synthesis of lactic acid isomers labelled with carbon-14; metabolism of lactate in tissues, tumors and intact animals; enzyme adaptations following physiol. stresses; effects of x-irradiation on nutrient value of foods; developed use of transketolase assay in tissues to assess thiamine deficiency. Home: 109 Sherbrook Rd., Syracuse 13224. Office: 750 E. Adams St., Syracuse, N.Y. 13210.*

BRINCKE, William Draper, Am. physician, pomologist; b. Kent County, Del., Feb. 9, 1798; s. John and Elizabeth (Gordon) B.; grad. Princeton, 1816; M.D., U. Pa., 1819; m. Sarah T. Physick, 1821; m. 2d, Elizabeth Bispham Reeves, 1832. Began practice of medicine, Wilmington, 1819, Phila., 1825; physician concerned with contagious diseases City Hosp., 1827-39; active in control of Asiatic cholera epidemic, 1832; a pomologist, developed numerous fruit varieties, worked primarily with small fruits and pears; published findings of strawberry expt. in Farmer's Cabinet, 1846; frequent contbr. to Horticulturist; a founder Am. Pomol. Soc.; retired from med. practice due to ill health, 1859. Died Dec. 16, 1862.

BRINCKERHOFF, Henry Morton, Am. elec. engr.; b. Fishkill-on-Hudson, N.Y., Apr. 20, 1868; s. Peter Remsen and Helen (Morton) B.; grad. Stevens Inst. Tech., 1890; m. Florence L. Fay, Jan. 20, 1903. Constrn. work Thomson-Houston Co. on West End Street Ry., Boston; asst. engr. in power house Utica Belt Line St. Ry., 1891-92; foreman in charge car equipment Gen. Electric Co., Boston, Coney Island and Bklyn. Ry.; asst. elec. engr. Intramural Ry., (first 3d rail elevated road in U. S.), World's Columbian Expn., Chgo., 1893; elec. engr. Met. West Side Elevated Ry. (1st large elevated road for city transportation equipped with electricity), 1894, asst. gen. mgr. and gen. mgr. 1898-1906; elec. asso. of William Barclay Parsons since 1906. Mem. Am. Soc. E.E., Western Soc. Engrs. Died Oct. 13, 1949.

BRINDLEY, James, English engr.; b. Derbyshire, Eng., 1716; built canal with fixed aqueduct over river Irwell (1st of its kind) for Duke of Bridgewater, 1759-61; connected Trent and Mersey rivers by means of Grand Trunk Canal; improved Newcomen steam engine; produced float for regulation of hot boiler, circa 1760. Died Turnhurst, Eng., Sept. 30, 1772.

BRINDLEY, Paul, Am. physician, pathologist; b. Maypearl, Tex., Dec. 27, 1896; s. George Goldthwaite and Mattie (Hanes) B.; student U. Tex., Austin, 1918-20; B.S., U. Tex. Med. Sch., Galveston, 1923, M.D., 1925; m. Anne Mae Ammons July 2, 1929. With U. of Tex., from 1925, prof. pathology, from 1929; pathologist Med. Br. Hosp., from 1929; cons. in pathology USPHS Hosp., from 1931. Fellow A.C.P., Coll. Am. Pathol., Am. Soc. of Clin. Pathology; mem. A.A.A.S., A.M.A., So. Med. Soc., Am. Assn. Pathologists and Bacteriologists, Internat. Assn Med. Museums. Contbr. articles to med. jours. Died Dec. 28, 1954.

BRINELL, Johann August, Swedish engr.; b. 1849; invented machine for measuring hardness of alloys and metals, 1900; Brinell no. pertains to indentation produced on metal by ball pressure test. Died 1925.

BRING, Erland Samuel, Swedish mathematician; b. Ausas skn, Kristianst, Sweden, Aug. 19, 1736; s. Jöns and Christina Elisabeth (Lagerlöf) B.; ed. U. Lund (Sweden); m. Ingrid Catharina Ringberg, 1791. Showed that gen. quintic equation can be transformed to trinomial form, 1786. Died Lund, May 20, 1798.

BRINK, David Maurice, physicist; b. Hobart, Tasmania, Australia, July 20, 1930; s. Maurice Ossian and Victoria (Finlayson) B.; B.Sc., U. Tasmania, 1951; D.Phil,. Magdalen Coll., Oxford (Eng.) U., 1955; m. Verena Wehrli, 1958; children—Anne-Katherine, Thomas David, Barbara Verena. Research lectr. Balliol. Coll., Oxford U., 1954-58, fellow, tutor, 1958—; instr. Mass. Inst. Tech., 1956-57. Rhodes scholar Oxford U., 1951-54. Author: (with G. R. Satchler) Angular Momentum, 1961; Nuclear Forces, 1965; also articles. Research on structure of atomic nuclei. Home: 9A Mansfield Rd., Oxford, Eng.*

BRINK, Frank, Jr., Am. biophysicist; b. Easton, Pa., Nov. 4, 1910; s. Frank and Lydia (Wilhelm) B.; B.S., Pa. State Coll., 1934; M.S., Cal. Inst. Tech., 1935; Ph.D., U. Pa., 1939; m. Marjory Gaylord, May 1, 1939; children—Patricia (Mrs Reinhard A. Mayer), David Warner. Faculty, Med. Coll. Cornell U., 1940-41, U. Pa., 1941-48, Johns Hopkins, 1948-53; prof., mem. Rockefeller Inst., N.Y.C., 1953—, acting dean grad. studies, 1954-58, dean, 1958—. Mem. div. com. biol. and med. sci. NSF, 1953-59; chmn. Pres.'s Com. on Nat. Medal Sci., 1963-64. Mem. Nat. Acad. Sci., Biophys. Soc., A.A.A.S., Am. Acad. Arts and Scis., Soc. Gen. Physiology, Am. Physiol. Soc., Am. Soc. Naturalists, Am. Inst. Biol. Sci. Editor: Biophys. Jour., 1960-63. Publs. on studies of nerve cells, using biochem. and electrochem. methods, role of calcium ions in nerve cells, action on nerve cells of narcotics and anesthetics. Home: Box 443, Route 1, Pleasant Valley Rd., Titusville, N.J. 08560. Office: Rockefeller U., N.Y.C. 10021.*

BRINK, Norman George, Am. chemist; b. Littleton, Colo., Aug. 31, 1920; s. Norman Rosell and Julia (Dunn) B.; A.B., Princeton, 1942, M.S., 1943, Ph.D., 1944; m. Elizabeth Yurenda, Jan. 12, 1947. Research chemist Merck & Co., Inc., Rahway, N.J., 1944-50, asst. dir. organic, biochem. research, 1950-56, dir. bio-organic chemistry, 1956-66, dir. univ. relations, 1966—. Proctor fellow Princeton, 1944; Merck Fgn. fellow Stockholm, 1952-53. Mem. A.A.A.S., Am. Chem. Soc., Am. Soc. Biol. Chemists, N.Y. Acad. Scis., Soc. Chemistry and Industry. Research, numerous publs. on isolation, chem. structure and biochem. function of variety of naturally occurring compounds of biol. or medicinal importance. Home: 2143 Buttonwood Lane, Westfield, N.J. 07090. Office: Rahway, N.J. 07065.*

BRINK, R(oyal) Alexander, biologist; b. Woodstock, Ont., Can., Sept. 16, 1897; s. Royal Wilson and Elizabeth Ann (Cuthbert) B.; B.S.A., Ont. Agrl. Coll., 1919; M.S., U. Ill., 1921; D.Sc., Harvard, 1923; m. Edith Margaret Whitelaw, Dec. 27, 1922 (dec. May 1962); children—Andrew Whitelaw, Margaret Alexandra (Mrs. Edward Chisholm Ingraham); m. 2d, Joyce Hickling, Oct. 19, 1963. Faculty, U. Wis., Madison, 1922—, prof. genetics, 1931—, chmn. dept. genetics, 1939-51. Mem. Nat. Acad. Scis., Am. Acad. Arts and Scis., A.A.A.S., Bot. Soc. Am., Genetics Soc. Am. (past pres.), Am. Soc. Naturalists (past pres.). Research, numerous publs. on role endosperm in seed devel., structural changes in maize chromosomes, unstable loci in maize, paramutation and control of gene expression in developing organisms. Home: 4237 Manitou Way, Madison, Wis. 53711.*

BRINKERHOFF, Lloyd Allen, Am. plant pathologist; b. Provo, Utah, July 9, 1915; s. Joseph and Phoebe (Allen) B.; B.S., U. Ariz., 1937, M.S., 1939; postgrad. Cornell U.; Ph.D., U. Minn., 1962; m. Regina Smith, Feb. 4, 1938; children—James Allen, Mary Anne. Plant pathologist U. S. Dept. Agr., 1937-39, 42—, prof. dept. botany and plant pathology Okla. State U., Stillwater, 1948—. Mem. Am. Phytopath. Soc., Am. Inst. Biol. Scis., Sigma Xi. Research, publs. on nature of infection and control of Phymatotrichum root rot of pecan, dissemination, overwintering and variability of cotton bacterial blight pathogen; methods of inoculation, sources of resistance, heritability of resistance and effect of temperature on expression of bacterial blight of cotton and Verticillium wilt. Home: 515 S. Stanley St., Stillwater, Okla. 74074.*

BRINKHOUS, K(enneth) M(erle), Am. pathologist; b. Clayton County, Ia., May 29, 1908; s. William and Ida (Voss) B.; student U. S. Mil. Acad., 1925; A.B., State U. of Ia., 1929, M.D., 1932; D.Sc. (hon.), University of Chicago, 1967; m. to Frances E. Benton, Sept. 5, 1936; children—William Kenneth, John Robert. Asst. in pathology State U. of Ia., 1932-33, instr., 1933-35, asso. in pathology, 1935-37, asst. prof., 1937-45, asso. prof., 1945-46; prof. pathology U. of N.C., 1946-62, alumni distinguished prof., 1962—. Member hematology study, Nat. Inst. of Health, U. S. Ph.S., 1948-52, chmn. 1959-62, chmn. pathology study section, 1957-59; mem. bd. editors Jour. Lab. and Clin. Medicine, 1948-53, Proceedings Soc. Exptl. Biology and Medicine, 1950-53, Blood, 1950-65. Mem. panel on blood coagulation Nat. Research Council, 1951-62, chairman, 1954-62; chairman med. advisory committee Hemophilia Foundation, 1954—; mem. Nat. Bd. Med. Examiners, 1957-59; mem. sci. adv. bd. Armed Forces Inst. of Pathology, 1957—, chmn., 1964-65; mem. sci. adv. bd. Am. Nat. Red Cross, 1965—; sec. gen. Internat. Com. Hemostasis and Thrombosis, 1965-67. Served from capt. to lt. col. M.C., U. S. Army, 1941-46; with 34th Division; at Darnell Gen. Hosp., and 8th Med. Lab in Australia, New Guinea, Netherlands East Indies and Philippine Islands; col., Med. Reserve Corps, since 1946. Co-recipient Ward Burdick Award, Am. Soc. Clin. Pathologists, 1940, recipient same, 1963; recipient O. Max Gardner award, 1961. Fellow A.A.A.S., Med. Soc. of N.C., College Am. Pathologists; Association Am. Physicians, Internat. Society Hematology, Internat. Acad. Pathology, A.M.A., Am. Association Pathologists and Bacteriologists, Am. Soc. Clin. Pathologists, Am. Soc. Exptl. Pathology (pres. 1965-66), Fedn. Am. Socs. Exptl. Biology (pres. 1966-67), Univs. Asso. Research and Education Pathology (president 1964—), Central Soc. Clin. Research, Society Experimental Biol. and Medicine, Am. Assn. U. Profs., Phi Beta Kappa, Phi Lambda Upsilon, Alpha Omega Alpha, Sigma Xi, Phi Chi. Bd. editors Thrombosis at Diathesis Haemorrhagica, 1961—. Basic studies in blood coagulation, hemorrhage, and thrombosis; laboratory diagnosis of bleeder states; plasma fractions for treatment of hemophilia; anticoagulant therapy for diffuse intravascular clotting and hemorrhage. Home: Chapel Hill, N.C.

BRINKHURST, Ralph Owen, zoologist; b. London, Eng., Mar. 13, 1933; B.Sc. with spl. honors in Zoology, U. London, 1955, Ph.D., 1958; m. Shirley Evelyn Withers, July 29, 1958; children—Sara Louise, Paul James. Faculty, U. Liverpool (Eng.), 1957-65, lectr., 1959-65; asso. prof. zoology, asso. Gt. Lakes Inst., asso. Royal Ont. Mus., U. Toronto (Ont., Can.), 1966—. Cons. in river pollution to various cities in Gt. Britain, in worm systematics to insts. and colls. throughout world. Mem. Internat. Limnological Assn., Freshwater Biol. Assn., Ecol. Soc., Am. Soc. Limnology and Oceanography. Research, publs. on biology of water-striders including population control and polymorphism, taxonomy and ecology of freshwater oligochaete worms and their value in pollution detection including identification keys to Europe, Africa, N.Am. Home: 1949 Ingledale Rd., Clarkson, Ont. Office: Dept. Zoology, U. Toronto, Toronto, Ont., Can.*

BRINKLEY, John, Brit. astronomer; b. Woodbridge, Suffolk, Eng., 1763; sr. wrangler, 1st Smith's prizeman Caius Coll., Cambridge (Eng.), u. 1788; M.A., 1791, D.D., 1806; became Andrews prof. astronomy Dublin (Ireland) U., also 1st astronomer royal for Ireland, 1792; named prebendary of Kilgaghlin, also rector Derrybrush, 1806; became bishop of Cloyne, 1826. Fellow Royal Soc., 1803 mem. French Acad. Sci., 1820, Royal Irish Acad. (pres. 1822-35), Royal Astron. Soc. (pres. 1831-33). Recipient Copley medal, 1824, Conyingham medal, 1817. Author: Elements of Astronomy, 1808; also articles. New theory of astron. refractions with tables for their calculations, 1814; wrote catalogue of 47 fundamental stars. Died Sept. 14, 1835.

BRINKMAN, August, Jr., Norwegian zoologist; b. Bergen, Norway, May 16, 1912; s. August and Therese (Stillmann) B.; Cand.mag., U. Oslo, 1937, Mag.-scient., 1939; Dr.philos., U. Bergen, 1952; m. Ingegerd Persson, Mar. 10, 1943; children—Inger Marie (Mrs. Christian Höy), Ian August. Asst. biologist U. Biol. Lab., Oslo, Norway, 1941; research biologist Gothenburg (Sweden) Mus. Natural History, 1941-43; lectr. zoology Bergens Mus., 1946-49; faculty U. Bergen, 1949—, prof. zoology, 1952—, head zool. lab., 1952—. Mem. Assn. for Advancement Sci. Bergen, Norwegian Acad. Sci. Author: The Microscope and Microscopical Preparation, 1951; also articles. Research on morphology, taxonomy and distbn. of worms including trematodes, nematodes, hirudineans, nemerteans, parasitology and hygiene, whale biology and history of their taxonomy. Home: Professorveien 8, Minde pr., Bergen, Norway.*

BRINKMAN, Hendrik, Dutch physicist; b. Kampen, Nov. 5, 1909; s. Jan Willem and Alida Willemina (Sonderman) B.; Ph.D. in Scis., U. Utrecht; m. E.C. van Markesteyn, Apr. 8, 1936; children—Madeleen, Annemarie Engeline, Boudewijn Matthijs. Sci. collaborator physics lab. Philips' Gloeilampenfabrieken N.V., Eindhoven; head research lab. Kema N.V., Arnhem; dir. physics lab. State U. Groningen, also prof. physics. Mem. Nederlandse Natuurkundige Vereniging, Am. Phys. Soc. Research and numerous publs. on atomic, nuclear and molecular physics, elec. charges. Home: Botanicuslaan 11, Haren. Office: Westersingel 34, Groningen, Netherlands.

BRINKMANN, August, Norwegian zoologist; b. Bergen, Norway, May 16, 1912; s. M. C. August and Theresia (Stillmann) B.; ed. U. Oslo, Bergen Mus.; M.A. in sci.; Ph.D.; m. Ingegerd Persson, Mar. 10, 1943; children—Inger-Marie, Jan August. Research asst. biology lab. U. Oslo; sci. collaborator Mus. Natural History, Gothenburg, 1941-43; ofcl. san. service Norwegian Armed Forces, London, 1943-45; reader zoology dept. Bergen Mus., 1946; prof. zoology Univ. Lab. Biology, Bergen, 1954. Mem. Norwegian Acad. Sci. Author: Contribution to our Knowledge of the Mon . . . Trematodes, 1940; The Microscope and Microscopical Preparations, 1950; Fish Trematodes from Norwegian Waters, 1952. Home: Professorveien 8, Minde pr. Office: Allegaten 49, Bergen, Norway.

BRINTON, Daniel Garrison, Am. anthropologist; b. Thornbury, Chester County, Pa., May 13, 1837; grad. Yale, 1858, A.M., LL.D.; grad. Jefferson Med. Coll., 1861; Sc.D., U. Pa. Asst. surgeon, surgeon and med. director 11th Army Corps, 1862-65; editor Med. and Surg. Reporter, 1867-87; prof. Am. archaeology and linguistics, U. Pa. Author: The Myths of the New World; The Religious Sentiment; American Hero-Myths; Chronicles of the Mayas; The Lenapé and Their Legends; Races and Peoples; The American Race; The Pursuit of Happiness; Nagualism; Lectures on the Religions of Primitive Peoples; Pioneer in N. Am. anthropology; gave 1st systematic classification of aboriginal langs. of Am.; 1st to recognize genetic relationships between certain Indian tribes of U. S.; made important contbns. to ethnol. studies of Am. Indian and to Am. archeology; aided in establishing a Mayan-Christian chronology. Died Phila., Pa., July 31, 1899.

BRINTON, Willard Cope, Am. mech. engr.; b. West Chester, Pa., Dec. 22, 1880; s. Samuel Lewis and Elizabeth Brinton (Smith) B.; grad. State Normal Sch., West Chester, 1900; S.B. in M.E. cum laude. Lawrence Scientific Sch. Harvard, 1907; m. Laura MacDonald Moses, Apr. 17, 1920. With H. Brinton & Co., Phila., 1900-03, Westinghouse Electric & Mfg. Co., East Pittsburgh, Pa., 1907-10; asst. to v.p. U. S. Motor Co., N.Y., 1910-12; mech. engr. Bush Terminal Co., N.Y., 1912-13; consulting practice from 1913, involving work in various parts of U. S., Europe, Japan and China; pres. and treas. Terminal Engring. Co., Lawrence Safety Brake Co. Mem. Am. Soc. M.E., Am. Statis. Assn. (v.p. 1919). Author: Graphic Methods for Presenting Facts, 1914; Graphic Presentation, 1939; also numerous articles. Died Nov. 29, 1957.

BRINTON, William, physician; b. Kidderminster, Eng., Nov. 20, 1823; M.B., London U., 1847; began med. studies King's Coll., London U.; M.D., 1848; 6 children. Lectr. forensic medicine St. Thomas's Hosp, later physician, lectr. physiology. Fellow Royal Soc., 1864, Coll. Physicians. Author: Pathology, Symptoms, and Treatment of Ulcer of the Stomach, 1857; Lectures on Diseases of the Stomach, 1859; On the Medical Selection of Lives for Assurance, 1856; On Food and its Digestion, being an Introduction to Dietetics; also contbr. to Todd's Cyclopaedia of Anatomy and Physiology; Peaks, Passes, and Glaciers. also articles; Research on disease of stomach, including localization of cancer of the stomach, 1857; hypertrophy of the submucous connective tissue of the stomach, linitas plastica, (Brinton's disease) 1859. Died 1867.

BRION, Abel Justin, French vet. scientist; b. Bazeilles, France, June 26, 1906; s. A. and J. (Bastogne) B.; Docteur Vétérinaire Vet. Coll. at Alfort (France), Faculty Medicine, Paris, France; Licence ès-Scis., Faculty Sci., Paris and Lyon; m. Larivière, Dec. 1, 1934; children—Alain, Anne-Marie (Mrs. Caine). Faculty, Nat. Vet. Coll., Alfort and Lyon, 1931—, head prof. dept. medicine, dir. research lab. on poultry diseases, dir. research lab. on equine viral disease, Alfort, 1955—. Decorated officier la Légion d'Honneur; comdr. of Mérite Agricole; officer Order des Palmes Académiques; croix Services Militaires Voloritaires. Mem. World Vet. Assn., (pres. 1950—), Nat. Coll. Vet. Surgeons (pres. 1966), Council World Poultry Sci. Assn., Vet. Acad., Royal Acad. Medicine Belgium (hon.), others. Author: Precis de Jurisprudence Vétérinaire, 1943, 3d edit. 1962; Vade. Mecum du Vétérinaire, 1945, 12th edit. 1965; (with H. Ey) Psychiatrie Animale, 1964. Research, publs. primarily on diseases of domestic animals; discoverer vaccines against Panileukopenia of cats, Newcastle Disease of fowls, Equine Influenza; discoverer virus causing Rhino-tonsilletis of dogs. Home: 4 Avenue Emile Pouvillon, Paris (Seine). Office: 7 Avenue du Général de Gaulle, Maisons-Alfort (Seine), France.*

BRIOSCHI, Francesco, Italian mathematician; b. Milan, Italy, Dec. 22, 1824; doctorate, U. Pavia (Italy), 1845. Prof. applied math. U. Pavia, 1852-61; gen. sec. Ministry Instrn., 1861-62; founder Milan Superior Tech. Inst., 1862, dir., prof. hydraulics and analysis until 1897; senator Italy. Mem. Royal Acad. Lincei Rome (pres. from 1884), French Acad. Scis., 1880. Author: La Teorica dei determinanti e le sue principale applicazioni, 1854; numerous papers. Contbd. to advance math. analysis in Italy; worked on theory of determinants and invariants, theory of algebraic forms, theory of elliptic functions with application to quintics, relations between groups and differential equations, symmetric functions of sums of powers of roots of equation, Abelian integrals, modular equations. Died Milan, Dec. 14, 1897.

BRIOT, Charles-Auguste-Albert, French mathematician; b. Sainte-Hippolyte, France, July 19, 1817; doctorate, École Normale, 1842. Became prof., Reims, France, 1842, Lycée d'Orleans, 1842, Faculty Lyons (France), 1845, Lycée Bonaparte, 1848, Lycée Sainte-Louis, 1851; became asst. Ecole Poyltechnique, 1850, entrance examiner, 1854-72; master confs. École Normale, 1857-82; apptd. titular prof. math. physics Sorbonne, 1870. Author: Lecons de mechanique, 1861; Essai sur la théorie mathematique de la lumière, 1864; Theorie méchanique de la chaleur, 1869; Traité des fonctions abéliennes, 1879; (with Bouquet) Theorie des fonctions doublement periodiques, 1859. Studied nature of integrals at singular point; investigated (with Bouquet) doubly periodic functions; researched problems in mechanics, astronomy, and analytic geometry. Died Bourg-d'Ault, France, Sept. 30, 1882.

BRIOT, Nicolas, inventor; b. Damblain, France, 1579; s. Didier Briot; m. Pauline Nisse (dec. 1608); m. 2d, Esther Petau, 1611. Engraver gen., govt. France, 1605-25; coin maker to Charles I of Eng.; chief engraver to English mint, 1633; master Scottish mint, 1635. Author: Raisons, moyens et propositions pour faire Toutes les monnoies du royaume a l'avenir univormes, et faire cesser Toutes falsifications, 1615. Improved German coining press and introduced it into Eng., circa 1630. Died Eng., 1646.

BRIQUET, Pierre, French physician; b. Châlons-sur-Marne, 1796; M.D., Paris, 1827. Intern, 1820; became instr. Faculty Medicine, Paris, 1831; physician Cochin Hosp., from 1836, Charité Hosp., from 1846. Mem. Acad. Medicine. Author: Traité thérapeutique du quinquina, 1853; Traité clinique et thérapeutique de l'hystérie, 1859. Described syndrome involving aphonia, shortness of breath, hysterical paralysis of diaphragm (Briquet's syndrome), 1859; 1st to describe ataxia analgiea hysterica, 1859; studied varicose veins, also gangrene of lungs, cholera. Died Paris, Nov. 25, 1881.

BRISBANE, Sir Thomas Makdougall, Scottish astronomer; b. Brisbane House, Largs, Ayrshire, Scotland, July 23, 1773; s. Thomas and Eleanor (Bruce) B.; ed. Acad. Kensington (Scotland), also Edinburgh (Scotland) U.; m. Anna Maria Makdougall, 1819; 2 sons, 2 daus. Commd. ensign, 1789, advanced through grades to gen., 1841; with Sir Ralph Abercromby in W.I., 1795-98; in Jamaica, 1800-08; studied astronomy at obs., Brisbane, Australia; named asst. adj.-gen., 1810; brig.-gen. under Wellington in Peninsula; in Can., 1813; gov. New S. Wales, Australia, 1821-25; built obs. at Paramatta, nr. Sydney, Australia, 1822, magnetic obs. at Makerstoun, Scotland, 1841; returned to Eng., 1825; to Scotland, 1826; col. 34th regt., 1826. Recipient Keith medal, 1848; Brisbane, Australia, named in his honor. Mem. Royal Soc. Edinburgh (pres. 1833-60), Royal Astron. Soc. (Gold medal 1828), French Acad. Scis., 1818. Fellow Royal Soc., 1810. Knighted, 1814. Author: A Method of Determining the Time with Accuracy from a Series of Altitudes of the Sun taken on the Same Side of the Meridian, 1811; On the Repeating Circle and a Method of Determining the Latitude by a Sextant or Circle with Simplicity and Accuracy from Circum-meridian observations taken at Noon, 1819; A Catalogue of 7,385 Stars, chiefly in the Southern Hemisphere, Prepared from Observations made 1822-26 at the Observatory of Paramatta, 1835; Died Largs, Jan. 27, 1860.

BRISCOE, Henry Vincent Aird, Brit. chemist; b. London, Eng., Sept. 24, 1888; ed. Imperial Coll., London. Prof. inorganic chemistry Armstrong Coll., Newcastle-on-Tyne, Eng., 1921; dir. inorganic and phys. chem. labs. Imperial Coll. Sci., 1932-34. Original work on boron, rhenium, selenium; determined atomic weights of vanadium, tin, boron, thallium. Died 1961.

BRISEID, Kjell, Norwegian pharmacologist; b. Oslo, Norway, Feb. 25, 1921; s. Tormod Gram and Brynhild (Briseid) Jensen; student Inst. Pharmacy, U. Oslo, 1940-46, Dr.philos., 1956; student Sch. Pharmacy, Copenhagen, Denmark, 1947-48; lic.pharm., Copenhagen, 1954; postgrad. dep. pharmacology U. Edinburgh, 1955-56; m. Anne Marie Kjernlie, May 19, 1945; children—Jon Audun, Tormod; m. 2d, Gudrun Karen Bieltvedt, Dec. 8, 1967. Postgrad. scholar Faculty Math. and Natural Sci., U. Oslo, 1951-55, faculty Inst. Pharmacy, 1957—, prof. pharmacodynamics, 1966—. Mem. com. on biol. standardization Scandinavian Pharmacopoiea Council, 1953—. Mem. Scandinavian, Norwegian pharmacological socs., Norwegian Physiol. Soc., Norwegian Pharm. Soc. Author: Papirkromatografisk-fluorimetrisk Bestemmelse Av Digitalisglykosider Som Grunnlag for En Styrkebestemmelse Av Folium Digitalis, 1955; also articles. Editorial bd. Acta Pharmacologica et Toxicologica. Research on phytochemistry and pharmacology of cardiac glycosides in digitalis purpurea, pharmacological and physiol. investigations on bradykinin and bradykinin-like polypeptides with biol. activity; methods for biol. standardization of drugs. Home: 45, Thereses gate, Oslo, Norway.*

BRISSAUD, Edouard, French neurologist; b. Besancon, France, Apr. 15, 1852; M.D., Paris Faculty Medicine, 1880. Asst. chief, then chief med. clinic Pitié; became preparator practical work in path. anatomy Faculty Paris, 1883, tchr. path. anatomy, 1887; became physician St. Antoine's Hosp., 1889, tchr., 1889-93, prof. neurol. pathology, from 1894; in charge Charcot's course on diseases of nervous system Salpêtrière, 1893-94. Mem. Acad. Medicine. Author: Recherches anatomo-pathologiques . . . , 1880; Anatomie du cerveau de l'homme, 1893; Lecons sur les maladies nerveuses, 1895-96; L'hygiène des asthmatiques, 1896; Nouvelle pratique médico-chirurgicale, 1911-12. Editor: (with Bouchard and Charcot) Traité de Medicine, 6 vols., 1891-94. Founder Neurol. Review. Classified (with Bourneville) various kinds mental degeneration as tuberous sclerosis of brain; described double innervation of face; gave clin. descriptions of torticollis, also tics, spasms; medico-legal expert on injuries relating to conversion hysteria. Died Paris, Dec. 19, 1909.

BRISSEAU, Pierre, surgeon; b. Paris, 1631; ed. Montpellier, France; M.D., Coll. Physicians Tournai (Belgium), 1677. Mil. physician, hosps., Tournai, also Mons, Belgium, Dounai, Belgium, from 1709. Author: Lettre touchant les hôpitaux de troupes, 1691; Traité des mouvements sympathiques, 1692; Méthode pour bien régler les hôpitaux, 1706; Lettre touchant les remèdes secrets, 1708; Traité de la cataracte et du glaucome, 1709. Pioneer in demonstrating that cataract depends on opacity of crystalline lens, 1706. Died Dounai, Sept. 10, 1717.

BRISSON, Mathurin Jacques, French naturalist; b. Fontenay-le-Comte, France, Apr. 3, 1723; s. Louise-Gabrielle (Jourdain) Brisson. Asst. to Reaumur; prof. physics Coll. de Navarre; prof. Ecoles centrales de Paris, from 1794. Mem. French Acad. Scis., 1759. Author: Le Regne animal . . . , 1756; Ornithologie . . . , 1760; Dictionnaire raisonné de physique . . . , 1781; Observations sur les nouveaux decouvert aerostatiques . . . , 1784; Pesanteur spécifique des corps, 1787. Proposed system based on natural unit of length (led to introduction meter). Died Seine-et-Oise, France, June 23, 1806.

BRISSOT, Pierre, physician; b. Fontenay-le-Comte, France, 1478; s. Jean Brissot; M.D., Paris, 1514; student of Villemor; prof. l'Ecole de Paris, from 1504. Author: Apologetica disceptatio qua docetur, per quae loca sanguis mitti debeat in viscerum inflammationibus praesertim in pleuritide, 1525. First to break with Arabo-Galenic medicine; revived Hippocrates' doctrine that patient should be bled sufficiently nr. lesion, also that in pleurisy in which vena cava drained region, patient may be bled on right or left side; revival resulted in research and discovery of valves of veins. Died Evora, Portugal, 1522.

BRISTOL, Charles Lawrence, Am. biologist; b. Ballston Spa, N.Y., Sept. 29, 1859; s. Lawrence W. and Caroline (Hawkins) B.; B.S., N.Y. U., 1883, M.S., 1888; Ph.D., 1895; m. Ellen Gallup, Jan. 28, 1890; children—Lawrence, Elisabeth (Mrs. W. E. Greenleaf). Prof. zoölogy, U. S.D., 1887-91; fellow Clark U., 1891-92, U. Chgo., 1892-93, prof. biology N.Y. U., 1894-1925, prof. biology on University World Cruise, 1926-27. Directed zoöl. expdns. to Bermuda and made successful expts. in transporting tropical marine animals alive to N.Y. Aquarium. Died Aug. 27, 1931.

BRISTOL, Leverett Dale, Am. hygienist; b. Chgo., June 2, 1880; s. Frank Milton and Nellie (Frisbie) B.; B.S., Wesleyan U., Middletown, Conn., 1903; M.D., Johns Hopkins, 1907; Dr. P.H., Harvard, 1917; m. Addie Louise Knox, June 27, 1907; children—Corabelle (Mrs. Richard Osborn Rice), Adelaide (Mrs. Livingston Lord Satterthwaite), Leverett Frisbie. Practiced in St. Paul, 1908-13; asso. prof. bacteriology Syracuse U., 1913-14; prof. bacteriology U. N.D., also dir. state health labs. 1914-16; fellow and asst. Harvard Med. Sch., 1916-17; state commr. health, Me., 1917-21; prof. preventive medicine and pub. health U. Minn., 1922-23; dir. N.Y. State Health Demonstrations, 1923-29; health dir. Am. Tel. & Tel. Co., 1929-44; exec. dir. Hosp. Council of Greater N.Y., 1944-45; commr. Me. Dept. Health and Welfare; 1945-47; cons. on health and indsl. medicine, 1947-48; chief, div. cancer control Pa. Dept. Health, 1948—. Mem., officer various health and med. orgns. and coms.; cons. on health to govt. and pvt. agys. Mem. A.M.A., Am. Pub. Health Assn., Nat. Tb Assn., Am. Assn. Indsl. Physicians and Surgeons, Indsl. Health Council, Internat. Soc. Med. Health Officers (hon.), Sigma Xi, Phi Beta Kappa. Author: Industrial Health Service; chpt. on Health of Office Workers in Handbook of Business Administration, 1931; chpt. on Industrial Medicine in Cyclopedia of Medicine, 1940; also articles. Died Feb. 20, 1957.

BRISTOL, William Henry, Am. mathematician; b. Waterbury, Conn., July 5, 1859; s. Benjamin Hiel and Pauline (Phelps) B.; M.E., Stevens Inst. Tech., 1884; m. J. Louise Wright, Sept. 8, 1885 (dec. 1888); m. 2d, Elise H. Myers, June 28, 1899. Organizer, tchr., manual tng. dept. Workingman's Sch., N.Y., 1882-86; instr. math. Stevens Inst. Tech., 1886-88, asst. prof., 1888-99, prof., 1899-1906; organizer Bristol Co., 1889, also pres. Recipient John Scott Legacy medal Franklin Inst., Phila., 1890, medal and diploma Chgo. Expn., 1893, Silver medal Paris Expn., 1900, Gold medal St. Louis expn., 1904, Grand prize Panama P.I. Expn., 1915. Inventor numerous recording instruments for pressure, temperature and electricity, steel belt-lacing, sound amplifying and talking motion picture apparatus. Died June 18, 1930.

BRISTOWE, John Syer, English physician; b. Camberwell, Eng., Jan. 19, 1827; s. John Syer and Mary (Chesshyre) B.; ed. St. Thomas's Hosp.; M.B., London, 1850, M.D., 1852; LL.D., Edinburgh, 1884; m. Miriam Isabelle Stearns, Oct. 9, 1856; 4 daus, including Beatrice M., 5 sons. Became house physician St. Thomas's Hosp., 1849, physician, 1860; lectr. on medicine, 1876-92; Croonian lectr., 1872, Lumleian lectr., 1879. Licentiate Soc. Apothecaties. Fellow Royal Soc., 1881, Royal Coll. Physicians; mem. Royal Coll. Surgeons, Med. Soc. London (pres. 1893). Author: Annual Report of the Medical Officer of Health . . . (1857-82); Theory and Practice of Medicine, 1876; Clinical Lectures and Essays on Diseases of the Nervous System, 1880. Editor: St. Thomas's Hospital Reports, 1870-76. Skilled in diagnosis and treatment of diseases of nervous system; gave a detailed description of trichina spiralis, a parasitic worm in the muscles of man; showed that cholesterol precipitation causes gallstones, 1887. Died Monmouth, Eng., Aug. 20, 1895.

BRITTAN, Martin Ralph, Am. zoologist; b. San Jose, Cal., Jan. 28, 1922; s. Ralph Hinton and Addie (Martin) B.; A.B., San Jose State Coll., 1946; Ph.D., Stanford, 1951; m. Ruth Marie Luebke, Aug. 10, 1947; children—Penelope, Pamela. Faculty, S.D. Sch. Mines and Tech., 1949-50, San Diego State Coll., 1950-51; faculty Sacramento State Coll., 1951—, prof., 1961—; fishery cons. Merced River Project, 1965-66. Mem. Am. Soc. Ichthyologists and Herpetologists, Ecol. Soc. Am., Am. Fisheries Soc., Soc. for Study Evolution, Soc. Animal Behavior, Soc. Systematic Zoology, Soc. Limnology and Oceanography, Wildlife Soc., Sigma Xi. Editor, Ichthyologica, 1966—. Research on systematics and ecology of fishes, especially tropical fresh waters. Home: 1820 Venus Dr., Sacramento 95825.*

BRITTON, Doyle, Am. chemist; b. Los Angeles, Mar. 6, 1930; s. John Joseph and Dorothy (Dearing) B.; B.S., U. Cal. at Los Angeles, 1951; Ph.D., Cal. Inst. Tech., 1955; m. Judith Gavin van Valkenburg, Oct. 27, 1962; children—Jennifer Dearing, David Frederick. Faculty, U. Minn., Mpls., 1955—, prof. chemistry, 1960—. Mem. Am. Crystallographic Assn., Am. Chem. Soc. Research, publs. on X-ray crystallographic determination of crystal and molecular structures with emphasis on intermolecular interactions. Office: Dept. Chemistry, U. Minn., Mpls. 55455.*

BRITTON, Edgar Clay, Am. chemist; b. Rockville, Ind., Oct. 25, 1891; s. Joseph and Bertha (Hirshbrunner) B.; A.B., U. Mich., 1915, Ph.D., 1918; m. Grace van Huss, 1916 (div. 1937); children—Harold E., Joseph H., Leunis G.; m. 2d, Mildred A. Proud, 1937; children—Linda Ann, Daniel E. Teaching asst. U. Mich., Ann Arbor, 1914-18, instr. in organic chemistry, 1918-20; chemist Dow Chem. Co., Midland, Mich., 1920-32, dir. organic research lab., 1932-56. With Nat. Defense Com., 1944. Recipient Perkin medal, 1956. Fellow Soc. Chem. Industry, Inst. Chemistry; mem. Chem. Soc. (pres. 1952), London (Eng.) Chem. Soc., A.A.A.S. Chem. patentee. Research synthetic organic chemistry in phenols, amines, halogen compounds, acetic acid, insecticides, fungicides, pharms., detergents, plastics; developed new catalysts to speed up conversion of oil refinery by-products into synthetic rubber; improved methods for producing carbolic, acetic acids. Died July 31, 1962.

BRITTON, Hubert Thomas Stanley, Brit. chemist; b. Apr. 22, 1892; s. Thomas Ernest and Clara B.; D.Sc., London (Eng.) U., 1926, Bristol (Eng.) U., 1934; diploma Imperial Coll.; also ed. Royal Coll. Sci., S. Kensington, Eng.; m. Edith Greenslade, 1916 (dec. 1959); 1 son, 1 dau. Chemist aero. inspection directorate Air Ministry, 1916-20; asst. lectr. King's Coll., London U., 1920-23; lectr. Norwood Tech. Coll., 1927-28, Univ. Coll., U. Exeter, 1928-35; prof. chemistry, dir. Washington Singer Labs., U. Exeter, 1935-57, ret., 1957, emeritus prof., 1957-60; Leverhulme Research fellow, 1957-59. Fellow Royal Inst. Chemistry, Chem. Soc.; mem. Soc. Analytical Chemistry, Soc. Chem. Industry. Author: Hydrogen Ions, their Determination and Importance in Pure and Industrial Chemistry, 1929, 4th edit., 1955; Conductometric Analysis, 1934; Chemistry, Life and Civilisation, 1930, 40. Contbr. papers on phys., inorganic chemistry. Died Dec. 30, 1960.

BRITTON, Nathaniel Lord, Am. botanist; b. New Dorp, S.I., N.Y., Jan. 15, 1859; E.M., Columbia, 1879, Ph.D., 1879, Sc.D. (hon.), 1904; LL.D., U. Pitts., 1912; m. Elizabeth Gertrude Knight, Aug. 27, 1885. Asst. in geology Columbia, 1879-86, instr. in geology and botany, 1886-90, also instr. in zoölogy, 1887-88, adj. prof. botany, 1891-96, prof. emeritus, 1896—; founder dir.-in-chief N.Y. Bot. Garden, 1896-1929, dir. emeritus, from 1929. With N.J. Geol. Survey, 5 yrs.; field asst. U. S. Geol. Survey, 1882. Mem. Bot. Soc. Am. (pres. 1896-98, 21), N.Y. Acad. Sci. (pres. 1905-07), N.Y. State Forestry Assn. (pres. 1913). Author: The Flora of New Jersey; Illustrated Flora of Northern United States and Canada, 3 vols., 1896-98, 2d edit., 1913 (with Addison Brown); Manual of the Flora of the Northern States and Canada; Flora of Bermuda, 1918; The Bahama Flora (with C. F. Millspaugh), 1920; The Cactaceae (with J. N. Rose), 4 vols., 1919-23. Died N.Y.C., June 25, 1934.

BRITTON, Wilton Everett, Am. entomologist; b. Marlboro, Mass., Sept. 18, 1868; s. Benjamin Howard and Emily Eliza (Wright) B.; B.S., U. N.H., 1893, D.Sc., 1930; postgrad. Cornell U., 1893-94; Ph.D., Yale, 1903; m. Bertha Madeline Perkins, Apr. 30, 1895. Horticulturist, Conn. Agrl. Expt. Sta., 1894-1901, entomologist, 1901—, also Conn. state entomologist, 1901—; lecturer Yale Sch. Forestry, 1901-05; supt. Conn. Geol. and Natural History Survey, 1925—. Chmn. Conn. Tree Protection Examining Bd. Author: Check-List of the Insects of Conn., 1920. Asso. editor Jour. Econ. Entomology, 1909-28. Died Feb. 15, 1939.

BRITVAN, Yakov Mironovich, Russian pathophysiologist; b. 1903; grad. Odessa Med. Inst., 1927; D.Med. Sci. Asst. dept. path. physiology Odessa Med. Inst., 1929-31, lectr., 1931-41; head therapeutic sect. aircraft factory hosp., 1941-45; prof., head dept. path. physiology Vinnitsa Med. Inst., 1945—. Mem. Ukrainian Soc. Pathophysiologists (bd. mem.). Editor, co-author: The Pathology of Respiration and Blood Circulation, 1958. Research and numerous publs. on problems of central nervous mechanisms, regulation of breathing and blood circulation in various path. states, proposed classification of disturbances of rhythm of breathing, theory of pathogenesis of periodic breathing, created various exptl. models, including neurosis of respiratory system, research on conditioned reflex disturbances of breathing, discovered various gen. biol. patterns in body reactions to hypoxic state. Address: Vinnitsa Med. Inst., ulitsa Pirogova 42, Vinnitsa, Ukraine SSR, USSR.

BRIXEY, John Clark, Am. mathematician, educator; b. Mounds, Okla., June 28, 1904; s. Albin Monroe and Ethel (Buchanan) B.; B.A., U. Okla., 1924, M.A., 1925; Ph.D., U. Chgo., 1936; m. Dorothea B. Morrison, Dec. 26, 1926; children—John Clark, Dorothy Jane (Mrs. George W. Ingels). Faculty math. U. Okla., Norman, since 1925—, professor since 1947—, consulting professor math. statistics in preventive medicine and pub. health, 1961—. Mem. Am. Math. Soc., Math. Assn. Am., Phi Beta Kappa, Sigma Xi, Pi Mu Epsilon. Author (with R. V. Andree) Fundamentals of College Mathematics, 1954, rev., 1961, Modern Trigonometry, 1955. Research in theory of numbers and math. statistics. Home: 927 S. Pickard St., Norman, Okla. 73069.*

BRIXNER, Berlyn, Am. optic engr.; b. El Paso, Tex., May 21, 1911; s. Hayes Rutherford and Dolores (Moss) B.; student Tex. Coll. Mines and Metallurgy, 1928-30, Tex. U., 1930-32; m. Betty Jo Williams, May 8, 1940; (div. Jan. 1947); children—Vivian Annette (Mrs. Barry Copeland), Mary Kathleen; m. 2d, Audrey Chew, June 16, 1958. Staff, W. W. Waite's Path. Lab., El Paso, 1932-35; regional photographer Soil Conservation Service, U. S. Dept. Agr., 1935-42, cartographic engr., 1942-43; staff Los Alamos Sci. Lab., U. Cal., Los Alamos, 1943—, optics group leader for devel. high-speed cameras for explosion research, 1945—. Recipient Robert Gordon Meml. award Soc. Photo-Optical Instrumentation Engrs., 1966. Mem. Soc. Motion Picture and Television Engrs. (editorial bd. 1965—, du Pont Gold medal 1966). Research, publs. on devel. high-speed rotating-mirror cameras, lens design system; designed spl-purpose lenses, many types high-speed cameras.

Patentee high speed camera, ultra high-speed light shutter, wide-angle condensing lens system. Home: 1342 46th St. Office: P.O. Box 1663, Los Alamos, 87544.*

BRIXNER, Lothar Heinrich, chemist; b. Karlsruhe, Germany, Dec. 30, 1928; s. Georg A. and Susanne (Wegele) B.; B.S., Technische Hochschule Fridericiana, Karlsruhe, 1950, M.S., 1952, Ph.D. summa cum laude, 1955; m. Marianne Wielandt, May 17, 1955; 1 dau., Diana Isabella. Came to U.S., 1956, naturalized, 1962. Instr., Technische Hochschule Karlsruhe, 1955-56; research chemist pigments dept. E. I. du Pont de Nemours & Co., Inc., Wilmington, Del., 1956-63, sr. research chemist, 1963-66, research asso. pigments dept. research labs., 1966——. Mem. Am. Chem. Soc. Research, publs. in solid state chemistry of complex oxides, ferroelectric, thermoelectric and semiconducting compounds, optical properties of oxo-metallates especially those useful as laser hosts; developer solid state laser operating in continuous wave output at room temperature. Home: 1631 E. Lafayette Dr., West Chester, Pa. 19380. Office: Exptl. Station, Wilmington, Del. 19898.*

BRIZUELA, Waldemar Idel Manlio, Argentinian biochemist, bacteriologist; b. Bragado, Argentina, May 4, 1921; s. Marcelino and Teresa (Cambiaso) B.; grad. in biochemistry Nat. U. Córdoba, 1947; m. Nélida Moreda, Aug. 21, 1925; children—Néstor Daniel, Vilma Beatrice, Marcelo Raul. Asst. prof. microbiology Med. Faculty, Cath. U. Córdoba (Argentina); with Dept. Pediatrics, Nat. U. Córdoba, Dept. Hygiene and Microbiology, Inst. Chem. Scis., Nat. U. Córdoba. Contbr. numerous articles to sci. publs. Home: 1.200 Av. Julio A. Roca-Cordoba, Argentina. Office: 177 Av. Figueroa Alcorta, Cordoba, Argentina.*

BROADBENT, Leonard, English biologist; b. Yorkshire, Eng., Mar. 27, 1916; s. Harry and Elisabeth (Broadbent) B.; B.Sc. with honors, King's Coll., London, Eng., 1939, Ph.D., 1948, D.Sc., 1957, Dip.Ed., 1940; m. Kathleen Mary Ives, Dec. 25, 1941; children—John Kenneth, Margaret Helen, Elizabeth Gilian. Asst. master Lady Manners Sch., Bakewell, Eng., 1940-41; asst. adv. entomologist Agrl. Adv. Service, Midland Province, Eng., 1941-45; plant pathologist Rothamsted Exptl. Sta., Harpenden, Eng., 1945-58; head plant pathology div. Glasshouse Crops Research Inst., Littlehampton, Eng., 1958-65; prof. biology, head Sch. Biol. Scis., Bath (Eng.) U. Tech., 1965——. Mem. Hort. Edn. Assn. (past pres.), Inst. Biology (past hon. sec., pres. elect), Assn. Applied Biologists (past hon. sec.), Biol. Council (past hon. sec.), Internat. Soc. for Hort. Sci. (chmn. commn. plant protection 1964——). Author: Virus Diseases of Brassica Crops, 1957; also numerous articles. Research on epidemiology and control of plant virus diseases in potato, Brassica, lettuce, tomato and other crops, biology and control of aphids. Home: The Dell, Church Lane, Widcombe, Bath. Office: Bath U. Tech., Claverton Down, Bath, Somerset, Eng.*

BROADBENT, William Henry, English physician; b. Lindley, Eng., Jan. 23, 1835; s. John and Esther (Butterworth) B.; ed. Owens Coll., Manchester, Eng., also Paris; M.B., London, 1858, M.D., 1860; numerous hon. degrees; m. Eliza Harpin, 1863; 2 sons, including John Francis Harpin; 3 daus., including Mary Ethel. Physician, London Fever Hosp., 1860-79; became pathologist St. Mary's Hosp., 1860, lectr. in comparative anatomy, 1861, physician, from 1865, lectr. in medicine, 1871-88. Lettsomian lectr. Med. Soc. London, 1074; mem. Royal Commn. on Fever Hosps., 1881; physician to royal family, from 1891. Fellow Royal Soc., Royal Coll. Physicians (sr. censor 8195, Croonian lectr. 1887, Lumleian lectr. 1891). Author: On Cancer, 1866; (treatise on heart diseases), 1897; Collected Papers (edited by son Walter), 1908. Investigated cancer, paralysis, heart disease, typhoid fever, and aphasia; described visible retraction in adherent pericardium (Broadbent's sign); advanced Broadbent's hypothesis of hemiplegia, 1866; worked on prevention and cure of Tb. Died July 10, 1907.

BROADHEAD, Garland Carr, Am. geologist; b. Albemarle County, Va., Oct. 30, 1827; s. Achilles and Mary Winston (Carr) B.; student U. Mo., 1850-51; M.S., 1873; Western Mil. Inst., Ky., 1851-52; m. Marion Wallace Wright, Dec. 1864 (dec. 1883); 2d, Victoria Regina Royall, June 1890. Civil engr. Pacific R.R. of Mo., 1852-57; asst. geologist Mo., 1857-61; U. S. dep. collector internal revenue, Mo., 1862-64; asst. engr. Mo. Pacific R.R., 1864-66; U. S. assessor 5th Dist., Mo., 1866; asst. geologist Ill., 1868, Mo., 1871-73; state geologist Mo., 1873-75; on surveys and constrn. of rys. in Kan., 1879-80; prof. geology U. Mo., 1887-97. Spl. agt. 10th Census, on quarry industry, for Mo. and Kan.; mem. Mo. River Commn., 1884-1902. Author several geol. reports of Mo. and Ill., other geol. publs. Died Dec. 1912.

BROBECK, John Raymond, Am. physiologist, educator; b. Steamboat Springs, Colo., Apr. 12, 1914; s. James Alexander and Ella (Johnson) B.; B.S., Wheaton Coll., 1936, LL.D., 1960; M.S., Northwestern U., 1937, Ph.D., 1939; M.D., Yale, 1943; m. Dorothy Winifred Kellogg, Aug. 24, 1940; children—Stephen James, Priscilla Kimball, Elisabeth Martha, John Thomas. Instr. physiology Yale, 1943-45, asst. prof., 1945-48, asso. prof., 1948-52, editor Yale Jour. Biology and Medicine, 1949-52; prof., chmn. dept. physiology U. Pa., Phila., 1952——. Sci. research adv. bd. Nat. Assn. Retarded Children. Mem. Am. Physiol. Soc., Am. Inst. Nutrition, Endocrine Soc., Am. Soc. Clin. Investigation, Phila. Coll. Physicians, Sigma Xi. Studies of temperature regulation, control of food and water intake, and control of neuroendocrine systems. Described (with Tepperman and Long) exptl. hyperphagia, 1943; described (with Anand) exptl. aphagia in animals with lesions in hypothalamus of brain, 1951. Home: 224 Vassar Av., Swarthmore, Pa. 19081. Office: Richards Bldg., U. Pa., Phila. 19104.*

BROCA, Pierre Paul, French surgeon, anthropologist; b. Sainte-Foy la Grande, Gironde, June 28, 1824; B.Sc., College of Sainte-Foy la Grande, 1840; M.D., Paris, 1849. Served successively as surgeon to St. Antoine, La Pirie, Hospital des Cliniques, Hospital Neeker; prof. external pathology, Paris; prof. clinical surgery, Paris. Founded Anthropology Lab., l'École des Hautes Études, 1858; Société d'Anthropologie, 1859; Revue d'Anthropologie, 1872. Mem., Acad. of Medicine; became Senator, 1880. Author: Atlas d'anatomie descriptive du corps humain (4 vols.), 1841-66; L'Ethnologie de la France, 1859; Mémoires d'anthropologie, (6 vols.), 1871-88. Founder of modern brain surgery; localized center of speech in the brain, 1861; expert on aphasia; investigated comparative morphology of brain, topography of skull and brain, prehistoric trepanations. Died Paris, France, July 9, 1880.

BROCARD, Henry (-Pierre-Jean-Baptiste), mathematician; b. Vignot, France, May 13, 1845; ed. Ecole Polytechnique. Numerous positions in govt. service; served to lt. col., territorial engrs.; dir. Meteorol. Insts., Algiers, from 1874; asso. with Meterol. Bull. Algeria, from 1875. Author: Analyse indeterminée du premier degré, 1896; Notes de bibliographie des courbes géométriques, 1897; Courbes géométriques remarquables, 1922. Studied geometry of triangle, including Brocard point and Brocard circle. Died Bar-le-Duc, France, Jan. 16, 1922.

BROCCHI, Giovanni Baptista, geologist; b. Bassano, Italy, Feb. 18, 1772; prof. botany, Brescia; became insp. Milan (Italy) Mines, 1809; engr. to viceroy of Egypt, 1822. Author: Trattato mineralogico e chemico sulle miniere di ferro del Mella, 1808; Memoria mineralogica sulla Valle di Fassa nel Tirolo, 1811; Conchiologia fossile subapennina con osservazioni geologiche, 1814; Dello stato fisico del suolo di Roma, 1820. Studied structure of Apenine mountain range and fossils of its strata; demonstrated Rome does not occupy site of extinct volcano, as previously held. Died Khartoum, Sudan, Sept. 26, 1826.

BROCH, Ole-Jacob, Norwegian mathematician; b. Frederistad, Norway, Jan. 14, 1818. Prof. math. U. Christiana, Norway; dir. Norwegian Nat. R.R., from 1855; senator, from 1869; dir. Internat. Bur. Weights and Measures, Breteuil, from 1872. Mem. French Acad. Scis., 1875. Author: Manuel de trignométrie, 1851; Traite de mécanique, 1854; Leçons sur les mathématiques supérieures, 1861. Died Breteuil, Feb. 5, 1889.

BROCHANT DE VILLIERS, Andre-Jean-Marie, French geologist; b. Villiers, France, Aug. 6, 1773; Engr., École polytechnique, 1800. Prof. mineralogy Sch. Pezay, then at Paris, from 1815; gen. insp. mines; dir. mirror manufacture, St.-Gobain, France; chief editor Jour. Mines. Mem. French Acad. Scis., 1816. Author: Traité élémentaire de minéralogie, 1818, Mémoire pour servir à une description géologique de la France, 1830-38; Explication de la carte geologique de France. Through work (with Élie de Beaumont and Armand Petit (Dufrénoy) in Alps and Pyrenees, confirmed existence of metamorphism and attributed it to action of water; investigations helped bring about geol. map of France. Died May 16, 1840.

BROCHIER, Hubert Louis, French economist; b. Coublevice, France, Oct. 12, 1923; s. Ernest and Jeanne Brochier; Licencie en Droit, Faculté de Droit, Grenoble, France, 1943; Docteur ès science economiques, Faculté de Droit, Paris, 1949; m. Marguerite-Marie Redoutey, Sept. 16, 1947; children—Chantal, Anne-Claire, Emmanuel, Jean-Christophe. Faculty, Faculté de Droit et de Science Economique, Grenoble, 1952——, prof., 1952-63; prof. Faculté de Droit et de Sciences Economique de Paris, 1963——. Expert monetary and fiscal question on Vietnam, UN, 1959. Mem. French Assn. Econ. Sci. (sec.-gen.), Internat. Assn. Econ. Sci. Author: Finances publiques et redistribution des revenus, 1950; Economie Finaciere (with P. Tabatoni), 1959; Le meracle economique japonais, 1965; also articles. Research in econ. politics, interrelationship of sociology, econs. and cybernetics. Home: 16 rue Cortambert, Paris 16, France.*

BROCHNER-MORTENSEN, Knud, Danish physician; b. Fredericia, July 4, 1906; s. Valdemar and Ingrid (Shäffer) B.-M.; M.D., U. Copenhagen; m. Else Stein, Apr. 16, 1932; children—Inge, Hans, Jens, Kirsten. Staff physician in internal medicine; prof. internal medicine, dir. med. dept. U. Copenhagen. Mem. Danish Soc. Internal Medicine, Am. Assn. Rheumatology, Argentine, Swedish socs. rheumatology, also other Danish sci. socs. Author: Uric Acid in Blood and Urine; Arthritis Urica, also numerous publs. on biochemistry, clin. research, med. edn. Home: 1, Frederik d. V's Vej, Copenhagen. Office: Medical Dept. A., Rigshospitalet, Blegdamsvej, 9, Denmark.

BROCK, Byron Britton, geologist; b. Kingston, Ont., Can., July 1, 1904; s. Reginald Walter and Mildred (Britton) B.; B.Applied Sci., U. B.C., 1926; postgrad. St. John's Coll., Cambridge, Eng., Queen's U., Kingston; Ph.D. in Geology, U.Wis., 1934; m. Barbara Grote, Stirling, Nov. 21, 1929; children—Elizabeth Mary (Mrs. G. F. Struan Robertson), Patrick Willet Grote. Asst. geologist Colony of Hong Kong, 1926-27; mineral exploration geologist Selection Trust, Ltd., Yugoslavia, 1927-28; with exploration dept. Britannia Mining co., 1929-30; mgr. Rhokana field div., No. Rhodesia, 1934-41; asst. to cons. geologist Anglo-Am. Corp., Johannesburg, South Africa, 1945-52, cons. geologist, 1952-65; hon. lectr. geology dept. U. Cape Town, Rondebosch, Cape, South Africa, 1965——. Fellow Royal Geog. Soc. Geol. Soc. (London, Eng.); mem. Geol. Soc. South Africa (past pres., mem. council 1950-66, Draper medal 1961), Soc. Econ. Geologists (regional v.p. for Africa 1960-63), Internat. Union Geol. Scis. (mem. world commn. on mineral distbn. 1963——), Soc. Econ. Geologists, Geol. Soc. Am., Am. Geophys. Union, Instn. Mining and Metallurgy, Canadian Inst. Mining and Metallurgy. Research and publs. on order as opposed to chaos in world patterns of mineral distbn. and crustal fragmentation, structural mosaics in a hierarchy, vertical tectonics, evolution mountain-bldg., lineament patterns and geol. cruxes, lunar and Martian tectonics. Home: Bosky Dell, Grant Av., Simon's Town, Cape. Office: Geology Dept., U. Cape Town, Rondebosch, Cape, South Africa.*

BROCK, John Fleming, South African physician; b. Port Elizabeth, South Africa, Mar. 27, 1905; s. John and Edith (Fleming) B.; B.A., U. Cape Town (S. Africa), 1925; postgrad. Rhodes Sch., Oxford, Eng.; M.A., Oxford U., 1931, D.M., 1935; m. Ruth Mary Lomberg, June 3, 1933; children—Virginia R. M. (Mrs. Dirk-jan Meerburg), David John, John Christopher Fleming, Bridget Edith (Mrs. Johann Lubbe). Med. 1st asst. Brit. Postgrad. Med. Sch., Hammersmith, London, 1933-35; asst. dir. research in medicine U. Cambridge (Eng.), 1935-38; prof. medicine U. Cape Town, 1938——; head dept. medicine Groote Schuur Hosp., 1938——. Cons. in nutrition WHO, 1949——. Fellow A.C.P. (hon.), Royal Coll. Physicians (London); mem. S. African Coll. Physicians, Surgeons and Gynecologists (council, pres. 1965——), Med. Assn. S. Africa (Research Silver medal 1962). Author: (with M. Autret) Kwashiorkor in Africa, 1952; Advances in Human Nutrition, 1961; also numerous articles. Research on pathophysiology of nutrition in relation to human health and disease, protein-calorie malnutrition. Home: Bendoran, Bishopscourt Dr., Bishopscourt, Cape, South Africa. Office: Dept. Medicine, Med. Sch. U. Cape Town, Cape Town, South Africa.*

BROCK, Norbert, German physician; b. Dorsten-Westf, May 26, 1912; s. Johannes and Franziska (Hunecke) B.; ed. univs. Friburg/Br., Münster, Graz, Düsseldorf; M.D.; m. Edith Hilia Priske, Dec. 22, 1944; children—Barbara-Annette, Gabriela, Jürgen, Ulrich. Reader in pharmacology; specialist in internal medicine, 1946; head dept. pharmacology Asta-Werke AG, Brackwede, 1949; titular prof. U. Münster, 1954. Mem. German Soc. Pharmacology, Soc. Research on Circulation, German Soc. Internal Medicine, Biometrical Soc. Research and publs. on pharmacology, cellular metabolism, gen. pharmacotherapy and chemotherapy of cancer. Home: Am Rehhagen 222, Uerentrub über Bielfeld. Office: Bielefeldstrasse 83, Brackwede i.W., Germany.

BROCK, Reginald Walter, Canadian geologist; b. Perth, Ont., Can., Jan. 10, 1874; s. Thomas and Marian (Jenkins) B.; student Ottowa Collegiate Inst., 1890, Heidelburg U., 1895-97; M.A., Toronto U., 1895; LL.B., Queens U., 1921, U. Hong Kong, 1933; m. Mildred Britton, Nov. 28, 1900; children—P. Willet, Byron Britton, David Hamilton, Thomas Leith, Phillip Holton Gilbert. Chmn. dept. geology and petrography Queens U., 1902-07; dir. geol. survey, 1907-14; dep. minister Dept. Mines, 1914; dean faculty applied sci. U. B.C. Attended Internat. Geog. Congress, Paris, 1931. Served to maj. Canadian armed forces, 1914-18. Decorated Jubilee medals King George V, 1935. Fellow Royal Soc. Can., Geol. Soc. London, Geol. Soc. Am. Author: The Physical Basis of Canada in Canada and Its Provinces. Discovered Nelson batholith robe in B.C.; proved Asian continent once included Fijian archipelago; conducted geol. investigation of Hong Kong. Died July 30, 1935.

BROCKLESBY, Richard, English physician; b. Somersetshire, Eng., 1722; s. Richard and Mary (Alloway) B.; ed. Edinburgh, Scotland; M.D., Leiden (Netherlands), 1745; degree U. Dublin, 1754; M.D., Cambridge, Eng., 1754. Practiced medicine, London, Eng., 1746-58, also after 1763; physician to army in Germany, 1758-63. Fellow Royal Soc., 1746; fellow, licentiate Coll. Physicians. Author: An Essay Concerning the Mortality among Horned Cattle, 8 vols., 1746; Economical and Medical Observations (1st book on army hygiene principles, camp diseases and treatments, regulations for field hosps.), 1764; also papers pub. in Philos. Trans. Suggested music be used to treat sick persons; recommended deep burial for diseased cattle. Died London, Dec. 11, 1797.

BROCKS, Karl Adolph Christian, German geophysicist; b. Kiel, Germany, Jan. 29, 1912; s. Carl Emil and Johanna (Mordhorst) B.; student U. Kiel,

1930-36; D., U. Berlin (Germany), 1939; m. Ursula Wiese, Dec. 16, 1942; children—Ute, Sabine, Dietrich. Research asst. German Research Found., 1939-40; asst. Meteorol. Inst., Berlin, 1942-45; faculty Hamburg U., 1948-53, prof., dir. Geophys. Inst., 1960—; faculty Hannover Tech. U., 1960; founder, dir. Radiometeorol. and Maritime Meteorol. Inst., Hamburg U., 1961—; coordinator, leader Atlantic Expdn., Meteor, 1965. Pres. sci. adv. council German Weather Service, 1967—. Mem. German Geophys. Soc. (pres.), German Meteorol. Soc., Nachrichtentechnische Gesellschaft. Research and publs. on devel. method to measure atmospheric stratification by light rays, devices to measure air-sea interaction and atmospheric turbulence above free ocean, methods to predict radar ranges above sea with simple meteorol. measurement; research on atmospheric turbulence and radio-wave propagation in atmosphere. Home: 3 Billeweg, 2057 Wentorf B. Hamburg, Germany.*

BROCQ, Louis Anne Jean, French dermatologist; b. Laroque-Timbaut, France, Feb. 1, 1856; ed. Paris, intern, 1879, docteur en medicine, 1882; practised St. Louis Hosp., Paris. Author: Traitement des maladies de peau, 1890; Précis élémentaire de dermatologie, 5 vols., 1893-96; Cliniques dermatologiques, 1924; (with Jacquet) Traitement des dermatoses par la petite chirurgie et les agents physiques, 1898. Specialized in cutaneous diseases, especially eczema, lichenization, fungoid mycosis, thrush; introduced beer lees for treatment of funuculosis; suggested term parapsoriasis for exfoliative dermatitis of unknown cause, 1902. Died, Dec. 18, 1928.

BROCQ-ROUSSEAU, Denis, French veterinarian; b. Paris, France, Oct. 17, 1869; ed. Alfort Sch., 1887, Mil. Sch. Saumur, 1891; Docteur es Sciences, U. Paris, 1907. Asst. at lab., Nancy, France; organized Mil. Lab. of Vet. Research, World War. I. Mem. French Acad. Scis., 1947, Acad. Medicine. Author: Le streptocoque gourmeux; Traité des foins; Le sérum normal, 1924-49. Studied role of toxalbumins in the alimentation of men and animals; prepared serum against horse strangles and anthrax. Died Paris, Jan. 22, 1950.

BRODAL, Alf, Norwegian neuro-anatomist; b. Oslo, Norway, Jan. 25, 1910; s. Peter and Helen (Obenauer) B.; grad. Med. Faculty, U. Oslo, 1937; Dr. hon. causa, U. Uppsala (Sweden), 1966; m. Inger Hannestad, July 10, 1935; children—Inger Helene (Mrs. Bodvar Vandvik), Anne Brit, Per. Asst. dept. anatomy Odont. High Sch. Norway, 1937-43; prosector anatomy U. Oslo, 1943-50, prof., 1950—, dean med. faculty, 1964-66, vice rector univ., 1967—. Chmn. div. for medicine Norwegian Research Council for Sci. and Humanities, 1960—. Rockefeller fellow U. Oxford (Eng.), 1946-47. Recipient Monrad-Krohn's prize, 1941, 60; Fridtjof Nansen's prize, 1952; Barany medal, 1963; La Jahre prize, 1966. Fellow Norwegian Acad. Scis., Deutsche Akad. der Naturforscher, A.A.A.S.; mem. Am. Neurol. Assn. (corr.), Nordic Neurol. Assn. (asso.), Royal Soc. Medicine London, Norwegian Neurol. Soc. (past chmn.), Norwegian Med. Assn. (editor jour. 1961). Author: Neurological Anatomy in Relation to Clinical Medicine, 1948; The Cranial Nerves, 1959; The Reticular Formation of the Brain Stem, 1957; (with J. Jansen) Aspects of Cerebellar Anatomy, 1954; (with O. Pompeiano, F. Walberg) The Vestibular Nuclei and their Connections, 1961; numerous articles, chpt. in book. Editor: (with R. Fänge) The Biology of Myxine, 1963. Research on structure of brain especially exptl. studies of connections between various parts and functional aspects. Home: 14 Prestasen, Blommenholm, Baerum, Norway. Office: 47 Karl Johansgt, Oslo, Norway.*

BRODBECK, Arthur Joseph, Am. social psychologist; b. Tarrytown, N.Y., June 7, 1922; s. Arthur J. and R. (Blessing) B.; B.S., U. Chgo., 1946; M.A., State U. Ia., 1947, Ph.D., 1949. Asst. prof. U. Ill., Urbana, 1948-51; asst. prof. grad. sch. Boston U., 1951-56; research asso. Psychoanalytic Inst., Beverly Hills, Cal., 1957-60, Yale Law Sch., New Haven, 1961-66, Center for Urban Edn., N.Y.C., 1966—; vis. prof. Wesleyan U., 1964-65, U. Pa., 1965-66. Fellow Center for Advanced Study of Behavioral Scis., 1955; top award for tv research 1961. Mem. N.Y. Acad. Sci., A.A.A.S., Am. Sociol. Assn., Psychol. Assn. Am. Author: (with Eugene Burdick) American Voting Behavior, 1958; (with H. D. Lasswell) Constitution and Character, 1967. Research, publs. in analysis of socialization and edn. as legal and power process, analysis of communication as a way to promote effective collaboration between specialists, devel. of theory of art in social process that shows feeling to be a value process operating in communities. Home: Woodstock Tower, Tudor City, N.Y.C. 10017. Office: 33 W. 42d St., N.Y.C. 10036.*

BRODE, Robert B(igham), Am. physicist; b. Walla Walla, Wash., June 12, 1900; s. Howard Stidham and Martha Catherine (Bigham) B.; B.S., Whitman Coll., 1921, D.Sc. (honorary), 1954; Ph.D., California Institute of Technology, 1924; student Oxford (Eng.) U. (Rhodes scholar), 1924-25, Gottingen (Germany) U. (Internat. Edn. Bd. fellowship), 1925-26, Princeton U. (Nat. Research Council fellow), 1926-27; m. Bernice Hedley Bidwell, Sept. 16, 1926; children—William (dec.), John Howard. Asso. physicist U. S. Bur. Standards, 1924; asst. prof. physics U. of Calif. at Berkeley, 1927-30, asso. prof., 1930-32, prof. since 1932; asso. dir. National Science Foundation, 1958-59; vis. prof. Mass. Inst. Tech., 1932; physicist dept. terrestrial magnetism Carnegie Inst., 1941, sect. T., applied physics lab. O.S.R.D. Johns Hopkins, 1942-43, supervising research and development on proximity fuse and fire control equipment; group leader Los Alamos (N.M.) Atomic Bomb Lab., 1943-46, Harvard, summer 1948. Guggenheim fellow London and Cambridge, 1934-35; Fulbright award, Manchester, 1951-52. Fellow Am. Phys. Soc., A.A.A.S.; mem. Am. Assn. Physics Teachers, Am. Assn. U. Profs. (2d v.p. 1960-61), Am. Assn. Rhodes Scholars, American Academy of Arts and Sciences, also Nat. Acad. Scis., Phi Beta Kappa, Sigma Xi. Author articles in physics publs. Specializes in cosmic rays and electronics; research on mass of meson; proximity fuse; absorption coefficients of slow electrons in metal vapors and gases. Home: 1471 Greenwood Terrace, Berkeley 8, Cal.

BRODE, Wallace R(eed), Am. chemist; b. Walla Walla, Wash., June 12, 1900; s. Howard S. and Catherine (Bigham) B.; B.S., Whitman Coll., 1921, D.Sc., 1955; M.S., U. Ill., 1922, Ph.D., 1925; D.Sc., Ohio State U., 1958, Ohio Wesleyan U., 1962; m. Ione Sundstrom, Mar. 19, 1941. Asst. chemistry U. Ill., 1921-24; chemist U. S. Nat. Bur. Standards, 1924-26, asso. dir., 1947-58; faculty Ohio State U., 1928-48, prof. 1939-48; head Paris Liaison Office, OSRD, 1944-45; head sci. dept. U. S. Naval Ordnance Test Sta., 1945-47; sci. adviser to U. S. Sec. State, 1958-60; dir. Barnes Engring. Co., Washington, 1960—; fgn. sec. Am. Chem. Soc., Washington, 1965—. Cons. to industry.; mem. Pres.'s Com. on Scientists and Engrs., 1958; co-chmn. People to Peoples Com. for Scientists and Engrs., 1957-58, 65—; mem. Sci. Manpower Commn., 1960—. Recipient Exceptional Service medal U. S. Dept. Commerce, 1958; Applied Spectroscopy medal Soc. Applied Spectroscopy, 1958; Guggenheim fellow, 1926-28; Marburg lectr. Am. Soc. for Testing Materials, 1950. Mem. Nat. Acad. Scis., Am. Chem. Soc. (dir. 1951-60, Priestley medal 1960), A.A.A.S. (pres. 1958, chmn. bd. dirs., 1959), Optical Soc. Am. (pres. 1961, editor, jour. 1950-60), Sci. Service (treas. 1958—), Sci. Research Soc. Am. (chmn. 1954-57), Sigma Xi (pres. 1961-62). Author several books including Chemical Spectroscopy, 1939; (with others) Laboratory Outlines of Organic Chemistry, 1940; (with others) The Roger Adams Symposium, 1955; (with others) Physical Methods in Chemical Analysis, 1960. Editor: Sci. in Progress, 1962—. Research and publs. in synthetic organic chemistry, dyes, spectrochemistry, studies in sci. edn., personnel, programs. Home: 3900 Connecticut Av., N.W., Washington 20008. Office: 1155 16th St., Washington 20036.*

BRODERIP, William John, Brit. naturalist; b. Bristol, Eng., Nov. 21, 1789; s. William Broderip; B.A., Oriel Coll., Oxford (Eng.) U., 1812. Called to bar Lincoln's Inn, 1817; lawyer western circuit, 1817-22; magistrate Thames police ct., 1822-46, Westminster Ct., 1846-56; elected bencher Gray's Inn, 1850, treas., 1851. Mem. Zool. Soc. (co-founder), Geol. Soc. (sec.), Royal Soc., 1828, Linnean Soc. Contbr. numerous articles on mammals, birds, reptiles, crustacea, mollusca, conchifera, cirrigrada. Described animal habits; collected natural objects; outstanding conchological collections. Died London, Feb. 27, 1859.

BRODERS, Albert C., Am. pathologist; b. Va., Aug. 8, 1885; M.D., Med. Coll. Va., 1910; Sc.D. (hon.), 1929; M.S., Minn. U., 1920; Sc.D (hon.), Washington and Lee, 1949; m. 1915; 3 children. Mem. staff Mayo Clinic, Rochester, Minn., from 1912, asst. surg. pathologist, 1912-19, asso., 1919-22, head sect. div. surg. pathology, 1922-35, 36-45, apptd. dir., cons., 1945; prof. surg. pathology, dir. cancer research Med. Coll. Va., 1935-36; prof. pathology Mayo Found., from 1936. Fellow Coll. Physicians; mem. A.M.A., Soc. Clin. Pathologists. Research on thyroid gland, chronic gastric ulcer and gastric carcinoma, epithelioma, Tb asso. with malignant neoplasia, grading of elinieopathology of cancer, tumor of extremities; gave classification of 4 types of malignant tumors based on undifferentiated state of cells, 1920. Died 1964.

BRODETSKY, Selig, mathematician; b. Olviopol, Russia, Feb. 10, 1888; s. Akiva and Ada Brodetsky; B.Sc., U. London; M.A., Cambridge U.; Ph.D., U. Leipzig; m. Mania Berenblum; 1 son, 1 dau. Isaac Newton research student, 1910-13; lectr. applied math. U. Bristol, 1914-19; reader U. Leeds, 1920-24, prof., 1924-48, prof. emeritus applied math., 1949-54. Fellow Royal Astron. Soc., Royal Aero. Soc., Inst. Physics. Author: A First Course in Nomography; The Mechanical Principals of the Aeroplane; Sir Isaac Newton, A Brief Account of His Life and Work; The Meaning of Mathematics, also articles on math., physics, aeronautics. Died May 18, 1954.

BRODIE, Arnold F., Am. biochemist; b. Boston, Dec. 31, 1923; s. Harry E. and Tessie (Getman) B.; B.A., Northeastern U., 1946; M.S., Boston U., 1947; postgrad. Mass. Inst. Tech.; Ph.D., U. Pa., 1952; m. June Wiskind, Nov. 24, 1948; children—Todd D., Leslie S. Research asst. Biochem. Lab., Mass. Gen. Hosp., Boston, 1952-54; research biochemist Leonard Wood Meml. Lab., Harvard Med. Sch., 1954-56, asso. dept. microbiology, 1956-60, asst. prof. dept. microbiology, 1960-63; prof. dept. microbiology U. So. Cal. Sch. Medicine, 1963—, chmn. physiology div. Am. Soc. Microbiology. Mem. Acad. Northeastern U., Am. Soc. Biol. Chemistry, A.A.A.S., Am. Soc. Microbiology, Am. Chem. Soc. Research and numerous publs. in means by which biol. systems obtain energy, electron transport pathways in bacteria, role of quinones in respiratory process as well as chemistry of naturally occurring quinones. Home: 1141 Romney Dr., Pasadena, Cal. 91105. Office: 2025 Zonal Av., Los Angeles 90033.*

BRODIE, Sir Benjamin Collins (the elder), English surgeon; b. Winterslow, Eng., June 9, 1783; s. Peter Bellinger and (Collins) B.; studied under Abernethy and Wilson in London; became pupil of Everard Home, St. George's Hosp., 1803; D.C.L., Oxford U.; m. Miss Sellon, 1816; eldest son, Benjamin Collins. Apptd. prof. comparative anatomy and physiology Royal Coll. Surgeons, 1816, pres., 1844; physician to George IV; sergeant-surgeon to William IV, 1832, then to Victoria. Recipient Copley medal, 1811. Fellow Royal Soc., 1810 (pres. 1858-61); mem. French Acad. Scis. Author: Diseases of the Joints, 1818. Credited with 1st surg. treatment of varicose veins, 1814; reduced number of amputations of joints through conservative treatment; described inflammation of head of tibia or Tb abscess (Brodie's abscess), 1832; gave possibly 1st description of intermittent claudication in man, 1846; opposed homeopathy. Died Broome Park, Surrey, Eng., Oct. 21, 1962.

BRODIE, Sir Benjamin Collins (the younger), English London, Eng., June 8, 1817; s. Benjamin Collins and (Sellon) B.; ed. Balliol. Coll., Oxford (Eng.); B.A., 1838; D.C.L., Oxford, 1872; m. dau. of John Vincent Thompson, 1844. Became prof. chemistry Oxford, 1855; Waynflete prof. chemistry, 1865-73. Fellow Royal Soc., 1849; mem. Chem. Soc. (pres. 1859-60). Research and publs. on allotropic states of carbon, beeswax; discovered arotic acid, graphitic acid. Died Oct. 8, 1880.

BRODIE, Bernard Beryl, pharmacologist; b. Liverpool, Eng., Aug. 7, 1909; s. Samuel and Esther (Ginsberg) B.; B.S., McGill U., 1931; Ph.D., N.Y.U., 1935; Ph.D. (hon.), U. Paris (France), 1963; D.Sc., Phila. Coll. Pharmacology and Sci., 1965; m. Anne Lois Smith, Aug. 30, 1950. Asst. prof. pharmacology N.Y. U. Med. Sch., 1943-47, asso. prof. biochemistry, 1947-50; chief lab. chem. pharmacology Nat. Heart Inst., NIH, 1950—; vis. prof. George Washington U. Med. Sch., also Georgetown U. Med. Sch., 1950—. Chmn. sci. adv. com. Nat. Council Alcoholism, 1963—. Served with Canadian Army, 1926-28. Recipient Distinguished Service award Dept. Health, Edn. and Welfare, 1958; Shionogi Commemoration lectr., Japan, 1962; Karl Beyer award, 1962; Julius Sturmer meml. lectr., 1962; Torald Sollmann award, Am. Pharmacology Soc., 1963; Distinguished Achievement award Modern Medicine medical journal, 1964. Mem. Nat. Acad. Scis., Internat. Pharm. Soc., Am. Coll. Neuropsychopharmacology (pres. 1965), Am. Soc. Biol. Chemists, Am. Soc. Pharmacology and Exptl. Therapeutics, Harvey Soc., N.Y., Washington acads. scis., Royal Soc. Medicine. Author: Metabolic Factors Controlling Duration of Drug Action, 1963; Drug Enzyme Interactions, 1964. Founder, U. S. editor Life Sciences, also Internat. Jour. Neuropharmacology. Author numerous articles in field. Research on drug metabolism and enzymes; membrane permeability; body water; neurochemistry; biochemistry of function; biochem. evolution; biol. control systems; mechanism of drug action. Home: 4977 Battery Lane, Bethesda 14. Office: Nat. Heart Inst., Nat. Institute Health, Bethesda, Md.

BRODIN, Harald, Swedish physician; b. Bjuraker, Sweden, Mar. 23, 1918; s. Ernst and Julia (Palm) B.; M.D., Lund U., 1955; m. Great Törnfelt, July 8, 1944. Asst. prof. physical medicine Lund U., 1962-64, Uppsala (Sweden) U., 1964-66; demonstrator phys. medicine U. Stockholm (Sweden), 1966—; head physician dept. orthopedic surgery, Boden, Sweden, 1957-59, dept. physiotherapy, Lund, 1959-64, dept. orthopedic surgery, Gävle, Sweden, 1964-66; head physician dept. physiotherapy Karolinska Sjukuset, Stockholm, 1966—; demonstrator Karolinska Institutet, Stockholm, 1966—. Mem. Scandinavian Soc. Manual Medicine (pres. Swedish sect.), Swedish Soc. Phys. Medicine, Rehab. and Social Medicine (v.p.), Società Italiana Med. Fis. E. Riabilitazione (hon.). Author: (with Bang, Bechgaard, Kaltenborn, Schlötz) Manipulation of the Spine, 1966; also textbooks on physiotherapy, 1963-64, articles. Research on longitudinal growth of bones, treatment of knee joint Tb, pedagogic problems in tng. of muscle power and in chronic lumbar pain; diagnosis of painful disorders in backs and joints. Home: 12 Tornvägen, Näsbypark, Sweden. Office: Karolinska Sjukhuset, Stockholm 60, Sweden.*

BRODSKY, Aleksandr Ilich, Russian phys. chemist; b. 1895; grad. Moscow U. Prof., Inst. Phys. Chemistry, Ukrainian Acad. Sci., 1926-39, dir., 1939—. Mem. USSR (corr.), Ukrainian acads. scis. Author: Studies on the Thermodynamics and Electrochemistry of Solutions, 1931; Modern Electrolyte Theory, 1934; Physical Chemistry, 6 edits., 1929-48; The Theory of Fine Fractionation and Separation of Mixtures by Thermodiffusion, 1948; co-author: The Use of the Isotope Method in Studying the Mechanism of Chemical Reactions, 1942; Chemistry of Isotopes, 1952, 57. Organized research on chem. reactions by isotope methods, 1934, 1st in USSR to prepare and use stable isotopes for this purpose. Address: Ukrainian Acad. Sci., Vladimirskaya ulitsa 54, Kiev, Ukraine SSR, USSR.

BRODSKY, Allen, Am. physicist; b. Balt., Nov. 5, 1928; s. Nathan M. and Gertrude (Silberman) B.; B. Engring. in Chem. Engring., Johns Hopkins, 1949, M.A. in Physics, 1960; Sc.D., U. Pitts., 1966; m. Paula Fishman, June 17, 1951; children—Richard S., Karen S., Jay R. Head health physics unit Naval Research Lab., Washington, 1950-52; pres. Health Physics Services, Inc., Balt., 1955-57; radiol. def. officer Fed. Civil Def. Adminstrn., Olney, Md., 1956-57; health physicist AEC, Washington, 1957-61; research asso. U. Pitts. Grad. Sch. Pub. Health, 1961-66, asso. prof. dept. occupational health, 1966—. Cons. to bus. firms in Pitts. Diplomate Am. Bd. Health Physics, Am. Acad. Indsl. Hygiene. Mem. Health Physics Soc. (mem. standards com. 1959—, nat. dir., pres. W.Pa. 1967—), Am. Phys. Soc., Internat. Radiation Protection Assn., Am. Indsl. Hygiene Assn., Am. Nuclear Soc., Am. Statis. Assn., Am. Pub. Health Assn., Am. Conf. Govt. Indsl. Hygienists. Editor, compiler: (with G. Victor Beard) Controlling Radiation Emergencies, 1960. Research and publs. on methods of measuring nuclear radiation for assessment radiation hazards, measurement pulsed neutrons and X-rays, and gamma radiation with photog. emulsions, methods of evaluating hazards from fallout or radioactivity, radioactivity in humans. Home: 2047 Brookfield Dr., Pitts. 15243.*

BRODY, Alfred Walter, Am. physician; b. N.Y.C., Feb. 20, 1920; s. Hyman and Sophia (Naidich) B.; B.A., Columbia, 1940, M.A. in Chemistry, 1941; M.D., L.I. Coll. Medicine, 1943; D.Sc., U. Pa., 1955; m. Shirley Bloom, Dec. 17, 1943; children—Betty Ann, Carolyn Gail, Frances Linda, Robert Seymour. Faculty, U. Pa., 1951-54; faculty Creighton U., Omaha, 1954—, prof. medicine, 1964—. Physiologist, dir. pulmonary function lab. Creighton Meml.-St. Joseph's Hosp., Omaha, 1955—. Mem. Am. Physiol. Soc., A.A.A.S., N.Y. Acad. Sci., A.M.A., Am. Coll. Clin. Pharmacology and Chemotherapy, Omaha Research Soc. (sec.-treas. 1962—), Neb. Heart Assn. (chmn. cardiovascular research 1957—) Omaha Tb Soc. (pres. 1965). Research and publs. on respiration, circulation, blood flow. Home: 5122 Capitol St., Omaha 68132.*

BRODY, Eugene Bloor, Am. psychiatrist; b. Columbia, Mo., June 17, 1921; s. Samuel and Sophie (Brody) B.; A.B., U. Mo., 1940, M.A., 1941; M.D., Harvard, 1944; grad. N.Y. Psychoanalytic Inst., 1957; m. Marian Holen, Sept. 23, 1944; children—Julie, James, John. Chief psychiatry Mil. Hosps., European Command, U. S. Army, 1946-48, cons. Internat. Mil. Tribunal, Nürnberg, Germany, 1947; instr. Yale U. Sch. Medicine, 1948-51, asst. prof., 1951-53, asso. prof., 1953-57; prof. psychiatry U. Md. Sch. Medicine, 1957—, chmn. dept., dir. Psychiat. Inst., 1959—; cons. editor Jour. Nervous and Mental Disease, 1960-67, editor-in-chief, 1967—; Mem. profl. adv. bd. VA, 1963-67; bd. dirs., profl. adv. bd. Nat. Assn. for Mental Health. Fellow Am. Psychiat. Assn., Am., Internat. psychoanalytic assns.; mem. Peruvian, Canadian psychiat. assns., Am. Psychopath. Assn., Am. Social Assn., Inter-Am. Council Psychiat. Assns., Inter-Am. Soc. Psychology. Author (with F. C. Redlich) Psychotherapy with Schizophrenics, 1952; The Minority Group Adolescent in the United States, 1967; (with G. O. Klee, R. R. Monroe) Psychiatric Epidemiology and Mental Health Planning, 1967. Contbr. numerous articles to sci. jours. Research on relationships of society; culture and mental illness; minority groups and prejudice. Home: Skyline Rd., Ruxton, Md. Office: 645 W. Redwood St., Balt. 21201.*

BRODY, Jacob Allan, Am. physician; b. Bklyn., May 5, 1931; s. Simon and Rosella (Mindell) B.; B.A., Williams Coll., 1952; M.D., State U. N.Y. Downstate Coll. Medicine, 1956. Epidemic intelligence officer Communicable Disease Center, Atlanta, 1957-59; med. officer Middle Am. Research Center, Panama C.Z., 1959-61; exchange scientist Inst. Poliomyelitis and Virus Encephalitis, Moscow, USSR, 1962; chief epidemiology sect. Arctic Health Research Center, Anchorage, 1963-65; chief Epidemiology br. Nat. Inst. Neurol. Diseases and Blindness, NIH, Bethesda, Md., 1966—. Fellow Am. Pub. Health Assn. (epidemiology sect., chmn. membership com.); mem. Nat. Multiple Sclerosis Soc. (med. adv. bd.), Am. Epidemiol. Soc., Am. Assn. Immunologists, Fedn. Am. Socs. Exptl. Biology, Alpha Omega Alpha. Research and numerous publs. on encephalitis epidemics and role of inapparent infections, effects of vaccination on epidemiological patterns of polio, methods for collection and testing blood specimens for field studies, natural history of rubella in a virgin population and showed that gamma globulin prevents infection; evaluated measles vaccines in highly susceptible Eskimo populations. Home: 3044 R St. N.W., Washington 20007. Office: 7550 Wisconsin Ave., Bethesda, Md. 20014.*

BRODY, Samuel, agrl. chemist; b. Garbatchi, nr. Baranowitz, Lithuania (now Poland), Feb. 8, 1890; s. Avram J. and S. Deborah (Malov) B.; came to U. S., 1906, naturalized, 1912; A.B., U. Cal., 1917, A.M., 1919; Ph.D., U. Chgo., 1928; Guggenheim fellow, Europe, 1930-31; m. Sophie Edith Dubosky, Aug. 15, 1920; children—Eugene Bloor, Dr. Arnold Jason. Asst. biochemist U. Cal. Med. Sch., 1919-20; asst. prof. dairy and agrl. chemistry U. Mo., 1920, asso. prof., 1925, chmn. inter-dept. com. on growth and metabolism, from 1926; chmn. Herman Frasch Found. research project in agrl. chemistry Mo. Coll. Agr., 1929-40, chmn. com. on muscular work in farm animals, from 1941, prof. dairy husbandry, 1945, chmn. com. on influence of climatic factors on physiol. reactions and productivity in farm animals, from 1946. Recipient Borden award, 1950. Fellow A.A.A.S.; mem. Am. Soc. Naturalists, NRC (mem. sub-com. on energy metabolism in relation to agr.; sub-com. on standardization), Am. Assn. Univ. Profs., Am. Chem. Soc. (mem. div. biol. chemistry), Am. Soc. Biol. Chemists, Am. Inst. Nutrition, Am. Dairy Sci. Assn., Am. Animal Sci. Prodn., Soc. for Devel. and Growth. Czechoslovak Acad. Agr., Sigma Xi, Gamma Sigma Delta. Author: Bioenergetics and Growth, 1945; also chapters in books and articles in jours. Died Aug. 6, 1956.

BRODY, Sylvia, Am. psychologist; b. N.Y.C., Oct. 17, 1914; d. Isidor and Gussie (Sperling) B.; B.A. in Psychology, New Sch. for Social Research, 1947; M.A., N.Y. U., 1948, Ph.D. in Clin. Psychology, 1954; m. Sidney Axelrad, May 11, 1949. Tchr. elementary and high schs., N.Y.C., 1935-44; ednl. cons., psychotherapist Caroline Zachary Inst. Human Devel., N.Y.C., 1944-48; psychologist Menninger Found., Topeka, 1948-50; individual practice psychoanalysis, N.Y.C., 1950—; counsellor parent edn. Child Study Assn., N.Y.C., 1950-54; sr. psychologist pediatric psychiatry service, dir. infant devel. research project Lenox Hill Hosp., N.Y.C., 1963—; dir. child devel. research project City U. N.Y. Author: Patterns of Mothering, 1956; Passivity, 1964. Research on types of maternal behaviour with infants, child analysis, child psychotherapy, infant devel. Home: 1148 Fifth Av., N.Y.C. 10028.*

BROEKHUYSEN, Gerrit Jeronoimo, zoologist; b. Java, Apr. 19, 1908; s. Gerardus (Johannus) and Elisabeth (Cornelia) B.; Candid. ex Biol., Leiden (Holland) U., 1929, Doctorate E., 1933, Ph.D., 1936; m. Marie Henriette Boots, Apr. 16, 1936; children—Paul, Jim Gerard Paul. Faculty, U. Capetown (South Africa), 1937-39, 46—, asso. prof. zoology, 1967—, fellow, 1960—; entomologist, Pretoria, South Africa, 1939-41; freshwater biologist, Java, 1941-46. Recipient Gill Meml. medal, 1963. Mem. Netherlands Ornithol. Union (corr.), Deutsche Ornithologen Gesellschaft (corr.), Cape Bird Club (hon. life), South African Ornithol. Soc. (editor The Ostrich 1954—). Author: The Birds Around Us, 1966; also numerous articles. Research in entomology, marine biology and ornithology. Home: 12 Balfour Av., Newlands, Cape, South Africa. Office: Dept. Zoology, U. Cape Town, Randebosch, South Africa.*

BROEMSER, Philipp, physiologist; b. Rüdesheim/Rhine, Germany, July 20, 1886; s. Georg Josef and Anna Friederike (Engel) B.; ed. Freiburg, also Munich, Berlin, Germany; doctorate Physiologisches Institut, Marburg, Germany, 1911; m. Maria Kunst, 1927. Asst. to O. Frank, Physiologisches Institut, Munich, several years, dir. from 1934; mil. physician during World War I; became mem. faculty Munich, 1918, asso. prof., 1922; apptd. prof., Basel, Switzerland, 1925; in Heidelburg, Germany, 1930-34. Author: Lehrbuch der Physiologie, 1934; papers pub. in German biol. and physiol. jours. Studied blood vessels; developed O. Frank's principles of registering blood circulation which made possible 1st undistorted picture of small changes in pressure and flow. Died Munich, Nov. 11, 1940.

BROER, Lambertus Johannes Folkert, Dutch physicist; b. Dordrecht, Netherlands, Jan. 17, 1916; s. Ime and Jansje (van der Meulen) B.; Degree in Theoretical Physics, U. Amsterdam (Netherlands), 1940, Doctorate in Physics, 1945; m. Gabriellina van Delden, Aug. 24, 1944; children—Christine, Margaretha, Ime. With U. Amsterdam, 1941-49, lectr. theoretical physics, 1946-49; prof. fluid dynamics Tech. U., Delft, Netherlands, 1949-61; prof. theoretical physics Tech. U., Eindhoven, Netherlands, 1962—. Cons. Inst. Plasma Physics, Jutphaas, 1963—. Mem. Netherlands Phys. Soc. (pres. 1963—), Royal Inst. Engrs., Netherlands Math. Soc. Editor: Applied Sci. Research, 1961—. Research, publs. on paramagnetic relaxation, theory of spectra, gas dynamics, effect of chem. reactions in flow problems, magneto hydrodynamics, theory of wave propagation. Home: 20 Panterlaan, Son, Netherlands. Office: 2 Insulindelaan, Eindhoven, Netherlands.*

BROGDEN, Wilfred John, psychologist; b. Sydney, Australia, May 6, 1912; s. John and Elsie (Taylor) B.; came to U. S., 1914, naturalized, 1926; student So. Meth. U., 1929-32; A.B., U. Ill., 1933, Ph.D., 1936; m. Elinor Taylor Davis, Sept. 8, 1935; children—Penelope Joel, Ann Abigail. NRC Postdoctoral fellow Johns Hopkins Sch. Medicine, 1936-37, research asst. psychiatry, 1937-39; faculty U. Wis., Madison, 1939—, prof. psychology, 1946—, asst. dean Grad. Sch., 1947-51, asso. dean, 1951-58. NDRC research grantee Brown U., 1942-43, U. Wis., 1944-45. Recipient Presdl. certificate appreciation, 1945. Mem. Am., Midwestern psychol. assns., Psychonomics Soc., Soc. Exptl. Psychologists, A.A.A.S., Sigma Xi. Research and publs. on motivation and incentive animal conditioning, role of angle from body in tracking proficiency, efficiency verbal learning and form serial position curve, study psychoacoustics; discovered sensory preconditioning. Home: 1429 Vilas Av., Madison, Wis. 53711.*

BRÖGGER, Waldemar Christopher, Norwegian mineralogist, geologist; b. Oslo, Norway, Nov. 10, 1851; studied at Bonn (Germany), Strasbourg (France); also studied in Italy, Eastern Baltic region; 2 sons. Became asst. to Geol. Survey of Norway, 1875; in 1878 taught in univ.; named prof. mineralogy U. Natural History, Stockholm, Sweden; became prof. mineralogy U. Oslo, 1890; elected 1st pres. U. Kristiania; lectr. Johns Hopkins, 1900. Mem. French Acad. Scis. Author: Über die Apatitgänge Norwegens, 1874; Die Miner Alien der Syenitregmatitgänge der Südnorwegugit und Nephelinsyenite; Mineralogie und Kristallographie, 1890; Die Minaralien der südnorwegischen Granitpegmatitgänge, 1906; Die eruptivgesteine des Kristianiagebiets, 1894-1924; also articles on paleontology and geology. Made original and detailed studies of Norwegian archipelago, including fossils and eruptive dykes; pioneer in Norway in defending natural evolution of species. Died Boekkelaget, Norway, Feb. 17, 1940.

BROGLIE, Louis Victor Pierre Raymond, Duc de, French physicist; b. Dieppe, France, Aug. 15, 1892; s. Duc Victor and Pauline (d'Armaillé) B.; ed. Sorbonne, Paris; license in history, 1910; license in science, 1913; D.Sc., 1924; Dr. honoris causa from numerous fgn. univs. Instructor at Sorbonne, from 1926; prof. in Faculty of Scis. there, 1932-62; 1st dir. Inst. Henri Poincaré, from 1932. Mem. sci. com. Commissariat of Atomic Energy, 1947. Recipient Nobel prize in physics, 1929, Kalinga prize, 1952. Mem. French Academy Scis., 1933 (permanent sec., from 1942); Fellow Royal Soc., 1953; mem. Académie Française, 1944; Fgn. assoc. Nat. Acad. Scis.; also mem. numerous fgn. assns. Author: Recherches sur la théorie des quanta; (with father Maurice) Introduction a l'étude des rayons X et gamma; Matière et lumière; Physique et micro-physique; Savants et découvertes; Nouvelles perspectives en microphysique; Sur les sentiers de la science, also numerous sci. works for layman. Proposed that matter and radiation have both wave and particle properties, thereby laying foundation of wave mechanics; has made great effort to find causal (as opposed to probabalistic) interpretation of wave mechanics. Address: 94, rue Perronet, Neuilly-sur-Seine, France.

BROGLIE, Maurice Duc de, French physicist; b. Paris, Apr. 27, 1875; s. Victor, Duc de Groglie, and Pauline d'Armaille; ed. Coll. Stanislas, École Navale; D. ès Scis., D. en Physique, D. hon. c., Oxford U.; m. Camille de Rochetaillée, 1904. Naval officer, 1895-1904; prof. Coll. de France, 1942-45; adminstr. for matériel téléphonique; founder pvt. obs. attached to École des Hautes Etudes. Mem. sci. com. Commn. Atomic Energy, 1957. Mem. French Acad. Scis., 1924, Royal Soc. London (fgn. mem., Hughes Gold medal), also Académie française, 1934, French Soc. Physics (pres. 1924), Acad. of Marine. Author: Atomes, radioactivité et transmutations; also papers on electricity, radioactivity, X rays, nuclear physics. Died July 14, 1960.

BROGREN, Gösta Egon, Swedish physicist; b. Orebro, Sweden, May 3, 1915; s. Gustaf Fredrik and Alma (Persson) B.; M.A., U. Uppsala (Sweden), 1939, Ph.D., 1943, Dr.Sc., 1949; m. Ragna Gunborg Grette, June 13, 1943; children—Eva Annika, Lena Monika. Asst. prof. physics U. Uppsala, 1949-55; faculty Chalmers U. Tech., Gothenburg, Sweden, 1956—, prof. atomic physics, 1963—. Fellow Brit. Phys. Soc.; mem. Swedish, Japanese phys. socs. Research, publs. on X-ray diffraction in perfect crystals, X-ray spectroscopy. Home: 4A, Föreningsgatan, Gothenburg, Sweden.*

BROIDA, Herbert Philip, Am. physicist; b. Aurora, Colo., Dec. 25, 1920; s. Theodore D. and Lucy (Shatz) B.; B.A., U. Colo., 1944; M.A., Harvard, 1945, Ph.D., 1949; postgrad Imperial Coll. (Eng.), U. Cambridge (Eng.), 1959-60; m. Ina Burnes, Aug. 8, 1948; children—John Phillip, Allison Burnes. Instr. physics Wesleyan U., Conn., 1944-45; research asso., teaching fellow Harvard, 1945-49; research asso. Georgetown U., Sch. of Med., Washington, 1957-63; physicist Nat. Bur. Standards, Washington, 1949-56, sect. chief, program coordinator, 1956-59, sr. research fellow, 1959-63; prof. physics U. Cal. at Santa Barbara, 1963—. Guggenheim fellow, 1952-53; NSF sr. postdoctoral fellow, 1959. Recipient Dept. Commerce gold medal, 1960; Arthur S. Flemming award, Washington Jr. C. of C., 1956. Fellow Am. Phys. Soc., A.A.A.S., Am. Optical Soc.; mem. Combustion Inst., Am. Assn. Physics Tchrs., Phi Beta Kappa, Sigma Xi. Study of upper atmosphere physics; isotope analysis; properties of condensed gases; spectroscopy of flames; solar spectra; diatomic spectra, trapped radicals, chem. kinetics. Home: 4071 Naranjo Dr., Santa Barbara, Cal. 93105.

BROIDA, Victor, engr.; b. Moscow, Russia, Dec. 25, 1907; s. Michel and Marie (Kamrasch) B.; ed. Poly. Sch. Grenoble (France); Ph.D. in Engring., U. Paris (France); m. Helene Sénégas, Nov. 6, 1937; 1 son, Max. Engr., dir. Thermique Industrielle, 1930-37, Automatisme industriel, 1938-40, Sci. Work Orgn., 1944-51; prof. Charleroi, France, 1954-56. Mem. Swiss Assn. Automation (corr.), Soc. Engrs. France, Internat. Fedn. Automatic control (v.p. Bulletin d'information). Author: Heat Inertia in Problems of Automatic Control of Temperature, 1940; Automatisme, Regulation automatique, Servomécanisme,

1957. Home: 13, rue de la France-Mutualiste, Boulogne-sur-Seine, France. Office: 56, rue de l'Eglise, Paris 15, France.

BROKAW, Richard S(pohn), Am. chemist; b. Orange, N.J., Mar. 26, 1923; s. Albert D. and Clara (Spohn) B.; B.A., Swarthmore Coll., 1943; A.M., Princeton, 1949, Ph.D., 1951; m. Frances Lydia Alford, June 27, 1947; children—James A. II, Richard Spohn, Frances C., Susan E. Aero. research scientist NASA Lewis Research Center, Cleve., 1952-55, asso. head combustion fundamentals sect., 1955-56, head, 1956-57, chief phys. chemistry br., 1957—. Mem. adv. com. on sci., engring., other specialized personnel Ohio Selective Service, 1959—. Mem. A.A.A.S., Am. Chem. Soc., Am. Inst. Aeros. and Astronautics, Combustion Inst., Soc. Engring. Sci., Sigma Xi. Research and numerous publs. on rates of chem. reactions in gases, heat conduction in gases, including influence of chem. reaction, combustion including ignition, flame speeds and burning rates of liquid fuels. Home: 17403 Edgewater Dr., Lakewood, O. 44107. Office: 21000 Brookpark Rd., Cleve. 44135.*

BROLLEY, John Edward, Jr., Am. physicist; b. Chgo., Jan. 15, 1919; s. John Edward and Marie (Schaeffer) B.; B.S., U. Chgo., 1940; M.S., Ind. U., 1948, Ph.D., 1949. Physicist, Los Alamos Sci. Lab., 1949—. Fellow Am. Phys. Soc., A.A.A.S.; mem. Archaeol. Inst. Am., Sigma Xi. Research and publs. in nuclear physics, especially neutron fission, proton-proton scattering, light nuclei interactions; theory of meson physics. Home: 719 41st St. Office: Los Alamos Sci. Lab., Los Alamos 87544.*

BROMFIELD, William, English surgeon; b. London, Eng., 1712; instrn. in surgery, London. Began practice surgery London, 1741; lectr. anatomy, surgery; founder (with Martin Madan) Lock Hosp. for Venereal Disease, also surgeon; apptd. surgeon St. George's Hosp., 1761; surgeon household of King George III. Author: An Account of English Nightshades, 1757; Narrative of a Physical Transaction with Mr. Aylet, Surgeon at Windsor, 1759; Chirurgical Cases and Observations, 2 vols., 1773; also works on methods of inoculating for small pox. Active in treatment of venereal disease. Died London, Nov. 24, 1792.

BROMLEY, D(avid) Allan, physicist, educator; b. Westmeath, Ont., Can., May 4, 1926; s. Milton Escourt and Susan Anne (Anderson) B.; B.Sc. in Engring. Physics, Queen's U., Kingston, Ont., 1948, M.Sc. in Physics, 1950; Ph.D. in Nuclear Physics, U. Rochester, 1952; M.A. (hon.), Yale, 1961; m. Patricia Jane Brassor, Aug. 30, 1949; children—David John, Karen Lynn. Came to U. S., 1949. Operating engr. Hydro Electric Power Commn. Ont., Niagara Falls, 1947; research officer Nat. Research Council of Can., Ottawa, Ont., 1948, fellow, 1952; faculty U. Rochester (N.Y.), 1952-55; sr. research officer, sect. head Atomic Energy Can. Ltd., Chalk River, Ont., 1955-60, installed 1st tandem Van de Graaff accelerator, 1959; asso. prof. physics, asso. dir. heavy ion accelerator lab. Yale, 1960-61, prof., 1961—, dir. nuclear structure lab., 1963—. Mem. panel on nuclear structure research NSF, 1960; chmn. com. nuclear scis. Nat. Acad. Scis., 1965—; cons. Los Alamos Sci. Lab., High Voltage Engring. Corp., Brookhaven Nat. Lab., Oak Ridge Nat. Lab., Acad. Press, Bell Telephone Labs.; dir. United Nuclear Corp. Recipient medal Gov. Gen. Can., 1948. Fellow Am. Phys. Soc. (council 1966—); mem. Canadian Assn. Physicists (vis. lectr. 1963), Internat. Union Pure and Applied Physics, Sigma Xi (pres. Yale chpt. 1962-63). Editor: (with E. W. Vogt) Proc. of Kingston Conf. on Nuclear Structure, 1960; cons. editor: Academic Paperbacks in Physics; asso. editor Phys. Rev., Nuclear News. Research and publs. on nuclear structure and nuclear reaction mechanisms; collaborated in construction of 1st variable energy cyclotron, 1st tandem electrostatic accelerator, 1st Emperor Van de Graaff Accelerator; developed 1st semiconductor nuclear detectors for nuclear reaction studies, pioneered in study of He3 induced reactions, reaction mechanism studies with tandem accelerator; spectroscopic studies with heavy ions; precision nuclear mass measurements via threshold reaction studies; discovered nuclear molecules. Home: 35 Tokeneke Dr., North Haven, Conn. 06473. Office: Wright Nuclear Structure Lab., Yale, 206 Whitney Av., New Haven 06520.

BROMWICH, Thomas John I'Anson, Brit. mathematician; b. 1875; D.Sc., Cambridge; fellow, praelector math. St. John's Coll., also univ. lectr. math. Cambridge; prof. math. Queen's Coll. (now U. Coll.), Galway, Ireland, 1902-07. Mem. London Math. Soc. (sec.). Author: Introduction to the Theory of Infinite Series; Table of Elementary Integrals; Quadratic Forms and their Classification by Means of Invariant Factors. Fellow Royal Soc. A. rehabilitator of Heaviside's math.; studied algebra. Died Aug. 24, 1929.

BRONDSTED, Holger Valdemar, Danish biologist; b. Ronne, Denmark, July 22, 1893; s. Karl Gustav and Clara (Thomsen) B.; Magister scientiarum, U. Copenhagen (Denmark), 1917, Dr.phil., 1936; m. Agnes Hansen, Apr. 12, 1919; children—Karen (Mrs. Knud Lund), Hans, Henrik. Asst., zool. lab. Vet. Highsch., Copenhagen, 1917-20; lectr. Birkerod State Gymnasium, 1920-41, headmaster, 1941-48; prof. gen. zoology U. Copenhagen, ret., 1964. Fellow Rockefeller Inst., Stockholm, Sweden, 1934-36; prop., lab. for physiology, 1936-48; mem. Konsistorium, U. Copenhagen. Fellow Internat. Inst. Embryology; mem. Soc. Natural History Copenhagen (chmn. 1950-52). Author: (in Danish) Textbook on General Zoology, 1956-59; The Atomic Age and Our Biological Future, 1957; Planarian Regeneration, 1968; also papers on systematics, exptl. research on cytology and regeneration of sponges and planarians. Former mem. editorial bd. Jour. Embryology and Exptl. Morphology. Address: 23 Stockholmsgade, Copenhagen 0, Denmark.*

BRONFENBRENNER, Martin, Am. economist; b. Pitts., Dec. 2, 1914; s. Jacques Jacob and Martha (Ornstein) B.; A.B., Washington U., St. Louis, 1934; Ph.D., U. Chgo., 1939; postgrad. Northwestern U. Law Sch.; m. Teruko Okuaki, Nov. 13, 1951; children—Kennth, June. Asst., U. Chgo., 1937-38; faculty Central YMCA Coll., Chgo., 1938-40; economist U. S. Treasury, 1940-41; financial economist Fed. Res. Bank, Chgo., 1941-42, 46-47; asso. prof. U. Wis., 1947-54, prof., 1954-57; prof. econs. Mich. State U., 1957-58, U. Minn., Mpls., 1958-62; prof. econs. Grad. Sch. Indsl. Adminstrn., Carnegie Inst. Tech., Pitts., 1962—, chmn. dept. econ., 1966—; vis. fellow Behavioral Scis. Center, Stanford, Cal., 1966-67. Fiscal economist, SCAP, Tokyo, Japan, 1949-50; economist UN, Econ. Commn. for Asia and Far East, Bangkok, Thailand, 1952. Fulbright lectr., Japan, 1962-63. Mem. Am. Econ. Assn., Am. Statis. Assn., Econometric Soc., Assn. for Asian Studies, Phi Beta Kappa. Author: Academic Encounter, 1961; also numerous articles. Research on theory of income distbn., monetary and fiscal policy especially theory of inflation, Japanese econ. devel. and prospects; reformulation of Marxian economies in contemporary terms. Home: 6612 Forward Av., pitts. 15217.*

BRONGERSMA, Leo Daniel, Dutch anatomist; b. Bloemendaal, May 17, 1907; s. S. H. and W. F. (Klinkhamer) B.; Ph.D. in Sci., U. Amsterdam; m. M. Sanders, 1934. Asst., Mus. Geology of Amsterdam, 1928-34; curator State Mus. Natural History, Leyde, 1932, asst., 1934-35, curator, 1936-47, prin. curator, 1948-58, dir., 1950, prin. dir., 1958; prof. vertebrate taxonomy at Leyde, 1938-50; head sci. expdn. to Star Mountains, Dutch New Guinea, 1959. Mem. Royal Netherlands Geog. Soc., Am. Soc. Ichtycology and Herpetology (hon. fgn.), Geology Soc. London (corr.), Royal Dutch Acad. Sci., Assn. Sci. Research in Tropics, Dutch Commn. for Internat. Protection of Nature. Office: State Museum of Natural History, Raamsteeg 2, Leiden, Netherlands.

BRONGERSMA, Margaretha Sanders, Dutch oceanographer; b. Kampen, Feb. 20, 1905; d. H. H. and G. (de Jong) B.; Ph.D. in Sci., U. Amsterdam; m. L. D. Brongersma, 1934. Asst., Zool. Mus. of Amsterdam, 1929; curator, 1930; asst. of Dubois collections, Leiden, 1933; research in South Africa, 1938; prof. zoogeography U. Leiden, 1947, prof. oceanography, 1956; sci., Sci., dept. Inst. Minerology and Zoology, Leiden, 1958. Mem. oceanography group UNESCO. Mem. Soc. Mineral Geology. Author: Die fossilen Fische der alttertiären Süsswasserablagerungen aus Mittel Sumatra; The Importance of Upwelling Water to Vertebrate Paleontology and Oilgeology; Mass Mortality in the Sea, also others. Address: Houtlaan 3, Leiden, Netherlands.

BRONGNIART, Adolphe-Theodore, French biologist; b. Paris, France, Jan. 14, 1801; s. Alexandre Theodore and Cecile Coquebert (de Montbret) B.; agrégé en medecine U. Paris, 1827, M.D., 1826. Tchr. medicine Faculty Medicine, U. Paris, 1828-30; named prof. botany Muséum d'histoire naturelle, 1833, placed in charge of instrn., acquisition and upkeep, 1854; named insp. gen. of univ., 1852; named mem. Imperial Council Pub. Instrn., 1866. Mem. Fr. Acad. Scis., Société botanique de France (founder, 1st pres.), Société d'agriculture. Author: Sur la génération et le développement de l'embryon des phanéroganes, 1827; Nouvelles recherches sur l'épiderme; Enumération des genres de plantes cultivées au Muséum d'histoire naturelle de Paris, 1843; Histoire des végétaux fossiles, 1828; Tableau des genres de végétaux fossiles, 1849. Founder (with Audouin, Dumas), Annales des scis. naturelles, 1824. Founder sci. of vegetable paleontology; developed classification system for bot. fossils; research on devel. of pollen. Died Paris, Feb. 18, 1876.

BRONGNIART, Alexandre, French mineralogist, geologist; b. Paris, Feb. 5, 1770; s. Alexandre Theodore and Anne-Louise (Degremont) B.; student chemistry with A. Lavoisier; degree in mining engineering, 1794; m. Cecile Coquebert; children—Adolphe, Herminie (Mrs. J. B. Dumas), Mathilde (Mrs. J. V. Audoeun). Prof., Société gymnastique, 1787-94, École Centrale, Collège des Quatre Nations, from 1797; mining engr., 1794-97; dir., porcelain factory, Sèvres, France, from 1800; chief engr. of mines, from 1818; prof. geology Mus. Natural History, from 1822. Mem. Société philomatique, French Acad. Scis. Author: Traité élémentaire de minéralogie avec des applications aux arts, 1807; La géographie mineralogique des environs de Paris, 1811; Sur les Caractères zoologiques des Formations, 1821; Tableau méthodique et caractéristique des principales espèces minérales, 1824; Classification et caracteres mineralogiques des roches homogenes et heterogenes, 1827; Tableau des terrains, qui composent l'ecorce du globe, 1829; Traité des arts céramique ou des poteries, 1844. Revived art of painting on glass; developed ceramic chemistry; studies of Paris Basin (with Cuvier) indicated that strata could be recognized by distinctive fossils (using this principle, separated tertiary strata North Central France into natural units arranged in chronological sequence); in zoology classified reptiles into orders of Surians, Batrachians, Chelonians, and Ophidians. Died Paris, Oct. 7, 1847.

BRONGNIART, Antoine-Louis, French chemist; b. Paris; prof. Coll. Pharmacy; apothecary to Louis XVI; prof. Paris Mus. Natural History. Author: Tableau analytique des combinaisons et des décompositions de différentes substances (chart of chem. decompositions), 1778. Died Paris, 1804.

BRONK, Detlev Wulf, Am. physiologist; b. N.Y.C., Aug. 13, 1897; s. Mitchell and Marie (Wulf) B.; A.B., Swarthmore, 1920; postgrad. U. Pa., 1921; M.S., U. Mich., 1922, Ph.D., 1926, recipient over 45 hon. degrees univs. and colls.; m. Helen A. Ramsey, Sept. 10, 1921; children—John Everton Ramsey, Adrian, Mitchell Herbert. Mem. faculty Swarthmore Coll., 1926-29, U. Pa., 1929-49, Cornell 1940-41; pres. Johns Hopkins U., 1949-53; pres. Rockefeller U., N.Y.C., 1953—. Coordinator of research. Air Surgeons Office, Hdqrs. Army Air Forces, 1942-46. Has held numerous endowed named lectureships, 1938—; dir. Johnson Research Found., U. Pa., 1929-49; chmn. NRC, 1946-50; pres. Nat. Acad. Scis., 1950-62; chmn. bd. NSF, 1956-64; mem. Pres.' Sci. Adv. Com., 1956-63; cons.-at-large, 1963—; chmn. Panel on Internat. Sci., 1957—. Trustee Atoms for Peace Awards, Inc., Oceanographic Inst., Woods Hole, Tulane U., U. Pa., Bucknell U., Marine Biol. Lab., Johns Hopkins U., Population Council, Protein Found., Rockefeller Bros. Fund, Sloan-Kettering Inst.; chmn. bd. trustees Renssalaer Poly. Inst., 1965—; vice chmn. N.Y. State Sci. and Tech. Found., 1965—. Recipient award for exceptional civilian service, 1946, Officer Brit. Empire, Longacre award, Aero. Med. Assn., 1948, Priestley award, Dickinson Coll., 1956; Gold medal Internat. Ben Franklin Soc., 1958; medal Soc. Promoting Internat. Sci. Relationships, 1959, Gold medal Holland Soc., 1961, medal Franklin Inst., 1962; Presdl. Medal of Freedom, 1964; Pub. Welfare medal Nat. Acad. Scis., 1964. Fellow A.A.A.S. (pres. 1952), mem. or hon. mem. many Am. and fgn. prof. socs., including Royal Soc., London, French Acad. Scis., sometime officer several. Asso. editor Jour. Gen. Physiology, 1954—. Research in infra-red spectroscopy, physiology of sense organs and nervous system, volume flow of blood, synaptic mechanisms; evolved methodology with aid of elec., optical and electro-microscopes to measure changes in nerve cells during passage of stimuli to brain; evaluated molecular structure of nerve cells, placement of atomic and ionic components in the cells, measurement of pressure that carries oxygen around body. Home: President's House, Rockefeller U., N.Y.C. 10021. also Hill House Farm, Media, Pa. Office: Rockefeller U., N.Y.C., 10021.

BRONK, J(ohn) Ramsey, Am. biochemist; b. Phila., Dec. 20, 1929; s. Detlev Wulf and Helen (Ramsey) B.; A.B., Princeton, 1952; Ph.D., Oxford U., 1955; m. Sylvia Smith, June 6, 1955; children—Richard, Christopher. Officer, USPHS, Nat. Heart Inst., 1956-58; asst. prof. zoology Columbia, 1958-60, asso. prof., 1960-65, prof., 1965-66; prof. biochemistry U. York (Eng.), 1966—. Rhodes scholar, 1952-55; Guggenheim fellow, 1964-65. Mem. Am. Soc. Biol. Chemists, Biochem. Soc., Soc. Gen. Physiologists. Research and publs. on conversion of nutrient energy in mitochondria, control of oxidative processes in mammalian cell by thyroid hormones, active transport of amino acids, nature and control of process of urea synthesis in liver. Home: 126 The Mount, York. Office: U. York, Heslington, York, Eng.*

BRONK, Otto von, see von Bronk, Otto.

BRONN, Heinrich Georg, German paleontologist; b. Ziegelhausen, Germany, Mar. 3, 1800; s. Georg Ernst and Elisabeth Margarete (Herzberger) B.; studied public affairs in Heidelberg, then natural scis. from 1817. Pvt. lectr. in Heidelberg, 1921; prof. natural history, 1822, then prof. zoology and dir. zool. collection, 1933; paleontolgeol. research trips to No. Italy and So. France, 1824-27; edited with C. V. Leonhard (later with son Georg) Jahrbuch für Mineralogie, Geognostik und Petrefaktenkunde, also editor other jours. Recipient Wollaston medal from London Geol. Soc., 1861. Author: Lethaea geognostica, 1834-38; Handbuch einer Geschichte der Natur, 3 vols., 1841-49; Index Paleontologicus; 1848-49; Allgemeine Zoologie, 1850; Die Klassen und Ordnungen des Thierreichs, 1859-63. Paved way for theory of evolution in paleontology; translated Darwin; work in area of gen. and systematic paleontology, paleozoology and stratigraphy; attempted to include extinct organisms in a complete system of zoology; several fossil types named after him; main contbn. was drawing nearly all gen. conclusions possible at that time from available empirical body of knowledge of paleontology and hist. geology. Died Heidelberg, Germany, July 5, 1862.

BRONOWSKI, Jacob, mathematician; b. Poland, Jan. 18, 1908; s. Abram and Celia B.; M.A., Ph.D.,

Cambridge (Eng.) U., 1933; m. Rita Coblentz (Colin), 1941; 4 daus. Naturalized Brit. subject. Sr. lectr. math. Univ. Coll., Hull, 1934-42; engaged in war-time research, 1942, sci. dep. to Brit. Chiefs of Staff Mission to Japan, 1945, wrote classical Brit. report The Effects of Atomic Bombs at Hiroshima and Nagasaki; worked with UNESCO, applying statis. research to econs. of industry, 1947-50; pres. Brit. Library Assn., 1957-58; Carnegie vis. prof. Mass. Inst. Tech., 1953. Author: (radioplay) Journey to Japan, 1950; The Common Sense of Science, 1951; Science and Human Values, 1958; also numerous papers in algebraic geometry and topology, math. statistics. Research and publs. on nature of sci. thinking, and what logical and mech. systems and machines can do to help explain it; concerned with the logic of expt. and with broad range of philos. problems bearing on anatomy of research and sci. method.

BRÖNSTED, Johannes Nicolaus, Danish chemist; b. Varde, Denmark Feb. 22, 1879; ed. Polytechnique Inst., Copenhagen, 1898; master's in chemistry, 1902; Ph.D., Copenhagen, Denmark, 1908. Assistant at U. Copenhagen chem. lab., 1905; became prof. chemistry, Copenhagen, 1908; dir. Physico-Chem. Inst., 1930-47. Investigated reaction kinetics; fundamental thermodynamics of solutions; catalysis, kinetic properties of ions; developer theory of acids and bases in solutions; worked on 2d law thermodynamics; (with Georg von Hevesy) worked on separation of mercury isotopes. Died Copenhagen, Dec. 17, 1947.

BROOK, Marx, Am. physicist; b. N.Y.C., July 12, 1920; s. Abraham and Esther (Shapiro) B.; B.S., U. N.M., 1944; M.A., U. Cal. at Los Angeles, 1949, Ph.D., 1953; m. Dorothy R. Strickland, Jan. 3, 1947; children—Janet, James, Georgia. Teaching asst., jr. geophysicist U. Cal. at Los Angeles, 1946-53; research physicist N.M. Inst. Mining and Tech., 1954-55, faculty, 1955—, prof. physics, sr. research physicist, 1959—; trustee N.M. Tech. Research Found., 1965—. Fellow Am. Phys. Soc.; mem. Am. Geophys. Union, A.A.A.S., Am. Meteorol. Soc., Am. Assn. U. Profs., Am., Assn. Physics Tchrs. Research and publs. in lightning and thunderstorms, disposition of electric charge in clouds, discharge mechanisms and processes, charge separation processes, growth of rain and relationship to electrical forces, radar studies of noise in clouds, thunder spectrum, electromagnetic radiation processes. Home: 1216 North Dr., Socorro, N.M. 87801.*

BROOKE, Charles, English surgeon; b. Eng., June 30, 1804; s. Henry James B.; ed. at St. John's Coll., Cambridge (Eng.) U.; B.M., 1828, M.A., 1853; studied medicine at St. Bartholomew's Hosp. Mem. surg. staff Met. Free Hosp., Westminster Hosp. Fellow Coll., Surgeons, Royal Soc., 1847; mem. Meteorol. Soc. (pres.), Royal Micros. Soc. (pres.). Author: A Synopsis of the Principal Formulae and Results of Pure Mathematics, 1829; Motion of Sound in Space, 1835; also articles on his inventions of self-recording meteorol. instruments. Edited and rewrote: Elements of Natural Philosophy (Golding Bird), 6th edit., 1867. Invented bead suture for treatment of deep wounds; research on theory of microscope, also invented aids for shifting of lenses; improved illumination of microscope, 1846-52. Died May 17, 1879.

BROOKE, John Mercer, Am. physicist; b. Tampa, Fla., Dec. 18, 1826; s. George Mercer and Lucy (Thomas) B.; ed. Kenyon Coll.; grad. Naval Acad., 1847; With Naval Obs., 1851-53; in charge astron. dept. N. Pacific surveying and exploring expdn. in sloop-of-war Vincennes, comd. Fennimore Cooper in survey of route between San Francisco, Sandwich Islands, Japan and China; resigned from U.S.N. 1861; made chief, bur. ordnance and hydrography under Sec. Stephen Mallory; prof. physics Va. Mil. Inst., 1866-99, emeritus, 1899. Recipient Gold medal Acad. Berlin. Invented Brooke gun; discovered utility of air-space in cannon; designed plans for iron-clad vessel with submerged ends (used in reconstrn. of Merrimac); drew directions for cruise of Shenandoah for destruction of whaling fleet; inventor deep sea sounding apparatus. Died 1906.

BROOKE, M(arion) M(urphy), Am. parasitologist; b. Atlanta, Dec. 6, 1913; s. Thomas Russell and Jessie (Jones) B.; A.B., Emory U., 1935, M.A., 1936; Sc.D., Johns Hopkins, 1942; m. Marian Dobler, Dec. 21, 1940; children—Thomas Russell, Martha Louise, Barbara. With lab. br. Dept. Health, Edn. and Welfare Communicable Disease Center, Atlanta, 1945—, chief lab. consultation and devel. sect., asst. chief tng. and consultation, 1962—; with USPHS, 1945—, sci. dir. (col.), 1958—; asso. prof. parasitology Med. Sch. Emory U., 1946—; asso. prof. parasitology U. N.C. Sch. Pub. Health, 1963—. Mem. expert adv. panel parasitic diseases WHO, Geneva, Switzerland, 1964—. Recipient Meritorious Service medal USPHS, 1965. Diplomate Am. Bd. Microbiology in Pub. Health Microbiology. Mem. Am. Soc. Tropical Medicine and Hygiene, Am. Pub. Health Assn., Am. Soc. Parasitologists, Sci. Research Soc. Am. (past br. pres.), Sigma Xi, Phi Sigma. Author: Amebiasis-Methods in Laboratory Diagnosis, 1958. Mem. editorial bd. Health Laboratory Science, 1964—; Public Health Laboratory, 1962—. Contbr. numerous articles to profl. jours. Co-developer PVA-fixative technique for preservation and shipment of specimens for lab. diagnosis of amebiasis, triton giemsa staining procedure to prevent this source of false positive diagnoses; co-discoverer transfer of malarial parasites between slides during mass staining of blood films; contbr. to diagnosis, epidemiology of amebiasis, malaria, enterobiasis, toxoplasmosis and other parasitic diseases. Home: 1343 Emory Rd. N.E., Atlanta 30306. Office: Communicable Disease Center, Atlanta 30333.*

BROOKHART, John Mills, Am. physiologist; b. Cleve., Dec. 1, 1913; s. Leslie Shellabarger and Anna Rose (Mills) B.; B.S., U. Mich., 1935, M.S., 1936, Ph.D., 1939; m. Anna Louise Simon, Aug. 26, 1939; children—Cornelia Mills (Mrs. Steven R. Frank), Constance Lee (Mrs. John L. Strawn), John Howard. Postdoctoral fellow in neurology Inst. Neurology Northwestern U. Med. Sch., 1939-40; faculty physiology Loyola U. Sch. Medicine, 1940-46; asst. prof. physiology U. Ill. Coll. Medicine, 1946-47; asst. prof. physiology Northwestern Med. Sch., 1947-49; asso. prof. physiology U. Ore. Med. Sch., 1949-51, prof., chmn. dept., 1952—; cons. USPHS, 1951-67, mem. adv. council on health research facilities, 1967—. Mem. American Physiological Society (president 1965-66), Federation American Socs. for Exptl. Biology (exec. com. 1964-67), A.A.A.S., Soc. Exptl. Biology and Medicine, Internat. Brain Research Orgn. (central council 1966), Ore. Neuropsychiat. Soc., Portland Acad. Medicine, Am. Acad. Arts and Scis., Sigma Xi, Delta Phi. Mem. editorial bds. Ann. Rev. Physiology, 1958-61, Jour. Neurophysiology, 1960-64, chief editor, 1964—. Research and publs. on physiology of central nervous system, neural control of respiration, mating behavior, mechanisms of generation of elec. activity by brain, mechanisms of transfer of activity between nerve cells, integration of control of posture. Home: 3126 N.E. 39th Av., Portland, Ore. 97212.*

BROOKS, Alfred Hulse, Am. geologist; b. Ann Arbor, Mich., July 18, 1871; s. Thomas Benton and Hannah (Hulse) B.; studied in Germany, 1890-91; B.S., Harvard, 1894; postgrad. in Paris; hon. D.Sc., Colgate U., 1920; m. Mabel W. Baker, Feb. 23, 1903. Asst. geologist U. S. Geol. Survey, working in various states, and 1898-1923 engaged in geol. and exploratory work in Alaska; chief Alaskan geologist. Vice-chmn. Alaska R.R. Commn., 1911-12. Recipient gold medals Am. Geog. Soc., Geog. Soc. Paris, 1913. Wrote The Geography and Geology of Alaska, and other papers on Alaska, and on mil. subjects. Died Wash., D.C., Nov. 22, 1924.

BROOKS, Chandler McCuskey, Am. physiologist; b. Waverley, W.Va., Dec. 18, 1905; s. Earle Amos and Mary (McCuskey) B.; A.B., Oberlin Coll., 1928; M.A., Princeton, 1929, Ph.D., 1931; m. Nelle Irene Graham, June 25, 1932. Teaching fellow, NRC fellow Harvard Med. Sch., 1931-33; faculty Johns Hopkins, 1933-48, L.I. Coll. Hosp. Med. Sch., 1948-50; chmn., prof. dept. physiology State U. N.Y. Downstate Med. Center, Bklyn., 1950—, dir. grad. edn., 1956-66, dean Grad. Sch., 1966—. Mem. sects. NIH. Recipient Citation, Internat. Physiol. Congress, 1965. Guggenheim fellow, 1946-48, Rockefellow fellow, 1950, China Med. Bd. N.Y. fellow, 1961-62. Mem. Harvey Soc. (pres. 1965), A.A.A.S. (council), Am. Heart Assn. (council), Am. Soc. Pharm. and Exptl. Therapeutics, Internat. Brain Research Orgn., Nat. Soc. Med. Research, N.Y. Acad. Scis., Soc. for Exptl. Biology and Medicine, Soc. for Study Internal Secretions, Soc. for Study Nervous and Mental Diseases, A.A.A.S., Am. Coll. Cardiology, Am. Coll. Pharmacology and Chemotherapy, Am. Inst. Biol. Scis., A.M.A. (spl. affiliate), Am. Physiology Soc., Phi Beta Kappa, Sigma Xi, Alpha Omega Alpha; hon. mem. Nat. Acad. Medicine Buenos Aires, Cardiology Soc. Argentina, biology socs. Montevideo, Uruguay, also Santiago, Chile, Inst. History of Medicine and Med. Research New Delhi, India. Author: (with others) Excitability of the Heart, 1955, Humors, Hormones, and Neurosecretions, 1962 (with Kiyomi Koizumi) Japanese Physiology, Past and Present, 1965. Editor: (with Paul F. Cranefield) The Historical Development of Physiological Thought, 1959; (with F. F. Kao, B. B. Lloyd) Cerebrospinal Fluid and the Regulation of Ventilation, 1965. Research and publs. on central control of autonomic system, function of the hypothalamus, motor cortex function in regulation of posture, activity in heart and nerve cells. Home: 623 2d St., Bklyn. 11215.*

BROOKS, Charles F., Am. meteorologist; b. St. Paul, Minn., May 2, 1891; s. Morgan and Frona Marie (Brooks) B.; student U. Ill., 1907-08; A.B., Harvard, 1911, as of 1912, A.M., 1912, Ph.D., 1914; m. Eleanor Merritt Stabler, June 4, 1914; children—Edward Morgan, Margaret, Sylvia, Barbara, Edith Herrick, Norman Herrick, Frona. Research asst. Blue Hill Obs., 1912-13; asst. in meteorology and physiography Harvard, 1913-14; asst. in physiography, Radcliffe College, 1914; asst., collaborator in farm mgmt. U. S. Dept. Agr., 1914-18; instr. geography Yale, 1915-18; instr. meteorology U. S. Signal Corps Sch. of Meteorology, College Station, Tex., World War, 1918; meteorologist U. S. Weather Bur., as editor Mo. Weather Rev., 1918-21; asso. prof. meteorology and climatology Clark U., 1921-26, prof., 1926-32, vis. lectr., 1941; prof. meteorology and dir. Blue Hill Obs., Harvard, 1931-57, emeritus, 1957—; vis. prof. U. Chgo., 1939; vis. lectr. U. S. Weather Bur., 1943-44; cons., 1948; pres. Mt. Washington Obs. Mem. internat. commns., Climatol., 1931-45, Snow and Glaciers, 1936-48, Instruments and Methods of Observation 1947-51 (pres. sub-committee on station instruments and methods 1947-53), Clouds and Hydrometeors 1947-53. Fellow A.A.A.S., Royal Meteorol. Soc., Am. Geog. Soc., Inst. Aero. Scis. (asso.); mem. Am. Acad. Arts and Scis., Am. Meteorol. Soc. (organizer, editor 1919-25, 27-36, 39; sec. 1919-54, hon. sec. 1954—), Assn. Am. Geographers (pres. 1947), Am. Geophys. Union (chmn. meteorol. sect. 1935-38), Phi Beta Kappa, Sigma Xi. Author: "Why the Weather?" Joint author: Climatology of N. Am. and West Indies (part of a 5 vol. series); Climatic Maps of North America; Eclipse Meteorology; Science from Shipboard; International Cloud Atlas. Died Jan. 8, 1958.

BROOKS, Edward Morgan, Am. meteorologist; b. New Haven, Mar. 19, 1916; s. Charles F. and Eleanor (Stabler) B.; A.B. cum laude, Harvard, 1937; M.S., Mass. Inst. Tech., 1939, Sc.D., 1945; m. Sarah Karena Bergh, Dec. 24, 1941; children—Karena (Mrs. George Poonen), Daniel, Roxana, Valerie, Marita, Bernard, Larimore, Galen. Map analyst Pan-Am. Airways, Pearl City, Hawaii, 1939-40, meteorology tutor, San Francisco, 1940, meteorology instr., Coral Gables, Fla., 1941-42; instr. research asso. Mass. Inst. Tech., 1942-46; faculty Inst. Tech. St. Louis U., 1946-61, prof., 1952-61; sr. meteorologist Allied Research Assos., Boston and Concord, Mass., 1961-63; staff scientist GCA Corp., Bedford, Mass., 1963-65, prin. scientist, 1965-67; profl. meteorologist W. E. Howell Assos., Bedford, Mass., 1967; prof. N.Y. State U. Coll. Arts and Scis., Plattsburgh, 1967—; cons. Mallinckrodt Chem. Works, 1949-61, McDonnell Aircraft Corp., 1959-60. Mem. Am. Meteorol. Soc. (chmn. St. Louis chpt. 1948-49), Am. Geophys. Union, A.A.A.S., Soc. Engring. Edn., Am. Astronom. Soc., Earth Sci. Technologies Assn., Amateur Telescope Makers of Boston, Sigma Xi, Pi Mu Epsilon. Contbr. numerous articles to sci. jours. Discovered, named "tornado cyclone" about ten miles in diameter in which tornadoes are located; discovered that seasons occur on only one side of the planet Venus; invented several simple, more accurate ways of determining the times when stars and planets are hidden by the moon; computed average drift currents for North Atlantic Ocean; discovered that sunspots are accompanied by cool, wet weather in Midwest. Home: 321 Kenrick St., Newton, Mass. 02158.*

BROOKS, Frank Pickering, Am. physiologist; b. Portsmouth, N.H., Jan. 2, 1920; s. Frank E. and Florence (Towle) B.; A.B. cum laude, Dartmouth, 1941; M.D., U. Pa., 1943, D.Sc., 1951; m. Emily Elizabeth Marden, July 5, 1942; children—William B., Sally E., Robert P. Fellow gastroenterology Lahey Clinic, Boston, 1948-50; USPHS fellow physiology Jefferson Med. Coll., 1951-52; faculty U. Pa., 1952—, asso. prof. medicine and physiology, 1959—, chief gastrointestinal sect. hosp., 1962—. Recipient Research Career Devel. award USPHS, 1963. Mem. Am. Physiol. Soc. (chmn. steering com. gastrointestinal sect.), Physiol. Soc. Gt. Britain, Am. Gastroent. Assn., A.C.P., Am. Clin. and Climatol. Assn. Research and publs. on central nervous control of gastric secretion and specificity of areas of hypothalamus. Home: 206 Almur Lane, Wynnewood, Pa. 19096. Office: 3400 Spruce St., Phila. 19104.*

BROOKS, Frederick A., Am. agrl. engr.; b. Mpls., May 1, 1895; s. Morgan and Frona (Brooks) B.; B.E.E., U. Ill., 1917, M.E., 1927; Sc.D., Mass. Inst. Tech., 1920; m. Margaret H. Ward, Sept. 7, 1929; children—Audrey M. (Mrs. Audrey Stolz), Emily F. (Mrs. Robert E. Lynde), Deborah A. (Mrs. Arthur Corra), Brenda D. (Mrs. Vinson Jester). Airplane designer Curtiss Aeroplane & Motor Corp., Buffalo, 1917-18; asst. engr. Dunlop Tire & Rubber Corp., Buffalo, 1920-21; sales engr. Hall-Scott Motor Car Co., Berkeley, Cal., 1923-25; chief engr. Johnson Gear Co., Berkeley, Cal., 1925-28; asst. to chief engr. Byron Jackson Co., Los Angeles and Berkeley, Cal., 1930-31; faculty U. Cal., Davis, 1931—, research agrl. engr., prof. emeritus, 1962—. U. S. del. UNESCO-Australia symposium on arid zone climatology, 1956; collaborator U. S. Forest Service div. Forest Fire Research, 1953—. Guggenheim fellow, 1959; recipient Cyrus Hall McCormick medal, 1960. Fellow Am. Soc. Agrl. Engrs., Am. Soc. M.E. (life); mem. Am. Meterol. Soc. (award in bioclimatology 1966), Am. Geophys. Union, Sigma Xi. Research and numerous publs. on parallel-beam measurement of atmospheric radiation over short paths with hohlraum chilled with liquid nitrogen, and through whole atmosphere simultaneously with spl. radiosonde flights; project leader in extensive field research in frost protection; coop. field expts. on eddy transfers of momentum, heat, moisture over 12-acre irrigated turf surface. Deceased.

BROOKS, Frederick Tom, botanist; b. Wells, Somerset, Dec. 17, 1882; s. Edward Brooks; ed. Emmanuel Coll., Cambridge; M.A. LL.D., St. Andrews; m. Emily Broderick, 1907. Reader in mycology, prof. botany Cambridge U., emeritus prof., 1948—. Mem. Agrl. Research Council, 1941-46, 49-52. Fellow Royal Soc., 1930, Bot. Soc. Edinburgh (hon.), Royal Soc. of Edinburgh; mem. Brit. Mycol. Soc. (hon., pres. 1922), Brit. Assn. (pres. bot. sect. 1935, gen. sec. 1935-46). Author: Plant Diseases; (with D. H. Scott) Flowering Plants and Flowerless Plants, also articles. Died Mar. 11, 1952.

BROOKS, Harvey, Am. physicist; b. Cleve., Aug. 5, 1915; s. Chester Kingsley and Elizabeth (Brown) B.; grad. Hill Sch., Pottstown, Pa., 1932; A.B., Yale, 1937, D.Sc., 1962; Ph.D., Harvard, 1940, D.Sc., 1963; D.Sc., Union College, 1962, Kenyon College, 1963; D.Sc., Brown University, 1964; m. Helen Gordon Lathrop, Oct. 20, 1945; children—Alice Lathrop, Katharine Gordon, Kingsley Chapin, Rosalind Hickox. Mem. Soc. Fellows, Harvard, 1940-42, staff Underwater Sound Lab., 1942-45, prof. applied physics, 1950—, dean engring. and applied physics, 1957—; asst. dir. Ordnance Research Lab., Pa. State U., 1945-46; asso. head Knolls Atomic Power Lab., Schenectady, 1946-50. Mem. Pres.'s Sci. Adv. Com., 1960-64, Nat. Sci. Bd., 1963—, Naval Research Adv. Com., 1963—. Recipient Ernest Orlando Lawrence award AEC, 1960. Fellow Am. Phys. Soc., Acoustical Soc., Am., Am. Philos. Soc.; mem. Nat. Acad. Scis., Am. Acad. Arts and Scis. Editor-in-chief, Physics and Chemistry Solids, 1956—. Participated in devel. of acoustic homing torpedo and scanning sonar in World War II; research and publs. in solid state theory, reactor theory. Home: 46 Brewster St., Cambridge, Mass. 02138.*

BROOKS, Morgan, Am. elec. engr.; b. Boston, Mar. 12, 1861; s. Francis A. and Frances (Butler) B.; Ph.B., Brown U., 1881; M.E., Stevens Inst. Tech., 1883; m. Frona Marie Brooks, Apr. 24, 1888; children—Henry M., Charles F., Frances (Mrs. Lincoln Colcord), Frederick A., Roger, Edith, Mrs. Frona B. Hughes, Dorothy Prescott (Mrs. Joseph M. Thomas). With Am. Bell Telephone Co., Boston, 1884-86; sec.-treas. St. Paul Gas Light Co., 1887-90; organizer Elec. Engring. Co. of Mpls., prof. elec. engring. U. Neb., 1898-1901, U. Ill., 1901-29, emeritus. Fellow Am. Inst. EE. (dir. 1907-10, v.p. 1910-12); mem. Am. Soc. M.E. (life), Illuminating Engring. Soc., Western Soc. Engrs., Sigma Xi. Contbr. to engring. mags. Patentee automatic telephone system, 1896. Died Apr. 25, 1947.

BROOKS, Norman Herrick, Am. civil engr.; b. Worcester, Mass., July 2, 1928; s. Charles Franklin and Eleanor (Stabler) B.; A.B., Harvard, 1949, S.M., 1950; Ph.D., Cal. Inst. Tech., 1954; m. Frederika Nelson, Dec. 22, 1948; children—Diana Stabler, Alexander Nelson, Laura Elizabeth. Faculty, Cal. Inst. Tech., 1953—, prof. civil engring., 1962—; vis. asso. prof. SEATO Grad. Sch. Engring., 1959-60; vis. prof. Mass. Inst. Tech., 1962-63. Hydraulic cons. Los Angeles County Sanitation Dists, 1955—; cons. on ocean disposal of sewage for projects in Los Angeles, San Diego, Seattle, 1955—. Chmn. com. on vis. scientists program in hydrology for Univs. Council on Water Resources and Am. Geophys. Union, 1964—. Mem. Am. Soc. C.E. (chmn. hydraulics group Los Angeles sect. 1962, Research prize, Cunningham and J. C. Stevens award 1959, Hering medal 1957, 62), Am. Geophys. Union, Internat. Assn. for Hydraulic Research, Am. Assn. U. Profs., Univs. Council on Water Resources, Phi Beta Kappa, Sigma Xi. Research and publs. on hydraulics especially alluvial streams, diffusion of sewage in oceans, turbulence studies, dispersion in streams and ground water flow. Home: 2521 Santa Anita, Altadena, Cal. 91001. Office: Cal. Inst. Tech., 1201 E. California Blvd., Pasadena, Cal. 91109.*

BROOKS, Robert Richard, chemist; b. Bristol, Eng., Apr. 9, 1926; s. William Henry and Andree (Briere) B.; B.Sc. with honors, Bristol U., 1952; Ph.D., U. Cape Town, 1960; m. Mary Yvonne Myatt, Aug. 26, 1950; 1 dau., Sarah Louise. With Imperial Smelting Corp., Avonmouth, Bristol, 1952-54, E. S. and A. Robinson, Bristol, 1954-56, Cape Times Ltd. (South Africa), 1957-58; lectr. chemistry U. Cape Town, 1959-60; reader chemistry Massey U., Palmerston North, New Zealand, 1960—. Vis. asso. prof. geology U. Cal. at Los Angeles, 1966-67. Mem. Royal (asso.), New Zealand (asso.) insts. chemistry. Contbr. articles to profl. jours. Research on trace analysis in geochemistry with particular reference to emission and atomic absorption spectroscopy, applied to problems in geochemistry and biogeochemistry in relation to biogeochem. prospecting and trace elements in sea water and marine organism. Home: 1 Seaton Ct., Palmerston North, New Zealand.*

BROOKS, Summer Cushing, Am. biologist; b. Sapporo, Japan, Aug. 17, 1888 (parents Am. citizens); s. William Penn and Eva Bancroft (Hall) B.; B.S., U. Mass., 1910; Ph.D., Harvard, 1916; m. Matilda Neuffer Moldenhauer, July 14, 1917. Asst. in botany Mass. Agrl. Expt. Sta., Amherst, 1910-11; bio-chemist Research Inst. Nat. Dental Assn., Cleve., 1916-17; Hanna research fellow in pathology Western Reserve U., 1917; research fellow tropical medicine Harvard, 1917-19; asso. prof. physiology and biochemistry Bryn Mawr Coll., 1919-20; biologist, hygienic lab. USPHS, Washington, 1920-26; prof. physiology Rutgers U., 1926-27; prof. of zoology U. Cal. at Berkeley, since 1927. Trustee Marine Biol. Lab., Woods Hole, Mass. Mem. A.A.A.S., Am. Bot. Soc., Am. Physiol. Soc., Am. Chem. Soc., Soc. Exptl. Biol. Medicine, Western Soc. Naturalists (pres. 1933), Phi Beta Kappa, Sigma Xi. Author: Permeability of Living Cells (with Matilda M. Brooks), 1941. Contbr. numerous articles to sci. periodicals. Died Apr. 23, 1948.

BROOKS, Vernon Bernard, physiologist, educator; b. Berlin, Germany, May 10, 1923; s. Martin B. and Margarete (Hahlo) B.; B.A., U. Toronto, 1946, Ph.D., 1952; M.Sc., U. Chgo., 1948; m. Nancy Fraser, June 29, 1950; children—Martin Fraser, Janet Mary, Nora Vivian. Lectr., asst. prof. McGill U., Montreal, Que., Can., 1950-56; asst., asso. prof. Rockefeller Inst., N.Y.C., 1956-64; prof., chmn. dept. physiology N.Y. Med. Coll., N.Y.C., 1964—. Vis. fellow Australian Nat. U., 1954-55. Mem. Canadian, Am. physiol. socs., Assn. Research in Nervous and Mental Disease (editorial bd. 1962-65), A.A.A.S., N.Y. Acad. Scis., N.Y. Acad. Medicine, Am. Assn. U. Profs., Sigma Xi. Research and publs. on interaction of nerve cells at single junctions and in complex systems, feedback systems enhancing contrast. Home: 98 Edgars Lane, Hastings-on-Hudson, N.Y. 10706. Office: 1 E. 106th St., N.Y.C. 10029.*

BROOKS, William Keith, Am. naturalist; b. Cleve., Mar. 25, 1848; grad. Williams Coll., 1870, LL.D., 1893; grad. Harvard, Ph.D., 1875; LL.D., Hobart Coll. U. Pa.; m. Amelia Schultz, June 1878 (died 1901). Asst., Boston Soc. Natural History, 1875-76; asso. prof. and prof. zoölogy, Johns Hopkins, from 1876, head biol. dept., from 1894. Recipient medal Soc. d'Acclamitation, Paris; challenger medal Edinburgh; medal St. Louis Expn., 1904. Mem. Nat. Acad. Scis., Am. Philos. Soc., Am. Acad. Arts and Scis. Author: Handbook of Invertebrate Zoölogy; The Stomatopoda of H. M. S. Challenger; Provisional Hypothesis of Pangenesis, 1877; Heredity, 1883; Monography of the Genus Salpa, 1893; Foundations of Zoology, 1899-1907; The Oyster, 1891. Research on morphology of marine animals, especially tunicates, crustaceans and the oyster. Died Balt., Nov. 12, 1908.

BROOKS, William Penn, Am. agriculturist; b. South Scituate, Mass., Nov. 19, 1851; s. Nathaniel and Rebecca Partridge (Cushing) B.; B.S., Mass. Agrl. Coll., 1875, postgrad., 1876; Ph.D., Halle, 1897; hon. degree Nogaku Hakushi, Japanese Dept. Edn., 1919; m. Eva Bancroft Hall, Mar. 29, 1882; children—Rachel Bancroft, Sumner Cushing. Prof. agr. Imperial Coll. of Agr., Japan, 1877-88, prof. botany, 1880-88, pres. ad interim, 1880-83, 86-87; prof. agr. Mass. Agrl. Coll., 1889-1908, lectr. on agr., 1908-18, pres. ad interim, 1903, 05-06; agriculturist Mass. Agrl. Exptl. Sta., 1889-1921, dir., 1906-18; cons. agriculturist, 1918-21. Contbr. to 2d, 3d and 4th and editor 5th and 6th ann. reports Imperial Coll. Agr. of Japan. Author: Agriculture, 3 vols., 1901; General Agriculture, Dairying and Poultry Farming. Died 1938.

BROOKS, William Robert, astronomer; b. Maidstone, Kent, Eng., June 11, 1844; s. William and Caroline (Wickings) B.; came to U. S., 1857; acad. edn., Marion, N.Y.; hon. A.M., Hobart, 1891; D.Sc., Hamilton, 1898; m. Mary E. Smith, Oct. 15, 1868. Founded Red House Observation, Phelps, N.Y., 1874, where he discovered 11 comets; in charge Smith Observation, Geneva, N.Y., 1888—; prof. astronomy, Hobart Coll.; lectr. Recipient Warner gold prizes for astron. discoveries; 10 medals from Astron. Soc. Pacific; Lalande medal, Paris Acad. Scis. spl. gold medal for list and photographs of cometary discoveries, St. Louis Expn., 1904; spl. gold medal and diploma Astron. Soc. of Mexico, for comet discoveries. Discovered 27 comets (many with use of telescope of own constrn.); pioneer in photography and its application to astronomy. Died May 3, 1921.

BROOM, Leonard, Am. sociologist; b. Boston, Nov. 8, 1911; B.S., Boston U., 1933, A.M., 1934; Ph.D., Duke, 1937; m. Gretchan Noel Cooke, Aug. 31, 1940; children—Karl Cooke, Dorothy Howard. Faculty, U. Cal. at Los Angeles, 1941-59, chmn. dept. anthropology and sociology, 1952-57, prof., 1953-59; prof., chmn. dept. sociology U. Tex., Austin, 1959-66, Ashbel Smith prof. sociology, 1963—; vis. prof. Inst. Advanced Studies, Australian Nat. U., 1964-65. Fulbright research fellow B.W.I., 1950-51, Faculty fellow Fund For Advancement Edn., 1953-54; Guggenheim fellow, 1958; hon. fellow Australian Nat. U., 1958; fellow Center for Advanced Study in Behavioral Scis., 1962-63. Fellow Am. Sociol. Assn., Am. (editor Rev. 1955-57, v.p. 1962-63), Royal anthrop. assns.; mem. Sociol. Research Assn., Southwestern, Australian, New Zealand sociol. assns., Population Assn., Internat. Inst. Differing Civilization. Author: (with Ruth Riemer) Removal and Return, 1949; (with Frank G. Speck) Cherokee Dance and Drama, 1951; (with Philip Selznick) Sociology, 1955; (with John I. Kitsuse) The Managed Casulty, 1956; (with Norval D. Glenn) Transformation of the Negro American, 1965. Research on social stratification and study of subordinate peoples. Home: 3010 Oakhurst Av., Austin, Tex. 78703.*

BROOM, Robert, morphologist; b. Paisley, Scotland, Nov. 20, 1866; s. John Broom; M.D., Glasgow, 1895, also C.M., D.Sc., LL.D.; several hon. doctorates; m. Mary Baird Baillie, 1893. Prof. geology and zoology Victoria Coll., Stellenbosch, S. Africa, 1903-10, keeper vertebrate paleontology and anthropology Transvaal Mus. Recipient Wollaston medal Geol. Soc., 1949. Fellow Royal Soc., 1920 (Croonian lectr. 1913, Royal medal 1928), Royal Soc. Edinburgh (hon.); corr. mem. Zool. Soc., Linnean Soc. New S. Wales, Paleontol. Soc. A., Geol. Soc. China, Am. Mus. N.Y.; pres. S. African Assn. for Advancement of Sci., 1933. Author: Origin of Human Skeleton, 1930; Mammal-like Reptiles of South Africa, 1932; The Coming of Man: was it accident of design? 1933; (with G. W. H. Schepers) Fossil Ape-Men of South Africa, 1946; Finding the Missing Link, 1940; (with J. T. Robinson) Sterkfontein Ape-man, 1950; also papers. Died Pretoria, S. Africa, Apr. 6, 1951.

BROPHY, James J., Am. physicist; b. Chgo., June 6, 1926; s. James J. and Ella Helen (Nerad) B.; B.S. in Elec. Engring., 1947; M.S., Ill. Inst. Tech., 1949, Ph.D. in Physics, 1951; m. Muriel Ann Johnson, Aug. 26, 1949; children—James J., John R., Thomas C. Research physicist Research Inst. (formerly Armour Research Found.), Ill. Inst. Tech., Chgo., 1951-53, supr. solid state physics, 1953-56, asst. dir. physics research div., 1956-61, dir. tech. devel., 1961-63, v.p., 1963—. Fellow Am. Phys. Soc.; mem. A.A.A.S., Sigma Xi. Author: (with L. V. Azaroff) Electronic Processes in Materials, 1963.; Semiconductor Devices, 1964; contbg. author: Fluctuation Phenomena in Solids, 1964. Editor: (with J. W. Buttrey) Organic Semiconductors, 1962. Home: 4901 Lawn Av., Western Springs, Ill. Office: 10 W. 35th St., Chgo. 60616.*

BROPHY, Truman William, Am. oral surgeon; b. Will County, Ill., Apr. 12, 1848; s. William and Amelia (Cleveland) B.; D.D.S., Pa. Coll. Dental Surgery, 1872; M.D., Rush Med. Coll., 1880; LL.D., Lake Forest U., 1889, Loyola U., 1924; Sc.D., U. Pa., 1914; m. Emma Jean Mason, May 8, 1873 (dec. 1899); children—Mrs. Jean Mason Barnes, Mrs. Florence Logan, Truman W., Mrs. Alberta L. Holloway; m. 2d, E. W. Strawbridge, Mar. 31, 1908. Pres., prof. oral surgery Chgo. Coll. Dental Surgery; oral surgeon Michael Reese, Frances Willard hosps. Pres. for U. S. of 13th Internat. Med. Congress, Madrid, Spain, 1903; del. 4th Internat. Dental Fedn. Decorated officer Legion of Honor, France; recipient Internat. Miller Memorial prize Internat. Dental Fedn., 1925. Author: Oral Surgery, 1915; Cleft Palate and Cleft Lip, 1923. Died Feb. 4, 1928.

BROSI, Albert Ralph, Am. chemist; b. Coatsburg, Ill., Nov. 12, 1907; s. Albert and Dollie (Gray) B.; B.S., U. Chgo., 1930, Ph.D., 1938; m. Pauline Potter, Oct. 23, 1937; children—Ruth Ellen (Mrs. Roger William Phillips), George Ralph. Chemist, Oak Ridge Nat. Lab., 1947—. Mem. Am. Chem. Soc., Am. Phys. Soc., A.A.A.S. Research on nuclear measurements, elucidation of decay schemes of radioactive isotopes, neutron cross sects. and beta-ray polarizations. Home: 105 W. Price Lane, Oak Ridge 37830. Office: P.O. Box X, Oak Ridge 37831.*

BROSSET, Cyrill, inorganic chemist; b. Lübeck, Sept. 20, 1908; s. Theodor and Tamara (Egoroff) B.; Dr. phil. U. Stockholm (Sweden), 1942; m. Carlotta Björnstjerna; children—Yvonne (Mrs. Yvonne Brosset-Lemkow), Marianne, Thomas; m. 2d, Carie Baumann. Head dept. inorganic chemistry Chalmers U. Tech., 1953—; dir. Swedish Inst. Silicate Research, 1956. Mem. Swedish Commn. Air Purification; tech. adviser Swedish Inst. Nat. Def., 1946. Home: Barrskogsgränd 11. Office: Inst. Oorganisk kemi, Chalmers t.h., Gibraltargatan 5-A, Gothenburg, Sweden.*

BROSSI, Arnold, biochemist; b. Winterthur, Switzerland, Dec. 19, 1923; s. Arnold and Ida (Krebs) B.; Ph.D., Eidgenössische Technischen Hochschule, Zurich, Switzerland, 1952; m. Hanni Leemann, Mar. 31, 1951; children—Mario, Angela, Franca. Group leader F. Hoffmann-La Roche & Co., Basle, Switzerland, 1952-62, dir. chem. research, Nutley, N.J., 1962-66, v.p., dir. chem. research, 1967—. Mem. Am., Swiss chem. socs., N.Y. Acad. Scis., Chem. Soc. London, Am. Soc. Pharmacognosy, A.A.A.S. Research and publs. on alkaloids and antibiotics; syntheses in field of medicinal chemistry, also patents. Home: 67 Harrison St., Verona N.J. 07044. Office: Hoffmann-La Roche Inc., Nutley, N.J. 07110.*

BROTZEN, Franz Richard, materials scientist; b. Berlin, July 4, 1915; s. Georg and Lena (Pacully) B.; B.S. in Metall. Engring., Case Inst. Tech., 1950, M.S. in Phys. Metallurgy, 1953, Ph.D. in Phys. Metallurgy, 1954; m. Frances B. Ridgway, Jan. 31, 1950; children—Franz Ridgway, Julie Ridgway. With A. Quimica Bayer, Ltda., 1934-40, J. Magnus e Cia, Ltda., 1940-41, A. G. McKee & Co., 1941-42, R. G. LeTourneau, Inc., 1947-48, U. S. Bur. Mines, 1950-51, Case Inst. Tech., 1951-54; faculty dept. mech. engring. Rice U., Houston, 1954—, prof., 1959—, dean engring., 1962-66. Vis. lectr. U. Brazil, 1963, 65. Guggenheim fellow, 1960-61. Mem. Am. Inst. Mining, Metall. and Petroleum Engrs., Am. Phys. Soc., Am. Soc. Metals, Am. Soc. Engring. Edn., Soc. Engring. Sci., Sigma Xi, Tau Beta Pi. Research, and publs. on lattice imperfection theory, electron emission phenomena, alloy theory. Home: 3612 Overbrook Lane, Houston 77027.*

BROUARDEL, Paul-Camille-Hippolyte, French physician; b. St. Quentin, France, Feb. 13, 1837; s. Auguste Brouardel; ed. Paris, under Marey; doctorate, 1865. Intern, Hosp. Cochin; physician Central Bur. Hosp., from 1869, also St. Antoine Hosp.; prof. legal medicine Paris Faculty Medicine, from 1874, dean, from 1887. Del., Internat. Congress against Tb, 1905. Mem. Acad. Medicine, French Acad. Scis., 1892, Soc. Legal Medicine (pres.). Author: Étude médico-légale sur la combustion du corps humain, 1878; Rapport sur les essais de vaccination cholérique, 1885;

Le secret médical, 1887; Hygiène des ouvriers employés dans les fabriques, 1889; La fièvre typhoide, 1895; La lutte contre la tuberculose, 1901; La profession médicale au commencement du XXe siecle, 1903; (with Gilbert) Traité de médecine et de thérapeutique, 10 vols., 1895-1902. Investigated Tb, also cancer of larynx, hygiene, rabies, variations of human temperature, elimination of urea during liver diseases. Died Paris, July 23, 1906.

BROUGH, Bennett H., Brit. metallurgist; m. Barabara Lloyd; 1 son, 1 dau.; instr. mine surveying, asso. Royal Sch. Mines; mem. council Instn. Mining Engrs., Chartered Inst. Secs., Internat. Testing Assn.; sec. Iron and Steel Inst.; fellow Geol. Soc., Royal Inst. Chemistry (council). Author: Treatise on Mine Surveying, 1888, 13th edit., 1908; also papers on mining and metallurgy. Died Oct. 3, 1908.

BROUGHTON, Levin Bowland, Am. chemist; b. Pocomoke City, Md., Mar. 29, 1886; s. William Thomas and Alice Mary (Bowland) B.; B.S., U. Md., 1908, M.S., 1911; Ph.D., Ohio State U., 1926; m. Laurise McDonnell, Dec. 27, 1911; children—Elinor C., Levin Barnett. Asst. chemist Md. Agrl. Expt. Sta., 1908-11; asst. prof. chemistry U. Md., 1911-14, asso. prof., 1914-18, prof. agrl. chemistry, 1918-29, prof. chemistry and state chemist since 1929, dean Coll. of Arts and Scis., since 1938. Mem. Am. Chem. Soc. (councillor 1937-38), Assn. Ofcl. Agrl. Chemists (v.p. 1939), Sigma Xi. Died Dec. 1943.

BROUN, John Allan, magnetician, meteorologist; b. Dumfries, Scotland, Sept. 21, 1817; ed. Edinburgh (Scotland) U.; m. Isaline Vallouy, 1850; 3 sons, 2 daus. Dir. magnetic obs., Makerstoun, Scotland, 1842-49; dir. Trevandrum Magnetic Obs., 1852; built obs. on Augustia Malley, (highest peak of Travancore Ghats) India, 1855; left India, 1865; went to Lausanne, Switzerland, Stuttgart, Germany, then to London, Eng., in 1873. Fellow Royal Soc., 1853 (Royal medal 1878). Author: Observations of Magnetic Declination made at Trevandrum and Augustia Malley . . . , 1873. Contbd. numerous papers to Royal Soc. One of founders of study of earth magnetism; instrumental in establishing its laws; discovered that earth loses or gains magnetic intensity as a whole showed earth to have magnetic period of 26 days (due to sun's rotation); defined annual period of magnetic intensity; showed that great magnetic disturbances proceed from particular solar meridians, 1857; invented a gravimeter. Died Nov. 22, 1879.

BROUN, Maurice, Am. ornithologist, naturalist; b. N.Y.C., Aug. 27, 1906; s. Jacob and Rebecca Broun; ed. pub. schs.; D.Sc. (hon.), Muhlenberg Coll., 1952; m. Irma Knowles Penniman, Jan. 15, 1934. Ornithol. work with Edward H. Forbush, 1927-29; pioneer devel. Pleasant Valley Sanctuary, Lenox, Mass., 1929-31; research asso. Austin Ornithol. Research Sta., S. Wellfleet, Mass., 1931-34; established nature trails near Rutland, Vt., summers 1939-42, also sanctuaries and nature trails in five other states; curator Hawk Mountain Sanctuary, Kempton, Pa., 1934-66. Recipient Certificate of Merit award Am. Motors-Nash Conservation Assn., 1953. Fellow Del. Valley Ornithol. Club; hon. mem. Rochester Acad. Sci.; elective mem. Am. Ornithologists Union; mem. Am. Fern Soc., Wilson Ornithol. Soc. Author: Index to North American Ferns, 1938; Hawks Aloft, The Story of Hawk Mountain, 1949; also annual newsletter Hawk Mountain Sanctuary Assn., 1939-66. Contbr.: Birds of Massachusetts and Other New England States, 3 vols., 1927-29; also numerous other books and profl. jours. Compilation of ecologic, geog., nomenclatural data N.Am. pteridophytes; studies of hawk migrations. Home: Strawberry Hill Farm, Route 1, New Ringgold, Pa. 17960. Office: Hawk Mountain Sanctuary, Kempton, Pa. 19529.*

BROUNCKER, William, mathematician; b. Castle Lyons, Ireland, circa 1620; s. William and Winefride (Leigh) B.; D. Physics, Oxford U., 1646. Lord viscount of Castle Lyons; a founder, 1st pres. Royal Soc.; pres. Gresham Coll. Author: Experiments of the Recoiling of Guns, 1702. Derived concept of continued fraction, 1658; discovered computation of logarithms by infinite series; gave 1st infinite series for area of equilateral hyperbola. Died Westminster, Eng., Apr. 5, 1684.

BROUSSAIS, François-Joseph Victor, French physician; b. St. Malo, France, Dec. 17, 1772; ed. Coll. Dinan (France), also Hosp. Port Malo, then under Billard and Duret at Brest, France, then at Paris. Surgeon in army and navy during wars of French Revolution; physician, Mcht. Marine; returned to Paris, 1814; became prof. Mil. Hosp., Val de Grâce, France, 1820; apptd. prof. gen. pathology Faculty Medicine Paris, 1831. Decorated cross Ordre de la Reunion, ribbon Legion of Honor. Mem. Inst. France. Author: Histoire des phlegmasies ou inflammations chroniques, 2 vols., 1808; Examen de la doctrine médicale généralement adoptée, 1816; Traité de physiologie, 2 vols., 1822-24; Annales de la médecine physiologique, 1822; Catéchisme de la médecine physiologique, 1824; De l'irritation et de la folie, 1828. Believed most disease was caused by gastro-intestinal irritation and should be treated with spl. diets, purges, bloodletting; thought all diseases originated from excessive stimulation of some bodily organ and spread from there to other parts of body; his views were largely discredited during cholera epidemic in Paris, 1832. Died Vitry-sur-Seine, France, Nov. 17, 1838.

BROUSSEAU, Kate, Am. psychologist; b. Ypsilanti, Mich.; d. Julius and Caroline Yakeley) B.; ed. State Normal Sch. also in Germany, Collège de France, École d'Anthropologie, Paris, U. Minn., U. Chgo. Law Sch.; Docteur de L'Universitéé Paris, 1904. Tchr., Los Angeles High Sch.; teacher psychology and pedagogy Los Angeles State Normal Sch.; prof. psychology Mills Coll., 1907-28; dir. psychol. service Inst. Family Relations, Los Angeles, 1929—. Made psychol. survey of Sonoma State Home for Feeble Minded, 1914-16, testing about 1,200 children; served in French Army, World War I, dir. Foyers du Soldat, on Lorraine Front; with French Army of Occupation in Germany and in devastated dists. of northern France; awarded commemorative medal by French Govt., 1920. Author: L'Éducation des Nègres aux États-Unis; Mongolism, Mental and Physical Characteristics of Mongolian Imbeciles. Died July 8, 1938.

BROUSSONET, (Pierre-Marie-) Auguste, French naturalist; b. Montpellier, France, Feb. 28, 1761; ed. in medicine; asst. to Louis Jean Marie Daubenton, Coll. de France; perpetual sec. Soc. Agr.; mem. Legislative Assembly, 1789; after fall of Girondists, fled to Spain. then practiced medicine, Morocco; became prof. botany, Montpellier, 1805; became mem. legislative body 1st Empire, 1796. Mem. French Acad. Scis., 1785; Royal Soc. (hon.), 1782. Author: Ichthyologiae decas I, 1782; Elenchus plantarum horti montispeliensis, 1805. Introduced Merino sheep, also Angora goat into France, carried out bot. researches in Spain and north Africa; pursued ichtyological research in England. Died Montpellier, July 27, 1807.

BROUWER, Dirk, astronomer; b. Rotterdam, Netherlands, Sept. 1, 1902; s. Martinus and Louisa (Van Wamelen) B.; Ph.D., U. Leiden (Netherlands), 1927; D.Sc., U. La Plata, 1959; m. Johanna A. M. deGraaf, Nov. 1, 1928; 1 son, James Martin. Came to U. S., 1928, naturalized, 1937. Fellow, Internat. Edn. Bd., 1927-28; faculty Yale, 1928-66, prof. astronomy, dir. Obs., chmn. dept. astronomy, 1941-66, Munson prof. natural philosophy and astronomy, 1944-66. Mem. adv. coms. to various govt. agys.; cons. U. S. Naval Research Lab., Jet Propulsion Lab. Mem. Nat. Acad. Scis., A.A.A.S., Royal (fgn. asso., Gold medal 1955), Am. astron. socs., Am. Acad. Arts and Sci., Royal Netherlands Acad. (corr.), Buenos Aires Acad., Internat. Astron. Union. Author: (with G. M. Clemence) Methods of Celestial Mechanics. Editor, Astron. Jour., 1941—. Research and numerous articles on celestial mechanics, planetary theory, lunar theory, theory artificial satellite motion, astron. constants, variable rotation earth. Home: 363 Willow St., New Haven 06511. Died 1966.

BROUWER, Ede, Dutch physiologist; b. Anlo, Oct. 3, 1893; s. Lammert and Ytje Brouwer; M.D., U. Groningen; m. M. Reitsma, Sept. 29, 1921; children—Hiltje, Ytje. Asst., Physiology Lab., Groningen, 1915-17; pediatrician, Groningen, 1919-21; physiologist Nat. Exptl. Sta. Agr., Hoorn, 1921-27; prof. animal physiology Agrl. U., Wageningen, 1939-64; ret., 1964. Mem. numerous agrl. assns. and orgns. Research and numerous publs. on animal nutrition, specifics on milk, herbs, hay. Address: Hinkeloordse weg 8, Wageningen, Netherlands.

BROUWER, Hendrik Aldert, Dutch geologist; b. Medemblik, Netherland, Sept. 20, 1886; s. Egbert Luitzen and Hendrika (Poutsma) B.; student Tech. U., Delft, Netherlands, 1903-09, D.Sc., 1910; postgrad. Sorbonne, also Mus. Natural History, Paris, 1909-10; m. Louis Betsy Van der Spil, June 2, 1909; children—L. E. Jan, Elisabeth (Mrs. T. Gerritze), M. Aldert. With mining dept. Batavia Netherlands E. Indies, 1910-17; prof. U. Delft (Netherlands), 1918-29; prof. U. Amsterdam, 1929-57, prof. emeritus, 1957—, dir. Geol. Inst., 1929-57; exchange prof. U. Mich., 1922; prof. U. Utrecht, 1925-50. Hon. mem. European, Am. and E. Asian socs.; mem. Royal Netherlands Acad. Scis. Author: Alkali Rocks in the Transvaal, 1910-17; Geology of the Netherlands East Indies, 1925; Geological Investigations in the Eastern East Indian Archipelago, 5 vols., 1922-27; Geological Expedition to the Less Surda Lunda Islands, 4 vols., 1940-42; Geological Explorations in Celebes, 1947; also numerous articles. Geol. and petrological investigations in S. Africa, Brazil, E. Indian Archipelago, Spain, Corsica, Scandinavia, especially structural geology, petrology and volcanology. Home: 90 Stadionweg, Amsterdam, Netherlands.*

BROUWER, Luitzen Egbertus J., Dutch mathematician; b. Netherlands, Feb. 27, 1881. Prof. U. Amsterdam, Netherlands. One of founders modern topology; proved that dimensionality of a Cartesian space is topological invariant; work in point sets considered by many most important since Cantor's; believed in intuitionism concerning numbers.

BROWER, Jacob Vradenberg, Am. archaeologist, explorer, b. nr. York, Mich., Jan. 21, 1844; s. Abraham Duryea and Mary R. B.; ed. pub. schs.; enlisted in vol. cav., 1862, in U. S. vol. navy, 1864; auditor Todd Co., Minn., 1867-73; mem. Minn. legislature, 1873; register U. S. land office, St. Cloud, Minn., 1874-79; Itasca State Park commr., Minn., 1891-95; charted source of Miss. river, 1889, of Mo. river, 1896; discovered mounds and ancient village site, Itasca Lake, 1894-95; rediscovered site of Quivira, 1897-98; discovered 1,125 ancient mounds, Mille Lac, Minn., 1900. Pres. Quivira Hist. Soc. (assn. of explorers, authors and ethnol. students). Author: The Mississippi River and Its Source, 1893; Prehistoric Man at the Head Waters of the Mississippi, 1895; The Missouri River and Its Utmost Source, 1896; Quivira, 1898; Harahey, 1899; Mille Lac, 1900; Kathio, 1901; Kakabikansing, 1902; Minnesota—Discovery of Its Area, 1541-1665, 1903; Kansas—Monumental Perpetuation of Its Earliest History, 1541-1896, 1903. Died 1905.

BROWMAN, Ludvig Gustav, Am. zoologist; b. De Kalb, Ill., Apr. 23, 1904; s. Andrew Henry and Anna Elizabeth (Valsas) B.; student No. Ill. U., 1922-24; B.A., U. Chgo., 1928, Ph.D., 1935; m. Audra Elizabeth Arnold, Mar. 25, 1934; children—Andrew Arnold, Audra Elizabeth Ann, David Ludvig, Catherine Phebe. Tchr., prin., jr. high sch. Shabbona Community Sch., 1924-26; tchr. biology high sch. Harvard (Ill.) Community Schs., 1928-31; faculty U. Mont., Missoula, 1937—, prof., 1946—, chmn. zoology, 1948-56, supr. health sci. animal facilities, 1960—. USPHS-NIH grantee, 1954, 56, 60—. Fellow A.A.-A.S.; mem. Am. Anat. Assn., Am. Inst. Biol. Scis., Am. Soc. Zoologists, Genetics Soc., Internat. Soc. for Study Biol. Rhythms, Mont. Acad. Sci., N.W. Sci. Assn., Soc. Exptl. Biology and Medicine, Soc. for Study Devel. and Growth, Sigma Xi, Phi Sigma, Phi Kappa Phi. Research and publs. on central optic system in microphthalmic rats. Home: 664 S. 6th East, Missoula, Mont. 59801.*

BROWN, Alexander Ephraim, Am. inventor; b. Cleve., May 14, 1852; s. Fayette and Cornelia C. (Curtis) B.; ed. pub. schs., Cleve.; spl. course civil engring. Bklyn. Poly. Inst., 1872; U. S. Geol. Survey, 1 yr.; m. Carrie M. Barnett, Nov. 14, 1877. Vice pres., gen. mgr. Brown Hoisting Machinery Co., 1880—. Inventor hoisting machinery, machines and appliances for iron, steel and coke works, ship building, process and material for reinforcing concrete, known as Fensinclave, also numerous patents. Died 1911.

BROWN, Anthony William Aldridge, entomologist; b. Horley, Eng., Nov. 18, 1911; s. William and Maisie (English) B.; B.Sc.F., U. Toronto (Ont., Can.), 1933, M.A., 1935, Ph.D., 1936; m. Jocelyn Evill, June 11, 1938; children—Hilary, Virginia, Kathryn. Asst. entomologist Can. Dept. Agr., Ottawa, Ont., 1937-42; dir. chem. warfare Canadian Army, 1942-47; biologist WHO, Geneva, Switzerland, 1956-58; prof., head dept. zoology U. Western Ont., London, 1948—. Cons. WHO, 1959—, USPHS, 1955. Decorated Order Brit. Empire. Fellow Royal Soc. Can.; mem. Entomol. Soc. Can. (past pres., Gold medal 1963), Am. Mosquito Control. Assn. (past pres.), Entomol. Soc. Am. (past president), Canadian Society Zoologists (president). Author: Insect Control by Chemicals, 1951; Insecticide Resistance in Arthropods, 1958; Entomology: Veterinary and Medical, 1954; also numerous articles. Research on nitrogenous metabolism in insects, Canadian forest insects, factors attracting mosquitoes to man and animals, aerial application insecticides, biochemistry and genetics of resistance to insecticides. Home: 275 Victoria St., London, Ont., Can.*

BROWN, Arnold Lanehart, Am. physician, pathologist; b. Wooster, O., Jan. 26, 1926; s. Arnold L. and Wilda (Woods) B.; student U. Richmond, 1943-45, Med. Coll. Va., 1945-49; postgrad. London Hosp., U. London (Eng.); m. Betty Simpson Brown, Oct. 2, 1949; children—Arnold Lanehart III, Anthony C., Allen W., Fletcher S., Lisa K. Postdoctoral fellow NIH, 1956-58; asst. pathologist Presbyn. and St. Luke's Hosp., Chgo., 1958-59; cons. Mayo Clinic, Rochester, Minn., 1959—; faculty U. Minn. Grad. Sch. Mpls., 1959—, asso. prof. pathology, 1967—; practice medicine specializing in pathology, Chgo., 1958-59, Rochester, 1959—. Mem. research and lab. com. Mayo Found., 1962—. Mem. Am. Soc. Exptl. Pathology, Am. Assn. Pathologists and Bacteriologists, Internat. Acad. Pathology, Electron Microscope Soc. Am., Am. Gastroenterology Assn. Research, numerous publs. on biochemistry and fine structure of normal and hypertrophic heart, electron microscopy of cardiac myosin molecule, various aspects of diseased heart, fine structure of abnormal blood cells, pathology gastrointestinal tract. Home: 814 10th St. S.W., Rochester, Minn. 55901.*

BROWN, Arthur, Am. virologist; b. N.Y.C., Feb. 12, 1922; s. Samuel S. and Ida (Hoffman) B.; B.A., Bklyn. Coll., 1943; postgrad. U. Ky., Ph.D., U. Chgo., 1950; m. Elaine Belaief, Dec. 24, 1947; children—Karen A., Kenneth M., Stephen S., David P. Instr., State U. N.Y. Coll. Medicine, Bklyn., 1951-55; supervisory microbiologist virus and rickettsia dept. Ft. Detrick, Frederick, Md., 1955—. Cons., co-dir. Ph.D. student George Washington U. Sch. Medicine, 1959—; lectr. dept. microbiology U. Md., 1957-60. Sr. fellow Inst. Molecular Biology, U. Geneva (Switzerland), 1963-64. Mem. Am. Acad. Microbiology, Am. Soc. for Microbiology (chmn. virology div.), Soc. Exptl. Biology and Medicine, Am. Assn. Immunologists, A.A.A.S., Tissue Culture Assn., Research Soc. Am., Sigma Xi. Adv. editor Jour. Infectious Disease, 1961-65. Research, publs. on interactions between host cells and viruses, viral genetics. Office: Virus and

Rickettsia Dept., U. S. Army BioLabs., Ft. Detrick, Frederick, Md. 21701.*

BROWN, Arthur Frederick, Brit. physicist; b. Castle-Douglas, Scotland, Sept. 2, 1920; s. James Carr and Jane (Torry) B.; M.A., in Math. and Natural Philosophy, Edinburgh (Scotland) U., 1942; Ph.D. in Physics, Birmingham (Eng.) U., 1947; m. Margaret McGowan Templeton, Feb. 21, 1945; children—Nicholas, Alison. With atomic energy project Oxford (Eng.) U., 1942-46; Brit. Iron and Steel Research Assn. fellow Cavendish Lab., Cambridge (Eng.) U., 1946-47, Dept. Sci. and Indsl. Research Sr. fellow, 1948-50, Imperial Chem. Industries fellow, 1950-52; faculty Edinburgh U., 1952-67, reader physics, 1963-67; prof. physics City U., London, Eng., 1967—. Mem. Scottish Examination Bd., 1965—. Fellow Royal Soc. Edinburgh (sec. to meetings 1963—), Inst. Physics, Instn. Metallurgists; mem. Inst. Physics and Phys. Soc. in Scotland (past chmn.). Author: Statistical Physics, 1967; also articles. Translator: Crystal Physics (G. S. Zhdanov), 1965. Contbr. to separation of uranium isotopes; research on mech. properties of metals especially using high resolution microscopy, radioactive tracers and elec. resistivity. Home: 36 Ling's Coppice, London S.E. 21. Cffice: City U., London E.C.1, Eng.

BROWN, Bancroft Huntington, Am. mathematician; b. Hyde Park, Mass., Nov. 11, 1894; s. Edward Waters and Ruth (Mowry) B.; A.B., Brown U., 1916, A.M., 1917; Ph.D., Harvard, 1922; A.M. (hon.), Dartmouth, 1931; LL.D., St. Anselm's Coll., 1962; m. Eleanor Pairman, Aug. 10, 1922; children—John Pairman, Barbara (Mrs. Thomas W. Streeter, Jr.), Joanna, Margaret. Instr. math. Harvard, 1919-21, Radcliffe Coll., 1920-21; faculty Dartmouth, Hanover, N.H., 1922—, prof. math., 1931-45, B.P. Cheney prof., 1945-62, B.P. Cheney prof. emeritus, 1962; instr. Mass. Inst. Tech., Cambridge, 1928-29, St. Anselm's Coll., Manchester, N.H., 1933-35. Mem. Math. Assn. Am. (v.p. 1938-39), N.H. Acad. Sci. (pres. 1937-38), Phi Beta Kappa, Sigma Xi. Study of cartography; equilong transformations of Euclidean space; geometry. Home: 7 Ripley Rd., Hanover, N.H.

BROWN, Charles, inventor; b. London. Employed by Sulzer, Winterthur, Switzerland, from 1851; built Zulzer steam engines until 1871; created valve gear permitting use of superheated steam, 1866; founder Swiss Locomotive Factory, 1871; constructed fast stamping machine; built alternating current motor, 1900. Died 1905.

BROWN, Charles Carroll, Am. civil engr.; b. Austinburg, O., Oct. 4, 1856; s. George Pliny and Mary Louise (Seymour) B.; studied engring. Cornell U., 1874-75; C.E., U. Mich., 1879, A.M. (hon.), 1913; m. Cora Stanton, Sept. 10, 1878 (dec.); children—Edith Stanton, Edwin Stanton, m. 2d, Eileen Finkle, Jan. 2, 1930. Prof. civil engring. Rose Poly. Institute, 1883-86, Union Coll., 1886-93; cons. engr. N.Y. State Bd. Health, 1888-93; city engr., Indpls., 1894-94; editor Municipal Engring., 1896-1917; with Ill. Div. of Hwys., 1918-19; prof. engring. Valparaiso U., 1919-21; engr. Dept. Pub. Works, St. Petersburg, Fla., 1921-23; city engr., Lakeland, Fla., 1923-27; prof. civil engring., U. Fla., 1927-33; chmn. Gainesville City Plan Bd.; cons. engr. Fla. Mapping Project. Fellow Fla. Engrs. Soc. (pres.); mem. Am. Soc. C.E. (life), Am. Pub. Works Assn. (pres., sec.), Am. Assn. Engrs., Sigma Xi. (2d nat. pres.). Author: Report on Croton Water Shed of the City of New York, 1889; Directory of American Cement Industries (5 edits.), 1901-09; Handbook for Cement Users (3 edits.), 1901-05. Died Nov. 26, 1949.

BROWN, Charles Howard, Am. physician; b. Bklyn., Apr. 4, 1913; s. Charles Harvey and Julia (Heath) B.; student Yale, 1933-34; A.B. with distinction, Wesleyan U., 1934; M.S., U. Chgo., 1936, M.D., 1938; m. Mary L. Laverty, June 10, 1938; children—Christopher Heath, Judith Harvey (Mrs. Jack Williams), Donald Howard. Instr. psychology U. Chgo., 1935-36; with Cleve. Clinic Found. and Hosp., 1948—, head dept. gastroenterology, prof., 1959—, bd. govs., 1963—. Diplomate Am. Bd. Internal Medicine, Am. Bd. Gastroenterology; fellow Am. Coll. Gastroenterology (hon.), A.C.P.; mem. A.M.A., Am. Fedn. Clin. Research, Am. Gastroenterology Assn., Am. Soc. Gastrointestinal Endoscopy (v.p. 1966-67), Acad. Psychosomatic Medicine, Am. Therapeutic Soc., Am. Assn. Study of Liver, Phi Beta Kappa. Author: Diagnostic Procedures in Gastroenterology, 1967. Research and numerous articles on gastroenterology, ulcer, colitis, liver disease. Home: 2808 Van Aken Blvd., Cleve. 44120. Office: 2020 E. 93d St., Cleve. 44106.*

BROWN, C(hristian) Henry, Am. oculist; b. Lancaster, Pa., May 8, 1857; s. Edwin H. and Sussn A. (Widmyer) B.; student Franklin and Marshall Coll., Pa.; M.D., U. of Pa., 1878. Phys. Phila. Hosp., 1878-80, Lancaster County Hosp., 1881-83; located in Phila., 1887; founder Phila. Optical Coll., 1889, also pres.; founder, sec. Bd. Health, Lancaster. Author: Optometric Record Book, 1891; The Optician's Manual, Vol. 1, 1895, Vol. 2, 1901; Clinics in Optometry, 1905; State Board Questions, 1909. Deceased.

BROWN, Clair Alan, Am. botanist; b. Port Allegany, Pa., Aug. 16, 1903; s. Charles Melvin and Jennie H. (Burrows) B.; B.S. cum laude, N.Y. State Coll. Forestry, 1925; M.A., U. Mich., 1926, Ph.D., 1934; m. Maude Nichols, Sept. 4, 1926 (dec. 1962); children—Dorcas Ellen (Mrs. Hubert Smith), Sarah Janet (dec. 1962). Faculty, La State U., Baton Rouge, 1926, 31—, prof. botany, 1944—. Bot. cons. to oil, paper and lumber industries. Guggenheim Meml. fellow, 1952; Edmund Niles Huyck fellow, summers 1958, 59; fellow Advanced Biology Inst., Costa Rica, summers 1961, 62. Named hon. pres. 78th extraordinary session Societe Botanique de France, 1959. Mem. Bot. Soc. Am., Torrey Bot. Club, Am. Soc. Taxonomists, Internat. Assn. Plant Taxonomy, Am. Fern Soc., Ecol. Soc. Am., La. Acad. Sci. Author: Ferns and Fern Allies of Louisiana (with D. C. Correll), 1942; Louisiana Trees and Shrubs, 1945; Vegetation of the Outer Banks of North Carolina, 1959; Palynological Techniques, 1960; Commercial Trees of Louisiana, 1956; Mississippi Trees, 1966; also numerous articles. Research on floras of La., Isle Royale, Mich., N.C., on pollen. Home: 1180 Stanford Av., Baton Rouge 70808.*

BROWN, (Alexander) Crum, chemist; b. Edinburgh, Scotland, Mar. 26, 1838; ed. Edinburgh, also London, Heidelberg, Marburg; Prof. of chemistry Edinburgh, U., 1869-1908. Fellow Royal Soc., 1879. Worked on theory of isomerism, also organic compounds of sulphur; made 1st studies (with Thomas Richard Fraser of relationship between chem. structure of drugs and physiologic action; investigated phlogiston theory, also Brodie's chem. notation system, application of math. to chemistry; devised modern structural formulas, 1864, formulated Crum Brown rule pertaining to substition in benzene derivatives. Died Oct. 28, 1922.

BROWN, David, Brit. chemist; b. nr. Leeds, Eng., July 3, 1936; s. James and Jane (Cowey) B.; B.Sc. with first class honors in chemistry, Sheffield U., 1957, Ph.D., 1960; m. Rita McNeil, July 8, 1961. Sci. officer U. K. Atomic Energy Authority, Harwell, Eng., 1960-64, sr. sci. officer, 1964—. Mem. Brit. Chem. Soc. Research, publs. on preparative chemistry of protactinium related elements. Home: Windrush, South Row, Chilton, Berks, Eng. Office: Atomic Energy Research Establishment, Harwell, Didcot, Berks, Eng.*

BROWN, David Henry, Am. biol. chemist; b. Ely, Nev., June 17, 1921; s. E. Bryce and Verna (Cryst) B.; B.S. with honors, Cal. Inst. Tech., 1942, Ph.D., 1948; m. Barbara A. Illingworth, Nov. 24, 1951; children—Philip, Christopher, Emily. Research asst. Cal. Inst. Tech., NDRC, 1942-45; co-dir. field analytical lab. div. 9 NDRC-Dugway Proving Ground, Chem. Warfare Service, Bushnell, Fla., 1944-45; Frasch Found. postdoctoral fellow plant biochemistry div. biology Cal. Inst. Tech., 1947-48; Merck fellow biol. chemistry Washington U. Sch. Medicine, St. Louis, 1948-50, faculty, 1950—, prof. biol. chemistry, 1962—. Mem. A.A.A.S., Am. Soc. Biol. Chemists, Biochem. Soc (Eng.), Sigma Xi. Research and publs. on enzymatic processes in carbohydrate metabolism in mammalian tissues.*

BROWN, D(enton) J(acobs), Am. chemist; b. Hampton, Pa., Apr. 13, 1882; s. A. David and Anna Mary (Jacobs) B.; A.B., U. Tex., 1910; Ph.D., U. Chgo., 1918; m. Sallie Sloan; 1 son, Denton Sloan. Instr. chemistry U. Tex., 1914-15, 17-18; asst. prof. chemistry Agrl. and Mech. Coll. of Tex., 1915-17; asso. prof. chemistry U. Neb., 1918-27, prof. 1927-50, emeritus since 1950. Mem. Am. Chem. Soc., Phi Beta Kappa, Sigma Xi. Research in electrochemistry and rates of reaction. Died Oct. 4, 1955.

BROWN, Donald D., Am. biologist. educator; b. Cin., Dec. 30, 1931; s. Albert Louis and Louise (Rauh) B.; student Dartmouth, 1949-52; M.D., M.S. in Biochemistry, U. Chgo., 1956; m. Linda Jane Weil, July 2, 1957; children—Deborah Lin, Christopher Charles, Sharon Elizabeth. Research asso. Nat. Inst. Mental Health, NIH, 1957-59; USPHS spl. fellow Pasteur Inst., Paris, France, 1959-60; spl. fellow Carnegie Instn. of Washington. 1960-62, staff mem. dept. embryology, 1962—; asst. prof. biology Johns Hopkins, Balt., 1964-66, asso. prof., 1966—. Mem. Am. Soc. Biol. Chemists, Soc. for Growth and Devel., A.A.A.S. Research and publs. on biochem. studies of embryos, RNA synthesis in eggs and embryos of amphibia. Home: 4818 Greenspring Av., Balt. 21209. Office: 115 W. University Pkwy., Balt. 21210.*

BROWN, Donald Robert, Am. psychologist; b. Albany, N.Y., Mar. 5, 1925; s. J. Edward and Natalie (Rosenberg) B.; A.B., Harvard, 1948; M.A., U. Cal. at Berkeley, 1951, Ph.D., 1951; m. June E. Gole, Aug. 14, 1945; children—Peter Douglas, Thomas Matthew, J. Noah. Faculty Bryn Mawr (Pa.) Coll., 1951-64; prof. dept. psychology U. Mich., Ann Arbor, 1964—; research cons., research asso. Mary Conover Mellon Found., Vassar Coll., 1952—. Cons. on Evaluation Commn. on Coll. Physics, 1963—. Fellow Center for Advanced Study in Behavioral Scis., 1960-61. Fellow Am. Psychol. Assn.; mem. Soc. for Study Social Issues, A.A.A.S., Am. Assn. U. Profs., Sigma Xi, Psi Chi. Contbg. author: American College, 1961; College and Character, 1964. Research and publs. on basic cognitive functions underlying social judgement, social psychology, higher edn. Home: 1315 Hill St., Ann Arbor, Mich. 48104.*

BROWN, Donald Wesley, Am. physician; b. Burlington, Colo., July 17, 1926; s. Joseph Donald and Seletha (Marrs) B.; B.A., U. Colo., 1949, M.D., 1952; m. Margaret Ruth Jacobson, June 21, 1952; children—Wesley, Benjamin, Andrew, Theodore. NIH fellow in cardiology U. Colo. Med. Center, Denver, 1955-56, dir. div. nuclear medicine, 1963—, asso. prof. radiology, 1966—; practice internal medicine, Honolulu, 1956-62; trainee in nuclear medicine AEC, 1962-63. Mem. Gov.'s Adv. com. on Radiation, 1964—. Fellow A.C.P.; mem. A.A.A.S., Soc. Nuclear Medicine. Contbr. articles to tech. jours. Developed method processing and displaying radioisotope scans in patients by digital computer; contbd. to devel. radioisotope diagnostic procedures. Home: 515 S. Garfield St., Denver 80209. Office: 4200 E. 9th Av., Denver 80220.*

BROWN, Ellen, Am. physician; b. San Francisco, Apr. 30, 1912; d. Warner and Jessie (Milliken) B.; student Stanford, 1930-31; B.A. with highest honors, U. Cal. at Berkeley, 1934; postgrad. Cornell U. Med. Coll.; M.D., U. Cal. Med. Sch., San Francisco, 1939. With U. Cal., Berkeley, 1940-42, asst. physician, 1940-42, research asst. Inst. Child Welfare, 1941-42; with U. Cal. Med. Sch., San Francisco, 1943-44, clin. instr. medicine, asst. to Dr. William J. Kerr, 1943-44, faculty, 1946—, prof. medicine, 1959—; research fellow physiology Harvard Med. Sch., 1944-46. Mem. San Francisco County Med. Soc., Assn. Am. Med. Colls., A.A.A.S., N.Y. Acad. Scis., Soc. Exptl. Biology and Medicine, Am. Physiol. Soc., Am. Fedn. Clin. Research, Western Soc. Clin. Research, San Francisco, Cal., Am. heart assns., Am. Coll. Angiology, Phi Beta Kappa, Sigma Xi, Iota Sigma Pi, Alpha Omega Alpha. Research and publs. in field capillary pressure and permeability; blood volume and vascular capacity, cardiac failure; cardiac complications of pregnancy, peripheral circulation in relation to pain syndromes and vascular diseases. Office: U. Cal. Med. Center, San Francisco 94122.*

BROWN, Elmer Burrell, Am. hematologist; b. N.Y.C., Apr. 1, 1926; s. Elmer Burrell and Pearl (Farris) B.; student U. Mo., 1943-44; A.B., Oberlin Coll., 1946; M.D., Washington U., 1950; m. Phyllis Stephanie Yace, May 15, 1954; children—Cynthia, Joanne, Catherine, Laura. USPHS trainee in hematology Washington U., 1954-55, NRC fellow in med. sci., 1955-57, instr., 1955-59; USPHS Spl. Research fellow Enzyme Sect. Lab., Cellular Physiology Nat. Heart Inst., 1957-59; asst. prof. medicine Washington U., 1959-64, asso. prof. medicine, 1964—; asst. physician Barnes Hosp., St. Louis, 1957—; apptd. chief hematology clinic Washington U. Clinics, 1961, chief div. hematology, 1964—; cons. hematology St. Louis City Hosp., Washington U. Clinics. Diplomate Am. Bd. Internal Medicine. Mem. Am. Fedn. Clin. Research, Internat., Am. socs. hematology, A.A.-A.S., Am., Central socs. clin. investigation, Central Clin. Research Club, Soc. Exptl. Biology and Medicine, Phi Beta Kappa, Sigma Xi, Alpha Omega Alpha, numerous others. Co-editor: Progress in Hematology, 1964-66. Research and publs. directed towards understanding mechanism and control of iron absorption; transfer of iron within the body; effects of excessive amounts of iron on body processes. Home: 86 Aberdeen Pl., St. Louis 63105.*

BROWN, Ernest Benton, Jr., Am. physiologist; b. Mortons Gap, Ky., July 13, 1914; s. Ernest Benton and Ida (Croft) B.; student U. Chgo., 1931-33; B.S., U. Ky., 1937, M.S., 1942; Ph.D., U. Minn., 1949; m. Helen Frances Carrier, Nov. 29, 1934; children—Sylvia Joy (Mrs. Robert E. Davis), Ernest Benton III. Faculty, U. Minn., 1946-61, prof., 1955-61; prof., chmn. physiology U. Kan. Med. Center, Kansas City, 1961—. Mem. A.A.A.S., Soc. Exptl. Biology and Medicine, Am. Physiol. Soc., Am. Heart Assn., Assn. Am. Med. Colls., N.Y. Acad. Scis., Phi Beta Kappa. Research, publs. on useful time of consciousness at high altitude, evaluation of increased sensitivity of respiratory control mechanisms to CO2 following prolonged hyperventilation, influence of high concentrations of CO2 on cardiovascular function, distbn. of blood flow, potassium metabolism and electrolyte distbn., determination of carbon dioxide buffer capacity of blood and tissues in acute and chronic exposures to elevated CO2. Home: 8005 Roe St., Prairie Village, Kan. 66208. Office: U. Kan. Med. Center, Kansas City, Kan. 66103.*

BROWN, Ernest William, mathematician; b. Hull, Eng., Nov. 29, 1866; s. William and Emma (Martin) B.; B.A., Christ's Coll., Cambridge, Eng., 1887, fellow, 1889-95, hon. fellow, 1911, M.A., 1891, Sc.D., 1897, Adams prize, 1907; hon. A.M., Yale U., 1907; D.Sc., Adelaide U., 1914, Yale, 1933, Columbia, 1934; LL.D., McGill U., 1935. Prof. math. Haverford (Pa.) Coll., 1891-1907, Yale, 1907-32. Recipient Bruce Medal, 1920. Mem. Am. Acad. Arts and Scis., Nat. Acad. Sci. (Watson medal 1937), Am. Astron. Soc. (v.p. 1923-25; pres. 1928-31). Author: Treatise on the Lunar Theory, 1896; A New Theory of the Moon's Motion, 1897-1905; Tables of the Motion of the Moon, 1920; Planetary Theory (with C. A. Shook), 1933. Research in lunar motion led to his suggestion of variable rate of rotation of earth. Died New Haven, Conn., July 22, 1938.

BROWN, Fay Cluff, Am. physicist; b. Washington Court House, O., Nov. 23, 1881; s. Argo A. and Jennie (Cluff) B.; A.B., Ind. U., 1904; A.M., U. Ill., 1906; Ph.D., Princeton, 1908, postgrad.; m. Dora Davidson, July 14, 1908; children—Herman C., Barbara E. (Mrs. Walter A. Radius), Robert B., Annabel,

(Mrs. Richard D. Olson). Instr., U. Ill., 1905-07; prof. U. Ia., 1909-17; asst., acting dir. Nat. Bur. Standards, Washington, 1918-27; organizing dir. Mus. Sci. and Industry, N.Y.C., 1927-32; dir. sci. exhibits A.A.A.S., 1932-37; curator Chgo. Mus. Sci. and Industry, 1937-40; mem. staff Radiation Lab., Mass. Inst. Tech., 1940-46. Chief examiner U. S. Civil Service, 1927; mem. Pres.'s Com. for Classification Civil Employees. Jacobus fellow Princeton, 1908. Fellow Am. Phys. Soc. (councillor 1923-26); mem. Inst. Physics, Am. Inst. City N.Y. (dir. 1928), Soc. Arts and Scis. (dir. 1928), Phi Beta Kappa, Sigma Xi. Editor: The Sea for Sam, 1935; And Thats Why, 1932. Research and publs. on photo conductivity and electron theory, method of determining bomb trajectories and improving bomb sights; Invented phonopticon for translating printed page into meaningful sound. Home: 3030 Newark St., Washington 20008.*

BROWN, Frank Arthur, Jr., Am. biologist; b. Beverly, Mass., Aug. 30, 1908; s. Frank Arthur and Arletta (Robinson) B.; A.B., Bowdoin Coll., 1929; M.A., Ph.D., Harvard, 1934; m. Jennie Wentworth Pettegrove, June 24, 1934; children—Charlotte (Mrs. William Mair Russell III), Frank Arthur III, Marilyn Diane (Mrs. Joseph Steven Maranchie). Faculty, Northwestern U., Evanston, Ill., 1937—, prof., 1946-57, chmn. biol. scis., 1949-57, Morrison prof. biology, 1957—; head dept. zoology Marine Biol. Lab., Woods Hole, Mass., 1945-49, trustee, 1946—; chmn. adv. panel Office Naval Research, 1952-57; dir. Gen. Biol. Supply House (Chgo.). Fellow A.A.A.S.; mem. Am. Soc. Zoology, Am. Soc. Naturalists, Soc. Gen. Physiology (past pres.), Union Am. Biol. Soc., Soc. Study Growth and Devel., Am. Physiol. Soc., Ecol. Soc. Am., Am. Soc. Plant Physiologists, Am. Geophys. Union, Soc. Exptl. Biology and Medicine, Am. Soc. Limnology and Oceanography, Am. Biol. Soc., Sigma Xi., others. Author: (with others) Comparative Animal Physiology, 1950; (with C. L. Prosser) Biological Clocks, 1962. Editor: Selected Invertebrate Types, 1950. Research and publs. primarily in field of animal modification of behavior, color changes, crustacean endocrinology, biol. rythms. Home: 906 Greenleaf Av., Wilmette, Ill. 60091. Office: Dept. Biology, Northwestern U., Evanston, Ill. 60201.*

BROWN, F(rank) E(merson), Am. chemist; b. nr. Cuba, Kan., Feb. 9, 1882; s. Prairie Frank and Marie Elizabeth (Barnhill) B.; A.B., Kan. State Teacher's Coll., 1911, B.S. U. Chgo., 1912-17, Ph.D., 1918; m. May Maria Holmes, Dec. 25, 1910; children—Frank Emerson, Holmes M.; m. 2d, Louise Jaggard, July 23, 1920; 1 dau., Louise Jaggard. Tchr., country schs., Republic County, Kan., 1899-1902, supt. of schs., prin. of high sch., Portis, Kan., 1905-07, Hill City, Kan., 1907-09; supt. of schs., Collinsville, Okla., 1911-1912; chemistry teacher Chgo. schs., Y.M.C.A. schs., 1912-18; chemistry teacher Fresno (Cal.) high sch. and jr. coll., 1914; asst. prof. of chemistry Ia. State Coll., Ames, 1917, asso. prof., 1918, prof., 1923-52, emeritus, 1952—. Dir. Ia. Sci. Talent Search, 1947—. Recipient Sci. Apparatus Makers award in Chem. Edn., Am. Chem. Soc., 1958. Mem. Am. Chem. Soc., Senate of Chem. Edn., (chmn. com. on membership affairs 1950-51); Ia. Acad. Sci. (chmn. com. sci. talent search 1950—; v.p. 1950-51, pres. 1951-52, dir. 1956—), A.A.A.S., Sigma Xi. Contbg. ed. Jour. Chem. Edn., 1928-40; editorial bd. Ia. State Coll. Jour. Sci. 1945—. Author: A Short Course in Qualitative Analysis, 1932; Qualitative Analysis Work Book, 1937; also articles on research and teaching of chemistry to chem. jours.; abstractor Chem. Abstracts, 1922—. Died Sept. 10, 1959.

BROWN, Fred, chemist; b. Manchester, Eng., Feb. 5, 1923; s. Frederick and Elsie (Armstrong) B.; B.A., Cambridge U., 1943, M.A., Ph.D., 1946; m. Eileen Hathway, Aug. 14, 1945; children—Allison, Jennifer, Anthony. Asst. lectr. U. North Wales, 1946-49; with Atomic Energy of Can. Lab., Chalk River, Ont., 1949—, now head research chemistry br. Mem. Chem. Inst. Can. Research and numerous publs. on mechanisms of chem. reactions, study of nuclear fission, penetration of energetic ions into solids. Home: 1 Tweedsmuir Pl., Deep River, Ont. Office: Atomic Energy of Can., Chalk River, Ont., Can.*

BROWN, Fred, English virologist; b. Clayton, Eng., Jan. 31, 1925; s. Fred and Jane (Fielding) B.; B.Sc., U. Manchester, 1944, M.Sc., 1946, Ph.D., 1948; m. Audrey Alice Doherty, May 1, 1948; children—Roger, David. Asst. lectr. chemistry Manchester U., 1946-48; lectr. Bristol U., 1948-50; sr. sci. officer Hannah Dairy Research Inst., Ayr, Scotland, 1950-53; sr. research officer Christie Hosp., Manchester, Eng., 1953-55; prin., sr. prin. sci. officer Animal Virus Research Inst., Pirbright, Surrey, Eng., 1955—. Mem. Biochem. Soc., Soc. for Gen. Microbiology. Research and numerous articles on structure of polysaccharides, virus structure and mechanisms of virus duplication. Home: Syndal, Glaziers Lane, Normandy, Surrey. Office: Animal Virus Research Institute, Pirbright, Surrey, Eng.*

BROWN, Frederick Martin, Am. entomologist; b. N.Y.C., Mar. 24, 1903; s. Frederick Wellington and Josephine A. L. (Rilke) B.; student Columbia Univ., 1920-1921; m. Hazel M. Heffron, September 1, 1928. Head sci. dept. Fountain Valley Sch., Colorado Springs, Colo., 1930—; lectr. Colo. Coll., 1935—; research asso. Carnegie Mus., Pitts., Am. Mus. Natural History, N.Y.C. Recipient Univ. medal U. Colo., 1960. Fellow A.A.A.S.; mem. Royal, Am. entomol. socs., Lepidopterists Soc. Author: America's Yesterday, 1935; Butterflies of Colorado, 1957; also numerous articles. Research on Andean fauna of Ecuador and fauna of Antilles, history of entomology, especially of field collectors in colonial time and during exploration of West. Address: Fountain Valley Sch., Colorado Springs, Colo. 80907.*

BROWN, George, Am. physician; b. Wilton, N.H., Oct. 11, 1823; s. Ephraim and Sarah (King) B.; attended U. Vt., Jefferson Med. Coll., Phila.; M.D., U. City N.Y.; m. Catherine Wood, Nov. 28, 1850, at least 1 son, Robert A. Began practice medicine, Barre, Mass., 1850; became supt. Elm Hill Sch. for Feeble-Minded Children, 1851, expanded sch. into largest pvt. instn. of its kind in U. S., (with wife) devoted his life to sch.; mem. Barre Sch. Bd. Died May 6, 1892.

BROWN, George, mathematician; b. 1650; named minister Parish of Kilmauro, Ayr, Scotland, 1680. Author: Rotula arithmetica, with an Account therof, 1700; A Specie Book serving at one View . . . , 1700; A Compendious, but a Compleat System of Decimal Arithmetik, . . . , 1701; Arithmetica Infinita, 1718. Invented instrument for teaching arithmetic (Rotula Arithmetica). Died 1730.

BROWN, George Bosworth, Am. biochemist; b. Birmingham, Ala., Apr. 18, 1914; s. Edwin S. and Phyllis (Bosworth) B.; B.S., Ill. Wesleyan U., 1934, D.Sc., 1960; M.S., U. Ill., 1936, Ph.D., 1938; m. Katherine Matthews, May 4, 1940. Faculty, Med. Coll., Cornell U., 1938-51, prof. biochemistry Sloan-Kettering Inst. div. Cornell U. Grad. Sch. Med. Scis., N.Y.C., 1952—; with Sloan-Kettering Inst. for Cancer Research, N.Y.C., 1946—, mem., 1948—. Cons. to govt. agys. Rockefeller travelling fellow, 1949. vis. prof. Fulbright, Australian Nat. U., 1965. Mem. Am. Chem. Soc. (chmn. N.Y. sect. 1966-67), A.A.A.S., Harvey Soc., Biochem. Soc. (London), Am. Soc. Biol. Chemists, Am. Assn. for Cancer Research. Editor: Biochemical Preparations, vol. X, 1963; editorial bd. Jour. Am. Chem. Soc., 1958-63, Jour. Biol. Chemistry, 1954-66. Research and publs. on nucleic acid metabolism, purines, anticancer agts., carcinogens, bioorganic chemistry. Home: 800 Grove St., Mamaroneck, N.Y. 10543. Office: 145 Boston Post Rd., Rye, N.Y. 10580.*

BROWN, George Harold, Am. electrical engineer; b. North Milwaukee, Wis., Oct. 14, 1908; s. James Clifford and Ida Louise (Siegert) B.; B.S., U. Wis., 1930, M.S., 1931, Ph.D., 1933, E.E., 1942; m. Julia Elizabeth Ward, Dec. 26, 1932; children—James Ward, George H. With Radio Corp. Am., 1933-37, 38—, successively research engr., Camden, N.J., research engr., Princeton, N.J., dir. Systems Research Lab., chief engr. Comml. Electronic Products div., Camden, N.J., chief engr. indsl. electronic products, 1933-59, v.p. engring., 1959-61, vice pres. of research and engineering, 1961-65, exec. v.p. research and engring., 1965—, also dir.; director RCA Communications, Incorporated; cons. engr., 1937-38. Exec. bd. George Washington council Boy Scouts. Fellow I.E.E.E. (Edison medal 1967), A.A.A.S.; mem. Nat. Acad. Engineering, Sigma Xi. Author: (with R. A. Bierwirth and C. N. Hoyler) Radio Frequency Heating, 1947; also articles sci. jours. Patentee in field. Research on antenna, radiation theory, electrical current use in drying penicillin and pasteurization, radio-frequency heating; showed that nonsinusoidal distribution of current caused performance of vertical radiators to vary from theory, 1934; devised Turnstile antenna, 1936; explained calculation of directional antenna patterns, 1937; formulated equation to solve problems of heat flow, etc., 1937-8; developed vestigal sideband filter for TV, 1938; developed test of UHF-TV. Home: 117 Hunt Dr. Office: RCA, Princeton, N.J.

BROWN, G(eorge) Malcolm, Canadian physician; b. Campbellford, Ont., Can., July 16, 1916; s. George A. and Elizabeth (Stewart) B.; M.D., Queen's U., 1938; Ph.D., Oxford U., 1940; m. Helen Louise Gatch, Sept. 20, 1950; children—Alison, Alexandra, Malcolm. Physiologist malarial research unit Royal Army Med. Corps, 1943-45; asso. prof. medicine Queen's U., Kingston, Can., 1946-51, prof. medicine, 1951-65; chmn. Med. Research Council Can., Ottawa, Ont., 1965—. Mem. med. adv. com. Def. Research Bd. Fellow Royal Coll. Physicians, A.C.P., Royal Soc. Can.; mem. NRC Can., Sci. Council Can., Royal Coll. Physicians and Surgeons Can. (pres. 1962-64), Coll. Physicians and Surgeons Ont. (pres. 1956-58). Research, numerous publs. in malaria, cold physiology, haematology and gastroenterology. Home: 477 Manor Av., Ottawa 2. Office: Med. Research Council, Ottawa 7, Ont., Can.*

BROWN, George Van Ingen, Am. surgeon; b. St. Paul, Jan. 15, 1862; s. Matthew Wilson and Emily (Lynch) B.; D.D.S., Pa. Coll. Dental Surgery, 1881; M.D., Milwaukee Med. Coll., 1895, C.M., 1895; A.B., No. Ill. Coll., 1898; M.D., Marquette U., 1909; m. Elizabeth Kathleen Selby Jones, Sept. 22, 1884; 1 son, Selby Van Ingen. Practiced oral surgery and dentistry, St. Paul and Duluth, Minn., 1881-98; specialized in plastic surgery, Milw., 1898—; prof. operative dentistry and oral surgery, dean dental dept. Milw. Med. Coll., 1898-1902; spl. lecturer on oral surgery, dental dept. U. of Ill., 1902-03, State U. of Ia., 1903-04, U. Tenn., 1904, Vanderbilt U., 1905, So. Dental Coll., Atlanta, 1909-15; prof. oral surgery and oral pathology State Univ. Ia., 1904-10; prof. oral and plastic surgery U. Wis., 1920-37; chief of staff St. Mary's Hosp., 1936; on surg. staff Columbia Hosp.; plastic surgeon State of Wis. Gen. Hosp., at U. Wis. also Wis. Orthopedic Hosp., Madison; cons. staff Milw. Children's Hosp., Wis. Meth. Hosp., Madison Gen. Hosp.; attending mem. surg. staff, Milw. County Hosp., Wauwatosa. Recipient Jarvie medal N.Y. State Dental Soc. Fellow A.C.S., Tri-State Med. Soc. (pres. 1920-21); mem. A.M.A. and affiliated assns., Pan-American Med. Assn., Am. Soc. Plastic and Reconstructive Surgery, Am. Bd. Plastic Surgery, many other Am. and fgn. socs. Author: The Surgery of Oral and Facial Diseases and Malformations, 4th edit., 1938; chapter in Ochsner's Surgical Diagnosis and Treatment, chapter in Sajous' Cyclopedia of Practical Medicine, also monographs, papers and occasional poems. Mem. advisory council Living Age (mag.). Died Apr. 2, 1948.

BROWN, Glenn Halstead, Am. chemist; b. Logan, O., Sept. 10, 1915; s. James E. and Nancy (Mohler) B.; B.S., Ohio U., 1939; M.S., Ohio State U., 1941; Ph.D., Ia. State U., 1951; m. Jessie Ruth Adcock, May 27, 1943; children—Larry H., Nancy K., Donald S., Barbara J. Faculty, U. Miss., 1941-48, 48-50, Ia. State U., 1945-48, U. Vt., 1950-52, U. Cin., 1952-60; prof., chmn. chemistry dept. Kent (O.) State U., 1960-65, dean research, 1963—, dir. Liquid Crystal Inst., 1965—. Chmn. Internat. Liquid Crystal Conf., 1965. Fellow Ohio Acad. Sci. (past pres.); mem. Am. Chem. Soc., A.A.A.S., Am. Inst. Chemists, Am. Crystallographic Assn., Sigma Xi, Phi Lambda Upsilon. Author: (with F. A. Anderson) Fundamentals of Chemistry, 1944; (with others) Laboratory Manual for Organic Chemistry, 1944; Record Book for Quantitative Analysis, 1954; (with E. M. Sallee) Quantitative Chemistry, 1963. Research and publs. on structure by x-ray techniques, structure and properties of liquid crystals, phenomenon of photochromism. Home: 470 Harvey Av., Kent, O. 44240.*

BROWN, Gordon Campbell, Am. virologist, educator; b. Zanesville, O., Sept. 28, 1912; s. Clarence M. and Helen (Gordon) B.; B.A., Ohio State U., 1934, M.A., 1935; Sc.D., Johns Hopkins, 1942; m. Marion Whitehead, Oct. 5, 1935; children—Gordon Kingsley, Susannah Elizabeth Fry. Research scientist Parke, Davis & Co., 1935-39; research fellow Yale, 1936; faculty dept. epidemiology U. Mich., Ann Arbor, 1942—, prof., 1954—. Cons., U. S. Dept. Health, Edn. Welfare, NIH, Mich. Health Dept., 1956—. Diplomate Am. Bd. Microbiology. Fellow Am. Assn. Immunologists, Am. Acad. Microbiologists, Central Soc. for Clin. Research; mem. Am. Epidemiological Soc., Am. Soc. Microbiology, Am. Pub. Health Assn., Soc. for Exptl. Biology and Medicine, Assn.. Tchrs. Preventive Medicine, Sigma Xi, Delta Omega. Research and publs. on epidemiology, virus vaccines, response of infants to combined vaccines, preventive medicine, diagnosis of virus diseases. Home: 2550 Devonshire Rd., Ann Arbor, Mich. 48104.*

BROWN, Harley Procter, Am. zoologist; b. Uniontown, Ala., Jan. 13, 1921; s. Harley Procter and Martha (McGinniss) B.; A.B., A.M., Miami U., Oxford, O., 1942; Ph.D. (scholar, fellow) Ohio State U., 1945; m. Laura Clifford Williams, June 1, 1942; 1 dau., Mary Hamilton (Mrs. Gary Wayne Catron). Instr., U. Ida., Moscow, 1945-47, Queens Coll., Flushing, N.Y., 1947-48; faculty U. Okla., Norman, 1948—, prof. zoology, 1962—, acting chmn. zoology dept., 1958-59. Recipient Bausch and Lomb Sci. award, 1938; Speakers' Bur. Silver award, 1942. Mem. A.A.A.S., Am. Inst. Biol. Sci., Am. Micros. Soc., Am. Soc. Zoologists, Am. Assn. U. Profs., Entomol. Soc. Am., Soc. Protozoologists, Soc. Systematic Zoology, Southwestern Assn. Naturalists, Okla. Acad. Sci., Okla. Ornithol. Soc., Phi Beta Kappa, Sigma Xi, Phi Eta Sigma, Phi Sigma, Delta Phi Alpha. Research and publs. on structure protozoan flagella, mechanics flagellar locomotion, life histories various kinds freshwater insects, pharians. Home: 529 Dakota St., Norman, Okla. 73069.*

BROWN, Harrison Scott, Am. geochemist; b. Sheridan, Wyo., Sept. 26, 1917; s. Harrison Harvey and Agatha (Scott) B.; B.S., U. Cal. at Berkeley, 1938; Ph.D., Johns Hopkins, 1941; LL.D., U. Alta., 1960; Sc.D., Rutgers U., 1964, Amherst U., 1966; m. Rudd Owen, Nov. 11, 1949; children—Eric, Scott. Instr. chemistry Johns Hopkins, Balt., 1941-42; research asso. in chemistry U. Chgo., 1942-43, asst. prof. Inst. Nuclear Studies, 1946-48, asso. prof., 1948-51; asst. dir. chem. div. Clinton Labs., Oak Ridge, Tenn., 1943-46; prof. geochemistry, science and government California Institute of Technology, Pasadena, Cal., 1951—. Vice pres. Internat. Council of Sci. Unions, 1963—. Recipient Lasker Found. award, 1958. Mem. Nat. Acad. Scis. (fgn. sec. 1962—), A.A.A.S. (award 1947), Am. Chem. Soc. (award in pure chemistry 1952), Geol. Soc. Am., Am. Geophys. Union, Am. Astron. Soc. Author: Must Destruction Be Our Destiny?, 1946; The Challenge of Man's Future, 1954; (with Bonner and Wier) The Next Hundred Years, 1957; also numerous articles. Co-discoverer methods of determining age of earth and meteorites; inventor technique for determining age of elements; discovered regularities in distbn. of certain trace elements in meteorites; research in chem. relationships between sun and planets; introduced concept of class I, class II, class III planets. Home: 623 E. California Blvd., Pasadena, Cal. 91109.*

BROWN, Harry Fletcher, Am. chemist; b. Natick, Mass., July 10, 1867; s. William H. and Maria F. (Osgood) B.; A.B., Harvard, 1890, A.M., 1892; D.Sc., U. Del., 1930; m. Florence M. Hammett, Oct. 26, 1897. Chief chemist U. S. Naval Torpedo Sta., Newport, R.I., 1893-1900; engaged in investigation and devel. of smokeless powder in Navy Dept.; gen. supt. Internat. Smokeless Powder & Chem. Co., 1900-04; dir. of mfr. smokeless powder E. I. du Pont de Nemours & Co., 1904-15, v.p. in charge smokeless powder dept., 1915-19, then v.p., dir.; Christiana Securities Co. Mem. Am. Chem. Soc., Nat. Edn. Assn. Died Feb. 28, 1944.

BROWN, Herbert Charles, chemist; b. London, Eng., May 22, 1912; s. Charles and Pearl (Stine) B.; A.S., Wright Jr. Coll., Chgo., 1935; B.S., U. Chgo., 1936, Ph.D., 1938; m. Sarah Baylen, Feb. 6, 1937; 1 son, Charles Allan. Prof. inorganic chemistry Purdue U., Lafayette, Ind., 1947-59, Richard B. Wetherill prof. chemistry, 1959, Richard B. Wetherill research prof., 1960—; vis. prof. U. Cal., Los Angeles, 1951, Ohio State U., 1952, U. Mexico, 1954, U. Cal., Berkeley, 1955, U. Colo., 1957, U. Heidelberg, Germany, 1963; chem. cons. Argonne Nat. Labs.; co-dir. research War Research Project and Manhattan Project U. Chgo. F. E. Clark lectr., 1953; Harrison Howe lectr., 1953; Freud-McCormick lectr., 1954; Centenary lectr. Eng., 1955; T. W. Talley lectr., 1956; Falk-Plaut lectr., 1957; Julius Stieglitz lectr., 1958; Max Tishler lectr., 1958; Kekule-Couper Centenary lectr., 1958; E. C. Franklin lectr., 1960; Ira Remsen lectr., 1961; Edgar Fahs Smith lectr., 1962; Seydel-Wooley lectr., 1966. Recipient Research award Purdue Sigma Xi, 1951; Nichols medal, 1959; Synthetic Organic Chem. Mfrs. Assn. medal, 1960; Herbert Newby McCoy award, 1965. Fellow Chem. Soc. (London); mem. Nat. Acad. Scis., Am. Acad. Arts and Scis., A.A.A.S., Am. Chem. Soc. (award 1960), Sigma Xi, Alpha Chi Sigma, Phi Lambda Upsilon. Author: Hydroboration, 1962. Contbr. numerous articles to profl. jours. Codiscoverer sodium borohydride and similar reducing agts. now used in pharm. industry; discoverer the hydroboration reaction; developed quantitative methods for study steric strains in molecules. Home: 1840 Garden St., West Lafayette, Ind. 47906.*

BROWN, Horace Manchester, Am. surgeon; b. New Bedford, Mass., Oct. 12, 1857; s. John Cheney and Jane Elisabeth (Taylor) B.; M.D., Univ. Med. Coll. (New York U.), 1880; postgrad., London, Paris, Munich, Brussels, Berlin; m. Fanny Louise Eldred, Oct. 25, 1882. Practiced in Milw., 1880—; surgeon C.,M.&St.P. Ry., 41 yrs.; builder, and owner Lakeside Hosp. 26 yrs. until its closure, 1915; later surgeon Columbia Hosp. Dir. Physicians' Radium Assn. Author: The Songs of Bilitis (English version), 1904. Translator: De Venenis of Abonus, Anathomia of Mondinus. An editor Annals of Med. History. Died Jan. 18, 1929.

BROWN, Ivan Willard, Jr., Am. thoracic surgeon; b. Newfane, N.J., July 6, 1915; s. Ivan Willard and Agnes (Clark) B.; student U. Rochester, 1933-36; B.S., M.D., Duke, 1936-40; m. Madeline Davis, Dec. 28, 1939; children—Sandra E. (Mrs. James A. Ling), Diane E., Ivan W. III. Instr. physiology Duke U. Med. Sch., Durham N.C., 1939-40, instr. pathology, 1940-42, asso. surgery, 1945-51, asst. prof. surgery, 1951-55, asso. prof., 1955-60, prof. surgery, 1960—. Markle scholar in med. sci., 1948-53; 1st award for research Am. Glycerine Products Assn., 1953. Mem. A.C.S., Am. Fedn. Clin. Research, Am., So. surg. assns., Am. Assn. Thoracic Surgery, Soc. Thoracic Surgery, Soc. Vascular Surgery, Internat. Cardiovascular Soc., Soc. Univ. Surgeons, Soc. Exptl. Biology, Medicine, Am. Soc. Clin. Investigation, Internat., Am. socs. hematology, Am. Soc. Artificial Internal Organs, Phi Beta Kappa, Sigma Xi, Alpha Omega Alpha. Research and publs. in cardio vascular surgery, hypothermia, hyberbaric oxygenation. Home: 1709 Vista St., Durham, N.C.*

BROWN, Jack Harold Upton, Am. sci. adminstr.; b. Nixon, Tex., Nov. 16, 1918; s. Gilmer W. and Thelma (Patton) B.; B.S., S.W. Tex. State Coll., 1939; postgrad. U. Tex.; Ph.D., Rutgers U., 1948; m. Jessie Carolyn Schulz, Apr. 14, 1943. Instr. physiology U. Tex., 1939-41; lectr. physics S.W. Tex. State Coll., 1943-44; instr. phys. chemistry Rutgers U., 1944-45, research asso. Bur. Biol. Research, 1944-48, fellow Gerard Swope Found., 1947-48; head biol. research, fellow Mellon Inst., Pitts., 1948-50; lectr. U. Pitts., 1948-50; asst. prof. physiology U. N.C., 1950-52; vis. scientist Oak Ridge Inst., 1951; faculty U. Sch. Medicine, 1952-60, prof., 1958-60, chmn. dept., 1959; with Div. Research Facilities and Resources, 1960-65, chief spl. research resources, 1962-64, acting chief gen. clin. research centers br., 1963-64, asst. div. chief, 1964-65; asst. dir. Nat. Insts. Gen. Med. Scis., NIH, Bethesda, Md., 1965-67, associate director, since 1967—; professional lecturer at Georgetown Med. Sch., Washington, 1961—. Recipient Research award Sigma Xi, 1961; Fulbright lectr., 1950; Gerard Swope fellow, 1947-48. Fellow A.A.A.S.; mem. I.E.E.E., Am. Chem. Soc., N.Y. Acad. Scis., Soc. Exptl. Biology and Medicine, Endocrine Soc., Am. Physiol. Soc., Pi Kappa Delta, Phi Lambda Upsilon, Alpha Chi. Author: (with S. B. Barker) Basic Endocrinology, 1962, 2d edit., 1966; Physiology of Man in Space, 1963. Editor sci. series Space Biology (C. C. Wunder), Blood and Body Functions (J. F. Ferguson), 1965. Research and publs. on devel. of cancer-inhibiting drug used to treat human carcinoma; inventor of several instruments for better lab. analysis, new techniques for analysis and determination of site of hormone action; devel. of field of biomed. engring. Home: 10667 Montrose Av. Office: Nat. Insts. Gen. Med. Scis., NIH, Bethesda, Md. 20014.*

BROWN, James Barrett, Am. physician; b. Hannibal, Mo., Sept. 20, 1899; s. Albert Sydney and Mary (Kerr) B.; student Washington U., St. Louis, 1917-19, M.D., 1923; children by previous marriage—James Barrett, Charles Sydney; m. 2d, Bertha Phillips Phillips, Sept. 30, 1946; adopted children—Jane Hamilton (Mrs. Robert D. Stanza), Frances Reith (Mrs. Harold Maxwell Stuhl). Faculty, Washington U. Sch. Medicine, 1930—, prof. clin. surgery, 1946—; chief reconstructive and maxillo-facial surgery and burns Valley Forge Gen. Hosp., Phoenixville, Pa., 1942—. Cons. VA; mem. Danforth Found. for Med. Edn., 1960—, Barrett Brown Found. for Research and Study in Reconstructive Surgery and Cancer. Recipient Am. Design award Lord & Taylor, 1944. Mem. A.C.S., Internat. Soc. Surgeons, Surgeons Club, Am. Surg. Assn., so., Western, Central surg. assns., Am. Soc. Plastic and Reconstructive Surgeons, Am. Assn. Plastic Surgeons (hon.), Am. Soc. for Surgery Hand, Soc. Head and Neck Surgeons. Author: (with Frank McDowell) Plastic Surgery of the Nose, 1951, 65; Skin Grafting, 1943, 49, 59; (with Minot P. Fryer, F. McDowell) Surgery of Face, Mouth and Jaws, 1954; Neck Dissections, 1954; (with M. Fryer) Postmortem homografts, 1960; (with Thomas Zaydon) Early Treatment of Facial Injuries, 1964; also numerous articles. Research on use split skin grafts, use synthetic prostheses, mgmt. burns, direct control war injuries, high energy electron burns, repair pure atomic burns, elongation cleft palates, simplified closure cleft lips, cancer head and neck, partoid tumors, rehab. of severe burns. Home: 710 S. Hanley Rd., St. Louis 63105.*

BROWN, James Campbell, Brit. chemist; b. Aberdeen, Scotland; s. George Brown; ed. Royal Coll. Chemistry and Sch. Mines; B.S. with 1st class honors, U. London, 1867, D.Sc., 1870; LL.D., Aberdeen U.; m. Ellen Fullarton Henderson, 1872; Teaching asst. chem. dept. U. Aberdeen, 1864-66; lectr. on exptl. sci. and toxicology Liverpool Royal Infirmary Sch. Medicine, 1867-83; head Lancashire County Lab., from 1875; prof. chemistry U. Coll., Liverpool, 1875-83, Victoria U., 1884-1903; Contbr. papers to sci. publs. Died Mar. 14, 1910.

BROWN, James Greenlief, Am. plant pathologist; b. St. Clair, Mich., Nov. 21, 1880; s. George Simeon and Ida Evelyn (Graham) B.; attended Ferris Inst., Big Rapids, Mich.; B.S., U. Chgo., 1916. M.S., 1917, Ph.D., 1925; m. Clara May MaNeil, June 5, 1912; 1 dau., Imogene (dec.), Lab. asst. in botany U. Chgo., 1907-08; research asst. Carnegie Desert Lab., Tucson, 1909-11; instr. biology U. Ariz., Tucson, 1909-15, asst. prof. biology, 1916-19, prof. plant pathology Agrl. Coll., and plant pathologist Agrl. Expt. Sta., 1920-52, also head dept. agr., from 1922. Fellow A.A.A.S. (mem. council 1941); mem. Am. Soc. Plant Physiologists, Am. Phytopathol. Soc. Bot. Soc. Am., Mycol. Soc. Am., Soc. Am. Bacteriologists, Sigma Xi, Phi Beta Kappa, Phi Kappa Phi. Author: Crown Gall on Coniefers; effect of penicillin on crown gall; also various papers on cotton, cactus and other plant diseases. Built machine used for delinting and surface sterilizing cotton seed with sulphuric acid. Died Apr. 1, 1954.

BROWN, James Salisbury, Am. inventor; b. Pawtucket, R.I., Dec. 23, 1802; s. Sylvanus and Ruth (Salisbury) B.; m.. Sarah Phillips Gridley, 1829, 2 daus., 1 son James. Pattern maker for David Wilkinson, Pawtucket, 1817-19; joined Pitcher & Gay, cotton machinery mfrs., 1819, became partner, 1823, gained control of firm, 1842, changed name to Brown Machine Works, continually expanded firm; patented improvement for lathe slide-rest (1st invented by his father), 1820; designed cutter for cutting bevel gears, 1830; patented spl. drilling machine, 1838; produced guns and gun making machines during Civil War, returned to mfg. cotton machinery after war. Died Dec. 29, 1879.

BROWN, Joe Robert, Am. neurologist; b. Mt. Pleasant, Ia., Nov. 24, 1911; s. James Smith and Olive (Smith) B.; B.A., U. Ia., 1933, M.D., 1937; M.S. in Neurology and Psychiatry, U. Minn., 1943; m. Rebecca Frisbee, Aug. 14, 1937; children—Hugh Frisbee, Carolyn Emily, Steven Robert. Chief neurol. service VA Hosp., Mpls., 1946-47, chief neurology, psychiatry service, 1947-49, area med. cons. VA, 1949—; with Mayo Grad. Sch. Medicine U. Minn., 1939-43, 49—, prof. neurology, 1963—; sr. cons. neurology Mayo Clinic, Rochester, Minn., 1966—; mem. neurology grad. tng. grant com. Nat Inst. Neurol. Diseases and Blindness, 1957-61, chmn., 1961. Mem. A.M.A., Am. Acad. Neurology (editorial bd. Jour. Neurology, 1957—, sec., treas., trustee), Nat. Multiple Sclerosis Soc. (mem. med. adv. bd. 1956—), Central Neuropsychiat. Assn. (pres. 1966), Acad. Aphasia (gov. bd. 1966), Assn. for Research in Nervous and Mental Disease, Phi Beta Kappa, Sigma Xi, Alpha Omega Alpha. Author: (with F. L. Darley, A. E. Aronson, R. E. Yoss) Motor Speech Disorder, 1968. Research and publs. in field mechanism and treatment of bulbar poliomyelitis, rehab. of neurologically handicapped, clin. aspects of cerebellar diseases, disorders of motor speech and language. Home: Mail Route 72, Rochester. Office: 200 1st St. S.W., Rochester, Minn. 55901.*

BROWN, John, instrument maker; flourished 1648-95; s. Thomas Brown(e). Designer, maker numerous precision instruments including barometers, baroscopes, dials; asso. Henry Sutton, Walter Hayes; once employed by Samuel Pepys. Author several texts on design and use of instruments he made.

BROWN, John, Brit. physician; b. Buncle, Berwickshire, Scotland, 1735; student Edinburgh (Scotland) U.; M.D., St. Andrews U., Scotland, 1779; m. Miss Lamond, 1765; 4 sons, including William Cullen, 4 daus. Pvt. tutor, Edinburgh; went to London, 1786. Author: Elementa medicinae, 1780; A Short Account of the Old Method of Cure and Outlines of the New Doctrine. Founder of Brunonian system of medicine which alleged most diseases were due to weakness; opposed blood-letting as a treatment. Died Oct. 17, 1788.

BROWN, John Lott, Am. psychologist; b. Phila., Dec. 3, 1924; s. John Lott and Carolyn (Francis) B.; B.S., Worcester Poly. Inst., 1945; M.A., Temple U., 1949; Ph.D., Columbia, 1952; m. Catharine Hertfelder, June 11, 1948; children—Patricia, Judith, Anderson, Barbara. Mem. sci. staff Columbia, 1951-54; head psychology div. Aviation Med. Acceleration Lab., Naval Air Devel Center, Johnsville, Pa., 1954-59, cons. Aero Computer Lab., 1959-64; asst. prof. physiology U. Pa., Phila., 1955-62, asso. prof., dir. grad. physiol. program, 1962-65; mem. research staff Neurol. Inst., U. Freiburg, Germany, 1961-62; dean Kan. State U. Grad. Sch., Manhattan, 1965-66, v.p. acad. affairs, 1966—. Cons. Gen. Electric Missile and Space Vehicle Dept., 1960-61. Fellow Am. Psychol. Assn., Optical Soc. Am.; mem. Am. Physiol. Soc., Psychonomic Soc., A.A.A.S., Sigma Xi, Psi Chi. Research and publs. on vision and neurol. aspects of visual system, human factors involved in space flight. Home: 2383 Grandview Terrace, Manhattan, Kan. 66502.*

BROWN, John Pinkney, Am. arboriculturist; b. Rising Sun, Ind., Jan. 19, 1842; s. Elbridge G. and Adaline (Style) B.; ed. Hanover Coll.; m. Mary E. Stephens, 1868. With Mississippi River service in U. S. steamers. Organizer, sec., treas. Internat. Soc. Arboriculture. Originated system of tree planting by rys. for future timber and tie supply; established model forest farm of 200 acres on which were planted 200,000 young forest trees. Author: Practical Arboriculture, 1906. Editor Arboriculture. Died 1915.

BROWN, Joseph Rogers, Am. inventor; b. Warren, R.I., Jan. 26, 1810; s. David and Patience (Rogers) B.; m. Caroline B. Niles, Sept. 18, 1837, 1 child. Perfected and built linear dividing engine, 1850; perfected vernier caliper reading to thousandths of an inch, 1851, applied vernier to protractors, 1852; became partner (with Lucian Sharpe) in firm J. R. Brown & Sharpe, 1853, incorporated as Brown & Sharpe Mfg. Co., 1868; micrometer caliper, 1867; invented precision gear cutter to make clock gears and to supply his jobbing customers with gears, 1855; greatest achievement was invention of universal grinding machine, patent issued, 1877 (after his death). Died Isles of Shoals, N.H., July 23, 1876.

BROWN, Laurie Mark, Am. physicist; b. Bklyn., Apr. 10, 1923; s. William and Elvira (Fleischman) B.; A.B., Cornell U., 1943, Ph.D., 1951; m. Judith Kobrin, Dec. 27, 1942 (dec. May 1963); children—Joana Lisa, Julie Elena. Faculty physics Northwestern U., 1950—, prof., 1961—. Mem. Inst. for Advanced Study (NSF fellow), Princeton, 1952-53; cons. Argonne Nat. Lab., 1960—.; vis. prof., Vienna, 1966, Rome, 1967. Fulbright Research scholar, Italy, 1958-60. Fellow Am. Phys. Soc. Research and publs. in passage of charged particles in matter, quantum electrodynamics, high energy nuclear physics, elementary particles. Home: 807 Milburn St., Evanston, Ill. 60201.*

BROWN, Lawrason, Am. physician; b. Balt., Sept. 29, 1871; s. William Judson and Mary Louise (Lawrason) B.; A.B., Johns Hopkins, 1895, M.D., 1900; Sc.D., Dartmouth, 1931; m. Martha Lewis Harris, Oct. 8, 1914. Asst. resident physician Trudeau Sanatorium, Saranac Lake, N.Y., 1900-01, resident physician, 1901-12, vis. physician, 1912-14, later cons. physician; cons. Waverly Hills Sanatorium, Louisville; instr. Trudeau Sch. Tb, from 1914. Trustee N.Y. State Hosp. for Incipient Tb, Ray Brook, Potts Meml. Hosp., Livingston; mem. adv. council Henry Phipps Inst., U. Pa., Milbank Found. First pres. Stevenson Soc. Am. Author: Rules for Recovery from Tuberculosis; (with H. L. Sampson) Intestinal Tuberculosis; (with Fred H. Heise) The Lung and Tuberculosis. Died Dec. 1937.

BROWN, Leland Arthur, Am. biologist; b. Marion, O., Nov. 18, 1897; s. Emmett Gibson and Alice (Boyd)

B.; B.S., Denison U., 1922; M.A., U. Pitts., 1924; Ph.D., Harvard, 1927; LL.D., Transylvania U., 1956; m. Lucille Cambier Melvin, Nov. 28, 1923; children—Leland Campbell, Peter Melvin. Instr. zoology U. Pitts., 1923-24; Austin teaching fellow Harvard, 1925-27; asst. prof. State U. Ia., 1927-29; asso. prof. George Washington U., 1929-32; prof. biology Transylvania U., Lexington, Ky., 1932—, dean, 1941-65, v.p., 1958-66, dir. instnl. research, 1965—, curator Sci. Mus., 1956—. Sci. cons. Richmond (Va.) Pub. Schs., 1967. Mem. Am. Soc. Zoologists, Ky. Acad. Sci. (pres. 1943), Assn. Instnl. Research, Am. Assn. Mus. Author: Early Philosophical Apparatus, 1959. Editor: Rafinesque Memorial Papers, 1942. Research and publs. on factors regulating prodn. of males in waterfleas; distbn. and growth in relation to temperature: early philos. apparatus. Home: 420 6th St., Lexington, Ky. 40508.*

BROWN, Milton Herbert, Canadian physician; b. Havelock, Ont., Can., Sept. 6, 1898; s. Johnson and Augusta (Hicks) B.; M.B., U. Toronto, 1924, B.Sc. in Medicine, 1925, M.D., 1928, D.P.H., 1939; m. Eleanor Graydon, June 1, 1929 (dec. 1944); m. 2d, Vera Constance Richardson, June 26, 1950; children—John Peter, Pamela Jane. Asso. dir. U. Toronto (Ont.) Sch. Hygiene, 1955—, prof., head dept. pub. health, 1955—; asst. dir. Connaught Med. Research Labs., Toronto, 1952—; mem. chest clinic Toronto Western Hosp., 1933—. Chmn. panel on infection and immunity def. research adv. com. Def. Research Bd., 1951-65; mem. pub. health research adv. com. Dept. Nat. Health and Welfare, 1960—. Decorated Order Brit. Empire. Diplomate Am. Bd. Preventive Medicine. Fellow Coll. Chest Physicians, Royal Soc. Medicine. Research and publs. on immunology and control communicable diseases and health adminstrn. Home: 42 McRae Dr., Toronto 17. Office: 150 College St., Toronto 5, Ont., Can.*

BROWN, Paul L., Am. soil scientist; b. Ash Grove, Kan., June 1, 1918; s. Fred W. and Antoinette (Lawson) B.; B.S., Kan. State U., 1941, M.S., 1949; Ph.D., Ia. State U., 1956; m. Mary Stewart, Jan. 25, 1946; children—R. Douglas, Richard G. Asst. soil surveyor U. S. Dept. Agr., Kan. State U., 1946-48; soil scientist U. S. Dept. Agr., Ft. Hays Br. Expt. Sta., Hays, Kan., 1948-56, Mont. State U., Bozeman, 1956—. Mem. Am. Soc. Agronomy, Soil Sci. Soc. Am., Canadian Soc. Soil Sci., Sigma Xi, Gamma Sigma Delta. Contbr. articles to sci. jours. Patentee soil moisture probe) pioneer in research on effect of plant populations, row spacings and soil moisture on grain sorghum prodn. in dryland areas; summarized effect of legumes and grasses in dryland cropping system in Great Plains. Home: 1220C N. 8th St. Office: Agr. Bldg., Mont. State U., Bozeman, Mont. 59715.*

BROWN, Paul Lopez, Am. zoologist; b. Anderson, S.C., Feb. 18, 1919; s. Ziney and Ola (Davis) B.; B.S., Knoxville Coll., 1941; M.S., U. Ill., 1948, Ph.D., 1955; m. Julia Mae Parham, Jan. 12, 1944 (dec. 1964); children—Pauletta Louise, Julian Ziney, Gloria Jean, Nanola Katherine. Faculty, So. U., Baton Rouge, 1948-58, asso. prof., 1956-58; prof. Fla. A. and M. U., 1958-59; prof., head dept. biology Va. State Coll., Norfolk Div., 1959—; asst. program dir. NSF, Washington, 1963-64, cons., 1964—. Mem. Am. Soc. Zoologists, Am. Inst. Biol. Scis., Va., Ill. acads. scis., Am.. Micros. Soc. Research and publs. on biology of crayfish. Home: 2825 Woodland Av., Norfolk, Va. 23504.*

BROWN, Percy, Am. röntgenologist; b. Cambridge, Mass., Nov. 24, 1875; s. Isaac Henry and Mary Elizabeth (Kennedy) B.; student Lawrence Sci. Sch. (Harvard), 1893-96; M.D., Harvard, 1900; m. Bernice Mayhew, Dec. 7, 1904. Röntgenologist Carney Hosp., 1903-10, St. Elizabeth's Hosp., 1905-11, L.I. Hosp., Boston, 1906-10; cons. röntgenologist Carney Hosp., 1911-13; instr. röntgenology Harvard Med. Sch., 1911-22; röntgenologist Boston Children's Hosp., Boston Infants' Hosp., 1903-06, 10-22, St. Luke's Hosp., N.Y.C., 1924-29; röntgenologist in chief Western Pa. Hosp., Pitts., 1923. Gold medalist Radiol. Soc. N.A., 1922. Ray Soc., 1923. Fellow A.C.P., Am. Coll. Radiology; mem. A.M.A., Mass. Med. Soc., Boston Soc. Med. Sciences, Am. Urol. Soc., Am. (pres. 1911, Caldwell lectr. 1923), Phila. (hon.) röntgen ray socs., Röntgen Soc. London, Deutsche Röntgen Gesellschaft, N.Y. Röntgen Soc., Röntgen Ray Soc. New Eng., Boston Sci. Soc., Boston Soc. Natural Hist. Author: American Martyrs to Science through the Roentgen Rays, 1935. Co-author of Science of Radiology for First Am. Congress of Radiology, Chicago, 1933. Contbr. to sci. jours., author of sundry monographs on subjects dealing with the X-Ray. Died Oct. 8, 1950.

BROWN, Philip King, Am. physician; b. Napa, Cal., June 24, 1869; s. Henry Adams and Charlotte Amanda (Blake) B.; A.B., Harvard, 1890, M.D., 1893; U. Berlin, 1895-96, Göttingen, 1896; m. Helen Adelaide Hillyer, Mar. 7, 1900; children—Hillyer Blake, Harrison Cabot, Phoebe Hearst, Bruce Worcester. Practiced medicine at San Francisco, from 1893; asst. in nervous diseases U. Cal., 1894, asso. prof. clin. medicine, 1896-98, instr. animal pathology, 1896-99; cons. physician Mt. Zion Hosp., San Francisco; cons. pathologist French Hosp., 1896-1901; asso. in medicine and instr. clin. pathology Cooper Med. Coll, 1899-1902; instr. clin. pathology and exptl. medicine, U. Cal.; med. dir. Southern Pacific Hosp.; attending physician City and County Hospital, 1905-17; founder, med. dir. Arequipa Sanatorium (for tubercular wage earning women), Manor, Cal. One of organizers San Francisco Settlement Assn. and San Francisco Boys' Club. Recorded (with W. Ophüls) a fatal case of severe leukopenia, 1901. Died Oct. 1940.

BROWN, Raymond Russell, Am. biochemist; b. Calgary, Alta., Can., Dec. 23, 1926; s. William Frederic and Anna (Kring-Wilson) B.; B.S., U. Alta., 1948, M.S., 1950; Ph.D. in Physiol. Chemistry, U. Wis., 1953; m. Eleanor J. Springer, Dec. 27, 1952; children—Laura J., Jeffrey F., Douglas E. Came to U. S., 1950, naturalized, 1955. Faculty, U. Alta., 1948-50; faculty U. Wis. Med. Sch., Madison, 1955—, prof. biochemistry, 1965—, acting dir. div. clin. oncology, 1962-64. Mem. Am. Chem. Soc., Am. Assn. Cancer Research, Am. Soc. Biol. Chemistry. Research and numerous publs. on amino acid metabolism in human patients, drug metabolism, vitamin B6 nutrition in man. Home: 2817 Van Hise Av., Madison, Wis. 53705.*

BROWN, Robert, Scottish botanist; b. Montrose, Scotland, Dec. 21, 1773; s. James Brown; ed. Marischal Coll., Aberdeen, also U. Edinburgh. Became army ofcl., Ireland, 1795, London, 1798; naturalist with Capt. Flinder's Australasian expdn. (brought back more than 4000 plant species), 1801-05; librarian Linnean Soc. also to Joseph Banks; apptd. curator bot. dept. Brit. Mus., 1827. Fellow Royal Soc., 1811 (Copley medal 1839); mem. Natural History Soc. Edinburgh, Acad. Scis. Inst. France (fgn. asso.). Author: Prodromus florae novae Hollandiae et insulas Van-Diemen, 1810; General Remarks, Geographical and Systematical on the Botany of Terra Australis, 1814; A Brief Account of Microscopical Observations . . . on the Particles Contained in the Pollen of Plants, . . . , 1828; papers contbd. to si. jours. First Brit. writer to use natural system of classification instead of Linnaean; studied plant physiology of over 4,000 species of plants; investigated impregnation of plants; discovered mode of fertilization and role of pollen, also movement of microscopic particles (Browning movement), 1827; discovered and named cell nucleus, 1831; determined relationships of flower to plant axis, also of flower parts to one another. Died London, Eng., June 10, 1858.

BROWN, Robert Harold, Am. geographer; b. Rochester, N.Y., Sept. 16, 1921; s. Harold Cecil and Marion (Johnson) B.; B.S., U. Minn., 1948, M.A., 1949; Ph.D., U. Chgo., 1957; m. Helene Adeline Zukey, Sept. 1, 1945; children—Suzanne Odette, Kurtis Johnson. Faculty, St. Cloud State Coll., 1949-64; prof., chmn. dept. geography U. Wyo., Laramie, 1964—. Mem. Assn. Am. Geographers (chmn. Gt. Plains-Rocky Mountain div.), Nat. Council for Geog. Edn., Regional Sci. Assn., Phi Delta Kappa. Author: Political Areal Functional Organization, 1957; (with Philip Tideman) Atlas of Minnesota Occupancy, 1961. Research on ways in which men spatially organize earth's surface, attempts to find methods of studying organized area that will lead to scientifically defensible predictions of future, dynamic cartography, hierarchical arrangement of society. Home: 1303 Mitchell St., Laramie, Wyo. 82070.*

BROWN, Robert K., Am. dentist; b. Sharpsburg, Pa., Sept. 22, 1893; s. William Richard and Bella Jane (Dyer) B.; student Indiana State Normal Sch. of Pa., 1908-11; D.D.S., U. of Mich., 1919, M.S., 1928; m. 2d, Inez Fredrica Rieger, Sept. 14, 1938; children—(1st marriage) Robert Benaway, Patricia Ann. Tchr. Pa. pub. schs., 1911-12; practiced dentistry, Cleve., 1919-21 Ann Arbor, Mich., 1935—instr. operative dentistry Sch. of Dentistry, U. Mich., 1921-23, asst. prof., 1923-28, prof. and dir. operative clinic, 1928-35. Fellow Am. Coll. Dentists; mem. Am. Dental Assn., Mich. State and Washtenaw dental socs. Mem. publ. com. The Dental Survey; contbg. editor Jour. of Mich. State Dental Soc. Contbr. articles to dental jours. Died Mar. 28, 1944.

BROWN, Roger William, Am. psychologist; b. Detroit, Apr. 14, 1925; s. Frank H. and Muriel (Graham) B.; B.A., U. Mich., 1947, M.A., 1948, Ph.D., 1952; M.A. (hon.), Harvard, 1962. Faculty, Harvard, 1952-57, 62—, prof. social psychology, 1962—, chmn. dept. social relations, 1967—; faculty Mass. Inst. Tech., 1957-62. Chmn. behavioral scis. study sect. Nat. Insts. Mental Health, 1961-62. Guggenheim fellow, 1965-66. Mem. Am. Acad. Arts and Scis., Am. Psychol. Assn. (pres. div. personality and social psychology 1965-66), New Eng. Psychol. Assn. (pres. 1965-66), Linguistic Soc. Am., Soc. Research on Child Devel. Author: Words and Things, 1958; (with others) Language, Thought, and Culture, 1958; A Study of Thinking, 1956; New Directions in Psychology, 1962; Social Psychology, 1965. Research and publs. in linguistics. Home: 100 Memorial Dr., Cambridge, Mass. 02142.*

BROWN, Ronald Frederick, Am. chemist; b. Washington, Apr. 14, 1910; s. Virgil Lee and Laura Lee (Hoover) B.; B.S., U. Md., 1932; A.M., Harvard, 1937, Ph.D., 1939; m. Allie M. Sandridge, Mar. 24, 1935; children—Karen Allgood, Stephen Mayo. Jr. chemist U. S. Geol. Survey, 1934-36; instr. chemistry Harvard, 1939-40, Purdue U., 1940-42; faculty U. So. Cal., Los Angeles, 1942—, prof. chemistry, 1955—, head dept., 1953-56, 57-63, chmn. faculty senate, 1960, exec. com., 1957-61. Vis. research asso. Cal. Inst. Tech., 1963-64. Fulbright fellow, 1956-57. Fellow A.A.A.S.; mem. Am. Chem. Soc., Am. Assn. U. Profs., Chem. Soc., Sigma Xi, Phi Kappa Phi, Phi Lambda Upsilon, Alpha Chi Sigma. Research and publs. in fields of rearrangements in organic chemistry, synthesis of possible anti-malarials, effects of substitution on ring closure reactions, synthesis of bromoalkyl amines, of dinitriles, effects of free radicals on phenoxyacetic acids, reactions of phenoxyacetic acids and esters, linear enthalpy entropy relation, electronegativity, polarizability. Home: 3827 S. Ridgeley Dr., Los Angeles 90008.*

BROWN, Russell Wilfrid, Am. bacteriologist; b. Gray, La., Jan. 17, 1905; s. John Daniel and Lizzie Elna (Saulsby) B.; B.S., Howard U., 1926; grad. student U. Chgo., 1927-29; M.S., Ia. State U., 1932, Ph.D., 1936; m. Mildred Marguerite McConnell, Oct. 31, 1932. Mem. faculty Tuskegee Inst., 1936—, prof. bacteriology, 1943—, chmn. div. natural scis., 1942-46, dir. Carver Research Found., 1944-57, dean research, 1957-62, v.p., dean grad. program, 1962—; research fellow Ia. State U., 1933-36, research asst., 1942-43; sr. postdoctoral fellow NSF, Yale Sch., Medicine, 1956-57; spl. research mammalian cell and tissue culture and virus host cell relationships. Bd. dirs. Oak Ridge Asso. Univs., 1963—; bd. trustees Stillman Coll., 1965—; trustee Carver Research Found. Recipient Alumni Merit award Ia. State U., 1961; Alumni Distinguished Postgrad. Achievement Sci. award Howard U., 1955. Mem. Am. Soc. Microbiology (pres. S.E. br. 1961-62), A.A.A.S., Nat. Inst. Sci. (pres. 1949-50), Am. Chem. Soc., N.Y. Acad. Scis., Tissue Culture Assn., Am. Soc. Cell Biology, Am. Acad. Microbiology, Sigma Xi, Phi Beta Kappa, Phi Kappa Phi, Beta Kappa Chi, Alpha Phi Alpha, Scabbard and Blade. Contbr. profl. jours. Home: P.O. Box 496, Tuskegee Inst., Ala. 36088.*

BROWN, Samuel, Am. surgeon; b. Rockbridge County, Va., Jan. 30, 1769; s. John and Margaret (Preston) B.; B.A., Dickinson Coll., 1789; med. degree U. Aberdeen (Scotland); m. Catherine Perry, 1809, 2 children. Introduced smallpox vaccination at Lexington, Ky., 1802; prof. theory and practice of medicine Transylvania U., 1819-25; started North Am. Med. and Surg. Jour., 1825; founder Kappa Lambda Soc. of Hippocrates (soc. of men pledged to profl. ideals). Died Huntsville, Ala., Jan. 12, 1830.

BROWN, Samuel Horton, Jr., Am. physician; b. Phila., Nov. 16, 1878; s. Samuel Horton and Cecelia Elizabeth (Greaney) B.; M.D., U. Pa., 1899; m. Margaret Julia Linnane, June 21, 1916; children—Samuel Horton, III, Mary Elizabeth, James Linnane, Franklin Luburg. Resident physician Howard Hosp., 1899-1900; asst. dermatologist Phila. and Univ. Hosps.; dermatologist No., So. dispensaries; cons. dermatologist M.E. Orphanage; asst. ophthalmologist Episcopal Hosp., St. Christopher's Hosp. for Children; ophthalmologist Mt. Sinai Hosp.; asst. ophthalmologist, outpatients department Pa. Hosp.; ophthalmologist 9th and 50th local exemption boards, Phila.; St. Luke's and Children's Hosp., Am. Hosp. for Diseases of Stomach. Author: Eczema, Its Causes, Diagnosis and Treatment, 1906; (with William C. Posey) History of Will's Hospital, Philadelphia, 1931. Asso. editor Am. Year-Book of Medicine and Surgery, 1904-05, Therapeutic Rev., 1904, Annals of Ophthalmology, 1904-06, Am. Medicine, 1906. Editor: Hughes' Practice of Medicine, 7th, 8th and 9th edits., 1904-06. Editor Weekly Roster and Med. Digest. Died June 12, 1940.

BROWN, Sanborn C., Am. physicist; b. Beirut, Lebanon, Jan. 19, 1913 (parents Am. citizens); s. Julius Arthur and Helen (Conner) B.; A.B., Dartmouth, 1935, M.A., 1937; Ph.D., Mass. Inst. Tech., 1944; m. Lois L. Wright, June 21, 1940; children—Peter M., Stanley W., Prudence E. Asst. physics Dartmouth, 1935-37; faculty Mass. Inst. Tech., 1937—, prof. physics, 1962—, asso. dean Grad. Sch., 1963—. Fellow Am. Acad. Arts and Scis. (chmn. Rumford com. 1955-58, chmn. com. ednl. activities 1957-61, sec. 1964-67), Am. Phys. Soc.; mem. Am. Assn. Physics Tchrs. (Distinguished Service citation 1962), Internat. Union Pure and Applied Physics (pres. commn. on physics edn.), Internat. Council Sci. Unions (mem. interunion commn. on teaching sci.), Sigma Xi. Author: Basic Data of Plasma Physics, 1959, rev. edit., 1961; Count Rumford, Physicist Extraordinary, 1962; Introduction to Electrical Discharges in Gases, 1966. Co-editor: International Education in Physics, 1960; Why Teach Physics, 1964; The Education of a Physicist, 1966; Count Rumford on the Nature of Heat, 1967; Basic Data of Plasma Physics, 1967. Editor: Electrons, Ions, and Waves—Selected Works of William Phelps Allis, 1967. Home: 37 Maple St., Lexington, Mass. 02173. Office: 77 Massachusetts Av., Cambridge, Mass. 02139.*

BROWN, Sanger (Monroe), physician; b. Bloomfield, Ont., Can., Feb. 16, 1852; s. Stewart and Catherine (Comer) B.; student Albert Coll., Belleville, Ont., 1872-73; M.D., Bellevue Hosp. Med. Coll. (New York U.), 1880; m. Bella Christy, July 9, 1885; 1 child—Christy. Asst. physician Hosp. for Insane, Ward's Island, N.Y., 1880-81, Danvers (Mass.) State

Hosp. for Insane, 1881; asst. physician Bloomingdale Asylum, N.Y., 1882-85, acting med. supt., 1886; prof. neurology Post-Grad. Med. Sch., Chgo., 1890-—; prof. med. jurisprudence and hygiene Rush Med. Coll., 1892-97; asso. prof. medicine and clin. medicine U. Ill., 1901-06, prof. clin. neurology, 1901. Conducted (with Prof. E. A. Schäfer) series of vivisection expts. on monkeys; described hereditary type spinocerebellar ataxia (Sanger Brown's ataxia). 1886-87. Died Apr. 1, 1928.

BROWN, Sidney George, elec. engr., inventor; b. Chgo., July 6, 1873; s. Sidney Brown; ed. Univ. Coll., London; m. Alice Stower, 1908. Prof. firm S. G. Brown, Ltd., Action; founder, chmn. Telegraph Condenser Co.; fellow Univ. Coll., London. Fellow Royal Soc., Inst. Physics; mem. Inst. Elec. Engrs. Contbr. papers to publs. of sci. socs. Inventor drum cable relay and magnetic shunt which relayed 1st messages over long submarine cables, 1899, also cable magnifying relay and thermo-electric relay; discovered (with Henry Hozier) 1st practical methods of directing Hertz waves (1st beam system), 1899; inventor single point iridium microphonic relay, 1908, granular carbon microphone relay, an improved telephone receiver useful in wireless telegraph (1st Reed telephone), 1910, 1st practical loud speaker for wireless; devised methods by which airships and airplanes could receive wireless messages, 1913; discovered stimulation of oxidation and other chem. actions by catalytic action of high frequency alternating currents, 1914; inventor gyroscopic compass used on board ship, 1914, gyroscopic compass for use in air, artificial horizon, automatic helmsman, gun directional compass, also turning indicator for airplanes, 1929, airplane speed indicator, 1930, electro-megaphone for sea, land, and air, 1931, spl. magnetic compass for airplanes, wireless receiving, 1932-33, liquid microphone sound devices. Died Aug. 7, 1948.

BROWN, Sidney Overton, Am. physiologist; b. Valera, Tex., Jan. 3, 1910; s. James Overton and Bela (Bonnett) B.; B.A., Tex. A. and M. Coll., 1932; Ph.D., U. Tex., 1936; postgrad. U. Chgo., 1940; m. Meta Suche, Aug. 28, 1938; 1 dau., Lorelei Bela. Faculty, Tex. A. and M. Coll., 1936-42, 46-—, prof. zoology, 1949-—, head radiation biology lab., 1959-—; research physiologist Tex. Engring. Sta., 1959-—. Mem. Tex. Acad. Sci. (pres.), Am. Assn. U. Profs. (past pres.), Sigma Xi, Phi Kappa Phi. Research and numerous publs. in physiology, nutrition, radiation biology. Home: 700 Gilchrist E. St., College Station, Tex. 77840.

BROWN, Stimson Joseph, Am. astronomer; b. Penn Yan, N.Y., Sept. 17, 1854; s. John Randolph Brown; student Cornell, 1871-72; grad. U. S. Naval Acad., 1876, at head of class; m. Alice Graham, Nov. 18, 1878; m. 2d, Elizabeth Sharp Pettit, Nov. 12, 1913. Served in U. S. Coast and Goed. Survey, 1879-81; prof. math. USN, 1883-—; astron. dir. U. S. Naval Obs., 1898-1901; dir. Nautical Almanac, 1900-01; on duty at U. S. Naval Acad., 1901, head dept. math. and mechanics June 1907, retired. Observed Catalogue of Stars for the Berliner Jahrbuch, at Annapolis, Md., 1885-87, also at obs. U. Wis., 1887-90; author text book on algebra, analytical geometry, trigonometry, also The Calculus (for midshipmen). Deceased.

BROWN, Sylvanus, Am. inventor; b. Valley Falls, R.I., June 4, 1747; s. Philip and Priscilla (Carpenter) B.; m. Ruth Salisbury, 1 son, James Salisbury. Learned millwright trade; served aboard Continental Navy vessel Alfred, at beginning of Am. Revolution; worked for State of R.I. arsenal; supervised constrn. several grist and saw mills, New Brunswick; made short trip to Europe, returned to Pawtucket, R.I., reestablished machine shop; assisted Samuel Slater in constructing Am.'s 1st practical power spinning wheel, 1790, credited with crucial part in turning Slater's memories of English spinning machines into working model; developed many machines essential to profitable constrn. of textile machinery; possibly 1st to use slide-crest lathe; superintended furnaces in cannon factory, Scituate, R.I., 1796-1801. Died Pawtucket, July 30, 1824.

BROWN, Thomas, Brit. philosopher, psychologist; b. manse of Kilmabreck, Kirkcudbright County, Scotland, Jan. 9, 1778; s. minister of Kilmabreck and (Kirkdale) B.; M.D., U. Edinburgh (Scotland), 1803. Disciple of Dugald Stewart; apptd. coadjutant to Stewart at U. Edinburgh, 1810. Author: Criticism of Erasmus Darwin's Zoonomia (pub. at age 20); also several vols. of poetry; Lectures on the Philosophy of the Mind (pub. posthumously), 1820. Postulated principle of suggestion to explain functions of mind; believed muscular sensation gave unity and significance to perception; first to consider in detail the secondary laws of association. Died Brompton, Kent, Eng., Apr. 2, 1820.

BROWN, Thomas Clachar, Am. geologist; b. Lunenburg, Mass., Mar. 28, 1882; s. James and Mary (Dunlop) B.; B.A., Amherst Coll., 1904; Ph.D., Columbia, 1909; m. Vesta Leland Sweezy, Apr. 12, 1907; children—Jessie Elizabeth, Helen Maron, Richard Leland, Florence Mary. Asst. prof. geology Middlebury (Vt.) Coll., 1909-11, Pa. State Coll., 1911-12; prof. geology Bryn Mawr (Pa.) Coll., 1912-17; farmer, tchr. Fitchburg (Mass.) High Sch., 1918-34. Recipient Walker prize Boston Soc. Natural History, 1932. Mem. Geol. Soc. Am. Made detailed studies of glacial and post glacial history of central Mass. from Nashua Valley to Conn. Valley. Died Feb. 28, 1934.

BROWN, Thomas Richardson, Am. physician; b. Balt., Sept. 11, 1872; s. Thomas R. and Harriet (Carrington) B.; grad. Balt. City Coll., 1889; A.B., Johns Hopkins, 1892, postgrad. 1892-93, M.D., 1897; m. Jean McComb Albert, Nov. 27, 1902; 1 dau., Eleanor Albert. Practiced in Balt., from 1899; asso. prof. medicine, Johns Hopkins; attending phys Union Meml. Hosp., Women's Hosp., Ch. Home and Infirmary. Trustee Johns Hopkins U.; chmn. med. adv. bd. Alfred I. du Pont Hosp. of Nemours Found. Mem. A.M.A., Md. Med. and Chirug. Faculty, Alpha Delta Phi, Assn. Am. Physicians, Am. Gastroenterol. Assn., Phi Beta Kappa. Demonstrated presence of eosinophilia in trichiniasis, 1898. Died Sept. 26, 1950.

BROWN, Wade Hampton, Am. pathologist; b. Sparta, Ga., Oct. 18, 1878; s. George Rives and Laura Virginia (Brown) B.; B.S., U. Nashville, 1899; postgrad. U. Chgo., 1902-03; M.D., Johns Hopkins, 1907; m. Beth Gillies, Oct. 29, 1908; children—Wade Gillies, Elspeth, Wade Hampton. Instr. pathology U. Va., 1907-08; instr. U. of Wis., 1908-10, asst. prof., 1910-11; prof. pathology U. N.C., 1911-13; with Rockefeller Inst. since 1913, asso. mem., 1914-22, mem. sci. staff for med. research since 1922. Mem. Assn. Am. Physicians, Am. Soc. Pathologists and Bacteriologists, Am. Soc. for Exptl. Pathology, Am. Soc. for Pharm. and Exptl. Therapeutics, Soc. for Exptl. Biology and Medicine, A.A.A.S., Sigma Xi. Contbr. papers to med. procs. and jours. Research in biology of syphilitic infections, constitutional factors and phys. environment in relation to heredity and disease. Died Aug. 4, 1942.

BROWN, W(ade) Lynn, Am. psychologist; b. Kennerd, Tex., Jan. 3, 1906; s. Charles Thomas and Emma (Miller) B.; B.A., U. Tex., 1932, Ph.D., 1937; m. Straus Evelena Berthaut, Aug. 30, 1933; 1 dau., Barbard Irene (Mrs. Dean Canty). Faculty, U. Tex., Austin, 1937-—, prof. psychology, 1958-—, head exptl. psychology, 1958-64, dir. radiobiol. lab., 1964. Mem. Am. Psychol. Assn. Contbg. author: Response of the Nervous System to Ionizing Radiation, 1961. Research and numerous articles on effects of ionizing radiation on behavior and perception of Macaca Mullata monkeys. Home: 1412 Gaston Av., Austin, Tex. 78703.*

BROWN, Walter Creighton, Am. biologist; b. Butte, Mont., Aug. 18, 1913; s. D. Franklin and Isabella (Creighton) B.; A.B., Coll. Puget Sound, 1935, M.A., 1938; postgrad. U. Rochester; Ph.D., Stanford, 1950; m. Jeanette B. Snyder, Aug. 20, 1950; children—Pamela H., James C., Julia E. Head, sci. dept. Clover Park High Sch., Tacoma, 1938-42; faculty Stanford, 1949-50, Northwestern U., 1950-53; faculty Menlo Coll., Menlo Park, Cal., 1955-—, dean sci. and math. div., 1960-65, dean instrn., 1966-—. Research asso., lectr. biology div. Stanford, 1955-—; Fulbright prof. Silliman U., P.I., 1954-55. Fellow A.A.A.S., Cal. Acad. Scis.; mem. Am. Soc. Zoologists, Am. Ecol. Soc., Internat. Soc. Tropical Ecology, Am. Soc. Ichthyologists and Herpetologists, Soc. Systematic Zoology, Am. Assn. U. Profs., Sigma Xi. Research and publs. on ecology, distbn., evolution of amphibians and reptiles in tropical islands of Pacific. Address: Menlo Coll., Menlo Park, Cal. 94025; also Div. Systematic Biology, Stanford U., Stanford, Cal. 94305.

BROWN, Walter L., Am. physicist; b. Charlottesville, Va., Oct. 11, 1924; s. Frederick L. and Maude (Towne) B.; student Davidson Coll., 1942-43; B.S. in Physics, Duke, 1945; M.A., Harvard, 1947, Ph.D., 1951; m. Lucie Mae Oakes, June 14, 1946; children—Stephen Lyons, Stuart Nelson, Virginia Marie, Keith Anderson. Mem. tech. staff Bell Telephone Labs., Murray Hill, N.J., 1950-—, head radiation physics research dept., 1957-—. Research on radiation produced defects in semicondrs., properties of semicondr. surfaces, semicondr. nuclear particle detectors, investigation of high energy particles trapped in space in earth's magnetic field. Home: 138 Cambridge Dr., Berkeley Heights, N.J. 07922. Office: Bell Telephone Lab, Murray Hill, N.J.

BROWN, William, Brit. biologist; b. Middlebie, Dumfriesshire, Scotland, Feb. 17, 1888; s. Gavin and Margaret (Broatch) B.; M.A. with 1st class honors, U. Edinburgh (Scotland) U., 1908, B.Sc., 1909; D.Sc., Imperial Coll. Sci. and Tech., London (Eng.) U., 1913; m. Lucy Doris Allen, Jan. 20, 1921; children—Lucy Margaret, Gavin William, Alison Grace Hilda (Mrs. John Beaumont), Elizabeth Mary (Mrs. Ian Merrylees). With Imperial Coll. Sci. and Tech., London U., 1912-53, prof. plant pathology, 1928-53, head bot. dept., 1938-53. Fellow Royal Soc. London; mem. Brit. Mycol. Soc. (past pres.), Assn. Applied Biologists (past pres.), Inst. Biology. Research and numerous publs. on physiology, growth and reprodn. in fungi, physiol. disease in plants caused by fungi and bacteria, specific diseases comml. crops. Home: 93 Church Rd., Hanwell, London W.7, Eng.*

BROWN, William Fuller, Jr., Am. physicist; b. Lyon Mountain, N.Y., Sept. 21, 1904; s. William Fuller and Mary (Williams) B.; B.A. with honors in English, Cornell U., 1925; Ph.D., Columbia, 1937; m. Nancy Shannon Johnson, Aug. 17, 1936; 1 son, Eric Ramsay. Tchr., Carolina Acad., Raleigh, N.C., 1925-27; lectr. physics Columbia, 1928-38; asst. prof. Princeton, 1938-43; staff Naval Ordnance Lab., 1941-45; research physicist Sun Oil Co., 1946-55; sr. research physicist Minn. Mining & Mfg. Co., St. Paul, 1955-57, cons., 1957-63, 65-—; prof. elec. engring. U. Minn., Mpls., 1957-—. Guest prof. Max-Planck Institut für Metallforschung, Stuttgart, Germany, 1963-64; mem. adv. com. on ferromagnetism Office Naval Research, 1949-55. Recipient Meritorious Civilian Service award U. S. Navy, 1945; Fulbright scholar Weizmann Inst. Sci., 1962. Fellow A.A.A.S., Am. Phys. Soc., N.Y. Acad. Scis.; mem. Am. Assn. Physics Tchrs. (mem. Coulomb's law com. 1944-50), Phi Beta Kappa, Sigma Xi, Phi Kappa Phi, Epsilon Chi. Author: Magnetostatic Principles in Ferromagnetism, 1962; Micromagnetics, 1963. Contbg. author: Handbook of Physics, 1958; Handbuch der Physik, Vol. 17, 1956. Contbr. articles to sci. jours. Research on electromagnetics; magnetomechanical effects and ferromagnetic domains; micromagnetics; plasticity and elasticity; dielectrics; applied theoretical physics; statistics. Home: 2033 Fremont Av., St. Paul 55119.

BROWN, William George, chemist; b. Newcastle-on-Tyne, Eng., Nov. 5, 1853; s. William Robert and Jane Gillie (Sanderson) B.; B.S. (Miller scholar), U. Va., 1877; Morgan fellow, Harvard, 1884, U. of Heidelberg (Germany), 1880-81; Ph.D. (hon.), U. of N.C., 1889; m. Isabelle White, Nov. 14, 1895. Pof. chemistry, instr. in geology and mineralogy, E. Tenn. U., 1877-79; prof. gen. and agrl. chemistry U. Tenn., 1879-80, chemistry and mineralogy, 1880-83; instr. chemistry U. Va., 1883-85; prof. chemistry and physics S.C. Mil. Acad., 1885-86; prof. chemistry Washington and Lee U., 1886-94; asst. chemist U. S. Dept. of Agr., 1894-96; prof. chemistry U. Mo., 1896-—, dir. of labs., 1905-10, prof. industrial chemistry, 1910-19. Elected dir. Tech. Sch., Newark, 1884 (declined); elected adj. prof. agr. U. S.C., 1885 (declined). Mem. U. S. Assay Commn., 1913. Editor U. Mo. Studies, 1904-11. Died Aug. 8, 1920.

BROWN, William Henry, Am. botanist; b. Richmond, Va., Oct. 6, 1884; s. John Henry and Julia (Wright) B.; B.S., Richmond Coll., 1906; Ph.D., Johns Hopkins, 1910; m. Mary Angus Blythe, June 3, 1927. Sci. asst. U. S. Fisheries Lab., Beaufort, N.C., 1908; grad. asst. Johns Hopkins, 1908, fellow, 1909-10. Bruce fellow, 1910; bot. investigation in Jamaica, 1910; asst., desert lab. Carnegie Instn., 1910; sci. asst. Mich. Agrl. Expt. Sta., also instr. plant physiology Mich. Agrl. Coll., 1910-11; plant physiologist Bur. Sci., Manila, 1911-23; asso. prof. botany U. Philippines, 1915-18, prof. and head of dept., 1919-24; chief, div. investigation Bur. Forestry, Manila, 1918-20; dir. Bur. Sci., Manila, 1924-33. Author: Vegetation of Philippine Mountains, 1919; A Textbook of General Botany, 1925; Laboratory Botany, 1925; The Plant Kingdom, 1935. Editor: Minor Products of Philippine Forests, Vol. 1, 1920, vols. 2 and 3, 1921. Editor in chief Philippine Jour. Sci. Manila, 1924-33. Died Nov. 9, 1939.

BROWN, William Louis, Jr., Am. biologist; b. Phila., June 1, 1922; s. William L. and Beulah (Brown) B.; B.S., Pa. State U., 1947; Ph.D., Harvard, 1950; m. Doris E. Rutherford, Apr. 6, 1946; children—Dorothy Ann, Alison E., Creighton T. Parker Traveling fellow Harvard, Australia, 1950-51; Fulbright Research fellow U. Melbourne (Australia), 1951-52; asst. curator insects Mus. Comparative Zoology, Harvard, 1952-54; asso. curator, 1954-60, asso. entomology, 1960-—; faculty Cornell U., Ithaca, N.Y., 1960-—, asso. prof. entomology, 1964-—. Recipient Donisthorpe prize Santschi Soc. 1963. Mem. Soc. Systematic Zoology, Soc. for Study Evolution, Am., Kan. entomol. socs., Cambridge Entomol. Club, Sigma Xi. Research and numerous publs. on dacetine ants, systematic revision generic groupings world ants; contbr. to gen. evolutionary theory on geog. variation animals, character displacement, centrifugal hypothesis speciation, role gen. adaption in evolution, evolutionary reduction. Home: 8 Oak Crest Rd., R.D. 1, Ithaca, N.Y. 14850.*

BROWN, William Oscar, Am. sociologist; b. Cistern, Tex., Oct. 18, 1899; s. William Robert and Josephine Irene (Darling) B.; A.B., U. Tex., 1921; M.A., B.D., So. Meth. U., 1924; Ph.D., U. Chgo., 1930; m. Ida Lonstein, Apr. 14, 1936; children—Stephen M., Keith M., Susan N. Faculty, U. Cin., 1925-35; research analyst WPA, 1935-37; asst. prof. Howard U., 1937-43; with OSS and State Dept., 1943-53, chief African br., OSS, 1944-46; dir. African studies, prof. sociology Boston U., 1953-65, prof. emeritus African studies, 1965. Cons. on African affairs State Dept.; consultative dir. Internat. African Inst. Mem. Am. Sociol. Assn., African Studies Assn., Am. Population Assn., Council on Fgn. Relations, N.Y. Acad. Scis., Acad. Polit. and Social Scis. Research and publs. on problems of race conflict, minority problems, use of social scis. in study of background, movements, devels. and problems of modern Africa. Home: 121 Colbourne Crescent, Brookline, Mass. 02146.*

BROWN-SÉQUARD, Charles Edward, French physician, physiologist; b. Port Louis, Mauritius, Apr. 8, 1817; s. Edward and Charlotte (Séquard) Brown; M.D., Paris, 1846; L.L.D., Cambridge (Eng.) U., 1881; m. 3 times; 1 son, 1 dau. Visited U. S., 1852;

returned to Paris, 1853; subdued cholera epidemic, Mauritius, 1854-55; prof. Va. Med. Coll., Richmond, 1855; sci. tchr., Paris, 1855-57; founder, pub. Jour. de Physiologie, 1858-64; physician at Nat. Hosp. for the Paralysed and Epileptic in London; prof. physiology and pathology nervous system, Harvard, 1863-68; prof. pathology, Paris, 1869-72; practiced medicine in N.Y., 1872-76; prof. physiology, Geneva, Switzerland, 1877; prof. exptl. medicine Coll. of France, 1878-94. Lectr. physiology and pathology, Eng.; named fellow Faculty Physicians and Surgeons, Glasgow, Scotland, 1859; founder (with Vulpian and Charcot) Archives de Physiologie, Paris, editor, 1889. Fellow Royal Soc., 1860, Royal Coll. Physicians; mem. Soc. Biology (sec. 1848, pres. 1886), French Acad. Scis., 1886. Contbr. papers to French, Brit., Am. sci. jours. Traced origin of sympathetic nerve fibers in spinal cord; 1st to show epilepsy could be produced in guinea pigs; discovered disease (named after him) characterized by lesion of one lateral half of spinal cord, causing paralysis on one side and sensation on other; studied adrenal glands; suggested use of testicular extracts for rejuvenation, 1889; studies of internal secretions led to successful treatment for myxedema. Died Sceaux, Apr. 1, 1894.

BROWNE, Denis, surgeon; b. Melbourne, Australia, Apr. 28, 1892; s. Sylvester and Anne Catherine (Stawell) B.; ed. Kings Sch., Sydney, Australia; M.D.; m. Moyra Ponsonby, Dec. 10, 1945; children—Clemence, Desmond, Rosemary. Recorder surgery, supt. hosp. for sick children, London, Eng., surgeon emeritus, 1958; Hunterian prof. Royal Coll. Surgeons, 1947, 49, 50, 51. Com. dir. Fountain and Carshalton Hosp. Recipient William E. Ladd Meml. medal in pediatric surgery, 1957. Mem. Royal Coll. Surgeons, Brit. Assn. Pediatric Surgery, French Urol. Soc., Internat. Coll. Surgeons (co-pres.). Contbr. articles to med. jours. Home: 16 Wilton St., London S.W.1. Office: 46 Harley St., London W.1, Eng.

BROWNE, John, Brit. physician; b. probably Norwich, Eng., 1642; ed. St. Thomas's Hosp., London. Became surgeon in ordinary to Charles II, London, 1677; surgeon St. Thomas's Hosp., 1683-91. Author: A Treatise of Preternatural Tumours, 8 vols., 1678. Gave 1st description of cirrhosis of liver, 1685. Died early 18th century.

BROWNE, Patrick, naturalist; b. Ireland, 1720; s. Edward Browne; student botany, phys. sci., Paris, France; M.D., Leyden, Netherlands, 1743. Research geology, botany, natural history Jamaica, B.W.I.; mapped Jamaica, 1755. Author: Civil and Natural History of Jamaica, 1756. Contbr. articles on birds and fishes of Ireland in Exshaw's London Mag, 1774. Noted naturalist. Died Aug. 29, 1790.

BROWNE, Ralph Cowan, Am. roentgenologist, inventor; b. Salem, Mass., Nov. 15, 1880; s. Josiah Hill and Katherine (Cowan) B.; grad. high sch., Salem, 1898; m. Mary Belle Moody, Jan. 15, 1908 (dec. Apr. 1952); m. 2d, Florence May Hart Cox, Nov. 21, 1952. Principally interested in research and invention; became pres. Brown Apparatus Co.; tech. expert L. E. Knott Apparatus Co.; Roentgenologist Salem Hosp. Inventor of elec. system and mechanism adopted by U. S. Govt. in North Sea mine barrage, World War I; inventor Browne portable X-ray apparatus, Browne air-lift mine pump, zincit chalcopyrite detector (used in wireless telegraphy), high resistance transmitters (used in telephony). Died Jan. 1, 1960.

BROWNE, Stanley George, English physician; b. London, Eng., Dec. 8, 1907; s. Arthur and Edith (Lillywhite) B.; M.B., B.S. with honors, London U., 1933, Diploma in Tropical Medicine, 1936, M.D., 1954; m. Ethel Marion Williamson, Nov. 15, 1940; children—Derek Stanley, Alastair Raymond, Christopher Harold. Med. missionary Bapt. Missionary Soc., Yakusu, Belgian Congo, 1936-59, prin. Sch. for Med. Assts., 1937-59, med. supt. Yalisombo Leprosarium, 1950-59; sr. specialist leprologist, dir. leprosy research unit Uzuakoli, Eastern Nigeria, 1959-66; asso. lectr. U. Ibadan, 1960-65; dir. Leprosy Study Centre, London, 1966—. Cons. adviser in leprosy Ministry of Health, 1966—; med. cons. Leprosy Mission, Homes of St. Giles, 1966—; lectr. leprosy, univs., med. schs., schs. tropical medicine. Decorated officer Order Brit. Empire, Chevalier de l'Ordre Royal du Lion, Officier de l'Ordre de Léopold, II; Murchison scholar Royal Coll. Physicians, 1934. Fellow Royal Coll. Surgeons, Royal Coll. Physicians, Royal Soc. Tropical Medicine and Hygiene, Royal Soc. Medicine, Internat. Soc. Tropical Dermatology, Internat. Leprosy Assn. (sec.-treas. 1966); mem. Lepra (med. sec. 1966—). Asso. editor Internat. Jour. Leprosy. Research and numerous publs. in clin. manifestations of filariasis in Congo and discovery of crab as intermediate host, pioneer studies of leprosy in Africa. Home: 16 Bridgefield Rd., Sutton Surrey, Eng. Office: 57A Wimpole St., London, W.1, Eng.*

BROWNE, Sir Thomas, Brit. physician; b. London, Eng., Oct. 19, 1605; s. Anna (Garraway) B.; ed. at Winchester Coll. and Broadgate Hall (now Pembroke Coll.), Oxford, Eng.; B.A., 1626; M.A., 1629; student medicine at Montpellier, France, Padua, Italy; M.D., Leyden, Netherlands, 1633; incorporated M.D., Oxford, 1637; m. Dorothy Mileham, 1641; 12 children, including Edward, Thomas, Elizabeth (Mrs. Lyttleton), Anne (Mrs. Edward Fairfax). Practiced medicine in Oxfordshire; moved to Norwich, Eng., 1637, and practiced there; remained a Royalist during civil war; created "socius honorarius" of Coll. of Physicians, 1664. Knighted, 1671. Author: Religio medici, 1642; Pseudodoxia epidemica: Enquiries into Vulgar Errors, 1646; Hydriotaphia: Urn Burial; The Garden of Cyprus, 1658; Collected Works (editor Simon Wilkin), 1835. Tried to reconcile skepticism of sci. with religious faith; carried out embryological expts.; believed in witches; refused to accept Copernican system of astronomy; made collections of plants and birds' eggs. Died Norwich, Oct. 19, 1682.

BROWNELL, Gordon Lee, Am. physicist; b. Duncan, Okla, Apr. 8, 1922; s. Roscoe David and Mable (Gourley) B.; B.S., Bucknell U., 1943; Ph.D., Mass. Inst. Tech., 1950; div.; children—Wendy, Peter, David, James. Applied physicist Mass. Gen. Hosp., Boston, 1961—; asso. prof. nuclear engring. Mass. Inst. Tech., 1961—. Bd. dirs. Neuroresearch Found.; trustee Retina Found. Fellow Am. Phys. Soc.; mem. A.A.A.S., Health Physics Soc., Biophysics Soc., Radiation Research Soc., Am. Nuclear Soc., Am. Assn. Physicists in Medicine. Author: (with J. Stanburg and D. Riggs) Endemic Goiter, 1956; also numerous articles. Editor: (with G. J. Hine) Radiation Dosimetry, 1958. Developer use of positron emitting isotopes for med. scanning, med. application of reactors; research in radiation dosimetry. Home: 100 Memorial Dr., Cambridge, Mass. 02142. Office: Physics Research Lab., Mass. Gen. Hosp., Boston 02114.

BROWNELL, Lloyd Earl, Am. chem. engr.; b. Potsdam, N.Y., Nov. 8, 1915; s. Earl Harvey and Ida (Kenyon) B.; B.Ch.E., Clarkson Coll., 1937, B. Mech. Engring., M. Chem. Engring., 1939; U. Mich., 1942, Ph.D., 1947; m. Janet Doris Emmons, Aug. 14, 1938; children—Gary Gene, Stephen Bruce, John Charles, Pamela Sue, Carol Cam. Instr. dept. chem. engring., U. Mich., 1942-47, asst. prof., 1947-51, asso. prof., 1951-54, prof., 1954—, supr. Fission Products Lab., 1951-60, dir. AEC-NSF Summer Insts., 1961-67. Cons., IAEA. Mem. Am. Inst. Chem. Engring., Am. Chem. Soc., Am. Nuclear Soc., A.A.A.S., N.Y. Acad. Scis., Am. Ordnance Assn. Author: Radiation Uses in Industry and Science, 1961; (with G. G. Brown) Unit Operations, 1950; (with E. H. Young) Process Equipment Design, 1959. Research and numerous articles on fluid flow in porous media; filtration; design of process equipment; gamma irradiation of food; ballistics. Patentee recovery process of sugar from sugar beets; centrifugal evaporator; automobile tire design. Home: 823 Barton Dr., Ann Arbor, Mich. 48105.*

BROWNING, Carl Hamilton, Scottish bacteriologist, pathologist; b. 1881; ed. Glasgow (Scotland) Acad. and U., M.B., Ch.B. with honors, 1903; M.D. (prizeman) with honors, Glasgow U., 1907, LL.D., 1952; D.P.H., Oxford U. (Eng.), 1913; LL.D., St. Andrews U. (Scotland), 1935. Ofcl. asst. to Prof. Ehrlich, Frankfurt/Main, Germany; dir. Bland-Sutton Inst. Pathology, Middlesex (Eng.) Hosp., 1914-19; prof. Bacteriology U. London (Eng.); Gardiner prof. bacteriology, Glasgow U. and Western Infirmary, 1919-51. Recipient Cameron prize, Edinburgh (Scotland) U., 1936. Carnegie Travelling fellow, pathology, 1904. Fellow Royal Soc., 1938, Royal Coll. Physicians, Gasgow, Coll. Pathologists; mem. Path. Soc. Author: Chemotherapy in Trypanosome Infections, 1908; (with others) Studies in Immunity, 1909, Recent Advances in the Diagnosis and Treatment of Syphilis, 1924, Textbook of Bacteriology, 11th edit. Muir and Ritchie's Manual, 1949; Applied Bacteriology, 1918; Immunochemical Studies, 1925; Chemotherapy with Antibacterial Dyestuffs in Experimental Chemotherapy, vol. 2, 1964. Research, publs. on chemotherapy, immunology.

BROWNING, Henry Charles, endocrinologist; b. Nottingham, Eng., Feb. 25, 1912; s. Frederick and Catherine (Durham) B.; B.Sc. with 1st class honors, U. Bristol (U.K.), 1938, Ed. Dip., 1939, Ph.D., 1942; m. Claire Harmon Wise, Oct. 17, 1961; children—Nicole Browning, Nadine Browning; stepchildren—Lucinda Wise, Margaret Wise. Came to U. S., 1946, naturalized, 1952. Lectr. zoology U. Sheffield (Eng.), 1941-43; USPHS sr. cancer research fellow Nat. Cancer Inst., Bethesda, Md., 1946-47; asst. prof. anatomy, research asso. pathology Yale Med. Sch., 1947-52; asso. prof. anatomy, U. Ind. Med. Sch., Bloomington, 1952-53, U. P.R., San Juan, 1953-54; faculty dental br. U. Tex., Houston, 1954—, prof. endocrinology, 1957—. chmn. dept. anatomy, 1958—; vis. prof. Baylor U. Med. Sch., Houston, 1954—, U. St. Thomas, Houston, 1958—. Mem. Am. Zool. Soc., Am. Inst. Biol. Scis., Endocrine Soc. Am. Genetics Assn., Am. Assn. U. Profs., Am. Assn. Anatomists, Soc. for Exptl. Biology and Medicine, Am. Assn. for Cancer Research, Tex. Acad. Scis., Sigma Xi. Research and numerous publs. on hormonal influence on cancer of endocrine organs; role luteotropic hormone in mammals; ovarian endocrinology. Home: 1201 Archley St., Houston 77025. Office: U. Texas, Dental Br., P.O. Box 20068, Houston 77025.

BROWNING, John M., Am. inventor; b. Ogden, Utah, Jan. 21, 1855; s. Jonathan and Elizabeth (Caroline) B. Made first gun at 13, of scrap iron in father's gunshop; patented breech-loading rifle, 1879, repeating rifle, 1884, box magazine, 1895; automatic guns adopted by European govts.; automatic pistol adopted by U. S. Govt., 1908, machine guns and machine rifle by U. S. Govt., 1918. Died Nov. 26, 1926.

BROWNING, Philip Embury, Am. chemist; b. Rhinebeck, N.Y., Sept. 9, 1866; s. William Garretson and Susanna Rebecca (Webb) B.; A.B, Yale, 1889, Ph.D., 1892; postgrad. U. Munich, 1893-94, Sorbonne, 1913-14; m. Elizabeth Sophia Bradley, Dec. 12, 1899. Asst. in chemistry Yale, 1889-93, instr., 1894-98, asst. prof., 1898-1929, asso. prof., 1929-32, curator chem. exhibit, from 1929. Author: An Introduction to the Rarer Elements, 1903, 08, 12; (with F. A. Gooch) Notes on Qualitative Analysis, 1898, Outlines of Qualitative Chemical Analysis, 1906. Died Jan. 2, 1937.

BROWNLEY, Floyd Irving, Jr., Am. chemist; b. Atlanta, Jan. 1, 1918; s. Floyd Irving and Ruth (Ballentine) B.; B.S., Wofford Coll., 1939; M.S. (fellow), Va. Poly. Inst., 1941; Ph.D. in Chemistry (fellow), Fla. State U., 1952; D.Sc., Wofford Coll., 1966; m. Martine Newlin Watson, July 17, 1943; children—Martine Watson, Karen Ruth. Faculty, Clemson (S.C.) Coll., 1939-40, 45—, prof. chemistry, 1952—, dept. head, 1959; chemist Hercules Powder Co., 1941-42. Sr. postdoctoral fellow Univ. Coll., Dublin, Ireland, 1962. Mem. Am. Inst. Chemists, Chem. Soc., Sigma Xi, Phi Kappa Phi. Author: Laboratory Exercises in General Chemistry, 1960. Research and publs. on analytical chemistry fluorine. Home: 100 Bradley St., Clemson, S.C. 29631.*

BROXON, James William, Am. physicist; b. nr. Jefferson Twp., Ind., July 12, 1897; s. William Chester and Victoria Ann (Gillespie) B.; A.B., Wabash Coll., 1919; M.A., U. Minn., 1920, Ph.D., 1926; m. Vera Maude Peacock, Dec. 16, 1922; children—William David, Patricia Jane. Asst. in physics Wabash Coll., Crawfordsville, Ind., 1917-18, instr. German, 1917-18, instr. math., 1918-19; teaching fellow in physics U. Minn., 1919-22, instr. physics, summer 1921; faculty U. Colo., Boulder, 1922—, prof. physics, 1929-63, prof. emeritus, 1963, chmn. dept., 1954-56; instr. Yale, 1924-25; research asso., metall. lab. U. Chgo., 1943-44. Recipient Robert L. Stearns award Asso. Alumni of U. Colo., 1964. Fellow Am. Phys. Soc.; mem. Colo.-Wyo. Acad. Sci. (pres. 1952-53), Am. Assn. Physics Tchrs., Am. Assn. U. Profs. (pres. U. Colo. 1935-36, 47-48), Phi Beta Kappa (pres. U. Colo. 1955-58), Sigma Xi, Sigma Pi Sigma, Lambda Chi Alpha. Author: (textbook) Mechanics, 1960. Research and publs. on cosmic rays; demonstrated relations of cosmic ray intensities and bursts to solar and geophys. variables; derived law of radial dependence of sunspot magnetic fields. Home: 945 14th St., Boulder, Colo. 80302.*

BROYLES, Carter D., Am. physicist; b. Eckman, W.Va., Nov. 1, 1924; s. William Allen and Martha (Pigg) B.; B.S. in Physics, U. Chattanooga, 1948; Ph.D., Vanderbilt U., 1952; m. Patricia Ridges, Apr. 30, 1955; children—Suzanne Marie, Stephen Carter. Research asso. Vanderbilt U., Nashville, 1952; research staff Sandia Corp., Albuquerque, 1952—, supr. radiation physics div., 1957-64, mgr. high altitude nuclear burst physics dept., 1964—; sci. dir. program 800, Def. Atomic Support Agy., Albuquerque, 1961-62. Recipient Meritorious Civilian Service award Dept. Def., 1964. Fellow Am. Physics Soc.; mem. Am. Assn. Physics Tchrs., Sigma Xi. Investigation of fluid dynamics; nuclear radiation measurements, plasma physics; magnetic spectrometer studies of radioactive isotopes.

BROZEK, Arthur, geneticist; b. Bohemia, Mar. 30, 1882; s. Hypolit and Mary (Loula) B.; student natural history, math. U. Prague (Czechoslovakia); m. Mary Jarmila Müldner, 1918. Prof. genetics Charles U., Prague. Mem. Czech Assn. Sci., Eugenic Soc. (Czechoslovakia), Bot. Soc., Am. Genetic Assn. Author: The History of Heredity, 1930; papers dealing with Mendelian heredity, especially on plant genus Mimulus. Address: Plant Physiol. Inst., Dept. Genetics, Charles U., Prague II, Banatecka str. 433, Czechoslovakia.

BROZEK, Josef (Maria), psychologist; b. Melnik, Bohemia, Aug. 14, 1913; s. Josef F. and Filomena (Sourek) B.; Ph.D., Charles U., Prague, Czechoslavakia, 1937; postgrad. U. Pa., U. Minn.; m. Eunice Marie Magnuson, Mar. 23, 1945; children—Josef Tomas, Margaret Maria, Peter Maria. Came to U. S., 1939, naturalized, 1945. Scientist, prof. lab. physiol. hygiene U. Minn., 1941-58; prof., head dept. psychology Lehigh U., Bethlehem, Pa., 1959-63, research prof., 1963—. Nat. lectr. Sigma Xi, 1962. Mem. Am. Psychol. Assn., Am. Assn. Phys. Anthropology, Am. Physiol. Soc., History of Sci. Soc. Author: (with A. Keys, others) The Biology of Human Starvation, 1950; author, editor Body Measurements and Human Nutrition, 1956; Symposium on Nutrition and Behavior, 1957; (with A. Henschel) Techniques for Measuring Body Composition, 1961; Body Composition, 1963; Human Body Composition, 1965; The Biology of Human Variation, 1966. Research and numerous publs. on characterization of psychol. changes in nutritional deficiencies; composition of human body, Soviet and East European (slavic) psychology, history of psychology. Home: 265 E. Market St., Bethlehem, Pa. 18018.*

BRUCE, Archibald, Am. physician, mineralogist; b. N.Y.C., Feb. 1777; s. William and Judith (Bayard) B.; A.B., Med. Faculty Columbia, 1797; M.D.,

U. Edinburgh (Scotland), 1800; m. 1803. Secured charter Coll. of Physicians and Surgeons, N.Y.C., 1807, mem. faculty (1st Am. prof. materia medica and mineralogy), 1807-11; founder Am. Mineralogy Jour., 1810; mem. faculty Queens Coll., 1812-18; mem. most leading scientific socs.; an original mem. N.Y. Hist. Soc.; bructie metal was named for him (a magnesium hydroxide which he discovered in N.J.); discovered deposits of zinc oxide, Sussex County, N.J. Died N.Y.C., Feb. 22, 1818.

BRUCE, David, Scottish physician; b. Scotland, flourished circa 1660; s. Andrew Bruce; M.A., St. Andrews, Scotland; student physics Paris, Montpellier, France; M.D., Valence, France, 1657. Became doctor of physics, Oxford, Eng., 1660; physician to Duke and Duchess of York; later settled in Edinburgh, Scotland. Mem. Coll. Physicians, orig. mem. Royal Soc., London.

BRUCE, David, physician, bacteriologist; b. Melbourne, Australia, May 29, 1855; s. David and Jane (Hamilton) B.; ed. U. Edinburgh (Scotland), 1876-81, M.B., C.M., M.D., 1881; m. Mary Elizabeth Steele, 1883. Asst. to Dr. Herbert Stone; entered Army Med. Service, 1883; served in Malta, 1884-88; worker Robert Koch's lab., Berlin, Germany, 1888; asst. prof. pathology Army Med. Sch., Netley, Eng., 1889-94; subsequently in Pietermaritzburg, Natal, then Zululand; returned to Eng., 1901; apptd. supr. of Royal Soc. commn. to investigate sleeping sickness in Uganda, 1903; head Malta fever commn., 1904; promoted surgeon-gen., 1912; comdt. and dir. research on trench fever and tetanus, Royal Army Med. Coll., Millbank, 1914-18. Knighted, 1908. Recipient Mary Kingsley medal Liverpool Sch. Tropical Medicine, 1905, Leeuwenhoek medal, 1915. Manson medal Royal Soc. Tropical medicine, Buchanan medal, 1922, Albert medal, 1923. Fellow Royal Soc., 1899 (Royal medal 1904); mem. French Acad. Scis., 1918. Discovered Brucella melitensis bacterium causing Malta fever and traced it to milk of Maltese goats, 1887; discovered cause of sleeping sickness and studied its transmission by tsetse fly, 1895; discovered cause of nagana (disease of horses and cattle in Central Africa), 1895; investigated Mediterranean fever, 1904-06. Died London, Eng., Nov. 27, 1931.

BRUCE, Everend Lester, Canadian geologist; b. Toledo, Ont., Can., 1884; B.Sc., Queens U., 1909; M.A., Columbia, 1912, Ph.D., 1915; m. Mrs. H. C. Horwood, 1923; children—Douglas, Geoffrey. With Geol. Survey of Can., 1913-19; prof. minerology Queens U., from 1919, head dept., from 1944. Fellow Geol. Soc. Am. (v.p. 1940, 42, pres. 1943), Royal Soc. Can. Author: Mineral Deposits of the Canadian Shield. Died 1949.

BRUCE, James, Scottish explorer; b. Kinnaird, Scotland, 1730; s. David and Marion (Graham) B.; ed. Edinburgh U.; m. Adriana Allen, 1754. Engaged in wine merchants' trade, 1754-61; studied ancient languages; visited Spain, Portugal, and Netherlands, 1755; Brit. consul, Algiers, also commd. to study Algierian ruins, 1763-65; travelled in Mediterranean countries. 1765-68, Africa, 1768-72; reached Gondar, Ethiopia, also source of Blue Nile in Ethiopia, 1770; returned to Eng., 1774. Author: Travels To Discover the Source of the Nile in the Years 1768-1773, 1790. Made 1st accurate drawings of Paestum; (with Luigi Balugani) sketched ancient ruins of Palmyra and Baalbek. Died Kinnaird, Apr. 27, 1794.

BRUCE, Sir John, Scottish surgeon; b. Dalkeith, Scotland, Mar. 6, 1905; s. John and Elizabeth (Clapperton) B.; M.B., U. Edinburgh, hon. Ch.B.; m. Mary Craig, July 6, 1935. Surgeon, Edinburgh Royal Infirmary; Regius prof. surgery clinic U. Edinburgh. Hon. fellow Royal Coll. Surgeons Gt. Britain, Glasgow, Ireland, A.C.S., Coll. Physicians, Surgeons and Gynaecologists of South Africa, Royal Coll. U. Surgeons of Denmark; mem. Royal Coll. Surgeons of Edinburgh (past pres.). Author: Manual of Surgical Anatomy, also articles. Home: 26 Moray Pl. Office: University Medical School, Teviot Pl., Edinburgh, Scotland.

BRUCE, William Robert, biophysicist; b. Hamhung, Korea, May 26, 1929; s. George Findlay and Ellen (Tate) B.; B.Sc., U. Alta., 1950; Ph.D., U. Sask., 1956; M.D., U. Chgo., 1958; m. Margaret MacFarlane, June 15, 1957; children—Graham Douglas, Lynda Jeanne. Faculty, U. Toronto (Ont., Can.), 1959—, prof. biophysics, 1966—; mem. physics sect. Ont. Cancer Inst., Toronto, 1959—. Mem. Am. Assn. for Cancer Research. Research and publs. on x-ray and gamma ray penetration, control of red blood cell prodn., action of anticancer agts. on normal marrow stem cells and leukemia stem cells. Home: 4 Marshfield Ct., Don Mills, Ont. Office: 500 Sherbourne St., Toronto 5, Ont., Can.*

BRUCE, William Speirs, Brit. naturalist, explorer; b. Aug. 1, 1867; s. Samuel Nobel Bruce; ed. U. Edinburgh; LL.D., Aberdeen; m. Jessie Nigg, 1901; 1 son, 1 dau. Lectr. geography Heriot Watt Coll., 1899-1901, from 1917; lectr. Ch. of Scotland Tng. Coll., Edinburgh, 1899-1901; George Heriot Research fellow U. Edinburgh, 1900-01; dir. Scottish Oceanographical Lab., Edinburgh; asst. Challenger Expdn. Commn.; asst. in zoology Sch. Medicine, Royal Coll., Edinburgh; naturalist Scottish Antarctic Expdn., 1892-03; in charge Ben Nevis Obs., 1895, 96; zoologist Jackson-Harmsworth, Polar Expdn., 1896, 97; naturalist Maj. Andrew Coats' Expdn. to Novaya Zemlya, Wiche Islands, also Barent's Sea, 1898; naturalist H.S.H. the Prince of Monaco's Expdn., Spitsbergen, 1898, 99, 1906; leader Scottish Nat. Antarctic Expdn. (S.Y. Scotia), 1902-04; discovered 150 miles of coast line of Antarctica (Coats' Land); bathymetrically surveyed S. Atlantic Ocean and Weddel Sea; explored and surveyed Prince Charles Foreland, other parts of Spitsbergen, 1906, 07, 09, 12, 14, 19, 20. Recipient Gold medal Royal Scottish Geog. Soc., 1904, Royal Gold medal (Patron's) Royal Geog. Soc., 1910, Livingstone Gold medal Am. Geog. Soc., 1920. Fellow Royal Soc. Edinburgh (Neill prize, Gold medal 1911, 13); mem. New Zealand Inst. (hon.), Institut Oceanographic (mem. com. perfectionnement Paris), Gesellschaft für Erdkunde zu Berlin (hon.). Author: Scientific Results of the Voyage of S.Y. Scots, 1902-04; Polar Exploration; also papers. Died Oct. 31, 1921.

BRUCH, Karl Wilhelm Ludwig, German anatomist, physiologist; b. Mainz, Germany, May 1, 1819; student, Giessen, and Berlin, Germany, doctors degree, 1842. Described lamina basalis chorioideae (Bruch's membrane), 1844. Died Jan. 4, 1884.

BRÜCHE, Ernst, German physician; b. Hamburg, Germany, Mar. 28, 1900; s. Franz and Elsie Brüche; ed. Technische Hochschule, Danzig; diploma in engring.; Ph.D.; m. Dorothee Lilienthan, 1929; children—Erika, Barbara, Hanni. Asst., prof. agrege Technische Hockschule, Danzig; head lab., later dir. AEG Forschungs Inst., dir. AEG Research Inst.; dir. Suddeutschen Lab., Mosbach; dir. lab. physics, dir. Physik-Verlag G.m.b.H.; hon. prof. U. Berlin, 1943, Technische Hochschule Karlsruch, 1961. Recipient Leibniz silver medal. Mem. numerous German and internat. socs. of physics. Editor: (with Scherzer) Geometrische Elektronenoptic; (with Recknagel) Elektronengeräte. Worked on 1at elec. electron microscope, perfected geometric electronic optic. Home: Baden Schmelzgarten 7. Office: Baden Pflazgaf Otto Strasse, Mosbach, Germany.

BRUCK, George, physicist; b. Budapest, Hungary, Oct. 20, 1904; s. Maximilian and Elizabeth (Himmler) B.; Fed. Degree in Applied Physics, Tech. U. Vienna (Austria), 1927; m. Edith Ann Voigt, Dec. 17, 1929. Came to U. S., 1939, naturalized, 1944. Asst. chief engr. Compagnie des Lamps, 1929-36; resident engr. Fabbr. Ital. Magn. Marelli, 1936-41; research engr. Crosley Corp., 1941-43; dir. spl. studies Nat. Union, 1943-44; chief engr. Hudson Am. Corp., 1944; research engr. Raytheon Mfg., 1944-46; chief photo engr. Spltys., Inc., Syossett, L.I., N.Y., 1946-53; pres. Bruck Industries, Inc., 1953-54; cons. Radiation, Inc., Melbourne, Fla., 1953-54; chief scientist Electronic div. AVCO Corp., Cin., 1954—. Fellow I.E.E.E.; mem. Am. Phys. Soc. (sr.). Research and numerous publs. on electro-optics, navigational devices, mil. electronics. Patentee electronic devices. Home: 1019 Brayton Av., Cin. 45215. Office: AVCO Corp., Cin. 45241.*

BRUCK, Henri D., physicist; b. Rosslau, Germany, May 19, 1909; s. Paul and Elisa (Metzger) B.; Doctor der Naturwissenschaften, U. Hamburg (Germany), 1935; Docteur ès Sciences, U. Paris, 1942; m. Francoise B. Laburie, Apr. 26, 1948. Compagnie generale, de T.S.F., 1942-48; head dept. Commissariat à l'Energie Atomique, France, 1948—; prof. Institut National des Sciences et Techniques Nucléares, 1962-Decorated Chevalier de l'Ordre National du Mérite. Mem. Société Française de Physique. Author: Accélérateurs Circulaires de Particules, 1966; also articles. Research on prototype of 1st French electron microscope, gen. design and tube first French Van de Graaff accelerator; gen. design and part proton synchrotron; design and bldg. electron positron storage ring. Home: 11bis Av., Carnot 92, Sceaux, France. Office: Boite Postale N°2, 91 Gif sur Yvette, France.*

BRÜCK, Hermann Alexander, astronomer; b. Berlin, Germany, Aug. 15, 1905; s. H. H. and Margaret (Weyland) B.; student U. Kiel (Germany), 1924-25, U. Bonn., 1925; D.Phil., U. Munich (Germany), 1928; D.Phil.Habil., U. Berlin, 1935; Ph.D., U. Cambridge (Eng.), 1941; m. Irma Waitzfelder, Sept. 8, 1936 (dec. 1950); children—Mary Cecilia, Peter Michael; m. 2d, Mary T. Conway, Nov. 21, 1951; children—Anne Margaret, Catherine Patricia, Andrew Alexander. Astronomer, Potsdam (Germany) Astrophys. Obs., 1928-36; research asso. Vatican Obs., 1936; observer Solar Physics Obs., Cambridge, 1937; John Couch Adams astronomer Cambridge U., 1943-46; asst. dir. Cambridge Obs., 1946-47, dir. Dunsink Obs., Dublin, Ireland, also prof. astronomy Dublin Inst. for Advanced Studies; 1947-57; astronomer royal for Scotland, Regius prof. astronomy U. Edinburgh, 1957—. Fellow Royal Astron. Soc.; Royal Soc. Edinburgh; mem. Royal Irish Acad., Edinburgh Math. Soc., Internat. Astron. Union, Astronomische Gesellschaft, Pontifical Acad. Scis. (mem. council 1964), Acad. Scis. Mainz (fgn.). Research and numerous articles on spectroscopy, physics of sun, structure of galaxy. Address: Royal Obs., Edinburgh 9, Scotland.*

BRUCK, Richard Hubert, mathematician; b. Pembroke, Ont., Can., Dec. 26, 1914; s. Joseph Hubert and Hellise (Workman) B.; B.A., U. Toronto (Ont.), 1937, M.A., 1938, Ph.D. (Univ. fellow), 1940; m. Helen Olive Glorine Troop, June 29, 1940. Came to U. S., 1940, naturalized, 1948. Instr., U. Ala., 1940-42; faculty math. U. Wis., Madison, 1942—, prof., 1952—, Guggenheim fellow, Univ. research fellow, 1946-47; vis. prof. statistics U. N.C., Chapel Hill, 1963-64. Cons. Rand Corp., 1961—; Fulbright lectr., Australian Nat. U., Canberra, 1963; research lectr. Can. Math. Congress, Saskatoon, 1963. NSF grantee, Germany, Eng., 1959-60. Mem. Am. (asso. sec. 1945-48, asst. editor Bull. 1945-47, Proc. 1955-57, mem. council 1959), London math. socs., Math Assn. Am. (Chauvenet prize 1956). Asso. editor Jour. Algebra, 1963—; asst. editor Canadian Math. Jour., 1964—. Author: A Survey of Binary Systems, 1958; various research papers in algebra and geometry. Home: 542 Charles Lane, Madison, Wis. 53711.*

BRÜCKE, Ernst Theodor von, see von Brücke, Ernst Theodor.

BRÜCKE, Ernst Wilhelm von, see von Brücke, Ernst Wilhelm.

BRUCKE, Franz, Austrian pharmacologist; b. Jan. 15, 1908; s. Ernst Theodore and Paula (Rölfs) B.; ed. univs. Innsbruck, Vienna, Berlin, Leipzig; M.D.; m. Gertraud Rittler, 1939. Asst. to profs. P. Rona and C. Neuberg, Berlin; med. asst. to O. Löwi, Graz, Austria, 1932-34; dir. asst. 1st med. clinic U. Vienna, 1934, asst. Inst. Pharmacology, 1934-46, dir., 1946, asso. prof. pharmacology, toxicology and prescription medicine, 1948, also dean. Rockefeller scholar, London, Cambridge, 1936-37. Mem. Vienna Health Council. Mem. Acad. Sci. Research of U. Vienna (corr.). Collaborator: Archives International, Pharmacological Rev. Address: Linke Wienzeile 12, Vienna VI, Austria.

BRÜCKE, Hans Gottfried von, Austrian surgeon; b. Leipzig, Dec. 31, 1905; s. Ernst and Paula (Roelfs) B.; ed. U. Innsbruck, U. Vienna; M.D.; m. Helga Hintner, Jan. 11, 1932; children—Ursula, Peter, Elisabeth. Asso., Johns Hopkins Hosp., Balt.; with surg. clinic, Vienna, Frauenklinik, Vienna, Innsbruck; Graz; with Surg. Inst. Graz, later 1st asst.; physician in chief Mürzzuschlag and Wagna hosps. Fellow Internat. Coll. Surgeons; mem. Internat., German, Austrian socs. surgery, Austrian Soc. Anesthesia. Author over 100 sci. works on surgery and physiology, including Die Chirurgie der Gallenwege, 1956. Home: Johann-Fux-Gasse 8, Graz, Austria. Office: Landeskrankenhaus, Wagna-Steiermark, Austria.

BRÜCKNER, Arthur Bernhard, ophthalmologist; b. Dorpat, Livonia (now Tartu, Estonia), Aug. 24, 1877; s. Alexander and Lucie (Schiele) B.; ed. univs. Heidelberg, Jena, Munich, Wurtzburg, Leipzig; M.D.; m. Hertha Teichmüller, May 5, 1901; m. 2d, Ruth Freudenberg, May 15, 1953; children—Wolfgang, Werner, Roland, Ekkehart. Prof. agrege, Wurtzburg, Königsberg, 1907, Berlin, 1912; full prof. at Königsberg, 1910; asso. prof. at Jena, 1921, Basel, 1923. Mem. Swiss Soc. Ophthalmology, Scottish Ophthal. Soc. Author: (with F. Schieck) Grundriss der Augenheilkunde; Cytologie des Auges, 1919; Grundzüge der Brillenlehre, 1924; Physiologic and Clinical Ophthalmologic Problems, 1932; Lehrbuch der Augenheikunde, 1948. Mem. (rev.) Ophthalmologica, 20 years. Address: Oberer Batterieweg 74/76, Basel, Switzerland.

BRÜCKNER, Eduard, geographer, climatologist; b. Jena, Germany, July 29, 1862; s. Alexander and Lucie (Schiele) B.; studied in Dorpat, Dresden and Munich, 1881-85, degree in 1885; m. Ernestine Stein, 1888; 1 dau. Taught phys. geography at Hamburg marine obs.; prof. at U. Bern, 1888, U. Halle, 1904, U. Vienna, 1906-27; founder jour. Zeitschrift für Gletscherkunde, 1906, editor until 1927. Mem. Internat. Glacier Commn. for Germany and Austria, Vienna Acad. Scis., Academia dei Lincei (Rome), Natural Sci. Orient Assn. (pres.), Austrian Adriakommission. Author: (with A. Penck) Die Alpen im Eiszeitalter, 1903-09; Klimaschwankungen seit 1700 nebst Bemerkungen über die Klimaschwankungen der Diluvialzeit (his most important book, pub. in Pencks geographische Abhandlungen 4, 1890). Contbr. numerous articles to profl. jours. Studied effect of ice age on earth's surface formations, particularly in Alps, also climate fluctuations; results from meteorological, economic-statistical, ethnographic and hygienic sources; Brückner Period (discovered 1887) was time of 35-year fluctuations of damp/cold and dry/warm epochs. Died Vienna, Austria, May 20, 1927.

BRUDEVOLD, Finn, dentist; b. Gjovik, Norway, June 12, 1910; s. Peder and Ingrid (Haugom) B.; L.D.S., State Dental Sch., Oslo, Norway, 1932, D. Honoris Causa Odontology, 1965; D.D.S., U. Minn., 1940; M.S., U. Rochester, 1954; A.M., Harvard, 1958; m. Esther Asher, June 27, 1941; children—Anne, Catherine, Christine. Came to U. S., 1939, naturalized, 1949. Practice dentistry, 1932-39; instr. State Dental Sch., Oslo, 1937-38, Tufts U. Sch. Dentistry, 1942; sr. dental surgeon Norwegian Pub. Health Service, 1944-45; faculty Tufts U. Sch. Dentistry, 1946-48, asst. prof., 1948-49; asst. prof. dental research U. Rochester, 1949-58, dir. research Eastman Dental Dispensary, 1949-58; chief preventive dentistry, prof. dentistry, dir. gen. clinic Forsyth Dental Center, Boston, 1958—; prof. dentistry Harvard Sch. Dental Medicine, 1958—. Hon. fellow Acad.

Dentura Prosthetics, 1951-52. Corr. mem. European Orgn. for Research on Fluorine and Dental Caries Prevention, Norwegian Dental Assn.; fellow A.A.A.S.; mem. Am. Dental Assn., Internat. Assn. Dental Research, Mass. Dental Soc., Omicron Kappa Epsilon. Research and numerous articles on composition human teeth, chemistry enamel and its relation to saliva and to composition enamel, ingested or topically applied fluoride to increase caries resistance. Home: 284 Woodward, Waban, Mass. 02168. Office: 140 Fenway, St., Boston 02115.*

BRUDUS, Dionysius, physician; b. Portugal, 1478; court physician to King Manuel; prof. U. Paris. Author criticism of Pierre Bissot for his opinion on blood letting; De Ratione Victus, 1544; work on diet and fevers. Died 1522.

BRUDZINSKI, Josef, Polish physician; b. Bolewo, Poland, Jan. 26, 1874; head rector Warsaw (Poland) U. Discovered 2 signs of meningitis (when neck of patient is bent, flexure movements of ankle, knee and hip occur, also when passive flexion of lower limb on one side is made, similar movement occurs in opposite limb, now called signs of Brudzinski). Died Warsaw, Dec. 18, 1917.

BRUECKNER, Keith A(llan), Am. theoretical physicist; b. Mpls., Mar. 19, 1924; s. Leo John and Agnes (Holland) B.; B.A., U. Minn. 1945; M.A., 1947; Ph.D., U. Cal. at Berkeley, 1950; m. Elsa Dekking, Aug. 12, 1960; children—Jan Keith, Anthony, Leslie. Prof. physics U. Pa., Phila., 1956-59; prof., chmn. dept. physics U. Cal. at La Jolla, 1959-61, dean letters and sci., 1963-65, dean grad. studies, 1965, prof. physics, dir. Inst. for Radiation Physics and Aerodynamics, 1965—; v.p., dir. research Inst. for Def. Analyses, Washington, 1961-63, cons., 1953—. Cons. to industry, govt. labs. Recipient Heineman prize for math. physics Heineman Phys. Soc., 1963. Editor: Advances in Theoretical Physics Series, Pure and Applied Science Series, 1965. Research in theory of many body systems, with particular application to nuclear structure, electron gap, and liquid and solid helium isotopes. Home: 7723 Ludington Pl., La Jolla, Cal. 92037. Office: U. Cal., San Diego, P.O. Box 109, La Jolla, Cal. 92037.*

BRUECKNER, (Johannes) Max, German mathematician; b. Harthau bei Zittau, Germany, Aug. 5, 1860; s. Johann Gottlieb and Natalie Hedwig (Radelli) B.; Ph.D., U. Leipzig, Germany, 1886. Tchr. realgymnasium, Zittau, 1887-97, gymnasium, Bautzen, Germany. Author: Elemente der vierdimensionalen Geometrie, 1894; Vielecko und Vielflache, 1900. Studied problems of multi-dimensional geometry; donated over 200 self-made math. models of polyhedrons to math. inst. U. Heidelberg, Germany. Died Bautzen, Nov. 1, 1934.

BRUECKNER, Rolf, German physicist; b. Urnshausen, Germany, Mar. 10, 1928; s. Fritz and Gertrud (Nennstiel) B.; Diploma-physicist, U. Wuerzburg (Germany), 1955; Doctor, Acad. Techn., Clausthal, Germany, 1961; m. Elisabeth Doepke, July 14, 1956; 1 son, Gerd. Scientist, Max-Planck Inst. for Silicate Sci., Wuerzburg, 1955—; faculty Acad. Tech., Clausthal, 1964—. Research and publs. on structure, phys. and phys.-chem. properties of glasses, glass-fibers, melts, and fluids, surface and interface phenomena, corrosion of refractories by glasses, melts and slags, rheological properties of melts and clay-water-systems, streaming and electro-optical birefringence, spectroscopy. Home: 7 Franz Schubertstr., Office: 2 Neunerplatz, 87 Wuerzburg, Germany.*

BRUELL, Jan Herbert, Am. exptl. psychologist; b. Bielsko, Poland, Dec. 27, 1920; s. Walter and Frieda (Schreinzer) B.; Diplompsychologe, U. Heidelberg (Germany), 1949; Ph.D. in Exptl. Psychology, Clark U., 1953; m. Tillie E. Dienst, Feb. 5, 1948; children—Susan, Peter, Steven. Came to U. S., 1949, naturalized, 1954. Faculty, Western Res. U., 1953-64, asso. prof., 1957-64; prof. psychology Case Western Res. U., Cleve., 1964—; research psychologist Highland View Cuyahoga County Hosp., 1954—. Cons. VA, 1964—. Recipient numerous research grants NSF, also USPHS. Mem. Am. Psychol. Assn., Am. Assn. U. Profs., A.A.-A.S., Am. Genetic Assn., Am. Soc. Human Genetics. Research, publs. on visual space perception in normal and brain injured individuals, behavior genetics inheritance of behavior; developed motor theory of visual localization. Home: 3026 Warrington St., Shaker Heights, O. 44120. Office: Case Western Res. U., Cleve. 44106.*

BRUES, Alice Mossie, Am. phys. anthropologist; b. Boston, Oct. 9, 1913; d. Charles Thomas and Beirne (Barrett) Brues; A.B., Bryn Mawr Coll., 1933; Ph.D., Radcliffe Coll., 1940. Faculty, U. Okla. Sch. Medicine, 1946-65, prof., 1960-65; vis. prof. anthropology U. Colo., Boulder, 1965-66, prof., 1966—; staff mem. Southwestern Homicide Investigators Seminar, 1954—; curator Stovall Mus., Norman, Okla., 1956-65. Fellow Am. Anthrop. Assn.; mem. Am. Assn. Phys. Anthropologists, Am. Assn. Anatomists, Am. Soc. Human Genetics, Soc. Study Evolution, Phi Beta Kappa, Sigma Xi. Asso. editor: Am. Jour. Phys. Anthropology, 1962—. Research and publs. on analysis of genetic factors, sex linkage, human eye color; study of disease and life expectancy as shown by skeletons of prehistoric populations, study of evolution in man, including math. analysis of natural selection in blood groups, math. models for predicting effects of selection and chance factors on population changes. Home: 4325 Prado Dr., Boulder, Colo. 80302.*

BRUES, Austin Moore, Am. physician; b. Milw., Apr. 25, 1906; s. Charles Thomas and Beirne (Barrett) B.; A.B., Harvard, 1926, M.D., 1930; m. Mildred Carter, June 1, 1930; children—Roger Austin, Nancy Carter (Mrs. Arnold Noven), Charles Thomas. Faculty, Harvard, 1937-45; faculty U. Chgo., 1945—, prof. medicine, 1952—; sr. biologist div. biol. and med. research Argonne (Ill.) Nat. Lab., 1946—, dir., 1946-62. Mem. U. S. delegation UN Com. on Effects of Atomic Energy, 1955—; mem. exchange mission in radiobiology USSR, 1959; cons. U. S. Dept. Def, AEC, NRC, Internat. Commn. on Radiol. Protection, WHO. Mem. Am. Assn. for Cancer Research (pres. 1954-55), Radiation Research Soc. (pres. 1955-56), Am. Acad. Arts and Scis., Am. Assn. Physicians, Am. Clin. and Climatol. Soc., Internat. Soc. Cell Biology. Editor: Low Level Irradiation, 1959; Concepts of Aging and Biological Organization, 1964. Research and numerous publs. on cancer, cell biology, effects of radiation, physiology and medicine. Home: 4907 Lee Av., Downers Grove, Ill. 60515. Office: 9700 S. Cass Av., Argonne, Ill. 60439.*

BRUES, Charles Thomas, Am. zoölogist; b. Wheeling, W.Va., June 20, 1879; s. Charles Thomas and Ada (Mossie) B.; B.S., U. Texas, 1901, M.S., 1902; postgrad. Columbia 1903-04; A.M., Harvard, 1942; m. Beirne Barrett, June 16, 1904; children—Austin Moore, Alice Mossie. Fellow in zoölogy Columbia, 1903-04; spl. field agt. U. S. Dept. Agr., 1904-05; curator invertebrate zoölogy Milw. Pub. Mus., 1905-09; instr. econ. entomology Harvard, 1909-12, asst. prof., 1912-26, asso. prof., 1926-35, prof. entomology, 1935-45, prof. emeritus, 1946—; asso. curator insects, Mus. Comparative Zoology. Biol. research West Indies, 1910, 12, 1926-27; Dutch East Indies, 1937; Philippines, 1949. Mem. NRC, 1917-19. Fellow Am. Acad. Arts and Scis., A.A.A.S., Entomol. Soc. Am. (pres. 1929); mem. Am. Assn. Econ. Entomologists, Cambridge Entomol. Club, Am. Soc. Naturalist, Boston Soc. Natural Hist., Chgo. (hon.), Fla. acads. sci., Fla. Entomol. Soc. (hon.), Sigma Xi. Author: (with A. L. Melander) A Key to the Families of North Amer. Insects, 1915, Insects and Human Welfare, 1920 (2d ed., 1947), Classification of Insects, 1931. Insect Dietary; An Account of the Food Habits of Insects, 1945. Editor: Psyche (jour. entomology) from 1909. Research and publs. on embryology and habits of insects, especially Hymenoptera and Diptera. Died July 22, 1955.

BRUEVICH, Nikolay Grigorevich, Russian mech. engr.; b. 1896; grad. Moscow U., 1923, Moscow Aviation Inst., 1930. Instr., Zhukovsky Air Force Engring. Acad., 1929—; asso. Inst. Mech. Engring., USSR Acad. Sci., 1951—. Decorated Order of Lenin (3). Mem. USSR Acad. Sci. Author: The Kinetostatics of Spatial Mechanisms, 1937; The Precision of Mechanisms, 1946; Calculating and Computing Devices, 1954; Basic Principles of Creating Control Devices for Electronic Computers, 1963. Specialist in theory of mechanisms, machines and precision mechanics. Address: Inst. Mech. Engring., USSR Acad. Sci., M. Kharitonevsky p. 4, Moscow, USSR.

BRUGGER, Heinrich, German physician; b. Ascheberg, Westphalia, Germany, Dec. 3, 1895; s. Franz and Anna (Lendermann) B.; ed. Münster, Giessen, Tübingen, Freiberg (all Germany) univs.; M.D.; m. Grete Cramer, June 1, 1926; children—Margaret, Annette. Asst. several hosps., Hamburg, Germany, 1923-28; dir. physician-in-chief Pediatric Clinic, Wangen, Germany, 1928—, hon. prof. U. Tubingen, 1949—. Recipient Cross of Merit, German Fed. Republic. Fellow Am. Coll. Chest physicians; mem. S. German, German, Austrian Tb assns. (hon.). Author: Tuberkulose des Kindes, 1948; Die Elektrochir. Behandl. der Tuberkulose, 1949. Studies on symptoms of pulmonary Tb in young children and adolescents, also other diseases of lungs, lymph glands, tumors and their surg. treatment. Address: Wangen-Allgaü, Kinderheilstatte, W. Germany.

BRUGGER, Robert Melvin, Am. physicist; b. Oklahoma City, Jan. 13, 1929; s. Melvin and Allene (McCully) B.; B.A., Colo. Coll., 1951; M.A., Rice U., 1953, Ph.D., 1955; m. Barbara Irene Lett, June 13, 1953; children—James Robert, Carolyn Irene. Head solid state physics Phillips Petroleum Co., Idaho Falls, Ida., 1955-66, assigned to U.K. Atomic Energy Research Establishment, Harwell, Eng., 1962-63; head solid state physics Ida. Nuclear Corp., Idaho Falls, 1966—. Fellow Am. Phys. Soc.; mem. Am. Nuclear Soc. Contbg. author: Slow Neutron Scattering, 1965. Research, publs. on neutron scattering from which are obtained a detailed description of forces holding together molecules, liquids and crystals. Home: 841 Claire View St. Office: P.O. Box 1845, Idaho Falls, Ida. 83401.*

BRUGIS, Thomas, Brit. surgeon; b. Eng., flourished circa 1640; M.D. Surgeon during Brit. civil wars, later settled at Rickmansworth, Hertfordshire, Eng. Author: The Marrow of Physicke, 1640; Vade Mecum, or a Companion for a Chirurgion, 1651.

BRUGNONE, Giovanni, Italian veterinarian; b. Ricaldone, Italy, Aug. 27, 1741; student at sch., Alfort; dir. Vet. Sch., Turin, Italy; became prof. comparative anatomy U. Turin, 1780. Mem. French Acad. Scis. Publs. on animal anatomy and vet. medicine. Died Turin, Mar. 3, 1818.

BRUGSCH, Heinrich Karl, German Egyptologist; b. Berlin, Prussia, Feb. 18, 1827; doctorate, U. Berlin. In Egypt for Prussian government, 1853-54; worked in Berlin Museum; sent to Persia, 1860-61; vice-consul, 1864-66; prof. of Egyptology, U. Göttingen, 1868; dir., Cairo School of Egyptology, 1870-79. Made pasha, 1881. Author: Grammaire demotique, 1855; Geographische Inschriften altägyptische Denkmäler (3 vol.), 1857-60; Reise der Königlichen Preussischer Gesandtschaft nach Persien (2 vol.), 1862-63; Hieroglyphisch-demotisches Wörterbuch (2 vol.), 1867-82; Geschichte Ägyptens unter den Pharaonen, 1877; Dictionnaire geographique de l'ancienne Egypte, 1877-80; Thesaurus inscriptionum ägyptiacarum, 1883; Religion und Mythologie der alten Ägypter, 1891. Pioneer in decipherment of demotic; published hieroglyphic-demotic dictionary. Died Charlottenburg, Germany, Sept. 9, 1894.

BRUGUIÈRES, Jean-Guillaume, naturalist; b. Montpellier, France, 1750; with Kerguélen's 2d expdn. to South Seas, 1773; studied coal fossils, France; discovered deposit of pozzolanae fossils on expdn. (with Olivier) to Middle East. Mem. French Acad. Scis., 1796. Author: Histoire naturelle des vers (completed by Lamarck); Choix de mémoires sur divers objets d'histoire naturelle. Gender of plants of onagraceae order from tropical Asia and Polynesia named in his honor. Died Ancona, Italy, Oct. 1, 1798.

BRUHAT, François, French mathematician; b. Paris, France, Apr. 8, 1929; s. Georges and Berthe (Hubert) B.; student École Normale Supérieure, 1948-52; Docteur ès Sciences Mathématiques, Paris, 1955. Staff, Centre National de la Recherche Scientifique, 1952-55; faculty, Faculté des Sciences, Nancy, France, 1955-61; prof., Faculté des Sciences, Paris, 1961—. Decorated Palmes Académiques. Research and publs. in lie groups and algebraic groups theory. Home: 80 Bd Pasteur, Paris, France.*

BRÜHL, Gustav (pseudonym Kara Giorg), physician; b. Herdorf, Germany, May 31, 1826; ed. Munich, Halle, Berlin, Germany; studied laryngology, Prague, Vienna. m. Margarete Reise; 2 sons, 1 dau. Came to U. S., 1848. Physician, St. Mary's Hosp., Cin., several years; lectr. laryngology Miami Med. Coll., Cin.; made anthrop. and ethnol. study trips to Central and S.Am; founder, editor Der Deutsche Pionier, Cin., 1869-70. Co-founder Peter Claver Soc. for Edn. Negro Children. Mem. numerous profl. socs. Author: Die Kulturvölker Alt-Amerikas, 1875; Aztlan-Chicomoztoc, 1879; travel, verse books; numerous med. articles. Tried to support thesis of pre-Columbian syphilis by archaeol. research. Died Cin., Feb. 16, 1903.

BRÜHL, Count John Maurice, astronomer, diplomat; b. Wiederau, Germany, Dec. 20, 1736; s. F. W., Graf von Brühl; ed. Leipzig; m. Alicia Maria, dowager countess of Egremont, 1776; 1 son, 1 dau.; m. 2d, Maria Chowne, 1796. Joined Saxon diplomatic service at Paris, 1755, Warsaw, Poland, 1759; ambassador to London, 1764-1809. Author: Three Registers of a Pocket Chronometer, 1785; Latitudes and Longitudes of several Places Ascertained, 1786; Nouveau journal du chronometre, 1790; On the Investigation of Astronomical Circles, 1794; A Register of Mr. Mudge's Timekeepers, 1794; also articles on astronomy. With others determined latitudes and longitudes of Brussels, Belgium, Frankfort, Germany, Dresden, Germany, Paris, France by using Hadley's sextant and Emery's chronometer; built obs. at villa in Harefield and set up astron. circle. Died June 8, 1809.

BRÜHL, Julius Wilhelm, German chemist; b. Warsaw, Poland, Feb. 13, 1850; s. Ludwig and Emma (Bamberg) B.; studied under Hofmann in Berlin, Germany; m. Lili Bamberger, 1880. Became asst. to Landolt, 1873; prof. in Lemberg and Freiburg (Germany), 1884-87; took over Bernthsen's pvt. lab. in Heidelberg, Germany, 1887-98. Editor German edition: Treatise on Chemistry (Roscoe, Schorlemmer). Research and numerous publs. on terpene and its derivatives, 1880, relationship between refractive index and constn. of a substance, refractivity of benzol, 1894, and tautomeric substance; 1899; gave the names "enol" and "keto" to the two forms of acetoacetic acid; developed apparatus for fractional distillation in vacuum; applied phys. chem. methods to organic chemistry. Died Heidelberg, Feb. 5, 1911.

BRUK, Isaak Semenovich, Russian elec. engr.; b. 1902; grad. Moscow Higher Tech. Sch., 1925. With Power Engring. Inst., USSR Acad. Sci., 1935—, dir. Inst. Electronic Control Machines, 1959—, also dir. Lab. Control Machines and Systems. Mem. USSR Acad. Scis. Author: A Machine for the Integration of Differential Equations, 1941; The Stability of Electrical Systems, 1945; An Electric Minimizer, 1948; The M-2 Electronic High-Speed Computer, 1956; Dynamic Patterns of Power Systems, 1962. Drafted design for 1st Soviet machine for integration ordinary differential equations, 1936-38; supr. design and constrn. alternating current computer for study elec. systems, 1945-47; designed M-1, M-2, M-3 electronic high-speed digital computers. Address: Inst. Electronic Control Machines, Leninsky prospect 14, Moscow, USSR.

BRUMBERGER, Harry, chemist; b. Vienna, Austria, Aug. 28, 1926; s. Leon and Rose (Kraft) B.; came to U. S., 1940, naturalized, 1948; B.S., Bklyn. Poly. Inst., 1949, M.S., 1952, Ph.D., 1955; m. Vilma Musry, June 21, 1950; children—Jesse, Eva. Research asso. dept. chemistry Cornell U., 1954-57; faculty Syracuse U., 1957-—, asso. prof. dept. chemistry, 1962-—; cons. IBM Mfg. Research Lab., Endicott, N.Y., 1964-65, Supplies Engring. Lab., Vestal, N.Y., 1962-65. Mem. Am. Chem. Soc., Am. Crystallographic Assn., A.A.A.S., Am. Assn. U. Profs., N.Y. Acad. Scis., Sigma Xi. Editor: Small-Angle X-Ray Scattering, 1967. Research and publs. on determination of surface area of colloidal systems by x-ray methods; exptl. observation of critical phenomena in liquid mixtures; investigation on polymer structure by x-ray methods; exptl. technique of small-angle x-ray scattering.

BRUMFIEL, Charles Francis, Am. mathematician; b. Matthews, Ind., Aug. 13, 1914; s. Walter Blaine and Amy (Troyer) B.; B.S., Ball State Coll., 1939; M.S., U. Chgo., 1944; Ph.D., Purdue U., 1954; m. Joanna Norene Findling; children—Gregory, Vincent. Instr., Ill. Inst. Tech., Chgo., 1942-44; indsl. engr. U. S. Steel Corp., Chgo., 1944-45; prof. math. Ball State Coll., Muncie, Ind., 1946-60, U. Mich., Ann Arbor, 1960-—. Author: (with Eicholz and Shanks) Algebra I, 1960, Geometry, 1960, Introduction to Mathematics, 1962, Fundamental Concepts of Elementary Mathematics, 1962; Algebra II, 1963; (with others) Arithmetic: Concepts and Skills, 1963; (with Eicholz, Shanks and O'Doffee) Principles of Arithmetic, 1963; (with Eicholz, Shanks and Fleenor) Pre-Calculus Mathematics, 1965; also others. Home: 1205 S. Forest St., Ann Arbor, Mich.*

BRUMMELKAMP, W. H., Dutch physician; b. Keboemen, Netherlands, Mar. 21, 1928; s. Reindert and Jans Brummelkamp; grad. U. Groningen (Netherlands) Med. Sch., 1952, Ph.D. in Pathology, 1956; postgrad. surgery U. Amsterdam (Netherlands), 1956-66; m. Tjadiene Dons., May 23, 1953. Faculty, U. Amsterdam, 1962-67; chief dept. surgery St. Lucas Hosp., Amsterdam, 1966-—. Recipient Tilanus medal in surgery, 1965; Cressy Morrisson award N.Y. Acad. Scis., 1965. Mem. Dutch Surg. Soc., Dutch Soc. for Thoracic Surgery, Am. Coll. Chest Physicians, N.Y. Acad. Scis. Author: On Meningeomas, 1956; Hyperbaric Oxygen Therapy in Clostridial Infections Type Welchii; also articles. Research on hyperberic oxygen treatment in anaerobic infections, mediastinal anomalies as cause of abdominal manifestation, thoracic trauma. Home: 5 Dopperkade. Office: St. Lucas Hosp., 164 Jan Toorop-str., Amsterdam, Holland.*

BRUMPT, Emile, biologist; b. Paris, France, Mar. 7, 1877; ed. Alsace, under Blanchard; doctorate in natural sci., 1901, in medicine, 1906. Mem. expdns. to Africa, Asia, Am.; became prof. parasitology and med. natural history, Paris, 1907; prof. parasitology U. Sao Paulo (Brazil), 1913-14; replaced Blanchard as prof. parasitology, Paris, 1919-48; founder Annals Human and Comparative Parasitology, 1933, also dir. Mem. Acad. Medicine (Prix Emile Brumpt founded in his honor 1948). Author: Precis de parasitologie, 1910; Travaux practiques de parasitologie. Identified many pathogenic species. Died July 7, 1951.

BRUMSHTEYN, Mikhail Solomonovich, Russian pathoanatomist; b. 1903; grad. Med. Faculty, 2d Moscow U., 1925; D.Med. Sci., 1952. Research physician Gorky City Hosp., 1925-30; asst. dept. path. anatomy Gorky Med. Inst., 1930-31, Pediatric Faculty, 2d Moscow Med. Inst., 1941; sr. asso. All-Union Inst. Exptl. Medicine and clinic of Inst. Roentgenology and Radiology, 1931-41; asso. path. anatomy lab. Main Mil. Med. Bd. of Soviet Army, 1946-48; prorector Astrakhan Med. Inst., prof., head chair path. anatomy, 1952-—. Mem. All-Union Soc. Pathoanatomists (bd. mem.). Author of some 60 works including Data on the Pathological Anatomy of Electric Shock, 1952. Address: Astrakhan Med. Inst., ulitsa Mechnikova 20, Astrakhan, RSFSR, USSR.

BRUN, Edmond Antoine, French engineer; b. Saint Cannat (B du R), France, Dec. 31, 1898; s. Antoine Marius and Marie (Villecrose) B.; B.Sc., U. Marseille, 1921, M.S. (fellowship nat. competitive test), 1923; Doctor of Sciences, Paris, 1934; m. Suzanne Vincent, September 8, 1923. Professor spl. courses Lycée Nice, 1925-30, Parisian Lycée, 1930-42; lectr., then prof. fluid mechanics Faculté des Sciences, 1942-—; dir. Laboratoire d'Aerothermique, prof. Ecole Nationale Superieure de l'Aéronautique, 1942-—. Mem. Armed Forces, 1917-19. Decorated chevalier Legion of Honor; officer Military Merit (Brazil); comdr. Palmes Académiques, Laureate Acad. Scis. Fellow Royal Area Soc., Am. Inst. Aeros. and Astronautics (hon.), American Astronautical Society; foreign fellow National Academy of Scis., Internat. Astronautics Acad., Société Francaise d'Astronautique (pres. 1960-62), Internat. Astronautical Fedn. (pres. 1962-64), Societé Francaise des Thermiciens (pres.). Author articles and books on aerodynamic heating convection, flight and icing of aircraft. Home: 8, pl. Commerce, Paris XV, France. Office: 4 ter route des Gardes, g2-Mendon, Hauts-de-Seine, France.

BRUN, Robert Michel, Swiss skin biochemist; b. Geneva, Switzerland, Feb. 27, 1926; s. Antonin and Elisa (Terrier) B.; student Coll. Geneva, 1938-45; Chem. Engr. degree, U. Geneva, 1950, Scis. Dr., 1954; m. Odile Pierre, Dec. 27, 1950; children—Nicole, Olivier. Head of lab. dept. dermatology U. Hosp. Geneva, 1950-—; privat-docent Faculty Medicine Geneva, 1961-—, in charge research, 1963-—; lectr. on investigative dermatology and allergology, 1961-—. Pres. scis. sect. Nat. Inst. Geneva. Mem. Swiss Soc. Dermatology, Swiss Soc. Allergy, Am. Chem. Soc. Research, publs. on physiology and pharmacodynamy of sweat and sebum secretion, pharmacodynamy of depigmentation of skin, researches on acanthosis and keratin; investigations on allergy, exptl. eczema and senzitisation with different substances; expt. on chem. X-rays protection. Home: 15 Athenee, 1206 Geneva. Office: Dept. Dermatology, Hosp. Cantonal, 1211 Geneva 4, Switzerland.*

BRUN, Viggo, Norwegian mathematician; b. Oct. 13, 1885; s. Soren and Lorentze (Petersen) B.; ed. Oslo, Norway, Gottingen, Germany, Paris, France univs.; Ph.D.; m. Laura Michelsen, July 22, 1940. Prof., Tech. Inst. Norway, 1924-46, U. Oslo, 1946-56. Recipient Gumerus medal; fellow U. Oslo, 1910-24. Mem. Oslo, Trondheim, Uppsala, Heisingfors acads. sci. Author: La serie $1/5+1/7+1/11+1/13$ . . . ou les dénominateurs sont 'nombres premiers jumeaux' est convergente ou finie, 1919; Le Crible d'Eratosthène et le théorème de Goldbach, 1920; Algorithmes euclidiens pour trois et quatre nombres, 1957. Home: Drobak, Norway. Office: U. Oslo, Norway.

BRUNCK, Heinrich von, see von Brunck, Heinrich.

BRUNCK, Otto, chemist; b. 1866; nephew of Heinrich von Brunck; prof., Freiberg, Germany, from 1896; worked in metal analysis; found practicable method for determining sulpher in coal, 1905; proved value of diazetyldioxim as agt. for quantitative determination of nickel and for separating nickel from other elements, 1907. Died 1946.

BRUNDAGE, Albert Harrison, Am. toxicologist; b. Candor, N.Y., Mar. 3, 1862; s. Amos H. and Sarah M. (Dimmick) B.; M.D., U. Med. Coll. (N.Y. U.), 1885; Ph.G., Bklyn. Coll. Pharmacy, 1892, Pharm.D., 1897; hon. A.M., U. Nashville, 1898; M.S., R.I. Coll. Pharmacy and Allied Scis., 1905; m. S. A. Holt, Sept. 26, 1888. Prof. toxicology and physiology Bklyn. Coll. Pharmacy, 1898-1903 (pres. 1893-94), and R. I. Coll. Pharmacy and Allied Scis., 1903-07; prof. emeritus toxicology and physiology, depts. medicine, dentistry and pharmacy Marquette U., from 1908. Toxicologist to Bushwick Hosp., 1904-21; lectr. on Tb, Bklyn. Com. for Prevention Tb, from 1908; lectr. for A.R.C., from 1918; pres. N.Y. State Bd. of Pharmacy, 1903, and Board examiner in toxicology and posology, 1901-04, chmn. state com. on poisons. Asso. in immuno-therapy Polhemus Clinic, L.I. Coll. Hosp. Founder of first open-air classes in Bklyn. pub. schs. Author: A Manual of Toxicology, 13th edit., 1921; Practical Points in Physiology, 2 edits., 1904. Died Mar. 12, 1936.

BRUNEL, Isambard Kingdom, English engr.; b. Portsmouth, Eng., Apr. 9, 1806; s. Marc Isambard Brunel; ed. Paris, France and in England. Built Clifton suspension bridge, started in 1836; also numerous other bridges in Eng., including Royal Albert, opened in 1859; became chief engr. Gt. Western Ry., 1833; originated broad gauge; designed Gt. Western (1st transatlantic steamship), 1838; also 1st screw steamship, Gt. Britain, 1845; Gt. Eastern (largest steamship of time), launched in 1858. Died Westminster, Eng., Sept. 15, 1859.

BRUNEL, Marc Isambard, engr., inventor; b. Hacqueville, France, Apr. 25, 1769; ed. seminary of St. Nicaise at Rouen; naval apprentice, France, 1786-93; because of the Terror, left France for U. S., 1793; surveyor in region bordering Lake Ontario; apptd. surveyor for canal connecting Lake Champlain and the Hudson; chief engr. N.Y.; evolved mech. method of making ships' blocks and sailed to Eng. to interest Brit. govt., 1799, apptd. supt. for erection machines Portsmouth (Eng.) dockyard (early example of completely mechanized prodn.); owner, sawmills, Battersea, London. Decorated Legion of Honor, 1829. Fellow Royal Soc., 1814 (v.p. 1832); mem. Royal Acad. Stockholm, French Acad. Scis., 1826. Built 1st underwater tunnel, 1825-43; inventor machines for sawing and bending timber, printing, knitting stockings, bootmaking; designed 1st floating landing stage, Liverpool, Eng., also suspension bridge Ile de Bourbon. Died London, Dec. 12, 1849.

BRUNELLESCHI, Filippo, Italian architect; b. Florence, Italy, 1377; s. Ser Brunellesco Lapi and Giuliana Spini; matriculated at Arte della seta (goldsmith's guild), 1404; apprentice to goldsmith; 1 adopted son, Andrea Buggiano. Began career as sculptor, then became architect; designed dome of Santa Maria del Fiore (1st major work of Renaissance architecture, prototype probably Persian), 1420-34, foundling hosp., 1421, Pazzi chapel, 1430; made discoveries in perspective (included in Leone Battista Alberti's Della Pittura); pioneer in concepts of controlled space; invented boat (badalone) to carry marble via river. Died Florence, Apr. 15, 1446.

BRUNER, Harry Davis, Am. pharmacologist; b. Jeffersonville, Ind., July 18, 1911; s. Henry and Florence (Davis) B.; B.S., U. Louisville, 1932, M.D., 1934, M.S., 1936; Ph.D., U. Chgo., 1939; m. Gladys Christine de Bord, Sept. 6, 1931; children—Robert H. A., Frederick D. Asst. prof. U. N.C., 1939-42, prof. pharmacology, 1947-49; research asso. dept. surgery U. Pa., 1942-45, asst. prof., 1945-47; chief scientist Oak Ridge Inst. Nuclear Studies, 1949-52; prof. physiology Emory U., 1952-56; chief med. research br. div. biology and medicine U. S. AEC, Washington, 1956-60, asst. dir. health and med. research, 1960-—. Recipient Jefferson Research medal S.C. Acad. Sci., 1939. Mem. A.A.A.S., Am. Soc. for Pharmacology and Exptl. Therapeutics, Am. Phys. Soc., Radiol. Research Soc., Soc. for Exptl. Biology and Medicine. Author: Methods in Medical Research, Vol. 8, 1961. Research and numerous articles on chem. agts., gen. toxology and physiology, factors in hematopoiesis, cardio-pulmonary blood flow, radiation damage from radioisotopes localized in body, effects radiation and treatment radiation sickness. Home: 9404 Byeforde Rd., Kensington, Md. 20795. Office: Div. Biology and Medicine, AEC, Washington 20255.*

BRUNER, Jerome Seymour, Am. psychologist; b. N.Y.C., Oct. 1, 1915; s. Herman and Rose (Glücksmann) B.; A.B., Duke, 1937; A.M., Harvard, 1939, Ph.D., 1941; D.H.L. (honorary), Lesley College, 1964; D.Sc. (honorary), Northwestern University, 1965; LL.D., Temple U., 1965; LL.D., U. Cin., 1966; m. Katherine Frost, Nov. 10, 1940 (div. 1956); children—Whitley, Jane; m. 2d, Blanche Marshall McLane, Jan. 16, 1960. Assisted establishment monitoring service German broadcasts for U. S. intelligence, 1941; asso. dir. Office Pub. Opinion Research, Princeton, 1942-44, mem. Inst. Advanced Study, 1951; govt. surveys pub. opinion on war problems, 1942-43; polit. intelligence, France, 1943; research Harvard, 1945-—. prof. psychology, 1952-—, dir. Center for Cognitive Studies, Harvard, 1961-—; lectr. Salzburg Seminar, 1952; cons. psychol. problems, 1948-—; Bacon professor University of Aix-en-Provence, France, 1965; education advisor book division Time, Incorporated. Chmn. curriculum study group Nat. Acad. Sciences, 1959-61; mem. White House Panel on Ednl. Research and Devel. Guggenheim fellow Cambridge U., 1955. Fellow American Psychol. Assn. (recipient Distinguished Sci. Contbn. award 1962, pres. 1964-65), Am. Acad. Arts and Sciences, Swiss Psychol. Society (hon.), Soc. Psychol. Study Social Issues (past pres.), Am. Assn. U. Profs., Puerto Rican Academy of Arts and Scis. (hon.). Author: Mandate from the People, pub. 1941; The Process of Education, pub. 1960; (with Smith and White) Opinions and Personality, 1956; (with Goodnow and Austin) A Study of Thinking, 1956; On Knowing: Essays for the Left Hand, 1962; Toward a Theory of Instruction, 1966; Studies in Cognitive Growth, 1966. Editor: Public Opinion Quar., 1943-44; syndic Harvard U. Press, 1962-63. Contbr. articles tech. profl. jours. Research on nature of perception, learning, thinking, as affected by motives, personality. Home: 6 Follen St., Cambridge, Mass.

BRUNETTI, Fausto, Italian physician; b. Roma, Italy, Aug. 2, 1910; s. Federico and Vaerini Brunetti; M.D., U. Padova (Italy), 1934; m. Caterina Bellino, Dec. 8, 1940. Head ear, nose and throat clinic U. Turin (Italy), 1948-—. Mem. Collegium O.R.L.-A.S., Société Francaise d'O.R.L., Internat. Coll. Surgeons, Società Italiana di ORL, Österreichische Otolaryngologische Gesellschaft, Deutsche Gesellschaft d. HND-Artze, Accademia di Medicina di Torino. Research, numerous publs. on ear, nose and throat, tumors, anesthesiology, vestibular and audiological ways, plastic surgery. Home: Corso G. Ferraris 61, Torino, Italy.*

BRUNETTI, Paolo, Italian physician; b. Florence, Italy, Apr. 1, 1932; s. Natale and Gina (Mazzanti) B.; Degree in Medicine, U. Pavia (Italy), 1956; m. Maria Teresa Quercioli, Aug. 3, 1963; 1 dau., Elena. Asst. dept. medicine U. Cagliari, 1956-59; postdoctoral fellow dept. biochemistry arthritis research unit U. Mich., Ann Arbor, 1959-60; prof. medicine dept. medicine U. Perugia (Italy), 1960-—. Research and publs. on hematology; investigated enzyme-deficiency haemolytic anemias and patterns of their hereditary transmission. Home: 10 Via Oddi Sforza, Perugia, Italy.*

BRUNFELS, Otto (Otho), humanist, physician, botanist; b. Mainz, Germany, circa 1488; dr. of medicine from Basel, Switzerland, 1530; m. Dorothea Heiligenhensin, 1524. Carthusian monk in Mainz and Strasbourg; became priest, 1514; came in contact with humanists (especially Hutten) and fled monastery, 1521; became Lutheran; tchr. Carmelite sch. in

Strasbourg until 1532; went to Bern as city physician, 1532. Author: Herbarium, vivae icones ad naturae imitationem, 1530; Contrafayt Kräuterbuch, 1532; Onomasticon s. lexicon medicinae, 1534. Called father of modern botany by Linné; his herbal of native plants of Germany and many new plants with very accurate drawings; also interested in edn., promoting study of natural scis. Died Bern, Switzerland, Nov. 25, 1534.

BRUNHES, Jean, French ecologist; b. Toulouse, France, 1869; ed. L'ecole Normale Superieure; passed secondary sch. tchrs. exam, 1892. Tchr. geography Fribourg, Lausanne univs. (Switz.); apptd. prof. human geography Coll. France, 1912; traveled to Orient and U. S. Author: L'Irrigation dans la peninsule Iberique et dans l'Afrique du Nord, 1902; La Geographie humaine (seminal work for a new study of relationships between man and his natural environment), 1910; Geographie humaine de la France, 1920. Died Boulogne-sur-la-seine, France, 1930.

BRUNHUBER, August David, German geologist; b. Burghausen, Germany, Jan. 23, 1851; s. Franx Xaver and Theresia (Koch) B.; ed. Munich, also Freiburg, Germany, 1869-77; m. Sofie Wilhelmine Reisenegger, 1908; practiced medicine, opened eye clinic, Regensburg, 1881. Mem. Regensburger Naturwissenschaftlicher Verein (chmn. 1900-20, geol. dir. to soc.). Investigated paleogeographic relationships, laid modern basis for geology of Regensburg 3-state corner, So. Germany. Died Regensburg, Feb. 15, 1928.

BRUNIG, Eberhard Friedrich Wilhelm Otto, German forester; b. Schoeningen, Germany, June 16, 1926; s. Otto Friedrich Wilhelm and Anneliese (Kuthe) Brünig; student U. Freiburg, Germany, 1950, Oxford (Eng.) U., 1951; M.A., U. Goettingen, Germany, 1952, D.Sc. in Forestry, 1953; m. Birgit Mahr, May 4, 1962; children—Stephanie Nicole, Alexandra Elisabeth. With Hesse State Forest Service, 1953; with Brit. Overseas Civil Service, 1954-63, sect. forest officer Bintulu, 1959-60, working plans officer Sarawak, 1960-63; inventory expert in Thailand, FAO, UN, 1963-64; sci. officer Inst. for World Forestry, Reinbek, Germany; lectr. tropical silviculture, forest productivity and mgmt. U. Hamburg. Cons. forester in silviculture and mgmt. to pvt. forest estates in Germany, 1965—. Mem. Commonwealth Forestry Assn., Brit. Ecol. Soc., Internat. Soc. Tropical Ecology, Soc. Forester Gt. Britain, Ecol. Soc. Am., Soc. Tropical Foresters, Flor.-Soziol. Arb.gem., Research and publs. on devel. of forest inventory and mgmt. methods for tropical forestry, study flora, ecology and vegetation history of heath forests of Sarawak and Brunei. Home: 3 Vor Den Hegen, Aumuehle, Germany. Office: Inst. for World Forestry, Reinbek, Germany.*

BRUNK, Hugh Daniel, Am. mathematician; b. Manteca, Cal., Aug. 22, 1919; s. Hugh Dennis and Velma Lee (Benson) B.; A.B., U. Cal. at Berkeley, 1940; M.S., Rice Inst., 1942, Ph.D., 1944; m. Jean Young, Oct. 17, 1942; children—Bridget, Gretchen, Heidi. Asst. prof. math. Rice Inst., Houston, 1946-51; mathematician Sandia Corp., Albuquerque, 1951-52; asso. prof. math. U. Mo., Columbia, 1952-57, prof., 1957-61, prof. statistics, chmn. dept., 1963—; prof. math. U. Cal. at Riverside, 1961-63. Fulbright lectr. Math. Inst., U. Copenhagen, Denmark, 1958-59; vis. sr. lectr. statistics Univ. Coll. Wales, 1966-67. Author: An Introduction to Mathematical Statistics, 1960; also articles, Statis. research in estimation, hypothesis testing, probability, distbns., limit theorems; math. research in inequalities for convex functions. Home: Route 8, Columbia, Mo. 65201.*

BRUNN, Heinrich von, see von Brunn, Heinrich.

BRUNN, Herman Karl, mathematician; b. Rome, Italy, Aug. 1, 1862; s. Heinrich and Ida (Bürkner) von B.; ed. under G. Bauer, A. Pringsheim, Munich, under K. Weierstrass, L. Kronecker, L. Fuchs, Berlin; m. Emma Ney, 1900; 1 son. Librarian, Technische Hockschule, Munich, Germany, also hon. prof. U. Munich, from 1906. Author: Über Ovale und Eiflächen (dissertation), 1887, Kurven ohne Wendepunkte (Habilitation paper, with dissertation was origin of theory of convex figures, especially Brunn-Minkowski theory used in modern calculus of variations), 1899; publs. on convex figures and topology. Died Munich, Sept. 20, 1939.

BRUNN, Lukas, mathematician; b. Annaberg, Germany, 16th century; ed. under Abraham Riese, Annaberg, then in Leipzig, Germany, 1598-1601, under Johann Praetorius, Altdorf, Switzerland, from 1607, under Johann Hauer, Nuremberg, Germany, 1612; became magister, 1611; insp. art mus., Dresden, Germany, from 1619. Author: Praxis perspectivae, d.i. von Verziehungen, 1615; Euclidis elementa practica, 1625. Invented precision micrometer. Died Dresden, Jan. 1, 1628.

BRUNN, Walter Albert Ferdinand von, see von Brunn, Walter Albert Ferdinand.

BRUNNER, Edmund de Schweintz, Am. sociologist; b. Bethlehem, Pa., Nov. 4, 1889; s. Franklin Henry and Nina (De Schweinitz) B.; B.A., Moravian Coll., 1909, M.A., 1912, Ph.D., 1914, L.H.D., 1935; M.A., U. Queensland, Australia, 1937; LL.D., U. Natal, South Africa, 1954; m. Mary Vogler, Dec. 16, 1912 (dec. 1947); children—Edmund de Schweintz, Wilfred Robert; m. 2d, Lousene Rousseau, 1948. Exec. sec. Country Ch. Commn. Moravian Ch., 1912-20; dir. town and country surveys Inst. Social and Religious Research, N.Y.C., 1921-30; prof. sociology Columbia, N.Y.C., 1930-55, chmn. Bur. Applied Social Research, 1951-63, prof. emeritus, 1955—; lectr. Drew U., 1921-29, Danburg State Coll., 1958—. Collaborator, adviser U. S. Dept. Agr., 1941-51; sr. cons., organizer Inst. Social Research, U. Natal, Durban, S. Africa, 1954. Fellow Am. Sociol. Assn.; mem. Rural Sociol. Soc. (past pres.), Fgn. Policy Assn. Author numerous books most important being: The Town and Country Church in the United States (with H. N. Morse), 1923; Surveying Your Community, 1925; (with G. Hughes, M. Patten) American Agricultural Villages, 1927; Village Communities, 1927; Rural Korea: a Social and Economic Survey, 1928; Immigrant Farmers and Their Children, 1929; (with J. H. Kolb) Rural Social Trends, 1933; A Study Rural Society, 1935; (with Irving Lorge) Rural Trends in Depression Years, 1937; (with Hsin-Pao Yang) Rural America and the Extension Service, 1949; American Society: Urban and Rural Patterns (with W. C. Hallenbeck), 1955; The Growth of a Science, 1957; (with David Wilder, Corinne Kirchner, John Newberry, Jr.) An Overview of Adult Education Research, 1959. Research and publs. on church as social instn., significant demographic difference between rural farm and rural nonfarm populations and communities; pioneered adaption sociol. research methods to underdeveloped areas. Address: 10 High Ridge Rd., Wilton, Conn. 06897.*

BRUNNER, Johann Conrad, anatomist; b. Diessenhofen, Switzerland, Jan. 16, 1653; s. Eduard Brunner; student medicine, Straussburg, France; M.D.; later student anatomy, Paris. Practice medicine, Diessenhofen, 10 years; became prof. medicine, Heidelberg, Germany, 1687; later returned to pvt. practice. Author: Experimenta nova circa pancreas, 1683; De glandulis in intestino duodeno hominis detectis, 1688. Research on glands of digestive system, effects of removing pancreas in dogs; discovered and described mucous glands in submucous layer of duodenum (Brunner's glands), 1687. Died Mannheim, Germany, Oct. 2, 1727.

BRUNNER, Johann Conrad von, see von Brunner, Johann Conrad.

BRUNNER, Richard Ludwig, Austrian chemist; b. Vienna, Austria, Jan. 5, 1900; s. Ludwig and Rosa (Buchneder) B.; student chemistry Technische Hochschule; engring. degree; Ph.D. in Chemistry; m. Marianne N. Asst. Tech. U. Vienna, 1920-29; mem. council adminstrn. Office Patents, Vienna, 1929; chemist in chief S. A. Brasserie Schwechat, Vienna, 1929-42; dir. Inst. Research and Devel. (brewery and malt industry), Prague, Czechoslovakia, 1942-45; chief chemist, sci. cons. Biochemie, 1946-63; dir. Inst. Biochemistry and Microbiology, Technische Hochschule, Vienna, 1963—. Mem. Assn. Austrian and German chemists, Am. Soc. Chemistry, Austrian Soc. Microbiology and Hygiene. Co-editor: Die Antibiotica, vol. I, 1962. Contbr. publs. concerning antibiotic biochemistry, breweries, also problems relating to penicillin (dissolution of basic salts, acidity in preparation. Home: Siebensterngasse 16, Vienna VII, Office: Getreidemarkt 9, Vienna VI, Austria.

BRÜNNOW, Franz Friedrich Ernst, astronomer; b. Berlin, Germany, Nov. 18, 1821; s. Johann and Wilhelmine (Weppler) B.; Ph.D., Berlin, 1843; m. Rebecca Lloyd Tappan, 1857; 1 son, Rudolf. Became head pvt. obs., Bilk, Germany, 1847; became asst. Berlin Obs., 1851; named prof., dir. obs., Ann Arbor, Mich., 1854; apptd. astronomer royal for Ireland, also dir. Dunsink Obs., 1865; ret., 1874. Recipient Gold medal Amsterdam Acad., 1848. Author: Lehrbuch der sphärischen Astronomie, 1851; numerous works on comets, small planets, fixed star parallel axes. Exponent of classical position astronomy. Died Heidelberg, Germany, Aug. 20, 1891.

BRUNO, Francesco, Italian botanist; b. Alimena, Palermo, Italy, Mar. 25, 1897; s. Santi and Tedesco (Rosa) B.; Ph.D. in Natural Scis., 1921, in Bot. Scis., 1927. Dir. Inst. and Bot. Gardens, Messina, Italy, 1936-39, Palermo, 1939—; dir. Phitopathologique, 1939-53; pres. sch. agr. U. Palermo, 1943-50, sch. pharmacy. Mem. Acad. Sci. and Letters Palermo, Acad. Natural and Econ. Scis. Palermo, Acad. Peloritana Messina, Acad. Gioenia Catana, Italian Bot. Soc. Author numerous works on vegetal physiology, biology, applied botany. Address: Orto Botanico, via Lincoln, Palermo, Italy.

BRUNO, Frank J(ohn), sociologist; b. Florence, Italy, June 1, 1874; s. Jerolomo and Zippora Elizabeth (Menchini) B.; brought by parents to U. S., 1876; A.B., Williams Coll., 1899; S.T.B., Yale, 1902; LL.D., Washington U., 1946; postgrad. Hartford (Conn.) Theol. Sem., 1905-06, N.Y. Sch. of Social Work, 1908, Columbia, 1913-15; m. Susan Grey L. Topham, May 28, 1902 (dec. May 17, 1950); 1 son, John Grey; m. 2d Joanna C. Colcord, Nov. 24, 1950. Ordained to ministry Congl. Ch., 1902; asst. pastor Waterbury, Conn., 1902-04; pastor S. Congl. Ch., Granby, Conn., 1904-05, S. Congl. Ch., Pueblo, Colo., 1905-07; gen. agt. Asso. Charities, Colorado Springs, Colo., 1907-11; supt. Charity Orgn. Soc., N.Y.C., 1911-14; gen. sec. Asso. Charities (later Family Welfare Assn.), Mpls., 1914-25; lectr. U. Minn., 1915-25, professorial lectr., 1919-25, acting chmn. dept. sociology and social work, 1919-22; prof. applied sociology and head dept. social work, Washington U., 1925-45; in British Mus. and London Sch. of Economics, 1933-34. Pres. Minn. Conf. of Social Work, 1916-17, Mo. Assn. for Social Welfare, 1938-39. Mem. Am. Sociol. Soc. Am. Assn. Social Workers (pres. 1927, 29, 42; mem. exec. com.), Internat. Assn. Schs. of Social Work (exec. com.), Internat. Conf. Social Work (chmn. Am. sect. 1934-38), Nat. Tb Assn. (bd. dirs. 1941-43), Nat. Soc. Crippled Children (bd. dirs. 1945), Am. Assn. Sch. of Social Work (bd. dirs. since 1940), Am. Social Hygiene Assn., Nat. Conf. of Social Work (pres. 1932-33), Family Welfare Assn. Am. Author: The Theory of Social Work, 1936; Trends in Social Work, 1946. Contbr. to Ency. Brit., The Survey, Social Sci. Abstracts, The Family, Social Work Year Book. Cons. editor Social Science Abstracts, Ency. of Social Scis. Died Aug. 7, 1955.

BRUNO, Giordano, philosopher, cosmologist; b. Nola, nr. Naples, Italy, Jan. or Feb. 1548; studied in Naples; M.A., U. Toulouse, 1580; entered Dominican order, 1565; ordained priest, 1572; accused of heresy and fled, 1576; wandered throughout Europe 1576-93; at Noli, Savona, Venice, Bergamo, Geneva, 1576-79; at Toulouse, 1579-81; at Sorbonne, Paris, 1581-83; in England, 1583-85; in Paris, 1586; prof. philosophy, U. W. Henberg, 1586-88; in Germany, 1588-92; excommunicated by Lutheran Church Superintendent, Helmstedt, 1589; taught art of memory to Giovanni Mocenigo in Venice, 1592-93; arrested by Inquisition, Venice, Italy, 1593; transferred to Rome, where he was imprisoned and tried, 1593-1600. Author: Ors reminiscendi, 1583; Spaccio della bestia trionfante; De la causa, principio, e uno; De l'infinito universo e mondi, La cena de la ceneri (all 1584); Figuratio Aristotelici physici auditus, 1586. Advocated pantheistic dynamism; posited an infinite God, who creates an infinite, knowable universe; derived his moral and metaphysical motives from Plotinus, Marsilio Ficino, and other Neoplatonists, and Nicholas of Cusa; drew his cosmology from Copernicus, Lucretius, and, to a lesser extent, Aristotle; conceived of the earth as revolving around a moving sun; believed stars were centers of other planetary systems; held earth and innumerable other planets to be noble and animated bodies moving in an infinite space; attempted to replace traditional reliance upon authority with interrogation of nature; in confuting Aristotelian physics and cosmology foreshadowed Galileo; his works mark change from medieval to modern sci. thought. Burned at stake, in Campo dei Giori, Rome, Feb. 17, 1600.

BRUNO OF LONGOBURGO, Italian surgeon; b. Longoburgo, Calabria; flourished mid 13th century; ed. Salerno. Practiced medicine in Padua and Verona. Author: Chirugia Magna, 1252, also abridged version Chirurgia Minor. In Chirugia Magna described ligatures of silk or linen; in wounds of the omantum he directed that the portion which have become either green or black must be removed and all veins and arteries ligated with silk thread; distinguished between arterial and venous bleeding and in bleeding from wounds of the extremities he recommended the member should be raised high so that the blood does not run out; warned against hairs, oil and salve between lips of a wound since these hinder healing; wounds, when possible, should be closed and kept dry, no moist medication being applied; early advocate of aseptic technique; moved the medicine of his day toward a more exptl. and observational direction; integrated more of the knowledge of Arabic medicine into the body of Western medicine.

BRUNS, (Ernst) Heinrich, astronomer; b. Berlin, Germany, Sept. 4, 1848; s. Christian Gerhard and Caroline Henriette (Haase) B.; ed. U. Berlin; m. Marie Wilhelmine Schleussner. Calculator, Pulkovo (Russia) Obs.; became observer Dorpat (now Tartu, Estonia) Obs., 1873; became prof. math. U. Berlin, 1876; apptd. prof. astronomy, also dir. univ. obs., Leipzig, Germany, 1882. Author: Über die Perioden der elliptischen Integrale erster und zweiter Gattung, 1875; Die Figur der Erde, 1878; Grundlinien des Wissenschaftichen Rechnens, 1903; Wahrscheinlichkeitsrechnung und Kollektivmasslehre, 1905; Das Gruppenschema für zufällige Ereignisse, 1906. Developed potential theory as exception to attraction principle; presented proof there are only 10 closed algebraic integrals; asserted that geodetic measurements give true figure of earth only when other types of measurements are considered. Died Leipzig, Sept. 23, 1919.

BRUNS, Henry Dickson, Am. physician; b. Charleston, S.C., 1859; s. John Dickson and Sarah (Dickson) B.; student U. Va., 1876-78, U. La.; resident student Charity Hosp., New Orleans; M.D., Jefferson Med. Coll., 1881; m.; children—T. M. Logan, John Dickson, James Henry, Thomas Nelson Carter. Pathologist, vis. oculist Charity Hosp.; surgeon-in-chief Eye, Ear, Nose and Throat Hosp., New Orleans; prof. diseases of eye New Orleans Polyclinic. Mem. adv. council Nat. Com. for Prevention of Blindness. Translator (from French): Mind Your Eyes (Francisque Sarcey), 1886. Editor New Orleans Med. and Surg. Jour. Died May 19, 1933.

BRUNS, Herbert A. O., German biologist; b. Wilhelmshaven, Germany, July 11, 1920; s. August and

Berta (Brost) B.; ed. univs. Berlin, Göttingen (both Germany); Ph.D. in Natural Scis.; m. Margarete, Apr. 26, 1949; 1 son, Gunter. Faculty Inst. Applied Zoology, U. Wurtzburg (Germany), 1951-56; dir. Lab. Ornithol. Protection, State of Hamburg, W. Germany, 1956-63. Mem. German Soc. Zoology, German Ornithol. Soc., Fedn. German Biologists, Internat. Union Applied Ornithology, Internat. Soc. Protective Ornithology, World Union for Protection of Life. Editor: Ornithologische Mitteilungen; Biologische Mitteilungen; Angewandte Ornithologie; Das Leben Zeitschrift für Biologie und Lebensschutz. Co-editor: Waldhygiene. Address: 2 Hamburg-Sasel, Ilsenweg 11, W. Germany.

BRUNS, V. F., Am. research agronomist; b. Herman W. and Bertha (Neuhaus) B.; B.S. in Agr., Ft. Hays State Coll., 1944; postgrad. Wash. State U., 1953; m. Bernadine L. Kalb, Feb. 18, 1944; children—Rodrick K., Buckley K., Vicki L. With U. S. Dept. Agr., 1936-—, asst. agronomy in charge bindweed exptl. project, Canton, Kan., 1944-47, research agronomist weed investigations Irrigated Agr. Research and Extension Center, Prosser, Wash., 1947-—, prin. Agrl. Research Service, 1965-—. Mem. Weed Soc. Am., A.A.A.S., Ecol. Soc. Am., Western Weed Control Conf., Wash. State Weed Conf., Sigma Xi. Research, publs. on devel. of use of aromatic solvents, acrolein and other herbicides for control aquatic weeds in irrigation systems; contbd. to devel. use of 2,4-D amitrole, dalapon and other herbicides for control weeds on ditchbanks and other noncrop lands; studies on tolerance of crops to herbicides in irrigation water and residues, effect fresh water storage on germination of weed seeds. Home: 1819 Benson Av. Office: Irrigated Agr. Research and Extension Center, Prosser, Wash. 99350.*

BRUNS, Viktor von, see von Bruns, Viktor.

BRUNSCHVICG, Leon, French philosopher; b. Paris, France, Nov. 10, 1869; ed. at École normale, 1888; agrégé in philosophy, 1891; doctor, 1897; taught at Rouen, Paris, Lycées Condorcet and Henri IV; became prof. Sorbonne, Paris, 1909; mem. Franch Acad. Scis., 1919, Société des amis de Spinoza (founder), Kant's Gesellschaft, Acad. Moral and Polit. Scis., French Soc. Philosophy (founder, pres. 1936). Mem. central com. for rights of man League of Nations. Author: Spinoza et ses contemporains, 1894; La modalité du judgment, 1897; Introduction à la vie de l'esprit, 1900; L'idéalisme comtemporain, 1805; Nature et liberté, le génie de Pascal, 1925; Les etapes de la philosophie mathematique, 1912; De la vraie et de la fausse conversion. Proponent of philosophy of critical idealism; extended teachings of Kant and Hegel; regarded math as highest level of thought yet reached. Died Aix-en-Provence, France, Feb. 10, 1944.

BRUNSCHWIG, Alexander, Am. surgeon, gynecologist; b. El Paso, Tex., Sept. 11, 1901; s. Felix and Pauline (Harris) B.; B.S., U. Chgo., 1923, M.S., 1924; M.D., Rush Med. Coll., 1927; postgrad. Strasbourg, and Paris, France; M.D., (hon.), Laval U. Que., Can.; Doctoris Honoris Causa, U. Strasbourg (France), 1959, univs. Bordeaux, Montpellier; m. Lea Naye, June 16, 1926; children—Louise (Mrs. Paul Sivak), Roxane (Mrs. Bruno Pavia). Practice Chgo., 1928-—; surg. staff U. Chgo. Clinics, became prof. surgery U. Chgo., 1940; attending surgeon, chief gynecol., dept. Meml. Hosp. for Treatment Cancer and Allied Diseases; prof. clin. surgery Cornell U. Coll. Med., 1947-—; cons. gynecologist, surgeon N.Y. Infirmary; cons. surgeon N.Y. Polyclinic Hosp. Mem. Unitarian Service Com. Med. Teaching Missions to Czechoslovakia, 1946, Austria, 1947. Recipient Medal Charles U. of Prague, 1946; medal U. Brussels (Belgium); Officier Legion of Honor (France); Gold medal Societie des Journees Medicales de Bruxelles; medal U. Bologna (Italy), Lucy Wortham James award, James Ewing medal. Fellow A.C.S., Internat. Coll. Surgs. (hon.), Am. Surg. Assn.; hon. or corr. mem. various fgn. med. and profl. socs.; mem. Soc. Exptl. Biology, Soc. Exptl. Pathology, Assn. for Cancer Research, Soc. Clinical Surgery, Soc. U. Surgeons, A.M.A., Soc. Pelvic Surgeons, N.Y. Gynecol. Soc. Author: The Surgery of Pancreatic Tumors; Radical Surgery in Advanced Cancer of the Abdomen; L'exenteration pelvienne, 1964; also articles med. jours. Studies in bone sarcoma; correlated histologic changes and clin. symptoms in irradiated Hodgkin's disease and lymphoblastoma of lymph nodes; studied epithelealization of chronic osteomyelitic cavities; contbns. in various areas of cancer pathology and gynecologic surgery. Home: 223 Monterey Av., Pelham 65, N.Y. Office: 444 E. 68th St., N.Y.C.

BRUNSCHWIG, Hieronymus, surgeon; b. Germany, circa 1450; Alsatian army surgeon; lived in Strasbourg, France. Author: Dis ist das buch der Cirurgia, 1497; Liber de arte distillandi, 1500; Liber pestilentialis, 1500; Haussapoteck, 1539; Hausarzneybüchlein, 1591. Gave 1st description gunshot wounds, 1497; work on distillation became standard 16th century treatise on distilled remedies. Died 1533.

BRUNSON, Joel Garrett, Am. pathologist; b. Greenville, S.C., Apr. 22, 1923; s. James Edwin and Leila (Ballenger) B.; student Furman U., 1940-43, Miss. State U., 1944; M.D., U. Buffalo, 1950. Faculty, U. Minn. Med. Sch., 1955-59, asst. prof. pathology, sr. research fellow USPHS, 1957-59; prof., chmn. dept. pathology U. Miss. Med. Center, Jackson, 1959-—. Chmn. pathology, study sect. USPHS, 1963-—; cons. VA Hosp., Jackson, 1959-—. Diplomate Nat. Bd. Med. Examiners, Am. Bd. Pathology. Mem. Am. Assn. Pathologists and Bacteriologists, Internat. Acad. Pathology, PanAm. Med. Assn., N.Y. Acad. Scis., Cryobiology Soc., Am. Soc. for Exptl. Pathology, Reticuloendothelial Soc., A.A.A.S., Am. Nuclear Soc., Am. Heart Assn., Am. Assn. U. Profs., Nat. Soc. for Standard Med. Vocabulary, Miss. Assn. Pathologists (pres. 1966-—), Sigma Xi. Editorial bd. Am. Jour. Pathology. Research and publs. on exptl. cardiovascular diseases especially rheumatic fever and renal diseases, effects damaging agts. on susceptibility to infective agts. and to immune reacting mechanisms, effects of adrenal gland hormones on immune response, detection of malignant cells by fluorescence. Home: R.F.D. 2, Terry, Miss. 39170. Office: U. Med. Center, Jackson, Miss. 39216.*

BRUNTON, Thomas Lauder, physician; b. Hiltonshill, Roxburghshire, Scotland, Mar. 14, 1844; s. James and Agnes (Stenhouse) B.; M.D., U. Edinburgh (Scotland), 1865, M.B., C.M., 1866; Baxter scholar Vienna (Austria); studied at Berlin, Germany, Amsterdam, Netherlands, Leipzig, Germany; m. Louisa Jane Stopford, 1879; 3 sons, 3 daus. House physician Edinburgh Infirmary, 1866-67; became lectr. medicine and pharmacology Middlesex Hosp., 1870, St. Bartholomew's, 1871; casualty physician, 1871-75; asst. physician, 1875-95; physician, 1895-1904. Lettsomian lectr., 1886; Goulstonian lectr., 1877; Croonian lectr., 1889; Harveian orator, 1804. Fellow Royal Soc., 1874. Author: Textbook of Pharmacology and Therapeutics, 1885. Research and numerous publs. on physiol. action of drugs, circulation, treatment of disease, digitalis, nitrates, inorganic salts and enzymes; used amyl nitrate for tension and pain of angina pectoris; a founder of modern pharmacology. Died London, Sept. 16, 1916.

BRUNTON, William, engr., inventor; b. Dalkeith, Scotland, May 26, 1777; s. Robert Brunton; student mechanics and engring. in father's shop; m. Anne Elizabeth Button, Oct. 30, 1810. Work in cotton mills New Lanark, Scotland, from 1790, under Boulton and Watt, from 1796; partner, mech. mgr. Eagle Foundry, Birmingham, Eng., from 1815; civil engr. London, Eng., 1820-35; part owner Cwm Avon Tin Works, Glamorganshire, S. Wales, Maesteg works, from 1838, brewery Neath, S. Wales. Mem. Inst. Engrs. Designer mech. works, foundries; builder copper smelting furnaces; adapted original methods of reducing and mfg. metals; developed steam engines used in navigation on Humber, Trent and Mersey rivers, 1st Liverpool ferry, 1814; attempted to improve ventilation for collieries; developed calciner. Died Camborne, Cornwall, Eng., Oct. 5, 1851.

BRUSH, Charles Francis, Am. inventor; b. Euclid, O., Mar. 17, 1849; s. Isaac Elbert and Delia Williams (Phillips) B.; M.E. U. Mich., 1869, M.S. (hon.), 1899, Sc. D., 1912; Ph.D., Western Res. U., 1880, LL.D., 1900; LL.D., Kenyon Coll., 1903; m. Mary E. Morris, Oct. 6, 1875; children—Charles Francis, Mrs. Edna Perkins, Helene. Analytical chemist, Cleve., 1870-73; iron and ore commn. mcht., 1873-77; founder Brush Electric Co.; pres. Cleveland Arcade Co., from 1887; founder, pres. Linde Air Products Co. Trustee Western Res. U.; corporator Case Sch. Applied Sci.; established C. F. Brush Found. for the Betterment of the Human Race, 1928, for research in eugenics. Recipient Rumford medal for practical devel. of electric arc lighting, Am. Acad. Arts and Scis., 1899; Edison medal, 1913. Fellow Am. Acad. Arts and Scis. Pioneer investigator of electric lighting; invented Brush electric arc light, 1878, patented over 50 other inventions, including the storage battery and other devices. Died Cleve., June 15, 1929.

BRUSH, George Jarvis, Am. mineralogist; b. Bklyn., Dec. 15, 1831; s. Jarvis and Sarah (Keeler) B.; Ph.B., Yale, 1852, A.M. (hon.), 1857; LL.D., Harvard, 1886; enrolled mining sch. Freiberg, Germany, 1853-55, Royal Sch. Mines, London, Eng.; m. Harriett Silliman Trumbull, Dec. 23, 1864. Asst. in chemistry U. Va., 1852-53; prof. metallurgy Yale, 1855-71, prof. mineralogy, 1864-98, prof. emeritus, 1898-1912, dir. Sheffield Sch. Sci., 1872-98, also trustee. Mem. Nat. Acad. Scis. Author: Manual of Determinitive Mineralogy, 1874. Contbr. to J. D. Dana's System of Mineralogy. Collector 15,000 specimens of mineral species donated to Sheffield Sch. Died Feb. 6, 1912.

BRUSON, Herman Alexander, Am. chemist; b. Middletown, O., July 20, 1901; s. Samuel J. and Rebecca (Arnovitz) B.; B.Sc., Mass. Inst. Tech., 1923; D.Sc., Federal Polytecknik Inst., Zurich, Switzerland, 1925; m. Virginia Haber, Mar. 30, 1929; children—Rita (Mrs. Howard Vactor), Dorothy, Barbara. Chemist, Goodyear Tire & Rubber Co., Akron, O., 1925-28; chemist, group leader Rohm & Haas Co., Phila., 1928-48; mgr. high polymer research Indsl. Rayon Co., Cleve., 1948-52; v.p. chem. div. Olin Mathieson Chem. Corp., New Haven, 1952-—; sect. editor Chem. Abstracts, 1947-60; asso. prof. organic chemistry Temple U., 1939-48. Mem. Am., Brit. chem. socs., A.A.A.S., Am. Inst. Chemists, Chemists Club N.Y. Contbr. to Organic Reactions, 1949. Publs. and numerous patents in plastics, synthetics, petroleum products, detergents, bactericides, insecticides, other chem. compounds. Home: Pleasant Hill Rd., Woodbridge, Conn. 06525. Office: Olin Mathieson Chemical Corp., New Haven.*

BRUUN, Egon, Danish allergologist; b. Copenhagen, Denmark, Feb. 1, 1909; s. Arthur and Gella B.; M.D., U. Copenhagen; student Tb treatment Forlanini Inst., Rome, Italy, 1940; m. Birte Dela, Mar. 7, 1936. Asst. to profs. Klinge and Rossle, univs. Münster, Berlin (both Germany), from 1938; asst. prof. E. B. Salem, Stockholm, Sweden, from 1943; staff Broussais Hosp. and Pasteur Inst., Paris, France, from 1950; physician-in-chief Danish Red Cross Asthma Sanitarium, Norway, allergy clinic of hosp. U. Copenhagen, 1955-—; reader, chair of allergy clinic U. Copenhagen, 1950-—. Vis. lectr. fgn. univs.; pres. 5th Congress Study Allergy, Copenhagen, 1962. Decorated Distinguished Service Order, Danish Red Cross. Mem. Danish, No. (pres.) socs. allergology, Am. Acad. Allergy, European Acad. Study Allergy (v.p.). Author: Experimental Investigations in Serum Allergy, with Reference to Aethiology of Rheumatic Joint Disease, 1940; other publs. on internal medicine, exptl. and clin. allergy. Editor: Who's Who in Allergology; Laerebog i Allergology. Home: Gersonweg 8, Hellerup. Office: Ostergade 18, Copenhagen K., Denmark.

BRUYLANTS, Albert Léon Gustave Marie, Belgian chemist; b. Brussels, Belgium, Aug. 22, 1915; s. Pierre Joseph and Lambertha Marie (Huyberechts) B.; student Groeningen (Netherlands) U.; Docteur en Scis., Catholic U. Louvain (Belgium), 1938; hon. doctorate U. Rennes, 1965; m. Marie Celine Galley, Nov. 14, 1942; children—Anne, Monique, Philippe, Pierre-Albert, Olivier. Research fellow Fonds Nat. de la Recherche Scientifique, 1941-42; faculty U. Louvain, 1942-—, prof. chemistry, 1946-—, doyen faculty scis., 1964-68. Vis. prof. U. Marseilles (France), 1958; advanced fellow Belgian Am. Ednl. Found., 1959-60. Mem. Royal Acad. Belgium, Soc. Indsl. Chemistry (hon.), chem. socs. Belgium (pres. 1961-62), France, U. S., Switzerland. Author: (with J. C. Jungers, J. Verhulst) Chimie générale, 3 vols.; also articles. Research in photochem. chlorination of aliphatic hydrocarbons, properties and synthesis of nitrogen compounds, correlations between structure and reactivity of organic compounds from point of view of phys. organic chemistry. Home: 78, chaussée de Bruxelles, Winksele-Veltem. Office: Laboratoire de Chimie Générale et Organique, 98, rue de Namur, Louvain, Belgium.*

BRUYN, Cornelis Adriaan Lobry de, educator; prof. Amsterdam, Netherlands; worked in field of organic chemistry; studied reactions of di-nitro compounds, action of ammonia on carbohydrates (which yielded osamines), action of alcohols as solvents (leading to isolation of free hydroxylamine and hydrazine). Died 1904.

BRYAN, Clause S., Am. veterinarian; b. Bedminster, Pa., June 5, 1908; s. Amos and Anne (Stever) B.; B.S., Pa. State Coll., 1930; M.S., Mich. State Coll., 1932, Ph.D., 1937, D.V.M., 1942; m. Jean Lenore Miller, June 9, 1933; children—Marjorie Ann, Nelda Jane. Instr. bacteriology and pub. health Mich. State Coll., 1931-41, asst. prof., 1941-42, asso. prof., 1942-43, prof. bacteriology and pub. health, 1943-44, prof., head dept. surgery and medicine, 1944-49, dean of vet. medicine, from 1947. Mem. Am. Vet. Med. Assn., U. S. Livestock San. Assn., Am. Pub. Health Assn., Conf. Research Workers of N.Am., Internat. Assn. Milk and Food Sanitarians, A.A.A.S., Am. Dairy Sci. Assn. Author: Dairy Bacteriology and Public Health; contbg. author Bovine Mastitis; also numerous articles. Died July 30, 1951; buried Lansing, Mich.

BRYAN, George Hartley, Brit. mathematician, aero. pioneer; b. Cambridge, Eng., Mar. 1, 1864; s. Robert Purdie Bryan; ed. Peterhouse, Cambridge; m. Mabel Williams, 1906; 1 dau. Recipient Smith's prize, 1889, Hopkins prize, 1920, Gold medal Inst. Naval Architects, 1901, Gold medal Royal Aero. Soc., 1914, Silver medal Prestito Littorio, 1926-27, spl. grant Research Dept., 1917-20. Fellow Royal Soc., 1895, Former pres. Cambridge Entomol. Soc., Postal Micros. Soc., Math. Assn., Inst. Aero. Engrs.; hon. mem. Calcutta Math. Soc. Author: The Longitudinal Stability of Aeroplane Gliders, 1902; Stability in Aviation, 1911; The Rigid Dynamics of Circling Flight; The Acoustics of Moving Sources with Application to Air Screws, 1921; The Canonical Forms of the Equations of Motion of an Airplane in Still or Gusty Air, 1921; Effects of Compressibility on Stream Line Motions, 1911-20; The Theory of Initial motions and Its Applications to the Aeroplane, 1922. Pioneer in application fundamental equation of rigid dynamics to problem of stability of airplanes; used model gliders in aero. expts. Died Oct. 13, 1928.

BRYAN, Kirk, Am. geologist; b. Albuquerque, July 22, 1888; s. Richard W. D. and Susie Hunter (Patten) B.; A.B., U. N.M., 1909; D.Sc. (hon., 1947; A.B., Yale U., 1910, Ph.D., 1920; A.M. (hon.), Harvard, 1942; m. Mary Catherine MacArthur, July 11, 1923; children—Richard Conger, Mary Catherine, Kirk, Margaret Stuart. With U. S. Geol. Survey, from 1912 sr. geologist, from 1927; instr. geology Yale, 1914-17; lectr. physiography Harvard, 1926-27, asst. prof., 1927-30, asso. prof., 1930-43, prof., from 1943. Served to 2d lt., engrs., U. S. Army, 1918-19, AEF in France. Fellow A.A.A.S. (v.p. and chmn.

Sect. E, 1939), Geol. Soc. Am. (v.p. 1948), mem. Am. Acad. Arts and Scis., Assn. Am. Geographers, Soc. Am. Mil. Engrs., Am. Geog. Soc., Am. Geophys. Union, Boston Geol. Soc. (pres. 1936), Geol. Soc. Wash., N.W. Sci. Sigma Xi, Pi Kappa Alpha. Geologist, Columbia Basin Project, 1923; Nat. Geo. Soc., Chaco Canyon Expdn., 1923-25, Middle Rio Grande Conservancy Dist., 1927, 34, 35, Mexican Govt., San Juan Project; research in problem of frozen ground, glacial chronology, application of physiography and geology to problems in archaeology; developed concept of pediments as a land form characteristic of erosion in arid and semiarid regions. Died Aug. 22, 1950.

BRYAN, Worcester Allen, Am. surgeon; b. Alexandria, Tenn., Sept. 1, 1873; s. Joshua Lester and Elizabeth Jane (Wood) B.; A.B., Cumberland U., 1893, A.M., 1897; M.D., Vanderbilt U., 1899; postgrad. N.Y. Polyclinic, 1902, U. Vienna, 1910; m. Emma Horatia Berry, Sept. 7, 1904; children—Anne Smith, Elizabeth Nelson, Worcester Allen, Emma Berry. Asst. to chair practice of medicine Vanderbilt U., 1899-1900, to chair surgery and demonstrator of surgery, 1900-07, lectr. on principles of surgery, 1902-06, adj. prof. surgery, 1906-09, prof. principles of surgery, 1909-11, prof. surgery and clin. surgery, 1911-25, prof. clin. surgery, 1925—; prof. oral surgery Vanderbilt Dental Dept., 1902-25; surgeon Protestant, Vanderbilt and Nashville gen. hosps., Watauga Sanitarium, Ridgetop, Tenn. Author: Principles of Surgery. Died Apr. 30, 1940.

BRYANT, Frank Augustus, Am. physician, surgeon; b. North Jackson, Pa., Oct. 18, 1851; s. Chauncey Elliott and Hannah (Corse) B.; M.D., Bellevue Hosp. Med. Coll. (New York U.), 1895; postgrad. Post-Grad. Med. Sch. and Hosp., New York; m. Sarah M. Mitchell, Nov. 28, 1871. Practiced in N.Y., 1895—; prin. of sch. for correction of speech disorders; resident physician Burke Found. for Convalescents. Author: Causes and Treatment of Stammering, 1907; Speech Disorders and Their Treatment, 1914; Manual of Free Gymnastics, 1884. Died Apr. 1921.

BRYANT, Jay Clark, Am. biochemist; b. Susquehanna, Pa., Feb. 6, 1905; s. Clayton Summer and Mary (Tallman) B.; B.S., Pa. State U., 1932; M.S., Cornell U., 1955; Ph.D., Georgetown U., 1963; m. Jean Barzhe, Oct. 30, 1940. Research biochemist tissue culture lab. Lab. Biology Nat. Cancer Inst., NIH, Bethesda, Md., 1947—. Mem. N.Y. Acad. Scis., Am. Chem. Soc., Am. Assn. Cancer Research, Soc. Cell Biology, Am. Soc. Exptl. Pathology, Tissue Culture Assn., Soc. Cryobiology, A.A.A.S. Author: Tissue Culture. Contbr. chpt. (with V. J. Evans) to advances in Tissue Culture at National Cancer Institute in U. S. A., 1965. Research and publs. on long-term cultivation of strains of mammalian cells in vitro and their relationship to cancer in vivo, chemistry of run-off water from soils. Home: 1511 Glenallan Av., Silver Spring, Md. 20902. Office: 9000 Wisconsin Av., Bethesda, Md. 20014.*

BRYANT, Joseph Decatur, Am. surgeon; b. East Troy, Wis., Mar. 12, 1845; s. Alonzo and Harriet (Adkins) B.; M.D., Bellevue Hosp. Med. Coll., 1868; LL.D., New York U., 1908; m. Annette Crum, 1874. Intern Bellevue Hosp., 1869-71; asst. to chair of anatomy Bellevue Coll., 1871, lectr. on surg. anatomy (summer course), 1871-74, asst. demonstrator anatomy, 1875-77, prof. gen. descriptive and surg. anatomy, 1877-83, asso. prof. orthopedic surgery, 1883-95; prof. principles and practice of surgery, operative and clin. surgery Univ. and Bellevue Hosp. Med. Coll., 1898—; san. insp. N.Y.C. Health Dept., 1873-79, commr., 1887-93; commr. N.Y. State Bd. Health, 1887-93; staffs of Governors Cleveland, Hill and Flower; cons. or vis. surgeon West Side Dispensary, 1872-75, Bur. for Med. and Surg. Relief, 1874-80, 82, Charity Hosp., 1881-82, Bellevue Hosp., 1882, N.Y.C. Insane Asylum, 1882—, St. Vincent's Hosp., 1887—, Northwestern Dispensary, Hosp. for Ruptured and Crippled, Woman's Hosp., St. Joseph's Hosp. (Yonkers), Hackensack Hosp.; physician to Grover Cleveland and family. Author med. monographs; Operative Surgery, 2 vols., 4th edit.; Bryant and Buck's American System of Surgery, 8 vols. Died Apr. 7, 1914.

BRYANT, W(illiam) Schier, Am. physician; b. Boston, May 15, 1861; s. Henry and Elizabeth Brimmer (Sohier) B.; A.B., cum laude, Harvard, 1884, A.M., M.D., 1888; m. Martha Lyman Cox, 1887 (dec.); children—Mrs. Mary Cleveland Blanchard, Elizabeth Sohier, Mrs. Alice de Vermandois Frank, Julia Cox, Gladys de Brion, William Sohier. Began practice in Boston; aural surgeon Boston Dispensary; asst. in anatomy and otology Harvard; sr. asst. surgeon Mass. Charitable Eye and Ear Infirmary; in N.Y., 1903; adj. prof. dept. diseases of ear N.Y. Post-Grad. Med. Sch. and Hosp.; cons. otolaryngologist Manhattan State Hosp.; sr. asst. surgeon aural dept. N.Y. Eye and Ear Infirmary; instr. otology Coll. Phys. and Surg. (Columbia); clin. asst. dept. otology Vanderbilt Clinic; asst. surgeon St. Bartholomew's Clinic; clin. instr. and attending surgeon, otol. dept. Cornell U. Med. Sch.; physician in class of nose, throat and ear diseases Presbyn. Hosp. Dispensary. Fellow A.C.S., Boylston Med. Soc.; mem. Boston Med. Library Assn., Mass. Med. Soc., Mass. Benevolent Med. Soc., A.M.A., (chmn. sect. laryngology and otology), Am. Otol. Soc. med. socs. State N.Y., County N.Y., Am. Bd. Otolaryngology, Am. Laryngol., Rhinol. and Otol. Soc., Assn. Mil. Surgeons of U. S. Author: Anatomy and Physiology of the Ear, and Tests of Hearing (in Burnett's System of Diseases of the Ear, Nose and Throat), 1893; Ear Section, Knight and Bryant's med. publs. Deceased.

BRYCE, James (Viscount of Dechmont), Brit. statesman, historian, polit. writer; b. May 10, 1838; s. James and Margaret (Young) B.; ed. U. Glasgow (Scotland), LL.D., 1886; B.A., Trinity Coll., Oxford (Eng.) U., 1862, D.C.L., 1870, Litt.D., 1914; numerous hon. degrees Brit., U. S., Australian, Canadian, European univs.; m. Elizabeth Marion Ashton, 1889. Called to bar Lincoln's Inn, 1867; regius prof. civil law Oxford (Eng.) U., 1870-93; M.P. for Tower Hamlets, from 1880, for Aberdeen (Scotland), 1885-1907; under-sec. state for fgn. affairs, from 1886; chancellor Duchy Lancaster (Eng.), from 1892; pres. Bd. Trade, from 1894; chief sec. for Ireland, 1905-07; A.E. and P. at Washington, 1907-13. Chmn. Royal Commn. on Secondary Edn., 1894; mem. senate London U. Fellow Royal Soc., 1893, Royal Geog. Soc. (hon.); mem. Inst. France (fgn.), royal acads. Turin (Italy), Brussels (Belgium), Naples (Italy), St. Petersburg (now Leningrad, USSR), Stockholm (Sweden), Brit. Acad. (pres.), others. Author: The Flora of the Island of Arran, 1859; The Holy Roman Empire, 1862; Report on the Condition of Education in Lancashire, 1867; The Trade Marks Registration Act with Introduction and Notes on Trade Mark Law, 1877; Transcaucasia and Ararat, 1877; The American Commonwealth (chief work, classic aimed at portraying U. S. polit. system in theory and practice), 1888; Impressions of South Africa, 1897; Studies in History and Jurisprudence, 1901; Studies in Contemporary Biography, 1903; The Hindrances to Good Citizenship, 1909; South America: Observations and Impressions, 1912; University and Historical Addresses, 1913; Essays and Addresses on War, 1918; Modern Democracies (compared prin. rep. govts. of world), 1921. Died Jan. 22, 1922.

BRYCE, Thomas Hastie, Scottish physician; b. Oct. 20, 1862; s. William Bryce; ed. U. Edinburgh; M.A., M.S., LL.D., Glasgow. Lectr. anatomy Queen Margaret Coll., Glasgow, 1890; lectr. anatomy U. Glasgow, 1892-1909, prof., 1909-35. Mem. Royal Commn. on Ancient Monuments, Scotland, 1908. Fellow Royal Soc., 1922, also Royal Soc. Edinburgh (Keith prize 1906); pres. Glasgow Archaeol. Soc., Soc. Antiquaries of Scotland. Author: The Book of Arran, Prehistoric Sepulchral Remains, 1910; also papers on anatomy, anthropology, archaeology. Editor Quain's Anatomy, 11th edit., vols. I, IV, parts 1 and 2. Demonstrated (with John Hammond Teacher) human embryo aged 13 to 14 days (Bryce-Teacher ovum), 1908; traced devel. of vascular relations of spleen. Died Oxford, Eng., May 16, 1946.

BRYCE-SMITH, Derek, Brit. chemist; b. London, Eng., Apr. 29, 1926; s. Charles Phillip and Amelia (Thick) B.-S.; B.Sc., U. London, 1945, B.Sc., 1948, Ph.D., 1951, D.Sc., 1961; m. Marjorie Mary Anne Stewart, Sept. 8, 1956 (dec. May 1966); children—Madeleine Anne, Duncan Charles, Hazel Virginia, David Alexander. Research chemist Powell Duffryn Research Ltd., London, 1945-46; research chemist Dufay-Chromex Ltd., Boreham, Wood, Hertfordshire, 1946-48; Imperial Chem. Industries postdoctoral fellow King's Coll., London, 1951-55, asst. lectr., 1955-56, lectr. chemistry U. Reading (Eng.), 1956-63, reader in chemistry, 1963-65, prof. organic chemistry, 1965—. Cons. Shell Research Ltd., 1959-61, Esso Research Ltd., 1961-64, E. I. duPont de Nemours & Co., Wilmington, Del., U. S., 1964—. Fellow Chem. Soc. (London), Am. Chem. Soc. Research, numerous publs. on photochemistry of aromatic compounds and p-quinones; discoverer, or co-discoverer photo-reactions of benzenes, isomerization to fulvenes; ring-contractive 1, 3-cycloaddition of olefins, polymerization, ring-expansion to cyclo-octatetraenes by addition of acetylenes, addition of maleic anhydride and maleimides; research on chemistry of unsolvated Grignard reagts. and alkylmagnesium alkoxides and organometallic compounds of potassium, sodium, lithium, calcium, silver; mechanistic studies on alkali-metallation of aromatic rings. Home: Orchard House, Grove Rd., Sonning Common, nr. Reading, Eng.*

BRYDON, James Emerson, Canadian mineralogist; b. Portage la Prairie, Man., Can., June 28, 1928; s. Frank A. and Cora (Grobb) B.; B.Sc. with honours, U. Man., 1951; M.Sc., U. Mo., 1954, Ph.D., 1956; m. Vera Warren, Nov. 22, 1951; children—Dianne, Jereleen, Franklin Leigh. With chemistry div. Can. Dept. Agr., 1951-58, research scientist Soil Research Inst., Ottawa, Ont., 1959—. Mem. Agrl. Inst. Can., Canadian Soc. Soil Sci. (sec., pres.-elect), Soil Sci. Soc. Am., Mineral. Assn. Can., Mineral Soc. Am., Clay Minerals Soc. Research and publs. on distbn. of clay minerals in Canadian soils and sediments, nature of alteration and decomposition of clay minerals under various natural and lab. conditions; co-discoverer of naturally occurring dioctahedral A1 chlorite.*

BRYGOO, Edouard Raoul, French biologist; b. Likke, France, Apr. 22, 1920; s. Robert J. and Yvonne (Daume) B.; Docteur en Médecine, Bordeaux, France, 1943; Ph.D., Paris, 1963; m. Janine Beauche, Aug. 10, 1943; children—Yves, Anne, Kate, Claude. With Mobile Hygiene Service, Cameroon, 1946-50; head Pasteur Inst. Lab., Saigon, Vietnam, 1950-52; asst. dir. Pasteur Inst., Tananarive, Madagascar, 1954-61, dir., 1962—. Head plague-bilharziases dept. Ministry Health of Madagascar. Mem. Madagascan Acad., Internat. Soc. Mycology, Soc. Exotic Pathology. Author: Parasitologie des Caméleons de Madagascar, 1963; also articles. Research in comparative and human parasitology, microbiology; studies on plague, bilharziases, flariases and mycoses of Madagascar. Home: 26, rue du XIV Juillet, Bergerac 24, France. Office: Institut Pasteur, BP 1224, Tananarive, Madagascar.*

BRYSON OF HERACLEA, Greek geometer, Pythagorean; b. circa 520 B.C.; student of Pythagoras (according to Iamblichus); contemporary of Antiphon; advanced problem of quadrature of the circle by circumscribing and inscribing polygons at same time; with Antiphon prepared elaboration of method of exhaustion. Died 450 B.C.

BRYSON, Reid Allen, Am. meteorologist; b. Detroit, June 7, 1920; s. William R. and Elma (Turner) B.; B.A., Denison U., 1941; postgrad. U. Wis.; Ph.D., U. Chgo., 1948; m. Frances E. Williamson, June 13, 1942; children—Anne, William, Robert, Thomas. Faculty, U. Wis., Madison, 1946-56, prof., 1957—, chmn. dept. meteorology, 1957-60; prof., head dept. meteorology U. Ariz., 1956-57. Mem. com. on geography Nat. Acad. Sci.-NRC, 1958-63, chmn., 1961-63. Mem. Wis. Phenological Soc. (past pres.), Phi Kappa Phi (hon.). Author: (with J. F. Lahey, H. A. Corzine, C. W. Hutchins) Atlas of 300 Millibar Wind Characteristics, 1960. Research and numerous articles on application of climatology to anthropol. and geog. problems, use of meteorol. techniques in phys. limnology, circulation patterns of lakes, identification of climatic change asso. with cultural change in Gt. Plains of 13th century. Office: Dept. Meteorology, U. Wis., Madison, Wis. 53706.*

BUACHE, Jean Nicolas, French geographer; b. La Neuville, France, Feb. 15, 1741; geographer to King of France, from 1782; prof. geography l'École Normale, from 1794; curator Naval Depository; mem. Bur. Longitudes; mem. French Acad. Scis., 1782. Author memoir on Solomon Islands. Reported on improvement of maps to French Acad. Scis., 1793. Died Nov. 21, 1825.

BUACHE (DE VERPONT), Philippe, French geographer; b. Paris, Feb. 7, 1700; studied architecture at Academie; m. dau. of Guillaume Delisle, 1729. Became 1st geographer to king, 1729; geography tutor to son of dauphin. Mem. French Acad. Scis., 1730. Author: Essai de géographie physique (divided seas and land masses into symmetrical basins separated by mountain chains), 1752; Considerations géographiques sur les nouvelles découvertes de la grande mer (suggested link between Asia and Am., also existence of Alaska and Aleutian Islands), 8 vols., 1753; Atlas physique, 1754; Parallele des fleunes et toutes les parties du monde. Died Jan. 27, 1773.

BUBNOFF, Serge von, see von Bubnoff, Serge.

BUCALOSSI, P., Italian physician; b. San Miniato, (Pisa), Italy, Aug. 9, 1905; s. Alfredo and Cosetti Maria Bucalossi; med. degree U. Pisa, Sch. Medicine, 1928; m. Eugenia Goisis, Jan. 12, 1938. Staff, Nat. Cancer Inst., Milan, Italy, 1935—, dir.-gen., 1957—; pres. Fatebenefratelli Hosp., Milan. Vice pres. for Europe, UICC; exec. com. Internat. Research Agy. for Cancer; mayor Milan, 1964—; hon. dep. Chamber of Deps. attached to work and Social Security sect. Mem. Italian League against Cancer (pres. Milanese sect.), Italian Cancer Soc., Italian Soc. Surgery, Italian Soc. Pathology, Assn. Fancaise pour l'étude due cancer. Dir., Jour. Tumori. Research, numerous publs. on cancer especially endocrinological and endocrinosurg. problems of cancer, pathology and clin. study of mammary cancer and malignant melanoma, etiological factors of most malignant tumors. Home: 15 Via Bigli. Office: 22 Piazzale Gorini, Milano, Italy.*

BUCH, Baron Christian Leopold von, see von Buch, Baron Christian Leopold.

BUCH, Kurt Karl Wilhelm, Finnish chemist; b. Helsinki, Finland, May 9, 1881; s. Mac and Anna (Stirtmann) B.; Ph.D., U. Helsinki; m. Mathild Lindegvist, Apr. 11, 1908; children—Anna-Martia, Nora-Margareta. With chem. sect. Inst. Research, Helsinki, from 1919; prof. chemistry Acad. Abo (Turku, Finland), from 1934; prof. U. Helsinki, 1942-51, retired, 1951. Mem. Assn. Finnish Chemists (hon.), numerous sci. socs. Finland. Contbd. numerous publs. on chemistry of sea and atmosphere and their counterbalances. Address: Rönnvägen 50, Helsinki, Finland.

BUCHAN, Alexander, Scottish meteorologist; b. Kinnesswood, Scotland, Apr. 11, 1829; ed. Edinburgh (Scotland) U. Tchr., 1848-60; became sec. Scottish Meteorol. Soc., 1860; curator mus. and library Royal Soc. Edinburgh, 1878-1906; a founder obs., Ben Nevis, Scotland, 1883. Fellow Royal Soc., 1898. Author: Handy Book of Meteorology, 1867; (papers)

Mean Pressure and Prevailing Winds of the Globe, 1869, Report on Atmospheric Circulation, 1889. Advanced sci. study of meteorology; established that pressure varies with longitude and latitude in world maps of isobars for months and year, pub. 1868-69; studied mean specific gravity of oceans, prevailing winds, and atmospheric pressure. Died Edinburgh, May 13, 1907.

BUCHAN, William, Brit. physician; b. Ancram, Eng., 1729; student divinity, medicine Edinburgh (Scotland) U.; practiced medicine Yorkshire, later Ackworth; physician to foundling hosp.; specialist diseases of children, Edinburgh, circa 1766, later London, Eng., circa 1778. Recipient Gold medal from Empress of Russia. Author: De Infantum Vita Conservanda, 1761; Domestic Medicine; or the Family Physician, 1769; Observations concerning the Prevention and Cure of the Veneral Disease, 1796. Died London, Feb. 25, 1805.

BUCHANAN, George, English physician; b. Myddelton Square Islington, Eng., Nov. 5, 1831; s. George Adam B.; B.A., U. Coll., London, Eng., 1851; M.D., London U., 1855; LL.D., U. Edinburgh (Scotland), 1893; m. Mary Murphy; m. 2d, Alice Mary Asmar Seaton; 2 sons, 4 daus. Physician, London Fever Hosp., 1861-68; became fellow U. Coll., 1864. cons. physician; Lettsomian lectr., 1867; apptd. permanent insp. med. dept. of privy council, 1869; prin. med. officer local govt. bd., 1879-92; censor, 1892-94. Chmn., Royal Commn. on Tb. Fellow Royal Coll. Physicians, Royal Soc., 1882. Research and publs. on typhus fever, control of cholera; reduced mortality from phthisis; did much to secure extinction of typhoid fever where formerly it was endemic; helped bring cholera and Tb under control; founded central pub. health dept. of state in Eng. Died London, May 5, 1895.

BUCHANAN, George, Brit. surgeon; b. 1827; M.A., M.D., LL.D.; civil surgeon Army in Crimea; surgeon Western Infirmary, Glasgow, Glasgow Royal Infirmary; prof. Clin. surgery Glasgow U., to 1900; pres. surg. sect. Brit. Med. Assn., 1888. Author: Camp Life in the Crimea; Clinical surgery (an inaugural address), 1874; On Lithotrity, with Cases, 1880; Radical Cure of Inguinal Hernia in Children, 1880; Talipes Varus, 1880; Faure's Storage Battery, and Electricity in Surgery, 1881. Editor: Glasgow Med. Jour.; Anatomists Vade Mecum, 10th edit. Died Apr. 19, 1906.

BUCHANAN, George Dale, Am. anatomist; b. Wichita Falls, Tex., Oct. 1, 1928; s. Arthur Odell and Ollie (Gertrude (Wilkins) B.; B.A., Rice Inst., 1950, M.A., 1954, Ph.D., 1956; m. Suzanne Freeman, Sept. 8, 1950; children—Richard Dale, Robin Sue. Faculty, Sam Houston State Coll., Huntsville, Tex., 1956-58, asso. prof. biology, 1956-58; research asso. Rice Inst., 1958-59; faculty U. Tenn. Med. Units, Memphis, 1959—, asso. prof. anatomy, 1966—; vis. prof. morphology and biology U. del Valle, Cali, Colombia, 1966. Mem. A.A.A.S., Am. Inst. Biol. Scis., Am. Assn. Anatomists, So. Soc. Anatomists, Am. Soc. Zoologists, Endocrine Soc., Soc. for Study Fertility (Eng.), Sigma Xi (Research award Rice chpt. 1954). Research and publs. on calcium and phosphate excretion in rats, mice, dogs, chickens as affected by parathyroid glands, delayed implantation in armadillos, natural history armadillos, reproductive physiology of ferret including histology and histochemistry of uterus during reproductive cycle. Address: U. Tenn. Med. Units, 62 S. Dunlap St., Memphis 38103.*

BUCHANAN, James William, Am. zoologist; b. Basil, O., Jan. 30, 1888; s. James Wilson and Almeda (Jenkins) B.; B.S., Ohio U., 1913; postgrad. U. London (Eng.), 1919; Ph.D., U. Chgo., 1921; m. Pearle Oliver, July 20, 1918; children—James O., William Ervine. Began as tchr., 1906; asso. prof. biology U. Miss., 1913-15; fellow in zoology U. Chgo., 1915-16; instr. zoology N.Y. U., 1916-17; sr. instr. 3d Army Post Schs., Coblenz, Germany, 1919; instr. and asst. prof. biology Yale, 1921-30; asso. prof. zoology Northwestern U., 1930-33, prof., 1933-49, chmn. dept., 1940-49; Morrison prof. zoölogy, 1945-49; acting dean Coll. Liberal Arts, 1945-46. Hancock prof. zool., dir. research Hancock Found., U. So. Cal., since 1949. Fellow A.A.A.S. (sec. Sect. F. 1940-48); mem. Am. Soc. Zoölogists, Am. Physiol. Soc., Am. Nature Assn., Chgo. Acad. Sci., (hon. life mem.), Phi Beta Kappa, Sigma Xi. Died June 27, 1952; buried Inglewood, Cal.

BUCHANAN, John Machlin, Am. biochemist; b. Winamac, Ind., Sept. 29, 1917; s. Harry James and Eunice (Miller) B.; B.S., DePauw U., 1938; M.S., U. Mich., 1939, D.Sc., 1961; Ph.D., Harvard, 1943; m. Elsa Nilsby, Dec. 11, 1948; children—Claire Louise, Stephen James, Lisa Renée, Peter Nilsson. Faculty, U. Pa., 1943-46, 48-49; NRC fellow Med. Nobel Institut, Stockholm, Sweden, 1946-48; faculty U. Pa., 1949-53, prof. biochemistry, 1950-53; prof., head div. biochemistry Mass. Inst. Tech., 1953-67, John and Dorothy Wilson prof., 1967—; vis. prof. biochemistry U. Cal. at Berkeley, 1959. Mem. Nat. Acad. Scis., Am. Soc. Biol. Chemists, Am. Chem. Soc. (Eli Lilly award in biol. chemistry 1951), Am. Acad. Arts and Scis. Jours. Research and numerous articles on biochemistry glycogen, oxidation ketones, fatty acids and acetoacetates, biosynthesis purine nucleotides, biosynthesis methionine and function tetrahydrofolic acid, vitamin B12, isolation and identification 5-methyltetrahydrofolic acid, metabolism E. coli B after infection with bacteriophage, enzymatic fixation nitrogen, determination structural site enzymes. Home: 56 Mariam St., Lexington, Mass. 02173. Office: Dept. Biology, Mass. Inst. Tech., Cambridge, Mass. 02139.*

BUCHANAN, Roberdeau, Am. astronomer; b. Phila., Nov. 22, 1839; s. McKean and F. Selina (Roberdeau) B.; B.S., Lawrence Sci. Sch., Harvard, 1861; m. Lyla M. Peters, 1888. Mathematician, Nautical Almanac Office, U. S. Naval Obs.; made calculations for Nautical Almanac, from 1879. Author: Observations on the Declaration of Independence, 1890; Treatise on the Projection of the Sphere, 1890; The Mathematical Theory of Eclipses; several genealogies, biography of Thomas McKeen. Died Dec. 18, 1916.

BUCHANAN, William, Am. polit. scientist; b. Richmond, Va., Dec. 25, 1918; s. Daniel Littleton and Cora (Briggs) B.; A.B., Washington and Lee U., 1941, M.A., 1941; M.A., Princeton, 1953, Ph.D., 1955; m. Vivian Landrum, Aug. 8, 1946; children—James Landrum, David Briggs, Mary Warrington. Asst. prof. govt. Miss. State U., 1952-56; asst. prof. polit. sci. U. So. Cal., Los Angeles, 1956-58, asso. prof., 1958-62; prof. polit. sci. U. Tenn., Knoxville, 1962-65; prof., chmn. polit. sci. dept. Washington and Lee U., Lexington, Va., 1965—; vis. research prof. legislative process U. Cal. at Berkeley, 1959-60. Mem. council Interuniv. Consortium for Polit. Research, 1964-66. Mem. Am. (mem. council 1966—), So. polit. sci. assns., Am. Assn. U. Profs. Author: (with Hadley Cantril) How Nations See Each Other, 1953; (with J. C. Wahlke, H. Eulau, L. C. Ferguson) The Legislative System, 1962; Legislative Partisanship, 1963; also articles. Research on pub. opinion and its impact on govt. through polit. parties, electoral system and legislative process. Home: 618 Ross Rd., Lexington, Va. 24450.*

BUCHER, Gordon Edwards, Canadian insect pathologist; b. Toronto, Ont., Can., Mar. 28, 1917; s. Otto Frank and Junietta (Edwards) B.; B.A., U. Toronto, 1937, M.A., 1939; Ph.D., Ohio State U., 1946; m. Marion Dearle, Sept. 10, 1949; 1 dau., Carolyn. Entomologist, Can. Dept. Agr., 1946—, officer-in-charge research lab., Kingston, Ont., 1949-55, insect pathologist Research Inst., Belleville, Ont., 1955—. Mem. Canadian, Ont., Am. entomol. socs., Canadian Soc. Microbiologists, Sigma Xi. Contbr. chpt. to Insect Pathology, 1963. Mem. editorial bd. Jour. Invertebrate Pathology, 1963-66. Research and publs. on entomophagous chalcids, physiology of house fly especially effect of cold on survival, population dynamics of forest insects, insect diseases and microorganisms of insects as grasshoppers, tent caterpillars, hornworms, cutworms; discovered new species of Clostridium; demonstrated pathogenicity of Pseudomonas and Torula. Home: 13 Rosewood Av. Office: 228 Dundas St. E., Belleville, Ont., Can.*

BUCHER, Karl, Swiss physician, pharmacologist; b. Basel, Switzerland, Sept. 22, 1912; s. Karl and Ida (Hauser) B.; M.D., U. Basel, 1936; m. Elisabeth Baumgartner, July 17, 1941; children—Urs-Michael, Katharine. Pharmacologist, Inst., U. Basel, 1936-40; research pharmacologist, CIBA Ltd., Basel, 1940-46; research fellow med. physics U. Cal. at Berkeley, 1947; chmn. Inst. Pharmacology, U. Basel, 1948—, prof., 1949—. Author: Die reflektorische Steurerung der Lungenatmung, 1952; also numerous articles. Research on regulation of respiration, cough, antitussives, gas exchange in lungs, elastic forces in lungs, interrelationships between left and right heart performance, antirheumatics. Home: 2 Nonnenweg, 4000, Basel, Switzerland CH.*

BUCHER, Otto Max, Swiss histologist, embryologist, physician; b. Lucerne, Switzerland, Mar. 1, 1913; ed. univs. Zurich (Switzerland), Berlin (Germany); M.D.; m. 1937; children—Edith, Max, HansJörg. Asst., Inst. Anatomy, U. Zurich, 1941-47; instr. anatomy, histology, embryology U. Zurich, 1941—, titular prof., 1947—; asso. prof. histology and embryology, dir. Inst. Histology and Embryology, Sch. Medicine, Lausanne, Switzerland, 1950—. Mem. Internat. Soc. Cell Biology, Tissue Culture Club, Assn. Anatomists, others. Author: Histologie und mikroskopische Anatomie des Menschen, mit Berucksichtigung der Histophysiologie und der mikroskopischen Diagnostik, 1959. Contbr. sci. jours. Home: 3, rue du Trabandan, Lausanne. Office: 9, rue du Bugnon, Lausanne, Switzerland.

BUCHER, Walter H(erman), Am. geologist; b. Akron, O., Mar. 12, 1888; s. August J. and Maria (Gebhardt) B.; Ph.D. U. Heidelberg (Germany), 1911; Sc.D. (hon.), Princeton, 1947, Columbia, 1957; m. Hannah E. Schmid; children—John Eric, Mary Dorothy, Margaret Louise, Robert Walter. With U. Cin., 1913-40, prof. geology, 1925-37, head dept., 1937-40; prof. structural geology Columbia, 1940-56, prof. emeritus, 1956, head dept., 1950-53. Recipient Bowie medal Am. Geophys. Union, 1954; L. v. Buch medal, Deutsche Geologische Gesellschaft, 1955. Fellow Am. Acad. Arts and Scis.; mem. Nat. Acad. Sci., NRC (div. chmn. 1940-43), Geol. Soc. Am. (pres. 1954-55), Geophys. Union (pres. 1948-53), Paleontol. Soc. Am., A.A.A.S., N.Y. Acad. Scis. (pres. 1944); hon. mem. Geol. de France, Deutsche Geol. Ges., Soc. Geol. Belgique. Author: The Deformation of the Earth's Crust, 1933; also articles sci. jours. Advanced theory on origin and geog. pattern of mountain chains on earth's surface, 1956; studied joints; cryptovolcanic structures; primary structures of sediments; tectonics. Died Houston, Feb. 17, 1965.

BUCHERER, Alfred Heinrich, physicist; b. Cologne, Germany, July 9, 1863; s. Heinrich and Eleanor Anne (Archibald) B.; student Technische Hochschule Hannover (Germany), 1884, Johns Hopkins, from 1885, under K. F. Braun, Strasbourg, France, after 1895; m. Camilla Hegeler, 1892. Prof. phys. chemistry, Bonn, Germany, 1899-1923. Author: Experimentelle Bestätigung des Relativitätsprinzips, 1908; Quantentheorie und Gravitationseffekt, 1924. Conducted expts. which confirmed Lorentz' and Einstein's theory of relativistic dynamics and aided in gen. recognition of theory of relativity, 1908-09. Died Bonn, Apr. 16, 1927.

BUCHERER, Hans Theodor, German chemist; b. Cologne-Ehrenfeld, Germany, May 19, 1869; s. Heinrich and Eleanor Anne (Archibald) B.; m. Heina Oppermann, 1895; 4 sons, 1 dau. Active in industry; prof. chem. tech. Technische Hochschule, Munich, Germany, 1926-34. Author: Die Teerfarbstoffe mit besondere Berucksichtigung der synthetischen Methode, 1904; Lehrbuch der Farbenchemie, 1921; (with R. Möhlau) Farbenchemische Praktik, 1920. Worked in dye chemistry and dyeing technique; Bucherer reaction of salts of sulphur acids on aromatic amines and phenols became technically important in naphthaline series. Died Benediktbeuern, Germany, May 29, 1949.

BÜCHERL, Emil Sebastian, German physiologist; b. Furth, Germany, Nov. 6, 1919; s. Alois and Burga; student univs. Munich, Rome, Heidelberg; M.D.; m. Rosemarie, May 29, 1957; 1 dau., Anja Sabine. Joined Surgery Clinic, U. Heidelberg, 1944; staff Amberg Municipal Hosp., also U. Inst. Pathology, Göttingen, Germany, 1948-51; Sabbastberg, 1951-52; with Surgery Clinic, Berlin U., 1957—. Mem. Internat., German, Berlin surgery socs., Berlin Med. Soc., Am. Artificial Organs Soc. Research and publs. on organ transplants, heart surgery, artificial heart, physio-pathology of blood vessels, treatment of operative shock. Home: Joachim Friedrich Strasse 53, West Berlin, West Germany. Office: Feie Universität, Chir. Klinik Neukölln, Berlin, West Germany.

BUCHHEIM, Rudolf Richard, pharmacologist; b. Bautzen, Germany, Mar. 1, 1820; s. Christian and Amalia (Bruchmann) B.; ed. Dresden, also Leipzig, Germany; doctorate, 1845; m. Minna Coelestine Pescheck, 1845; 3 sons, 3 daus. Apptd. asso. prof. pharmacology, dietetics, history medicine, med. lit., Dorpat (now Tartu, Estonia), 1847, established univ. inst. pharmacology (prototype found.; pub. more than 100 exptl. works under his direction); prof. pharmacology, Giessen, Germany, from 1867. Author: Über pharmakologische Untersuchungen, 1856; Lehrbuch der Arzneimittellehre, 1856. Introduced exptl. chem. and physiol. methods into pharmacology. Died Giessen, Dec. 25, 1879.

BUCHHOLZ, John Theodore, Am. botanist; b. Polk County, Neb., July 14, 1888; s. Conrad C. and Christine (Weber) B.; B.S., Ia. Wesleyan Coll., 1909; A.B., State U. Ia., 1909; M.S., U. Chgo., 1914, fellow, 1916-17, Ph.D., 1917; studied Ia. Lakeside Lab. and Cold Spring Harbor, N.Y.; m. Olive Peterson, Aug. 15, 1912; children—Olive Miriam, Christine, Ruth Elizabeth. Mem. faculty Ark. State Normal Sch., Conway, 1909-18; prof. biology, West Tex. State Normal Coll., Canyon City, 1918-19; prof. botany, head dept. U. Ark., 1919-26; prof. botany U. Tex., 1926-29; prof. botany U. Ill., from 1929, head dept., 1938-42. Visiting investigator Carnegie Instn. dept. of genetics, Cold Spring Harbor, summers, 1921-41. Fellow A.A.A.S.; mem. Bot. Soc. Am. (pres. 1941), Am. Soc. Naturalists, Genetics Soc. of America, Am. Assn. Univ. Profs., Torrey Bot. Club, Sigma Xi. Research and publs. on morphology and embryology of conifers; on the genetics of Datura, especially the role of pollen-tube growth in heredity of polyploids, on plants with extra chromosomes and genes affecting pollen-tube growth. Died July 1, 1951; buried Urbana, Ill.

BUCHI, George Hermann, Am. chemist; b. Baden, Switzerland, Aug. 1, 1921; s. George J. and Martha (Muller) B.; D.Sc., Fed. Inst. Tech., Zurich, 1947; m. Anne Westfall Barkman, Aug. 20, 1955. Came to U. S., 1948, naturalized, 1955. Postdoctoral fellow U. Chgo., 1948-51; prof. chemistry Mass. Inst. Tech., Cambridge, 1951—. Recipient Ruzicka award, 1958, Fritzsche award, 1958. Mem. Nat., Am. acads. scis. Research and numerous publs. on chemistry of natural products, organic photochemistry. Home: 100 Memorial Dr., Cambridge, Mass. 02142.*

BUCHNER, Eduard, chemist; b. Munich, Germany, May 20, 1860; s. Ernst and Friederike (Martin) B.; ed. Tech. U., also U. Munich; Ph.D., 1888; m. Lotte Stahl; 2 sons, 1 dau. Asst. to von Baeyer, 1890; became lectr., Munich, 1891, later in Kiel, Germany; asso. prof., Kiel, 1895; became asso. prof. analytical and pharm. chemistry, Tübingen, Germany, 1896; apptd. prof. chemistry U. Berlin (Germany), 1898, U. Breslau (now Wroclaw, Poland), 1909; U. Würzburg (Germany), 1911. Recipient Nobel prize, 1907. Author: (with H. Buchner and M. Hahn) Die Zymasegärung, 1903. Synthesized di-iodo-acetamid, 1888, pyrozole, 1889; showed that alcoholic fermentation is caused by action of enzymes rather than physiol. processes in yeast cells, 1897; discovered zymase, 1st enzyme to be isolated, 1897. Died Focsani, Rumania, Aug. 13, 1817.

BUCHNER, Franz, German pathologist; b. Boppard, Rhine, Jan. 20, 1895; grad. Giessen, 1925. Prof. in Freiburg im Breisgau. Author: Pathogenese der peptischen Veränderungen, 1931; Koronarinfarkt, 1935; Koronarinsuffizienz, 1939; Allgemeine Pathologie, 1950; Spezielle Pathologie, 1955. Research on origin of duodenal and stomach ulcers; coronary insufficiency.

BUCHNER, (Karl) Georg, naturalist; b. Goddelau, Germany, Oct. 17, 1813; s. Ernst and Luise Caroline (Reuss) B.; ed. Strasbourg, France, 1831-33, 35; studied medicine, Giessen, Germany, 1833-35; Ph.D., U. Zurich (Switzerland), 1836. Engaged in revolutionary activities as student in Giessen, founder Gesellschaft für Menschenrechte; became lectr. natural history U. Zurich, 1836. Mem. Soc. Natural History Strasbourg. Author: (plays) Dantons Tod, 1835, Leonce und Lena, 1836, Woyzeck, 1836; Lenz (novella fragment), 1836; Mémoire sur le systeme nerveux du barbeau (dissertation on brain nerves of fish). Died Zurich, Feb. 19, 1837.

BUCHNER, Hans Ernst Angass, German bacteriologist; b. Munich, Germany, Dec. 16, 1850; s. Ernst and Friederike (Martin) B.; ed. U. Munich (Germany), Leipzig, Germany; M.D., 1874; m. Augusta Stutz, 1882. Became physician Bavarian Army, 1875; became asso. prof. hygiene, Munich, 1880, prof., 1894, apptd. dir. Hygienisches Institut, 1894. Author: Atiologische Therapie und Prophylaxis der Tuberkulose, 1883; Die neueren Gesichtspunkte in der Immunitätsfrage, 1892. Pioneer in research on gamma globulins; research on immunity; devised methods of studying anaerobic bacteria. Died Munich, Apr. 5, 1902.

BUCHNER, Johann Andreas, German chemist, pharmacist; b. Munich, Germany, Apr. 6, 1783; s. Johannes Buchner; grad. Erfurt Akademie, Germany; M.D. (hon.), U. Bonn, Germany, 1819; 1 son, Ludwig Andreas. Chief pharmacist, central pharmacy of pub. hosps., Munich, 1807-18; asso. prof. U. Landshut, 1818-22, prof. pharmacology, toxology (1st in Germany separate from chemistry), dir. pharm. inst., from 1822; established pharm. inst. which became part of U. Munich, 1840; mem. Bayerische Akademie der Wissenschaften. Author: Erster Entwurf eines Systems der chemischen Wissenschaft und Kunst, 1815; Würdigung der Pharmazie in staatswissenschaftlicher Beziehung, 1818; Über der Trennung der Pharmazie von der Heilkunst, 1819; Vollständiger Inbegriff der Pharmazie, 6 vols., 1821-36; Lehrbuch der analytischen Chemie und stöchiometrie, 1836. Chem.-pharm. work mainly in chemistry of plant products; discovered number of glucosides; 1st to isolate paraffin from Tegernsee rock oil; helped establish pharmacy as an accepted science. Died June 5, 1852.

BUCHNER, Ludwig Friedrich Karl Christian, German physician, philosopher; b. Darmstadt, Germany, Mar. 29, 1824; s. Ernst and Luise Caroline (Reuss) B.; student philosophy, medicine, Giessen, (Germany), Strasbourg (France), Würzburg (Germany), Vienna (Austria); doctorate Giessen, 1848; m. Sophie Thomas, 1860; 4 children including George. Apptd. lectr. U. Tübingen (Germany), 1854-55; physician, writer, Darmstadt; lectr., U. S., 1872-73, later in Germany; founder: German League of Free Thinkers, 1881. Author: Kraft und Stoff, 1855; Natur und Geist, 1857; 6 Vorlesungen über die Darwinische Theorie, 1868; Der Mensch und seine Stellung in der Natur, 1869; Der Gottesbegriff und seine Bedeutung in der Gegenwart, 1874; Aus dem Geistesleben der Tiere, 1876; Das künftige Leben und die moderne Wissenschaft, 1889; Am. Sterbelager des Jahrhunderts, 1894; Darwinismus und Sozialismus, 1894; Im Dienste der Wahrheit, 1891. Exponent of materialism; used Darwin's theory in his ateleological philosophy. Died Darmstadt, Apr. 1, 1899.

BUCHNER, Max Franz Christian, chemist; b. Bamberg, Germany, July 10, 1866; s. Carl Christian and Elise (Seilböck) B.; ed. tech. high schs. Munich and Wurzburg, Germany; doctorate under A. Hautzsch, 1898; hon. dr. engring. Tech. High Sch. Karlsruhe (Germany); m. Sophie Rottenhöfer, 1888; 5 daus., including Irma (Mrs. Wilhelm Eduard Bachmann). Co-owner C. C. Buchner Verlag Bamberg pub. house and bookstore, 1886-92; active in chem. industry, often in Norway, 1898-1924; organizer ACHEMA exhbn. and conf., 1920, founder ann., 1925; founder Soc. for Chem. Apparati (DECHEMA), 1926, also Chemische Fabrik (mag.). Recipient Liebig medal. Contbd. numerous papers to chem. jours. Worked on electrolysis of organic compounds also prodn. pure aluminum oxide from German clays and lesser bauxites; creator carundum ceramics: pioneer in devel. chem. apparatus; patentee numerous inventions. Died Mahle/Hanover, Germany, Apr. 10, 1934.

BUCHNER, Paul Ernst Christof, zoologist; b. Nuremberg, Germany, Apr. 12, 1886; s. Wilhelm and Julie (Mengin) B.; Ph.D. in Natural Scis. honoris causa, in Biol. Scis. honoris causa; M.D. honoris causa; m. Miliana Coppa, Mar. 12, 1913; 1 son, Giorgio. Titular prof. chair zoology, univs. Griefswald, Breslau, Leipzig, Munich (all Germany); pres. Center of Studies of Isle of Ischia. Mem. German Soc. Zoology, numerous acads. Contbd. numerous publs. on endosymbiosis of animals with micro-organisms, also natural history studies and bathing life on island of Ischia. Address: via S. Alessandro 15, Porto d'Ischia, Naples, Italy.

BUCHSBAUM, Solomon Jan, physicist; b. Stryj, Poland, Dec. 4, 1929; s. Jacob and Bertha (Rudoerfer) B.; B.S., McGill U., 1952, M.S., 1953; Ph.D., Mass. Inst. Tech., 1957; m. Phyllis Norma Isenman, July 3, 1955; children—Rachel Joy, David Joel, Adam Louis. Staff mem. Mass. Inst. Tech., 1957-58; mem. research staff Bell Telephone Labs., Murray Hill, N.J., 1958—, head solid state and plasma physics dept., 1960-65, dir. electronics research lab., 1965—. Cons. Convair, 1957-58, U. S. AEC, 1965—; chmn. div. plasma physics Am. Phys. Soc., 1958. Recipient Anne Molson Gold medal, 1953. Fellow Am. Phys. Soc. Author: (with W. P. Allis, A. Bers) Waves in Anistropic Plasmas, 1963. Asso. editor: Physics of Fluids, 1962-64. Research, numerous publs. in plasma physics both gaseous and solid state, gaseous electronics, quantum electronics.*

BUCHTHAL, Fritz, physician; b. Witten, Germany, Aug. 19, 1907; s. Sally and Hedwig (Weyl) B.; student U. Freiburg, 1925, Stanford, 1926-28; M.D., Berlin U., 1931; hon. degrees U. Münster, 1961, U. Zurich (Switzerland), 1965, m. Margaret Agnes Lennox, Aug. 19, 1955. Staff physiology, U. Berlin (Germany), 1930-32; staff lab. for physiology of work U. Copenhagen (Denmark), 1933-43, dir. Inst. Neurophysiology, 1946—, prof. neurophysiology, 1955—; staff Inst. Physiology, U. Lund (Sweden), 1943-45; head dept. clin. neurophysiology U. Hosp., Copenhagen, 1945—; vis. Horovitz prof. N.Y. U., 1965. Mem. Royal Danish Acad. Sci., Acad. Tech. Scis.; hon. mem. French Neurol. Soc.; corr. mem. German Neurol. Soc., German Physiol. Soc. Author: (with Kaiser, P. Rosenfalck) Rheology of Muscle, 1951; Clinical Electromyography, 1957; (with A. Rosenfalck) Sensory Evoked Potentials in Man, 1966; also articles. Research on mechanism of muscular contraction, application of electrophysiol. methods to diagnosis of nerve and muscle disease, relation of anticonvulsants in blood to clin. and EEG seizures. Home: 24 Söbredden, Gentofte, Denmark. Office: 36 Juliane Mariesvej, Copenhagen Ö, Denmark.*

BUCHWALD, Nathaniel Avrom, Am. neurophysiologist; b. Bklyn., July 19, 1924; s. S.S. and Nellie (Miller) B.; B.S., U. Miami, 1946; Ph.D., U. Minn., 1953; m. Jennifer Sullivan, Dec. 30, 1952; children —Katherine, Scott, Elizabeth. Instr. anatomy Tulane U., New Orleans, 1953-57; faculty U. Cal. at Los Angeles, 1957—, prof. anatomy, 1965—, mem. Brain Research Inst., 1961—. Cons. in neurophysiology VA Hosp., Long Beach, Cal., 1957-64. Sr. fellow, recipient Career Devel. award USPHS, 1958-67. Mem. Am. Anat. Soc., Am. Physiol. Soc., A.A.A.S., Radiation Research Soc., Am. Acad. Neurology (asso.), Am. Assn. U. Profs. Research, publs. on relationship of brain elec. activity to behavior, particularly in problems of learning and performance, behavioral and elec. responses to small doses of ionizing radiation. Home: 4550 Estrondo Dr., Encino, Cal. 91344.

BUCHWALD, Niels Fabritius, Danish plant pathologist; b. Aalborg, Denmark, Aug. 10, 1898; s. Niels Bredahl and Astrid (Fabritius) B.; candidatus magic terii, U. Copenhagen, 1924; m. Karen Mikkelsen, May 22, 1928; children—Niels Vagn, Jorgen Steen. Faculty, Royal Vet. and Agrl. Coll. Copenhagen, 1925—, prof., chief dept. plant pathology, 1944—; resident dr. Cornell U., 1930-31. Recipient Ridder of Dannebrog I. Mem. Danish Mycol. Soc. (pres. 1944), Acad. Tech. Scis. Denmark (chmn. virus disease com. 1948-60), Norwegian Acad. Scis., Swedish Royal Agrl. and Forestry Acad., Swedish Bot. Soc., Finnish Plant Pathology Soc. Author: Fysiogenic Plant Diseases I-II; Fungi imperfecti (Deuteromycetes); Diseases of Horticultural Plants; several textbooks on plant diseases. Editor: Naturhistorie Sidende, 1937-66. Research, publs., contbns. to Am. Ency. on mushrooms, polypores and other kinds of fungi. Home: 68 Dalgas Blvd., Copenhagen, Denmark.*

BUCK, Carl Darling, Am. philologist; b. Orland, Me., Oct. 2, 1866; s. Edward and Emeline (Darling) B.; A.B., Yale, 1886, Ph.D., 1889; mem. Am. Sch. Classical Studies, Athens, 1887-89; studied in Leipzig, 1889-92; Ph.D. (hon), U. Athens (Greece), 1912; Litt. D., Princeton, 1935; m. Clarinda Darling Swazey, Sept. 10, 1889; children—Carl Edward, Howard Swazey, Clarinda Darling. Asst. prof. Sanskrit and Indo-European comparative philology U. Chgo., 1892-94, asso. prof., 1894-1900, prof., 1900-03, prof. and head dept., 1903-33, Martin A. Ryerson distinguished service prof., 1930-33, prof. emeritus, 1933—; ann. prof. Am. Sch. Classical Studies, Athens, 1923-24. Mem. Am. Philos. Soc., Am. Acad. Arts and Scis., Am. Philol. Assn. (pres. 1915-16), Am. Linguistic Soc. (pres. 1927, 37), Phi Beta Kappa, Delta Kappa Epsilon. Author: Vocalismus der oskischen Sprache Leipzig, 1892; Hale-Buck Latin Grammar (with William G. Hale), 1903; Grammar of Oscan and Umbrian, 1904; Sketch of the Linguistic Conditions in Chicago, 1903; Introduction to the Study of the Greek Dialects, 1909, rev. edit., 1955; Comparative Grammar of Greek and Latin, 1933; Reserve Index of Greek Nouns and Adjectives (with W. Petersen), 1945; Dictionary of Selected Synonyms in the Principal Indo-European Languages, 1949; also articles in philol. jours. Mem. bd. editors Classical Philology. Died Feb. 8, 1955.

BUCK, Edward Clark, engr.; b. 1873; s. James Buck. Did extensive engring. work, explorations, surveys in Australia, New Zealand, South Africa; waterworks, dep., acting city engr., Pretoria, Transvaal, 1903-08; cons. engr., London, Trinidad, 1908-13; a pioneer Trinidad oilfields; dir. pub. works, Brit. Guiana; explored to sources of Orinoco and Amazon rivers; ret., 1921; cons. engr., petroleum mining expert. Fellow Royal Geog. Soc.; mem. Instn. Mech. Engrs. (1st chmn. Caribbean br. 1938, mem. council), Instn. Civil Engrs., Instn. Structural Engrs. Research and publs. on oil fuels; inventor, patentee improved method of elec. ry. safety signalling, patent astron. abney level, incendiary and armor piercing shells, others. Died Nov. 20, 1950.

BUCK, Gurdon, Am. surgeon; b. N.Y.C., May 4, 1807; s. Gurdon and Susannah (Manwaring), B., M.D., Coll. Phys. and Surg., N.Y.C., 1830; studied in Paris (France), Berlin (Germany), Vienna (Austria), 1832-34; m. Henriette Wolff, July 27, 1836, 1 son, Albert H. Vis. surgeon N.Y. Hosp., 1837, St. Luke's Hosp., 1846; asso. with N.Y. Eye and Ear Infirmary, 1852-62; vis surgeon Presbyn. Hosp., 1872; among chief contbns. was Buck's extension (a treatment of thigh fractures by weights and pulleys), 1860; pioneer in plastic face surgery. Author: Description of an Improved Extension Apparatus for the Treatment of Fracture of the Thigh, 1867; Contributions to Reparative Surgery, 1876. Died N.Y.C., Mar. 6, 1877.

BUCK, Harold Winthrop, Am. elec. engr.; b. N.Y.C., May 7, 1873; s. Albert Henry and Laura S. (Abbott) B.; Ph.B., Yale, 1894; E.E., Columbia Sch. Mines, 1895; m. Charlotte R. Porter, 1902; children—Winthrop Porter, Gurdon; m. 2d, Mary Perry, 1941. Entered Schenectady works Gen. Electric Co., 1895, student, later asst. engr.; elec. engr. Niagara Falls Power Co., from 1900. Fellow Am. Inst. E.E. (pres. 1916-17); mem. Franklin Inst., Engring. Inst. Can. Patentee mech. and elec. devices, also process for making corundum in elec. furnace. Died Aug. 5, 1958; buried Greenwich, Conn.

BUCK, Jirah Dewey, Am. physician; b. Fredonia, N.Y., Nov. 20, 1838; s. Reuben and Fanny B.; ed. at Belvidere, Ill.; M.D., Cleveland Homeo. Coll., 1864; m. Melissa M. Clough, Oct. 3, 1865. Prof. physiology Cleveland Homeo. Coll., 1866-71; prof. therapeutics, dean Pulte Med. Coll., 1880—. Pres. Theosophical Soc. in Am. 1892-94, Am. Inst. Homoeopathy, 1890, Ohio State Homeo. Med. Soc., 1875. Author: A Study of Man and the Way to Health, 1888; Mystic Masonry, 1896; The Soul and Sex in Education. Deceased.

BUCK, John Bonner, Am. biologist; b. Hartford, Conn., Sept. 26, 1912; s. George Sumner and Carrie Elizabeth (Bonner) B.; A.B., Johns Hopkins, 1933, Ph.D., 1936; m. Elisabeth Tennent Mast, Dec. 22, 1939; children—Peter, Susan, Judith, Alan. Asst. zoology Johns Hopkins, 1933-36; NRC fellow Cal. Inst. Tech., 1936-37; research asso. Carnegie Instn., 1937-39; asst. prof. zoology U. Rochester, 1939-45; physiologist NIH, 1945—, chief, lab. phys. biology, 1962—; mem. Johns Hopkins expdns. to Jamaica, 1936, 41, 62; vis. prof. U. Wash., 1951, Cal. Inst. Tech., 1953; guest Cambridge (Eng.) U., 1963-64; instr. Marine Biol. Lab., Woods Hole, Mass., 1942-44, 57-59, trustee, 1959—. Mem. Soc. Gen. Physiologists (pres. 1960), Am. Soc. Zoologists (v.p. 1956). Contbr. articles to sci. jours. Spl. research in chromosome structure, insect respiration, firefly physiology. Mem. editorial bd. Biol. Bull., 1957-61, 65—, Jour. Morphology, 1964—.*

BUCK, John Henry, physicist; b. London, Eng., Sept. 22, 1912; s. Henry G. and Ann (Crooke) B.; came to U. S., 1935, naturalized, 1942; B.Sc., U. Sask.; 1934, M.Sc., 1935; Ph.D., U. Rochester, 1938; m. Carolyn F. Wilcox, July 31, 1940; children—John A., Barbara C. Instr., U. Rochester, 1935-38, Mass. Inst. Tech., 1938-41; supr. physics div. Socony Vacuum Oil Co., Paulsboro, N.J., 1946-50; sr. physicist, project engr. Union Carbide & Chem. Co., Oak Ridge, Tenn., 1950-53; v.p., gen. mgr. Wells Surveys, Inc., Tulsa, 1953-58; v.p. engring B.J. Electronics Santa Ana, Cal., 1958-59; v.p., gen. mgr. instruments div. budd Co., Phoenixville, Pa., 1959—. Recipient Army/Navy certificate appreciation, 1946. Fellow Am. Phys. Soc., A.A.A.S. Study of nuclear physics; nuclear reactors; cyclotrons; radar navigation and bombing. Home: 499 Bair Rd., Berwyn, Pa. Office: Box 245, Phoenixville, Pa. 19460.

BUCK, Peter Henry (Maori name: Te Rangi Hiroa), ethnologist; b. Urenui, Taranaki, New Zealand, Aug. 15, 1880; s. William Henry and Nga-Rongo B.; student Te Aute Coll., New Zealand, 1896-98; M.B., Otago Med. Sch., U. New Zealand, 1904, Ch.B., 1904, M.D., 1910, D.Sc., 1937; M.A., Yale, 1936; D.Sc., U. Rochester, 1939; m. Margaret Wilson, Aug. 14, 1905. Med. officer of health, New Zealand, 1905-08; mem. Parliament, New Zealand, 1904-14; dir. Maori hygiene, New Zealand, 1919-27; ethnologist, Bishop Museum, Honolulu, 1927-32, 1934-36; vis. prof. Yale U., 1932-34; dir. Bishop Mus., prof. anthropology Yale, 1936-48; pres. bd. trustees, Bernice P. Bishop from 1948. Recipient Hector medal New Zealand, Rivers Meml. medal Royal Anthrop. Inst., London. Fellow Royal Soc. of New Zealand, Royal Anthrop. Inst., A.A.A.S.; mem. Am. Anthrop. Assn., Polynesian Soc., Sigma Xi. Author (under Maori name of Te Rangi Hiroa): Evolution of Maori Clothing, 1926; Material Culture of the Cook Islands, 1927; Samoan Material Culture, 1930; Ethnology of Tongareva, 1932; Ethnology of Manihiki and Rakahanga, 1932; Mangaian Society, 1934; Ethnology of Mangareva, 1938; Mangaian Society, 1934; Ethnology of Mangareva, 1938; Vikings of the Sunrise, 1938; Anthropology and Religion, 1939; Arts and Crafts of Cook Islands, 1943; Introduction to Polynesian Anthropology, 1945; The Coming of the Maori, 1948; also articles. Theorized about migration routes of Polynesians, Died Honolulu, Hawaii, Dec. 1, 1951.

BUCK, R(obert) Creighton; Am. mathematician; b. Cin., Aug. 30, 1920; s. Robert Jirah and Martha (Creighton) B.; A.B., U. Cin., 1941, M.A., 1942; Ph.D., Harvard, 1947; m. Ellen Fedder, Dec. 28, 1944; children—Nancy Elizabeth, Donald Paul. Mem. Harvard Soc. Fellows, Cambridge, Mass., 1942-43, 45-47; asst. prof. Brown U., Providence, 1947-50; asso. prof., U. Wis. Madison, 1950-54, prof., 1954—, chmn. dept., 1964—. Mem. project FOCUS, Inst. Def. Analyses, 1959-60; chmn. Com. Undergrad. Program, 1959-63; mem. film panel Com. Ednl. media, 1963-66; mem. programmed learning panel Sch. Math. Study Group, 1960-64; mem. U. S. Commn. Math. Instrn., 1963—; exec. com. dir. math. NRC, 1963-65; math panel Nat. Security Agy. Sci. Adv. Bd., 1963—. Guggenheim fellow, 1958-59. Mem. Am. Math. Soc. (mem. council 1959—, exec. com. 1960, editor Proc. 1964—), Math. Assn. Am. (bd. govs. 1960-63). Luth. Acad. Scholarship, Phi Beta Kappa, Sigma Xi. Author: Advanced Calculus, 2d edit., 1965; Polynomial Expansions (with R. P. Boas), 1958; also articles. Editor: Studies in Modern Analysis, 1962; Modern Analysis Series, 1962. Research, publs. on representation of analytic functions, structure of gen. function algebras, and topology. Home: 3601 Sunset Dr., Madison, Wis. 53705.*

BUCKENS, Félix, Belgian engr., mathematician; b. Tchen-Tcheou, June 3, 1916; s. Camille Fernand and Jeanne (Matsuo) B.; Ph.D. in Applied Scis., U. Louvain (Belgium); m. Suzanne Goosens, June 3, 1956; children—Paul, Marie-Madeleine, Pierre, Cecile. Instr. U. Louvain, 1948-50, prof., from 1950; dir. IRSAC center, Congo, 1952-55. Recipient laureate of L. Empain inter-faculty prize in sci., 1941, laureate alumni prize U. Found.; research fellow Cal. Inst. Tech., 1947. Mem. Assn. Engrs. and Industrialists. Contbr. articles mechanics, applied math. to publs. Home: 3, avenue des Hetres, Heverlee. Office: Inst. Méc. et Math. appl., rue des Célestins, Heverlee, Belgium.

BUCKHAM, Sir Arthur McDougall, Brit. engr.; b. July 8, 1879; s. Fred E. Duckham; ed. Blackheath Sch.; m. Maud Peppercorn; 1 son, 2 daus. Worked with George Livesay in engring. workshops; dir.-gen. aircraft prodn.; mem. Air Council; mem. council Ministry of Munitions. Author: Atlantic Letters of World Affairs, 1932. Research, patentee in areas of furnace work, coal carbonization, engring. as related to chem. devels. Died Feb. 14, 1932.

BUCKINGHAM, Edgar, Am. physicist; b. Phila., July 8, 1867; s. Lucius Henry and Angelina Bradley (Hyde) B.; A.B., Harvard, 1887, postgrad., 1887-89; postgrad. U. Strasbourg (now in France), 1889-90; Ph.D., Leipzig, Germany, 1893; m. Elizabeth Branton Holstein, July 15, 1901; children—Katharine, Stephen Alvord. Asst. in physics Harvard, 1888-89, 91-92, Strassburg, 1889-90; mem. faculty Bryn Mawr Coll., 1893-99; instr. physics U. Wis., 1901-02; asst. physicist, bur. soils U. S. Dept. Agr., 1902-05; with Bur. of Standards, 1905-37, serving as physicist. Lectr. on thermodynamics Grad. Sch. U. S. Naval Acad., 1910-12; asso. sci. attaché U. S. Embassy, Rome, Italy, 1918. Author: An Outline of the Theory of Thermodynamics, 1900. Died Apr. 29, 1940.

BUCKLAND, Cyrus, Am. inventor; b. Manchester, Conn., Aug. 10, 1799; s. George and Elizabeth B.; m. Mary Locke, May 18, 1824, 3 children. Instrumental in manufacture of eccentric bit and auger used in cutting lock, guard plate, side plate, breech plate, rod spring and barrels to gunstocks; designed and patented rifling machine to cut groove of regularly decreasing depth from breech to muzzle (sec. of war purchased U. S. Govt. rights to invention). Died Springfield, Mass., Feb. 26, 1891.

BUCKLAND, Francis Trevelyan, English naturalist; b. Oxford, Eng., Dec. 17, 1826; s. William and Mary (Morland) B.; B.A., Christ Ch., Oxford, 1848; student surgery St. George's Hosp., London, 1848-51. Became army surgeon, London, 1854; insp. salmon fisheries, 1867-80. Founder, Land and Water, 1866. Author: Curiosities of Natural History, 1857-72; Fish Hatching, 1863; Logbook of a Fisherman and Zoologist, 1875; Natural History of British Fishes, 1881; Notes and Jottings from Animal Life, 1882; ann. reports on salmon fisheries in Quarterly Review; reports on Scottish salmon fisheries, 1871, Norfolk fisheries, 1875, crab and lobster fisheries, 1877, Scottish herring fisheries, 1878, sea fisheries, 1879; articles in Field. Editor: Natural History of Selborne (White), 1876. Authority on pisciculture; investigated econ. questions affecting artificial supply of salmon; devised ladders to assist salmon in reaching spawning grounds; helped secure internat. agreement to prevent extermination of N. Atlantic fur seal. Died Oxford, Dec. 19, 1880.

BUCKLAND, William, English geologist; b. Axminster, Eng., 1784; s. Charles Buckland; B.A., Corpus Christi Coll., Oxford, 1805; m. Mary Morland, 1825; 1 son, Francis Trevelyan. Ordained priest, 1808; apptd. prof. mineralogy, Oxford, Eng. 1813, reader in geology, 1819; became canon Christ Ch., Oxford, 1825; dean Westminster, 1845-56. Fellow Corpus Christi Coll., 1808-25; made geol. tour S.W. Eng., 1808-12. Fellow Royal Soc., 1818 (Wollaston medal 1848); mem. Geol. Soc. (pres. 1824, 40); French Acad. Scis., 1839. Author: Reliquiae diluvianae, 1823; Geology and Mineralogy Considered with Reference to Natural Theology, 2 vols., 1836; Bridgewater treatise (compendium of previous geol. and paleontol. sci.), 1836; numerous papers. Described secondary paleontol. formations and their fauna; 1st in England to note action of glacial ice on rocks; tried to reconcile new geol. and paleontol. discoveries with Cuvier's anti-evolution theory and Bible. Died Aug. 15, 1856.

BUCKLER, Thomas Hepburn, physician; b Balt., Jan. 4, 1812; s. William and Anne (Hepburn) B.; attended St. Mary's Coll., Balt.; M.D., U. Md., 1835; studied in clinics, Paris, France, 1830; m. Anne Fuller, 1831; m. 2d, Eliza Ridgely, Nov. 21, 1865. Physician of Balt. City and County Almshouse, several years; physician to pub. figures including Chief Justice Roger B. Taney, Pres. James Buchanan, Gen. Robert E. Lee, 1850-55; Southern sympathizer during Civil War; practiced medicine under license of French Govt., Paris, 1866-90; advocated use of ammonium phosphate in treating rheumatism, laparotomy for intestinal obstruction, rest and open-air treatment for Tb. Author: A History of Epidemic Cholera, as it appeared at the Baltimore City and County Almshouse, in the Summer of 1849, with Some Remarks on the Medical Topography and Diseases of this Region, 1851; On the Etiology, Pathology and Treatment of Fibro-bronchitis and Rheumatic Pneumonia, 1853. Died Apr. 20, 1901.

BUCKLER, William Hepburn, Am. archeologist; b. Paris, France, Feb. 1, 1867 (parents Am. citizens); s. Thomas Hepburn and Eliza (Ridgely) B.; M.A., Trinity Coll., Cambridge U., Eng., 1890; LL.B., Cambridge U., 1891; postgrad. Law Dept., U. Md., 1893-94; M.A., Oxford, 1925, D.Litt., 1937; LL.D., Aberdeen, 1935, Johns Hopkins, 1940; m. Georgina Grenfell Walrond, May 25, 1892; children—Lucy Ridgely (Mrs. Vivian Seymer), Barbara Isabel (Mrs. Charles Wrinch). Practiced law, Balt., 1894-1902; sec. emergency com. after Baltimore fire, Feb. 1904; sec. U. S. special embassy to King's wedding, Spain, 1906; sec. U. S. Legation Madrid, Spain, 1907-09; mem. staff Am. Expdn. to Sardis, Asia Minor, 1910-14; special agt. Dept. of State, Embassy, London, 1914-18; attached Am. Comm. to Negotiate Peace, Paris, June-Dec. 1919; made journeys in Asia Minor, 1924, 26, 30, 33; asso. All Souls Coll., Oxford, Eng., 1924-25. Mem. council Am. Soc. for Archeol. in Asia Minor; v.p. Soc. for Hellenic Studies, Soc. for Roman Studies; fellow Brit. Acad. Author: History of Contract in Roman Law (Yorke Prize, Cambridge), 1894; Relation of Roman Law to Other Historical Sciences (Vol. II, Proc. Internat. Congress Arts and Sciences, St. Louis), 1904; Chapter VI, in Hollander and Barnett's Studies in American Trade Unionism, 1906; Chapter XXII in Ripley's Railway Problems, 1907; Lydian Inscriptions, 1924; Sardis-Greek and Latin Inscriptions (with David M. Robinson), 1932; Monuments of Western Phrygia . . . (with W. M. Calder, C. M. Cox and K. Guthrie), 1933, 39. Died Mar. 2, 1952.

BUCKLEY, Albert Coulson, Am. neuropsychiatrist; b. Phila., Aug. 6, 1873; s. William Coulson and Lucy Ann (Davis) B.; A.B., Central High Sch., Phila., 1894; M.O., Medico-Chirurg. Coll., Phila., 1897; m. Harriet Ellis Baily, 1904. Asst. neurologist Medico-Chirurg. Hosp., 1897-1906; asso. prof. normal histology Medico-Chirurg. Coll., 1899-1908, asso. prof. psychiatry, 1906-12, hon. cons. psychiatrist, from 1931; asst. phys. Friends Hospital, 1906-18, med. supt., from 1918; alienist Phila. Orthopedic Hosp., 1912-21; prof. psychiatry Grad. Sch. of Medicine, U. Pa., from 1919, asso. prof. psychiatry School of Medicine, from 1930; neurologist Frankford Hosp., 1930. Diplomate Am. Board Psychiatry and Neurology. Fellow Coll. Physicians of Phila., A.M.A. Author: The Basis of Psychiatry, 1920; Nursing Mental and Nervous Diseases, 1927. Died Aug. 17, 1939.

BUCKLEY, Joseph Paul, Am. pharmacologist; b. Bridgeport, Conn., Jan. 12, 1924; s. Morris F. and Rose (Graff) B.; B.S., U. Conn., 1949; M.S., Purdue U., 1951- Ph.D., 1952; m. Shirley E. Shipman, Aug. 16, 1947. Faculty, U. Pitts., 1952—, prof., chmn. dept. pharmacology, 1958—, prof. dept. pharmacology Sch. Dentistry, 1963—; research investigator Haskell Labs. E. I. Dupont, Newark, Del., 1954; staff pharmacologist St. John's Gen. Hosp., Pitts. Cons. pharmacologist U. S. Vitamin Pharm. Corp., A. H. Robbins Co., Eaton Labs. Recipient Angiology Research Found. Honors Achievement award, 1965. Fellow Am. Coll. Angiology; mem. A.A.A.S. (sec. pharm. scis. sect.), Am. Pharm. Assn. (chmn. pharmacology symposium com. 1964-65, Found. award in pharmacodynamics 1966), Acad. Pharm. Scis. (chmn. sect. pharmacology and biochemistry 1964-66) N.Y. Acad. Sci., Sigma Xi (pres. Pitts. chpt.). Research, publs. on central mechanism of angiotensin II pressor activity; hypotensive drugs; coronary dilators. Home: 1264 Arrowood Dr., Pitts. 15243.*

BUCKLEY, Oliver Ellsworth, Am. physicist; b. Sloan, Ia., Aug. 8, 1887; s. William Doubleday and Sarah (Jeffrey) B.; B.S., Grinnell Coll., 1909; Ph.D., Cornell U., 1914; D.Engring., Case Inst. Tech., 1948; Sc.D., Columbia, 1948; m. Clara Louise Lane, Oct. 1914; children—Katherine, Barbara, Juliet, William. Instr. physics Grinnell Coll., 1909-10, Cornell U., 1910-14; research physicist Western Electric Co., 1914-25; with Bell Telephone Labs. from 1925, asst. dir. research, 1927-33, dir., 1933-36, 37—, exec. v.p., 1936-40, pres., 1940-51, chmn. bd., 51-52. Chmn. sci. adv. com. to Office for Def. Moblzn., 1951. Recipient Medal for Merit, Presdl. citation, 1946. Mem. Nat. Acad. Sci., Am. Philos. Soc., Am. Acad. Arts and Scis., N.Y. Acad. Scis., Am. Phys. Soc., Am. Inst. E.E., Acoustical Soc. Am., A.A.A.S., Franklin Inst. Contbr. articles to profl. lit. Invented mercury vapor diffusion lamp, ionization manometer; was motivating source and research head for devel. under-water cables for telephone and telegraph, mainly trans-oceanic cable; patentee preparation vacuum tubes, submarine cable systems, conductors, signaling systems, other inventions. Home: 13 Fairview Terrace, Maplewood, N.J.

BUCKLEY, William, English mathematician; b. Lillieshal, Salop, Eng., 1519; student King's Coll., Cambridge (Eng.) U., 1537-45, tchr. arithmetic and geometry, 1548-49, also served at court, math. tutor to Edward VI, 1545-48; tutor to royal pages, 1550-51. Author: Arithmetica memorativa (1st work on theory of combinations), circa 1550. Designed sundials. Died 1571.

BUCKMANN, Adolf L., German marine biologist; b. Elze Krs. Alfeld, Germany, Jan. 17, 1900; s. Rudolf H. and Elisabeth (Kreusler) B.; student U. Leipzig (Germany), 1919-20; Dr.rer.nat., U. Hamburg (Germany), 1923; m. Hildegard Thomae, Oct. 13, 1924; children—Walter, Detlef, Hildegard, Irmgard, (Mrs. Karl Heinrich Goelnitz). Asst., Deutsche Wissenschaftliche Kommission fur Meeresforschung, Helgoland, 1924-36; head dept. fishery biology Biologische Anstalt Helgoland, 1936-46; dept. chief fishery biology Max-Planck-Inst. Meeresbiologie Wilhelmshaven, 1948-53; prof. hydrobiology and fisheries U. Hamburg, dir. Institut für Hydrobiologie und Fischereiwissenschaft, 1953-66, ret., 1966. Sci. chmn. Deutsche wiss. Kommission für Meeresforschung, 1952-59; dir. Biolog. Anstalt Helgoland, 1953-60. Author: Das Problem der optimalen Befischung, 1963. Research and publs. on marine biology, dynamics of food, availability and ecology of fish. Home: 39, Heilwig-Str., Hamburg 20, Germany.*

BUCKTON, George Bowdler, English chemist, entomologist; b. Hornsey, Eng., May 24, 1818; s. George and Eliza (Merricks) B.; m. Mary Ann Odling, 1865; 1 son, 5 daus. Worked in chemistry, also entomology, astronomy. Fellow Royal Soc., 1857. Author: Monograph of British Aphids, 1876-83; Monograph of British Cicadae or Teltigridae, 1890-91; Natural History of Eristalsis Tenax or the Drone Fly, 1895. Discovered, isolated mercuric methyl, 1852-65; studied parthenogenesis of aphids. Died Sept. 25, 1905.

BUCKWALTER, Tracy V., Am. inventor; b. Jersey Shore, Pa., Apr. 28, 1880; s. David B. and Ellen Virginia (Harmen) B.; grad. high sch.; m. Hattie Mae Emmons, Oct. 22, 1902 (dec. May 1941); children—Lawrence E., Emory T., Theodore J., Eugene P., Norman R., Tracy V.; m. 2d, Sara Porter Gregory, Nov. 18, 1941. Began as apprentice with elec. contractor, Phila., 1896; with Pa. R.R., Altoona, Pa., as machinist, 1900, asst. foreman, 1901, draftsman, 1906, foreman motive power engring. dept., 1911; with Timken Roller Bearing Co., Canton, O., as chief engr., 1916-23, v.p., 1922-46; dir. Spun Steel Corp., developed electric trucks, gas-electric locomotives, while at Altoona roller bearing steel mills, roller bearing machine tools Timken locomotive, inexpensive bearing for automotive vehicles; supervised tests of locomotive and railroad car axles under auspices Assn. Am. Railroads since 1933, steam locomotive balancing since 1936. Recipient Modern Pioneer award Nat. Assn. Mfrs., 1939, Henderson medal Franklin Inst., 1946. Mem. Soc. Automotive Engrs., Am. Soc. M.E., Am. Welding Soc., A.A.A.S., Am. Geog. Soc., Ohio Forestry, Am. Mus. Natural History, Princeton Engring. Assn. (Bracket mem.). Author: The Railroad; also tech. booklets on axle testing, roller bearings, locomotives, automotive equipment. Died Mar. 14, 1948; buried Ft. Lauderdale.

BUCKY, Gustav, radiologist; b. Leipzig, Germany, Sept. 3, 1880; ed. State U. Geneva (Switzerland); M.D., State U. Saxony, Leipzig, 1907. Came to U. S. Clin. prof. Albert Einstein Coll. Medicine, N.Y.C., 1956—. Recipient Golden Key of Merit, Am. Congress Physiotherapy, 1933. Contbd. numerous med., sci. publs., 3 books. Inventor Bucky diaphragm, Bucky camera, (with Albert Einstein) automatic exposure device; originated Grenz ray therapy.

BUCQUET, Jean-Baptiste-Marie, French chemist; b. Paris, France, Feb. 18, 1746; s. Antoine-Joseph and Marthe (Marotin) B.; studied physics and chemistry; tchr. pharmacy Paris Faculty of Medicine, 1775; became full prof., 1776; royal censor; mem. Royal Soc. Medicine, French Acad. Scis., 1777. Author: Introduction a l'etude des corps naturels tirés du règne minéral, 1773; Introduction a l'étude des corps tirés du règne végétal (most complete and methodical table of vegetal analysis of the time), 1773; Mémoire sur la manière dont les animaux sont affectés par les différents fluides aériformes méphitiques, 1778. Died Paris, Jan. 24, 1780.

BUCQUOY, Jules, French physician; b. Peronne, France, Aug. 14, 1829; studied in Paris; interne, 1851; Docteur en Medicine, 1855; aggregation, 1863; prof. Hôtel-Dieu, 1867-68; physician Cochin Hosp., 1870-85; became mem. Acad. Medicine, 1882, pres., 1908; founder Soc. Therapeutics. Author: Leçons clinique sur les maladies du coeur, 1869. First (with Dujardin-Beaumetz) to use strophanthus as a cardiac stimulant 1885; research on heart disease, pulmonary gangrene. Died June 30, 1920.

BUDD, William, English physician; b. North Taunton, Eng., 1811; studied in Paris, London, Edinburgh; M.D. Mem. staff of hosp., Bristol, Eng. Fellow Royal Soc., 1871. Author: Typhoid Feber; Its Nature, Mode of Spreading, and Prevention, 1873. Advocated disinfection to prevent the spread of contagious disease; proposed methods of ridding Bristol of Asiatic cholera, and Eng. of rinderpest, 1866; proved that typhoid fever is contagious. Died Jan. 9, 1880.

BUDDE, Emil Arnold, German physicist; b. Geldern, Germany, July 28, 1842; Dr. Phil., U. Bonn (Germany), 1864; Dr. Ing. Author: Zur Kosmologie der Gegenwart, 1872; Mechanik der Punkte und starren Systeme, 1890. Pioneer in constrn. practical metal-filament electric lamps. Died Feldafing, Aug. 19, 1921.

BUDDEN, Kenneth George, physicist; b. June 23, 1915; s. George Easthope and Gertrude Homer (Rea) B.; M.A., Ph.D., St. John's Coll., Cambridge (Eng.) U.; m. Nicolette Ann Lydia de Longesdon, 1947. Staff, Telecommunications Research Establishment, 1938-41, Brit. Air Command, Washington, 1941-44, Air Command, S.E. Asia, 1945; fellow, dir. studies, lectr. physics St. John's Coll., Cambridge U., 1965—. Fellow Royal Soc., 1966; mem. Inst. Physics, Instn. E.E.'s Cambridge Philos. Soc. Author: Radio Waves in the Ionosphere, 1961; The Wave-Guide Mode Theory of Wave Propagation, 1961; Lectures on Magnetoionic Theory, 1964. Research, publs. on theory of propagation of radio waves especially in magnetosphere, ionosphere, also around earth. Home: 15 Adams Rd., Cambridge. Office: Cavendish Lab., Free Sch. Lane, Cambridge, Eng.

BUDDINGH, G. John, med. microbiologist; b. Netherlands, Aug. 7, 1904; s. Cornelis and Dirkje (Karel) B.; came to U. S., 1907, naturalized, 1919; A.B., Calvin Coll., 1929; M.D., Vanderbilt U., 1935; m. Alice E. Mitchell, May 28, 1941. Practice medicine, specializing in microbiology, New Orleans, 1948—; prof., head microbiology La. State U. Sch. Medicine, 1948—. Mem. A.A.A.S., Am. Soc. Pathologists and Bacteriologists, Am. Assn. Exptl. Pathology, Am. Pub. Health Assn., Am. Acad. Microbiology. Research, numerous publs. on devel. of chick embryo technic for virus culture, viral and bacterial infections. Home: 3541 Carondelet St., New Orleans 70115.*

BUDDINGTON, Arthur Francis, Am. geologist; b. Wilmington, Del., Nov. 29, 1890; s. Osmer Gilbert and Mary (Wheeler) B.; Ph.B., Brown U., 1912, M.S., 1913; Ph.D., Princeton, 1916; Sc.D. (hon.), Brown U., 1942; LL.D., Franklin and Marshall Coll., 1958; Sc.D. (honorary) University Liège (Belgium), 1967; m. Jene Elizabeth Muntz, Sept. 10, 1924; 1 dau., Elizabeth (Mrs. L. E. Branagan). Instr., Brown U., 1917-19; with geophys. lab. Carnegie Instn., 1919-20; faculty Princeton, 1920—, Blair prof. emeritus, 1959—. Geologist, U. S. Geol. Survey, 1943-61. Recipient Andre H. Dumont medal Geol. Soc. Belgium, 1960. Mem. Nat. Acad. Scis., Geol. Soc. Am. (Penrose medal 1954), Mineral. Soc. Am. (Roebling medal 1958; pres. 1942), Mineral. Soc. Gt. Britain (hon.), Am. Philos. Soc., Am. Acad. Arts and Scis., Soc. Econ. Geologists, Am. Geophys. Union). Asso. editor Am. Jour. Sci. Research and numerous publs. on geology and mineral deposits of S.E. Alaska and Adirondacks; magnetite iron ore deposits of N.Y., N.J., Pa.; role of iron-titanium oxide minerals in reverse remanent magnetism, their use as a geothermometer and oxygen barometer. Home: 185 Prospect Av., Princeton, N.J. 08540.*

BUDÉ, Guillaume, French chemist; b. Paris, France, 1467; s. Dreux Budé; studied law U. Orleans (France); 7 children. Participated in founding Collegium Trilingue (now Coll. France); ambassador of Louis XII to Pope Leo X; apptd. maître des requêtes, 1522; several times provost of guilds. Author: De Asse et partibus, 1514. Described use of nitric acid for separation of gold from silver. Died Amsterdam, Netherlands, Aug. 23, 1540.

BUDEANU, Constantin, Rumanian electrotech. engr.; b. 1886; prof. Bucharest (Rumania) Polytechnic Inst.; mem. Acad. Rumanian People's Republic, several internat. sci. socs. Author: Puissances réactives et fictives, 1927; Problem of Electrification in Rumania, 1944; Practical General System of Sizes and Units, 1956; other works on elec. engring., deformed régime, reactive power, rationalization of unity system. Died 1959.

BUDESINSKY, Brestislav, Czechoslovakian chemist; b. Prague, Czechoslovakia, Apr. 10, 1928; s. Zde-Zdenek and Barbora (Cervenková) B.; Dipl. Ing., Tech. U., Prague, 1951; C.Sc., Chem. U. Pardubiee, 1959; m. Zdenka Krchová, June 2, 1953; children—Tomás. Chemist, Spofa United Pharm. Works, Prague, 1954-56. vice head analytical chemistry dept. 1956-59; staff Nuclear Research Inst., Czechoslovak Acad. Sci., Prague, 1960—, head analytical chemistry dept., 1963—. Mem. analytical div. IAEA, Vienna, Austria, 1963—. Recipient prize Czechoslovak Acad. Scis., 1965. Mem. Czechoslovak Chem. Soc. (mem. analytical group 1959—). Author: XO and MTB as Chromogenic Reagents; also numerous articles. Prins. and examples of applications of chelatometry in organic analysis; various new methods in organic analysis via functional groups; new theoretical devels. on complex formation in solution; some new chromogenic and extractive reagents. Home: 349/16 Pakomerická, Prague 8. Office: Nuclear Research Inst., Czechoslovak Acad. Sci., Rez u Prague, Czechoslovakia.*

BUDGE, Julius, German physiologist; b. Wetzlar, Germany, 1811; physician; tchr. anatomy and physiology, Bonn, Germany, later Greifswald, Germany. Author (in German): Research on the Nervous System, 1841-42; General Pathology, 1843; Compendium of Physiology, 1864. Discovered origin of various capillary canals of liver, also origin of sympathetic nerve in spinal cord; found that specific region of brain controls urinary and genital organs. Died Greifswald, 1888.

BUDINGTON, Robert Allyn, Am. zoologist; b. Leyden, Mass., Oct. 22, 1872; s. Stephen Buckland and Ereda (Baker) B.; B.A., Williams Coll., 1896, M.A., 1899, Sc.D., 1929; postgrad. Columbia, 1899-1903; m. Mabel Frances Stone, Dec. 27, 1906; children—Robert Allyn, William Stone. Instr. sci. and math. Dow Acad., Franconia, N.H., 1896-98; asst. in biology Williams Coll., 1898-99; asst. demonstrator physiology Coll. Phys. and Surg. Columbia, 1900-02; instr. physiology and zoology Mt. Hermon (Mass.) Sch., 1903-05; instr. biology Wesleyan U., Conn., 1905-08; asso. prof. zoology Oberlin Coll., 1908-13, prof. zoology, 1913-40, prof. emeritus, 1940—, head dept., 1913-36; Marine Lab., Woods Hole, Mass., 1902-10, instr. embryology, 1912-19. Fellow A.A.A.S.; mem. Am. Soc. Zoölogists, Am. Naturalists, Ohio Acad. Sci. Author: (with H. W. Conn) Advanced Physiology and Hygiene, 1909; Physiology and Human Life, 1927. Contbr. to profl. jours. Died Oct. 23, 1954.

BUDINI, Paolo, Italian physicist; b. Lussingrande, Italy, Aug. 28, 1916; m. Ambra Vidich, Apr. 7, 1949; children—Marco, Piero. Prof. theoretical physics Trieste U., 1954-64; dep. dir. Internat. Center for Theoretical Physics, Trieste, 1964—. Mem. Italian Phys. Soc., Am. Inst. Physics, Venetian Acad. Sci. Research, publs. on cosmic rays, ionization, quantum electrodynamics, limits of validity of dynamical groups. Home: 6/1 Salita Contovello. Office: 6 Piazza Oberdan, Trieste, Italy.*

BUDKER, Gersh Itskovich, Russian physicist; b. 1918; grad. Moscow U., 1941. Asso., Inst. Atomic Energy, USSR Acad. Sci., 1946-57, dir. Inst. Nuclear Physics, Siberian dept., 1957—; prof. Moscow Engring. and Physics Inst., 1956-57. Del., Internat. Conf. Physicists, Geneva, 1956. Mem. USSR Acad. Sci. (Presidium mem. Siberian dept. 1958—). Author: En Route to Anti-Matter, 1963. Research and publs. on theory of heterogeneous uranium graphite reactors, kinetic theory, regulation of atomic reactors and theory and calculation of cyclic charged particle accelerators. Address: Inst. Nuclear Physics, Siberian Dept., USSR Acad. Sci., Norosibirsk, RSFSR, USSR.

BUDLE, John, English mining engr.; b. Kyo, nr. Tanfield in Durham, Eng., 1773; taught by father; became mgr. Wallsend colliery, 1806. Author: A Synopsis of the Newcastle Coalfield, 1838; also article describing method of ventilating mines and underground lighting, 1813. Introduced improved method of coal working by which most of a vein could be removed instead of much remaining to support mine roof; assisted Davy on expts. with Davy Lamp. Died Oct. 10, 1843.

BUDNER, Stanley, Am. psychologist; b. N.Y.C., Nov. 12, 1933; s. Louis and Sara (Ladir) Budnitzky; B.A., Coll. City N.Y., 1955; Ph.D., Columbia, 1960; m. Nancy A. Salisbury, June 20, 1961; children—Anmiryam, Thaddeus Arnold. Research asst. Bur. Applied Social Research, 1956-58; sr. research psychologist N.Y. Psychiat. Inst., 1959-61; research asso. Sch. Social Work, Columbia, 1961-67; dir. research and demonstration center Fordham Univ. Sch. Social Service, New York City, 1968—; research director Mental Retardation Clinic, N.Y. Med. Coll., 1962—; research asso. N.Y.C. Bd. Edn., 1962-64; lectr. Hunter Coll., 1960-67, Columbia U. Sch. Nursing, 1961—. Cons. Retarded Infants Service, 1964—. Mem. Am. Psychol. Assn., A.A.A.S., Am. Sociol. Assn., N.Y. Acad. Scis., Phi Beta Kappa. Research on relation personality to behavior and ideology, personality and intellectual functions. Home: 680 West End Av., N.Y.C. 10025.*

BUDYKO, Mikhail Ivanovich, Russian meteorologist, geophysicist; b. 1920; grad. Leningrad Poly. Inst., 1942; D.Physico-Math. Sci., 1951. With Voeykov Main Geophys. Obs., 1942—, dir., 1954—; prof., 1953—. Recipient Lenin prize, 1958. Author: The Thermal Balance of the Earth's Surface, 1959; The Changing Climate and Ways of Converting It, 1962; co-author: The Thermal and Water Balance of the Earth's Surface, 1959; Seasonal Changes in Climatic Factors of Geographical Zonality, 1962. Editor: The Thermal Balance of the Earth's Surface, 1960. Address: Voeykov Main Geophys. Obs., Malaya Spasskaya 7, Leningrad, USSR.

BUDZANOWSKI, Andrzej Kazimierz, Polish physicist; b. Lwów, Poland, Mar. 13, 1933; s. Kazimierz and Zofia (Arnold) B.; M.Sc., Jagellonian U., 1955, Ph.D., 1961, Docent habil., 1966; m. Elzbieta Latala, Nov. 12, 1959; 1 son, Maciej. With Jagellonian U., 1954—, adj., 1962—; with Inst. Nuclear Physics, Cracow, Poland, 1958—, sr. research officer, 1962-65, leader fast neutron reactions lab., 1965—. Fellow Phys. Soc. (Gt. Britain); mem. Polish Phys. Soc. Research, publs. in nuclear reactions, polarization phenomena in stripping of deuterons and in inelastic scattering of protons, interaction of deuterons and alpha particles with atomic nuclei, measurements of total reaction cross-section and angular distbns. of elastic scattering, structure studies using inelastic alpha-scattering. Home: 3a B.Prusa, Kraków. Office: 152 Radzikowskiego, Kraków 23, Poland.*

BUEDING, Ernest, Am. biochemist; b. Frankfurt am Main, Germany, Aug. 19, 1910; s. Frederick and Katia (Margoulieff) B.; B.A., Goethe Coll., Frankfurt, 1928; M.D., U. Paris, 1936; m. Raya Palzeff, Apr. 3, 1940; 1 son, Robert. Came to U. S., 1939, naturalized, 1944. Prof. pharmacology, chmn. dept. La. State U. Sch. Medicine, 1954-60; prof. pathobiology Sch. Hygiene and Pub. Health, Johns Hopkins, Balt., 1960—, prof. dept. pharmacology and exptl. therapeutics Sch. Medicine, 1966—. Fulbright prof. pharmacology U. Oxford, Eng., 1959, Guggenheim fellow, 1963; mem. Commn. on Parasitology Armed Forces Epidemiological Bd., 1953—; mem. expert panel on parasitic diseases WHO, 1963—; mem. parasitology panel U. S.-Japan Med. Sci. Exchange Program, 1965—. Mem. Am. Soc. Biol. Chemists, Am. Soc. Pharmacology and Exptl. Therapeutics, Am. Chem. Soc., A.A.A.S., Brit. Biochem. Soc., Brit. Pharmacol. Soc., Phi Beta Kappa. Contbr. chpts. to books, numerous articles to profl. jours. Research on intermediary carbohydrate metabolism, comparative biochemistry of parasites and of intestinal smooth muscle, exptl. chemotherapy of parasitic infections, characterization of glycogen. Home: 4001 Roundtop Rd., Balt. 21218. Office: 615 N. Wolfe St., Balt. 21205.*

BUEHLER, Calvin A., Am. chemist; b. Stone Creek, O., Nov. 29, 1896; s. Philip E. and Mary (Frohlich) B.; B.Chem. Engring., Ohio State U., 1918, M.S., 1920, Ph.D. (DuPont fellow), 1922; postgrad. U. Chgo.; m. Katherine Boies McCallen, Dec. 20, 1963. Chemist, Barrett Co., Phila., 1918-19; faculty U. Tenn., Knoxville, 1922-67, prof. chemistry, 1928-67, head dept. chemistry, 1940-63, Alumni Distinguished Service prof., 1963-67. Recipient So. Chemist award, 1950. Mem. Am. Chem. Soc., A.A.A.S., Am. Assn. U. Profs. Research, numerous publs. on methods for identification organic compounds, organic syntheses, 1, 2-enediols and explanation of their stability, synthesis potential medicinals. Home: 925 Keowee Av., Knoxville, Tenn. 37919.*

BUEHLER, Robert Joseph, Am. math. statistician; b. Alma, Wis., May 1, 1925; s. Theodore and Ruth (Gobar) B.; B.S., U. Wis., 1948, M.S., 1949, Ph.D., 1952; m. Barbara Jean Martin, June 20, 1964. Mem. staff Sandia Corp., Albuquerque, 1951-55; instr. math. U. Wis., Madison, 1955-57; asst. prof. statistics Ia. State U., Ames, 1957-60, asso. prof., 1960-63; prof. statistics U. Minn., Mpls., 1963—. Study of statistical inference. Home: 901 W. County Rd. G2, St. Paul 55112. U. Minn., Mpls. 55455.*

BUEKER, Elmer D., Am. anatomist; b. Hartsburg, Mo., July 30, 1903; s. Julius Herman and Meta (Uthlaut) B.; B.S., Mo. State Coll., 1927; M.A., Colo. U., 1929; Ph.D., Wash. U., 1942; m. Elizabeth Cantrell Aber, Aug. 26, 1930; 1 dau., Anne (Mrs. Barry Butcher). Asso. prof. anatomy Georgetown U., 1947-50, Coll. Medicine U. Mo., 1950-55; prof. anatomy N.Y. U., 1955—, mem. grad. faculty of arts and sci., 1956—, acting chmn. dept., 1967—. Fellow N.Y. Acad. Sci., A.A.A.S.; mem. Harvey Soc., Soc. for Growth and Devel., Am. Assn. Anatomists, Am. Assn. U. Profs., Soc. Exptl. Biology and Medicine. Am. Assn. Zoologists, Sigma Xi. Research, on numerous publs. on nerve growth stimulating properties of mouse sarcomas which led to isolation of nerve growth factor; further studies proved that nerve growth factor is a protein of two peptides, A and C, which are active only when combined as AC.*

BUELL, Abel, Am. inventor, silversmith; b. Killingworth, Conn., Feb. 1742; son of John Buell; m. Mary Parker, 1762; m. 2d, Aletta Devoe, 1771; m. 3d, Mrs. Rebecca Parkman, 1779; m. 4th, Sarah. Apprenticed to silversmith Ebenezer Chittenden, Killingworth; opened own shop, 1762; 1st signs of his ability were some Conn. 5 shilling bank notes which he artfully improved to 5 pound notes (this indiscretion cost him some months in jail plus branding and confiscation of property); upon release from prison, constructed lapidary machine for cutting and finishing precious stones; learned craft of typefounding, produced 1st known example of Am. typefounding, 1769; granted 100 pounds by Conn. Assembly to aid in establishing type-foundry at New Haven, 1769; began copperplate engraving, 1770; produced map of territories of U. S. according to Peace of 1783 (his chief engraving work), 1784; remained in New Haven where his business operations extended to operating packet boats; developed marble quarry; owned 2 privateersmen; fashioned silver; cast type; practiced engraving; a diffuse and rarely profitable businessman; constructed money coining machine, 1785; traveled to England, 1789; worked at N.Y. cotton mfg. plant, 1793; returned to Hartford (Conn.) and continued silversmithing and engraving, 1799; silversmith in Stockbridge, Mass., 1805; a believer in Thomas Paine's doctrines. Died Alms House, New Haven, Mar. 10, 1822.

BUELL, Mary Van Rensselaer, Am. biochemist; b. Madison, Wis., June 14, 1893; d. Charles Edwin and Martha (Merry) Buell; B.A., U. Wis., 1914, M.A., 1915, Ph.D., 1919. Prof. biochemistry U. Chgo., 1951-60; project dir. Enzyme Inst. U. Wis., Madison, 1960—, Mem. Am. Soc. Biol. Chemists. Research, numerous publs. on chemistry of nucleic acids and nucleotides, relation of hormones to metabolism of carbohydrates, devel. of ultramicro. procedures for analysis of enzyme activity. Home: 2620 Arbor Dr., Madison 53711.*

BUERGER, Martin Julian, Am. crystallographer; b. Detroit, Apr. 8, 1903; s. Martin John and Julie (Weber) B.; S.B., Mass. Inst. Tech., 1925, S.M., 1927, Ph.D., 1929; D.h.c., U. Bern (Switzerland), 1958; m. Lila MacAskill, July 5, 1938; children—Maria (Mrs. Louis Friedrich), Laura (Mrs. John Sawyer), Janet, Dorothy (Mrs. Bruce MacLeod, Patricia. Faculty, Mass. Inst. Tech., 1929—, prof., 1956—. Fellow Mineral. Soc. Am. (Roebling medal 1958); mem. Nat. Acad. Scis., Am. Crystallographic Assn., Geol. Soc. Am. (Arthur L. Day medal 1951); hon. mem. Academia Brasileira de Ciencias, Accademia delle Scienze di Torino, Deutsche Mineralogische Gesellschaft, Accademia Nazionale dei Lincei, Bayerische Adademie der Wissenschaften (corr.), Osterreichische Akademie der Wissenschaften (corr.), Real Sociedad Espanola de Historia Natural Madrid. Author: X-ray Crystallography, 1942; Elementary Crystallography, 1956; (with Leonid V. Azaroff) The Powder Method, 1958; Vector Space, 1959; Crystal-Structure Analysis, 1960; The Precession Method, 1964; Contemporary Crystallography, 1968; also numerous articles. Co-editor, Internat. Tables for Crystallography, 1946—, Zeitschrift für Kristallographie, 1953—. Research on arrangements of atoms in crystals. Home Weston Rd., Lincoln, Mass. 01773. Office: Dept. Geology and Geophysics, Mass. Inst. Tech., Cambridge, Mass. 02139.*

BUERKLE, Jack V., Am. sociologist; b. West Frankfort, Ill., Aug. 9, 1923; s. Henry Adam and Clemence (Henderson) B.; B.A., U. Ill., 1948 M.A., 1949; Ph.D., U. Ia., 1954; m. Marla Louise Edwards, June 1946; children—Stephen Vincent, Melanie Lake. Asst. prof. Lake Forest Coll., 1954-55; asst. prof. Yale, 1955-60; asso. prof. Temple U., Phila., 1960-63, prof. chmn. dept. sociology, 1963—; vis. prof. Der Wirtschaftschochschule, Mannheim, West Germany, 1966-67. Mem. Am. Sociol. Assn., Am. Psychol. Assn., Eastern Sociol. Soc., Sigma Xi. Contbr. articles to sci. jours. Research in the social psychol. areas of interpersonal perception and social motivation, including family and deliquency; co-inventor of Yale Marital Interaction Battery. Home: 526 Revere Rd., Merion Station Pa. 19066. Office: Dept. Sociology, Temple U., Phila. 19122.*

BUESCHER, Edward Louis, Am. virologist; b. Cin., July 24, 1925; s. Edwin B. and Geneva (Summe) B.; B.S., U. Dayton, 1945; M.D., U. Cin., 1948; m. Elizabeth L. Fincel, June 19, 1947; children—M. Christine, E. Stephen, Michael D., Monica A., Teresa M. Research asso. Children's Hosp. Research Found., Cin., 1948, 49-50; virologist, dept. virus and rickettsial diseases Army Med. Dept., Research and Grad. Sch., Washington, 1950-51; chief dept. virus diseases Far East Med. Research Unit, U. S. Army, Tokyo, 1951-54; asst. chief dept. virus and rickettsial diseases Walter Reed Army Inst. Research, Washington, 1954-56, chief dept. virus diseases, 1956—, dir. div. communicable disease, immunology, 1967—; clin. asso. prof. pediatrics Georgetown U. Med. Sch., 1964—; mem. commn. on virus infections Armed Forces Epidemiol. Bd., 1965—. Recipient Gorgas medal, 1965. Mem. Am. Assn. Immunologists, Am. Soc. Microbiology, Am. Fedn. Clin. Research, Am. Epidemiological Soc., Infectious Disease Soc. Am. Research, numerous publs. on description of natural history of Japanese encephalitis virus; co-discoverer Rubella virus; ecology of human respiratory viruses. Home: 9213 Midwood Rd., Silver Spring, Md. 20910. Office: Dept. Virus Diseases, Walter Reed Army Inst. Research, Washington 20012.*

BUESGEN, Moritz Heinrich Wilhelm Albert Emil, German botanist; b. Weilburg/Lahn, Germany, July 24, 1858; s. Moritz and Emilie (Buderus) B.; studied in Bonn, Berlin, Germany, Strasbourg, Germany (now France); doctorate 1882; m. Martha Vollert, 1889; 1 son, 2 daus. Asst. to de Bary at zool. sta. in Naples, Italy; lectr. at Jena, Germany; became prof. botany, Eisenach, Germany, 1893, Hanover/Münden, Germany, 1901; research expdns. to Dutch E. Indies, 1902, 03, to Cameron and Togo, 1908. Author: (with E. Münch) Bau und Leben der Waldbäume (standard work), 1897. Research on woodland trees; proved that plants produce honeydew as a defense when irritated by parasitic fungi or leaf lice. Died Hanover/Münden, June 12, 1921.

BUETTNER, Konrad Johannes Karl, geophysicist; b. Westendorf, Germany, Oct. 6, 1903; s. Johannes S. and Elisabeth (Kreuser) B.; Dr. phil., Gottingen U., 1926; m. Lucie S.A. Fischer, Aug. 18, 1933; 1 son, Michael J. F. Came to U. S., 1947, naturalized, 1955. Researcher obs., Potsdam, 1927-31; researcher, asst. prof. U. Kiel, 1931-47; researcher U. S. Sch. Aviation Medicine, Randolph Field, Tex., 1947-53; faculty U. Wash., Seattle, 1953—, prof. atmospheric scis. 1956—. Cons., Rand Corp., Santa Monica, Cal. 1960—. Mem. Am. Meteorol. Soc. (Bioclimatology award 1960), Am. Geophys. Union, Am. Physiol. Soc., Internat. Soc. Biometeorology. Author: Physikalische Bioklimatologie, 1938. Discoverer that half of radiant energy in cosmos is in form of cosmic rays; pioneer in cosmic ray and solar radiation measurement in airplane; studies in physics of human heat exchange, water entry into human skin. Home: 906 37th Av., Seattle 98122. Office: U. Washington, Seattle 98105.*

BUETTNER-JANUSCH, John, Am. anthropologist, geneticist; b. Chgo., Dec. 7, 1924; s. Frederick William and Gertrude (Buettner) J.; Ph.B., U. Chgo., 1948, S.B., 1949, A.M., 1953; Ph.D., U. Mich., 1957; m. Vina Mallowitz, Sept. 22, 1950. Instr., research asst. U. Utah, 1953-55; instr. Wayne U., 1956; research asst. U. Mich., 1956-58; asst. prof. Yale, 1958-62, asso. prof., 1962-65; faculty Duke, Durham, N.C., 1965—, prof. anatomy, zoology, 1967—. Member American Association for Advancement Science, Am. Assn. Phys. Anthropologists, Am. Soc. Human Genetics, Internat. Primatological Soc., Am. Anthrop. Assn. Author: Origins of Man, 1966; also articles. Research on human and nonhuman primate evolution by study hemoglobin molecule primates, population distbn. human biochem. traits. Home: 804 Berkeley St., Durham, N.C. 27705.*

BUFF, Heinrich Ludwig, German chemist, physicist; b. Rödelheim, Germany, May 23, 1805; studied under Liebig in Giessen, Germany. Chemist, Kestner's Factory in Thann, Alsace; worked in Gay-Lussac's Lab., Paris, France; named prof. physics, Giessen, 1839. Discovered (with Wöhler) silicon hydride; verified proportionality between quantity of electricity used and amount of decomposition produced. Died Giessen, Dec. 23, 1878.

BUFFON, Georges Louis Leclerc, Comte de, French naturalist; b. Montbard, near Burgundy, France, Sept. 7, 1707; s. Benjamin François Leclerc; ed. in law, Jesuit coll. in Dijon, France; law degree, 1726; studied medicine at Angers; m. Marie-Francois de Saint-Belin, 1752; 1 son. Traveled to Italy and England; translated several English books into French; inherited property from his mother and then devoted his life to science; apptd. supt., Jardin du Roi and Royal Mus., 1739. Fellow Royal Soc., 1740; mem. French Acad. Scis., 1733; elected to Acad. française, 1753. Author: Histoire naturelle, générale et particulière avec la description du cabinet du roi (15 vols., eventually 44 vols.), 1749-1804; Theorie de la terre, 1749; Histoire des oiseaux (9 vols.), 1770-83; Histoire des mineraux (5 vols.), 1783-88; Les époques de la nature, 1780; Translator into French: Treatise on Fluxions (Newton), Vegetable Staticks (S. Hale). In his natural history attempted to treat intelligibly, under one heading, all the facts of nature, previously available only in isolated and disconnected sources; formulated theory of slow causes, by which one interprets the unknown of the past by the known of the present; showed the earth had had its own special ages, changes, epochs (in short, a history of its own); created a new geological chronology; discussed degeneracy of animals and limits set for each species by climates, mountains, and seas; anticipated some aspects of modern genetics and calculus of probabilities. Died Paris, Apr. 15, 1788.

BUGBEE, Henry Greenwood, Am. urologist; b. Waterbury, Conn., 1881; s. Walter T. and Flora (Greenwood) B.; M.D., Columbia U. Coll. of Phys. and Surg., 1903; m. Della Searles, 1906; children—Dorothy B. Kelsey, Eleanor Clift, Henry G., Jr. Surgeon-in-chief Vassar Bros. Hosp., 1906-10; urologist Woman's and Lawrence hosps. (pres. med. bd.); urologist St. Luke's Hosp., N.Y.; cons. urologist Vassar Bros. Hosp., Poughkeepsie, Muhlenberg Hosp., Plainfield, Mountainside Hosp., Montclair, Mather Hosp., Port Jefferson. Fellow A.C.S.; mem. N.Y. Acad. of Medicine, A.M.A. (chmn. sect. of urology), Am. Urol. Assn. (pres.), Am. Assn. Genito-Urinary Surgeons (pres., sec.), Am. Bd. Urology, N.Y. Urol. Assn. (pres.), Clin. Soc. of Genito-Urinary Surgeons (sec., pres.), Quiz, Royal (Budapest) med. socs., Société d'Urologie Française, Société Internationale d'Urologie (del. for U. S.; pres. of Congress 1939). Author of sect. in Lewis' Surgery. Mem. editorial bd. Jour. Urology. Contbr. articles to jours. Died Jan. 17, 1945.

BUGG, D. V., Brit. physicist; b. Ipswich, Eng., July 16, 1935; s. Edward Herbert and Maud (Keeble) B.; B.A., Emmanuel Coll., Cambridge U., 1957, M.A., Ph.D., 1960; m. Rosemary V. Chambers, Mar. 29, 1962; children—Nicholas, Timothy. Demonstrator physics Cambridge U., ofcl. fellow Emmanuel Coll., Cambridge, 1960-63; sr. sci. officer Rutherford Lab., Chilton, Berks., Eng., 1963—. Research, publs. in elementary particle physics with bubble chamber and electronic techniques chiefly applied to nucleon-nucleon interactions. Home: Old Bakehouse, Astron Tirrold, Berks. Office: Rutherford Lab., Chilton, Berks., Eng.*

BUGGE, Thomas, Danish mathematician, astronomer; b. Copenhagen, Denmark, Oct. 12, 1740; became prof. astronomy and mathematics U. Copenhagen, 1777. Mem. French Acad. Scis., 1804. Author: Premiers principes de l'astronomie sphérique et théorique, 1790; mathématiques pures ou abstraites, 1813-14. Made several maps of Denmark; studied theoretical and spherical astronomy, pure mathematics. Died Copenhagen, Jan. 15, 1815.

BUGGS, Charles Wesley, Am. microbiologist; b. Brunswick, Ga., Aug. 6, 1906; John Wesley and Leonora Victoria (Clark) B.; A.B., Morehouse Coll., Atlanta, 1928; M.S. (Rosenwald scholar 1931-34), U. Minn., 1932, Ph.D. (Shevlin fellow medicine 1933), 1934; m. Maggie Lee Bennett, Dec. 27, 1927; 1 dau., Margaret Leonora. Prof. biology, chmn. div. scis. Dillard U., 1934-43, 49-56; from instr. to asso. prof. bacteriology Sch. Medicine, Wayne State U., 1943-49; prof. microbiology Sch. Medicine, Howard U., 1956—, chmn. dept., 1958—. Mem. A.A.A.S., Am. Soc. Microbiology, N.Y. Acad. Scis., Soc. Exptl. Biology and Medicine, Sigma Xi, Alpha Phi Alpha, Sigma Pi Phi. Author: Premedical Education for Negroes, 1949. Spl. research on resistance bacteria to antibiotics. Home: 4103 18th Pl. N.E., Washington 20018.*

BUGLIARELLO, George, engr.; b. Trieste, Italy, May 20, 1927; s. Federico and Spera (Gefter-Wondrich) B.; Dr. Ing., U. Padova, 1951; M.S., U. Minn., 1954; Sc.D., Mass. Inst. Tech., 1959; m. Virginia Upton Harding, Jan. 28, 1960; children—Federico David, Nicholas Luigi. Research engr. U. Padova, 1951, cons. engr., asst. to chmn. hydraulic structures, 1954-55; research asst., research asso. hydrodynamics lab. Mass. Inst. Tech., 1956-59; asst. prof. civil engring. Carnegie Inst. Tech., Pitts., 1959-63, asso. prof., 1963-66, chmn. Interdisciplinary Biotech. Program, 1965—, prof. biotech., civil engring., 1966—. Partner Enviroconsult, 1966—; spl. cons. NIH, 1966—; lectr. NATO Advanced Study Inst. on Surface Hydrodynamics, 1966; chmn. Engring. Found. Research Conf. on Team Approaches to Complex Problem Solving, 1967. Mem. Am. Soc. C.E. (chmn. Fluid dynamics com.), Soc. Rheology, Am. Soc. Engring. Edn., Internat. Soc. Hemorheology (sec.), Internat. Assn. Hydraulic Research, Natural Philosophy Soc. Contbr. numerous articles to profl. jours. Research on fluid mechanics, hemodynamics, engring. creativity techniques, water resources engring., engring. implications biol. flows, devel. of bioengring. edn. Home: 1154 Murray Hill Av., Pitts. 15217.*

BUGYI, Balázs, Hungarian physician; b. Kolozsvar, Hungary, May 21, 1921; s. József and Anna (Temesváry) B.; Physician, U. Szeged (Hungary), 1935, Physico-chemist, 1937; Candidate Med. Scis., Hungarian Acad. Scis., Budapest, 1962. Fellow, U. Berlin, also U. Munich, 1938-41; Brit. Council fellow U. Birmingham (Eng.), 1947; asst. prof. U. Szeged, 1937, U. Budapest, 1938-40, 45-47, U. Kolozsvár, 1940-44; med. insp. industry Health Ministry, 1947-52; asso. prof. High Sch. Phys. Edn. in Sport Physiology, Budapest, 1946-49; chief radiologist County Hosp., Szolnok, 1952-56; chief radiologist Ganz Maraly Polyclinics, Budapest, 1956—. Author: (with Kováts) Occupational Mycotic Diseases of the Lung, 1968; also numerous articles. Developed paleopath. bone diagnosis using nutture, hygienic and endocrinological factors; used human paleonendocrinology to study behaviour of hist. nation; studied osteal growth and musculature. Home: 18 Ferenczy Istvan utca, Budapest V. Office: Ganz-Mávag Policlinic, Budapest. VIII, Kőbányai ut 21/23.*

BUHL, Karl Wolfgang Claus, German botanist; b. Neisse, Germany, Dec. 31, 1908; s. Konrad and Magda (Gerhard) B.; ed. univs. Breslau, Halle, Kiel (all Germany); Ph.D. in Natural Scis.; m. Ingeborg Bauer, Oct. 17, 1937; children—Karin, Klaus. Asst. Institut Pfanzenkrankheiten, U. Bonn (Germany), 1935-37; with Pflanzenschutzamt, Breslau (now Wroclaw, Poland), from 1938, Biol. Bundesanstalt, Institüt für Getreide-, Oelfrucht- und Futterpflanzenkrankheiten, from 1939, dir, 1954—. Mem. Ges fur angewandte Botanik und Angewandte Entomologie, Nederlandse Pläntziektenkundige Vereniging, Vereinigung Deutscher Pflanzenarzte. Co-author: Sorauer, Handbuch der Pflanzenkrankheiten; Kirchner, Bestimmungsbuch für Krankheiten und Schädlinge bei Getreidekulturen. Contbr. sci. works on damages caused by animals in farmlands. Address: Kiel-Kitzeberg, Schonkamp 23, W. Germany.

BUHL, Ludwig von, see von Buhl, Ludwig.

BUISSIERE (or BUSSIERE), Paul, anat. writer, surgeon; b. France; surgeon Orange, France; became Huguenot exile to Copenhagen, Denmark; naturalized English citizen, 1688; surgeon London, Eng. Mem. French Acad. Scis., Fellow Royal Soc. (mem. council 1719). Author: Lettre à M. Bourdelin pour servir de réponse au sieur Méry sur l'usage du Trou oval dans le Foetus, 1700; Nouvelle Déscription Anatomique du Coeur des Tortues terrestres de l'Amérique et de ses Vaisseaux, 1713. Contbr. to Memoires of Acad. Scis., Acta Eruditorum, Philosophical Transactions. Pioneer in introducing course of lectures on anatomy and physiology into Eng. Died Jan. 22, 1739.

BUISSON, Henri, French physicist, astronomer; b. Paris, France, July 15, 1873; prof. physics Faculty Scis., Marseilles, France; mem. French Acad. Scis., 1932. Verified (with Fabry) Doppler effect experimentally. Died Jan. 6, 1944.

BUIST, Archibald Johnston, Am. surgeon, gynecologist; b. Charleston, S.C., Feb. 7, 1872; s. John Somers and Margaret Sinclair (Johnston) B.; A.B., Princeton, 1893; M.D., Med. Coll. State of S.C., 1896; m. Alice Stock Mitchell, 1899; 1 son, Archibald Johnston; m. 2d, Elizabeth Roller Gestefeld, 1916. Began practice at Charleston, 1896; prof. gen. surgery Med. Coll. State of S.C., 1904-10, prof. abdominal surgery and gynecology, 1910-12, prof. gynecology, 1912-39, emeritus since 1939; vis. gynecologist Roper Hosp. Diplomate Am. Bd. Surgery. Fellow A.C.S., So. Surg. Soc., Southeastern Surg. Congress; mem. Am., So. med. assns. Died Sept. 12, 1943.

BUKANTZ, Samuel C., Am. physician; b. N.Y.C., Sept. 12, 1911; s. Barnett and Bertha (Stelsom) B.; B.S., Washington Sq. Coll., N.Y. U., 1930; M.D., N.Y. U., 1934; m. A. Jewell Williams, Apr. 5, 1941; children—Jessica, Dorothy. Staff, Barnes Hosp., 1948-58, Jewish Hosp., St. Louis, 1954-58; med. and research dir. Jewish Nat. Home Asthmatic Children and Children's Asthma Research Inst. and Hosp., Denver, 1958-63; staff Colo. Gen. Hosp., 1958-63, Children's Hosp., Denver, 1962-63; asso. physician Bellevue Hosp., N.Y.C., 1964—; sr. clin. investigator Schering Corp., 1963-64, asso. med. dir., clin. investigator, 1964-65; dir. dept. clin. research Hoffmann-La Roche, Inc., Nutley, N.J., 1965—; faculty Washington U., St. Louis, 1946-58, asst. dean, 1948-54; asso. prof. clin. medicine U. Colo. Sch. Medicine, Denver, 1958-63; asso. prof. clin. medicine N.Y. U. Sch. Medicine, N.Y.C., 1964—. Mem. A.M.A., Am. Acad. Allergy, A.C.P., Am. Coll. Chest Physicians, Royal Soc. Medicine. Home: 412 Redmond Rd., South Orange, N.J. 07079. Office: Hoffmann-La Roche, Inc., Nutley, N.J. 07110.*

BUKOVAC, Martin John, Am. hort. physiologist; b. Johnston City, Ill., Nov. 12, 1929; s. John and Sadie (Fake) B.; B.S., Mich. State U., 1951, M.S., 1954, Ph.D., 1957; m. Judith Ann Kelley, Sept. 5, 1956; 1 dau., Janice Louise. Faculty, Mich. State U., East Lansing, 1957—, prof., 1963—; vis. lectr. Japan Atomic Energy Radioisotope Sch., 1958; adviser Internat. Atomic Energy Agy., Vienna, 1961. NSF Sr. Postdoctoral fellow U. Oxford, Bristol, 1965-66. Mem. Am. Soc. Plant Physiologists, A.A.A.S., Am. Soc. Hort. Sci., Bot. Soc. Am., Soc. Devel. Biologists, Sigma Xi, Alpha Zeta. Numerous publs. on modification of plant growth and devel. using growth-regulating chems. Home: 2015 Brentwood Av., East Lansing, Mich. 48823.*

BUKOWSKI VON STOLZENBURG, Geiza, geologist; b. Bochnia, Galicia (now in Poland), Nov. 25, 1858; studied geology under E. Suess, paleontology under M. Neumayr, Vienna, Austria; doctorate 1887; m. Katherine Wehrmann, 1904; became mem. staff state geol. inst. in Vienna, 1889, head geologist; until 1918; dir. geol. inst., Warsaw, Poland, from 1918. Made stratigraphic-paleontol. studies in Greece and Turkey; pioneer in geol. research in unstudied areas of Austria, especially Galicia and Dalmatia. Died Bochnia, Feb. 1, 1937.

BULBROOK, Richard David, English biochemist; b. London, Eng., Apr. 29, 1926; s. Frederick and Violet (Butler) B.; B.Sc., London U., 1951, M.Sc., 1952, Ph.D., 1953; m. Joan Alice Swithinbank, Oct. 4, 1952; children—Susan, Stephen, Paul. Starling-Bayliss scholar Univ. Coll., London, 1952; staff Imperial Cancer Research Fund, London, 1953—, head sect. clin. biochemistry, 1965—. Mem. Soc. for Endocrinology (joint sec. 1966—). Publs. on investigation of hormones in human cancer, predictions of clin. course of human breast cancer, and hormonal mgmt. of disease; (with J. L. Hayward) research on precancerous endocrine status in normal women. Home: Moleside, Marden, Kent, Gt. Britain. Office: Imperial Cancer Research Fund, Lincoln's Inn Fields, London, W.C. 2, Eng.*

BULENGER (or BOULENGER), Julius Caesar, French theologian; b. Loudun, France, 1558; s. Pierre Bulenger; M. in Canon Law; joined Sec. of Jesus left order 1594, rejoined 1614; tchr. Coll. Harcourt, Paris, France, Toulouse, France, Pisa, Italy; Author: Deo theatro ludisque scenicis 1603. Opuscula, 1621. Published Hippocrates' Aphorisms, 1599. Made qualified attack on astrology. Died Cahors, France, 1628.

BULGARELLI, Rolando, Italian pediatrician; b. Modena, Italy, July 7, 1917; s. Eliseo and Teresa (Rossi) B.; M.D.; prof. pediatrics clinic, since 1958; dir. chair of child care U. Genoa (Italy). Author: I tumori endotodacici del bambino, 1952; Compendio de Puericultura, 1958-62; also monographs, numerous articles on child care. Home: via Don Minzoni 14, Genoa. Office: Institutio Gaslini, Geova Quarto, via 5 Maggio 39, Genoa, Italy.

BULKLEY, L(ucius) Duncan, Am. dermatologist; b. N.Y., Jan. 12, 1845; s. Henry D. and Juliana (Barnes) B.; A.B., Yale, 1866, A.M., 1869; M.D., Coll. Phys. and Surg., Columbia, 1869; studied dermatology abroad; m. Katherine La Rue Mellick, May 28, 1872. Attending phys. N.Y. Skin and Cancer Hosp., 1882—; cons. phys. N.Y. Hosp., 1894—; cons. dermatologist Randall's Island Hosp., Hosp. for Ruptured and Crippled, Manhattan Eye and Ear Hosp. Author: Acne and Its Treatment, 1885; Syphilis in the Innocent, 1894; Manual of Diseases of the Skin, 1898; Eczema and Its Management, 1901; Compendium of Diseases of the Skin, 1912; Diet and Hygiene in Diseases of the Skin, 1913; Cancer, Its Cause and Treatment, Vol. I, 1915, Vol. II, 1917; The Medical Treatment of Cancer, 1919. Home: New York, N.Y. Died July 20, 1928.

BULKTON, George Bowdler, Brit. chemist; b. Oakfield, Eng., May 24, 1818; lived in Haslemere, Eng. Fellow Royal Soc., 1857. Discovered mercury ethyl; prepared (with Hofmann) acetamide and disulphoethotic acid. Died Haslemere, Sept. 25, 1905.

BULL, Carroll Gideon, Am. immunologist; b. Knoxville, Tenn., June 22, 1884; s. William G. and Emma (White) B.; B.S., Peabody Coll. (U. Nashville), 1907; M.D., U. Nashville, 1910; student U. Chgo., U. Mich., Harvard; m. Zelma Smith, 1914; children—Nancy, Carrollyn. Tchr. bacteriology and pathology Lincoln Meml. U., 1910-12; fellow Nelson Morris Inst., Chicago, 1912-13; asst. and asso. in pathology Rockefeller Inst., N.Y.C., 1913-17; asso. prof. immunology, Sch. of Hygiene and Pub. Health, Johns Hopkins, 1918, prof., 1921—. Research in agglutination of bacteria in vivo and toxin and anti-toxin for B. Welchi. Died May 31, 1931.

BULL, Colin Bruce Bradley, geophysicist; b. Birmingham, Eng., June 13, 1928; s. George Ernest and Alice (Collier) B.; B.Sc. with honours, U. Birmingham (Eng.), 1948, Ph.D., 1951; postgrad. U. Cambridge (Eng.); m. Diana Gillian Garrett, June 16, 1956; children—Nicholas, Rebecca, Andrew. Shell student U. Cambridge, 1951-52; geophysicist Brit. N. Greenland Expdn., 1952-54, chief scientist, 1953-54; Imperial Chem. Industries research fellow U. Birmingham, 1955-56; sr. lectr. U. Wellington (New Zealand), 1956-61; faculty Ohio State U., Columbus, 1961—, prof. geology, 1965—, dir. Inst. Polar Studies, 1965—. Recipient H.M. Queen Elizabeth's Polar medal, 1954. Mem. Glaciological Soc., Am. Geophys. Union, Arctic Inst. N.Am., Sigma Xi. Research, publs. on geog., geophys. and glaciological exploration of Greenland, and Antarctica. Home: 4187 Olentangy Blvd., Columbus, O. 43214.*

BULL, Ephraim Wales, Am. horticulturist, legislator; b. Boston, Mar. 4, 1806; s. Epaphras and Esther (Wales) B.; m. Mary Walker, Sept. 10, 1826. Developed and exhibited Concord grape, 1853; mem. Mass. Ho. of Reps., 1855, chmn. com. on agr.; chmn. agr. Mass. Senate, 1856; mem. Mass. Bd. Agr. 1856-58. Developed numerous new strains of grapes. Died Sept. 26, 1895.

BULL, Henry Bolivar, Am. biochemist; b. Stateburg, S.C., June 16, 1905; s. Deasaussure and Caroline (Reese) B.; B.S., U. S.C., 1927; M.S., U. Minn., 1928, Ph.D., 1930; postgrad. U. Rochester; m. Fredrica Alway, Feb. 13, 1935; 1 dau., Jean. Instr., U. Minn., 1929-31; NRC fellow U. Berlin (Germany), 1931-32; asst. prof. U. Minn., St. Paul, 1932-36; postdoctoral U. Vienna (Austria), 1934; faculty Northwestern U. Med. Sch., Chgo., 1936-52, prof., 1945-52; vis. prof. Cal. Inst. Tech., Pasadena, Cal., 1943; prof., head dept. U. Ia., Iowa City, 1952-63, research prof. biochemistry, 1963—. Mem. Am. Chem. Soc., Am. Soc. Biol. Chemists, Soc. Gen. Physiologists, Biophys. Soc. Author: Biochemistry of the Lipids, 1937; Physical Biochemistry, 1943; Introduction to Physical Biochemistry, 1964; also numerous articles. Research on phys. chemistry of proteins, surface chemistry of proteins, titration of proetins. Home: 309 Sunset St., Iowa City 52240.*

BULL, Ludlow (Sequine), Egyptologist; b. N.Y.C., Jan. 10, 1886; s. Charles Stedman and Mary Eunice (Kingsbury) B.; A.B., Yale, 1907; LL.B., Harvard, 1910; Ph.D., U. Chgo., 1922; m. Katharine Davis Exton, Nov. 25, 1924; children—Frederick Kingsbury, 2d, Roger Ludlow, Agnes Davis. Admitted to N.Y. bar, 1911, practiced in N.Y.C., 1910-15; fellow in Semitics, U. Chgo., 1920-22, mem. Oriental Inst., expdns. to Egypt, Mesopotamia, Syria, 1919-20, Egypt, 1923; asst. curator Egyptian dept. Met. Mus., N.Y.C., 1922-28, asso. curator, 1928—; lectr. in Egyptology, Yale, 1925-36, research asso. with rank of prof., 1936—; curator Yale Egyptian Collection, 1925—. Asso. fellow Davenport Coll. Mem. Com. Mediterranean Antiquities, Am. Council Learned Socs., 1930-36. Fellow Am. Geog. Soc.; mem. Am. Oriental Soc. (del. to centenary Royal Asiatic Soc. 1923; pres. 1939-40), Archeol. Inst. Am., Soc. Bibl. Lit., Egypt Exploration Soc. (London), Am. Research Center in Egypt, Antiquarian and Landmarks Soc. Conn., Palestine Exploration Soc. (Jerusalem) Soc. Egyptol. Reine Elisabeth (Brussels). Author: The Rhind Mathematical Papyrus, Vol. II (with A. B. Chace, and H. P. Manning), 1929; Inscriptions at Deir el Hagar, in H. E. Winlock, Ed Dakhleh Oasis, 1936. Editor of 6 vols. Publs. of Egyptian Expdn., 6 vols. Publs. of the Egyptian dept. Met. Mus. N.Y.C. Mem. editorial board Metropolitan Museum Studies, 1928-34. Contbr. articles to profl. jours. Died July 1, 1954; buried Litchfield, Conn.

BULL, Nina, Am. research psychologist; b. Buffalo, Nov. 4, 1880; d. Ansley and Cornelia (Rumsey) Wilcox; student pvt. sch.; m. Henry Adsit Bull, Dec. 7, 1901 (div.); children—Katherine, Harry, Marian (Mrs. Eames). Hon. pres. Brain Research Found., Inc., N.Y.C., 1930-40; research asso. in psychiatry Coll. Phys. and Surg., Columbia, N.Y.C., 1940-50; dir. research project for study motor attitudes N.Y. State Psychiat. Inst., N.Y.C., 1950-60. Fellow N.Y. Acad. Scis.; mem. Am. Psychol. Assn., A.A.A.S. Author: The Attitude Theory of Emotion, 1951; The Body and Its Mind, An Introduction to Attitude Psychology, 1962. Formulation, publs. on attitude theory and its place physiology and psychology; goal orientation, frustration and depression studies. Home: 310 E. 55th St., N.Y.C. 10022.*

BULL, William Tillinghast, Am. surgeon; b. Newport, R.I., May 18, 1849; s. Henry B.; A.B., Harvard, 1869, A.M.; M.D., Coll. Phys. and Surg., N.Y., 1872; studied with Dr. Sands, also at Bellevue Hosp., and in Vienna, Berlin, Paris, and London; m. Mary Nevins Blaine, May 30, 1893. Practiced medicine specializing in surgery, N.Y.C., from 1875; in charge N.Y. Dispensary, 1875-77, Chambers St. Hosp., 1877-88; vis. surgeon N.Y. Hosp., from 1883; cons. surgeon St. Luke's Hosp., from 1883, Hosp. for Ruptured and Crippled, N.Y., State Emigrants' Hosp.; demonstrator anatomy Coll. Phys. and Surg., Columbia, 1879-80, prof. practice of surgery, 1889-1904. Fellow Am. Surg. Assn. One of 1st surgeons in Am. to adopt antisepsis; developed procedures for treatment abdominal gunshot wounds, hernia, cancer of breast. Died N.Y.C., Feb. 22, 1909.

BULLARD, Sir Edward Crisp, English geophysicist; b. Norwich, Eng., Sept. 21, 1907; s. Edward John and Eleanor (Crisp) B.; B.A., Cambridge (Eng.) U., 1929, Ph.D., 1932, Sc.D., 1948; m. Margaret Ellen Thomas, July 23, 1931; children—Belinda, Emily (Mrs. Gregory Stewart), Henrietta, Polly. With Cambridge U., 1931-39, 45-48, 57—, prof. geophysics, 1964—, Smithson Research fellow Royal Soc., 1936-43, asst. dir. research dept. geodesy and geophysics, 1956-60; prof. physics Toronto (Ont., Can.), U., 1948-49; dir. Nat. Phys. Lab., Teddington, Eng., 1950-55. Dir. Bullard & Sons Ltd., Norwich, IBM (U.K.). Recipient Chree medal Phys. Soc., 1956, Gold medal Royal Astron. Soc., 1965; Wollaston medal Geol. Soc. London, 1967. Fellow Royal Soc. (Hughes medal), Geol. Soc. Am. (hon., Day Medal); mem. U. S. Nat. Acad. Sci. (fgn. asso., Agassiz medal 1965), Am. Acad. Arts and Scis. (fgn. hon.). Research, numerous publs. on physics of solid earth especially flow of heat through ocean floor, origin of earth's magnetic field, measurement of gravity. Home: 19 Clarkson Rd. Office: Madingley Rise, Madingley Rd., Cambridge, Eng.*

BULLARD, Fred Mason, Am. geologist; b. McLoud, Okla., July 20, 1901; s. Ezra Grant and Alice (Mason) B.; B.S., U. Okla., 1921, M.S., 1922; Ph.D., U. Mich., 1928; m. Bess Mills, July 28, 1923; children—Fredda Jean, Peggy Rae (Mrs. John L. Marshall). Field geologist Okla. Geol. Survey, 1919-23; faculty U. Tex., Austin, 1924—, prof. geology, 1939—, chmn. dept., 1929-37, grad. adviser for geology, 1947-49; vis. prof. U. Mich., summers 1933, 37, Nat. U. Mexico, summers 1943-46, Vassar Coll., 1949, Columbia, summers 1949-51. Fulbright research scholar, Italy, 1952-53; Fulbright lectr. U. de San Agustin, Peru, 1959; chief of party U. S. AID-U. Tex. group U. Baghdad, Iraq, 1962-64. Fellow Geol. Soc. Am., Mineral. Soc. Am.; mem. Am. Assn. Petroleum Geologists (Distinguished lectr. 1945-54), Am. Geophys. Union, Phi Beta Kappa, Sigma Xi (pres. Tex. 1934-36), Sigma Gamma Epsilon (nat. pres. 1953-57). Author: Volcanoes, In History, In Theory, In Eruption, 1962; also articles. Research on Cretaceous stratigraphy of Tex., Okla. and adjacent areas, studies of meteorites, of use of heavy minerals to identify source of sediments; research in volcanology, particularly Parícutin Volcano, Mexico; studies in establishing volcanic cycles, cycles of lava flows, other means of tracing cyclic behavior of volcanoes; studies on ignimbrites. Home: 903 W. 30th St., Austin, Tex. 78705.*

BULLEIN, William, Brit. physician; b. Isle of Ely, Eng., early in the reign of Henry VIII; m. widow of Sir Thomas Hilton; studied medicine on continent. Rector of Blaxhall, Suffolk, Eng., 1550-53. Author: Bulwarke against Sickness (includes an early English herbal), 1562-63; A comfortable Regiment and a very wholesome order against the moste perilous Pleurisie, whereof many doe daily die within this Citee of London and other places, 1562; A Dialogue against the Fever Pestilence, 1564-65; A briefe and short discourse of the Vertue and Operation of Balsame. With an Instruction for those that have their Health to Preserve the Same, pub. 1585. Died 1576.

BULLEN, Francis Peter, Australian physicist; b. Melbourne, Australia, May, 9, 1929; s. Rogerson and Elizabeth (O'Donnell) B.; B.Sc., U. Melbourne (Australia), 1951, M.Sc. with honors, 1953; Ph.D., U. Sheffield (Eng.), 1961; m. Aileen Mary Dalton, Dec. 11, 1954; children—Stephen Peter, Margaret Mary, Paul Raymond, Helen Mary, Rogerson Luke, Mark Vincent, Catherine Mary. Sci. officer Aero. Research Labs., Dept. Supply, Melbourne, 1953-60, sr. research scientist, 1961-67, prin. research scientist, 1967—. Research, publs. on phys. mechanisms of plastic deformation, hardening and fracture of metals; codiscoverer irreversible effect of hydrostatic pressure on yielding and brittleness in metals of bodycentered cubic structure. Home: 32 Toorak Av., Croydon, Victoria, Australia. Office: Box 4331, G.P.O., Melbourne, Australia.*

BULLEN, Keith Edward, applied mathematician; b. Auckland, New Zealand, June 29, 1906; s. George Sherrar and Maud (Burfoot) B.; student Auckland U. Coll., 1923-25, D.Sc. (hon.), 1963; M.A., U. New Zealand, 1927, B.Sc., 1930; Ph.D., Cambridge (Eng.) U., 1936, Sc.D., 1945; M.A., Melbourne (Australia) U., 1945; m. Florence Mary Pressley, 1935; children —John Edward, Anne. Master, Auckland Grammar Sch., 1926-27; lectr. math. Auckland U. Coll., 1927-39; spl. lectr. Hull U. Coll., Eng., 1933; Strathcona exhibitioner St. John's Coll., Cambridge, 1932-33; sr. lectr. U. Melbourne, 1940-45; prof. applied maths. U. Sydney, (Australia), 1945—. Vice pres. Internat. Sci. Com. for Antarctic Research, 1958-62; convener Australian Nat. Com. for Antarctic Research, 1958-62; chmn. Australian Nat. Com. for Internat. Geophys. Year, 1955-60. Recipient Lyle medal Australian NRC, 1949; Walter Burfitt prize and medal Royal Soc. New South Wales, 1953; William Bowie medal Am. Geophys. Union, 1961; Research medal Royal Soc. Victoria, 1965. Fellow Royal Soc. London, Am. Geophys. Union, Royal Soc. New Zealand (hon. Hector medal 1952;, Australian Acad. Sci. (Found.); mem. Internat. Assn. Seismology and Physics of Interior of Earth (past pres.), Internat. Union Geodesy and Geophysics (v.p. 1963-67), U. S. Nat. Acad. Sci. (fgn. asso.), Am. Acad. Arts and Sci. (Fgn. hon.), Geol. Soc. Am. (corr., Day medal 1963). Author: Introduction to the Theory of Seismology, 1947, 3d edit., 1965; Seismology, 1954; Introduction to the Theory of Mechanics, 1949, 7th edit., 1965; also numerous articles. Determined values of density, pressure, gravitation intensity, compressiblity and rigidity throughout interior of Earth; research on Earth's inner core, structure terrestrial planets, Mars, Venus and Mercury, origin of Moon, seismic investigation of nuclear explosions, origin times of early hydrogenbomb explosions, earthquake travel-times, free Earth oscillations. Address; Dept. Applied Maths., U. Sydney, Sydney, N.S.W., Australia.*

BULLER, Arthur Henry Reginald, botanist, mycologist; b. Birmingham, Eng., Aug. 19, 1874; s. Alban Gardner and Mary Jane (Huggins) B.; student Queen's Coll., Mason Coll., 1892, also Munich, Germany; B.Sc., London, 1896; D.Sc., Birmingham, Eng., 1903; Ph.D., Leipzig, Germany; L.L.D. U. Man. (Can.), 1924. Lectr. botany, Birmingham, 1901-04; prof. botany U. Man., Winnipeg, Can., 1904-36, prof. emeritus, 1936. Pres. sect. for mycology and bacteriology 6th Internat. Bot. Congress, Amsterdam, Netherlands, 1938; Fellow Royal Soc., 1929 (Royal medal 1937); pres. Brit. Mycol. Soc., 1913, Canadian Phytopathological Soc., 1920, Royal Soc. Can. (Flovell medal 1929), 1927-28, Bot. Soc. Am., 1928; sec. sect. k Brit. Assn. Author: Researches on Fungi, Vols. I-VI, 1909-34. Essays on Wheat, 1919; Practical Botany, 1929; (with G. R. Bisby and J. Dearness) The Fungi of Manitoba, 1929. Collaborator: The Fungi of Manitoba and Saskatchewan, 1938. Editor: (with C. L. Shear) Selecta fungorum carpologia (Tulasne, English transl. by W. B. Grove), 3 vols., 1931. Increased knowledge of gen. biology, also of sexuality of fungi, of how common fungi adapt to environment, of relation of structure of fungi parts to function. Died Winnipeg, July 3, 1944.

BULLER, Walter Lawry, New Zealand ornithologist; b. Newark, Bay of Islands, New Zealand, Oct. 9, 1838; s. James Buller; ed. Wesley Coll., Auckland, New Zealand; Sc.D. (hon.), Cambridge (Eng.) U., 1900, Tubingen, Germany. Founder weekly Maori newspaper, editor-in-chief Maori Messenger, 1861; resident magistrate, 1862; judge, native land ct., 1865; sec. to agt.-gen. for New Zealand, Eng., 1871; called to bar at Inner Temple, 1874; practice law in Supreme Ct., New Zealand, until 1886. Decorated Legion of Honor, mansion house com. Paris Exhbn., 1889. Fellow Royal Soc., 1879, Linnean Soc., Geol. Soc. Author: History of the Birds of New Zealand, 1873; Manual of the Birds of New Zealand, 1882; numerous papers. Made thorough study of ornithology of New Zealand. Died July 19, 1906.

BULLIARD, Pierre, French botanist; b. Aube-Pierre, France, Nov. 24, 1752; s. Francois and Elisabeth Tripier; studied botany and anatomy at Clairvaux Abbey; studied engraving with Francois Martinet in Paris, France; m. Angélique de Voulges; m. 2d, Marie Lemeicier. Author: Flora parisiensis, 1776; Aviceptologie francaise, 1778; Herbier de la France, 1780; Dictionnaire élémentaire de botanique, 1783; Histoire des plantes vénéneuses, 1784; Histoire des champignons de la France, 1791. Research on Parisian flora, mushrooms, spontaneous generation; discovered cause of wheat rust. Died Sept. 1793.

BULLIS, Harvey Raymond, Jr., Am. marine fisheries scientist; b. Milw., June 14, 1924; s. Harvey Raymond and Isabelle (Johnson) B.; B.S., U. Wis., 1949; M.S., U. Miami, 1951; m. Lois May Fischer, Apr. 7, 1944; children—Betty Susan (Mrs. Alan Grant Burton), Margaret Louise, Jeanmarie, Harvey Raymond III. Fishery scientist Bur. Comml. Fisheries U. S. Fish and Wildlife Service, 1950—, dir. Exploratory Fishing and Gear Research Base, Pascagoula, Miss., 1955—; mem. U. S. AID Fishery Survey Team to W. Africa, 1961; mem. Nat. Acad. Scis.; com. sci. technologic base of P.R. economy; chief scientist fishing expdns. Gulf Mexico, Caribbean Sea. Mem. Am. Fisheries Soc., Am. Soc. Ichthyologist and Herpetologists, Soc. Systematic Zoology, Am. Malacological Union. Research, numerous publs. on marine fauna and fishery resources of tropical western Atlantic region; dir. research leading to discovery of several large new fishery resources, foremost extensive internat. shrimp fishery off northeastern S.Am.; dir. collection of over two million research specimens which have yielded over 200 new species. Home: 101 Hague St. Office: Bur. Comml. Fisheries Exploratory Fishing Base, Pascagoula, Miss. 39567.*

BULLOCH, William, Brit. bacteriologist; b. Aberdeen, Scotland, Aug. 19, 1868; s. John Bulloch; ed. Aberdeen U., LL.D., 1920; studied in Leipzig, Germany, Vienna, Austria, Paris, France, also King's Coll., London, Eng.; M.D.; m. Irene Adelaide Baker, 1923; Asst. prof. pathology U. Coll., London; bacteriologist Lister Inst. Preventive Medicine, also chmn. governing body; emeritus prof. bacteriology U. London; examiner U. London, U. Leeds (Eng.), U. Liverpool (Eng.), Royal Coll. Vet. Surgeons; Tyndall lectr. Royal Instn., 1922; mem. govt. foot and mouth disease com., med. Research Council under Nat. Ins. Act. Grocers' Co. Research scholar in preventive medicine. Fellow Royal Soc., Royal Soc. Medicine (hon., pres. path. sect.). Contbns. on bacteriological, path. subjects. Died Feb. 11, 1941.

BULLOCK, Roberts Cozart, Am. mathematician; b. Oxford, N.C.; s. James Dudley and Dena (Roberts) B.; A.B., U. N.C., 1926, M.A., 1928; Ph.D., U. Chgo., 1932; m. Pauline Cox, June 30, 1935; 1 dau., Rebecca Roberts. Instr. math. U. N.C., Chapel Hill, 1928-30; head dept. Lambuth Coll., Jackson, Tenn., 1932-34; head math. dept. Ark. Tech. Coll., Russellville, 1934-35; faculty N.C. State U., Raleigh, 1935—, prof., 1947—; asso. dir. Rocket Research Project, 1952-61, dir., 1961—. Mem. Am. Math. Soc., Math. Assn. Am., Am. Soc. Engring. Edn., Phi Beta Kappa, Sigma Xi, Phi Kappa Phi, Lambda Chi Alpha. Research in theory of motion of arty. rockets. Home: 1415 Dixie Trail, Raleigh, N.C.*

BULLOCK, Theodore Holmes, Am. zoologist, educator; b. Nanking, China, May 16, 1915 (parents Am. citizens); s. A. Archibald and Ruth (Beckwith) B.; A.B., U. Cal. at Berkeley, 1936, Ph.D., 1940; m. Martha Runquist, May 30, 1937; children—Christine, Stephen. Sterling fellow Yale, 1940-41, Rockefeller fellow, 1941-42, research asso., instr. neuroanatomy, 1942-44; asst. prof. anatomy U. Mo. Sch. Medicine, 1944-46; faculty U. Cal. at Los Angeles, 1946-66, prof. zoology, 1955-66; prof. neuroscis. U. Cal. at San Diego, 1966—. Mem. Am. Physiol. Soc., A.A.-A.S., Soc. Gen. Physiology, Am. Soc. Zoologists (pres., 1964-65), Am. Acad. Arts and Scis., Nat. Acad. Scis., Phi Beta Kappa, Sigma Xi. Author: Structure and Function in the Nervous Systems of Invertebrates, 1965. Contbr. numerous articles in field to tech. jours.*

BULLOCK, William A., Am. inventor; b. Greenville, N.Y., 1813; 1 dau. Apprentice to iron founder and machinist, Catskill, N.Y., 1821-34; owner machine shop (developed shingle-cutting machine), Plattsville, N.Y., 1836-38; established unsuccessful shingle mfg. firm, Savannah, Ga.; then opened shop making hay and cotton presses of his own design, also artificial legs, N.Y.C.; opened patent agy., machine shop (3 original designs came from this shop, grain drill, seed planter, lath cutting machine), Phila., 1849; printed daily newspaper The Banner of The Union, 1849-53, became interested in printing machinery, devoted rest of life to devel. and eventual patenting of Bullock press which revolutionized printing by printing on both sides of the paper, printing from continuous roll of paper and cutting newsprint either before or after printing, 1863. Died Phila., Apr. 12, 1867.

BULLOUGH, William Sydney, Brit. biologist; b. London, Eng., Apr. 6, 1914; s. Fred Sydney and Letitia (Cooper) B.; B.Sc., U. Leeds, Eng., 1935, Ph.D., 1937, D.Sc., 1944; m. Helena Florence Gibbs, Mar. 23, 1942; children—Helena Anne, William Arthur. Lectr. zoology, U. Leeds, 1937-44; lectr. zoology McGill U., Montreal, Que., Can., 1944-46; research fellow Royal Soc. London, 1946-51; research fellow Brit. Empire Cancer Campaign, 1951-52; prof., chmn. dept. zoology Birkbeck Coll., U. London, 1952—. Recipient U. Brussels, U. Helsinki medals. Fellow Zool. Soc. London, Inst. Biology; mem. numerous socs., hon. mem. Soc. Investigative Dermatology. Author: Practical Invertebrate Anatomy, 1950; Vertebrate Sexual Cycles, 1951; The Evolution of Differentiation, 1967. several children's books. Research, numerous publs. on control of reproductive cycles, sex hormones led to problem of mitotic control in adult mammals; described mechanisms that control mitotic and functional homeostasis, extraction of a new group of chems. now called chalones which influence these mechanisms, clearer understanding of way in which cancer develops when these mechanisms break down. Home: New Cottage, Uplands Rd., Kenley, Surrey, Eng. Office: Birkbeck Coll., Malet St., London W.C.1, Eng.*

BULLOWA, Jesse G. M., Am. physician; b. N.Y.C., Oct. 19, 1879; s. Moritz and Mary (Grunhut) B.; A.B., Coll. City of N.Y., 1899; M.D., Coll. Phys. and Surg. (Columbia), 1903; m. Sadie Nones, Sept. 24, 1907; children—Margaret, James, Elizabeth, Jean, Anne. Vis. physician Harlem and Willard Parker hospitals; clin. professor medicine N.Y. U., Coll. Medicine, since 1928; cons. physician N.Y. Infirmary for Women and Children, Norwalk (Conn.) Gen. Hosp.; cons. serologist at Long Beach (N.Y.) Hosp. Fellow N.Y. Acad. of Medicine, A.A.A.S., N.Y. Acad. Scis.; mem. A.M.A., N.Y. Pathol. Soc., Soc. for Exptl. Biology and Medicine, Assn. for Study of Internal Secretions, Nat. Tb Assn., Am. Assn. Immunologists, Am. Trudeau Soc., Phi Beta Kappa. Translator: Bechold's Colloids in Biology and Medicine, 1919. Author: The Management of the Pneumonias, 1937; The Specific Therapy of the Pneumonias, 1939. Contbr. articles to med. jours. Died Nov. 10, 1943.

BULLRICH, Kurt, German physicist; b. Honneroth, Germany, Feb. 9, 1920; s. Peter and Hilde (Püngeler) B.; student U. Frankfurt, Germany, 1939-40, U. Munich, Germany, 1940-41; Ph.D., U. Frankfurt, 1942. Univ. dozent U. Mainz, Germany, 1963—. Research, publs. in field atmospheric physics; light scattering in the turbid atmosphere. Home: 7 Ehrhard St., Mainz Fed. Rep. Germany. Office: U. Mainz, Fed. Rep. Germany.*

BULMAN, Oliver Meredith Boone, Brit. geologist; b. London, May 20, 1902; s. H. H. and B. E. Boone; ed. Cambridge U.; Ph.D., Sc.D., London U.; m. Marguerite Fearnsides, 1938; children—Jane, Louisa, Charlotte, William. Geology planner, Imperial Co., London, also at Cambridge; instr. paleozoology Cambridge U., prof. geology, since 1955. Fellow Royal Soc.; mem. Geol. Fören Stockholm (corr.), Kgl. Fysiogr. Sällsk. Lund (fgn.), Paleontology Soc. India. Contbr. articles to jours. Address: Sedgwick Museum, Dowing Street, Cambridge, England.

BULOW, Kurd von, see von Bulow, Kurd.

BULWER, John, Brit. physician; b. Eng.; flourished 1654; s. Thomas B. Author: Chirologia, 1644; Philocophus, or the Deafe and Dumbe Man's Friend, 1648; Pathomyotomia, 1649; Anthropometamorphosis, 1650; Vultispex criticus seu phiognomia medici; Glossiatrus; Otiatrus. Discovered methods of communication with deaf and dumb by use of sign lang., gestures, lip reading, articulation; discovered deaf enjoy music through medium of teeth.

BUMKE, Oswald, psychiatrist, neurologist; b. Stolp (Pomerania) Sept. 25, 1877; s. Albert and Emma (Westphal) B.; m. Hedwig Burkart, 1910; 1 son, 1 dau. Prof., Rostock, 1914, Breslau, 1916, Leipzig, 1929; prof. psychiatry U. Munich, from 1924, rector, 1928-29; dir. Munich Neurol. Clinic, for 22 years (all Germany). Sec. of Archiv fur Psychiatrie; chmn. editorial collegium Munchner· Medizinische Wochenschrift. Author: Über nervose Entartung, 1912; Lehrbuch der Geisteskrankheiten, 1919; Kultur und Entartung, 1922; Das Uterbewusstsein, 1922; Die gegenwartige stromungen in der Psychiatrie, 1928; Die Psychoanalyse, Eine Kritik, 1931; Die Psychoanalyse Seele, 1941. Editor: Handbuch der Geisteskrankheiten, 1928-33; Handbuch der Neurologie, 1935-37. Major studies on pupil of eye in mental disease; discovered peculiar dilation of pupil which results from certain psychic stimuli (known as Bumke's pupil), 1904; studies on psycho-phy. relationships in brain; also known as physician (attended Lenin in Moscow). Died Munich, Jan. 5, 1950.

BUMM, Ernst, gynecologist; b. Würzburg, Germany, Apr. 15, 1858; s. Kaspar and Barbara (Gutbrod) B.; studied medicine in Würzburg; doctorate, 1880; m. Lillie Leube, 1897; 3 sons, 1 dau. Became lectr. in gynecology and obstetrics, Würzburg, 1885; became prof., Basel, Switzerland, 1894, Halle, Germany, 1900, in Berlin, Germany, 1904; dir. gynecol. clinic. Author: Grundriss zum Studium der Geburtshilfe, 1902. First to apply bacteriology systematically to obstetrics and gynecology, especially in puerperal fever and surg.

infections; improved surg. techniques; his textbook became a model for later med. textbooks. Died Munich, Germany, Jan. 2, 1925.

BUMSTEAD, Henry Andrews, Am. physicist; b. Pekin, Ill., Mar. 12, 1870; s. Samuel Josiah and Sarah Ellen (Seiwell) B.; A.B., Johns Hopkins, 1891; Ph.D., Yale, 1897; m. Luetta Ullrich, Aug. 18, 1896. Asst. in physics Johns Hopkins, 1891-93; instr. in physics Sheffield Sci. Sch., Yale, 1893-1900; asst. prof., 1900-06, prof., Yale Coll., and dir. Sloane Phys. Lab., 1906-—, Yale U. Sci. attaché Am. Embassy, London, 1918. Fellow Am. Acad. Arts and Scis. Investigated properties of delta rays emitted by metals under influence of alpha rays, 1911-20; showed that fast-moving electrons are produced by alpha particles colliding with gas molecules, 1916. Contbr. numerous articles to sci. jours. on exptl. and theoretical physics, especially electromagnetic phenomena. Died Dec. 31, 1920.

BUNAU-VARILLA, Philippe Jean, French engr.; b. Paris, France, July 26, 1860; ed. Ecole Polytechnique; engr., Ecole des Ponts et Chaussées, 1883. Became asst. engr. to de Lesseps, Panama Canal, 1884, chief engr., 1885; named plenipotentiary minister from Panama to U. S. (negotiated Hay-Herran treaty, 1903, also sale of French rights in Panama to U. S.), 1903. Decorated Grand Croix, Legion of Honor. Author: Panama; le passé, le présent, l'avenir, 1902; Panama; le traffic, 1902; Panama, 1913; La grande aventure de Panama, 1920; L'autojavellisation imperceptible, 1926; Radiolyse chimique, 1927; From Panama to Verdun, 1940. Organized insurrection in Panama that resulted in its independence from Columbia; negotiated treaty giving U. S. control of the Canal Zone; devised new system for water purification; supervised engineering projects in Europe and Africa. Died Paris, May 18, 1940.

BUNCE, Donald F(airbairn) M(acDougal), II, Am. anatomist, educator; b. Harrisburg, Pa., July 15, 1920; s. Wesley Hibbard and Jean (Fairbairn) B.; B.S., U. Miami, 1951; M.S., U. Ill., 1959, Ph.D., 1960; m. Lorraine Pelch, May 1, 1954; children—Gregory Alan Chip, Graham Alison Dale. Pres., Bunce Sch. Lab. Technique, Coral Gables, Fla., 1945-48; physiologist Armour Labs., Chgo., 1953-56; dir. research Chgo. Pharmacal Co., 1956-57; instr. anatomy Tulane U. Sch. Medicine, 1960-62; research prof. physiology, Coll. Osteopathic Medicine and Surgery, Des Moines, 1962-67, dir. grad. sch., 1962-—, chmn. depts. physiology and pathology, 1967-—. Vis. fellow Inst. Exptl. Surgery, Copenhagen, Denmark, 1962; vis. prof. Karolinska Inst., Stockholm, Sweden, 1965. Fellow Am. Coll. Angiology, A.A.A.S., N.Y., Ia. acads. sci.; mem. A.M.A., Am. Assn. Anatomists, Anat. Soc. Gt. Britain, So. Soc. Anatomists (exec. sec. 1960-62), Path. Soc. Gt. Britain, Soc. Exptl. Biology and Medicine, Am. Assn. U. Profs., L'Union Internationale d'Angéiologie, Société Français d'Angéiologie et d'Histopathologie, Sigma Xi, Sigma Alpha Epsilon. Editorial bd. Angéiologie, Paris, France, 1960-—. Author: Laboratory Guide to Microscopic Anatomy, edit. 2, 1966; also articles. Research in anatomy and diseases of blood vessels, physiology and pathology of fetus and infant; invented Bunce double hemostat used to remove surgically blood vessels and other tissues so that they may be studied in their living condition. Home: 717 61st St., Des Moines 50312.*

BUNCE, Stanley Chalmers, Am. chemist; b. Bayonne, N.J., Aug. 21, 1917; s. Arthur Chalmers and Elizabeth (Sticht) B.; B.S. in Chemistry, Lehigh U., 1938, M.A. in Edn., 1942, Ph.D. in Organic Chemistry, Rensselaer Poly Inst., 1951; m. Lillis Adele Jackson, Oct. 2, 1943; children—Gale Elizabeth (Mrs. Andrew Schmidt), Judith Preston, James Arthur. High sch. sci. tchr., Hershey, Pa., 1939-41, Bound Brook, N.J., 1941-43; research chemist Johns-Manville Corp., Manville, N.J., 1943-46; faculty Rensselaer Poly. Inst., Troy, N.Y., 1946-—, prof., 1958-—. Fellow A.A.A.S.; mem. Am. Chem. Soc., New Eng. Assn. Chemistry Tchrs., Fedn. Am. Scientists. Author: (with others) Principles of Chemistry, 1966; also articles. Research on reactions of organic compounds in which 3-membered rings are present, especially steps by which molecular rearrangements take place; prepared new compounds. Home: Box 68 Taconic Lake Rd., Grafton, N.Y. 12082. Office: Dept. Chemistry, Rensselaer Poly. Inst., Troy, N.Y. 12181.*

BUNCEL, Erwin, chemist; b. Presov, Czechoslovakia, May 31, 1931; s. Ignacz and Irene (Sharman) B.; came to Eng., 1946, Brit. citizen, 1955; B.Sc., U. London, 1954, Ph.D., 1957; m. Penny Bienenfeld, Dec. 16, 1956; children—Irene, Jacqueline. Research asso. U. N.C., Chapel Hill, 1957-58; postdoctoral fellow NRC Can., Hamilton, 1958-61; research chemist Am. Cyanamid Co., Stamford, Conn., 1961-62; faculty Queen's U., Kingston, Ont., Can., 1962-—, asso. prof. chemistry, 1966-—. Collaborator chem. kinetics data project Nat. Acad. Sci.-Nat. Bur. Standards, 1959-64. Mem. Am. Chem. Soc., Canadian Inst. Chemistry, Chem. Soc. London, Am. Acad. Arts and Scis., N.Y. Acad. Scis. Author, co-author Tables of Chemical Kinetics, Homogeneous Reactions, vol. 1, 1962, vol. 2, 1964; also articles, revs. Research in organic, specializing in phys. organic, chemistry on topics including acid-base catalysis, hydrogen-deuterium exchange processes and isotope effects, nucleophilic substitution in carbohydrates and in aromatic polynitro compounds.*

BUNDE, Carl Albert, Am. clin. pharmacologist; b. Ashland County, Wis., Apr. 22, 1907; s. Karl and Alivina (Brozinske) B.; A.B., U. Wis., 1933, A.M., 1934, Ph.D., 1937; M.D., Southwestern Med. Coll., 1948; m. Wilma Elizabeth Bell, May 30, 1930. Fellow, U. Okla. Sch. Medicine, 1937-38, faculty, 1938-42, asst. prof., 1941-42; asst. prof. Baylor U. Coll. Medicine, 1942-43; asso. prof. physiology and pharmacology Southwestern Med. Coll., 1943-49; v.p. research Pitman-Moore Co., Indpls., 1949-60; dir. med. research William S. Merrell Co., Cin., 1960-—. Fellow Am. Coll. Clin. Pharmacology; mem. Soc. Zoologists, Endocrine Soc., Am. Physiol. Soc., Soc. for Exptl. Biology and Medicine, A.M.A., Am. Fedn. for Clin. Research, Am. Therapeutic Soc., Am. Soc. for Study Sterility. Research, publs. on endocrinology, circulation, metabolism, therapeutics. Home: 3738 Donegal Dr., Cin. 45236. Office: William S. Merrell Co., Cin. 45215.*

BUNDY, Francis Pettit, Am. physicist; b. Columbus, O., Sept. 1, 1910; s. Lyman E. and Edith C. (Scott) B.; M.Sc., Otterbein Coll., 1931, D.Sc., 1959; Ph.D., Ohio State U., 1937; m. Hazel V. Forwood, Oct. 24, 1936; children—John, Freda (Mrs. John Hofland), Susanne (Mrs. George Moffat), David. Faculty, Ohio U., 1937-42; physicist Underwater Sound Lab., Harvard, 1942-45, USN Underwater Sound Lab, New London, Conn., 1945-46; physicist Gen. Electric Research and Devel. Center, Schenectady, 1946-—. Recipient Naval Ordnance Devel. award, 1945; Army-Navy certificate of appreciation, 1947. Mem. A.A.A.S., Am. Phys. Soc., Acoustical Soc. Am. Editor: (with W. R. Hibbard, H. M. Strong) Progress in Very High Pressure Research, 1961. Research, publs. on spectroscopy of hot high velocity flames, vacuum type self-supporting thermal insulation, very high pressure phenomena.*

BUNGE, Alexander von, see von Bunge, Alexander.

BUNGE, Nicolai Andreevich, Russian chemist; b. Russia, 1842; s. Andrei Bunge; prof., U. Kiev (Russia); studied laws of organic and inorganic electrolysis. Died 1914.

BUNIAKOVSKI, Victor Jakovlevich, mathematician; b. Russia, Dec. 15, 1804; s. Jakov Buniakovski; D-es-s., Faculty Sci., Paris, 1825; became prof. math. U. St. Petersburg, Russia, 1845. Mem. Petersburg Acad. Sci. (pres. 1864-89). Author: Principles of Mathematical Theory of Possibilities, 1846. Invented a theory of inequalities. Died St. Petersburg, Nov. 30, 1889.

BUNIM, Joseph J., physician; b. Wolozin, Russia, Nov. 5, 1906; s. Moses and Minnie (Joselowsky) B.; brought to U. S., 1910, naturalized, 1914; B.S., Coll. City N.Y., 1926; M.D., N.Y. U., 1930, Dr.Med Scis., 1938; m. Miriam Schild, Dec. 30, 1934; children—Lesley Schild (Mrs. Morton Heafitz), Elizabeth Rose (Mrs. Harvey Karten), Michael Ben. With N.Y. U., Coll. Medicine, 1936-52, asso. prof. medicine, 1949-52; chief arthritis and rheumatism br. Nat. Inst. Arthritis and Metabolic Diseases, NIH, Bethesda, Md., 1952-—, chief clin. investigations, 1953-—; asso. prof. Johns Hopkins, Balt., 1953-—; cons. physician Balt. City Hosps., 1953-60; attending staff physician Walter Reed Gen. Hosp., 1957-—; clin. prof. medicine Georgetown U. Sch. Medicine. Mem. NRC, 1960-63; mem. adv. panel on chronic degenerative diseases WHO, 1963-—; leader group rheumatolcgists on cultural exchange visit for U. S. A., Russia and Scandinavian countries, 1964. Grover Powers fellow infectious diseases Yale Coll. Medicine, 1932; recipient Heberden medal Brit. Empire Rheumatism Soc.; Presdl. citation N.Y. U. Sch. Medicine, posthumous 1965; Joseph J. Bunim Meml. Fund for Joseph J. Bunim Meml. Lecture, established in his honor. Diplomate Am. Bd. Internal Medicine. Fellow N.Y. Acad. Medicine, Washington Acad. Scis., A.C.P.; mem. Am Rheumatism Assn. (past pres.), N.Y. Rheumatism Assn., N.Y. Heart Assn., Harvey Soc., Alpha Omega Alpha, Sigma Xi; hon. mem. Heberden Soc. (medal), Canadian Rheumatism Assn., Argentine, Mexican, Italian rheumatology socs. Research and numerous publs. in immunology, collagen diseases, especially nature diagnosis and treatment of rheumatic diseases. Died July 8, 1964.*

BUNIN, Konstantin Petrovich, Russian metallurgist; b. Yakaterinodar (now Krasnodar), RSFSR, 1910; grad. Dneptropetrovsk Metall. Inst., 1932. Head dept. metal sci. Inst. Ferrous Metallurgy, Ukrainian Acad. Sci., 1948-—. Mem. Ukrainian Acad. Sci. (corr.). Author over 100 works including 5 monographs: Chilled Cast Iron, 1947; Iron Carbide Alloys, 1949; The Structure of Cast Iron, 1952; co-author: Spheroidal Graphitic Cast Iron, 1955. Research on phase transitions of metals, crystallization mechanism of cast iron and its phase and structural changes in solid state. Address: Ukrainian Acad. Sci., Vladimirskaya ulitsa 54, Kiev, Ukraine SSR, USSR.

BUNIN, Konstantin Vladimirovich, Russian infectionist; b. 1912; D. Med. Sci. Prof., head chair infectious diseases 1st Moscow Med. Inst., 1954-—. Mem. Learned Med. Council, RSFSR Ministry Health; mem. Higher Certification Commn. Mem. Moscow (bd. mem.), All-Russian (bd. mem.), All-Union (bd. mem.) socs. infectionists, microbiologists and epidemiologists. Author over 120 works including The Cardio-Vascular System in Infectious Diseases; Early Differential Diagnosis of Infectious Diseases; Infectionist's Pocket Manual, 1960; Infectious Diseases, 2d edit., 1961. Address: 1st Moscow Med. Inst., B. Pirogovskaya ulitsa 2-6, Moscow, USSR.

BUNIVA, (Michele) Francesco, Italian physician; b. Pignerol, Italy, May 15, 1761. Prof. medicine U. Turin, then prof. pathology, 1801-14; dir. Turin Vet. Sch. Mem. French Acad. Sci. 1819. Author works on comparative pathology. Died Pincina, Italy, Oct. 26 1834.

BUNKER, Don Louis, Am. chemist; b. San Fernando, Cal., July 12, 1931; s. Jack and Mabel (Lesser) B.; B.S., Antioch Coll., 1953; Ph.D., Cal. Inst. Tech., 1957; m. Joyce Reppard, June 20, 1952 (div. Apr. 1959); 1 dau., Susan; m. 2d, Susan Swartz, July 20, 1961; children—David, Brian. Staff mem. Los Alamos Sci. Lab., 1957-65; asso. prof. chemistry U. Cal. at Irvine, 1965-—. Mem. Am. Chem. Soc., Am. Phys. Soc., A.A.A.S., Sigma Xi. Author: Theory of Elementary Gas Reaction Rates, 1965; also articles. Research on rates at which chem. reactions occur. Home: 2324 Arbutus St., Newport Beach, Cal. 92660. Office: Dept. Chemistry, U. Cal. at Irvine, Cal., 92664.*

BUNKER, John Philip, Am. physician; b. Boston, Feb. 13, 1920; s. Philip H. and Emily L. (Glover) B.; A.B., Harvard, 1942, M.D., 1945; m. Mary Franklin Bush, Aug. 12, 1944; children—Jane Williams, Katherine Ford, John Philip, Emily Lane. Practice medicine specializing in anesthesia, Boston, 1950-60, Stanford, Cal., 1960-—; asst. in anesthetics Mass. Gen. Hosp., 1950-53, asso. anesthetist, 1953-60; asso. in anesthetics Harvard Med. Sch., 1952-55, asst. clin. prof. anesthetics, 1955-60; prof. anesthetics Stanford, Palo Alto, Cal., 1960-—. Mem. com. on anesthetics NRC, 1959-—, chmn. subcom. on nat. halothane study, 1963-—; mem. pharmacology tng. com. NIH, 1961-66. Study of metabolic effects of blood transfusions; pharmacology of anesthesia. Home: 672 Foothill, Stanford, Cal. 94305. Office: Stanford Med. Center, Palo Alto, Cal. 94304.

BUNKER, Merle Eugene, Am. physicist; b. Kansas City, Mo., Feb. 8, 1923; s. Noah F. and Ada (Spitler) B.; student N.C. State Coll., 1943-44; B.S. in Mech. Engring. Purdue U., 1946; M.S. in Physics, Ind. U., 1948, Ph.D., 1950; m. Wilma Jean Horine, May 10, 1943; children—Bruce C., Paul G. Exptl. nuclear physicist Los Alamos Sci. Lab., U. Cal., 1950-—. NSF sr. postdoctoral fellow Inst. for Theoretical Physics, Copenhagen, Denmark, 1964-65. With AUS, 1943-46; Manhattan Project, 1944-46. Fellow Am. Phys. Soc.; mem. Sigma Xi. Contbr. articles on research in nuclear physics, reactor tech. to tech. jours. Research on radioactive decay and low-energy nuclear structure. Home: 2218 46th St. Office: Box 1663 Los Alamos 87544.*

BUNNETT, Joseph Frederick, Am. chemist; b. Portland, Ore., Nov. 26, 1921; s. Joseph and Louise (Boulan) B.; B.A., Reed Coll., 1942; Ph.D., U. Rochester, 1945; m. Sara Anne Telfer, Aug. 22, 1942; children—Alfred Boulan, David Telfer, Peter Sylvester. Research chemist Western Pine Assn., Portland, 1945-46; faculty Reed Coll., 1946-52, asst. prof., 1948-52; faculty U. N.C., 1952-58, asso. prof., 1953-58; faculty Brown U., Providence, 1958-66, prof., 1959-66, chmn. chemistry dept., 1961-64; prof. chemistry U. Cal. at Santa Cruz, 1966-—. Fulbright Research scholar U. Coll., London, Eng., 1949-50; Guggenheim fellow, Fulbright Research scholar U. Munich, Germany, 1960-61, Mem. Am. Acad. Arts and Scis., A.A.A.S., Am. Chem. Soc., Chem. Soc. (London), Gesellschaft deutscher Chemiker. Editor: Accounts of Chemical Research. Research, articles on mechanism organic reactions, organic synthesis. Home: 608 Arroyo Seco, Santa Cruz, Cal. 95060.*

BÜNNING, Erwin, German biologist; b. Hamburg, Germany, Jan. 23, 1906; s. Hinrich and Hermine (Winkler) B.; student U. Berlin, U. Göttingen, 1925-28; Dr., U. Berlin, 1929; m. Eleonore Walter, Jan. 30, 1935; children—Ilse (Mrs. Richard M. Franklin), Otto, Ingrid. Research asst., U. Frankfurt/Main, 1928-29; asst. a. lectr. U. Jena (Germany), 1930-35; prof. U. Königsberg (Germany), 1935-42, U. Strassburg, 1942-45, U. Tübingen (Germany), 1945-—. Mem. Deutsche Forschungsgemeinschaft, Wissenschaftrat., Leopoldina-Acad., Heidelberg, N.Y. acads. scis., Nat. Acad. Sci. (fgn. asso.), Japanese (hon.), Am. (corr.) bot. socs.; corr. mem. Berlin, Göttingen, Munich acads. sci. Author several books. Research, numerous publs. on bot., other biol. topics. Home: 20 Waldhäuser Str. Office: Institute for Biology, 1 Morgenstelle, Tübingen, Germany.*

BUNSEN, Robert Wilhelm, German chemist; b. Göttingen, Germany, Mar. 31, 1811; s. Christian and Aug. Friederike (Quensel) B.; Ph.D., U. Göttingen, 1830; postgrad. (state scholar) Paris, Switzerland, Salzburg, Venice, 1832;33. Privatdocent, U. Göttingen, 1834-36; tchr. trade sch., Kassel, 1836-38; prof. U. Marburg, 1838-51, U. Breslau, 1851-52, U. Heidelberg, 1852-89. Honored by Deutsche Elektrochemische Gesellschaft when it changed its name to Deutsche Bunsen Gesellschaft, 1901. Fellow Royal Soc., 1858. Corr. chemistry sect. French Acad. Scis., 1853, fgn. asso., 1882. Author: Gasometrische Methoden, 1857; Chemische Analyse durch Spektralbeobachtung, 1860; Flammenreaktionen, 1866. Developed spectrum analysis (with Kirchoff), 1859; outlined principles of chem. analysis by spectral methods; showed that under certain conditions each element emits definite and characteristic spectrum; used spectral analysis to discover the new elements, caesium, 1860, rubidium, 1861; prepared analytically many pure compounds; produced magnesium in quantity and demonstrated the brilliance of its flame when burned in air; developed methods of gas analysis, 1853, and flame tests; isolated and investigated free cacodyl, (thus furthering study of organic radicals) 1837-42; Investigated geysers of Iceland, 1846; isolated lithium in quantity, 1855; conducted (with Roscoe) extensive investigations on measurements of light intensities and developed normal light unit, 1857; experimented on variation of melting points with pressure; made accurate determination of expansion of water on solidification; designed new types of lab. equipment, including gas burner, 1855, zinc-carbon galvanic battery, filter pump, 1868, greasespot photometer, 1844, various calorimeters (all bear his name). Died Heidelberg, Germany, Aug. 16, 1899.

BUNT, Lucas Nicolas Hendrik, Dutch mathematician; b. Edam, June 10, 1905; s. Hendrik and Jannetje Elizabeth Maria (Klinkert) B.; Ph.D. in science U. Amsterdam; m. E. C. Huisman, Jan. 4, 1932; children—Lydia, Hendrik. Math. tchr., 1929-46; prof. agrégé mathematics State U. Utrecht and Groningen, from 1946, dir. science dept. Inst. Pedagogic Formation; vis. prof. U. S., 1956-57. Mem. Math. Assn. Am. Author: Bijdrage tot de theorie der convexe puntverzamelingen, 1934; Van Ahmes tot Euclides, 1954; Statistiek voor het Voorbereidend Hoger en Middelbaar Onderwijs, 1956, also reports to Commn. Instrn. Math. Home: J. P. Thijsselaam, 89, Utrecht. Office: Pedagogisch Instituut, Afd. Leraarsopleiding, Lucas Bolwerk 11, Utrecht, Netherlands.

BUNTE, Hans Hugo Christian, German chemist; b. Wunsiedel, Germany, Dec. 25, 1848; s. Carl and Fanny (Geiger) B.; studied chemistry at Polytechnikum in Stuttgart, Heidelberg, and Erlangen (all Germany); hon. dr.engring., Technische Hochschulen in Vienna, Austria, Munich and Hanover, Germany; m. Wilhelmine Stölzel, 1877; 2 sons including Karl, 3 daus. Lectr., Technische Hochschule, Munich, 1872; became gen. sec. Deutscher Verien von Gas und Wasserfachmännern (German Assn. Gas and Water Specialists) in 1884; prof. chem. tech. Technische Hochschule Karlsruhe (Germany), 1887-1919, founder, dir. Gas Inst., 1907; (with N. H. Schilling) pub. Jour. für Gasbeleuchtung, 1876-1920. Author: Fortschritt und chemische Technik, 1897. Fundamental research on effect of steam supply in generators; 1st dependable determinations of heat values; developed formula for gross efficiency of heating; built essential basis for heat econs.; recognized significance of carbon dioxide content of flue-gases; introduced sci. methods into study of gas, fuel and heating. Died Karlsruhe, Aug. 17, 1925.

BUNTING, Arthur Hugh, applied botanist; b. Sept. 7, 1917; s. S. P. and R. Bunting; M.Sc. (Rhodes scholar for Transvaal), U. Witwatersrand, Johannesburg, S. Africa, 1938; Ph.D., Oriel Coll., Oxford (Eng.) U., 1941; m. Elsie Muriel Reynard, 1941; 3 sons. Asst. chemist Rothamsted Exptl. Sta., 1941-45; mem. Human Nutrition Research unit Med. Research Council, 1945-47; chief sci. officer Overseas Food Corp., 1947-51; sr. research officer Sudan Ministry Agr., 1951-56; prof. agrl. botany Reading (Eng.) U., 1956——. Fellow Inst. Biology; mem. Assn. Applied Biologists (pres. 1963-64), Soc. Exptl. Biology, Brit. Ecol. Soc., Nutrition Soc., Inst. Biology. Co-editor: Jour. Applied Ecology. Research, publs. on tropical agr., history econ. plants, relation between environment, physiology, morphology crop plants. Home: 27 The Mount, Caversham, Reading. Office: Reading U., Reading, Eng.

BUNYATYAN, Grachiya Khachaturovich, Russian biochemist; b. 1907; grad. Med. Faculty, Yerevan U., 1930. Instr., Yerevan U., 1930-37, prof., 1937——; rector, 1942-46; head biochemistry sect. Inst. Physiology, Armenian Acad. Sci., 1944——, acad. sec. dept. biol. sci., 1947-56. Mem. Armenian Acad. Sci. (v.p. 1962——). Research and publs. on role of unsaturated phosphatides and choline in oxydizing processes, oxidation and stblzn. of vitamin C and cortical control of metabolism. Address: Yerevan University, ulitsa Abovyana 104, Yerevan, Armenia SSR, USSR.

BUNZEL, Ruth L., anthropologist; b. N.Y.C., Apr. 18, 1898; d. Jonas and Hattie (Bernheim) Bunzel; A.B., Barnard Coll., 1918; Ph.D (Social Sci. Research Council and Rockefeller Found. fellows), Columbia, 1929. Lectr., Barnard Coll., 1929-30, Columbia, 1933-40, OWI, 1942-45; asso. dir. Research in Contemporary Cultures, 1947-51; adj. prof. anthropology Columbia, N.Y.C., 1952——. Field researcher N.M., Ariz., Guatemala, Mexico, Chinese community N.Y., 1924-32, 36-37, 39, 47-51; cons. Columbia-Presbyn. Sch. Nursing, 1963-66; research asso. Inst. Intercultural Studies, 1953——. Mem. Am. Anthrop. Assn., Am. Ethnol. Soc., Soc. Applied Anthropology, A.A.A.S. Author: The Pueblo Potter, 1929; also articles, revs., monographs. Editor: (with M. Mead) Golden Age of Anthropology, 1960. Home: 62 Perry St., N.Y.C. 10014.*

BUOGO, Giulio, Italian chemist; b. Palermo, Italy, May 24, 1892; s. Onorato and Tereza (Mazzola) B.; Ph.D. in Chemistry; m. Adele Raja, Apr. 2, 1921; 1 child, Silvia (adopted). Univ. prof. Mem. Italian Soc. Pharmacology. Author: Scienza dell'alimentazione, 1942; Techologia chimica e chimica industriale, 1947; Eroi del Persiero nella Scienza e nella Technica; Alimentazione e Salute nelle Ricerche Biochimiche e nella Igiene Sociale; 11 Chimico nella Vita Moderna. Address: Lungomare Sauro 25, Bari, Italy.

BUONANNI (or BONANNI), Filippo, Italian naturalist; b. Rome, 1638. Mem. Soc. of Jesus; librarian Roman Coll. Author: Numismata pontificum romanorum, 2 vols., 1699; Ordinum religisorum catalogus, 3 vols., 1706-10; Museum Kircherianum, 1709; Trattato sopra la vernice, 1720; Gabinetto Armonica, 1716. Wrote work on shellfish (1681) in which he argued that remora could not stop ship but that a third cause such as adverse current might both stop ship and cause remora to cling; denied reproductive intercourse in shellfish as well as egg laying; supported theory of spontaneous generation; discussed why shellfish fatter at full moon. Died 1725.

BUONO, Pietro, Italian alchemist; flourished 1330; received med. edn.; physician of Ferrara, Italy. Author: The Precious Pearl, 1330. First to mention ceramical glaze (combination of calcined lead and tin).

BURALI-FORTI, Cesare, Italian mathematician; b. Arezzo, Italy, 1861; tchr. Royal Acad. Artillery and Engineering, Turin, 1903-31; assisted in preparation of Formulaire de mathématiques (Peano), 5 vols., 1895-1905. Originated paradox that well-ordered series of all ordinal numbers defines a new ordinal number which is not one of the all (led to attack on Cantor's theory of infinite number theory). Died Turin, Italy, 1931.

BURBANCK, William Dudley, Am. biologist; b. Indpls., Aug. 20, 1913; s. George G. and Flora (Kokemiller) B.; A.B., Earlham Coll., 1935; M.S., Haverford Coll., 1936; Ph.D., U. Chgo., 1941; m. Madeline Palmer, Sept. 7, 1940; children—Melinda Ann (Mrs. John M. Miller), George Palmer. Faculty, Earlham Coll., Richmond, Ind., 1936-38, Coll. City N.Y., 1941-42; faculty, chmn. dept. biology, Drury Coll., Springfield, Mo., 1942-49; instr. invertebrate zoology Marine Biol. Lab., Woods Hole, Mass., 1943-50; vis. prof. Emory U., Atlanta, 1949-50, prof. biology, 1950-—, chmn. dept., 1952-57, dir. Lullwater Field Lab., 1961-—. Bd. mgrs. Highlands Biol. Sta., 1958-61, trustee, 1961-——; mem. Marine Biol. Lab. Corp., 1943-——. NSF research grantee, 1959-——. Mem. Assn. Southeastern Biologists (Best Research Paper award 1961, exec. com. 1958-61, 63-66, pres. 1964-65), Ecol. Soc. Am., Am. Soc. Limnology and Oceanography, Soc. Protozoology, Am. Inst. Biol. Scis., Atlantic Eastern Research Soc., Societas Internationalis Limnologiae, Ga. Ornithol. Soc., Ga. Acad. Sci., Marine Assn. U.K., Sigma Xi. Author: chpt. Selected Invertebrate Types, 1950; also articles. Research on genus Cyathura, studies of estuarine species Cyathura polita including taxonomic classification, geog. distbn. along U. S. Eastern coast, morphol. variation, physiol. tolerances; ecol. protozoology especially effect of nutrition on symbiotic relationships of species of ciliates. Home: 1164 Clifton Rd. N.E., Atlanta 30307.*

BURBANK, Luther, Am. naturalist; b. Lancaster, Mass., Mar. 7, 1849; s. Samuel Walton and Olive (Ross) B.; ed. Lancaster Acad.; Sc.D., Tufts, 1905; m. Elizabeth J. Waters, Dec. 21, 1916. Always devoted to study of nature, especially plant life; moved to Santa Rosa, Cal., 1875; conducted Burbank's Expt. Farms. Originator Burbank potato, rapid-growing edible thornless opuntias (cactus), Gold, Wickson, Apple, October, Chalco, America, Santa Rosa, Formosa, Beauty, Eldorado, and Climax plums; Giant Splendor, Sugar, Standard and Stoneless prunes; a new fruit, the Plumcot; Burbank and Abundance cherries; Peachblow, Burbank and Santa Rosa roses, gigantic forms of amaryllis, tigridias, the Shasta Daisy, Giant and Fragrance callas; and various new apples, peaches, nuts, berries and other valuable trees, fruits, flowers, grasses, grains and vegetables. Spl. lectr. on evolution, Leland Stanford Jr. U. Author: Training of the Human Plant; Methods and Discoveries, 12 vols.; How Plants Are Trained to Work for Man, 8 vols., 1921; also numerous mag. articles. Died Santo Rosa, Cal., Apr. 11, 1926.

BURBANK, Wilbur Swett, geologist; b. Amesbury, Mass., Mar. 30, 1898; s. Wilbur Augustus and Emma Elizabeth (Swett) B.; S.B., Mass. Inst. Tech., 1919, S.M., 1920, post grad., 1924-25; m. Beryl Frances Loughlin, Apr. 1, 1933; children—John Francis, Phillip Augustus. With U. S. Geol. Survey, 1920-——, successively mineral resource investigations, gen. geologic mapping. Republic of Haiti, geologic mapping, Mont., studies copper desposits of Mich., fed. and state coop. study geology and mineral resources of Colo., 1926-39, adminstrv. com. work geologic programs, 1920-58. Soc. Econ. Geologists rep. earth scis. div. NRC, 1934-36, exec. com. div. earth scis., 1955-58. Recipient distinguished service award Dept. Interior, 1958. Mem. Am. Geophys. Union (v.p. volcanology 1935-38, pres. 1947-50, sec. tectonophysics sect 1947-53), Soc. Econ. Geologists, Geol. Soc. Am., Mineral. Soc. Am., Geochem. Soc., No. New Eng. Acad.

Scis, Colo. Sci. Soc., Geol. Soc. Washington. Contbr. numerous tech. articles profl. publs. Research on geology of West Indes, Haiti; discovered volcanic calderas of San Juan mountains; investigated factors controlling formation of mineral deposits in shallow volcanic environments, as related to chemical and thermodynamic principles. Home: 9 Bayberry Lane, Exeter, N.H. 03833. Office: U. S. Geological Survey, Washington 20242.

BURBRIDGE, Thomas Nathaniel, Am. pharmacologist, educator; b. New Orleans, July 12, 1921; s. Leonidas Tullius and Marie E. (Harrison) B.; A.B., Talladega Coll., 1941; M.D., U. Cal. at San Francisco, 1948, Ph.D., 1956; m. Yvonne Mason, Oct. 22, 1945; children—Thomas Nathaniel, Lynne, Terry-Anne, Nancy Marie, Leigh Mason. Vis. prof. U. Indonesia Med. Sch., 1952-55; asso. prof. pharmacology U. Cal. Med. Center at San Francisco, 1956-——. Cons. Cal. state hosps. for mentally ill, 1957-——. Recipient Lederle Med. Faculty award, 1959. Mem. Am. Soc. Pharmacology and Exptl. Therapeutics, Soc. Clin. Pharmacology and Therapeutics, A.A.A.S. Research, publs. on drugs affecting central nervous system, tranquilizers, psychic energizers, alcohol, narcotics. Home: 1676 9th Av., San Francisco 94122.*

BURBURY, Samuel Hawksley, English mathematician; b. Kenilworth, Eng., May 18, 1831; s. Samuel and Helen B.; B.A., 1854, M.A., 1857 St. John's Coll., Cambridge; m. Alice Ann Taylor, Apr. 12, 1860; 4 sons, 2 daus. Became lawyer Lincoln's Inn, 1858; practiced at parliamentary bar from 1860, then entered chamber practice, ret., 1908. Fellow Royal Soc., 1890. Author: (with Henry W. Watson) The Application of Generalised Co-ordinates to the Kinetics of a Material System, 1879; (with Watson) The Mathematical Theory of Electricity and Magnetism, 1885-89; various papers to Philosophical Magazine. Carried on work of Clerk Maxwell; tried to place electrostatics and electromagnetism on more formal math. basis. Died Aug. 18, 1911.

BURCH, Cecil Reginald, Brit. physicist; b. Oxford, Eng., May 12, 1901; B.A., Ph.D., Cambridge (Eng.) U. Research Met. Vickers Co., Manchester, Eng., 1923-33; research asso. H. H. Wills Physics Lab., Bristol (Eng.) U., 1936-44, research fellow, 1944-48; Warren research fellow, after 1948. Leverhulme fellow in optics Imperial Coll. Sci. and Tech., 1933-35. Fellow Royal Soc. (Rumford medal 1954), 1944. Working with vacuum distillation, developed oil condensation pump which made possible distillation of such oils as olive oil, arachis oil, beeswax, others; prin. of oil condensation pump later used to produce high energy particles; built 1st Schwartzschild-aplanaticaspheric reflecting microscope.

BURCH, George James, Brit. physicist; b. 1852; s. George and Jane (Hicks) B.; M.A., D.Sc. (Oxon), Cheshunt Coll., Oxford (Eng.) U.; m. Constance Emily Jeffries, 1884; 3 sons, 2 daus. Prof. physics U. Coll., Reading, Eng.; univ. extension staff lectr. Oxford. U. Fellow Royal Soc. Author: The Capillary Electrometer; Electrical Science; The Pronunciation of English by Foreigners; Practical Physiological Optics. Contbr. numerous papers to sci. publs. Research electricity, light; inventor improved methods for use capillary electrometer for recording, measuring rapid changes of electromotive force (photographed currents produced by speaking into telephone); created temporary color-blindness by exposing eye to sunlight in focus of burning glass using glasses of various colors. Died Mar. 17, 1914.

BURCH, Guy Irving, Am. population analyst; b. Clayton, N.M., May 24, 1899; s. Fred Irving and Minnie Boone (Boggs) B.; ed. Culver Military Acad., 1914-16, Pawling Sch., 1917-18, Cavalry Officers Tng. Sch., Leon Springs, Tex., 1918, Columbia U., 1919-23, 25; m. A. Wilhelmine Taylor, June 24, 1920; children—Sally Ann, Caroline Sue. Founder, dir. Population Reference Bur., since 1929; mem. Council on Population Policy, 1935-36; supplied Joint Army and Navy Selective Service Com. with age group data for original draft bill, 1940; chmn. Population-Resources Roundtable, 1947. Fellow Population Assn. Am., (charter and organizing mem.); mem. Am. Eugenics Soc. (dir. 1932-47, sec. 1933-36). Author: Population Roads to Peace or War (with E. Pendell), 1945, rev. edit. Human Breeding and Survival, 1947. Contbg. editor Eugenics, 1931. Editor: Human Facts in Science, 1935-36; Population Bulletin, since 1940. Contbr. articles to mags. Died Jan. 13, 1951.

BURCH, Philip Robert, English physicist, biologist; b. Huntingdon, Eng., Apr. 15, 1920; s. Robert and Gertrude (Barnard) B.; B.A., Cambridge (Eng.) U., 1948, M.A., 1952; Ph.D., U. Leeds (Eng.), 1952; m. Jane Elizabeth Ramsey, July 21, 1949; children—Stephen Frank, Belinda Jane, Matthew Robert. Research asst. dept. med. physics U. Leeds 1949-54, external sci. staff Med. Research Council, med. Research Council fellow, 1954-59, dep. dir. Med. Research Council environmental radiation research unit, 1959-——; vis. scientist health physics div. Oak Ridge Nat. Lab., 1961-62; hon. reader dept. med. physics, U. Leeds, 1964-——. Recipient Roentgen award Brit. Inst. Radiology, 1958. Research, numerous publs. on measurements of natural radioactivity in man, measurement cosmic radiation, theory of cavity ionization chambers, embryogenesis, cellular differentiation, mitotic- and growth-control, classical immune response, carcinogenesis, autoimmunity and aging. Home: 48 Henconner Lane, Leeds 7, Yorkshire, Eng.*

BURCHAM, Paul B(aker), Am. mathematician, educator; b. Fayette, Mo., Feb. 22, 1916; s. Frank Eli and Bula (Richardson) B.; B.A., Central Coll., 1935; M.A., Northwestern U., 1938, Ph.D., 1941; m. Helen Spencer, Ann Leonard. Instr. math. Central Coll., Fayette, 1935-36; staff Southwestern Bell Telephone Co., Webster Groves, Mo., 1936-37; faculty U. Mo. Inst. in Math., Columbia, Mo., 1946—, chmn. dept. math., 1948-66, prof., 1954—. Recipient Distinguished Alumni award Central Meth. Coll., 1963. Rotarian. Author: (with Betz and Ewing) Differential Equations with Applications. Study of summability of series; several inclusion relations in domain of Hausdorff matrices. Home: 401 Westmount St., Columbia, Mo.

BURCHAM, William Ernest, English physicist; b. Norfolk, Eng., Aug. 1, 1913; s. Ernest Barnard and Edith (Pitcher) B.; B.A., Cambridge (Eng.) U., 1934, Ph.D., 1937; m. Isabella Mary Todd, Jan. 3, 1942; children—Sheila Mary (Mrs. George Barker), Margaret Ann. Stokes student Pembroke Coll., Cambridge, 1937-39; sci. officer Ministry of Supply, 1939-44; fellow, lectr., dean Selwyn Coll., Cambridge, 1944-51; univ. lectr. in physics Cambridge U., 1946-51; Oliver Lodge prof. physics U. Birmingham (Eng.), 1951—. Cons. U.K. Atomic Energy Authority, 1963—. Fellow Royal Soc., 1957; mem. Inst. Physics and Phys. Soc., Am. Phys. Soc. Author: Nuclear Physics, An Introduction, 1963; also articles. Research on nuclear reactions using accelerated particles. Home: 95 Witherford Way, Birmingham 29, Eng.*

BURCHARD OF MT. SION (or Burchardus de Monte Sion), geographer; b. Barby, nr. Magdeburg, Germany; mem. Dominican order; traveled through Holy Land, also Syria, Armenia, Egypt, Cyprus, possibly as early as 1272, until 1282; reached Cairo, Egypt, 1285, probably as leader Rudolf of Hapsburg's delegation to sultan of Egypt. Author: Descriptio terrae sanctae (geog. manual on Near East based on empirical observations and organized into 13 chpts., for centuries used as model for travel writers and cartographers), written 1271-91. Died after 1285.

BURCHELL, William John, English naturalist; b. Fulham, Eng., circa 1782; s. Matthew Burchell; hon. D.C.L., Oxford (Eng.) U., 1834. Botanist at St. Helena, 1805-10; studied Cape Dutch at Cape Town, S. Africa, 1810; travelled through S. Africa collecting natural history specimens, 1811-15; went to Lisbon, Portugal, 1825; in Rio de Janeiro, Brazil, 1825-26; travelled through Brazilian forest collecting over 15,000 natural species of plants and insects, 1826-29. Fellow Linnean Soc. London. Author of two vols. on African travels, 1822; also zool. articles. Discovered many new animal and plant species some of which were named in his honor; astron. and meteorol. observations; extensive drawings and collections of plants. Died Mar. 23, 1863.

BURCKHARDT, Carl Emanuel, geologist, paleontologist; b. Basel, Switzerland, Mar. 26, 1869; s. Wilhelm and Maria Caroline (Sarasin) B.; studied botany and geology in Geneva, Basel, Zurich (all Switzerland), 1888-93; doctorate 1893. Worked with Swiss geol. commn.; with E. Suess and W. Waagen in Vienna, K.A. v. Zittel in Munich, Germany, 1894-95; began at the state geol. inst. in La Plata, Argentina, 1896; studied the Cordilleras in Argentina and Chile, 1897-98; went to Munich in 1901; began as asst. at Bayerische Obergamt in Germany, 1903; head geologist Instituto geologico in Mexico, 1904-15; afterward did pvt. work. Studied distbn., stratigraphy and paleontology of Triassic, Jurassic and chalk formations in Central Am.; fundamental work on paleogeography, climatic conditions, ammonite fauna and paths of Mexican mesozoicum. Died Mexico, Aug. 26, 1935.

BURCKHARDT, Johann Heinrich, German botanist; b. Sulzbach, Germany, Aug. 5, 1676; s. Georg Christoph and Innocentia Rosina (Schaetz) B.; M.D., 1700; m. Anna Sophia Overlach, 1711; city physician in Wolfenbüttel, Germany. Author: Epistola ad Godfredum Leibnitzium, qua characterem plantarum naturalem, nec a radicibus, nec ab aliis partibus plantarum minus essentiabibus peti posse ostendit (letter to Leibniz which suggests classifying plants according to their sexual organs); considered a forerunner of Linnaeus. Died Wolfenbüttel, May 3, 1738.

BURCKHARDT, Johan Jakob, Swiss mathematician; b. Basel, Switzerland, July 13, 1903; s. Wilhelm and Eleonor (Vischer) B.; Dr.Phil., U. Zürich (Switzerland), 1928; m. Helen Alice Grossmann, July 24, 1934; children—Christoph, Brigitte (Mrs. Utelli), Heinrich, Elisabeth (Mrs. Ernst), Johann-Jakob. Faculty, U. Zürich, 1933—, prof., 1942—. Mem. Swiss Math. Soc. (pres.). Author: Die Bewegungsgruppen der Kristallographie, 1966; also hist. and biographical articles on crystallography, astronomy, math. Editor jour. Commentarii Mathematici Helvetici. New formulation of math. crystallography. Home: 4 Bergheimstrasse, Zürich CH-8032, Switzerland.*

BURCKHARDT, Johann Karl, mathematician, astronomer; b. Leipzig, Germany, Apr. 30, 1773; studied practical astronomy with Baron Franz Xaver von Zach at the Baron's observatory on the Seeberg, nr. Gotha; assisted von Zach in observing right ascension of stars, 1795-97; von Zach recommended him to Lalande at Paris; went to Paris, 1797, adjunct astronomer, bd. of longitude; naturalized, 1799; replaced Lalande at obs. of École militaire. Mem. Bur. Longitudes, French Acad. Scis., 1804. Author: Tables de la lune, 1812; Tables axillaires, 1814; Traité sur la comete de 1770, 1806. Translated into German 1st 2 vols. of Mécanique céleste (Laplace). His astron. tables were the basic works of his time; calculated orbits of comets; assisted Lalande. Died Paris, June 22, 1825.

BURDACH, Karl Friedrich, German anatomist; physiologist; b. Leipzig, Germany, June 12, 1776; prof. anatomy and physiology at Dorpat, Estonia, Königsberg (now Kaliningrad, USSR). Author: Physiologie als Erfahrungswissenschaft, 1826; Vom Baue und Leben des Gehirns und Rückenmarks, 3 vols., 1819-25. Gave 1st complete description of cuneate fasciculus of spina cord (fasciculus cunneatus or Burdach's column), 1806; research on anatomy of nervous system. Died Königsberg, July 16, 1847.

BURDEN, Henry, ironmaster, inventor; b. Dunblane, Scotland, Apr. 20, 1791; s. Peter and Elizabeth (Abercrombie) B.; m. Helen McQuat, Jan. 17, 1821. Came to Am., 1819; made variety of labor-saving machines, including a threshing machine. Patented machine for making wrought iron spikes, 1825; invented improved plow, 1st cultivator made in U. S., self-acting machine for rolling iron into bars; 1st patented horseshoe machine (his most widely known invention), 1835. Died Troy, N.Y., Jan. 19, 1871; buried Albany, N.Y.

BURDEN, Robert Prentice, Am. civil engr.; b. Somerville, Mass., Jan. 26, 1918; s. Harry P. and Lunetta (McPhetres) B.; B.S. in Civil Engring., Tufts U., 1939; M.S., Harvard, 1940, Ph.D., 1948; m. Rhoda Lee Davis, June 16, 1941; children—Christopher Lee, Leslie Prentice, Robert Thaddeus. With Rockefeller Found., N.Y.C., 1948-60; asst. to dean (Grad. Sch. Pub. Health, co-dir. water resources program div. engring. and applied sci. Harvard, Cambridge, Mass., 1960—. Cons. to White House, numerous fgn. govts., cos., WHO, Nat. Acad. Scis., 1960—. Mem. Am. Soc. C.E., Am. Pub. Health Assn., Instn. Civil Engrs. (London), Sigma Psi, Delta Omega. Author: (with Leslie Banks) Technological Development of Africa, 1954. Research and numerous publs. on environmental systems, design, innovation in health and devel. practices in emergent countries, adaptation of computer studies for water resource devel. Home: 5 Scott St., Cambridge, Mass. 02138.*

BURDETTE, Walter James, Am. surgeon, geneticist; b. Hillsboro, Tex., Feb. 5, 1915; s. James S. and Ovazene (Weatherred) B.; A.B., Baylor U., 1935; M.A., U. Tex., 1936, Ph.D., 1938; M.D., Yale, 1942; m. Kathryn Lynch, Apr. 9, 1947; children—Susan, William J. Faculty, La. State U., 1946-55, U. Mo., 1955-56, St. Louis U. Sch. Medicine, 1956-57, U. Utah, 1957-65; prof. surgery, asso. dir., chief thoracic surgery U. Tex. M.D. Anderson Hosp. and Tumor Inst., Houston, 1966—. Cons. VA hosps., NIH, Nat. Cancer Inst., Am. Cancer Soc., Nat. Acad. Sci., Oak Ridge Inst. Nuclear Studies Hosp., Nat. Heart Inst.; Gibson lectr. Oxford U. (Eng.), 1966; vis. investigator Max Planck Inst. für Biochimie, Tübingen, Germany, 1966; Mem. Nat. Adv. Cancer, Heart councils, Nat. Adv. Commn. on Smoking and Health. Rockefeller travel fellow, 1958. Diplomate Am. Bd. Surgery, Am. Bd. Thoracic Surgery. Fellow A.C.S.; mem. Soc. Surgery of Alimentary Tract, Am. Assn. Cancer Research (dir.), Am. Cancer Soc., Am., Western, So. surg. assns., Soc. Clin. Surgery, A.M.A., Genetics Soc. Am., A.A.A.S., Soc. Exptl. Biology and Medicine, Societe Internationale de Chirurgie, others. Editor: Etiology and Treatment of Leukemia, 1957; Methodology in Human Genetics, 1962; Methodology in Mammalian Genetics, 1963; Methodology in Basic Genetics, 1963; Primary Hepatoma, 1965; Carcinoma of the Alimentary Tract, 1965; Viruses Inducing Cancer, 1966. Research, publs. on gen. and thoracic surgery, genetics of cancer; invertebrate hormones, fine structure and metabolism of human cardiac muscle in vitro, shock, human and insect chromosomes, oncogenic viruses. Home: 239 Chimney Rock Rd., Houston 77024. Office: U. Tex. M. D. Anderson Hosp., Houston 77025.*

BURDICK, C. Lalor, Am. chemist; b. Denver, Apr. 14, 1892; B.S., Drake U., 1911; M.S., Mass. Inst. Tech., 1913; Ph.D., U. Basel (Switzerland), 1915; m. Allison Ward, 1938; children—Lalor, Cynthia (Mrs. John Brill). Research asso. Cal. Inst. Tech., 1917; metall. engr. Guggenheim Bros., N.Y. and Chile, 1918-24; v.p., cons. engr. Anglo Chilean Consol. Nitrate Corp., 1925-28; with E. I. du Pont De Nemours & Co., Wilmington, Del., 1929—. Trustee emeritus Marine Biol. Lab., Woods Hole, Mass.; cons. Longwood Found. Mem. N.Y. Acad. Sci., A.A.A.S., Am. Chem. Soc., Inst. Am. Chem. Engrs., Am. Soc. Study of Sterility, Soc. Study Fertility, Internat. Planned Parenthood Fedn., Planned Parenthood Fedn. Am. (dir., Mem. exec. com.), Phi Beta Kappa. Home: 4400 Lancaster Pike. Office: 9137 DuPont Bldg., Wilmington, Del. 19805.*

BURDIN, Claude, mining engr.; b. Lépin, Savoy, May 18, 1778; ed. École Polytechnique, 1807; 1 adopted son, Achille Frandin Burdin. Prof. Sch. Mines of St. Etienne, France; head of mines for Cantala, from 1822; engr., dir. French mines, from 1847. Mem. French Acad., Scis., 1842. Author: Mémoires sur les roues hydrauliques où l'eau agit par sa réaction contre des palettes ou des canaux mobiles, 1824. Built a reaction turbine which was set up in Pontgibaud; devised a locomotive which carried its own movable rails, 1830; studied submarine navigation. Died Nov. 12, 1873.

BURDON, David Joseph, geologist; b. Buttevant, Ireland, Feb. 27, 1914; s. William and Helen (O'Connell) B.; B.E., B.Sc., Nat. U. Ireland, 1935, M.Sc., 1949, D.Sc., 1963; Ph.D., London U., 1949; m. Kathleen O'Reilly, June 12, 1949; children—Barbara Anne, David John, Gerald Desmond, Andrew Nicholas, Paul Christopher. Geologist, surveyor Kolar Gold Fields, Mysore State, South India, 1938-47; asst. water engr. Colonial Service, Cyprus, 1949-52; hydrogeologist FAO, Syria, 1952-60, Greece, 1960-63, Egypt, 1963-64, Rome hdqrs., 1964—. Fellow Geol. Soc. London, Royal Geol. Soc. Cornwall; mem. Soc. Econ. Geology, Internat. Assn. Hydrogeologists, Société Geologie de France, Geologist Assn.; asso. mem. Instn. Mining and Metallurgy; profl. asso. Inst. Water Engrs. Author: Handbook of the Geology of Jordan, 1959; (with N. Papakis) Handbook of Karst Hydrogeology, 1963; also articles, profl. reports. Research in application of geology to mining, civil engring., water devel., specializing in groundwater. Home: Rathclare House, Buttevant, Ireland. Office: FAO, UN, Viale delle Terme di Caracalla, Rome, Italy.*

BURDON, James, Brit. chemist; b. Stoke-on-Trent, Eng., Sept. 6, 1932; s. Edward and Minnie (Bailey) B.; B.Sc. with first class honors, U. Birmingham, 1953, Ph.D., 1956, D.Sc., 1966; m. Joan Christine Bishop, July 14, 1956; children—Kevin James, Elaine Christine, Susan Elisabeth. Research fellow U. Birmingham, Eng., 1956-58, lectr. in chemistry, 1958-67, sr. lectr., 1967—. Rockefeller Found. fellow, vis. asso., Cal. Inst. Tech., 1961-62. Mem. Chem. Soc. London (Eng.), Am. Chem. Soc. Research, publs. on organic chemistry of fluorine containing compounds. Home: 66 Chesterwood Rd., Kings Heath, Birmingham 13. Office: Chemistry Dept., U. of Birmingham, Edgbaston, Birmingham, 15, Eng.*

BURDON-SANDERSON, Sir John Scott, English physiologist; b. Jesmond, Eng., 1828; s. Richard and Elizabeth (Sanderson) B.; studied at U. Edinburgh (Scotland), 1847-51, received M.D.; M.A., Oxford (Eng.) U., 1883, D.M., 1895; m. Ghetal Herschell, Aug. 9, 1853. Apptd. med. registrar St. Mary's Hosp., London, Eng., 1853, became lectr., 1854; named med. officer health for Paddington, 1857; apptd. insp. of privy council, 1860; named physician Brompton Hosp. for Consumption, also Middlesex Hosp., 1860; ret., 1870; prof. medicine Oxford; became Jodrell prof. physiology U. Coll., 1874. Mem. 3 royal commns.; Harveian orator Coll. Physicians, 1878. Fellow Royal Soc., Royal Coll. Physicians. Recipient Baley medal, 1880. Research and publs. on origin of infectious diseases; suspected relationship of micro-organisms to disease; investigated (with F. J. M. Page) elec. phenomena in excitatory process of frog and tortoise hearts, 1883; measured speed of nerve impulse with photography, 1890. Died Oxford, Nov. 23, 1905.

BUREAU, Florent J., Belgium mathematician; b. Jemeppe sur Sambre, Belgium, Dec. 17, 1906; student U. Liège (Belgium), 1925-29, U. Paris, 1929-30, U. Berlin, 1930, U. Copenhagen, 1931, U. Rome, 1931-36; m. Odette Frenay, 1942; children—Nicole, Anne-Marie. Now prof. U. Liège; prof. U. Chgo., 1953-54, Duke, Durham, N.C., 1955-56, U. Del., Newark, 1968. Author: Calcul Vectoriel et Calcul Tensoriel; also numerous articles. Research on theory of functions, ordinary differential equations, partial differential equations. Home: 5 Place d'Italie, Liège, Belgium.*

BURES, Jan, Czechoslovakian neurophysiologist; b. Ctyri Dvory, Czechoslovakia, June 13, 1926; s. Rudolf and Marie (Pislová) B.; M.D., Charles U., Prague, Czechoslovakia, 1950; candidate scis. Czechoslovak Acad. Scis., 1955, D.Sc., 1963; m. Olga Komorádová, June 11, 1949; 1 dau., Olga. Research asst. Inst. Physiology, Charles U. Sch. Medicine, Prague, 1949-50; research asst. Inst. Physiology, Czechoslovak Acad. Scis., Prague, 1950-58, head lab. physiology, 1958—. Mem. Czechoslovak Physiol. Soc., Soc. Higher Nervous Activity, Internat. Brain Research Orgn. (charter, mem. central council 1964—). Author: (with M. Petrán, J. Zachar) Electrophysiological Methods in Biological Research, 1960; also numerous articles. Research on pathophysiology of epilepsy, mechanism of spreading EEG depression, application of spreading depression as functional ablation procedure in behavorial studies (with O. Buresová); studies on functional orgn. of brain, neural plasticity at single neuron level. Home: 18 Lipová, Prague 2. Office: 1083 Budejovická, Prague 4-Krc, Czechoslovakia.*

BURG, Anton Behme, Am. educator; b. Dallas City, Ill., Oct. 18, 1904; s. Frank W. and Sadie (Quinton) B.; B.Sc. U. Chgo., 1927, M.Sc., 1928, Ph.D., 1931. Research, Kimberley-Clark Corp., Neenah, Wis., 1928-29; research asst., instr. chemistry U. Chgo., 1929-39; faculty U. So. Cal., Los Angeles, 1939—, prof., 1943—, dept. head, 1940-50. Recipient Tolman medal, 1961; University Assos. award for Creative Scholarship, 1963-64. Mem. Am. Chem. Soc., A.A.A.S.,

Am. Assn. U. Profs., Phi Beta Kappa, Sigma Xi. Pioneer work boron hydrides useful for rocket fuels; research, numerous publs. leading to new inorganic plastics, fluorine chemistry. Office: Univ. of Southern Cal., Los Angeles 90007.

BÜRG, Johann Tobias, astronomer; b. Vienna, Austria, Dec. 24, 1766; prof. astronomy, Klagenfurt, Austria; later became astronomer, obs., Vienna. Recipient prize for solution of a problem of moon's motions from French Inst.; Mem. French Acad. Scis., 1812. Produced lunar tables. Died Wiesenau, Austria, Nov. 25, 1834.

BÜRGEL, Bruno Hans, German astron. writer; b. Berlin, Germany, Nov. 14, 1875; s. Adolf and Emilie (Sommer) Trendelenburg; m. Franziska Sophie Sobek, 1901; 1 son. With Urania Obs., Berlin, from 1895; Author: Aus fernen Welten, 1910; Vom Arbeiter zum Astronomen, 1919; Du und das Weltall, Die seltsamen Geschichten des Dr. Ulebuhle, 1920; Weltall und Naturgefuhl, 1926; Das Weltbild des modernen Menschen, 1932; Saat und Evnte, 1942; others. Contbr. articles to publs. Popularized sci., astronomy for layman; advocated adult edn. Died Potsdam, Germany, July 8, 1948.

BURGER, Alwyn Johannes, S. African geochronologist; b. S. Africa, Dec. 19, 1926; s. Paulus Andreas and Anna (Burger) B.; M.Sc. in Physics, U. Stellenbosch, 1950; Ph.D., in Geochronology, U. Cape Town, 1959; m. Johanna Etresia Pretorius, Apr. 17, 1954; children—Herman, Marieta, Chrisna. Staff, Nat. Phys. Research Lab., Council for Sci. and Indsl. Research, Pretoria, S. Africa, 1951—, head geochronology div., 1958-66, chief research officer, 1966—. Mem. S. African Inst. Physics (found.). Research, publs. on radiometric dating of S. African granite and acid lavas using uranium-lead and thorium-lead decay schemes; modification of mineral separating equipment for minute quantities of radioactive minerals, application of solid source mass spectrometry to isotope geology. Home: 202 Carinus. Office: P.O. Box 395, Scientiae, Pretoria, Transvaal, S. Africa.*

BURGERS, Anton Cornelis Jacobus, Dutch endocrinologist; b. Arnhem, Netherlands, Aug. 21, 1925; s. Anton C. J. and Cecile (Hendriks) B.; B.Sc., Rijks U. Utrecht, Netherlands, 1948, M.Sc., 1952, Ph.D., 1956. Staff Rijks U. Utrecht, 1947-62, research asso., 1962—; Dutch Orgn. Pure Research research fellow Med. Sch., Capetown, S. Africa, 1953-55; WOSUNA fellow Caribbean Marine Biol. Inst., Carmabi, Curacao, Antilles, 1956-57; Rockefeller fellow Hormone Research Lab., U. Cal. at Berkeley, 1959-60; program specialist UNESCO dept. advancement scis., Paris, France, 1965—. Lectr., Far E., India, Thailand, Taiwan, Japan. Marine Biol. Research Lab. grantee, Naples, Italy, 1958-62, 63-65. Mem. Japanese Zool. Soc., Dutch Pharmacological and Physiol. Soc., Anat. Soc. Author: Investigations into the Action of Certain Hormones and Other Substances on the Melanophores of the South African Clawed Toad, 1956; also numerous articles. Research on frog tests for pregnancy, structure activity relationships of amines and polypeptides; discovery several MSH types in a single pituitary gland. Home: 103 Vollenhovenlaan, Utrecht, Netherlands. Office: UNESCO, Pl. de Fontenoy, Paris, France.*

BURGERS, Johannes Martinus, physicist, educator; b. Arnhem, Netherlands, Jan. 13, 1895; s. Johannes Martinus and Johanna Hendrika (Romijn) B.; Dr. Phys. and Math. Scis., U. Leiden (Netherlands), 1918; honorary degrees Free U. Brussels (Belgium), 1948, U. Poitiers (France), 1950; m. Jeannette D. Roosenschoon, July 30, 1919 (dec. Aug. 1939); children—Anna Charlotte, Jan Herman, Jeannette Marion Veronica; m. 2d, Anna Margretha Verhoeven, Aug. 20, 1941. Prof., Tech. U., Delft, Netherlands, 1918-55; research prof. Inst. Fluid Dynamics and Applied Math., U. Md., College Park, 1955—. Decorated knight Order Netherlands Lion. Mem. Am. Inst. Physics, Am. Geophys. Union, Internat. Union Theoretical and Applied Mechanics (gen. sec. 1946-52), Am. Acad. Arts and Scis., Philos. Soc. Washington, N.Y., Washington acads. scis., Royal Netherlands Acad. Scis., Acad. Scis. Torino. Author: Experience and Conceptual Activity, 1965. Research on fluid dynamics, rheology, plasma physics, theory of turbulence. Home: 4622 Knox Rd., College Park, Md. 20740.*

BURGERS, Willy Gerard, Dutch chemist; Arnhem, Netherlands, Aug. 16, 1897; s. Johannus and Johanna H. (Romijn) B.; ed. Groningen, Leyde univs.; Ph.D. in chemistry, 1928; m. Mathilde Kraus, Jan. 7, 1926. Collaborator (under dir. Sir William Bragg; Davy Faraday Research Lab., London, 1924-27; with physics lab. Philips Soc., Eindhoven, 1927-40; prof. phys. chemistry Tech. U. Delft, since 1940; prof. Purdue U., 1949-50. Recipient Luigi Losana Gold medal Italian Assn. Metallurgy, 1959; Heyn medal German Metallurgy Soc., Gustav Trasenster medal, Liege, 1961. Mem. Royal Dutch Acad. Amsterdam, French Soc. Metallurgy (hon.). Author: Handbuch der Metallphysik; Recrisallisation des métaux. Home: Huis te Hoornkade 22, Ryswijk/Z.H. Office: Lab. voor Metallkunde, Rotterdamseweg 137, Delft, Netherlands.

BURGESS, Alan, English physicist; b. Oldham, Eng., Nov. 9, 1933; s. James and Annie (Walker) B.; B.Sc., U. Coll., U. London (Eng.), 1955, Ph.D., 1959; M.A., U. Cambridge (Eng.), 1966; m. Lore Fish, Apr. 4, 1961; children—Gina, Neil. Research fellow physics dept. U. Coll., U. London, 1958-62, lectr., 1964-66; sr. research scientist N.Y. U., 1962-63; mem. Inst. for Advanced Study, Princeton, N.J., 1963-64; univ. lectr. dept. applied math. and theoretical physics, fellow U. Coll., Cambridge U., 1966—. Cons. Atomic Energy Research Establishment, Harwell, Eng., intermittently 1960—, Culham Lab., 1964—. Fellow Roya. Astron. Soc. Research, publs. on basic atomic collision and radiation processes with applications to lab. and astrophys. plasmas especially solar corona; discovered importance of dielectronic recombination in high temperature plasmas. Home: 5 Lowbury Crescent, Oakington, Cambridge, Eng.*

BURGESS, Albert Franklin, Am. entomologist; b. Rockland, Mass., Oct. 2, 1873; s. Emory and Mary Ann (Lewis) B.; B.S., Mass. Agrl. Coll., 1895, M.S., 1897; m. Mary Emily Dwight, June 20, 1904; children—Emory Dwight, Albert; m. 2d, Josie Fay Goodell, Jan. 1, 1945. Asst. entomologist Mass. State Bd. Agr., 1895-99; asst. in entomology U., Ill., 1899-1900; asst. insp. nurseries and orchards, Ohio, 1900-02, chief insp., 1902-07; expert in charge of breeding expts., U. S. Dept. Agr., 1907-43; worked with control of San Jose scale insect, gypsy moth problem, circa 1907; head Bur. Entomology, 1913-28; prin. entomologist U. S. Plant Quarantine and Control Administration, 1928-34, U. S. Bur. of Entomology and Plant Quarantine, 1933-43. research U. Mass., 1934-36. Fellow A.A.A.S., Entomol. Soc. Am.; mem. Am. Assn. Econ. Entomologists, Sigma Xi. Author: History of the Gypsy Moth, 1953. Asso. editor Jour. Econ. Entomology. Combatted gypsy moth and San Jose scale insect by importing natural enemies and by using lime sulphur. Died Feb. 23, 1953.

BURGESS, Alexander Manlius, Am. physician; b. St. Albans, Vt., May 4, 1885; s. Thomas and Mary T. (Sargent) B.; A.B., Brown U., 1906, Sc.D., 1954; M.D., Harvard, 1910; Ed.D., R.I. Coll., 1959; m. Abby Bullock, June 8, 1910; children—Alexander Manlius, Samuel B., Robert S., Abby (Mrs. John A. Rockett). Pathologist, asst. prof. pathology McGill U. Med. Sch., Montreal, Que., Can., 1912-13; practice medicine, Providence, 1914-49; faculty Brown U., 1922-50, prof. health and hygiene, 1944-50, chmn. div. univ. health, 1924-44; physician-in-chief Charles V. Chapin Hosp., Providence, 1926-40, cons. physician, 1940—; physician-in-chief R.I. Hosp., Providence, 1945-48; dir. med. edn. Miriam Hosp., Providence, 1955—, Newport Hosp. (R.I.), 1955-58, Meml. Hosp., Pawtucket, R.I., 1957-63. Nat. chmn. Nat. Com. for Resettlement Fgn. Physicians, 1949-58, vice chmn., 1958-64; area sect. chief medicine VA., 1949-55, area cons., 1955; Chapin orator R.I. Med. Soc., 1961. Fellow A.C.P. (Alfred Stengel award 1958, past gov. R.I., past mem. bd. regents, past 2d v.p.), Providence Med. Assn. (past pres.), New Eng. Diabetes Assn. (past pres.). Research, numerous publs. on diabetes and hypertension, oxygen therapy with description of new method of application; descriptions of various malignant and hematological conditions. Home: 107 Bowen St., Providence 02906. Office: Miriam Hosp., 164 Summit Av., Providence 02906.*

BURGESS, C(ecil) Edmund, Am. mathematician, educator; b. Happy, Tex., Jan. 21, 1920; s. John Wesley and Sallie (Crawford) B.; B.S., W. Tex. State Univ., 1941; Ph.D., U. Tex., 1951; m. Charlotte June Stevenson, Feb. 20, 1948; children—Grant Lewis, Carol Jean. Instr. math. U. Tex., 1941-42, 46-51; faculty math. U. Utah, Salt Lake City, 1951—, prof., 1961—, also chairman department of mathematics. Visiting lecturer U. Wis., Madison, 1956-57; vis. mem. Inst. for Advanced Study, Princeton, N.J., 1962-63. Mem. Am. Math. Soc., Math. Assn. Am. Research on structure of continua, embeddings of surfaces in Euclidean three-space. Home: 2236 Logan Av., Salt Lake City 84108.*

BURGESS, Charles Frederick, Am. chem. engr.; b. Oshkosh, Wis., Jan. 5, 1873; s. Frederick and Anna A. (Heckman) B.; B.S., U. Wis., 1895, E.E., 1897, Ph.D., 1926; D.Engring., Ill. Inst. Tech., 1944; m. Ida M. Jackson, June 25, 1903; children—Betty, Jackson. Instr. and asst. prof. elec. engring. U. Wis., 1895-1900, prof. applied electro-chemistry and chem. engring., 1900-13; engr. for Wis. R.R. Commn., 1908-13; later pres., chmn. bd. C. F. Burgess Lab.; dir. Burgess-Parr Co. and Burgess-Manning Co. Perkin medal of Chemical Societies, 1932; Edward Goodrich Acheson, award, 1942. Mem. Am. Electrochem. Soc. (pres.), Soc. Chem. Industry, Am. Chem. Soc. Am. Gas Inst., Western Soc. Engrs. (Octave Chanute medal 1911), Am. Electroplaters' Soc., Royal Instn. Gt. Britain, Beta Theta Pi, Tau Beta Pi, Alpha Chi Sigma. Inventor process for electrolytic purification of iron; numerous other inventions include various iron alloys, improvements in dry cells, over 400 patents. Died Feb. 13, 1945.

BURGESS, Edward Sandford, Am. botanist; b. Little Valley, N.Y., Jan. 19, 1855; s. Chalon and Emma (Johnston) B.; A.B., Hamilton Coll., 1879, A.M., 1882, Sc.D., 1904; fellow Johns Hopkins, 1880-81; Ph.D., Columbia, 1899; m. Irene S. Hamilton, Dec. 30, 1884. Tchr. botany, Washington, 1881-95, Martha's Vineyard Summer Inst., 1880-95, Johns Hopkins U., 1885; prof. natural sci. Hunter Coll., 1895—, acting pres., 1908. Pres. Torrey Bot. Club, 1912-13. Author: History of Pre-Clusian Botany, 1902; Species and Variations of Biotian Asters, 1906. Died Feb. 23, 1928.

BURGESS, George Kimball, Am. physicist; b. Newton, Mass., Jan. 4, 1874; s. Charles A. and Addie L. (Kimball) B.; S.B., M.I.T., 1896; D.Sc., U. Paris, 1901; hon. D. Engr., Case Sch. Applied Sci., 1923, Lehigh U., 1925; m. Suzanne Babut, Jan. 5, 1901. Taught physics Mass. Inst. Tech., U. of Mich., 1900-01, and U. Cal.; asso. physicist, Nat. Bur. Standards, Washington, 1903-13, physicist, chief div. of metallurgy, 1913-23; dir. Bur. Standards, from 1923, mem. various coms. connected with it; engaged in pyrometric and metall. researches. U. S. del. 7th Internat. Conf. on Weights and Measures, Paris, 1927, World Engring. Congress, Tokyo, 1929; pres. Annual Conf. on Weights and Measures. Mem. fgn. service and engring. coms. NRC; chmn. Fed. Specifications Bd.; mem. NACA. Fellow Am. Phys. Soc., A.A.A.S. Author: Recherches sur la constante de Gravitation, 1901; Experimental Physics—Freshman course, 1902; The Measurement of High Temperatures (with H. LeChâtelier), 1912. Active in devel. and adoption of internat. temperature scale; supported program to redetermine absolute values of elec. units; suggested (with Waidner) the black body radiation of hollow tube immersed in molten platinum as natural reproducible standard of light, 1908 (experimentally realized in late 1920's). Died July 2, 1932.

BURGESS, Robert Lewis, Am. ecologist; b. Kalamazoo, Sept. 12, 1931; s. James Lewis and Hazel (Warren) B.; B.S., U. Wis., 1957, M.S., 1959, Ph.D., 1961; m. Vera Thiel Ballegoin, July 30, 1955; children—Karen Elaine, Steven Robert, Susan Patricia, Ellen Kaihan. Research assistant Univ. Wisconsin, 1957-60; assistant professor of botany Ariz. State U., 1960-63; asst. prof., asso. prof. botany N.D. State U., Fargo, 1963—; dir. Desert Inst., Tempe, Ariz., 1963; vis. asso. prof. botany Pahlavi U., Shiraz, Iran, 1965-66. Mem. A.A.A.S., Am. Inst. Biol. Scis., Ecol. Soc. Am., Brit. Ecol. Soc., Nature Conservancy, Internat. Assn. Tropical Ecology, S.W. Assn. Naturalists, Sigma Xi. Author: Cells: Their Structure and Function, 1962; Bibliography of the Natural History of Iran, 1966. Research, publs. in structure and function of plant communities; contbns. to math. relations between organisms and environment, and spatial distbn. of environmental factors. Address: N.D. State U., Fargo, N.D. 58102.*

BURGESS, W(illiam) Starling, Am. naval architect and engr.; b. Boston, Dec. 25, 1878; s. Edward and Caroline Louisa (Sullivant) B.; A.B., Harvard, 1901; m. Helene Adams Willard, Oct. 1901 (dec. 1902)); m. 2d, Rosamond Tudor, Oct. 13, 1904; children— Edward, Frederick Tudor, Starling; m. 3d, Elsie Janet Foos, Nov. 1925; children—Ann, Diana. Established W. Starling Burgess Co., builders of yachts and comml. vessels, 1904; established Burgess Co. and Curtis, aeroplane builders, Marblehead, Mass., 1910; entered from Burgess & Morgan Ltd., naval architects, N.Y. 1926; mem. Burgess & Donaldson, Inc., since 1931. Designed Enterprise, defender America's cup against Sir Thomas Lipton's Shamrock V, 1930; also designer exptl. aluminum ship, 1935. Recipient Collier prize, 1915, for greatest progress in aviation during preceding year. Died Mar. 19, 1947.

BURGHELE, Teodor, Rumanian surgeon; b. 1905; rector Medico-Pharm. Inst., Bucharest, Rumania mem. Acad. Socialist Republic Rumania, many fgn. sci. socs. Author: (with N. Hortolomei) Über Hämaturien bei Nierenoeckenentleerungsstörungen, 1935; (with others) Emergency Cases in Urology, The Shock Kidney Trial Copy, 1963, Vesical Troubles in Medullary Traumatisms, 1963. Specialist in urology, esp. renal Tb; contbr. to surg. treatment of mitral stenosis and pericarditis.

BÜRGL, Emil, Swiss chemist; b. Bern, Switzerland, Apr. 19, 1872; s. Friedrich and Magdalene (von Känel) B.; m. Pauline Lucie Bandi, 1899; 2 sons. Prof. pharmacology, med. chemistry, from 1906; physician. Author: Chlorophyll als Pharmakonz, 1932; Die Arzneimittelkombinationen, 1938; Die Durchläsigkeit der Haut für Arzneien und Gifte, 1942. Noted for theory of quantitative effect of combination of 2 drugs with same end effect; studied effect of chlorophyll, penetrability of skin by medicines, poisons. Died Bern, Jan. 30, 1947.

BÜRGL, Jost, mathematician; b. Lichtensteig, Switzerland, Feb. 28, 1552; m. Miss Bramer; m. 2d, Catherine Braun. Clockmaker to ct. of Landgrave Wilhelm IV, Hessia, 1579-92; imperial clockmaker, Prague, Czechoslovakia, from 1604. Author: Arithmetische und Geometrische Progresstabulen, 1620. Conceived idea of logarithms before 1610; found proof for prostaphaeresis (used for solution of all triangles, sectioning of circle, constrn. of heptagon); inventor double compass, triangular instrument, instrument for perspective drawing. Died Kassel, Germany, Jan. 31, 1632.

BURGOS, Mario Hector, Argentine anatomist; b. Buenos Aires, Argentina, June 21, 1921; s. Remigio S. and Leticia (Asborno) B.; B.D., Nat. Coll. Manuel Belgrano, 1940; M.D., Med. Sch., U. Buenos Aires, 1948; m. Aída E. Mónaco, Nov. 24, 1949; children—Patricia, Claudio, Virginia, Mariano, Leticia, Octavio.

Instr., Inst. Microscopic Anatomy and Embryology, Med. Sch., U. Buenos Aires, 1943-46; research fellow Inst. Biology and Exptl. Medicine, Buenos Aires, 1946-53, dept. anatomy Harvard Med. Sch., Boston, 1953-55; research asso. Cornell Med. Coll., N.Y.C., 1955-57; prof., dir. Inst. Histology and Embryology, Med. Sch., Cuyo U., Mendoza, Argentina, 1957—; vis. prof. dept. anatomy U. Wash., Seattle, 1959-60. Recipient Gold medal Pub. Assistance, 1949. Mem. Soc. Argent. Biología, Asociación Latin, Am. Ciencias Fisiol., Internat. Fertility Assn., Am. Assn. Anatomists, Histochem. Cytochem. Assn., Internat. Soc. Cell Biologists, Internat. Soc. Comparative Endocrinology, Pan Am. Med. Assn. Research and publs. on male and female reproductive tract, cell biology, mitochondria as biol. pumps and active transport. Home: 2167 Martines de Rosas, Mendoza, Argentina.*

BURHOP, Eric Henry Stoneley, physicist; b. Jan. 31, 1911; s. Henry A. and Bertha B.; B.A., Melbourne U., 1932, M.Sc., 1933; Ph.D., Trinity Coll., Cambridge (Eng.) U., 1937; m. Winifred Stevens, 1936; 2 sons, 1 dau. Research, Cavendish Lab., Cambridge U., 1933-35; research physicist, lectr. Melbourne U., 1935-45, dep. dir. Radio Research Lab., 1942-44; lectr. math. U. Coll., London, Eng., 1945-49, reader in math., 1949-50, reader in physics, 1950-60, prof. physics, 1960—. Tech. officer, mission to Berkeley, Cal., 1944-45. Recipient Joliot-Curie medal, 1966. Fellow Royal Soc., 1963; mem. World Fedn. Sci. Workers (asst. sec.), Phys. Soc. London. Author: The Challenge of Atomic Energy, 1951; The Auger Effect, 1953; (with H. S. W. Massey) Electronic and Ionic Impact Phenomena, 1953. Research, publs. on atomic, nuclear physics, application of elementary particle physics to nuclear physics studies, also K meson interaction processes, atomic collision processes, theory of X-radiation and Auger effect. Home: 39 Templemere, Oatlands Dr., Weybridge, Eng. Office: Dept. Physics, U. Coll., Gower St., London W.C. 1, Eng.

BURIAN, Hermann Martin, ophthalmologist; b. Naples, Italy, Jan. 14, 1906; s. Richard Anton and Marie (Drucker) B.; M.D., U. Belgrade (Yugoslavia), 1930; D.Sc., Colby Coll., 1945; m. Gladys Simmons Hart, Sept. 19, 1940; children—Richard, Peter. Came to U. S., 1936, naturalized, 1940. Asst. prof. ophthalmology Dartmouth, 1936-45; practice medicine specializing in ophthalmology; prof. ophthalmology, Iowa City, 1951—. Chmn., Am. Com. on Optics and Visual Physiology, 1964—; mem. Am. Orthoptic Council, pres., 1956-58; mem. visual com. Armed-Forces-NRC. Fellow A.M.A. (Hector Gold medal 1963); mem. Am. Acad. Ophthalmology and Otolaryngology, Am. Opthalmol. Soc., Biophys. Soc., Optical Soc. Am., A.A.A.S., Soc. for Exptl. Biology and Medicine, Internat. Soc. for Clin. Electroretinography (sec. Western hemisphere), Sigma Xi. Research, numerous publs. on binocular visual act in normal and abnormal states neuromuscular system eye, physiology and electrophysiology retina in normal states, structure angle of anterior chamber in normal and diseased states. Home: 430 Lee St., Iowa City 52240.*

BURIDAN, Jean, French astronomer; b. Bethune, France, circa 1300; maître ès arts, 1329; lectr. Paris U., rector, 1327, 40. Author commentaries on Aristotle's works, including Physics, Metaphysics, De anima, Parva naturalia, Meteorologica, De coelo et mundo, De generatione et corruptione, Nicomachean Ethics, Politics, Forerunner of modern dynamics; disciple of William of Ockham, stressed doctrine of nominalism; anticipated Newton's 1st law of motion; pioneer in suggesting extension of mechanics to heavens. Died circa 1385.

BURK, Dean, Am. chemist; b. Oakland, Cal., Mar. 21, 1904; s. Frederic and Caroline (Frear) B.; B.S., U. Cal. at Berkeley, 1923, Ph.D., 1927; postgrad. U. London (Eng.), Kaiser Wilhelm Inst. for Biology, Harvard; m. Mildred Alice Chaundy, Jan. 28, 1929; children—Diana, Wendy, Frederic Chaundy. Chemist, Fixed Nitrogen Research Lab., U. S. Dept. Agr., 1929-39; asso. prof. Cornell U. Med. Coll., 1939-41; chemist Nat. Cancer Inst., NIH, N.Y.C., 1939-41, Bethesda, Md., 1941—, supervisory research chemist, 1959—; research master, grad. faculty George Washington U., Washington, 1947—. Guest research worker U.S.S.R. Acad. Sci., Moscow, 1935; spl. cancer research fellow Kaiser Wilhelm Inst. for Cell Physiology, Berlin, Germany, 1950; fgn. mem. Max Planck Inst. for Cell Physiology, Berlin, 1953—. Bd. dirs. Sci. Resources Found., Boston. Recipient Gerhard Domagk prize for cancer research Domagk Found., U. Muenster, Germany, 1965. Mem. Am. Chem. Soc. (Hillebrand award 1952), Am. Soc Biol. Chemists, Am. Assn. Cancer Research, Am. Soc. Plant Physiology, N.Y., Washington acads. scis., Harvey Soc., Soc. for Exptl. Biology and Medicine, A.A.A.S., Max Planck Assn. for Advancement Sci. (Goettingen, Germany), Royal Soc. Medicine (London, Eng.). Author-editor: Cancer, 1945; Approaches to Tumor Chemotherapy, 1947; Cell Chemistry, 1953; Contbr. numerous articles to tech. jours. Research on quantum efficiency in photosynthesis by green plants, nature specific biochemistry cancer cells, formation simple complexes oxygen gas with cobalt-amino compounds under physiol. conditions simulating those with hemoglobin, elucidation kinetics rates enzyme reactions; discovered vitamins and anti-vitamins; research on energy and mechanims biol. nitrogen fixation. Home: 4719 44th St., Washington 20016. Office: Nat. Cancer Inst., Bethesda, Md. 20014.*

BURKARD, Otto Michael, Austrian physicist; b. Graz, Austria, Nov. 24, 1908; s. Otto Benno and Bertha (Petzl-Hirschmann) B.; Ph.D., U. Graz, 1933; m. Herta Emma Waidbacher, Mar. 28, 1942; children—Rainer Ernst, Hildegund Herta, Helmut Otto, Hans Michael. Prof. physics and math. Fed. Trade Sch. for Machine and Elec. Engring., Graz-Goesting, 1938-49; prof., head Inst. Meteorology and Geophysics, U. Graz, 1949—. Corr. mem. Austrian Acad. Scis. Co-editor: Gerlands Beiträge zur Geophysik. Research, numerous publs. on ionosphere and space. Home: 32 Grillparzerstrasse, Graz 8010, Austria. Office: 1 Halbärthgasse, Graz A-8010, Austria.

BURKE, Arthur Devries, Am. dairy technologist; b. Wheeling, W.Va., Jan. 1893; s. Thomas Carrol and Anna (Little) B.; B.S., U. Wis., 1916; M.S., Ohio State U., 1920, postgrad. 1927-28; m. Marguerite Outcalt. Feb. 1, 1921. Dairy inspr. Huntington, W.Va., 1916-17; instr. dairy dept. Ohio State U., 1919-20; asst. prof. of dairying Okla. A. and M. Coll., 1920-22, asso. prof., 1922-29; prof. and head dairy dept. Ala. Poly. Inst., 1929-46. Mem. adv. council Sealtest, Inc., 1935-48. Mem. Am. Dairy Sci. Assn. (sec., v.p., pres. so. sect.), Internat. Assn. Ice Cream Mfrs. (mem. statis. research com. 1939-47), Ala. Dairy Products Assn. (pres., v.p., sec.; exec. sec.). Author: Practical Ice Cream Making, 1933, revised 1945; Practical Dairy Tests, 1935; Practical Manufacture of Cultured Milks and Kindred Products, 1938. Tech. editor Milk Dealer, 1920—; Ice Cream Rev., 1929—. Contbr. tech. jours. Died Aug. 16, 1950.

BURKE, Edmund, Brit. statesman, polit. writer; b. Dublin, Ireland, Jan. 12, 1729; s. Richard Burke; student Trinity Coll., Dublin, 1743-48; studied law, Middle Temple, London, 1750; m. Jane Nugent, 1756; 2 sons. Failure to practice law led to father's withdrawal of support; depended on lit. for livelihood from 1755; launched Ann. Register (yearbook of polit. and econ. information), 1756; pvt. sec. to Willimm Gerard Hamilton, 1759; went to Scotland when Hamilton was apptd. sec. to Earl of Halifax, 1761; returned to Eng., 1763; pvt. sec. to prime minister Lord Rockingham, 1765; mem. Parliament, 1765-94; apptd. agt. to province of N.Y., 1771; apptd. paymaster of forces, 1782; served on select com. on affairs of E. India Co., circa 1783; lord rector of Glasgow, 1784; engaged in impeachment of Warren Hastings of govt. of India, 1787-94. Author: A Vindication of Natural Society, 1756; Philosophical Inquiry into the Origin of our Ideas on the Sublime and Beautiful, 1756; An Account of the European Settlements in America, 1757; Speech on Conciliation, 1775; Reflection of the Revolution in France and on the Proceedings in Certain Societies in London Relative to That Event, 1790; also many other works. Considered one of world's greatest orators; possessed unusual ability to apply broad philos. concepts to specific problems of govt.; an outstanding leader of Whig party; noted for struggle for sanity and liberalism in treatment of Am. colonies; criticized French Revolution; basic tenets included belief in orderly constitutional govt., belief that action of govt. should conform with basic permanent elements of institutional life of the people, belief that circumstances must determine application of principles, that aristocracy is law of nature, belief in divine intent that rules human soc.; rejected many reforms in belief that they often go amiss; rejected doctrine of inherent, absolute human rights; believed that freedom can be attained only be reconciling man to the state; considered polit., social and religious instns. representative of wisdom of the ages; called the intellectual father of modern conservatism; a resolute critic of totalitarian doctrines. Died July 8, 1787.

BURKE, Joseph Eldrid, Am. metallurgist, ceramist; b. Berkeley, Cal., Sept. 1, 1914; s. Charles Eldrid and Ruth (Hadcock) B.; B.A., McMaster U., Hamilton, Ont., Can., 1935; Ph.D., Cornell U., 1940; m. Kathleen Mary Wilson, Sept. 16, 1939; children—Charles R., Margaret E. Group leader Los Alamos Lab., Manhatten Dist., 1943-46; asso. prof. U. Chgo., Inst. Metals, 1946-49; mgr. metallurgy Knolls Atomic Power Lab., Gen. Electric Co., Schenectady, 1949-54, mgr. ceramics Research and Devel. Center, 1954—. Fellow Am. Ceramic Soc. (past chmn. basic sci. div.); mem. Am. Soc. Metals, Am. Inst. Mining and Metall. Engrs., Brit. Ceramic Soc., A.A.A.S. Author: (with A. U. Seybolt) Procedures in Experimental Metallurgy, 1953; also articles. Editor: The Metal Beryllium (with D. W. White), 1955; Progress in Ceramic Science, vols. 1-4, 1961, 62, 63, 66. Research on recrystallization, grain growth and sintering (fundamental phenomena of processing) in metals and oxide ceramics, properties and behavior of metals and ceramics used for prodn. nuclear energy. Home: 33 Forest Rd., Burnt Hills, N.Y. 12027. Office: Gen. Electric Co., Research and Devel. Center, Schenectady 12301.*

BURKE, Kevin Charles Antony, geologist; b. London, Eng., Nov. 13, 1929; s. Charles Henry and Kathleen (Daly) B.; B.Sc., U. Coll. London, Ph.D., 1953; m. Angela Marion Phipps, Jan. 23, 1960; children—Nicholas, Matthew, Jane. Faculty, U. Ghana, 1953-56; with atomic energy div. Geol. Survey Great Britain, 1956-60; with Internat. Atomic Energy Agy., Korea, 1960; faculty U. W.I. Jamaica, 1961-65; prof. geology U. Ibadan, Nigeria, 1965—. Mem. Geol. Soc. London. Research, publs. on structural and regional geology of Ireland, Caribbean, Africa; nuclear raw materials. Address: U. Ibadan, Nigeria.*

BURKE, Philip George, English physicist; b. London, Eng., Oct. 18, 1932; s. Henry and Frances (Sprague) B.; B.Sc. in Physics with 1st class honors, U. Exeter (Eng.), 1953; Ph.D., U. Coll., London, 1956; m. Valerie M. Martin, Aug. 29, 1959; children—Helen Frances, Susan Valerie, Pamela Jean. Research asso. U. Coll. London, 1956-57; lectr. U. London Inst. for Computer Sci., 1957-59; research physicist Lawrence Radiation Lab., Berkeley, Cal., 1959-62; prin. sci. officer U.K. Atomic Energy Authority, Harwell, Eng., 1962—; vis. lectr. dept. astrophysics Princeton, 1966. Cons. Lockheed Missiles & Space Co., Palo Alto, Cal., 1965—. NATO fellow Nat. Bur. Standards, Washington, 1965. Fellow Phys. Soc. London; mem. Am. Phys. Soc. Research, publs. on theory scattering of nucleons by light nuclei, bubble chamber expts., resonances in elementary particle physics, applied dispersion relations to elementary particle scattering; discovered theoretically narrow resonances in electron atom scattering. Home: 1 Main Rd., E. Hagbourne, Didcot, Berkshire. Office: Theoretical Physics div. A.E.R.E., Harwell, Berkshire, Eng.*

BURKE, T(homas) Finley, Am. physicist; b. N.Y.C., June 28, 1918; s. Thomas F. and Elizabeth M. (Boyd) B.; A.B., N.Y. U., 1939; postgrad. Ind. U. Mass. Inst. Tech.; m. Barbara Howe, Aug. 1, 1942; children—Marjorie Lynn, Judith Howe. Physicist, U. Cal. Div. War Research, 1942-46; project engr., sect. head, asst. chief engr. Melpar, Inc., 1949-55; asst. dir. central research lab., Westinghouse Air Brake Co., 1952-53; mem. sr. staff Ramo-Wooldridge Corp., Los Angeles, 1955-57, Rand Corp., Santa Monica, Cal., 1957—. Mem. spl. team NRC, summer 1951. Recipient S.F.B. Morse medal for physics. Fellow Am. Phys. Soc. Research on particle accelerators; high vacuum systems; weapon system analysis; acoustics; electromechanical systems. Home: 7004 Fernhill Dr., Malibu, Cal. Office: Rand Corp., Santa Monica, Cal.

BURKE, Victor, Am. bacteriologist; b. Kent, Wash., Mar. 12, 1882; s. Harry and Sarah Eugenia (Jones) B.; A.B., Stanford, 1907, A.M., 1908, Ph.D., 1911; grad. study Columbia, 1910; m. Georgina B. Spooner, Oct. 19, 1912. Asst. naturalist Albatross expdn. to Japan, 1906; instr. bacteriology Stanford, 1918-21; asst. prof. bacteriology Wash. State Coll., 1921-23, asso. prof., 1923-25, prof. and head dept., 1925-49; made state prof. of bacteriology and public health, 1947. Mem. Soc. Am. Bacteriologists, Sigma Xi, Phi Kappa Phi. Author: The Cyclogasteridae, 1930; also many articles in field of bacteriology. Bacteriol. editor Webster's New Internat. Dictionary. Died Oct. 7, 1958.

BURKE-GAFFNEY, Michael Walter, Can. astronomer; b. Dublin, Ireland, Dec. 17, 1896; s. Thomas and Joan (O'Donnell) B-G.; B.E., Dublin U., 1917; Ph.D., Georgetown U., 1935; Dr. Engring. N.S. Tech. Coll., 1955. Came to Can., 1920, naturalized, 1925. Engr., War Office, London, Eng., 1917-18, Air Ministry, 1918-20; lectr. astronomy Regis Coll., Toronto, Ont., Can., 1935-39; dean engring. St. Mary's U., Halifax, N.S., Can., 1940-48, prof. astronomy, 1948-65, prof. emeritus, 1965—. Mem. Canadian com. Internat. Astron. Union, 1964-67, cons. to commn. 41, 1964—. Mem. Royal Astron. Soc. Can., Am. Astron. Soc., Canadian Aeros. and Space Inst., Profl. Engrs. Assn. N.S. Author: Kepler and Jesuits, 1944; Daniel Seghers, 1961; also numerous articles. Research in celestial mechanics, possible orbits of artificial satellites; tracked and computed orbits of Sputnik I and II, 1957. Home: 923 Robie St., Halifax, N.S., Can.*

BÜRKER, Karl Jakob Sebastian, German physiologist; b. Aug. 10, 1872; student U. Tübingen (Germany), 1892-98, U. Heidelberg (Germany); Dr. sc.nat., 1897; M.D., 1900; M.D. h.c., U. Geissen (Germany), 1932. Became pvt. docent U. Tübingen, 1901, prof. physiology, 1904; prof., from 1947; prof., dir. physiol. inst. U. Giessen, 1917. Author: Der Muskel und das Gesetz von der Erhaltung der Kraft, 1902; Die Lebensvorgänge des menschlichen Körpers, 1925. Research on surface and shape of blood corpuscles, 1922. Died 1957.

BURKHOLDER, Donald Lyman, Am. mathematician, educator; b. Octavia, Neb., Jan. 19, 1927; s. Elmer and Susie (Rothrock) B.; B.A., Earlham Coll., 1950; M.S., U. Wis., 1953; Ph.D., U. N.C., 1955; m. Jean Annette Fox, June 17, 1950; children—Kathleen, James Peter, William. Faculty math. U. Ill., Urbana, 1955—, prof., 1964—. Fellow Inst. Math. Statistics; mem. Am. Math. Soc., Inst. Math. Statistics, Am. Statis. Assn. Editor: The Annals of Mathematical Statistics, 1964—. Research on stochastic approximation, statis. sufficiency, conditional expectations, ergodic theory, probability theory. Home: 506 W. Oregon St., Urbana, Ill. 61801.*

BURKHOLDER, Paul R(ufus), Am. microbiologist; b. Orrstown, Pa., Feb. 1, 1903; s. Wm. Rankin and Mary Ellen (Schubert) B.; A.B., Dickinson Coll., 1924; Ph.D., Cornell U., 1929; Nat. Research Council fellow in botany, Harvard, 1932-33, Columbia, 1933-34; M.A. hon., Yale, 1944; Sc.D., hon., Dickinson Coll., 1949; m. Lillian Miller, Feb. 4, 1930; children—Franz M., Peter M., Karl M. Instr. botany Cornell U., 1924-28; biol. curator Buffalo Mus. of Sci., 1929-32; asst. prof. Conn. Coll., 1934-37, asso. prof., 1937-38; asso. prof. U. of Mo., 1938-40; asso. prof. Yale, 1940-43, Eaton prof. botany 1944-53; chairman of department of Plant Science 1950-53; head dept. bacteriology U. of Georgia, 1953-56; dir. research Bklyn. Botanic Garden, 1956-61; chmn. marine biology programs Lamont Geol. Obs. Columbia, 1961—. Mem. Am. Assn. Advancement Sci., Nat. Acad. Sci., Bot. Soc. Am. (sec. 1940-45), Am. Soc. Naturalists (pres. 1948), Am. Soc. Microbiologists, Soc. Protozool., Soc. Gen. Microbiol., Torrey Bot. Club, Sigma Xi. Author papers in field. Research in microbiology; marine biology; study of antibiotics; algae; chloromycetin; growth substances. Office: Lamont Geol. Obs., Palisades, N.Y.

BÜRKNER, Hans Friedrich Hermann, German shipbuilder; b. Dresden, Germany, Jan. 11, 1864; s. Friedrich Hermann Ludwig and Helene Camilla (Jordan) B.; student math. and scis. Heidelberg, Leipzig, Germany, Polytechnikum, Dresden; student shipbldg. Technische Hochschule, Berlin, Germany, D. Engring. (hon.); m. Adolphine Emma Ernestine Anna Kaben; children—5 daus. Entered German navy, named shipbldg. master, 1894, adminstrv. officer, 1898-1905, with constrn. dept., 1905-18, retired, 1919. Recipient Golden Meml. medal Soc. Tech. Shipbldg. Author: Erinnerungen und Gedanken eines alten Kriegsschiffbauers, 1940. Worked on bldg. ships resistant to enemy fire (dividing ship into watertight compartments, armor plating, increasing maneuverability, stability, reducing danger of fire and shattering and others). Died Ringelheim, Germany, Oct. 29, 1943.

BURKS, Ardath Walter, Am. polit. scientist; b. Covington, Ky., May 1, 1915; s. Alonzo Edwin and Clara (McCracken) B.; A.B., U. Cin., 1939; M.A., U. Minn., 1941; postgrad. U. Colo.; Ph.D., Sch. Advanced Internat. Studies, Johns Hopkins, 1949; m. Jane Virginia Lyle, Nov. 15, 1941; 1 son, Riki Stephen. Intelligence specialist Army Air Corps, War Dept., 1941-43, Joint Army-Navy Air Intelligence Div., Office Naval Intelligence, Washington, 1946; faculty Rutgers-the State U., New Brunswick, N.J., 1948—, prof. polit. sci., 1959—, chmn. dept., 1962-65, dir. internat. programs, 1966—; dir. Inst. on Asian Studies, U. Hawaii, 1966. Social Sci. Research Council fellow, Japan, 1952-53; Fulbright prof., Japan, 1958-59; Ford Found. grantee, 1962; Rutgers Research Council fellow, Japan, 1965. Mem. Am. Polit. Sci. Assn., Assn. Asian Studies, Asia Soc. N.Y., Japan Soc. N.Y., Am. Assn. U. Profs. (pres. Rutgers chpt. 1960-61), Internat. House Japan. Author: The Government of Japan, 1964; (with Djang Chu) Far Eastern Governments and Politics, 1956. Contbr. articles to profl. jours., encys. Application of principles of comparative politics to Eastern Asia, especially Japan; studies in modernization, including urban phenomena, especially in Japan. Home: 30 W. Maple Av., Bound Brook, N.J. 08805.*

BURLAGE, Henry Matthew, Am. pharmacist, educator; b. Rensselaer, Ind., May 23, 1897; s. Max and Mary (Linzbach) B.; A.B., Ind. U., 1919; M.A. (Austin teaching fellow), Harvard, 1921; Ph.G, B.S., Purdue U., 1924, D.Sc., 1961; Ph.D., U. Wash., 1929; m. Alleda Virginia Robb, Dec. 29, 1925; 1 son, Robb Kendrick. Faculty, Purdue, 1921-24, asso. prof., 1929-31; instr. U. Wash., 1924-27; asso. prof. drug analysis Ore. State Coll., 1927-29; instr. Ore. Bd. Pharmacy, 1927-29; prof. U. N.C., 1931-47; prof. pharmacy and pharm. chemistry U. Tex., Austin, 1947—, dean Coll. Pharmacy, 1947-62. Fellow Am. Coll. Apothecaries, A.A.A.S.; mem. Am. (mem. council 1961-64), Tex. pharm. assns., Am. Chem. Soc., Am. Assn. Colls. Pharmacy (exec. com. 1959-61, pres. 1960-61), Acacia, Phi Beta Kappa, Sigma Xi, Phi Kappa Phi, Rho Chi, Kappa Psi, Alpha Chi Sigma, Phi Lambda Upsilon. Author: Study Guide to Essential Literature of Pharmacy, 1962, rev., 1964, 65, 66; (with M. L. Jacobs) Index to the Plants of North Carolina with Reputed Medicinal Uses, 1959; (C. O. Lee, L. W. Rising) Orientation to Pharmacy, 1959; also numerous articles. Editor: Physical and Technical Pharmacy, 1963. Research in pharm. techniques, plant analysis, assay procedures. Home: 702 E. 43d St., Austin, Tex. 78751.*

BURLET, Claude, French physician; b. Bourges, France, 1664; m. Miss Dodart. Asso. with botanist Dodart; 1st physician to Philip V of Spain, 1700-18, to French Royal family, 1727. Mem. French Acad. Scis., 1699. Made studies on Spanish pharmaceutics, hydrology. Died Paris, France, Aug. 10, 1731.

BURLEY, Gordon, chemist; b. Giessen, Germany, Feb. 15, 1925; s. Bernard L. and Eleanor (Bensinger) B.; A.B., Temple U., 1948; M.S., U. Md., 1950; Ph.D., Georgetown U., 1962; m. Jaylee Montague, Nov. 28, 1957. Came to U. S., 1939, naturalized, 1948. Research asso., geophys. lab. Carnegie Inst., Washington, 1950-52; phys. chemist Nat. Bur. Standards, Washington, 1952-67; phys. chemist div. reactor licencing AEC, Washington, 1967—; instr. Nat. Bur. Standards Grad. Sch., 1962-63, Howard U., 1967. Mem Am. Chem. Soc., Am. Crystallographic Assn., Mineral. Soc. Am. Research and publs. on crystallography of silver iodide and its role in cloud seeding; exptl. X-ray atomic scattering factors; theoretical calculations in lattice dynamics; fission product release and propagation in nuclear reactors. Home: 4701 Willard Av., Chevy Chase, Md. 20015. Office: AEC, Div. Reactor Licencing, Washington 20545.*

BURLEY, Walter, Brit. philosopher; b. circa 1275; studied at Merton Coll., Oxford (Eng.) U., under Duns Scotus at U. Paris. Tchr. at Oxford, Paris, Toulouse, France; almoner to Philippa, consort of Edward III, 1327; became envoy to Pope, 1327, 1330; reputed tutor to Black Prince, circa 1342. Author: Liber de vita et moribus philosophorum et poetarum; also numerous treatises on Aristotle. His theory of knowledge resembled that of St. Thomas Aquinas; disagreed with Aristotle's argument for unmoved 1st mover. Died circa 1345.

BURLING, Temple, Am. psychiatrist; b. Chgo., Mar. 22, 1896; s. James Perkins and Terese (Temple) B.; student Grinnell Coll., 1913-15; B.S., U. Chgo., 1921; M.D., Rush Med. Coll., 1923; m. Katherine White, Jan. 30, 1924; children—Robbins, James Perkins, Helen Temple (Mrs. Malcolm Kenyon Ottaway). Practice medicine, Mpls., 1925-26; field epidemiologist Minn. Bd. Health, 1926-28; with Inst. for Juvenile Research, Chgo., 1931-34, Winnetka (Ill.) Pub. Schs., 1934-37; psychiatrist R. H. Macy & Co., 1937-40; dir. Providence Child Guidance Clinic, 1940-47; research vocational rehab. Nat. Com. for Mental Hygiene, 1947-48; prof. Sch. Indsl. and Labor Relations, Cornell U., Ithaca, N.Y., 1948-64, prof. emeritus, 1964—; vis. prof. Middle E. Tech. U., Ankara, Turkey, 1962-63. Mem. A.M.A., Am. Psychiat. Assn., Am. Orthopsychol. Assn., A.A.A.S., Soc. for Gen. Systems Research. Author: (with T.A.C. Rennie, Luther Woodward) Vocational Rehabilitation in Psychiatric Patients, 1950; The Give and Take in Hospitals (with Edith Lentz, Robert Wilson), 1956; also articles. Research in vocational rehab. recovered psychotics, human relations in hosps., human relations in retailing. Home: R.D. 1, Trumansburg, N.Y. 14886. Office: Ives Hall, Cornell U., Ithaca, N.Y. 14850.*

BURLINGAME, C. Charles, Am. psychiatrist, univ. prof.; b. Rockford, Ill., Oct. 27, 1885; s. Charles Henry Camlin and Ella S. F. (Dagwell) B.; M.D., Ill. Gen. Med. Coll., 1908; m. Ruth Beardsley Parsons, Dec. 31, 1912. Asst. phys. Westboro (Mass.) State Hosp., 1908-12; med. dir., asst. supt., acting supt. Fergus Falls (Minn.) State Hosp., 1912-15; indsl. psychiatrist Cheney Bros., Manchester, Conn., 1915-17, 19-21; exec. officer joint adminstrv. bd. Columbia U.-Presbyn. Hosp. Med. Center, N.Y.C., 1921-28; exec. v.p. Presbyn. Hosp., N.Y.C., 1923-25; in private psychiatric practice, N.Y.C., 1925-31; hosp. cons. govt. of Uruguay and other S. Am. and European countries from 1925; psychiatrist in chief Inst. of Living, 1931-50; asso. in psychiatry Columbia from 1932; prof. psychiatry and mental hygiene Yale, 1936-38; cons. in psychiatry U. S. Vets. Hosp., Newington, Conn., from 1931, Neurol. Inst., N.Y.C.; 1932-39; attending psychiatrist Vanderbilt Clinic, N.Y. City, 1932-34, asso. attending psychiatrist since 1934; attending neuropsychiatrist Vet. Home & Hosp. Commn., Conn.; cons. in psychiatry St. Francis Hosp., Hartford, Conn., from 1933; cons. in psychiatry Charlotte Hungerford Hosp., Torrington, Conn., since 1934; clin. cons. Hosp. St. Raphael, New Haven, from 1943; cons. psychiatry Hartford Hosp., 1945; cons. psychiatry and neurology Meriden Hosp.; sr. psychiatry U. S. Vets Hosp., Northampton; cons. psychiatry to sec. war, 1944-45. Diplomate Am. Bd. Psychiatry and Neurology. Mem. A.C.P., Am., So. psychiat. assns., Assn., N.Y. Acad. Medicine, Am. Soc. of Research Psychosomatic Problems, Am. Psychiat. Found., A.M.A., Assn. Research in Nervous and Mental Diseases, Assn. Study Internal Secretions. N.E. Soc. Psychiatry. Pan-Am. Med. Assn., N.E. Conf. Sociedad Cubana de Neurologia and Psiquiatria, many other med. socs. Editor Digest of Neurology and Psychiatry; asso. editor Am. Jour. Psychiatry. Died July 22, 1950; buried West Hartford, Conn.

BURMAN, Jan, Dutch botanist; b. Amsterdam, Netherlands, 1707; tchr. botany in Amsterdam, from 1738. Author: Phytanhoza, 1736; Thesaurus Zeylanicus, exhibens plantas in Insula Zeylana Nascentes, 1737; Rariorum Africanarum Plantarum ad vivum delineatarum, 1738, 39; Herbarium amboinense, 1741-50; Auctuarium, 1755; Vachendorfia, 1757; De ferrariae charactere, 1757; Flora malabarici, 1769; The Flora of the Sunda Islands, 7 vols. Studied flora of Tunisia and Algeria. Died 1780.

BURMAN, Nikolaus Laurens, Dutch botanist; b. Amsterdam, Holland, 1734; s. Jan Burman; became prof. botany, succeeding his father, Amsterdam, Author bot. works including Flora Indica (completed by J. G. Koenig), 1768. Established genera Erodium and Pelargonium, 1759. Died 1793.

BURMEISTER, Hermann Carl Conrad, German entomologist; b. Straslund, Germany, Jan. 15, 1807; studied at Greifswald; M.D., Halle, Germany, 1829. Named prof. natural history, Cologne, Germany, 1842; prof. zoology, Halle; became mem. Nat. Assembly, Frankfort, Germany, 1848. Author: Lehrbuch der Naturgeschichte, 1830; Handbuch der Entomologie (compilation of everything then known about entomology), 5 vols., 1832-55; Grundriss der Naturgeschichte, 1833, 51; Zoologischer Handatlas, 1835-43; Handbuch der Naturgeschichte, 1837; Geschichte der Shoepfung, 1843, 51; Die Organisation der Trilobiten, 1843; Die Labyrinthodonten, 1849-50; Geologische Bilder zur Geschichte der Erde und ihrer Bewohner, 1851; Systematische Übersicht der Tiere Brasiliens, 1854-56. Died May 2, 1892.

BURMESTER, Ludwig Ernst Hans, German mathematician; b. Othmarschen, Holstein, Germany, May 5, 1840; s. Gottfried and Wilhelmine (Weigel) B.; ed. at univs. of Dresden, Göttingen, Heidelberg (all Germany); hon. dr. Technische Hochschule Hanover; m. Gabriele Schalowetz, 1868; 3 sons, 1 dau. Tchr. in Lodz, Poland; moved to Dresden, 1870; lectr., 1871; became prof. descriptive geometry, Dresden, 1872; with Technische Hochschule, Munich (Germany), from 1887, prof. emeritus, 1912. Mem. Bavarian Acad. Sci., Deutsche Akademie der Naturforscher. Author: Theorie und Darstellung der Beleuchtung gesetzmässig gestalteter Flächen, 1871; Grundzüge der Reliefperspektive, 1883; Lehrbuch der Kinematik I, 1888; also numerous articles in math. and tech. jours. Research on theory of lines of equal lighting intensity on a surface, relief perspective and its effect on stage figures; improved acad. standard of teaching in applied geometry; worked on theory of movement of rigid and changing images; research on light beam optics; offered explanation for geometric optical illusions. Died Munich, Apr. 20, 1927.

BURN, Joshua Harold, English pharmacologist; b. Barnard Castle, Eng., Mar. 6, 1892; s. John George and Josephine (Howson) B.; M.D., Emmanuel Coll., Cambridge U., 1913; D.Sc., Yale, 1957; M.D. (hon.), Mainz U., 1964; Dr. hon., U. Paris, 1965; m. Katharine Pemberton, Oct. 4, 1928; children—George, Isabel (Mrs. David Caddy), Josephine (Mrs. Morris Oliver), Margaret (Mrs. Michael Gerrard), Frances (Mrs. Stephen Gray), Robert. Dir. pharmacol. lab. Pharm. Soc., 1926-37; prof. pharmacology Oxford U., 1937-59; vis. prof. pharmacology Washington U., St. Louis, 1959—. Recipient Gairdner Internat. award Gairdner Found. Toronto, 1959. Fellow Royal Soc., Nat. Inst. Scis. of India (hon.), Author: Drugs, Medicines and Man, 1962; Our Most Interesting Diseases, 1964; The Autonomic Nervous System, 1963. Research, publs. on biol. standardization of drugs, hormones, mechanism of release of noradrenaline, causes of fibrillation in the heart. Home: 2 Capel Close, Oxford, Eng.*

BURNAM, Curtis Field, Am. surgeon, radiologist; b. Richmond, Ky., Jan. 17, 1877; s. Anthony Rollins and Margaret (Summers) B.; A.B., Central U., Ky., 1895; M.D., Johns Hopkins, 1900; m. Florence Overall, Oct. 10, 1908; 1 dau., Mary (Mrs. Howard C. Smith). Resident gynecologist Johns Hopkins Hosp., 1900-05, later asso. gynecologist, also vis. physician in ray therapy; asso. in gynecology Johns Hopkins Med. Sch. until 1912, asso. prof. surgery; radiologist Howard A. Kelly Hosp., Balt. Fellow A.C.S.; mem. A.M.A., Med. and Chirur. Faculty of Md., Am. Gynecol. Soc., Am. Urol. Assn., So. Surg. Assn., So. Med. Assn., Am. Radium Soc. (Janeway medal 1936), Am. Coll. Radiology. Author: (with Howard Atwood Kelly) Diseases of the Kidneys, Ureters and Bladder, 1914, also many med. papers, including those relating to work with Dr. Samuel Crowe of Johns Hopkins, on prevention and cure of deafness. Died Nov. 29, 1947.

BURNET, Sir Frank Macfarlane, Australian immunologist; b. Traralgon, Australia, Sept. 3, 1899; s. Frank and Hadassah (Mackay) B.; M.D., Geelong Coll., Melbourne U., 1922; Ph.D., London U., 1928; m. Edith Linda Marston Druce, July 10, 1928; children—Elizabeth (Mrs. Paul Dexter), Ian, Deborah (Mrs. John Giddy). Beit meml. fellow Lister Inst., London, 1926-27; asst. dir. Walter and Eliza Hall Inst. Med. Research, 1928-31, 34-44, dir., also prof. exptl. medicine U. Melbourne, 1944-65, prof. emeritus, 1966—, Rowden White research fellow Sch. Microbiology, 1966—. Chmn. Med. Research Adv. Com., Papua, New Guinea, 1962. Recipient Lasker Found. award, 1952; Von Behring prize Marburg U. 1952; Nobel prize in physiology and medicine (with P. Medawar), 1960; knighted, 1951. Fellow Royal Soc.; mem. Australian Acad. Sci. (pres. 1965—), Australian Med. Assn., Pathol. Soc. U.K. Author: Virus as Organism, 1945; (with F. J. Fenner) The Production of Antibodies, 1949; Viruses and Man, 1955; The Clonal Selection Theory of Acquired Immunity, 1959; Principles of Animal Virology, 1960; Natural History of Infectious Diseases, 3rd edit., 1962; Integrity of the Body, 1962; (with I. R. Mackay) Autoimmune Diseases, 1963. Research, publs. on cultivation of viruses in chick embryo; genetics of influenza virus; clin. studies on autoimmune diseases. Home: 13 Edward St., Kew, Melbourne, Victoria, Australia.*

BURNETT, Charles Henry, Am. otologist; b. Phila., May 28, 1842; s. Eli Seal and Hannah (Mustin) B.; grad. Yale, 1864; M.D., U. Pa., 1867; studied otology in Europe, 1870-72; m. Anna Davis, June 18, 1874. Practiced medicine specializing in otology, Phila., 1872-1902; prof. otology Phila. Polyclinic; pres. Am. Otol. Soc. Author: The Ear; Its Anatomy, Physiology, and Diseases, 1877; Hearing and How To Keep It, 1879. Editor: System of Diseases of the Ear, Nose, and Throat, 1893; Textbook of Diseases of the Ear, Nose and Throat, 1901. Developed operation for relief of progressive deafness and vertigo by performing tympanotomy and removing incus. Died Bryn Mawr, Pa., Jan. 30, 1902.

BURNETT, Charles Theodore, Am. psychologist; b. Springfield, Mass., June 24, 1873; s. Charles Martin and Alice (Munyan) B.; A.B., Amherst, 1895, L.H.D., 1930; L.H.D., Bowdoin, 1944; Ph.D., Harvard, 1903; m. Sue Winchell, Dec. 16, 1914; children—David Winchell, Bettina and Audrey (twins). Began as instr. in psychology Bowdoin Coll., 1904, prof. psychology since 1909. Fellow A.A.A.S.; mem. Am. Psychol. Assn., Phi Beta Kappa. Author: Splitting the Mind, 1925; Hyde of Bowdoin. 1931. Died Jan. 31, 1946.

BURNETT, Edwin Clark, Am. surgeon; b. Mansfield, O., Jan. 19, 1854; s. Dwight and Mary Ann (Bristol) B.; M.D., St. Louis Med. Coll. (now med. dept. Washington U.), 1883. Practiced medicine, Olney, Ill., 1883-84; returned to St. Louis to make spl. study of genito-urinary diseases in clinics at St. Louis Med. Coll.; chief of clinic, 1890, lectr. on syphilis, 1893; clin. prof. genito-urinary diseases Washington U., 1904-11. Deceased.

BURNETT, John Harrison, English botanist; b. Jan. 21, 1922; s. T. Harrison Burnett; B.A., M.A. (Kitchener scholar), Merton Coll., Oxford (Eng.) U., 1947; Ph.D. (Christopher Welch scholar), 1953; m. E. Margaret Bishop, 1945; 2 sons. Lectr. Lincoln Coll., 1948-49; fellow Magdalen Coll., 1949-53; lectr., demonstrator Oxford U., 1949-53; lectr. Liverpool (Eng.) U., 1954-55; prof. botany St. Andrews (Scotland) U., 1955-60, dean faculty sci., 1958-60; prof. botany King's Coll., Newcastle, Tyne, Eng., 1961-63, dean Faculty Sci., 1966—; prof. botany U. Newcastle/Tyne, 1963—. Fellow Royal Soc. Edinburgh; mem. Scottish Hort. Research Inst. (chmn. 1959—), Eugenics Soc., Inst. Biology, Bot. Soc. Brit. Isles Author: General Mycology, 1967. Editor: New Phytologist (also trustee); The Vegetation of Scotland (also contbr.). Research, publs. on gymnosperms, population genetics of plants, conservation, physiology of genetics and fungi, especially mating systems. Home: 54 Moorside S., Newcastle/Tyne. Office: Dept. Botany, the univ., Newcastle/Tyne 1, Eng.

BURNETT, Sir William, Scottish surgeon; b. Montrose, Scotland, Jan., 1779; surgeon's mate, asst. surgeon on ships; served at Battle of Nile and Trafalgar; after Trafalgar in charge hosps., for prisoners at Portsmouth and Forton; physician, insp. hosps. Mediterranean fleet, from 1810; med. charge of Russian fleet in Medway, from 1813; pvt. physician Chichester, Eng., until 1822; later physician gen. of navy. Fellow Royal Soc., 1833, Royal Coll. Physicians; created knight, 1831. Author: An Account of the Bilious Remittent in the Mediterranean Fleet in 1810-13, 1814; Official Report on the Fever in H.M.S. Bann on the Coast of Africa and Amongst the Royal Marines in the Island of Ascension, 1824; An Account of a Contagious Fever Prevailing Amongst the Prisoners of War at Chatham, 1831. Prepared a disinfectant solution composed mainly of zinc chloride, also a variation containing a small amount of ferrous chloride, 1857; responsible for reforms in treatment and reporting of diseases aboard ship; introduced med. reforms in navy, including better treatment for naval mental patients; patented a disinfectant, Burke's fluid, for preservation of timber, canvas, cordage and other materials. Died Chichester, Eng., Feb. 16, 1861.

BURNEY, Leroy E., Am. physician; b. Burney, Ind., Dec. 31, 1906; B.S., Ind. U., 1928, M.D., 1930; M.P.H., Johns Hopkins, 1932; Sc.D., Jefferson Med. Coll., 1957, DePauw U., 1958, Ind. U., 1959, Woman's Med. Coll. Pa., 1960; LL.D., Seton Hall U., 1957; m. Mildred Hewins, Feb. 20, 1932; children—Robert G., Kay S. (Mrs. Rhett W. Butler). Regional officer, Kansas City, Mo., 1940-43; asst. surgeon gen., dep. chief Bur. State Service, Washington, 1954-56; surgeon gen., 1956-61; ret., 1961; v.p. for health scis. Temple U., Phila., 1961—. Diplomate Am. Bd. Preventive Medicine. Mem. A.C.P., A.M.A., Assn. Am. Med. Colls., Assn. Mil. Surgeons U. S., Coll. Physicians Phila., Nat. Sanitation Found., Philadelphia County Med. Soc. Research publs. on pub. health and health care adminstrn. Home: 901 Rock Creek Rd., Bryn Mawr, Pa. Office: 3420 N. Broad St., Phila.*

BURNHAM, Charles Russel, Am. cytogeneticist; b. Hebron, Wis., Jan. 13, 1904; s. Warren and Isabelle (Black) B.; student U. Minn., 1920-22; B.A., U. Wis., 1924, M.S., 1925, Ph.D., 1929; m. Mabel Lucile Strickland, July 18, 1936; children—Sarah Lucile (Mrs. David B. Mertz), Barbara Jean (Mrs. Arthur Lee). NRC fellow Cornell U., Bussey Instn. at Harvard, Cal. Inst. Tech., 1929-31; teaching fellow Cal. Inst. Tech., 1931-32; faculty W.Va. U., 1934-38; faculty U. Minn., St. Paul, 1938—, prof. cytogenetics, 1943—. Gosney fellow Cal. Inst. Tech., 1947-48. Mem. Genetics Soc. Am., Am. Genetics Assn., Am. Naturalists, Soc. for Study Evolution, Minn. Human Genetics League, Am. Soc. Agronomy. Author: Discussions in Cytogenetics, 1962; also articles. Research on use of chromosomal aberrations in studies of inheritance and chromosome behavior in crop plants, planning and synthesis of spl. stocks for these studies. Home: 1539 Branston St., St. Paul 55108.*

BURNHAM, Frederick Russell, Am. explorer; b. Tivoli, Minn., May 11, 1861; s. Edwin O. and Rebecca (Russell) B.; m. Blanche Blick, 1884 (dec. Dec. 1939); m. 2d, Ilo K. Willits, Oct. 28, 1943. Cowboy, scout, guide, miner, dep. sheriff in West; went to Africa, 1893; scout in Matabele War in Rhodesia; discovered in granite ruins of an ancient civilization of Rhodesia a buried treasure of gold and gold ornaments dating before Christian era; led expdn. to explore Barotzeland preparatory to bldg. of Cape to Cairo Ry.; took active part in 2d Matabele War on staff of Sir Frederick Carrington; commd. to capture or kill the Matabele God M'Limo and succeeded in entering his cave in Matopa Mts. and killing him; operated gold mines in Klondike and Alaska, 1898-1900; sent for by Lord Roberts, Jan. 1900, to go to S. Africa for service in the Boer War, made chief of scouts of British Army in field; and invalided home, 1901; made surveys of Volta River, W. Africa, 1902, exploring parts of French Nigeria Hinterland of Gold Coast Colony, and took active part in native troubles of that time; comd. an exploration of magnitude from Lake Rudolph to German E. Africa, covering a vast region along Congo basin and head of the Nile, 1903-04; discovered a lake of 49 sq. miles composed almost entirely of pure carbonate of soda of unknown depth; made archaeol. discovery of Maya civilization extending into Yaqui country, as shown by stone carvings and writings, 1908; engaged with John Hays Hammond in diverting Yaqui River through a system of canals into delta containing 700 sq. miles of land; v.p. Dominguez Oilfields Co. Author: Scouting on Two Continents; Taking Chances. Died Sept. 1, 1947; buried at Three Rivers, Cal.

BURNHAM, Sherburne Wesley, Am. astronomer; b. Thetford, Vt., Dec. 12, 1838; hon. A.M., Yale, 1878; Sc.D., Northwestern U., 1915. Took up study of astronomy as an amateur, and made many discoveries, especially of double stars, with a 6-inch refractor; observer in pvt. obs., Chgo., 1870-77, Dearborn Obs., Chgo., 1877-81, 82-84, Washburn Obs., Madison, Wis., 1881-82; astronomer Lick Obs., 1888-92; prof. practical astronomy and astronomer Yerkes Obs., U. Chgo., from 1893. Discovered 1,274 new double stars; expert commr. to test the seeing on Mt. Hamilton, Cal., resulting in locating the Lick Obs. there, 1879. Recipient Gold medal Royal Astron. Soc. (for discovery and measurement of double stars), 1894; Lalande prize in astronomy Paris Acad. Scis., 1904. Fellow, 1874, asso. 1898, Royal Astron. Soc.; asso. fellow Am. Acad. Arts and Scis. Author of Vol. I, publs. of Yerkes Obs. (gen. catalogue of stars discovered by him), 1900; also a gen. catalogue of all known double stars visible in Northern Hemisphere, for Carnegie Instn., Washington, 1907; Measures of Proper-Motion Stars, 1912. Died Chgo., Ill., Mar. 11, 1921.

BURNS, Allan, Scottish surgeon; b. Glasgow, Sept. 18, 1781; s. John Burns; in med. service with army, 1804; lectr. on anatomy and surgery, Glasgow. Author: Observations on Diseases of the Heart, 1809; Observations on the Surgical Anatomy of the Head and Neck, 1812. Pioneer in recognition of mitral stenosis, also described heart murmur mechanism, 1809; gave probably 1st description of chloroma (or chlorosarcoma, green cancer), also described fascia lata falciform process, 1811. Died 1813.

BURNS, B(enedict) Delisle, neurophysiologist; b. London, Eng., Feb. 22, 1915; s. Cecil Delisle and Margaret (Hannay) B.; student Tubingen (Germany) U., 1931-32; B.A., Cambridge U., 1936; M.R.C.S., L.R.C.P., U. Coll. Hosp., London, 1939; children—Martin, Nicholas, Julian, Gale; m. 2d, Monika Kasputis, Sept. 25, 1954; 1 dau., Ramune. Mem. research staff Med. Research Council Gt. Britain, London, 1945-50; faculty McGill U., Montreal, Que., Can., 1950-66, prof. physiology, 1958-66, chmn. dept., 1965-66; sci. adviser to dept. vets. affairs Queen Mary Hosp., Montreal, 1956-66; mem. visual panel Canadian Def. Research Bd., 1956-66; head, div. physiology and pharmacology Nat. Inst. for Med. Research, London, Eng., 1966—. Author: The Mammalian Cerebral Cortex, 1958; The Uncertain Nervous System, 1968. Home: Fir Island, Mill Hill. Office: Nat. Inst. for Med. Research, Mill Hill, London, N.W. 7, Eng.*

BURNS, Edward Leroy, Am. pathologist; b. Kansas City, Mo., June 28, 1905; s. Merle C. and Mary Louise (Arnold) B.; A.A., Kansas City Jr. Coll., 1924; M.D., Washington U., St. Louis, 1928; m. Eloise Simpson Garland, Dec. 8, 1934; children—Elizabeth, Thomas. Mem. faculty Washington U., St. Louis, 1928-30, 33-37, 44, La. State U. Sch. Medicine, 1937-45; vis. pathologist Path. Meml. Hosp., Fremont, O., 1948-57; head pathology Mercy Hosp., Toledo, 1945-65; dir. Inst. Med. Research, Toledo Hosp., 1965-67; pres. Cancer Cytology Research Fund, 1967—. Mem. Am. Soc. Clin. Pathologists (past pres.), Am. Assn. Pathologists and Bacteriologists, Am. Soc. Cytology, Am. Soc. Exptl. Pathology, Internat. Acad. Pathology, Coll. Am. Pathologists, Am. Cancer Soc., A.M.A., Soc. Exptl. Biology and Medicine, Am. Assn. for Cancer Research. Contbn. in field of cancer, producing cancers in mice and developing a community uterine cancer detection program by means of Papanicolaou tests; contbns. in field of edn. through initiation and devel. of continuing edn. program of Am. Soc. Clin. Pathologists. Home: 2630 Westchester Rd., Toledo 43615. Office: 2313 Madison Av., Toledo 43624.*

BURNS, James MacGregor, Am. polit. scientist; b. Melrose, Mass., Aug. 3, 1918; s. Robert Arthur and Mildred (Bunce) B.; B.A., Williams Coll., 1939; M.A., Ph.D., Harvard, 1947; postgrad. London (Eng.) Sch. Econs.; m. Janet Dismorr Thompson, Mar. 23, 1942; children—David MacGregor, Timothy Stewart, Deborah Edwards, Margaret. Faculty, Williams Coll., Williamstown, Mass., 1941—, now prof. polit. sci., James P. Baxter, II prof. history and publ. affairs. Exec. sec. nonferrous metals commn. Nat. War Labor Bd., 1942-43; staff Hoover Commn., Washington, 1948. Recipient Woodrow Wilson award, 1957. Mem. Phi Beta Kappa. Author: Congress on Trial, 1949; (with others) Government by the People, 1950; Roosevelt: The Lion and the Fox, 1956; Kennedy: a Political Profile, 1959; The Deadlock of Democracy: Four Party Politics in America, 1963; Presidential Government: The Crucible of Leadership, 1966. Home: High Mowing Bee Hill, Williamstown, Mass. 01267.*

BURNS, John, Scottish physician; b. Glasgow, Scotland, Nov. 12, 1775; s. John Burns; studied medicine Glasgow U.; surgeon's clk., Glasgow Infirmary, 1792; became lectr. on anatomy, 1792, later on midwifery; apptd. surgeon to Royal Infirmary, 1800; became Regius prof. surgery U. Glasgow, 1815. Fellow Royal Soc., 1830; mem. Inst. France. Author: Anatomy of the Gravid Uterus, 1799; Dissertation on Inflammation, 1800; Principles of Midwifery, 1809; Popular Directions for the Treatment of the Diseases of Women and Children; Principles of Surgery, 2 vols., 1828-38; Principles of Christian Philosophy, 1828, 6th edit., 1846. Died on board a steamer, nr. Portpatrick, June 18, 1850.

BURNS, John J., pharmacologist; b. Flushing, N.Y., Oct. 8, 1920; s. Thomas F. and Katherine (Kane) B.; B.S., Queens Coll., 1942; M.A., Columbia, 1948, Ph.D., 1950. With Lab. of Chem. Pharmacology, Nat. Heart Inst., Bethesda, Md., 1950-60, dept. chief, 1957-60; dir. research, pharmacodynamics div. Wellcome Research Labs., Burroughs Wellcome & Co., Inc., Tuckahoe, N.Y., 1960-65; v.p. for research Hoffmann-La-Roche Inc., Nutley, N.J., 1966—. Vis. prof. pharmacology Albert Einstein Coll. Medicine, 1961. Mem. Am. Soc. Pharmacology and Exptl. Therapeutics, Am. Soc. Biol. Chemists, N.Y. Acad. Scis. Numerous publs. on metabolism of drugs, vitamins and carbohydrates; elucidated pathway of ascorbic acidbiosynthesis in the rat; discovered missing enzymatic step in guinea pig, monkey and man which prevents synthesis of Vitamin C; metabolic data which led to introduction of new drugs for treatment of arthritis and gout; studies on drug interactions. Home: 415 E. 52d St., N.Y.C. 10022. Office: Kingsland Av., Nutley, N.J. 07110.*

BURNS, Kevin, physicist; b. Pleasant Ridge, N.B., Can., Mar. 1, 1881; s. John and Gertrude (Campbell) B.; brought to U. S., 1885; A.B., U. Minn., 1903, Ph.D., 1910; postgrad., Europe, 1911-12; D.Sc., St. Bonaventure Coll., 1947; m. Hazel Bunney, 1911 (dec. 1917); 1 son, Kevin; m. 2d, Ruth Buchanan, 1926; children—John Buchanan, George Campbell. Asst. at Lick Obs., Cal., 1904-07, U. Minn., 1907-10; Martin Kellogg fellow Lick Obs., residence in Europe, 1911-12; asst. physicist Bur. Standards, Washington, 1913-17, asso., 1917, physicist, 1917-19; astronomer Allegheny Obs., Pitts., 1920-51, asst. dir., 1930-51. Mem. Philos. Soc. Washington, A.A.A.S., Internat. Astron. Union, Sigma Xi. Specialist in spectroscopy, pioneer in spectrochem. analysis: measured standard wavelengths; determined stellar velocities and distances. Died Apr. 30, 1958.

BURNS, Robert Kyle, Am. embryologist, anatomist; b. Hillsboro, W.Va., July 26, 1896; s. William McLauren and Sarah Elizabeth (White) B.; A.B., Bridgewater Coll., 1916, Sc.D., 1953; student U. Va., summer 1915; Ph.D., Yale, 1924; m. Emily Lucile Moore, June 21, 1924; children—Robert Kyle, William Moore, John McLauren. Instructor biology, Bridgewater Coll., 1916-17; fellow in zoölogy, Yale, 1920-24; instructor zoölogy, University of Cincinnati, 1924-25, assistant professor, 1925-28; asst. prof. anatomy, Univ. of Rochester, 1928-30; asso. prof. 1930-40; staff member Carnegie Institution of Washington, Balt., 1940-62; interim prof. zoology Bridgewater (Va.) Coll., 1962—; lectr. zoology U. Cal., Santa Barbara, 1965; staff, biological station, U. of Va., 1940-46; hon. prof. zoology, Johns Hopkins University, 1945—; visiting professor U. Fla., 1955; exchange prof. Sorbonne, U. Paris, 1955-56. Pvt. USMC 1918-19. Mem. A.A.A.S., Nat. Acad. Scis., Internat. Inst. Embryology, Am. Soc. Zoölogists, Am. Assn. Anatomists, Am. Soc. Naturalists Soc. for Growth and Devel., Sigma Xi. Contbr. sci. jours., various books. Investigation of parabiosis and experimental transformation of sex in amphibians; interrelations of reproductive tract and hypophysis; growth and development after heteroplastic grafting; reproduction in natural population of opposum; sex reversal in gonads and genital tract of opposum embryos induced experimentally by hormones; pronephoric duct and development of mesonephros. Home: 102 N. 2d St., Bridgewater, Va.

BURNS, Robert Martin, Am. chemist; b. Liongmont, Colo., Jan. 9, 1890; s. Thomas and Mattie (Ash) B.; A.B., U. Colo., 1915, A.M., 1916, Sc.D. (hon.), 1945;

Ph.D., Princeton, 1921; m. Ada Kneale, Sept. 11, 1924; 1 dau. Madja (Mrs. John Hovey Gould. With Barrett Co., 1921-22, Western Electric Co., 1922-25; with Bell Telephone Labs., 1925-55, chem. dir., 1945-55; sci. adviser Stanford Research Inst., Menlo Park, Cal., 1955—. Cons. Sprague Electric Co., North Adams., Mass., 1955—. Recipient Perkin medal Soc. Chem. Ind., 1952, Whitney award Nat. Assn. Corrosion Engrs., 1953. Mem. Electrochem. Soc. (past pres., Acheson medal 1956), Am. Chem. Soc. (past chmn. N.Y. sect.), A.A.A.S., Sigma Xi. Author: (with W. W. Bradley) Protective Coatings for Metals, 3d edit., 1967; also numerous articles. Research on mechanism corrosion metals. Home: 1587 Cañada Lane, Woodside, Cal. 94062. Office: Stanford Research Inst., Menlo Park, Cal. 94025.*

BURNS, William George, Brit. chemist; b. Glasgow, Scotland, Nov. 18, 1925; s. William Millott and Isabella (Vine) B.; B.A., St. Catharine's Coll., Cambridge, Eng., 1946, Ph.D., 1949; m. Mary Pugh, July 6, 1963; children—Paul Justin, Sian Teresa. Postdoctoral research fellow National Research Council of Canada, Ottawa, Ontario, 1949-50; research chemist Hilger & Watts Ltd., London, Eng., 1950-54; Atomic Energy Research Establishment, Harwell, Berks, Eng., 1954—. Mem. Faraday Soc., Chem. Soc. Research, publs. on chem. effects following interaction radiation with chem. substances, including steps of chain reactions produced by ultraviolet radiation and effects different types of ionizing radiation on organic substances. Home: Fern Hill, 33 Lockstile Way, Goring, Reading, Berks. Office: Chemistry Div., Atomic Energy Research Establishment, Harwell, Berks., Eng.*

BURNSIDE, Orvin Charles, Am. agronomist; b. Hawley, Minn., June 9, 1932; s. John J. and Sena (Dwyre) B.; B.S., N.D. State U., 1954; M.S., U. Minn., 1958, Ph.D., 1959; m. Delores S. Schattschneider, Dec. 22, 1954; children—Bruce D., Kristi L. Prof. agronomy U. Neb., Lincoln, 1959—. Mem. Weed Sci. Soc. Am. Contbr.: Principles of Weed Control, 1967. Research and publs. on devel. of mech., cultural and chem. weed control systems for agronomic crops; persistence and dissipation of herbicides; determination of influence of environment on herbicide toxicity. Home: 6111 Lexington St., Lincoln, Neb. 68505.*

BURNSIDE, William, Brit. mathematician; b. London, Eng., July 2, 1852; s. William and Emma (Knight) B.; student St. John's, Cambridge, Eng., 1871-73; B.S., Pembroke Coll., 1875; Sc.D., Dublin, Ireland; LL.D., Edinburgh, Scotland; m. Alexandrina Urquhart, 1886; 2 sons, 3 daus. Fellow lectr. Pembroke Coll., 1875-86, named hon. fellow, 1900; prof. math. Royal Naval Coll., Greenwich, Eng., 1885-1919. Fellow Royal Soc., 1893 (council 1901-03, Royal medal 1904); mem. London Math. Soc. (pres. 1906-08, De Morgan medal 1899). Author: Theory of Groups, 1897; Theory of Probability, 1928; numerous papers. Worked on elliptic functions, also differential geometry, gen. theory of discontinuous groups of finite order, probability theory, theory of automorphic functions, hydrodynamics, theory of potential. Died West Wickham, Eng., Aug. 21, 1927.

BURNSTOCK, Geoffrey, physiologist; b. London, Eng., May 10, 1929; s. James Hyman and Nancy (Green) B.; B.Sc., King's Coll., U. London, 1953, Ph.D., 1957; M.Sc. (hon.), U. Melbourne (Australia), 1962; m. Nomi Hirschfeld, Apr. 9, 1957; children—Aviva, Tamara, Dina. Med. Research Council Research fellow Nat. Inst. for Med. Research, Millhill, London, 1956-57, dept. pharmacology Oxford (Eng.) U., 1957-59; Rockefeller Travelling fellow dept. physiology U. Ill., 1959; faculty U. Melbourne, 1959—, prof., chmn. dept. zoology, 1964—. Mem. Brit., Australian physiol. socs., European Soc. for Biochem. Pharmacology, Anat. Soc. Australia and New Zealand, Australian Soc. for Med. Research. Research, publs. on cellular structure and electrophysiology of smooth muscle and autonomic nervous system, mechanism of transmission from autonomic nerves to smooth muscle, evolution and devel. vertebrate autonomic nervous system. Home: 24 Redesdale Rd., Ivanhoe, Melbourne, Victoria, Australia.*

BURR, Arthur Albert, Am. engineer, educator; b. Manor, Sask., Can., Aug. 23, 1913; s. Charles A. and Mary (McCartney) B.; B.A., U. Sask., 1938, M.A., 1940; Ph.D. in Physics, Pa. State U., 1943; m. Leslie Mae Dickin, July 1, 1941; children—Janet, Leslie, Leonard Charles. Came to U. S., 1940, naturalized, 1951. Research physicist Armstrong Cork Co., Lancaster, Pa., 1943-46; faculty Rensselaer Poly. Inst., Troy, N.Y., 1946—, dean Sch. Engring., 1962—. Cons. to govt. Mem. Am. Soc. Engring. Edn., Am. Phys. Soc., Am. Soc. for Metals (Outstanding Tchrs. award, 1952), Soc. for Nondestructive Testing, Am. Inst. Mining, Metall. and Petroleum Engrs., Electrochem. Soc., N.Y. Acad. Scis., Sigma Xi. Research, publs. thermodynamics solids, alloys metallic systems. Home: 983 Spring Av., Troy, N.Y. 12180.*

BURR, C(olonel) B(ell), Am. physician; b. Lansing, Mich., Nov. 3, 1856; s. Allen R. and Catharine (Foote) B.; M.D., Coll. Phys. and Surg. (Columbia), 1878. Specialist in nervous and mental diseases; asst. phys. Eastern Mich. Asylum at Pontiac, 1878-85, asst. med. supt., 1885-89, med. supt., 1889-94; med. dir. Oak Grove Hosp. for nervous and mental diseases, Flint, Mich., 1894-1920. Fellow A.C.P. Author: Practical Psychology and Psychiatry, 1921. Compiler (for Mich. Med. Soc.) Medical History of Michigan, 1929. Died Apr. 11, 1931.

BURR, George Oswald, Am. biochemist; b. Conway, Ark., Oct. 6, 1896; s. George Hutchinson and Ruth (Herndon) B.; A.B., Hendrix Coll., 1916, LL.D., 1936; M.A., Ark. U., 1920; Ph.D., Minn. U., 1924; postgrad. U. Ill., U. Chgo.; m. Mildred Lawson, Aug. 22, 1925 (dec. 1962); m. 2d, Violet Karlstad, Aug. 23, 1964. Prin., Crossett Ark. High Sch., 1916-17; prof. Ky. Wesleyan Coll., 1917-18; research asso. U. Cal. at Berkeley, 1924-28; faculty U. Minn., 1928-46, prof., 1936-40, dir. div. physiol. chemistry, 1940-46; head dept. physiology and biochemistry Hawaii Sugar Expt. Sta., 1946-61, cons., 1961-65; research adviser Taiwan Sugar Corp., Tainan, 1965—. Mem. industry adv. com. Sugar Research Found., Inc., 1948-61; mem. NRC, 1944-48. Guggenheim fellow for European study, 1934; recipient Outstanding Achievement award U. Minn., 1955; Spl. award Am. Oil Chemists Soc. Mem. Am. Chem. Soc., Am. Soc. Biol. Chemists, Am. Inst. Nutrition, Soc. Plant Physiologists, Sigma Xi. Author: (with E. S. Miller) Plant Metabolism, 1934; also numerous articles. Editor: (with Rosendahl, Gortner) Physico-Chemical Properties of Plant Saps, 1934, J. Arthur Harris; Botanist and Biometrician, 1936. Asso. editor Jour. Nutrition, 1941-46, Jour. Phys. Chemistry, 1943-44, Archives Biochemistry, 1943-53. Co-discoverer polyunsaturate essential fatty acids; elucidated chem. nature vitamin E; developed new methods for study fats; applied radioisotopes to problems in field and factory; developed comml. control flowering in sugar fields; invented method for making drying oils for paint, gamma-ray scale used in sugar mills to weigh cane and bagasse, sunlight integrator, temperature integrator. Home: 112 Niuiki Circle, Honolulu 96821. Office: 1527 Keeaumoku, Honolulu 96822; also Expt. Sta., Taiwan Sugar Corp., Tainan, Taiwan.*

BURR, William Hubert, Am. civil engr.; b. Watertown, Conn., July 14, 1851; s. George William and Marion Foote (Scovill) B.; C.E., Rensselaer Poly. Inst., 1872; m. Caroline Kent Seelye, 1876 (dec. 1894); children—Mrs. Marion Elisabeth Mars, William Fairfield, George Lindsley; m. 2d, Gertrude Gold Shipman, 1900; 1 dau., Mrs. Anne Louisa Colgate. Began practice as civ. engr., 1872; prof. rational and tech. mechanics Rensselaer Poly. Inst., 1876-84; asst. to chief engr., later gen. mgr. Phoenix Bridge Co., 1884-91; prof. engring. Harvard, 1892-93; prof. civil engring. Columbia, 1893-1916, emeritus; civil engr. and cons. engr., N.Y.C., 1916—. Consulting engr. to dept. pub. works, 1893-95, parks, 1895-97, of docks, 1895-97, and then dept. of bridges and bd. of water supply (all N.Y.); mem. bd. engrs. to investigate feasibility of proposed bridge across North River, 1894; mem. bd. to locate deep water harbor on coast of southern Cal., 1896; mem. Isthmian Canal Commn. to examine and report on route for interoceanic canal across Central Am. Isthmus, 1902; mem. Isthmian Canal Commn., bd. cons. engrs., 1905—; mem. adv. bd. of engrs. for constrn. Barge Canal by State N.Y., 1911; mem. bd. cons. engrs. by commns. states of N.Y. and N.J. for constructing vehicular tunnel under Hudson river at N.Y.C., 1919; cons. engr. N.Y. State Transit Commn., 1923-24, Port N.Y. Authority, 1925—, for constrn. of Ft. Washington Suspension Bridge across Hudson River, and for other bridges being built for Port. Awarded 1st place in national competition for proposed memorial bridge across Potomac at Washington, 1900. Fellow Am. Acad. Arts and Sciences. Author: The Stresses in Bridge and Roof Trusses, 1881; Elasticity and Resistance of the Materials of Engineering, 1883; Ancient and Modern Engineering and the Isthmian Canal, 1902; (with M. S. Falk) The Graphic Method by Influence Lines for Bridge and Roof Computation, 1905, The Design and Construction of Metallic Bridges, 1912; Suspension Bridges, Arch Ribs and Cantilevers, 1913. Died Dec. 13, 1934.

BURRAGE, Walter Lincoln, Am. physician; b. Boston, Oct. 21, 1860; s. Alvah Augustus and Elizabeth Amelia (Smith) B.; A.B., Harvard, 1883, A.M., 1888; M.D., Harvard Med. Sch., 1888; m. Sally Swan, Oct. 3, 1894. House physician Boston City Hosp., 1886-88, Woman's Hosp., N.Y., 1888-90; vis. gynecologist Carney, St. Elizabeth hosps., Boston, 1890-1903; surgeon to out-patients Free Hosp. for Women, 1890-1901; clin. instr. in gynecology Harvard, 1893-95. Hon. fellow Am. Gynecol. Soc. Author: Gynecological Diagnosis, 1910; A History of the Massachusetts Medical Society, 1781-1922, 1923. Collaborator: Dr. Howard A. Kelly's Am. Medical Biography, 2 vols., 1912, also rewrote Kelly's Appendicitis, and assisted with his Medical Gynecology, 1910, Biographies, 1920, Dictionary of American Medical Biography (2049 biographies), 1928; Medicine in Massachusetts, 1930; Catalogue of Honorary, Past and Present fellows, Massachusetts Medical Society, 1781-1931 (11,126 names) 1931. Died Jan. 26. 1935.

BURRELL, Edward Parker, Am. inventor, mech. engr.; b. Hall, N.Y., Feb. 11, 1871; s. Edward and Elizabeth (Parker) B.; M.E., Cornell U., 1898, M.M.E., 1899; m. Katharine Ward, Dec. 8, 1904. Successively designing engr., works engr., works mgr., Warner and Swasey Co., 1900-24, dir. engring., 1924—; directed design of large telescopes built by this co., including 72-inch reflecting telescope for Dominion Astrophys. Obs., Victoria, B.C., 69-inch reflecting telescope for Ohio Wesleyan U., 20-inch refractor for Chabot Obs., Oakland, Calif. Designed and constructed model of proposed 200-inch telescope for Mt. Wilson Obs.; patentee many inventions. Died Mar. 21, 1937.

BURRELL, George Arthur, Am. chemist; b. Cleve., Jan. 23, 1882; s. Alexander A. and Jane (Penny) B.; student Ohio State U., 1902-04, Chem. E., 1918; Sc.D., Wesleyan U., 1919; m. Mary L. Schafer, 1906; 1 dau., Dorothy May; m. 2d, Naomi L. Schafer, June 16, 1914. Chemist U. S. Geol. Survey, 1904-08; in charge research work, gas mine gas, and natural gas and gasoline investigations U. S. Bur. Mines, Pitts., 1908-16; cons. engr. petroleum and natural gas work, 1916-43; asst. to dir. Bur. Mines, 1917; located supply of helium gas in Tex. and initiated govt. helium program in charge constrn. of refineries for Island and Raritan Refining Cos., N.Y.C., 1919-20; v.p., pres. Island Refining Co.; gen. mgr. Raritan Refining Co.; pres. Burrell Corp., 1923-52, became chmn. bd.; pres. Atlantic States Gas Co., 1936-54; v.p. Commonwealth Gas Corp., 1942-54; retained by Russian govt. to modernize natural gas industry, 1930-31. Recipient Lamme medal for achievements in engring. Ohio State U., 1935; Hanlon award, Nat. Gasoline Assn. Am., 1948. Mem. Am. Petroleum Inst., Am. Chem. Soc., Am. Inst. Chem. Engrs., Am. Inst. Chemists, Sigma Xi. Author: Handbook of Gasoline, 1917; Recovery of Gasoline from Natural Gas, 1925; An American Engineer Looks at Russia, 1932; and also many papers and govt. publs. on gas, gasoline, petroleum and allied subjects. Inventor Burrell gas detector, Burrell gas analysis apparatus; co-inventor Burrell-Oberfell process of extracting gasoline from natural gas by charcoal methods; designed and built many natural gas refineries. Died Aug. 16, 1957.

BURRELL, H(erbert) Cayford, Am. geologist; b. Boston, Dec. 26, 1903; s. Herbert Leslie and Caroline White (Cayford) B.; student Middlesex Sch., Concord, Mass., 1915-21; S.B., Harvard, 1928, A.M., 1920, Ph.D., 1946; postgrad. U. Wis., 1932-33; m. Mary Josephine Runkel, July 25, 1929; children—Frederick R., Patricia R. Geologist Cerro de Pasco Copper Corp., Morococha, Peru, 1929-32; 1st resident geologist Zinc Corp., Ltd., Broken Hill, N.S.W., Australia, 1934-37; with Central Geog. Survey, 1937-39; lab. research on Broken Hill ores Harvard, 1939-42; with coordinator Interam. Affairs, Bd. Econ. Warfare, and Fgn. Econ. Adminstrn. in connection with devel. operations and purchase manganese ores for U. S. Govt., 1942-45; staff geologist Oliver Iron Mining Co. (U. S. Steel Corp. subsidiary) in cos. exploration for iron ore in Venezuela, 1946-49; trans. to raw materials dept. U. S. Steel Corp. of Del., Pitts., 1949, served as geologist; mgr. raw materials devel. Columbia-Geneva Steel div. U. S. Steel. Mem. Am. Inst. Mining and Metall. Engrs., Mining and Metall. Soc. Am. Author: Geology of the Broken Hill Ore Deposit, Broken Hill, N.S.W., Australia (with J. K. Gustafson, M. D. Garretty), 1950. Died Nov. 9, 1953; buried Salem, Mass.

BURRETT, Claude Adelbert, Am. physician; b. Monroe County, N.Y., July 13, 1878; s. Cyrus A. and Ida I. (Sage) B.; Ph.B., Syracuse U., 1902; postgrad., 1902-03; M.D., Cleve. Homeo. Med. Coll., 1905; m. Clara Virginia Partridge, July 13, 1905; children—Adelbert Partridge, Helen Louise, John Barton, Virginia. Dir. pathogenic lab. and instr. in toxicology U. Mich., 1905-08, asst. prof. genito-urinary surgery, dermatology and electrotherapeutics, 1908-13, prof. surgery and genito-urinary surgery, 1913-14; prof. surgery, acting dean Coll. of Homeo. Medicine, Ohio State U., 1914-15, dean coll., 1915-22; dean, med. dir. N.Y. Med. Coll., Flower and Fifth Av. hosps., 1925—, pres., 1939—. Fellow A.C.S. Died Mar. 3, 1941.

BURRI, Conrad Robert, Swiss petrologist; b. Zurich, Switzerland, May 22, 1900; s. Robert Josef and Emilie (Walser) B.; student Swiss Fed. Inst. Tech., U. Berne; Ph.D., U. Zürich, 1926; hon. D.Sc., U. Madrid (Spain), 1967; m. Mariette Kircheisen, June 3, 1930. Asst., U. Freiburg i. Br., 1926-29; joined faculty Swiss Fed. Inst. Tech., 1929, became prof. petrology, 1954; now prof. petrology U. Zurich. Pres., Found. Vulkaninstitut Immanuel Friedlaender, Zurich. Fellow Mineral. Soc. Gt. Britain, Mineral. Soc. Am.; hon. mem. Real Sociedad de Historia Natural (Madrid); corr. mem. Instituto Lucas Mallada Investigaciones geológicas (Madrid), Soc. géol. de Belgique, acads. of Padova, Naples. Author: (with P. Niggli) Die junge Eruptivgesteine des mediterranen Orogens, 2 vols., 1945; Das Polarisationsmikroskop, 1950; Petrochemische Berechnungsmethoden, 1959; (with R. L. Parker, E. Wenk) Die optische Orientierung der Plagioklase, 1967. Research on petrology especially chem., volcanic rocks, optics of plagioclases. Home: 15 Eichhalde 8053, Zurich, Switzerland.*

BURRIEL MARTI, Fernando, Spanish analytical chemist; b. Valencia, Spain, May 29, 1905; s. Juan and Teresa M.; ed. Valencia, Madrid, Brussels univs.; Ph.D. in chemistry; m. Puana Maria Barcelo, May 24, 1934; children—Fernando, Maria, Josefa, Juan Manuel, Maria Teresa. Asst. prof. U. Madrid, later prof., from 1945; prof. chemistry Superior Tech. Sch. for Engring. of Telecommunications; prof. analytical chemistry U. Grenad. Mem. Midland Soc. Gt. Britain (hon.), Soc. Analysts London. Author: Quimica Analitica Cualitativa; Flame Photometry. Home: calle Isaac Peral, 1 6°, Madrid 15. Office: Ciudad universitaria, Madrid, Spain.

BURRILL, Meredith Frederic, Am. geographer; b. Houlton, Me., Dec. 23, 1902; s. Fred Wilson and Carrie Louise (Odiorne) B.; A.B., Bates Coll., 1925; M.A., Clark U., 1926, Ph.D., 1930; D.Sc., Bates Coll, 1960; m. Sarah Bannister, May 30, 1927; children—Robert Meredith, Elizabeth Ellen (Mrs. David Henry Allard). Instr. geography schs. and colleges, 1926-30; asst. prof. geography, Okla. A.&M. Coll., 1930-31, asso. prof., 1931-37, prof., 1937-40; research adviser McGill U., 1931; unit head and acting regional chief Land Use Planning Sect., Region 8, Resettlement Adminstrn., 1935-37; econ. geographer Gen. Land Office, Dept. of Interior, 1940-42, chief Div. of Research and Analysis, same, 1942-43; dir. U. S. Bd. on Geog. Names, 1943-48, exec. sec., also dir. Office of Geography, Dept. of Interior, 1948—. Chmn. Nat. Research Council com. on Social Geography, 1948-51, vice chmn. adv. com. to office Naval Research, geog. br., 1949-52. Del. internat. geog. congresses, 1949, 52, 60, 64; chmn. UN Group of Experts on Geog. Names, 1960, 66. Member board of trustees Bates College, 1966—. Decorated Order of Alfonso X, El Sabio; recipient So. Regional Grant-in-Aid, Social Sci. Research Council, 1932. Fellow Royal Geog. Soc.; corr. mem. Geog. Soc. of Lima; mem. Assn. of Am. Geographers (pres. 1966), Am. Name Soc. (pres. 1955), Am. Dialect Soc., Am. Geog. Soc., A.A.A.S., Am. Geophys. Union, Phi Beta Kappa. Author: A Socio-Economic Atlas of Okla., 1936; Water for Industry, 1956. Research on and official standardization of geographic names in U. S.; work on international cooperation in this field; work in development of toponymy and geography. Home: 5503 Grove St., Chevy Chase, Md. 20015. Office: Dept. Interior, Washington.*

BURRIS, Robert H(arza), Am. biochemist; b. Brookings, S.D., Apr. 13, 1914; s. Edward Thomas and Mable (Harza) B.; B.A., S.D. State Coll., 1936; D.Sc., 1966; M.S., U. Wis., 1938, Ph.D., 1940; m. Katherine Irene Brusse, Sept. 12, 1945; children—Jean Carol, John Edward, Ellen Louise. NRC fellow Columbia, 1940-41; faculty U. Wis., Madison, 1941—, prof. biochemistry, 1951—, chmn. dept., 1958—. U. S. State dept. cons. in India, 1961; cons. NSF, 1953-57, NIH, 1961—. Guggenheim fellow, Helsinki, Finland, Cambridge, Eng., 1954. Mem. Nat. Acad. Sci. (mem. subcom. on sci. and pub. policy 1964—), Am. Soc. Microbiologists, Am. Soc. Biol. Chemists, Soc. Gen. Physiologists, Am. Chem. Soc., A.A.A.S., Am. Inst. Biol. Scis., Biochem. Soc. (Eng.), Scandinavian, Am. (past pres.), socs. plant physiologists. Author: (with W. W. Umbreit, J. F. Stauffer) Manometric Methods, 1945; also numerous articles. Research on plant and bacterial respiration and metabolism nitrogen and organic acids; pioneered applications stable isotope N15 to studies biol. nitrogen fixation showing ammonia is an early product nitrogen fixation and compound which reacts to form organic nitrogenous compounds in nitrogen fixing organisms. Home: 1015 University Bay Dr., Madison, Wis. 53705.*

BURROUGHS, William Seward, Am. inventor; b. Auburn, N.Y., Jan. 28, 1855; s. Edmund and Ellen Burroughs; m. Ida Selover, 1879; children—Jennie, Horace, Mortimer, Helen. Worker in father's shop making models for castings and new inventions, St. Louis, 1881; employed by Future Great Mgf. Co., St. Louis, 1881-84; invented machine to solve arithmetical problems, 1844-45 (not commercially practical); organized Am. Arithmometer Co. to produce machines for solving arithmetical problems, St. Louis, 1885; granted patent for 1st practical machine, 1892; awarded John Scott medal of Franklin Inst. for his invention, 1897. Died Citronelle, Ala., Sept. 15, 1898.

BURROUGHS, Wise, Am. animal scientist; b. Tipton, Ia., Dec. 19, 1911; s. Ernest Otho and Flora (Stout) B.; student Blackburn Coll., 1930-32; B.S., U. Ill., 1934, Ph.D., 1939; m. Helen Stevenson, Aug. 21, 1937; children—Phyllis M., Charles W., John R. Asst., Ohio Agr. Expt. Sta., 1939-45; asso. prof. Ohio State U., 1945-50; prof. animal sci. Ia. State U., Ames, 1951—. Recipient Am. Feed Mfrs. award, 1954; John Scott medal award City of Phila., 1958; Ford Found. Efficiency Agrl. award, 1960. Mem. Am. Soc. Animal Sci. (Morrison award 1966), Am. Soc. Exptl. Biology and Medicine, Sigma Xi, Phi Kappa Phi, Alpha Zeta, Gamma Sigma Delta. Author: Physiology of Digestion in the Ruminant, 1965; also numerous articles. Research in digestive processes in ruminants, stilbestrol feeding to beef cattle, urea addition to beef cattle rations. Patentee stilbestrol feeding to beef cattle. Home: 3003 West St., Ames, Ia. 50010.*

BURROW, Trigant, Am. phylobiologist, psychiatrist; b. Norfolk, Va., Sept. 7, 1875; s. John W. and Anastasia (Devereux) B.; A.B., Fordham, 1895; M.D., U. Va., 1899, Ph.D., Johns Hopkins, 1909; postgrad. U. Va., Munich, Vienna, Johns Hopkins and Zurich, 1900-10; m. Emily Sherwood Bryan, Aug. 9, 1904; children—John D., Emily Sherwood (Mrs. Hans Syz). Demonstrator in biology U. Va., 1899-1900; asst. physician U. Frauenklinik, Munich, 1900; asst. in exptl. psychology Johns Hopkins, 1906-09, in clin. psychiatry Johns Hopkins Hosp., 1911-27; practice and research in psychiatry and psychoanalysis, 1911-23, social psychiatry and the group method of analysis, 1923-28; sci. dir. Lifwynn Found. for Lab. Research in Analytic and Social Psychiatry, Westport, Conn., 1927—. Participant 2d Internat. Symposium on Feelings and Emotions. Moosehart, Ill., 1948. Mem. A.M.A., A.A.A.S., Med. and Chirurg. Faculty of Md., Am. Psychopath. Assn., Am. Psychiatric Assn., Am. Psychol. Assn., Am. Anthropol. Soc., Human Genetics Soc. Am., N.Y. Acad. Scis., So. Soc. for Philosophy and Psychology, Phi Beta Kappa, Phi Delta Theta. Author: The Social Basis of Consciousness, 1927; The Structure of Insanity, 1932; The Biology of Human Conflict, 1937; The Neurosis of Man—An Introduction to a Science of Human Behavior, 1949; Science and Man's Behavior, The Contribution of Phylobiology, 1953. A Search for Man's Sanity, The Selected Letters of Trigant Burrow, 1958; also articles in field of medicine, exptl. psychology, psychoanalysis, individual and social psychiatry and phylopathology. Research in phylopathology, or in modifications of behavior induced through adjusting organism's internal tensional pattern, also in instrumental recording of these physiol. changes, from 1928. Died May 24, 1950.

BURROWS, Charles R., Am. engineer; b. Detroit, Mich., June 21, 1902; s. Charles W. and Maud F. (Murdock) B.; B.S.E., U. Michigan, 1924; A.M., Columbia U., 1927; E.E., Michigan, 1935; Ph.D., Columbia, 1938; m. Lola May Schwingel (d. 1964); m. 2d. Marcelle, Sept. 10, 1966; children: C. Robert, Donna M. Mem. Technical Staff, Bell Telephone Labs, New York, 1924-45; prof., dir., School of Electrical Engineering, Cornell U., 1945-56; assoc. chief scientist, G.E. Advanced Elec. Center, Cornell; senior scientist, Radio Engineering Lab., Dynamics Corp. Vice-pres. Ford Instrument, Div. of Sperry-Rand; Radiation Inc.; Page Communication Engineers; Datronics Engineers Inc. Fellow, Institute of Electrical & Electronic Engineers; Sigma Xi; Internat. Council Scientific Unions; Internat. Scientific Radio Union; Am. Physical Soc.; Am. Astronomical Soc.; Am. Institute of Astronautics and Aeronautics; Am. Geophysical Union. Awarded Presidential Certificate of Merit, 1948. Author numerous publications. Specialist in radio wave propagation; research on long and short wave transatlantic transmission; proximity fuse; ultrashort wave propagation; radio astronomy, ionosphere, vacuum tubes. Home: 4721 Old Dominion Dr., Arlington, Va. 22207. Office: 725 23rd St., N.W., Washington, D.C., 20006.

BURROWS, Sir George, Brit. physician; b. London, Eng., Nov. 28, 1801; s. George Man Burrows; B.A., Ealing and Caius Coll., Cambridge; 1825, M.B., 1826, M.L., 1829, M.D., 1831; LL.D., 1881; D.C.L., Oxford, 1872; at least 1 son, F. A. Burrows. Jr. fellow Ealing and Caius Coll., 1825-35, hon. fellow, 1880, coll. rep. Gen. Med. Council, 1860; became joint lectr. med. jurisprudence St. Bartholomew's Hosp., 1873, sole lectr., 1834, joint lectr. medicine, 1836, sole lectr., 1841, 1st asst. physician, 1834, physician, 1841-63; physician Christ's Hosp., 1860-71; named physician extraordinary to queen, 1870, physician in ordinary, 1873-87. Licentiate, fellow Royal Coll. Physicians (Gulstonian lectr., 1834, Croonian lectr., 1835-36, Lumleian lectr., 1843-44, censor, 1839, 40, 43, 46, councillor, 15 years, treas., 1860-63, pres. 1871-76); fellow Royal Soc.; pres. Brit. Med. Assn., 1862, Royal Med. and Chirurg. Soc., 1869-71. Author: On Disorders of the Cerebral Circulation, 1846; On the Connection Between Affections of the Brain and Diseases of the Heart, 1846. Died London, Dec. 12, 1887.

BURROWS, Montrose Thomas, Am. surgeon; b. Halstead, Kan., Oct. 31, 1884; s. Thomas Forbes and Carolina Melvina (Richards) b.; A.B., U. Kan., 1905; M.D., Johns Hopkins, 1909; m. Flora Barbara Hege, Sept. 4, 1918; children—Bette Burrows Tanner, Helen Eugenia Ferrey, Zelta Reynolds, Loy Montrose. Fellow and asst. Rockefeller Inst. Med. Research, 1909-11; instr. anatomy, Cornell U. Med. Sch., 1911-15; asso. in pathology and resident pathologist Johns Hopkins Med. Sch. and Hosp., 1915-17; acting prof. pathology Washington U., 1917-20; asso. prof. surgery, same, and dir. research labs. Barnard Free Skin and Cancer Hosp., St. Louis, Mo., 1920-28; later specialized in cancer treatment and research, Pasadena, Cal. Mem. A.M.A., A.A.A.S., Am. Assn. Anatomists, Am. Soc. Exptl. Pathology, Soc. Exptl. Biology and Medicine, Sigma Xi. Contbr. to biol. revs. Research in tissue culture, vitamin theory of cancer, heart muscle contraction poliomyelitis, cancer, focal infections; devised (with Alexis Carrel) method for cultivating tissue in vitro, 1910; introduced term, tissue culture. Died Aug. 21, 1947.

BURROWS, William, Am. microbiologist; b. New Haven, Mar. 6, 1908; s. William and Winnefred E. (Johnson) B.; B.S., Purdue U., 1928; M.S., U. Ill., 1930; Ph.D., U. Chgo., 1932; postgrad. (fellow) Johns Hopkins; m. Margaret Pound., June 24, 1931; children—Mary (Mrs. Carl Freeland Eveleigh). Research asso. U. Chgo., 1932-35, faculty, 1937—, prof. microbiology, 1947—; Gen. Edn. Bd. fellow, 1935-37. Responsible investigator OSRD, 1942-46, NIH, 1946-52, 55—, Commn. on enteric infections Armed Forces Epidemiological Bd., 1962—; cons. to pvt. labs., govt. agys. Fellow Inst. Medicine Chgo., Am. Pub. Health Assn., A.A.A.S., American Academy of Microbiology; member American Society Microbiologists, Soc. for Exptl. Biology and Medicine, Am. Assn. Immunologists, Ill. Soc. for Microbiology (past pres.), Sigma Xi. Author: Textbook of Microbiology, 19th edit., 1968; also numerous articles. Editorial bd. Jour. Infectious Diseases, 1937—, Dorland Med. Dictionary, 1960. Research on immunity Asiatic cholera, antigenic analysis cholera vibroios, nature cholera toxin, quantitative assay cholera toxin and antitoxin; described copronatibody; developed cholera vaccines. Home: 5805 Dorchester Av., Chgo. 60637.*

BURRUS, John T., Am. surgeon; b. Surry County, N.C., July 13, 1876; s. John G. and Bettie (Reece) B.; M.D., Davidson Coll., 1899, Balt. Med. Coll., 1900, Grant U., 1901; grad. N.Y. Skin and Cancer Hosp.; postgrad. N.Y. Polyclinic, N.Y. Post-Grad. and Univ. Coll. of Medicine, also abroad; m. Mary B. Atkins, Apr. 2, 1899; 1 dau., Iris. Began practice at Jonesville, 1899; moved to High Point, 1904; owner and surgeon High Point Hosp.; vis. surgeon Western N.C. Hosp.; chief surgeon Carolina & Yadkin River R.R.; surgeon Thomasville Bapt. Orphanage; vis. lectr. in clin. surgery Duke U. Pres. N.C. State Bd. Health. Fellow A.C.S. Died June 8, 1938.

BURSTEIN, Elias, Am. physicist; b. N.Y.C., Sept. 30, 1917; s. Samuel and Sarah (Plotkin) B.; A.B. Bklyn. Coll., 1938; M.A., Kan. U., 1941; postgrad. Mass. Inst. Tech., Cath. U.; m. Rena B. Benson, Sept. 19, 1943; children—Joanna Bliss, Sandra Joy, Miriam Stephanie. Asst. instr. chemistry dept. U. Kan., 1939-41; research asst. chemistry dept. Mass. Inst. Tech., 1941-43, research asso., 1943-44; projects engr. White Research Assos., Boston, 1944-45; with USN Research Lab., Washington, 1945-58, physicist 1945-48, head physics sect., crystal br., 1948-58, head semiconductor br., 1958; prof. physics U. Pa., Phila., 1958—. Asso. mem. working group on semicondr. devices of adv. group on electron tubes Dept. of Def., 1955-58; Navy rep. panel on semicondr. materials Materials Adv. Bd., NRC-Nat. Acad. Scis., 1956, solid state adv. panel, 1956—; mem. Office Naval Research Com. on thermoelectric devices, 1958, materials com. Air Force Study Group, 1958; cons. Philco Applied Research Lab., Blue Bell, Pa., 1962—, Libbey Owens Ford Glass Research Lab., Toledo, 1964—, semicondr. br. USN Research Lab. 1960—; sec. bd. editors Solid State Communications, 1963—. Recipient Wash. Acad. Scis. ann. phys. scis. award, 1956, Brooklyn Coll. Chemistry dept. gold medal, 1958, Navy civilian meritorious service award, 1957, Research Soc. Am. pure sci. award, 1958. Fellow Am. Phys. Soc. (sec.-treas. solid state physics div. 1956-61), Optical Soc. Am.; mem. Phi Beta Kappa. Discovered infrared photo conductivity in semicondrs containing impurities, and excitation and photo ionization spectra of semiconductors containing impurities, 1950, cyclotron resonance in semicondrs. at infrared frequencies, 1957, Theory of Lattice Vibration infrared spectra of homopolar crystals, 1955, interband magneto-optical effects in semicondrs. 1957, infrared detector based on supercondr. tunneling junctions, 1962. Home: 729 Arlington Rd., Narberth, Pa. 19072.*

BURSTYN, Gunther, Austrian mil. engr.; b. Aussee, Austria, July 6, 1879; s. Adolf and Juliane (Hoffmann) B.; ed. mil. engring. sch. Hainburg, Austria, 1895-99; Ph.D. (hon.), Technische Hochschule, Vienna, Austria, 1944; m. Gabriele Wagner, 1910; children—1 son, 1 dau. Officer, C.E., charge bldg., reconstructing railroads and bridges, World War I; engring. adviser Austrian Ministry Mil. Affairs, 1926-33. 1st to conceive idea of tank for mil. purposes, circa 1912 (ideas not implemented); inventor collapsible tank block, clinometer, 1920.

BURT, Cyril Lodowic, Brit. psychologist; b. London, Eng., Mar. 3, 1883; s. Cecil Barrow and Martha (Evans) B.; student Jesus Coll., Oxford, Eng., 1902-06, M.A., D.Sc.; postgrad. Würzburg (Germany) U.; LL.D., Aberdeen (Scotland); Litt.D., U. Reading (Eng.); m. Joyce Muriel Woods, May 13, 1931. Lectr. exptl. psychology U. Liverpool (Eng.), 1908-13; lectr. psychology U. Cambridge (Eng.), 1913-15; psychologists London County Council, 1913-30; prof. edn. U. London, 1924-31, prof. psychology, 1931-50. Mem. various govt. coms. on edn., indsl. fatigue, selection mil. personnel. Hon. fellow Jesus Coll., Oxford; created knight. Fellow Brit. Acad., Brit. Psychol. Soc. Author: Distribution of Educational Abilities, 1917; Mental and Scholastic Tests, 1921; The Young Delinquent, 1925; The Subnormal Mind, 1935; How the Mind Works, 1933; The Backward Child, 1937; The Factors of the Mind, 1940; The Causes and Treatment of Backwardness, 1952; The Psychological Study of Typography, 1959. Research on psychol. tests, factorial techniques, gifted child. Home: 9, Elsworthy Rd., London N.W.3, Eng.*

BURT, Edward Angus, Am. botanist; b. Athens, Pa., Apr. 9, 1859; s. Howard Fuller and Miranda (Forsyth) B.; grad. N.Y. State Normal Sch., Albany, 1881; A.B., Harvard, 1893; A.M., 1894, Ph.D., 1895; m. Clara M. Briggs. Aug. 21, 1884; children—Angus Edward, Albert Forsyth, Farlow, Howard. Teacher, Albany Boys' Acad., 1880-85; prof. natural sci. State Normal Sch., Albany, 1885-91; prof. botany Middlebury (Vt.) Coll., 1895-1913; asso. prof. botany Washington U., 1913-18, prof., 1918-25; librarian and mycologist Mo. Bot. Garden, 1913-25. Died Apr. 27, 1939.

BURT, Horace Greeley, Am. engr.; b. Jan. 1849; C.E., Univ. of Mich., 1872. Began ry. service, 1868; resident engr. C. & N.W. Ry., 1873-81, div. supt., 1881-87, chief engr., 1887-1888, 3d v.p., 1896-97; gen. mgr. Fremont, Elkhorn & Mo. Valley and Sioux R.R. cos., 1888-96, C., St.P., M.&O. Ry., 1896; pres. U.P. R.R., 1898-1904; traveled around world, 1904-05; cons. engr., 1905-09; receiver C.G.W. Ry., 1909; cons. engr., 1909-11; chief engr. of com. of investigation smoke abatement and electrification of ry. terminals, 1911—. Died May 19, 1913.

BURT, John, Am. inventor; b. Wales, N.Y., Apr. 18, 1814; s. William and Phoebe (Coles) B.; m. Julia Calkins, Dec. 3, 1835, 3 children. Dep. surveyor Mich., 1841; began constrn. railroad from Marquette to Lake Superior, completed in 1857; 1st supt. Saulte St. Marie Canal; devised number of improvements for manufacture of pig and wrought iron involving methods of carbonization (patented 1869); patented type of canal lock 1867, put into use, 1881; pres. Lake Superior & Peninsula Iron Co., Burt Freestone Co. Died Detroit, Aug. 16, 1886.

BURT, Stephen Smith, Am. physician; b. Oneida, N.Y., Nov. 1, 1850; s. Oliver T. and Rebecca (Johnston) B.; student Cornell U., 1869-70; M.D., Coll. Phys. and Surg., Columbia, 1875, Roosevelt Hosp., 1877; A.M. (hon.); Yale, 1890. Instr. medicine and phys. diagnosis, 1882-84 N.Y. Post-Grad. Med. Sch. and Hosp., prof. medicine, 1884-1908, also mem. corp.; prof. thoracic diseases U. Vt., 1884-85; attending phys. N.Y. Post-Grad. Hosp. Author: Exploration of the chest in Health and Disease, 1889. Died Mar. 26, 1932.

BURT, Wayne Vincent, oceanographer; b. South Shore, S.D., May 10, 1917; s. John David and Mary (McDuffee) B.; B.S., Pacific Coll., 1939; certificate in meteorology U. S. Naval Acad., 1944; M.S., Scripps Instn. Oceanography U. Cal. at Los Angeles, 1948, Ph.D., 1952; Sc.D., George Fox Coll., 1963; m. Grace Louise DuBois, Jan. 15, 1941; children—John Alan, Christine Louise (Mrs. James F. Young), Laurence Wayne, Darcy Jean. Material engr. Kaiser Co., Inc., Washington, 1942; instr. U. Ore., 1946; asst. Scripps Instn. Oceanography, 1946-48; asso. oceanographer, 1948-49; asst. prof. oceanography, research oceanographer Chesapeake Bay Inst., Johns Hopkins, 1949-53, asst. dir., 1953; research oceanographer U. Wash., Seattle, 1953-54; asso. prof. Ore. State U., 1954-59, prof., chmn. oceanographic dept., 1959—, dir. Marine Sci. Center, 1964—. Mem. com. on oceanography Nat. Acad. Sci., 1965—. Mem. Marine Tech. Soc. (dir. 1965—), Am. Geophys. Union (past pres. oceanographic sect.), Am. Soc. Limnology and Oceanography (past pres. Pacific sect.), Am. Meteorology Soc. Marine Biol. Assn. U.K. Contbr. articles to tech. jours. Research on transmission light in natural waters, air-sea interaction, phys. oceanography. Home: 1615 Hillcrest St., Corvallis, Ore. 97330.*

BURT, William Austin, Am. inventor; b. Worcester, Mass., June 13, 1792; s. Alvin and Wealthy (Austin) B.; m. Phoebe Cole 1813; 5 children, including John. Served as justice of peace, postmaster, county surveyor Detroit; invented the typographer (predecessor of typewriter), patented, 1829; elected surveyor Macomb County, 1841; apptd. dist. surveyor in Mich; asso. judge Mich. Circuit Ct., 1833; postmaster Mt. Vernon (Mich.); U. S. dep. surveyor Washington D.C. with Gen. Land Office, 1833-55; surveyed parts of Iowa, Wisconsin, and Upper Peninsula of Michigan; discovered iron ore in Mich. Peninsula. Constructed solar compass, patented, 1836, equatorial sextant, patented, 1856. Recipient Scott medal Franklin Inst., 1840. Author: Key to the Solar Compass, 8th edit., 1909. Died Detroit, Aug. 18, 1858.

BURT, William Henry, Am. mammalogist; b. Haddam, Kan., Jan. 22, 1903; s. Frank P. and Hattie (Carlson) B.; A.B., U. Kan., 1926, A.M., 1927; Ph.D., U. Cal. at Berkeley, 1930; m. Leona Suzan Galutia, Sept. 15, 1928. Research fellow Cal. Inst. Tech., 1930-35; faculty U. Mich., Ann Arbor, 1935—, prof. zoology, 1949—, asst. curator, 1935-38, curator mammals, 1938—. Mem. Am. Soc. Mammalogists (past dir., past pres.), Soc. Systematic Zoology (past mem. council), Izaak Walton League (past pres. Mich. chpt.), NRC, Soc. for Study Evolution, Wildlife Soc., Ecol. Soc., Australian Mammal Soc., Wilson, Cooper ornithol. socs. Author: Mammals of Michigan, 1946, rev., 1948; A Field Guide to the Mammals, 1951, rev., 1964; Mammals of the Great Lakes, 1957; also numerous articles. Research on anatomy birds and mammals, populations, home range, territorial behavior in mammals, faunal studies U. S. and Mexico; described several new species mammals. Home: 2365 Ayrshire St., Ann Arbor, Mich. 48105.*

BURTON, Alan Chadburn, physiologist; b. London, Eng., 1904; s. Frank and Annie (Tyrrell) B.; B.Sc., U. Coll., U. London (Eng.), 1925; M.A., U. Toronto (Ont., Can.), 1928, Ph.D. in Physics, 1932; LL.D., U. Alta., 1964; m. Clara Burton, Aug. 3, 1933; 1 son, Peter Francis. Demonstrator in physics U. Coll., U. London, 1925-26; sci. master Liverpool (Eng.) Collegiate Sch., 1926-27; research asst. NRC Can., scholar, student fellow U. Toronto, 1927-32; research fellow U. Rochester, N.Y., 1932-34; Rockefeller Tng. fellow Johnson Found. Med. Physics, U. Pa., 1934-40; research fellow Banting Inst. Toronto, 1940-45; faculty U. Western Ont., London, 1945—, now prof., head dept. biophysics, 1947—. Named to Order Brit. Empire. Fellow Royal Soc. Can., Royal Soc. Arts (U.K.), A.S.E.B. (past chmn. fed. bd.); mem. Canadian Assn. Physicists, Canadian (past pres.), Am. (past pres.) physiol. socs., Biophys. Soc. (pres. 1965—). Author: (with O. G. Edholm) Man in a Cold Environment, 1953; Physiology and Biophysics of the Circulation, 1965; also numerous articles. Research in high-frequency heating electrolytes, physiology body temperature regulation, heart exchanges with animals, applied physiology clotting and protective equipment, hemodynamics and peripheral circulation, biophysics red cells blood, steady state in biology, atomic weights, theory growth in cancer. Home: 243 Epworth Av., Office: Dept. Biophysics, U. Western Ont., London, Ont., Can.*

BURTON, Donald, Brit. chemist; b. Morley, Yorks, July 29, 1892; s. Arthur Angell and Annie Kate (Risegarne) B.; D.Sc., U. Leeds; m. Ellen Alexandra Valentina Short, 1926; 1 son, Donald Arthur Prudence. Chief chemist Bolton, 1925-51; prof. leather industries U. Leeds, 1951-59, prof. emeritus 1959; hon. dir. Procter Internat. Research Lab., 1955-59. Fellow Royal Inst. Chemists; mem. Soc. Leather Trades Chemists (hon. treas.), Am. Leather Chemists Assn., City and Guilds London Inst., Internat. Union Socs. of Leather Chemists (hon. treas.). Author: (with Robertshaw) Sulphated Oils and Allied Products, 1939. Research, numerous publs. on tech. of leather, oils and sulphated oils. Address; 1 Blacklands Close, Saffron Walden, Essex, England.

BURTON, Eli Franklin, Canadian physicist; b. Feb. 14, 1879; s. George and Eliza (Barclay) B.; B.A., Emmanuel Coll., Cambridge (Eng.) U.; Ph.D., U. Toronto; m. Fannie May Wicher, 1906; 1 son, 1 dau. Demonstrator physics U. Toronto, 1906-11, asso. prof., 1911-24, prof., 1923-32, head dept. physics, dir. McLennan Lab., from 1932. Mem. NRC of Can., 1937-46. Fellow Royal Soc. Can.; pres. Royal Canadian Inst., Toronto, 1931-32. Author: Physical Properties of Colloidal Solutions, 1916, 3d edit., 1938; Superconductivity, 1934; (with C. A. Chant) College Physics, 1933; (with H. G. Smith, J. O. Wilhelm) Phenomena at the Temperature of Liquid Helium, 1940; (with W. H. Kohl) The Electron Microscope, 1942, 2d edit., 1946; also papers. Research on liquid helium, colloids, cosmic rays, and radar. Died Toronto, July 6, 1948.

BURTON, George Dexter, Am. inventor; b. Temple, N.H., Oct. 26, 1855; s. Dexter L. and Emily F. B.; ed. Appleton Acad., Comer's Comml. Coll., Boston; m. Frances C. Jones, Jan. 1894. Editor and pub. New England Star, New Ipswich, N.H., 1873-77. Pres., Am. Electric Forge Co., Electrochem. Pulp and Paper Co., Reno (Nev.) Reduction Works, The Burton Co., mills at Clinton and Holliston, Mass. Lectr. on heating and working metals by electricity. Inventor Burton stock car, liquid process of heating and welding metals by electric current, process of unhairing and tanning animal skins and hides by electricity; process for degumming and separating vegetable fibres by electricity; numerous other discoveries. Died Jan. 7, 1918.

BURTON, George Joseph, Am. med. entomologist; b. N.Y.C., Mar. 20, 1919; s. Samson De Linco and Jeanne (Rutborg) B.; B.S., Coll. City N.Y., 1939, M.S., 1940; Ph.D., N.Y. U., 1946; m. Ellen Maria Carey, Jan. 6, 1962. Plant quarantine insp. U. S. Dept. Agr., 1945-47; prof. biology and entomology State Tchrs. Coll., East Stroudsburg, Pa., 1947-50; commd. col. U. S. Public Health Service, 1953, malariologist U. S. Operations Mission, Monrovia, Liberia, 1953-55, Kathmandu, Nepal, 1955-58; entomologist Tech. Cooperation Mission, New Delhi and Ernakulum, India, 1958-61; vector-borne disease specialist in filariasis AID, Georgetown, Brit. Guiana, 1961-63, U. S. officer-in-charge NIH-Ghana Med. Research joint research program, Accra, also chief med. entomologist, 1963-65, sci. dir. for med. entomology, etiology div. Nat. Cancer Inst., Bethesda, Md., 1965—. Entomologist USN, 1943-45, AUS, 1950-53. Member International Filariasis Assn., A.A.A.S., Am. Soc. Tropical Medicine and Hygiene, Royal Soc. Tropical Medicine and Hygiene, Am. Soc. Parasitologists, Am. Mosquito Control Assn., Entomol. Soc. Washington. Research, publs. on bionomics of mosquitoes transmitting filariasis, blackflies transmitting onchocerciasis, mosquitoes transmitting malaria; developed, produced films on epidemiology of filariasis and onchocerciasis; developed apparatus for collecting mosquitoes and blackflies, ground equipment for dispersing insecticides; photomicrography of med. important arthropods; research and control of vectors of malaria, filariasis and onchocerciasis. Home: 4977 Battery Lane, Bethesda 20014. Office: 7550 Wisconsin Av., Bethesda, Md. 20014.*

BURTON, Glenn Willard, Am. geneticist; b. Clatonia, Neb., May 5, 1910; s. Joseph Fearn and Nellie May (Rittenburg) B.; B.Sc., U. Neb., 1932; M.Sc., Rutgers U., 1933, Ph.D., 1936, D.Sc., 1955; D.Sc., U. Neb., 1962; m. Helen Maruine Jeffryes, Dec. 16, 1934; children—Elizabeth (Mrs. John Edward Fowler), Robert Glenn, Thomas Jeffryes, Joseph William, Richard Bennett. Chmn. div. agronomy U. Ga., Athens, 1950-64, U. Found. prof., 1957—; prin. geneticist. crops research div. agrl. research service (formerly div. forage crops and diseases) U. S. Dept. Agr., Ga. Coastal Plain Expt. Sta., Tifton, Ga., 1936—. Guest lectr. Cornell U., U. Minn., N.C. State Coll., Ohio State U., Ore. State Coll., U. Mo., Tex. Inst. Techl, Va. Poly Inst. Recipient Ann. Agr. award So. Seedsmen Assn., 1950; Superior Service award U. S. Dept. Agr., 1955; Ford Almanac Research award, 1962; U. S. Golf Assn. Green Sect. award, 1965; Gold medal Men's Garden Clubs Am., 1965, Golden medallion award Am. Grassland Council, 1965. Mem. Am. Soc. Agronomy Stevenson award 1949, John Scott award 1957, (past chmn. crops div., past. pres.), A.A.A.S., Bot. Soc. Am., Am. Soc. Range Mgmt., Genetics Soc. Am., Sigma Xi, Alpha Zeta. Research, numerous publs. on improvement pasture and turf grasses for Southeastern U. S. Home: 421 W. 10th St. Office: Ga. Coastal Plain Expt. Sta., Tifton, Ga. 31794.*

BURTON, Harry Edward, Am. astronomer; b. Onawa, Ia., June 11, 1878; s. William and Sarah Martha (Van Dorn) B.; A.B., U. Ia., 1901, M.S., 1903; fellow in math. State U. Ia., 1902-03; m. Ina Burroughs Robinson, Aug. 22, 1911. Apptd. computer U. S. Naval Obs., 1909; advanced through grades to prin. astronomer; head of equatorial div. from 1929. Mem. Am. Astron. Soc., Internat. Astron. Union, Sigma Xi. Discovered the separation of Comet Mellish into 2 components, 1915; derived new elements of orbits of satellites of Mars and redetermined position of equator of Mars, 1929. Wrote introduction to Observations of Double Stars, 1928-44. Co-author of Publications, U. S. Naval Observatory, Second Series, Vol. XII. Editor Manual of Field Astronomy, for Naval Officers detailed to Hydrographic Surveys. Contbr. to Astron. Jour. Died July 19, 1948.

BURTON, Joseph Ashby, Am. physicist; b. Onley, Va., Aug. 22, 1914; s. Vernon S. and Loleta (Boggs) B.; B.S., Wash. & Lee U., 1934; Ph.D. in Phys. Chemistry, Johns Hopkins, 1938; m. Denison Laws, Aug. 29, 1936; children—Delano Burton (Mrs. Leroy M. May), W. Butler, John D. Mem. tech. staff Bell Telephone Labs., Murray Hill, N.J., 1938—, dir. chem. phys. research, 1958—. Bd. visitors Bartol Research Found., Swarthmore, Pa., 1960-66. Fellow Am. Phys. Soc.; mem. Phi Beta Kappa. Research on photoelectric surfaces; low voltage neutron generator; bombardment induced conductivity; phys. chemistry of semiconductors; thermionic emission; impurity effects in germanium and silicon; luminescent materials. Home: 22 Linden Lane, Chatham, N.J. Office: Bell Telephone Labs., Murray Hill, N.J.

BURTON, Milton, Am. chemist; b. Stapleton, N.Y., Mar. 4, 1902; B.S. in Chem. Engring., New York U., 1922, M.S., 1923, Ph.D., 1925; m. Frances Louise Paperno, May 19, 1934 (died 1944); 1 son, James; m. 2d Sarah Holt Foust, May 18, 1946; 1 s. Thomas. Teaching fellow New York U., 1922-24, U. fellow, 1924-25; various industrial positions, 1925-35; with N.Y. U., 1935-36, 38-42, Univ. of Cal., 1937-38; chief of radiation chem. section, atomic energy project, Metall. Lab., U. of Chicago, 1942-45, Clinton Labs., Oak Ridge, Tenn., 1945-46; mem. Radiol. Safety Sect. and consultant to Office of Q.M. Gen., Operation Crossroads, Bikini, 1946; prof. of chemistry U. of Notre Dame since 1945, director radiation lab., 1947—. Fulbright lecturer, guest prof., Guggenheim Fellow, Göttingen, 1955-56. Fellow Am. Inst. Chemists (sec.-treas. N.Y. chapter 1941-42); mem. Radiation Research Soc. (councillor-at-large 1956-57, pres. 1958-59), Société de Chimie Physique, France, Am. Chem. Soc., A.A.A.S., Ind. Acad. Sci. (W. A. Noyes award 1952), Faraday Soc. United Kingdom, Fedn. Am. Scientists (councillor-at-large, 1957-59), Am. Assn. U. Profs., Sigma Xi. Author: (with G. K. Rollefson) Photochemistry and the Mechanism of Chemical Reactions, published in 1939. Co-editor of: Comparative Effects of Radiation. Contributor articles professional jours. Mem. editorial bd. Jour. Phys. Chemistry, 1949-54; asso. editor Jour. Chem. Physics, 1965—. Photochemical studies; research in effects of electrical discharge and high-energy radiation; research on ultra-high-velocity processes and details of energy transfer. Home: 730 Indiana Av., Mishawaka, Ind. 46544. Office: Radiation Lab., U. Notre Dame, Notre Dame, Ind.*

BURTON-OPITZ, Russell, Am. physician, physiologist; b. Ft. Wayne, Ind., Oct. 25, 1875; s. Charles and Anna B.; M.D., Rush Med. Coll., 1895; S.B., U. Chgo., 1897, postgrad., 1897-98. S.M., 1902, Ph.D., 1905; U. Vienna, 1898; m. Jeanette Jonassen, 1909 (dec. 1930); 1 dau., Arlyn; m. 2d, Elizabeth Elliot Phillips Cordts, 1932. Asst. in physiology U. Breslau, 1898-1901; investigator Marine Biol. Sta., Naples 1901; asst. in physiology Harvard, 1901-02; asst. Columbia, 1902-03, instr., 1903-04, adj. prof., 1904-10, asso. prof. physiology, 1909-23, head dept. of physiology, 1909-11, lectr. in physiology, 1923—; cons. physician Cumberland Hosp.; cons. diseases of heart Lenox Hill, Englewood, North Hudson, Holy Name, Christ, Hackensack hosps. (all N.Y.C.). Fellow A.A.A.S.; mem. A.M.A., Am. Physiol. Soc., Soc. Exptl. Medicine and Biology, Am. Soc. Naturalists, Deutsche Physiol. Gesellschaft, Am. Soc. Biol. Chemists, Med. Soc. State N.Y., N.Y. County Med. Soc., Am. Soc. Pharm. and Exptl. Therapy, N.Y. Cardiol. Soc. (pres.), Sigma Xi, Alpha Omega Alpha (pres.). Contbr. to Am. and fgn. physiol. and med. jours. Author: Text Book of Physiology, 1920; Advanced Lessons in Practical Physiology, 1920; Elementary Manual of Physiology. Died Nov. 18, 1954.

BURTT, Davy Joseph, botanist; b. Mar. 7, 1870; s. Dennis Davy; M.A., D.Phil., Oxford U.; Ph.D., Cambridge U.; student U. Cal.; research student Cornell U., Cambridge U.; m. Alice Bolton, 1896; Asst. to dir. Royal Botanic Gardens, Kew, Eng., 1891-92; lectr. tropical forest botany, imperial forestry inst. Oxford U., univ. demonstrator forestry; asst., dept. botany U. Cal., 1893-96, instr., 1901-02; botanist Cal. Agrl. Expt. Sta., 1896-1901; asst. curator U. S. Dept. Agr., Washington, 1902-03; govt. agrostologist, botanist, chief div. botany Dept. Agr., Pretoria, S. Africa, 1903-13. Fellow Royal Geol. Soc., Linnean Soc., Royal Soc. S. Africa, Royal Soc. Arts; mem. Soc. Bot. France, Linnean socs. Lyons, Bordeaux. Author: A First Check-list of the Flowering Plants and Ferns of the Transvaal and Swaziland; Notes on the Genus Ficus; The Native Vegetation of Alkali Lands; The Stock-ranges of North-Western California; The Native Vegetation and Crops of the Colorado Delta; Observations on the Inheritance of Characters in the Maize Plant; New or Noteworthy South African Plants; A Manual of the Flowering Plants and Ferns of the Transvaal with Swaziland, South Africa; The Collection and Preparation of Herbarium Specimens; Notes on the Forest Flora of Northern Rhodesia; The Taxonomic Position of the Pentacyclic-Sympetalae; The Classification of Coniferae; On the Primary Groups of Dicotyledons; The Classification of Tropical Woody Vegetation-Types. Died Aug. 20, 1940.

BURWASH, Lachlan Taylor, Canadian explorer; b. Cobourg, Ont., Can., Sept. 5, 1874; s. Nathaneal and Margaret (Procter) B.; student Albert Coll., Victoria Coll. (both Can.); M.E., U. Toronto, Can., 1912. Engr. Can. govt., Klondike, Can., 1892-1912; govt. explorer, Canadian Arctic; mining developer Mackenzie River, Great Bear Lake regions, Can., from 1932; pres. Burwash Yellow Knife Gold Mines Ltd. Author: Canada's Western Arctic, 1931. Followed route of 1845 Franklin Expdn., 1925-26, 30; flew to magnetic pole, 1929. Died Cobourg, Dec. 21, 1940.

BURWELL, Arthur Warner, Am. chemist; b. Rock Island, Ill., Aug. 26, 1867; s. Charles A. and Cornelia P. (Bonnell) B.; student Kaiser Wilhelm Univ., Strassburg, Germany (Ph.D.); m. Bertha Schade, Dec. 22, 1898; children—Richard Bonnell, Oliver Peckham, Cornelia. Chemist Standard Oil Co., 1893-98; cons. chemist to 1922; research in oxidation of petroleum hydrocarbons, 1922-26, practical oxidation of petroleum hydrocarbons since 1926; v.p. and tech. dir. Alox Corp., Niagara Falls, N.Y. Recipient Gold medal Western N.Y. Sec. of Am. Chem. Soc. for work in producing and utilizing fatty acids and other chemicals from petroleum, 1941. Mem. Am. German Chem. Socs. Automotive Engrs., Am. Soc. Testing Materials, Electrochem. Soc., Am. Petroleum Inst., Am. Inst. Chemistry. Author: Oiliness, 1935; also many articles in tech. jours. on decomposition of petroleum hydrocarbons, lubricants. Patentee in mfr. of lubricants. Died May 24, 1946.

BURWELL, Robert Lemmon, Jr., Am. chemist; b. Balt., May 6, 1912; s. Robert L. and Anne (Lewis) B.; B.S., St. John's Coll., 1932; Ph.D., Princeton, 1936; m. Elise Frank, Dec. 23, 1939; children—Polly, Augusta Somervell. Procter postdoctoral fellow Princeton, 1935-36; instr. Trinity Coll., Hartford, Conn., 1936-39; faculty dept. chemistry Northwestern U., Evanston, 1939—, prof., 1952—, chmn. dept., 1952-57. Mem. com. on phys. chemistry Nat. Acad. Sci.-NRC, 1962-65. Mem. Am. Chem. Soc. (chmn. div. phys. chemistry 1958-59), Catalysis Soc. (dir. 1965—, Faraday Soc., Chem. Soc. London, Internat. Congress on Catalysis (dir. 1955-64). Research, publs. on mechanisms chem. reactions, particularly reactions on surfaces of solids. Home: 2759 Girard Av., Evanston, Ill. 60201.*

BUSBECK, Augier (Bousbecq, or Augerius Ghislenias Busbequius), Flemish diplomat, botanist; b. Comines, Flanders, 1522; served under Charles V, then his bro. Ferdinand I of Austria; sent as ambassador to Solyman II by emperor Ferdinand, Constantinople, 1555; made 2d visit, 1556; imprisoned; returned 1562; held various positions imperial ct.; spent last years as treas. to Elizabeth of Austria at French Ct. Author: Legationis turcicae epistolae quatuor, 1589; Letters from France to the Emperor Rudolph. Introduced various plants from the Levant, including lilac, tulip, Indian chestnut tree to Europe; 1st to study Gothic lang. of Crimea region; discovered Monumentum ancyranum. Died St. Germain, France, Oct. 28, 1592.

BUSCAINO, Vito Maria, Italian neurologist, psychiatrist; b. Trapani, Italy, Dec. 1, 1887; s. Giuseppe and Maria (Cernigliaro) B.; studied medicine and surgery; M.D.; m. Nada Pinti, Sept. 23, 1920; children—Geuseppe Andrea, Maria Grazia, Rosalba. Asst. clinics, Florence; dir. neurology clinic, Cana, then at Naples. Recipient Gold medal Pres. of Republic Culture and Science, also of Republic for Health. Mem. Italian Soc. Neurology (hon. pres.), Italian Soc. Psychiatry (hon. pres.), Acad. Med. and Surg. Sciences (pres. 1963 at Soc. nazion. di scienze, lettere et arti), A.A.A.S. (corr.), Soc. Biochem. Psychopharmacology Paris (corr.), Am. Acad. Neurology (hon. corr.), also others. Research, numerous publs. on neurobiology, normal and path. perceptions, emotions, conscience, language, epilepsy, schizophrenia. Home: parco Grifeo 38, Naples. Office: piazza S. Pasquale a Chiaia 19, Naples, Italy.

BUSCH, August Ludwig, astronomer; b. Danzig, Poland, 1804; prof. math.; later mem. staff Obs., Königsberg, Germany, became dir., 1846. Author: Vorschule der darstellenden Geometrie, 1846; The Works of Bessel, 1849; also collection of observations made by J. Bradley at Kew and Wanstead, 1838. Observed total eclipse of sun, July 28, 1851. Died Königsberg, 1855.

BUSCH, Daryle H., Am. chemist; b. Carterville, Ill., Mar. 30, 1928; s. Dwight H. and Ione (Bauman) B.; B.A., So. Ill. U., 1951; M.S., U. Ill., 1952, Ph.D. (Bersworth fellow), 1954; m. Geraldine Barnes, Mar. 11, 1951; children—Derek H., Michael C., Steven J., Cheryl Ann, Kristina Marie. Faculty, Ohio State U., Columbus, 1954—, prof., 1963—, head div. inorganic chemistry, 1959-66; cons. E. I. duPont, 1956—, NIH, 1961-65, NSF, 1965—, Beaunit Fibers, 1966—; cons. editor Allyn & Bacon; plenary lectr. 10th Internat. Conf. Coordination Chemistry, St. Moritz, Switzerland, 1966. Unrestricted grantee Research Corp., 1960-63; recipient Am. Chem. Soc. award inorganic chemistry, 1963. Mem. Am. Chem. Soc., A.A.A.S., Sigma Xi, Pi Mu Epsilon, Phi Lambda Upsilon, Sigma Pi Sigma. Mem. editorial bd. Chem. Revs., 1967—, Internat. Letters, Inorganic and Nuclear Chemistry, 1965—, Am. Chem. Soc. Monographs, 1964—. Contbr. numerous articles to profl. jours. Studies in transition metal chemistry, application of phys. measurements to study of their compounds, nature of interactions between metal ions and groups bound to them, function of metal ions in the control of chem. reactions. Home: 1930 Cambridge Blvd., Columbus 43212. Office: 88 W. 18th Av., Columbus, O. 43210.*

BUSCH, Eduard Axel Valdemar, Danish neuro-surgeon; b. Copenhagen, Denmark, Sept. 9, 1899; s. Valdemar and Johanne (Becher) B.; ed. Copenhagen U., also hosps. of Copenhagen, Stockholm, Baltimore, N.Y.; M.D., m. Rigmor Schaffer, Mar. 27, 1937; Children—Johanne, Axel. Dir. dept. neuro-surgery hosp. Copenhagen U., also prof. neuro-surgery. Mem. numerous sci. assns. in Europe and U. S. Research, publs. on neuro-surgery. Home: Rormosehus, Raagelije. Office: University of Copenhagen, Copenhagen, Denmark.

BUSCH, Friedrich, German meteorologist; b. Recklinghausen, Germany, July 4, 1851; s. Josef Friedrich and Agnes (Schulte) B.; studied math. and natural scis. in Münster, Germany, hon. Ph.D., Münster, 1925; m. Theresia Schipper, 1881; 3 sons, including Fritz, Max, 3 daus., including Maria (Mrs. Aloyd Timpe). Became tchr., Paulinisches Gymnasium (secondary sch.), Münster, 1876, Gymnasium Laurentiunum, Arnsberg, Westphalia, Germany, 1880. Author: (with Ch. Jensen) Tatsachen und Theorien der atmosphärischen Polarisation, 1911; also numerous articles in meteorol. jours. Research on meteorol. optics; studied the Bishop's Ring which was particularly visible after the Krakatoa eruption and disappeared after 1886; studied Babinet, Arago and Brewster points (neutral points of atmospheric polarization) and recognized that they occur simultaneously with solar phenomena. Died Arnsberg, Aug. 30, 1931.

BUSCH, Georg Adolf, Swiss physicist; b. Zurich, Switzerland, Sept. 12, 1908; s. Franz Paul and Elisa (Staub) B.; diploma Swiss Fed. Inst. Tech., Zurich, 1933, Dr.sc.nat., 1938; m. Margarete Klemm, Apr. 8, 1940; children—Dora, Hansjurg, Franziska, Annamaria. Prof. physics Swiss Fed. Inst. Tech., 1948—, dir. Lab. for Solid State Physics, 1955 ; vis. prof. Carnegie Inst. Tech., Pitts., 1952. Mem. council Swiss Natural Sci. Foundel, 1962. Mem. Finnish Acad. Sci., Swiss, German, Am. phys. socs., Sigma Xi. Research, numerous publs. on discovery of new ferroelectrics and antiferroelectrics, free-electron magnetism in semicondrs., new magneto-optic effects in ferromagnetic semicondrs. Home: 119 Hadlaub, Zurich, Switzerland.*

BUSCH, Harris, Am. med. scientist; b. Chgo., May 23, 1923; s. Maurice Ralph and Rose (Feigenholtz) B.; B.S., U. Ill., 1943, M.S., 1946; M.S., U. Wis., 1950, Ph.D., 1952; m. Rose Klora, June 16, 1945; children—Daniel Avery, Laura Anne, Gerald Irwin, Fredric Neal. Faculty, Yale, 1952-55; faculty U. Ill., 1955-60, prof. pharmacology, 1959-60; prof. pharmacology, chmn. dept. Baylor U. Coll. Medicine, Houston, 1960—, prof. biochemistry, chmn. dept., 1960-62. Cons. to pvt. cos., govt. agys. USPHS fellow, Baldwin scholar, Yale, 1952-55; Am. Cancer Soc. scholar, 1955. Mem. Am. Assn. for Cancer Research (dir., 1963—, past pres. Southwestern sect.), Tex. Med. Center Research Soc. (past pres.), Am. Soc. Biol. Chemists, Am. Soc. for Pharmacology and Exptl. Therapeutics, Soc. for Growth, Biochem. Soc. (Gt. Britain), Soc. for Exptl. Biology and Medicine. Author: Chemistry of Pancreatic Diseases, 1958; An Introduction to the Biochemistry of the Cancer Cell, 1963; The Nucleus of the Cancer Cell, 1963; Frontiers in Medicine, 1963; Histones and other Nuclear Proteins, 1965; Chemotherapy, 1966; also numerous articles. Editor, Methods in Cancer Research, vol. 1, 1966, vols. II, III, 1967. Research on nucleus, isolation nuclei cancer cells, demonstration nucleolar nucleic acids cancer cells different from those of some normal cells, methods for isolation mass quantities nuclear and nucleolar RNA, nuclear proteins, histones, acidic proteins, nucleolar enzymes for synthesis, interconversion and destruction nucleolar RNA, effects of drugs on nuclear proteins and nucleic acids tumor cells. Home: 4966 Dumfries Dr., Houston, 77035.*

BUSCH, Max Gustav Reinhold, German chemist; b. Hochneukirch, nr. Dusseldorf, Germany, Aug. 16, 1865; s. Peter and Lisette (Lindgens) B.; student Technische Hochschule Charlottenburg; doctorate from Erlangen, 1889; m. Frieda Leuze, 1894; 3 daus. Prof. pharmacy, applied chemistry and chem. tech., from 1912. Research and numerous publs. on nitrogen-carbon compounds; synthetized numerous heterocyclic substances; discovered desmotropism of oxim and hydroxylamine compounds, intramolecular rearrangements and molecular compounds of oppositely constructed substances; developed protective method of catalytic hydrogenation and dehalogenation. Died Erlangen, Aug. 26, 1941.

BUSEMANN, Herbert, mathematician; b. Berlin, Germany, May 12, 1905; s. Alfred and Olga (Wetzlar) B.; student U. Munchen, Germany, 1925-26; U. Paris, U. Gottingen, 1926-27, 28-30, Ph.D., 1931; 1927-28; m. Ruth Elsa Helbig, July 2, 1939. Came to U. S., 1936, naturalized, 1943. Asst. Inst. for Advanced Study, Princeton, N.J., 1936-39; faculty Johns Hopkins U., 1939-40, Ill. Inst. Tech., 1940-45, Smith Coll., 1945-47; prof. math. U. So. Cal., 1947-64, distinguished prof., 1964—; Fulbright research fellow, 1952-53. Mem. Royal Danish Acad. Scis., Am. Math. Soc., Math. Assn. Am. Author: The Geometry of Geodesics, 1955; Convex Surfaces, 1958; (with P. J. Kelly) Projective Geometry and Projective Metrics, 1953. Research and publs. in systematic investigation of those parts of geometry in which the existence of shortest connections plays a role; novel types of problems in higher-dimensional convexity. Home: 5532 Overdale Dr., Los Angeles 90043. Office: U. So. Cal., Los Angeles 90007.*

BUSEY, Samuel Clagett, Am. physician; b. Montgomery County, Md., July 23, 1828; M.D., Univ. of Pa., 1848; LL.D., St. Mary's U., Balt. Practiced medicine, Washington. Prof. materia medica, diseases of infancy and childhood, theory and practice of medicine, then emeritus prof. theory and practice of medicine Georgetown Univ., Washington. Pres. med. Soc. of D.C., 1877, 94-98. Author: Occlusion and Dilatation of Lymph Channels, Acquired forms; Lymph Channels. Died 1901.

BUSH, George Leonard, Am. chemist, educator; b. New Haven, W.Va., Dec. 18, 1897; s. George Fletcher and Mary (Oliver) B.; B.S. in Edn., Ohio State U., 1922, M.A. in Chemistry, 1923; D.Ed. in Sci. Edn., Columbia, 1940; m. Alma A. Millonig, June 13, 1931; children—Richard Wayne, Charles Arthur. Tchr. prin. pub. high schs., 1916-46; faculty St. Cloud (Minn.) Tchrs. Coll., 1946-47; faculty Kent (O.) State U., 1947—, prof. chemistry, 1951—, prof. emeritus, 1966—. Mem. Am. Chem. Soc., A.A.A.S., Ohio Acad. Sci., Sigma Xi, Phi Lambda Upsilon, Phi Delta Kappa, Kappa Delta Pi. Author: (with others) Dynamic Chemistry, 1936, Senior Science, 1937, A Biology of Familiar Things, 1939, Chemistry Today, 1954, New Senior Science, 1957, Worlds of Science, 1959, Investigations in Chemistry, 1962, The Atom, 1961; Science Education in Consumer Buying, 1941. Contbr. over 500 articles to sci. jours. Home: 7802 Birchwood Dr., Kent, O. 44240.*

BUSH, Katharine Jeannette, Am. zoölogist; b. Scranton, Pa., Dec. 30, 1855; d. William Henry and Eliza Ann (Clark) B.; studied many yrs. under Prof. A. E. Verrill; Ph.D., Yale, 1901. Asst. zoöl. dept. Yale U. Mus., 1879—; on U. S. Fish Commn. several yrs.; assisted in revision of Webster's Dictionary, resulting in Webster's Internat. Dictionary, 1890. Author: The Tubicolous Annelids of the Tribes Sabellides and Serpulides—Harriman Alaska Expedition, Vol. XII, 1905. Died 1937.

BUSH, Kenneth Arthur, Am. mathematician; b. Oneonta, N.Y., Feb. 24, 1914; s. Ivan DeWitt and Julia (Cary) B.; B.A., Columbia, 1936, M.A., 1939; Ph.D., U. N.C., 1950; m. Julia M. Huddleston, Jan. 17, 1942; children—Cary Wilson, Sarah Stephenie, William DeWitt. Faculty, U. S. Naval Acad., Annapolis, Md., 1941-46, State U. N.Y., Plattsburgh, 1946-48, 50-52, U. Ill., Urbana, 1952-54; prof. math., head dept. U. Ida., Moscow, 1954-61; prof. math. Wash. State U., Pullman, 1961—. Cons. Project Whirlwind, Mass. Inst. Tech., 1951, Project SCAMP, U. Cal. at Los Angeles, 1953; dir. NSF Summer Inst., U. Ida., 1960-61; vis. lectr. Soc. Indsl. and Applied Math., 1962. Mem. Math. Assn. Am. (vice chmn. Pacific N.W. sect. 1958, chmn. 1959), Econometric Soc., Operations Research Soc. Am., Inst. Math. Statistics. Research in combinatorial problems, matrices, classical analysis, math. statistics. Home: 900 Derby St., Pullman, Wash. 99163.*

BUSH, Laurens Earle, Am. mathematician; b. Martins, S.C., Mar. 24, 1900; s. Laurens Ashley and Mary Elizabeth (Oswald) B.; B.S., The Citadel, 1919; S.M., U. N.C., 1926; student U. Chgo., 1929; Ph.D., Ohio State U., 1931; m. Winnie Davis Kearse, June 23, 1920; children—Patricia Earle (Mrs. Jerome Long), Edward Ashley. Instr. U. N.C., 1926-30, Ohio State U., 1931-33; prof., chmn. dept. math. Coll. St. Thomas St. Paul, 1933-53; professor of mathematics Kent State University, 1953—, head department of mathematics, 1953-64; vis. prof. North Tex. State Coll., summers 1938-41. Dir. William Lowell Putnam math. competition for Math. Assn. Am., 1948-65. Mem. Math. Assn. Am. (bd. govs. 1947-49, past chmn. Minn. sect., chmn. Ohio sect. 1958-59), Am. Math. Soc., Sigma Xi. Bd. editors Nat. Math. mag., 1935-45. Contbr. articles profl. jours. Contributions to structure of linear associative algebra; definition of an algebra; theory of integers. Home: 408 Burr Oak Dr., Kent, O. 44240.

BUSH, Spencer Harrison, Am. metallurgist; b. Flint, Mich., Apr. 4, 1920; s. Edward C. and Rachel (Roser) B.; student Flint Jr. Coll., 1938-40, Ohio State U., 1943-44; B.S., Chem.E., Univ. Mich., 1948, B.S. in Metall. Engring., 1948, M.S., 1950, Ph.D., 1953; m. Roberta Lee Warren, Aug. 28, 1948; children—David S., Carl E. Asst. chemist Dow Chem. Co., 1940-42; instr. dental materials U. Mich., 1951-53; with Hanford Atomic Products Operation, Gen. Electric Co., Richland, Wash., 1953-65, supr. phys. metallurgy, 1954-57, supr. fuels fabrication devel., 1957-60, metall. specialist, 1960-63, cons. metallurgist, 1963-65; cons. to dir. Battelle Meml. Inst., Pacific N.W. Lab., Richland, 1965—. Lectr. metall. engring. Center for Grad. Study, U. Wash., 1953-67, affiliate prof., 1967—; mem. adv. com. on reactor safeguards U.S. AEC, 1966—. Am. Foundrymen's Soc. fellow, 1948-50. Mem. Am. Soc. for Metals (chmn. program council 1966-67, trustee 1967—), Am. Inst. Mining, Metall. and Petroleum Engrs. (past chmn. nuclear metallurgy com., chmn. ann. speaker com. 1967-68), Am. Nuclear Soc. (mem. adv. editorial bd. Nuclear Applications 1965—), Am. Welding Soc. (mem. Pressure vessel research com. 1967—), Am. Soc. for Testing and Materials, Brit. Inst. Metals, Sigma Xi. Research, publs. on phys. and mech. metallurgy of nuclear materials, temper embrittlement of steels, surface hardening of pearlitic malleable irons, gold base and chrome-cobalt base dental alloys, kinetics, oxidation in zirconium alloys, effect of fabrication structural materials, stress corrosion, reactor safety, Home: 630 Cedar St. Office: P.O. Box 999, Richland, Wash. 99352.*

BUSH, Vannevar, Am. elec. engr.; b. Everett, Mass., Mar. 11, 1890; s. Richard Perry (D.D.) and Emma Linwood (Paine) B.; B.S., M.S., Tufts, 1913, Sc.D., (hon.), 1932; Eng. D., Mass. Inst. Tech., Harvard, 1916; LL.D., Brown U., Middlebury Coll., 1939, Johns Hopkins, 1940, U. Pa., Yale 1942, Washington U., U. Buffalo, 1946, Princeton, 1947, Colby Coll., 1951; Eng.D., Poly. Inst. of Bklyn., 1941, Rutgers Coll., 1942. Carnegie Inst. Tech., 1948; Sc.D., Harvard. Williams Coll., 1941, Stevens Inst. Tech., 1943, Trinity Coll., 1946, W.Va. U., 1947, Columbia, 1947, U. Cambridge (Eng.), 1950, Boston U., 1959; m. Phoebe Davis, Sept. 5, 1916; children—Richard Davis, John Hathaway. With test dept. Gen. Electric Co., 1913; with inspection dept. USN, 1914; instr. math. Tufts, 1914-15, asst. prof. elec. engring., 1916-17; research on submarine detection with spl. bd. on submarine devices USN, 1917-18; asso. prof. elec. power transmission Mass. Inst. Tech., 1919-23, prof., 1923-32; v.p., dean engring., 1932-38; pres. Carnegie Instn. of Washington. 1935-55., trustee, 1958—. Trustee Carnegie Corp. of N.Y., 1939-55, trustee emeritus; dir. Metals and Controls Corp., Am. Tel. & Tel. Co.; chmn. Merck & Co., Inc. Mem. adv. com. Nat. Security Resources Bd. Trustee Tufts, Johns Hopkins, 1943-55; life mem. corp. Mass. Inst. Tech., hon. chmn. corp., 1959—; chmn. bd. Graphic Arts Research Found.; trustee George Putnam Found.; regent Smithsonian Instn., 1943-55. Lamme medal Am. Inst. E. E., 1935, Edison medal, 1943; Research Corp. award, 1939; Ballou medal, Tufts, 1941; Holley medal Am. Soc. M.E.; John Scott medal Phila. City Trusts, 1943; gold medal Nat. Inst. Social Scis., 1945; Marcellus Hartley award Nat. Acad. Scis.; Wash. award Western Soc. Engrs., 1946; Hoover medal Asso. Engring. Nat. Acad. Scis.; Scientists, 1947; D.S.M., Roosevelt Meml. Assn., 1945, Tufts Coll., 1947; John Fritz medal 1951; award Am. Inst. Cons. Engrs., 1953; John J. Carty medal and award, Nat. Acad. Scis., William Proctor prize, Sci. Research Soc. Am., 1954. Mem. Nat. Adv. Com. for Aeronautics (chmn. 1939-41), Bus. Adv. Council of Dept. Commerce, 1939-41; chmn. Nat. Def. Research Com., 1940-41; dir. OSRD, 1941-46; chmn. Joint Com. New Weapons and Equipment of Joint U. S. Chiefs of Staff, 1942-46; chmn. Joint Research and Devel. Bd. 1946-47; chmn. Research and Devel. Bd., Nat. Mil. Establishment, 1947-48. Fellow Am. Inst. E.E., Am. Phys. Soc.; mem. Am. Acad. Arts and Sci., Nat. Acad. A.A.A.S. Am. Soc. Engring. Edn., (Lamme award 1955), Am. Soc. M.E. (hon.), Soc. Naval Architects and Marine Engrs. (hon.), Franklin Inst. (hon.; Levy medal 1928), Am. Philos. Soc., Am. Math. Soc., Sigma Xi, Tau Beta Pi, Phi Beta Kappa. Author: (with W. H. Timbie) Principles of Electrical Engineering, 1922; Operational Circuit Analysis, 1929, Endless Horizons, 1946; Modern Arms and Free Men, 1949. Builder of differential analyzer (machine for solving differential equations), 1942; research on transmission line transients, operational circuit analysis, gaseous conduction apparatus, analytical devices, principles of elec. engring. Home: 304 Marsh St., Belmont, Mass. 02178. Mass. Inst. Technology, Cambridge, Mass. 02139.*

BUSHKE, Abraham, dermatologist; b. Nakel/Netze, Poland, Sept. 27, 1868; s. Julius and Eva (Bernstein) B.; M.D., 1891; m. Erna Fränkel, 1900; at least 3 children—Albrecht, Franz, Wilhelm. Worked with Rudolf Virchow and Robert Koch in Berlin, Germany; asst. to H. Helferich at surg. clinic, Greifswald, until 1894; asst. to Albert Neisser dermatol. clinic, Breslau, Germany, 1895-97, to Edmund Lesser in Berlin until 1904; dir. dermatol. dept. of hosp. Am Urban, Berlin, 1904-06; dir. dermatol. dept. Rudolf-Virchow Hosp., Berlin; became titular prof., 1908, asso. prof., 1920; forced to retire, 1933. Author: (with M. Gumpert) Geschlechtskrankheiten bei Kindern, 1926; (with A. Joseph) Lehrbuch der Gonorrhoe, 1926; over 250 other publs. First (with Otto Busse) to report on European blastomycosis (Busse-Buschke disease), 1895; research on treatment of syphilis, inherited lues (syphilis), fungoid disease of skin; pointed out tonsils as entryways for infection, 1894. Died Theresienstad concentration camp, Feb. 25, 1943.

BUSHNELL, David, Am. inventor; b. Saybrook, Conn., 1742; grad. Yale, 1775. Completed man-propelled submarine boat, 1775; originator modern submarine warfare; lt. Continental Army, 1779; capt., 1781; commander Corps Engrs., West Point, 1783. Died Warrenton, Ga. 1824.

BUSHNELL, George Ensign, Am. surgeon; b. Worcester, Mass., Sept. 10, 1853; s. George and Mary Elizabeth (Blake) B.; A.B., Yale, 1876, M.D., 1880; m. Adra Holmes, 1881 (dec. 1896); 2d, Ethel M. Barnard, Dec. 24, 1902. Apptd. asst. surgeon U. S. Army, 1881, capt. asst. surgeon, 1886, maj. chief surgeon vols., 1898, maj. surgeon, 1898, maj. M.C., hon. discharge from vol. service, 1899; commd. lt. col. U. S. Army, 1908, col., 1911; comd. U. S. Army Comd. U. S. A. Gen. Hosp., Ft. Bayard, N.M., about 14 yrs.; ret. by operation of law, 1917. Author: A Study in the Epidemiology of Tuberculosis, 1920. Deceased.

BUSHNELL, John Horace, Jr., Am. biologist, educator; b. Grand Rapids, Mich., Mar. 17, 1925; s. John Horace and Margaret (Wagemaker) B.; B.A., Vanderbilt U., 1948; postgrad. U. Mich.; M.S., Mich. State U., 1956, Ph.D., 1961; m. Judith Mary Clark, Mar. 3, 1951; children—John Palmer, Jeffery Stevens, Romney Clark, Heather S. Officer, Chief Naval Operations Sino Am. Coop. Orgn., USN, China, 1943-46; educator, diplomatic service U. S. State Dept. and Afghan Ministry of Edn., Kabul, Afghanistan, 1948-50; asst. sec., sales mgr. Wagemaker Boat Corp., U. S. Molded Shapes, Grand Rapids, 1950-55; instr. zoology Mich. State U., 1959-60; asst. prof. biology Washington and Jefferson Coll., 1960-64; faculty U. Colo., 1964—, asso. prof. biology, 1966—; guest lectr. Harvard Biol. Labs., 1966. Recipient NIH Summer Research award Marine Biol. Labs., Woods Hole, Mass., 1962, Edwin S. Linton fellow, 1962-64. Mem. Am. Ecol. Soc., Am. Soc. Zoologists, Am. Micros. Soc., Animal Behavior Soc., Marine Biol. Assn. U.K., Sigma Xi. Asso. editor Am. Midland Naturalist, 1967—. Contbns. to knowledge of ecology, physiol. ecology, behavioural ecology of aquatic invertebrate animals; studies, publs. on growth, form, longevity, ecology, systematics, zoogeography, evolution of certain invertebrates, especially Ectoprocta, Coelenterata, Porifera, Acarina. Home: 2825 6th St., Boulder, Colo. 80302.*

BUSIGNIES, Henri Gaston, electronic-communications engr.; b. Sceaux, France, Dec. 29, 1905; s. Henri and Juliette (Benoit) B.; degree in elec. engring., Paris, 1926; D.Sc. (honorary), Newark Coll. Engineering, 1958; m. Cecile Phaeton, July 15, 1931; 1 dau., Monique (Mrs. Henri Charles Honeck). Came to U. S., 1940, naturalized, 1953. Research, development engr. Les Laboratoires, Le Materiel Telephonique, Paris labs. Internat. Tel. & Tel. Corp., 1928-35, dept. head, 1935-38, head project on direction finders, radar, instrument landing, receivers, antennas, 1938-41; lab. head Fed. Telecommunication Labs., Internat. Tel. & Tel. Corp., Nutley, N.J., 1941-46, dir., 1946-48, tech. dir., 1948-54, exec. v.p., 1954-56, pres., 1956—, president I. T. & T. labs. div. 1958—, v.p., 1960-65, gen. tech. dir. corp., 1960—, sr. v.p., 1965—; dir. Internat. Standard Electric Corp. (N.Y.C.). Recipient Lakhovsky award Radio Club of France, 1926; certificate commendation for outstanding service USN, 1947; Presdl. Certificate of Merit, 1948; Pioneer award air navigation I.R.E., 1959; David Sarnoff award I.E.E.E., 1964. Fellow I.E.E.E.; member of French Engrs. in U. S., Soc. French Civil Engrs., French Society Advancement of Sciences, National Academy of Engineering. Contbr. articles profl., tech. publs. Patentee in field. Home: 71 Melrose Pl., Montclair, N.J. Office: 320 Park Av., N.Y.C. 22.

BUSINCO, Lino, Italian physician; b. Montecreto, Modena, July 7, 1908; s. Ettore and Amabile (Bagatti) B.; M.D.; m. Potenzia Oppo, Aug. 10, 1937; children—Elena, Luisa, Salvatore. Instr. gen. pathology Sch. Medicine, U. Rome; dir. Center Bio'l. Studies Sovereign and Mil. Order of Malta. (recipient award). Mem. Internat. Commn. Allergy, Allergy socs. Ialy, France, Spain, Portugal, Brazil, Germany. Research, numerous publs. on allergies. Home: via G. B. De Rossi 15A, Rome. Office: via Arigento, 6, Rome, Italy.

BUSK, George, microscopist; b. St. Petersburg, Russia, Aug. 12, 1807; s. Robert and Jane (Wesly) B.; articled student Coll. Surgeons, 6 years; student St. Thomas, St. Bartholomew's hosps.; m. Ellen Busk, Aug. 12, 1843. Served as surgeon in navy; settled in London, 1855; treas. Royal Instn.; Hunterian prof.; trustee Hunterian Mus. Fellow Royal Coll. Surgeons (became pres. 1871), Royal Soc., 1850, Linnean Soc., Zool. Soc., Micros. Soc. (a founder; pres. 1848-49); mem. Geol. Soc. Recipient Lyell medal, 1878, Wollaston medal, 1885; Bryozoa genus, Buskia, named in his honor. Author: Catalogue of Marine Polyzoa in the British Museum, 1852; A Monograph of the Fossil Polyzoa of the Crag, 1859; Report on the Polyzoa Collected by H.M.S. Challenger, 1884-86; also numerous articles and translations. Microscopic research on lower forms of life; 1st to formulate sci. arrangement for Bryozoa, 1856. Died 1886.

BUSKIRK, Elsworth Robert, Am. physiologist; b. Beloit, Wis., Aug. 11, 1925; s. Ellsworth Fred and Laura (Parman) B.; student U. Wis., 1943; B.A., St. Olaf Coll., 1950; M.A., U. Minn., 1951, Ph.D., 1954; m. Mable Heen, Aug. 28, 1948; children—Laurel Ann, Kristine Janet. Research asso. Lab. for Physiol. Hygiene, U. Minn., Mpls., 1954; physiologist Research and Engring. Center, Natick, Mass., 1954-57, Inst. Arthritis and Metabolic Diseases, NIH, Bethesda, Md., 1957-63; prof. applied physiology Inst. for Sci. and Engring., Pa. State U., University Park, 1963—, dir. Lab. for Human Performance Research 1963—. Cons. calorie sub com. NRC. Mem. Am. Physiol. Soc., Gerontology Soc., Am. Coll. Sports Medicine (past pres.), Aerospace Med. Soc., Ergonomics Soc. Contbg. author Science and Medicine of Exercise and Sports, 1960. Contbr. articles to sci. jours. Research in human stress physiology, study cardiovascular and respiratory function, metabolism, thermal regulation. Home: 216 Hunter Av., State College, Pa. 16801.*

BUSS, Arnold Herbert, Am. psychologist; b. Bklyn., Aug. 7, 1924; s. Harry and Esta (Agree) B.; B.A., N.Y. U., 1947; M.A., Ind. U., 1949, Ph.D., 1952; m. Edith H. Nolte, June 12, 1948; children—Arnold, David, Laura. Faculty, State U. Ia., 1951-52, U. Pitts., 1957-65; chief psychologist Carter State Hosp., Indpls., 1952-57; prof. psychology Rutgers U., New Brunswick, N.J., 1965—. Mem. Am. Psychol. Assn., A.A.A.S. Author: The Psychology of Aggression, 1961, Psychopathology, 1966. Research, publs. on techniques for study human aggression in lab. Home: 22 Chandler Rd., Edison, N.J. 08817. Office: 88 College Av., New Brunswick, N.J.*

BUSSE, Ewald William, Am. psychiatrist; b. St. Louis, Aug. 18, 1917; s. Frederick Ewald and Emily (Stroh) B.; B.A., Westminster Coll., 1938, Sc.D., 1960; M.D., Washington U., St. Louis, 1942; m. Ortrude Helen Schnaedelbach, July 18, 1941; children—Ortrude Susan (Mrs. William Parry Dinsmoor White), Barbara Ann, Ewald Richard, Deborah Emily. Faculty, U. Colo. Sch. Medicine, Denver, 1946-53; prof. psychiatry, chmn. dept., Duke, Durham, N.C., 1953—, dir. Center for Study Aging and Human Devel., 1957—, J. P. Gibbons prof. psychiatry, 1965—. Cons. USN, U. S. Army, VA., NIH. Mem. Am. Bd. Psychiatry and Neurology (sec.-treas.). Research, numerous publs. on psychosomatic disorders, problems aging and aged, electroencephalographic studies. Home: 1132 Woodburn Rd., Durham, N.C. 27705.*

BUSSE, Otto Emil Franz Ulrich, pathologist; b. Gühlitz, nr. Perleberg, Germany, Dec. 6, 1867; s. Hermann and Ottilie (Hansen) B.; doctorate from Greifswald, 1892; m. Lotte Grawitz; m. 2d, Elisabeth Engeling, 1913; children—Paul Busse-Grawitz, Ernst, 2 daus. Became prof., 1902; became head dept. Hygienic Inst., Poznan, Poland, 1904; prof. path. anatomy, Zurich, Switzerland; dir. Path. Inst. until 1912. Author: (dissertation) Über die Heilungsvorgänge an Schnittwunden der Haut, 1892; Das Obduktionsprotokoll, 1903; also parts of med. publs., manuals. Research on histology of infection, saccharomycosis; discovered 1st pathogenic yeasts, 1894; 1st (with A. Buschke) to report on European blastomycosis (Busse-Buschke disease), 1895. Died Zurich, Feb. 3, 1922.

BUSSE, Warren Froemming, Am. phys. chemist; b. Chgo., Dec. 2, 1900; s. Gustave M. and Emma (Froemming) B.; Ph.B., U. Wis., 1924, Ph.D., 1927; m. Fritzi Cook, Dec. 25, 1932; 1 son, David M. From physicist to dir. phys. research B. F. Goodrich Co., Akron, O., 1928-42; asso. dir. research Gen. Aniline & Film Co., Eston, Pa., 1942-48; tech. dir. Inst. Textile Tech., Charlottesville, Va., 1948-50; research fellow E.I. du Pont de Nemours & Co., Inc., Wilmington, Del., 1950-66. Chmn. textile div. Gordon Research Conf.; tchr. semantics U. Del., Newark, 1962—. Vis. fellow Yale, 1966. Fellow Am. Phys. Soc.; mem. Am. Chem. Soc., Rheology Soc., Phi Beta Kappa, Sigma Xi, Phi Kappa Tau. Contbr. articles to profl. jours. Patentee in field. Research on mechanism of rubber elasticity, fatigue resitance textiles, rheology plastics. Home: 803 Greenwood Rd., Westover Hills, Wilmington, 19807.*

BUSSIERE, see Buissiere.

BUSSY, Antoine Alexandre Brutus, French chemist; b. Marsielles, France, May 21, 1794; ed. École polytechnique, 1813; docteur es médecine; became lectr. Paris Sch. Pharmacy, 1821, prof., 1836, dir., 1844-73. Mem. Acad. Medicine (pres.), French Acad. Scis., 1850, Hygiene Council for Seine, Soc. Pharmacy (pres.), Sci. Union Pharmacists of France (founder). Author: Traité des moyens de reconnaitre les falsifications des drogues, 1829. Discovered method of preparing magnesium by heating magnesium chloride and potassium together, 1831; liquified several gases previously considered inert; prepared (independently of Wöhler) metallic beryllium, 1828. Died Paris, Feb. 1, 1882.

BUSTAD, Leo K., Am. biologist; b. Stanwood, Wash., Jan. 10, 1920; s. Rasmus and Thora (Larson) B.; B.S., Wash. State U., 1941, M.S., 1948, D.V.M., 1949; Ph.D., U. Wash., 1960; m. Signe E. Byrd, June 13, 1942; children—Leo B., Karen A., Rebecca L. Mgr. exptl. animal farm Hanford Labs., Gen. Electric Co., 1949-65; dir. radiobiology lab., prof. radiation biology U. Cal., Davis, 1965—, dir. NIH Tng. Program in Lab. Animal Medicine, 1966—. Mem. various coms. Nat. Acad. Scis.-NRC; mem. exec. com. Regional Primate Research Center U. Wash., 1962—, mem. adv. com., 1965—; guest lectr. Coll. Vet. Medicine Wash. State U., 1955—. Mem. Am. Physiol. Soc., Radiation Research Soc., Am. Vet. Med. Assn., A.A.A.S., Soc. Exptl. Biology and Medicine, Sigma Xi. Editor: Biology of Radioiodine, 1963; (with R. O. McClellan) Swine in Biomedical Research, 1966. Research, numerous publs. on biol. effects of important radionuclides including radioiodine, radiostrontium and radium as well as late effects of beta irradiation of skin, metabolic behavior in domestic animals of a variety of radionuclides important in nuclear medicine, nuclear industry and fallout. Home: Route 1, 42 Walnut Lane, Davis, Cal. 95616.*

BUTANI, Dhamo Kessowdas, entomologist; b. Sind, Pakistan, Feb. 1, 1923; s. Kessowdas M. and Bhambhibai (Israni) B.; B.Sc. in Agr., Coll. Agr. Poona (India), 1944, M.Sc., 1947; m. Mohni Hassaram Shahani, May 12, 1944; children—Prakash, Neelum, Shabnum, Sonia, Rajesh. Research scholar Indian Council Agr. Research, 1947-50; quarantine insp. Govt. of India plant prot. dept., 1950-51; field entomologist research and extention, 1951-58; asst. prof. entomology and zoology Agr. Coll., Sabour, Bihar, 1958-60; entomologist Regional Research Centre, I.A.R.I., 1961—. Fellow Royal Entomol. Soc. London, Entomology Soc. India (life, counciller); mem. Bihar Acad. Agrl. Sci., Sci. and Technol. Soc. India. Research, numerous publs. on morphology, anatomy and control of various insect pests, mites and rats especially sugarcane, cotton, oilseeds and millet pests; worked out insecticidal schedules of treatments against pest complex of various crops. Home: Pir wali gali, SIRSA, Dist. Hissar, India. Office: Regional Research Center, I.A.R.I., Sirsa, Dist., Hissar, India.*

BUTCHER, James Walter, Am. entomologist, educator; b. Ramsaytown, Pa., Feb. 14, 1917; s. Louis and Mary (Yeloushan) B.; B.S., U. Pitts., 1943; M.S., U. Minn., 1949, Ph.D., 1951; m. Mary K. Culley, June 18, 1944; children—Craig, Mary Helen. With Gulf Research & Devel. Corp., 1945-46, U. S. Dept. Agr., 1951-52, Minn. Dept. Agr., 1954-57; faculty dept. entomology Mich. State U., East Lansing, 1957—, prof., 1965—. Fulbright scholar U. Vienna, 1966-67. Mem. Entomol. Soc. Am., Ecol. Soc. Am., Canadian Entomol. Soc., Canadian Soc. Zoologists, Mich., Ind., Man., Ont. entomol. socs. Research, numerous publs. on forest entomology, insect ecology and behavior, pesticide side effects, soil zoology. Home: 3666 E. Hiawatha St., Okemos, Mich. 48823. Office: Dept. Entomology, Mich. State U., East Lansing, Mich. 48823.*

BUTCHER, William Dean, Brit. dermatologist; b. Oct. 17, 1846; s. William and Rachel (Deane) B.; ed. U. London, St. Bartholomew's Hosp.; m. Fanny Bazett, 1877; 2 daus., 3 sons. House surgeon to Sir James Paget, 1868; practiced in Calcutta, India, 1870-74; surgeon in S.S. Gothic, 1894; cons. surgeon London Skin Hosp. Electrotherapeutic Soc. rep. Congress Physiotherapy, Liege, 1905, Berlin Röntgen congress, 1906, Röntgen Soc. rep. Internat. Congress Electrology and Radiology, Amsterdam, 1908, electrotherapeutic sect. Royal Soc. Medicine rep. Congress Physiotherapy, Paris, 1910. Fellow Philos. Soc. mem. Royal Coll. Surgeons, Röntgen Soc. (pres. 1908-09), Royal Soc. Medicine (pres. electrotherapeutic 1909-10), Ealing Sci. and Microscopic Soc. (pres.). Translator: Radiotherapy in Skin Diseases (Belot); Medical Electricity (Guilleninot); Mechanism of Life (Leduc). Research and publs. on radium, x-rays, osmotic growth, phys. basis of edn. Introduced 1st Röntgen installation in Eng. at London Skin Hosp., 1898. Died Jan. 10, 1919.

BUTENANDT, Adolf Friedrich Johann, German biochemist; b. Bremerhaven-Lehe, Germany, Mar. 24, 1903; s. Otto Louis Max and Wilhelmine (Thomfohrde) B.; ed. Oberrealschule, Bremerhaven-Lehe, Univ. of Marburg, Germany; Ph.D., U. of Goettingen, 1927; Ph.D. (hon.), U. Graz; M.D. honoris causa, U. Tübingen, Dr. rer. nat. (hon.); D.V.M., U. Munich; D.Sc. (honoris causa), U. Leeds, Cambridge, St. Louis U.; M.D. (hon.), U. Thessaloniki, Wien: Dr. rer. nat. h.c. U. of Madrid; Dr. Ing. e.h., Berlin; m. Erika von Ziegner, Feb. 28, 1931; children—Ina, Otfrid, Heide, Eckart, Anke, Imme, Maike. Privatdozent in biochem. U. of Goettingen, 1931-33; prof. of organic chemistry Technische Hochschule, Danzig, 1933-36; dir. Kaiser-Wilhelm Inst. for Biochemistry, Berlin, also hon. prof. U. of Berlin; 1936-45; dir. Max-Planck Inst. for Biochemistry, Tübingen, also prof. of physiol. chemistry U. of Tübingen, 1945-56; dir. Max-Planck Inst. for Biochemistry, Munich, also prof. physiol. chemistry U. Munich, 1956—; pres. Max-Planck Soc., 1960—. Recipient Nobel prize for chemistry, 1939. Mem. N.Y. Acad. Scis., Bavarian Acad. Sci. Goettingen Acad. of Sci., and other sci. socs. Research on plant insecticides; hormones. Address: München 1, Postfach 647, Marsopstr. 5, Germany.

BUTEO, Jean, French mathematician; b. Charpey, France, 1492; joined Order St. Antoine; later became gen. Author: Delphinatici opera geometrica, 1554; Géométrie; Logistique; (earliest work on elementary geometry in French), 1559; Réfutation de la prétundue quadrature du cercle faite par Orance finée. Worked on duplication of cube. Died Caam, France, circa 1565.

BUTLER, Allan Macy, Am. physician; b. Yonkers, N.Y., Apr. 3, 1894; s. George Prentiss and Ellen (Mudge) B.; Litt.B., Princeton, 1916; M.D., Harvard, 1926; m. Mabel H. Churchill, June 24, 1921; children—Margaret B., Allan C., Beverley A. Asst., Rockefeller Inst., 1926-28; faculty Harvard, 1929—, prof. pediatrics, 1945-60, prof. emeritus, 1960—; chief Children's Med. Service, Mass. Gen. Hosp., Boston, 1941-60; dir. clin. services, chief pediatrics Met. Hosp., Detroit, 1960-62; spl. cons. Cal. State Dept. Pub. Health, 1962-64. Lectr. dept. pediatrics Stanford Med. Sch., 1964-65; investigator OSRD, 1942-45; cons. AID, 1965, Head Start, Office Econ. Opportunity, 1965—. Recipient Ernst Boas award for contbns. to social medicine, 1964. Mem. Am., (past pres.), New Eng. (past pres.) pediatric socs., Physician Council (past pres.), Physicians Forum (past chmn.), Am. Acad. Arts and Scis., Assn. Am. Physicians. Research, numerous publs. on biol. chemistry, clin. medicine, endocrinology and nutrition. Address: Vineyard Haven, Mass. 02568.*

BUTLER, Amos William, Am. zoölogist; b. Brookville, Ind., Oct. 1, 1860; s. William Wallace and Hannah (Wright) B.; A.B., Ind. U., 1894, A.M., 1900, LL.D., 1922; LL.D., Hanover Coll., 1915; m. Mary I. Reynolds, June 2, 1880; children—Mrs. Carrie Hannah Watts, Mrs. Alice Kaylor (dec.), Wm. Reynolds, Gwyn Foster, Mrs. Anne Harrison, Hadley Butler (dec.). Ornithologist, dept. of geology and resources of Ind., 1896-97; sec. Ind. Bd. State Charities, 1897-1923. Mem. White House Children's Conf., 1909. Lectr. econs. Purdue U., 1905. Pres. Nat. Conf. Charities and Corrections, 1906-07; chmn. Am. com. on Internat. Prison Congress, Washington, 1910; mem. Am. Prison Assn.; v.p. Internat. Prison Congress; del. from U. S. to Internat. Prison Congress, London (v.p. sect. 2), 1925, Prague, 1930; pres. Ind. Conf. Charities and Correction, 1915; fellow A.A.A.S.; chmn. exec. com. Ind. Soc. for Mental Hygiene, 1918-25, pres., 1925-30; sr. sociologist U. S. Bur. Efficiency, 1928-29; founder Internat. Com. on Mental Hygiene, Washington, 1930; mem. adv. com. to Nat. Commn. on Law Observance and Enforcement, 1929-31; mem. exec. bd. Am. Inst. of Criminal Law and Criminology, 1935—. a founder Am. Anthrop. Soc. and Am. Assn. Mammalogists. Author: Birds of Indiana; Also Indiana—A Century of Progress, The Development of Public Charities and Corrections. Died Aug. 5, 1937.

BUTLER, Charles, naturalist; b. High Wycombe, Eng., 1559; entered Magdalen Hall, Oxford (Eng.); bible clk. Magdalen Coll.; B.A., 1583-84, M.A., 1587; master free sch. at Basingstoke, Hampshire, Eng., 1587-94; officiated at poor vicarage of Laurence-Wotton, 1594-1642. Author: The Feminine Monarchy, or the History of Bees, 1609; The Principles of Music, 1636. Proposed an original system of spelling. Died Laurence-Wotton, Mar. 29, 1647.

BUTLER, Charles St. John, Am. physician; b. Bristol, Tenn., Mar. 1, 1875; s. Matthew Moore and Mary Taylor (Dulaney) B.; student King Coll.; A.B., Emory and Henry Coll., 1895, LL.D., 1932; M.D., U. Va., 1897; m. Ingeborg Maria Nordqvist, July 4, 1899; children—Maria Nordqvist (Mrs. Harry L. Brockmann), Martha Amanda (Mrs. Erik W. Ehn), Ruth Elizabeth. Began practice at Bristol, 1899; entered Med. Corps, U. S. Navy, as lt., j.g., 1900, and advanced through grades to rear adm., 1935; ret., 1939; instr. bacteriology and tropical medicine, U. S. Naval Med. Sch., 1907-21; lectr. George Washington U., Jefferson Med. College, 1923-24; comdg. officer Naval Med. Sch., Washington, 1921-24, 27-32; comdg. officer U. S. Naval Hosp., Bklyn., 1932-35; dir. gen. pub. health Republic of Haiti, 1924-27; comdg. officer U. S. Naval Med. Supply Depot, Bklyn., 1935-36, U. S. Navy Med. Center, Washington, 1936-38. Mem. med. bd. NRC, 1924-26. Diplomate Am. Bd. Internal Medicine. Fellow A.M.A., A.C.S., A.C.P., N.Y Acad. Medicine; mem. A.A.A.S., Am. Acad. Tropical Medicine (pres. 1940), N.Y. Soc. Tropical Medicine (pres. 1935), Washington Acad. Scis., Am. Soc. Tropical Medicine (pres. 1927), Am. Soc. Clin Pathologists, Mil. Surgeons of U. S., Soc. of Medicine of Haiti (hon.). Author: Syphilis Sive Morbus Humanus, 1936, coll. edit. (2d), 1939; also numerous papers dealing with tropical medicine, seasickness. Died Oct. 7, 1944.

BUTLER, Clifford Charles, English scientist; b. May 20, 1922; s. C.H.J. and O. Butler; B.Sc., Reading (Eng.) U., 1942, Ph.D., 1946; m. Kathleen Betty Collins, 1947; 2 daus. Demonstrator physics Reading U., 1942-45; from asst. lectr. to lectr. physics Manchester (Eng.) U., 1945-53; reader in physics Imperial Coll., London U., 1953-57, prof. physics, 1957-63, asst. dir. Physics Dept., 1955-62, prof. physics, head Physics Dept., 1963—; dean Royal Coll. Sci., 1966—. Recipient Charles Vernon Boys prize London Phys. Soc., 1956. Fellow Royal Soc., 1961; sec. Gen. Internat. Union Pure and Applied Physics. Research, publs. on electron diffraction, cosmic ray, elementary particle physics. Home: The Gatehouse, Fulmer Rd., Gerrards Cross, Buckinghamshire, Eng. Office: Physics Dept., Imperial Coll. Sci. and Tech., Prince Consort Rd., London S.W. 7, Eng.

BUTLER, Elmer Grimshaw, Am. biologist; b. Parish, N.Y., Feb. 13, 1900; s. Frank Alexander and Elizabeth (Grimshaw) B.; A.B., Syracuse U., 1921, D.Sc., 1941; M.A., Princeton, 1925, Ph.D., 1926; m. Eleanor Brill, June 30, 1927. Instr., U. Vt., 1921-23; faculty Princeton, 1926—, Class of 1877 prof. zoology, 1937-60, Henry Fairfield Osborn prof. biology, 1960—, chmn. dept. biology, 1933-48. Trustee Marine Biol. Lab., Woods Hole, Mass. 1949—. John Simon Guggenheim fellow, 1950. Fellow A.A.A.S.; mem. Am. Soc. Zoologists (past pres.), Am. Philos. Soc., Soc. for Devel. and Growth (past pres.), Am. Assn. Anatomists, Internat. Inst. Embryology, Internat. Soc. for Cell Biology. Research, numerous publs. on normal and exptl. embryology and regeneration, particular cellular differentiation and establishment structure organisms, relation nerves to regenerative activities, devel. supernumerary structures, regeneration central nervous system. Home: 19 Lake Lane, Princeton, N.J. 08540.*

BUTLER, George Frank, Am. physician; b. Moravia, N.Y., Mar. 15, 1857; s. Isaac and Asenath (Chase) B.; M.D., Rush Med. Coll., Chicago, 1889; A.M. (hon.), Valparaiso U., 1908; m. Nannie Blanche Porter, Mar. 21, 1882. Pharmacist, Pittsfield, Mass., 1874-78; in sheep and drug bus., Kan., 1878-86; lectr. med. pharmacy and materia medica Rush Med. Coll., 1889-92; prof. materia medica, therapeutics and clin. medicine Northwestern U. Women's Med. Sch., 1890-96, Coll. Physicians and Surgeons, Chgo., 1892-1906; prof. medicine Dearborn Med. Coll., 1905-06; prof. internal medicine Chgo. Post-Grad. Med. Sch., 1905-07; med. supt. Alma Springs Sanitarium, Alma, Mich., 1900-05; prof. and head dept. therapeutics, prof. clin. and preventive medicine Chicago Coll. Medicine and Surgery, 1906-15, emeritus, 1915—; pres. faculty and prof. diseases of kidneys and nervous system Practitioners' Coll., Chgo. 1910-12; phys. of Cook Co., Ill., 1911-13; med. dir. N. Shore Health Resort, Winnetka, Ill. Author: Textbook of Materia Medica, Therapeutics and Pharmacology, 1896. Died June 22, 1921.

BUTLER, Gordon Cecil, Canadian biochemist; b. Ingersoll, Ont., Can., Sept. 4, 1913; s. Irvin and Edna (Harris) B.; B.A., U. Toronto (Ont.), 1935, Ph.D., 1938; m. Jean Meeke, July 3, 1937; children—Judith, Stephen, Gregory, Susan. Researcher, U. London (Eng.), 1938-40; research chemist Charles E. Frosst Co., Montreal, Que., Can., 1940-42; staff NRC atomic energy project, 1945-47; prof. biochemistry U. Toronto, 1947-57; dir. biology and health physics div. Atomic Energy of Can., Ltd., 1957-65; dir. div. radiation biology NRC, Ottawa, Ont., 1965—. Fellow Royal Soc. Can.; A.A.A.S.; mem. Canadian Soc. for Cell Biology, Canadian Biochem. Soc., Canadian Physiol. Soc., Am. Soc. Biol. Chemists, Health Physics Soc. Research, publs. biochemistry, radiation chemistry and health physics. Home: 209 Melrose Av., Ottawa 3, Ont. Office: NRC, Ottawa 2, Ont., Can.*

BUTLER, Howard Crosby, Am. archeologist; b. Croton Falls, N.Y., Mar. 7, 1872; s. Edward Marchant and Helen Belden (Crosby) B.; A.B., Princeton, 1892, A.M., 1893; postgrad. Columbia Sch. Architecture, Am. Schs. Classical Studies, Rome, Athens; univ. fellow in archeology, Princeton, 1893, 97; fellow Am. Sch. Classical Studies, Rome, 1897-98. Organized and conducted archaeol. expdns. in Syria, 1899-1900, 04-05, 09; prof. history of architecture, Princeton, from 1905, also resident master Grad. Coll.; dir. Am. Excavations at Sardis in Asia Minor. Recipient Drexel Gold medal, 1910. Mem. Archaeol. Inst. Am. (hon.), A.I.A., Am. Acad. Arts and Letters. Author: Scotland's Ruined Abbeys, 1900; The Story of Athens, 1902; Archaeology and Other Arts, 1903; Sardis, 1922. Revealed previously unknown civilizations and recovered many scientifically valuable artifacts in his later expdns. to Syria, 1910-22. Died Am. Hosp., Neuilly, France, Aug. 15, 1922.

BUTLER, John Alfred Valentine, English chemist; b. Winchcombe, Eng., Feb. 14, 1899; s. Alfred and Mary Ann (Powell) B.; M.Sc., Birmingham (Eng.) U., 1922, D.Sc., 1927; m. Margaret Lois Hope, July 2, 1929; children—Gavin, Elizabeth (Mrs. E. Warrington), William. Lectr., U. Coll. Swansea (Wales), 1922-26, U. Edinburgh (Scotland), 1926-39; Rockefeller fellow Rockefeller Inst., Princeton, N.J.,1939-41; exec. officer Brit. Sci. Office, Washington, 1941-44; phys. chemist Chester Beatty Research Inst., 1949; prof. phys. chemistry U. London (Eng.), 1954-66. Recipient Meldola medal, 1929. Fellow Royal Soc., 1956; mem. Biochem. Soc., Biophys. Soc., Faraday Soc. Author: Chemical Thermodynamics, 1934; Science and Human Life, 1957; Inside the Living Cell, 1959; Life of the Cell, 1964; also numerous articles. Research on thermodynamics of solutions and electrochemistry, proteins, nucleic acids, effects of radiation. Home: Nightingale Corner, Rickmansworth, Herts, Eng.

BUTLER, John M., Am. psychologist; b. Hector, Minn., May 12, 1917; s. John Patrick and Myrtle (Schroeder) B.; B.S., U. Minn., 1939, Ph.D., 1949; m. Helen Machat, July 12, 1956; children—James T., Steven. Faculty, U. Chgo., 1947-—, prof. psychology, 1960-—. Mem. Am., Midwestern psychol. assns., A.A.A.S., Soc. Multivariate Exptl. Psychologists, Sigma Xi. Author: (with L. N. Rice, A. K. Wagstaff) Quantitative Naturalistic Research, 1963. Studies on process and outcome of psychotherapy, theory of factor analysis. Home: 18510 Perth, Homewood, Ill. Office: 5848 University Av., Chgo. 60637.*

BUTLER, Philip A., Am. biologist; b. Upper Montclair, N.J., May 15, 1914; s. Marshall A. and Cora (Lewis) B.; B.S., Northwestern U., 1935, Ph.D., 1940; m. Ona Cunningham, Aug. 29, 1941; children—Patricia Ann (Mrs. William Kwachka), Penelope (Mrs. Ira Arlook). Fishery biologist U. S. Fish and Wildlife Service, Annapolis, Md., 1946-48, fishery biologist, lab. dir. Bur. Comml. Fisheries, U. S. Fish and Wildlife Service, Gulf Breeze, Fla., 1948-—; mem. nat. tech. adv. com. on water quality criteria, 1967. Mem. Nat. Shellfisheries Assn. (pres. 1961-62, 1962-63), A.A.A.S., Am. Fisheries Soc., Am. Soc. Limnology and Oceanography, Sigma Xi. Contbr. articles to sci. jours. Demonstrated effects of flood waters on oyster prodn.; developed techniques for monitoring effects of cyclic hydrographic changes on oysters and other sedentary animals; documented contamination of estuarine environment with pesticides and developed special techniques pesticide bioassay tests; initiated nationwide program to monitor pesticides in the marine environment. Home: Sabine Island, Gulf Breeze. Office: Bur. Comml. Fisheries, Biol. Lab., Sabine Island, Gulf Breeze, Fla. 32561.*

BUTLER, Ralph, Am. otolaryngologist; b. Loag, Pa.; s. James and Rachel M. (James) B.; B.E., West Chester State Normal Sch. (now West Chester Tchrs. Coll.), 1893; M.D., U. Pa., 1900; studied diseases of ear, nose and throat, Vienna, 1901-02, Berlin, 1906; m. Ida Shaw, Dec. 18, 1905. Resident St. Joseph's Hosp., Phila., 1900-01; asst. aural surgeon U. Pa., 1902-06, instr. in otology, 1907-16, asst. prof. otology, 1916-24, prof. laryngology and vice dean of otolaryngology, grad. school of medicine, 1918-46, emeritus prof. laryngology, from 1946; prof. diseases nose and throat Phila. Polyclinic and Coll. for Grads. in Medicine, 1912-18; cons. otolaryngology Lankenau Hosp., Drexel Home and Women's Hosp. of Phila. Fellow A.C.S.; mem. A.M.A., Am. Otol. Soc., Am. Laryngol. Assn., Am. Laryngol., Rhinol and Otol. Soc., Med. Soc. State Pa., Phila. County Med. Soc., Alpha Kappa Kappa. Died Apr. 1954.

BUTLER, Thomas Cullom, Am. pharmacologist, educator; b. Phoenix, Apr. 2, 1910; s. Thomas Berry and Marian (Cullom) B.; A.B., Vanderbilt U., 1930, M.D., 1934; m. Pauline Campbell, Aug. 10, 1936; children—Barbara, Thomas Campbell, Marian. Faculty dept. pharmacology Vanderbilt U., 1934-44; tech. aide malaria program OSRD, 1944-46; faculty Johns Hopkins, 1946-50; prof. pharmacology U. N.C., Chapel Hill, 1950-—, dir. Center for Research in Pharmacology and Toxicology, 1965-—. Mem. Am. Soc. for Pharmacology and Exptl. Therapeutics, Soc. for Exptl. Biology and Medicine. Research, numerous publs. on anesthetics and hypnotics, chem. alterations of drugs in body, distbn. of drugs in tissues of body, excretion of drugs by kidney, studies of intracellular pH. Home: Route 1, Box 337, Chapel Hill, N.C. 27515.*

BUTLER, William John, vet. surgeon; b. Bowling, Scotland, May 31, 1881; s. Hugh and Isabella (Fenwick) B.; brought to U. S., 1889; D.V.S., N.Y. U., 1903; m. J. Ozella Cato, May 24, 1921; children—Hugh Cato, Cato Kay. Began practice in Mont., 1903, also live stock insp. and rancher; mining in Mexico, 1910-13, also rep. Am. stock men, at Mazatlan; state vet. surgeon of Mont. since 1913; dir. labs. Mont. Live Stock Sanitary Bd.; adminstr. of relief, Montana, 1934-35. Del. from U. S., to 13th Internat. Congress of Veterinary Medicine, Zurich, Switzerland, 1938; dir., mem. Vet. Bd. Internat. Live Stock Expn., Chgo.; dir. Mont. State Fair; pres. Mont. Stallion Registry Bd.; chmn. Mont. Milk Control Bd., from 1935. Mem. U. S. Live Stock San. Assn. (pres. 1922-23), Mont. Bd. Entomology, Am., Mont. (pres. 1938-39) vet. med. assns., Western States L.S.S. Assn. (pres. 1924-25), Am. Pub. Health Assn., Internat. Assn. Dairy and Milk Insps., A.A.A.S. Author of pamphlets and articles on diagnosis and control of disease in domestic and wild animal life, production of clean milk and its care, and range conditions affecting live stock. Died Oct. 29, 1948.

BUTLER, William Morris, Am. neurologist; b. Maine, N.Y., Mar. 26, 1850; s. William and Nancy (Smith) B.; A.B., Hamilton Coll., 1870, A.M., 1874; M.D., Coll. Phys. and Surg. (Columbia), 1873; postgrad. in nervous diseases Bellevue Med. Coll., L.I. Med. Coll., École de Médecine, Paris, and under Charcot and Déjerine, La Salpêtrière, Paris; m. Mary Bradford, Oct. 9, 1874; 1 son, Bradford; m. 2d, Anna Clarke; m. 3d, Ann L. Pell. Began practice, Montclair, N.J., 1873; specialist in nervous and mental diseases, Bklyn., 1883-—; 1st asst. physician Middletown State Homoe. Hosp., 1874-83, later cons. alienist; prof. mental diseases N.Y. Homoe. Coll. and Flower Hosp., 1904-17. Mem. N.Y. State Bd. Med. Examiners, 1892-1903. Hon. mem. Western N.Y. Homoe. Soc. Author: Mental Diseases and Their Homoeopathic Treatment, 1910. Died June 23, 1940.

BUTLEROV, Alexander Michailovich, Russian chemist; b. Christopol, Russia, Aug. 25, 1828; M. Chemistry, U. Kazan (Russia), 1851; D. Chemistry and Physics, U. Moscow, 1854; m. Nadezhda Mikailovna Glumilina, 1851; children—Michail, Vladimir. Asso. prof. chemistry, Kazan, 1854-57, apptd. prof., 1857, rector, 1860-61, 62-63; with U. St. Petersburg, 1867-80. Mem. Russian Acad. Scis. Author: Vvedenie k polnomu izuceniju organiceskoj chimii, 1864; numerous articles. Invented term chem. structure; made formula correspond to substance; discovered tertiary alcohols; 1st to prepare formaldehyde; recognized character of tautomerism; isolated and synthesized compounds in petroleum. Died near Kazan, Aug. 5, 1886.

BUTLIN, Sir Henry Trentham, English surgeon; b. Camborne, Eng., Oct. 25, 1845; s. William Wright and Julia Crowther (Trentham) B.; ed. St. Bartholomew's Hosp., from 1864; qualified Dr., 1867; D.C.L. (hon.), Durham; m. 1st, Annie Tipping Balderson, 1873 (dec. 1916). House surgeon to Sir James Paget, after 1867; asst. surgeon St. Bartholomew's Hosp., 1881, full surgeon, 1892, cons. surgeon, became lectr. hosp. sch., 1897; 1st dean faculty medicine of reconstructed U. London; gov. Rugby Sch.; councillor, treas., pres. Brit. Med. Assn. Fellow Royal Coll. Surgeons (pres. 1909). Author: Diseases of the Tongue, 1885. Contbr. to devel. of Brit. laryngology; research on pathology of carcinoma and sarcoma. Died London, Jan. 24, 1912.

BÜTSCHLI, Otto, German zoologist; b. Frankfort/Main, Germany, May 3, 1848; s. Friedrich and Phil. Caroline (Culmann) B.; studied mineralogy, chemistry and zoology in Karlsruhe, Heidelberg, and Leipzig (all Germany); hon. dir. Jena and Cambridge; m. Hedwig Sophie Wilhelmine Hoffmann, 1 dau.; m. 2d, Mathilde Lange, 2 daus. Asst. to K. Moebius, U. Kiel (Germany); faculty, Karlsruhe; prof. Heidelberg, from 1878. Mem. Berlin, Göttingen, Munich, St. Petersburg, Brussels acads. Author: Studien über die ersten Entwicklungsvorgänge der Eizelle, der Zellteilung und der Conjugation der Infusorien, 1876; Untersuchungen über Strukturen, insbesondere über Strukturen nicht zelliger Erzeugnisse des Organismus, 1898; Untersuchungen über Mikrostrukturen des estarrten Schwefels, 1900; Untersuchungen über organische Kalkgebilde nebst Bemerkungen über organische Kieselgebilde, 1908; Protozoen, 1880-89; Mechanismus und Vitalismus, 1901; Vorlesungen über vergleichende Anatomie, 1910; Untersuchungen über mikroskopische Schäume und die Protoplasma, 1892; Über den Bau quellbarer Körper, 1896. Pioneer research in cytology; research on protozoa (1878-85), evolution of invertebrates, nuclear and cell division, protoplasmic structure, bacteria suggested protoplasm is foamlike or alveolar; helped establish its fluid nature; in his invertebrate studies, advanced understanding of devel. of insects, gastropods; investigated structure of nematode worms. Died Heidelberg, Feb. 3, 1920.

BUTT, Hugh Roland, Am. physician; b. Belhaven, N.C., Jan. 8, 1910; s. Harry Roland and Maybelle (Jarvis) B.; student Va. Polytech. Inst., 1927-29; M.D., U. Va., 1933; Mayo Found. U. Minn., 1937; m. Mary Dempwolf, Apr. 8, 1939; children—Selby, Lucy, Charles, Frances. Intern St. Luke's Hosp., Bethlehem, Pa., 1933-34; fellow medicine Mayo Found., 1934-37, 1st asst., 1937-38, instr., 1938-43, asst. prof., 1943-47, asso. prof., 1947-52, prof., 1952-—; cons. physician Mayo Clinic, St. Mary's Hosp., 1938-—. Chmn. sci. counselors Nat. Cancer Inst., 1961-62; member National Advisory Cancer Council, 1966-—. Recipient John Horsley Meml. prize U. Va., 1938. Diplomate American Board Internal Medicine (member board; subspecialty gastroenterology). Fellow A.C.P.; member American Society Clinical Investigation, American Gastroenterological Association, also Central Society, Clin. Research, Association of American Physicians American Med. Assn. Author: Vitamin K (with Snell), 1941. Author papers, monographs. Research on gastroenterology; (with Snell) used vitamin K and bile to treat hemorrhagie diathesis in jaundice, 1938. Home: 1014 7th St. S.W., Rochester, Minn.

BUTT, Wilfrid Roger, Brit. biochemist; b. Southampton, Eng., May 2, 1922; s. Albert Thomas and Charlotte (Coffin) B.; B.Sc., London U., 1945; Ph.D., Birmingham U., 1954; m. Patricia Doreen Sharp, Apr. 25, 1951; children—Susan Elizabeth, John Anthony. Exptl. asst. Ministry of Supply London, 1939-45; research asst., endocrine unit London Hosp., 1946-49; cons. biochemist, dept. clin. endocrinology Women's Hosp., Birmingham, Eng., 1949-—; hon. research fellow U. Birmingham, 1959-65, hon. lectr., 1965-—. Fellow Royal Inst. Chemistry; mem. Assn. Clin. Biochemistry (chmn. Midland region 1959-60), Chem. Soc., Biochem. Soc., Soc. of Endocrinology. Author: Hormone Chemistry, 1966; Chemistry of Gonadotrophins, 1966. Research, publs. on devel. of methods for purification and assay of steroids and pituitary hormones; purification of gonadotrophins from human pituitary glands and from urine for chem., biol., clin. applications including chem. analysis of hormones, immunological assay and use of hormones in treatment of male and female fertility. Home: 383 Tilehouse Lane, Tidbury Green, Solihull, Warwickshire, Eng. Office: Women's Hosp., Showell Green Lane, Sparkhill, Birmingham, Eng.*

BUTTENBERG, Dietrich Theodor Walter, German physician; b. Plauen, Germany, Mar. 27, 1924; s. Otto Albin Oskar and Luise (Merkel) B.; Faculty, U. Heidelberg (Germany), 1949-—, prof. medicine, 1965-—; chief Klinik Wuhr, Rosenheim, Germany, 1965-—. Author: Mammography, 1964; Gelbkörperhormontherapie, 1966; also articles. Research on mammography, endocrinology, vitamins, techniques of operations. Home: 3 Münchener, Office: 4 Droste-Hülshoff, Rosenheim, Germany.*

BUTTENSTEDT, Carl, German aero. pioneer; b. Volkstedt, Saxony, Germany, July 29, 1845; s. Andreas and Auguste Elisabeth (Schuppe) B.; m. Rosa Beier, 1881; 4 sons, 2 daus. Tchr., mil. acad. at Weiseenfels/Saale, Germany; then in mining adminstrn; worked solely on sociol. and aero. research after 1904. Corr. mem. Royal Acad. La Stella d'Italia. Author: Das Flugprinzip, 1889; Das Geheimnis des Flugmediums, 1894; Naturstudien zur einer neuen Segeltheorie, 1894; also articles in aero. mags. Theoretical father of glider; suggested birds beat wings to accelerate forward motion, but are held up in air by constant displacement of supporting column of air by gliding. Died Berlin-Friedrichshagen, Germany, Sept. 20, 1910.

BUTTERFIELD, William John Hughes, English physician; b. Birmingham, Eng., Mar. 28, 1920; s. William Hughes and Doris (Pritchard) B.; B.A., Oxford (Eng.) U., 1942, B.M., B.Ch., 1945, M.A., 1946; M.D., Johns Hopkins, 1951; m. Ann West Sanders, Feb. 1946 (dec. Oct. 1948); 1 son, Jonathan West Hughes; m. 2d, Isabel Ann Foster Kennedy, Mar. 16, 1950; children—Sarah Harriet Ann, Jeremy John Nicholas, Toby Michael John. Sci. staff Med. Research Council, 1947-58; prof. exptl. medicine U. London at Guy's Hosp., 1958-63, prof. medicine, 1963-—; staff U.K. Atomic Weapons Trials, 1952, 56. Chmn. Bedford Diabetic Survey, 1962; cons. expert com. on diabetes WHO, 1964-—; chmn. liaison com. Woolwich-Erith New Town Integrated Med. Care Com., 1965-—; mem. med. sub-com. U. Grants Com., 1966-—. Decorated Order Brit. Empire. Fellow Royal Coll. Physicians London, N.Y. Acad. Scis. (hon.); mem. European Soc. for Study Diabetics. Author: (with others) on Burns, 1953, Shorter Textbook of Medicine, 1962, On diabetes, 1966; also numerous articles. Pioneered investigations on long range and clin. course of flash burns and difficulties of mass treatment in nuclear def.; showed that insulin does not pass from blood into tissues as easily in diabetics as in normals suggesting peripheral block to hormone action; organized 1st mass screening for undetected diabetics in Bedford, Eng. Home: 3, Maids of Honour Row, Richmond, Surrey, Eng. Office: Dept. Medicine, Guys Hosp., London, S.E., Eng.*

BÜTTNER, Christian Wilhelm, natural scientist; b. Wolfenbüttel, Germany, Feb. 27, 1716; s. Johann Christian and Christiana (Ulrich) B.; trained as pharmacist; named magister U. Göttingen (Germany), 1755; asso. prof., 1758; prof., 1763; privy councilor to Saxon-Weimar court, Jena, Germany, 1783; asso. mem. Societät der Wissenschaften Göttingen from 1762. Author: Vergleichungstafeln der Schriftarten verschiedener Völker in den vergangenen und gegenwärtigen Zeiten, 2 vols, 1771-81. Suggested that races of mankind developed from a single basic form; influenced J. F. Blumenbach and others; gave his naturalia collection to Göttingen, 1773. Died Jena, Oct. 8, 1801.

BUTYLIN, Aleksey Grigorevich, Russian obstetrician, gynecologist; b. Chistopol, Kazan Guberniya, 1886; grad. Med. Faculty, Kazan U., 1914; D.Med. Sci., 1937. Intern, Irkutsk Med. Inst., 1920-24, asst., later lectr. Obstetrics and Gynecology Clinic, 1924-38; prof., chair obstetrics and gynecology Alma-Ata Med. Inst. (during evacuation Kursk Med. Inst. to Alma-Ata), 1941-43; prof., head chair obstetrics and gynecology Kursk Med. Inst., 1938-—, also chmn. State Exam. Commn. Chmn., Kursk City and Oblast, USSR and RSFSR Ministry Health Commns. on Obstetrics. Mem. Kursk Obstet. and Gynecol. Soc. (chmn.). Author some 50 works on obstetrics and gynecology. Mem. editorial council Obstetrics and Gynecology. Address: Kursk Med. Inst., ulitsa Karla Marksa 3, Kursk, RSFSR, USSR.

BUXBUAM, Johannes Christian, botanist; b. Merseburg, Germany; christened, Oct. 5, 1693; s. Andreas and Maria Dorothea (Bretnitz) B.; studied in Leipzig, Jena and Halle (all Germany); accompanied Count Romanzov to Constantinople, afterwards traveling through Asia Minor; Peter the Great called him to St. Petersburg, Russia, 1724; participated in founding Acad. Scis.; prof. Imperial U. Author: Enumeratio plantarum in agro Hallensi crescentium, 1721; Centuriae plantarum minus cognitarum circa Byzantium et in Oriente observatarum, 1728-40. First to describe plants of coastal lands of Black Sea, Asia Minor and Armenia; moss genus Buxbaumia named after him. Died Merseburg, July 17, 1730.

BUXTON, Charles Lee, physician; b. Superior, Wis., Oct. 14, 1904; s. Edward Timothy and Lucinda (Lee) B.; B.S., Princeton, 1927; M.D., Columbia, 1932, Med. Sc.D., 1940; M.A., Yale, 1954; m. Helen Morgan Rotch, Sept. 3, 1938 (div. 1967); children—Timothy, Anthony, Edward, Lucinda; m. Margaret Mithoefer, November 24, 1967. Director endocrine clinic Coll. Physicians and Surgeons, N.Y.C., 1938-54, asso. prof., 1947, asso. attending in obstetrics, 1947-54; prof. obstetrics and gynecology Yale Med. Sch., 1954—, chmn. dept. obstetrics, gynecology, 1954-66, dir. infertility clinic, 1954—; cons. obstetrics, gynecology Wm. W. Backus Hosp., Norwich, Charlotte Hungerford Hosp., Torrington, Meriden, New Britain Gen., Stamford, Hartford hosps. (all Conn.). Recipient Lasker award, 1965. Fellow Am. Coll. Obstetrics and Gynecology; mem. A.M.A., Am. Endocrine Soc., Am. Fertility Soc., Am. Assn. Obstetrics and Gynecology, N.Y., New Haven, New Eng. obstetrics socs., Am. Gynecol. Soc., Soc. Gynecol. Investigation, Soc. de Obstetrics and Gynecology de Brasil, Soc. Royal Belge de Gynecologie et a'obstetrique; Brit. Soc. Study of Fertility. Author: (with E. T. Engle) Diagnosis and Therapy of Gynecological Endocrine Disorders, 1949; (with A. Southam) Human Infertility, 1958; Psychophysical Methods for Relief of Childbirth Pain, 1962; also articles. Research in obstet. and gynecol. problems. Home: 24 Cliff St., New Haven 06511. Office: 333 Cedar St., New Haven 06510.*

BUXTON, James Basil, Brit. veterinarian; ed. Royal Vet. Coll., London, Eng., Liverpool U.; M.A.; Diploma in Vet. Hygiene. Lectr. vet. hygiene Royal Vet. Coll., 1911-12, prin., dean, 1936-54; vet. supt. Wellcome Physiol. Research Labs., 1912-22; dir. farm labs. Med. Research Council, 1922-23; prof. animal pathology Cambridge U., 1923-36, dir. Inst. Animal Pathology. Recipient John Henry Steel Gold medal, 1934. Fellow Royal Coll. Vet. Surgeons (pres. 1932-33, mem. council from 1920), Queen's Coll; mem. numerous vet. socs. Died May 25, 1954.

BUXTON, Leonard Halford Dudley, Brit. archeologist, anthropologist; b. Apr. 18, 1889; s. Dudley W. Buxton; ed. Exeter Coll., Oxford; M.A., D.Sc.; m. Marie Louise Montgomery, 1923; 1 dau. Bursar, Exeter Coll.; univ. reader in phys. anthropology Oxford U., from 1928, proctor, 1929. Albert Kahn Travelling fellow, 1921. Fellow Royal Geog. Soc., Soc. Antiquaries. Author: Peoples of Asia, 1925; China, 1930; (with S. Gisbon) Oxford University Ceremonies, 1935; also papers. Noted for studies of China and Japan; made extensive excavations in Sudan and Cyprus. Died Mar. 5, 1939.

BUXTON, Patrick Alfred, Brit. biologist, entomologist; b. London, Eng., Mar. 24, 1892; s. Alfred Fowell and Violet (Jex-Blake) B.; ed. Trinity Coll., Cambridge (Eng.) U., St. George's Hosp., London; m. Muryell Gladys Rice; 1 son, 4 daus. Med. entomologist to Iraq, 1914-18, to govt. Palestine, 1921-24; dir. expdn. to Samoa, 1924-26; dir. dept. entomology London Sch. Hygiene and Tropical Medicine, 1926-55; prof. med. entomology U. London, 1933-55. Recipient Linnean Gold medal, 1953. Fellow Royal Soc., 1943, Royal Entomol. Soc. London. Author: Animal Life in Deserts, 1924; Researches in Polynesia and Melenesia; The Louse, 1947; The Natural History of Tsetse Flies. Expert on relation of insects to problems of preventive medicine. Died Dec. 13, 1955.

BUYS-BALLOT, Christopher Hendrik Dirk, Dutch meteorologist; b. Kloetinge, Zealand, Netherlands, Oct. 10, 1817; ed. U. Utrecht (Netherlands). Became prof. math. U. Utrecht, 1847, prof. of exptl. physics, 1870; apptd. dir. Royal Dutch Meteorol. Inst., Utrecht, 1854-87. Author: Changements périodiques de la témpérature, 1847; Eenige reglen voor te wachten van weerverandering in Nederland, 1860; Suggestions of a Uniform System of Meteorological Observations, 1872-73. Established system of storm signals in Europe; invented aeroklinoscope; formulated a meteorol. law of storms (named for him) which asserts that if one stands with his back to the wind, the area of low pressure is to his left; advocated uniform system of meteorol. observations. Died Utrecht, Feb. 3, 1890.

BUZZARD, Sir Edward Farquhar, Brit. physician; b. London, Dec. 20, 1871; s. Thomas Buzzard; ed. Magdalen Coll., Oxford, St. Thomas's Hosp.; M.D., M.A.; hon. doctorates, Manitoba, Belfast. Regius prof. medicine Oxford U., 1928-43, emeritus, from 1943; physician in ordinary to King, 1932-36, extra physician, from 1937; v.p. Radcliffe Infirmary; cons. physician St. Thomas's Hosp., Ministry of Pensions; physician, lectr. med. pathology Royal Free Hosp.; physician Belgrave Hosp. for Children, Nat. Hosp. for Paralyzed and Epileptic; hon. fellow Magdalen Coll., from 1928; hon student Christ Ch., 1943. Recipient Osler Meml. medal, 1940. Fellow Royal Coll. Physicians (Goulstonian lectr. acute infective and toxic conditions of nervous system 1907, Harveian orator 1941); Royal Soc. Medicine (pres. sects. clin. medicine, neurology and psychiatry); fellow or mem. Brit. Med. Assn. (pres. 1936), Assn. Brit. Neurologists (pres. 1938), Assn. Physicians, (pres. 1941); hon. freeman Soc. Apothecaries. Co-author: Pathology of the Nervous System. Contbr. author: System of Medicine (Allbutt), Modern Medicine (Osler), Index of Differential Diagnosis (French). Contbr. papers to jours. Died Dec. 17, 1945.

BUZZATI-TRAVERSO, Adriano Antonio, Italian geneticist; b. Milan, Italy, Apr. 6, 1913; s. Giulio Cesare and Alba (Mantovani) Buzzati; D.Sc., U. Milan, 1936; M.D. (hon.), U. Louvain, 1962. Asst. instr. zoology U. Pavia, 1937-44, prof. genetics, 1948—, dir. Inst. Genetics, 1948-62; dir. div. genetics Instituto Italiano di Idrobiologia, Pallanza, 1944-48; vis. prof. U. Cal., 1951-52; prof. biology Scripps Instn. Oceanography, U. Cal. at La Jolla, 1953-59, research asso. 1961—; dir. sci. div. Italian Nat. Com. for Nuclear Energy, 1958-60; dir. Internat. Lab. Genetics and Biophysics, Naples, 1962—. Mem. exec. coms. Internat. Cell Research Orgn., European Molecular Biology Orgn. Mem. A.A.A.S. (life), Genetics Soc., Soc. for Study Evolution, Biometric Soc., others. Author: (with L. Cavalli) Teoria dell'Urto, 1948; Perspectives in Marine Biology, 1958; Immediate and Low Level Effects of Ionizing Radiations, 1960; also numerous articles. Demonstrated that polygenic mutations can be induced by mutagenic agts. such as X-rays and that new genetic variability produced may increase rate of phenotype change under natural selection pressure. Home: 26 Via Posillipo, 80123 Naples. Office: 10 Via Marconi, 80125 Naples, Italy.*

BUZZI, Alfredo Patricio, Argentine cardiologist; b. Buenos Aires, Argentina, Oct. 11, 1930; s. Alfredo and Maria (Matthews) B.; M.D., Buenos Aires Sch. Medicine, 1957; m. Matilde Ana Lamberti, May 28, 1962; children—Alfredo Eugenio, Patricio Carlos. Asst. instr. Sala 4, Hosp. de Clinicas, Buenos Aires, 1957-58, instr. medicine, 1958-64; docent Facultad de Medicina, Buenos Aires, 1964—; became cons. cardiovascular disease Centro Nacional de Rehabilitacion, 1959, chief dept. angiology, 1963; became cardiologist Navy Hosp., Bueons Aires, 1961. Recipient Premio Pedro Schneidewind Bahia, 1951, Premio Ana Malenky, 1963. Mem. Sociedad Argentina de Historia de la Medicina (pres. 1965-66), Sociedad Argentina de Cardiología, Sociedad de Angiologia, Sociedad de Medicina Interna. Research and publs. on mechanism of blood flow redistribution following vasodilatation; nature of peripheral vascular response during arteriography; abnormalities of digital pulse vol. in Raynaud's syndrome; plethysmographic grading of severity of atherosclerosis obliterans. Home: 1949 Obligado. Office: 615 Pueyrredón, Buenos Aires, Argentina.*

BYERLY, Perry, Am. seismologist; b. Clarinda, Ia., May 28, 1897; B.A., M.A., Ph.D., U. of Calif., 1920-21, graduate school, 1921-24; Sr. Fulbright scholar U. Cambridge, Eng., 1960-61; LL.D., Univ. California at Berkeley, 1966; 3 sons, Perry Edward, David, Donald; m. 2d Lillian Lizee, 1941. Instructor physics, University of Nevada, 1924-25; successively instr., asst. prof., asso. prof., prof. seismology U. Cal., asst. prof., asso. prof., prof. seismology U. Cal., 1925-64, chmn. dept geol. scis., 1949-54, now dir. emeritus seismographic stas., prof. emeritus. Guggenheim fellow, 1929, 52; Smith-Mundt Act lectr. Nat. U. Mexico, 1954; Condon lectr. Univs. Ore. Fellow Royal Astron. Soc., Geol. Soc. Am., Am. Geophys. Union, Internat. Assn. Seismology and Physics of Interior of Earth (pres., 1960-63); mem. Am. Acad. Arts and Scis., Nat. Research Council, Wash. Acad. Scis., Nat. Acad. Scis. (chmn. sect. on geophysics 1957-60, chmn. panel seismology and gravity of Internat. Geophys. Yr., 1956-58), Earthquake Engring. Research Inst., Seismol. Soc. Am. (hon.). Author: Seismology, 1942; contbr. articles on research to Bull, of Seismol. Soc. of Am., etc. Study of nature of forces at source of earthquakes; energy in earthquake waves; California seismology; roots of mountains; earth structure. Home: 6037 Contra Costa Rd., Oakland, Cal. 94618. Office: Earth Scis. Bldg., U. Cal., Berkeley, Cal. 94720.

BYERLY, Theodore Carroll, Am. biologist; b. Melbourne, Ia., May 3, 1902; s. William Henry and Lulu May (Crook) B.; B.A., State U. Ia., 1923, M.S., 1925, Ph.D., 1926; m. Helen Frances Freeman, May 31, 1929; children—Carroll (Mrs. Norman Holcomb), David S., Nora (Mrs. Thomas D. Bolita). Instr., U. Mich., 1926-28, Hunter Coll., 1928-29; physiologist div. animal husbandry, bur. animal industry U. S. Dept. Agr., 1929-37; poultry husbandman in charge poultry husbandry div., 1941-53; chief animal and poultry husbandry research br. Agrl. Research Service, 1953-55, asst. dir. livestock research, 1955-57, dep. adminstr. Farm Research, 1957-62; adminstr. Cooper State Research Service, Washington, 1962—; prof. Md. U., 1937-41. Recipient Borden award for contbn. to poultry sci. Poultry Sci. Assn., 1943; Distinguished Service award U. S. Dept. Agr., 1965. Author: Livestock and Livestock Products, 1964; also numerous articles. Research on efficiency of feed conversion and reproductive physiology. Home: 6-J Ridge Rd., Greenbelt, Md. 20770. Office: U. S. Dept. Agr., Coop. State Research Service, Washington 20250.*

BYERLY, William Elwood, Am. mathematician; b. Phila., Dec. 13, 1849; s. Elwood B.; A.B., Harvard, 1871, Ph.D., 1873; m. Alice Worcester Persons, May 28, 1885; children—Robert Wayne, Francis Parkman; m. 2d, Anne Carter Wickham Renshaw, July 23, 1921. Asst. prof. math. Cornell, 1873-76; asso. prof. math. Harvard, 1876-81, prof., 1881-1913. Fellow Am. Acad. Arts and Scis. Author: Elements of Differential Calculus, 1879; Elements of Integral Calculus, 1881; An Elementary Treatise on Fourier's Series and Spherical, Cylindrical and Ellipsoidal Harmonics, 1893; Problems in Differential Calculus, 1895; Generalized Coördinates, 1916; Introduction to the Calculus of Variations, 1917. Died Swarthmore, Pa., Dec. 20, 1935.

BYERRUM, Richard Uglow, Am. biochemist; b. Aurora, Ill., Sept. 22, 1920; s. Earl Edward and Florence (Uglow) B.; A.B., Wabash Coll., 1942; Ph.D., U. Ill., 1947; m. Claire Somers, Apr. 3, 1945; children—Elizabeth, Robert, Mary. Research asso. U. S. Chem. Corps., toxicity dept. U. Chgo., 1944-47; faculty Mich. State U., East Lansing, 1947—, prof. biochemistry, 1957—, acting dir. Inst. Biology and Medicine, 1961-62, dean Coll. Natural Sci., 1962—. Traveling scientist Am. Chem. Soc., 1960-64. Recipient Travel award Internat. Congress Biochemistry, Vienna, Austria, 1958, Montreal, Que., Can., 1959. Mem. A.A.A.S., Am. Soc. Plant Physiologists, Am. Soc. Biol. Chemists, Soc. for Exptl. Biology and Medicine, Mich. Acad. Arts., Sci. and Letters, Am. Chem. Soc., Federated Biol. Soc., Mich. Assn. Professions, Ingham County Med. Soc., Sigma Xi (Jr. Research award 1958), Phi Beta Kappa, Phi Lambda Upsilon, Alpha Chi Sigma, Beta Theta Pi, Gamma Alpha, Phi Kappa Phi. Author: (with Ball, Fairley, Lillevik) Experimental Biochemistry, 1956; also numerous articles, revs. Research on biosynthesis alkaloids. Patentee cancer tumor inhibiting materials. Home: 602 Wildwood Dr., East Lansing, Mich. 48823.*

BYERS, George William, Am. entomologist; b. Washington, May 16, 1923; s. George and Helen (Kessler) B.; B.Sc., Purdue U., 1947; M.Sc., U. Mich., 1948, Ph.D., 1952; m. Gloria Wong, Dec. 16, 1955. Faculty, U. Kan., Lawrence, 1956—, prof. entomology, curator Snow Entomol. Mus., 1965—. Rackham fellow, U. Mich., 1952-53. Mem. Entomol. Soc. Am., Soc. Systematic Zoology, Soc. for Study Evolution, Kan. Entomol. Soc., Sigma Xi. Contbr. articles to tech. jours. Editor: Systematic Zoology, 1963-67. Research on biology and classification crane flies and scorpion-flies. Home: 2215 Princeton Blvd., Lawrence, Kan. 66044.*

BYERS, Horace Robert, Am. meteorologist; b. Seattle, Mar. 12, 1906; s. Charles Hopkins and Harriet (Ensminger), B.; A.B., U. Cal. at Berkeley, 1929; S.M., Mass. Inst. Tech., 1932, Sc.D., 1935; m. Frances Isabel Clark, Oct. 6, 1927; 1 dau., Henrietta (Mrs. T. W. Bilhorn). Various research positions, 1927-35; research meteorologist U. S. Weather Bur., 1935-40; faculty U. Chgo., 1940-65, prof. meteorology, 1944-65, chmn. dept., 1948-60; dean geoscis., distinguished prof. meteorology Tex. A. and M. U., College Station, 1965—. Dir. U. S. Govt. Thunderstorm Project, 1945-49. Trustee Univ. Corp. for Atmospheric Research, 1960—, chmn. bd., 1962-64. Recipient award of merit Chgo. Tech. Socs. Council, 1959. Mem. Am. Inst. Aeros. and Astronautics (Losey award 1941), Am. Meteorol. Soc. (Brooks award 1960, pres. 1951-53), Am. Geophys. Union (sect. pres. 1946-47), Internat. Assn. Meteorology and Atmospheric Physics (pres. 1960-63), Royal Meteorol. Soc., Nat. Acad. Scis., A.A.A.S., Am. Geog. Soc., Sigma Xi. Author: General Meteorology, 1937, 44, 59; Elements of Cloud Physics, 1965; (with R. R. Braham) The Thunderstorm, 1949; also numerous articles. Research on structure and circulation of thunderstorms, physics of clouds and precipitation. Home: 305 Brookside Dr., Bryan, Tex. 77801.*

BYFORD, Henry Turman, Am. gynecologist; b. Evansville, Ind., Nov. 12, 1853; s. William H. and Anne (Holland) B.; grad. Williston Sem., 1870; M.D., Chicago Med. Coll. (Northwestern U.), 1873; m. Lucy Larned, Nov. 8, 1882. Practiced medicine, prof. gynecology Coll. Medicine, U. Ill., Chgo., 1892-1913, emeritus; cons. gynecologist St. Luke's and Chgo. Lying-In hosps. Fellow A.C.S. Author: Manual of Gynecology; To Panama and Back, 1908; (with William Heath Byford) Diseases of Women. Co-author: American Text Book of Gynecology; Keating and Coe's Clinical Gynecology; Kelly and Noble's Operative Gynecology. Died June 5, 1938.

BYKOV, Boris Aleksandrovich, Russian geog. botanist; b. 1910; grad. Biol. Faculty, Kazakhstan U., 1939; D.Biol. Sci., 1955. Sr. asso. Bot. Inst, Kazakhstan Acad. Sci. Mem. Kazakhstan Acad. Sci. (corr.). Author: The History of the Spruce Woods of Tyan Shan, 1947; Geobotany, 1957; co-author: The Trees and Bushes of Alma-Ata, 1941, also others. Specialist in floristic systematics. Address: Bot. Inst., Kazakhstan Acad. Sci., Alma-Ata, Kazakhstan SSR, USSR.

BYRD, Richard Evelyn, Am. explorer; b. Winchester, Va., Oct. 25, 1888; s. Richard Evelyn and Eleanor Bolling (Flood) B.; ed. Shenandoah Valley Mil. Acad., Va. Mil. Inst., U. of Va.; grad. U. S. Naval Acad., 1912; numerous hon. degrees; m. Marie D. Ames, Jan. 20, 1915. Ensign U. S. Navy, 1912, advanced through grades to lt. comdr.; promoted to comdr. after north polar flight, 1926, to rear adm., 1930. Entered Aviation Service 1917; comdr. U. S. Air Forces of Can., July 1918, until Armistice; comdr. aviation unit of Navy-MacMillan Polar Expdn., June-Oct. 1925; made flight in airplane with Floyd Bennett over North Pole and back to base at Kings Bay, Spitzbergen, May 9, 1926, covering distance of 1,360 miles in 15½ hours; made trans-Atlantic flight with 3 companions, from N.Y. to France, distance of 4,200 miles, flight lasting

42 hours, June 29-July 1, 1927; flew over South Pole, Nov. 29, 1929; made 1st expdn. to Antarctic, 1928-30, 2d expdn., 1933-May 10, 1935; on both expdns. made important discoveries, among them being Edsel Ford Mountains and Marie Byrd Land; spent 5 mos. of winter night alone at scientific work in shadow of South Pole. In 1939 was made comdr. of U. S. Antarctic Service, an expdn. sent to the Antarctic by Govt.; made four noteworthy flights resulting in discovery of five new mountain ranges, five islands, more than 100,000 square miles of area, a large peninsula, and 700 miles of hitherto unknown stretches of antarctic coast. During World War II served with Fleet Admiral King in Washington and Fleet Admiral Nimitz in Pacific; overseas 4 times (3 times in Pacific, once, Western front in Europe); apptd. comdg. officer U. S. Navy Antarctic Expdn., 1946. Advisor Dept. Defense, Polar defense and strategy. Presented by President Coolidge with Hubbard gold medal, 1926, "for valor in exploration"; recipient Congressional Medal of Honor, 1926, Spl. Congressional medals, 1930, 37, 46; Congressional Life Saving Medal of Honor, Navy D.S.M., Navy Cross, Navy Flying Cross. Patron's medal of Royal Geog. Soc. (British, 1931), and gold medal Reale Societa Geografica (Italy, 1931); Elisha Kent Kane medal Phila. Geog. Soc.; Langley medal of aerodromics Smithsonian Instn., David Livingstone Centenary medal Am. Geog. Soc., D.S.M. of State of N.Y. presented by Gov. Franklin D. Roosevelt; also 65 other medals; received from President Roosevelt, 1940, gold star in recognition of services as comdr. of U. S. Antarctic Service Expdn., 1939-41; 22 citations from Navy Dept.; twice awarded Legion of Merit medal; also many awards fgn. govts. and learned socs. Mem. Phi Beta Kappa, Kappa Alpha, and about 200 other orgns. Author: Skyward, 1928; Little America, 1930; Discovery, 1935; Exploring with Byrd; Alone, 1938. Died Mar. 11, 1957.

BYRGIUS, Justus, Swiss mathematician; b. Lichtensteig, Switzerland, Feb. 28, 1552; supt. obs. Hesse, Germany; inventor astron. instruments and clockmaker for William IV of Hesse, from 1579; entered service of Emperor Rudolf II in Prague, 1604; later returned to Kassel in Hesse. Constructed a number of instruments, including a celestial globe and a sector; compiled table of logarithms (independently of J. Napier), between 1603 and 1611. Died Kassel, Germany, Jan. 31, 1632.

BYRNE, Donn, Am. psychologist; b. Austin, Tex., Dec. 19, 1931; s. Bernard Devine and Rebecca (Singleton) B.; student Stanford, 1949-51, Ph.D., 1958; B.A., Fresno State Coll., 1953, M.A., 1956; m. Lois Ann Pugsley, Sept. 12, 1953; children—Keven Singleton, Robin Lynn. Instr., San Francisco State Coll., 1957-59; faculty U. Tex., Austin, 1959-—, prof., 1966-—, asst. chmn. dept. psychology, 1964-66; vis. prof. Stanford, 1966-67. Mem. Am., Southwestern psychol. assns., Psychonomic Soc., A.A.A.S., N.Y. Acad. Scis., Am. Assn. U. Profs. Author: (with H. C. Lindgren) Psychology: An Introduction to the Study of Human Behavior, 1961; (with H. C. Lindgren, L. Petrinovich) Psychology: An Introduction to a Behavioral Science, 1966; Personality: a Research Approach, 1966; (with Hamilton) Personality Research: A Book of Readings, 1966; also articles, chpts. in books. Editor: (with P. Worchel) Personality Change, 1964. Research in interpersonal attraction, devel. repression-sensitization scale. Home: 6705 Lexington Rd., Austin, Tex. 78731.*

BYRNE, John, physician; b. Kilkeel, Ireland, Oct. 13, 1825; s. Stephen and Elizabeth (Sloane) B.; M.D., U. Edinburgh (Scotland), 1846; grad. N.Y. Med. Coll., 1853. Came to Am., 1848; practiced medicine, Bklyn., 1848-1902; mem. exec. bd., also clin. prof. uterine surgery L.I. Coll. Hosp.; surgeon-in-chief St. Mary's Hosp., N.Y.C., 1858-1902; devised means of using electric cautery-knife in surgery of malignant disease of uterus. Author: Clinical Notes on the Electric Cautery in Uterine Surgery, 1872. Died Montreux, Switzerland, Oct. 1, 1902.

BYRNE, John V(incent), Am. oceanographer; b. Hempstead, N.Y., May 9, 1928; s. Frank E. and Kathleen (Barry) B.; A.B., Hamilton Coll., 1951; M.A., Columbia, 1953; Ph.D., U. So. Cal., 1957; m. Shirley O'Connor, Nov. 26, 1954; children—Donna Marie, Lisa Kay, Karen Lynn, Steven John. Research geologist Humble Oil & Refining Co., Houston, 1957-60; faculty dept. oceanography Ore. State U., Corvallis, 1960-—, prof., 1965-—. Marine cons. Ore. Dept. Geology and Mineral Industries, 1963-—; program director oceanography NSF, 1965-67. Recipient Carter award for teaching excellence Ore. State U., 1964. Mem. Geol. Soc. Am., Am. Geophys. Union, A.A.A.S., Am. Assn. Petroleum Geologists. Research, publs. on recent sedimentation in Gulf of Cal., Chenier Plain of La., ocean off Ore., geomorphology of Ore. coast. Home: 1501 N. 12th St., Corvallis, Ore. 97330.*

BYRNE, Joseph (J. Grandson), physiologist; b. Ireland, Mar. 21, 1870; s. Patrick and Margaret (O'Neill) B.; B.A., Carlow Coll. (Royal U., Ireland), 1890; M.A., St. Francis Xavier's Coll., N.Y., 1893; M.D., Coll. Phys. and Surg. (Columbia), 1895; LL.B., N.Y. Law Sch., 1900; LL.D., Fordham U., 1921. Came to U. S., 1891, naturalized, 1897. Practiced at N.Y., since 1897; dean Fordham U. Med. Sch., 1917-21; pres. med. bd. Central and Neurol. Hosp., Welfare Island, N.Y.; cons. neurologist City, Fordham, and Neurol. hosps.; mem. adv. bd. Health Dept., N.Y.C. Fellow N.Y. Acad. Medicine, A.C.P.; mem. Royal Coll. of Surgeons (Eng.), Soc. for Exptl. Study of Biology and Medicine, N.Y. Neurol. Soc. Author: Physiology of the Semi-circular Canals, and Their Relation to Seasickness, 1912; Seasickness and Health, 1912; The Mechanism of Pain, 1918; Sensory-Psychic Integration, 1924; The Pupils in Visceral and Somatic Disorders, 1926; Studies on the Physiology of the Eye, Still Reaction, Sleep, Dreams, Hibernation, Repression, Hypnosis, Narcosis, Coma, and Allied Conditions, 1933; Clinical Studies on the Physiology of the Eye, 1934; Studies on the Physiology of the Middle Ear, 1938; The Effect of Stimulation of the Cortex Cerebri Upon the Mechanisms which Mediate Movements of the Iris and the Membrana Tympani, 1937; The Mechanism of Sensation and Emotion, 1941; The Physiological Basis of Perception, 1941. Made extensive exptl. and clin. studies, contbd. articles on mechanism of sensation. Died May 13, 1945.

BYRNES, Charles Metcalfe, Am. neurologist; b. Claiborne County, Miss., Nov. 4, 1881; s. Charles Ralston and Helen (Metcalfe) B.; student La. State U.; B.S., U. N.C., 1902; M.D., Johns Hopkins, 1906; m. Louise Alexander McCosh, Sept. 1, 1920; 1 dau., Louise Metcalfe. Demonstrator in anatomy Johns Hopkins, 1902-03 instr. clin. neurology, 1909, asso., from 1918; adj. prof. anatomy U. Va., 1906-09; lectr. in neuropathology U. N.C., 1935; neurologist Ch. Home and Infirmary, Balt. Contbr. to Forchheimer's Therapeusis of Internal Disease, 1913, Tice's Practice of Medicine, 1920. Died Nov. 29, 1936.

BYRNES, Ralph Leonidas, physician; b. Walcott, Ia., Mar. 30, 1878; s. Thomas (M.D.) and Jennie (Allen) B.; B.Sc., State U. Ia., 1902, M.Sc. and M.D., 1906; studied Harvard, 1911; spl. work in bacteriology, New Haven, 1918, in pulmonary tb, Yale Army Med. Sch.; m. Edith Whitney Merritt, Oct. 6, 1908. Hosp. service, 1906-08; practice at Avoca, Ia., 1908-10; at Pottenger's Sanatorium, Monrovia, Cal., 1910; prof. bacteriology and pathology U. Utah, dir. State Bd. Health and Lab., 1911-15; prof. pathology, bacteriology and clin. microscopy U. So. Cal., 1915-16; prof. diseases of the chest Coll. Med. Evangelists, Los Angeles, 1919-23; established first endocrine and mental hygiene clinic, Belvedere Health Center, Los Angeles County Health Dept., 1930. Diplomate Am. Bd. Internal Medicine. Mem. A.M.A., Am. Pub. Health Assn., A.C.P., Pi Kappa Alpha, Phi Beta Pi. Died Feb. 16, 1943; buried Walcott, Ia.

BYRON, William Glenn, Am. geographer; b. Denver, Jan. 28, 1923; s. Walter and Susan (Fulenwider) B.; A.B., U. Cal. at Los Angeles, 1948, M.A., 1951; Ph.D., Syracuse U., 1954; m. Grechen Kumnick, Aug. 28, 1949; children—Mary, Beverly, Robert, Susan. Blue printer Vega Aircraft Co., 1941-42; high sch. tchr., Porterville, Cal., 1949-51; mem. faculty Cal. State Coll. at Los Angeles, 1954-—, prof. geography, 1962-—, chmn. dept., 1958-64; mgr. Meridian Co., Monterey Park, Cal., 1958-—. Named Outstanding Prof., Cal. State Coll., Los Angeles, 1965-66. Grantee NSF, 1961, 63, 64. Mem. Cal. Council Geography Tchrs. (sec.-treas. -1957-59), Nat. Council Geog. Edn. (Cal. coordinator 1959-62), Assn. Am. Geographers, Assn. Pacific Coast Geographers, Pacific Coast Council Latin Am. Studies. Co-editor: Patterns on the Land, 1958. Pioneer research on feasibility of relating photometric measurements to cartography; identified and explored new archaeol. zone in coastal marshlands of Nayarit, Mexico. Home: 18807 E. Leadora Av., Glendora, Cal. Office: 5151 State College Dr., Los Angeles.*

BYWATERS, Eric George Lapthorne, English physician; b. London, Eng., June 1, 1910; s. George Ernest and Ethel (Penney) B.; M.B., B.S., London U., 1935; m. Betty Euan-Thomas, Aug. 10, 1935; children—Caroline (Mrs. Barry McCullagh), Elizabeth (Mrs. Jasper Jewett), Jane. Rockefeller Research Travelling fellow, 1937; research fellow Harvard, 1938; Beit Meml. Research fellow, Empire Rheumatism Council Research fellow, 1939; physician Hammersmith Hosp., London, 1940-46, hon. physician, 1947-—; physician Brit. Postgrad. Med. Sch., 1940-46, Med. Research Council investigator, lectr. medicine, 1945-46, sr. lectr., 1947-—; investigator Med. Research Council, 1943, actind dir. shock unit, 1944, hon. dir. rheumatism research unit, Taplow, Eng., 1958-—; hon. physician Canadian Red Cross Meml. Hosp., Taplow, 1947-—; prof. rheumatology U. London, 1958-—. Mem. expert coms. on rheumatic fever and connective tissue diseases WHO, 1966-—. Recipient Gairdner Found. award, 1963. Fellow Royal Coll. Physicians; mem. Assn. Physicians Gt. Britain and Ireland, Brit. Med. Assn., Renal Soc., Heberden Soc. (past pres.); hon. mem. Am., Dutch, Argentinian, French, German, Czechoslovak, Yugoslavian rheumatism assns. Editorial bd. Annals Rheumatic Diseases. Research, numerous publs. on mechanisms of shock in air raid victims leading to renal fialure, arthritis. Home: 53 Burkes Rd., Beaconsfield, Bucks., Eng. Office: M.R.C. Rheumatism Research Unit, Canadian Red Cross Meml. Hosp., Taplo, Maidenhead, Berks., Eng.*

C

CABANES, Jean Louis, German ornithologist; b. Berlin, Germany, Mar. 8, 1816; s. Benoit-Jean and Marie Luise (Fahland) C.; studied natural sci. in Berlin; m. Jeanne Renaldi, 1849. Became mus. curator, Charleston, S.C., 1839; became sci. helper, Berlin, 1841, named asst., 1846; 1st curator Zool. Mus., Berlin U., 1850-93. Mem. German Ornithol. Soc. (founder). Founder, Jour. für Ornithologie, 1853; numerous articles. Author: Museum Heinaenum, 4 vols., 1855-63. Systematic study of birds especially those of So. and middle Am.; instrumental in establishing natural classification in ornithology; noted several instances of gynandromorphism (having both male and female characteristics) in birds, 1874. Died Friedrichshagen nr. Berlin, Feb. 20, 1906.

CABANIS, Pierre Jean Georges, French physician, philosopher; b. Cosnac, Charente-Maritime, France, June 5, 1757; s. Jean Baptiste C.; ed. College of Brives; studied medicine, Paris, France, 1777-83; m. Charlotte de Grouchy, May 14, 1796; children—Aminthe-Geneviève, Annette-Paméla. Prof., Acad. Warsaw (Poland); sec. to Prince-Bishop of Poland at Vilnius; became adminstr. Paris hosps., 1789; prof. hygiene, Paris, 1795-99, became prof. legal medicine and history of medicine, 1799; physician to Mirabeau; friend of Condorcet; mem. Council of Five Hundred; revolutionist in French revolution. Author: Observations sur les hôpitaux, 1789; Journal de la maladie et de la mort d'Honoré G.-u.-R Mirabeaus, 1791; Du degré de certitude de la médecine, 1798; Rapports du physique et du moral de l'homme, 1802; Coup d'oeil sur les révolutions et sur la réforme de la médecine, 1804; Courtes observations sur les affections catarrhales, 1807. Pioneered physiol. psychology; developed theory of coenesthesis; regarded soul as faculty of body, but later accepted concept of immortality; advocated med. reforms; tried to rebuild French med. edn. on clin. grounds. Died Meulan, France, May 5, 1808.

CABANNES, Henri, French mech. engr.; b. Montpellier, France, Jan. 21, 1923; s. Jena and Marie (Fabry) C.; ed. École normale supérieure; Ph.D. in Sci.; m. Madeleine Lebon, Sept. 10, 1948; children—Jean-Pierre, Jean-Paul, André, Hélène, Benoit. Prof., Sch. of Sci., Marseille, France; vis. prof. U. Quebec (Can.); prof. mechanics Sch. of Sci., Paris. Laureat, Acad. Scis. Mem. Nat. Council Research Sci. Author: Théorie des ondes de choc; Cours de mécanique générale, 1961; Problèmes de Mécanique Générale, 1965; Cours de Magnetodynamique des Fluides, 1965. Research on shock waves, especially curvature and singularities in axisymmetrical case, kinetic theory of ionized gases and magneto-fluid-dynamics, problem of wedge. Home: 23, allée de Trevise, Sceaux (Seine), France. Office: 9, quai St.-Bernard; Paris 5, France.*

CABANNES, Jean, French physicist; b. Marseille, France, Aug. 12, 1885; prof. physics Sch. Sci. of Montpellier and Paris; corr. mem., section gen. physics French Acad. Scis., 1932, then mem., 1946; mem. Com. Longitudes. Contbd. to understanding of molecular diffusion phenomena; succeeded in spectroscoping light of night sky.

CABASSO, V. J., virologist; b. Egypt, June 21, 1915; s. Jacques and Fortunée (Bouskela) C.; Bachelier en Philosophie, Lycee Francais, Alexandria, Egypt, 1933; M.S., Hebrew U., Jerusalem, Israel, 1938; postgrad. Sorbonne, Paris; Sc.D., U. Algiers (Algeria), 1941; m. Anna Sara Cooper, June 26, 1948; children—Jacqueline Louise, Phillip Joseph. Came to U. S., 1946, naturalized, 1958. Fellow Pasteur Inst. Paris, 1939; research fellow Pasteur Inst., Tunis, Tunisia, 1940-44; head dept. bacteriology and labs. UNRRA, Middle E. and Greece Missions, 1944-46; research virologist Lederle Labs. div. Am. Cyanamid Co., Pearl River, N.Y., 1946-58, head virus immunological research dept., 1958-—. Diplomate Am. Bd. Microbiology. Fellow N.Y. Acad. Scis., Am. Acad. Microbiology, N.Y. Acad. Medicine; mem. Pharm. Mfrs. Assn. (chmn. div. human virus products 1966-—), U. S. Livestock San. Assn. (mem. Western Hemisphere com. on non-primate animal virus characterization 1964-—), Nat. Acad. Scis. (mem. com. on standard methods for vet. microbiology 1966-—), N.Y. Soc. Tropical Medicine, Am., N.Y. State pub. health assns., Am. Soc. for Microbiology, A.A.A.S., Am. Forestry Assn. Research, numerous publs. on adaptation of canine distemper virus to chicken embryo with devel. serological procedures and effective vaccine; early demonstration of biol. fuel cell; 1st cultivation of infectious canine hepatitis virus in tissue culture with devel. serological procedures and effective vaccine; contbd. to devel. vaccine against mumps, poliomyelitis (oral, live), measles, Bluetongue, rabies, hog cholera, bovine rhinotracheitis. Home: 66 Lt. Cox Dr. Office: Lederle Labs., Pearl River, N.Y. 10965.*

CABEO, Niccolo, Italian physicist; b. Ferrara, Italy, 1585; Jesuit priest; prof. moral philosophy and mathematics at Parma; tchr. mathematics at Genoa. Author: Philosophia magnetica, 1629; commentary on Aristotle's Meteorologica. Research on electrification of bodies by friction; noted elec. repulsion; provided an innovation in hydrostatics by his concept of apparatus to measure velocity of a fluid current;

repeated W. Gilbert's expts. on magnetism, but interpreted results differently; opposed Galileo's ideas on tides and motion of projectiles. Died Genoa, Italy, June 30, 1650.

CABLE, Joe Wood, Am. physicist; b. Murray, Ky., Feb. 17, 1931; s. Thomas Ray and Hettie (McHood) C.; A.B., Murray State Coll., 1952; Ph.D., Fla. State U., 1955; m. Wanda L. McReynolds, Apr. 9, 1950; children—Joe Mac, Jeffrey Wood, Michael Andrew. Physicist, Oak Ridge Nat. Lab., 1955——. Mem. Am. Phys. Soc. Contbr. numerous articles to profl. jours. Research on neutron scattering techniques to study magnetic and electronic properties of solids. Home: 120 Albany Rd. Office: P.O. Box X, Oak Ridge, 37830.*

CABOT, Arthur Tracy, Am. surgeon; b. Boston, Jan. 5, 1852; s. Samuel and Hannah Lowell (Jackson) C.; A.B., Harvard, 1872, A.M., 1878, M.D., 1876; postgrad. Vienna and Berlin; m. Susan Shattuck, Aug. 16, 1882. Practiced medicine, Boston, 1878-86; vis. surgeon Mass. Gen. Hosp., 1886-1902; cons. surgeon Children's Hosp., N.E. Hosp. for Women and Children; instr. oral pathology and surgery Harvard, 1878-80, clin. instr. genito-urinary surgery, 1885-96, fellow, from 1896. chmn. Mass. Commn. on Hospitals for Consumptives. Fellow Am. Surg. Assn., Am. Acad. Arts and Scis. Contbg. author: Surgery; Its Principles and Practice (W. W. Keen), Vol. IV, 1908. Contbr. to jours. Leading genito-urinary surgeon of New Eng.; promoted pub. health and campaign against Tb. Died Nov. 4, 1912.

CABOT, Hugh, Am. surgeon; b. Beverly Farms, Mass., Aug. 11, 1872; s. James Elliot and Elizabeth (Dwight) C.; A.B., Harvard, 1894, M.D., 1898; LL.D., Queen's U., Belfast, 1925; m. Mary Anderson Bolt, Sept. 1902 (dec.); children—Hugh Cabot, Mary Anderson (dec.), John Bolt, Arthur Tracy; m. 2d, Elizabeth Cole Amory, Oct. 1938. Began practice, Boston, 1900; asst. surgeon and surgeon, Mass. Gen. Hosp., 1902-19; surgeon Bapt. Hosp., 1900-19; asst. prof. surgery Harvard Med. Sch., 1910-18, clin. prof. 1919; prof. surgery, U. Mich., 1919-30, dean Med. Sch., 1921-30; prof. surgery Grad. Sch., U. Minn., 1930-39; surgeon Mayo Clinic, 1930-39. Fellow A.C.S., Royal Soc. Medicine (London) (hon.); mem. Am. Surg. Assn., A.M.A., Am. Assn. Genito-Urinary Surgeons, Assn. Français d'Urologie, Société Internationale d'Urologie, Phi Beta Kappa, Sigma Xi, Alpha Omega Alpha. Author: Surgical Nursing, 4 edits., 1924, 30, 37, 40; The Doctor's Bill, 1936; The Patient's Dilemma, 1940. Editor: Modern Urology, 1918, 3d edit., 1936. Contbr. articles in med. jours. Died Aug. 14, 1945.

CABOT, John, explorer; b. Genoa, Italy, 1450; 3 sons—Lewes, Sebastian, Santius. Engaged in trading activities in early life; became interested in finding Western water route to Orient; moved with family to Bristol, Eng., circa 1484; calculated that he would try to reach island of Brazil or Seven Cities (of fable) and then continue to Asia; attempted trips for several years (all unsuccessful); when news of Columbus' discovery came (1493) he gave up idea of reaching islands, decided instead to push on to Asia; by ofcl. permission of King Henry VII (1496) finally set sail in ship Mathew with 18 men, 1497; sailed N.W. from Ireland and reached Cape Breton Island, 1497, claimed it for Henry VII, convinced he had reached Asia; sailed North, naming several more islands on his way back, reached Bristol, 1497; convinced Henry VII that another voyage could reach Japan and thus enrich throne; granted new patents by king, 1498; left Bristol with intentions to reach Greenland (or Asia as he thought) and follow coast South, 1498; reached Greenland in early June, named it Labrador's Land for 1 of his sailors, Joao Fernandes, (called Llavrador); sailed North encountering many icebergs; crew mutinied and refused to proceed further North, 1498; reached modern Baffin Land, in 66° latitude, thought it to be part of Asia and proceeded South; passed Newfoundland, N.S. and New Eng. still searching for Japan as far South as 38th parallel; finding no signs of civilization, returned to Eng. late 1498. Died 1498.

CABOT, Richard Clarke, Am. physician; b. Brookline, Mass., May 21, 1868; s. James Elliot and Elizabeth (Dwight) C.; A.B., Harvard, 1889, M.D., 1892; LL.D., U. Rochester, 1930; L.H.D., Syracuse U., 1934; m. Ella Lyman, Oct. 26, 1894. Physician to outpatients Mass. Gen. Hosp., 1898-1908, asst. vis. physician, 1908-12, chief med. staff, 1912-21; asst., Harvard, 1899-1903, faculty, 1903-34, prof. clin. medicine, 1919-33, lectr. philosophy Prof. Josiah Royce's Harvard Sem. course in logic, 1903-04, prof. social ethics, 1920-34. Pres. Nat. Conf. Social Work, 1931. Recipient Gold medal Nat. Inst. Social Scis., 1931. Fellow Am. Acad. Arts and Scis. Author: Clinical Examination of the Blood, 1896; Physical Diagnosis, 10 edits., 1901-30; Social Service and the Art of Healing, 2 edits., 1909-28; Differential Diagnosis, Vol. I, 4 edits., 1911-19, Vol. II, 3 edits., 1915-24; What Men Live By, 1914; Laymen's Handbook of Medicine, 1916, rev. edit., 1937; Social Work, 1919; Facts on the Heart, 1926; Adventures on the Borderlands of Ethics, 1926; The Meaning of Right and Wrong, 1933; The Art of Ministering to the Sick (with Russell L. Dicks), 1936; Christianity and Sex, 1937. Discovered Cabot's ring bodies in stained red blood cells in some cases of anemia; advocate of group medicine; inaugurated social service system of med. care; emphasized autopsy correlations; initiated case-history method in med. edn. Died Cambridge, Mass., May 8, 1939.

CABOT, Sebastian, explorer, cartographer; b. Bristol, Eng., circa 1476; s. John Cabot. Accompanied his father in service of Henry VII on explorations that resulted in discovery of Labrador or Newfoundland, 1497 (1st Brit. claim in New World); made map of Glascony and Guienne for Henry VIII, 1512; capt. Spanish Navy, 1512-16; mem. Spanish council of New Indies, 1518; comd. Spanish expdn. to Brazil which failed to yield expected wealth, 1525-29, was banished to Oran, 1530; reinstated as pilot major, 1533; made colored world map, 1544; re-entered Brit. service, 1548; received pension from Edward VI, 1549; became gov. Mcht. Adventurers Co. (later Muscovy Co.), 1553; organized expdn. to search for northeast passage to Far East, which contbd. to devel. trade with Russia. Died London, 1557.

CABRERA, Benjamin David, Philippine parasitologist; b. Tarlac, Philippines, Mar. 18, 1920; s. Eusebio R. and Macaria (David) C.; M.D., U. Philippines, 1945; M.P.H., Tulane U., 1950; m. Melanie Gregorio, Sept. 28, 1945; children—Eleanora, Edgardo, Benjamin Gerardo. Parasitologist, 3d Med. Lab., U. S. Army, Rizal, Philippines, 1946-47; faculty U. Philippines, 1947——, prof., chmn. parasitology Inst. Hygiene, 1964——, head pest control and other services, 1959——. Guest lectr. Johns Hopkins, Balt., 1961-62; mem. Philippine NRC. Recipient Manila Med. Soc. Research award, 1960, Abbott-PMA Research Contest award, 1961, Philippine Legion of Honor for research in filariasis from Pres. Ferdinand Marcos, 1966. Fellow Philippine Soc. Pathologists; mem. Philippine Pub. Health Assn., Philippine Soc. Parasitologists (sec.), Philippine Entomol. Soc., Manila Med. Soc., Philippine Med. Assn., Phi Sigma, Phi Kappa Phi. Author: Manualin Parasitology, (with Serafin Juliano, Angelina Latonio), 1965; also articles. Discovered prin. vector of bancroftian filariasis in Sorsogon, Philippines; surveyed for filariasis in Philippines and discovered Anopheles minimus flavirostris as its prin. vector in remote areas of country; discovered existence of Brugia malayi in Palawan with its mosquito vector; studied egg laying habits of Enterobius vermicularis among sch. children, reinfection rates of ascariasis among treated sch. children in Philippines, snail which is intermediate host of Paragonimus in Philippines. Home: 73 Mindanao Av., Quezon City, Philippines. Office: 625 Herran St., Ermita, Manila, Philippines.*

CACERES, Cesar Augusto, physician; b. Puerto Cortes, Honduras, Apr. 9, 1927; s. Julian Rios and Mariana (Culotta) C.; B.S., Georgetown U., 1949, M.D., 1953. Practiced medicine, specializing in internal medicine, Washington, D.C., 1956——; spl. med. research fellow George Washington U., 1956-60, asst. prof. medicine Sch. Medicine, 1963-65, professorial lectr. bio-med. engrng. Sch. Engrng., 1964——, asso. prof. medicine Sch. Medicine, 1965——; chief Instrumentation Field Sta., heart disease control program div. chronic disease USPHS, Dept. Health, Edn. and Welfare, 1960——. Recipient Superior Service Honor medal Dept. Health, Edn. and Welfare, 1963, Superior Service Group award, 1966. Mem. Am. Heart Assn., N.Y. Acad. Sci., Am. Fedn. Clin. Research, Assn. Computing Machinery, Am. Pub. Health Assn., Assn. Advancement Med. Instrumentation, Fed. Profl. Assn. (pres.-elect). Author: (with O. Schmitt) Computer Assisted Studies of Biomedical Problems, 1964; Biomedical Telemetry, 1966; (with L. W. Perry) Innocent Murmurs, 1967. Research and numerous articles on formulation and demonstration of automated med. data and signal processing, multidimensional statis. studies, analysis of acoustic properties of heart sounds and stethoscopes, devel. of automatic fluorescent antibody screening devices for streptococci, investigations into feasibility of ultrasound for detection of heart disease. Home: 1722 Q St. N.W., Washington 20009. Office: 2121 K St. N.W., Washington 20037.*

CADE, Roger, math. physicist; b. Portmouth, Eng., Feb. 20, 1924; s. George William and May (Graddon) C.; B.Sc. U. London (Eng.), 1944, M.Sc., 1947, Ph.D., 1952; m. Adelene Sophia Jane Fettus, Dec. 24, 1948; children—Sylvia, Angela Lorraine. Lectr. physics S.-E. Essex Tech. Coll., Dagenham, Eng., 1945-47; asst. lectr. math. Queen's U., Belfast, No. Ireland, 1947-49; faculty U. W.I., Jamaica, 1949-64, sr. lectr. math., 1959-64; prof. math. U. P.R., Mayaguez, 1964-66, U. Zambia, Lusaka, 1966——. Research and publs. on electrostatics, solution of boundary-value problems, theory of electrostriction, gen. theory electrostatic force in continuum with spl. reference to electric double layer. Address: U. Zambia, P.O. Box 2379, Lusaka, Zambia, Africa.*

CADET DE GASSICOURT, (Charles-Les-Jules-) Ernest, French physician; b. Paris, France, Oct. 31, 1826; s. Charles-Louis and Clementine (DuBois) C. de G.; M.D., Paris, 1857; joined service of childhood diseases, Trousseau Hosp., Paris, taught at Ste. Eugénie Hosp. Mem. Acad. Medicine (sec.). Author: Traité clinique des maladies de l'enfance, 1880-84. Founder, dir. Revue mensuelle des maladies d'enfance, 1883-98. Discovered symptom of exanthema in scarlet fever; studied diseases of infancy and childhood. Died June 10, 1900.

CADLE, Richard Dunbar, Am. chemist; b. Painesville, O., Sept. 23, 1914; s. Thomas Priday and Margaret (Dunbar) C.; B.A., Western Res. U., 1936; Ph.D., U. Wash., 1940; m. Edna Fay Saer, May 2, 1940; children—Donald Richard, Steven Howard, Gary Alan. Chemist, Procter & Gamble Co., Cin., 1940-47; sr. chemist Naval Ordnance Test Sta., China Lake, Cal., 1947-48; sr. chemist Stanford Research Inst., Menlo Park, Cal., 1948-55, chmn. atmospheric chem. physics dept., 1955-63; program scientist Nat. Center Atmospheric Research, Boulder, Colo., 1963——. Gen. chmn. Internat. Symposium on Chem. Reactions in Lower and Upper Atmosphere, 1961. Mem. Am. Chem. Soc., Am. Assn. Contamination Control (nat. dir. 1961-63), Am. Meteorol. Soc., Am. Geophys. Union, Phi Beta Kappa, Sioma Xi, Phi Lambda Upsilon. Author: Particle Size Determination, 1955; Particle Size, 1965; Particles in the Atmosphere and Space, 1966; also articles. Investigation in chem. reactions and nature of aerosol particles in natural and in contaminated atmospheres, using both lab. and field techniques, autoxidation of organic substances and inventor series of rocket propellants. Home: 4415 Chippewa Dr. Office: Nat. Center Atmospheric Research, Boulder, Colo. 80302.*

CADWALADER, Thomas, Am. surgeon; b. Phila., 1708; s. John and Martha (Jones) C.; m. Hannah Lambert, 1738. A founder (with Benjamin Franklin) Phila. Library, 1731; performed earliest recorded autopsies in Am. 1742; subscribed to founding Pa. Hosp., 1751. Mem. Am. Philos. Soc. One of most noted 18th century Am. physicians. Died Trenton, N.J., Nov. 14, 1799.

CAELIUS AURELIANUS, physician; flourished Sicca, Numidia, 5th century; author: Liber celerum vel acutarum passionarum; De morbis acutis; De morbis chronicis; wrote on chronic and acute diseases using works of Soranus of Ephesus; compiled compendium of med. scis. Discussed nomenclature, etiology, symptomatology, pathology, treatment and diagnosis of various diseases; described pthisis, gout, motor and sensory paralyses, encephalitis, epilepsy, stammering and speech defects; differentiated epileptic seizures from hysterical attacks; recommended humane treatment of the insane.

CAESALPINUS, see Cesalpino.

CAGLE, Fred Ray, Am. herpetologist; b. Marion, Ill., Oct. 9, 1915; s. Fred and Agnes (Guiney) C.; student U. Ia., 1935; B.Ed., U. So. Ill., 1937; M.S., U. Mich., 1938, Ph.D. (Rackham fellow), 1943; m. Josephine Alexander, June 18, 1938; children—Fred Ray, Mary Jo (Mrs. Moragne). Faculty, U. So. Ill., 1938-41; faculty Tulane U., New Orleans, 1946——, prof. zoology, 1949——, coordinator research, 1959-63, v.p., 1963-65, v.p. instnl. devel., 1965——. Chmn. com. sponsored research Am. Council Edn., 1965——; mem. sci. and tech. com. Library Congress, 1962——; mem. U. S. Nat. Commn., com. natural scis. UNESCO, 1961-63; cons. brs. govt., univs. Bd. dirs. Gulf South Research Inst. Recipient Darwin medal USSR Acad. Scis., 1959; Alumni Achievement award U. So. Ill., 1965. Fellow Herpetologists League, A.A.A.S.; mem. Am. Inst. Biol. Scis., Am. Soc. Ichthyologists and Herpetologists, Soc. Systematic Zoology, La. Acad. Scis., Soc. for Study Evolution, Southwestern Assn. Naturalists, Assn. Southeastern Biologists, Netherlands Assn. Herpetology, Am. Soc. Mammalogists, Brit. Herpetological Soc., Wildlife Soc., Conf. Biol. Editors, Ecol. Soc. Am., Am. Assn. U. Profs., Am. Soc. Naturalists, Nat. Council U. Research Adminstrs., Fedn. Internationale de Documentation, Phi Sigma, others. Author: (with others) Vertebrates of the United States, 1957. Editor-in-chief Copeia, 1955-59; sect. editor Biol. Abstracts, 1957-63; asso. editor Am. Midland Naturalist, 1959-63. Research, publs. on habits, physiology of various turtles in Ill., bats, raccoons, reptiles in U. S. Home: 6320 Story St., New Orleans, 70118.*

CAGLIOSTRO, Conte Alessandro, see Balsamo, Giuseppe.

CAGNIARD, Louis, French geophysicist, b. Le Havre, France, Dec. 9, 1900; s. Louis and Virginie (Desperrois) C.; Agrégation de Physique, Ecole Normale Supérieure, 1924; D.Physics, 1928, D. in math., 1938; m. Georgette Pesty, Dec. 30, 1924 (dec. May 1949). Asst., Lille (France) U., 1924-28; dir. geophys. exploration cos., 1928-38; prof. Strasbourg (France) U., 1928-38; chief sci. sect. French F.I.A.T., Germany, 1945-47; prof. applied geophysics Paris (France) U., 1947——; dir. Centre de Recherches Géophysiques du Centre Nat. de la Recherche Scientifique, 1957——. Named chevalier de la Légion d'Honneur, comdr. des Palmes Académiques. Author: Réflexion et réfraction des ondes séismiques progressives, 1940; La Prospection Géophysique, 1950; (with C. H. Dix, Edward A. Flinn) Reflection and Refraction of Progressive Seismic Waves, 1962; also numerous articles. Research on gravimetric, magnetic, electric, electromagnetic and seismic methods of geophys. prospection, geophys. study ice and subglacial layers, mil. applications of certain methods of underwater detection, secular variation of terrestrial magnetic field. Patentee prin. magnetotelluric prospection method. Home: 397 Rue de Vaugirard, 75 Paris 15°, France.*

CAGNIARD DE LA TOUR, Charles, French biologist, physicist, inventor; b. Paris, May 31, 1777; ed. Rebais Mil. Sch., École polytechnique, Paris, École des ingenieurs-geographes. Mem. French Acad. Scis., 1851. Invented siren for determining number of vibrations corr. to sounds of various pitches, 1819, acoustic pyrometer, portable mill, blowing machine (new application Archimedes' screw known as carnardelle), machine for studying flight of birds, chronometer balance for summing up engine power; showed degrees of heat at which liquids under pressure were resolved into gases; determined temperatures and pressures corr. to critical state for various liquids, 1822; observed critical point of ether and other substances; investigated resonances, vibration of spheres, sound vibrations in liquids, formation of sound by vibrating strings, 1834-40; created gas lighting, Paris St.-Louis Hosp., 1819; discovered (independently of Theodor Schwann) that yeast is composed of living cells. Died Paris, July 5, 1859.

CAHILL, George Francis, Jr., Am. physician; b. N.Y.C., July 7, 1927; s. George Francis and Eva (Wagner) C.; B.S., Yale, 1949; M.D., Columbia, 1953; M.A. (hon.), Harvard, 1966; m. Sarah Townsend duPont, Dec. 20, 1949; children—Colleen T., Peter duPont, George Francis III, Sarah, Eva, Elizabeth. Practice medicine specializing in metabolic disease, Boston, 1955—; faculty Harvard Med. Sch., Boston, 1955—, asso. prof. medicine, 1965—, dir. Elliott P. Joslin Research Lab., 1963—. Mem. metabolism study sect. NIH, 1966—; cons. VA. Diplomate Am. Bd. Internal Medicine. Mem. Am. Diabetes Assn. (Lilly Award, 1965, dir. 1966—), Endocrine Soc. (Oppenheimer award 1963), Am. Physiol. Soc., Am. Fedn. Clin. Research, A.M.A., A.A.A.S., Am. Clin. and Climatol. Assn. Author: (with A. Renold) Adipose Tissue, 1965. Editor: (with A. Marble) Chemistry and Chemotherapy of Diabetes Mellitus, 1962; asso. editor Jour. Clin. Investigation, 1962—; mem. editorial bd. Jour. Lipid Research, 1964—; Jour. Clin. Endocrinology and Metabolism, 1965—; Diabetes, 1967—. Research, publs. on mechanisms controlling concentration of glucose in blood, its prodn., removal and factors altering rate of each of these. Home: 150 Winding River Rd., Needham, Mass. 02192. Office: 170 Pilgrim Rd., Boston 02215.*

CAHILL, Laurence James, Jr., Am. physicist; b. Frankfort, Me., Sept. 21, 1924; s. Laurence J. and Wilma (Lord) C.; student U. Me., 1942-43; B.S., U. Chgo., 1950; B.S., U. S. Mil. Acad., 1946; M.S., U. Ia., 1956, Ph.D., 1959; m. Alice Adeline Krieger, Sept. 10, 1949; children—Laurence James III, Thomas F., Daniel A. Staff, U. Ia., 1954-59, research asso., 1954-59; faculty U. N.H., Durham, 1959—, prof. physics, 1965—, dir. Space Scis. Center, 1966—; chief physics NASA Hdqrs., Washington, 1962-63, cons., 1962-65; vis. prof. U. Cal. at San Diego, 1965-66. Cons. NSF, 1965—. Recipient NASA award for sustained superior performance, 1963. Mem. Am. Phys. Soc., Am. Geophys. Union, A.A.A.S., Sigma Xi. Research and publs. on measurement by rocket-borne magnetometer of elec. currents in ionosphere, measurement boundary between earth's magnetic field and interplanetary medium, ring current of charged particles encircling earth and causing magnetic storms. Home: 9 Woodman Rd., Durham, N.H. 03824.*

CAHILL, Thaddeus, Am. inventor; b. Mount Zion, Ia., 1867; s. Timothy and Ellen (Harrington) C.; studied Oberlin Acad., 1884-85, and in labs.; LL.B., Columbian (now George Washington) U., 1892, LL.M., 1893, D.C.L., 1900; unmarried. Admitted to bar, 1894, practiced several yrs. Invented elec. typewriter; invented process of producing music electrically, known as telharmony; pioneer in U. S. of art of distbg. music electrically from a central sta. to receiving telephones on premises of subscribers; also inventions in composing machines, heat engines, wireless telephony and wired-wireless; removed lab. from Washington to Holyoke, Mass., 1902, and to N.Y., 1911. Died N.Y.C., Apr. 12, 1934.

CAHN, J., French pharmacologist; b. Paris, Mar. 25, 1923; s. Achille and Fanny (Bender) C.; Doctor es Sciences, U. Paris, 1961; m. Gisele Sibille Cahn, Sept. 19, 1946; children—Martine, Gilles, Remy, Florence, Didier. Dir. Center Exptl. Therapy, Hosp. la Pitié, Paris, 1953—; research asso., dept. cardiology, Phila. Gen. Hosp., Pa., 1955; head neurochem. lab. Sch. Medicine, Paris, 1961—; lectr. psychophysiology Faculty Scis., Paris, 1965—. Recipient French Nat. Acad. Medicine award, 1960. Mem. French socs. Pharmacology, Anesthesiology, Biochemistry, Neurosurgery, Psychiatry, Italian Soc. Pharmacology, Internat. Soc. Biochem. Pharmacology, N.Y. Acad. Scis., numerous others. Research, numerous publs. on cardiology, brain, anesthesiology, pain pharmacology. Home: 18 rue J. M. de Heredia, Paris 7. Office: 83 Bvd. de l'Hopital, Paris 13, France.*

CAHOURS, Auguste Andrée Thomas, French chemist; b. Paris, France, Oct. 2, 1813; student of Chevreul; warden of Paris mint; became prof. École polytechnique, Paris, 1871. Mem. French Acad. Scis. Author: Leçons de chimie générale élémentaire, 2 vols., 1825-56. Discovered several compounds including xylene, anisol, amyl alcohol, methyl salicylate, allyl alcohol, tin tetraethyl, also sulfonium bases independently; determined (with Bineau) density of steams under varying temperatures; prepared acid chlorides from phosphorus pentachlorides. Died Paris, Mar. 17, 1891.

CAIANIELLO, Eduardo Renato, Italian physicist; b. Naples, Italy, June 25, 1921; s. Giuseppe and Sammartino (Lidia) C.; Ph.D., Naples; m. Persico Carla, 1946; children—Dora, Eva, Orietta, Silvia. Dir. Instituto di Fisica Teorica, also Scula di Perfezionamente in Fisica Teorica e Nucleare, U. Naples. Mem. Italian, Am. phys. socs., Accademia Pontaniana, Nat. Soc. Sci., Lit. and Arts, N.Y. Acad. Sci. Home: via Manzoni 63, Naples. Office: Istituto di Fisica Teorica, Monstra d'Oltremare Pad. 19, Naples, Italy.

CAILLETET, Louis Paul, French physicist; b. Châtillon-sur-Seine, France, Sept. 21, 1832; mem. French Acad. Scis.; 1st (about same time as Pictet) to liquefy nitrogen, oxygen, carbon monoxide using compression cooling, and sudden expansion, 1877; built pump which could maintain pressure of several hundred atmospheres; postulated critical point at which distinction between liquid and vapor disappears. Died Châtillon-sur-Seine, Jan. 5, 1913.

CAIN, James Clarence, Am. physician; b. Kosse, Tex., Mar. 19, 1913; s. Thomas Marshall and Aileen (Jackson) C.; B.A., U. Tex., 1937, M.D., 1937; M.Medicine, U. Minn., 1948; m. Ida May Wirtz, June 6, 1938; children—Stephanie Cannon (Mrs. Karl Van D'Eldon), Mary Lucinda (Mrs. William Carlton Moore), Katherine May (Mrs. Jerry Wayne Snider), James Alvin. Faculty, U. Tex. Med. Sch., 1939-40; fellow Mayo Clinic, Rochester, Minn., 1940-48, staff, 1948—, cons. gastroenterology, 1948—; personal physician Pres. L. B. Johnson, 1937—; clin. prof. U. Minn., Rochester, 1964—. Mem. Nat. Adv. Commn. on Med. Manpower, 1966—; chmn. Nat. Adv. Com. to Selective Service for Selection of Doctors, Dentists and Allied Med. Personnel, 1965—; mem. Minn. Bd. Med. Examiners, 1957—. Mem. A.M.A. (Billings gold medal 1963, council on nat. security 1965—), A.C.P. (gov. for mem. council on nat. security 1965—), A.C.P. (gov. for Minn. 1965—), Am. Gastroent. Assn., Assn. for Study Liver Diseases, Am. Fedn. Clin. Research, Sigma Xi. Research and numerous publs. on liver and gastrointestinal tract, gastric ulcer. Home: Cain's Mesa, Route 4, Rochester 55901. Office: Mayo Clinic, Rochester, Minn. 55901.*

CAINELLI, G., Italian chemist; b. Trent, Italy, June 8, 1932; s. Giulio and Tullia (Molinari) C.; Diplom.Ing.Chem., Eidgenoessische Technische Hochschule, Zürich, 1955, Dr.tech.wiss., 1958; m. Gabriella Penna, Oct. 11, 1958; children—Giulio, Michele, Asst., Eidgenoessische Technische Hochschule, 1958-60; with Politecnico di Milano (Italy), 1960—, prof. chemistry, 1963—. Mem. Societa chimica Italiana, Am. Chem. Soc. Research, publs. on elucidation of structure of natural products (terpenes and terpenoids), functionalization of non-activated methyl groups with lead tetra-acetate including partial synthesis of steroidal hormone aldosterone, organoboron chemistry including synthesis of olefins from enolderivatives of chetones and aldehydes, geminal dimetallic organic compounds, including new olefination reaction for carbonyl compounds. Home: 4 Sirtori Cinisello-Milano, Italia.*

CAIRNES, John Elliott, Irish economist; b. Castle Bellingham, W. Luth, Ireland, Dec. 26, 1823; s. William and Mary Anne (Wolsey) C.; B.A., Trinity Coll., Dublin, Ireland, 1848, M.A., 1854; m. Eliza Charlotte Alexander, 1860; 3 children. Mem. staff Engrs. Office, Galway; Whately prof. polit. economy, Dublin, 1856-61; prof. polit. economy and jurisprudence Queen's Coll., Galway, Ireland, 1861-66, U. Coll., London, Eng., 1866-72; admitted to Irish bar, 1857. Author: The Slave Power, 1862; also treatises on polit. economy. Defended wage-fund theory; known for theory of non-competing groups; belonged to classical sch. Died July 8, 1875.

CAIRNS, John, Jr., Am. limnologist; b. Conshohocken, Pa., May 8, 1923; s. John and Eunice (Fesmire) C.; A.B., Swarthmore Coll., 1947; M.S., U. Pa., 1949, Ph.D., 1953; m. Jean Barbara Ogden, Aug. 5, 1944; children—Karen, Stefan, Duncan, Heather. With Acad. Natural Scis., Phila., 1948-66, acting chmn. limnology dept., 1961-63, asst. chmn., 1964-66; prof. zoology U. Kan., Lawrence, 1966—; faculty Rocky Mountain Biol. Lab., Crested Butte, Colo., 1961-63, Temple U., 1962-63, U. Mich. Biol. Sta, Pellston, 1964-66. Mem. Am. Micros. Soc., Soc. Protozoologists, Am. Fisheries Soc., Soc. Limnology and Oceanography, Am. Inst. Biol. Sci., A.A.A.S., Pa. Acad. Sci., Assn. Southeastern Biologists, Biol. Soc. Washington. Author: Population Dynamics, 1965. Contbr. numerous articles to profl. jours. Studies on ecol. relationships of aquatic organisms with emphasis on mgmt. of water resources, factors determining structure of fresh-water protozoan populations and effects of pesticides, detergents, indsl. wastes upon aquatic organisms. Home: 1530 Learnard Av., Lawrence, Kan. 66044.*

CAIRNS, Stewart Scott, Am. mathematician; b. Franklin, N.H., May 8, 1904; s. James George and Laure (Dorion) C.; A.B., Harvard, 1926. A.M., 1927, Ph.D., 1931; m. Kathleen Hand, June 20, 1928; children—James Donald, Charles Edward. Instr. Harvard, 1927-28, Yale, 1929-31; faculty Lehigh U., Bethlehem, Pa., 1933-38; asst. prof. Queens Coll., Flushing, N.Y., 1938-46; prof., chmn. dept. Syracuse (N.Y.) U., 1946-48; prof. U. Ill., Urbana, 1948—, head dept. 1948-58; mem. Inst. for Advanced Study, 1936-37, 59-60, 62-63; Fulbright lectr. Strasbourg (France) U., 1954-55. Cons. Dept. Def., 1950—, Rand Corp., Santa Monica, Cal., 1949—. Mem. Am. Math. Soc., A.A.A.S., Phi Beta Kappa, Sigma Xi, Pi Mu Epsilon. Author: Introductory Topology, 1961. Contbr. articles in topology, analysis and combinatorics to tech. jours. Home: 607 W. Mich. Av., Urbana, Ill. 61801.*

CAIUS, John (also Key or Kaye), English physician; b. Norwich, Eng., Oct. 6, 1510; s. Robert and Alice (Wodanell) C.; grad. (fellow) Gonville Hall, Cambridge (Eng.) U., 1533, M.A., 1535; studied under Montanus and Vesalius at Padua (Italy); M.D., 1541. Lectured on Aristotle, Padua; lectr. anatomy London, Eng., 1544-64; became physician in London, 1547; physician to Edward VI, Queen Mary, Queen Elizabeth I until 1568; refounded and endowed Gonville and Caius Coll., 1557, master, 1558-1573. Fellow Coll. Physicians in London (pres.). Author (under pseudonym Londonensis): A Boke or Counseill against the Disease commonly called the Sweate or Sweatyng Sicknesse, 1552; De Medendi methodo, 1554, 56; Annals of the College from 1555 to 1572; Hippocrates de medicamentis; De ratione victus, 1556; Account of the Sweating Sickness in England (original title was De Ephemera Britannica), 1556, reprinted 1721; History of the University of Cambridge, 1568; De thermis Britannicis (probably never printed), Of Some Rare Plants and Animals, 1570; De Canibus Britannicis libellus 1570, reprinted 1729; De pronunciatione Graecae et Latinae Linguae, 1574. Promoted study of anatomy by obtaining the bodies of prisoners for demonstrations; studied sweating sickness. Died Cambridge, July 29, 1573.

CAJGFINGER, Henri, French surgeon; b. Piennes, France, Dec. 9, 1926; s. Zelman and Marthe (Rosezweig) C.; grad. Faculté de Medecine, Lyons, France, 1945, Docteur en Medecine, 1958; m. Ida Berglas, Jan. 26, 1953; children—Marianne, Edith, Laurence. Chief clinic ear nose and throat Lyons, U., 1959—; 1st asst. in ear nose and throat dept. Croix Rousse Hosp., Lyons, 1963—. Mem. French Ear Nose and Throat Soc., European Rhinologic Soc. Author: (with H. Martin, A. Persillon) Surgery of Stapedo-Vest.'s Ankylosis Syndrome, 1958; also numerous articles. Research on surgery of stape's Ankylosis syndrome, difficult decanulations of tracheotomized children, epistaxis in pregnancy, ear nose and throat manifestations in myasthenis, oesophagoscopy, ear and rubella, deafness, and anoxia of newborn infant, cervical chondroma, sudden deafness, pharyngo-laryngeal disorders in dermatomyositis, radiotherapeutic treatment of glomus jugularis tumors, amygdalectomy and nephritis, Melkerson Rosenthal's syndrome, laryngeal von Recklinghausen's disease, treatment of troubles of endolabyrinthical fluid's pressure. Address: 36 Rue Victor Hugo, Lyons, Rhone 69 France.*

CAJORI, Florian, mathematician; b. St. Aignan, nr. Thusis, Switzerland, Feb. 28, 1859; s. George and Catherina (Camenisch) C.; came to U. S., 1875; B.S., U. Wis., 1883, M.S, 1886; student Johns Hopkins; Ph.D., Tulane, 1894; LL.D., U. Colo., 1912, Colo. Coll., 1913; Sc.D., U. Wis., 1913; m. Elizabeth G. Edwards, Sept. 3, 1890; 1 son, Florian Anton. Asst. prof. math. 1885-87; prof. applied math., 1887-88, Tulane U.; prof. physics Colo. Coll., 1889-98, math., 1898-1918, dean dept. of engring., 1903-18; prof. history of math., U. Cal. at Berkeley, from 1918. Mem. Am. Math. Soc., Math. Assn. America (pres. 1917), Deutsche Mathematiker-Vereinigung, A.A.A.S., Math. Assn. (Eng.), History of Science Soc. (v.p. 1924, 25); fellow Am Acad. Arts and Sciences, Comité intern. d'histoire d. sciences (v.p. 1929). Author: The Teaching and History of Mathematics in the United States, 1890; A History of Mathematics, 2d edit., 1919; A History of Elementary Mathematics, 2d edit., 1917; A History of Physics, 1899; Introduction to the Modern Theory of Equations, 1904; Early Mathematical Sciences in North and South America, 1928; Mathematics in Liberal Education, 1928; Career of F. R. Hassler, 1929; History of Mathematical Notations, 2 vols., 1928-29. Contbr. to math. jours. Died 1930.

CALABI, Eugenio, mathematician; b. Milan, Italy, May 11, 1923; s. Giuseppe and Maria (Bassani) C.; came to U. S., 1939; B.S. in Chem. Engring., Mass. Inst. Tech., 1946; M.A., U. Ill., 1947; Ph.D., Princeton, 1950; m. Giuliana Segré, Sept. 3, 1952; children—Nora J., Joseph A. Asst. prof. math. La. State U., 1951-54; vis. asst. prof. math. Cal. Inst. Tech., 1954-55; asst. prof. math. U. Minn., 1955-57, asso. prof., 1957-60, prof., 1960-64; prof. math., U. Pa., Phila., 1964—. Served with AUS, 1943-46. Study of differential geometry of complex manifolds. Home: 516 Sabine Circle, Wynnewood, Pa.*

CALCAGNINI, Celio, Italian philosopher, poet, astronomer; b. Ferrara, Italy, 1479; became prof. belles-lettres U. Ferrara, 1520; served in armies of Maximilian and Julius II; sent on various diplomatic missions to Rome and Hungary. Author numerous works including: Quaestionum epistolicarum libri III, 1608. Suggested the earth moves around the sun. Died Ferrara, 1541.

CALDANI, Leopoldo Marco Antonio, Italian anatomist, physiologist; b. Bologna, Italy, Nov. 21, 1725; M.D., Bologna, 1750; became prof. practical medicine Bologna, 1755; prof. anatomy Padua, Italy, 1771-

1805. Fellow Royal Soc., 1772. Author: Institutiones pathologicae, 1772; Institutiones physiologicae, 1773; (with a nephew, Floriano) Icones anatomicae, 4 vols., 1801-13/14; Explicatio iconum anatomicarum, 5 vols., 1802-14. Research on function of spinal cord, effect of electricity on nerves, heart and muscle. Died Padua, Dec. 20, 1813.

CALDAROLA, Leonardo, Italian surgeon; b. Bari, Italy, Nov. 24, 1921; s. Francesco and Beatrice (Aldergo) C.; Degree with Honours in Medicine and Surgery, U. Bari, 1945; m. Carla Papa, Oct. 2, 1957; children—Beatrice, Francesco. Faculty U. Turin (Italy), 1946—, univ. lectr. in surg. therapy, 1957-61, tchr. oncological surg. techniques Faculty of medicine, 1966—; head surgeon Oncol. Inst. Turin, 1961—. Mem. Italian Inst. Cancerology, Italian Inst. Surgery, Italian Inst. Pathology, Italian Inst. Cytology, Italian Inst. Nuclear Medicine, Internat. Coll. Surgeons. Research and numerous publs. on endarterial blood transfusion, realization of artificial heart-lung device with related exptl. and clin. applications, realization of artificial kidney with exptl. and clin. applications, treatment of malignant tumors with radioactive isotopes. Home: 79 Corso Re Umberto. Office: 31 Via Cavour, Turin, Italy.*

CALDECOTT, John, Brit. astronomer, meteorologist; b. Eng., 1800; comml. agt. for rajah of Travancore at Allepey, 1832-36; dir. of rajah's obs. at Trivandrum, India, 1837-49; fellow Royal Astron. Soc., Royal Soc., 1840. Research and publs. on ground temperature at various depths (1st to do this in tropics), showing that earth there is 5° to 6°F. hotter than air, 1842-45, bi-annual inversion of law of variation near the magnetic equator which he attributed to monsoon; observed and computed elements for great comet of 1843. Died Trivandrum, India, Dec. 16, 1849.

CALDERWOOD, W. L., Brit. naturalist; b. Glasgow, Scotland, 1865; s. Henry Calderwood; ed. Edinburgh (Scotland) U., also Naples, Germany; m. Lily de Courcy Pullen; 1 son, 2 daus. Dir., Marine Biol. Assn. Lab., Plymouth, Eng., 1889-93; insp. salmon fisheries Scotland, 1898-1930; demonstrator Edinburgh U.; naturalist, staff Fisheries Bd. Scotland; fisheries adviser to N. Scotland Hydro-Electric Bd., 1943—. Fellow Royal Soc. Edinburgh. Author: Mussel Culture and Bait Supply, 1895; The Life of Salmon, 1907; Salmon Rivers and Lochs' of Scotland, 2d edit., 1921; Salmon and Sea Trout, 1930; Salmon Hatching, 1931; Salmon! Experiences and Reflections, 1938; also numerous sci. papers and official reports. Reported on salmon fisheries of E. Can., Newfoundland; adviser on salmon conservation in Brit. hydro-electric projects. Died May 2, 1950.

CALDWELL, Charles, Am. surgeon; b. Caswell County, N.C., May 14, 1772; s. Charles and Miss (Murray) C.; M.D., U. Pa., 1796; m. Eliza Leaming, 1799. Prof. faculty phys. scis. U. Pa.; surgeon Whisky Insurrection, 1794. founder med. dept. Transylvania U., 1819, prof. insts. medicine and clin. practice, 1819-37; 1st prof. U. Louisville (Ky.) 1837-49; asso. with U. Nashville, after 1849. Author: Physiology Vindicated in a Critique on Liebig's Animal Chemistry, (pamphlet attacking Liebig), 1843; also more than 200 books and papers. Editor Port Folio, Phila., 1812. Introduced med. science into Mississippi Valley; opposed use of chemistry in medicine, physiology, pathology. Died Louisville. July 9, 1853.

CALDWELL, Eugene Wilson, Am. physician; b. Savannah, Mo., Dec. 3, 1870; s. W. W. and Camilla (Kellogg) C.; B.S., U. Kan., 1892; M.D., U. and Bellevue Hosp. Med. Coll. (N.Y. U.), 1905; spl. student Coll. Phys. and Surgs., Columbia, 1898-99; m. Elizabeth Perkins, 1913. Engaged in expts. in wireless telephony (with L. I. Blake) for U. S. Lighthouse Establishment, 1893-95; asst. engring. dept. N.Y. Telephone Co., 1895-97; from 1897 has devoted nearly all time to exptl. work with Röntgen rays and to their practical application in diagnosis. Inventor Caldwell Liquid Interrupter, spl. forms of Röntgen ray tubes for therapeutic uses and many other appliances used with Röntgen rays; prof. Röntgenology. Coll. Phys. and Surg., Columbia, 1917-18. Author: The Röntgen Rays in Therapeutics and Diagnosis (with William A. Pusey), 1903. Died June 23, 1918.

CALDWELL, Otis William, Am. botanist; b. Lebanon, Ind., Dec. 18, 1869; s. Theodore Robert and Belle C.; B.S., Franklin Coll., 1894, LL.D., 1917; Ph.D., U. Chgo., 1898; m. Cora Burke, Aug. 25, 1897. Prof. biology Eastern Ill. State Normal Sch., 1899-1907; asso. prof. botany, 1907-13, prof. botany and dean Univ. Coll., 1913-17, U. Chgo.; prof. edn. Tchrs. Coll., Columbia, and dir. Lincoln Exptl. Sch., 1917-27; dir. Div. Sch. Experimentation of Inst. Ednl. Research, 1920-27; dir. Inst. Sch. Experimentation, 1927-35; ret. as emeritus prof. Vis. prof. U. Cal., 1931, Atlanta U., 1937-38. Fellow A.A.A.S. (chmn. com. on place sci. in edn. 1924-40, gen. sec., 1933-47). Author: Laboratory and Field Manual of Botany, 1901; Plant Morphology, 1903; Practical Botany (with J. Y. Bergen), 1911; Introduction to Botany, 1914; Elements of General Science, 1914; Laboratory Manual of General Science (with others), 1915; Then and Now in Education, 1923; Biology in the Public Press, 1923; Open Doors to Science, 1925; Introduction to Science (with F. D. Curtis), 1929; Biological Foundations of Education (with C. C. Skinner and J. W. Tietz), 1931; Biology for Today (with F. D. Curtis and N. H. Sherman), 1933; Do You Believe It (with G. E. Lundeen), 1934; Everyday Biology (with F. D. Curtis and N. H. Sherman), 1940; Everyday Science (with F. D. Curtis), 1943. Editor: Science Remaking the World (with E. E. Slosson), 1923. Contbr. to science and ednl. jours. Died July 5, 1947.

CALDWELL, Ralph Merrill, Am. plant pathologist; b. Brookings, S.D., June 27, 1903; s. Peter and Margaret (Christie) C.; B.S., South Dakota State Coll., 1925; M.S., U. of Wis., 1927, Ph.D., 1929; m. Margaret Dunlap, Sept. 12, 1931; 1 dau., Janet Harriet (now Mrs. Ralph W. Storts). Assistant in the department of botany, U. of Wis., 1925-28; state leader in barberry eradication in Wis., U. S. Dept. Agr., 1928-30; agent, plant pathology, U. S. Dept. Agr., Purdue U., 1930-31, asso. pathologist, 1931-37, chief, dept. botany and plant pathology, Purdue U. Agrl. Expt. Station, 1937-50, head of dept. of agrl. botany, Purdue U., 1943-50, head dept. botany and plant pathology, 1950-54, professor plant pathology, 1937—. Fellow A.A.A.S., Am. Soc. Agronomy; mem. Am. Phytopath. Soc. (treas.; bus. mgr. Phytopathology 1944-46, nat. councillor-at-large 1964-66), Indiana Acad. Scis., Sigma Xi. Editor Phytopathology, 1954-57. Contbr. sci. papers to profl. jours. Research on nature and control of diseases of cereal crops; gave elucidation of genetics of disease and insect resistant wheat, oats, barley; developed new varieties winter wheat, oats, barley. Home: 628 Terry Lane West Lafayette, Ind. 49706. Office: Dept. Botany and Plant Pathology, Purdue U., Lafayette, Ind. 49707.*

CALDWELL, William E(dgar), Am. obstetrician; b. Northfield, O., Feb. 23, 1880; s. Milton Etsil and Susanna Adams C.; M.D., N.Y. Univ. and Bellevue Hosp. Med. Sch., 1904. Adj. asst. attending physician, Bellevue Hosp., also instr. in obstetrics N.Y. U. Med. Sch., 1909-14, asst. prof. and asst. attending physician Bellevue Hosp. Obstet. Service, 1914-20; asso. prof. Columbia U., 1920-27, prof. clin. obstetrics and gynecology, from 1927, asso. dir. Sloane Hosp., from 1920. Fellow A.C.S.; mem. A.M.A., med. socs. State of N.Y., County N.Y., N.Y. Obstet. Soc., N.Y. Acad. Medicine, Am. Gynecol. Soc., Am. Gynecol. Club, Sigma Xi. Contbr. articles to jours. Proposed (with H. C. Moloy) Caldwell-Moloy classification of female pelvis, with regard to form and measurements, 1933. Died Apr. 1, 1943.

CALESNICK, Benjamin, Am. pharmacologist; b. Phila., Dec. 27, 1915; s. Samuel and Ida (Lichtenstein) C.; B.S., St. Joseph's Coll., 1938; A.M., Temple U., 1941; M.D., Hahnemann Med. Coll., 1944; m. Sophie Adele Brenner, Jan. 27, 1945; 1 son, Jay Lee. Faculty Hahnemann Med. Coll., Phila., 1946—, prof. pharmacology, 1966—; dir. sect. human pharmacology Hahnemann Med. Coll. and Hosp. Phila., 1957—; chief Hypertension Clinic, St. Joseph's Hosp., Phila., 1956—; lectr. pharmacology Woman's Med. Coll., 1950-52; vis. instr. pharmacology U. Pa., 1949-50. Recipient Honors Achievement award Angiology Research Found., 1964-65. Fellow Royal Soc. Health; mem. A.M.A., Am. Soc. for Pharmacology and Exptl. Therapeutics, Soc. for Exptl. Biology and Medicine, Soc. Toxicology, Soc. Nuclear Medicine, Am. Coll. Clin. Pharmacology and Chemotherapy, Alpha Omega Alpha. Contbr. to med. textbooks. Pioneer in study of effects of various compounds and drugs in man, including techniques for studying antitussive agts. and various pulmonary functions. Home: 1036 S. 54th St., Phila. 19143. Office: 235 N. 15th St., Phila. 19102.*

CALHOUN, Fred Harvey Hall, Am. geologist; b. Auburn, N.Y., June 27, 1873; s. John Hamilton and Ellen (Hall) C.; B.S., U. Chgo., 1898, Ph.D., 1902; m. Grace B. Ward, June 9, 1904; children—John Ward, Fred. Asst. geol. dept. U. Chgo., 1899-1902; asst. prof. geology and physics Ill. Coll., 1902-04; prof. geology and mineralogy Clemson (S.C.) Coll., 1904-59, also dir. agrl. dept., 1915-33, dean sch. chemistry and geology, 1933-50; cons. geologist S.A.L. R.R.; asst. geologist U. S. Geol. Survey, 1903-15. Fellow Geol. Soc. Am., A.A.A.S.; mem. S.C. Acad. Science (pres.), Alpha Nu, Alpha Chi Sigma, Phi Kappa Phi. Author of geol. monographs. Died May 2, 1959.

CALHOUN, John Bumpass, Am. ecologist; b. nr. Elkton, Tenn., May 11, 1917; s. James B. and Fern (Madole) C.; B.S., U. Va., 1939; Ph.D., Northwestern U., 1943; m. Edith Delight Gressley, Aug. 15, 1942; children—Catherine Carson, Cheryl Hause. Faculty, Emory U., 1943-44, Ohio State U., 1944-46, Johns Hopkins, 1946-49; spl. fellow Jackson Lab., Bar Harbor, Me., 1949-51; research psychologist Walter Reed Army Med. Center, Washington, 1951-54, Nat. Inst. Mental Health, NIH, Bethesda, Md., 1954—; fellow Center for Advanced Study in Behavioral Scis., Palo Alto, Cal., 1962-63. Mem. com. on uncommon animals NRC, 1966—, Fed. Interagy. Com. on Internat. Biol. Program, 1966—. Fellow A.A.A.S.; mem. Ecol. Soc., Soc. for Gen. Systems Research, Am. Naturalists, Am. Soc. Mammalogists, Wildlife Soc. Author: The Ecology and Sociology of the Norway Rat, 1963; also articles. Editor N. Am. Census Small Mammals, 1948-56. Showed evolution to culminate in twelve as optimum group size for mammals; formulated theory brain function; research on impact of population on mental health and quality of life of man. Home: 5705 Cheshire Dr. Office: Nat. Inst. Mental Health, Bethesda, Md. 20014.*

CALHOUN, John Caldwell, Am. polit. theorist; b. Abbeville Dist., S.C., Mar. 18, 1782; s. Patrick and Martha (Caldwell) C.; grad. Yale, 1804, Litchfield (Conn.) Law Sch., 1806; m. Floride Bouneau, Jan. 1811, 9 children. Admitted to S.C. bar, 1807; mem. U. S. Ho. of Reps. S.C., 1811-17, acting chmn. house com. on fgn. affairs, 1811, one of group called War Hawks, presented resolution recommending declaration of war on Eng., 1812; resigned U. S. Ho. of Reps. to become U. S. sec. of war, 1817-25; vice pres. U. S. under John Q. Adams, 1825-29, under Andrew Jackson, 1829-32 (resigned because of dispute with Jackson over states rights and nullification); leader and polit. theoretician of states rights point of view; formulated S.C.'s policy during nullification crisis, declaring that a state can nullify laws it considers unconstl., 1832-33; mem. U. S. Senate from S.C., 1832-43, 45-50, leading senatorial champion of slavery and the So. cause of states rights under his philosophy of "concurrent majorities" or mutual checks whereby each sect. of the country was to share equally in fed. power; U. S. sec. of state under Tyler, 1843-45. R. K. Crallé published compilation of Calhoun's works, 6 vols., 1851-55. Died Washington, D.C., Mar. 31, 1850.

CALKINS, Gary Nathan, Am. zoologist; b. Valparaiso, Ind., Jan. 18, 1869; s. John W. and Emma F. (Smith) C.; S.B., Mass. Inst. Tech., 1890; Ph.D., Columbia, 1897, Sc.D., 1929; m. Anne Marshall Smith, June 28, 1894; m. 2d, Helen Richards Colton, 1909; children—Gary Nathan, Samuel Williston. Asst. biologist Mass. Bd. Health, also lectr. biology Mass. Inst. Tech., 1890-93; faculty Columbia, from 1894, prof., 1904-06, prof. protozoology, 1906-39, emeritus, from 1939. Biologist New York State Cancer Lab., 1904-08; clk. of corp. Marine Biol. Lab. Fellow A.A.A.S., N.Y. Zool. Soc.; mem. Am. Soc. Naturalists, Am. Morphol. Soc., Soc. Exptl. Biology and Medicine, Nat. Acad. Scis., Am. Soc. Cancer Research (pres. 1913-14), Soc. Exptl. Biology and Medicine (pres. 1919-21). Author: The Protozoa (Vol. VI, Columbia U. Biol. Series), 1901; Protozoölogy, 1908; Biology, 1914; Biology of the Protozoa; 1926; also papers. Died Jan. 4, 1943.

CALLA, Étienne, French engr.; b. Paris, 1760; pupil of Vaucanson; built spinning machine factory, 1788; constructed steam boiler for Fulton's steamship, 1800; produced machine tools for all revolutions, 1830. Died 1835.

CALLAN, Edwin J(oseph), Am. physicist; b. Floral Park, N.Y., Nov. 29, 1922; s. James F. and Katherine E. (Huntington) C.; B.S. (scholar), Manhattan Coll., 1943; M.S., Ohio State U., 1960; scholar Dublin Inst. Advanced Studies, 1964-65; m. June M. Frost, Jan. 31, 1946; children—Patricia A., Susan J. Physicist, chief thermal research sect., concrete research div. Waterways Expt. Sta., Jackson, Miss., 1946-53; USAF asst. for programming directorate research Wright Air Devel. Center, Wright-Patterson AFB, O., 1953-56, dir. plans, analysis Aerospace Research Labs., 1956—; cons. Interservice Group on Flight Vehicle Power, 1960-62. Mem. Am. Phys. Soc., Am. Chem. Soc., Am. Inst. Aeros. and Astronautics. Research on thermal, elastic properties of concrete, concrete for radiation shielding; discovered quantized exponential mass spectrum of elementary particles, calculations of Auger and Coster-Kronig transition rates and fluorescence yields, matrix methods for computing hypergeometric functions. Home: 541 Chaucer Rd., Dayton, O. 45431. Office: Aerospace Research Lab.-ARB, B450, Wright-Patterson AFB, O. 45433.

CALLANDREAU, Pierre Jean Octave, French mathematician, astronomer; b. Angouême, France, Sept. 18, 1852; studied at École polytechnique, Paris, France, under Leverrier at Sch. Astronomy, Paris Obs. Astronomer at Paris Obs.; prof. astronomy l'École polytechnique. Mem. French Acad. Scis. Author: (with Tisserand) Traité de mécanique céleste. Research on figures of equilibrium of celestial bodies, periodic comets of Jupiter including their capture, relation to shooting stars, the small planets and the breaks of the ring; calculated orbits. Died Paris, Feb. 13, 1904.

CALLARD DE LA DUCQUERIE, J(ean)-B(aptiste), French physician, botanist; b. Caen, France, 1628; prof. medicine U. Caen; responsible for 1st bot. garden at Caen. Mem. Acad. Caen, French Acad. Scis. Author: Lexicon medicum etymologicum (contains 11,000 etymologies of bot., surg., and med. terms), 1673; Catalogus plantarum in locis paludosis, 1714. Died Caen, Feb. 26, 1718.

CALLEN, Earl Robert, Am. physicist; b. Phila., Aug. 28, 1925; s. Abraham and Mildred (Goldfarb) C.; A.B., U. Pa., 1948, M.A., 1951; Ph.D., Mass. Inst. Tech., 1954; m. Anita Blatt, Dec. 16, 1949; children—Liza, Melany, Jane, Jody. Physicist, Nat. Security Agy., 1955-59; physicist U. S. Naval Ordnance Labs., Silver Spring, Md., 1959—; research prof. Catholic U., 1961—. Fellow Am. Phys. Soc., Washington Philos. Soc. Contbr. chpt. to Reinhold Ency. of Physics, 1966. Research on temperature dependence of magnetic anisotropy and magnetostriction; statis. mechanics of ferro and antiferromagnetism; interaction of helicons and magnons. Home: 9110 LeVelle Ct.,

Chevy Chase, Md. 20015. Office: U. S. Naval Ordnance Labs., Silver Spring, Md. 20910.*

CALLEN, Herbert Bernard, Am. physicist; b. Phila., July 1, 1919; s. Abraham and Mildred (Goldfarb) C.; B.S., Temple U., 1941, M.A., 1942; Ph.D., Mass. Inst. Tech., 1947; m. Sara Smith, Jan. 21, 1945; children—Jill, Jed. Physicist, Kellex Corp., Manhattan Project, N.Y.C., 1944-45; physicist Bumblebee project Princeton, 1945; research asso. Mass. Inst. Tech., 1947-48; mem. faculty U. Pa., Phila., 1948—, prof. physics, 1956—. Cons. Univac div. Sperry Rand Corp., 1950—; mem. adv. com. Nat. Magnet Lab., Mass. Inst. Tech., 1963—. Fellow Am. Phys. Soc. (exec. com., solid state physics div. 1964-65). Author: Thermodynamics, 1960. Research and papers on solid state physics and statis. mechanics, especially fluctuation-dissipation theorem, spin systems. Home: 23 Derwen Rd., Bala-Cynwyd, Pa. 19004.*

CALLENDAR, Hugh Longbourne, English physicist; b. Hatherop, Eng., Apr. 18, 1863; s. Hugh C.; ed. at Marlborough and Cambridge (Eng.); became fellow, 1886; M.A., 1888; LL.D., 1898; m. Victoria Mary Stewart, 1894; 3 sons, 1 dau.; prof. physics McGill U., Montreal, Que., Can., 1893, U. Coll., London, Eng., 1898-02; named prof. Imperial Coll. Sci., 1902. Recipient Rumford medal Royal Soc., 1906, 1st Duddell Meml. medal Phys. Soc., 1924; named Comdr. Order Brit. Empire, 1920. Fellow Royal Soc., 1894, Royal Soc. Can. Author: Callendar Steam Tables, 1915; Properties of Steam and Thermodynamic Theory of Turbines, 1920; also articles. Research on steam engines; prepared Callendar steam equations; developed method of measuring total heat of steam at high pressures; invented electric resistance thermometer, compensated air thermometer; research (with Barnes) on methods of measuring specific heats, 1899. Died London, Jan. 21, 1930.

CALLENDER, George William, Brit. surgeon; b. Clifton, Eng., 1830; became student at St. Bartholomew's Hosp., 1849; several children; house surgeon at St. Bartholomew's Hosp., asst. surgeon, 1861-71, surgeon from 1871, became registrar, 1854; demonstrator anatomy; lectr. on comparative anatomy and anatomy; became lectr. on surgery, 1873; treas. of med. sch. Fellow Royal Coll. Surgeons, Royal Soc., 1871. Author book on anatomy of parts involved in femoral rupture, 1863; also articles, anat. treatises. Died on ship in Atlantic, Oct. 20, 1878.

CALLET, Jean-Charles, French mathematician, b. Versailles, France, Aug. 25, 1744; became prof. hydrography, Vannes, 1788; apptd. prof. math. U. Paris, 1792. Author: Tables de Gardiner (most exact table of logarithms then known), 1783; Tables de logarithmes (trigonometric logarithms derived from decimal div. of angles), 1795. Died Nov. 14, 1799.

CALLEY (or CAWLEY), John, Brit. inventor; b. Dartmouth, Eng.; mfr. window glass, Dartmouth; with Newcomen developed 1st really functional indsl. steam engine, 1712; also developed other machines. Died 1725.

CALLIHAN, Dixon, Am. physicist; b. Scarbro, W.Va., July 20, 1908; s. Alfred D. and Janie (Dixon) C.; A.B., Marshall U., 1928, D.Sc., 1964; M.A., Duke, 1931; Ph.D., N.Y.U., 1933; m. Alva Stroh, Nov. 13, 1936. Faculty Coll. City N.Y., 1934-48; staff mem. div. war research Columbia, 1942-45; staff mem. Union Carbide Corp., Oak Ridge Nat. Lab., 1945—. Pres., Oak Ridge Community Playhouse, 1952-55, bd. dirs., 1963—. Fellow Am. Phys. Soc., Am. Nuclear Soc. (mem. standards com. 1959—), mem. U.S.A. Standards Inst. (standards com. N16 1967—). Editor: Nuclear Sci. and Engring., 1965—. Research on critical conditions for nuclear chain reactions; developed methods for separation uranium isotopes. Home: 102 Oak Lane. Office: P.O. Box Y, Oak Ridge 37830.*

CALLINICUS, Egyptian alchemist; b. Heliopolus, Egypt; flourished 7th century; invented Greek fire which was a mixture consisting of inflammable petroleum and potassium nitrate to supply oxygen, quicklime; since it burned on water it could be used against a wooden fleet.

CALLIPPOS OF CYZICOS, Greek astronomer; b. circa 370 B.C.; went with Polemarchus to Athens where he stayed with Aristotle; studied under Eudoxus; corrected and improved Eudoxus' theory of concentric spheres to account for the movements of sun, moon and planets; originated Callipic cycle of 76 years which was equal to 4 Metonic cycles minus a day; known for planetary theories; most of his works have been lost. Died circa 300 B.C.

CALLISTHENES OF OLYNTHES, astronomer; b. circa 360 B.C.; student of Aristotle; studied at Sch. at Assos; historian on the expdns. of Alexander the Gt. Wrote on Alexander's expdn., a history of Greece from the peace of Antalcides to the Phocian War, history of the Phocian War. (none of his writings have survived). Sent a complete list of eclipses observed in Babylon over a period 1,900 years to his uncle, Aristotle; assisted in forming Aristotle's pneumatic theory of earthquakes which stated they were caused by underground pockets of compressed air. Executed for treason, circa 328 B.C.

CALLOW, John Michael, mining engr., metallurgist; b. Northrepps, Norwich, Norfolk, Eng., July 7, 1867; s. Michael John and Emily (Neave) C.; ed. in Eng.; came to U. S., 1890; m. Roberta More, 1893; children—Bessie Roberta, Margaret Roper More, Francis Marie, Michael John. Engr., draftsman Stearns, Roger Mfg. Co., 1892-93; engr. Metallic Extraction Co., 1894; operating mines and mills in San Juan Co., Colo., 1895-96; on engring. staff Samuel Newhouse, 1898-1901; pvt. practice, 1901-06, inventing Callow settling tank and Callow traveling belt screen; pres. and mgr. Gen. Engring. Co., N.Y., 1906—; designed and built 500-ton plant for Nat. Copper Co., Mullan, Ida., installing pneumatic flotation cells, 1912; designing and constrn. engr., Mt. Isa Mines, Australia, 1928-31. Originator of pneumatic flotation in treatment of ores; awarded 18 patents. Awarded Douglas medal by Am. Inst. Mining and Metall. Engrs., for achievements in non-ferrous metallurgy, 1925. Died July 27, 1940.

CALLOWAY, Doris Howes, Am. educator, nutritionist; b. Canton, O., Feb. 14, 1923; d. Earl John and Lillian (Roberts) Howes; B.S., Ohio State U., 1943; postgrad. Johns Hopkins Hosp., 1944; Ph.D., U. Chgo., 1947; m. Nathaniel O. Calloway, Feb. 14, 1946 (div. Sept. 1956); children—David Karl, Candace Mary. Research dietitian dept. medicine U. Ill., Chgo., 1945; cons. therapeutic nutrition Med. Assos. Chgo., 1948-51; nutritionist, head metabolism lab. and chief nutrition br. Armed Forces Food and Container Inst., Chgo., 1951-61; chmn. dept. food sci. and nutrition Stanford Research Inst., Menlo Park, Cal., 1961-64; prof. nutrition U. Cal., Berkeley, 1963—. Mem. subcom. on ascorbic acid and pantothenic acid com. on dietary allowances Food and Nutrition Bd., NRC, Nat. Acad. Sci., 1965—, mem. panel atmosphere regeneration Space Sci. Bd., 1966—. Recipient Meritorious Civilian Service award Dept. Army, 1959. Mem. Am. Inst. Nutrition, Am. Dietetic Assn., Sigma Xi. Author: (with Igel) Nutrition, a Programmed Text, 1964; (with Bogert and Briggs) Nutrition and Physical Fitness, 1966. Editor: Human Ecology in Space Flight, 3 vols., 1966—; asso. editor Nutrition Reviews, 1961—. Research in relationships between nutrition and stress, including study of problems of space flight; survival, and energy and protein metabolism; radiation injury; mil. feeding. Home: 25 Forest Lane, Berkeley, Cal. 94708.*

CALLOWAY, Jean Mitchener, Am. mathematician; b. Indianola, Miss., Dec. 18, 1923; s. James Earl and Mittie Lou (Mitchener) C.; B.A., Millsaps Coll., 1944; M.A., U. Pa., 1949, Ph.D., 1952; m. Anne Marie Whitney, June 21, 1952; children—Nancy Lou, Catherine Anne. Asst. instr. math. Millsaps Coll., 1944; instr. math. McCallie Sch., Chattanooga, 1944-47; asst. instr. math. U. Pa., 1947-52; asst. prof. math. Carleton Coll., 1952-58, asso. prof., 1958-60, acting chmn. dept., 1953-54, 58-59; mem. Inst. for Advanced Study, Princeton, N.J., 1958-59; Olney prof. math., chmn. dept. Kalamazoo (Mich.) Coll., 1960—. Mem. writing team Sch. Math. Study Group, summers 1959, 60, mem. panel on supplementary publs., 1961-63. Mem. Math. Assn. Am. (chmn. Mich. sect. 1963-64), Pi Kappa Alpha. Episcopalian. Author: Fundamentals of Modern Mathematics, 1964. Home: 1229 Knollwood Av., Kalamazoo 49007.*

CALMAN, William Thomas, Brit. zoologist; b. Dundee, Scotland, 1871; B.Sc., St. Andrews U., 1895, D.Sc., 1900, LL.D.; m. Alice Jean Donaldson, 1906; 1 son, 1 dau. Asst. lectr., demonstrator in zoology U. Coll., Dundee, 1895-1903; asst. zool. dept. Brit. Mus., 1904-21, dep. keeper zoology, 1921-27, keeper zoology, 1927-36. Lectr. zoology St. Andrews U., 1940-46. Fellow Royal Soc., 1921, Linnean Soc. (pres. 1934-37, Gold medal 1949), Royal Soc. Edinburgh (hon.); mem. Brit. Assn. (pres. sect. D 1930), Ray Soc. (sec. 1919-46). Author: The Life of Crustacea, 1911; The Classification of Animals, 1949; articles on crustacea, others Ency. Britannica, 11th edit., other sci. publs. Died Sept. 29, 1952.

CALMETTE, Albert Léon Charles, French physician, bacteriologist; b. Nice, France, July 12, 1863; M.D., U. Paris, 1886; LL.D., Cambridge U.; m. Mlle de la Salle. Physician, French Navy, 1883-90; studied bubonic plague, Oporto, Portugal, 1889; founder, 1st dir. Pasteur Inst., Saigon, 1891-93, Lille, France, 1896-1919; prof. in Lille; acting adminstrv. head Pasteur Inst., Paris, 1917-33. Hon. fellow Royal Soc., 1921, Royal Soc. Medicine; mem. Acad. Scis. (sect. medicine and surgery), Acad. Medicine, Acad. Colonial Scis. Author: L'ankylostomiase, maladie sociale des exploitations houillères, 1905; Recherches sur l'épuration biologique des eaux d'égout, 8 vols., 1905-14; Len Venins, les animaux venimeux et la sérothérapie antivenimeuse, 1907; Recherches expérimentales sur la tuberculose, 1907-14; L'infection bacillaire et la Tuberculose chez l'homme et chez les animaux, 2d edit., 1924 (translated into English); La vaccination preventive coutre la Tuberculose par B.C.G., 1927. Discovered (with Yersin and Borrel) serum for snakebite, successfully inoculated animals with anti-plague serum; introduced Calmette's reaction (tuberculin conjunctival test); developed (with Guérin) vaccine Bacille Calmette Guérin for protection from Tb (still widely used). Died Paris, France, Oct. 29, 1933.

CALNE, Roy Yorke, Brit. surgeon; b. London, Eng., Dec. 30, 1930; s. Joseph Robert and Eileen (Gubbay) C.; student Lancing Coll., 1944-47; M.B., B.S., Guys Hosp. Med. Sch., London, 1953, M.S., 1961; m. Patricia Doreen Whelan, Mar. 2, 1956; children—Jane, Sarah, Deborah, Suzanne, Russell. Harkness research fellow in surgery Harvard Med. Sch., also asst. in surgery Peter Bent Brigham Hosp., Boston, 1960-61; sr. lectr. in surgery Westminster Hosp., London, 1962-65; prof. surgery, head dept. Cambridge (Eng.) U., 1965—, fellow Trinity Hall. Recipient Hallett prize Royal Coll. Surgeons, 1958, Jacksonian prize, 1961, Hunterian Prof., 1963, Cecil Joll prize, 1965. Fellow Assn. of Surgeons of Gt. Britain, Royal Coll Surgeons. Author: Renal Transplantation, 1963; (with H. Ellis) Lecture Notes in Surgery, 1965; also articles. Research on mechanisms of rejection of kidney transplants, methods of prevention of graft destruction. Home: 22 Barrow Rd., Cambridge, Eng.*

CALOI, Pietro, Italian geophysicist; b. Verona, Italy, Feb. 22, 1907; s. Bernardo and Vittoria (Tiranti) C.; Ph.D. in Math., U. Padua (Italy), 1929; m. Antonietta Borriero, Nov. 3, 1938; children—Francesco, Vittoria, Lucia, Rita, Benedetto. Asst. geophysics Istituto Geofisico Trieste, 1931-37; prin. geophysicist U. Rome (Italy), 1937-41, chief geophysicist, 1941-49, prof. geophysics, dir. Obs., 1949—. Cons. to Italian govt. agys. Recipient prize for physics Accademia d'Italia, 1941, prize Feltrinelli for geodesy and geophysics Accademia Nazionale dei Lincei, 1956. Mem. Accademia Nazionale dei Lincei, Am. Geophys. Union, Commn. Seismologique Européenne (past pres.). Author: Attività-Sismica in Italia, 1930-40; Acc. d'Italia, 1942; On the Upper Mantle, 1967. Research and publs. on epicentral coordinates, determination, seismic waves, earthquakes, Rayleigh waves, dynamical properties of dams, hydrodynamical properties of dams, hydrodynamical properties of lakes and seas. Home: 29 Mario Fascetti, Rome, Italy.*

CALVERT, Frederick Grace, Brit. chemist; b. London, Eng., Nov. 14, 1819; s. Col. Calvert; studied under Gerardin at Rouen (France), Sorbonne, Collége de France, École de médecine (all Paris, France). Practised and studied chemistry, France, 1835-46; chem. tchr., specialist, mfr., Manchester, Eng., 1846-73; prof. chemistry Royal Instn., Manchester. Fellow Royal Soc., 1859. Author: Dyeing and Calico Printing; also articles. Research on tanning, desulphurization of coal, protection of iron ships from rust, iron puddling, manufacture of coal-tar products expecially phenic or carbolic acid (1st to manufacture it in a pure state in Eng.). Died Vienna, Austria, Oct. 24, 1873.

CALVERT, Jack George, Am. phys. chemist; b. Inglewood, Cal., May 9, 1923; s. John George and Emma (Eschstruth) C.; B.S. in Chemistry, U. Cal. at Los Angeles, 1944, Ph.D., 1949; m. Doris Arlene Breimon, Nov. 8, 1946; children—Richard John, Mark Steven. Mem. faculty Ohio State U., 1950—, prof. chemistry, 1960—, chmn. dept., 1964—. Cons. air pollution tng. com. USPHS, 1964—. Named Honor Prof. of Year, Coll. Arts and Scis., Ohio State U., 1957, recipient Alumni award distinguished teaching, 1961. NRC Can. fellow, 1949. Fellow Ohio Acad. Sci.; mem. Am. Chem. Soc., Am. Assn. U. Profs., Air Pollution Control Assn., Phi Beta Kappa, Sigma Xi, Pi Mu Epsilon, Phi Lambda Upsilon, Kappa Alpha, Alpha Chi Sigma. Author: (with J. N. Pitts, Jr.) Photochemistry, 1965; also articles. Research in photochemistry, reaction kinetic, mechanisms free radical reactions. Home: 2535 McVey Blvd., W., Worthington, O. 43085. Office: 88 W. 18th Av., Columbus, O., 43210.*

CALVERT, Lauriston Derwent, crystallographer; b. Zeehan, Tasmania, Dec. 10, 1924; s. Frederick Clifford and Ivy (Clark) C.; B.Sc., U. New Zealand, 1946, B.A., 1948, M.Sc., 1949, Ph.D., 1952; m. Margery McKay, May 5, 1954; children—Alistair Graham, Margaret Joan. Postdoctoral fellow NRC, Ottawa, Ont., Can., 1952-54, crystallographer div. applied chemistry, 1954—. Mem. Am. Crystallographic Assn. Co-editor Structure Reports, 1962—. Research and publs. on X-ray crystal structures, apparatus for high-temperature and low-temperature X-ray diffraction. Home: R.R. 2, Cumberland, Ont. Office: Div. Applied Chemistry, Nat. Research Council, Ottawa, Ont., Can.*

CALVERY, Herbert Orion, Am. pharmacologist; b. Eddy, Tex., Dec. 9, 1897; s. Luther and Theresa Irene (Marricle) C.; ed. Peniel Coll.; B.S., Greenville Coll., Ill., 1919; A.B., U. Ill., 1921; M.S., 1923, Ph.D., 1924; m. Gertrude V. Lane, June 2, 1925; children—Catherine Ann, George Herbert. Tchr., 1919; asst. chemistry U. Ill., 1920-22, fellowship 1922-24; asst. prof. U. Louisville, 1924-25; instr. Johns Hopkins Med. Sch., 1925-27; asst. prof. physiol. chemistry U. Mich. Med. Sch., 1927-35; senior pharmacologist FDA, 1935-36, chief div. pharmacology, 1936-43. Fellow A.A.A.S.; mem. Am. Chem. Soc. (chmn. biol. div. 1939-40), Am. Soc. Biol. Chemists, Am. Soc. Pharmacology and Exptl. Therapeutics, Soc. Exptl. Biology and Medicine, Am. Pub. Health Assn., Sigma Xi, Alpha Chi Sigma, Phi Lambda Upsilon, Phi Sigma. Contbr. to jours., texts, encys. on biochemistry and toxicology. Died Sep. 23, 1945.

CALVET, Jean Paul Emile, French physician; b. Chaumont, France, Dec. 29, 1908; s. Louis and Marthe (Baur) C.; M.D., Sch. Medicine, Paris; m. Renée Cottineau, June 29, 1932; children—Marie-Francine, Sylvie, Marie-Noëlle, Isabelle. Extern, Paris hosps., 1927-28, intern, 1929-30, became surgeon, 1943; named chief of clinic, aid in anatomy Sch. of Medicine, 1934, later prof.; apptd. chief of surgery service Fernand Widal Hosp., 1954; chief of surgery service Lariboisière Hosp., 1956——. Mem. Surgery Acad. (titular), Orthopedic and Traumatology Soc. Author: Publications diverses sur des sujets chirurgicaux, 1936; Précis de physiologie applique à l'education physique. Home: 12, rue les Pelibes, Paris. Office: 16, rue Remy de Gourmont, Paris, France.

CALVIN, John, theologian and political theorist; b. Noyon, France, 1509; s. Gerard Caulvin (or Cauvin); studied Latin, College de la Marche, Paris, 1523-28; studied law at Orleans U. and Greek at Bourges; m. Idelette de Bure. Began to preach the reformed doctrines while at Bourges and avowed himself a disciple of the reformation, circa 1528; preached at Noyon, 1529; went to Paris but was forced to leave because of a sermon which he delivered on justification by faith; forced because of religious persecution to go to Basel, then Ferrara, Paris, Geneva and Strasbourg; recalled to Geneva where he established theocratic rule under his own direction; established academy at Geneva, 1559, where he taught theology. Author: Institutes of the Christian Religion, 1535. Intense biblicism and resolute theocentricity are hallmarks of his theology; his was most dynamic political thought of reformation; taught predestination in frank and uncompromising way; taught natural order is foundation of all legal and moral relations between men; Christian Church and state are created by God but are designed for different purposes and must be kept autonomous and distinct; jurisdiction of state is temporal and must be confined to physical and external existence of men; Church jurisdiction is spiritual and its authority should include no element of secular concern; princes must be obeyed, even while they oppress; oppressive rulers cannot without sin be resisted forcibly by private citizens, although sometimes divine authority may be claimed for action, guided by lower magistrates, against a heretical or infidel sovereign. Max Weber argued that Calvinism, by its emphasis on such virtues as thrift, industry, sobriety and responsibility, created a capitalistic outlook and the worldly asceticism of modern business; this view much controverted. Died Geneva, Switzerland, May 27, 1564.

CALVIN, Lyle David, Am. statistician, educator; b. Dannebrog, Neb., Apr. 12, 1923; s. David A. and Muriel (Harvey) C.; grad. Parsons (Kan.) Jr. Coll., 1943; B.S. in Meteorology, U. Chgo., 1948; B.S., N.C. State Coll., 1947, Ph.D., 1953; m. Shirley Jeanne Schmidt, Apr. 19, 1952; children—James Arthur, Ronald David, Janet Lee. Biometrician, G. D. Searle & Co., Chgo., 1950-52; asst. statistician N.C. State Coll., Raleigh, 1952-53; statistician Agrl. Expt. Sta., asso. prof. Ore. State U., 1953-57, prof., 1957——, chmn. dept. statistics, 1962——; vis. prof. U. Edinburgh, 1967. Served from pvt. to 1st lt. USAAF, 1943-46. Mem. Am. Statis. Assn., Inst. Math. Statis., Biometric Soc. (pres. WNAR 1964-65), Royal Statis. Soc., Sigma Xi. Research on sampling methods; experimental design and analysis. Home: 3463 Crest Dr., Corvallis, Ore. 97330.

CALVIN, Melvin, Am. chemist; b. St. Paul, Minn., Apr. 8, 1911; s. Elias and Rose (Hervitz) C.; B.S., Mich. Coll. Mining and Tech., 1931; Ph.D., U. Minn., 1935; D.Sc., Mich. Coll., 1955, U. Nottingham, 1958, Oxford University, 1959, Northwestern University, 1961, University of Notre Dame, 1965; m. Marie Genevieve Jemtegaard, Oct. 4, 1942; children—Elin B., Karole R., Noel M. Faculty U. Cal. at Berkeley, 1937——, prof. chemistry 1947——, dir. chem. biodynamics group, Lawrence Radiation Lab., 1946——; dir. laboratory of chem. biodynamics, 1963——, asso. dir. Lawrence Radiation Laboratory, 1967——. Recipient Nobel prize in chemistry, 1961. Fellow Royal Soc., 1959 (Davy medal 1964); mem. Nat. Acad. Scis., Am. Philos. Soc., Am. Acad. Arts and Scis., numerous other profl. assns. Author: (with G. E. K. Branch) The theory of Organic Chemistry, 1941; (with others) Isotopic Carbon, 1949; (with A. E. Martell) The chemistry of the Metal Chelate Compounds, 1952; (with J. A. Bassham) the Path of Carbon in Photosynthesis, 1957; Chemical Evolution, 1961; (with J. A. Basshan) the Photosynthesis of Carbon Compounds, 1962. Research, publs. on photosynthesis, chem. evolution, biophysics; new concepts about behavior of all types of organic molecules under variety of conditions, laying foundation for modern theoretical organic chemistry; prin. interest is application of basic physics and chemistry of molecules to some of fundamental biodynamic problems of biology. Home: 2683 Buena Vista Way, Berkeley, Cal. 94708. Office: Lab. Chem. Biodynamics, U. Cal., Berkeley, Cal. 94720.

CALVISIUS, Sethus, German music theoretician; b. Gorsleben, Germany, Feb. 21, 1556; s. Jakob and Elisabeth (Kruse) Kallwitz; enrolled U. Leipzig (Germany), 1576; student Helmstedt, 1579-80; studied only music after 1580; m. Magdalene Junge, Feb. 10, 1595; 2 sons, 1 dau. In Magdeburg, after 1572; took various positions as cantor, music tchr., astronomer and linguist at Leipzig, beginning in 1581; taught pvt. students. Author numerous publs. including: Compendium musicae, 1594; Exercitatio musica tertia, 1609; Exercitatione, musicae duae, 1600; Meloporia, sive melodiae . . . , 1592; Harmoniae cantionum ecclesiasticarum, 1597; Opus chronologicum, 1685. Wrote extensively on music theory of his time. Died Leipzig, Nov. 24, 1615.

CALW, Ulrich Rülein von, see von Calw, Ulrich Rülein.

CAMA, H. R., Indian biochemist; b. Bombay, India, Apr. 15, 1921; s. Ratansha F. and Mithan (Marshall) C.; B.Sc. with honors, Bombay U., 1944, B.A. with honors, 1945, M.Sc., 1946; Ph.D., U. Liverpool (Eng.), 1949, D.Sc., 1959. Biochemist, Royal United Hosps., Liverpool; staff dept. biochemistry U. Liverpool; prof. biochemistry, dean faculty sci. Indian Inst. Sci., Bangalore; investigator-in-charge several OSIR schemes Govt. of India. Fellow Royal Inst. Chemistry (U.K.); mem. Biochem. Soc. (U.K.), Soc. Biol. Chemists (life), Internat. Congress B.C., Internat. Congress Nutrition. Research, numerous publs. on carotenoids of Indian flowers and fruits, distbn. of vitamins A1 and A2 in Indian marine and freshwater fish liver oils and their biochem. and nutritional significance, metabolic derivatives of vitamins A1 and A2, nutritional work on solvent-extracted vegetable oils and proteins, cholesterol metabolism. Office: Dept. Biochemistry, Indian Inst. Sci., Bangalore-12, India.*

CAMAIN, Robert, French physician; b. Rion Landes, France, May 7, 1915; s. Henri and Madeleine (Court) C.; M.D., U. Bordeaux, 1939; postgrad. Institut Pasteur, Paris, 1945-46; m. Raymonde Giabicani Camain, May 7, 1946; children—Lise, Nadine, Marie-France. Staff, Institut Pasteur, Cayenne, French Guyana, 1947-48, Inst. Pasteur, Dakar, Senegal, 1950-—; faculty U. Dakar Sch. Medicine, 1950-—, prof. histology and embryology, dean Sch. Medicine, 1966-—. Decorated Legion d'Honneur; Merite Senegalais. Mem. Internat. Acad. Pathology, Soc. Belge de Medicine Tropicale, Soc. de Biologie, Assn. des Anatomistes de Langue francais. Research, numerous publs. on tropical pathology, including bilharziasis, fungus diseases, viral hepatitis and sequellae in tropical regions, cirrhosis, kwashiorkor, leprosy; cancer in Affrica, including primary carcinoma of liver, kapusi sarcoma, Burkitt's disease, tumor of orbitae in Senegal. Home and office: BP220 Institut Pasteur, Dakar, Senegal.*

CAMBEL, Ali Bulent, mech.; aerospace engr.; b. Merano, Italy, Sept. 4, 1923; s. H. Cemil and Remziye (Hakki) C.; came to U. S., 1943, naturalized, 1951; B.S., Robert Coll., 1942; postgrad. U. Istanbul, (Turkey), Mass. Inst. Tech.; M.S., Cal. Inst. Tech., 1946; Ph.D., State U. Ia., 1950; m. Marion de Paar, Dec. 20, 1946; children—Metin, Emel, Leyla, Sarah. Faculty, State U. Ia., 1947-53, asst. prof., 1950-53; staff thermodynamicist Canadian Def. Research Bd., 1953; asso. prof. Northwestern U., Evanston, Ill., 1953-56, dir. Gas Dynamics Lab., 1953-—, prof., 1956-60, Walter P. Murphy Distinguished Prof., 1960-—, chmn. dept. mech. engring. and astron. scis., 1957-—. Staff dir. Pres.'s Interdepartmental Energy Study, White House, 1963; cons. to govt. agys., pvt. cos.; Sigma Xi nat. lectr., 1961-62. Recipient Leadership award YMCA, Service award Immigrants Service League, 1965. Mem. Am. Soc. M.E., Am. Inst. Aeronautics and Astronautics (Edward Pendray award 1959), Am. Soc. Engring. Edn. (McGraw award 1960), Am. Assn. U. Profs., Am. Phys. Soc. Author: (with B. H. Jennings) Gas Dynamics, 1958; (with D. Duclos, T. P. Anderson) Real Gases, 1963; Plasma Physics and Magnetofluidmechanics, 1963; (with Marion Cambel) Plasma Physics, 1965; also numerous articles. Research on aerothermochemistry, plasma physics, magnetofluidmechanics, energy conversions originated method keeping stable flame burning in high speed jet aircraft. Home: 2027 Orrington Av., Evanston, Ill. 60201.*

CAMBEY, Henri Prudence, French inventor; b. Troyes, France, 1787; inventor dividing machine, also a lathe; improved theodolite and heliostat; constructed instruments for Paris (France) Obs. including great equatorial; discovered attenuation of declination of magnet needle by a copper plate. Died 1847.

CAMEL, George Joseph, botanist; b. Brunn, Moravia (now Czechoslovakia), Apr. 21, 1661; ed. at U. Brünn; joined the Soc. of Jesus, Brünn, 1682; became pharmacy clk., Neuhaus, Czechoslovakia, 1685; became pharmacist at Krumman, 1686; became physician, pharmacist, 1687; went to Manila, P.I., 1688; opened pharmacy for the poor in Manila. Linnaeus changed name Thea japonica (chinensis) to Camelia in his honor. Author: Gazophylacium, 1767; Medicinal Plants; also articles on plants and natural history. Observations on zoology, native pharmacy, botany including med. herbs. Died Manila, May 2, 1706.

CAMERARIUS, Elias, German alchemist; b. Tübingen, Württemberg, Feb. 17, 1673. Prof., Tübingen; physician, councillor to Duke of Württemberg. Mem. Academia Naturae Curiosorum, 1692. Author several treatises, including; Spirituum animalium statum naturalem . . . ; Spiritum D. Boylii fumantem obviaque circa ipsum phaenomena; Usum et abusum potuum thee et caffe in his regionibus. Research, expts. with ammonium sulphide. Died Tübingen, Feb. 6, 1734.

CAMERARIUS, Joachim, German physician, botanist; b. Nuremberg, Germany, 1534; s. Joachim Camerarius. Practiced medicine, Nuremberg; obtained possession bot. library of Gesner with 1500 wood engravings. Author: Hortus medicus; Epitome mathioli de plantis, 1586; other bot. works. Cultivated rare plants. Died 1598.

CAMERARIUS, Rudolf Jakob, German botanist, physician; b. Tübingen, Germany, Feb. 12, 1665; s. Elias Rudolf and Regina Barbara (Neuffer) C.; ed. at Tübingen; studied under his father; m. Christine Magdalene Crafft, 1689; 1 son, Alexander. Prof. physics Tübingen, 1688-95; became prof. medicine, dir. botanic garden Tübingen, 1695. Author: De sexu plantarum, 1695; Opuscula botanica, pub. 1797; also articles. Pioneered work in exptl. genetics; helped establish sexual theory in plants; defined stamen as male and ovary as female parts of plants, 1694; after experiments and observations, described roles of stamen, pistil, and pollen in fertilization. Died Tübingen, Sept. 11, 1721.

CAMERER, Johann Friedrich Wilhelm, physician, physiologist; b. Stuttgart, Germany, Oct. 17, 1842; s. Johann Wilhelm and Julie (Hirzel) C.; began study math., physics, chemistry Polytechnikum Stuttgart, 1861; began study medicine Tübingen, 1861; Dr.h.c., Tübingen; m. Hedwig Gugler, 1867; 1 son, Johann Wilhelm; 4 daus. Dr. med., 1865. Research staff Physiol. Inst., Tübingen; became physician, Crailsheim, 1867; chief army physician in war 1870-71; later moved to Ulm, Germany, Riedlingen, and finally to Urach, Germany, 1883. Mem. German Soc. Pediatrics (hon.), 1894. Research on child physiology, pediatrics; a founder of modern techniques of infant care. Died Urach, Mar. 25, 1910.

CAMERON, D(onald) Ewen, psychiatrist; b. Bridge of Allan, Scotland, Dec. 24, 1901; s. Duncan and Margaret Isabel (Conacher) C.; M.B., Ch.B., U. Glasgow, 1924, M.D. with distinction, 1936; m. Jean Carruthers Rankine, Aug. 5, 1933; children—Duncan Hume, Airlie A. C., D. Stuart, James R. Intern, resident surgeon Glasgow Mental Infirmary, 1924-25; asst. physician Glasgow Royal Mental Hosp., 1925-26; Henderson research scholar in psychiatry Johns Hopkins, 1926-28; volontairarzt Burgholzli Clinic, Zurich, Switzerland, 1928-29; physician charge reception unit Province Mental Hosp., Brandon, Man., Can., 1929-36; sr. research psychiatrist Worcester (Mass.) State Hosp., 1936-37, resident dir. research, 1937-38; prof. neurology and psychiatry Albany Med. Coll., also neurologist and psychiatrist in chief Albany Hosp., 1938-43; prof., chmn. dept. psychiatry McGill U., also psychiatrist in chief Royal Victoria Hosp., dir. Allan Meml. Inst. Psychiatry, Montreal, 1943-—; cons. psychiatrist Montreal Gen. Hosp.; dir. psychiatry and aging research labs. VA Hosp., Albany, N.Y., 1964-—; research professor of psychiatry Albany Med. Coll., 1964 -—. Mem. expert adv. panel mental health WHO, 1952-—; mem. bd. examiners Am. Psychiat. Assn., 1938-39, bd. psychiat. examiners State N.Y., 1941-43. Hon. president Manfred Sakel Found. Diplomate Am. Bd. Psychiatry and Neurology. Fellow N.Y. Acad. Medicine, Am. Psychiat. Assn. (pres. 1952-53), Royal Coll. Physicians and Surgeons Can.; hon. fellow Am. Geriatrics Soc.; member A.M.A., Brit., Que. Indsl., Canadian medical assns., World Fedn. Mental Health, Royal Medico-Psychol. Assn. (hon.), A.A.A.S., Assn. Research Nervous and Mental Disease, Que. Psychiat. Assn. (pres.), American Psychopath. Assn. (pres. 1962-63), World (pres. 1961-66), Canadian, Que. (pres. 1956-57) psychiat. assns., Soc. Biol. Psychiatry (pres. 1965-66). Author: Objective and Experimental Psychiatry, 1935; (with H. G. Ross) Human Behavior and its Relation to Industry, 1944, Studies in Supervision, 1945; Remembering, 1947; Life is for Living, 1948; General Psychotherapy; Dynamics and Procedures, 1950. Contbr. profl. jours. Introduced insulin treatment for schizophrenia in N. Am., 1938; established transcultural psychiatry, 1952, RNA as substrate for memory, 1956. Home: Mt. Whitney Rd., Lake Placid, N.Y. Office: VA Hosp., Albany, N.Y.*

CAMERON, Dale Corbin, Am. psychiatrist, hosp. ofcl.; b. Hendley, Neb., July 10, 1912; s. Joseph Robert and Veda (Corbin) C.; A.B., U. Neb., 1933, M.D., 1936; M.P.H., Johns Hopkins, 1951; m. Irma Christine Lippold, June 10, 1936; children—Robert William, Marsha Kay (Mrs. John A. Soucheray). Intern, USPHS Hosp., San Francisco, 1936-37; resident in psychiatry U. Colo., 1939-40; clin. dir. USPHS Hosp., Ft. Worth, 1940-41, exec. officer, 1941-43; asst. chief Nat. Inst. Mental Health, 1945-50; med. cons. health emergency planning Office Surgeon Gen., 1953-54; med. dir. Minn. Dept. Pub. Welfare, 1954-60; clin. prof. psychiatry and neurology U. Minn.,

Mpls., 1954-60; asst. supt. St. Elizabeth's Hosp., Washington, 1960-62, supt., 1962-67; chief drug dependence unit WHO, Geneva, Switzerland, 1967-—; apptd. clin. prof. psychiatry George Washington U., Washington, 1960. Chmn. com. on drug addiction and narcotics NRC, 1961-67. Asst. surgeon-med. dir. USPHS, 1936-54, med. dir., asst. surgeon gen., 1960-67. Mem. A.M.A. (chmn. com. on narcotic addiction 1962-63, chmn. com. on alcoholism and addiction 1963-67), Am. Psychiat. Assn. (treas. 1963-68), Washington Psychiat. Soc. (chmn. liaison with standing com. Jud. Conf. D.C. 1961-67), Phi Beta Kappa, Alpha Omega Alpha, Phi Beta Pi. Publs. on adminstrn. mental health programs, forensic psychiatry, drug dependence. Home: 50 Rue de Moillebeau, 1202 Geneva. Office: World Health Orgn., Av.-Appia, 1211 Geneva 27, Switzerland.

CAMERON, Donald Malcolm, Am. astronomer; b. Jan. 10, 1912; B.S., Case U., 1934; M.Sc., Ohio State U., 1936, Ph.D. in physics, 1941. Research asst. astronomer Warner and Swasey Obs., Case U., since 1945; mem. Astron. Soc.; research on infra-red spectrum of hydrogen selenide and on M-type stars. Address: 4400 Tamalgra Dr., South Euclid, Ohio.

CAMERON, Edwin J(ohn), Am. chemist; b. Cambridge, Mass., Sept. 17, 1895; s. John J. and Annie (Ellis) C.; B.S., Mass. Inst. Tech., 1920; Ph.D., George Washington U., 1927; m. Dorothy Ellouise Pray, Sept. 21, 1921; 1 son, John Pray. Bacteriologist, Comml. Solvents Corp., Terre Haute, Ind., 1920-23; Nat. Canners Assn., Washington, 1923-36, asst. dir. research lab., 1936-39, dir. research lab., 1939-55, dir. all research labs., 1955-—. Mem. Soc. Am. Bacteriol., Am. Chem. Soc., Inst. Food Technologists. Contbr. articles on thermophilic bacteria, sources of thermophilic bacteria and relations to canned food spoilage; thermophilic contamination of beet and cane sugar. Died Mar. 21, 1955.

CAMERON, Frank Kenneth, Am. chemist; b. Balt., Feb. 2, 1869; s. John Malcolm and Elizabeth (Fitz-Patrick) C.; A.B., Johns Hopkins, 1891, Ph.D., 1894; m. Katherine Boyle, Sept. 14, 1899 (dec. Nov. 25, 1903); m. 2d, Virginia B. Newton, Oct. 6, 1908 (dec. Nov. 15, 1954); children—Francis, Katherine. Research fellow, instr. chemistry Cornell U., 1894-95; asso. prof. chemistry Catholic U. Am., 1895-97; research asst., instr. phys. chemistry Cornell U., 1897-98; expert, U. S. Dept. Agr., 1898, chemist Div. of Soils, 1899; in charge Lab. Soil Chemistry, 1899-1915; prof. chemistry U. N.C., 1926-46, emeritus prof., 1946-58; cons. practice. Recipient Herty medal, 1939. Mem. various sci. socs., including A.A.A.S., Am. Chem. Soc., Soc. Electrochemistry, Elisha Mitchell Sci. Soc. Author: Les Constituants Mineraux des Solutions des Sols (with James M. Bell), 1907; The Soil Solution, 1911; also over 200 research papers. Asst. editor Zeitschrift Kolloide Chemistry, 1910-14, Jour. Phys. Chemistry, 1910-23, 31-33, Jour. Indsl. and Engring. Chemistry, 1912-21. Died Aug. 18, 1958.

CAMERON, John Roderick, Am. med. physicist; b. Chippewa Falls, Wis., Apr. 21, 1922; s. Duncan and Mary (O'Connell) C.; B.S., U. Chgo., 1947; M.S., U. Wis., 1949, Ph.D. in Nuclear Physics, 1952; m. Lavonda Donovan, Aug. 2, 1947; children—Anne, Carol. Asst. prof. nuclear physics U. Sao Paulo (Brazil), 1952-54; project asso. in nuclear physics U. Wis., Madison, 1954-55, faculty radiology, physics, 1958-—, prof., 1965-—; asst. prof. nuclear physics U. Pitts., 1955-58. Cons. VA Hosp., Madison; AEC, Washington; Internat. Atomic Energy Agy., Vienna, Austria. Mem. Am. Assn. Physicists in Medicine (pres. 1968), Soc. Nuclear Medicine (trustee 1964-67), Am. Phys. Soc., Biophys. Soc., Radiation Research Soc., A.A.A.S. Research, publs. in radiation dosimetry using thermoluminescence, bone mineral measurement. Home: 6620 Wood Circle East, Middleton, Wis. 53562. Office: U. Wis., Madison, Wis. 53706.*

CAMERON, Sir (Gordon) Roy, pathologist; b. Victoria, Australia, June 30, 1899; s. G. Cameron; ed. Melbourne (Australia), Freiburg (Germany) univs.; LL.D., Edinburgh U., 1956, Melbourne U., 1962. Asst. dir. Walter and Eliza Hall Inst. Research, Melbourne, 1926-28; reader in morbid anatomy U. Coll. Med. Sch., U. London (Eng.), 1935-37, prof., 1937-64, hon. cons. pathologist, 1964-—; asst. editor Jour. Pathology and Bacteriology, 1935-55. Mem. Agrl. Research Council, 1948-56, Med. Research Council, 1952-56, adv. council, Beit Meml. Fellowship for Med. Research, council, Imperial Cancer Research Fund, 1948-—; found. pres. Coll. Pathologists, 1962-—. Fellow Royal Soc. (Royal medal 1960), Royal Coll. Physicians London; mem. Royal Coll. Physicians and Surgeons Glasgow (hon.). Author: Pathology of the Cell, 1952; (with W. G. Spector) Pathology of the Injured Cell, 1961; (with W. C. Hov) Biliary Cirrhosis, 1962; numerous articles Jour. Pathology and Bacteriology. Initiated new discipline of micropathology from studies of liver; improved methods for keeping blood supply from damaged liver and preventing bile outflow. Home: 56 Camlet Way, Hadley Wood, Barnet, U.K.

CAMERON, Thomas Wright Moir, parasitologist; b. Glasgow, Scotland, Apr. 29, 1894; s. Hugh Cameron; M.A., Ph.D., Edinburgh (Scotland) U.; m. Stella Blanche; 1 dau. Sr. research asst. Inst. Agrl. Parasitology, London, Eng., 1923-25; lectr., Milner fellow Dept. Helminthology, London Sch. Hygiene and Tropical Medicine, 1925-29; lectr. helminthology U. Edinburgh, 1929-32; now prof. emeritus parasitology McGill U., Montreal, Que., Can. Mem. Royal Soc. Medicine (sec. tropical disease and parasitology sect., v.p. comparative medicine sect.), Am. Soc. Parasitologists (pres. 1949), Royal Soc. Can. (pres. sect. V 1949-50), Canadian Soc. Biology (pres. 1959), Canadian Soc. Zoology (pres. 1960), World Fedn. Parasitologists (pres. 1965-—). Author: Animal Diseases in Relation to Principles of Man; Internal Parasites of Domestic Animals; Principles of Parasite Control; Parasites of Man in Temperate Climates; Early History of Caribe Islands; Parasites and Parasitism. Research on parasitic helminths, diseases of animals in relation to man. Office: P.O. Box 110, Ste. Anne de Bellevue, Que., Can.

CAMISHION, Rudolph Carmen, Am. surgeon; b. Riverside, N.J., July 16, 1927; s. Anthony and Florence (Marino) C.; B.S., St. Joseph's Coll., 1950; M.D., Jefferson Med. Coll., 1954; m. Nancy I. Muzzarelli, June 28, 1952; children—Germaine, Sandra and Mary (twins), Lisa, Nancy, Janice. Faculty, Jefferson Coll. Medicine, Phila., 1959-—, asso. prof. surgery, 1964-—; attending surgeon Jefferson Med. Coll. Hosp.; cons. thoracic surgery VA Hosp., Phila. Diplomate Am. Bd. Surgery, Am. Bd. Thoracic Surgery. Mem. A.C.S., A.M.A., A.A.A.S., Am. Assn. for Thoracic Surgery, Am. Heart Assn., Am. Thoracic Soc., Assn. Am. Med. Colls., Internat. Cardiovascular Soc., Laennec Soc., N.Y. Acad. Scis., Soc. U. Surgeons, Soc. for Vascular Surgery, numerous others, Alpha Omega Alpha. Author: Basic Medical Electronics, 1964. Contbr. articles to sci. jours. Research on heart, lungs, vascular physiology, surgery. Home: 1101 Cherry Lane, Riverton, N.J. 08077. Office: 1025 Walnut St., Phila. 19107.*

CAMP, Ezra John, Am. mathematician, b. Roanoke, Ill., Nov. 26, 1906; s. John J. and Anna (Schertz) C.; B.A., Goshen (Ind.) Coll., 1928; M.S., U. Chgo., 1932, Ph.D., 1935; m. Lois Lapp, Sept. 4, 1930; children—Evelyn, Richard, John (dec.), Larry. Instr. math. Goshen Coll., 1929-31; Kidder (Mo.) Coll., 1932-33; acting prof. math. Yankton (S.D.) Coll., 1935-37; asso. prof. math. Macalester Coll., St. Paul, 1937-43, prof., 1943-—. Mem. Am. Math. Soc., Math. Assn. Am., Sigma Xi. Author: Mathematical Analysis, 1956. Study of math. analysis; applied mathematics. Home: 3470 Siems Ct., St. Paul 55112.*

CAMPACCI, C. A., Brazilian agronomic engr.; b. Piracicaba, Brazil, Nov. 7, 1919; s. Luciano and Maria B. Campacci; degree Escola Superior de Agricultura Luiz de Queiroz, 1943; m. Maria C. Motta, Dec. 21, 1949; children—Paulo Motta, Luciano Motta. Phytopathology asst. Agrl. Sch. Luiz de Queiroz; phytopathology asst. Instituto Biológico de Sao Paulo (Brazil), chief sect. fungicides, 1962-—. Mem. Sociedade Paulista de Agronomia, Sociedad Brasileira de Defensivos para a Lavoura e Pecuária, Sociedad Brasileira de Olericultura, Associacão Latinoamericana de Fitotecnia. Research, numerous publs. on fungicides and pesticides in states of São Paulo and Brazil. Home: 719 Gaspar Fernandes. Office: 1252 Av. Cons. Rodrigues Alves, São Paulo, São Paulo, Brazil.*

CAMPAIGNE, Ernest Edwin, Am. chemist; b. Chgo., Feb. 13, 1914; s. John Herbert and Nellie (Daufel) C.; B.S., Northwestern U., 1936, M.S., 1938, Ph.D. in Biochemistry, 1941; m. Jean Hill White, Jan. 1, 1941; children—David Alan, Claudia Jean (Mrs. Ronald L. Buskirk), Barbara Naomi. Instr., Bowdoin Coll., 1940-41; research asso. Northwestern U., 1941-42; asso. biochemist M.D. Anderson Hosp. for Cancer Research, Galveston, Tex., 1942-43; faculty dept. chemistry Ind. U., Bloomington, 1943-—, prof., 1953-—. Vis. prof. U. Cal. at Los Angeles, 1954-55; cons. NIH, 1960-64, 66-—; indsl. chem. cons., 1952-—. Fellow N.Y. Acad. Scis., Ind. Acad. Sci.; mem. Am. Chem. Soc., Chem. Soc. (London), A.A.A.S. Author: (with J. C. Muhler and C. H. Rohrer) Introduction to Chemistry, 1960; Elementary Organic Chemistry, 1962. Research and publs. on synthesis of drugs, antihistamines, anticonvulsant drugs for treatment of epilepsy, molecular dimensions of drugs to interpret their optimum dimensions and nature of receptor sites. Home: 1240 E. Wylie St., Bloomington, Ind. 47401.*

CAMPANELLA, Tommaso (Giovan Domenico), Italian philosopher, geologist; b. Stilo, Calabria, Italy, Sept. 5, 1568; educated at Morgentia and Cosenza; joined Dominican order, 1583; lived and wrote in Naples and Rome, Italy; returned to Stilo, 1598; imprisoned during the Spanish Inquisition for being revolutionary, 1599; released from Naples prison 1626 through the efforts of Pope Urban VIII; rearrested then regained freedom, Rome, 1628; went to Paris where he was received by Cardinal Richelieu, 1634. Author numerous books including: Prodromus philosophiae instaurandae . . . , 1617; De sensu rerum et magia, 1620; Apologia pro Galileo, 1622; Realis philosophiae epilogisticae partes IV, 1623; Civitas solis; Atheismus triumphatus, 1631; Philosophia rationalis, 1637; Universalis philosophiae seu metaphysicarum, 1637; De monarchia Hispanica, 1640. Supported views of Nicolas of Cusa and Telesio; believed in philosophy based on perception and individual consciousness to which he ascribed power, will, and knowledge; anticipated Descartes in proclaiming self-consciousness to be the basic principle of knowledge and certitude; believed in planes of being and religion; attempted to reconcile in a new synthesis Christian doctrines and the naturalistic tendencies of his time; advocated exptl. method; proposed that heavy and light bodies would fall at the same rate of speed in a vacuum; an astrologer. Died Paris, May 21, 1639.

CAMPANI, Guiseppe, Italian astronomer; b. Rome, Italy, 17th century; lived in Rome; expert in lens grinding; worked with bro. Matteo Campani-Alimensis; built lens for the astronomer Cassini Guiseppe ordered by King Louis XIV; at his death Pope Benedict IV bought all his instruments and presented them to U. Bologna (Italy). Built power telescopes which possessed focal distance of 26-42 m.; observed the spots on Jupiter; publs. on Saturn.

CAMPBELL, Alan Keith, Am. polit. scientist; b. Elgin, Neb., May 31, 1923; s. Charles E. and Anna (Schneckloth) C.; A.B., Whitman Coll., 1947; M.P.A., Wayne U., 1949; Ph.D., Harvard, 1952; m. Linna Jane Owen, Mar. 9, 1945; children—Kimberly Ann, Charles Duncan. Asst. dir. Harvard Summer Sch., instr., vis. lectr., 1950-57; mem. staff N.Y. Met. Region Study, 1957-58; research cons. N.Y. State Commn. on Govtl. Operations of N.Y.C., 1959; prof., chmn. polit. sci. dept. Hofstra Coll., 1954-60; dep. comptroller for adminstrn. State of N.Y., 1960-61; vis. prof. Columbia U., 1961-62; prof. polit. sci., dir. met. studies program Maxwell Grad. Sch. Syracuse U., 1961-—; mem. faculty Salzburg (Austria) Seminar, 1965. Sheldon Traveling fellow; Volker fellow. Mem. Am. Polit. Sci. Assn., Am. Soc. Pub. Adminstrn., Internat. Soc. for Community Devel., Am. Council on Edn. (mem. com. on instrn., evaluation), Phi Beta Kappa, Phi Epsilon Pi. Author: (with Seymour Sacks) Metropolitan America: Governmental Systems and Fiscal Patterns, 1967; (with others) The Negro in Syracuse, 1964; co-editor Case Studies in American Government, 1962. Contbr. articles in field to profl., sci. jours. Research on relations of socio-economic characteristics of communities to their polit., govtl., fiscal behavior with emphasis on differences between suburbs and cities; findings demonstrate importance of income, govtl. systems, flow of intergovtl. funds on such behavior; research on determinants of ednl. performance; findings demonstrate importance of socio-econ. background of family. Home: Ten Eyck Av., R.D., Cazenovia, N.Y. 13035.*

CAMPBELL, Alastair Heriot, Australian physician; b. Melbourne, Australia Jan. 30, 1917; s. Frederick Harper and Agnes (Docker) C.; M.B.B.S., Melbourne U., 1940, M.D., 1961; m. Wilma Mary Dobbs, May 27, 1943; children—Andrew John, Jennifer Ann. Staff, Grenvale Sanatorium, Victoria, Australia, 1946, Repatriation Gen. Hosp., Melbourne, 1947-51; Wunderley Traveling scholar, 1952-53; jr. specialist chest diseases Repatriation Chest Clinic, Melbourne, 1954; specialist chest diseases Repatriation Dept., Queensland, 1955-59, Victoria, 1959-64; cons. chest diseases Central Office, Australian Repatriation Dept., Melbourne, 1964-—. Mem. Australian Commonwealth Tb. Adv. Council, 1963-—. Recipient Darcy Cowan prize, 1965. Fellow Australasian Coll. Physicians; mem. Thoracic Soc. Australia, (pres. 1965-66), Australian Anthropology Soc. (pres. 1965-67), Repatriation Med. Officers Assn. (pres. 1962-—), Australian Med. Assn., Royal Soc. of Victoria. Author: (with others) The Aborigines and Torres Islanders of Queensland, 1958; also articles. Research on chest diseases and possible causal relationship between lung cancer and both Tb. and chronic bronchitis, obstructive airway disease. Home: 105 Banksiast Heidelberg, Melbourne. Office: Repatriation Dept., St. Kilda Rd., Melbourne, Victoria, Australia.*

CAMPBELL, Andrew, Am. inventor; b. Trenton, N.J., June 14, 1821; m. 1848, 4 children. Brushmaker, St. Louis, 1842-50; built 1st St. Louis omnibus; patented printing machine, 1858, began Campbell Country Press; erected plant, Bkly., 1866; developed 2 revolution picture press, 1867, large press for fine illustrations, 1868; made 1st press which printed, inserted, pasted, folded and cut in 1 continuous operation. Died Bklyn., Apr. 13, 1890.

CAMPBELL, Sir Archibald Campbell, phys. scientist; b. Florence, Scotland, Feb. 22, 1835; s. Archibald Douglas and Caroline Agnes (Dick) C.; LL.D., U. Glasgow, 1907; m. Augusta Clementina Carrington, July 7, 1864; joined army, 1854, ret., 1868; afterwards active in politics, the aux. forces and sci.; polit. party organizer. Fellow Royal Soc., 1907. Research on astron. and phys. scis.; pioneered in mechanics of aerial propulsion; developed numerous precision instruments; obtained photographic impressions through various opaque objects before Röntgen's announcement of same in 1895; came close to discovering X-rays. Died July 8, 1908.

CAMPBELL, Berry, Am. neurologist; b. St. Paul, Mar. 21, 1912; s. E. Paul and Fan (Berry) C.; A.B. in Zoology, U. Cal. at Los Angeles, 1932; Ph.D. in Anatomy, Johns Hopkins, 1935; m. Irene Wilson, June 26, 1933; children—Carolyn Lyell (Mrs. Robert William Dickerman), John Howland, Richard Dana, Cathryn Alice. Asst. prof. anatomy U. Okla. Med. Sch., 1937-40; asst. prof. anatomy U. Tenn. Med.

Sch., 1942; fellow neurology Columbia, 1942-43; faculty U. Minn., 1943-58; research prof. neurosurgery Loma Linda U., Los Angeles, 1959—; prof. physiology, chmn. dept. Cal. Coll. Medicine, U. Cal., 1965—. NRC fellow in med. scis., 1935-36, 1936-37; John Simon Guggenheim fellow., 1940-41, 1941-42; hon. research fellow U. Coll., London, Eng., 1953-54. Mem. A.A.A.S., Am. Soc. Herpetologists and Ichthyologists, Am. Neurol. Assn., Am. Acad. Neurology, Assn. for Research in Nervous and Mental Diseases, Am. Assn. Anatomists, Am. Soc. Mammalogists, Assn. for History of Medicine, Am. Physiol. Soc., Am. Assn. Phys. Anthropologists, Harvey Soc., Los Angeles Acad. Medicine. Research and numerous articles on natural history, comparative anatomy of vertebrates, correlation of structure and function in brain, allergic reactions of brain, milk antibodies, serum therapy. Home: 444 N. Alta Vista Av., Monrovia, Cal. 91016. Office: 1200 N. State St., Los Angeles 90033.*

CAMPBELL, Dan Hampton, Am. biochemist; b. Fremont, O., Jan. 18, 1907; s. Ralph Edward and Edna (Moses) C.; A.B., Wabash Coll., 1930; M.S., Washington U., St. Louis, 1932; Ph.D., U. Chgo., 1935; Sc.D., Wabash Coll., 1960; m. Margaret Kathryn Dorr, May 12, 1930; 1 son, John Hampton. Instr., Washington U., 1930-33; Logan fellow U. Chgo., 1933-36, instr., 1936-39, asst. prof., 1939-42; mem. faculty Cal. Inst. Tech., Pasadena, 1942—, prof. biochemistry, 1950—. Cons., Office Med. Research, 1942-47, NIH, 1948—. Chmn. research and devel. bd., dir. Internat. Chem. and Nuclear Corp., 1960—. Recipient Merit award City of Hope, Nat. Med. Center, 1955, 57; award USPHS, 1962. Mem. A.A.A.S., Am. Chem. Soc., Am. Assn. Immunologists, Soc. Exptl. Biology and Medicine, Brit. Soc. Immunology, Am. Acad. Allergy, Arctic Inst. of N.Am., Am. Assn. U. Profs., Internat. Soc. Hematology. Author: (with John E. Cushing) Principles of Immunology, 1953; (with J. S. Garvey, N. E. Cremer and D. H. Sussdorf) Methods in Immunology, 1963; also numerous articles. Research in chemistry of factors involved in immunity to diseases and allergic reactions, devel. of precise methods to characterize vaccines and active components in serum of immunized animals; patentee substitutes for human blood plasma, specific insoluble adsorbent for removing antibodies from body fluids. Home: 1154 Mt. Lowe Dr., Altadena, Cal. 91001.*

CAMPBELL, Douglas Houghton, Am. botanist; b. Detroit, Dec. 16, 1859; s. James Valentine and Cornelia (Hotchkiss) C.; Ph.M., U. Mich., 1882, Ph.D., 1886, LL.D., 1932; postgrad., Bonn, Tübingen and Berlin. Instr. biology Detroit High Sch., 1882-86; prof. botany Ind. U., 1888-91, Leland Stanford U., 1891-1925, emeritus from 1925. Fellow Am. Acad. Arts and Scis., Royal Soc. Edinburgh; mem. Linnaean Soc. (fgn.), Nat. Acad. Scis., Am. Philos. Soc., various other socs., Am. and European. Author: Elements of Structural and Systematic Botany, 1890; Structure and Development of Mosses and Ferns, 1895, 3d edit., 1918; Lectures on Evolution of Plants, 1899; A University Text Book of Botany, 1902, 2d edit., 1907; Plant Life and Evolution, 1911; Outline of Plant Geography, 1926; Evolution of the Land Plants, 1940. Continental Drift and Plant Distribution, 1943; also monographs and other papers. Died Feb. 23, 1953.

CAMPBELL, Edward De Mille, Am. chemist; b. Detroit, Sept. 9, 1863; s. James Valentine and Cornelia (Hotchkiss) C.; B.S. in Chemistry, U. Mich., 1886; m. Jennie M. Ives, 1888. Chemist, Ohio Iron Co., Zanesville, 1886-87, Sharon (Pa.) Iron Co., 1887-88, Dayton (Tenn.) Coal & Iron Co., 1888-90; asst. prof. metallurgy U. Mich., 1890, jr. prof. metallurgy and metall. chemistry, 1893, jr. prof. analytical chemistry, 1896, prof. chem. engring. and analytical chemistry, 1902, dir. chem. lab. and prof. chem. engring. and analytical chemistry, 1905, prof. chemistry and dir. chem. lab., 1914, prof. chemistry and metallurgy and dir. chem. lab., 1920-25. Died Sept. 18, 1925.

CAMPBELL, Edward James Moran, English physician, physiologist; b. Yorkshire, Eng., Aug. 31, 1925; s. Edward Gordon and Clare Irene (O'Callaghan) C.; B.S., U. London, 1946, M.D., 1951, Ph.D., 1954; m. Diana M. E. Green, Dec. 18, 1954; children—Fiona, Susan, Robert, Jessica. House staff Middlesex Hosp., 1949-50, lectr., 1950-52, registrar, 1953-54, 1955-61, prof. medicine; fellow Johns Hopkins, 1955; lectr. medicine Postgrad. Med. Sch. London (Eng.), 1962—; physician Hammersmith Hosp., London, 1962—. Fellow Royal Coll. Physicians; mem. Assn. Physicians, Physiol. Soc., Thoracic Soc., Med. Research Soc., Royal Soc. Medicine. Author: Respiratory Muscles and the Mechanics of Breathing, 1958; (with Dickinson, Slater) Clinical Physiology, 1963. Introduced (with Howell) inappropriateness theory of dyspnea, developed rebreathing method for blood pCO_2; developed methods of treating respiratory failure. Home: 9 Acacia Gardens, London N.W. 8, Eng.*

CAMPBELL, Frank Leslie, Am. entomologist; b. Phila., Sept. 5, 1898; s. Andrew Jackson and Esther (Williams) C.; B.S., U. Pa., 1921; M.S., Rutgers U., 1924; D.Sc., Harvard, 1926; m. Elizabeth Mildred Boyd, Oct. 8, 1921; children—Drew Boyd, Lucile Esther (Mrs. Dallas M. Cooper); m. 2d, Ina Mae Lee, Apr. 17, 1947. Asst. prof. N.Y. U., 1926-27; entomologist U. S. Dept. Agr., 1927-36; faculty Ohio State U., 1936-42; editor Sci. Monthly, Washington, 1943-48; with CIA, Washington, 1948-53; exec. sec. div. biology and agr. Nat. Acad. Scis.-NRC, Washington, 1953-64. Mem. numerous sci. socs. Contbr. articles to profl. jours. A pioneer in application of quantitative methods to entomol. studies; research on insect structure, growth, devel., toxicity of insecticides, insect physiology. Home: 2475 Virginia Av., N.W., Washington 20037.*

CAMPBELL, George Ashley, Am. telephone research engr.; b. Hastings, Minn., Nov. 27, 1870; s. Cassius Samuel and Lydia Lorraine (Ashley) C; B.S., Mass. Inst. Tech., 1891; A.B., Harvard, 1892, A.M., 1893, Ph.D., 1901; studied Göttingen, Vienna and Paris; m. Caroline Gillis Sawyer, 1913; children—Alexander Hovey (dec.), Ashley Sawyer. With Am. Tel. & Tel. Co., 1897-1934; with Bell Telephone Labs., 1934-35. Distinguished Service medal I.R.E., 1936; Elliott Cresson medal Franklin Inst., 1939. Edison medal Am. Inst. E.E., 1940. Mem. Am. Acad. Arts and Scis., Math. Soc., Math. Assn. Am., Am. Phys. Soc., A.A.A.S. Recipient Pioneering research in connection with loading, crosstalk, 4-wire repeater circuits, sidetone reduction, electric wave filters, inductive interference, antenna arrays, maximum output networks, Fourier integrals and electrical units. Author: Collected Papers, 1937. Died Nov. 10, 1954; buried Easthampton, Mass.

CAMPBELL, Gilbert Sadler, Am. surgeon; b. Toronto, Ont., Can., Jan. 4, 1924; (parents Am. citizens); s. Gilbert S. and Ellen (Thorson) C.; student Hampden-Sydney Coll., 1939-40; A.B., U. Va., 1943, M.D., 1946; M.S., U. Minn., 1949, Ph.D., 1954; m. Dorothy Nugent, Sept. 18, 1947 (div. Mar. 1960); children—Kathryn Ellen, Rebecca Sadler, Thomas Kim, William Riley; m. 2d, Joan Louise Hancock, Sept. 28, 1961; children—Lisa Hancock, John Gilbert. Faculty, U. Minn., 1954-58, asst. prof., 1955-58; prof. U. Okla. Med. Center, also chief surgery VA Hosp., Oklahoma City., 1958-65; prof., chmn. dept. surgery U. Ark. Med. Center, Little Rock, 1965—. Recipient Horsley prize U. Va., 1954; Markle scholar in med. scis., 1954-59. Mem. Soc. for Exptl. Biology and Medicine, Soc. U. Surgeons, Am. Assn. Thoracic Surgery, Am. Surg. Assn., Soc. for Vascular Surgery, Am. Physiol. Soc., Am. Soc. for Artificial Internal Organs. Research, numerous publs. on heart and lung surgery, blood vessel surgery, circulation of lung. Home: 7201 Kingwood St., Little Rock 72205.*

CAMPBELL, Henry Fraser, Am. surgeon; b. Augusta, Ga., Feb. 10, 1824; s. James and Mary (Eve) C.; M.D., Med. Coll. Ga., 1842; m. Sarah (Sibley) Bosworth, June 17, 1844. Began practice medicine, Augusta, 1842; prof. comparative and micros. anatomy Med. Coll. Ga., 1854-57, prof. anatomy, 1857-66, prof. orthopedic surgery and gynecology, 1868-91; prof. anatomy, surgery New Orleans Sch. Medicine, 1866-68; surgeon Confederate Army, 1861, Bd. Med. Examiners; med. dir. Gen. Mil. Hosp. Richmond (Va.); mem. Ga. Bd. Health, 1875; recipient prize for investigation of excreto-secretory system, A.M.A., 1857; corr. Acad. Natural Scis. Phila.; mem. Imperial Acad. Medicine St. Petersburg; pres. Ga. Med. Assn., 1871; a founder Am. Gynecol. Soc., 1876; pres. A.M.A., 1885; co-editor So. Med. and Surg. Jour., 1857-61. Pioneer in preventive medicine; made observations on abortive treatment of gonorrhea, 1845, epidemic dengue fever, 1851, typhoidal fever, 1853, dysentery; proposed theory that sympathetic nervous system is the excito-secretory (his term) mechanism which activates glands, 1857; Investigated sympathetic nerves in reflex phenomena; asserted portability of atmospheric germs, also noncontagious nature of yellow fever, 1880. Died Augusta, Ga., Dec. 15, 1891.

CAMPBELL, Howard Ernest, Am. mathematician; b. Detroit, Sept. 20, 1925; s. Howard E. and Marie (Brown) C.; student Stevens Inst. Tech., 1943-44; B.S. in Elec. Engring., U. Wis., 1946, M.S. in Math., 1947, Ph.D. in Math., 1949; m. Ruth Mary Noland, June 27, 1950; children—Tanaquil Ruth, Howard Blaine, Thane George, Lowell Lete. Grad. asst. U. Wis., Madison, 1946-49; instr. U. Pa., Phila., 1949-51; asst. prof. Emory U., Atlanta, 1951-56; asst. prof. math. Mich. State U., East Lansing, 1956-59, asso. prof., 1959-63; prof. math., head dept. U. Ida., Moscow, 1963—. Served with USNR, 1943-46. Mem. Am. Math. Soc., Math. Assn. Am. (Consultants Bur. of com. on undergrad. program in math.), Sigma Xi, Pi Mu Epsilon. Contbr. articles on associative and non-associative algebras to profl. publs. Home: 205 S. Garfield St., Moscow, Ida. 83843.*

CAMPBELL, Ian, Am. geologist; b. Bismarck, N.D., Oct. 17, 1899; s. Dugald and Agnes (Gilkison) C.; A.B., U. Ore., 1922, A.M., 1924; postgrad. Northwestern U., 1923-24; Ph.D., Harvard, 1931; m. Catherine Robbins Chase, Sept. 16, 1930; 1 son, Dugald Robbins. Asst. prof. geology La. State U., Baton Rouge, 1925-28; instr. Harvard 1928--31; geologist U. S. Geol. Survey, 1929-35, 41-50; faculty Cal. Inst. Tech., 1931-59, prof. petrology, 1946-59; field engr. U. S. Navy Radio and Sound Lab., 1944-46; state geologist, chief Cal. State Div. Mines and Geology, San Francisco, 1959—. Cons. geologist various cos., 1925-59. Mem. Am. Inst. Mining Engrs. (past chmn. indsl. minerals div.), A.A.A.S. (past pres. Pacific div.), Cal. Acad. Scis. (trustee 1960—), Am. Geol. Inst. (past pres.), Mineral. Soc. Am. (past pres.), Geol. Society of America (associate editor 1962—, president 1968), also member Association of American State Geologists (past pres.), Soc. Mining Engrs. (Western vice chmn. 1966—), Am. Assn. Petroleum Geologists, Am. Inst. Mining, Metall. and Petroleum Engrs., (Hardinge award 1962), Am. Assn. U. Profs., Am. Geophys. Union, Assn. Engring. Geologists, Geochem. Soc., Geol. Soc. Am., Mineral. Soc. Am., Soc. Econ. Geologists Am. Inst. Profl. Geologists, Contbr. articles to tech. jours. Research on indsl. minerals, engring. geology, petrology. Home: 1333 Jones St., San Francisco 94109. Office: Ferry Bldg., San Francisco 94111.*

CAMPBELL, James, physiologist; b. Glasgow, Scotland, July 18, 1907; s. James and Robina (Wylie) C.; B.A., U. Toronto (Ont., Can.), 1930, M.A., 1932, Ph.D., 1938; m. Mary Louise Allen, Apr. 17, 1954; children—James, Elizabeth Mary, Christine Louise, Sheila Wylie. Research asst. physiology McGill U., 1933; faculty U. Toronto, 1934-41, 52—, prof. physiology, 1962—. Mem. Canadian, Am. physiol. socs., Canadian Biochem. Soc., Biochem. Soc. (U.K.), Am. Chem. Soc., Am. Diabetes Assn., Nutrition Soc. Can. Contbg. author Physiological Basis of Medical Practice, 7th edit. 1961; Hormonal Factors in Carbohydrate Metabolism, 1953; Symposium on Experimental Diabetes, 1954; Growth Hormone, Nature and Actions, 1955. Research and publs. on effects anterior pituitary gland on fat metabolism, effects growth hormone in prodn. diabetes; developed process for canning potable water. Home: 54 Summerhill Gardens, Toronto 7, Ont., Can.*

CAMPBELL, James LeRoy, Am. physician; b. Fulton County, Ga., July 15, 1870; s. Thomas Jefferson II and Mary Jane (Brown) C.; M.D., Atlanta Med. Coll.; m. Mary Jones, Sept. 20, 1899; children—Lula Grove (Mrs. George M. Ivey), James LeRoy. Prof. surg. anatomy, clin. surgery, Emory U. Sch. Medicine, 1905-20; prof. clin. surgery, 1920-40; chief, surg. service, Emory U. Div. of Grady Hosp., 1921-30; emeritus prof. clin. surgery, Emory U. Sch. Medicine, 1940-48. Author of present Ga. state law for control of cancer; established state aid cancer clinics for treatment of indigent cancer victims in Ga.; secured amendment to constn. of Ga., exempting edni. instns. from taxation, 1918. Ga. state chmn. Am. Soc. for Control Cancer, 1920-29; chmn. cancer commn. Med. Assn. Ga., 1918-48; chmn. exec. com. Ga. Div. of Women's Field Army of Am. Soc. for Control of Cancer, 1937-47; bd. dirs. Am. Soc. for Control Cancer, 1939-46. Fellow A.M.A., A.C.S.; mem. So. Med. Assn., Fulton Co. Med. Soc., Med. Assn. Ga., Phi Beta Kappa, Omicron Delta Kappa, Alpha Omega Alpha. Author of numerous articles in profl. jours. Died June 11, 1948.

CAMPBELL, Capt. John, Scottish navigator; b. Kirkcudbrightshire, Scotland, circa 1720; s. John Campbell; passed examination for naval lt., 1744; m. Apprenticed to master of coasting vessel; became lt., then comdr. under Sir Edward Hawke; became vice adm., 1778; apptd. gov., comdr.-in-chief Newfoundland; ret. 1787. Suggested Hadley's quadrants be enlarged to a sextant which could measure angles up to 120°, 1747. Died London, Dec. 16, 1790.

CAMPBELL, John Alexander, Am. physician, educator; b. Cin., June 29, 1914; s. Archibald and Elizabeth M. (Harris) C.; B.S., U. Cinn., 1935, M.D., 1937; m. Willie Dickins, July 9, 1938; children—Nancy, Duncan, Jamie. Intern, Detroit Receiving Hosp., 1937-38; postgrad. tng. radiology Henry Ford Hosp., Detroit, 1938-41; asst. radiologist St. Joseph Hosp., Ann Arbor, Mich., 1941; instr. radiology USPHS, 1942-45; chief radiologist Indpls. City Hosp., 1944-45; faculty radiology Ind. U. Med. Center, 1941—, prof., 1956—, chmn. dept., 1957—; dir. acad. radiol. services Univ., Marion County Gen., Indpls. VA hosps., Ind. U. Student Health Service; cons. radiation adv. control commn. Ind. Bd. Health; radiol. cons. Ind. Civil Def. Bd.; mem. com. radiology, Nat. Acad. Scis., Nat. Research Council. Mem. Marion County Civil Def. Bd. Dir. Indpls. Council World Affairs, Doctors Dixieland Band, Indpls. Civic Theatre; mgr. Pony League Baseball; mem. Indpls. Mayor's Citizens Adv. Com., 1965—. Diplomate Am. Bd. Radiology in radiology and nuclear medicine. Fellow Am. Coll. Radiology (councilor, commn. tech. affairs); mem. Am. Roentgen Ray Society, Radiological Society North Am. (1st v.p.), Council Acad. Socs., Assn. Am. Med. Colls. (member of the council of academic societies), Assn. of Univ. Radiologists (president 1964-65). Author civil def. bull. Research in cineradiography of congenital heart disease; improved equipment and methods. Home: 5201 Grandview Dr., Indpls. 8. Office: 1100 W. Michigan St., Indpls. 7.*

CAMPBELL, John TenBrook, Am. surveyor; b. nr. Montezuma, Ind., May 21, 1833; s. Joseph and Rachel (TenBrook) C.; one term Western Manual Labor Acad.; m. Annie Butterfield, Dec. 15, 1864. Capt. Co. H. 21st Ind. Vols., 1861; right leg permanently crippled in battle, Baton Rouge, 1862; resigned, Oct. 29, 1862; asst. provost-marshal 7th Congl. Dist. of Ind., June 24-Nov. 4, 1863; treas. Parke Co., Ind., 2 terms; asst. assessor internal revenue, 1870; Republican; later Greenbacker; returned to Rep. party, 1890; defeated for state senate as Greenbacker, 1870; journal clerk, Ind. Senate, 1878; 1st asst., Ind. Bur. of Statistics, 1878-83; co. surveyor 10 yrs. Devised 5 new problems in trigonometry as applied to surveying; also 2 new problems in curve work, with formula for application in practice. Deceased.

CAMPBELL, Kenneth Nielsen, Am. chemist; b. Hillsdale, Mich., May 31, 1905; s. R. Burke and Vivian (Nielsen) C.; student Kalamazoo Coll., 1924-26; B.S. in Chemistry, U. Chgo., 1928, Ph.D., 1932; m. Barbara H. Knapp, June 17, 1933. Fellow, Pa. State U., 1933-35, U. Ill., 1935-36; faculty U. Notre Dame, 1936-53, prof. chemistry, 1945-53; dir. medicinal chemistry Mead Johnson Research Center, Evansville, Ind., 1953——. Cons. USPHS, 1949-52. Recipient Certificate of Merit, Ind. Tech. U., 1958, Centennial award U. Notre Dame, 1965. Mem. Am. Chem. Soc. (past chmn. div. medicinal chemistry), A.A.A.S., N.Y. Acad. Scis., Chem. Soc. (London, Eng.). Contbr. numerous articles to tech. jours. Research in chemotherapy, petroleum chemistry, synthetic drugs. Home: 8216 Petersburg Rd., Evansville 47711. Office: Mead Johnson Research Center, Evansville, Ind. 47721.*

CAMPBELL, Leon, Am. astronomer; b. Cambridge, Mass., Jan. 20, 1881; s. William J. and Leonora (Rawding) C.; student pub. schs., Cambridge, and spl. instrn. Harvard Obs.; m. Fredrica J. Thompson, June 15, 1905; children—Leon, Florence May, Malcolm Fredric, Ruth Evelyn, Eleanor Beatrice. Asst. at Harvard Obs., 1899-1911, in charge of Arequipa (Peru) sta., 1911-15, variable star investigator, and astronomer, from 1915; instr. in astronomy Harvard, 1928, Pickering Meml. astronomer, from 1931. Mem. Am. Astron. Soc., Am. Assn. Variable Star Observers (pres. 1919-22), Internat. Astron. Union, Lima Geog. Soc. (hon). Contbr. to annals of Harvard Obs. and astron. mags.; co-author, the Story of Variable Stars, 1941. Studied (with T. E. Sterne) periodic variation of stars, 1937, found that the changes of period were purely statis. fluctuations in 377 stars they observed, and the changes of period well established for only 5 stars. Died May 10, 1951.

CAMPBELL, L(inzy) Leon, Jr., Am. microbiologist; b. Panhandle, Tex., Feb. 10, 1927; s. Linzy Leon and Eula (McSpaaden) C.; B.A., U. Tex., 1949, M.A., 1950, Ph.D., 1952; m. Alice Pauline Dauksa, Feb. 7, 1953. Faculty, Wash. State U., Pullman, 1954-59, Western Res. U., Cleve., 1959-62; prof. microbiology U. Ill., Urbana, 1962——, head, dept., 1963——. Mem. microbiology tng. grant panel Nat. Inst. Gen. Med. Scis., 1964——. Nat. Microbiol. Inst. fellow, 1952-54. Fellow Am. Acad. Microbiology; mem. Am. Soc. for Microbiology (councilor at large 1962-64, chmn. publs. bd. 1965——), Soc. for Gen. Microbiology, Am. Soc. Biol. Chemists, Am. Chem. Soc., A.A.A.S. Editor: (with H. Orin Halvorson) Spores III, 1965; mem. editorial bd. Jour. Bacteriology, 1961-64, editor, 1964-65, editor-in-chief, 1965——. Research, publs. on microbial metabolism; enzymes; fermentations; food microbiology; thermophilic microorganisms; thermophilic bacteriophage; dissimilatory sulfate reduction. Home: 401 W. Delaware St. Office: 127 Burrill Hall, Urbana, Ill. 61801.

CAMPBELL, Maurice, English cardiologist; b. Oxford, Eng., Dec. 3, 1891; s. John E. and Sarah (Hardman) C.; scholar New Coll. Oxford, 1910-14; B.M. B.Ch., Oscon, 1916, D.M., 1921; m. Mary Chrimes, Aug. 24, 1924; children—Heather (Mrs. Hynes), Patricia (Mrs. John Ambrose), Clare (Mrs. Max Reese), Christopher, Donald. Staff, Guy's Hosp., London, Eng., 1920-26, physician in charge cardiac dept., 1926-56; physician Nat. Heart Hosp., London, 1929-58; med. supt. Orfington Hosp., 1939-44. Decorated officer Brit. Empire, 1918. Fellow Royal Coll. Physicians, London; mem. Brit. Cardiac Soc. (pres. 1956-60) Brit. Heart Found. (chmn. council 1960-65), Sherlock Holmes Soc. (chmn. 1953-55). Editor: Brit. Heart Jour., 1938-58. Latest research, publs. on genetic and environmental causes of malformations of heart. Home: 47 Arkwright Rd., Hampstead, London NW 3, Eng.*

CAMPBELL, Paul Andrew, Am. physician; b. Frankfort, Ind. Dec. 25, 1902; s. Walter William and Elizabeth (Kempf) C.; B.Sc., U. Chgo., 1924, M.D., 1929; m. Eleanor Studebaker Carlisle, Dec. 21, 1933; 1 dau., Eleanor Elizabeth (Mrs. Richard Thomas). Surgeon Culver (Ind.) Mil. Acad., 1930-34; pvt. practice medicine, specializing in ear, nose and throat, Chgo., 1935-40; faculty Rush Med. Coll. U. Chgo., 1935-40; dir. research USAAF Sch. Aviation Medicine, 1942-45; pvt. practice medicine, Chgo., 1946-50; faculty U. Ill., 1946-50; head research USAF Sch. Aviation Medicine, 1950-53; mem. staff Am. Embassy, London, Eng., 1953-56; asst. to comdr. Air Force Office Sci. Research, Washington, 1956-58; chief space medicine div. USAF Sch. Aerospace Medicine, 1958-60, head advanced studies group, 1960-62, comdr., 1962-63, ret., 1963. Recipient Bauer Space Medicine award. Mem. A.C.S., Internat. Acad. Astronautics, A.M.A., Sigma Xi, numerous others. Author: Medical and Biological Aspects of the Energies of Space, 1961; Earthman-Spaceman-Universal Man, 1965. Research, publs. in aviation and space medicine, otolaryngology. Home and office: 321 Park Hill Dr., San Antonio 78212.*

CAMPBELL, Robert Dale, Am. geographer; b. Omaha, Dec. 2, 1914; s. Robert Ward and Emma (Klempnauer) C.; B.A., U. Colo., 1938, M.A., 1940; Ph.D., Clark U., 1949; m. Marian McAnelly, Dec. 30, 1962; 1 son, Duncan; step-children—Diane Tucker, Ken Tucker, Margaret Tucker. Cartographer, OSS, Washington, 1942-43; climatologist research and devel. br. Office Q.M. Gen., Washington, 1946-47; prof., chmn. dept. geography and regional sci. George Washington U., 1947-66; regional planner Ford Found., Calcutta, India, 1964-66; v.p., dir. area operation Matrix Corp., Washington, 1966——. Vis. lectr. Alexandria U. (Egypt), 1952-53, Peshawar U., Pakistan, 1957-58, Oxford U., 1955; prin. investigator, dir. 10 maj. research projects, 1948-64. Mem. Am. Assn. U. Profs., Assn. Am. Geographers, Regional Sci. Assn., Phi Beta Kappa, Sigma Xi, Pi Gamma Mu. Author: Pakistan: Emerging Democracy, 1963; (with Eric Fischer, Eldon Miller) A Queston of Place, 1967; (with V. Nath) Regional Planning for India, 1967. Developer information system for urban planning, systems concepts in regional and econ. devel. planning, computer-based mgmt. information systems; studies in psychol. geography, culture transfer problems, nat. personality. Home: 8219 Lilly Stone Dr., Bethesda, Md. 20034. Office: 421 King St., Alexandria, Va.*

CAMPBELL, William, metallurgist; b. Gateshead-on-Tyne, Eng., June 24, 1876; s. Thomas and Franceska (Albrecht) C; grad. Civil Service Dept., King's Coll., London, 1892; St. Kenelms Coll., Oxford, 1892-94; Durham U. Coll. Sci., 1894-97, A.Sc., 1896, B.Sc., 1897, M.Sc., 1903, Sc.D., 1905; Royal Sch. Mines, London, 1899-1901; Ph.D., Columbia, 1903, A.M., 1905, Sc.D., 1928; m. Estelle M. Campbell. Demonstrator in metallurgy and lecturer in geology Durham Coll. Sci., 1898-99; Royal Exhbn. of 1851 research scholar Royal Sch. Mines, London, 1899-1901; univ. fellow Columbia, 1902, Barnard fellow, 1903; lectr. European geology Sch. Mines, Columbia, 1903-06, instr. metallurgy, 1904-07, adj. prof., 1907-12, asso. prof., 1912-14, prof., 1914-24, Howe prof., 1924——. Metallographer tech. br. U. S. Geol. Survey, 1907-11, metallographer Bur. Mines 1911-21; lectr. metallurgy U. S. Naval Acad. Post-Grad. Sch., 1913. Editor Sch. Mines Quar., 1910; asst. editor Internat. Jour. Metallography, Jour. Indsl. and Engineering Chemistry. Fellow Geol. Soc. London, N.Y. Acad. Scis., (v.p. 1911). Recipient Saville Shaw medal, Soc. Chem. Industry, 1903; Carnegie scholar Iron and Steel Inst., 1903; mem. com. on alloy steel NRC; mem. advisory com. U. S. Bur. Standard. Died Dec. 16, 1936.

CAMPBELL, William, bacteriologist; b. Leith, Scotland, June 9, 1889; s. John Kennedy and Anne (Tait) C.; ed. univs. Edinburgh, London and Paris, M.B., Ch.B.; m. Gladys Frankland Dogshun, 1923; 2 sons. Med. resident French Hosp., London, Eng., 1911; sr. resident med. officer London Sch. Tropical Medicine's Hosp., 1912; pathologist Royal Hosp. for Sick Children, Edinburgh; asst. to prof. pub. health in connection with bacteriological work, Edinburgh, 1912-14; dep. supt. Infectious Diseases Hosp., Bradford, 1913-14; asst. to prof. pathology U. Edinburgh, 1914-18; army specialist in bacteriology Royal Army Med. Corps, 1915-19; pathologist to Pellagra Commn., 1918; cons. pathologist Alex. Dist. E.E.F., 1916-19; pathologist, Bradford, 1919-24; bacteriologist St. Luke's Hosp., Bradford, 1919-24; control bacteriologist Govt. Wool Disinfecting Sta., Liverpool, 1920-24; lectr. indsl. pathology and medicine Tech. Coll. Bradford, 1922-24; extern examiner Govt. Sch. Medicine, Cairo, Egypt, 1918-19; examiner U. Witwatersrand (S. Africa) 1924-41; mem. S. African nat. com. of Internat. Union Biol. Scis., 1927-32; mem. health com. investigating health conditions of S.W. Africa, 1945-46; pres. Cape Western br. Med. Assn. S. Africa, 1933; mem. council Royal Soc. S. Africa, 1931; m. Cape Hosp. Bd., 1933-34; mem. fed. council Med. Assn. S. Africa; prof. bacteriology U. Cape Town until 1943; hon. bacteriologist Groote Schuur Gen. Hosp., Capetown until 1943; med. officer to Kaokoveld and Ovambozand; mandated territories for adminstrn. of S. Africa from 1943; cons. physician, bacteriologist Rondebosch, Mowbray and Woodstock hosps. Fellow Royal Soc. Contbr. med. jours. Research on causes of morbidity and mortality at Tsumeb and Abenab, S.W. Africa. Died Mar. 18, 1953.

CAMPBELL, William Wallace, Am. astronomer; b. Hancock County, O., Apr. 11, 1862; s. Robert Wilson and Harriet (Welch) C.; B.S., U. Mich., 1886; M.S. (hon.), 1899; Sc.D., Western U. Pa., 1900, U. Mich., 1905, U. Western Australia, 1922, Cambridge U., 1925, Columbia U., 1928, U. Chgo., 1931; LL.D., U. Wis., 1902, U. Cal., 1932; m. Elizabeth Ballard Thompson, Dec. 28, 1892; children—Wallace, Douglas, Kenneth. Prof. math. U. Colo., 1886-88; instr. astronomy U. Mich., 1888-91; astronomer Lick Obs., 1891-1930 (emeritus), acting dir., 1900, dir., 1901-30, emeritus; pres. U. Cal., 1923-30, emeritus. In charge Lick Obs. eclipse expdn. to India, Jan. 1898, Ga., May 1900, Spain, Aug. 1905, Flint Island, Jan. 1908, Kiev, Russia, Aug. 1914, Goldendale, Wash., June 1918, Wallal, Western Australia, Sept. 1922; mem. expdn. to Lower Cal., Mexico, Sept. 1923; Silliman lectr. Yale, 1909-10; William Ellery Hale lectr. Nat. Acad. Scis., 1914; Halley lectr. Oxford, 1925. Lalande prize (Gold medal), Paris Acad. Scis., 1903, Janssen prize (Gold medal), 1910; Gold medal Royal Astron. Soc., 1906; Draper Gold medal Nat. Acad. Scis. 1906; Bruce Gold medal, 1915. Mem. Am. Acad. Arts and Scis., French Acad. Scis. Author: The Elements of Practical Astronomy, 1899; Stellar Motions, 1913; Stellar Radial Velocities (with J. H. Moore), 1928. Made observations on nova auriga, nebulae, Wolf-Rayet stars, comets, varius bright line stars; showed that Mars has only 1/4 as much atmosphere as the earth; computed astron. calendars. Died Berkeley, Cal., June 14, 1938.

CAMPBELL, Willis Cohoon, Am. orthopedic surgeon; b. Jackson, Miss., Dec. 18, 1880; s. Charles C. and Lula (Cohoon) C.; M.D., U. Va., 1904; postgrad. Royal Orthopedic Hosp., London; m. Elizabeth Yerger, June 30, 1908; children—Louise, Willis, Elizabeth, George. Practiced at Memphis, Tenn., from 1906; prof. orthopedic surgery U. Tenn. Coll. Medicine, from 1910; cons. orthopedic surgery, Bapt. Meml. and St. Joseph's hosps., U. S. Marine Hosp. No. 12; chief staff Dr. Willis C. Campbell Clinic (founder 1920), Crippled Children's Hosp., Hosp. for Crippled Adults (founder 1923); helped organize Crippled Children's Hosp. Sch., 1918; attending orthopedic surgeon Methodist Hosp. Fellow A.C.S.; Author: Orthopedic Surgery, 1930; Orthopedics of Childhood (monograph), 1927; Operative Orthopaedics, 1939. Developed new method of treatment for reconstrn. of stiff joints and bone grafting to promote healing of fractures; noted for massive onlay bone-graft; pioneer in devel. of arthroplastics; began use of sulfanilamide in bone surgery cases. Died May 4, 1941.

CAMPER, Pieter (Petrus), Dutch anatomist, naturalist; b. Leyden, Netherlands, May 11, 1722; s. Florent Camper; grad. in philosophy and medicine U. Leiden, 1746; studied natural history in Eng., 1748-49; visited Paris, France, Lyons, France, Geneva, Switzerland; prof. philosophy, medicine and surgery U. Franeker, Netherlands, 1750-52; in Eng., 1752-55; held chair of anatomy and surgery the Athenaeum, Amsterdam, Netherlands, 1755-60, ret.; prof. medicine, surgery and anatomy U. Groningen (Netherlands), 1763-73; became dep. Assembly of Friesland, 1762; held seat in the Council of State, 1787-89. Founder agrl. soc., also socs. for advancement of progress in med. sci. Fellow Royal Soc., 1750; mem. French Acad. Scis.; mem. most sci. orgns. of Europe. Author: Dissertatio de quibusdam oculi partibus, 1746; Demonstrationum anatomicopathologicarum, 1760-62; De admirabili analogia inter stirpes et animalia, 1764; Sur les différences que présente le visage dans les races humaines, 1781; Oeuvres de P. Camper . . . , 3 vols., 1803. Research on human and comparative anatomy including the measurement of the facial angle of the forehead, observations of racial differences; believed difference of cast and form of intelligence depends on facial angle; described superficial layer of superficial fascia of lower abdomen (Camper's fascia); discovered the air content of bird bones, fibrous structure of the crystalline lens; advocated symphysiotomy in difficult labor. Died The Hague, Netherlands, Apr. 7, 1789.

CAMPOS, Paulo Campana, Philippine physician; b. Dasmariñas, Cavite, P.I., July 27, 1921; s. Jose S. and Luisa (Campana) C.; M.D., U. Philippines, 1945; m. Lourdes Espiritu, Dec. 9, 1951; children—Jose Paulo Romualdo, Paulo Jose, Enrique Placido Filemon. Physician in charge radioisotope lab. Philippine Gen. Hosp., Manila, 1956——, chmn. dept. medicine, 1959——, attending physician, head dept. medicine Research Lab., 1960——; prof. medicine Coll. Medicine U. Phillippines, 1962——, head dept. medicine, 1960——, dir. comprehensive community health program, 1966——. Cons. medicine St. Luke's Hosp., Vets. Meml. Hosp., 1958——. Fulbright grantee, 1952; IAEA grantee, 1965. China Med. Bd. N.Y. grantee, 1961; Rockefeller grantee, 1966. Fellow Am. Coll. Cardiology Philippine Heart Assn. (pres. 1966——), Am. Nuclear Soc. Research, publs. on use of radioisotopes in investigation of goiter. Home: 245 Vilaruel St., Pasay City, Philippines. Office: Philippine Gen. Hosp., Taft Av., Manila, P.I.*

CAMSELL, Charles, geologist, explorer; b. Ft. Liard, N.W.T., Can, Feb. 8, 1876; s. Julian S. and Sarah (Foulds) C.; ed. St. John's Coll., U. Man., Queen's U., Kingston, Harvard, Mass. Inst. Tech.; B.A.; LL.D., Queen's U, 1922, U. Alta., 1929, U. Man., 1936; m. Isabel Doucie, 1905; 1 son, 2 daus. Explored in northwestern Can., 1894-1900; geologist Algoma Central Ry., 1901; geologist Canadian No. Ry., 1902-03, Geol. Survey of Can., 1904; geol. investigator, explorer B.C., No. Can., 1904-20; responsible for original exploration and mapping of some of larger rivers of northwestern Can.; in charge B.C. and Yukon br. of Geol. Survey, 1918; dep. minister mines and resources for Can., ret. 1946. Hon. fellow St. Johns Coll., Winnipeg, 1938. Fellow Royal Soc. Can. (pres. 1930), Geog. Soc. Am. (v.p. 1937), Royal Geog. Soc. (hon.); mem. Engring. Inst. Can. (pres. 1932), Am. Inst. Mining, Metall. and Petroleum Engrs. (hon., dir. 1939-45), Canadian Inst. Mining and Metallurgy (hon., pres., Gold medal 1931), Canadian Geog. Soc. (1st pres. 1930-41). Author: (autobiography) Son of the North, also numerous articles on geology and geography of B.C. and No. Can. Died Dec. 19, 1958.

CAMUS, Charles Étienne Louis, French mathematician; b. Crecy-en-Brie, France, Aug. 25, 1699; s. Anne (Poirier) C.; ed. Navarre Coll.; examiner Sch. Engring., Paris, France, later prof. math.; went with Maupertuis and Clairaut to measure a degree of meridian in Lapland, 1736; named royal architect, 1739. Recipient prize for ship work, 1727. Fellow Royal Soc., 1764; mem. Acad. Architecture (perpetual sec.),

French Acad. Scis. (became sub-dir. 1749, 60, dir., 1750, 61). Author: Cours de mathématiques, 1766. His measurements of latitude showed oblateness of earth at poles; invented rudder machine; research on forms of cogs with cogged wheels and pinions. Died Paris, France, May 4, 1768.

CANAL-FEIJOO, Enrique José, Argentinian parasitologist; b. Santiago del Estero, Argentina, Mar. 6, 1899; s. Enrique and Ermilia (Corbalán) Canal-Feijoo. Doctor en Medicina, Facultad de Medicina, Buenos Aires, 1925; postgrad. Institut für Schiffs und Tropenkrankheiten, Hamburg, Germany, Facultad de Medicina de Paris; m. Mariá Cármen C. Jiménez Aleorta, June 19, 1935; children—Dámaso Enrique José, Carmen Ermilia (Mrs. Juan Carlos Salinas), Bernardo Rafael, José Eduardo Napoleón. Head lab. Dirección Regional de Paludisano, Santiago del Estero, 1927-47; chief service Hosp. Mixto de Santiago del Estero, 1930-47; prof. parasitology Universidad Nacional de Tucuman, Argentina, 1941—; prof. parasitology Facultad de Bioquímica, 1941-64; prof. parasitology Facultad de Medicina, Universidad Nacional de Tucumán, 1950—. Mem Pub. Health Council Santiago del Estero (became pres. 1944), Royal Soc. Tropical Medicine and Hygiene, Sociedad Argentina de Patologicá Infecciosa y Epidemiología. Author: Tratamiento de la Enfermedad de Chagas, 1950; also numerous articles. Described malaria syndrome for 1st time, 1932; discovered 2d and 3d clinical cases of Coccidiodes immitis in Argentina; one of 1st to describe black-granuled mycetoma in Madurella in Argentina. Home: 74 Cordoba, S.M. Tucumán, Argentina.*

CANALS, Étienne Joseph, French biol. physicist and chemist; b. Palau-del-Vidre, Jan. 18, 1888; s. Etienne and Philomene (Maurell) C.; ed. Montpellier Sch. of Sci. and Pharmacology; Ph.D. in Phys. Sci., also in Pharmacology; m. Louise Ferrière, Apr. 2, 1918. Titular prof. physics Sch. Pharmacy, Montpellier, also dean, 1945-58; mem. Cons. Com. on Advanced Teaching, 1946-58; insp. of pharmacies, 1948-58. Decorated officer Legion of Honor; laureat Inst. of France, Acad. Physicians. Mem. acads. pharmacy of Barcelona (hon.), France (corr.), Acad. Medicine of France (corr.), Royal Acad. Pharmacy of Madrid (corr). Author: Le platre chirurgical; Le rôle physiologique du magnésium chez les végétaux et l'emploi des radioisotopes en pharmacie; also papers on physics applied to biology and chemistry, including questions of diffused light, Raman effect, radioactivity, radioactive indicators, diffusion across membranes. Address: Palau-del-Vidre (P.-O.), France.

CANANO, Giovanni Battista (or Giambattista Canani), Italian anatomist; b. Ferrara, Italy, 1515; s. Ludovico and Lucrezia (Brancaleoni) C.; pupil of Marcantonio della Torre. Prof. anatomy, Ferrara; chief physician to Pope Juliua II, to 1555; with pub. health service, Ferrara, 1555-79. Author: Musculorum humani corporis picturata dissectio, circa 1572; Anatomia, 1574. Discovered palmaris brevis muscle which tenses palm of hand, also valves of venal, iliac, azygos veins. Died Jan. 29, 1579.

CANAPE, Jean, French physician; flourished 16th Century; physician to Francis I, 1542; lectr. to surgeons, Lyons, France. Author: le Guidon pour les barbiers et les chirurgiens, 1538; Opuscules de divers auteurs médecins, 1552. Translator: Surgery (Guy de Chauliac), 1542; also major Latin works including several of Galen's works. First to teach surgery in French.

CANAVAL, Richard, Austrian geologist; b. Klagenfurt, Austria, Mar. 25, 1855; s. Josef Leodegar and Ottilie Edle (von Rostorn) C.; Dr.phil., U. and Tech. U. Graz (Austria); studied at mining Acad. Leoben, Dr. mont. h.c.; m. Maria Frelin Thinn von Thinnfeld, 1895. Entered state service in mining, 1886; held managerial position in Klagenfurt, 1907-18. Research and numerous publs. in geology, history of mining industry especially Carinthia and the S.E. Alps. Died Klagenfurt, July 31, 1939.

CANBY, Joel Shackelford, Am. anthropologist; b. Denver, Aug. 1, 1919; s. Henry S. and Eleanor (Shackelford) C.; B.A., Colo. Coll., 1941; M.A., Harvard, 1948, Ph.D., 1950; m. Marion Beeley, Oct. 28, 1961; children—Christine Downing, Barbara Downing, Laura G. Dir. Harvard-Honduras expdn., 1948; dir. Center for Advanced Anthrop. Studies, Guatemala, 1949-50; research specialist Mid East Inst., Iraq, 1950-52; prof., chmn. dept. social relations U. Baghdad (Iraq), 1951-54; adviser Minister of Social Affairs, Iraq, 1952-54; prof. anthropology and coordinated behavioral scis. State U. N.Y. at New Paltz, 1955-57; head tng. devel. sect. Systems Devel. Corp., Santa Monica, Cal., 1958-62; research specialist, supr. Apollo crew simulation, space and information div. N.Am. Aviation, Inc., Downey, Cal., 1962—. Lectr. San Fernando State Coll., 1958-59, U. Cal. at Los Angeles, 1959-60; postdoctoral fellow Viking Fund, 1949-50, Middle East Inst. of Washington, 1950-52. Mem. Am. Anthrop. Assn., Soc. for Applied Anthropology, Human Factors Soc., Am. Acad. Polit. and Social Sci. Author: Civilizations of Ancient America, 1951; also articles. Research in cultural beginnings in Central Am., effects of agr. on formerly nomadic peoples (Kurds and Bedouin in Iraq), application of principles of behavioral sci. to advanced systems. Home: 16812 Livorno Dr., Pacific Palisades, Cal. 90272. Office: 12214 Lakewood Blvd., Downey, Cal. 90241.*

CANDALE, François de Foix (Comte de), French mathematician; bishop in southern France; founder chair of math. U. Bordeaux (France); called the French Euclid, he was interested in a better transl. of Euclid's Elements (but made no contbn. to gen. theory of geometry). Died 1594.

CANDAU, Marcolino Gomes, physician; b. Rio de Janeiro, May 30, 1911; s. Julio and Augusta (Gomes) C.; M.D., M.P.H., sch. medicine State of Rio de Janeiro, 1933, student U. Brazil, 1937, Johns Hopkins, 1940-41; LL.D., U. Mich., Johns Hopkins, U. Edinburgh (Scotland); M.D., U. Geneva (Switzerland); Sc.D., Bates Coll.; m. Ena de Carvalho, May 16, 1936; children—Marcos de Carvalho, Nelson de Carvalho. Charge various health services State of Rio de Janeiro, 1934-38; participated in mosquito eradication campaign waged by Northeast Malaria Service, 1939; chief health services State of Rio de Janeiro, 1941-42, asst. chief med. services, 1943; asst. supt., later supt. Co-op. Health Services, Brazilian Govt., 1944-50; asst. prof. hygiene, sch. medicine, State of Rio de Janeiro, 1938-50; dir. Div. of Orgn. of Health Services, WHO, Geneva, Switzerland, 1950, asst. dir.-gen. Dept. Adv. Services, 1951; asst. dir. Pan Am. San. Bur., Washington, 1952-53, also dept. dir. Regional Office for the Americas, WHO; dir.-gen. WHO 1953—. Recipient Bronfman prize Am. Pub. Health Assn., 1961, Eduardo Liceaga medal Govt. of Mexico. Hon. fellow Argentine Med. Assn. (fgn.), Am. Public Health Assn., Royal Society Promotion Health (Gt. Britain), Nat. Acad. Med. (Peru), Peruvian Pub. Health Assn., Royal Acad. Med. (Ireland), Royal Soc. Med. (Lon.), fellow Royal Coll. Physicians (London); mem. Geneva Med. Soc. (hon.), Am. Hosp. Assn. (hon.), Am. Venereal Disease Assn., (hon.). Brazilian Soc. Hygiene, Soc. Medicine and Surgery (Rio de Janeiro), Royal Soc. Tropical Medicine and Hygiene (Gt. Britain), Inter-Am. Assn. San Engring. (hon.), Canadian Pub. Health Assn. (hon.). Expert in epidemiology, emphasizing modern methods of sanitation, adequate use of nutritional sci. and agrl. methods to improve world-wide level of health. Home: 14 av. Dumas, Geneva. Office: WHO, Palais des Nations, Geneva, Switzerland.

CANDÈZE, Ernest Charles Auguste, Belgian entomologist; b. Liège, Belgium, Feb. 27, 1827; student under Jean Theodore Lacordair at Liège; studied medicine in Liège and Paris (France); asso. first with Lacrodair, later with M. F. Chapuis, Selys de Longchamps, Robert m'Lachlan of Eng. Commr., Natural History Mus., Brussels, Belgium. Mem. Belgian Acad., Order of Leopold (officer), Belgian Entomol. Soc. (a founder), Acad. Scis. Liège, also many fgn. entomol. socs. Author: The Adventures of a Cricket; The Doryphora in Belgium; also monograph of 4 vols. on Elateridae, 1857-60. Research on the coleopterous family, Elateridae; named many species of Elateridae. Died Glain, Belgium, June 30, 1898.

CANDOLLE, Alphonse Louis-Pierre Pyrame de, naturalist; b. Paris, France, Oct. 27, 1806; s. Augustin Pyrame de Candolle; 1 son, Anne Casimir Pyrame. Received chair of botany at Geneva, Switzerland, 1842. Called assembly of botanists to resolve problems of nomenclature, 1867. Mem. French Acad. Scis., 1851. Author: Géographie botanique raisonnée, 1855; Lois de la nomenclature botanique, 1867; (with son) Monographiae phanerogamarum, 1879-91; Origine des plantes cultivées, 1883; also continued father's work, Prodromus systematis naturalis regni vegetabilis, 1824-74. Research on variables involved in disappearance of family names; introduced a morphologic system for plant classification which showed a discrepancy between morphologic relationships and physiologic species; thought contractile spongioles at root tips produced the rise of sap and absorption of water; tried to discover laws of plant distbn. based on temperature. Died Geneva, Apr. 4, 1893.

CANDOLLE, Anne Casimir Pyrame de, Swiss botanist; b. Geneva, Switzerland, 1836; s. Alphonse Louis-Pierre Pyrame de Candolle. Author study on tropical plant family Piperaceae; (with father Alphonse Louis-Pierre Pyrame de C.) Monographiae phanerogamarum, 1879-91; Considérations sur l'étude de la phyllotaxie, 1881; Anatomie comparée des feuilles chez les familles de dicotylédones. Died 1918.

CANDOLLE, Augustin Pyrame de, Swiss botanist; b. Geneva, Switzerland, Feb. 4, 1778; began studies under Vaucher at Coll. Geneva; went to Paris, 1796; M.D., Paris, 1804; m., 1 son, Aphonse Louis-Pierre Pyrame. Became dep. of Cuvier at Collège de France, 1802; made bot. and agrl. survey of French kingdom, summers 1806-12; apptd. prof. botany, dir. bot. garden, 1807; received chair of botany U. Montpellier (France), 1810; held chair natural history, dir. (with son) of bot. garden at Geneva, 1817-34. Mem. French Acad. Scis., 1810; Soc. Agr. Author: Historia plantarum succulentarum, 1799; Astragalogia, 1802; Flore française, 1803-15; Propriétés medicales des plantes, 1804; Rapports sur ses voyages botaniques et agronomiques, 1813; Théorie élémentaire de la botanique, 1813; Regni vegetabilis systema naturale, 2 vols., 1819, 21; Prodromus systematis regni vegetabilis, 7 vols., (completed by son), 1824-39; Organographie végétale, 3 vols., 1827; Physiologie végétale, 3 vols., 1832. Introduced, elaborated and perfected in the plant kingdom Cuvier's system of classification which was based on natural characteristics; introduced the term taxonomy to describe the sci. of classification, 1813. Died Geneva, Sept. 9, 1841.

CANEPARIO (or Caneparius, Canepari), **Pietro** (or Petrus) **Maria,** Italian physician; flourished 1619; prof. medicine, Venice, Italy. Author: De atramentis cujuscunque generis, 1619; book on inks (describing pyrites, cadmia, magnesia, marcasite, preparation of inks and ink powders, preparation and uses of oil of vitriol, mineral colors and pigments, spagyril or chem. medicines prepared by alchemy).

CANGADEVA, Hindu mathematician; flourished 1st half 12th century; founded sch. for study of Siddhántásiromani, 1205-06. Was last important mathematician of medieval India.

CANISTRIS, Opicinus de, see de Canistris.

CANN, John Rusweiler, Am. biophys. chemist; b. Bethlehem, Pa., Dec. 11, 1920; s. John Henry and Anna L. (Rusweiler) C.; B.S., Moravian Coll., 1942; M.S., Lehigh U., 1943; M.A., Princeton, 1945, Ph.D., 1946; m. Minerva Elda Butz, Sept. 7, 1946; children—Susan Elaine, Richard Louis, David Claude. Research asst. Manhattan Project, Princeton, 1943-46; research asso. Cornell U., Ithaca, N.Y., 1946-47; research fellow Cal. Inst. Tech., Pasadena, 1947-48, sr. research fellow, 1948-51; asst. prof. biophysics U. Colo. Med. Center, Denver, 1951-57, asso. prof., 1957-63, prof., 1963—. Mem. planning group for biophys. materials Nat. Inst. Gen. Med. Studies, 1965; mem. panel molecular biology Nat. Sci. Found., 1967—. Spl. research fellow USPHS, 1960-61. Mem. Am. Chem. Soc. (adv. com. colloid div. 1956-58, nat. colloid symposium com. 1959-65), A.A.A.S., Am. Soc. Biol. Chemists, Am. Assn. Immunologists, N.Y. Acad. Sci., Biophys. Soc., Sigma Xi. Author: (with D. W. Talmage) The Chemistry of Immunity in Health and Disease, 1961; also numerous articles. Mem. editorial bd. Immunochemistry, 1964—; abstractor Excerpta Medica, 1952—. Research in phys. chemistry and immunology of biologically important macromolecules such as antibodies, enzymes, serum proteins, myoglobin, electrophoresis, sedimentation; fractionation of blood proteins, studies on interaction of proteins with each other and with small molecules. Home: 6341 E. Eastman Av., Denver 80222.*

CANNAVA, Alberto, Italian physician; b. Catania, Apr. 28, 1906; s. Vincenzo and Rosalie (Nicolosi) C.; M.D., circa 1928; diploma of honor with highest distinction; m. Eleonora Barcellona, Sept. 20, 1935; children—Lia, Nadia. Asst., aide U. Catania, then at Genoa; successively instr. pharmacology, prof. U. Sassari, then at Catania; former pres. Sch. Pharmacy, Sassari. Mem. numerous sci. socs. Contbr. publs. on various expts. Home: via Castanzo 15, Catania. Office: via Androne 81, Catania, Italy.

CANNIZZARO, Stanislao, Italian chemist; b. Palermo, Italy, July 13, 1826; mother's maiden name was Di Benedetto; studied in Palermo, Naples, Pisa, Turin (all Italy). Asst. to Rafaelle Piria, 1845-46; participated in Sicilian Revolution, 1848; fled to Paris, France, 1849; prof. phys. chemistry Nat. Coll. Alexandria (Egypt), 1851-55; prof. chemistry Genoa (Italy), 1855-61; joined Garibaldi in Palermo, 1860; prof. inorganic and organic chemistry, dir. lab. annex Palermo, 1861-71, held chair chemistry Rome, Italy, 1871; became senator, 1871; v.p. Italian Senate; founder, became dir. Italian Inst. Chemistry, U. Rome, 1871. Fellow Royal Soc., 1889 (Copley medal 1891); mem. French Acad. Scis. Author: Sunto di un corso di filosofia chemica, 1858; Scritti interno alla teoria molecolare e atomica, 1896; also articles. Discovered cyanamide dibenzoyl (with Ross), Cannizzaro reaction of aromatic aldehydes; rediscovered Avogadro's hypothesis and used it in determination of molecular and atomic weights with hydrogen and specific heats as standards; proved the unity of inorganic and organic chemistry. Died Rome, May 10, 1910.

CANNON, Annie Jump, Am. astronomer; b. Dover, Del., Dec. 11, 1863; d. Wilson Lee and Mary Elizabeth (Jump) C.; B.S., Wellesley, 1884, M.A., 1907; spl. work in astronomy, Radcliffe Coll.; D.Sc., U. Del., 1918; Dr. Astronomy, U. Groningen, Holland, 1921; LL.D., Wellesley Coll., 1925; D.Sc., Oxford U., 1925, Oglethorpe, 1935; Mt. Holyoke, 1937. Asst. Coll. Obs., 1897-1911, curator astron. photographs, 1911-38, William Cranch Bond astronomer and curator, from 1938. Recipient Henry Draper medal, for investigations in astron. physics, 1931; Ellen Richards research prize, 1932. Author various Harvard Coll. Obs. Annals. In course of photographic work discovered 300 variable stars, 5 new stars, 1 spectroscopic binary and numerous stars having bright lines or variable spectra; completed a catalogue of almost 300,000 stellar spectra which fills ten quarto volumes of the annals, all of which are published; made an extension to the catalogue, giving the spectra of fainter stars. Died Apr. 1941.

CANNON, Jack A., Am. surgeon; b. Salina, Kan., July 17, 1919; s. Charles Heaton and Bess May (Beadle) C.; B.A., U. Cal. at Los Angeles, 1940; M.D., Harvard, 1943; m. Helen Stacia Sineszko, Feb. 15, 1949; children—Susan Gail, Patricia Bess, Jack

Charles, Michael Gerald, Deborah Christine. From intern to asst. resident surgery Mass. Gen. Hosp., Boston, 1944-46; admitting room physician, then surg. resident Los Angeles County Gen. Hosp., 1948-49, sr. surg. resident, 1949-50; instr. surgery U. So. Cal. Med. Sch., 1919-50; mem. faculty U. Cal. at Los Angeles Med. Center, 1950—, prof. surgery, 1963—, also mem. staff Center for Health Scis.; head physician surgery Harbor Gen. Hosp., Torrance, Cal., 1951-52; staff surgeon Wadsworth VA Hosp., 1950-51, 52-60; mem. staff St. John's, Santa Monica hosps. (both Santa Monica). Grantee USPHS, 1954—. Served to capt. M.C., AUS, 1946-47; with M.C., USAF, 1953. Diplomate Nat. Bd. Med. Examiners, Am. Bd. Surgery. Fellow A.C.S.; mem. Am., Cal., Los Angeles County (bd. dirs. Bay dist.) med. assns., Am. Surg. Assn, Soc. Univ. Surgeons, Soc. Vascular Surgery, Internat. Cardiovascular Soc., Pan-Pacific, Pacific Coast, Western surg. assns., Bay, (bd. dirs., past pres.), Los Angeles surg. socs., Soc. Grad. Surgeons Los Angeles County Gen. Hosp., Alpha Omega Alpha. Spl. research homotransplantability of tissues by host alteration, studies in reconstructive arterial surgery. Home: 121 Udine Way, Los Angeles 90024.*

CANNON, Walter Bradford, Am. physiologist; b. Prairie du Chien, Wis., Oct. 19, 1871; s. Colbert Hanchett and Wilma (Denio) C.; A.B., Harvard U., 1896, A.M., 1897, M.D., 1900, Sc.D., 1937; hon. degrees Yale, univs. Liege, Strasbourg, Paris, Madrid, other univs.; m. Cornelia James, June 25, 1901; children—Bradford, Wilma Denio, Linda, Marian, Helen. Faculty Harvard, from 1899, George Higginson prof., 1906-42, emeritus prof., from 1942. Harvard exchange prof. to France, 1929-30; Linacre lectr., Cambridge U., 1930; vis. prof. Peiping Union Med. Sch., 1935; hon. fellow Stanford, 1941. Recipient Friedenwald medal. Am. Gastroentol. Assn., 1941. Fellow A.A.A.S. (pres. 1939), Am. Acad. Arts and Scis., mem. Nat. Acad. Scis. (fgn. sec.), Acad Scis. USSR (hon.), Am. Philos. Soc., Am. Physiol. Soc., Assn. Am. Physicians, Soc. Exptl. Biology and Medicine, A.M.A., Mass. Med. Soc., Phi Beta Kappa; fgn., hon. or corr. mem. Société de Biologie (Paris), R. Academia delle Scientze, Bologna, Sociedad de Biologia, Buenos Aires, Société Belge de Biologie (Brussels), Royal Soc. Medicine (Budapest), Royal Soc. London (Croonian lectr. 1918), Royal Swedish Acad. Sci., nat. acads. medicine Spain, Mexico, Royal Soc. Edinburgh, British Physiol. Soc., Academia de Medicina (Barcelona), Académie Royale de Medecine de Belgique (Belgium). Author: A Laboratory Course in Physiology, 1910; The Mechanical Factors of Digestion, 1911; Bodily Changes in Pain, Hunger, Fear and Rage, 1915, revised edit., 1929; Traumatic Shock, 1923; The Wisdom of the Body, 1932, revised edition was published in 1939; Digestion and Health, 1936; Autonomic Neuro-effector Systems (with A. Rosenblueth), 1937; The Way of an Investigator, 1945. Contbr. articles describing movements of stomach and intestines, effects of emotional excitement, organic stabilization, chem. mediation of nerve impulses, also papers on med. edn. and in defense of med. research; introduced bismuth meal for roentgenographic studies of digestive functions; demonstrated fight-flight reaction of body to stress; discovered epinephrine-like hormone sympathin; conducted studies in traumatic shock; investigated ductless glands, especially adrenals; showed how secretions of hormones prepare body to meet emergencies; developed concept of homeostasis (the condition of stability maintained by organism's regulatory mechanisms). Died Oct. 1, 1945.

CANNON, William Austin, Am. botanist; b. Washington, Mich., Sept. 23, 1870; s. George Henry and Lucy Marie (Cole) C.; U. Mich.; A.B., 1899, A.M., 1900, Stanford U.; univ. fellow, Columbia, 1900-02, Ph.D., 1902; m. 2d, Ella Shaw Varney. Asst., N.Y. Bot. Gardens, 1902-04; resident investigator Desert Lab., 1904-06; staff mem. dept. bot. research, Carnegie Inst. Washington, 1906-24, acting dir. 1911-12, research asso., 1925. Spl. research, Save the Redwoods League, 1925-26. Lectr. in botany, Stanford U., 1926—. Mem. A.A.A.S., other socs. Author: Studies in Heredity as Illustrated by the Trichomes of Species and Hybirds of Juglans, Oenothera, Papaver, and Solanum, 1909; Root Habits of Desert Plants, 1911; Conditions of Parasitism in Plants, 1910; Botanical Features of the Algerian Sahara, 1913; Plant Habits and Habitats in the Arid Portions of South Australia, 1921; Vegetation of South Africa, 1924; Physiological Features of Roots, 1925. Research on cytology of plant hybrids; plant anatomy; anatomy of plant parasites; morphology and physiology of arid region plants, and of root systems.

CANO, see Del Cano.

CANONICUS, Joannes (John the Canon, Marbres), Brit. natural philosopher; flourished 1329; student, Paris; D.D., Oxford, Eng.; studied under Duns Scotus, Paris. Franciscan monk; tchr. theology, Oxford, until his death. Author: In Aristoleis, Lib. VII, 1481; Lecturae magistrales; Quaestiones disputatae; Quaestiones dialectices, 1492. Discussed vacuum; stated vacuum could not exist; studied Aristotelian philosophy, civil and canon law. Died circa 1340.

CANTACUZENE, Jean, Rumanian microbiologist; b. Bucharest, Rumania, Nov. 25, 1863; mem. staff Pasteur Inst. Labs.; prof. medicine Faculty of Medicine, Bucharest. Mem. French Acad. Scis. (corr.). Research on cholera and spirochetes in birds, immunity of invertebrates; prepared BCG vaccine for children. Died Bucharest, Jan. 14, 1932.

CANTANI, Arnoldo, physician; b. Hainsbach, Bohemia, Feb. 15, 1837; studied medicine at Prague, Czechoslovakia; prof. toxicology and pharmacology in Pavia, Italy, 1864-67; dir. clin. inst., Milan, 1867; dir. clin. inst., Naples, from 1868; founder 1st bacteriol. lab. in Italy, 1885, 1st antirabic inst. and lab. for exptl. pathology; senator of Italy, 1889. Author: Patologia e terapia del ricambio materiale, 1875-83; Spezielle Pathologie und Therapie der Stoffwechselkrankheiten, 1880-84; Manuale di farmacologia clinica, 1884-92; Pro sylvia, 1893. Translator: Lehrbuch der speziellen Pathologie und Therapie (Niemeyer), 1858. Research on nutritional disorders, typhus, malaria, and tb; suggested Cantani's diet for diabetics, also enteroclysis for treatment of endogenous intoxication (named for him); introduced German med. methods into Italy. Died Naples, Apr. 30, 1893.

CANTELL, Kari Juhani, Finnish virologist; b. Mäntyharju, Finland, Aug. 23, 1932; s. Lauri Oskar Immanuel and Ester (Haapakoski) C.; Candidate medicine, U. Helsinki, 1952, Licenciate medicine, 1956, M.Sc.D., 1959, Docent, 1963; m. Aila Onnia Kokko, June 18, 1958; children—Heikki Juhani, Lauri Tapio, Aaro Sakari. Asst. physician U. Helsinki, 1956-58; staff Nat. Serum Inst., Helsinki, 1959—, head virus dept., 1962-65, prof., head virus div., 1965—. Sec. gen. Finnish NRC for Med. Scis., 1965-67. Postdoctoral fellow research dept. Children's Hosp. Phila., 1960-62; recipient Ann. med. expt. Fenn. award, 1965. Mem. Finnish Soc. for Pathology. Research, publs. on interferon, viral interference, mumps virus. Home: 30 F. Koivikkotie, Helsinki 63. Office: 166 Mannerheimintie, Helsinki, 28, Finland.*

CANTILLON, Richard, Irish economist; b. County Kerry, Ireland, 1680; m. dau. of Mr. Omani; children—3 daus., including Henrietta. Merchant banking house, London also Paris; became head of banking house, Paris; probably assisted John Law with floating of paper money, France. Author: Essai sur la nature du commerce en général (first treatise on subject, Jevons regarded book as beginning of econs. as a sci.), 1755. His work often called cradle of polit. economy has been quoted by Quesnay (his fundamental doctrine based on Cantillon), also Adam Smith and Condillac. Murdered by his cook, London, May 14, 1734.

CANTLE, Sir James, Scottish surgeon; b. Keithmore, Banffshire, Scotland, Jan. 17, 1851; s. William Cantle; ed. Milne's Inst., Charing Cross Hosp.; M.A. in Natural Sci. with honors, Aberdeen (Scotland) U., LL.D., 1919; M.B., C.M.; D.P.H., London (Eng.) U.; m. Mabel Barclay, 1884; 4 sons. Demonstrator anatomy Charing Cross Hosp., 1872-87, asst. surgeon, from 1877, surgeon, from 1887; went to China, 1887, dean Coll. Medicine for Chinese, 1889-96. Examiner U. Abderdeen; surgeon comdt. U.M.S.C., London, 1885-88; plague officer London County Council; cons. surgeon. Fellow Royal Coll. Surgeons Eng.: mem. Caledonian Soc. (pres. 1902-03), Royal Soc. Tropical Medicine and Hygiene (founder, pres. 1921-23). Author: Plague: Degeneration amongst Londoners, 1885; Physical Efficiency, 1906. Founder, co-editor Jour. Tropical Medicine. Made cholera expdn. to Egypt, 1883, investigated other tropical diseases. Died May 28, 1926.

CANTON, John, English physicist; b. Stroud, Eng., July 31, 1718; apprenticed to weaver, then sent to London, 1737; a son, William. Schoolmaster, London; head pvt. sch. in Spital Sq., London. Fellow Royal Soc., 1749 (Copley medal 1751). Author: Method of Making Artificial Magnets without the Use of Natural Ones, 1750. Demonstrated electrification of air; showed that water is compressible, 1762; invented electroscope, electrometer; 1st to make powerful artificial magnets, also to observe magnetic storms; discovered Canton's phosphorus, phosphorescent calcium sulfide; suggested that decomposing animal matter caused luminescence of sea; 1st in Eng. to verify Franklin's hypothesis on nature of electricity. Died London, Mar. 22, 1772.

CANTONI, Giulio L., biochemist; b. Milan, Italy, Sept. 29, 1915; s. Umberto and Nella (Pesaro) C.; M.D., U. Milan, 1939; m. Gabriella Sobrero, May 29, 1965; 1 dau., Allegra. Came to U. S., 1941, naturalized, 1947. Research asst. U. Milan, 1934-39, U. Oxford, Eng., 1940-42, U. Mich., 1942-43; instr. N.Y. U., 1943-45; asst. prof. pharmacology L.I. Coll. Medicine, 1945-48; sr. fellow Am. Cancer Soc., 1948-50; asso. prof. pharmacology Western Res. U., 1950-54; chief Lab. of Gen. and Comparative Biochemistry, Intramural Research, Nat. Inst. Mental Health, Bethesda, Md., 1954—. Mem. Am. Soc. Biol. Chemists, Biochem. Soc. Eng., Am. Chem. Soc., Harvey Soc., Marine Biol. Lab. Editor: (with D. R. Davies) Procedures in Nucleic Acid Research, 1966. Research and numerous publs. pharmacology of automatic nervous systems and bacterial toxins, intermediary metabolism and enzyme chemistry, transmethylation reactions, metabolism of amino acids, nucleic acid chemistry and biology. Home: 6938 Blaisdell Rd., Bethesda 20034. Office: NIH, Bldg. 10, Bethesda, Md. 20014.*

CANTOR, Georg Ferdinand Ludwig Philipp, German mathematician; b. St. Petersburg, Russia, Mar. 3, 1845; s. George Woldemar and Maria (Böhm) C.; ed. univs. Zurich (Switzerland), Göttingen (Germany), Frankfort (Germany); Ph.D., U. Berlin (Germany), 1867. Became lectr. U. Halle (Germany), 1869, named asso. prof., 1872, prof. math., 1879-1913. Founder, Internat. Mathematicians Congresses, 1897. Author: Grundlagen einer allgemeinen Mannigfaltigkeitslehre, 1883; Beiträge zur Begründung der transfiniten Mengenlehr, 2 vols., 1895-97; Gesammelte Abhandlungen, 1932; also contbr. to Founding of the Theory of Transfinite Numbers, 1915. Developed theory of sets of points, theory of irrational numbers; defined (with Dedkind and others) real numbers and proposed Cantor-Dedkind axiom of one-to-one correspondence; introduced transfinite numbers, originated theory of systems of numbers, including finite, infinite, or different orders of infinite sets; this devel. of set theory is basis of modern analysis; his new approach to concept of infinity brought new critical examination of foundations of mathematics. Died Halle, Jan. 6, 1918.

CANTOR, Moritz Benedikt, German mathematician; b. Mannheim, Germany, Aug. 23, 1829; s. Isaak Benedict and Nelly (Schnapper) C.; studied at Heidelberg, Germany, Berlin, Germany; Ph.D., Göttingen, Germany, 1851; m. Thelly Gerotwohl; 1 son, 1 dau. Became asst. prof., Heidelberg, 1853, asso. prof., 1863, hon. prof., 1877, prof., 1908-13. Author: Mathematische Beiträge zum Kulturleben der Völker, 1861; Vorlesungen über Geschichte der Mathematike, 4 vols., 1880-1908. Contbd. to internat. math. history and hist. research. Died Heidelberg, Apr. 10, 1920.

CANTOR, Nathaniel, Am. sociologist, criminologist; b. Indpls. Nov. 26, 1898; B.A., Columbia U., 1921, Ph.D., 1925; LL.B., U. Buffalo, 1929. Asso. prof. U. Buffalo, 1928-32, prof. 1932-57, chmn. dept., 1942-57; vis. prof. Columbia 1951-52, vis. prof. edn. U. Cal. at Los Angeles, 1957; cons. U. S. Fgn. Operations Adminstrn. for European Productivity Agy., 1954-55. Mem. Am. Prison Assn., Am. Sociol. Soc. Author: Crime, Criminals and Criminal Justice, 1932; Crime, 1935; Crime and Society, 1939; The Dynamics of Learning, 1943; Employee Counseling: A New Approach to Industrial Psychology, 1945; Learning Through Discussion, 1951; The Teaching-Learning Process, 1953, other. Contbr. sociol., psychiat. and criminal jours. of U. S. and abroad. Special work in causes of crime, treatment of offenders, European development in crime research, social work and correctional field, mental hygiene and edn. Died Dec. 5, 1957.

CANTRIL, Hadley, Am. psychologist; b. Hyrum, Utah, June 16, 1906; s. Albert Hadley and Edna (Meyer) C.; B.S., Dartmouth, 1928, Sc.D., 1960; Ph.D., Harvard, 1931; LL.D., Washington and Lee U., 1949; m. Mavis Katherine Lyman, June 18, 1932; children—Albert Hadley, Mavis Ann (Mrs. Donald M. Jansky). Instr., Dartmouth, 1931-32, Harvard, 1932-35; asst. prof. Tchrs. Coll., Columbia, 1935-36; faculty Princeton, 1936—, prof. psychology, 1945-53, Stuart prof., chmn. dept. psychology, 1953-55, research asso., 1955—. Chmn. bd., sr. counsellor Inst. for Internat. Social Research, 1955—, dir. Office Pub. Opinion Research, 1940-55; pres. Research Council, Princeton, 1940—; cons. exec. br. White House, 1955-56. Guggenheim fellow, 1949. Mem. Am., Eastern (past pres.) psychol. assns., Soc. for Psychol. Study Social Issues (past pres.), Phi Beta Kappa. Author numerous books the most recent being: The "Why" of Man's Experience, 1950; (with William Buchanan) How Nations See Each Other, 1953; The Politics of Despair, 1958; (with Charles H. Bumstead) Reflections on the Human Venture, 1960; Soviet Leaders and Mastery over Man, 1960; Human Nature and Political Systems, 1961; The Pattern of Human Concerns, 1965; The Human Dimension: Experiences in Policy Research, 1967; (with Lloyd Free) The Political Beliefs of Americans, 1967; also numerous articles. Editor: Tensions that Cause Wars, 1950; Public Opinion, 1935-46, 1951; The Morning Notes of Adelbert Ames, Jr., 1960. Research on attitude, opinion, and perception, social and mass movements. Home: 124 Mercer St., Princetown, N.J. 08540.*

CAPALNA, S., Rumanian biochemist; b. Barsau-Mare-Dej, Rumania, July 3, 1928; s. Simion A. and Paulina (Prodan) C.; doctorate in biochemistry U. Bucharest (Rumania), 1962; m. Ana Popescu, Sept. 28, 1953; children—Gabriela-Mariana, Simion-Adrian. Asst., chair of biochemistry Faculty Medicine, Cluj, Rumania, 1953-59; asst., chair biochemistry Faculty Medicine, Bucharest, 1959-61, sr. lectr. biochemistry, 1961-64; prin. researcher, inst. physiology RSR Acad., Bucharest, 1960—. Fellow or mem. Unions of Soc. of Med. Scis., Bucharest, A.A.A.S. Contbr. papers to sci. publs. Research on biochemistry of atherosclerosis, metallic ions in enzyme activity, radiobiology and radioprotection, glycine effect after irradiation. Home: Bloc P8, Scara 2, Mihai Bravu, Bucharest, 10, Rumania. Office: Bd. 1 Mai 11, Bucharest 2, Rumania.*

CAPE, John Anthony, Am. physicist; b. Helena, Mont., Nov. 2, 1929; s. Joseph Battiste and Carmela (Orsatti) C.; A.B., Carroll Coll., 1951; M.S., Mont.

State U., 1953; Ph.D., U. Notre Dame, 1958; m. Judith Lee Bennett, May 12, 1962; children—Christa Anthony, Randal. Prof., Carroll Coll., 1955-56; instr. U. Notre Dame, 1957-58; research asso. U. Ill., 1958-60; research specialist Atomics Internat., 1960-63; mem. tech. staff N. Am. Aviation Sci. Center, Thousand Oaks, Cal., 1963—. Mem. Am. Phys. Soc., Sigma Xi. Research on thermo-phys. properties of metallic oxides, nitrides; optical and thermal properties of defects in alkali-halides; magnetic properties of superconducting alloys; localized impurity states in metals and alloys; structure and interactions in ferro and antiferromagnetic compounds.

CAPEL, Charles Edward, Am. mathematician; b. Troy, N.Y., Dec. 26, 1922; s. Charles Edward and Frances (Albert) C.; A.B., N.Y. State Tchrs. Coll., 1947; M.A., U. Rochester, 1950; Ph.D., Tulane U., 1953; m. June Arlene Semple, Sept. 22, 1945; children—Janice Roberta, Gail Ann. Instr. math. Geneseo (N.Y.) State Tchrs. Coll., 1947-49, Tulane U., 1950-51; asst. prof. math. U. Miami (Fla.), 1953-58; research mathematician Westinghouse Research Labs., Pitts., 1958-60; prof. math. Miami (O.) U., 1960—. Served with AUS, 1943-45. Mem. Am. Math. Soc., Math. Assn. Am. (chmn. Ohio sect. 1963-64), Sigma Xi, Pi Mu Epsilon. Author: (with others) Universal Mathematics I, 1954, Elementary Mathematics of Sets, 1955; also research papers on topology. Home: 5 Patrick Dr., Oxford, O. 45056.*

CAPELLA, Martianus Minneus Felix (Felix Capella), encyclopedist; b. Carthage or Madaura, Africa, circa 420; practiced law at Carthage; thought to be proconsul. Author: Satyricon or Nuptials of Philology and Mercury (an ency.), circa 470. First to specify 7 liberal arts (grammar, dialectic, and rhetoric; geometry, astronomy, arithmetic, and music); stated Mercury and Venus revolve around sun, while earth is at system's center (sometimes presumed to be basis for Copernicus' work).

CAPELLO, Carlo Felice, Italian geographer; b. Moncalieri, Turin, Italy, Sept. 10, 1905; s. Giovanni and Maria Teresa (Briatore) C.; Ph.D. U. Turin, 1929; m. Palma Sanguinetti, Sept. 27, 1934; 1 dau., Elena. Prof. geography, sch. letters U. Turin. Mem. Italian Geog. Soc., Glaciology Soc., Cambridge, other socs. Research in phys. geography, cartography, meteorology, glaciology, geomorphology, hist. geography. Home: Corso Unione Sovietica 248, Turin. Office: Università, via Carlo Alberto 10, Turin, Italy.

CAPITAN, Joseph Louis, French biologist, anthropologist; b. Paris, France, Apr. 19, 1854; studied under Claude Bernard; doctorate, 1883. Head med. clinic Hôtel de Dieu, Paris, 1885; became physician of mil. hosp. Bégin, Vincennes, France, 1914; apptd. lectr. path. anthropology Sch. Anthropology, 1892; became prof. med. geology, 1898; apptd. prof. Am. antiquities College France, 1908; founder Internat. Inst. Anthropology. Mem. Acad. Medicine, Soc. Biology (sec., v.p.), Soc. Anthropology. Author: la Préhistoire, 1923; l'Humanité préhistorique dans la vallée de la Vézère, 1924; le Travail en Amerique avant Colomb, 1914. Authority on European pre-history and pre-Columbian Am.; discovered rupestrine painting (made it possible to identify various prehistoric races); important archaeol. diggings in Brittany and Algeria. Died Paris, Sept. 1, 1929.

CAPLOW, Theodore, Am. sociologist; N.Y.C., May 1, 1920; s. Samuel Nathaniel and Florence (Israel) C.; student Columbia, 1936-38; A.B., U. Chgo., 1939; M.A., U. Minn., 1941, Ph.D., 1946. Faculty, U. Minn., 1946-60; prof. sociology grad. faculty Columbia, N.Y.C., 1961—. Pres., Mendota Research Group, Englewood, N.J., 1957-65. Vis. prof. various fgn. univs., 1950-62. Author: Sociology of Work, 1954; (with R. McGee) Academic Marketplace, 1957; Principles of Organization, 1964. Research, publs. on codification of principles of human orgn. in terms ind. of particular times, places, and cultures. Home: 50 Central Park West, N.Y.C. 10023.*

CAPPELL, Daniel Fowler, Scottish physician; b. Glasgow, Scotland, Feb. 28, 1900; s. Robert and Anabella (Fowler) C.; ed. U. Glasgow; hon. M.D.; m. Isabella Griffin, 1927. Prof. pathology U. St. Andrews, 1931-45; named dean Sch. Medicine, 1939; prof. pathology U. Glasgow, 1945—; mem. Gen. Council Physicians, 1940. Fellow Royal Coll. Physicians, Coll. Pathologists; mem. Assn. Clin. Pathologists (former assn. pres., pres. council), Path. Soc. Medicine. Published Muir's Textbook of Pathology. Home: 50 Dowanside Rd., Glasgow. Office: Pathology Dept. The Univ. and Western Infirmary, Glasgow, Scotland, Gt. Britain.

CAPPELLANI, Francesco, Italian physicist; b. Catania, Italy, Aug. 27, 1935; s. Santi and Lilia (Carvso) C.; Degree in Nuclear Physics, Inst. Physics, Torino (Italy) U., 1958; m. Luisa Gabardini, Sept. 26, 1966. With radioelements group C.N.R.N., Milano, Italy, 1958-60, Ispra, Italy, 1960-61; mem. nuclear chemistry group chemistry dept. C.C.R. Euratom,-ISPRA, Varese, Italy, 1961—. Research, publs. on nuclear spectroscopy and solid state semiconductor detectors, decay schemes of excited nuclei and thermal neutron cross sects. determination, constrn., properties and applications of silicon and germanium detectors. Home: 15 Mazzorin Luvinate, Varese. Office: C.C.R. Euratom, Ispra, Varese, Italy.*

CAPPELLETTI, Carlo, Italian botanist; b. Verona, Italy, July 12, 1900; s. Antonia and Lina (Povegliotti) C.; Ph.D. in Natural Scis., U. Verona; hon. doctorate U. Lille (France); m. Margherita Bassi, Aug. 27, 1928; children—Antonio, Elsa. Dir. Bot. Inst. of Torino, 1932-48; dir. Bot. Inst. of Padua (Italy), 1948—. Mem. Acad. dei Lincei, Rome, Agrl. Acad. Torino, acads. Scis. Verona, Padua, Istituto Veneto, Venice, Italian Soc. Botany, other socs. Author: Trattato di botanica; many other publs. on botany. Home: via Orto Botanico 15. Office: via Orto Botanico 13, Padua, Italy.

CAPPS, Richard Huntley, Am. physicist; b. Wichita, Kan., July 1, 1928; s. Charles M. and Anna (Palmer) C.; B.A., U. Kan., 1950; M.A., U. Wis., 1952, Ph.D., 1955; m. Joan P. Salatino, June 18, 1955. Research asso. U. Cal. at Berkeley, 1955-57; faculty U. Wash., 1957-58; research asso. Cornell U., 1958-60; faculty Northwestern U., Evanston, Ill., 1960—, prof., 1965-67; prof. Purdue University, Lafayette, Ind., 1967—. Cons. Argonne Nat. Lab., 1960—. Mem. Am. Phys. Soc. Contbr. numerous articles to profl. jours. Research on theoretical investigation of basic laws of fundamental sub-atomic particles. Home: 135 Indian Rock Dr., West Lafayette, Ind. 47906. Office: Purdue Univ., Lafayette, Ind., 47907.*

CAPRA, Baldassare (or Balthasar), Italian astronomer, philosopher; b. Milan, Italy, 1580. Author: Tyrocinia astronomica, 1606; Usus et fabrica circini . . . , 1655. Challenged Galileo; declared himself inventor compass of proportion. Died 1626.

CAPRARO, Vittorio, Italian physiologist; b. Trento, Italy, Dec. 10, 1911; s. Renato and Maria (Masera) C.; M.D., U. Rome (Italy), 1936; postgrad. (fellow in physiology), U. Pavia (Italy), 1938, U. Milan (Italy), 1939-50; m. Cordelia Frugoni, Sept. 5, 1939; children—Sabina (Mrs. Colantuoni), Patrizia (Mrs. Lovo), Luca, Brigida. Asst., U. Rome, 1937, U. Pavia, 1938; faculty U. Milan, 1939-51, prof., chmn. gen. physiology, 1960—; prof., chmn. gen. physiology U. Urbino, 1951-54, U. Parma, 1954-60; co-dir. Inst. C Erba for Therapeutical Research Milan, 1951-61; med. dir. Columbus Hosp., Milan, 1964—. Mem. A.A.A.S., N.Y. Acad. Scis., Società Ital. di Fisiol., Società Italiana di Biologia Sperimentale. Author: Tecnica Chimica Biologica, 1945; also numerous articles. Research on permeability of biol. membranes, mechanism of action of posthypophyseal hormones on water passage, mechanism of intestinal absorption of vitamin B12 of sugars and amino acids. Home: 8 Via Ponchielli 20129 Milan, Italy.*

CAPRON, A. R. C., French parasitologist; b. Lens, France, Dec. 30, 1930; s. E. G. and M. (Curt) C.; Docteur en Médecine, U. Lille (France), 1958, Licence ès Sciences, 1959; m. J. Marot, May 10, 1955; 1 dau., Isabelle. With Lille Faculty Medicine, 1955-60, prof., 1961—; asst. Institut Pasteur de Madagascar, 1958-60; lectr. Paris Faculty Medicine, 1961—; chief parasitological service Centre Hospitalier et Universitaire Lille, 1965—. Lauréat de l'Académie Nationale de Médecine. Mem. Société de Pathologie exotique, Société de Zoologie, Société de Pathologie thoracique, Société Internationale de Mycologie. Research, numerous publs. on comparative helminthology, epidemiology of bilharziasis, biology of Schistosoma, antigenetic structure structur of helminths, relationships between host and parasite, diagnostic possibilities of immunoelectrophoresis, perfection of new methods of immunological investigation of human helminthiases. Home: 132 Avenue de la Republique-La Madeleine, Nord France. Office: Faculte de Medecine, Lille 59, France.*

CAPRON, Paul Corneille, Belgian radiochemist; b. Ixelles, Belgium, Apr. 5, 1905; s. Adolphe and Romaine (Vindevogel) C.; Docteur en Sciences Chimiques with distinction, U. Louvain (Belgium), 1929; m. Marie Stevens, Aug. 6, 1929; children—Anne-Marie (Mrs. Guy Debry), Francoise, Claire (Mrs. Jacques Massion), Marie-Agnès (Mrs. Henry Hugel). With Lab. de Recherche, Union Chim. Belge, 1929-32; with F.N.-R.S., 1932-42, asso., 1937-42; faculty U. Louvain (Belgium), 1942—, prof., 1943—, became sci. councilor, 1962, also mem. adminstrn. bd. Chmn. sci. com. Institut Interuniversitaire des Sciences Nucleaires. Mem. Société de Chimie-Physique de France. Co-editor, Industrie Chimique Belge. Research and publs. on radiation chemistry, hot atom chemistry, including isotope effects in chem. effects of radiative neutron capture, nucleogenesis of recoil atom and its chem. fate, solid state chemistry; among first to discover primary formation of radicals on irradiation of hydrogen. Home: Les Bruyères, Nethen, Belgium. Office: Centre de Physique Nucléaire, Parc d'Arenberg, Heverlee, Louvain, Belgium.*

CAPSTAFF, John George, color photography authority; b. Gateshead-on-Tyne, Eng., Feb. 24, 1879; s. John Squire and Elizabeth (Hogg) C.; ed. Heaton Sch. of Science and Art and Armstrong Coll., Durham; m. Alice Grace Wallace, of Newcastle-on-Tyne, Sept. 23, 1912; children—Phyllis Mary, Elizabeth. Came to U. S., 1913. Research in photography Eastman Kodak Co., Rochester, N.Y., 1913-60; pioneered in 16 millimeter motion pictures; responsible for application of reversal process to motion pictures; holds many photog. patents; authority on color photography. Recipient Modern Pioneer award, 1940; Progress medal of Soc. Motion Picture Engrs., 1944. Fellow Royal Photog. Soc. (hon., Progress Medal 1947), Photog. Soc. Am. (hon.). Died Jan. 31, 1960.

CAPUTO, Antonio, Italian biochemist; b. Naples, Italy, Feb. 4, 1927; s. Giuseppe and Teresa C.; Med. Degree, U. Naples, 1951; m. Marzia Donnini, June 22, 1953; children—Stefano, Fiorella, Francesca. Staff, Sci. dept. Istituto Regina Elena for Cancer Research, Rome, Italy, 1951—, head biochem. unit, 1954-59, dir. sci. dept., 1960—; prof. gen. pathology, dir. Inst. Gen. Pathology, U. Perugia (Italy), 1966—. Recipient 1st prize C. Erba for young scientists, 1954. Co.-author: Advances in Protein Chemistry, vol. 19, 1964. Research and numerous publs. on chem. carcinogenesis, relations between function and structure of protein molecules, tumor constituents, composition of path. fluids.*

CAPUTTO, Ranwel, Argentinian biol. chemist; b. Buenos Aires, Argentina, Jan. 1, 1914; s. Salvador and Vincenta (De Rosa) C.; M.D., Universidad Nacional de Córdoba, 1940; m. Joaquina Dora Prieto, June 9, 1943; children—Ranwel, Dora, Beatriz, Lilia. Research asso. Cambridge U., Eng., 1941-42; instr. biochemistry Universidad Nacional de Córdoba, 1943-45; research asso. Instituto de Investigaciones Bioquímicas Fundación Campomar, 1945-50; research biochemist dept. medicine U. Okla., 1950-59; head biochem. sect. Okla. Med. Research Inst., 1959-63; chmn. Departamento Química Biológica, Instituto de Ciencias Químicas, Universidad Nacional de Córdoba, Argentina, 1963; dir. Instituto de Ciencias Químicas, Universidad Nacional de Córdoba, 1964. Mem. Am. Chem. Soc., Biochem. Soc. London, Am. Soc. Biol. Chemists, Sigma Xi, Alpha Omega Alpha. Research and numerous publs. on isolation and identification of glucose 1-6 diphosphate and uridine-diphosphoglucose, isolation of neuraminlactose, characterization of several new enzymatic activities, biochem. function of vitamin E. Home: Astrada y Salsacate (Alto Verde), Córdoba, Argentina.*

CAQUOT, Albert Irénée, French engr.; b. Bouziers, France, July 1, 1881; ed. École Polytechnique; m. Jeanne Lecomte, July 1, 1881; 1 dau., Suzanne (Mme. Jean Leheron-Kerisel). Constrn. engr. Société Pelnard-Considère; prof. French Nat. Sch. Mines; gen. dir. Ministry of Air, 1918, 28, 38; mem. Superior Council Sci. Research and Tech. Progress; pres. Soc. for Encouragement of Nat. Industry; pres. Gen. Commn. Sci. Orgn.; pres. French Assn. Normalization, mem. French Nat. Com. Productivity. Mem. French Acad. Scis., 1934, pres., 1952; mem. French Soc. Econ. Geography. Specialist in reinforced concrete; intensive studies in aeronautics; dir. tech. dept. of mil. aviation during World War I; made observation balloons for warships. Address: 1, rue Beethoven, Paris 16, France.

CARABELLI, Georg, Hungarian dentist; b. Hungary, 1787. Described a tubercle or 5th cusp on the lingual surface of upper molar teeth, 1844; said to be hereditary; known as Carabelli's cusp or tubercle. Died 1842.

CARACI, Giuseppe, Italian geographer; b. Florence, Italy, Dec. 23, 1893; s. Biagio and Domenica (Longo) C.; Ph.D. in Letters; m. Iolanda Marcellino, Mar. 3, 1933; children—Domenica, Giovanni, Ilaria, Maria. Titular prof. geography, dir. inst. geog. scis. and cartology U. Rome. Mem. many Italian and other sci. socs. Author various books; contbr. to jours. Home: via Giacinto Carini 32, Rome. Office: piazza della Republica 10, Rome, Italy.

CARATHÉODORY, Constantin, German mathematician; b. Berlin, Germany, Sept. 13, 1873; s. Stephanes and Despina (Petrocochino) C.; ed. Belgian Mil. Acad., 1891-95; student math. Berlin, 1900-02, Göttingen, Germany, 1902-04; Ph.D. With Brit. Engring. Corps, Egypt, 1898-1900; prof. Hanover, Germany, 1909, Breslau, Germany (now Wroclaw, Poland), from 1910, Göttingen, from 1913, Berlin, from 1918, Singrha U., 1920-22, Munich, Germany, from 1924; rector Reorganized U. Athens (Greece), from 1939. Vis. Carl Schurz prof. U. Wis., 1936-37. Mem. numerous acads. Author: Vorlesungen über reelle Funktionen, 1918; Conformal Representation, 1932; Geometric optics, 1937; Funktionen theorie, 1950; Mass und Integral und Ihre Algebraisierung, 1955. Major contbn. in calculus of variations (added theory of discontinuous solutions, created new constn. of field theory which he connected with partial differentiation calculus); advanced study of function theory, functions of several complex variables, meromorphic functions and theory of several integrals; contbd. numerous works in geometric optics (mirror telescope, projection apparatus), mechanics, thermodynamics. Died Munich, Feb. 2, 1950.

CARBON, Max William, Am. physicist; b. Monon, Ind., Jan. 19, 1922; s. Joseph William and Mary Olive (Goole) C.; B.S. in Mech. Engring., Purdue U., 1943, M.S., 1947, Ph.D., 1949; m. Phyllis Camille Myers, Apr. 13, 1944; children—Ronald Allen, Jean Ann, Susan Jane, David William, Janet Elaine. With Hanford works Gen. Electric Co., 1949-55, head heat transfer unit, 1951-55; with research and advanced devel. div. Avco Mfg. Corp., 1955-58, chief thermodynamics sect., 1956-58; professor and chmn. nuclear engring. department U. Wis. Coll. Engring., 1958—. Served to capt., ordnance dept., AUS, 1943-46. Mem.

Am. Nuclear Soc., A.A.A.S., Am. Soc. Engring. Edn., Am. Soc. M.E., Am Assn. U. Profs., Sigma Xi, Tau Beta Pi. Investigation of direct energy conversion; heat transfer; nuclear power. Home: Mech. Engring. Bldg., U. Wis., Madison, Wis. 53706.*

CARCAVI, Pierre de, see de Carcavi.

CARDANI, Cesare, Italian chemist; b. Milan, Italy, May 9, 1922; s. Pierino and Virginia (Provera) C.; Ph.D. in Chem. Scis.; m. Albertina Caglieris, May 25, 1954; children—Marco, Silvia. Prof. organic chemistry Poly. Sch. of Milan. Mem. Am., Italian chem. socs. Research and publs. on natural organic substances. Home: via Bocconi 9, Milan. Office: Piazza Leonardo de Vinci 32, Milan, Italy.

CARDANO, Girolamo (or Geronimo), Italian mathematician, physician; b. Pavia, Sept. 24, 1501; illegitimate s. Facio Cardano (Milan jurist); studied medicine U. Pavia, U. Padua, 1523-24; m., several sons. Denied admittance to Coll. of Physicians in Milan because of his illegitimacy, but was finally admitted after establishing reputation as able physician; pub. lectr. in math., Milan, 1534-36; practiced medicine, taught math. and medicine, Pavia, also Milan, 1543-61; summoned to Scotland as med. adviser to Archbishop Hamilton of St. Andrews, 1552; shortly after, proceeded to court of Edward VI in Eng.; his reputation and med. practice waned after execution of his son (convicted of poisoning his unfaithful wife), 1560; banished from Milan shortly thereafter on some unspecified accusation, decided to leave the city although the decree was rescinded; taught medicine in Bologna, 1562-70; moved to Rome, was arrested for heresey or debt (possibly both), 1570; upon release, accepted into Coll. of Physicians, Rome, and pensioned by Pope Gregory XIII, but prohibited from teaching and publishing further works. Author: Practica arithmeticae, 1539; Ars magna, 1545; De subtilitate rerum, 1550; De varietate rerum, 1557; De vita propria liber; also works on math., astronomy, astrology, medicine, alchemy. Wrote 1st great Latin treatise devoted to algebra, 1545; recognized negative roots; studied particular cases of bi-quadratic equations; his name is attached to solution of the cubic (Cardan's Rule) although he obtained it from Tartaglia (its discoverer) and published it without permission; considered problems in probability, including what later became known as the Petersburg Problem; invented Cardan suspension and Cardan shaft; devised method for teaching deaf and dumb to read and write with Braille-like code; gave 1st clear account of typhus fever, 1536; accepted only 3 of the four elements, since he regarded fire as form of motion; distinguished between chem. combination, solution, and mech. mixture. Died Rome, Sept. 20, 1576.

CARDER, Dean Samuel, Am. seismologist; b. Medford, Ore., Aug. 11, 1897; s. Eli Woodworth and Cora (Redden) C.; B.S. in Mining Engring., Ore. State U., 1921; M.S., U. Ida., 1925; Ph.D., U. Cal. at Berkeley, 1933; m. Ellen Mitchell, May 29, 1923 (div. June 1939); children—Thelma M., Bruce M.; m. 2d, Harriet B. Burg, Aug. 1, 1942; children—Alan Burg, Laura Deanne. With U. S. Coast and Geodetic Survey, 1933-65, chief seismologist, 1953-63, research seismologist Office Research and Devel., 1963-65; seismologist Environmental Sci. Service Adminstr., U. S. Dept. Commerce, Earthquake Mechanism Lab., San Francisco, 1965-67, emeritus, 1967—. U. S. mem. seismology com. Pan Am. Inst. Geography and History, Buenos Aires, 1961, Guatemala, 1965. Recipient Colbert medal Soc. Am. Mil. Engrs., 1956. Fellow Geol. Soc. Am., A.A.A.S., Washington Acad. Sci.; mem. Seismol. Soc. Am. (past pres.), Am. Geophys. Union, Soc. Exploration Geophysicists, Earthquake Engring. Research Inst. Author: Earthquake Investigation in the Western United States, 1965; also articles. Designed and developed displacement elements of a strong motion seismograph; established correlation between water load in a large reservoir and small earthquakes; confirmed existence of maj. discontinuities at depth of 400 and 650 kilometers; helped develop formulas to predict ground motion and accelerations from large underground explosions; assisted in devel. travel times from surface foci in distance ranges up to 100 degrees. Home: 5828 Fresno Av., Richmond, Cal. 94804. Office: 390 Main St., San Francisco 94105.*

CARDIN, Augusto, Italian chemist; b. Galerina, Trevise, Feb. 13, 1891; former titular prof. physiology Sch. Math., Phys. and Natural Scis., U. Camerino, also rector; now prof. physiology, dean Sch. Math., Phys. and Natural Scis., U. Sassari (Italy); instr., inst. physics Sch. Pharmacology. Address: via Muroni 29, Sassari, Italy.

CARDIS, Fernand, Swiss pathologist; b. Lausanne, Switzerland, Nov. 7, 1898; s. Jules and Marie (Buffat) C.; ed. Lausanne, also Basel, Switzerland, Berlin, Germany; M.D.; m. Jeanne Zurcher, June 2, 1926; children—Raymond, Jeanne-Marie, Francois, Jacqueline. Asst. path. anatomy; physician-in-chief Leysin sanatorium; instr., then prof. agrégé physiology U. Lausanne Sch. Medicine. Mem. Med. Soc., Paris Hosps., Nat. Acad. Medicine, Paris, other socs. Research and publs. on Tb, other aspects of respiratory pathology. Home: 2, rue Montbenon, Lausanne. Office: 10, rue Beau Sejour, Lausanne, Switzerland.

CARDON, Philip Vincent, Am. agronomist; b. Logan, Utah, Apr. 25, 1889; s. Thomas Barthelmy and Lucy (Smith) C.; B.S., Utah State Agrl. Coll., 1909, LL.D., 1948; M.S., U. Cal., 1933; m. Leah Ivins, Sept. 17, 1913; children—Lucy Elizabeth (Mrs. Calvin L. Rampton), Margaret Ivins (Mrs. Gerald Wessler), Philippe. Spl. agt., cereal investigations, U. S. Dept. Agr., Washington, 1910-11, sci. asst., 1911-13, asst. agronomist, 1913-14, in cotton investigations, 1914-18, dry land agr., 1918-19; prof. agronomy and agron., Mont. State Agrl. Coll. and Expt. Sta., 1919-21; dir. so. br. Utah State Agrl. Coll., 1921-22; editor, Utah Farmer, 1922-25; farm economist, Utah Agrl. Expt. Sta., 1926-27, dir. 1928-35, regional dir. land policy sect., Agrl. Adjustment Adminstrn., U. S. Dept. Agr., 1934-35; prin. agronomist, in charge forage crops and diseases, bur. of plant industry, U. S. Dept. Agr., 1935-39, asst. chief, 1939-42, asst. research adminstr., agrl. research adminstrn., 1942-45, research adminstr. agrl. research adminstrn., 1942-45, research adminstr. 1945-46, spl. asst. to chief, bur. plant industry, soils and agr. engring., 1946-48, research adminstr., 1948-51, dir. grad. sch., 1952-53; dir. gen. FAO, 1953-56; cons. Southern Regional Edn. Bd. Chmn. U. S. delegation to 4th Internat. Grassland Congress, Great Britain, 1937; mem. tech. secretariat UN Conf., on Food and Agr., 1943, chmn. standing adv. com. agr., food and agr. orgn.; mem. Mexican-U. S. Agr. Commn., 1945-46; mem. U. S. del. 4th Conf., Food and Agr. Orgn., 1948, U. S. del. 5th Conf. Food and Agr., 1949; chmn. 9th Congress Internat. Seed Testing Assn., 1950; chmn. sect. Agr. and Forestry, Alaskan Sci. Conf., 1950; alternate U. S. del. 4th Inter-Am. Conf. on Agr., Montevideo, 1950, 6th Conf. Food and Agr. Orgn., 1951 (mem. coordinating com. 1951-53); pres. 6th Internat. Grassland Congress, 1952. Recipient Distinguished Service Award, U. S. Dept. Agr., 1948. Fellow A.A.A.S., Am. Soc. Agronomy; mem. Am. Farm Econ. Assn., Wash. Bot. Soc., Phi Kappa Phi, Alpha Zeta. Contbr. bulls. and tech. papers in agronomy and farm econs. Research on field crops, farm econs., land utilization. Died Oct. 13, 1965.

CARDWELL, Alvin Boyd, Am. physicist; b. Lenoir City, Tenn., Oct. 16, 1902; s. John Wesley and Martha (Duff) C.; B.S., U. Chattanooga, 1925, D.Sc., 1961; M.S., U. Wis., 1927, Ph.D. (fellow in physics), 1930; m. Edna Evangaline Zirkle, Dec. 27, 1930; children—Edward (dec.), Nancy, Charles Evan. Faculty, Tulane U., 1930-36; prof. physics Kan. State U., Manhattan, 1936—, head dept. physics, 1937-53, 56—, physicist in charge Engring. Expt. Sta., 1946-53, 56—, physicist in charge Agrl. Expt. Sta., 1936-53, 56—, asso. dean Sch. Arts and Scis., 1953-55, dir. Bur. Gen. Research, 1954—; research physicist Manhattan Project, 1944-46. Recipient Distinguished Prof. award Kan. State U., 1957. Fellow Am. Phys. Soc., A.A.A.S.; mem. Kan. Acad. Sci., Am. Assn. Physics Tchrs., Sigma Xi, Phi Kappa Phi, Gamma Alpha, Gamma Sigma Delta, Alpha Phi Eta, Sigma. Contbr. research articles on phys. electronics and solid state physics to tech. jours. Home: 1502 N. 10th St., Manhattan, Kan. 66502.*

CARELL, Erich, German polit. scientist; b. Ponzan, June 12, 1905; s. Paul and Lina (Muller) C.; Ph.D. in Polit. Sci., U. Munich; m. Ilse Scheiber, Feb. 8, 1949; children—Marion, Christoph, Barbara, Suzanne. Became prof., Munich, 1931; prof. humanitarian scis., econs. and statistics U. Würzburg, 1943. Mem. Ifv-Instituts (Munich), Assn. for Social Politics, Soc. for Econ. and Social Scis. Author: Sozialokonomische Theorie und Knojunkturproblem, 1931; Bodenknappheit und Grundrentenbildung, 1948; Unternehmergewinn und Arbeitslohn, 1950; Grundlagen der Preisbilding, 1951; Allgemeine Volkswirtschaftslehre, 1963. Home: Millt. Dullenbergweg 5. Office: Dornerschulstrasse 10, Würzberg, Germany.

CAREY, Eben James, Am. physician; b. Chgo., July 31, 1889; s. Frank White and Mary Anne (Curran) C.; pre-med. studies, U. Cal., 1909-11; 1st and 2d yrs. in medicine, 1911-13; B.S., Creighton U., 1916, M.S., 1918, D.Sc., 1920; M.D., Rush Med. Coll., 1925; m. Helene Lichnovsky, Sept. 3, 1919; 1 dau., Mary Anne. Instr. and asst. prof. anatomy Creighton U., 1914-20; prof. and dir. dept. of anatomy, Marquette U. Sch. Medicine, Milw., 1920-26, also dean med. students, 1921-26, acting dean, 1926, dean and prof. anatomy, 1933-47; med. dir. Marquette Free Dispensary, 1924; chief staff Marquette U. Hosp., 1926. Recipient Silver medal for exhibit illustrating original investigation on intrinsic wave mechanics of nervous and muscular systems, A.M.A., 1937. Hon. fellowship Am. Coll. of Dentistry, 1939. Fellow A.A.A.S., A.M.A., N.Y. Acad. Scis.; mem. Wis., Milwaukee County (pres. 1942) med. socs., Milw. Acad. Medicine (pres. 1939-40), Chgo. Inst. Medicine, Alpha Omega Alpha, Pi Kappa Epsilon, Phi Chi (chmn. exec. trustees, editor Quar. 1932-47). Author: Studies in Anatomy, 1924; also many tech. articles. Silver medal for sci. exhibit on exptl. bone origin and pathology, A.M.A. 1928, citation of merit for sci. exhibit of continued exptl. studies on bone origin, 1929; citation of merit, 1930, 31; gold medal for studies on origin of muscle and bone, Radiol. Soc. N.Am., 1933; gold medal for exhibit on motor nerve endings, A.M.A., 1942. Died June 5, 1947.

CAREY, Henry Charles, Am. economist, publisher; b. Phila., Dec. 15, 1793; s. Matthew Carey. Leading partner Carey, Lea & Carey (important Am. publishing house, Am. publisher for Thomas Carlyle, Washington Irving, Sir Walter Scott); del. Pa. Constl. Conv., 1872. Author: The Rates of Wages (essay), 1835; Principles of Political Economy (internat. acclaimed; translated into Italian and Swedish), 3 vols., 1837-40; Commercial Associations in France and England, 1845; Past, Present and Future (announced attitude as foe of free trade system), 1848; Harmony of Interests: Manufacturing and Commercial, 1851; Slave Trade, Domestic and Foreign, 1853; Letters to the President, 1858; The Principles of Social Science, 3 vols., 1858-59; Unity of Law, 1872; most works translated into German, French, Italian, Russian, Spanish. Leader of Am. nationalist sch. polit. economy; converted from laissez-faire to protectionist doctrine, 1844; critic of English classical economists and socialist econ. thought. Died Phila., Oct. 13, 1879.

CAREY, Mathew, economist, publisher; b. Dublin, Ireland, Jan. 28, 1760, s. Christopher Carey; m. Bridget Flahavan, 1791, 9 children including Henry. Editor, Freeman's Journal, Dublin, 1780-83, Volunteer's Journal, 1783-84; worked in printing office set up by Benjamin Franklin at Passy, France; came to Am., 1784; founder, editor Pa. Herald, 1785, Columbian Magazine, 1786; published Am. Museum mag., 1789; dir. Bank of Pa., 1802; mem. Phila. Soc. for Promotion of Nat. Industry; founder Hibernian Soc. for relief of Irish immigrants; his writings constitute a major influence in hist. devel. and direction of Am. nationalist sch. of econ. thought. Author: Autobiographical Sketches, 1829; Miscellaneous Essays, 1830. Advocated soil conservation, predicted major advances in agrl. tech.; believed (in contrast to Malthus) that means of subsistence in indsl. society would increase more quickly than population; advocated integrated, balanced productive system as boon to common good as well as to nat. wealth; urged fed. constrn., subsidization of internal improvements. Died Phila., Sept. 16, 1839.

CAREY, S. Warren, Australian geologist; b. Campbelltown, New South Wales, Australia, Nov. 1, 1911; s. Tasman George and Hannah (Harley) C.; B.Sc., U. Sydney (Australia), 1932, M.Sc., 1934, D.Sc., 1939; m. Austral Mary Robson, June 15, 1940; children—Alice Elspeth, Harley Roberts, Robyn Anne, David, Warren C. Research scholar U. Sydney, 1932-34; petroleum geologist New Guinea, Oil Search Ltd., 1934-38, Australasian Petroleum Co., 1938-42; chief govt. geologist Tasmania, 1944-46; found. prof. geology U. Tasmania, Hobart, Australia, 1946—; vis. prof. Yale 1959-60. Econ. cons. Hydro-Electric Commn., Tasmania, 1946-60; geol. cons. for pub. projects and instns., pvt. cons.; tech. expert UN, Israel, 1963-64. Mem. Am. Assn. Petroleum Geologists, Am. Geophys. Union, Seismol. Soc. Am., Australian Assn. Mining and Metallurgy, geol. socs. Australia, France, India, Alta. Soc. Petroleum Geologists. Contbg. author, editor: Continental Draft—A Symposium, 1958; Dolerite—A Symposium, 1957. Publs. on def. continental drift hypothesis; originated Tethyan Shear System concept, theory on rotation of So. hemisphere eastward with respect to the No., orocline concept in tectonics; (with others) developed theory that earth currently expanding. Home: 24 Richardson Av., Hobart, Tasmania, Australia.

CARHART, Henry Smith, Am. physicist; b. Coeymans, N.Y., Mar. 27, 1844; s. Daniel S. and Margaret (Martin) C.; A.B., Wesleyan, Conn., 1869, A.M., 1872, LL.D., 1893; student Yale, 1871-72, Harvard, 1876, U. Berlin, 1881-82; LL.D., U. Mich., 1912; Sc.D., Northwestern, 1912; m. Ellen M. Soule, Aug. 30, 1876. Prof. physics and chemistry Northwestern U., 1872-86; prof. physics U. Mich., 1886-1909, emeritus prof., from 1909. Mem. Internat. Jury of Awards, Paris Expn., of Electricity, 1881; pres. Bd. Judges, Dept. Electricity, Chgo., Expn., 1893; mem. Jury of Awards, Buffalo Expn., 1901; U. S. del. Internat. Elec. Congress, Chgo., 1893, St. Louis, 1904; guest of Brit. Assn. Adv. Science to S. Africa, 1905; mem. preliminary conf. on elec. units and standards, Berlin, 1905, U. S. del., London, 1908; del. U. Mich. to Darwin Centennial Celebration, Cambridge, Eng., 1909. Author: University Physics, 1894-96; Electrical Measurements (with G. W. Patterson), 1895; College Physics, 1910; (with H. N. Chute) High School Physics, 1901; 1910, First Principles of Physics, 1912, Physics with Applications, 1917. Research in electricity on standard cells and primary batteries; devised Carhart-Clark cell. Died Feb. 13, 1920.

CARIDROIT, Fernand, French biologist; b. France, 1895; aggregation in sci., M.D., Paris; became dir. physiol. lab. Coll. France; investigated sexual endocrinology of birds, genetics of gallinaceans and other birds. Died 1950.

CARION, Johannes (Nägelin, Gewürznägelein, or Caryophyllus), German astronomer, astrologer; b. Bietigheim, Germany, Mar. 22, 1499; student math. and astronomy, Tübingen, Germany; Dr.h.c., 1535. Became Magister; named ct. astronomer to Prince Joachim I, Brandenburg, 1522; assigned to diplomatic missions abroad. Author: Practica Nürnberg (astrological calendars), 1531; Brief World Chronicle (edited and completed with timetable by Melanchton).

Made prophecies, including prediction of Berlin flood of 1524. Died Berlin, circa 1538.

CARL, Johann Samuel, German chemist, physician; b. Öhringen, Württemberg, 1676; s. Johann Ernest Carl; student medicine U. Halle (Germany), licence, 1699; pupil of Stahl. Physician to Count of Wittgenstein, Count of Isenburg-Stolberg, then to King of Denmark, 1736. Author: Lapis Lydius Philosophico-pyrotechnicus ad ossium fossilium docimasiam analytice demonstrandam adhibitus . . . , 1703. Editor: Stahl's lectures. Showed identity of natural, artificial, antimonial cinnabar, and gave its composition as 6 parts mercury to 1 part sulphur. Died Melldorf, Holstein, June 13, 1757.

CARLANDER, Kenneth Dixon, fishery biologist; b. Gary, Ind., May 25, 1915; s. Lester William and Ruth Emelia (Larson) C.; B.A. cum laude, U. Minn., 1936, M.S., 1938, Ph.D., 1943; m. Harriet Coleman Bell, June 23, 1939. Ornithologist, Panhandle Plains Hist. Soc. Mus., Canyon, Tex., 1933; lab. technician U. Minn., 1936-38; aquatic biologist Minn. Dept. Conservation, 1938-46; asst. prof. Ia. State U., 1946-48, asso. prof., 1948-57, prof., 1957—; leader Ia. Coop. Fishery Research Unit, 1946-66. Research grantee Atomic Energy Commn., Nat. Sci. Found. Exec. bd. 15th Internat. Congress Limnology, 1960-62; editorial referee Fisheries Research Bd. of Can., 1957; mem. Ia. Natural Resources Council, 1961, Iowa Water Resources Research Institute, 1964—; mem. panel of fishery experts FAO of UN; Ford Foundation assignment, Egypt, 1965-66. Fellow American Inst. Fishery Research Biologists (bd. control 1957-61), A.A.A.S. (council); mem. Am. Institute of Biological Sciences, Am. Soc. Ichthyologists and Herpetologists (bd. govs. 1952-56), Am. Soc. Limnology and Oceanography, Wildlife Soc., Am. Fisheries Soc. (pres. 1960-61), Biometric Soc., Ecol. Soc. Am., Internat. Assn. Theoretical and Applied Limnology (member central committee 1962—), American Society of Naturalists, National Assn. Biology Tchrs., Am. Assn. U. Profs., Izaak Walton League, United World Federalists (chpt. chmn.), Ia. Acad. Sci. (dir. 1959-64, v.p.), Ia. Zool. Soc. (1957-59), Nature Conservancy, Japanese Society of Population Ecology, Sigma Xi (served as chpt. pres. 1963), Phi Kappa Phi, Gamma Sigma Delta. Author: Handbook of Freshwater Fishery Biology, rev. edit., 1953. Editorial bd.: Progressive Fish Culturist, 1951-52, Jour. Wildlife Mgmt., 1952-53, Am. Fishery Soc. Trans., 1956-59, Proc. Ia. Acad. Sci., 1957. Age and growth studies of fish; investigation of relationship of standing crop of fish to ecological factors; evaluation of fish population changes in Clear Lake, Iowa. Home: 2322 Knappa St., Ames, Ia.*

CARLESON, Lennart Axel Edvard, Swedish mathematician; b. Stockholm, Sweden, Mar. 18, 1928; s. Bengt A. F. and Aina (Elfverson) C.; Ph.D., U. Uppsala (Sweden), 1950; m. Inga Britta M. Wall, Oct. 21, 1953; children—Caspar, Beatrice. Prof., U. Stockholm, 1954-55; prof. U. Uppsala, 1955-67, research prof., 1967—. Mem. Swedish Acad. Sci., Am. Acad. Arts and Scis. Editor, Acta Mathematica, 1955—. Research, publs. in real and complex analysis; proved corona conjecture in function algebras, 1962, that Fourier series of square integrable functions converge outside sets of measure zero, 1966. Home: Döbelnsgatan 30D, Uppsala, Sweden.*

CARLETON, George, English bishop; b. Norham Castle, Northumberland, Eng., 1559; studied under Bernard Gilpin, Apostle of North; M.A., Oxford, Eng., 1579; became fellow Merton Coll., 1580; a son, Henry. Vicar of Mayfield, Sussex, Eng., 1589-1605; first prebendary of Llandaff, later apptd. bishop, 1618; sent to Synod of Dort by James I, 1618; apptd. bishop of Chichester, 1619. Recipient medal (for learning and piety at Synod of Dort), Dutch States. Author: Astrologomania, the Madnesse of Astrologers: or, An Examination of Sir Christopher Heydon's Booke, Entituled A Defense of Judiciarie Astrologie (published almost 20 years after it was written), 1624; also works on theology. Believed astrology was invention of devil spread by Zoroaster, that it was inseparable from magic and not part of natural philosophy. Died 1628.

CARLETON, Mark Alfred, Am. plant pathologist; b. Jerusalem, O., Mar. 7, 1866; s. Lewis D. and Lydia Jane (Mann) C.; B.Sc., Kan. Agrl. Coll., 1887, M.Sc., 1893; m. Amanda Elizabeth Faught, Dec. 29, 1897. Cerealist, U. S. Dept. Agr., from 1894; agrl. explorer for U. S. Govt. in Russia and Siberia, 1898-99; cereal expert, then plant pathologist Cuyamel (Honduras) Fruit Co. In charge U. S. grain exhibit, mem. jury of awards Paris Expn., 1900; chmn. group 84, jury of awards St. Louis Expn., 1904; on leave of absence to conduct work of Pa. Chestnut Tree Blight Commn., 1912-13. Decorated order Mérite Agricole, French Govt. Research and publs. on grain rusts; introduced many new wheat varieties into U. S., notably Russian Kubanka durum and red Kharkov winter wheats. Died 1925.

CARLETON, Nathaniel Phillips, Am. physicist; b. Burlington, Vt., Mar. 16, 1929; s. Phillips Dean and Katharine (Pease) C.; B.A. in Physics, Harvard, 1951, M.A., 1952, Ph.D., 1955; m. Kathryn Eleanor Lerch, June 12, 1951; children—Sarah Louise, Nathaniel Phillips, Mary Johanna, Jane Elizabeth, Glen Bennett. Researcher Avco Research Lab., Everett, Mass., 1958, Air Force Research Labs., Bedford, Mass., 1959-62; research asso. Nat. Bur. Standards, Boulder, Colo., 1960-62; researcher Smithsonian Astrophys. Obs., Cambridge, Mass., 1962—. Tchr., researcher Harvard, 1951—. Mem. Am. Phys. Soc., Am. Geophys. Union. Contbr. articles to profl. jours. Research in interpretation phenomena upper atmosphere. Home: West St., Carlisle, Mass. 01741. Office: Pierce Hall, Harvard U., Cambridge, Mass.

CARLEVARO, Enzo, Italian engr.; b. Voghera, Pavia, Italy, Dec. 24, 1894; s. Ettore and Caterina (Troncone) C.; Ph.D. in Engring.; m. Maria Sorrentini, Mar. 19, 1921; children—Mario, Lea, Aldo. Former adminstr. Azienda Tranviaria, Soc. Electronics and Aeronautics; past dean Sch. Engring.; now prof. U. Naples. Mem. Italian Electrotech. Assn. (past v.p. Naples sect.), Acad. Pontanizna, Thermotech. Assn. (past pres.), Internat. Inst. of Cold, Radiosci. Union. Research and publs. on tech. physics, electrotechnology. Home: via Girolamo Santacroce 5, Naples. Office: via Mezzocannone 16, Universitz, Naples, Italy.

CARLISLE, Sir Anthony, English anatomist, surgeon; b. Stillington, Durham, Eng., Feb. 15, 1768; studied medicine at York, Eng., also under Mr. Green at Durham; resident pupil of Mr. Henry Watson, surgeon to Westminster Hosp.; completed edn. at London, Eng. Surgeon to Westminster Hosp., 1793-1840; prof. anatomy and surgery Royal Acad. Art, London, 1808-24; surgeon extraordinary to Prince of Wales (George IV). Fellow Royal Soc., 1804; mem. Coll. Surgeons (mem. council beginning in 1815, examiner 1825-40, became pres. 1829, 39). Research and publs. on anatomy and surgery; discovered (with Nicholson) electroylytic separation of water into hydrogen and oxygen using voltaic pile; introduced an amputating knife for surgery. Died London, Nov. 2, 1840.

CARLITZ, Leonard, Am. mathematician; b. Phila., Dec. 26, 1907; s. Michael and Anna (Schneyer) C.; A.B., U. Pa., 1927, M.A., 1928, Ph.D., 1930; m. Clara Skaler, Sept. 1, 1931; children—Michael, Robert. Asst. prof. math. Duke U., Durham, N.C., 1932-37, asso. prof., 1937-44, prof., 1944—; Nat. Research fellow in math. Cal. Inst. Tech., also Cambridge U., 1930-32; mem. Inst. for Advanced Study, Princeton, N.J., 1935-36. Research and publs. on theory of numbers, power series, combinatorial analysis, spl. functions. Home: 2303 Cranford Rd., Durham, N.C. 27706.*

CARLSEN, Tage Guido, Danish physicist; b. Varde, Jan. 21, 1914; s. Emil and Johanne (Clemmesen) C.; D.Sc., U. Copenhagen, 1958; m. Dora Christensen, July 19, 1944. Placed in charge research Royal Danish Inst. of Pharmacy, 1943, apptd. lectr., 1950; prof. physics Tech. U. Denmark, 1957—. Research on reproducibility of mass determinations; constrn. of modified form of Christiansen-filter; light-sensitivity of chemicals, especially storage of medicaments; high precision comparison of mass- and length-prototypes. Author: On the Spectral Distribution of the Light Sensitivity of Some Chemicals, 1958. Home: Malmmoseve; 3A, 2840 Holte, Denmark. Office: Lundtofte, 2800 Lyngby, Denmark.*

CARLSON, Anton Julius, physiologist; b. Bohuslan, Sweden, Jan. 29, 1875; s. Carl and Hedwig (Anderson) Jacobson; A.B., Augustana Coll., 1889, A.M., 1899; Ph.D., Stanford U., 1903; hon. degrees from 8 univs. and colls.; m. Esther Shegren, Sept. 26, 1905; children—Robert Bernard, Alice Esther, Alvin Julius. Came to U. S., 1891. Research asso. Carnegie Instn., 1903-04; asso., asst. prof., prof. and chmn. dept. of physiology, U. Chgo., 1904-40, later Frank P. Hixon Distinguished Service prof. emeritus. Cons. U. S. Food and Drug Adminstrn., USPHS; lectr. in China under auspices of Rockefeller Found., 1935; with Am. relief expdn. in Europe, 1918-19; mem. Internat. Congresses of Physiology in Vienna, 1909, Groningen, 1913, Edinburgh, 1923, Stockholm, 1927, Boston, 1930, Leningrad and Moscow, 1935, Copenhagen, 1950, Montreal, 1953; mem. med. and research coms. Nat. Found. of Infantile Paralysis. Recipient Gold medal A.M.A., Distinguished Service citation Minn. Med. Assn.; named Humanist of Year, 1953. Fellow A.A.A.S. (past pres.); pres. Nat. Soc. for Med. Research, Research Council on Problems of Alcohol, Chgo. Com. on Alcoholism; pres. Am. Biol. Soc., Am. Physiol. Soc., Fedn. of Am. Socs. for Exptl. Biology, Inst. of Medicine, Am. Assn. Univ. Profs.; mem. Am. Gerontological Soc. (pres.), Nat. Acad. Sci., NRC, A.M.A. Am. Inst. Nutrition, Am. Inst. Chemists, also biological and med. socs. of France, Germany, Sweden, China and Argentina. Author (books): Control of Hunger in Health and Disease; The Machinery of the Body; also some 200 research reports. Research and publs. on heart and circulation, lymph, saliva, immune bodies, metabolism, various glands, gastric secretions and movements. Died Chgo., Sept. 2, 1956.

CARLSON, Clarence Selmer, Am. mathematician; b. Dalesburg, S.D., Jan. 6, 1901; s. Erick August and Bertha (Hovde) C.; A.B., St. Olaf Coll., 1926; M.S., U. Ia., 1928; postgrad. State U. Ia., 1930-31, 32-33; m. Lorraine Eloise Baumann, July 17, 1931; 1 son, Erik Edvard. With Carlson Dept. Store, Beresford, S.D., 1919-22; asst. in instrn. St. Olaf Coll., 1925-26; grad. asst. State U. of Ia., 1926-28; from instr. to prof., St. Olaf Coll., Northfield, Minn., 1928—, chmn. dept. mathematics, 1931-68. Mem. Math. Assn. Am., Am. Math. Soc., Nat. Council Tchrs. Mathematics. Study of geometry; Cremona and birational transformations. Home: 1 Walden Pl., R. 2, Northfield, Minn. 55057.*

CARLSON, Elizabeth, Am. mathematician; b. Mpls. Oct. 2, 1896; s. Carl Emil and Alice (Johnson) C.; B.A., U. Minn., 1917, M.A., 1918, Ph.D., 1924. Instr. math. and physics Knox Coll., Galesburg, Ill., 1919-20; faculty U. Minn., Mpls., 1924—, now prof. math. emeritus, 1965—; prof. math. Macalester Coll., 1965, now ret. Mem. Phi Beta Kappa, Sigma Xi. Study of projective geometry; math. analysis: approximation theory. Home: 3024 14th Av. S., Mpls. 55407.

CARLSON, James Gordon, Am. zoologist; b. Port Allegany, Pa., Jan. 24, 1908; s. James August and Mabel (Johns) C.; A.B., U. Pa., 1930, Ph.D., 1935; m. Elizabeth Shirley, Dec. 24, 1936; children—Shirley Johns, Bette Walker (Mrs. Robert Larry Schrader), James Marvin. Demonstrator biology Bryn Mawr Coll., 1930-31, instr., 1931-35; faculty U. Ala., 1935-46, asso. prof., 1945-46; sr. biologist indsl. hygiene research lab. USPHS, Bethesda, Md., 1946-47; prof. zoology dept. zoology and entomology U. Tenn., Knoxville, 1947—, head dept., 1947-67, dir. Inst. Radiation Biology, 1956—, Alumni Distinguished Service prof., 1962—. Cons. biology div. Oak Ridge Nat. Lab., 1947—. Rockefeller fellow in nat. scis., 1940-41; Pub. Health Service Spl. fellow Heidelberg (Germany) U., 1964-65. Mem. A.A.A.S., Am. Assn. U. Profs., Am. Inst. Biol. Scis., Am. Soc. Cell Biology, Am. Soc. Naturalists, Am. Soc. Zoologists, Assn. Southeastern Biologists, Radiation Research Soc., N.Y., Tenn. acads. sci., Phi Beta Kappa, Sigma Xi, Phi Kappa Phi. Research, publs. on cell morphology and cell division, including effects of chem. agts., ultraviolet radiations and ionizing radiations. Home: 2134 Island Home Blvd., Knoxville, Tenn. 37920.*

CARLSON, Lars A., Swedish physician; b. Stockholm, Sweden, Nov. 14, 1928; s. Fritz D. and Marie-Louise (Ljungberger) C.; M.D., Karolinska Institutet, Stockholm, 1956; m. Kerstin I. Rudin, Mar. 18, 1953; children—Björn Einar Fritz, Mats Lars, Pia Maria. Asso. prof. internal medicine Karolinska Institutet, 1961-62, asso. research prof., 1962—. European editor Jour. Atherosclerosis Research, 1964—. Numerous publs. on devel. specific method for determination of triglycerides in tissues; elevation of blood plasma lipid levels in coronary heart diseases; effect of exercise on blood lipids, lipid moblzn. Home: Författarvägen 27, Bromma, Sweden. Office: King Gustaf V Research Inst., Stockholm 60, Sweden.

CARLSON, Loren Daniel, Am. physiologist; b. Davenport, Ia., May 5, 1915; s. Frank Daniel and Esther (Lind) C.; B.S., St. Ambrose Coll., 1937; Ph.D., U. Ia., 1941; m. Marion Gross, June 7, 1941; children—Eric, Christopher, Allen, Katherine. Faculty, U. Wash., 1945-60, prof. physiology, biophysics, 1955-60; prof., chmn. dept. physiology, biophysics U. Ky., 1960-66; prof., chmn. physiology, chmn. div. scis. basic to medicine U. Cal., Davis, 1966—. Fellow A.A.A.S., Am. Inst. Aeros. and Astronautics (asso.); mem. Am. Phys. Soc., Am. Soc. Zoologists, Soc. for Exptl. Biology and Medicine, Aerospace Med. Assn., Internat. Acad. Astronautics, Sigma Xi. Author: Man in the Cold, 1954; also numerous articles. Established requirements, design and test of oxygen equipment for Air Force and comml. air lines; field, lab. studies of acclimatization to cold environments in man and animals; studies of temperature regulation, aerospace physiology. Home: 22 Meadowbrook Dr., Davis, Cal. 95616.*

CARLSON, Lucile, Am. geographer, educator; b. Minot, N.D., Jan. 23, 1904; d. John August and Anna C. (Clambey) Carlson; B.A., State Coll., Minot, 1931; M.A., U. Wash., 1942, Ph.D., 1947. Mem. faculty U. Wash., 1947-48; faculty Western Res. U., Cleve., 1948—, asso. prof. geography, 1960—, chmn. dept., 1961—. Recipient Whitbeck prize Jour. Geography, 1954. Mem. Assn. Am. Geographers, African Studies Assn., Am. Geog. Soc., Am. Assn. U. Profs., Sigma Xi, Phi Beta Kappa. Author: Geography and World Politics, 1958, Africa's Lands and Nations, 1967; co-author World Geography, 1958. Research, publs. on European Arctic and Africa. Home: 2833 Derbyshire Rd., Cleve. 44118.*

CARLSON, Oscar Norman, Am. metallurgist; b. Mitchell, S.D., Dec. 21, 1920; s. Oscar and Ruth Belle (Gammill) C.; B.A., Yankton Coll., 1943; Ph.D., Ia. State U., 1950; m. Virginia Jyleen Forsberg, July 30, 1946; children—Gregory Norman, Richard Norman, Karen Virginia. Mem. faculty Ia. State U., 1943—, prof., sr. metallurgist Ames Lab., 1960—, chmn. dept. metallurgy, chief, metallurgy div. Ames Lab., 1962-66; spl. research nuclear metals and alloys, phase studies binary alloy systems, brittle-ductile behavior metals and alloys; developed process preparing and purifying yttrium, vanadium, zirconium, calcium, hafnium metals. Mem. Am. Soc. Metals (chmn. Des Moines 1957-58), Am. Chem. Soc., Am. Inst. Metall. Engrs., Am. Soc. Engring. Edn., Ia. Acad. Scis., Sigma Xi, Phi Kappa Phi, Phi Lambda Upsilon. Home: 811 Ridgewood, Ames, Ia. 50010.*

CARLSON, Thomas Arthur, Am. chemist; b. Waterbury, Conn., Apr. 1, 1928; s. Arthur L. and Jenny (Sandstrom) C.; B.S., Trinity Coll., Conn., 1950; M.A., Johns Hopkins, 1951, Ph.D., 1954; m. Effie E. Bradley, Dec. 27, 1950. Sr. staff mem. chemistry div. Oak Ridge Nat. Lab., 1954—. Guggenheim fellow Research Inst. for Physics, Stockholm, 1966-67. Mem. Am. Phys. Soc., Research Engring. Soc. Am. Research in field of atomic and molecular consequences of nuclear decay, basic nature of various processes involving multiple ionization. Home: 114 Dixie Lane. Office: Oak Ridge Nat. Lab., Oak Ridge 37830.*

CARLSON, William Dwight, Am. radiation research adminstr.; b. Denver, Nov. 5, 1928; s. Dwight I. and Irene (Gilkison) C.; D.V.M., Colo. State U., 1952, M.S., 1956; Ph.D. in Radiology (Am. Vet. Med. Assn. fellow), U. Colo., 1958; m. Beverley Ann Bradshaw, June 12, 1950; children—Earl Dwight, Susan Elaine. Practice vet. medicine, Littleton, Colo., 1952-53; faculty Colo. State U., Ft. Collins, 1953-55, 57-68, prof. radiology, 1962-68, chmn. dept. radiology and radiation biology, 1964—, chmn. bd. trustees Research Found., 1964-68, acting dir., 1966-68; v.p. Heath Engring. Labs., 1966-68; president University of Wyoming, Laramie, 1968—. Consultant, mem. adv. coms. various govt. agys.; sr. vet., USPHS res. officer, 1962—. Fellow A.A.A.S.; mem. Am. Vet. Med. Assn., Colo. Vet. Med. Soc., Am. Assn. Acad. Clinicians, Am. Vet. Radiology Soc. (pres. 1965), Educators in Vet. Radiol. Sci. (founder), Am. Bd. Vet. Radiology (organizing mem., v.p. 1966—), Nuclear Medicine Soc. Am. (nat. trustee 1964—), Radiation Research Soc., Colo. Nuclear Medicine Soc.; asso. mem. Colo. Radiol. Soc., Colo., Denver med. socs.; mem. Sigma Xi, Phi Zeta, Phi Kappa Phi. Author: Veterinary Radiology, 1961. Editor: Proc. of Internat. Symposium on the Effects of Ionizing Radiation on the Reproductive System, 1964. Contbr. to numerous profl. textbooks. A pioneer in modern vet. radiology; prin. investigator or dir. numerous radiol. projects. Home: 1306 Ivinson, Laramie, Wyo. 82070.*

CARLSSON, Arvid Per Emil, Swedish pharmacologist; b. Uppsala, Sweden, Jan. 25, 1923; s. Gottfrid O. H. and Lizzie (Steffenburg) C.; M.D., U. Lund (Sweden), 1951; m. Ulla-Lisa Maria Christoffersson, Dec. 29, 1945; children—Bo, Lena, Hans, Maria, Magnus. Faculty, U. Lund, 1951-59, asso. prof., 1956-59; prof. pharmacology U. Göteborg (Sweden), 1959—. Research and numerous articles on role of biogenic amines as neurohumoral transmitters, mode of action of psychotropic drugs. Home: 50 Torlld Wulffsgatan, Göteborg SV, Sweden.*

CARLSTEN, Arne Claes, Swedish clin. physiologist; b. Malmö, Sweden, June 18, 1921; s. Claes Ivar and Elsa (Billsten) C.; M.D., U. Lund, 1950; postgrad. U. Göteborg, 1950-51, Karolinska Institutet, Stockholm, Sweden, 1951-54; m. Gertrud Alsi Hedvig Petren, Dec. 27, 1947; children—Claes, Johan, Anders, Hans, Marie-Louise. Asst. prof. physiology U. Lund (Sweden), 1950-51; faculty U. Göteborg (Sweden), 1951-54, prof. physiology, 1955—; reader clin. physiology Karolinska Inst., Stockholm, 1954-55; head hosp. dept. clin. physiology Sahlgren's Hosp., Göteborg, 1955—, med. dir., 1960—. Mem. Scandinavian Soc. Clin. Chemistry and Clin. Physiology (mem. bd.), Swedish Soc. Clin. Physiology, Swedish Soc. Lab. Drs. Author: The Circulatory Response to Muscular Exercise in Man, 1966; also numerous articles. Research on source and significance of histamine destroying enzyme in blood and lymph, functional changes in ECG, effect of positive pressure ventilation on central circulation; interpretation of auricular arrhythmia in total A-V block; disturbances in gen. hemodynamics at radical neck surgery, pathophysiol. consequences of varicose veins at rest, at different body postures and at exercise, metabolic aspects of human heart muscles in normal persons and with different cardiovascular diseases. Home: 9 Änggardsplatsen, Göteborg, SV, Sweden.*

CARMALT, William Henry, Am. surgeon; b. Friendsville, Pa., Aug. 3, 1836; s. Caleb and Sarah (Price) C.; ed. boarding schs. in Pa., N.J. and Va.; M.D., Coll. Phys. and Surg. (Columbus), 1861; hon. A.M., Yale, 1881; m. Laura Woolsey Johnson, Dec. 8, 1863; children—Ethel, Laurance Johnson, Geraldine Woolsey. In practice, N.Y., 1861-69; studied in Germany, 1869-74; lectr. ophthalmology and otology, Yale, 1876-79, prof. 1879-81, prof. principles and practice of surgery, 1881-1907; cons. surgeon, New Haven Hosp.; pres. Gen. Hosp. Soc. of Conn. Fellow Am. Surg. Assn. (pres. 1907); mem. Am. Ophthalmol. Soc., Am. Otol. Soc. Author: Heredity and Crime, 1909; Epidemiology of Abortion in Cows, 1868; articles. Contbd. to knowledge of skin cancer, ameboid movements of malignant cells, testing for color blindness; improved instruments, including aural forceps, curved hemostat, tongue forceps. Died July 17, 1929.

CARMAN, Eric Hawstone, physicist; b. Melbourne, Australia, Feb. 6, 1922; s. George Hewstone and Beatrice Louise (Mathers) C.; B.Sc., U. Melbourne, 1950, M.Sc., 1952, Ph.D., 1954; m. Mollie Chennell, May 31, 1947; 1 son, Gregory John. Head physics dept. B.S.A. Group Research Centre, Sheffield and Birmingham, Eng., 1954-57; lectr. physics Imperial Coll., London, 1957-60; head physics dept. U. Coll. Townsville, Australia, 1960-64; reader physics, head dept. U. Botswana, Lesotho and Swaziland, S. Africa, 1964—. Under U. S. Govt. contract established joint radio research projects between U. Botswana, Lesotho and Swaziland, Athens Ionospheric Inst., U. Athens, 1966. Mem. Australian Inst. Physics (asso.), Inst. Physics (asso.), Phys. Soc. Eng. Publs. on measurements (with Walter Kannaluik) of heat transfer coefficients of air and rare gases at various temperatures; developed new methods for preparing fine iron powders with permanent magnet properties; developed drilling machine based on spark erosion principles; founder Upper Atmosphere Labs., Townsville for investigation of freak radio transmission at very high frequencies across equator. Address: U. Botswana, Lesotho and Swaziland, P.O. Roma, Lesotho, So. Africa.*

CARMICHAEL, Emmett Bryan, Am. biochemist; b. Shelbyville, Mo., Sept. 4, 1895; s. George Frank and Amelia Grant (Tingle) C.; student Central Meth. Coll., Fayette, Mo., 1914-16; B.A., U. Colo., 1918, M.S., 1922; Ph.D., U. Cin., 1927; m. Lelah Marie Van Hook, Nov. 23, 1921. Instr., U. Colo., 1919-24, U. Cin., 1924-26; bacteriologist William S. Merrell Co., 1926-27; faculty U. Ala., 1927-60, prof. biochemistry, 1932-60, chmn. dept., 1927-60, prof. emeritus, 1966—; asst. dean Med. Coll. Ala. and Sch. Dentistry, Birmingham, 1959-66. Chmn. Gorgas Scholarship Found., 1957—. Cited by Central Coll., 1954, U. Ala., 1966; Phi Beta Man of Year, 1954; recipient So. Chemist award, 1965; Gorgas award Ala. Med. Assn., 1966. Mem. Ala. Acad. Sci. (past pres.), A.A.A.S. (past nat. chmn. acad. conf.), Am. Chem. Soc. (past chmn. Ala. sect.), Soc. Exptl. Biology and Medicine (past vice chmn. So. sect.), So. Med. Assn. (past acting sec. and mgr.), Am. Soc. Biol. Chemists, Am. Physiol. Soc., Am. Assn. Clin. Chemists, Internat. Soc. Toxinology, Am. Assn. Hist. Medicine, Am. Inst. Chemists (pres. 1967—), Sigma Xi (past pres. Ala. chpt.), Alpha Epsilon Delta (past nat. pres., Distinguished Service award 1966), Phi Beta Pi (past nat. pres.). Author: Laboratory Manual of Physiological Chemistry, 1932, 7th-9th editions (with W. W. Carlson), 1949-53, 10th edition (with W. J. Wingo, J. W. Woods), 1957; also numerous articles. Editor Ala. Jour. Med. Scis., 1964—. Research on toxins, natural and therapeutic, pepsin, glycogen, biog. sketches. Home: 3501 Redmont Rd., Birmingham, Ala. 35213.*

CARMICHAEL, Henry, Am. chemist; b. Bklyn., Mar. 5, 1846; s. Daniel and Eliza C.; A.B., Amherst Coll., 1867, A.M., 1870; Ph.D., U. Göttingen, 1871; m. Annie Darling Cole. Prof. chemistry Ia. Coll., 1871, Bowdoin Coll., 1872-86; lecturer Me. Med. Sch., 1872-86; state assayer of Me., 1872-86; lecturer M.I.T., 1899-1901. Moved to Boston, 1886; inventor of processes for manufacture of fibreware, of soda and bleach by electrolysis, and many others; expert in patent causes; assayer, metallurgist and inventor metall. processes. Died 1924.

CARMICHAEL, Hugh, physicist; b. Farr, Scotland, Nov. 10, 1906; s. Dugald and Agnes (Macaulay) C.; B.Sc., Edinburgh (Scotland) U., 1929; Ph.D., Cambridge (Eng.) U., 1936, M.A., 1939; m. Margaret Elizabeth May Maclennan, Oct. 23, 1937; children—Dugald Macaulay, Margaret Lind (Mrs. Wilson Stuart), Elizabeth Agnes (Mrs. G. A. Cooper), Hugh Alexander Lorne. Sr. exptl. officer Ministry of Supply, U.K., 1939-40, prin. sci. officer, mission to Can., 1944-50; prin. research officer Atomic Energy of Can., head gen. physics br., Chalk River, Ont., 1950—; vis. sci. U. S. Nat. Acad. Sci., 1965. Fellow Royal Soc. Can.; mem. Can. Assn. Physicists, Am. Phys. Soc., Am. Geophys. Union. Contbr. articles to profl. publs. Designed and developed quartz fibre electrometers, fused silica microbalances, meteorol. radiosonde apparatus, nuclear reactor control ion chambers, cosmic ray neutron monitors; investigations of cosmic ray ionization bursts, time variation, latitute and altitude variations, studies of solar flares and their effects on cosmic radiation and planetary space. Home: 9 Beach Av., Deep River, Ont. Office: Atomic Energy of Can. Ltd., Chalk River, Ont., Can.*

CARMICHAEL, Hugh Thompson, psychiatrist; b. Peterborough, Ont., Can., Feb. 10, 1898; s. Duncan Nevin and Jessie Eliza (Bolster) C.; M.D., C.M., Queen's U., Kingston, Ont., 1923; M.S., U. Minn., 1931; m. Charlotte Louise Greenwood, June 27, 1923 (dec. 1935); children—William Greenwood, Donald Keith, Charlotte Elizabeth (Mrs. Michael Goble); m. 2d, Gladys Carolyn Ruhland, July 10, 1937; children—Sandra Jennifer, Peter John Hugh. Came to U. S., 1926, naturalized, 1932. Faculty, Albany Med. Coll., 1930-31; psychiatrist Worcester (Mass.) State Hosp., 1931-35; faculty U. Chgo., 1935-43, 60-61; faculty U. Ill., Coll. of Medicine, 1943-61, prof. psychiatry, 1947-66; tng. analyst Inst. for Psychoanalysis, Chgo., 1947-67, supervising analyst, 1949-67. Cons. brs. govt.; mem. com. USPHS. Diplomate Am. Bd. Psychiatry and Neurology (dir. 1957-65, v.p. 1963). Fellow Am. Psychiat. Assn. (life), Inst. Medicine Chgo., Am. Orthopsychiat. Assn., Am. Coll. Psychiatrists; mem. A.M.A. (rep. joint commn. accreditation of hosps. 1959-67, treas. 1963-67), Group for Advancement Psychiatry, Am. Psychoanalytic Assn. (exec. com. 1956), Am. Psychosomatic Soc. (charter), World Fedn. Mental Health, World Med. Assn., Internat. Psychoanalytic Assn., Am. Assn. U. Profs., Canadian Psych. Assn., Sigma Xi, others. Research, publs. on role of endocrines in mental disorders, psychosomatic medicine, psychoanalytic therapy, psychiat. edn.*

CARMICHAEL, Leonard, Am. psychologist; b. Germantown, Philadelphia on November 9, 1898; s. Thomas Harrison and Emily Henrietta (Leonard) C.; grad. Germantown Friends Sch., 1917; B.S., Tufts Coll., 1921, Sc.D., 1937; Ph.D., Harvard, 1924, grad. study U. Berlin, 1924; recipient numerous honorary degrees; m. Pearl Kidston, June 30, 1932; one daughter, Martha (Mrs. S. Parker Oliphant). instr. biology, part time, Tufts Coll., 1923-24; instr. psychology Princeton, 1924-26, asst. prof., 1926-27; asso. prof. psychology Brown U., 1927-28, prof., 1928-36, also dir. psychol. lab., 1927-36, and dir. lab. sensory physiology, 1934-36; chmn. dept. psychology, dean faculty of arts and sci., U. Rochester, 1936-38; pres. Tufts Coll. and dir. lab. sensory psychology and physiology, 1938-52; sec. (the seventh) Smithsonian Institution. 1953-64; v.p. for research and expln. Nat. Geog. Soc.; lectr. Harvard, summers 1927-31; vis. prof. exptl. psychology Clark U., 1931-32; vis. prof. psychology Harvard, 1935; vis. prof. Radcliffe College, 1935, University of Washington, 1940; Fellow American Acad. Arts and Scis., A.A.A.S.; mem. Am. Philos. Soc., Nat. Acad. Sci. (chmn. sect. psychology 1950-53), Nat. Research Council, Soc. Exptl. Psychologists. Soc. Research in Child Development, Nat. Geog. Soc. (mem. bd. trustees), Am. Psychol. Assn. (pres. 1939-40), Soc. Exptl. Biology and Medicine, International Union of Biological Sciences (pres. sect. exptl. psychology and animal behavior 1961), Soc. of the Cin., S.A.R., Am. Legion, Newcomen Soc., Lit. Soc., Phi Beta Kappa, Sigma Xi; hon. mem. Ergonomics Research Soc. Eng., Soc. Francaise de Psychologie. Author: (with H. C. Warren) Elements of Human Psychology, 1930; (with W. F. Dearborn) Reading and Visual Fatigue, 1947. Editor, part-author: Manual of Child Psychology, 2d edition, 1954. Co-editor: The Selection of Military Manpower, 1952; Basic Psychology, 1957. Associate editor Jour. Genetic Psychology, Genetic Psychology Monographs. Brit. Jour. Ednl. Psychology. Editor Houghton Mifflin Co. series of books on psychology. Contbr. psychol. jours. Research on relations of sense organs to behavior; (with H. H. Jasper) 1st in U. S. to record electroencephalograms, 1935; Developed new techniques for electronic recording of eye movements; studied visual fatigue in reading. Home: 4520 Hoban Rd., Washington 7. Office: Nat. Geog. Soc., 17th and M Sts. N.W., Washington.*

CARMICHAEL, Richard, Brit. surgeon; b. Dublin, Ireland, Feb. 6, 1779; s. Hugh Carmichael; apprentice to Peile, surgeon; Asst. surgeon to Wexford Militia, 1795-1802; began practice surgery, Dublin, 1803; became surgeon St. George's Hosp. and Dispensary, 1803, Lock Hosp., 1810; surgeon Richmond, Whitworth, Hardwicke hosps., 1816-36; founder (with Adams, McDowell) Richmond Hosp. Sch. Medicine (later became Carmichael Sch.), 1826. Mem. Irish Coll. Surgeons, Royal Acad. Medicine France (corr.). Author: An Essay on the Effects of Carbonate of Iron upon Cancer, 1806, 2d edit., 1809; An Essay on the Nature of Scrofula, 1810; An Essay on the Venereal Diseases which have been confounded with Syphilis and the Symptoms which arise exclusively from that Poison, 1814; also numerous articles. Improved treatment of venereal diseases, including mercury adminstrn.; advocated separation of pharmacy from medicine and surgery. Died June 8, 1849.

CARMICHAEL, Robert Daniel, Am. mathematician; b. Goodwater, Ala., Mar. 1, 1879; A.B., Lineville, 1898; Ph.D., Princeton, 1911; m., 1901; 4 children. Prof. math. Ala. Presbyn. Coll., 1906-09; asst. prof. Ind. U., 1911-12, asso. prof., 1912-15; faculty U. Ill., 1915—, prof., 1920-47, prof. emeritus, 1947—, head. dept., 1929-34, acting dean grad. sch., 1933-34, dean, 34-47, dean emeritus, 1947—. Mem. exec. com., div. phys. scis. NRC, 1929-32. Mem. Math. Soc. (v.p. 1922, editor Trans. 1931-36), Math. Assn. (pres. 1923), Philos. Assn. Editor-in-chief Am. Math. Monthly, 1918-19. Research and publs. on analytic solutions of linear difference equations, theory of numbers, theory of relativity, theory of groups, philosophy of math. Address: P.O. Box 335, Griggsville, Ill.

CARMODY, Thomas Edward, Am. surgeon; b. Shiawassee County, Mich., May 22, 1875; s. Thomas and Mary Ann (Gorman) C.; D.D.S., U. Mich., 1897, D.D.Sc., 1898; grad. Sch. Medicine, U. Colo., 1903; m. Mary Jane McBride, Nov. 7, 1899; children—David, Ruth P. (Mrs. William G. Summers), Mary Alice (Mrs. Howard D. Cobb). In practice as physician and surgeon 1903-46, specializing in otorhinolaryngology, bronchoesophagology, oral and plastic surgery; prof. bacteriology and histology Dental Coll., U. Denver, 1898-1905, prof. oral surgery and rhinology, 1905-32; asst. in laryngology and otology Med. Sch., U. Colo. 1905-33, chief otolaryngology, child research council, research dept., 1928-36. Surgeon general of Colo., 1909-11. Fellow A.C.S., Am. Coll. Dentists, Internat. Coll. Surgeons; mem. Denver County Med. Soc. (sec., 1904, pres., 1923), Denver Dental Soc. (pres., 1907), Colo. Otolaryngol. Soc. (1st pres.), Colo. Soc. for Crippled Children (1st pres.), Am. Acad. Opthal. and Otolaryn. (pres., 1923), Am. Bronchoesophagological Soc., Am. Laryn., Rhinol. and Otol. Soc. (pres., 1936), Am. Laryn. Assn. (pres., 1941), Am. Otol. Assn., Am. Soc. Oral and Plastic Surgeons, Am. Soc. Plastic and Reconstructive Surgery, A.M.A. (chmn. otolaryn. sec., 1931); mem. 1st Internat. Otolaryn. Congress, Copenhagen, Denmark, 1929; bd. dirs. Nat. Soc. for Crippled Children. Died Aug. 30, 1946.

CARMON, James Lavern, Am. statistician; b. Mount Airy, Ga., May 7, 1926; s. L. Whitt and Annie Mae (Funk) C.; B.S., U. Ga., 1948; M.S., U. Md., 1950; Ph.D., N.C. State Coll., 1954; m. Elizabeth Joan Lee, Feb. 15, 1946; 1 dau., Lee Anne. Faculty, U. Ga., Athens, 1950—, statistician, dir. Computer Center, 1962—. Mem. Assn. Computing Machinery, Am. Statis. Assn., Biometrics Soc., Genetics Assn., Sigma Xi, Phi Kappa Phi. Research, publs. in biostatics, genetics, computer applications to biol. research. Home: 304 Greencrest Dr., Athens, Ga. 30601.*

CARNAHAN, Howard Leon, Am. geneticist; b. Erie, Kan., Mar. 25, 1920; s. Lee and Olive (Houghton) C.; A.A., Parsons (Kan.) Jr. Coll., 1939; B.S., Kan. State U., 1942; M.S., U. Minn., 1947, Ph.D., 1949; m. Shirley J. Wallace, Feb. 8, 1945; children—Lew Wallace, Timothy Leon. Research asst. U. Minn., 1942, 1946-48; asst. prof. agronomy Pa. State U., 1949-52; research agronomist U. S. Regional Pasture Research Lab. U. S. Dept. Agr., State College, Pa., 1953-60, agronomist in charge, 1957-60, research agronomist Alfalfa investigations USDA, Reno, Nev., 1960-65; dir. research Arnold-Thomas Seed Service, Fresno, Cal., 1965—; asso. editor Agronomy Jour., 1964-65; asso. editor Crop Science, 1964-65. Mem. Am. Genetic Assn., Am. Soc. Agronomy, Crop Sci. Soc., Sigma Xi. Contbr. numerous articles in field to sci. jours. Developed improved varieties of red clover, orchard grass and alfalfa; contbd. to knowledge of genetics of white clover and to knowledge of cytogenetics of bromegrass, reed canarygrass and intergeneric Lolium-Festuca hybrids; basic studies in plant breeding methodology including breeding for disease and insect resistance. Home: 6756 N. Lafayette St., Fresno 93705. Office: P.O. Box 2345, Fresno, Cal. 93723.*

CARNALL, Rudolf von, German geologist; b. Glatz, Germany (now Klodzko, Poland), Feb. 9, 1804; s. Arvid Conrad and Mathilde (le Cointe) C.; Dr.phil. h.c., Berlin, Germany, 1855; m. Emilie Büttner; 2 daus. Practical and ofcl. govt. employee 1844-61; re-established mining sch., Tarnowitz, 1839; docent mining engring. Berlin U., 1849-55 supt. of mines and mining office dir., Breslau, Germany (now Wroclaw, Poland); councilor in mines and mining sect., Prussian Ministry of Commerce, 1855-61. Mem. German Geol. Soc. (co-founder 1848). Author: Die Sprünge in Steinkohlengebirge, 1835; Geognostische Karte von Oberschlesien, 1844; Geognostische Karte von den Erzlagerstätten bei Tarnowitz und Beuthen in Oberschlesien, 1844; Bergmännisches Taschenbuch . . . Oberschlesiens, Years 1-4, 1844-47; Die Bergwerke in Preussen und deren Besteuerung, 1850; Die Bergwerksverhältnisse im Preussischen Staate, 1856. Founder ofcl. jour. (in existence until 1945). Discovered chlorkalium-chlormagnesium salts (Carnallit); played important part in opening up salt deposits nr. Stassfurt. Died Breslau, Nov. 17, 1874.

CARNAP, Rudolf, philosopher; b. Wuppertal, Germany, May 18, 1891; s. Johannes S. and Anna (Dörpfeld) C.; student U. Freiburg Baden, Jena; Ph.D., U. Jena, 1921; Sc.D. (hon.), Harvard, 1936; LL.D., U. Cal. at Los Angeles, 1963; H.L.D., U. Mich., 1965; m. Elizabeth Ina von Stöger, 1933 (dec. 1964). Came to U. S., 1935, naturalized, 1941. Instr. philosophy U. Vienna (Austria), 1926-31; prof. natural philosophy German U., Prague, Czechoslovakia, 1931-35; prof. philosophy U. Chgo., 1936-52; prof. philosophy U. Cal. at Los Angeles, 1954-62, research philosopher, 1962—; vis. prof. Harvard, 1940-41. Fellow Am. Acad. Arts and Scis., Brit. Acad. (corr.); mem. Am. Philos. Assn., Assn. Symbolic Logic, Philos. Sci. Assn. Author: Der Raum, 1922; Physikal. Begriffsbildung, 1926; Der Logische Aufbau der Welt, 1928; Scheinprobleme der Philosophie, 1928; Abriss der Logistik, 1929; The Unity of Science, 1934; Logische Syntax der Sprache, 1934; Die Aufgabe der Wissenschaftslogik, 1934; Philosophy and Logical Syntax, 1935; Logical Syntax of Language, English translation, 1937; Foundations of Logic and Mathematics, 1939; Introduction to Semantics, 1942; Formalization of Logic, 1943; Meaning and Necessity, 1947; Logical Foundations of Probability, 1950; The Continuum of Inductive Methods, 1951; Einfuhrung in die symbolische Logik, 1954; Introduction to Symbolic Logic and its Applications, 1958; Induktive Logik und Wahrscheinlichkeit, 1958; also book chpts., articles. Writings collected in The Philosophy of Rudolf Carnap (editor Paul A. Schilpp), 1963. A founder of logical positivism; made important contbns. to logic, semantics and philosophy of sci.; early rejected almost all traditional philosophy, but later modified this view; all propositions were held to be tautological, scientific, or non-sensical; one of most influential of contemporary philosophers. Home: 11728 Dorothy St., Los Angeles 90049.*

CARNE, Joseph, Brit. geologist; b. Truro, Eng., Apr. 17, 1782; s. William Carne; ed. Wesleyan Sch., Keynsham, nr. Bristol, Eng.; m. Mary Thomas, Mar. 23, 1808; became mgr. Hayle Copper Works, 1810, Penzance Bank, 1820. Fellow Royal Soc., 1818; mem. Cambridge Philos. Soc. (hon.), Cornwall Geol. Soc. (treas.). Publs. on Cornish geology, 1816-51, also on mining. Died Oct. 12, 1858.

CARNELL, Paul Herbert, Am. chemist; b. Oakfield, Wis., May 27, 1917; s. Herbert Clyde and Fannie (Carstens) C.; A.B. magna cum laude, Albion Coll., 1939; Ph.D., Western Res. U., 1943; m. Phyllis Wipple, June 21, 1942; children—Nancy (Mrs. Harvey Hoeltzel), Mike, Cheryl, Beth, Cara. Research chemist Phillips Petroleum Co., Bartlesville, Okla., 1943-47; dir. research Leonard Refineries, Alma, Mich., 1947-48; asst. prof. Marietta Coll., 1948-49; faculty Albion (Mich.) Coll., 1949-66, prof., chmn. dept. chemistry, 1952-66; research asso. Yale, 1959-60; vis. scientist Mich. Acad. Sci., Arts and Letters, 1960-66; with U. S. Office Edn., 1966—. Mem. Am. Chem. Soc., N.Y. Acad. Scis., Yale Chemists' Assn., Sigma Xi. Author: Molecular Equilibrium, 1963. Research, publs. in petroleum chemistry; synthetic membranes; lubricating oils; fuel oils; fuel oil stability; treating with hydrofluoric acid; viscosity. Home: 1209 Highland Dr., Silver Spring, Md. 20910. Office: Div. of Coll. Support, U. S. Office of Edn., Washington 20202.*

CARNES, William Henry, Jr., Am. pathologist; b. Ft. Worth, Nov. 2, 1909; s. William Henry and Sarah (Thompson) C.; A.B., Columbia, 1932; M.D., Johns Hopkins, 1936; m. Elizabeth Ann Irwin, June 27, 1950. Asso. in pathology Johns Hopkins, 1938-39; instr. pathology Columbia, 1939-41; faculty Stanford, 1941-47, 51-56, asso. prof. pathology, 1945-47, prof. pathology, 1955-56; faculty Johns Hopkins U., 1947-51, asso. prof. pathology, 1950-51; prof. pathology, head dept. U. Utah, 1956—; asst. pathologist The Presbyn. Hosp., N.Y., 1939-41; pathologist Johns Hopkins Hosp., Balt., 1947-51, Salt Lake County Gen. Hosp., 1956-65; chief cons. pathology VA Hosp., Salt Lake City, 1956—; pathologist-in-chief Univ. Hosp., Salt Lake City. Mem. A.A.A.S., Am. Assn. Pathologists and Bacteriologists, A.M.A., Am. Assn. Cancer Research, Coll. Am. Pathologists, Am. Soc. Exptl. Pathology, Histochem. Soc., Internat. Acad. Pathology, Johns Hopkins Med. and Surg. Assn., Salt Lake County Med. Soc., Utah Soc. Pathologists, Utah State Med. Assn., Sigma Xi. Research and publs. in exptl. pathology, including carcinogenesis and cardiovascular disease. Home: 2226 Hubbard Av., Salt Lake City 84108.*

CARNEY, Frank, Am. geologist; b. Watkins, N.Y., Mar. 15, 1868; s. Hugh and Esther R. (Beahan) C.; A.B., Cornell U., 1895, Ph.D., 1909; m. Mary E. Keegan, June 26, 1890; children—Esther L. (Mrs. H. H. Martin), Ewart Gladstone, Harry Beahan, Mary F. (Mrs. J. W. Cunnick), Frances E. (Mrs. P. A. Knoedler). Instr., Starkey Sem., 1887-90, prin., 1894-95; instr. Keuka Inst., 1895-1900; asst. in geology Cornell U., 1901, instr. Summer Sch. of Geography, 1901-04, Summer Sch., 1914-16; vice prin. Ithaca High Sch., 1901-04; prof. geology Denison U., 1904-14, prof. geology and geography, 1915-17; chief geologist Nat. Refining Co., from 1917, in charge Land Dept., 1923-29; prof. geology and geography Baylor U., from 1929. Lectr. geography U. Va., summer 1909-11; prof. geology U. Chgo., summer 1912; acting prof. geology U. Mich., 1912-13; asst. geologist Ohio Geol. Survey, 1907-17. Research on glacial deposits of N.Y. State; discovered evidence of more than one major advance of ice; believed that life changed progressively during geologic time. Died Dec. 13, 1934.

CARNOCHRAN, John Murray, Am. surgeon; b. Savannah, Ga., July 4, 1817; s. John and Harriet (Putnam) A.; grad. U. Edinburgh (Scotland), 1834; M.D., Coll. Physicians and Surgeons, N.Y.C., 1836; studied medicine, Paris, France, 1836-42, London, Eng., 1842-47; m. Estelle Morris. Practiced medicine, N.Y.C., 1847-87; surgeon-in-chief N.Y. State Emigrant Hosp. Ward's Island, 1851; prof. surgery N.Y. Med. Coll., 1851-62; health officer port of N.Y., 1870-71; mem. N.Y. Medico-Legal Soc., 1871-87. Author: Etiology, Pathology, and Treatment of Congenital Dislocation of the Head of the Femur, 1850; Contributions to Operative Surgery and Surgical Pathology, 1858; Cerebral Localization in Relation to Insanity, 1884; many case reports on surg. procedures. First to excise maxillary nerve for relief of neuralgia of face, 1858; introduced ligation of carotids in treatment elephantiasis, 1867; other work included 5 cases of amputation at hip joint, removal of entire diseased lower jaw, exsection of entire ulna with preservation of arm function. Died Oct. 28, 1887.

CARNOT, Lazare Nicolas Marguerite, French mil. engr., mathematician; b. Nolay, Burgundy, France, May 13, 1753; s. Claude and Marguerite (Pothier) C.; student l'École de Mézières, 1771-73; m. 1791; children—Nicolas Léonhard Sadi, Lazare Hippolyte. Became officer of Corps de Génie, 1773; made capt., 1783; became revolutionary and elected dep. for Pas de Calais; voted for execution of king; stationed at Aire, 1791; commr. mil. matters, 1792-93; returned to Paris and became mem. Com. Pub. Safety; capt., 1793-94; became maj. in engrs., 1795; one of 5 dirs., 1795-96; a founder of the Inst., became mem., 1796, reelected, 1800; pres. Directory, 1796-97; minister of war, 1800; ret., 1801-14; named gen. of div., gov. of Antwerp, Belgium, by Napoleon, 1814; joined Napoleon during Hundred Days; named minister interior and peer of France; proscribed during 2d restoration; later lived in Magdeburg, Germany. Mem. French Acad. Scis. Author: Essai sur les machines en général, 1783; Réflexions sur la métaphysique du calcul infinitésimal, 1797; Oeuvre mathématiques, 1797; De la corrélation des figures de géométrie, 1801; Géométrie de position, 1803; Principes fondamentaux de l'equilibre et du mouvement, 1803; Essai sur les transversales, 1806; De la défense de places fortes, 1810; De la stabilité des corps flottants, 1814. Research on projective geometry, including derivation of classical theorems using negative magnitudes; earliest proof that kinetic energy is lost in collision of bodies; studied principles of active defense, including Carnot's wall. Died Magdeburg, Aug. 2, 1823.

CARNOT, Marie-Adolphe, French chemist; b. Paris, France, Jan. 27, 1839; s. Hippolyte Carnot; ed. Paris polytechnic, École des Mines. Gen. insp. mines; prof. mineralogy and analytic chemistry; dir. École des Mines. Mem. French Acad. Scis., Acad. Agr. Research and publs. on relation of sci. and agr.; improved sanitation of Paris; made geol. and agronomic map of region of Indre. Died June 21, 1920.

CARNOT, Nicolas Léonard Sadi, French physicist; b. Paris, France, June 1, 1796; s. Lazare Nicolas Marguerite Carnot; studied at Lycée Charlemagne, also École polytechnique, Paris, 1812-14. Officer, Engrs. Corps, 1814-19, lt. Staff Corps, 1819-27, capt. in Engrs., 1827, left service, 1828. Mem. Reunion polytechnicienne industrielle, Assn. polytechnique. Author: Réflexions sur la puissance motrice du feu et sur les machines propres à développer cette puissance, 1824. Founder of modern thermodynamics; Carnot's principles were developed into 2d law of thermodynamics; Carnot's cycle showed relationship between heat and mech. energy, also contains germinal ideas of entropy; his work anticipated that of Helmholtz, Joule, Colding, Kelvin, and Mayer. Died Paris, Aug. 24, 1832.

CARNOY, Jean Baptiste, biologist; b. Hainaut, Belgium, Jan. 22, 1836; received doctorate; govt. grantee; ordained; worked in Leipzig, Berlin, Bonn (all Germany), Vienna, Austria, also with Francesco Castracane in Rome, Italy; became vicar at Celles, nr. Tournai, Belgium; tchr. microscopy, founder sch. cytology, Louvain, Belgium; founder jour. la Cellule, 1884. Research and publs. on anatomy, nucleus, cell segmentation, albuminoid membrane. Died Schuls, Switzerland, Sept. 6, 1899.

CARO, Heinrich, German chemist; b. Posen, Germany (now Poznan, Poland), Feb. 13, 1834; s. Simon C. and Amalie (Schnitzler) C.; studied at Royal Trade Inst., also U. Berlin (Germany), 1852-55; Ph.D., Munich, Germany, 1877; postgrad., Heidelberg, Germany, from 1904; hon. degrees from several univs.; m. Edoth Sarah Eaton, 1866; 3 sons, 4 daus. Worked for dye firm, Mülheim; chemist Roberts, Dale & Co. (dye firm), Manchester, Eng., later co-owner; returned to Germany, 1866; an organizer, developer Badische Anilin und Soda-Fabrik, Mannheim, Germany, dir., 1868-89. Hon. mem. VDI (co-founder German engring. assn. 1856, chmn. 1892-93), also mem. an assn. to protect interests of German chemists in industry (founder 1877, chmn. 1898-1900). Discovered several dyes used in textile industry and biol. research including methylene blues, azo-dyes, also peroxymonosulfuric acid (Caro's acid). Died Dresden, Germany, Sept. 11, 1910.

CARO, Nikodem, chemist; b. Lodz, Poland, May 23, 1871; s. Albert and Rosa (Rubinstein) C.; studied chemistry, Berlin, Germany; Dr.Ing. h.c.; Dr.Agr. h.c.; m. Else Friedmann; 1 dau., Vera. Founded chem. lab.; cons. to chem. industry. Mem. Acad. Leningrad. Research and numerous publs. on catalytic oxidation of ammonia, partial condensation of gases, gasifying and gas prodn. of turf; developed cyanamide process for nitrogen fixation. Died June 27, 1935.

CAROCHEZ, Noel-Simon, optician; b. France; became mem. Bur. Longitudes, 1795; worked 20 years to perfect optical instruments; specialist on constrn. of achromatic spectacles and telescopes. Died 1814.

CAROTHERS, Wallace H(ume), Am. chemist; b. Burlington, Ia., Apr. 27, 1896; s. Ira Hume and Mary Elizabeth (McMullen) C.; B.S., Tarkio Coll., 1920; M.S., U. Ill., 1921, Ph.D., 1924; postgrad. U. Chgo.; m. Helen E. Sweetman, Feb. 21, 1936. Instr. chemistry, Tarkio Coll., 1918-20, U. S.D., 1921-22, U. Ill., 1924-26, Harvard, 1926-28; research chemist du Pont Co., Wilmington, Del., 1928-37; co-inventor of important new synthetic rubbers; expert on plastics; inventor nylon, 1935; worked with diamines and dicarboxylic acids; developed neoprene. Editor: Organic Syntheses, 1933; asso. editor Jour. Am. Chem. Soc., 1930-37. Died Apr. 29, 1937.

CAROUGE, Bertrand-Augustin, French astronomer; b. Dol, France, Oct. 8, 1741; s. Bertrand and Marguerite (Blanchard) C.; studied under Lalande, Paris, France; apptd. to adminstrv. position in post office, 1798. Contbr. to Connaissance des temps; Globe céleste (both Lamarche). Sci. calculations for Lalande's l'Astronomie. Made tables of phases of moon, formulas of parallaxes. Died Paris, Mar. 29, 1798.

CARPENTER, Arthur Howe, Am. metallurgist; b. Georgetown, Colo., Oct. 19, 1877; s. Franklin Reuben and Annette Fuller (Howe) C.; student Ohio U., 1894, A.M., 1914; student Northwestern U., 2 yrs.; m. Margaret Lucile Evans, June 5, 1901; children—Franklin Dafydd, Margaret Annette (Mrs. D. M. Dutton), Mary Elizabeth (Mrs. S. L. McCarthy). Assayer, research chemist, 1894-96, asst. supt. Deadwood and Delaware Smelting & Refining Co., Deadwood, S.C., 1898-99; jr. partner firm Carpenter & Caprenter, Denver, 1900; supt. Clear Creek Mining & Reduction Co., Golden, Colo., 1901-03; research work, Calumet, Mich., 1904; gen. mgr. Takilma (Ore.) Mining & Smelting Co., 1905; chief chemist Am. Smelting & Refining Co., Leadville and Denver, 1906-08; prospecting, Nev., 1910-11; research metallurgist, Am. Vanadium Co., 1912-18; gen. mgr. Colo. Vanadium Corp., 1918-20; asst. prof. metallurgy, Armour Inst. Tech. (Ill. Inst. Tech.), Chgo., 1920-28, asso. prof., 1929, head metallurgy div., lectr. geology, astronomy, meteorology; emeritus prof., 1944-56. Staff cons. metallurgy Armour Research Found.; cons. practice. Fellow A.A.A.S.; mem. Am. Inst. Mining and Metall. Engrs., Am. Soc. Testing Materials, Soc. Promotion Engring. Edn., Astron. Soc. of Pacific, Am. Assn. Variable Star Observers, Phi Lambda Upsilon, Alpha Chi Sigma, Pi Gamma Mu, Delta Tau Delta. Contbr. mining and metall. publs. Inventor methods of covering pipe with lead. Made 20½ telescope used by Elgin Obs. for daily Arcturus ceremony at Century of Progress Expn., Chgo., 1933, 34. Died Mar. 20, 1956.

CARPENTER, Arthur Whiting, Am. chem. engr.; b. Wellsville, N.Y., Mar. 30, 1890; s. Samuel and Clara (Whiting) C.; S.B., Mass. Inst. Tech., 1913, M.S. in Chem. Engring.; m. Irma Coon, May 8, 1948. City chemist, Alliance, O., 1914; chemist water purification plant, Akron, O., 1915-17; tech. service engr. Goodyear Tire & Rubber Co., Akron, 1919-21, devel. engr., 1923-26; supt. Holtite Mfg. Co., Balt., 1922; devel. engr. B. F. Goodrich Co., Akron, 1927, mgr. testing labs., 1928-55; cons. chem. engr., rubber technologist, 1955—. Loaned by B. F. Goodrich Co. to Nat. Resources Security Bd., Sept 1948; asst. dir. raw materials OPM, to 1949; prin. indsl. specialist and cons. rubber conservation div. WPB, 1941-42. Served from 1st lt. to capt., San. Corps, U. S. Army, 1918-19. Recipient Charles Goodyear medal Am. Chem. Soc., 1957. Fellow Am. Inst. Chemists; mem. Am. Soc. Testing Materials (pres. 1946-47; hon. mem 1955—), Am. Inst. Chem. Engring., Nat. Soc. Profl. Engrs., Am. Chem. Soc. Contbr. chpt. on Physical Testing and Specifications in Am. Chem. Soc. Monograph No. 74, The Chemistry and Technology of Rubber, 1937. Obtained authors and edited books pub. by Am. Soc. for Testing Materials; Symposium on Rubber, 1932; Symposium on the Applications of Synthetic Rubbers, 1944. Contbr. numerous tech. papers on testing and instruments to sci. and trade jours. Contbr. to devel. standard specifications and methods of test for rubber products. Home: 943 Genesee Rd., Akron, O. 44303.*

CARPENTER, Charles Congden, Am. zoologist; b. Denison, Ia., June 2, 1921; s. Harry Alonzo and Myrtle Ruth (Barber) C.; A.B., N. State Tchrs. Coll., Marquette, Mich., 1943; postgrad. Wayne State U., 1945; M.S., U. Mich., 1947, Ph.D., 1951; m. Mary Francis Pitynski, Sept. 2, 1947; children—Janet Eleanor, Caryn Sue, Geoffrey Congden. Teaching fellow U. Mich., 1947-51, instr., 1951-52; asst. prof. Wayne State U., 1952; faculty U. Okla., also U. Okla. Biol. Sta., Norman, 1952—, prof., 1966—, curator reptiles Stovall Mus. Sci. and History, 1956—; scientist Galapagos Internat. Sci. Project, 1964. N.Y. Zool. Soc. Research grantee, Jackson Hole (Wyo.) Research Sta., 1951; NSF Research grantee, 1956—, Mexico, 1962-66, Galapagos Islands, 1962-64. Fellow Okla. Acad. Sci., Herpetologists League; mem. Animal Behavior Soc. (sec. 1966—), Grassland Research Found. (past bd. govs., past sec.-treas.), S.W. Assn. Naturalists (gov. 1964—), Ecol. Soc. Am., Brit. Ecol. Soc., Am. Ornithol. Union, Wilson Ornithol. Soc., Am. Soc. Ichthyologists and Herpetologists, Herpetologist League, Am. Soc. Mammalogists, Am. Soc. Zoologists, Am. Inst. Biol. Scis., Okla. Acad. Sci., Wilderness Soc., Nature Conservancy. Contbr. articles to tech. jours. Research on specific aggressive display patterns of lizards, behavior taxonomy reptiles, ecol. niche determinations of reptiles, ethograms of small mammals. Home: 1218 Cruce St., Norman, Okla. 73069.*

CARPENTER, Clarence Ray, Am. psychologist, anthropologist; b. nr. Cherryville, N.C., Nov. 28, 1905; s. Clarence E. and Gaddie Lee (Harrelson) C.; A.B., Duke, 1928, M.A., 1929; Ph.D., Stanford, 1932; postgrad. Yale, 1931-34; m. Mariana Evans, July 16, 1932 (dec.); children—Richard Lee, Lane Evans; m. 2d, Ruth E. Jones Chamblee, Oct. 8, 1966. Asst. prof., Bard Coll., 1934-38, Coll. Phys. and Surg., 1938-40; faculty Pa. State U., 1940—, prof. psychology, 1948-64, research prof. psychology and anthropology, 1965—, head dept. psychology, 1952-58, dir. div. acad. research and services, 1957-62, dir. div. instructional services, 1961-63, asst. to pres. Milton S. Hershey Med. Center, 1963-64; Ford Found. Distinguished vis. prof. behavioral scis. U. N.C., Chapel Hill., 1964-65. Cons. to pvt. cos., ednl. insts. Cited for distinguished contbn. to edn. in Pa., Dept. Pub. Instrn., 1961. Fellow N.Y. Zool. Soc.; mem. N.E.A. (pres. American Association for Higher Education dept. 1965-66), Nat. Acad. Sci. (mem. primate com. 1958—), Internat. Primatological Soc. (sec. Western hemisphere), Am. Psychol. Assn., A.A.A.S., Assn. Higher Edn., Am. Assn. Anthropologists, Am. Sci. Film Assn., Am. Soc. Animal Behavior, Soc. for Religion in Higher Edn. Author: Naturalistic Behavior of Non-human Primates, 1965; also numerous articles. Research on behavioral effects of hormones on birds, studies in non-human primates, analysis, evaluation of films and ednl. TV. Home: 258 Twigs Lane, State College, Pa. 16801. Office: 214 Burrowes Bldg., University Park, Pa. 16802.*

CARPENTER, Esther, Am. zoologist; b. Meriden, Conn., June 4, 1903; d. Ernest Charles and Nettie Jane (Hale) Carpenter; B.A., Ohio Wesleyan U., 1925, D.Sc. (hon.), 1956; M.S., U. Wis., 1927; fellow biology, Bryn Mawr Coll., 1927-28; Ph.D., Yale, 1932. Research asst. embryology Carnegie Inst. Tech., 1932-33; part-time instr. Albertus Magnus Coll., 1933-34; mem. faculty Smith Coll., 1933—, prof. zoology, 1953—; chmn. dept., 1955-60, Myra M. Sampson prof. zoology, 1963—. Research, Strangeways Research Lab., Cambridge, Eng., 1953-54, 61. Howald scholar Ohio State U., 1942-43. Mem. Tissue Culture Assn., Growth Soc., Am. Soc. Cell Biology, Am. Soc. Zoologists, Am. Assn. Anatomists, Sigma Xi. Contbr. articles thyroid, pituitary, vitamin A. Research in differentiation and beginning of function in embryonic thyroids of chick and rat; effect of excess vitamin A in diet of young rats on thyroid gland and on population of thyroid-stimulating cells in the pituitary glands. Home: 55 Prospect St., Northampton, Mass. 01060.*

CARPENTER, Eugene R., Am. brain surgeon; b. Knobnoster, Mo., Oct. 5, 1873; s. William D. and Emma (Shanks) C.; student U. Mich., 1894-97; M.D., Jefferson Med. Coll., 1898 (winner de Schweinitz medal on ophthalmology, Dercums neurol. prize, and otol. prize); m. Lucile Snyder, July 20, 1916. Intern, Kings Co. Hosp., 1898-99; postgrad. Manhattan Eye and Ear Hosp., N.Y. Eye Infirmary, also Vienna and London, 1907-08; practiced in Dallas, 1921-34. Fellow A.C.S. Designer of numerous surg. instruments. Died Oct. 11, 1934.

CARPENTER, Ford Ashman, Am. meteorologist, aeronaut; b. Chgo., Mar. 25, 1868; s. Lebbaeus Ross and Charlotte (Eaton) C.; ed. Dilworth Acad.; Carson Astron. Obs.; U. S. Balloon and Airship Schs., etc.; LL.D., Whittier Coll., 1913; Sc.D., Occidental Coll., 1921. With U. S. Weather Service various stas., 1888-1919; spl. observer, 1940-41; mgr. dept. meteorology and aeronautics, Los Angeles C. of C., 1919-41. Hon. lectr. U. Cal., summers 1914-16, 1939-41; lectr. U. S. Army Aviation School, San Diego, 1915, Monterey Mil. Encampment, 1916-17; faculty (lectr. meteorology) So. br. U. Cal., 1919-30; lectr. meteorology AS, War Dept., 1915-44, Babson Inst., 1921-35, Columbia, Cornell and Northwestern, 1923-38. N.Y. U., U. S. Mil. Acad., U. S. Naval Acad., Poly. Inst. Bklyn., Carnegie Inst. Pitts., Field Mus., 1925-38, Goodyear-Zeppelin Co., 1926-29, War Coll. (Washington), etc., Meteorological adviser Palos Verdes Estates, 1914-20, Pauba Rancho, 1921-31, TWA, 1927-30, Santa Fe Ry. Co., 1922-35, Am. Airways, TWA, United Airlines, 1927-38, Los Angeles Municipal Airport, 1927, Hollywood Bowl, 1928, Amer.-Hawaiian S.S. Co., 1934-40. Climatol. adviser to Frank A. Vanderlip, 1927-37. Selected and surveyed L.A. Municipal Airport, 1927. Meteorol. observer of aerial bombing of former German battleships, 1921. Internat. balloon pilot Fédération Aeronautique Internationale since 1921. Meteorological and aero. adviser to naval affairs com. of 72d Congress, 1930; mem. 8th Internat. Geog. Congress, Washington, 1904, Internat. Congress Tb, Washington, 1908; mem. photog. com. standards, U. S. Dept. Agr., 1908. Climatol. commr. Seattle Expn. (gold medal for meteorol. exhibit), 1909; first photographed red snow in natural colors, 1911; asst. in U. S. Weather Bur. meteorograph ascents into stratosphere, alt. 108,000 ft., 1913; mem. Pan Am. Med. Congress, 1915, 1st Internat. Aero Congress (v.p.), Omaha, 1921. Past fellow A.A.A.S., Royal Meteorol. and Geog. Socs. (London), Am. Seismol. Soc., Am. Assn. U. Profs., fellow San Diego Soc. Nat. History, So. Cal. Acad. Sci. (pres. 1929-31, v.p. 1932-39), Am. Climatol. and Clin. Assn., Nat. Aero. Assn., Assn. Mil. Engrs., Sigma Xi, Phi Beta Kappa. Author of monographs, pamphlets, articles, etc., including Climate and Weather of San Diego; Influence of the College Spirit; Aviator and Weather Bureau; Meteorological Methods; Aerial Pathways; Roadbeds of the Air; Weather and Flight; Aids to Air Pilots; Climatic Comparisons; Old Probabilities mate; Commercial Climatology; Gen. "Billy" Mitchell As I Knew Him; Sailing Around America's Shores of Two Oceans; Climatology of a Block of Ice, 1945. Contbr. Atlantic Monthly, Sci. Am., Nation's Bus., etc. Editor, Meteorology and Aeronautics, 1919-41. Inventor of anemometric scale, hythergraph, televentscope and ventograph. Died Nov. 1947.

CARPENTER, Frank Gilbert, Am. chemist; b. Washington, Mar. 26, 1920; s. Charles Gilbert and Clara May (Barlow) C.; B.S., U. Md., 1942; M. Chem. Engring., U. Del., 1946, Ph.D., 1949; m. Angela Briefs, Aug. 11, 1945; children—Louise Maria, Christel Mary Anna, Bernard J., Therese Margot, Pauline Clara, Joel Francis. Research asso. on bone char project Nat. Bur. Standards, Washington, 1948-63; dir. cane sugar refining research project U. S. Dept. Agr., So. Regional Lab., New Orleans, 1963—. Mem. Am. Chem. Soc. Research and numerous publs. on phys. chemistry, processing problems in sugar refining industry especially bone char decolorizing process. Home: 29 Crane St., New Orleans 70124. Office: 1100 Robert E. Lee Blvd., New Orleans 70119.*

CARPENTER, Frederick Hiltman, Am. biochemist; b. Cortez, Colo., June 8, 1918; s. Nathaniel Elliott and Clara (Kelley) C.; A.B., Stanford, 1940, M.A., 1941, Ph.D., 1944; m. Elizabeth Louise Card, Jan. 24, 1943; children—Carol Elizabeth, Nathaniel Elliott, Sarah Ellen, Arthur Niles. Faculty, U. Cal. at Berkeley, 1949—, prof. biochemistry, 1962—. Guggenheim fellow U. Paris, Orsay, 1964-65. Mem. Am. Soc. Biol. Chemists, Am. Chem. Soc., Harvey Soc., A.A.A.S., Sigma Xi. Contbr. numerous articles to profl. jours. Research, publs. on action of ribonuclease on RNA, action of mustard gases, synthesis of penicillin, peptide synthesis, protein chemistry, relationship of structure to biol. activity of insulin. Home: 1958 Thousand Oaks Blvd., Berkeley, Cal. 94707.*

CARPENTER, Geoffrey Douglas Hale, Brit. physician; b. Eton College, Oct. 26, 1882; s. P. Herbert and Emma (Hale) C.; ed. St. Catherine's, Oxford, St. George's Hosp.; certificate London Sch. Tropical Medicine, 1910; D.M., Oxford U.; m. Amy Frances Thomas, 1919. Entered Colonial Med. Service, 1910, ret., 1930; apptd. by Royal Soc. to Spl. Sleeping Sickness Commn., Uganda, 1910-14; active service E. African campaign, 1914-18; specialist officer for sleeping sickness control in Uganda, 1920-30; spl. investigator into tse-tse fly in Ngamiland for Sec. of State, 1930-31; Hoep prof. zoology (entomology) Oxford U., 1938-48. Licentiate Royal Coll. Physicians. Mem. Royal Coll. Surgeons; pres. S. Eastern Union Sci. Socs., 1936-37, S. Royal Entomol. Soc. London, 1945-46; v.p. Linnean Soc. London, 1935-36. Author: A Naturalist on Lake Victoria, 1920; A Naturalist in East Africa, 1925; Mimicry, 1933; reports on tse-tse fly in Reports of the Sleeping Sickness Commission of the Royal Society, from 1913; also tech. contbns. to publs. of sci. socs. Died Jan. 30, 1953.

CARPENTER, Sir (Henry Cort) Harold, Brit. metallurgist; b. Bristol, Eng., Feb. 6, 1875; s. William Lant and Annie Grace (Viret) C.; M.A., Oxford (Eng.) U.; Ph.D., Leipzig (Germany) U.; D.Sc. (hon.), U. Wales; D. Metallurgy (hon.), U. Sheffield (Eng.); also ed. U. Manchester (Eng.); m. Ethel Mary Lomas, 1905. Research fellow (under Perkins), demonstrator Owens Coll., Manchester, 1898-1901; head chem., metall. depts. Nat. Phys. Lab.; prof. metallurgy Victoria U., Manchester, 1906-13, Royal Sch. Mines, Imperial Coll. Sci. and Tech., S. Kensington, Eng., from 1913. Mem. Adv. Council Dept. Sci. and Indsl. Research, also chmn. Metallurgy Research Bd.; James Forrest lectr. Instn. C.E.'s, 1927. Recipient Thomas Turner gold medal, 1929, Bessemer gold medal, 1931; named Carl Lueg laureate Verein Deutscher Eisenhüttenleute, 1937, Honda laureate Nippon Kinyoku Gakkai, 1940. Fellow Royal Soc., 1918; asso. Royal Sch. Mines; mem. Swedish Royal Acad. Sci. and Indsl. Research (corr.), Am. Inst. Mining and Metall. Engrs. (hon.), Société Nationale pour l'encouragement de l'Industrie (hon.), Inst. Metals (pres. 1918-20, platinum medal 1939), Iron and Steel Inst. (Carnegie gold medal 1905, pres. 1935-37), Instn. Mining and Metallurgy (gold medal 1932, pres. 1934). Author: (with Robertson) Metals, 2 vols., 1939. Studies, publs. on alloys of iron and carbon at high temperatures, tool steels, copper, aluminum; worked on recrystallization of metals, described changes in carbon steel structure during temperature shifts in critical range. Died Swansea, Wales, Sept. 13, 1940.

CARPENTER, Horace Francis, Am. mineralogist, conchologist; b. Pawtucket, R.I., Oct. 19, 1842; s. Horace and Charlotte C.; spl. course in analyt. chemistry Brown U., 1860-61; m. Jennie Hastings, Feb. 23, 1895. Began in gold and silver refining business, 1860; ret. from firm of H. F. Carpenter & Son, 1912; discoverer of process of extracting gold and silver from photog. waste; also discovered process of obtaining chemically pure gold for comml. purposes. Treas. N.E. Mfg. Jewelers and Silversmiths' Assn. 18 yrs. Regarded as leading authority in R.I. on mollusks; presented City of Providence his library on natural history (237 vols.) and his collection of 1,200 species and varieties of minerals and 4,000 species of shells, consisting of 75,000 specimens; discovered 3 new shell-bearing mollusks. Died Feb. 28, 1937.

CARPENTER, John Melvin, Am. zoologist; b. Terre Haute, Ind., May 21, 1910; s. John Roscoe and Maude (Batchelder) C.; B.A., U. Tex., 1936, M.A., 1940, Ph.D., 1946; m. Mary Wilkes Meadors, July 2, 1946; children—Mary M., Gene B., John H. Research asso. Clayton Found. Research, 1938-45; faculty U. Tex., 1942-46; asst. prof. U. Tenn., 1946-53; prof. head zoology dept. U. Ky., 1953-63, chmn., 1963-65, chmn. biol. sci. div., 1954-55, dir. NSF undergrad. research program in zoology, 1959-61. Fellow A.A.-A.S.; mem. Am. Inst. Biol. Sci., Am. Soc. Zoologists, Ecol. Soc. Am., Soc. Study Evolution, Am. Assn. U. Profs., Assn. Higher Edn., Am. Genetics Assn., Sigma Xi. Author: (with R. W. Barbour, J. M. Edney) Laboratory Studies in Principles of Zoology, 1955, rev., 1958; (with A. W. Jones) Microtechnique, 1952, rev. 1957; also articles. Research on nutrition of Drosophila, reproductive potential of various insects, effects of radioisotopes on insects, seasonal population cycles in insects, mutation in Drosophila melanogaster. Home: 208 Tahoma Rd., Lexington, Ky. 40503.*

CARPENTER, Louis George, Am. cons. engr.; b. Orion, Mich., Mar. 28, 1861; s. Charles K. and Jennette (Coryell) C.; B.S., Mich. Agrl. Coll., 1879, M.S., 1883, D.Eng., 1927; U. Mich., winters, 1881-82, 1883-84; Johns Hopkins, 1885-86, 1887-88; m. Mary J. C. Merrell, Feb. 17, 1887 (dec. 1921); children—Charles L., Jeannette (Mrs. Roe Emery); m. 2d, Katherine M. Warren, Sept. 30, 1922. Asst. and asst. prof. math. Mich. Agrl. Coll., 1881-88; prof. engring. and physics, Colo. Agrl. Coll., 1888-1911. Irrigation engr., Colo. Expt. Station, later dir. of sta., 1899-1910; irrigation expert U. S. Dept. Agr.; organized 1st systematic instrn. in irrigation engring. and investigation in that line, 1888; spl. agt. and field geologist U. S. artesian wells investigation, 1890; expert in irrigation litigation, U. S. vs. Rio Grande dam, Elephant Butte Internat. Case; state engr., Colo., 1903-05; cons. engr. and irrigation expert for state in suit of state of Kan. against Colo., of Wyo. against Colo., etc.; cons. engr. many important dams, irrigation and hydraulic enterprises, etc.; mem. bd. arbitration selected by both parties as referee and chmn. to settle electric lighting controversy at Colorado Springs, 1907; mem. Irrigation Commn. of B.C., 1907-08, to determine foundations for new water code (adopted by B.C. Parliament, 1908); arbitrator chosen by both sides in dispute over waters of North Platte, U. S. vs. Wyo., 1920; expert for Pueblo, Canon City and Salida in stopping pollution of Arkansas River by mining debris; cons. engr. preparing case for Colo. in Wyo. vs. Colo., etc. Chevalier du Mérite Agricole, France, 1895; gold medals, Paris and Portland expns. Died Denver, Colo., Sept. 12, 1935.

CARPENTER, Malcolm Breckenridge, Am. neuroanatomist; b. Montrose, Colo., July 7, 1921; s. Grover B. and Haidee (Moritz) C.; B.A., Columbia, 1943; M.D., L.I. Coll. Medicine, 1947; m. Carolyn Ivins Sloan, July 19, 1949; children—Duncan B., Gregory S., Rustin I. Faculty, Columbia, N.Y.C., 1953—, prof. anatomy, 1962—. Mem., chmn. neurol. sci. tng. com. Nat. Inst. Neurol. Diseases and Blindness, 1962-66; mem. adv. bd. Parkinson Disease Found., 1962—. Markle scholar Columbia, 1953-58. Mem. Am. Assn. Anatomy, N.Y. Acad. Medicine, Assn. Research in Nervous and Mental Diseases, Harvey Soc., Am. Neurol. Assn., Internat. Brain Research Orgn., Cushing Soc., Sigma Xi. Author: (with Truex) Human Neuroanatomy, 1964. Editorial bd. Neurology, 1963—. Neuroanat. research related to basal ganglia, cerebellum, vestibular system. Home: 185 Delhi Rd., Scarsdale, N.Y. 10583. Office: 630 W. 168th St., N.Y.C. 10032.*

CARPENTER, Nathanael, Brit. mathematician, philosopher; b. Devonshire, Eng., 1589; s. John Carpenter; student Oxford (Eng.) U., 1605-07; elected Devonshire fellow Exeter Coll., 1607; B.A., 1610; M.A., 1613; B.D., 1620; D.D., 1626. Tchr., Chelsea, Eng.; schoolmaster King's Wards, Dublin, Ireland. Author (under name of N. C. Cosmopolitanus): Philosophia Libera triplici exercitationum decade proposita, 1621; two books on geography, 1625; also sermons. Studied math., geography, philosophy; attacked Aristotelian system. Died 1635.

CARPENTER, Rolla Clinton, Am. engr.; b. Orion, Mich., June 26, 1852; s. Charles K. and Jennette (Coryell) C.; B.S., Mich. Agrl. Coll., 1873, M.S.; C.E., U. of Mich., 1875; M.M.E., Cornell, 1888; LL.D., Mich. Agrl. Coll., 1906; m. Marion Dewey, 1876. Instr. and prof. math. and civil engring. Mich. Agrl. Coll., 1875-90; asso. prof. engring., 1890-95, prof. exptl. engring., 1895-1917, Cornell U. (emeritus). Cons. engr. for Helderburg, Cayuga Lake, Quaker Portland, Gt. No., Belleville Portland, Cal. and Atlas Portland cement cos. Constructed numerous power stas. for elec. rys.; patent expert in several important cases; engr. for high pressure fire system, City of Balt., 1911; N.Y.C. for high pressure pumping engines, 1911-12; engr. for Kopper's Co., Bklyn. pumping engines, 1914; lighting and heating of city bldgs. 1913-16. Judge machinery and transp. Chgo. Expn., 1893, Buffalo Expn., 1901, Jamestown Expn., 1907. Author: Experimental Engineering, 1890, 1902; Heating and Ventilating, 1898, 1910; (with Prof. Diedrichs) The Gas Engine; Heating and Ventilation (New Internat. Ency. and Kidder's Archtl. Pocket Book). Died Jan. 19, 1919.

CARPENTER, Thorne Martin, Am. physiol. chemist; b. Malden, N.Y., Sept. 18, 1878; s. John Reed and Sarah Twilla (Fisher) C.; B.S., Mass. Agrl. Coll., 1902; B.S., Boston U., 1902; Ph.D., Harvard, 1915; Sc.D., Mass. State Coll., 1946; m. Kathrine Evelyn Murphy, Aug. 30, 1920. Research chemist Nutrition Lab. Carnegie Instn., Boston, 1907-37, acting dir., 1937-42, dir., 1943-45; ret., 1946; research asso. Sch. Pub. Health Harvard, 1946-48; Am. rep. Com. Standardization of Certain Methods Used in Making Dietary Studies, Health Orgn. League Nations, Rome, 1932; cons. mem. Evans Meml. Service Mass. Meml. Hosp., 1934-35. Recipient Army and Navy certificate of appreciation for work in Office Research and Sci. Devel. in World War II, 1947. Fellow Am. Acad. Arts and Scis., A.A.A.S., Am. Inst. Nutrition (past pres.); mem. Am. Chem. Soc., Sigma Xi, Phi Kappa Phi. Author: (with Francis G. Benedict) Respiration Calorimeters for Studying the Respiratory Exchange and Energy Transformations of Man, 1910, The Metabolism and Energy Transformations of Healthy Man During Rest, 1910, Food Ingestion and Energy Transformations With Special Reference to the Stimulating Effect of Nutrients, 1918; A Comparison of Methods For Determining the Respiratory Exchange of Man, 1915; Tables, Factors, and Formulas for Computing Respiratory Exchange and Biological Transformations of Energy, 1924; Human Metabolism With Enemata of Alcohol, Dextrose and Levulose, 1925. Mem. editorial bd. Jour. Nutrition, 1930-36, Quarterly Jour. Studies on Alcohol, 1940-46. Contbr. numerous articles to profl. jours. Research, publs. on exchange of carbon dioxide and oxygen; devel. of apparatus for analysis of atmospheric and room air, metabolism of alcohol, effect of sugars on biol. gas exchange. Home: 959 Washington St., Norwood, Mass. 02062.*

CARPENTER, William Benjamin, English naturalist, physiologist; b. Exeter, Eng., Oct. 29, 1813; s. Lant Carpenter; apprentice to a physician; studied medicine, U. Coll. London, 1833-35, Edinburgh, Scotland, 1835-39, M.D., 1839; LL.D., Edinburgh, 1871; 5 sons including William Lant, Philip Herbert. Lectr., Bristol (Eng.) Med. Sch.; became Fullerian prof. physiology Royal Inst., London, 1844; prof. forensic medicine U. Coll., London; registrar U. London, 1856-79; leader (with Wyville Thomson) deep sea expdn., Lightning and Challenger, north of Ireland, 1868-70. Recipient Lyell medal Geol. Soc., 1883. Fellow Royal Soc., 1844; mem. French Acad. Scis. (corr.), Brit. Assn. (pres. 1872), Inst. France. Author: The Principles of General and Comparative Physiology, 1839; Popular Cyclopaedia of Science, 1843; Principles of Human Physiology, 1846; the Microscope and Its Revelations, 1868; Nature and Man, 1877. Editor, Brit. and Fgn. Medico-Chirurg. Rev., 1847-52. Research in zoology, botany, microscopy, comparative and mental physiology, 1843-71; advocated theory of vertical circulation of ocean currents. Died London, Nov. 19, 1885.

CARPENTIER, Jules Adrien Marie Louise, French engr.; b. Paris, France, Aug. 30, 1851; ed. Paris Polytechnic Sch., 1871; mem. French Acad. Scis., Bur. Longitudes. Research in electric and telegraphic apparatus; inventor: melograph, melotrope, submarine periscopes, automatic sight anti-aircraft guns; perfected electric clock, automatic photog. equipment. Died June 30, 1921.

CARPINO, Louis A., Am. chemist; b. Des Moines, Dec. 13, 1927; s. Pete and Angela (Ortale) C.; B.S., Ia. State Coll., 1950; M.S., U. Ill., 1951, Ph.D., 1953; m. Barbara Pepe, Aug. 30, 1958; children—Philip, Alexandra, Nicholas, Christine. Faculty, U. Mass., Amherst, 1954—, prof. chemistry, 1967—. Mem. Am. German chem. socs., Chem. Soc. London. Research, publs. on devel. new synthetic techniques in field of organic chemistry. Home: 11 Mount Pleasant St., Amherst, Mass. 01002.*

CARPUE, Joseph Constantine, Brit. surgeon, anatomist; b. London, Eng., May 4, 1764; ed. Jesuit Coll., Dovay, also St. George's Hosp.; staff surgeon Duke of York's Hosp., Chelsea, Eng., for 12 years; surgeon Nat. Vaccine Instn. until 1846. Taught anatomy privately, 1800-32; cons. surgeon St. Pancras Infirmary. Fellow Royal Soc., 1817. Author: Description of the Muscles of the Human Body, 1801; Account of two Successful Operations for Restoring a Lost Nose . . . , 1816; History of the High Operation for the Stone . . . , 1819; An Introduction to Electricity and Galvanism with Cases Showing Their Effects in the Cure of Disease, 1803. Intorduced the Indian method of rhinoplasty; emphasized suprapubic operation for stone. Died Jan. 30, 1846.

CARR, Charles Jelleff, Am. pharmacologist; b. Balt., Mar. 27, 1910; s. Joshua Barney and Pearl (Jelleff) C.; B.S., U. Md., 1933, M.S., 1934, Ph.D., 1937; D.Sc., Purdue U.. 1964; m. Mary Agnes McGrath, June 14, 1932; children—Daniel J., Noel Edward, Joseph Barney. Faculty U. Md., 1933-55, prof. pharmacology, 1950-55; prof., head dept. pharmacology Purdue U., 1955-57; chief pharmacology unit Psychopharmacology Service Center, Nat. Inst. Mental Health, Bethesda, Md., 1957-63; chief sci. analysis br. Life Sci. div. Army Res. Office, Arlington, Va., 1963-67; dir. life scis. research office Fedn. American Societes for Exptl. Biology, Bethesda, 1967—. Member Am. Chem. Soc., Soc. Pharmacology and Exptl. Therapeutics, N.Y. Acad. Scis. Author: Pharmacologic Principles of Medical Practice, 1965. Contbr. numerous articles in field to profl. jours. Home: 12 River Meadow Dr., Ellicott City, Md. 21043. Office: 9650 Rockville Pike, Bethesda, Md. 20014.*

CARR, Edward Albert, Jr., Am. physician; b. Cranston, R.I., Mar. 3, 1922; s. Edward A. and Florence (Hodge) C.; A.B. summa cum laude, Brown U., 1942; M.D., cum laude, Harvard, 1945; m. Nancy Albosta, Dec. 27, 1952; children—Sharon L., Cynthia F. Research fellow, instr. pharmacology Harvard, 1948-51; exchange fellow St. Bartholomew's Hosp., London, Eng., 1952-53; faculty U. Mich., Ann Arbor, 1953—, prof. pharmacology, 1962—, asso. prof. internal medicine, 1967—, dir. program in investigative clin. pharmacology, 1962—; sr. staff U. Hosp., Ann Arbor, 1957—; dir. Upjohn Center for Clin. Pharmacology, 1967—. Cons., Ann Arbor VA Hosp., 1954—. Mem. Central Soc. Clin. Research, Central Clin. Research Club, American Thyroid Assn., Soc. Nuclear Medicine, A.A.A.S., Endocrine Soc., Am. Fedn. Clin. Research, Am. Soc. Pharmacology and Exptl. Therapeutics, Royal Soc. Medicine (fgn. affiliate), Phi Beta Kappa, Sigma Xi, Alpha Omega Alpha. Author: (with A. V. Wegst, R. Ganatra) Radioisotopes in Biology and Medicine, 1964; also numerous articles, abstracts, revs. Research on devel. radio-isotopic method for visualizing normal and diseased hearts, adverse reactions to drugs; devel. tng. and research in clin. pharmacology. Home: 3050 Foxcroft, Ann Arbor, Mich. 48104.*

CARR, Emma Perry, Am. chemist; b. Holmesville, O., July 23, 1880; d. Edmund Cone and Anna Mary (Jack) Carr; B.S., U. Chgo., 1905, Ph.D., 1910; postgrad. Queen's U., Belfast, Ireland, 1919, U. Zurich, (Switzerland) 1925, 1929-30; D.Sc., Allegheny Coll., 1939; Russell Sage Coll., 1941, Mount Holyoke Coll., 1952, Hood Coll., 1957. Asst. dept. chemistry Mount Holyoke Coll., South Hadley, Mass., 1901-04, instr. 1905-08, asso. prof., 1910-13, prof., chmn. dept., 1913-46, prof. emeritus, 1946—; vis. prof. Inst. Chemistry, Nat. Autonomous U. Mexico, 1944. Voting del. Internat. Union Pure and Applied Chemistry, Washington, 1926, Bucharest, 1925, Lucerne, 1936. Alice Freeman Palmer fellow Am. Assn. U. Women, 1929-30. Fellow, Am. Phys. Soc.; mem. Am. Chem. Soc. (Garvan medal 1937; James Flack Norris award N.E. sect. 1957), Am. Optical Soc., A.A.A.S., Phi Beta Kappa, Sigma Xi, Sigma Delta Epsilon (hon.), Iota Sigma Pi (hon.). Research has concerned application of phys. chemistry to organic problems with spl. reference to absorption spectra of organic compounds in the vacuum ultraviolet. Home: 1 Calvin Circle, Evanston, Ill. 60201.*

CARR, Francis Howard, English chemist; b. Croydon, Eng., Mar. 13, 1874; s. Henry Carr; ed. City and Ghilds Coll., London; D.Sc., Manchester, Eng.; m. Hilda Mary Sykes, 1898; 3 daus. Salter's research fellow, 1894-98; chief mfg. chemist Burroughs, Wellcome & Co., 1898-1914; dir., chief chemist Boots' Pure Drug Co., 1914-19; chmn. Brit. Drug Houses, Ltd., London; fellow Imperial Coll. Sci. Past pres. Soc. Chem. Industry, Assn. Brit. Chem. Mfrs. Co-author: Organic Medicinal Chemicals. Contbr. to sci. and tech. publs. Isolated (with G. Barger and H. H. Dale) ergotoxine. Address: The White House, Petersfield, Hants, Eng.

CARR, Harvey, Am. psychologist; b. Morris, Ill., 1873; s. Hamilton and Bell (Garden) C.; student DePauw U., 1893-95; B.Sc., U. Colo., 1901, M.Sc., 1902; Ph.D., U. Chgo., 1905; m. Antoinette Cox, Dec. 30, 1908; children—Frances Garden, Laurence Hamilton, Virginia Thurston. Instr. psychology Pratt Inst., Bklyn., 1906-08; asst. prof. psychology U. Chgo., 1908-16, asso. prof., 1916-23, prof., 1923-38, chmn. of dept., 1926-38, prof. emeritus, from 1938. Adv. editor Journal of Gen. Psychology; coöperating editor Comp. Psychology Monographs. Mem. Am. Psychol. Assn. (pres. 1926), Sigma Xi, Sigma Nu. Author: Textbook of Psychology, 1925; An Introduction to Space Perception, 1935. Contbns. on comparative psychology, visual space perception, ednl. theory. Died Culver, Ind., June 27, 1954.

CARR, Howard Earl, Am. physicist; b. Headland, Ala., Sept. 16, 1915; s. Samuel Tilden and Annie (Freeman) C.; B.S., Auburn U., 1936; M.A., U. Va., 1939, Ph.D., 1941; m. Carolyn Audrey Taylor, June 25, 1939; children—Howard Earl, Carolyn Ann. Faculty U. S.C., 1941-43, U. S. Naval Acad., 1946-48; faculty Auburn (Ala.) U., 1948—, head prof. physics, 1953—. Cons. Oak Ridge Nat. Lab., 1951-58, USAF, 1958-60. Fellow Am. Phys. Soc. (sec. Southeastern sect. 1956—) A.A.A.S.; mem. Am. Assn. Physics Tchrs., Am. Inst. Physics (Ala. regional counselor 1960—), Ala. Acad. Sci. (pres. 1957), Sigma Xi, Phi Kappa Phi. Research on separation of isotopes; negative ion formation; kinetics of ions and molecules; devel. techniques for teaching physics. Home: 342 Payne St., Auburn, Ala. 36830.*

CARR, Malcolm Wallace, Am. oral surgeon; b. N.Y.C., Oct. 8, 1899; s. Stephen Gould and Mary Helen (Cordts) C.; D.D.S. with honors, U. Pa., 1922; postgrad. Columbia Coll. Phys. and Surg., 1923-24; m. Helen Ruth Houston, Mar. 29, 1942 (dec. Oct. 1960). Asst. attending oral surgeon Fifth Av. Hosp., 1923-31, asso. attending oral surgeon, 1931-36; instr. oral surgery, clin. asst. vis. oral surgeon Vanderbilt Clinic, 1924-25; attending oral surgeon N.Y. Polyclinic Med. Sch. and Hosp., 1929-55, prof. oral surgery, 1949-55, cons. oral surgeon, 1955;—; asso. oral surgery St. Mary's Hosp. for Children, 1933-55; dir. oral surgery Met. Hosp., 1934-55, mem. med. bd., vis. oral surgeon, 1934-55, cons. oral surgeon, 1955—; courtesy staff oral surgery Doctors Hosp., 1935—; guest lectr. grad courses oral surgery Grad. Sch. Medicine, U. Pa., 1941—, lectr. dept. oral and maxillo-facial surgery, Sch. Dentistry, 1941-43; asso. prof. oral surgery, asso. oral surgeon Flower and Fifth Av. Hosps., N.Y. Med. Coll., 1936—; attending oral surgeon, dir. oral surgery, mem. med. bd. St. Luke's Hospital, 1940-62, cons. oral surgeon, 1962—; cons. oral surgeon Knickerbocker, Roosevelt, St. Barnabas hosps., Bird S. Coler Meml. Hosp. and Home; cons., instr. oral surgery U. S. Naval Hosp., St. Albans, L.I., N.Y., 1948—. Mem. mayor's adv. council health and hosps., City N.Y., 1954—; panel med. experts Supreme Ct. N.Y., 1953—. Recipient award merit Am. Dental Assn., 1940, N.Y. Acad. Dentistry, 1942, Brit. Ministry War, 1945, Federation Dentaire Internationale 11th Internat. Congress, London, 1952, U. Pa., 1956, Henry Spenadle medal and award First Dist. Dental Soc., 1965. Diplomate Am. Bd. Oral Surgeons (founder mem.). Fellow dental surgery Royal Coll. Surgeons (Eng., hon.); fellow Am. Coll. Dentists (pres. 1945-46), N.Y. Acad. Dentistry (pres. 1941-42), Am. Soc. Anesthesiologists, A.A.A.S.; mem. Am. Dental Assn., Am. Soc. Oral Surgeons, Acad. Internat. Medicine and Dentistry (pres. 1960—), Internat. Assn., Dental Research, A.M.A., Federation Dentaire Internationale; hon. mem. Am. Dental Soc. Europe, Colegio de Circujanos Dentistas de Puerto Rico, Federacion Odontologica Latino Americana, also mem. Japanese Stomatological Assn., Sigma Xi. Author: Dentistry: An Agency of Health Service, 1946; Acute Infections of the Face and Neck, 1948; Collected Papers, 1922-52. Asso. editor Jour. Dental Research, 1926-33; contbg. editor, mem. editorial bd. Annals of Dentistry, 1936—; asso. editor, mem. editorial bd. Jour. Oral Surgery, 1950—. Research and publs. on oral, maxillo-facial surgery; infections of dental origin; role of dental services in hosps. Home: 530 Park Av. Office: 52 E. 61st St., N.Y.C. 10021.*

CARR, Thomas Deaderick, Am. astronomer; b. Ft. Worth, Jan. 2, 1917; s. Archibald Fairly and Louise (Deaderick) C.; B.S., U. Fla., 1937, M.S., 1939, Ph.D., 1958; postgrad. Duke, 1939-40, U. Chgo., 1946-47; m. Glenna Dodson, May 20, 1961; 1 dau., Susan Catherine. Physicist, Ballistic Research Lab., Aberdeen Proving Ground, Md., 1940-45; civilian scientist Bur. Ordnance Navy Dept. Bikini A-Bomb Tests, 1946; physicist Air Force Missile Test Center, Patrick AFB, Fla., 1950-56; faculty U. Fla., Gainesville, 1958—, prof. dept. physics, astronomy, 1966—. Recipient citation for exceptional civilian service U. S. War Dept., 1945. Mem. Am. Phys. Soc., Am. Astron. Soc., Internat. Astron. Union, Am. Geophys. Union, Internat. Radio Sci. Union, Phi Beta Kappa, Sigma Xi. Author: (with A. G. Smith) Radio Exploration of the Planetary System, 1964. Research and publs. primarily in field of planet Jupiter. Home: 1546 S.W. 35th Pl., Gainesville, Fla. 32603.*

CARRADORI, Gioachino, Italian physicist; b. Prato, Tuscany, 1758; ed. Pisa, Italy; conducted research on absorption, adhesion, heat, electricity. Author: Teoria del calore, 1787; On Fertility of Land, circa 1802; a treatise on galvanism; also sci. articles in jours. of Milan and Pavia. Died 1818.

CARRARA, Nello, Italian physicist; b. Florence, Italy, Feb. 19, 1900; prof. magnetic waves, sci. sch. U. Florence; full prof., 1940—; named dir. center of studies for physics of microwaves Nat. Council Research; mem. Nat. Coms. for Physics and Tech., also Nat. Inst. for Space Research; mem. Internat. Radiosci. Union Spacewarn Contact for Italy, for 3d group of study Com. of Space Research. Address: Università degli Studi, Florence, Italy.

CARRÉ, Ferdinand, French inventor; b. France, 1824. Built refrigeration machine with ammonia solution, 1859-62; perfected compression machine for use in refrigeration, 1863; constructed a heat interchanger, evaporators, condensors, refrigerating vessels. Died 1900.

CARRÉ, Louis, French mathematician; b. Seine-et-Marne, France, July 26, 1663; studied theology seminare de S. Magloire, Collège de Provins; sec. to Malebranche, 1687-93. Mem. French Acad. Scis. Author: Méthode pour la mesure des surfaces, 1700; also wrote on gen. theories of sound and different mus. harmonies. Died Paris, France, Apr. 11, 1711.

CARREL, Alexis, surgeon, physiologist; b. Sainte-Foy-lès-Lyon, France, June 28, 1873; L.B., U. Lyons, (France), 1890, M.D., 1900; m. Anne de La Motte. Prosector, U. Lyons, 1900-02; came to Can., 1904, to U. S., 1905; staff U. Chgo., 1905; mem. staff Rockefeller Inst. for Med. Research, 1906-12, mem. 1912-39, emeritus, 1939-44; mem. spl. mission for French Ministry Pub. Health, beginning in 1939. Decorated Legion d'Honneur (France); recipient Nobel prize in medicine and physiology, 1912. Mem. French Acad. Scis., 1927. Author: Treatment of Infected Wounds, 1917; Man the Unknown, 1935; (with Lindbergh) The Culture of Organs, 1938; Voyage to Lourdes, 1950 (posthumous). Originated methods for suturing blood vessels in the transplantation of organs, also method of organ perfusion; (with Lindbergh) developed artificial heart; (with Dakin) developed a sodium hypochlorite antiseptic which was used during World War I; research on cultivation of tissues. Died Paris, France, Nov. 5, 1955.

CARRELLI, Antonio, Italian physicist; b. Naples, Italy, July 1, 1900; s. Raffaele and Silvia (Scardaccione) C.; m. Eleonora Laliccia, Dec. 4, 1937 (dec.); m. 2d, Elisa Pellerano, Feb. 14, 1965; children—Fabrizia, Claudio, Paolo. Prof. exptl. physics U. Catania, 1930-32; prof. U. Naples, 1932—, also dir. inst. exptl. physics. Vice pres. RAI-TV, 1946-55, pres., 1955-59; v.p. Commn. European Community Atomic Energy (EURATOM). Pres. Nat. Com. on Physics, 1962-64; pres. sci. com., nat. commn. UNESCO. Mem. Nat. Acad. Lincei, Rome, Nat. Acad. Scis., Arts and Letters, Naples, Microlambda Soc. (pres. 1948-52). Research, publs. on spectral dispersion, fluorescence, thermodynamics, particle theory, ultrasonics, viscosity of liquids, nuclear physics. Home: 6 Piazza d'Ovidio, Napoli, Italia. Office: 51-53 rue Belliard, Bruxelles, Belgique.*

CARRÉRE (or CARRERA), François, physician; b. Perpignan, France, Mar. 1622; M.D., U. Barcelona (Spain), 1654; elected rector U. Perpignan (not admitted to post), 1666; apptd. chief medicine Spanish Armies, 1676. Author: Vario omnique falso astrologiae conceptu, 1657; Salute militum tuenda, 1679; med. works in Catalan lang. Died Barcelona, May 11, 1695.

CARRICHTER, Bartholomäus, physician, astrologer; b. Reckingen, Netherlands, early 16th century; personal physician to emperors Ferdinand I and Maximilian II; author: Kräuterbuch, . . . , 1573; Kräuter und Artzneybuch, 1631; Practica, . . . , 1579; Buch von der Harmoney . . . , 1686. Probably believed in humor theory of disease. Died before 1574.

CARRICK, Robert, zoologist; b. Glasgow, Scotland, Apr. 4, 1911; s. John and Isabella (Bates) C.; B.Sc. with 1st class honors in Zoology, U. Glasgow, 1933; Ph.D., U. Edinburgh (Scotland), 1937; m. Christina Mackenzie, Apr. 2, 1938. Asst., U. Aberdeen (Scotland), 1936-37, sr. lectr., 1948-52; lectr. U. Leeds (Eng.), 1937-48; sr. prin. research scientist Commonwealth Sci. and Indsl. Research Orgn., Canberra, Australia, 1952—; head Antarctic biology Mawson Inst. for Antarctic research, U. Adelaide (South Australia), 1967—. Chmn. biology working group Sci. Com. for Antarctic Research, 1960-63. Mem. Internat. Union for Conservation Nature (exec. mem. for Australia 1963—). Mem. Brit. Ornithologists Union, Royal Australasian Ornithologists Union, Australian Mammal Soc., Scottish Ornithologists Club, Brit. Trust for Ornithology. Editor: (with M. W. Holdgate, J. Prevost) Antarctic Biology, 1964. Research and publs. on population ecology of birds and seals especially factors controlling breeding; proved that social competition greatly inhibits breeding, and prevents population explosions and crashes, spacing protects against disease; recommendations on conservation of Australian and Antarctic fauna. Office: Mawson Inst. for Antarctic Research, U. Adelaide, South Australia.*

CARRIER, George Francis, Am. engr.; b. Millinocket, Me., May 4, 1918; s. Charles Mosher and Mary (Marco) C.; M.E., Cornell, 1939, Ph.D., 1944; m. Mary Casey, June 30, 1946; children—Kenneth, Robert, Mark. Research asso. Harvard, 1944-46. Gordon MacKay prof. mech. engring., 1952—; asst. prof. Brown U., 1946-47, asso. prof., 1947-48, professor, 1948-52; asso. mng. editor Quar. Applied Mathematics since 1952. Fellow Am. Acad. Scis.; mem. A.S.-M.E., Sigma Xi, Nat. Acad. Scis. Editor: The Foundations of High Speed Aerodynamics, 1951. Contbr. articles tech. jours. Study of hydrodynamics and applied math. Home: Rice Spring Lane, Wayland. Office: Harvard U., Cambridge, Mass.

CARRIKER, Melbourne Romaine, Am. marine biologist; b. Santa Marta, Colombia, Feb. 25, 1915 (parents Am. citizens); s. Melbourne Armstrong and Myrtle Carmelita (Flye) C.; B.S., Rutgers U., 1939; Ph.M., U. Wis., 1940, Ph.D., 1943; m. Meriel Roosevelt McAlister, Nov. 17, 1943; children—Eric B., Bruce L., Neal A., Robert R. Mem. faculty Rutgers U., 1946-54; asso. prof. zoology U. N.C., 1954-61; supervisory fishery research biologist Biol. Lab. of U. S. Bur. Comml. Fisheries, Oxford, Md., 1961-62; dir. systematics ecology program Marine Biol. Lab., Woods Hole, Mass., 1962—. Fellow A.A.A.S.; mem. Am. Malacological Union, Am. Inst. Biol. Scis., Am. Soc. Zoologists, Atlantic Estuarine Research Soc. (pres. 1961), Ecol. Soc. Am. (mem. coms.), European Malacological Union, Nat. Shellfisheries Assn. (editor Proceedings 1953-55, pres. 1957-59), Inst. Malacology (pres., 1964, editor 1962—), Marine Biol. Assn. of U.K., Soc. Limnology and Oceanography (mem. coms.), Soc. Systematic Zoology. Research and numerous publs. on life history and environmental relationships of oysters, quahogs, fresh water pond snails; shell boring and demineralization by boring gastropods; estuarine ecology. Home: 243 Elm Rd., Falmouth, Mass. 02540. Office: Marine Biol. Lab., Woods Hole, Mass. 02543.*

CARRINGTON, Richard Christopher, English astronomer; b. Chelsea, London, Eng., May 26, 1826; s. Richard Carrington; B.A., Trinity Coll., Cambridge, Eng., 1848. Observer, Durham (Eng.) U., 1849-52; built pvt. obs. at Redhill, Surrey, Eng., 1853-61; mgr. Brentford brewery, 1858-65. Fellow Royal Soc., 1860; mem. Royal Astron. Soc. (sec. 1857-62, Gold medal 1859). Author: Results of Astronomical Observations Made at the Observatory of the University, Durham, from October 1849 to April 1852, 1855; A Catalogue of 3,753 Circumpolar Stars Observed at Redhill . . . , 1857; Observations of the Spots on the Sun . . . , 1863; On the Distribution of the Perihelia of the Parabolic and Hyperbolic Comets . . . , 1860. Research on motions of sun spots; determined elements of sun's rotation; discovered systematic drift of photosphere which produced inequal rates of rotation of earth around sun; 1st observation of solar flare. Died Chrut, Surrey, Nov. 27, 1875.

CARRION, Daniel, Peruvian med. student; b. 1850. Established identity of Oroya fever, verruga peruana, named Carrión's disease in his honor. Died 1885.

CARRO AMIGO, Santiago, Spanish physician, surgeon; Santiago de Compostela, b. Dec. 15, 1931; s. Antonia and Mercedes (Amigo) Carro; ed. U. Santiago de Compostela, Central U. Madrid; M.D. Instr. surgery Sch. Medicine, Santiago de Compostela, 1960-61; entered Mil. Health Service, 1962. Recipient Jose Miguel Guitarte award, 1955-56, 58-59; fellowship Internat. U. Menendez Pelayo Ag., 1956; Gomez Ulla prize Medecine et chirurgie de guerre (rev.), 1962. Mem. Soc. Clin. Hypnology Santiago de Compostela, Soc. History of Spanish Medicine, Soc. Surgery of Galice. Translator (from English): Como hacer la historia clínica n cirurgia, 1962. Med. sci. editor, works on path. surgery, psychomomatic illness, history of medicine. Home: rua del Vilar, 12, Santiago de Compostela. Office: Hospital Clinico, Santiago de Compostela, Spain.

CARROLL, James, physician, surgeon; b. Eng., June 5, 1854; s. James and Harriet (Chiverton) C.; grad. med. dept. U. Md., 1891; postgrad. Johns Hopkins Hosp.; m. Jennie M. George Lucas, May 1888. Asso. with Walter Reed, in study of Sanarelli's supposed yellow fever bacillus, 1897-1902, U. S. and Cuba; 1st lt., asst. surgeon U. S. Army; prof. bacteriology and clin. microscopy Army Med. Sch.; prof. bacteriology Washington Post-Grad. Sch.; prof. bacteriology and pathology, med. dept. George Washington U.; curator Army Med. Mus. prof. bacteriology and pathology Columbia, also Army Med. Sch., 1902. Permitted himself to be infected by mosquito bite, thus proving the mosquito the definitive carrier of yellow fever, also proved agt. Is a filterable virus. Died 1907.

CARROLL, John Bissell, Am. psychologist; b. Hartford, Conn., June 5, 1916; s. William J. and Helen M. (Bissell) C.; B.A., Wesleyan U., Middletown, Conn., 1937; Ph.D., U. Minn., 1941; M.A., (hon.), Harvard, 1953; m. Mary Elizabeth Searle, Sept. 6, 1941; 1 dau., Melissa. Instr., Mt. Holyoke Coll., 1940-42, Ind. U., 1942-43; lectr. U. Chgo., 1943-44; research psychologist U. S. Dept. Army, 1946-49; faculty Grad. Sch. Edn., Harvard, 1949-67, prof. edn., 1957-67, dir. Lab. for Research in Instrn., 1957-67; sr. psychologist Ednl. Testing Service, Princeton, N.J., 1967—. Mem. com. aptitude examiners Coll. Entrance Examination Bd., 1952-65; mem. adv. com. Commr. of Edn. on New Edn. Media, 1961-64. Fellow Am. Psychol. Assn., A.A.A.S.; mem. Social Sci. Research Council (mem. com. on linguistics and psychology 1952-61), Am. Council Learned Socs. (mem. com. on lang. program 1951-55), Psychometric Soc. (past pres.), Linguistic Soc. Am., Am. Edn. Research Assn., Modern Lang. Assn., Psychonomic Soc., Nat. Acad. Edn. (founding). Author: Study of Language, 1953; (with S. M. Sapon) Modern Language Aptitude Test, 1958; Language and Thought, 1964; also numerous articles. Editor: Language, Thought and Reality, 1956. Research on basic dimensions individual differences in lang. abilities, factor analysis and psychol. test methodology, methodology fgn. lang. teaching, measurement lang. phenomena; developed practical tests fgn. lang. aptitude. Office: Ednl. Testing Service; Princeton, N.J. 08540.*

CARROLL, Kenneth Kitchener, Canadian biochemist; b. Carrolls, N.B., Can., Mar. 9, 1923; s. Lawrence and Sarah (Estey) C.; B.Sc., U. N.B., 1943, M.Sc., 1946; M.A., U. Toronto, 1946; postgrad. McGill U.; Ph.D., U. Western Ont., 1949, postgrad. Canadian Life Ins. Officers Assn. fellow; postgrad. Cambridge U., 1952-54; m. Margaret Aileen Ronson, Aug. 26, 1950; children—Douglas, Stephen, James. Faculty, U. Western Ont., London, Can., 1954—, prof., acting head, dept. med. research, 1965—. Merck fellow, 1952-53; Agrl. Research Council fellow, 1953-54. Mem. Canadian Biochem. Soc., Canadian Physiol. Soc., Canadian Nutrition Soc. (sec. 1965—), Chem. Inst. Can. (councillor 1963-66), A.A.A.S., Am. Chem. Soc., Am. Soc. Biol. Chemists, Am. Oil Chemists Soc., Biochem. Soc. Britain, Canadian Fedn. Biol. Socs. (hon. asso. sec. 1966—). Editor, Canadian Fedn. News, 1962—. Studies on metabolism of fatty acids and cholesterol; chem. carcinogens; bacterial lipids; methods for separation and analysis of lipids. Home: 561 St. George St., London, Ont., Can.*

CARROLL, Lewis, see Dodgson, Charles Lutwidge.

CARROLL, Robert, Am. mathematician; b. Chgo., May 10, 1930; s. Walter S. and Dorothy (LeMonnier) C.; B.S., U. Wis., 1952; Ph.D., U. Md., 1959; m. Berenice Jacobs, Sept. 7, 1957; 1 son, David. Aero. research scientist NASA, 1952-54; NSF fellow, Nancy, France, 1959-60; with Rutgers U., 1960-64, asso. prof., 1963-64; asso. prof. U. Ill., 1964-67, prof., 1967—. Mem. Am. Math. Soc., Soc. Math. France. Contbr. articles in field to sci. jours. Contbns. to abstract theory of singular and degenerate Cauchy problems, abstract evolution equations and Cauchy problems, spectral theory of Green's operator. Home: 412 W. Oregon St., Urbana, Ill. 61801.*

CARRON DU VILLARDS, Charles-Joseph, ophthalmologist; b. Annecy, Italy, 1800; docteur de l'université Turin (Italy); surgeon-oculist to king of Sardinia; practiced at Annecy; went to Paris, 1832; founder 1st dispensary for diseases of eye in Paris,

France, 1835. insp. gen. mil. surgery, Brazil. Mem. Acad. Scis. Turin, Soc. Medicine Paris, Imperial Acad. Rio de Janeiro. Author: Guide pratique pour l'étude des maladies des yeux, 1838; Sur l'opération de la cataracte. Performed delicate eye operations. Died Rio de Janeiro, Brazil, Feb. 2, 1860.

CARRUTHERS, John Bennett, botanist; b. Jan. 19, 1869; s. W. Carruthers; ed. Greifswald (Prussia) U.; m. Frances Helen Louise Inglis. Demonstrator botany Royal Vet. Coll., London, 1893; prof. botany Coll. Agr., Downton, Hants., 1895; on spl. mission by Colonial Govt. and Ceylon Planters Assn. to investigate disease of cacao tree, 1890 (discovered fungus causing disease and gave measures for its prevention and cure); govt. mycologist, asst. dir. Royal Botanic Gardens, Ceylon, 1900; dir. agr. and govt. botanist Federated Malay States, 1905-10. Fellow Royal Soc. Edinburgh, Linnean Soc. Research and numerous articles on algae, mycology, plant pathology. Died July 17, 1910.

CARRUTHERS, Peter Ambler, Am. physicist; b. Lafayette, Ind., Oct. 7, 1935; s. Maurice Earl and Nila (Ambler) C.; B.S., Carnegie Inst. Tech., 1957, M.S., 1957; Ph.D., Cornell U., 1960; m. Jean Ann Breitenbecher, Feb. 26, 1955; children—Peter, Debra, Kathryn. Asst. prof. Cornell U., Ithaca, N.Y., 1961-63, asso. prof., 1963-67. Vis. asso. prof. Cal. Inst. Tech., 1965. Alfred P. Sloan Research fellow, 1963-65; NSF senior postdoctoral fellow U. Rome, 1967-68. Mem. Am., Italian phys. socs. Author: (with R. Brout) Lectures on the Many-Electron Problem, 1963; Introduction to Unitary Symmetry, 1966. Research on theory of strong interactions of elementary particles, especially explanation of excited states of nucleon and pion-nucleon interactions, theory of heat transport in insulating crystals, theory of irreversible processes. Home: 149 Pine Tree Rd., Ithaca, N.Y. 14850.*

CARRUTHERS, William, Brit. chemist; b. Glasgow, Scotland; B.Sc., U. Glasgow, 1942, Ph.D., 1949. Mem. sci. staff Med. Research Council, 1949-66; lectr. in chemistry U. Exeter, 1966—. Recipient Joseph Black Medal in chemistry U. Glasgow, 1943. Editor: (with James Cook) Progress in Organic Chemistry, vols. 5, 6, 7. Research, publs. into cancer producing substances, identification of automatic hydrocarbons in petroleum. Office: Chemistry Dept., U. Exeter, Eng.*

CARSON, Gordon Bloom, Am. mech. engr.; b. High Bridge, N.J., Aug. 1, 1911; s. Whifield R. and Emily (Bloom) C.; B.S., Case Inst. Tech., 1931, D.Eng., 1957; M.S., Yale, 1932, M.E., 1938; m. Beth Lacy, June 19, 1937; children—Richard W., Emily E. (Mrs. Lee Allen Duffus), Alice L. (Mrs. William P. Allman), Jean H. Dean of engineering, director engineering experiment station Ohio State U., Columbus, 1953-58, v.p. Research Found., 1958—; chmn. Ohio State Adv. Com. Sci., Tech. and Specialized Personnel. Dir. Indsl. Nucleonics Corp., Knowledge Communication Fund. Recipient Tech. Man of Year award Columbus, 1957; citation Ohio Soc. Profl. Engrs., 1961. Fellow Am. Soc. M.E., A.A.A.S.; mem. Nat. Soc. Profl. Engrs., Am. Inst. Indsl. Engrs. (nat. pres. 1957), Am. Soc. Engring. Edn., Am. Ordnance Assn. Editor: Prodn. Handbook, 1958. Research on machine tool automation, mfg. and automation for consumer goods industries, remote listening center concept for univs. Home: 2125 Elgin Rd., Columbus 43221. Office: 190 N. Oval Dr., Columbus, O. 43210.*

CARSON, Hampton Lawrence, Am. geneticist; b. Phila., Nov. 5, 1914; s. Joseph and Edith (Bruen) C.; A.B., U. Pa., 1936, Ph.D., 1943; m. Meredith Shelton, Aug. 14, 1937; children—Joseph II, Edward Bruen. Instr., U. Pa., 1938-42; faculty Washington U., St. Louis, 1943—, prof. genetics, 1956—; vis. prof. biology U. Sao Paulo (Brazil), 1951. Fulbright Research scholar zoology dept. U. Melbourne (Australia), 1961. Mem. Genetics Soc., Soc. for Study Evolution, Am. Soc. Naturalists, Soc. Zool. Population Genetics, A.A.A.S., Phi Beta Kappa, Sigma Xi. Author: Heredity and Human Life, 1963; also numerous articles. Research in causes hereditary variation in populations. Home: 240 N. New Ballas Rd., St. Louis 63141.*

CARSON, James, Scottish physician; b. Scotland, 1772; M.D., U. Edinburgh (Scotland), 1799; practiced medicine in Liverpool, Eng.; gave evidence at the Angus trial, 1808. Fellow Royal Soc., 1837. Author: Remarks on a Late Publication . . . , 1808; Reasons for Colonizing the Island of Newfoundland, 1813; A Letter to the Members of Parliament . . . , 1813; An Enquiry into the Causes of the Motion of the Blood, 1815, 2d rev. edit. An Inquiry into the Causes of Respiration . . . , 1833; A New Method of Slaughtering Animals for Human Food, 1839. Suggested artificial pneumothorax could be used in the treatment of pulmonary Tb, 1821. Died Sutton, Surrey, Eng., Aug. 12, 1843.

CARSON, John Renshaw, Am. research engr.; b. Pitts., June 28, 1887; s. John D. and Ada R. (Johnston) C.; B.S., Princeton, 1907, E.E., 1909, M.S., 1912; postgrad. Mass. Inst. Tech., 1907-08; D.Sc., Brooklyn Poly. Inst., 1936; m. Frances Atwell, July 22, 1913; 1 son, John R. Instr. in physics Princeton, 1912-14; engr. transmission theory devel. Am. Tel. & Tel. Co., 1914-34; research mathematician, Bell Telephone Labs., N.Y.C., from 1934. Recipient Liebmann Meml. prize for invention in radio and contbns. to math. theory of electric circuits, I.R.E., 1924; Elliott Cresson medal Franklin Inst., for contbns. to art of elec. communication, 1939. Author: Electric Circuit Theory and the Operational Calculus, 1927; Elektrische Ausgleichvorgänge und Operatorenrechnung, 1929. Installed 1st carrier current system for telephones; invented sideband system for multiple messages. Died Oct. 31, 1940.

CARSON, Rachel L(ouise), Am. sci. writer; b. Springdale, Pa., May 27, 1907; d. Robert Warden and Maria Frazier (McLean) Carson; A.B., Pa. Coll. for Women, 1929, D.Litt., 1952; A.M., Johns Hopkins, 1932; D.Sc., Oberlin Coll., 1952; Doctor of Letters (honorary) Drexel Institute Tech., 1952; D. Litt. (honorary) Smith College, 1953; spl. studies Marine Biol. Lab., various summers since 1929. Mem. zoology staff U. Md., 1931-36, Johns Hopkins, summer schs., 1930-36; joined staff Bur. Fisheries (now Fish and Wildlife Service) as biologist, 1936, editor-in-chief Fish and Wildlife Service 1949-52. Eugene Saxton Meml. fellowship, 1949; Guggenheim fellowship, 1951-52. Recipient George Westinghouse A.A.A.S. Sci. Writing award, 1950; John Burroughs medal, 1952; Henry G. Bryant Gold medal, 1952; Nat. Book award (non-fiction), 1951; Page-One award, 1952; Frances K. Hutchinson medal, 1952; Gold Medal, N.Y. Zoological Soc.; Silver Jubilee Medal, Ltd. Editions Club, 1954; Book award, Nat. Council Women U. S., 1956, Achievement, Am. Assn. U. Woman, 1956; Schweitzer medal Animal Welfare Inst. 1962; Constance Lindsay Skinner award Women's Nat. Book Assn., 1963; New Eng. Outdoor Writers Assn. Award, 1963; Conservationist of the Year award Nat. Wildlife Fedn., 1963; Achievement award Einstein College Medicine (women's division), 1963; special citations Garden Club of America, Pa. Federation Women's Clubs, Izaak Walton League America, 1963. Fellow Royal Society Literature; mem. National Institute of Arts and Letters, Audubon Soc., Soc. Women Geographers. Author: Under the Sea Wind, 1941; The Sea Around Us, 1951; The Edge of the Sea, 1956; Silent Spring, 1962; also articles. Well-known writer on biol. observations, research in marine biology, imprudent use of insecticides and resultant hazards. Died Apr. 14, 1964.

CARSON, Stanley Frederick, Am. microbiologist; b. San Francisco, Oct. 4, 1912; s. David S. and Erma (Simon) C.; A.B., Stanford U., 1934, Ph.D., 1941; m. Dorothy M. Wieser, Apr. 22, 1944; children—Suzanne, Scott, Wendy. Haskins Lab. fellow Stanford, 1941-42; sr. microbiologist Merck & Co., 1942-45; group leader Wyeth Inst., Phila., 1945-46; sr. biologist biology div. Oak Ridge Nat. Lab., 1947, asst. dir. biology division, 1948-62, associate director, 1963-67, deputy director, 1967—. Mem. Nat. Sci. Found., adv. panel on molecular biology, 1955-57, adv. panel on spl. facilities and programs, 1961—; v.p. microbial metabolism div. Internat. Congress on Microbiology, Rome, 1953. E. R. Squibb lectr. Rutgers U., 1958; hon. research prof. Univ. of Ga., 1960—; professor biomed. sciences U. Tenn., 1967—. Fellow A.A.A.S.; mem. Am. Chem. Soc., Am. Soc. Microbiology, Am. Soc. Biol. Chemists. Asso. editor Jour. Bacteriology, 1951-56, Bacteriol. Revs., 1958-64; editorial bd. Ann. Revs. Microbiology, 1965—. Contbr. numerous articles to profl. jours. Pioneer research with radioactive isotopes on role of carbon dioxide in cellular biochemistry; patentee methods of producing streptomycin and penicillin; research on microbial metabolism, mechanism of enzyme action. Home: 109 Pleasant Rd., Oak Ridge 37830. Office: Biology Div., Oak Ridge Nat. Lab., P.O. Box Y, Oak Ridge 37831.*

CARSTAIRS, George Morrison, psychiatrist; b. Mussoorie, India, June 18, 1916; s. George and Elizabeth H. C.; ed. George Watson's Coll., Edinburgh, Scotland; M.A., M.D., Edinburgh U.; m. Vera Hunt, 1950; 2 sons, 1 dau. Asst. physician Royal Edinburgh Hosp., 1942; Commonwealth fellow U. S., 1948-49; Rockefeller research fellow, 1950-51; Henderson research scholar, 1951-52; sr. registrar Maudsley Hosp., 1953; sci. staff Med. Research Council, 1954-60; prof. psychol. medicine U. Edinburgh, 1961—. Dir. unit for research on epidemiology of psychiatric illness, 1960—; Reith lectr., 1962; pres. World Fedn. Mental Health, 1967-68. Fellow Royal Coll. Physicians Edinburgh; mem. Royal Soc. Medicine, Internat. Epidemiological Assn., Royal Medico-Psychol. Assn. Author: The Twice Born, 1957; This Island Now, 1963. Study cultural factors in personality devel.; research on epidemiology psychiatric illness. Address: Univ. Dept. Psychiatry, Royal Edinburgh Hosp., Morningside Park, Edinburgh 10, Scotland.

CARSTEN, Mary E. (Mrs. Don Marlin), biochemist, physiologist; b. Berlin, Germany, Mar. 2, 1922; c. Paul and Frida (Born) Carsten; came to U. S., 1940, naturalized, 1946; B.A., N.Y. U., 1946, M.S., 1948, Ph.D., 1951; m. Don Marlin, Apr. 23, 1964. Instr., N.Y. U., 1951-53; research asso. dep. microbiology, Coll. Phys. and Surg., Columbia, N.Y.C., 1953-55; research asst. physiol. chemist Sch. Medicine, U. Cal. at Los Angeles, 1956-61, asso. research biochemist, 1961-63, asso. prof. depts. physiology and gynecology, 1963—. Established investigator Los Angeles County Heart Assn., 1961-64. Nat. Found. Infantile Paralysis fellow, 1954-55; Am. Cancer Soc. fellow, 1955-57; recipient Los Angeles County Heart Assn. award for cardio-vascular research, 1962, 63, 64, Research Career Devel. award USPHS, 1964—. Mem. Am. Soc. Biol. Chemists, Am. Chem. Soc., N.Y. Acad. Scis., Am. Physiol. Soc., Am. Soc. for Microbiology, Sigma Xi. Contbg. author Ion Exchangers in Organic and Biochemistry, 1957; Thyrotropin, 1963; Biology of Gestation, 1968. also numerous articles. Research on ion exchange chromatography amino acids, characterization various antibodies and blood group substances, purification and characterization contractile proteins of skeletal, heart and smooth muscles and species differences, isolation and purification cardiac sarcotubular vesicles, Home: 624 N. Highland Av., Los Angeles 90036.*

CARSWELL, Robert, physician, pathologist; b. Paisley, Scotland, Feb. 3, 1793; ed. U. Glasgow (Scotland); studied at hosps. of Paris (France), Lyons (France); M.D., Marischal Coll., Aberdeen, Scotland, 1826; studied morbid anatomy at Paris, 1827; m. Marguerite Chardenot. Worked on a series of 2000 water color drawings of diseased structures, 1828-31; became prof. path. anatomy U. Coll., London, Eng., 1831; physician U. Coll. Hosp; began pvt. practice medicine, circa 1836; resigned professorship, 1840 and accepted position as physician to the King of Belgium. Author: Illustrations of the Elementary Forms of Diseases, 1837; contbg. author: Cyclopaedia of Practical Medicine; also articles on induration, melanosis, mortification, perforation, scirrhus, tubercle. Died June 15, 1857.

CARTAILHAC, Edouard Philippe Emile, French anthropologist; b. Marseille, France, 1845; prof. anthropology Toulouse, France; asst. curator Toulouse Mus.; dir. rev. Materiaux pour l'histoire naturelle et primitive de l'homme. Author: L'age de pierre dans les souvenirs et superstitions populaires, 1878; L'age de pierre en Asie, 1880; Histoire de la science; les premiers travaux sur les monuments megalithiques, 1886; La France prehistorique, 1889. Contbns. to palentology. Died Geneva, Switzerland, 1921.

CARTAN, Élie Joseph, French mathematician; b. Dolomieu, Isère, France, Apr. 9, 1869; s. Joseph and Anne (Cottaz) C.; ed. Ecole normale supérieure; Sc.D., 1894; D. honoris causa, U. Liege (Belgium), 1934, Harvard, 1936, Free U. Berlin (Germany), 1947, U. Bucharest (Rumania), Catholic U. Louvain (Belgium), 1947, U. Pisa (Italy), 1948; m. Marie-Louise Bianconi, 1903; 1 son, Henri Paul, 1 dau. Lectr. U. Montpellier (France), 1894-96, U. Lyons (France), 1896-1903; prof. U. Nancy (France), 1903-09; lectr. U. Paris (France), 1909-12, prof., 1912-40, prof. Sch. Physics and Chemistry, 1910-41, hon. prof., 1940-51. Mem. Bureau des longitudes. Fellow Royal Soc., 1947; mem. French (v.p. 1945, pres. 1946), Norwegian acads. scis., Lincei, other fgn. acads. Author: Thèse sur la structure des groupes de transformations finis et continus, 1894; Leçons sur les invariants intégraux, 1922; Leçons sur la géométrie des espaces de Riemann, 1st edit., 1928, 2d edit., 1946; Leçons sur la géométrie projective complexe, 1921; Sur las espaces de Finsler, 1933; Les espaces metriques fondés sur la notion d'aire, 1933; La topologie des groupes de Lie, 1936; Leçons sur la théorie des espaces à connexion projective, 1937; La théorie des groupes finis et continus et la géométrie différentielle, 1937; Leçons sur la théorie des spineurs, vol. I, II, 1938; Selecta, 1939; Les systemes differentiels extérieurs et leurs applications geometriques, 1945. Applied Lie's theory of continuous groups to obtain classification of hypercomplex number systems; chief areas of work were math. physics, differential geometry, group theory. Died Paris, May 6, 1951.

CARTAN, Henri Paul, French mathematician; b. Nancy, France, July 8, 1904; s. Élie Joseph and Marie-Louise (Bianconi) C.; student École Normale Supérieure, Paris, 1923-26; Docteur ès scis., Paris, 1928; hon. doctorates U. Münster (Germany), 1952, Eidg. Techn. Hochschule, Zurich, 1955; m. Nicole Antoinette Weiss, Sept. 14, 1935; children—Jean, Francoise, Etienne, Mireille, Suzanne. Prof. Lycée Caen, 1928-39; chargé de cours U. Lille (France), 1929-31; maitre de confs., then prof. U. Strasbourg (France), 1931-40; prof. U. Paris, 1940—. Faculty École Normale Supérieure 1940-65. Mem. Math. Soc. France, Am., London (hon.) math. socs. Author: (with S. Eilenberg) Homological Algebra, 1956; Analytic Functions, 1961; also articles. Research on analytic functions of one or several variables, theory of sheaves, algebraic topology, homological algebra, potential theory. Home: 95, blvd. Jourdan, Paris 14, France.*

CARTER, George, English physicist; b. Liverpool, Eng., Feb. 16, 1934; s. Edward Mark and Louisa (Evans) C.; ed. Liverpool Coll. Tech., Wigan and Dist. Coll. Tech.; B.Sc., 1954, spl. B.Sc., London, 1956; Ph.D., Liverpool U., 1959; m. Hilda May, Apr. 19, 1958; 1 dau., Susan. Tech. asst., research physicist Dunlop Rubber Co., Liverpool, 1951-56; postdoctorate fellow Liverpool U., 1959-60, lectr. elec. engring., 1960-64, sr. lectr., 1964—. Fellow Chem. Soc.; mem. Inst. Physics (asso.), Phys. Soc. (asso.). Research and publs. on bombardment of solids, thermal desorption from and diffusion in solids, mass spectrometric investigation of electron impact phenomena in gases, ultra high vacuum technology. Home: 76 Winifred Lane, Aughton, Lancashire. Office: Liverpool Univ., Ashton St., Liverpool, Lancashire, U.K.*

CARTER, George Stuart, Brit. zoologist; b. Eastbourne, Eng., Sept. 15, 1893; s. George Charles and Hilda (Keane) C.; ed. Marlborough Coll., 1907-12, Caius Coll., Cambridge (Eng.) U., 1912-14, 1919-23. Lectr. zoology Glasgow (Scotland) U., 1924-30; fellow Corpus Christi Coll., Cambridge U., 1930—, sr. tutor, 1942-45, dir. studies in biol. scis., 1933-60; mem. biol. expdns. to Paraguay, 1926, Brit. Guiana, 1933, Uganda, 1952. Fellow Linnean Soc. London (past v.p., gold medal 1966), Zool. Soc. London, Royal Soc. Edinburgh, Inst. Biology. Author: A General Zoology of the Invertebrates, 1940; Animal Evolution, 1957; A Hundred Years of Evolution, 1957; The Papyrus Swamps of Uganda, 1956; Structure and Habit in Vertebrate Evolution, 1967; also numerous articles. Research on life of fishes in tropical swamps especially air-breathing fishes, swamps as an environment for life in gen., theory of evolution in present day form. Home: 40 Grange Rd., Cambridge, Eng.*

CARTER, Henry Rose, Am. sanitarian; b. Clifton Plantation, Caroline County, Va., Aug. 25, 1852; s. Henry Rose and Emma Caroline (Coleman) C.; C.E., U. Va., 1873, postgrad. in math. and applied chemistry, 1874-75; M.D., U. Md., 1879; m. Laura Hook; Sept. 29, 1880. Entered Marine Hosp. Service (now USPHS), 1879, apptd. by spl. act of Congress, 1915; mem. Rockefeller Yellow Fever Commn., Central and S. Am., 1916; in charge malarial work for USPHS, 1917-18, with spl. reference to anti-malarial measures for cantonments; yellow fever work, Peru, and sanitary adviser Peruvian Govt., 1920-21. Hon. chmn., vice permanent sec. Nat. Malaria Com.; mem. Yellow Fever Council, Internat. Health Bd., Rockefeller Found. Devoted attention mainly to sanitation in connection with yellow fever and malaria, beginning at the Ship Island Quarantine Sta. on Gulf of Mexico, 1888; had charge in control of several yellow fever epidemics in Southern states; determined incubation period of yellow fever; discoverer extrinsic incubation of yellow fever, 1900-01; this announcement leading to later discovery of Walter Reed, that mosquitoes are carriers of yellow fever; inaugurated quarantine system in Cuba, 1899, 1900. Died Washington, D.C., Sept. 14, 1925.

CARTER, Henry Vandyke, physician; b. Eng., 1831; ed. Univ. Coll., St. George's Hosp.; joined Bombay (India) Med. Service, 1858; prof. anatomy and physiology Grant Med. Coll.; returned to Europe to study leprosy in Norway, Italy, Greece, Algiers, Crete, India. Discovered the cause of rat bite fever was the bacterium Spirillum minus, 1861-74; studied the Asiatic form of relapsing fever using monkeys, 1882-87, also mycetoma, surra, malaria. Died 1897.

CARTER, Herbert Edmund, Am. biochemist; b. Mooresville, Ind., Sept. 25, 1910; s. George Benjamin and Edna (Pidgeon) C.; A.B., DePauw U., 1930, Sc.D., 1952; A.M., U. Ill., 1931, Ph.D., 1934; m. Elizabeth Winifred DeWees, Aug. 30, 1933, children—Anne Winsett, Jean Elizabeth. Instr. in chemistry U. Ill., 1933-35, asso., 1935-37, asst. prof. 1937-43, asso. prof. 1943-45, prof., 1945—, head dept. chemistry and chem. engring., 1954-67, v.p. for acad. affairs, 1967—. Mem. Pres.'s Com. on Nat. Medal of Sci., 1963—; mem. nat. sci. bd. NSF, 1963—. Awarded Rector Scholarship, Rector Fellowship, DePauw University; Eli Lilly & Co. annual award, 1943, award in lipid chemistry Am. Oil Chemists Soc., 1966. Mem. exec. com., div. chemistry and chem. tech. NRC, 1949-55, 57-58. Mem. Am. Chem. Soc. (dir., asso. editor Bio-Chemistry 1961—; recipient William H. Nichols medal N.Y. sect.), Am. Inst. Nutrition (sec. 1945-47), Am. Soc. Biol. Chemists (editorial bd. 1951-60, editorial com. 1963—, pres. 1956-57), Nat. Acad. Scis. (chmn. sect. biochemistry 1963—), Sigma Xi, Phi Beta Kappa, Phi Eta Sigma, Blue Key, Gamma Alpha, Alpha Chi Sigma. Mem. editorial bd. Bio Chem. Preparations, editor-in-chief, Vol. I. Contbr. tech. publs. Work on biochemistry of amino acids, isolation and structure determination of antibiotics (especially streptomycin), and of complex lipids; discovered group of complex glycolipids in plants which he named phytoglycolipid. Home: 704 W. Delaware St., Urbana, Ill.

CARTER, John Charles, med. apparatus co. exec.; b. Paris, Tenn., Mar. 12, 1911; s. Glenn Owen and Suzanne (Sweeney) C.; student Mount Union Coll., 1930-32, Columbia U., 1934-35; m. Anne Ellis McKay, Nov. 26, 1936; children—Janne Lee, Glenn McKay. With Union Carbide Corp., 1933—, tech. writer, motion picture prodn., 1934-38, with oxygen therapy dept., 1938-54, gen. mgr. oxygen therapy dept., N.Y.C., 1954-64, manager med. products dept., 1964—. Chmn. med. gases com. Compressed Gas Assn., 1964; v.p. Hosp. Industries Asso., 1964. Dir. Darien Family Counselling Service. Mem. Nat. Acad. Scis. (mem. hyperbaric oxygen com. 1962-65), Am. Assn. Inhalation Therapy (founder), Sigma Nu. Coauthor: Hyperbaric Oxygen Medicine, 1964. Research on utilization of gases in medicine; devel. portable liquid oxygen systems for use by pulmonary and cardiac cripples; hyberbaric oxygen systems; filter for producing bacteria free oxygen and other gases. Home: 277 Davenport Ridge Rd., New Canaan, Conn. Office: 270 Park Av., N.Y.C.

CARTER, John Robert, Am. physician, educator; b. Buffalo, Apr. 21, 1917; s. John Harvey and Gertrude Ann (Buckpitt) C.; B.S., Hamilton Coll., 1939; M.D., U. Rochester, 1943; m. Adelaide Briggs, May 8, 1943; children—Marilyn Anne, Jeanne Catherine. Faculty dept. pathology State U. Ia., Iowa City, 1944-59; prof., chmn. dept. pathology and oncology U. Kan. Med. Center, Kansas City, 1960-66; dir. Inst. Pathology, prof., chmn. dept. pathology, Western Res. U. Sch. Medicine, Cleve., 1966—. Dir., Univs. Asso. for Research and Edn. in Pathology, 1965—. Mem. A.A.A.S., Am. Assn. Pathologists and Bacteriologists, Internat. Acad. Pathology, Am. Soc. Clin. Pathologists, Am. Soc. Exptl. Pathology, Soc. Exptl. Biology and Medicine, Coll. Am. Pathologists, Phi Beta Kappa, Sigma Xi, Alpha Omega Alpha. Contbr. articles to profl. jours. Research on blood coagulation. Home: 36570 Ridge Rd., Willoughby, O. 44094. Office: Institute of Pathology, Western Reserve University, Cleve. 44106.*

CARTER, Robert Clifton, Am. geneticist; b. Gate City, Va., June 29, 1909; s. Charles Chester and Mary (Moore) C.; B.S., Va. Poly. Inst., 1931; M.S., Ia. State U., 1939, Ph.D., 1956; m. Alyce Wylde Rose, Nov. 20, 1940. Asst. co. agrl. agt. Va. Agrl. Extension Service, Blacksburg, 1931-33, co. agrl. agt., 1933-37, 39-40, 42-47; sec. Va. Beef Producers Assn., Richmond, 1947-48; prof. animal genetics Va. Poly. Inst., Blacksburg, 1948—. Mem. rev. com. Livestock Research Task Force Study, U. S. Dept. Agr., 1964-65. Mem. A.A.A.S., Am. Soc. Animal Sci. (chmn. breeding and genetics com. 1956, sec. So. sect. 1966, v.p. 1967), Am. Genetics Assn. (council 1965—, sec. 1967), Genetics Soc. Am., Biometrics Soc., Am. Soc. Zoologists, N.Y., Va. acads. sci., Am. Inst. Biol. Sci., Sigma Xi, Phi Kappa Phi. Editorial bd. Jour. Heredity, 1964—. Research, publs. in genetics of juvenile growth rate and body conformation in cattle and sheep; prodn. systems and grazing mgmt. of beef cattle and sheep. Home: 405 Hemlock Dr., Blacksburg, Va. 24060.*

CARTERET, Philip, explorer; lt. of Dolphin, Byron's voyage, 1764-66, comdr., 1766; became comdr. Swallow on voyage (with S. Wallis) from Plymouth, Eng. to S. Pole, 1766; separated from Wallis at Straits of Magellan, then explored along equatorial Pacific, Soloman Islands, Canal of St. George between New Ireland and New Britten, Island of New Hanover; apptd. to post ranks of navy, 1771; became comdr. Druit frigate, 1777; apptd. to Endymion, 1779; ret., 1794. Author: An Account of the Voyages undertaken for making Discoveries in the Southern Hemisphere, 3 vols.; The Inhabitants of the Coast of Patagonia; Account of Camelopardolis found at the Cape of Good Hope. Discovered and named Sandwich, Byron, New Hanover, Duke of Portland's Admiralty Islands; discovered Charlott Islands, Gloucester Islands, Pitcairn Island, 1767; named straits between New Britain and New Ireland, St. Georges Channel; corrected Dampier's survey of Mindanao. Died Southampton, Eng., July 21, 1796.

CARTERON, Jean, French telecommunications engr.; b. Paris, Mar. 8, 1926; s. Léon and Marthe (Dauvergne) C.; telecommunications engring. degree Poly. Sch. Paris; m. Nicole Mounier, Sept. 30, 1949; children—Dominique, Isabelle, Florence, Pau. Chief of service, math. and nuclear studies Electricité de France, 1959-63; prof. Superior Sch. Electricity, 1956—; named dir. S.A.C.S., 1964. Recipient Blondel medal. Mem. French Assn. Calculations and Treatment of Information (pres.). Home: 8, Hameau des Bergeronnettes, La Celle St.-Cloud. Office: 35, boulevard Brune, Paris 14, France.

CARTHEUSER, Johann Friedrich, German chemist; b. Hayne, Stolberg, Sept. 29, 1704; a son, Friedrich August. M.D., Halle. Prof. chemistry, pharmacy and materia medica, later of anatomy, botany, therapeutics and pathology U. Frankfort/Oder (Germany), 1740-59. Mem. Berlin Acad. Author: Specimen amoenitiatum naturae . . . , 1733; Amoenitatum naturae, sive historiae naturalis . . . , 1735; Elementa chemiae medicae . . . , 1736; Dissertatio chymico-physica . . . , 1754; Dissertationes physico-chymico-medico-medica . . . , 1774, 73, 75. Influenced the devel. pharmacy; tried to determine active prins. of plants. Died Frankfort/Oder, June 22, 1777.

CARTIER, Jacques, French navigator, explorer; b. St. Malo, Brittany, 1491. Sent by Francis I of France as leader expdn. to discover N.W. passage to East, 1534; reached Newfoundland, 1534, named Brion Island off coast; mistook Magdalen and Prince Edward islands for main coast; stayed 10 days in Gaspé Harbour making friends with Huron-Iroquois Indians from Que.; returned to St. Malo, 1534; set out again, 1536, anchored in Pillage Bay (named it Bay of St. Lawrence); after reaching island of Orleans was told by Indians he was in Canada (Indian word for village); reached Indian village of Hochelaga (now site of Montreal), 1536; returned to St. Malo with 12 Indians as proof of his discovery of what he thought was Northern Mexico, 1537; set sail again, 1541, made his hdqrs. nr. Que. at Cap Rouge; his reports of the warmer, more fertile areas of Can. stimulated French settlement there; gave tech. advice in naturical matters, also acted as Portuguese interpreter, 1544-57. Author: Brief Récit et succinte narration, 1545. First explorer of Gulf of St. Lawrence; discovered and sailed up St. Lawrence River, 1534; ascertained that Newfoundland was an island; never discovered the Northwest Passage and natural resources he was looking for. Died St. Malo, Sept. 1, 1557.

CARTWRIGHT, Edmund, Brit. inventor; b. Marnham, Nottinghamshire, Eng., Apr. 24, 1743; s. William Cartwright; B.A., U. Oxford (Eng.), 1764, M.A., 1766; elected fellow Magdalen Coll., 1764; several children including Edmund, Elizabeth (Mrs. John Penrose), Frances Dorothy, Mary (Mrs. Henry E. Strickland). Incumbent of Brampton Yorkshire, Eng.; rector of Goadby Marwood, Leicestershire, Eng., 1779; visited cotton-spinning mills nr. Matlock, 1784; became prebendary of Lincoln, 1786; conceived idea of a weaving mill, 1787; patented wool-combing machine, 1789-92; sold Doncaster factory, 1793; moved to London, Eng.; patented alcohol engine, 1797; agrl. experimenter for the dukes of Bedford and Woburn, Berfordshire, Eng., 1800-07; bought farm in Sevenoaks, Kent, Eng. Received cash award from Parliament, 1809. Fellow Royal Soc., 1821; mem. Soc. Arts. Contbr. to Monthly Rev.; also wrote poems. Invented 1st power loom, which wove wide cloth for practical purposes, 1785; patented rope-making machine, 1792; assisted Robert Fulton in steamboat experiments. Died Hastings, Eng., Oct. 30, 1823.

CARTWRIGHT, George Eastman, Am. physician; b. Lancaster, Wis., Dec. 1, 1917; s. Walter C. and Vera (Eastman) C.; B.A., U. Wis., 1939; M.D., Johns Hopkins, 1943; m. Helene N. Cleare, 1948; children—Margaret Ann, Jane Ann, Christine Ann, Candace Ann, Peter Edmund. Practice medicine, specializing in internal medicine, Salt Lake City, 1947—; faculty Coll. Medicine U. Utah, 1947—, prof. medicine, 1958—, head of the department of medicine, 1967—. James Waring Meml. lectr. Univ. of Colo., 1957; George R. Minot lectr. A.M.A., San Francisco, 1964; mem. various coms. NRC, Nat. Cancer Inst., VA, NIH. Diplomate Am. Bd. Internal Medicine, Am. Bd. Nutrition. Mem. Assn. Am. Physicians, Am. Soc. Clin. Investigation, Western Soc. Clin. Research, Am. Inst. Nutrition, Soc. Exptl. Biology and Medicine, Western Assn. Physicians, A.C.P., Am. Soc. Clin. Nutrition, A.M.A., Phi Beta Kappa, Sigma Xi. Author: Diagnostic Laboratory Hematology, 1954. Contbg. editor Principles of Internal Medicine, 1954-66. Contbr. numerous articles to profl. jours. Research in diseases of blood. Home: 2870 Floribunda Dr., Salt Lake City 84117. Office: U. Med. Center, Salt Lake City 84112.*

CARTWRIGHT, Mary Lucy, English mathematician; b. 1900; d. W. D. Cartwright; ed. St. Hugh's Coll., Oxford (Eng.) U., M.A., Ph.D.; M.A., Cambridge (Eng.) U., Sc.D, 1949; asst. mistress Alice Ottley Sch., Worcester, Eng., 1923-24; faculty Wycombe Abbey Sch., Buckinghamshire, Eng., 1924-27; Yarrow research fellow Girton Coll., Cambridge U., 1930-34, fellow 1934-49, lectr. math., 1935-59, mistress Girton Coll., 1949—, reader in theory of functions, 1959—. Cons. USN math. research projects Stanford, Princeton, 1949. Fellow Cambridge Philos. Soc., Royal Soc. (Sylvester medal 1964), 1947; mem. Math. Assn. (v.p.), London Math. Soc. (pres. 1961-63), Math. Assn. (1951-52). Research, publs. on integral functions, theory of functions of complex variable, theory of ordinary differential equations especially nonlinear equations, topological aspects of theory. Address: Girton Coll., Cambridge, Eng.

CARTWRIGHT, Oscar Ling, Am. taxonomic entomologist; b. Sharpsville, Pa., Apr. 12, 1900; s. William Robert and Lydia Blanche (McDowell) C.; B.S., Allegheny Coll., 1923; M.Sc., Ohio State U., 1925; m. Sara Marie Richbourg, Dec. 18, 1928. Research entomologist S.C. Agr. Expt. Sta., 1925-45, 47; sanitarian USPHS, 1945-46; asso. curator div. insects U. S. Nat. Mus., Smithsonian Instn., 1948-63, curator coleoptera, dept. entomology, 1963—; spl. taxonomic research in Scarabaeidae; field studies in Costa Rica, 1951, El Salvador, 1958, Tex., 1951, Ariz., 1956, Fla., 1959. Recipient Smithsonian Service award, 1954. Fellow Entomol. Soc. Am.; mem. Entomol. Soc. Washington, Biol. Soc. Washington, Soc. Systematic Zoologists, Alpha Chi Rhi. Phi Beta Phi, Gamma Alpha. Contbr. numerous reports and articles. Completed detailed life history of So. corn stalk borer, estimating annual loss and suggesting controls; taxonomic studies on the scarab species, described many scarab species new to sci. Home: 2110 W. Greenwich St., Falls Church, Va. 22043. Office: Dept. Entomology, Smithsonian Instn., Washington 20560.*

CARTY, John J., Am. elec. engr.; b. Cambridge, Mass., Apr. 14, 1861; ed. Cambridge Latin Sch.; D.Engring., Stevens Inst. Tech., 1915, N.Y. U., 1922; D.Sc., U. Chgo., Bowdoin, 1916, Tufts, 1919, Yale, 1922, Princeton, 1923; LL.D., McGill, 1917, U. Pa., 1924; m. Marion Mount Russell, Aug. 8, 1891; 1 son, John Russell. Began with Bell System in Boston, 1879, served in various positions, including chief engr. N.Y. Telephone Co., 1889-1907, chief engr. Am. Tel. & Tel. Co., 1907-19, v.p., 1919-30. Directed 1st transcontinental telephone wire line, (completed 1915). Connected radio stations by telephone circuits, (forerunner of, chain broadcasting, 1924. Trustee Carnegie Instn., Washington, Carnegie Corp., N.Y. Recipient John Fritz medal, engring. socs., 1928. Fellow Am. Acad. Arts and Scis., Am. Inst. E.E. (pres. 1915-16, Edison medal 1918); hon. mem. Franklin Inst. (Longstreth medal 1903, Franklin medal 1916). Pioneer in devel. of telephone. Applied metallic circuit to commercial use of phone, 1881. Laid basis for battery phone system. Invented battery switchboard, other improvements (24 patents, 1883-90). Proved electrical induction between parallel circuits mostly electrostatic, 1899. Died Dec. 27, 1932.

CARUS, Carl Gustav, German physiologist, psychologist, philosopher; b. Leipzig, Germany, Jan. 3, 1789; s. August Gottlob and Christiana Elisabeth (Jäger) C.; M.D., U. Leipzig, 1811; m. Caroline Carus, 1811; 6 sons, 5 daus. Prof. gynecology Med.-Surg. Acad., Dresden, Germany, from 1814; dir. Inst. Obstetrics, Dresden; as personal physician to King of Prussia, traveled to Italy, Switzerland, 1827; visited Eng., Scotland, 1844. Recipient medal for research on circulation of larva of nevrotera Acad. Medicine, Paris, France, 1832. Pres., Leopoldina Halle, Germany, 1862. Author: Lehrbuch der Zootomie, 1818; Lehrbuch der Gynäkologie, 2 vols., 1820; Zur Lehre von der Schwangerschaft u Geburt, 1822-24; Über den Blutkrieslauf der Insekten, 1827; Grundzüge der vergleichenden Anatomie und Physiologie, 3 vols., 1828; Vorlesungen über Psychologie, 1831; Neun Briefe über Landschaftsmalerei, 1831; Grundzüge einer neuen Kranioskopie, 1841; Psyche, 1846; System der Physiologie, 2 vols., 1847-49; Symbolik der menschlischen Gestalt, 1853. Described 6 vertebrae in human skull, 1828, also longitudinal axis of pelvic canal as a curved line with symphysis pubis as its center (curve of Carus); attempted to outline nervous system; adherent of Schelling in philosophy; sought to explain consciousness and devel. of soul. Died Dresden, July 28, 1869.

CARUS, Julius Victor, German zoologist; b. Leipzig, Germany, Oct. 23, 1823; s. Ernst A. and Agnes (Küster) C.; studied medicine and sci. at Leipzig and Dorpat, Estonia, 1841-46; also studied at univs. of Würzburg, (Germany) and Freiburg (Germany); m. Sophie C. Hasse, 1853; 3 daus.; m. 2d, Alexandra Petroff, 1886; 1 son. Became asst. physician in Leipzig, 1846, doctor of medicine and surgery, 1849; apptd. conservator Anat. Inst., Oxford, Eng., 1849; named prof. comparative anatomy, Leipzig, 1851; apptd. dir. Zool. Collection at Leipzig, 1853; asso. in zool. research with Prof. William Thomson of Edinburgh, 1872-74. Author: Zur näheren kenntniss des generationwechsel, 1849; System der thierischen Morphologie, 1853; Scones zootomicae, 1857; Über die Werthbestimmung zoologischer Merkmale, 1854; Geschichte der Zoologie, 1872. (with Gerstäcker) Handbuch der Zoologie, 2 vols., 1863. Founder, Zoologischer Angzeiger, 1878. Translated works of Darwin, Spencer, Lewe, Beals, Aristotle into German. Died Leipzig, 1903.

CARUS, Paul, philosopher; b. Ilsenburg, Germany, July 18, 1852; s. Gustav and Laura (Krueger) C.; ed. Gymnasium at Stettin, U. Strassburg; Ph.D., U. Tübingen, 1876; m. Mary Hegeler, Mar. 29, 1888. Editor, The Open Court, and The Monist, Chgo. Author: The Ethical Problem; Primer of Philosophy; Religion of Science; Karma; Nirvana; Eros and Psyche; Crown of Thorns; The Chief's Daughter; Godward; Nature of the State; Surd of Metaphysics; Sacred Tunes; Amitabha; Rise of Man; Story of Samson; Bride of Christ; God, an Enquiry Into Man's Highest Ideals; The Pleroma, an Essay on the Origin of Christianity; Truth on Trial; Personality and the Interpersonal; Nietzsche and other Exponents of Individualism; Mechanistic Principle and the Non-Mechanical; Principle of Relativity from the Standpoint of the Philosophy of Science; Truth and Other Poems; K'ung Fu Tze (dramatic poem); Goethe; The Venus of Milo. Searched for scientific basis for religion. Died Feb. 11, 1919.

CARVALHO, Alcides, Brazilian agronomist; b. Piracicaba, Sao Paulo, Brazil, Sept. 20, 1913; s. Aristides P. and Clementina F. Carvalho; student Sch. Agronomy, U. Sao Paulo, 1930-34; m. Antoniete C.e S. Carvalho, Feb. 18, 1962; children—Mariana, Marta. Asst. genetic dept. Instituto Agronomico, Campinas, Sao Paulo, 1935-42, with Dep. I. Agronomica, 1942-—, chief genetics, 1966-—. Mem. Sociedad Brasileira Para o Progresso da Ciencia, Sociedade Brasileira de Genética, Am. Genetics Assn., A.A.A.S., Soc. for Study Evolution, Soc. Latino-Americana de Fitotecnia. Contbg. author: Genética—Aspectos modernos da genética pura e aplicada. Research, numerous publs. on plant breeding, taxonomy, flower biology, genetics of Coffea grabica. Home: 1249 Eco. Glicerio, Campinas, Sao Paulo, Brazil.*

CARVER, Charles Ellsworth, Jr., Am. civil engr.; b. Burlington, Vt., May 27, 1922; s. Charles Ellsworth and Etta (Chapman) C.; B.S. in Civil Engring., U. Vt., 1947; M.S., Mass. Inst. Tech., 1949, Sc.D., 1955; m. Florence Elizabeth Mallory, Sept. 23, 1950; children—Beth Anne, Lynne Mallory, Charles Ellsworth III. Instr., U. Mass., Amherst, 1949-51; sr. hydrodynamics engr. Glenn L. Martin Co., Balt., 1955-56; lectr. applied mechanics Johns Hopkins, Balt., 1955-56; research asso. hydrodynamics Woods Hole (Mass.) Oceanographic Instn., 1956-58; faculty U. Mass., 1958-—, prof. civil engring., 1962-—. Cons. to pvt. cos. Mem. Am. Soc. C.E., Am. Geophys. Union, Am. Soc. Engring. Edn., Phi Beta Kappa, Tau Beta Pi, Sigma Xi. Contg. author Biological Waste Treatment, 1956; also articles. Hydrodynamic research undersea weapons systems; research on dynamics helicopter weapons systems, absorption oxygen for air bubbles and water droplets; photomicroscopic studies of flow of non-Newtonian fluids. Home: 15 Sunset Av., Amherst, Mass. 01002.*

CARVER, George Washington, Am. agrl. chemist, botanist; b. of slave parents, nr. Diamond Grove, Mo., circa 1864; in infancy lost father, and was stolen and carried into Ark. with mother, who was never heard of again; was bought from captors for a race horse valued at $300, and returned to former home in Mo.; worked way through high sch., Mpls., Kan., and later through coll.; B.S.Agr., Ia. State Coll. A. and M. Arts, 1894, M.S. Agr. 1896; D.Sc., Simpson Coll., 1928. Elected mem. faculty, Ia. State Coll. A. and M. Arts, placed in charge greenhouse, devoting spl. attention to bacterial lab. work in systematic botany; tchr. Tuskegee Inst. from 1896, also dir. Dept. Agrl. Research. Apptd. collaborator Bur. Plant Industry, U. S. Dept. Agr., div. mycology and disease survey, 1935. Recipient Spingarn medal, 1923, Roosevelt medal, 1939; plantation on which he was born made a nat. monument, 1953. Mem. Royal Soc. Arts, London. Developed 300 types of synthetic material from peanuts, including dyes, soap, cheese, milk substitutes; developed 118 by-products from sweet potatoes; made dyes from local clays; taught soil improvement; urged diversification of crops. Died Jan. 5, 1943.

CARWRIGHT, Mary Lucy, Brit. mathematician; b. Aynho, Northants, Eng., Dec. 17, 1900; d. William Digby and Lucy (Bury) Cartwright; B.A., St. Hugh's Coll., Oxford (Eng.) U., 1919, M.A., 1923, Ph.D., 1930; Sc.D., Cambridge U., 1949; Math. mistress Alice Ottley Sch., Worcester, Eng., 1923-24, Wycombe Abbey Sch., Bucks, Eng., 1924-27; fellow Girton Coll., Cambridge, Eng., 1930-—, mistress, 1949-—; faculty Cambridge U., 1933-—, reader in theory of functions, 1959-—. Cons., Office of Naval Research, 1949. Named Comdr. Order of Dannebrog, 1961; recipient Sylvester medal Royal Soc., 1964. Mem. Math. Assn., London Math. Soc., Cambridge Philos. Soc., Inst. Math. and its Applications, Am. Math. Soc. Author: Integral Functions, 1956; also numerous articles. Research on theory analytic functions and series, theory of nonlinear differential equations especially consideration of equations connected with radio and stability problems of automatic control. Address: Girton Coll., Cambridge, Eng.*

CARY, William, inventor; b. 1759; studied under Ramsden; owner ind. bus., 1790-1825. Mem. Astron. Soc. (charter). Contbr. Meteorol. Diary to Gentleman's Mag., for several years. Built 1st transit circle made in Eng. (2 feet in diameter with microscope for reading off), 2 ½ foot altitude and azimuth instrument, also sextants, microscopes, reflecting and refracting telescopes. Died Nov. 16, 1825.

CARYOPHYLLUS, see Carion, Johannes.

CASAGRANDE, Joseph Bartholomew, Am. anthropologist, educator; b. Cin., Feb. 14, 1915; s. Louis Bartholomew and Alma (Hausske) C.; B.A., U. Wis., 1938, Wis. scholar, 1938-39; Ph.D. (Univ. fellow 1940-41), Columbia, 1951; m. Mary Deveney, Aug. 15, 1945; children—Louis Bartholomew, Mary Leonora, Laurie Jean, Katherine Alma. Instr. anthropology Queens Coll., summer 1949, U. Rochester, 1949-50; staff Social Sci. Research Council, Washington and N.Y.C., 1950-60; prof., head dept. anthropology U. Ill., Urbana, 1960-—. Field trips to Comanche, Okla., 1940, Ojibwa, Wis., 1941, Navaho, Ariz., 1956, Ecuador, 1962, 63, 64; exec. sec. 29th Internat. Congress Americanists, N.Y.C., 1949; adj. asso. prof. Am. U., 1953-56; lectr. Fgn. Service Inst., 1956. Recipient Demoblzn. award Social Sci. Research Council, 1946-47, 48-49. Fellow Am. Anthrop. Assn. (exec. bd. 1961-63), Royal Anthrop. Soc. Gt. Britain and Ireland; mem. Am. Ethnol. Soc. (v.p., pres. 1961-63), Soc. Applied Anthropology (regional v.p. 1960-61), Anthrop. Soc. Wash. (sec. 1953-56), Linguistic Soc. Am., Phi Beta Kappa. Author: Comanche Linguistic Acculturation, 1955. Editor: In the Company of Man: Twenty Portraits by Anthropologists, 1960. Contbr. numerous articles to profl. jours. Home: 302 W. Florida Av., Urbana, Ill. 61803.*

CASAL, Gaspar, Spanish physician; b. Oviedo, Spain, 1679; ed. Alcala de Henares; practiced in Madrid, Spain and Oviedo; physician to the king; protomedico of Castilla; author: Historia natural y medica de el principado de Asturias, 1762. Known as Spanish Hippocrates; earliest description of pellagra which was pub. after his death, 1762; he called it mal de la rosa; pellagral lessions on the back of the neck are called Casal's collar or necklace. Died Madrid, 1759.

CASANOVA COLAS, José, Spanish physicist; b. Valencia, Spain, Nov. 12, 1923; s. Jose and Dolores (Colas) Casanova; Ph.D., U. Valencia; m. Mercedes Roque, 1 Sept. 12, 1946; children—Jose Luis, Carlos, Mercedes. Mem. Faculty Scis., Valladolid, Spain; advanced from instr. to prof. Mem. Royal Spanish Soc. Physics and Chemistry. Author: Microcapillary Technic for the Incorporation of Nuclear Targets in Mass of Photographic Emulsions, 1958; Nueva técnica de trabajo con emulsiones fotonucleares, 1960; Problemas de física, 1962. Home: calle Puente Colgante, 2, Valladolid. Office: Facultad de Ciencias, Valladolid, Spain.

CASARETT, George William, Am. pathologist; b. Rochester, N.Y., Aug. 17, 1920; s. George William and Caroline (Rocco) C.; student U. Toronto, 1938-41; Ph.D., U. Rochester, 1952; m. Marion I. Wells, June 12, 1944; children—Leslie G., Vicki W. Scientist, sect. head Atomic Energy Project, AEC, Med. Sch. U. Rochester, 1947-—, faculty dept. radiation biology and biophysics, 1953-—, prof., 1963-—, dir. radiobiol. research, research asso. radiation therapy div., radiology, 1959-—; cons. USPHS, UN, Colo. State U., U. Tenn. Mem. Nat. Council Radiation Protection. Recipient Silver medal for Manhattan Project War Dept.; 1st award for research Am. Roentgen Ray Soc., 1959; 1st award cum laude for fundamental research Radiol. Soc. N.Am., 1959, 64. Fellow N.Y. Acad. Scis.; mem. A.A.A.S., Am. Assn. Anatomists, Am. Assn. U. Profs., Am. Inst. Biol. Scis., Am. Soc. Exptl. Pathology, Gerontol. Soc., Radiation Research Soc., Soc. Exptl. Biology and Medicine, Sigma Xi. Asso. editor Radiation Research Jour., 1960-64. Contbr. numerous articles to profl. jours. Discovered indirect mechanisms of pathologic effects of ionizing radiation from external and internal sources on mammalian body cells, tissues and organs throughout life, greater effectiveness of certain modes of temporal protraction of radiation dose for causing male sterility and for destroying certain organs and cancers, radiation induction of premature changes like aging changes in connective tissue and small blood vessels. Home: 107 French Rd., Rochester 14618. Office: 260 Crittenden Blvd., Rochester, N.Y. 14620.*

CASE, Ermine Cowles, Am. geologist, paleontologist; b. Kansas City, Mo., Sept. 11, 1871; s. Theodore S. and Julia (Lykins) C.; B.A., M.A., Kan. U., 1893; M.S., Cornell U., 1895; Ph.D., U. Chgo., 1896; m. Mary Margaret Snow, 1898; children—Theodore J., Francis H. Lab. asst. chemistry Kan. U., 1893-94; mem. field party in Wyo., Kan. U., 1899; prof. geology and phys. geography Wis. State Normal Sch., Milw., 1897-1907; became asst. prof. hist. geology and paleontology U. Mich., 1907, named jr. prof., 1909, became curator paleont. collections, 1911, named prof., 1912; named dir. Mus. Geology (now Paleontology), 1928; chmn. dept. geology, 1934-41; leader field parties to W. and S.W. Mem. Am. Soc. Naturalists, Soc. Vertebrate Paleontologists, Washington Acad. Sci., Mich. Acad. Sci. (became pres. 1912), Paleontol. Soc., Geol. Soc. Am., Am. Philos. Soc., Am. Soc. Mammalogists, A.A.A.S. Author: Wisconsin, its geology and Physical Geography, 1907; also numerous articles, monographs. Research on vertebrate fossils, Pelycosauria, and the zygomatic arch, Permian reptiles and amphibians of Tex.; interpreted the environment of extinct vertebrates, 1912-26. Died 1953.

CASES QUERALT, José, Spanish agronomical engr.; b. Lerida, Spain, Aug. 7, 1909; s. Florencio and Fermina Queralt Biosca; ed. Superior Tech. Sch. Agronomical Engrs.; m. Mathilde Mendez de Vigo, May 17, 1941; children—José Ignacio, María Teresa, Fernando, Juan Antonio, Florencio Luis. Chief insp. 3d zone; engring. dir. Phytosan. Center of Barcelona (Spain); dir. Sch. Agrl. Experts; prof. gen. phytotechnic and agrl. industries. Mem. Royal Acad. Sci. and Arts of Barcelona. Author: Estudio sobre la fitoprofilaxis agrícola en España, 1963. Address: calle Bruch, 42, Barcelona 10, Spain.

CASEY, Albert Eugene, Am. pathologist; b. N.Y.C., Mar. 13, 1903; s. Eugene Joseph and Anna Alma (Powell) C.; A.B., Spring Hill Coll., 1922; M.D., St. Louis U., 1927; m. Bourdon Eason Veazey, Apr. 19, 1928; children—Anna Elizabeth (Mrs. Charles L. Payne), Albert Eugene; m. 2d, Joanne Gunn, Nov. 8, 1952; 1 son, Paul Travis. Asst. in anatomy St. Louis U., 1924-27, asso. prof. pathology, 1936-38; asst., asso. Rockefeller Inst., 1927-34; asso. prof. U. Va., 1934-36; sr. asst. prof. pathology and bacteriology La. State U., 1938-42; sr. pathologist Charity Hosp. of La., 1938-42; pathologist, dir. labs. Birmingham (Ala.) Bapt. Hosps., 1942-—; prof. pathology U. Ala., 1953-—; cons. pathologist Childrens, Holy Family, Peoples, Univ. hosps. Diplomate Am. Bd. Pathology. Mem. Coll. Am. Pathologists (chmn. S.E. regional com. 1954-57), Ala. Assn. Pathologists (pres. 1947), Soc. Exptl. Biology and Medicine (council 1941-42), Am. Soc. Clin. Pathology (counselor 1947-50), Am. Soc. Exptl. Pathology, Am. Assn. Anatomists, Am. Assn. Cancer Research, N.Y. Acad. Sci., A.A.A.S., Am. Assn. Blood Banks (Ala. rep. 1959-—), Am. Assn. Pathology and Bacteriology, Am. Pub. Health Assn., A.M.A., Internat. Acad. Pathology, Am. Assn. Phys. Anthropologists, Harvey Soc., Sigma Xi, others. Author: Encyclopedia of Pathologists of Southern U. S., 1963; Host Reactions and Cancer, 1962; also numerous articles. Research on hematology, exptl. cancer, viruses, phys. anthropology, pathology, and med. history, especially syphilis and cytology. Home: 2011 Southwood Rd., Birmingham 35216. Office: 924 S. 18th Dr., Birmingham, Ala. 35205.*

CASEY, John, Irish mathematician; b. Kilkenny, Cork, Ireland, May 12, 1820; ed. Kilkenny and Mitchelstown; scholar Trinity Coll.; hon. degrees from U. Dublin (Ireland), Royal U. Ireland; headmaster central sch. Kilkenny, taught in Kingstown, 1862-73; mem. faculty Catholic U. until 1881; then joined faculty of U. Coll. Honored by Norwegian Govt. Fellow Royal Soc., 1875; mem. Royal Irish Acad. (mem. council, Recipient Cunningham medal); also sci. socs. in Eng., Belgium, France. Author books on plane geometry, trigonometry. Research on cubic transformations, conic sects., Euclidian problems; solved Poncelet's theorem geometrically. Died Dublin, Jan. 3, 1891.

CASHMAN, Robert Joseph, Am. physicist; b. Wilmington, O., Sept. 27, 1906; s. John C. and Corinna (Smithson) C.; A.B., Bethany Coll., 1928, Sc.D. (hon.), 1953; A.M., Northwestern U., 1930, Ph.D. in Physics, 1935; m. Agnes Jones, June 8, 1940; children—Linda Lloyd (Mrs. Franklyn Modine), John Elliott. Asst. physics Northwestern U., Evanston, Ill., 1928-30, instr., 1930-36, asst. prof., 1936-40, asso. prof., 1941-47, prof., 1947—; physicist Nat. Def. Research Com., U. Mich., 1941, dir., at Northwestern U., 1941-45; USN contracts, 1945—, U. S. Army contracts, 1962-64, USAF contracts, 1962—. Cons., Gen. Electric Co., Lynn, Mass., Farnsworth Electronics Co., Fort Wayne, Ind., Internat. Tel. & Tel., Ft. Wayne, Dupont Corp., Wilmington, Del., U. Mich., Ann Arbor, Mich., Baldwin Piano Co., Cin. Recipient certificates of Commendation U. S. Army, USN, 1947. Fellow Am. Phys. Soc., Am. Optical Soc., mem. Kappa Alpha. Patentee in field photoelectric sensors. Important work includes research in photoelectricity, photoconductivity, thallous sulfide, lead sulfide, lead selenide photoconductive cells, temperature measurement. Home: 830 Indian Rd., Glenview, Ill. Office: Dept. Physics, Northwestern U., Evanston, Ill.*

CASIDA, Lester Earl, Am. reproductive physiologist; b. Chula, Mo., Apr. 9, 1904; s. Edward Ellsworth and Minnie (Molloy) C.; B.S. in Edn., N.E. Mo. State Tchrs. Coll., 1926; A.M., U. Mo., 1927, Ph.D., 1932; m. Ruth Barnes, Aug. 2, 1927; children—Lester Earl, John Edward, Betty Ruth (Mrs. Robert Allen Damerau). Tchr. rural sch., Grandy County, Mo., 1922-24; acting prof. biol. sci. Ariz. State Tchrs. Coll., 1929-30; acting prof. agr. Ark. State Tchrs. Coll., 1930-31; NRC fellow U. Wis., 1932-34, faculty, 1934—, prof. genetics, 1946—. Cons., NIH, 1962; cons. reproductive physiology Ford Found., India, 1964. Recipient Am. Dairy Sci. Assn. Borden award, 1954, Am. Soc. Animal Sci. Morrison award, 1959, Animal Physiology and Endocrinology Research award Am. Soc. Animal Sci., 1965, Master and Pioneer medal 5th Internat. Congress Animal Reproduction, 1964, Order Cavalier Ufficiale, Republic of Italy, 1966; cited for merit U. Mo. Alumni, 1965. Mem. Endocrine Soc., Soc. Exptl. Biology and Medicine, Am. Soc. Animal Sci., Am. Dairy Sci. Assn., Soc. for Study Fertility, Am. Inst. Biol. Scis., Sigma Xi. Research and numerous publs. on causes of infertility in female mammal, synchronization of estrus in groups of females, physiology of corpus luteum, formation and regression changes in reproductive organs of postpartum female. Home: 4229 Mandan Ct., Madison, Wis. 53711.*

CASIDA, Lester Earl, Jr., Am. microbiologist; b. Columbia, Mo., Aug. 25, 1928; s. Lester Earl and Ruth (Barnes) C.; B.S., U. Wis., 1950, M.S., 1951, Ph.D., 1953; m. Mardelle E. Baumgartner, Aug. 23, 1953; children—Nancy Ann, Sharon Ann. Microbiologist, Abbott Labs., North Chicago, Ill., 1951, Pabst Labs., Milw., 1953-54; fermentation biochemist Pfizer and Co., Bklyn., 1954-57; faculty Pa. State U., University Park, 1957—, prof. microbiology, 1966—. Cons., Texaco Corp., Beacon, N.Y., 1964—. Mem. Am. Soc. Microbiology, Soc. Indsl. Microbiology, Am. Chem. Soc., Mycology Soc. Am., Soc. for Gen. Microbiology. Research and publs. on new indsl. fermentation, microbial soil ecology including discovery new group soil microorganisms. Patentee microbial process for indsl. prodn L-lysine. Home: 203 E. Mitchell Av., State Coll., Pa. 16801. Office. Dept. Microbiology, Pa. State U., University Park, Pa. 16802.*

CASIMIR, Hendrik B. G., Dutch physicist; b. Den Haag, Netherlands, July 15, 1909; s. Rommert and Teunsina (Borgman) C.; student U. Copenhagen, 1929, 30; D.Sc., Leyden (Netherlands) U., 1931; Dr.honoris causa Technol. U., Aachen, 1966, U. Louvain, 1966, Technol. U., Copenhagen, 1965, U. Edinburgh, 1966; m. Josina Maria Jonker, Aug. 11, 1938; children—Elisabeth M. (Mrs. Peter Gruys), Rommert J., Anna M. (Mrs. Rein Saariste), Teunsina D. (Mrs. Mark van Hasselt), Gerda J. Asst., Leyden U., 1931-32, asso. prof., 1945—, asst. E.T.H., Zürich, Switzerland, 1932-33, Leyden U., 1933-42; with Philips Electr., Eindhoven, Netherlands, 1942—, became codir. for research labs., 1946, mem. bd. mgmt., 1956—. Recipient Gold medal Teylers Found., 1936; Alfred Ewing medal Inst. Civil Engrs., London, 1966. Mem. Royal Acad. Scis. Amsterdam (v.p. 1966), European Indsl. Research Mgmt. Assn. (pres. 1966), Royal Flemish, Am. acads. arts and scis., Royal Inst. Engrs. (Netherlands), Dutch Phys. Soc. Author: Magnetism and very Low Temperatures, 1940; On the Interaction between Atomic Nuclei and Electrons, 1936; also numerous articles. Research on math. formalism of quantum mechanics, thermodynamics of superconductors, paramagnetism at very low temperatures and theory of paramagnetic relaxation, theory of hyper fine structure (quadrupole moments, perturbations), irreversible thermodynamics (extension of Onsager's relations), influence of retardation on Van der Waals forces. Home: 7 De Zegge Heeze, N.Br., Netherlands. Office: N.V. Philips' Gloeilampenfabrieken, Eindhoven, Netherlands.*

CASLARI, David ben Abraham, physician, translator; flourished 13th century; 1 son, Abraham ben David. Lived in Narbonne, 1284. Translator (from Latin into Hebrew) Galen's treatise, De inaequali intemperencie or De malitia complexionis diversae (Sefer ro'a mezeg mithalef).

CASO Y ANDRADE, Alfonso, Mexican anthropologist; b. Mexico City, Mexico, Feb. 1, 1896; s. Antonio and María (Andrade) C.; U. Mexico, 1929; doctor honoris causa U. Nat. Mexico, U. Mérida, U. Albuquerque, U. Morelia; m. María Lombardo, Aug. 21, 1922; children—Beatriz (wife of Carlos Solorzano), Andrés, Alejandro Eugenia (wife of Luis Rius); m. 2d, Aida Lombardo. Prof. faculty philopophy and letters, 1918-40, Escuela de Layes, 1919-29; director Escuela Nacional Preparatoira, 1928; head dept. archeology Museo Nacional, 1930-33, dir., 1933-34; dir. explorations, Monte Albán, Oaxaca, 1931-44; dir. higher learning and sci. investigation, 1944; rector U. Nacional Mexico, 1944-45; sec. nat. properties and adminstrv. inspection, 1946-48; dir. Inst. Nacional Indigenista, Revista Estudios Antropologicos. First Medal, Viking Fund for Archaeology, 1952. Fellow Royal Anthrop. Inst. Great Britain and Ireland (hon.); mem. Nat. Acad. Scis. Antonio Alzate (pres.), Nat. Coll., Soc. Geography and Statistics, N.Y. Acad. Sciences, Washington Academy Scis., Archaeol. Inst. Am. (hon.), Am. Philos. Soc., Royal Anthropological Institute (hon.), Soc. Am. Paris (hon.), Am. Anthrop. Assn., Soc. Geography and History Guatemala, Am. Assn. Tchrs. Spanish, Deutsche Gesellschaft für Volkerkunde. Author numerous books including Urnas de Ooxaca, 1950; The People of the Sun, 1952; Codex Bodley, 1960; Codex Selden, 1964. Dir., founder of the Boletin Bibliog. de Antropologia Am. Home: Avenue Central 234, Tlacopac, Sn. Angel, Mexico 20, D.F. Office: Instituto Nacional Indigenista, Av. Revolución 1279, Mexico City 20, Mexico.

CASON, Hulsey, Am. psychologist; b. Lexington, Ga., Feb. 21, 1893; s. Emory Hugh and Jessie (Jones) C.; A.B., Mercer U., 1913; A.M., Columbia, 1920, Ph.D., 1922; m. Eloise May Boeker, Sept. 6, 1923; children—Roger Lee, Jean; m. 2d Marion Conrad, Aug. 30, 1939; 1 son, Emory Conrad. Asst. prof. psychology Syracuse U., 1923-26; faculty U. Rochester, 1926-27, prof. psychology, 1927-30; prof. psychology U. Wis., 1930-40, U. Miami, from 1948; research psychologist, USPHS, 1940-48. Lectr., U.N.C., 1923, Columbia, 1927, 29, U. Wis., 1930. Diplomate in Clin. Psychology, Am. Bd. Profl. Psychology, 1948. Fellow Am. Psychol. Assn., A.A.A.S.; mem. Am. Assn. U. Profs., So. Soc. for Philosophy and Psychology, Midwestern Psychol. Assn., Sigma Xi, Phi Delta Kappa, Psi Chi. Author: Conditioned Pupillary and Eyelid Reactions, 1922; Laws of Exercise and Effect, 1924; Common Annoyances, 1930; Pleasant and Unpleasant Activities, 1932; Nightmare Dream, 1935; Psychopathic Personality, 1942; Concept and Symptoms of the Psychopath, 1948 (also in Spanish, 1950, 51). Mem. editorial bd. jour. of Psychology, from 1935. Died May 1, 1950.

CASORATI, Felice, Italian mathematician; b. Pavia, Italy, Dec. 17, 1835; apptd. asso. prof. U. Pavia, 1857, prof., 1861. Author: Teorica delle funzioni di variabili complesse, 1868. Advanced study of analysis in Italy; worked on infinitesimal geometry and theory of functions of complex variables. Died Casteggio, Italy, Sept. 11, 1890.

CASPARI, Ernest Wolfgang, Am. geneticist; b. Berlin, Germany, Oct. 24, 1909; s. Wilhelm and Gertrud (Gerschel) C.; Ph.D., Göttingen (Germany) U., 1933; M.A., Conn. Wesleyan U., 1950; m. Hermine Berta Abraham, Aug. 16, 1938. Came to U. S., 1938, naturalized, 1944. Asst. zoology U. Göttingen, 1933-35; asst. microbiology U. Istanbul (Turkey) Med. Sch., 1935-38; fellow, asst. prof. biology Lafayette Coll., 1938-44; asst. prof. zoology, research asso. U. Rochester (N.Y.), 1944-46, prof. biology, 1960—, chmn. dept., 1960-65; asso. prof. biology Conn. Wesleyan U., 1946-47, prof. biology, 1949-60; research asso. dept. genetics Carnegie Instn. Washington, Cold Spring Harbor, N.Y., 1947-49. Mem. Am. Acad. Arts and Scis., Genetics Soc. Am. (treas. 1951-53, v.p. 1965, pres. 1966), A.A.A.S. (v.p., chmn. sect. F zool. science 1963), American Society of Naturalists (v.p. 1961), Am. Soc. Zoologists, Soc. for Study Evolution, Soc. Developmental Biology, Sigma Xi. Editor: Advances in Genetics, 1959—, Genetics, 1968—. Research, numerous publs. on developmental and biochem. genetics of eve color in moth Ephestia; genetic control of permeability and mitochondria; evolutionary importance of cytoplasmic inheritance; genetic basis of behavior; genetic changes in somatic cells. Home: 80 Penarrow Rd., Rochester, N.Y. 14618.*

CASPARI, Max Edward, Am. physicist; b. Frankfurt/Main, Germany, Mar. 17, 1923; s. Wilhelm Louis and Gertrud (Gerschel) C.; student London (Eng.) U., 1941-45; A.B., (Conn.) Wesleyan U., 1948; Ph.D., Mass. Inst. Tech., 1954; m. Sarah Bockoven, Dec. 28, 1951; children—Rachel Elizabeth, Matthew Ernst, Alexander Paul. Came to U. S. 1947, naturalized 1952. Research asst. Lab. for Insulation Research, Mass. Inst. Tech., 1948-54; instr. to prof. physics U. Pa., 1954—, prof. 1964—. Cons. Frankford Arsenal, Phila., 1961—. Fellow Am. Phys. Soc.; mem. Am. Assn. U. Profs., Sigma Xi. Angular correlation measurements in magnetic materials; dielectric breakdown; ferroelectricity. Home: 235 S. 15th St., Phila. 19102.*

CASPER, Johann Ludwig, German physician; b. Berlin, Germany, 1796; asst. prof. U. Berlin, from 1825, prof., from 1839, dir. med. sch., from 1841; editor Weekly Jour. Medicine, from 1833. Author: Essays on Medical Statistics and Official Medicine, 1825-37; On Probable Duration of Human Life, 1843; Prakisches Handbuch der Gerichtlichen, 2 vols., 1856. One of most renowned practicing physicians in Germany; reformed legal medicine to maintain its close relation with sci. medicine in gen. Died Berlin 1864.

CASPERSSON, Tobjörn Oskar, Swedish physician; b. Motala, Sweden, Oct. 15, 1910; s. Oskar Fredrik and Eva (Kalen) C.; M.D., Stockholm U., 1936; m. Siv. Cunnarson, 1937; children—Gunnel, Kestin, Lena. Dir. Med. Nobel Inst. for Med. Cell Research and Wallenberg Lab. for Exptl. Cell Research, Stockholm; prof. med. cell research and genetics, 1944—. Mem. natural sci. research council of Sweden, 1946-51, Swedish Anticancer Soc. research council, 1951-57, Swedish UNESCO Council, 1946—; cons. for UNESCO in Latin Am., 1950. Mem. council Swedish Inst. for Internat. Cultural Relations; mem. bd. Swedish-Am. Found.; mem. UN Radiation Com. Hon. lectureships various U. S. and European univs. Mem. Royal Swedish Acad. Sci., Royal Soc. Sci., Belgian, Brazilian acads. sci. Work in med. physics, especially research in cellular physics. Home: Scheelegatan 17, Stockholm. Office: Karolinska Institut, Stockholm 60, Sweden.

CASS, Leo Joseph, Am. physician; b. Medford, Mass., Dec. 17, 1908; s. John W. and Eleanor (Baldwin) C.; A.B., Harvard, 1933, M.D., 1938; m. Victoria Maxwell, Aug. 11, 1941; children—Leo M., Victoria B. Practice medicine, Boston, 1942—; physician Univ. Health Service, Harvard, 1942—, dir. health services Law Sch., 1947—, acting chief medicine, asso. dir., 1960; pres. Cass Research Assos., Boston, 1959—, San Jorge Research Inst., San Juan, P.R., 1960—; co-dir. Chandler Hovey Unit for Research and Treatment Multiple Sclerosis, Brookline, Mass., 1961—; hon. pres. staff L.I. Hosp., Boston; hon. chief medicine Brooks Hosp., Brookline. Corecipient Hull award Phila. Clin. Conv., A.M.A., 1965. Mem. A.M.A., Mass. Med. Soc., Am. Therapeutic Soc., Am. Coll. Chest Physicians, N.Y. Acad. Scis., Am. Coll. Health Assn., Aerospace Med. Assn., Mass. Orchid Soc., Pan-Am. Soc., Drug Information Assn. (founding mem., v.p. 1965), Internat. Acad. Law and Sci. (pres. 1965-66). Asso. editor Current Therapeutic Research, 1960-66. Therapeutic research and publs. on methods testing drugs for efficacy and toxicity, steroid studies. Home: 4 Myopia Hill Rd., Winchester, Mass. 01890. Office: Austin Hall, Harvard Law Sch., Cambridge, Mass. 02138.*

CASSEGRAIN, Giovanni D., French physician, astronomer; b. 1625; became prof. Collège de Chartres; wrote a treatise on megaphones; invented reflecting telescope which used a small convex mirror to reflect the light rays onto the reflector, behind which was the eyepiece, 1672. Died 1712.

CASSEL, Gustav, economist; b. Oct. 20, 1866; s. Oscar and Leontine (Dahlstrand) C.; ed. Stockholm, Sweden; doctorate in math., Uppsala, 1895; hon. doctorates, univs. Cologne, Munich, Innsbruck, Athens, Stockholm; m. Johanna Bjornson-Möller, 1895; 2 sons, 2 daus. Became prof. polit. economy, Stockholm, 1904; Rhodes meml. lectr. Oxford U., 1932; Swedish del. World Monetary and Econ. Conf., London, 1933; lectr. Internat. Inst., Geneva, Switzerland. Corr. for Sweden Royal Econ. Soc.; mem., pres. Royal Swedish Acad. Sci.; hon. mem. Am. Econ. Assn. Author: Nature and Necessity of Interest, 1902; Theoretische Sozilaokonomie, 1918; The World's Monetary Problems, Two Memoranda to the League of Nations, 1921; Money and Foreign Exchange after 1914, 1922; Fundamental Thoughts in Economics, 1925; Post-War Monetary Stabilisation, 1928; The Crisis in the World's Monetary System, 1932; Quantitative Thinking in Economics, 1935; The Downfall of the Gold Standard, 1936. Died Jan. 14, 1945.

CASSELLA, Joseph, Italian astronomer, mathematician; b. Cusano, Italy, 1755. Tchr. sci., astronomy. Made observations on occultation of stars. Died Naples, Italy, 1808.

CASSELS, James Macdonald, English physicist; b. Sept. 9, 1924; s. Alastair Macdonald and Ada White (Scott) C.; ed. St. Lawrence Coll., Ramsgate, Eng., Trinity Coll., Cambridge (Eng.) U.; B.A., M.A., Ph.D.; m. Jane Helen Thera Lawrence, 1947; 1 son, 1 dau. Harwell fellow, prin. sci. officer Atomic Energy Research Establishment, Harwell, Eng., 1949-53; lectr., 1953; from sr. lectr. to prof. exptl. physics U. Liverpool (Eng.), 1956-59, Lyon Jones prof. physics, 1960—. Vis. prof. Cornell U., Ithaca, N.Y., 1959-60. Fellow Royal Soc., 1959; mem. Inst. Physics, Phys. Soc., Am. Phys. Soc. Research, publs. on atomic, nuclear, elementary particle physics. Home: Overstrand, The Esplande, Cressington Park, Liverpool 19. Office: Chadwick Lab., U. Liverpool, Liverpool 3, Lancashire, Eng.

CASSELS, John William Scott, Brit. mathematician; b. Durham City, Eng., July 11, 1922; s. John William and Muriel S. (Lobjoit) C.; M.A., Edinburgh (Scotland) U., 1943; Ph.D., Cambridge (Eng.) U., 1949; D.hon. causa, U. Lille (France), 1965; m. Constance Mabel Merritt, Aug. 6, 1949; children—Patricia Mary, John William Merritt. Lectr., Manchester (Eng.), U., 1949-50; faculty Cambridge U., 1950—, Sadleirian prof. pure math., 1967—; fellow Trinity Coll., Cambridge U., 1949—. Fellow Royal

Soc., 1963; mem. London Math. Soc., Cambridge Philos. Soc. Author: Diophantine Approximation, 1957; Geometry of Numbers, 1959; also articles. Research on arithmetic especially theory of diophantine approximation, geometry of numbers and diophantine equations. Home: 3 Luard Close, Cambridge, U.K.*

CASSERIO (or CASSERI), Guilio, Italian anatomist, physician; b. Piacenza, Italy, 1561; studied under Fabricius ab Aquapendente; M.D., Padua, Italy. Prof. anatomy and surgery U. Padua, 1609-16. Author: De vocis auditusque organis hist. anatomica, 1600; Pentaesthesion, 1609; Tabulae anatomicae, 1627; Tabulae de formato foetu, 1645. Discovered anterior ligament of malleus (middle ear); one of earliest to describe maxillary sinus and posterior fontanel; 1st to describe musculocutaneous nerve, coracobrachialis muscle (previously recognized by Arantius, 1594), larynx and laryngotomy; compared organs of hearing in adult human, human foetus and several animals; noted glands located between tarsus and conjunctiva of eyelid, 1609; illustrated anatomy of organs of speech and hearing. Died Padua, 1616.

CASSIDY, Harold Gomes, chemist; b. Havana, Cuba, Oct. 17, 1906; s. Walter Clarence and Camilla G. (Casseres) C.; student Akron U., 1923; A.B., Oberlin Coll., 1930, A.M., 1932; Ph.D., Yale, 1939; m. Kathryn Myra Childs, May 19, 1934. Research fellow Oberlin Coll., 1932-33; research chemist William S. Merrell Co., Cin., 1933-36; instr. Oberlin Coll., 1936-37; faculty Yale, New Haven, 1938—, prof. chemistry, 1958—. Research chemist Manhattan Engring., other O.S.R.D. project, 1942-45; seminar leader Danforth Workshop on Liberal Arts Edn., 1962-65; regional lectr. Am. Chem. Soc., 1952, 56, 59; sr. fellow Center for Advanced Studies, Wesleyan U., 1965-66. Fellow A.A.A.S., N.Y. Acad. Scis.; mem. Am. Chem. Soc., Fedn. Am. Soc. Exptl. Biology, Conn. Acad. Arts and Scis., Sigma Xi. Author: (with J. English) Principles of Organic Chemistry, 1949; Adsorption and Chromatography, 1951; Fundamentals of Chromatography, 1957; The Sciences and The Arts, 1962; (with K. A. Kun) Oxidation-Reduction Polymers, 1965; also numerous articles. Pioneer in field oxidation-reduction polymers. Home: 163 East Rock Rd., New Haven 06511.*

CASSIN, John, Am. ornithologist; b. Delaware County, Pa., Sept. 6, 1813. Became mcht., Phila., 1834, later established engraving and lithographing bus.; mem. Acad. Natural Scis.; identified and arranged 26,000 ornithol. specimens in collection of Dr. Thomas B. Wilson. (largest collection then in existence); studied taxonomy, synonomy and nomenclature of ornithology; contbr. sci. publs. and ornithological sections of U. S. govt. publs. on Japan, on route from the Mississippi to Pacific Ocean, and on southern hemisphere. Author: Illustrations of the Birds of California, Texas, Oregon, British, and Russian America (which added 50 new species to Audubon's Birds of America), 1856. Died Phila., Jan. 10, 1869.

CASSINI, Alexandre Henri Gabriel (vicomte de), French botanist; b. Paris, May 9, 1784; s. Jacques (or Jean) Dominique and Claude-Marie-Louise (de la Myre-Mory) C. Studied botany; became lawyer; worked for civil tribunal, from 1810; elected mem. Chamber of Peers, 1830. Mem. French Acad. Scis. Author: Opuscules phytologiques, 3 vols., 1826; also articles in Dictionnaire des scis. naturelles, Jour. de physique. Studied astronomy, then turned to bot. research specializing in Compositae. Died Paris, Apr. 16, 1832.

CASSINI, Giovanni Domenico (or Jean Dominique), astronomer; b. Perinaldo, Italy, June 8, 1625; s. Jacobo and Julia (Crovese) C.; ed. Vallebone, also Genoa, Italy; m. Geneviève Delaistre, 1673; children—Jean-Baptiste, Jacques. Helped Marquis Malvasia build obs. at Bologna, Italy, 1644; became prof. astronomy U. Bologna, 1650; mem. papal commn. to end dispute between Ferrara and Bologna over nav. and river rights; became insp. waterways; went to Paris, France, 1669; became French subject, 1673; directed constrn. of Paris Obs., became its dir.; trained astronomers, including Jesuit scholars sent to China. Fellow Royal Soc., 1672. Author: Ephemerides bonomienses mediceorum siderum, 1668; Connaissance des temps, 1679; Divers ouvrages d'astronomie, 1731. Invented quartic curve (Cassini's oval or gen. lemniscate); developed theory on motion of comets; showed cause of moon's libration; observed and measured rotation periods of Jupiter, Mars, Venus, 1665-67; discovered 4 satellites of Saturn, 1671-84; observed dark div. in Saturn's ring (Cassini's div.); established accurate tables for periods of Jupiter's satellites; made earliest systematic observation of zodiacal light; cooperated in determining solar parallax; determined obliquity of ecliptic and eccentricity of earth's orbit; showed earth is prolate spheroid; estimated distance of Mars from sun (7 per cent in error), 1679; found method for determining ship's position at sea; improved gnomon and meridan for purpose of fixing solstices and reforming calendar. Died Paris, Sept. 11, 1712.

CASSINI, Jacques, French astronomer; b. Paris, Feb. 8, 1677; s. Giovanni Domenico and Geneviève (Delaitre) C.; m. Suzanne-François Charpentier de Charmois; children—César François, Dominique Jean, Dominique Joseph. Traveled to Bologna, Italy, 1695, then to Holland and Eng.; succeeded father as dir. Paris Obs.; as mem. commn. Paris Acad. Scis., helped establish velocity of sound in air at 1,106 feet per second. Fellow Royal Soc., 1698, mem. French Acad. Scis., Acad. Berlin (Germany), Inst. Bologna. Author: De la grandeur et de la figure de la terre, 1720; Éléments d'astronomie et tables astronomiques, 1740. Engraved Planisphère terrestr, 1694, 96; continued father's work on Saturn; devoted prin. research to determination of figure of earth; helped father measure meridian to Canigou, France, 1701, and continued it to Dunkerque, 1718; completed measuring arc of meridian between Paris and Collioures; on basis of his measurements of arcs of meridian, which gave diminishing figure of earth as it approached poles, he surmised polar diameter of earth is greater than equatorial; built clock and concave mirror for use in melting metals, 1747. Died Thury, France, Apr. 18, 1756.

CASSINI, Jacques (or Jean) Dominique (comte de), French mathematician, geographer; b. Paris, June 30, 1748; s. César François and Charlotte Drouin (de Vandeuil) Cassini de Thury; ed. Juilly; m. Claude-Marie-Louise de la Myre-Mory, Apr. 7, 1773; 1 son, Alexandre Henri Gabriel. Traveled to Am., Africa, Italy, 1775; succeeded father as dir. Paris Obs., 1784. Named senator and comte by Napoleon I. Fellow Royal Soc., 1789; mem. French Acad. Scis., 1770. Author: Manuel de l'étranger qui voyage en Italie, 1778; Exposé des opérations faites en France en 1787 pour la jonction des observatoires de Paris et de Greenwich; Mémoire sur une nouvelle boussole; Extrait d'observations astronomiques . . . ; Déclinaison de l'aiguille aimantée, 1791; De l'influence de l'équinox, 1794; Précis sur la prison des Bénédictins anglais; Mémoires pour servir a l'histoire des sciences . . . , 1810; Extraits des observations astronomiques et physiques, 1810; Les veillées du village, 1830; Entretiens sur la religion, 1862. Completed topog. map of Paris begun by father (pub. 1793), also his Carte topographique de la France, 1744-87; active in dividing France into sects.; reduced mech. errors in astron. observations by use of repeating circle. Died Thury, Oct. 18, 1845.

CASSINI DE THURY, César François, French astronomer, geographer; b. Paris, June 17, 1714; s. Jacques (or Jean) Dominique and Suzanne-Francoise Charpentier (de Charmois) C.; m. Charlotte Drouin de Vandeuil; 1 son, Jacques (or Jean) Dominique. Succeeded father as dir. Paris Obs. Fellow Royal Soc., 1750. Attempted (with Maraldi and Lacaille) to measure sound exactly, but failed to notice influence of temperature; specialized in geodesy; carried out geometrical surveys of France and Flanders; checked father's measurement of a degree in France, 1740; commd. to prepare topog. map of France based on geodesic triangulation (completed by son in 1789), 1744. Author: Méridienne de l'Observatoire de Paris, 1744; Description géométrique de la Terre, 1775; Description géométrique de la France, 1784. Died Thury, France, Sept. 4, 1784.

CASSINIS, Roberto, Italian geophysicist; b. Roma, Italy, Oct. 31, 1921; s. Gino and Sofia (Pericoli) C.; Dr.in Engring., Politecnico di Milano (Italy); m. Maria M. Maviglia, Oct. 14, 1950; children—Riccardo, Carlo, Alessandro. With Fondazione Lerici del Politecnico, Milan, Italy, 1947-64, asst. prof., 1959-64, chief geophysicist, 1958-60, asst. gen. mgr., 1961-64; prof. applied geophysics U. Palermo (Italy), 1964—; cons. Fondazione Lerici, Milan. Mem. Soc. Expln. Geophysicists, European Assn. Expln. Geophysicists. Research and publs. on geophys. expln. especially with seismic methods. Home: 30 Lombardia, Milano, Italy. Office: 260 Calatafimi, Palermo, Italy.*

CASSIODORUS, Flavius Magnus Aurelius, Roman mathematician, statesman; b. Calabria, Italy, circa 487; gov. of Sicily; in service of Odoacer, Ostrogoth King of Italy; sec. to Theodoric and treasurer and councilor under him; consul 514-537; ret. from public life, circa 550; founded monastery of Vivarium, and probably monastery at Castellum; at Vivarium taught monks to transcribe manuscripts from Greek texts. Author: Chronik; De rebus gestis Gothorum; Historia ecclesiastica; Introduction to Divine and Human Readings; also wrote on music; compiled math. works; founder of trivium and quadrivium, medieval basis for edn.; made numerous mech. devices; 1st to use terms rational and irrational in modern math. sense. Died circa 583.

CASSIOS DIONYSIOS, Greek biologist, botanist; b. Utica; flourished circa 88 B.C.; translated into Greek the work on agr. of the Carthaginian Mago, reduced it from 28 to 20 books, added extracts from other Greek authors (work now lost); author work on roots; credited with an illustrated pharmacopoeia.

CASSIRER, Ernst Alfred, philosopher; b. Breslau, (now Wroclaw, Poland), July 28, 1874; s. Eduard and Jenny Cassiter; ed. univs. of Berlin, Leipzig, Heidelberg, Marburg (all Germany) m. Toni Bondy, 1902; 3 children. Joined U. Hamburg (Germany), 1919, later became full prof., apptd rector, 1930; with Oxford (Eng.), U., 1933-35, U. Göteberg (Sweden), 1935-41, Yale, 1941-44, Columbia U., N.Y.C., 1944-45. Author: Substance and Function and Einstein's Theory of Relativity, 1923; An Essay on Man, 1944; Rousseau, Kant, Goethe, 1945; Language and Myth, 1946; Myth of the State, 1946; The Problem of Knowledge, 1950; The Philosophy of the Enlightenment, 1951; The Platonic Renaissance in Italy, 1953; The Philosophy of Symbolic Forms, I, 1953, II, 1955; Substance and Function, and Einstein's Theory of Relativity, 1953; The Question of Jean-Jacques Rousseau, 1954. Studied epistemology and methodology of history; proponent of Neo-Kantian Marburg Sch. philosophy; worked with theories of math, physics, and chemistry. Died N.Y.C., Apr. 13, 1945.

CASSIRER, Richard, German neurologist; b. Breslau, Germany (now Wroclaw, Poland), Apr. 23, 1868; s. Louis and Emilie (Schiffer) C.; studied medicine at Berlin, Freiburg (both Germany); M.D., 1891; also student Vienna (Austria); m. Hedwig; 2 sons, 1 dau. Asst. at nerve clinic, Breslau; became asst. nerve polyclinic, Vienna, 1895, named co-dir., 1919; became lectr. on tabes and tabetic psychosis, 1903; apptd. prof. 1912. Author: Die vasomotorisch-trophischen Neurosen, 1901, 12; (with H. Oppenheim) Die Vasomotorisch trophischen Erkrankungen, 1907; also articles in psychological and neurol. jours. Research on localized neurology, brain surgery. Died Berlin, Germany, Aug. 20, 1925.

CASSIUS, Andreas, physician, chemist; b. Schleswig, Denmark, circa 1605; s. Christian Andreas and Sophia (Vesteria) C.; M.D., Leiden (Netherlands), 1632; m. Catharina Willers, July 8, 1637; m. 2d, Gertrud Staphorst, 1642; 2 sons, 2 daus. Practiced medicine in Hamburg, Germany; personal physician to Prince Johann von Holstein, also bishop of Lübeck. Author: Disputatio inauguralis de miscellaneis medicis. Known as alchemist; discovered purple pigment used in glass, enamel and chinaware (purple of Cassius), 1687. Died Hamburg, May 27, 1673.

CASTEL, Louis Bertrand, French physicist, mathematician; b. Montpellier, France, Nov. 11, 1688; s. André-Guillaume Castel; ed. Sch. Jesuits Toulouse, France, Coll. Louis-le-Grand. Joined Soc. of Jesus, 1703. Mem. Royal Soc., 1730, Acad. Bordeaux, Acad. Rouen, Soc. Royale de Lyon, Author: Traité de la pesanteur universelle, 1724; Le Clavecin doculaire, 1725; le Plan d'une mathématique abrégée, 1727; la Mathématique universelle, 1728; Mémoires de Trévoux, 1735; Résponse à M. d'Anville sur le pays de Kamtchatka et de Jeco, 1737; Optique des couleurs, 1740; also account of Newton's system, 1743. Believed everything could be explained using principle of gravity of bodies and action of spirits; built chromatic harpsichord to illustrate his treatise on melody of colors. Died Paris, France, Jan. 11, 1757.

CASTELLANI, Marquis Aldo (Conte di Chisimaio), physician; b. Florence, Italy, Sept. 8, 1879; s. Ettore and Violante (Giuliani) C.; ed. U. Florence, also Bonn, Germany, Lister Inst., London, Eng., London Sch. Tropical Medicine; M.D.; m. Josephine Ambler Stead, Nov. 19, 1910; 1 dau., Jacqueline (Lady Killearn). Prof. tropical medicine, physician to King Umberto of Italy; mem. faculties Ceylon Med. Coll., U. Naples, Tulane U., New Orleans, La. State U., Ross Inst., London, U. Rome; prof. honoris causa Instituto de Medicina Tropical de Lisboa (Portugal). Recipient numerous decorations and awards. Fellow Accademia dei Lincei, Pontifical Acad. Scis.; mem. Circolog della Caccia, Rome, Am. Acad. Dermatology, French Soc. Microbiology (hon.), French Soc. Dermatology (hon.), Deutsche Dermatologische Gesellschaft (hon.). Author: Scoperta della etiologia della Malattie del Sonno; Framboesia tropicale; (with A. J. Chalmers) Manual of Tropical Medicine; Fungi and Fungous Diseases, Climate and Acclimatization, (with Jacono) Manuale di Clinica Tropicale, 1937; Malattie dell' Africa, 1946; Little Known Tropical Diseases, 1954; Ulcerations de la Jambe, 1958; Microbes, Men and Monarchs, 1960, 63, 68. Devised new serological methods (Castellani's absorption test, others); demonstrated cause of sleeping sickness and how it is transmitted (organism named for him) discovered spirochete of yaws; investigated parasitic diseases of skin; described brochospirochetosis (Castellani's bronchitis). Home: Villa Azzura, Rua Conde Ferreira 13, Cascais. Office: Instituto Medicina Tropical, Rua Junqueira, Lisbon, Portugal.

CASTELLANI, Maria, mathematician; b. Milan, Italy; d. Paolo and Eloisa (Namias) Castellani; Ph.D. in Math., U. Rome (Italy), 1919; postgrad. (fellow) Girton Coll., Cambridge (Eng.) U., (fellow) Bryn Mawr Coll. Came to U. S., 1945, naturalized, 1953. Actuary, statistician Nat. Security Fund, Rome, 1921-30, chief actuary, head statis. dept., 1932-46; actuary Internat. Labour Office, Geneva, Switzerland, 1930-32; privatdocent, lectr. U. Geneva, 1932-40, U. Rome, 1940; Lena Haag prof. math., chmn. dept. U. Kansas City, 1946-59; prof. math. Fairleigh Dickinson U., Teaneck, N.J., 1959—, chmn. dept., 1959-63. Cons. actuary Vatican, 1936-37, Internat. Agrl. Inst., FAO, 1936. Mem. Internat. Fedn. Bus. and Profl. Women (past v.p.), Am. Math. Soc., Inst. Math. Statistics, Econometric Soc., Am. Statis. Assn. Author: Sulla Frequenza della Invalinita, 1925; La Teoria dei Campioni, 1954. Research and publs. on algebraic geometry, actuarial and statis. math. and probability. Home: Butler Hall, 400 W. 119th St., N.Y.C. 10027. Office: Fairleigh Dickinson U., Teaneck, N.J. 07666.

CASTELLI, Benedetto, Italian mathematician, physicist; b. Perugia, Italy, 1577; studied under Galileo at Padua, Italy; studied at Florence, Italy;

joined Benedictine order, Monte Cassino, 1595; later abbot there; instr. math., Pisa, Italy, also at Coll. della Sapienza, Rome, Italy, 1615-25; employed to prevent flooding in Valley of Pisa between river Verchio and Fiome Morto by Ferdinand, grand duke of Tuscany, 1623; named papal mathematician by Urban VII, 1623; prof. U. Rome; apptd. to assist in drainage of Tuscany by Pope Urban VIII, 1625. Author: Della misura dell'acque correnti (compilation of Leonardo's notes), 1638; De la mesure des eaux courantes, pub. 1664; Studied law of continuity for steady flow of water (Castelli's law); invented helioscope; originated sound hydraulic tech.; taught Torricelli publs. on drainage, flood control, and currents. Died Rome, 1644.

CASTELLI, Pietro, Italian botanist; b. Rome, Italy, 1574; studied under Andrea Cesalpino; tchr. at Rome and Messina, Sicily; laid out bot. gardens, Messina, 1635; chemist and surgeon. Author: Treatise on the Odoriferous Hyena, 1638; De abusu circa dierum criticorum enumerationem, 1642; numerous pamphlets on sci. especially use of medicinal plants; helped compile catalogue of Messina gardens. Advocated study of human body through dissection. Died Messina, 1662.

CASTELLO BLANCO, Joao Rodriguez de, see Amatus Lusitanus.

CASTELNUOVO, Guido, Italian mathematician; b. Venice, Italy, Aug. 14, 1864; s. Enrico Castelnuovo; studied at U. Padua (Italy); prof. analytic and projective geometry U. Rome (Italy), 1891-35; became mem. Nat. Research Council, 1944. Mem. Acad. of the Lincei (pres. until 1952, apptd. senator for life 1950), French Acad. Scis. Author: Geometria analitica e proiettíva, 1903; Calcolo della probabilitá, 1919; Spazio e tempo secondo le veduti di Einstein, 1923. Research on algebraic geometry including invariant theory of surfaces, rationality of plane involutions; characterized rational surfaces using certain invariants; studied calculus of probability, natural philosophy especially determinism and chance, causality and indeterminancy. Died Rome, Apr. 27, 1952.

CASTER, Kenneth E(dward), Am. paleobiologist; b. New Albany, Pa., Jan. 26, 1908; A.B., Cornell U., 1929, M.S., 1930, Ph.D., 1933; m. Anneliese Schloh, June 18, 1933. Instr. geology Cornell U., 1929-35, Geneseo State Coll., 1936; instr.-curator U. Cin., 1936-39, prof. geology follow Grad. Sch., 1948—; vis. prof. U. Cal. at Berkeley, 1960, U. So. Cal., 1965; Fulbright vis. prof. U. Tasmania, 1957, U. Cologne (Germany), 1964. Guggenheim fellow Brazil, Colombia, 1947-48, S. Africa, Australia, 1956-57. Mem. Paleontol. Research Inst. (pres. 1967—), Ohio Acad. Sci. (pres. 1961), Paleontol. Soc. (pres. 1960), Geol. Soc. Am., Paleont. Gesellschaft, Soc. Econ. Paleontologists and Mineraologists, A.A.A.S., geol. socs. Australia, Brasil, France, Norway, Sweden, Paleont. Soc. Brasil, Paleont. Assn. (Gt. Britain), Soc. Vertebrate Paleontology, Evolution Soc. Author: (with Von Engeln) Geology, 1952; also numerous articles. Research on systematic paleontology (invertebrate), problematica, Devonian stratigraphy, So. Hemisphere geology, continental drift. Home: 425 Riddle Rd., Cin. 45220.*

CASTIGLIONI, Arturo, physician; b. Trieste, Italy, Apr. 10, 1874; s. Victor and Enrichetta (Bolaffio) C.; M.D., Vienna Univ., 1896; m. Marcella Sanguinetti, Apr. 26, 1903; children—Laura Luzzatto, Victor F. Came to U. S., 1939, naturalized 1946. Clin. asst. Vienna, 1896-98; head sanitary service, Lloyd Triestino, 1899-1918, of Italian Line, 1918-36; prof. history medicine, U. Padua, 1922-38; mem. high council Pub. Health, Rome, 1922-29; prof. U. for Foreigners, Perugia, 1934-38; lectr. history medicine Univs. of Sao Paulo, Rio de Janeiro, Buenos Aires, Santiago, Chile, 1930; Hideyo-Noguchi lectr. Johns Hopkins, Balt., 1933; research asso., lectr. history medicine Yale, 1939, prof. 1943-47. Fellow N.Y. Acad. Med.; hon. mem. Royal Soc. Medicine (London), Academia Nacional de Medicina (Buenos Aires, Gesellschaft der Aerzte (Vienna), Am. Assn. History Medicine. Author: Il Volto de Ippocrate, 1925; Storia della Medicina, 1927 (3d edit., 1949), Italian Med. 1932; History of Tuberculosis, 1933; The Renaissance of Medicine in Italy, 1934; L'orto della Sanita, 1934; Adventures of the Mind, 1946 (1st pub. in Italy as Incantesimo e Magia). Contbr. articles and essays to English, Italian, French, Dutch, Spanish, German and Am. med. jours. Original work in medico-hist. investigation. Died Jan. 21, 1952; buried Milan, Italy.

CASTILLO, Leopoldo Sanchez, Philippine animal scientist; b. Clarin, Philippines, Nov. 15, 1921; s. Adriano and Gertrudes (Sanchez) C.; B.S. in Agr., U. Philippines, 1948; B.S. in Edn., Colegio de San Jose, 1950; M.S., Cornell U., 1955, Ph.D., 1960; m. Gelia O. Tagumpay, June 16, 1954; children—Evello, Gertrudes, Nina. Instr., St. Paul's Acad., Inabanga, Bohol, Philippines, 1948-51; faculty Coll. Agr., U. Philippines, College, Laguna, 1951—, research asso. prof., 1962—, in charge dairy husbandry dept. animal husbandry, 1951-54, head dept. animal husbandry, 1957-58, in charge animal nutrition div., 1962—, in charge nutrition div. Dairy Tng. and Research Inst., 1962—; vis. asso. prof. Cornell U., Ithaca, N.Y., 1966-67. Mem. Philippine Nat. Food and Agr. Orgn. Com., 1963-66; cons. to U. S. and Philippine govt. agys.; mem. Philippine NRC. Fgn. Adminstrn. Operations-Nat. Econ. Council tng. grantee, 1954-55; Rockefeller Found. fellow, 1958-60. Mem. Soc. for Advancement Research, A.A.A.S., Philippine Assn. for Advancement Sci., Am. Dairy Sci. Assn., Am. Soc. Animal Sci., Philippine Soc. Animal Sci. (past v.p., Past pres.), Sigma Xi, Phi Kappa Phi, Gamma Sigma Delta, Phi Sigma. Cons. editor Philippine Agriculturist, 1965—; editor Philippine Jour. Animal Sci., 1967—. Research, publs. on livestock and poultry feeding and mgmt., utilization of farm by-products as well as forages and leaf meals from leguminous and non-leguminous plants as feeds. Address: College, Laguna, Philippines.*

CASTILLO NAJERA, Francisco, Mexican physician, diplomat; b. Durango, Mexico; s. Romualdo Castillo and Rosa (Najera) C.; A.B., Instituto Juarez, Durango, 1904; M.D., Facultad Nacional de Medicina, Mexico City, 1912, surgeon, 1913; m. Eugenia Davila, Mar. 14, 1917; children—Francisco, Rosa Eugenia, Luis, Guillermo. Began as physician, 1913; dir. mil. hosps., Leon y Torreon, 1915-18, Juarez Hosp., Mexico City, 1918-20, Med. Mil. Sch. of Mexico, 1920; minister from Mexico to China, 1922-24, to Belgium, 1927-30, to Holland, 1930-32, to France, 1932-35; became ambassador from Mexico to U. S., 1935; later chmn. Mexican delegation to UN. Author: The Campaign Against Yellow Fever in Mexico, 1923, others. Contr. in study of yellow fever. Died Dec. 20, 1954.

CASTILLON, Jean (Salvemini, Giovanni), mathematician; b. Castiglione, (Toscana), Jan. 15, 1708; s. Giuseppe and Maria Maddalena Lucia (Braccesi) Salvemini; doctor juris, Pisa, Italy; Dr.phil., Utrecht, Netherland, 1754; m. Elisabeth du Frèsne, 1745; m. 2d, Madeleine Ravène, 1759. Came to Switzerland, 1729; became dir. humanistic sch., 1737; moved to Lausanne, Switzerland, 1745; named rector U. Utrecht, 1758, prof. math. and astronomy, 1751; prof. math., Berlin, Germany; became 1st astronomer Berlin Obs., 1765, dir. math. dept., 1787. Mem. Berlin Acad. Scis., Acad. Göttingen, Royal Soc., London, acads. of Bologna, Mannheim, Padua, Prague. Publs. include: Opuscula of Newtown 1744; Correspondence of Leibniz and Johann Bernoulli, 1745. Editor: Introductio in analysin infinitorum (Euler), 1748. Translator: (into French) Locke's Elements of Natural Philosophy, Some Thoughts Concerning Reading; Amsterdam and Leipzig, 1757; (into Italian), Alexander Pope's Essay on Man. First to solve problem of inscribing a triangle, whose sides pass through 3 given points, into a circle; studied arty., writings of Newton and Euler; math. work on Kardiode, a curve named after him. Died Berlin, Germany, Oct. 11, 1791.

CASTLE, John Granville, Jr., Am. physicist; b. Buffalo, Sept. 9, 1924; s. John Granville and Grace (Draffan) C.; B.A. magna cum laude, U. Buffalo, 1947; Ph.D., Yale, 1950; m. Margaret Breese, Sept. 7, 1946; children—William B., Leslie Ann, Clytie M., John Lemoyne, Bruce Allen. Instr., U. Buffalo, 1950-51, research asso., 1952-55; research physicist Cornell Aero. Lab., 1951-52; physicist Westinghouse Research Labs., Pitts., 1955-67; prof. elec. engring. U. Pitts., 1967—. Cons. to Nat. Bur. Standards, Redstone Arsenal. Fellow Am. Phys. Soc.; mem. A.A.-A.S., Phi Beta Kappa, Sigma Xi. Author: (with others) Science By Degrees: Temperature from Zero to Zero, 1965; also articles. Research on magnetic and collision and flow properties of gases, magnetic resonance properties of solids; devel. of vortex free air thermometer, ion collector, solid state maser, talking typewriter. Home: 2059 Hampstead Dr., Pitts. 15235.*

CASTLE, Raymond Nielson, Am. chemist; b. Boise, Ida., June 24, 1916; s. Ray Newell and Lula (Nielson) C.; student Boise Jr. Coll., 1934-35; B.S., U. Ida., 1939; M.A., U. Colo., 1941, Ph.D., 1944; m. Ada Necia Van Orden, June 24, 1937; children—Raymond Norman, Dean Lowell, David Elliott, George Leonard, Elizabeth Anne, Edith Eilene, Christian Daniel, Lyle William. Faculty, U. N.M., Albuquerque, 1946—, prof. chemistry, 1956—, chmn. dept. chemistry, 1963—; research asso. Smith, Kline & French Labs., Phila., summer 1955; research fellow Tech. U., Copenhagen, Denmark, 1961-62. Mem. Am. Chem. Soc., Pharm. Soc. Japan, N.M. Acad. Sci., Chem. Soc. (London). Contbr. numerous articles to profl. jours. Editor, Jour. Heterocyclic Chemistry, 1963—. Research on synthesis of nitrogen heterocycles to potential physiol. activity, notably anticancer, antimalarial, diuretic, heterocyclic chemistry in ring systems with two nitrogen atoms adjacent in six-membered ring, namely pyridazines and condensed ring pyridazines, optical crystallography of organic compounds; discovered novel displacement reaction of halogen from activated positions in nitrogen heterocycles with phosphorus pentasulfide. Home: 1316 Avenida Manana N.E., Albuquerque 87110.*

CASTLE, William Bosworth, Am. physician; b. Cambridge, Mass., Oct. 21, 1897; s. William Ernest and Clara Sears (Bosworth) C.; student Harvard, 1914-17, M.D., 1921, D.Sc., 1964; M.S., Yale, 1933; M.D., U. Utrecht (Netherlands) 1936; D.Sc., U. Chgo., 1952, U. Pa., 1966; LL.D., Jefferson Med. Coll., Phila., 1964; L.H.D., Boston Coll., 1966; m. Louise Miller, July 1, 1933; children—William Rogers, Anne Louise. Asst. medicine Harvard Med. Sch., 1925-27, alumni asst., 1927-28, asst., 1928-29, instr. to asso. prof., 1929-37, prof., 1937-57, George Richard Minot prof. medicine, 1957-63, Francis Weld Peabody faculty prof. medicine, 1963-68; dir. Thorndike Meml. Lab., Boston City Hosp., 1948-63, hon. dir., 1963-68, jr. vis. physician Boston City Hosp., 1933-48, asst. vis. physician, 1948-55, vis. physician, 1956-63, dir. II and IV Med. Services, 1940-63, cons. physician, 1963—; dir. Rockefeller Found. Commn. for Study Anemia in P.R., 1931-32; sr. cons. hematology Lemuel Shattuck Hosp., 1955—; cons. medicine Beth Israel Hosp., 1956—; cons. staff in internal medicine Mt. Auburn Hosp., 1965—. Recipient John Phillips Meml. prize, A.C.P., 1932; William Procter Jr. internat. award for distinguished service in scis., Phila. Coll. Pharmacy and Sci., 1935; Walter Reed medal, Am. Soc. Tropical Medicine, 1939; Mead Johnson & Co. award for research on vitamin B complex, 1950; Gordon Wilson medal Am. Clin. and Climatol. Assn., 1961; George W. Kober medal Assn. Am. Physicians, 1962; John M. Russell award of Markle Scholars, 1964; ann. hon. Lecture award Albany Medical Coll., 1964; Distinguished Lecture award Coll. Medicine U. Ky., 1965; Oscar B. Hunter award Am. Therapeutic Soc., 1965, Joseph Goldberger award for clin. nutrition A.M.A. and Nutrition Found., 1966. Fellow Am. Acad. Arts and Scis., Royal Coll. Physicians, A.C.P. (master 1957—); hon. fellow Royal Coll. Phys. and Surg. of Can., Royal Australasian Coll. Physicians; mem. Am. Acad. Tropical Medicine, A.A.A.S., A.M.A., Am. Philos. Soc., Am. Soc. Clin. Investigation (pres. 1940-41, emeritus 1943—), Am. Soc. Exptl. Pathology, Am. Soc. Tropical Medicine and Hygiene, Am. Fedn. Clin. Research (sr.), Assn. Am. Physicians (pres. 1959-60, emeritus 1960—), Nat. Acad. Scis., Royal Coll. Physicians (Edinburgh), Phi Beta Kappa, Alpha Omega Alpha; corr. mem Société Internationale Europeenne Hematologie, l'Academie royale de Medecine de Belgique, Brit. Med. Assn.; hon. mem. numerous others. Demonstrated lifesaving effects of parenteral therapy with crude liver extract in advanced cases of sprue, and iron-therapy cure for anemia of hookworm disease, 1931-32; major research has been in pathologic physiology of nutritional and hemolytic anemias, relation between achylia gastrica and pernicious anemia, concept of erythrostasis as phenomenon underlying various types of accelerated red cell destruction. Home: 22 Irving St., Brookline, Mass. 02146. Office: Boston City Hospital, Boston 02118.*

CASTLE, William Ernest, Am. zoölogist; b. Alexandria, O., Oct. 25, 1867; s. William Augustus and Sarah (Fassett) C.; A.B., Denison U., 1889; A.B., Harvard, 1893, A.M., 1894, Ph.D., 1895; LL.D., Denison, 1921; Sc.D., Wisconsin, 1921; m. Clara Sears Bosworth, Aug. 18, 1896 (died May 22, 1940); children—William Bosworth, Henry Fassett (dec.), Edward Sears. Taught Latin, Ottawa (Kan.) U., 1889-92; instr. vertebrate anatomy U. Wis., 1895-96; instr. biology, Knox Coll., 1896-97; instr. zoölogy, Harvard, 1897-1903, asst. prof., 1903-08, prof., 1908-36, now emeritus; research asso. in genetics, U. Cal., Berkeley, 1936—; also research asso. Carnegie Instn. Washington. Recipient Kimber award, genetics, Nat. Acad. Scis., 1955. Fellow Am. Acad. Arts and Scis.; mem. Nat. Acad. Scis., Am. Philos. Soc., Boston Soc. Natural History, A.A.A.S., Am. Soc. Naturalists (pres. 1918), Am. Soc. Zoölogists (pres. Eastern Branch 1905-06), Phi Beta Kappa, Sigma Xi. Author: Heredity in Relation to Evolution and Animal Breeding, 1911; Genetics and Eugenics, 1916, 4th rev. edit., 1930; Genetics of Domestic Rabbits, 1930; Mammaliam Genetics, 1940; also several publs. on subjects of heredity and evolution. Performed significant expts. on natural selection (using piebald hooded rats) which illustrate importance of mutation in alteration of species. Died, Berkeley, Cal., June 3, 1962.

CASTLEMAN, Benjamin, Am. physician; b. Everett, Mass., May 17, 1906; s. Samuel and Rose (Michaelson) C.; B.A., Harvard, 1927; M.D., Yale, 1931; M.D. (hon.), U. Goteborg (Sweden), 1966; m. Anna Alice Segal, Dec. 22, 1935; children—Ruth (Mrs. Emery E. Griffin, Jr.), Jean (Mrs. Lewis Chase), Paul Arnold. House officer pathology Mass. Gen. Hosp., 1931-32, resident pathologist, 1932-35, mem. staff, 1935—, chief dept. pathology, 1953—, editor case records, 1951—; mem. faculty Harvard Med. Sch., 1935—, prof. pathology, 1962—; cons. pathologist Emerson (Mass.) Hosp., Worcester (Mass.) Meml. Hosp., Mass. Eye and Ear Infirmary, Mass. Soldiers Home; vis. prof. All India Inst. Med. Scis., New Delhi, 1961-62, Scandinavian univs., 1964; lecture tours fgn. med. schs. Mem. part III com. Nat. Bd. Med. Examiners, 1964—; chmn. allocations com. Mass. Heart Assn., 1962-65; mem. expert com. med. edn. WHO, 1958-62, cons. cancer unit, 1963. Mem. Am., Mass. med. assns., Am. Acad. Arts and Scis., Am. Assn. Pathologists and Bacteriologists, Internat. Acad. Pathology (pres. 1962-63), Am. Soc. Exptl. Pathology, New Eng. Soc. Pathologists, New Eng. Cancer Soc. Research on the thymus and parathyroid glands; pulmonary disease; hypertension; myasthenia gravis. Home: 203 Clinton Rd., Brookline, Mass. 02146. Office: Dept. Pathology, Mass. Gen. Hosp., Boston 02114.*

CASTNER, Hamilton Young, chemist; b. Bklyn., 1859; ed. Bklyn. Poly. Inst., Columbia Coll. Sch. of Mines; went to Eng. as analytical chemist; invented electrolytic method of isolating sodium and chlorine from brine and founded (with Kellner) a co. for working it. Died 1899.

CASTRACANE, Francesco, Italian microscopist; b. Fano, Italy, July 19, 1817; s. Leonardo and Laurentina (Caleotti della Zecca) C.; ed. Jesuit sch., 1826-37; ordained priest, 1840. Research, publs. on methods of illumination, diatoms; discovered structure of valves and method of reproduction. Died Mar. 27, 1899.

CASTRÉN, J(orma) A(dolf), Finnish physician, ophthalmologist; b. Tornio, Finland, July 2, 1919; s. Karl Adolf and Sylvia (Tallus) C.; M.D., Helsinki (Finland) U., 1947, M.Sc.D., 1955; Competency prof. ophthalmology, Oulu (Finland) U., 1965; m. Irma Tellervo Nordman, June 27, 1947; children—Anita Irmeli, Leena Synnöve. Asst., Eye Clinic, Karolinian Hosp., Stockholm, Sweden, 1948-49, Birmingham, Ala., 1958; staff Helsinki U. Eye Hosp., 1952—, teaching specialist physician, 1966—; vice prof. ophthalmology Oulu U., 1965; ophthalmologist Lahti Central Mil. Hosp., 1949-51. Sec., Eye Found., 1960-61. Mem. Soc. Ophthalmologists Finland (past sec.), Duodecim (Finland), Assn. Physicians Finland, Soc. Ophthalmologists Finland. Author: Significance of Prematurity on the Eye, 1955; (with others) Recommendations of Light in Hospitals, 1965; also numerous articles. Research on eyes of premature infants, relating fibroplasia to oxygen therapy in incubators, methods of eye surgery. Home and office: 5 A Merikatu, Helsinki, Finland.*

CASTRO, see Rodrigues de Castro, Estevan.

CASWELL, Alexis, Am. astronomer; b. Taunton, Mass., Jan. 29, 1799; s. Samuel and Polly (Seaver) C.; grad. Brown U., 1822, D.D. (hon.), 1841, LL.D., 1865; m. Esther Thompson, 1830; (dec. 1850); m. 2d, Elizabeth Edmands, 1855; 6 children. Prof. ancient langs. Columbian U., Washington, 1825-27; entered ministry as pastor of a Baptist Ch., Halifax, N.S., Can., 1827; prof. math. and natural philosophy Brown U., 1828-63, pres., 1868-72, trustee, 1873-75, regent, 1840; became asso. fellow Am. Acad. Arts and Scis., 1850; v.p. A.A.A.S., 1855; chosen by U. S. Govt. as one of 50 incorporators Nat. Acad. Scis., 1863; became pres. Nat. Exchange Bank, also Am. Screw Co. (both Providence), 1863; pres. R.I. Hosp., 1875-77; dir., v.p. Providence Athenaeum. Author: Smithsonian Contributions to Knowledge, 1860; other works include: Lectures on Astronomy, 1858, Memoirs of John Barstow, 1864. Made precise measurements on temperature and pressure and observations on winds, clouds, moisture, rain, storms, and the aurora from College Hill in Providence; partially through his efforts meteorology came to be recognized as a sci. Died Providence, Jan. 8, 1877.

CASWELL, John, Brit. mathematician; b. Somerset, Eng., 1655; grad. Wadham Coll., Oxford, Eng., 1671; tchr. math. Oxford U. from 1677, elected Savilian prof. astronomy, 1709. Author: Exercitatio geometrica de dimensione figuarum, 1684; Catoptricae et dioptricae sphaericae elementa, 1695; Astronomiae physicae et geometriae elementa, 1702; also a work on trigonometry, 1685. Made new measurements of densities of 29 metallic materials; experiments on measuring height with mercury barometer. Died 1712.

CASWELL, Randall Smith, Am. physicist; b. Eugene, Ore., Feb. 7, 1924; s. Albert Edward and Constance (Edwards) C.; student U. Ore., 1940-42; S.B., Mass. Inst. Tech., 1947, Ph.D., 1951; m. Jean Marden Miller, June 14, 1945; children—William Edward, Virginia Lee, Anne Marden, Ellen Sue, Wendy Jean, Julia Constance. Asso. prof. physics U. Ky., 1950-52; physicist neutron physics Nat. Bur. Standards, Washington, 1952—; vis. physicist Centre d'Etudes Nucléaires de Saclay, France, 1963-64; adj. prof. physics Am. U., Washington, 1957—. Fellow Am. Phys. Soc. Exptl. measurement of Beta-ray average energies, neutron penetration in water, response of neutron instruments, theoretical calculations of neutron cross sections using nuclear optical model and statis. models. Home: 2209 Salisbury Rd., Silver Spring, Md. 20910. Office: Nat. Bur. Standards, Washington 20234.*

CATALA, Joaquin, Spanish physicist; b. Manresa, Barcelona, Spain, Sept. 14, 1911; s. José and Maria Dolores (De Alemany) C.; M.A., U. Barcelona, 1932; Phil.Sc., U. Madrid (Spain), 1940; m. Adela Torrente, Jan. 17, 1940; 1 dau. Mari Carmen (Mrs. Dilla). Lectr. physics U. Barcelona, 1933-39; U. Madrid, 1940-42; prof. physics U. Valencia (Spain), 1943—; chief Meteorol. Centre Levante, Valencia, 1963—; dir. Inst. Corpuscular Physics, 1950—. Recipient 1st medal physics, 1959, Alfonso el Sabio plate. Mem. Acad. Sci. Madrid and Barcelona, Inst. Nuclear Studies (vice chmn. 1966). Author: Fisica General, 1958; also numerous articles. Research in nuclear physics, both low and high energy, photographic emulsion technique. Home: 4, Cerdan de Tallada, Valencia, Spain.*

CATALAN, Eugène Charles, Belgian mathematician; b. Bruges, Belgium; 1814; studied at l'École polytechnique, Paris, France; Docteur ès Science math., 1841; Aggregation Scis., 1846. Resigned position with Dept. Bridges and Hwys. to teach math.; prof. math. Collège de Chalons-sur Marne; math. tchr. Collège Charlemagne; taught math Lycée S.-Louis, beginning in 1848; participated in revolution of 1848; became prof. analysis U. Liège (Belgium), 1865. Author: Élements de geometrié, 1843; Notions d'astronomie, 1860; also numerous publs. on multiple integrals, gen. theory of surfaces, math. analysis, calculus of probability, geometry, superior arithmetic. Developed circulants, spl. forms of determinants; research on spherical harmonics, analysis of differential equations, transformation of variables in multiple integrals, theory of continuous fractions, uncertain series, indefinite products. Died Liège, 1894.

CATALANO, Giuseppe, Italian botanist; b. Palermo, Italy, Dec. 8, 1888; s. Emanuele and Carolina (Cocchiara) C.; Ph.D. in Natural Scis. Former dir. bot. inst. Agronomical Sch. of Naples (Italy); dir. Exptl. Sta. for Medicinal Plants, Naples; prof. Botany, dir. inst. and bot. garden, sch. of sci. U. Naples. Mem. Superior Council of Agr. and Forests; mem. undercom. for medicinal plants Nat. Council Research. Mem. Acad. Phys. and Math. Sci., Naples, Acad. Sci., Letters and Arts, Palermo, Acad. Gioenia (corr. Catania), Acad. Pontaniana, Naples, other socs. Author: Teoria generale della foglia, 1946; Botanica agraria, 1948; Ricordi di filosofia agraria, 1948-52; Teoria biologica della musica, 1963. Address: via Luigia Sanfelice 5, Naples, Italy.

CATALDI, Pietro Antonio, Italian mathematician; b. Bologna, Italy, Apr. 15, 1552; studied at Acad. Design, Florence, Italy; apptd. prof. math. and astronomy Florence, 1563, Perugia, Italy, 1572; then founder early math. acad. at Bologna; prof. U. Bologna, 1584. Author: Pratica arithmetica, 4 parts, 1602, 06, 16, 17; Trattato dei numeri perfetti, 1603; Opusculum de lineis rectis, 1603; Trattato del mondo brevissimo di trovare la radice quadra delli numeri, 1613; Regola della quantità, o cosa di cosa, 1618; L'Algebra discorsiva, 1618; La nuova algebra proporzionale, 1619; L'Algebra applicata, 1622; Editor 1st 6 books Euclid's Elements, 1620. Devoted many years to numerical calculations; pioneer in use of continued fractions; invented common form of continued fractions, 1613; developed 1st symbolism for continued fractions, using standard notation; used continued fractions systematically to find sq. roots of numbers; contbd. to study of quadrature of circle, 1612, theory of roots, 1613. Died Bologna, Feb. 11, 1626.

CATANI, Renato Amilcare, Brazilian chemist; b. Sao Paulo, Brazil, Sept. 15, 1916; s. Alberto and Maria (Catani) C.; ed. E.S.A. Luiz de Queiroz, U. Sao Paulo; Engenheiroagronomo; Doctor in Agronomy; m. Maria Jose Mendes, June 13, 1942; children—Marilia, Afranio. Head, dept. soil fertility Instituto Agronomico, Sao Paulo, 1942-55; prof. chemistry E.S.A. Luiz de Queiroz, U. Sao Paulo, 1955—. Mem. Brazilian Soc. for Progress of Sci., Brazilian Soc. Chemistry, Brazilian Soc. Soils, Brazilian Soc. Metals. Contbr. papers to sci. publs. Research on chem. composition of coffee, sugar cane, cotton, other tropical plants, also on chem. composition and fertility of soils of state of Sao Paulo. Home: 1141 Av. Brasil, Piracicaba, Sao Paulo, Brazil.*

CATELAN (or CATALAN), Laurens, French pharmacist; b. Montpellier, France, 16th century; apothecary, lectr. U. Montpellier; lectured on med. herbs, demonstrated pharmacy, 1605. Author: Descours sur la theriaque, 1614; Rare et curieux discours de la plante appelle mandragore (description of mandrake root), 1638. Died after 1639.

CATES, Louis Shattuck, Am. mining engr.; b. Boston, Dec. 20, 1881; s. Edwin Wallace and Emily Allen (Johnson) C.; S.B., Mass. Inst. Tech., 1902; Dr. Engring., Mich. Coll. Mining and Tech., 1936; D.Sc., U. Ariz., 1946, Columbia, 1947; m. 3d. Ethel Chesbrough Lewis, May 12, 1951. Mine operator in Mexico, 1902; asst. to pres. Nat. Steel & Wire Co., N.Y., 1903-04; in charge of constrn. and devel. at Bingham Canyon. Utah, for Boston Consolidated Mining Co., 1904-08. gen. mgr., 1909; with Ray (Ariz.) Consol. Copper Co., 1910-22; asst. gen. mgr. Utah Copper Co., 1919-22, gen. mgr., 1922-30, v.p., 1923-30; asst. gen. mgr., Bingham & Garfield Ry. Co., 1919-22, gen. mgr., 1922-30, v.p., 1923-30; pres. Phelps Dodge Corp., 1930-47, chmn. bd., 1947-59; dir. Phelps Dodge Copper Products Corp., L.I. R.R. Co., Phelps Dodge Refining Corp., Niagara Fire Ins. Co., So. Peru Copper Corp. Mem. div. engring. and indsl. research NRC, 1939-41. Recipient Saunders Gold medal Am. Inst. Mining and Metall. Engr., 1939; Gold medal Mining and Metall. Soc. Am., 1956. Mem. Copper and Brass Research Assn. (v.p., dir. 1931-32), Am. Mining Congress (pres. 1925), Mining and Metall. Soc. Am. (pres. 1931), Am. Inst. Mining and Metall. Engrs. (v.p. 1934-37, dir. 1919-21, 1931-34, 1936-39, pres. 1946), Nat. Inst. Social Scis. (life). Developed system making it possible, by underground methods, to mine low grade ores. Died Oct. 29, 1959.

CATESBY, Mark, English naturalist; b. Castle Hedington, Essex, Eng., Apr. 3, 1683; s. John and Elizabeth (Jekyll) C.; studied natural sci., London; m. Elizabeth Rowland, Oct. 2, 1747; children—Mark, Ann. Came to Am., 1712; studied floral and fauna in Va., 1712-19, S.C., Ga., Fla., 1722-25, Bahamas, 1726. Fellow Royal Soc., 1733. Author: The Natural History of Carolina, Florida, and the Bahama Islands (containing 200 engraved zool. plates), Vol. 1, 1731, Vol. 2, 1743, appendix, 1748; Hortus Britanno-Americanus: or, A Curious Collection of Trees and Shrubs, 1763, also papers on ornithology and anthropology. Laid found. for future devel. of Am. ornithology by combining illustrations (showing birds in natural environments) with reasonably accurate textual descriptions; introduced new theory of bird migrations; made earliest tech. identification of a vertebrate fossil (mammoth) in Am. circa 1724; studied fishes, reptiles, and insects of Isle of Providence; theorized that Am. Indians were of Asian origin; possibly was 1st to make illustrations of Am. insects. Died London, Eng., Dec. 23, 1749.

CATHCART, Edward Provan, Brit. physician, surgeon; b. Apr. 1877; s. Edward Moore Cathcart; M.B., Ch.B., Glasgow (Scotland) U., 1900, M.D., 1904, S D.Sc., 1906, LL.D., 1948; postgrad. univs. Munich, Berlin, Leningrad; LL.D., St. Andrews U., 1928; m. Gertrude Dorman Bostock; 3 daus. Research scholar Lister Inst., London, 1902-04, asst., 1904-05; Grieve lectr. chem. physiology U. Glasgow, 1905-15, Gardiner prof., 1919-28, Regius prof. physiology, 1928-47; prof. physiology London Hosp. Med. Sch., 1915-19. Hon. research asso. Carnegie Inst. Instn. Washington, 1911. Mem. Med. Adv. Com., Scotland; mem. Agrl. Research Council; mem Med. Research Council; mem. Army Hygiene Adv. Com.; mem. Gen. Med. Council; chmn. Indsl. Health Research Bd.; mem. adv. com. agrl. sci. Devel. Commn.; chmn. Scottish Health Services Com.; mem. adv. com. nutrition Ministry Health. Fellow Royal Soc., 1920, Royal Soc. Edinburgh. Author: The Physiology of Protein Metabolism; The Human Factor in Industry; also papers. Died Feb. 18, 1951.

CATHELIN, Fernard, French surgeon; b. France, 1873. Author: Manuel pratique de la lithotritie, 1911; La circulation du liquide cephalo-rachidien avec applications a la therapeutique, 1912; Cinq années de pratique et d'enseignement a l'Hôpital d'urologie et de chirurgie urinaire, 1913; Chirurgie urinaire de guerre, 1919; Les migrations de oiseaux, 1920; (with A. Grandjean) L'infection gonococcique et ses complications, 1928. Editor: Travaux annuels de l'Hôpital d'urologie et de chirurgie urinaire, 1919-26. Credited with introducing caudal anesthesia by injecting anesthetic into epidural space through sacrococcygeal ligament, 1901; advocated caudal anesthesia in 2d stage of labor, 1905. Died 1945.

CATLIN, Charles Albert, Am. chemist; b. Burlington, Vt., May 10, 1849; s. Henry Wadhams and Mary Cobb (Mayo) C.; B.S., U. Vt., 1872, Ph.B., 1873, Sc.D., 1913; spl. course Mass. Inst. Tech., 1894, 95; m. Frances L. Herrick, June 20, 1877. Chemist, Rumford Chem. Works, Providence, 1873-75, 1878-1916. Author: Baking Powders, 1899. Inventor and patentee of chem. processes and applications, many of which relate to mfr. of phosphates for dietetic purposes. Died Apr. 2, 1916.

CATLIN, George, Am. ethnologist, anthropologist, artist; b. Wilkes-Barre, Pa., July 26, 1796; s. Putman and Polly (Sutton) C.; student, law sch. of Reeves and Gould, Litchfield, Conn., 1817-18; studied painting, 1820-23; academician Pa. Acad. Fine Arts, 1824; m. Clara B. Gregory, May 10, 1828; children—Elizabeth Wing, Clara Gregory, Louise Victoria, George. Accompanied Gov. Clark to meeting of Winnebagoes, Menomonees, Shawnees, Sacs, Foxes, 1830-31; painted portraits of Black Hawk and his braves, traveled up Missouri to mouth of Yellowstone River, painting 10 tribes, including the Mandans, 1832; ascended Platte to Ft. Laramie, painting Pawnees, Omahas, Otoes, Arapahoes, Cheyennes, on way to Great Salt Lake, 1833; visited Comanches, other tribes of S.W., 1834; ascended Mississippi to Falls of St. Anthony, visiting Miss. Sioux, Objibaways, Saukes (Sacs), returned 1836; visited Seminoles and Euchees in Fla., 1837, Osceola and other Seminole chiefs imprisoned in Charleston, 1837; visited France, 1845-48; traveled in S.Am., met von Humboldt. Author: Letters and Notes on the Manners, Customs, and Conditions of the North American Indians, 1841; The Breath of Life, or Mal-Respiration, 1861; Life Amongst the Indians, 1867; Last Rambles Amongst the Indians of the Rocky Mountains and the Andes, 1867; The Lifted and Subsided Rocks of America, with their influences on the oceanic, atmospheric, and land currents, and the Distribution of the Races, 1870; The O-kee-pa, a religious ceremony of the Mandans; Catlin's Notes in Europe, 2 vols.; also over 1200 oil paintings, etchings, drawings, mostly of Indians in native surroundings in N. Am., Uruguay, Yucayali, on Amazon. Suggested reservation of pub. lands for nat. park, 1832; selected Yellowstone region for park; made valuable contbn. to anthropology of Am. Indians; discovered red pipestone. Died Jersey City, Dec. 23, 1871.

CATO, Marcus Porcius (The Censor), Roman biologist, statesman; b. Tusculum, Italy, 234 B.C.; quaester under Scipio, from 204 B.C.; became consul, 195 B.C.; gen. army, served in war, Spain, 194 B.C., war against Antiochus, 191 B.C.; became censor in 184 B.C. and ambassador to Carthage, 150 B.C. Advocated 3d Punic war, sought to restore morality and simplicity of early Roman republic. Author: De agricultura (or De re rustica), a farm manual, (edited by Keil, 1882); Origines (extant in fragments). Recommended cabbage poultice as cure for all human diseases including cancer; believed black bile and swollen spleen were results of malarial fever. Died 149 B.C.

CATON, Richard, Brit. surgeon; b. 1842; M.D.; LL.D., univs. Edinburgh, Padua, Liverpool; m. Annie Ivory, 1942; 2 daus. Former pro-chancellor, emeritus prof. U. Liverpool; past pres., cons. physician Liverpool Royal Infirmary; mem. Gen. Med. Council, 1904-26; vice chmn. Liverpool Sch. Tropical Medicine; past pres. Liverpool Med. Instn. Fellow Royal Coll. Physicians (past mem. council, Harveian Orator); mem. Hellenic Soc. Author: Temples and Ritual of Asklepios; Prevention of Valvular Disease of the Heart; Iemhotep and Ancient Egyptian Medicine; How To Live. Was 1st in succeeding in leading off and recording action currents of brains of animals (led to devel. of electroencephalograph, 1875); research on pathology of pneumonia and pleuro-pneumonia, pathology and treatment of chorea, treatment and symptoms of enteric fever. Died Jan. 2, 1926.

CATRON, Damon V(on), Am. nutritionist; b. Russiaville, Ind., Sept. 12, 1915; s. Andrew E. and Lucille (Avery) C.; B.S. with distinction (4-H Club scholar), Purdue U., 1938; M.S., U. Ill., 1945; Ph.D., Ia. State U., 1948; m. 2d, Geraldine Wright Rogers, Apr. 10, 1965; children by previous marriage—Damon Dwayne, Patricia Ann (Mrs. Ross Christensen), Carmen Rosalee (Mrs. Vincent Pezzullo), Kenneth Randolph Rogers. Instr. high sch. New Castle, Ind., 1938-40; jr. livestock specialist Purdue U., 1940-41; poultry specialist Ralston-Purina Co., St. Louis, 1942-43; faculty U. Ill., 1943-44; nutritionist Honeggers' & Co., Forrest, Ill., 1944-45; faculty animal sci. dept. Ia. State U., 1945-60, prof., 1953-60; v.p. Walnut Grove Products Co., Inc., Atlantic, Ia., 1960-64; dir. life scis. research, research div. W. R. Grace & Co., Washington Research Center, Clarksville, Md., 1964-66; coordinator food sci. and nutrition U. Mo., Columbia, 1966—. Recipient Outstanding Research award Am. Feed Mfrs., 1953, Hon. award Farm House Frat., 1964, Distinguished Nutritionist award Distillers Research Council, 1964. Fellow Am. Inst. Chemists, A.A.A.S.; mem. Am. Soc. Animal Sci., Am. Inst. Nutrition, Soc. Exptl. Biology and Medicine, Fedn. Am. Socs. Exptl. Biology, N.Y. Acad. Scis., Am. Dairy Sci. Assn., Inst. Food Technologists, Alpha Zeta, Kappa Delta Pi, Phi Sigma Biol. Soc., Gamma Sigma Delta, Phi Kappa Phi, Sigma Xi, Ceres, others. Research and publs. on nutritional requirements of animals; growth supplements and antibiotics in animal diets; protein, enzyme studies; nutritional aspects of disease. Home: 1407 Ridgemont Ct., Columbia, Mo. 65201.*

CATTELL, Raymond Bernard, psychologist; b. West Bromwick, Eng., Mar. 20, 1905; s. Alfred Ernest and Mary (Field) C.; B.Sc. with 1st honors, King's Coll., London, Eng., 1926; Ph.D., U. Coll., U. London, 1929, D.Sc., 1939; m. Monica Rogers, Dec. 8, 1930 (div. 1934); 1 son, Hereward; m. 2d, Alberta Karen Schuettler, Apr. 2, 1946; children—Mary, Heather, Roderic, Devon. Came to U. S., 1937, naturalized, 1947. Asst. lectr. U. S.W. Eng., 1927-30; dir. Leicester Child Guidance Clinic, 1931-37; research asso. to E. L. Thorndike, 1937-38; G. Stanley Hall Prof., Clark U., 1938-41; lectr., fellow Harvard, 1941-44; cons. Adj. Gen.'s Office, 1944-45; Distinguished Research prof., dir. Personality Lab., U. Ill., Urbana, 1945—. Chief cons. Inst. Personality and Ability Testing, Champaign, Ill., 1949—. Darwin Research fellow, 1935-37. Recipient Wenner Gren prize N.Y. Acad. Sci., 1950. Mem. Am. Psychol. Assn., Brit. Psychol. Soc., Soc. Multivariate Exptl. Psychology, Psychometric Soc. Author numerous books most important being: Description and Measurement of Personality, 1946; Personality, 1950; Personal and Motivational Structure and Measurement, 1957; Meaning and Measurement of Anxiety and Neuroticism (with I. Scheier), 1961; Personality and Social Psychology, 1964; The Scientific Analysis of Personality, 1965; Objective Personality and Motivation Tests, 1967; also many articles. Editor: Handbook of Multivariate Experimental Psychology, 1966. Research on determination main functional unities in personality by factor analysis, evaluation of genetic and cultural influences on personality, Mava analysis psychol. inheritance; objective human motivation and conflict measurement with math. model; developed objective measurements for small group and nat. culture pattern classification; discovered evidence for fluid ability concept and culture fair intelligence tests; improved factor analytic and taxonomic methodology. Home: 615 Kirby Rd., Champaign, Ill. Office: Laboratory Personal Analysis, U. Ill., Urbana, Ill. 61820.*

CAUCHOIX, Robert-Aglaé, optician; b. France, Apr. 24, 1776; built optical and other sci. instruments; inventor achromatic glass with long focal length. Died Feb. 5, 1845.

CAUCHY, Baron Augustin Louis, French mathematician; b. Paris, Aug. 21, 1789; s. Louis François and Marie Madeleine (Desestre) C.; ed. École Polytechnique, also École ces Ponts et Chaussées; m. Aloise de Bure, 1818; 2 daus. Engr., port of Cherbourg, France; instructorships in algebra Faculty Scis., in math. physics Coll. France, in mechanics École Polytechnique (all Paris), until 1830; forced into exile because of his adherence to Catholic faith and French monarchy, 1830; prof. math. physics U. Turin (Italy), 1830-37; apptd. tutor to Duke of Bordeaux, Prague, Czechoslovakia, 1832; returned to Paris, 1837; named prof. math. astronomy Faculty Scis. Paris, 1848, reinstated 1854; also prof. École Polytechnique. Fellow Royal Soc., 1832; mem. French Acad. Scis. Author: Méthode pour déterminer à priori le nombre des racines réelles, 1813; Théorie des ondes, 1815; Cours d'analyse de l'École Polytechnique, 1821; Le calcul infinitésimal, 1823; Leçons sur les applications du calcul infinitésimal à la géométrie, 2 vols., 1826-28; Mémoire sur l'application du calcul des residus à la solution des problèmes de physique mathematique, 1827; Mémoires sur la resolution des équations numeriques et la théorie de l'elimination, 1829; Mémoire sur la dispersion de la lumière, 1836; Note sur la développement des fonctions en series ordonnées suivant les puissance ascendantes des variables, 1846; also over 500 articles. Concentrated on principles of calculus, functions of complex variable, number theory, group theory; algebraic analysis; developed rigorous methods of analysis; 1st to rigorously prove Taylor's theorem; helped originate theory of permutation groups; invented calculus of residues; gave integral calculus a logical basis; applied math. to physics; 1st to attempt finding a math. basis for ether; formulated refractive index in terms of wave length; investigated reflections in metals, 1842, also wave surfaces in 2 axis crystals; pioneer in molecular mechanics; substituted idea of continuity of geometrical displacements for principle of continuity of matter; in optics, developed wave theory and simple dispersion formulae; originated theory of stress in elasticity; studied hydrodynamics and astronomy; accounted for motion of Pallas. Died Sceaux, France, May 23, 1857.

CAUDELL, Andrew Nelson, Am. entomologist; b. 1872; custodian Orthoptea, Nat. Mus.; wrote many articles based on museum material; described many new species. Died 1936.

CAUER, Wilhelm, German mathematician; b. Berlin, Germany, June 24, 1900; s. Wilhelm and Marie (Koch) C.; ed. Tech. U. Berlin, U. Berlin, U. Bonn (all Germany); m. Karoline Cauer, 1925; 2 sons, 4 daus. Asst. lectr. Math. Inst. Göttingen, Germany; Rockefeller fellow Mass. Inst. Tech., Cambridge, Mass., 1930-31; prof. applied math. Göttingen; engr. with aircraft co., Kassel, 1935-36; head Mix & Genest AG. Labs., also prof. applied math. Tech. U. Berlin, 1936-45. Author: Ein reaktanztheorem, 1931; Theorie der linearen Wechselstromschaltungen, Vol. 1, 1941. Founded network syntheses, a spl. branch of electrotechnics dealing especially with transmission techniques. Died Berlin, Apr. 22, 1945.

CAUJOLLE, Fernand, French pharmacologist; toxicologist; b. Toulouse, France, July 14, 1901; s. Etienne and Marie-Elisabeth (Bonnefont) C.; Chem. Engr., Ecole Nationale Supérieure de Chimie de Toulouse, 1922, Dispensing Chemist, 1925, Docteur en Médecine, 1929, Docteur ès Sciences, 1939; m. Odette Courtade, Dec. 28, 1931 (dec.); children—Anne-Marie, Jacqueline, Raymond; m. 2d, Denise Caujolle, Aug. 6, 1960. Faculty, U. Alger, 1923-28; faculty Faculty Pharmacy, Toulouse, 1939—, prof. chem. pharmacy and pharmacodynamics, 1941—; prof. indsl. hygiene Ecole Nationale Supérieure de Chimie de Toulouse, 1958—; dir. Centre de Recherches sur les Toxicités, C.N.R.S., Toulouse, 1960—. Expert toxicology and pharmacology French Ministry Health, 1960—. Decorated officier Légion d'honneur, Commandeur des Palmes Académiques. Mem. French Académie de Médecine et de Pharmacie, French Commn. du Codex (corr.), French Ordre des Pharmaciens, Académie des Sciences (pres.), Inscriptions et Belles Lettres. Research on exptl. toxicology and pharmacodynamics, especially thermal sulfur, boron, cobalt, stain, germanium derivatives. Address: 205 route de Narbonne, Toulouse 31, France.*

CAULLERY, Maurice, French naturalist; b. Bergues, France, Sept. 5, 1868; D.Natural Scis., 1895; instr., Lyons and Marseille, France; succeeded Giard as head dept. evolution living organisms Faculty Scis., Paris, France, 1909; improved Zool. Sta. of Wimereux, Lab. of Evolution of Living Organisms. Recipient Linnaean gold medal, 1947. Mem. French Acad. Scis., 1928. Research, publs. on parasitism, sexuality, influence of environment, and invertebrate zoology. Died Paris, July 13, 1958.

CAUQUIL, Germaine Anne, French chemist; b. Montpellier, France, Nov. 1, 1897; d. Germain and Augustine (Banau) C.; Ph.D. in Phys. Sci. and Chem. Engring., U. Montpellier; lectr., then prof. chemistry. Laureate, Sch. of Science; recipient phys. sci. prize Taupie, Le Bel prize; laureate, Acad. Scis. for Berthelot prize, named Berthelot physician. Mem. chem. socs. France (laureat), London, Soc. Phys. Chemistry, Am. Chem. Soc. Address: 25, bd. du. Jeu-de-Paume, Montpellier (Herault), France.

CAUS, Salomon de, see de Caus.

CAUSEY, Nell Bevel (Mrs. David Causey), Am. invertebrate zoologist; b. Trenton, Tenn., Dec. 8, 1910; d. Harvey M. and Nettie (Hester) Bevel; B.S., Coll. Ozarks, 1931; M.A., U. Ark., 1937; Ph.D., Duke, 1940; m. David Causey, Aug. 2, 1938. Tchr. high schs. Ark., 1931-36; instr. zoology U. Ark., 1940-44; faculty La. State U., Baton Rouge, asso. prof. zoology, 1966—. Mem. Soc. Systematic Zoology, Soc. Study Evolution, Nat. Speleological Soc., Phi Beta Kappa, Sigma Xi. Editor, Proc. La. Acad. Scis., 1964—. Contbr. numerous articles to tech. jours. Research in taxonomy and life history millipeds N.Am. and Mexico, both epigean and cave species. Home: 1110 Magnolia Woods, Baton Rouge 70806.*

CAUTLEY, Sir Proby Thomas, Brit. engr.; b. Stratford St. Mary's, Eng., 1802; col. of Engrs., India; served in Bhurtpore, 1825; asst. in reconstrn. of Doab canal, circa 1824-30, dir., 1831-43; made plans for Ganges canal (censored by Arthus Cotton), supt. of constrn., 1843-45, 48-54; in Eng., 1845-48; mem. Council of India, 1858-68. Recipient (with Hugh Falconer) Wollaston medal Geol. Soc. Fellow Royal Soc., 1846. Explored geology of Siwalik range; made large collections of fossils (in Brit. Mus.); wrote on canals and fossils of India; responsible for 300 mile canal to Cawnpore, India, also 500 mile Ganges canal across Solani River (includes aqueduct 300 yards long and 50 yards wide). Died Sydenham, London, Eng., Jan. 25, 1871.

CAVA, Michael P., Am. chemist; b. Bklyn., Feb. 13, 1926; s. Michael R. and Catherine (Lombardo) C.; B.S., Harvard, 1946; M.S., U. Mich., Ph.D., 1951; m. Esther Laden, June 11, 1951; 1 son, John M. Postdoctoral fellow Harvard, 1951-53; from asst. prof. to prof. Ohio State U., 1953-65; prof. Wayne State U., Detroit, 1965—. Cons., Smith, Kline and French, Phila., 1965—; mem. study sect. NIH, 1966—. Alfred P. Sloan Found. fellow. Mem. Am. Chem. Soc., Am. Soc. Pharmacognosy. Author: (with M. J. Mitchell) Cyclobutadiene and Related Compounds; also numerous articles. Research on chemistry of 4-membered ring organic compounds; isolation and chemistry of alkaloids of tropical plants. Home: 8120 E. Jefferson St., Detroit.

CAVALIERI, Francesco Bonaventura, Italian mathematician; b. Milan, Italy, 1598; studied under Galileo's disciple Benedetto Castelli, Pisa; joined Congregation of Hieronymites, circa 1613; entered Jesuit Order, 1613; prof. math., Bologna, Italy, 1629-47. Author: Geometrie indivisibilum continuorum nova, 1635; Trigonometria plana et sphaerica, 1635; Appendice della nova practicca astrologica, 1640; Exercitationes geometricae, 1647. Used Kepler's idea of infinitely small geometrical quantites to develop Archimedes method of exhaustion into a method of indivisibles (forerunner of integral calculus), 1629; provided rigorous proof of Pappus theorem relating to volume of solid of revolution; research on vol. of solids of revolution, burning mirrors; gave almost perfect proof that sum of angles of spherical triangle is greater than 180° and less than 450°; one of 1st to recognize and popularize value of logarithms; determined focal lengths of glass lenses, 1647; believed gravity a force caused by external action. Died Bologna, Nov. 30, 1647.

CAVALIERI, Ugo, Italian physician; b. Ancona, Italy, Feb. 3, 1919; s. Ugo and Maria (Duranti) C.; M.D., U. Bologna, 1942; m. Maria Luisa Moneta, Apr. 12, 1944; children—Enrico, Maria Gloria. Asst. physician Hosp. of Ancona City, 1942-45; dir. med. sect. Hosp. of Cingoli, 1945-48; mem. Workers Geriatric Center, Clinica del Lavoro, U. Milan, 1949-54; dir. Geriatric Hosp., Pia Casa, Abbiategrasso, Milan, 1954—; cons. physician Mental Hosp. Paolo Pini, Milan. Mem. Italian Gerontol. Soc. (hon.), Italian Gerontol. Assn. (physician pres.). Author: Patologia della Vecchiaia, 1957; Gerontologia e Igiene, 1958; La Cura del Paziente Anziano, 1961. Publs. on a theory of pathogenesis of osteoporosis; pharmacology of ethanol in human beings; definition of immobilization pathology in aged. Home: 11 Piazza Grandi, Milan. Office: Pia Casa, Abbiategrasso, Milan, Italy.*

CAVALLERO, Cesare, Italian physician; b. Alexandria, Italy, Aug. 23, 1913; prof., 1956—; prof. anatomy and path. histology Sch. Medicine and Surgery, dir. Sch. Specialization in Legal Medicine, U. Pavia (Italy); dir. Inst. Anatomy and Path. Histology; mem. Italian Soc. Endocrinology (dir.), Deutsche Gesellschaft für Endokrinologie, Italian Soc. Pathology, Allergography and Biol. Experimentation, Italian Soc. for Study of Nutritional Diseases, Internat. Soc. Neurovegetal System, Lombardy Soc. Medico-Biol. Sci., Italian League against Tumors. Address: via Castelmorrone 30, Milan, Italy.

CAVALLI, Giancarlo, Italian physician; b. Bologna, Italy, Nov. 5, 1928; s. Amilcare and Maria (Maccaferri) C.; D. Medicine and Surgery, U. Bologna, 1952; m. Luciana Bianchedi, Apr. 12, 1924; children—Andrea, Chiara, Daniela. Faculty, Inst. Med. Pathology, U. Bologna, 1952-53, Inst. Clin. Medicine, 1953—, prof. histology and gen. embryology, 1959—, prof. med. pathology, 1962—, teaching prof. Sch. Specialisation in gastroenterology, 1964—. Recipient Selezione Marzotto, Soc. Internal Medicine Italy, 1959. Mem. N.Y. Acad. Scis., Italian Soc. Gastroenterology, Italian Soc. Histochemistry. Author: (with C. Cacciari, E. Pisi) Splenoportografia e Splenomanometria, 1957; also numerous articles. Research on behavior of desoxyribonucleic acid in course of spermatogenesis, in mitosis and various physio-path. diseases of liver, exptl. and clin. studies in intestinal absorption, and physio-pathology of portal area. Home: 13, Viale XII Giugno, Bologna, Italy.*

CAVALLI-SFORZA, Luigi Luca, Italian geneticist; b. Genova, Italy, Jan. 25, 1922; s. Pio and Attilia (Manacorda) C-S.; degree in medicine U. Pavia (Italy), 1944; M.A., U. Cambridge (Eng.), 1950; m. Alba Maria Ramazzotti, Jan. 12, 1946; children—Matteo, Francesco, Tomajo, Violetta. Asst. in research U. Cambridge, 1948-50; dir. research in microbiology Istituto Sieroterapico Milanese, 1950-57; prof. genetics U. Parma (Italy), 1958-62; prof., chmn. genetics dept. U. Pavia, 1962—; dir. internat. lab. of genetics and biophysics, Naples, Italy, 1964, vice dir., dir. Pavia sect., 1965—. Mem. Biometric Soc. (pres. 1966—). Author: (with A. Buzzati) Target Theory and Elementary Biological Units, 1948; Statistical Methods in Biology, 1958; also numerous articles. Research on origin of resistance to antibiotics, sex and sex determining factors in bacteria, role of chance in determining human evolution, methods of phylogenetic analysis with spl. interest in human origin. Home: 18 Fatebenesorelle, Milano, Italy. Office: 14 S. Epifanio, Pavia, Italy.*

CAVALLO, Tiberius, philosopher, physicist; b. Naples, Italy, Mar. 30, 1749; went to London, Eng., 1771. Fellow Royal Soc., 1779; mem. Royal Acad. Scis. Naples. Author: On Medical Electricity, 1780; A Treatise on the Nature and Properties of Air and other Permanently Elastic Fluids, 1781; Complete Treatise on Electricity, 1786; A Treatise on Magnetism in Theory and Practice, 1787; The Medicinal Properties of Facitious Air, 1798; Elements of Natural and Experimental Philosophy, 1803; also articles. Invented various instruments including a micrometer, electrometer, condenser for measuring quantity and force of electricity; research on chemistry, magnetism, influence of air and light on plant growth; his research paved way for discoveries in organic life. Died London, Eng., Dec. 21, 1809.

CAVANILLES, Antonio José, Spanish botanist; b. Valencia, Spain, Jan. 16, 1745; prof. philosophy Murcia, Spain; preceptor of children of Duke del Infantado, Spanish ambassador to Paris, France, 1777-89; student botany Paris; became dir., Royal Bot. Gardens, Madrid; under commission from King of Spain studied plants and natural history of Spain; correspondant l'Institut de France. Author: Monadelphiae Classis Dissertationes decem, 1785-86; Icones et Descriptiones plantarum Hispaniae, 1791-99; Observiones sobre el Cultivo del Arroz en el . . . de Valencia y su Influencia en la Salud Publica, 1795; Annales de Historia Natural, 1800. Described numerous new plants; proponent Linnaean system of botany. Died Madrid, Spain, May 4, 1804.

CAVÉ, François, French inventor; b. Mesnil, France, Sept. 12, 1794; carpenter, foreman, cashmere scarf factory; machine builder, Paris. Invented an adjustable expansion control with movable slide valve, oscillating steam engine for manufacture of cloth (used widely as improved by John Penn for navigation), also a wheel for ship propulsion; constructed 1st steam hammer, 1836, also steam engines, locomotives, ship hulls, mills, rolling mills, machine tools; studied boilers, ship propellers, transmission shaft torsion. Died Paris, Mar. 10, 1875.

CAVEN, Robert Martin, Brit. chemist; b. Freemantle, Eng., Apr. 26, 1870; s. R. Caven; ed. Univ. Coll., Nottingham; D.Sc., U. London; m. Ethel Mary Willson, 1906; 1 dau. Asst. city analyst, Birmingham, Eng., 1893-95; lectr. chemistry Univ. Coll., Nottingham, 1895-1918; prin. Tech. Coll., Darlington, 1919-20; prof. inorganic and analytical chemistry Royal Tech. Coll., Glasgow, Scotland, from 1920. Fellow Royal Inst. Chemistry, Chem. Soc.; mem. Soc. Chem. Industry. Author: (with D. G. D. Lander) Systematic Inorganic Chemistry, 1906; Systematic Qualitative Analysis, 1909; A Short System of Qualitative Analysis, 1917; Carbon and its Allies (Textbook of Inorganic Chemistry, Vol. V, edited by J. Newton Friend), 1918; The Foundations of Chemical Theory, 1920; Quantitative Chemical Analysis, Part I, 1923, Part II, 1925; Gas and Gases (Home University Library), 1926; (with J. A. Cranston) Symbols and Formulae in Chemistry, 1928; also papers. Gen. editor Blackie's Manuals of Pure and Applied Chemistry. Died July 16, 1934.

CAVENDISH, Henry, Engish physicist, chemist; b. Nice, France, Oct. 10, 1731; s. Lord Charles Cavendish and Lady Anne Grey; student Peterhouse Coll., Cambridge U., 1749-53. Spent most of his life as recluse in London, devoting himself entirely to sci. research; wrote numerous papers and sci. memoirs. Fellow Royal Soc. London, 1760; fgn. asso. French Acad. Scis., 1803. Cavendish Phys. Lab. at Cambridge named in his honor. Discovered properties of hydrogen, 1766, and determined its specific gravity; determined composition of water by showing that it can be produced synthetically by combustion of hydrogen; determined composition of atmosphere, 1781; speculated that air contains a small amount of an inert gas (later discovered to be argon); worked out composition of nitric acid and synthesized it; determined specific gravity of carbon dioxide; experimented in electricity in early 1770's, and anticipated much of the work of next half century, though he published nothing on that subject; one of 1st to apply calculus to theory of electricity; measured Newton's gravitational constant through what is now known as the Cavendish Expt., 1798, thus was able to calculate the mass and mean density of earth with considerable accuracy; believed that heat is caused by internal motion of particles. Died London, Eng., Feb. 24, 1810.

CAVENDISH, Richard, English mathematician; b. Trimley, Eng.; flourished 1539-1600; ed. Oxford; M.A., Cambridge, 1573; 1 dau., Douglas (Mrs. Richard Hakluyt). Surveyor, hydrographer, chart-maker for King; employed to carry letters to Mary Queen of Scots by Duke of Norfolk, 1568-69; became mem. Parliament for Denbigh, 1572, 85; apptd. to law office by Elizabeth, but excluded by judges, 1587. Wrote theol. tract; translated Euclid into English. Died circa 1601.

CAVENTOU, Eugène, French chemist; b. Paris, 1824; s. Joseph Bienaimé and Marie-Joseph (Labre) C.; mem. Acad. Medicine; studied surface of cailcedra and carapa in order to extract a new febrifuge, the touboucorinin, 1857-59; made (with Bouchardat) gen. study of Algerian wines; studied brominated ethyl, transformation from alcohol to glycol; isolated new carbides crotonylin and hexoylene. Died Feb. 13, 1912.

CAVENTOU, Joseph Bienaimé, French chemist, pharmacologist; b. Saint-Omer, France, June 30, 1795; son, Eugéne; prof. toxicology l'École supérieure de pharmacie de Paris (France). Bronze statue erected in his honor, Paris. Author: Nouvelle nomenclature chimique, 1816; (with Pelletier), Recherches sur l'action qu'exerce l'acide nitrique sur la nature nacrée des calculs biliares, 1817; Traité élémentaire de pharmacie théorique, . . . , 1819; Recherches chimiques sur quelques matières animales saines et morbides, 1843. Research on alkaloids; discovered (with Pelletier) strychnine, 1818, brucine, and quinine (from cinchona bark), 1820, also (with Meissner) veratrine, 1818, colchicine, 1819, cinchonine, 1820; introduced the term, chlorophyll, 1818. Died Paris, May 5, 1877.

CAVERT, Henry Mead, Am. physician; b. Mpls., Mar. 30, 1922; s. William L. and Mary (Mead) C.; B.S., U. Minn., 1942, M.D., 1951, Ph.D., 1952; m. June L. Sederstrom, Jan. 27, 1946; children—John Mead, Harlan McCrea, Winston Peter. Research fellow Am. Heart Assn., 1951-54, investigator, 1954-57; faculty U. Minn., Mpls., 1953—, professor dept. physiology, 1967—, asst. dean Coll. Med. Scis., 1957-64, asso. dean Med. Sch., 1964—; vis. prof. U. Edinburgh (Scotland), 1961-62. Mem. heart program project com. B, Nat. Heart Inst., NIH, 1966—. Nat. Heart Inst. Spl. Research fellow, 1961-62. Mem. Assn. Am. Med. Colls. (chmn. com. on student aspects internat. med. edn. 1968—), Am. Physiol. Soc., Minn. Acad. Sci., A.A.A.S., Sigma Xi, Alpha Omega Alpha, Phi Lambda Upsilon. Author: (with others) The Machinery of the Body, 1937, 5th edit., 1961; also articles. Research on metabolic fuel substances used as sources of energy for heart and skeletal muscle contraction, role of cardiac and skeletal muscle glycogen stores metabolic processes, distbn. water to cells and tissues isolated heart and skeletal muscle, role insulin in regulating accessibility of blood glucose to intracellular fluid heart and muscle cells. Home: 3328 48th Av. S., Mpls. 55406.*

CAVOLINI, Filippo, Italian naturalist; b. Naples, Italy, 1756; prof. natural history Royal U. Naples; jurisconsul; mem. French Acad. Scis.; author: Mémoires; studied zophyta, plants, polypi. Died 1810.

CAWLEY, Sir Charles Mills, Brit. chemist; b. Gillingham, Kent, Eng., May 17, 1907; s. John and Emily (Hall) C.; B.Sc., Imperial Coll. Sci. and Tech., London, Eng., 1928, M.Sc., 1929, Ph.D., 1932, D.Sc., 1950; m. Florence Mary Ellaline Shepherd, July 14, 1934; children—Penelope Ann (Mrs. John Boyd Potter). Chemist, Fuel Research Sta., Greenwich, London, 1929-53; dir. Hdqrs., Dept. Sci. and Indsl. Research, London, 1953-59; chief scientist Ministry of Power, London, 1959-67; civil service commr., sci. and engring. adviser to commn., 1967—. Chmn., Admiralty Fuels and Lubricants Adv. Com., 1957-63. Decorated comdr. Order Brit. Empire. Fellow Imperial Coll. Sci. and Tech., Inst. Petroleum, Inst. Fuel (past v.p.), Royal Inst. Chemistry, Royal Soc. Arts. Research and publs. on oil from coal, hydrogenation of coal and coal products, flame warfare Home: 4 Longlands, Worthing, Sussex, Eng. Office: Civil Service Commn., 23 Savile Row, London, W. 1, Eng.*

CAXTON, William, Brit. inventor; b. Kent, Eng., 1422; learned printing in Cologne, Germany, 1471-72; a dau., Elizabeth (Mrs. Gerald Croppe); apprenticed to Robert Large, a London silk mcht., 1438; mcht. at Bruges, 1446-70; mem. Co. of Mercers in London; gov. Mcht. Adventurers in Low Countries, Bruges, 1463-69; negotiated comml. treaties with dukes of Burgundy; set up press in partnership with Colard Mansion; returned to Eng., 1476; established press at Westminster. Translator: The Recuyell of the Historyes of Troye, 1469-71, (printed 1474-75); The Game and Playe of the Chesse; wrote prefaces and epilogues. Printed nearly 100 publs. and translated many of them. Issued 1st known piece of printing from Caxton press in Eng., 1477; 1st dated book printed in Eng. (The Dictes and Sayings of the Philosophers [Earl Rivers], a translation from the French). Contbd. an 8th book to Higden's Polychronicon. Died 1491.

CAYER, Jean Ignace, French astronomist, physicist; b. Lyons, France, 1704; became canon of Fourviere, 1724. Mem. Royal Soc. Fine Arts (now Acad. Lyons); author: Calculs astronomiques et des réflexions sur pusage de sonner les cloches dans les temps d'orage, 1745. Studied math., effects of lightning. Died 1754.

CAYEUX, Lucien, French geologist; b. Semousies, France, Mar. 26, 1864; studied under J. Gosselet and Barrois, Faculté des Sciences de Lille (France). Asst., Faculté des Sciences, Lille, 1887-91; asst. l'École des Mines, Paris, France, titular prof. 1907-12; became prof. College France, 1912; named prof. applied geology and agr. Nat. Agronomic Inst., 1901. Recipient Prix Valliant, French Acad. Scis., 1897, Grand Prix, 1923. Mem. French Acad. Scis. Author: l'Étude micrographique des terrains sédimentaires, 1897; Introduction à l'etude petrographique du roches sédimentaires, 1916. Research on lithology of sedimentary rocks in France; used phys., microchem. and chromatic methods to analyze sedimentary rocks; studied siliceous rock and chalk of Bassin of Paris. Died Mauvessur-Loire, Nov. 1, 1944.

CAYLEY, Arthur, Brit. mathematician; b. Richmond, Eng., Aug. 16, 1821; s. Henry and Maria Antonia (Doughty) C.; ed. (sr. wrangler, 1st Smith's prize 1842), Trinity Coll., Cambridge (Eng.) U.; student law Lincoln's Inn; m. Susan Moline, 1863. Practiced law, 1849-63; fellow Cambridge, 1842-52, 1875-95, Sadlerian prof. pure math., 1863-95, became hon. fellow, 1872. Recipient Copley medal 1882. Fellow Royal Soc., 1852 (Royal medal 1859); mem. Brit. Assn. (became pres. 1883). Author: A Treatise on Elliptic Functions, 1876; Single and Double Theta Functions, 1881; also numerous articles which were collected and issued in 13 vols., 1889-98. Developed theory of algebraic invariants, theory of matrices, theory of groups, geometry of n-dimensional space, geometry of "absolute," abstract geometry; studied higher singularities of curves and surfaces, classification of cubic curves, additions to theories of rational transformation and correspondence, theory of elliptic functions, secular acceleration of moon's mean motion, attraction of ellipsoids, hypergeometry, invariance, algebra of matrices (used by Heisenberg in quantum mechanics), theoretical dynamics, spherical and phys. astronomy. Died Cambridge, Jan. 26, 1895.

CAYLEY, Sir George, English inventor, aero. engr. b. Scarborough, Yorkshire, Eng., Dec. 27, 1773; studied at York, Nottingham (both Eng.); studied chemistry and electricity under George C. Morgan. A founder Regent Street Polytechnic, London, Eng. Author: On Aerial Navigation (laid founds. of aerodynamics); publs. on mech. flight, 1809-54. Research and publs. on air travel; built large glider, 1809; studied properties of air and power for flight; suggested symmetrically angled wings and a double decker; patented hot air engine, 1837; invented new type of telescope, artificial limbs, caterpillar tractor, tension wheel; studied agrl. reclamation methods; built heliocopter, 1843; 1st to attempt math. explanation of basic principles of flight; father of Brit. aeronautics. Died Scarborough, Dec. 15, 1857.

CAYLUS, Anne Claude Philippe, French archaeologist; b. Paris, Oct. 31, 1692; s. Comte de and Marie Marguerite (le Valois de Villette de Murcay) C. Count, Marquis of Esternay; served in Spanish War of Succession; travelled in Italy and Levant; returned to Paris, 1717. Mem. French Acad. Inscriptions. Author: Recueil d'antiquités egyptiennes, étrusques, grecques, romaines, et gauloises, 7 vols., 1752-67; Tableaux drawn from the Iliad, Odyssey and Aenid, 1757; Collection of Ancient Paintings after the Designs of Bartoli; Fairy Tales and other fictional works; also treatises on antiquities. Studied ancient method of encaustic painting; studied art and engraving. Died nr. Paris, Sept. 5, 1765.

CAYREL, Roger, French astronomer; b. Bordeaux, France, Dec. 4, 1925; s. Jean and Jeanne (Liotard) C.; student Ecole Normale Superieure, Paris; Dr., U. Paris, 1957; m. Giusa de Strobel, Aug. 12, 1956; children—Lora, Marguerite, Francoise. Research fellow Paris Astrophys. Inst., 1951-59, 61-62, Cal. Inst. Tech., Pasadena, 1959-60; asst. astronomer Paris Obs., 1962-66, astronomer, 1966—. Mem. Inst. Advanced Study, Princeton, N.J., 1966-67; bd. dirs. Paris Astrophys. Inst. Recipient Palmes Academiques. Mem. Internat. Astron. Union, Am., Royal astron. socs. Research, publs. on stellar atmosphere theory, chem. composition of stars, line formation in stellar spectra, physics of solar chromosphere. Home: 6 rue du Douanier Rousseau, Paris 14. Office: Paris Obs., 92 Meudon, France.*

CAZAUX, Pierre Lucien Maurice, French pharmacologist; b. Cazaubon, Gers, Aug. 11, 1902; s. Charles Bernard and Nady (Laporte) C.; ed. Sch. Pharmacy, Paris, Sch. Sci., Bordeaux, France; m. Marie-Thérèse Pery, Dec. 6, 1928; 1 son. Bernard-André. Mem. Soc. Pharmacy Bordeaux (titular), Hydrology Soc. France (hon.), Internat. Soc. Hydrology, Acad. Pharmacy (corr), Inst. Coimbra, Royal Acad. Pharmacy Madrid. Research and publs. on physio-chem. character, chem. composition, biol. properties, pharmacodynamics, and metaplasia of mineral waters. Home: 40, allées de Tourny, Bordeaux. Office: 8, place de la Victoire, Bordeaux, France.

CAZENAVE, Pierre-Louis-Alphée, French dermatologist; b. Paris, May 5, 1802; studied under Biett, St. Louis Hosp.; docteur en médecine, 1827; aggregation, 1836. Placed in charge choleraic br. St. Louis Hosp., 1831, replaced Alibert as tchr. of therapeutics, 1838, tchr. diseases of skin Sch. Medicine, from 1841, resigned 1866. Author: Abrégé pratique des maladies de la peau, 1828; Lecons sur les maladies de la peau, 1841-44, 45-56; Traité des maladies du cuir chevelu, 1850; Pathol. des mal. de la peau et de la syph., 1868-69; Bibl. med. les gourmes, 1873. Described pemphigus foliaceus, a form of pemphigus accompanied by flaccid scabby bullae, 1844, lupus erythematosus (Cazenave's lupus), 1850; returned dermat. studies to sci. form. Died Garches, France, 1877.

CAZIN, Achille Auguste, French physicist; b. Perpignan, France, 1832; prof. and researcher. Author: Essai sur la détente et la compression des Gaz, 1862; Théorie élémentaire des machines à air chaud, 1865; la Chaleur, 1866; les Forces physiques, 1869; l'Etincelle électrique, 1877; la Spectroscopie, 1878; Traité théorique des piles électriques, 1881. Research on electricity, magnetism, thermodynamics, thermal properties of gases and vapors. Died Paris, France, 1877.

CAZPSKI, Siegfried, chemist, physician; b. Obica, Posen, Poland, May 28, 1861; s. Simon and Rosalie (Goldenring) C.; student under H. von. Helmholtz; m. Margarethe Koch, 1887; 3 sons, 3 daus. including Elizabeth (Mrs. Wilhelm August Flitner). Became sci. co-worker Optical Works Carl Zeiss, Jena, Germany, 1884, mem. directory, 1891, head, 1903. Author: Theorie der Optischen Instrumente, 1893; Optica Geometrisch. Verified qualitatively (with Gocke) Helmholtz equation. Died Weimar, Germany, June 19, 1907.

CEBYSEV, Pafnutij L'vovic, Russian mathematician; b. Oktatovo, Russia, May 26, 1821; diploma U. Moscow (Russia); 1840. prof. math. U. St. Petersburg (Russia), 1847-82, taught analytic geometry, higher algebra, theory of numbers, integral calculus. Mem. Inst. France (asso. fgn.), Royal Soc. London (fgn.). Collaborated with Boumiakovsky to pub. 2 vols. of collected works of Euler, 1849. Established the existence of limits within which must comprise the sum of logarithums of primes inferior to given number, 1850; research and numerous publs. on the theory of dispersion, theory of numbers, theory of least squares, interpolation theory, infinite series, calculus of variations, theory of probability, primes. Died Aug. 12, 1894.

CECCO D'ASCOLI (or Stabili, Francesco degli), Italian astrologer, astronomer; b. Ascoli Piceno nr. Marche, Italy, 1269; apptd. physician, astrologer to Carlo, duke of Calabria, 1326; named prof. astrology and math. U. Bologna (Italy), 1322. Author: l'Acerba, (poetic work on astronomy, meteorology, stellar influence and physiognomy, minerals), 4 books, 1272; also commentary on Tractatus de sphaera mundi (Johannes de Sacrobosco). Tried for heresy because of his criticism of the Divine Comedy of Dante, condemned and burned at the stake with his writings, Florence, Italy, Sept. 16, 1327.

CECI, Antonio, Italian orthopedic surgeon; b. Italy, 1852; pupil of Durante; surgeon, Pisa, Italy. In surgery performed 1st kineplasty (amputation allowing muscles to be retained for motor use in artificial limbs), 1896; also performed early laryngotomies, splenectomies; popularized use local anesthesia; skilled as plastic surgeon. Died 1920.

CECIONI, Francesco, Italian mathematician; b. Leghorn, Italy, Dec. 1, 1884; s. Olderigo and Isolina (Cantinelli) C.; Ph.D. in Math.; m. Maria Franchini, Aug. 24, 1911; children—Giovanni, Luigi, Giacomo, Anna, Raffaele, Giovanna, Clara. Univ. asst., then prof. in a secondary sch., 1907-21; prof. Naval Acad., 1921-26; univ. prof., 1926-60; now prof. emeritus U. Pisa (Italy). Mem. Toscan Soc. Sci. and Letters, La Colombaria, Florence. Author: Lezioni sui fondamenti della matematica. Home: via Trieste 55, Livorno, Italy. Office: Istituto Matematica, Universita di Pisa, Italy.

CEDRO, Victorio Carmelo Federico, Argentinian bacteriologist; b. Martinez, Argentina, May 25, 1916; s. Federico and Angela (Paviglianiti) C.; D.V.M., U. Buenos Aires, 1938, M.D., 1957; m. Agustina Ernesta Caselli, July 11, 1942; children—Ercilia Angela, Elda María, Victor Agustín. Expert infectious diseases div. Inst. Zoonosis, Agr. Dept., 1944-47, chief Brucellosis div., 1948-57; dir. Nat. Centre Agrl. and Animal Research, 1959-61; dir. Inst. Zoonosis, Nat. Inst. Agrl. and Animal Tech., Buenos Aires, Argentina, 1957-—, Brucellosis coordinator, 1962-—. Mem. Argentine Microbiology Assn., World Vet. Assn., Argentine Soc. Transmissible Diseases (expert com.). Contbr. numerous articles to profl. publs. Research on microbiology, zoonosis (Tb, leptospirosis, virus, brucellosis), carriers and reservoirs of Argentine hemorrhagic fever. Discoverer vaccination method for swine brucellosis, antigen for brucellosis diagnostic in blood. Home: 2167 Maestro Alvarez, San Fernando, Buenos Aires. Office: 134 Chorroarín, Buenos Aires. Argentina.*

CEFFONS (or DE CEFFONA), Pierre, French natural philosopher; Cistercian monk in Clairvaux. Author: Commentary on the Sentences (treats mech. and astron. theories, questions and denies commonly accepted opinions such as finiteness and uniqueness of world, also rotation of celestial spheres by intelligences). Died circa 1351.

CEI, José Miguel, biologist, herpetologist; b. Pisa, Italy, Mar. 23, 1918; s. Luigi and Ginette (Piallini) C.; B., Classic Lyceum, 1936; student U. Florence (Italy), 1937-39; Doctorate Degree, U. Pisa (Italy), 1940; Doctorate Natural Scis. (hon.), U. Tucuman (Argentina), 1952; m. Sylvana Silvi, Apr. 10, 1940; children—Roberto Luigi, Vera Catalina, Luis Marcos. Asst. prof. U. Florence, 1942-47; prof. Inst. Zoology, U. Nacional Tucuman, 1948-54, dir. Inst. Gen. Biology, 1952-54; prof. dept. sci. research Nat. U. Cuyo, Mendoza, Argentina, 1955-58, prof. Inst. Biology, 1958-—, dir., 1958-—, head dept. morphology Med. Sch., 1965-66; head dept. ecology U. Chile, Santiago, 1957. Mem. Assn. Latino-Americana Herpetologos-Ictiologos (founder), A.A.A.S., Am. Soc. Ichthyologists and Herpetologists, Herpetologists League, Sociedad Biología Argentina, Sociedad de Biología de Chile. Author: Gli Animali sulle Terre e Oceani, 1946; Darwin—Antology, 1947; Biologia General, 1949; Batracios de Chile, 1962; also numerous articles. Research on sex-cycle in amphibians, regressive evolution of eye, neotropical amphibians; discovered physiol. bipolarity in internal rhythm of the Anurans, sibling-species in neotropical Anurans. Home: Casilla Correo 327, Mendoza. Office: Instituto Biológica, Nat. U. Cuyo, Mendoza, Argentina.*

CÉIDIGH, Pádraig Sean O., Irish zoologist; b. Dublin, Ireland, July 14, 1933; s. Patrick John and Eileen (MacMahon) C.; B.Sc. with honors, U. Coll., Dublin, 1955; M.Sc., U. Coll., Galway, Ireland, 1956, Ph.D., 1960; m. Mairin Chonaire, Oct. 2, 1956; children—Mairtin, Muris. Biologist, Salmon Research Trust Ireland, Newport, 1956; asst. zoology dept. U. Coll. Galway, Ireland, 1956-62, prof., 1962-—. Cons. on biology; mem. adv. com. on fisheries Fisheries br. Irish Dept. Agr. Fellow Zool. Soc. London; mem. Royal Dublin Soc., Marine Biol. Assn. U.K. Research on adult Decapoda and larval Decapoda off W. coast Ireland, plankton Irish coast. Home: 61 Ard na Mara, Galway, Ireland.*

CEJP, Karel, Czechoslovakian botanist, mycologist; b. Rokycany, Czechoslovakia, Feb. 22, 1900; s. Karel and Barbora Cejp; RNDr, U. Charles, Praha, Czechoslovakia, 1923, Dr.Sc., 1956; m. Darija Ottisová, Sept. 12, 1951. Asst., Bot. Inst., Charles U., 1923, prof. systematic botany, 1952-—. Mem. Czechoslavak Sci. Soc. Mycology, Bot. Soc. Czechoslovakia, Soc. Linneene de Lyon, Am. Mycol. Soc. Research, numerous publs. and monographs on morphology of plants, mycology, fungi, oomycetes. Home: Rokycany 1-79, Srbova 2, Czechoslovakia. Office: Praha 2, Benátská 2, Czechoslovakia.*

CELLI, Angelo, Italian physician; b. Cugli, Italy, 1857; pro. hygiene, Palermo, Italy, 1886; established Pasteur Inst., Palermo; became asst. prof. hygiene U. Rome, 1887, prof., 1890. Author: Manuale dell' uppiziale sanitario, 1899; Manuale d' igienista, 1906-07; Die Malaria im ihrer Bedeutung für die Geschichte Roms und der römischen Campagna (monograph), 1929. Observed meningococcus (later studied by Weichselbaum), 1884; apparently saw dysentery bacillus before Shiga; differentiated (with Marchiafava) the tertian, quartan, and aestive-autumnal types of malarial parasites, 1889; active in malarial control; freed Roman Campagna, Italy, from malaria. Died 1914.

CELLINI, Benvenuto, Italian artist; b. Florence, Italy, 1500; s. Giovanni C.; apprenticed to goldsmiths; employed in cutting dies, medalling and enamelling, seal engraving; made gold metals, coins for pope; engr. for Pope Clement VII during war with France; created Perseus in bronze for grand square of Florence, 1545-54. Wrote on jewelry, enamelling, coinings, art of making gold and silver bases, casting of statues in bronze, quality of marble for statuary, art of design; established theory of perspective as aid to painter. Died circa 1571.

CELMER, Walter Daniel, Am. chemist; b. Plymouth, Pa., Sept. 13, 1925; s. Peter Thomas and Katherine (Zolnierowicz) C.; B.S. cum laude, Bucknell U., 1947; Ph.D., U. Ill., 1950; m. Florence Theresa Mack, June 15, 1946; children—Leslie, Janet, Robert, Ann. With Charles Pfizer & Co., Inc., Bklyn., 1950-58, Groton, Conn., 1959-—, research supr., 1954-61, research mgr., 1961-—. Upjohn fellow, 1948-50. Mem. Am. Soc. Biol. Chemists, Am. Chem. Soc., N.Y. Acad. Scis. Research, numerous publs. in chemistry antibiotics, natural products and related compounds; discovered structures of mycomycin, thiolutin and oleandomycin; stereochemistry, biogenesis and mode-of-action of macrolide antibiotics; invented triacetyloleandomycin and related therapeutically-useful drugs. Home: 14 Greenway Rd., New London, Conn. 06320. Office: Med. Research Labs., Charles Pfizer & Co., Inc., Groton, Conn. 06320.*

CELORIA, Giovanni, astronomer; b. Monferrato, Italy, 1842; engr., U. Turin (Italy), 1863; with obs. of Milan, Italy, Berlin and Bonn, Germany; became aide-astronomer Brera Obs., 1866, 2d astronomer 1873; named prof. theoretical geodesy Tech. Inst. Milan, 1875; named 1st astronomer, dir. Milan Obs., 1900. Apptd. pres. Royal Italian Commn. on Geodesy, 1902. Author: (monographs) The Moon; The Comets; Manual of Popular Astronomy; Atlas of Astronomy; Cosmography; The Earth and the Universe; (stellar catalogue) Mean Positions for 1870 of 1118 Stars; also articles. Showed individual stars increase rapidly in density from galactic pole to plane of Galaxy, that the fainter the stars the more abruptly the Galaxy commences while a gradually increasing condensation characterizes the brighter stars; concluded the Milky Way is a region of stellar condensation; star soundings at north galactic pole indicated further increase of optical power would not increase number of stars visible in that direction, thus demonstrating that the universe is limited in extent; studied eclipses from B.C. 322 to 99; computed orbits of double stars, 1888. Died Jan. 29, 1920.

CELS, Jacques Philippe Martin, French botanist, horticulturist; b. Versailles, France, June 15, 1740; became mem. l'Institut, 1795; participated in drafting rural code to form nursery, nr. Montrouge, France. Mem. Soc. Agr. Acclimatized large number of exotic plants. Died Montrouge, May 15, 1806.

CELSIUS, Anders, Swedish astronomer; b. Uppsala, Sweden, Nov. 27, 1701; s. Nils C. Prof. astronomy U. Uppsala, 1730-44; builder and dir. Uppsala Obs., 1740; travelled in Germany, Italy, and France; on French expdn. to measure degree of meridian in Lapland, 1736. Fellow Royal Soc., 1735; mem. Stockholm, Berlin acads., Inst. Bologna, Royal Soc. Uppsala (sec.). Author: Nova methodus distantium solis a terra determinandi, 1730; De observationibus pro figura telluris determinanda, 1738. Regarded as founder of Swedish astronomy; determined (with Zanotti) earth's ecliptic obliquity; verified Newton's theory that earth is flattened at poles; recorded observations of Aurora Borealis (made 1716-32), 1733; studied daily variation of magnetic declination; advocated use of Gregorian calendar; invented centigrade thermometer, 1742. Died Uppsala, Apr. 25, 1744.

CELSIUS, Magnus Nils, Swedish mathematician, archeologist; b. Alfta, 1621; children—Olof, Anders. Prof. math. U. Uppsala (Sweden); author treatise on fish, another on plants of Uppsala; 1st to decipher Runes of the Hensingland. Died 1679.

CELSUS, Aulus Cornelius; flourished under Tiberius, 14-37. Author: De medecina (8 books), ency. on agr., mil. sci., rhetoric; also wrote on plastic operations, removal of tonsils, dental practices, hygiene; studied urinary and stomach diseases, skin diseases; gave 1st indications of how neuropsychiat. disorders were grouped relative to other med. problems of that period, 1st to mention heart dieases and insanity, also gave 1st account on use of ligature. Known as Roman Hippocrates because of his imitation of the Greek physician; introduced Hippocratic system into Rome.

CENDROWSKI, Wojciech Stefan, Polish physician; b. Warsaw, Poland, Nov. 11, 1928; s. Stefan and Wladyslawa (Ukleja) C.; grad. Med. Acad. Warsaw, U. Warsaw, 1952, M.D., 1963; m. Hanna Monika Grzeda, Jan. 25, 1961; children—Piotr, Tomasz. Asst. neurology Dist. Hosp., Bydgoszcz, Poland, 1953-56; research fellow, adj. in neurology Polish Psychoneurol. Inst., Pruszkow, Warsaw, 1956-—; scholar in demyelinating diseases dept. neurology research group in demyelinating diseases Med. Research Council, U. Newcastle (Eng.), 1966 Mem. internat. panel corr. neurologists Nat. Multiple Sclerosis Soc., N.Y.C., 1964-—. Mem. Am. Acad. Neurology, Commn. Geog. Neurology, World Fedn. Neurology (corr.). Author: Multiple Sclerosis, 1966; also numerous articles. Research on geog. distbn. of sporadic and familial multiple sclerosis in Poland, epidemiology and genetics of multiple sclerosis, supression of exptl. allergic encephalomyelitis; discovered increased content of serum haptoglobin protein in multiple sclerosis. Home: 82 26 Woloska, Warsaw 12, Poland. Office: Psychoneurol. Inst., Pruszkow, Poland.*

CENSORINUS, Roman mathematician, astrologer; flourished 3d century (during reigns of Alexander Severus and Gordian); wrote on natural history of man, also music, astronomy, religious rites, doctrines of Greek philosophers. Author: De accentibus (now lost); De die natali (on astrology and horoscopes, useful for its chronology), 238.

CENTERS, Richard, Am. psychologist; b. Paducah, Ky., Oct. 16, 1912; s. Cager and Eula (Heath) C.; A.B., U. Ky., 1944; M.A., Princeton, 1946, Ph.D., 1947. Instr. psychology Rutgers U., New Brunswick, N.J., 1946-48; faculty U. Cal. at Los Angeles, 1948-—, prof. psychology, 1965-—. Mem. Am. Psychol. Assn., Soc. for Psychol. Study Social Issues. Author: The Psychology of Social Classes, 1949; contbr. numerous articles to tech. jours. Research on class consciousness in Am., relationship class consciousness to polit. and social behavior, psychology attitudes, beliefs, motivation, personality. Home: 11557 Wyandotte St., North, Hollywood, Cal. 91605.*

CERISE (or CERISI), Laurent Alexandre Phillibert, physician; b. Aosta, Italy, Feb. 2, 1809; M.D., U. Turin, 1828; practiced medicine specializing in physiology, Paris, France; mem. Acad. Medicine. Author: Des fonctions et des maladies nerveuses dans leurs rapports ave l'èducation sociale et privée, morale et physique, 1842; Mélanges mèdicopsychologiques, 1872. Follower of Buchez. Died Paris, Oct. 6, 1869.

CERLETTI, Ugo, Italian neuropsychiatrist; b. Conegliano, Trevisa, Sept. 12, 1877; M.D.; hon. doctorate Sorbonne; named prof. neuropsychiatry U. Bari, 1925; founder Neuropsychiat. Clinic of Rome, 1928; dir. inst. research Psychiat. Hosp. of Monbello. Recipient Susca prize Italian Acad. for sci. discoveries and research in physics, chemistry and medicine. Author: Lezioni di malattie nervose e mentali, 1949; many other sci. publs. Address: via Savoia 37, Rome, Italy.

CERMAK, Jack Edward, Am. engr. b. Hastings, Colo., Sept. 8, 1922; s. Joseph and Helen (Herman) C.; B.S., Colo. State U., 1947, M.S., 1948; Ph.D., Cornell U., 1959; postgrad. (NATO postdoctoral fellow) Cambridge (Eng.) U., 1961-62; m. Helen Jane Carlson, Dec. 17, 1949; children—Douglas Karl, Jonathan Joel. Faculty Colo. State U., Ft. Collins, 1948—, prof.-in-charge fluid mechanics program, 1960—, chmn. engring. sci., 1963—; dir., v.p. Research Foundation, 1965—. Visiting lecturer University Tex., Cambridge (England) University, U. Hokkaido, Technion; cons. Metronics, Inc., 1963, 66, Worthington, Skilling, Helle, Jackson, 1965, Cornell Aero. Labs., 1966-67; mem. bd. mems. Univ. Corp. for Atmospheric Research, 1966-67; pres., chmn. 10th Midwestern Mechanics Conference, 1966-67. Member of American Society Engineering Education (chairman of mechanics div. 1965—), Am. Inst. Aeros. and Astronautics, Am. Meteorol. Soc., Am. Geophys. Union, Am. Soc. C.E., Am. Soc. M.E., (fluid mechanics com.), Sigma Xi. Contbr. articles to profl. pubs. Research in lab. simulation of atmospheric phenomena, electrokinetics.*

CERMAK, Vladimír, Czechoslovakian phys. chemist; b. Ceské Budejovice, Czechoslovakia, July 5, 1920; s. Ludvík and Anna (Guthová) C.; Dr. rer. nat., Charles U., Prague, Czechoslovakia, 1949; m. Vera Görlichová, Mar. 30, 1945. Sci. asst. Charles U., Inst. Phys. Chemistry, 1946-49; postdoctoral fellow Czechoslovak Acad. Scis., Prague, 1949-53, staff Inst. Phys. Chemistry, 1952—, head of mass spectrometry dept., 1954—; vis. fellow Joint Inst. for Lab. Astrophysics, Boulder, Colo., 1964-65. Recipient State award for chemistry, 1954. Research and publs. on application of mass spectrometer to study of reactions and detection of excited particles; charge transfer and Penning ionization. Home: 28 Dvorecká, Prague 4. Office: 7 Máchova, Prague 2, Czechoslovakia.*

CERNUSCHI, Felix, physicist, astrophysicist; b. Montevideo, Uruguay, May 17, 1907; s. Felix and Teresa (Torterolo) C.; C.E., U. Buenos Aires (Argentina), 1932; Ph.D., Cambridge (Eng.) U., 1938; postgrad. Institut Henri Poincare, Paris, France, Poly. Inst. Zurich (Switzerland), Princeton. m. Zulema Frias, June 24, 1947; children—Felix, Bruno. Astrophysics research Cordoba Obs., 1937-39; prof. Tucuman (Argentina) U., 1939-43; research asso. Harvard Obs., 1945-47; sci. adviser UNESCO, Paris, 1947-48; vis. prof. U. P.R., 1948-49; prof. astronomy and physics U. Montevideo, 1949—, dir. astronomy dept., 1955—; dir. physics dept., faculty engring. U Buenos Aires, 1957—, prof., 1962—. Asociacion Argentina para el Progreso de las Ciencias fellow Princeton, 1938-39; Guggenheim fellow Harvard Obs., 1941-46. Fellow Am. Phys. Soc., A.A.A.S.; mem. Nat. Acad. Scis. (Argentina), Asociacion Fisica Argentina, Asociacion Argentina para el Progreso de las Ciencias, Asociacion Uruguay para el Progresso de las Ciencias, Internat. Union Theoretical and Applied Mechanics, others, Sigma Xi. Author: Sur les Neutrons, 1937; An Elementary Theory of Condensation, 1939; Super-Novae in the Newton-core Stars, 1939; Tentative theory of the Origin of Cosmic Rays, 1939; On the Behavior of Matter at Extremely High Temperatures and Pressures, 1939. Panorama Universitario Tucuman, 1941; La Ciencia en la Educación Intelectual, 1945; Como debre Ensenarse la Ciencia, 1965; others. Contbr. articles to profl. jours. Contributed to theories of liquids, critical phenomena, dielectric breakdown, adsorption, origin of chem. elements, cosmic grains, polarization of stars' light, evolution of cosmic clouds. Home: Reconquista 398, Montevideo. Office: Cerrito 73, Montevideo, Uruguay; also Paseo Colon 850, Buenos Aires, Argentina.*

CERRAI, Enrico, Italian chemist; b. Livorno, Italy, Apr. 7, 1924; s. Guido and Bianca (Betti) C.; Dr.-Inorganic and Phys. Chemistry, U. Pisa, Italy, 1949; m. Maria Teresa Soriani, Apr. 12, 1953; children—Federico, Guido. Researcher, Centro Informazioni Studi Esperienze Labs., Segrete (Milan) Italy, 1949—, head chemistry labs., 1956—; prof. separations of isotopes Poly. Sch., Milan, 1950-59, prof. nuclear materials tech., 1957—. Mem. Societa di Chinica Italiana, Gruppo di Studio delle Acque, Soc. for Analytical Chemistry. Research, publs. on deuterium enrichment, isotope exchange, chem. separations (reversed-phase partition chromatography) and analytical methods, treatment of liquid radioactive waste, radioactivity determinations in natural samples, chemistry of watercooled nuclear reactors with particular reference to water radiolysis, nuclear materials corrosion. Home: Via Griziotti 8. Office: Casella Postale 3986, Milan, Italy.*

CERRETELLI, P., Italian physiologist; b. Milano, Italy, Oct. 21, 1932; s. Berto and Jolanda (Croci) C.; M.D., U. Milano, 1956, Ph.D., Sport Medicine, 1959, Libero Docente in Human Physiology, 1962; m. Sogliani Maria Grazia, Oct. 26, 1964; 1 dau., Silvia. Faculty, U. Milano Sch. Medicine, 1956-62, asso. prof. physiology, 1966—. NATO fellow U. N.Y. at Buffalo, 1963-64, vis. asst. prof. physiology, 1964-65. Mem. Italian Physiol. Soc., Associazione Ergonomica Italiana. Research, publs. on energetics of muscular work, oxydative and anaerobic energy sources in man and animals in various physiol. conditions and in isolated mammalian muscle, physiology of cardiovascular system, respiratory system. Home: 8, Via Pirandello, Milano, Italy.*

CERUTI, Arturo, Italian botanist; b. Turin, Italy, Sept. 15, 1911; univ. prof., 1959—; prof. botany, sch. math., phys. and natural sci. U. Turin. Address: Università degli Studi, via Mattioli 25, Torino (306), Italy.

CESALPINO, Andrea, Italian botanist, physician; b. Tuscany, Italy, June 6, 1519; studied medicine and botany under Colombo and Ghini at U. Pisa (Italy); taught medicine, botany and philosophy at Pisa; became 2d dir. Pisan Bot. Gardens, 1554; prof. materia medica, Pisa, from 1555; physician to Pope Clement VIII in Rome, 1592; prepared herbarium for Bishop Tornabono, circa 1550. Author: De plantis libri, 1583; Tractationum philosophicorum tomus unus, 1588; Quaestionum medicarum, 1593; De metallicus libri tres Andrea Caesalpino auctore, 1596; Understood form of blood circulation before Harvey; described pulmonary circulation; used term circulation by 1559; described manufacture of alum, extraction of mercury from cinnabar; anticipated Linnean system of classification, by classifying plants according to fruits and flowers; stated minerals form regular geometrical forms when crystallized; knew of fossils; influenced in philosophy by Aristotle and Averroës. Died Rome, Feb. 23, 1603.

CESARI, Lamberto, mathematician; b. Bologna, Italy, Sept. 23, 1910; s. Cesare and Amelia (Giannizzeri) C.; Ph.D. in Math., U. Pisa (Italy), 1933; postgrad. (fellow) Munich U., Scuola Normale Sup. Pisa; m. Isotta Hornauer. Came to U. S., 1948. Asst. prof. math. Consiglio Nazionale Ricerche, Roma, 1935-39, U. Roma, 1938-39; asso. prof. math. U. Pisa, 1939-42; asso. prof. math. U. Bologna, 1942-46, prof., 1947-49; vis. prof. math. U. Cal. at Berkeley, 1949, U. Wis., Madison, 1950, Purdue U., Lafayette, Ind., 1950-52; prof. math. Purdue U., 1952-60; prof. math. U. Mich., Ann Arbor, 1960—. Mem. Am. Math. Soc., Math. Assn. Am., Am. Assn. U. Profs., Am. Civil Liberties Union, Unione Mathematica Italiana, Accademia delle Scienze di Bologna (corr. mem.), Sigma Xi. Author: Surface Area, 1956; Asymptotic Behavior and Stability Problems in Ordinary Differential Equations, 1959. Contbr. papers in fields of real functions, calculus of variations, surface area theory, optimal control theory, ordinary and partial differential equations, numerical analysis and applied math. Home: 719 W. Madison St., Ann Arbor, Mich. 48104.*

CESARIS DEMEL, Venceslao, Italian anatomist; b. Turin, Italy, Mar. 14, 1897; s. Antonio and Amalia (Mantovani) C.; M.D.; m. Luisa Terni, Sept. 2, 1943. Prof. anatomy and path. histology U. Siena (Italy). Mem. Italian Soc. Pathology, Italian Soc. Path. Anatomy, Acad. Physiocrats, Siena. Author: Trattazione di argomenti di patologia neoplastics; also numerous articles. Home: via Curatone 9, Siena. Office: via Laterino 8, Siena, Italy.

CESARO, Ernesto, Italian mathematician; b. Naples, Italy, Mar. 12, 1859; studied under Catalan at Liège, Belgium. Prof. at Naples. Author: Introduzione alla teoria mathematica della elasticita, 1895; Geometria intrinseca, 1896; Fondamento intrinseco della pangeometria, 1905. Research in intrinsic geometry, br. geometry which he greatly enlarged, also worked on divergent series; examined fundamental concepts of local probability. Died Torre Annunziata, Italy, Sept. 12, 1906.

CESI (or CAESIUS), Bernard, Italian theologian, mineralogist; b. Mantua, circa 1581; Jesuit prof. theology, Modena and Parma, Italy. Author: Mineralogia, sive naturalis philosophiae thesauri, 1636. First to use term mineralogy in modern sense. Died Modena, Sept. 4, 1630.

CESTONI, Diacinto, naturalist; b. Santa Maria in Giorgio nr. Ancona, Italy, May 10, 1637; s. Vittorio and Settimia C.; studied in Leghorn, Italy and Rome, Italy, 1648-56; studied with vice chancellor of Coll. Physicians of Pontifical State; chief pharmacist in ct. of grandduke of Tuscany, Leghorn, 1656—. Research and publs. on the verification of works of F. Redi and A. Vallisnieri in view of traditional ideas on natural philosophy especially on life cycles of parasites, fleas, and reproductive cycle of algae. Died June 29, 1718.

ÇETIN, Enver Tali, Turkish microbiologist; b. Van, Turkey, June 18, 1925; s. Arif Hikmet and Inayet (Bulak) Ç.; M.D., Med. Faculty, Istanbul (Turkey) U., 1950, Specialist in Microbiology, 1954. Researcher, Inst. Microbiology and Contagious Diseases, Med. Faculty, Istanbul U., 1951-54, 1956-58, asst. prof., 1958-65, prof. microbiology, 1965—; researcher London Sch. Hygiene and Tropical Medicine, 1955-56, Inst. Pasteur, Paris, France, 1958-59. Turkish rep. Internat. Subcom. on Phage-Typing of Staphylococci, 1962—. Mem. Turkish Soc. Biology, Turkish Soc. Microbiology, Turkish Med. Assn., Société Française Microbiologie, N.Y. Acad. Scis. Author: General Virology, 1962; Viral and Rickettsial Diseases, 1963; Practical Microbiology, 1965; also numerous articles. Demonstrated presence of heat labile and heat stabile factors and antibodies in normal and immune guinea pig sera against T-system bacteriophages; research on antibiotic susceptibility of different microorganisms, haemolysin inhibiting substance in Staphylococcus aureus strains. Home: 8/5 Fakülte ap. Börekci Sokak, Beyazit, Istanbul, Turkey.*

CEVA, Giovanni, Italian mathematician; b. circa 1647, Milan, Italy; hydraulic engr.; employed by Duke of Mantua; commr. of Archducal Chamber of Mantua, Italy. Author: De lineis rectis se invicem secantibus constructio statica, 1678; Geometriae motus; De mundi fabrica . . . , De re nummaria geometrice tractata. Discovered theorems on theory of transversals; 1st clear math. writings on econs.; gave static and geometric proofs for concurrency of straight lines through vertices of triangles (Ceva's theorem). Died 1734.

CEVA, Tommaso, Italian mathematician; b. Milan, Italy, Dec. 20, 1648; ed. as Jesuit. Taught mathematics at Jesuit College, Milan. Author: Opuscula Mathematica, 1699. Composed a poem in Latin on Descartes' system; described an instrument for mechanically executing bisection of angles; discussed the cycloid. Died Milan, Feb. 3, 1736.

CHABANEU, François, chemist; b. Charente, France, Apr. 21, 1754; studied theology (expelled from sch. due to his views on metaphysics); became prof. math. at Passy (at age 17); later prof. mineralogy, physics and chemistry, Madrid, Spain. Research on platinum; produced ingot of malleable platinum, 1783. Died Nontron, Jan. 1842.

CHABBAL, Robert, French physicist; b. Nimes, France, Feb. 6, 1927; s. Jean and Lucie Chabbal; Doctorat es Sciences, Ecole Normale Superieure, 1957; children—Jean, Sylvie. Prof. Faculty of Scis., Paris, 1959-65, Faculty of Scis., Orsay, France, 1965—; dir. Lab. Aime-Cotton, CRNS, 1962—. Mem. French Phys. Soc., Am. Optical Soc. Research in atomic and optical spectroscopy. Address: Lab. Aime-Cotton, C.N.T.S. II, Campus D'Orsay, 91 Orsay, France.*

CHABBERT, Yves Achille, French bacteriologist; b. Bordeaux, France, Apr. 1, 1921; M.D., 1946; m. Marie Vialard, Aug. 28, 1947; children—Veronique, Sophie, Jean-Christophe. Staff, Pasteur Inst., Paris, 1948—, chief lab. antibiotherapy. Mem. Société Francaise de Microbiologie. Research, numerous publs. on bacterial resistance to antibiotics, sensitivity testing, treatment of infectious diseases with antibiotics Home: 20 Rue de Turenne, Paris (4). Office: Institut Pasteur, Rue du Dr. Roux, Paris, France.*

CHABERT, Philabert, French veterinary surgeon, bacteriologist; b. Lyons, France, Jan. 6, 1737; dir. Veterinary Sch., Alfort, France; mem. French Acad. Scis., Soc. Agriculture; editor The Veterinary Almanac, 1782-90. Author: Traité des maladies vermineuses, 1782; Traité du charbon ou anthrax dans les animaux, 1783. Described geanders, 1779. Died Alfort, Sept. 8, 1814.

CHABOT, Georges, French geographer; b. Besançon, France, Apr. 5, 1890; s. Charles and Marthe (Scohy) C.; ed. École normale supérieure; agrégé in history and geography; D.Litt.; Dr. Honoris causa, U. Uppsala; m. Marcelle Durand, Apr. 4, 1914; children—Suzei (Mrs. A. Trautsolt), Jean-Rene. Prof., dean Sch. Letters, Dijon, 1928-45; prof., dir. Inst. Geography, Sorbonne, Paris, until 1960; pres. Nat. Center Sci. Research, also French Com. Internat. Geog. Union, since 1960. Hon. mem. geog. socs. of Finland, Belgium, Serbia, Norway, Germany. Author: les Plateaux du Jura central, 1927; la Cotiere orientale de la Dombes, 1927; la Bourgogne, 1942; les Villes, 1948; la Finlande et les Pays scandinaves, 1958. Address: 38 rue Oger, Bourg-la-Reine (Seine), France.

CHABRIER DE LA SAULNIÈRE, Pierre, French chemist; b. Neuvy sur Loire, France, Oct. 25, 1912; s. Philippe and Emilienne (Desserprix) C.; received degree in chem. engr., U. of Lyon, 1936; doctorat, Paris, 1942; m. Rose Bruneteau, 1941; 2 sons, 2 daus. Affiliated with Centre National de la Recherche Scientifique, from 1940; dir. of research, since 1962; mem. of permanent commission on French pharmocopoeia; officer of Palmes Académiques; Chevalier de la Santé Publique; Berthelot medal; editor in chief of series Les industries, leurs productions, leurs nuisances; author of more than 350 scientific articles. Research on sulphur derivatives; anticonvulsants; excitants of the central nervous system; analgesics; local anesthetics; anti-thyroid preparations; since 1957, special researches on organophosphores; reaction mechanisms; enzyme action; insecticides; industrial hygiene. Home: 242 Boulevard Saint-Germain 75, Paris 7, France. Office: Laboratoire du C.N.R.S. 45 Orleans la Source, France.*

CHABRY, Laurent, French biologist; b. Roanne, France, 1855; M.D., 1881; doctor en science, 1887; asst. at the sch. hautes études; adj. dir. Concarneau Lab.; master of confs. at Faculty of Sci., Lyons, France. Publs. on flying of birds, double position of equilibrium in wings of Coleoptera; research on Tb, exptl. embryology. Died Riorges, France, Nov. 23, 1894.

CHACE, Fenner Albert, Jr., Am. zoologist; b. Fall River, Mass., Oct. 5, 1908; s. Fenner Albert and Mary (Buffinton) C.; A.B., Harvard, 1930, A.M., 1931, Ph.D., 1934; m. Janice Dexter Grinnell, Sept. 1, 1934; 1 dau., Linda Dee (Mrs. Richard W. Mayo, Jr.). Began career as assistant curator marine invertebrates Mus. Comparative Zoology, Harvard, 1934-42, tutor in biology, 1940-41; curator crustacea, 1942-46, curator div. marine invertebrates Smithsonian Instn., Washington, 1946-63, sr. zoologist, invertebrate zoology, 1963-—, supr. Life in the Sea Hall, 1961-65. Mem. Bd. U. S. Civil Service Examiners. Alexander Agassiz fellow in oceanography Mus. Comparative Zoology, Harvard, 1935-40. Mem. Am. Soc. Limnology and Oceanography, Biol. Soc. Washington, Soc. Systematic Zoology, Sigma Xi. Research and publs. on morphology, taxonomy, classification, and distbn. of shrimps, lobsters, crabs, and related groups of decapod and stomatopod crustaceans. Home: 6 Leland Ct., Chevy Chase, Md. 20015. Office: Smithsonian Instn., Washington 20560.*

CHACHAVA, Mikhail Konstantinovich, Russian surgeon; b. 1902; grad. Med. Faculty, Tbilisi U., 1928, postgrad., 1928-30; D.Med. Sci., 1937. Asst., later asst. lectr. Tbilisi Med. Inst., 1931-44, prof., 1944-47, head gen. chair hosp. surgery, 1948-50, head chair hosp. surgery Pediatric and Health Hygiene Faculty, 1951-—. Mem. All-Union (bd. mem.), Tbilisi (dep. chmn.) surg. socs. Author over 100 works including Surgical Treatment of Ulcers, 1951; Burns and Constriction of the Esophagus, 1952; Skin Grafts, 1963; Tuberculosis of the Knee Joint; Manual of Specific Surgery, 2 vols. Co-editor: Large Med. Ency., 2d edit. Address: Tbilisi Med. Inst., ulitsa Melikishvili 16, Tbilisi, Gruz. SSR, USSR.

CHACHIAN, Pavlaki, Armenian physician; b. Constantinople, Turkey, 1806; s. Emmanuel Chachian; studied medicine, Padua, Italy, Paris, France; introduced auscultation in Constantinople. Died Nancy, France, 1887.

CHACON, Dionysio (Denis) Daca (or Daza), Spanish physician; b. Valladolid, Spain, 1503; wound surgeon to Charles V; served as mil. surgeon in Flanders; personal surgeon to Archduke Maximilian; surgeon to Don Juan of Austria, to Spanish fleet, to Valladolid Hosp. Author: Practica y teorica de cirurgia, en romancy en latin, (most exhaustive book in Spanish on surgery at time), 1600. Developed new method of operation for aneurysm; 1st to suggest ligation of nasal polyps; disagreed with then current belief that gunshot wounds were poisonous. Died circa 1503.

CHACORNAC, Jean, French astronomer; b. Lyons, France, June 21, 1823; studied at Obs., Marseille, France; asst. astronomer Obs., Paris, France. Recipient Prix, French Acad. Scis. Author: Atlas écliptique, 1856. Discovered 6 asteroids; research on sun-spots. Died Villeurbane, France, Sept. 23, 1873.

CHADWICK, Claude Simpson, Am. biologist; b. Carthage, Tex., Jan. 2, 1907; s. Daniel Davis and Anna K. (Westmoreland) C.; B.S., Centenary Coll., 1927; M.S., Vanderbilt U., 1928, Ph.D., 1936; m. Mary Evelyn Jenkins, Aug. 22, 1931; children—Elizabeth Ann (Mrs. Lee Martin Russell), Caroline Danielle (Mrs. James Hiram Beasley, Jr.). Faculty, Vanderbilt U., 1928-51; prof. biology George Peabody Coll., 1951-54, head biology dept., 1954-63, chmn. sci. div., 1956-63; head dept. biology Emory and Henry Coll., Emory, Va., 1963-—. Mem. com. ednl. policies div. biology and agr. Nat. Acad. Scis.-NRC, 1955-58; mem. adv. com. Biology Data Book, 1960-64. Mem. Am. Soc. Zoologists, Am. Inst. Biol. Scis., Soc. Exptl. Biology and Medicine, Assn. S.E. Biologists, Va. Acad. Sci., Sigma Xi. Author: Information Program in Principles of Biology, 1961, rev., 1963. Research, publs. on relationship of pituitary gland to the water drive, pituitary-thyroid-skin gland relationship in induction of molting, high mitosis-inducing effect of pituitary hormone, effect of prolactin on skin of certain Amphibia; cyclic changes in cytology of mammalian pituitary in relation to reproductive cycle. Address: P.O. Box KK, Emory, Va. 24327.*

CHADWICK, Sir Edwin, English physician; b. Longsight, Eng., Jan. 24, 1800; educated for the bar; became lit. asst. to Jeremy Bentham; investigator Royal Commn. on Poor Laws, from 1832, became full mem., 1833; apptd. sec. to poor law commrs., 1834 (Commn. dissolved in 1846); commr. Bd. of Health for improving water supply, drainage and cleansing of gt. towns, 1848-54. Contbd. essays to Westminster Rev.; wote report of 1834 which procured passage of poor laws. Advocated using experts in certain depts. of local affairs instead of elected ofcls. Died Surrey, Eng., 1890.

CHADWICK, George Halcott, Am. geologist; b. Catskill, N.Y., May 27, 1876; s. Nathaniel Kimball and Celia Serena (Halcott) C.; Ph.B., U. Rochester, 1904, M.S., 1907; Sc.D., St Lawrence U., 1940; m. Bertha Elisabeth Ellwanger, Feb. 22, 1908 (dec. Mar. 1937); children—Elizabeth Ellwanger, George Halcott; m. 2d, Irene Brugger, Sept. 30, 1944. With Ward's Natural Sci. Establishment, 1896-1906, N.Y. State Mus., 1906-07, St. Lawrence U., 1907-14, U. Rochester, 1914-23, Empire Gas & Fuel Co., Okla., 1923-25, Williams Coll., 1925-26, Vassar, 1929, U. Newark, 1943-44, Natural Sci. Assn. of the Catskills, Inc., from 1944; with various surveys: N.Y. State, Pa. Topographic and Geologic, 1926-31, Cities Service Group, 1930-31, Nat. Park Service, 1935-37. Fellow Geol. Soc. Am.; mem. Phi Beta Kappa, Sigma Xi, Alpha Delta Phi. Contbr. papers on Devonian Stratigraphy, geology of Mt. Desert Island, Me. Died Aug. 1953.

CHADWICK, Sir James, Brit. physicist; b. Oct. 20, 1891; s. J. J. Chadwick; ed. Manchester Secondary Sch., Univs. of Manchester and Berlin; Ph.D., Cambridge U.; M.Sc., Victoria; D.Sc. (hon.) Oxford and Birmingham, Reading, Dublin, Leeds and McGill, Exeter; LL.D., Liverpool and Edinburgh; m. Aileen Stewart-Brown, 1925; twin daus. Lectr., asst. dir. radioactive research. Cavendish Lab. (Cambridge); master, Gonville and Caius Coll., Cambridge, 1948-58, now fellow. Lyon Jones prof. physics U. Liverpool, 1935-43. Knighted, 1945. Fellow Royal Soc., 1927, fgn. mem. Royal Acad., Brussels; Royal Danish Acad. of Sci.; Royal Acad. Sci., Amsterdam; corr. mem. Sachsische Akademie der Wissenschaften, Acad. of Sci., Leipzig. Recipient Hughes medal Royal Soc., 1932, Copley Medal, 1950; Nobel prize (physics), 1935; U. S. Medal for Merit, 1946; Faraday Medal, 1950, Franklin Medal, 1951; elected to Pontificia Academia Scientiarum, 1961. Co-author: Radiations from Radioactive Substances (with Lord Rutherford and C. D. Ellis), 1930; papers on radioactivity. Proved existence of neutron by bombarding beryllium with alpha particles, 1932; worked on generation of chain reaction and on nuclear disintegration; head of British delegation to U. S., 1943; actively engaged in experiments that led to devel. of atomic bomb. Home: Wynne's Parc, Denbigh, North Wales, Eng.

CHADWICK, Leigh E(dward), physiologist; b. Washington, Aug. 9, 1904; s. DeWitt Clinton and Charlotte Alice Rothwell (Pechel) C.; B.S., Haverford Coll., 1925; fellow Am.-German student exchange, Phillips-Universitat, Marburg, Germany, 1927-28; M.A., U. Pa., 1929; M.A., Harvard, 1938, Ph.D., 1939; m. Maria Beatrice Kievenaar, Nov. 17, 1941. Instr., French and German, 1925-32; instr. biology Pueblo (Colo.) Jr. Coll., 1939-41; asst. physiology U. Rochester Sch. Medicine and Dentistry 1941-42, instr., 1942-44; chief entomology br. med. labs. Army Chem. Center, Md., 1944-56; prof. entomology U. Ill., 1956-—, former head dept. Mem. Am. Acad. Arts and Sci., A.A.A.S., Am. Physiol. Society, Entomological Society of America, Society of General Physiologists, Sigma Xi, Phi Beta Kappa. Research on morphology and physiology of insects. Home: 504 South James Street, Champaign, Ill. Office: U. Ill., Dept. Entomology, Urbana, Ill.

CHADWICK, Wallace Lacey, Am. engr.; b. Loring, Kan., Dec. 4, 1897; ed. U. Redlands; hon. D.Eng.S., 1965. Mem. staff Southern Cal. Edison Co., for 35 years, became v.p., 1951, ret., 1962; recipient Sprague award Instrument Soc. Am., 1963. Worked on design and constrn. of Cal. river aqueduct, Big Creek 2A project, also others related to devel. of water power; contbr. to devel. of first digital-computer control system used for start-up and control of thermal power plants.

CHAFFEE, Emory Leon, Am. physicist; b. Sommerville, Mass., Apr. 15, 1885; s. Emory Franklin and Belle Genevieve (Carter) C.; S.B., Mass. Inst. Tech., 1907; A.M., Harvard, 1908, Ph.D., 1911 (Bowdoin Prize), S.D., 1944; D. Engring. (hon.), Case Inst. Tech., 1935; m. Alice Hampson, March 25, 1924. Instr. physics and elec. engring. Harvard, 1911-17, asst. prof. physics, 1917-23, asso. prof., 1923-26, prof., 1926-35, Gordon McKay prof. physics and communication engring. 1935-46, Gordon McKay prof. applied physics, 1946-53, prof. emeritus, 1953-—, Rumford prof. physics, 1940-53, dir. Cruft Lab. 1940-46, co-dir. Lyman Lab. of Physics 1947-53, dir. Labs. of Engring. Sci. and Applied Physics, 1948-53. Fellow Am. Acad. Arts and Scis., Am. Phys. Soc., I.R.E. (v.p. 1925); mem. Tau Beta Pi, Sigma Xi. Author Physical Laboratory Manual, 1914; Theory of Thermionic Vacuum Tubes, 1933; also numerous articles in physics and engring. jours. Co-author: Electronic Circuits and Tubes, 1948. Home: 130 Goden St., Belmont, Mass. 02178. Office: Cruft Laboratory, Harvard University, Cambridge, Mass. 02138.*

CHAGAS, Carlos, Brazilian physician; b. 1879; discovered trypanosoma cruzi, causative agent of Brazilian or Am. trypanosomiasis, known as Chagas' disease, 1909. Died 1934.

CHAIN, Ernst Boris, biochemist; b. Berlin, Germany, June 19, 1906; s. Michael Chain; grad. in chemistry Freidrich-Wilhelm U., Berlin, 1930; hon. degrees univs. Liège, Bordeaux, Turin, Paris, La Plata, Cordova, Brazil, Montevideo; m. Anne Beloff, 1948; children—Benjamin, Daniel, Judith. Went to Eng., 1933. Staff, Charité Hosp., Berlin, 1930-33; asst. Sch. Biochemistry, Cambridge (Eng.) U., 1933-35; demonstrator, lectr. chem. pathology Sir William Dunn Sch. Pathology, Oxford (Eng.) U., 1935-48; sci. dir. Internat. Research Centre for Chem. Microbiology, Instituto Superiore di Sanità, Rome, Italy, 1948-61; prof. biochemistry Imperial Coll., U. London, 1961-—. Recipient (with Fleming and Florey) Nobel prize in medicine, 1945, Silver Berzelius medal Swedish Med. Soc., 1946, Pasteur medal Institut Pasteur and Societé de Chimie Biologique, prize Harmsworth Meml. Fund, Paul Ehrlich Centenary prize, 1954, Gold medal for therapeutics Worshipful Soc. Apothecaries, London, 1957; decorated comdr. Legion of Honor, Grand Ufficiale al Merito della Repubblica Italiana. Fellow Royal Soc., 1949; mem. Academie de Medicine, Academie des Sciences (both Paris), Real Academia de Ciencias (Madrid), Weizmann Inst. Sci. (Israel), Nat. Inst. Scis. (India), Sicieta Chimica Italia (Marotta medal 1962), N.Y. Acad. Medicine, Finnish Biochem. Soc., others. Contbr. papers, monographs on penicillin and antibiotics. Research on tumor metabolism, mechanism of lysozyme action, methods of biochem. microanalysis, 1935-39; systematic study of antibacterial substances produced by micro-organisms (with H. W. Florey) leading to reinvestigation, properties, isolation, chem. structure of penicillin and other antiobiotics; investigation of carbohydrate-amino acid relationship in nervous tissue; studied mode of action of insulin, fermentation tech., 6-amino-penicillanic acid and penicillinase-stable penicillius, lysergic acid prodn. in submerged culture and isolation of new fungal metabolites (all since 1948).

CHAKLIN, Vasiliy Dmitrievich, Russian orthopedist, traumatologist; b. 1892; grad. Med. Faculty, Kharkov U., 1917; D.Med. Sci., 1924. Intern, asst., lectr. dept. faculty surgery and operative surgery Kharkov Med. Inst., asso. Kharkov Orthopedic Inst., 1917-30; prof., 1931-—; founder, dir. Urals (now Sverdlovsk) Research Inst. Traumatology and Orthopedics, head chair orthopedics, traumatology and field surgery Sverdlovsk Med. Inst., 1931-44; dir. Moscow Inst. Prosthetics, then sci. dir. Moscow Orthopedic Hosp., 1945-55; head clinic Central Inst. Traumatology and Orthopedics, USSR Ministry Health, 1956-—; chief orthopedist and traumatologist Moscow City, 1959-—. Mem. learned council RSFSR Ministry Health. Mem. USSR Acad. Med. Sci. (corr.), All-Union Soc. Traumatologists and Orthopedists (dep. chmn.), Purkinje Czech Med. Soc. (hon.), Bulgarian Soc. Orthopedists and Traumatologists (hon.). Author 130 works including Infectious Diseases of the Bones, Joints and Cartilages; Fractures; Traumatology and Orthopedics; Operative Orthopedics; Orthopedics, 2 vols., 1957; Principles of Operative Traumatology and Orthopedics, 1962. Address: Central Inst. Traumatology and Orthopedics, USSR Ministry Health, Teply p. 16, Moscow, USSR.

CHAKO, Nicholas, mathematician, physicist; b. Hotova, Epirus, Albania, Nov. 11, 1910; s. Kyriacus (Qako) Demetrius and Victoria John (Tako) C.; came to U. S., 1929, naturalized, 1935; B.S., U. Paris-Marseille (France), 1928; Ph.D., Johns Hopkins, 1934; Dr-ès-Sciences, Sorbonne, 1966; attended Harvard, 1933-34, also Mass. Inst. Tech.; postdoctoral fellow University of Michigan, 1933; fellow by courtesy Univ. Chicago, 1942-44; m. Bernardine van Looy, July 16, 1952; 1 son, Alexander Constantine. Prof., head math. and physics sects. state Gymnasium, Scutari, Albania, 1936-37; mem. staff Cruft Lab., also tutor Harvard, 1938-40; staff spectroscopic lab. Mass. Inst. Tech., 1940-41; lectr. Ill. Inst. Tech., 1941-42; asso. prof. Kan. State Coll., 1946-47, Ala. Poly. Inst., 1947-49; Fulbright exchange prof. Utrecht (Holland) U., 1950-51; guest prof. Chalmers U. Tech., Gothenburg, Sweden, 1951-52; vis. lectr. Lund (Sweden) U., spring 1952; lectr. Brown U., summer 1944; research asso. Inst. Math. Scis., N.Y.U., 1953-56; prof. physics Adelphi Coll., 1955-56; dept. math. Queens Coll., 1956-—; vis. prof. Inst. Space Studies, N.Y.C., summer 1961; invited lectr. Polish Acad. Scis., also Athens (Greece) U., summer 1965; invited prof. French Atomic Energy Center, Saclay, also exchange prof. U. Paris, 1965-66. Tech. adviser Signal Corps, 6th Service Command, 1942; cons. OSS, 1942-45; mathematician, research engr. Russell Electric Co., Chgo., 1942-46. Recipient ann. prize Royal Soc. Engrs. and Chalmers Alumni Assn., 1952. Fellow Inst. Math. and Applications, Gt. Britain; mem. Am., Dutch, Polish phys. socs., Am., French math. socs., A.A.A.S., N.Y. Acad. Scis. Author: Anglo-Saxon Universities, 1937-39; Emperor Anastasius I, 1940; Contribution à la Théorie de la Diffraction, 1966. Research and publs. on absorption of light by organic compounds, geometrical and electron optics, diffraction theory, foundations of quantum mechanics, asymptotic integration, spl. functions, partial differential equations. Home: 138-10 Franklin Av., Flushing, N.Y. 11355.*

CHAKRAVARTI, Debi Mukherji, Indian chemist; b. Calcutta, India, Apr. 2, 1924; d. Sachindra Kumar and Kamalabala (Banerjee) Mukherji; B.Sc., Calcutta U., 1941; M.Sc., U. Coll. Sci., Calcutta, 1943; D.Phil., Oxford U., 1949; m. Ram Narayan Chakravarti, May 18, 1950; 1 son, Deb Narayan. Faculty, Bethune Coll., Calcutta, 1944-47, 50-—, prof. chemistry, 1965-—. W. Bengal state scholar Dyson Perrins Lab., Oxford U., 1947-49. Premchand Roychand scholar Calcutta U., 1950; recipient Mouat medal Calcutta U., 1952. Fellow Chem. Soc. (London); mem. Instn. Chemists (India). Editorial bd. Jour. and Proc. Instn. Chemists. Research and publs. on structures of alkaloids; surveyed Indian jungle beans and jungle yams for sterols and steroid sapogenins and discovered (with R. N. Chakravarti) Dioscorea prazeri and Dioscorea deltoidea as rich Indian sources of diosgenin; studied synthetic monoterpenes and natural higher terpenes. Home: Piii, Block F, New Alipore, Calcutta-53, India.*

CHAKRAVARTI, Robindra Nath, Indian pathologist; b. Agra, India, Aug. 2, 1925; s. Hem Chandra and Kiran Bala C.; I.Sc., Agra Coll., 1943; M.B., B.S., Lucknow (India) U., 1948; D.T.M. and H., Calcutta (India) U., 1957, D.Phil., 1961; m. Arati Sanyal, Aug. 14, 1950; children—Ranjan, Manas, Aloke, Seema. House physician King George's Med. Coll. Hosp., Lucknow, 1948; demonstrator pathology, 1949-52; staff Central Drug Research Inst., Lucknow, 1952-62, sr. sci. officer, 1956-62; asst. prof. exptl. medicine Inst. Post-Grad. Med. Edn. and Research, Chandigarh, India, 1962—; hon. lectr. anatomy U. B.C.; Vancouver, Can., 1959-60; tchr., examiner postgrad. med. studies Punjab (India) U., 1963—; examiner Bombay (India) U., 1965—. Colombo Plan fellow, Can., U. S. A., 1959-60; recipient certificate in exptl. pathology and electron microscopy Govt. of Can., 1960. Mem. Indian Assn. Pathologists, Soc. for Exptl. Biology and Medicine, Gerontological Soc., Indian Sci. Congress Assn. (sect. com. 1965—). Research, publs. on exptl. medicine, exptl. atherosclerosis in rabbits, use thyroid hormones and estrogens to ameliorate atherosclerosis. Home: 50/6JB, Sector 24 A, Chandigarh, Punjab, India.*

CHALCIDIUS, Roman natural philosopher; flourished 300-350; Latin translator and commentator of the Timaeos; neoplatonist, until end of 12th century Plato was known almost exclusively in west through his work.

CHALKE, Herbert Davis, Brit. physician; b. Porth, S. Wales, June 15, 1897; s. Richard David and Naomi (Davies) C.; ed. univs. Wales, Cambridge, London; M.A., Cambridge, 1925; D.P.H., 1930; m. Kathleen Freeman; one son, David John. Med. officer Met. Asylum's Bd., 1926-30; Tb physician, Wales, Dorset, Hampstead, 1930-39; dep. med. officer Health Borough of Hampstead, 1936-39, 46-48; hygiene officer Royal Army M.C., 1939-45; officer Comdg. Hygiene Co. (territorial Army), 1947-54; div. med. officer London County Council, 1947-57; med. officer health Met. Borough, Camberwell, 1954-64. Lectrs. panel Faculty Med. History, Worshipful Soc. Apothecaries; lectr. at colls., hosps., profl. socs. Licentiate, mem. Royal Coll. Physicians London. Fellow Royal Soc. Health, Brit. Med. Assn.; mem. Soc. Med. Officers Health (past pres., chmn. council), Soc. Authors, Brit. Soc. for Internat. Health Edn., Soc. for Study of Addiction, Med. Council on Alcoholism (hon. sec.). Author: Hygiene and Public Health, 1962, 63, 65. Co-author: Preliminary Hygiene, 1967. Co-author, editor: Radiation and Health, 1962. Contbr. articles and editorials to profl. jours. Hon. editor Bull. on Alcoholism. Research on prevention and control of Tb, smallpox, typhus, food infections, home accidents, also on loss of sense of smell as contributory factor in gas poisoning of the elderly, needs and problems of the elderly, role of social workers, health knowledge among the public, effects of health edn., smoking habits of children and results of spl. anti-smoking edn., alcohol edn. and prevalence of alcoholism, impact of TV advt. on the public, radiation and health. Address: No. 5, Egerton Gardens, London N.W. 4, Eng.*

CHALLIS, James, Brit. astronomer; b. Essex, Eng., Dec. 12, 1803; s. John C.; ed. Trinity Coll., Cambridge (Eng.) U., 1821; became sr. wrangler, 1825; fellow, 1826-31; rector Papworth Everard, Cambridgshire, 1830-52; Plumian prof. astronomy, 1836-82; dir. Cambridge Obs., 1836-61. Mem. Royal Astron. Soc., Royal Soc., 1848. Author: Notes on the Principles of Pure and Applied Calculation; and Applications of Mathematical Principles to Theories of the Physical Forces, 1869; astron. lectures, 1879; also publs. on scriptures, edn. Looked for unknown planet (Neptune) using Leverrier's computation and observed it without knowing it, 1846; invented meteorscope, 1848, transit reducer, 1849, collimating eyepiece, 1850; gave theory of phys. forces which states phys. forces are modes of pressure of same ethereal medium. Died 1882.

CHALOUPKA, Jiri, Czechoslovakian microbiologist; b. Malsice, Czechoslovakia, Mar. 25, 1926; s. Vojtěch and Bozena (Hejdová) C.; M.S., Charles U., Prague, Czechoslovakia, 1952; Ph.D., Czechoslovakian Acad. Sci., 1955; m. Bohumila Peroutková, Feb. 18, 1950; 1 dau., Alena. Staff Inst. Microbiology, Czechoslovakian Acad. Sci., Prague, 1952—, head dept. gen. microbiology, 1956—; external lectr. in molecular biology Charles U., 1964—; vis. staff Chester Beatty Research Inst., London, Gt. Britain, 1959; research asso. dept. bacteriology U. Wis., Madison, 1962. Mem. Czechoslovakian Soc. Microbiology (head sect. gen. microbiology 1965—), Czechoslovakian Soc. Biochemistry. Research and publs. on regulation of bacterial growth and enzyme synthesis; discovered regulation synthesis of extracellular bacterial proteolytic enzymes by amino acids, degradation of cell wall in bacilli during growth, relationship between synthesis of structural components of cell and growth. Home: 2534 Hlavní, Prague 4. Office: Czechoslovak Acad. Sci., 1083 Budejovická, Prague 4, Czechoslovakia.*

CHAMBERLAIN, Francis Le Conte, Am. physician; b. Santa Cruz, Cal., 1905; M.D., U. Cal., 1934; D.Sc. in Internal Medicine, Columbia, 1937. Resident cardiologist Mass. Gen. Hosp., Boston, 1937-38; cardiovascular research fellow Harvard, 1938-39; now clin. prof. medicine U. Cal. Served as surgeon USPHS; cons. VA Hosp., Ft. Miley. Mem. cardiovascular certification exam. bd. Diplomate Am. Bd. Internal Medicine, specializing cardiovascular. Mem. Am. Heart Assn. (past pres.). Mem. 1st Am. group to explore interior of heart with catheter (important to devel. heart surgery); wrote article Intravenous Saline Infusion as clinical Test for Right Heart and Left Heart Failure, others. Address: 87 Farnsworth Lane, San Francisco 17.

CHAMBERLAIN, Joseph Wyan, Am. astronomer; b. Boonville, Mo., Aug. 24, 1928; s. Gilbert Lee and Jessie (Wyan) C.; A.B., U. Mo., 1948, A.M., 1949; M.S., U. Mich., 1951, Ph.D., 1952; m. Marilyn Jean Roesler, Sept. 10, 1949; children—Joy Anne, David Wyan, Jeffrey Scott. Project scientist aurora and airglow USAF Cambridge Research Center, Bedford, Mass., 1951-53; research asso. to prof. Yerkes Obs., Chgo., 1953-62; asso. dir. space div. Kitt Peak Nat. Obs., Tucson, 1962—. Cons. Pres.'s Sci. Adv. Com. Space Sci. and Tech. Panel, Los Alamos Sci. Lab., Environmental Sci. Services Adminstrn. Alfred P. Sloan fellow, 1961-63. Fellow Royal Astron Soc., A.A.A.S., Am. Geophys. Union; mem. Am. Phys. Soc., Am. Astron. Soc. (Warner prize 1961), Internat. Astron. Union, Internat. Union Geodesy Geophysics, Internat. Sci. Radio Union, Nat. Acad. Sci. Author: Physics of the Aurora and Airglow, 1961; also articles. Mem. editorial bd. Sci., 1964—; Planetary Space Sci., 1963—; Jour. Atmospheric Sci., 1961—. Observational and theoretical research on aurora and airglow; theoretical research on solar wind, atmospheres of Mars and Venus. Office: P.O. Box 4130, 950 N. Cherry Av., Tucson 85717.*

CHAMBERLAIN, Owen, Am. nuclear physicist; b. San Francisco, July 10, 1920; A.B., Dartmouth, 1941, Cramer fellow, 1941; Ph.D., U. Chgo., 1949; m. 1943; 4 children. Instr. physics U. Cal. at Berkeley, 1948-50, asst. prof., 1950-54, asso. prof., 1954-58, prof., 1958—; civilian physicist Manhattan Dist., Berkeley, Los Alamos, 1942-46; Guggenheim fellow, 1957-58; Loeb lecturer at Harvard University, 1959. Recipient Nobel prize (with Emilio Segré) for physics, for discovery anti-proton, 1959. Fellow Am. Phys. Soc.; mem. Nat. Acad. Scis. Research in high-energy physics including triple-scattering and polarized-target expts; also research using photog. emulsions to produce visual examples of an anti-proton and proton or neutron in which the particles die simultaneously. Address: Physics Dept., U. Cal., Berkeley 4, Cal.

CHAMBERLAIN, Richard Hall, Am. physician; b. Jacksonville, Fla., May 25, 1915; s. William Douglas and Lucille (McKowen) C.; A.B., Centre Coll., 1934; M.D., U. Louisville, 1939; m. Merle Johnson, Aug. 23, 1941. Faculty, U. Pa., Phila., 1940—, prof. radiology Sch. Medicine and Grad. Sch. Medicine, 1952—, chmn. dept. Sch. Medicine, 1961—. Mem. Nat. Council on Radiation Protection, 1952—, ofcl. del. UN Sci. Com. on Effects Atomic Radiation, 1963—; cons., mem. numerous coms. USPHS, WHO, U. S. Dept. State, VA. Fellow Am. Coll. Radiology (v.p. 1964-65); mem. A.M.A., Am. Radium Soc., Am. Roentgen Ray Soc., Am. Cancer Soc., Radiol. Soc. N.Am. (1st v.p. 1957), Sigma Xi, Alpha Omega Alpha, Alpha Kappa Kappa. Research, publs. on radiation therapy, radioisotopes in clin. medicine, radioisotope scanning methods, fluoroscopic image amplification, radiation protection in medicine and pub. health. Home: 8327 Germantown Av., Phila. 19118.*

CHAMBERLAND, Charles Édouard, French bacteriologist; b. Chilly-le-Vignoble, France, Mar. 12, 1851; s. Auguste and Appoline (Philibert) C.; studied phys. scis. l'École normale, 1871-74; became asst. to Pasteur at physiol. chemistry lab. on rue d'Ulm, Paris, 1875; joined Pasteur Inst., 1888, became under-dir., 1904. Mem. French Acad. Scis. Author: Recherches sur l'origine et le développement des organismes microscopiques, 1879; le Charbon et la vaccination charbonneuse, d'après les travaux récents de M. Pasteur, 1883; les Eaux d'alimentation dans l'hygiène et les maladies épedémiques, 1885. Pioneered in bacteriology; with Pasteur opposed the idea of spontaneous generation; using his research on resistance of small organisms to high temperatures, he developed rules for sterilization of cultures for several organisms; (with L. Pasteur, P. Roux) proved rabies virus is present in the blood and nerve tissue of rabid animals; invented an unglazed porcelain filter; worked to improve pub. hygiene; advocated a vaccination made from coal products. Died 1908.

CHAMBERLEN, Peter, Brit. obstetrician; b. Southampton, Eng., 1572; s. William Chamberlen; m. Sara de Launc. Received licence to practice midwifery from the bishop of London, Eng.; mem. Coll. Physicians. Attempted to organize midwifery of London (rejected by Coll. Physicians) 1616; invented 1st practical obstetric forceps (kept a family secret for almost a century). Died 1626.

CHAMBERLEN, Peter, physician; b. 1601; s. Peter Chamberlen; ed. Merchant Taylors' Sch., also Emmanuel Coll., Cambridge (Eng.) U.; M.D., U. Padua (Italy), 1619; Lectr. anatomy to Barber Surgeons; physician to king. Fellow Coll. Physicians. Author: A Voice in Rhama, or The Cry of the Women and Children, echoed forth in the Compassions of Peter Chamberlen, 1647; A Vindication of Public Artificial Baths, 1648; also pub. scheme of politics, method for propelling carriages by wind; several theol. schemes. Patentee new way of writing and printing true English, 1672. Advocated incorporation of midwives; petitioned parliament to institute a system of hydro-therapeutics to prevent plague, 1648. Died May 8, 1683.

CHAMBERS, Leslie Addison, Am. biophysicist; b. Mystic, Ia., Oct. 11, 1905; s. Clarence Edwin and Bertha Alice (Lang) C.; B.S., Tex. Christian U., 1927, M.S., 1928; Ph.D., Princeton, 1930; m. Ione Lee Way, May 30, 1930; children—Sabra Lee (Mrs. Herbert H. Henke), John Leslie, Linda Jean (Mrs. Victor Caldwell), Alys Ione (Mrs. Paul Moss). Dir. research R.A. Taft San. Engring. Center USPHS, 1951-56; dir. research Los Angeles Air Pollution Control Dist., 1956-60; dir. Allan Hancock Found. U. So. Cal., Los Angeles, 1960—; mem. Nat. Air Conservation Commn.; mem. adv. panel spl. biol. facilities NSF; chmn. bd. Inst. Urban Ecology; cons. govt. agys. Fellow A.A.A.S., Am. Pub. Health Assn., N.Y. Acad. Scis.; mem. Am. Physiol. Soc., Am. Soc. Microbiology, Electron Microscope Soc. Am., Research Soc. Am., Soc. Pediatric Research, Soc. Immunologists, Air Pollution Control Assn., Am. Chem. Soc., Soc. Am. Bacteriologists, Sigma Xi, Phi Sigma. Contbr. numerous articles to profl. jours. Publs. on biol. and chem. effects of ultra-sounds; properties of viruses, rickettsiae, and bacteria; environmental health, air pollution; aerobiology. Home: 1195 S. Los Robles Av., Pasadena, Cal. 91106. Office: Allan Hancock Found., U. So. Cal., Los Angeles 90007.*

CHAMBERS, Llewelyn Gwyn, Welsh mathematician; b. Hornchurch, Eng., Feb. 11, 1924; s. Llewelyn and Sarah Gwen (Rowlands) C.; B.Sc. with first class honors in Math., U. Coll. N. Wales, 1944, M.Sc., 1947; m. Mona Rees Owen, Aug. 25, 1955; 1 son, Huw Gwyn. Sci. officer Royal Naval Sci. Service, 1944-51; professorial staff Royal Mil. Coll. Sci., 1951-54; lectr. math. U. Coll. N. Wales, Bangor, 1954-65, sr. lectr., 1965—. Fellow Inst. Math. and its Applications; mem. Edinburgh Math. Soc. Editorial bd. Y Gwyddonydd, 1962—. Research and publs. on cosmology including possibility of charge creation and properties of geons, aerials and asso. electromagnetic problems, oceanic waves.

CHAMBERS, Randall M(arion), Am. psychologist; b. Vincennes, Ind., Feb. 26, 1927; s. Frank Earl and Elizabeth (Randall) C.; A.B., Ind. U., 1948; M.A., U. Mo., 1951; Ph.D., Western Res. U., 1954; m. Mary Jane Fisher, Sept. 7, 1949; children—Mark Randall, Craig Franklin. Research asso. Western Res. U., 1951-53; physiol. psychologist Jackson Meml. Lab., Bar Harbor, Me., 1953-55; research psychologist USAF Research & Devel. Command, Tex., 1955-57; asso. prof. psychology Rutgers U., 1957-58; head Human Factors Engring. br. Aviation Med. Acceleration Lab. U. S. Naval Air Devel. Center, Johnsville, Pa., 1958-64, head psychology div., 1964—; asso. psychology and physiology Sch. Medicine U. Pa., Phila., 1958—. Recipient Faculty Citation award, 1962, Distinguished Service award, 1966, Incentive Superior Accomplishment award, 1963. Fellow Am. Psychol. Assn., Am. Sociol. Assn.; mem. Am. Inst. Aeros. and Astronautics, Aerospace Med. Assn., N.Y. Acad. Scis., Am. Inst. Biol. Scis., Human Factors Soc., Soc. For Psychophysiol. Research, Nat. Acad. Scis., Sigma Xi. Author: (with others) Unusual Environments and Human Behavior, 1963. Contbr. numerous articles to profl. jours. Publs. on studies of effects of high acceleration stress on performance and limitations of astronauts and test pilots; devel. of dynamic flight simulation and acceleration tng. using the human certrifuge. Home: 39 Home Rd., Hatboro, Pa. 19040. Office: U. S. Naval Air Devel. Center, Johnsville, Pa. 18974.*

CHAMBERS, Robert, biologist; b. Erzerum, Turkey, Oct. 23, 1881; s. Robert and Elizabeth (Lawson) C.; B.A., Robert Coll., Constantinople, Turkey, 1900; M.A., Queens U., Kingston, Can., 1902, LL.D., 1944; Ph.D., U. Munich (Germany), 1908; research student, Columbia, 1911-12; m. Bertha Inez Smith, June 15, 1910; children—Robert (dec. 1941), William Nesbit, Edward Lucas, Bradford; m. 2d, Eloise Parkhurst, Dec. 3, 1954. With A.B.C.F.M. in Turkey until 1909; lectr. biology Toronto (Can.) U., 1909-11; asst. prof. histology U. Cin., 1912-15; instr. anatomy, Cornell U. Med. Coll., N.Y.C., 1915-19, asst. prof., 1919-23, prof. microscopic anatomy, 1923-28; research prof. biology, Washington Sq. Coll. (N.Y. U.), 1928-48, prof. emeritus continuing research through NIH grant. Trustee Marine Biol. Corp., Woods Hole, Biol. Lab., Cold Spring Harbor. Fellow A.A.A.S.; mem. Marine Biol. Lab., Am. Soc. Zoologists (pres. 1948-49), Am. Soc. Naturalists, Am. Assn. Anatomists (v.p. 1939), Am. Assn. Cancer Research, Soc. Exptl. Biology and Medicine, Harvey Soc. (pres. 1944-45, 1945-46), Union Am. Biol. Soc. (pres. 1946), British Biol. Corp., Am. Physiol. Soc., Am. Soc. Bot., Soc. Gen. Physiology; asso. fellow N.Y. Acad. Scis., fgn. corr. mem. Nat. Acad. Med., France. Recipient Traill medal Linnean Soc., London, for researches on physical nature of protoplasm, 1925, John Scott medal City of Phila., for invention of devices for microdissection of living cells, 1925. Contbr. to Ency. Brit., also numerous research articles to American and European jours. Co-editor: Protoplasma (jour.). Made investigations in cytology and on phys. nature of protoplasm and the constituents of the living cell with micromanipulator he devised for dissecting and injecting living cells under highest magnification of compound microscope, cellular physiology, cancer in tissue culture, and of blood capillary circulation in traumatic shock. Died July 22, 1957.

CHAMBERS, Robert, naturalist; b. Peebles, Scotland, July 10, 1802; s. Jean (Gibson) C.; student at classical acad. in Edinburgh, Scotland; left sch. in 1816; m. twice; survived by 3 sons, 6 daus. Taught at Portobello, beginning in 1818; bookseller, Edinburgh, 1818-32; founder pub. firm W. & R. Chambers; visited U. S., 1860. Mem. Royal Soc. Edinburgh. Author: Illustrations of the Author of Waverly, 1822; Traditions of Edinburgh, 1823; Picture of Scotland, 2 vols., 1826; Life of James I, 1830; A Gazeteer of Scotland, 1835; Vestiges of the Natural History of Creation, 1844; Book of Days, 2 vols., 1862-64. Studied Scottish character, dialect and countryside; Darwin said his book Vestiges cleared way fcr new view of world's creation. Died Mar. 17, 1871.

CHAMBON, Joseph, French physician; b. Grignan, Provence, France, 1647. Doctor of Faculty at Avignon; then went to Marsailles but also travelled in Italy, Germany and Poland; physician to King John Sobieski; went to Holland and Eng., then returned to France; licentiateship at Paris and practiced medicine there; imprisoned for 2 years in Bastille because of memoir in def. of a Neapolitan in the Bastille; upon release returned to Marsailles; physician to the galleys; retired, 1705. Author: Principes de physique rapportes à la médecine pratique, 3 parts, 1711, 1714, 1716; Traité des Métaux et des Minéreaux et des Rémèdes qu'on en peut tirer, 1714.

CHAMBON DE MONTAUX, Nicolas, French physician; b. Brévannes, France, Sept. 21, 1748; s. Jean Baptiste and Marguerite (Froussard) Chambon; m. Augustine. Head physician Salpetrière Hosp.; 1st physician Army, also insp. mil. hosps.; Jacobin, during French Revolution; mayor of Paris, 1792-93 (presided at judgement and execution of Louis XVI). Mem. Royal Soc. Medicine. Author: Des maladies des femmes, 1785; Des maladies de la grossesse, 1789; Traité de la fièvre maligne, 1787. Research on medicine, especially effect of mineral water and electricity in treatment of certain diseases. Died Nov. 1, 1826.

CHAMBRELENT, François Jules Hilaire, engr., agronomist; b. St.-Pierre, Martinique Islands, Feb. 17, 1817; ed. at Paris, Sorèze Coll., École polytechnique, École des ponts et chaussées, 1836. Named engr. Dept. Civil Engring., 1841, chief engr., 1865, gen. insp., 1879, in charge drainage, irrigation and drying of grounds, region of Gironde. Mem. Nat. Soc. Agr., French Acad. Scis. Research on soil and sand; improved sandy moors of Gascogne by drainage and drying, later planted pine trees in region. Died Paris, Nov. 13, 1893.

CHAMPETIER, Georges Hippolyte, French chemist; b. Paris, France, Feb. 3, 1905; s. Hippolyte and Berthe (Constant) C.; Ingénieur Ecole supérieure de Physique et de Chimie, 1925, Docteur ès sciences physiques, 1933; m. Eugénie Haussonville, Apr. 19, 1927; children—Marc, Yves, Monique (Mrs. Damman), Anne-Marie (Mrs. de Philip). Asst. gen. chemistry Institut de biologie physico-chimique, 1928-37; research chief Institut de chimie de Paris, 1937-47; prof. macromolecular chemistry Faculté des Sciences, Paris, 1953—; dir. studies Ecole supérieure de Physique et de Chimie, Paris, 1950—; adj. dir. Centre National de la Recherche Scientifique, 1951-56; became pres. Institut Pasteur de Paris, 1966. Pres., Comité de chimie macromoléculaire de la Délégation générale à la Recherche Scientifique et Technique, 1961—. Author: Eléments de la chimie, 1947; Chimie générale, 1945; Molecules géantes et leurs applications, 1948; Dérivés cellulosiques, 1954; Chimie macromoléculaire, 1957; Fibres textiles, 1959; also numerous articles. Pioneered in macromolecular chemistry; research and numerous publs. on textile fibres, plastomers elastomers, cellulose and celluloid derivatives, polyamides; preparations of giant molecules by polymerization and polycondensation. Address: 10, rue Vauquelin, Paris 5ème (Seine) France.*

CHAMPIER, Symphorien, French physician; b. Saint-Symphorien-le-Châtel, France, 1472; ed. Paris (France) Med. Sch. at Montpellier, France; M.D., 1492. Physician successively to Charles V, Louis XII, Duke of Lorraine; physician at Lyons, 1502-03; participated in French mil. campaigns, Italy, 1507-15; began practice medicine again in Lyons, 1515; placed in charge of hygiene in Lyons; alderman in Lyons, 1520-31; magistrate in Lyons, 1531-33; a founder of Trinity Coll. in Lyons. Author numerous books including: De claris medicinae scriptoribus, 1506; Vocabulorum medicinalium, 1508; Rosa Gallica, 1512; Symphonia Platonis, 1516; Symphonia Galeni ad Hippocratem, Celsi ad Avicinnam, 1528; also a life of Arnald de Villanova and another of Mesuë. His med. texts resurrected the doctrines of Hippocrates and Galen in France; tried to reconcile the med. views of Hippocrates, Galen, Celsus and Avicenna. Died circa 1539.

CHAMPION, Frank Clive, Brit. physicist; b. London, Eng., Nov. 2, 1907; s. Frank Charles and Alice (Killick) C.; M.A., St. John's Coll., Cambridge (Eng.) U., Ph.D., 1932; Asst. lectr. Univ. Coll., Nottingham, Eng., 1932-34; lectr. King's Coll., London U., 1934-46, reader, 1946-59, prof. exptl. physics, 1959—. Fellow Inst. Physics (London). Author: (with N. Davy) Properties of Matter, 1936, 3d edit., 1959; Electronic Properties of Diamonds, 1963. Contbr. numerous articles to profl. jours. Research on nuclear physics, scattering of alpha and beta particles by matter, direct test of spl. relativity momentum and energy relations from cloud chamber photographs of electron-electron collisions, solid state physics, optical and electronic properties of diamond, sulphur and magnesium oxide, diamond as radiation counter and register of radiation damage. Home: 279, Chiswick Village, London, Eng.*

CHAMPION, Kenneth Stanley Warner, physicist; b. Sydney, Australia, Dec. 7, 1923; s. Cecil A. B. and Ellen C. (Moxham) C.; B.Sc., U. Sydney, 1944; Ph.D., U. Birmingham (Eng.), 1951; m. Mavis Audrey Hinckley, Nov. 27, 1948; children—Annette Helen, Gwendalyn Ruth, Geoffrey Phillip, Sandra Beryl. Research asso. Mass. Inst. Tech., Cambridge, 1952-54; asst. prof. Tufts U., Medford, Mass., 1954-59; sect. chief Space Physics Lab., Air Force Cambridge Research Labs., Bedford, Mass., 1959-63, chief atmospheric structure br., 1963—; research asso. Mass. Inst. Tech., 1956-59; vis. prof. U. Adelaide (Australia), 1964. Fellow Phys. Soc. London; mem. Am. Phys. Soc., Am. Geophys. Union. Contbg. author: Cospar International Reference Atmosphere, 1965; also articles. Co-editor U. S. Standard Atmosphere, 1962. Research on plasma physics including intense magnetic fields, atmospheric density, temperature and other properties from satellite and rocket measurements. Home: 6 Rolfe Rd., Lexington, Mass. 02173. Office: Atmospheric Structure Br., Air Force Cambridge Research Labs., L. G. Hanscom Field, Bedford, Mass. 01730.*

CHAMPLAIN, Samuel de, French explorer; b. Brouage, France, 1567; s. Antoine de Champlain. Under officer in service of Henry IV, Brittany, 1598, Cadiz, Spain, 1601; visited Mexico, 1602; made voyage to Can., 1603; colonized Que., Can., 1608; promoted fur trade Gt. Lakes region, 1610-20; 1st gov. Can., from 1620; exile in Eng., 1629-32, returned to Que., 1632. Author: Voyages de la Nouvelle France, 1632. Discovered Lake Champlain, 1610; explored St. Lawrence region of Can.; extended French explorations as far W. as Wis. in U. S.; knowledgeable about Indians of Can., especially Hurons; persuaded Cardinal Richlieu to found Co. of 100 Assos. to run New France (in Can.). Died Que., Dec. 25, 1635.

CHAMPOLLION-FIGEAC, Jean-François, French Egyptologist; b. Figeac, France, Dec. 23, 1790; s. Jacques Champollion-Figeac; ed. l'École des langes Orientales, Coll. de France; m. Rosine Blanc; became mem. teaching staff, lyceé, Grenoble, France, 1807; named adj. prof. history Faculty Letters, Grenoble, 1809; returned to Paris to study Egyptian antiquities; became supt. dept. Egyptian antiquities Louvre, 1822, conservator Egyptian mus., 1826, head, course on Egyptian archeology, 1827; sent by Charles X to see Egyptian relics in Italian museums, 1824; dir. sci. expdn. to Egypt, 1828; chair of Egyptian archaeology founded for him at Coll. de France, 1831. Mem. Acad. Inscriptions. Author: Précis du système hiéroglyphique des Anciens Egyptiens, 1924; Grammaire Egyptienne, 1836-41; Dictionnaire hiéroglyphique. Deciphered Egyptian hieroglyphs through study of Rosetta stone, which led him to discovery to their figurative, ideographic, and alphabetic nature; made Egyptian script readable by regarding it phonetically. Died Paris, Mar. 4, 1832.

CHAMPOLLION-FIGEAC, Jean Jacques, French archaeologist; b. Figeac, France, Oct. 5, 1778; s. Jacques Champollion-Figeac; studied at Grenoble, France; conservator of library, also prof. to dean of Greek lit. Grenoble, 1809-16; aided Napoleon in Hundred Days return, 1815; conservator manuscripts Bibliothèque nationale, Paris, France, 1828; prof. paleography l'École des chartes; librarian at palace of Fontainbleau, 1849. Mem. French Acad. Scis. (corr.). Author, Antiquités de Grenoble, 1807; Nouvelles recherches sur les patois ou idioms vulgaires de la France, 1809; Annales des lagides ou chronologie des rois grecs d'Egypte, 1819; Nouvelles recherches sur la ville gauloise d'Uxellodunum, 1820; Résume complet de chronologie, 1827; l'Egypt ancienne et moderne, 1850. Research on ancient Egyptian chronology, 1820-27; worked (with his bro. Jean-François) to import Egyptian archeol. finds. Died May 9, 1867.

CHANCA, Diego Alvarez, physician; b. 15th century. Ct. physician to Ferdinand and Isabella of Spain; accompanied 2d expdn. of Columbus to Am., 1493, saved life of Columbus and others suffering malaria; selected site for 1st permanent settlement of Isabella, Haita. Author of detailed and accurate account of New World as he saw it during 3-month stay; also wrote study of treatment of pleurisy, pub. 1506, after his return to Spain.

CHANCE, Britton, Am. biochemist; b. Wilkes-Barre, Pa., July 24, 1913; s. Edwin M. and Eleanor (Kent) C.; B.S., U. Pa., 1935, M.S., 1936, Ph.D., 1940; Ph.D., Cambridge U., 1942, D.Sc., 1952; M.D. (hon.), Karolinska Institutets, 1962; m. Jane Earle, 1938; children—Eleanor C. Swett, Britton, Jan, Peter; m. 2d Lilian Streeter Lucas, 1956; Ann Lucas, Gerald Lucas, A. Brooke Lucas, William Lucas, Margaret, Lilian, Benjamin. Asst. prof. biophysics U. Pa., Phila., 1940-48, prof., 1949—, acting dir. Johnson Research Found., 1940-41, chmn. dept. biophysics Sch. Medicine, 1949—, dir. Eldridge Reeves Johnson Found., 1949—, E. R. Johnson prof. biophysics, 1964; staff mem. Mass. Inst. Tech., 1941-46; investigator OSRD, 1941; Guggenheim fellow Molteno Inst., Cambridge, Nobel Inst., Stockholm, 1946-48; mem. vis. com. Mass. Inst. Tech. and Bartol Research Found., 1955-59; mem. President's Sci. Adv. Com., 1959-60; cons. NSF, 1951-56. Harvey lectr., 1954, Phillips lectr., 1956, Pepper lectr., 1957, Hackett lectr., 1966; Exchange scholar U. S.-USSR, 1963. Recipient Paul Lewis award in enzyme chemistry, 1950, Presdl. certificate of merit, 1950, Morlock award I.R.E., 1961; Genootschapps medaille Dutch Academy Sci., 1965; Keilin medal, 1966; Franklin medal, 1966. Mem. Nat. Acad. Scis., Am. Acad. Arts and Scis., Am. Soc. Biol. Chemists, Harvey Soc., Am. Chem. Soc. (Howe award 1966), Am. Physiol. Soc., Am. Philos. Soc., Biophys. Soc., Royal Soc. Arts (London), Biochem. Soc., Sigma Xi, numerous others. Editor: Energy-Linked Functions of Mitochondria, 1963. Co-editor: Waveforms, 1949; Electronic Time Measurements, 1949; Rapid Mixing and Sampling Techniques in Biochemistry, 1964; Control of Energy Metabolism, 1965. Research, publs. on studies of mechanisms of enzyme action, especially in intermediate metabolism, discovering enzyme-substrate compounds in multi-engine systems; in vivo measurement of cell metabolism in various physiol. states; applied photometric techniques to biol. problems; inventor automatic steering, radar nav. and bombsight devices. Home: 4014 Pine St., Phila. 19104.

CHANCE, Sir James Timmins, Brit. optician; b. Mar. 22, 1814; s. William Chance; ed. London u. (now Univ. Coll.); B.A. (scholar), Trinity Coll., Cambridge, 1838; M.A., 1841; M.A. ad eundem, Oxford, 1848; m. Elizabeth Ferguson, 1845, (dec. 1887). Partner, Chance Bros. & Co., Spon Lane, Oldbury, nr. Birmingham, from 1839, head of firm, 1865; worked on manufacture and improvement of dioptric illuminating apparatus for lighthouses, from 1859; worked to correct existing errors and deficiencies in island lighthouses with Royal Commn. of 1859. Mem. Instn. Civil Engrs. (asso., Telford gold medal and premium for paper on optical apparatus used in lighthouses 1867). Introduced numerous improvements and designed lights for lighthouses in many parts of world. Died Jan. 8, 1902.

CHANCEL, Gustave-Charles-Bonaventure, French chemist; b. Loriol, France, Jan. 18, 1822; ed. École centrale, 1840; docteur ès sciences, 1848. Worked with Pelouze; became asst. chemist École des mines, 1846; named titular prof. chemistry Faculty of U. Montpellier (France), 1852; became dean Faculty of Sci., U. Montpellier, 1865; apptd. rector Acad. Montpellier, 1879. Recipient le prix Jecker, 1884. Mem. French Acad. Scis. (corr.). Author: Cours élémentaire d'analyse chimique, 1851; (with Gehardt) Précis d' analyse chimique, 2 vols., 1873-74; also numerous articles. Research on aliphatic and aromatic organic compounds, acetones. Died Montpellier, Aug. 5, 1890.

CHANCOURTOIS, Alexandre Émile Béguyer de, see de Chancourtois.

CHANDLER, Asa Crawford, Am. biologist; b. Newark, Feb. 19, 1891; s. Frank Thomas and Augusta (Jappé) C.; A.B., Cornell U., 1911; M.S., U. Cal., 1912, Ph.D., 1914; m. Belle Clarke, June 1, 1914 (dec.); 1 dau., Dorothy Belle; m. 2d, Ina Henrietta Sands, July 9, 1921 (dec.); children—Frank Sands, Emily Alice; m. 3d, Mrs. Lillie Moore Laughlin, Dec. 23, 1944, Grad. asst. zoöl. dept. U. Cal., 1911-14; instr. zoölogy, later asst. prof. Ore. State Agrl. Coll., 1914-18; instr. biology Rice Inst., Houston, 1919-24; charge hookworm research lab. Sch. Tropical Medicine, Calcutta, India, 1924-27; prof. biology Rice Inst., 1927-58; spl. cons. USPHS, 1942-47. Fellow A.A.A.S.; mem. Am. Soc. Parasitologists, Am. Soc. Tropical Medicine and Hygiene, Am. Micros. Soc., Am. Soc. Naturalists, Sigma Xi. Author: Animal Parasites and Human Disease, 1918; Anthelmintics and Their Uses (with R. N. Chopra), 1928; Hookworm Disease, 1929; Introduction to Parasitology, 9th edit. 1955; The Eater's Digest, 1941. Made helminthological survey of India, 1924-27. Died Aug. 23, 1958.

CHANDLER, David Culbertson, Am. zoologist; b. Walnut Grove, Minn., July 11, 1906; s. Frank Orlando and Amanda (Culbertson) C.; A.B., Greenville Coll., 1929; A.M., U. Mich., 1930, Ph.D., 1934; m. Pearl Carlson, Jan. 1, 1935; children—Robert, Candace. Faculty, U. Ark., 1934-38; chmn. sci. dept. McMurray Coll., 1935-36; faculty Ohio State U., 1938-49, prof. limnology, 1947-49; prof. limnology Cornell U., 1949-53; prof. zoology U. Mich., Ann Arbor, 1953—, dir. Gt. Lakes research div., 1960—. Fellow A.A.A.S.; mem. Am. Microscop. Soc. (past pres.), Am. Soc. Limnology and Oceanography (past pres.), Ecol. Soc. Am., Internat. Assn. Theoretical and Applied Limnology, Mich. Acad. Sci., Sigma Xi, Phi Sigma, Phi Kappa Phi. Contbr. articles to tech. jours. Developed limnological program, Gt. Lakes research. Home: 500 Highland Rd., Ann Arbor, Mich. 48104.*

CHANDLER, Seth Carlo, Am. astronomer; b. Boston, Sept. 17, 1846; s. Seth and Mary (Cheever) C.; ed. Boston English High Sch.; LL.D., DePauw U., 1891; m. Carrie M. Herman, Oct. 20, 1870. Aid, U. S. Coast and Geod. Survey, 1864-70; life ins. actuary, 1870-85; editor Astron. Jour., from 1896. Recipient Watson Gold medal, Gold medal Royal Astron. Soc. Eng. Fellow Am. Acad. Arts and Scis. Showed period of precession of earth around axis of figure to be 428 days, on basis of observations made during 1884-85. Died Dec. 31, 1913.

CHANDRA, Purna, soil microbiologist; b. Khatauli, India, June 23, 1929; s. Dalip Singh and Gokalia (Devi) C.; M.Sc., Agra U., 1951; Ph.D., Ore. State U., 1958; m. Shashi Bala Bhutani, Aug. 23, 1964; 1 son, Neleen Vivaik. With Govt. Agrl. Coll., Kanpur, India, 1951-55, research scholar, 1953-55; research fellow Ore. State U., 1955-59, jr. bacteriologist, 1958-59; lectr. U. Baghdad, Iraq, 1959-60; microbiologist Can. Dept. Agr., Swift Current (Sask.) Exptl. Sta., 1961-63; asst. prof. Mt. Allison U., 1963—. Mem. A.A.A.S., Canadian Soc. Microbiology, Internat. Soil Sci. Soc. Research, studies, publs. on use of Gibberellic acid and trace elements in composting, in nitrification and sulfur oxication process; effect of herbicides, fungicides and insecticides on soil microorganisms; nitrogen transformation in soils when different fertilizers and organic residues are applied; study of the microflora of ancient soils of Iraq; biosynthesis of volucrosporin; effect of freezing and thawing on soil microflora. Home: 52 Squire St., Sackville, N.B., Can. Office: Mt. Allison U., Sackville, N.B. Can.

CHANDRASEKHAR, Subrahmanyan, theoretical astrophysicist; b. Lahore, India, Oct. 19, 1910; M.A., Presidency Coll., Madras, 1930; Ph.D., Trinity Coll., Cambridge, 1933, Sc.D., 1942; m. Lalitha, Madras India, Sept. 1936. Came to U. S., 1936; naturalized, 1953. Govt. India scholar in theoretical physics, Cambridge, 1930-34; fellow, Trinity Coll., Cambridge, 1933-37; research asso. Yerkes Obs., Williams Bay and U. of Chicago, 1937, asst. prof., 1938-41, asso. prof. 1942-43, prof., 1944-47, distinguished service prof., 1947—, Morton D. Hull Distinguished Service prof., 1952—. Recipient Bruce medal Astron. Soc. Pacific, 1952, gold medal Royal Astron. Soc. (London), 1953; Rumford medal Am. Acad. Arts and Scis., 1957, Nat. Medal of Sci. (U. S.), 1966. Fellow Royal Soc., 1944 (Royal medal 1962); mem. Nat. Acad. Scis., Am. Philos. Soc., Am. Acad. Arts and Scis., Cambridge Philos. Soc., Am., Royal astron. socs. Author: An Introduction to the Study of Stellar Structure, 1939; Principles of Stellar Dynamics, 1942; Radiative Transfer, 1950; Hydrodynamic and Hydromagnetic Stability, 1961. Mng. editor The Astrophysical Jour., 1952—. Contbr. various sci. periodicals. Developed a theory of white dwarf stars which posits a limit of mass of dwarf star (Chandrasekhar's limit) and an inverse relationship of its mass and radius; his theory explains final stages of stellar evolution; other work in dynamics of stellar systems, theory of stellar atmospheres, radiative transfer, hydrodynamics and hydromagnetic relativity. Address: Lab. for Astrophysics and Space Research, 933 E. 56th St., Chgo. 60637.

CHANEY, Ralph W(orks), Am. geologist, paleobotanist; b. Chgo., Aug. 24, 1890; s. Fred A. and Laura Jeanette (Works) C.; B.S., U. Chgo., 1912, Ph.D., 1919; D.Sc., U. Ore., 1944; m. Marguerite Seeley, June 1, 1917; children—Richard Works, Ellen, David Hinman. Asso. with Francis W. Parker Sch., Chgo., 1914-17; instr. U. Ia., 1917-19, asst. prof., 1919-22; research asso. Carnegie Inst., Washington, 1922-56; prof. paleontology, curator, Museum of Paleontology, U. Cal., 1931-56, emeritus, 1957—, asst. dir. radiation lab., 1944-45, cons., 1953—; vis. prof. Stanford, 1957. Field work in Cenozoic paleobotany in Westen America since 1916, Central America and South America, 1930, 32, 35, 40, Mongolia, China and P.I., 1925. China, Manchuria, Chosen and Japan, 1933, 37, China, 1948, Japan, 1950—; collaborator National Park Service, 1955—. Dir., counselor Save-the-Redwoods League, pres., 1961; mem. adv. bd. Nat. Park Service, 1943-54. Mem. Paleontology Soc. Japan, Nat. Acad. Sci., Am. Philos. Soc. (council 1951-54). Fellow Geol. Soc. Am. (v.p. 1940), Paleontol. Soc. Am. (pres. 1939), Cal. Acad. Sci.; mem. bot. socs. Am., Japan (fgn.), Gamma Alpha, Theta Tau. Author of numerous books; editor vols. 22-27, Bull. Dept. Geol. Sciences, U. of Calif. Pubs. Contbr. to tech. publs. Work in tertiary paleobotany, geology of western Am. and eastern Asia; developed a dynamic approach to tertiary forests based on close resemblance of fossil and living plants. Home: 1129 Keith Av., Berkeley 8, Cal.

CHANG, Chen Chung, mathematician; b. Tientsin, China, Oct. 13, 1927; s. P. C. and S. T. (Tsai) C.; came to U. S., 1944, naturalized, 1959; B.A., Harvard, 1949; M.A., U. Cal. at Berkeley, 1950, Ph.D., 1955; m. Marjorie Galvan, Aug. 1, 1951; children—Ann, Alice, Peter, Julia. Instr., Cornell U., 1955-56; asst. prof. U. So. Cal., 1956-58; asst. prof. math. U. Cal. at Los Angeles, 1958-61, asso. prof., 1961-64, prof., 1964—; cons. to industry. Sr. postdoctoral fellow NSF, 1962-63; sr. Fulbright research scholar Oxford (Eng.) U., 1966-67, also vis. fellow All Souls Coll. Mem. Am. Math. Soc., Assn. Symbolic Logic, Phi Beta Kappa, Sigma Xi, Pi Mu Epsilon. Contbr. articles profl. jours. Study of abstract algebra; set theory; logic. Home: 521 Veteran Av., Los Angeles 90024.

CHANG, Jeffrey Peh-I, biologist; b. Changteh, Hunan, China, Oct. 10, 1917; s. S. Y. and C. L. (Chiu) C.; B.S., Nat. Central U., 1941; M.S., U. Ill., 1946, Ph.D., 1949; m. Jeannette A. Klemola, Aug. 21, 1948; children—JoAnn, Peter. Came to U. S., 1945, naturalized, 1953. Research asso. U. Ill., U. Kan., U. Tex. M.D. Anderson Hosp. and Tumor Inst., 1949-59; asso. biologist, asso. prof. U. Tex. M.D. Anderson Hosp. and Tumor Inst., Houston, 1959-64, biologist, prof., 1964—; cons. NIH Nat. Cancer Inst., 1958-61, Sch. Aerospace Medicine USAF, 1962-64. Contbr. numerous articles to profl. jours. Developed techniques for staining mitochondria in frozen-dried tissues, concentrating sputum cells by millipore filtration, staining auto-radiographs of non-diffusible isotopes; devised open-top cryostat for pathologic diagnosis and research; developed sect. freeze-substitution technique for histochem. research; research in cell biology, specifically histochemistry, electron microscopy, chem. carcinogenesis. Home: 4512 Valerie St., Bellaire, Tex. 77401. Office: 6723 Bertner St., Houston 77025.*

CHANG, Min Chuch, biologist; b. Taiyuan, Shansi, China, Oct. 10, 1908; s. Gen Shu and Shih (Laing) C.; B.Sc., Tsing Hua U., Peking, China, 1933; Ph.D., Cambridge U. (Eng.), 1941; m. Isabelle Chin, May 28, 1948; children—Francis Hugh, Claudia, Pamela. Came to U. S., 1945. Research asst. Tsing Hua U., 1933-38; research worker Cambridge U., 1941-45; with Worcester Found., Shrewsbury, Mass., 1945—, sr. scientist, 1956—; research prof. Boston U., 1961—. Recipient The Ortho award, 1950, Albert Lasker award, 1954, Ortho award and medal, 1961. Mem. Am. Phys. Soc., Am. Assn. Anatomists, Am. Acad. Arts and Scis. Contbr. numerous articles in field to sci. jours. Studies, publs. on animal reprodn., artificial insemination, mammalian fertilization, transplantation of mammalian eggs, early embryonic devel., oral contraceptives. Home: 15 Fiske St. Office: 222 Maple Av., Shrewsbury, Mass. 01545.*

CHANG Ts'ang, Chinese mathematician; b. circa 250 B.C. Author: Arithmetic Rules in Nine Sections (greatest arithmetic classic of China). Gave area of segment of a circle as $1/2(c + a)a$, where c is the chord and a the altitude of the segment. Died circa 152 B.C.

CHANNING, Walter, Am. physician; b. Newport, R.I., Apr. 15, 1786; s. William and Lucy (Ellery) C.; B.A., Harvard, 1808, M.D., 1812; M.D., U. Pa., 1809; studied in London and Edinburgh; m. Barbara Higginson Perkins, 1815; m. 2d, Elizabeth Wainwright, 1831. First prof. obstetrics and med. jurisprudence Harvard Med. Sch., 1815-circa 1844, dean, 1819-47; co-editor Boston Med. and Surg. Jour., 1828; librarian Mass. Med. Soc., 1822-25, treas., 1828-40; a founder Boston Lying-In Hosp., 1832; mem. Am. Acad. Arts and Scis. Author: Six cases of Inhalation of Ether in Labor, 1847; Treatise on Etherization in Childbirth, 1848; Professional Reminiscences of Foreign Travel; New and Old, 1851; A Physician's Vacation, 1856. 1st to use ether in childbirth cases, 1847; 1st to describe pernicious anemia in pregnancy, 1842; gave account of puerperal fever in Europe, 1852, and ether-chloroform controversy. Died Boston, July 27, 1876.

CHANORIER, Jean, French agronomist; b. Lyons, France, Nov. 15, 1746; worked at Croissy, France; dir. finance; elected dep. to Council of 500; mem. French Acad. Scis. (non-resident). Wrote dissertation on sheep and experimented with wool; introduced sheep (brought from Spain) to France. Died May 29, 1806.

CHANT, Clarence Augustus, Canadian astronomer; b. nr. Toronto, Ont., Can., May 31, 1865; B.A., U. Toronto, 1890, M.A., 1900; postgrad. U. Leipzig, 1898; Ph.D., Harvard, 1901; m. 1894; 2 daus. Tutorial fellow U. Toronto, 1891, lectr., 1892, asso. prof., 1907, prof., 1918, dir. David Dunlap Obs., 1933, emeritus prof. astrophysics, 1935-56. Fellow Am. Phys. Soc., Royal Astron. Soc., Royal Astron. Soc. Can. (pres. 1904-07, Royal Soc. Can.; mem. Internat. Astron. Union, Am. Stron. Soc. Author: Our Wonderful Universe, 1928; (with E. F. Burton) Text Book of College Physics, 1933; The Teaching of Astronomy in the University of Toronto, 1954. Editor: Jour. Royal Astron. Soc. Can., 1907-56. Organized expdn. to Australia to observe solar eclipse to text Einstein theory, 1922. Died Nov. 18, 1956.

CHANT, Donald Alfred, Am. entomologist; b. Toronto, Ont., Can., Sept. 30, 1928; s. Sperrin Noah and Nellie (Cooper) C.; B.A., U. B.C., 1950, M.A., 1952; Ph.D., U. London (Eng.), 1956; m. Christina Margaret Rutherford, Feb. 5, 1959; children—Patrick Colin, Jeffrey Alan. Research officer Biol. Control Lab., Can. Dept. Agr., Vancouver, B.C., 1950-56, Biol. Control Research Inst., Belleville, Ont., 1956-60, head sect, 1959-60; dir. Research Lab., Vineland-St. Cathrine, Ont., 1960-64; apptd. entomologist, prof. biol. control, chmn. dept. biol. control U. Cal. at Riverside 1964; chairman dept. zoology Univ. Toronto. Fellow Royal Entomological Society London; mem. Am. Inst. Biol. Scis., Ecol. Soc. Am., Entomol. Soc. Am., Entomol. Soc. Can., Canadian Soc. Zoologists, Brit. Ecol. Soc. Contbr. numerous articles to tech. jours. Research on systematics mite family Phytoseiidae, generic concepts in group, biology and ecology predacious mites, predation, use of predacious mites in biol. control. Home: 17 Old Mill Terrace, Toronto, Ont., Can.*

CHANTEMESSE, André, French bacteriologist; b. Puy, France, Oct. 13, 1851; M.D., Paris, France, 1884; became asst. exptl. and comparative pathology Faculty of Medicine, Paris, 1883; opened inst., Paris, 1889; fought typhus epidemic, Lille, France, 1893; prof. respiratory diseases Faculty Medicine, Paris, 1891-92; became prof. exptl. and comparative pathology, Paris, 1896, named prof. hygiene, 1898; apptd. insp. gen. hygienic services, 1903. Author: Pathologie générale et expérimentale: les processus généraux (with Podwyssotsky), 2 vols., 1901-05; (with Borel) Moustiques et fièvre, 1906; Hygiene internationale frontieres et prophylaxie, 1907; Traité d'hygiène maritime, 1909. Mem. French Acad. Medicine. Research on epidemic dysentery, pneumoenteritis of pigs, microbial contamination of water, typhoid fever; discovered dysentery microbe. Died Paris, Feb. 25, 1919.

CHANTRE, Ernest, French anthropologist; b. Lyons, France, Jan. 13, 1843; Doctorate of Scis.; became tchr. ethnography Faculty of Letters, Lyons, 1892; traveled in Orient; dir. Mus. of Lyons, 1877-1910; tchr. anthropology Faculty of Scis., Lyons, 1881-1908. Research on geology, paleontology, glaciology; invented mensuration instruments; advocated conservation of monuments and famous sites. Author: Recherches anthropologiques dans le Caucase, 5 vols., 1895-97; Etudes paléo-ethnologiques dans le bassin du Rhône, 4 vols., 1875-80; Paléontologie humaine, 1901. Died Ecully, Rhône, France, Nov. 24, 1924.

CHANTRY, George William, chem. physicist; b. Wallasey, Cheshire, Eng., Apr. 13, 1933; s. George William and Sophia (Johnston) C.; B.A., Oxford U., 1954, D.Phil., 1959, M.A., 1960; m. Diana Margaret Rhodes Martin, July 15, 1956; children—Richard Alexander Rhodes, Catherine Jane Rhodes. Research asso. Cornell U., Ithaca, N.Y., 1958-60; sr. sci. officer Nat. Phys. Lab., Teddington, Middlesex, Eng., 1960—. Research, publs. on theory of Raman intensities, understanding of structure of simple free radicals, instrumentation and application of far infra-red spectroscopy. Home: 2 Walnut Tree Rd., Shepperton, Middlesex. Office: Nat. Phys. Lab., Teddington, Middlesex, Eng.*

CHANUTIN, Alfred, Am. biochemist, educator; b. New Haven, June 3, 1897; s. Morris and Bertha (Hillquit) C.; Ph.B., Yale, 1917, Ph.D., 1923; m. Beverley Doswell Anderson, Dec. 20, 1927; 1 dau., Martha Langhorne (Mrs. Hilton H. Dier, Jr.). Faculty U. Pa., 1920-21, Cornell U., 1921-22, U.Ill., 1923-24; asso. prof. biochemistry U. Va., Charlottesville, 1924-29, prof., 1929-67, emeritus, 1967—. Mem. hematology panel NIH, mem. com. on sterilization blood plasma and red cells NRC. Mem. Am. Chem. Soc., Am. Soc. Biology Chemistry, Radiation Research Soc., Sigma Xi, Alpha Omega Alpha. Contbr. numerous articles to sci. jours. Research on creatine metabolism, renal insufficiency, effect of chem. warfare agts., plasma and erythrocyte proteins, electrophoresis, metabolism of red blood cells during storage of blood.*

CHANY, Charles, physician; b. Budapest, Hungary, Sept. 1, 1920; s. Nicolas and Irene (Lobi) Csanyi; M.D., U. Paris, 1954; Asst. virologist Pasteur Inst., Paris, 1954-60; Rockefeller fellow NIH, Bethesda, Md., 1957; head virus research unit St. Vincent de Paul Hosp. and Cancer Inst., Villejuif, France, 1961—; prof. microbiology Paris U. Med. Sch.; dir. viral research unit Inst. Nat. Health Research, St. Vincent de Paul Hosp. Decorated Nat. Order of Merit, France. Mem. French, Brit., Am. socs. microbiology and medicine. Publs. on 1st description of adenoviral pneumonia in children; research on para-influenza virus; interferon and its antagonists. Home: 12 Theophraste Renaudot, Paris 15. Office: Hosp. of St. Vincent De Paul, Paris, France.*

CHAO, Jowett, zoologist; b. Meikee, Chekiang, China, Nov. 16, 1915; a. Chien Chung and Mien (Tsao) C.; B.S., West China Union U., 1939; M.S., Nat. Central U., 1942; Ph.D., Cornell U., 1949; m. Cherie P. H. Chow, July 11, 1939; 1 dau., Victoria Cherie. Came to U. S., 1945, naturalized, 1954. Biology tchr. Nankei Sch., Chungkiang, China, 1937-38; biology instr. West China Union U., 1939-40; entomologist Ministry of Agr. and Forestry, China, 1942-44; chemistry tchr. Putney Sch., Vt., 1947-48; chemist Vitaminerals, Inc., Glendale, Cal., 1948-54; asst. research zoologist U. Cal., Los Angeles, 1954-59, asso. research zoologist, 1959-65, research zoologist, 1965—. Mem. Am. Soc. Parasitologists, Am. Soc. Trop. Medicine and Hygiene, Am. Mosquito Control Assn., Entomol. Soc. Am., Soc. Protozoologists, Sigma Xi. Research, publs. on host-parasite-vector relationship studies and in vitro culture of malaria and other blood parasites; biology and internal microorganisms of mosquitos. Home: 228 S. Palm Dr., Beverly Hills, Cal. 90212.*

CHAPANIS, Alphonse, Am. psychologist; b. Meriden, Conn., Mar. 17, 1917; s. Anicatas and Mary (Barkevich) C.; B.A., U. Conn., 1937; M.A., Yale, 1942, Ph.D., 1943; m. Marion Amelia Rowe, Aug. 23, 1941 (div. 1960); children—Roger and Linda (twins); m. 2d, Natalia Potanin, Mar. 25, 1960. Psychologist, Tenn. Dept. Pub. Health, Franklin, 1939-40; asst. psychologist Aero Med. Lab., Wright Air Devel. Center, Dayton, O., 1942-43; with Johns Hopkins, 1946-53, asso. prof., 1949-56, research contract dir., 1952-53, prof. psychology and indsl. engring., 1956-63, prof. psychology, 1963—; tech. staff Bell Telephone Labs., 1953-54; liaison scientist Office Naval Research br. office U. S. Embassy, London, Eng., 1960-61. Mem. sci. adv. council def. and space group Chrysler Corp., Detroit, 1962—; cons. Aerospace Corp, 1964—. Recipient Franklin V. Taylor award Soc. Engring. Psychologists, 1963. Fellow Am. Psychol. Assn.; mem. A.A.A.S. (life), Optical Soc. Am., Human Factors Soc. (pres. 1963-64), Ergonomics Research Soc., Soc. Exptl. Psychologists, Eastern Psychol. Assn., Sigma Xi (nat. lectr. 1965). Author: (with Garner, Morgan) Applied Experimental Psychology, 1949; Research Techniques in Human Engineering, 1959; Man-Machine Engineering, 1965; also numerous articles. Pioneered field human factors engring.; research on gen. exptl. psychology emphasizing psychophysiology vision. Home: 6316 Falkirk Rd., Balt. 21212. Office: Dept. Psychology, Johns Hopkins, Balt. 21218.*

CHAPIN, Henry Dwight, Am. physician; b. Steubenville, O., Feb. 4, 1857; s. Henry Barton and Harriet Ann (Smith) C.; A.B., Princeton, 1877, A.M., 1885; M.D., Columbia, 1881; m. Alice Delafield, June 1, 1907. Prof. emeritus, diseases of children, N.Y. Post-Grad. Med. Sch. and Hosp. Former pres. Hosp. Social Service Assn. of N.Y.; dir. N.Y. Post-Grad. Med. Sch. and Hosp. Mem. Am. Pediatric Soc. (pres. 1910-11), N.Y. Acad. Medicine, N.Y. County Med. Soc. Author: Theory and Practice of Infant Feeding, 1902 (3 edits.); Vital Questions, 1905; A General Treatise on Diseases of Children, 1909 (6 edits.); Health First—The Fine Art of Living, 1917; Heredity and Child Culture, 1922. Also sociol. and med. articles in various mags. Chmn. pub. health div. of N.Y. State Reconstruction Commn. Awarded Columbia U. medal for outstanding contributions to problems relating to the care of children and as a pioneer in hospital social service, 1933. Died June 27, 1942.

CHAPIN, John Bassett, Am. physician; b. N.Y., Dec. 4, 1829; s. William and Elizabeth H. (Bassett) C.; A.B., Williams Coll., 1850; M.D., Jefferson Med. Coll., 1853; LL.D., Jefferson, and Williams; m. Harriet E. Preston, Mar. 18, 1858. Med. supt. State Hosp., Willard, N.Y., 1869-84; phys.-in-chief, Pa. Hosp. for Insane, Phila., 1884-1911. In a public communication, 1862, recommended changes in asylum construction which would provide for segregation of insane in detached blocks, according to classes and conditions; views were adopted in Willard Hosp. Author: Compendium of Insanity for Physicians and Students, 1899. Ret., 1911. Died Jan. 18, 1918.

CHAPIN, S. F., Am. entomologist; b. Mich., Mar. 12, 1839; M.D., Columbia; m. Maria Endicott, 1866; asst. surgeon under Gen. Sheridan during Civil War; moved to Auburn, Cal., 1866; practiced medicine, San Jose, Cal., 1878-85; became mem. Cal. Bd. State Hort. Commn., 1881; insp. fruit pests, 1883-85. Promoted control of insect pests; conducted many experiments with pesticides of day; noted coccids or scale insects. Died nr. Auburn, Mar. 14, 1889.

CHAPMAN, Alfred Chaston, Brit. chemist; b. 1869; ed. Univ. Coll., London; student of Williamson, Charles Graham. Sr. demonstrator applied chemistry Univ. Coll., London; Pub. Analyst, City of St. Alban, from 1903; hon. prof. Inst. Superieur des Fermentations, Ghent, Belgium; examiner Inst. Gen. Chemistry; mem. bd. studies in chemistry U. London; mem. ct. U. Leeds; mem. consultative com. on chemistry London County Council; mem. adv. council plant and animal products Imperial Inst.; mem. sci. panel apptd. under safeguarding of Industries Act, Bd. Trade; mem. Royal Commn. Awards to Inventors, from 1921; mem. Govt. Com. Ethyl Petrol, 1928-30; mem. Forest Products Research Bd., 1921-26; mem. chem. research bd. Dept. Sci. and Inssl. Research, from 1931. Fellow or mem. Soc. Pub. Analysists and other Analytical Chemists (pres. 1914-16), Inst. Brewing (pres. 1911-18), Chem. Soc. (council), Inst. Chemistry Gt. Britain and Ireland (pres. 1921-24), Royal Micros. Soc. (pres. 1924-26), Medico-Legal Soc. London, Fed. Council for Pure and Applied Chemistry, Royal Inst. Gt. Britain (mgr. 1930-32, v.p. 1931), Fellow Royal Soc., 1920, Royal Inst. Chemistry, Soc. de Zymologie Pure et Appliquée, Belgium. Author: The Hop and Its Constituents, 1905; (with F. G. S. Baker) An Atlas of the Saacharomycetes, 1906; Brewing, 1912; also papers. Died Oct. 17, 1932.

CHAPMAN, Alvan Wentworth, Am. botanist, physician b. Southampton, Mass., Sept. 28, 1809; s. Paul and Ruth (Pomeroy) C.; A.B., Amherst Coll., 1830; m. Mary Ann (Simmons) Hancock, Nov. 1839. Practiced medicine, Quincy, Marianna and Apalachicola, Fla.; became leading botanist in South; friend and correspondent of Asa Gray; genus Chapmania named in his honor; helped Union soldiers escape from prison nr. his home at Apalachicola during Civil War. Author: Flora of the Southern States (only manual for So. botany until after turn of century), 1860. Pioneered research on semi-tropical vegetation of western Fla. Died Apalachicola, Apr. 6, 1899.

CHAPMAN, Arthur Barclay, Am. geneticist; b. Windermere, Eng., Oct. 25, 1908; s. William Daniel and Nora (Moss) C.; came to U. S., 1925, naturalized, 1936; B.S., Wash. State U., 1930; M.S., Ia. State U., 1931; Ph.D., U. Wis., 1935; m. Winifred Mary Rollin, Sept. 1, 1934; children—Barbara Nora (Mrs. William Joseph Vaughan), Winifred Nell (Mrs. Douglas James Baker), Mary Jane. NRC fellow Ia. State U., Ames, 1936, U. Chgo., 1937; faculty U. Wis., Madison, 1936—, prof. genetics, 1947—. Rockefeller Found. award for research and teaching in Poland, 1960; Fulbright fellow, 1966-67; Guggenheim Meml. Found. fellow, 1966-67. Mem. Am. Soc. Animal Sci. (pres. 1965, dir. 1966), Am. Dairy Sci. Assn., Genetics Soc. Am., Am. Genetic Assn., Brit. Soc. Animal Prodn., Biometric Soc., A.A.A.S., Sigma Xi, Alpha Zeta, Gamma Alpha, Phi Sigma. Asso. editor Jour. Animal Sci., 1958-60, editor, 1961-63. Research, numerous publs. on relation of selection procedures and mating systems to genetic improvement livestock, kind and amount genetic and environmental variation influencing econ. traits livestock, genetic effects of cumulative irradiation in rats. Home: 1117 Risser Rd.. Madison, Wis. 53705.*

CHAPMAN, Carleton Burke, Am. physician; b. Sycamore, Ala., June 11, 1915; s. John G. and Mary (Anderson) C.; A.B., Davidson Coll., 1939; B.A. (Rhodes scholar), Oxford U., 1938, M.A., 1950; M.D., Harvard, 1941, M.P.H., 1944; m. Ruth Horine, Aug. 30, 1940; children—Nancy C. (Mrs. Jack A. Collins), John G., Mary A. Faculty, U. Minn., 1947-53; prof. medicine U. Tex. Southwestern Med. Sch., Dallas, 1953-66, dir. Weinberger Lab. Cardiovascular Research; dean Dartmouth Med. Sch., Hanover, N.H., 1966—. Recipient Distinguished Achievement award Modern Medicine, 1966. Mem. Am. Soc. Clin. Investigation, Am. Fedn. Clin. Research, Central Soc. Clin. Research, Assn. Am. Physicians, A.C.P., Am. Coll. Cardiology, Am. Heart Assn. (pres. 1964-65). Author: (with Jere H. Mitchell) Starling on the Heart, 1965. Research and publs. on physiology of exercise, circulatory control systems and adaptation to stress, bases of circulatory physiology in 19th century, clin. cardiology. Home: Thetford, Vt. 05074.*

CHAPMAN, David Leonard, Brit. chemist; b. Norfolk, Eng., Dec. 8, 1869; s. David and Maria (Wells) C.; ed. Christ Church, Oxford, Eng.; m. Muriel Catharine Channing Holmes, 1918; 1 dau., Ruth Holmes (Mrs. Anthony Tomkinson). Schoolmaster, then asst. in Dixon's Chem. Dept., Manchester, Eng., 1897-1907; became fellow Jesus Coll., Oxford, 1907, dir. Sir Leoline Jenkins Labs. until 1944, vice prin., 1935-44. Fellow Royal Soc., 1913. Research and publs. on chemistry and chem. physics; Introduced a theory of the detonation wave, 1899; (with C. H. Burgess) confirmed non-existence of photochem. extinction. Died Oxford, Jan. 17, 1958.

CHAPMAN, Dean Roden, Am. aerodynamicist; b. Fort Sumner, N.M., Mar. 8, 1922; s. James Perry and Mary (Zweifel) C.; A.A., Los Angeles City Coll., 1941; B.S., Cal. Inst. Tech., 1944, M.S., 1944, Ph.D., 1948; m. Marguerite Cole, Dec. 25, 1944; children—Anita, Donald Cole. Instr., Cal. Inst. Tech., 1943-44; research engr. Jet Propulsion Lab., Pasadena, Cal., 1947; research sci. NACA, Moffett Field, Cal., 1946, 48-57, staff sci. NASA, Moffett Field, 1957—. Recipient Sperry award Inst. Aero. Scis., 1952; Rockefeller Pub. Service award, 1959; award for exceptional sci. achievement NASA, 1964. Mem. Am. Inst. Aeros. and Astronautics. Research and publs. in supersonic aerodynamics, atmosphere entry into planetary atmospheres, properties distrbn. and origin of tektites; demonstrated reprodn. of tektite shapes by aerodynamic ablation at hypervelocities. Home: 251 Whitclem Ct., Palo Alto, Cal. 94306. Office: NASA Ames Research Center, Moffett Field, Cal. 94035.*

CHAPMAN, Frank Michier, Am. ornithologist; b. Englewood, N.J., June 12, 1864; s. Lebbeus and Mary A. (Parkhurst) C.; acad. edn.; Sc.D. (hon.), Brown, 1913; m. Fannie Bates Embury, 1898; l son, Frank Michler. Asso. curator ornithology and mammalogy Am. Mus. Natural History, 1888-1908, curator ornithology, from 1908, originator there of habitat of bird groups and seasonal bird exhibits. Zool. explorations in temperate and tropical Am., from 1887; dir. bur. publs. A.R.C., 1917-18, commr. Latin Am. 1918-19. Recipient medal Roosevelt Meml. Assn., 1928. Fellow Am. Ornithologists' Union (pres. 1911); hon. mem. N.Y. Zool. Soc., Brit. Ornithologists' Union, Sociedad Ornithologica del Plata, Deutschen Ornithologischen Gesellschaft; mem. Nat. Acad. Scis. (1st Elliot medal 1918), Am. Philos. Soc., Burroughs Meml. Assn. (pres. 1921-25, medal 1929), Explorers Club (v.p. 1910-18), Linnaean Soc. N.Y. (elected pres. 1897, 1st Linnean Medal 1912). Author: Handbook of Birds of Eastern North America, 1895; Bird-Life, a Guide to the Study of Our Common Birds, 1897; Bird Studies with a Camera, 1900; A Color Key to North American Birds, 1903; The Economic Value of Birds to the State, 1903; The Warblers of North America, 1907; Camps and Cruises of an Ornithologist, 1908; The Travels of Birds, 1916; The Distribution of Bird-Life in Colombia, 1917; Our Winter Birds, 1918; What Bird Is That?, 1920; Birds of Urubamba Valley, Peru, 1921; The Distribution of Bird-Life in Ecuador, 1926; My Tropical Air Castle, 1929; Autobiography of a Bird-Lover, 1933; Life in an Air Castle, 1938; The Post-Glacial History of Zonotrichia capensis; Birds and Man; also numerous papers on birds and mammals. Editor and founder Bird-Lore. Innovated museum exhibits in which birds are shown in simulated natural habitats. Died Nov. 15, 1945.

CHAPMAN, Frederick, Brit. paleontologist; b. Camden Town, London, Feb. 13, 1864; s. Robert and Eleanor (Dinsey) C.; m. Eleanor Dinsey, 1890; 1 son, 1 dau. Asst. to Prof. Judd, geol. dept. Royal Coll. Sci., 1881-1902; paleontologist State Nat. Mus., Melbourne, Australia, 1902-27; commonwealth paleontologist, Australia, 1927-35; lectr. paleontology U. Melbourne, 1920-32; mem. Australian NRC, from 1922; hon. curator Naroanoa Native Garden, Al Balwyn.; hon. paleontologist Victoria Geol. Survey. Recipient Syme prize and medal for research, 1920. Fellow Royal Micros. Soc., Royal Soc. S. Australia (Clarke medal for research 1932); mem. royal socs. N.S. Wales New Zealand; asso. Linnean Soc., London; pres. Micros. Soc., Victoria, 1919-20, Field Naturalist Club, Victoria, 1920, Royal Soc. Victoria, 1929-30. Author: The Foraminifera, 1902; New or Little-Known Victorian Fossils in National Museum, 30 parts, 1902-27; Australiasian Fossils, 1914; Silurian Bivalved Mollusca of Victoria, 1908; Succession and Homotaxial Relationships of the Australian Cainozoic System, 1914; Antarctic Results (Shackleton and Mawson Expeditions, 1907-09 and 1911-14), 1917, 37; Tertiary and Cretaceous Fish Remains of New Zealand, 1918; Foraminifera and Ostracoda of New Zealand, 1926; The Sorrento Bore, 1928; Open-air Studies in Australia, 1929; Guide to Fossil Collections, National Museum, Melbourne, 1929; The Book of Fossils, 1934; also papers. Died Dec. 10, 1943.

CHAPMAN, Henry George, Brit. physician; b. Ealing, Eng., Jan. 13, 1879; s. Henry Chapman; ed. Ormond Coll., U. Melbourrne; M.D., B.S.; m. Julia E. A. Cox, 1903; 1 son, 2 daus. Acting prof. physics U. Adelaide, 1901; demonstrator pathology U. Melbourne, 1902; became lectr. demonstrator physiology U. Sydney (Australia), 1903, asst. prof., 1913, prof. pharmacology, 1917, physiology, 1921, also dir. cancer research; lectr. Tech. Coll., Sydney, 1908; hon. path. chemist Royal Prince Alfred Hosp., Sydney. Chmn. Silicosis Joint Com. Recipient Syme prize for sci. research in Australia, 1909. Pres. Linnean Soc. New S. Wales, 1917-18. Research and publs. in biochemistry. Died May 25, 1934.

CHAPMAN, John, physician; b. Nottingham, Eng., 1822; ed., Paris, 1844, St. George's Hosp., London; M.D., St. Andrews (Scotland), 1857; apprentice to watchmaker, Worksop, Scotland; went to Adelaide, Australia as watchmaker and optician; returned to Europe, 1844; became bookseller, London, 1849; editor, propr. Westminster Rev., from 1851; practiced medicine, from 1857; moved to Paris, 1874. Author: Chloroform and other Anaesthetics, 1864; Functional Disorders of the Stomach, 1864; Diarrhoea and Cholera, 1865; Seasickness, 1869; Medical Institutions of the United Kingdom, 1870; Prostitution, 1870; Neuralgia, 1873; Medical Charity, 1874. Developed an elongated ice bag for application to spine (used as a remedy for cholera and sea-sickness), circa 1865. Died Nov. 25, 1894.

CHAPMAN, John Stewart, Am. physician; b. Sweetwater, Tex., Jan. 30, 1908; s. Alfred A. and Ollie (Johnson) C.; B.A., B.S., So. Meth. U., 1927, M.A., 1928; M.D., U. Tex., 1932; m. Marianne Ryan, Nov. 12, 1932; 1 dau., Carolyn. Practice medicine specializing in pulmonary disease, 1943-52; prof. internal medicine, asst. dean U. Tex. Southwestern Med. Sch., Dallas, 1952—. Cons. to local and area VA hosps., pvt. hosps. in Dallas, also govt. agys. Mem. Adv. Com. on Study Profl. Communications on Radiol. Health, 1966-67, Nat. Adv. Com. on Radiation, 1966-67. Diplomate Am. Bd. Internal Medicine. Fellow A.C.P.; mem. A.M.A. (council environmental and public health), Am. Thoracic Soc. (pres. 1966-67). Contbg. author Principles of Internal Medicine, 1950-58. Editor: Anonymous Mycobacteria in Human Disease, 1960. Research and publs. on atypical mycobacteria, sarcoidosis, eosinophilic leucocyte, epidemiology of Tb. Home: 3606 Lovers Lane, Dallas 75225.*

CHAPMAN, Loring F., Am. psychologist; b. Cal., Oct. 4, 1929; s. Lee E. and Elinore (Gundry) Scott; B.S., U. Nev., 1950; Ph.D., U. Chgo., 1955; m. Toy Farrar, June 14, 1954; children—Robert, Antony, Pandora. Faculty, Cornell U. Med. Coll., N.Y.C., 1955-61, U. Cal. at Los Angeles, 1961-65; clin. research prof. med. psychology U. Ore. Med. Sch., Portland, 1965—. Dir. research Fairview Hosp. and Tng. Center, Salem, Ore., 1965—; asst. research psychologist N.Y. Hosp., 1955-61; cons. Pacific State Hosp., Pomona, Cal., 1962-65. Recipient Thornton Wilson prize for research in psychiatry, 1958; USPHS Career Devel. award, 1964. Mem. Am. Physiol. Soc., Am. Psychol. Assn., Am. Acad. Neurology, Am. Assn. Mental Deficiency, Royal Soc. Medicine London, Am. Neurol. Assn., Sigma Xi. Author: Pain and Suffering, 1965, also articles. Research on behavioral consequences of surg. removal of brain tissue, functions of specific regions in human brain, neural mechanisms underlying psychosomatic disease, definition of neuroanat. and biochem. features of reaction of human skin to injury, computer analysis of surface and deep brain activity during learning, perception, emotion, research in pain, epilepsy, schizophrenia, mental retardation. Office: 2250 Strong Rd., Salem, Ore. 93710.*

CHAPMAN, Robert Hollister, Am. topographic engr.; b. New Haven, July 29, 1868; s. Charles Wesley and Etta (Sperry) C.; ed. Corcoran Sci. Sch., Washington; m. Frances Beardsley Andrews, June 1, 1907. With U. S. Geol. Survey, 1880-1920; made topographic surveys and explorations in prin. Western and So. states, maps of portion of Death Valley and adjacent deserts and high sierras, Cal.; sent to Ottawa to introduce U. S. topographic methods in Geol. Survey of Can., 1909; in charge field work, on Vancouver Island, 1909, 10, 11; acting supt. Glacier Nat. Park, Mont., 1912; returned to U. S. Geol. Survey, 1913. Author numerous bulls. pub. by govts. U. S. and Can. Personal explorations in No. Selkirks, B.C., 1915, and No. Canadian Rockies, 1919. Died Jan. 11, 1920.

CHAPMAN, Sir Robert William, Australian engr.; b. Stony Stratford, Bucks, Dec. 27, 1866; ed. Wesley Coll., U. Melbourne (Australia); m. Eva Maud Hall, 1889; 6 sons, 2 daus. Worked in ry. and bridge constrn.; apptd. lectr. Adelaide (Australia) U., 1889, prof. engring., 1907-37; pres. S. Australian Sch. Mines, from 1939. Pres. Australian Inst. Mining and Metallurgy, 1920-21, Instrn. Engrs., Australia, 1922, S. Australian Inst. Surveyors, 1914-28. Author: Astronomy for Surveyors; Reinforced Concrete; pamphlets on Australian tides; also papers on S. Australian timbers, wire ropes, reinforced concrete, various aspects of structural engring. Died Feb. 27, 1942.

CHAPMAN, Seville, Am. physicist; b. Seville, Spain, Nov. 4, 1912 (parents Am. citizens); s. Charles Edward and Elizabeth (Russell) C.; A.B. in Physics, U. Cal. at Berkeley, 1934, Ph.D. in Physics, 1938; m. Mary Bried, Aug. 1, 1942; children—Clark Russell, Diane Adams, Ralph Lyon. Faculty, U. Kan., Lawrence, 1938-41, Stanford (Cal.), 1941-48; prin. physicist Cornell Aero. Lab., Inc., Buffalo, 1948-50, asst. dept. head, 1950-51, dept. head, 1951-57, dir. physics div., 1957—. Mem. A.A.A.S., Am. Assn. Physics Tchrs., Am. Astronaut. Soc., Am. Geophys. Union, Am. Inst. Aeros. and Astronautics, Am. Meteorol. Soc., Am. Phys. Soc., I.E.E.E., Sigma Xi. Author: How to Study Physics, 1949. Research and publs. on atmospheric physics, electrification of thunderclouds, devel. of missile and space systems and radars, arms control, water resources. Home: 94 Harper Rd. Buffalo 14226.*

CHAPMAN, Sydney, mathematician, geophysicist; b. Eccles, Lancashire, Eng., Jan. 29, 1888; s. Joseph and Sarah Louise (Gray) C.; B.Sc. in Engring., Manchester U., 1907, M.Sc., 1908, D.Sc., 1912; B.A. in Math., Cambridge U., 1911, M.A., 1914, Sc.D., 1958; M.A., Oxford U., 1946; Sc.D., U. Alaska, 1958, U. Mich., 1960, U. Colo., 1962, U. Paris, 1962, U. Exeter, 1963, U. Newcastle, 1965, U. Sheffield, 1968; D.Tech., U. Brunel, 1968; m. Katharine Nora Steinthal, Mar. 23, 1922; children—Cecil Hall, Robert Gray, Mary Milnes (Mrs. Ian McAlley), Richard Joseph Ernest. Chief asst. Royal Obs., Greenwich, 1910-14, 1916-18; lectr. math. Trinity Coll., Cambridge, Eng., 1914-19; prof. Manchester U., 1919-24, Imperial Coll. Sci. and Tech., London, 1924-46; prof. natural philosophy Oxford U. 1946-53; research asso. Cal. Inst. Tech., 1950-51; prof. geophysics U. Alaska, 1951—; research staff High Altitude Obs., Boulder, Colo., 1955—; sr. research scientist Inst. Sci. and Tech., U. Mich., 1959-62. Pres. Spl. Commn. Geophys. year, 1953-59. Hon. fellow Trinity Coll., Cambridge, 1957, Queen's Coll., Oxford, Imperial Coll., London; recipient Antonio Feltrinelli internat. prize Academie Nazionale dei Lincei, Rome, 1956, Bowie medal Am. Geophys. Union, 1962, Symons Meml. Gold Medal Royal Meteorol. Soc., 1965, Hodgkins medal Smithsonian Instn., 1965. Fellow Royal Soc., 1919 (Copley medal 1964); mem., past pres. London Math. Soc., Royal Meteorol. Soc., Royal Astron. Soc., London Phys. Soc., Internat. Union Geodesy and Geophysics, Internat. Assn. Meteorology, Internat. Assn. Terrestrial Magnetism and Electricity; mem. Am. Phys. Soc., Nat. Acad. Scis.; hon. mem. many learned socs. Author: Earth's Magnetism, 1936; (with T. G. Cowling) Mathematical Theory of Non-Uniform Gases, 1939; (with J. Bartels) Geomagnetism, 1940; IGY: Year of Discovery (Edison prize as best 1959 sci. book for youth), 1959; Solar Plasma, Geomagnetism and Aurora, 1964. Modified the accurate kinetic theory of gases proposed by Maxwell in 1867, thereby discovering gaseous thermal diffusion and confirming it experimentally, 1912-17; later work on auroras, magnetic disturbances in and composition of ionosphere, atmospherical tides, etc.; demonstrated power of thermal diffusion in highly ionized gases (as in solar corona), 1958. Address: High Altitude Observatory, Boulder, Colo. 80302.*

CHAPMAN, Valentine Jackson, botanist; b. Alcester, U.K., Feb. 14, 1910; s. Thomas J. and Amy (Wynn) C.; student Marlborough Coll., Wilshire, Eng., 1925-29; B.A., Cambridge (Eng.) U., 1932, M.A., 1935, Ph.D., 1935; m. Phyllis Claire Parks, June 22, 1938; children—David James, Michael John, Peter Jackson. Henry fellow Harvard, 1935-36; lectr. botany, Manchester U., 1936-37; fellow Gonville and Cains Coll., Cambridge, Eng., 1937-44; u. demonstrator botany U. Cambridge, 1937-46; prof. botany Auckland (New Zealand) U., 1946—. Named Hon. City Botanist, Auckland City Council, 1946. Fellow Linnean Soc. London, World Acad. Arts and Sci.; mem. Am. Ecol. Soc., Internat. Phycological Soc., Torrey Bot. Club, New Zealand Ecol. Soc., Australian Marine Sci. Soc., New Zealand Marine Sci. Soc. Author: The Algae, 1948, rev., 1962; Seaweeds and Their Uses, 1950; Salt Marshes and Salt Deserts of the World, 1960; Coastal Vegetation, 1964; also numerous articles. Research in ecology and physiology maritime plants. Home: 5 Coronation Rd., Epsom, Auckland S.E. 3, New Zealand.*

CHAPOT, H., plant geneticist; b. Paris, France, Mar. 27, 1922; s. Rene and Marie (Guillaume) C.; Ingenieur agricole, Nat. Sch. Agr., Grignon, France, 1944; grad. in plant genetics Orstom, 1945; m. Simone Bonnenfant, Nov. 5, 1946; children—Monique Marie Jeanne, Gerard Marcel Rene. Head citrus sect. Hort. Service, Dept. Agr. of Morocco, 1946-57; head citrus and subtropical fruits sect. Directorate for agrl. Research and Edn., Ministry of Agr., 1957-63; chief hort. service Nat. Inst. for Agrl. Research Morocco, Rabat, 1963-66; prof. fruit culture Nat. Sch. for Agr., Meknes, Morocco, 1963-66; citrus prodn. expert FAO, Antalya, Turkey, 1967—. Mem. European Assn. for Research on Plant Breeding, Internat. Orgn. Citrus Virologists. Author: (with L. Blanc, G. Cuenot) Agrumes et Fruits Subtropicaux aux U.S.A., 1952; Les Agrumes au Liban, 1954; (with V. L. Delucchi) Maladies, Troubles et Predateurs des Agrumes au Maroc, 1964; also articles. Research on citrus virus diseases in Mediterranean area, taxonomy and pomology of citrus. Home: 40-Dair. Bl. Apartman, Fener Cadd., Antalya. Office: FAO Project, P.K. 35, Antalya, Turkey.*

CHAPPE, Abbé Claude, French engr.; b. Brûlon, France, Dec. 25, 1763; s. Ignace Chappe and Marie-Renée de Vernoy (de Vert) Chappe d'Auteroche; ed. at Rouen, France; also sem. of La Fliche. Joined religious order as monk; his benefice was suppressed by the legislature in 1789; returned to Brulon; with his bro., set up shop to work on telegraph, 1791; named telegraphic engr., 1793; given job of establishing semaphore telegraphs from Paris to Lille, 1793, completed in 1794; when others claimed credit for his invention, he committed suicide, 1805. Statue erected in his honor, Paris, 1893. Author: Lettres sur le nouveau télégraphe, 1798. Created aerial telegraph, semaphore; extensive research on electric telegraph, 1790, synchronized electric system, 1791, shutter system, 1791, semaphore system, 1792. Died Paris, Jan. 23, 1805.

CHAPPE D'AUTEROCHE, Jean Baptiste, French astronomer; b. Mauriac, Auvergne, France, Mar. 2, 1722; s. Jean, and Madelaine (de La Farge) C. d'A.; studied sci. at Collége Louis-le-Grand; ordained priest; worked as astronomer; assisted Cassini in delineating gen. map of France; became adj. astronomer replacing La Lande, 1759. Mem. French Acad. Scis. Author: Memoire du passage de Venus sur le soleil, . . . , 1762; Antidote ou examen du mauvaus livre initulé: Voyage en Sibére, 1768. Translator and editor: Tables (Halley), 1752. Observed the passage of Venus in front of the sun, Tobolsk, Siberia, 1761, Cal., 1769. Died San José, Cal., Aug. 1, 1769.

CHAPPUIS, James, French radiologist; b. Bessancon, France, Nov. 5, 1854; D.Sc.; prof. physics, Montauban, 1877, Pointiers, 1878, Paris, 1881, Ecole Central Arts et Manufact. Author: Lecons de physique, 1892; Cours de physique, 1898. With H. Chauvel, recorded 1st roentgenographic studies of biliary concretions, 1896.

CHAPTAL, Jean Antoine Claude, French chemist; b. Nogaret, Lozérer, France, June 4, 1756; s. Antoine and Françoise (Brunel) C.; studied at College de Rodez; received doctorate in medicine U. Montpellier, (France), 1777; received chair of chemistry U. Montpellier, 1781; prof. organic chemistry Polytechnic Inst., Paris, France, during French Revolution; arrested during Revolution; released and placed in charge of Grenelle powder mills by committee of public safety; returned to Montpellier after the fall of Robespierre and reorganized univ.; under Bonaparte became senator and later treas. of senate, supervisor of nat. edn. in council of state, minister of interior; named to Chamber of Peers, 1819. Created comte de Chanteloup. Founded sch. of arts, conservatory, chambers of commerce and soc. for encouragement of industry. Mem. French Acad. Scis., 1796 (became pres. 1825); Fellow Royal Soc., 1825. Author: Eléments de chymie, 3 vols., 1790-1803; Chimie appliquée aux arts, 4 vols., 1807. Established works for the prodn. of acids, alum, and soda; improved the manufacture of sulphuric acid, beet-root sugar, dyes, saltpeter for gunpowder; applied chemistry to industry and agr.; introduced the metric system to France; reorganized prisons and hospitals; established model farm and system of distbn. of agrl. seeds; organized plans for extension of Louvre; extended network of French highways. Died Paris, July 30, 1832.

CHAPUT, Victor Alexandre Henri, French surgeon; b. Tonnerre, France, 1857; M.D., 1885; surgeon Brussels Hosp., Boucicaut Hosp., Lariboisière Hosp.; author works on intestinal surgery, grafts, asepsis; suggested universal use of rubber gloves by physicians. Died Paris, France, 1919.

CHARACHON, Robert Louis, French physician; b. Lyons, France, Sept. 14, 1932; s. Jean and Marguerite (Roche) C.; M.D., U. Lyons, 1961; m. Claude Blanc, Sept. 3, 1957; children—Delphine, Remi, Guillaume. Head clinic, asst. Hosps. Lyon, 1961-66; asst. prof. ear nose and throat U. Grenoble (France), 1966—. Mem. Société Français d'Oto rhino-laryngologie, Author: Anatomy of the Arteria auditiva interna, 1961; (with D. Rebattu) The Heavy Epistaxis, 1961; (with D. Gignoux, Labayle) Ethmoid Tumors, 1964; also numerous articles. Research on internal auditory artery in the human by epoxy resin injection allowing study of anastamosis in external wall of cochlea; studies on carcinology and laryngo-bronchology. Home: 4 place Bir-Hakeim, Grenoble, France.*

CHARAKA, Hindu surgeon; flourished 100 A.D.; b. Pancanada, Kashmir; probably physician to King Kaniska, A.D. 100. Author: Caraka samhita (represents Atreya's system of medicine).

CHARAS, Moyse, physician, chemist; b. Uzès, France, Apr. 2, 1619; s. Moise and Marguerite (Folchier) C.; studied at the pharmacies of Montpellier, Blois (both France); M.D., U. London (Eng.); m. Suzanne Felix; m. 2d, Madeleine Hadancourt. 1666; 11 children. Practiced pharmacy Orange, Netherlands, then Paris; became apothecary to Duke of Orleans, 1663; prof. chemistry Jardin du roi, circa 1671-85; moved to Eng., 1685 because of revocation of Edict of Nantes; pharmacist to King Charles II; practiced medicine in Spain; prosecuted by Inquisition; returned to France, and was converted to Roman Catholicism. Mem. French Acad. Scis. Author: Theriaque D'Andromachus, 1668; Nouvelles experiences sur la vipere, 1669; Traité sur la thériaque, 1667; Pharmacopée royale galenique et chymique, 2 vols., 1672. Observed the heat produced when oil of vitriol was mixed with water, 1692; thought this reaction caused the heat in hot springs; classified salts into volatile, fixed and acid. Died Paris, Jan. 21, 1698.

CHARCOT, Jean-Baptiste, French physician, explorer; b. Neuilly-sur-Seine, France, July 15, 1867; student l'École alsacienne, Salpêtriere, 1876-85; m. Jean Hugo, 1896; m. 2d, Marguerite Cléry, Jan. 24, 1907. Became clin. chief for Prof. Raymond at Salpêtriere, 1897; studied cancer at l'Institut Pasteur; traveled with the American, Vanderbilt to the Mediterranean and down the Nile; traveled to Ireland, 1900; mem. expdns. to the Arctic, 1901, Antarctic, 1903-05, 08-10, Hebrides, 1915; research voyages to the Atlantic and Mediterranean, 1920-23; explored Greenland and No. polar regions, 1930-33. Mem. French Acad. Scis., Author: la Navigation mise à la portée de tous, 1901; le Porquoi Pas? dans l'Antarctique, 1910; la Mer de Groenland, 1929. Research on ethnography, geophysics, hydrography, meteorology, microbiology, bacteriology, terrestrial magnetism and glaciology; nautical studies; made new maps of the Antarctic and observed atmospheric electricity there. Died Reykjavik, Iceland, Sept. 16, 1936.

CHARCOT, Jean Martin, French physician; b. Paris, Nov. 29, 1825; s. Martin Charcot; student Lycée St. Louis, until 1844; M.D., U. Paris, 1853; m. Augustine-Victoire Richard, 1864; children—Jeanne, Jean. Apptd. physician Central Hosp. Bur., 1856; asso. with Salpetrière Hosp., from 1862, organized mus. and lab. path. anatomy, 1879; prof. path. anatomy Med. Faculty Paris, from 1860. Founder (with others) several med. jours. Recipient Montyon prize, 1880; subject of monument at Salpetrière, 1889, also at Lamalou in Herault. Mem. Acad. Medicine, Acad. Scis., 1883. Author: Leçons cliniques sur les maladies des vieillards et les maladies chroniques, 1868; Leçons sur les maladies du systeme nerveux, 1873; Leçons sur les maladies du foie, des voies biliares et des reins, 1877; Iconographie photographique de la Salpetrière, 1878-81; Leçons sur les localisations dans les maladies du cerveau et de la moelle épinière, 1880; (with P. Richer) Les Démoniaques dans l'art, 1887 (with P. Richer) Les Difformes et les Malades dans l'art, 1889. Developed at Salpetrière greatest clinic of his time for diseases of nervous system; exerted gt. influence on devel. of sci. of neurology; described characteristics of condition known as tabes dorsalis; differentiated multiple sclerosis from paralysis agitans; using hypnotism, investigated nature of hysteria; credited by Sigmund Freud (who was his pupil) with contbns. to early psychoanalytic formulations on hysteria. Died Lac des Settons, Nièvre, France, Aug. 16, 1893.

CHARDIN, Jean, explorer; b. Paris, France; baptized Nov. 26, 1643; s. Daniel and Jeanne (Guiselin) C.; went to Indies and Persia, 1665, returned to Paris, 1670; in 1671 left for Constantinople, Caucasus, Persia, and India; became jeweler to Charles II of Eng.; named Brit. plenipotentiary to State of Holland, 1683. Fellow Royal Soc., 1682. Author: Recit du couronnement du roi de Perse Solimen III, 1671; Voyage en Perse et aux Indes orientales, 1686; Journal du voyage du Chevalier Chardin, 4 vols., 1686-1711. Research and publs. on Persian manners, customs, law, adminstrn., monuments including mores and customs of ancient Persians; studied Turks and Arabs. Died London, Eng., 1713.

CHARDON DE COURCELLES, Étienne, French physician; b. Amagne, France, Mar. 9, 1705; ed. Faculty of Medicine, Reims, France; bachelor, 1741. Doctor at Faculty of Reims; 2d physician to the navy at Brest; prof. surgery at Brest; mem. staff Sch. of Astronomy for 30 years, later dir. Mem. Académie de marine. Author: Traite d'anatomie. 1750; Mémoire sur le régime végétal des gens de mers, pub. 1781. Research on the removal of salt from sea water; fought typhoid epidemic of 1757 by burning beds and clothing of victims, disinfection of utensils. Died July 5, 1775.

CHARDONNENS, Louis Jules Marie, Swiss chemist; b. Fribourg, Switzerland, Jan. 11, 1898; s. Auguste and Therese (Monney) C.; B.A., U. Fribourg, 1917, Dr. Chemistry, 1924; postgrad. U. Munich; Dr. honoris causa, U. Laval, 1952; m. Helene Castella, Sept. 16, 1947. Faculty, U. Fribourg, 1928—, extraordinary prof. chemistry, 1934-39, ordinary prof., 1939—, dir. Inst. Inorganic and Analytical Chemistry, 1962—, dean Faculty Scis., 1941-42, 57-58, univ. rector, 1950-52. Mem. Swiss, French, London chem. socs. Contbr. articles to profl. jours. Research on reactivity of some organic compounds, synthetic organic chemistry. Home: 73 Chemin Ritter, Fribourg, Switzerland.*

CHARDONNET, Jean Charles, French geographer, economist; b. La Rochelle, Sept. 9, 1913; s. Charles and Pauline (Danjean) C.; ed. Ecole normale supérieure, U. Paris; Ph.D. in letters; m. Jacqueline Largeault, Mar. 1, 1943; children—Jacques, Bruno, Alain, Patrick. Asst., Sorbonne, 1940-45; prof. Sch. Letters, Nancy, 1945-47, later at Dijon; prof. Inst. Polit. Studies of Paris, Nat. Sch. Geog. Sci., 1947—, Center Ind. Studies, Geneva, 1950—. Author: Le relief des Alpes du Sud, 1947; L'économie française, 1958; Les sources d'énergie, 1962. Home: 5, rue Henri Regnault, Saint-Cloud. Office: 27, Saint-Guillaume, Paris 7, France.

CHARDONNET, Louis Marie Hilaire Bernigaud, French chemist; b. Besançon, France, May 1, 1839; ed. Ecole Polytechnique (under Louis Pasteur), 1859, Ecole des ponts et chaussees; m. Mlle. de Ruolz-Montchal, 1865. Engr., Ponts et Chaussees. Author: Don Carlos, 1874. Patentee rayon, 1884 (Perkin medal 1914); inventor actineograph (to measure dark rays of sun) for aviation; investigated effect of ultraviolet rays on organisms, made glass that allowed passage of ultra-violet rays. Died Paris, France, Mar. 12, 1924.

CHARGAFF, Erwin, chemist; b. Czernowitz, Austria, Aug. 11, 1905; s. Hermann and Rosa Chargaff; Dr.phil. in chemistry, U. Vienna (Austria), 1928; m. Vera Broido, 1929; 1 son, Thomas H. Came to U. S., 1928, naturalized, 1940. Research fellow Yale, 1928-30; asst. U. Berlin (Germany), 1930-33; research asso. Institut Pasteur, Paris, France, 1933-34; faculty Columbia, N.Y.C., 1935—, prof. biochemistry, 1952—; vis. prof., Sweden, 1949, Japan, 1958. Brazil, 1959; Einstein vis. prof. Collège de France, Paris, 1965; with OSRD, 1942-44. Harvey Soc. lectr., 1956; Plenary lectr. Internat. Congress Biochemistry, Vienna, 1958. Guggenheim fellow, 1949, 58; recipient Pasteur medal Soc. chim. biol., Paris, 1949, Neuberg medal, N.Y.C., 1959; Soc. chim.biol. medal, Paris, 1961; Charles Leopold Mayer prize Académie des Sciences, Paris, 1963, Dr. H. P. Heineken prize Royal Netherlands Acad. Scis., Amsterdam, 1964, Bertner Found. award, Houston, 1965. Fellow Am. Acad. Arts and Scis.; mem. Nat. Acad. Scis., Royal Swedish Physiographic Soc. (fgn. mem.), Am. Chem. Soc., A.A.A.S., Soc. Biol. Chemists, Harvey Soc., Am. Soc. Cell Biology, Chem. Soc. (London), Brit. Biochem. Soc., Schweiz. Chem. Gesellschaft. Author: Essays on Nucleic Acids, 1963; also numerous articles. Editor: (with J. N. Davidson) The Nucleic Acids, 3 vols., 1955, 60. Demonstrated existence different deoxyribonucleic acids in different biol. species; discovered base-pairing regularities in DNA; research on methods for quantitative analysis nucleic acid composition. Home: 350 Central Park W., N.Y.C. 10025.

CHARGEYSHVILI, Archil Kornelevich, Soviet otorhinolaryngologist; b. Villageo of Zanati, Georgia, 1903; grad. Med. Faculty, Tbilisi Med. U., 1928; Cand. Med. Sci., 1939; D.Med. Sci., 1951. Otorhinolaryngologist in Far East, 1932-34; asst. dept. otorhinolaryngology Tbilisi Med. Inst., 1934-52, prof., head chair otorhinolaryngology, 1952—. Mem. Georgian Soc. Otorhinolaryngologists (founder Abkhazian, Adzhar, Kutaisi and Tbilisi brs.). Author of 60 works including Manual on Ear, Nose and Throat Diseases, 6 edits., 1937-61; co-author, editor Symposium of Proceedings of the Department of Otorhinolaryngology of Tbilisi Medical Institute, 1958; co-author 4 vol. manual or otorhinolaryngology, 1960. Address: Tbilisi Med. Inst., ulitsa Melikishvili 16, Tbilisi, Gruz. SSR, USSR.

CHARLES, Don C(laude), Am. psychologist; b. St. Joseph, Mo., Apr. 22, 1918; s. Claude C. and Helen (Miller) C.; B.A., State Coll. Ia., 1941; M.A., U. Neb., 1947, Ph.D., 1951; m. Anne Palmer Malonee, Aug. 12, 1947; children—Linda, Christopher, Laura, Andrew. High sch. tchr., Denmark, Ia., 1941-42; instr., guidance cons., U. Neb., 1948-51; prof. psychology Ia. State U., Ames, 1951—. Fellow Am. Psychol. Assn., Gerontol. Soc.; mem. Midwestern, Ia. psychol. assns. Author: (with W. R. Baller) Psychology of Human Growth and Development, 1961; Psychology of the Child in the Classroom, 1964; also articles. Research in human psychol. devel., longitudinal studies intelligence in mature years, characteristics behavioral changes in mentally retarded persons through life span. Home: Woodview, Route 4, Ames, Ia. 50010.*

CHARLES, Guy de, see de Charles.

CHARLES, Jacques Alexandre César, French physicist; b. Loiret, France, Nov. 12, 1746; held minor office in French govt.; later became pof. physics Conservatoire des Arts et Metiers, Paris; tchr. Sorbonne, Paris; given pension and apartment in Louvre by Louis XVI. Mem. French Acad. Scis., 1795. Popularized Franklin's one-fluid theory on electricity; 1st to use hydrogen for balloon inflation, 1783 (his balloon ascents led to aero. craze); research on expansion of gases; stated effect of changes in temperature on gas volume (Charles law, which anticipated Gay-Lussac's law); invented a megascope, thermometric hydrometer; improved Gravesande's heliostat. Died Paris, Apr. 7, 1823.

CHARLES, Robert George, Am. chemist; b. Pitts., Oct. 19, 1926; s. George M. and Ethel (Thompson) C.; B.S., U. Pitts., 1948, M.S., 1950, Ph.D., 1952. Chemist, Westinghouse Research Labs., Pitts., 1952—. Mem. Am. Chem. Soc., Chem. Soc. London, A.A.A.S. Author: (with others) The Binding Force, 1966; also articles. Research in coordination chemistry, thermodynamics, lasers. Home: 4873 Christopher Dr., Allison Park, Pa. 15101. Office: Westinghouse Research Labs., Pitts. 15235.*

CHARLESBY, Arthur, Brit. physicist; b. London, Eng., Oct. 12, 1915; s. Samuel and Ethel (Cogan) C.; B.Sc. (Royal scholar), Imperial Coll., U. London, 1937, Ph.D. in Physics, 1939; m. Irene Goulding, Aug. 29, 1958. Prin. sci. Officer radiation effects sect. Atomic Energy Research Establishment, Harwell, Eng., 1949-55; head nuclear and radiation dept. T.I. Research Lab., Hinxton, nr. Cambridge, Eng., 1955-57; prof. physics, head dept. Royal Mil. Coll. Sci., Shrivenham, Eng., 1957—. Vis. prof. U. Brussels, 1966; lectr., cons. polymer and radiation effects, 1950—; mem. various sci. coms. on materials and utilization of radiation. Mem. Brit. Rheology Soc. (pres.), Royal Instn., Biophys. Soc., Assn. for Radiation Research, others. Author: Atomic Radiation and Polymers, 1960. Editor: Internat. Series Radiation Effects in Materials. Contbr. over 150 articles to profl. jours. Research on polymers and elastomers, reinforcement, effects of nuclear radiation in polymers, analogies with simple systems, effect of radiation conditions and environment, radiation protection, radiosensitization, comparison of different types of radiation, thermoluminescence of polymers and organic materials, effect of glow discharges, oxidation of metals, behavior of materials in a nuclear environment.*

CHARLESWORTH, Edward, Brit. physician; b. Ossington, Nottinghamshire, Eng., 1783; s. John Charlesworth; studied under Dr. E. Harrison of Horncastle; M.D., Edinburgh, Scotland U., 1807; m. Miss Rockcliffe; practiced at Lincoln; became vis. physician to the asylum, 1820; physician to Lincoln County Hosp. Obtained opportunities for patients in insane asylum to exercise outside; classified the insane; instrumental in abolishing mech. restraints for the insane; improved structure and arrangement of asylum. Died Feb. 20, 1853.

CHARLESWORTH, John Kaye, Brit. geologist; b. Leeds, Eng., Jan. 3, 1889; s. George and Louise (Kaye) C.; student U. Leeds, 1907-10, U. London, 1910-11; D.Sc., U. Leeds, 1920; Ph.D., U. Breslau, Germany (now Wroclaw, Poland), 1914; D.Sc. (hon.), Queens U., Belfast, Ireland, 1957; m. Janet Cumming Gibson, Mar. 14, 1922; children—Marian Louise Kaye (Mrs. Derek Collier Lyddon), Henry Alexander Kaye. Sr. lectr. Victoria U., Manchester, Eng., 1919-21; prof. geology Queen's U., 1921-54, prof. emeritus, 1954—; Decorated Comdr. Brit. Empire; recipient Neill prize Royal Soc. Edinburgh, 1954, Prestwich medal Ecol. Soc. London. Fellow Royal Soc. Edinburgh, Geol. Soc. London, Royal Geog. Soc.; mem. Royal Irish Acad. Author: The Geology of Ireland; An Introduction, 1953; The Quarternary Era, 2 vols., 1957; Historical Geology of Ireland, 1963; also articles. Research on mode of disappearance of ice in Brit. Isles at close of Glacial Period. Home: Corrig, Ballycastle, County Antrim, N. Ireland.*

CHARLETON, Rice, Brit. physician; b. Eng., 1710; ed. at Oxford (Eng.); M.A., 1747; M.B.; M.D., 1757; practiced medicine, Bath, Eng.; physician Bath Gen. Hosp., 1757-81. Fellow Royal Soc., 1748; mem. London Coll. Physicians. Author: Chemical Analysis of Bath Waters, 1750; An Inquiry into the Efficacy of Bath Waters in Palsies, 1774; Tract the Third, containing Cases of Patients Admitted into the Hospital at Bath under the Care of the Late Dr. Oliver. Proved that many palsy cures attributed to waters of Bath, were actually cases of lead poisoning (caused by drinking cider from lead-lined casks) and cured by removal of cause and frequent bathing. Died 1789.

CHARLETON, Walter, Brit. physician; b. Shepton Mallet, Somerset, Eng., Feb. 2, 1619; s. of rector of Shepton Mallet in Somerset; ed. Magdalen Hall, Oxford, Eng., 1635; M.D. by king's mandate, 1643. Nominally physician to Charles I and Charles II; practised medicine, London, Eng., 1650-92; went to Nantwick, returned to London before 1698; sr. censor Coll. Physicians, 1698-1706; Harveian orations, 1702, 06; became Harveian librarian, 1706. Fellow Royal Soc., 1663. Coll. Physicians. Author several books including: Spiritus gorgonicus . . . , 1650; Physiologia Epicurio-Gassendo-Charletoniana . . . , 1654; Eoconomia animalis novis . . . , 1659; Exercitationes physico-anatomicae . . . , 1659; Hypothesibus superstructura et mechanice explicata, 1659; Exercitationes pathologicae . . . , 1661; also numerous med., philos. and antiquarian tracts. Research and publs. on atomic theory, barometer, magnitude of atoms, gravity and levity, heat and cold, fluidity and viscosity, lodestone and motion, physics, theology, natural history, and archaeology; his works show transition from scholastic way of writing to newer method of recording observations and drawing conclusions; tried to prove Stonehenge was built by Danes. Died Apr. 24, 1707.

CHARLEVOIX, Pierre François Xavier de, French explorer, historian, botanist; b. St. Quentin, France, Oct. 24, 1682; s. Francois and Antoinette (Forestier) de C.; student College Louis le Grand, Paris. France, 1701-04; ordained priest Soc. Jesus, 1698; prof. grammar, Can., 1705-09; prof. rhetoric Jesuit coll., Que., Can., 1705-09; tchr. College Louis le Grand, 1709-19, became prefect. messenger of the regent of France to Can., 1720; tchr., France, 1723-33; tchr. of Voltaire. Wrote numerous descriptive histories of France, Am., Japan, 1715-44, including: Histoire de Saint Domingue, 1730; Histoire du Japon, 1736; Histoire du Paraguay. 1756; Description des plantes principales de l'Amérique équinoxiale. Research on the flora of Japan; observed geography of interior Am., also its flora, fauna, customs and manners of Indians. Died Feb. 1, 1761.

CHARLIER, Carl Wilhelm Ludwig, Swedish astronomer; b. Östersund, Sweden Apr. 1, 1862; prof., Lund, Sweden; research on stellar clusters and related celestial phenomena, astrophysics, calculation of probabilities, theoretical optics. Died Lund, Nov. 5, 1934.

CHARLOIS, Auguste Honoré Pierre, French astronomer; b. La Cardière, France, Nov. 26, 1864; ed. Sch of Bros.; student (under J. Perrotin) Obs. of Mont-Gros, nr. Nice, France, from 1880; astronomer Obs. Nice, until his death. Recipient Janssen prize, 1889. Determined longitude of Nice-Milan, 1882; made trip to Spain to observe transit of Venus, 1882; one of first to use photography in research; discovered 77 asteroids and 104 small planets; calculated elliptical elements of comet orbits; measured 338 double stars; research on Mercury and Eros; rediscovered Andromochus. Died Nice, Mar. 26, 1910.

CHARLTON, Thomas Malcolm, Brit. engr.; b. South Normanton, Derbyshire, Eng., Sept. 1, 1923; s. William and Emily (Wallbank) C.; ed. U. Nottingham, Cambridge U.; B.Sc. in engring., M.A.; m. Valerie McCulloch, Sept. 18, 1950; children—Richard, Edward. Asst. to minister aero. prodn., T.R.E. at Malvern, 1943-46; engr. at Merz and McLellan, Newcastle/Tyne, 1946-54; lectr. mech. constrn., mem. corp. Sidney Sussex Coll., Cambridge, 1954-63; prof., head Sch. Civil Engring., U. Belfast, 1963—. Mem. London Inst. Civil Engrs. Author: Model Analysis of Structures, 1954; Energy Principles in Applied Statics, 1959; Analysis of Statically Indeterminate Frameworks, 1961. Home: 6 Mt. Pleasant, Belfast 9. Office: David Keir Bldg., University of Belfast, Belfast 9, N. Ireland.

CHARM, Stanley E., Am. chem. engr.; b. Boston, Oct. 18, 1926; s. Louis David and Anna (Gorfinkle) C.; B.S., U. Mass., 1950; M.S., Wash. State Coll., 1952; B.S. in chem. engring. Mass. Inst. Tech., 1952, Sc.D., 1957; m. Shirley Zitaner, June 15, 1952; children—Anne Sima, Susan Rachel, Elizabeth Martha. Research chemist Am. Home Foods, Rochester, N.Y., 1952-53; research asst. Mass. Inst. Tech., 1953-55, instr., 1955-57, asst. prof., 1957-63; asso. prof., sci. dir. New England Enzyme Center Tufts Med. Sch., Boston, 1963—; cons. Bur. Comml. Fisheries, Gloucester, Mass., 1963—. Mem. Inst. Food Technologists, Am. Chem. Soc., Am. Inst. Chem. Engrs., Soc. Rheology, Soc. Cryobiology. Author: Fundamentals of Food Engineering. Research and publs. in rheology of food, heat transfer and flow of food materials, blood rheology and flow in microcirculations, large scale purification of proteins, large scale manufacture of enzymes. Home: 21 Concolor Av., Newton, Mass. Office: 136 Harrison Av., Boston 02111.*

CHARMOT, Guy Denis, French physician; b. Toulon, France, Oct. 9, 1914; s. Ulysse J. and Claire (Esmieu) C.; Docteur en Medecine, Faculté de Lyon, 1938; m. Edith Dubuisson, Nov. 30, 1948; 1 child, Dominique. Physician, Corps de Santé des Troyes de Marine; staff Services de Santé Publique hosps. in Cameroons, Ivory Coast, Chad, Senegal, Madagascar, Belgian Congo; prof. l'Ecole du Service de Santé des Troyes de Marine. Mem. Soc. Tropical Pathology, Med. Soc. Marseilles Hosp., Med. Soc. Black Africa, Madagascar Med. Soc. Author: (with Martin, Mafart, Bertrand) Médecine Tropicale; also numerous articles. Research on blood disorders, liver diseases, spleenomegaly, parasitic diseases. Home: 72 Bd. de Reuilly, Paris (12). Office: 28 Cours Albert 1e, Paris, France.

CHARNEY, Jule Gregory, Am. meteorologist; b. San Francisco, Jan. 1, 1917; s. Ely and Stella (Litman) C.; A.B., U. Cal. at Los Angeles, 1938, M.A., 1940, Ph.D., 1946; m. Elinor Marie Kesting, Feb. 2, 1946; children—Nicolas H., Nora K., Peter E. Instr., lectr. in physics meteorology U. Cal. at Los Angeles, 1942-46; research asso. in meteorology U. Chgo., 1946-47; Nat. Research Council fellow U. Oslo (Norway), 1947-48; mem., dir. theoretical meteorology group Inst. for Advanced Study, Princeton, N.J., 1948-56; prof. meteorology Mass. Inst. Tech., Cambridge, 1956—; lectr. U. Chgo., summers 1950-55; Assos. lectr. Woods Hole Oceanographic Inst., summer 1954; lectr. Imperial Coll., London, Eng., summer 1957. Recipient Robert M. Losey award Inst. Aero. Scis., 1957; Symons Meml. Gold medal Royal Meteorol. Soc., 1961. Mem. Am. Meteorol. Soc. (Meisinger award 1949, Carl-Gustav Rossby research medal 1964). Fellow Am. Acad. Arts and Scis., Am. Geophys. Union; mem. Royal Swedish Acad. Scis. (fgn.), Am. Civil Liberties Union, Am. Math. Soc., Nat. Acad. Scis., Phi Beta Kappa, Sigma Xi. Originated theory of geostrophic scaling and baroclinic instability; pioneered weather prediction by high speed computers; developed theory of Gulf Stream and Equatorial undercurrents as inertial boundary layer phenomenon; contbns. to cloud physics. Home: 288 Prince St., West Newton, Mass. 02165.*

CHARNOCK, Thomas, Brit. alchemist; b. Kent, Eng., 1526; studied under chemist James S. from Salisbury; m. Agnes Norden, 1562. Served at Calais, 1557; set up pvt. lab., and performed expts., Somerset, Eng., until his death. Author: The Breviary of Naturall Philosophy, 1557; Aenigma ad alchimiam, 1572; A Booke of Philosophie, 1566. Known as alchemical writer who searched for philosopher's stone (James S. is said to have bequeathed to him secret of stone). Died Apr. 1581.

CHARON, Jean Octave Emile, French physicist; b. Paris, France, Feb. 25, 1920; s. Jules and Yolande (Charles) C.; ed. Ecole supérieure de physique et chimie; degree in engring. m. Jacqueline Lengelle, Sept. 15, 1942; children—Catherine, Irène, Elisabeth, Jean Jacques. Sci. attache French embassy, Washington, 1945-50; in charge research Commissariat of Atomic Energy, Saclay, 1955-63. Recipient Crown of Nautlus prize, Seuil, 1961, 63. Author: La connaissance de l'univers, 1961; Du temps de l'espace et des hommes, 1962; Elements d'une theorie unitaire d'univers, 1962; L'homme à sa découverte, 1963; Relativité générale, 1963. Home: 30, av. Saint-Laurent, Orsay (S.-&-O.). Office: 23-25, av. Mac-Mahon, Paris 17, France.

CHARPENTIER, Jacques, French physician, philosopher; b. Clermonten-Beauvaisis, France, 1521; ed. at Paris (France) and Coll. of Bourgogne; received license in medicine. Prof. philosophy Coll. Bourgogne; became prof. math. College de France, 1566; physician to Charles, IX; rector Académie de Paris until his death. Author numerous books including: Platonis cum Aristotele in universa philosophia comparatio, 1573; Universae artis disserendi descriptio, 1564; Libri quatuordecim qui Aristotelis esse dicuntur de secretiore parte divinae sapientiae secundum Aegyptos, 1572. Supported fanatically the beliefs of Aristotle and attacked Ramus; accused of causing the Saint Bartholomew massacre, 1572. Died Paris, Feb. 1, 1574.

CHARPENTIER, Johann H., geologist; b. Freiberg, Germany, Dec. 8, 1786; s. Johann F.W. and Dorothea Wilhelmina (Zobel) C.; ed. Mining Acad. Freiberg; m. F.v.d. Gablenz; 1 dau. Indsl. geologist, first in Silesia, later in France; discovered Bex Salt Mines, 1813, mgr., 1813-55. Author: Essai sur la constitution geologique des Pyrénées, 1823; Sur la cause probable du transport de bloc erratique de la Suisse, 1835; Essai sur les glaciers du bassin du Rhône, 1841. Studied glaciers in Pyrenees, 1808-12; demonstrated that valley glaciers in Alps once extended to lower levels, thus indicating glaciers in recent past geologic ages were more extensive than at present, 1834; researches in geomorphology. Died 1855.

CHARPENTIER, Leon Joseph, Am. entomologist; b. Houma, La., Mar. 28, 1910; s. E. and Aurelie (Fanguy) C.; student La. State U., 1930-34, 66; m. Eleanor Chauvin, Nov. 25, 1937; children—Kathleen (Mrs. Walker Holt), Kenneth. With Entomology Research div. Agrl. Research Service, U. S. Dept. Agr., Houma, 1931—, asso. entomologist, 1956-61, entomologist, 1961-64, sr. research entomologist, 1964—. Recipient Certificate of Merit awards Farm Bur., 1966, 67. Mem. Entomol. Soc. Am., Am., Internat. socs. sugar cane technologists. Contbr. numerous articles to profl. jours. Established efficient tropical parasite of sugarcane borer in La.; developed control programs for sugarcane borer and soil arthropods; discovered vectors of sugarcane mosaic disease. Home: Route 3, Box 117. Office: U. S. Dept. Agr., Box 387, Houma, La. 70360.*

CHARPENTIER, Pierre Marie Augustine, French physician; b. Argenton-sur-Creuse, France, 1852; qualified as instr. Faculty Medicine, 1878; prof. med. physics Faculty of Nancy (France); mem. Acad. Medicine. Research on nerve physiology (especially sensorial and emotional excitation on heart), photometry, persistance of impressions, auditory sensations, speed of nervous reactions; analyzed retinal function; discovered 2 modes of sensibility to light in eye, also nervous oscilations caused by excitations. Died Argenton-sur-Creuse, 1916.

CHARPENTIER COSSIGNY, Joseph François, French engr.; b. Port-Louis, Ile de France, 1730; s. Jean-Francois Charpentier Cossigny; m. Marie-Francoise Menassier. Visited China and Benegal, India, 1753; joined mil. commissary, 1757; named insp. hosps., 1758; became royal engr. Ile de France, 1760; made polit. mission to Java, 1761-62; returned to France, 1770; began work on black wood plantation, Port Louis, 1776; became dep. to French Constl. Assembly, 1789; worked in gun powder factories, Ile de France, 1800, France, 1801-09; dir. gun powder manufacture, Ile de France, 1800. Mem. French Acad. Scis. (corr.), 1774. Author: Essae sur la fabrique de l'indigo, 1779; Voyage au Benegale Voyage à Canton, 1800; Recherches physiques et chimiques sur la fabrication de la poudre à canon, 1807. Studied culture of coffee, sugar cane, manufacture of sugar and rum, maritime commerce and geography. Died Paris, 1809.

CHARPIN, Jacques, French physician; b. Aix en Provence, France, Mar. 20, 1921; s. Eysee and Jeanne (Latig) C.; m. Huguette Cavaglione, June 28, 1948; children—Jean Michel, Denis, Marion, Dany. Faculty, U. Aix Marseilles (France), 1955—, prof. medicine, 1962—. Author: Allergy, 1961. Research, numerous publs. on atmospheric allergens, indsl. and natural pollutants and molds in France. Home: 8 Bd. d'Athenes, Marseilles, France.*

CHARPY, Adrien, French physician, anatomist; b. Caluire, France, 1848; ed. at med. sch. Lyons, France; doctorate, 1873; prof. anatomy Faculty of Lyons; became prof. anatomy, Toulouse France, 1889; Author: Studies of Anatomy, 1891; (with Poirier Nicholas) Traité d'anatomie humaine, 5 vols., 1892-1904; also articles on voluntary muscles, the position of the uterus, capacity of cecum, brachial plexis, peeling of skin, pancreatic canals, rectal veins. Died Toulouse, July 1911.

CHARPY, Georges, French chemist, metallurgist; b. Rhone, France, Sept. 1, 1865; prof. École polytechnique, also École des mines. Mem. French Acad. Scis., 1913. Research on metall. and chem. reactions, influence of corrosion on mech. properties, influence of time on fast deformation of metals, tensile strength of metals in shocks; built a drop pendulum. Died Paris, Nov. 25, 1945.

CHARRIERE, Joseph-Frederic Benoit, surgeon, instrument maker; b. Cerniat, Switzerland, 1803; devised guillotine-type instrument for excising tonsils, circa 1842, also scale for measuring size of urethral sounds and catheters in which consecutive numbers differ by 1/3 millimeter in diameter, recorded circa 1850; improved instruments for lithotripsy, for treatment of urethral stricture, also for vaginal speculum.

CHARRIN, Albert (Benoit-Jérôme), French physiologist, pathologist; b. Condrieu, France, Nov. 25, 1857; student medicine, Lyons, France, from 1879; doctorate, 1885; lab. dir. Institut Pasteur, 1885-86; named auditor Com. French Hygiene, 1887; apptd. physician of hosps., 1890; became asst. prof. arsonval at Collége France, 1892; mem. missions to study cholera, Spain and Italy, recipient gold medal; became prof. in dept. gen. pathology College France, 1896. Author: The Poisons of the Organism (Urine, Tube Digestion, Tissues), 3 vols., 1893-97; Clinique médicale de l'Hôtel-Dieu, 1896; Course of Applied Pathology, 1897; Les défenses naturelles de l'organisme, 1898, Pathologie générale infectieuse, 1898. Research on microbes, poisons, exptl. paralysis, infections, effect of toxins on circulation, natural defense in the new-born, transmission of path. wastes, rabies, mucus. Died Paris, May 17, 1907.

CHARTIER, Jean, French physician; b. Paris, France, 1610; s. Rene Chartier; prof. Paris Faculty of Medicine; physician to French king; physician, prof. medicine Royal Coll. France. Author: La science du plomb sacré des sages ou de l'antimoine, 1651; translator Treatise of Fevers (Palladius), 1646. Died 1662.

CHARTIER, René, French physician; b. Vendôme, France, 1572; student Coll. of Angers, studied math., Bordeaux, France, and rhetoric at Bayonne, attended Med. Sch. in Paris, France; B.A., 1606; M.D., 1608; children—Jean, Philippe. Became doctor to the wives of Henry IV, 1612; named physician to Louis XIII, 1613; named prof. surgery Royal Coll., 1617; traveled through Spain, Eng., and Italy. Translated Galen and Hippocrates: Hippocratis Coi et Claudii Galeni Pergameni archiatron opera, 13 vols., 1639-79. Died Oct. 29, 1654.

CHARTON, André, French biologist; b. Ossey, France, Sept. 28, 1911; s. Charles and Charlotte Charton; Doctorat Véterinaire; Agregation des Ecoles Véterinaires, 1943; Diplome, Institut Pasteur; Licence in Droit; m. Magdelaine Quinet, Jan. 27, 1936; children—Geneviève, Francoise. Named research chief Ecole Véterinaire, Alfort, France, 1937, dir., 1964; became lectr Ecole d'Alfort, 1950, prof. pathology, 1953; Decorated Chevalier de la Légion d'Honneur, Officier des Palmes Académiques, officer du Mérite Agricole. Titular mem. Vet. Acad. France. Author: Nutrition des Mammiféres domestiques; Bases physiologiques; also numerous articles. Research on nutritional pathology of domestic animals. Home: 41 bis Avenue des Charmes, Fontenay aux Bois, 94, France. Office: Ecole Nationale Veterinaire, Alfort 94, France.*

CHARY, Chintamanny Ragdonatha, Indian astronomer; with Madras (India) Obs. for 40 years, 1st asst. for 17 years; in command parties to observe total eclipses of sun, Wunpurthy, 1868, Avenaski, Coinbatore Dist., 1871. Author: Transit of Venus, 1874; also articles. Editor native calendar. Made observations with transit circle for star catalogue, 1862; discovered 2 new variable stars. Died Madras, Feb. 5, 1880.

CHASE, David Marion, Am. physicist; b. Denver, Jan. 20, 1930; s. Leo David and Margaret (Moffett) C.; A.M., Princeton U., 1953, Ph.D., 1955; B.A., U. Colo., 1951; m. Juliann Carol Miller, May 4, 1963. Physicist Naval Research Lab., Washington, 1950-53; staff physicist Los Alamos Sci. Lab., 1954-57; vis. asst. prof. physics Ia. State U., 1956; sr. physicist TRG, Inc., Melville, N.Y., 1957—. Mem. Am. Phys. Soc., Am. Geophys. Union. Contbr. articles in field to sci. jours. Derived motion of charged test particles in gen. relativity; formulated adiabatic approximation and treated scattering of nucleons by nuclei with collection motion; contbr. to application of unified model of nuclear structure; analyzed coherence of light propagating in turbulent media; applied kinematics and similarity to space-time correlations in turbulence. Home: 1 Henhawk Lane, Huntington, N.Y. 11743. Office: Rt. 110, Melville, N.Y. 11749.*

CHASE, Frederick Lincoln, Am. astronomer; b. Boulder, Colo., June 28, 1865; s. George Franklin and Augusta Ann (Staples) C.; A.B., U. Colo., 1886; Ph.D., Yale, 1891. Asst. in Yale Obs., 1890; asst. astronomer, 1891-1911, acting dir. same, 1910-13. Author: Heliometer Triangulation of the Victoria Comparison Stars (Annals of the Cape Observatory, 1897); Triangulation of the Principal Stars of the Cluster in Coma Berenices (trans. Yale Univ. Obs., Vol. I, Part V, 1896); Parallax Investigations on 163 Stars, mainly of Large Proper Motion (trans. Yale Univ. Obs., Vol. II, Part I, 1906); Parallax Investigations on 35 Selected Stars, Vol. II, Part II, 1910; Parallax of 41 Southern Stars, Vol. II, Part III, 1912; Catalogue of Yale Parallax Results, Vol. II, Part IV, 1912. Died Nov. 8, 1933.

CHASE, Herman Burleigh, Am. biologist; b. New Hampton, N.H., May 7, 1913; s. Edwin Elmer and Ada Mabel (Burleigh) C.; A.B., Dartmouth, 1934; Ph.D., U. Chgo., 1938; m. Elizabeth Studley Brown, Aug. 30, 1937; children—Elizabeth Studley, Burleigh Brown (dec.), Anne Louise, Catharine Marie. Teaching asst. zoology U. Chgo., 1935-38; instr. zoology U. Ill., 1938-44, asst. prof., 1944-48; asso. prof. biology Brown U. 1948-52, prof., 1952—, R. P. Brown, professor biology, 1960—, chairman biology department, 1963-65, director Inst. for Health Sciences, 1967—. Mem. American Soc. Zoologists, Genetics Soc. Am., Am. Soc. Human Genetics, A.A.-A.S., Soc. Study Evolution, Soc. Study Growth and Development, Am. Soc. Naturalists, Radiation Research Soc., N.Y. Acad. Sci., Am. Assn. U. Profs., Sigma Xi, Phi Beta Kappa. Contbr. papers to profl. jours. on physiol. genetics, mouse genetics, dermatology. Home: 175 Ferris Av., Rumford 16, R.I. Office: Biology Dept., Brown U., Providence 02912.*

CHASE, Isaac McKim, Am. mech. engr.; b. Balt., May 27, 1837; s. Alexander and Mary Ann (Cruser) C.; grad. Md. Inst.; m. Emeline Hall, Apr. 1, 1878. Entered Washington Navy Yard, 1868, becoming master mechanic; expert in marine propellers and propulsion. Author: Screw Propellers and Marine Propulsion, 1895; Art of Pattern Making, 1903. Died 1903.

CHASE, Pliny Earle, Am. scientist; b. Worcester, Mass., Aug. 18, 1820; s. Anthony and Lydia (Earle) C.; grad. Harvard, 1839, M.A. (hon.), 1844; LL.D. (hon.), Haverford Coll., 1844; m. Elizabeth Brown Oliver, 1843, 6 children. Prof. natural scis. Haverford Coll., 1871, prof. philosophy and logic, 1875, acting pres., 1886; lectr. psychology and logic Bryn-Mawr Coll.; mem. Am. Philos. Soc. (v.p., sec., Magellanic medal for paper Numerical Relations between Gravity and Magnetism); linguist; mgr. Franklin Inst.; fellow A.A.A.S., 1874. Author textbooks including: Elements of Arithmetic, 1844; The Common School Arithmetic, 1848; Elements of Meteorology for Schools and Households, 1884. Died Haverford, Pa., Dec. 17, 1886.

CHASE, Robert Arthur, Am. surgeon; b. Keene, N.H., Jan. 6, 1923; s. Albert Henry and Georgia Beulah (Bump) C.; B.S. cum laude, U. N.H., 1945; M.D., Yale, 1947; m. Ann Crosby Parker, Feb. 3, 1946; children—Deborah Lee, Nancy Jo, Robert N. Mem. faculty Yale Sch. Medicine, 1948-54, 59-62, asst. prof. surgery, 1959-62; mem. faculty U. Pitts., 1957-59, resident plastic surgeon, also teaching fellow, 1957-59; attending surgeon VA Hosp., W. Haven, Conn., 1959-62, also Grace New Haven Community Hosp., 1959-63; prof., exec. dept. surgery Stanford Sch. Medicine, 1963—; cons. plastic surgery Christian Med. Coll. and Hosp., Vellore, S. India, 1962. Recipient Francis Gilman Blake award Yale Sch. Medicine, 1962. Diplomate Am. Bd. Surgery Am. Bd. Plastic Surgery, Am. Bd. Med. Examiners. Fellow A.C.S.; mem. Am. Soc. Plastic Surgery, G.E. Lindskog Surg. Soc., Am. Soc. Surgery Hand, Am. Soc.

Cleft Palate Rehab., Am. Assn. Surgery Trauma, Plastic Surgery Research Council, New Eng. Soc. Plastic and Reconstructive Surgery, Found. Am. Soc. Plastic and Reconstructive Surgery (bd. dirs.), Soc. Univ. Surgeons, Sigma Xi. Contbr. articles profl. jours. Home: 839 Northampton Dr., Palo Alto, Cal.*

CHASE, Sherret Spaulding, Am. botanist; b. Toledo, June 30, 1918; s. Clement Edwards and Helen (Kelsey) C.; student U. Ariz., 1935-36; B.S., Yale, 1939; Ph.D., Cornell U., 1947; m. Catherine Compton, Nov. 27, 1943; children—Catherine (Mrs. Stephen Jepson), Helen, Sherret, W. Compton, Alice. Faculty Ia. State U., 1947-53; with DeKalb (Ill.) Agrl. Assn., Inc., 1954-66, research geneticist, 1954-66, dir. internat. seed operations, 1964-66, cons., 1966—; Bullard fellow Harvard U., 1966-67; cons. seed research and devel., 1966—; dir. Semillas Agricolas S.A., Madrid, Spain, 1965-66; dir. DeKalb-Italiana SPA, Venezia Mestre, Italy, 1964-66; dir. Agrl. Assn. Ltd., New Delhi, India, 1965-66. Fellow A.A.A.S.; mem. Am. Soc. Agronomy, Genetics Soc. Am., Am. Genetic Assn., Botan. Soc. Am., Ill. Acad. Sci., N.Y. Acad. Sci., Am. Forestry Assn., Am. Inst. Biol. Scis., Sigma Xi, Phi Kappa Phi, Gamma Sigma Delta. Contbr. articles in field to sci. jours. Research in field of cytotaxonomy of Najas; parthenogenesis in maize; development of "monoploid method" for development of inbred lines in maize; maize breeding and breeding methods; forest tree genetics and breeding. Home: 612 S. Main St., Sycamore, Ill. 60178. Office: Botanical Museum, Harvard U., Oxford St., Cambridge, Mass. 02138.*

CHASLES, Michel, French mathematician; b. Epernon, France, Nov. 15, 1793; entered l'École polytechnique, Paris, France, 1812; received master's degree in engring., 1815. Participated in def. of Paris, 1814; studied mathematics, Chartres, 1815; became prof. geodesy and applied mechanics École Polytechnique, 1841; named prof. advanced geometry Sorbonne, Paris, 1846, a founder Faculty Scis., chmn. advanced geometry div. Fellow Royal Soc., 1854; mem. French Acad. Scis., 1851 (pres. 1860). Author: Apercus historique sur l'origine et le développement des méthodes en géométrie, 1837; Traité de géometrie superieure, 1852; Trois livres de porismes d'Euclide, 1863; Traité des sections coniques, 1865; Rapport sur les progrès de la géométrie, 1871. Developed projective geometry (independently of J. Steiner); solved problem of attraction of ellipsoid on an external point, 1846; developed gen. theory of homography and duality (reciprocity); introduced the term, harmonic ratio; used method in which five curves from which all others can be projected are symmetrical with respect to a center for reduction of cubis; developed new theory of conical sections and surfaces of 2d degree. Died Paris. France, Dec. 18, 1880.

CHASON, Jacob Leon, Am. physician; b. Monroe, Mich., May 12, 1915; s. Ben and Ida (Beiser) C.; A.B. with distinction, U. Mich., 1937, M.D., 1940; m. Helen Pelok, May 19, 1942; children—Steven, Ellen, David. Vis. fellow neuropathology Mayo Clinic, 1952; faculty Wayne State U., Detroit, 1949—, prof. neuropathology, 1958—, dir. resident tng., 1961-64, acting chmn. dept. pathology, 1964—. Asso. pathologist William Beaumont Hosp., Detroit, 1963-64; acting chief pathology Detroit Gen. Hosp., 1964—; cons. to hosps. NIH sr. fellow Hammersmith Hosp., London, Eng., 1959-60. Mem. Am. Acad. Neurology, A.M.A., Am. Soc. Clin. Pathologists, Coll. Am. Pathologists, Internat. Acad. Pathology, Sigma Xi. Contbg. author Text on Systemic Pathology, 1958. Research and publs. on mechanisms of responses of nerve cell to mech. injury imparted through intact skull. Home: 4862 Keithdale St., Bloomfield Twp., Mich. 48013. Office: 1400 Chrysler Freeway, Detroit 48207.*

CHASSAIGNAC, Pierre Marie Édouard, French surgeon; b. Nantes, France, 1804; studied medicine in Nantes; became prof., U. Paris. Mem. Acad. de Médecine. Author: Etudes d'anatomie et de pathologie chirurgicale, 1851; Traité de l'écrasement linéaire, 1856; Traité pratique de la suppuration et du drainage chirurgical, 1859; Traité clinique et pratique des operations chirurgicales, 1861-62. Developed effective technique for surg. drainage involving use of rubber tube drains, 1859; described nodule which could be either of large anterior tubercles of 6th cervical vertebra, known as Chassaignac's tubercle, circa 1845. Died Versailles, France, Aug. 26, 1879.

CHATELET, Gabrielle-Émilie (Marquise du), French mathematician; b. Paris, Dec. 17, 1706; d. Baron de Breteuil; student math. under Clairaut, Koenig and Maupertuis; m. Marquis du Chatelet-Lomont, 1725. Lady-in-waiting to the queen; Voltaire's mistress. Author: Institutions de physique, 1740; Dissertation sur la natur et la propagation de feu, 1744; Traduction des principes de Newton, 1756. Helped introduce Newtonian theory in France; research on fire; supported Leibnizian theory of conservation of living forces. Died Luneville, France, Sept. 10, 1749.

CHATIN, Gaspard Adolphe, French botanist; b. nr. Tullins, France, Nov. 30, 1813; D. of Sci., 1839; D. of Pharmacy, 1840; D. Medicine, U. Paris (France), 1844; 1 dau. Joannès. Became pharmacist hosp. Beaujon, France, 1841; became prof. botany l'École supérieure pharmacy, 1848; dir. Sch. Pharmacy, 1873-86. Mem. French Acad. Scis., 1874, Acad. Medicine. Author: Comparative Anatomy of Vegetables, 2 vols., 1856-92; The Water Cress, 1866; The Trufflé, 1892. Research on organic structure of plants, composition of plant sap, respiration of fruits, assimilation of iodine in plants and animals; showed iodine can prevent devel. of endemic goiter and cretinism; classified vegetables by anatomic characteristics. Died Essarts-le-Roi, France, Jan. 13, 1901.

CHATIN, Joannès, French anatomist, zoologist; b. Paris, Aug. 19, 1847; s. Gaspard Adolphe Chatin; Licencié ès scis. naturelles, 1869; Docteur en médecine, École superieure de pharmacie de Paris, 1871, Docteur es scis., 1872; Agrégation, 1874. Tchr. botany École des hautes études, 1874-76; became lectr. anat. zoology Faculty Scis., U. Paris, 1877, later prof. and chmn. anatomy, comparative physiology, prof. histology; head micrographic lab., Havre, France; apptd. insp. meats (for trichina) from Am., 1881. Mem. French Acad. Scis., 1900, Acad. Medicine, Hygienic Council. Author: Les organes des sens dans la série animale, 1880; Contributions experimentales à l'étude de la chromatopsie, 1881; La trichine et la trichinose, 1883; La cellule nerveuse, 1890; La cellule animale, 1892; Les organes de relation chez les invertébrés (les vertébrés), 2 vols., 1894; Les organes de nutrition et de reproduction chez les invertébrés (les vertébrés), 2 vols., 1894; La machoire des insectes, 1897. Studied various aspects of animal anatomy, including snout of mole, proboscis of Lepidoptera. Died Essarts-le-Roi, France, July, 1912.

CHATT, Joseph, Brit. chemist; b. Horden, Eng., Nov. 6, 1914; s. Joseph and Marjory Elsie (Parker) C.; B.A., Emmanuel Coll., Cambridge, Eng., 1937; Ph.D., 1940, M.A., 1941, Sc.D., 1956; m. Ethel Williams, May 31, 1947; children—Elizabeth Mary, Joseph. Research chemist Woolwich Arsenal, 1941-42; with Peter Spence, & Sons Ltd., Widnes, Eng., 1942-46, chief chemist; Imperial Chem. Industries Research fellow Imperial Coll., London, 1946-47; with Imperial Chem. Industries, 1947-62, Akers group mgr. research dept. heavy organic chems. div., 1960-62; dir. unit nitrogen fixation Agrl. Research Council, 1962—; prof. inorganic chemistry Queen Mary Coll., London, 1963-64; prof. chemistry U. Sussex, Brighton, Eng., 1964—; distinguished vis. prof. chemistry Pa. State University, 1960; visiting professor Yale, 1962; Royal Society Leverhulme vis. prof. Univ. Rajasthan, Jaipur, India, 1966-67. Fellow Royal Soc., 1961; mem. Chem. Soc. (past hon. sec., past mem. council), Royal Inst. Chemistry, Soc. Chem. Industry, Am. Chem. Soc. Contbg. author Chemistry of Carbon Compounds, 1956; Organo-Metallic Chemistry, 1960. Research and numerous publs. on organo-transition metal chemistry, bonding ligands to metal atom in co-ordination compounds. Home: 28 Tongdean Av., Hove, Sussex, BN 3 6TN, Eng. Office: Chem. Lab, U. Sussex, Brighton, BN 1 9QJ, Eng.*

CHATTERJA, Jyoti Bhusan, Indian physician; b. Calcutta, India, Feb. 16, 1919; s. N.N. and Kamala (Mukherjee) C.; M.B., B.S., U. Calcutta, 1942, M.D., 1949; m. Irabati Bhattacharjee, Feb. 27, 1943; 1 dau., Bharati. Research officer Indian Council Med. Research, Sch. of Trop. Medicine, Calcutta, 1944-54, prof. haematology, head dept., 1955—, officer-in-charge haematol. unit, 1955—, dir. Sch. Trop. Medicine, 1966—, tchr. U. Coll. Medicine, 1957—. Councillor to Internat. Socs. Haematology, Blood Transfusion and Reticuloendothelial; pres. med. and vet. sect. Indian Sci. Congress, 1964-65. Recipient Coates medal, 1958; Barclay medal, 1964; Amirchand prize, 1964; Minto medal, 1965. Fellow Nat. Inst. India, Indian Acad. Med. Scis., Coll. Pathologists (London); corr. fellow A.C.P.; mem. Indian Soc. Haematology (pres. 1963-64), Indian Assn. Pathologists (pres. 1968). Research and numerous publs. on characterization of nutritional anaemias in India with reference to deficiency of iron, folic acid and vitamin B12; discovered Hb.E. in Bengalees, various types of genetic interaction between Hb.E. and other abnormal haemoglobins; pattern of deficiency of gluco-6-phosphate dehydrogenase in India. Home: No. 116, B Block, Lake Town, Calcutta 55, India.*

CHATTERJEA, S(anti) K(umar), Indian mathematician; b. Howrah, India, Nov. 5, 1930; s. Profulla K. and Saila (Banerjea) C.; M.Sc. in Pure Math., Calcutta (India) U., 1953; Ph.D., Jadavpur U., 1964; m. Momota Banerjea, Mar. 3, 1957; children—Rekha, Dipa, Soma. Prof. math. Bangabasi Coll., Calcutta, 1954—. Recipient 1st award U. Grants Commn., New Delhi, India, 1966-67. Mem. Calcutta Math. Soc., Indian Sci. Congress, Bangia Bijanan Parishad, Unione Matematica Italiana. Author: Pre-University Mathematicis Part I (Algebra), Part II (Trigonometry), 1960; (with M. C. Bagchi) Advanced Algebra, 1961; Lectures in Mathematics, vol. I, 1964; also articles. Staff reviewers Am. Math. Soc., 1961—, Zentralblatt für Mathematik, 1962—. Research on various properties of spl. functions—non-linear recurrence relations and their characterizational nature, definite integrals, Turan and analogous expressions, calculus, generating functions, integral representation for product of 2 functions of same kind, operational formulas and their applications, various generalizations of Laguerre and Bessel polynomials, new techniques of operational methods. Home: 28/1B Serpentine Lane, Calcutta-14, W. Bengal, India.*

CHATTERJEE, Hemendra Nath, Indian physician; b. Chapra, Bihar, India, Aug. 11, 1904; s. Bhabataran and Umasashi (Bannerjee) C.; student Visvavarati U., India, 1912-14, City Coll., Calcutta, India, 1920-22, R. G. Kar Med. Coll., 1922-28; M.B., Calcutta U., 1928, M.D., 1948; m. Nalini Bannersee, Feb. 14, 1927; children—Arun Kumar, Partha Sarathi. Lectr. pathology and bacteriology R. G. Kar Med. Coll., 1933-42; research scientist Standard Pharm. Works, Ltd., 1942-45; vis. physician Calcutta Nat. Med. Coll., 1946-58; sr. physician E.S.I.S. Hosp., Belur, nr. Calcutta, 1966—; officer in charge inquiry on bone marrow, also inquiry on secondary anemias Govt. India Research Fund Assn., 1936-39. Durbhanga Research scholar Calcutta U., 1936-39, Capt. Kalyan Kumar Mukherjee Research fellow, 1939-41; Wathumull Found. Am. scholar, 1948-49; Ghosh Travelling fellow U. Cal., 1948-49; State scholar Govt. W. Bengal, 1948-49, Griffith prize Calcutta U. Fellow Royal Soc. Tropical Medicine and Hygiene; mem. Fedn. Am. Socs. Exptl. Biologists, Am. Assn. Immunologists, Am. Soc. Microbiology, Indian Med. Assn. Author: Studies in the Dimensions of Erythrolytes of Man, 1940; also numerous other books, articles. Developed new treatment of cholera which greatly reduced mortality in India, new treatment of uremia in cholera, new treatment of hyperpyremia in cholera; studies of pathology and biochemistry of cholera, erythrocytes as means of diagnosis, biochemistry and treatment of anemia of pregnancy, bone marrow in different diseases. Home: 9 Romes Mitter Rd., Calcutta, West Bengal. Office: E.S.I.S. Hosp., Beur, Dist. Howrah, West Bengal, India.*

CHAUCER, Geoffrey, English poet; b. London, Eng., circa 1340; s. John C.; m. Philippa de Roet, circa 1366; children—Thomas, Elizabeth, Agnes, Lewis. Became member household of Prince Lionel, from 1357; joined the army of Edward III, 1359; captured by the French and ransomed; made diplomatic missions to the continent at least 9 times, 1368-87; comptroller of customs on wool, skins, hides, London, 1374-86; clk. of king's works, 1389-91. Elected knight of the shire of Kent, 1386. Author: Book of the Duchess, 1369; Treatise on Astrolabe, circa 1391; The Legend of Good Women; Troilus and Criseyde; Canterbury Tales; many other works. Best known English poet of the middle ages; knew contemporary astronomical and astrological doctrine; prepared treatise on astrolabe. Died Westminster, Eng., Oct. 25, 1400.

CHAUFFARD, Anatole Marie Emile, French physician; b. Avignon, France, Aug. 22, 1855; ed. at Med. Sch., Paris, France; doctor medicine, 1882; apptd. physician of hosps., 1883; became instr., 1886; doctor Paris Hosp.; became chief of service St. Perine, Broussais, 1889, Lenac, 1894, Cochin, 1895 (all France); prof. clin. medicine Faculty of Paris; prof. history of medicine, 1909. Recipient gold medal for internship, 1881. Author: Crises in Illness, 1886. Research on illness of liver and cardio-renal apparatus, gastric determinations of typhoid fever, metabolism; described familial hemolytic jaundice (now known as Chauffard-Minkowski disease), 1899. Died Paris, Nov. 1, 1932.

CHAUFFARD, Marie Denis Etienne Hyacinthe, French physician; b. Avignon, France, Dec. 26, 1796; s. Denis Francois Agricol and Rose (Chapuis) C.; studied at Hosp. of Nîmes; doctorate Montpellier, France, 1818; 1 son, Paul Emile. Practiced medicine in Avignon; chief doctor at hosps. and prisons of Avignon; pres. hygiene council; prof. anatomy; mayor of Avignon, 1847-48. Publs. include: Treatise of Internal Inflammations, 2 vols., 1831; Resumé of Practical Medicine, 1833. Mem. Acad. Medicine (became corr. 1835 and nat. asso. 1871). Died Dec. 14, 1880.

CHAUFFARD, Paul Emile, French physician; b. Avignon, France, May 18, 1823; s. Marie Denis Etienne Hyacinthe C.; studied medicine, Paris, France; doctorate, 1846. Became physician in chief hosps. of Avignon, 1856; named physician hosps. of Paris, 1861; apptd. insp. gen. pub. instrn., 1874; received chair pathology and gen. therapeutics Faculty of Paris, 1871; founder Mixed Faculty of Lyons (France), 1877. Recipient Gold medal for internship. Mem. Acad. Medicine, 1874. Author: Principles of General Pathology, 1862; Of Spontaneity and Specifics of Illness, 1867; Life, 1873; also writings on history and med. philosophy. Advocate of vitalism. Died Feb. 6, 1879.

CHAULIAC, Guy de, see de Chauliac.

CHAULNES, Louis Marie Joseph Romain d'Albert d'Ailly (Duc de Picquigny); French chemist; b. 1741; s. Michel Ferdinand Chaulnes; apptd. camp marshal, 1756. Fellow Royal Soc., 1764. Author Mémoire sur la véritable entrée du monument qui se trouve non loin du Caire, pres de Sakhara, 1777. Research on carbonic acids and salts; discovered way of obtaining acidulated water, also method for extraction and purification of salts in urine. Died Oct. 24, 1792.

CHAULNES, Michel-Ferdinand d'Albert d'Ailly, French phys. scientist; b. Dec. 31, 1714; s. Louis-Auguste and Marie-Anne-Romaine (de Beamanou de Lavarden) C.; m. Anne Josephe Bonnier de la Mosson, 1734; 1 son, Marie Joseph d'Albert d'Ailly. Profl. natural history; invented electric machine; made obsur le soleil; also other publs. Built improved instruments for astron. observations; research on physics, soldier; named peer of France, 1745; apptd. lt. gen., gov. of Picardie, 1752. Mem. French Acad. Scis. (hon.), 1743. Author: Observations du passage de Vénus servations using microscope. Died Paris, France, Sept. 23, 1769.

CHAUMETON, François Pierre, French physician; b. Changé-sur-Loire, Sept. 20, 1775; dir. Dictionnaire des sciences Médicales for several years; named surgeon 3d class in army, western France, 1793; left army to study surgery in mil. hosp., Metz; rejoined army as pharmacist on coast of Cherbourg; later studied natural scis., history of medicine and ancient langs.; doctor to French troops from Holland. Author: Essai d'entomologie médicale, 1805; Flore du dictionnaire des sciences médicales, 1813-20; also other publs. Contbg. author: Universal Biography; Dictionary of Medical Sciences. Died Aug. 10, 1819.

CHAUMETTE, Antoine, French surgeon; b. Vergesac, France; flourished 1560; studied medicine, Montpellier, France, also Paris; practiced surgery in Vergesac. Author: Enchiridion chirurgicum, externorum morborum remedia complectens (summary of surgery), 1564.

CHAUSSIER, François, French physician; b. Dijon, France, July 2, 1746; s. Jean Baptiste and Catherine (Mortenne) C.; M.D., Besancon, 1780. Reorganized (with Fourcroy) med. instrn. during French Revolution; prof. Paris Faculty Medicine; named 1st physician Maternity Hosp., 1804; prof. chemistry École Polytechnique, Paris; surgeon prisons, physician hosps., Dijon; pub. prof. anatomy, Dijon. Mem. French Acad. Scis., 1796. Author: Consultation Médico-Légale sur une accusation d'enfanticide, 1785; Exposition sommaire des muscles suivant la classification et la nomeclature methodiques adoptées as cours d'anatomie de Dijon, 1789; Observations chirurgico-légales sur un point important de la juresprudence criminelle, 1790; Exposition sommaire de la structure et des differentes parties de l'encéphale ou cerveau, 1807; Manuel médico-lègal des poisons, 1824. Described median raphe of corpus callosum; described hydrogen expts. to Dijon Acad., 1777; ignited hydrogen with elect. spark; reported carbon dioxide is formed by heating wood or gum, 1780; discovered sodium thiosulphate by action of sulphur dioxide on sodium sulphide solution, 1799. Died Paris, June 19, 1828.

CHAUVEL, Jules-Fidèle Marie, French surgeon; b. Quintin, France, June 9, 1841; studied medicine at Rennes, France; student sch. in mil. service at Strasbourg, France; doctorate, 1862. Employed by l'École du Val-de-Grâce; dir. health service of Mil. Govt. of Paris; physician at infirmary of Invalides; participated in Franco-German War. 1870; became agregate prof. med. operations, 1873; prof. operations and formal preparations at Val-de-Grâce, 1880-90, introduced ophthalmology; named dir. govt. and mil. health service of Paris, also mem. tech. health and attendance service, 1895-1903. Mem. Soc. Surgery, Ophthalmol. Soc., Acad. Medicine. Author: Precise Surgical Operations, 1877; The Precise Theory and Practise of the Eye Examination; Ophthalmological Studies; Treatise on Surgical Practise in the Army, 1890. Studied war and medicine including infection, amputation, effect of contaminated air. Died Paris, 1908.

CHAUVENET, Édouard Jérôme, French chemist; b. Bompas (Pry.-Or.), France, Oct. 13, 1881; s. Auguste and Marie (Vidal) C.; ed. U. Montpellier, U. Paris; Ph.D. in sci.; m. Marie Sarrbayrouze, Oct. 10, 1905; 1 son, Robert. Inst., Sch. Sci. of Besançon; prof. at Caen; dir. Inst. Chemistry and Tech.; dean Sch. Sci. at Caen; founder Inst. Chemistry of Caen. Recipient gold medal from acads. of Genoa and Argentina. Mem. Chem. Soc. France (pres. Normandy sect.); hon. mem. acads. of Genoa, Milan, Bari, Audbreusio. Author book on phys. chemistry. Research and numerous publs. on phys. and organic chemistry, plastics, milk and fruits. Address: 4, place Monseigneur-des-Hameaux, Caen (Calvados), France.

CHAUVENET, William, Am. astronomer, mathematician; b. Milford, Pa., May 24, 1820; s. William Marc and Mary B. (Kerr) C.; grad. with high honors Yale, 1840; m. Catherine Hemple, 1842. Prof. math. Naval Asylum, Phila., 1841, serving on U.S.S. Mississippi; influenced orgn. of U. S. Naval Acad., 1845, chmn. dept. math. and astronomy, 1853-55; prof. Washington U., St. Louis, 1855-69, chancellor, 1862-69; elected mem. Am. Philos. Soc., Am. Acad. Arts and Scis.; an incorporator Nat. Acad. Scis.; pres. A.A.A.S. Author: A Treatise on Plane and Spherical Trigonometry, 1850; Manual of Spherical and Practical Astronomy, 1863; A Treatise on Elementary Geometry with Appendices Containing a Collection of Exercises for Students and an Introduction to Modern Geometry, 1870. Died St. Paul, Minn., Dec. 13, 1870.

CHAVANNE, Joseph, geographer; b. Graz, Austria, Aug. 7, 1846; studied in Prague, Czechoslovakia, also Graz. Travelled through N. and Central Am., 1867-69, Africa, from 1870; later employed by Central Meteorol. Bur., Vienna, Austria; became journalist Vienna Geog. Soc. periodical, 1875; with Rundschau für Geographie und Statistik, Germany, 1881. Author: Die Sahara, 1878; Afghanistan, 1879; Afrikas Ströme und Flüsse, 1881; Geographische Charakter-Bilder, 1886; Carte générale de l'Afrique Oesterreich-Ungarn, 1882-86. Made charts and maps of the Sahara, Congo and Morroco. Died Buenos Aires, Argentina, Dec. 7, 1902.

CHAVEUA, (Jean Baptiste) Auguste, French physiologist; b. Villeneuve-la-Guyard, France, Nov. 21, 1827; ed. L'Ecole Veterinaire, Alfort, France; M.D., Lyons, France, 1875; became chief of service at a sch. in Lyons, 1848; named prof. anatomy and physiology L'Ecole Veterinaire, 1864; placed in charge of study of bovine pest, 1865; named insp. gen. of vet. schs., 1866; became dir. L'Ecole Veterinaire of Lyons, 1875, also prof.; prof. exptl. pathology Mus. Natural History; named prof. exptl. and comparative medicine, Lyons, 1875. Mem. French Acad. Scis., 1878, also Acad. Medicine. Author: Treatise on Comparative Anatomy of Domestic Animals, 1855; The Muscular Work and Its Representative Energy, 1891; The Life and Energy within the Animal, Introduction to the Study of Sources of the Transformations of the Directional Force while Laboring in Physiological Work, 1894. Founder Rev. of Tb, Jour. Gen. Physiology and Pathology. Research on comparative anatomy, infectious diseases, digestive Tb; pioneer in support of pasteurization; studied (with Marey) phenomena of circulation; investigated problems of gen. physiology. Died Paris, Jan. 5, 1917.

CHAVIN, Walter, Am. biologist; b. N.Y.C., Dec. 6, 1925; s. Isidor and Fanny (Kesch) C.; B.S., Coll. City N.Y., 1946; M.S., N.Y. U., 1949, Ph.D., 1954. Instr. dept. biology Coll. City N.Y., 1946-47; research asst. N.Y. Aquarium, N.Y. Zool. Soc., N.Y.C., 1948-49; instr. dept. zoology U. Ariz., Tucson, 1949-51; research specialist Am. Mus. Natural History, N.Y.C., 1951-53; research asso., cons. Argonne Nat. Lab., AEC, Chgo., 1955-58; prof. endocrinology and radiation biology dept. biology Wayne State U., Detroit, 1953——. Chmn. radiation biology dirs. AEC; cons. Am. Inst. Biol. Scis.; panelist, referee NSF. Grantee, Am. Cancer Soc., AEC., Damon Runyon Fund, NSF, USPHS, Pharm. Industry; NSF sr. postdoctoral fellow, Sorbonne, 1960-61. Fellow A.A.A.S.; mem. Am. Soc. Zoologists (mem. exec. com. div. comparative endocrinology), Inst. in Radiation Biology (dir. 1958——), Mich. Nucleonic Soc. (pres. 1965——, mem. council 1961——), Am. Inst. Biol. Scis., Am. Soc. Zoologists. Endocrine Soc., Soc. Endocrinologique de France, Mich. Acad. Arts and Sci., Radiation Research Soc., Soc. Exptl. Biology and Medicine, Aerospace Med. Assn., Am. Physiol. Soc., Sigma Xi. Author: Anatomy and Physiology Laboratory, 1960; also numerous articles. Research on pigment cell physiology, radiation effects and endocrine physiology of fish. Home: 1368 Joliet Pl., Detroit 48207.*

CHAVIRA, Enrique Navarrete, Mexican astronomer; b. Mexico City, Mexico, June 5, 1925; s. Alvaro Chavira and Mercedes (Navarrete) Liceaga. Bachiller, Nat. Autonomous U. Mexico, 1949; m. Luz Ma. Martinez Suarez, Jan. 28, 1956; children—Elsa, Elizabeth, Anabella, Denev. With Nat. Astron. Obs., Nat. Autonomous U. Mexico, Mexico D.F., 1940——, investigator, 1948——; staff Nat. Astrophys. Obs. Pub. Edn., Puebla, Mexico, 1952——, sec., 1956——. Mem. Internat. Astron. Union, Am. Astron. Soc., Astron. Soc. Pacific. Research, publs. on observation of OB stars, blue stars, variable stars, flare stars, discoveries of supernovae and infrared objects. Home: 2128 20 Sur, Puebla, Puebla, Mexico.

CHAWLA, S., Indian physician; b. India, Oct. 25, 1927; d. Ishwar Das and Kaushalya (Nirula) Grover; M.B.B.S., Lady Hardinge Med. Coll., 1950; D.M.R., Bernard Inst. Radiology, 1957; m. T.P.S. Chawla, July 11, 1960. Head, Palam Maternity and Child Welfare Center, Delhi, 1952; asst. Lady Hardinge Med. Coll., New Delhi, 1953-55, lectr., 1957-59, prof., head radiology dept., 1962——; asst. prof. All India Inst. Medicine, New Delhi, 1959-62. Mem. Indian Acad. Med. Scis., Indian Soc. Gastroenterology, Indian Assn. Radiology, Assn. Med. Women in India. Research, publs. on intestinal Tb, rare congenital and fetal abnormalities. Address: 24 Lady Hardinge Rd., New Delhi 1, India.*

CHAYKA, Vergeniy Ivanovich, Russian pathoanatomist; b. 1902; grad. Kiev Med. Inst., 1927; D.Med. Sci., 1940. Head, Central Lab. and Prosectorium, October Revolution Hosp., Kiev, 1932-40; lectr. dept. path. anatomy Kiev Med. Inst., 1934-40, prof., head chair path. anatomy, 1940——, dep. dir. for sci. work and studies, 1941-44. Cons., Kiev Oblast Dept. Health; mem. learned med. council Ukrainian Ministry Health. Mem. Kiev Soc. Pathoanatomists (chmn.). Author of 50 works including The Connective Tissue Skeleton of the Heart and its Importance in Pathology, 1940. Address: Kiev Med. Inst., b. Tarasa Shevchenko 13, Kiev, Ukrainian SSR, USSR.

CHAZALLON, Antoine-Marie-Remi, hydrographer, cartographer; b. Desaignes, France, Jan. 17, 1802; s. Etienne and Marguerite (Patouillard) C.; ed. Poly. Sch., 1822; hydrographical engr., 1824. Began cartographic work on Atlantic coastline of Europe, Beatemps-Beaupré, 1824; named 1st class engr., 1842; elected to Constituent Assembly, Ardèche; cartographic work at Havre, France, 1853-56. Decorated Legion of Honor, 1861. Mem. French Acad. Scis., 1869. Author: Pilote française. Research on tides and secondary tides; calculation of declination; invented a tide chart. Died Desaignes, Dec. 23, 1872.

CHAZELLES, Jean Matthieu de, see de Chazelles.

CHAZY, Jean, French mathematician; b. Villefranche, France, Aug. 15, 1882; s. Claude Chazy; began studies at l'École normale, 1902; received doctorate, 1910; m. Miss Veillard, Mar. 26, 1919; children—Hélène, Claude, Jeanne. Lectr., Grenoble, France, then Lille, France; prof. differential and integral calculus at Lille; prof. rational mechanics Faculty of Sci., Sorbonne, Paris, from 1934; prof. analytical math. Central Sch. Arts and Manufacture, Paris; examiner mechanics École Polytechnique. Decorated Legion of Honor. Mem. French Acad. Scis., 1937 (3 times Laureate), Math. Soc. (became pres. 1934). Author: The Theory of Relativity and Celestial Mechanics, 1928; Course of Rational Mechanics, 1933; numerous publs. on differential equations. Research on peripheral movement of Mercury, application of theory of relativity to celestial mechanics, differential equations, problem of 3 bodies, displacement of perihelion of Mercury. Died Paris, 1955.

CHEADLE, Vernon Irvin, Am. botanist; b. Salem, S.D., Feb. 6, 1910; s. Henry Melvin and Inez (Engleman) C.; A.B., Miami U., 1932; M.A., Harvard U., 1934, Ph.D., 1936; student S.D. State Coll., 1927-28; m. Mary Jenkins Low, Dec. 23, 1939; 1 son, William Gerald. Field supr. barberry eradication U. S. Dept. Agr., S.D., 1931; Austin teaching fellow Harvard, 1933-36; instr. botany R.I. State Coll., 1936-41, asst. prof., 1941-42, prof., head dept., 1942-52, dir. grad. div., 1943-52; prof. botany U. Cal., Davis, 1952-62, chmn. dept., 1952-60, acting vice chancellor, 1961-62; chancellor U. Cal., Santa Barbara, 1962——. Fulbright fellow, 1959. Mem. Am. Acad. Arts and Scis., Am. Assn. U. Profs., Am. Inst. Biol. Scis., Am. Soc. Naturalists, Am. Soc. Plant Taxonomists, Internat. Soc. Plant Morphologists, Soc. Study Evolution, Soc. Study Growth and Devel., Torrey Bot. Club, Phi Beta Kappa, Sigma Xi. Author (with E. A. Palmatier) Discussion Manuals, Botany I & II, 1949-50, Discussion Manual Biology I, 1952. Research, publs. in anatomy of conducting tissue in wood and bark. Home: 543 Channel Islands Rd., U. Cal., Santa Barbara, Cal. 93106.*

CHEADLE, Walter Butler, English physician; b. Colne, Eng., Oct. 15, 1835; s. James and Eliza (Butler) C.; B.A., Caius Coll., Cambridge (Eng.) U., 1859, M.B., 1861; m. Anne Murgatroyd, Jan. 31, 1866; m. 2d, Emily Mansel, Aug. 4, 1892; 4 sons. Staff, St. Mary's Hosp., 1869-92, dean med. sch., 1869-73. Fellow Royal Coll. Physicians. Author: The North-West Passage by Land (based on explorations of western Can. 1862-64), 1865; On the Principles and Exact Conditions to be Observed in the Artificial Feeding of Infants, and the Diseases which arise from Faults of Diet in Early Life, 1899. First to define nature of infantile scurvy, 1877; wrote on infantile rheumatism, 1899. Died Mar. 25, 1910.

CHEBYSHEV, Pafnuti Lvovich, Russian mathematician; b. Borovks, Russia, 1821; ed. U. Moscow (Russia). Became prof. math. U. St. Petersburg (Russia), 1859-80. Mem. Inst. France (corr. 1860, asso. fgn. mem. 1874), Royal Soc. London (fgn.). Research and publs. on convergence of Taylor series, prime numbers, problem of obtaining rectilinear motion by linkage, probability theory, quadratic forms, integral theory, gearings, constrn. geog. maps; devised a straight line motion; proved Bertrand's postulate. Died 1894.

CHEEK, Donald Brook, pediatrician; b. Adelaide, South Australia, Apr. 12, 1924; s. Roydon Arthur and Olive (Brook) C.; M.B., B.S., Adelaide U., 1947, M.D., 1953; D.Sc., U. Cin., 1955; m. Mary Ellen Whitmore, June 20, 1952; children—Suzanne Valerie, Cecelia Anne. Practice medicine, specializing in pediatrics, Cin., 1953-56, Dallas, 1957-59, Melbourne, Australia, 1959-62, Balt., 1962——; research fellow, sr. research asso. Childrens Hosp. Research Found., 1953-56; faculty U. Tex., 1957-58; dir. research unit Royal Melbourne Childrens Hosp., 1959-62; asso. prof. pediatrics Johns Hopkins, 1962——; pediatrician, dir. div. growth Johns Hopkins Hosp.; prin. investigator research grant NIH, cons. nutrition study sect. Recipient Borden award Am. Acad. Pediatrics, 1968. Diplomate Am. Bd. Pediatrics. Mem. Soc. Pediatric Research, Australian Physiol. Soc., Australian Med. Assn., Australian Pediatric Assn., Johns Hopkins Med. Soc. Author: Human Growth, Body Composition, Cell Growth, Energy and Intelligence, 1968. Publs. on studies of homeostatic changes in acrodynia, measurement of tissue chloride and extracellular fluid in children, blood adrenalin in infants; muscle cell size in various conditions. Home: Box 71, Gibson Island, Md. 21056. Office: Johns Hopkins Hosp., Balt. 21205.*

CHELTON, Dudley Boyd, Am. cryogenic engr.; b. Balt., July 17, 1928; s. Louis G. and Anna (Shaw) C.; B. Mech. Engring., Ohio State U., 1948; M.S., Mass. Inst. Tech., 1949; m. Jean N. Hays, May 31, 1947; children—Dudley Boyd, Cynthia Anne. Research asso. Los Alamos Sci. Lab., 1950-51; chief cryogenic systems sect. Nat. Bur. Standards Inst. for Materials Research, Boulder, Colo., 1951——. Cons. in cryogenic engring., cons. to aerospace activities. Recipient Gold medal U. S. Dept. Commerce, 1953. Mem. Am. Inst. Aeros. and Astronautics, Sci. Research Soc. Am.,

Sigma Xi, Tau Beta Pi. Contbns. to field of cryogenic engring. concentrating on systems and processes; cryostats, liquid hydrogen bubble chambers, liquefaction of gases, cryogenic refrigeration systems, liquid hydrogen tech. and safety. Home: 806 Brooklawn Dr. Office: Nat. Bur. Standards, Cryogenics Div., Boulder, Colo. 80302.*

CHEN, Di, physicist; b. Chekiang, China, Mar. 15, 1929; s. Hsun-yu and Chien (Wang) C.; came to U. S., 1954, naturalized, 1965; B.S. in Elec. Engring., Nat. Taiwan U., Taipei, 1949-52; M.S., U. Minn., 1956; Ph.D., Stanford, 1959; m. Lynn C. Wang, June 14, 1958; children—Andrew An-ju, Daniel Ting-yuan. Asst. prof. elec. engring. dept. U. Minn., Pls., 1959-62; sr. prin. research scientist Honeywell Corp. Research Center, Hopkins, Minn., 1962—. Mem. I.E.E.E., Am. Inst. Physics, Sigma Xi, Eta Kappa Nu. Research and publs. on emitting-sole magnetron amplifier, magneto-optic and magnetic properties of thin films of manganese bismuth; invented absorption type laser modulator. Home: 5731 Woodland Rd. Office: 500 Washington Av. S., Hopkins, Minn. 55343.*

CHEN, Francis F., physicist; b. Canton, China, Nov. 18, 1929; s. M. Conrad and Evelyn (Chu) C.; came to U. S., 1936, naturalized, 1953; A.B., Harvard, 1950, M.A., 1951, Ph.D., 1954; m. Edna C. S. Lau, Mar. 31, 1956; children—Sheryl, Patricia Ann, Robert. Staff mem. Plasma Physics Lab., Princeton, 1954-64, research physicist, 1964—; vis. physicist Centre d'Etudes Nucleaires, Fontenay-aux-Roses, France, 1962-63; vis. prof. City Coll. N.Y., 1966-67. Mem. Am. Phys. Soc., Phi Beta Kappa, Sigma Xi. Contbr. chpt. to Plasma Diagnostic Techniques, 1965. Research on theory and expt. relating to instabilities of a plasma in a magnetic field and escape of plasma from magnetic bottles; devel. of method of Langmuir probes for plasma measurements. Home: Autumn Hill Rd. Office: Plasma Physics Lab., P.O. Box 451, Princeton, N.J. 08540.*

CHEN, Ko Kuei, pharmacologist; b. Shanghai, China., Feb. 26, 1898; s. Jui Chien and Ta Chi (Chou) C.; B.S., U. Wis., 1920, Ph.D., 1923, Sc.D., 1952; M.D., Johns Hopkins, 1927; Sc.D., Phila. Coll. Pharmacy, 1946; m. Amy Ling, July 15, 1929; children—Thomas T., Mei (Mrs. Grant V. Welland). Came to U. S., 1918, naturalized, 1947. Sr. asst. Peking Union Med. Coll., 1923-25; asso. Johns Hopkins, 1927-29; dir. pharmacologic research Eli Lilly & Co., Indpls., 1929-63; prof. pharmacology Ind. U., Indpls., 1937—. Recipient Remington Honor medal N.Y. br. Am. Pharmacological Assn., 1965. Mem. A.M.A., Am. Soc. Pharmacology and Exptl. Therapy (past pres.), Am. Phys. Soc., A.A.A.S., German Pharmacological Soc., Swiss Chem. Soc., Academica Sinica. Author: (with C. F. Schmidt) Ephedrine and Related Substances, 1930; also numerous articles. Research on ephedrine, new synthetic analgesics, vasoconstrictors, digitalis, cardiotonic glycosides; developed cyanide poison antidotes. Home: 7975 Hillcrest Rd., Indpls. 46240. Office: 1100 W. Michigan St., Indpls. 46207.

CHEN, Yu Why, mathematician; b. Nantungchow, China, Apr. 1, 1910; s. Yin-ko and Wu (Chen); Ph.D., U. Gottingen (Germany), 1934; m. Cho-Tchin Tsiang, Oct. 21, 1938; 1 son, Victor. Came to U. S., 1945, naturalized, 1951. Prof. math. Nat. U., also S.W. U., Peking, China, 1936-45; research fellow Courant Inst. Math. Sci., Inst. for Advanced Study, Princeton, N.J., 1946-50; asso. prof. U. Okla., Norman, 1950-51; prof. math. Wayne State U., Detroit, 1952—. Mem. Am. Math. Soc., Math. Assn. Am., A.A.A.S. Research, publs. in theory and applications of partial differential equations. Home: 16207 Ohio St., Detroit 48221.*

CHEN, Yung Ming, Chinese mathematician; b. Peking, China, June 22, 1924; s. Chun Sen and Mo Tak (So) Chen; B.Sc. in Aero. Engring., Nat. Chiao Tung U., Shanghai, 1947; D.Sc., Hokkaido U., 1961. Tchr. math. Pin Ching Middle Sch., Kowloon, Hong Kong, 1948-56; lectr. math. Nanyang U., Singapore, 1956-58; tutor math. U. Hong Kong, 1958-60, asst. lectr., 1960-63, lectr., 1963—. Reviewer Zentralblatt für Mathematik, 1962—. Generalized Hardy-Littlewood maximal theorem and Riesz theorem on conjugate functions; devised spl. asymptotic approximation method useful for generalizing some theorems concerning functions in lower form. Home: A4, Eastbourne Ct., Eastbourne Rd., Kowloon, Hong Kong.*

CHENEVIX, Richard, chemist, mineralogist; b. Ireland, 1774; s. Col. Chenevix; student Dublin (Ireland) U.; m. Countess of Ronault, June 4, 1812. Resided in Paris, France, 1808. Recipient Copley Gold medal. Fellow Royal Soc., 1801, Royal Soc. Edinburgh, Irish Acad. Contbr. articles to French chem. jours., from 1798, English jours., from 1800. Principal research on analysis of metals; formed an alloy of palladium and mercury. Died Apr. 5, 1830.

CHENEY, Clarence Orien, Am. psychiatrist; b. Poughkeepsie, N.Y., 1887; s. Albert Orion and Caroline (Adriance) C.; A.B., Columbia, 1908, M.D., 1911; m. Josephine Scott, June 7, 1915; 1 son, Robert Scott. Asst. physician and pathologist Manhattan State Hosp., 1912-17; asst. dir. N.Y. State Psychiat. Inst., 1917-22; asst. supt. Utica State Hosp., 1922-26; supt. Hudson River State Hosp., Poughkeepsie, 1926-31; dir. N.Y. State Psychiat. Inst. and Hosp., 1931-36; dir. N.Y. Hosp., Westchester Div., 1936-46, emeritus, 1946-47; instr. psychiatry Cornell Med. Sch., 1917-22, Syracuse Med. Sch., 1922-26; prof. clin. psychiatry, Columbia, 1931-33, prof. psychiatry, 1933-36; prof. clin. psychiatry, Cornell Med. Sch., 1936-47. Fellow A.C.P., N.Y. Acad. Medicine, Am. Psychiat. Assn. (sec.-treas. 1928-33; chmn. bd. examiners 1933-38; pres. 1935-36); mem. N.Y. Psychiat. Soc. (pres. 1943-44), N.Y. Soc. for Clin. Psychiatry (pres. 1934-35), N.Y. Neurol. Soc., A.M.A., N.Y. State Med. Soc. (exchmn. sect. on neurology and psychiatry), Westchester County Med Soc., Assn. for Research in Nervous and Mental Diseases, Royal Medico-Psychol. Assn. (Eng.) (corr. mem.). Sigma Xi, Alpha Chi Rho, Alpha Omega Alpha. Editor: Outlines of Psychiatric Examinations, 1934. Asso. editor of Psychiat. Quar. and Am. Jour. Psychiatry. Died Nov. 4, 1947.

CHENEY, Ralph Holt, Am. biologist; b. Maynard, Mass., June 3, 1896; s. Levi Raymnd and Mary E. (Billington) C.; B.S., Boston U., 1918, A.M., 1919; D.Sc., L.I. U., 1962; M.S., Harvard, 1922, Sc.D., 1923; m. Agnes Lyford Gray, June 11, 1924. Instr., Western Res. U., Cleve., 1923; asst. prof. N.Y. U., 1924-29; prof., chmn. biology dept. L.I. U., Bklyn., N.Y., 1929-46, prof. biol. scis. Coll. Pharmacy, 1932-66, seminar professor, adviser biology students Brooklyn center of Long Island University, 1966—; investigator physiology and pharmacology Marine Biol. Lab., Woods Hole, Mass., 1929—; resident in econ. plants Bklyn. Bot. Garden, 1930-51, hon. asso., 1951—; asso. prof. biology Bklyn. Coll., 1946-48, prof. biology, 1949-66, emeritus, 1966—. Cons. to beverage industry, 1925—, Applied Physiology and Pharmacology, 1930—. Fellow A.A.A.S., N.Y. Acad. Scis.; mem. Soc. Gen. Physiologists, Soc. Cell Biology, Am. Soc. Pharmacology and Exptl. Therapeutics, American Society of Zoologists, Bot. Soc. Am., Torrey Botany Club (president 1967), Phi Beta Kappa, Sigma Xi. Author: Coffee, 1925; also numerous articles. Adv. editor Econ. Botany, 1947-57; Am. collaborator Acta Phytotherapeutic, 1952—, Quar. Jour. for Crude Drug Research, 1961—. Research on effects caffein and caffein-containing beverages on muscles, and nerves of frog, rat, and man, also comparative effects caffeine, theobromine, and theophylline on gametes, effects gamma and ultraviolet irradiation, and radiomimetic chems. on sperm and eggs and early devel. of Arbacia, medicinal plants. Home: 612 Ocean Av., Bklyn. 11226; (summer) Marine Biol. Lab., Woods Hole, Mass. 02543. Office: L.I. U., Bklyn. Center Campus, Bklyn. 11201.*

CHENG, Chia Chung, chemist; b. Tientsin, China, May 5, 1925; s. K. L. and Chui-yuen (Chien) C.; B.S., Nat. U. Chekiang, 1948; M.A., U. Tex., 1951, Ph.D., 1954; m. Katherine K. S. Cheng, May 30, 1953; children—Amy Yuwei, Anna Yumin, Alice Yuray. Came to U. S., 1949. Faculty dept. chemistry N.M. Highland U., Las Vegas, 1954-57; research asso. dept. chemistry Princeton, 1957-59; head med. chemistry sect. Midwest Research Inst., Kansas City, Mo., 1959—. Mem. A.A.A.S., Am. Chem. Soc., N.Y. Acad. Scis., Sci. Research Soc. Am., Sigma Xi, Phi Lambda Upsilon. Contbr. numerous articles to profl. jours. Research on medicinal and organic chemistry programs involving the synthesis and evaluation of antibiotics, antimetabolites, alkylating agts., vitamin analogs, natural products, anticancer and antimalarial compounds. Home: 6028 Walnut St., Kansas City 64113. Office: 425 Volker Blvd., Kansas City, Mo. 64110.*

CHENG, David K., engineer; b. Kiangsu, China, Jan. 10, 1918; s. Han J. Cheng and Ying Hsui; B.S.E.E., National Chiao-Tung University, 1938; S.M., 1944, Sc.D., 1946, Harvard Univ.; m. Enid Kwok, Mar. 27, 1948; son, Eugene Alan; came to U. S., 1943. Research engr., Central Radio Corp., National Resources Commission of China, 1938-43; electronics engr. and project engr., U.S.A.F. Communication Lab., Cambridge, Mass., 1946-48; asst. prof. elec. engrg., Syracuse Univ., 1948-51; assoc. prof., Syracuse, 1951-55; prof., 1955—; research project dir., Syracuse Univ. Research Institute, 1949—. Consultant, IBM Corp., 1952-53, General Electric Co., 1957-61; consulting editor, Addison-Wesley Publishing Co., 1961—. Fellow, Inst. of Elec. and Electronics Engrs.; A.A.A.S.; mem, N.Y. Acad. Scis.; Sigma Xi; Am. Soc. Engrg. Education; Am. Assoc. Univ. Prof. Recipient numerous scholarships, awards. Author: Analysis of Linear Systems, 1959. Investigation of Z-transform theory of antenna arrays; phase-error effects in microwave reflectors; signal-processing array synthesis; gain and signal-to-noise ratio optimization techniques; electromagnetic problems in plasma environments; diffraction field measurements with light-modulated scattering techniques. Home: 102 Harpers Court, DeWitt, New York, 13214. Office: Electrical Engineering Dept., Syracuse Univ., Syracuse, N.Y., 13210.*

CHENG, Ping-Yao, chemist; b. Amoy, China, Sept. 15, 1921; s. Foo-sing and Bin (Chang) C.; B.S., Amoy U., 1945; M.S., U. Ore., 1950; Ph.D. (Rosenberg fellow) U. Cal., Berkeley, 1953; m. Yueh-Lai Tsai, Sept. 20, 1952; children—Jye-Shern, Myra. Came to U. S., 1949, naturalized, 1963. Mem. research faculty U. Cal., Berkeley, 1951-54; mem. staff Rockefeller Found., N.Y.C., 1956-64; biophysicist Lawrence Radiation Lab., U. Cal., Livermore, 1964—. Nat. Found. Infantile Paralysis fellow Cal. Inst. Tech., 1954-56. Mem. Am. Phys. Soc., Am. Chem. Soc., Am. Soc. Biol. Chemists, Biophys. Soc. Research in sedimentation velocity, viscosity, diffusion and optical rotatory dispersion, size, structure and functions of viruses, enzymes and nucleic acids. Home: 888 Caliente Av., Livermore, Cal. 94550.*

CHENG, Sin-I, mech. engr.; b. Changchow, China, Dec. 28, 1921; s. Mou-yi and Shou-jen (Hsu) C.; came to U. S., 1948, naturalized, 1963; B.S., Chiaotung U., Shanghai, China, 1943; M.S., U. Mich., 1949; M.A., Princeton U., 1952, Ph.D., 1952; m. Jean Sing, Oct. 25, 1950; children—Andrew, Thomas, Irene. Faculty Princeton (N.J.) U., 1951—, prof. aero. engring. 1960—. Cons. to industry. Asso. fellow Am. Inst. Aeros. and Astronautics. Author: (with L. Corcco) Combustion Instability in Rocket Motors, 1956; also articles. Research in gas dynamics and propulsion scis., nature and origin of sudden failure of rocket motors in operation, probable remedies. Home: 379 Prospect Av., Princeton, N.J. 08540.*

CHENG, Thomas Clement, biologist; b. Nanking, China, Nov. 5, 1930; s. James T. and Dorothy (Lee) C.; came to U. S., 1946, naturalized, 1954; student W.Va. U., 1947-49; A.B., Wayne State U., 1952; M.S., U. Va., 1956, Ph.D., 1958; m. Barbara Ann Schimmel, May 31, 1956; children—Thomas Clement, James Bradford, Allison Ellen. Lab. asst. Wayne State U., 1952; instr. U. Va., Charlottesville, 1957-58; asst. prof. biology U. Md., Balt., 1958-59; asst. prof. Lafayette Coll., Easton, Pa., 1959-62, asso. prof., 1962-64; chief parasitology and immunology sect. N.E. Research Center, USPHS, Narragansett, R.I., 1964-65; asso. prof. zoology, U. Hawaii, Honolulu, 1965—. Mem. Surgeon Gen's Com. on Food Protection, USPHS, 1964-65. Recipient Andrew Fleming award U. Va., 1958; Darbaker award Pa. Acad. Sci., 1961; Roy and Ira Jones award for superior teaching Lafayette Coll., 1962. Mem. Am. Microscopical Soc., Soc. Protozoologists, Am. Soc. Zoologists, Am. Soc. Parasitologists, A.A.-A.S., Helminthological Soc. Washington, Am. Inst. Biol. Scis., N.Y. Acad. Scis., Soc. Exptl. Biology and Medicine, Am. Soc. Limnology and Oceanography, Nat. Shellfisheries Assn., Sigma Xi, Phi Sigma. Author: The Biology of Animal Parasites, 1964; Marine Molluscs as Hosts for Symbioses, 1967; also numerous articles. Editor: Some Immunological and Biochemical Aspects of Host-Parasite Relationships, 1963. Mem. editorial com. Pa. Acad. Sci., 1960-65. Research in devel. of parasitic flatworms, physiology of animal parasites, nutritional requirements of parasites, cellular and humoral def. mechanisms of molluscs, exptl. invertebrate pathology, interrelationships among invertebrate animals. Home: 3480 Alani Dr., Honolulu 96822.*

CHENU, Jean Charles, French physician, naturalist; b. Metz, France, Aug. 30, 1808; studied medicine, Paris, France; became mil. physician and surgeon, 1829; served in cavalry, war of the Orient, 1834; took care of collections of Benjamin Delessert; named maj. Aid to Seine police, 1845; participated in the Crimean War. Vice pres. Soc. Aid for Wounded Mil. (wrote report for soc. on the med.-surg. services of ambulances and hosps. during Franco-Prussian War 1870-71). Author: Illustrations conchyliologique, 1847; also publs. on water minerals. Drafted Encyclopédie d'histoire naturelle, 35 vols., 1850-61. Died Paris, Nov. 19, 1879.

CHEPIKOV, Konstantin Romanovich, Russian geologist; b. Jan. 6, 1901; grad. Moscow Mining Acad., 1929. Former oil prospector, Kerch Peninsula and Siberian Ural-Volga region; head Lab. Oil Geology, Inst. Geol. Sci., 1947-54; dep. dir. Petroleum Inst., USSR Acad. Sci., 1954—. Recipient Stalin prize, 1946. Mem. USSR Acad. Sci. (corr.). Author: A Brief Outline of the Geological Structure and the Oil Deposits of the Kerch Peninsula, 1930; The Classification of Upper Permian Redrock by Tetrapod, 1946; The Age of the Ufa Deposits, 1948; co-author: New Findings on Organic Micro-Residues in Oils from the European USSR, 1963, also others. Address: Petroleum Inst., USSR Acad. Sci., Leninsky prospect 29, Moscow, USSR.

CHERASKIN, Emanuel, Am. physician; b. Phila., June 9, 1916; s. Herman and Celia (Homes) C.; A.B., U. Ala., 1939, M.A., 1941, D.M.D., 1952; M.D., U. Cin., 1943; Douter Honoris Causa, Universidade de Sao Paulo; m. Carol Elizabeth Elwood, Sept. 23, 1944; 1 dau., Lisa. Gen. practice medicine, Moundville, Ala., 1947-48, specializing in oral medicine, Birmingham, Ala., 1952—; prof., chmn. div. oral surgery, oral medicine Sch. Dentistry U. Ala., 1956-62, prof., chmn. dept. oral medicine, 1962—; asst. prof. dept. medicine Med. Coll. Ala., 1959—; cons. VA Hosp., Birmingham, Tuskegee, Ala., 1952—, Predictive Medicine Program, Hollywood, Cal., 1965—. Recipient Samuel Charles Miller Meml. Lectr. award Am. Acad. Dental Medicine, 1964. Mem. A.M.A., Am. Dental Assn., Am. Acad. Dental Medicine. Author: (with L. L. Langley) The Physiological Foundation of Dental Practice, 1956, The Physiology of Man, 1965; Diagnostic Stomatology, 1961. Research and publs. primarily in field of health and proneness profiles. Home: 1316 S. 27th Pl., Birmingham 35205. Office: 1919 7th Av. S., Birmingham, Ala. 35233.*

CHERBULIEZ, Emile, chemist; b. Mulhouse, France, Jan. 22, 1891; s. Emile and Emma (Koeckert) C.; student U. Geneva (Switzerland), 1908-09; ing.-

chem., Swiss Fed. Tech. Inst., Zurich, 1913, Dr. rer.-nat., 1917; Dr.phil., U. Munich (Germany), 1915; m. Jeanne Stephani, July 16, 1925; children—Cecile (Mrs. DeHaas), Isaline, Theodore, André. Faculty U. Geneva, 1922—, became prof. pharm. chemistry, 1928, prof. organic and pharm. chemistry, 1952, ret., 1966. Pres., Com. Atomic Sci., 1959-62; v.p. Nat. Council Research, Switzerland; pres. Com. Chemistry, Switzerland. Mem. Soc. helvetique Sc. Nat.; Swiss Chem. Soc. (hon.), Biochem. Soc., London. Pres. bd. editors Helvetica Chimica Acta. Contbr. papers to sci. publs. Research on fractionation of caseinogen, microdetermination of peptide sequences with tracer technique, formation and transformation of esters of inorganic and organic acids of phosphorus, flourinated alcohols. Home: 48 Fossard, 1211 Conches nr. Geneva, Switzerland.*

CHEREAU, Achille, French physician; b. Bar-sur-Sein, France, Aug. 23, 1817; studied medicine, Paris, France; M.D., 1841; practiced medicine, Paris; became librarian Faculty Medicine, Paris. 1877. Decorated Legion of Honor, 1871; recipient Civrieux prize French Acad. Medicine, 1848. Mem. Fr. Acad. Medicine. Author: Memoire servir a l'étude des maladies des ovaires, 1844; La vérité sur la morte de Jean-Jacques Rousseau, 1866; Essai sur les origines du journalisme médical francais, 1867; Guillotin et la guillotine, 1871; Le Parnasse médical francais ou dictionnaire des médecins-poètes de la France, 1874. Research, publs. on syphilis, med. journalism, maladies of women, also insects, especially ants. Died Paris, 1885.

CHERENKOV, Pavel Aleksandrovich, Russian physicist; b. Voronezh Guberniya, July 15, 1904; grad. Voronezh U., 1928; postgrad. Physics Inst., USSR Acad. Sci., 1930-35; Cand. Natural Sci., 1935; D.Natural Sci., 1940. Prof., asso. Inst. Phys. Problems, USSR Acad. Sci. Decorated Order of Lenin; recipient Nobel prize in physics (with I. M. Frank and I. E. Tamm), 1958. Mem. USSR Acad. Sci. (corr.). Research and publs. on composition of matter; discovered Cherenkov-Vavilov effect in photogenesis by liquids subjected to gamma-radiation whose charged particles have velocity greater than that of light in same medium (basis for devel. various methods of recording charged particles and designing Cherenkov counters), 1934; his research opened up new studies of cosmic rays and high-energy subatomic particles; generation of radio waves by means of Cherenkov effect used in devel. latest Soviet rockets. Address: Inst. Physical Problems, USSR Acad. Sci., Kaluzhskoe sh. 32, Moscow, USSR.

CHEREPAKHIN, Gavriil Klimentevich, Russian obstetrician and gynecologist; b. 1888; grad. Med. Faculty, Kharkov U., 1911; D.Med. Sci. Asst. dept. obstetrics and gynecology Don U., Rostov, 1920-29, lectr., 1929-31; prof., head chair obstetrics and gynecology Gorky Med. Inst., 1931—; chief obstetrician and gynecologist Gorky City. Chmn., Gorky City Obstetrics Commn.; mem. Commns. for Obstetrics and Gynecol. Care, USSR and RSFSR Ministry Health. Decorated Order of Lenin. Mem. All-Russian Soc. Obstetricians and Gynecologists (chmn. Gorky br. 1931—). Author of 120 works including Experience in Dealing with Breakdowns of Higher Nervous Activity by Means of Certain Pharmaco-Dynamic Substances; co-author: Hysterokymography and its Use in Obstetrics and Gynecology. Address: Gorky Med. Inst., pl. Minina i Pozharskogo 10, Gorky RSFSR, USSR.

CHERN, Shiing Shen, mathematician; b. Kashing, China, Oct. 26, 1911; s. Lien C. and Mei (Han) C.; B.Sc., Nankai U., Tientsin, China, 1930; M.Sc., Tsing Hua U., Peiping, China, 1934; D.Sc., U. Hamburg, (Germany), 1936; m. Shih Ning Tsen, July 28, 1939; children—Paul, May. Came to U. S., 1949, naturalized, 1961. Prof. math. Tsing Hua U., 1937-43; mem. Inst. for Advanced Study, Princeton, N.J., 1943-45; acting dir. Inst. Math., Academia Sinica, Nanking, China, 1946-48; prof. math. U. Chgo., 1949-59; prof. math. U. Cal. at Berkeley, 1960—. Mem. Am. Math. Soc. (v.p. 1963-64), Nat. Acad. Sci., Am. Acad. Arts and Scis. Editor Proceedings Am. Math. Soc., 1955-56, Transactions Am. Math. Soc., 1957-58, Ill. Jour. Math., 1960-65. Research in differential geometry in the large; developed Chern characteristic classes in fiber spaces and algebraic geometry. Home: 8336 Kent Ct., El Cerrito, Cal. Office: Dept. Math., U. Cal., Berkeley, Cal. 94720.*

CHERNENKO, Semion Fedorivich, Russian biologist; b. Sept. 13, 1877. Mem. staff Michuring Central Genetic Lab., from 1926; became prof., 1935; also on staff Fruit and Vegtable Inst., Michurinsk. Recipient Stalin prize, 1947. Author: Half-century of Work in the Garden, 1957; also articles on genetics. Bred series of new apple and pear types with different ripening periods.

CHERNIAEV, Ilia Ilyich, Russian chemist; b. Jan. 20, 1893; degree Petrograd U., 1915; student of L. A. Chugaev; mem. staff platinum inst. USSR Acad. Scis., 1918—, dir. chemistry inst., 1941—; prof. Leningrad U., 1932—, Moscow Oil Inst., 1935-41, U. Moscow, 1945—. Recipient Stalin prize, 1946, 52. Mem. USSR Acad. Scis. Author: Problems of Complex Compounds Chemistry, 1936. Research on complex compounds in inorganic chemistry; studied reductions of irridium compounds, also oxidations reactions of complex platinum compounds; discovered phenomenon of change in rotation sign of light polarization plane in optically active amino compounds of tetravalent platinum when converted into amide or imide compounds. Address: Institut obsheney inorganicheskoy khimii, V-71 Leminsk prosp. 31, Moscow, USSR.

CHERNICK, Jack, Am. physicist; b. Newark, Nov. 27, 1911; s. Isadore and Sarah (Tessler) C.; student Rutgers U., 1929-32; B.S. in Math., U. Chgo., 1938; M.S. in Math., Bklyn. Coll., 1941; m. Norma Leonia Weiner, Feb. 19, 1939; children—Irene Alice, Michael Ross, Julian August. Mathematician, Ballistics Research Lab., Aberdeen Proving Ground, Md., 1941-47; physicist Brookhaven Nat. Lab., Upton, L.I., N.Y., 1947—. Mem. adv. com. for reactor physics U. S. AEC, 1951—. Fellow Am. Nuclear Soc. (Spl. award 1966). Editorial com. Nuclear Sci. and Engring. Jour., 1955—. Research and publs. on theory of reactor kinetics and resonance absorption of neutrons in reactors; invented high flux beam reactor. Home: 495 S. Country Rd., East Patchogue, N.Y. 11772. Office: Brookhaven Nat. Lab., Upton, L.I., N.Y. 11973.*

CHERNIGOVSKY, Vladimir Nikolaevich, Russian physiologist; b. Mar. 1, 1907; grad. Med. Faculty, Perm. U., 1930; D.Med. Sci. Asst. dept. physiology Orenburg Vet. Inst., 1930-32, Sverdlovsk Med. Inst., 1932-37; sr. asso. dept. gen. physiology Leningrad br. All-Union Inst. Exptl. Medicine, 1937-41; prof., later head chair physiology Naval Med. Acad., Leningrad, lab. head Pavlov Inst. Physiology, USSR Acad. Sci., 1942-53, dir. Pavlov Inst. Physiology, 1960—, acad. sec. dept. physiology, 1963—, bur. mem. dept. biol. sci., until 1963; dir. Inst. Normal and Path. Physiology, USSR Acad. Med. Sci., 1953-59. Sci. cons. Leningrad Tb Inst. Recipient Pavlov prize, 1944. Mem. USSR Acad. Med. Sci. (v.p. 1953-57), USSR Acad. Sci., All-Union Soc. Physiologists, Pharmacologists and Biochemists (chmn. Moscow dept., bd. mem.). Author over 100 works including Interoceptors, 1949; co-author: Problems of the Nervous Regulation of the Blood, 1953; Afferent Systems of the Internal Organs, 1943. Editor: Bull. Exptl. Biology and Medicine. Research on enteroceptive reflexes and functional interrelationships between internal organs and cephalic brain cortex; studied reflex control in blood system and role of nervous system in pathogenesis of several diseases. Address: Inst. of Physiology, USSR Acad. Sci., n. Makarova 6, Leningrad, USSR.

CHERNOFF, Amoz Immanuel, Am. physician; b. Malden, Mass., Mar. 17, 1923; s. Isaiah and Celia (Margolin) C.; B.S., Yale, 1943, M.D., 1947; m. Renate Rosemarie Fisher, Jan. 25, 1953; children—David F., Susan N., Judith A. Research fellow hematology, asst. dir. hematology research labs. Michael Reese Hosp., Chgo., 1949-51; A.C.P. fellow medicine Washington U., St. Louis, 1951-52, Spl. USPHS fellow, 1952-53, faculty, 1951-56, asst. prof., 1954-56; asso. prof. medicine Duke, 1956-58; research prof. dir. Meml. Research Center, U. Tenn., Knoxville, 1958—. Recipient USPHS Research Career award, 1962—. Diplomate Am. Bd. Internal Medicine. Fellow A.C.P.; mem. Am. Soc. Clin. Investigation, So. Soc. Clin. Investigation, Central Soc. Clin. Research, Am. Fedn. Clin. Research, Am. Soc. Human Genetics, Am. Soc. Hematology, Internat. Soc. Hematology. Research and numerous publs. on hereditary hemolytic anemias especially abnormal hemoglobins, discovered several variants; developed procedures for study of these compounds. Home: 8017 Bennington Dr., Knoxville, Tenn. 37919.*

CHERNOFF, Herman, Am. statistician; b. N.Y.C., July 1, 1923; s. Max and Pauline (Markowitz) C.; B.S., Coll. City N.Y., 1943; Sc.M., Brown U., 1945, Ph.D., 1948; m. Judith Ullman, Sept. 7, 1947; children—Ellen Sue, Miriam Cheryl. Research asso. Cowles commn. for research in econs. U. Chgo., 1948-49; asst. prof., U. Ill., Urbana, 1949-51, asso. prof., 1951-52; asso. prof. Stanford (Cal.) U., 1952-56, prof., 1956—. Mem. Inst. Math. Statistics (pres.). Author: With (L. E. Moses) Elementary Decision Theory, 1959. Research in large sample theory, optimal design of experiments. Home: 825 San Francisco Ct., Stanford, Cal.*

CHERNOGOROV, Ivan Alekseevich, Russian internist; b. 1893; grad. Med. Faculty, Kharkov U., 1918; D.Med. Sci., 1941. Physician various Moscow polyclinics, 1922-30; asst. dept. propedeutics of internal diseases 1st Moscow Med. Inst., sr. asso. Moscow Research Inst. Functional Diagnostics and Therapy, 1930-41; head chair propedeutics internal diseases Kursk Med. Inst., 1941-45; dep. dir. sci. work Inst. Therapy, USSR Acad. Med. Sci., 1945-50, sci. dir., head dept. cardiovascular diseases, 1959—; prof., 1941—; head chair internal diseases Moscow Med. Stomatological Inst., 1950-59. Mem. All-Russian (Presidium mem.), Moscow (bd. mem., dep. chmn. cardiological sect.) socs. therapeutists, All-Union Znanie Soc., All-Union (bd. mem.), All-Russian (dep. chmn.) socs. cardiologists. Author of 50 works including Changes in the Electrocardiogram in Cases of Damage to the Myocardium; The Physiological Nature of Heart Blockages, 1946; Angina Pectoris; Disturbances of the Cardiac Rhythm, 1962. Address: Inst. Therapy, USSR Acad. Med. Sci., B. Serpukhovskaya ulitsa 27, Moscow, USSR.

CHERNYAEV, Ilya Ilich, Russian inorganic chemist; b. Jan. 20, 1893; grad. Physico-Math. Faculty, Petrograd U., 1915. With Inst. for Study Platinum, USSR Acad. Sci., 1918—, asso. Inst. Gen. and Inorganic Chemistry, 1934-41, dir., 1941-62, head dept. simple and complex inorganic compounds, 1962—; asst. chair inorganic chemistry Leningrad U., 1918-32, prof., head, 1932-35; prof. Moscow Petroleum Inst., 1935-41, Moscow U., 1945—. Decorated Order of Lenin (3); recipient Stalin prize (2). Mem. USSR Acad. Sci. Author: Problems of Complex Compounds Chemistry, 1936; Experimental Basis of the Law of Transinfluence, 1954; The Tasks of Complex Compounds Chemistry, 1959. Research and publs. on complex compounds, oxidation reactions of complex platinum compounds, reduction reactions of iridium compounds by means of various sugars (proved that nitro group bonded with platinum via nitrogen); discovered phenomenon of change in rotation sign of light polarization plane in optically active amino compounds of tetravalent platinum upon their conversion into amido or imido compounds. Address: Inst. Gen. and Inorganic Chemistry, USSR Acad. Sci., Leninsky prospect 31, Moscow, USSR.

CHERNYAK, Valentin Zakharevich, Russian vet surgeon; b. 1893; grad. Warsaw Vet. Inst., 1917. D.Vet. Sci. Asst. lectr. path. anatomy infectious diseases Leningrad Vet. Inst., 1927-29, head chair path. anatomy, 1933—; prof., 1929—; head chair path. anatomy Voronezh Vet. Inst., 1929-33; founder, dir. Path. Anatomy Lab., Leningrad Vet. Research Inst., 1933-38. Author: Pathoanatomical Diagnosis of the Main Infectious Diseases of Farm Animals, 1957. Address: Leningrad Vet. Inst., Chernigovskaya ulitsa 5, Leningrad, USSR.

CHERREAU, Antoine, French chemist; b. Paris, France, 1776; studied useful cryptogams, paregoric elixers, esculines, opium; introduced system for naming medicines. Died Paris, 1848.

CHERRIE, George Kruck, Am. naturalist; b. Knoxville, Ia., Aug. 22, 1865; s. Martin and Agnes (Breckenridge) C.; ed. Knoxville, Ia., and State Agrl. Coll., 1880; m. Stella M. Bruere, Dec. 1, 1895; m. 2d, Esther Atwell, Jan. 12, 1934. Asst. taxidermist, U. S. Nat. Mus., 1888, accepting position in Am. Mus., N.Y., later; apptd. taxidermist and curator of birds, mammals and reptiles, Nat. Mus., Costa Rica, 1889; asst. curator ornithology, in charge of dept., Field Mus. Natural History, Chgo., 1894-97; conducted explorations in Valley of the Orinoco for Lord Rothschild, 1897-99; ornithol. explorations, for Tring Mus., in French Guiana, 1902-03; explorations in Trinidad and Valley of Orinoco, for Bklyn. Inst. Mus., 1905 and 1907; curator ornithology and mammalogy, Bklyn. Inst. Arts and Scis., 1899-1911; mem. Am. Mus. expdn. to Colombia, S.A., valley of the Magdalena River and the high interior, 1913; rep. Am. Mus. Natural History on Roosevelt expdn. through S. Am., 1913-14; naturalist Collins-Day S. Am. expdn., 1914-15; dir. Cherrie-Roosevelt S. Am. expdn. for Am. Mus. Natural History, 1916-17; S. Am. service for U. S. Bur. Naval Intelligence, 1918-19; a leader Anthony-Cherrie expdn. to Ecuador for Am. Mus. Natural History, 1920-21; Cherrie expdn. to Ecuador for same, 1921; Mazaroni River diamond fields, Brit. Guiana, 1922; expdn. to Central Brazil, 1924; naturalist Roosevelt-Simpson Asiatic expdn. for Field Mus., Chinese Turkestan, 1925-26; leader Marshall Field Brazilian expdn., 1926. Mem. Am. Ornithologists' Union; hon. fellow Am. Mus. Natural History, 1921. Wrote on Central American Birds with descriptions of new species, etc., in The Auk and Proc. U. S. Nat. Mus., 1890-96; also The Ornithology of Santo Domingo (Publs. Field Mus. of Natural History, Chgo.), 1896; New Birds of the Orinoco Region and of Trinidad, 1909; Dark Trails, Adventures of a Naturalist, 1930. Died Jan. 20, 1948.

CHERRY, Thomas MacFarland, mathematician; b. Melbourne, Australia, May 21, 1898; s. Thomas and Edith (Gladman) C.; student Scotch Coll., Melbourne, 1908-14; B.A., U. Melbourne, 1918; B.A., Cambridge (Eng.) U., 1922, Ph.D., 1924, Sc.D., 1950; D.Sc. (hon.), Australian Nat. U., U. Western Australia; m. Olive Ellen Wright, Jan. 24, 1931; 1 dau., Gillian Eleanor (Mrs. John David Stowell). Fellow, Trinity Coll., Cambridge, 1924-28; asso. prof. Manchester U., 1924-25, Edinburgh U., 1927 (both Eng.); prof. Math. Melbourne U., 1929-63. Found. fellow Australian Acad. Sci. (sec. 1956-59, pres. 1961-65); Fellow Royal Soc., 1954; mem. London, Am., Australian math. socs. Research, publs. on differential equations, expansions, asymptotic series, and applications to phys. sci. Home: 7 Mountain Grove, Kew, Victoria. Australia. Died Nov. 21, 1966.

CHERVIN, Nicolas, French physician; b. S.-Laurent-d'Oingt, France, Oct. 6, 1783; studied medicine in Paris, France, 1812; received doctorate; studied typhoid at mil. hosp. in Mayence, Germany, 1813; sent to Am. to study yellow fever, 1814; continued research in Spain, 1822. Mem. French Acad. Medicine. Recipient le prix Montyon Acad. Medicine, 1832. Author: L'examen critique des prétendues preuves de contagion de la fievre jaune, 2 vols., 1828; De l'identité de nature des fièvres d'origine paludéene, 1842. Investigated tropical diseases in Am. by visiting ports, slave ships, leper colonies, dissecting corpses, examining fish and foods; recognized yellow fever is not contagious. Died Bourbonne-les-Bains, France, Aug. 14, 1843.

CHERWELL, 1st Viscount, Frederick Alexander Lindemann, Brit. physicist; b. Sidmouth, 1886; s. A. P. Lindemann; ed. Blairlodge, Darmstadt, Paris; Ph.D., Berlin U. Prof. exptl. philosophy Oxford U., 1919, 53-56, elected fellow Wadham Coll., 1919, hon. fellow, 1956, emeritus Student of the House Christ Ch., from 1956; exptl. pilot, dir. phys. lab. RAF, Farnborough, personal asst. to Churchill, 1940. Fellow Royal Soc. Author: The Physical Significance of the Quantum Theory; also papers. Research on quantum theory, conditions in upper atmosphere; advocate of saturation bombing. Died July 3, 1957.

CHESLDEN, William, Brit. surgeon, anatomist; b. Lincolnshire, Eng., Oct. 19, 1688; s. George and Deborah (Hubbert) C.; apprentice to Mr. Wilkes, surgeon of Leicester; studied under William Cowper in London, Eng.; apprentice to Mr. Ferne, St. Thomas Hosp.; m. Deborah Knight; 1 dau., Deborah Wilhelmina (Mrs. Charles Cotes). Admitted to Barber Surgeons' Co., 1710; became lectr. anatomy, London, 1711; surgeon St. Thomas's Hosp., 1719-38, St. George's Hosp., 1734-37, Chelsea Hosp., 1737-52; named head surgeon for Queen Caroline, 1727. Fellow Royal Soc., 1712; mem. French Acad. Scis., 1729, Royal Acad. Surgery Paris (1st fgn. asso.). Author: Syllavus sive anatomicus in usum theatri anatomici Wilhelmi Cheseldem chirugi autoris impensis, 1711; Anatomy of the Human Body, 1713; Osteographia, 1733; Treatise on the High Operation for the Stone, 1723. One of first surgeons to perform lateral operation for stone (lithotomy); produced artificial pupils to cure certain types of blindness (iridotomy), 1727; designed Fulham Bridge, 1729. Died Bath, Eng., Apr. 10, 1752.

CHESNEY, Cummings C., Am. elec. engr.; b. Selinsgrove, Pa., Oct. 28, 1863; s. John C. and Jane (McFall) C.; B.S., Pa. State Coll., 1885; m. Elizabeth Cutler, 1891; children—Malcolm, M., Elizabeth, Margaret, Katherine, Barbara. Tchr. math., chemistry Doylestown Sem., also Pa. State Coll., 1885-88; joined William Stanley's lab. force, Great Barrington, Mass., 1888; with U. S. Elec. Lighting Co., Newark, 1889-90; an incorporator Stanley Electric Mfg. Co., Pittsfield, Mass., 1890, became 1st v.p. and chief engr., 1904; 1st v.p. and chief engr. Stanley-G. I. Electric Mfg. Co.; mgr. Pittsfield works Gen. Electric Co., 1906-27, v.p., chmn. mfg. com. pres. Pittsfield Coal Gas Co.; dir. Boston & Albany R.R. Co.; dir., officer various banks, ins. cos. Mem. Am. Inst. E.E. (Edison medal 1922, pres. 1926-27), Soc. Arts, London, Eng. Pioneer in elec. improvements; laid out 1st polyphase power transmission plant put into successful operation in Am.; pioneer in designing alternating current generators for high voltages. Died Nov. 27, 1947.

CHESSHER, Robert, Brit. surgeon; b. Hinckley, Eng., 1750; stepfather, Mr. Whalley; ed. Bosworth Sch.; apprenticed to stepfather; studied under Dr. Denman. House surgeon to Middlesex Hosp.; later returned to Hinkley. Used double-inclined plane to support fractured legs, 1790; invented several instruments for supporting weak spines and head, for application friction to paralyzed limbs or muscles. Died Jan. 31, 1831.

CHESSIN, Henry, Am. physicist; b. Cleve., Dec. 8, 1919; s. Sam and Sarah (Glass) C.; B.S., Western Res. U., 1947; M.S., Purdue U., 1950; Ph.D., Poly. Inst. Bklyn., 1957; m. Sylvia Bray, Dec. 25, 1950. Instr. physics Poly. Inst. Bklyn., 1951-57; sr. scientist U. S. Steel Co., Pitts., 1957-64; prof. physics State U. N.Y., Albany, 1964—; cons. Dudley Obs., E. F. Fullam, Inc. Mem. Am. Phys. Soc., Am. Crystallographic Assn., Inst. Metals, Sigma Xi. Research in x-ray crystallography, transformations and lattice dynamics of metals and metallic solid solutions. Home: 875 Birchwood Lane, Schenectady 12309. Office: 1223 Western Av., Albany, N.Y. 12203.*

CHESTER, Frederick Dixon, bacteriologist; b. San Domingo, Haiti, Oct. 8, 1861; s. Edwin Smith and Elizabeth (Walthall) C.; B.S., Cornell, 1882, M.S., 1885; m. Emma L. Sherwood, June, 1883. Prof. geology and botany, Del. Coll., 1882-9; mycologist and bacteriologist, Del. Agrl. Expt. Sta., bacteriologist in charge state bacteriol. and pathol. lab., 1899-1906; in bus., from 1906. Fellow A.A.A.S.; mem. Am. Geol. Soc., Soc. for Promotion of Agrl. Sci., Am. Bacteriol. Soc. Author: Manual of Determinative Bacteriology, 1901; also monographs on geology of Del. and Eastern Md., papers and reports on mycology and bacteriology. Died Jan. 1, 1943.

CHESTER, Robert, Brit. astronomer, alchemist; b. Chester, Eng., flourished 1182; author numerous manuscripts including: De astrolabio; Morienus romanus, 1182; De diversitate annorum ex Roberto Cestrensi super tabulas toletanas, posthumously circa, 1370. Translated sci. works from Arabic to Latin.

CHESTER JONES, Ian, zoologist; b. Montreal, Que., Can., Jan. 3, 1916; s. Hugh Creswell and Ada Ann (Hollowood) J.; B.Sc., U. Liverpool (Eng.), 1938, Ph.D., 1941, D.Sc., 1958; m. Nansi Ellis Williams, Feb. 28, 1943; children—Hugh Chester, Gareth Chester, Ruth Chester. Faculty, U. Liverpool, 1946-47, 49-58, sr. lectr. zoology, 1955-58; research fellow Harvard Sch. Dental Medicine, 1947-49; prof. zoology, chmn. U. Sheffield (eng.), 1958—, dean Faculty Pure Sci., 1965—. Mem. Brit. Soc. Endocrinology. (pres. 1966—). Author: The Adrenal Cortex, 1957; also numerous articles. Research on nature and mode of action of glands secreting hormones in vertebrate animals. Home: 98 Ashdell, Sheffield S10 2TN, Eng.*

CHESTNUT, Alphonse F., Am. marine biologist; b. Stoughton, Mass., Nov. 20, 1917; s. Kleopas and Josephine (Wolan) C.; B.Sc., Coll. William and Mary, 1941; M.Sc., Rutgers U., 1943, Ph.D., 1949; m. Janet Hamilton Wood, Oct. 16, 1943; children—Alfred Page, John William, Biologist, Chesapeake Corp., Va., 1941-42; research asso. oyster research lab. Rutgers U., 1943-48; dir. Inst. Fisheries Research U. N.C., Morehead City, 1955—, faculty, 1948—, prof. marine biology, 1959—. Mem. N.C. Mus. Adv. Commn., N.C. Comml. Fisheries Study Commn.; mem. sci. adv. bd. Fla. Conservation Dept., 1962—; adviser N.C. Conservation Devel., 19——. Fellow Am. Inst. Fisheries Research, mem. Nat. Shellfish Assn. (past pres., editor proc.), N.C. Acad. Sci., Atlantic Estuarine Soc. (past sec.), Am. Soc. Limnology and Oceanography, Ecol. Soc., Am. Soc. Zoologists, Sigma Xi. Contbg. author Survey of Marine Fisheries of North Carolina, 1951; also articles on oysters, clams, scallops. Home: 110 Holly Lane, Morehead City, N.C. 28557.*

CHESTON, Warren Bruce, Am. physicist; b. Rochester, N.Y., Mar. 15, 1926; s. George L. and Clara (Hoesterey) C.; B.S., Harvard, 1947; postgraduate Columbia, 1947-48; Ph.D., U. Rochester, 1951; m. Roberta J. Bohrod, Nov. 1, 1950; children—Stephen, Rebecca, Dena, Nicolas. Asst. prof. Washington U., St. Louis, 1951-53; asst. prof. U. Minn., Mpls., asso. prof., 1957-62, prof. 1962—, dir. Space Sci. Center, 1965-67, dean Inst. Tech., 1967—. Fulbright lectr. U. Utrecht, Netherlands, 1958-59; sci. attache Dept. State, London, Eng., 1963-65. Fellow, Am. Phys. Soc. Theoretical studies of interaction of mesons with light nuclei; hyperon-nucleus interaction; many-body problems. Home: 3850 Richfield Rd., Mpls. 55410. Office: Inst. of Technology, U. Minn., Mpls. 55455.

CHETVERUKHIN, Nikolay Fedorovich, Russian mathematician; b. 1891. Prof. Mem. RSFSR Acad. Pedagogical Sci. Author: An Introduction to Advanced Geometry, 1934; Problems of the Methodology and Methods of Geometrical Structures in a School Course on Geometry, 1946; A Course of Descriptive Geometry, 1963. Editor: Collection of Problems in Descriptive Geometry, 1963. Research on theory of geometrical structures, descriptive geometry, principles of geometry. Address: RSFSR Acad. Pedagogical Sci., B. Polyanka 58, Moscow, USSR.

CHEVALIER, (Louis Marie) Arthur, French optician; b. Paris, Mar. 15, 1830; s. Charles Louis Chevalier; optician in Paris; author: Hygiène de la vue, 1861; l'Art de l'opticien, 1863; l'Etudiant micrographe, traité théorique et pratique du microscope et des préparations, 1865; publs. on hygiene of sight, maladies of the eyes, use of the microscope, photography in medicine; perfected the universal visometer, ophthalmoscope, axometer, pupilometer. Died Paris, 1874.

CHEVALIER, Auguste Jean Baptiste, French botanist; b. Domfront, France, June 23, 1873; D.Natural Sci.; traveled in Senegal and Sudan studying geog. botany and econ. flora; organizer bot. services Colonial Lab. of Mus., France; investigated Ubangi of Lake Tchad, 1902-03; returned from Occidental Africa, 1905; gathered bot. material in Indochina; prof. colonial agronomy Mus. Natural History. Mem. French Acad. Scis., 1937 (editor Revue d'agriculture et de géographie botanique). Author: les Végétaux utiles de l'Afrique tropicale francaise, 1905; Traité de geographie physique, 1923. Research in plant geography; assembled bot. information from various countries including listings of useful plants such as coffee, tea, cotton, wood. Died Paris, France, June 4, 1956.

CHEVALIER, Charles Louis, French optician; b. Paris, France, Apr. 18, 1804; s. Louis Vincent C.; 1 son, Louis Marie Arthur. Installed Valois gallery, Palais-Royal; assisted French Soc. for Advancement Nat. Industry (Recipient medal 1834). Author: Notice sur l'usage des chambres obscures et des chambres claires, 1839; Notes rectificatives pour servir à l'histoire des microscopes, 1835; Des microscopes et de leur usage, 1839; Nouveau manuel complet du physician préparateur, 1853. Built (with his father) microscope; built first achromatic microscope in France, 1823, pneumatic machine, 1846; perfected solar microscope; invented new instruments concerned with diaphragm and lens of camera. Died Paris, Nov. 21, 1859.

CHEVALIER, Jean Baptiste Alphonse, French chemist; b. Langres, France, 1793; prof. L'ecole de pharmacie; mem. Acad. Medicine. Author: (with Payen) Traité des reactifs chimiques, 1822; (with A. Richard and Guillemin) Dictionnaire des drogues simples et composées. 1826-29; Dictionnaire des alterations et des falsifications des substance alimentaires, médicamenteuses et commerciales, 1850-52; Recherches chronologiques sur les moyens appliqués à la conservation des substances alimentaires, 1858; Traité des désinfectants, sous le rapport de l'hygiene publique, 1862. Died Paris, 1879.

CHEVALIER, Louis Henri, French sociologist, anthropologist; b. L'Aiguillon-sur-Mer (Vendée), France, May 29, 1911. Prof., Ecole normale supérieur, Inst. Polit. Studies, Nat. Inst. Demographic Studies, Collège de France. Author: Le problème démographique nord-africain; Madagascar: population et resources; Démographie générale; La formation de la population parisienne au XIX siècle; Classes laborieuses et classes dangereuses à Paris pendant la première moitié du XIX siècle. Address: Collège de France, Paris, France.

CHEVALLIER, François, French mathematician; b. Ste.-Maur, France; studied under Abbé Galloys; math. tchr. École du roi et des pages de la petite ecuries; prof. math. Collège de France; mem. French Acad. Scis. Author: Des effets de la poudre à canon, principalement dan les mines, 1707. Died Jan. 1748.

CHEVALLIER, Jean-Gabriel-Augustin, French optician; b. Mautes-sur-Seine, France, 1778; 1 son, Charles-Victorin-Réaumur; optician, Paris, France; ct. optician to king; optical engr. to Kings of Westphalia, also to Louis XVIII. Recipient First Class Gold medal Emperor of Russia. Mem. numerous socs., including Acad. Scis. St. Petersburg. Author: Essai sur l'art de l'ingénieur en instruments de physique expérimentale, 1819. Invented mech. barometer, 1801, balance for weighing grains, 1819, galactrometer, 1823, hydrometrical instrument and binoculars; built 1st good microscope objective; kept records of Paris climate for over 30 years. Died 1848.

CHEVANDIER DE VALDROME, Jean-Pierre-Eugène-Napoléon (Chevandier, Eugène), French agronomist; b. Saint Quirin, France, Aug. 18, 1810; engring. diploma École centrale, 1831. Head lab. École centrale; became engr., asst. dir. for mfg. plant; Circey, 1834; mem. adminstrv. council of St. Gobain; mem. Gen. Council Agr. and Commerce, 1850; elected dep. from Meurthe to French Assembly, 1859, 63, 69; became v.p. Legislative Corps, 1869; became minister interior Parliamentary Cabinet, 1870. Mem. French Acad. Scis. (corr.), Central Soc. Agr. Research and observations on reforestation, plant irrigation, effect of water on wood, specific weights, uses of wood; studied timber, Eng. and Germany. Died Paris, Dec. 1, 1878.

CHEVENARD, Pierre, French matallurgist; b. Thizy, France, Dec. 31, 1888; metallurgist Société de Commentry, Fourchambault et Decazeville; mem. French Acad. Scis., 1946; invented dilatometer which showed transformation point of alloys and recorded diagram of balance of phases; his research on stress limits led to prodn. of steels which could withstand high temperatures; built mechanism to show cycles of elastic hysteresis of textiles drawn with constant speed. Died 1960.

CHEVERS, Norman, Brit. physician; b. Greenhithe, Eng., 1818; grad. U. Glasgow (Scotland), 1839; M.D.; practiced in London, Eng., until 1848; civil surgeon, India, until 1876; prin., prof. medicine Calcutta (India) Med. Coll. Contbr. numerous articles to profl. jours. Gave 1st description of chronic constrictive pericarditis, 1842. Died 1886.

CHEVREUL, Michel Eugène, French chemist; gerontologist; b. Angers, France, Aug. 31, 1786; s. Michel and Etinette-Madeleine (Bachelier) C.; pupil of Vauquelin at Coll. de France, 1803; doctorate, Harvard, 1886. Became chemist in factory directed by Vauquelin, 1804; became asst. in pvt. instn. founded by Fourcory, 1809; named prof. physics Lycée Charlemagne, 1813; apptd. dir. dye works Gobelins by Louis XVIII; became prof. chemistry Mus. Natural History, 1830, dir., 1864-79. Fellow Royal Soc., 1826. Author: Recherches chimiques sur les corps gras d'origine animale, 1823; Considérations générales sur l'analyse et sur ses applications, 1824; Considérations sur l'analyse organique, 1924; Leçons de chimie appliquée à la teinture, 1828-31; De la loi du contraste simultané des couleurs et de l'assortiment des objects colorés (book led to elaborate study dyestuffs, use of benzol), 1839; De la baguette divinatoire, du pendule dit explorateur et des tables tournantes, 1854; Lettres adresées à M. Villemain sur la méthode en général et sur la definition du mot fait, relativement aux sciences, aux lettres, aux beauxarts, 1856; Des couleurs et des leurs application aux arts, 1864; Histoire des connaissances chimiques, 1866; De la méthode a posteriori expérimentale et de la généralité de ses applications, 1870; Guano du Perous, 1874; Sur l'affinité capillaire, 1876; Sur la combinaison du chlorhydrate d'ammoniaque avec les chlorures de sodium et de potassium, 1877; Sur les draps de laine teints en noir bleuâtre, 1879; Considerations générales sur les méthodes scientifiques, 1883. Pioneer in analysis of organic substances; improved soap- and candlemaking industry through investigations of oils, fats, soap formation; patentee (with Gay-Lussac) candle manufacture; a founder of modern organic chemistry; discovered and named stearic and oleic acids (margarine); defined chem. species, also noted existence of isomerism, advanced theories of molecular motion and esters, isolated creatine from muscle, 1832; discovered quercitin and cetyl alcohol; found sugar in diabetec urine to be glucose; studied physics and psychology of color; pioneer in gerontology; when in his 90s, he studied psychol. effects of old age. Died Paris, Apr. 9, 1889.

CHEW, Geoffrey Foucar, Am. theoretical physicist, educator; b. Washington, June 5, 1924; s. Arthur Percy and Pauline (Foucar) C.; B.S. in Physics, George Washington U., 1944; Ph.D., U. Chgo., 1948; m. Ruth Elva Wright, June 16, 1945; children—Berkeley Arthur and Beverly Randall (twins). Research physicist, Los Alamos Sci. Lab., 1944-46; physicist Lawrence Radiation Lab., Cal., 1948-49; asst. prof. physics, U. Cal. at Berkeley, 1949-50; asst. prof. U. Ill., 1950-51; asso. prof. U. Cal. at Berkeley, 1951-55, prof. 1955-57, prof. physics, 1957—. Fulbright lectr. Les Houches, France, 1953, 60, 65; Hughes prize Am. Phys. Soc., 1962; Overseas fellow Churchill Coll., Cambridge, Eng., 1962-63. Mem. Nat. Acad. Scis., Am. Phys. Soc. Author: S-Matrix Theory of Strong Interactions, 1961. Developed bootstrap theory of nuclear particles based on analytic S-matrix; studies on Regge pole theory of high-energy nuclear reactions. Home: 10 Maybeck Twin Dr., Berkeley, Cal. 94708.

CHEYMOL, Jean Henri, French physician, pharmacologist; b. Castillon, June 24, 1896; s. Justin and Marie (Mauriac) C.; ed. Collège Saint-Eugène at Aurillac; M.D., Ph.D. in sci., Ph.D. in pharmacology; m. Yvonne Grand, July 13, 1926; children—Georges, Claude. With health service, 1939-40, Eugene Gley at Collège de France; asst., lectr., now prof., dir. Inst. Pharmacology, Sch. Medicine, Paris. Lectr. in Europe. Recipient Montyon prize in physiology, 1948, Parkin prize Acad. Sci., 1960, Acad. Medicine prize. Mem. Nat. Acad. Medicine. Author numerous books. Research and publs. on respiratory and emetic centers; numerous new synthetic molecules found to be alive pharmacologically. Home: 1, parvis Notre-Dame, Paris. Office: Faculte de medicine, Paris, France.

CHEYNE, George, Brit. physician; b. Methlick, Scotland, 1671; studied under Dr. Archibald Pitcairn; M.D.; hon. diploma Edinburgh (Scotland) Coll., 1724; m. Margaret Middleton; several children including John. Began practice of medicine, London, 1702. Fellow Royal Soc., 1701. Author: New Theory of Fevers, 1702; Fluxionum methodus inversa, 1703; Rudimentarium methodl fluxiorum inversae specimina, 1705; The English Malady or a Treatise of Nervous Diseases of all Kinds, 1735; The Natural Method of Curing the Diseases of the Body, 1742. Proposed that hypochondria is caused by excess moisture in the air and variability of weather, 1733; recommended strict adherence to milk and vegetable diet; research on muscle contractions connected with nervous diseases; his Philosophical Principles of Natural Religion (1705) probably prompted Newton to add new Quaeries on methodological, epistemological, and metaphysical problems to the 1706 Latin edition of his Opticks. Died Bath, Eng., Apr. 13, 1743.

CHEYNE, John, physician; b. Leith, Scotland, Feb. 3, 1777; med. degree, U. Edinburgh (Scotland), 1795. Asst. surgeon, surgeon to regt. arty in Eng. and Ireland, 1795-99; in charge ordnance hosp., Leith Ft., 1799-1809; settled in Dublin, Ireland, 1809; named physician-gen. to forces in Ireland, 1820; physician Meath Hosp., also prof. practice physic Coll. Surgeons, 1811-15; named physician to House of Industry, 1815; ret. and left Dublin, 1831. Author: Essays on the Diseases of Children, 1801-02; On Hydrocephalus Acutus (possibly 1st description of acute hydrocephalus), 1808; The Pathology of the Membrane of the Larynx and Bronchia, 1809; Cases of Apoplexy and Lethargy, 1812. Studied diseases of children, also acute and epidemic diseases; gave 1st description of a breathing irregularity (Cheyne-Stokes respiration), 1818. Died Edinburgh, Jan. 31, 1836.

CHEZY, Antoine de, see de Chezy.

CHIANG, Cheng-Wu, hydronamic engr., educator; b. Jehol, China, Mar. 15, 1907; s. Nei Heng and Lan (Ling) C.; B.S., Tongshan Engring. Coll., Nat. Chiao Tung U., 1930; m. Hou Chiu Li, Jan. 7, 1926; children—Pai Chung, Ko Chung, Chih Chung, Cheng Chung, Chi Chung. Municipal and san. engr., 1930; sr. engr. Yung Hwa Engring. Co., Tientsin, 1931-33, Taiyuan Flood Prevention Commn., 1933-34; chief cons. engr. Chin-sui Mil. Gen Hdqrs., 1934-40; head designing dept. Szu-chuan Hwy. Bur., 1940-42; chief engr., dist. office Nat. Salt Adminstrn., 1942-54; asst. to chief engr. Shihmen Reservoir Devel. Commn., 1954-58; prof. hydrodynamics Cheng Kung U., Tawan, Republic of China, 1958—; part-time prof. Navy Engring. Inst., Republic of China, 1959—; dir. Tainan (Taiwan) Hydraulic Lab., Ministry of Econ. Affairs; cons. engr. Agrl. Machinery, Operation and Maintenance Office, Taiwan Sugar Co.; research fellow Engring. Research Center of Academia Sinica. Hon. fellow U. Minn., 1963. Mem. Am. Soc. C.E., Am. Geophys. Union, Am. Water Resources Assn., Chinese Engrs. Assn., Chinese Inst. Civil Engring., Chinese Inst. San. Engring., Chinese Inst. Municipal Engring., Chinese Inst. Hydraulic Engring. Research and numerous publs. in devel. of formulars in hydrodynamics. Home: 149 N. Park Rd., Tainan. Office: Civil Engring. Dept., Provincial Cheng Kung U., Tainan, Taiwan, Republic of China.*

CHIANG, Huai Chang, biologist; b. Sungkiang, China, Feb. 15, 1915; s. Wentse and Hsiu (Hsiu) C.; B.S., Tsing Hua U., Peiping, China, 1938; M.S., U. Minn., 1946, Ph.D., 1948; m. Zoh Ing Shen, Sept. 8, 1946; children—Jeanne, Katherine, Robert. Came to U. S., 1945, naturalized, 1958. Research asst. Tsing Hua U., 1938-44; faculty U. Minn., St. Paul, 1945—, prof. entomology, 1960—. Guggenheim fellow, 1956. Mem. Entomol. Soc. Am., Ecol. Soc. Am., Canadian Entomol. Soc., Royal Ent. Soc. London, Minn. Acad. Scis., Japanese Soc. Population Ecology, Soc. Animal Behavior, Sigma Xi. Contbr. numerous articles to tech. jours. Research on population ecology, relations insects, crop-growth and yield, behavior, activity and tolerance to phys. factors insects. Home: 1896 Carl St., St. Paul 55113.*

CHIAPPA, Sergio, Italian radiologist; b. Milan, Italy, July 19, 1927; s. Mario and Beatrice (Cima) C.; grad. in medicine with full marks and honours, U. Milan, 1950; specialist in radiology U. Pavia (Italy), 1952; m. Eugenia Levi, May 12, 1957; children—Fabrizia, Valeria. Asst., Radiol. Inst., U. Pavia, 1950-53; staff Radiol. Institut, Serafimerlazarettet, Stockholm, Sodersjukhuset, also Radiumhemmet, Stockholm, Sweden, 1955-58; asst. prof. Radiol. Inst., U. Milan, 1958—; chief radiol. dept. Fatebenefratelli Hosp., Milan, 1965—. Mem. Italian Soc. Radiology and Nuclear Medicine. Author: La neuroradiologia delle regioni encefaliche-Trattato di Medicina; Interna, 1960; Indagini cliniche e radiologiche sulla tunzionalità esocrina del pancreas, 1959; Vascular Roentgenology, 1964; Linfoadenografia e radioterapia endolinfatica nelle metastasi da neoplasie della sfera urologica, 1965; also numerous articles. Discovered radioactive contrast medium and its use in endolymphatic radiotherapy for detection of tumors. Home: 8 Piazza Aquileia, Milano, Italy.*

CHIARAMONTI, Scipion, Italian astronomer, philosopher; b. Cesena, Italy, 1565; instr. philosophy, Pisa, Italy; founder l'Academie des Offuscati. Author: Anti-Tycho, 1621; Caesenoe historia; works on astronomy, math., history. Opposed Tycho Brahe on comets; confuted by Kepler and Galileo. Died 1652.

CHIARI, Herman, Austrian pathologist, physician; b. Vienna, Austria, Dec. 6, 1897; s. Hans and Oellacher Chiari; M.D., U. Vienna; m. Elisabeth Spohn, 1924. Asst., Inst. Gen. and Exptl. Pathology, U. Vienna, Inst. Anat. Pathology, instr., 1931, dir. Inst. Anat. Pathology, 1936, dean Sch. Medicine, full prof. anat. pathology, 1951-54. Mem. Austrian Acad. Sci., Nat. Acad. Leopoldina, Med. Soc. Newcastle upon Tyne (hon.), Med. Soc. Vienna, German Soc. Pathology. Author: Pathologische Anatomie des akuten Gelenkrheumastimus; Herz- und Gefässkrankheiten; Diebetes mellitus; Gelenkerkrankungen, others. Address: Frankgasse 6, Vienna IX, Austria.

CHIAROTTI, Gianfranco, Italian physicist; b. Coma, Italy July 6, 1928; s. Luigi and Maria (Paltrinieri) C.; Ph.D. in physics; m. Aida Repanai, Oct. 26, 1959; children—Guido, Laura. Dir., Inst. Physics, U. Messina (Italy), also prof. physics. Mem. Italian, Lombarde socs. physics, Am. Phys. Soc., Peloritana Acad. Scis. and Letters. Research and publs. on solid state physics. Home: via Regina Margherita. Office: Istituto di Fisica Università, Messina, Italy.

CHIARUGI, Vincenzo, psychiatrist, dermatologist; b. Empoli, Italy, Feb. 20, 1759; grad. medicine Pisa, Italy, 1780; practiced medicine, Florence, Italy; doctor to Achispedoli of St. Maria Nuova, Florence; dir. constrn. of new mental hosp., 1785, became head; named prof. skin and mental diseases, Santa Maria Nuova, 1802, apptd. supt., 1818. Author: Lettere sopra un caso ul mal venero, 1783; Della pazia in genere e inspecie, 1793-94; Suggio teorico-pratico sulle malattie cutanee sordide, 1799; la Fisica dell'uomo, 1811-12; Saggio di ticerche sulla pellagra, 1814; Soptra una supposta specie di ermafrovitismo, 1819. Considered father of Italian clin. psychiatry; research on pellegra; 1st in Europe to stop using chains and fetters in treating insane, 1793; developed new methods and reforms for treating insane. Died Dec. 20, 1820.

CHIBISOV, Konstantin Vladimirovich, Russian sci. photographer; b. Mar. 1, 1897; grad. Moscow U., 1922. With Air Force Research and Testing Inst. Aerial Photography, 1918-30; an organizer, asso. All-Union Cinematography Research Inst., 1930—; chmn. Commn. for Sci. Photography and Cinematography, USSR Acad. Sci., 1948—; prof. Moscow U., 1950—. Decorated Order of Lenin; recipient Stalin prize, 1950. Mem. USSR Acad. Sci. (corr.). Author: The Theory of Photographic Processes, 1935; Theory of Photographic Emulsion Synthesis, 1937; The Nature of Photographic Sensitivity, 1948; The Nature of Light Sensitivity Nuclei in Photographic Emulsions, 1953; Research on the Nature of Photographic Sensitivity, 1957. Research and publs. on photographic sensitometry, theory of photographic emulsion synthesis, nature of photographic sensitivity. Home: B. Bronnaya 25/III, Moscow. Office: Moscow University, Leninskie gory, Moscow, USSR.

CHICHESTER, Clinton Oscar, Am. food technologist; b. Bklyn., Feb. 11, 1925; s. Clinton O. and Anna M. (Hartmann) C.; B.S., Mass. Inst. Tech., 1949; M.S., U. Cal. at Berkeley, 1951, Ph.D., 1954; m. Nancy J. Case, August 1966; children—Clinton, Catherine, Christine, Samuel, Cecile. Faculty, U. Cal. at Davis, 1954—, prof., chmn. food tech., 1962—. Consultant to governmental agencies. Recipient Bernardo O'Higgins medal Chile. Member of the American Chem. Soc., Inst. Food Technologists, Soc. for Exptl. Stress Analysis, Assn. Automotive Medicine, N.Y. Acad. Scis., Spectroscopy Soc., Am. Soc. Biol. Chemistry, Am. Inst. Nutrition, Am. Pub. Health Assn., Am. Optical Soc., Am. Inst. Chem. Engrs. Editor: (with L. W. Slanetz, A. R. Gaufin, Z. J. Ordal) Microbiological Quality of Foods, 1963; Research in Pesticides, 1965; (with E. M. Mrak, G. F. Stewart) Advances in Food Research series; also numerous articles. Research in biosynthesis carotenoids, food sci. and tech. Home: 1314 Pine Lane, Davis, Cal. 95616.*

CHICHIBABIN, Alexei Eugenivich, Russian chemist; b. Kusemino, nr. Poltava, Russia, Mar. 29, 1871; became asst. U. Moscow, 1900; apptd. prof. Tech. High Sch., Moscow, 1909; contbr. papers on tervalent carbon and other topics in organic chemistry to jours. Developed Chichibabin amination reaction, 1914, also Chichibabin pyridine synthesis. Died Paris, Aug. 15, 1945.

CHICKERING, Arthur Merton, Am. arachnologist; b. North Danville, Vt., Mar. 23, 1887; s. Orville Elmore and Alice (Finley) C.; Ph.B., Yale, 1913; M.S., U. Wis., 1916; Ph.D., U. Mich., 1927; m. Mabel Adele Kehler, Feb. 13, 1909; children—Alice Pope (Mrs. Orville Crays), Orville M., Donald Hugh. Asst., instr. Beloit Coll., 1913-18; prof., chmn. dept. biology Albion (Mich.) Coll., 1918-57, chmn. div. sci. and math., 1931-56; instr. zoology U. Mich., 1925-26; asso. in arachnology Mus. Comparative Zoology, Harvard, 1953—, cons., 1962-63. Cited by Mich. State Sentate, 1943; Guggenheim fellow, 1957-58; NSF grantee, 1963—. Fellow A.A.A.S.; mem. Mich. Acad. Sci., Arts, Letters, (past pres.), Am. Microscopical Soc. (past pres., editorial bd. 1955-66), Chgo. Acad. Sci. (hon. life), Soc. Systematic Zoology, Entomol. Soc. Am., Phi Beta Kappa, Sigma Xi. Numerous publs. on collections, descriptions new species spiders; spermatogenesis in Belostomatidae and Nepidae. Home: 65 Hammond St., Cambridge 02138. Office: Mus. of Comparative Zoology, Harvard U., Cambridge, Mass. 02138.*

CHICOYNEAU, François, French physician; b. Montpellier, France, 1672; s. Michael and Catherine (Pichotti) C.; M.D., 1693; m. Catherine Fournier Chirac; m. 2d; children include: Aimé François, Jean François, Jean Joseph. Chancellor, prof. anatomy and botany Montpellier; intendant Jardin royal; became physician of King Louis XV, 1731. Mem. French Acad. Sci. (asso.), 1732. Worked during the plague of Marseille, France, 1720; research on mercurial friction. Died Versailles, France, Apr. 13, 1752.

CHIEN, Shih-Liang, chemist; b. Honan, China, Jan. 9, 1908; s. Hung-Yeh and Jen-Kung (Hsu) C.; B.S., Nat. Tsing Hua U., 1931; M.S., U. Ill., 1932, Ph.D., 1934; m. Wan-Tu Chang, Jan. 16, 1927; children—Robert Chun, Shu, Frederick Fu. Prof. chemistry Nat. U. Peking, 1934-37, 1946-49, Nat. S.W. Asso. U., 1938-40; research chemist New Asiatic Chem. and Pharm. Research Inst., 1941-45; prof. chemistry Nat. Taiwan U., 1949-51, pres., 1951—. Mem. Joint Bd. Dirs., Palace Mus. and Nat. Central Mus., 1951—; trustee China Found. for Promotion Edn. and Culture, 1952—; Joint Com. on Entrance Examinations for Univs. and Colls. in Taiwan, 1954—; mem. Council Academia Sinica, 1957—, Nat. Council on Sci. Devel., 1959—, Atomic Energy Council, 1960—. Mem. Chinese Chem. Soc., Chinese Soc. Advancement Natural Scis., Philosophy of Sci. Assn., Chinese Soc. Advancement Sci. Research, publs. in theoretical and synthetic organic chemistry; derived gen. formula for calculation of number of stereoisomers theoretically possible for compounds containing pairs of similar asymmetric carbon atoms. Home: 20 Foochow St., Tapei, Taiwan.*

CHIEN, Shu, physiologist; b. Peiping, China, June 23, 1931; s. Shih-liang and Wan-tu (Chang) C.; student Nat. Peking U., 1947-48; B.M., Nat. Taiwan U., 1953; Ph.D., Columbia, 1957; m. Kuang-Chung Hu, Apr. 7, 1957; children—May, Ann. Came to U. S., 1954. Faculty, Columbia, N.Y.C., 1954—, asso. prof. physiology, 1964—. Mem. Am. Physiol. Soc., A.A.A.S., Harvey Soc., Instrument Soc. Am., N.Y. Acad. Scis., Radiation Research Soc., Soc. Exptl. Biology and Medicine, Sigma Xi. Research and publs. on analysis of factors concerned with regulation of blood volume and blood viscosity; evaluation of roles played by sympathetic nervous system and endocrine systems in compensatory responses to hemorrhage; study of circulatory disturbances in shock due to hemorrhage, bacterial toxins, histamine and cardiac dysfunction. Home: 277 Van Nostrand Av., Englewood, N.J. 07631. Office: 630 W. 168th St., N.Y.C. 10032.*

CHIGOT, Paul Louis, French surgeon; b. Etaples, Feb. 21, 1906; s. Eugène and Marthe (Colle) C.; M.D.; m. Pauline Loyson, 1933; children—Jean-Paul, Catherine. Intern, hosps. of Paris, later surgeon; chief service Trousseau Hosp.; prof. Sch. Medicine, Paris. Mem. Acad. Sci., Royal London Soc. Medicine (hon.). Author: Traité de traumatologie infantile. Address: 17, rue de Bourgogne, Paris, France.

CHIHARA, Theodore Seio, Am. mathematician; b. Seattle, Mar. 14, 1929; s. George I. and Nobue (Fushiki) C.; B.S., Seattle U., 1951; M.S., Purdue U., 1953, Ph.D., 1955; m. Amy E. Hirabayashi, Jan. 28, 1956; children—Laura, Lisa, Linda, Jerome, Gregg. Teaching asst. Purdue U., 1952-55; faculty in math. Seattle U., 1954—, head dept., 1958-66, prof., 1965—; summer faculty asso. Boeing Airplane Co., summers 1957, 58. Mem. Wash. State Adv. Com. on Math., since 19——. Mem. Am. Assn. U. Profs., Am. Math. Soc., Math. Assn. Am., Sigma Xi. Research and publs. on spl. functions, orthogonal polynomials, moment problems. Home: 3854 Cascadia Av. S., Seattle 98118.*

CHILD, Charles G. 3d, Am. surgeon; b. N.Y.C., Feb. 1, 1908; s. Charles G. and Helen (Francis) C.; grad. Philips Exeter Acad., 1926; A.B., Yale, 1930; M.D., Cornell Med. Coll., 1934; m. Margaret MacC Austin, June 14, 1941; children—Caroline W., Margaret F., Helen D., Cleland G., Charles A., Elizabeth M. Intern, asst. resident and resident surgeon N.Y. Hosp., 1934-42; instr., asst. clin. prof. surgery, asso. clin. prof. surgery Cornell Med. Coll., 1942-52; surgeon-in-chief New Eng. Center Hosp., Boston, 1952-54; prof. surgery, chmn. dept. Tufts U. Med. Sch., 1952-58; dir. first surg. service Boston City Hosp., 1954-58; prof. surgery, chmn. dept. University of Michigan Medical Sch., Ann Arbor, 1959—; dir. surgery Wayne County Gen. Hosp., Eloise. Mich. Served as lieutenant M.C., USNR, 1944-46. Fellow A.C.S.; mem. Am., New Eng. surg. socs., Soc. U. Surgeons, Soc. Clin. Surgery. Author articles profl. jours. Editor Jour. Surg. Research. Research on general surgery; liver diseases and neoplasia and pancreatico duodenal physiology; portal hypertension. Home: 3202 N. Maple Rd. Office: U. Mich. Med. Center, Ann Arbor, Mich.

CHILD, Charles Manning, Am. biologist; b. Ypsilanti, Mich., Feb. 2, 1869; s. Charles Chauncey and Mary Elizabeth (Manning) C.; Ph.B., Wesleyan U., Conn., 1890, M.S., 1892, hon. D.Sc., 1928; Ph.D., U. Leipzig (Germany), 1894; m. Lydia Van Meter, Aug. 15, 1899; 1 dau. Jeannette Manning. Grad. asst. Wesleyan U., 1890-92, Naples Zool. Sta., 1894, 1902; asst. in zoology U. Chgo., 1895-96, successively asso. instr., asst. prof., asso. prof., and prof., 1916-34, emeritus, 1934; vis. prof. Duke, 1930; Rockefeller Found. vis. prof., Tohoku Imperial U., Sendai, Japan, 1930-31; later lectr. in biology Stanford. Fellow A.A.A.S.; mem. Acad. Scis., Am. Soc. Zoologists, Am. Naturalists, Am. Physiol. Soc., Am. Assn. Anatomists, Linnean Soc. London (fgn. mem.), Société Royale Zoologique de Belgique (hon.), Phi Beta Kappa. Author: Die physiologische Isolation von Teilen des Organismus (Leipzig), 1911; Senescence and Rejuvenescence, 1915; Individuality in Organisms, 1915; The Orgin and Development of the Nervous System from a Physiological Viewpoint, 1921; Physiological Foundations of Behavior, 1924; Patterns and Problems of Development, 1941; also numerous articles giving results of researches, in Am. and European jours. Died Dec. 19, 1954.

CHILDE, Henry Langdon, Brit. inventor; b. 1781; exhibited "dissolving views" at Adelphi Theater in the 1830's and 40's; gave magic lantern demonstrations in Manchester, Eng., and other large provincial towns. Invented chromatrope; used achromatic lens and an improved oil lamp to increase the brilliancy of pictures. Died 1874.

CHILDE, Vere Gordon, archeologist; b. Sydney, Australia, Apr. 14, 1892; s. S. H. Childe; ed. U. Sydney, Queen's Coll., Oxford; B.Litt., 1916; D.Litt.; D.Sc.; LL.D., Edinburgh U., 1957. Librarian Royal Anthrop. Inst., 1925-27; prof. prehistoric archaeology U. Edinburgh, prof. prehistoric European archaeology, dir. Inst. Archaeology, U. London, 1946-56. Fellow Brit. Acad., Royal Anthrop. Inst., Soc. Antiquities (Scotland); hon. mem. German Archeol. Inst., Italian Palentol. Inst., Estonian Learned Soc., Swiss Prehistory Soc., Royal Irish Acad., Soc. Antiquaries of North, K. Nederlandsche Akademie van Wetenschappen det Norske Videnskaps—Akaemi. Author: The Dawn of European Civilisation, rev. edit., 1950; The Aryans, 1926; The Most Ancient East, 1928; The Danube in Prehistory, 1929; The Bronze Age, 1931; Skara Brae, 1931; New Light on the Most Ancient East, 1934; The Prehistory of Scotland, 1935; Man Makes Himself, 1937; What Happened in History, 1942; Progress and Archaeology, 1944; Prehistoric Communities of the British Isles, 1947; Piecing together the Past, 1956; A Short Introduction to Archaeology, 1956; Society and Knowledge, 1956; The Prehistory of European Society (pub. posthumously), 1958. Noted for his studies of European prehistory. Died Oct. 19, 1957.

CHILDREN, John George, Brit. scientist; b. Tunbridge, Eng., May 18, 1777; s. George C.; ed. Queen's Coll., Cambridge, Eng.; until 1798; m. Miss Holwell; m. 2d, Caroline Wise, 1809; m. 3d, Mrs. Towers, May 31, 1819. mem. staff Brit. Mus., 1816-40. Recipient Royal Instn. medal, 1828. Fellow Royal Soc., 1807 (sec. 1826-27, 30-37). Contbr. numerous articles on minerals to tech. jours.; pub. notes on electricity, 1808-15; translated chem. tracts, 1819-22. Constructed largest galvanic battery in existence at time, 1813; discovered method of extracting silver without using mercury. Died Jan. 1, 1852.

CHILDS, Ernest Carr, physicist; b. London, Eng., Nov. 6, 1907; s. William and Elizabeth (Carr) C.; B.Sc., U. London King's Coll., 1927, Ph.D., 1931; Ph.D., Clare Coll., Cambridge (Eng.) U., 1934, Sc.D., 1945; m. Marion Schmidt, Dec. 14, 1935; children—Thomas Henry Carr, Brian Matthew Colin. Research physicist Cambridge Instrument Co. Ltd., London, 1928-29; staff U. Cambridge, 1934—, asst. dir. research, 1945-62, reader, 1962—; vis. prof. U. Ill., also Kearney Found. lectr. U. Cal., 1958; hon. dir. unit soil physics Agrl. Research Council, 1951—; Fellow Darwin Coll., Cambridge. Fellow Cambridge Philos. Soc.; mem. Brit. Soc. Soil Sci., Am. Geophys. Union, Nat. Inst. Agrl. Engring. (gov.), Nat. Coll. Agr. Engring (gov.). Author: (with A. Campbell) Measurement of Inductunce, Capacitance and Frequency, 1935; also numerous articles, chpt. in book. Research on phys. basis of movement of water in porous matter both saturated and unsaturated especially applications to irrigation and drainage of soils. Home: 243 Hinton Way, Gt. Shelford, Cambridge, Eng.*

CHILDS, Henry E. Jr., Am. vertebrate ecologist; b. Providence, Jan. 8, 1924; s. Henry E. and Judith (Hopkins) C.; A.A., U. Cal. at Berkeley, 1947, A.B., 1949, M.A., 1951; Ph.D., 1959; m. Mary E. Baeck, Mar. 14. 1947; children—Stephanie H., Eric L. Research biologist U. Cal. at Davis, 1951-54; instr. Compton Coll., 1954-57; instr. Cerritos Coll., Norwalk, Cal., 1957-63, chmn. div. life sci., 1964—. NSF Faculty fellow, 1959-60; Louis Agassiz Fuertes Research grantee, 1950. Mem. So. Cal. Acad. Sci. (dir. 1964—, v.p. 1965), Cooper Ornithol. Soc., A.A.A.S., Am. Inst. Biol. Sci., Am. Soc. Mammalogists, Wildlife Soc., Ecol. Soc., Soc. Animal Behavior, Wildlife Disease Assn. Author: (with L. L. Cramer), Introductory Life Science, 1964; also articles. Research on bird and mammal biology, effects chronic irradiation on wild small mammals. Home: 15053 Neartree Rd., La Mirada, Cal. 90638. Office: 11110 E. Alondra Blvd., Norwalk, Cal. 90650.*

CHILIKIN, Mikhail Grigorevich, Russian elec. engr.; b. 1909; grad. Moscow Power Engring. Inst., 1935, postgrad., 1935-38; Cand. Tech. Sci., 1938; D.Tech. Sci., 1954. Instr., Moscow Power Engring. Inst., 1935-41, head chair elec. equipment for indsl. enterprises, 1951—, dir. bd. studies, until 1943, dep. dir. for studies. 1943-52, dir., 1952—; prof., 1951—. Chmn., sci. and Tech. Com. for Automated Elec. Drives and Use Elec. Machines; chmn. sect. for power equipment tech. council USSR Gosplan. Author: A General Course on Electric Drives; a High Voltage Electric Drive. Mem. editorial bd. Elecktrichestvo, Vestnik vysshey shkoly. Research on electrodrive systems particularly systems with asynchronous gliding sleeves with controlled mercury vapor rectifiers. Address: Moscow Power Engring. Inst., Krasnokazarmennaya 17, Moscow, USSR.

CHILLINGWORTH, John, Brit. astronomer, mathematician; b. Northunberland, Eng.; fellow Merton Coll., Oxford (Eng.) U., became prin. St. John's Hall, 1440; named jr. proctor of univ., 1441. Author: Algorismus; Canones et tabulae astronomicae; De Judiciis astronomiae; Arithmeticum opus. Founded sch. of zealous promoters of math. inquiry, Oxford. Died May 17, 1445.

CHILTON, Charles, biologist, surgeon; b. Leominster, Eng., Sept. 27, 1860; s. Thomas and Jane (Price) C.; ed. Canterbury Coll., Otago (New Zealand) U.; M.D., C.M., Edinburgh (Scotland) U.; M.A., D.Sc., New Zealand; postgrad. in medicine, Edinburgh, Heidelberg, Vienna, London, 1895-1900; LL.D., Aberdeen, Scotland. Tchr., Christchurch (New Zealand) schs., 1881-86; tutor Dunedin (New Zealand) Tng. Coll., 1886-88; rector Port Chalmers (New Zealand) Dist. High Sch., 1888-95; practice medicine specializing in ophthalmic surgery, Christchurch, 1900-02; prof. biology Canterbury Coll., U. New Zealand, 1902-29, rector Canterbury Coll., 1921-28, prof. emeritus, from 1929. Mem. Bd. Edn., New Zealand; active in ednl. and sci. work. Fellow Linnean Soc., London, New Zealand Inst. (pres. 1913-14); mem. Zool. Soc. (corr. London), Royal Soc. New S. Wales (hon.), Am. Mus. Natural History (hon.). Editor: The Subantarctic Islands of New Zealand, 1909. Research and publs. on Crustacea of New Zealand, Australia, India, subterranean Crustacea of New Zealand subantarctic islands, Amphipoda of Scottish nat. antarctic expdn. Died Oct. 25, 1929.

CHILTON, Thomas Hamilton, Am. chem. engr.; b. Greensboro, Ala., Aug. 14, 1899; s. Claudius L. and Mabel (Pierce) C.; student U. Ala., 1915-16; Chem. Engr., Columbia U., 1922; D.Sc., U. Del., 1943; m. Cherridah McLemore, June 29, 1926; children—Thomas McLemore, Daniel Tanner. Chemist, F. J. Carman, 1922-25; with E. I. Du Pont de Nemours & Co., Wilmington, Del., 1925-59, tech. dir. engring. dept., 1946-58; Regents prof. U. Cal. at Berkeley, 1959-60; Fulbright lectr. univs. Kyoto and Nagoya (Japan), Nancy and Toulouse (France), 1960-61; vis. prof. U. New South Wales, Sydney, Australia, 1960; Neely vis. prof. chem. engring. Ga. Inst. Tech., Atlanta, 1962-63; vis. prof. chem. engring. U. Del., 1963-64; vis. prof. chem. engring. Cal. Inst. Tech., Pasadena, 1965; vis. prof. chem. engring. U. Va., Charlottesville, 1965-66. Recipient Egleston medal Columbia Engring. Alumni Assn., 1934; Chandler medal Columbia U., 1939, Univ. medal, 1951. Mem. Am. Inst. Chem. Engrs. (pres. 1951, Founders award 1958), Am. Chem. Soc., A.A.A.S., Am. Assn. Engring. Edn., Nat. Acad. Engring. Contbr. numerous articles to profl. jours. Research and devel. process for prodn. nitric acid operating under pressure, simplified correlation of data on pressure loss suffered by fluids flowing through beds of granular packing, method of predicting rates of absorption of vapors in solvents from measurements on pressure loss, simple method of expressing performance data on absorption in differential contacting. Home: Yorklyn Rd., Hockessin, Del. 19707.*

CHIN, Tom Doon Yuen, physician, epidemiologist; b. Taishan, Kwangtung, China, May 29, 1922; s. Yoke Poy and Ken Jin (Yee) C.; M.D., U. Mich., 1946; M.P.H., Tulane U., 1950; m. Rita Chung, June 8, 1950; children—Donna, Robert. Came to U. S., 1934, naturalized, 1959. Pub. health physician La. State Dept. Health, 1950-51; with Communicable Disease Center, USPHS, Kansas City, 1954—, asst. chief, 1956-64, chief, 1964—; faculty U. Kan. Sch. Medicine, Kansas City, 1955—, clin. prof. microbiology, 1964—. Diplomate Am. Bd. Preventive Medicine, Am. Bd. Microbiology. Fellow A.A.A.S., Am. Pub. Health Assn.; mem. Am. Epidemiol. Soc., Infectious Diseases Soc. Am., Am. Soc. Microbiology, Am. Soc. Tropical Medicine and Hygiene, N.Y. Acad. Scis., A.M.A., Pan Am. Med. Assn., Reticuloendothelial Soc., Delta Omega. Asso. editor Am. Jour. Epidemiology. Research and publs. on epidemiology of enterovirus infections, acute respiratory infections, histoplasmosis, leukemia, epidemiology of poliomyelitis following introduction vaccines, defined importance of various viruses in alimentary tract and respiratory tract to human disease. Home: 9910 Fontana Lane, Overland Park, Kan. 66207. Office: 2002 W. 39th St., Kansas City, Kan. 66103.*

CHINAKAL, Nikolai Andreevich, Russian mining engr.; b. Nov. 19, 1888; grad. Dnepropetrovsk Mining Engrs. Inst., 1912; Ph.D. in Tech. Scis. (hon.); became asso. with Donbas Mines, 1912; apptd. asst. to authorized rep. Central Adminstrn. of Coal Industry in Makeevskii Region, 1920; prof., dir. advanced mining constrn. Kirov Poly. Inst., Tomsk, 1940-44; dir. Siberian Br. Inst. Mining, USSR Acad. Scis., 1957—. Mem. Council of Labor and Def. (developed plan for restoring Donbas); mem. Cen central com. All-Russian Union Miners, 1921, also chief econ. sect., central com. Recipient Stalin prize. Mem. USSR Acad. Scis. Author: System of Exploitation with Shield Reinforcements, 1943; Shield Method of Exploitation, Progressive Method Using Systems of Exploitation in the Kuzbas, 1957; (with N. V. Marevich) Shield Method of Exploitation with Gravity Filling of Worked-out Space, Progressive Method Using Systems of Exploitation in the Kuzbas, 1957. Contbr. to sci. publs. Developed and improved systems of utilization and mechanization of coal deposits; formulated scheme of shield reinforcement for exploitation of thick strata of precipitous slopes. Office: Inst. of Mining of Siberian Dept. of USSR, Acad. of Scis., Irkutsk, Siberia, USSR.

CHING, Te May Tsou (Mrs. Kim Kwong Ching), physiologist; b. Soochow, China, Jan. 9, 1923; d. S. W. and S. (Tsao) Tsou; B.S., Nat. Central U., Nanking, China, 1944; M.S., Mich. State U., 1950; Ph.D., Mich. State U., 1954; m. Kim Kwong Ching, Apr. 6, 1946; children—Franklin Gee-tan, Annabel Gee-Lan. Came to U. S., 1948, naturalized, 1964. Research asst. Mich. State U., East Lansing, Mich., 1952-54, research instr., 1954-56; faculty Ore. State U., Corvallis, 1956—, asso. prof., 1962—. Mem. Genetic Soc. Am., Am. Soc. Plant Physiologist, Am. Soc. Agronomy, Am. Soc. Oil Chemists, Canadian Soc. Plant Physiologists, Societas Physiologia Plantarum, A.A.A.S., Sigma Xi, Xi Sigma Pi, Sigma Delta Epsilon. Research and publs. on induced and simultaneous aberrations of mitosis in plants, safe storage conditions for forage seeds, freeze-drying coniferous pollen for preservation, biology of seed devel. and germination, lipid metabolism in coniferous seeds. Home: 3240 Elmwood Way, Corvallis, Ore. 97330.*

CHIPLONKAR, Vasant Trimbak, Indian physicist; b. Belgaum, India, Aug. 23, 1912; s. T. L. and Bhagirthi (Kelkar) C.; B.Sc., Royal Inst. Sci. Bombay (India), 1932; M.Sc., Banaras Hindu U., 1934, Ph.D., 1947, research fellow, 1934-37; m. Mandakini Lele, Dec. 15, 1941; children—Ashok, Hemant, Ajit, Varsha. Asst. lectr. Royal Inst. Sci. Bombay, 1937-40, asso. prof., 1954-60, prof. physics, 1960—; demonstrator Banaras Hindu U., 1940-46, prof. physics M.N. Coll., Visnagar, India, 1946-54. Mem. Indian Phys. Soc. (v.p. 1962-63), Indian Conv. Spectroscopists (sec. 1964). Research, publs. on gaseous electronics, rectification in cold cathode glow discharge constn. current generator gas device, impedance of normal glow discharge, distbn. of current density on cathode surface; plasma physics radial proficles of electron density and energy in diffusion controlled plasma, thermionic plasma converters; ion-impact, light excitation, angular distbn. of sputtering from metal targets. Home: 13-D Ranade Rd., Bombay 28. Office: Inst. Sci., MME Cama Rd., Bombay 1, India.*

CHIRAC, Pierre, French physician; b. Conques, France, 1650; M.D., Montpellier, France, 1683; became prof. theology U. Montpellier, 1678, prof. medicine. 1684; named chief physician to army of Roussillou in Catalogue, 1692; 1st physician to Duke of Orleans, from 1715; dir. Royal Garden, from 1718; named 1st physician to Louis XV, 1731. Mem. French Acad. Scis., 1699. Author: Treatise on Malignant Fevers, 1742; Dissertations et Consultations medicinales, 3 vols., 1744-55. Studied circulatory system, also heart-beat, treatment of battle wounds, structure of hair and of liver; believed in existence of neuro-lymphatic capillaries; distinguished for work in plague of Rochefort (Mal de Siam), 1692. Died Marly, France, Mar. 1, 1732.

CHIRIGOS, Michael Anthony, Am. biochemist; b. Weirton, W.Va., Sept. 14, 1924; s. Anthony and Mary

(Michallos) C.; A.A., Balt. Jr. Coll., 1949; B.S., Western Md. Coll., 1952; M.S., U. Del., 1954; Ph.D., Rutgers U., 1957; m. Mary Eleftherious Lazapoulos, Sept. 19, 1954; children—Fevronia, Michael Anthony, Marianthia. Research fellow Rutgers Inst. Microbiology, 1954-55, research fellow dept. agrl. biochemistry, 1955-57; research fellow Nat. Heart Inst., NIH, Bethesda, Md., 1957-59, biochemist lab. chem. pharmacology Nat. Cancer Inst., 1959-63, head viral chemotherapy sect. drug evaluation br. Cancer Chemotherapy Nat. Service Center, 1963-66, head virus, disease modification sect. leukemia and lymphoma br. Etiology, Viral Oncology, 1966—; panel mem. Bd. U. S. Civil Service Examiners. Mem. Soc. Exptl. Biology and Medicine, Am. Assn. Cancer Research, A.A.A.S., Sigma Xi. Contbr. numerous articles to profl. jours. Developed meningeal leukemia in animals to trace devel. of meningeal leukemia occurring in humans with the aim of ascertaining mechanism of leukemia spread to brain and most efficient therapy to control this syndrome; studies on oncogenic viruses in vitro and in vivo, mechanisms of controlling oncogenic viruses and their induced diseases which include drug therapy, immuno-therapy and vaccines. Home: 318 Linthcum Rd., Rockville, Md. 20851. Office: 7550 Wisconsin Av., Bethesda, Md. 20014.*

CHISHOLM, William, Sr., inventor, mfr.; b. Scotland, Aug. 12, 1825; was a sailor, then, 1847-52, builder at Montreal; removed, 1852, to Cleve.; became mgr. Cleve. Rolling Mills, then mfr. of spikes, bolts; discovered practical method for mfr. of screws from Bessemer steel; organized Union Steel Co.; invented new method for mfg. steel shovels and spades, also new method of steam hoisting and pumping engines, conveyors for coal and ore. Died 1908.

CHISINI, Oscar, Italian mathematician; b. Bergame, Mar. 14, 1889; Ph.D. in math. Prof. analytical, projective and descriptive geometry univs. Bologna, Modena, Cagliari, Milan; dir. Periodico di matematiche. Mem. dei Lincei Acad., acads. of Bologna, Milan, Turin, Royal Soc. Sci. of Liege. Research on algebraic geometry. Address: University of Milan, Milan, Italy.

CHISTOVICH, Aleksey Nikolaevich, Russian pathoanatomist; b. 1905; grad. Leningrad Mil. Med. Acad., 1926; D.Med. Sci., 1935. Asst. dept. path. anatomy Leningrad Mil. Med. Acad., 1926-35, lectr., 1935-39, dep. head chair path. anatomy, 1942-45, prof., head chair, 1946—; prof. chair path. anatomy Kuybyshev Mil. Med. Acad., 1940-42; cons. path. anatomy dept. Leningrad Tb Research Inst. Mem. USSR Acad. Med. Sci. (corr.). Author over 80 works including: co-author Outline of Pathological Anatomy and the Clinical Aspects of Tuberculosis, 1949. Research and publs. on path. anatomy, anatomy and pathogenesis of Tb, especially bone Tb, blood diseases, gunshot wounds of lungs and other organs, infected wounds, effects of antibiotics on morphological symptoms of infections. Address: Leningrad Mil. Med. Acad., ulitsa Lebedeva 6, Leningrad, USSR.

CHISTOVICH, Andrey Sergeevich, Russian psychiatrist; b. St. Petersburg, Russia, 1897; grad. Petrograd. Mil. Med. Acad., 1922; D.Med. Sci. Head chair psychiatry Novosibirsk Med. Inst., 1938-46; head psychiat. clinic Pavlov Inst. Evolutionary Physiology, head psychiat. sect. Inst. Physiology, USSR Acad. Sci., 1946-52; head chair psychiatry Naval Med. Acad., 1952-56, Leningrad Mil. Med. Acad., 1956—. Mem. Czechoslovak. Med. Soc. Co-editor Psychiatry sect. Large Med. Ency., 2d edit. Research and publs. on path. physiology and pathogenesis of delirium, infectious psychoses (tularemia, brucellosis, rheumatic psychoses), etiological role of latent infection in genesis of psychoses. Address: Leningrad Mil. Med. Acad., ulitsa Lebedeva 6, Leningrad, USSR.

CHITANI, Toshizo, Japanese chemist; b. Tokyo, Japan, 1901; grad. Tokyo U., 1901; studied in Germany; became prof. Osaka U., then prof. Tokyo Municipal U., since 1953. Recipient Hattori Hoko-kai prize. Author: Heavy Hydrogen and Heavy Water; also contrb. articles to jours. Authority on ionized water.

CHITTENDEN, Frank Hurlbut, Am. entomologist; b. Cleve., Nov. 3, 1858; s. S. King and Harriet M. C.; licentiate Cornell U., 1881; Sc.D., Western U. of Pa., 1904; Asst. entomologist, U. S. Dept. Agr., from 1891; entomologist, truck crop insect investigations, from 1917. Author: Insects Injurious to Vegetables; also bulls, and other papers on econ. and tech. entomology for U. S. Dept. Agr. Died Sept. 15, 1929.

CHITTENDEN, Russell Henry, Am. physiol. chemist; b. New Haven, Conn., Feb. 18, 1856; s. Horace Horatio and Emily Eliza (Doane) C.; Ph.B., Sheffield Scientific Sch., Yale, 1875, Ph.D., 1880, LL.D., 1922; Heidelberg U., 1878-79; LL.D., U. Toronto, 1903, U. Birmingham, 1911, Wash. U., 1915; Sc.D., U. Pa., 1904; m. Gertrude Louise Baldwin, June 20, 1877; children—Edith Russell, Alfred Knight, Lilla Millard (Mrs. Harry Gray Barbour). Asst., Yale, 1874-77, instr., 1877-78, 1879-82, prof. physiol. chemistry, 1882-1922, dir. Sheffield Sci. Sch., 1898-1922; lectr. physiol. chemistry, Columbia, 1898-1903. Mem. referee bd. of cons. sci. experts to Sec. of Agr.; mem. exec. com. NRC, 1917; U. S. rep. Inter-Allied Sci. Food Commn., London, Paris and Rome, 1918. Fellow Am. Acad. Arts and Scis., N.Y. Acad. Medicine (hon.); mem. Nat. Acad. Sciences, Am. Philos. Soc. Société des Sciences Médicales et Naturelles de Bruxelles, Am. Soc. Naturalists (pres. 1893), Am. Physiol. Soc. (pres. 1895-1904), Am. Soc. Biol. Chemists (pres. 1907); corr. mem. Société de Biologie, Paris. Editor: Studies in Physiological Chemistry, 4 vols., 1884, 1901. Author: Digestive Proteolysis, 1895; Physiological Economy in Nutrition, 1905; Nutrition of Man, 1907; History of the Sheffield Scientific School (2 vols.), 1928; Development of Physiological Chemistry in the United States, 1930; also many papers on physiol. subjects in Am. and foreign jours. Probably founded 1st physiol. chemistry lab. in U. S.; pioneer in research on biochemistry of nutrition and digestion; 1st to isolate glycocoll and glycogen. Died New Haven, Dec. 26, 1943.

CHITWOOD, Benjamin Goodwin, Am. zoologist; b. Chgo., Dec. 21, 1907; s. Richard Mortimer and Eleanor Mary (McCombie) C.; B.A. with honors in biology, Rice Inst., 1928; M.S., George Washington U., 1929, Ph.D., 1931; m. May Belle Hutson, Apr. 17, 1927 (div. Apr. 1952); children—Marie Diana (Mrs. Karl Bushong), Edward Richard; m. 2d, Lily B. Devries, Mar. 2, 1962; 1 dau., Haunani-Lynn Devries (adopted). Jr. nematologist div. nematology U. S. Bur. Plant Industry, 1928-29, field asst. div. small fruit, 1930, asst. zoologist div. zoology, 1931-37, asso. nematologist div. nematology, 1937-50; asso. prof. zoology Cath. U. Am., 1950-52; chief nematologist Fla. State Plant Bd., Gainesville, 1955-58; sr. scientist Lab. Comparative Biology, Kaiser Found. Research Inst., Berkeley, Richmond, Cal., 1958-61; sr. pathologist U. Hawaii Agr. Expt. Sta., Lihue, Kauei, 1961-62; lectr. biology Western Wash. State Coll., Bellingham, 1963-65; clk. Tacoma Pub. Library, 1965-66, sr. exptl. aide dept. entomology Western Wash. Research and Extension Center, Puyallup, 1967—. Mem. A.A.A.S. (mem. com. on classification animal kingdom 1936—), Helminth Soc. Wash. (past pres.). Author: An Introduction to Nematology, 1937, 38, 40, 50; also numerous articles. Research on soil fumigation, plant disease control, Systematics Nemata, Nematomorpha, Kinorhyncha, Tardigrada. Address: Expt. Sta. Lab., Pioneer Av.; Puyallup, Wash. 98371.*

CHIU, Wan Cheng, meteorologist; b. Meihsien, Kwangtung, China, Nov. 1, 1919; s. Dow Nan and Ah-Jun (Lim) C.; came to U. S., 1945, naturalized, 1962; B.S., Nat. Central U., Chungking, China, 1941; student U. Chgo., 1945-46; M.S., N.Y. U., 1947, Ph.D., 1951; m. Margaret C. Y. Liu, Feb. 6, 1954; children—Linda, Ellen, Elaine Amy. Staff, N.Y. U., 1951-61, meteorologist, 1955-60, research scientist, 1960-61; prof. meteorology U. Hawaii, Honolulu, 1961—. Mem. Am. Meteorl. Soc., Am. Geophys. Union, Royal Meteorol. Soc. Eng. Research and publs. on atmospheric tides, large scale atmospheric wind systems, energies asso. with various scales of atmospheric motions. Home: 216 Kalalau St., Honolulu 96821.*

CHIU, Ying Nan, chemist; b. Canton, China, Nov. 25, 1933; s. Nien Tai and Yun-Dan (Liang) C.; student Nat. Taiwan U., 1950-52; B.A., Berea Coll., 1955; M.S., Yale, 1956, Ph.D., 1960; m. Lue-Yung Chow, June 18, 1960; children—Han-Seh, Kong-Seh. Post-doctoral fellow Columbia University, N.Y.C., 1960-62; research associate University Chicago, 1962-64; asst. prof. Cath. U. Am., 1964-66, asso. prof., 1966—. Cons. Melpar, Inc., Falls Church, Va., 1966; lectr. Summer Sci. Inst., Academia Sinica, Nat. Taiwan U. Nat. Tsing-Hua U. Monsanto Chem. Co. fellow, 1958; China Internat. Found. scholar, 1953. Mem. Am. Chem. Soc., Am. Phys. Soc., Sigma Xi, Phi Kappa Phi. Research, publs. on application of angular momentum and irreducible tensor methods in molecular quantum mechanics and molecular spectroscopy, expansion of irregular solid spherical harmonics, theory of higher multipole radiation in molecules. Home: 9298 Edmonston Rd., Greenbelt, Md. 20770.*

CHIZHIKOV, David Mikhaylovich, Russian metall. engr.; b. 1895; grad. Moscow Mining Acad., 1924; D.Tech. Sci., 1935. With Moscow Copper Electrolyte Plant and Vladikavkaz Lead and Zinc Plant, until 1927, Konstantinovka Zinc Plant, 1927-30; an organizer, 1st dir. Non-Ferrous Metall. Research Inst., 1930; prof. Moscow Inst. Non-Ferrous Metals and Gold, 1933-41; asso. Inst. Metallurgy, USSR Acad. Sci., 1939—. Chmn., Sci. and Tech. Council on Non-Ferrous Metals and Gold; in charge tech. consultations on numerous non-ferrous metall. enterprises. Recipient Stalin prize, 1942, 50. Mem. USSR Acad. Sci. (corr.). Author: The Metallurgy of Heavy Non-Ferrous Materials, 1948; The Metallurgy of Zinc, 1938; The Metallurgy of Lead, 1944; Non-Ferrous Metallurgy, 1958, also textbooks on zinc and lead. Address: Inst. Metallurgy, USSR Acad. Sci., Leningradsky prospect 49, Moscow, USSR.

CHLADNI, Ernst Florence Friedrich, German physicist; b. Wittenberg, Germany, Nov. 30, 1756; s. Ernst Martin and Johanna Sophia C.; educated in law; grad. U. Leipzig (Germany), 1782; physicist at Wittenberg; gave mus. performances and sci. lectures. Author: Entdeckungen über die Theorie des Klanges, 1787; Die Akustik, 1802; Über die Hervorbringung der menschliche Sprachlaute, 1824; Neue Beiträge zur Akustik, 1817. A founder of sci. of acoustics; discovered longitudinal vibrations in string or rod, 1787, rotating oscillation of rotating waves, the cause of consonance and dissonance in tuning forks; studied vibration of plates by sprinkling them with sand and bowing the edges (figures produced are called Chladni's figures); showed that communication of vibrations in material bodies is subject to constant math. laws; measured velocity of sound in gases other than air; invented the euphonium. 1789; discovered boundary of high tones to be 22,000 oscillations per second; held the theory of aerolites, 1794; one of 1st to demonstrate that meteorites could not have originated on earth. Died Breslau, Germany (now Wroclaw, Poland), Apr. 3, 1827.

CHODAT, Robert, botanist; b. Moutier-Grandval, Switzerland, Apr. 6, 1865; ed. Bienne, Berne, Bale, Geneva; Bachelier es Scis., Geneva, Docteur es Scis. (Naturelles), 1887; hon. doctorates Victoria U., 1902, Cambridge U. Became asso. prof. systematic botany Geneva U., 1889, prof., 1891, prof. botany, 1900, doyen Faculty Scis., 1898-1906, rector, 1908-10; dir. Inst. Botany; dir. Jardin alpin LaLinneae, also Lab. Alpine Biology, Bourg-St.-Pierre, Grand St. Bernard; became dir. Herbier Boissier, 1918. Mem. Linnean Soc. (fgn. mem., Gold medal 1933), Brit. Assn. (asso.), Duetsche Akademie der Naturforscher Halle, Royal Acad. Belgium (asso.), Inst. France (corr.), Belgian Soc. Biology, (corr.), acads. scis. Turin, Russia, Modena (all corr.), Am. Bot. Soc. (corr.). Author: Monographia polygalacearum; Distribution et origine de l'espèce et des groupes chez les Polygalacées; Sur l'origine des tubecribles dans les bois; Sur le polymorphisme du Scenedesmus acutus; Scenedes mus, 1926; La Feuille des Iridées; Algnes vertes de la Suisse; La niege rouge; Principles de botanique, 1906; Biologie des Plantes, 1917; Sur le polymorphisme des Algues, 1908; Végétation du Paraguay, 1916-21; La Théorie de la mutation généralisée, chez le Chlorella rubenscens, 1929; Monographies d'Algues en culture pure—les Pteridopsides des temps paléozoiques; also articles. Died Apr. 1934.

CHODOROW, Marvin, Am. physicist; b. Buffalo, July 16, 1913; s. Isidor and Lena (Cohen) C.; B.A., U. Buffalo, 1934; Ph.D., Mass. Inst. Tech., 1939; m. Leah Ruth Turitz, Sept. 19, 1937; children—Nancy Julia, Joan Elizabeth. Research asso. Pa. State Coll., 1940-41; instr. physics Coll. City N.Y., 1941-43; sr. project engr. Sperry Gyroscope Co., 1943-47; mem. faculty Stanford, 1947—, prof. applied physics and elec. engring., 1954—, dir. microwave lab., 1959—, exec. head div. applied physics, 1962—; cons. Def. Dept., also Rand Corp.; vis. lectr. Ecole Normale Superieure, Paris, France, 1955-56; designed 1st klystron for microwave relay systems, 1946, 1st megawatt klystron, 1949, 1st pulsed high power traveling-wave tubes, 1952-57. Fulbright fellow, Cambridge (Eng.) U., 1962-63. Fellow I.E.E.E. (W. R. G. Baker award 1962), Am. Phys. Soc.; mem. Nat. Acad. Engring., A.A.A.S., Am. Assn. Physics Tchrs., Phi Beta Kappa, Sigma Xi. Co-author: Fundamentals of Microwave Electronics, 1964; also articles. Home: 247 La Cuesta Dr., Ladera, Menlo Park, Cal. Office: Microwave Lab., Stanford Univ., Stanford Cal. 94305.*

CHOE, Sang, Korean marine biologist; b. Annui, Korea, Jan. 21. 1927; s. Young Joo and Ah Ji (Rhee) C.; B.S. in Agr., U. Tokyo (Japan), 1952, postgrad., 1952-57, Dr.Agri., 1961; m. Kyoung JaWooh, Nov. 6, 1955; children—Hyon Jae, Hyon Whoo. Research asst. dept. fisheries U. Tokyo, Japan, 1957-62; asso. prof. marine biology Fisheries Coll., Pusan Nat. U., Korea, 1962-63; sr. biologist biology div. Atomic Energy Research Inst., Seoul, Korea, 1963—; instr. dept. zoology Seoul Nat. U., 1964—. Chmn. spl. com. for fisheries Korea, FAO, 1964—; chmn. Korean oceanographic com. Korean Nat. Commn., UNESCO, since 1965—. Recipient Distinguished Publ. award Pusan Nat. U., 1963. Mem. Japanese Soc. Sci. Fisheries. Ecol. Soc. Japan, Malcological Soc. Japan, Zool. Soc. Korea, Plankton Soc. Japan, Am. Soc. Limnology and Oceanography. Author: Biology of the Japanese Common Sea Cucumber, 1963; also articles. Research on ecology of Tapes japonica, life histories of Loripes pisidium and Diopatra neapolitana, biol. and ecol. bases for multiplication of Stichopus japonicus, phytoplankton, measurement marine primary productivity of Korean sea waters; successfully raised from hatching certain species of cuttlefishes and squids. Home: 99-5 Chongam-dong. Office: 217, Kongnouung, dong-Soungbuk-ku, Seoul, Korea.*

CHOKUYEN, Ajima, Japanese mathematician; worked on diophantine analysis, Malfatti problem (to inscribe 3 circles in a triangle, each tangent to other 2), treated special cases of problem to determine number of figures in repetition in circulating decimals; improved Yenri theory by taking equal divs. of chord instead of arc. Died 1798.

CHOMBART DE LAUWE, Paul-Henry, French sociologist; b. Cambrai, France, Apr. 8, 1913; s. Henry and Marie (Blanchemain) C. de L.; Licencé-ès-lettres, U. Paris, 1936, Doctorat-ès-lettres, 1955; ethnol. studies in Africa, 1936. Placed in charge research Nat. Center Sci. Research, 1945; founder group of social ethnologists Mus. of Man, 1949; became dir. studies Sch. Advanced Studies, Sorbonne, 1960; named dir. Center Social Ethnology and Psychosociology, 1965. Mem. Soc. Sociology, Soc. Psychology, Soc. Polit. Scis., Soc. Econs. Author: La vie quotidienne des familles ouvrières, 1956; (with others) La femme dans la société, 1963; (with others) Famille et habitation, 1960; also papers. Research in urban demography and sociology, with spl. attention to role of women. Home: 10, rue de Chateaudun, Ivry (94), France. Office: 1, rue du Onze Novembre, Montrouge (92), France.*

CHOMEL, Antoine, French astronomer; b. Moulins, France, Mar. 20, 1668; s. Antoine C.; studied rhetoric and grammar; joined Soc. of Jesus, 1683, took first vows, 1685; traveled to Canton, China via Turkey and Persia, 1698-1701. Mem. French Acad. Sci., 1699. Translated Latin poetry into French. Made astron. observations in China; introduced Western astronomy to Orient. Died Canton, China, May 12, 1702.

CHOMEL, August François, French physician; b. Paris, France, Apr. 13, 1788; M.D., École pratique, 1813; chief intern, physician Charité Hosp., Paris, 1814-23; became prof. clin. medicine, 1826; later prof. clin. medicine l'Hôtel-Dieu; participated in reform of med. sch., Paris. 1823; became lab. asst. to Fouquier, Paris, 1823; physician to King Louis-Philippe and royal family. Author: Éléments de pathologie général, 1817; De l'existence des fièvres; Des fièvres et maladies pestilentielles, 1821; Lecons de clinique, 3 vols., 1834-40; Des dyspepsies, 1857. Mem. French Acad. Medicine. Described peripheral neuritis, 1828; studied pathology. Died Chateau de Morsang, Apr. 9, 1858.

CHOMEL, Pierre-Jean-Baptiste, French botanist; b. Paris, France, Sept. 2, 1671; s. Jean-Baptiste and Anne François (Le Breton) C.; studied under his uncle Noël at U. Lyons, (France); studied botany under Tournefort at U. Paris; M.D., 1697; m. Denise Corps; m. 2d, Marie Jeanne Jobert; children—Jean Baptiste Louise, Louis Denis, Charles Jean. Named royal physician, 1706; created bot. gardin l'École de pharmacie; named prof. botany Med. Faculty at Paris. 1705; physician at Hôtel-Dieu, 1710-39. Mem. French Acad. Scis., 1720. Author: Abrégé de l'histoire des plantes usuelles, 1712. Participated in arguments of physicians against surgeons. Died Paris, July 4, 1740.

CHOMIAK, Marian, Polish neuroanatomist; b. Nadolce, Poland, Dec. 18, 1912; s. Kacper and Zofia (Wisniewska) C.; MVD, Faculty Vet. Medicine, Warsaw (Poland) U., 1946; m. Celina Chomiak, Apr. 17, 1947; children—Maria, Agnieszka. With Faculty Vet. Medicine, Mariae Curie-Skodowska U., Lublin, Poland, 1944-48, dep. dean, 1951-54, dean Faculty Vet. Medicine, 1954-55; with Faculty Vet. Medicine, Agrl. Coll., Lublin, 1948—, prof. neuranatomy, 1963—, dep. rector, 1955-62, rector, 1965. Decorated Gold Cross of Merit, medal of X-Anniversary Polisch People's Republic, Knight's Order of Poland's Regeneration Order. Mem. Internat. Assn. Vet. Anatomists, Polish Anatomy Assn., Polish Zool. Assn., Polish Assn. Vet. Sci. Research on devel. and microstructure of central nervous system of domestic animals; innervation individual or-organs of domestic animals. Home: 36/11 Al.PKWN, Lublin, Poland.*

CHOMPRÉ, Nicolas Maurice, French mathematician, physicist; b. Paris, France, Sept. 23, 1750; s. Pierre C.; Named dir. Bur. Mines and Agr., 1771; dir. gen. corr., 1777-86; bur. chief at Tresor royal; sec. Govt. of Boulonnaise; named geometer Ministry of Interior, 1794; chief Bur. for Fgn. Affairs; named consul in Malaga, 1795; mem. Council for Mitigating Civil Disputes, 1806-14. Author: Élemens d'arithmetique, d'algèbre de géométrie, 1777. Publs. on physics especially galvanism. Died Ivry sur Seine, France, July 24, 1825.

CHOPART, François, French surgeon; b. Paris, France, Oct. 30, 1743; s. Francois Turlure and Marie-Anne c.; M.A., Hôtel-Dieu, 1761; M.D., 1770; studied surgery under Antoine Petit. Became prof. surgery l'École pratique, circa 1770; named demonstrator physiology Paris, 1782; later prof. surgery, surgeon; head Hosp. Chanty, Paris; named prof. pathology, 1789; apptd. surgeon, chief Sch. Medicine, 1790; apptd. physician to Dauphin Louis XVII, 1793. Author: les Lésions à la tête, 1771; (with Desault) Traité des maladies chirurgicales, 1789; Traité des maladies des voies urinaires, 1791. First to perform mediotarsal disarticulation (named in his honor); founder (with Desault) urinary surgery. Died Paris, June 9, 1795.

CHOPPIN, Arthur Richard, Am. chem. engr.; b. Alexandira, La., Dec. 13, 1897; s. Dr. Arthur R. and Margaret (Whittington) C.; B.A., La State U., 1918, M.S., 1925; Ph.D., Ohio State U., 1929; m. Eunice D. Bolin, Mar. 29, 1919; children—Arthur Richard, George Purnell. Teacher, Rapides Parish (La.) Sch. Bd., 1919-20; asst. principal, Vermillion (La.) Sch. Bd., 1920-23; grad. asst., La State U., 1924-25, asst. chemistry, 1925-26, instr., 1926-28, asst. prof., 1929-31, asso. prof., 1931-36, prof., since 1936, prof. and mgr. lab. stores, 1939-43; asst. to dean of Coll. of Chemistry and Physics, 1944, dean of the college, and professor of chemistry since 1944. Served as 2d lt., Machine Gun Corps, World War I. La. State Civilian Gas Consultant mem. State Emergency Relief Bd.; dir. Pelican Boys' and Girls' States; past pres. Baton Rouge (La.) Rotary Internat. Club; advisory mem. Econ. Development Com. of La.; adjutant Univ. Post of Am. Legion; state comdr. Am. Legion of La., 1946, nat. vice comdr., 1965-66; bd. dirs. Gulf South Research Inst.; dir. radiol. and special weapons div. La. Civil Def.; vice chmn. Louisian bd. Nuclear Energy; chmn. Community Recreation Council, Selective Service com. of La. State U.; sec. La. Wildlife Assn., 1932. Fellow Am. Inst. Chemists, Am. Inst. Chem. Engrs.; mem. Am. Chem. Soc., Am. Men of Sci. A.A.A.S., Sigma Xi. Holder of several United States patents. Contbr. of articles to profl. jours. Research in chem. kinetics, spectroscopy, by-product utilization, pollution control. Home: 722 S. Lakeview Dr., Baton Rouge. 70810. Office: La. State U., Baton Rouge, 70803.*

CHOPPIN, Gregory Robert, Am. chemist; b. Eagle Lake, Tex., Nov. 9, 1927; s. Gilbert P. and Nellie (Guidroz) C.; B.S., Loyola U., New Orleans, 1949; Ph.D., U. Tex., 1953; m. Ann M. Warner, June 9, 1951; children—Denise, Suzanne, Paul, Nadine. Began career as research scientist Lawrence Radiation Laboratory, University of California at Berkeley, 1953-56; faculty Fla. State U., Tallahassee, 1956—, prof. chemistry, 1963—; vis. scientist Centre D'Etude Nucleaires, Mol-Donk, Belgium, 1962-63. Chmn. subcom. on radiochemistry com. on nuclear sci. Nat. Acad. Sci.-NRC, 1966—. Named Young Man of Year, El Cerrito (Cal) Jr. C. of C., 1956. Mem. Am. Chem. Soc., A.A.A.S., Phi Beta Kappa, Sigma Xi., Author Experimental Nuclear Chemistry, 1961; Nuclei and Radioactivity, 1964; (with B. Jaffe) Chemistry The Science of Matter, Energy, And Change, 1965; also numerous articles. Co-discoverer of element 101, mendelevium; elucidation of factors important in complexation of lanthanide and actinide ions in aqueous solutions; provided spectroscopic evidence on structure in water. Home: 840 Watt Dr., Tallahassee 32303.*

CHOQUET, Gustave, French mathematician; b. Solesmes, France, Mar. 1, 1915; s. Gustave and Marie (Fosse) C.; Sc.D., U. Paris, 1946; m. Marie Pihan, 1942; children—Bernard, Christian, Claire, Daniel; m. 2d, Yvonne Bruhat, 1961. Prof. math. French Inst. Poland, 1946; lectr. U. Grenoble, 1947-49; faculty U. Paris, 1949—, now prof. math.; prof. Poly. Sch. Paris, 1965—. Decorated Officier d'Academie; Chevalier de la Legion d'Honneur. Mem. Am., French math. socs. Numerous publs. on theory of real functions, potential theory, functional analysis; theory of capacities and integral representations. Home: 16 Avenue d'Alembert, Antony, France. Office: Inst. Henri Poincare, 11 rue Pierre Curie, Paris 5, France.*

CHORUS, Alphons Maria Joseph, Dutch psychologist; b. Munstergeleen, Apr. 18, 1909; s. Arnold and Pauline (Pijls) C.; ed. U. Nimegue, U. Utrecht; Ph.D.; m., 1905; 9 children. Psychologist, Pedological Inst.; reader U. Nimegue; prof. psychology U. Leiden, 1947—, founder, dir. Psychol. Inst.; founder psychology service of Netherlands, The Hague, 1950. Author: Sociale psychologie; Psychologie van de menselijke levensloop; Het beeld van de mens in de oude biografie. Home: Rijingeesterstraatweg 11 Oegstgeest. Office: Rynsburgerweg 169, Leiden, Netherlands.

CHOU, Shelley Nien-chun, physician; b. Chekiang, China, Feb. 6, 1924; s. Shelley P. and Tse-tsun (Chao) C.; B.S., St. John's U., Shanghai, China, 1946; M.D., U. Utah, 1949; M.S., U. Minn., 1954, Ph.D., 1964; m. Jolene Johnson, Nov. 24, 1956; children—Shelley T., Dana, Kerry. Resident, U. Minn. Hosps., 1950-55; practice medicine, specializing in neurosurgery, Salt Lake City, 1955-58, Bethesda, Md., 1959, Mpls., 1960—; clin. asst. Coll. Medicine U. Utah. 1956-58; vis. scientist Nat. Insts. Neurol. Diseases and Blindness NIH, 1959; faculty U. Minn., 1960—, asso. prof. neurosurgery, 1965—. Diplomate Am. Bd. Neurosurgery. Mem. A.M.A., Congress Neurol. Surgery, Soc. Nuclear Medicine, Harvey Cushing Soc., Neurosurg. Soc. N.Am., Forum U. Neurosurgeons, A.A.A.S., Phi Rho Sigma. Contbr. numerous articles to profl. jours. Publs. on studies of intracranial lesions using radioactive angiography techniques; malformations of cerebral vasculature; neurol. dysfunctions of urinary bladder. Home: 2 Otter Lane, North Oaks, Minn. 55110. Office: B-590 Mayo Meml., 412 S.E. Union St., Mpls. 55455.*

CHOUARD, Claude Henri, French physician; b. Paris, July 7, 1931; s. Pierre and Denise (Petit-Dutaillis) C.; ed. Paris Faculty Scis., Faculty Medicine; m. Isabelle Sainflou, July 7, 1959; children—Christophe, Matthieu, Clément. Intern, Paris hosps., 1955; became asst. in neuroanat. research, 1960; head Neuroantomic Clinic, Paris, 1962—; aggregate prof. otorhinolarynxology Paris Faculty Medicine. Mem. French Soc. Surgeons, Soc. Pathology, Internat. Soc. Plastic Surgeons. Author: (with others Chirurgie Cervicale-faciale, 1965; (with others) Encyclopédie Medico-Chirurgicale; also papers. Neuroanat. research on origins of facial nerve, pneumogastric nerve, constn. solitary bundle, mechanism of facial paralyses. Address: 10, Bd. Flandrin, Paris 16, France.*

CHOUKAS, Michael Eugene, sociologist; b. Samos Island, Greece, Nov. 16, 1901; s. Nicholas and Calliope (Doukas) C.; A.B., Dartmouth Coll., 1927; A.M., Columbia, 1928, Ph.D., 1934; m. Gertrude Spitz, June 20, 1927; 1 son, Michael E. Came to U. S., 1916, naturalized, 1924. Prof. of sociology, Dartmouth Coll. since 1929, chmn. dept. 1947-51; 1962-64; prof. sociology, Bates Coll., summers, 1939-40; Jorns Hopkins U., 1950. Mem. N.H. Com. on Morale, 1941; Office of Strategic Services, 1944-45; dir. of survey on alcoholism in N.H., 1946. Member American Sociological Soc., Byzantine Soc., Phi Beta Kappa, Alpha Tau Omega. Mem. Greek Orthodox Church. Author: Black Angels of Athos, 1934; Propaganda Comes of Age, 1965. Contbr. articles to Sociological Rev., and other jours. and periodicals. Defined propaganda and its differentiation from all other forms of promotional activities; made a sociological study of monastic life. Home: Hanover, N.H.

CHOULANT, Johann Ludwig, German physician; b. Dresden, Nov. 12, 1791; M.D., Leipzig, 1818. Physician, Royal Hospital, Friedrichstadt, 1821-27; prof. theoretical medicine and dir. of polyclinic, Univ. Dresden, 1923; prof. practical medicine and dir. clinic, 1827; dir. academy, 1842. Author: Tafeln zur Geschichte der Medizin, 1822; Handbüch der Bücherkunde für die älteren Medizin, 1828; Bibliotheca medico-historica, 1842; Die anatomische Abbildung des XV und XVI Jahrhunderts, 1843; Geschichte und Bibliographie der anatomischen Abbildung, 1852; Geschichte und Bibliographie der illustr. Inkunabeln aus den Gebiete der Medizin und Naturwissenschaften, 1858. Composed chronological tables and 2 bibliographies indispensable to study of medical history, especially on anatomical illustration. Died Dresden, Germany, July 18, 1861.

CHOVNICK, Arthur, Am. geneticist; b. N.Y.C., Aug. 2, 1927; s. Herman N. and Fannie (Hutkin) C.; A.B., Ind. U., 1949, M.A., 1950; Ph.D., Ohio State U., 1953; m. Elinor Joy Mosher, June 7, 1949; children—Lisa Beaumont, Benjamin Paul. Faculty, U. Conn., 1953-59, asst. dir. Biol. Lab., Cold Spring Harbor, N.Y., 1959-60, lab. dir., 1960-62; prof. genetics, head dept. U. Conn., 1962—. Mem. Genetics Soc. Am., Am. Genetics Assn., A.A.A.S., Am. Inst. Biol. Sci., Am. Soc. Naturalists, Sigma Xi. Research and publs. on structural and functional orgn. genetic material in higher organisms. Home: P.O. Box 123, North Windham, Conn. 06256. Office: Dept. Genetics, U. Conn., Storrs, 06268.*

CHOW, Kao Liang, neuropsychologist; b. Tientsin, China. Apr. 21, 1918; s. Su Tau and Tau Yu (Tsau) C.; B.S., Yenching U., China, 1943; Ph.D., Harvard, 1950; m. Margaret W. C. Zee, May 2, 1964. Came to U. S., 1946, naturalized, 1963. Staff, Yerkes Lab. Primate Biology, Orange Park, Fla., 1947-54, research asso., 1947-54; faculty U. Chgo., 1954-61; faculty Stanford Med. Sch., Palo Alto, Cal., 1961—, prof. neurology Mem. Internat. Brain Research Orgn., N.Y. Acad. Scis., Am. Physiol. Assn., A.A.A.S., Sigma Xi. Research and publs. on establishment of connections from thalamus to cortex, effects light deprivation on behavior and anatomy of eye in new born animals, neural basis of learning and memory. Home: 101 Alma St., Palo Alto, Cal. 94301.*

CHOW, Ven Te, civil engr.; b. Hangchow, China, Aug. 14, 1919; s. Chun Tan and Chin (Yu) C.; B.S., Nat. Chiao Tung U., 1940; M.S., Pa. State U., 1948; Ph.D., U. Ill., 1950; m. Lora (Shu), June 3, 1961; children—Margot, Marana. Came to U. S., 1943, naturalized, 1962. Head instr., engr. Great China Sch., 1940-41; lectr. U. Utopia, 1941-43; faculty China Agrl. and Textile Engring. Coll., 1943-45; civil engr. Bur. Pub. Works, Formosa, 1945-46; faculty Nat. Chiao Tung U., 1946-47; hydrologist Ill. State Water Survey, Champaign. 1950-51; faculty U. Ill., Urbana, 1951—, prof. hydraulic engring., 1958—. Cons., lectr. various countries, hon. cons. UNESCO, 1966—; mem. U. S. Nat. Com. for Internat. Hydrological Decade, 1964—. Recipient Epstein award Epstein and Sons, 1955, Achievement award Chinese Inst. Engrs., 1965. Mem. Am. Geophys. Union (vis. scientist 1965—), A.A.A.S., Am. Soc. Engring. Edn., Am. Soc. C.E. (Research prize 1962), Internat. Assn. for Hydraulic Research, Internat. Commn. on Irrigation and Drainage, Sigma Xi, Phi Tau Phi. Phi Kappa Phi. Author: Open-Channel Hydraulics, 1959. Editor-in-chief Handbook of Applied Hydrology 1964; editor Advances in Hydrosci., 1964—; asso. editor Jour. Water Resources Research, 1964—; cons. editor McGraw-Hill Book Co. Devel. stochastic methods for hydrologic analysis and design and engring planning for drainage, flood control and other water resource projects; devel. sci. of watershed hydraulics; publs. in hydrology, hydraulics, water resources, and hydrosci. Home: 2014 S. Anderson St., Urbana, Ill. 61801.*

CHRESTIEN, André-Jean, French physician; b. Sommières, France, June 2, 1758; studied, U. Montpellier (France); M.D., 1779; sec. to Dr. Lamure; municipal officer Montpellier, also head mil. hosp., Montpellier, during French Revolution. Mem. Acad. Medicine. Author: Mémoire sur l'épidemie de variole qui sevit à Montpellier en l'an VI; Traité de médecine iatroleptique, 1811; Originated treatments of milk diets, snails for Tb, also Kina resin, chick-pea coffee for various maladies; advocated innoculation, use of gold in medicine; invented iatroliptice method. Died Montpellier, Mar. 11, 1840.

CHRETIEN, Henry, French mathematician, astronomer; b. Paris, France, Feb. 1, 1870; studied at École supérieure d'électricité; D.Sc., U. Paris, 1906; m. Madeline Rose Marie; 1 dau., Mrs. Arthur C. Neeseman. Astronomer, Obs., Meudon, 1901-06; staff astrophys. service Obs., Nice, France; worked with mil. aero. sect., during World War I; named hon. prof. Sorbonne, Paris Optical Inst., 1920; designed telescope U. S. Naval Obs., Washington, 1932. Recipient Acad. award, 1954. Author: Traité du calcul des combinaisons optiques. Invented anamorphic lens for cinemascope, 1931. Died Washington, Feb. 6, 1956.

CHRÉTIEN, Jacques, French pneumophysiologist; b. Ciré d'Aunis, Jan. 23, 1922; s. Joseph and Isabelle (Deverts) C.; Docteur en Médecine, Faculté de Médecine, Paris, 1953; Pneumophtisiologue qualifié, 1953; m. Paule Lequitte, July 31, 1945; children—Jean, Claire, Laure, Paul-Henri, Pierre. Head clinic, 1953-54; became prof. agrégé Faculté de Médecine, Paris, 1961; staff Hôpitaux de Paris, 1948, became physician, 1963. Author: (with P. Renault) Tuberculose des Bronches; (with A. Meyer) Les Hemoptysies traceobronchiques; (with G. Roussel, G. Brouet) Les Chorio-Epitheliomes; (with G. Mathe, G. Richet) Traité de Sémiologie medicale; also numerous articles. Research on broncho-pulmonary pathology and clin. anatomy. Home: 12 Boulevard Jean Mermoz 92, Neuilly/Seine, France. Office: Hôpital Cochin Rue du Faubourg Saint Jacques, Paris XIV°, France.*

CHRIEN, Robert Edward, Am. physicist; b. Cleve., Apr. 15, 1930; s. Friederich and Anna (Goros) C.; B.S. in Physics, Rensselaer Poly. Inst., 1952; M.S., Case Inst. Tech., 1955, Ph.D., 1958; m. Susan Varga, June 6, 1953; children—Robert Edward, Katherine Suzanne, Elizabeth Jane, Thomas George. Staff, Brookhaven Nat. Lab., Upton, L.I., N.Y., 1957—. physicist, 1965—. Staff, joint U. S.-Greece coop. program NRC Democritus, Athens, 1966; Mem. Nuclear cross sect. adv. group U. S. AEC. Mem. Am. Phys. soc., A.A.A.S., Sigma Xi. Patentee on fluorescent lamp phosphors. Research on properties of neutroninduced reactions including de-excitation of levels by gamma-ray emission, applications of on-line computer techniques to nuclear physics research. Home: 51 S. Country Rd., Bellport, N.Y. 11713. Office: Dept. Physics, Brookhaven Nat. Lab., Upton, N.Y. 11973.*

CHRISMAN, Oscar, Am. paidologist; b. Gosport, Ind., Nov. 16, 1855; s. Benjamin and Eliza (Bastian) C.; grad. Ind. State Normal Sch., 1887; A.B., Ind. U., 1888, A.M., 1893; fellow Clark U., 1892-94; student in Jena and Berlin, 1894-96; Ph.D., U. Jena, 1896; m. Drusilla Lukenbill, Oct. 6, 1883; children—Chrisman, Oscie Drusilla (Mrs. R. D. Gladding) (dec.). Tchr. pub. schs., 1876-83, prin., 1883-85; prin. Longfellow Sch., Houston, 1888-89; supt. schs., Gonzales, Tex., 1889-92; prof. Kan. State Normal Sch., 1896-1901; prof. paidology Ohio U., 1902-26, prof. psychology, 1904-22. Formulated idea of sci. of child, 1893. originated term paidology, made this his life's work. Author: The Historical Child, 1920. Died Feb. 2, 1929.

CHRIST, Johann Friedrich, archeologist; b. Coburg, Germany, Apr. 26, 1700; s. Johann Sebastian and Anna Euphrosyna (Drechsel) C.; ed. at Jena, also Halle, Germany; sec., tutor, instr. to noble and wealthy families; became asso. prof. history Leipzig, Germany, 1731; traveled in Eng., France, Italy; named prof. poetics (also lectured on archeology and art history), 1739; taught Lessing and Winckelmann; expert on gems, antiques; collected monograms. Author: Noctes Academicae, 1727-29; Abhandlungen über die Literatur und Kunstwerke, Vornehmlich des Altertums, 1776. Died Leipzig, Sept. 3, 1756.

CHRISTANUS PRACHATICENSIS, see Christian of Prague.

CHRISTENSEN, Bert Einar, Am. chemist; b. Duluth, Minn., Oct. 20, 1904; s. Olaf B. and Lena (Hanson) C.; B.S., Wash. State Coll., 1927; Ph.D., U. Wash., 1932; m. Emelyn Marie Burke, Aug. 31, 1932; children—Gerald Roger, Robert Leonard, Joyce Emelyn (Mrs. James Harris). Research chemist Atmospheric Nitrogen Corp., Syracuse, N.Y., 1927-28; faculty Ore. Corvallis, 1932—, prof. chemistry, 1944—, chmn. dept. chemistry, 1956—. Cons., Krishell Labs., Portland, Ore., 1948—. Mem. Am. Chem. Soc. (vis. asso. 1956—), Am. Assn. U. Profs., A.A.A.S., Sigma Xi (award 1957). Research and publs. on synthesis organic compounds biol. interest including pyrimidine, quinazoline, purine derivatives, certain amino acids; chemistry and synthesis psoralene and certain of its derivatives. Home: 337 N. 23d St., Corvallis, Ore. 97330.*

CHRISTENSEN, Borge Christian, Danish physician; b. Copenhagen, Denmark, July 12, 1911; s. Soren Christian and Mary (Jeppesen) C.; M.D., U. Copenhagen, 1937, D.Sc., 1943; m. Sigbrit Rosenstand, Sept. 18, 1940; children—Britta (Mrs. Gunnar Diemer), Knud Christian, Soren Christian. Practice medicine specializing in internal medicine and health physics; head dept. radioisotopic research and treatment Finsen Inst., Copenhagen, 1951—; examiner internal medicine U. Copenhagen, 1951—. Adviser, Danish govt. agys.; mem. exec. com. Comité Internat. de Photobiology, 1956—, v.p., 1957—; mem. expert adv. panel on radiation WHO, 1957—; mem. Internat. Commn. on Radiol. Protection, 1959—, Internat. Commn. on Radiol. Units and Measurements. Mem. Health Physics Soc. (standards com. 1961—), Radiation Research Soc., N.Y. Acad. Scis. Author: (with Bent Buchmann) Progress in Photobiology, 1961; (with Michael Schroder) The Atomic Age and its Radiation, 1961; also articles. Hon. editorial bd. Health Physics, 1964—. Research, evaluation of radioisotopes in biol. research and med. treatment, cellular level-oxygen metabolism, transport of inorganic ions through cell membranes; radiation protection and evaluation of risks from ionizing radiation. Home: 30 Trorodvej, Vedbaek, Denmark. Office: 49, Strandboulevard, Copenhagen, O, Denmark.*

CHRISTENSEN, C. M., Am. mycologist; b. Sturgeon Bay, Wis., Aug. 8, 1905; s. Peter Karl and Christine C.; B.S., U. Minn., 1929, M.S., 1930, Ph.D., 1937; m. Katherine Wallace Barry, Sept. 27, 1935; children—Sarah Ellen (Mrs. Wm. R. Nelson), Melanie Barry, Jane Martin (Mrs. Gary T. Vance). With U. Minn., 1930—, prof., 1948—; mem. Mexican agrl. program Rockefeller Found., 1959-60, 63, 64, 65. Mem. Am. Phytopathol. Soc., Am. Soc. Microbiology, Am. Microbiol. Soc., Mycol. Soc. Am., Am. Assn. Cereal Chemists, Brit. Mycolog. Soc. Author: Common Edible Mushroom, 1943, The Molds and Man, 1953, Common Fleshy Fungi, 1946. Research and publs. on relation of fungi to deterioration of stored grains and seeds; isolated mycotoxins and linked them to specific diseases whose causes had been unknown. Home: 2350 Carter Av., St. Paul, Minn. 55108.*

CHRISTENSEN, Halvor Niels, Am. biochemist; b. Cozad, Neb., Oct. 24, 1915; s. Niels Peter and Matena (Smidt) C.; B.S., Kearney State Coll., 1935; M.S., Purdue U., 1937; Ph.D., Harvard, 1940; m. Mayme O. Matthews, Aug. 28, 1939; children—Haldan N., Carl E., Karen A. Asso., Harvard, instr. Med. Sch., 1942-44; dir. chemistry labs. Mary I. Bassett Hosp., Cooperstown, N.Y., 1944-47; dir. chem. research Childrens Hosp., Boston, 1947-49; faculty Tufts U. Sch. Medicine, 1949-55; prof., chmn. dept. biol. chemistry U. Mich., Ann Arbor, 1955—. Cons., NIH, 1961—, mem. various coms. Guggenheim fellow Carlsberg Lab., Copenhagen, Denmark, 1952. Fellow Am. Acad. Arts and Scis., mem. Am. Chem. Soc., Am. Inst. Nutrition, Am. Soc. Biol. Chemists, Biophys. Soc. Author: Body Fluids and the Acid Base Balance, 1964; Biological Transport, 1962; (with G. Palmer) Enzyme Kinectics, 1966; also numerous articles. Demonstrated origin peptide wastage, in intravenous nutrition, absence random collection peptides in tissues, alkali-metal dependence transport, alkali metal fluxes as result transport organic solutes; systematized amino acid transport for various cells and tissues; research on structure peptide antibiotics. Home: 2200 Devonshire Rd., Ann Arbor, Mich. 48104.*

CHRISTENSEN, Harold T(aylor), Am. sociologist; b. Preston, Ida., Mar. 10, 1909; s. Henry Oswald and Nettie (Taylor) C.; student Ricks Coll., 1926-27, 28-29; B.S., Brigham Young U., 1935, M.S., 1937; Ph.D., U. Wis., 1941; m. Alice Spencer, June 5, 1935; children—Carl, Boyd, Janice (Mrs. Francis Hedquist), Larry, Gayle. Tchr. elementary sch., Hibbard, Ida., 1927-28; faculty Brigham Young U., 1935-47; leader div. farm population and rural life bur. agrl. econs. Northeastern region U. S. Dept. Agr., 1944-45; prof. head dept. sociology Purdue U., Lafayette, Ind., 1947-62, prof. sociology, 1962—. Fulbright Research scholar, Denmark, 1957-58. Fellow Am. Sociol. Assn.; mem. Sociol. Research Assn., Nat. Council on Family Relations (past pres.), Population Assn. Am., Author: Marriage Analysis, 1950, rev., 1958. Editor: Handbook of Marriage and the Family, 1964. Research and publs. on marriage and family relationships, particularly courtship behavior, child spacing, family size as these phenomena relate to mental health individual and overall stability family. Home: 2167 Tecumseh Park Lane, West Lafayette, Ind. 47906. Office: Purdue U., Lafayette, Ind. 47907.*

CHRISTENSEN, Leo Martin, Am. chemist; b. Dow City, Ia., 1898; B.Sc., Ia. State Coll., 1923, Ph.D. in Biophys. Chemistry (fellow), 1926; m. 1926; 4 children. Apptd. instr. chemistry Ia. State Coll., 1923, asst. prof., 1931-35 research biochemist Commml. Solvents Corp., 1926-31; cons. Chem. Found., Inc., N.Y., 1935-37; sec.-treas. Kan. Co. sect., 1937-39; head dept. agrl. chemistry, Ida., 1939-41; research exec. chemurgy project, Neb., 1941-45; pvt. cons., Neb., 1945-48; v.p. Indsl. Found., 1950. Mem. Indsl. Devel. Council.; cons. WPB. Fellow Inst. Chemistry; mem. Am. Chem. Soc. Research on use of castor plants for insecticides, prodn. of fermentation chems. and vegetable oils, alcohol motor fuels and engines, synthetic rubber, thermo-setting plastics, indl. plant location. Died 1955.

CHRISTENSEN, Niels Anton, engr.; b. Toerring, Denmark, Aug. 16, 1865; s. Christian and Anne Marie (Nielsen) Jensen; apprentice in shipbuilding and marine engring.; grad. Tech. Inst. Copenhagen; m. Mathielde Thomesen, Aug. 19, 1894; 1 dau., Esther Marie (Mrs. Charles Jacob Young). Came to U. S. 1891, naturalized. Successively in charge machinery for waterworks of Calcutta, India, then charge refining nitrate of soda, Chile; asso. Fraser-Chalmers, machinery mfrs., 1892-93; asst. to supt., chief engr. Edward P. Allis Co., Milw., 1894-96; founded Christensen Engring. Co., for mfr. ari brakes on cable cars, 1896, with co. until 1903. Active during World War I, World War II developing spl. hydraulic and compressed air equipment for aircraft. Held over 200 patents. Died Oct. 1952.

CHRISTENSEN, Ralph Jorgensen, Am. physicist; b. Provo, Utah, Aug. 2, 1906; s. Christen Carl and Ellen S. (Jorgensen) C.; B.S., in Physics Brigham Young U., 1928; M.A., U. Cal. at Berkeley, 1930, Ph.D., 1932; m. Clara Clyde, Aug. 14, 1928; children—Eleanor (Mrs. Charles L. Brandt), Ralph C. Teaching fellow U. Cal., Berkeley, 1928-32, research asso. agr. Expt. Sta., 1932-33, research physicist div. war research, 1941-45; physics tchr. San Mateo (Cal.) City Coll., 1932-41; with U. S. Navy Electronics Lab., San Diego, 1945-67, asso. tech. dir., 1958-61, tech. dir., 1961-67; tech. dir. Naval Electronics Lab. Center, 1967—. Mem. USN Bd. Civil Service Examiners for Profls. in Naval Labs. in Cal., 1952-55, chmn., 1954-55; mem. USN Anti-Submarine Warfare Research and Devel. Planning Council, 1961—, chmn., 1964-65; mem. USN Bur. Ships Research and Devel. Council, 1961-66. Fellow Am. Phys. Soc.; mem. I.E.E.E. (sr.), Sigma Xi. Co-discoverer phototropic vertically migrating acoustic-scattering layer in ocean, later shown to be complex of organisms; research in ocean acoustics, x-ray generation. Home: 3580 Charles St., San Diego 92106. Office: U. S. Navy Electronics Lab., Center, San Diego 92152.*

CHRISTIAN, Gary Dale, Am. chemist; b. Eugene, Ore., Nov. 25, 1937; s. Roy and Edna (Trout) C.; B.S., U. Ore., 1959; M.S., U. Md., 1962, Ph.D., 1964; m. Suanne Byrd Coulbourne, June 17, 1961; children—Dale Brian, Carol Jean. Research asst. U. Md., College Park, 1961, asst. prof., 1965-66; research analytical chemist Walter Reed Army Inst. Research, Washington, 1961-65, 66-67; asst. prof. U. Ky., Lexington, 1967—. Vice pres. ChrisFeld Precision Instruments Corp., Beltsville, Md., 1964—. Mem. Am. Chem. Soc., A.A.A.S., Ky. Acad. Sci., Am. Assn. Univ. Profs., Sigma Xi. Research and publs. on analytical methods for trace and ultra-trace determination of metals in biol. matherials; devel. ultramicro clin. analyses by electroanalytical methods; determined formation of complexes between metals of body and biologically active compounds; invented a constant current coulometer. Home: 3413 Crimson King Ct., Lexington, Ky. 40502.*

CHRISTIAN, Henry Asbury, Am. physician; b. Lynchburg, Va., Feb. 17, 1876; s. Camillus and Mary Elizabeth (Davis) C.; A.B., A.M., Randolph-Macon Coll., 1895, LL.D., 1923; M.D., Johns Hopkins, 1900; A.M., Harvard, 1903; Sc.D., Jefferson Med. Coll., 1928, U. Mich., 1938; LL.D., Western Res. U., 1931, U. Western Ont., 1938; m. Elizabeth Sears Seabury, June 30, 1921. Asst. pathologist Boston City Hosp., 1900-02, asst. vis. pathologist, 1902-05; asst. pathologist Children's Hosp., 1903-07; instr. pathology, Harvard, 1903-05, theory and practice of physic, 1905-07, asst. prof., 1907-08, dean Faculty of Medicine and Med. Sch., 1907-12, Hersey prof., 1908-39, emeritus from 1939, recalled to active teaching, 1942-46; clin. prof. medicine, Tufts Coll. Med. Sch., 1943-46; dean Faculty of Medicine and Med. Sch., Boston, 1908-12; physician-in-chief Carney Hosp., 1907-12, Peter Bent Brigham Hosp., 1910-39; vis. physician, Beth Israel Hosp., Boston, 1942-46. Resident chmn. division medical scis., NRC, 1919-20. Recipient Distinguished Service medal A.M.A., 1947. Fellow Am. Acad. Arts and Scis., Am. Coll. Phys.; hon. fellow Royal Coll. Phys. (Can.); mem. Assn. Am. Phys. (pres. 1935), Am. Assn. Pathologists and Bacteriologists, A.M.A., Am. Soc. for Clin. Investigation (pres. 1919), A.A.A.S., Am. Soc. Exptl. Pathology, Interstate Post Grad. Med. Assn. N.Am. (pres. 1931), Sigma Chi, Phi Beta Kappa (asso.), Alpha Omega Alpha, Sigma Xi. Author: Diagnosis and Treatment of Heart Disease; Principles and Practice of Medicine (4 editions); Bright's Disease; Purpuras; Non-valvular Heart Disease; also papers on pathological and clinical med. subjects. Editor (for Oxford University Press) Oxford Medicine and Oxford Monographs. Described condition known as chronic idiopathic xanthomatosis (Hand-Schuller-Christian disease), 1919, nodular nonsuppurative panniculitis (Christian-Weber disease), 1928. Died Aug. 24, 1951.

CHRISTIAN, John Edward, Am. chemist; b. Indpls., July 12, 1917; s. George Edward and Okel (Waltz) C.; B.S., Purdue U., 1939, Ph.D, 1944; postgrad. U. Cal. at San Francisco, 1947; m. Catherine Ellen Spponer, July 23, 1948; 1 dau., Linda Kay. Retail pharmacist, 1935-39; control chemist Upjohn Co., Kalamazoo, 1939-40; research fellow Purdue U., Lafayette, 1940-44, faculty, 1945—, coordinator bionucleonics research, 1948—, prof. pharm. chemistry, 1950-59, prof. bionucleonics, 1959—, head dept. radiol. control, 1956-59, head bionucleonics dept., 1959—, dir. Inst. for Environmental Health Studies, 1965—. Cons., Smith, Kline and French Labs., Phila., 1946—. Recipient Chilean Iodine Ednl. Bur. award, 1956, Julius Sturmer award Phila. Coll. Pharmacy and Sci., 1958, Fellow A.A.A.S., Ind. Acad. Sci.; mem. Am. Pharm. Assn. (Justin L. Powers Research Achievement award 1962, Ebert medal 1957), Am. Soc. Bacteriologists, Am. Chem. Soc., Am. Assn. U. Profs., Am. Assn. Coll. Pharmacists, Health Physics Soc., Conf. on Radiol. Health, Am. Pub. Health Assn., Sigma Xi, Rho Chi, Phi Lambda Upsilon, Sigma Pi Sigma, Author: (with G. L. Jenkins, G. P. Hager) Quantitative Pharmaceutical Analysis, 1953; (with S. C. Rothman) Constructive Uses of Atomic Energy, 1949; (with L. M. Parks, L. E. Harris, P. J. Jannke) Inorganic Chemistry in Pharmacy, 1949; (with A. Adelmann) Radioactivity for Pharmaceutical and Allied Research Laboratories, 1960; (with S. Rothchild, B. D. Rupe, W. F. Bousquet) Advances in Tracer Methodology, Vol. 1, 1963; (with W. F. Bousquet) Radioisotopes and Nuclear Techniques in the Pharmaceutical and Allied Industries, 1960; also numerous articles. Research on analytical applications of radioisotopes; pharm. quality control using radioisotope assays; radioactivity in man and animals. Home: 1301 Woodland Av., West Lafayette, Ind. 47906. Office: Purdue U., Lafayette, Ind. 47907.*

CHRISTIAN, John Jermyn, Am. endocrinologist; b. Scranton, Pa., Apr. 12, 1917; s. John Oren and Margaret (Jermyn) C.; A.B., Princeton, 1939; postgrad. Columbia, U. Mich., U. Pa.; Sc.D., Johns Hopkins, 1954; m. Patricia Hart, Nov. 8, 1958; children—John Jermyn, Patricia Jean. Research asst. Pa. Game Com., 1946-47; research pharmacologist Wyeth, Inc., 1948-51; physiologist Naval Med. Research Inst., 1951-59; research asso. Johns Hopkins Sch. Hygiene and Pub. Health, 1954-59; asso. dir. Penrose Research Labs., Phila. Zool. Soc., 1959-62; asso. prof. comparative pathology U. Pa., Phila., 1959-63; asso. mem. Albert Einstein Med. Center Research Labs., Phila., 1963—. Mem. endocrinology study sect. NIH, 1958-63; com. comparative pathology Nat. Acad. Scis.-NRC, 1963—; cons. State Dept. in Thailand, 1960. Mem. Ecol. Soc. (Mercer award 1957), Phila. Endocrine Soc. (dir. 1965-67), Wildlife Disease Assn. (pres. 1965-67), Am. Inst. Biol. Scis., Soc. Exptl. Biology and Medicine, Am. Soc. Mammalogists, N.Y. Acad. Scis. Research, numerous publs. on theory of endocrine-behavioral feed-back regulation of population growth. Home: 1718 Sheffield Dr., Norristown, Pa. 19403.*

CHRISTIAN, John Wyrill, English metallurgist; b. Scarborough, Eng., Apr. 9, 1926; s. John William and Louisa (Crawford) C.; B.A., Queen's Coll., Oxford U., 1945, D.Phil., 1949, M.A., 1950; m. Maureen Lena Smith, July 26, 1949; children—Louise Hilda, John William, Timothy James. Demonstrator, Oxford U., 1948-51, Pressed Steel Co. research fellow, 1951-55, lectr., 1955-58, George Kelley reader in metallurgy, 1958-67, prof. phys. metallurgy, 1967—, professorial fellow St. Edmund Hall, 1963—. Vis. scientist NRC, Ottawa, Can., 1956; vis. prof. U. Ill., 1954, 63; Republic Steel distinguished vis. prof. Case Inst. Tech., 1962-63. Fellow Inst. Physics and Phys. Soc., Instn. Metallurgists, Inst. Metals. Author: (with W. Hume-Rothery, W. B. Pearson) Metallurgical Equilibrium Diagrams, 1952; Theory of Transformation in Metals and Alloys, 1965. Research, publs. on phase diagrams, defects in crystalline solids, theory of transformations in metals and alloys, deformation processes in refractory metals, especially at sub-zero temperatures. Home: 20 Linkside Av., Oxford, Eng.*

CHRISTIAN OF PRAGUE (Christanus Prachaticensis), physician; b. Prague, Czechoslovakia, 1368; astronomer to the emperor and king of Bohemia; dean, rector Philos. Faculty of Prague, 1403; became vicar of St. Michael, Prague, 1405; visited John Husin Constanee, 1415; exiled from Prague, 1427-29; became rector of univ., 1434. Author: Astrolabium Ptolemaec algonsmus prosaycus; book on herbal medicine, bloodletting, the plague, 1409; also wrote on arithmetic and iatromathematics. Died 1439.

CHRISTIANSEN, Christian, Danish physicist; b. Lorborg, Jutland, Denmark, Oct. 9, 1843; prof. physics, Copenhagen, Denmark, 1886-1917. Author: Indledning til den mathematiske fysik, 1887-89, 2 vols.; improved water-jet pump. verified Sellmeier's theory of anomalous dispersion using refractive index of fuchsine (magenta). Died Copenhagen, Nov. 28, 1917.

CHRISTIANSEN, Jens Anton, Danish chemist; b. Vejle, Sept. 6, 1888; s. Frederick and Helvig (Nielsen) C.; Ph.D.; m. Marie Inger, Feb. 24, 1914. Asst., lab. of Carlsberg, 1911-15; asst. chem. lab. of univ., 1915; lectr. chemistry U. Copenhagen, 1921, prof., 1931-59, vice rector, 1955-56, dir. Inst. Phys. Chemistry, 1948-59, now prof. emeritus; dir. Inst. Chemistry, Technol. U., 1931-48. Mem. Danish Acad. Sci. and Letters, Norwegian Acad. Sci., acads. sci. of Goteborg, Stockholm and Lund, Polish (hon.), Danish (hon.), London assns. chemistry, N.Y. Acad. Scis. Research and numerous articles on kinetics, mechanism of chem. reactions (chain reactions, mechanism of enzyme reactions, receptor mechanism of internal ear). Address: Sundvaenget 9, Hellerup, Denmark.*

CHRISTIE, Arthur Carlisle, Am. physician, radiologist; b. West Sunbury, Butler County, Pa., Dec. 29, 1879; s. Milton Hughes and Harriet Josephine (Rhodes) C.; M.D., Ohio Wesleyan U., 1904, M.S., Ohio 1919; D.Sc., Am. U., 1942; grad. Army Med. Sch., Washington, 1907; m. Maude Irene Hopkins, June 1, 1904; children—Mrs. Geneva Irene Morris, Carlisle Van Dyke, Milton Arthur, Harriet Inez Beck. Practiced Clymer, N.Y., 1904-06; joined Med. Corps., U. S. Army, 1906; prof. operative surgery and Roentgenology, Army Med. Sch., 1912-16; resigned as capt.; gen. charge X-ray work for Army, 1917-Aug. 1918; prof. radiology George Washington U. Med. Coll.; specialized in Roentgenology; prof. clinical radiology, Georgetown U. Med. Sch.; cons. USPHS. Pres. Fifth Internat. Congress Radiology, Chgo., 1937; chmn. com. on radiology NRC. Mem. editorial bd. Am. Jour. Roentology and Radium Therapy. Fellow A.M.A., A.C.P., Internat. Coll. Surgeons, Brit. Faculty Radiologists (hon.); mem. Roentgen Ray Soc. (pres.). Radiol. Soc. N.A., Am. Coll. Radiology (pres.). Author: Manual of X-ray Technique, 1913, 17; Roentgen Diagnosis and Therapy, 1924; Economic Problems of Medicine, 1935. Died June 22, 1956.

CHRISTIE, Dan Edwin, Am. mathematician; b. Dover-Foxcroft, Me., Oct. 11, 1915; s. Dan Foss and Blanche Ellen (Hamlin) C.; A.B. summa cum laude, Bowdoin Coll., 1937; postgrad. (Henry fellow) St. John's Coll., Cambridge, Eng., 1937-38; A.M., Princeton, 1940, Ph.D., 1942; m. Eleanor Wilson, Aug. 31, 1940; 1 son, Mark Edwin. Instr., Bowdoin Coll., Brunswick, Me., 1942—, prof. math. and physics, 1955—, chmn. dept. math., 1964—. Lectr. basic pre-meteorol. program U. S. Air Force, 1943-44, basic pre-radar sch. U. S. Navy, 1944-46. Mem. Am. Math. Soc., Math. Assn. Am. (chmn. Northeastern sect. 1960-61, com. undergraduate program in math. 1963-66), Am. Assn. Physics Tchrs., A.A.A.S., Phi Beta Kappa, Sigma Xi. Author: Intermediate College Mechanics, 1952; Vector Mechanics, 1964. Research and publs. on homotopy group theory, math. aspects of thermodynamics, dimensional analysis. Home: 4 Atwood Lane, Brunswick, Me. 04011.*

CHRISTIE, John McDougall, geologist; b. Calcutta, India, Dec. 4, 1931; s. John and Anne (Logie) C.; B.Sc., U. Edinburgh, 1953, Ph.D., 1956; m. Helen Clark Herd, Aug. 24, 1957; children—Catherine M., John D., Ann M. Came to U. S., 1956. Instr. Pomona Coll., 1956-58; asst. prof. U. Cal., Los Angeles, 1958-64, asso. prof., 1964—. Guggenheim fellow, 1964-65. Fellow Geol. Soc. Am.; mem. Am. Geophys. Union, A.A.A.S., Sigma Xi. Contbr. articles in field to sci. jours. Research on mechanisms of deformation of minerals, especially quartz; contbns. to structural geology and structural petrology; deformation in mylonite zones and metamorphic rocks. Home: 1027 Hartzell St., Pacific Palisades, Cal. 90272. Office: Dept. Geology, U. Cal., Los Angeles 90024.*

CHRISTIE, Samuel Hunter, Brit. mathematician; b. London, Eng., Mar. 22, 1784; s. James C.; ed. Trinity Coll., Cambridge, Eng.; B.A., 1805; m. Elizabeth Theodora Claydon, May 12, 1808; 1 son, W. H.; m. 2d, Margaret Ellen Malcolm, Oct. 15, 1844. Tchr. math., prof. Woowich Mil. Acad., 1806-54. Mem. Compass Com. Fellow Royal Soc., 1826 (sec. 1837-54). Author: An Elementary Course of Mathematics for the Use of the Royal Military Academy and for Students in General, 1845; also articles. Made 1st observation of how slow rotation of iron produces polarity; showed conductivity of several metals varies directly as sq. of diameter of conducting wire and inversely as length; proved direct effect of solar rays; improved horizontal and dipping needle of compass; suggested balance of 4 resistances (leading to Wheatstone's invention). Died Twickenham, Eng., Jan. 24, 1865.

CHRISTIE, Thomas, Brit. physician; b. Lanarkshire, Eng., 1773; ed. U. Aberdeen (Scotland); M.D., Aberdeen, 1810; surgeon E. India Co., Ceylon, 1797-1810; practiced medicine in Cheltenham, 1810-29; named physician extraordinary to prince regent, 1812. Author: An Account of the Introduction, Progress, and Success of Vaccination in Ceylon, 1811. Introduced vaccination to Ceylon, 1802. Died Oct. 11, 1829.

CHRISTIE, Sir William Henry Mahoney, Brit. astronomer; b. Woolwich, Eng., Oct. 1, 1845; s. Samuel Hunter and Margaret Ellen (Mahoney) C.; ed. King's Coll. London, Eng., Trinity Coll., Cambridge (Eng.) U.; (hon. D.Sc., Oxford (Eng.) U.; m. Violette Mary Hickman, 1881; 2 sons. Began career in astronomy at Royal Obs., Greenwich, Eng., apptd. chief asst.; astronomer royal, 1881; elected fellow Trinity Coll., Cambridge U., 1869; Fellow Royal Soc., 1881; mem. Royal Astron. Soc. (pres. 1890-92); corr. mem. Acad. Scis. Paris, 1896, Acad. Sci. St. Petersburg. Author: Manual of Elementary Astronomy. Pub. paper on movements of stars in 1877; research on sunspot movements and movement of stars; modified G. B. Airy's transit circle; tried to determine radial velocities of brighter stars; (with E. W. Mauder) photographed sun daily. Died Jan. 22, 1922.

CHRISTISON, J(ohn) Sanderson, physician; b. Brechin, Forfarshire, Scotland, Mar. 13, 1856; s. Robert and Martha (Sanderson) C.; ed. schs. at Glasgow, Wigan, Upholland, Edinburgh and London; came to U. S. with parents, 1875; M.D., U. of City of N.Y., 1877. Was asst. physician N.Y.C. Lunatic Asylum, acting surgeon to Workhouse (prison) Blackwell's Island; asst. physician N.Y.C. Insane Asylum; acting physician Bellevue Hosp., outdoor dept.; later asst. physician Wis. State Hosp. for Insane; then in pvt. practice. Sec. Criminological Soc. of Chicago. Author: Crime and Criminals, 1898, 2d edit., 1901; Brain in Relation to Mind, 1899; Farmer Kilroy, 1902; The Tragedy of Chicago—A Study in Hypnotism, 1906. Died 1908.

CHRISTISON, Sir Robert, Scottish toxicologist; b. Edinburgh, Scotland, July 18, 1797; s. Alexander Christison; M.D., Edinburgh, Scotland, 1819, LL.D. (hon.); m. Miss Brown, 1827; 3 sons. House physician to Edinburgh Infirmary, 1817-20, physician, from 1827; prof. med. jurisprudence, Edinburgh, 1822-32; prof. medicine and therapeutics, Edinburgh, med. adviser to Brit. crown, 1829-66. Created baronet, 1871. Mem. Edinburgh Coll. Physicians (pres. 1839, 48), Royal Soc. Edinburgh (pres. 1868-73), Brit. Assn. (pres. 1875-76). Author: Treatise on Poisons, 1829; Granular Degeneration of the Kidneys, 1839; Commentary on the Pharmacopoeias of Great Britain, 1842. 1832-77; Research on renal pathology; toxicology; formulated instrns. for exams. of cadavers for legal purposes which became accepted guide. Died Edinburgh, Jan. 23, 1882.

CHRISTMANN, Jakob, German mathematician, philologist; b. nr. Mentz, Germany, 1554; successively prof. Hebrew, logic, Arabic, Coll. of Heidelberg, 1592-1613. Author: Solar Observations; Theory of Moon, 1611 (both on chronology). Died 1613.

CHRISTOFFEL, Elwin Bruno, mathematician; b. Monschau, Germany, Nov. 10, 1829; s. Franz Carl and Maria Helena (Engels) C.; Ph.D. in Math., U. Berlin, 1856. Became asst. and lecturing prof. math., Berlin, 1859, Polytechnikum Zurich, 1862, Trade-Acad. (part of Tech. U.), Berlin, 1869-72; became one of 1st profs. Strasbourg (France) U., 1872. Author: Geschichte mathematische Abhandlung, 2 vols., pub. 1910. Studied algebraic and Abelian functions, differential geometry, potential theory, invariants theory, problem of equivalence of quadratic forms; Christoffel symbols in theory of quadratic differential forms named for him. developed Riemann's geometry; invented process called covariant differentiation; 1st discussion of mapping of polygons; worked in math. physics. Died Strasbourg, Mar. 15, 1900.

CHRISTOV, Christo Jankov, Bulgarian physicist; b. Varna, Bulgaria, June 12, 1915; s. Ianko Kalchev and Ekaterina (Baeva) C.; Physicist, U. Sofia, 1938; m. Rossitsa Tihomirova, Dec. 28, 1941; children—Evgeni, Ekaterina, Vladimir. Staff, Faculty Scis., U. Sofia, 1942—, prof., 1951—, head chair atomic physics, 1957—; head divs. theory elementary particles Bulgarian Acad. Scis., Sofia, 1947-63, head div. cosmic ray physics, 1957-63, dir. Inst. Physics, 1963—. Recipient Dimitrov's award 2d degree. Mem. Union Sci. Collaborators. Author: Textbook on Electrodynamics and on Mathematical Methods in Physics; also articles. Research on crystal physics, binding energy, transition of electromagnetic waves, statis. physics, stochoastic processes in gas mixtures and cosmic rays, axiomatic approach in classical mechanics, theory of relativity. Home: 23a Shipka St., Sofia, Bulgaria.*

CHRISTY, David, Am. geologist; b. 1802. Journalist, 1824-35; apptd. agt. Am. Colonization Soc. in Ohio, 1848; delivered messages before Ohio Ho. of Reps., 1849, 50, published as pamphlet "On the Present Relations of Free Labor to Slave in Tropical and Semi-Tropical Countries"; published The Republic of Liberia: Facts for Thinking Men, 1852, Cotton is King: Or the Economical Relations of Slavery, 1855, pamphlet "Ethiopia: Her Gloom and Glory," 1857; made geol. observations reported in series of letters 1st published in Cin. Gazette; geologist Nanthala & Tuckasege Land & Mineral Co. of N.C.; engaged in writing Geology Attesting Christianity, 1867. Made geol. observations of eastern and middle U. S.; probably 1st to draw approximate sects. of strata from Atlantic to Ia., from Lake Erie to Gulf of Mexico. Died circa 1868.

CHRISTY, Nicholas Pierson, Am. physician; b. Morristown, N.J., June 18, 1923; s. Leroy and Elizabeth (Baker) C.; A.B., Yale, 1945; M.D., Columbia U., 1951; m. Beverly Vairin Morris, June 21, 1947; children—Nicholas Pierson, Martha Vairin. Asst. vis. physician Delafield Hosp., N.Y.C., 1955—, 1st med. div. Bellevue Hosp., N.Y.C., 1958—; asso. attending physician Presbyn. Hosp., N.Y.C., 1962—; faculty Columbia Coll. Phys. and Surg., N.Y.C., 1956—, asso. prof. medicine, 1962—, asso. clin. prof., 1965—; chmn. dept. medicine Roosevelt Hosp., N.Y.C., 1965—. Recipient Borden award, Joseph Mather Smith prize Columbia U. John and Mary R. Markle scholar. Diplomate Am. Bd. Internal Medicine. Mem. Harvey Soc., A.A.A.S., Soc. for Exptl. Biology and Medicine, Am. Soc. for Clin. Investigation, Am. Fedn. for Clin. Research, N.Y. Acad. Scis., Laurentian Hormone Conf., Am. Physiol. Soc., Royal Soc. Medicine (affiliate), Am. Assn. for Study Liver Diseases, Endocrine Soc. Editor-in-chief Jour. Clin. Endocrinology, and Metabolism. Home: 8 Peter Cooper Rd., N.Y.C. 10010. Office: Roosevelt Hosp., N.Y.C. 10019.*

CHRISTY, Robert F(rederick), physicist; b. Vancouver, B.C., Can., May 14, 1916; s. Moise Jacques and Hattie Alberta (McKay) C.; A.B., U. of B.C., 1935, A.M., 1937; Ph.D., U. of Calif., 1941; m. Dagmar Elizabeth Lieven, May 31, 1941; children—Thomas E., Peter R. Came to U. S., 1938, naturalized, 1945. Research, metall. lab., U. of Chicago, 1942-43, at Los Alamos, N.M., 1943-45; asst. prof. physics, U. of Chicago, 1946; asso. prof., and prof. physics, Calif. Inst. Tech. since 1946. Fellow Am. Phys. Soc.; mem. Internat. Astron. Union, Nat. Acad. Scis. Study of cosmic rays; research in the fields of astrophysics; theoretical physics; nuclear physics. Home: 1330 S. Euclid Av., Pasadena, Cal. 91106.

CHRISTY, Robert W(entworth), Am. physicist; b. Chgo., Nov. 2, 1922; s. Walter Christian and Ruth Adele (Seifried) C.; A.A., U. Chgo., 1942, M.S., 1949, Ph.D., 1953; m. Cynthia Park, May 7, 1955. Mem. faculty Dartmouth, 1953—, prof. physics, 1962—, chmn. dept. physics and astronomy, 1963—; cons. Motorola, Inc., 1952-53, TRW Space Tech. Labs., 1958—. Mem. Am. Phys. Soc., Am. Assn. Physics Tchrs., A.A.A.S., Sigma Xi, Alpha Delta Phi. Author: (with A. Pytte) Structure of Matter, 1965.

Investigations of structural imperfections in crystals through mech., elec., and optical properties; studies of thin-film insulators; effects of plasma wave propagation in metals. Home: Bragg Hill Rd., Norwich, Vt. 05055. Office: Wilder Physics Lab., Dartmouth Coll., Hanover, N.H. 03755.*

CHROBAK, Rudolf, Austrian gynecologist; b. Troppau, Austria, July 8, 1843; s. Joseph and Magdalene (Eitelberger) C.; studied medicine, Vienna, Austria, 1861-66; M.D.; m. Helen Lumpe, 1868; at least 3 daus. Became lectr., 1871; named prof. extraordinary, 1880, became prof. U. Vienna, 1889, dir. U. Clinic, until 1908. Author: (with A. Rosthorn) Die Erkrankung der weiblichen Geschlechtsorgane, Vol. 1, 1896; also numerous articles in med. jours., textbooks, manuals. Improved surg. techniques; research on uterus. Died Vienna, Oct. 1, 1910.

CHRUSCIEL, Tadeusz Leslaw, Polish physician, pharmacologist; b. Lwów, Poland, Jan. 30, 1926; s. Stanislaw and Bronislawa (Markowska) C.; M.D., Jagiellonian U., Kraków, Poland, 1951, Candidate Med. Scis., 1958; m. Maria Sliz, Apr. 11, 1955; children—Magdalene, Peter, Wojciech. Staff, Jagiellonian U., Kraków, 1950-58; chief dept. pharmacology Silesian Acad. Medicine, Zabrze-Rokitnica, Poland, 1955—, prof., 1966—. Clin. pharmacology cons., 1961—; rector's dep. for sci. film, 1964—. Recipient medal for exemplary work in health service, 1953; medal for work in health service Silesian County, 1965; Rockefeller Found. fellow dept. pharmacology Oxford U., 1959. Mem. Polish Acad. Sci. (mem. com. for physiol. scis. 1963—), Polish Pharmacological Soc. (vice pres. 1966—), Polish Physiol. Soc. (past v.p.), Royal Soc. Medicine (affiliate), Internat. Union for Biochem. Pharmacology (local sec. for Poland 1966—), Polish Med. Soc., Polish Pharm. Soc., Copernican (Polish) Soc. Scis. of Nature, Polish Soc. Animal Care, Polish Soc. Sci. Film. Contbg. author: Physiology of Animals, 1964; Research and numerous publs. on exptl. atherosclerosis, role of autonomic nervous system in hyperlipemia induced by surface active agts., adrenergic mechanisms, analysis of role (with W. Blaschko) of precursors of adrenaline in central nervous system, exptl. toxoplasmosis, exptl. collagenosis; 1st description (with others) of atherosclerosis in pigeons; discovered antiatherosclerotic action of pharnesic acid analogs; postulated (with M. Chrusciel) hypothesis on role of inhibition of decarboxylases in etiopathology of lupus erythematosus disseminatus. Home: 9/5 Dworcowa, Zabrze, Poland. Office: 38 Marx's Zabrze-Rokitnica, Poland.*

CHRYSIPPOS OF SOLI, Greek philosopher; b. Soli, Cicilia, circa 280 B.C.; studied under Zeno in Athens, Greece, Cleanthis, Asclepiades; 3rd Greek leader of Stoics of Sali in Cicilia; succeeded Cleanthes as head of acad., Athens; wrote numerous treatises (extant in fragments); systematized Stoicism and reconciled various factions; found possible combinations of ten axioms to be more than 1,000,000. Died circa 207 B.C.

CHRYSOLOGUE, Noël André (le Pere), French astronomer; b. Gy, France, Dec. 8, 1728; studied under Lemonnier, Paris, France; Capucian monk. Author: Mappemonde projetée sur l'horizon de Paris, 1774; Théorie de la surface actuelle de la terre, 1806. Editor: Abré d'astronomie pour l'usage des planisphere, 1778. Measured altitudes of Vosges, Jura and Alps; perfected Toricelli's barometer; research in geology. Died Gy, Sept. 8, 1808.

CHRYSOSPATHIS, Panayotis, Greek gen. surgeon; b. Athens, Greece, Mar. 1915; s. John George and Mary (Patsiadou) C.; grad. Med. Sch., Athens, 1938. Faculty, Med. Sch. Athens, asso. prof. surgery, 1954—; vis. faculty Chiba U., Japan, 1962; staff Aretaion Hosp., Athens; head surg. dept. Pammacaristos Hosp., Athens. Fellow Greek Surg. Soc., A.C.S.; mem. Internat. Coll. Surgeons, Societie Internat. de Chirugie, Soc. Med. Internat. de Photographi Cinema et Television Endoscopic. Research, numerous publs. on personal modification type gastrectomy billroth to prevent dumping syndrome, personal modification temporary cervical colostomy combining replacement of esophagus using large bowel. Home: 46 Gizzis St. Psychicho. Office: Pammacaristos Hosp., 43, Iacovaton St., Athens 906, Greece.*

CHRYSTAL, George, Scottish mathematician; b. Mill of Kingoodie, Aberdeenshire, Scotland, Mar. 8, 1851; s. William and Margaret (Burr) C.; entered Aberdeen (Scotland) U., 1867; studied at Peterhouse, Cambridge (Eng.) U., 1872-75; B.A., 1875; M.A., Corpus Christi Coll., Cambridge U., 1878; LL.D., U. Aberdeen, 1887, U. Glasgow (Scotland), 1911; m. Margaret Anne Balfour, Sept. 22, 1903; 4 sons, 2 daus. Lectr., Corpus Christi Coll.; named prof. math. St. Andrews (Scotland) U., 1877; became prof. math. U. Edinburgh (Scotland), 1879, apptd. dean faculty of arts, 1891. Recipient Royal medal Royal Soc. London, 1911. Mem. Royal Soc. Edinburgh (Keith prize). Author: Algebra, An Elementary Textbook for the Higher Classes of Secondary Schools, 2 parts, 1886-89; Introduction to Algebra, 1898; Non Euclidean Geometry. Research in exptl. physics under Clerk Maxwell at Cambridge; verified Ohm's law using the relation between current and electromotive force in a wire, 1876; discovered properties of lenses and doublets; research on oscillations in lakes, 1911; invented various instruments. Died Edinburgh, Nov. 3, 1911.

CHRZCZONOWICZ, Stanislaw, Polish chemist; b. Dzwinsk, Poland, Jan. 23, 1911; s. Karol and Seweryna (Stankiewicz) C.; magister, U. Stefan Batory, Vilno, Poland, 1936; dr.tech.scis., Tech. U., Lódz (Poland), 1957; m. Janina Grochowska, July 15, 1939; 1 dau., Barbara. Asst. lectr. dept. organic chemistry U. Stefan Batory, 1936-39; faculty dept. organic tech. Tech. U. Lódz, 1945—, prof., 1959—, dean Faculty Chemistry, 1961—, dir. Inst. Polymers, 1966—. Pres. council Inst. Synthetic Fibers, 1959—; mem. council Inst. Paints and Varnishes, also Inst. Cellulose and Paper, Decorated Krzyz Kawalerski Orderu Odrodzenia Polski, Polonia Restituta. Mem. Polski Towarzystwo Chemiczne. Author: (with W. Albrecht, W. Czternastek, M. Wlodarczyk, A. Ziabicki) Poliamidy, 1964; also articles. Research on chemistry of polyamides, especially catalytic methods of synthesis, chemistry of organosilicon compounds, polycondensation of monomers to polysiloxanes. Patentee synthesis of polyamide-6, antistatic preparation of synthetic fibers, preparation of heat-resistant materials from silicone resins. Home: 40 Kilinskiego, Lódz, Poland.*

CHU, Ju-Chin, chem. engr.; b. Taitsang, Kiangsu, China, Dec. 14, 1919; s. C. N. and C. (Wang) C.; B.Sc. in Chemistry, Nat. Tsing Hua U., 1940; Sc.D., Mass. Inst. Tech., 1946; m. Ching Chen Chu, Feb. 25, 1945; children—Gilbert, Steven, Morgan. Sr. chem. engr. Shell Chem. Corp., 1946; faculty Washington U., St. Louis, 1946-49, faculty Poly. Inst. Bklyn., 1949-66, prof., 1955-66; tech. adviser autonetics N. Am. Rockwell Corp., Anaheim, Cal., 1966—; tech. dir. Chem. Constrn. Corp., N.Y.C., 1956-57; cons. to indsl. firms, 1948—. Recipient Achievement award Chinese Inst. Engrs., 1961, Gold medal and scroll Republic of China, 1961, medal of honor U. Liege (Belgium), 1961. Mem. Academica Sinica (life), Am. Inst. Chem. Engrs., Am. Chem. Soc., Am. Soc. Elec. Engrs., Am. Inst. Aeros. and Astronautics, Sigma Xi, Tau Beta Pi, Phi Upsilon, Phi Tau Phi, Iota Tau. Adv. editor Consultants Bur., 1966—. Research, publs. and patents in propulsion, energy conversion fuel cells, distillation, nuclear tech., applied kinetics and thermodynamics, heat and mass transfer, fluidization drying, extraction. Home: 2816 E. Marywood Lane, Orange, Cal. 92667. Office: 3370 Miraloma Av., Anaheim, Cal. 92803.*

CHU SHIH-CHIEH, Chinese mathematician; b. Yenshan, China; flourished circa 1280-1303. Tchr. math. China; extensive travels; settled in Kuanglin, China, continued teaching, scholarship. Author: Introduction to Mathematical Studies (contained nothing new; explained addition, subtraction, multiplication, div., measuring areas and solids, methods of excess and deficiency, solution of simple equations, rules of signs for algebraic addition and multiplication; important because it brought Chinese algebra to Japan, 1598, thus greatly influencing Japanese math.), 1299; Precious Mirror of the Four Elements (explains method for solution of 4 linear equations containing 4 unknowns—method of solving simultaneous equations very nearly same as method of elimination and substitution except that concept of determinant is missing; also contains method for solution of numerical equations of any degree, diagram for finding binomial coefficients up to 8th power—equivalent to Pascal's triangle; also contains discussion of summation of integral finite series), 1303. Second book places him among greatest mathematicians of all time.

CHU, Ting Li, chemist; b. Peiping, China, Dec. 26, 1924; s. C. H. and S. (Tsui) C.; B.S., Catholic U. Peking (China), 1945, M.A., 1948; Ph.D., Washington U., St. Louis, 1952; m. Shirley Shan-Chi Yu, Sept. 4, 1954; children—Dennis, Dora, Daniel. Came to U. S., 1948, naturalized, 1961. Faculty, Duquesne U., Pitts., 1952-56, asso. prof. chemistry, 1955-56; staff Westinghouse Research Labs., Pitts., 1956-67, fellow scientist, 1963-64, mgr. epitaxy growth, 1964-67; prof. elec. engring. and electronic scis., So. Methodist U., Dallas, 1967—. Mem. Am. Chem. Soc., Electrochem. Soc., I.E.E.E. Research and publs. on preparation techniques for crystals and thin films of electronic materials, crystal growth process and defect formation. Home: 6211 W. Northwest Hwy., Dallas 75225.*

CHUBB, Lewis Warrington, Am. elec. engr.; b. Ft. Yates, N.D., Oct. 22, 1882; s. Charles St. John and Sarah L. (Eaton) C.; M.E. in Elec. Engring., Ohio State U., 1905; Sc.D., Allegheny Coll., 1933; D.Sc., U. Pitts., 1933; m. Mary Porter Everson, Mar. 28, 1910 (dec. 1919); children—Lewis Warrington, John Everson, Morris Wistar; m. 2d, Ora Lee (Dias) McGregor, May 10, 1926; 1 dau., Vivian McGregor. With Westinghouse Elec. & Mfg. Co., 1905-30, in research and devel. magnetic materials, 1906-18, in charge research sect.; with elec. devel. sect., materials and process engring. dept., 1916-20; mgr. radio engring. dept., 1920-30; asst. v.p. charge engring. RCA Victor Co., Camden, N.J., 1930; dir. Westinghouse Research Labs., East Pittsburgh, 1930-48. Mem. spl. uranium com. Nat. Acad. Scis., 1941-42; mem. spl. planning bd. on atomic energy OSRD, 1941-42, cons. to dir., 1941-43. Recipient Lamme medal Ohio State U., 1934, John Fritz medal for 1947. Fellow Am. Inst. E.E., I.R.E.; mem. Phys. Soc., Sigma Xi, Tau Beta Pi. Awarded more than 120 patents as result of elec., mech. and chem. researches; Contbr. many articles and papers. Invented submarine detection methods, electric torpedoes, numerous control systems for tanks, planes, etc., polarized auto headlight, gearless clock, 1932. Died Wilkinsburg, Pa., Apr. 2, 1952.

CHUBB, Talbot Albert, physicist; b. Pitts., Nov. 5, 1923; s. Charles F. and Mary Clare (Albert) C.; A.B., Princeton, 1944; Ph.D., U. N.C., 1950; m. Martha Capps, Oct. 24, 1947; children—Mary Carroll, Nancy Henderson, Talbot Spence, Constance Lamont. Physicist, U. S. Naval Research Lab., Washington, 1950-58, head Upper Air Physics br., 1958—. Recipient Elisha Mitchell Soc. award U. N.C., 1951, E. O. Hulbert award Naval Research Lab., 1953. Fellow Am. Geophys. Union, Am. Phys. Soc., Am. Astron. Soc. Studies, publs. on sun and stars from rockets which show that x-rays are primary variable portion of sun's emission, that during solar flares x-rays are cause of radio fadeout. Located about 20 x-ray stars including extragalactic nebulae, showing that the x-ray star at center of Crab nebula is not neutron star, but is about 1 light year in diameter. Home: 5023 N. 38th St., Arlington, Va. 22207. Office: U. S. Naval Research Lab., Washington 20390.*

CHUBB, Walston, Am. materials engr.; b. Washington, July 23, 1923; s. Robert Walton and Irene (Sylvester) C.; A.B., Harvard, 1944; B.S., Mo. Sch. Mines, 1948, M.S. (Ludlow-Saylor Wire Co. fellow), 1949; m. Carolyn Elizabeth Carpenter, June 16, 1951; children—Walston, Catherine Louise. Asst. engr. Brush Beryllium Co., Luckey, O., 1949-51; prin. engr. Battelle Meml. Inst., Columbus, O., 1951-57, fellow materials tech. div., 1957-66, fellow materials devel. fuels and materials for nuclear power plants, devel. uranium carbide and its alloys, zirconium engring. dept., 1966—. Mem. Am. Soc. Metals, Am. Inst. Mining, Metall. and Petroleum Engr., Am. Ceramic Soc., Sigma Xi. Research and publs. on alloys, zirconium-uranium alloys, iron-chronium-aluminum alloys, hot-hardness equipment for use to 2000° C.*

CHUDAKOV, Aleksandr Evgenievich, Russian physicist; b. 1921; Ph.D. in physico-math. scis. Moscow U., 1948. Mem. staff Lab. Cosmic Rays, Lebedev Physics Inst., since 1948; mem. Acad. Sci. USSR, 1948; recipient Lenin prize, 1960, for discovery and investigation of outer radiational belt of earth and moon.

CHUDOBA, Karl Franz, German mineralogist; b. Writzow, Sept. 10, 1898; s. Bartholomäus and Rosa (Kessler) B.; Ph.D.; m. Charlotte Mager, Aug. 15, 1930. Asst., U. Vienna, Mus. Natural History, Vienna, U. Friburg-en-Brisgau; prof. U. Bonn. Mem. Soc. Friends Mineralogy and Geology (co-founder, hon.), Commn. Study Precious Stones (hon.). Author: Allgemeine und Spezielle Mineralogie; Taschenbuch der Schmuck und Edelsteine. Address: Thomas-Mann Strasse 7, Göttingen, Germany.

CHUFAROV, Grigoriy Ivanovich, Russian phys. chemist; b. Nov. 2, 1900; grad. Urals Poly. Inst., 1928; with Urals Physicochem. Inst., 1931-36, Urals Physicotech. Inst., 1936-39; dir. Inst. Chemistry, Urals br. USSR Acad. Sci., 1939-46; rector Urals U., 1946-56; asso. Urals br. USSR Acad. Sci., 1956—. Decorated Order of Lenin (2). Mem. USSR Acad. Sci. (corr.). Research and publs. on mechanism and kinetics of dissociation and reduction of metal oxides; corrosion of metals in acids and action of inhibitors and hot tinning, zincing, and decarbonization of ferrosilicon steel. Address: Urals Branch, USSR Acad. Sci., Sverdlovsk, RSFSR, USSR.

CHUGAEV, Leo Alexandrovich, Russian chemist; b. Moscow, Russia, Oct. 4, 1873; prof. inorganic chemistry, U. Leningrad, Russia; research on platinum metals, inner complex salts; worked on optical activity and organic chemistry; discovered some glyoxime compounds and characterized them. Died Vologda, U.S.S.R., Sept., 1922.

CHUKHROV, Fedor Vasilevich, Russian mineralogist, geochemist; b. July 15, 1908; grad. Moscow Geol. Survey Inst., 1932. Asso., Inst. Geol. Sci., USSR Acad. Sci., 1936-50, dep. dir., 1950-55, bur. mem. dept. geology and geog. sci., 1960—, dir. Inst. Geology of Ore Deposits, Petrography, Mineralogy and Geochemistry, 1955—. Del., Internat. Mineral. Assn., Washington, 1963. Decorated Order of Lenin; recipient Stalin prize, 1950. Mem. USSR Acad. Sci. (corr.). Research and publs. on mineral. and geochem. study of ore deposits in Kazakhstan, study of colloids in earth's crust, mineralogy of oxidation zone. Address: Inst. of Geology of Ore Deposits, Petrography, Mineralogy and Geochemistry, USSR Acad. Sci., Staromonetny p. 35, Moscow V-17, USSR.

CHUMNOS, Nicephoros, Byzantine theologian, philosopher; b. circa 1250; studied under George (Gregory) of Cyprus at Constantinople; 1 dau., Irene. Pvt. sec. to Andronicos II, also army comdr.; sent on mission to Persia, 1272-75; prefect of secretariat; became gov. of Thessalonica, 1309; entered monastery and assumed name Nathanael, 1320. Author numerous publs. including: Treatise on Matter; Treatise on Meteorology (including cooling of air by winds, hail, the origin of winds). Died Jan. 16, 1327.

CHUN, Carl, German zoologist; b. Höchst nr. Frankfurt, Germany, Oct. 1, 1852; s. Gustav and (Urich) C.; studied at U. Göttingen, U. Leipzig (both Germany); grad. 1874; Dr.med.h.c., U. Oslo (Norway), 1911; m. Lily Vogt, 1884; 2 daus. Became lectr. U. Leipzig, 1878, named prof., 1898, rector, 1907-08; became prof. U. Königsberg (Germany), 1883, U. Breslau (now Wroclaw, Poland), 1891; worked in Naples, Italy, Canary Islands, 1887-88; head Valdivia Expdn., 1898-99. Author: Aus den Tiefen des Weltmeeres, 1900. Research on deep sea fauna, oceanography, botany. Died Leipzig, Apr. 11, 1914.

CHUPKA, William Andrew, Am. phys. chemist; b. Pittston, Pa., Feb. 12, 1923; s. William and Antoinette (Valinchus) C.; B.S., U. Scranton, 1943; M.S., U. Chgo., 1949, Ph.D., 1951; m. Olive A. Pirani, May 21, 1955; children—Jocelyn T., Marc W. Instr. chemistry Harvard, 1951-54; asso. physicist Argonne (Ill.) National Laboratory, 1954-67, senior physicist, 1967—. Mem. committee on high temperature chem. phenomena Nat. Research Council, 1961—; chmn. Gordon Research Conf. on High Temperature Chemistry, 1964. Guggenheim fellow, 1961-62. Developed application of mass spectrometry to study of high temperature chemistry, 1955—; exptl. tests of statis. theory of mass spectra, 1959—; photoionization mass spectroscopy, 1965—. Home: 1136 Cardinal Lane, Naperville, Ill. Office: 9700 S. Cass Av., Argonne, Ill. 60440.*

CHUQUET, Nicholas, French mathematician; b. Paris, 1445; M.D., U. Paris; lived in Lyons. Author: Triparty en la science des nombres, drafted, 1484, 1st pub., 1880. Known for treatise on rational and irrational numbers and theory of equations; offered clear explanation of role of zero symbol; employed system of exponential notation similar to that used today; was 1st to use radical sign with an index to indicate roots; devised method of extracting square and cube roots which was far superior to any other known in 15th century; made systematic use of positive and negative numbers and gave rules for working with them; clearly defined arithmetical and geometrical progression and related them to each other; discussed what was later found to be the germinal idea of logarithms. Died 1500.

CHURCH, Arthur Henry, Brit. botanist; b. Eng., 1865; ed. Jesus Coll., Oxford, Eng.; M.A., D.Sc.; univ. lectr. botany; ret., 1930. Fellow Royal Soc., 1921. Author: Plant Life of the Oxford District. Illustrator: Types of Floral Mechanism (accurate delineations of 12 early spring flowers). Editor Oxford Bot. Memoires, 1919-25. Died Apr. 24, 1937.

CHURCH, James Marion, Am. chem. engr.; b. Worcester, Mass., Sept. 25, 1903; s. Frank L. and Mary (Andrews) C.; A.B., William Jewell Coll., 1925; Ph.D., Pa. State U., 1933; m. Dorothy Borden, June 3, 1933; children—Larry B., Thomas M., James A. Instr., Allegheny Coll., Meadville, Pa., 1927-28, Drexel Inst. Tech., Phila., 1928-30; research chemist, devel. engr. Monsanto Chem. Co., St. Louis, 1932-40; prof. chem. engring. Columbia, 1940—. Recipient Distinguished Alumni award William Jewell Coll., 1955, Distinguished Nat. Service award, Phi Lambda Upsilon 1962. Mem. Am. Chem. Soc. (N.Y. sect. councillor, past chmn. div.) Am. Inst. Chem. Engring., Soc. Plastics Engrs., Ph Lambda Upsilon (nat. pres. 1954-57, nat. sec. 1945-54), Sigma Xi, Alpha Xi Sigma. Author: (with H. R. Simonds) Encyclopedia of Basic Materials for Plastics, 1967, Concise Guide to Plastics, 1962. Research and publs. on organic chem. process, polymerization, plastics processes, flammability of textiles. Home: 6 Kenwood Rd., Tenafly, N.J. 07670. Office: 377 Eng. Terrace, Columbia U., N.Y.C. 10027.*

CHURCH, Lloyd Eugene, Am. anatomist, dentist; b. Littleton, W.Va., Sept. 25, 1919; s. Howard and Mary (Henderson) C.; B.S., W.Va. U., 1942; D.D.S., U. Md., 1944; M.S., George Washington U., 1951, Ph.D., 1959; m. Hildegard Cascio, Apr. 1, 1964; 1 dau., Pamela Gail. Instr. oral surgery and anesthesia Dental and Med. Sch., Med. Coll. Va., Richmond, 1945-46; clin. instr. anatomy George Washington U. Sch. Medicine, Washington, 1952-62, lectr., 1962-64, asst. professorial lectr. anatomy, 1964-67, asso. research prof. anatomy, 1967—; fellow, staff dept. oral pathology Armed Forces Inst. Pathology, Washington, 1959-62, vis. scientist, 1962-63; sr. research sci. Nat. Biomed. Research Found., Washington, 1963-66; vis. prof. Bangalore (India) U., 1966; attending oral surgeon Suburban Hosp., Bethesda, Md., 1965-66. Mem. President's Com. on Employment of Physically Handicapped, 1961—; Md. Gov.'s Com. on Employment Physically Handicapped, 1965—; mem. other adv. profl. coms., cons. nat. orgns., govt. Recipient Distinguished Citizenship award State of Md., Sword of Hope, Am. Cancer Soc. Fellow A.A.A.S., Royal Microscope Soc., Am. Coll. Dentistry, Internat. Assn. Oral Surgery; mem. Am. Assn. Anatomists, Am. Dental Assn., Am. Soc. Exptl. Biology and Medicine, Am. Acad. Oral Pathology, Internat. Assn. Dental Research, N.Y. Acad. Scis., Royal Soc. Health, Sigma Xi, Omicron Kappa Upsilon. Research and numerous publs. on bone, especially embryology and pathology of Avian long bones; devel. and growth of face specifically mandible; X-ray studies of mandible; comparative studies of eruption of teeth, growth of head in relation to skeletal age; ethnic studies of facial growth; growth and devel. of temporomandibular joint. Home: 4890 Battery Lane. Office: 8218 Wisconsin Av., Bethesda, Md. 20014.*

CHURCH, Sir William Selby, Brit. physician; b. Dec. 4, 1837; s. John and Isabella (Selby) C.; ed. Univ. Coll., Oxford, M.S., Sc.D.; LL.D., Glasgow; Sc.D Victoria U.; D.C.L., Durham U.; m. Sybil Constance, 1875; 1 son, 1 dau. Lee's reader anatomy Christ Ch, Oxford; lectr. comparative anatomy, cons. physician St. Bartholomew's Hosp.; cons. physician City of London Hosp. for Diseases of Chest, Royal Gen. Dispensary; hon. fellow Univ. Coll., Oxford. Univ. rep. Gen. Med. Council, 1889-99; mem. Royal Commn. on Care and Treatment of Sick and Wounded, S. African Campaign, 1900, Royal Commn. on Arsenical Poisons, 1901, on Vivisection, 1906. Pres., Royal Soc. Medicine, Royal Coll. Physicians. Contbr. to Allbutt's System of Medicine, also to jours. Research on embolic aneurisms, acute rheumatism. Died Apr. 27, 1928.

CHURCHILL, Edward Perry, Am. zoologist; b. Allerton, Ia., July 19, 1882; s. Edward Payson and Flavilla (Kellogg) C.; A.B., State U. Ia., 1907; Ph.D., Johns Hopkins, 1916; m. Nellie Arlene Lewis, Aug. 12, 1908; children—Marjorie Ann (Mrs. Robert J. Fibikar). Prin. high schs. Ia., 1907-12; asst. to commr. fisheries U. S. Bur. Fisheries, Washington, 1916-20; faculty U. S.D., Vermillion, 1920—, prof. zoology, 1926—, head dept., 1926-47, prof. emeritus, 1967—. Mem. Am. Soc. Zoologists, Am. Assn. Allied Scis., S.D. Acad. Scis. (past pres.), Phi Beta Kappa, Sigma Xi. Author: Fishes of South Dakota, 1933; (with W. H. Over) The Life History of the Blue Crab, 1919; I'd Rather be a Professor, 1955; Three Thousand Coyotes and I, 1961. Research and publs. on food and feeding mechanism Miss. River clams, life cycle migrations edible crab Chesapeake Bay, fishes S.D., oysters and oyster industry Atlantic and Gulf Coast, larvae edible crab Chesapeake Bay. Home: 415 S. University St., Vermillion, S.D. 57069.*

CHURCHILL, Ruel Vance, Am. mathematician; b. Akron, Ind., Dec. 12, 1899; s. Abner Cain and Meldora (Friend) C.; B.S., U. Chgo., 1922; M.S., U. Mich., 1925, Ph.D., 1929; m. Ruby Frances Sicks, Aug. 19, 1922; children—Betty (Mrs. Paul R. McMurray), Eugene S. Faculty, U. Mich., Ann Arbor, 1922—, prof. math., 1942—, supr. research projects, 1930-31, 46-49, 53-54, prof. emeritus, 1965—; vis. lectr. U. Wis., 1941; research U. Freiburg, Germany, 1936, USAAF, 1944, Cal. Inst. Tech., 1949. Mem. Am. Math. Soc. (mem. council 1946-49, editorial com. 1949), Math. Assn. Am. (v.p. 1956-57, bd. govs. 1959-61), Soc. for Indsl. and Applied Math., Phi Beta Kappa, Sigma Xi. Devel. and application of operational calculus. Reviewer, Math. Revs., 1940—. Author: Complex Variables and Applications, 2d edit., 1960; Fourier Series and Boundary Value Problems, 2d edit., 1963; Operational Mathematics, 2d edit., 1958. Research, publs. on advanced math. analysis, especially integral transforms, and applications to physics and engring. Home: 1704 Hermitage Rd., Ann Arbor, Mich. 48104.*

CHURCHMAN, C(harles) West, Am. educator; b. Phila., Aug. 29, 1913; s. Clarke Wharton and Norah (Fassitt) C.; A.B., U. Pa., 1935, M.A., 1936, Ph.D., 1938; m. Gloria King, Sept. 27, 1950; 1 son, Daniel Wharton. Instr. philosophy U. Pa., 1937-42, head dept., asst. prof. philosophy, 1945-48; asso. prof. philosophy Wayne U., 1948-51; prof. engring. adminstrn. Case Inst. Tech., 1951-58; prof. bus. adminstrn. U. Cal. at Berkeley, 1958—, also asso. director space sciences laboratory; head mathematics sect. U. S. Ordnance Lab. Frankford Arsenal, 1942-45; co-founder Inst. Exptl. Method (research city planning), Philadelphia, 1945-48; director research System Devel. Corp., 1962-63. Fellow A.A.A.S., Operations Research Soc. Am.; mem. Inst. Mgmt. Scis. (pres. 1962), Philosophy of Sci. Assn. (governing bd), Am. Philos. Assn., Assn. Symbolic Logic, American Statistical Association. Author: Elements of Logic, 1940; Theory of Experimental Inference, 1948; Methods of Inquiry (with R. L. Ackoff), 1950; Prediction and Optimal Decision, 1961. Co-editor: Measurement of Consumer Interest, 1946; Introduction to Operations Research, 1957; Measurement: Theories and Definitions, 1959. Editor of Philosophy of Science (quarterly), 1948—. Management Science (quarterly), 1954—. Work on foundations of decision theory, 1948; definition of psychological and sociological concepts in decision theory framework, 1946; application of decision theory to the meaning of scientific method, 1948, 1961; developed science of values and ethics, 1948, 1961. Home: 639 Belvedere St., San Francisco. Office: Bus. Adminstrn. Dept., U. Cal., Berkeley 4, Cal.*

CHURCHMAN, John Woolman, Am. bacteriologist; b. Burlington, N.J., Jan. 8, 1877; s. Horace and Edith Anna (Woolman) C.; A.B., Princeton, 1898, A.M., 1901; M.D., Johns Hopkins, 1902; A.M., Yale, 1915; m. Martha Bertrand Jaramillo, Oct. 27, 1923. Volunteer asst. Surg. Clinic, Breslau, Germany, 1905-06; asst. resident surgeon and resident surgeon Johns Hopkins Hosp., 1906-11; instr. surgery, Johns Hopkins, 1909-11; asst. prof. surgery Yale, 1912-14, prof, from 1914, also acting head of dept.; attending surgeon and acting surgeon in chief, New Haven Hosp. and Dispensary; prof. exptl. therapeutics and dir. Lab. of Exptl. Therapeutics, Cornell U. Med. Sch. Fellow A.C.S., N.Y. Acad. Sci. Officier de l'Instruction publique, and Officier d'Académie, France. Recipient Alvarenga prize, Coll. Physicians of Phila., 1921. Discovered specific bactericidal action of gentian violet with regard to staphylococci, 1912. Died July 12, 1937.

CHURG, Jacob, physician; b. Dolhinow, Poland, July 16, 1910; s. Wolf and Gita (Ravich) C.; M.B., U. Wilno, Poland, 1933, M.D., 1936; m. Vivian Gelb, Oct. 18, 1942; children—Andrew Marc, Warren Bernard. Came to U. S., 1936, naturalized, 1943. Practiced medicine, specializing in pathology, Teaneck, N.J., 1946—; research asso. Mt. Sinai Hosp., N.Y.C., 1946—, asso. attending pathologist, 1962—; pathologist Barnert Meml. Hosp., 1946—; lectr. Fairleigh Dickinson U. Sch. Dentistry, Teaneck, N.J., 1957—; asso. clin. prof. pathology N.J. Coll. Medicine and Dentistry, 1958—; clin. prof. pathology Mt. Sinai Sch. Medicine, 1966—; cons. pathologist VA Hosp., East Orange, N.J., 1966—. Diplomate Am. Bd. Pathology. Fellow Coll. Am. Pathologists; mem. A.M.A., Am. Assn. Pathologists and Bacteriologists, N.Y. Acad. Scis., N.Y. Acad. Medicine, Harvey Soc., Am. Soc. Exptl. Pathology, Internat. Acad. Pathology. Research, numerous publs. in syndrome of allergic inflammation of blood vessels and connective tissue, arterial lesions due to various types of kidney damage, ultrastructure of kidney in various renal and systemic diseases, effects of dusts upon ultrastructure of the lung. Address: 711 Ogden Av., Teaneck, N.J. 07666.*

CHUSID, Joseph George, Am. neurologist; b. Newark, Aug. 23, 1914; s. Morris and Kate (Kellner) C.; A.B., U. Pa., 1934, M.D., 1938, m. Kathryn McCool, Mar. 5, 1942; children—Joseph M., Joanna. Practice medicine, specializing in neurology, N.Y.C., 1940; mem. staffs St. Vincent's Hosp., 1947—, Columbia-Presbyn. Med. Center, 1965—; faculty U. Pa. Med. Sch., 1941-45, Seton Hall Coll. Medicine, 1961-65; asso. clin. prof. neurology Columbia, N.Y.C., 1965—. Diplomate Am. Bd. Psychiatry & Neurology. Fellow Am. Acad. Neurology; mem. Am. Physiol. Soc., Soc. for Exptl. Biology & Medicine, Am. Electroencephalographic Soc., Assn. for Research in Nervous and Mental Diseases, Alpha Omega Alpha. Author: (with J. J. McDonald) Correlative Neuroanatomy & Functional Neurology, 13th edit., 1967. Research in cortical connections of monkey's brain; exptl. epilepsy in monkeys and rats; effects of drugs and metals on brain of monkeys and rats; electroencephalographic studies of man and animals; pathogenesis of epilepsy. Home: 3390 Emeric Av., Wantagh, N.Y. 11793. Office: 153 W. 11th St., N.Y.C. 10011.*

CHUZEN, Murai, Japanese mathematician; flourished 1780; pupil of Nakane Genjun; author: Kaisho Tempei Sampo (on numerical higher equations), 1765; Sampo Doshi-mon, 1781; used Pascal triangle in expressing coefficients of terms in expansion of a binomial, 1781.

CHVOSTEK, Franz, Austrian internist; b. Vienna, Austria, Oct. 3, 1864; s. Franz and L. Katherina F. (Aloisia) C.; finished med. studies Vienna, 1888; m. Wilhelmine Kruder; 1 son. Became intern Clinic of Heinrich Bramberger, 1886; asst. Psychiat. Clinic under T. Meynert, 1890-92; later intern Clinic of Otto Kahler; became head intern div. Sophienspitales, 1897; prof. Med. Clinic, Vienna, 1909-33. Author: Über alimentäre Glykosurie bei Morbus Basedowii, 1892; Über des Wesen der paraoxysmalen Hämoglobinurie, 1894; Zur Ätiologischen der alsuten Gelenkrheumatismus, 1898. Research on neurology, internal medicine, disease of ductless glands; knew significance of tonsils; pancreatic anemia named for him; founder genetics and hereditary pathology. Died Gropenstein, Austria, Apr. 17, 1944.

CHVOSTEK, Franz, Austrian surgeon; b. May 21, 1835; s. Vinzenz C.; grad. in medicine Josephs Acad., 1861; m. L. Katherina F. Aloisia; 1 son, Franz. Leader, Duchek Clinic; intern at Garnisonspitales. Discovered tapping face of patient with tetany causes spasm of the facial muscles, 1876; research on pathology and therapy of spinal and nervous diseases; used elec. therapy. Died Nov. 16, 1884.

CHYUNG, Moon Ki, Korean ichthyologist; b. Soonchon City, Korea, Sept. 19, 1898; B.S., Tokyo Imperial U., 1929; D.Sc., Pusan Fisheries Coll., 1962; m. Han So Bang; children—Bang Chin (Mrs. Chung Heng Lee), Suck Cho, Myung Cho. Chief, fisheries research stas., Pyungan-Bukto and Mokpo, 1930-47; pres. Nat. Pusan Fisheries Coll., 1947; dir. fisheries bur. Korean Ministry Agr., 1947-51; chief Central Fisheries Inspection Sta., 1951-60; chief Heung-Han Fisheries Research Inst., Seoul, 1963—. Mem. monetary bd. Bank of Korea, 1962—. Recipient several culture and writing awards. Mem. Nat. Acad. Sci. (committeeman 1954—), Korean Fisheries Assn. (pres. 1952), Central Fisheries Cooperation (steering com. 1962—), Korean Fisheries Tech. Assn. (chmn. 1965—). Author: Names of Korean Fish, 1934; Korean Laver, 1937; Korean Alaska Pollack, 1937; A Catalogue of the Fishes in the Yalu River, 1954; The List of the Marine Algae of Korea, 1955; The Aspects of the Korean Seas and the Distribution of Marine Life Therein, 1959; Illustrated Encyclopedia The Fauna of Korea (2) Fishes, 1961; The Historical Study on the Investigation of Fishes in Korea, 1961. Extensive research on fishes of Korea with descriptions of new species. Home: 110 Nesudong Chongroku. Office: Heunghan Fisheries Research Inst., IPO 1121, Seoul, Korea.*

CIAMICIAN, Giacomo Luigi, Italian chemist; b. Trieste, Aug. 25, 1857; became prof., Padua, Italy 1887; apptd. prof. gen. chemistry Bologna, Italy, 1889. Mem. French Acad. Scis. Research on pyrrole, 1880-88, essential oils, 1889-99, chem. action of light in inducing oxidation of aldehydes and ketones, 1900-15, pyrroline, plant alkaloids, effect of chem. compounds on devel. plants, synthesis of amino acids. Died Bologna, Jan. 1, 1922.

CIASSI, Giovanni Maria, Italian botanist; b. Treviso, Italy, 1554; author: Mediationes de natura plantarum, 1677; studied germination, sap circulation, plant irritability. Died 1679.

CICERO, Marcus Tullius, Roman polit. theorist, statesman; b. Arpinum (now Arpino), Jan. 3, 106 B.C.; s. Marcus Tullius and Helvia Cicero; student of Archias the Poet, also Phaedrus the Epicurean, Diocotus the Stoic, Molo (n) of Rhodes; studied law under Q. Mucius Scaevola and his nephew of the same name; m. Terentia, circa 77 B.C. (div. 46 B.C.); children—Marcus, Tullia; m. 2d, Pubilia (div.). Served in mil. campaign, 89 B.C.; became legal pleader in Forum, Rome, 81 B.C.; traveled east because of poor health, studied philosophy in Athens; returned to Rome, 77 B.C.; elected quaestor (paymaster), 76 B.C., served in Sicily; elected aedile, 70 B.C., praetor, 67 B.C., consul, 64 B.C.; prosecuted Catiline on conspiracy charge, 63 B.C.; accused by tribune Clodius of violating due process of law, went into exile in Greece, 58 B.C.; returned triumphantly to Rome, 57 B.C.; apptd. proconsul (gov.) of Cilicia, 52 B.C.; joined Pompey against Julius Caesar, but submitted to Caesar's victory; after Caesar's death denounced Marc Antony in series of Philippics, 44-43 B.C., sought to recreate the republic; proscribed by ruling Triumvirate (which included Antony), 43 B.C. Author: De republica; De legibus; De officiis; also many other writings on philosophy, rhetoric. Devoted much effort to making Greek philosophy accessible in Latin literary form; a conservative; argued that while the state belongs to the people, their will must be expressed by sovereign authority; advanced an influential doctrine of natural law, arguing that true law is right reason in agreement with nature. Slain by Antony's soldiers at Formia, Dec. 7, 43 B.C.

CIDENAS (Kidinnu), Babylonian astronomer; b. circa 343 B.C.; head astron. sch. at Sippra; discovered precession of equinoxes.

CID PALACIOS, Rafael, Spanish astronomer, meteorologist; b. Vigo, Oct. 22, 1918; s. Jaime and Maria Palacios (Burgos) Cid P.; Ph.D. in sci.; m. Maria Eva Castro Carbo, Apr. 10, 1949; children—Maria Eva, Rafael, Alberto, Fernando, Maria Jose. Collaborator, Obs. of Santiago; meteorologist Nat. Service in Canary Islands and Saragossa Sea; asst. prof. U. Saragossa; prof. astronomy and geodesy Sch. Sci. of Saragossa. Mem. Acad. Sci. of Saragossa, Assn. for Progress of Sci. Author: Orbitas de estrellas dobles; Eclipse de sol; Problema de cuerpos; Mecania teorica. Home: calle Corona de Aragon, 50. Office: Faculty of Science, University of Saragossa, Saragossa, Spain.

CIERMAUS, Johann, mathematician; b. Bois-le-Duc, Low Countries (now Netherlands), circa 1600; became a Jesuit, 1619; prof. Jesuit Coll., Louvain, Belgium, later, Antwerp, Belgium. Author: Annus positionum mathematicarum, 1640. First to suggest an automatic calculating machine, 1640. Died Portugal, 1648.

CIFERRI, Raffaele M., Italian botanist; b. Fermo, May 30, 1897; s. Giuseppe and Eugenia (Ciccioli) C.; Ph.D. in Agrl. Scis., Tech. Agrl. Inst.; m. Angela Maria Borgna, Oct. 25, 1925; children—Fiorella, Orio, Donatella. Asst., U. Pavia, Florence, later instr.; vice dir. Cryptogamical Lab., Pavia; prof. Sch. Agr., U. Florence. Research and numerous publs. on botany. Died Pavia, Feb. 14, 1964.

CIGANEK, Leodegar, Czechoslovakian physician, electrophysiologist; b. Bratislava, Czechoslovakia, Feb. 10, 1925; s. Leodegar and Francisca (Wagner) C.; Medicinae Universae Doctor, Comenius U., Bratislava, 1950, Candidatus Scientiarum, 1960, Docent Neurologiae, 1965; m. Viera Dingova, Apr. 20, 1953; children—Leodegar, Lea. Physician, Neuropsychiat. Clinic, Comenius U., 1950-54; faculty Med. Faculty, 1966—; sci. worker Inst. Normal and Path. Physiology (formerly Inst. Exptl. Medicine), 1954—. Mem. Royal Soc. Medicine (affiliate), Czechoslovak Med. Soc. J. E. Purkynie. Author: Die Elektroencephalographische Lichtreizantwort der menschlichen Hirnrinde, 1961; also articles. Research on elec. responses of brain to various stimuli, surg. treatment of epilepsies. Home: 6/c Belehradska St. Office: 1, Sienkiewicz St., Bratislava, Czechoslovakia.*

CIGNA, Giovanni Francesco, Italian, physician, chemist; b. Mondovi, Italy, July 2, 1734; grad. U. Turin (Italy), 1757; became prof. anatomy U. Turin, 1770. Fellow Royal Soc., 1764; mem. Acad. Scis. Turin (a founder). Author: De caussa extinctionis flammae d animatium in aëre inclosorum, 1750; Sull'uso dell'electricita nella medicine e sulla irritabilita Halleriana, 1757. Research on electricity and evaporation, cause of extinction of flame and animal life in enclosed space; discovered burning phosophorous decreases the elasticity of air. Died Turin, July 16, 1790.

CIMMINO, Gianfranco, Italian mathematician; b. Naples, Italy, Mar. 12, 1908; s. Francesco and Olimpia Gibellini (Tornielli-Boniperti) C.; Ph.D. in math.; m. Maria Martinez, Mar. 26, 1940; children—Francesco, Olimpia, Giovanni, Oretnsia, Giuseppe, Beatrice, Pietro. Asst., instr. U. Naples, 1928-38; titular chair math. analysis U. Cagliari, 1938, Bologna, 1940. Mem. Acad. Sci. of Bologna, Societa nazionale of Naples (corr.), Accademia nazionale of Modena. Research and publs. on math. Home: via Albertazzi 6/5. Office: via Trombetti 4, Bologna, Italy.

CINADER, Bernhard, immunologist; b. Vienna, Austria, Mar. 30, 1919; s. Leon and Adele (Schwarz) C.; B.S., U. London, 1945, Ph.D., 1948, D.Sc., 1958; 1 dau., Agatha. Prin. sci. officer Inst. Animal Physiology, Babraham, Cambridge, 1956-59; head of subdiv. immunochemistry Ont. Cancer Inst., Toronto, 1958—; professor dept. med. biophysics, pathol. chemistry U. Toronto, 1958—; organizer, chmn. symposia N.Y. Acad. Scis., 1962, Fedn. European Biochem. Socs., Vienna, 1965, Toronto, 1966. Enrique E. Ecker lectr. in exptl. pathology Western Res. U., 1964. Fellow N.Y. Acad. Scis.; mem. Biochem. Soc., Brit., Canadian socs. immunology, Am. Assn. Immunologists, Soc. Exptl. Biology and Medicine. Editor: Antibodies to Biologically Active Molecules, 1967; The Regulation of the Antibody Response, 1967. Established that specificity and inheritance of antibody can be regulated by immunological unresponsiveness to the body's own molecules; coined term eniotypy for polymorphism in which some individuals of a species possess a molecule which others lack and developed criteria by which this type of deficiency can be recognized; has shown that inhibition of enzymes by antibody depends on steric hindrance of access to catalytic site of enzyme and antibody-imposed shape changes in enzyme; has elucidated structure and shape of antigen-antibody aggregates; proposed regulation of antibody response through heterogeneous receptors on precursors of antibody-forming cells and through steering mechanism. Home: 34 Rowanwood Av., Toronto. Office: 500 Sherbourne St., Toronto 5, Ont., Can.*

CINQUINI, Maria dei Conti Cibrario, Italian mathematician; b. Genoa, Italy, Sept. 6, 1905; d. Giullo and Cristina (Botto) dei Conti Cibrario; Ph.D. in Math., U. Turin; m. Silvio Cinquini, July 14, 1938; children—Giuseppe, Vittoria, Carlo. Prof., U. Pavia (Italy). Mem. Unione Matematica Italiana. Lombarde Inst. (corr.), Acad. Sci. and Letters. Contbr. numerous articles to math. revs. Home: Corso Cairoli 96. Office: Mathematics Inst., University of Pavia, Pavia, Italy.

CINQUINI, Silvio, Italian mathematician; b. Spezia, Italy, Sept. 4, 1906; s. Enea and Teodolinda (Miotti) C.; Ph.D., U. Bologna, 1929; postgrad. Scuola Normale Superiore at Pisa, 1931; m. Maria Cibrario, July 14, 1938; children—Giuseppe, Vittoria, Carlo. Prof., Scuola Normale Superiore, Pisa, 1936-38; faculty U. Pisa, 1931-38; prof. math. analysis U. Pavia, 1939—, dean faculty math., phys. and natural scis., 1957—. Mem. Sci. Commn., Italian Math. Union, 1958-61. Decorated comdr. Order of Merit, Italian Republic; recipient Gold medal for achievements in sch., culture and arts. Mem. Lombardian Acad. Scis. and Letters. Author: Funzioni quasi periodiche, 1950; Elementi di Teoria delle funzioni di variabile reale, 1960; (with Maria Cinquini Cibrario) Equazioni a derivate parziali di tipo iperbolico, 1964; Lezioni di Analisi Matematica, 1965; also articles. Research on theory of functions of real variables, approximations of functions, analytic functions, calculus of variations, semiperiodic functions, ordinary differential equations, equations of partial derivatives. Home: 96 Corso Cairoli, Pavia, Italy.*

CIOLEK, Erazm. (Vitello, Erasmus), Polish mathematician; b. 1210; author: Vitellionis perspectivae libri decem (collection of his works pub. posthumously 1533); also wrote on philosophy, sci. of celestial movements. Compared axioms, theorems and hypotheses of Euclid, Ptolemy, Apollonios, Theodore, Menalaus, Theon, Pappus, Probus and Al-Hazem. Died 1285.

CIRUELO, Pedro Sanchez, mathematician; b. Daroca, Spain, circa 1470; student Salamanca, Spain; Ph.D., Paris, France; prof. U. Paris; author: Cursus quatuor mathematicamum artium liberalium, 1516. Died circa 1560.

CISIN, Ira Hubert, Am. statistician; b. N.Y.C., Sept. 1, 1919; s. Bertram R. and Esther (Berlin) C.; B.S., N.Y. U., 1939; M.A., Am. U., 1951, Ph.D., 1957; m. Pattsie V. Miller, Nov. 21, 1946; children—Daniel, Frederick, Tobi. Chief profl. staff Attitude Research br. Def. Dept., 1946-52; asst. dir. research design Human Resources Research Office, Washington, 1952-59; project dir. Cal. Dept. Pub. Health, Berkeley, 1959-62; dir. social research group George Washington U., 1962—, prof. sociology, 1964—. Cons. CBS, Mich. Bell Telephone Co., USPHS, Cal. Dept. Pub. Health, Bur. Social Sci. Research, Am. U., Nat. Safety Council. Mem. A.A.A.S., Am. Statis. Assn., Am. Sociol. Assn., Inst. Math. Statistics, Am. Assn. For Pub. Opinion Research. Research in quantitative methods in social research, including sampling, data collection and scaling. Home: 6828 Wilson Lane, Bethesda, Md. 20034. Office: 818 18th St. N.W., Washington 20006.*

CIST, Jacob, Am. naturalist, inventor; b. Phila., Mar. 13, 1782; s. Charles and Mary (Weiss) C.; m. Sarah Hollenback, Aug. 25, 1807, at least 2 daus. Postmaster, Wilkes-Barre, Pa.; mined anthracite coal, Mauch Chunk, Pa., 1815-21; collected fossil plants and flora, described coal formations, 1815-21; founder, corr. sec. Luzerne County Agrl. Soc.; invented artist's paint mixing mill, 1803, printer's ink from anthracite coal, 1808, anthracite coal burning stove, 1817. Died Dec. 30, 1825.

CITOYS, François, French physician; b. Poitiers, France, 1572; M.D., Poitiers, 1598; m. Madelaine Joulain; several children. Practiced medicine, Poitiers, from 1608, Poitou, from 1620; became physician to king, 1628; courier in service of Richelieu, 1635-38; returned to Poitou, 1638; prof. Faculty Medicine, Poitiers, until 1652. Author: Opuscula medica, 1639. Described lead colic (Poitou colic), 1616. Died 1652.

CITROEN, André, French automotive engr.; b. 1878; gear, automobile mfr.; introduced to France all-steel body (automobile) and assembly line constrn., 1919; invented front wheel drive, 1934; developed advt. methods including skywriting. Died 1935.

CITTERIO, Carlo, Italian neuropsychiatrist; b. Cernusco, Italy, Oct. 11, 1927; s. Gaetano and Maria (De Micheli) C.; grad. Facoltà di Medicina e Chirurgia U. Bologna, 1954; Specializzazione in Clinica Neuropsichiatrica, U. Parma, 1961. Became instr. psychiatry, 1963, instr. criminal anthropology, 1965; became asst. Ospedale Psichiatrico, Verona, Italy, 1962, Ospedale Psichiatrico, Rome, 1964. Named cons. neurology INAM, 1965. Mem. Italian Psychiatry Soc., Italian Sect. of Psychopathology Forum. Research and publs. in psychopharmacology, endocrine psychiatry, gen. psychopathology, alcoholism and symptomatic psychoneurosis. Home: 34 Piazzale Medaglie d'Oro, Rome. Office: Ospedale Psichiatrico Provinciale, Rome, Italy.*

CIUCA, A., Rumanian geriatrician; b. Slatina, Rumania, Oct. 5, 1920; s. Dumitru and Dumitra Ciuca; grad. Med. U. Bucharest, 1947; m. Constanta Bilcu, Oct. 12, 1948; 1 dau., Svetlana. Country physician, 1947-48; chief physician Dist. Severin, 1948-50; chief physician Region Severin, 1950-51, health state insp., 1951-52; chief physician County Banat, 1953-56; gen. dir. med. assistance Ministry of Health, Bucharest, Rumania, 1956-58; chief sect. health orgn. Center for Health Protection, 1956-58; chief sect. sci. research Inst. Geriatrics, Bucharest, 1958—, also mem. sci. counsel; chmn. sect. for social medicine, 1966. Recipient distinctions for san. work, Red Cross Work; Work Decoration. Author: Health State of the Population Bucharest, 1958; Outpatient Care (Dispensarisation) of elderly, Bucharest, 1964. Research, publs. on demographic studies, longevity, biol. age, adjustment to old age, health status, med. social care of population; these studies have approached areas of fundamental as well as applicable research in social medicine and gerontology. Address: Inst. for Geriatrics, Bucharest, Romania.*

CIUSSA, Riccardo, Italian chemist; b. Sassari, Apr. 27, 1877; s. Ignazio and Giovanna (Pinna) C.; Ph.D. in chemistry. Prof. pharm. chemistry, 1924; prof. agrarian chemistry U. Bari, 1942, dean Sch. Pharmacy, 1927—, prof. emeritus, 1952. Mem. Pugliese Acad. Sci. (pres. 1942-47). Author: Chimica organica, 1946; Lezioni di chimica generale e inorganica, 1952, numerous others. Address: via Cavour 40, Bari, Italy.

CIVIALE, Jean, French physician; b. Thieza, France, July 5, 1792; s. Pierre and Jeanne (Usse) C.; studied medicine, Paris, from 1815; became student of Dupuytren, 1817; M.D., 1820; m. Marie-Dix-Août-Faugère. Physician at Brioude; returned to Paris, performed kidney stone operation; apptd. physician Neckar Hosp., 1829; Recipient Grand prize for surgery, 1827; decorated Legion d'honneur, 1867. Mem. French Acad. Medicine, French Acad. Sci., 1847. Author: Traité pratique sur les maladies des organes génito-urinaires, 3 vols., 1837-42, Du traitement médicale et preservatif de la pierre et de la gravelle, 1840; De l'urétrotomie, 1849. Research on genito-urinary diseases; devised best lithotrite of time, 1826; popularized lithotripsy. Died 1867.

CLAASSEN, Howard H., Am. physicist; b. Hillsboro, Kan., Apr. 10, 1918; s. John W. and Elizabeth (Hiebert) C.; A.A., Tabor Coll., 1938; B.S., Bethel Coll., Kan., 1941; M.S., U. Okla., 1943, Ph.D., 1949; m. Esther E. Wiebe, Aug. 29, 1941; children—John, Eileen, Nancy. With Phillips Petroleum Co., 1943-47; asst. prof. U. Okla., 1949-52; faculty Wheaton (Ill.) Coll., 1952-66, prof. physics, 1959-66; sr. physicist, chemistry div. Argonne (Ill.) Nat. Lab., 1966——. Recipient Rosenberger medal U. Chgo., 1964. Mem. A.A.A.S., Am. Phys. Soc., Soc. Applied Spectroscopy, Sigma Xi. Author: The Noble Gases, 1966; also articles. Made 1st preparations of higher fluorides including platinum hexafluoride, ruthenium hexafluoride, rhodium hexaflouride, xenon tetrafluoride; determination of vibrational frequencies and molecular potential functions of numerous fluorides and oxide fluorides. Home: 1N018 Ethel St., Wheaton, Ill. 60187. Office: Argonne Nat. Lab., Argonne, Ill. 60439.*

CLAASSEN, Peter Walter, Am. biologist; b. Hillsboro, Kan., Mar. 17, 1886; s. Deitrich and Elizabeth (Wall) C.; grad. normal course, McPherson (Kan.) Coll., 1909; A.B., U. Kan., 1913, A.M., 1915; Ph.D., Cornell U., 1918; m. Evelyn Strong, Dec. 22, 1917; children—Sarah Evelyn, Richard Strong. Prin. pub. sch., Hillsboro, 1909-11; asst. entomologist State of Kan., 1913-15; instr. biology, Cornell U., 1915-16, asst. prof., 1918-26, prof., 1926-37; asst. prof. entomology, U. Kan., 1916-17; prof. biology (leave of absence), Tsing Hua Coll., Peking, China, 1924-25. Specialized in study of plecoptera. Author: Laboratory Text in General Biology; Plecoptera Nymphs of North America. Died Aug. 16, 1937.

CLADIS, John Baros, Am. physicist; b. Dawson, N.M., June 21, 1922; s. George William and Mary (Chousis) Baros; B.S., Colo. U., 1944; Ph.D. in Nuclear Physics, U. Cal. at Berkeley, 1952; m. Genevieve Liradjis, June 15, 1947; children—Mary Katherine, Christine Diane, George Baros, Mark Sydney. Research physicist Lawrence Radiation Lab., Berkeley, Cal., 1948-52; staff Los Alamos Sci. Lab., 1952-55; mgr. space physics research Research Labs., Lockheed Missiles & Space Co., Palo Alto, Cal., 1955——. Session chmn. Advanced Study Inst. on Radiation Belts, NATO, Bergen, Norway, 1965. Mem. Am. Phys. Soc., Am. Geophys. Union, Sigma Xi, Tau Beta Pi, Pi Mu Epsilon, Sigma Pi Sigma, Alpha Chi Sigma. Contbr. articles to tech. jours. Research on high-energy proton and neutron scattering, nuclear weapons, satellite and rocket investigations geomagnetically-trapped particles, interaction solar wind with geomagnetic field, theory acceleration geomagnetically trapped particles by fluctuation of ionospheric currents; invented interferometer radar antenna. Home: 15 Bear Gulch Dr., Portola Valley, Cal. 94026. Office: Lockheed Missiles & Space Co., Palo Alto, Cal. 94304.*

CLADO, Sprio, gynecologist; b. Smyrna, Turkey, 1862; became naturalized French citizen, 1886; studied medicine U. Paris (France); M.D., 1887. Became intern, 1882, physician, 1887; named lab. chief at Pitiê, 1887; apptd. clin. chief Hôtel-Dieu, 1889. Author: Etude sur une bactérie septique de la vessie, 1895; Traité d'hystéroscopie, 1898; Diagnostic gynécologique, 1902. Research on gynecology, appendicitis, urinary infection, cranial cerebral topography; described suspensory ligament of ovary covered with peritoneum (Clado's band), circa 1892. Died 1920.

CLAESSON, Stig Melker, Swedish phys. chemist; b. Kumla, Sweden, Feb. 5, 1917; s. Wiktor and Ruth (Johansson) C.; matriculation, Uppsala (Sweden) U., 1937, fil.kand., 1939, fil. lic., 1942, laureate, 1946; m. Ingrid Elisabet Moring, Jan. 5, 1944; children—Eva Charlotte, Ingrid Marianne, Caroline Elisabet. With U. Uppsala, 1946——, prof. phys. chemistry, dir. Inst. Phys. Chemistry, 1949——; Distinguished vis. prof. U. Del., 1961-62. Recipient Wallmark award Royal Acad. Scis., 1946; Lindström Minnesfond medal, 1958. Rockefeller fellow Cal. Inst. Tech., Pasadena, U. Cal. at Berkeley, 1947. Fellow N.Y. Acad. Scis.; mem. Royal Acad. Scis., Royal Acad. Engring. Scis., Royal Sci. Soc., Swedish Nat. Com. Chemistry, Swedish Nat. Com. Optics, Am. Phys. Soc., Am. Chem. Soc., Sigma Xi. Editor: Fifth Internatl Symposium Free Radicals, 1961; Nobel Symposium 5 Fast Reactions and Primary Processes in Chemical Kinetics, 1967. Research, publs. on adsorption chromatography of solutions, vapor phase adsorption chromatography, chromatography of macromolecules, dilute solutions of macromolecules, photochemistry and fast reactions using flashphotlysis, ultrafast reactions using multiple flashes; developed instruments for high precision measurements, especially using optical methods. Home: 14 C Ö. Slottsgatan, Uppsala, Sweden.*

CLAGETT, O(scar) Theron, Am. surgeon; b. Jamesport, Mo., Oct. 19, 1908; s. Oscar Frederic and Effie (Stevens) C.; M.D., U. Colo., 1933, Doctor of Science (honorary), 1962; M.S., U. Minn., 1938, grad. student, 1935-40; m. Alicia M. Eames, Nov. 3, 1934; children—Mary Alice, Nancy Jane, Barbara Joan, Martha Eleanor, James Stevens, Robert Scott. Intern Colo. Gen. Hosp., Denver, 1933-34; practice of medicine, Glenwood Springs, Colo., 1934-35; asst. surgeon Mayo Clinic, 1938-40; head sect., div. of surgery, 1940, prof. surgery Mayo Found. Grad. Sch. U. Minn., 1951——; vis. prof. surgery Johns Hopkins Hosp., 1957. Awarded Norlin medal, U. Colo., 1947. Secretary American Board Thoracic Surgery. Fellow A.C.S., mem. Am. Med. Assn., American Association of Thoracic Surgery (past president), American Central surg. assns., Western Surg. Soc., Am. Heart Assn., Mexican Nat. Acad. Surgery; hon. mem. Royal Australasian College Surgeons, Royal College Surgeons (Ireland), Thoracic Soc. (Eng.), Soc. Thoracic Surgeons Gt. Britain and Ireland. Research on cardiovascular diseases. Home: 1611 Merrihills Dr. Office: Mayo Clinic, Rochester, Minn.

CLAIR, Alexander, French engr.; b. St. Etienne, France; mfr. precision instruments, Paris, France; made precision dynamometer for steam engines, 1854, also recording dynamometer, rotation recording dynamometer to determine work performance of drive shaft. Died 1886.

CLAIRAUT, Alexis Claude, French mathematician; b. Paris, May 7, 1713; s. Jean-Baptiste Clairaut; tutored by father. Mem. Maupertuis's expdn. to Lapland to measure a degree of meridian, 1736. Recipient prize for Théorie de la lune, 1765, also prizes St. Petersburg Acad. Scis., 1752, 1761. Mem. French Acad. Scis., 1731 (asso. dir., 1743); Fellow Royal Soc., 1737. Author: Recherches sur les courbes à double courbure (1st complete book on solid analytical geometry), 1731; Éléments de géométrie, 1741; Théorie de la figure de la terre, 1743; Eléments d'algèbre, 1746; Théorie de la lune, 1752; Mémoire sur l'orbite apparente du soleil autour de la terre, en ayant égard aux perturbations produites par la lune et par les principales planétes, 1757; Théorie du mouvement des comètes, 1760; also numerous papers. Specialized in celestial mechanics; correctly determined shape of earth; connected gravity of points on surface of rotating ellipsoid with compression and centrifugal force at equator (Clairaut's theorem), 1743; gave a solution for astron. 3 body problem, 1750; applied Newton's theories to estimation of perturbating influence of planets and comets; explained motion of moon, 1752; calculated effects of Venus' gravitational pull on earth, as compared with that of moon (completed Euler's and d'Alembert's work on this subject); computed change in gravity at high altitude; calculated path of Uranus before its discovery; 1st to obtain reasonable figure for mass of Venus, 1757; calculated perihelion of Halley's comet and predicted its return, 1759; investigated oscillation of conical pendulums; wrote on achromatic telescopes; designed and produced 1st aplanatic objective; designed circular slide rule; advanced theory of capillarity based on attraction between parts of fluid itself; found solutions in 1st order differential equations, as well as in higher degrees. Died Paris, May 17, 1765.

CLAIRMONT (or KLARBERG), Paul, surgeon; b. Vienna, Austria, Jan. 10, 1875; s. Wilhelm and Ottilie (von Pickler) C.; studied medicine under Eiselsberg at U. Vienna; m. Emmy Koller, 1921; 1 son, 1 dau. Became asst. Serological Therapeutics Inst., Vienna, 1898; vol. doctor with A. Eiselsberg, Königsberg, Germany, 1900; asst. to Eiselsberg, Vienna, 1903; named titular prof., Vienna, 1911, apptd. 1st physician in Rudolf-Spital sect. of univ., 1912; surgeon in Balkan War; prof. surgery U. Zürich (Switzerland), 1918-41. Author: Die Bedeutung der Magenradiologie für der Chirurgie, 1911; Operationen an anus und Rectum . . . , 1923; Uber den experimentell erzeugte Ulcus ventriculi und seine Heilung durch dem Gastroenterotomie, 1918. Serological-bacteriological studies; research on physiology and pathology; improved gastrointestinal surgery; tried to ameliorate recurrent dislocation of shoulder by strengthening ligaments, 1917. Died St. Prex, Switzerland, Jan. 1, 1942.

CLAISEN, Ludwig Rainer, German chemist; b. Cologne, Germany, Jan. 14, 1851; s. Heinrich Wilhelm and Emilia (Berghaus) C.; ed. Bonn and Göttingen, Germany; grad. 1875; lectr., Manchester, Eng., Munich and Bonn, Germany, from 1878, prof. Aachen, 1890-97, Kiel, 1897-1904; hon. prof. Berlin, 1904-07 (all Germany); pvt. research in chem. analysis and synthesis, from 1907. Originated Claisen reaction for synthesis of aromatic ketonic ester; proved (with W. Wislicenus) the essentials of tautomerism; introduced use of sodamide. Died Bad Godesberg, Germany, Jan. 5, 1930.

CLANNY, William Reid, inventor; b. Bangor, Ireland, 1776; ed. Edinburgh, Scotland; M.D., 1803; asst. surgeon in navy; practised medicine at Bishopwearmouth, Sunderland, until his death. Recipient gold and silver medals Soc. Arts, 1816, 17. Contbr. articles to sci. jours. Invented safety lamp for miners, 1812; strived to reduce fatalities resulting from mine explosions. Died Jan. 10, 1850.

CLAPARÈDE, Édouard, Swiss psychologist; b. Geneva, Switzerland, Mar. 24, 1873; ed. Geneva, Leipzig, Paris univs.; M.D., Prof. exptl. psychology Geneva U., from 1915; founder Inst. J. J. Rousseau for study of educational psychology; dir. Archives de Psychologie; psychol. cons. Egyptian govt. sch. reform, 1928-29; sec. Com. Internat. Congresses Psychology. Mem. Brit. Psychol. Soc. (hon.). Author: L'association des Idées, 1903; Esquisse d'une théorie biologique du sommeil, 1905; Psychologie de l'enfant, 1905; L'ecole sur mesur, 1920; Comment diagnostiquer les aptitudes des écoliers, 1925; L'education fonctionelle, 1931; La genèse de l'hypothèse, 1934; Le développement mental, 1946. Used functional approach to study of child psychology. Died Geneva, Sept. 29, 1940.

CLAPARÈDE, Jean Louis René Antoine Édouard, naturalist; b. Geneva, Switzerland, Apr. 24, 1832; prof. comparative anatomy, U. Geneva, from 1862; traveled in Norway, England, Italy. Author: Etudes sur les infusoires et les rhizopodes, 1858-60; Recherches sur l'évolution des araignées, 1862; Recherches anatomiques sur les oligochètes, 1862; les Annélides chétopodes du golfe de Naples, 1868; Recherches sur la structure des annélides sédentaires, 1873. Studies infusoria, rhizopods, spiders, oligochaetes, and annelids. Died Sienna, Italy, May 31, 1871.

CLAPEYRON, Benoit-Paul-Émile, French engr.; b. Paris, Feb. 26, 1799; ed. École Polytechnique, also École des Mines. With G. Lamé in Russia, 1820-30; lectr., St. Petersburg, Russia; with No. France ry. line; in charge French and English locomotives; studied Bordeaux-Site and Bordeaux-Bayonne lines; had bridge of Asnières and bridge over Garonne River built; prof., later chief engr. École des ponts et chaussées. Mem. French Acad. Scis. 1858. Author: Mémoire sur la force motrice de la chaleur, 1843; (with Lamé) Vues politiques et pratiques sur les travaux publics en France, 1832, Plan d'écoles générales et speciales pour l'agriculture, l'industrie manufacturière, le commerce et l'administration, 1833. Contbd. to math. theory of elasticity of solid bodies; gave graphic presentation of Carnot theorem; found relation between conversion heat, steam pressure, vol. change; investigated (with Lamé) equilibrium of firm, homogeneous bodies; applied these studies to constrn. locomotives in patent on cam driving and overlapping of valves, 1842, also to bridge bldg. and in plans for bridges and arches; investigated atmospheric steam ry. from Paris to St.-Germain; built traction engine for Paris-Versaille R.R. Died Paris, Jan. 28, 1864.

CLAPHAM, Arthur Roy, English botanist; b. Norwich, Eng., May 24, 1904; s. George and Dora (Harvey) C.; B.A. (Found. scholar) Downing Coll., Cambridge, Eng., 1925, Ph.D. (Frank Smart student), 1929, M.A., 1930; m. Brenda North Stoessiger, Mar. 27, 1933; children—Elizabeth Harvey (Mrs. Humphrey Rang), Jennifer Margaret (Mrs. David Newton), David Harvey. Crop physiologist Rothamsted Exptl. Sta., 1928-30; demonstrator Oxford (Eng.) U., 1930-44; prof. botany Sheffield (Eng.) U., 1944——, pro-vice-chancellor, 1954-58, acting vice-chancellor, 1965-66. Mem. Field Studies Council, 1943——; chmn. Brit. nat. com. Internat. Biol. Programme, 1964——. Fellow Royal Soc., 1959, Linnean Soc.; mem. Brit. Ecol. Soc., Bot. Soc. Brit. Isles, Soc. Exptl. Biologists, Nature Conservancy (chmn. sci. policy com. 1963——). Author: (with W. O. James) The Biology of Flowers, 1935; (with T. G. Tutin, E. F. Warburg) Flora of the British Isles, 1952, 62; (with Tutin, Warburg) Excursion Flora of the British Isles, 1959; also articles. Research on flora and vegetation of Britain. Home: 8 Claremont Pl., Sheffield 10, Eng.*

CLAPIÉS, Jean de, see de Clapiés.

CLAPP, Frederick Gardner, Am. geologist; b. Boston, July 20, 1879; s. Edward Blake and Mary Frances (Jones) C.; S.B. in Geology, Mass. Inst. Tech., 1901; m. Helen Drew Ripley, Dec. 28, 1908; children—Clara Frances, Edward Gardner, Priscilla. Instr. geology Mass. Inst. Tech., 1901-03; geologist U. S. Geol. Survey, 1902-08; cons. geologist and petroleum engr., specializing in reports on oil and gas properties from 1908; mng. geologist. The Associated Geol. Engrs., 1912-18; chief geologist The Associated Petroleum Engrs. from 1919; also petroleum and natural gas expert Canadian Dept. Mines, 1911-15; in charge geol. explorations in China, 1913-15; in charge investigations in Australia and New Zealand, 1923-25; petroleum adviser Imperial Govt. of Iran, 1927, 28, 33; explorations in Iran and Afghanistan, 1934-38; Okla. oil operations, 1939-43; spl. lectr. oil geology, Harvard, 1921; reported to com. of U. S. Senate in Teapot Dome Case, 1923. Fellow Geol. Society America, A.A.A.S., Am., Royal geog. socs.; mem. Am. Inst. Mining and Metall. Engrs., Geol. Soc. Washington, Am. Asiatic Assn., Am. Geophys. Union, Am. Petroleum Inst., Am. Assn. Petroleum Geologists, Inst. of Petroleum (London), Soc. Economic Geologists, Royal Central Asian Soc., Am. Inst. for Iranian Art and Archeology, Société Géologique de France. Author books and papers on travel, geology, petroleum, natural gas, geography and water supply. Died Feb. 18, 1944.

CLAPROTH, Justus, German inventor; b. Kassel, Germany, Dec. 28, 1728; s. Johannes Kaspar and Kunigunde (Schneider) C.; Dr.juris, U. Göttingen (Germany), 1754; m. Catharine Böse, 1764. City sec., 1752; became auditor, judge, assessor, 1753; named prof. law U. Göttingen, 1761, received chair in Lang. Coll., 1771. Became mem. Mfg. Bd. at Göttingen, 1754. Author: Kurze Vorstellung von Lauf des Processes, 1757; Ohnmassgeblicher Entwurf eines Gesetzbuches, I and II, 1774; Einleitung in den Prozess Gottingen, I, 1779, II, 1780. Invented process to regenerate waste paper; improved book binding. Died Göttingen, Feb. 20, 1805.

CLAR, Eric (Julius), chemist; b. Herrnskretschen, Austria, Aug. 23, 1902; s. Carl Clar and Sophie (Hofmann) C.; Dipl.Ing., Tech. U., Dresden, Germany, 1926, Dr.Ing., 1927; Dr.Ing.Habil., Prague, Czechoslovakia, 1942; m. Louisa Keilitz, July 6, 1943; children—Douglas, Frederick. Asst., Tech. U., Dresden, 1927-29; chief chemist Istituto Ronzoni Milano, Italy, 1930-33; Privat Lab. Herrnskretschen, 1933-46; dozent Tech. U., Prague, 1942-45; faculty U. Glasgow (Scotland), 1946——; prof., 1966——. Recipient Golden Kekulé medal German Chem. Soc., 1965. Mem.

Chem. Soc. London, Faraday Soc. Author: Aromatische Kohleenwasserstoffe, 1941, 52; Polycyclic Hydrocarbons, 1964; also numerous articles. Syntheses of polycyclic hydrocarbons and investigation of their phys. properties; classification of absorption bands, annelation principle, use of aromatic sextet as exact symbol for description of electronic fine structures. Home: 30 Lilybank Gardens, Glasgow W.2, Scotland, U.K.*

CLARA, Max, physician; b. Völs, Feb. 12, 1899; s. Josef and Therese (Edenstraser) C.; ed. Bzano Gymnasium, U. Innsbruck; M.D.; m. Henny Maurer, Sept. 8, 1926; children—Helga, Max Josef, Bernhard Oswald. Asst. histological embryologist at inst. U. Innsbruck; agrégé at Rome; full prof. at Leipzig and Munich; prof. at Istanbul (Turkey). Mem. Leopoldina Halle (Sarles), Internat. Orgn. Brain Research. Author: Das Nervensystem des Menschen, 3 vols.; Entwicklugsgeschichte des Menschen, 5 vols.; Lipide. Handbuck der Histochemie. Home: Bonnerstrasse 24, Munich, Germany. Office: Beyazit, Universite, Istanbul, Turkey.

CLARK, Alfred Joseph, Brit. pharmacologist; b. Glastonbury, Somerset, Aug. 19, 1885; s. Francis Joseph C.; B.A., 1907, M.B., 1910, Cambridge Univ.; M.D., St. Bartholomew's Hosp., 1914; m. Beatrice Powell Hazell, 1919; 2 sons, 2 daughters. Lecturer in Pharmacology, Guys Hospital, London, 1913; prof., Univ. of Cape Town, 1919; prof., Univ. College, Univ. of London, 1920-26; prof. of Materia Medica, Edinburgh Univ., 1926-41. Mem. Medical Research Council, 1934-41. Mem. Royal College Physicians, 1912; Fellow, Royal College Physicians, 1921; Mem. Royal College Physicians, Edinburgh, 1926; Fellow Royal Socs., London, 1931, Edinburgh, 1927. Author: Applied Pharmacology; Mode of Action of Drugs on Cells; Comparative Physiology of the Heart; General Pharmacology, 1937. Died July 30, 1941.

CLARK, Alva Benson, Am. elec. engr.; b. Clay Center, O., Feb. 15, 1890; s. George Frederick and Nellie Judith (McIntyre) C.; B.E.E., U. Mich., 1911; m. Anna C. Harper, Nov. 25, 1920 (dec.); children—Judith H., Patricia H. (dec.); m. 2d, Helen Kerstetter, Aug. 13, 1938. Elec. engr. Am. Tel. & Tel. Co., 1911-34; with Bell Telephone Labs., 1934-55, dir. transmission devel., 1935-40, dir. systems devel., 1940-44, v.p., 1944-55. Del., Comité Consultatif International Telephonique and Internat. Electro-Tech. Comm. meetings, Belgium and Scandinavia, 1930. Cons., OSRD, 1941-45; dir. research and devel. NSA, 1954-—. Fellow Am. Inst. E.E., Acoustical Soc. Am., I.R.E. (sr.), Sigma Xi, Tau Beta Pi. Author several papers; holds 44 patents on elec. communication devices. Died Nov. 14, 1955.

CLARK, Alvan, Am. astronomer, lens maker; b. Ashfield, Mass., Mar. 8, 1804; s. Alvan and Mary (Bassett) C.; A.M. (hon.), Amherst Coll., Chgo. U., Princeton, Harvard; m. Maria Pease, Mar. 25, 1826, children include Alvan Graham, George Bassett. Began career as engraver and portrait painter; started firm Alvan Clark & Sons, Cambridge, Mass., 1846; produced world's largest telescopes; produced lenses for Vienna Observatory, Wesleyan U., Middletown, Conn., Lick Obs., Cal., 18 inch lens for Dearborn Obs., Evanston, Ill., 23 inch for Princeton, 26 inch telescopes for Naval Observatory and U. Va., 30 inch telescope for Pulkova Observatory, Russia, 1879; his firm made 1st achromatic lenses in U. S.; made 36 inch lens for Lick telescope. Recipient Rumford medal Am. Acad. Arts and Scis. Discovered (with son Alvan G.) companion star of Sirius, 1861. Died Cambridge, Mass., Aug. 19, 1887.

CLARK, Alvan Graham, Am. astronomer, lens maker; b. Fall River, Mass., July 10, 1832; s. Alvan and Maria (Pease) C.; m. Mary Willard, Jan. 2, 1865. Discovered (with father) companion star of Sirius, 1861, for which he was awarded Lalande gold medal French Acad. Sci.; discovered 16 double stars; made 40 inch lenses of Yerkes telescope (then world's largest) for U. Chgo. at Williams Bay, Wis., 1897. made 30 inch refractor for Imperial Observatory, St. Petersburg, Russia; mem. total eclipse expdn. to Spain, 1870, Wyo. Mission, 1878. Fellow Am. Acad. Arts and Scis., A.A.A.S. Died Cambridge, Mass., June 9, 1897.

CLARK, Austin Hobart, Am. biologist; b. Wellesley, Mass., Dec. 17, 1880; s. Theodore Minot and Jeanette (French) C.; A.B., Harvard, 1903, postgrad., 1904; m. Mary Wendell Upham, Mar. 6, 1906 (dec. Dec. 1931); children—Austin Bryant Jackson, Sarah Wendell, Hugh Upham, Anne Bradstreet, Mary Holmes; m. 2, Leila Gay Forbes, Sept. 23, 1933. Organizer expdn. to Margarita Island, Venezuela, 1901; in zoöl. research, Lesser Antilles, 1903-05; acting chief sci. staff U. S. Fisheries S.S. Albatross, 1906-07; mem. staff Smithsonian Inst., 1908-50, ret., dir. press service, A.A.A.S.; vice-chmn. Am. Geophys. Union; pres. Washington Acad. Scis.; pres. Entomol. Soc. Washington; mem. exec. com. and long range planning com. So. Assn. Sci. and Industry; mem. adv. bd. Va. Fisheries Lab.; research in oceanography, marine biology, ornithology and entomology; exec. com. bd. govs. Nature Conservancy. Fellow Royal Geog. Soc. Author: Animals of Land and Sea, 1925; Nature Narratives, Vol. I, 1929, Vol. II, 1931; The New Evolution, 1930; Animals Alive, 1948; also monographs and bulls. and 600 sci. articles in English, French, Italian, German and Russian. Died Oct. 28, 1954.

CLARK, Avon Maxwell, Australian geneticist; b. Melbourne, Australia, Apr. 18, 1918; s. Reg. A. and Ida J. (Stow) C.; B.Sc., Melbourne U., 1940, M.Sc., 1941; Ph.D., Cambridge (Eng.) U., 1949; m. Elis Gray Anderson, Jan. 17, 1942; children—Janina, Julian Maxwell. Lectr. histology Melbourne Med. Sch., 1943-44; lectr. zoology Melbourne U., 1945-47, sr. lectr., 1950-55, reader genetics, 1955-59; sr. research scholar Cambridge U., 1947-49; prof. U. Tasmania, 1960-63; prof. biology, chmn. Sch. Biol. Sci., Flinders U., Bedford Park, Australia, 1964-—, pro-vice chancellor, 1966-—. Mem. Nat. Radiation Adv. Com., 1959-—, Radiation Health Com., 1962-—, Nat. Health and Med. Research Council, 1961-—. Carnegie Corp. N.Y. Research scholar, 1965. Mem. A.A.A.S., Australian and New Zealand Assn. for Advancement Sci., Genetics Soc., Biochem. Soc. Australia. Research, publs. on physiology of regeneration and of behaviour in Protozoa, enzyme activities of blood corpuscles, enzyme activity in developing embryos, radiation and chem. induced mutation, relation between carcinogenesis and mutagenesis; orgn. of grad. and undergrad. schs. in biol. scis. Office: Flinders U., Bedford Park, S.A., Australia.*

CLARK, Byron Bryant, Am. pharmacologist; b. Temple, Tex., Apr. 5, 1908; s. Oscar Wilder and Ida (Hansen) C.; B.A., Baylor U., 1930; M.S., State U. Ia., 1932, Ph.D., 1934; m. Gladys Lawson, Jan. 26, 1931; children—Barbara Gene (Mrs. Edward Riter), Jack Bryant, Kenneth Edward. Research asso. in path. chemistry Gen. Hosp., State U. Ia., 1931-36; faculty Albany Med. Coll., 1936-47; prof., chmn. dept. pharmacology Tufts U. Sch. Medicine, Boston, 1947-57; dir. pharmacology Mead Johnson Research Div., Evansville, Ind., 1957-58, dir. pharmacology and chemotherapy Mead Johnson Research Center, Evansville, 1958-62, v.p., 1962-—. Cons. drug cos., hosps. Fellow A.A.A.S.; mem. Am. Soc. Pharmacology and Exptl. Therapeutics, Soc. Exptl. Biology and Medicine, Soc. Toxicology, Sigma Xi. Research, numerous publs. Exptl. Therapeutics, Soc. Exptl. Biology and Medicine, Soc. Toxicology, Sigma Xi. Research, numerous publs. on insulin and carbohydrate metabolism, ethyl and methyl alcohol pharmacology and metabolism, drugs on blood and hemoglobin, gastric secretion, antacids, antispasmodics, autonomic drugs, cardiac antiarrythmic drugs, analgesic and antipyretic drugs, pharmacology and physiol. disposition of quaternary ammonium compounds. Home: 701 S. Boeke Rd., Evansville 47714. Office: 939 Bond St., Evansville, Ind. 47721.*

CLARK, Charles Alexander, Brit. dental radiologist; b. London, Eng., 1860; ed. Nat. Dental Hosp.; L.D.S.; m.; 1 son, 2 daus. Ret. dental radiologist; cons. radiologist Royal Dental Hosp.; dental surgeon Lucifer Match Factory, Ex-officio mem. Dental Bd., 1918-19. Fellow Royal Soc. Medicine (radiology sect., phys. med. sects.); mem. British Dental Assn. (pres. Met. br. 1919), Rcentgen Soc. (past mem. council). Author: Dental Radiography, 1926; co-author: Griffith's Metallurgy, also articles on dental radiology. Demonstrator value x-rays in dental surgery and technique. Died Mar. 11, 1939.

CLARK, Charles Lester, Am. mathematician; b. San Jose, Cal., Nov. 17, 1917; s. Charles James and Minnie Bethiah (Lester) C.; student San Jose State Coll., 1935-38; A.B., Stanford, 1939, A.M., 1940; Ph.D., U. Va., 1944; m. Jean Show, Sept. 8, 1940; children—Charles Dennis, Robert Keith, Jeffrey Craig. Instr. math. U. Va., 1942-44, vis. prof., 1955-56; from asst. prof. to prof. Ore. State Coll., 1944-57; prof. math. Cal. State Coll. at Los Angeles, 1957-—, head math. dept., 1957-64, dir. instl. research, 1964-—, dir. coll. computing center, 1961-—; cons. edn., industry, 1958-—. Royall Victor fellow Stanford, 1940-41; du Pont fellow U. Va., 1941-42. Mem. Am. Math. Soc., Math. Assn. Am., A.A.A.S., Assn. Computing Machinery, Am. Assn. U. Profs., Phi Beta Kappa, Sigma Xi. Study of analysis; topology; arc reversing transformations. Home: 3780 Greenhill Rd., Pasadena, Cal. Office: Dept. Mathematics, Cal. State Coll., Los Angeles 90032.

CLARK, David Delano, Am. physicist; b. Austin, Tex., Feb. 10, 1924; s. David Lee and Grace (Delano) C.; student U. Tex., 1941-42, 46; A.B. in Physics, U. Cal. at Berkeley, 1948, Ph.D. (AEC predoctoral fellow 1950-52), 1953; m. Gladys Braunstein, Dec. 27, 1949; children—Marcia Susan, Gordon Richard, Janet Mirella. Research asso. Brookhaven Nat. Lab., 1953-55, vis. scientist summers, 1957, 58; mem. faculty engring. physics Cornell U., 1955-—, asso. professor, 1958-65, professor, since 1965-—; director of Nuclear Reactor Laboratory, 1960-—; cons. Gen. Atomic Co., summer 1959. Served with USAAF, 1942-46. Euratom fellow Euratom Research Center, Ispra, Italy, 1962. Mem. Am. Phys. Soc., Am. Nuclear Soc., Fedn. Am. Scientists, Phi Beta Kappa, Sigma Xi. Contbr. articles profl. jours. Basic exptl. research on interactions of particles and nuclei, nuclear structure, applications of nuclear physics, especially in nuclear reactors. Home: 105 Needham Pl., Ithaca, 14850. Office: Nuclear Reactor Lab, Cornell U., Ithaca, N.Y. 14850.*

CLARK, Eugenie, ichthyologist; b. N.Y.C., May 4, 1922; d. Charles and Yumico (Mitomi) Clark; B.A., Hunter Coll., 1942; M.A., N.Y. U., 1946, Ph.D., 1950; m. Ilias Konstantinu, Nov. 4, 1950 (div. Apr. 1966); children—Hera, Iris, Themistokles, Nikolas; m. 2d, Chandler Brossard, May 10, 1966. Research asst. Scripps Inst. Oceanography, 1946-47; research asst. Am. Mus. Natural History, 1948-50, research asso. dept. animal behavior, 1954-—, exec. dir. Cape Haze Marine Lab., 1955-67; asso. prof. N.Y. U., 1966-—; adj. prof. New Eng. Inst. for Med. Research, 1966-—, research asso., 1956-—. Leader Am. del. Israel S. Red Sea Expdn., 1962; cons. biology br. Oceanic Scis. Office Naval Research, 1962-—; cons. oceanographic facilities NSF, 1963-—; cons. Shark research panel Am. Inst. Biol. Scis., 1960-—. Mem. Am. Soc. Ichthyologists and Herpetologists, Soc. Women Geographers. Author: Lady With a Spear, 1953. Contbr. numerous articles to sci. jours. Research on poisonous properties, sexual behavior, conditioned responses, morphology, physiology and systematics of fishes in Micronesia, Japan, Red Sea, Caribbean, Gulf of Mexico, Mediterranean, combining lab. studies and expts. with field observations, field behavior of sharks. Home: 7356 Point of Rocks. Office: Cape Haze Marine Lab., Blind Pass, Sarasota, Fla. 33581.*

CLARK, Francis Eugene, Am. microbiologist; b. Fairbury, Neb., Mar. 4, 1910; s. Walter Maxwell and Sylvia (Wookey) C.; B.A., U. Colo., 1932, B.D.E., 1933, M.A., 1933, Ph.D., 1936; m. Evelyn Irene Larson, Aug. 15, 1933; children—Lowell Eugene, Linda Jean. With Agrl. Research Service, U. S. Dept. Agr., 1936-—, chief microbiologist, 1961-—; mem. grad. faculty Colo. State U., 1957-—; mem. plant chemistry div. Govt. New Zealand, 1962-63. Mem. adv. com. on microbiology UNESCO, 1961-—; lectr., cons. soils div. CSIRO, Australia, 1963; lectr., cons. Czechoslovak Acad. Sci., 1964; mem. NRC com. soil biology, 1956-63. Mem. Internat. Soc. Soil Sci. Am. Soc. Agronomy (asso. editor 1963-65), Soil Sci. Soc. Am. (asso. editor 1951-66), Am. Soc. Microbiology, Geochem. Soc. Am., Wash. Acad. Sci. Editor: Ecology of Soilborne Pathogens, 1965, co-editor Soil Nitrogen, 1965, Methods of Soil Analysis, 1965, Aerobic Sporeforming Bacteria, 1946; asso. editor Agronomy Jour., 1963-66, Geoderma, 1966-—. Research, numerous publs. field taxonomy of soil bacteria, ecology of soilborne plant pathogens, transformations of soil nitrogen, microbiology of rhizosphere, taxonomy of cellulolytic corynebacterial and putrefactive anaerobes, methods of soil analysis, decomposition of organic materials in soil. Home: 1901 Country Club Rd. Office: P. O. Box 998, Ft. Collins, Colo., 80521.*

CLARK, Frank Eugene, Am. mathematician; b. St. Louis, Sept. 16, 1919; s. Arthur Henry and Persis (Kidder) C.; A.B., Dartmouth, 1941; M.A., Duke, 1946, Ph.D., 1948; m. Esther Lucille Sommer, June 12, 1948; children—David, Daniel, Carolyn, Margaret. Vis. instr. math. Duke, 1947-48; instr. Tulane U., New Orleans, 1948-50; faculty math. Rutgers U., New Brunswick, N.J., 1950-—, prof., 1960-—, chmn. univ. coll. math. dept., 1959-—. Mem. Am. Math. Soc., Math. Assn. Am., Soc. for Indsl. and Applied Math., Phi Beta Kappa, Sigma Xi. Study of linear programming; algebraic inequalities. Home: 820 Shadowlawn Dr., Westfield, N.J.*

CLARK, George Bromley, Am. mining engr.; b. Pleasant Grove, Utah, June 5, 1912; s. William Edward and Cora (Bromley) C.; B.S., U. Utah, 1935, M.S., 1946; Ph.D. in Mining Engring., U. Ill., 1952; m. Barbara Hall, July 9, 1954; children—Elizabeth, Margaret, George Hall, Melinda, John Hall, David Hall, Richard Hall. Jr. engr. Tintic Standard Mining Co., 1935-38, 39-40; instr. U. Utah, 1938-39; sampling foreman U. S. Bur. Mines, 1940-41; faculty U. Ill., Urbana, 1946-54, prof., 1951-54; prof. mining engring., chmn. dept. U. Mo., Rolla, 1954-61, asso. dir. research, 1961-—, dir. Rock Mechanics Research Center, 1963-—. Cons. in explosives, blasting, nuclear weapons effects. Mem. Am. Inst. Mining Engrs., A.A.-A.S., Am. Soc. Testing and Materials, Am. Geophys. Union, Am. Soc. for Engring. Edn., Am. Soc. C.E., Sigma Xi, Sigma Gamma Epsilon. Author: (with Robert S. Lewis) Elements of Mining, 1964; also articles. Research in ammonium nitrate-fuel oil blasting agts., nuclear weapons effects, static and dynamic rock mechanics, viscoelastic waves, waves in earth materials; inventor multitube manometer. Home: 815 Soest Rd., Rolla, Mo. 65401.*

CLARK, George Lindenberg, Am. chemist; b. Anderson, Ind., Sept. 6, 1892; s. Ralph Bliven and Olive (Burnett) C.; B.A., DePauw U., 1914; M.S., U. Chgo., 1914, Ph.D., 1918; Sc.D., DePauw U., 1937; m. Mary Mason Johnson, June 19, 1919; children—Mary Ann (dec.), Ralph Burnett, George Mason, Jean, Carolyn Johnson. Instr. in chemistry DePauw U., 1914-16; asso. prof. chemistry Vanderbilt, 1919-21; nat. research fellow, Harvard, 1922-24; asst. prof. applied chem. research Mass. Inst. Tech., 1924-27 (installed and directed 1st indsl. X-ray research lab.); prof. chemistry U. Ill., 1927-53; research prof. analytical chemistry, 1953-60, emeritus 1960, vis. prof. DePauw U., 1961. Recipient Grasselli medal, 1932; Mehl medal, 1944; Orton lectureship award, 1946. Fellow A.A.-A.S.; mem. Am. Chem. Soc., (Mobile-Pensacola sect. award, 1952), Am. Crystallographic Assn., Am. Inst. Chemists, Radiol. Soc. Am., Am. Soc. Testing Materials (Marburg Meml. lectr. 1927), Phi Beta Kappa, Phi Beta Kappa Assos., Sigma Xi, Phi Lambda Upsilon, Alpha Chi Sigma; founder, Electron Microscope Soc. Am., 1943. Author: Applied X-Rays, 1926,

4th edit., 1955; also many papers in chem. and physical jours. Editor-in-chief Ency. of Chemistry, 1957, Supplement, 1958, 2d edit., 1966. Editor Ency. Spectroscopy, 1960, Microscopy, 1961; Ency. X-Rays and Gamma Rays, 1963. Research on X-ray spectroscopy and radiography, diffraction, electron microscopy, spectro-photometry, photochemistry, all instrumental analytical methods. Home: 406 Penn Av., Urbana, Ill.

CLARK, George Thomas, Brit. archaeologist; b. London, Eng., May 26, 1809; s. George and Clara (Dicey) C.; ed. at Charterhouse; m. Ann Price Lewis, Apr. 3, 1850; children—Godfrey Lewis; also 1 dau. Constructed two divs. of Gt. Western Ry.; a founder Royal Archaeol. Inst. Mem. Cambrian Archaeol. Assn. (trustee), Brit. Iron Trade Assn. (pres. 1876), Archaeol. Assn. Author: History and Description of the Great Western Railway, 1839; Medieval Military Architecture in England, 2 vols.; also articles. Research on mil. and hist. importance of the earthworks of Wales; showed use of the mounds of Norman times. Died Jan. 31, 1898.

CLARK, Henry James, Am. botanist, zoologist; b. Easton, Mass., June 22, 1826; s. Henry Porter and Abigail (Orton) C.; grad. U. City N.Y., 1848, B.S., Harvard, 1854; m. Mary Young Holbrook, 1854, 8 children. Pvt. asst. to Louis Agassiz, 1854-65; asst. prof. zoology Scientific Sch. at Harvard, 1860-65; prof. botany, zoology and geology Pa. State Coll., 1866-69, U. Ky., 1869-72; prof. Mass. Agrl. Coll., 1872-73. Lectr., Mus. Comparative Zoology. Author: (pamphlet) A Claim for Scientific Property, 1863; Mind in Nature; or the Origin of Life, and the Mode of Development of Animals, 1865. Discovered flagellated cells of living sponges and demonstrated their animal nature; studied Protozoa; cell theory; advanced knowledge of histology of Coelenterata and Porifera. Died July 1, 1873.

CLARK, Hubert Lyman, Am. zoologist; b. Amherst, Mass., Jan. 9, 1870; s. William Smith and Harriet Kapuolani Richards (Williston) C.; A.B., Amherst Coll., 1892; Ph.D., Johns Hopkins, 1897; hon. Sc.D. Olivet (Mich.) Coll., 1927; m. Fannie Lee Snell, Apr. 4, 1899; children—William Smith, Stirrat, Janet Stirrat, Edith. Prof. biology Olivet Coll., 1899-1905; asst. Museum Comparative Zoology, Harvard, 1905-12, curator of echinoderms, 1912-27, became curator of marine invertebrates, and asso. prof. zoology, 1927, emeritus prof., also curator; acting prof. Williams Coll., 1920-21; acting asso. prof. Stanford U., 1936; research asso. Hancock Found., U. So. Cal., 1946-47. Made sci. investigations in Jamaica, Tobago, Bermuda, Galapagos Islands and Australia. Recipient Clarke Meml. medal Royal Soc. of New South Wales, Australia, for research in Australian sci., 1947. Wrote numerous sci. papers, including 3 monographs on Australian Echinoderms, 1921, 38, 46. Spl. editor for echinoderms Webster's New Internat. Dictionary, 2d edit. Died Cambridge, Mass., July 31, 1947.

CLARK, Sir James, physician; b. Cullen, Scotland; Dec. 14, 1788; M.A., Aberdeen, Scotland; ed. Coll. Surgeons, Edinburgh, Scotland; M.D., U. Edinburgh, 1817; m. Barbara Stephen, circa 1860; a son, J. F. Clark. Practiced in Rome, Italy, 1816-26; went abroad to study effect of climate on pthisis; army physician, 6 years; moved to London, Eng., circa 1828; named physician to Duchess of Kent, 1835; physician to Victoria, 1835-37, physician in ordinary, from 1837, mem. several royal commns.; mem. senate London U., 1838-65; mem. Gen. Med. Council, 1858-60. Fellow Royal Soc., 1832. Author: The Sanative Influence of Climate, 1829; The Influence of Climate in Chronic Diseases, 1829; Treatise on Pulmonary Consumption and Scrofulous Diseases, 1835; Treatise on Pulmonary Consumption, 1835. Advocated stricter examinations and requirements for med. students; studied ways to conceal nauseous flavor of some drugs. Died Bagshot Park, London, June 29, 1870.

CLARK, John Bates, Am. polit. economist; b. Providence, Jan. 26, 1847; s. John Hezekiah and Charlotte (Huntington) C.; student Brown U.; A.B., Amherst, 1872, A.M., 1878, Ph.D., 1890; postgrad. univs. Heidelberg and Zürich; LL.D., Princeton, 1896, Amherst, 1897, U. Christiania, Norway, 1911, Columbia U., 1929; D. Polit. Sci., U. Tubingen (Germany), 1928; m. Myra A. Smith, Sept. 28, 1875; children—Frederick Huntington, Alden Hyde, John Maurice, Helen Converse. Prof. polit. economy and history Carleton Coll., 1877-81; prof. history and polit. sci. Smith Coll., 1882-93; prof. polit. economy Amherst, 1892-95; prof. polit. economy Columbia, 1895-1923. Lectr. polit. economy Johns Hopkins, 1892-95; dir. div. econs. and history, Carnegie Endowment for Internat. Peace, 1911-23. Author: The Philosophy of Wealth, 1885; The Distribution of Wealth, 1899; The Control of Trusts, 1901; The Problem of Monopoly, 1904; Essentials of Economic Theory, 1907; The Modern Distributive Process (with F. H. Giddings); Control of Trusts (enlarged edit., with J. M. Clark). Editor Polit. Sci. Quar., 1895-1911. Died Mar. 21, 1938.

CLARK, John Charles, Am. physicist, govt. ofcl.; b. Shelby, O., Sept. 16, 1903; s. Clinton J. and Elizabeth J. (Knight) C.; A.B., Ohio Wesleyan U., 1925; M.S., U. Chgo., 1927; Ph.D., Stanford, 1935. Instr. physics Lehigh U., 1927-29; Stanford, 1930-35; asso. prof. physics Mich. State U., 1936-41; asso. div. leader Los Alamos Sci. Labs., 1946-57; research cons. Gen. Dynamics/Astronautics, San Diego, 1957-62; sci. cons. U. S. Govt., Gen. Dynamics, 1962-63; sci. attache U. S. Dept. State, Washington, 1963-66. Cons. to industry and govt., 1960-63. Recipient Civilian Meritorious Service award U. S. Army, 1953. Mem. Am. Phys. Soc., Gamma Alpha. Author numerous research reports on X-rays, high energy physics, detonation phenomena. Address: P.O. Box 2165, La Jolla, Cal. 92037.*

CLARK, John Graham Douglas, Brit. anthropologist; b. Shortlands, Kent, Eng., July 28, 1907; s. Charles Douglas and Maud (Shaw) C.; B.A., Cambridge U., 1930, M.A., 1933, Ph.D., 1933, Sc.D., 1953; m. Gwladys Maude White, July 18, 1936; children—Margaret Helen (Mrs. John G. Nandris), William G. D., Philip C. L. Research, 1930-35; lectr. Cambridge (Eng.) U., 1935-41, 46-52, Disney prof. archaeology, 1952—. Vis. lectr. Edinburgh U., 1949, U. Glasgow, 1955, Harvard, 1957, Otago U., 1964, Bristol U., 1967. Recipient Hodgkins gold medal Smithsonian Instn., 1967. Fellow Brit. Acad.; fellow or hon. fgn. mem. Royal Soc. Antiquaries of Copenhagen, Swiss Prehistoric Soc., German Archaeol. Inst., Royal Irish Acad., Finnish Archaeol. Soc., Am., Royal Danish acads. arts and scis., Royal Netherlands Acad. Scis.; mem. Prehistoric Soc. (editor Proc. 1935—, pres. 1958-62), Soc. Antiquaries of London (v.p. 1959-62). Author: Archaeology and Society, 1939; Prehistoric Europe, The Economic Basis, 1952; World Prehistory, 1961; (with Stuart Piggott) Prehistoric Societies, 1967; also numerous others. Research, numerous publs. on Mesolithic Stone Age hunting cultures of N.W. Europe since end of Ice Age in context of environmental changes, including excavation of Mesolithic site of Star Carr, Eng.; evolution of econ. life in prehistoric Europe, pattern of world prehistory. Home: 19, Wilberforce Rd. Office: Univ. Mus. Archaeology and Ethnology, Downing St., Cambridge, Eng.*

CLARK, John Wesley, Am. physicist; b. Williamsburg, Va., Apr. 1, 1915; s. Wesley Plummer and Anna (Henderson) C.; B.A., Mont. State U., 1935; M.S., U. Ill., 1937, Ph.D., 1939; m. Ruth Jeannette Hartley, Nov. 24, 1937; children—John H., Stephen W., Robert D., William K. Mem. tech. staff Bell Telephone Labs., N.Y.C., 1939-46; sect. head Collins Radio Co., Cedar Rapids, Ia., 1946-50; gen. sales mgr. Varian Assos., Palo Alto, Cal., 1950-54; head nuclear electron labs. Litton Industries, Beverly Hills, Cal., 1954-55; mgr. nucleonics div. Hughes Aircraft Co., Culver City, Cal., 1955-63; sr. research asso. Battelle Meml. Inst., Columbus, O., 1963-65; cons. ocean engring., Santa Monica, 1965—. Fellow I.E.E.E.; mem. Am. Inst. Aeros. and Astronautics, Marine Tech. Soc., Am. Phys. Soc., Am. Nuclear Soc., A.A.A.S., N.Y. Acad. Sci., Sigma Xi. Developed microwave gas switching tubes for radar, comml. microwave communication systems, electronic remote controlled systems for nuclear, underwater environments; discovered effects transient nuclear radiation on electronic systems. Address: 422 23d St., Santa Monica, Cal. 90402.*

CLARK, Josiah Latimer, Brit. civil engr.; b. Gt. Marlow, Mar. 10, 1822. Engr., Britannia Tubular Bridge, Wales, 1847; asst. engr., engr.-in-chief Electric and internat. Telegraph Co., 1850; partner Clark and Stanfield, floating dock and hydraulic canal life engrs.; with partners (Forde and Taylor) supervised manufacture and laying of more than 100,000 miles of submarine cables in various parts of world. Fellow Royal Soc., 1889; mem. Inst. Elec. engrs. (4th pres.), Instn. Civil Engrs. Author: Description of Britannia and Conway Tubular Bridges, 1850; Treatise on Electrical Measurement, 1868; (with Sabine) Electrical Tables and Formula, 1871; Dictionary of Electrical Measures, 1891. Invented double bell insulators for telegraph wires, system of protection for steel telegraph cable, pneumatic system of transmission of telegrams through pipes, Clark's standard cell, numerous telegraphic improvements; proposed names and system of volts, ohms, farads, 1861. Died Oct. 30, 1898.

CLARK, Kenneth Edwin, Am. psychologist; b. New Madison, O., Dec. 18, 1914; s. Harry H. and Nellie (Tremps) C.; B.Sc., Ohio State U., 1935, M.A., 1937, Ph.D., 1940; m. Helen Titelmaier, June 29, 1942; children—Patricia (Mrs. Todd Storm), Virginia, Joyce. Tchr., Ashtabula (O.) County Schs., 1935-37; faculty U. Minn., 1940-42, 46-60, prof., 1956-60, chmn. dept. psychology, 1957-60, asso. dean Grad. Sch., 1960; dean U. Colo. Coll. Arts and Scis., 1961-63; dean Coll. Arts and Scis., U. Rochester, 1963—. Mem. Pres.'s Com. on Nat. Medal of Sci., 1962-65. Trustee Am. Bd. Examiners in Profl. Psychology (past pres.). Mem. Am. Psychol. Assn., A.A.A.S., Am. Statis. Assn., Phi Beta Kappa, Psi Chi. Author: America's Psychologists, 1957; Vocational Interests of Nonprofessional Men, 1961; also articles. Editor, Jour. Applied Psychology, 1961—. Research on factors contbg. to research success psychologists, devel. measures preferences for work activities for use in counseling young persons going into occupations not requiring coll. Home: 57 Southern Pkwy., Rochester, N.Y. 14618.*

CLARK, Leland Charles, Jr., Am. physiologist; b. Rochester, N.Y., Dec. 4, 1918; s. Leland Charles and Helen (Foote) C.; B.S. with honors in Sci., Antioch Coll., 1941; Ph.D. (NRS fellow) U. Rochester Sch. Medicine and Dentistry, 1944; m. Eleanor W. Wyckoff, Dec. 28, 1939; children—Susan Lee (Mrs. Wayne Wooley), Joan, Linda, Rebecca. Faculty, Antioch Coll., Yellow Springs, O., 1944-55, 56-58, Fels Research Inst., Yellow Springs, 1944-58, Coll. Medicine U. Cin., 1955-58; faculty U. Ala., University, 1958—, prof. biochemistry dept. surgery Med. Coll., 1963—. Cons., Surgeon Gen., NIH, 1961-65. Recipient Research Career award NIH, 1962. Fellow N.Y. Acad. Scis. Fedn. Am. Socs. for Exptl. Biology; mem. Am. Chem. Soc., A.A.A.S., Am. Physiol. Soc., Am. Soc. Artificial Internal Organs, Am. Assn. U. Profs., Sigma Xi. Contbr. to med. textbooks. Research, publs. in physiology, biochemistry, polarography, pharmacology. Patentee bubble-defoam heart lung machine, telethermometer, oxygen electrode, hydrogen diagnostic electrodes. Home: 1312 Willoughby Rd., Birmingham, Ala. 35216.*

CLARK, Leonard Bertrand, Am. zoologist; b. Oakland, Cal., Mar. 18, 1902; s. William Alfred and Adelaide (Bertrand) C.; student U. Man., Can., 1923-26; m. Averill Ambes Zimmerman, Aug. 28, 1929; children—Averill Joan (Mrs. Donald White), Susan Diane. Faculty, Union Coll., Schenectady, 1931—, chmn. dept. biol. scis., 1945—, prof. biology, 1946—; vis. prof. zoology U. St. Andrews (Scotland), 1957-58; Fulbright-Hays vis. prof. Am. U., Beirut, 1964-65. Mem. Soc. Am. Zoologists, A.A.A.S., Soc. Gen. Physiologists, Corp. Marine Biol. Lab., Am. Assn. U. Profs., Radiation Research Soc., Phi Beta Kappa, Sigma Xi. Research, numerous publs. on animal behavior, sensory physiology, biol. effects of visible light, radiobiology.*

CLARK, Melville, Am. physicist; b. Syracuse, N.Y., Dec. 19, 1921; s. Melville and Dorothy (Speich) C.; S.B., Mass. Inst. Tech., 1943, postgrad., 1943-44; postgrad. U. N.M., 1945-46, Princeton, 1946; A.M., Harvard, 1947, Ph.D., 1949. Physicist, Radiation lab., Mass. Inst. Tech., Cambridge, 1942-45, Manhattan Dist., Los Alamos, 1945-46, Brookhaven Nat. Lab., Upton, N.Y., 1949-53, Radiation Lab., U. Cal. at Livermore, 1953-55; asso. prof. Mass. Inst. Tech., 1955-62; sr. engr. specialist Sylvania Electric Products, Waltham, Mass., 1962-64; sr. cons. scientist Avco Corp., Wilmington, Mass., 1964-67; sr. scientist NASA, Cambridge, 1967—; pres. Melville Clark Assos., Cochituate, Mass., 1955—, Meldor Enterprises Corp., 1960-66; v.p., dir. Clark Music Co., Syracuse, N.Y., 1960. Cons. to pvt. cos. Mem. Am. Phys. Soc., Am. Inst. Physics, A.A.A.S., Acoustical Soc. Am., Sigma Xi. Author: (with D. Rose) Plasmas and Controlled Fusion, 1961; (with Hansen) Numerical Methods of Reactor Analysis, 1964; also numerous articles. Invention, design, devel. of microwave components devel. theories for neutron and gamma-ray transport through media, for nuclear reactor calculations, for cross sect. measurements, for atomic and hydrogen bomb design; invention, design, devel. new musical instruments; research in factors important to auditors in musical instrument sounds, test equipment for mus. acoustics. Home: 8 Richard Rd., Cochituate, Mass. 01778.*

CLARK, Robert Arthur, Am. mathematician; b. Melrose, Mass., May 3, 1923; s. Arthur Henry and Persis (Kidder) C.; student Colo. Coll., 1940-42; A.B., Duke, 1944; M.S., Mass. Inst. Tech., 1946, Ph.D., 1949. Faculty, Case Inst. Tech., Cleve., 1950—, prof. math., 1964—; mem. U. S. Army Math. Research Center, U. Wis., Madison, 1961-62. Mem. Phi Beta Kappa, Sigma Xi, Sigma Chi. Study of shell theory; asymptotic theory of differential equations; elasticity.

CLARK, Robert Bernard, English zoologist; b. London, Eng., Oct. 13, 1923; s. Joseph Lawrence and Dorothy (Halden) C.; B.Sc., Chelsea Poly., 1944 B.Sc., U. Exeter, 1950; Ph.D., U. Glasgow, 1956; D.Sc., U. London, 1965; m. Mary Eleanor Laurence, July 11, 1956. With Road Research Lab. (DSIR), 1944-47; asst. lectr. U. Glasgow, 1950-53; asst. prof. U. Cal. at Berkeley, 1953-55; lectr. U. Bristol (Eng.), 1956-65; prof. U. Newcastle upon Tyne (Eng.), 1965—. Mem. Soc. for Exptl. Biology, Marine Biol. Assn., zool. socs. London, France, other sci. socs. Author: Dynamics in Metazoan Evolution, 1964; Practical Course in Experimental Zoology, 1966; also papers. Research on structure and integrative function of brain of polychaete worms, its role as source of hormones controlling growth and sexual maturation and as control and memory-storage center in learning, also research on musculature, locomotion and evolution of polychaete worms. Address: Dept. of Zoology, Univ. Newcastle upon Tyne, Newcastle upon Tyne, 1, Eng.*

CLARK, Robin Jon Hawes, chemist; b. Rangiora, New Zealand, Feb. 16, 1935; s. Reginald Hawes and Marjorie (Thomas) C.; B.Sc., U. Canterbury, Christchurch, New Zealand, 1955, M.Sc. with 1st class honors, 1957; Ph.D., U. Coll., London, Eng., 1961; m. Beatrice Rawdin Brown, May 30, 1964. Demonstrator, U. Canterbury, 1956-57; research and teaching fellow U. Otago (New Zealand), 1958; demonstrator, asst. lectr. U. Coll., London, 1959-62, lectr., 1963—. Vis. asso. prof. Columbia, N.Y.C., 1965; vis. prof. U. Western Ont. (Can.), 1968; lectr. univs. in Europe, Eng., U. S. A., 1961—. Mem. Chem. Soc. (London), Royal Inst. Chemistry. Author: The Chemistry of Titanium and Vanadium, 1968. Research, publs. on chemistry of early transition elements, particularly titanium and vanadium, stabilization of high coordination numbers and low oxidation states, electronic spectra, diffuse reflectance spectra and magnetism of transition metal complexes, application of low frequency infrared spectra to study of structures of inorganic complexes. Home: 5 Farm Rd., Edgware, Middlesex, Eng. Office: William Ramsay and Ralph Forster Labs., U. Coll., Gower St., London W.C.1, Eng.*

CLARK, Thomas, Scottish chemist; b. Ayr, Scotland, Mar. 31, 1801; entered Glasgow (Scotland) U., 1827; M.D., 1831; with counting house of Macintosh (the waterproofer), Glasgow, Scotland; later joined St. Rollox Chem. Works; became apothecary to infirmary, 1829; named prof. chemistry Marischal Coll. and U., Aberdeen, Scotland, 1833-60; became lectr. chemistry Glasgow Mechanics Instn., 1836. Contbr. articles on pharmaceutics to tech. jours. Discovered pyrophosphate of soda, 1836; invented soap-test for testing hardness of water, process for softening chalk waters; introduced calcium hydroxide to soften water, use of water as constituent of some acids. Died Glasgow, Nov. 27, 1867.

CLARK, Victor Malcolm, Brit. molecular scientist; b. Ilkeston, Eng., Dec. 31, 1925; s. Henry and Eva (Baker) C.; Ph.D. (Exhibition, scholar, fellow), Gonville and Caius Coll., Cambridge (Eng.) U., 1950; m. Geseke Brandis, July 2, 1960; children—Duncan, Andrew, Gordon. Rockefeller Found. fellow Harvard, 1951, research fellow, 1952; univ. demonstrator, ofcl. fellow Gonville and Caius Coll., Cambridge U., 1953-58, lectr., Cambridge U., 1958-64; prof. molecular scis. U. Warwick, Coventry, Eng., 1964—. Mem. Brit., German, Am. chem. socs. Research, publs. on acylations reactions, especially phosphorylation, in realm between chemistry and biochemistry. Home: 118 Coventry Rd., Warwick, Eng. Office: Sch. Molecular Scis., U. Warwick, Coventry, Warwickshire, Eng.*

CLARK, Sir Wilfrid Le Gros, Brit. anatomist; b. Hemel, Hempstead, Eng., June 5, 1895; s. Edward Travers and Ethel (Clapton) C.; D.Sc., U. London, 1928; M.A., Oxford U., 1934; hon. D.Sc., U. Durham (Eng.), 1949, U. Manchester (Eng.), 1962, U. Witwatersrand, 1965, U. Edinburgh, 1966; M.D. (hon.), U. Melbourne (Australia), 1952, U. Oslo (Norway), 1961; LL.D., U. Malaya, 1953; m. Freda Constance, Dec. 20, 1924; children—Joan Le Gros (Mrs. Peter Uzzell), Pauline Le Gros (Mrs. Miles Hardie). Prin. med. officer, Sarawak, 1920-23; prof. anatomy London U., 1924-34; prof. Oxford U., 1934-62. Recipient Hunterian medal, 1929, Viking medal, 1955, Huxley medal, 1958. Fellow Royal Soc., 1935; mem. Anat. Soc., Physiol. Soc., Royal Anthrop. Inst., Nat. Acad. Sci. Washington, N.Y. Acad. Sci. Author: Tissues of the Body, 5th edit., 1965; Antecedents of Man, 2d edit., 1962; Fossil Evidence for Human Evolution, 2d edit., 1964; also numerous articles. Research on connections in brain, especially sensory systems, growth and regeneration of muscular tissues, fossil apes from Miocene deposits of E. Africa, australopithecines of S. and E. Africa, comparative anatomy of tree shrews of S.E. Asia showing their affinities with primates. Home: Fawler's End, Fawler, Charlbury, Oxford, Eng.*

CLARK, William, Brit. anatomist; b. Newcastle-on-Tyne, Eng., Apr. 5, 1788; s. John C.; B.A., Trinity Coll., Cambridge Eng., 1808; fellow, 1809-27; studied medicine in London, Eng.; M.D., 1827; m. Mary Willis, 1827; 1 son. Prof. anatomy Cambridge, 1817-66; rector of Guiseley, Yorkshire, Eng., 1826-59. Fellow Royal Soc., 1836. Author: Analysis of a Course of Lectures on the Anatomy and Physiology of the Human Body, 1822; also articles. Translator from Dutch: A Handbook of Zoology (J. Vander Hoeven), 1856-58. Responsible for acquisition of extensive museum of comparative anatomy, 1832; laid founds. for sch. biol. sci. at Cambridge. Died Sept. 15, 1869.

CLARK, William Mansfield, Am. chemist; b. Tivoli, N.Y., Aug. 17, 1884; s. William Mansfield and Caroline (Hobson) C.; B.A., 1907, M.A., 1908, Williams College; Ph.D., Johns Hopkins, 1910; hon. Sc.D., Williams, 1933; m. Rose Willard Goddard, Sept. 14, 1910; children: Harriet A., Miriam G. Chemist with Bureau of Dairy Industry, Dept. of Agriculture, 1910-20; prof. of chemistry, Hygiene Laboratory, Public Health Service, 1920-27; dir., Dept. of physiological Chemistry, Johns Hopkins, 1927-52; Professor emeritus, 1952-66; research prof., Dept. of Chemistry, Johns Hopkins Univ., 1952-64. Ed. board, Journal of Biological Chemistry, 1933-51; chairman, Division of Chemistry and Chem. Technology, National Research Council during WWII. Nichols Medal, Am. Chem. Soc., 1936; Borden Award, 1944; President's Certificate of Merit, 1945; Passano Award, 1957. Mem., Nat. Acad. Scis.; Am. Phil. Soc.; Am. Soc. of Biological Chemists (pres. 1933-34); Soc. of Am. Bacteriologists (pres. 1933). Author: The Determination of Hydrogen Ions, 1920; Topics in Physical Chemistry, 1948; Oxidation-Reduction Potentials of Organic Systems, 1960. Investigation of physical methods in control of pH and oxidation-reduction; contribution to study of life processes. Died Jan. 19, 1964.

CLARKE, Alexander Ross, Brit. geodesist; b. Sutherlandshire, Scotland, Dec. 16, 1828; ed. Woolwich, Eng.; m. Frances Dixon, 1853; army engr.; in charge trigonometrical operations Ordnance Survey, 1854-81. Fellow Royal Soc., 1862 (Royal medal 1887), Royal Soc. Edinburgh (hon.); corr. mem. Imperial Acad. Sci. St. Petersburg. Author: Geodesy (standard work), 1880. Contbd. to prin. triangulation of Brit. Isles, also correlated work with that in other countries; reduced, pub. observations on figure of earth, 1858, on standards of length, 1866; proposed Clarke ellipsoids as bases of reference, 1880. Died Feb. 11, 1914.

CLARKE, Alfred Carpenter, Am. sociologist; b. Milford, Conn., June 26, 1921; s. Stanley Newton and Madge (Haviland) C.; A.B. cum laude, Marietta Coll., 1948; M.A., Ohio State U., 1950, Ph.D., 1955; m. Daisy Jackson, Aug. 29, 1948; children—Kenneth Carpenter, James Alfred. Research asst. dept. sociology Ohio State U., 1949-50, faculty, 1951—, prof., 1964—, vice-chmn., 1967—, dir. research projects Grad. Sch., 1958—. Fellow Am. Sociol. Assn.; mem. Am. Assn. U. Profs. (pres. 1959-60), Nat., Ohio (pres. 1965-66) councils on family relations, Ohio Valley Sociol. Soc., Ohio Acad. Sci., Alpha Kappa Delta, Beta Gamma Sigma. Author: (with R. Dynes, S. Dinitz, I. Ishino) Social Problems: Dissensus and Deviation in an Industrial Society, 1964; also articles. Series editor Appleton-Century-Crofts Series in Sociology, 1965—. Demonstrated significant differences between social stratification and patterns of work and leisure, occupational aspirations, mate-selection, adaptability of Air Force pilot trainees, social adaptation of aged to later life demands. Home: 4016 Windermere Rd., Columbus, O. 43221. Office: Dept. Sociology, Ohio State U., Columbus, O. 43210.*

CLARKE, Arthur Haddleton, Jr., Am. malacologist; b. Danvers, Mass., July 12, 1926; s. Arthur H. and Irene (Baker) C.; A.B., Boston U., 1952; M.Sc., Cornell U., 1958; Ph.D., Harvard, 1960; m. Louise Carol Robinson, July 3, 1948; 1 son, Arthur R. Marine biologist Lamont Geol. Obs., Palisades, N.Y., 1957-59; curator mollusks Nat. Mus. Can., Ottawa, Ont., 1959—. Inst. Malacology, Malcological Soc. London, Conchological Soc. Gt. Britain and Ireland, Am. Malacological Union (pres. 1967-68). Author: Annotated List and Bibliography of the Abyssal Marine Mollusks of the World, 1962; also articles. Editor: Malacologia, 1962—; asso. editor Canadian Field-Naturalist, 1960—. Research on arctic marine mollusks, abyssal marine mollusks, N.Am. freshwater mollusks. Home: 1778 Gilbert Av., Ottawa, Ont., Can.*

CLARKE, Beverly Leonidas, Am. chemist; b. Nashville, Sept. 30, 1900; s. Thomas Hopkins and Ida Clyde (Gallegher) C.; B.S. in Chemistry cum laude, George Washington U., 1921; M.A., Columbia, 1923, Ph.D., 1924; m. Ruth Adeline Johnston, Sept. 2, 1930; children—Thomas Beverly, James Johnston. Am.-Scandinavian Found. fellow Nobel Inst., Stockholm, Sweden, 1921-22; fellow in research Carnegie Instn. Washington, Carmel, Cal., Tucson, 1924-25; NRC fellow chemistry Stanford, 1925-27; instr. Columbia, also Coll. City N.Y., 1922-24; tech. asst. to U. S. comml. attache, Paris, London, Eng., summer 1923; materials chemist Bell Telephone Labs., N.Y.C., Murray Hill, N.J., 1927-45; dir. chem. control Merck & Co., Inc., Rahway, N.J., 1945-60; research officer Fedn. Am. Socs. Exptl. Biology, Washington, 1963—. Cons. NDRC, Dept. State, Dept. Def., Office Censorship, 1942-64. Recipient Army-Navy certificate of appreciation for civilian work, 1947. Fellow N.Y. Acad. Scis. (life); mem. Am. Chem. Soc. (cons. to office exec. sec. 1960-63), A.A.A.S., Fedn. Internat. de Documentation. Author: The Romance of Reality, 1927; Marvels of Modern Chemistry, 1932; (with M. Edward Maretn) The Doctor Looks at Murder, 1937; also numerous articles, stories to New Yorker. Research on application analytical chemistry to industry, microchemistry, application statis. methods to analytical chemisry, application statis. methods to sci. communications, oral and written, intensive exam. multiple-authorship trends in sci. papers. Home: 4600 Connecticutt Av. N.W., Washington 20008. Office: 9650 Rockville Pike, Bethesda, Md. 20014.*

CLARKE, Charles Baron, botanist; b. Andover, Hampshire, Eng., June 17, 1832; s. Turner Poulter and Elizabeth (Parker) C.; ed. Kings Coll. Sch., London, Eng., 1846-52; Trinity Coll., 1952-56, B.A.; M.A., Lincoln's Inn, 1859. Tchr., Queens Coll., 1956-65; staff Presidency Coll., Calcutta, India; supt. Calcutta Bot. Gardens, 1869-71. Fellow Royal Soc., 1882; mem. Linnean Soc., Geol. Soc., Geologists Assn. Author: Commelinacae, 1881; Crytandracae, 1883. Collected plants from Andover, Eng. and Bengal, India; studied flora of Kashmir and India; classified numerous species of plants in Eng. and India. Died Aug. 25, 1906.

CLARKE, Charles Kirk, physician; b. Can.; s. Charles and Emma (Kent) C.; M.D., LL.D., U. Toronto; m. Emma De Veber Andrews (dec. 1902); m. 2d, Theresa Gallagher; 4 sons, 2 daus. Asst. supt. Hamilton and Rockwood Hosps. for Insane, later supt. Rockwood Hosp. for Insane, Toronto Hosp. for Insane, until 1911, Toronto Gen. Hosp., 1911; dean Med. Faculty, U. Toronto, 1907-20, also prof. psychiatry; med. dir. Canadian Nat. Com. for Mental Hygiene, 1918-24. Editor: U. Toronto Med. Bull.; an editor Jour. Insanity, Canadian Jour. Mental Hygiene. Research and numerous publs. on psychiat. and medico-legal subjects, ornithology. Died Jan. 1924.

CLARKE, Donald Alston, Am. pharmacologist; b. Mansfield, Pa., Aug. 27, 1915; s. Carl Dean and Genevieve (Clark) C.; B.S., Phila. Coll. Pharmacy & Sci., 1937; A.M. (research fellow), Cornell U. Med. Coll., 1946, Ph.D., 1950; m. Dolores B. DeRevuelta, May 19, 1951; children—Philip Carl, Dolores, Brian. Apothecary-in-chief Cornell Med. Center, N.Y. Hosp., 1937-49, mem. various coms., 1956-62; faculty Sloan-Kettering Inst. Cancer Research, Cornell, 1951-62, head solid tumor screening sect., 1951-62; v.p., treas., dir. operations CANCIRCO, Inc., Rye, N.Y., 1962-64; med. dir. Clairol, Inc., N.Y., Conn., 1963-67, sci. dir. biomed. research, 1967—. Mem. faculties, insts. on hosp. pharmacy, Ann Arbor, Mich., 1946, Chgo., 1947, Princeton, 1948, Storrs, Conn., 1954; mem. pharm. survey com. Am. Council on Edn., 1946-51; mem. pharmacology com. drug evaluation panel Cancer Chemotherapy Nat. Service Center, 1959-63. Bd. trustees Columbia Coll. Pharmacy; bd. dirs. Westchester div. Am. Cancer Soc. Mem. Am. Pharm. Assn. (past 1st v.p.), Am. Soc. Hosp. Pharmacists (past vice chmn.), Assn. Advancement Profl. Pharmacy (now Am. Coll. Apothecaries), A.A.A.S., N.Y. Acad. Scis., Am. Soc. Pharmacology and Exptl. Therapeutics, Am. Therapeutics Soc., Am. Assn. Cancer Research, Soc. Exptl. Biology and Medicine, Am. Inst. Biol. Scis., Sigma Xi. Former editor Hosp. Pharmacy Forum, hosp. pharmacy sect. Interprofl. Bull.; contbg. editor El Farmaceutico. Research and publs. on application of digitalis glycosides in treatment of congestive heart failure, also use of mercurial diuretics; devel. useful agts. in field of cancer chemotherapy; studies in biochemistry and cancer biology; application of new methods for toxicological evaluation of substances to be used in and on humans. Home: 25 Geneva Rd., R.D. 2. Office: Clairol Research Lab., Tindall and New Canaan Avs., Norwalk, Conn. 06850.

CLARKE, Edward Daniel, Brit. mineralogist; b. Willingdon, Eng., June 5, 1769; s. Edward and Anne (Grenfield) C.; began studies at Jesus Coll., Cambridge Eng., 1786; B.A., 1790, M.A., 1794; LL.D., Cambridge 1803; m. Angelica Rush, 1806; 2 sons. Became tutor to Duke of Dorset's nephew, 1790; toured Gt. Britain; companion to Lord Berwick in France, Switzerland, Italy, 1792; tutor to family of Sir Roger Mostyn, 1794-96; elected fellow Jesus Coll., 1798; undertook (with Mr. Gipps) 3-year tour of No. Europe, Asia Minor, Syria, Palestine, Turkey, Greece, Egypt, 1799; ordained, 1805; became vicar of Harlton; rector of Yeldham, Essex, Eng., 1809-22; prof. mineralogy Cambridge U., 1808-21, elected librarian, 1817. Cambridge founded a chair of mineralogy in his honor, 1808. Author: The Mineral Kingdom, 1808; Travels in Various Parts of Europe, Asia and Africa, 1810. Studied mineralogy and natural scis.; collected vases, coins, minerals in Italy, in 1790's; also minerals throughout world; built a balloon. Died Pall Mall, Eng., Mar. 9, 1822.

CLARKE, Frank Wigglesworth, Am. geologist; b. Boston, Mar. 19, 1847; s. Henry Ware and Abby Mason (Fisher) C.; S.B., Lawrence Scientific Sch., Harvard, 1867; D.Sc., Columbian, 1891, Victoria, Manchester, 1903; LL.D., Aberdeen U., 1906, U. Cin., 1914; m. Mary P. Olmsted, Sept. 9, 1874. Instr. Cornell, 1869; prof. chemistry Howard U., Washington, 1873-74; prof. chemistry and physics U. Cin., 1874-83; chief chemist U. S. Geol. Survey and hon. curator minerals U. S. Nat. Museum from 1883. Hon. pres. Internat. Com. on Chem. Elements. Chevalier de la Légion d'Honneur; Wilde medalist Manchester (Eng.) Lit. and Philos. Soc., 1903. Mem. Internat. Jury Awards, Paris Expn., 1900. Author: Weights, Measures and Money of All Nations; Elements of Chemistry; Constants of Nature; Report on the Teaching of Chemistry and Physics in the United States, 1881; (with Louis M. Dennis) Elementary Chemistry, 1902, Laboratory Manual of Elementary Chemistry, 1902; Recalculation of Atomic Weights, 3d ed., 1910. Investigated constitution of natural silicates; made geochem. calculations on proportions of shales, sandstones and limestones in sedimentary deposits. Died May 23, 1931.

CLARKE, Frederick James Patrick, English physicist, ceramist; b. Harrow, Eng., Apr. 19, 1929; s. Patrick Theobald and Evelyn (Ediss) C.; student Prior Park Coll., Bath, Eng., 1939-47; B.Sc. in Gen. Math., U. London (Eng.), 1953, B.Sc. with spl. physics honors, 1954, Ph.D., 1957; m. Nathalie Pamela Constance Cullen, May 22, 1956; children—Audrey, Michael, Stephen. Sr. prin. scientist Atomic Energy Research Establishment, Harwell, Eng., 1954—. Fellow Inst. Physics (chmn. subcom. 1966—), mem. Brit. Ceramic Soc. (basic sci. council 1966—). Research, publs. on deformation and fracture of ceramics, effects of reactor irradiation on properties of reactor materials. Home: Walsingham, East Hendred, Berkshire, Eng. Office: Atomic Energy Research Establishment, Harwell, Eng.*

CLARKE, George Leonard, Am. biologist; b. Providence, Aug. 7, 1905; s. Prescott O. and Mary (Chase) C.; A.B., Harvard, 1927, A.M., 1928, Ph.D., 1931; m. Marion Sherman Butcher, Sept. 5, 1929; children—David Butcher, Prescott, William Leonard. Faculty, Harvard, 1930—, prof. biology, 1963—, chmn. com. on oceanography, 1954-63; with Woods Hole Oceanographic Inst., 1931—, marine biologist, 1940—; lectr. UNESCO, India, 1957. vis. investigator Inst. Oceanography, Monaco, 1958; participant various oceanographic expdns. Recipient Fulbright award, 1951. Mem. Limnol. Soc. Am. (pres. 1942-46), Am. Soc.

Limnology and Oceanography (sec.-treas. 1946-49), Am. Acad. Arts and Sci., Am. Ornith. Union, Am. Soc. Zoology, Ecol. Soc. Am., Am. Geophys. Union, A.A.A.S. Am. Inst. Biol. Sci., Sigma Xi, numerous others. Author: Elements of Ecology, 1954, rev., 1965. Contbr. articles to profl. jours. Developer Clarke-Bumpus plankton sampler, bathy-photometer; research in productivity of sea, behavior of animals, ecol. effects of daylight, meaning of bioluminescence in sea. Home: 44 Juniper Rd., Belmont, Mass. 02178.* Office: Biol. Labs., Harvard, Cambridge, Mass. 02138.*

CLARKE, Jacob Augustus Lockhart, Brit. anatomist; b. Pimlico, Eng., 1817; ed. Guy's and St. Thomas hosps.; diploma Apothecaries Soc.; M.D., St. Andrews (Scotland). Practiced medicine at Pimlico; physician Hosp. for Epilepsy and Paralysis, during 1870's. Fellow Royal Soc., 1854 (Royal medal 1864); mem. London Coll. Physicians. Contbr. numerous articles to med. jours. Concentrated on micros. research of brain and nervous system, also applied new methods which revolutionized histological research; introduced Canada balsam to mount histologic specimens; discovered and described intermediolateral column of cells in spinal cord (Dorsal nucleus of Clarke), 1851; did research in pathology; gave (with John H. Jackson) 1st complete description of syringomyelia, 1867; also wrote on muscular afflictions, nervous disorders and locomotor ataxia. Died London, Eng., June 25, 1880.

CLARKE, John, Brit. physician; b. Northamptonshire, Eng., 1761; s. John C.; ed. St. Paul's Sch., St. George's Hosp. Practiced medicine in Chancery Lane; license in midwifery Coll. Physicians, 1787; chief midwifery practitioner, London; lectr. midwifery St. Bartholomew's Hosp. Author: Essay on the Epidemic Disease of Lying-in Woman, 1787; Practical Essays on Pregnancy and Labour, and the Diseases of Lying-in Women, 1793; Commentaries on Some of the Most Important Diseases of Children. First to give exact description of laryngismus stridulus; described infantile tetany; also wrote on tumor of placenta. Died Aug., 1815.

CLARKE, John, English mathematician; b. Norwich, Eng., 1682; s. Edward and Hannah (Parmeter) C.; ed. Gonville and Caius Colleges, Cambridge U., B.A. 1703, M.A. 1707, D.D. by royal command, 1717; m.; a son. Royal chaplain and canon of Canterbury, 1721; Dean of Salisbury, 1728. Author: An Enquiry into the Cause and Origin of Evil, 2 vols., 1720; A Demonstration of some of the principal sections of Sir Isaac Newton's Principles of Natural Philosophy, 1730. Translated Grotius' De Veritate, 1711; published English translation of Rohault's Traité de Physique and System of Natural Philosophy; edited 2d edition of Ditton's An Institution of Fluxions, 1726; wrote notes on Wollaston's Religion of Nature, 1722. Died Salisbury, Eng., Feb. 10, 1757.

CLARKE, J(ohn) F(rederick) Gates, entomologist; b. Victoria, B.C., Can., Feb. 22, 1905; s. Robert Wilson and Ida Charlotte (Gates) C.; came to U. S., 1916, naturalized, 1934; student U. Wash., 1923-24; Ph.C. in Pharmacy, Wash. State U., 1926, B.S. in Zoology, 1930; M.S. in Entomology, Cornell U., 1931, postgrad., 1935-36; postgrad. U. Paris (France), 1945-46; Ph.D. in Entomology, U. London (Eng.), 1953; m. Thelma Blanche Canterbury Miesen, June 14, 1929; children—John Frederick Gates, Carol Canterbury. Pharmacist, Offerman Drug Co., Bellingham, Wash., 1926-29; instr. biology Wash. State Coll., 1931-35; entomologist Dept. Agr., 1936-54; curator div. insects U. S. Nat. Museum, Smithsonian Instn., 1954-62, chmn. dept. entomology, 1963-65, sr. entomologist, 1965—; field expdns. to Pacific Islands, Micronesia, 1952-53, W.I., V.I., 1956, 58, Eng., Austria, 1957, 60, S.Am., 1958-59, Yucatan, 1960, S. Pacific, Polynesia 1961, Tahiti, Raivavae, Tubuai, Rapa, Fiji, 1963, Dominica, 1965, Sarawak, Ceylon, Australia, Kenya, 1967, Marquesas Islands, 1968. Del. Dept. Agr. to 8th Entomol. Conf., Stockholm, Sweden, 1948; presented papers numerous internat. confs. Grantee NRC, 1934, 35, Smithsonian Instn., 1947, 67, Am. Philos. Soc., 1950, NSF, 1958, 61, Office Naval Research, 1961, 62. Fellow Royal Entomol. Soc., Entomol. Soc. Am. (chmn. com. proposal establish Nat. Inst. Entomology 1957); mem. Entomol. Soc. Washington (chmn. program com. 1955-58, pres. mem. exec. com. 1960), Soc. Brit. Entomology, Lepidopterists Soc. (exec. council 1967), Sigma Xi, Phi Kappa Phi, Rho Chi, Phi Sigma; hon. mem. Soc. Crucena de Ciencias Naturales (Bolivia), Entomol. Soc. Peru, Am. Entomol. Soc. Author: Giant Golden Book on Butterflies, 1963; also numerous articles. Inventor philatelic tool. Home: 5115 72d Av., Glenridge, Hyattsville, Md. 20784. Office: Dept. Entomology, U. S. Nat. Mus., Smithsonian Instn., Washington 20560.*

CLARKE, John Mason, Am. paleontologist; b. Canandaigua, N.Y., Apr. 15, 1857; s. Noah T. and Laura Mason (Merrill) C.; A.B., Amherst, 1877, A.M., 1882; U. of Göttingen, 1882-84; Ph.D., Marburg, 1898; LL.D., Amherst, 1902, Johns Hopkins, 1915; Sc.D., Colgate, 1909, U. Chgo., 1916, Princeton, 1919; m. Fannie V. Bosler, 1895. Prof. geology and mineralogy Smith Coll., 1881-84; lectr. geology Mass. State Agrl. Coll., 1885-86; asst. paleontologist State of N.Y., 1886; prof. geology and mineralogy Rensselaer Poly. Inst., from 1894; asst. state geologist, 1894, state paleontologist, 1898-1904, state geologist and paleontologist and dir. State Mus. and Science Dept., U. State of N.Y., from 1904. Chmn. geol. com. NRC, 1917. Recipient Hayden gold medal, 1908; gold medal Permanent Wild Life Protection Fund, 1920; Spindiaroff prize Internat. Congress of Geologists, Stockholm, 1910. Fellow Am. Acad. Arts and Scis., Geol. Soc. Am. (pres. 1916-17), Paleontol. Soc. (1st pres.), Nat. Parks Assn. (v.p.). Author: Sketches of Gaspé; The Magdalen Islands; Heart of Gaspé; The Life of James Hall; Organic Dependence and Disease; L'Ile Percée; New Devonian Crustacea, 1882; Early Devonic History of New York and Eastern North America, 1908; and over 150 scientific papers. Authority on Devonian period. Died Albany, N.Y., May 29, 1925.

CLARKE, William, English alchemist, physician; b. Swainswyke, England, ca. 1640; B.A., 1661, M.A., 1662, Fellow, 1663, M.D., Merton College, Oxford. Practiced medicine in Bath; moved to Stepney, London. Author: The Natural History of Nitre: or, a Philosophical Discourse of the Nature, Generation, Place, and Artificial Extraction of Nitre, with its Vertues and Uses, 1670. Discussed nitro-aerial particles, distillation of nitre; believed nitre was cause of meteors. Died Stepney, London, Apr. 24, 1684.

CLARKE, William Branwhite, geologist; b. East Bergholt, Suffolk, Eng., June 2, 1798; B.A., Cambridge (Eng.) U., M.A., 1824; curate of Ramsholt, Suffolk; made numerous geol. trips to continent; Anglican clergyman in New South Wales, Australia, 1840-70; visited Tasmania, 1856, 60; became fellow St. Paul's Coll., 1853. Trustee Australian Mus. Fellow Royal Soc., 1876; mem. Philos. Soc. New South Wales (v.p.). Contbr. numerous sci. articles to tech. jours.; also poems. Research on Australian coal measures; discovered gold in New South Wales, 1841, tin, 1849, diamonds, 1859. Died June 17, 1878.

CLARKE, William Eagle, ornithologist; b. Leeds, Mar. 16, 1853; s. William Clarke; ed. Yorkshire Coll.; LL.D.; m., 2 daus. Hon. supr. bird collection Royal Scottish Mus., Edinburgh, keeper natural history dept., 1906-21. Mem. Wild Birds Adv. Com. for Scotland. Fellow Royal Soc. Edinburgh; mem. Brit. (pres. 1917-20, Godwin-Salvin medal), Am. (hon.) ornithologists unions, Zool. Soc. Scotland (v.p.), Yorkshire Naturalists Union (pres. 1905), Internat. Com. for Bird Protection, Vienna, Budapest ornithol. socs. Author: (with others) Handbook of Yorkshire Vertebrata, 1881, Bartholomew's Physical Atlas, vol. v, Zoogeography, 1911; Studies in Bird Migration, 2 vols., 1912. Co-editor: Scottish Naturalists, Anns. Scottish Natural History; editor Saunders British Birds, 3d edit., 1927, also numerous articles on vertebrates. Died May 10, 1938.

CLARKE, William James, Am. pathologist; b. Seattle, July 16, 1919; s. William J. and Gladys (Kearney) C.; B.S. in Vet. Sci., State Coll. Wash., 1943, D.V.M., 1944, Ph.D. in Research Pathology, 1962; m. Betty Louise Morgan, Sept. 22, 1942; 1 dau., Judith Ann. Asst. chief veterinarian Research Labs., Swift & Co., Chgo., 1944; dep. veterinarian State of Wash., Olympia, 1944-46; practice vet. medicine, Seattle, 1946-53, 55-56; sr. scientist Exptl. Animal Farm, Biology Lab., Hanford Labs., Gen. Electric Co., Richland, Wash., 1956-58, mgr., 1959-60, sr. scientist research pathology Biology Lab., 1962-65; NIH fellow research pathology Wash. State U., Pullman, 1960-62; mgr. pathology sect. biology dept. Battelle-N.W., Richland, 1965—. Lectr., Wash. State U., 1958-59, U. Wash. Grad. Sch., 1962, U. Munich (Germany), 1963. Mem. Am. Coll. Vet. Toxicologists (past chmn. com. on radiation toxicology), Am. Wash. State vet. med. assns., Am. Soc. Vet. Physiology and Pharmacology, Conf. Research Workers in Animal Diseases, Pacific N.W. Soc. Pathologists, Health Physics Soc., Sigma Xi, Alpha Psi. Research, numerous publs. in radiobiology and radiation pathology, cellular and tissue pathology; devel. and extension histological, histochem. and autoradiographic techniques for assessment cellular, tissue and organ damage. Home: 1618 Birch St. Office: Biology Dept., Battelle-N.W., Richland, Wash. 99352.*

CLASSEN, Alexander, German chemist; b. Aachen, Germany, Apr. 13, 1843; s. Mathias Hubert and Anna Katharine (Schoenen) C.; studied at Giessen, Germany, Berlin, Germany; Dr. Ing. h.c.; m. Johann Bischof, 1883; 2 sons, 1 dau. Asst. in Rostock, Germany; became asso. prof. analytical chemistry Technische U. Aachen, 1870, named prof. inorganic chemistry, 1882. Author: (with H. Danneel) Quantitative Analyse durch Elektrolyse, 1927; Handbuch der analytischen Chemie, 2 vols., 1922-24; Theorie und Praxis der Massanalyse; (with Sir Henry Roscoe) Lehrbuch der anorganischen Chemie, 2 vols., 1895. Used electrolysis in analytical chemistry; considered father of electroanalysis. Died Aachen, Jan. 28, 1934.

CLAU, Christopher, see Clavius, Christopher.

CLAUDE, Georges, French chemist; b. Paris, France, Sept. 24, 1870; entered Sch. Physics and Chemistry, 1886; chief lab., elec. powerhouse, Halles; engr. Thomson-Houston French Co.; imprisoned as collaborator in World War II, 1945; freed, 1949. Mem. French Acad. Scis., 1924. Author: l'Electricité à la portée de tout le monde, 1904; Air liquide, oxygène, azote, 1909. Proved acetylene dissolved in acetone can be transported, 1897; liquefied air by expansion method, 1902; manufactured oxygen, separated rare gases; produced liquid chlorine for use in poison gas attacks during World War I; suggested use of neon lamps for lighting, 1910; invented electromagnetic wave measurement, neon lighting signs; tried to use inequalities of ocean temperature to produce power; invented new method of ammonia synthesis, 1917. Died May 23, 1960.

CLAUDE, Inis Lothair, Jr., Am. polit. scientist; b. Yellville, Ark., Sept. 3, 1922; s. Inis Lothair and Parilla Jane (Pledger) C.; B.A. with High Honors, Hendrix Coll., 1942; M.A., Harvard, 1947, Ph.D. (Chase prize 1949), 1949; m. Marie Stapleton, Aug. 1, 1943; children—Susan, Robert Burr, Cathy. Instr., then asst. prof. govt. Harvard, 1949-56; asso. prof. polit. sci. U. Del., 1956-57; asso. prof. polit. sci. U. Mich., 1957-60, prof., 1960—; vis. research scholar Carnegie Endowment Internat. Peace, 1960-61; faculty chmn. course internat. relations Inst. Social Studies, The Hague, 1964-65. Mem. exec. com. Center Research Conflict Resolution, 1959-63; chmn. com. internat. orgn. Social Sci. Research Council, 1962—; mem. research group UN financial problems Brookings Instn., 1962-63, adv. com. UN policy study program, 1963—, occasional lectr. Army and Navy war colls., also UN Fgn. Service Tng. Seminar; cons. Dept. State, 1962—. Faculty fellow Fund Advancement Edn., 1951-52; Rockefeller research grantee, 1958-59; Horace H. Rackham Grad. Sch. research grantee, 1963; Guggenheim fellow, 1964-65; Fulbright research grantee, 1964-65. Mem. Am. (Woodrow Wilson Found. award 1963), Internat. (rapporteur gen. internat. orgn. 1964) polit. sci. assns., Am. Soc. Internat. Law, Commn. Study Orgn. Peace (exec. com. 1957—), Am. Assn. U. Profs., Am. Civil Liberties Union. Author: National Minorities: An International Problem, 1955; Swords Into Plowshares: The Problems and Progress of International Organization, 3d edit., 1964; Power and International Relations, 1962. Bd. editors Internat. Orgn., 1960—; chmn. bd. editors Jour. Conflict Resolution, 1961-63; cons. editor internat. affairs Random House, Inc., 1962—. Research in internat. orgn., particularly on polit. devel. of UN, power factor in internat. politics. Home: 2124 Chaucer Dr., Ann Arbor, Mich. 48103.*

CLAUDER, Gabriel, physician; b. Altenburg, Saxe-Altenburg, Oct. 18, 1633; s. Joseph C.; ed. Univ. Jena; M.D., Leipzig; pupil of Rolfinck. Traveled in Holland, Germany, Italy; practiced medicine in Altenburg; physician to various Saxon princes. Mem. Academia Naturae Curiosorum. Author: Tinctura Universali, 1678; Dissertatio de Methodus balsamandi corpora humana, 1679; De invento cinnabarino; several papers under name, Theseus. Wrote about thermoscope, gunshot wounds, mirrors and natural history; works show knowledge of recent experiments on air (possibly by Mayow); discussed volatile salt of air; based writings on own experience and historical evidence. Died Oct. 10, 1691.

CLAUDET, Antoine François Jean, photographer, inventor; b. Lyons, France, Aug. 12, 1797; Employed by bankers office from 1818; joined glass works of Choisy-le-Roi, became dir.; came to London, Eng., became glass warehouseman, 1829; daguerreotype photographer, from 1840; photographer in ordinary to queen. Recipient medal Soc. Arts, 1853, council medal Gt. Exhbn. of 1851; named Chevalier of Legion of Honor. Fellow Royal Soc., 1853. Contbr. articles to tech. jours. Invented glass-cutting machine, 1833; one of 1st to use collodion process in photography; invented numerous photog. devices including dynactinometer, photographometer, focimeter, stereomonscope, system of unity of measure for focusing enlargements, system of photosculpture. Died Dec. 27, 1867.

CLAUDIUS, Friedrich Matthias, German anatomist; b. Marburg, Germany, June 1, 1822; studied medicine and natural sci., Jena, Göttingen and Kiel (all Germany) univs.; Dr.Phil., Göttingen, 1844; Dr.Med., Kiel, 1852. Conservator, Zool. Mus., Kiel, 1849-52; field physician Schleswig-Holstein Army; became prof. anatomy Marburg, 1859. Author: Physiologische Bemerkungen über das Gehörorgan der Cetaceen und das Labyrinth der Säugethiere, 1858. Described polyhedral cells of cochlear duct, 1852, ovarian fossa; research on hearing ability and apparatus of various animals. Died Kiel, Jan. 20, 1869.

CLAUS, Adolf Karl Ludwig, German chemist; b. Cassel, Germany, June 6, 1838; s. Heinrich and Charlotte (Richter) C.; studied medicine, Marburg, Germany, chemistry, Göttingen, Germany; m. Alice Warder, 1867; 4 children. Asso. prof. chemistry and chem. tech., Freiburg, Germany. Research and publs. on quinoline derivatives, aliphatic aromatic ketones; studied benzol, 1867; discovered phenazine; suggested structure for benzene ring. Died Horheim nr. Freiburg, May 4, 1900.

CLAUS, Carl Carlovich, see Klaus, Karl Karlovich.

CLAUS, Carl Ernst, Russian chemist, pharmacist; B. Dorpat, Estonia, Jan. 11, 1796; m. Luise Dorothea Bathe, 1821; 1 son, 1 dau. Began work in pharmacy at age 14; became pharmacist, Saratow, Russia, 1816; owner pharmacy Kazan, to 1826; became asst. chemistry Chem. Inst., Dorpat, 1831; expdn., Asia, 1834; named asso. prof. chemistry, Kazan, 1837, prof., 1843; named prof. pharmacy, dir. Pharm. Inst., Dorpat, 1852. Mem. Berlin Acad. Scis. Recipient Demidov prize. Author: Grundrüge der analytischen Physochemie, 1837; Reise in den Steppen . . . , 2 vols., 1837-38; also numerous articles. Explored (with Goebel) steppes between Urals and Volga; isolated and discovered ruthenium, 1845; studied botany, platinum residues. Died Dorpat, Mar. 24, 1864.

CLAUSBERG, Christlieb, mathematician; b. Danzig, Poland, Dec. 27, 1689; student Altdorf, Switzerland; m. Anna Margareta Fleimann. Tchr. arithmetic and Hebrew, Danzig, before 1729, arithmetic and computation in Hamburg, Lübeck, Leipzig, (all Germany), 1730-33; in service of Danish Royal Court, until 1746. Author: Demonstrative Rechenkunst, 1732. Reformed math. edn.; research on the math. of accounting, problems of rebates and discount, forensic practices. Died Copenhagen, Denmark, June 6, 1751.

CLAUSEN, Harry John, Am. anatomist; b. Clinton, Ia., Oct. 18, 1904; s. Claus and Mary (Claussen) C.; A.B., U. Ia., 1926, M.S., 1930; Ph.D., N.Y. U., 1932; m. Irene Krouth, July 3, 1928; 1 son, Philip J. Asst. curator dept. exptl. biology Am. Mus. Natural History, N.Y.C., 1932-37; faculty U. Colo. Sch. Medicine, Denver, 1937-46, asso. prof., 1945-46; prof., chmn. dept. anatomy Loyola U. Sch. Dentistry, New Orleans, 1946-48; asso. prof. anatomy U. Ark., Little Rock, 1948-50, 57-58, U. S.D. Sch. Medicine, Vermillion, S.D., 1950-57; scientist adminstr. NIH, USPHS, Bethesda, Md., 1958——, exec. sec. anatomy, physiology, pharmacology and clin. research coms. career devel. rev. br. div. research grants, 1959-61, supervisory exec. sec. all sects., 1961——. Fellow A.A.A.S.; mem. Am. Assn. Anatomists, Am. Soc. Zoologists, Soc. Exptl. Biology and Medicine, Soc. Growth and Devel. Contbr. articles to tech. jours. Research in human devel. involving endocrine diseases, influence drugs on growth and devel., congenital anomolies in human; described new species bacteria responsible for tumor formations in some animals. Home: 8383 Colesville Rd., Silver Spring, Md. 20910. Office: 5300 Westbard Av., Bethesda, Md. 20014.*

CLAUSEN, Jens (Christian), botanist; b. N. Eskilstrup, Denmark, Mar. 11, 1891; s. Christen Augustinus and Christine (Christensen) C.; cand. mag. U. Copenhagen, Denmark, 1920, Dr. Philosophy, 1926; Dr. Agriculture, Uppsala, Sweden, 1957; m. Anna Hansen, Oct. 28, 1921 (dec. 1956). Came to U. S., 1931, naturalized, 1943. Staff mem., dept. plant biology Carnegie Instn. of Washington, Stanford, Cal., 1931——, emeritus, 1956——; prof. biology Stanford, 1951——. Lectr., vis. prof. various colls. Recipient Mary Soper Pope medal of botany, 1948. Mem. Royal Danish, Royal Swedish acads., Am. Acad. Arts and Scis., U. S. Nat. Acad Sci., Internat. Assn. Biosystematists, Soc. Study of Evolution (pres. 1955), Am. Soc. Naturalists, Am. Soc. Agronomists. Author: Stages in the Evolution of Plant Species, 1951. Publs. on co-discovery of ecol. races within species; genetic determination of species differences; heredity and environmental factors; world tree lines. Home: 2461 Romona St., Palo Alto, Cal. 94301.

CLAUSEN, John Adam, Am. sociologist; b. N.Y.C., Dec. 20, 1914; s. Adam Peter and Mary (Blum) C.; A.B., Cornell U., 1936, M.A., 1939; Ph.D., U. Chgo., 1949; m. Suzanne Ravage, June 2, 1939; children—Christopher John, Peter Anthony, Bruce Martin, Michael Allen. Research asso. Va. Planning Bd. 1941-42; study dir. troop attitude research War Dept., 1942-46; asst. prof. sociology Cornell U., 1946-48; social sci. research cons. Nat. Inst. Mental Health, Bethesda, Md., 1948-51, chief lab. socio-environmental studies, 1951-60, research career award com., 1960-64; mem. com. on internat. centers Nat. Insts. Health, 1961-63; prof. sociology, dir. Inst. Human Devel. U. Cal. at Berkeley, 1960-65, research sociologist, prof. sociology, 1965——. Mem. com. social sci., psychiatry Social Sci. Research Council, 1952-57, chmn. com. social structure and socialization, also director; member committee disaster studies NRC, 1954-57. Research adv. com. Cal. Dept. Mental Hygiene, 1964-66; research com. Cal. Dept. Corrections, 1962-65; mem. spl. med. adv. group VA, 1965——; mem. Adv. Com. on Abuse of Depressant and Stimulant Drugs, Food and Drug Adminstrn., 1965——; mem. Scottish Rite Joint Com. Schizophrenia Research, 1965——. Council Assn. Aid Crippled Children. Recipient Center Advanced Study Behavioral Scis. fellow, 1957-58. Fellow Am. Sociol. Assn. (exec. com., council 1957-61, chmn. com. on certification in social psychology), Am. Orthopsychiat. Assn., A.A.A.S.; member Am. Psychopathological Association, Soc. Research in Child Devel., Gerontol. Society, American Public Health Association (member committee on public health and behavioral scis. 1958-60). Unitarian. Author: Sociology in the Field of Mental Health, 1956; (with others) Measurement and Prediction, 1950, Explorations in Social Psychiatry, 1957; also articles profl. jours. Asso. editor: Sociometry, 1956-59, editor, 1959-62; editor (with R. Strauss) Medicine and Society, 1963; editorial bd.; Am. Jour. Ortho-psychiatry, 1956-59; Psychosomatic Medicine, 1962——, Jour. Psychiat. Research, 1963——; regional editor Social Sci. and Medicine, 1966——. Contributions to social psychiatry: formulation of the relationships social, cultural, and psychological factors in drug use and addiction; demonstration of relationship between family structure and occurrence of schizophrenia; research on social psychological antecedents of smoking; formulation and documentation of the process of socialization in the family. Home: 1963 Yosemite Rd. Office: Inst. Human Development, U. Cal., Berkeley, Cal.

CLAUSEN, Robert Theodore, Am. biologist; b. N.Y.C., Dec. 26, 1911; s. Adam Peter and Mary (Blum) C.; A.B., Cornell U., 1933, A.M., 1934, Ph.D., 1937; m. Edna Wadleigh Rublee, Jan. 31, 1942; children—Eric Neil, Joanna Margaret, Thomas Paul, Heidi Elizabeth. Faculty, Cornell U., Ithaca, N.Y., 1933——, prof., 1949——; collaborator U. S. Dept. Agr., 1943. Mem. A.A.A.S., Ecol. Soc., Soc. Naturalists, Soc. Study Evolution, Bot. Soc., Soc. Plant Taxonomy, Torrey Bot. Club, Am. Fern Soc. (past pres.). Author: Sedum of the Trans-Mexican Volcanic Belt, 1959. Research, numerous publs. in flowering plants and ferns, especially Sedum, Pachyrrhizus, Najas, Ophioglossaceae; ecol. surveys, exptl. studies, description of new species. Home: 1421 Slaterville Rd., Ithaca, N.Y. 14850.*

CLAUSIUS, Rudolf Julius Emanuel, German math. physicist; b. Köslin, Prussia (now Koszalin, Poland), Jan. 2, 1822; s. Carl E. and Charlotte Wilhelmine (Schultze) Gottlieb; degree U. Berlin (Germany), 1848; also ed. Halle, Germany; m. Adelhaide Rumpan, 1859; 2 sons, 4 daus.; m. 2d, Sophie Sack, 1886. Became lectr. on physics Arty. and Engring. Sch. Berlin, also docent U. Berlin, 1850; prof. physics Tech. U. Zurich (Switzerland), 1855-67, Würzburg, Germany, 1867-69, Bonn, Germany, from 1869. Fellow Royal Soc., 1868; mem. French Acad. Scis., 1865, numerous other sci. socs. Author: Über das Wesen der Wärme verglichen mit Licht und Schall, 1857; Die Potentialfunktion und das Potential, 1859; Die mechanische Wärmetheorie, 2 vols., 1864-67 and later edits.; Der zweite Hauptsatz der Wärmetheorie, 1867; Über die Energievorräte in der Natur . . . , 1885. Reconciled Carnot's theory of heat to equivalence of heat and work, and thus formulated 2d law thermodynamics, or principle of entropy, 1850; mathematized principle of conservation of energy; related pressure and temperature in working changes of state (expressed in Clausius-Clapeyron equation); established kinetic theory of gases, on the basis of which he showed that oxygen molecule is diatomic; found inner energy of ideal gas ind. of vol.; deduced formula for mean velocity of gas molecules; also worked in phys. chemistry, electricity; contributed to theory of electrolysis; pointed out that Ohm's law applied to electrolytes. Died Bonn, Aug. 24, 1888.

CLAUSS, Alexander Richard Christian, chemist; b. Chemnitz, Germany, Jan. 26, 1929; s. Victor Eugen and Dora-Liddy (Reinecker) C.; Dipl.-Chem., Hochscule Regensburg, 1952; Dr.rer.nat., Tech. Hochschule Darmstadt, Germany, 1954; m. Lea Van Bever, Nov. 24, 1960; children—Karin, Isabelle, Christian. Asst., Techn. Hochschule Darmstadt, 1954-56; research asst. European Research Assos., Brussels, Belgium, 1956-58; group leader, sub-dir. lab. Société d'Etudes de Recherches et d'Applications pour l'Industrie, Brussels, 1959——. Research, publs. on structure determination of graphitic acid, surface chemistry of carbon, Donnan equilibrium and membrane potential investigations of membranes of graphitic acid; detection of sheets of C monolayers by electron microscopic technique; preparation of acetylene metal carbonyl compounds, in-pile diffusion studies fission products in uranium oxide, high and low temperature fuel cells. Home: 50 Onze Lieve Vrouwstrat, Alsemberg, Belgium. Office: 1091, chaussée d'Alsemberg, Brussels, 18, Belgium.*

CLAUSS, Gustav Ernst Friedrich, German geodesist; b. Landau, Germany, Mar. 19, 1871; s. Robert Nikolaus and Elisabeth Marie (Pister) C.; studied geodetics Tech. U. Munich (Germany); m. Marie Juliane Hoeraut, 1907; 4 children. In charge triangulation and leveling in Bavaria for Bavarian Land Survey Office; became prof. Tech. U. Munich, 1926, named hon. prof., 1930. Mem. Bavarian Acad. Scis. (mem. geodesic commn. 1919). Research and publs. on connecting Bavarian and Prussian survey systems and establishing uniform geog. coordinates; wrote German Standard Sheet DIN 3025 on geodesy. Died Munich, Oct. 6, 1938.

CLAVASIO, Dominicus de, see de Clavasio.

CLAVEUS, see Duclo, Gaston.

CLAVIUS (or CLAU), Christopher, astronomer, mathematician; b. Bamberg, Bavaria, Germany, 1537; studied at Coimbra and Rome, Italy; joined Soc. of Jesus, 1555; became lectr. astronomy and math., Jesuit Coll., Rome, 1565. Author: Euclidis elementorum libri XV, 1574; Epitome arithmeticae practicae, 1583; Calendarii romani Gregoriani explicatio, 1603; Algebra Christophori Clavii Bombergensis, 1608; Opera mathematica, 5 vols., 1611-12; Geometria practica, 1604; also wrote on the Spheric of Theodosius and the Sphaera of Johannes de Sacro Bosco. Participated in Gregorian calendar reform, 1582; introduced German algebraic symbolism to Italy; research on trigonometry, geometry, astronomy, astrolabe, constrn. of sundials; originated way of dividing a measuring scale into divisions of any size which predated Vernier's measure; introduced the decimal point, 1593. Died Rome, Feb. 12, 1612.

CLAXTON, Thomas Folkes, English astronomer; b. London, Apr. 29, 1874; s. J. D. and Ellen (Folkes) C.; m. Dora Curwen Westray, 1899. Mem. staff Royal Obs., Greenwich, 1890-93; apptd. asst. dir. Royal Alfred Obs., Mauritius, 1895, dir., 1896; dir. Royal Obs., Hong-Kong, 1912-32. Mem., hon. sec. bd. dirs. Mauritius Inst. and Mus., 1903-11; mem. Internat. Commns. for Exploration of Upper Air, Maritime Meteorology, Terrestrial Magnetism, till 1932; pres. Conf. Dirs. of Far Eastern Weather Services, Hong Kong, 1930. Fellow Royal Astron. Soc.; mem. Mauritius Meteorol. Soc. (hon. sec.), 1896-1911. Author: Magnetic and Meteorological Observations, Mauritius (ann.), 1896-1910; Seismological Observations, 1898-1910; The Climate of Hong-Kong; The Winds of Hong-Kong; Isotyphs for the Far East, 1932; also papers on cyclones, other meteorol. topics. Died July 12, 1952.

CLAY, Albert Tobias, Am. orientalist; b. Hanover, Pa., Dec. 4, 1866; s. John Martin and Mary Barbara C.; A.B., Franklin and Marshall Coll., 1889; grad. Mt. Airy Theol. Sem., 1892; Ph.D., U. Pa., 1894; A.M. (hon.), Yale, 1910; LL.D., Pa. Coll., 1913, Litt.D., Muhlenberg Coll., Pa. 1918; m. Elizabeth Sommerville McCafferty, June 11, 1895. Assyrian fellow, 1892-93, instr. Hebrew, 1892-95, U. Pa.; instr. O.T. theology, Chgo. Luth. Sem., 1895-99; instr. Hebrew, Mt. Airy Theol. Sem., 1905-10; lectr. Hebrew, Assyrian, and Semitic archaeology, 1899-1903, asst. curator Babylonian and Semitic antiquities, 1899-1910, asst. prof. Semitic philology and archaeology, 1903-09, prof., 1909-10, U. Pa.; Laffan prof. Assyriology and Babylonian lit., Yale, 1910——. Curator Yale Babylonian Collection, 1912——; Reinicker lectr. Episcopal Theol. Sem., Alexandria, Va., 1908; ann. prof., Am. Sch. Oriental Research in Jerusalem, 1919-20. Mem. Archaeol. Inst. Am. (v.p., 1917-25; apptd. to visit Bagdad and arrange for establishment Sch. Oriental Research, 1920, 23), Am. Oriental Soc. (librarian 1912-25, treas.). Author: Business Documents of Murashu Sons, dated in the reign of Darius II (2 vols.); Documents from Temple Archives of Nippur, dated in the region of Cassite Rulers (3 vols.); Legal and Commercial Transactions, dated in the Assyrian and Neo-Babylonian and Persian Periods; Aramaic Endorsements, O.T. and Semitic Studies, in memory of W. R. Harper; Amurru, the Home of the Northern Semites; Personal Names of the Cassite Period; Business Transactions of the First Millennium B.C.; Legal Documents from Erech, Seleucid Era, Epics, Hymns, Omens and Other Texts, vols. I, II and IV, and II of Babylonian Records in the Library of J. Pierpont Morgan; Miscellaneous Inscriptions in the Yale Babylonian Collection; The Empire of the Amorites; Neo-Babylonian Letters from Erech; A Hebrew Deluge Story in Cuneiform; The Origin of Biblical Traditions; (with H. V. Hilprecht) Business Documents of Murashu Sons of Nippur, dated in the reign of Artaxerxes I; (with Morris Jastrow, Jr.) An Old Babylonian Version of the Gilgamesh Epic. Died Sept. 14, 1925.

CLAY, Charles, Brit. surgeon; b. Arden Mills, Eng., Dec. 27, 1801; s. Joseph Clay; apprentice to Kinder Wood, surgeon; attended practice Royal Infirmary, Manchester, Eng.; entered Edinburgh (Scotland) U., 1821; m. Vaudrey, 1823 (dec. before 1839); 3 children (dec. before 1839); m. 2d, Boreham. Practiced at Ashton-under-Lyne, England; med. officer health Audenshaw, England; sr. officer, lectr. mid-wifery St. Mary's Hosp., Manchester, England. Licentiate Royal Coll. Surgeons Edinburgh; extra-licentiate Royal Coll. Physicians London. Mem. Manchester Med. Soc. (pres.), Manchester Numis. Soc. (editor, pres.). Author: The British Record of Obstetric Medicine and Surgery for 1848 and 1849; The Results of all the Operations for the Extirpation of Diseased Ovaria by the large Incision from September 13, 1842 to the present Time, 8 vols., 1848; The Complete Handbook of Obstetric Surgery, 1856, 3d edit., 8 vols., 1874; Geological Sketches and Observations on Fossil Vegetable Remains . . . , from the great South Lancashire Coal Field, 8 vols., 1839; History of the Currency of the Isle of Man, 8 vols., 1849. One of 1st to successfully remove ovarian tumors, 1848; study and practice of ovariotomy; 1st in Eng. to cure varicose vein with Vienna paste; invented speculum for operation of squint. Died Poulton-le-Fylde, Sept. 19, 1893.

CLAY, Jacob, Dutch physicist; b. Berkhout, Netherlands, 1882; studied under Kamerlingh Onnes at Leiden, Netherlands; prof., lycée at Delft, Netherlands, U. Amsterdam, from 1929. Author: History of Law of Nature in Philosophy, 1915. Pioneer in study of cosmic rays; discovered (independently of Compton) variation of intensity of rays with latitude; discovered transverse effect, thus proving that radiation consists of electrically charged particles. Died Amsterdam, 1955.

CLAYTON, Carlyle Newton, Am. plant pathologist; b. Liberty, S.C., Dec. 20, 1912; s. William Carter and Ruth (Newton) C.; B.S., Clemson U., 1934; Ph.D., U. Wis., 1940; m. Zelma Adelaide Shumate, Aug. 31, 1935; children—Jo Ann (Mrs. W. A. Leist), Carlyle Andrew. Research asst. U. Wis., 1935-40; faculty Clemson U., 1940-45; asso. prof. plant pathology N.C. State U., 1945-50, prof. plant pathology, 1950——. Mem. Am. Phytopathol. Soc., N.C. Acad. Sci., A.A.-A.S., Am. Inst. Biol. Sci., Sigma Xi, Gamma Sigma Delta. Research and publs. on fruit diseases which led to improved understanding and control of them. Home: 2607 Van Dyke Av., Raleigh, N.C. 27607. Office: 2419 Gardner Hall, Raleigh, N.C. 27607.*

CLAYTON, Donald Delbert, Am. astrophysicist, nuclear physicist; b. Shenandoah, Ia., Mar. 18, 1935; s. Delbert Homer and Avis (Kembery) C.; B.S., So. Meth. U., 1956; Ph.D., Cal. Inst. Tech., 1962; m. Mary Lou Keesee, Dec. 18, 1954; children—Donald Douglas, Devon Charles. Research fellow in physics Cal. Inst. Tech., 1961-63; staff scientist Aerospace Corp., El Segundo, Cal., part time 1961-63; faculty Rice U., Houston, 1963—, asso. prof. physics and space sci., 1965—; vis. asso. physics Cal. Inst. Tech., 1966-67. Mem. Am. Phys. Soc., Am. Astron. Soc., Am. Geophys. Union, A.A.A.S., Phi Beta Kappa, Sigma Xi. Contbr. articles to profl. lit. Research on formation of elements in stars, age of elements, age of galaxy, evolution of stars. Office: Dept. Space Sci., Rice U., Houston 77001.*

CLAYTON, Frances Elizabeth, Am. geneticist; b. Texarkana, Tex., Nov. 6, 1922; d. Carl C. and Louise (Heath) C.; A.A., Texarkana Coll., 1942; B.S., Tex. Womens U., 1944; M.A., U. Tex., 1947, Ph.D., 1951. Instr., U. Ark., Fayetteville, 1950-51; instr. U. Tex., Austin, 1951-52, Rosalie B. Hite postdoctoral fellow, 1952-53, research scientist, 1953-54; faculty U. Ark., 1954—, prof. zoology, 1961—; vis. colleague in genetics U. Hawaii, Honolulu, 1963-64. Mem. Genetics Soc., Am. Genetic Assn., Evolution Soc., A.A.A.S., Soc. Exptl. Biology and Medicine, Am. Soc. Zoologists, Am. Micros. Soc. Contbr. articles to tech. jours. Research on devel. eye mutants in Drosophila, genetic irradiation effects during different stages cell division and devel., cytology species Drosophila. Home: 665 Lindell, Fayetteville, Ark. 72701.*

CLAYTON, Henry Helm, Am. meteorologist; b. Murfreesboro, Tenn., Mar. 12, 1861; s. Henry Holmes and Maria L. (Helm) C.; ed. pvt. schs., 1869-77; m. Frances Fawn Coman, Sept. 21, 1892; children—Henry Comyn (dec.), Lawrence Locke, Frances Lindley. Asst., Astron. Obs., Ann Arbor, Mich., 1884-85; asst. Harvard Astronomical Obs., 1885-86; observer Blue Hill Meteorol. Obs., 1886-91; local forecast ofcl. U. S. Weather Bur., 1891-93; meteorologist, Blue Hill Meteorol. Obs., 1894-1909; dean Sch. Aeros. Assn. Inst., Boston, 1909-10. Employed by Oficina Meteorologica Argentina to study methods of weather forecasting in Argentine Republic and also to inaugurate nr. Cordoba a sta. for exploring upper air by means of kites, Mar.-Oct. 1910; forecast ofcl. Oficina Meteorologica Argentina, Buenos Aires, 1913-22; researches, in coöperation with Smithsonian Instn., in regard to relation of world weather changes to observe conditions on sun, 1923-25; pvt. weather service and cons. meteorologist for bus. orgns., 1925-42; research asso. Harvard, 1943-44. In charge, 1905, of Tiesseren de Bort-Rotch expdn. for exploring atmosphere over Atlantic ocean with kites and sounding balloons; accompanied Oscar Erbsloh in German balloon Pommern, Oct. 1907, when record-making balloon voyage was made from St. Louis to Asbury Park, N.J. Inaugurated new system of weather forecasting, based on solar heat changes in Argentina, 1918. Cons. expert in Cloud Atlas prepared for Hydrographic Office, under Capt. Sigsbee, U. S. Navy; for anemometers, Blue Hill box kite, etc. Del. Pan-Am. Sci. Congress, Washington, 1915; del. Argentine Weather Service to 6th Internat. Meteorol. Conf., Holland, 1923. Fellow Am. Acad. Arts and Scis. Author: World Weather, 1923; Solar Radiation and Weather, 1925; Solar Activity, 1926; World Weather Records (pub. by Smithsonian Instn.), 1927 and 1934; Atmosphere and the Sun, 1930; Solar Relations to Weather, 1943; also numerous papers on meteorol. subjects (1939-41); studies of periodic changes in solar activity pub. by Smithsonian Instn. and in Jour. Atmosphere, Electricity and Terrestrial Magnetism of Carnegie Instn. Helped develop science of aerology; through his studies of clouds, provided basis for Clayton Egnell law (velocity increases with height); invented attachment for anemometers, Blue Hill box kite. Died Norwood, Mass., Oct. 27, 1946.

CLAYTON, John, scientist; b. Eng. circa 1650; educated as a theologian; 1 son, John; emigrated to Va. with his son in 1705; atty. gen. Va.; research on stained glass; discovered gas could be distilled from crude coal and stored.

CLAYTON, John, botanist; b. Fulham, Eng., circa 1685. Came to Am., 1705. Asst. clk. Gloucester County, Va., 1705-22, 1st clk., 1722-73; collected bot. specimens in Middle Tidewater dists. of Va.; his specimens identified and categorized in John Frederick Gronovius' book Flora Virginia, 1739, 43; part of his collection now in Nat. Herbarium in England. Mem. several learned socs. Europe. A genus of plants popularly known as spring beauty also called Claytonias named in his honor. Died Virginia, Dec. 15, 1773.

CLAYTON, Robert Norman, chemist; b. Hamilton, Ont., Can., Mar. 20, 1930; s. Norman and Gwenda (Twist) C.; B.Sc., Queens U., 1951, M.Sc., 1952; Ph.D., Cal. Inst. Tech., 1955. Research fellow Cal. Inst. Tech., 1955-56; faculty Pa. State U., 1956-58; faculty U. Chgo., 1958—, prof. chemistry, 1966—. Mem. Am. Geophys. Union, A.A.A.S., Sigma Xi. Research, publs. on distbn. of stable isotopes of light elements in nature; application to problems in geology. Home: 5532 South Shore Dr., Chgo. 60637.*

CLAYTON, Roderick Keener, biophysicist; b. Tallinn, Estonia, Mar. 29, 1922; s. John H. and Helena (Mullerstein) C.; B.S., Cal. Inst. Tech., 1947, Ph.D., 1951; m. Betty Jean Compton, June 28, 1944; children—Roderick Dale, Ann Keener. Faculty, U. S. Naval Postgrad. Sch., 1952-57; NSF Sr. Postdoctoral fellow U. Oxford (Eng.), Norwegian Inst. Tech., Trondheim, Oak Ridge Nat. Lab., 1957-58; sr. biophysicist biology div. Oak Ridge Nat. Lab., 1958-62; vis. prof. microbiology Dartmouth, 1962-63; sr. investigator C. F. Kettering Research Lab., Yellow Springs, O., 1963-66; prof. biology, biophysics Cornell U., Ithaca, N.Y., 1966—; mem. staff instrn. Marine Biol. Lab., Woods Hole, Mass. Fellow A.A.A.S.; mem. Biophys. Soc., Am. Soc. Biol. Chemists, Soc. Gen. Physiologists, Am. Assn. Plant Physiologists. Author: Molecular Physics in Photosynthesis, 1965. Contbr. numerous articles to profl. jours. Research in mechanisms by which light energy is converted to chem. energy in plants, nature of photochem. reaction centers in photosynthetic bacteria.*

CLEAVELAND, Parker, Am. geologist; b. Byefield, Mass., Jan. 15, 1780; s. Parker and Elizabeth (Jackman) C.; M.A., Harvard, 1802; M.D., Dartmouth Coll., 1823; LL.D., Bowdoin Coll., 1824; m. Martha Bush, Sept. 9, 1806. Taught sch., York, Me., 1799-1802; tutor math. and natural philosophy Harvard, 1803-05; prof. math. and natural philosophy Bowdoin Coll., Brunswick, Me., 1805-58; prof. materia medica Med. Sch. of Me., Brunswick, 1820-58. Mem. Am. Acad. Arts and Scis., Am. Philos. Soc. Author: Elementary Treatise on Mineralogy and Geology (1st Am. work on subject, classified minerals according to chem. composition, rather than crystal form); 1816; Agricultural Queries, 1827. The mineral Cleavelandite named for him. Died Brunswick, Me., Oct. 15, 1858.

CLEBSCH, Rudolf Friedrich Alfred, German mathematician; b. Königsberg, Prussia, Germany, Jan. 19, 1833; s. Ernst C.; studied at Königsberg, 1850-54; state examination in math. and physics; Ph.D., 1854; m. Elise Heinel, 1859; 4 sons; m. 2d, Minna Rays, 1867; Became prof. theoretical mechanics Karlsruhe (Germany), 1858; named prof. math. Giessen (Germany), 1863; apptd. prof. Göttingen (Germany), 1868. Author: Theorie der Elastizität Körper, 1862; Theorie der Abelschen Funktion, (with W. P. Gordan), 1866; Theorie der binären algebraischen Formen, 1872; also numerous articles. Founder (with C. Neumann) Mathematischen Annalen, 1868. Research on theory of invariants, use of higher transcendentals in geometry, math. physics, calculus of variations, partial differential equations of the first order, gen. theory of curves; proved theorems on the pentahedron set up by J. J. Sylvester and J. Steiner; used deficiency as a fundamental principle to classify algebraic curves; application of elliptic theory and Abelian functions to geometry and the study of rational and elliptic curves; used geometry in Abelian functions; introduced determinants into new areas of math. Died Göttingen, July 11, 1872.

CLEGG, Arthur Bradbury, Brit. physicist; b. Heswall, Eng., Mar. 15, 1929; s. Frederick Bradbury and Beatrice (Andrew) C.; B.A., Cambridge (Eng.) U., 1952, Ph.D., 1955, M.A., 1956; m. Marguerite Ingeborg Davis, Oct. 23, 1956; children—Karen Marguerite, Peter David. Research fellow Synchrotron Lab., Cal. Inst. Tech., Pasadena, 1955-58; sr. research officer dept. nuclear physics U. Oxford (Eng.), 1958-66, fellow Jesus Coll., 1964-66; prof. nuclear physics U. Lancaster (Eng.), 1966—. Fellow Inst. Physics; mem. Am. Phys. Soc. Author: High Energy Nuclear Reactions, 1965; also articles. Research on reactions induced in nuclei by nucleons of 100-150 Me V including nuclei structure; properties and reactions of elementary particle states. Home: Strawberry Bank, Cannon Hill, Westbourne Rd., Lancaster, Eng.*

CLEGG, Samuel, English inventor, engr.; b. Manchester, Eng., Mar. 2, 1781; studied under Dr. Dalton; apprenticed to Boulton and Watt. Employed by Henry Lodge to adapt new lighting in cotton mill; installed gas light for Mr. Rudolph Ackermann, 1813; became engr. Chartered Gas Co., 1814; engr. Portuguese Govt.; employed by Brit. Govt., 1838. Mem. Instn. Civil Engrs. Wrote treatise on mfr. of coal gas. Invented lime purifiers for gas; patented water meter, 1815, 18; invented wet gas meter, 1816, also telescopic gas container, revolving retort, retort for continuous carbonization. Died Middlesex, Eng., Jan. 8, 1861.

CLEGHORN, Robert Allen, psychiatrist; b. Cambridge, Mass., Oct. 6, 1904; s. Allen Mackenzie and Edna (Gartshore) C.; M.D., U. Toronto (Ont., Can.), 1928; D.Sc. in Physiology, Marischal Coll., Aberdeen, Scotland, 1932; m. Sheena Marnoch, Apr. 2, 1932; children—Mhairi, Jane (Mrs. Z. M. Santiago), Ailie M. (Mrs. B. D. Fletcher), John M. Dir. lab. exptl. therapeutics Allan Meml. Inst., Montreal, Que., Can., 1946-64, dir. inst., 1964—; chmn. dept. psychiatry McGill U., Montreal, Que., Can., 1964—, prof. psychiatry, 1960—. Fellow Royal Coll. Physicians and Surgeons Can., Am. Coll. Psychiatrists; mem. Am. Psychosomatic Soc. (past pres.), Que. Psychiat. Assn. Research and numerous publs. on physiology of adrenal glands and autonomic nervous system, wound shock in battle casualties, clin. psychiatry. Home: 3160 St. Sulpice Rd., Montreal, Que., Can.*

CLELAND, Archibald, Brit. surgeon; b. Eng., 1709; surgeon to Wade's Regiment of Horse; proposed theory that loud noises such as thunderclaps and gun shots could cause deafness because air pressure of disturbances bends ear drum and ear bones inward; devised a catheter tube for the nose, 1731 (discovered that this treatment could sometimes lead to various nervous disorders). Died 1774.

CLELAND, Herdman Fitzgerald, Am. geologist; b. Milan, Ill., July 13, 1869; s. David James and Margaret (Betty) C.; student Gates Coll., Neligh, Neb.; B.A., Oberlin Coll., 1894; Ph.D., Yale, 1900; postgrad. U. Neb., 1895, U. Chgo., 1896, Cornell U., 1901; m. Helen Williams Davison, Aug. 9, 1910 (dec. 1923); children—Margaret Jane, Elizabeth Davison; m. 2d, Emily Leonard Wadsworth, Oct. 1, 1925; children—Eunice Louise, Cynthia Herdman. Asst. in geology, Cornell U., 1900-01; instr. in geology and botany, 1901-04; asst. prof. geology and mineralogy Williams Coll., 1905-07, prof. geology from 1907; instr., Cornell Summer Sch., 1900-02, U. of Tenn. Summer Sch., 1904. Del., Internat. Geol. Congress, Mexico, 1906, Belgium, 1922, Spain, 1926, Internat. Geog. Congress, Geneva, Switzerland, 1908. Fellow Am. Acad. Arts and Scis., A.A.A.S.; mem. Geol. Soc. Am., Am. Geog. Soc., Seismol. Soc. Am., Paleontol. Soc., Am. Inst. Mining and Metall. Engrs., Am. Archaeol. Soc., N.Y. Acad. Sci., Phi Beta Kappa, Sigma Xi. Author: Fauna of the Hamilton Formation, 1901; Fossils and Stratigraphy of the Middle Devonic of Wisconsin, 1911; Physical and Historical Geology (2 vols.), 1917; Practical Applications of Geology and Physiography, 1920; Our Prehistoric Ancestors, 1928; Why Be an Evolutionist, 1930. Noted for writings on geology and evolution. Died in S.S. Mohawk, off coast of N.J., Jan. 24, 1935.

CLELAND, John, physician; b. Perth, Scotland, June 15, 1835; ed. U. Edinburgh (Scotland); M.D., LL.D.; m. Ada Marion Spottiswoode Balfour, 1888; 1 son. Prof. anatomy and physiology Queen's Coll., Galway, Eire, 1863-77; Regius prof. anatomy U. Glasgow (Scotland), 1877-1909. Licentiate Royal Soc. Coll. Surgeons, Edinburgh. Fellow Royal Soc., 1872. Author: The Mechanism of the Gubernaculum, 1856; Animal Physiology, 1874; Directory for Dissection, 1876; Evolution, Expression, and Sensation, Cell Life and Pathology, 1881; Scala Naturae and other Poems, 1887; also numerous memoirs, lectures, addresses. Coauthor: Memoirs and Memoranda in Anatomy, 1889; Cleland and Mackay's Human Anatomy, 1896. Died Mar. 5, 1924.

CLELAND, Ralph Erskine, Am. botanist; b. Le Claire, Ia., Oct. 20, 1892; s. Charles Samuel and Edith (Collins) C.; A.B., U. Pa., 1915, M.S., 1916, Ph.D., 1919; LL.D., Hanover Coll., 1957; Sc.D., U. Pa., 1958; m. Elizabeth P. Shoyer, June 11, 1927; children—William Wallace, Robert Erskine, Charles Frederick. Faculty biology dept. Goucher Coll., 1919-38, chmn. dept., 1937-38; prof. Ind. U., Bloomington, 1938-58, Distinguished Service prof., 1958-63, emeritus, 1963—, chmn. botany dept., 1938-63, dean Grad. Sch., 1950-58, co-dir. Aerospace Research Applications Center, 1963—. Chmn. div. biology and agr. NRC, 1948-51; mem. U. S. Nat. Commn. for UNESCO, 1958-60. Guggenheim fellow, 1927-28. Mem. Am. Acad. Arts and Scis., Internat. Union Biol. Scis. (past v.p.), Assn. Grad. Schs. (past sec.-treas.), Am. Philos. Soc. (John F. Lewis award 1937, v.p. 1965—), Bot. Soc. Am. (award merit 1956, past pres.), Nat. Acad. Scis., Genetics Soc. Am. (past pres.), Bot. Soc. Korea (hon. life), Genetics Soc. Japan, (hon. life), German Bot. Soc. (corr.). Research, publs. on chromosome behavior evening primrose; analyzed evolutionary history evening primroses. Home: 1300 E. 1st St., Bloomington, Ind. 47401.*

CLEMENCE, Gerald Maurice, Am. astronomer; b. Smithfield, R.I., Aug. 16, 1908; s. Richard R. and Lora (Oatley) C.; Ph.B., Brown U., 1930; Sc.D., Case Inst. Tech., 1954; Dr., Universidad de Cuyo (Argentina), 1961; m. Edith Melvina Vail, Aug. 17, 1929; children—Gerald Vail, Theodore Grinnell. Astronomer, U. S. Naval Obs., Washington, 1930-63, dir. nautical almanac, 1945-58, sci. dir., 1958-63; sr. research associate, lecturer dept. astronomy, Yale, 1963-66, professor of astronomy, 1966—. Recipient Conrad medal USN, 1962; USN Distinguished Service award and medal, 1963. Fellow Am. Acad. Arts and Scis.; mem. Nat. Acad. Scis., Royal (asso., Gold medal 1965), Royal Canadian (hon.) astron. socs., Bureau des Longitudes (France). Author: (with Dirk Brouwer) Methods of Celestial Mechanics, 1961; (with E. W. Woolard) Spherical Astronomy, 1965; also numerous articles. Research on motions of planets; introduced measurement of time by means of orbital motions of moon and earth; defined second as fraction of year instead of day. Home: 123 York St., New Haven, Conn. 06511.*

CLEMENCE, Richard Vernon, Am. economist; b. Greenville, R.I., Oct. 13, 1910; s. Richard R. and Lora (Oatley) C.; Ph.B. magna cum laude in Econs., Brown

U., 1934, A.M., 1937; M.A., Harvard, 1940, Ph.D., 1948; m. Eleanor Prescott, Dec. 5, 1942; 1 dau., Melissa. Instr., Boston U., 1937-38, Mass. Inst. Tech., 1939-42; research asso. Nat. Bur. Econ. Research, 1942; faculty Wellesley (Mass.) Coll., 1947—, chmn. dept. econs., 1949—, prof. econs., 1958—, A. Barton Hepburn prof., 1965—; dir. planning div. Paine Brooke Assos., 1959—; dir. Benjamin Chase Co., Derry, N.H. Cons. New Eng. Econ. Edn. Council, 1956—, Kingston Ins., 1960—, Paine Brook Fund, 1964—. Initiator, bd. dirs. Wellesley Tutuorial. Life fellow Nat. Assn. Bus. Economists (plaque 1966), Royal Econ. Soc. (Eng.); mem. Am. Econ. Assn., Econometric Soc., Econ. History Assn., Phi Beta Kappa. Author: Income Analysis, 1951; Economics of Defense, 1952; (with Doody) The Schumpeterian System, 1950; Economic Change in America (with Lambie), 1951; (with Wronski) Modern Economics, 1965; also articles. Developed new technique of income analysis. Home: 61 Beverly Rd., Wellesley, Mass. 02181.*

CLEMENT, Anthony Calhoun, Am. biologist; b. Spartanburg, S.C., Nov. 24, 1909; s. Thomas Newton and Garrie (Miles) C.; B.S., U. S.C., 1930; postgrad. N.Y. U.; A.M., Princeton, 1933, Ph.D., 1935; m. Octavia Cooper, Dec. 20, 1959. Faculty, Coll. Charleston (S.C.), 1935-49, prof., 1945-49; faculty Emory U., Atlanta, 1949—, prof. biology, 1956—. Program dir. for developmental biology NSF, Washington, 1958-59, cons., 1959-62. Trustee Marine Biol. Lab., Woods Hole, Mass. Guggenheim fellow, 1954-55. Fellow A.A.A.S.; mem. Am. Soc. Zoologists, Soc. for Developmental Biology, Internat. Inst. Embryology, Sigma Xi, Phi Beta Kappa. Research, publs. on effects centrifugal force on devel. mollusks, cytochemistry, origin body axes and symmetry, cytoplasmic composition and cell differentiation. Home: 1543 Knob Hill Dr. N.E., Atlanta 30329.*

CLEMENT, Jacqueline, French biochemist; b. Epernay, Marne, France, Nov. 3, 1916; d. Lucien Henri and Laure (Roche) Champougny; Licence ès Sciences, 1939, Thèse de Doctorat d'Etat, 1947; m. Clement, May 24, 1947; children—Yves, Marie, Laure. Researcher, Nat. Center for Sci. Research, 1943-53, dep., 1953-59, chief, 1959—; faculty Faculty Sci., Inst. Cancer Research, U. Paris (France), 1940-59; faculty Faculty Sci., Lab. Animal Physiology and Nutrition, U. Dijon (France), 1960—. Recipient Prix Essec, 1957; named Chevalier de l'ordre des Palmes Academiques, 1963. Mem. Internat. Confs. on Problemes Lipids, Soc. Biol. Chemistry. Research, numerous publs. on metabolism of lipids, dehydrogenase of fatty heavy acids, lipase, cholesterolesterase, phospholipase A and B of pancreatic juice, steatosis, phospholipids of intestinal mucosa processes of digestion and resorption and of stocking.*

CLÉMENT-DESORMES, Nicolas, French chemist; b. Dijon, France, Jan. 12, 1779; studied in libraries, later student of chemistry Montgolfier and Guyton de Morveau; m. dau. of Desormes; asst. to Guyton, Ecole Polytechnique; prof. Conservatoire des arts et métiers; prof. applied chemistry at Paris; with Desormes owned chem. works; gen. mgr. Gohain & Co., 1820; produced sugar from beets at Rambouillet; research on iodine and its compounds, catalysts; improved method of producing sulphuric acid; determined true nature of carbon monoxide; direct measurement of specific heats. Died Paris, Nov. 21, 1841.

CLEMENTE, Carmine Domenic, Am. anatomist; b. Pennsgrove, N.J., Apr. 29, 1928; s. Ermanno and Caroline (Friozzi) C.; A.B., U. Pa., 1948, M.S., 1950, Ph.D., 1952; m. Dorothy Warren, Dec. 19, 1955. Faculty, U. Cal., Los Angeles, 1952—, prof., chmn. dept. anatomy, 1963—; hon. research asso. U. Coll., London, Eng., 1953-54. Vis. scientist Nat. Inst. Neurol. Diseases and Blindness, Bethesda, Md., 1958; cons. Sepulveda VA Hosp., 1956—; mem. anat. tng. grants com. Nat. Inst. Gen. Med. Scis. NIH, 1964—; fellow Bank Am.-Giannini Found., 1953-54, mem. med. adv. panel, 1964—. Mem. Am. Assn. Anatomists, Am. Physiol. Soc., Am. Acad. Neurology, Am. Acad. Cerebral Palsy, Biol. Stain Commn., Internat. Brain Research Orgn., Sigma Xi. Research, numerous publs. on brain mechanisms responsible for behavior, effects of ionizing radiations on nervous system, nerve degeneration and regeneration, neurophysiology of sleep and wakefulness, sexual behavior and internal inhibition. Home: 11341 Cashmere St., Los Angeles 90049.*

CLEMENTI, E., chemist; b. Trent, Italy, Nov. 19, 1931; s. Ambrogio and Laura (Marzari) C.; Liceo Classico, U. Trento; Dr.Pure Chemistry, U. Pavia (Italy); m. Hildegard Cornelius, Aug. 13, 1956. Postdoctoral staff Politecnico di Milano (Italy) Instituto di Chimica Industriale, 1955, Fla. State U., Tallahassee, 1956-57, U. Cal. at Berkeley, 1958-60, U. Chgo., 1960-61; group leader research lab. IBM, San Jose, Cal., 1961-65; vis. prof. U. Chgo., 1966; mgr. large scale sci. computation dept. IBM Research Lab., San Jose, 1967—. Mem. Phys. Soc. Asst. editor Jour. Chem. Physics. Research, publs. on sublet-triplet transition mechanisms, carbon vapor thermodynamical properties, molecular orbital computation in linear molecules; systematic study in atomic structure, especially relativistic and correlation effect; constrn. of large scale programs for molecular computation; reaction kinetic study by theoretical techniques. Home: 23240 Deerfield Rd., Los Gatos, Cal., 95030. Office: Monterey and Cottle Rds., San Jose, Cal. 95114.*

CLEMENTS, Frederic Edward, Am. plant ecologist; b. Lincoln, Neb., Sept. 16, 1874; s. Ephraim G. and Mary (Scoggin) C.; B.Sc., U. Neb. 1894, M.A., 1896, Ph.D., 1898, LL.D., 1940; m. Edith Schwartz, May 30, 1899. Instr. and asso. prof. botany U. Neb., 1894-1906, prof. plant physiology, 1906-07; prof. and head dept. botany, U. Minn., 1907-17; state botanist, dir. bot. survey of Minn.; in charge ecol. research, Carnegie Instn., Washington, 1917-41; in charge Lab. Ecol. Research from 1941; collaborator U. S. Soil Conservation Service from 1934; cons. Nat. Hwy. Research Bd., 1935. Fellow A.A.A.S., (gen. sec., 1910); mem. Neb. Acad. of Science, Am. Microscop. Soc., Bot. Soc. of Am. (v.p. 1905; councillor, 1906-10), Am. Geographers Assn., Am. Nomenclature Commn., Internat. Nomenclature Commn., Internat. Congress Science, Am. Nature Study Soc. (dir. 1905), Am. Breeders Assn., Ecol. Soc. of Am., Paleontol. Soc., Soc. Am. Foresters, Am. Meteorol. Soc., Am. Soc. Mammalogists, Am. Soc. of Naturalists, Am. Soc. Plant Physiologists, Societas Phytogeographica Suecana of Sweden (hon.), Reale Academia Agricultura of Italy (hon.), Sigma Xi, Phi Beta Kappa. Author: The Phytogeography of Nebraska (with Dr. Pound), 1898, 2d edit., 1900; Histogenesis of Caryophyllales, 1899; Greek and Latin in Biological Nomenclature, 1902; Herbaria Formationum Coloradensium, 1902; Development and Structure of Vegetation, 1904; Research Methods in Ecology, 1905; Plant Physiology and Ecology, 1907; Cryptogamae Formationum Coloradensium, 1908; Genera of Fungi, 1909; Minnesota Mushrooms, 1910; Rocky Mountain Flowers, 1913 (with Dr. Edith Clements); Plant Succession, 1916; Plant Indicators, 1920; Aeration and Air-Content, 1921; The Phylogenetic Method in Taxonomy (with Dr. Hall), 1923; Experimental Pollination (with Dr. Long), 1923; Experimental Vegetation (with Dr. Weaver), 1924; Phytometer Method in Ecology (with Dr. Goldsmith), 1924; Plant Succession and Indicators, 1928; Flower Families and Ancestors (with Dr. Edith Clements), 1928. Known for ecol. research, soil conservation work. Died July 26, 1945.

CLEMENTS, Frederick William Arthur, Australian pediatrician; b. New South Wales, Australia, Aug. 9, 1904; s. William Ernest and Ida Ruth (Brown) C.; M.B., B.S., U. Sydney, 1928, M.D., 1937, D.P.H., 1933, D.T.M., 1935; m. Muriel Ellen Willis, Dec. 12, 1931; children—Pamela (Mrs. David H. Cureton), Rosemary Joan (Mrs. John E. Jefferis). Lectr., U. Sydney, Sch. Pub. Health and Tropical Medicine, 1931-38, sr. lectr. child health and nutrition, 1951—. Dir. Inst. Anatomy, Canberra, dir. Commonwealth Nutrition Unit, 1938-49; chief nutrition sect. WHO, Geneva, 1949-51; cons. pediatrician Karitane Mothercraft Home. Author: Infant Nutrition, 1949; Child Health, Its Origins and Promotion (with B. D. McCloskey) 1964; (with J. F. Rogers) You and Your Food, 1967, Diet in Health and Disease, 1966; also numerous articles. Research on nutrition problems in Australia, New Guinea, including endemic goitre and deficiency diseases. Home: 41 Cocupara Av., Lindfield, New South Wales, Australia.*

CLEMM, August, German chemist; b. Giessen, Germany, Dec. 8, 1837; s. Friedrich and Luise (Müller) C.; student chemistry at Giessen; m. Fanny Heyer, 1863; 4 children. With Anilinfabrik Dyckerhoff, Clemm & Co., Mannheim Germany, 1862, became coowner, 1863; head Badische Anilin-und Soda Fabrik Ludungshafen plant, 1866-82; mem. Bavarian Landtag, 1883-99, v.p., 1893-97, pres., 1897-99; named Hon. citizen of Ludwigshafen; decorated Royal Barvarian Ritterkreuzes. Leader in anilin manufacture. Died Schoss Hardt nr. Neustadt, Germany, Oct. 28, 1910.

CLEMM, Carl Friedrich, German chemist; b. Giessen, Germany, Aug. 16, 1836; s. Friedrich and Luise (Müller) C.; ed. Polytechnikum, Karlsruhe, Germany; studied chemistry in Giessen; m. Marie Hoff, 1863; 6 children. Began career in chem. factory of his uncle Karl Clemm-Lennig; founded own company Dyckerhoff, Clemm & Co. in 1861; founder cellulose plant, Waldhof nr. Mannheim, Germany; nat. liberal mem. Reichstages, 1887-93. Named hon. citizen of Ludwigshafen, Germany. Mem. Dutch East Africa Soc. Important in cellulose manufacture. Died Ludwigshafen, Feb. 20, 1899.

CLEMM, Hans, German chemist; b. Ludwigshafen, Germany, Dec. 18, 1872; s. Carl Friedrich and Helene (Bassermann) C.; studied chemistry in Kiel Germany; m. Maria Clemm, 1903; after training in cellulose plants Waldhof and Permnan (Estonia), became tech. dir. Waldhof plant in 1902 (under his direction became greatest such factory in Europe); later established connections with U. S. and Scandinavia. Named hon. senator U. Heidelberg (Germany). Mem. Assn. Cellulose and Paper Chemists and Engrs. (pres.) Leader in cellulose and paper chem. industry. Died Darmstadt, Germany, Oct. 29, 1927.

CLEMMESEN, Johannes, Danish med. pathologist; b. Copenhagen, Denmark, Nov. 14, 1908; s. Johan and Marie (Gran) C.; student Metropolitan Sch., Copenhagen, 1919-26; Medico-surg. B., U. Copenhagen, 1933, D.Med. Sci., 1938, Specialist Clin.Medic., 1943, Specialist Pathology, 1947. Research asst. radium dept. U. Hosp., Aarhus, Denmark, 1933-37; Oxley Cancer Research fellow exptl. pathology U. Leeds (Eng.), 1937-38; asst. hosps. Copenhagen, 1938-50; pathologist Old Peoples Town, Copenhagen, 1950-55; chief pathologist Finsen Inst., Copenhagen, 1955—, Norre Hosp.; dir. Danish Cancer Registry, 1942; asso. prof. path. anatomy Royal Dental Coll., 1950-55. Mem. subcom. on cancer registration WHO, 1950-57, mem. panel on statistics, 1950-63, panel on cancer, 1963. Mem. Internat. Cancer Union (various coms. on geog. pathology), Am. Assn. Cancer Edn. (hon.), Am. Assn. Cancer Research (corr.). Author: X-Radiation and Heterotransplantation, 1938; Statistical Studies in the Aetiology of Malignant Neoplasms, 1965; also articles. Research on effect total-body irradiation on transplantation immunity; a rev. of cancer surveys compared with concepts of causes. Office: 49 Strandboulevarden, Copenhagen O., Denmark.*

CLEMSON, Thomas Green, Am. mineralogist; b. Phila., July 1, 1807; studied practical lab., Paris, at Sorbonne under Thénard, Gay-Lussac and DuLong; student lab. of Robiquet, from 1827, École des Mines Royale, 1828-32; m. Anna Maria Calhoun, 1838 (dec. 1850). Cons. engr. in Paris, France, Phila., Washington, D.C., 1832-39; asso. with father-in-law, John C. Calhoun, in so. agr. and gold mining, from 1839; U. S. charge d'affaires in Belgium, 1844-51; engaged in planting, assaying, also ofcl. rep. Belgian govt., 1853-61; a founder Md. Agrl. Coll., 1856; U. S. supt. agr., 1859-61; supr. mines and metal works Trans-Mississippi Dept., Confederate States Am., 1865; endowment in his will led to founding Clemson (S.C.) Agrl. Coll. Mem. many learned socs. Research and numerous articles on sci. agr., agrl. edn., mining, mineralogy. Died Ft. Hill, S.C., Apr. 6, 1888.

CLEOMEDES; Greek astronomer, 1st century A.D., disciple of Posidonios. Author of The Circular Theory of the Heavenly Bodies (summarizes the astron. sci. of Stoics). maintained that earth is spherical in form, that number of fixed stars is infinite, and that moon's rotation on its axis occurs simultaneously with its synodical revolution. Noted refraction of light.

CLEOSTRATOS OF TENEDOS, astronomer; b. Tenedos; flourished 538-432 B.C.; possibly a successor to Thales. Wrote an astron. poem. Introduced signs of the zodiac and 8 year cycle of intercalations; according to Hygin, he pointed out 1st of 2 stars in constellation of chariot (Chenreaux).

CLÉRAMBAULT, see de Clérambault.

CLERK, Sir Dugald, Brit. engr.; b. Glasgow, Scotland, Mar. 31, 1854; s. Donald Clerk; ed. W. of Scotland Tech. Coll., Andersonian Coll., Young Chair Tech. Chemistry, Glasgow, Yorkshire Coll. Sci., Leeds, Eng.; D.Sc., Manchester, Eng., also Leeds; LL.D., St. Andrews U., also Glasgow; D. Eng., Liverpool, Eng.; m. Margaret Hannay, 1883. Dir. engring. research Admiralty, 1916-17; cons. engr.; dir. Nat. Gas Engine Co. Ltd. Judge, Automobile Club Trials, Richmond Show, 1899, 1000 Miles Trials, 1900, Glasgow Reliability Trials, Glasgow Exhbn., 1901, Crystal Palace 1000 Miles Reliability Trials, 1903; Forrest lectr. Instn. Civil Engrs., 1904, 20; chmn. Delegacy City and Guilds Engring Coll., South Kensington, Eng., 1918-19; mem. adv. com. for aeros. also chmn. internal combustion engine com. Air Ministry, 1916-19; mem. Panel Bd. Invention and Research, 1915-19; mem. Air Inventions Com., 1918-19; mem. Univ. Grants Com.; mem. Court, Goldsmiths Co.; prime warden, 1926-27. Recipient President's medal Gas Inst., 1902; created knight comdr. Order Brit. Empire, 1917. Fellow Royal Soc., 1908 (Royal medal), Chem. Soc.; mem. Instn. Civil Engrs. (v.p., Watt medal 1882, Telford prize 1882, 86, Telford Gold medal 1907), Instn. Mech. Engrs., Royal Instn. (v.p.), Royal Soc. Arts (chmn. council 1915-17, Albert medal 1922); pres. Jr. Inst. Engrs., London, 1905-06, Soc. Brit. Gas Industries, 1906-08, Inc. Instn. Automobile Engrs., 1908-09, engring. sect. Brit. Assn., Dublin, Ireland, 1908, Instn. Gas Engrs., 1919-20. Author: The Gas, Petrol, and Oil Engine, 1909-13. also sci. papers and lectures. Inventor internal combustion engines, also Clerk cycle gas engine. Died Nov 12, 1932.

CLERSELIER, Claude, French physicist; b. 1614; lawyer at Parlement, 1647; treas. for Dept. of Auvergne; collected and published posthumous writings of Descartes. Author: Objections contre les méditations de Descartes, 1647. Editor: Lettres, 1667; Traité du homme; le Traité de la formation du foetus; le Traité de la lumière et le traité du monde, 1667 (all by Descartes). Died Paris, Apr. 3, 1684.

CLESSIN, Stephan, German paleontologist; b. Würzburg, Germany, Nov. 13, 1833; s. Joseph and Apollonia (Vornberger) C.; m. Ida Erhard, 1862; 2 sons, 1 dau. First became army officer; later employed by Bavarian R.R. service. Recipient Golden Linné medal Swedish Acad. Scis. Author: Deutsche Exkursions-hollusken-Frauna, 1877; Die Mollusken Fauna Osterreich, 1884; Ungarns und der Schweiz, 1887; also numerous articles. Pub., Malakaroologische Blätter, 1877-91. Studied molluscs, prehistoric shells, malacology, conchybiology. Died Regensburg, Germany, Dec. 21, 1911.

CLÈVE, Per Theodor, Swedish chemist, geologist; b. Stockholm, Sweden, Feb. 10, 1840; studied under Svanberg at Uppsala (Sweden) U., also studied in Paris, France; worked with Wurtz in Paris, then returned to Uppsala; adj. chemistry Stockholm Technol. Inst., 1870-74; prof. Uppsala. Chmn. of the com. that selected Nobel prize winners in chemistry. The mineral cleveite was named in his honor; recipient Davy medal, 1894. Author: Memoir on Ammoniacal Platinum Bases, 1872; Dictionnaire de Chimie, 1883; Analyse chimique qualitative, 1885; Scheele, 1886; (with Wurtz) Dictionnaire de chimie; (with Frémy) Encyclopedie chimique. Studied complex ammoniacal-platinum compounds and discovered their isomers, 1874; showed tetravalence of thorium and the properties of cerium, lanthanium, and erbium; predicted didymium might contain another element (neodymium, which was discovered by Welsbach, 1885); discovered thulium and holmium in erbia; research on scandium, North Sea plankton. Died Uppsala, June 18, 1905.

CLEVELAND, Forrest Fenton, Am. physicist, educator; b. Falmouth, Ky., Jan. 10, 1906; s. George Edward and Flora (Wyatt) C.; A.B., Transylvania Coll., 1927; M.S., U. Ky., 1930, Ph.D., 1934; postgrad. U. Chgo., U. Mich.; m. Marguerite Ewalt, Aug. 17, 1929. Tchr. pub. schs. Ky., 1927-30, 34-35; instr. physics U. Ky., 1930-34; prof. physics, head dept. Lynchburg Coll., 1935-39; asst. prof. physics Armour Inst. Tech., 1939-40; asso. prof. physics Ill. Inst. Tech., Chgo., 1940-43, prof., 1943——. Vis. prof. physics Ohio State U., summer 1947. Recipient research prize Va. Acad. Sci., 1939. Fellow A.A.A.S., Am. Phys. Soc.; mem. Am. Assn. Physics Tchrs., Am. Assn. U. Profs., Sigma Xi, Sigma Pi Sigma. Author: Experiments in Physics, A Laboratory Manual, 1938; (with others) General Physics Laboratory Manual, 1940. Asso. editor Jour. Chem. Physics, 1950-53; editor, pub. Spectroscopia Molecular, 1952——. Research, numerous publs. on absorption and scattering of light by molecules and deduction of geometrical arrangement of atoms, forces required to move atoms, strengths of bonds holding atoms, energies obtained when atoms come together to form molecules. Home: 1633 E. Hyde Park Blvd., Chgo. 60615.*

CLEVELAND, Sidney Earl, Am. psychologist; b. Boston, Jan. 22, 1919; s. Herbert Carlos and Edith (Willey) C.; A.B., Brown U., 1941; M.A., U. Neb., 1942; Ph.D., U. Mich., 1950; m. Marjorie Lucille Spacht, Nov. 27, 1942; children—John Alvin, Carol Tennant, Mark Eastman, Sarah Downing. With VA Hosp., Houston, 1950——, chief psychology service, 1962——; clin. prof. psychology Baylor U. Coll. Medicine, 1957——; pvt. practice clin. psychology, Houston, 1961——. Mem. Am., Southwestern, Interam., Tex. psychol. assns. Author: (with Seymour Fisher) Body Image and Personality, 1958; The Body Percept, 1965. Studies, numerous publs. investigation of body image and personality, role played by these variables in area of psychosomatic medicine. Home: 12021 Tall Oaks Lane, Houston 77024. Office: 2002 Holcombe Blvd., Houston 77031.*

CLEVENGER, Shobal Vail, neurologist; b. Florence, Italy, Mar. 24, 1843; s. Shobal Vail and Elizabeth (Wright) C.; brought to U. S., 1843; grad. Chgo. Med. Coll., 1879; m. Mariana Knapp (dec. 1910). Spl. pathologist Cook County Insane Asylum, Dunning, Ill., 1883; lectr. Chgo. Art Inst., Sch. Pharmacy, Law Sch.; supt. Ill. Eastern Hosp. for Insane, Kankakee, 1893. Author: Comparative Physiology and Psychology, 1884; Spinal Concussion, 1889; Medical Jurisprudence of Insanity, 1898; The Evolution of Man and His Mind, 1903; Therapeutics, Materia Medica, and the Practice of Medicine, 1905. Gave 1st description of inferior temporal sulcus of cerebral hemisphere, circa 1880. Died Mar. 24, 1920.

CLEWE, Thomas Hailey, Am. med. scientist; b. San Francisco, Sept. 9, 1925; s. Ernest George and Genevieve (Hailey) C.; B.S., Stanford, 1949, M.D., 1955; m. Geralyn Lambson, July 23, 1961; children—Jane Elizabeth, Carol Ann, Craig Thomas. Research asso. dept. obstetrics, gynecology Sch. Medicine Stanford, 1954-56; med. research fellow Population Council, dept. anatomy Sch. Medicine Yale, 1956-57, instr. anatomy, 1957-58; asst. prof. research human reprodn. U. Kan., 1958-61; asst. research prof. dept. obstetrics, gynecology Sch. Medicine Vanderbilt U., 1961-66; research asso. div. genetic and developmental disorders Delta Regional Primate Center, asso. prof. anatomy Sch. Medicine, Tulane U., New Orleans, 1966——. Recipient Rubin award Am. Soc. Study Sterility, 1959. Mem. Am. Assn. Anatomists, Am. Soc. Zoologists, Internat. Primatological Soc., Am. Fertility Soc., Internat. Fertility Assn., Tissue Culture Assn., N.Y. Acad. Scis., Soc. Sci. Study Sex, Soc. Gynecologic Investigation, Phi Beta Kappa, Sigma Xi. Publs. on devel. of techniques for study of secretory, ovum transport functions of fallopian tube; early stages of reprodn. Home: Bocage Royale Apts., Covington, La. 70433. Office: Delta Regional Primate Research Center, Covington, La. 70433.*

CLIFCORN, LaVerne Edward, Am. chemist; b. Rockford, Ill., Aug. 28, 1906; s. Edward F. and Wilhelmina (Mulder) C.; B.A., U. Wis., 1928, M.A., 1930, Ph.D., 1934; m. Bernice E. Dengel, Sept. 30, 1930; children—Joan B., (Mrs. Thomas Tephly), Charlene L. (Mrs. Elmer Ciesiel). Chemist, State of Wis., 1928-36; with Continental Can Co., 1936-55, dir. fundamental research div., 1950-55, asso. dir. research metal div., 1955; supr. container utilization research div. Am. Can Co., Barrington, Ill., 1955-58; mgr. research dept. Nat. Can Corp., 1958-61; dir. research Crown Cork & Seal Co., Phila., 1961——. Cons., U. S. Dept. Def., 1962——; chmn. gen. com. on foods NRC, 1960-66, Mem. Agrl. Research Inst. (v.p. 1954-55), Inst. Food Technologists (pres. 1954-56), Am. Chem. Soc., Contbr. articles to tech. jours., chpts. to books. Patentee in field. Developed stabilized iodized salt; co-inventor end over end agitating processes; research on nutritional value preserved foods, pub. health aspects metal container packaging. Home: 115 N. Ela Rd., Barrington, Ill. 60010. Office: 3501 W. 31st., Chgo. 60623.

CLIFFORD, Peter, surgeon; b. Dublin, Ireland, June 28, 1921; s. Cornelius and May (Tuomey) C.; M.B., B.Ch., B.A.O., U. Coll., Dublin, 1945, Diploma Laryngology and Otology, R.C.P. and S., Inst. Laryngology London, (Eng.), 1951; M.Ch., (N.U.I.), 1957; m. Jayne Elizabeth Hughes, Nov. 12, 1955; children—Sarah Jayne, Blaise Patricia, Senan Cornelius Ballard, Piers Patrick Ballard. House surgeon Coventry and Warwickshire Hosp., 1945-46, Royal Infirmary, Gloucester, Eng., 1946-47; med. officer, Tanganyika, 1949-50, Kenya, 1954-56; surg. register Royal Victory Infirmary, Newcastle-upon-Tyne, Eng., 1951-54; cons. surgeon dept. head and neck surgery Kenyatta Nat. Hosp., Nairobi, Kenya, 1958——; hon. lectr. surgery U. East Africa, 1964——; hon. asso. scientist Sloan Kettering Inst. for Cancer Research, N.Y.C., 1963——. Kenya assessor E. African Med. Research Council, 1963——. WHO fellow in cancer geog. pathology, 1960. Fellow Assn. Surgeons E. Africa (past pres.), Royal Soc. Medicine (Harrison prize 1963); mem. Am. Assn. for Cancer Research (corr.). Research, publs. on treatment of malignant disease of head and neck, surg. mgmt., biologic and chemotherapeutic treatment, epidemiological studies including Burkitts lymphoma and nasopharyngeal carcinoma. Address: Dept. Head and Neck Surgery, Kenyatta Nat. Hosp., Box 30024, Nairobi, Kenya, E. Africa.*

CLIFFORD, William Kingdon, English mathematician; b. Exeter, Eng., May 4, 1845; s. William and Mrs. (Kingdon) C.; ed. King's Coll., London, Eng., Trinity Coll., Cambridge Eng.; became fellow, 1868; m. Lucy Lane, Apr. 7, 1875; 2 surviving daus.; apptd. prof. math. U. London, 1871-79. Fellow Royal Soc., 1874. Author: Elements of Dynamics, 1879-87; Seeing and Thinking, 1879; Lectures and Essays, 1879; Mathematical Papers (edited by R. Tucker), 1882; The Common Sense of Exact Sciences (completed by Karl Pearson), 1885; also articles. Research on non-Euclidean geometry, theory of functions, theory of graphs, elliptical functions, universal algebra, biquaternions; generalized (with Reye) theory of polars, of curves, and of surfaces; classified loci; developed concept of consciousness as mind stuff. Died Madeira, Portugal, Mar. 3, 1879.

CLIFFORT, George, Dutch botanist; b. Amsterdam, Netherlands, 1685; established bot. garden, zoo, and mus. (with Linnaeus as dir.) at Hartecamp. Honored by Linnaeus in plant name cliffortie, also in publ. l'Hortus Cliffortianus, 1737. Died Amsterdam, 1750.

CLIFT, William, Brit. naturalist; b. Burcombe, Cornwall, Eng., Feb. 14, 1775; s. Robert C.; studied at sch., Bodmin, Cornwall; apprentice to John Hunter, physician; m. Caroline Amelia Pope, 1799; children—William Home, Caroline Amelia (Mrs. Richard Owen). Sec. to John Hunter, 1792-93; caretaker Hunter's collections, 1793-1844; osteologist and med. draughtsman. Fellow Royal Soc., 1823; mem. Chem. Soc. Compiled catalogue of osteology in Hunterian Mus; contbd. papers to Geol. Soc.'s Trans., 1829, 35, Philos. Trans., 1815, 23, Edinburgh New Philos. Jour. 1831; contbd. drawings to sci. books including ones of Matthew Baillie and Sir Everard Home. Died June 20, 1849.

CLIFTON, Charles Egolf, Am. microbiologist; b. Etna, O., Mar. 23, 1904; s. Allen Benton and Lulu (Egolf) C.; B.A., Ohio State U., 1925, M.Sc., 1926; Ph.D., U. Minn., 1928; m. Esther Ora Carlson, Sept. 7, 1932; children—Charles Egolf, John Peter. Fellow, Mayo Found., 1927-28; research chemist Eastman Kodak Co., 1929; faculty Stanford (Cal.), 1929——, prof. microbiology, 1940——; Gen. Edn. Bd. fellow Cambridge (Eng.) U., 1936-37, Technische Hogeschool, Delft, Holland, 1936-37. Mem. Am. Soc. Microbiology, A.A.A.S., Soc. for Exptl. Biology and Medicine, Soc. Gen. Physiologists, Sigma Xi. Author: Introduction to the Bacteria, 1950; Introduction to Bacterial Physiology, 1957; also numerous articles. Editor, Ann. Rev. Microbiology, 1947, 67. Research on size filtrable viruses, growth bacteria, metabolism anaerobic bacteria, aging bacteria. Home: 421 Van Buren St., Los Altos, Cal. 94022. Office: Dept. Med. Microbiology, Stanford, Stanford, Cal. 94305.*

CLIFTON, Kelly Hardenbrook, Am. radiobiologist; b. Spokane, Wash., July 22, 1927; s. John Minton and Nora Marie (Toole) C.; B.A., U. Mont., 1950; M.S., U. Wis., 1951, Ph.D., 1955; m. Mayre-Lee Harris, Aug. 27, 1949; children—Kelly Hardenbrook, William Harris, Brice Minton. Am. Cancer Soc. fellow Children's Cancer Research Found., Boston, 1955-56, research asso., 1956-59; research fellow pathology dept. Med. Sch. Harvard, 1957-59; faculty Med. Sch. U. Wis., Madison, 1959——, prof. radiobiology, 1967——, chmn. undergrad. biology core curriculum com., 1967——. Mem. Am. Assn. Cancer Research, Soc. Exptl. Biology and Medicine (past sect. chmn.), Radiation Research Soc., Am. Soc. Exptl. Pathologists, Am. Soc. Zoologists. Editorial adv. bd. Cancer Research, 1965——. Described cell type of origin and physiol. mechanism of induction of mammotropin-secreting pituitary tumors resulting from long-term estrogen treatment, radiation effects on model solid carcinoma systems in mice. Home: 1218 University Bay Dr., Madison, Wis. 53705.*

CLINARD, Marshall Barron, Am. sociologist; b. Boston, Nov. 12, 1911; s. Andrew Marshall and Gladys (Barron) C.; B.A., Stanford, 1932, M.A., 1934; Ph.D., U. Chgo., 1941; m. Ruth Blackburn, Aug. 28, 1937; children—Marsha (Mrs. Richard Schacht), Stephen A. Faculty, U. Ia., 1937-41, Vanderbilt U., 1945-46; with U. S. Bur. Census, 1941-43, OPA, 1943-45; faculty U. Wis., Madison, 1946——, prof. sociology, 1951——. Fulbright research prof., Sweden, 1954-55; cons. urban community devel. Ford Found., India, 1958-60, 62-63; cons. U. S. Dept. Labor, 1966-67. Mem. Am. Sociol. Assn. (past mem. council), Midwest (pres. 1965-66), Indian social. socs., Soc. for Study Social Problems (past pres.), Sociol. Research Assn., Internat. Criminological Soc. Author: The Black Market, 1952; Sociology of Deviant Behavior, 1957, rev., 1962, 68; Anomie and Deviant Behavior, 1964; Slums and Community Development, 1966; (with Richard Quinney) Criminal Behavior Systems, 1967. Research, publs. on analysis and explanation of deviant behavior, occupational crime, sociology of slum; constructed typol. analysis for criminal behavior. Home: 6022 Green Tree Rd., Madison, Wis. 53711.*

CLINE, Isaac Monroe, Am. meteorologist; b. Madisonville, Tenn., Oct. 13, 1861; s. Jacob Leander and Mary Isabel (Wilson) C.; A.B., Hiwassee College, 1882, A.M., 1885; M.D., U. Ark., 1885; Ph.D., Tex. Christian U., 1896; Sc.D., Tulane U., 1934; m. Cora M. Ballew, Mar. 17, 1887 (dec. Sept. 1900); children—Allie May (Mrs. Ernest E. B. Drake), Rosemary (Mrs. Vora Williams), Esther Ballew (Mrs. Albert Allen Jones). Entered U. S. Weather Service (then Signal Corps, U. S. A.), 1882; asst. observer, Little Rock, 1883-85; in charge of observation sta., Abilene, Tex., 1885-89, Galveston, Tex., 1889-91; local forecaster and sect. dir. Tex. Sect. Climatol. Service, Weather Bur. of U. S. Dept. of Agr., 1891-1901; at New Orleans, 1901-35, in charge forecast center embracing Tex., Okla., Ark. and La.; also in charge coöperation between Mexican Weather Service and U. S. Weather Bur.; prin. meteorologist U. S. Weather Bur., ret., 1935. Instr. climatology U. Tex., 1897-1901. Del. 2d Pan-Am. Sci. Congress, Washington, 1915; mem. Union Goedesique et Geophysique, Commission pour l'Etude des Raz de Maree. Fellow Am. Meteorol. Soc. (pres. 1934-35), Am. Geog. Soc., A.A.A.S.; mem. Nat. Inst. Social Scis. Author of many bulls. and articles on climate of S.W., its effect on health and on agr., including Relation of Storm Tides to the Center and Movement of Tropical Hurricanes (a contbn. to knowledge and forecasting of hurricanes), Tropical Cyclones (Introducing the integration method for 1st time in study of storms, and presenting new conclusions which define and describe cyclone characteristics), 1926; author Characteristics of Tropical Cyclones. Died Aug 3, 1955.

CLOËTTA, Max, pharmacologist; b. Zurich, Switzerland, July 21, 1868; s. Arnold Leonhard and Maria (Locher) C.; studied medicine at Geneva, Switzerland, Zurich, Strasbourg, France; m. Julie Lucy Ann Spöndling, 1899; 3 sons, 2 daus. Asst., Strasbourg; asso. prof. pharmacology at Zurich, 1901, prof., 1907; rector, 1914; wrote numerous articles and essays many of which appeared in Archiv für experimentelle Pathologie und Pharmakologie; isolated diagalen (digitalis preparation suitable for injection), 1904; discovered uroprotein acid; research on drug addiction, dynamics of blood circulation in lungs, exptl. pharmacology. Died Zurich, June 23, 1940.

CLOIZEAUX, Jacques Marie Henri le Grand des, French physicist; b. Evreux, July 9, 1928; s. Roger and Marie (de Maupeau) C.; ed. Ecole normale superieure, Paris; Ph.D. in sci.; m. Nicole de Cremoux, Oct. 21, 1957; children—Cecile, Etienne, Francois. Engr., French Commisariat of Atomic Energy, Center Nuclear Studies, Saclay. Mem. French Soc. Physics, Am. Phys. Soc. Contbr. articles to physics publs. Home: 186, Bat M., av. Aristide-Briand, Antony (Seine). Office: Cen-Saclay, boite postale 2, Gif-sur-Yvette (S.-et-O.), France.

CLOOS, Ernst, geologist; b. Saarbrücken, Germany, May 17, 1898; s. Ulrich and Elisabeth (Heckel) C.; student univs. of Freiburg and Göttingen; Ph.D., U. of Breslau, 1923; m. Margaret Spemann, Dec. 27, 1923; children—Gisela, Veronica. Came to U. S., 1929, naturalized 1938. Geologist, Seismos Co., Hannover, Germany, 1924-29, in charge geophysical exploration in Tex., 1924-26, Eng., 1927, Germany and Iraq, 1927-28; conducted investigation Sierra Nevada granites, 1929-30; lecturer Johns Hopkins U., 1931-37, asso. prof., 1937-41, prof. since 1941. Chmn. Commn. Md. Dept. Geology, Mines and Water Resources, 1962. Guggenheim fellow, 1956-57, Fellow Geological Soc. of London; member Nat. Academy of Sciences, Geological Society Am. Geophys. Union A.A.A.S., Am. Philos. Soc., Geol. Soc. London, Geol. Soc. Finland, Geologists Assn., Phi Beta Kappa, Sigma Xi. Condr. research expdns. Mesopotamia, Persia, Cal. and Tex. Research in structural geology of Appalachians and Petrofabrics. Home: 610 Fairway Dr., Towson 4, Md. Office: Johns Hopkins University, Balt. 18.

CLOOS, Hans, German geologist; b. Magdeburg, Germany, Nov. 8, 1885; s. Ulrich and Elisabeth (Heckel) C.; studied natural scis. at Bonn, Jena, Freiburg (all Germany); m. Elli Gruters; 4 children. Oil geologist on expdn. to S.W. Africa, 1909, Netherlands Indies, 1911-13; prof. Breslau, Germany, 1919, Bonn, 1926; trips to Scandinavia, 1925-26, U. S., 1927, 33, 48, South Africa, 1929, 36. Recipient Penrose medal, 1948; Leopold von Buch plaque. Mem. Geol. Assn. (chmn. 1931), numerous acads. Author: Gespräch mit der Erde, 1947; Das Mechanismus tiefvulkanischer Vorgänge, 1921; also numerous articles. Chief editor Geologische Rundschau, from 1923. Pioneer in granite tectonics; one of 1st to use true scale models in deciphering mechanics of faulting; studied structure and devel. of continents. Died Bonn, Sept. 26, 1951.

CLOQUET, Hippolyte, French physician; b. Paris, France, Mar. 10, 1787; studied under Cuvier; took internship, 1810; defended doctorate thesis, 1815; aggregation, 1821. Prof., Royal Athenaeum; distinguished prof. anatomy. Mem. Acad. Medicine. Author: l'Encyclopédie méthodique, le systeme anatomique, 4 vols.; Mémoire sur les ganglions nerveux des fossess nasales, 1818; Osphrésiologie, ou traité des odeurs, 1821; (with Vicq d'Azyr) Faune des medecins, 6 vols., 1822-28. Described swelling of nasopalatine nerve in the anterior palatine canal (Cloquet's ganglion), 1818, deviation of nasal septum, 1821, vasomotor rhinitis, 1821. Died Paris, 1843.

CLOQUET, Jules Germain, French physician, surgeon; b. Paris, France, Dec. 28, 1790; s. Jean-Baptiste-Antoine C.; studied medicine at Rouen, France, later at Val de Grâce Hosp., Paris; docteur en médecine, 1817; aggregation in surgery, 1824. Became substitute prof. for A. Dubois, Faculty of Paris, 1827, named prof. anatomy, 1831, apptd. prof. of clin. surgery, 1834, and later of surg. pathology; operated at St. Antoine Hosp.; surgeon to emperor. Mem. French Acad. Scis., 1855. Author: Anatomie de l'homme, 5 vols., 1821-31; Anatomie des vers intestinaux, 1824; Manuel d'anatomie descriptive, 3 vols., 1825; Pathologie chirurgicale, 1831. Gave 1st description of the central canal of vitreous (hyaloid canal), circa 1821; studied treatment for rupture of perineum, harelip operations, hernias, capillary vessels of iris. Died Paris, Feb. 24, 1883.

CLOS, Dominique, French botanist; b. Sorèze, France, May 25, 1821; s. Jean-Antoine C.; licencié ès sciences and docteur en médecine (with thesis The paratids), 1845; prof. botany Faculty of Toulouse; dir. Garden of Plants, Toulouse, 1853; ret. in 1889; mem. French Acad. Scis., 1881. Author: Les plantes au point de vue litteraire, 1868; Essai de tératologie taxinomique, 1871; Des stipules et de leur role a l'inflorescence et dans la fleur, 1878; Nouvel apercu sur la théorie des inflorescences; also numerous articles on bot. theory, various parts of plants, vegetal teratology, Mediterranean bot. zones, sci. agr., botany of ancients. Died Aug. 19, 1908.

CLOSE, Sir Charles (Frederick), see Arden-Close, Sir Charles (Frederick).

CLOSE, Maxwell Henry, Irish geologist; b. Merrion Square, Dublin, Ireland, Oct. 23, 1822; s. Henry Samuel and Jane (Waring) C.; ed. at Weymouth, Trinity Coll.; B.A., 1846, M.A., 1847; began as cleric; later became geologist; pres. Royal Geol. Soc. Ireland, 1878-79; treas. Royal Irish Acad., 1878-1903; mem. council Royal Dublin Soc. Author (under assumed names): Ausa dynamica: Force, Impulsion, and Energy (John O'Toole), 1884; A Few Chapters in Astronomy (Claudius Kennedy), 1894; Notes on General Glaciation in Ireland; The General Glaciation of Iar-Connaught. Studied glaciation and its effects on Ireland. Died Sept. 12, 1903.

CLOT-BEY (Antoine Barthelemy), French physician; b. Grenoble, France, Nov. 5, 1793; s. Louis and Marie (Bérard) C.; docteur en medecine Faculty Medicine, Montpellier, France, 1817; docteur en chirurgie, 1823. Became asst. physician Hosp. Charity, Marseilles, France, 1822; called to Egypt, 1825, where he became physician for pasha of Egypt, organized health council and san. service there; Hosp. of Abou-Zabel, nr. Cairo, Egypt, built for him, organized sch. medicine there; transferred this sch. to Cairo, 1833; served as physician during bubonic plague and cholera epidemics; an organizer of pub. instrn. in Egypt; after death of Méhémet-Ali returned to Marseilles, contbg. to state collection of Egyptian antiques he brought with him, 1852. Author many publs. including: De la peste observée in Egypte, 1840; Reorganisation de service medicale civil et militaire en Egypte en 1856, 1862; De l'ophtalmie et de la vaccination, coup d'oeil sur la peste, 1840; Mémoires, edited in 1949. Introduced smallpox vaccination to Egypt, 1832. Died Aug. 20, 1868.

CLOUD, Marshall Morgan, Am. surgeon; b. Carroll County, Va., Oct. 9, 1868; s. Columbus Henry and Mary Emily (Parker) C.; M.D., honor medalist, U. Kan., 1892; grad. U. S. A. Med. Sch., 1897; B.S., U. So. Cal., 1904, A.M., 1906; grad. study Stanford, 1905, U. Chgo., 1906; m. Mary Frances Moore, June 19, 1894; children—Dorothy (Mrs. Frederick B. Pinkus), Marguerite (Mrs. Allison J. Wallace, Jr.), Mary, Frances. Asst. supt. Kan. State Hosp., Topeka, 1893-95; clin. prof. ophthalmology, U. So. Cal., 1910-13, prof. mil. medicine, 1920-23; ophthalmologist Nat. Soldiers Home, Sawtelle, Cal., 1910-13, Santa Fe Ry., 1910-21; examining surgeon, U. S. Pension Bur.; on staff Hollywood Hosp. and Gen. Hosp., Los Angeles. Fellow A.C.S. Author: Sanitary Analysis of Water, 1905; Curing Our Nerves, 1934; Facts About Alcoholic Drinks, 1934. Inventor, with M. F. Volkman, of horizontal-base range finder for artillery fire, 1904. Died 1937.

CLOUD, Preston E., Jr., Am. geologist, educator; b. West Upton, Mass., Sept. 26, 1912; s. Preston E. and Pauline (Wiedemann) C.; B.S., George Washington U., 1938; Ph.D., Yale, 1940; m. Frances Webster, June 1, 1952; children—Karen Frances, Lisa Anne, Kevin Bruce. Instr. geology Mo. Sch. Mines, 1940-41; research fellow Yale, 1941-42; geologist U. S. Geol. Survey, 1942-46, 48-51; asst. prof., curator invertebrate paleontology Harvard, 1946-48; prof. geology U. Minn., 1961-65, chmn. dept. geology, 1961-63, head sch. earth scis., 1962-64; prof. geology U. Cal., Los Angeles, 1965—. Recipient A. Cressey Morrison award, 1940, Rockefeller Pub. Service award, 1956, Distinguished Service award U. S. Dept. Interior, 1959. Mem. Nat. Acad. Scis., Soc. Study Evolution, Geol. Soc. Am., Paleontol. Soc. Am., Phi Beta Kappa, Sigma Xi, numerous others. Author: Terebratuloid Brachiopods of the Silurian and Devonian, 1942; (with V. E. Barnes) The Ellenburger Group of Central Texas, 1948; (with R. G. Schmidt, H. W. Burke et al) Geology of Saipan, Mariana Islands, 1959; (with others) Environment of Calcium Carbonate Deposition West of Andros Island, Bahamas, 1962. Contbr. numerous articles to profl. jours. Research into patterns of evolution among fossil invertebrates, brachiopod phylogeny, regional geology, deposition and geochemistry of modern carbonate sediments, reconstrn. past biol. environments, past geography of marine organisms, origin and earliest records of life, evolution of atmosphere and lithosphere. Office: Dept. Geology, U. Cal., Los Angeles 90024.*

CLOUDSLEY-THOMPSON, John Leonard, zoologist; b. Murree, India, May 23, 1921; s. Ashley G. G. and Muriel E. (Griffiths) Thompson; B.A., Pembroke Coll., Cambridge U., 1946, M.A., 1948, Ph.D., 1950; D.Sc., U. London, 1960; m. Jessie Anne Cloudsley, 1944; children—John Hugh, Timothy, Peter Leslie. Lectr. zoology King's Coll., U. London, 1950-60; prof. zoology U. Khartoum (Sudan), 1960—; keeper Sudan Natural History Mus. Mem. Cambridge Iceland Expdn., 1947, Expdn. to So. Tunisia, 1954, expdns. to various parts Central Africa, 1960—. Named Hon. capt., freeman City of London, also Liveryman, Worshipful Co. of Skinners. Fellow Linnean Soc., Royal Entomol. Soc. London, Zool. Soc., Inst. Biology, World Acad. Art and Sci.; mem. Biol. Council (past hon. sec.), Soc. Expt. Biology (past mem. council), Royal Soc. (past mem. council), London Natural History Soc. (past mem. council), Internat. Soc. Biometeorology (past mem. exec. bd.), Soc. for Exptl. Biology. Author: Biology of Deserts, 1954; Spiders, Scorpions, Centipedes and Mites, 1958; Animal Behaviour, 1960; (with John Sankey) Land Invertebrates, 1961; (with M. J. Chadwick) Life in Deserts, 1964; Animal Conflict and Adaptation, 1965; Desert Life, 1965; Animal Twilight, 1967. Research on sensory physiology, ecology and behaviour of terrestrial arthropods, especially in relation to water relations and diurnal rhythms of activity, adaptations of desert animals to their environments. Home: Glendooen, 10, Lower Green Rd., Esher, Surrey, Eng. Office: Dept. Zoology, U. Khartoum, Sudan.*

CLOUET, Jean-François (-Louis), French chemist, metallurgist; b. Singly, France, Nov. 11, 1751; ed. mil. sch., also pupil of Monge at Mezières, France. Founded factory, Singly; then became tchr. chemistry, Mezières; in charge founderies of Chouvancy, Givonne, Villancy, Daigny, during French Revolution; immigrated to Cayenne, French Guiana, 1799; prof. Sch. Engring., Cayenne. Mem. French Acad. Scis., 1796. Contbd. to Journal de physique, Annales des mines, Annales de Chimie. Developed method for fabricating cast steel or forged iron (used in Daigny factory to furnish arms at arsenals of Douai and Metz for nat. def.); showed that steel is formed by heating wrought iron and diamond powder together, 1799; worked (with Lavoisier) on saltpeter; confirmed Scheele's observation that passing ammonia gas over strongly heated charcoal causes prussic acid to form; worked on composition of enamels; wrote on manufacture of laminated figures; also interested in botany and effects of different climates. Died on voyage to Cayenne, June 4, 1801.

CLOUSTON, Sir Thomas Smith, physician; b. Orkney, Apr. 22, 1840; s. Robert Clouston; ed. Aberdeen U.; M.D., LL.D.; LL.D., Edinburgh U.; m. Harriet Segur Storer; 2 sons, 1 dau. Physician supt. Royal Asylum, Morningside, Edinburgh; lectr. mental diseases Edinburgh U.; med. supt. Cumberland and Westmorland Asylum. Mem. Med. Psychol. Assn. (past pres.), Medico-Chirurg. Soc. Edinburgh (past pres.), Royal Coll. Physicians Edinburgh (pres. 1902-03). Author: Clinical Lecturers on Mental Diseases; The Neuroses of Development; An Asylum or Hospital House with Plans; The Hygiene of Mind; Unsoundness of Mind. First to demonstrate relationship between congenital syphilis and dementia paralytica, 1877. Died Apr. 19, 1915.

CLOUTIER, Gilles Georges, physicist; b. Quebec, Can., June 27, 1928; s. Phileas and Valeda (Nadeau) C.; B.A., Laval U., 1949, B.A.Sc., 1953; M.Sc., McGill U., 1956, Ph.D., 1959; m. Colette Michaud, May 1, 1954; children—Hélène, Suzanne, Pierre. Research officer C.A.R.D.E., Valcartier, Que., Can., 1953-54; sr. mem. sci. staff RCA Research Labs., Montreal, 1959-63; asso. prof. physics U. Montreal, 1963-67, prof. physics, 1967—. Recipient Can. Centennial medal. Mem. Canadian Assn. Physicists, Am. Phys. Soc., I.E.E.E. Author articles. Research on microwave optics, electromagnetic waves propagation, plasma physics. Home: 485 Valery, Brossard, Que. Office: Dept. Physics, Univ. Montrea, Montreal, Que., Can.*

CLOWER, Robert Wayne, Am. economist; b. Pullman, Wash., Feb. 13, 1926; s. Fay Walter and Mary (Gilchrist) C.; B.A. in Econs., Wash. State U., 1948, M.A., 1949; B.Litt in Econs., Oxford (Eng.) U., 1952; m. Frances Hepburn, Jan. 7, 1946; children—Ailsa, Leslie, Robert, Stephanie, Valerie. Instr., Wash. State U., 1947-49, asst. prof., 1952-57; asst. tutor New Coll., Oxford U., 1949-52; vis. prof. econs. U. Punjab, Lahore, West Pakistan, 1954-56; prof. econs. Northwestern U., Evanston, Ill., 1957—; dir. econ. survey Liberia, Monrovia, 1961-62; Michaeimas Term vis. prof. Cambridge (Eng.) U., 1962; John Maynard Keynes vis. prof. U. Essex, Colchester, Eng., 1965-66. Cons., Commonwealth P.R. Ports of Authority, 1964; chmn. dept. econs. Northwestern U., 1958-64. Rhodes scholar, 1949-52; Nuffield student, 1950-52; Ford Found. Faculty Research fellow, 1960-61; Guggenheim Found. fellow, 1965-66. Mem. Am. Econ. Assn., Royal Econ. Soc., Econometric Soc., Am. Statis. Assn. Author: (with D. W. Bushaw) Introduction to Mathematical Economics, 1957; (with J. F. Due) Intermediate Economic Analysis, 1961, 66; (with Dalton, Walters, Harwitz) Growth without Development, 1966; also articles. Editor: Am. Econ. Rev., 1962-65. Research on theory of money and prices especially stock-flow aspects of market behavior on micro founds. Keynsian econs., pure theory of money economy, theory of savs. Home: 2138 Orrington St., Evanston, Ill. 60201; also Harvey's Farm, Peldon, Essex, Eng.*

CLOWES, Frank, chemist; b. Bradford, Yorkshire, 1848; s. Francis and Mary (Low) C.; ed. Royal Sch. Mines, London, Royal Coll. Sci., Dublin, U. Würzburg; D.Sc., U. London; m. Mabel H. Waters, 1877. Lectr. chemistry and physics with lab. Queenwood Coll., Hants.; prof. chemistry Univ. Coll., Nottingham, 1881-97, also 1st prin., emeritus prof. chemistry and metallurgy; dir. London County council Chem. Staff and Labs., 1897-1913; gov. Dulwich Coll.; expert adviser gas supply Corp. of London; visitor Royal Inst., 1903-04, 19-20. Fellow Royal Inst. Chemists (v.p. 1901-03, 11-13), Chem. Soc.; mem. Brit. Assn. (v.p. chemistry sect. 1893-1914), Soc. Chem. Industry (pres. 1879-98), Instn. Civil Engrs. (Forest lectr. 1901), Midland Counties Inst. Engring. (hon.), Chester Soc. Sci. (hon.), Royal Instn. Author: Textbook on Qualitative Analysis, 9th edit., 1874; Textbook on Quantitative Analysis, 12th edit., 1891; Elementary Practical Chemistry, Part I, 7th edit., 1896, Part II, 9th edit.; The Detection and Estimation of Inflammable Gases and Vapours in the Air, 1896, also numerous articles. Died Dec. 18, 1923.

CLOWES, George Henry Alexander, chemist; b. Ipswich, Eng., Aug. 27, 1877; s. Josiah Pratt and Ellen (Seppings) C.; student Royal Coll. Sci., London, Göttingen U., Berlin U., Pasteur Inst. (Paris); Ph.D., Göttingen, 1899; D.Sc., Butler U., 1931; LL.D., Wabash Coll., 1938; m. Edith Whitehill Hinkel, June 1910; children—Alexander Temple (dec.), George H. A., Allen Whitehill. Came to U. S., 1900, naturalized, 1921. With Eli Lilly & Co., Indpls., 1918-58, dir. research lab., Indpls., Woods Hole, Mass., 1920-46. Recipient Banting Medal, 1947. Mem. Am. Chem. Soc. for Exptl. Biology and Medicine, Biochem. Soc., Cancer Assn., Immunologists and Pathologists, Chem. Soc. (Eng.). Research in cancer; coöperated in devel. of insulin, liver extract, penicillin, etc. Died Aug. 25, 1958.

CLOWES THE ELDER, William, Brit. surgeon; b. Kingsbury, Warwickshire, Eng., circa, 1540; s. Thomas C.; surgery apprentice to George Keble (London surgeon); army surgeon in France, 1563; naval surgeon, 1563-69, 88; practiced surgery, London, 1569; surgeon St. Bartholomew's Hosp., 1581-85, also Christ's Hosp.; army surgeon in Low Countries, 1585-87; returned to practice in London; published surg. treatises, 1579-1602; surgeon to queen; asst. in ct. of Barber-Surgeons Co., 1588; his books (best surg. works of Elizabethan Eng.) are De Morbo Gallico, 1579; Proved Practise for all young Chirurgians, 1591; Treatise of the French or Spanish Pocks by John Almenar, 1591; A Profitable and Necessary Book of Observation, 1596; A Right Frutefull and Approved Treatise for the Artificiall Cure of the Struma or Evill, cured by the Kinges and Queenes of England, 1602. Died 1604.

CLUFF, Leighton Eggertsen, Am. physician; b. Salt Lake City, June 10, 1923; s. Lehi Eggertsen and Lottie (Brain) C.; B.S., U. Utah, 1944; M.D., George Washington U., 1949; m. Beth Allen, Aug. 19, 1944; children—Claudia Beth, Patricia Leigh. With Johns Hopkins Hosp., 1949-52; vis. investigator, asst. physician Rockefeller Inst. for Med. Research, 1952-54;

Instr., resident physician dept. medicine Johns Hopkins U. and Johns Hopkins Hosp., 1954-55, asst. prof., 1955-57, asso. prof., 1957-64, prof., 1964-66, head div. allergy and infectious diseases, 1964-66, physician Hosp., 1955—; prof., head dept. medicine U. Fla. Sch. Medicine, Gainesville, 1966—. Cons. Army Chem. Corps Biol. Lab.; mem. Armed Forces Epidemiology Bd. Commn. on Streptococcal and Staphylococcal diseases; chmn. tng. grant com. Nat. Inst. Allergy and Infectious Diseases NIH; drug research bd. NRC-Nat. Acad. Scis. Markle scholar in med. scis. Mem. Am. Soc. Clin. Investigation, Assn. Am. Physicians, Soc. Exptl. Biology and Medicine, Am. Assn. Immunologists, So. Soc. Clin. Investigation, A.C.P. Contbg. author: (with J. H. Mulholland) Symposium on Endotoxin, 1964; (with F. R. Fekety) Maxcy's Principles of Public Health and Epidemiology, 1965; (with J. C. Allen) Prevention of Infection, 1965; (with J. E. Johnson) Section III Hypersensitivity Responses to Infectious Agents, Immunological Diseases, 1965; (with P. S. Norman) Drugs of Choice, 1965-66; others. Studies, publs. on mechanisms of fever and action of bacterial toxins; source, control and mechanisms of staphylococcal disease; frequency, types and causes of untoward reactions to drugs; mechanisms of resistance to infections; effects of psychol. factors in infections. Home: 3217 N.W. 18th Av. Office: U. Fla. Coll. Medicine, Gainesville, Fla. 32601.*

CLUSIUS, Carolus, (or Charles de l'Ecluse), botanist; b. Arras, France, Feb. 19, 1526; s. Michel and Guillemette (Quincaailt) de l'Ecluse; studied law, Louvain, Belgium, 1546-48; licensed jurist, 1548; studied medicine, Marburg, Germany, 1548-49; also student in Wittenberg, Germany; M.D., Montpellier, France, 1555; postgrad., Paris, France, 1560-62. Lived in S. Netherlands, 1555-60; visited Spain and Portugal, 1564, 65, Austria, Hungary, then Eng., 1561, 71, 79, 80; dir. emperor's garden, Vienna, Austria, 1573-87; apptd. prof. botany, Leyden, Netherlands, 1593. Author: Rariorum aliquot stirpium per Hispanas observatarum historia, 1576; Rariorum aliquot stirpium per Pannoniam, Austriam et vicinas historia, 1583; Rariorum plantarum historia, 1601; Fungorum historia, 1601; Exoticorum libri decem, 1605; also many transls., maps, commentaries on history and geography. Helped found sci. botany; established, preserved and developed garden cultures; cultivated new, fng. and exotic plants, including chestnut, 1576, potato, 1588, Primulae, others. Died Leyden, Apr. 4, 1609.

CLUVER, Philip (also Cluverius or Cluwer), geographer; b. Danzig (now Poland), 1580; ed. Leyden, Netherlands; travelled to Eng., France, Germany, Poland, Italy in pursuit of information for writings; settled Netherlands, pensioner Leyden Acad., after 1616. Author: Germania Antiqua, 1616; Siciliae Antiquae libri duo, Sardinia et Corsica Antiqua, 1619; Italia Antiqua (pub. posthumously), 1624; others. Recognized by some as founder of hist. geography. Died Leyden, 1623.

CLYMER, George E., Am. inventor; b. Bucks County, Pa., 1754; m. Margaret Backhouse, 3 daus. A carpenter by trade; lived in Phila.; spent last years in Eng. distributing his "Columbian" printing press. Devised pump to clear cofferdams for 1st permanent bridge across Schuylkill River; invented Columbian (an improved printing press; 1st outstanding Am. invention in printing field), 1816; recipient gold medal for invention from King of Netherlands. Died London, Eng., Aug. 27, 1834.

COACHMAN, Lawrence Keyes, Am. oceanographer; b. Rochester, N.Y., Apr. 25, 1926; s. Kendrick Powell and Grace (Lewis) C.; B.N.S., Tufts Coll., 1945; A.B., Dartmouth, 1948; M.F., Yale, 1951; Ph.D., U. Wash., 1962; m. Nancy Bosworth Park, Oct. 24, 1953; children—Martha Lincoln, Katherine Atwood, Mary Lawrence (dec.), John Kendrick. Oceanographer, hydrographer Blue Dolphin Labrador Expdns., Hanover, N.H., 1950-55; sr. scientist Zoofysiologisk Inst., Oslo, Norway, 1955-57; research, teaching asst. dept. oceanography U. Wash., Seattle, 1957-62, asst. prof., 1962-66, asso. prof., 1966—; chmn. U. S. Oceanographic Exchange Delegation to USSR, 1964. Fellow Arctic Inst. N.Am.; mem. Am. Geophys. Union, A.A.A.S., Am. Soc. Limnology and Oceanography. Constructed gen. model water masses and circulation Arctic Ocean; research on methods for extraction and analysis gases from glacier ice, age determination glacier ice. Home: 3815 46th Av. N.E., Seattle 98105.*

COANDA, Henri-Marie, aero. engr.; b. Bucharest, Rumania, June 6, 1885; s. Constantin Coanda; ed. Charlottenburg Tech. Highsch., Berlin, Liège (Belgium) U., École Supérieure électricité, Montfiore; studied sculpture under Rodin; Ph.D. in Engring., other degrees in aero-engring., elec. engring., refrigerator engring. Chief engr. Bristol Aircraft; turned to research after various bus. ventures; worked in Paris, 1933-37. Conducted air flow expts. by riding cowcatcher of a locomotive at night, 1906; designer, pilot of 1st jet plane, 1910; originated Coanda effect (used in prodn. of saucer-type aircraft), 1937-38; developed machine for desalinating sea water.

COATES, Donald Robert, Am. geologist; b. Grand Island, Neb., July 23, 1922; s. Frank Jefferson and Harriet (Ferris) C.; B.A., Coll. Wooster, 1944; M.A., Columbia, 1948, Ph.D., 1956; m. Jeanne Louise Grandison, Mar. 18, 1944; children—Cheryl D., Donald Eric, Lark J. Faculty Earlham Coll., Richmond, Ind., 1948-51; geologist, project chief U. S. Geol. Survey, Tucson, Ariz., 1951-54; faculty Harpur Coll. (now State U. of N.Y.), Binghamton, 1954—, chmn. dept. geology, 1954-63, prof., 1963—. Soil scientist Charles Kettering Found., Richmond, Ind., 1948-51; research geologist U. S. Geol. Survey, Vestal, N.Y., 1958-61; vis. geoscientist Am. Geol. Inst., 1963-65; cons. attys., Binghamton, N.Y., 1955—, NSF, 1964—, N.Y. State Atty. Gen., 1965—. Recipient NSF award for Sustained Superior Performance, 1964. U. S. Geol. Survey research grantee, 1958-61. Fellow A.A.A.S., Geol. Soc. Am.; mem. Am. Geophys. Union, Nat. Assn. Geology Tchrs. (pres. Eastern sect. 1962), Phi Beta Kappa. Editor: Geology of South-Central New York, 1963. Contbr. to Science—A Process Approach, 1965, also articles, reports. Pioneer in modern statis. geomorphology; devel. new mapping techniques for geology field measurements; analog system for computation of pre-glacial topography in N.Y.; multicycle theory for geomorphic evolution of Finger Lakes region; produced 1st math. computerized program to analyze system of related variables of hydrology, geomorphology, and geology. Home: 212 Edgewood Rd., Vestal, N.Y. 13850. Office: Dept. of Geology, State U. of N.Y. at Binghamton, Binghamton, N.Y. 13901.

COATNEY, G(eorge) Robert, Am. pharmacologist, malariologist; b. Falls City, Neb., May 3, 1902; s. Edward Ernest and Ida Virginia (Banner) C.; A.B., Grand Island (Neb.) Coll., 1925; A.M., U. Neb., 1926, Sc.D. (hon.), 1963; Ph.D., Ia. State U., 1932; Sc.D., Bowling Green (Ohio) State U., 1958; Sc.D. (hon.), University of Nebraska, 1963; m. Eva Mae Rice, May 2, 1929; children—Cathryn Ann, John Edward. Faculty Grand Island Coll., 1926-30, Neb. Wesleyan U., 1931-33, Peru (Neb.) State Tchrs. Coll., 1933-38; with NIH, 1938—, sr. protozoologist, 1942-45, sec. malaria study sect., 1946-48, head chemotherapy sect. Lab. Tropical Disease, 1947-57, asst. chief, 1955-58, acting chief, 1958; chief of Lab. Parasite Chemotherapy sect. Lab. Tropical Disease, 1947-57, asst. chief, 1955-58, acting chief, 1958; chief of Lab. Parasite Chemotherapy, 1959—; lectr. tropical pub. health Harvard, 1955—; vis. prof. preventive medicine and pub. health Howard U., 1955—; prof. pharmacology La. State Univ. Med. School, New Orleans, Louisiana, 1966—. Commissioned scientist (lieutenant comdr.) regular corps USPHS, 1945, sr. scientist, 1947, scientist dir., 1949. U. S. del. to VI Internat. Congress Tropical Medicine and Malaria, Lisbon, 1958; mem. expert com. malaria WHO, 1948, 49, 60, 65, list of experts on malaria, 1949—; mem. Commn. on Malaria Armed Forces Epidemiology Bd., 1964, asso. dir. commn. malaria, 1965—, mem. commn. parasitic diseases, 1965—; mem. Sci. Adv. Council and Liberian Institute of American Foundation Tropical Medicine. Recipient Army-Navy Certificate of Appreciation, 1947; Dept. of Army Certificate of Appreciation, 1953; Darling Found. Medal and Prize, 1954, Gorgas Medal, 1954; Alumni Merit award Ia. State U., 1956; Distinguished Service medal La. State U., 1966. Hon. fellow Royal Sanitary Inst. Gt. Britain; fellow of New York Academy of Science; member A.A.A.S., Am. Soc. Parasitiologists, Royal Am. (pres. 1962) socs. tropical medicine and hygiene, Tropical Medicine Assn. Washington (pres. 1961), Am. Acad. Tropical Med., Am. Acad. Microbiology, Am. Pub. Health Assn., Assn. Mil. Surgeons U. S., Acacia, Sigma Xi, Phi Kappa Phi. Author: (with others) Survey of Antimalarial Agents, 1952, Chemotherapy of Malaria, 1955. Contbr. sci. papers tech. jours. Studies on the biology of malaria and other blood inhabiting protozoa, mainly in birds, monkeys, and man; made a systematic study of antimalarial agents for control and cure of the disease in man, began in 1942; established 1st facility for evaluation of antimalarial drugs in human subjects (prisoner volunteers), 1944; made 1st systematic study of simian (monkey) malarias to show that many of them are infectious to man. Home: 307 Gatehouse Dr. W., Gatehouse Apts., Metairie, La. 70001. Office: La. State U., Sch. Medicine, 1542 Tulane Av., New Orleans 70112.*

COBB, Collier, Am. geologist; b. Mt. Auburn Plantation, Wayne County, N.C., Mar. 21, 1862; s. Needham B. and Martha Louisa (Cobb) C.; student Wake Forest Coll., Sc. D. (hon.), 1917; student U. N.C.; A.B., Harvard, 1889, A.M., 1894; m. Mary L. Battle, Jan. 27, 1891 (dec. 1900); children—William Battle, Collier, Mary Louisa; m. 2d, Lucy P. Battle, Apr. 6, 1904 (dec. 1905); m. 3d, Mary Knox Gatlin, Oct. 27, 1910. Tchr., pub. schs. of N.C.; lectr. in State normal schs.; asst. Harvard, 1888-90; instr. Mass. Inst. Tech., 1890-92; prof. geology, U. N.C., 1892; Kenan research prof., 1920-21, studying shore lines N. Pacific, Gulf of Mexico, and Caribbean Sea. Tchr. geology, summer, Harvard, 1891, Knoxville, 1902, 09, Biltmore Forest Sch., 1905-12, Cornell, 1928, U. of N.C. Mem. Balt. Conf. on China-Am. Relations, 1925. Fellow Geol. Soc. Am., A.A.-A.S., Assn. Am. Geographers. Edited and published small illustrated paper, 1871-75; pub. map of North Carolina, 1879, 6 editions. Author: Where the Wind Does the Work; Human Habitations; Landes and Dunes of Gascony; Pocket Dictionary of Common Rock and Rock Minerals, 2d edit., 1915; Geography of North Carolina, 1880, 5th edit., 1915. Discoverer of Enfield horse, in early Pleistocene deposits of N.C. Studied moving sands, coast lines, soils. Died Nov. 28, 1934.

COBB, Stanley, Am. neuropathologist; b. Brookline, Mass., Dec. 10, 1887; A.B., Harvard, 1910, M.D., 1914; M.D., U. Munich (Germany), 1928. Surg. house officer Peter Bent Brigham Hosp., 1914-15; vol., physiol. research lab. Johns Hopkins, 1915-16, asst. physiology, later in psychiatry, 1916-17, asst. psychiatrist, 1917-18, asso. psychiatrist, 1918-19; faculty Harvard, 1919-23, 26—, Bullard prof. neuropathology, 1926-54, prof. emeritus, 1954—. Asst. neurologist Mass. Gen. Hosp., Boston, 1919-20, psychiatrist-in-chief, 1934-54; vis. neurologist; Boston City Hosp., 1925-34; with Office Sci. Research and Devel., 1944. Fellow A.M.A., Psychiat. Assn.; mem. Neurol. Assn. (pres. 1949), Assn. Research Nervous Mental Diseases (pres. 1940), Psychosomatic Soc. (pres. 1950), Soc. Clin. Investigation, Ornithol. Union. Research on physiology of cerebral circulation, physiol. and clin. epilepsy, effects of anoxia on brain, electroencephalography, language, psychosomatic medicine.

COBB, W(illiam) Montague, Am. anatomist, phys. anthropologist; b. Washington, Oct. 12, 1904; s. William Elmer and Alexzine E. (Montague) C.; A.B., Amherst Coll., 1925; M.D., Howard U., 1929; Ph.D., Western Res. U., 1932; Sc.D., Amherst Coll., 1955; LL.D., Morgan State Coll., 1964; m. Hilda Bradley Smith, June 26, 1929; children—Carolyn Elizabeth (Mrs. Robert S. Wilkinson, Jr.), Hilda Amelia (Mrs. Leander C. Gray). Faculty, Howard U., 1928—, prof., 1942—, head dept., 1947—; lectr. phys. anthropology Catholic U. Am., Washington, 1957-58; fellow anatomy Western Res. U., 1933-39, asso. anatomy, 1942-44. Recipient Selective Service medal U. S. Congress, Meritorious Pub. Service award Govt. of D.C., 1961. Mem. Anthrop. Soc. Washington (past pres.), Am. Assn. Phys. Anthropologists (past pres.), Medico-Chirurg. Soc. D.C. (past pres.), John A. Andrew Clin. Soc. (past pres.), Nat. Med. Assn. (past pres. Distinguished Service medal 1955), A.A.A.S. (past v.p., chmn. sect. H), Am. Assn. Anatomists, Anat. Soc. Gt. Britain and Ireland, Am. Soc. Mammalogists, Am. Soc. Mammalogists, Am. Assn. History Medicine, Am. Eugenics Soc. (dir.), Assn. for Study Negro Life and History. Numerous publs. on devel. of graphic method of learning anatomy; demonstrated craniofacial union in mammals and man; phys. anthropology of Am. Negro; studies in suprasternal bones in man, age changes in adult human skeleton, med. care and plight of Negro; description of dentition of walrus. Home: 1219 Girard St. N.W., Washington 20009.*

COBBOLD, Thomas Spencer, Brit. parasitologist; b. Ipswich, Eng., 1828; s. Richard Cobbold; M.D., Edinburgh (Scotland) U., 1851. Curator, Edinburgh Anat. Mus., 1851-56; prof. comparative anatomy, Edinburgh; lectr. on botany and zoology, London, Eng., 1857-84; became curator, prof. geology Brit. Mus., 1868; prof. botany Royal Vet. Coll., from 1873; spl. prof. helminthology. Fellow Royal Soc., 1864. Author: Entozoa, an introduction to the study of Helminthology with reference more particular to the internal parasites of man, 1864, Entozoa (supplement), 1869; Catalogue of the Specimens of Entozoa in the Museum of the Royal College of Surgeons of England, 1866; Tapeworms, 1866; Our Food-producing Ruminants and the Parasites which reside in them, 1871; The Grouse Disease, 1873; Worms, 1872; The Internal Parasites of our Domesticated Animals, 1873; Parasites, 1879; Human Parasites, 1882; Parasites of Meat and Prepared Flesh Food, 1884; contbns. to Cyclo. of Anatomy and Physiology (Todd), 1858, Dictionary of Medicine (Quain); also many memoirs. Specialist on parasitic worms in man and animals; introduced names Bilharzia haemotobia, 1864, Filaria bancrofti (in honor of Joseph Bancroft), 1877; also worked in geology. Died Mar. 20, 1886.

COBINE, James Dillon, Am. physicist; b. Oklahoma City, May 10, 1905; s. Kenner Rice and Mary (Cockrell) C.; B.S., U. Wis., 1931; M.S., Cal. Inst. Tech., 1932, Ph.D., 1934; postgrad. Harvard. With Wis. Telephone Co., Eau Claire, Wis. summer 1930; grad. asst. Cal. Inst. Tech., 1933-34; instr. Harvard, 1935-38, faculty instr., 1938-41, asst. prof., 1941-45; physicist Gen. Electric Co. Research Lab., Schenectady, 1945—. Recipient certificate of appreciation for World War II research U. S. War and Navy Depts., 1947. Fellow Am. Phys. Soc., I.E.E.E.; mem. Am. Soc. Engring. Edn., Sigma Xi, Tau Beta Pi, Eta Kappa Nu. Author: Gaseous Conducters, 1941; (with others) Electronic Circuits and Tubes, 1947; contbg. author: Smithsonian Physical Tables, rev., 1951; High Temperature Technology, 1956, Ency. Electrochemistry, 1964; McGraw Hill Ency. Sci. and Tech., 1960; Engineering Aspects of Magnetohydrodynamics, 1964; McGraw Hill Yearbook of Sci. and Tech., 1963. Patentee in field. Research on elec. discharges in gases, and applications to radar; studies of magnetic materials and photoelectric phenomena. Patentee in field. Home: R.D. 1, Rexford, N.Y. 12148. Office: G.E. Research Lab., The Knolls, Schenectady.*

COBLENTZ, Virgil, Am. chemist; b. Springfield, O., Mar. 12, 1862; s. John Philip and Susan (Zitzer) C.; grad. Wittenberg Coll.; Ph.G. and Pharm.M., Phila. Coll. Pharmacy, 1882; Ph.D., U. Berlin, 1891; M.D., U. Wuerzburg, 1895; m. Anna Bauer, Mar. 7, 1889.

Prof. materia medica and toxicology Cin. Coll. Pharmacy, 1884-87; prof. chemistry N.Y. Coll. Pharmacy, Columbia, 1891-1911; chief chemist, E. R. Squibb & Sons, N.Y.C., 1911-17; pathologist Hazard Hosp., Long Branch, N.J. Chmn. N.Y. sect. Soc. Chem. Industry of Gt. Britain, 1902-04, N.Y. sect. Verein Deutscher Chemiker, 1909-14; mem. com. revision U. S. Pharmacopoeia, 1900-20. Author: Handbook of Pharmacy, 1895; The Newer Remedies, 1897; Sadtler and Coblentz Medical and Pharmaceutical Chemistry, 1899; Volumetric Analysis. Died 1932.

COBLENTZ, William Weber, Am. physicist, astronomer; b. North Lima, O., Nov. 20, 1873; B.S., Case Inst. Tech., 1900, Sc.D., 1930; A.M., Cornell U., 1901, Ph.D., 1930; m. 1924, 2 children. Carnegie research asst., hon. fellow Cornell U., 1903-05; physicist Bur. of Standards, 1904-45; cons. physicist, 1945——. Recipient Potts medal Franklin Inst., 1910; Janssen medal Paris Acad. Sci., 1920; Scott medal and premium, Phila., 1924; Gold key Am. Congress Phys. Therapy, 1934; Rumford Gold medal Am. Acad. Arts and Scis., 1937; Soc. Applied Spectroscopy medal, 1953. Mem. Nat. Acad. Scis., Am. Phys. Soc., Optical Soc. (Ives medal 1945), Am. Astron. Soc., Soc. Psychical Research, A.M.A.; corr. mem. Soc. francaise de physique, Union Geodesy and Geophysics. Noted for work on infra-red, ultra-violet, planetary and stellar radiation; further work on phys. study of firefly, ice formation on plants, photoelectric properties of materials, stellar radiometry, planetary temperatures. Died 1962.

COCCHI, Antonio, Italian physician; b. Benevento, Italy, Aug. 3, 1695; s. Giacinto and Beatrice (Bianchi) C.; studied in Florence, Italy, later at U. Pisa (Italy); received degree, 1716; studied under Tommaso Puccini, Florence; m. Gaetana Debi; m. 2d, Teresa Orsola Piombanti; children—Beatrice (Mrs. Angelo Tavanti), Raimondo. Elected to Med. Coll. Florence, 1735; became instr. anatomy and philosophy, 1736; practiced medicine in Florence. Co-founder renovated Bot. Soc., 1734. Fellow Royal Soc., 1736. Author: A Treatise on the Use of Cold Water Not Only for Preserving Sanity but also for the Cure of Other Ills; Trattato dei Bagni di Pisa. Translator: Xenophontis ephesii ephesialorum libri V de amoribus anthiae et abrocomae, 1726; Del matromonio ragionamento di un filosofo mugellano, 1762; The Pythagorean Diet of Vegetables only Conducive to the Preservations of the Health and the Cure of Disease, 1745; The Life of Asclepiades, the Celebrated Founder of the Asclepiadic Sect in Physic. Died Jan. 1, 1758.

COCCONI, Giuseppe, physicist; b. Como, Italy, Oct. 3, 1914; s. Giovanni and Ida (Cachat) C.; Ph.D., U. Milan (Italy), 1937; m. Vanna Tongiorgi, Aug. 9, 1945; children—Anna, Alan. Prof. physics U. Milan, 1940-42; prof. physics U. Catamia, Italy, 1942-47; prof. physics Cornell U., 1947-63; research physicist Centre Européen Recherche Nucléaire, Geneva, Switzerland, 1963——. Studies of strong interactions, particularly cosmic rays; origin of cosmic radiation; radio astronomy; studies of interactions produced by protons accelerated in synchrotrons. Home: 12 Crets de Champel. Office: CERN, Geneva, Switzerland.*

COCHIN, Denis-Claude, French botanist; b. 1698; introduced many rare plants to Châtillon, France. Author: Jardin des curieux ou catalogue raisonné des plantes les plus belles et les plus rares (catalog of all plants cultivated in exptl. garden at Châtillon), 1777. Died Aug. 19, 1786.

COCHRAN, Donald Gordon, Am. entomologist; b. New Hampton, Ia., July 5, 1927; s. Charles Douglas and Clara (Borlang) C.; B.S., Ia. State U., 1950; M.S., Va. Poly. Inst., 1952; Ph.D., Rutgers U., 1954; m. Frances Beale Craig, Sept. 5, 1952; children—Victoria Ann, Megan Lee, Christopher Craig. Research entomologist U. S. Army Edgewood (Md.) Arsenal, 1955-57; asso. prof. Va. Poly. Inst., Blacksburg, 1957-64, prof. entomology, 1964——. Mem. A.A.A.S., Am. Inst. Biol. Scis., Entomol. Soc., Am. Genetics Assn., Sigma Xi. Research, publs. on insect resistance to insecticides; biochemistry of insect muscle and its comparison with muscle from other animals. Home: 1406 Hillcrest Dr., Blacksburg, Va. 24060.*

COCHRAN, George Wilson, Am. plant virologist; b. Topeka, July 7, 1919; s. Charles Vernon and Faith (Hammaker) C.; B.S., Kan. State U., 1941, M.S., 1942; Ph.D., Cornell U., 1947; m. Phyllis Louise Cole, Oct. 23, 1943; children—Patricia Elaine, Lois Beverly. Postdoctoral fellow Rockefeller Inst. for Med. Research, Princeton, N.J., 1946-47; asso. prof. Utah State U., Logan, 1948-55, prof. plant virology, 1955——, faculty honor lectr., 1963. Fulbright lectr. Netherlands, 1957-58; guest scholar lectr. Grad. Sch., Kan. State U., 1964. Mem. Am. Phytopath. Soc. (past councilor), Am. Soc. for Hort. Sci., Am. Pomological Soc., Am. Chem. Soc., Electron Microscopy Soc. Am., Am. Soc. for Microbiology, Am. Inst. Biol. Scientists, A.A.A.S., N.Y. Acad. Scis., Utah Acad. Scis., Arts and Letters, Sigma Xi, Phi Kappa Phi, Gamma Sigma Delta, Alpha Zeta. Contbr. numerous articles to tech. jours. Developed chromatographic procedures for virus isolation identification and purification; detected naturally occurring infective forms viral RNA; demonstrated cell-free synthesis infectious RNA western equine encephalitis virus, cell-free synthesis whole tobacco mosaic virus by isolated primed normal plant chloroplasts; isolated and purified viral RNA replicative mechanism tobacco mosaic virus. Home: 991 Sumac Dr., Logan, Utah, 84321.*

COCHRAN, James Harvey, Am. entomologist; b. Abbeville, S.C., Sept. 14, 1913; s. Harvey Nickles and Leona (Greene) C.; B.S., Clemson U., 1935; M.S., Ia. State U., 1936, Ph.D., 1946; m. Mildred Viola Batson, Aug. 28, 1940; children—Andrew H., Sandra Eve, Jennifer Greene. Research entomologist E. I. duPont de Nemours & Co., Wilmington, Del., 1938-42, 46-47; asso. entomologist S.C. Expt. Sta., 1947-53; prof., head entomology and zoology dept., state entomologist Clemson U., 1953——. Mem. Entomol. Soc. Am., S.C. Entomol. Soc., S.C. Acad. Sci., Sigma Xi, Gamma Sigma Delta (pres. Clemson chpt. 1961), Alpha Zeta, Phi Kappa Phi. Contbr. articles to sci. jours. Basic, applied research in entomology; insect toxicology. Home: 434 Pendleton Rd., Clemson, S.C. 29631.

COCHRAN, Lloyd Curtis, Am. plant pathologist; b. Frankfort, Ind., Apr. 5, 1906; s. Morris Edgar and Bertha E. (Kellar) C.; B.S., Purdue, 1928; M.S., Mich. State U., 1930, Ph.D., 1936; m. Maud V. Tague, Aug. 21, 1933; children—Lettie L. (Mrs. James Mortimer), John M. With Citrus Expt. Sta., U. Cal. at Riverside, 1936-41; with U. S. Dept. Agr., 1941——, chief fruit and tree nut br., crop div. Agrl. Research Service, Beltsville, Md., 1959——. Mem. Am. Phytopath. Soc., Washington Bot. Soc., Fla., Internat. hort. socs. Research, publs. on virus diseases of deciduous fruits, virus diseases of citrus. Home: 10805 Bornedale Dr., Adelphi, Md. Office: Plant Industry Sta., Beltsville, Md.*

COCHRAN, William, Scottish physicist; b. July 30, 1922; s. James and Margaret Watson (Baird) C.; M.A, Ph.D., Edinburgh (Scotland) U., 1943-46; m. Ingegerd Wall, 1953; 1 son, 2 daus. Asst. lectr. Edinburgh U., 1943-46, prof. physics, 1964——; demonstrator, lectr. Cambridge (Eng.) U., 1948-62, fellow Trinity Hall, 1951-64, reader in physics, 1962-64. Research fellow abroad, 1950-51, 58-59. Fellow Royal Soc., 1962; mem. Royal Soc. Edinburgh, Inst. Physics and Phys. Soc. (Guthrie medal 1966). Author: (with H. Lipson) Vol. III of the Crystalline State, 1954. Research, publs. on physics, chemistry of crystals. Address: Dept. Natural Philosophy, Edinburgh U., Drummond St., Edinburgh, Scotland.

COCHRAN, William G(emmell), statistician; b. Rutherglen, Scotland, July 15, 1909; s. Thomas and Jeanie W. (Gemmell) C.; M.A., Glasgow U., 1931; M.A., Cambridge U., 1938; m. Betty Mitchell, July 17, 1937; children—Elizabeth, Alexander, Teresa. Came to U. S., 1939, naturalized, 1946. Statistician Rothamsted Exptl. Sta., 1934-39; prof. math. statistics Ia. State Coll., 1939-46; asso. dir. Inst. Statistics, U. N.C., 1946-48; prof. biostatistics, sch. hygiene Johns Hopkins, 1948-57; professor statistics Harvard, 1957——. Member American Statistics Association (pres. 1953), Inst. Math. Statistics (pres. 1946), Internat. Statis. Inst. Author: Experimental Designs (with G. M. Cox), 1950; Sampling Techniques, 1953. Work on more effective collection of data by experiments and surveys, so that informative results are obtained. Home: 2 Ardley Pl., Winchester, Mass. Office: Harvard U., Cambridge, Mass.

COCHRANE, Robert, physician; b. Peitaho, N. China, Aug. 11, 1899; s. Thomas and Grace Hamilton (Greenhill) C.; ed. Eltham Coll., London, Eng.; M.B., Ch.B., M.D.; m. Ivy Gladys Nunn, Aug. 1927; children—Robert Cameron, Ian Hamilton, Grace Hamilton. Med. dir. Lady Willingdon Leprosarium, Chingleput, 1935-44; dir., prof. medicine Christian Med. Coll., Vellores, India, 1944-48; cons. med. technician Am. Leprosy Missions; cons. leprologist Ministry Health, London, 1951-64. Mem. Royal Soc. Medicine, Royal Soc. Tropical Medicine, Internat. Leprosy Assn. (v.p.). Author: Practical Textbook of Leprosy. Editor: Leprosy in Theory and Practice; Joan Wright, 1956; Treatment of Leprosy, also articles. Home: 419a Anchors Ditchling Rd., Brighton, Sussex, Eng. Office: 57a Wimpole St., London W.1, Eng.

COCKAYNE, Leonard, botanist; b. Derbyshire, 1855; s. W. Cockayne; ed. Owens Coll., Manchester; Ph.D.; D.Sc., U. New Zealand, 1932; m. Maria Maude Blakeley. Tchr., 1881-85; with pvt. exptl. sta., 1887-1904; researcher, 1888-1934; explored Chatham and sub-Antarctic lands; econ. study forests New Zealand State Forest Service, 1919-34, later hon. botanist. Mem. Royal Commn. on Forestry, 1913, Cawthorn Commn., 1919. Recipient Hector medal and prize, 1912, Hutton Meml. medal, 1914, Mueller Meml. medal and prize, 1928, Veitch Meml. medal, 1932. Fellow Royal Soc. (Darwin medal 1928), 1912, Linnaean Soc., New Zealand Inst. (pres. 1918-19); mem. Bot. Soc. Am. (corr.), Mass. Hort. Soc., Bot. Soc. Germany, Forest Soc. Finland, Plant-Geog. Soc. Sweden, Royal Sci. Soc. (Gothenburg) (fgn.). Author: New Zealand Plants and Their Story; The Vegetation of New Zealand; The Cultivation of New Zealand Plants, also numerous memoirs on ecology, botany, evolution. Died July 8, 1934.

COCKBURN, William, Brit. physician; b. Eng., 1669; s. Sir William Cockburn; M.A., Edinburgh, Scotland; med. student at Leiden, Netherlands, 1691; m. Mary de Baudisson, 1698; m. 2d, Lady Mary Fielding, Apr. 5, 1729. Physician to the fleet, 1694; practiced medicine in London, Eng. before 1710; physician Greenwich Hosp., 1731. Licentiate London Coll. Physicians; Fellow Royal Soc., 1696. Author: Lues venerea; Symptoms, Nature and Cure of a Gonorrhea; Account of the Nature and Cure of Looseness, 1710; contbd. Operation of a Blister to the Philos. Trans., 1699. Discovered medication for dysentery and supplied it to army and navy. Died Nov. 1739.

COCKCROFT, Sir John Douglas, English nuclear physicist; b. Todmorden, Yorks, Eng., May 27, 1897; s. J. A. and A. M. (Fielden) C.; ed. U. Manchester, St. John's Coll., Cambridge U., M.A., Ph.D., M.Sc. (Tech.); D.Sc. (hon.), Oxford, 1949, London U., 1950, Cambridge, 1954, Dublin, 1952, Canberra, 1952, Leicester, 1958; LL.D., U. Toronto, Glasgow U., 1951, Dublin, 1952, Sydney, 1952, Manchester, 1953, Birmingham, 1954, Melbourne, 1952, St. Andrews, 1955; m. to E. Elizabeth Crabtree, 1925; children—1 son, 4 daus. Fellow St. John's Coll., 1928-46; Jacksonian prof. natural philosophy U. Cambridge, 1939-46. Chief supt. Air Defense Research and Dev. Establishment, Ministry of Supply, 1941-44; dir. atomic energy div. Nat. Research Council of Can., 1944-46; dir. Atomic Energy Research Establishment, 1946-58; mem. Atomic Energy Authority, 1958-1967; master Churchill Coll., Cambridge (Eng.) U., chancellor Australian Nat. U., 1961-65; pres. of Manchester U. Inst. Sci. and Technology. Chairman Def. Research Policy Com., 1952-54. Fellow Royal Soc., 1936; mem. Inst. Elec. Engrs. Created Knight, 1948. Hon. fellow St. John's Coll., Cambridge. Recipient (with Dr. Ernest Thomas Sinton Walton) Nobel prize in physics, 1951; given hon. freedom of Todmorden, 1946; awarded Hughes medal Royal Society, Medal of Freedom with golden palms (U. S.), 1947, J. A. Ewing medal Instn. Civil Engrs., 1948, Chevalier Legion of Honor (France), 1950; Royal Medal Royal Soc., 1954; Kelvin Medal, Institute of Civil Engineers, 1956; recipient Faraday medal, 1955; decorated Order of Merit, 1957; Atoms for Peace prize, 1961. Mem. Brit. Assn. for Advancement Science (pres. 1961-62), Am. Acad. Arts and Scis.; fgn. member of Royal Danish Acad., Royal Swedish Acad. Author papers in nuclear physics Proc. of Royal Soc., also papers on tech. subjects Jour. Inst. Elec. Engrs. With Walton, built particle accelerator and pioneered in transmutation of atomic nuclei by artificially accelerated atomic particles; originally used protons to bombard lithium atoms (1932); but continued research on transmutation of nuclei using protons and deuterium. Died Sept. 18, 1967.

COCKER, Wesley, chemist; b. Lancashire, Eng., Jan. 31, 1908; s. William and Elizabeth Ellen (Buckley) C.; B.Sc. with 1st class honors, U. Manchester (Eng.), 1928, M.Sc., 1929, Ph.D., 1931, D.Sc., 1958; M.A., Sc.D., U. Dublin (Ireland), 1948; m. Eleanor Gertrude Garstang, Sept. 1, 1938; 1 dau., Katharyn Eleanor (Mrs. William Colvin Given). Grad. research prizeman, scholar U. Manchester, 1928-29, Dalton Chem. scholar, 1929-30, Beyer fellow, 1931-32; research chemist dyestuffs group I.C.I. Ltd., 1932-34; chief chemist, works mgr. Cocker Chem. Co., Ltd., 1934-37; asst. lectr. U. Coll., Exeter, Eng., 1937-39; lectr. King's Coll., Newcastle, Eng., 1939-46; univ. prof. chemistry, head dept. chemistry Trinity Coll., U. Dublin, 1946——. Cons. Cocker Chem. Co., Ltd., 1937——. Fellow Trinity Coll., 1958——. Fellow Royal Inst. Chemistry (v.p., council), Inst. Chemistry Ireland; Chem. Soc. (past mem. council); mem. Soc. Chem. Industry (past mem. council), Royal Irish Acad. (mem. council 1965——, sec. com. of sci.), Biochem. Soc. Research, numerous articles on syntheses alpha-amino acids, mechanisms of some organic reactions; elucidated structure and investigated chemistry organic compounds extracted from plants, trees. Joint patentee synthesis methacrylic esters. Home: 176 Orwell Rd., Dublin 14, Republic of Ireland.*

COCKERELL, Theodore Dru Alison, zoologist; b. Norwood, Eng., Aug. 22, 1866; s. Sydney John and Alice Elizabeth (Bennett) C.; ed. pvt. schs. in Eng. and Middlesex Hosp. Med. Sch.; m. Annie S. Fenn, 1891 (dec. 1893); children—Austin, Martin; m. 2d, Wilmatte Porter, 1900. Resided in Colo., 1887-90, studying entomology, botany, etc.; curator pub. mus., Kingston, Jamaica, 1891-93; prof. entomology N.M. Agrl. Coll., 1893-96, 1898-1900; entomologist N.M. Agrl. Expt. Sta., 1893-1901; cons. entomologist Ariz. Agrl. Expt. Sta., 1900-09; tchr. biology, N.M. Normal U., 1900-03; curator Colo. Coll. Mus., 1903-04; lectr. on biology, Colo. State Prep. Sch., 1904-1927; lectr. on entomology U. Colo., 1904-06, prof. systematic zoology, 1906-12; prof. zoology, 1912-34, emeritus, 1934. Fellow A.A.A.S., Am. Mus. Natural History (hon.); corr. mem. Phila. Acad. Natural Scis., Am. Entomol. Soc.; mem. Am. Philos. Soc. Author: Zoology; Zoology of Colorado; also over 3,000 articles and notes in scientific publs., principally on mollusca, insects, fishes, palaeontology, geographic distbn. of life, fossil insects, natural history of various regions, and subjects connected with evolution. Made explorations in Siberia, Japan, South America, Madeira Islands, Russia, Australia, Morocco, Central and South Africa. Died Jan. 26, 1948.

COCKERILL, William, inventor; b. Lancashire, Eng., 1759; children—William, Charles James, John,

Nancy (Mrs. James Hodson). Mechanic in Lancashire; worked in St. Petersburg, Russia, beginning in 1794 and in Sweden beginning in 1796; began manufacturing spinning and weaving machinery at Verriers, Belgium, 1799, Liège, Belgium, 1807-12. Invented spinning and weaving machines. Died Liège, 1832.

COCKLE, Sir James, lawyer, mathematician; b. 1819; s. James Cockle; B.A., Trinity Coll., Cambridge (Eng.) U., 1842, M.A., 1845; m. Adelaide Wildin, Aug. 22, 1855; 8 children. First chief justice Queensland, Australia, 1863-79. Created knight, 1869. Fellow Royal Astron. Soc., Royal Soc., 1865, London Math. Soc. Contbr. to theory of differential equations; original study of linear differential equations of the order n-1; worked on solution to gen. equation of 5th degree; 1st to discover and develop properties of functions called criticoids or differential invariants. Died 1895.

CODACCIONI, J. L., French physician; b. Marseilles, France, Apr. 29, 1929; s. A.M. and M. (Torelli) C.; M.D., U. Aix-Marseilles, 1954; m. A.M. Plenet, July 13, 1965. Dir. clinic Faculté de Médecine, Marseille, 1956-60; med. asst. hosps., 1960-64; became lectr., agrégé, physician, 1965; now asso. prof. clin. endocrinology Faculté de Medecine, Aix-Marseilles U. Cons. endocrinology and metabolic diseases French Rys., S.N.C.F.-Mediterranée. Mem. French Endocrinology Soc., Royal Soc. Medicine, Assn. French Speaking Diabetologist, European Assn. for Study Diabetes. Author: (with J. Vague, J. Nicolino, H. Serment, R. Simonin) Notions d'Endocrinologie Clinique; also numerous articles. Research on relationship of obesity and diabetes, pathogenesis and treatment of goiter. Home: 5 Montée du Plateau, 13 Marseilles 7. Office: Hôpital de la Conception 144 rue St. Pierre, 13 Marseilles 5, France.*

CODDINGTON, Earl Alexander, Am. mathematician; b. Washington, Dec. 16, 1920; s. Cyrus A. and Lillian (Dezarn) C.; Ph.D., Johns Hopkins, 1948; m. Susan Klaber, Nov. 17, 1945; children—Alan A., Robert H., Claire H. Instr. math. Johns Hopkins, 1948-49, Mass. Inst. Tech., 1949-52; faculty math. U. Cal. at Los Angeles, 1952—, prof., 1959—. Fulbright lectr. U. Copenhagen (Denmark), 1955-56. Mem. Am. Math. Soc., Math. Assn. Am., Phi Beta Kappa, Sigma Xi. Author: (with N. Levinson) Theory of Ordinary Differential Equations, 1955; An Introduction to Ordinary Differential Equations, 1958. Research in differential equations and analysis. Home: 764 Wildomar St., Pacific Palisades, Cal. 90272. Office: Math. Dept., U. Cal., Los Angeles 90024.

CODDINGTON, Henry, mathematician; b. circa 1800; ed. Cambridge (Eng.) U. Fellow Royal Soc., 1829. Wrote on optics; recommended microscopes with grooved sphere lens to avoid marginal blurring caused by spherical aberration (Coddington lens, first described by Brewster). Died Rome, Italy, Mar. 3, 1845.

CODE, Arthur Dodd, Am. astronomer; b. Bklyn., Aug. 13, 1923; s. Lorne Arthur and Jessica (Dodd) C.; M.S., U. Chgo., 1947; Ph.D. in Astronomy and Astrophysics, 1950; m. Mary Ella Guild, Oct. 9, 1943; children—Alan D., Douglas Merritt, Edith Louis, David Arthur. Research asst. Yerkes Obs., 1946-49; instr. astron. U. Va., 1949-50; faculty U. Wis., 1951-56, 58—, prof. astronomy, chmn. dept., 1958—, dir. Washburn Obs. of univ., 1958—; asso. prof. astronomy Cal. Inst. Tech., also mem. staff Mt. Wilson and Palomar Obs., 1956-58. Mem. Am. Astron. Soc., Sigma Xi. Spl. research photoelectric spectrophotometry of stars and galaxies, devel. of instruments. Home: 2813 Mason St., Madison 5, Wis. 53705.*

CODE, Charles Frederick, Am. physician; b. Winnipeg, Man., Can., Feb. 1, 1910; s. Abraham and Gertrude (Drewry) C.; M.D., B.Sc. in Medicine, U. Man., 1934; Ph.D., U. Minn., 1939; m. Gwendolyn Irene Bond, Dec. 30, 1935; children—Wendy Drewry (Mrs. Gene Goodrow), Carla Radford, Charles Allan. Came to U. S., 1938, naturalized, 1953. With Mayo Found., Rochester, Minn., 1934-35, 1937-38, 1940—, prof. physiology, 1942—, dir. for med. edn. and research, 1966—; lectr. physiology U. Coll., London, Eng., 1936-37; instr. U. Minn., 1938-39; staff Mayo Clinic, 1940-42, cons. in physiology, 1942—; chmn. sect. physiology Mayo Clinic and Mayo Found., 1942—. Recipient Physiology prize, also Gold medal for research U. Man., 1930; Theobald Smith award and medal for research A.A.A.S., 1938; Bayliss-Starling Meml. scholar U. Coll., London, 1936. Mem. Am. Physiol. Soc. (editor Handbook on the Physiology of the Alimentary Canal 1965-66), Am. Gastroent. Assn. (governing bd. 1963—), Fedn. Am. Socs. for Exptl. Biology (chmn. gastroenterology symposium com. 1963—). Author: (with Brian Creamer, Jerry F. Schlegel) An Atlas of Esophageal Motility in Health and Disease, 1959; also numerous articles. Research in role histamine in allergy, motor action esophagus in health and disease, motor action alimentary canal. Home: 803 8th St. S.W., Rochester 55901. Office: Mayo Found., Rochester, Minn. 55902.*

CODEGONE, Cesare, Italian civil engr.; b. Novare, Mar. 16, 1904; s. Pietro and Teresa (Biglieri) C.; ed. U. Turin; civil engr.; m. Camilla Valsania, Dec. 8, 1943; children—Contardo, Marco. Prof., Poly. Sch. Turin; instr. civil engring. Mem. Acad. Sci. Turin (corr.), Associazione Termotecnica Italianna (sec.-gen.). Author: Fisica Tecnica, 5 vols., 1962. Home: via Tadini 23. Office: Politecnico, Turin, Italy.

CODERE, Helen Frances, Am. anthropologist; b. Winnipeg, Man., Can., Sept. 10, 1917; d. Charles Francis and Mabelle (Prosser) C.; came to U. S., 1919, naturalized, 1925; B.A. summa cum laude, U. Minn., 1939; Ph.D., Columbia, 1950. Instr., Vassar Coll., 1946-50, asst. prof., 1951-53, asso. prof., 1955-57, prof., 1958-63; faculty Bennington Coll., 1963-64; prof. anthropology Brandeis U., Waltham, Mass., 1964—; vis. lectr. U. B.C., Vancouver, Can., 1954-55. Fellow Am. Anthropol. Assn. (exec. council 1966—), A.A.A.S., African Studies Assn., N.Y. Acad. Scis., Phi Beta Kappa. Author: Fighting with Property, a Study of Kwakiutl Potlatching and Warfare, 1792-30, 1950; also articles. Research on Kwakiutl ethnohistory and instns., econ. anthropology and study cultural change.*

CODINGTON, John Fort, Am. chemist; b. Macon, Ga., Feb. 9, 1920; s. Arthur H. and Kate Haynes (Fort) C.; A.B., Emory U., 1941, M.A., 1942; Ph.D., U. Va., 1945; m. Celia Unger, Dec. 20, 1952; children—William, Ida Kate, Dorothy. Staff chemotherapy sect. NIH, Bethesda, Md., 1945-49; with Faculté de Médecine de Paris (France), 1945-49, Columbia, 1951-55; staff Sloan Kettering Inst. for Cancer Research, N.Y.C., 1955—, asso., 1960-67; asst. biochemist Mass. Gen. Hosp., also research asso. Harvard Med. Sch., 1967—. Fellow A.A.A.S.; mem. Am. Soc. Biol. Chemists, Am. Chem. Soc., Sigma Xi. Research, publs. on synthesis of potential antimalarial compounds, relationship between structure and activity in peptide antibiotics, peptide synthesis, synthesis and biol. properties of analogs of thiamine, synthesis and biol. properties of nucleic acid analogs, carbohydrates. Home: 1725 Commonwealth Av., West Newton, Mass. 02165. Office: Lab for Carbohydrate Research, Mass. Gen. Hosp., Boston 02114.*

CODRONCHI, Giovanni Battista, Italian physician; b. Imola, Italy, 1547; practiced medicine, Imola; wrote earliest existing work entirely devoted to larynx, 1597, also 1st important work on legal medicine, 1597. Author: De Christiana actuta medendi ratione libriduo . . . , 1591; De morbis veneficis ac veneficiis libri quatuor, 1595; De vitiis vocis libri duo . . . , 1597; De morbis qui Imolae et alii communite. hoc anno 1602 . . . , 1603; De rabie, hydrophobia communitei dicta, libri duo, de sale absynthii . . . , 1610; De annis climactericis necnon de ratione vitandi eorum pericula itemque de modis vitam producendi, 1623. Died 1628.

CODY, George Dewey, Am. physicist; b. Flushing, N.Y., May 16, 1930; s. George Dewey and Anne (Lynch) C.; B.A. summa cum laude, Harvard, 1952, M.A., 1954, Ph.D., 1957; m. Barbara Gorman, Nov. 7, 1959; children—George, Elizabeth, Monica. Mem. staff RCA Labs., Princeton, N.J., 1957-66, group head, 1966—. J. Parker fellow, 1957-58; recipient David Sarnoff award RCA, 1960, 64. Mem. Am. Phys. Soc. Research in superconductivity and high temperature thermal conductivity. Home: 14 Southern Way, Office: care RCA Labs., Princeton, N.J. 08540.*

COE, Michael Douglas, Am. anthropologist; b. N.Y.C., May 14, 1929; s. William Rogers and Clover (Simonton) C.; A.B., Harvard, 1950; Ph.D., 1959; m. Sophie Dobzhansky, June 5, 1955; children—Nicholas, Andrew, Sarah, Peter. Asst. prof. U. Tenn., 1958-60; faculty Yale, 1960—, professor anthropology, 1968—. Adviser, Robert Woods Bliss Collection Pre-Columbian Art, Dumbarton Oaks, Harvard, 1963—. Fellow Royal Anthrop. Soc., Am. Anthropol. Assn. A.A.A.S.; mem. Mexican Soc. Anthropology, Soc. for Am. Archaeology. Author: La Victoria, An Early Site on the Pacific Coast of Guatemala, 1961; Mexico, 1962; The Jaguar's Children: Pre-Classic Art of Central Mexico, 1965; The Maya, 1966; (with Kent V. Flannery) Early Cultures and Human Ecology in South Coastal Guatemala; 1967; also numerous articles. Research on beginning of village life in Mesoamerica, devel. high civilization in So. Mexico and Central Am. particularly Olmec culture of Veracruz. Home: 376 St. Ronan St., New Haven 06511.*

COELHO, Eduardo, Portuguese physician; b. Santo Tirso, Portugal, Dec. 7, 1895; s. António C. Araújo and M. (Araújo) C.; M.D., U. Lisbon (Portugal), 1923; m. Macieiro Coelho, July 14, 1925; 5 children. Staff, U. Lisbon, 1924—, prof. internal medicine, 1949—, dir. dept. internal medicine, 1949-50, dir. Center Cardiology, Instituto de Alta Cultura, 1950—; dir. dept. cardiology Santa Maria Hosp., Lisbon, 1954—. Author: L'infactus du myocarde, 1934; A Patologia na circulaçao coronária, 1937; Liçoes de Cardiologia, 5 vols., 1954-65; Fisiopatologia e diagnóstico das cardiopatias cong., 1966; Da Filosofia da Medecina, 1961; also numerous articles. Research in cardiology, especially acute chronic constrictive pericarditis, interpretation of electrocardiographic alterations; hemodynamic studies, including acquired and congenital cardiopathies, mitral valvular diseases, paramiloidose, hemosiderose, primary myocardiopathies, chronic cor pulmonale, angiocardiographie, hemodymaic and vectorcardiographic study of Complex Fallot Type. Home: Rua Viriato, 1, Lisbon, Portugal.*

COESTER, Fritz, physicist; b. Berlin, Germany, Oct. 16, 1921; s. Robert and Helen (Pfaff) C.; Ph.D., U. Zurich (Switzerland), 1944; m. Elizabeth Morgan Garrett, Dec. 27, 1952; children—Janet Carol, William Robert, Hans Carl, Michael Fritz, Susan Elizabeth, Thomas Matthew. Came to the United States of Am., 1947, naturalized, 1952. Physicist, Sulzer Bros., Inc., Winterthur, Switzerland, 1944-46; research asst. U. Geneva (Switzerland), 1946-47; faculty State U. Ia., Iowa City, 1947-63, prof., 1960-63; mem. Inst. for Advanced Study, Princeton, N.J., 1953-54; sr. physicist Argonne (Ill.) Nat. Lab., 1963—. Research in theoretical physics, theoretical nuclear physics; quantum theory of fields. Home: Rural Route 2, Iowa City. Office: Argonne Nat. Lab., Argonne, Ill. 60440.*

COFER, Charles Norval, Am. psychologist; b. Cape Girardeau, Mo., June 1, 1916; s. Charles Norval and Ernestine (Osterloh) C.; A.B., S.E. Mo. State Coll., 1937; M.A., U. Ia., 1937; Ph.D., Brown U., 1940; m. Justine Marie Donnelly, Aug. 3, 1940; children—Thomas Michael, Jonathan Charles. Faculty, George Washington U., Washington, 1941-47, asst. prof., 1946-47; faculty U. Md., College Park, 1947-59, prof. psychology, 1951-59; prof., dir. grad. studies in psychology N.Y. U., 1959-63; vis. prof. U. Cal. at Berkeley, 1962-63; prof. Pa. State U., University Park, 1963-67; prof. chmn. dept. psychology U. Md., College Park, 1967—. Member American District of Columbia (past president), also Maryland (past pres.), Eastern (past pres., dir. 1965—) psychol. assns., Psychonomic Soc. (bd. govs. 1965—), Sigma Xi. Author: (with M. H. Appley) Motivation: Theory and Research, 1964; also numerous articles. Research in verbal learning and verbal behavior, basic associative processes, structure lang. as factor in relation to problem solving and thinking. Home: 805 Sligo Creek Pkwy., Takoma Park, Md. 20012. Office: Univ. of Md., College Park, Md. 20742.*

COFER, Leland Eggleston, Am. indsl. hygienist; b. Richmond, Va., Nov. 16, 1869; s. Nathan Pliny and Effie (Mountcastle) C.; student Richmond Coll., 1884-87; M.D., Richmond Med. Coll., 1889; m. Clara Drake-Smith, Aug. 25, 1893 (dec. 1906); m. 2d, Luisita Leland, Oct. 15, 1919. Served in U. S. Marine hosps. at Boston, Buffalo, Norfolk, Savannah, Mobile, San Diego and San Francisco, 1889-1900; organized campaign for eradication of plague from Seattle and N.W.; chief quarantine officer H. I., 1900-08; organized quarantine system there and was pres. Hawaiian Bd. Health, having charge of Leprosy Sta.; appointed asst. surgeon gen. USPHS, 1908, in charge div. of maritime quarantine of U. S. and its possessions and dependencies, also of med. function of immigration; loaned by Govt. as health officer, Port of N.Y., 1916; dir. indsl. hygiene State of N.Y. Mem. A.M.A., Am. Assn. Mil. Surgeons, Am. Assn. Tropical Medicine. Died Feb. 17, 1948.

COFFEY, Robert Calvin, Am. surgeon; b. Caldwell County, N.C., Oct. 20, 1869; s. Patterson Vance and Nancy Martitia (Estes) C.; prep. edn., Globe (N.C.) Acad.; M.D., Ky. Sch. Medicine, Louisville, 1892; m. Clarissa Ellen Coffey, Aug. 9, 1893; children—Jay Russell, Wilson Bryan, Robert Mayo. Began practice Moscow, Ida., 1892; moved to Portland, Ore., 1900; owner and chief surgeon Dr. Robert C. Coffey Clinic and Hosp. Fellow A.C.S. Devised method of treatment of gastro-enteroptosis, entitled the "hammock operation." Wrote monograph on Gastroptosis; chapter on Diseases of the Pancreas, in Binnie's Regional Surgery. Died Nov. 9, 1933.

COFFIN, James Henry, Am. mathematician, meteorologist; b. Williamsburg, Mass., Sept. 6, 1806; s. Matthew and Betsy (Allen) C.; grad. Amherst Coll., 1828, M.A. (hon.), 1831; LL.D., Rutgers, 1859; m. Aurelia Jennings, 1833; m. 2d, Abby Young, 1851. Opened pvt. sch. for boys, Greenfield, Mass., 1829, added manual labor dept., 1830 (1st sch. of its kind in U. S., was beginning of Fellenberg Manual Labor Instn.); taught at Ogensburg, N.Y., 1837-39; tutor Williams Coll., 1840-43; prof. math. and natural philosophy Lafayette Coll., 1846-73; collaborated in work of Smithsonian Instn., pub. results of meteorol. studies under auspices Smithsonian Instn. as: Winds of the Northern Hemisphere, 1853; Psychrometrical Tables, 1856; The Orbit and Phenomena of a Meteoric Fire Ball, 1869; The Winds of the Globe, or the Laws of the Atmospheric Circulation over the Surface of the Earth, pub. posthumously, 1876; author textbooks Solar and Lunar Eclipses, 1845; Analytical Geometry, 1849; Conic Sections, 1850. Research in meteorology; established observatory on Mt. Greylock, 4,000 feet above sea level; advanced his theory of atmospheric circulation, which included Buys-Ballot's law. Died Easton, Pa., Feb. 6, 1873.

COFFIN, John Huntington Crane, Am. mathematician; b. Wiscasset, Me., Sept. 14, 1815; grad. Bowdoin Coll., 1834, LL.D., 1884; married, 5 children. Prof. math. on various vessels, U. S. Navy, 1836-43; with U. S. Naval Obs., Washington, 1843-53; prof. math., astronomy and navigation U. S. Naval Acad., 1853-66; chief editor Nautical Almanac, 1866-77. Mem. Am. Acad. Scis., Am. Philos. Soc., Nat. Acad. Scis. Author: Observations with the Mural Circle, with Formulas, Tables, and Discussions, 1845-49; The Compass, 1863; Navigation and Nautical Astronomy, 1868; Observations of the Total Eclipse of the Sun,

August 1869, 1884; also wrote various math. papers. Died Washington, D.C., Jan. 8, 1890.

COFFIN, Selden Jennings, Am. astronomer; b. Ogdensburg, N.Y., Aug. 3, 1838; s. James Henry and Aurelia M. (Jennings) C.; A.B., Lafayette Coll., 1858, A.M., 1861; grad. Princeton Theol. Sem., 1864; Ph.D., Hanover Coll., 1876; m. Mary A. Angle, N.J., Dec. 22, 1875 (dec. 1889); 2d, Emma F. Angle, Dec. 23, 1891. Tutor and adj. prof. math. Lafayette Coll., 1864-76, prof., 1876-86, registrar, 1886-1904, prof. astronomy, 1873-1915. Author: Conic Sections, 1878; Record of the Men of La Fayette, 1891. Compiled his father's The Winds of the Globe, 1876; revised Olmsted's Astronomy, 1882. Died Mar. 15, 1915.

COFFIN, Walter Harris, dentist; b. Portland, Me., 1853; s. C. R. Coffin; ed. Royal Coll. Sci., Univ. Coll., St. Thomas's Hosp., Royal Sch. Mines, Royal Dental Hosp.; m. Sophia Lydia Walters. Former lectr., examiner Nat. Dental Hosp. Fellow Chem. Soc., Royal Soc. Medicine, Phys. Soc. London; mem. Brit. Assn., Brit. Dental Assn. (past pres. Met. br.), Soc. for Psychical Research (founder), Fabian Soc., also hon. and corr. mem. various Am. and fgn. sci. orgns. Editor: Trans. Odontological Soc., Brit. Dental Jour., 13 years. Research and publs. on use of peroxide of hydrogen, generalized treatment of irregularities, gutta percha impressions in orthodontia, primary batteries. Died Apr. 10, 1916.

COFFIN, William Carey, Am. engr., architect; b. Pitts., Sept. 7, 1862; s. William Carey and Jane McCormick (Osborne) C.; C.E., Western U. Pa. (now U. Pitts.), 1883; D.Sc., U. Pitts., 1936; m. Vida Hurst, 1889. With Keystone Bridge Co., 1883; chief engr. Ft. Pitt Boiler Works 1883-85; with Riter-Conley Mfg. Co., 1885-1908, v.p. from incorporation, 1898; asst. gen. sales agent Jones & Laughlin Steel Co., 1909-15; v.p. Blaw-Knox Co., 1915-23; engr. and architect, 1923-27, ret. Designed and built some of largest blast furnaces, steel plants and oil refineries in U. S. and Can. and secured many large contracts in U. S. and fgn. countries, including electric power houses in Dublin, Ireland, Glasgow, Scotland, and Bristol, Eng.; later designed and built several residences at Miami Beach, Fla. Am. Iron and Steel Inst. Author: Governmental Regulation of Cooperation in Trade, Seeds of Progress and Success, New Approach to Spiritual Revival, The Place of Big Business in a Democracy, Enduring Faith; also studies of social economic conditions in Europe, and S. Am. Died Dec. 4, 1944.

COFFON, J(ean) B(aptiste) G., physician; b. Cerdon, France 1658; studied at Lyons, France; M.D., Montpellier, France. Chief physician Army of Alps; followed Marshal Tessé to Spain; returned to France, settled in Lyons. Author: Résponse aux observations de Chicoyneau, Verney et Soulier sur la nature, les événements et le traitement de la peste à Marseille, 1721; Relation et dissertation sur la bête de Gévaudon, 1722. His sanitary measures in Lyons protected many inhabitants from contagious diseases; blamed plague on insects in blood. Died Lyons, Sept. 30, 1730.

COFMAN-NICORESTI, Carol Adolph, biochemist; b. Nicoresti, Moldavia, Sept. 1881; s. Alter Moise and Gisella (Witling) C.-N.; ed. Lycée Cordeanu, London South Coll., King's Coll., St. Thomas Med. Sch.; B.èsLett.ès.Sc.; m. Beatrice Winchcombe, 1914; 4 daus. Came to London, 1899; studied chemistry, pharmacy, medicine; mng. dir. Solidol Chem. Ltd.; dir. Société Française du Lysol Solidol S/A., Solidol G.m.b.H. Mem. trade marks and poison law coms. Brit. Chem. Mfrs. Fellow Royal Soc. Medicine; mem. Pharm. Soc. Research and publs. on energy, constitution of matter. Patentee for solidification of Lysol, volatile oils; turpentine, various med., chem. and hygienic appliances, preservation of mushrooms. Died Apr. 27, 1938.

COGAN, David Glendenning, Am. physician; b. Fall River, Mass., Feb. 14, 1908; s. James Joseph and Edith (Ives) C.; B.A., Dartmouth, 1929; M.D., Harvard, 1932; m. Frances Capps, July 14, 1934; children—Frances (Mrs. William W. Parson), Ann, Priscilla. Faculty, Harvard Med. Sch., 1937—, prof. ophthalmology, 1955—, dir. Howe Lab. of Ophthalmology, 1940—, editor-in-chief Archives of Ophthalmology, 1960-66, Henry Willard Williams prof. ophthalmology, chmn. dept., 1963—. Mem. Am. Acad. Arts and Scis., Am. Ophthalmol. Soc. Author: Neurology of Ocular Motor System, 1948; Neurology of the Visual System, 1965. Publs. on studies of clin. physiology and pathology of eye. Home: 30 Clark St., Belmont, Mass. 02178. Office: 243 Charles St., Boston 02114.*

COGAN, Thomas, Brit. physician; b. Chard, Eng., circa 1545; B.A., Oxford (Eng.) U., 1563, M.A., 1566, M.B., 1574; m. Ellen Trafford, before 1586. Master, Manchester (Eng.) Grammar Sch., 1575; practice medicine Manchester; family physician Sir Richard Shuttleworth, 1591-93. Author: The Well of Wisdom, 1577; The Haven of Health, 1584; Epistolarum familiarium M.T. Ciceronia epitome; Epistolae item aliae familiares Ciceronis; Orationes aliquot faciliores Ciceronis. Died 1607.

COGGESHALL, Arthur Sterry, Am. paleontologist; b. Bridgeport, Conn., July 17, 1873; s. Sterry Israel and Harriet Ellen (Jeffries) C.; D.Sc., Occidental Coll., 1950; m. Jennie Louise Smith, Apr. 24, 1895; children—Ethyl Adele (Mrs. Elmer P. Kuhn), Mildred Olive (Mrs. Benedict Kristoff), Hazel Eloise (Mrs. Rigby Pogmore); m. 2d, Adelaide Arneson, Oct. 28, 1946. With Am. Mus. Natural History, N.Y.C., 1896-99, curator pub. edn., preparator-in-chief dept. paleontology, Carnegie Mus., Pitts., 1899-1929; designed and perfected cast steel method of mounting large Dinosaurus, 1904; dir. St. Paul Inst., 1929-31; chief Ill. State Mus., Springfield, 1931-37; dir. Santa Barbara (Cal.) Mus. Natural History, 1937-58. Specialized work, Dinosaurs, Brit. Mus. Natural History Br., 1905; Natural History Mus., Jardin des Plantes, Paris, France, 1901; Mus. Fur Nature Künde, Berlin, 1908; Royal Mus. Natural Hist., Vienna, Austria, 1909; Musco Geologico, Bologna, Italy, 1909; Imperial Mus., St. Petersburg, Russia, 1911; Nat. Mus. of Argentina, La Plata, Argentina, 1912; Nat. Mus. Spain, Madrid, 1913. Mem. forest adv. com., sec. Western Mus. Conf. Fellow A.A.A.S.; mem. Am. Assn. Museums (council). Lectr. natural history subjects and travels. Died Aug. 13, 1958.

COGGESHALL, George Whiteley, Am. chemist; b. Des Moines, Dec. 21, 1867; s. John M. and Mary J. (Whiteley) C.; B.S., Grinnell Coll., 1890; postgrad. Harvard, 1891-92; Ph.D., Leipzig, 1895; m. Anna Torrey, Sept. 6, 1900; children—Elizabeth (Mrs. John C. West), Mary (Mrs. John H. Hollands), Dorothy (Mrs. Walter P. Wilson). Instr. phys. chemistry, Harvard, 1895-97; developed new chem. products, 1898-1910; chief chem. engr. Inst. Indsl. Research, Washington, 1911-23; dir. Indsl. Research Labs., 1923-29; dir. research, S. D. Warren Co., 1925-40. In charge surfacing concrete vessels, Emergency Fleet Corp., 1918. Fellow A.A.A.S., Am. Acad. Arts and Scis.; mem. Am. Chem. Soc., Am. Electrochem. Soc., Am. Inst. Mining and Metall. Engrs., Soc. Chem. Industry, Am. Inst. Chem. Engrs., Washington Acad. Scis., Am. Genetic Assn., Am. Geog. Soc. Contbr. papers on standard calomel, electrodes, titanium mordants, potash from feldspathic rocks. Developed various chem. processes, including Portland cement, rare metal compounds, gasoline from heavy oils, phosphates, paper pulp, paper specialties. Died Nov. 18, 1944.

COGGESHALL, Norman David, Am. physicist; b. Ridgefarm, Ill., May 15, 1916; s. Lester B. and Grace (Blaisdell) C.; B.A., U. Ill., 1937, M.S., 1938, Ph.D. in Physics, 1942; m. Margaret J. Danner, Aug. 22, 1940; children—Nancy, David, Gwen, Philip. Instr. physics U. Ill., 1942-43; research physicist Gulf Research & Devel. Co. Pitts., 1943-50, asst. dir. physics div., 1950-55, dir. analytical scis. div., 1955-61, dir. phys. scis. div., 1961-66, dir. process scis. dept., 1966—. Mem. adv. panel Nat. Bur. Standards. Fellow Am. Phys. Soc.; mem. Am. Chem. Soc., Am. Soc. Testing Materials (chmn. E-14 mass spectrometry), Corporate Assos., Am. Inst. Physics, Am. Petroleum Inst. (adv. coms.). Research, patents, publs. in molecular spectroscopy, chem. physics and process instrumentation. Home: 1322 Barbara Dr., Verona, Pa. 15147. Office: P.O. Drawer 2038, Pitts. 15147.*

COGHILL, George Ellett, Am. anatomist; b. Beaucoup, Ill., Mar. 17, 1872; s. John Waller and Elisabeth (Tucker) C.; student Shurtleff Coll., Alton, Ill., 1891-94; A.B., Brown, 1896, Ph.D., 1902, D.Sc., 1934; M.S., U. of N.M., 1899; Sc.D., Pittsburgh, 1931, Denison, 1933; m. Muriel Anderson, Sept. 13, 1900; children—Robert De Wolf, James Tucker, Louis Waller, Muriel, Benjamin Anderson. Asst. prof. biology U. N.M., 1899-1900; prof. biology Pacific U., Forest Grove, Ore., 1902-06; prof. biology and embryology, Willamette U., Salem, Ore., 1906-07; prof. zoölogy, Denison U., Granville, O., 1907-13; asso. prof. anatomy U. Kan., 1913-15, prof., 1915-25, head dept., 1918-25, sec. Sch. Medicine, 1918-24; prof. comparative anatomy, Wistar Inst. Anatomy and Biology, 1925-27, mem. bd. advisers, Wistar Inst., from 1926, mem. inst., 1927-36; mng. editor Jour. Comparative Neurology, 1927-33; vis. lectr. advanced anatomy, Univ. Coll., London, 1928. Recipient Daniel Giraud Elliot gold medal Nat. Acad. Scis., 1934. Author: Anatomy and the Problem of Behavior, 1929. Research on nervous system of Amphibia; worked on correlation of structure and function in devel. nervous system in relation to behavioral devel. Died July 23, 1941.

COGHILL, Robert DeWolf, Am. chemist; b. Providence, Oct. 5, 1901; s. George Ellett and Muriel (Anderson) C.; A.B., U. Kan., 1921, M.S., 1922; Ph.D., Yale, 1924; D.Sc., Bradley U., 1952; m. Margaret Butcher, June 10, 1925 (div. Aug. 1936); children—William Sheridan, Mary Ann (Mrs. Gerald P. Murphy); m. 2d, Caroline Hale Arnold, Sept. 17, 1936 (dec. Aug. 1946); children—Jane Griswold (Mrs. James T. Elliott), George Ellett; m. 3d, Marjorie Moburg, Nov. 27, 1947. Nat. Tb Assn. Research fellow Yale, 1924-26, instr., 1927-28, asst. prof., 1928-39; chief fermentation div. No. Regional Research Lab., U. S. Dept. Agr., Peoria, Ill., 1939-45; dir. research Abbott Labs., North Chicago, Ill., 1945-57; spl. asst. to dir. Nat. Cancer Inst., NIH, Bethesda, Md., 1957-65; sci. adviser life scis. area Stanford Research Inst., Menlo Park, Cal., 1965—; v.p. Indsl. Research Inst., N.Y.C., 1956-57, pres., 1957. Trustee Biol. Abstracts. Cited for Distinguished Service, U. Kan., 1947; recipient U. S. Pres.'s medal for merit, 1948. Mem. Am. Chem. Soc. (Midwest award St. Louis sect. 1949), Soc. Am. Bacteriologists, N.Y. Acad. Scis., A.A.A.S., Am. Soc. Biol. Chemists, Sigma Xi. Author: Organic Compounds (with Julian M. Sturtevant), 1936; also numerous articles. Patentee in field. Research on improvement penicillin prodn., new drugs, removal horse-specific antigen from antitoxin, rendering it safe for sensitized persons. Address: 3630 Tahoma Pl. W., Tacoma 98466.*

COHART, Edward Maurice, Am. physician; b. N.Y.C., Dec. 8, 1909; s. Maurice A. and Emma (Chess) C.; A.B., Columbia, 1928, M.D., 1933, M.P.H., 1947; M.A. (hon.), Yale, 1956; m. Mary Schleifer, Oct. 19, 1933; children—Paul, Diana (Mrs. David A. Reisen). Practice medicine specializing in internal medicine, N.Y.C., 1935-42; faculty Yale Sch. Medicine, 1948—, prof. pub. health, 1956—, chmn. dept., 1966—. Dep. commr. N.Y.C. Dept. Health, 1955-56. Diplomate Am. Bd. Preventive Medicine. WHO Travelling fellow, 1963, 65. Fellow Am. Coll. Preventive Medicine; mem. A.A.A.S., Am., Conn. (past pres.) pub. health assns., A.M.A., Conn. Acad. Preventive Medicine (past pres.), Pub. Health Cancer Assn. (past pres.), New Eng. Pub. Health Assn. (pres. 1966), Assn. Schs. Pub. Health (sec.-treas. 1965—), Assn. Tchrs. Preventive Medicine. Contbr. articles to tech. jours., chpts. to books. Epidemiological research in cancer; operations research in publ. health practice. Home: 625 Ellsworth Av., New Haven 06511.*

COHEE, George Vincent, Am. geologist; b. Indpls., Feb. 4, 1907; s. Frank Lloyd and Stella (Holsapple) C.; B.S., U. Ill., 1931, M.S., 1932, B.S., 1933, Ph.D., 1937; m. Vera French, Aug. 14, 1930. Asst. geologist oil and gas div. Ill. Geol. Survey, 1936-42; asst. state geologist Ind. Geol. Survey, 1942-43; petroleum analyst Petroleum Adminstrn. for War, 1943; research asso. U. Mich., 1947-49; chmn. dept. geology U. Ark., 1951-52; with U. S. Geol. Survey, 1943-51, 1952—, chmn. geologic names com., also chief geologic names rev. staff, Washington, 1952—. Vice chmn., sec. Am. Commn. on Stratigraphic Nomenclature, 1954-55, 1963-64; v.p. N.Am. Sub-commn. for Tectonic Map of World, Internat. Geol. Congress, 1960-65. Fellow Geol. Soc. Am.; mem. Soc. Econ. Paleontologists and Mineralogists, Am. Assn. Petroleum Geologists (hon. mem., sec.-treas. 1960-62), Geol. Soc. Washington (sec. 1956-57, pres. 1965), Geochem. Soc., Sigma Xi. Numerous publs. on stratigraphy especially of Mich. structural basin. Home: 5508 Namakagan Rd., Washington 20016. Office: U. S. Geol. Survey, Washington 20242.*

COHEN, Alan Seymour, Am. physician; b. Boston, Apr. 9, 1926; s. George I. and Jennie (Laskin) C.; A.B. magna cum laude, Harvard, 1947; M.D. magna cum laude, Boston U., 1952; m. Joan Elizabeth Prince, Sept. 12, 1954; children—Evan Bruce, Andrew Hollis, Robert Adam. Sr. asst. surgeon USPHS, 1953-55; research fellow Mass. Gen. Hosp.-Harvard Med. Sch., 1956-58; instr. medicine Harvard Med. Sch., 1959; faculty Boston U. Sch. Medicine, 1960—, asso. prof. medicine, 1962-65, asso. prof. medicine and pharmacology, 1965—; dir. arthritis and connective tissue sect. U. Hosp., Boston U. Med. Center, 1960—; vis. physician, mem. Evans Meml. U. Hosp., Boston, 1963—. Cons. VA, Boston, 1960—, Commonwealth of Mass., Nat. Found., N.Y.C., 1962—; mem. med. research com., arthritis tng. grants com. Nat. Inst. Arthritis and Metabolic Diseases. Recipient Maimonides award Greater Boston Med. Soc., 1952. Mem. Am. Soc. for Clin. Investigation, Am. Fedn. Clin. Research, Electron Microscope Soc. Am., Soc. for Exptl. Biology and Medicine, Am. Soc. for Cell Biology, Am. Soc. for Exptl. Pathology, N.Y. Acad. Scis., Am. Rheumatism Assn. (exec. com. 1962—), New Eng. Rheumatism Soc. (pres.), A.M.A., Mass. Med. Soc., Arthritis Found. (exec. bd., medicine and scis. com. Mass. chpt. 1965—). Editor: Laboratory Diagnostic Procedures in the Rheumatic Diseases, 1967. Research, numerous publs. on fine structure, chemistry of amyloid fibril and its high resolution ultrastructure, chloroquine in treatment of rheumatoid arthritis; first to describe specific fibrous appearance of amyloid in electron microscope. Home: 54 Winston Rd., Newton Center, Mass. 02159. Office: Arthritis and Connective Tissue Disease Sect., U. Hosp., 750 Harrison Av., Boston 02118.*

COHEN, Albert Kircidel, Am. sociologist; b. Boston, June 15, 1918; s. Morris and Clara (Scolnick) C.; A.B., Harvard, 1939, Ph.D., 1951; M.A., Ind. U., 1942; m. Naty B. Manguerra, Dec. 12, 1948. Dir. orientation Ind. Boys Sch., Plainfield, 1942; faculty Ind. U., 1947-65; prof. sociology U. Conn., Storrs, 1965—. Vis. asso. prof. U. Cal., Berkeley, 1960-61; fellow Center for Advanced Study in Behavioral Scis., Stanford, Cal., 1961-62. Mem. Am., Eastern sociol. socs., Soc. for Study Social Problems, Internat., Am. socs. criminology. Author: Delinquent Boys: The Culture of the Gang, 1955; Deviance and Control, 1966. Editor: (with Karl F. Schuessler and Alfred R. Lindesmith) The Sutherland Papers, 1956. Devel. gen. theory of deviant behavior and its control. Home: Route 3, Box 151, Storrs, Conn. 06268.*

COHEN, Alvin Jerome, Am. geochemist; b. Louisville, July 21, 1918; s. Herman H. and Lillie (Levy) C.; B.S., U. Fla., 1940; Ph.D., U. Ill., 1949; postgrad. California Institute of Technology, Tech.; m.; children—Amanda Ellen, Anne Jessica, David Jefferson. With phys. chemistry div. U. S. Naval Ordnance Test Sta., 1951-53; fellow in glass research Mellon

Inst., Pitts., 1953-58, sr. fellow, 1958-62; prof. geochemistry, dept. earth and planetary scis. Space Research Coordination Center, U. Pitts., 1963——. Cons. to glass industry, to industry on problems relating to lunar surface. Mem. A.A.A.S., Am. Chem. Soc., Am. Astron. Soc., Am. Phys. Soc., Am. Ceramic Soc., Geochem. Soc., Chem. Soc. London, Soc. Glass Technology, Am. Geophys. Union, Mineralog. Soc. Research, numerous publs. in field of theory of terrestrial origin of tektites by impact of asteroids or comets producing impact explosion crater, radiation effects in quartz and silicate glasses, oxidation, reduction reactions in solid state, shock deformation in minerals by meteorite impact, lunar surface; inventor variable transmission glass. Office: 506 Langley Hall, Univ. Pitts., Pitts. 15213.*

COHEN, Bernard L., Am. physicist; b. Pitts., June 14, 1924; s. Samuel and Mollie (Friedman) C.; B.S., Case Inst. Tech., 1944; M.S., U. Pitts., 1948; D.Sc., Carnegie Inst. Tech., 1950; m. Anna Foner, Mar. 30, 1950; children—Donald, Judith, Frederick, Ernest. With Oak Ridge Nat. Lab., 1950-58; prof. U. Pitts., 1958, dir. Schaife Nuclear Physics Lab., 1965——. Fellow Am. Phys. Soc. Research in various fields of nuclear physics including nuclear reactions and nuclear structure. Home: 5414 Albemarle Av., Pitts. 15217.*

COHEN, Bertram, dentist; b. Johannesburg, South Africa, Aug. 27, 1918; s. Morris and Pauline (Solomon) C.; B.D.S., U. Witwatersrand, 1942, D.D.S., 1959; M.S.D., Northwestern U., Chgo., 1948; m. Hazel Alice Evans, Oct. 3, 1950. Lectr., U. Witwatersrand, 1948-54; C. J. Adams research fellow U. London, Postgrad. Med. Sch., 1954-55; sr. Leverhulme research fellow in oral pathology Royal Coll. Surgeons, London, 1956-60, Nuffield prof., dir. dept. dental sci., 1960——. Chmn. dental com. Med. Research Council. Fellow Am. Coll. Dentists, Royal Soc. Medicine, Internat. Acad. Pathology, Pathology Soc. Great Britain. Mem. editorial bds. Brit. Dental Jour., Internat. Dental Jour., Dental Abstracts, Oral Research Abstracts. Research, publs. on devel. and morphological pecularities of tissues between teeth and their significance in devel. of periodontal disease; study of oropharyngeal tumors. Home: 3 Northwick Terrace, London N.W. 8. Office: Royal Coll. Surgeons, London W.C. 2, Eng.*

COHEN, David, linguist, anthropologist; b. Tunis, Tunisia, July 24, 1922; s. Albert and Mathilde (Koskas) C.; Licences, Sorbonne, Paris; Diplômes, École Nationale des Langues Orientales; Doctorate Hautes Études; m. Christiane Buret, Oct. 29, 1946; children—Danielle, Jean, Jacqueline, Elisabeth. Staff, Centre National de la Recherche Scientifique, also 4th sect. École des Hautes Études. Cons. for Afro-Asiatic langs. and cultures Conseil de l'Europe. Mem. Société de Linguistique, Assn. pour la Traduction Automatique, Société de Sociologie, Société Asiatique. Author: Dialecte arabe de Mauritanie, 1963; Parler des Juifs de Tunis, 1965; Études de Linguistique Semitique et Arabe; Dictionnaire des Racines Semitiques; also articles. Research on comparative grammar of semitic langs.; comparative dialectology of Arabic; automatic analysis of classical Arabic; contributed to critical reappraisement of aggrammatism and socalled verbal deafness in lang. pathology. Home: 6 Louvois Viroflay (Sein and Oise) 78, France. Office: 47 des Ecoles Paris 5e, France.*

COHEN, D(avid) Walter, Am. periodontologist; b. Phila., Dec. 15, 1926; s. Abram and Goldie (Schlein) C.; student U. Pa., 1943-45, D.D.S., 1950; research fellow Beth Israel Hosp., Boston, 1950-51; m. Betty Ann Axelrod, Dec. 19, 1948; children—Jane Ellen, Amy Sue, Joanne Louise. Faculty, U. Pa., Phila., 1951——, prof. periodontics, chmn. dept., Sch. Dental Medicine, 1963——; vis. lectr. Temple U., 1953——; vis. prof. U. Ill., 1962, Boston U., 1964——; chief periodontics Phila. Gen. Hosp., 1951——; also cons. to surgeon gen. USPHS, USAF, VA, Army hosps. Diplomate Am. Bd. Periodontology. Fellow Am. Acad. Oral Pathology, A.A.A.S., Am. Coll. Dentists, mem. Am. Soc. Periodontists (pres. 1965), Am. Dental Assn., Internat. Assn. Dental Research, N.Y. Acad. Scis., Acad. Stomatology, Am. Acad. Periodontology, Omicron Kappa Upsilon, Sigma Xi, others. Co-author 4 books on periodontia, dental anatomy, current therapy. Sci. editor Alpha Omegan, 1961-64; editorial staff Current Contents. Research and publs. on pathology of the soft tissues of the mouth and oral skeleton; periodontal diseases in children; pain killers. Home: 124 Colwyn Lane, Bala Cynwyd, Pa. 19004. Office: 4001 Spruce St., Phila. 19104.*

COHEN, Emil Wilhelm, geologist; b. Aakjär nr. Horsens, Jutland, Denmark, Oct. 12, 1842; s. Joseph Gerson and Bertha Emilie (Haln) C.; studied chemistry and physics in Berlin, Germany, Heidelberg, Germany, beginning 1863; m. Lina Häusser, 1875. Became lectr. at Heidelberg, 1871; traveled through gold and diamond fields of S. Africa, 1873-74; became asso. prof. petrography, Strasbourg, France, 1878; adminstrv. mem. geol. commn., dir. geol. Land Survey of Alsace-Lorraine; became prof., 1884; apptd. prof. mineralogy and geology, Greifswald, Germany, 1885. Author: Sammlung von Mikrophotographien zur Veranschaulichung der mikroskopischen Struktur von Mineralien und Gesteinen, 1880-84; Meteoreisenstudien, 1891-1900; Meteoritenkunde, 1894-1905. A founder of modern petrography; studied cosmic petrography; made analytic and microscopic studies of meteor iron and its minerals; leader in meteorology. Died Greifswald, Apr. 13, 1905.

COHEN, Ernst Julius, Dutch chemist; b. Amsterdam, Netherlands, Mar. 7, 1869; Dr. degree U. Amsterdam, 1893; asst. to van 't Hoff U. Amsterdam, 1893; prof. phys. chemistry U. Utrecht (Netherlands), from 1902. Fellow Royal Soc., 1926; mem. Royal Dutch Acad. Scis. Author: J. H. van 't Hoff, 1899; Das Lachgas Eine Chemisch-kulturhistorische Studie, 1902; Uit het land van Benj. Franklin, 1928. Research and publs. on allotropes of tin and antimony, metastability, electrochemistry, piezo chemistry. Died Auschwitz, Poland, Mar. 5, 1944.

COHEN, Ezechiel Godert David, physicist; b. Amsterdam, Holland, Jan. 16, 1923; s. David Ezechiel and Sophie Louise (de Sterke) C.; B.S. in Math., Physics and Astronomy, U. Amsterdam, 1947, Ph.D., 1957; m. Maria Arnoldina Linnekamp, Apr. 19, 1950; children—Michael Benjamin, Andrea Margaret. Came to U. S., 1963. First asst. U. Amsterdam, 1950-61, asso. prof., 1961-63; research asso. U. Mich., 1957-58, Johns Hopkins, 1958-59; prof., mem. Rockefeller Inst., 1963——. Mem. Am., Netherlands phys. socs. Editor: Fundamental Problems in Statistical Mechanics, I, 1961, II, 1968. Research in statis. mechanics and its applications in various brs. of physics. Home: 144-15 Charter Rd., Jamaica, N.Y. 11435. Office: Rockefeller U., York Av. and 66th St., N.Y.C. 10021.*

COHEN, Jacob (da Silva) Sòlis, Am. physician; b. N.Y., Feb. 28, 1838; s. Myer David and Judith Simirah (da Silva Solis) C.; M.D., U. Pa., 1860; LL.D., Jefferson Med. Coll., Temple U., 1912; m. Miriam Bingswanger, Feb. 10, 1874. Mil. surgeon, 1861-65; practiced medicine specializing in diseases of throat and chest, Phila., from 1866; hon. prof. laryngology Jefferson Med. Coll., Phila.; emeritus prof. diseases of throat Phila. Polyclinic and Coll. for Graduates in Medicine; cons. phys. Hosp. for Insane, S.E. Dist. Pa. Author: A Treatise on Inhalation; Diseases of the Throat and Nasal Passages, 2d ed., 1879; Croup in Its Relations to Tracheotomy; The Throat and the Voice, 1880. Pioneered in devel. of laryngology; credited with performing 1st successful operation for cancer of larynx, recorded 1867. Died Phila., Dec. 22, 1927.

COHEN, Jerome Bernard, Am. metallurgist; b. Bklyn., July 16, 1932; s. David Israel and Shirley (Silverman) C.; S.B., Mass. Inst. Tech., 1954, Sc.D., 1957; m. Lois Nesson, Sept. 15, 1957; children—Elissa, Andrew. Sr. scientist AVCO Corp., Wilmington, Mass., 1958-59; faculty Northwestern U., Evanston, Ill., 1960——, prof. materials sci., 1965——; liaison scientist Office Naval Research, London, Eng., 1966-67. Fulbright scholar, Paris, 1957-58. Mem. Am. Inst. Mining, Metall. and Petroleum Engrs. (Hardy medal 1960), Am. Assn. U. Profs., Am. Crystallographic Assn., A.A.A.S., Sigma Xi, Tau Beta Pi, Phi Lambda Upsilon. Author: Diffraction Methods of Materials Science, 1966; also articles. Application of X-ray diffraction to study of deformation and local atomic arrangements in alloys, ceramics, structure of compounds, thermodynamics of alloys. Home: 362 Jackson Av., Glencoe, Ill. 60022. Office: Materials Sci. Tech., Northwestern U., Evanston, Ill. 60201.*

COHEN, Julius Berend, English chemist; b. Manchester, Eng., May 6, 1859; s. Sigismund and Zarah Zara Cohen; student Owens Coll., Manchester, Eng.; B.Sc., Victoria; Ph.D., U. Munich (Germany); D.Sc. (hon.), Leeds, Eng.; LL.D., Glasgow; m. Hilda Hughes, 1892; 1 son, 2 daus. With Clayton Aniline Col., 1880-82; demonstrator Victoria U., Manchester, 1884-90; lectr. organic chemistry Yorkshire Coll., 1890-1904; prof. organic chemistry Leeds U., 1904-24. Mem. chemotherapy com. Med. Research Com., 1924-32. Fellow Royal Soc., 1913. Author: Researches on Organic Chemistry; Theoretical Organic Chemistry; Practical Organic Chemistry; Organic Chemistry for Advanced Students, 3 parts; Class-book of Organic Chemistry, 2 vols.; Smoke, a Study of Town Air. Research on laws of aromatic substitution and optical activity; authority on smoke abatement. Died June 14, 1935.

COHEN, Leonard Arlin, Am. neurophysiologist; b. Bklyn., May 4, 1924; s. Abraham and Alice (Spieller) C.; B.A., U. Conn., 1948; Ph.D., Yale, 1952; m. I. Sheila Gold, June 25, 1950; children—Seth David, Jared Ben, Ellyn Beth. Faculty, Sch. Medicine U. Pitts., 1951-61, asso. prof., 1959-61; head dept. physiology Albert Einstein Med. Center, Phila., 1961——; adj. asso. prof. dept. physiology Sch. Medicine U. Pa., 1961——; prof. dept. phys. medicine and rehab. (physiology) Temple U. Health Scis. Center, Phila., Pa., 1967——. Postdoctoral fellow lab. physiology Oxford U., Eng., 1956-58. Recipient James Hudson Brown award Yale, 1950. Fulbright fellow NSF, 1956-58. Mem. Am. Physiol. Soc., A.A.A.S., Am. Assn. U. Profs., Royal Soc. Medicine, Aerospace Med. Assn., Am. Inst. Aeros. and Astronautics. Established that neck sensation plays a major role in balance and orientation of living body and is at least as important as semicircular canals in inner ear; protection of astronauts against high sustained acceleration by rotating them around their longitudinal axis at right angles to the direction of acceleration. Home: 224 Old Lancaster Rd., Merion, Pa. 19066. Office: Albert Einstein Med. Center, York and Tabor Rds., Phila. 19141.*

COHEN, Louis, physicist; b. Kiev, Russia, Dec. 16, 1876; s. Abraham and Nattie (Resnik) C.; B.Sc., Armour Inst. Tech., 1901; student U. Chgo., 1902; Ph.D., Columbia, 1905; m. Ethel Slavin, Jan. 3, 1904; 1 dau., Mrs. Louis P. Sissman. With scientific staff Bur. of Standards, Washington, 1905-09; with Elec. Signaling Co., 1909-12; cons. practice from 1913; prof. elec. engring., George Washington U., 1916-29; cons. engr., War Dept., 1920-24; lectr. Bur. of Standards, from 1928; U. S. del. Provisional Tech. Com. Internat. Conf. on Elec. Communication, Paris, France, 1921; mem. adv. tech. bd. Conf. on Limitation of Armament, Washington, 1921-22; tech. expert German-Austrian claim commn., 1929-31. Fellow A.A.A.S., Am. Inst. E.E., Am. Inst. Radio Engrs., Am. Phys. Soc. Author: Formulae and Tables for Calculation of Alternating Current Problems, 1913; Heaviside's Electrical Circuit Theory, 1928; also many papers in scientific and tech. jours. Inventor of many devices in radio and cable telegraphy; research on alternating current, theory of elec. circuit. Died Sept. 28, 1948.

COHEN, Marshall Harris, Am. radio astronomer, educator; b. Manchester, N.H., July 5, 1926; s. Solomon and Mollie (Epstein) C.; B.Elec. Engring., Ohio State U., 1948, M.S., 1949, Ph.D., 1952; m. Shirley Kekst, Sept. 19, 1948; children—Thelma, Linda, Sara. Research asso. Ohio State U., 1950-54; faculty Cornell U., 1954-66; prof. applied electrophysics U. Cal. at San Diego, 1966——. Vis. asso. prof. radio astronomy Cal. Inst. Tech., 1965. Guggenheim fellow Paris Obs., 1960-61. Fellow A.A.A.S.; mem. Am. Astron. Soc., Am. Geophys. Union, Internat. Astron. Union, Internat. Sci. Radio Union. Research, publs. on radio bursts from sun, scintillation of radio sources, diameters of radio sources. Home: 8587 La Jolla Scenic Dr., La Jolla, Cal. 92037. Office: Dept. Applied Electrophysics, U. Cal., San Diego 92037.*

COHEN, Morris, Canadian chemist; b. Regina, Sask., Can., July 10, 1915; s. Israel and Esther (Brownstone) C.; B.A., Brandon Coll., 1934; M.A., U. Toronto, 1935, Ph.D., 1939; m. Beatrice Sohn, July 15, 1940; children—Matthew, Andrew. With NRC Can., Ottawa, Ont., 1943——, head metallic corrosion and oxidation, 1946——; vis. prof. U. New S. Wales, 1966. Recipient W. R. Whitney award Nat. Assn. Corrosion Engrs., 1960. Fellow Chem. Inst. Can.; mem. Electrochem. Soc., Am. Chem. Soc., A.A.A.S. Research, numerous publs. on reactions of metal surfaces using techniques of phys. chemistry, electrochemistry and electron optics, passivity, oxidation and electron diffraction, clean metal surfaces and reaction products on such surfaces. Home: 504 Mansfield St., Ottawa 13. Office: NRC, Ottawa, Ont., Can.*

COHEN, Paul Joseph, Am. mathematician; b. Long Branch, N.J., Apr. 2, 1934; s. Abraham and Minnie (Kaplan) C.; student Bklyn. Coll., 1950-53; M.S., U. Chgo., 1954, Ph.D., 1958; m. Christina Martha Karls, 1963; children—Eric, Steven. Instr., U. Rochester, 1957-58, Mass. Inst. Tech., 1958-59; fellow Inst. Advanced Study, Princeton, 1959-61; mem. faculty Stanford, 1961——, prof. math., 1964——. Recipient Fields medal Internat. Math. Union, 1966; Bocher prize Am. Math Soc., 1964; Research Corp. award Research Corp., 1964. Mem. Nat. Acad. Scis. Proved impossibility of demonstrating continuum hypothesis from the axioms of set theory; has studied partial differential equations and harmonic analysis. Home: 755 Santa Ynez St., Stanford 94305.

COHEN, Philip Pacy, Am. biochemist; b. Derry, N.H., Sept. 26, 1908; s. David Harris and Ada (Cottler) C.; B.S., Tufts Coll., 1930; Ph.D., U. Wis., 1937, M.D., 1938; m. Rubye H. Tepper, June 15, 1935; children—Philip T., David B., Julie A., Milton T. Faculty, U. Wis., Madison, 1941——, prof., chmn. dept. physiol. chemistry, 1948——, acting dean Sch. Medicine, 1961-63. Chmn. com. on growth NRC, 1954-56; mem. bd. sci. counsellors Nat. Cancer Inst., 1957-61; cons. U. S. State Dept., U. Mexico, 1958; mem. physiol. chemistry study sect. NIH, 1959-61, mem. nat. adv. cancer council, 1963-67, mem. adv. com. to dir., 1966——; mem. adv. com. biology and medicine U. S. AEC, 1963——. Recipient Commonwealth Fund award Oxford U., 1958; hon. mem. faculty U. Chile Sch. Medicine, 1965. Fellow A.A.A.S.; mem. Am. Chem. Soc., Am. Soc. Biol. Chemists, Biochem. Soc. (Eng.), Harvey Soc., Argentina Biochemical Soc., Sigma Xi. Research, numerous publs. on intermediary nitrogen metabolism, enzymology, transaminases, enzymes of urea biosynthesis, comparative and evolutionary biochemistry, mechanism of thyroxine action, embryogenesis and differentiation. Home: 1117 Oak Way, Madison, Wis. 53705. Office: 1215 Linden Dr., Madison, Wis. 53705.*

COHEN, Robert, Am. physicist; b. Indpls., Oct. 15, 1924; s. Daniel and Bertha (Schmidt) C.; B.S. in Chemistry, Wayne U., 1947; M.S., U. Mich., 1948; postgrad. Purdue U., 1948-51; Ph.D., Cornell U., 1956; m. Carolyn Christie Campbell, July 21, 1963; children—Clark Emerson, Clara King. Research asso. Cornell U., 1951-52; physicist Inst. for Telecommunication Scis. and Aeronomy, ESSA, Boulder, Colo., 1956——. Cons. NASA, 1966——. Mem. I.E.E.E. (wave propagation com.), Research Soc. Am. (Boulder Scientist award 1964), Am. Geophys. Union, Internat. Union Radio Sci. (mem. commns. 3 and 4), Sigma Xi, Kappa Mu Epsilon, Sigma Pi Sigma. Contbg. author: Ionospheric Sporadic E, 1962; Physics of Geomagnetic phenomena, 1967. Research, publs. on very high frequency ionospheric forward scatter, ionospheric irregularities in equatorial electrojet, ionospheric F-region irregularities (Spread F), equatorial aeronomy using incoherent scatter radar, plasma instabilities in ionosphere. Home: Sunshine Canyon. Office: Dept. Commerce Bldg., Boulder, Colo. 80302.*

COHEN, Robert Abraham, Am. physician; b. Chgo., Nov. 13, 1909; s. Ezra Harry and Catherine (Kurzon) C.; S.B., U. Chgo., 1930, Ph.D., M.D., 1935; m. Mabel Jean Blake, Mar. 21, 1933; children—Donald Edward, Margery Jean. Sr. fellow Inst. for Juvenile Research Chgo., 1939-40; clin. dir. Chestnut Lodge, Rookville, Md., 1946-52; dir. clin. investigations, NIH Nat. Inst. Mental Health, Bethesda, Md., 1952-. Fellow Am. Psychiat. Assn., Washington Sch. Psychiatry; mem. Am., Washington (past pres.) psychoanalytic assns., Washington Psychiat. Assn. (past pres.), Washington Acad. Medicine. Research publs. on manic-depressive psychosis, psychotherapeutic process social psychiatry, research adminstrn. Home: 4514 Dorset Av., Chevy Chase, Md. 20015. Office Nat. Inst. Mental Health, Bethesda, Md. 20014.*

COHEN, Robert Sonné, physicist; b. N.Y.C., Feb. 18, 1923; s. Mordecai M. and Mabel (Reinschreiber) C.; B.A., Wesleyan U., Middletown, Conn., 1943; M.S., Yale, 1943, Ph.D. (NRC fellow); 1948; m. Robin Gertrude Hirshhorn, June 18, 1944; children—Michael, Daniel, Deborah. Instructor physics Yale University, New Haven, Conn., 1943-44, instr. philosophy, 1949-51; sci. staff, div. war research Columbia, 1944-46; asst. prof. physics, philosophy Wesleyan U., 1949-57; asso. prof. physics Boston U., 1957-59, prof., chmn. dept., 1959-; chairman of Boston Colloquium of Philos. Science; vis. lectr. humanities and philosophy of sci. Mass. Inst. Tech., 1958-59, 61-62; vis. prof. history of ideas Brandeis U., 1959-60; Inst. History Philosophy and Science lecturer, American University, Washington, summers 1958-65; vis. prof. Yugoslav Acad. Sci., 1963, Hungarian Acad. Sci., 1964. Am. Council Learned Soc. fellow philosophy and sci., 1948-49; Ford faculty fellow, Cambridge, Eng., 1955-56; Inst. Philosophy and Sociology vis. Fellow Polish Acad. Scis., Warsaw, 1962. Mem. A.A.A.S., American Physical Society, American Association Physics Tchrs., Am. Philos. Assn., History Sci. Soc., Philosophy Sci. Assn., Am. Assn. U. Profs., Emergency Civil Liberties Com., Am. Civil Liberties Union, Am. Inst. Marxist Studies (chmn.), N.A.A.C.P. Author articles, books in field. Study of the foundations of thermodynamics; logical empiricism and natural science; concept and theory formation in the physical sciences; dialectical materialism and science; kinematic and other relativity theories; action-at-a-distance and field theories. Home: 44 Adams Av., Watertown, Mass. 02172. Office: 111 Cummington St., Boston 02215.

COHEN, Sanford Irwin, Am. psychiatrist, educator; b. N.Y.C., Sept. 5, 1928; s. George A. and Gertrude (Slater) C.; A.B. magna cum laude, N.Y. U., 1948; M.B., M.D., Chgo. Med. Sch., 1952; m. Jean Steinbruecker, Nov. 30, 1952; children—Jeffrey, Debra, John, Robert. Faculty dept. psychiatry Duke Med. Center, Durham, N.C., 1956-, prof., 1964-, head div. psychosomatic medicine and psycho-physiol. research dept. psychiatry, 1964-, lectr. dept. psychology, 1960-. Instr., Washington Psychoanalytic Inst., 1964-; cons. VA Hosp., Durham, 1957-; USPHS, Nat. Inst. Mental Health, 1963-66. Markle scholar in med. sci., 1957-62; recipient Robert Morse award for excellence in sci. writing, 1964. Fellow Am. Psychiat. Assn., Am. Coll. Clin. Pharmacology; mem. A.A.A.S., Am. Psychosomatic Soc. Contbr. numerous articles to profl. jours., chpts. to books. Research on psychophysiol. effects of altered sensory environments, central nervous influences on cardiovascular functions and disease. Home: 1334 Welcome Circle, Durham, N.C. 27705.*

COHEN, Saul G(erald), Am. chemist; b. Boston, May 10, 1916; s. Barnet Mayer, and Ida (Levine) C.; A.B., Harvard, 1937, M.A., 1938, Ph.D., 1940; m. Doris Brewer, Nov. 27, 1941; children—Jonathan B., Elisabeth J. Pvt. asst., postdoctoral fellow, instr. Nat. Def. Research Com., Harvard, 1939-43; Nat. Research fellow, instr. U. Cal. at Los Angeles, 1943-44; with Pitts. Plate Glass Co., 1944-45; chief chemist Polaroid Corp., Cambridge, Mass., 1945-50, cons. 1950-; faculty Brandeis U., Waltham, Mass., 1950-, prof., 1952-, chmn. Sch. Sci., 1950-55, dean faculty, 1955-59, chmn. chemistry dept., 1959-66; vis. prof. Harvard Med. Sch., 1965. NSF European Travel grantee, 1957; Guggenheim fellow, Fulbright Sr. Research scholar, U.K., 1958-59. Fellow Am. Acad. Arts and Scis., A.A.A.S.; mem. Am. Chem. Soc., Chem. Soc. London, Phi Beta Kappa, Sigma Xi. Research, publs. in mechanisms of organic reactions, photochemistry, free radicals, vinyl polymerization, high polymers, photog. processes, stereochemistry, enzyme reactions. Home: 39 Moon Hill Rd., Lexington, Mass. 02173. Office: Dept. Chemistry, Brandeis U., Waltham, Mass. 02154.*

COHEN, Seymour Stanley, Am. biochemist; b. N.Y.C., Apr. 30, 1917; s. Herman and Lena (Tanz) C.; B.S., Coll. City N.Y., 1936; Ph.D. in Biol. Chemistry, Columbia, 1941; m. Elaine Pear, July 12, 1940; children—Michael, Sara. NRC fellow Rockefeller Inst., 1941-42; mem. faculty U. Pa., 1943-, prof. biochemistry in pediatrics, 1954-, Charles Hayden Found.-Am. Cancer Soc. prof. biochemistry, 1957-, Hartzell prof., chmn. dept. therapeutic research, Sch. Medicine, 1963-; Guggenheim fellow Pasteur Inst., Paris, France, 1947-48. Trustee Marine Biol. Lab., Woods Hole, Mass.; bd. sci. cons. Sloan-Kettering Inst., mem. adv. com. Aspen Biol. Inst. Recipient certificate war research OSRD, 1945, War Manpower Commn., 1945; War Research medal Columbia, 1943; Eli Lilly award and medal Am. Soc. Bacteriology, Immunology and Pathology, 1951; 1st Mead Johnson award Am. Acad. Pediatrics, 1952; medal Soc. de Chimie Biologique (France), 1964. Fellow A.A.A.S. (Newcomb Cleveland award 1955), Nat. Acad. Scis., Am. Acad. Arts and Scis., Soc. Gen Physiologists (councilor). Editorial bd. Virology, 1954-59, Jour. Biol. Chemistry, 1959-65, Jour. Cell Comp. Physiology, 1966-, Current Topics in Developmental Biology, 1966-. Biochem. studies of virus-infected cells; discovered origin of ribose; contbns. to knowledge of modes of action of antitumor and antibacterial drugs. Home: 43 Rockglen Rd., Phila. 19151.*

COHEN, Solly Gabriel, physicist; b. London, Eng., Nov. 17, 1920; s. Lewis and Fay Cohen; B.A., Cambridge U., 1942, Ph.D., 1949; m. Dalia Carmi, May 1951; children—Haran, Yuul. Research physicist Cavendish Lab., Cambridge, 1942-44, Brit. Govt. Mission to Montreal, Que., Can., 1944-46; research fellow Cambridge U., 1946-49; faculty Hebrew U., Jerusalem, Israel, 1945-, prof. physics, 1961-, dean Faculty Sci., 1965-; vis. prof. Princeton, 1958-59. Mem. Israeli Acad. Scis. (life), Nat. Israeli Council for Research and Devel. (exec. council), Am. Phys. Soc. Research, publs. on low energy nuclear physics and spectroscopy; application to solid state problems; hyperfine interactions in solids. Home: 13 Rashba, Jerusalem, Israel.*

COHEN, Theodore, Am. chemist; b. Arlington, Mass., May 11, 1929; s. Gideon and Lillian Sarah (Leight) C.; B.S., Tufts Coll., 1951; Ph.D., U. So. Cal., 1955; postgrad. (Fulbright grantee, Ramsay Meml. fellow), U. Glasgow (Scotland), 1955-56; m. Pearl Bernice Silverman, July 22, 1954; children—Bret Ronald, Rima Jean. Faculty chemistry dept. U. Pitts., 1956-, prof. chemistry, 1966-. Mem. Am. Chem. Soc., Chem. Soc. (London, Eng.). Research, publs. on reactions of alicyclic and aromatic diazonium ions and pyridine N-oxides, nuclear magnetic resonance studies of aggregation of organic molecules in aqueous solution; discovered organocopper intermediates in copper promoted organic reactions. Home: 2825 Shady Av., Pitts. 15217.*

COHEN, Yves Yvan Charles, biologist, pharmacologist; b. Tunis, Tunisia, Oct. 6, 1927; s. Henri and Mathilde (Nataf) C.; D.Sc., U. Paris (France), 1955; m. Nicole Monnot, July 28, 1951; children—Henri, Anne, Jean-Michel. Attaché de recherches CNRS, 1951-54; prof. École Nationale de Medecine et de Pharmacie, Rouen, France, 1955-61; prof. Faculty Pharmacy, U. Paris, 1962-; chief laboratoire de Controle pharmaceutique des radioéléments Centre D'Etudes Nucleaires de Saclay, 1955—. Corr. mem. Commn. Permanente de La Pharmacopée Francaise, 1960—; expert pharmacologist Ministère des Affaires Sociales, 1960—. Decorated Chevalier dans L'Ordre des Palmes Academiques. Mem. Assn. des Physiologistes, de Therapeutique et Pharmacodynamie, Chimique de France, Société de Biologie. Research, numerous publs. on radio pharmacology, physico-chem. and biol. studies of radioactive isotopes used in medicine, distbn., fate and mode action drugs using radioisotopes and labelled compounds including membrane permeability and receptors binding. Office: 4 Av. de L'Observatoire, 75 Paris 6e, France.*

COHN, Edwin Joseph, Am. biochemist; b. N.Y.C., Dec. 17, 1892; s. Abraham and Maimie (Einstein) C.; student Amherst Coll., 1910-13, D.S., 1944; B.S., U. Chgo., 1914, Ph.D., 1917; postgrad. Harvard, 1915-17, M.A. (hon.), 1943, D.S.; 1945; D.S., Columbia, 1945; M.D. (hon.) Geneva, 1946, Berne, 1947; m. Marianne Brettauer, July 30, 1917 (dec.); children—Edwin J., Alfred; m. 2d Rebekah Higginson, June 15, 1948. NRC fellow in chemistry, 1919-22; studied at Carlsberg Lab., Copenhagen and Cambridge (Eng.) U., asst. prof. phys. chemistry Harvard, 1922-28, asso. prof., 1928-35; prof. biol. chemistry, head dept. phys. chemistry, Harvard Med. Sch., 1935-49, Higgins U. prof., from 1949, chmn. div. med. scis., 1936-49, dir. U. Lab. Phys. Chemistry related to med. and pub. health, from 1949, chmn. dept. biophys. chemistry from 1950. Recipient Alvarenga prize Coll. Physicians Phila., 1942, Passano award for distinguished service to Am. clin. medicine, 1945; John Scott medal, Phila., 1946; John Phillips Meml. medal, 1946, Theodore William Richards medal, 1948 (both A.C.P.); Medal of Science, Free U. Belgium, 1947; Medal of Merit, U. S. Govt., 1948; French Legion Honor, 1952. Silliman lectr. Yale, 1946; Am. Swiss Found. lectr., 1947; Belgian Am. Ednl. Found. lectr., 1947; Julius Streglitz meml. lectr., A.C.S., 1949. Felow A.A.A.S., Am. Acad. Arts and Scis., N.Y. Acad. Sci.; mem. Nat. Acad. Sci., Am. Philos.

Soc., Am. Chem. Soc. Am. Soc. Biol. Chem., Am. Physiol. Soc., Sigma Xi. Contbr. articles on chemistry of natural products and systems, liver fractions, plasma fractions, physical chem. of proteins, blood, and other tissues to profl. jours. Author: (with J. T. Edsall) Proteins, Amino Acids and Peptides, 1943; Research in Medical Sciences, March of Medicine, 1946. Developed new methods for separating component parts of blood; isolated fibrinogen, serum albumin, gamma globulin; extracted substance controlling pernicious anemia from liver, 1927. Died Boston, Oct. 1, 1953.

COHN, Emil, physicist; b. Neustrelitz, Germany, Sept. 28, 1854; s. August and Charlotte (Hahn) C.; studied at Leipzig, Heidelberg, (both Germany), Strasbourg (France); grad. 1878; m. Marie Goldchmidt, 1885; 2 daus. Lectr. theoretical physics at Strasbourg, 1884, then asso. staff prof. 1918; became hon. prof. Rostock, Germany, 1918, Freiburg, Germany, 1920; ret. in Heidelberg, 1935; emigrated to Switzerland, 1939. Author: Das elektromagnetische Feld. Vorlesungen über die Maxwellsche Theorie, 1900. Research on electrodynamics of moving media, modified magnetic fields; determined dielectric properties from the wave length in wire, in water, and in ice, 1888-92; determined electromagnetic field equalization in moving bodies, 1902, 27; measured dielectric constants in crystals, 1931. Died Ringgenberg, Switzerland, Jan. 28, 1944.

COHN, Ferdinand Julius, German botanist; b. Breslau, Jan. 24, 1828; s. Isaak and Amaha (Nissen) C.; ed. Breslau, also Berlin; m. Pauline Reichenbach; became asso. prof. botany, Breslau, 1859, prof., 1971; founder Inst. Plant Physiology, Breslau, 1866, dir., 1872. Mem. Royal Soc., 1897, Linnean Soc. (Gold medal 1895). Author: Über blutahnliche Farbungen, 1850; Die Wunder des Blutes, 1854; Organismen der Pockenlymphe, 1872; Untersuchungen über Bacterien, 1872-75. Significantly improved the microscope; contbd. to cell theory; discovered cilia in zoospores and analyzed their movement; studied morphology and life history of lower algae and fungi; demonstrated that bacteria are plants; discovered proteid crystals in potato; showed that protoplasm of animal and plant cells is essentially the same, thus that there is but one phys. basis of life; studied nature of zoogloea, formation and germination of true spores; bacteria in air and water; considered the founder of bacteriology; assisted Koch in publishing his work on anthrax. Died Breslau, June 25, 1898.

COHN, Fritz, German astronomer; b. Königsberg, Germany (now Kaliningrad, Russia), May 12, 1866; s. Callmann and Henriette (Rosenberg) C.; educated Königsberg and Berlin, Germany; degree in astronomy, 1888; m. dau. of C. F. W. Peters, 1897. With Königsberg Obs., 1891-1909; observator, from 1900; prof. astronomy Berlin, 1909; also dir. Recheninstitute Berlin; asso. Royal Astron. Soc. London. Essays and other publs. in astron. and sci. handbooks, jours.; made meridian observations; developed very accurate new method for determination of planetoid paths. Died Berlin, Dec. 14, 1921.

COHN, Harvey, Am. mathematician; b. N.Y.C., Dec. 27, 1923; s. Morris and Leah (Spielmann) C.; B.S. Coll. City N.Y., 1942; M.S., N.Y. U., 1943; Ph.D., Harvard, 1948; m. Bernice Blaufarb, Mar. 8, 1951; children—Anthony, Susan. Teaching fellow math. Harvard, 1947-48; asst. prof. Wayne U., 1948-54, asso. prof., 1954-56; vis. asso. prof. Stanford, 1954-55; faculty Washington U., St. Louis, 1956-58, prof. math., head Computer Center, 1957-58; prof., head math dept. U. Ariz., Tucson, 1958—. Cons., Argonne (Ill.) Nat. Labs., 1958—; mem. adv. com. Autonomous U. Guadalajara, Mexico, 1963—; mem. com. on regional devel. math. NRC, 1962—. William Lowell Putnam prize fellow Harvard, 1946-47; U. S. Army Office Ordnance Research grantee, 1952-54, NSF grantee, 1959—, Cottrell Research Corp. grantee, 1959-61. Mem. Am. Math. Soc., Assn. Computing Machinery, Math. Assn. Am., Phi Beta Kappa, Sigma Xi. Reviewer, Math. Revs., 1956—, Zentablatt für Mathematik, 1961—. Research on applications of high speed computing to pure math., algebraic and geometric number theory, modular functions of several variables. Home: Box 827, Route 5, Tucson 85718.*

COHN, Isidore, Jr., Am. surgeon; b. New Orleans, Sept. 25, 1921; s. Isidore and Elsie (Waldhorn) C.; B.S., Tulane U., 1942; M.D., U. Pa., 1945, M.Sc., 1952, D.Sc., 1955; m. Jacqueline Heymann, July 4, 1944; children—Ian Jeffrey, Lauren Kerry. With La. State U. Sch. Medicine, 1952-, prof. surgery, 1959-, chmn. dept. surgery, 1962-; sr. vis. surgeon, surgeon-in-chief Charity Hosp. of La., New Orleans, 1962-; cons. VA Hosp., New Orleans; cons. Touro Infirmary, New Orleans. Mem. Am. Bd. Surgery, A.C.S., Am., So. surg. assns., Soc. U. Surgeons, Internat. Surg. Soc., Am. Gastroent. Assn., A.M.A., Assn. for Acad. Surgery, Sigma Xi. Author: Strangulation Obstruction, 1961. Publs. on bacterial and antibiotic studies in intestinal obstruction and large bowel surgery, intestinal antisepsis, cancer transplantation in intestinal surgery. Home: 510 Iona St., Metairie, La. 70005. Office: 1542 Tulane Av., New Orleans 70112.*

COHN, Lassar, German chemist; b. Hamburg, Germany, Sept. 6, 1858; s. Jacob Marcus and Hanna (Hewe) C.; studied at Heidelberg, Bonn, Königsberg (all Germany); grad. 1880; m. Martha Samuel; 1 son, 1 dau. Prof. at Königsberg, 1894-97, 1902-09; Munich, Germany, 1897-98; later asso. with various indsl. cos. Author: Arbeitsmethoden für organischchemische Laboratorien, 1891; Die Chemie in täglichen Leben, 1896; Praxis der Harnanalyse, 1897; Einführung in die Chemie in leichtfasslicher Form, 1899. Research in pure and applied organic chemistry, electrolysis; improved nitrometer and other lab. instruments. Died Königsberg, Oct. 9, 1922.

COHN, Richard M., Am. mathematician; b. N.Y.C., Sept. 2, 1919; s. Hugo and Elsie (Rothschild) C.; B.A., Columbia, 1939, M.A., 1941, Ph.D., 1947. Faculty, Rutgers, The State U., New Brunswick, N.J., 1948—, prof. math., 1959—. Mem. Am. Math. Soc., Math. Assn. Am. Author: Difference Algebra, 1965; also articles. Algebra of difference equations; study of criteria for nonrandomness; properties of impedance function of an elec. network. Home: 203 S. Adelaide Av., Highland Park, N.J. 08904. Office: Rutgers, The State U., New Brunswick, N.J. 08903.*

COHN, Waldo E., Am. biochemist; b. San Francisco, June 28, 1910; s. Max Samuel and Rae (Marcus) C.; B.S., U. Cal. at Berkeley, 1931, M.S., 1932, Ph.D., 1938; m. Charmian Edlin, Apr. 14, 1943; children—Marcus Tufts, Dunell Edlin. Research asso. Harvard Med. Sch., 1939-42; tutor Harvard, 1939-42; chemist Oak Ridge Nat. Lab., Manhattan Project, 1943-47, biochemist, 1948—; vis. prof. U. Paris, 1963; vis. prof. Rockefeller U., 1967; Fulbright scholar, 1955-56; dir. Office Biochem. nomenclature, Nat. Acad. Sci., 1965—. Guggenheim fellow, 1955-56, 62-63. Fellow A.A.A.S., Am. Acad. Arts and Scis.; mem. Am. Soc. Biol. Chemists, Am. Chem. Soc. (Chromaography award 1962), Biochem. Soc. (England). Inventor (with E. R. Tompkins, J. X. Khym) of ion-exchange elution chromatography, with spl. ref. to separation of rare earths, fission products, biochem. substances, nucleic acid constituents. Home: 102 Plymouth St., Oak Ridge 37830. Office: Oak Ridge Nat. Lab., Box Y, Oak Ridge 37630.*

COHNHEIM, Julius Friedrich, German physician pathologist; b. Demmin, Pomerania, Germany, July 20, 1839; studied medicine at Berlin, Würzburg, and elsewhere; prof. in Kiel (Germany), 1868-72; Breslau (now Wroclaw, Poland), 1872-78; and Leipzig (Germany), from 1878. Author: Untersuchungen über die embolischen Prozesse, 1872; Neue Untersuchungen über die Entzündung, 1873; Vorlesungen über allgemeinen Pathologie, 2 vols., 1877-80; Die Tuberkulose vom Standpunkte der Infectionslehre, 1881; Geschichtlicher Abhandlungen, 1885. Described and named pseudoleukemia, 1865; showed pus is composed mainly of leukocytes which have passed through capillary walls, 1867; produced Tb in rabbits, 1877; 1st to freeze tissue for microscopic studies; research on inflammation and embolism; originated "embryonic" theory of cancer. Died Leipzig, Germany, Aug. 15, 1884.

COHRS, Paul, German vet. pathologist; b. Cederan, Saxony, Germany, Mar. 22, 1897; s. Christoph and Liddy (Schilling) C.; Dr.med.vet., Sch. Vet. Medicine, Dresden, 1923; Dr.med.vet.h.c., U. Berlin-E., 1959; m. Martha Dorothea Schoene, Oct. 28, 1928; children—Jens, Hans-Uwe. Faculty, U. Leipzig, Sch. Vet. Medicine, 1924-37, asso. prof., head Inst. Histology and Embryology, 1928-37; prof., dir. Path. Inst., Sch. Vet. Medicine, Hanover, Germany, 1937-65, emeritus, 1965—. Mem. German Assn. Pathologists, World Assn. Vet. Pathologists, Assn. Neuropathologists. Author: (with Nieberle) Lehrbuch der speziellen Pathologie der Haustiere, 4th edit., 1961; (with O. Seifried, R. Baumann) Lehrbuch der allgemeinen Pathologie für Tierärzte, 7th edit., 1950; (with R. Jaffé, H. Meessen) Pathologie der Laboratoriumstiere, 1958; also numerous articles. Address: 8 Gerlach Strasse, Hanover, West Germany.*

COHRS, Wilhelm Christof Paul, German veterinarian; b. Oederan (saxe), Mar. 22, 1897; s. Christoph and Liddy (Schilling) C.; D.V.M., U. Dresden; m. Dorothea Schoene, Oct. 25, 1928; children—Jens, Hans-Uwe. Asst., Inst. Vet. Pathology, U. Leipzig, instr., prof., dir. Inst. Histological and Embryological Vet.; full prof., dir. Inst. Pathology, Superior Vet. Sch., Hanover, Germany. Mem. German Soc. Pathology, World Assn. Vet. Pathologists, Italian Soc. Vet. Sci. (hon.). Author: (with Jaffé and Meesen) Pathologie der Laboratoriumstiere, 1958; (with Nieberle) Lehrbuch der Pathologischen Anatomie der Haustiere, 1961; (with Seifried and Baumann) Kitt's Lehrbuch der Allgemeinen Pathologie der Haustiere. Home: Hans Boecklerallee 16, Hanover, Germany.

COINDET, Jean François, Swiss physician; b. Geneva, Switzerland, July 12, 1774; ed. Edinburgh, Scotland; physician in Geneva; in 1820 introduced iodine for treatment of thyroid disorders. Died Nice, France, Feb. 11, 1834.

COITER, Volcher, anatomist; b. Groningen, Netherlands, 1534; studied under Fallopius and Eustachius in Padua, Italy, also under Aldrovandi; became student in Montpellier, France, 1555, Bologna, Italy, 1559-62, 63, also Rome, Perugia; m. Helena. City physician, Nurnberg, Germany, 1569-75; then prof., Bologna. Author: De ossibus et cartilaginibus corporis humanis tabulae, 1566; Diversorum animalium sceletorum explicationes, 1575. Pioneer in study osteogenesis; a founder path. anatomy; studied embryology of birds and man, 1564; discovered 2 superior muscles of nose. Died Brienne, France, June 2, 1576.

COKER, Ernest George, engr.; s. G. Coker; M.A., Cambridge U.; D.Sc., Sidney, Louvain; D.Sc., U. Edinburgh; m. Alice Mary King, 1899 (dec. 1941). Asso. prof. civil engring. McGill U., Montreal, Que., Can.; prof. civil and mech. engring. City and Guilds of London Tech. Coll., Finsbury; prof. emeritus civil and mech. engring. U. London, 1934-46. Recipient Howard N. Potts medal Franklin Inst., 1922, Levy medal, 1926. Fellow Royal Soc., 1916 (Rumford medal 1936), Royal Soc. Edinburgh; mem. Inst. Civil Engrs. (Telford medal 1921, Howard Quinquennial medal 1932), Instn. Mech. Engrs. (council, Thomas Hawksley medal 1922), Royal Instn., Inst. Naval Architects (asso., Gold medal 1911). Author: (with Filon) A Treatise on Photo-Elasticity, also articles. Died Apr. 9, 1946.

COKER, Robert Ervin, Am. biologist; b. Society Hill, S.C., June 4, 1876; s. William Caleb and Mary (McIver) C.; B.S., U. N.C., 1896, LL.D., 1959; Ph.D., Johns Hopkins, 1906; D.Sc., U. S.C., 1947; m. Jennie Louise Coit, Oct. 11, 1910; children—Robert Ervin II, Coit McLean. Custodian, acting dir. U. S. Fisheries Lab., 1902-04; spl. investigator Marine Fisheries and Guano Industry, Peru, 1906-08; dir. U. S. Bur. Fisheries, 1910-15, chief div. sci. inquiry, 1915-22; prof. zoology U. N.C., 1922-39, Kenan prof. zoology, 1953—, emeritus, 1953—; chmn. dept. zoology, div. nat. scis., 1935-43; vis. prof., cons. marine biology U. P.R., 1955—, emerito, 1963—. Recipient Mayflower award, 1947; Gardner award, 1950. Fellow Chgo. Acad. Sci.; mem. Am. Inst. Biol. Scis., Ecol. Soc. Am. Limnology Soc. Am., Am. Soc. Zoology, Elisha Mitchell Soc., N.C. Acad. Sci., A.A.A.S. Author: This Great and Wide Sea, 1944; Streams, Lakes, Ponds, 1954; also numerous articles. Home: 40 Oakwood Dr., Chapel Hill, N.C. 27514.*

COKER, William Chambers, Am. botanist; b. Hartsville, S.C., Oct. 24, 1872; s. James Lide and Susan Armstrong (Stout) C.; B.S., S.C. Coll., 1894; Ph.D., Johns Hopkins, 1901; studied at Bonn, Germany, 1901-02; LL.D., U. S.C., 1925; D.Sc., U. of N.C., 1947; m. Louise Venable, Oct. 28, 1934. Asso. prof. botany U. N.C., 1902-07, prof., 1907-20, Kenan prof. botany, 1920-44, Kanan research prof. botany, 1944-45, emeritus since 1945; also dir. Coker Arboretum. Chief bot. staff Bahama expdn. Geog. Soc. Balt., 1903. Fellow A.A.A.S.; mem. Bot. Soc. Am. (chmn. Southeastern sect.), Am. Soc. Naturalists, Am. Mycol. Soc., Elisha Mitchell Sci. Soc., Am. Forestry Assn., Phi Beta Kappa, Sigma Xi. Editor; Jour. Elisha Mitchell Sci. Soc., 1904-45; hon. curator botany Charleston Mus., 1943-53. Author: Vegetation of the Bahama Islands, 1905; The Plant Life of Hartsville, S.C., 1912; The Trees of North Carolina (with H. R. Totten), 1916; The Saprolegniaceae, 1923; The Clavarias of the United States and Canada, 1923; The Gasteromycetes of the Eastern United States and Canada (with J. N. Couch), 1928; Trees of the Southeastern States (with H. R. Totten), 1934; The Boletaceae of North Carolina (with A. H. Beers), 1943; The Stipitate Hydnums of the Eastern United States (with A. H. Beers), 1951. Contbr. numerous articles on morphological botany, particularly on gymnosperms and fungi. Died June 27, 1953.

COLBERG, Christoph, geodesist; b. Woldegk, Prussia, Germany, July 7, 1776; s. Julius Daniel F. and Charlotte Dorothea (Fuchs) C.; Dr.phil., U. Warsaw (Poland); m. Carolina Fleurietta Wilhelmina Merker, 1806; 6 children. Entered ofcl. state service 1st in Prussia, then Poland and finally in Russia; prof. applied math. U. Warsaw until 1930. Author: Post und Reise Karte von Polen und Posen, 1817; Beschreibung eines Planimeters etc., 1820. Invented measuring instruments; designed 1st postcards and travel maps in Polish lang. Died Warsaw, Sept. 5, 1831.

COLBERG, Marshall Rudolph, Am. economist; b. Chgo., June 11, 1913; s. Rudolph E. and Elvira (Wester) C.; A.B., U. Chgo., 1934, A.M., 1938; Ph.D., U. Mich., 1950; m. Sarah Naomi McCoy, Sept. 17, 1966; children (by previous marriage)—Marsha, Daniel. With WPB, Washington, 1940-43, Civilian Prodn. Adminstrn., Washington, 1946-47; faculty Fla. State U., Tallahassee, 1950—, prof., 1953—, chmn. dept. econs., 1954-67. Cons. U. S. Office Edn., since 1966—. Mem. Am. Econ. Assn. (com. on edn. 1963—), So. Econ. Assn. (pres. 1962), Mt. Pelerin Soc. Author: Human Capital in Southern Development, 1965; (with Allen & Buchanan) Prices, Income & Public Policy, 1959; (with Greenhut) Factors in the Location of Florida Industry, 1962; (with Forbush and Whitaker) Business Economics, 1964. Devised theory of joint costs under monopoly; research on industry, particularly in the South and Fla.; developed theory of human capital movement within a country; predicted econ. consequences of sch. integration; theory and measurement of life earnings profiles; research on minimum wage, area redevel. Home: 2308 Don Andres Av., Tallahassee, Fla. 32304.*

COLBERT, Edwin Harris, Am. paleontologist; b. Clarinda, Ia., Sept. 28, 1905; s. George Harris and Mary (Adamson) C.; student N.W. Mo. State Coll., 1926; A.B., U. Neb., 1928; A.M., Columbia, 1930, Ph.D., 1935; m. Margaret Mary Matthew, July 8, 1933; children—George Matthew, David William, Philip Valentine, Daniel Lee, Charles Diller. Staff, Am. Mus. Natural History, 1930—, curator, 1943—; faculty Columbia, N.Y.C., 1945—, prof. vertebrate paleontology, 1945—; asso. curator Acad. Natural Sci. Phila., 1937-48; research asso. No. Ariz. Soc. Sci. and Art, 1949—. Recipient John Strong Newberry prize Columbia, 1931. Fellow Geol. Soc. Am., N.Y. Zool. Soc.; mem. Soc. Vertebrate Paleontology (past pres.), Soc. for Study Evolution (past pres.), Palontol. Soc. (past v.p.), Nat. Acad. Sci. U. S. (Daniel Giraud Elliot medal 1935), Paleontol. Soc. India, Acad. Nac. de Ciencias Argentina, A.A.A.S., Soc. Ichthyologists and Herpetologists, Am. Soc. Mammalogists, Soc. Systematic Zoology, Sigma Xi. Author: Evolution of the Vertebrates, 1955; Dinosaurs, 1961; (with Marshall Kay) Stratigraphy and Life History, 1965; The Age of Reptiles, 1965; also numerous paleontol. papers, monographs. Research on extinct amphibians, reptiles and mammals especially late Tertiary mammals and early Mesozoic reptiles, fossil faunas from N. and S.Am., Asia, and Africa. Home: 497 Park Av., Leonia, N.J. 07605. Office: Am. Mus. Natural History, N.Y.C. 10024.*

COLDEN, Cadwallader, botanist; b. Dunse, Scotland, Feb. 17, 1688; s. Alexander Colden; M.D., U. Edinburgh, 1705; m. Alice Christie, Nov. 11, 1715, children—Jane, Elizabeth, David. Came to Am., 1710, N.Y.C., 1718; surveyor gen. N.Y., 1720; apptd. to N.Y. Gov.'s Council, 1721; lt. gov. colony N.Y., 1761-76; supported Britain in Revolutionary War; signed original charters for Marine Soc. of N.Y., N.Y. State C. of C.; 1st advocate of cooling regimen in treating fevers; proposed cure for cancer. Author: History of the Five Indian Nations Depending upon New York 1727; Account of Diseases Prevalent in America, 1736; Essay on the Cause and Remedy of the Yellow Fever so Fatal at New York, 1743; An Explication of the First Causes of Action in Matter, and, of the Cause of Gravitation, 1745; Principles of Action in Matter, 1752. Introduced Linnaean system of classification into Am. only months after its publ. in Europe; sent descriptions of 300 to 400 Am. plants to Linnaeus. Died L.I., N.Y., Sept. 28, 1776.

COLDING, Ludwig August, Danish physicist; b. Arnakke nr. Holbock, Denmark, July 13, 1815; ed. Polytechnic Sch., 1837; became insp. of streets, 1845, of water, 1847; became engr. for Village of Copenhagen, 1858; prof. Polytechnic Sch., 1865; mem. Copenhagen Soc. Scis. Author: les Cyclones tropicaux, 1871; des Mouvemont de l'eau souterraine, 1872; les tempetes et les irruptions de la mer en 1872, 1881; Developed ideas of a mech. equivalent heat without knowing of the works R. Mayer, Joule and Helmoltz. Died Mar. 21, 1888.

COLE, Aaron Hodgman, Am. biologist; b. Greenwich, N.Y., Oct. 21, 1856; s. Morgan C. and Lydia Ann (Hodgman) C.; A.B., Colgate U., 1884, A.M. 1887; grad. student Johns Hopkins, 1889, U. Chgo. 1893, 1896, 1898; m. Emma Sarah Mason, Dec. 29 1885. Instr. natural scis. Peddie Inst., 1884-88 lectr. zoölogy and geology, Colgate U., 1888-92 instr. zoölogy, Cold Spring Harbor Biol. Lab., 1893 lectr. biology, U. Chgo. extension div., 1895-1906 instr. biology, Chgo. Tchrs. Coll., 1906-13; instr technique of biol. projection and anesthesia of animals, U. Chgo., 1901. Popular lectr. on bacteriology 1895-1913, delivering popular lectures on vita phenomena of lower animals and plants. Inventor o sci. apparatus, eye-shields, and methods of highl magnifying on screen images of microscopic animal and plants; demonstrated method of teaching biolog from living plants and animals with a projectio microscope, 1905; discovered successful method o culture for amoeba, and method of showing move ment of sap in leaves of plants. Asso. editor Unite Editors Ency. and Dictionary and author of article on The Projection Microscope and Its Use and An esthesia of Animals and Plants. Author and pub. Manual of Biological Projection and Anesthesia o Animals, 1907. Died Dec. 31, 1913.

COLE, Alfred Dodge, Am. physicist; b. Rutland, Vt Dec. 18, 1861; s. Israel D. and Alice (Ware) C.; A.B Brown, 1884, A.M., 1887; studied Johns Hopkins 1884-85, Summer Sch., Harvard, 1888, Berlin, 1894 95, Summer Sch., Cornell, 1897, U. Chgo., summers 1898, 1899, 1900, 1904; m. Emily Downer, June 18 1889. Instr. chemistry and physics Denison U., 1885 87, acting prof., 1887-88, prof., 1888-1901; prof physics, Ohio State U., 1901-07, Vassar Coll., 1907 08; prof. physics and head dept., Ohio State U., 1908 26; prof. physics, from 1926. Research guest, Na Bur. of Standards, Washington, and U. Berlin, 1912 13. War work on electron radio receivers U. S. Nav lab., Washington, summer 1917, and U. S. Bu Standards, summer 1918. Investigated infrared radia tion; electric waves; electric oscillations, impa excitation, capillary electrometer. Died Columbus, O Dec. 1, 1928.

COLE, Arthur, Am. physicist; b. Evanston, Ill., Oc 28, 1929; s. Harry and Bertha (Slavitt) C.; B.A., Ri Inst., 1951, M.A., 1953, postgrad.; Ph.D., U. Tex

1956; m. Allyce Tinsley, Sept. 4, 1951; children—Aron Christopher, Amantha Erema. Electronic, radiol. physicist U. Tex. M. D. Anderson Hosp. and Tumor Inst., 1951-54, asst. physicist, 1956-57, 58-61, asso. physicist, 1961-65, physicist, 1965—, instr. physics Postgrad. Sch. Medicine, 1952-54, instr. radiation physics Med. br., Galveston, 1954, fellow biophysics Austin, Houston, 1954-56, asst. prof. physics 1957-60, asst. prof. biophysics, 1960-61, asso. prof. biophysics, 1961—, asso. grad. faculty Grad. Sch. Biomed. Scis., 1964—. Nat. Acad. Scis. Research fellow Radiol. Research Unit, Harwell, Eng., 1957-58. Mem. Biophys. Soc., N.Y. Acad. Scis., Radiation Research Soc. Editor: Theoretical and Experimental Biophysics, 1967. Research, publs. on design and constrn. of various manual and automatic radiation measuring devices; determination of phys. factors related to treatment in radiation therapy machines; utilized radiation particle beams of low penetration to determine location of radiation sensitive sites within cells and microorganisms; study of molecular and archtl. arrangement of chromosomes from mammalian cells using electron microscopy. Home: 8874 Chatsworth St., Houston. Office: U. Tex. M. D. Anderson Hosp. and Tumor Clinic, 6723 Bertner Av., Houston 77025.*

COLE, Elbert Charles, Am. biologist; b. Northampton, Mass., Apr. 2, 1891; s. George Elbert and Alice L. (French) C.; A.B. cum laude, Middlebury Coll., 1915; A.M., Trinity Coll., 1918; Ph.D., Harvard, 1924; m. Ida Nell Ainsworth, Oct. 28, 1916 (dec. Apr. 1932); children—Gerald A., Elbert Charles, Phyllis Anne (Mrs. William Acker Rice Deming); m. 2d, Margaret Caldwell Grierson, July 20, 1933. Instr., Hartford (Conn.) High Sch., 1915-22, Trinity Coll., 1918-19; Austin Teaching fellow Harvard, 1922-24; faculty Williams Coll., Williamstown, Mass., 1924—, prof., 1932-45, Samuel Fessenden Clarke prof. biology, 1945-56, prof. emeritus, 1956, chmn. dept. biology, 1931-51. Fellow A.A.A.S.; mem. Am. Assn. Anatomists, Am. Soc. Zoologists, Biol. Stain Commn., Marine Biol. Corp., Phi Beta Kappa (Alumni plaque 1961), Gamma Alpha. Author: Introduction to Biology, 1933; General Biology, 1939; a Textbook of Comparative Histology, 1941; also articles. Research on use vital methylene blue in studies on nervous tissues, on dye hematoxylin, use dyes and other chems. in study structures and functions animals; developed lab. methods and apparatus. Address: 2 Chipman Park, Middlebury, Vt. 05753.*

COLE, Fay-Cooper, Am. anthropologist; b. Plainwell, Mich., Aug. 8, 1881; s. George LaMont and Ida Josephine (Upright) C.; B.S., Northwestern U., 1903, Sc.D. (hon.), 1928; postgrad. U. Chgo., U. Berlin (Germany); Ph.D., Columbia, 1914; LL.D., Beloit Coll., 1944; m. Mabel Elizabeth Cook, Oct. 20, 1906; 1 son. Ethnologist Field Mus. Natural History, also spl. investigator Philippine Bur. Sci., 1906; asst. curator Malayan ethnology and phys. anthropology Field Mus. Natural History, 1912-23; lectr. anthropology Northwestern U., 1921-23; faculty U. Chgo., 1924—, prof., 1927—, chmn. dept. anthropology, 1927—. Mem. Social Sci. Research Council, 1925-30; vice-chmn. div. anthropology and psychology NRC, 1927-30, chmn., 1929-30, mem. exec. com., 1942—. Recipient Gold medal Geog. Soc. Chgo. Author: Chinese Pottery in the Philippines, 1912; Wild Tribes of Davao Gulf, Mindanao, 1913; Traditions of the Tinguian, 1922; The Long Road, 1933; Rediscovering Illinois, 1937; The Story of Man, 1937; Peoples of Malaysia, 1945. Co-author: The Nature of the World and of Man, 1926; Making Mankind, 1929; This World and Man, 1937. Office: Univ. of Chgo., Chgo. 60637.

COLE, Francis Joseph, Brit. zoologist; b. Feb. 3, 1872; s. William and Elizabeth Cole; ed. Jesus Coll., Oxford, Eng.; D.Sc., U. Oxford; m. Annie Clow Menzies; 1 son. Lectr. in zoology U. Liverpool (Eng.), 1895-1906; prof. zoology Univ. Coll. (now Univ.) Reading (Eng.), 1906-39. Recipient Rolleston prize Oxford, 1902, Neill Gold medal Royal Soc. Edinburgh (Scotland), 1908. Fellow Royal Soc., 1926. Author: History of Anatomical Injections, 1921; History of Protozoology, 1926; Early Theories of Sexual Generation, 1930; Leeuwenhoek's Zoological Researches, 1937; Observationes Anatomicae Selectiores Amstelodamensium, 1938; Microscopic Science in Holland in the Seventeenth Century, 1938; Bibliographical Reflections of a Biologist, 1939; History of Comparative Anatomy from Aristotle to the Eighteenth Century, 1944, 49; Dr. William Croone on Generation, 1947; History of Micro-dissection, 1951; The History of Albert Durer's Rhinoceros in Zoological Literature, 1953; Bell's Law, 1955; Henry Power and the Circulation of the Blood, 1957; Harvey's Animals, 1957; Obiter Dicta Bibliographica, 1958; also numerous papers in sci. jours. Died Jan. 27, 1959.

COLE, Frank Nelson, Am. mathematician; b. Ashland, Mass., Sept. 20, 1861; s. Otis and Frances Maria (Pond) C.; A.B., Harvard, 1882, A.M., Ph.D., 1886; m. Martha Marie Streiff, July 26, 1888. Lectr. math. Harvard, 1885-87; instr. and asst. prof. math. U. Mich., 1888-95; prof. math. Columbia, from 1895. Contbd. to study of finite groups using combinatorial analysis; treated sextic equation, 1886. Died May 27, 1926.

COLE, Grenville Arthur James, geologist; b. London, Eng., Oct. 21, 1859; s. J. J. Cole; ed. City of London Sch., Royal Sch. Mines; m. Blanche Vernon; 1 son. Prof. geology Royal Coll. Sci. for Ireland, from 1890; dir. Geol. Survey Ireland, from 1905; examiner London, Cambridge, Oxford, other univs. Recipient Murchison medal Geol. Soc., 1909. Fellow Royal Soc., 1917; pres. geol. sect. Brit. Assn., 1915. Geog. Assn., 1919, Irish Geog. Assn., 1919-22. Author: Aids in Practical Geology, 1891; Open-air Studies, 1895; The Changeful Earth, 1911; Rocks and their Origins, 1912; Outlines of Mineralogy for Geological Students, 1913; The Growth of Europe, 1914; Ireland, the Land and the Landscape, 1915; Ireland the Outpost, 1919; Common Stones, 1921; Memoir and Map of Irish mineral localities (Geol. Survey), 1922; (with Blanche Cole) As We Ride, 1902; sect. on Europe in The World We Live in, 1917; also sci. papers. Died Apr. 20, 1924.

COLE, Harold Harrison, Am. physiologist; b. Waterloo, Wis., Feb. 11, 1897; s. Clarence Elliot and Hattie (White) C.; B.S., U. Wis., 1920; M.S., U. Cal. at Berkeley, 1925, LL.D., 1965; Ph.D., U. Minn., 1928; m. Jessie Clift, 1929 (dec. 1952); 1 son, Cliff H.; m. 2d Cynthia Clegg, Mar. 2, 1955; 1 dau., Nancy. Faculty, U. Cal., Davis, 1928—, faculty research lectr., 1943, prof., 1943-64, chmn. dept. animal husbandry, 1952-60, emeritus prof., 1964—. Chmn. subcom. on bloat Com. Animal Health NRC, 1944-54, Com. on Beef, 1959-60; mem. Com. Range Land Utilization, adviser to Cal. Bd. Forestry. Recipient Morrison award, 1952, Animal Physiology and Endocrine award, 1963. Mem. Am. Assn. Anatomists, Am. Soc. Animal Sci., A.A.A.S., Soc. Exptl. Biology and Medicine, Endocrine Soc., Am. Dairy Sci. Assn., Soc. Study Fertility, Sigma Xi. Editor: (with P. T. Cupps) Reproduction in Domestic Animals, 1959; Introduction to Livestock Production, 1962; Gonadutropins: Their Chemical and Biological Properties and Secretory Control, 1964. Asso. editor Jour. Animal Sci., 1943-45. Contbr. numerous articles to profl. jours. Discovered that blood serum of pregnant mares is a rich source of a hormone stimulating gonads, practical means of preventing milk fever in dairy cattle. Home: 528 Miller Dr., Davis, Cal. 95616.*

COLE, Jack Westley, Am. physician; b. Portland, Ore., Aug. 28, 1920; s. Alva Warren and Louise (Shafer) C.; A.B., U. Ore., 1941; M.D., Wash. U., 1944; M.A., Yale U., 1966; m. Ruth Adele Kraft, Dec. 22, 1943; children—Deborah, Linda, Douglas, John. Mem. faculty Western Res. U. Sch. Medicine, 1952-63; prof., chmn. dept. surgery Hahnemann Med. Coll. and Hosp., 1963-66; Ensign prof., chmn. dept. surgery Yale U. Sch. Medicine, 1966—; cons. various hosps. Eleanor Roosevelt Internat. Cancer Research fellow, 1962. Mem. Am. Surg. Assn., Halsted Soc., Soc. Surgery of Alimentary Tract, Am. Soc. Cell Biology, Soc. Cryobiology. Research and publs. on histochemistry, cytochemistry, carcinogenesis; studies dealing with cellular kinetics in normal and abnormal intestinal epithelium. Home: Prospect Court, Woodbridge, Conn. 06525. Office: 333 Cedar St., New Haven, Conn. 06510.*

COLE, Keith David, Australian physicist; b. Cairns, Australia, Mar. 2, 1929; s. John Copson and Jessie (White) C.; B.Sc., U. Queensland, Brisbane, Australia, 1950, B.Sc. (with honors), 1952, Dip.Ed., 1953, M.Sc., 1954, D.Sc., 1967; m. Ailsa Anne Moore, Dec. 29, 1956; children—Maxine, James, David. Secondary sch. tchr., Cairns, Sydney, 1953-55; mem. Australian Nat. Antarctic Research Expdn., Macquarie Island, 1956; physicist Antarctic div. Dept. External Affairs, Melbourne, Australia, 1957-63; physicist, seconded to upper atmosphere sect. Commonwealth Sci. and Indsl. Research Orgn., Australia, 1962; research asso. U. Chgo., 1963-64; research fellow Nat. Centre for Atmospheric Research, Environmental Scis. and Services Adminstrn., Boulder, Colo., 1965; research asso. U. Colo., 1966; prof. Sch. Phys. Scis., La Trobe U., Bundoora, Australia, 1966—. Fellow Australian Inst. Physics; mem. Am. Geophys. Union. Research and publs. on theoretical physics of upper atmosphere, magnetosphere and interplanetary space; theory of geomagnetic storms particularly of ring current and stable auroral red arcs. Home: 3 Lavender Park Rd., Eltham, Victoria. Office: Sch. Phys. Scis., La Trobe U., Bundoora, Victoria, Australia.*

COLE, Kenneth Stewart, Am. biophysicist, govt. ofcl.; b. Ithaca, N.Y., July 10, 1900; s. Charles Nelson and Mabel (Stewart) C.; A.B., Oberlin Coll., 1922, Sc.D., 1954; Ph.D., Cornell U., 1926; D.Sc., U. Chgo., 1967; M.D., U. Uppsala (Sweden), 1967; m. Elizabeth Evans Roberts, June 29, 1932 (dec. Mar. 1966); children—Roger Braley, Sarah Roberts (Mrs. Walter Main). Research fellow Harvard, Cambridge, Mass., 1926-28, U. Leipzig (Germany), 1928-29; asst. prof. physiology Columbia, N.Y.C., 1929-37, asso. prof., 1937-46; prof. biophysics and physiology U. Chgo., 1946-49, prin. biophysicist metall. lab., 1942-46; tech. dir. Naval Med. Research Inst., Bethesda, Md., 1949-54; chief lab. biophysics Nat. Inst. Neurol. Diseases and Blindness, Nat. Insts. Health, Bethesda, 1954-66, sr. research biophysicist, 1966—; prof. biophysics U. Cal. at Berkeley, 1965—. Priestley lectr. Pa. State U., 1939; lectr. U. Pa., Phila., 1956, Yale, 1962; regents prof. U. Cal. at Berkeley, 1963-64; mem. bd. Biol. Lab., Cold Spring Harbor, L.I., N.Y., 1940-45; mem. founding bd. Found. for Advanced Edn. in Scis. Bd., 1958-65. Guggenheim fellow Inst. for Advanced Study, Princeton, N.J., 1941-42; recipient National Medal of Science, 1967. Fellow Am. Phys. Soc., A.A.A.S., N.Y. Acad. Science; mem. Nat. Acad. Scis., Am. Acad. Arts and Scis., Am. Physiol. Soc. (council 1963-65), Société Philomatique de Paris, Soc. Gen. Physiologists, Biophys. Soc. (founding com. 1957-58, council 1958-62), pres. 1963-64), Sociedad Brasileira de Biologia (hon.), Sigma Xi, Alpha Epsilon Delta (hon.), Epsilon Chi (hon.). Research, publs. on conductance of living cell membranes, ion permeabilities responsible for impulse propagation in giant squid nerve fiber. Office: NIH, Bethesda, Md. 20014.*

COLE, LaMont Cook, Am. ecologist; b. Chgo., July 15, 1916; s. Fay-Cooper and Mabel (Cook) C.; S.B., U. Chgo., 1938; M.S., U. Utah, 1940; Ph.D., U. Chgo., 1944; m. Ann Louise Schuster, Aug. 27, 1940; children—John LaMont, George Frederic. Faculty, Ind. U., 1946-48; faculty Cornell U., Ithaca, N.Y., 1948—, prof. ecology, 1953—, chmn. dept. zoology, 1964-65, chmn. sect. on ecology and systematics div. biol. scis., 1965-67. Mem. adv. com. Environmental Biology NSF, 1958-61; chmn. spl. rev. team math. and statistics NIH, 1963—. Mem. Ecol. Soc. Am. (pres. elect 1966-67), Am. Inst. Biol. Scis. (gov. 1965—, exec. com. 1966—, v.p. 1968), A.A.A.S., Biometric Soc., Am. Statis. Assn., Am. Soc. Zoologists, Am. Soc. Naturalists, Editor Ecology, 1958-63. Research, numerous publs. on statis. and math. study populations, analyzation spatial distbn. population mems., methods analyzing population growth; discovered relationship between population fluctuations and theoretical random fluctuations; related features life histories to effects on population characteristics. Home: R.D. 3, Trumansburg, N.Y. 14886. Office: Langmuir Lab., Cornell U., Ithaca, N.Y. 14850.*

COLE, Lawrence Wooster, Am. psychologist; b. Toledo, May 15, 1870; s. Irving and Josephine (Webb) C.; A.B., U. Okla., 1899; A.M., Harvard, 1904, Ph.D., 1910; m. Fannie B. Cooksey, June 28, 1900; children—Elizabeth J., Margaret, Mary Louise. Supt. schs., El Reno, Okla., 1897-1900; instr. and asso. prof. psychology, 1900-04, prof., 1904-08, U. Okla.; instr. exptl. psychology Wellesley Coll., 1908-09; prof. psychology and edn., U. Colo., 1910-38, prof. emeritus, 1938-46, dir. Sch. Social Service, 1911-20. Fellow A.A.A.S.; mem. Am. Psychol. Assn., Phi Beta Kappa, Sigma Xi. Contbr. various reports of exptl. investigation in comparative and human psychology. Author: Factors of Human Psychology. Translator of Durprat's Psychologie Sociale. Died Mar. 19, 1946.

COLE, Leon J(acob); Am. zoologist; b. Allegany, N.Y., June 1, 1877; s. Elisha Kelley and Helen Marion (Newton) C.; student Mich. Agrl. Coll., 1894-95, 1897-98; A.B., U. Mich., 1901; Ph.D., Harvard, 1906; Sc.D., Mich. State Coll., 1945; m. Margaret Belcher Goodenow, Aug. 28, 1906; children—Margaret Valeria, Edward Goodenow. Asst. in zoölogy U. Mich., 1898-1902, teaching fellow Harvard U., 1902-06, chief Div. Animal Breeding and Pathology, Agrl. Expt. Sta., R.I., 1906-07; instr. zoölogy, Sheffield Sci. Sch. (Yale), 1907-10; asso. prof. exptl. breeding U. Wis., 1910-14, prof., 1914-18, prof. genetics, 1918-47; on leave as chief Animal Husbandry Div., Bur. Animal Industry, U. S. Dept. Agr., 1923-24; chmn. Div. Biology and Agr., NRC, 1926-27. Mem. Harriman Alaska Expdn., 1899; zoöl. expdn. to Yucatan, 1904. Investigator, U. S. Bur. Fisheries, summers 1901-06. Fellow A.A.A.S. (v.p. Sect. F., 1940); Poultry Sci. Assn.; mem. Am. Soc. Zoölogists, Am. Genetic Assn.; Genetic Soc. Am. (v.p. 1937, pres. 1940), Eugenics Soc., Am. Soc. Naturalists (v.p. 1917; sec. 1927-31), Am. Ornithologists' Union, Am. Soc. Mammalogists, Soc. Animal Prodn., Wis. Acad. Sci. (pres. 1924-27), Sigma Xi (pres. Wis. Chapter 1917-18; nat. exec. com. 1932-34), Phi Kappa Phi, Gamma Sigma Delta, Phi Sigma (hon. nat. pres., 1940-46); corr. mem. Czechoslovak Acad. Agr. Contbr. on zoölogy, animal behavior, genetics and animal breeding. Organizer Am. Bird Banding Assn., 1909. Died Feb. 17, 1948.

COLE, Leonard Jay, Am. physiologist; b. N.Y.C., July 31, 1916; s. Joseph and Sarah (Liebson) C.; B.S., Coll. City N.Y., 1939; M.S., N.Y. U., 1941; postgrad. U. Cal. at Berkeley; m. Joyce Antonette Carlson, Aug. 20, 1946; children—Mark David, Sharon Linda. Chemist, Mare Island Naval Shipyard Lab., Cal., 1942-48; research specialist U. Cal. Exptl. Sta., Berkeley, 1948-50; biochemist U. S. Vets. Hosp., Oakland, Cal., 1950-51; research radiobiologist U. S. Naval Radiol. Def. Lab., San Francisco, 1951-55, head exptl. pathology br., 1955—. Fellow A.A.A.S., mem. Am. Physiol. Soc., Am. Chem. Soc., Radiation Research Soc., N.Y. Acad. Sci., Transplantation Soc., Am. Soc. Microbiology, Cryobiology Soc., Soc. Exptl. Biol. Medicine, Sigma Xi. Asso. editor Radiation Research, 1959-62. Contbr. numerous articles to profl., sci. jours. Pioneer in marrow transplantation and immunology in x-irradiated animals, use cell markers for detecting donor cells in radiation chimeras, radiation carcinogenesis. Home: 242 Polhemus Av., Atherton, Cal. 94025. Office: U. S. Naval Radiol. Def. Lab., San Francisco 94135.*

COLE, Robert Hugh, Am. chemist; b. Oberlin, O., Oct. 26, 1914; s. Charles Nelson and Mabel (Stewart) C.; A.B., 1935; A.M., Harvard, 1936, Ph.D., 1940;

m. Elisabeth French, Apr., 24, 1943. Research supr. Woods Hole Oceanographic Inst., NDRC, 1940-46; asso prof. physics, U. Mo., 1946-47; asso. prof. chemistry, Brown U., Providence, 1947-51, prof., 1951-59, Metcalf prof., 1959—, chmn. dept., 1949-61. Author: Underwater Explosions, 1947; (with J. S. Coles) Physical Principles of Chemistry, 1964. Study of intermolecular forces; dielectric properties of matter. Home: 45 Humboldt Av., Providence 02906.

COLE, Roger David, Am. biochemist; b. Berkeley, Cal., Nov. 17, 1924; s. Naylor Elmer and Frances (Slankard) C.; B.S., U. Cal. at Berkeley, 1948, Ph.D., 1954; m. Thelma Bennett, July 11, 1944; children—David Naylor, Miriam Faith, Janice Joy. Jr. research biochemist U. Cal., Berkeley, 1954-55, faculty, 1958—, prof. biochemistry, 1965—; Nat. Found. Infantile Paralysis fellow Nat. Inst. Med. Research, London, Eng., 1955-56; research asso. Rockefeller Inst. Med. Research, N.Y.C., 1956-58; Guggenheim Meml. fellow Lab. for Molecular Biology, Cambridge, Eng., 1966-67. Com. mem., sect. chmn., lectr. sci. confs., 1955—; cons. numerous pubs., 1960—. Mem. A.A.A.S., Am. Soc. Biol. Chemists, Am. Chem. Soc., Sigma Xi. Contbr. numerous articles to profl. jours. Editorial bd. Archives Biochemistry and Biophysics; Biochimica Biophysica Acta. Research, publs. on devel. new methods for purification of hormones; techniques for determination of molecular structure of proteins; structure of hemoglobin. Home: 109 Villa Ct., Lafayette, Cal. 94549. Office: Biochemistry Bldg., U. Cal., Berkeley, Cal. 94720.*

COLE, W. Storrs, Am. micropaleontologist; b. Albany, N.Y., July 16, 1902; s. Frederick Willard and Edna (Storrs) C.; B.S., Cornell U., 1925, M.S., 1928, Ph.D., 1930; m. Gladys Florine Watt, June 3, 1926. Paleontologist, Huasteca Petroleum Co., Mexico, 1926-27, Sun Oil Co., Dallas, 1930-31; faculty Ohio State U., Columbus, 1931-46, prof. geomorphology, 1945-46; prof. geology, chmn. dept. Cornell U., Ithaca, N.Y., 1946-62, prof., 1962—. Research asso. in paleontology Scripps Instn. Oceanography, 1931, 35; cons. paleontologist Fla. Geol. Survey, 1929-47; geologist U. S. Geol. Survey, 1947—; research affiliate U. Hawaii, summer 1965, mem. geology and geography sect. NRC, 1944-47; adv. council N.Y. State Mus., 1947-62. Fellow Geol. Soc. Am. (3d v.p. 1959); mem. Paleontol. Soc. (pres. 1953), Am. Assn. Geographers, Paleontol. Research Inst., Cushman Found. for Foraminiferal Research. Research, numerous publs. in identification, biology, ecology, stratigraphic range and geog. distbrn. larger Foraminifera, especially of Ams. and Indo-Pacific region, correlation by means of these organisms of sediments encountered in wells drilled on Bikini and Eniwetok atolls, geomorphic studies of Appalachian highlands. Home: 310 Fall Creek Dr., Ithaca, N.Y. 14850.*

COLE, Warren Henry, Am. physician; b. Clay Center, Kan., July 24, 1898; s. George Henry and Mary (John) C.; B.S., U. Kan., 1918; M.D., Wash. U., 1920, D.Sc., 1967; m. Clara Lund, June 13, 1942. Mem. dept. surgery Wash. U. Sch. Medicine, 1926-36; prof. surgery, head dept. U. Ill. Coll. Medicine, Chgo., 1936-66. Recipient Leonard Research prize Am. Roentgen Ray Soc., 1926; Certificate of Merit, St. Louis Med. Sch., 1927; Distinguished Service award U. Kan., 1949, Wash. U., 1955. Mem. Am. Surg. Assn. (pres. 1959), Soc. U. Surgeons (pres. 1940), Am. Cancer Soc. (pres. 1960), A.C.S. (pres. 1955), Am. Geriatric Soc. (pres. 1958), Am. Assn. Colon Surgeons (pres. 1960), Am. Assn. Surgeons of Trauma (pres. 1955), A.M.A. (Distinguished Service Award 1965), Chgo. Surg. Soc. (pres. 1942), Chgo. Med. Soc. (pres. 1950), Western Surg. Assn. (pres. 1950), Am. Goiter Assn. (pres. 1958), Royal Coll. Surgeons Edinburg, Royal Coll. Surgeons Eng. Author: (with Robert Zollinger) Textbook of Surgery, 8th edit., 1964; (with Tilden Everson) Spontaneous Regression of Cancer, 1966. Contbr. numerous articles to sci. jours. Developed cholecystography, a method of visualizing the gall bladder; research, publs. on etiology and treatment of goiter, gallbladder disease; clin. research on dissemination of cancer with spl. reference to prevention; treatment of cancer of colon and adjuvant chemotherapy of cancer of breast. Home: 8 W. Kensington Rd., Asheville, N.C. 28804.*

COLE, William, Brit. physician; b. 1635; M.B., Gloucester Hall, Oxford (Eng.), 1660, M.D., 1666; practiced at Worcester, 1666-92; in London, Eng., from 1692; fellow Coll. Physicians. Author: Cogitata de secretione animali, 1674; Physico-Med. Essay Concerning the Late Frequency of Apoplexies, 1689; Novae hypotheseos, ad explicanda febrium intermittentium symptomata hypotyposis, 1694; Disquisitio de perspiratione insensibili, 1702. Belonged to mech. sch. medicine; research on apoplexies; adopted Sydenham's treatment for small-pox; advocated use of Peruvian bark (quinine) for treatment of fever. Died June 12, 1716.

COLEMAN, Arthur Philemon, Canadian geologist; b. 1852; s. Francis and Emmeline (Adams) C.; ed. Victoria U., Cobourg, Can., Breslau (now Wroclaw, Poland) U.; M.A., Ph.D. Prof. geology and natural history Victoria U., Cobourg, 1881-90; prof. geology U. Toronto (Ont., Can.), 1890-1922, emeritus, from 1922; geologist Bur. Mines Ont., 1893-1909. Recipient Victoria medal Royal Geog. Soc., 1932. Fellow Royal Soc., 1910; mem. Geol. Soc. Am. (pres. 1915), Canadian Inst. (pres. 2 years), Brit. Assn. (pres. sect. C. 1910, Murchison medal 1910), Royal Soc. Can. (pres. 1921). Author: The Canadian Rockies, New and Old Trails; Ice Ages, Recent and Ancient, 1926; also reports on econ. geology of Ont. in govt. publs., papers in sci. jours. Explored and mapped areas of Canadian Rockies; made geol. survey and mapped Sudbury Dist., also Lake Iroquois shore, Labrador, Toronto Interglacial Beds, other areas in Ont. Died Feb. 27, 1939.

COLEMAN, Bernard D., Am. mathematician; b. N.Y.C., July 5, 1930; s. Nathaniel and Pauline (Listain) C.; B.S. cum laude, Ind. U., 1951; M.S., Yale, 1952; Ph.D., 1954; m. Jane Winthrop Candia, Mar. 27, 1965; 1 son, David Augustus. Research chemist E.I. duPont de Nemours & Co., Wilmington, Del., 1954-57; sr. fellow Mellon Inst., Pitts., 1957—; prof. math. Carnegie-Mellon U., 1967—; visitor Istituto Matematico, Università di Bologna (Italy), 1960-61; vis. prof. Johns Hopkins, 1962-63; adj. prof. U. Pitts., 1964-65; vis. lectr. U. Manchester (Eng.), 1965, Università di Pisa (Italy), 1966. Mem. Soc. For Natural Philosophy. Author: (with H. Markovitz, W. Noll) Viscometric Flows of Non-Newtonian Fluids, 1965. Mem. editorial bd. The Archive for Rational Mechanics and Analysis, 1962—. Editor-in-chief Springer Tracts in Natural Philosophy, 1967—. Research and numerous publs. in hydrodynamics of non-classical fluids, founds. of thermodynamics, theory of wave propagation in materials with memory; developed gen. theory of fading memory in non-linear materials. Office: 4400 5th Av., Pitts. 15213.*

COLEMAN, Howard S., Am. educator; b. Everett, Pa., Jan. 10, 1917; s. Howard Solomon and Amy (Ritchey) C.; B.S., Pa. State U., 1938, M.S., 1939, Ph.D., 1942; m. Madeline Maria Gloria Fiorillo, Sept. 6, 1942; children—Michael Howard, Madeline Frances, Thomas Robert, Carl William. Faculty, Pa. State U., 1934-47, also dir. optical inspection lab.; asso. prof., dir. optical research lab. U. Tex., 1947-51; with Bausch & Lomb, Inc., Rochester, N.Y., 1951-62, mgr., v.p. research and engring., 1954-62; head physics research dept., tech. asst. to v.p. in charge research Melpar, Inc., Falls Church, Va., 1962-64; prof. elec. engring., dean Coll. Engring., U. Ariz., 1964—; cons. Melpar, Inc., Burr-Brown Research Corp., Tucson, Electro-Optical Systems, Incorporated, Pasadena, California, Kollsman Instrument Corporation. Recipient Joint Services Merit award. Fellow Optical Soc. Am.; mem. Phys. Soc., Meteorol. Soc., Inst. Aero. Soc., Am. Assn. Physics Tchrs., Am. Soc. Metals, Internat. Com. Optics, Am. Geophys. Union, Am. Inst. Physics, Am. Soc. Engring. Edn., Nat. Soc. Profl. Engrs. Research, devel. of optical inspection devices. Home: 5511 E. Burns St., Tucson 85711.*

COLEMAN, James Smoot, Am. polit. scientist; b. Provo, Utah, Feb. 4, 1919; s. Jacob and Allie (Smoot) C.; B.A., Brigham Young U., 1947; M.A., Harvard, 1948, Ph.D., 1953; m. Ursula Maria Finken, June 23, 1965; children—S. James Smoot, Robert L. Teaching fellow dept. govt. Harvard, 1949-50, 52-53; from instr. to prof. dept. polit. sci. U. Cal. at Los Angeles, 1953-65, dir. African Studies center, 1960-65; dir. polit. sci. research program East African Inst. Social Research, Kampala, Uganda, 1965-67; asso. dir. Rockefeller Found., rep. Rockefeller Found. to U. East Africa, Nairobi, Kenya, 1967—; dir. social scis. div. Inst. for Devel. Studies, Univ. Coll., Nairobi, 1967—. Bd. dirs. Social Sci. Research Council, 1962-64. Recipient Woodrow Wilson prize Am. Polit. Sci. Assn., 1958. Mem. African Studies Assn. (U. S., pres. 1963-64), Internat. Congress Africanists (v.p. 1962-67), Am. Polit. Sci. Assn., Internat. African Inst., African Studies Assn., Soc. for Internat. Devel. Author: Nigeria: Background to Nationalism, 1958. Editor (with Gabriel Almond) Politics of the Developing Areas, 1960; (with Carl Rosberg) Political Parties and National Integration in Tropical Africa, 1964; Education and Political Development, 1965. Research and publs. on new modes of analysis of polit. systems of developing areas. Home: Black Rock Av. Office: Rockefeller Foundation, P.O. Box 7543, Nairobi, Kenya.*

COLEMAN, Leslie Charles, Canadian botanist, entomologist; b. Ont., Can.; ed. U. Toronto, U. Gottingen (Germany); mycologist, entomologist Govt. Mysore, India, 1908-13, dir. agrl., 1913-26, 1927; prof. U. Toronto, Ont., 1926-27; decorated companion of Order of Indian Empire. Author bulls., papers on plant pathology, entomology, agr. Died Sept. 14, 1954.

COLEMAN, Nathaniel Terry, Am. soil scientist; b. Halifax, Va., Nov. 24, 1919; s. Oliver Crump and Elizabeth (Terry) C.; B.A., Knox Coll., 1940; Ph.D., N.C. State Coll., 1949; m. Martha Elizabeth Peerman, Sept. 19, 1944; 1 son, Robert Peerman. Asst. dir. soil testing N.C. Dept. Agr., 1948-49; postdoctoral fellow U. Cal. at Berkeley, 1949-50; asso. prof. N.C. State Coll., Raleigh, 1950-54, prof., 1954-60; dir. agrl. sect. IAEA, Vienna, 1960-61; prof. soils and plant nutrition U. Cal. at Riverside, 1961—, chmn. dept., 1961-65, asso. dir. Dry Lands Research Inst., 1964-66. Fellow A.A.A.S., Am. Soc. Agronomy; mem. Soil Sci. Soc. Am. (editorial bd. proc. 1959-65), Am. Geophys. Union, Clay Minerals Soc. Editor: Natural Resources: Air, Land, Water, 1964. Mem. editorial bd. Soil Sci. Research and numerous publs. on interrelations between mineralogy and ion-exchange properties of soils; established existence of large liquid junction potentials in ion-exchange systems, demonstrated their importance in electrometric measurements of ion activites; defined conditions under which exchangeable aluminum ions and adsorbed hydroxy-aluminum polymers occur in soils. Home: 5442 Glenhaven Av., Riverside, Cal. 92506.*

COLEY, William Bradley, Am. surgeon; b. Westport, Conn., Jan. 12, 1862; s. Horace Bradley and Clarine Bradley (Wakeman) C.; B.A., Yale, 1884; M.D., Harvard, 1888, A.M. (hon.), 1911; postgrad. N.Y. Hosp., 1890; A.M. (hon.), Yale, 1910; m. Alice Lancaster, June 4, 1891; children—Bradley Lancaster, Malcolm, Helen Lancaster (Mrs. William Boone Nauts). In practice as physician, from 1888; staff N.Y. Hosp., 1890; cons. surgeon Meml. Hosp.; surgeon in chief Hosp. for Ruptured and Crippled, Mary McClellan Hosp. (Cambridge, N.Y.); cons. surgeon Fifth Avenue Hosp., Sharon (Conn.) Hosp. Fellow Am. Surg. Assn. A.C.S.; hon. mem. Assn. Surgeons Gt. Britain and Ireland. Contbg. author: Twentieth Century Practice of Medicine—part on Cancer (Vol. XVII), 1897; Hernia (in Dennis' System of Surgery), 1896; Hernia (in Warren and Gould's International Text Book of Surgery), 1898; chpt. on hernia in Progressive Medicine, from 1898. Hernia (in Keen's Surgery), 1907. Tried method of using erysipelas and prodigiosus toxins in treatment of sarcoma, circa 1893. Died Apr. 16, 1936.

COLGATE, Stirling Auchincloss, Am. physicist; b. N.Y.C., Nov. 14, 1925; s. Henry Auchincloss and Jeanette Thurber (Prwyn) C.; B.A. in Physics, Cornell U., 1948, Ph.D., 1952; m. Rosemary Williamson, July 12, 1947; children—Henry Auchincloss II, Mary Sarah, Arthur Stirling. With Lawrence Radiation Lab., at Berkeley, Cal., 1951-52, at Livermore, Cal., 1952-64; engaged in electron and accelerator physics work, devel. nuclear weapons and tests in Eniwetok, 1955; with Controlled Thermonuclear Fusion project, 1955-64; del. 2d Internat. Conf. Peaceful Uses of Atomic Energy, Geneva, Switzerland, 1958; tech. adviser Conf. Discontinuance Nuclear Weapons Tests, Geneva, 1959; part-time lectr. elec. engring. U. Cal. at Berkeley, 1960-64; pres. N.M. Inst. Mining and Tech., 1965—; partner Richard M. Colgate, patent devel., 1958—. Cons. AEC, 1960—; mem. nuclear panel Sci. Adv. Bd., 1959-61; adv. com. fluid mechanics NASA, 1960-62; cons. ballistic missile div. USAF, 1960-62; cons. Def. Atomic Support Agy., 1962-64. Fellow Am. Phys. Soc.; mem. Am. Astron. Soc., N.Y. Acad. Sci., Sigma Xi. Mem. editorial bd. Nuclear Fusion, 1959-64. Research on controlled thermonuclear fusion with toroidal pinches and plasma confinement astrophysics; gamma ray absorption measurement, annihilation of positrons in flight; origin of cosmic ray theories; quasi-stellar source theories. Address: New Mexico Inst. Mining and Tech., Socorro, N.M. 87801.*

COLHOUN, John, botanist; b. Castlederg, Ireland, May 15, 1913; s. James and Rebecca (Lecky) C.; B.Sc., Queen's U., Belfast, Ireland, 1933, B.Agr. with 1st class honors, 1934, M.Agr., 1937; D.I.C., Imperial Coll. Sci., U. London, 1937, Ph.D., 1940, D.Sc., 1955; M.Sc., Manchester U., 1964; m. Margaret Waterhouse, July 30, 1949; children—Lucy Margaret, Mary Elizabeth, Jacqueline Helen. Faculty, Queen's U., Belfast, 1940-60, reader mycology and plant pathology, 1954-60; staff Ministry Agr. for N. Ireland, 1939-60, prin. sci. officer plant pathology div., 1951-60; Barker prof. cryptogamie botany U. Manchester (Eng.), 1960—. Fellow Inst. Biology, Linnean Soc. London; mem. Brit. Mycol. Soc. (past pres.), Assn. Applied Biologists, Am., Netherlands phytopath. socs., Bot. Soc. Brit. Isles, Author: (with A. E. Muskett) Diseases of the Flax Plant, 1947; Club Root Diseases of Crucifers, 1958; also articles. Research on influence of environment on incidence of crop diseases, devel. techniques for testing of efficiency of fungicides used for control of plant diseases, fine structure and function of asexual reproductive organs of fungi, biology of fungi pathogenic to plants. Home: 12 Southdown Crescent, Cheadle Hulme, Cheshire, Eng. Office: The Univ., Manchester, 13, Eng.*

COLIN, Father Elie, French astronomer; b. Graulhet, France, Nov. 28, 1852; Jesuit novice Sem. of At Alain, Lavaur, France; ordained priest, 1885; apprentice in astronomy at Stonyhurst, Eng., 1887; sent to mission at Madagascar in 1888 and there founded Obs. of Tanarive; named laureate of French Acad. Scis. for work on cyclones of Indian Ocean, chevalier Legion of Honor. Author: Observations météorologiques, 24 vols., 1888-1913; also article Le magnétisme de Madagascar in Histoire de Madagascar (Grandidier), 1932. Died Apr. 10, 1923.

COLIN, Henri Ernest, French botanist; b. Bains-les-Baines, France, Nov. 1, 1880; ed. Sem. of S-Dié; Licencié ès sciences, 1905; Docteur ès sciences, 1911; ordained priest, 1905; named lectr. Catholic insts. of Paris and Lille, (France), 1911; prof. vegetal physiology at Paris, 1921; dean Free Faculty Sci., Paris, 1932; became deaf in 1914, but continued his sci. work. Mem. French Acad. Scis., 1937 (le prix Montagne 1912, le prix Vaillant 1934). Author: Les transformations expérimentales des etres vivants, 1923; De la matière a la vie, 1929; Les diastases, 1931. Research on nature, genesis and transformations of plant glucides, sugars of algae and mushrooms, inulin compositions, monocotyledons, grafted hybrids, structure of pectic matter including its deterioration and flocculation. Died Mar. 21, 1943.

COLIN, Jean Jacques, French chemist; b. Riom, France, Dec. 16, 1784; learned chemistry from circle of Guy-Lassac; prof. chemistry at Dijon, Colin, France; prof. chemistry at St. Cyr, France 1818-57. Author: Cours de chimie, 1827; Considérations élémentaires sur les propotions chimiques, les équivalents et les atomes, 1841; also (with Milne-Edwards) articles on vegetable physiology especially germination of cereals and chlorophyll respiration. Research on fabrication of soaps, fermentation of sugar, colored substances of Polygonium tinctorium and madder; (with Ganthier de Chambry) discovered action of iodine on amylacous substances in 1814. Died Mar. 9, 1865.

COLIN, Léon, French physician; b. Saint-Quérin, Paris, France, Apr. 16, 1830; studied in Strasbourg, then in Val-de-Grâce; named prof. Sch. Application in 1859, asst. dir., 1879-1881; succeeded Laveran in chair of epidemiology of armies at Val-de-Grâce in 1867; employed in mil. hosp. of small pox, Bicetre, France, during war of 1870; dir. health services of mil. of Paris; insp. gen., pres. tech. com., 1888-1895; mem. council hygiene of Paris; mem. cons. com. of hygiene of France. Mem. Acad. Medicine (pres. 1905) Author: Traité des fièves intermittentes, 1870; La Variole au point de vue epidemiologique et prophylactique, 1873; la Fièvre typhoide dans l'armée (Lacaze prize 1878); Traité des maladies épidémiques, 1882; Paris: sa tapographie, son hygiène, ses maladies (Montyon prize 1885). Died Feb. 22, 1906.

COLIN, Sébastien, French physician; b. 1519; s. Raoul Colin; m. Mlle. Bonnet; 1 son, Adam. Author under name of Lisset Benancio: La Déclaration des abuz et tromperies que font les apoticaires, 1553; L'ordre et régime qu'on doit garder et tenir en la cure des fièvres, 1558; translated Traicité de la peste, 1566. Pub. one of 1st med. texts in French. Died circa 1578.

COLLADON, (Jean-) Daniel, Swiss engr.; b. Geneva, Switzerland, Dec. 15, 1802; became prof. mechanics at l'École centrale, Paris, 1829, at U. Geneva, 1834; mem. French Acad. Scis., 1876. Research (with Sturms) on compressibility of liquids, 1834; determined speed of sound in water, 1837-41; developed method for measuring the performance of a steam engine using the power of a paddle wheel boat, 1841-44; invented apparatus used in the Mont Anis and St. Gothard tunnels. Died Geneva, July 3, 1893.

COLLAR, Arthur Roderick, English mathematician, physicist; b. Feb. 22, 1908; s. Arthur and Louie C.; M.A., D.Sc., Cambridge (Eng.) U.; m. Winifred Margaret Moorman, 1934; 2 sons. With aerodynamics dept. Nat. Phys. Lab., 1929-41, structural and mech. engrng. dept. Royal Aircraft Establishment, 1941-45; Sir George White prof. aero. engrng. U. Bristol (Eng.), 1945—, pro vice chancellor, 1967. Chmn. Aero. Research Council, 1964—. Recipient George Taylor Gold medal, 1947, Orville Wright prize, 1958, J. E. Hodgson prize, 1960. Fellow Royal Soc., Royal Aero. Soc. (pres. 1964, Gold medal 1966). Author: (with others) Elementary Matrices, 1938. Editor: (with others) Hypersonic Flow. Research, publs. on aeroelasticity, matrix analysis and differential equations, fluid dynamics, wind tunnel design, stability of accelerated motion. Address: Queen's Bldg., Bristol U., Bristol 8, Eng.

COLLATZ, Lothar, German mathematician; b. July 6, 1910; s. Carl and Maske C.; ed. Göttingen, Munich, Berlin (all Germany) univs.; Ph.D., D. honoris causa; m. Martha Togny, 1940. Prof., dir. Inst. for Applied Math. and Computer Centre, U. Hamburg (Germany), 1952—. Mem. Gesellschaft für Angewandte Mathematik und Mechanik; Deutsch Mathematiker-Vereinigung. Author: Eigenwertaufgaben mit technische Anwendungen; Numerische Behandlung von Differentialgleichungen; Differential gleichungen für Ingenieure. Work in functional analysis, differential and integral equations, numerical analysis, matrices, other topics. Address: Institut für Angewandte Mathematik, Rothenbaumchaussee 67-69, 2 Hamburg 13, Germany.

COLLES, Abraham, surgeon; b. Milmount nr. Kilkenny, Eire, 1773; studied at Dublin (Ireland) U., Edinburgh, Scotland, London, Eng.; diploma Irish Coll. Surgeons, 1795; M.D., 1796; m. Sophia Cope, 1807; at least one son, William. Asst. to Astley Cooper in London; also attended London hosps.; practised medicine, Dublin, 1797-99; resident surgeon, 1799-1813; vis surgeon Steven's Hosp., Dublin, 1813-41; prof. anatomy and surgery, 1804-36. Mem. Irish Coll. Surgeons (pres. 1802, 30). Author: Treatise on Surgical Anatomy, 1811. Ligated subclavian artery, 1811, 13; described space under perineal fascia containing ischiocavernosus, transverse perineal muscle (Colles' space), 1811; discovered Colles' fracture of radius, 1814; 1st European to successfully tie innominate artery. Died Nov. 16, 1843.

COLLET, André Jean, French pathologist; b. Lyon, France, Oct. 2, 1920; s. Louis Félix and Jeanne (Marcel) C.; Doctorat en Médecine, Faculty Medicine Lyon, 1949; m. Jeanne Zundel, Mar. 1, 1947; children—Jean-Charles, Catherine, Marie-Laure. Various positions with Inst. physiology Lyon Faculty Medicine, also Inst. Cardiologic Research, Royat, France; head research Inst. Histology, Lyon; with Center Coal Mining Research, Creil, France, 1950—; lectr. Ameins Nat. Sch. Medicine and Pharmacy, 1962—. Mem. French Soc. Respiratory Pathology, French Soc. Electronic Microscopy, French Soc. Occupational Medicine. Author: (with A. Policard) Physiologie du tissue conjonctif normal et pathologique, 1961. Research on normal and path. physiology of circulatory system, connective tissue, morphology of normal and path. lung (exptl. and human), pathology of pulmonary diseases caused by dust, also on fibrogenesi, including conventional and infra-structural morphology, metabolism, pharmacodynamism. Home: 7, rue Henry Le Chatelier, Verneuil en Halatte, Oise. Office: Cerchar, B.P. no. 27, Creil, Oise, France.*

COLLET DES COSTILS, Hippolyte-Victor, French engr.; b. Caen, France, Nov. 20, 1773; s. Jean Collet Des Costils; studied chemistry with Vauquelin at Paris, later student Ecole des mines; accompanied Monge and Bertholet on Egyptian expdn.; after his return to Paris in charge lab. Ecole des mines, became chief engr., 1809, provisional dir., 1815; mem. geol. expdn. to Italy, 1813. Research on platinum, iron ores, decomposition of galena, hardening of quiet lime; discovered iridium in 1803. Died Dec. 6, 1815.

COLLETT, Armand Rene, Am. chemist; b. Hartford City, Ind., Feb. 3, 1895; s. August and Eugenie (Loriaux) C.; A.B., W.Va. U., 1918; Ph.D., Yale, 1923; m. Dorothy May Atwood, Aug. 7, 1922; children—Dorothy May (Mrs. Otto Vernon Vande Linde Jr.), Florence Marie (Mrs. Clifford Shreve). Instr. chemistry Yale, 1923-24; faculty chemistry W.Va. U., 1924-65, prof., dean emeritus, 1965—, chmn. dept. chemistry, 1947-49, 61-62, asst. dean Coll. Arts and Scis., 1949-51, dean, 1951-61. Mem. Am. Chem. Soc., W.Va. Acad. Sci., Phi Beta Kappa, Sigma Xi, Gamma Alpha, Alpha Chi Sigma, Phi Lambda Upsilon. Author: (with Samuel Morris) Experimental Procedures in General Chemistry, 1938. Research, publs. on kinetics of organic reactions, solubility relations of disubstituted benzene derivatives, electrolytic oxidation studies on coal and humic acids, studies of inert electrode systems. Home: 806 Des Moines Av., Morgantown, W.Va. 26505.*

COLLETT, John, Am. geologist and farmer; b. Eugene, Ind., Jan. 6, 1828; grad. Wabash Coll., 1847, A.M., 1850; Ph.D., 1879; M.D., Central Med. Coll. Asst. state geologist, 1870-78; chief bur. statistics and geology, 1879-80; state geologist, 1881-85. Has published numerous geol. reports, 6 vols., 110 papers, 22 maps and gen. sections, nearly 2,000 figures, most of which were since copied by Mo. and Pa. Died 1899.

COLLIE, George Lucius, Am. geologist; b. Delavan, Wis., Aug. 11, 1857; s. Joseph and Ann Elizabeth (Foote) C.; B.S., Beloit Coll., 1881; A.M., Harvard, 1891, Ph.D., 1893; m. Katharine E. Burrows of Chicago, March 26, 1896 (dec.); children—Helen Tannisse (dec.), Kenneth Gordon. Asst. prin., prin. Delavan High Sch., 1885-90; Morgan fellow, Harvard, 1891-92; prof. geology, 1892-1923, prof. anthropology, 1923-31, curator Logan Mus. Archaeology Beloit Coll., 1893-1931, dean 1899-1931, acting pres., 1902-03, 05-08, also dir. Logan Mus. Sch. for Prehistoric Research, Les Eyzies, France, Tebersa, Algeria; Asst. Wis. Geol. Survey, 1898. Fellow Geol. Soc. Am., A.A.A.S.; mem. Am. Anthrop. Assn., Phi Beta Kappa. Traveled around the world, 1910-11, on geol. trip covering 40,000 miles. Writer on geol. and ednl. topics. Engaged in research for early man in France and Algeria, 1926-28. Died Dec. 28, 1954.

COLLIE, John Norman, Brit. chemist; b. Alderley Edge, Eng., Sept. 10, 1859; s. John and Selina Mary (Winkworth) C.; D. Philosophy, Würzburg, Germany, 1883; D.Sc., LL.D. Asst., Queen's coll., Belfast, Ireland, 1880-82; lectr. Cheltenham Ladies Coll., 1883; asst. prof. Univ. Coll., London, Eng., 1887, prof. organic chemistry, 1902-28, dir. chem. lab., 1912-28, emeritus prof., from 1928; prof., Pharm. Soc., 1896. Fellow Royal Soc., 1896, Royal Soc. Edinburgh, Royal Geog. Soc. (v.p.); mem. Alpine Club (pres.). Author: Climbing on the Himalaya, 1902; (with H. E. M. Stutfield) Climbs and Exploration in the Canadian Rockies, 1903; publs. on organic chemistry. Investigated (with Ramsay) properties of inert gases; studied various organic substances. Died Sligachan, Isle of Skye, Nov. 1, 1942.

COLLIER, Herbert Bruce, Canadian biochemist; b. Toronto, Ont., Can., Oct. 10, 1905; s. Robert Victor and Eliza (Caswell) C.; B.A., U. Toronto, 1927, M.A., 1929, Ph.D., 1930; m. Mary A. Rodger, July 24, 1930; children—Carol (Mrs. William Vine), Rodger. Head dept. biochemistry West China U., Chengtu, 1930-39; biochemist Inst. Parasitology McGill U., Montreal, Que., Can., 1939-42; faculty Dalhousie U., Halifax, N.S., Can., 1942-46, asso. prof. biochemistry, 1943-46; prof. head biochemistry U. Sask., Saskatoon, Can., 1946-49; faculty U. Alta., Edmonton, Can., 1949—, prof. biochemistry, 1949-64, prof. biochemistry in pathology, 1964—. Fellow Royal Soc. Can., Internat. Soc. Hematology; mem. Am. Chem. Soc., Am. Soc. Biol. Chemists, Canadian Biochem. Soc., Canadian Soc. Clin. Chemists, Sigma. Xi. Research, numerous publs. on metabolism of phenothiazine as an anthelmintic in domestic animals, phenothiazine derivatives as enzyme inhibitors, mechanisms of drug anemias; red-cell enzymes and membrane properties. Home: 13107 Churchill Crescent. Office: U. Alta. Hosp., Edmonton, Alta., Can.*

COLLIMITIUS, Georgius (Lycoripensis), mathematician; b. Rain am Lech, Germany, 1482; studied math., U. Ingolstadt, (Germany), 1496-97; magister, 1520; m. Martha; 1 son, 2 daus. Became lectr. math. Vienna (Austria), 1503; named physician Vienna Imperial Ct., 1510; cartographer, ofcl. envoy; published astrological calendars, ephemerides, almanacs. Author: De romain calendarin correctione consilium, 1515; Regiment für die lauff der pestilenz, 1521; Artificum de applicatione astrologiae ad medicinam, 1531. Studied astronomy, math., relationship of weather and health; recommended calendar improvements. Died Vienna, Mar. 25, 1535.

COLLINGS, Clyde Wilson, Am. urological surgeon; b. Vancouver, Wash., Feb. 28, 1892; s. Dellbert A. and Emma May (McCafferty) C.; student U. Wash., 1913-15; M.D., Ore. U., 1919; m. Martha Monigle, 1944; 1 son, Anthony; children (by previous marriage)—Clyde Wilson, Amzell Iona. Urol. interne and resident urologist Bellevue Hosp., N.Y.C., 1919-21, asst. attending urologist, 1921-36; sr. instr. urologic surgery, chief urologic clinic N.Y.U. Med. Coll., 1921-36; asst. vis. urologist St. Vincent Hosp., 1921-26; cons. urologist St. Joseph's Hosp., Far Rockaway, L.I., 1921-52; founder mem. med. staff Doctors Hosp., 1929-52; sr. surgeon attending staff Los Angeles County Gen. Hosp.; asso. prof. surgery (urology) Sch. Medicine, Coll. Med. Evangelists; chief urologic clinic and sr. surgeon attending urologic staff White Meml. Hosp. Fellow Royal Soc. Medicine Eng., A.C.S.; mem. A.M.A., Am., French urol. assns., Delta Tau Delta. Contbr. to med. jours. of U. S., England and France. Inventor of radio electric knife for transurethral surgery, 1923; devised 1st operation with cutting high frequency current and knife electrode through urethroscope, 1923. Died July 4, 1952.

COLLINGWOOD, Cedric Alex, English entomologist; b. London, Eng., Jan. 20, 1919; s. Lawrance Arthur and Anna (Koenig) C.; 1st class certificate plant pathology, Chelsea Poly., 1940; B.Sc. with 2d class honors, 1941; m. Helen Scott Rae, Nov. 6, 1943 (dec. Jan. 1966). Entomol. asst. Harker Adams Agrl. Coll., 1942-46; adviser entomologist Nat. Agr. Adv. Service, Everham Area Lab., 1947-53, dep. regional adv. entomologist, East Midlands, 1954-60, S.W. Eng., 1960-63, regional entomologist, S.E. Eng., 1963—. Entomol. cons. (hon.) Amateur Entomologists Soc. Fellow Royal Entomol. Soc. London; mem. Inst. Biology, Assn. Applied Biologists (council 1963-66), Soc Brit. Entomology, European Soc. Nematology, Internat. Soc. Biometeorology. Cons. ed., Entomologists Record. Research, publs. primarily on economic entomology; infestation and damage assessment methods; ecology, chemical control of black currant gall mite; taxonomy, distribution of ants. Home: Oxford House, Rotten Row, Bradfield, Berkshire. Office: Coley Park, Reading, Berkshire, Eng.*

COLLINGWOOD, Sir Edward Foyle, English mathematician; b. Jan. 17, 1900; s. C. G. and Lilburn (Tower) C.; ed. Royal Naval Coll., Osborne, also Dartmouth (both Eng.); M.A., Ph.D. Sc.D., Trinity Coll., Cambridge (Eng.) U. Steward, Trinity Coll., lectr. Faculty Math., Cambridge U., 1930-38; dir. sci. research with admiralty del., Washington, 1942, officer in charge sweeping div., 1943, chief scientist admiralty design dept., 1943-45; chmn. council Durham (Eng.) U., 1963—. Gibson lectr. U. Glasgow (Scotland), 1961; chmn. Newcastle (Eng.) Regional Hosp. Bd., 1953—, Central Health Services Council, 1963—; mem. Med. Research Council, 1960—, treas., 1960-67, Royal Commn. on Med. Edn., 1965—. Fellow Royal Soc., 1965, Royal Soc. Edinburgh; mem. London Math. Soc. (pres. 1959-60), Internat. Hosp. Fedn. (v.p. 1959—). Author: (with A. J. Lohwater) The Theory of Cluster Sets, 1966. Research, publs. on theory of value distbn., boundary properties of functions of complex variable, math. analysis. Address: Lilburn Tower, Alnwick, Northumberland, Eng.

COLLINS, Conrad Green, Am. physician; b. New Orleans, Apr. 23, 1907; s. Charles and Amelia (Haydel) C.; B.S., Tulane U., 1922, M.D., 1928, M.S. in Gynecology, 1931; m. Louise Morris Carroll, Oct. 9, 1935; children—Louise Carroll (Ms. Oliver Hypolite Dabezies), Conrad Green, Claudia Elizabeth. Faculty, Tulane U., New Orleans, 1931—, prof., chmn. dept. gynecology and obstetrics, 1950—; obstetrician, gynecologist-in-chief Tulane unit Charity Hosp. La., 1950—. Nat. cons. Surgeon Gen. USAF, 1952-62. Mem. A.C.S. (regent 1962—), Am. Gynecol. Club, Am. Gynecol. Soc., Am. Assn. Obstetricians and Gynecologists, Am. Coll. Obstetricians and Gynecologists. Contbg. author: Management of the Patient with Cancer; also numerous articles. Research on ligation vena cava and ovarian vessels in treatment thrombophlebitis, cortisone in treatment ligneous cellulitis in female pelvis, early repair vesicovaginal fistula. Home: 1423 State St., New Orleans 70118.*

COLLINS, Edward Treacher, Brit. surgeon; b. London, Eng., 1862; s. W. J. Collins; ed. Univ. Coll. Sch., Middlesex Hosp.; m. Hetty Emily Herrick, 1894; 1 son, 1 dau. House surgeon Royal London Ophthalmic Hosp., 1884-87, pathologist, 1887-94; Hunterian prof. pathology and surgery Royal Coll. Surgeons, 1893-94, Erasmus Wilson lectr., 1899-1900, Bowman lectr., 1921; Montgomery lectr. Royal Coll. Surgeons in Ireland, 1924; vis. ophthalmic surgeon to ophthalmic schs. Met. Asylums Bd., 24 years; cons. surgeon Royal London Ophthalmic Hosp.; cons. ophthalmic surgeon Charing Cross Hosp. Recipient William Mackenzie medal, 1931. Licentiate Royal Coll. Physicians London, Soc. Apothocaries. Fellow Royal Coll. Surgeons Eng.; mem. Internat. Ophthal. Council (hon. chmn.), A.M.A. (hon.), Copenhagen Med. Soc. (fgn.), Hungarian Royal Soc. Medicine (hon.), Hungarian Ophthal. Soc. (hon.), Egyptian Ophthal. Soc. (hon.), Ophthal. Soc. U.K. (pres. 1917-18, 24), Council Brit. Ophthalmologists (pres. 1923-26). Author: Researches into the Anatomy and Pathology of the Eye, 1896; In the Kingdom of the Shah, 1896; Pathology and Bacteriology of the Eye, 1911; Arboreal Life and the Evolution of the Human Eye, 1922; History and Traditions of the Moorfields Eye Hospital, 1929; also articles on ophthalmology. Died Dec. 13, 1932.

COLLINS, George B., Am. physicist; b. Washington, Jan. 3, 1906; s. Guy N. and Christine (Schmidt) C.; grad. U. Md., 1928; Ph.D., Johns Hopkins, 1932; m. Elsa Leser, June 6, 1934; children—Peter, Lucy, Robert. Faculty, U. Notre Dame, South Bend, Ind., 1933-40; with radiation lab. Mass. Inst. Tech., Cambridge, 1940-46; faculty, chmn. physics dept. U. Rochester (N.Y.), 1946-50; with Brookhaven Nat. Lab., Upton, N.Y., 1950—, dir. cosmotron, 1950-60, mem. council State U. N.Y. at Stony Brook, 1956—. Recipient Medal of Merit for radar work. Fellow Am. Phys. Soc. Dep. editor, author Mass. Inst. Tech. Radiation Lab. series. Early high energy interactions research with electrons, Cerenkov Effect, electron disintegration of nucleus, high energy research of strong interactions. Home: 46 Bellport Lane, Bellport, N.Y. Office: Brookhaven Nat. Lab., Upton, N.Y.*

COLLINS, George Lewis, Am. physician; b. Providence, Feb. 10, 1852; s. George L. and Laura S. (Capron) C.; Ph.B., Brown U., 1873; M.D., Harvard, 1879; postgrad. U. Leipzig, Paris, Vienna and Berlin. Vis. physician and surgeon to various R. I. hosps. and charitable orgns.; cons. physician Providence Lying-In Hosp., R.I. R.C. Orphan Asylum, R.I. State Sanatorium; cons. surgeon R.I. and St. Joseph's hosps.; dir. USPHS. Worked to eradicate trachoma in Am. South. Died June 17, 1940.

COLLINS, Guy N., Am. botanist; b. Mertensia, N.Y., Aug. 9, 1872; s. George and Maria Anne (Hathaway) C.; student Syracuse U., 1890-91; m. Christine Hudson, Aug. 3; 1903; children—George Briggs, Perez Hathaway, Asst. botanist Bur. Plant Industry, U. S. Dept. Agr., 1901-10, botanist, 1910-20, botanist in charge biophys. investigations, 1920—. Conducted explorations in Liberia for N.Y. Colonization Soc., 1891-97; visited Mexico, Guatemala, Costa Rica, P.R. and Haiti for U. S. Dept. Agr. Author: Economic Plants of Puerto Rico (with O. F. Cook), 1903; also numerous articles on tropical agr. and genetics of maize. Died Aug. 14, 1938.

COLLINS, Henry B(ascom), Am. anthropologist; b. Geneva, Ala., Apr. 9, 1899; s. Henry Bascom and Sophie (Neville) C.; A.B., Millsaps Coll., 1922, Sc.D., 1940; A.M., George Washington U., 1925; m. Carolyn A. Walker, Nov. 26, 1931; 1 dau., Judith Ann (Mrs. A. L. Pagani). Sr. ethnologist Bur. Am. Ethnology, 1939-51, sr. anthropologist, 1951-64, acting dir., 1964-65; sr. scientist Smithsonian Office Anthropology, Washington, 1965—. Hon. v.p. 32d Internat. Congress Americanists, Copenhagen, 1956; v.p. 2d Internat. Congress Anthrop. and Ethnol. Scis., Copenhapen, 1938, 7th Internat. Congress, Moscow, 1964. Recipient Gold medal Royal Danish Acad. Scis. and Letters, 1936. Fellow A.A.A.S., Am. Anthrop. Assn., Washington Acad. Scis., Artic Inst. N.Am. (bd. govs., com. chmn.); mem. Am. Assn. Phys. Anthropologists, Am. Soc. Archaeology, Sigma Xi, Sigma Upsilon, Pi Kappa Alpha. Author: Preshistoric Art of the Alaskan Eskimo, 1929; Archeology of St. Lawrence Island, Alaska, 1937; Outline of Eskimo Prehistory, 1940; (with others) The Aleutian Islands, Their People and Natural History, 1946; Arctic Area (Program of the History of America), 1953. Editor: Science in Alaska, 1952. Contbr. numerous articles to profl. jours. Publs. on Smithsonian expdns. to Arctic regions; prehistoric Eskimo culture and art, their origins and relationships to cultures in Eurasia and Am. Home: 2557 36th St. N.W., Washington 20007. Office: Smithsonian Instn., Washington 20560.*

COLLINS, Heron S., Am. mathematician; b. Charlotte, N.C., Nov. 17, 1922; s. Thomas Gordan and Freda (Huskey) C.; B.S., Wofford Coll., 1948; M.S., Tulane U., 1950, Ph.D., 1952; m. Sarah Joyce Harkey, Aug. 23, 1952; children—James Heron, Thomas Welton, Jill Louise. Asst. prof. U. Md., 1952-53, U. S.C., 1953-54; faculty La. State U., Baton Rouge, 1954—, prof., 1962—. Mem. Am. Math. Soc., Phi Beta Kappa, Sigma Xi. Contbr. numerous articles to math. jours. on measure and integration theory; analytic function theory; Banach spaces and algebras. Home: 508 Magnolia Woods St., Baton Rouge.

COLLINS, John, Brit. mathematician; b. Wood Eaton, Oxfordshire, Eng., Mar. 5, 1625; student math. of John Marr; bookseller apprentice in Oxford, Eng.; clk. Prince Charles's Kitchen; served at sea, Crete, 1642-49; mathematical tchr. London, Eng.; govt. clk., 1660-72. Fellow Royal Soc., 1667. Author: Doctrine of Decimal Arithmetick, pub. 1685; Arithmetic in Whole Numbers and Fractions, both Vulgar and Decimal with Tables for the Forbearance and Rebate of Money, pub. 1688; Commercium epistolicum (collection of letters related to discussion between Leibniz and Newton about invention of differential and integral calculus), pub. 1712, also papers, trade pamphlets. Collected and printed sci. information. Died Nov. 10, 1683.

COLLINS, Samuel Cornette, Am. mech. engr.; b. Democrat, Ky., Sept. 28, 1898; s. John W. and Rachel (Caudill) C.; B.S., U. Tenn., 1920, M.S., 1924; Ph.D. in Chemistry, U. N.C., 1927, D.Sc., 1957; m. Lena Arbragine Masterson, Sept. 4, 1929. Prof., Carson-Newman Coll., Jefferson City, Tenn., 1925-26; asso. prof. U. Tenn., Martin, 1927-28; prof. Tenn. State Tchrs. Coll., Johnson City, 1928-30; faculty Mass. Inst. Tech., 1930—, prof. mech. engring., 1949-64, prof. emeritus, 1964—; sr. engring staff Arthur D. Little, Inc., Cambridge, Mass., 1964—; cons. engr., Cambridge, 1945-64. Recipient Kamerlingh Onnes Gold medal Netherlands Refrigeration Soc., 1958. Fellow Am. Acad. Arts and Scis. (Rumford medal 1965); mem. Am. Soc. M.E. Author: (with R. W. Cannaday) Expansion Machines for Low Temperatures, 1958; also numerous articles. Patentee in field, designed and developed cryogenic equipment: developed pump-oxygenator system for open heart surgery. Home: 28 Village Hill Rd., Belmont, Mass. 02178.*

COLLINS, William Edward, Am. psychologist ; b. Bklyn., May 16, 1932; s. William Edward and Loretta (Brasier) C.; B.S., St. Peter's Coll., 1954; M.A., Fordham U., 1956, Ph.D., 1959. With Fordham U., 1954-59, grad. instr., 1958-59; research psychologist U. S. Army Med. Research Lab., Ft. Knox, Ky., 1959-61; with Civil Aeromed. Inst., FAA, Oklahoma City, 1961—, chief psychology lab., 1965—. Mem. Armed Forces NRC Com. on Vision and NAS-NRC Com. on Hearing, Bioacoustics and Biomechanics, 1963—; adj. asso. prof. U. Okla., 1963—. Mem. Am., Eastern, Okla. psychol. assns., A.A.A.S., Aerospace Med. Assn., Psychonomic Soc., N.Y. Acad. Sci., Sigma Xi. Research, publs. on visual thresholds of color-blind; function of vestibular system, visual-vestibular interaction. Office: AC-118 CAMI-FAA, Box 25082, Oklahoma City 73125.*

COLLINS, William Francis, Am. physician; b. New Haven, Jan. 20, 1924; s. William Francis and Jane (Shanley) C.; B.S., Yale, 1944, M.D., 1947; m. Gwendolyn Ruth Davis, Dec. 12, 1950; children—William Francis III, Peter Davis, Ruth Ellen. Nat. Found. for Infantile Paralysis fellow Washington U. Med. Sch., St. Louis, 1953-54; faculty Western Res. U. Sch. Medicine, Cleve., 1954-63, asso. prof. neurosurgery, 1960-63; prof., chmn. div. neurol. surgery, neurosurgeon-in-chief Med. Coll. Va., Richmond, 1963—. Research, publs. on orgn. of sensory pathways of central nervous system. Home: 5105 Cary St. Rd., Richmond, Va. 23226.*

COLLINSON, Peter, English biologist; b. Windermere, Eng., 1694; m. Mary Bushell; 1 son, 1 dau. As partner with brother, improved father's bus. and began trade with Am. colonies; Established botanic garden at Mill Hill. Fellow Royal Soc., 1728; founding mem. Soc. Antiquaries. Author: Hortus Collinsonianus; Plants Cultivated by Peter Collinson, 1843; also articles. Improved English system of horticulture; research on naturalization of plants and trees; advocated cultivation of flax, hemp, silk, grape in Am. Died Essex, Eng., 1768.

COLLIP, James Bertram, Canadian biochemist, physician; b. Belleville, Ont., Nov. 20, 1892; s. James Dennis and May F. (Vance) C.; B.A., U. Toronto, 1912. M.A., 1913, Ph.D., 1916, D.Sc., 1952; Rockefeller fellow U. Alta., 1921-22, D.Sc., 1924, M.D., 1926, LL.D., 1946; LL.D., U. Man., 1935; D.Sc., Harvard 1936, McGill U., 1955, U.B.C., 1950, Western Ont. U., 1964; LL.D., Queen's U., 1941, Oxford U., 1946, U. London, 1948; M.D. (hon.), Laval U., 1947; D. Med. Sci., U. Montreal, 1959; m. Vivian Ray, 1915; children—Margaret (Mrs. McBride), Barbara (Mrs. Wyatt), John. From lectr. to prof. U. Alta., 1915-28; asst. prof. U. Toronto, 1922; prof. biochemistry, head dept. McGill U., 1928-41, research prof., dir. research inst. endocrinology, 1941-47; dean faculty medicine Western Ont. U., 1947-61, prof. med. research, 1947—. Recipient awards including Cameron prize U. Edinburgh, 1937; Medal Freedom with silver palm (U. S.), 1947; named comdr. Brit. Empire, 1943. Fellow Royal Soc. Can., A.A.A.S., Royal Soc., 1933; N.Y. Acad. Medicine, Montreal Medico-Chirurgical Soc., A.C.P., Royal Coll. Physicians; mem. Nat. Research Council Can. (dir. div. med. research 1946-57), Am. Soc. Biol. Chemists, Am. Assn. Study Internal Secretions, Biochem. Soc. Gt. Britain, Brit. Physiol. Soc. Blood Chemistry, Am., Canadian, Brit. physiol. socs. Soc. Exptl. Biology and Medicine, Canadian Med. Assn., others; Sigma Xi (past pres.), Alpha Omega Alpha. Research and publs. in field. Isolated parathormone from parathyroid gland and introduced it in treatment of tetany, 1925; first to isolate pure insulin, 1921. Died June 19, 1965.

COLLIS, Ronald Thomas, meteorologist; b. London, Eng., July 22, 1920; s. Thomas James and Dora (Coombes) C.; B.A. with honors, Oxford U., 1949, M.A., 1951; m. Ethyl Maxine Sirstins, March 31, 1951; 1 son, Gregory. Came to U. S., 1958, naturalized, 1964. Meteorologist, Decca Radar Ltd., London, 1955-58; mgr. aerophysics lab. Stanford Research Inst., 1958—. Fellow Royal Meteorol. Soc. (London), Am. Meteorol. Soc.; mem. Research Soc. Am. Research and publs. on radar, radio applications in meteorol. instrumentation, including the develop. of original techniques for measuring rainfall over extended areas by radar; played maj. role in developing new methods of probing atmosphere using laser energy in instruments called Lidars; identified various hitherto unsuspected structures in clear air; studies of problems of detecting clear air turbulence in upper air. Home: 181 N. Castanya Way, Menlo Park, Cal. 94025.*

COLLITZ, Hermann, philologist; b. Bleckede, Hanover, Feb. 4, 1855; grad. Gymnasium Johanneum at Lüneburg, 1875; student philology, Göttingen, 1875-79, Berlin, 1879-82; A.M., Ph.D., Göttingen, 1878; L.H.D., U. Chgo., 1916; m. Klara Hechtenberg, Aug. 13, 1904. Asst. librarian, 1883-86, instr. Sanskrit and comparative philology, U. Halle, 1885-86; asso. prof. German, 1886-97, prof. comparative philology and German, 1897-1907, Bryn Mawr Coll.; prof. Germanic philology, Johns Hopkins, 1907-27 (emeritus); lectr. Linguistic Institute, 1928, 31. Author: Die Verwantschaftsverhältnisse der Griechischen Dialekte, 1885; Die neueste Sprachforschung, 1886; Das Schwache Praeteritum und seine Vorgeschichte, 1912. Editor: Sammlung der Griechischen Dialektinschriften, 4 vols., 1884-1915; Bauer's Waldeck Dialect Dictionary, Leipzig, 1902; Hesperia (Schriften zur Germanischen Philologie), 1912—. Died May 13, 1935.

COLLOMB, Edouard, French Geologist; b. Mar. 8, 1801; Licencié ès sciences; accompanied Agassiz on his voyages. Author: La Carte géologique de l'Espagne, 1864; pub. 1st geol. map of Spain. Research on age, area, flow and moraine of European glaciers, eruptions of Mt. Etna. Died Paris, France, May 28, 1875.

COLMAN, Norman Jay, Am. agriculturist; b. nr. Richfield Springs, N.Y., May 16, 1827; s. Hamilton and Nancy (Sprague) C.; LL.B., Louisville Law Sch., 1851; LL.D., U. Mo., 1905; D.Agr., U. Ill., 1905; m. Clara Porter, 1851 (dec. 1863); m. 2d, Catherine Wright, 1866. Practiced law, New Albany, Ind., 1850-52; elected dist. atty., 1852; removed to St. Louis, 1852; established Colman's Rural World, became its editor; dean agrl. editors in U. S. Alderman St. Louis, 1855-56; served in Civil War; mem. Mo. Ho. of Reps., 1865-66; defeated for lt. gov., 1868, elected for term 1875-77; U. S. commr. agr., 1885-89, when dept. was elevated to exec. branch of govt., became 1st sec. of agr. of U. S., 1889. Issued call and presided over conv. of dels. from agrl. colls. in U. S., 1885, urged adoption of laws upon Congress creating new system of expt. stas. in connection with agrl. colls. in U. S.; selected by commn. to head govt. horse-breeding farm, Ft. Collins, Colo., for establishment of a breed of Am. trotting-bred carriage horses. Decorated officer Merite Agricole, France, 1889. Died 1911.

COLOMBAT, Marc, physician; b. Vienna, Austria, June 28, 1797; s. Marc and Julie (Gay) C.; studied under Dr. Prunelle, U. Montpellier (France); thesis Faculty Medicine, Paris, France, 1828; 1 dau., Emile. Founder, Inst. Orthophonics, Paris, 1829; developed orthophonic method for treating speech impairments; studied the voice. Died June 20, 1851.

COLOMBO, Matteo Realdo, Italian anatomist; b. Cremona, Italy, 1516; studied medicine under Vesalius at Padua (Italy); became asst., then successor of Vesalius at Padua; 1st prof. anatomy at Pisa (Italy), 1545-48; held chair of anatomy at Papal U., Rome, Italy, until 1559. Author: De re anatomica, 1559. Discovered pulmonary circulation of blood, 1558; 1st to use living animals in lab. expts. especially in function of heart and lungs; 1st to mention the folding of the peritoneum; 1st to describe mediastinum. Died Rome, 1559.

COLON, José Angel, Puerto Rican meteorologist; b. Coamo, P.R., Nov. 24, 1921; s. Angel M. and Carmen (Pérez) C.; B.A., U. P.R., 1944; M.S., U. Chgo., 1950, Ph.D., 1960; m. Lydia Vélez, May 20, 1951; children—Dorothy (Mrs. Edwin R. Rivera), Olga, Jose. Faculty, U. P.R., 1945-49; instr. U. Chgo., 1951-52, research asso., 1952-54; research forecaster U. S. Weather Bur., San Juan, P.R., 1954-58, chief, 1964—; supervisory research meteorologist Nat. Hurricane Research Project, Miami, Fla., 1959-62, 64; cons. research meteorologist Internat. Meteorol. Center, Internat. Indian Ocean Expdn., Bombay, India, 1962-64. Mem. Nat., Royal (Eng.) meteorol. socs., Am. Geophys. Union, Sigma Xi. Contbr. articles to tech. jours. Research on structure, formation, motion and evolution tropical hurricanes, air-sea interactions, redistbn. energy from ocean to atmosphere, formation and structure Indian cyclones, Indian monsoon circulations. Home: Weather Bur. Bldg., Ponce de Leon Av., Stop 4, San Juan 00906. Office: Weather Bur., Internat. Airport, San Juan, P.R. 00913.*

COLONNETTI, Gustav, Italian engr.; b. Turin, Italy, Nov. 8, 1886; s. Alcibiade and Paola (Callegaris) C.;

ed. Poly. Sch. of Turin; Ph.D. in Engring. and Sci., U. Turin Dr. honoris causa univs. Toulouse, Lausanne, Poitiers, Liege; m. Laura Badini-Confalonieri, 1927; children—Elena, Pier-Giorgio, Lia, Alberto, Silvia, Margherita. Prof. constrn. scis.; rector Poly. Sch. Turin and Sch. Engrs. of Pisa.; now emeritus prof. Pres., Council of Research, 1945-56. Mem. Nat. Acad. dei Lincei, Pontificia Acad. Scientiarum, Turin Acad. Sci. Author: Scienze delle Costruzioni; L'equilibre des corps deformables; Elastoplastica. Home: corso Moncalieri 62. Office: Isituto Dinamometrico Italiano, vir Onorato Vigliani 104, Turin, Italy.

COLOT, German, French surgeon; flourished 1480. Practiced lithotomy; performed 1st operation for removal of gall stones, 1470.

COLOT, Laurent, French surgeon; b. near Troyes, France; flourished around 1550. Court surgeon in the reign of Henry II, 1547-59; lithotomist at Hotel-Diey; for 3 generations the secret of lithotomy as practised by him was kept in the family. Died Paris.

COLPA, Johannes Pieter, Dutch chemist; b. Arnhem, Jan. 26, 1926; s. Johannes and Maria Johanna (Oostingh) C.; ed. U. Amsterdam, Cambridge U.; Ph.D. in natural scis.; m. J. P. Boonstra, Dec. 20, 1951; 1 son, Alex. Collaborator lab. phys. chemistry U. Amsteram, until 1957; chemist research lab. Shell, 1957—; prof. theoretical chemistry U. Amsterdam, 1963—. Editor: (rev.) Molecular Physics. Research and numerous publs. on molecular spectroscopy and ondulatory molecular mechanics. Home: Jacob Obrechtstraat 82. Office: Nieuwe Prinsengracht 126, Amsterdam, Netherlands.

COLT, Samuel, Am. inventor, firearms mfr.; b. Hartford, Conn., July 19, 1814; s. Christopher and Sarah (Caldwell) C.; m. Elizabeth H. Jarvis, June 5, 1856. Invented multi-shot firearm of revolving barrell type, constructed wooden model; constructed 2 pistols, 1831, sent description to U. S. Patent Office, 1832, received 1st U. S. patent for 1st practical revolving firearm (widely used in opening of Am. West), 1836; introduced electricity as agt. for igniting gun powder; operated mfg. plant, Hartford; built an underwater telegraph, 1843. Died Hartford, Jan. 10, 1862.

COLTMAN, John Wesley, Am. physicist; b. Cleve., July 19, 1915; s. Robert W. and Louise (Tyroler) C.; B.S., Case Inst. Tech., 1937; M.S., U. Ill., 1939, Ph.D., 1941; m. Charlotte Waters Beard, June 10, 1941; children—Sally Louise, Nancy Jean. Research scientist Westinghouse Research Lab., Pitts., 1941-43, sect. mgr., 1943-49, dept. mgr. 1949-60, asso. dir., 1960-64, dir. math. and radiation, 1964—. Mem. adv. group on electron devices Dept. Def., N.Y.C., 1961-65; mem. adv. com. electrophysics NASA, Washington, 1964-66; chmn. Forest Hills Borough Planning Commn., 1958-61. Recipient Longstreth medal Franklin Inst., 1961; named Man of Year in Sci., Pitts., 1960. Fellow, Am. Phys. Soc., I.E.E.E., mem. Sigma Alpha Epsilon. Co-inventor Scintillation Counter; developed first X-ray image amplifier, research in noise in imaging. Home: 3319 Scathelocke Rd. Office: Westinghouse Research Lab., Pitts. 15235.*

COLTON, Gardner Quincy, Am. dentist, inventor; b. Georgia, Vt., Feb. 7, 1814; s. Walter and Thankful (Cobb) C.; studied medicine under Dr. Willard Parker, N.Y.C., 1842. Gave exhbn. of effects of nitrous oxide (laughing gas) when inhaled, N.Y.C., 1844, toured other cities with the gas; invented an electric motor, 1847; participated in Cal. gold rush, 1849; justice of peace San Francisco; reporter Boston Transcript, 1860; 1st (with Horace Wells) to use nitrous oxide for extraction of a tooth, 1844, tech. problems prevented widespread use for some time, reintroduced it as dental anesthetic, 1869. Died Rotterdam, Holland, Aug. 9, 1898.

COLTON, Harold Sellers, Am. zoologist, archaeologist; b. Phila., Aug. 29, 1881; s. Sabin Woolworth and Jessie (Sellers) C.; B.S., U. Pa., 1904, M.S., 1906, Ph.D., 1908; LL.D., U. Ariz., 1955; D.Sc., No. Ariz. U., 1959; m. Mary Russell Ferrell, May 23, 1912; children—J. Ferrell, Sabin Woolworth IV (dec.). With Carnegie Instn. Table. Stazions Zoologica, Napoli, Italy, 1908-09; faculty U. Pa., Phila., 1909-54, prof., 1926-54; dir. Mus. No. Ariz., Flagstaff, 1928-60. Recipient City of Flagstaff award for pub. service, 1963; U. S. Dept. Interior Conservation Service award, 1959. Mem. Am., Nat. geog. socs., Am. Soc. Zoologists, Am. Soc. Naturalists, Am. Anthrop. Soc., Ecol. Soc. Am., Am. Soc. Mammalogists, Acad. Nat. Sci. Phila., Ariz. Acad. Sci., Am. Micros. Soc., Soc. for Am. Archeologists. Author: (with Frank C. Baxter) Days in Painted Desert, 1932; (with L. L. Hargrave) Handbook of Northern Arizona Pottery Wares, 1937; Hopi Kachina Dolls, 1948; Potsherds, 1953; Black Sand, 1960; North of Market Street, 1961; Prehistoric Dogs of the Southwest, 1966; others; also numerous articles. Research on ecology mollusks, physiology ascidians, anatomy Lac insect, geography prehistoric Indians no. Ariz., geology cinder cones in San Francisco volcanic field. Home: Box 699, Flagstaff, Ariz. 86002.*

COLUMBO, Matteo Realdo, see Colombo, Matteo Realdo.

COLUMBUS, Christopher, Italian navigator, explorer; b. Genoa, Italy, date unknown; s. Domenico and Suzanna (Fontanarossa) Columbo; studied at U. Pavia (Italy); m. Felipe Moñiz de Perestrello, circa 1478; 2 illegitimate sons (by Beatriz Enriquez), Diego, Ferdinand. Map and chart maker, Santo Porto (Madeira Islands), circa 1479; became convinced from this work that the East could be reached by sailing West; used 3 sources to determine this theory: Travels (Marco Polo), Imago Mundi (Pierre d'Ally), Cosmographia (Aeneas Sylvius Piccolomini, Pope Pius II); misjudged in his calculations the amount of land comprising the surface of earth and the distance to India; tried to gain help for venture from John II of Portugal, failed and went to Spain; received little attention at Spanish court; had friends plead his case in Eng. and France; happened to meet Father Juan Perez (Queen Isabella's confessor), which led to support by Spanish throne; prepared for his 1st voyage with 3 ships, Santa Maria, Pinta (commanded by Martin Pinzon), Nina (commanded by Vincente Yonez); set sail, Aug. 3, 1492; put in at Teneriffe (Canary Islands) when Pinta lost its rudder 3 days after embarking; set sail again in 3 days; sighted New World (he believed it to be "Indies," as Asia was called, Oct. 12, 1492; discovered islands of Santa Maria de la Concepcion (Bahama Islands), Fernandina (Amelia Island), Isabella (Bahama Islands), Cuba, Hispaniola; built fort La Navidad on Santa Maria Island, left 44 men there; returned to Spain, Mar. 14, 1493, was received with great fanfare; set out on 2d voyage with 1500 men (including 12 missionaries) and 17 ships, Sept. 25, 1493; discovered islands of Dominica (Windward Islands), Marigalante, Guadalupe, Monserrat, Antigua (Leeward Islands), San Martin, Santa Cruz, Virgin Islands, Puerto Rico; established mining camp in Central Hispaniola; shipped 1st slaves into New World on 2d voyage; started on 3d voyage, May 30, 1498, discovered island of Trinidad and coast of S.Am.; sailed as far South as Honduras, went to Jamaica; retuned to Cuba, was arrested and put in chains by royal gov. Francisco de Bobadilla (had been sent by King Ferdinand, who never favored Columbus, to rule Columbus' discoveries and investigate charges of cruelty); released in Spain, made provisions for 4th voyage, leaving on May 9, 1502; still hoped to find passage to the East on this voyage; was refused entry into port of Santo Domingo; lost 2 ships in a gale and 2 others in wreck off Jamaica, where some of his men revolted and settled; managed to get supplies and repair his ships, returned to Spain, Sept. 12, 1504; his voyages must be observed as great navigational feats; his use of only basic navigational instruments may rate him as 1 of best navigators of all time. First European to observe the variation of the magnetic declination. Died Vallodolid, Spain, May 20, 1506; buried Santa Maria de las Cuevas, Seville, Spain; reinterred Cathedral of San Domingo, Hispaniola; reinterred Havana, Cuba; reinterred Cathedral, Seville.

COLUMNA, Fabius (or Fabio Colonna), Italian botanist; b. Naples, Italy, 1567; received classical education. Assisted in founding Academia de Lincei, Rome, ca. 1616. Author: Phytobasanos, 1592; Ecpharasis, 1606. Works remarkable for accuracy of descriptions and beauty of figures; 1st to use copper plates for botanical figures; recognized and used distinction of genera; developed principles of botany; tried to identify medicinal plants described by ancients; coined term, petal; discovered many new plants; wrote compilation of then known knowledge of botany. Died Naples, 1647.

COLWELL, Arthur Ralph, Sr., Am. physician; b. Chgo., July 8, 1897; s. Lewis W. and Grace (Stryker) C.; S.B., U. Chgo., 1919; M.D., Rush Med. Coll., 1921; m. Jeane Haskins, Sept. 1, 1921; children—Arthur Ralph, John A., Mary Ann (Mrs. Charles D. Nitchie). NRC fellow Harvard Med. Sch., 1927-29; Asst. prof. Rush Med. Coll., Chgo., 1923-33; faculty Northwestern U. Med. Sch., 1933—, prof. medicine, chmn. dept., 1950-65, emeritus, 1965—; practice medicine specializing in diabetes, Chgo., 1923—. Trustee Sprague Memorial Institute. Member of the American Diabetes Association (past pres.), Assn. Am. Physicians Central Soc. for Clin. Research. Author: Diabetes Mellitus, 1947; Types of Diabetes Mellitus and Their Treatment, 1950; also numerous articles. Research on timing insulin preparations, course diabetes mellitus, vascular complications diabetes. Home: 260 E. Chestnut St., Chgo. 60611. Office: 707 Fairbanks Ct., Chgo. 60611.*

COLWELL, Robert Neil, Am. photogrammetrist; b. Star, Ida., Feb. 4, 1918; s. Arthur Ludlow and Mary (Kinnison) C.; B.S., U. Cal. at Berkeley, 1938, Ph.D., 1942; m. Betty Louise Larson, Apr. 14, 1942; children—Arthur Edwin, John Eric, Nancy Louise, Robert Richard. Asst. prof. forestry U. Cal. at Berkeley, 1947-50; chief of research Navy Photo Interpretation Center, 1951-52; asso. prof. U. Cal. at Berkeley, 1953-57, prof. forestry, 1957—. Chmn., NRC Com. on Crop Geography and Vegetation Analysis, 1951-53. Mem. Internat. Soc. Photogrammetry (pres. commn. on photo interpretation 1952-56), Am. Soc. Photogrammetry (Distinguished lectr. 1965-66, Fairchild award 1956, Abrams award 1954, Photo Interpretation award 1963), Soc. Am. Foresters, Sigma Xi (nat. lectr. 1963-64). Editor: Manual of Photo Interpretation, 1960. Research publs. on disease detection on agrl. crops by aerial photography, inventory of livestock by aerial photography, assessment of trafficability conditions by aerial photography, determination of distbn. patterns of sprays, dusts and spores by means of radioactive tagging with isotopes, determination of rate of transport of organic solutes in plants by tagging with radio isotopes. Home: 1300 Juanita Dr., Walnut Creek, Cal. 94529.*

COLWIN, Arthur Lentz, biologist; b. Sydney, Australia, Jan. 26, 1911; B.Sc., McGill U., 1933, M.Sc., 1934, Ph.D. (Nat. Research Council Can. fellow 1935-36; Moyse Travelling fellow Cambridge (Eng.) U., 1934-35; Seessel fellow Yale, 1936-37, Royal Soc. Can. fellow, 1937-38; m. Laura North Hunter, June 15, 1940. Came to U. S., 1936, naturalized, 1942. Mem. faculty Queens Coll., 1940—, prof., 1957—; Fulbright research fellow Tokyo (Japan) U., 1953-54; vis. scientist Nat. Inst. Med. Research, London, Eng., 1960; spl. research fertilization, devel. biology, cell contacts and assn., membrane structure and behavior. Trustee Marine Biol. Lab., Woods Hole, Mass., 1962—. Served to capt. USAAF, 1943-46. Fellow N.Y. Acad. Scis.; mem. Internat. Inst. Embryology, Internat. Soc. Cell Biology, Am. Soc. Zoologists, Soc. Study Devel. and Growth, Electron Microscope Soc. Am. Contbr. profl. jours. Asso. editor Jour. Exptl. Zoology, 1964—, Jour. Morphology, 1964—. Research on normal and experimental embryology; sperm-egg association; fertilization; egg cortical changes; cell division and differentiation. Office: Dept. Biology, Queens Coll., Flushing, N.Y. 11367.

COMAN, Dale Rex, Am. pathologist; b. Hartford, Conn., Feb. 22, 1906; s. Edward L. and Florence (Rex) C.; student U. R.I., 1924-26; B.A., U. Mich., 1928; M.D., McGill U., 1933; m. Mona Charity Segal, Dec. 22, 1937; children—Michael Dale, Charity Beth. Practice medicine, specializing in pathology, Phila., 1937—; faculty U. Pa., 1937—, prof., chmn. dept. pathology, 1954—. Diplomate Am. Bd. Pathology. Fellow A.A.A.S.; mem. Am. Assn. Cancer Research, Am. Assn. Pathologists and Bacteriologists, Internat., Am. socs. cell biology, Am. Soc. Exptl. Pathology, Tissue Culture Assn., Sigma Xi, Alpha Omega Alpha. Author: The Technique of Postmortem Examination, 1934. Demonstrated reduced mut. adhesiveness of cancer cells and established our current concepts of invasiveness of cancer; studies on cell surface structure and chem. factors affecting cellular adhesion, cell motility and chemotaxis, mechanisms of metastasis and the distbn. of metastatic tumors. Home: Garden Ct. Apts., 47th and Pine Sts., Phila. 19143. Office: Dept. Pathology, U. Pa. Med. Sch., Phila. 19104.*

COMAR, Cyril Lewis, Am. biologist; b. Dudley, Eng., Mar. 28, 1914; s. David and Bertha (Simon) C.; naturalized U. S. citizen, 1941; B.S., U. Cal. at Berkeley, 1936, postgrad.; Ph.D., Purdue U., 1941; m. Mildred Cashin, Aug. 7, 1939; children—Anne Patricia, Thomas Allan, Louise Elaine. Asst. chemist Fibrebd. Products, Antioch, Cal., 1936-37; research asso. Mich. State U., East Lansing, 1941-43; prof., biochemist U. Fla., Gainesville, 1943-48; prof. biophysics, lab. dir. U. Tenn., Knoxville, 1948-53; chief biomed. research, 1954-57; prof., head dept. phys. biology Cornell U., 1957—. Mem. Nat. Com. on Radiation Protection, 1966—, Nat. Adv. Com. on Radiation, 1962—; mem. Fed. Radiation Council, 1964—, chmn. adv. com., 1964—; cons. to govt. agys., research insts. Recipient Medal of Honor, City of Paris, 1957. Mem. Am. Chem. Soc., Am. Inst. Nutrition, Am. Inst. Biol. Chemists, Soc. for Exptl. Biology and Medicine, Nat. Acad. Scis. (food protection com. 1961—, A.A.A.S., Radiation Research Soc., Health Physics Soc., N.Y. State Soc. for Med. Research, Am. Vet. Med. Assn. (hon.), Sigma Xi, Phi Lambda Epsilon, Gamma Alpha, Gamma Sigma Epsilon, Phi Kappa Phi, Phi Zeta Alpha. Author: Radioisotopes in Biology and Agriculture, 1955; also numerous articles. Editor: Mineral Metabolism—an Advanced Treatise, 1960. Research on passage radioactive fission products through food chain into human population, metabolism alkaline earths, phys. techniques and concepts in biol. research. Home: 8 Highland Park, Ithaca, N.Y.*

COMARR, A. Estin, Am. urologist; b. Chgo., July 15, 1915; s. Louis and Esther (Massel) C.; A.B., U. So. Cal., 1937; M.B., Chgo. Med. Sch., 1940, M.D., 1941; m. Ruth Harriette Litchman, Sept. 4, 1938; 1 dau., Cynthia Ester. Chief neurol.-urology, asst. chief spinal cord injury service VA Hosp., Long Beach, Cal., 1946—; faculty U. Cal. at Los Angeles, 1953-55; chief urology service, cons. spinal cord injuries Rancho Los Amigos Hosp., Downey, Cal., 1955—, pres., mem. bd., 1964—; clin. prof. urology Loma Linda U. Sch. Medicine, 1966—; clin. prof. surgery U. So. Cal., 1966—; staff Los Angeles County Gen. Hosp., 1955—. Decorated Bronze Star medal; Meritorious Service plaque; Belgian Fourregere; recipient numerous citations of merit, including Am. Legion, 1952, D.A.V., 1952, So. Cal. chpt. Nat. Multiple Sclerosis Soc., 1959, Nat. Soc. Crippled Children and Adults, 1960, El Hosp. Colonia, Mexico City, 1961; 1st Sci. Meritorious award Chgo. Med. Sch. Alumni Assn., 1963; Tutelary St. award W.H. and C. Gottsche Rehab. Center, 1966; Citation of Commendation, Paralyzed Vets. Am., 1966. Diplomate Am. Bd. Urology. Mem. A.M.A., Am. Urol. Assn., Am. Paraplegia Soc. (founding and past pres.), A.C.S., Internat. Coll. Surgeons, Internat. Med. Soc. Paraplegia, World Med.

Assn., VA Urol. Assn., Kappa Zeta, Tau Epsilon Phi, Phi Delta Epsilon, others. Contbr. numerous articles to med. jours. Pioneered in care of spinal cord injuries. Office: VA Hosp., Long Beach, Cal. 90801.*

COMBASLE, Mathieude, see de Combasle.

COMBEROUSSE, Charles-Jules-Felix de, see de Comberousse.

COMBES, Charles Pierre Mathieu Combes, French engr.; b. Cahors, France, Dec. 26, 1802; tchr. Sch. Mines of Saint-Etienne; dir. Firminy mines; apptd. prof., dir. Sch. Mines, Paris, France 1857; insp. gen. of mines. Mem. French Acad. Scis. (became pres. 1853), French Soc. Agr. Patentee universal machine (hydraulic turbine). Invented anemometer; research and publs. on ventilation of coal mines, ventilators and turbines; designed theodolite for underground surveying, also steam machine regulator. Died Paris, June 11, 1872.

COMBS, C(larence) Murphy, Am. anatomist; b. Louisville, Apr. 13, 1925; s. C. H. and Mary (Murphy) C.; A.B., Transylvania Coll., 1946; M.S. Northwestern U., 1948, Ph.D., 1950; m. Virginia Lee Thompson, Aug. 24, 1946 (div. Oct. 1964); children—Jeanne Marie, Stephen Murphy, Nancy Clare. Instr., W.Va. U. Med. Sch., 1948; Ward fellow Northwestern U. Med. Sch., 1946-50, faculty, 1950-66, prof. anatomy, 1963-66; prof., chmn. dept. anatomy Chgo. Med. Sch., 1966—; asso. prof. U. P.R. Med. Sch., 1958-60; chief sect. perinatal physiology Nat. Inst. Neurol. Disease and Blindness, San Juan, P.R., 1958-60, spl. cons., 1958. Spl. lectr. Ill. State Psychopathic Inst., 1954-58. USPHS Sr. Research fellow, 1959-61; recipient Research Career Devel. award USPHS, 1961-64. Mem. Am. Assn. Anatomists, Am. Acad. Neurology, Internat. Brain Research Orgn., Am. Assn. U. profs., A.A.A.S., Biol. Stain Commn., Sigma Xi. Editorial bd. Dorland's Med. Dictionary, 1965. Research, publs. on relationships between cerebellum and other parts of central and peripheral nervous systems, interconnections between cerebral cortex and diencephalon, gross structure of spinal cord segments. Home: 1706 Washington St., Evanston, Ill. 60602. Office: 2020 W. Ogden Av., Chgo. 60612.*

COMBS, George Ernest, Jr., Am. animal nutritionist; b. Arcadia, Fla., Feb. 21, 1927; s. George Ernest and Blanche (Hall) C.; B.S. with high honors, U. Fla., 1951, M.S., 1953; Ph.D., Ia. State U., 1955; m. Christine Walker, Feb. 19, 1948; children—Stuart Keith, Brian Scott, Gary Glen. Faculty U. Fla., Gainesville, 1951-52, 55—, prof. animal nutrition, 1967—. Mem. Am. Inst. Nutrition, Am. Soc. Animal Sci., Sigma Xi, Gamma Sigma Delta (Jr. Faculty Research award 1965). Research, numerous publs. on mineral, amino acid, vitamin requirements and interrelationships. Home: 630 N.W. 36th St., Gainesville, Fla. 32601.*

COMBS, Gerald Fuson, Am. nutritionist; b. Olney, Ill., Feb. 23, 1920; s. Lloyd R. and Ina (Fuson) C.; B.S. in Agr., U. Ill., 1940; Ph.D., in Animal Nutrition, Cornell U., 1948; m. Lily McMaster Ijams, Mar. 26, 1943; children—Gerald F., Lawrence L., John W., Gregory L. Faculty nutrition U. Md., 1948-65, 1966—; asst. head nutrition sect. Office Internat. Research, NIH, Bethesda, Md., 1965-66. Recipient Research award Am. Feed Mfrs. Assn., 1953; teaching and research award in agr. Agr., 1963. Mem. Nat. Acad. Sci. Internat. Union Nutritional Sci. (U. S. nat. com. 1962—), Am. Inst. Nutrition, Poultry Sci. Assn., Worlds Poultry Sci. Assn., Soc. Exptl. Biology and Medicine, N.Y. Acad. Scis., Sigma Xi, Phi Kappa Phi. Research, numerous publs. on poultry and human nutrition including energy, protein, amino acids, vitamins, minerals and fat, nutritional interrelationships; human nutrition researchers. Home: 10754 Kinloch Rd., Silver Spring, Md. 20703. Office: U. Md., College Park, Md.*

COMBY, Jules, French physician; b. Pompadour, France, 1853; studied at Faculty of Medicine, Paris, France; student of Bouchard; specialized in infantile medicine; established dispensary of Villette, France; chief of services Hosp. Trousseau, France, also Hosp. Childhood Diseases. Author: Dictionnaire d'hygiène des enfants, 1885; Traité des maladies de l'enfance. Founder, Archives de médecine des enfants. Promoted sterilized milk for babies. Died May, 1947.

COMFORT, Mandred Whitset, Am. physician; b. Hillsboro, Tex., June 10, 1895; s. Edgar Whitset and Eulah (Stoud) C.; A.B., Austin Coll., 1916, LL.D., 1954; M.D., U. Tex., 1921; M.S. in Neurology, U. Minn., 1926; m. Aurella Jones, Mar. 14, 1931. Adj. prof. anatomy U. Tex., 1921-23; fellow Mayo Found., Rochester, Minn., 1923; asso. in medicine Mayo Clinic, 1928, prof. medicine Mayo Found. Grad. Sch., 1946-57; cons. physician St. Mary's Hosp., 1928-57. Cons., Nat. Cancer Inst., 1946-50, and 1952-57. Diplomate Am. Bd. Internal Medicine. Fellow A.M.A., A.C.P., mem. Assn. Am. Physicians, Am. Gastro-ent. Assn. (pres. 1957), Central Soc. for Clin. Research, Sigma Xi, Alpha Kappa Kappa, Alpha Omega Alpha. Contbr. numerous articles on gastroenterology to profl. jours. Died Aug. 7, 1957.

COMINGS, Edward Walter, Am. chem. engr.; b. Phillipsburg, N.J., Feb. 24, 1908; s. Robert Morrow and Nellie (Breen) C.; B.S., U. Ill., 1930; Sc.D., Mass. Inst. Tech., 1934; m. Jeannette Florentine Rice, Nov. 14, 1931; children—Gordon Robert, David Edward, Miriam Louise (Mrs. John Lobell). Asst. prof. N.C. State Coll., 1935-36; faculty U. Ill., 1936-51; prof. chem. engring., 1947-51; head Purdue U. Sch. Chem. and Metall. Engring. 1951-59; dean engring. U. Del., Newark, 1959—; chmn. Gordon Conf. High Pressure Research, 1956. Fulbright lectr. Delft Tech. Inst., Netherlands, 1957. Guggenheim fellow, 1957. Recipient Naval Ordnance Devel. award for work on smoke munitions, 1945. Fellow A.A.A.S.; mem. Am. Chem. Soc. (past div. chmn.), Delaware Academy of Scis. (president 1966), American Inst. Chem. Engrs. (William H. Walker award 1956), Am. Soc. Engring. Edn., Sigma Xi, Phi Lambda Upsilon, Alpha Chi Sigma, Omega Chi Epsilon, Tau Beta Pi, Phi Eta Sigma, Tau Kappa Epsilon. Author: High Pressure Technology, 1956. Contbr. chpt. to The Applications of Chemical Engineering, 1940. Research, numerous publs. on drying, extraction and thickening of slurries; high pressure viscosity and thermal conductivity, high pressure chem. reactions, turbulent flow in air jets and smoke munitions, high velocity jet spray dryer. Home: 509 Windsor Dr., Newark, Del. 19711.*

COMMANDINO, Federigo, Italian mathematician; b. Urbino, Italy, 1509; studied medicine in Padua, Italy, 1535; physician, mathematician to Duke of Urbino at Verona, Italy. Translated works of ancient mathematicians and geometricians including Archimedes, Appllonius, Euclid. Applied computation methods of Archimedes to the center of gravity. Died Urbino, 1575.

COMMELIN (or COMMELYN), Caspar (or Gaspard), Dutch botanist; b. Amsterdam, Netherlands, Oct. 14, 1668; Succeeded his uncle Jean Commelin as prof. botany, dir. Amsterdam Bot. Garden; mem. French Acad. Scis., L'Academie des Curieux de la Nature. Author: Flora Malabarica, 1696; Horti medici Amstelodamensis plantarum usualium catalogus, 1697. Died Amsterdam, Dec. 25, 1731.

COMMELIN, Jean (Jerome), Dutch botanist; b. Amsterdam, Netherlands, 1629; dir., prof. Bot. Garden of Amsterdam. Author: Catalogus plantarum indigenarum Hollandie, 1869; Les Hespérides des pays-Bas, 1676; Description of Plants of the Botanical Gardens of Amsterdam, 1697. Studied flora of Holland, edible plants in Indies. Died Amsterdam, 1692.

COMMERSON, Philibert, French botanist; b. Châtillon-les-Dombes, France, 1727; received doctorate, Montpellier, France, 1755; founder botanic garden, Châtillon-les-Dombes; moved to Paris, France, 1764; became mem. expdn. of Bougainville, 1766. Collected plants around world; described fishes of Mediterranean (at request of Linnaeus); made drawings and collected specimens of plants and animals during Bougainville's expdn. Died Island of Mauritius, 1773.

COMMON, Andrew Ainslie, English astronomer; b. Newcastle/Tyne, Eng., Aug. 7, 1841; m. Ann Matthews, 1867; 1 son, 3 daus.; LL.D., U. St. Andrews (Scotland), 1891; mem. firm Matthew Hall & Co., san. engrs.; bd. visitors Royal Obs., 1894-1903. Fellow Royal Soc., 1885; mem. Royal Astron. Soc. (Gold medal 1884, pres. 1895-97). Observed satellites of Mars and Saturn, comet of 1882; 1st to apply photography to the study of nebulae; improved the sightings of guns; made large reflecting equatorial telescopes, small mirrors for observing eclipses for Royal Soc., South Kensington and Greenwich Obs. Died Ealing, Middlesex, June 2, 1903.

COMMON, Robert Haddon, agrl. chemist; b. Larne, North Ireland, Feb. 25, 1907; s. Robert Hall and Alice (Magill) C.; B.Sc., Queen's U., Belfast, 1928, B.Agr., 1929, M.Agr., 1931, D.Sc., 1957; Ph.D., London U., 1935, D.Sc., 1944; m. Renate Liselotte Gueterbock, Oct. 16, 1935; children—Edith (Mrs. D. S. Layne), Christine (Mrs. Jagat Singh), Robert, Sonia, Jennifer, Andrew. Faculty, Queen's U., Belfast, 1929-47; prof. agrl. chemistry, chmn. dept. Macdonald Coll., McGill U., Que., Can., 1947—. Fellow Royal Inst. Chemistry, Agrl. Inst. Can., Royal Soc. Can.; mem. Chem. Inst. Can., Can. Inst. Food Tech., Sigma Xi. Research, publs. on mineral metabolism and steroid estrogens of domestic fowl, effects of estrogens on blood and tissues of fowl. Address: Box 223, Macdonald Coll. P.O., Que., Can.*

COMMONER, Barry, Am. biologist; b. Bklyn., May 28, 1917; s. Isidore and Goldie (Yarmolinsky) C.; A.B. with honors, Columbia, 1937; M.A., Harvard, 1938, Ph.D., 1941; D.Sc., Hahnemann Med. Coll., 1963; m. Gloria C. Gordon, Dec. 1, 1946; children—Lucy Allison, Frederic Gordon. Asst. biology Harvard, 1938-40; instr. biology Queens Coll., 1940-42; asso. editor Sci. Illustrated, 1946-47; asso. prof. plant physiology Washington U., St. Louis, 1947-53, prof., 1953—, chmn. dept. botany, 1965—, dir. Center for Biology Natural Systems, 1966—. Vice pres St. Louis Com. for Nuclear Information, 1958-65, pres., 1965—; bd. dirs. Scientists Inst. Pub. Information, 1963—, Am. Inst. Biol. Sci., 1964—. Recipient Newcomb Cleveland prize A.A.A.S., 1953. Fellow A.A.A.S. (chmn. com. on sci. in promotion of human welfare 1958-66); mem. Soc. Biol. Chemists, Soc. Gen. Physiologists, Am. Soc. Naturalists, Am. Soc. Plant Physiologists Am. Assn. U. Profs. (pres. chpt.), Phi Beta Kappa, Sigma Xi, Hon. adv. panel Problems of Virology, 1957; editorial bd. Internat. Review of Cytology, 1957-65, Theoretical Biology, 1960-64. Research on chem. and phys. basis of processes characteristic of functional living cells; studies of biosynthesis of tobacco mosaic virus in living tobacco tissue; electro spin resonance technique to study roles of free radicals in a variety of biol. processes. Home: 25 Crestwood Dr., Calyton, Mo. 63105. Office: Washington U., St. Louis 63130.*

COMOLET, Raymond René, French engr.; b. Beaupouyet (Dordogne), Apr. 28, 1920; s. Victor and Anne Marie (de Bony) C.; ed. Ecole normale supérieure; Ph.D. in sci.; m. Marie Rose Leveque, Dec. 16, 1944; children—Marie Helen, Pierre, Thierry, Arnaud. Prof., Faidherbe Lycée at Lille, Sch. Sci. of U. Nancy, then Paris. Mem. U. Assn. Fluid Mechanics. Author: Etude sur les butees a air; Cours de mécanique expérimentale des fluides; Recueil de problemes corriges de mécanique expérimentale des fluides. Home: 9, av. de Longchamps, Boulogne-Billancourt (Seine). Office: 11, rue Pierre-Curie, Paris 5, France.

COMPERE, Clinton Lee, Am. physician, orthopaedic surgeon; b. Greenville, Tex., Feb. 17, 1911; s. Edward L. and Clara (Davison) C.; B.S., U. Chgo., 1936, M.D., 1937; m. Katharine Gram, Mar. 31, 1949; children—Clinton Lee, Mary Katharine. Individual practice orthopaedic surgery, Chgo., 1946—; mem. sr. attending staff Chgo. Wesley Meml. Hosp., 1949—, chief staff, 1964—; acad. dir. Prosthetic Research Center, Chgo., 1955—, Prosthetic-Orthotic Edn., Chgo., 1958—; cons. 5th Army Hdqrs., 1947—; cons. amputee clinics Regional Office VA, 1947—; asso. prof. orthopaedic surgery Northwestern U. Med. Sch., 1954-65, prof., 1965—; mem. coms. NRC-Nat. Acad. Sci., Dept. Health, Edn. and Welfare. Vice pres. med. affairs, bd. dirs. Rehab. Inst. Chgo.; mem. med. adv. com. Ill. Div. Vocational Rehab. Recipient citation President's Com. Employment Physically Handicapped, 1959. Diplomate Am. Bd. Orthopaedic Surgery. Mem. Am. Acad. Orthopaedic Surgeons (pres. 1963-64), A.M.A., A.C.S., Am., 20th Century orthopaedic assns., Internat. Soc. Orthopaedic Surgery and Traumatology, Alpha Omega Alpha. Co-author: Fracture Treatment, 1937; also articles. Research and edn. in field of extremity prosthetics, bone pathology, rehab. Home: 2600 Orrington Av., Evanston, Ill. 60201. Office: 737 N. Michigan Av., Chgo. 60611.*

COMPTON, Arthur (Holly), Am. physicist, univ. chancellor; b. Wooster, O., Sept. 10, 1892; s. Elias and Otelia Catherine (Augspurger) C.; B.S., Coll. Wooster, 1913, Sc.D., 1927; M.A., Princeton, 1914, Ph.D. (Porter Ogden Jacobus fellow), 1916, Sc.D., 1934; postgrad. (Nat. Research fellow) Cambridge (Eng.) U., 1919-20; Sc.D., Ohio State U., 1929, Yale, 1929, Brown U., 1935, Harvard, 1936; M.A., Oxford U., 1934; numerous other hon. degrees; m. Betty Charity McCloskey, June 28, 1916; children—Arthur Alan, John Joseph. Instr. physics U. Minn., 1916-17; research engr. Westinghouse Lamp Co., 1917-19, also civilian asso. U. S. Signal Corps, developing airplane instruments, 1917-18; prof. physics, head dept. Washington U., 1920-23; prof. physics U. Chgo., 1923-29, Charles H. Swift distinguished service prof., 1929-45, chmn. dept. physics, dean div. phys. scis., 1940-45, dir. Metall. Atomic Project, 1942-45; chancellor Washington U., St. Louis, 1945-53, prof. natural history, 1953-61. Chmn. com. on X-rays and radioactivity NRC, 1922-25; research asso. Carnegie Instn., 1931-41; mem. Nat. Cancer Adv. Bd., 1937-44; mem. Presdl. Commn. on Higher Edn., 1946; U. S. del. UNESCO, Paris, 1946. Regent, Smithsonian Instn., from 1938. Recipient Rumford Gold medal Am. Acad. Arts and Scis., 1927; Nobel prize for physics (with C. T. R. Wilson), 1927; Gold medal Radiol. Soc. N.Am., 1928; Matteucci Gold medal Italian Acad. Scis., 1930; Hughes medal Royal Soc. London, 1940; Franklin medal Franklin Inst., 1940; Medal for Merit, U. S. Govt., 1946. Guggenheim fellow, 1926-27; fellow Balliol Coll., Oxford U., 1934-35. Fellow Am. Phys. Soc. (pres. 1934), A.A.A.S. (pres. 1942), Western Soc. Engrs. (hon.); mem. Am. Philos. Soc. (Franklin medal 1946), Nat. Acad. Scis. (chmn. physics sect. 1938-41), Phi Beta Kappa, Sigma Xi, Alpha Tau Omega, Gamma Alpha; fgn. or hon. mem. numerous European and other sci. acads. Author: (monograph) Secondary Radiations Produced by X-rays, 1922; X-rays and Electrons, 1926; The Freedom of Man, 1935; (with S. K. Allison) X-rays in Theory and Experiment, 1935; Human Meaning of Science, 1940, also numerous sci. articles; co-author: On Going to College, 1938. Discoverer change in wave-length of X-rays when scattered (Compton effect); total reflection of X-rays; (with C. H. Hagenow) complete polarization of X-rays; (with R. L. Doan) X-rays spectra from ruled gratings; elec. character of cosmic ray; directed world cosmic ray survey, 1931-34; directed work resulting in 1st atomic chain reaction. Died Berkeley, Cal., Mar. 15, 1962.

COMPTON, Karl Taylor, Am. physicist; b. Wooster, O., Sept. 14, 1887; s. Elias and Otelia (Augspurger) C.; Ph.B., Coll. of Wooster, 1908, M.S., 1909, D.Sc., 1923; Ph.D., Princeton, 1912, D.Sc., 1930; D.Sc., Stevens Inst. Tech., 1931, Boston U., 1932, Columbia, 1940, N.Y. U., 1946, Cambridge U., 1952, Israel Inst. Tech., 1954; LL.D., Harvard, 1930, U. Wis., 1934, Williams Coll., 1936, Johns Hopkins, 1937, U. Cal., 1941, Northwestern U., 1942, Tufts Coll., 1943, Coll. William and Mary, 1947, D. Eng., Bklyn. Poly. Inst., 1930, Case Sch. Applied Sci., 1931, Rutgers U., 1941, Worcester Poly. Inst., 1946;

Dr. Applied Sci., Ecole Polytechnique, Montreal, 1944; numerous other hon. degrees; m. Rowena Rayman (dec.); 1 dau., Mary Evelyn (Mrs. Russell Alderman) m. 2d, Margaret Hutchinson; children—Jean Corrin (Mrs. C. W. Boyes), Charles Arthur. Instr. chemistry Coll. Wooster, 1909-10; instr. physics Reed Coll., Portland, Ore., 1913-15; asst. prof. physics Princeton, 1915-19, prof., 1919-30, chmn. dept. physics, 1929-30; pres. Mass. Inst. Tech., 1930-48, chmn. of corp., 1948-54. Chmn. research and development bd. Nat. Mil. Establishment, 1948-49; cons. physicist Dept. Agr. and Gen. Electric Co., 1924-30; mem. vis. com. U. S. Bur. Standards, 1931-41; mem. bus. adv. and planning council Dept. Commerce, 1933-36; mem. adv. com. U. S. Weather Bur., 1935-48, nat. defense research com. Office Sci. Research and Development, 1940-47, chief Office of Field Service, 1943-45, dir. Pacific br., 1945; mem. Sci. Intelligence Mission to Japan, 1945; chmn. U. S. Radar Mission to U.K., 1943; mem. sec. war's spl. adv. com. on atomic bomb, 1945; chmn. Research Bd. Nat. Security, 1945-46, Joint Chiefs of Staff Evaluation Bd. on Atomic Bomb Tests, 1946, Research and Development Bd., 1948-49, Naval Research Adv. Com., 1946-48, Nat. Security Tng. Commn., 1951-54. Trustee Edison Found., 1946-53, Ford Found., 1946-51, Meml. Found. Neuro-Endocrine Research, 1932-47, New Eng. Indsl. Research Found., 1941-48, Rockefeller Found. and Gen. Edn. Bd., 1940-53 (exec. coin. 1941-42, nominating com. 1941-44), Sloan Found., 1942-54, Sloan-Kettering Inst., 1947-54, Brookings Inst., 1940-50, Princeton (charter 1952-54), Population Council, 1953-54; mem. adv. bd. Bartol Research Found., 1927-36, chmn. governing bd. Am. Inst. Physics, 1931-36; Cons. Brit. Parliamentary and Sci. Com., 1954. Pilgrim Trust lectr. Royal Soc. London, 1943. Fellow Am. Phys. Soc. (councillor; v.p. 1925-27; pres. 1927-29), Optical Soc. of Am.; mem. A.A.A.S. (pres. 1935-36; exec. com. 1931-40), Am. Philos. Soc., Am. Chem. Soc., Franklin Inst., Am. Soc. M.E., Am. Acad. Arts and Scis., Am. Inst. N.Y., Am. Soc. Engring. Edn. (v.p 1937; pres 1938-39), Phi Beta Kappa, Sigma Xi, Alpha Tau Omega, Tau Beta Pi. Recipient Rumford medal Am. Acad. Arts and Scis., 1931, Medal for Merit, 1946, Washington Award, 1947, Marcellus Hartley medal Nat. Acad. Sci., 1947, Lamme medal Am. Soc. Engring. Edn., 1949, William Procter prize for sci. achievement Sci. Research Soc. of Am., 1950, Hoover medal Founder Engring. Socs., 1951, Joseph Priestley award Dickinson Coll., 1954. Research on contact differences of potential and Peltier effect, structure of crystals by X-ray photography, photoelectric effect, ionization, thermionic emission. Died June 22, 1954.

COMPTON, W(alter) Dale, physicist; b. Chrisman, Ill., Jan. 7, 1929; s. Roy L. and Marcia (Wood) D.; B.A., Wabash Coll., 1949; M.S., U. Okla., 1951; Ph.D., U. Ill., 1955; m. Jeanne C. Parker, Oct. 14, 1951; children—Gayle Corinne, Donald Leonard, Duane Arthur. Physicist, U. S. Naval Ordnance Test Sta., China Lake, Cal.; physicist, U. S. Naval Research Lab., Washington, 1955-61; prof. dept. physics U. Ill., Urbana, 1961—, dir. coordinated sci. lab., 1965—. Fellow Am. Phys. Soc.; mem. Washington Acad. Scis., Research Soc. Am. Author: (with J. H. Schulman) Color Centers in Solids, 1962. Research in radiation damage to materials, color centers in solids, elec. transport in solids. Home: 404 Eliot Dr., Urbana, Ill. 61801.*

COMROE, Julius H(iram), Jr., Am. physiologist, pharmacologist; b. York, Pa., Mar. 13, 1911; s. Julius Hiram and Mollie (Levy) C.. A.B., U. of Pa., 1931; M.D., 1934 Commonwealth Fund Fellow, Nat. Inst. Med. Research, London, 1939; m. Jeanette Wolfson, June 30, 1936; 1 dau., Joan Catherine. Interne, Hosp. of U. of Pa., 1934-36; instr. in pharmacology U. of Pa. Med. Sch., 1936-40, asso. in pharmacology, 1940-42, asst. prof., 1942-46; prof. physiology and pharmacology U. of Pa. Grad. Sch. of Medicine, 1946-57; clin. physiologist Hospital of U. of Pa., 1946-57; prof. of physiology, dir. Cardiovascular Research Inst., U. Cal. Med. Center, San Francisco, 1957—; chmn. 1st Teaching Inst. (1953), Asso. American Medical Colleges, also chairman 1961 Institute; mem. U. S. Pub. Health Service Pharmacology Study Sect., 1953-55, chairman Physiology Study Section 1955-58; board sci. counselors Nat. Heart Inst., 1957-61; mem. Nat. Adv. Mental Health Council, 1958-62; member of The National Advisory Heart Council, 1963-67; mem. U. S. national committee I.U.P.S., 1962—. Cons. med. research div. Chem. Warfare Service, 1944-46. Recipient Am. Physiol. Soc. Travel award, 1938; Distinguished achievement award, Modern Medicine, 1961. Fellow A.C.P., A.A.-A.S.; mem. Am. Physiol. Soc. (pres. 1960-61), Am. Soc. for Pharmacology and Exptl. Therapeutics (councillor 1953-56), Nat. Acad. Scis., Am. Acad. Arts and Sciences, Harvey Society (honorary member), American Society for Clinical Investigation, Phi Beta Kappa, Alpha Omega Alpha, Sigma Xi. Author: Physiological Basis for 02 Therapy, 1950; Methods in Medical Research, Vol. 2, 1950; The Lung: Clinical Physiology and Pulmonary Function Tests, 1955, 62; Physiology of Respiration, 1964. Editor: Physiology for Physicians, 1963-66, Circulation Research, 1966—. Cons. editor Am. Jour. Physiology, 1954-62, Pharmacological Revs., 1949-51; ed. bd. Jour. Med. Edn., 1955-60, Am. Rev. of Respiratory Diseases, 1959-63. Research on neuromuscular transmission; aortic and carotid bodies; pulmonary function; regulation of respiration; autonomic drugs. Home: 555 Laurent Rd., Hillsborough, Cal. Office: Cardiovascular Research Inst., U. Cal. Med. Center, San Francisco 94122.

COMSTOCK, Anna Botsford, Am. entomologist, artist; b. Otto, N.Y., Sept. 1, 1854; d. Marvin and Phebe (Irish) Botsford; grad. Chamberlain Inst., Randolph, N.Y., 1873. Cornell, 1878; studied art at Cooper Union and under John P. Davis; m. John Henry Comstock, Oct. 7, 1878. Exhibited Chicago Expn., 1893, Paris Expn., 1900; bronze medal for wood engraving, Buffalo Expn., 1901; Asst. prof. Cornell extension work in Nature Study, 1899; lectr. Stanford extension work, 1899-1900; asst. prof. Cornell U., 1913-20; prof. nature study, 1920-22. Asso. dir. of Am. Nature Assn. Named One of 12 Greatest Living Am. Women by Nat. League Women Voters, 1923. Author: Ways of the Six-Footed; 1903; How to Keep Bees; Handbook of Nature-Study, 1911; The Pet Book, 1914; Bird, Animal, Tree and Plant Notebooks, 1914; (with husband) How to know the Butterflies, 1904. Made more than 600 wood engravings pub. in Comstock's Manual for the study of Insects, 1895. Editor Nature Study Review, 1917-23. Died Aug. 24, 1930.

COMSTOCK, Cyrus Ballou, Am. engr.; b. West Wrentham, Mass., Feb. 3, 1831; s. Nathan and Betsey (Cook) C.; grad. U. S. Mil. Acad., 1855; m. Elizabeth Blair, Feb. 3, 1869. Bvt. 2d lt. engrs., 1855; advanced through grades to col., 1888; Chief engr. Army of Potomac, 1862-63; sr. engr. at siege of Vicksburg; chief engr. in assault and capture of Ft. Fisher, N.C.; sr. a.-d.-c. to Grant, 1864-66; col. a.-d.-c. to Grant and Sherman, 1866-70. Mem. and pres. Miss. River Commn.; mem. permanent Bd. Engrs. for Fortifications; ret., 1895; advanced to rank of brig. gen. ret. by act of 1904. Author: Primary Triangulation of the U. S. Lake Survey. Contbd. to flood control of Mississippi River; specialist in river and harbor improvement. Died 1910.

COMSTOCK, Daniel Frost, Am. physicist, engineer; b. Newport, R.I., Aug. 14, 1883; s. Ezra Young and Nellie Preston (Barr) C.; S.B., Mass. Inst. Tech., 1904; studied U. of Berlin, 1905, U. of Zürich, 1905-06, U. of Basel, 1906, Ph.D., studied U. of Cambridge, Eng., 1906-07, under J. J. Thomson; m. Joan Barton, June 30, 1925; children—Daniel Frost, Charles Barton. Apptd. to teaching staff, Mass. Inst. Tech., 1904, instr. in theoretical physics, 1905-10, asst. prof., 1910-15, asso. prof., 1915-17; directed scientific work on development of means for detection of hostile submarines, World War I, president Comstock & Wescott, Inc., Engrs. 1912-14; v.p. Kalmus, Comstock & Wescott, Inc., engrs., 1914-25; dir. of scientific work on, and principal inventor of the process for producing Motion Pictures in natural color known as the Technicolor process, developed by Kalmus, Comstock & Wescott, Inc., for the Technicolor Motion Picture Corp., during 1914-25; v.p. Technicolor Motion Picture Corp., 1918-25; pres. Comstock & Wescott, Inc., since 1925; engaged almost exclusively with research and development on war projects during World War II; now engaged in industrial research. Pres., treas., dir. Stator Co., co-inventor of new refrigeration process owned by same. Member National Advisory Council to Committee on Patents, Ho. of Reps., 1939. Fellow Am. Academy Arts and Sci., Am. Physical Society; mem. Am. Chemical Society. Club: St. Botolph (Boston). Author: Nature of Matter and Electricity (with L. T. Troland), 1917. Contbr. original research articles on modern theory of electricity and optics to Am. and English tech. jours. Research on relativity theory; theory of matter; theoretical physics; magnetic properties of saturated iron; Stator system of refrigeration; color motion picture photography; electromagnetic theory. Home: Lincoln, Mass. Office: 765 Concord Av., Cambridge 38, Mass.

COMSTOCK, George Cary, Am. astronomer; b. Madison, Wis., Feb. 12, 1855; s. Charles Henry and Mercy (Bronson) C.; Ph.B., U. Mich., 1877, Sc.D., 1907; LL.B., U. Wis., 1883; LL.D., U. Ill., 1907; admitted to bar, but never practiced; m. Esther Cecile Everett, June 12, 1894; 1 dau., Mary (Mrs. George Carey). Recorder and asst. engr. U. S. Lake Survey, 1874-78; asst. engr. on improvement Miss. River; asst. astronomer Washburn Obs.; computer Nautical Almanac Office; prof. math. and astronomy, Ohio State U., 1885-87; prof. astronomy, 1887-1922, dir. Washburn Obs., 1889-1922, dir. and dean Grad. Sch., 1906-20, U. Wis., ret., 1922. Fellow Am. Acad. Arts and Scis. Author: Method of Least Squares, 1890; Text-Book of Astronomy, 1900; Field Astronomy for Engineers, 1902; The Sumner Line as an Aid to Navigation, 1919. Noted for work on double stars. Died May 21, 1934.

COMSTOCK, Gregory Jamieson, Am. metallurgist; b. Cleve., June 29, 1893; s. John Belden and Roseanna (Cook) C.; Ph.B., Yale, 1917; M. Engring. (hon.), Stevens Inst. Tech.; m. Elizabeth Singleton, Oct. 2, 1922; children—John Belden, Nancy Rutledge. With Am. Hardware Corp., 1912-21, Internat. Silver Co., 1921-26, Firth-Sterling Steel Co., 1926-31, Handy & Harmon, 1931-38; pres. metall. cons. firms Gregory Comstock Inc., Essex, Conn., Comstock Inc., Higganum, Conn., 1938—; faculty Stevens Inst. Tech., 1938-61, prof. powder metallurgy, 1941-61, now emeritus; Mem. Am. Soc. Metals, Am. Inst. Mining and Metall. Engrs., Am. Ordnance Assn., Soc. Carbide Engrs., Metal Sci. Club, MPIF (Pioneer award 1964), Sigma Xi. Research, publs. in area of hard cemented carbides, a metallic material compounded from powders which has been used by govt. and bus. Patentee in field. Home 69 Main St. Office: 47 Main St., Essex, Conn. 06426.*

COMSTOCK, John Henry, Am. entomologist; b. Janesville, Wis., Feb. 24, 1849; s. Ebenezer and Susan (Allen) C.; B.S., Cornell, 1874; postgrad. Yale, U. Leipzig; m. Anna Botsford, Oct. 7, 1878. Instr. entomology Cornell, 1875-77, asst. prof., 1877-78, prof. entomology, invertebrate zoölogy, 1882-1914; U. S. entomologist, Washington, 1879-81. Lectr. Vassar Coll., 1877; non-resident prof. entomology, Leland Stanford Jr. U., 1891-1900. Author: Insect Life; Notes on Entomology; Introduction to Entomology; How to Know the Butterflies (with his wife); The Spider Book; The Wings of Insects. Research on coccidae provided basis for determining systematic relationship of these insects, also provided basis for subsequent systematic work on coccidae in U. S.; research on wing venation of insects, relationship and descent of lepidoptera. Died Mar. 20, 1931.

COMSTOCK, Ralph Ernest, Am. geneticist; b. Spring Valley, Minn., July 19, 1912; s. Charles Roy and Bertha (Drewes) C.; B.S., U. Minn., 1934, M.S., 1936, Ph.D., 1938; m. Helen Agnes Bartel, Aug. 8, 1936; children—Cynthia Pam (dec.), Mary Sue, John Alan. Faculty, U. Minn., St. Paul, 1937-43, 1957—, prof. animal husbandry, 1957-65, prof., head dept. genetics, 1965—; asso. prof. N.C. State Coll., 1943-46, prof., 1947-57; head animal husbandry P.R. Expt. Sta., 1946-47. Mem. Genetics Soc. Am., Am. Soc. Animal Sci., Biometrics Soc., Soc. for Study of Evolution, Agronomy Soc. Am., Soc. Am. Naturalists, Sigma Xi, Gamma Sigma Delta. Research, numerous publs. in quantitative and population genetics, clarification of relative efficiencies of systems in plant and animal breeding, level of dominance in gene action with reference to effects on quantitative characters, problems of obtaining definitive information on level of dominance. Home: 1958 Roselawn Av. W., St. Paul 55113.*

COMTE, Auguste (Isidore Auguste Marie Francois Xavier), French sociologist, philosopher; b. Montpellier, France, Jan. 19, 1798; s. Louis Xavier and Rosalie (Boyer) C.; ed. École Polytechnique, Paris, 1814-16; m. Caroline Massin, 1825. Took part in defense of Paris, 1814; contemplated career in U. S., but was dissuaded by friend's report of materialistic spirit there; tutor in Paris; became disciple and sec. of social philosopher Saint-Simon, 1817; lectr., from 1826; proposed chair history of phys. scis. Coll. France; apptd. tutor École Polytechnique. 1832, examiner for applicants, 1836; also filled engagements as math. tchr.; founded Positivist Soc., 1848; moved toward mysticism inspired by memories of his friend, Clotilde de Vaux, who died 1845; subsisted in final years on subsidy provided by J. S. Mill and other admirers. Author: Plan des travaux scientifiques nécessaires pour réorganiser la société, 1822; Considérations philosophiques sur les sciences et les savants, 1829; Cours de philosophie positive, 1851-54; Catéchisme positiviste, ou sommaire exposition de la religion universelle, 1852; Synthèse Subjective, 1856; publ. series of articles as asso. of Saint-Simon, until their breach in 1922; also contbr. articles to Le Producteur. Believed man's intellectual devel. was covered by Law of 3 Stages: theological, metaphysical, and positive (latter calls for explanation of phenomena through observation, experiment, and hypotheses); classified scis. on basis of complexity and degree of specialty; held that all scis. contribute to sociology (a term he coined), and that positive philosophy is only solid found. for social reorganization; his positivism and sociology have been important seminal influence. Died Paris, Sept. 5, 1857.

COMTE, Edouard Charles Pierre, French geophysicist; b. Paris, France, July 8, 1903; s. Albert and Henriette (Taschereau) C.; ed. Ecole normale supérieure, Princeton; Ph.D. in sci.; m. Nadia Vergniaud, Feb. 9, 1939; children—Dominique, Patrice; Master of research Nat. Center Sci. Research; instr. Sorbonne; chief service Atomic Energy Commn.; dir. research Jacob-Holtzer; later pres., dir.-gen.; v.p., dir. Cie Ateliers et Forges de la Loire; adminstr. Cie Acieries de la Mazine-Firming. Mem. Association des cadres dirigeants de l'industrie. Research and publs. on geology and geophysics. Home: 76 bis, rue Cassini. Office: 12, rue de la Rochefoucauld, Paris, France.

COMTE, Joseph-Achille, French naturalist; b. Grenoble, France, Sept. 29, 1802; s. Jean Baptiste and Marie Philippine (Robinot) C.; student medicine, intern Paris (France) hosps., 1823; m. Aglaé de Boucauville. Tchr. natural history Charlemagne Coll.; head clk. French Ministry Edn., from 1848; dir. Ecole preparatoire, Nantes, France, for 10 years. Author: Recherches anatomiques et physiologiques relatives à la prédominance du bras droit sur le bras gauche, 1828; Règne animal de Cuvier, disposé en tableaux méthodiques, 1832-41; (with Milne-Edwards) Cahiers d'histoire naturelle, 1836-45; Oeuvres complètes de Buffon avec les suites, 1846; Traité complet d'histoire naturelle, 1844-48; Musée d'histoire naturelle, 1854. Died Jan. 17, 1866.

CONANT, Alban Jasper, Am. archeologist; b. Chelsea, Vt., Sept. 24, 1821; s. Caleb and Sarah (Barnes)

C.; grad. Gouverneur Wesleyan Sem., 1844; studied art, N.Y.C.; A.M. (hon.), Madison U., U. Mo., 1871; m. Sarah M. Howes, 1845 (dec. 1867); m. 2d, Brianna C. Bryan, 1869. Curator agr. and mining Mo. State U.; chmn. commn. under U. S. land grant; founded sch. mines and metallurgy, supr., 3 yrs.; helped found Western Acad. Art, 1860; Délégué Correspondant Institution Ethnographique, Paris. Painted portraits of Lincoln, Sherman, other leading Americans. Author: The Archaeology of the Missouri Valley (republished in many European transls.), 1876; Footprints of Vanished Races in the Mississippi Valley, 1879; My Acquaintance with Abraham Lincoln. Pioneer in archeol. research in U. S. Died Feb. 3, 1915.

CONANT, Hezekiah, Am. inventor; b. Dudley, Mass., July 28, 1827; s. Hervey and Dolly (Healy) C.; m. Sarah Williams Learned, Oct. 4, 1853; m. 2d, Harriet Knight Learned, Nov. 1859; m. 3d, Mary Eaton Knight, Dec. 6, 1865; 2 children. Patented pair of lasting pinchers for use in shoe mfg., 1852; journeyman machinist, Boston, Worcester, Mass., 1852-55; employed at Colt Firearm Co., Hartford, Conn., 1855-circa 1859; invented gas check for breech loading firearms, 1856; invented machines for dressing sewing thread and winding thread on spools automatically, 1859, sold patent to Willimantic Linen Co. (Conn.), 1860, mech. expert, 1860-68; organized Conant Thread Co., Pawtucket, R.I., 1868, merged with J. & P. Coates Co. of Paisley, Scotland (largest mfrs. in Europe), 1869, named changed from Conant Thread Co. to J. & P. Coates Co., Ltd., 1893. Died Jan. 22, 1902.

CONANT, James Bryant, Am. chemist; b. Dorchester, Mass., Mar. 26, 1893; s. James Scott and Jennett Orr (Bryant) C.; A.B., Harvard, 1913, Ph.D., 1916; also numerous hon. degrees; m. Grace Thayer Richards, April 17, 1921; children—James Richards, Theodore Richards. Prof. chemistry Harvard, 1919-33, pres., 1933-53; U. S. high commr. for Germany, 1953-55; U. S. ambassador to Fed. Republic Germany, 1955-57; condr. study on the American Pub. High Sch., 1957-62, study of the Edn. of Am. Tchrs., 1962—(both under grant from Carnegie Corp., N.Y. chmn.). Nat. Def. Research Com., 1941-46; dep. dir. OSRD, 1941-46; mem. gen. adv. com. AEC, 1947-52; ednl. adviser to Ford Found. in Berlin, Germany, 1963-65. Mem. Nat. Acad. Scis. Author: On Understanding Science, 1947; Education in a Divided World, 1948; Science and Common Sense, 1951; Modern Science and Modern Man, 1952; Education and Liberty, 1953; The Citadel of Learning, 1956; Germany and Freedom, 1958; The American High School Today, 1959; The Child, The Parent, and The State, 1959; Education in the Junior High School Years, 1960; Slums and Suburbs: A Commentary on Schools in Metropolitan Areas, 1961; also The Education of American Teachers, 1963; Two Modes of Thought, 1964; Shaping Educational Policy, 1964. (with N. H. Black) a high sch. chemistry textbook; also other textbooks on organic chemistry. Editor: Harvard Case Histories in Experimental Science, 2 vols., 1957. Research in organic chemistry, especially on free radicals, hemoglobin, superacid solutions and chlorophyll; contbd. to organizing of U. S. atomic energy projects during World War II; conducted important researches in Am. edn. Home: 200 E. 66th St., N.Y.C. 10021. Office: 730 3d Av., N.Y.C. 10017.*

CONARD, Henry Shoemaker, Am. botanist; b. Phila, Sept. 12, 1874; s. Thomas P. and Rebecca (Baldwin) C.; B.S., Haverford Coll., 1894, M.A., 1895, D.Sc., 1945; Ph.D., U. Pa., 1901; (Johnston scholar) Johns Hopkins, 1905-06; Sc.D., Grinnell Coll., 1944; m. E. Laetitia Moon, Apr. 13, 1900 (dec. Nov. 1946); children—Elizabeth (Mrs. Harold Corkey), Rebecca (Mrs. Porter French), Alfred F.; m. 2d, Louisa Sargent, Dec. 20, 1950. Tchr. sci. Westtown Sch., Pa., 1895-99; instr. U. Pa., 1904; prof. botany Grinnell Coll., 1906-44; vis. research prof. State U. Ia., 1944-52. Named Eminent Ecologist, Ecol. Soc. Am., 1954. Fellow Ia. Acad. Sci. (hon. life), A.A.A.S.; mem. Fla. Acad. Sci., Am. Bryological Soc., Bot. Soc. Am., Brit. Bryological Soc., Am. Assn. U. Profs., Am. Mycological Soc. Author: Waterlilies, 1905; Hay-Scented Fern, 1908; How to Know the Mosses, 1944; also numerous articles. Translator: Plant Sociology (J. Braun-Blanquet), 1905; Plant Life Danube Basin (Kerner), 1951. Home: Lake Hamilton, Fla. 33851.*

CONARD, Robert Allen, Am. physician; b. Jacksonville, Fla., July 29, 1913; s. Robert Allen and Aline (Rucker) C.; B.S., U. S.C., 1936; M.D., Med. Coll. S.C., 1941; m. Marion L. Laue, June 18, 1948; children—Katrina, Christopher, Randall, Lisa. Commd. lt. (j.g.) USN, 1941, advanced through grades to rear adm., 1965; project officer nuclear med. research and radiol. safety BUMED Project, 1947-50; with Naval Radiol. Def. Lab., 1947-50; med. researcher Naval Med. Research Inst., Bethesda, Md., 1950-56; ret., 1956; sr. scientist Brookhaven Nat. Lab., Upton, L.I., N.Y., 1956—; head med. surveys Marshallese people exposed to fallout, 1954. Mem. Space Sci. Bd. Panel, 1964—, panel WHO, 1965. Mem. Radiation Research Soc., Nuclear Med. Soc., N.Y. Acad. Scis., A.A.A.S., Tissue Culture Soc., Conf. on Radiol. Health. Research, numerous publs. on effects radiation on physiology of gastrointestinal tract, effects radiation on bone growth. Home: 32 Ivy Lane, Setauket, L.I., N.Y. 11785. Office: Brookhaven Nat. Lab., Upton, N.Y. 11973.*

CONCATO, Luigi Maria, Italian physician; b. Padua, Italy, 1825; prof. medicine Bologna, Italy, from 1860, Turin, Italy, from 1878; author: La Percussione nella diagnosi delle cardiopatie, 1868; La diagnosi generale dei tumori addominale, 1881; described Tb inflammation of serous membranes, 1881; skilled in use of microscope. Died Turin, 1882.

CONDAMINE, Charles Marie de la, see La Condamine, Charles Marie de.

CONDILLIAC, Étienne Bonnot de, French philosopher, psychologist; b. Grenoble, France, Sept. 30, 1715; entered Seminary of St. Sulpice; student theology Sorbonne, Paris. Took holy orders, became abbé de Mureaux, 1740; lived in Paris, became friend of Diderot, later of J. J. Rousseau; preceptor of Duke of Parma, 1755-68; spent later years in retirement at Flux, nr. Beaugency; mem. Acad. Scis., 1768. Author: Essai sur l'origine des connaissances humaines, 1746; Traité des systèmes, 1749; Traité des sensations, 1754; Traité des animaux, 1755; Logique, 1780. Devoted most of life to speculation and writing; most important contbns. in psychology and philosophy, also wrote on politics, history, economics, logic, grammar; introduced philosophy of Locke to France; important in history of French philosophy for his theory of sensationism, which states that all human knowledge, judgement and volition is derived exclusively from sensation and transformed sensation. Died Flux, Aug. 3, 1780.

CONDIT, Paul Taylor, Am. physician; b. N.Y.C., Oct. 27, 1918; s. Kenneth H. and Marjorie W. (Brown) C.; A.B. cum laude, Princeton, 1940, A.M., 1941, Ph.D., 1958; M.D., Johns Hopkins, 1950; m. Mary Louise Winterode, Apr. 18, 1942; children—Nancy P., Paul Taylor, Philip H. II. Sr. investigator Nat. Cancer Inst., Bethesda, Md., 1951-53, 55-58; fellow in medicine Johns Hopkins, 1953-55; asso. head cancer sect. Okla. Med. Research Found., Oklahoma City, 1958-64, head, 1964—; asso. prof. research medicine, biochemistry, radiology U. Okla. Sch. Medicine, 1958—. Fellow A.C.P., Am. Coll. Clin. Pharmacology and Chemotherapy; mem. Am. Chem. Soc., A.M.A., Am. Assn. Cancer Research, A.A.A.S., Am. Soc. Pharm. and Exptl. Therapeutics, N.Y. Acad. Sci. Research, publs. in cancer chemotherapy, mechanism of folic acid antagonists, combination of radiation therapy and chemotherapy. Home: 1605 Norwood Pl., Oklahoma City 73120. Office: 825 N.E. 13th St., Oklahoma City 73104.*

CONDON, Edward Uhler, Am. physicist; b. Alamogordo, N.M., Mar. 2, 1902; s. William Edward and Caroline (Uhler) C.; A.B., U. Cal. at Berkeley, 1924, Ph.D., 1926; D.Sc., Delhi (India), N.M. Inst. Mining and Tech., 1950, Am. U., also Alfred U., 1951; m. Emilie Honzik, Nov. 9, 1922; children—Marie (Mrs. Wayne Thornton, Jr.), Paul Edward, Joseph Henry. Asso. prof. physics Princeton, 1928-37; asso. dir. research Westinghouse Electric Corp., 1937-45; dir. Nat. Bur. Standards, Washington, 1945-51; dir. research Corning Glass Works, Corning, N.Y., 1951-54; Wayman Crow prof. physics Washington U., St. Louis, 1956-63; prof. physics, fellow Joint Inst. for Lab. Astrophysics, U. Colo., Boulder, 1963—. Sci. adviser to Spl. Senate Com. on Atomic Energy, 1945-46; mem. NACA, 1945-51; mem. Pres.'s Evaluation Bd., Naval Atomic Bomb Tests, 1946; mem. S-1 Com. on Mil. Use of Atomic Energy, 1941. Mem. Am. Phys. Soc. (pres. 1946), A.A.A.S. (pres. 1953), Am. Assn. Physics Tchrs. (pres. 1964), Société Française de Physique (hon.), Royal Swedish Acad. Engring. Scis. (hon.), Nat. Acad. Scis., Am. Philos. Soc., Am. Acad. Arts and Sci. Author: Franck-Condon Principle in Molecular Spectroscopy, 1926; (with Ronald Gurney) Quantum Theory of Radio-active Disintegration, 1928; Theory of Atomic Spectra (with George H. Shortley), 1935; (with Hugh Odishaw) Handbook of Physics, 1958. Vice pres. Ann. Revs., 1963—; editor Revs. of Modern Physics, 1957—; chmn. editorial bd. Internat. Sci. and Tech., 1961—. Research, publs. on quantum mechanics of nuclear physics and of atomic and molecular spectra. Home: 761 Cascade Av., Boulder, Colo. 80302.

CONDORCET, Marie-Jean-Antoine-Nicolas Caritat, Marquis de, French mathematician, social thinker; b. Ribemont, Picardy, France; Sept. 17, 1743; ed. Jesuit Coll., Rheims, France, also Paris; m. Sophie de Grouchy, 1786. Became geometer; dep. from Paris, sec. French Legislative Assembly, 1791, pres., 1792. Recipient Berlin prize for theory of comets. Mem. French Acad. Scis. 1769 (permanent sec. from 1785); French Acad., 1782. Author: Eloges des academiciens morts avant, 1699; Essai d'analyse, 1768; Vie de Voltaire, 1787; Esquisse d'un tableau historique des progrès de l'esprit humain, 1794; numerous sci. memoirs. Did work on integral calculus and problem of 3 bodies. Contributed to theory of probability; advocated concept of progress in human hist. devel.; thought that death might ultimately be conquered; considered a founder of social scis. as area of study. A physiocrat, he advocated free trade and elimination of corvée; advocated enfranchisement of women and nat. system of public edn.; wrote a Girondist constitution (defeated by Jacobins, who proscribed and imprisoned him in April 1794). Died in jail, Bourg-la-Reine, France, Apr. 8, 1794.

CONDORELLI, Luigi, Italian pathologist, cardiologist; b. Rome, Italy, May 28, 1899; s. Mario and Adele (De Fiore) C.; ed. in medicine, surgery, biochemistry, med. pathology U. Rome; m. Gerolama Giancio, May 26, 1928; children—Adele, Ines, Mario, Raffaella, Salvatore, Marialuisa, Giuliana. Prof. gen. med. pathology U. Cagliari, 1933; full prof. med. pathology U. Bari, 1935; prof. med. clinic U. Catana, 1938; prof. med. pathology and methodology clinics U. Rome, also dir. Sch. Cardiology and Med. Clinic. Mem. Italian (pres.), European (pres.), French socs. cardiology; hon. mem. Deutsche Gesellschaft für Kreislaufforschaft, Swiss Acad. Med. Sci., Brit. Cardiological Soc., Am. Coll. Chest Physicians, Nat. Acad. Medicine of Mexico, Nat. French Acad. Medicine, numerous others. Author: over 250 sci. publs., including Le neurassiti primarie endemiche in Italia di natura presumibilmente virale, e loro rapporti con le neurassiti virali ad zeiologia accerata e nosograficamente definité, 1955; Fisiopatologia clinica e terapa delle arteriopatie obiliteranti, 1958; Exploration de la circulation veineuse peripherique, 1960; Baric Peripheric Vascular Reflex with Local Effect, 1961; La regolazione coronarica al vaglio della sperimentazzione clinica, 1961; Sulla sensibilita vascolare periferica, 1962. Address: via Gasperia 10, Rome, Italy.

CONDRAU, Gion, Swiss psychotherapist; b. Disentis, Switzerland, Sept. 1, 1919; s. Leo and Creszentia (Müheim) C.; student U. Fribourg, 1938-39; M.D., U. Berne (Switzerland), 1943; Ph.D., Zurich (Switzerland), 1949; m. Irmgard Buschhausen, Mar. 30, 1959; children—Isabella Maria, Gion Fidel, Claudius Duri. Practice medicine, specializing in psychotherapy, Zurich, 1952—; cons. psychiatrist Frauenklinik, U. Zürich, 1962—, prof. psychotherapy, psychosomatic medicine and med. psychology, U. Fribourg and U. Zürich (Switzerland), 1967—. Mem. Swiss Soc. Psychosomatic Medicine (founder, pres.). Author: Angst und Schuld als Grundprobleme der Psychotherapie, 1962; Daseinsanalytische Psychotherapie, 1963; Einführung in die Psychotheranie, 1964; Die Daseinsanalyse von Medard Boss und ihre Bedeutung für die Psychiatrie, 1965; Psychosomatik der Frauenheilkunde, 1965; also articles. Application of existential philosophy in psychotherapy and psychosomatic medicine; psychosomatic research in gynecology and obstetrics. Address: 770 Strehlgasse 8704 Herrliberg, Switzerland.*

CONE, Thomas Edward, Jr., Am. physician; b. Bklyn., Aug. 15, 1915; s. Thomas Edward and Helen (McMahon) C.; B.A., Columbia, 1935, M.D., 1939; m. Barbara Kirkland Cross, Dec. 5, 1939 (div. 1967); children—Thomas Edward III, Elizabeth Gray (Mrs. Robert Christian Smallridge), Mary Cross; m. Marjorie Ann. Davis, 1967. Commissioned Commd. Lt. (j.g.) USN, 1941, advanced through grades to capt., 1955-63, chief pediatric service U. S. Naval Hosp., Phila., 1951-53, chief pediatric and dependent services Nat. Naval Med. Service, Bethesda, Md., 1953-63; ret., 1963; cons. pediatrics Nat. Inst. Neurologic Diseases and Blindness, Bethesda, 1958-63; asso. clin. prof. pediatrics Georgetown U. Sch. Med., 1953-63; lectr. pediatrics Johns Hopkins Med. Sch., 1959-63; asso. clin. prof. pediatrics Harvard Med. Sch., 1963-66; prof. pediatrics Southwestern Med. Sch., 1966—. Mem. A.M.A., New Eng., Am. pediatric socs., Benjamin Waterhouse Med. History Soc., Alpha Omega Alpha. Studies on incidence and nature of high blood pressure in children produced by pheochromocytoma; causes for accelerative growth trends in children in U. S., Western Europe. Home: 4206 Larchmont Av., Dallas 75205.*

CONGDON, Charles C., Am. pathologist; b. Dunkirk, N.Y., Dec. 13, 1920; s. Charles C. and Jessie (Cooper) C.; A.B., U. Mich., 1942, M.D., 1944, M.S., 1950; m. Margaret Louise Ribble, Apr. 1947; children—Peggie Dune, Mary Dawn, Claudia, Kyle E., Lara Paige. Med. officer, pathologist Nat. Cancer Inst., 1952-55; biologist Oak Ridge Nat. Lab., 1955—; prof. biology U. Tenn., 1966—. Coordinator Bone Marrow Confs., 1957—; mem. com. on tissue transplantation Nat. Acad. Sci.-NRC, 1957-66; mem. acute leukemia task force Nat. Cancer Inst.-NIH, 1962—. Mem. A.A.A.S., Am. Assn. Cancer Research, Am. Assn. Pathologists and Bacteriologists, A.M.A., Am. Soc. Exptl. Pathology, Am. Soc. Clin. Investigation, Am., Internat. socs. hematology, Radiation Research Soc., Soc. Exptl. Biology and Medicine, Sci. Research Soc. Am. Editor: Experimental Hematology, 1957—; editorial bd. Transplantation, 1963—. Research, numerous publs. on exptl. treatment of radiation injury by bone marrow transplantation; structural changes in germinal centers of lymphatic tissues during an immune response. Home: P.O. Box 683, Oak Ridge 37830. Office: P.O. Box Y, Oak Ridge 37831.*

CONGER, Alan Douglas, Am. biologist; b. Muskegon, Mich., Mar. 23, 1917; s. Louis H. and Hazel (Snyder) C.; A.B. with honors, Harvard, 1940, M.A., Ph.D., 1947; m. Priscilla Nash, Aug. 25, 1944; children—Lucy L., Alan D., David R., Priscilla C. Sr. biologist Oak Ridge Nat. Lab., 1947-52; Fulbright sr. research scholar Postgrad. Med. Coll., London, 1952-53; prof. biology U. Fla., Gainesville, 1958-64; prof., head radiation biology Temple U. Med. Sch., Phila., 1964—. Cons. Oak Ridge, Brookhaven nat. labs. Asso. editor Radiation Research, 1959-63, Radiation Botany, 1960—, Mutation Research, 1961—. Research, publs. on effects of ionizing radiations on living cells, primarily effects on chromosomes. Home: 506 Old Gulph Rd., Penn Valley, Narberth, Pa. 19072. Office: 3420 N. Broad St., Phila. 19140.*

CONGREVE, Sir William, Brit. inventor; b. Woolwich, Eng., May 20, 1772; s. Sir William and Rebecca (Elmstone) C.; ed. Royal Acad. at Woolwich, Cambridge, Eng.; studied law; became officer of royal arty. attached to Royal Lab., 1791; comptroller, 1814-28; directed to form 2 rocket cos., 1809; lieutenant colonel in Hanoverian arty., 1811; served with rocket co., Leipzig, Germany, 1813, South France, 1814; succeeded to his father's baronetcy, 1814; mem. Parliament, 1812-28. Fellow Royal Soc., 1811. Author: A Concise Account of the Origin and Progress of the Rocket System, 1807; An Elementary Treatise on the Mounting of Naval Ordnance, 1812; Description of the Hydro-pneumatic Lock invented by Colonel Congreve, 1814; A Short Account of a Patent lately Taken Out by Sir William Congreve for a New Principle of Steam Engine, 1819; Systems of Currency, 1819; A Treatise on the General Principles, Powers and Facility of Application of the Congreve Rocket System as Compared with Artillery, 1827. Invented Congreve rocket, 1808, detonator used during Napoleonic war, 1805, also gun-recoil mounting, worked on perpetual motion machine, steam engine, color-printing process, rocket for killing whales; suggested steam in steam engine be superheated in long tube of small diameter, 1821. Died Toulouse, France, May 16, 1828.

CONIA, Jean Marie, chemist; b. Laventie, France, Feb. 27, 1921; s. Andre Edouard and Isabelle (Cortyl) C.; student U. Lille, 1939-40, U. Montpellier, 1941-43; D.Sc., U. Paris, 1950; m. Mary Francoise Cortyl, May 29, 1950; children—Christian, Patrick, Isabelle. With Nat. Center Sci. Research, Ecole Normale Superieure, Paris, France, 1946-57; chef de Travaux à la Sorbonne, Paris, France, 1955-57; maitre de confs. U. Caen, France Faculty of Scis., 1957-60, prof. organic chemistry, 1961—. Mem. Société Chimique de France, Am. Chem. Soc. Research, publs. on alkylation of saturated and non-saturated ketones, small ring compounds and cyclobutanones, thermolysis and photolysis of non-saturated carbonyle compounds, rearrangements and thermal cyclisations. Home: 31 rue des Jardins, 14-Caen, France.*

CONKLIN, Edmund Smith, Am. psychologist; b. New Britain, Conn., Apr. 19, 1884; s. Edmund Sidney and Catherine Annette (Smith) C.; B.H., Springfield (Mass.) YMCA Coll., 1908; M.A., Clark U., 1909, fellow in psychology, 1909-11, Ph.D., 1911, Sc.D., 1939; m. Helen Corey Holbrook, June 27, 1915; children—Marietta Muir, Edmund Holbrook. Asst. prof. psychology and acting head dept. U. Ore., 1911-13, prof. and head dept., 1913-29, prof. and chmn. dept., 1929-34, acting dean grad. sch., 1922-23; prof. and head dept. psychology, Ind. U., 1934-42; vis. prof. U. Chgo. Div. Sch., winter 1931 summers 1930-33; Am. Psychol. Assn. rep. NRC, 1939-41. Mem. editorial bd. Jour. Genetic Psychology, Genetic Psychology Monographs, Psychological Record. Fellow A.A.A.S.; mem. Am. Psychol. Assn., Am. Assn. U. Profs., Midwestern Psychol. Assn. (pres. 1938-39), Sigma Xi, Phi Gamma Delta. Author: Principles of Abnormal Psychology, 1927 (rev. edit. 1935); Psychology of Religious Adjustment, 1929. Principles of Adolescent Psychology, 1935; (with F. S. Freeman) Introductory Psychology for Students of Education, 1939. Died Oct. 6, 1942.

CONKLIN, Edwin Grant, Am. biologist; b. Waldo, O., Nov. 24, 1863; s. Abram V. and Maria (Hull) C.; S.B., Ohio Wesleyan U., 1885, A.B., 1886, A.M., 1889, Sc.D., 1910; Ph.D., Johns Hopkins, 1891, LL.D., 1940; Sc.D., U. Pa., 1908, Yale, 1930; LL.D., Western Res. U., 1925, U. Pa., 1943, Princeton, 1945; m. Belle Adkinson, June 13, 1889 (dec. Mar. 1940); children—Paul, Mary (Mrs. Samuel Masland), Isabel. Prof. biology Ohio Wesleyan, 1891-94; prof. zoölogy Northwestern U., 1894-96, U. Pa., 1896-1908; prof. biology Princeton, 1908-33, emeritus, from 1933, spl. lectr. in biology. Trustee Woods Hole Lab., from 1897, also Woods Hole Oceanographic Instn.; pres. Bermuda Biol. Station, 1926-36; mem. adv. bd. Wistar Inst.; pres. Sci. Service, 1936-45. Recipient John J. Carty Gold medal and award, 1942-43; Gold medal Nat. Inst. Social Sci., 1943. Fellow A.A.A.S. (pres. 1936), Am. Acad. Arts and Scis.; mem. Nat. Acad. Scis., Am. Soc. Zoölogists (pres. 1899), Assn. Am. Anatomists, Am. Soc. Naturalists (pres. 1912), Am. Philos. Soc. (pres. 1942-45, 1948—), Phi Beta Kappa, Sigma Xi. Co-editor Biol. Bull., Jour. Exptl. Zoölogy, Genetics. Author: Heredity and Environment; Mechanism of Evolution; Direction of Human Evolution; Synopsis of General Morphology; Future of Evolution; Revolt Against Darwinism; Science and the Faith of the Modern; Embryology and Evolution; Problems of Development; Biology and Democracy; Freedom and Responsibility; What Is Man?; Man, Real and Ideal. Research in embryology and cytology on fertilization and orgn. of the egg, cell lineages, early devel. of mollusks, brachiopods, ascidians. Tried to determine to what extent adult structures could be traced back to egg cell, and in what form they appear. Authority on environment and heredity, critic of racist theories of nazism and fascism; advocate of internat. cooperation in sci. research. Died Princeton, N.J., Nov. 21, 1952.

CONKLIN, Harold Colyer, Am. anthropologist; b. Easton, Pa., Apr. 27, 1926; s. Howard S. and May W. (Colyer) C.; A.B., U. Cal. at Berkeley, 1950; Ph.D., Yale, 1955; m. Jean M. Morisuye, June 11, 1954; children—Bruce Robert, Mark William. From instr. to asso. prof. anthropology Columbia, 1954-62; lectr. anthropology Rockefeller Inst., 1961-62; prof. anthropology Yale, 1962—, chmn. dept., 1964—; field research in Philippines, 1945-47, 52-54, 55, 57-58, 61, 62-63, 64, 65, Malaya and Indonesia, 1948, 57, Cal. and N.Y., 1951, 52, Guatemala, 1959. Bd. dirs., mem. plans and policy com. Social Sci. Research Council, 1963—; spl. cons. Internat. Rice Research Inst., Los Banos, Philippines, 1962—; book rev. editor Am. Anthropologist, 1960-62; mem. Pacific sci. bd. Nat. Acad. Scis.-NRC, 1962-66. Served with AUS, 1944-46. Fellow Am. Anthrop. Assn. (exec. bd. 1965—), Royal Anthrop. Soc. N.Y. Acad. Scis. (sec. sect. anthropology 1956), Sigma Xi; mem. Am. Ethnol. Soc. (councilor 1960-62), Koninklijk Inst. voor Taal- Land- en Volkenkunde, Linguistic Soc. Am., Soc. Am. Archaeology, Kroeber Anthrop. Soc., Phila. Anthrop. Assn., Am. Geog. Soc., Am. Oriental Soc., Assn. Asian Studies, Classification Soc., La Société de Linguistique de Paris, Far Eastern Prehistory Soc., Asia Soc., Am. Folklore Soc., Soc. Ethnomusicology, Soc. Econ. Botany. Study of the integral system of shifting cultivation in the Philippines; linguistics; ethnography; Malaysian studies.

CONKWRIGHT, Nelson Bush, Am. mathematician; b. Winchester, Ky., Dec. 5, 1898; s. John B. and Nora (Bush) C.; B.A., U. Ky., 1922; Ph.D., U. Ill., 1926; m. Ruth Freeman, Dec. 24, 1929. Faculty, U. Ia., Iowa City, 1926—, prof. math., 1955—. Mem. Phi Beta Kappa, Sigma Xi, Gamma Alpha. Asso. editor Am. Math. Monthly, 1942-51. Author: Differential Equations, 1934; Introduction to the Theory of Equations, 1941; also papers to math. jours. Study of differential equations; theory of equations. Home: 335 Beldon Av., Iowa City, Ia. 52241.*

CONLEY, Robert Thomas, Am. chemist; b. Summit, N.J., Dec. 27, 1931; s. Thomas M. and Pauline (Mente) C.; B.S., Seton Hall U., 1953; M.A., Princeton, 1955, Ph.D., 1957; m. L. Doris Rutscher, May 7, 1955; children—Debra J., Bryan R., Robin E. Asst. prof. Canisius Coll., Buffalo, 1956-61; faculty Seton Hall U., South Orange, N.J., 1961-67, prof. chemistry, 1964-67; professor chemistry, chairman department Wright State University, Dayton, O., 1967—. Cons. in organic, polymer and analytical chemistry, 1958—. Newman Club fellow, 1954-55; E. I. Du-Pont de Nemours fellow, 1955-56. Mem. Am. Chem. Soc., Soc. Applied Spectroscopy, A.A.A.S., N.J. Acad. Sci., Sigma Xi. Author: Infrared Spectroscopy, 1966; also numerous articles. Research on fundamental reactions of polymers at elevated temperatures in oxygen containing atmospheres, chem. transformations involved in structural rearrangements from carbon to nitrogen, preparation and chem. reaction of small ring compounds. Home: 33 Woods Dr., Dayton, O. 45432.*

CONN, George Keith Thurburn, Brit. physicist; b. Aberdeen, Scotland, Oct. 18, 1911; s. John Thurburn and Margaret Jane (Keith) C.; M.A., U. Aberdeen, 1933; B.A., U. Cambridge (Eng.), 1935, Ph.D., 1938; m. Violet Lindsay Henderson, Apr. 10, 1939; children—David Lindsay Thurburn, Alasdair Keith Thurburn, Neil Graham Thurburn. Faculty, reader physics U. Sheffield (Eng.), 1938-56; prof., head dept. physics U. Exeter (Eng.), 1956—, dean Faculty Sci., 1958-61, dir. Norman Lockyer Obs., 1961—. Cons. MOR Cos., 1949-56. Fellow Inst. Physics, Royal Astron. Soc., Cambridge Philos. Soc.; mem. Brit. Iron and Steel Research Assn. (past mem. research com.). Author: Nature of the Atom, 1939; Wave Nature of the Electron, 1938; (with D. E. Avery) Infrared Methods, 1960; (with H. D. Turner) Evolution of the Nuclear Atom, 1966; also articles. Editor: Research, 1956-62. Research on optics, including infrared and far infrared regions of spectrum, vacuum physics, mechanism of conduction in solids, optical properties of metals and semi-conductors. Home: 6, W. Garth Rd., Exeter, Devon, Eng.*

CONN, Herbert William, Am. biologist; b. Fitchburg, Mass., Jan. 10, 1859; s. Reuben Rice and Harriet E. (Harding) C.; A.B., Boston U., 1881, A.M.; Ph.D., Johns Hopkins, 1884; m. Julia M. Joel, Aug. 5, 1885; children—Bertha (Mrs. Van Tuyl Hart Bien), Harold Joel. Instr. biology Wesleyan U., Conn., 1884-86, asst. prof., 1886-88, prof., from 1889. Dir. Cold Spring Harbor Biol. Lab., 1889-97; bacteriologist Storrs Expt. Sta., 1890-1906; dir. lab. Conn. Bd. Health, from 1905; lectr. Trinity Coll., 1887-89. Author: The Story of the Living Machine, 1899; The Method of Evolution, 1900; An Elementary Physiology and Hygiene for Use in Schools, 1903; practical Dairy Bacteriology, 1907; Introductory Physiology and Hygiene, rev. edit. 1908; Biology (introductory study for use in colls.), 1912; Elementary Physiology and Hygiene (for upper grammar grades), 1913; Social Heredity and Social Evolution, 1914; Bacteria, Yeasts, and Molds in the Home, 1917; Physiology and Health, rev. edit., 1924. Specialist in bacteriology of dairy products; made extensive expts. in bacteriology of cream ripening; introduced to Am. dairymen use of pure cultures of bacteria for artificial inoculation of cream for butter making, 1893; suggested prodn. of sanitary butter through pasturization of cream, use of pure cultures of latic acid bacteria; among 1st to show close relationship between unsanitary conditions on dairy farms and bacteria in milk; 1st to prove that typhoid fever may be carried by oysters. Died Apr. 18, 1917.

CONN, Jerome W., Am. physician; b. N.Y.C., Sept. 24, 1907; s. Joseph and Dora (Kobrin) C.; student Rutgers U., 1925-28; M.D., U. of Mich., 1932; D.Sc., Rutgers U., 1964; m. Elizabeth Stern, June 17, 1932; children—Phyllis, J. William. Instr. medicine, U. Hosp., Ann Arbor, Mich., 1934-38, asst. prof. medicine, 1938-43, asso. prof., 1943-50, prof. internal medicine, 1950-58, distinguished Univ. prof., 1958—, also dir. div. metabolism and endocrinology, 1943—; cons. to surgeon gen. U. S. Army, 1945-48, surgeon gen. USPHS, 1947-48; Banting Meml. lectr. Am. Diabetes Assn., 1958; Henry Russell lectr. U. Mich., 1961, Charles L. Brown Meml. lectr. Temple U., 1961; William Middleton lectr. Wis. Soc. Internal Medicine, numerous others; Asso. editor Jour. Lab. and Clin. Medicine, 1946-51. Mem. com. on metabolism in trauma U. S. Army Dept. Research and Devel.; mem. med. fellowship bd. NRC, 1953—. Recipient Modern Medicine award for achievement in endocrinology, 1957; citation for contbn. to medicine Mich. Med. Soc., 1962, Ricketts award U. Chgo., 1967, numerous others. Diplomate Am. Bd. Internal Medicine. Fellow A.C.P., A.C.S. (hon.); mem. Am. Soc. Clin. Investigation (council 1950-53), Central Soc. Clin. Research (v.p. 1953, pres. 1954), Am. Inst. Nutrition, Am. Diabetes Assn. (pres. 1962), A.M.A. Assn. Am. Physicians, Endocrine Soc., Sigma Xi, Alpha Omega Alpha, Phi Kappa Phi. Contbr. articles to med. jours., sects. to books. Studies in human nutrition, normal human metabolism, diseases of metabolism, endocrinology. Home: 200 Orchard Hill Dr. Office: University Hospital, Ann Arbor, Mich.*

CONNELL, George Edward, Am. biochemist; b. Saskatoon, Sask., Can., June 20, 1930; s. James Lorne and Mabel (Killins) C.; B.A., U. Toronto (Ont.), 1951, Ph.D., 1955; postgrad. N.Y. U.; m. Sheila Harriet Horan, Dec. 27, 1955; children—James, Benjamin, Caroline Jane, Thomas Geoffrey, George, Margaret Mary. Faculty dept. biochemistry U. Toronto, 1957—, prof., chmn. dept., 1965—. Mem. Med. Research Council Can., 1966—. NRC Can. fellow, Ottawa, 1955-56. Mem. Canadian Arthritis and Rheumatism Soc. (dir. 1966—). Research, publs on protein structure and function, structure of enzymes, antibodies and other plasma proteins. Home: 105 Lytton Blvd., Toronto 12, Ont., Can.*

CONNELL, Karl, Am. surgeon; b. Omaha, circa 1879; s. William J. and Mattie (Chadwick) C.; M.D., Coll. Phys. and Surg., Columbia, 1900; m. Frank Hovey-Roof, Sept. 28, 1922; children—Barbara, Karl, Francis Chadwick, Asso. in surgery Columbia, 1910-14; asst. attending surgeon Roosevelt Hosp., N.Y., 1908-19; prof. surgery Creighton U., 1919-24. pres. Presbyn. Hosp., Omaha, 1920-24. pres. Connell Apparatus from 1930. Decorated D.S.M. Fellow A.C.S. Author: Surgical Therapeusis. Inventor anesthetic apparatus. Died Oct. 18, 1941.

CONNEY, Allan Howard, Am. biochemist; b. Chgo., Mar. 23, 1930; s. Leo Y. and Celia (Gasway) C.; B.S., U. Wis., 1952, M.S., 1954, Ph.D., 1956; m. Diana Locke, Sept. 5, 1954; children—Michael, Steven. Pharmacologist, Nat. Heart Inst., NIH, Bethesda, Md., 1957-60; head biochem. pharmacology dept. Wellcome Research Lab., Burroughs Wellcome & Co., Inc., Tuckahoe, N.Y., 1960—. Field editor Jour. Pharmacology and Exptl. Therapeutics. Mem. Am. Soc. Biol. Chemists, A.A.A.S., Am. Assn. Cancer Research, Soc. Toxicology, Am. Pharm. Assn., Am. Soc. Pharmacology Exptl. Therapeutics. Research, publs. on biochem. effects of drugs, carcinogens and steroid hormones, metabolism of drugs, carcinogens and steroid hormones, induced enzyme synthesis and ascorbic acid biosynthesis. Home: 27 Hickory Hill Dr., Dobbs Ferry, N.Y. 10522. Office: 1 Scarsdale Rd., Tuckahoe, N.Y. 10707.*

CONNICK, Robert Elwell, Am. chemist; b. Eureka, Cal., July 29, 1917; s. Arthur Elwell and Florence (Robertson) C.; B.S., U. Cal. at Berkeley, 1939, Ph.D., 1942; m. Frances Spieth, Dec. 19, 1952; children—Mary Catherine, Elizabeth, Arthur, Megan, Sarah, William. Faculty, U. Cal. at Berkeley, 1942—, prof. chemistry, 1952—, chmn. dept. chemistry, 1958-60, dean Coll. Chemistry, 1960-65, vice chancellor, 1965-67. Mem. Nat. Acad. Scis., Am. Chem. Soc., Phi Beta Kappa, Sigma Xi. Research, numerous publs. in inorganic chemistry, reaction mechanisms and complex-ion formation in aqueous solutions, lifetime water molecules in first coordination sphere metal ions measured by nuclear magnetic resonance. Home: 50 Marguerita Rd., Berkeley, Cal. 94707.*

CONNOLLY, Joseph Peter, Am. geologist; b. Cleve., Nov. 15, 1890; s. Peter Albert and Bertha Elizabeth (Orwig) C.; A.B., Oberlin Coll., 1912; A.M., U. Mo., 1915; A.M., Harvard 1916, Ph.D., 1927; spl. studies in geology Mass. Inst. Tech., 1917; m. Anna Ruth

Lewis, Dec. 22, 1924; children—Lewis Peter, Thomas Joseph. Geol. fieldworker Mo. Bur. Mines and Geology, 1914-15; instr. in geology U. Mo., 1919; prof. mineralogy and petrography, S.D. State Sch. Mines, from 1919, also v.p., 1926-35, pres. from 1935; cons. geologist. Leader Nat. Geog. Soc.-S.D. Sch. of Mines, Joint Paleontol. Expdn., 1940. Fellow Geol. Soc. Am., Mineral Soc. Am., A.A.A.S.; mem. Soc. Econ. Geologists, Am. Inst. Mining and Metall. Engrs., Mining and Metall. Soc. Am., Am. Soc. for Engring. Edn., Sigma Xi, Gamma Alpha, Sigma Gamma Epsilon. Editor: Black Hills Engr., from 1935. Writer of bulletins on mineralogy and econ. geology of Black Hills; contbr. to profl. jours. Died Oct. 7, 1947.

CONNOR, Ralph, Am. chemist; b. Newton, Ill., July 12, 1907; s. Stephen A. and Minnie (Ross) C.; B.S., U. Ill., 1929; Ph.D., U. Wis., 1932; D.Sc., Phila. Coll. Pharmacy and Sci., 1954, University Pennsylvania, 1959, Brooklyn Polytechnic Inst., 1967; LL.D., Lehigh U., 1966; m. Margaret Raef, Sept. 1, 1931; 1 son, Stephen R. Tech. aide nat. def. research com. OSRD, 1941-45, chief div. 8, 1943-45; faculty U. Pa., 1935-44; asso. dir. research Rohm & Haas Co., Phila., 1945-48, v.p. in charge of research, 1948—, mem. exec. com., 1948—, chmn. bd., 1960—. Mem. div. chemistry and chem. tech. NRC, 1953-57, mem. tech. adv. panel on biol. and chem. warfare, 1954-61. Decorated King's medal for service in cause freedom (Eng.); recipient medal for merit U. S. A., Naval Ordnance Devel. award; Gold medal Am. Inst. Chemists, 1963; Chem. Industry medal Soc. Chem. Industry, 1965. Mem. Am. Chem. Soc. (chmn. bd. 1956-58, Priestley medal 1967). Author: (with Fuson, Price, Snyder) A Brief Course in Organic Chemistry, 1941; also articles. Research on organic sulfur compounds, high pressure catalysis, reaction mechanisms. Home: 234 N. Bent Rd., Wyncote, Pa. 19095. Office: Independence Mall W., Phila. 19105.*

CONOLLY, John, Brit. physician; b. Market Rasen, Eng., May 27, 1794; s. M.D., Edinburgh, Scotland, 1821; D.C.L., Oxford, Eng.; m. dau. of Sir John Collins, 1816. Served in army, 1812-15; prof. medicine U. Coll., London, Eng., 1823-44; dir. insane asylum, Hanwell, Eng., 1839-44; vis. physician Middlesex (Eng.) Asylum; a founder Asylum for Insane, Eralswood, Eng. Mem. Brit. Med. Assn. (charter). Author: The Treatment of the Insane without Mechanical Restraints, 1856; treatises on the insane and asylum methods, 1847-56; articles in med. jours. Introduced no restraint system to asylums in Eng. Died Hanwell, Mar. 5, 1866.

CONON OF SAMOS, Greek astronomer, mathematician; flourished 250 B.C.; traveled Western part of Greek world; settled in Alexandria, Egypt; received at ct. of Ptolemy IV Evergete. Author 7 books on astronomy. Invented curve called spiral of Archimedes; investigated conic sections; developed a calendar based on rise and setting of stars; discovered constellation Coma Berenices; collected observations of solar eclipses in Egypt.

CONRAD, Frank, Am. elec. engr; b. Pitts., May 4, 1874; s. Herbert M. and Sadie (Cassidy) C.; ed. pub. schs., Pitts.; D.Sc. (hon.), U. Pitts., 1928; m. Flora Selheimer, June 18, 1902; children—Francis H., Crawford J., Jane L. Began as bench-hand Westinghouse Elec. & Mfg. Co., Pitts., 1890, became gen. engr., 1914, asst. chief engr., from 1921. Recipient Edison medal Am. Inst. E.E., 1931, Lamme medal, 1936, John Scott medal City of Phila., 1933; Morris Liebman prize Inst. Radio Engrs., 1936. Early experimenter in radio, using phonograph records for broadcasting, 1919; built radio transmitter for Westinghouse Co., 1920; instrumental in developing Sta. KDKA, Pitts., made broadcast regarded as birth of pub. broadcasting, Nov. 2, 1920; patentee over 200 inventions, including round watt-hour electric meter, pantagraph trolley for electric trains, electric clock, devices for automobile ignition. Died Miami, Fla., Dec. 11, 1941.

CONRAD, Marcel Edward, Jr., Am. physician; b. N.Y.C., Aug. 15, 1928; s. Marcel Edward and Lulu (Geraghty) C.; B.S., Georgetown U., 1949, M.D., 1953; m. Patricia Jane Hutchon, Jan. 16, 1948; children—Marcel Edward III, Mark E., Carol J. Asst. chief hematology service Walter Reed Gen. Hosp., Washington, 1958-60, chief, 1965—, comdg. officer 43d Surg. Hosp., 1960-61, chief dept. gastroenterology Walter Reed Army Inst. Research, Washington, 1962, asst. chief dept. hematology, 1963-65, chief dept. hematology, 1965—; clin. asst. prof. medicine Georgetown Med. Sch., 1965—; hematologic cons. Surgeon Gen. U. S. Army, 1965—. Fellow A.C.P., Internat. Soc. Hematology; mem. Am. Soc. Hematology, Soc. Exptl. Biology and Medicine, A.A.A.S., Am. Fedn. Clin. Research. Research, numerous publs. in hematology and gastroenterology, iron metabolism, hemolytic disorders, infectious hepatitis. Home: 9702 Sutherland Rd., Silver Spring, Md. 20901. Office: Dept. Hematology, Walter Reed Army Inst. Research, Washington 20012.*

CONRAD, Timothy Abbot, Am. geologist, paleontologist; b. nr. Trenton, N.J., June 21, 1803; s. Solomon White Conrad; studied natural history and science privately. Assisted his father in printing bus. until father's death, 1831; became especially interested in conchology; became mem. Phila. Acad. Scis., 1831; state geologist and paleontologist of N.Y., 1837-42; mem. Am. Philos. Soc., Phila. Acad. Scis. His work is not well known because of his opposition to Darwin's theory of evolution; contbr. articles to various jours.; drew many of plates for his own works. Author: American Marine Conchology; or, Descriptions and Colored Figures of the Shells of the Atlantic Coast, 1831; Fossil Shells of the Tertiary Formations of North America, 1832; Paleontology of the State of New York (1838-40). Died Trenton, Aug. 9, 1877.

CONRADI, Johann Wilhelm Heinrich, German physician; b. Marburg/Lahn, Germany, Sept. 22, 1780; s. Johann Ludwig and Anna Florentine (Seip) C.; studied medicine in Marburg, 1797-1802; doctorate, 1802; m. Johanna Friederike Schreidt; 5 sons, 1 dau.; m. Marie Ernestine Schulze, 1824; 2 sons. Became asso. prof., Marburg, 1803, prof., 1805; named dir. of polyclinic, 1809, stationary clinic, 1812; sent to Heidelberg, Germany, to establish med. clinic, 1814; named dir. polyclinic Göttingen, Germany, 1823; dir. stationary clinic, 1837-53; ret. from directorship, 1853, but continued teaching. Author textbooks on gen. and specialized pathology, therapy. One of 1st in Germany to oppose ideas of Broussais on gastroenteritis and bloodletting. Died Göttingen, June 17, 1861.

CONRADTY, Conrad, German chemist; b. Münchaurauch, Germany, July 15, 1827; s. Johann Thomas and Eva Barbara (Dittler) C.; m. Katharina Götz, 1854; 3 sons, 1 dau. Salesman and pencil mfr.; founded pencil factory, Nürnberg, Germany, 1855, later extended prodn. to paints, carbon-sticks for arc lamps, electrodes. Studied electro-carbon. Died Nürnberg, June 17, 1901.

CONRING, Hermann, German physician; b. Norden, East Friesland, Nov. 9, 1606; ed. Univs. Helmstädt; Leyden; Ph.D., M.D., 1636. Chair of Natural Philosophy, 1632, chair of Medicine, chair of Law, Univ. Helmstädt; physician and councillor to Princess Regent of Friesland, 1649. Author: De Hermetica Aegyptiorum vetere et nova Paracelsicorum Medicina, 1648; In universam artem medicam . . . Introductio, 1654; De Morborum remediis Magicis et Unguento Armario. Wrote on medicine, philosophy, natural science, history, law, poetry, antiquities, chemistry; an Aristotelian and Galenist, opposed Paracelsists, did not acknowledge existence of Egyptian Hermes or Egyptian medicine and chemistry; against introduction of chemistry into medicine; made laudanum from opium. Died Dec. 12, 1681.

CONRINGIUS (or CONRING), Hermann, physician; b. Norden, E. Friesland, Netherlands, Nov. 9, 1606; s. Hermann and Galatea (Copin) C.; studied at U. Helmstedt (Netherlands), U. Leiden (Netherlands); m. Anna Marie Stucke, Apr. 21, 1636; 11 children including: Herman, Anna Maria (Mrs. D. John Saubertus), Marie Sophie, Johanne. Pursued professions of divinity, law, and medicine; prof. medicine, and other positions U. Helmstedt; named physician to Christina of Sweden, 1650. Named councillor of state by the Province of E. Friesland, also cts. of Brunswick, Denmark, and Sweden; received ann. pension from King of France. Author: Germanicorum corpurum habitus antique et novi causis. 1645; Introductio in universam artem medicam, 1654; Treatise de Hermetica Aegyptiorum medicina; also numerous hist. publs. Supported Harvey's theory of circulation of blood; helped improve German public law. Died Helmstadt, Dec. 12, 1681.

CONSIDÈRE, Armand Gabriel, French engr.; b. Port-sur-Soône, France, June 8, 1841; ed. Ecole Polytechnique; chief engr. Ponts et Chaussées. Laureate, Soc. for Encouraging Industry. Mem. French Acad. Scis., 1892. Stated rules for calculation of structures of reinforced concrete; constructed lighthouse at Eckmühl. Died Paris, France, Aug. 3, 1914.

CONSTAM, Emil Joseph, chemist; b. N.Y., Feb. 19, 1858; s. Joseph and Karolina (Grünberg) C.; studied at Polytechnikum of Switzerland; grad. U. Zurich (Switzerland), 1882; m. Johanna Ida Stierlin, 1885; 3 sons, 1 dau. Became pvt. lectr. Polytech. U., Zurich, 1885, prof., 1899; became dir. Fed. Fuel Testing Inst., 1907. Author: Elektrolytische Darstellung einer neuen Klasse oxydierender Substanzen, 1897; Einfluss der Festigkeit von Steinkohlenbriketts . . . , 1904; also articles in chem. jours. Research on fuels and economy, testing of fuels. Died Zurich, Feb. 11, 1917.

CONSTANCE, Lincoln, Am. botanist; b. Eugene, Ore., Feb. 16, 1909; s. Lewis Llewellyn and Ella (Clifford) C.; A.B., U. Ore., 1930, M.A., 1932; Ph.D., U. Cal. at Berkeley, 1934; m. Sara Luten, July 12, 1936; one son, William Clifford. Faculty, State U. Wash., 1934-37; faculty U. Cal. at Berkeley, 1937—, prof. botany, 1949—, curator, dir. Herbarium, 1962—, dean Coll. Letters and Sci., 1955-62, vice chancellor, 1962-65; vis. lectr., acting dir. Gray Herbarium, Harvard, 1947-48. Guggenheim fellow, 1953-54. Fellow Cal. Acad. Sci.; mem. Am. Acad. Arts and Scis., A.A.A.S., Bot. Soc. Am. (certificate of merit 1959), Cal. Bot. Soc., Am. Inst. Biol. Scis., Soc. Argentina Bot. (corr.), Société de Biogéographie (Paris), Soc. for Study Evolution, Internat. Soc. Plant Taxonomists, Am. Taxonomy Soc., Am. Soc. Naturalists. Contbr. numerous articles to tech. jours., monographs. Research on classification flowering plants with respect to evolution, migration, distbn. Home: 47 Alamo Av., Berkeley, Cal. 94708.*

CONSTANT, F(rank) Woodbridge, Am. physicist; b. Mpls., June 1, 1904; s. Frank H. and Annette (Woodbridge) C.; grad. Lawrenceville Sch., 1921; B.S., Princeton, 1925; Ph.D., Yale, 1928; m. Elizabeth Bass, Jan. 25, 1940. Nat. research fellow Cal. Inst. Tech., Pasadena, Cal., 1928-30; faculty Duke, Durham, N.C., 1930-46, asso. prof. physics, 1942-46; prof. physics, chmn. dept. Trinity Coll., Hartford, Conn., 1946—. With OSRD, 1942-46. Mem. Phi Beta Kappa, Sigma Xi. Author: Theoretical Physics: Mechanics, 1954; Theoretical Physics and Electromagnetism, 1958; Fundamental Laws of Physics, 1963. Research in mechanics, theoretical physics, fundamental laws of physics; ferromagnetism, electromagnetism. Home: Gun Mill Rd., Bloomfield, Conn.*

CONSTANT, Hayne, engr.; b. Sept. 26, 1904; s. Frederick Charles and Mary Theresa C.; M.A., M.I. Mech.E., Queen's Coll., Cambridge (Eng.) U. Lectr. Imperial Coll. Sci., 1934-36; work on gas turbine project Royal Aircraft Establishment, from 1936, head engine dept., from 1941, head research dept power jets, from 1944; dept. dir. Nat. Gas Turbine Establishment, from 1946, dir., 1948-60. Recipient Clayton prize Inst. M.E.'s, 1947. Fellow Royal Soc., 1948, Royal Aero. Soc. (Busk Meml. prize 1932, Gold medal 1963). Author: Gas Turbines and their Problems, 1948. Contbr. to govt. papers, tech. jours. Address: Ripelhyrst, Kiln Way, Grayshott, Hindhead, Sly, Surrey, Eng.

CONSTANTIN, Julien Noël, French botanist; b. Paris, France, Aug. 16, 1857; licence es sciences mathématiques, 1879; aggregation des sciences physiques, 1880; docteur. Prof. organography, paleontologic botany Mus. Natural History, 1901-20; prof. Ecole normale superieure; lectr. Faculty Sci., Bordeaux, France. Mem. French Acad. Scis., 1912, Acad. Agr., Mycol. Soc., Bot. Soc. Author: Les végétaux et les milieux cosmiques; Les Plantes, 1922; works on mushrooms and influence of environment on plants. Investigated medicinal value of white ringed mushroom; (with Lucet) examined path. effects of mushrooms. Died Paris, Nov. 17, 1936.

CONSTANTIN, Robert, botanist, linguist; b. Caen, France, 1530; grad. physician U. Caen, 1564; student langs. and letters under G. Clutin. Practiced medicine Montaban, France; also tchr., translator Greek. Author: Lexicon Graeco-Latinum, 1562; Theophrasti de historia plantarum, 1584; Treatise on the Medical Plants of Provence, 1597. Died Germany, 1605.

CONSTANTINE THE AFRICAN (Constantinus Africanus), physician, translator; b. Carthage, Italy, 1015; was mcht. in Carthage; sec. to Robert Guiscard; Travelled, gathered numerous med. works, studied med. techniques, Babylon and India also Ethiopia, Egypt, for 39 years. Became Benedictine monk, retired to abbey of Monte Cassino, Italy. Translator: Kitab-al-malaki (Cali b. Al-Abbas); Aphorisms (Hippocrates), Prognosticon; Megatechne (Galen), De locis, Commentaires; also writings on Aristotelian and other Greek and Arab philosophies. Through his translations (used as texts at Sch. Salerno), he renewed study of Greek medicine in Italy and Western Europe, introduced Arab med. techniques to Italy, revived interest in sci. in Western Europe. Died 1087.

CONSTANTINESCU, Liviu, Rumanian geophysicist; b. Ighisul Vechi, Rumania, Nov. 26, 1914; s. Romul and Ann (Dancasiu) C.; B.Sc., in Physics and Chemistry, U. Bucharest (Rumania), 1935, Ph.D., 1941; D.Sc., Higher Commn. for Diplomas, 1959; m. Sofia Stoicoviciu, Dec. 28, 1939; 1 son, Dan-Horia. Asst., U. Bucharest, 1937-47; geophysicist Surlari Geophys. Obs., Geol. Inst. Rumania, 1943-58; reader geophys. potential fields Inst. Mines, Bucharest, 1950-57; prof. geophysics Inst. for Oil, Gas and Geology, Bucharest, 1957—; sr. research worker Inst. Geology, Acad. head dept. applied geophysics, 1958-61, Center for Geophys. Research Acad., 1961—; head dept. geophysics Inst. for Oil, Gas and Geology, 1960—. Sec. gen. Rumanian Nat. Com. for Geodesy and Geophysics, 1966—. Decorated Rumanian Order of Labour. Mem. Acad. Rumania (corr.), Assn. Hungarian Geophysicists (hon.), Am. Geophys. Union, Soc. Exploration Geophysicists, Internat. Assn. Geomagnetism and Aeronomy (past mem. com. on observatories), European Seismol. Commn. Author: Prospectiuni Geofizice, vol. I, 1964, editor, co-author, vol. II, 1965. Editor: Editura Technica. Research, articles on geomagnetism, gen. morphology and spl. features of magnetic storms, normal distbn. and secular variation of main field, processing and interpretation of gravity data downward analytical continuation of potential fields, gravity tidal effects, focal mechanism of earthquakes, seismicity, seismotectonics. Home: 67-73, Mihai Bravu, Bucharest 39. Office: 21 Bd., Ana Ipatescu, Bucharest 22, Rumania.*

CONTI, Roberto, Italian mathematician; b. Florence, Italy, Apr. 29, 1923; s. Bruno and Carla (Lanza) C.; student Scuola Normale Superiore, Pisa, Italy, 1941-43, 44-45, 46-47; m. Nuccia Liverani, May 28, 1949; children—Cecilia, Raffaello. Faculty U. Florence, 1947-56, 58—, prof., 1958—; prof. U. Catania (Italy), 1956-58. Sec., Centro Internazionale Matematico Estive, 1954—; mem. Mem. Italian Math. Union (sci. commn.), Am. Math. Soc., Math. Soc.

France. Author: (with G. Sansone) Nonlinear Differential Equations, 1964; also articles. Research on ordinary differential equations, calculus of variations, optimum control theory. Home: 14/A G.B. Amici, Firenze, Italia 50131.*

CONTRERAS, Raul, Mexican pathologist; b. Puebla, Mexico, Apr. 7, 1921; s. Raul and Antonia (Rodriguez) C.; M.D., U. Puebla, 1947; M.S., Nat. U. Mexico, 1961; m. Ana Maria Artime, Oct. 13, 1951; children—Raul, Jesus Ramon. Pathologist, Gen. Hosp., Puebla, 1951-53; dir. Inst. Biology and Exptl. Medicine, U. Puebla, 1950-53, prof. pathology, 1949-53; research Nat. Inst. Cardiology, Mexico City, 1953-64, chief morbid anatomy service, dept. pathology, 1965; prof. pathology Nat. U. Mexico, Mexico City, 1962—; prof. pathology Nat. Poly. Inst., Mexico City, 1953—. Recipient Gedeon Richter award, 1948; Francisco Marin award, 1947; Centennial medal Armed Forces Inst. Pathology, 1962. Mem. Mexican Assn. Pathologists, Mexican Council Pathologists, N.Y. Acad. Scis., Internat. Acad. Pathology, Latin Am. Soc. Pathology. Research, publs. on bacterial endocarditis, cardiovascular syphilis, isolated myocarditis, myocardial sarcoidosis, morphological bases of arterial hypertension, arteritis of pulseless disease. Home: 35 Bonampak, Mexico D.F. 13. Office: 300 Cuauhtemoc, Mexico D.F. 7, Mexico.*

CONVERSE, Phillip E., Am. social scientist; b. Concord, N.H., Nov. 17, 1928; s. Ernest L. and Evelyn (Eaton) C.; B.A., Denison U., 1949; M.A. in English Lit., State U. Ia., 1950; Certificate des Études Francaises, U. Paris, 1954; M.A. in Sociology, U. Mich., 1956, Ph.D. in Social Psychology, 1958; m. Jean G. McDonnell, Aug. 25, 1951; children—Peter, Timothy. Staff, U. Mich., Ann Arbor, 1956—, prof. sociology and polit. sci., 1965—, program dir. Survey Research Center, 1965—, dir. Inter-U. Consortium for Polit. Research, 1962—. Horance H. Rackham fellow, 1955-56; Social Sci. Research Council grantee, 1961-62; NSF sr. postdoctoral fellow, 1967-68. Recipient Fulbright award, Paris, Bordeaux, France, 1959-60. Mem. Am. Sociol. Assn. (past mem. com. on data retrieval), Social Sci. Research Council (com. on comparative politics 1967—), Nat. Acad. Scis. (com. on information in Behavior Scis.), Am. Polit. Sci. Assn., Phi Beta Kappa. Author: (with T. Newcomb, Ralph Turner) Social Psychology: The Study of Human Interaction, 1965; (with A. Campbell, W. E. Miller, D. E. Stokes), The American Voter, 1960, Elections and the Political Order, 1966; also articles. Research on empirical theory of mass voting processes, structure and devel. of pub. opinion in polit. areas, orgn. of large-scale archives of machine-readable data for social scis. Home: 1312 Cambridge St., Ann Arbor, Mich. 48104.

CONVERSI, Marcello, Italian physicist; b. Tivoli, Italy, Aug. 25, 1917; s. Alessandro and Amina (Radiciotti) C.; student Liceo Scientifico C. Cavour, Rome, 1932-36; D.Physics, U. Rome, 1940; m. Elena LaCava, Apr. 3, 1953; children—Daniele, Alessandra, Mauro. Asst., U. Rome, 1940-47, prof. advanced physics, 1958—, dir. physics dept., 1961, 63—; postdoctoral fellow U. Chgo., 1947-49; prof. exptl. physics U. Pisa, Italy, 1950-58, dir. physics dept. Premio Ministro Publica Istruzione per la Fisica, 1964. Fellow Am. Phys. Soc.; mem. Accademia Nazionale dei Lincei. Discovered with E. Pancini and O. Piccioni that cosmic ray mesons were not agt. of nuclear forces; devel. with A. Gozzini 1st pulsed track detector. Home: 4 Via Vajna, Roma. Office: U. Rome, Roma, Italy.*

CONVEY, John, physicist; b. Durham, Eng., Mar. 29, 1910; s. John and Mary (Catterson) C.; B.Sc., U. Alta., Can., 1933, M.Sc., 1936; Ph.D., U. Toronto, 1940; D.Sc., McMaster U., 1959; m. Annette Therese Lemieux, Dec. 30, 1939; children—Annette (Mrs. R. M. Spillane), Jacqueline (Mrs. Trent Gow), John, Nicole. Asso. prof. physics U. Toronto, 1946-48, chief phys. metallurgy div. Mines br., 1948-51, dir. Mines br. Dept. Energy, Mines and Resources, Ottawa, Ont., Can., 1951—; chmn. sub-com. Canadian Nat. Productivity Council, 1961-63; mem. adv. com. sci. and medicine rev. group Expo 67; asso. com. mem. Nat. Museum of Sci. and Engring. Recipient Sorby prize, 1942. Fellow Inst. Physics of Great Britain; mem. Can. Inst. Mining and Metallurgy Blaylock medal, Am. Inst. Mining, Metall. and Petroleum Engrs., Am. Phys. Soc., Canadian Standards Assn. (dir. 1963—), Inst. Mining and Metallurgy Gt. Britain, Canadian Welding Bur. (mem. adminstrv. bd.), Electrochem. Soc., Am. Faraday Soc., Alta. and Northwest Chamber of Mines and Resources, Am. Soc. for Metals (pres. 1966-67). Home: 9 Bayswater Pl., Ottawa. Office: 555 Booth St., Ottawa, Ont., Can.*

CONWAY, Arthur William, math. physicist; b. Wexford, Eire, Oct. 2, 1875; s. Myles Conway; ed. Univ. Coll., Dublin, Ireland, Corpus Christi Coll., Oxford, Eng.; M.A., Oxford; M.A., Royal U. Ireland, also D.Sc. (hon.), 1908; Sc.D., Dublin, 1938; LL.D., St. Andrews, 1938; m. Daphne Bingham; 1 son, 3 daus. Registrar, prof. math. physics Univ. Coll., Dublin, 1909-40, pres., 1940-47; fellow Royal U. Ireland; hon. fellow Corpus Christi Coll.; senator Nat. U. Ireland, 1909-47, vice-chancellor, 1942-45. Fellow Royal Soc. (council 1935-36). pres. Royal Irish Acad., Roya Dublin Soc. Editor: Mathematical Works of Sir William Rowan Hamilton. Contbr. memoirs on math. physics. Died July 11, 1950.

CONWAY, Edward Joseph, Irish biochemist, biophysicist; b. Nenagh, Ireland, July 3, 1894; s. William Francis and Mary (McCready) C.; grad. in physiology and chemistry with 1st class honors U. Coll., Dublin, Ireland, 1916, M.Sc., 1917, grad. in Medicine, 1921, D.Sc., 1927; D.Sc., Trinity Coll., Dublin, 1952; M.D., Nat. U. Ireland, Dublin, 1952; m. Mabel Edith Hughes, Aug. 7, 1934; children—Dorothy, Stephanie, Maureen, Beatrice. Research staff U. Frankfurt, Germany, 1928-29; prof. biochemistry and pharmacology U. Coll., Dublin, 1932-65; extern examiner biochemistry Cambridge (Eng.) U., 1952; dir. unit cell metabolism Med. Research Council, Dublin, 1959—. Chmn. council Dublin Inst. Advanced Studies, 1960—; mem. AEC, Republic of Ireland, 1957—; U. Coll. Dublin rep. Med. Research Council Ireland, 1957—; mem. Irish nat. commn. UNESCO, 1962—. Fellow Royal Irish Acad., Royal Soc., Royal Soc. Arts., Royal Coll. Physicians Ireland (hon.), Royal Inst. Chemistry, Inst. Chemistry Ireland (hon.), Inst. Biology (London, Eng.); mem. Biochem. Soc., Physiol. Soc., Lister Inst., Academie Septrionale, Chem. Soc., Royal Coll. Physicians (hon.), Royal Irish Acad. (mem. council 1962), Pontifical Acad. Scis. Author: Microdiffusion Analysis and Volumetric Error, 1939; The Biochemistry of Gastric Acid Secretion, 1957; also numerous articles. Research on microdiffusion analysis, gastric prodn. hydrochloric acid by redox pump, transport of ions by same process, chem. evolution of ocean; invented micro-diffusion unit, micro burette. Home: Woodbank, Killiney, Co. Dublin, Office: Merville House, Foster A., Blackrock, Dublin, Ireland.*

CONWAY, Harry Donald, mech. engr.; b. Chatham, Kent, Eng., Dec. 3, 1917; s. John and Ada Frances (Young) C.; B.Sc. (Sir John Johnson scholar, Sir Joseph Whitworth scholar), London U., 1942, Ph.D., 1945, D.Sc., 1949; M.A., U. Cambridge, 1946; m. Dorothy D. Adams, Aug. 24, 1946; children—Geoffrey S., Peter S. Sci. officer Nat. Phys. Lab., Teddington, Eng., 1942-45; univ. demonstrator in engring. Cambridge (Eng.) U., also asst. dir. studies St. Catharines U., 1946-47; asso. prof. mechanics Cornell U., Ithaca, N.Y., 1947-48, prof. applied mechanics, 1948—; John Simon Guggenheim fellow, vis. prof. Imperial Coll., London (Eng.) U., 1953-54; Julius F. Stone vis. prof. Ohio State U., 1958-59. NSF Sr. Postdoctoral fellow, 1961-62. Author: Aircraft Strength of Materials, 1947; Mechanics of Materials, 1950; also numerous articles. Research in applied elasticity, mechanics materials, plates, shells, vibrations. Office: Thurston Hall, Cornell Univ., Ithaca, N.Y. 14850.*

CONWAY, Herbert, Am. plastic surgeon; b. Ft. Wayne, Ind., June 25, 1904; s. James Francis and Irene (McCarthy) C.; student Miami U., 1921-24; B.S., U. Cin., 1929, M.S., 1932, M.B., 1928, M.D. (Taft fellow in surgery 1928-29), 1929; m. Frances Gallagher, Nov. 7, 1936; children—Karen, Richard William, Catherine Lanning. Attending surgeon N.Y. Hosp., 1935-42, attending surgeon, attending surgeon in charge plastic surgery, 1945—; faculty Cornell U. Med. Coll., 1932—, asso. prof. clin. surgery, 1945-55, prof. clin. surgery, 1955—; cons. plastic surgeon VA Hosp., Bronx, N.Y., 1945—; cons. plastic surgery White Plains (N.Y.) Hosp., Bellevue Hosp., Hosp. Spl. Surgery; cons. plastic and reconstructive surgery Health Center, Inc., N.Y.C., 1950—; cons., lectr. plastic surgery St. Albans (N.Y.) Naval Hosp., 1957—; 1st vis. prof. Found. Am. Soc. Plastic and Reconstructive Surgery, Brazil and Peru, 1965. Rep., Acad. Surgery of Peru to A.C.S., 1950—; pres., trustee Found. Am. Soc. Plastic and Reconstructive Surgery, 1960-61; spl. State Dept. lectr. India, Saudi Arabia, Pakistan, Lebanon, 1962-63; mem. com. tissue transplantation NRC-Nat. Acad. Scis., 1958-63. Recipient Colles medal Royal Coll. Surgeons Ireland. Diplomate Am. Bd. Plastic Surgery (chmn.), Am. Bd. Surgery. Mem. A.M.A., N.Y. Surg. Soc., Internat. Soc. Surgery, Soc. Med. Cons. World War II, Brit. Assn. Plastic Surgeons, A.C.S., Pan Am. Med. Assn. (chmn. plastic surgery sect. 1952-59), Am. Assn. Plastic Surgeons (trustee 1954-57, pres. 1960-61); hon. mem. Acad. Surgery of Chile, Chilean, Spanish, Japanese socs. plastic and reconstructive surgery, Acad. Surgery of Peru, Argentine Med. Assn., French, Argentine assns. plastic surgery, Surg. Soc. South Africa, Academia de Ciencias Medicas (Barcelona), Hellenic Surg. Soc. Author: (with Richard B. Stark) Plastic Surgery 100 Years Ago, 1953; Tumors of the Skin, 1956; (with Norman E. Hugo, John Tulenko) Surgery of Tumors of the Skin, 1966; also articles. Editor Transplantation Bull.; U. S. editor Revista Latino Americana de Cirurgia Plastica; asso. editor Journal of Surgery. Home: 580 Park Av., N.Y.C. 10021. Office: 232 E. 66th St., also 525 E. 68th St., N.Y.C. 10021.

CONWAY, William Gaylord, Am. zoologist; b. St. Louis, Nov. 20, 1929; s. Frederick Eldridge and Alice Harriet (Gaylord) C.; A.B., Washington U., 1951. Curator birds St. Louis Zool. Park, 1950-56; curator birds N.Y. Zool. Park, N.Y.C., 1956-60, asso. dir., 1960-61, curator birds, 1956—, general director, 1966—. Director American Committee International Wildlife Protection. Fellow New York Zool. Soc.; mem. Brit. Avicultural Soc., Am. Ornithologists Union, Cooper Ornithol. Soc., Internat. Wild Waterfowl Assn. (dir.), Wilson Ornithol. Club, Wild Animal Propagation Trust (pres.), Am. Assn. Zool. Parks and Aquariums (president), Internat. Survival Service Commn., Am. Soc. Ichthyologists and Herpetologists. Contbr. articles profl. jours. Expeditions to Trinidad, Argentina, Bolivia. Conservation and educational exhibition of wildlife. Office: N.Y. Zoological Park, N.Y.C. 60.

CONWELL, Esther Marly, Am. physicist; b. N.Y.C., May 23, 1922; d. Charles and Ida (Korn) Conwell, B.A., Bklyn. Coll., 1942; M.S., U. Rochester, 1945; Ph.D., U. Chgo., 1948; m. Abraham Rothberg, Sept. 30, 1945; 1 son, Lewis. Instr., Bklyn. Coll., 1946-51; tech. staff Bell Telephone Labs., Murray Hill, N.J., 1951-52; physicist, program mgr. Gen. Telephone & Electronics Labs., Bayside, N.Y., 1952—; vis. prof. Ecole Normale Superieure, Paris, France, 1962-63. Fellow Am. Phys. Soc., A.A.A.S.; mem. Research Engring. Soc. Am., Phi Beta Kappa. Author: High-Field Transport in Semiconductors, 1967. Contbr. numerous articles to sci. jours. Research in solid state physics, particularly in field of transport in semiconductors. Office: Gen. Telephone & Electronics Labs., Bayside, N.Y. 11360.*

CONWENTZ, Hugo Wilhelm, botanist; b. Danzig (now Poland), Jan. 20, 1855; s. Albert Wilhelm Conwentz; studied natural scis. Breslau, Germany (now Wroclaw, Poland), Göttingen, Germany; m. Greta Ekelöf, 1919. Became asst. botany in Breslau, 1876; head W. Prussian Mus., Danzig, 1879-1909. Author: Flora des Bernsteins, 1886; Moorbrücken im Tal der Sorge, 1891; Die Eibe in Westpreussen, 1892; Seltene Waldbäume im Westpreussen, 1895; Die Gefährung der Naturdenkmäler und Vorschläge zu ihrer Erhaltung, 1904; Heimatkunde in der Schule, 1904; also numerous others. Research and publs. on paleobotany, plant geography; pioneered nature preservation. Died Berlin-Schöneberg, Germany, May 12, 1922.

CONYBEARE, William Daniel, English geologist, paleontologist; b. St.-Botolph, June 1787; ed. Westminster, Christ Ch., Oxford; M.A., 1811; m., 1814; at least 1 son, William John. Vicar of Axminster, Devonshire, 1836-44; dean of Llandaff, 1845-57; founder (with Sir Henry de la Beche) Bristol Philos. Instn. and Mus. Fellow Royal Soc., 1832, Geol. Soc. London; mem. French Acad. Scis., 1827. Author: On the Origin of a remarkable Class of Organic Impressions occurring in Nodules of Flint, 1814; Hydrographical Basin of the Thames; Theory of Mountain Chains; Ichthyosaurus; Report on the Progress, Actual State and Ulterior Prospects of Geological Science. Gave 1st description of ichthyosaurus. Died Itchenstoke, nr. Portsmouth, Eng., Aug. 12, 1857.

CONZE, Alexander Christian Leopold, German archeologist; b. Hanover, Germany, Dec. 10, 1831; s. Georg and Sophie (Hewig) C.; studied at U. Göttingen, U. Berlin (both Germany); degree in archeology, 1855; m. Elise Eramann, 1861; 4 sons, 2 daus. Became asso. prof. Halle, Germany, 1863; apptd. prof. U. Vienna (Austria), 1869; dir. Deutsches Archäogisches Institut, 1887-1905. Mem. Acad. Vienna. Author: Beiträge zur Geschichte der griechischen Plastik, 1868; Vorlegeblätter für archeologische Übungen, I-VIII, 1869-86; Röhnische Bildwerke eiuheun schen Fundoet im Osterrech, 1872; Albertrümer von Pergamon, 1885; also articles. Archeol. research in Greece, Italy, Asia Minor; made excavation in Pergamon; discovered Samothrake. Died Berlin, Germany, July 19, 1914.

CONZETT, Homer Eugene, Am. physicist; b. Dubuque, Ia., Oct. 16, 1920; s. Andrew and Christina (Accola) C.; B.A., U. Dubuque, 1942; Ph.D., U. Cal. at Berkeley, 1956; m. Nancy Carolyn Tapscott, Apr. 30, 1960; children—Andrew Churchill, Rebecca Ann. Research physicist Lawrence Radiation Lab. U. Cal., Berkeley, 1956—; Fulbright lectr. U. Tokyo, 1957-58; vis. research prof. U. Grenoble (France), 1966-67. Mem. Am. Phys. Soc. Research on analysis and theory of scattering of particles by nuclei, determinations and analysis of polarization of protons scattered from few-nucleon systems, investigations of neutron interaction through studies involving three particles in final state of nuclear reaction. Home: 318 Vassar Av., Berkeley, Cal. 94708.*

COODE, Sir John, Brit. engr.; b. Bodmin, Eng., Nov. 11, 1816; s. Charles and Anne (Bennett) C.; ed. Bodmin Grammar Sch.; m. Jane Ruce, 1842. Cons. engr., Westminster, Eng., 1844-47; resident engr. in charge works at Portland Harbour, 1847, engr. in chief, 1856-72. Mem. Instn. Civil Engrs. (pres. 1889-91). Contbr. papers on harbors (Chesil Bank, 1852, especially valuable). Probably most distinguished harbor engr. of century, asso. with works at Colombo, Ceylon, 1874-85. Died 1892.

COOK, Adrian Maxwell, Brit. microbiologist; b. Hull, Eng., Sept. 1, 1915; s. Harold and Emily (Greendale) C.; student U. Nottingham (Eng.); B.Pharm., U. London, Ph.D., Dip.Bact.; m. Kathleen May Howard, Mar. 16, 1946. Faculty, U. London Sch. Pharmacy, 1946—, prof. pharm. microbiology. Chmn. research U.K. Panel on Gamma and Electron Sterilization. Mem. Soc. Chem. Industry (chmn. microbiology group), Royal Soc. Health (chmn. pharm. group), Internat. Assn. Microbiol. Socs., Federatione Internat. Pharmaceutique, Pharm. Soc., Royal Inst. Chemistry, Soc. Gen. Microbiology, Soc. Applied Bacteriology, Inst. Food Sci. and Tech. Research, publs. on dynamics and evaluation of bacteriocidal activity of chems., heat, and ionizing radiations. Home: Probyn House, 42, Coram St., London W.C.1., Eng.*

COOK, Albert John, Am. naturalist; b. Owosso, Mich., Aug. 30, 1842; s. Ezekiel and Barbara Ann (Hodge) C.; B.S., Mich. Agrl. Coll., 1862, M.S., 1865, D.Sc., 1905; studied at Harvard, 1867-88; m. Mary H. Baldwin, June 30, 1870; m. 2d, Sarah J. Eldredge, July 3, 1897. Instr. math. Mich. Agrl. Coll., 1867-69, prof. zoölogy and entomology, 1868-93, curator Gen. Mus., 1875-93, entomologist Expt. Sta., 1888-91; prof. biology Pomona Coll., Cal., 1893-1911; state commr. of horticulture Cal., 1911-16. Condr. extension work in agr., U. Cal., 1894-1905. Author: Manual of the Apiary; Injurious Insects of Michigan; Silo and Silage; Maple Sugar and the Sugar Bush; Birds of Michigan; California Citrus Culture, 1913. Died Sept. 29, 1916.

COOK, Arthur Bernard, archeologist; b. Oct. 22, 1868; s. William Henry and Harriet (Bickersteth) C.; ed. (Major scholar, Craven U. scholar) Trinity Coll., Cambridge, Eng.; Litt.D.; m. Emily Maddox, 1894; 1 dau. Prof. Greek, Bedford Coll., London, Eng., 1892-1907; reader in classical archeology U. Cambridge, 1908-31, Laurence prof. classical archeology, 1931-34. Fellow Trinity Coll., 1893-99; examiner in classical tripos, 1898-1904, 1908. Fellow Brit. Acad.; mem. Socs. for Promotion of Hellenic and Roman Studies, Classical Assn., Folk-Lore Soc., Cambridge Antiquarian Soc., Cambridge, Am. philol. socs., German Archeol. Inst. Author: The Metaphysical Basis of Plato's Ethics, 1895; Zeus, a Study in Ancient Religion, Vol. I, 1914, Vol. II, 1925, Vol. III, 1940; The Rise and Progress of Classical Archaeology, 1931; also many articles. Gen. editor Methuen's Handbooks of Archaeology, 1928-49. Died Cambridge, Apr. 26, 1952.

COOK, Arthur Herbert, English chemist; b. London, Eng., July 10, 1911; s. Arthur Bert and Catherine (Oliver) C.; B.Sc., Imperial Coll., U. London, 1932, Ph.D., Diploma, 1934, D.Sc., 1947; postgrad. Kaiser Wilhelm Inst. für Medizinische Forschung. Chemist, Imperial Chem. Industries, Ltd., 1936; faculty Imperial Coll. 1937-49, asst. prof., 1947-49; asst. dir. Brewing Industry Research Found., Nutfield, Surrey, Eng., 1949-57, dir., 1957—. Bd. dirs. A. D. Little Inst., Musselburgh, Scotland. Fellow Royal Soc., Royal Inst. Chemistry. Author: (with F. Mayer) Chemistry of Natural Coloring Matters, 1943; also numerous articles. Editor: Chemistry and Biology of Yeasts, 1958; Barley and Malt, 1962; editor-in-chief Dictionary of Organic Compounds, 5 vols., 1965. Research on chemistry and biochemistry of natural materials including pigments, penicillin, brewing and fermentation. Address: Lyttel Hall, Nutfield, Surrey, Gt. Britain.*

COOK, Charles Davenport, Am. pediatrician; b. Mpls., Nov. 30, 1919; s. Henry Wireman and Ellen (Davenport) C.; B.A. summa cum laude, Princeton, 1941; M.D. cum laude Harvard, 1944; M.A. (hon), Yale, 1964; m. Sheila Gamble, Mar. 10, 1945; children—Andrew D., Sheila D., Peter G., Charles Davenport. Faculty, Harvard Med. Sch., 1949-64; mem. staff Children's Med. Center, Boston, 1952-64, physician, 1958-64; prof. pediatrics, chmn. dept. Yale, 1964—. Diplomate Am. Bd. Pediatrics. Mem. Am. Acad. Pediatrics, Am. Pediatric Soc., Am. Physiologic Soc., Am. Thoracic Soc., Assn. Pediatric Ambulatory Services, Soc. Pediatric Research. Mem. editorial bd. Jour. Pediatrics, 1960, trustee ednl. found., 1961-63, chmn., 1963. Research, publs. on pulmonary physiology and diseases in children; fetal physiology and the devel. of lung. Home: 176 Armory St., New Haven 06511. Office: 333 Cedar St., New Haven 06510.*

COOK, C(harles) Wayne, Am. range scientist; b. Gove, Kan., Oct. 28, 1914; s. Carl F. and Mabel (Davenport) C.; B.S. in Botany, Ft. Hays Kan. State Coll., 1940; M.S. in Range Sci., Utah State U., 1942; Ph.D., Tex. A. and M. U., 1950; m. Eula Lewis, June 8, 1940; 1 son, Charles Randall. With Soil Conservation Service, 1942-43; faculty Utah State U., Logan, 1946—, asst. dean Coll. Natural Resources, 1965—. Recipient Hoblitzelle Nat. award in agrl. scis., 1953. Mem. Am. Soc. Range Mgmt. (pres. 1967—), Am. Soc. Animal Sci., Soil Conservation Soc. Am., Sigma Xi, Phi Kappa Phi, Delta Epsilon, Xi Sigma Pi, Alpha Zeta. Author: Basic Problems and Techniques in Range Research, 1962; contbg. author Pasture and Range Research Techniques, 1962; Forages, 1962. Research, numerous publs. on nutritional qualities of all range types of intermountain area, methods and techniques in use of herbicides and mech. means of controlling noxious range weeds, prins. in mgmt. methods for rehabilitating and using range lands. Home: 1373 Maple Dr., Logan, Utah 84321.*

COOK, Clarence Sharp, Am. physicist, govt. ofcl.; St. Louis, Crossing, Ind., Aug. 18, 1918; s. Clarence C. and Musa (Sharp) C.; A.B., DePauw U., 1940; M.A. in Physics, Ind. U., 1942, Ph.D. in Physics, 1948; m. Marian N. Waring, June 19, 1943; children—Sherma Louise, Wayne William. Teaching asst. physics Ind. U., 1940-42, research asst., 1946-48; asst. prof. physics Washington U., St. Louis, 1948-53; head nuclear radiation br. U. S. Naval Radiol. Def. Lab., San Francisco, 1953-60, head nucleonics div., 1960-61, physics cons. to sci. dir., 1962—. Mem. bd. Civil Service Examiners for Scientists and Engrs., Pasa-

dena, Cal., 1955-58, chmn., 1957-58. Fulbright research scholar Aarhus (Denmark) U., 1961-62. Fellow Am. Phys. Soc., Cal. Acad. Scis.; mem. Am. Assn. Physics Tchrs.; Am. Geophys. Union, A.A.A.S., Phi Beta Kappa, Sigma Xi. Author: Modern Atomic and Nuclear Physics, 1961; Structure of Atomic Nuclei, 1964. Contbr. articles on nuclear physics and geophysics to sci. jours. Home: 2711 Monterey St., San Mateo, Cal. 94403. Office: U. S. Naval Radiol. Def. Lab., San Francisco 94135.*

COOK, Clinton Dana, Am. chemist, univ. ofcl.; b. St. Johnsburg, Vt., Feb. 20, 1921; s. Clinton Dana and Anna (Kubauec) C.; S.B., Mass. Inst. Tech., 1942; M.S., U. Vt., 1948; Ph.D., Ohio State U., 1951; m. Alice MacLaren Fisher, May 21, 1944; children—Dana, Allison Anne, Polly M., Timothy F., Cynthia C. Group leader, gen. chmn. devel. West Lynn Works, Gen. Electric Co., 1942-46; faculty U. Vt., Burlington, 1946-48, 1952—, prof. chemistry, 1959—, chmn. dept. chemistry, 1960-63, dean faculties, 1963-65, v.p. for acad. affairs, 65—; asst. instr., Abbott fellow Ohio State U., 1948-51; supervising chemist liquid dielectric sect. Pittsfield Works, Gen. Electric Co., 1951-52. Cons. organic chemistry U. S. Rubber, 1956-59, instnl. grants div. NSF, 1965—. Mem. Am. Chem. Soc., Sigma Xi (pres. Vt. chpt. 1956-57). Contbr. articles to tech. jours. Research, publs. in phenoxy radicals and mechanism autoxidation inhibition, chemistry quinone methides. Home: 64 Bilodeau Ct., Burlington, Vt. 05401.*

COOK, Earl Ferguson, Am. geologist; b. Bellingham, Wash., May 24, 1920; s. Earl Ferguson and Helen (Royer) C.; B.S. in Mining Engring., U. Wash., 1943, M.S. in Geology, 1947, Ph.D. in Geology, 1954; student U. Paris (France), 1945-46; U. Geneva (Switzerland), 1948-49; m. Jean E. Wiltse, June 21, 1947 (div. 1964); children—Jeanette, Randall, and Cynthia. Instructor of geology at U. Wash., 1947-48, Stanford, 1948; geologist, Geophoto Services, Denver, 1949-51; mem. faculty U. Ida., 1951-64, dean Coll. Mines, prof. geology, 1957-64, also dir. Ida. Bur. Mines and Geology, 1957-64; exec. sec. div. earth scis. Nat. Acad. Scis.-Nat. Research Council, Washington, 1964—; geologist Gulf Oil Corp., summers 1955, 56. Served with AUS, 1943-46; ETO. Decorated Purple Heart, 3 battle stars; recipient of Award of Merit from Austrian Province of Burgenland, 1966. Member of Geological Society of America, American Inst. Mining, Metall. and Petroleum Engrs., Am. Assn. Petroleum Geologists, Soc. Econ. Geologists, Am. Geophys. Union, Am. Inst. Profl. Geologists, N.W. Sci. Assn. Demonstrated the great extent (up to 10,000 sq. miles) and usefulness in interpretation of geologic history and geologic structures of ignimbrites—individual sheets of volcanic rock formed by the rapid spreading of hot, turbulent, subaerial volcanic density currents. Home: 1330 New Hampshire Av. N.W., Washington 20036. Office: Nat. Acad. Scis., 2101 Constitution Av. N.W., Washington 20418.

COOK, Edward Noble, Am. physician; b. St. Paul, Aug. 21, 1905; s. Edward and Jessie Gertrude (Noble) C.; B.A. U. Minn., 1926, B.S., 1927, B. Medicine, 1928, M.D., 1929, M.S. in Urology 1935; m. Jean Elizabeth Moore, June 14, 1934; children—Margaret (Mrs. C. M. Berndt, Jr.), Edward Noble, Nancy. Intern Kings County Hosp., Bklyn., 1929-30; mem. staff Mayo Found., U. Minn., 1930—, prof. urology, 1958—; staff Mayo Clinic, 1935—, cons. urology, 1935—; cons. urology Meth., St. Mary's hosps.; spl. research infections urinary tract, transurethral surgery. Served as lt. M.C., USN, 1939-47. Member of American Urol. Assn., A.M.A. (sec. sect. urology 1946-49, chmn. sect. 1949-50), Minn. Med. Assn., Olmsted County Med. Soc., Sigma Xi, Delta Upsilon, Alpha Kappa Kappa. Clubs: Country, University. Contbr. profl. jours. Particular interest in study of urinary tract infections and transurethral prostatic surgery. Home: Crocus Hill, Salem Rd., Rochester. Office: 200 1st St. S.W., Rochester, Minn.

COOK, Elton Straus, Am. chemist; b. Oberlin, O., Dec. 24, 1909; s. Edward Monroe and Bertha (Straus) C.; B.A., summa cum laude, Oberlin Coll., 1930; Ph.D., Yale, 1933; m. Elizabeth Luck, June 1, 1935; children—Edward Mark, David Charles. Fellow, Yale, 1933-34; research chemist, head dept. organic prodn. William S. Merrell Co., 1934-37; prof., head div. chemistry and biochemistry Institutum Divi Thomae, Cin., 1937—, asst. dir., 1943-45, dean, 1945—, v.p., trustee, 1955—. Recipient Diploma of Honor, Pan Am. Cancer Cytology Congress, 1957; award Gordon Research Conf., 1959. Fellow Am. Inst. Chemists (dist. dir. 1953—), Chem. Soc. London, A.A.A.S.; mem. Am. Chem. Soc. (Cin. Chemist award 1964), History of Sci. Soc., Soc. for Exptl. Biology and Medicine, Am. Assn. Cancer Research, N.Y. Acad. Sci., Internat. Cancer Congresses. Research, numerous publs. chemistry and pharmacology synthetic drugs, mechanism drug action in terms metabolism and enzymes, tissue injury in terms chem. factors affecting growth and metabolism with resulting therapeutic application, growth and metabolism cancer in terms accelerating and inhibiting factors, chemistry antibacterial substance in tissue; co-developer potassium bromide pellet procedure for infrared analysis. Home: 6503 Park Lane, Cin. 45227. Office: 1842 Madison Rd., Cin. 45206.*

COOK, Frederick Albert, Am. physician, explorer; b. Callicoon Depot, N.Y., June 10, 1865; s. Theodore Albert and Magdalene C.; M.D., U. N.Y., 1890; m. Mary Fidell Hunt, June 10, 1902. Surgeon, Peary Arctic expdn., 1891-92; surgeon Belgium Antarctic expdn., 1897-99, led expdns. to explore and climb Mt. McKinley, 1903-06. Recipient Gold medal Royal Soc., Belgium; Silver medal Royal Geog. Soc., Belgium. Mem. Am. Nat., Phila. geog. Socs., Kings County Med. Soc. Author: Through the First Antarctic Night, 1900; To the Top of the Continent, 1907; My Attainment of the Pole, 1909. His claim that he reached N. Pole Apr. 21, 1900, rejected by scientists in Copenhagen on grounds of insufficient evidence. Died Aug. 5, 1940.

COOK, Gerald Bernard, English chemist; b. London, Eng., Jan. 10, 1921; s. William George and Florence (Harris) C.; B.Sc. with 1st class honors, Birmingham (Eng.) U., 1941; Ph.D., Cambridge (Eng.) U., 1951; m. Nancy Combe Jackson, Oct. 11, 1943; children—Gillian Margaret, Deborah Frances, Fione Catrine. Staff, Cavendish Lab., Ministry of Supply, Cambridge, Eng., 1941-44; staff Joint Canadian-U.K. Atomic Energy Project, Montreal, Que., Can., also Chalk River, Ont., Can., 1944-47; staff U.K. Atomic Energy Authority, Harwell, 1947-60; head lab. IAEA, Vienna, Austria, 1960—. Author: (with J. F. Duncan) Modern Radiochemical Practice, 1952, Isotopes in Chemistry, 1967; also articles. Research on fission products of uranium and plutonium, indsl., sci. and med. uses of radioisotopes; analytical chemistry of trace inorganic elements in uranium. Home: 1 Homer Markt, Vienna I. Office: 11 Kärntnerring, Vienna I, Austria.*

COOK, Gilbert Richard, Am. physicist; b. Washington, Apr. 7, 1916; s. Gilbert Richard and Doris A. (Frederick) C.; B.S., Carnegie Inst. Tech., 1940; postgrad. U. Cal. at Los Angeles, M.S., U. So. Cal., 1954, Ph.D., 1959; m. Virginia Beatrice Blackmun, Dec. 11, 1964; stepchildren—Elin, Esther, Richard. Mem. engring. staff Western Electric Co., Kearny, N.J., 1940-41, Pacific Tel.&Tel. Co., Los Angeles, 1945-50; research asst. U. So. Cal., 1954-56, research asso., 1956-59; tech. staff TRW Space Tech. Labs., Los Angeles, 1959-60, Aerospace Corp., Los Angeles, 1960—. Mem. Am. Phys. Soc., Optical Soc. Am., Am. Geophys. Union, N.Y. Acad. Scis. Research, publs. on vacuum ultra-violet radiation physics, cross sects. for photo-absorption and photo-ionization processes, upper atmosphere physics. Home: 5679 Beaumont Av., La Jolla, Cal. 92037. Office: P.O. Box 95085, Los Angeles 90045.*

COOK, James, Brit. navigator; b. Marton village, Cleveland, Yorkshire, Eng., Oct. 28, 1728. Apprentice haberdasher, Staithes, Eng., 1740; apprentice, later mate Walker firm of shipowners, to 1755; with Royal Navy, from 1755, master, from 1759; marine surveyor, coast Newfoundland, Labrador, 1763; made 3 exploratory voyages into Pacific, 1768, 1772-75, 1776-79. Recipient medal Royal Soc. Fellow Royal Soc., 1776. Discovered Admiralty Islands, Society Islands; rediscovered Hawaiian Islands (which he named Sandwich Islands); observed solar eclipse at Burges Island, 1766, transit of Venus from Tahiti, 1769; explored coasts of New Zealand, Australia; circumnavigated globe. Died Kealakekua Bay, Hawaii, Feb. 14, 1779.

COOK, James Wilfred, chemist; b. London, Eng., Dec. 10, 1900; s. Charles William and Frances (Wall) C.; Ph.D., Univ. Coll., London, 1923, D.Sc., 1925; Sc.D., U. Dublin (Ireland), U. Nigeria; D., U. Rennes (France); LL.D., U. Exeter, 1967; m. Elsie Winifred Griffiths, Sept. 27, 1930 (dec. Apr. 1966); children—Geoffrey Michael, Laurence Martin, Robert James; m. Vera Elizabeth Ford, Mar. 25, 1967. Lectr. organic chemistry Sir John Cass Inst., 1920-28, chemist dept. sci. and indsl. research, 1928-29; research chemist Royal Cancer Hosp., London, 1929-39; prof. chemistry U. London, 1935-39, U. Glasgow (Scotland), 1939-54; prin. U. Coll., Exeter, Eng., 1954-55; vice chancellor U. Exeter, 1955-66; vice chancellor U. East Africa, Kampala, Uganda, 1966—. Chmn., mem. numerous U.K. Govt. coms. Decorated Officier de l'Ordre de Léopold (Belgium); named Knight Bachelor, 1963. Fellow Royal Soc., 1938; mem. Chem. Soc. (past v.p.), Royal Inst. Chemistry (past pres.). Editor: Progress in Organic Chemistry, vols. I-VII. Research, numerous publs. on discovery synthetic carcinogens, research on biologically active organic chems. Home: 3 Borup Av., P.O. Box 7110, Kampala, Uganda.*

COOK, John Call, geophysicist; b. Afton, Wyo., Apr. 7, 1918; s. Carl and Ella (Call) C.; B.S. in Physics with honors, U. Utah, 1942; M.S., Pa. State U., 1947, Ph.D. in Geophysics, 1951; m. Violet Kathryn Myers, July 29, 1949; children—Carl A., Eliot V. Staff, Mass. Inst. Tech. Radiation Lab., 1942-45; teaching asst., research asso. Pa. State U., 1945-51; sr. physicist S.W. Research Inst., San Antonio, 1951-64; chief geophysicist, dept. mgr. Geotech div. Teledyne Industries, Inc., Dallas, 1964—; Antarctic seismologist Arctic Inst. N.Am., Internat. Geophys. Year, 1957-58; asst. prof. physics Trinity U., San Antonio, 1958-59; asso. geophysicist Scripps Oceanographic Inst., 1961; vis. investigator Woods Hole (Mass.) Oceanographic Inst., 1964. Mem. Am. Phys. Soc., Am. Geophys. Union, Soc. Exploration Geophysicists, Sigma Xi, Sigma Pi Sigma. Contbr. articles on applied physics and geophysics to tech. jours. Research on natural gamma-radiation field, unorthodox geophys. prospecting methods, remote sensing by electromagnetic, acoustic and seismic, thermal and optical means, soil magnetism, high temperature methods and equipment, lasers, gravity instruments; invented electrostatic crevasse detector. Home: 9109 Leaside Dr., Dallas 75238. Office: 3401 Shiloh Rd., Garland, Tex. 75041.*

COOK, Maurice, English metallurgist; b. Hartlepool, Eng., Dec. 24, 1897; s. Andrew and Mary (Ridley) C.; B.Sc., Manchester (Eng.) U., 1919, M.Sc., 1920, D.Sc., 1940; Ph.D., Kings Coll., Cambridge (Eng.) U., 1923; m. Irene Wells, Aug. 23, 1921. Lectr., Manchester U., 1919-21; metallurgist C. A. Parsons & Co., 1924-26; with Imperial Chem. Industries Ltd., 1926-59, research dir., 1942-51, mng. dir., 1951-57, chmn., 1957-59; chmn. Steatite & Porcelain Products Ltd., 1946-57. Decorated Comdr. Order Brit. Empire; recipient Manchester U. Leblanc medal, 1919; Hon. City and Guilds London Insignia award, 1961; named hon. asso. Coll. Tech., Birmingham, Eng., 1950. Fellow Inst. Metals; mem. Instn. Metallurgists (founder, fellow, past pres.), Inst. Metals (past pres. Platinum medal 1957, Robertson medal 1959), Brit. Non-Ferrous Metals Research Assn. (past chmn.), Am. Inst. Mining and Metall. Engrs. (hon. life), Am. Soc. Metals (Distinguished life). Research, numerous publs. on alloy systems, phys. and mech. properties of non-ferrous metals and alloys, metal melting and casting, deformation, recrystallization, grain-growth and annealing of metals, testing of materials, welding. Home: 48 Rocky Lane, Perry Barr, Birmingham. 22B, Eng.*

COOK, Melville Thurston, Am. botanist; b. Coffeen, Ill., Sept. 20, 1869; s. William Harvey (M.D.) and Elizabeth Frances (Robinson) C.; student, DePauw U., 1888-89, 1891-93, A.M., 1902, Sc.D., 1940; A.B., Leland Stanford Jr. U., 1894; Marine Biol. Lab., Woods Hole, Mass., 1896, 99, 1900; U. Chgo., summers, 1897- 98; Ohio State Lab., Sandusky, summers, 1902, 03; Ph.D., Ohio State U., 1904; Sc.D., U. P.R., 1940; fellow New York Bot. Garden, 1906-07; m. Dora Reavill, Sept. 8, 1897; children—Harvey Reavill, Harold Thurston, Elizabeth (Mrs. Harry A. Ross). Prin. high sch. Vandalia, Ill., 1894-95; instr. in biology De Pauw U., 1895-97, prof. 1897-1904; chief dept. plant pathology and econ. entomology Estación Central Agronómica, Santiago de las Vegas, Cuba, 1904-06; plant pathologist Del. Agrl. Expt. Sta., Newark, 1907-11; state plant pathologist of N.J., and prof. plant pathology Rutgers Coll., 1911-23; plant pathologist N.J. Agrl. Expt. Sta., 1911-23, Insular Expt. Station, Rio Piedras, P.R., 1923-40, retired 1940; vis. prof. botany La. State U., 1944; editor Journal of Dept. Agr. of P.R., 1928-40, research lectr. sch. tropical medicine, 1926; lectr. embryology, Central Coll. Phys. and Surg., Indpls., 1902-03; on comparative anatomy Med. Coll. Ind., 1903-04. Fellow A.A.A.S., (v.p. 1921, chmn. bot. sect. 1921), Bot. Soc. Am.; mem. Am. Phytopathol. Soc. (pres. 1917), Ecol. Soc. Am., Am. Assn. Econ. Entomologists, Entomol. Soc. Am., Sigma Xi, Phi Beta Kappa, Pi Gamma Mu, Gamma Sigma Delta. Author: Diseases of Tropical Plants, 1912; Applied Economic Botany, 1919; College Botany, 1920; Los Enfermedades de las Plantas Económicas de las Antillas 1939; Los virusos de las plantas, 1943; Viruses and Virus Diseases of Plants, 1947. Contbr. to bot. jours. on plant pathology and econ. botany. Died Washington, D.C., Aug. 11, 1952.

COOK, Melvin Alonzo, Am. phys. chemist; b. Garden City, Utah, Oct. 10, 1911; s. Alonzo Laker and Alice Maude (Osmond) C.; B.A., U. Utah, 1933, M.A., 1934; Ph.D. (Loomis fellow), Yale, 1937; m. Wanda Garfield, June 19, 1935; children—Barbara Jean (Mrs. Stanley Kieth Petersen), Melvin Garfield, Virginia (Mrs. Gill Oldroyd Sanders), Merrill Alonzo, Krehl Osmond. Began career as research chemist Easter Lab., E. I. duPont de Nemours & Co., Gibbstown, N.J., 1937-47; prof. metallurgy U. Utah, Salt Lake City, 1947—, dir. Inst. Metals and Explosives Research, 1951-64, prof. mech. engring., 1965—; pres. IRECO Chems. and subsidiaries, Salt Lake City, 1958—. Cons. to numerous pvt. cos.; mem. adv. council Picatinny Arsenal, 1962-64; Reynolds lectr. U. Utah, 1952. Mem. Am. Chem. Soc. (Utah award 1961, E. V. Murphree award 1968), Am. Phys. Soc., N.Y. acads. scis., Utah Acad. (Sci. award 1954). Author: The Science of High Explosives, 1958; Prehistory and Earth Models, 1965; Science and Mormonism, 1967. also numerous articles. Patentee explosives. Research on theory detonation, adsorption gases on solids, soap chemistry, chem. bonds, solid state, quantum mechanics, continental drift, geol. chronometry; designed explosive devices, explosives; formulated free acid-base theory flotation. Home: 631 16th Av., Salt Lake City 84103.*

COOK, Orator Fuller, Jr., Am. botanist; b. Clyde, N.Y., May 28, 1867; s. Orator Fuller and Eliza (Hookway) C.; Ph.B., Syracuse U., 1890, D.Sc. (hon.), 1930; m. Alice Carter, Oct. 11, 1892. In charge dept. biology Syracuse U., 1890-91; made (1891-97) extended visits to Liberia for exploration and investigation as agt. N.Y. State Colonization Soc.; prof. natural scis. in Liberia Coll., 1891-97, pres., 1896-97; secured extensive collection of plants and animals for investigation in U. S. Nat. Mus., custodian and asst. curator U. S. Nat. Mus., from 1898. Spl. agt. in charge plant importation U. S. Dept. Agr., 1898-1900, in charge of investigation in tropical agr., from 1900, visited P.R., Guatemala, Mexico, Costa Rica; prof. botany George Washington U., 1904; botanist in charge palm classification Fairchild Tropical Garden, Cocoanut Grove, Fla., 1937. Research and publs. on Liberia and Africa colonization, P.R., tropical agr., botany, zoölogy, evolution, history of cultivated plants, especially on breeding, acclimatization, and cultural improvement of cotton and rubber plants, also on classification of palms and millipeds; as botanist representing U. S. Dept. Agr. with Bingham expdn. to Peru under auspices Nat. Geog. Soc. and Yale U., 1915, investigated plants used by the Incas; expdn. to Haiti on agrl. exploration and study for improvement of agrl. conditions., 1917; expdn. to China for study of agrl. conditions, 1919; Carnegie Instn. expdn. to C.Am. to study ancient Maya civilization, 1922; expdns. to Haiti, Panama, Mexico, Colombia, Ecuador, to investigate native cottons and sources of rubber, 1923-31; contbd. improvements in cotton cultivation; helped discover new species of rubber trees; showed that they can be grown in Fla., south of Miami. Died Lanham, Md., Apr. 23, 1949.

COOK, Robert Manuel, Brit. archeologist; b. Sheffield, Eng., July 4, 1909; s. Charles Robert and Mary (Arnold) C.; B.A., Cambridge U.; m. Kathleen Porter, July 11, 1938. Asst. lectr. classics, U. Manchester, 1934-38, lectr., 1938-45; Laurence reader, archeology Cambridge U., 1945-62, Laurence prof. archeology and ancient art, 1962—. Mem. German Archeology Inst. Author: Greek Painted Pottery; The Greeks till Alexander. Home: 15 Wilberforce Rd. Office: Museum of Classical Archeology, Little St. Mary's Lane, Cambridge, Eng.

COOK, Thomas Bratton, Jr., Am. physicist; b. Richmond, Ky., Aug. 28, 1926; s. Thomas Bratton and Willie Ethel (Wilson) C.; B.S., Western Ky. State Coll., 1947; M.S., Vanderbilt U., 1949, Ph.D., 1951; m. Virginia Page Preston, Dec. 19, 1947; children—Thomas Bratton III, Shelley Lynn. Mem. staff weapons effects dept. Sandia Corp., Albuquerque, 1951-55, sec. supr., vulnerability studies sect., 1955-56, div. supr. nuclear burst studies div., 1956-59, mgr. nuclear burst physics dept., 1959-62, dir. nuclear burst physics and math. research, 1962-67, vice president of research; 1967—. Consultant to the Office of Dir. Def. Research and Engring., Aerospace Corp., Ballistic Systems div. USAF, Office of chief of research and devel. Dept. Army, Def. Atomic Support Agy.; mem. USAF Sci. Adv. Bd.; AEC chmn. Joint AEC Dept. Def. Weapons Effects Group. Fellow Am. Phys. Soc.; mem. Am. Ordnance Assn., Sigma Xi. Research in physics of high altitude nuclear explosions, with emphasis on effects of neutrons, gamma-rays, and X-rays.*

COOKE, C(harles) Montague, Jr., Am. zoologist; b. Honolulu, Dec. 20, 1874; s. Charles Montague and Anna Charlotte (Rice) C.; A.B., Yale, 1897, Ph.D., 1901; Sc.D., U. Hawaii, 1936; m. Eliza Lefferts, Apr. 25, 1901; children—Carolene Alexander, Charles Montague. Studied nearly all large collections of Hawaiian land shells in museums of Europe and Am.; malacologist with Bishop Mus., Honolulu. Mem. Washington Acad. Scis., Acad. Natural Scis., Phila., Malacological Soc., London. Died Oct. 29, 1948.

COOKE, Josiah Parsons, Am. chemist, educator; b. Boston, Oct. 12, 1827; s. Josiah and Mary (Pratt) C.; grad. Harvard, 1848, LL.D., 1889; LL.D. Cambridge (Eng.) U., 1882; m. Mary Huntington, 1860. Erving prof. chemistry and mineralogy Harvard, 1850-94. Mem. Nat. Acad. Scis., Am. Acad. Arts and Scis. (corr. sec. 1873-92, pres. 1892-94). Author: Elements of Chemical Physics, 1860; First Principles of Chemical Philosophy, 1868; The New Chemistry, 1872, rev. edit., 1884; contributions from the chemical Laboratory of Harvard College, 2 vols., 1877-1889; Chemical and Physical Researches, 1881. First coll. instr. to use lab. in undergrad. course; made 1st attempt to classify elements according to atomic weight; determined atomic weight of antimony, 1877-81; studied (with Richards) ratio of atomic weights of hydrogen and oxygen, 1887-89. Died Newport, R.I., Sept. 3, 1894.

COOKE, Kenneth Lloyd, Am. mathematician; b. Kansas City, Mo., Aug. 13, 1925; s. Sidney Kenneth and Mildred Blanche (Brown) C.; B.A., Pomona Coll., 1947; M.S., Stanford, 1949, Ph.D., 1952; m. Margaret Sarah Burgess, Aug. 18, 1950; children—Catherine Sarah, Robert Kenneth. Instr., then asst. prof. math. State Coll. Wash., Pullman, 1950-57; mem. faculty Pomona Coll., 1957—, Joseph N. Fiske prof. math., 1963—, chmn. dept., 1961—; cons. RAND Corp., 1956-65; mathematician Research Inst. Advanced Studies, Balt., 1963-64. NSF faculty fellow, 1966-67. Mem. Am. Math. Soc., Math. Assn. Am., Soc. Indsl. and Applied Math., Am. Assn. U. Profs., Phi Beta Kappa, Sigma Xi. Author: (with Richard Bellman) Differential-Difference Equations, 1963. Devel. theory of differential difference equations, or functional differential equations, and emphasizing their importance. Home: 654 Northwestern Dr., Claremont, Cal. 91711.*

COOKE, Lewis Henry, Brit. metallurgist; ed. Royal Sch. Mines, London, Eng.; prof. mine surveying Imperial Coll. Sci.; mem. bd. studies in mining and metallurgy U. London; asso. Royal Sch. Mines. Fellow Geol. Soc.; mem. Instn. Mining and Metallurgy (mem. council, recipient Consol. Goldfields premium for improvements in mine surveying methods and instruments 1913, Consol. Goldfields Gold medal 1925). Revised Ore and Stone Mining (C. Le Nove Foster). Research and publs. on mine surveying in coal mining, duration of Brit. coal output, determination of specific gravity of minerals and rocks in field, theory of conglomerate formation. Died Aug. 23, 1929.

COOKE, Mordecai Cubitt, botanist; b. Horning, Eng., July 12, 1825; M.A. (hon.) St. Lawrence U, 1870; M.A., Yale, 1873; LL.D., N.Y., 1874; m. 1846; 3 sons, 2 daus. Apprentice wholesale drapery trade; usher boys' sch.; lawyer's clerk; tchr. in a nat. sch.; with India Mus., 1860-80; with Royal Bot. Gardens, Kew, Eng., from 1880. Recipient Victoria Medal of Honor, Royal Hort. Soc. Asso. Linnean Soc. (Linnean Gold medal 1903). Author: Illustrations of British Fungi, 8 vols., 1881-91; Introduction to the Study of Fungi, 1895; Fungoid Pests of the Flower Gardens, 1906; 40 other bot. works. Authority on fungi; specialized in Cryptogamia. Died Oct. 27, 1913.

COOKE, Robert Edmond, Am. physician; b. Attleboro, Mass., Nov. 13, 1920; s. Ronald Melbourne and Renee (Wuillumier) C.; B.S., Sheffield Sci. Sch., 1941; M.D., Yale, 1944; m. Nancy Perry, Sept. 3, 1965; children—Christopher C., W. Robert, Kim Mc., Robyn M., Wendy W. Faculty, Yale, 1950-56, asso. prof. pediatrics, physiology, 1954-56; asso. pediatrician Grace-New Haven Hosp., 1951-56; prof. pediatrics Johns Hopkins, 1956—; pediatrician-in-chief Johns Hopkins Hosp., 1956—, Given Found. prof. pediatrics 1962—; Grover Powers prof. pediatrics Nat. Assn. Retarded Children, 1957-59. Cons. pediatrician Childrens' Hosp., Balt., Balt. City Hosps.; asso. cons. staff gen. pediatrics Sinai Hosp. Balt.; cons. NIH Clin. Center; mem. Pres.'s Panel on Mental Retardation, 1961-62; Pres.'s Com. on Mental Retardation, 1966—; mem. bd. visitors John F. Kennedy Child Devel. Center U. Colo. Med. Center; chmn. nat. steering com. Project Head Start, Office Economic Opportunity, 1965—; vis. prof. pediatrics Howard U., 1966—. Recipient Ramsay Meml. Scholarship prize, 1942, Perkins Scholarship prize, 1943, Parker prize, 1944, Campbell Gold medal, 1944, E. Mead Johnson award, 1954. Mem. Am. Acad. Pediatrics, Am. Assn. on Mental Deficiency, A.M.A., Am. Soc. Clin. Investigation, Am. Acad. Pediatrics (joint com. pediatric research, edn. and practice), Nat. Assn. Retarded Children, Soc. Pediatric Research (pres. 1965-66), Aurelian Honor Soc., Phi Beta Kappa, Sigma Xi, Alpha Omega Alpha. Author (with S. Osler) Biosocial Basis of Mental Retardation, 1966. Research, publs. on chemistry of body especially salt and water, mental retardation. Home: 401 Somerset Rd., Balt. 21210. Office: Children's Med. Surg. Center, Johns Hopkins Hosp., Balt. 21205.*

COOKE, Thomas, Brit. optician; b. Allerthorpe, Yorkshire, Eng., Mar. 8, 1807; student nav. Tchr., Allerthorpe, Eng., from 1823, York, Eng., 1829-36; builder astron. telescopes, from 1851; mfr. turret clocks. Mem. Royal Astron. Soc. Author: On a New Driving-clock for Equatorials, 1859. Perfected astron. clock; helped restore Brit. prestige in practical optics; invented appliances for facilitating telescopic observations; brought system of equatorial mounting nr. perfection; applied steam to grinding and polishing of lenses (prodn. made easy and cheap, insured uniform quality). Died Oct. 19, 1868.

COOKE, William Bridge, Am. mycologist; b. Foster, O., July 16, 1908; s. William Thomas Hunter and Katharine May (Bridge) C.; B.A., U. Cin., 1937; M.S., Ore. State U., 1939; Ph.D., Wash. State U., 1950; m. Vivian Greenwald, June 12, 1942. Research asso. dept. plant pathology Wash. State U., Pullman, 1950-51; mycologist in charge fungus studies Robert A. Taft San. Engring. Lab., U. S. Dept. Health Edn. and Welfare, Cin., 1952-66; mycologist Cin. Water Research Lab., Fed. Water Pollution Control Adminstrn., Dept. Health Edn. and Welfare, 1965-66, U. S. Dept. Interior, Cin., 1966—. Fellow A.A.A.S., Ohio Acad. Sci., Am. Acad. Microbiology (charter); mem. Mycol. Soc. Am., Ecol. Soc. Am., Am. Inst. Biol. Scis., Bot. Soc. Am., Am. Soc. Plant Taxonomy, Am. Soc. Agronomy, Internat. Assn. Plant Taxonomists, Cal. Bot. Soc., Brit. Mycol. Soc., Internat. Soc. Human and Animal Micol. Author: A Laboratory Guide to Culture and Identification of Sewage Fungi, 1963. Research, publs. list of flora and fungi of Mt. Shasta, Cal. Developed techniques for demonstrating presence of fungus populations in sewage and polluted waters. Home: 1135 Wilshire Ct., Cin. 45230. Office: 4676 Columbia Pkwy., Cin. 45226.

COOKE, William Ernest, Australian astronomer, meteorologist; b. July 25, 1863; s. Ebenezer Cooke; M.A., Adelaide (Australia) U., 1889; m. Jessie Greayer, 1887; 3 sons, 2 daus. Became 1st asst. Adelaide Obs., 1883; govt. astronomer Western Australia, 1896-1912, New S. Wales, 1912-26; prof. astronomy Sydney (Australia) U.; ret., 1926. Author: Climate of Western Australia; Annual Meteorological Reports of Western Australia; Perth Catalogue of Standard Stars, 31°-41° S.; Meridian Observations, vols. 2-5; Sydney Catalogue of Intermediate Stars; also astrographic catalogues, numerous papers. Died Nov. 7, 1947.

COOKE, Sir William Fothergill, Brit. elec. engr.; b. Ealing, Middlesex, Eng., 1806; ed. Durham (Eng.) U., U. Edinburgh (Scotland), also studied medicine

Paris, Heidelberg; Brit. Army officer, India, 1826-31. Recipient Albert Gold medal Soc. Arts, 1867. Was shown principle of electric telegraphy by prof. Muncke, 1836; (with Sir Charles Wheatstone) patented telegraphic apparatus, 1837, produced workable instrument, 1845 (soon adopted by all ry. lines in Eng.). Died June 25, 1879.

COOKE, William Trevor, English physician; b. Walsall, Eng.; s. John George and Eleanor (Farrington) C.; B.A., Sidney Sussex Coll., Cambridge U., 1932, M.B., B.Chirurg., 1935, M.D., 1940; M.D. with honors, U. Birmingham (Eng.), 1955; m. Margaret Annie Michael Foxell, Sept. 14, 1944; children—Ann May (Mrs. Campbell Curtis), Alison Jane. Med. registrar Queens Hosp., Birmingham, 1936-37, 40-45, asst. dir. med. research, 1945-46; research fellow Harvard, Caroline Harold fellow Birmingham U., 1938-39; reader medicine Birmingham U., 1946—; cons. physician Birmingham United Hosp., 1946—. Fellow Royal Coll. Physicians; mem. Assn. Physicians, Brit. Soc. Gastroenterology, Physiol. Soc., Med. Research Soc., Australian Gastroenterology Soc. (hon.), Brazilian Soc. Gastroenterology and Nutrition. Research, numerous publs. on renal problems of water and electrolytic excretion, small and large intestines, especially prevention and delineation of adult celiac disease and its protean secondary manifestations in hematology, bone metabolism, nutritional deficiency. Home: 33 St. Marys Rd., Birmingham 17. Office: Gen. Hosp., Birmingham, Eng.*

COOKE-YARBOROUGH, Edmund Harry, English physicist; b. Campsall, Yorkshire, Eng., Dec. 25, 1918; s. George Eustace and Daphne (Wrinch) C.-Y.; M.A., Oxford (Eng.) U., 1946; m. Anthea Dixon, June 7, 1952; children—Anthony, Jane. Staff, Telecommunications Research Establishment, 1940-49; with Atomic Energy Research Establishment, Harwell, Eng., 1949—, head electronics and applied physics div., 1957—. Fellow Instn. Elec. Engrs., Inst. Physics; mem. I.E.E.E. Author: An Introduction to Transistor Circuits, 1957; also articles. Research on non-linear circuits, on-line data processing, nuclear instrumentation, direct conversion of heat to electricity. Office: Electronics and Applied Physics Div., Atomic Energy Research Establishment, Harwell, Didcot, Berks., Eng.*

COOKSON, Isabel Clifton, Australian paleobotanist; b. Melbourne, Australia, Dec. 25, 1893; d. John and Elizabeth (Summers) C.; B.Sc., U. Melbourne, 1916, D.Sc., 1932; Cotton Research scholar U. Manchester, 1926-27. Demonstrator in botany U. Melbourne, 1916-23, lectr., 1930-47, research fellow, 1948-59, part time research fellow, 1960—. Mem. Am. Bot. Soc. (corr.), Paleontol. Soc. India (hon.), Royal Soc. Victoria (life). Publs. on Victorian early Paleozoic plants, Tertiary plants from Victorian brown coal deposits; Australian Tertiary and Mesozoic pollen, spores and microplankton. Home: 154 Power St., Hawthorn 3122, Victoria, Australia.*

COOLEY, Charles Horton, Am. sociologist; b. Ann Arbor, Mich., Aug. 17, 1864; s. Thomas M. and Mary E. (Horton) C.; A.B., U. Mich., 1887, Ph.D., 1894; m. Elsie Jones, July 24, 1890; children—Rutger Horton, Margaret Horton (Mrs. James A. Kennedy, dec.). Mary Elizabeth. Statis. work in Interstate Commerce and Census Bur., Washington, 1889-91; asst. polit. economy, U. Mich., 1892-95, instr. sociology, 1895-99, asst. prof., 1899-1904, prof. from 1904. Mem. Am. Sociol. Soc. (pres. 1918). Author: Human Nature and the Social Order, 1902; Social Organization, 1909; Social Process, 1918; Life and the Student, 1927. Pioneer in field of sociology; saw human society as an organism; his views on reciprocal influence of the individual on society and of society on the individual remain basic to sociol. sci. Died May 8, 1929.

COOLEY, Denton A., Am. physician; b. Houston, Aug. 22, 1920; s. Ralph C. and Mary (Fraley) C.; B.A., U. Tex., 1941; M.D., Johns Hopkins, 1944; m. Louise Goldsborough Thomas, Jan. 15, 1949; children—Mary Fraley, Susan Mathias, Louise Goldsborough, Florence Talbot, Helen Thomas. Sr. surg. registrar Brompton Hosp., London, England, 1950-51; with Baylor U. Coll. Medicine, Houston, 1951—, prof., 1963—; chief cardiovascular surgery service St. Luke's Tex. Children's hosps.; attending surgeon Meth., Ben Taub Gen. hosps.; cons. in surgery Hermann, VA hosps. Recipient Distinguished Service certificate Houston Heart Assn., 1959. Mem. A.C.S. (gov.), Am. Bd. Thoracic Surgery (dir.), Am. Pan-Pacific, So., Western surg. assns., Tex., Houston surg. socs., Am. Coll. Cardiology, Am. Coll. Chest Physicians, Am. Assn. Thoracic Surgery, So., Tex., Am. med. assns., Tex. Acad. Sci., numerous others. Publs. on devel. disposable heart lung machine; one of first surgeons to institute hemodilution technique for open heart surgery; techniques of cardiovascular surgery. Home: 3014 Del Monte Dr., Houston 77019.*

COOLEY, Thomas Benton, Am. pediatrist; b. Ann Arbor, Mich., June 23, 1871; s. Thomas McIntyre and Mary Elizabeth (Horton) C.; A.B., U. Mich., 1891, M.D., 1895; D.Sc., 1940; postgrad. Harvard, Germany; m. Abigail Hubbard, Dec. 21, 1903; children—Emily Holland, Thomas McIntyre. Instr. med. dept. U. Mich., 1898-1900, asst. prof. hygiene, 1903-05; in pvt. practice, specializing in diseases of children, Detroit, from 1905; asst. chief Children's Bur., A.R.C., France, 1918-19; chief of pediatric service and chmn. staff Children's Hosp. of Mich., 1921-41; prof. and head dept. pediatrics Wayne U. Coll. Medicine, 1936-41, emeritus from 1941, Mem. Am. Inst. Nutrition, Soc. for Research in Child Devel. Soc. for Pediatric Research, Mich. State, Wayne County med. socs., Am. Acad. Pediatrics (pres.), Am. Pediatric Soc. (pres.), Detroit Acad. Medicine, Sigma Xi. Contbr. to Abt's Pediatrics, Am. Jour. Diseases Children, Brennemann's Pediatrics, Jour. Pediatrics. Reported (with E. R. Witwer and O. P. Lee) familial erythroblastic anemia (thalassemia, or Cooley's anemia), in children in Mediterranean area, 1927. Died Oct. 13, 1945.

COOLIDGE, Julian Lowell, Am. mathematician; b. Brookline, Mass., Sept. 28, 1873; s. Joseph Randolph and Julia (Gardner) C.; A.B., Harvard, 1895, LL.D., 1940; B.Sc., Oxford, 1897; Ph.D., U. Bonn (Germany), 1904; D.Sc., Lehigh U., 1938; m. Theresa Reynolds, Jan. 17, 1901; children—Jane Revere, Julian Gardner, Archibald Cary, Margaret Wendell, Elizabeth Peabody, Rachel Revere, John Phillips, Theresa Reynolds, Tchr. math. Groton (Mass.) Sch., 1897-99; instr. math. Harvard, 1900, asst. prof., 1908, prof., 1918-40, prof. emeritus, 1940. Fellow Am. Acad. Arts and Sciences; mem. Am. Math. Soc. (past v.p.), Math. Assn. Am. (pres. 1925), Assn. Math. Teachers in N.E. (past pres.), Phi Beta Kappa. Author: Elements of Non-Euclidean Geometry, 1909; Treatise on the Circle and the Sphere, 1916; Geometry of the Complex Domain, 1924; Introduction to Mathematical Probability, 1925; Algebraic Plane Curves, 1931; History of Geometrical Methods, 1940; History of the Conic Sections, 1943. Died Mar. 5, 1954.

COOLIDGE, William David, Am. physicist; b. Hudson, Mass., Oct. 23, 1873; s. Albert Edward and Martha Alice (Shattuck) C.; B.S., Mass. Inst. Tech., 1896; Ph.D., U. Leipsic, 1899; D.Sc., Union Coll., 1927, Lehigh U., 1927; M.D., U. Zurich, 1937; LL.D., Ursinus Coll., 1942; m. Ethel Woodard, Dec. 30, 1908 (dec. Feb. 1915); children—Elizabeth Belknap, Lawrence David; m. 2d, Dorothy Elizabeth MacHaffie, Feb. 29, 1916. Asst. prof. physics Mass. Inst. Tech., 1904-05; with Gen. Electric Co., 1905-61, v.p., dir. research, 1940-44, X-ray cons., 1945-61. Mem. vis. com. U. S. Bur. Standards, 1935-49; mem. Nat. Inventors Council, 1940-57, Nat. Def. Research Commn., World War II, Nat. Acad. Sci. Com. on Atom Bomb Project, 1941. Recipient Potts medal Franklin Inst., 1926, Franklin medal 1944; Hughes medal Royal Soc. London, 1927; Faraday medal Inst. Elec. Engrs. of Eng., 1939; Modern Pioneer award Nat. Mfg. Assn., 1940; Duddell medal Phys. Soc. Eng., 1942; Roentgen medal Friends of Germ. R. Mus., 1963. Mem. Nat. Acad. Scis., Am. Acad. Arts and Scis. (Rumford medal 1914), A.A.A.S., Am. Chem. Soc., Am. Inst. E.E. (Edison medal 1927), Am. Phys. Soc., Sigma Xi, Eta Kappa Nu. Research, numerous publs. on elec. conductivity of aqueous solutions at high temperatures; discovery, devel. method to make tungsten strong and ductile, and applications; inventor new form of X-ray tube (known as Coolidge tube), 1913. Home: 1480 Lenox Rd., Schenectady, 12308.

COOMBS, Howard Abbott, Am. geologist; b. Dallas, Apr. 10, 1906; s. Horace Milton and Anola (Sigerfoose) C.; B.S., M.S., U. Wash., Ph.D., 1936; m. Leila Ewing, Jan. 1, 1936; 1 dau., Carol Leigh. Faculty, U. Wash., Seattle, 1935—, prof., 1950—, chmn. geology dept., 1954—. Collaborator, U. S. Coast Survey State Wash. on earthquakes, 1952—; cons. numerous dams in U.S., Japan, Can. Fellow Geol. Soc. Am. Research, publs. on active volcanoes of world, Wash. earthquakes. Home: 3856 46th St. N.E., Seattle 98105.*

COON, Carleton Stevens, Am. anthropologist; b. Wakefield, Mass., June 23, 1904; s. John Lewis and Bessie (Carleton) C.; grad. Phillips Acad., 1921; B.A. Magna cum laude Harvard, 1925, M.A., Ph.D., 1928; m. Mary Goodale, 1926 (div. 1944); children—Carleton S., Charles A.; m. 2d, Lisa Dougherty Geddes, Mar. 31, 1945. Faculty, Harvard, 1928-42, prof. anthropology, 1945-48; prof., curator U. Pa., Phila., 1948-63; research curator U. Mus., Phila., 1963—. Cons. Scott Foresman & Co., 1956-66; mem. com. on edn. NSF, 1958-63; mem. Smithsonian Com. on use Soft Funds, 1965—. Recipient Viking medal Wenner-Gren Found., 1952; Gold medal Phila. Athenaeum, 1963. Mem. Nat. Acad. Scis., Am. Acad. Sci., N.Y. Acad. Scis., Am. Anthropol. Soc., Am. Acad. Phys. Anthropology. Author numerous books including: Caravan, 1951, 58; The Story of Man, 1954, 62; The Seven Caves, 1957; The Origin of Races, 1962; The Living Races of Man, 1965; also numerous articles. Research on cultures Berbers N. Africa, Albanians, peoples Middle E. from Morocco to Pakistan, also archaeology same region, racial history and racial classification of peoples of world, human evolution and modern racial classification, cave excavations in W. Africa. Address: 207 Concord St., Gloucester, Mass. 01930.*

COON, James Huntington, Am. physicist; b. Liberty, Mo., Nov. 9, 1914; s. Raymond H. and Mayme (Bryan) C.; A.B., U. Ind., 1937; Ph.D., U. Chgo., 1942; m. Joan Worth Newman, July 1, 1955; children—Leslie James, Ansley Bryan, Hilary Huntington. Research staff U. Wis., 1947; staff Los Alamos Sci. Lab., 1943-46, 1948—, group leader, 1950—. Mem. Am. Phys. Soc., Am. Geophys. Union. Research, publs. on low energy nuclear physics, nuclear weapons testing, nuclear test detection, space physics. Home: 116 Venado St. Office: Los Alamos Sci. Lab., Los Alamos 87544.*

COON, Julius Mosher, Am. pharmacologist, educator; b. Liberty, Mo., Oct. 29, 1910; s. Raymond Huntington and Mayme (Bryan) C.; A.B. in Chemistry, Ind. U., 1932; Ph.D. in Pharmacology, U. Chgo., 1938; M.D., U. Ill., 1945; m. Mary Elizabeth Bond, July 26, 1947; children—James Starner, Margaret Bryan. Faculty, U. Chgo., 1939-45, 1946-53; with FDA, 1946; prof., head dept. pharmacology Jefferson Med. Coll., Phila., 1953—. Mem. food protection com. Nat. Acad. Sci.-NRC, 1952—; cons. USPHS, NIH, 1958-64, Food and Drug Adminstrn., 1966—, Walter Reed Army Inst. Research, 1966—; mem. expert adv. panel on food additives WHO, 1966—. Member of the American Society Pharmacology and Exptl. Therapeutics, Soc. Toxicology, Soc. Exptl. Biology and Medicine, Am. Inds. Hygiene Assn., Radiation Research Soc., N.Y. Acad. Scis., Coll. Physicians Phila. Editorial bd. Jour. Pharmacology, 1954-57, Clin. Medicine, 1963—, Toxicology and Applied Pharmacology, 1967—. Research on nicotine axon reflexes in skin, bioassay of pituitary, chem. protection against radiation, toxicologic interactions of insecticides, food toxicology. Home: 130 Summit Av., Jenkintown, Pa. 19046. Office: 1025 Walnut St., Phila. 19107.*

COON, Minor J., Am. biochemist; b. Englewood, Colo., July 29, 1921; s. Minor D. and Mary (Jesser) C.; B.A. in Chemistry, U. Colo., 1943; Ph.D. in Biochemistry, U. Ill., 1946; m. Mary Lou Newburn, June 27, 1948; children—Lawrence, Susan. Instr., U. Pa., 1947-49, asst. prof., 1952-53, asso. prof., 1955, prof. physiol. chemistry, 1955; prof. biochemistry U. Mich., Ann Arbor, 1955—; USPHS Research fellow Nat. Cancer Inst., 1952-53. Mem. adv. council Life Ins. Med. Research Fund., 1961-65; mem. biochemistry study sect. NIH, 1962-65. Recipient Paul Lewis award in enzyme chemistry, 1959. Mem. Am. Soc. Biol. Chemists, Phi Beta Kappa, Sigma Xi. Editor: Biochemical Preparations, vol. 9, 1962. Research, numerous publs. on mechanism action enzymes involved in amino acid metabolism, carbondioxide fixation, hydroxylation reactions. Home: 1901 Austin Av., Ann Arbor, Mich. 48104.*

COONS, Albert Hewett, Am. immunologist; b. Gloversville, N.Y., June 28, 1912; s. Albert Selmser and Marion (Hewett) C.; A.B., Williams Coll., 1933, Sc.D. (hon.), 1960; M.D., Harvard, 1937; Sc.D. (hon.), Yale, 1961; m. Phyllis Watts, Dec. 27, 1947; children—Elizabeth Schuyler, Susan Wakefield, Hilary Watts, Wendy, Albert. Research and teaching faculties dept. bacteriology and immunology Harvard Med. Sch., Boston, 1940-42, 46—, vis. prof. bacteriology and immunology, also career investigator Am. Heart Assn., 1953—. Mem. study sect. USPHS, 1946-60; sci. counselor Nat. Inst. Allergy and Infectious Disease, 1960-62; asso. mem. Commn. on Immunization, Armed Forces Epidemiology Bd., 1950-59, mem., 1959—. Recipient Kimble Methodology award Conf. State and Provincial Pub. Health Dirs., 1958, Lasker award Am. Pub. Health Assn., 1959, Paul Ehrlich award West Germany, 1961, Passano Found. award 1962, T. Duckett Jones Meml. award Helen Hay Whitney Found., 1962, Gairdner Found. ann. award, 1962, Emil von Behring prize U. Marburg, W. German, 1966. Harvey lectr., 1957. Fellow Am. Soc. Arts & Scis., A.A.A.S.; mem. Am. Assn. Immunologists (pres. 1960-61), Histochem. Soc. (pres. 1964-65), Soc. Exptl. Biology and Medicine, Brit. Soc. Immunology, Am. Soc. Cell Biology, Internat. Soc. Cell Biology, Nat. Acad. Scis. Research and studies in antibody formation, especially of the cells responsible for their synthesis; devel. method for microscopic localization of large biol. molecules by means of specific antibodies directed against them. Home: 132 High St., Brookline, Mass. 02146.*

COOPER, Alfred, Brit. surgeon; b. Norwich, Eng., Dec. 28, 1838; s. William and Anna (Marsh) C.; ed. at Merchant Taylor's Sch., St. Bartholomew's Hosp.; m. Agnes Cecil Emmeline Duff, 1882; 3 daus., 1 son. Went with Thomas Smith to study anatomy in Paris; apptd. prosector to examiners of Royal Coll. of Surgeons on his return to Eng.; studied syphilis at Lock Hosp., Soho, London, Eng.; Fellow Royal Coll. Surgeons Eng. Author: Syphilis and Pseudo Syphilis, 1884; A Practical Treatise on Disease of the Rectum, 1887; (with F. Swinford Edwards) Diseases of the Rectum and Anus, 1892. Research on rectal diseases and syphilis. Died Mar. 3, 1908.

COOPER, Sir Astley Paston, Brit. surgeon; b. Brooke-Hall, Norfolk, Eng., Aug. 23, 1768; s. Samuel C.; studied at Edinburgh (Scotland) Med. Sch., 1787-88; m. Anne Cock, Dec. 1791 (dec. 1827); m. 2d, C. Jones, July 1828. Apptd. demonstrator St. Thomas' Hosp., London, Eng., 1789; lectr. anatomy Royal Coll. Surgeons, London, 1793-96, became prof. comparative anatomy, 1813, named examiner, 1822, elected pres. in 1827, and in 1836; apptd. surgeon Guy's Hosp., 1800; surgeon to King George IV and William IV. Created baronet after 1820. Fellow Royal Soc., 1802; mem. Medico-Chirurg. Soc. (a founder), French Acad. Scis., 1833. Author: The Anatomy and Surgical

Treatment of Hernia, 1807; On Dislocations and Fractures of the Joints, 1822; Lectures on the Principles and Practice of Surgery, 3 vols., 1824-27. Performed 1st sub-clavicle ligature, 1806; ligated external lilac artery, 1808; best known for operation for aneurysm (performed without antiseptics, by tying abdominal aorta) 1817; described deafness due to obstruction of auditory tube; 1st description of chronic cystic mastitis, 1831. Died London, Feb. 12, 1841.

COOPER, Delmer C(lair), Am. botanist; b. Sutherland, Ia., Apr. 10, 1896; s. Henry Ernest and Alvaretta (Frush) C.; A.B., Morningside Coll., 1916, D.Sc. (hon.), 1948; M.S., Purdue U., 1926; Ph.D., U. Wis., 1930; m. Lillian Margaret Scheuber, Aug. 30, 1933; children—Delmer Chapin, Elizabeth Jean (Mrs. Victor Wiggert), Robert Ernest. Faculty, Morningside Coll., 1917-19, Purdue U., 1926-28; faculty U. Wis., Madison, 1928—, prof. genetics, 1947-66, prof. emeritus, 1966—, research, supr. grad. students, 1930—. Mem. A.A.A.S., Bot. Soc. Am., Am. Assn. U. Profs., genetics socs. Am., Can., Potato Assn. Am., Am. Soc. Naturalists, Internat. Soc. Plant Morphologists, Wis. Acad. Sci., Arts and Letters, Sigma Xi, Phi Sigma. Research, publs. on reprodn.—sterility, cross pollination, devel. gametophytes in various plants. Home: 2749 Kendall Av., Madison, Wis. 53705.*

COOPER, Edward Joshua, astronomer; b. Stephen's Green, Dublin, Ireland, May 1798; s. Edward Synge and Anne (Verelat) C.; student Christ Ch., Oxford, 2 years; m. Miss L'Estrange; m. 2d, Sarah Frances Wynne; 5 daus. Travelled in Europe and the East; became mgr. uncle's estates, Markree, Sligo, 1830, succeeded to the estates, 1837; mem. Parliament for Sligo county, 1930-41, 57-59; built obs., Markree. Recipient Cunningham Gold medal Royal Irish Acad., 1858. Fellow Royal Soc., 1853. Author: Catalogue of Stars (from observations at Markree 1851-56, included more than 50,000 stars within 3 degrees of the ecliptic not catalogued before); Cometic Orbits, 1852; also contbns. to Royal Astron. Soc., 1853. Made astron. and meteorol. observations, 1833-63. Died Apr. 23, 1863.

COOPER, Edwin Lavern, Am. zoologist; b. Utica, Mich., Aug. 31, 1919; s. George E. and Ada (Dentel) C.; B.S., U. Mich., 1940, M.S., 1947, Ph.D., 1949; m. Margaret E. Simmons, Dec. 20, 1941; children—Marilyn, John. Research asso. Mich. Dept. Conservation, 1949-52; chief fishery biologist Wis. Conservation Dept., 1952-56; prof. zoology Pa. State U., State College, 1956—. Mem. A.A.A.S., Am. Inst. Biol. Scis., Ecol. Soc. Am., Am. Fisheries Soc., Am. Soc. Limnology and Oceanography, Am. Inst. Fisheries (regional dir. research biologists 1964-66). Research, publs. on regulation of wild fish populations for mgmt. sport fisheries. Home: 1282 Penfield Rd., State College, Pa. 16801.*

COOPER, Elias Samuel, Am. surgeon; b. nr. Somerville, O., Nov. 25, 1820; s. Jacob and Elizabeth (Walls) C.; M.D., St. Louis U., 1841. An organizer Med. Soc. Cal.; founded 1st med. coll. on Pacific coast (became part of Stanford 1908), San Francisco, 1858; with med. dept. U. Pacific. Contbr. articles to Jour. Northwestern Med. Sch., Cal. State Jour. Successfully removed uterine myoma suprapubically; sterilized instruments with alcohol; advocated use of silver wire in unnited fractures; developed cure for aneurysm by cutting down on sac and sewing it up from outside; conducted correctional operations for club foot. Died San Francisco, Oct. 13, 1862.

COOPER, Gustav Arthur, Am. paleontologist; b. College Point, N.Y., Feb. 9, 1902; s. Gustav August and Lucy (English) C.; B.S., Colgate U., 1924, M.S., 1926, Sc.D., 1953; Ph.D., Yale, 1929; m. Josephine Phelps Wells, July 21, 1930; children—Arthur Wells, Anne (Mrs. George R. Gay). Asst. curator stratigraphic paleontology Smithsonian Instn., Washington, 1930-43, curator invertebrate paleontology and paleobotany, 1943-56, head curator geology, 1956-64, chairman of the department of paleobiology, 1964-67, senior scientist, 1967—. Recipient of the Mary Clarke Thompson medal Nat. Acad., 1958. Fellow Paleontol. Soc. (medal 1964), Geol. Soc. Am. Author: Ozarkian and Canadian Brachiopoda; (with E. O. Ulrich) Chazyan and Related Brachiopods; also numerous articles. Research on Ordovician, Devonian and Permian stratigraphy, brachiopods. Home: 3425 Porter St., Washington 20016. Office: U. S. Nat. Mus., Washington 20560.*

COOPER, Irving S., Am. neurosurgeon; b. Atlantic City, July 15, 1922; s. Louis and Eleanor Lillian C.; B.A., George Washington U., 1942, M.D., 1945; M.S., Ph.D., U. Minn., 1951; m. Mary Dan Frost, Dec. 15, 1944; children—Daniel Alan, Douglas Paul, Lisa Frost. Intern U. S. Naval Hosp., St. Albans, N.Y., 1945-46; fellow neurosurgery Mayo Found., 1948-51; mem. faculty N.Y. U. Med. Sch., 1961-64, prof. clin. neurosurgery, 1954-64; research prof. neuroanatomy N.Y. Med. Coll.; director of neurosurgery St. Barnabas Hosp., N.Y.C., 1954—; spl. research devel., practice, teaching specialized brain operations for treatment Parkinsonism, related diseases, devel. cryogenic surgery. Served to lt. (j.g.), M.C., USNR, 1946-48. Recipient Lewis Harvey Taylor award Am. Therapeutic Soc., 1957; St. Barnabas Hosp. award, 1959; Modern Medicine award 1960; alumni Achievement award George Washington U., 1960; award in medicine N.Y. Philanthropic League, 1960; civic award in medicine Bronx Bd. Trade, 1961; Humanitarian award Nat. Cystic Fibrosis Found., 1962; Eliza Savage vis. prof. Australia, 1962. Diplomate in neurology Am. Bd. Neurology and Psychiatry, Am. Bd. Neurol. Surgery. Fellow A.C.S., Am. Geriatric Soc., N.Y. Acad. Medicine, N.Y. Acad. Sci.; mem. Harvey Cushing Neurosurg. Soc., A.M.A. (Hektoen Bronze Medal award 1957, 58, Certificate of Merit 1961), Neurosurg. Soc. Am., Am. Acad. Neurology, Am. Fedn. Clin. Research, Pan Am. Soc. U. S., Am. Congress Phys. Medicine and Rehab., Scandinavian Neurosurg. Soc., Med. Honor Soc., Sigma Xi (hon.), Alpha Omega Alpha; hon. mem. Neurol. and Neurosurg. Soc. Argentina, Soc. Neurology and Neurosurgery Cuba. Author: The Neurosurgical Alleviation of Parkinsonism, 1956; Parkinsonism: Its Medical and Surgical Therapy, 1961. Home: 76 Mount Tom Rd., Pelham Manor, N.Y. Office: 4422 3d Av., Bronx 57, N.Y.

COOPER, James Graham, Am. entomologist; b. N.Y.C., June 19, 1830; s. William and Frances (Graham) C.; grad. Coll. Phys. and Surg., N.Y.C., 1851; m. Rosa Wells, Jan. 9, 1866. Physician Pacific R.R. Survey Expdn., 1853-55; contract surgeon to U. S. Army; zoologist Geol. Survey of Cal.; became expert on geog. and biologic aspects of Pacific coast regions; wrote chapter on zoology for Natural Wealth of California (T. F. Cronise), 1868; practiced medicine, Santa Cruz, Cal., 1866-71; lived in Ventura County, Cal., 1871-75, Oakland, Cal., 1871-1902. One of first to collect materials and write about natural history of Cal. and Ore. Died July 19, 1902.

COOPER, John Allen Dicks, Am. physician; b. El Paso, Tex., Dec. 22, 1918; s. John Allen Dicks and Cora (Walker) C.; B.S. in Chemistry, N.M. State U., 1939; Ph.D. in Biochemistry, 1943, M.B., 1950, M.D., 1951; Doctor Honoris Causa, U. Brazil, 1958; m. Mary Jane Stratton, June 17, 1944; children—Margaret Ann, John Allen Dicks, Patricia Alison, Randolph Arend Stratton. Faculty, Northwestern U., Evanston, Ill., 1943—, prof. biochemistry, 1957—, asso. dean Med. Sch., 1959-63, dir. integrated program in med. edn., 1960-63, dean scis., 1963—, asso. dean faculties, 1963—; staff Passavant Meml. Hosp., Chgo., 1955—; dir. radioisotope service VA Research Hosp., Chgo., 1954-65; vis. prof. biophysics U. Brasil, 1956, U. Buenos Aires (Argentina), 1958. Mem. Coms., cons. various govt. agys. John and Mary Markle scholar, 1951-56; recipient Outstanding Alumni award N.M. State U., 1960. Mem. Am. Hosp. Assn. (mem. adv. com. on hosp. research and edn. trust 1964-66), Am. Cancer Soc. (mem. adv. com. on personnel for research 1962-66), Asso. Midwest Univs. (pres. 1965-66), Pan Am. Fedn., Nat. Assn. Med. Schs. (treas. adminstrv. com. 1964—), A.A.A.S., A.M.A., Am. Med. Writers Assn., Am. Assn. Biol. Chemists, Assn. Am. Med. Colls., Central Soc. for Clin. Research, Inst. Medicine, Sigma Xi, Alpha Omega Alpha. Research, numerous publs. on biochemistry and biochem. aspects of disease and use of radioactive isotopes in elucidating metabolic pathways. Home: 2243 Orrington Av., Evanston, Ill. 60201.*

COOPER, John Montgomery, Am. anthropologist; b. Rockville, Md., Oct. 28, 1881; s. James Joseph and Emma Lillie (Tolou) C.; student St. Charles Coll., Md., 1897-99; Ph.D., Am. Coll., Rome, Italy, 1902, S.T.D., 1905. Asst. pastor St. Matthew's Ch., Washington, D.C., 1905-18; instr. Catholic U. Am. 1909-23, asso. prof., 1923-28, prof. anthropology, 1928-49. Pres. Anthrop. Soc. of Washington, 1930-32, Am. Anthrop. Assn., 1940; sec., treas. Catholic Anthrop. Conf. 1926-49; v.p. Am. Folklore Soc., 1943. Fellow A.A.A.S.; mem. Washington Acad. Scis. Recipient Mendel medal, 1939. Author: Analytical and Critical Bibliography of Tribes of Tierra del Fuego, 1917; Birth Control, 1923; Play Fair, 1923; Content of Advanced Religion Course, 1924; Religion Outlines for Colleges, 4 vols., 1924-30; Children's Institutions, 1931; Northern Algonquian Supreme Being, 1934; Snares and Deadfalls of Northern Algonquians, 1938; Temporal Sequence and Marginal Cultures, 1941. Editor: Primitive Man. Died May 22, 1949.

COOPER, John N., Am. physicist; b. Kalamazoo, Feb. 4, 1914; s. Bert H. and Anna (Niessink) C.; A.B., Kalamazoo Coll., 1935; Ph.D., Cornell U., 1940; m. Elaine Norton, Sept. 5, 1936. Instr. physics U. So. Cal., 1940-43; asst. prof. physics U. Okla., 1943-46; physicist Lawrence Radiation Lab., U. Cal., Berkeley, 1944-45; faculty Ohio State U., 1946-56; prof. physics Naval Postgrad. Sch., Monterey, Cal., 1956—; cons. Ramo-Wooldridge Corp., TRW Systems, 1955-66. Fellow Am. Phys. Soc.; mem. Inst. Elec. and Electronic Engrs., Am. Assn. Physics Tchrs., Sigma Xi. Author (with A. W. Smith) Elements of Physics, 1957, 1964. Research and publs. in auger transitions in x-rays; low energy nuclear physics; stopping powers of elements for protons; superconductive thin films. Home: San Luis Av., Rt. 3, Box 637, Carmel, Cal. 93921. Office: Dept. Physics, Naval Postgrad. Sch., Monterey, Cal. 93940.*

COOPER, John Thomas, Brit. chemist; b. Greenwich, Eng., June 29, 1790; lectr. chemistry Russell Instn., later Aldersgate St. Sch. Medicine; mfr. chems. and apparatus. Described compounds of platinum, separation of lime and magnesia, analysis of zinc ores, ancient ruby glass, refractive index apparatus, catchuic acid; invented Cooper's tube for collecting gases over mercury, also spirit-lamp furnace and (with Carey) microscope illuminated with limelight. Died London, Sept. 24, 1854.

COOPER, Joseph Bonar, Am. psychologist; b. Indpls., Dec. 10, 1912; s. William Hand and Eva (Bonar) C.; B.A., U. Cal. at Los Angeles, 1936, Ph.D., 1940; m. Hazel Lucille Gasch, Nov. 16, 1949; children—Barbara, Gretchen, Gwendolyn, Mary. Faculty, San Jose (Cal.) State Coll., 1940—, prof. psychology, 1952—. Mem. Am. Psychol. Assn., A.A.A.S., Sigma Xi. Author: (with J. L. McGaugh) Integrating Principles of Social Psychology, 1963; also articles. Research on attitude theory, measurement emotion in prejudice by use galvanic skin response, variables responsible for devel. and perpetuation prejudice in individual, social behavior in various species. Home: 14036 Saratoga Hills Rd., Saratoga, Cal. 95070. Office: San Jose State Coll., San Jose, Cal. 95114.*

COOPER, Kenneth Ernest, English bacteriologist; b. Doncaster, Eng., July 8, 1903; s. Ernest and Elizabeth (Scott) C.; B.Sc., Ph.D., Leeds U.; m. Jessie Griffiths, Aug. 9, 1930. Research asst. color chemistry Huddersfield Tech. Coll., 1927-28; research asst. chemotherapy Leeds Med. Sch., 1928-33, lectr. bacteriology, 1935-38; faculty Bristol U., 1938—, head emergency pub. health lab., 1940-50, reader bacteriology, 1946-50, prof. bacteriology, 1950—. Licentiate Royal Coll. Physicians. Mem. Royal Coll. Surgeons, Soc. for Gen. Microbiology (gen. sec. 1954-60, treas. 1961—), Path. Soc. Gt. Britain, Brit. Med. Assn. Contbg. author: Analytical Microbiology, edited by Fred Kavanagh. Contbr. papers to jours. Research on antiseptics and antibiotics, methods of lab. diagnosis of bacterial infections, epidemiological studies of diphtheria, entinofevers, food poisoning, venereal diseases, pub. health standards for milk and ice cream, use of antibiotics and antiseptics in control of hosp. infections. Home: Fairfield, Tickenham, Clevedon, Somerset, Eng. Office: Bacteriology Dept., Medical Sch., University Walk, Bristol, Eng.*

COOPER, Leslie Hugh Norman, English oceanographer; b. Broadstairs, Kent, Eng., June 17, 1905; s. Charles Herbert and Annie (Silk) C.; B.Sc. with 1st class honors in Chemistry (Eyton Williams scholar), U. Coll. N. Wales, 1924, Ph.D., 1927, D.Sc., 1938; m. Gwynedd Daloni Seth Hughes, Sept. 4, 1935; children—Alastair Lane, Graham Charles Lane, Roger Lane, Justin Seth Lane, Veronica Lane. Chemist, Rubber Research Assn., Croydon, Surrey, Eng., 1927-29, Imperial Chem. Industries, Manchester, Huddersfield, 1929-30; chemist, oceanographer Marine Biol. Assn. U.K., Lab., Citadel Hill, Plymouth, Eng., 1930—. Fellow Royal Inst. Chemistry, Inst. Biology, Royal Soc., 1964; mem. Chem. Soc., Faraday Soc., Challenger Soc., Royal Meteorol. Soc., Geol. Soc., Geologists Assn., Ussher Soc., Devonshire Assn. Research, numerous publs. on chem. nutrients in English Channel, structure waters of Celtic Sea and deep Atlantic, phys. chemistry of sea water. Home: 2 Queens Gate Villas, Lipson, Plymouth. Office: Marine Biol. Lab., Citadel Hill, Plymouth, Devon, Eng.*

COOPER, Theodore, Am. physician; b. Trenton, N.J., Dec. 28, 1928; s. Victor and Dora (Popkin) C.; B.S., Georgetown U., 1949; M.D., St. Louis U., 1954, Ph.D., 1956; m. Vivian Cecilia Evans, June 16, 1956; children—Michael Harris, Mary Katherine, Victoria Susan, Frank Victor. USPHS fellow St. Louis U. Dept. Physiology, 1955-56; clin. asso. surgery br. Nat. Heart Inst., Bethesda, Md., 1956-58, chief animal lab., surgery br., 1959-60; faculty St. Louis U., 1960-66, prof. surgery, 1964-66; prof., chmn. dept. pharmacology U. N.M., Albuquerque, 1966—, on leave, 1967-69; asso. dir. artificial heart, myocardial infarction programs Nat. Heart Inst., Bethesda, Md., 1967—. With the U. S. Pub. Health Service Pharmacology and Experimental Therapeutics Study sect., 1964-67. Recipient Borden award, 1954. Mem. Am. Soc. Pharmacology and Exptl. Therapeutics, Am. Physiol. Soc., Soc. Exptl. Biology and Medicine, Am. Soc. Clin. Investigation, Am. Fedn. Clin. Research, Am. Soc. Artificial Internal Organs, Internat. Cardiovascular Soc., Am. Coll. Chest Physicians, Am., Mo. med. assns., St. Louis Med. Soc., St. Louis Heart Assn., A.A.A.S., Sigma Xi. Author (with others) Nervous Control of the Heart, 1965. Editorial bd. Jour. Pharmacology and Exptl. Therapeutics, 1965—, Circulation Research, 1966—; editor Supplements to Circulation, 1966—; co-editor for circulation Am. Jour. Physiology, Jour. Applied Physiology, 1967—. Contbr. numerous articles to med. jours. Discoverer new techniques of denervating heart which have helped delineate role of nerves in heart on its ability to function under a wide variety of circumstances and on its ability to respond to drugs. Home: 6711 Greyswood Rd., Bethesda 20034. Office: Bldg. 31, NIH, Bethesda, Md. 20014.*

COOPER, Thomas, chemist; b. Westminster, Eng., Oct. 22, 1759; s. Thomas Cooper; entered Oxford (Eng.) U., 1779; studied medicine London and Manchester; LL.D. (hon.), U.S.C., 1834, m. Alice Greenwood; m. 2d, Elizabeth Hemming, 1811; 8 children. Came to Am. in reaction to English conservative policies, 1794; convicted, sentenced and fined under Sedition Act, 1800; commr. in Luzerne County, Pa., 1801-04; state judge Pa., 1804-11; prof. chemistry Dickinson Coll., 1811-15; prof. applied chemistry and mineralogy U. Pa., 1816-19; prof. chemistry U. S.C., 1820, pres., 1821-34. Mem. Am. Philos. Soc. Author: On the Constitution, 1826; Lectures on Political Economy, 1826. Editor: Statutes at Large of South Carolina, 5 vols, 1836-39; Thomson System of Chemistry, 4 vols., 1818. Editor: Emporium of Arts and Sciences, 1813-14. Influential in establishing 1st sch. of medicine and 1st insane asylum in S.C. Died Columbia, S.C., May 11, 1839.

COOPER, William, see Cowper, William.

COOPS, Jan, chemist; b. Amsterdam, Holland, May 27, 1894; s. Jan. and Wilhelmina (Desselkoen) C.; Chem. Engr., Tech. U. Delft (Holland), 1919, Dr. Tech. Scis. cum laude, 1924. Asst. organic chemistry Tech. U. Delft, 1917-18; asst. prof. Netherlands' Sch. Econs., Rotterdam, 1918-29; prof. chemistry Free U., Amsterdam, 1929-64, founder, dir. chem. lab., 1932-64. Mem. bd. Central Inst. Physico-Chem. Constants, 1953—; mem. thermochem. commn. Union Internal. de Chimie Pure et Appliquée; mem. Commn. Chem. Thermodynamics; mem. Netherlands Chem. Council, 1954—; mem. Commn. Marine Soil Research; mem. adv. bd. RVO/TNO. Decorated knight Order Nederlandse Loeuw. Mem. Royal Netherlands Chem. Soc. (pres. sect. organic chemistry 1946-49, nat. pres. 1947-50). Author: Stero-isometry of Tartaric Acad., 1924; Calibration of Calorimeters for Reactions in a Bomb at Constant Volumn, 1950. Research and publn. on devel. high precision bomb calorimeter for very accurate measurements of heats of combustion of several series organic compounds; thermochem. investigations on heats of solution, vaporization and melting; purity control by means of melting curve techniques, constn. high-precision distillation-rate osmometer; studies on carbon-carbon energies of substituted tetra-phenyl-ethanes. Address: Arisotelslaan 40, Zetst, The Netherlands.*

COOVER, John Edgar, Am. psychologist; b. Remington, Ind., Mar. 16, 1872; s. John Calvin and Elizabeth Hadessa (Keller) C.; Ped.B., Colo. State Normal Sch., 1898; A.B., Stanford, 1904, A.M., 1905, Ph.D., 1912; m. Margaret Evelyn Brooks, June 3, 1905; 1 son, Calvin Clay. Various lines of business, and prin. schs. Colo. and Cal., until 1910; asst. dept. psychology Stanford, 1910, asso. prof., 1921-30, prof., 1930-33; asso. prof. psychology Johns Hopkins, 1927-28. Author: Formal Discipline from the Standpoint of Experimental Psychology, 1916; Experiments in Psychical Research at Leland Stanford, Jr. University, 1917; Metapsychics and the Incredulity of Psychologists (Chapt. XI, "The Case For and Against Psychial Belief"), 1927; The Quantitative Measurement of Higher Mental Processes in the Pioneer Studies of H. Ebbinghaus (Analysis 51, in "Methods in Social Science"), 1931. Joint Author: The Weise-Coover Kinaesthetic Method of Typing, 1924. Died Feb. 19, 1938.

COP, William, physician; b. Basel, Switzerland; physician to Louis XII and Francis I of France; credited with restoring med. art in France; excellent transls. of works of Greek physicians. Died 1532.

COPAUX, Hippolyte, French chemist; b. Paris, France, Mar. 7, 1872; investigated complex molybdates and tungstates and compounds of cobalt and nickel. Died Étampes, France, Aug. 28, 1934.

COPE, Cuthbert Leslie, English physician; b. London, Eng., May 21, 1903; s. Gilbert Edgar and Florence (Dell) C.; B.A. with 1st class honors in physiology, U. Oxford, 1924; B.M., B.Ch., U. Coll. Hosp. Med. Sch. London, 1927; D.M. 1934; m. Eileen Gertrude Putt, Aug. 9, 1937; children—Jonathan Leslie, David Robert. Beit Meml. Med. research fellow, 1929-36; first asst. med. unit Radcliffe Infirmary, Oxford U., 1938; dir. research human problems Nat. Coal Bd., 1947-48; physician Postgrad. Med. Sch., London, 1948—. Recipient Radcliffe prize, Oxford, 1935. Fellow Royal Coll. Physicians (Lumleian lectr., examiner), Royal Soc. Medicine (past pres. endocrine sect.); mem. Assn. Physicians, Med. Research Soc. Author: Adrenal Steroids and Disease, 1965. Research, publs. on metabolic disturbances in human disease, including devel. of new methods, especially use of isotopes to measure cortisol and aldosterone secretion rates. Home: 25, Birkdale Rd., Ealing, London W. 5. Office: Dept. Medicine, Postgrad. Med. Sch., London W. 12, Eng.*

COPE, Edward Drinker, Am. paleontologist; b. Phila., July 28, 1840; s. Alfred and Hanna (Edge) C.; ed. U. Pa., Phila. Acad. Scis., Smithsonian Instn.; A.M. (hon.), Haverford Coll., 1870; Ph.D. (hon.), Heidelberg (Germany) U., 1885; m. Annie Pim, Aug. 14, 1865, 1 child. Prof. comparative zoology and botany Haverford Coll., 1864-67; mem. Phila. Acad. Natural Scis., 1861, curator, 1865, mem. council, 1879; paleontologist U. S. Geol. Survey, 1870, discovered about 1000 new species extinct vertebrata; prof. geology and mineralogy U. Pa., 1889-95, prof. zoology and comparative anatomy, 1895-97. mem. Nat. Acad. Scis., A.A.A.S. (pres. 1896). Author: Synopsis of the Extinct Cetacea of the United States, 1867-68; Systematic Arrangement of the Extinct Batrachia, Reptilia and Aves of North America, 1869-70; Relation of Man to Tertiary Mammalia, 1875. Collected fossils in western U. S.; worked out evolutionary history of horse; theorized that natural movements of animals aided in alteration and devel. of moving parts (kinetogenesis). Died Phila., Apr. 12, 1897.

COPE, Freeman Widener, Am. phys. biochemist, physiologist; b. Peekskill, N.Y., 1930; s. T. Freeman and Frances (Thorndike) C.; A.B., Harvard, 1951; M.D., Johns Hopkins, 1955. Phys. biochemist, physiologist Aerospace Med. Research Lab., Johnsville, Warminster, Pa., 1957—. Mem. A.A.A.S., Am. Chem. Soc., Am. Physiol. Soc., Biophys. Soc., Aerospace Med. Assn. Co-discoverer photogenerated free radicals in melanin of eye; 1st to apply to biology in quantitative way principles of solid state physics, to demonstrate complexing of sodium ions in muscle by nuclear magnetic resonance. Home: 238 W. Court St., Doylestown, Pa. 18901. Office: Biochemistry Div., Aerospace Med. Research Lab., Johnsville, Warminster, Pa. 18974.*

COPE, Oliver, Am. surgeon; b. Germantown, Pa., Aug. 15, 1902; s. Walter and Eliza Middleton (Kane) C.; A.B., Harvard, 1923, M.D., 1928; Dr. Honoris Causa, U. Toulouse, 1950; m. Alice DeNormandie, Dec. 28, 1932; children—Robert DeNormandie, Eliza Middleton. Intern, resident surgery Mass. Gen. Hosp., 1928-32, asst. to asso. surgeon, 1934-46, vis. surgeon since 1946; Moseley travelling fellow Harvard, 1933, instr., later asso., asst. prof. surgery, med. sch., 1934-38, associate professor surgery, 1948-63, professor surgery, 1963—; chief of staff Boston unit of Shriners Burn Institute, since 1964. Responsible investigator Office Sci. Research and Development, 1942-45; mem. subcom. Burns Nat. Research Council, 1943-45; dir. research under contract with Office Naval Research, 1947—. Diplomate Am. Bd. Surgery. Fellow A.C.S.; member of American Surgical Association (president 1962-63), N.E. Surg. Soc., Internat. Soc. Surgery, A.A.A.S., Society Clinical Surgery, Soc. Clin. Investigation, Soc. U. Surgeons, A.M.A., Mass. Med. Soc., Am. Acad. Arts and Scis., Boston Surg. Soc. (pres. 1965). Research on experimental and surgical endocrinology; burns; surgical metabolism; medical education. Home: 20 Hubbard Park, Cambridge 38, Mass. Office: Mass. Gen. Hosp., Boston 14.

COPELAND, Arthur H., Am. mathematician; b. Rochester, N.Y., June 22, 1898; s. Albert Edward and Jenny (Morris) C.; B.A., Amherst Coll., 1921; M.A., Ph.D., Harvard, 1926; m. Dorothy West, June 16, 1925; 1 son, Arthur H. Faculty, Rice Inst. (now Rice U.), 1924-28, Buffalo U., 1928-29; faculty U. Mich., Ann Arbor, 1929—, prof. math., 1942—. Vis. lectr. Author: Geometry, Algebra and Trigonometry by Vector Methods, 1962. Research on founds. of probability and statistics, epistemology, inductive logic, deductive logic, mechanics, Fourier series. Home: 805 Oakdale St., Barton Hills, Ann Arbor, Mich. 48105.*

COPELAND, Donald Eugene, Am. biologist; b. Mendon, O., Feb. 6, 1912; s. Arland Murlin and Chloe (Severns) C.; A.B., Rochester U., 1935; M.A., Amherst Coll., 1937; Ph.D., Harvard, 1941; m. Marjorie Groves, June 20, 1941; children—Sandra Kay, Jane Hance, Diana Sue. Instr. zoology U. N.C., 1941-42; asst., then asso. prof. zoology Brown U., 1946-51; chief aviation physiologist Office Surgeon Gen., USAF, 1951-53; profl. asso. Nat. Acad. Scis.-NRC, 1953-56; secretary National Insts. Health, 1956-59; professor of zoology, Tulane University, 1959—; mem. Marine Biol. Lab., 1948—; Mem. morphology and genetics study sect., physiology study sect. Nat. Insts. Health, 1952-53. Served to capt. USAAF, 1942-46. Mem. Am. Assn. Anatomists, Am. Soc. Zoologists, Soc. Study Devel. and Growth, Assn. Southeastern Biologists. Research on electron microscopy and histophysiology of inorganic ion transport systems; oxygen in fish swim bladder; carbon monoxide in Portuguese-Man-of-War float; salt movement in fish gill, crab gill, shrimp gill, anal papillae of mosquito larvae; function of the glandular pseudobranch in fish. Home: 2808 Calhoun St., New Orleans 70118.

COPELAND, Murray M., Am. surgeon; b. McDonough, Ga., June 23, 1902; s. Edward Meadows and Mary Elizabeth (Speer) C.; A.B., Oglethorpe U., 1923, D.Sc., 1955; M.D., Johns Hopkins, 1927; m. Jean Brown, June 20, 1931. Instr. surgery Johns Hopkins, Balt., 1937-46; faculty U. Md., 1937-46, asst. prof. surgery, 1945-46; chief of surgery Kennedy VA Hosp., Memphis, 1946-47; prof., chmn. dept. oncology Georgetown U. Med. Center, Washington, 1947-60, prof. emeritus, 1960—; faculty M.D. Anderson Hosp. and Tumor Inst., U. Tex., Houston, 1960-67, asso. dir. edn., prof. surgery (oncology), 1963-67; v.p. Univ. Cancer Foundation (Internat.), 1967. Secretary-general Tenth International Union Against Cancer, Houston, 1967-70; consultant in surgery, also in cancer therapy numerous hosps. Diplomate Am. Bd. Surgery. Fellow N.Y. Acad. Scis.; mem. Am. Acad. Orthopaedic Surgeons, A.A.A.S., Am. Assn. Cancer Research, Am. Coll. Surgeons, A.M.A., Am. Radium Soc., Am. Assn. Cancer Insts., Assn. Hosp. Dirs. Med. Edn., Coordinators Cancer Teaching, Inst. Environmental Scis., Internat. Union Against Cancer, Soc. Head and Neck Surgeons, Southeastern Surg. Congress (past pres.), So. Med. Assn., So. Surg. Assn., Am. Cancer Soc. (pres. 1964-65, dir.-at-large 1957—), Phi Delta Kappa. Mem. editorial bd. Yearbook of Cancer, Ca—A Cancer Jour. for Clinicians, The Am. Surgeon, Med. Tribune, Surg. Digest. Research and publs. on histogenesis of bone neoplasms, breast, oral cavity, staging and treatment of cancers. Home: 1600 Holcombe Blvd., Houston 77025.*

COPELAND, Ralph, Brit. astronomer; b. Woodplumpton, Eng., Sept. 3, 1837; s. Robert and Elizabeth (Milner) C.; entered U. Göttingen, Germany, 1865; Ph.D., 1869; m. Susannah Milner, 1859; 1 son, 1 dau.; m. 2d, Theodora Benfey, Dec. 1871; 3 daus., 1 son. Went to Australia, 1853; worker on sheep run and at gold diggings in Victoria, 5 years; vol. apprentice Beyer, Peacock & Co., locomotive engrs., Manchester, Eng.; set up obs. in Manchester; mem. German Arctic Expn., 1869; became asst. astronomer at Lord Rosse's obs., Birr Castle, Parsontown, 1871; became asst. Dublin U., Dunsink, 1874; in charge obs. at Dun Echt, Aberdeen, Scotland; named astronomer royal for Scotland, 1889; prof., Edinburgh, Scotland. Mem. Royal Astron. Soc. Author: First Göttingen Catalogue of Stars, 1869; Copernicus, a Journal of Astronomy, 3 vols., 1891-1904. Observed Donati's comet, 1858, non-instantaneous occultation of K Concri by moon, also positions of all stars down to 9th magnitude in zone 2 degrees wide immediately south of celestial equator, 1863, transit of Venus from Mauritius, 1869 and from Jamaica, 1882; discovered great tree fern at Trinidad, 1874; discovered spectrum of temporary star Nova Cygni had been reduced to bright line, 1877; research on geodetics. Died Edinburgh, Scotland, Oct. 27, 1905.

COPELAND, Thomas, English surgeon; b. Byfield, Eng., May 1781; s. William Copeland; studied under Mr. Denham, Chigwell, Essex, under Edward Ford, London, also at St. Bartholomew's Hosp.; qualified as surgeon, 1804; army surgeon, Spain, 1809; practiced in London; apptd. surgeon extraordinary to Queen Victoria, 1837. Fellow Royal Soc., 1834, Royal Coll. Surgeons (council). Author: Observations on some of the principal Diseases of the Rectum, 1810; Observations on the Symptoms and Treatment of the Diseases Spine, more particularly relating to the Incipient Stages, 1815; also papers, including History of a Case in which a Calculus was voided from a Tumour in the Groin. Editor: Observations on the Diseases of the Hipjoint (E. Ford), 1810. Founder of rectum surgery; 1st to suggest removal of septum narium by forceps, in cases where its oblique positions obstructed passage of air through nostrils. Died Nov. 19, 1855.

COPEMAN, Sydney A. Monckton, Brit. physician; b. Feb. 21, 1862; s. Canon Copeman; ed. Corpus Christi Coll., Cambridge, St. Thomas's Hosp., London; M.A., M.D., Cambridge; Diploma in pub. health.; m. Ethel Margaret Boord (dec. 1944); 1 son, 2 daus. Ret. Med. officer Ministry Health; former sr. med. insp. His Majesty's Local Govt. Bd.; mem. Faculty Medicine, chmn. bd. studies in hygiene U. London, also examiner in state medicine; emeritus lectr. pub. health Westminster hosp.; Chadwick lectr. hygiene, 1914; former examiner in pub. health and forensic medicine and toxicology U. Bristol, in pub. health Royal Coll. Physicians, in hygiene and pub. health U. Leeds, in pub. health Royal Coll. Surgeons; research scholar, spl. commr. Brit. Med. Assn. Recipient Cameron prize U. Edinburgh, 1899, Fothergillian gold medal Med. Soc. London, 1899, Gold medal Internat. Faculty Scis., 1938. Fellow Royal Soc. (Buchanan gold medal 1902), 1903, Royal Coll. Physicians (London) (past mem. council, Milroy lectr. 1898), Zool. Soc. (past mem. council), Hunterian Soc. (hon.); mem. Royal Soc. Medicine (past pres. epidemiological sect., Jenner medal 1925), Med. Research Club (joint founder). Author: Vaccination, Its Natural History and Pathology (Milroy Lectures), 1898, also numerous articles on pub. health. Inventor glycerinated lymph, officially adopted in 1898 (in gen. use for anti-smallpox vaccination). Died Apr. 11, 1947.

COPENHAVER, John Harrison, Jr., Am. biologist; b. Ralston, Neb., Dec. 21, 1922; s. John Harrison and Dora (Tallman) C.; B.A., Dartmouth, 1946; M.S., U. Wis., 1949, Ph.D., 1950; m. Marion Lamson, June 30, 1946; children—John Harrison III, Margaret Ilse, Christine, Eric Charles, Lisa Carol. Asst. prof. pharmacology U. Tex., 1951-53; mem. faculty Dartmouth, 1953—, prof. zoology, 1960-61, prof. biology, chmn. dept. biol. scis., 1961—: NSF fellow U. Cal. at Berkeley, 1960-61. Mem. Am. Chem. Soc., Am. Soc. Biol. Chemists, Phi Beta Kappa, Sigma Xi. Research on enzyme chemistry of renal function; oxidative phosphorylation; cell transport mechanisms. Home: 21 Lyme Rd., Hanover, N.H.*

COPERNICUS, Nicolaus (Mikolaj Kopernik or Niklas Koppernigk), Polish astronomer; b. Thorn (Torun), Poland, Feb. 19, 1473, as subject of king of Poland; s. Niklas and Barbara (Watzenrode or Watzelrode) C.; studied classics and math. under W. Brudzewski, U. Cracow (Poland), 1491-94, law and astronomy under Domenico Maria da Navara, Bologna, Italy, 1496-1500, law and Greek, Padua, Italy, 1500-03, law, Ferrara, doctorate in canon law, 1503, medicine, Padua, 1503-06. Elected canon Cathedral of Frauenburg (Frombork), 1497; on leave of absence to lecture on astronomy and math., Rome, 1500; settled in Heilsberg, Poland, 1506; physician and pvt. sec. to his uncle and patron, Lucas Watzenrode, Bishop of Ermland, 1506-12; returned to Frauenburg, practiced medicine, also various polit., eccles. positions, 1512-20; adminstr. of Frauenburg, circa 1520; dep.

counselor finance, Prussia, circa 1522-29. Author: Commentariolus, 1530; De revolutionibus orbium coelestium libri VI, completed circa 1530, pub., 1543. Considered greatest astronomer since Ptolemy; work precipitated sci. revolution that culminated in researches of Newton; generally regarded as founder modern astronomy; 1st to work out in math. detail heliostatic theory solar system (states that planets revolve in circular orbits about sun, that earth itself is just another planet, that earth completes its revolution around sun in 1 year, that it rotates daily about its axis); maintained (to account for parallax not being observed) that radius of earth's orbit around sun was as but small point compared to radius of sphere of fixed stars (therefore stars, unlike planets, do not reflect motion of earth); demonstrated how heliostatic conception made it possible to calculate accurately planetary positions and planetary distances from sun, also to explain seasons, precession of equinoxes, and retrograde motion of planets. Died Frauenburg, May 24, 1543.

COPHER, Glover Hancock, Am. surgeon; b. Troy, Mo., Oct. 27, 1893; s. William Harrison and Sally (Duff) C.; A.B., U. Mo., 1916; M.D., Washington U., St. Louis, 1918; m. Marjorie Hulsizer, Jan. 12, 1924 (dec. May 1935); 1 dau., Marjorie Copher White. Practice surgery, St. Louis; staff Barnes Hosp., 1918—; faculty Washington U. Med. Sch., 1918—, clin. prof. surgery, 1950—. Recipient medal Am. Radiol. Soc., 1924. Mem. A.C.S., Am. Surg Assn., Clin. Soc. Surgery, Internat. Soc. Surgeons, A.M.A. Author: Methods of surgery, 1925; (with W. Cole, E. Graham) Diagnosis and Treatment of Diseases of the Biliary Tract, 1928; also articles. Co-discoverer cholecystography. Home: 5281 Westminster Pl., St. Louis 63108. Office: Barnes Hosp. Plaza, St. Louis 63110.*

COPHO OF SALERNO, anatomist; flourished 12th century; author De modo medendi; credited with 1st Salernian anat. treatise, Anatomia porci, which introduced anat. expts. in animals and reflects Arabic influence in terminology.

COPLAND, James, Brit. physician; b. Deerness, Orkney Islands, Nov. 1791; M.D., Edinburgh, Scotland, 1815. Fellow Royal Soc., 1833. Compiler Dictionary of Practical Medicine, 1832; contbr. profl. jours.; prolific med. writer. Died Kilburn, July 12, 1870.

COPLEY, Alfred L(ewin), physiologist; b. Germany, June 6, 1910; student U. Freiburg, 1930, U. Berlin, 1930-31, U. Königsberg, 1931, U. Frankfurt, 1931-32, U. Würzburg, 1932-33 (all Germany); M.D., U. Heidelberg (Germany), 1935, U. Basel (Switzerland), 1936; m. Nina Tryggvadottir, 1949; 1 dau., Una Dora. Came to U. S., 1937, naturalized, 1943. Research asso. Hixon Lab. for Med. Research, U. Kan. Med. Sch., 1940-42; fellow surgery, head lab. exptl. surgery U. Va. Med. Sch., 1943-44, research asso. in preventive medicine and bacteriology, 1944-45; vis. research asso. biology dept. N.Y. U., 1947-48, research asso. lab. cellular physiology, 1945-49; dir. project AEC, 1949-52; chief investigator NSF project, 1955-59, Nat. Heart Inst., NIH, 1960-65, Office Naval Research projects, 1950-52, 65—; research asso. hematology Mt. Sinai Hosp., N.Y.C., 1948-49; instr. Columbia Coll. Phys. and Surg., 1948-49; faculty N.Y. Med. Coll., 1949-52, 60—, research prof. physiology, 1962-64, research prof. pharmacology, 1965—, head hemorheological lab., 1960-65; adj. research prof. Newark Coll. Engring., 1967—; chargé de recherches, head lab. blood and vascular physiol. research lab. Internat. Children's Center, Paris, France, 1952-55; chargé de recherches Nat. Inst. Hygiène, head research lab. microcirculation Nat. Center Blood Transfusion, Paris, 1955-57; dir. exptl. research vascular diseases med. research labs. Charing Cross Hosp., London, 1957-59; vis. prof. pathology Royal Coll. Surgeons Eng., 1959; chief spl. hemorrhage and thrombosis research labs. VA Hosp., East Orange, N.J., 1965—, asso. chief staff, 1965-66. Mem. med. adv. bd. Nat. Blood Research Found., 1954—; mem. corp. Marine Biol. Lab., Woods Hole, Mass., 1954—. Fellow A.A.A.S., N.Y. Acad. Scis., Royal Soc. Medicine, Internat. Soc. Hematology; mem. Am. Physiol. Soc., Am. Soc. Hematology, Soc. for Exptl. Biology and Medicine, Assn. History of Medicine, Brit. Soc. Rheology, Research Assn. des Physiologistes (France), Sigma Xi. Editor: (with G. Stainsby) Flow Properties of Blood and other Biological Systems, 1960; (with E. H. Lee) Proc. 4th Internat. Congress on Rheology, part I, 1965; Symposium on Biorheology, 1965; Hemorheology, Proc. 1st Internat. Conf., 1967; editor-in-chief (with G. W. Scott Blair) Biorheology, 1965, an Internat. Jour., 1962—. Research, numerous publs. on blood clotting, mechanisms of trombosis, hemorrhage and hemostasis, blood vessel wall, comparative hematology, blood platelets, physiology of spleen, immunology, brain and liver metabolism, cholinesterase, radiobiology, biorheology of hair, exptl. Tb, microcirculation, hemorheology. Home: 50 Central Park W., N.Y.C. 10023. Office: Hemorrhage Thrombosis Research Labs., VA Hosp., East Orange, N.J. 07019.

COPPEE, George Ed Paul, Belgian physician; b. Mont-sur-Marchienne, June 30, 1909; s. George and Jeanne (Robert) C.; M.D., U. Liège; m. Mari-Henriette Bolly, July 6, 1939; children—Georges, Françoise, Paul. Asst., prof. U. Liège; physician-dir. Ernest Malvoz Inst. Pres. com. on hygiene and medicine Communauté européenne du charbon et de l'acier. Research and publs. on ergonomy. Home: 10, ave. du Hetre, Sclessin. Office: 4, quai du Barbour, Liège, Belgium.

COPPEN, Alec James, Brit. psychiatrist; b. London, Eng., Jan. 29, 1923; s. Herbert John and Margarete (Henshaw) C.; student Dulwich Coll., 1935-39, U. Bristol, 1947-53, Inst. Psychiatry, 1953-60; M.B., Ch.B., 1953, M.D., 1958, D.P.M., 1957; m. Gunhild Andersson, Aug. 9, 1952; 1 son, Michael. Research asst. to Aubrey Lewis at Inst. of Psychiatry, 1958-60; cons. psychiatrist West Park Hosp. and Med. Research Council, Neuropsychiat. Research Unit; hon. cons. St. George's Hosp., London. Mem. Royal Medico-Psychol. Assn. (mem. council, sec. research and clin. sects.), Royal Soc. Medicine, Brit. Med. Assn. Contbr. articles to profl. lit. Conducted biochem. investigations into causes of mental illness, especially manic depressive illness; investigations into psychosomatic aspects of menstrual disorder, gynecological conditions, into constl. factors in psychiatry. Home: 5 Walnut Close, Epsom, Eng. Office: Medical Research Council, Neuropsychiatric Research Unit, Green Bank, West Park Hospital, Epsom, Eng.

COPPENS, Rene, French geologist; b. Auray, France, Dec. 20, 1910; s. Jean Marie Frederic and Henriette (Denis) C.; ed. Faculty Scis., Caen, Rennes, Paris; D.Sc., Paris, 1949; m. Andree Dagorne (dec. 1957); children—Yves, Odile (Mme. Costantini), Sylvie; m. 2d, Madeline Rollinger, June 13, 1960. Prof., Coll. Clermont, Lycée de Vannes; lectr. Faculty of Scis., Nancy, 1955—, titular prof., 1958—, adj. dir. Center Radiol. Research, U. Nancy. Author: La Radioactivité des roches, 1957; (with A. Roubault) Précis de Géologie, 1966; also many articles. Using autoradiography, worked on radioactivity of certain rocks, focusing on relationship of this radioactivity and eventual presence of uranium deposits; other work with lead isotopes and carbon-14. Home: 35 Bd Jean Jaures. Office: BP 452, Nancy 54, France.*

COPPET, Louis Casimir de, see de Coppet.

COPPI, Germano, Italian chemist; b. Tortona, Italy, Sept. 22, 1935; s. Renato and Yolanda (Prigione) C.; degree Chemistry, U. Pavia (Italy), 1958; m. Nelda Paulin, July 6, 1963. Vice head microbiol. lab. Lab. Research C. Erba, Milan, Italy; head biochem. and microbiol. labs. Research Lab. De Angeli, Milan; head Radioisotopical Lab., Milan. Mem. Soc. Ital. Sci. Farmac. Applic., Soc. Ital. Biol. Sper., Soc. Ital. Microb. Research, publs. on chemotherapeutic agts., including mebinol, tetralysal, neomycinpamoate; pharmacological agts., including Ac. Nafcaproico (choleretic drug), naphthypramide (anti-inflammatory drug), nafiverine (Spasmolytic drug). Home: 46/9 L. Muratori. Office: 15 Sevio, Milan, Italy.*

COPSON, Edward Thomas, Scottish mathematician; b. Aug. 21, 1901; s. T. C. Copson; M.A., St. John's Coll., Oxford (Eng.) U.; D.Sc., U. Edinburgh (Scotland); m. Beatrice Mary Whittaker; 2 daus. Lectr. U. Edinburgh, 1922-29; lectr. St. Andrews (Scotland) U., 1930-34, regius prof. math., from 1950, master St. Salvator's Coll., 1954-57; asst. prof. Royal Naval Coll., Greenwich, Eng., 1934-35; prof. math. U. Coll., Dundee, Scotland, 1935-50. Fellow Royal Soc. Edinburgh (Keith prize 1939-41); mem. London, Edinburgh math. socs., Math. Assn., Inst. Math. and its Application. Author: The Theory of Functions of a Complex Variable, 1935; (with Bevan Baker) The Mathematical Theory of Huygens' Principle, 1939; Asymptotic Expansions, 1965; Metric Spaces, 1968. Research, publs. in classical analysis. Home: 42 Buchanan Gardens. Office: Dept. Math., St. Andrews U., St. Andrews, Fife, Scotland.

COQUEBERT DE MONTBRET, Charles Étienne (baron), French geographer; b. Paris, July 3, 1755; ed. Coll. Plessis; studied under Abbe Nollet, also Valmont de Bomare. Sec. to Bur. Consulates, Versaille, France, 1775; sent as marine commr. to Hamburg, Germany, 1776; apptd. consul gen. Hanseatic cities, 1777; returned to Paris, 1786; succeeded father as conseiller-correcteur, court of count; sent as agt. for commerce and navy, Dublin, Ireland, 1790; named by Com. on Pub. Safety to organize new system of weights and measures, after 1793; tchr. phys. geography at a republican lycée; named sec. mines; became commr. comml. relations, Amsterdam, Netherlands, after 18 Brumaire; in London, Eng., between peace of Amiens and break with Eng.; then became master of petitions State Council, also chief interior ministry Div. Statistics; chief of customs, after union of Holland and France; sec.-gen. Ministry Commerce, until 1814; prof. geology École des Mines. Mem. French Acad. Scis., 1816, Soc. Geography (founding mem.), Soc. Natural History, Soc. Agr. Author: La géographie physique statistique et commerciale de l'Europe; Éclaircissements préliminaires; also various articles. Editor articles on botany and rural economy Dictionary Natural Scis. Died Paris, Apr. 9, 1831.

COQUILLETT, Daniel William, Am. entomologist; b. McHenry County, Ill., Jan. 23, 1856; s. Francis Marquis Lafayette and Sarah Ann (Cokelet) C.; ed. dist. sch.; m. Tchr. dist. sch., 1876; prepared descriptive paper on caterpillars of U. S., with analytical keys to groups and species, for 10th Ann. Report, state entomologist of Ill., 1886; contbr. on applied entomology Germantown Telegraph, 2 yrs.; asst. state entomologist of Ill., 1881, wrote major portion of 11th Ann. Report, removed to Southern Calif., 1882; assisted Matthew Cooke on two works on entomology, 1883; investigated for U. S. Dept. Agr., outbreak of destructive grasshoppers in Central Cal., 1885, perfected method for destroying them by use of poisoned mash, also investigated cottony cushion scale insect, infesting citrus trees in So. Cal., 1886; experimented poisonous gases for destroying these and other noxious insects on trees and plants, inaugurating the hydrocyanic gas treatment; in employ U. S. Dept. Agr., from 1887, asst. entomologist, from 1893; removed to Washington, 1893; apptd. hon. custodian diptera U. S. Nat. Mus., 1896. Died 1911.

CORBELLE GUASCH, George Lawrence, Argentinian surgeon; b. Buenos Aires, Argentine, Apr. 24, 1926; s. Joseph and Catherine (Guasch) Corbelle; M.D., Buenos Aires U., 1953; m. Vicenta Ferrer, Dec. 21, 1960; 1 son, George Louis. Surgeon of 4th chair of surgery Durand Hosp., also surgeon medicine service and urgent surgery, 1951—; prof. surgery U. Buenos Aires, 1961—, instr. residents at 4' chair of surgery, 1964—. Recipient 1st prize Soc. Surgery Buenos Aires, 1961—, instr. residents at 4th chair of Hosp. Surgery Service, Paris, 1962-63; scholar Specia Labs., Paris, 1963. Mem. Argentine Med. Assn., Argentine Soc. Surgeons, Argentine Assn. Surgery, Soc. Surgery Buenos Aires (asso.), Argentine Soc. Cancerology, Faculte de Medicine de Paris (asso. fgn. mem.), Argentine Coll. Surgeons. Author: The Lymphography—Its Clinical and Surgical Value, 1967; also articles. Research on gastroent. surgery including the hepatic cirrhosis, exploration of thoracic canal by lymphography in cirrhosis, hemorrhages after porta; caval shunt, intrahepatic thrombosis after portacaval shunt, lymphography in cancer and in blood diseases and lymphedemas, pancreatic necrosis. Home: 1042 Uruguay, Buenos Aires. Office: 5044 Diaz Velez Av., Buenos Aires, Argentine.*

CORBEN, Herbert Charles, Am. physicist; b. Dorset, Eng., Apr. 18, 1914; s. Harold Frederick and Margaret (Hart) C.; B.A., U. Melbourne (Australia), 1934, B.Sc., 1935, M.A.M.Sc., 1936; Ph.D., Cambridge (Eng.) U., 1939; m. Mulaika Barclay, June 7, 1941 (div. Aug 1955); children—Deirdre (Mrs. John Shaw Sabine IV), Sharon (Mrs. Allen Golden), Gregory; m. 2d, Beverly Balkum, Oct. 25, 1957. Came to U. S., 1946, naturalized, 1950. Lectr., New Eng. U. Coll., Armidale, Australia, 1941; lectr. physics and math. U. Melbourne, 1942-46; acting dean Trinity Coll., Melbourne, 1942-45; asso. prof. physics Carnegie Inst. Tech., Pitts., 1946-51, prof., 1953-56; mem. tech. staff Ramo-Wooldridge Corp., Westchester, Los Angeles, 1956-57, asso. dir. Research Lab., Canoga Park, Cal., 1957-61; dir. quantum physics lab. TRW, Inc., Redondo Beach, Cal., 1961—. Lectr., U. Pitts., 1947-48, U. So. Cal., 1956-58; Fulbright lectr. univs. Genoa, Milan, Bologna, Padua, 1951-53; vis. scientist dept. math. Imperial Coll., London, Eng., 1962-63. Exhbn. 1851 scholar, 1936-39; Rouse Ball research fellow, 1936-39; Commonwealth Fund fellow, 1938-41. Fellow Am. Phys. Soc.; mem. Am. Assn. Physics Tchrs. Author: (with P. Stehle) Classical Mechanics, 1950, 60; Classical and Quantum Theories of Spinning Particles, 1968. Contbr. to books and dictionaries. Theoretical studies in basic and applied physics—relativity theory, field theory, classical and quantum mechanics, passage of electrons through matter, cosmic ray showers, theory of nuclear forces, nuclear reactor theory; author theory of elementary particles based on relativistic quantum theory of spinning top. Home: 247 34th St., Hermosa Beach, Cal. 90254. Office: 1 Space Park, Redondo Beach, Cal. 90278.*

CORBIN, Kendall B(rooks), Am. physician; b. Oak Park, Ill., Dec. 31, 1907; s. William Sherman and Emma (Heacock) C.; A.B., Stanford, 1931, M.D., 1935; m. Eryl Portia Wallace, Jan. 2, 1932; children—Kendall Wallace, Edwin Malcolm. Instr. in anatomy, Stanford, 1935-38; Nat. Research Council fellow in medicine, Neurology Inst., Northwestern U., 1937-38; asso. prof. anatomy, Tenn. U., then prof. and chief div. anatomy, 1938-46, in charge neurology, 1943-46; prof. neurology Mayo Found., Minn. U., cons. in neurology, Mayo Clinic, 1946—, head sect. neurology, 1956-63, sr. cons. neurology, 1963—, pres. staff, 1968; also asso. dir. Mayo Found. for Med. Edn. and Research, Grad. Sch. of U. Minn., 1950-54. Chmn. of Rochester Com. on Higher Edn. Diplomate Am. Bd. Psychiatry and Neurology. Mem. Amer. Neurol. Assn., Amer. Acad. Neurology, A.M.A., Am. Assn. Anatomists, Am. Physiol. Soc., Soc. Exptl. Biology and Medicine, Minn. Med. Assn., Central Neuropsychiatric Assn., Minn. Soc. Neurology and Psychiatry, Phi Beta Kappa, Sigma Xi, Alpha Omega Alpha. Contbr. articles on nervous system to med. jours. Introduced use of trihexyphenidyl in treatment of Parkinson's disease, 1949. Home: Crocus Hill, Route 2, Rochester. Office: Mayo Clinic, Rochester, Minn.*

CORCORAN, Eugene Francis, Am. chem. oceanographer; b. Arthur, N.D., Nov. 28, 1916; s. Harold Alyn and Katherine (Zieg) C.; B.S., N.D. Agrl. Coll., 1940; student San Diego State Coll., 1947-49; Ph.D., Scripps Instn. Oceanography U. Cal., 1958; m. Aldoris Caroline Brevig, Aug. 29, 1940; children—Kathryn Ann (Mrs. Oliver Dixon Edwards), Sheryl Jean. Instr. math. San Diego State Coll., 1946-49; research asst. Scripps Instn. Oceanography, 1949-50, grad. research biochemist, 1952-57; asst. prof. marine biochemistry Inst. Marine Scis. U. Miami, 1957-65, asso. prof., 1965—; asso. program dir. facilities and spl. programs Biol. and Med. Scis., NSF, 1967—; adv. com. Duke U. Oceanographic Program. Mem. Marine Biol. Assn. U.K., Am. Chem. Soc., Am. Soc. Limnology and Oceanography, Geochem. Soc., A.A.A.S., Sigma Xi. Research and publs. with primary organic prodn. in the sea and factors which influence it; studies of minor constituents in sea water (transition elements and organics) as well as major nutrients, (phosphates, silicates and nitrogen compounds); hydrographic conditions, (salinity, temperature, light). Home: 5990 S.W. 85th St., South Miami, Fla. 33143. Office: 1 Rickenbacker Causeway, Miami, Fla. 33149.*

CORDA, August Karl Joseph, botanist; b. Reichenberg, Germany, Sept. 22, 1809; self-educated; curator Nat. Mus., Prague, Czechoslovakia; made sci. tour of Tex. Mem. Acad. Scis. Vienna (corr.). Author: Monographia Rhizospermorum et Hepatiearum, 1929; Icones fungorum hucusque cognitorum, 6 vols, 1837-54; Prachtflora europaischen Schimmelbildung, 1839; Beiträge zur Flora der Vorwelft, 1845. One of first to study morphology of fossil plants; research on plant anatomy, cryptogamic botany. Died at sea, Sept. 1849.

CORDEN, Malcolm Ernest, plant pathologist; b. Portland, Ore., Nov. 8, 1927; s. Sidney E. and Catherine (Andrews) C.; student Reed Coll., 1945, U. Wash., 1945-46, 47, U. Ore., 1948-49; B.S., Ore. State U., 1952, Ph.D., 1955; m. Josephine Marie Willimont, Aug. 26, 1948; children—Jeffry Lynn, David Allen, John William, Cathy Jo, Steven Whitney, Karen Lynn. Asst. prof. plant pathology Conn. Agrl. Expt. Sta., New Haven, 1955-58; faculty Ore. State U., Corvallis, 1958—, prof. plant pathology, 1966—. Dir. botany and plant pathology NSF Undergrad. Research Participation Program, 1966—. Mem. Am. Phytopath. Soc., Am. Soc. Plant Physiologists, Am. Inst. Biol. Scis., Sigma Xi, Phi Kappa Phi. Research on physiology of wilt diseases of plants; characterization of fungal pectic enzymes, mode of action of fungicides. Home: 915 Merrie Dr., Corvallis. Ore. 97330.*

CORDIER, Pierre Louis Antoine, French mining engr., geologist; b. Abbeville, France, Mar. 31, 1777; ed. Sch. Mines, Paris; mem. expdn. to Egypt, 1798; became engr.-in-chief, 1808; apptd. prof. geology Mus. Natural History, Paris, 1819; gen. insp. of mines. Mem. French Acad. Scis., 1822. Author: Essay on the Internal Temperature of the Earth, 1827; various sci. memoirs. Authority on metallurgy and mining; enlarged geol. collection of Mus. of Natural History; investigated volcanic eruption. Died Paris, Mar. 30, 1861.

CORDON, Faustino, Spanish biologist; b. Madrid, Spain, Jan. 22, 1909; s. Antonio and Elena (Bonet) C.; Ph.D., in Pharmacy, U. Madrid; m. Maria Vergara, Dec. 7, 1948; children—Maria, Teresa, Elena, Ines. Researcher, Zeltia S.A. Labs.; head of service Inst. Biology and Serotherapy. Author: Immunidad y automultiplicación proteica, 1954; Significación de la fotosintesis y respiración celular, 1957; Introducción al origen y evolución de la vida, 1958; Generalización de los principios teoricos del darwinismo, 1961; La actividad cientifica y su medio social, 1962. Home: calle Alcala, 91, Madrid 9. Office: calle Bravo Murillo, 53, Madrid, Spain.

CORDUS, Euricius, German physician; b. Simtshausen, Hesse, 1486; s. Urban Solden C.; studied at U. Erfurt; m. Kunigunde Dunnwald, 1514; 5 sons and 3 daus. including Valerius, Philipp, Augustus, Juliane, Regine. Disciple of Luther; became prof. Marburg (Germany), 1527; named city physician, Bremen, Germany, 1534. Author: Botanologicon sur colloquium de herbis, 1534; Liber de Urinis, (opposed med. superstitions) 1543; Botanologicon (1st attempt to establish a sci. basis for botany in Germany), 1534; also 1st pharmacopoeia in Germany, 1535. First to produce sulphuric ether; gave an early description of acute infectious disease with symptoms of fever, sweating and skin lesions which was known as maliary fever and sweating sickness, 1529. Died Bremen, Dec. 24, 1535.

CORDUS, Valerius, physician, botanist; b. Simmershausen, Hesse, Feb. 18, 1515; s. Euricius and Kunnigund (Dunnward) C.; ed. by father; studied medicine Univs. Erfurt, Marburg, Wittenberg. Traveled throughout Germany to see mines and collect plants; lectured on pharmacy at Wittenberg; studied botany in Italy, 1542. Author: Annotationes in Dioscorides de materia medica; Sylva rerum fossilium, historia stirpium; De Halosantho seu Spermate Ceti vulgo dicto; Pharmacorum omnium quae in usu potiss. sunt componendorum ratio, 1535. One of greatest botanists of his time; his Pharmacopeia first work of its kind on pharmacy; described preparation of ether by distilling alcohol with concentrated sulphuric acid. Died Rome, Italy, Sept. 25, 1544.

CORE, Earl Lemley, Am. botanist; b. Core, W.Va., Jan. 20, 1902; s. Harry M. and Clara (Lemley) C.; A.B., W.Va. U., 1926, M.A., 1928; Ph.D., Columbia, 1936; D.Sc., Waynesburg Coll., 1957; m. Freda Bess Garrison, June 8, 1925; children—Ruth (Mrs. Harry Miller), Merle, Harry, David. Faculty, W.Va. U., Morgantown, 1928—, prof. botany, 1942—, chmn. dept. biology, 1948—; botanist Colombian Chincona Mission, Fgn. Econ. Adminstrn., Bogota, Colombia, 1943-45. A founder W.Va. U. Arboretum, 1948, Terra Alta (W.Va.) Biol. Sta., 1962. Mem. Bot. Soc. Am., A.A.-A.S., Internat. Soc. Plant Taxonomists, Sigma Xi. Author: Plant Taxonomy, 1955; (with Nelle Ammons) Woody Plants in Winter, 1958; (with P. D. Strausbaugh, B. R. Weimer) General Biology, 1961; (with P. D. Strausbaugh) Flora of West Virginia, 4 vols., 1952-64; also numerous articles. Investigations of Am. species of Scleria; taxonomy of vascular plants of eastern U. S. Home: 460 Brockway Av., Morgantown, W.Va. 26505.*

COREY, Elias James, Am. chemist; b. Methuen, Mass., July 12, 1928; s. Elias and Tina (Hashem) C.; S.B., Mass. Inst. Tech., 1948, Ph.D., 1951; A.M. (hon.), Harvard, 1959; m. Claire Higham, Sept. 14, 1961; children—David, and John. From instructor to assistant professor at the University of Ill., 1951-55, prof., 1955-59; prof. chemistry Harvard, 1959—; sci. adviser Chas. Pfizer Co., Union Carbide Co.; mem. sci. bd. Itek Corp. Cons. U. S. Office of Science and Technology. Recipient Swiss-Am. exchange fellowship, 1957; Guggenheim fellow, 1957-58; Alfred P. Sloan Found. fellow, 1956-59. Mem. Am. Acad. Arts and Scis., A.A.A.S., Am. (award pure chemistry 1959), Swiss, German chem. socs., Nat. Acad. Sci., Sigma Xi. Author numerous sci., publs. Research in theoretical, structural, and synthetic organic chemistry; stereochemistry. Home: 20 Avon Hill St., Cambridge, Mass.

CORFIELD, William Henry, Brit. physician; b. Shrewsbury, Dec. 14, 1843; s. Thomas Corfield; ed. Magdalen Coll., Oxford, Univ. Coll., London, med. schs. in Paris, Lyons; M.A., M.D., Oxford; m. Emily Madelina Pike. Med. fellow Pembroke Coll.; Burdett-Coutts geol. scholar, Radcliffe travelling fellow Oxford U.; examiner for honours Natural Sci. Sch., Oxford, 1868; prof. hygiene and pub. health Univ. Coll., London, also hon. san. adviser, founded 1st hygienic lab. San. adviser to His Majesty's Office Works, 1899-1903. Recipient Bronze medal Royal Soc. Pub. Mediciene, Belgium, 1901. Fellow Royal Coll. Physicians; mem. Epidemiol. Soc. London (pres.), Soc. Med. Officers of Health (past pres). Author: A Digest of Facts relating to the Treatment and Utilisation of Sewage, 3d edit., 1887; Lectures to the Royal Engineers on Water Supply Sewerage and Sewage Utilisation, 1874; Laws of Health, 9th edit., 1896; Dwelling-houses, their Sanitary Construction and Arrangements, 4th edit., 1898; Disease and Defective House Sanitation, 1896; Sanitary Knowledge, 1900; The Etiology of Typhoid Fever and Its Prevention, 1902. Discovered existence of lithodomus borings in Aymestry limestone of Silurian formation (showed that boring bivalves existed earlier than previously believed). Died Aug. 28, 1903.

CORI, Carl Ferdinand, biochemist; b. Prague, Czechoslovakia, Dec. 5, 1896; s. Carl I. and Maria (Lippich) C.; student Gymnasium, Trieste, Austria, 1906-14; M.D., German U. of Prague, 1920; Sc.D., Yale and Western Res. univs., 1947, Boston U., 1948, Cambridge (Eng.) U.; m. Gerty Theresa Radnitz, Aug. 5, 1920 (dec. Oct. 26, 1957); 1 son, Carl Thomas; m. 2d, Anne McK. Jones, Mar. 23, 1960. Came to U. S., 1922; naturalized, 1928. Asst. in pharmacology, U. Graz, Austria, 1920-21; biochem. State Inst. for Study Malignant Disease, Buffalo, 1922-31; prof. pharmacology and biochemistry, Washington U. Sch. Medicine, St. Louis, 1931—. Recipient (with G. T. Cori and B. A. Houssay) Nobel Prize in medicine and physiology, 1947; Willard Gibbs Medal, Am. Chem. Soc., 1948; Sugar Research Found. award, 1947, 50; Lasker award, 1946, Squibb award, 1947. Fellow Royal Soc., 1950; mem. Nat. Acad. Scis.; hon. mem. Harvey Soc.; mem. Am. Soc. Biol. Chemists, Am. Chem. Soc. (Mid-West award, 1946), A.A.A.S., Am. Philos. Soc., Sigma Xi. Contbr. articles, chiefly on carbohydrate metabolism and enzymes of animal tissues, to Am. sci. jours. Mem. editorial Bd. Jour. of Biol. Chemistry, Biochimica et Biophysica Acta. Research (with wife) into carbohydrate metabolism, especially discovery of glycolysis in live tumors and isolation (1936) of glucose-l-phosphate (Cori ester), leading to discovery of the course of catalytic conversion of glycogen to glucose by means of phosphorylase; (Nobel prize 1947); other work on influence of ovariectomy on tumor incidence, carbohydrate metabolism of malignant tumors; sugar in animal body; intestinal absorption; action of epinephrin on metabolism; phosphate changes in muscle; isolation of crystalline enzymes, mechanism of action of insulin. Home: 909 Lay Rd., St. Louis 24.

CORI, Carl Isidor, zoologist; b. Brüx, Germany, Feb. 24, 1865; s. Eduard and Rosina (Trinks) C.; Dr. phil., U. Leipzig (Germany), 1889; Dr. med., U. Prague (Czechoslovakia), 1891; m. Maria Lippich, 1892; 1 son, 2 daus. Asso. prof., head zool. sta., Trieste, from 1898; prof. in ordinary, Vienna, Austria, from 1908; prof. German U., Prague, 1919-35. Author: (with B. Hatschek) Elementarkurs der Zoolomie in 15 Vorlesungen, 1896; Der Naturfreund am Meeresstrande, 1928; Biologie der Tiere, 1935. Research on anatomy of Kamotozoas, Phoronideae; initiator, organizer Adriatic Sea research. Died Vienna, Aug. 31, 1954.

CORI, Gerty Theresa Radnitz, biochemist; b. Prague, Czechoslovakia, Aug. 15, 1896; d. Otto and Martha (Neustadt) Radnitz; grad. Realgymnasium of Tetschen, Czechoslovakia, 1914; M.D., German U. Prague Med. Sch., 1920, Sc.D., Boston U., 1948, Smith Coll., 1949, Yale, 1951, Columbia, 1954; m. Carl F. Cori, August 5, 1920; 1 son, C. Thomas. Came to U. S., 1922, naturalized, 1928. Asst., Children's Hosp., Vienna, 1920-22; asst. biochemist State Inst. for Study Malignant Diseases, Buffalo, 1922-31; research asso. Washington U. Med. Sch., 1931-47, prof. biol. chemistry, 1947-57. Mem. adv. bd. NSF. Recipient Mid-west award Am. Chem. Soc., 1946; Squibb award in endocrinology, 1947; Nobel prize in medicine and physiology (with C. F. Cori and B. A. Houssay), 1947; Garvan medal, 1948; Sugar research prize Nat. Acad. Scis., 1950; Borden award Assn. Med. Colls. Mem. Nat. Acad. Sci., Am. Philos. Soc., Am. Soc. Biol. Chemists, Harvey Soc., Am. Chem. Soc., Sigma Xi. Research (with husband) in carbohydrate metabolism, especially discovery of glycolysis in live tumors, isolation (1936) of glucose-l-phosphate (Cori ester), which led to discovery of course of catalytic conversion of glycogen to glucose by means of phosphorylase (Nobel prize 1947); also immunological study of complement of human serum; research into mechanism of action of hormones (particularly pituitary). Died St. Louis, Oct. 26, 1957.

CORIAT, Isador Henry, Am. psychiatrist, neurologist; b. Phila., Dec. 10, 1875; s. Harry and Clara (Einstein) C.; M.D., Tufts Coll. Med. Sch., 1900; spl. student in philosophy, Harvard, 1909-10; m. Etta Dann, Feb. 1, 1904. Asst. and 1st asst. phys. Worcester Insane Hosp., 1900-05; neurol. staff Boston City Hosp., 1905-19; neurologist, Mt. Sinai Hosp., 1905-14; cons. neurologist Chelsea Meml., Beth-Israel hosps., 1919-28; neuropsychiatrist Forsyth Dental Infirmary, 1913-29; instr. neurology Tufts Coll. Med. Sch., 1914-16; instr., mem. tng. com. and tng. analyst, Boston Psychoanalytic Inst. Neurologist, Med. Adv. Bd., World War. Fellow A.M.A., Mass. Med. Soc., Boston Med. Library, Am. Psychiat. Assn.; mem. Am. Psycho-Pathol. Assn. (v.p. 1931-32), N.E. Soc. Psychiatry, Internat. Psychoanalytic Assn. (v.p. 1936-37), Boston Med. History Club, Mass. Psychiat. Soc., Am. Psychoanalytic Assn. (pres. 1924-25, 36-37), Boston Soc. Neurology and Psychiatry, Boston Psychoanalytic Soc. (pres. 1930-32 and 1941-42), Jewish Acad. Arts and Scis., Am. Bd. Psychiatry and Neurology; hon. mem. Tau Epsilon Phi. Collaborating editor Jour. Abnormal Psychology, 1906-26, Psychoanalytic Rev.; collaborator for psychoanalytic terms for Dictionary of Psychology. Author: A Laboratory Manual of Clinical and Physiological Chemistry (with Dr. A. E. Austin), 1898; Religion and Medicine (with Drs. Worcester and McCoomb), 1908; Abnormal Psychology, 1910, 2d edit., 1914; The Hysteria of Lady Macbeth, 1912, 2d edit., 1919; The Meaning of Dreams, 1915; What is Psychoanalysis?, 1917; Repressed Emotions, 1920; Stammering, 1928; also monographs and articles on nervous and mental diseases, psychopathology and psychoanalysis. Died May 26, 1943.

CORIOLIS, Gaspard Gustave de, see de Coriolis.

CORIOU, Henri Étienne, French metallurgist; b. St. Goazec, France, Nov. 1, 1925; s. Yves Michel and Marie (Daniel) C.; licence ès sciences physiques U. Paris, 1947; m. Agnes Amiot, July 26, 1954. With French AEC, 1947—, head of service for aqueous corrosion and electrochemistry studies, Saclay, 1964—, lectr. Inst. Sci. and Nuclear Tech., 1967—. Pres., Commn. on Electrochemistry at High Temperatures, 1961-65; pres. group for nuclear materials corrosion European Fedn. Corrosion, 1967; sci. adviser Belgian Center for Corrosion Studies, 1967. Recipient Rist prize French Soc. Metallurgy, 1959. Mem. Centre français de la corrosion, Comité International de Thermodynamique et de Cinetique Electrochimiques, Société de Chimie Industrielle. Research, publs. on corrosion of high purity metals, alloys; discovery of spl. steels for use in highly oxidizing nitric media; discovery of a reference electrode for use in molten fluorides. Home: 15 rue du Progress, Le Plessis Robinson 92, France. Office: Dept. Chemistry, SECAE-CEN Saclay, B.P. n.2, 91 Gif-sur-Yvette, France.*

CORK, James M., Am. physicist; b. Yale, Mich., July 9, 1894; s. George M. and Jennie (Lee) C.; B.S., U. Mich., 1916, M.S., 1917, Ph.D., 1921; m. Laurie Kaufmann, 1918; children—Janet Lee (Mrs. John C. Wahr), James A. Asst. physicist U. S. Bur. Standards, 1919; asst. prof. physics, Pa. State Coll., 1919-20; instr. physics, U. Mich., 1920-25, asst. prof. 1926-31, asso. prof. 1932-37, prof., 1937-57; exchange prof. Victoria U., Manchester, Eng., 1926-27; cons. Argonne Nat. Lab., 1950-57. Mem. NDRC, 1942-45. Fellow Am. Phys. Soc., A.A.A.S.; mem. Washington Philos. Soc., Sigma Xi, Gamma Alpha. Author: Pyrometry (with W. P. Wood), 1927, 1941; Heat, 1933, 1942; Radioactivity and Nuclear Physics, 1946, 1957. Contbr. articles to tech. jours. on heat, X-rays and radioactivity. Died Nov. 27, 1957.

CORLETT, William Thomas, Am. physician; b. Orange, O., Apr. 15, 1854; s. William and Ann (Avery) C.; ed. Oberlin Coll., 1870-73; M.D., Wooster U., 1877; student and intern London Hosp., 1879-81, Hôpital St. Louis, Paris, France, 1881; diploma Royal Coll. Phys., London, 1881; later studied in Vienna, Berlin and Breslau; m. Amanda Marie Leisy, June 26, 1895; children—Christine L. (Mrs. Horace F. Henriques), Ann E. (Mrs. Daniel B. Ford), Helen A., Edward L. Prof. diseases of skin and genito-urinary diseases Wooster U., 1883-85; prof. dermatology and syphilology Western Res. U., 1885-1914, sr. prof., 1914-24, emeritus prof., 1924-48. Fellow Royal Soc. Medicine (Gt. Britain), A.A.A.S., A.M.A.; mem. Am. Dermatol. Assn. (hon., pres. 1905), Am. Acad. Dermatology and Syphilology (hon.); corr. mem. Brit. Assn. Dermatology and Syphilology. Author: Treatise on the Acute Infectious Exanthemata, 1901; The American Tropics, 1908; The People of Orrisdale and Others, 1918; Early Reminiscences, 1920. Wrote: The Scaly Diseases of the Skin (Vol. III, Morrow's System of Dermatology, etc.), 1894; The Vegetable Parasitic Diseases of the Skin (in Bangs and Hardaway's American Text-Book of Genito-Urinary Diseases, etc.), 1898; Purpura, Pompholyx and Pellagra, in Reference Handbook of the Medical Sciences, 1903; also on Lichen, Lentigo, Granuloma, Annulare in 1915 edition; The Medicine-Man of the American Indian and His Cultural Background, 1935. Also articles on diseases of skin in Am. and fgn. jours. Died June 11, 1948.

CORMACK, Allan MacLeod, Am. physicist; b. Johannesburg, South Africa, Feb. 23, 1924; s. George and Amelia (MacLeod) C.; B.S., U. Cape Town, 1944, M.S., 1945; postgrad. Cambridge U., 1947-49; m. Barbara Jeanne Seavey, Jan. 6, 1950; children—Margaret Jean, Jean Barbara, Robert Allan Seavey. Came to U. S., 1956, naturalized, 1966. Lectr., U. Cape Town, 1950-56; research fellow Harvard, 1956-57; faculty Tufts U., Medford, Mass., 1957—, prof., 1964—; visitor Lawrence Radiation Lab. U. Cal. at Berkeley, 1966-67. Fellow Am. Phys. Soc.; mem. South African Inst. Physics, Sigma Xi. Research on medium energy nucleon-nucleon and nucleon-nucleus scattering. Home: 18 Harrison St., Winchester, Mass. 01890. Office: Physics Dept., Tufts U., Medford, Mass. 02155.*

CORMONTAINGNE, Louis de, French mil. engr.; b. Strasbourg, France, 1695; dir. sieges of Menin, Ypres, Tournay, Belgium, 1734-45; author: Mémorial pour l'attaque des places; Mémorial pour la fortification permanente et passagère, 1805-09 (posthumous publ.); improved art of fortification; planned additions to fortifications of Metz and Thionville during reign of Louis XV of France. Died 1752.

CORNATZER, William Eugene, Am. biochemist; b. Mocksville, N.C., Sept. 23, 1918; s. William P. and Stella Augusta (Vogler) C.; student Mars Hill Coll., 1935-37; B.S., Wake Forest Coll., 1939; M.S., U. N.C., 1941, Ph.D., 1944; M.D., Bowman Gray Sch. Medicine, 1951; m. Margaret Virginia Freeman, Mar. 30, 1946; children—Nancy Freeman, William Eugene. Fels Research fellow U. N.C., 1941-45; asst. prof. biochemistry Bowman Gray Sch. Medicine, 1946-51; prof., head dept. biochemistry, dir. Ireland Research Lab., U. N.D. Sch. Medicine, Grand Forks, 1951—. Mem. research adv. com. human nutrition and consumer use U. S. Dept. Agr., 1965—. Recipient Frank Billing award for original investigation A.M.A., 1951. Diplomate Am. Bd. Clin. Chemistry (dir. 1960-66), Nat. Bd. Med. Examiners (mem. biochemistry test com. 1962-66). Fellow A.A.A.S., N.Y. Acad. Scis., A.C.P.; mem. Am. Assn. Clin. Chemistry (past mem. exec. com.), N.D. Acad. Sci. (past pres.), Am. Soc. Biol. Chemists, Am. Inst. Nutrition, Am. Chem. Soc., Am. Fedn. for Clin. Research, Am. Assn. for Study Liver Disease, Am. Assn. for Cancer Research, Soc. for Exptl. Biology and Medicine, Soc. for Clin. Research, Radiation Research Soc., N.C. Acad. Sci., Royal Soc. Medicine (affiliate), Sigma Xi, Alpha Omega Alpha. Research and numerous publs. on quinine metabolism and absorption, radiation effects and toxicity of isotopes, lipotropic agents, phosopholipid metabolism. Home: 307 Park Av., Grand Forks, N.D. 58201.*

CORNEJO, Mariano Harlan, Peruvian sociologist, politician; b. Arequipa, Peru, Oct. 29, 1870; s. Mariano Cornejo; ed. Ph.D., U. San Marcos, Lima, Peru, 1887; Doctorate in jurisprudence, 1889; m. Clorinda Cano; 1 son, 1 dau. Mem. Peruvian Chamber Deps., 1893-1904, pres. 1901-03; prof. U. Lima, 1896; senator, 1911-20. Mem. Ct. Internat. Justice, The Hague, Netherlands, 1916; pres. 10th Internat. Congress Sociology Geneva, Switzerland, 1930. Named officier d'Instruction Publique. Author: Sociologia general, 1908; many sci. articles. Mem. Spanish Acad., acads. Jurisprudence, History, Geography, Acad. Scis. Madrid (Spain), Internat. Inst. Sociology, Institut de France, Academie des Sciences morales et Politiques. Noted for comprehensive theory of group behavior; gave psychol. interpretation of factors which influence devel. soc. and instns. Died 1942.

CORNELIUS, Charles Edward, Am. physiologist; b. Huntington Park, Cal., Dec. 19, 1927; s. Samuel Paul and Alberta (Johnson) C.; B.S., U. Cal. at Davis, 1949, B.S. in Vet. Sci., 1951, D.V.M., 1953, Ph.D., 1957; m. Bette Jean Watt, Sept. 2, 1948; children—Stephen, Clifford, John, Aimee. Asst. prof. U. Cal., Davis, 1957-62, asso. prof., 1962-66; prof., dean Kan. State U. Coll. Vet. Medicine, 1966—. Mem. study sect. NIH Gen. Medicine, 1965—. Mem. Am. Physiol. Soc., Soc. for Exptl. Biology and Medicine, Am. Vet. Med. Assn., Am. Soc. Zoologists. Author: with Jiro J. Kaneko) Clinical Biochemistry of Domestic Animals, 1963. Co-editor Advances in Veterinary Science, 1966. Research, numerous publs. on biochem. metabolism important to urolithiasis, organic anion transport by the liver, liver function. Home: 2060 Hunting St., Manhattan, Kan. 66502.*

CORNELL, Ezra, Am. telegraph magnate; b. Westchester, N.Y., Jan. 11, 1807; s. Elijah and Eunice (Barnard) C.; m. Mary Ann Wood, Mar. 19, 1831, 1 son, Alonzo B. Worked in flour and plaster mills of J. S. Beebe, Ithaca, N.Y., 1828-41, became gen. mgr.; with Samuel F. B. Morse devised means for insulating telegraph wires on poles and helped erect line from Balt. to Washington; owned cos. bldg. lines between many major cities in East and Midwest, including Magnetic Telegraph Co., Erie & Mich. Telegraph Co., N.Y. & Erie Telegraph Co., merged with competing lines to form Western Union Telegraph Co., dir., 1855-74, largest stockholder until 1870; pres. N.Y. State Agrl. Soc., 1862; founded, endowed Cornell U., 1868, provided for edn. of women and poor students in liberal and mech. arts. Died Dec. 9, 1874.

CORNER, Edred John Henry, Brit. botanist; b. London, Eng., Jan. 12, 1906; s. Edred Moss and Henrietta (Henderson) C.; M.A., Cambridge (Eng.) U., 1929; m. Helga Dinesen Sondergoord, Apr. 9, 1953; children—John Kavanagh, Stephanie Christine, Dorothy Lindsay Helga. Asst. dir. gardens dept. Straits Settlements, 1929-46; field sci. co-operation officer UNESCO, Latin Am., 1947-48; faculty U. Cambridge, 1949—, prof. tropical botany, 1966—, fellow Sidney Sussex Coll., 1950—. Fellow Royal Soc. London (Darwin medal 1960); mem. Royal Geog. Soc. (Patron's medal 1966), Linnean Soc. London, Brit., Am. mycol. socs., Société Mycologique de France, Mycol. Soc. Japan, Internat. Union for Conservation of Nature. Author: Wayside Trees of Malaya, 1940; Clavaria and Allied Genera, 1950; Life of Plants, 1964; Cantharelloid Fungi, 1966; Natural History of Palms, 1966; also numerous articles. Research on fungal fructifications, tropical trees, evolution tropical forest, constrn. flowers, fruits and seeds. Home: 91 Hinton Way, Great Shelford, Cambridge, Eng.*

CORNER, George Washington, Am. med. biologist; b. Balt., Dec. 12, 1889; s. George Washington and Florence (Evans) C.; A.B., Johns Hopkins, 1909, M.D., 1913; Dr. honoris causa, Cath. U. Chile, 1942; D.Sc., U. Rochester, 1944, Boston U., 1948, Chgo., 1958; LL.D., Tulane, 1955, Temple 1956; M.D.S. (hon.), Women's Med. Coll., Phila., 1958; D.Sc., Oxford U., 1950, M.A., 1952; Litt.D., U. Pa., 1965; m. Betsy Lyon Copping, Dec. 28, 1915; children—George Washington, Hester Ann (dec.). Med. asst. Grenfell Labrador Mission, summers, 1912-13; asst. in anatomy, Johns Hopkins, 1913-14; resident house officer, Johns Hopkins Hosp., 1914-15; asst. prof. anatomy U. Cal. 1915-19; asso. prof. anatomy, Johns Hopkins, 1919-23, now professor emeritus of embryology; prof. anatomy U. Rochester, 1923-40; also curator Med. Library, 1938-40; dir. dept. of embryology, Carnegie Inst. of Washington, 1940-56; historian Rockefeller Inst., 1956-60, vis. prof., 1961—; exec. officer Am. Philos. Soc., 1960—; George Eastman vis. prof. Oxford U., 1952-53. Vicary lecturer, Royal Coll. Surgeons, London, 1936, Vanuxem lectr. Princeton U., 1942; Terry lectr. Yale, 1944; research prof. Commonwealth Fund, Univ. of Louisville Med. School, 1946. U. S. del. Internat. Congress of Endocrinology, 1941, pres. congress, 1964. Mng. editor, Am. Journal Anatomy, 1939-41. Recipient Squibb award Soc. Study of Internal Secretions, 1940; Presdl. Certificate of Merit, 1948; Passano Found. award, 1958. Hon. fellow Balliol Coll., Oxford, 1952-53. Hon. fellow Royal Soc. Edinburgh, Royal Coll. Obstetrics and Gynecology London; fellow Internat. Inst. Embryology; mem. Royal Soc. London (fgn.), Am. Assn. Anatomists (sec. 1930-38, pres. 1946-48), Soc. Exptl. Biology and Medicine, Am. Philos. Soc. (v.p. 1953-56) Nat. Acad. Scis. (v.p. 1953-57), Anat. Soc. Gt. Britain (hon.), Am. Assn. History Medicine, Phi Beta Kappa, Sigma Xi, fgn. corr. mem. numerous socs.; hon. mem. and fellow Am. and fgn. socs. Author books, 1927—; also numerous papers in field. Anatomical studies of ovarian and uterine cycle, which led to understanding of anatomical details of menstrual cycle and the functions of estrogen and progesterone; also contbd. to fetal and neonatal physiology. Office: 104 S. 5th St., Phila. 19106.*

CORNET, George, German bacteriologist, physician; b. Eichstätt, Germany, July 27, 1858; student U. Munich (Germany); visited sanatoriums of France and Italy; med. asst. at Görbersdorf; worked under Robert Koch, Berlin, Germany; resort physician, Reichenhall, Germany; named hon. prof. U. Berlin. Author numerous publs. including: Über Tuberkulose, 1890; Die Tuberkulose, 1896; Die Schwindsucht, 1900. Proved the existence of Tb bacillus and need for hygiene to reduce contagion. Died Berlin, Mar. 26, 1915.

CORNETTE, Claude-Melchior, French physician; b. Besançon, France, Mar. 1, 1744; s. Pierre-Claude and Claude (Sauvin) C.; student Jesuit sch., Besançon, 1760, pharmacy and chemistry Paris, France, 1763. Worked under Lassone (1st royal physician), 1772; dir. Marly Lab., from 1773; apptd. Médecin des épidémies, 1782; 1st royal physician, physician to Count of Artois, from 1784, emigrated from France, 1791. Mem. French Acad. Scis., 1778, Soc. Medicine. Studied decomposition of salts by muriatic acid, effects of acids on fatty substances, properties of ipecac, cinchona bark and opium; in charge sanitation Versailles, France. Died May 11, 1794.

CORNFIELD, Jerome, Am. statistician; b. New York, N.Y., Oct. 30, 1912; s. Samuel and Jenny (Haren) C.; B.S., New York Univ., 1933; Columbia Univ., U. S. Dept. Agric. Graduate School; m. Ruth June Bittler, July 6, 1937; children: Ann, Ellen. Statistician, U. S. Dept. Labor, 1934-47; U. S. National Institute of Health, 1947-58; 1960-67; prof. Johns Hopkins Univ., 1958-60; prof. Univ. Pittsburgh, 1967—. Taught at Columbia, American U., Univ. of Mich., Stanford U., Univ. of Oslo. Awarded Superior Science Award, U. S. Dept. Health, Education and Welfare, 1967. Mem., Am. Statistical Assn.; Institute Mathematical Statistics, A.A.A.S., Am. Epidem. Soc., Biometric Soc., Washington Acad. Sci., Virginia Acad. Sci. Author numerous articles. Developed statistical methods that contribute to sampling of finite populations, randomized clinical trial, biological assay, analysis of epidemiological studies, analysis of linear multiple compartment systems, analysis of variance, input-output systems. Home: 9600 Arnon Chapel Road, Great Falls, Virginia.*

CORNFORTH, John Warcup, chemist; b. Sydney, Australia, Sept. 7, 1917; s. John William and Hilda (Eipper) C.; B.Sc., U. Sydney, 1937, M.Sc., 1938; D. Phil., Oxford U., 1941; m. Rita H. Harradence, Sept. 27, 1941; children—Brenda, John, Philippa. Mem. sci. staff Med. Research Council, London, Eng., 1946-62; director research Milstead Lab. Chem. Enzymology, Shell Research Ltd., Sittingbourne, Kent, Eng., 1962—; asso. prof. Sch. Molelular Scis., U. Warwick, 1965—. Recipient Stouffer prize, 1967. Fellow Royal Soc., 1953, A.A.A.S.; mem. Chem. Soc. (Corday-Morgan medal 1953, Flintoff medal 1966), Biochem. Soc. (CIBA medal 1966), Am. Soc. Biol. Chemists (hon.), Am. Chem. Soc. Publs. on contbns. to chemistry of penicillin, total synthesis of steroids and other biol. active natural products, chemistry of heterocyclic compounds; biosynthesis of steroids; enzyme chemistry. Home: 187 Ufton Lane. Office: Broad Oak Rd., Sittingbourne, Kent, Eng.*

CORNIL, Andre Victor, French bacteriologist; b. Cusset, France, June 17, 1837; ed. in medicine Paris, France, doctorate thesis, 1864. Staff Lourcine Hosp., from 1864; med. faculty Parigi, from 1869, prof. path. anatomy, from 1882; dep. of Republican party, senator. Mem. Acad. Medicine, Anat. Soc. (pres. 1885). Author: (with Raver) Manuel d'histologie pathologique, 1869-76; Les bacteries et leur role dans l'anatomie, 1885. Contbr. articles on histology, bacteriology. Founder histology lab.; as senator worked to pass law regulating practice of medicine and pharmacy in France, also responsible for bills on sanitation of Paris, other large cities, and on water systems of Vigne and Verneuil (both France). Died Apr. 14, 1908.

CORNING, Hanson Kelly, anatomist; b. N.Y.C., Nov. 19, 1860; ed. Germany; student of Gegenbair, Waldeyer, Rabl; prof. Basel, Switzerland; author, editor textbooks on topog. anatomy, 1907, hist. devel. of anatomy, 1921. Died N.Y.C., Feb. 7, 1951.

CORNING, J(ames) Leonard, Am. neurologist; b. Stamford, Conn., Aug. 26, 1855; s. James Leonard and Sarah Ellen (Deming) C.; student U. Heidelberg; M.D., U. Würzburg (Germany), 1878; A.M. (hon.), Williams Coll., 1888; LL.D.; m. Julia Crane, May 12, 1883. Cons. in nervous and mental diseases to various hosps., N.Y. Author: Carotid Compression, 1882; Brain Rest, 1883, Brain Exhaustion; Local Anaesthesia, 1886; Hysteria and Epilepsy, 1888; A Treatise on Headache and Neuralgia, 1888; Pain in Its Neuro-Pathological and Neuro-Therapeutic Relations, 1894. Discoverer spinal anesthesia, 1885; demonstrated that action of certain medicinal substances, notably stimulants and sedatives, may be increased and prolonged while subject remains in compressed air; 1st to inject liquid paraffin into tissues and solidify it in situ. Died Aug. 24, 1923.

CORNISH, Vaughan, Brit. geographer; b. Suffolk, Eng., Dec. 22, 1862; s. C. J. Cornish; ed. Victoria U., Manchester, Eng.; D.Sc., Manchester; m. Ellen Agnes Provis, 1891 (dec. 1911); m. 2d, Mary Watson Floyer, 1913. Dir. tech. edn. Hampshire County Council, until 1895; engaged in geog. research, also in movement for preservation scenery of rural Eng., from 1895. Recipient Grand Prix for sci. photography Franco-Brit. Exhbn., 1908. Fellow Royal Geog. Soc. (Gill Meml. award 1900); pres. Geog. sect. Brit. Assn., 1923, 1928. Conf. Dels. of Corr. Socs., 1928; pres. Geog. Assn. Author: The Panama Canal and its Makers; Waves of the Sea and other Water Waves; The Travels of Ellen Cornish; Waves of Sand and Snow; A Geography of Imperial Defence; A Strategical Atlas of the Oceans; The Great Capitals; National Parks and the Heritage of Scenery, 1930; The Poetic Impression of Natural Scenery, 1931; The Scenery of England, 1932; Ocean Waves and Kindred Geophysical Phenomena, 1934; Scenery and the Sense of Sight, 1935; Borderlands of Language in Europe, 1937; The Scenery of Sidmouth, 1940; Historic Thorn Trees in the British Isles, 1941; A Family of Devon, 1942; The Beauties of Scenery, 1943; The Churchyard Yew, an Emblem of Immortality, 1964; Geographical Essays, 1946; Photography of Scenery, 1946; also various papers. Died May 1, 1948.

CORNMAN, Ivor, Am. biologist; b. Cleve., May 22, 1914; s. Robert Stevenson and Anna May (Evans) C.; A.B., Oberlin Coll., 1936; M.S., N.Y. U., 1939; Ph.D., U. Mich., 1949; m. Margaret Evans, June 21, 1947. Research fellow Sloan-Kettering Inst. Cancer Research,

1946-49; asst. research prof. George Washington U. Med. Sch., 1949-55, dir. research Cancer Clinic, 1954-55; dept. chief Hazleton Labs., Falls Church, Va., 1955-59, asst. dir. research, 1959-64; biol. cons., Woods Hole, Mass., Kingston, Jamaica, 1964—. Mem. Am. Soc. Pharmacognosy, Internat. Soc. Cell Biology, Soc. Gen. Physiology, Am. Assn. Cancer Research, Am. Assn. Anatomists, Am. Physiol. Soc., Am. Soc. Zoology, Tissue Culture Assn., Am. Soc. Cell Biology, Soc. Devel. Biology, Sigma Xi. Contbr. numerous articles to sci. jours. Research on mechanism of cell division; devising and operating novel simplified systems for finding new drugs in marine organisms. Home: 10A Orchard St., Woods Hole, Mass. 02543. Office: 15 Glenview Terrace, Kingston 6, Jamaica.*

CORNU, (Marie) Alfred, French physicist; b. Orléans, France, Mar. 6, 1841; studied at École polytechnique, later at École des Mines; docteur ès sciences. Head engr. of mines; prof. physics École polytechnique; worked on installation of Nice (France) Obs. Recipient Lacaze prize, 1878. Fellow Royal Soc., 1884 (Rumford medal 1884); mem. French Acad. Scis., 1878 (pres. 1896), French Soc. Physics (pres.), French Soc. for Advancement Scis., Bur. Longitudes. Author: Un Nouveau polarimètre, 1870; Sur le spectre ce l'aurore boréale du février, 1872; Détermination de la vitesse de la lumière, 1877. Research and numerous articles in optics and spectroscopy including deformation of solid bodies withk optical methods, ultraviolet rays; improved Fizeau method for determination of speed of light; proved a math. theory of electro-magnetic phenomena could be deduced from Laplace's Law; demonstrated form of surface of light waves (in isotropic atmosphere) placed in uniform magnetic fields; measured earth's density, 1873; research on tone intervals, quartz prism free of double refraction; eponym for Cornu spiral, curve used in calculating light intensities in Fresnel diffraction. Died Romorantin, France, Apr. 12, 1902.

CORNU, Maxime, French naturalist; b. Orléans, France, July 16, 1843; Licence es sciences, Ecole Normale Superieure, 1868; doctorate Faculty of Science, Paris, France, 1872. Asst. naturalist Garden of Plants, from 1874, lectr. organography, plant physiology, 1876, later prof. field cultivation; insp. gen. agr., from 1881. Contbr. numerous sci. articles. Work with cryptogams, mildew (1 of 1st to point out its dangers), phylloxera, wheat rust, Bremia Lactucae, in a gen. way, all plant parasites. Died May 1901.

CORNUT (or CORNUTI), Jacques Phillipe, physician, botanist; b. Paris, France, 1606. Author: Canadensium plantarum aliarumque nondum editarum historia (description of 79 Canadian plants with numerous figures), 1635; also one of 1st floras of area around Paris, 1653. Died Aug. 23, 1651.

CORNWELL, George William, Am. biologist; b. Benton Harbor, Mich., Dec. 4, 1929; s. Truman G. and Dorothy (Moore) C.; B.S., Mich. State U., 1955; M.S., U. Utah, 1960; Ph.D. (NSF, Rackham fellow), U. Mich., 1966; m. Meryl A. Parren, May 31, 1952; 1 dau., Laura Lynn. Research aide Fish and Wildlife Service, Fairbanks, Alaska, 1955; ranger, naturalist Nat. Park Service, Olympic Nat. Park, Wash., 1959; asst. prof. wildlife dept. forestry and wildlife Va. Poly. Inst., Blacksburg, 1963-66; asso. prof. wildlife Sch. Forestry, U. Fla., Gainesville, 1967—. Mem. Am., Brit. ornithol. unions, Wildlife Soc., Animal Behavior Soc., Ecol. Soc. Am., Cooper, Wilson ornithol. socs., Wildlife Disease Assn. Research in waterfowl biology, animal behavior, avian diseases, extension edn. work in areas of outdoor recreation and wildlife mgmt. Home: care J. A. Parren, R.F.D. 1, Sawyer, Mich. 49125. Office: Sch. of Forestry, U. of Fla., Gainesville, Fla. 32601.*

CORONELLI, Marco Vincenzo, Italian geographer, astronomer; b. Venice, 1650; studied mathematics and geography. Went to Paris at invitation of Cardinal of Estrées, 1681; returned to Venice, 1684; named cosmographer of Venetian Republic, 1685. Author: Description of the Morea, 1685; Atlanté Veneto, 1690; Corso geografico, 1694; Isolario . . . (2 vol.), 1696-8; Biblioteca universale (7 vol.), 1701-09; Cronologia universale, 1707; Ancient and Modern Rome, 1716; Calendaris perpetuo profano. Built, for Louis XIV, two Globes of Marly, four meters in diameter, 1683. Died Venice, 1718.

CORRADI, Giuseppe, Italian surgeon; b. Bevagua, Italy, 1830; prof. clin. surgery Rome (Italy) U.; author: Etudes cliniques sur le tetrecissement de l'urethre sur la taille et sur les Fistules de la vagine, 1870; Compendio di terapeutica chirurgica, 1876; inventor instruments and apparatuses. Died 1907.

CORREIA, J(osé) Pinto, Portuguese physician; b. Santarem, Portugal, Apr. 22, 1931; s. Jacob Pinto and M. Amelia (Duarte) C.; grad. Med. Sch. Lisbon (Portugal), 1956, M.D., 1964; postgrad. London (Eng.) Med. Sch., 1958-59, 61; m. M. Adelaide Vasconcelos, Sept. 12, 1957; children—Rosario, Clara Teresa, Margarida. Brit. Council scholar in gastroenterology, London, 1958-59, 61; 1st asst. U. Hosp., reader Med. Sch. Lisbon, 1965—; chief cons. gastroenterology Mil. Hosp. Luanda (Angola), 1963-65. Mem. Soc. Med. Scis. Lisbon, Soc. Gastroenterology, Soc. Biochemistry, Soc. Biology. Research and publs. places of absorption, malabsorption in gastrectomized and hepatic cirrhosis, clin. studies in hepatic cirrhosis, intestinal flora and hepatic histochemistry. Home: 142 Av. E.U.A., Lisbon 5, Portugal.*

CORRELL, Malcolm, Am. physicist, educator; b. Linton, Ind., May 3, 1914; s. John Thomas and Lista (Morgan) C.; A.B. in Physics, Ind. U., 1935; Ph.D., U. Chgo., 1948; m. Ruth Armstrong, July 16, 1938; children—Elizabeth, Timothy, Mark. Physicist, Elec. Sorting Machine Co., Grand Rapids, Mich., 1936-40; faculty U. Chgo., 1942-48, 51-52, Okla. A. and M. Coll., 1948-51, DePauw U., 1952-61; prof. physics, dir. gen. edn. U. Colo., Boulder, 1961—, now dir. Div. Integrated Studies. Chmn. com. on physics achievement test Coll. Entrance Exam. Bd., 1961—. Fellow A.A.A.S. (committeeman-at-large 1964—); mem. Am. Assn. Physics Tchrs. (pres. 1961), Am. Phys. Soc., Am. Inst. Physics (gov. 1962-65), Assn. for Gen. and Liberal Studies, Colo.-Wyo. Acad. Sci., Am. Assn. U. Profs. Research and publs. on cosmic radiation, solar magnetic fields. Home: 320 20th St., Boulder, Colo. 80302.*

CORRELL, Werner Wilhelm, German psychologist; b. Wasseralfingen, Wurt., June 29, 1928; s. Friedrich and Katharina (Bantel) C.; Ph.D. in Psychology, Pedagogy and Philosophy, U. Tübingen; m. Traute Rubert, Nov. 23, 1962. Teacher, 1949-53; prof. in U. S., 1951-52; asst. at Frankfurt, 1957-58; instr. psychology, 1958-60; prof. psychology Central Sch. Pedagogy, Flensburg, Germany, 1961—. Mem. sci. mission Harvard, 1963-64. Mem. German Soc. Psychology, Soc. Psychometry, Soc. Pedagogical Research. Author: Lernpsychologie, 1961; Lernstörungen, 1962; Reform des Erziehungsdenkens, 1963; Pädagogische Psychologie auf verhaltenspsychologischer Grundlange, 1964. Address: Pädagogische Hochschule, Mürwikerstrasse 77, 239, Flensburg, Germany.

CORRENS, Carl Wilhelm, German mineralogist; b. Tübingen, May 19, 1893; s. Carl Erich and Elisabeth Correns; ed. Schiller Gymnasium, Münster, Germany, U. Münster, U. Berlin; Ph.D., 1920; D.honoris cause U. Tübingen; m. Agnes Ballowitz, Dec. 30, 1921; children—Dietrich, Agnes Elisabeth. Prof. mineralogy U. Berlin; asso. prof. U. Rostock, 1927, full prof., 1930; full prof. U. Göttingen, 1938, prof. emeritus, 1961. Corr. mem. Royal Acad. Sci. of Sweden, Acad. Bologna, Soc. Econ. Palaeontologists and Mineralogists, Geol. Föreningen (Stockholm), Geol. Soc. Am.; hon. mem. Deutschen Miner. Gesellschaft. Author: Die Sedimente des äquatorialen Atlant. Oceans, 1935-37; (with T. Barth and P. Eskola) Die Entstehung der Gesteine, 1939; Einführung in die Mineralogy, 1949. Address: Richard Zsigmondy weg 9, Göttingen, Germany.

CORRENS, Karl Erich, German botanist; b. Munich, Germany, Sept. 19, 1864; s. Erich and Emilie Köchlin) C.; ed. at univs. of Munich, Graz, Austria, Berlin, Germany, Leipzig, Germany; grad. Munich, 1889; m. Elisabeth Widmer, 1892; 2 sons, 1 dau. Became lectr. Tübingen, Germany, 1891; named asso. prof., Leipzig, 1902; apptd. prof. at Münster/Westphal, 1909; dir. Kaiser Wilhelm-Institut for Biology, Berlin-Dahlem, 1914-33; prof. philosophy, 1920-24. Author numerous books including: Die neuen Vererbungsgesetze, 1905; (with R. Goldschmidt) Die Vererbung und Bestimmung des Geschlechts, 1913; also essays, articles. Research on genetics leading to rediscovery of Mendel's law of inheritance, 1900. Died Berlin, Feb. 14, 1933.

CORRIGAN, Dominic John, Irish physician; b. Dublin, Ireland, Dec. 1, 1802; s. John Corrigan; student Maymooth Coll.; M.D., Edinburgh, 1825. Practiced medicine, Dublin, from 1830; became lectr. medicine Carmichael Sch., 1833; physician House of Industry hosps., 1840-66; physician in ordinary to queen of Ireland; mem. Parliament from Dublin, 1868-74. Pres. Irish Coll. Physicians. Author: Ten Days in Athens (entertainment); also various lectures, pamphlet on medicine. Credited with discovery of aortic regurgitation, also with 1st description of peculiar pulse which accompanies the disease; pointed out that hypertrophy of heart in aortic disease is compensatory mechanism or condition rather than disease in itself, 1832; introduced term water-hammer (also called Corrigan pulse) for pulse in aortic insufficiency, 1832; supported distinction between typhus and typhoid fever, 1853. Died Feb. 1, 1880.

CORSO, John Fiermonte, Am. exptl. psychologist; b. Oswego, N.Y., Dec. 1, 1919; s. Onofrio Curro and Santa (Alcerillo) C.; B.Ed., State U. N.Y., Oswego, 1942; M.A., State U. Ia., 1948, Ph.D., 1950; m. Josephine Ann Solazzo, Feb. 8, 1943; children—Gregory Michael, Douglas Jerome, Christine Ann. chief sound and vibration sect. psychology br. Army Med. Research Lab., Ft. Knox, Ky., 1950-51; chief, human factors office Rome Air Devel. Center, Griffiss AFB, Rome, N.Y., 1951-52; prof. Pa. State U., 1952-62; prof., dir. dept. psychology St. Louis U., 1962-63; prof., chmn. dept. psychology State U. N.Y., Cortland, 1963—. Cons. psychology, 1952—. Recipient Hon. citation Am. Inst. Research, 1961; numerous research grants from govt. agys., corps., univs. Mem. Am., Midwestern, Eastern, N.Y., Pa. psychol. assns., Acoustical Soc. Am., A.A.A.S., Am. Assn. U. Profs., Pa., N.Y. acads. scis., Internat. Soc. Cybernetic Medicine, Psychonomic Soc., Human Factors Soc., Am. Inst. Physics, Sigma Xi. Contbg. author: Contemporary Approaches to Psychology, 1967; author Experimental Psychology of Sensory Behavior, 1967; contbr., reviewer Handbook of Environmental Biology, 1966. Research leading to revision of Am. standards for normal hearing; studies on deterioration of hearing as function of aging in men and women from 18 to 65 years; discovery of new techniques for measuring hearing by means of bone-conduction of head. Home: Cosmos Hill Rd., Cortland, N.Y. 13045.*

CORSON, Dale Raymond, Am. physicist, educator; b. Pittsburg, Kan., Apr. 5, 1914; s. Harry Raymond and Alta (Hill) C.; A.B., Coll. Emporia, 1934; M.A., U. Kan., 1935; postgrad. Ohio State U.; Ph.D., U. Cal. at Berkeley, 1938; m. Nellie Elizabeth Griswold, June 17, 1938; children—David, Bruce, Richard, Janet. Instr., research fellow U. Cal. at Berkeley, 1938-40; asst. prof. U. Mo., 1940-43, asso. prof., 1943-45; staff Los Alamos Sci. Lab., 1945-46; faculty physics Cornell U., Ithaca, N.Y., 1946—, prof., 1952—, chmn. dept. physics, 1956-59, dean coll. engring., 1959-63, provost, 1963—. Staff, Radiation Lab., Mass. Inst. Tech., 1941-43; tech. adviser War Dept., 1943-45. Recipient Presdl. certificate merit, 1948. Fellow Am. Phys. Soc.; mem. Phi Beta Kappa, Sigma Xi, Tau Beta Pi. Author: (with P. Lorrain, W. H. Freeman) Introduction to Electromagnetic Fields and Waves, 1962; also articles. Research in nuclear physics; cosmic rays; engring. Home: 144 Northview Rd., Ithaca, N.Y. 14850.

CORSSEN, Guenter, physician; b. Bremen, Germany, Feb. 6, 1916; s. Johannes and Margarete (Thiele) C.; M.D., U. Hamburg, 1950; m. Eva Schmidt, Nov. 17, 1955. Practice medicine specializing in anesthesiology, Galveston, Tex., 1952-60, Ann Arbor, Mich., 1961—; faculty Med. br. U. Tex., 1955-60; faculty U. Mich. Med. Center, 1961—, prof. anesthesiology, 1965—. Mem. N.Y. Acad. Scis., Internat. Anesthesiology Research Soc., Am. Soc. Anesthesiologists (1st Sci. award 1963), A.M.A. (Hektoen Med. award 1963, certificate of merit 1964). Research and publs. on drug dependence and tolerance in cultured human cancer cells, role of acetylcholine in mechanism of ciliary motion of human respiratory epithelium in vitro, visualization of smooth muscle elements in alveolar walls of human lung, exploration, clin. introduction phencyclidine derivative CI-581 for use in short lasting surg. procedures in humans. Home: 3490 Miller Rd. Office: 1405 E. Ann St., Univ. Hosp., Ann Arbor, Mich. 48104.*

CORT, Joseph Henry, Am. physiologist; b. Boston, Dec. 27, 1927; s. Boris and Eva (Kolidicka) C.; student Harvard, 1944-46; M.D., Yale, 1951; Ph.D. (Henry fellow) Cambridge (Eng.), U., 1953; D.Sc., Acad. Sci., Prague, Czechoslovakia, 1963; m. Ruth Mathilde Leyendecker, Aug. 29, 1949; children—Nicholas, Susannah, Alexis and Deborah (twins). Research fellow dept. exptl. medicine, Cambridge U., 1951-53; sr. lectr. dept. physiology U. Birmingham (Eng.), 1953-54; sr. sci. research officer Inst. Cardiovascular Research, Prague, 1954-65; prof. pharmacology and therapeutics U. Man. (Can.), Winnipeg, 1965—. Mem. Renal Assn., N.Y. Acad. Scis. Author: Physiologie der Koerperfluessigkeiten, 1957: (with V. Fencl) Electrolytes, Fluid Dynamics and the Nervous System, 1965; also articles. Research on relationship between nervous system, regulation of balance of salt in body. Home: 928 Renfrew Bay, Winnipeg 9, Man., Can.*

CORTE, Arturo Eduardo, Argentinian earth scientist; b. Salta, Argentina, Oct. 27, 1919; s. Francisco and Mansueta (Fontana) C.; Bachiller, Nat. Colegio, Salta, 1940; Dr. in Natural Scis., Cordoba (Argentina) U., 1945; postgrad. Stockholm (Sweden) U., 1948-50; m. Rosa Figueres, Dec. 21, 1946; children—Arturo, Elizabeth. Geologist, Hwy. Dept., Cordoba, 1945; chief geologist ground water div. Mendoza Geol. Survey, 1945-48; sec. Water Instituto, Universidad Nacional Cuyo, Mendoza, Argentina, 1946-48; chief scientist Antarctic Argentine Expdn., 1952-53; contract scientist U. S. Army Cold Regions Research and Engring. Lab., Hanover, N.H., 1955-63; dir. geology dept. Universidad Nacional del Sur, Bahia Blanca, Argentina, 1963—. Argentine Assn. for Advancement Scis. scholar, 1948-50. Mem. Geol. Soc. Am., A.A.A.S., Internat. Geoog. Union, Am. Geophys. Union, Soc. Cryobiology, Asociacion Geologica Argentina, Asociacion Argentina para el Progreso de las Ciencias. Contbg. author Review Articles in Engineering Geology, Vol. II, 1967. Research and publs. on geology of very cold regions of earth especially particle segregation by freezing, underground ice, effects of long cycles of freezing and thawing. Home: 741 Avenida de los Constituyentes, El Palihue, Bahia Blanca, Argentina.*

CORTEZ, Hernando (Hernán Cortés), Spanish explorer; b. Medellin, Estremadura, Spain, 1485; s. Martin and Catalina (Altamirano) C.; attended U. Salamanca, 1499-1501; m. Catalina Juarez, circa 1515. Soldier in San Domingo, 1504-11; accompanied Diego Valasquez in conquest of Cuba, 1511; became mayor of Santiago; given command of Spanish effort to explore and conquer Mexico, sailed with troops, 1519; founded town of Vera Cruz, 1519; heard reports of rich Kingdom of Aztecs; burned his fleet to cut off escape route, and so he could use the sailors as soldiers; marched on Mexico City; skillfully exploited Indian superstitions and internal malcontents of Aztec empire; admitted into Mexico City as relative of the Sun God, captured emperor Montezuma; Velasquez, who had tried to recall Cortez previously, sent troops under Narvaez to force his return to Cuba; Cortez left Mexico City, 1520, defeated Narvaez, returned to Mexico City and found armed natives blocking his way; defeated Aztec army, July 1520, recaptured Mexico City, 1521; created capt. gen. of Mexican Troops, also marques of Oaxaca by Spanish king; extended conquests in Mexico peninsula, 1521-26; visited Spain, 1828-30; discovered Lower Cal., 1536; returned to Spain, 1540; died in obscurity because of his failure to keep his power, property and positions in Spanish colonial Mexico. Died nr. Seville, Spain, Dec. 2, 1547.

CORTI, Alfonso, Italian anatomist; b. Gambarana, Pavia, Italy, June 15, 1822. Histologist in Vienna, Berlin, Utrecht and Turin; prosector (under Hyrtl), Vienna. Author: Recherches sur l'organe de l'ouie des Mammifères, 1851. Discovered complex organ of ear (organ of Corti or tunnel of Corti) by which sound is directly perceived (rodlike bodies arranged in a double row so as to form spiral tunnel), 1851; made important studies on structure of retina. Died Rome, Feb. 19, 1888.

CORTI, Bonaventura, Italian botanist; b. Viano, Italy, Feb. 26, 1729; s. Domenico and Vittoria (Bondioli) C.; studied in Reggio Emilia; ordained priest at Accademia degli Iponcondriaci; tchr. metaphysics and botany Coll. San Nazario. Reggio Emilia, 1754-67; became rector Coll. San Carlo, Modena Italy, 1767-1806; dir. Modena Bot. Garden; prof. botany and agr. U. Bologna (Italy), 1805-09; ret. from teaching and returned to Reggio Emilia, 1809. Author: Osservazioni microscopiche sulla tremella e gulla circolazione della fluido in una pianta acquajuola, 1774; Sulua circolazione del fluido scoperta in uarie piante, 1776; Mezzi ped distruggere i vermi che rodono il grano in erba, 1777. Research on plant tropisms or movements and their reactions to various solutions; in cellular anatomy and physiology discovered protoplasmic flow in plants. Died Reggio Emilia, Italy, Feb. 3, 1813.

CORTIE, Aloysius Laurence, Brit. astronomer; b. London, Eng., Apr. 22, 1859; student Stonyhurst, St. Beuno. Joined Soc. Jesus, 1878, ordained to priesthood Roman Catholic Ch., 1892; dir. Stonyhurst (Eng.) Obs., 1881; prof. math. Stonyhurst, for 27 years, music, for 19 years. Fellow Brit. Astron. Soc. (dir. solar sect. 1900-10); mem. Manchester Astron. Soc. (pres.). Reported on eclipses of 1905, 11, 14. Died May 17, 1925.

CORVISART, Lucien, French physician; b. Thonne la Long, France, June 9, 1824; s. Jean-Baptiste-Rene and Antoinette (Cliquot) C.; Docteur en medecine, 1852; intern Paris, France hosps. Personal physician to Napoleon III. Author: Dy spesie et consomption. Introduced 1st use of pepsin in therapeutics, also term tetany (tetanie), 1852. Died Dec. 24, 1882.

CORVISART DES MARETS, Baron Jean Nicolas, French physician; b. Dricourt, Champagne, France, 1755; studied law College of St. Barbe; student of Desault, Hôtel Dieu; doctor of medicine, U. Paris, France, 1782. Apptd. physician Charité Hosp., Paris, 1783; prof. med., 1786; held 1st chair of internal medicine, Hospice de l'Unité; named prof. practical medicine Collège de France, 1796-1804; mem. French Acad. Scis., 1811; personal physician to Napoleon. Author: Essai sur les maladies et les lesions organiques du coeur et des gros vaisseaux, 1806; Nouvelle méthode pour reconnaitre les maladies internes, 1808. Translated and added comments to Inventum novum (Auenbruggers). First description of dyspnea of effort, 1806; distinguished between organic and functional heart disease, 1806, between hypertrophy and dilatation of heart; discovered and described chronic hypertrophic myocarditis; classified heart disease using anat. structures; used term organic lesions in heart disease; differentiated between right and left heart failure; described mitral aortic valvular lesions, tricuspid stenosis. Died Sept. 18, 1821.

CORWIN, Alsoph Henry, Am. chemist; b. Marietta, O., Jan. 11, 1908; s. Clifford Egbert and Elizabeth Gillet (Stimson) C.; A.B., Marietta College, 1928, D.Sc. (honorary), Marietta College, 1953; Ph.D., Harvard, 1932; m. Irene Marguerite Davis, Aug. 6, 1938. Asso. in chemistry, Johns Hopkins Univ., 1932-39, asso. prof., 1939-44, prof. since 1944, chmn. Dept. of Chemistry, 1944-47. Official investigator Nat. Defense Research Com., 1942-45; consultant to Army Chem. Corps., 1944, 49-59, official investigator, 1945-50; cons. metrology, Nat. Bur. Standards, 1948-52; chief investigator, Office Naval Research, 1951-53; investigator Nat. Aeros. and Space Adminstrn., 1962-65. Mem. NRC panel adv. Nat. Bur. Standards, 1960—. Mem. Am. Chem. Soc. (recipient award of merit Maryland section 1965, com. adv. to Chem. Corps), Phi Beta Kappa, Sigma Xi, Delta Upsilon. Contributor papers to scientific jour. Investigation of relationships between chemical structure and biological activity of hemoglobin, chlorophyll and the castor bean toxin, ricin; chemistry of allergens; microchemical manipulations and precision weighing. Home: 2903 Overland Av., Baltimore 14. Office: Dept. of Chemistry, Johns Hopkins U., Balt. 18.

CORY, Charles Barney, Am. ornithologist; b. Boston, Jan 31, 1857; s. Barney and Eliza A. B. C.; ed. Boston schs. and Lawrence Sci. Sch. (Harvard), 1879; m. Harriet W. Peterson, May 29, 1883. Hon. curator ornithology Boston Soc. Natural History until 1905; hon. curator ornithology, 1895-1906, prof. and hon. curator dept. zoölogy, 1906-21, Field Mus. Natural History, Chgo. Pres. Am. Ornithologists' Union, 1904-05; fellow Zoöl. and Linnean socs. London. Author: Catalogue of West Indian Birds; Hunting and Fishing in Florida; The Birds of Eastern North America; How to Know the Shore Birds of North America; How to Know the Ducks, Geese and Swans of North America; The Birds of the West Indies; Key to the Water Birds of Florida; Hunting and Fishing in Florida; Key to the Birds of Eastern North America; The Birds of Illinois and Wisconsin. Died July 29, 1921.

COSANDEY, Florian, Swiss botanist; b. Sainte-Croix, June 16, 1897; s. Paul and Bertha (Vogt) C.; ed. in Lausanne, Innsbruck, Liège, Paris; Ph.D. in sci.; m. Lucy Germond, July 14, 1926. Prof. botany U. Lausanne, also dean Sch. Sci., rector. Research and publs. on algology. Home: 4, rue Chandolin. Office: University of Lausanne, Lausanne, Switzerland.

COSBY, Richard Sheridan, physician; b. N.Y.C., July 6, 1913; s. Charles R. and Mildred (Freeman) C.; A.B., Harvard, 1934, M.D., 1938; m. Mary Mayo, May 8, 1948. Fellow in physiology Western Sch. Medicine, 1939; research cardiologist Mass. Gen. Hosp., Boston, 1940-41; instr. medicine Harvard Med. Sch., 1942; practice medicine specializing in cardiology, Pasadena, Cal., 1946—; asso. clin. prof. medicine U. Southern California Sch. Medicine, 1954-67, clin. prof. of medicine, 1967—; dir. Pasadena Cardiovascular Research Found., 1964—; cons. cardiology Pacific Mut. Life Ins. Co., Los Angeles, No. Inyo Hosp., Bishop, Cal., Mono Med. Center, Bridgeport, Cal. Mem. Los Angeles County Heart Assn. (sec. 1957-58, dir. 1960—), A.C.P., Am. Coll. Chest Physicians, Am. Coll. Cardiology, Am. Heart Assn. (council on clin. cardiology), Am., Cal. med assns. to sci. jours. Research and publs. on cardiac physiology, cardiac catheterization, electrocardiography, pacemakers, data retrieval, clin. cardiology. Home: 2575 Lombardy Rd., San Marino, Cal. 91108. Office: 111 Congress St., Pasadena, Cal. 91105.*

COSGROVE, Gerald Edward, Jr., Am. pathologist; b. Dubuque, Ia., July 13, 1920; s. Gerald Edward and Madeline (Casutt) C.; B.S., U. Notre Dame, 1945; M.D., U. Mich., 1944; m. Marion Emma Thompson, Dec. 24, 1943; children—Nancy (Mrs. Don Smith), David, Mary, Judith, John, Paul, Christine. Pathologist, chief lab. service Gorgas Hosp., Ancon, C.Z., 1955-57; research pathologist biology div. Oak Ridge Nat. Lab., 1957—. Diplomate Am. Bd. Pathology. Mem. Am. Soc. Exptl. Pathology, Radiation Research Soc., Am. Soc. Parasitology, Am. Inst. Biol. Scis., Am. Soc. Ichthyologists and Herpetologists, Wildlife Disease Assn., Tenn. Acad. Sci. Research and publs. on pathology of radiation injury in mammals, diseases and parasites of lab. animals and wildlife. Home: 188 Waddell Circle. Office: Biology Div., Oak Ridge Nat. Lab., Oak Ridge 37830.*

COSGROVE, William Burnham, Am. biologist; b. N.Y.C., June 11, 1920; s. William L. and Evelyn (Burnham) C.; A.B., Cornell U. 1941; M.S., N.Y. U., 1947, Ph.D., 1949; m. Dolores C. Mangual, Aug. 6, 1949; children—Karen L., Bruce B. Grad. asst. N.Y. U., 1941-42, 46-49; asso. in zoology U. Ia., 1949-51, asst. prof., 1951-56, asso. prof., 1956-57; asso. prof. zoology U. Ga., Athens, 1957-64, prof., head dept., 1964—. Recipient Michael award for Research, 1967. Mem. Am. Soc. Zoologists, Soc. Protozoologists, Assn. Southeastern Biologists, A.A.A.S., Am. Assn. U. Profs., Sigma Xi, Phi Kappa Phi. Research, publs. on nutrition and chem. activities of unicellular animals and function of certain of their intracellular structures, properties and function of blood pigments of invertebrates. Home: 140 Lynwood Ct., Athens, Ga. 30601.*

COSSALI, Pietro, Italian mathematician; b. Verona, Italy, June 29, 1748; ed. at Jesuit college. Prof. of natural philosophy, 1787, prof. of astronomy, 1791, Univ. Parma; prof. mathematics, Univ. Padua, 1806-15. Inspector general of waters. Mem. Italian Institute, 1811. Author: Origine, trasporto in Italia e primi progressi in essa dell'Algebra, Storia critica (2 vol.), 1779-99; Ephemerides astronomiques, annually 1791-1806. Known for his history of algebra. Died Padua, Dec. 20, 1815.

COSSERAT, Eugene-Maurice-Pierre, French mathematician, astronomer; b. Amiens, France, Mar. 4, 1866; Aggregarion in math., Ecole normale superieure. Prof. astronomy, differential calculus Faculty Sci., Toulouse, France, from 1888; dir. obs., Toulouse, from 1908. Mem. French Acad. Scis., 1919, Bureau des longitudes. Author: Théorie des crops déformables (detailed elucidation of theory of elasticity), 1909. Studied rings and satellites of Saturn, planets, comets. Died May 31, 1931.

COSSIGNY, Joseph-François Charpentier, see Charpentier Cossigny, Joseph François.

COSSLETT, Vernon Ellis, English physicist; b. Cirencester, Eng., June 16, 1908; s. Edgar William and Anne (Williams) C.; B.Sc., Bristol U., 1929, Ph.D., 1932; Sc.D., Cambridge U., 1963; D.Sc. (hon.), U. Tübingen, Germany, 1963; postgrad. Kaiser Wilhelm Inst., Berlin, Germany, 1930-31, U. Coll., London, Eng., 1931-32; m. Anna Joanna Wischin, Oct. 3, 1940; children—Stephen Rhys, Anna Therese. Research fellow U. Bristol, 1932-35, Birkbeck Coll., London, 1939-40; lectr. Faraday House Elec. Engring. Coll., London, 1935-39; departmental lectr., elec. lab. Oxford U., 1940-46; ICI research fellow Cavendish Lab., Cambridge (Eng.) U., 1946-49, univ. lectr., 1949-65, reader in electron physics, 1965—. Fellow Corpus Christi Coll. Cambridge; mem. Assn. U. Tchrs., Gt. Britain (pres. 1953), Royal Micros. Soc. (pres. 1961-64), Inst. Physics and Phys. Soc. (v.p. 1961-65), Am. Phys. Soc.; hon. mem. Am. Electron Microscopy Soc., French Soc. Electron Microscopy. Author: Electron Optics, 1946, 50; Practical Electron Microscopy, 1951; (with W. C. Nixon) X-Ray Microscopy, 1960; Modern Microscopy, 1966. Research, publs. on devel. of electron microscopy, x-ray microscopy and electron microprobe analysis; inventor (with W. C. Nixon) projection X-ray microscope, (with P. Duncumb) scanning electron probe microanalyser; builder (with K. C. A. Smith) high voltage electron microscope (750 kilowatts). Home: 1, Long Rd., Cambridge, Eng.*

COSSMANN, Maurice, French engr., botanist; b. Paris, France, 1850; s. Herman-Maurice Cossmann. Head engr. French Nat. Railroads; devoted prime of life to botany. Author: Applications de l'électricité au material des chemins de fer, 1882; Iconographie complète des coquilles fossilles de l'ėocene des environs de Paris, 1904; others. Studied fossil pelecypodes and gasteropodes; explored sandy regions at Fontainebleau and areas surrounding Paris; founded Revue critique de Paléontologie. Died May 17, 1924.

COSSON, Ernest-Saint-Charles, French botanist; b. Paris, France, July 22, 1819; doctorate in medicine, 1847. Made bot. expdn. to Algeria, 1852-58. Mem. French Acad. Scis., 1873, Bot. Soc. France (founder, archivist, pres.). Author: Itineraire d'un voyage botanique en Algerie, 1857. Known for work on flora of Algeria, Paris; opposed (successfully) project to build an artificial sea by flooding So. Tunisia. Died Dec. 31, 1889.

COSTA, Aurelio, Italian physician; b. Brescia, Italy, June 24, 1903; s. Quinto and Elvira (Baltera) C.; degree in medicine, Turin (Italy) U., 1928; m. Teresa Ferraris, July 5, 1937; children—Elvira (Mrs. Ninni Fadda), Adele (Mrs. Gastone Fara), Paolo, Vittorio, Quinto, Emma. Staff, Mauriziano Hosp., Turin, 1932—, head physician, 1944—, dir. Inst. for Endocrine and Metabolic Diseases, 1942—, dir. service nuclear medicine, 1951—. Mem. endocrinology enterprize Nat. Council Researches, 1964. Fulbright scholar Thyroid Clinic, Mass. Hosp., 1957. Named comdr. Italian Order Chivalry, 1958. Mem. Med. Acad. Turin, Royal Soc. Medicine (affiliate). Author: Le malattie della tiroide, 1958; (with others) Diagnostica e Terapia con i radioisotpoi, 1962. Research and publs. on endemic cretinism and deaf-mutism, thyroid function during intrauterine life in human being. T.S.H. activity in fetal-mother symbiosis, endemic goiter and iodine metabolism in different regions of Italy, heart-catheterism. Home: 51 corso Galileo Ferraris, Turin. Office: Mauriziano Hosp., 46 corso Turati, Italy.*

COSTA, Cristóvao da, see Acosta, Cristobal.

COSTA, Erminio, Am. pharmacologist, neurologist; b. Cagliari, Italy, Mar. 9, 1924; s. Oreste and Viginia (Murgia) C.; M.D., U. Padua and Cagliari, Italy, 1947; Libera Docenza in Pharmacology, U. Caliari, 1954; m. Anna Marrazzi, June 10, 1950; children—Massimo, Robert H., Michael J. Came to U. S., 1956, naturalized, 1961. Faculty, U. Cagliari, Italy, 1948-56; med. research asso. Thudichum Psychiat. Research Lab., Galesburg (Ill.) State Hosp., 1956-60; dep. chief, head, sect. on clin. pharmacology Lab. Chem. Pharmacology, Nat. Heart Inst., Bethesda, Md., 1960-65; asso. prof. pharmacology Columbia, Coll. Phys. & Surg., N.Y.C., 1965—. Mem. Am. Physiol. Soc., Am. Soc. for Pharmacology and Exptl. Therapeutics, Am. Acad. Neurology, Soc. Biol. Psychiatry, Assn. for Research in Nervous and Mental Diseases, Am. Coll. Neuropsychopharmacology. Research on mechanisms of neuronal transmission; studies in physiol. role of serotonin and norepinephrine in central nervous system and peripheral transmission of nerve impulse. Home: 1 Wayside Lane, Scarsdale, N.Y. 10583. Office: 630 W. 168th St., N.Y.C. 10032.*

COSTA, Giovanni, Italian physicist; b. Rovigo, Italy, Sept. 11, 1930; s. Teobaldo and Teresa (Bianchi) C.; Dr. in Physics U. Padua, Italy, 1954; Ph.D., Mass. Inst. Tech., 1959; L.D., Rome U., 1962. Research asst. U. Padua, Italy, 1954-56; research asst. Mass. Inst. Tech. Lab. for Nuclear Sci., 1957-59; asst. prof. U. Bari, Italy, 1959-60; asst. prof. U. Padua, Italy, 1960-61; research asso. CERN, Geneva, Switzerland, 1962-63; asso. prof. U. Padua, Italy, 1963-66, 68—; research asso. Rutherford High Energy Lab., Didcot, Eng., 1967. Mem. Italian Phys. Soc. Research, publs. on physics of elementary particles, proposals for expts., prediction of Beta decay of hyperons, methods for determining relative parities of strange particles, phenomenological analysis of strange particle interactions. Home: 37 Porta Adige, Rovigo, Italy. Office: 8 Marzolo St., Padova, Italy.*

COSTACHEL, Octav, Rumanian physician; b. Zatreni Vilcea, Romania, Sept. 6, 1911; s. Haralamb Vasile and Maria (Craciunescu) C.; att., faculty Medicine, Iassy, Rumania, 1935, M.D., 1936, spec. radiol., 1939, spec. internal med., 1940; spec. med. isotopes Moscow (USSR) U., 1957, D.Sc., 1959; m. Chitulescu Irina, Mar. 8, 1959; children—Corina, Nadia. Asst. radiology Med. Faculty, Jassy, Rumania, 1940-49; dir. Oncological Inst., Bucharest, Rumania, 1949—; prof. oncology Med. Faculty, Bucharest, Rumania, 1952—; sec. gen. Ministry Health Rumania, 1948-49. Cons. oncology Bucharest hosps. Granted sci. merit order, 1966, medal for distinction in labor, 1954, N.I. Pirogov medal, 1966. Mem. Internat. Union Against Cancer (v.p. 1966—), Rumanian Oncological Soc. (pres. 1957), Royal Soc. Medicine (London), Internat. Soc. Chemotherapy N.Y., Internat. Soc. Transplantation. Contbg. author: Clinical and Experimental Oncological Research, 1958; General Oncology, 1961; Early Diagnosis of Malignant Tumours, 1964; Complex Treatment of Cancer, 1965. Research in polypassage of tumors at short intervals, new method of mass consultation for early detection of cancer. Address: Oncological Institute, 11 Bul. 1 Mai, P.O. Box 5916, Bucharest 62, Rumania.*

COSTAZ, Baron Louis, French engr.; b. Belley, France, Mar. 17, 1767; s. Claude and Claudine (Goujon) C.; student U. Valence, U. Paris (both France). Gen. dir. Dept. Civil Engring.; sec. Egyptian Inst., Mem. French Acad. Scis., 1831, Soc. Geography (a founder, pres. 1829). Went to Egypt with Napoleon, explored Nile River; discovered ancient canal (dug by Ptolemy II) between Red Sea and Nile. Died Fontainebleau, France, Feb. 15, 1842.

COSTE, Jacques-Marie-Cyprien-Victor, French naturalist; b. Castrles, France, May 10, 1807; s. Jacques and Marguerite (Julien) C.; med. studies U. Montpellier (France); student of Delpech (studied cholera epidemic in Eng., Scotland with him, 1832). Prof. ovology, anatomy l'Ecole Pratique, Paris, France, from 1836; prof. embryology College de France, from 1844; insp. maritime fishing, from 1862. Mem. Acad. Medicine, French Acad. Scis. (pres. 1871), 1851. Author: Memoire sur l'ovologie humaine, 1835; Instructions pratique sur la pisiculture, 1853. Studied fisheries, culture of oysters, Italy, 1854; interested in pisiculture; established center of fish breeding at Huningue; studied embryology; discovered and described germinal spot of animal ovum, 1837. Died Sept. 19, 1873.

COSTE, Jean-François, French physician; b. Villes, France, June 14, 1741; student of Antoine Petit, Paris, France; 1st French physician to take part in Am. Revolution; chief physician Invalides Hosp., from 1796, Grande Armée, 1803-07; author books on medicinal botany. Died Nov. 8, 1819.

COSTELLO, David Francis, Am. ecologist; b. Norfolk, Neb., Sept. 1, 1904; s. Thomas and Mary (Mallory) C.; A.B., Neb. State Tchrs. Coll., 1925; M.S., U. Chgo., 1926, Ph.D., 1934; m. Cecilia C. Waldkirch, June 12, 1929; children—Barbara M. (Mrs. Virgil M. McDougle), David K., Donald R. Instr., Marquette U., 1926-32; forest ecologist Rocky Mountain Forest and Range Expt. Sta., Ft. Collins, Colo., 1934-37, chief div. range research, 1937-53; chief div. range mgmt. wildlife habitat, recreation research Pacific N.W. Forest and Range Expt. Sta., Portland, Ore., 1953-64. Spl. lectr. Colo. State U., 1943-53; guest speaker U. Coll. North Wales, 1962. Commended for superior service U. S. Dept. Agr., 1963. Fellow A.A.A.S.; mem. Am. Soc. Range Mgmt. (dir. 1953-55), Ecol. Soc. Am., Audubon Soc., Outdoor Photographers League, Authors Guild. Author: Range Ecology, 1939; The World of the Porcupine, 1966; The World of the Ant, 1968. also numerous tech. and popular articles. Asso. editor Jour. Range Mgmt., 1950-53, Ecology, 1946-48. Developed bot. survey methods; research in livestock prodn. methods, renewal natural resources disturbed by man; discovered plant species new to sci. Home: 4211 N.E. 79th., Av., Portland, Ore. 97218.*

COSTER, Dirk, Dutch physicist; b. Amsterdam, Netherlands, Oct. 5, 1889; s. Barend and Aafje (van der Mik) C.; ed. U. Lund (Sweden), 1920-22; Ph.D., U. Leyden, 1922; U. Copenhagen (Sweden), 1922-23; m. Lina Maria Wijsman, Feb. 26, 1919; 4 children. Asst. in physics Tech. High Sch., Delft, 1916-20; prof. exptl. physics U. Groningen, since 1924; mem. Koninklyke Akademie van Wetenschappen, Amsterdam. Discovered (with von Hevesy) the element Hafnium. Address: Natuurkundig Laboratorium, Westersingel 34, Groningen, Netherlands.

COSTILOW, Ralph Norman, Am. microbiologist; b. Oxford, W.Va., Oct. 23, 1922; s. Berkeley and Mamie (Cline) C.; B.S., W.Va. U., 1948; M.S., N.C. State Coll., 1950; Ph.D., Mich. State Coll., 1953; m. Ann Beasley, Dec. 9, 1950; children—James B., Susan Ann. Research instr. N.C. State Coll., 1949-51; faculty Mich. State U., East Lansing, 1953—, prof. microbiology, 1960—. Mem. A.A.A.S., Am. Soc. for Microbiology. Research and publs. on food microbiology especially microbial physiology, natural lactic acid fermentations, physiology sporulation process in anaerobes, physiology insect pathogens, intermediary metabolism of amino acids by anaerobic bacteria. Home: 2012 Osage Dr., Okemos, Mich. 48864. Office: Dept. Microbiology, Mich. State U., East Lansing, Mich. 48823.

COTES, Roger, English mathematician; b. Burbage, Leicestershire, Eng., July 10, 1682; s. Robert and Grace (Farmer) C.; B.A., Trinity Coll., Cambridge U., 1702, M.A., 1706. Plumian prof. astronomy Trinity Coll., Cambridge U., from 1706; asst. to Isaac Newton in preparation of 2d edition of Principia, 1709-13; took holy orders, 1713. Fellow Royal Soc., 1711. Author: Logometria, 1713; Harmonia Mensurarum (earliest work in which decided progress was made in application of logarithms and of properties of circle to calculus of fluents); 1722; Hydro-Statical and Pneumatic Lectures, 1738. Studied calculus of infinite differences; revamped Flamsteed's and Cassini's solar and planetary tables; undertook to construct tables of moon on Newtonian principles; description of heliostat-telescope furnished with mirror revolving by clockwork showed that he had already in 1708 (independently of Hooke's project of 1674) anticipated system of equatorial mounting. Died Cambridge, Eng., June 5, 1716.

COTLAR, Mischa, mathematician; b. Kiev, Russia, Aug. 1, 1913; Ph.D., U. Chgo., 1953; m. Janny Frenkel. Research prof. dept. math. La Plata U., 1946-47; research prof. math. U. Buenos Aires (Argentina), 1948-50, prof. math., 1957—; Guggenheim fellow, research asso. U. Chgo., 1951-53; chmn. inst. math. U. Cuyo, Mendoza, Argentina, 1953-56. Vis. prof. Washington U., St. Louis, 1958. Recipient award Acad. Scis., Madrid, 1950. Mem. Acad. Scis., Lima, Peru, Am. Math. Soc., Sci. Soc. Argentina, Inst. Parapsychology of Buenos Aires (a founder). Editor jour. Unión Matemática Argentina. Contbr. to math. jours. Participant in expts. in psychokinesis, telepathy, mediumships; research on relationship of parapsychology to practice of yoga.

COTTA, Bernhard von, see von Cotta, Bernhard.

COTTA (or COTTEY), John, Brit. physician; b. Warwickshire, Eng., circa 1575; B.A., Trinity Coll., Cambridge (Eng.) U., 1595, M.A., Corpus Christi Coll., 1596; M.D., 1603. Practice medicine, Northampton, Eng., 1603-23 (or possibly 1650). Author: A Short Discoverie of the Unobserved Dangers of Severall Sorts of Ignorant and Inconsiderate Practisers of Physicke in England, 1612; The Triall of Witch-craft, 1616; Cotta contra Antonium . . . , 1623; The Infallible, True and Assured Witch, 1624. Attacked quack med. practitioners, including astrological medicine. Died circa 1650.

COTTAM, Clarence, Am. biologist; b. St. George, Utah, Jan. 1, 1899; s. Thomas P. and Emmaline (Jarvis) C.; A.B., Brigham Young U., 1926, M.S., 1927; Ph.D., George Washington U., 1936; postgrad. U. Utah, Am. U.; m. Margery Brown, May 20, 1920; children—Glenna Claire (Mrs. Ivan L. Sanderson), Margery B. (Mrs. Grant Osborn), Josephine (Mrs. Douglas F. Day), Carolyn (Mrs. Dwayne Stevenson). With U. S. Biol. Survey (name now U. S. Fish and Wildlife Service), Washington, 1929-54, asst. dir., 1946-54; dean Coll. Biology and Agr., prof. biology Brigham Young U., 1954-58. Dir. Rob and Bessie Welder Wildlife Found., Sinton, Tex., 1955—; cons. to Dept. Interior, other govtl. agys. Recipient Poage Humanitarian award, 1962; Frances K. Hutchenson medal Garden Clubs Am., 1962. Fellow Utah, Tex. acads. sci., A.A.A.S., Am. Ornithol. Union; mem. Wildlife Soc. (pres. 1949-50, Leopold medal 1955), Nat. Parks Assn. (chmn. trustees 1963—), Ecol. Soc. Am., Nat. Audubon Soc. (Distinguished Service medal 1961), Wildlife Mgmt. Inst., Am. Ornithol. Union, Am. Fisheries Soc., Audubon Naturalist Soc., Internat. Assn. Game Fish and Conservation Commnrs., Sigma Xi, others. Author: (with Herbert Zim) Insects, 1951. Research and publs. on wildlife mgmt., conservation, ecology, control of pests by environmental modifications. Address: P.O. Box 1396, Sinton, Tex. 78387.*

COTTE, Louis, French meteorologist; b. Laun, France, Oct. 20, 1740; s. Elisabeth Le Nain; ed. Coll. Oratory-Soissons; m. Antoinette-Marie-Madeleine du Coudray. Prof. philosophy, theology, Montmorency; curate of Montmorency, from 1767; renounced priesthood; curator St. Genevieve Library, Paris, France, from 1800; mem. Acad. Sci., Imperial Soc. Agr., Oratory. Author: Traité de météorologie, 1774; Memoires sur la météorologie. Discovered Enghein spring which contained sulphurous water, 1766; attempted to apply meteorol. information to medicine and agr.; considered 1 of founders of meteorology. Died Oct. 4, 1815.

COTTE, Maurice, French physicist; b. Grenoble, France, Jan. 19, 1908; s. Joseph and Marie (Maurice) C.; Licence ès Sciences Mathématiques et Physique, Agrégation de Physique, École Normale Supérieure, Paris, 1931, Doctor's Degree in Phys. Sci., 1938; m. Raymonde Richet, Aug. 11, 1947; 1 dau., Jacqueline. Asst., Faculty Sci., U. Grenoble (France), 1932-36; with Sci. Research Nat. Found., 1936-38; prof. Faculty Sci., U. Poitiers (France), 1942-57; prof. Faculty Sci., U. Paris, 1957—. Sci. counsellor Centre d'Energie Atomique, 1962; mem. reading com. Cahiers de Physique. Mem. Société Francaise de Physique, Am. Phys. Soc. Contbg. author: Principes de l'Électricité, 1966; also articles. Research on electron optics including Gaussian approximation, aberrations, space charge; wave propagation including piezoelectric media, ionosphere, concentric cables, lines, wave guides, group velocity, transmission tests, influence of dielectric losses; electronics, including filters, branching networks, ring modulator; electrostatics and magnetostatics, including elliptic aperature in a conducting plane, ironless windings. Home: 33, rue du Chateau, 92, Boulogne, France. Office: Faculté des Sciences 9 quai, St.-Bernard, Paris 5e, France.*

COTTEY, John, see Cotta, John.

COTTEAU, Gustave-Honoré, French zoologist, paleontologist; b. Auxerre, France, Dec. 17, 1818; licence in law, 1840; asst. judge, Auxerre, from 1846; judge, Coulommiers, France, from 1853, Auxerre, from 1862; mem. Geol. Soc. (pres. 1874), French Acad. Scis., 1887, Geol. Soc. London; specialized in study of paleontology, particularly echinoderms; quite well known in his day. Died Aug. 10, 1894.

COTTEREAU DU CLOS, Samuel, see Duclos, Samuel Cottereau.

COTTERILL, James Henry, Brit. engr.; b. Norfolk, Eng., Nov. 2, 1836; s. Joseph Cotterill; student Brighton Coll.; M.A., St. John's Coll., Cambridge, 1866; apprentice Messrs. Fairbairn and Co., engrs.; became lectr. Royal Sch. Naval Architecture, South Kensington, Eng., 1866, vice-prin., 1870; named prof. applied mechanics Royal Naval Coll., Greenwich, Eng., 1878; ret., 1897. Fellow Royal Soc., 1878; hon. v.p. Instn. Naval Architects. Author: Steam Engine considered as a Thermodynamic Machine, 1878, 3d edit., 1895; Applied Mechanics, 1884, 5th edit., 1900. Died Jan. 8, 1922.

COTTING, John Ruggles, Am. geologist; b. Acton, Mass., 1783; attended Harvard, Dartmouth Med. Sch. Ordained to ministry Congregational Ch., circa 1810; devoted most of his life to sci. pursuits; state geologist of Ga., 1835-37, made 1st geol. survey of state; gathered valuable collection of plants, minerals and fossils (divided among various colls. at his death); a copy of his state geol. report (1836) was requested by Czar of Russia for Royal Library. Author: Introduction to Chemistry, 1822; Synopsis of Lectures on Geology, 1825. Died Milledgeville, Ga., Oct. 13, 1867.

COTTON, Frank Albert, Am. chemist; b. Phila., Apr. 9, 1930; s. Albert and Helen (Taylor) C.; student Drexel Inst. Tech., 1947-49; A.B., Temple U., 1951, D.Sc., 1963; Ph.D., Harvard, 1955; m. Diane Dornacher, June 13, 1959; children—Jennifer Helen, Jane Myrna. Faculty, Mass. Inst. Tech., Cambridge, 1955—, prof. chemistry, 1961—. Fellow National Academy of Sciences, member A.A.A.S., American Chemical Soc. (award in inorganic chemistry 1962, Baekeland medal N.J. sect. 1963), Am. Crystallographic Assn., Am. Acad. Arts and Scis. Author: Chemical Applications of Group Theory, 1963; (with G. Wilkinson) Advanced Inorganic Chemistry, 1962, 2d edit., 1966. Studies in relationship of molecular and electronic structure to chem. properties.*

COTTON, Frederic Jay, Am. surgeon; b. Prescott, Wis., Sept. 24, 1869; s. Joseph Potter and Isabella (Cole) C.; A.B., Harvard, 1890, A.M., 1894, M.D., 1894; postgrad. Coll. Phys. and Surg., N.Y.C., U. Vienna; m. Jane Baldwin, Feb. 8, 1902. Practiced, Boston, 1897-38; asst. surgeon Children's Hosp., 1897-1902; surgeon Boston City Hosp., 1902-1931, Beth Israel Hosp., 1923-27; asst. in surgery, Harvard Med. Sch., 1903-04, later lectr. in surgery; asst. prof. surgery, Tufts Coll. Med. Sch., 1906-10. Cons. in surgery USPHS, from 1919. Fellow Am. Surg. Assn., A.C.S. (a founder). Author: Dislocations and Joint Fractures, 1910. Devised (with W. M. Boothby) apparatus for administering nitrous oxide-oxygen-ether anesthesia, 1912. Died Apr. 14, 1938.

COTTON, Richard T., Am. entomologist; b. Lydney, Glos., Eng., Jan. 27, 1893; s. William James and Agnes (James) C.; came to U. S., 1906, naturalized, 1912; B.S., Cornell U., 1914, M.S., 1918; Ph.D., George Washington U., 1924; m. Emily Willey, Dec. 18, 1917; 1 son, Robert T. Entomologist, Insular Expt. Sta., Rio Piedras, P.R., 1915-17, U. S. Dept. Agr., Fla., Washington, Kan., 1919-57; ret., 1957. Author: Pests of Stored Grain and Grain Products, 1956; (with H. H. Shepard et al) Methods of Testing Chemicals on Insects, 1958; The Chemistry and Technology of Cereals as Food and Feed, 1959; research and numerous publs. on protection of stored food products in U. S., resulting in devel. many new fumigants now in use for treating stored products; perfecting of procedures for protecting grain supplies acquired by govt. in connection with Loan Program; control of insects in flour mills, warehouses. Home: 2141 N.E. 27th Dr., Ft. Lauderdale, Fla. 33306.*

COTTON, Robert Henry, Am. chemist; b. Newton, Mass., Nov. 17, 1914; s. Leonard M. and Helen (Patenaude) C.; B.S., Bowdoin Coll., 1937; M.S., Mass. Inst. Tech., 1939; Ph.D., Pa. State U., 1944; m. Mildred Woodward Smith, Jan. 1, 1948; children—Dorothy (Mrs. Clarence Kirkwood), Leonard Wright, Thomas Carroll. Dir. Plymouth (Fla.) div. Nat. Research Corp., 1945-47; director research Holly Sugar Corp., 1948-53, Huron Milling Co., Harbor Beach, Mich., 1954-58; dir. research Continental Baking Co., Rye, N.Y., 1958—, v.p., 1965—. Chmn. tech. liaison com. U. S. Dept. Agr.-Am. Bakers Assn., 1959—; mem. vis. com. dept. nutrition and food sci. Mass.

Inst. Tech. Corp., 1966—; chmn. com. NRC, U. S. Army Natick labs. for cereal and general products. Member Am. Assn. for Advancement Science, Assn. Research Dirs. (past pres.), Inst. Food Tech., Am. Chem. Soc., Am. Assn. Cereal Chemists (pres. 1965-66), Am. Inst. Baking. Contbr. chpts. to Beet Sugar Technology, 1951; Sugar Beet Economics, 1952; Research and numerous publs. on purification of clays by acid electrodialysis; devel. of 1st successful frozen orange juice concentrate; pioneer of diffusion technique of rapid dehydration of vegetable materials; studies on wheat proteins and bread making process. Home: 56 Intervale Pl. Office: P.O. Box 731, Rye, N.Y. 10580.*

COTTRELL, Alan Howard, English metallurgist; b. July 17, 1919; s. Albert and Elizabeth Cottrell; B.Sc., U. Birmingham (Eng.), 1939, Ph.D., 1942, D.Sc.; D.Sc. (hon.) Columbia, 1965, Newcastle (Eng.) U., 1967; m. Jean Elizabeth Harber, 1944; 1 son. Lectr. metallurgy U. Birmingham, 1943-49, prof. phys. metallurgy, 1949-55; dep. head metallurgy div. Atomic Energy Research Establishment, Harwell, Eng., 1955-58; fellow, Goldsmiths' prof. Metallurgy Christ Coll. Cambridge (Eng.) U., 1958-65; deputy chief adviser studies sec. state for defense, Gt. Britain, 1965-67, chief adviser, 1967—. Recipient Rosenhain medal Inst. Metals, Platinum medal, 1965; Réamur medal Société Française de Metallurgie. Fellow Royal Soc., 1955 (v.p. 1964, Hughes medallist, 1961), Royal Swedish Acad. Scis.; mem. (fgn. hon.) Am. Acad. Arts and Scis. Author: Theoretical Structural Metallurgy, 1948, 2d edit. 1955; Dislocations and Plastic Flow in Crystals, 1953; The Mechanical Properties of Crystals, 1953, The Mechanical Properties of Matter, 1964; Theory of Crystal Dislocations, 1964; An Introduction to Metallurgy, 1967. Research, publs. on metallurgy, theoretical physics, operational analysis, tech. policy. Home: 19 Madingley Rd., Cambridge, Eng. Office: Ministry of Defense, Whitehall, London S.W. 1, Eng.

COTTRELL, Calvert Byron, Am. inventor; b. Westerley, R.I., Aug. 10, 1821; s. Lebbeus and Lydia (Maxson) C.; m. Lydia W. Perkins, May 4, 1849, 5 children. Machinist and employing contractor for Levalley, Lanphear & Co., Phoenix, R.I., 1840-55; in partnership (with Nathan Babcock) firm Cottrell & Babcock, mfrs. printing presses, Westerley, 1855-80, name changed to C. B. Cottrell & Sons, 1880; invented air spring for reversing bed of press, tapeless sheet delivery to drum cylinder, rotary color printing press, shifting tympan for a web perfecting press. Died Westerley, June 12, 1893.

COTTRELL, Frederick Gardner, Am. chemist; b. Oakland, Cal., Jan. 10, 1877; s. Henry and Cynthia L. (Durfee) C.; B.S., U. Cal., 1896, LL.D., 1927; postgrad. U. Berlin, 1901; Ph.D., U. Leipzig, 1902; m. Jessie M. Fulton, Jan. 1, 1904; 2 children. Le Conte fellow U. Cal., 1896-97, instr. phys. chemistry, 1902-06, asst. prof., 1906-11; chem. tchr. Oakland High Sch., 1897-1900, cons. chemist U. S. Bur. Mines, 1911, chief phys. chemist (field duty), 1911-14, chief chemist, 1914-15, chief metallurgist, 1916-19, asst. dir., 1919, 20, dir., 1920; chmn. div. chemistry and chem. tech. NRC, 1921-22; dir. Fixed Nitrogen Research Lab., U. S. Dept. Agr. 1922-27, chief div. fertilizer and fixed nitrogen investigation Bur. Chemistry and Soils, 1927-30, cons. chemist Bur. Chemistry and Soils, 1930-40, Bur. Plant Industry, 1940-43; pres. Research Associates, Inc., 1935-38; tech. cons. Smithsonian Instn., 1928-29, Research Corp., N.Y.C., from 1930. Mem. Am. Chem. Soc., Am. Inst. Mining Engrs., Am. Electrochem. Soc., Nat. Acad. Scis., Am. Philos. Soc., Société de Chimie Industrielle (hon.), Sigma Xi, Phi Beta Kappa. Research on nitrogen fixation, liquefaction of gases, recovery of helium; inventor Cottrell precipitator for precipitation of particles from gases; built device for prodn. of positive ion rays, 1930. Died Nov. 16, 1948.

COTTRELL, William Frederick, Am. sociologist; b. Idaho Falls, Ida., Aug. 19, 1903; s. William Franklin and Alice (Catlin) C.; A.A., Westminster Coll., Utah, 1922; A.B., Occidental Coll., 1925, A.M. 1929; postgrad. U. Utah, U. Cal. at Berkeley; Ph.D., Stanford, 1930; m. Annice Gertrude Lyman, Dec. 29, 1925; children—William, Robert Lyman, Barbara Colleen (Mrs. Frank R. Nelson, Jr.). Tchr., Milford (Utah) High Sch., 1925-26, Westminster Coll. (Utah), 1926-28; faculty Miami U., Oxford, O., 1930—, prof. govt., 1946—, chmn. dept. sociology and anthropology, 1960—, dir. Scripps Found. for Population Research, 1964—. Mem. numerous state commns. on mental health, aging, child welfare, alcoholism; cons. to industry. Co-recipient Essay prize Inst. for Social Research, Oslo, Norway, 1954. Mem. Am., Ohio Valley sociol. assns., Am. Polit. Sci. Assn., Population Assn. Am., Gerontological Soc., Soc. Study Social Problems, Phi Beta Kappa, Alpha Kappa Delta, Pi Sigma Alpha. Author: The Railroader, 1940, Men Cry Peace, 1954; Energy and Society, 1955. Contbr. to Handbook of Applied Psychology, 1950; The City in Mid-Century, 1957; Handbook of Social Gerontology, 1960; Aging and Leisure, 1961; Explorations in Social Change, 1964; Social Change in Developing Areas, 1965. Primary research on way modern man and society have been affected by technol. change. Home: 210 N. Campus Av., Oxford, O. 45056.*

COTUGNO (or COTUNNIUS), Domenico, Italian physician, anatomist; b. Ruvo, Naples, Italy, Jan. 29, 1736; s. Michele and Chiara (Assalemme) C.; ed. U. Naples. Surgeon Hosp., of Incurables, Naples; apptd. prof. anatomy U. Naples, 1766; physician to royal family. Mem. French Acad. Scis., 1810. Author: De aquaeductibus auris humani internae, 1761; De ischiade nervosa, 1764; Sciatica, 1765; De sedibus variolarium, 1769. Described nasopalatine nerve, 1760, sciatica Cotugno's disease), 1764; gave 1st good description of aqueduct nerve; 1st to describe cerebrospinal fluid, 1764; discovered albumin in urine; investigated internal ear, also chemistry of and effect on body of fluid of ear; cotunnite, nerve of Cotunnius, liquor Cotunnii all named in his honor. Died Naples, Oct. 6, 1822.

COUCH, James Russell, Am. biochemist; b. Grandview, Tex., June 10, 1909; s. James Roy and Mamie (Edwards) C.; B.S., Tex. Agrl. and Mech. U., 1931, M.S., 1934; Ph.D., U. Wis., 1948; m. Velma E. Holland, Oct. 5, 1934; children—Sandra Jean (Mrs. Theodore E. O'Connor), James Russell II, Robert Andrew, Robin Ann. Asst. poultry husbandman Tex. Agrl. Expt. Sta., College Station, 1931-34, asso. poultry husbandman, 1934-36, poultry husbandman, 1936-41; Rockefeller Research Found. fellow, 1946-48; prof. Tex. Agrl. and Mech. U., 1948—. Mem. Poultry Sci. Assn., Am. Chem. Soc., A.A.A.S., Soc. for Exptl. Biology and Medicine, Tex. Acad. Sci., Am. Inst. Nutrition, Am. Soc. Exptl. Biology, Am. Soc. Biol. Chemists, Am. Soc. Bacteriologists, Sigma Xi, Phi Kappa Phi, Gamma Alpha. Studies, publs. on vitamins, metabolism, antibiotics in poultry nutrition; vitamin E and Exudative Diathesis; function of folic acid and Vitamin B12; protein requirements for domestic fowl; fatty acids and poultry nutrition. Home: 204 Pershing St., College Sta., Tex. 77841.*

COUCH, John Nathaniel, Am. botanist; b. Prince Edward County, Va., Oct. 12, 1896; s. John Henry and Sallie Love (Terry) C.; student Trinity Coll. (Duke U.), 1914-17; A.B., U. of N.C., 1919, A.M., 1922, Ph.D., 1924; student L'Université de Nancy, spring 1919, U. of Wis., summer, 1923; m. Else Dorothy Ruprecht, May 28, 1927; children—John Philip, Sally Louise. Instr. botany, U. of N.C., 1917-18; science teacher high sch., Chapel Hill, N.C., 1919-20, Charlotte, N.C., 1920-21; instr. botany, U. of N.C., 1922-25, asst. prof., 1927-28, asso. prof., 1928-32, prof., 1932-45, Kenan prof. since 1945; Nat. Research Council fellow in botany, Carnegie Instn., 1925-26, Mo. Botanical Garden, 1926-27; with Johns Hopkins Bot. Exploration, Jamaica, B.W.I., summer, 1926; visiting prof. Johns Hopkins U., winters, 1933-35, U. of Va., summer, 1933; Cultural exchange specialist Dept. State, India, 1961; mem. N.C. Gov.'s Sci. Adv. Com., 1961-64. Spl. adviser to chmn. OSRD, 1944. Recipient Walker grand prize Boston Soc. Natural History, 1939; Meritorious Teachers award Assn. Southeastern Biologists, 1954; certificate of merit Bot. Soc. Am., 1956; first North Carolina award in science, 1964. Fellow A.A.A.S. (v.p., chmn. botany sect. 1962); mem. Nat Acad. Sci. India (hon. fgn. mem.), Soc. American Bacteriologists, Botanical Soc. of America (chairman Southeastern section, 1951), American Mycological Society (pres. 1943), Am. Soc. Microbiology, Society General Microbiology, American Mosquito Control Association, Nat. Acad. Sci., N.C. Acad. Sci. (pres. 1946-47, Jefferson award, Poteat medal 1937), Internat. Assn. Plant Taxonomists, Am. Soc. Plant Taxonomists, Indian Phytopath. Soc., Elisha Mitchell Sci. Soc. (pres. 1937-38), Sigma Xi. Author: The Gasteromycetes of the Eastern United States and Canada (with W. C. Coker), 1928, The Genus Septobasidium, 1938. Asso. editor Mycologa, 1937-39; editor Jour. Elisha Mitchell Sci. Soc., 1946-60. Contbr. articles on bot. subjects to professional jours. Research on sexuality, culture, and ciliary structure of water fungi; symbiosis between scale insects and fungi; parasitic fungi in mosquito larvae; Actinomy-cetales. Home: Chapel Hill, N.C.

COUCH, Jonathan, Brit. naturalist; b. Polperro, nr. Fowey, Eng., Mar. 15, 1789; s. Richard and Philippa C.; student medicine London; m. twice; children by 2d wife—Richard Quiller, Thomas Quiller, John Quiller. Practiced medicine Polperro, 1809-70; trained fishermen as asst. naturalists. Author: Cornish Fauna, 1834, 41; A History of the Fishes of the British Islands, 1860-65; History of Polperro, 1871; Illustrations of Instinct (storehouse of information, carefully collected and sifted, on habits of fishes, also illustrations giving unique representations of vivid natural colors of fishes while alive or immediately after death), deduced from the Habits of British Animals, 1847; other manuscripts now in library of Royal Instn. Cornwall; numerous papers in Brit. sci. publs. Aided Thomas Bewick and William Yarell in study nature; prin. work in ichthyology.

COUDER, André, French astronomer; b. Alençon, France, Nov. 27, 1897; Mem. staff Obs. Paris, from 1925, apptd. chief optical labs., 1926, then adjunct astronomer, 1937, became astronomer, 1943, titular astronomer. Mem. Royal Astron. Soc., Internat. Astron. Union (v.p.), Bur. des longitudes (pres. 1951), French Acad: Sci., 1954. Author: Récherches sur les déformations des grands miroirs employés aux observations astronomiques, 1932; (with A. Danjon) Lunettes et Télescopes, 1935. Research in astron. instruments and modern astrophysics; developed technique of polishing parabolic surfaces by previous thermic deformation. Home: 11, rue Bobierre, Bourg-la-Reine, Paris, France.

COUÉ, Emile, French physician, pharmacist; b. Troyes, France, Feb. 26, 1857; trained as pharmacist; studied hypnotism and suggestion under H. Bernheim, and A. Liebault, beginning in 1901; apothecary at Troyes, 1882-1910; founded free clinic for practice of his psychotherapeutic method, Nancy, France, 1910; taught in Europe and U. S. Author: Self Mastery by Conscious Autosuggestion, 1922; My System, 1923. Research on certain nervous troubles using hypnosis, suggestion and autosuggestion; believed to have introduced term, autosuggestion, 1922; emphasized power of imagination in healing of disease; claimed to bring about organic changes with suggestion. Died Nancy, July 2, 1926.

COUERBE, J. P., French chemist; b. Vertheuil, France, 1807; student toxicology under Lesueur, Paris, France, 1829; worked in Pelletier's lab., ret., after 1840. Author: Chimie du sulfure de carbone, 1838; Présence de l'arsenic dans le corps humain en putréfáction; Note sur des assements humains, 1859. First to prepare pure codeine and thebaine; investigated narcotine, narceine, xanthic acid and brain substance, also meconine (a pure substance of opium); original work in trying to determine bone age through chem. composition. Died Vertheuil, Oct. 9, 1867.

COUES, Elliott, Am. ornithologist; b. Portsmouth, N.H., Sept. 9, 1842; s. Samuel Elliott and Charlotte (Havenladd) C.; grad. Columbian U., 1861, M.D., 1863, hon. A.M., 1862, Ph.D., 1869; m. Jane Augusta McKenny, May 3, 1867; m. 2d, Mary Emily Bates, Oct. 25, 1887. Prof. zoology and comparative anatomy Norwich U., 1869-73; and naturalist U. S. Northern Boundary Commn., 1873-76; sec. and naturalist U. S. Geol. and Geog. Survey of the Territories, 1876-80; lectr. anatomy Med. Sch., Columbian U., 1877-83, prof. anatomy, 1883-87; long connected with Smithsonian Instn. Mem. Nat. Acad. Scis., Am. Ornithologists' Union (a founder, v.p.), Theosophical Soc. India (pres. Am. bd. control), Am. Soc. for Psychical Research (founder), numerous other socs. Author: Key to North American Birds, 1872; Field Ornithology, 1874; Birds of the Northwest; Fur-Bearing Animals, 1877; Monographs of North American Rodentia; Birds of the Colorado Valley; New England Birds, 1881; Dictionary of North American Birds, 1882; Biogen, a Speculation on the Origin and Nature of Life; The Daemon of Darwin; Kuthumi; Can Matter Think; Buddhist Catechism; A Woman in the Case; Signs of the Times; Citizen Bird; others. Introduced "Key" system into zoology (previously had been used only in bot. methods); discovered several bird species; his descriotions of birds are among most accurate known; did much to encourage systematic research in Am. ornithology; also made valuable contbns. in mammalogy. Died Balt., Dec. 25, 1899.

COUGHLIN, Richard James, Jr., Am. sociologist; b. Buffalo, Dec. 12, 1917; s. Richard James and Mary (Eardley) C.; B.S., Buffalo State U., 1941; M.A., Yale, 1950, Ph.D., 1953; m. Margaret Morgan, Feb. 7, 1946; children—Kenneth Morgan, Elizabeth Troth. U. S. vice consul, Saigon, 1946-48; research analyst U. S. State Dept., Washington, 1948-49; Fulbright Found. and Social Sci. Research Council grantee, Thailand, 1951-52; instr. Yale, 1953-55, asst. prof., 1955-60; rep. Asia Found., Hong Kong, 1957-59; asso. prof. York U., Toronto, 1960-63; prof. U. Va., 1963—, acting chmn. dept. sociology, anthropology, 1965-66, chmn., 1967—. Dir. Va. Council on Family Relations, 1964-67. Mem. Am., So. sociol. assns., Soc. Study Social Problems, Population Assn. Am., Assn. for Asian Studies. Author: Double Identity, The Chinese in Modern Thailand, 1960; (with Donn Hart, Phya Anuman) Southeast Asian Birth Customs, 1965. Research, publs. on factors affecting assimilation of non-Western minority in a non-Western society, urbanization in Asia, sociology of human reprodn. Home: 1204 Blue Ridge Rd., Charlottesville, Va. 22903.*

COULOMB, Charles Augustin de, French physicist; b. Angouleme, France, June 14, 1736; s. Henri and Catherine (Bajet) C.; ed. Collegium des Quatre-Nations, Paris, also sch. for mil. engrs., Mézières, France. Served as capt. Royal Engrs. Corps, Martinique, W.I., 1764-72; served on island of Aix, then Cherbourg, France, then Paris, 1781; named intendant French Water Commn., 1784; Conservator Map Commn.; apptd. commissary for orgn. of edn. by Napoleon, 1802, insp. gen. edn., 1805. Mem. French Acad. Scis., 1774. Author: Mémoire sur la statique des voutes, 1776; Recherches sur les moyens d'exécuter sous l'eau toutes sortes de travaux hydrauliques, 1779; Théorie des machines simples, 1781; Recherches théoriques et expérimentales sur la force de torsion et sur l'élasticité des fils de metal, 1784; Sur l'électricité et le magnétisme, 7 vols., 1785-89. Research on electricity and magnetism; invented torsion balance; applied Newton's inverse square law to electricity; demonstrated that force of elec. attraction or repulsion between 2 spheres is proportional to product of charges on each sphere and inversely proportional to square of distance between centers of spheres (Coulomb's law). Died Paris, France, Aug. 23, 1806.

COULOMB, Jean Marie François Joseph, meteorologist; b. Blida, Algeria, Nov. 7, 1904; s. Charles and Blanche (D'Izalguier) C.; ed. École normale supérieure, Paris; doctorate in sci.; m. Alice Gaydier, Sept. 29, 1928; children—Pierre, René, Marie-Blanche, Geneviève. Asst., Collège de France, 1928-31; physician Obs. of Puy-de-Dome, 1931-37; dir. Inst. Meteorology and Global Physics, Algeria, 1937-41; Paris, 1941-56; prof. Sch. Sci., Paris, 1941—; dir. Nat. Center Sci. Research, 1956-62; pres. Nat. Center Space Studies, 1962—. Mem., French Acad. Scis; Roy. Soc., Liege; Author: La physique des nuages, 1940; La constitution physique de la terre, 1952 (English edit. The Physical Constitution of the Earth, 1963). Home: 24, rue Desnouettes, Paris 15. Office: 129, rue de l'Université, Paris 7, France.

COULOMBRE, Alfred Joseph, Am. embryologist; b. Boston, Aug. 15, 1922; s. Charles Alfred and Cecelia Mary (Greene) C.; B.S., Cath. U. Am., 1947, M.S., 1949; Ph.D. (Adam T. Bruce fellow 1952-53), Johns Hopkins, 1953; student, summer investigator, Jackson Lab., 1947-50, 57; m. Jane Louise Lacy, June 26, 1948. Instr. Wabash Coll., 1948; instr. anatomy Yale Sch. Medicine, 1953-56, asst. prof., 1956-61; head sec. exptl. embryology, lab. neuro-anatomical scis. NIH, 1961—, chief, lab., 1962-67, sci. dir. Nat. Inst. Childhood Diseases, 1967—. Mem. adv. panel devel. biology NSF, 1958-61; mem. child health and human devel. tng. com. NIH, 1962—; commr. Sci. Manpower Commn., 1963—. Mem. Am. Assn. Anatomists, A.A.A.S., Assn. Research Ophthalmology, Soc. Developmental Biology (treas., 1965—), Am. Inst. Biol. Scis. (dir., 1965—), Md, Acad. Scis., Teratology Soc., Am. Bryological Soc., Phi Beta Kappa, Sigma Xi. Editorial bds. Devel. Biology, Bio-Sci., Jour. Exptl. Zoology, Am. Inst. Biol. Scis. Exptl. analysis of tissue interactions which control size, shape and orientation of each part of the embryonic eye throughout devel. Home: 8315 N. Brook Lane, Bethesda 20014. Office: Nat. Insts. Health, Bethesda, Md. 20014.*

COULSON, Charles Alfred, Brit. mathematician; b. Dudley, Worcestershire, Eng., Dec. 13, 1910; s. Alfred and Annie (Hancock) C.; B.A., Trinity Coll., Cambridge, Eng., 1931, M.A., 1934, Ph.D., 1935, D.Sc., 1945; m. Eileen Florence Burrett, Aug. 28, 1938; children—Andrew Charles, Martin Geoffrey, Janet Eileen, Wendy Ann. Fellow, Trinity Coll., Cambridge, 1934-38; lectr. math. U. Coll., Dundee, Scotland, 1938-45; Imperial Chem. Industries, fellow Oxford (Eng.) U., coll. lectr. U. Coll., 1945-47; Wheatstone prof. theoretical physics King's Coll., U. London (Eng.), 1947-52; Rouse Ball prof. math., fellow Wadham Coll., 1952—. Fellow Royal Soc., 1950, Royal Soc. Edinburgh; hon. mem. Am. Acad. Arts and Scis. Author: Waves, 1941; Electricity, 1948; Valence, 1952; Science and Christian Belief, 1955; Science, Technology and the Christian, 1960; (with A. Streitwieser, Jr.) Dictionary of pi-electron Calculations, 1965. Research and publs. on wave mechanics and relations between exptl. chemistry and math. Home: 64 Old Rd., Oxford, Headington, Oxford. Office: Oxford U., Math. Inst., 24-29, St. Giles, Oxford, Eng.*

COULSON, Charles Barrie, Brit. biochemist; b. Blidworth, Eng., Feb. 26, 1928; s. Alfred John and Doris May (West) C.; B.Sc. with Honors, Sheffield (Eng.) U., 1949, M.Sc., 1955; Dr.Nat.Sc., Charles U., Prague, Czechoslovakia, 1952; Ph.D., U. Wales, 1962; m. Nov. 13, 1948; children—Susan Theresa, David John. Postdoctoral research fellow Inst. Seaweed Research, Scotland, 1952-54; Agr. Research Council Research asst., Imperial Chem. Industries Research fellow, lectr. U. Coll., N. Wales, U. Wales, Bangor, 1955-60; project leader, head biochemistry and biophysics div. Arthur D. Little Research Inst., Edinburgh, Scotland, 1960-63; prof., head dept. biochemistry nutrition and food sci. U. Ghana, Legon, Accra, 1963-67; project dir. UNDP/FAO Food and nutrition project in Poland, Warsaw, 1965—. Fellow U.K. Royal Inst. Chemistry; mem. Biochem. Soc., Biophysics Soc., Brit. Soc. for Soil Sci., Royal Inst. Chemistry, Soc. for Chem. Industry, A.A.A.S. Joint editor Chromotographic Data, Jour. Chromatography, 1960-63; editorial bd. Jour. Chromatography, 1963—. Research and publs. on amino-acids, proteins of land plants and marine algae, gluten proteins, lucerne (alfalfa) triterpenoidal saponin including control of their growth depression with cholesterol and essential fatty acids; studies on soil humic acid characteristics, role of leaf polyphenols in these and soil profile devel., electrothermal studies of gluten hydration and alteration of rheological properties, lysine-lactose browning systems. Home: 123 Newhailes Crescent, Musselburgh, Midlothian, Scotland, U.K. Office: UNDP/FAO Project Office, ul. Mokotowska 14, Warsaw, Poland.*

COULSON, Roland A., Am. biochemist, educator; b. Rolla, Kan., Dec. 20, 1915; s. Emmett and Josephine (Armstrong) C.; A.B., U. Wichita, 1937; M.Sc., La. State U., 1939; Ph.D., U. London, 1944; m. Nancy Margaret Kelley, Nov. 16, 1944; children—Thomas Duncan, Carol Melissa. Faculty dept. biochemistry La. State U. Sch. Medicine, New Orleans, 1944—, prof., 1954—, asso. dean grad. studies, 1965—. Diplomate Am. Bd. Clin. Chemists. Fellow Am. Soc. Clin. Chemists; mem. Am. Soc. Biol. Chemists, Soc. Exptl. Biology and Medicine, Biochem. Soc. (Gt. Britain). Author: (with T. Hernandez) Biochemistry of the Alligator, 1964. Research and numerous publs. on metabolism in slow motion as observed in alligator and turtle, intermediary metabolism in living animal. Home: 2347 Mendez St., New Orleans 70112.*

COULTER, John Merle, botanist; b. Ningpo, China, Nov. 20, 1851; s. Moses Stanley and Caroline E. (Crowe) C.; A.B., Hanover Coll., 1870, A.M., 1873, Ph.D., 1882; Ph.D., Ind. U., 1884; m. Georgie M. Gaylord, Jan. 1, 1874. Botanist, U. S. Geol. Survey in Rocky Mountains, 1872-73; prof. natural scis. Hanover Coll., 1874-79; prof. biology Wabash Coll., 1879-91; pres. and prof. botany Ind. U., 1891-93; pres. Lake Forest U., 1893-96; prof. and head dept. botany U. Chgo., 1896-1925; adviser Boyce Thompson Inst. Plant Research, Yonkers, N.Y., 1925—. Prin. Bay View Summer U., 1893-96, Winona Summer Sch., 1895-98; founder and editor Bot. Gazette, 1875—. Fellow Am. Acad. Arts and Scis., A.A.A.S. (gen. sec. 1901, pres. 1918); mem. NRC, 1923-28. Author: Manual of Rocky Mountain Botany, 1885; Botany of Western Texas, 1891-94; Plant Relations, 1899; Plant Structures, 1899; Morphology of Gymnosperms (with Charles J. Chamberlain), 1901; Plant Studies, 1902; Morphology of Angiosperms (with Chamberlain), 1903; New Manual of Botany of the Central Rocky Mountains (with A. Nelson), 1909; A text-book of Botany, 1913; Elementary Studies in Botany, 1913; Fundamentals of Plant Breeding, 1914; Evolution of Sex in Plants, 1914; Plant Genetics, 1918. Died Yonkers, N.Y., Dec. 23, 1923.

COULVIER-GRAVIER, Remi-Armand, French astronomer; b. Reims, France, Feb. 26, 1803; worked in parents' winery; established obs., Paris, France, 1849. Author: Recherches sur les étoiles filantes, 1847; Précis des recherches sur les meteures et sur les lois qui les régissent, 1863. Prin. observations on shooting stars; tried to establish correlation between wind currents and atmospheric precipitation and motion of shooting stars; made futile attempts to have observation posts established throughout France to predict weather. Died Paris, Feb. 12, 1868.

COUMOULOS, George, chemist; b. N.Y.C., May 30, 1910; s. Dematrīus and Anna (Karamanis) C.; ed. U. Athens, Cambridge U.; Dr ès sc., Ph.D.; m. Helen Oekonomopoulos, Apr. 2, 1936; children—Demetrius, Marianna. Inds. cons. Am. Econ. Mission to Greece; head research Scalistiri Mining Enterprises; asst. dir. Greek Bank for Indsl. Devel., Athens.; prof. Nat. Tech. U. Athens. Mem. London Inst. Mines and Metallurgy, London Physics Instn., Chem. Soc. London, Am. Chem. Soc. Home: Odos Dryadon 8, Athens. Office: Odos Amalias 20, Athens, Greece.

COUNCILMAN, William Thomas, Am. pathologist; b. Pikesville, Md., Jan. 1, 1854; s. John T. and Christiana Drummond (Mitchell) C.; ed. St. John's Coll., Annapolis, Md.; M.D., U. Md., 1878, LL.D., 1907; student univs. Vienna and Leipzig; A.M. (hon.), Harvard, 1899, Johns Hopkins, 1902; LL.D., McGill, 1911; m. Isabella Coolidge, Dec. 17, 1894; children—Isabella Coolidge (Mrs. Frank Wigglesworth), Christiana Drummond (Mrs. William Otho Potwin Morgan), Elizabeth Lydia. Asso., and asso. prof. pathology, Johns Hopkins, 1886-91; Shattuck prof. path. anatomy Harvard, from 1892. Mem. Hamilton Rice expdn. to Amazon. Fellow Am. Acad. Arts and Scis. Phila. Acad. Medicine; pres. Assn. Am. Physicians. Author: A Study of the Bacteriology and Pathology of Two Hundred and Twenty Fatal Cases of Diphtheria, 1901; Pathology, 1902; Pathology: A Manual for Teachers and Students, 1912; Disease and Its Causes, 1913. Research on dysentery, cerebrospinal meningitis, diphtheria, smallpox; announced his discovery of probable etiology of smallpox. Died York Village, Me., May 26, 1933.

COUPER, Archibald Scott, chemist; b. Kirkintilloch, Scotland, Mar. 31, 1831; student U. Glasgow (Scotland), later U. Edinburgh (Scotland), 1852, also in Paris, France, (under Wurtz), and Germany. Became asst. in Edinburgh, 1858. Developed theory of molecular structure (linking of carbon atoms) similar to Kekulé's, but delay in publ. lost Couper priority; developed formulae nearer to those of modern notation than did Kekulé; introduced term structure; 1st to publish structural formula that showed linking of every atom for aromatic compounds, salicylic acid; 1st to publish ring formula; illustrated linking of atoms with each other by symbol of valence line, 1858; 1st to synthesize bromobenzene and p-dibromobenzene; distinguished elective affinity and degree of affinity. Died Kirkintilloch, Scotland, Mar. 11, 1892.

COUPLAND, Rex Ernest, anatomist; b. Jan. 30, 1924; s. Ernest Coupland; M.B., U. Leeds (Eng.), Ch.B. with honors, 1947, M.D. with distinction, 1952, Ph.D., 1954. With Leeds Gen. Infirmary, 1947; demonstrator, lectr. anatomy U. Leeds, 1948, 50-58; asst. prof. anatomy U. Minn. (U. S.), 1955-56; prof. anatomy Queen's Coll. St. Andrews U., Dundee, Scotland, 1958-67; prof. human morphology U. Nottingham (Eng.), 1967—. Mem. bds. biol. research and non-ionizing radiations Med. Research Council. Fellow Royal Soc. Edinburgh, 1960; mem. Inst. Biology, Anat. Soc. (mem. council), Anat. Soc. Gt. Britain and Ireland, Am. Assn. Anatomist, Soc. Exptl. Biology. Author: The Natural History of the Chromaffin Cell, 1965. Research, publs. on anatomy of human kidney, renal disease; specialist cell biology including anatomy, physiology and pharmacology of endocrine and nervous systems, functional and structural relationship between adrenal cortex and medulla, storage and release of biogenic amines; also of catechol amines by chromaffin cells. Home: East Riding, Fairfield Rd., Broughty Ferry, North Dundee, Scotland. Office: U. Nottingham, Nottinghamshire, Eng.

COUPLAND, Robert Thomas, Canadian educator; b. Winnipeg, Man., Can., Jan. 24, 1920; s. Thomas John and Gertrude (McLeod) C.; B.S.A., U. Man., 1946; Ph.D., U. Neb., 1949; m. Irene Gladys Butterworth, Nov. 16, 1945; 1 dau., Lorraine Dawn. Student asst., grad. asst. grassland research Exptl. Farm, Swift Current, Sask., Can., 1941-47, in charge Forest Exptl. Sta., Wasagaming, Man., 1946; faculty, head dept. plant ecology U. Sask., Saskatoon, Can., 1948—, prof., 1957—. Mem. Canadian subcom., also dir. Canadian grasslands project, Internat. Biol. Program. Recipient Contennial medal Can. Mem. Brit. Grassland Soc., Sask. Inst. Agrologists. Brit. Ecol. Soc., Ecol. Soc. Am., Weed Soc. Am., Am. Soc. Range Mgmt., Agrl. Inst. Can. Asso. editor Canadian Jour. Botany and Weeds. Research and publs. on classification of native grasslands; biology of weeds. Home: 515 Copland Crescent, Saskatoon, Sask., Can.*

COUPLET, Claude-Antoine, French hydraulic engr.; b. Paris, France, Apr. 20, 1642; s. Antoine Couplet; student math. (under Buhot) Ecole des pages. Prof. math. Ecole des pages, from 1670; guardian of machines, Obs., Paris, from 1670. Mem. Acad. Sci. (treas. 1696). Worked on canal system bringing water to garden of Versailles, France. Died July 25, 1722.

COURANT, Ernest David, physicist; b. Göttingen, Germany, Mar. 26, 1920; s. Richard and Nina (Runge) C.; came to U. S., 1934, naturalized, 1940; B.A., Swarthmore Coll., 1940; M.S., U. Rochester, 1942, Ph.D., 1943; m. Sara Paul, Dec. 9, 1944; children—Paul, Carl. Physicist, Can. Atomic Energy Project, Montreal, Que., 1943-46; research asso. Cornell U., 1946-48; asso. physicist Brookhaven Nat. Lab., Upton, N.Y., 1948-53, physicist, 1953-60, sr. physicist, 1960—; participated in design and constrn. of alternating-gradient synchrotron. Vis. prof. Yale, 1961-62, prof. physics, part-time 1962-66; prof. physics, part-time State U. N.Y., Stony Brook, 1967—; vis. asst. prof. Princeton, 1950-51. Fulbright Research fellow, Cambridge, Eng., 1956. Contbg. author: Handbuch der Physik, vol. 44, 1959. Co-discoverer principle alternating-gradient (strong-focusing) synchrotron. Home: 109 Bay Av., Bayport, N.Y. 11705. Office: Brookhaven Nat. Lab., Upton, N.Y. 11973.*

COURANT, Richard, mathematician; b. Lublinitz, Poland, January 8, 1888; s. Siegmund and Martha (Freund) C.; student U. Breslau (Germany), U. Zurich (Switzerland); Ph.D., U. Goettingen (Germany), 1910; E.D., Technische Hochschule (Darmstadt); Sc.D., Case Inst. Tech., 1958, N.Y.U., 1958; D.E., Technische Hochschule, Aachen, Germany, 1958; m. Nerina Runge, Jan. 1919; children—Ernest David, Gertrude A. Elizabeth, Hans Wolfgang Julius, Marianne Leonore. Came to U. S., 1934, naturalized, 1940. Asst. and instr. math. U. Goettingen, 1910-14; prof. math. U. Muenster, 1919-20; prof. math., dir. Math. Inst., Goettingen, 1920-33; lectr. U. Cambridge, Eng., 1933-34; prof. math., head math. dept. N.Y.U., 1934-58, prof. emeritus, science adviser, 1958—; also dir. Inst. Mathematics Scis. Recipient Navy Distinguished Pub. Service award, 1958. Mem. Am. Math. Soc., A.A.A.S., Am. Phys. Soc., Nat. Acad. Scis, N.Y. Acad. Scis., Math. Assn. Am., Am. Philos. Soc., Accademia Nazionale dei Lincei, Royal Netherlands, Acad. Sciences and Letters, Akademie der Wissenschaften, Royal Danish Acad. Sci. and Letters, Sigma Xi. Club: Cosmos. Author textbooks including: Vorlesungen über differential und Integralrechnung, 2 vols., 1927-29; Supersonic Flow and Shock Wave, (with K. O. Friedrichs), 1948; Methods of Mathematical Physics (with D. Hilbert), vol. I, 1953, vol. II, 1962. Studies in math. analysis and physics, theory of functions, calculus of variations. Home: 142 Calton Rd., New Rochelle, N.Y. Address: N.Y.U., 251 Mercer St., N.Y.C. 10012.*

COURMONT, Jules, French bacteriologist; b. Lyons, France, Jan. 26, 1865; licence es sciences naturelles, 1884; doctorate, 1891; aggregation, 1892; studied under Chauveau; Prof. hygiene, faculties of medicine, pharmacy, Lyons; practiced medicine, hosps., Lyons; founder Bacteriological Inst. Lyons, 1900, Mus. Hygiene, U. Lyons, 1910. Mem. Acad. Medicine. Author: Précis de bactériologie; (with Panisset) Précis de microbiologie des maladies infectieuses des animaux; (with Lesieur and Rochaix) Précis d'hygiene. Known chiefly as hygienist; one of 1st to link certain bacteria with illnesses in animals. Died Lyons, Feb. 24, 1917.

COURNAND, André F., physiologist; b. Paris, France, Sept. 24, 1895; s. Jules and Marguerite (Weber) C.; B.A., Sorbonne, Paris, 1913, P.C.B. in Sci., 1914; M.D., U. Paris, 1930; Dr. honoris causa, U. Strasbourg, 1957, U. Lyon, 1958, U. Brussels, 1959, U. Pisa, 1960; D.Sc., U. Birmingham, 1961; Gustavus Adolphus Coll., 1963, Columbia, 1965; m. Sibylle Blumer (dec. 1959); children—Muriel, Marie-Eve, Marie Claire; m. 2d, Ruth Fabian, 1963. Came to U. S. 1930, naturalized 1941. Prof. emeritus medicine Coll. Phys. & Surg., Columbia. Laureate (silver medal), faculty medicine U. Paris; recipient Andreas

Retzius silver medal Swedish Soc. Internal Medicine; Lasker award USPHS; (with D. W. Richards and Werner Forssman) Nobel Prize in medicine and physiology, 1956, Fellow Royal Soc. Medicine; mem. Nat. Acad. Scis. U. S. A., de l'Academie Nationale de Medecine (France), Academie Royal de Medecine de Belgique, Am. Physiol. Soc., Assn. Am. Physicians, Brit. Cardiac Soc., Swedish Soc. Internal Medicine, Soc. Medicale Hopitaux de Paris, Academie des Sciences, Institut de France (fgn. mem.). Author: Cardiac Catheterization in Congenital Heart Disease, 1949. First to use clinically technique of cardiac catheterization; many other contbns. to understanding of cardiocirculatory and pulmonary functions in normal and diseased humans. Home: 1361 Madison Av., N.Y.C. 28.

COURNOT, Antoine Augustin, French mathematician, economist; b. Gray, Haute Saône, France, Aug. 28, 1801; grad. École normale superieure, 1820. Asst. prof. Acad. Paris, France, 1831-34; prof. math. Acad. Lyons, France, 1834-35; rector, ednl. adminstr. Acad. Grenoble, France, 1835, insp. gen., 1838; insp. gen. Acad. Dijon, France, 1854-62. Author: Researches Into the Mathematical Principles of the Theory of Wealth, 1838; Exposition de la théorie des chances et des probabilités, 1843; Essai sur les fondements des nos connaissances et sur les caractires de la critique philosophique, 1851; Materialisme, vitalisme, rationalisme, 1875; Traité de l'enchaînement des ideés fondementales dans les sciences et dans l'histoires, 1861; Principes de la théorie des richesses, 1863; Considerations sur la marche des idées et des évenements dans les temps modernes, 1872; Revue sommaire des doctrines économiques, 1877. Considered founder of math. economics; applied math. probability to econ. theory; 1st to apply supply and demand functions to econ. analysis; made earliest known statement of modern theory of monopoly; introduced concepts of demand curve, duopoly, perfect competition. Died Paris, Mar. 30, 1877.

COURRIER, Marie Jules Constant Robert, French endocrinologist; b. Saxon-Sion, France, Oct. 6, 1895; s. Jules and Elmire (Anthoine) C.; ed. U. Nancy, Collège de Pont-à-Mousson, Facultè de Mèdecin, Strasbourg, France; dr.honoris causa, Rio de Janiero, Guiluc, Brussels, Liège, Geneva, Istanbul; m. Juliette Desmots, Aug. 1923; 2 daus. Successively asst. Faculty Médecine, Strasbourg, 1924; agrégé, d'Alger, 1926, became prof., 1935; named prof. College de France, Paris, 1938. Mem. French Acad. Medicine, French Acad. Scis. (life sec.), Royal Soc. London. Author: Endocrinologie de la gestation; also numerous articles. Endocrinological research on male, female, and thyroid glands. Home: 3 rue Mazarine, Paris, France.*

COURT, Arnold, Am. climatologist; b. Seattle, June 20, 1914; s. Nathan Altshiller and Sophie (Ravitch) C.; B.A., U. Okla., 1934; M.S., U. Wash., 1949; Ph.D., U. Cal. at Berkeley, 1956; m. Corinne H. Feibelman, May 27, 1941; children—David, Lois, Ellen. Reporter, city editor Duncan (Okla.) Banner, 1935-38; observer, meteorologist U. S. Weather Bur., 1938-43; climatologist Office Q.M. Gen., U. S. Army, 1946-51; research meteorologist U. Cal. at Berkeley, 1951-54, lectr., 1956-57; meteorologist U. S. Forest Service, Berkeley, 1956-60; chief applied climatol. br. Cambridge Research Labs., USAF, Bedford, Mass., 1960-62; sr. research scientist Lockheed-Cal. Co., Burbank, 1962-65; prof. climatology San Fernando Valley State Coll., Northridge, Cal., 1962—. Cons. to pvt. cos. Recipient spl. congl. medal U. S. Antarctic Service, 1939-41. Fellow A.A.A.S.; mem. Am. Geog. Soc., Am. Geophys. Union, Am., Royal (London Eng.) meteorol. socs., Am. Statis. Assn., Arctic Inst. N.Am., Assn. Am. Geographers, Glaciological Soc. Contbr. numerous articles to tech. jours. Discovered tendency for tropopause over Antarctic to be poorly-defined or disappear in late winter; introduced statis. theory extreme values into climatology; developed regression and correlation procedures for wind vectors. Home: 17168 Septo St., Northridge, Cal. 91324.*

COURTICE, Frederick Colin, Australian physiologist; b. Bundaberg, Australia, Mar. 26, 1911; s. Frederick and Mary (Pegg) C.; B.Sc., U. Sydney, 1933, D.Sc., 1946; D.Phil., Oxford U., Eng., 1935, M.A., 1945; m. Joyce Mary Seaton, Dec. 18, 1937; children—Rosemary Anne (Mrs. Ronald Ernest Murray), Anthony Colin, Susan Mary, Gillian Phyllis. Rhodes scholar, 1933-35; Beit fellow for med. research, Oxford U., 1937-38, reader in human physiology, 1945-48; sr. exptl. officer Chem. Def. Research Establishment, Porton, Eng., 1940-45; dir. Kanematsu Inst., Sydney, Hosp., 1948-58; prof. exptl. pathology Australian Nat. U., Canberra, 1958—. Chmn., Australian Nat. Scis. Com. of UNESCO, 1961-65; chmn. Nat. Radiation Adv. Com., 1965—; mem. adv. council Life Ins. Med. Research Fund, 1954—; mem. med. and sci. adv. com. Nat. Heart Found., 1960—. Fellow Australian Acad. Sci., Royal Australasian Coll. Physicians, Royal Australasian Coll. Surgeons; mem. Coll. of Pathologists Australia, Physiol. Soc. Eng., Australian Physiol. Soc. Author: (with J. M. Yoffey) Lymphatics, Lymph and Lymphoid Tissue, 1956. Research and numerous publs. on physiology of muscular exercise; tissue injuries, especially in burns and edema of lungs; function of lymphatic system; origin and devel. of atherosclerosis. Home: 35 State Circle, Canberra, Australia.*

COURTIVRON, Gaspard Le Compasseur, French mathematician; b. Courtivron, France, Feb. 28, 1715; s. Jean and Marie (de Clermont-Tonnere) C.; m. Cornette de Cely; m. 2d, Mademoiselle de Fussey, 1759. Mem. French Acad. Scis., 1744. Attempted to treat bovine diseases through inoculation, controlling contamination; propounded own theory of light; used blacksmith techniques to show mixture of minerals gave better quality iron. Died Oct. 5, 1785.

COURTOIS, Bernard, French industrialist; b. Dijon, France, Feb. 8, 1777; s. Jean-Baptiste C.; studied pharmacy at Auxerre; studied chemistry under Fourcroy in Paris, under Thénard at École polytechnique, Paris; ran his father's factory for mfg. soda from seaweed ash; worked in lab. of Count Antoine François de Fourcroy in the Ecole Polytechnique, to 1799; inducted into army, 1799; became mem. staff at Thénard Lab., 1802; manufactured saltpeter, sodium nitrate, and soda from kelp, beginning 1804. Recipient prize for discovery of iodine French Acad. Scis., 1831. Discovered detonating powder made by action of ammonia on iodine, 1811; discovered iodine while studying products of mother liquors produced from leaching ashes of burnt seaweed, 1811; isolated morphine (1st known alkaloid) while working with Guyton de Morveau on opium. Died Paris, Sept. 27, 1838.

COURTY, Clement Raymond François, French chemist; b. Bages, France, Jan. 23, 1902; s. Francois Baptiste and Marie (Azalague) C.; ed. Perpignon College; Montpellier Fac. of Sci., licence in physical sci.; D.Sc., Sorbonne; m. Simone Ferry; children: Janine, Monique, Philippe. Prof. of secondary ed., 1924-42; prof. of higher education, Caen, 1942-47; Lyon, since 1947. Titular prof. physical chemistry, Faculty of Sci., U. of Lyon. Awarded Palms of Acad.; Order of Public Health; mem., Legion of Honor; French Physics Soc., Lyon Pharmacy Soc. Author: Charbons Activés, 1952. Research in magnetochemistry, spectrographic adsorption, electrochemistry, quantitative mechanics; discovery of magnetic dosage of iron traces; research on amount of tin in tin plate, theoretic calculus on diamagnetism and the dimension of some atoms. Home: 5 rue des Chartreux, Lyon (1) 69, France. Office: Faculté des Sciences, 43 boulevard du 11 Novembre 1918, Villeurbanne 69, France.*

COURVILLE, Cyril Brian, Am. physician; b. Traverse City, Mich., Feb. 19, 1900; s. Philip Albert and Emma Amelia (Kroupa) C.; student Cedar Lake (Mich.) Acad., 1915-16; A.B., Emmanuel Missionary Coll., Berrien Springs, Mich., 1921; M.D., Coll. of Med. Evangelists, Loma Linda and Los Angeles, Calif., 1925; M.Sc. in embryology U. of So. Calif., 1930; D.M.S., Teijo University, Tokyo, Japan; married Margaret Louise Farnsworth, June 10, 1939. Instr. in anatomy Loma Linda University, Loma Linda, 1926-29, prof. neurology since 1934, head sect. nervous diseases; vol. asst. neurosurg. clinic Dr. Harvey Cushing, 1927; resident neurology and neurosurgery Los Angeles Co. Gen. Hosp., 1929-33, dir. Cajal Lab. Neuropathology, 1934—; consultant neuropathology, Los Angeles County coroner White Meml. Hosp., Los Angeles. Founder, organizer med. cadet corps, 1935, organized 47th Gen. Hosp., 1937, Diplomate Nat. Bd. Med. Examiners; fellow Am. Acad. Neurology; mem. Cal., Los Angeles County med. assns., Am. Neurol. Assn., Am. Acad. Forensic Scis., Am. Acad. Cerebral Palsy, Los Angeles Neurological Society, Am. Assn. Neuropathologists, British Anthropological Association. Republican. Adventist. Author sci. books, including Commotio Cerebri, 1953; Effects of Alcohol in the Nervous System of Man; contributor to the Study of Cerebral Anoxia, 1953, Forensic Neuropathology, 1964, others. Author and editor, Medical Cadet Corps Training Manual, 1942, 43. Author essays and articles profl. jours. Research effects of lack of oxygen in brain; brain concussion; characteristics of human brain tumors; histology of brain injury; effects of asphyxia at birth; brain poisons. Home: 1000 Oxford Way, Flintridge, Pasadena 3, Cal. Office: Los Angeles County Hosp., 1200 N. State St., Los Angeles.*

COURVOISIER, Ludwig Georg, Swiss surgeon; b. Basel, Switzerland, Nov. 10, 1843; mil. surgeon during Franco-Prussian War; became privat dozent surgery U. Basel, 1880, named prof., 1888. Author: Die häusliche Krankenpflege, 1874; Die Neurome, 1886; Kasuistisch-Statistische Beiträge; Zur Pathologie und Chirugie der Gallenwege, 1890. Research on gallbladder surgery; developed laws for malignant diseases of pancreas and gall bladder; originated an operation for relief of pyloric obstruction. Died Basel, Apr. 8, 1918.

COURY, Charles, French physician; b. Oct. 10, 1916; s. Alfred and Annette (Debbane) C.; ed. Faculté de Médecine, Paris; m. Jeanin Wiriot, July 11, 1952; children—Florence, Béatrice, Laure. Head med. and Tb. clinic Faculté de Paris, 1944; became physicians hosps. Paris, 1955; faculty Faculté de Medecine, 1958—, prof., 1965—. Mem. directary council Comité National de Défense contre la Tuberculose; med. expert for Tribunals. Mem. French Soc. for History Medicine. Author: Les Kystes dermoides intrathoraciques, 1945; (with M. Bariety). Le Dépistage radiologique systematique des affections du thorax, 1952, Le Médiastin et sa pathologie, 1958, Histoire de la médecine, 1963; l'Hippocratisme digital, l'osteo-arthropathie, 1950; also numerous articles. Research on tubercular and non-tubercular infections of thorax, distance retention of thoracic infections, mediastin pathology, tuberculin reactions, systematic radiol. detection, history of medicine and surgery. Home: 3 rue de Lasteyrie, Paris 16e, France. Office: Hôtel-Dieu, Place du Parvis, Notre-Dame, Paris 4e, France.*

COUSIN, Jacques-Antoine-Joseph, French mathematician; b. Paris, France, Jan. 28, 1739; asst. prof. physics Royal Coll., 1766-69; prof. math. École Militaire, from 1769; pub. adminstr., hosps., from 1790; mem. French Acad. Scis., 1722. Author: Introduction a l'étude de l'astronomie physique, 1787; Traité elementaire de physique, 1794. Abolished practice of putting 4 patients in one bed (ending scabies epidemic). Died Paris, Dec. 29, 1800.

COUSTEAU, Jacques-Yves, French oceanographer; b. St. Andre-de-Cubzac, France, 1911; ed. naval sch., Brest, France. With French underground during World War II; helped clear portions of French Mediterranean coast of mines, after World War II; pres. Compagnies Oceanographiques Francaises, comdr. research ship Calypso, 1950—; dir. Musée Océanographique de Monaco, 1957—. Decorated Legion of Honor, Croix de Guerre. Fgn. asso., Nat. Acad. Scis., U. S. Author: Par 18 Metres de Fond, 1946; La Plongée en Scaphandre, 1948; The Silent World, 1953; The Living Sea, 1963; The Voyage of the Calypso. Inventor (with French engr.) aqualung, 1943; 1st man to take color pictures underwater; inventor of process to use TV under water, 1945; designed "flying saucer" for underwater explorations capable of descending to over 1600 feet and remaining under water for 20 hours, 1959; founded undersea research group for French Navy. Home: 233 Foubourg St. Honore, Paris 8e, France.

COUTY, Louis, physician; b. 1835; Docteur en médicine; prof. indsl. biology Polytechnique Sch., Rio de Janeiro, Brazil; founder lab. Rio de Janeiro; research on curare and snake venom. Died Rio de Janeiro, 1885.

COVELLO, Mario, Italian toxicological chemist; b. Alvito, Italy, Oct. 27, 1901; s. Salvatore and Giovanna (Campana) C.; D.Pharmaceutical Chemistry, U. Naples (Italy), 1924; D.Honoris Causa, U. Paris, 1958, U. Nantes (France), 1966; m. Rita Gabrieli, Apr. 26, 1930; children—Lucio, Aldo. Faculty, Bari's U., Naples, 1933—, prof. pharm. and tossicological chemistry, 1948—, dean Faculty Sci., 1948-51, dir. Inst. Pharm. and Toxicological Chemistry, dean Faculty Pharmacy, 1951—. Recipient Gold medal for meritorious service in culture arts and sci., 1966, Gold Medal for Pub. Health, 1966. Mem. Società Chimica Italiana (chmn. Sezione Campana), Socitè Italiana di Scienze Farmaceutiche (mem. consiglio centrale). Research and numerous publs. on pharm. organic chemistry, analytical chemistry, biochemistry, especially synthetic jodorganics (medicines obtained from Jod-salycilics acids 3- and 5). Home: Via Parco Grifeo N. 38, Napoli, Italy.*

COVER, Morris Seifert, Am. veterinarian; b. Harrisburg, Pa., July 16, 1916; s. Elwood A. and Elizabeth (Seifert) C.; D.V.M., U. Pa., 1938; M.S., Kan. State U., 1942; Ph.D., U. Ill., 1952; m. June E. Minnich, Sept. 8, 1938 (dec. Apr. 1961); children—Charles Elwood, Wende Elizabeth; m. 2d, Janet C. Coblentz, May 28, 1966. Asst. poultry pathologist U. N.H., 1938-40; asst. prof. anatomy, histology Kan. State U. Vet. Coll., 1940-46; asso. prof. anatomy, histology U. Ill. Vet. Coll., 1946-52, prof. animal sci., asso. dean Agrl. Coll., dir. Agrl. Expt. Sta., 1952-67; staff veterinarian Ralston Purina Co., St. Louis, 1967—; pub. health veterinarian. Fellow A.A.A.S.; mem. Am. Vet. Med. Assn., Am. Assn. Animal Pathology, Poultry Sci. Assn. Research, numerous publs. in animal pathology; delineation of infectious synovitis treatment; isolation of viruses causing poultry diseases; devel. of treatments and control measures for chronic respiratory disease. Home: 131 Glenn Cove, Chesterfield, Mo. 63017. Office: Checkerboard Sq., St. Louis 63199.*

COVIAN, Miguel Rolando, neurophysiologist; b. Rufino, Santa Fé, Argentina, Sept. 7, 1913; s. Miguel Covian and Maria Bello de Covian; grad. U. Buenos Aires (Argentina), Sch. Medicine, 1943. Instr. dept. physiology U. Buenos Aires, 1937-42, mem. Instituto de Biologia y Medicina Experimental, 1943-55; prof., chief dept. physiology Sch. Medicine, Ribeirao Preto, Brazil, 1955—; vis. prof. dept. physiology Sch. Medicine, Porto Alegre, Brazil, 1954. Rockefeller Found. fellow Johns Hopkins, Balt., 1958-51. Mem. Latin Am. Assn. Physiol. Scis. (pres. 1966—), Internat. Brain Research Orgn., Biol. Soc. Buenos Aires, Biol. Soc. Ribeirao Preto, Brazilian Soc. Electroencephalography and Clin. Neurophysiology. Contbg. author: Textbook of Physiology, 1968. Central representation (with V. B. Mountcastle, C. R. Harrison) of some forms of deep sensibility; (with H. E. J. Houssay) arterial hypertension in hemidecorticate rats; (with R. F. Marseillan) interaction between specific visual system and reticular formation; (with J. Antunes-Rodrigues) specific alterations in sodium chloride intake after hypothalamic lesions in rat; physiology of septal area. Address: Sch. Medicine Dept. Physiology, Ribeirao Preto, S. P., Brazil.*

COVILLE, Frederick V(ernon), Am. botanist; b. Preston, N.Y., Mar. 23, 1867; s. Joseph Addison and Lydia (More) C.; A.B., Cornell U., 1887; D.Sc., George Washington U., 1921; m. Elizabeth Harwood Boynton, Oct. 4, 1890; children—Arthur Boynton (dec.), Stanley, Katharine, Cabot, Frederick. Instr. botany. Cornell U., 1887-88; asst. botanist U. S. Dept. Agr., 1888-93, botanist, 1893-1937; curator U. S. Nat. Herbarium, 1893-1937; acting dir. Nat. Arboretum, 1929-37. Procured Found. of Desert Lab. by Carnegie Instn. Fellow A.A.A.S. (v.p. botany. 1902). Author: Botany of the Death Valley Expdn.; Standardized Plant Names (joint author); and many bot. papers. George Robert White medal of honor. Mass. Hort. Soc., 1931. Died Washington, D.C, Jan. 9, 1937.

COWAN, Frederick Pierce, Am. health physicist; b. Bar Harbor, Me., July 3, 1906; s. Fred Herbert and Lena (Pierce) C.; B.A., Bowdoin Coll., 1928; M.A., Harvard, 1931, Ph.D., 1935; m. Eva Rachel Taylor, June 21, 1934. Faculty, Bowdoin Coll., 1928-29, Harvard, 1929-35, Radcliffe Coll., 1933-35, Rensselaer Poly. Inst., 1936-43; research asso. Radio Research Lab., Cambridge, Mass., 1943-45, Chrysler Corp., Detroit, 1946-47; health physicist Brookhaven Nat. Lab., Upton, N.Y., 1947—. Cons. Fed. Civil Def. Agy., 1951-55; chmn. Am. Bd. Health Physics, 1963-65; mem. Internat. Comm. on Radiation Units and Measurements, 1965—, Nat. Council on Radiation Protection, 1967—. Mem. Health Physics Soc. (pres. 1957-58), Am. Phys. Soc., Radiation Research Soc., I.E.E.E., A.A.A.S. Research, publs. on bioassay techniques, fallout studies, propagation of gamma rays, accelerator dosimetry and reactor safety; activities on standardization programs. Home: 22 Livingston Rd., Bellport, N.Y. 11713. Office: Brookhaven Nat. Lab., Upton, N.Y. 11973.*

COWAN, George Arthur, Am. chemist; b. Worcester, Mass., Feb. 15, 1920; s. Louis Abraham and Anna (Listic) C.; B.S. in Chemistry, Worcester Poly. Inst., 1941; D.Sc. in Chemistry, Carnegie Inst. Tech., 1950; m. Helen Siegel Dunham, Sept. 7, 1946. Research asst. Palmer phys. lab. Princeton, 1941; research asst., metall. lab. U. Chgo., 1942-45; research asst. Pupin phys. lab. Columbia, 1945; chemist Los Alamos Sci. Lab., 1945-46, staff mem., 1949—, group leader, radiochemistry, 1955—, asso. div. leader of test div., 1956—; teaching asst. Carnegie Inst. Tech., 1949. Chmn. bd. Los Alamos Nat. Bank, 1963—. Recipient E. O. Lawrence award AEC, 1965. Fellow Am. Phys. Soc., A.A.A.S.; mem. Fedn. Atomic Scis. (chmn. Los Alamos sect. 1955), Am. chem. soc. Research in nuclear energy and weapon devel., prodn. of heaviest elements; study of fission process, neutron spectroscopy. Home: 721 42d St. Office: P.O. Box 1663, Los Alamos, N.M. 85744.*

COWAN, John Charles, Am. chemist; b. Danville, Ill., Oct. 25, 1911; s. Charles C. and Ella (Davis) C.; student U. Colo., 1929-33; A.B., U. Ill., 1934, Ph.D., 1938; m. Lucile D. Chenoweth, Aug. 20, 1938; children—Karen L. (Mrs. Richard I. Ford), Colleen. Chemist, head oilseed sect. No. Utilization Research and Devel. div. Agrl. Research Sta., U. S. Dept. Agr., Peoria, Ill., 1944-59, chief oilseed crops lab., 1959—. Recipient Superior Service award U. S. Dept. Agr., 1948. Mem. Am. Chem. Soc., Am. Oil Chemists Soc. (A. E. Bailey award 1961), Federated Paint Socs., Sigma Xi, Phi Lambda Upsilon, Alpha Chi Sigma. Contbr. chpts. to Gilman's Organic Chemistry, 1953; Holman's Progress in Chemistry and Technology of Fats and Oils, 1958; Lundberg's Autoxidation and Antioxidants, 1962; Gaylord's Polyethers, 1963. Research, numerous publs. on preparation properties and use of soybeans, flax and their derived products including polyester rubbers, polyamide coatings and adhesives, flavor stability of edible soybean oil, orgn. vegetable oils, emulsion paints from linseed oil, polymerization, cyclization and other chem. modifications of oils and proteins. Home: 1614 W. Margaret St. Office: 1815 N. University St., Peoria, Ill. 61604.*

COWARD, William, Brit. physician; b. Winchester, Eng., 1656; B.A., Oxford (Eng.) U., 1677, M.A., 1683, M.B., 1685, M.D., 1687. Practice medicine Northampton, London, Epswich, Eng. Author: De Fermento volatili nutrivito conjectura rationis, 1695; Alcali Vindicatum, 1698; Remediorum Medicinalum Tabula, 1704; Opthalmoiatra, 1706. Ridiculed Cartesian notion of immaterial soul residing in pineal gland. Died Ipswich, 1725.

COWDRY, Edmund Vincent, anatomist; b. MacLeod, Alta., Can., July 18, 1888; s. Nathaniel H. and Anna (Ingham) C.; came to U. S., 1909; naturalized 1930; A.B., U. Toronto, 1909; Ph.D., U. Chgo., 1913; D.Sc., Institutum Divi Thomae, Cin., 1957; m. Alice Hanford Smith, Dec. 20, 1916; children—Edmund Vincent, Alice Moira (Mrs. Drew W. Luten, Jr.), Margaret Hanford (Mrs. Howard F. Park III). Faculty, Johns Hopkins, 1913-19, Peking (China) Union Med. Coll., 1917-21; asso. Mem. Rockefeller Inst., 1921-28; faculty Washington U., St. Louis, 1928-60, prof. anatomy, 1941-50, prof. emeritus, 1960—, dir. Wernse Cancer Research Lab., 1950-60; dir. research Barnard Free Skin and Cancer Hosp., St. Louis, 1936-48, Sci. Assos., Inc., St. Louis, 1964; also lectr. Cons., adviser to brs. of U. S. govt., instns. and govts. of India, China, others. Recipient numerous awards, citations including Bobst award, 1954; Gerontological Research Found. award, 1958; medal Japanese Assn. Anatomy, 1959; Minister of Edn. medal Republic of China, 1959; Hamdi Suat Aknar medal, Ankara, Turkey, 1959; award Bertner Found., 1960. Hon. fellow Royal Micros Soc. London; hon. mem. Soc. de Cancerologia de Guadalajara, Argentin Gerontol. Soc., Soc. Philomanthique de Paris, Nat. Geriatrics Soc., Academica Sinica (corr.), others. Author: Histology, 1943, 5th edit. (with John C. Finerty), 1960; Cancer Cells 1955; Editor: (with Madison Bentley) Problem of Nervous Disorder, 1934; (with S. Seno) Intracellular Membraneous Structure, 1965; Microscopic Technique in Biology and Medicine, 1943 (5th edit. with V. Emmel 1964); numerous books including Problems of Ageing, 1938; Care of Geriatric Patient, 1958, 63. Research, publs. on Am. and African yellow fever and their identity; discovered Cowdria ruminantium, causative agt. of heartwater, also (with Dr. Ham) the life cycle of agt. of East Coast fever in Kenya; exptl. work on promin. treatment of leprosy, (with Chonai Ruangsiri) adopted as standard practice until replaced. Home: 4961 Laclede Av., St. Louis 63108.*

COWELL, Philip Herbert, English astronomer; b. Alipore, Calcutta, India, Aug. 7, 1870; hon. D.Sc., Oxford (Eng.) U.; m. Phyllis Chaplin, 1901. Fellow Trinity Coll., Cambridge, Eng.; chief asst. Royal Obs., Greenwich, Eng.; supt. Nautical Almanac Office, 1910-30. Fellow Royal Soc., 1906. Investigated retardation of earth's rotation period (led to calculation that day is lengthening about 1 minute every thousand years). Died Aldeburgh, Suffolk, Eng., June 6, 1949.

COWEN, David, Am. neuropathologist; b. N.Y.C., July 29, 1907; s. David and Miriam (Goodman) C.; A.B., Columbia, 1928, M.D., 1932. Faculty, Columbia, 1937—, prof. neuropathology, 1963—; with Columbia Presbyn. Med. Center, 1937—, attending neuropathologist, 1963—; cons. neuropathologist U. S. VA Hosp., East Orange, N.J., Lenox Hill Hosp., N.Y.C. Mem. Am. Assn. Neuropathologists (pres. 1961-62), Am. Neurol. Assn., Am. Acad. Neurology, Am. Assn. Exptl. Pathology. Research, publs. on human congenital toxoplasmosis, exptl. congenital toxoplasmosis in mouse, atrophies and encephalomalacias in infancy and childhood, etiology and consequences of perinatal brain damage, perinatal infections of the nervous system, infantile neuroaxonal dystrophy, effects of anti-metabolite, 6-aminonicotinamide on nervous system. Office: 630 W. 168th St., N.Y.C. 10032.*

COWGILL, George Raymond, Am. physiologist, biochemist; b. St. Paul, Minn., Feb. 8, 1893; s. Frank Brooks and Ida Lillian (Hall) C.; student Hamline U., 1911-13, Sc.D., 1955; A.B., Stanford, 1916; Ph.D., Yale, 1921; Sc.D., U. So. Cal., 1947; m. Alice Mae Fesler, Sept. 7, 1922 (dec. 1957); 1 dau. Barbara Jean (Mrs. Allen R. Perrins); m. 2d, Grace Deuel, Mar. 31, 1959. Faculty, Yale, 1921—, prof. nutrition, 1944-60, prof. emeritus nutrition, 1960—; adj. prof. biochemistry and nutrition U. So. Cal., Los Angeles, 1960—. Mem. food and nutrition bd. NRC, 1941-47. Recipient Ann. Sci. award for research Grocery Mfrs., 1947. Fellow Am. Pub. Health Assn., N.Y. Acad. Scis.; mem. Am. Physiol. Soc., Am. Soc. Biol. Chemist, Soc. for Exptl. Biology and Medicine, Med. and Surg. Soc. Rio de Janeiro (hon.), Am. Inst. Nutrition (Mead Johnson award for research on vitamin B complex 1942, Osborne-Mendel award for research in basic sci. nutrition 1957). Author: Vitamin B Requirement of Man, 1934; also numerous articles. Editor: (with W. L. Marxer) The Art of Predictive Medicine, 1967. Editor Jour. Nutrition, 1938-58. Research on feeding expts. with dogs, vitamin B, food enrichment in U. S., other countries. Home: 1225 Charles St., Pasadena, Cal. 91103.*

COWIE, Alfred Tennant, Brit. endocrinologist; b. Daviot, Scotland, Aug. 16, 1916; s. Alfred William and Anna (Tennant) C.; M.R.C.V.S., B.Sc., Royal (Dick) Vet. Coll., Edinburgh U., 1938; Ph.D., Reading U., 1947, D.Sc., 1959; m. F. M. Valda Coppen, Sept. 18, 1943; children—Anna, George. Centenary fellow Royal Vet. Coll., 1939-41; staff Nat. Inst. for Research in Dairying, Reading, Eng., 1941—. Mem. Soc. for Endocrinology, Physiol. Soc., Brit. Vet. Assn., Royal Soc. Medicine. Author: Pregnancy Diagnosis Tests, 1947; also numerous articles. Editor: Milk—The Mammary Gland and Its Secretion, 1961. Research on hormonal mechanisms regulating growth of mammary gland and secretion of milk. Home: 6 Maiden Erlegh Dr., Earley, Reading. Office: Nat. Inst. for Research in Dairying, Reading, Eng.*

COWIE, Dean Bruce, Am. biophysicist; b. Vancouver, B.C., Can., Mar. 6, 1913; s. Louis James and Cora (Graf) C. (parents Am. citizens); student U. Cal. at Berkeley, 1935-38, Swarthmore Coll., 1938-39; m. Jean Vardeman Cockrell, Aug. 18, 1946; children—Katharine Morris, Dean Bruce, Carol Vardeman, Douglas Randolph. Research fellow Nat. Cancer Inst., USPHS, Bethesda, Md., 1939-43; mem. staff dept. terrestrial magnetism Carnegie Instn. Washington, 1943—. Hon. staff mem. Instituto de Biofisca, Universidade do Brazil, Rio de Janeiro, 1961—. Fellow Am. Phys. Soc., A.A.A.S. (council 1963—), Am. Acad. Microbiology, Washington Acad. Sci.; mem. Biophys. Soc. (council 1960-63, exec. council 1962, pres. elect 1966-67), Tex. Radiol. Soc. (hon.). Author: (with others) Studies of Biosynthesis in Escherichia Coli, 1955, Macromolecular Biosynthesis, 1964. Research, publs. on amino acid analogs, enzymes and enzyme kinetics, genetic relations between viral and host cell DNA. Home: 5401 Blackistone Rd., Washington 20016. Office: 5421 Broad Branch Rd., Washington 20015.*

COWLES, Edward Spencer, Am. neurologist, psychiatrist; b. Williamsburg, Va., Sept. 22, 1889; s. John Bertram and Harriet (Spencer) C.; student Coll. William and Mary; M.D., U. Coll. Medicine, Richmond, Va., 1907; postgrad. Harvard Med. Sch.; m. Florence Jaquith (dec.); children—Virginia, Mary; m. 2d. Nona de Mohrenschildt McAdoo, Aug. 1927; m. 3d. Lorraine Posey. Dir. Psychopathic Sanitarium, Portsmouth, N.H., 1910-16; head dept. psychopathology N.Y. Poly. Med. Sch. and Hosp., 1916-17; head war examining bd., N.Y.C. 1918; dir. Park Av. Hosp., N.Y.C., from 1916; dir. Body and Mind Found., Inc., N.Y.C.; med. mem. Joint Commn. to Investigate Healing for P.E. Ch. of Am. Fellow A.A.A.S. Author: Industrial Education, 1918; Psychopathology, 1918; Religion and Medicine in the Church, 1925; New Aspects of Chronic Alcoholism, 1937; A New Approach to the Pathology and Treatment of the Psychoneurosis and of the Melancholia Mania Psychosis, 1939; Don't Be Afraid, 1941; Alcoholism Can be Cured, 1950; Conquest of Fatigue and Fear, 1954. Popularized psychiatry through lectures and mag. articles. Died Nov. 16, 1954.

COWLES, Henry Chandler, Am. botanist; b. Kensington, Conn., Feb. 27, 1869; s. Henry Martyn and Eliza (Whittlesey) C.; A.B., Oberlin Coll., 1893; Ph.D., U. Chgo., 1898; Sc.D., Oberlin Coll., 1923; m. Elizabeth Waller, June 25, 1900; 1 dau., Harriet Elizabeth. Prof. natural sciences, Gates Coll., Neb., 1894-95; spl. field asst. U. S. Geol. Survey, summer, 1895; instr. botany U. Chgo., 1902-07, asst. prof., 1907-11, asso. prof., 1911-15, prof., from 1915, chmn. dept. of botany, 1925-34. Pres. ecology sect., Internat. Bot. Congress, 1930. Mem. Ecol. Soc. Am. (an organizer 1915, pres. 1918). Author: Vegetation of Sand Dunes of Lake Michigan, 1899; Plant Societies of Chicago 1901; Text-book of Plant Ecology, 1911; Plant Societies of Chicago and Vicinity, 1913. Editor Botanical Gazette, 1925-34. A founder of ecology, specializing in research on physiographic and comparative ecology, ecol. relations of dune vegetation, trees as indicators of past topog. conditions, and floristics of Chgo. region. Died Chgo., Sept. 12, 1939.

COWLES, Phillip Bishop, Am. bacteriologist; b. Wallingford, Conn., May 26, 1899; s. Frederic Morgan and Charity Sprague (Bishop) C.; grad. Choate Sch., 1917; B.A., Yale, 1921, Ph.D., 1929; m. Mabel Wiles Smith, June 19, 1923; children—Helen Patricia (Mrs. Philippe Verdier), Philip Bishop. Instr. mathematics and sci. The Choate Sch., 1921-27; instr. Yale Med. Sch., 1929-32, asst. prof., 1932-41, asso. prof., 1941-51, professor of microbiology, 1951-64, professor microbiology emeritus, 1964—, fellow Davenport Coll., 1936—, acting master, 1962-63. Mem. Nat. Bd. Med. Exam., 1950-55. Mem. Soc. Am. Bacteriologists. Research on bacteriostasis and disinfection; tetanus immunization; bacteriophage. Home: 224 Edgehill Rd., New Haven.

COWLEY, R. Adams, Am. surgeon; b. Layton, Utah, July 25, 1917; s. William Wallace and Alta (Adams) C.; student U. Utah, 1936-40; M.D., U. Md., 1944. Asst. dir. exptl. surgery dept. surgery U. Md., 1951-63, dir. Cardiopulmonary Lab., 1952-63, prof., head div. thoracic, cardiovascular surgery, 1961—. Mem. com. on shock, com. on hyperbaric oxygenation NRC. Mem. Soc. Thoracic Surgeons, A.C.S., Soc. Vascular Surgeons, Am. Assn. Thoracic Surgery, Am. Coll. Chest Physicians, A.A.A.S., Internat. Cardiovascular Soc., Soc. Artificial and Internal Organs. Contbr. numerous articles to sci. jours. Research in interdisciplinary approach to study of shock and methods of resuscitation. Home: 1010 St. Paul St. Office: U. Md. Hosp., Balt. 21201.*

COWLING, Thomas George, English mathematician; b. June 17, 1906; s. George and Edith Eliza Cowling; M.A., Ph.D., Oxford (Eng.) U., hon. fellow Brasenose Coll., 1966; m. Doris Moffatt, 1935; 3 children. Tchr. math. Imperial Coll. Sci., Eng., U. Col. Swansea (Wales), U. Coll. Dundee (Scotland), Manchester (Eng.) U.; tchr. U. Coll. Bangor (Wales), prof. math. 1945-48; prof. applied math. Leeds (Eng.) U., 1948—. Fellow Royal Soc., 1947, Royal Astron. Soc. (gold medallist 1956, pres. 1965-67), London Math. Soc. Author: (with S. Chapman) The Mathematical Theory of Non-Uniform Gases, 1938; Molecules in Motion, 1950; Magnetohydrodynamics, 1957. Research, publs. on astron. and atmospheric scis. including magnetohydrodynamics, theoretical astrophysics, kinetic theory gases. Address: 19 Hollin Gardens, Leeds 16, Yorkshire, Eng.

COWPER, Edward, inventor, printer; partner (with Augustus Applegath; in printing bus.; partner with brother) E. & E. Cowper Machines; prof. mfg. art and mechanics Kings Coll., London. Patentee curved stereotype plates and method of printing on long rolls of paper, 1816. Inventor improved method of distbg. ink, 1818, machine which prints both sides of paper simultaneously, (with brother) cylinder cardprinting press, (with Applegath) 4-cylinder machine, 1827.

COWPER (or COOPER), William, English surgeon, anatomist; b. Petersfield, Sussex, Eng., Mar. 8, 1666; s. Richard Cowper; apprenticed to surgeon, London, Eng., 1682; qualified as surgeon, 1691. Practiced in London; defended himself against Godefridus Bidloo, a Leyden prof., 1701. Fellow Royal Soc., 1696; mem. French Acad. Scis., 1699. Author: Myotomia reformata; or, a New Administration of the Muscles of the Humane Bodies, 1694; The Anatomy of Humane Bodies, 1698; Glandularum quarundam nuper detectarum ductuumque earum excretionum descriptio cum figuris, 1702; also med. articles. Discovered a pair of racemose glands located beneath the anterior part of the membranous part of the urethra (known as Cowper's glands in the male), 1698; presented classic description of aortic insufficiency; studied muscles. Died London, Mar. 8, 1709.

COWPER-COLES, Sherard Osborn, Brit. chemist, engr.; b. Cowper-Coles; m. Constaance Hamilton Watts, 1919; 3 sons, Fellow Royal Instn.; mem. Instn. Elec. Engrs., Instn. Mech. Engrs., Am. Iron and Steel Inst., Am. Inst. E.E., Am. Electro-Chem. Soc., Soc. Chem. Industry; founder Faraday Soc. Contbr. papers on sci. topics. Inventor Sheradizing process for rendering iron and steel rustless, process for inlaying and decorating metallic surafces, electolytic process for making parabolic reflectors, weak line of cleavage process for making copper wire, electrolytic process for recovery of zinc from its ores, electrolytic process for producing points and cutting edges and to supersede grinding; founder electrolytic galvanizing industry and chromium plating industry; developed electrolytic process for making copper sheets and tubes. Died Sept. 9, 1936.

COX, Allan V., Am. geophysicist; b. Santa Ana, Cal., Dec. 17, 1926; s. Vernon D. and Hilda (Schultz) C.; B.S., U. Cal., 1955, M.A., 1957, Ph.D., 1959. Geol. field asst. U. S. Geol. Survey, Alaska, 1950, 51, 54, geophysicist, Menlo Park, Cal., 1959——. Research asso. Stanford, 1962——. Mem. Geol. Soc. Am., Seismol. Soc. Am., Am. Geophys. Union, A.A.A.S., Soc. Terrestrial Magnetism and Electricity Japan, Inst. Nat. Scis. Ecuador, Sigma Xi. Contbd. to history of earth's magnetic field by analyzing magnetic traces imprinted on rocks by ancient magnetic fields; established chronology for North, South changes in direction of earth's magnetic field. Home: Star Route, Box 68, Redwood City, Cal. 94061. Office: Dept. Geophysics, Stanford U., Stanford, Cal. 94305.*

COX, David Roxbee, English theoretical statistician; b. July 15, 1924; s. S. R. Cox; M.A., St. John's Coll., Cambridge (Eng.) U.; Ph.D., U. Leeds (Eng.), 1949; m. Joyce Drummond, 1948; 4 children. With Wool Industries Research Assn., Eng., 1946-50; statis. lab. Cambridge U., 1950-55; vis. prof. U. N.C. (U. S.), 1955-56; reading prof. statistics Imperial Coll. Sci. and Tech., London, 1961-66, prof., 1966——. Mem. Royal Statis. Soc., Royal Biometric Soc., Operational Research Soc., Cambridge Philos. Soc. Author: (with others) Statistical Methods in the Textile Industry, 1949, Queues, 1961, Theory of Stochastic Processes, 1965, Statistical Analysis of Series of Events, 1966; Planning of Experiments, 1958, Renewal Theory, 1962. Editor Biometrika, 1966——; asso. editor Jour. Applied Probability. Research, publs. on application probability and statis. theory, operational research theory and methods. Address: Dept. Math., Imperial Coll. Sci. and Tech., 53 Princes Gate, Exhibition Rd., London S.W. 7, Eng.

COX, Doak Carey, Am. hydrologist-geologist; b. Wailuku, Maui, Hawaii, Jan. 16, 1917; s. Joel B. and Helen (Horton) C.; B.S., U. Hawaii, 1938; M.A. Harvard, 1941, Ph.D., 1965; m. Majorie L. Greiner, June 9, 1941; children—Catharine Elisabeth (Mrs. Stephenson Langmuir), Nancy Ann, Marion Louise, Charles Edwin, Helen Amelia. Geologist, U. S. Geol. Survey, 1941-45; geologist Expt. Sta., Hawaiian Sugar Planters Assn., Honolulu, 1946-56, prin. geophysicist, 1956-60; prof. geology U. Hawaii, Honolulu, 1960——, head tsunami div. Hawaii Inst. Geophysics, 1960-65, dir. Water Resources Research Center, 1964-——. Hydrologist, Pacific sci. bd. NRC, Arno Atoll, Marshall Islands, 1950; vis. prof. geology Stanford, 1960; cons. geologist Hawaiian and other Pacific Islands, 1946-64; sec. tsunami com. Internal Union Geodesy and Geophysics, 1960——; chmn. oceanography panel com. Alaskan earthquake Nat. Acad. Scis., 1964——; sec. Hilo Tech. Tsunami Adv. Council, 1962-66. Fellow A.A.A.S., Geol. Soc. Am.; mem. Hawaiian Acad. Sci. (past pres.), Am. Geophys. Union, Am. Meteorol. Soc., Am. Inst. Profl. Geologists, Assn. Engring. Geologists, Seismol. Soc. Am., Am. Inst. Mining, Meterol. and Petroleum Engrs. Author: (with F. P. Shepard, G. A. MacDonald) The Tsunami of April 1, 1946, 1950; (with G. A. MacDonald, D. A. Davis) Geology and Ground-Water Resources of Island of Kauai, 1960; also articles. Outlined geologic controls localizing fluospar desposits in Western U. S.; described tsunamis and effects, analysis their generation, propagation and shoreline behavior; described Hawaiian and other island ground-water bodies. Home: 1929 Kakela Dr., Honolulu 96822.*

COX, Edwin, Am. chemist; b. Richmond, Va., Sept. 20, 1902; s. Edwin Piper and Sally Bland (Clarke) C.; B.S. in chem. engring., Va. Mil. Inst., 1920, M.S. in chemistry, 1920; m. Virginia Bagby DeMott, May 19, 1927; 1 son, Edwin. With Va.-Carolina Chem. Corp., 1920-57, v.p., 1949-57; pvt. practice chem. engr., chemist, Aylett, Va., 1958——; pres. Tobacco By-Products and Chem. Corp.; partner Edwin Cox Assos., Richmond, Va.; dir. Commonwealth Lab., Inc. (Richmond). Trustee Virginia Inst. Sci. Research; mem. Va. Library Bd.; chmn. adv. com. Va. Selective Service. Personal Aide five Va. Govs., 1926-50. Served as lt. col. to col. Inf., U. S. Army, 1940-46. comdg. officer 176th Inf., Joint Chiefs of Staff Secretariat, SHAEF, operational control I. G. Farben, 1945; brig. gen. Va. N.G., ret. Dec. Legion of Merit, Bronze Star, Army Comm. Ribbon with oak leaf cluster; Distinguished Service award, Am. Chem. Soc., Va. sect., 1951; Gold medal award, Am. Institute Chemists, 1965. Registered engineer, Virginia and New York. Fellow A.A.A.S., American Inst. Chemists; mem. Am. Chem. Soc., Am. Inst. Chem. Engrs., Soc. Chem. Industry, Va. Acad. Sci., Mil. Order World Wars, Am. Legion, Va. (mem. bd.), Richmond hist. socs., Soc. Colonial Wars, Nat. Soc. Profl. Engrs., S. R., Jamestown Society (past gov.). Democrat (past chmn. 3d dist. com., past sec. Virginia State Committee). Protestant Episcopalian (lay reader, trustee parish). Knight of Malta. Clubs: Commonwealth, Richmond (Virginia); Chemists (N.Y.C.); Cosmos (Wash.); Golden Horseshoe. Author articles in chemical journals. Issued 14 patents in field. Research on agricultural chemicals; biological phosphate chemistry; phosphorous chemistry; proteins; sulphur; nicotine; air and surface pollution; process engineering; nonmetallic metallurgy and mining. Address: Holly Hill, Aylett, Va. 23009.

COX, Ernest Gordon, English chemist; b. Bath, Eng., Apr. 24, 1906; s. Ernest Henry and Rosina (Ring) C.; B.Sc., U. Bristol (Eng.), 1927, D.Sc., 1936; D.Sc., U. Newcastle (Eng.), 1964, Birmingham (Eng.) U., 1964; m. Lucie Grace Baker, Apr. 2, 1929; children—Patricia Ann, Keith Gordon. Research asst. Royal Instn., London, Eng., 1927-29; faculty U. Birmingham, 1929-45, reader chemistry, 1945; prof. structural chemistry U. Leeds (Eng.), 1945-60; sec. Agrl. Research Council, London, 1960——. Mem. Agrl. Research Council, 1957-60. Decorated Territorial Decoration, knight comdr. Order Brit. Empire. Fellow Royal Soc.; mem. Royal Inst. Chemistry, Inst. Physics. Research, numerous publs. on application of X-ray crystallography to study molecular structures of co-ordination compounds and carbohydrates. Home: 117 Hampstead Way, London, N.W. 11. Office: 160, Great Portland St., London, W. 1, Eng.*

COX, Guy Henry, Am. geologist; b. Lehigh, Ia., May 4, 1882; s. Edward Henry and Ada (Wilson) C.; B.S., Northwestern U., 1905; studied Sch. Mines, U. Cal., 1906; M.A., U. Wis., 1908, Ph.D., 1911; E.M., Mo. Sch. Mines and Metallurgy, U. Mo., Rolla, 1914; m. Kittie May Gates, Dec. 27, 1909. Geol. work in Wis., Wyo., Ill., summers 1904-11; instr. geology, U. Cal., 1909; asst. prof. mineralogy and petrology, 1910-11, prof. geology and mineralogy, 1911-20, Mo. Sch. Mines and Metallurgy. Civilian supervision govt. tng. camp, sects. A and B, Rolla, Mo., 1917-18. Mem. Cox & Radcliffe, cons. engrs., 1917-19; chief geologist, Jersey Oil Co., from 1920. Author: Field Methods in Petroleum Geology, 1920. Died 1922.

COX, Henry Joseph, Am. meteorologist; b. Newton, Mass., Apr. 5, 1863; s. Thomas and Hannah M. (Perkins) C.; A.B., Harvard, 1884; A.M. (hon.), Norwich U., 1887, Sc.D., 1914; m. Mary Cavanagh, Sept. 8, 1887; children—Henry Perkins, Arthur Cavanagh, Paul Greenwood (dec.). Prof. Norwich U., 1886-88; in weather service from 1884; sr. meteorologist, from 1924, became sr. in charge North Central Forecast Dist., Chgo.; also in charge corn and wheat region service of Weather Bur., and in charge spl. researches in agrl. meteorology. Fellow Am. Meteorol. Soc. (councilor). Author: Weather Bureau Records in Court; Lantern Slides in Teaching of Meteorology (Bull. 3, Geog. Soc. Chicago; Recent Advances in Meteorology; Notes of a Meteorologist in Europe; The Weather Bureau and the Cranberry Industry; Weather and Climate of Chicago; Weather Forecasting in United States (joint author); Influence of Great Lakes upon Movement of Storms; Thermal Belts and Fruit Growing in North Carolina Mountain Region. Died Jan. 7, 1930.

COX, Herald R(ea), Am. virologist; b. Rosedale, Ind. Feb. 28, 1907; s. Leo Robert and Pauline (Rea) C.; A.B., Ind. State U., 1928, Sc.D., 1964; Sc.D., Johns Hopkins, 1931, (hon.), U. Mont., 1942, m. Marion Alice Curry, June 19, 1932; children—Jane Earle (Mrs. Richard Alan Nozell), George Robert, Gordon Lee. Instr. immunology Johns Hopkins, 1931-32; asst. bacteriology and pathology Rockefeller Inst. for Med. Research, N.Y.C., 1932-36; asso. bacteriologist USPHS, Hamilton, Mont., 1936-40, prin. bacteriologist, 1940-42; asso. dir. viral research Lederle Labs., Am. Cyanamid Co., Pearl River, N.Y., 1942-44, dir. viral research, 1944——; vis. lectr. microbiology Harvard Sch. Pub. Health, 1951——. Mem. adv. council U. S. Army Chem. Corps, 1952——. Typhus Commn. medal U. S. Sec. of War, 1946; Ricketts award in med. scis. U. Chgo., 1951. Mem. A.A.A.S. (Theobald Smith award in med. sci. 1941), Am. Soc. for Microbiology (past pres.), Am. Soc. Microbiologists, Am. Soc. Immunologists, Am. Soc. for Tropical Medicine and Hygiene, Am. Pub. Health Assn., N.Y. Acad. Medicine. Contbr. numerous articles to tech. jours. Developed chicken embryo, yolk-sac vaccines against Rocky Mountain spotted fever, epidemic typhus fever; co-discovered Q fever; research and devel. on numerous living vaccines including rabies, hog cholera, Newcastle disease of poultry, trivalent oral poliomyelitis vaccines; developed first killed vaccines in field virology for Eastern, Western equine encephalitis viruses. Home: 150 Haverstraw Rd., Suffern, N.Y. 10901. Office: Lederle Labs., Pearl River, N.Y. 10965.*

COX, Jacques François, Belgium astronomer; b. Antwerp, Belgium, Aug. 16, 1898; s. Frans and Eugenie (Christiaens) C.; Ph.D. in Phys. Scis. and Math., Brussels U.; D. honoris causa univs. Besancon, Caen, Lille; m. Pauline Godfroie, Aug. 5, 1935; 1 son, Lucienne. Asst., instr., full prof. basic astronomy, hon. dean Brussels U. Mem. Acad. Royale de Belgique (hon. perpetual sec.), Inst. Navigation, L'Instituto de Coimbra (Portugal) (corr.). Author: Fluctuations Saisonnières de la Potation de la Terre; Répartition des Petites Planètes; Probabilité de la Découverte des Comètes; Représentation Conforme de la Terre Entière dans un Triangle. Address: 351 av. Louise, Brussels 5, Belgium.

COX, John, physicist; b. London, Eng., Nov. 20, 1851; M.A., Cambridge (Eng.) U., 1877. Warden, Cavendish Coll., 1877-87; MacDonald prof. physics, Cambridge, prof. McGill U., Montreal, Can., 1890-1901, dir. physics lab. from 1901; gave 1st clin. x-ray report of patient with bullet in leg (with Robert Charles Kirkpatrick), 1896. Died May 13, 1923.

COX, John Paul, Am. astrophysicist; b. Ft. Myers, Fla., Nov. 4, 1926; s. James B. and Bess L. (Tollette) C.; A.B., Ind. U., 1949, M.S., 1950, Ph.D., 1954. Faculty, Cornell U., 1954-62; vis. scientist Courant Inst. Math. Scis. N.Y.U., 1962-63; vis. fellow Joint Inst. Lab. Astrophysics, Boulder, Colo., 1963; asso. prof. astrophysics U. Colo., Boulder, 1963-65, prof., 1965——; cons. Smithsonian Astrophys. Obs., Cambridge, Mass., 1957, 59, 60, Los Alamos Sci. Lab., 1960——; asst. engr. Pratt & Whitney Aircraft Corp., East Hartford, Conn., 1958. Mem. Am. Phys. Soc., Am. Astron. Soc., N.Y. Acad. Scis., Internat. Astron. Union, Sigma Xi, Phi Eta Sigma. Research on theory of origin of stellar pulsations as being due to second helium ionization in the stellar envelope, nature and properties of helium stars. Home: 827 16th St., Boulder, Colo. 80302.*

COX, Leslie Reginald, Brit. paleontologist; b. London, Eng., Nov. 22, 1897; s. Walter and Jessie Lucy (Witte) C.; M.A., Sc.D. in Pateontology, Queens Coll. Cambridge; m. Hilda Cecilia Lewis, Sept. 17, 1925; children—Hilda, John. With dept. paleontology Brit. Mus., aux. curator, 1922, prin. sci. officer, 1946, asst. curator, 1961, ret., 1963. Mem. Royal Soc., London Geol. Soc. (Gyell medal), Geologists Assn. (pres. 1954-56), Paleontol. Assn. (pres. 1964), Malacol. Soc. London (pres. 1957-60). Author: Neogene and Quaternary Mollusca from the Zanzibar Protectorate, 1927; Fossil Fauna of the Samana Range, Mollusca of the Hangu Shales, 1930; The Jurassic Lamellibrance Fauna of Cutch, 1940-52; New Light on William Smith and His Work, 1942; contbr. sect. on mollusca to Treatise of Invertebrate Paleontology, 1960. Home: 30 Haslmere Av., Hendon, London N.W. 4. Office: British Museum, Cromwell Rd., London S.W.7, Eng.

COX, Oliver Cromwell, sociologist; b. Port-of-Spain, Trinidad, W.I., Aug. 25, 1901; s. William R and Virginia (Austin) C.; came to U. S., 1919, naturalized, 1926; B.S. in Law, Northwestern U., 1928; M.A. in Econs., U. Chgo., 1932, Ph.D. in Sociology, 1938; LL.D., Wiley Coll., 1945. Sr. statistician Chgo. Park Dist., 1936-37; asso. prof. sociology Louisville Municipal Coll., 1937-38; prof. econs. Wiley Coll., 1938-44; prof., head dept. social scis. Tuskegee Inst., 1944-49; prof., chmn. dept. sociology Lincoln U., Jefferson City, Mo., 1949——. Recipient George Washington Carver award Doubleday & Co., 1948. Fellow Am. Sociol. Assn.; mem. Mo. Acad. Sci., Am. Econ. Assn., Midwest Sociol. Soc. Author: Caste, Class and Race, 1948; Foundations of Capitalism, 1959; Capitalism and American Leadership, 1962; Capitalism as a System, 1964; also articles. Elucidated structure and theory modern capitalism; analyzed situations of race relations in various parts of world. Address: Lincoln U., Jefferson City, Mo. 65101.*

COX, Rachel Dunaway, Am. psychologist; b. Murray, Ky., Jan. 20, 1904; d. Enoch T. and Khadra (Fergeson) Dunaway; B.A., U. Tex., 1925; M.A., Columbia, 1930; Ph.D., U. Pa., 1943; m. Reavis Cox, Feb. 18, 1928; children—David Jackson, Rosemary Dunaway (Mrs. Jon J. Masters). Reporter, mem. editorial staff N.Y. Herald-Tribune, N.Y.C., 1927-29; dir. edn. West Side YMCA, N.Y.C., 1930-35; hosp. case worker A.R.C., Walter Reed Hosp., 1944; faculty psychology and child devel. Bryn Mawr (Pa.) Coll., 1944——, prof., 1955——, dir. Child Study Inst., 1944——, chmn. dept. edn. and child devel., 1944——. Recipient Service to Youth Guidance award B'nai B'rith, 1960. Mem. Acad. Certified Social Workers, Am., Eastern psychol. assns., Nat. Assn. Social Workers, Am. Personnel and Guidance Assn., Soc. for Projective Techniques (sec. 1951-53), Phi Beta Kappa, Sigma Xi. Author: Counselors and Their Work, 1945; also articles. Pioneer in program of bringing psychiatry and social case work to the field of guidance in elementary edn. Home: 503 Walnut Lane, Swarthmore, Pa. 19081. Office: Child Study Inst., Bryn Mawr Coll., Bryn Mawr, Pa.*

COX, William Sands, Brit. surgeon; b. Birmingham, Eng., 1802; s. E. T. Cox; student Guy's, St. Thomas' hosps., London, Eng., 1821-23, École de Médecine, Paris, France, 1824. Lectr. anatomy Birmingham, 1825; co-founder Birmingham Sch. Medicine, 1828, Queen's Hosp., 1841, Queen's Coll., Birmingham, 1843. Fellow Royal Soc., 1836. Author: A Synopsis of the Bones, Ligaments, and Muscles, Blood-vessels, and Nerves of the Human Body, 1831; A Memoir on Amputation of the Thigh at the Hip Joint, 1845; also articles in London Med. Gazette. Died Dec. 23, 1875.

COXE, Eckley Brinton, Am. mining engr., inventor; b. Phila., June 4, 1839; s. Charles Sidney and Ann (Brinton) C.; grad. U. Pa., 1858; studied mining in Europe, 1860-63; m. Sophia G. Fisher, June 27, 1868. Organized Coxe Bros. Co., 1865; pres. Cross Creek Co.; pres. Del., Susquehanna & Schuylkill R.R., 1890; a founder Am. Inst. Mining Engrs., pres., 1878-89; pres. Am. Soc. M.E., 1892-94; invented automatic slate-picking machine, mech. stoker, gyrating screens, steel measuring tapes. Died May 13, 1895.

COXE, John Redman, Am. physician; b. Trenton, N.J., Sept. 16, 1773; s. Daniel and Sarah (Redman) C.; studied under Benjamin Rush; M.D., U. Pa., 1794; postgrad. hosps. in London, Edinburgh, Paris; m. Sarah Cox, 6 children. Practiced medicine in Phila.; prof. chemistry U. Pa., 1809-19, prof. materia medica and pharmacy Med. Dept., 1819-35; physician to Pa. Hosp., Phila. Almshouse; resident physician Yellow Fever hosp., Bush Hill, during epidemic of 1797, physician of port during epidemic of 1798. Editor Med. Museum, 1805-11. Performed 1st smallpox vaccination in Phila. on his infant son (with virus obtained from Jenner), circa 1802; introduced Jalap plant into U. S. Died Phila., Mar. 22, 1864.

COXE, Warren Winfred, Am. ednl. psychologist; b. Belvidere, Ill., July 19, 1886; s. Herbert J. and Nellie M. (Warren) C.; B.S., Dakota Wesleyan U., 1911; postgrad. U. Chgo.; Ph.D., Ohio State U., 1923; m. Gertrude E. Burton, July 13, 1935; 1 dau., Sara Elizabeth. High sch. tchr., Rushford, Minn., 1911-13; prin., Fountain, Minn., 1913-15; asst. dir. Vocation Bur., Cin., 1916-21; psychologist Gen. Hosp., Cin., 1916-17; chief ednl. measurements bur. N.Y. Dept. Edn., 1923-29, dir. div. research, from 1929. Instr. summer schs. Ohio State U., 1922, 23, State Coll. for Tchrs., Albany, N.Y., 1924, 25, U. Wis., 1928; lectr. U. Tex., summer 1930; pres. Bethlehem Community Coll., 1946-47. Fellow Am. Statis. Assn. (pres. Albany chpt. 1935-37), A.A.A.S.; mem. Ednl. Research Assn. N.Y. (pres. 1930), N.Y. State Assn. Applied Psychologists (pres. 1937), Am. Ednl. Research Assn., Nat. Soc. for Study Edn., Am. Psychol. Assn., N.E.A., Am. Assn. Sch. Adminstrs. Author: The Influence of Latin on English Spelling, 1924; (with Ethel L. Cornell) Cornell-Coxe Performance Ability Scale (non-verbal intelligence test). Home: 120 Salisbury Rd., Delmar, N.Y. Office: State Dept. of Edn., Albany, N.Y. 12201.

COXETER, Harold Scott MacDonald, mathematician; b. London, Eng., Feb. 9, 1907; s. Harold Samuel and Lucy (Gee) Coxeter; B.A., Univ. of Cambridge (Eng.), 1929, Ph.D., 1931; LL.D., U. Alta. (Can.), Edmonton, 1957; m. Hendrina J. Brouwer, Aug. 20, 1936; children—Edgar, Susan (Mrs. Alfred D. Thomas). Rockefeller fellow Princeton, 1932-33, Procter fellow, 1934-35; faculty U. Toronto (Ont., Can.), 1936—, prof. math., 1948—; vis. prof. Dartmouth, 1964. Fellow Royal Soc. London, Royal Soc. Can.; mem. London (life), Edinburgh (hon.), Am. math. socs., Canadian Math. Congress (pres. 1965-67), Math. Assn. Am. Author: The Real Projective Plane, 1955; Introduction to Geometry, 1961; Regular Polytopes, 1963; Projective Geometry, 1964; (with W. O. J. Moser) Generations and Relations, 1965; Non-Euclidean Geometry, 1965; (with S. L. Greitzer), Geometry Revisited, 1967; also numerous articles. Generalization of kaleidoscope in Euclidean and non-Euclidean spaces; presentation finite groups, geometrical interpretation for integral octaves; application projective geometry to relativity theory. Home: 67 Roxborough Dr., Toronto 5, Ont., Can.*

COZE, Léon, French bacteriologist; b. Strasbourg, France, Oct. 13, 1819; s. Jean-Baptiste Coze; prof. pharmacy Strasbourg, later Nancy, France, from 1857; pioneer germ theory of disease, 1866-70. Died Nov. 4, 1896.

COZZI, Danila, Italian chemist; b. Pisa, Italy, Sept. 3, 1914; s. Averardo and Celide (Grazzini) C.; Ph.D. in chemistry; m. Giuliana Rivolta; 1 son, Gualtiero. Mem. Soc. Chim. Ital., Assn. Elettrotecnica Ital., Soc. Ital. de Mineralogie. Research and numerous publs. on dental products, analytical chemistry, electrochemistry. Home: via D. Buonvicini 28, Florence, Italy. Office: via Risorgimento 35, Pisa, Italy.

CRABBÉ, Jean, Belgian physician; b. Brussels, Belgium, Aug. 12, 1927; s. François and Simone (Doutreligne) C.; M.D., U. Louvain (Belgium), 1951, Agrégé de l'Enseignement supérieur, 1962; m. Marie De Guchteneere, Aug. 10, 1954; children—Isabelle, François, Christine, Bruno, Nicolas. Asst. in medicine Universitätsspital, Zurich, Switzerland, 1951-52; research asst. Hôpital cantonal, Geneva, Switzerland, 1954-55; research fellow in medicine Peter Bent Brigham Hosp., Harvard Med. Sch., Boston, 1955-58; research fellow Mass. Gen. Hosp., Harvard Med. Sch., 1959-60; asst. in medicine U. Clinics, St. Raphael, Louvain, 1960-62; cons. endocrinology U. Clinics St. Pierre, Louvain, 1962—, dir. endocrine research unit, 1962—; asst. prof. renal and endocrine physiology University of Lourain Medical School, 1966-67, associate professor, 1967—. Recipient of the René Beckers award, Belgium, 1955; John J. Larkin Meml. grantee, 1958; recipient Alumni award, Belgium, 1963. Mem. Société Belge d'Endocrinologie, Société de Physiologie, Société Belge de Médecine Interne, Société Belge de Recherches cliniques, Royal Soc. Medicine (U.K.), Endocrine Soc. (U. S. A.), Am. Fedn. for Clin. Research. Author: The Sodium Retaining Action of Aldosterone, 1963; also numerous articles. Research on role adrenal cortical secretion in immunity mechanisms, significance of aldosterone secretion in normal and path. states, mode of action of aldosterone on sodium transport in the kidney. Home: 55 Leopold Beosierlaan, Kessel-Lo, Belgium. Office: Univ. Clinics St. Pierre, Louvain, Belgium.*

CRABTREE, William, Brit. astronomer; b. Broughton, Eng., 1610; s. John Crabtree; m. Elizabeth Pendleton, Sept. 14, 1633; 3 children; Cloth merchant; made 1st observation (with Jeremiah Horrox) of transit of Venus, 1639. Died circa 1644.

CRACHERODE, Clayton Mordaunt, Brit. naturalist; b. Taplow, Eng., 1730; s. Mordaunt and Mary (Moris) C.; B.A., Oxford (Eng.) U., 1750, M.A., 1753; Curate, Bimsey, Eng.; trustee Brit. Mus., from 1784. Fellow Royal Soc., 1785. Collected fossils; bequeathed 4500 books to Brit. Mus. Died Taplow, Apr. 5, 1799.

CRADOCK, Edward, English alchemist; b. Staffordshire, Eng.; flourished 1571; B.A., Christ Church, Oxford (Eng.), 1555-56, M.A., 1558-59; D.D., 1565. Lady Margaret prof. divinity, 1565-94. Author: Documentium et practica, 1408; Tractus de lapide philosphico, 1415; A Treatise of the Philosopher's Stone, 1445; The Shippe of Assured Safetie, 1571.

CRAFT, Clarence Christian, Am. physician, magnetician; b. Gaston, S.C., Sept. 28, 1880; s. David Elmore and Mary Louisa (Richter) C.; B.S., S.C. Mil. Acad., 1902; M.D., George Washington U., 1909; m. Charlotte Maye Thomas, Dec. 6, 1911; children—Hume Richter, Warren Frederick. Aide, U. S. Coast and Geodesic Survey, 1903-05, on U. S. S. Patterson assisting in making soundings for Army and Navy cable, from Cape Flattery to Sitka, Alaska, summer 1903; surveying harbor of Kiska, Aleutian Islands, summer 1904, and survey of H.I., 1904-05; magnetic observer in various states, 1905-07; computer Dept. Terrestrial Magnetism, Carnegie Instn. Washington, to 1909; magnetic observer, Comdr. Peary's auxiliary ship Erik, 1908, going as far north as Etah, Greenland; surgeon and magnetic observer aboard yacht Carnegie from 1909-11; in pub. health work, Florence, S.C., 1914-16. 21-35. Died May 25, 1935.

CRAFT, Johann Daniel, chemist; b. Wertheim, Germany, Sept. 28, 1624; s. Iust and Susanna C.; began practice medicine, Zellerfeld, circa 1662; traveled throughout N.Am. and Europe; founded textile, glass and other factories; probably invented milk glass. Died Amsterdam, Netherlands, Apr. 16, 1697.

CRAFTS, A. S., Am. plant physiologist; b. Ft. Collins, Colo., June 25, 1897; s. Henry Alanzo and Elizabeth (Bleakley) C.; B.S., U. Cal. at Berkeley, 1927, Ph.D., 1930; LL.D., U. Cal. at Davis, 1966; m. Alice E. Hardisty, June 25, 1926; children—Harold Springer, Helen Elizabeth (Mrs. Charles E. Hedges). NRC fellow Cornell U., 1930-31; asst. botanist Cal. Agrl. Expt. Sta., Davis, 1931-36; asst. botanist, asst. prof. botany U. Cal. at Davis, 1936-39, asso. botanist, asso. prof., 1939-46, botanist, prof. botany, 1946-64, prof. botany emeritus, 1964—; vis. prof. P.R. Agrl. Expt. Sta., 1947-48. Mem. Am. Soc. Plant Physiology (pres. 1955-56), Weed Soc. Am. (pres. 1958), Zool.-Bot. Soc. Vienna, Am. Inst. Biol. Sci., Cal. Bot. Soc., Weed Sci. Soc. Am. Author: (with H. B. Currier, C. R. Stocking) Water in the Physiology of Plants, 1949; Translocation in Plants, 1961; The Chemistry and Mode of Action of Herbicides, 1961; (with W. W. Robbins) Weed Control, 1962; (with S. Yamaguchi) The Autoradiography of Plant Materials, 1964; also numerous articles. Confirmation of mass flow theory of solute translocation in plants; demonstration, proof that herbicides applied to foliage may move with foods in plants; explanation of uptake of lipid-soluble and water-soluble herbicides into plant foliage. Home: 626 B St., Davis, Cal. 95616.*

CRAFTS, James Mason, Am. chemist; b. Boston, Mar. 8, 1839; s. R. A. and Marian (Mason) C.; S.B., Lawrence Scientific Sch. (Harvard), 1858, LL.D., 1898; studied chemistry at Bergacademie, Freiberg, also univs. Heidelberg and Paris, 1850-55; m. Clemence Haggerty, June 13, 1868. Examined mines in Mex., 1866-67; prof. chemistry and dean chem. faculty, Cornell, 1868-71; prof. chemistry, Mass. Inst. Tech., 1871-80, organic chemistry, 1892-97, pres., 1898-1900; engaged in chem. research, Boston, from 1900. Recipient Jecker prize Paris Acad. Scis., 1885; Rumford medal, 1911; chevalier Legion of Honor, France, 1885. Fellow Am. Acad. Arts and Sciences; mem. Nat. Acad. Sciences; hon. mem. Royal Inst. Great Britain. Author: Qualitative Chemical Analysis, 1870. Also published Researches Upon Silicic Compounds, 1865; Arsenic Ethers, a Method of Synthesis by Means of Chloride of Aluminum, 1879; Studies in Thermometry, 1880; Catalysis in Concentrated Solutions, 1908; Thermometry, 1913-15. Discoverer (with Charles Friedel; of Friedel-Crafts reaction (synthetic reaction with anhydrous aluminum chloride acting as catalyst in producing hundreds of new carbon compounds). Died Ridgefield, Conn., June 20, 1917.

CRAFTS, Roger Conant, Am. anatomist; b. Lewiston, Me., Jan. 26, 1911; s. Seldon T. and Alice (Conant) C.; B.S., Bates Coll., 1933; Ph.D., Columbia, 1941; m. Margaret D. Findley, Aug. 10, 1938; children—Roger Conant, Susan D. Faculty, Boston U. Sch. Medicine, 1941-50; Francis Brunning prof., head dept. anatomy U. Cin. Coll. Medicine, 1950—. Fellow A.A.A.S.; mem. Am. Assn. Anatomists, Am. Assn. U. Profs., Soc. Exptl. Biology and Medicine, Assn. Am. Med. Colls., Phi Beta Kappa, Sigma Xi. Author: A Guide to the Regional Dissection of the Human Body, 1943; Textbook of Human Anatomy, 1966. Research, publs. on mechanics and effects of hormones on blood formation. Home: 3230 Daytona Av., Cin. 45211.*

CRAGG, James Birkett, Brit. biologist; b. North Shields, Eng., Nov. 8, 1910; s. Arthur William and Jane (Hall) C.; M.Sc. in Biology, U. Durham. Former prof. zoology U. Durham; dir. Research Sta. of Merlewood (nature protection) at Grange over Sands, Lancashire. Mem. Soc. for Exptl. Biology, Brit. Ecol. Soc., Inst. Biology. Co-editor: Jour. Animal Ecology; editor advances in Ecological Research. Home: Charney Rd., Grange over Sands. Office: Merlewood Research Sta., Grange over Sands, Eng.

CRAGGS, John Drummond, English physicist; b. Huddersfield, Eng., May 17, 1915; s. Thomas Lawson and Elsie (Roberts) C.; B.Sc., U. London (Eng.), 1937, M.Sc., 1938, Ph.D., 1941; m. Dorothy Ellen Margaret Garfitt, Aug. 2, 1941; children—Susan, Sarah. Research staff Kings Coll., U. London, 1937-38, Met. Vickers High Voltage Lab., 1938-47; faculty dept. elec. engring. and electronics U. Liverpool (Eng.), 1947—, prof. electronic engring., 1955—; vis. faculty U. Cal. at Berkeley, 1944-45. Fellow Inst. Physics, Phys. Soc. Author: (with S. C. Curran) Counter Tubes, 1949; (with J. M. Meek) Electrical Breakdown of Gases, 1953, High Voltage Laboratory Techniques, 1954; also numerous articles. Research on gaseous electronics, recombination processes in decaying plasmas, ionization processes in gases especially electron attachment ot molecules, electron detachment. Home: Stone Cottage, Newton-cum Larton, West Kirby, Cheshire, Eng. Office: U. Liverpool, Liverpool 3, Eng.*

CRAIG, Allen Thornton, Am. math. statistician; b. Marion, Ala., Aug. 5, 1904; s. Edgar Montgomery and Vivia (Roark) C.; A.B., U. Fla., 1927, M.A., 1928; Ph.D. in Math., U. Ia., 1931. Faculty, U. Ia., Iowa City, 1931—, prof. math., 1945—. Mem. research group M, NDRC, Washington, 1942. Fellow Inst. Math. Statistics (v.p. 1939-41); mem. Am. Math. Soc., Math. Assn. Am. Author: (with Robert V. Hogg) Introduction to Math. Statistics, 1959, 2d edit., 1965. Editorial bd. Annals Math. Statistics, 1938-49. Study of sampling theory and independence problems.

CRAIG, Cecil C., Am. mathematician; b. Otwell, Ind., Apr. 14, 1898; s. Harley E. and Lula (Abbott) C.; A.B., Ind. U., 1920, A.M., 1922; postgrad. U. Lund (Sweden); Ph.D., U. Mich., 1927; m. Ruth Swan, Sept. 3, 1927; 1 dau., Mary Elizabeth. Instr. math. Ind. U., 1920-22; faculty U. Mich., 1922—, prof., 1942—, dir. statis. research lab., 1946—. Am. Scandinavian Found. fellow, 1924-25; NRC fellow, Princeton, 1929-30, Stanford, 1930-31; Rockefeller Found. fellow, 1937-38. Recipient Shewhart medal, 1957. Fellow Inst. Math. Statis., Am. Statis. Assn., Am. Soc. Quality Control; mem. Am. Math. Soc., Math. Assn. Am., Biometric Soc., Phi Beta Kappa, Sigma Xi. Research in math. statistics and statis. quality control. Home: 1410 Iroquois Dr., Ann Arbor, Mich. 48104.*

CRAIG, Charles Franklin, Am. pathologist; b. Danbury, Conn., July 4, 1872; s. William Edward and Maria Hamlin (Payne) C.; M.D., Yale, 1894, hon. M.A., 1914; D.Sc., Tulane U., 1945; m. Lillian Osmun, July 7, 1893; children—Marjorie Lillian, Edward Arthur. Acting asst. Surgeon U. S. Army, 1898-1903; advanced through grades from 1st lt. to col. M.C., 1918; pathologist and bacteriologist Sternberg U. S. Army Gen. Hosp., Ga., 1898, Simpson Gen. Hosp., Va., 1898-99, Camp Columbia Hosp., Havana, 1899, U. S. Army Gen. Hosp., Presidio, Cal., 1899-1905, Div. Hosp., Manila, 1906; mem. U. S. Army Bd. for Study Tropical Diseases, Manila, 1906-07; lab. Ft. Leavenworth, Kan., 1907-09; attending surgeon N.Y.C., 1909; asst. curator Army Med. Museum, 1909-13, curator, 1919-20; asst. prof. bacteriology and clin. diagnosis Army Med. Sch., Washington, 1909-13, prof. bacteriology, parasitology and preventive medicine, also dir. labs., 1920-22, comdt. and dir. clin. pathology and preventive medicine, 1926-30; asso. prof. bacteriology, med. dept. George Washington U., 1910-11; comdg. officer Central Dept. Lab., Ft. Leavenworth, 1913-16, Dept. Lab. No 2, So. Dept., El Paso, Tex., 1916-17, Ft. Leavenworth, 1917-18; organized and comd. Yale Army Lab. Sch., 1918-19; med. insp. Hawaiian Dept., 1922-26; asst. comdt. Army Med. Center, Washington, 1930-31; comdt. Army Med. Center, Washington, 1930-31; prof. tropical medicine and dir. dept., sch. medicine Tulane U., 1931-38, emeritus prof. tropical medicine, from 1939. Fellow A.C.S., A.C.P., Assn. Mil. Surgeons U. S. (life, Founders medal 1948), A.M.A., Am. Pub. Health Assn.; mem. Am. Soc. Tropical Medicine (pres. 1914-15), Royal Soc. Tropical Medicine and Hygiene, Internat. Leprosy Assn., Internat. Soc. Tropical Medicine, Am. Soc. Parasitologists (pres. 1934-35), Am. Acad. Tropical Medicine (pres. 1935, Gold medal 1943), Am. Soc. Clin. Pathologists, Sigma Xi. Author: The AEstivo-Autumnal Malarial Fevers, 1901; The Malarial Fevers, Haemoglobinuric Fever and the Blood Protozoa of Man, 1909; The Parasitic Amoebae of Man, 1911; The Wassermann Test, 1918, 21; A Manual of the Parasitic Protozoa of Man, 1925; Amebiasis and Amebic Dysentery, 1935; Clinical Parasitology (with Faust), 1937; The Laboratory Diagnosis of Protozoan Diseases, 1941; The Etiology, Diagnosis and Treatment of Amebiasis, 1944; also chpts. in numerous med. texts. Asso. editor Am. Jour. Parasitology; editor Am. Jour. Tropical Medicine. Known for work on malarial parasites; described genus of flagellate protozoans (2 species of which inhabit intestine and cause dysentery-like symptoms) now called Craigia, 1906; demonstrated (with P. M. Ashburn) that a filterable virus causes dengue, 1907. Died Dec. 9, 1950.

CRAIG, Harmon, Am. geochemist; b. N.Y.C., Mar. 15, 1926; s. John Richard and Virginia (Stanley) C.; M.S., U. Chgo., 1950, Ph.D., 1951; m. Valerie Kopecky, Sept. 27, 1947; children—Claudia Campbell, Cynthia Camilla, Karen Constance. Research asso. Enrico Fermi Inst. for Nuclear Studies, 1951-55; prof. geochemistry, chmn. dept. earth scis. Scripps Instn. Oceanography and Dept. Earth Scis., U. Cal., San Diego, 1956—, scientist, expdn. leader Expdns. Monsoon, Antarctic and Pacific, 1961, Zephyrus, Mediterranean, Red Sea, 1962, Carrousel, South Pacific, 1964, Nova, Pacific, 1967. Guggenheim fellow Laboratorio di Geologia Nucleare U. Pisa (Italy), 1962-63. Mem. Am. Geophys. Union, Geol. Soc. Am. Editor (with S. L. Miller, G. J. Wasserburg) Isotopic and Cosmic Chemistry, 1964. Contbr. numerous articles to sci. jours. Investigations of origin and chemistry of volcanic and geothermal gases, steam, water; studies of chemistry and origin of meteorites, geochem. studies using isotopes of hydrogen, carbon, oxygen; studies of formation and mixing of deep ocean water masses; marine and atmospheric chemistry. Home: 8553 LaJolla Shores Dr., La Jolla, Cal. 92037.*

CRAIG, Homer Vincent, Am. mathematician; b. Denver, Aug. 26, 1900; s. Clarence Henry and Louisa (Harvey) C.; B.A., U. Colo., 1924; Ph.D., U. Wis., 1929; m. Janie Cecile Moore, Aug. 27, 1925; 1 dau., Jane Homer. Asst. prof. Colo. State U., 1925-26; asst. U. Wis., 1926-27, instr., 1927-29; asst. prof. U. Tex., 1929-39, asso. prof., chmn. applied math. dept., 1939-41, prof., 1942-43, grad. prof., 1943—, grad. adviser, 1947-60. Research specialist Boeing Airplane Co., summers 1950-61; cons. Gen Dynamics, Ft. Worth, 1961-63, 64—. Recipient Teaching Excellence award U. Tex. Students' Assn., 1964. Fellow A.A.A.S.; mem. Am. Math. Soc., Math. Assn. Am., Tensor Soc. (Am. rep. 1952—), Circolo Matematico di Palermo, Research Assn. for Applied Geometry, N.Y. Acad. Sci., Sigma Xi, Sigma Pi Sigma, Phi Kappa Psi. Author: Vector and Tensor Analysis, 1943; (with others) Dictionary of Mathematics, 1949, 59. Editorial staff Math. Mag., 1947-62, Tensor, 1951—. Contbr. articles to profl. publs. Originated theory of extensors, 1937; research on tensor and extensor analysis, generalized differential geometries, applications extensor theory, relativity. Home: 3104 Grandview St., Austin, Tex. 78705.*

CRAIG, John, mathematician; flourished circa 1693; prebend of Durnford in Cathedral of Salisbury (Eng.), 1708-26; prebend of Gillingham Major, 1726-31. Fellow Royal Soc., 1711. Author: Methodus Figurarum, 1685; Tractatus Mathematicus, 1703; De Calculo Fluentium, 1718; Theologiae Christianae Principia Mathematica, 1699. Developed theory of fluxions; made differential calculus known in Eng. Died London, Eng., Oct. 31, 1731.

CRAIG, John Merrill, Am. pathologist; b. Pasadena, Cal., Oct. 14, 1913; s. Volney M. and Elinor (Merrill) C.; A.B., U. Cal. at Berkeley, 1936, M.A., 1938; M.D. cum laude, Harvard, 1941; m. Elsa Fay Hartshorne, May 31, 1949; children—Robert C. Hartshorne, Marianna Hartshorne (Mrs. Charles Small), Caroline D. Hartshorne. Practice medicine specializing in pathology, Boston, 1950-59, 60—, Pitts., 1959-60; asst., asso., pathologist Childrens Hosp. Med. Centre, 1950-59; dir. labs. E.S. Magee Hosp., 1959-60; pathologist-in-chief Boston Lying-in Hosp., Free Hosp. For Women, 1960—; asst. prof. pathology Med. Sch. Harvard, 1952-59, clin. prof., 1960—; prof. pathology U. Pitts., 1959-60; mem. pathology study sect. USPHS, 1959-64. Mem. New Eng. Soc. Pathologists (pres.), A.M.A., Soc. Pediatric Research, Am. Pediatric Soc., Assn. Pathologists and Bacteriologists, Am. Soc. Exptl. Pathology, Histochem. Soc. Asso. editor: Am. Jour. Pathology, 1965—. Research, numerous publs. on pathology of respiratory distress syndrone in newborns, congenital metabolic disorders of children, cystic fibrosis of pancreas, agammaglobulinemia. Home: 194 Brattle St., Cambridge, Mass. 02138. Office: 221 Longwood Av., Boston 02115.*

CRAIG, Lyman Creighton, Am. chemist; b. Palmyra, Ia., June 12, 1906; s. William McCoy and Anna (Kitchell) C.; B.S., Ia. State U., 1928, Ph.D., 1931; NRC fellow Johns Hopkins, 1931-33; m. Rachel Parker, Nov. 25, 1937; children—Anna (Mrs. David Miller), David Lindley, Mary Elizabeth. Faculty, Rockefeller U., N.Y.C., 1933—, prof., 1949—. Recipient Albert Lasker Med. Research award in basic sci., 1963; Fisher award in analytical chemistry, 1965. Mem. Nat. Acad. Sci., Am. Acad. Arts and Scis., Am. Chem. Soc., A.A.A.S. Research, numerous publs. on chem. structure of various natural products; devel. separation methods, fractional distillation, fractional extraction and fractional dialysis. Home: 151 Rodney St., Glen Rock, N.J. 07452. Office: 66th and York Av., N.Y.C. 10021.*

CRAIG, Richard Ansel, Am. meteorologist; b. Abington, Mass., Mar. 23, 1922; s. Ansel W. and Rose (Beatty) C.; A.B., Harvard, 1942; M.S., Mass. Inst. Tech., 1944, Sc.D., 1948; m. Constance Arnold, Aug. 12, 1944; children—Malcolm A., Joyce E., Lee A., Janet S., Ronda C. With Radiation Lab., Mass. Inst. Tech., 1944-45; staff Woods Hole (Mass.) Oceanographic Instn., 1946; research fellow Harvard Coll. Obs., 1948-51; meteorologist, physicist Air Force Cambridge Research Center, Bedford, Mass., 1951-58; prof. meteorology Fla. State U., Tallahassee, 1958—. Cons. White Sands Missile Range, 1964—. Mem. Am. Meteorol. Soc. (chmn. bd. on profl. ethics 1961-64), Am. Geophys. Union. Author: The Upper Atmosphere, Meteorology and Physics, 1965; also articles. Asso. editor Jour. Meteorology, 1955-61; Jour. Geophys. Research, 1964-65. Theoretical and observational studies atmospheric ozone; research on solar weather relationships, transfer heat and water between ocean and air, atmospheric dynamics, stratospheric warmings. Home: 2326 Amelia Circle, Tallahassee 32304.*

CRAIG, Thomas, mathematician; b. Ayrshire, Scotland, Dec. 20, 1885; s. Alexander and Mary (Hall) C.; grad. Lafayette coll., 1875; Ph.D., Johns Hopkins, 1878; m. Louise Alvord, May 4, 1880. Lectr., Johns Hopkins, 1877-79, fellow, later asso. prof. math., 1881-92, prof., 1892-1900; mathematician U. S. Coast and Geodetic Survey, 1879-81; editor Am. Jour. Math., 1894-99. Author: Elements of the Mathematical Theory of Fluid Motion, 1879; A Treatise on Projections, 1882; A Treatise on Linear Differential Equations, 1889. Died May 8, 1900.

CRAIG, Winchell McKendree, Am. neurosurgeon; b. Washington Ct. House, O., Apr. 27, 1892; s. Thomas Henry and Eliza Orlena (Pine) C.; student Culver Mil. Acad., 1911; A.B., O. Wesleyan U., 1915; D.Sc., 1937; M.D., Johns Hopkins Med. Sch., 1919; M.S. in Surgery, U. Minn., 1930; m. Jean Katherine Fitzgerald, Feb. 16, 1928; children—Winchell McKendree, James Stewart, Jean Mary Patricia, Graham Fitzgerald. Fellow Mayo Foundation, Grad. Sch. U. Minn., 1921-24, instr., 1925, prof. neurosurgery, 1937; senior cons. Surgery Mayo Clinic, 1946—; neurol. surgeon, cons. Mayo Clinic, St. Mary's, Methodist hosps. Fellow A.M.A., A.C.S.; mem. Italian Soc. Neurosurgery (hon.), Neurosurgery French Lang. (hon.), Am. Surg. Assn., Am. Neurol. Assn., Western Surg. Assn., So. Surg. Assn., Minn. Soc. Neurology and Psychiatry, Internat. Surg. Soc., Central Neuropsy. Assn., Soc. Neurol. Surgeons. Central Soc. Clin. Research, Harvey Cushing Soc., Am. Acad. Neurol. Surgery (hon.), Johns Hopkins Alumnae Assn., Minn. Surg. Soc., Internat. Neurol. Assn., Assn. Mil. Surgeons U. S. (pres. 1953), Sigma Xi, Phi Beta Pi. Author 300 med. papers and chapters in monographs or systems of medicine. Editorial bd. Jour. Neurosurgery. Died Feb. 12, 1960.

CRAIGIE, David, Scottish physician; b. near Edinburgh, Scotland, 1793; M.D., U. Edinburgh, 1816; physician to Edinburgh Infirmary; owner, editor Edinburgh Med. and Surg. Jour.; fellow Edinburgh Coll. Physicians. Author: Elements of Morbid Anatomy, 1828; reputed to have introduced name relapsing fever in describing Edinburgh epidemic of 1843. Died 1866.

CRAIGMILES, Julian Pryor, Am. agronomist; b. Thomasville, Ga., Jan. 17, 1921; s. Joe E. and Grace (Beverly) C.; B.S., U. Ga., 1942, M.S., 1948; Ph.D., Cornell U., 1953; m. Mary Dixon, Dec. 12, 1948; children—Julian Pryor, Mary, Lee. Asso. agronomist U. Ga., 1952-56, agronomist, 1956-64, grad. faculty, 1957-64; supt. Rice-Pasture Expt. Sta., Beaumont, Tex., 1964—; grad. faculty Tex. A and M. U., 1964—. Recipient Ga. Plant Food Edn. Assn. award, 1957; Sears Roebuck award, 1958. Mem. Am. Soc. Agronomists, A.A.A.S., Crop Sci. Soc. Am., Soil Sci. Soc. Am. Contbr. numerous articles to profl. jours. Developed 1st male sterile sudangrass leading to release of 1st hybrid sudangrass, breeding lines of fescue, white clover and smooth brome grass. Home: 5925 Honeysuckle Dr. Office: Route 5, Box 366, Beaumont, Tex. 77706.*

CRAIK, Robert, Canadian physician, surgeon; b. Montreal, Que., Can., Apr. 22, 1829; s. Robert and Jean (Dickson) C.; M.D., McGill U., 1854, LL.D., 1895; m. Alice Symmers, 1856. House surgeon Gen. Hosp., 1854-60, later gov., cons., mem. mng. com.; demonstrator in charge practical anatomy McGill U., 1856-60, prof. clin. surgery, 1860-67, prof. chemistry, 1867-79, registrar, 1869-77, treas., 1875-1901, dean faculty medicine, prof. hygiene and pub. health, 1889-1901. Gov., v.p. Provincial Med. Bd.; govt. mem. Provincial Bd. Health; gov., cons. physician, sr. mem. med. bd. Royal Victoria Hosp. Mem. Royal Inst. for Advancement of Learning; pres. Medico-Chirurg. Soc. Author: Nature of Morbid Poisons and Germ Theory of Disease, 1854; Strychnia in Cholera, 1854; Papers on Purpura and Tetanus, 1855; Hyoscyamus Poisoning, 1858; Antisepsis in Successful Ovariotomies and Compound Joint Injuries, 1869-71; Medical Education, 1890. Died June 28, 1906.

CRAINZ, Franco, Italian gynecologist; b. Roma, Italy, May 18, 1913; s. Silvio and Ada (Fanelli) C.; D.M., U. Roma, 1936. Dir. Sch. for Midwives, Novara, U. Torino (Italy), 1956-64; prof. obstetrics and gynecology, dir. obstet. and gynecol. dept. U. Cagliari, 1964-66, U. Messina (Italy), 1966-67, U. Bari (Italy), 1968—. Research, publs. on anatomy and physiology of endo cervix, history of obstetrics and gynecology. Address: Clinica Ostetrica e Ginecologica dell' Università, Policlinico, 70124, Bari, Italy.*

CRAM, Donald James, Am. chemist; b. Chester, Vt., April 22, 1919; son of William Moffet and Joanna (Shelley) C.; B.S., Rollins College (Florida), 1941; M.S., U. Neb., 1942; Ph.D., Harvard, Nat. Research Fellow, 1945-47, 1947; m. Jean Turner, Dec. 22, 1941. Research chemist Merck and Co., 1942-45; asst. prof. chemistry U. Cal. at Los Angeles, 1947-50, asso. prof., 1950-56, prof. 1956—; chem. cons. Upjohn Co., 1952—, Union Carbide Company, 1960—; United States State Department exchange fellow to Instituto de Quimica, Universidad Nacional de Mexico, summer 1956; guest prof. U. Heidelberg (Germany), summer 1958. American Chemical Society fellow, 1947-58, Guggenheim fellow, 1954-55; California Junior C. of C. Young Man of Year, 1954; award for creative work in synthetic organic chemistry American Chemical Society, 1965; Herbert Newby McCoy award, 1965; award for creative research organic chemistry Synthetic Organic Chemical Mfrs. Association, 1965. Member American Chem. Soc., Nat. Acad. Scis., Swiss Chem. Soc., Chem. Soc. (Eng.), Sigma Xi. Author: (with G. S. Hammond) Organic Chemistry (textbook), 1959; Fundamentals of Carbanian Chemistry, 1965; (with John H. Richards and G. S. Hammond) Elements of Organic Chemistry, 1967. Contbr. chapts. in Steric Effects in Organic Chemistry. Research in field of stereochemistry, mold metabolites, large ring chemistry course of Neber rearrangement, synthesis of tropolone, asymmetric induction, conformational analysis, electrophilic substitution at saturated carbon. Home: 1250 Roscomare Rd., Los Angeles 90024.

CRAMBLETT, Henry Gaylord, Am. pediatrician, virologist; b. Scio, O., Feb. 8, 1929; s. Carl Smith and Olive (Fulton) C.; B.S., Mt. Union Coll., 1950; M.D., U. Cin., 1953; m. Donna Jean Reese, June 16, 1960; children—Deborah Kaye, Betsy Diane. Clin. research asso. Nat. Inst. Allergy and Infectious Diseases, Clin. Center, Bethesda, Md., 1955-57; faculty State U. Ia., 1957-60, asst. prof., 1958-60; faculty Bowman Gray Sch. Medicine, 1960-64, prof. pediatrics, 1963-64, dir. virology lab., 1960-64; prof. pediatrics Ohio State U., Columbus, 1964—, exec. dir. Children's Hosp. Research Found., 1964—, chmn. dept. med. microbiology, 1966—. Recipient Hoffheimer prize U. Cin., 1953, Eben J. Carey award in anatomy, 1950. Diplomate Am. Bd. Pediatrics, Am. Bd. Microbiology. Fellow Am. Acad. Microbiology, A.A.A.S.; mem. Infectious Diseases Society of America, Southern Society for Pediatric Research (past. president), Soc. for Pediatric Research, Am. Pediatric Soc., Am. Acad. Pediatrics, Midwest Soc. for Pediatric Research, Ohio Soc. Bacteriologists, Soc. for Exptl. Biology and Medicine, Am. Soc. for Microbiology, A.M.A., Alpha Omega Alpha. Research, publs. on etiologic assn. virus infections in illnesses of infants and children, estimation of importance of various viruses in morbidity and mortality in pediatric age group. Home: 2480 Sheringham Rd., Columbus, O. 43221.*

CRAMER, (Johann Baptist Joseph) August, psychiatrist; b. St. Gallen, Switzerland, Nov. 10, 1860; s. Heinrich and Emma (Deninger) C.; studied in Munich and Marburg, Germany; state exam, 1886; doctorate, 1887; postgrad. in psychiatry at Marburg, Freiburg, Germany; m. Amalie Johanna Luise Marie Schwaner; a son, Heinrich. Asst. to August Zinner in Eberswalde; became dep. dir. Göttingen Provinzialanstalt, 1895; named titular prof., 1897; became asso. prof. psychiatry, dir. inst., Göttingen, Germany, 1900; founder polyclinic (later became neurol. clinic); became privy councillor, 1907; founder Rasemühle sanatorium for nervous diseases, also an asylum, ednl. healing instn. for psychopathic welfare children. Author: Die Halluzinationen im Muskelsinn bein Geisteskranken, 1889; Gerichtliche Psychiatrie, 1897; Pubertät und Schule, 1911; Handbuch der Nervenkrankheiten im Kindesalter, 1912. Pioneer of modern mental instns. Died Göttingen, Sept. 5, 1912.

CRAMER, Carl Eduard, Swiss botanist; b. Zurich, Switzerland, Mar. 4, 1831; s. Salomon and Anna Magdalene (Burkhard) C.; studied in Zurich, Freiburg, Germany; doctorate 1855; studied under C. Nägeli; m. Aline Kesselring, 1860; 1 son and 2 daus., including Melanie Magdalene (Mrs. Ulrich Meister). Became lectr., 1857; prof. botany Eidgenössische Technische Hochschule, Zurich, 1861-1901, also established inst. for plant physiology; prof. U. Zurich, 1880-83, dir. bot. garden, 1882-93. Author: (with C. Nägeli), Pflanzen physiologische Untersuchungen, 1855-57; also articles. Founder plant architectonics; observed generalogy of a single cell up to last descendants of its division; research on cryptogam flora of Switzerland, bacteriology, devel. deviations of plant organs. Died Zurich, Nov. 24, 1901.

CRAMER, Caspar, German physician; b. Bautzen, Germany, 1648; prof. medicine U. Erfurt, Germany; pub. lectures, Collegium Chymicum, 1688; theorized that air is rarefied form of matter. Died Aug. 8, 1682.

CRAMER, Friedrich Detmar, German chemist; b. Breslau (now Wroclaw, Poland), Sept. 20, 1923; s. Johannes E. F. and Ilse (Kriebitzsch) C.; student Breslau U., 1944; Dr.rer.nat., Heidelberg (Germany), 1949; m. Marie-Luise Erdel, Nov. 29, 1947; children—Mattias, Johannes, Wigand, Martin, Daniel, Bettina, Franz. Asst., Heidelberg U., 1950-53, lectr., 1954-59; research fellow Cambridge (Eng.) U., 1953-54; prof. Darmstadt (Germany) Tech. U., 1959-62; dir. Max-Planck Inst. for Exptl. Medicine, Göttingen, Germany, 1962—; hon. prof. Braunschweig (Germany) U., 1965—. Author: Papierchromatographie, 1952; Einschlussverbindungen, 1954; also numerous articles. Research on inclusion compounds, including loose binding between molecules, mechanism of enzyme action, including substrate binding, enzyme models, structure and synthesis of nucleic acids, including chem. syntheses, biosynthesis, chem. modifications. Home: 18, Jakob-Henle-St., Göttingen 34, Germany.*

CRAMER, Gabriel, Swiss physicist; b. Geneva, Switzerland, July 31, 1704; apptd. prof. philosophy, Geneva, 1750. Author: Introduction à l'analyse des lignes courbes algebriques, 4 vols., 1750. Editor works of Johann Bernoulli, 4 vols., 1742, also correspondence between Bernoulli and Leibniz. Fellow Royal Soc., 1749. Studied quartic curves; formally introduced Y-axis; amplified rule for solving simultaneous linear equations, 1750; discovered Cramer's paradox involving curves of nth order; gave classification of quintic curves; found that Newton's theory of sound propagation did not always accord with practical experiences. Died Bagnoles, France, Jan. 4, 1752.

CRAMÉR, Harald, Swedish mathematician; b. Stockholm, Sweden, Sept. 25, 1893; s. Carl and Emelie (Cramér) C.; Ph.D., Stockholm U., 1917, Dr. Law (hon.), 1964; D.Sc. (hon.), Princeton, 1947; Ph.D. (hon.), Copenhagen (Denmark), 1950; m. Marta Hansson, June 20, 1918; children—Marie-Louise (Mrs. Christian Rasmussen), Thomas, Kim. Cons. actuary Swedish Life Ins. Cos., 1920-58; prof. math. statistics Stockholm U., 1929-58, pres., 1950-58; chancellor all Swedish univs., 1958-62; vis. prof. Princeton, 1946, Yale, 1947, 65, Columbia, 1963, U. Cal. at Berkeley, 1953, 66. Decorated comdr. first class Polar Star Order (Sweden); officer Legion of Honor (France). Mem. Royal Swedish, Norwegian acads. sci., Am. Acad. Arts and Scis., Internat. Statis. Inst. (hon.), Royal Statis. Soc. (hon., London), Swedish Actuarial Soc. (past pres., hon. pres. 1965—). Author: Random Variables and Probability Distributions, 1937; Mathematical Methods of Statistics, 1946; Elements of Probability Theory, 1954; (with M. R. Leadbetter) Stationary and Related Stochastic Processes, 1966. Research on modern math. probability theory, math. theory of statis. methods, including limit theorems, ruin problems, spectral representation of stochastic processes. Home: 33 Skärviksvägen, 18261 Djursholm, Sweden.*

CRAMER, Harold Leslie, Am. psychologist; b. Boston, Oct. 10, 1926; s. Harold Henry and Helen (Madden) C.; B.A. Ohio Wesleyan U., 1952; Ed.M., Harvard 1961, Ed.D., 1968; m. Marion Louise Skinner, Sept. 16, 1950 (div. Nov. 1964); children—Cynthia Ruth, Kathryn Louise, Joan Elizabeth, Martha Marion; m. 2d, Roxanne Herrick, Apr. 20, 1967. Tchr., Needham, Mass., 1952-53; pres., treas. Cramer Oil Inc., Needham, 1953-60; research asst. Harvard, 1961-65, IBM fellow, 1965-66; research dir. Brooks Found. East, Phila., 1966-67; asst. prof. Northeastern U., Boston, 1967—; research asso. Am. Center for Research in Blindness and Rehab. Newton, Mass., 1965-66. Mem. adv. bd. Center for Rate Controlled Recs., Louisville; cons. Library of Congress, Div. for Blind, Washington. Mem. Am. Psychol. Assn., Am. Ednl. Research Assn., Acoustical Soc. Am., Am. Inst. Physics, Kappa Delta Pi. Contbr. articles to sci. jours. Discovered presentation of auditory material with inter-aural time difference of 7½ milliseconds increases intelligibility binaural redundancy effect); discovered presentation of time-compressed speech by earphone is 20-25% more intelligible than by loudspeaker; developed computer technique for time-compressing speech (braided speech) that does not discard any speech and which works by overlapping segments of speech. Home: 156 Line St., Cambridge, Mass. 02139. Office: Northeastern U. College of Education, 102 Fenway, Boston 02115.*

CRAMER, Harrison Emery, Am. meteorologist; b. Johnstown, Pa., May 27, 1919; s. Frank W. and Ella (Emery) C.; A.B., Amherst Coll., 1941; S.M., Mass. Inst. Tech., 1943, Sc.D., 1948; m. Virginia M. Viets, Dec. 22, 1942; children—Anne, Dorothy, Nancy, William. Research asso. Mass. Inst. Tech., Cambridge, 1946-48, research meteorologist, 1949-65; group scientist dir., lab. environmental scis. GCA Corp. Technical Division, Bedford, Mass., also Salt Lake City, Utah, 1965—. Cons. meteorologist. Mem. Am. Geophys. Union, Am. Meteorol. Soc. (mem. com. on atmospheric turbulence and air pollution 1960—, Royal Meteorol. Soc., A.A.A.S., N.Y. Acad. Scis., Phi Beta Kappa, Sigma Xi. Research, publs. in measurements of atmospheric turbulence and diffusion, studies of energy exchange processes in atmospheric surface layer; devel. of semi-empirical techniques for predicting dispersal of windborne materials used in air pollution studies, toxic hazard evaluations; design of meteorol. data acquisition systems and applications. Home: 1581 Millbrook Rd., Salt Lake City 84106. Office: 75 E. Stratford Av., Salt Lake City 84115.*

CRAMER, Johann Andreas, German metallurgist; b. Quedlinburg, Dec. 14, 1710; taught assaying at Leyden and London; Councillor for mines and metallurgy, Blankenburg, 1743-73. Author: Docimasia, 1736; Elementa Artis Docimasticae, (2 vol.) 1739. Considered foremost assayer of his time. Died Berggiesshubel, Dresden, Dec. 6, 1777.

CRAMP, William, Brit. engr.; b. 1876; s. James and Mary Catherine (Robinson) C.; M.Sc., U. Birmingham; D.Sc., Victoria U.; M.Sc. in Tech., U. Manchester; m. May Marion Hartog, 1902; 2 daus. Apprentice, Rotherham and Sons, Coventry, 1892-97; engr. Ferranti Ltd., Hollingwood, 1897-1901; lectr. elec. design Central Tech. Coll., South Kensington, 1901-06; cons. engr., London, Manchester; lectr. U. Manchester, 1906-19; prof. elec. engr. Univ., Edgbaston, Birmingham, 1919-39. Faraday lectr., 1931. Recipient Silver medal for research on pneumatic conveying Royal Soc. Arts, 1922. Mem. Inst. Elec. Engrs. (chmn. N.W. sect. 1911, West Midlands sect. 1921, awarded premium 4 times for papers on elec. subjects). Inventor single phase motor (self-exciting alternator), apparatus for prodn. nitric acid from air, (with N. H. Searby) new system of coal-face lighting. Author: Armature Windings; Continuous Current Machine Design; Faraday and His Contemporaries; (with C. F. Smith) Vectors and Vector Diagrams, also numerous articles on elec. subjects, pneumatic conveying, flour treatment. Died Apr. 20, 1939.

CRAMPTON, Henry Edward, Am. zoölogist; b. N.Y.C., Jan. 5, 1875; s. Henry Edward (M.D.; and Dorcas Matilda (Miller) C.; student Coll. City N.Y., 1889-92; A.B., Columbia, 1893, Ph.D., 1899; m. Marion M. Tully, Oct. 27, 1896. Asst. in biology Columbia, 1893-95; instr. Mass. Inst. Tech., 1895-96; fellow, lecturer, tutor, and instr. in zoölogy Columbia, 1896-1901, adj. prof., 1901-04, prof., 1904-43, emeritus prof., 1943—; instr. embryology Marine Biol. Lab., Woods Hole, Mass., 1895-1902; in charge of embryology, Cold Spring Harbor, 1903-06; asso. Carnegie Inst.; curator invertebrate zoölogy Am. Mus. Natural History, 1909-20; asso. Bishop Museum, Honolulu. Scientific expdns. Islands of South Pacific Ocean, 1906, 07, 08, 09, British Guiana and interior of Brazil, 1911; expdns. to Bahamas, 1912, Puerto Rico, 1913-14, 14-15, South Seas, 1919, 23-24, Western Pacific, Asia, Malaysia, Australia, 1920-21, South Seas and Asia, 1928-29, Hawaiian Islands, 1929, 30, 31, 35; research asso., Am. Mus. Natural History, 1943—. Fellow N.Y. Acad. Scis. (pres.), A.A.A.S., Washington Acad. Scis.; mem. Am. Soc. Naturalists (v.p. 1921), Am. Soc. Zoölogists (v.p. 1911), Eugenics Soc. U. S. (sec.-treas.), Phi Beta Kappa, Sigma Xi. Author: The Doctrine of Evolution, 1911; also various monographs on evolution, embryology, exptl. zoölogy. Research in developmental mechanics; egg fertilization; embryology of gastropods; variation, selection and inheritance in Lepidoptera (Saturniidae); exptl. evolution; variation evolution and geog. distbn. of Polynesian mollusca; exptl. heredity of Hawaiian and Am. mollusca. Died Feb. 26, 1956.

CRAMPTON, Sir Philip, Irish surgeon; b. Dublin, Ireland, June 7, 1777; student medicine, Dublin; grad. medicine, Glasgow, Scotland, 1800. Became surgeon Meath Hosp., Dublin, 1798; pvt. tchr. anatomy, from 1800; surgeon to gen. forces in Ireland, for many years; surgeon in ordinary to Queen; mem. senate Queen's U. Fellow Royal Soc., 1812; mem. Royal Zool. Soc. Ireland (a founder), Dublin Coll. Surgeons (pres. 3 times). Described accommodation of eyes of birds to different distances. Died June 10, 1858.

CRAMPTON, Thomas Russell, Brit. engr.; b. London, Eng., 1816; devised 100 mile per hour locomotive with large driving wheels behind boiler; built undersea telegraph cable from Dover to Cape Gris-Nez. Died 1888.

CRANBERG, Lawrence, Am. physicist; b. N.Y.C., July 4, 1917; s. Hyman and Fanny (Rubinstein) C.; B.S., M.S. in Edn., Coll. City N.Y., 1937; A.M., Harvard, 1940; Ph.D., U. Pa., 1949; m. Charlotte Mount, Oct. 31, 1953; children—Alexis Mount, Nicole. Sr. physicist, group leader Signal Corps Engring. Labs., Ft. Monmouth, N.J., 1940-46; staff mem. Los Alamos Sci. Lab., 1950-63; prof. physics, U. Va., Charlottesville, 1963—. Cons., Tex. Nuclear Corp., Austin, 1957-65. Guggenheim fellow, 1958. Important work includes research into nuclear physics, neutron spectroscopy, ethical problems of scis., devel. of blind-aid devices; research in initiation of elec. breakdown in vacuum. Home: 1934 Blue Ridge Rd., Charlottesville, Va.*

CRANDALL, Stephen Harry, mech. engr., educator; b. Cebu, P.I., Dec. 2, 1920; s. William Harry and Julia J. (Kuenemann) C.; M.E., Steven Inst. Tech., 1942; Ph.D., Mass. Inst. Tech., 1946; m. Patricia E. Stickel, Jan. 21, 1949; children—Jane S., William S. Staff mem. Radiation Lab., Mass. Inst. Tech., 1942-44, instr. math. dept., 1944-46, instr. mech. engring. dept., 1946-47, asst. prof., mech. engring. dept., 1947-50, asso. prof., 1950-58, prof., 1958—, head div. applied mechanics, mech. engring. dept., 1957-59, 1961—; vis. lectr. Imperial Coll., London, 1949; vis. prof. Faculty of Sci., Marseille, 1960, vis. scholar U. Cal. Berkeley, 1964—. Fulbright fellow, 1949; NSF faculty sci. fellow, 1964. Fellow Am. Acad. Arts and Scis., Am. Acoustical Soc., A.A.A.S.; mem. Am. Soc. M.E., Am. Soc. Engring. Edn., Am. Math. Soc., Soc. Indsl. and Applied Math. Author: Engineering Analysis, 1956; (with W. D. Mark) Random Vibration in Mechanical Systems, 1963. Editor: Random Vibration, vol. 1, 1958; Mechanics of Solids, 1959; Random Vibration, vol. 2, 1963. Contbr. articles to tech. jours. on numerical analysis and applied mechanics. Home: Tabor Hill Rd., Lincoln, Mass. 01773.*

CRANE, Charles Kittredge, Am. narcotics researcher; b. Dalton, Mass., Aug. 28, 1881; s. Zenas and Ellen Judith (Kittredge) C.; Ph.B., Sheffield Sci. Sch., Yale, 1903; m. Margaret Diana Wilson, 1914 (div. 1926); 1 son, Peter. With Z. and W. M. Crane, paper mfrs., Dalton, 1903-11; asst. hon. sec. Lord Knutsford's Shell Shock Hosps., London, 1914-17; research on internat. narcotics problem, 1925—. Died Jan. 24, 1932.

CRANE, Frederick Loring, Am. biochemist; b. Montague, Mass., Dec. 3, 1925; s. Frederick Turner and Gertrude (Stange) C.; B.S., U. Mich., 1950, M.S., 1951, Ph.D., 1953; m. Helen Marguerite Eggerth, Apr. 7, 1950; children—Richard, Katherine, Eleanor, Thomas. Trainee, Inst. For Enzyme Research U. Wis., 1953-57, asst. prof., 1957-59; asst. prof. chemistry U. Tex., Austin, 1959-60; asso. prof. biology Purdue U., Lafayette, Ind., 1960-62, prof., 1962—; NSF postdoctoral fellow U. Stockholm, Sweden, 1963-64. Mem. Am. Soc. Biol. Chemists, Am. Chem. Soc. (Eli Lilly award in biol. chemistry 1961), Am. Soc. For Cell Biology, Am. Soc. Plant Physiologists, Scandinavian Soc. Plant Physiology, Bot. Soc. Am., A.A.A.S., Sigma Xi. Contbr. numerous articles to profl. jours. Research, publs. on biol. oxidations; discovery of coenzyme Q and electron-transferring flavoprotein in fatty acid oxidation; demonstration of requirement for plastoquinones and vitamin E in photosynthesis. Home: 1936 Indian Trail Dr., West Lafayette, Ind. 47906.*

CRANE, Horace Richard, Am. physicist; b. Turlock, Calif., Nov. 4, 1907; s. Horace Stephen and Mary Alice (Roselle) C.; B.S., Calif. Inst. Tech., 1930, Ph.D., 1934; m. Florence Rohmer LeBaron, Dec. 30, 1934; children—Carol Ann, Janet (dec.), George Richard. Research fellow, Calif. Inst. Tech., 1934-35; mem. faculty U. of Mich., since 1935, prof. physics, 1946—, chairman of department of physics, 1965—; research asso. (radar) Mass. Inst. Tech., 1940-41; physicist Carnegie Inst. of Washington, 1941, project dir., proximity fuze project, U. of Mich., 1941-43; atomic energy project, 1943-45; cons. NDRC, 1941-45. Vice pres. Midwestern Univs. Research Assn., 1956-57, pres., 1957-60; member policy bd. Argonne National Lab., 1957—. Fellow American Physical Society, A.A.A.S.; member National Academy Sciences, American Assn. Physics Teachers (president 1965), Sigma Xi. Contbr. scientific articles in profl. mags. Inventor of Race Track, a modified form of synchrotron for nuclear studies, 1946; made early discoveries in field of artificially produced radioactive atoms, 1934-39; research covering nuclear physics; biophysics; electronics; radio-carbon dating; high energy accelerators. Home: 830 Avon Rd., Ann Arbor, Mich.

CRANE, Robert Kellogg, Am. biochemist; b. Palmyra, N.J., Dec. 20, 1919; s. Wilbur Fiske, Jr. and Mary Elizabeth (McHale) C.; B.S., Washington Coll., 1942; Ph.D., Harvard, 1950; m. Mildred Ellen Price, July 19, 1941 (div. 1962); children—Barbara Joan, Jonathan Townley. Asst. biochemist Mass. Gen. Hosp., Boston, 1949-50; instr., asso. prof. biochemistry Med. Sch. Washington U., St. Louis, 1950-62; prof., chmn. dept. biochemistry Chgo. Med. Sch., 1962-66; prof., chmn. dept. physiology Med. Sch. Rutgers U., New Brunswick, N.J., 1966—. Mem. biochemistry test com. Nat. Bd. Med. Examiners; mem. steering com. Gastroenterology Research Group. Mem. Am. Chem. Soc., Am. Soc. Biol. Chemistry, A.A.A.S., Am. Soc.

Cell Biology, N.Y. Acad. Scis., Am. Gastroent. Assn., Biophys. Soc. Mem. editorial bd. Gastroenterology, 1963—; Biochimica et Biophysica Acta, 1965—. Research, numerous publs. primarily in field of cell metabolism. Office: Med. Sch., Rutgers U., New Brunswick, N.J. 08903.*

CRANEFIELD, Paul Frederic, Am. physiologist; b. Madison, Wis., Apr. 28, 1925; s. Paul Frederic and Edna (Rothnick) C.; Ph.B., U. Wis., 1946, Ph.D., 1951; M.D., Albert Einstein Coll. Medicine, 1964. Faculty, State U. N.Y. Downstate Med. Center, 1953-62; sr. research fellow dept. psychiatry Albert Einstein Coll. Medicine, 1960-64; asso. prof. pharmacology Coll. Phys. and Surg. Columbia, N.Y.C., 1964—; exec. sec. com. on publs. and med. information N.Y. Acad. Medicine, 1963-66; asso. prof. Rockefeller U., N.Y.C., 1966—. Mem. history of life scis. study sect. NIH. Recipient Schumann Med. Hist. Essay prize, 1956. Mem. Am. Physiol. Soc., Biophys. Soc., Cardiac Muscle Club, Am. Assn. History Medicine. Author: (with C. McC. Brooks) Historical Development of Physiological Thought, 1959; (with B. F. Hoffman) The Electrophysiology of the Heart, 1960. Editor: Bull. N.Y. Acad. Medicine, 1963-66; mem. editorial bd. Am. Jour. Physiology, 1963—, Jour. Applied Physiology, 1963—, Circulation Research, 1963—; editor Jour. Gen. Physiology, 1966—. Contbr. articles to profl. jours. Studies on elec. activity of single cells of heart, mechanism and nature of atrio-ventricular nodal delay, nature of anodal excitation of heart muscle, studies of nature and cause of ventricular fibrillation, studies of history of physiology, history of medicine, history of mental deficiency. Home: 48 W. 12th St., N.Y.C. 10011. Office: York Av. and 66th St., N.Y.C. 10021.*

CRANSTON, John, physician; b. Eng., 1625; s. James Cranston; m. Mary Clarke, June 3, 1658, 1 son, Samuel. Atty. gen. R.I., 1654-56; commr. R.I. Gen. Assembly from Newport, 1655-66; dep. gov. R.I., 1672-78; gov. R.I., 1678-80. Discovered element protactinium (Mendeleev's predicted eka-tantalum); collaborated with Frederick Soddy in important research on radioactivity. Died R.I., Mar. 12, 1680.

CRANTZ, Heinrich Johann Nepomuk von, see von Crantz, Heinrich Johann Nepomuk.

CRANZ, Carl Julius, German physicist; b. Hohebach, Germany, Jan. 2, 1858; s. Carl Hermann and Marie A. Magdalena (Kraus) C.; studied theology and philology at Tübingen, Germany; studied math. and natural scis. at Berlin, Germany and Tübingen, from, 1879; grad. 1883; Dr. Ing. h.c.; m. Klara Grub, 1891; 1 son, Hermann. Lectr. math. and mechanics Tech. U. Stuttgart (Germany); ins. mathematician; became asso. prof., head ballistic lab. Tech. Acad. Berlin, Charlottenburg, Germany, 1903; prof. tech. physics Tech. U. Berlin; sci. cons. to Govt. of China, Nankin, 1935-37. Author: Lehrbuch der Balistics, I, 1910 (with others, II, 1926, III, 1913, complementary vol., 1936). Extended the measuring methods of exptl. physics to ballistics; (with H. Schardin) created 1st perfect and fastest kinematographic pictures of projectiles; founded sci. ballistics. Died Esslingen, Germany, Dec. 11, 1945.

CRAPLET, Camille Charles, French biologist; b. Bagnolet, France, Jan. 6, 1920; s. Gaston and Marie Louise (Vitry) C.; Baccalauriat Math. and Philosophy, Collège Chaptal, Paris, France, 1937; D. Vetérinaire, École Nationale Vétérinaire d'Alfort, Paris, 1943; Licence in Sci., Faculté de Science, Paris, 1947; m. Josette Meunier, July 18, 1945; children—Michel, Alain, Pascal. Rural vet., 1942-46; lectr. Agrl. Sch. Grignon (France), 1948-54, prof. animal breeding, 1954—; also cons. animal feeding. Author: numerous books latest including: The Pig, 1961; The Veal, 1963. Home: 33 Sadi Cathot, Bagnolet 93, France. Office: École d'Agriculture, Grignon 78, France.*

CRARY, Albert Paddock, Am. geophysicist; b. Pierrepont, N.Y., July 25, 1911; s. Frank J. and Ella (Paddock) C.; B.S. magna cum laude, St. Lawrence U., 1931; M.S., Lehigh U., 1933. Oil prospector Ind. Exploration Co., Colombia, 1938-40; in Eng., Venezuela, Bahrein Island for United Geophy. Co., 1942-46; geophysicist Cambridge Research Center USAF, 1946-60; chief scientist U. S. Antarctic Research Program Nat. Sci. Found., 1959-67, dep. dir. div. environmental scis., 1967—; deputy chief scientist Internat. Geophy. Year Program in Antarctica, 1957-59. Recipient Navy Distinguished Public Service Award, 1959; American Geographical Soc.'s Cullum Geog. medal, 1960; Department Def. distinguished service award, 1959. Mem. A.A.A.S., American Geophys. Union, American Meteorol. Soc., Am. Geog. Soc., Arctic Inst. N.A., Seismol. Soc. Am., Soc. Exptl. Geophysics, Phi Beta Kappa. Author articles in field. Geophysical explorations in arctic and antarctic regions. Office: Nat. Sci. Found., Washington 25.

CRARY, Donald, Am. geologist; b. Pierre Point, N.Y, 1917; B.S. in Mining Engring., Lehigh U.; m. Miss Mapel; children—Martha Elaine, Donald J., John Albert, Sarah Ella. From geophysicist to supt. exploration Sohio Petroleum Co., 1945-52, mgr. N.W. div., 1952-55, gen. supt. exploration, 1955-57. Mem. Am. Assn. Petroleum Geologists, Soc. Econ. Geologists, Geol. Soc. Am. Skilled interpreter structures related to salt domes on Gulf Coast; studies resulting in discovery important oil and gas reserves, devel. new oil and gas fields La. region. Died 1957.

CRARY, Douglas Dunham, Am. geographer; b. Warren, Pa., Sept. 1, 1910; s. Clare J. and Irene (Horton) C.; A.B., U. Mich., 1933, M.A., 1934, Ph.D., 1947; postgrad. Harvard, 1938-40; m. Margaret Irene Fead, Sept. 15, 1934; children—Martha D., Thomas H., Elizabeth A., Rachel L. Asst. geography Harvard, 1938-40; faculty U. Mich., Ann Arbor, 1943—, prof. geography, 1963—; prof. African studies program McMaster U., Salisbury, Rhodesia, 1964. Lectr., Am. specialist program Office Ednl. and Cultural Exchange, Dept. State, Middle E. and N.E. Africa, 1960-61. Mem. Assn. Am. Geographers, Am. Geog. Soc., African Studies Assn., Mich. Acad. Sci., Arts and Letters, Sigma Xi, Phi Kappa Phi. Research, publs. on geography, Middle E. and Africa. Home: 1842 Cambridge Rd., Ann Arbor, Mich. 48104.*

CRASEMAN, Edgar, Swiss scientist; b. Strattlingen, Bern, Switzerland, Jan. 29, 1896; s. Edouard and Sophie (Hügli) C.; ed. Gymnasium, Gern, Fed. Poly. Sch., Zurich; Ph.D. in tech. scis.; D. honoris causa in Med. Vet., Faculty Medicine, U. Zurich, 1960; D. honoris causa in Sci. Agr., Superior Sch. Agr. at Stuttgart-Hohenheim, 1963; m. Elisabeth Bucher-Lussy; 1 son, Claes-Christian. Prof. agrl. chemistry, 1936; dean Faculty Agr., 1946-50. Recipient Henneberg-Lehmann prize George Auguste U., Göttingen, Germany, 1963. Mem. Italian Soc. for Progress of Zootechnics (corr.), Inst. Research (Brunswick, Germany) (corr.), Econ.-Agrarian Acad. Georgic Sons (corr.). Research and numerous publs. on physiology of nutrition and practical fodder for animals, conservation of forages. Home: Seestrasse 777, Meilen CH. Office: Universitätrasse 2, Zurich, Switzerland.

CRATEUAS, pharmacologist, flourished circa 100 to 60 B.C.; pharmacologist at ct. of Mithridates the Gt. (named plant Mithridatia for him); author comprehensive sci. work on pharmacology (at least 3 vols.), also author-illustrator book of colored pictures of plants with explanations of med. use; work influenced Sextius Niger, Dioscorides and all later pharmacology, medicine.

CRATO VON CRAFFTHEIM, Johannes (or Krafft, Johann), physician; b. Breslau (now Wroclaw, Poland), Nov. 1519; studied theology at Wittenberg, Germany; magister in classical langs.; M.D., Padua, Italy; m. Scharf von Weid, 1550; 1 son, 2 daus. Traveled through Italy; practiced medicine in Verona, Italy; returned to Breslau, 1550; city physician; personal physician to emperors Ferdinand and Maximilian II, Vienna, Austria, until 1581. Decorated Conies palatinus. Author: Idea doctrinae Hoppocraticae, 1554; Methodus therapeutica . . . , 1555; Ordnung oder Preservation zur Zeit der Pest, 1555; Isogoge medicinae, 1560; Perioche methodica in libros Gallin, 1563; De morbo gallico commentarium, 1594. Supported Galenic sch. medicine and opposed Paracelsian sch.; one of 1st to study contagious nature of some diseases. Died Breslau, Oct. 19, 1585 or 87.

CRAVER, Bradford North, Am. pharmacologist; b. Geneva, N.Y., Aug. 21, 1910; s. David H. and Eva (North) C.; A.B., Cornell U., 1932; M.S., Boston U., 1936, Ph.D., 1941; M.D., Wayne State U., 1944; m. Elena Borikova, Aug. 17, 1933; children—David, Frederick, Charles. Faculty, Am. Coll., Sofia, Bulgaria, 1932-34, Boston U., 1937-41, Wayne State U. Coll. Medicine, 1941-44; research asso. Manhattan Engring. Project, U. Rochester, 1944-45; sr. pharmacologist CIBA Pharm. Products, Inc., 1945-51; dir. Johnson & Johnson Research Found., New Brunswick, N.J., 1951-54; dir. pharmacology div. Squibb Inst. for Med. Research, New Brunswick, 1954—. Fellow A.A.A.S., N.Y. Acad. Scis.; mem. Am. Soc. for Pharmacology and Exptl. Therapeutics, Soc. for Exptl. Biology and Medicine, Am. Inst. Biol. Scis., Soc. Toxicology, Am. Coll. Clin. Pharmacology and Chemotherapy, A.M.A. (affiliate), Am. Assn. U. Profs. Contbr. numerous articles to profl. jours. Developed new drugs for mental, allergic, cardiovascular, renal and other diseases; co-designer heart perfusion apparatus. Home: 805 Mountain Av., Westfield, N.J. 07090. Office: George's Rd., New Brunswick, N.J. 08903.*

CRAVIOTO, Joaquin, Mexican med. nutritionist; b. Mexico City, Mexico, Sept. 12, 1923; s. Rafael G. and Carmen (Munoz) C.; M.D., Army Med. Sch.; Mexico, 1945; M.P.H., Mexican Sch. Pub. Health, 1947; postgrad. Pediat U.; postdoctoral fellow Bellevue Med. Center, U. N.Y.; research fellow U. Göteborg, Sweden; m. Maria Cristina Quintana O'Farril, Oct. 27, 1955; children—Alejandro, Patricia. Dep. chief nutrition ward Army Central Hosp., Mexico, 1948-49; head nutrition research lab. Children's Hosp., Mexico 1953-59; head nutrition dept., 1964; acting head Group for Research on Infantile Malnutrition, Mexico, 1959-60, head, 1961-62; nutrition officer Latin Am. area FAO, 1960; asso. dir. Inst. Nutrition of Central Am. and Panama, WHO, 1962-64; prof. nutrition Mexican Sch. Pub. Health, 1953-57, prof. postgrad. pediatrics, 1964—. Recipient Nat. Award Scis., Nat. Acad. Sci. Research of Mexico, 1962. Mem. Mexican, Am. socs. pediatric research, Mexican Soc. Biochemists, Am. Inst. Nutrition, Am. Acad. Pediatrics, N.Y. Acad. Scis. Publs. on establishment of diagnosis, prevention of infant malnutrition and its study in various social groups; effect on central nervous system. Office: 162 Dr. Marquez, Mexico D.F., Mexico.*

CRAW, Alexander, entomologist; b. Ayr, Scotland, 1850; came to U. S., 1873; in charge of J. W. Wolfskill orange grove, Cal., from 1875; asst. to D. W. Coquillet in investigation of control of cottony cushion scale, 1887-88; quarantine insp. Port San Francisco, 1890-1904; supt. entomology, insp. Hawaiian Bd. Agr. and Forestry, Honolulu, 1904-08. Exhibited insects Los Angeles Citrus Fair, 1881; 1st to suggest use of natural enemies for subjugation of pests; conceived and applied modern principles of hort. quarantine; named Physokermes insignicola Craw, Aspidiotiphagus citrinus (Craw), both common Cal. insects. Died 1908.

CRAWFORD, Adair, physician; b. Antrim, Ireland, 1748; student St. George's Hosp., Glasgow, Scotland; M.D.; prof. chemistry Mil. Acad. Woolwich; physician St. Thomas's Hosp. Fellow Royal Soc., 1786. Author: Experiments and Observations on Animal Heat and the Inflammation of Combustible Bodies, 1779; On the Matter of Cancer and on the Aerial Fluids, 1790; An Experimental Inquiry into the Effects of Tonics and other Medicinal Substances on the Cohesion of Animal Fibre, 1817. First to recognize presence of strontium; research on animal heat, specific and latent heats; discussed chemistry of respiration. Died Lymington, Hampshire, Eng., July 1795.

CRAWFORD, Bryce (Low), Jr., Am. chemist; b. New Orleans, Nov. 27, 1914; s. Bryce Low and Clara (Hall) C.; A.B., Stanford, 1934; A.M., 1935, Ph.D., 1937; m. Ruth Raney, Dec. 21, 1940; children—Bryce Low, III, Craig Llewellyn, Shery Ann. NRC research fellow Harvard, 1937-39; instr. Yale, 1939-40; faculty U. Minn., Mpls., 1940—, prof. chemistry, 1946—, chmn. chemistry dept., 1955-60, dean Grad. Sch., 1960—; Fulbright prof. Oxford U., Eng., 1951, Japan, 1966. Recipient Certificate of Merit, Pres. Truman, 1946; Guggenheim Found. fellow, 1950-51. Fellow Am. Phys. Soc.; mem. Nat. Acad. Scis., A.A.A.S., Optical Soc. Am., Am. Assn. U. Profs., Am. Chem. Soc., Am. Phys. Soc., Coblentz Soc., Sigma Xi, Phi Beta Kappa, Phi Lambda Upsilon. Contbr. numerous articles to tech. jours. Research on spectroscopy and structure of molecules, especially factors governing vibrational distortions of molecules. Home: 1545 Branston St., St. Paul 55108. Office: U. Minn., Mpls. 55455.*

CRAWFORD, Franzo Hazlett, Am. physicist; b. Dickinson, N.D., July 5, 1900; s. Lewis Ferandus and Cora Belle (Hazlett) C.; B.S., U. of N.D., 1920; B.A. (honors), Oxford U., 1923; Ph.D., Harvard U., 1928; m. Marie Edna Timberlake, June 18, 1928. Instr. physics, Northwestern U., 1923-24, Harvard U., 1927-28; faculty instr. Harvard U., 1928-30, asst. prof., tutor and chmn. tutorial bd. of natural sciences, 1930-36; visiting prof. physics, Williams Coll., 1936-37, Thomas T. Read professor of physics, 1937-66, professor emeritus, 1966—, former chairman of dept.; special research associate, Radio Research Lab., Harvard U., 1943-46; vis. prof. physics Columbia, summer 1951. Mem. Am. Com. Internat. Union Pure and Applied Physics, 1950-53. Rhodes scholar from N.D., 1920-23. Fellow Am. Phys. Soc., Am. Optical Soc., Am. Acad. Arts and Scis., N.Y. Acad. Sci.; mem. Phi Beta Kappa, Sigma Xi. Author: Heat, Thermodynamics and Statistical Physics, 1963. Study of the magnetic effect on the light from diatomic molecules; method for use of Jacobians in thermodynamics; several patents. Home: Kenwood Rd., Interlachen, Fla. 32048.

CRAWFORD, Frederick William, plasma physicist; b. Birmingham, Eng., July 28, 1931; s. William and Maud (Careless) C.; B.S., U. London, 1952; M.S., 1958; Ph.D., U. Liverpool, 1955, diploma in Edn., 1956, D.Eng., 1965; m. Beatrice Hutter, Oct. 21, 1963; 1 dau., Isabelle. Scientist research dept. Nat. Coal Bd., Isleworth, 1956-57; sr. lectr. elec. engring. dept. Coll. Advanced Tech., Birmingham, 1958-59; sr. research asso. Inst. For Plasma Research Stanford, 1959—; vis. scientist Atomic Energy Research Center, Saclay, France, 1961-62. Fellow Am. Phys. Soc., Inst. Physics, I.E.E.E.; mem. Sigma Xi. Research, numerous publs. on plasma wave phenomena, basic plasma processes and diagnostic techniques. Home: 859 Cedro Way. Office: Inst. For Plasma Research, Stanford, Via Crespi, Stanford, Cal. 94305.*

CRAWFORD, Jean Veghte, Am. chemist, coll. dean; b. Buffalo, Mar. 13, 1919; d. William J., Jr. and Mildred C. (Veghte) Crawford; A.B., Mt. Holyoke Coll., 1940; A.M., Oberlin Coll., 1942; Ph.D., U. Ill., 1950. Instr., Mt. Holyoke Coll., 1942-45; chemist Eastman Kodak Co., 1945-47; adj. prof., Randolph-Macon Woman's Coll., 1950-51; mem. faculty Wellesley Coll., 1951—, prof. chemistry, 1963—, chmn. dept., 1961-64, dean of students, 1966—. Mem. Am. Chem. Soc., Am. Assn. U. Profs., Phi Beta Kappa, Sigma Xi, Iota Sigma Pi, Sigma Delta Epsilon. Research on mechanism of organic reactions; heterocyclic nitrogen compounds. Home: 3 Hallowell House, Wellesley, Mass. 02181.

CRAWFORD, John, physician; b. May 3, 1746; med. degree, U. Leyden (Holland). Set forth theory of infection or contagion (his most useful contbn. to med. sci.); helped to found Soc. for Promotion of Useful Knowledge, Balt. Dispensary. Died Balt., May 9, 1813.

CRAWFORD, Lawrence, mathematician; b. Glasgow, Scotland, 1867; s. John Crawford; ed. U. Glasgow, King's Coll., Cambridge; M.A., D.Sc., LL.D.; m. Annie M. Spilhaus, 1903; 3 sons, 2 daus. Lectr. math. Mason Coll., Birmingham, Eng., 1893-98; prof. pure math. S. African Coll., Cape Town, 1899-1918; prof. U. Cape Town, 1918-38, vice chancellor, acting prin., 1931. Fellow Royal Soc. Edinburgh, royal socs. Edinburgh, S. Africa; pres. S. African Assn. for Advancement Sci., 1915-16. Contbr. papers to math. and sci. jours. Died Apr. 5, 1951.

CRAWFORD, Osbert Guy Stanhope, Brit. archaeologist; b. 1886; s. C.E.G. and Alice Luscombe (Mackenzie) C.; ed. Marlborough, Oxford; Litt.D., Cambridge, 1952, Southampton, 1955. Jr. demonstrator Sch. Geography, Oxford, 1921; excavated in Sudan under H. S. Wellcome, 1913-14; Rhind lectr. U. Edinburgh, 1943. Mem. Royal Commn. on Ancient and Hist. Monuments Eng., 1938-46. Recipient Victoria medal Royal Geog. Soc., 1940. Fellow Brit. Acad., Soc. Antiquaries (hon.); mem. Southeastern Union Scis. Socs. (pres. 1930), German Archaeol. Inst. (corr.), Austrian Anthrop. Soc. (corr.), Vienna Prehistoric Soc. (hon.); Am. Geog. Soc. (hon. corr.), Prehistoric Soc. (pres. 1938). Author: Man and His Past, 1921; The Andover District, 1922; Air Survey and Archaeology, 1924; Long Barrows and Stone Circles of the Cotswolds and the Welsh Marches, 1925; (with Alexander Keiller) Wessex from the Air, 1928; Air Photography for Archaeologists, 1929; Field Archaeology, 1932; Topography of Roman Scotland, 1949; (with F. Addison) Abu Geili Excavations Report, 1950; The Fung Kingdom of Sennar, 1951; Archaeology in the Field, 1953; Castles and Churches in the Middle Nile Region, 1953; (autobiography) Said and Done, 1955; The Eye Goddess, 1957. Founder, editor: Antiquity, 1927. Died Nov. 29, 1957.

CRAWFORD, Robert James, Canadian chemist; b. Edmonton, Alta., Can., July 8, 1929; s. Robert James and Margaret Alice (Pickles) C.; B.Sc., U. Alta., 1952, M.Sc., 1954; Ph.D., U. Ill., 1956; m. Agnes Joan Fisher, Oct. 6, 1956; children—Janet Maureen, Anne Katherine, Linda Joan, Eric Anthony. Asst. prof. chemistry U. Alta., Edmonton, 1956-62, asso. prof., 1962-67, prof., 1967—. Mem. Canadian Assn. U. Tchrs., Am. Chem. Soc., Chem. Soc. London, Chem. Inst. Can. (chmn. Edmonton sect. 1967-68). Research, articles on evidence for trimethylene species from pyrolysis of cyclic azo compounds; possible reaction mechanism for cyclopropane thermolysis; contbn. to mechanism of para-Claisen rearrangement; synthesis and studies of pyrazolines. Home: 7411 119th St., Edmonton, Alta., Can.*

CRAWFORD, Russell Tracy, Am. astronomer; b. Davis, Cal., Mar. 26, 1876; s. Frederick Gustavus and Mary Lanette (Foster) C.; B.S., U. Cal., 1897, Ph.D., 1901; postgrad. U. Berlin, winter 1911; m. Mary Crooke McCleave, Oct. 20, 1902 (died Apr. 21, 1903); m. 2d, Helen Alice Young, May 22, 1913. Faculty, U. Cal., 1903-58, instr. astronomy until 1906, asst. prof., 1906-10, asso. prof., 1910-19, prof., 1919-46, emeritus, 1946-58, chmn. dept., 1938-41, 42-58. Dir. Students Obs., 1939-46, dir. emeritus, 1946-58. Fellow A.A.A.S.; mem. Astron. Soc. Am., Astron. Soc. Pacific (pres. 1914), Astronomische Gesellschaft, Phi Beta Kappa, Sigma Xi. Investigated orbits of many comets and two satellites; computed gen. perturbations of several asteroids. Author: The Determination of Orbits of Comets and Asteroids. Editor: Cajori's Newton's Principia, A Revision of Motte's Translation. Died Dec. 21, 1958.

CRAYA, Antoine Edouard, physicist; b. Salonica, Greece, July 27, 1911; s. Paul and Thérèse (Nozza) C.; student Ecole Polytechnique, Paris, France, 1930-32; Docteur ès Sciences, Grenoble, France, 1957; m. Suzanne Million, Sept. 25, 1951. Research scientist Neyrpic Labs., Grenoble, 1936-51; vis. asso. prof. State U. Ia., 1951, Columbia U., N.Y.C., 1952; faculty U. Grenoble, 1953—, prof., 1958—, head dept. mechanics, 1967—, asso. dir. fluid mechanics labs., 1953—. Mem. Nat. Com. Sci. Research, 1967—; sci. cons. SOGREAH Labs., 1953—, Grenoble Center Nuclear Studies, 1956—. Laureate, Acad. Scis., 1963. Mem. Fluid Mechanics Assn. French Univs. (mem. council). Author: Contribution to the Analysis of Turbulence Associated with Mean Velocities, 1958; also articles. Research on wave propagation and instability in channels, flows with density stratification, structure of homogeneous turbulence asso. with a mean velocity field, confined jets, thermal and diffusive properties of swirling flows, flows of liquids of high magnetic diffusivity, instability in two phase flows. Home: 9, Rue Charles Péguy, Grenoble, France.*

CREDÉ, Carl Siegmund Franz, German gynecologist; b. Berlin, Germany, Dec. 23, 1819; s. Wilhelm Ludwig Credé; ed. at Berlin, Heidelberg (Germany); grad., 1842; M.D.; m. Cecilie von Cebrow, 1846; 8 children including Benno Carl. Asst. obstet. clinic, Berlin, 1843-48; lectr., active in obstetric research and lit. Berlin, 1850-56; became dir. Berlin Clinic, 1852; dir. Leipzig (Germany) clinic, 1856-87; pub. monthly Monatschrift für Geburtz-Kunde and Frauen Krankheiten, 1859-69. Author: Klinische Vorträge zur Geburtshilfe, 1853-54; also numerous articles. Originated a manual method for expression of the placenta (Credé's method), 1854, more accurate description pub., 1860; introduced the use of silver nitrate solution to prevent ophthalmia neonatorum in the newborn, 1884; invented incubator for premature infants (Credé's Incubator), 1888. Died Leipzig, Mar. 14, 1892.

CREDNER, Karl Friedrich Heinrich, German geologist; b. 1809; children—Karl Hermann, Rudolf. Privy councilor, dir. mining Halle region, Germany, from 1868; mineral crednerite named in his honor. Died 1876.

CREDNER, (Karl) Hermann (Georg), German geologist; b. Gotha, Germany, Oct. 1, 1841; s. Heinrich and Anna (Vey); studied geology and paleontology, Clusthal, Germany, Breslau (now Wroclaw, Poland), Gottingen, Germany; doctorate from Göttingen; hon. dr., Cambridge, Eng.; m. Marie Riebeck, 1872; 6 daus., including Hedwig (Mrs. Felix Berber). Traveled extensively in N.Am.; became asso. prof., Leipzig, Germany, 1870, prof. geology and paleontology, 1895; apptd. dir. geol. survey of Saxony, 1872. H. C. Found. Advancement Geology established in his honor on his 70th birthday; a glacier and a mountain named in his honor. Hon. mem. several sci. socs. Author: Beschreibung von Mineralvorkommen in Nordamerika, 1866/67; Die Gliederung der eozoischen Formationsgruppe Nordamerikas, 1869; Die vorsilurischen Gebilde der Oberen Halbinsel von Michigan in Nordamerika, 1869; Geognosie und Mineralreichtum des Alleghany-Systems, 1871; Elemente der Geologie, 1871; Über Lössablagerungen an der Zschopau und Freiberger Mulde, . . . , 1876; Der rote Gneis des Sächsischen Erzgebirges seine Verbandverhältnisse und genetischen Beziehungen zu der archäischen Schichtenreihe, 1877; Das Oligozän des Leipziger Kreises . . . , 1878; Über Glazialerscheinungen in Sachsen . . . , 1880; Die Stegocephalen Sächsische Graulitgebirge und seine Umgebung, 1889; Über die erzgebirgische Gneisformation und die Sächsische Königreich Sachsens, 1908. Research on glaciers, lower Permian sandstone nr. Dresden, formation of loess, earthquakes in Saxony; publs. on N.Am. Died Leipzig, July 22, 1913.

CREDNER, Rudolf, German geographer; b. Gotha, Germany, 1850; s. Heinrich and Anna (Vey) C.; m. Helene Ziervogel; 1 son, Wilhelm; became prof. geography, Griefswald, Germany, 1881. Author: Die Deltas, 1878; Die Reliktenseen, 1887-89; Rügen, 1893. Research on geomorphology, delta formation, glacial morphology. Died Greifswald, 1908.

CREECH, Hugh John, chemist; b. Exeter, Ont., Can., June 27, 1910; s. Richard Newton and Edith (Sanders) C.; B.A. U. Western Ont., 1933, M.A., 1935; Ph.D., U. Toronto (Ont.), 1938; postgrad. Harvard; m. Edna Marie Hearne, July 10, 1937; children—Richard Hearne, Joan Marie. Came to U. S., 1938, naturalized, 1945. Faculty, U. Md., College Park, Md., 1941-45, asso. prof., 1943-45; immunochemist Inst. for Cancer Research, Phila., 1945-47, head dept. chemotherapy, chmn. adminstrv. com., 1947-57, chmn. div. chemotherapy, 1957—; lectr. Bryn Mawr (Pa.) Coll., 1945-47; Mem. U. S. A. Nat. Com. on Internat. Union Against Cancer, 1957-60. Recipient awards Am. Cancer Soc., 1947-55, NIH, 1956-66. Mem. Am. Assn. for Cancer Research (dir., sec.-treas. 1952—), Am. Chem. Soc., Phila. Organic Chemists Club, A.A.A.S., Sigma Xi. Research, publs. on chems. to combat malaria, carcinogen-protein conjugates to immunize against cancer-producing chems., polysaccharide complexes and nitrogen mustard analogs to destroy tumors and produce mutation, fluorescent antibodies to detect microorganisms in tissues. Home: 702 Preston Rd., Erdenheim Phila. 19118. Office: 7701 Burholme Av., Phila. 19111.*

CREECH, Oscar, Jr., Am. surgeon; b. Nashville, N.C., Nov. 14, 1916; s. Oscar and Martha (Gulley) C.; B.S., Wake Forest Coll., 1937, D.Sc. (hon.), 1966; M.D., Jefferson Med. Coll., 1941; m. Dorothy Browne, Sept. 7, 1937; children—Oscar III, Diana, Archibald, Martha. Faculty, Tulane U., New Orleans, 1946-49, 56—, William Henderson prof., chmn. dept. surgery, 1956—; faculty Baylor U., 1949-56. Recipient J. Torrance Stewart Gold medal Jefferson Med. Coll., 1941; Distinguished Service citation Wake Forest Coll., 1960; Modern Medicine award for Distinguished Achievement, 1965. Diplomate Am. Bd. Surgery, Bd. Thoracic Surgery. Mem. A.C.S., A.M.A. (Hektoen Gold medal 1954, 59), Am. (pres. 1966—), So. surg. assns., Soc. U. Surgeons, Soc. Vascular Surgery (treas.), Am. Assn. Thoracic Surgery, Soc. for Exptl. Biology and Medicine, A.A.A.S., Internat. Cardiovascular Soc. (v.p. N.Am. chpt. 1966, treas. gen. group), Am. Heart Assn. (Contbns. to Research citation 1961), Internat. Soc. Surgery, Soc. Clin. Surgery (v.p. 1966), Internat. Surg. Group, Surg. Biology Club, Sigma Xi. Contbr. articles to med. jours. Investigation of cardiovascular physiology; organ transplantation; pathology and surgical techniques; perfusion chemotherapy for cancer. Home: 1120 State St., New Orleans 70118.

CREIGHTON, Elmer Ellsworth Farmer, Am. elec. engr.; b. Vallejo, Cal., 1873; A.B., Leland Stanford Jr. U., 1895, E.E., 1897; Sorbonne, Paris, 1898; École Supérieure de l'Électricité Paris, 1898-1900; unmarried. Engr. exptl. dept. Stanley Elec. Mfg. Co., Pittsfield, Mass., 1902-04; protective apparatus devels. and research, Gen. Elec. Co., 1904; asst. prof. elec. engring. Union U., 1904-06; cons. engr. Gen. Electric Co., also inventor and developer elec. protective apparatus, 1912—. Author many tech. papers pub. in Trans. A.I.E.E. from 1906. Died Jan. 12, 1929.

CREIGHTON, Henry Jermain Maude, chemist; b. Dartmouth, N.S., Mar. 2, 1886; s. Henry Dolby and Helen James (Robson) C.; B.A., Dalhousie U., 1906, M.A., 1907, LL.D., 1948; M.Sc., U. Birmingham (Eng.), 1909; student U. of Heidelberg (Germany 1909-10); D.Sc., Fed. Polytech., Zurich, Switzerland, 1911; Sc.D., Swarthmore (Pa.) College, 1957; married to Jean Hamilton Walker, June 21, 1916; children—Robert, Rosamond (Mrs. Syemour J. Ettman). Came to U. S., 1912. Lectr. on chem., Dalhousie U., 1911; instr. in chemistry, Swarthmore Coll., 1912-13, asst. prof., 1913-24, asso. prof., 1924-27, prof. 1927-52, now emeritus prof., head chemistry dept., 1927-49; consultant on electrochemistry; on leave of absence; with Manhattan Project at Columbia Univ., 1943-46. Columbia University War Research Medallion, 1946. Recipient Longstreth Medal, 1918, Potts Gold Medal, 1939, Modern Pioneer Award, 1940, Acheson Gold Medal and Prize, 1946. Representative, Electrochemical Soc. on Research Committee Am. Electroplaters' Soc., 1948-51. Mem. A.A.A.S., American Chemical Soc. (chairman Phila. sect., 1940-41), Electrochemical Soc. (vice pres. 1936-38, pres. 1939-40), Chem. Soc. (London), Soc. of Chemical Industry, Faraday Soc., Franklin Inst. (chmn. com. on science and arts, 1918; mem. Franklin medal committee, 1922-52, chmn. 1933), Nova Scotian Institute of Science, New York Academy Science, Chem. Inst. Can., Phi Beta Kappa (hon.), Sigma Xi (chmn. nat. lectureship com., 1938-43). Clubs: Chemists, Men's Faculty of Columbia U. (N.Y.); University (Montreal); Halifax, Royal Nova Scotia Yacht Squadron (Halifax, N.S.); English-Speaking Union, Royal Commonwealth Society, Over-Seas League (London, England). Author: Principles of Electrochemistry, 4th edit. 1943. Research on electrolysis, corrosion; conduction, organic electrochemical reduction. Home: Glen Margaret, N.S., Can.

CRELIN, Edmund Slocum, Am. anatomist; b. Long Branch, N.J., Apr. 26, 1923; s. Edmund Slocum and Agatha (Bublin) C.; B.A. cum laude, Central Coll. Ia., 1947; Ph.D., Yale, 1951; m. Marjorie Joyce McCain, Sept. 11, 1948; children—Sheryl, Edmund III, Robert, Carole. Faculty, Yale, 1951—, asso. prof. anatomy, 1961—. Cons. drug cos. Recipient F. G. Blake award Yale, 1961. Mem. Am. Assn. Anatomists, A.A.A.S., N.Y. Acad. Sci., Soc. Exptl. Biology and Medicine, A.M.A., Sigma Xi. Editor; Gray's Anatomy, 1967—. Discovered male hormone to be chief factor in causing differences in bony pelvis between males and females; proved cartilage cells change into bone cells during devel. of bones; analyzed changes in pelvic joints during pregnancy. Home: 124 Sunset Hill Dr., Branford, Conn. 06405.*

CRELL, Lorenz Florenz Friedrich von, see von Crell.

CRELLE, August Leopold, German mathematician; b. Eichwerder, Prussia, Germany, Mar. 11, 1780; Ph.D., U. Heidelberg (Germany), 1815. Constructed most of the Prussian roads built 1816-26; built 1st railroad in Germany; planned Berlin-Potsdam R.R.; mem. staff at Tech. Inst. until 1828; then joined Ministry of Ecclesiastical Affairs and Pub. Edn. Mem. Acad. Scis. Berlin. Author: Essay on a General Theory of Analytic Functions, 1826; Manual of Geometry, 1827. Founder, Journal des mathématiques pures et appliquées, 1826, Journal d'architecture, 1728, Crelle's Jour. Pub., Legendre's Geometry, 1823; Mathematical Works (LaGrange), 1823-24. His math. tables are widely used. Died Berlin, Germany, Oct. 6, 1855.

CREMER, Erika, Austrian phys. chemist; b. Munich, Germany, May 20, 1900; d. Max and Elsbeth (Rothmund) C.; Dr.phil., U. Berlin, 1927; Dr. phil.-habil., 1938; Dr.rer.nat.h.c., Berlin Tech. U., 1965. Research fellow, univs. Berlin, Freiburg im Breisgau, several Kaiser Wilhelm insts., 1927-40; faculty U. Innsbruck (Austria), 1940—, became dir. Phys. Chem. Inst., prof., 1959—. head Inst. Bioclimatic Research, Westerland/Sylt (Germany), 1936, Recipient Wilhelm Exner medal, Osterreicher Gewerbeverein, 1958, Prechtl medal, Technische Hochschule Wien, 1965. Mem. German Bunsen Soc., German Chem. Soc., Assn. Austrian Chemists; corr. mem. Austrian Acad. Scis. Author: (with Keulemans) Gas Chromatographie, 1959; (with Pahl) Kinetik der Gasreaktionen, 1961; also numerous articles. Research on kinetics in gas phase, solid state and on surfaces; developed first formula for explosion by branching of reaction chains; introduced gas chromatography for microanalysis and determination of phys. constants, 1947. Address: 20 Reithmannstrasse, 6020 Innsbruck, Austria.*

CREMONA, Luigi, Italian mathematician; b. Pavia, Italy, Dec. 7, 1830; student of F. Brioschi, U. Pavia; became prof. higher geometry, Bologna, Italy, 1860; named prof. geometry, graphical statics, Milan, Italy, 1866; became prof. higher math., dir. Engring. Sch., Rome, 1873. Recipient (with R. Sturm) Steiner prize Berlin Acad. Fellow Royal Soc., 1879; mem. French Acad. Scis., 1898. Author: Corso di static grafica,

1867; Le figure reciproche nella statica grafica, 1872; Elements of Projective Geometry, Eng. trans., 1885; Opera matematiche di Luigi Cremona, 1914, 15; also wrote on birational transformations (one case is Cremona transformation). Research in projective geometry and graphical statics; extended theory of transformation of curves and correspondence of points on curves to 3 dimensions; discovered (with G. Veronese) new properties of mystic hexagon; Died Rome, June 10, 1903.

CRESSMAN, George Parmley, Am. meteorologist; b. West Chester, Pa., Oct. 7, 1919; s. George Righter and Martha (Parmley) C.; B.Sc., in Physics, Pa. State Coll., 1941; M.Sc. in Meteorology, N.Y. U., 1942; Ph.D., U. Chgo., 1949; m. Nelia Marion Hazard, Feb. 28, 1942; children—Ruth, George, Catherine, Florence. Research asst. U. Chgo., 1946-49; cons. Air Weather Service, Andrews AFB, Washington, 1949-54; founder joint numerical weather prediction unit U. S. Weather Bur., Washington, 1954, 1962-64; dir. Office Nat. Meteorol. Services, Weather Bur., Silver Spring, Md., 1964-65, dir. Weather Bur., 1965—. Recipient USAF Meritorious Civilian Service award, 1956; U. S. Dept. Commerce Gold medal, 1961; Robert M. Losey award Am. Inst. Aero. Scis., 1965. Mem. Am. Meteorol. Soc., Washington Acad. Scis., Am. Geophys. Union, World Meteorol. Orgn. (mem. commn. for aerology 1953—, mem. adv. com. 1962—). Research, publs. on synoptic meteorology, automatic processing and analysis meteorol. data and numerical weather prediction. Home: 9 Old Stage Ct., Rockville, Md. 20852. Office: 8060 13th St., Silver Springs, Md. 20910.*

CRESSWELL, Eric, Brit. animal husbandman; b. Tynemouth, Eng., Sept. 25, 1926; s. George Edward and Nora (Sharp) C.; B.Sc., Durham (Eng.) U., 1946; M.Sc. (Cockshutt scholar), Toronto (Can.) U., 1955; Ph.D. (sr. research fellow), U. New Zealand, 1958; m. Rosalinde Conwy Moody, Dec. 10, 1959; children—Jeremy Wynne, Deryn Jane, Erica Lucy Penelope, Simon. Agrl. adviser Farmers Marketing & Supply Co., 1948; tech. adviser Brit. Glues & chems. Ltd., 1949; research officer, animal health div. Agrl. Research Council, 1949-53; vis. prof. animal husbandry Utah State U., Logan, 1958-59; prin. sci. officer, head sheep research sect. Roswell Research Inst., 1959—. Govt. adviser sheep prodn., Colombia, 1965-67; FAO adviser, Saudi Arabia, 1967—. Research, publs. on sheep husbandry, including skeletal studies, dentition, wintering, other fields. Address: Kenilworth, Gannock Rd., Deganwy, N. Wales, Gt. Britain.*

CRESTANI, Giuseppe, Italian meteorologist; b. Montebello, Vicence, May 14, 1897; s. Venceslao and Lucia (Bortolaso) C.; Ph.D. in phys. scis. Prof. meteorology; dir. G. Magrini Meteorologic Obs., Padua, Italy. Author numerous publs., including Meteorologia Aeronautica, 1920; Climatologia Ipogea, 1937. Address: via Sammicheli 47, Padua, Italy.

CRETCHER, Leonard Harry, Am. chemist; b. Degraff, O., July 25, 1888; s. Harry Donelly and Nancy (Black) C.; A.B., U. Mich., 1912; Ph.D. in Chemistry, Yale, 1916; m. Frances L. Hickok, July 10, 1920. Instr. chemistry U. Tenn., 1913-14; fellow organic chemistry Rockefeller Inst., 1916-17, asst., 1917-18; asst. chief chemistry Nat. Aniline & Chem. Co., 1919-20, research chemist, 1920-22; sr. fellow pure research Mellon Inst., Pitts., 1922-26, head dept., 1926-52, asst. dir., 1931-52, dir. research, 1952-54. Fellow A.A.A.S.; mem. Am. Chem. Soc. (Pitts. award Pitts. sect. 1945), Soc. Biol. Chemistry, Sigma Xi, Phi Lambda Upsilon, Theta Chi. Editorial bd. Jour. Organic Chemistry. Research, publs., patents in pyrimidine aldehydres, oxidation tertiary hydrocarbons, glycol ethers and esters, organic boron compounds, barbituric acids, chlorethers, alginic acid, sugar acids, equilibria in binary liquid systems, chemotherapy of pneumonia and malaria. Home: 144 N. Dithridge St., Pitts. 15213. Office: Mellon Inst., 4400 5th Av., Pitts. 15213.*

CRETZSCHMAR, Philipp Jakob, German anatomist, zoologist; b. Sulzbach/Taunus, Germany, June 11, 1786; s. Otto and Anna C. (Eberhardt) C.; studied natural philosophy in Würzburg, Germany, then medicine in Halle, Germany; continued studies in Würzburg, 1807, and received degree; m. C. Josephina Müller, 1815. Began practice medicine, Frankfort/Main, Germany, 1808; doctor in French army; worked in German hosps.; practiced surgery in Paris, France, also Vienna, Austria; with Catalonian Army in Spain; resumed practice medicine, Frankfort/Main; set up courses for surg. assistants in mil. hosp.; taught anatomy, later also zoology Senckenberg Med. Inst. Founder, Senckenbergische Naturforschende Gesellschaft and Fund, 1817, dir. for 30 years, also founder medal for outstanding research. Laid found. for natural sci. collection at Senckenberg Mus., Frankfort. Died Frankfort/Main, May 4, 1845.

CREUTZ, E(dward), Am. physicist; b. Beaver Dam, Wis., Jan. 23, 1913; s. Lester Raymond and Grace (Smith) C.; B.S., U. Wis., 1936, Ph.D., 1939; m. Lela Rollefson, Sept. 13, 1937; children—Michael John, Carl Eugene, Ann Jo Carmel. Research asst., instr. Princeton, 1939-42; physicist U. Chgo., 1942-44; group leader Manhattan project, Los Alamos, 1944-46; faculty Carnegie Inst. Tech., Pitts., 1946-56, prof., head dept. physics, dir. nuclear research Center, 1949-56; dir. John Jay Hopkins Lab. for Pure and Applied Sci., Gulf Gen. Atomic (Gen. Atomic division Gen. Dynamics Corporation until 1967), San Diego, 1955—, director of research, 1955-59, v.p. research and devel., 1959—. Cons., NSF, 1950—, AEC and its labs., 1946—, NASA, 1963—. Fellow Am. Phys. Soc., Am. Nuclear Soc. (dir. 1958-61), San Diego Natural History Soc. (pres. fellows 1965-66, mem. board of directors 1964—); member of the American Institute Physics (dir.-at-large gov. bd. 1965—). Research, numerous articles in scattering protons from Lithium and protons, resonance neutrons in uranium, angular distbn. neutrons from (D, n) reactions, metallurgy uranium and beryllium, fluid flow; inventor in field nuclear reactors and tech. Home: P.O. Box 765, Rancho Sante Fe, Cal. 92067. Office: P.O. Box 608, San Diego 92112.*

CREUTZFELDT, Otto Detlef, German physician; b. Berlin, Apr. 1, 1927; s. Hans Gerhard and Claire (Sombart) C.; student univs. Kiel, Heidelberg, Freiburg/Breisgau (all Germany); M.D., Freiburg/Br., 1953. Research asst. Brain Research Inst., U. Cal. at Los Angeles, 1960-61; head dept. neurophysiology, cons. neurologist, Max Planck Inst. Psychiatry, Munich, 1962—; docent U. Munich, 1963—. Recipient Hans Berger prize, German EEG Soc., 1965. Mem. Max Planck Soc., German Physiol. Soc., Internat. Brain Research Orgn. Cons, editor Jour. EEG Clin. Neurophysiology, 1966—, Jour. Exptl. Brain Research, 1965—, Kybernetik, 1964—. Research and numerous publs. on electrogenesis of electroencephalograms; information transmission in central nervous system, neuronal basis of visual perception; applied neurophysiology. Address: Max Planck Institute for Psychiatry, 2 Kraepelinstrasse, 23 Munich, West Germany.*

CREUTZFELDT, T. Werner Otto N. Carl, German physician; b. Kiel, Germany, May 11, 1924; s. Hans Gerhard and Clare (Sombart) C.; student U. Freiburg i. Br., 1943-44, U. Tuebingen, 1948; M.D., U. Kiel, 1950; m. Cora Elisabeth Glees, Dec. 1, 1962; children—Hans Nikolaus, Cornelius Werner. Research student, dept. anatomy U. Kiel, 1947-48; research asst., dept. pathology U. Freiburg, 1950-52, pvt. docent, 1957, prof., 1962; resident, dept. internal medicine, 1952, chief resident, 1958, physician-in-chief, 1960-64, prof., 1962; research fellow, dept. pathology Washington U., St. Louis, 1959, Baker Research Lab., New Eng. Deaconess Hosp., Boston, 1959; dir., chmn. dept. internal medicine U. Goettingen, 1964—. Mem. German socs. internal medicine, endocrinology, gastroenterology and diabetology, European Assn. for Study of Diabetes, European Soc. for Clin. Investigation, Royal Soc. Medicine, London, N.Y. Acad. Scis. Author: Oral Treatment of Diabetes, 1961; also articles. Research on morphology and function of islets of Langerhans, insulin secretion, pathogenesis of diabetes mellitus, treatment of diabetes, mechanism of action of oral antidiabetic drugs, pathogenesis and treatment of hepatitis and liver cirrhosis, pathogenesis, diagnosis and treatment of acute and chronic pancreatitis. Home: 41 Senderstrasse, Göttingen-Nikolausberg. Office: 1 Humboldtallee, Göttingen, W. Germany.*

CREVASSE, Lamar Earle, Jr., Am. physician; b. West Palm Beach, Fla., June 23, 1926; s. Lamar Earle and Ethyl (Venable) C.; A.B., Duke, 1950, M.D., 1954; m. Carol Lynn Austin, June 24, 1951; children—J. Lamar, Austin, Candace. NIH fellow cardiology Emory U., 1957-58; practice medicine, specializing in cardiology, Gainesville, Fla.; faculty U. Fla. Coll. Medicine, 1948—, asso. prof. medicine, 1963—. Markle scholar in med. sci. Mem. Am. Fedn. Clin. Research, Am. Heart Assn., A.C.P., A.M.A., So. Soc. Clin. Investigation. Research, publs. on mechanisms of action of cardiovascular drugs using radioactive isotopes; devel. of molecular model for muscle contraction. Home: 1920 N.W. 12th Rd., Gainesville, Fla. 32601.*

CRÉVECOEUR, see de Crévecoeur.

CREW, Henry, Am. physicist; b. Richmond, O., June 4, 1859; s. Wm. Henry and Deborah A. C.; A.B., Princeton, 1882; Ph.D., Johns Hopkins, 1887; m. Helen C. Coale, July 17, 1890; children—Alice H., Mildred, William H. Instr. physics, Haverford Coll., 1888-91; astronomer Lick Obs., 1891-92; prof. physics Northwestern U., 1892-1930; chief o fdiv. of basic scis. Century of Progress Expn., Chgo., 1930. Collaborator Astrophysical Journal, 1892-1942; del. Congress of Physicists, Paris, 1900. Fellow Am. Acad. Arts and Scis., Phi Beta Kappa; mem. Nat. Acad. Sciences, Am. Phys. Soc. (pres. 1909), Am. Philos. Soc., Am. Assn. U. Profs. (pres. 1929-30), History of Science Soc. (pres. 1930). Recipient Oersted medal Am. Assn. Physics Tchrs. 1941. Author: Principles of Mechanics, 1908; General Physics, 1908; Rise of Modern Physics, 1928; also numerous articles. Translator: Maurolycus' Optics, 1940. Did early spectroscopic research. Died Feb. 17, 1953.

CREW, William Henry, Am. physicist; b. Evanston, Ill., Aug. 24, 1899; s. Henry II and Helen C. (Coale) C.; B.S., U. S. Naval Acad., 1922; M.S., Johns Hopkins, 1924, Ph.D. in Physics, 1926; m. Dorothy I. Staines, July 4, 1931; 1 son, Henry III. Asst. physicist Naval Research Lab., Washington, 1925-28; faculty U. S. Naval Postgrad. Sch., 1928-29, N.Y. U., 1929-45; tech. aide OSRD, 1941-45; asst. sec. I.R.E. (now I.E.E.E.), also sec. Radio Tech. Planning Bd., 1945-46; asst. dean students Rensselaer Polytech. Inst., 1946-48; dean Coll. Engring. Scis., USAF Inst. Tech., 1948-50; asst. dir., sci. personnel Los Alamos Sci. Lab., 1950-65, vis. scholar U. Ariz., 1966—. Cons. prof. U. N.M., 1957-65; mem. N.M. adv. com. SSS, 1953-66. Fellow Am. Phys. Soc., A.A.A.S.; mem. I.E.E.E. (sr.), Am. Soc. Engring. Edn., U. S. Naval Inst. (asso.), Am. Assn. U. Profs. (asso.), Sigma Xi. Research and publs. in fields of photoelectricity and spectroscopy, mil. applications of optics. Home: 815 Camino de Fray Marcos, Tucson 85718.*

CREWE, Albert Victor, physicist; b. Bradford, Eng., Feb. 18, 1927; s. Wilfred and Edith (Lawrence) C.; B.Sc., U. Liverpool (Eng.), 1947, Ph.D., 1951; m. Doreen Patricia Blunsdon, Apr. 11, 1949; children—Jennifer Elisabeth, Sarah Jane, Elizabeth Jane, David Albert. Came to U. S., 1955, naturalized, 1961. Lectr., U. Liverpool, 1951-55; faculty U. Chgo., 1955—, prof. physics, 1962—, prof. Enrico Fermi Inst., 1962—; dir. particle accelerator div. Argonne (Ill.) Nat. Lab., 1958-61, laboratory director, 1961-67. Mem. Ill. Air Pollution Control Bd.; mem. Gov.'s Sci. Adv. Committee, State of Ill. Named Chicagoan of Year in Sci., Chgo. Jr. C. of C., 1962; Outstanding New Citizen, Citizenship Council Chgo., 1962; recipient Immigrants Service League's Ann. award for outstanding achievement in field sci., 1962. Fellow Am. Phys. Soc., Am. Nuclear Soc.; mem. Electron Microscope Soc. Author: (with J. J. Katz) Research U. S. A., 1963; also articles. Accelerator devel. including beam extraction systems for cyclotrons, design and constrn. of a synchrotron; research in high energy physics especially meson prodn.; electron microscope devel. Home: 63, Old Creek Rd., Palos Park, Ill. 60464. Office: Fermi Inst., 5630 Ellis Av., U. Chgo., Chgo. 60637.*

CREYSSEL, Jean Pierre Louis Marie, French surgeon; b. Marseille, France, Jan. 27, 1898; s. Jacques Adolphe and Marie (Marguery) C.; ed. Ampere Sch.; Ph.D. in Medicine, Faculty of Medicine, Lyons; M.D., 1925; m. Suzanne Chapuis; children—Roger, Jacqueline, Pierre. Intern, Hosps. of Lyons, 1921, surgeon, 1938; prof. agrege surgery Faculty of Medicine, Lyons, 1933; prof. traumatology, 1956; prof. orhtopedic surgery and traumatology, 1957. Mem. Acad. Surgeons, Soc. Surgeons of Lyons (past pres.), Internat. Soc. Surgeons. Author: Précis d'anatomie médico-chirurgicale; Traité de thérapeutic chirurgicale, 1957; Cancer des glandes salivaires; Shock traumatique; Fractures transcotyloïdiennes du bassin. Home: 37, place Bellecour. Office: Hopital Ed.-Herriot, Lyons, France.

CRICK, Francis Harry Compton, Brit. biologist; b. Northampton, Eng., June 8, 1916; s. Harry and Annie Elizabeth (Wilkins) C.; B.Sc., Univ. Coll., London, 1938, postgrad., 1937-39; Ph.D., Caius Coll., Cambridge, 1954; m. Ruth Doreen Dodd, 1940 (div. 1947); 1 son, Michael F. C.; m. 2d, Odile Speed, 1949; children—Gabrielle A., Jaqueline M. T. Scientist, Brit. Admiralty, 1939-47; with Strangeways Research Lab., Cambridge, 1947-49; mem. Med. Research Council Unit, 1949—, Med. Research Council Lab. on Molecular Biology, 1962—; with protein structure project Bklyn. Poly., 1953-54; vis. prof. Harvard; fellow Churchill Coll., Cambridge, 1959-61; elected fellow Univ. Coll., London, 1962; fellow Salk Inst. for Biol. Studies, San Diego, Cal. Recipient Prix Charles Leopold Meyer, French Acad. Scis., 1961; (with J. D. Watson and M. H. F. Wilkins) Nobel prize in medicine and physiology, 1962; Research Corp. award; Award of Merit, Gairdner Found., 1962; Warren Triennial Prize lectr., 1959. Fellow Royal Soc. London, 1959; fgn. hon. mem. Am. Acad. Arts and Scis. Proposed (with James D. Watson) model for double-helical structure of DNA which answers question of how hereditary material duplicates itself (considered most important single devel. in biology of 20th century), 1953.

CRILE, Dennis Rider Wood, Am. surgeon; b. Baltic, O., May 27, 1891; s. Austin D. and Winifred Augusta (Wood) C.; B.S., U. Wis., 1914; M.D., Harvard, 1917; m. Mary Dorothea Webb, Jan. 1, 1919; children—Dennis Michael, Dorothea Mary. Asst. prof. surgery, U. Ill. Coll. Medicine. Contbr. to med. jours., and chpts. dealing with compound fractures in Ochsner's Surgical Diagnosis and Treatment. Originator of method of resuscitation of dying persons by injection of adrenalin into heart. Died Mar. 21, 1937.

CRILE, George (Washington), Am. surgeon; b. Chili, O., Nov. 11, 1864; s. Michael and Margaret (Dietz) C.; B.S., Ohio No. U., 1885, A.M., 1888; M.D., Wooster U. (now Western Res. U.), 1887, A.M., 1894, LL.D., 1916; postgrad. Vienna, London, Paris; Ph.D. (hon), Hiram Coll., 1901; M.Ch., U. Dublin, 1925; LL.D., U. Glasgow, 1928; hon. doctorate U. Guatemala, 1939; m. Grace McBride, Feb. 7, 1900; children—Margaret (Mrs. Hiram Garretson), Elisabeth (Mrs. J. A. Crisler, Jr.), George Washington, Robert. Lectr. and demonstrator histology, Wooster U., 1889-90, prof. physiology, 1890-93, prof. principles and practice of surgery, 1893-1900; prof. clin. surgery Western Res. U., 1900-11, surgery, 1911-24; vis. surgeon Lakeside Hosp., 1911-24; a founder Cleve. Clinic Found., became dir. research Profl. dir. U. S. Army Base Hosp. No. 4, Lakeside Unit (B.E.F., No. 9), France, 1917-18; sr. cons. in surg. research, 1918-19. Decorated Chevalier Legion Honor (French), 1922;

recipient Alvarenga prize Coll. Physicians, Phila., 1901; Cartwright prize Columbia, 1897, 1903; Senn prize A.M.A., 1898; Am. med. medal for service to humanity, 1914; Nat. Inst. Soc. Sci. medal, 1917; Trimble Lecture medal, 1921; 3d laureate of Lannelongue Found. (Lannelongue Internat. medal of surgery presented by Société Internationale de Chirurgie de Paris), 1925; Cleve. medal for public service, 1931; Distinguished Service Gold Key Am. Assn. Anatomists, A.A.A.S., Am. Surg. Assn. (pres. 1923), vice, 1931; Distinguished Service Gold Key Am. Assn. A.C.S. (pres. 1916, chmn. bd. regents 1917-39), A.M.A. Am. Physiol. Soc., Am. Assn. Obstetricians, Gynecologists and Abdominal Surgeons, So. Med. Assn., Am. Philos. Soc.; mem. Assn. Am. Pathologists and Bacteriologists, Am. Soc. Clin. Surgery, Nat. Inst. Social Scis., NRC, Assn. Study Internal Secretions, Am. Heart Assn., other med. socs.; hon. or corr. fellow or mem. many Am. and Euopean scis. Author: Surgical Shock, 1897; Surgery of Respiratory System, 1899; Certain Problems Relating to Surgical Operations, 1901; On the Blood Pressure in Surgery, 1903; Hemorrhage and Transfusion, 1909; Anemia and Resuscitation, 1914; Anoci-Association (with Lower), 1914, 2d edit., title, Surgical Shock and the Shockless Operation through Anoci-Association, 1920; Origin and Nature of the Emotions, 1915; A Mechanistic View of War and Peace, 1915; Man, An Adaptive Mechanism, 1916; The Kinetic Drive, 1916; The Fallacy of the German State Philosophy, 1918; A Physical Interpretation of Shock Exhaustion and Restoration, 1921; The Thyroid Gland (with others), 1922; Notes on Military Surgery, 1924; A Bipolar Theory of Living Processes, 1926; Problems in Surgery, 1928; Diagnosis and Treatment of Diseases of the Thyroid Gland (with others), 1932; Diseases Peculiar to Civilized Man, 1934; The Phenomena of Life, 1936; The Surgical Treatment of Hypertension, 1938; Intelligence Power and Personality, 1941. Introduced theory that shock results from exhaustion of vasomotor center, 1901; developed block anesthesia (method of blocking nerve trunks with local anesthesia), circa 1901. Died Jan. 7, 1943.

CRIQUI, Fernand, French sci. writer; b. Strasbourg, France, May 14, 1921; s. Fernand and Alice (Sigle) C.; ed. in phys. edn. U. Strasbourg; m. M. Westermann. Former instr. Internat. Center Advanced Journalism Teaching, U. Strasbourg. Mem. Assn. Sci. Writers of France, Soc. Writers of Alsace and Lorraine. Contbr. over 200 articles, monographs on sci. and medicine to jours. and revs. in Europe and Am. Address: 35, rue Dietterlin, Strasbourg-Meinau, France.

CRISMON, Jefferson Martineau, Am. physiologist; b. Phila., Feb. 4, 1908; s. Kenneth Allen and Alley (Martineau) C.; A.B., Stanford, 1931, M.D., 1938; m. Cathrine Alice Stanton, June 12, 1937; children—Patricia, Daniel. Faculty, Stanford, 1936—, prof., 1951—, exec. head dept. physiology, 1951-63; hon. fellow research dept. pediatrics Yale Sch. Medicine, 1940-41. Mem. confs. on cold injury Josiah Macy Jr. Found., 1951-58; cons. Army Surgeon Gen., 1958—. Guggenheim fellow, 1957-58. Fellow A.A.A.S.; mem. Am. Heart Assn. (past chmn. research com.), Am. Physiol. Soc., Western Soc. for Clin. Research, Am. Fedn. for Clin. Research, Microcirculatory Soc., Sigma Xi. Asso. editor, Ann. Rev. physiology, 1946-53; editorial bd. Circulation Research, 1953-57, Circulation, 1957—. Research, publs. on distbn. of water and electrolytes as influenced by heat, very low temperatures, lack of blood supply, injury by heat, cold or other phys. changes, regulation of capillary blood flow, distbn. of blood flow in human skin especially during responses to inflammation causing agts. Home: 1805 Guinda St., Palo Alto, Cal. 94303. Offic: Dept. Physiology, Stanford, Stanford, Cal. 94305.*

CRISTOL, Stanley Jerome, Am. chemist; b. Chgo., June 14, 1916; s. Myer J. and Lillian (Young) C.; B.S., Northwestern U., 1937; M.A., U. Cal. at Los Angeles, 1939, Ph.D., 1943; m. Barbara Wright Swingle, June 1957; children—Marjorie Jo, Jeffrey Tod. Served as research chemist The Standard Oil Company of California, 1938-41; research fellow U. Ill., 1943-44; research chemist Dept. Agr., 1944-46; asst. prof., then asso. prof. U. Colo., 1946-55, prof., 1955—, chmn. dept. chemistry, 1960-62; vis. prof. Stanford, summer 1961. With OSRD, 1944-46; adv. panels NSF, 1957-63; cons. E. I. duPont de Nemours & Co., Inc., Esso Research & Engring. Co., Guggenheim fellow, 1955-56. Fellow, A.A.A.S. Chem. Soc. London; mem. Am. Chem. Soc. (chmn. organic chemistry div. 1961-62, adv. bd. petroleum research fund 1963-66), Am. Assn. U. Profs., Colo.-Wyo. Acad. Sci., Phi Beta Kappa, Sigma Xi, Phi Lambda Upsilon. Author: (with L. D. Smith, Jr.) Organic Chemistry, 1966. Editorial bd. Chem. Reviews, 1957-59; Jour. Organic Chemistry, 1964—. Contbr. research articles in sci. jours. Research in fields of organic reaction mechanisms and organic synthesis. Home: 2918 3d St., Boulder, Colo.

CRITCHFIELD, Charles Louis, Am. physicist; b. Shreve, O., June 7, 1910; s. R. Roy and Clara Mae (Prince) C.; B.S., George Washington U., 1934, M.A., 1936, Ph.D., 1939; m. Jean LaZelle Anderson, Aug. 31, 1935; children—Lewis, Robert, Barbara, Douglas. With Nat. Bur. Standards, Washington, 1930-37; instr. U. Rochester, N.Y., 1939-40; nat. research fellow Princeton, Inst. for Advanced Study, 1940-41; instr. Harvard, 1941-42; physicist Geophys. Lab., Washington, 1942-43; group leader Los Alamos, 1943-46; asso. prof. George Washington U., Washington, 1946; physicist Monsanto Chem. Co., Oak Ridge, 1946-47; professor physics Univ. of Minnesota, 1947-55; dir. sci. research Convair div. Gen. Dynamics Corp., San Diego, 1955-60; v.p. research Telecomputing Corp., San Diego, 1960-61; asso. div. leader theoretical physics div. Los Alamos Sci. Lab., 1961—. Recipient Alumni Achievement award George Washington U., 1964. Fellow Am. Phys. Soc. Editorial bd. Ann. Rev. Nuclear Sci., 1957-61; asso. editor Jour. of Franklin Inst., 1956-62, Phys. Rev., 1951-54. Author: Theory of Atomic Nucleus and Nuclear Energy Sources (with G. Gamow), 1949. Research papers in math. physics, especially nuclear theory, astro-physics, cosmic rays, and particle theory including H-H reaction for stellar energy (with Hans Bethe). Patents in field of weaponry and plastic balloons. Home: 391 El Conejo St., Los Alamos 87544. Office: Theoretical Div., Los Alamos Sci. Lab., Los Alamos, N.M.*

CRITCHLEY, Macdonald, Brit. neurologist; b. Bristol, Eng., Feb. 2, 1900; student Christian Bros. Coll., Clifton, Eng., Bristol U. Physician, Nat. Hosp., London, 1927—; neurologist King's Coll. Hosp., London, 1927—; former dean Inst. Neurology, London; cons. neurologist Royal Navy, 1939—. Decorated comdr. Brit. Empire. Mem. World Fedn. Neurology (pres. 1965—). Author: Parietal Lobes, 1953; Developmental Dyslexia, 1964; also numerous articles. Research on parietal lobe symphlomalology, disorders of lang., also reading retardation in children. Office: Nat. Hosp., Queen Sq., London, W.C.1, Eng.*

CRITTENDEN, Eugene Casson, Jr., Am. physicist; b. Washington, Dec. 25, 1914; s. Eugene C. and Norma (Snyder) C.; student George Washington U., 1930-33; A.B., Cornell U., 1934, Ph.D., 1939; m. Josephine Woolfolk, Nov. 8, 1942; children—Elizabeth, Robert. Asst. physics Cornell, Ithaca, N.Y., 1934-38; instr. Case Inst. Tech., Cleve., 1938-42, asst. prof., 1942-44, asso. prof., 1946-48, prof., 1948-52; research physicist Radiation Lab., U. Cal. at Berkeley, 1944-46; head, solid state physics Atomics Internat., Los Angeles, 1952-53; prof. U. S. Naval Postgrad. Sch., Monterey Cal., 1953—, chmn. physics dept., 1964—. Fellow Am. Phys. Soc., Sigma Xi. Important work includes research in nuclear physics, solid state physics, ferromagnetism, and superconductivity.

CROASDALE, Stuart, Am. mining engr.; b. Delaware Water Gap, Pa., Nov. 21, 1866; s. Evan Thomas and Ellen (Andre) C.; B.S. in chemistry, Lafayette Coll., 1888, M.S., Ph.D., 1891; m. Elma G. Shaw, 1891; children—Dorothy, Ernest Shaw, Evan Thomas. Was chief chemist, Holden Lixiviation Works, Aspen, Colo., 1891-93. Gillette (Colo.) Reduction Works, 1894-95, Globe Smelting & Refining Co., Denver, 1896-1900; cons. practice, 1900—; cons. engr. Anaconda Copper Co., Mont., 1903. Burro Mt. Copper Co., 1903-07, Calumet & Ariz. Copper Co., 1912-13, and many other cos., including Utah Copper Co., Nipissing Mines Co., etc.; pres. Alma Gold Corp., Denver. Pioneer in smelter smoke investigation, and comml. leaching of copper ores; inventor, volatilization process for treatment of ores, improved process for concentration of ores, hydometallurgical process for treatment of mercury ores, etc. Died Sept. 30, 1934.

CROATTO, Ugo, Italian chemist; b. Trieste, Italy, Nov. 2, 1914; s. Paolo and Mercede (Sauli) C.; Dr.-Chemistry, U. Padua, 1936; m. Costanza Bruno, July 10, 1943. Prof. chemistry U. Modena, 1949-54, U. Padova, 1954—; dir. Inst. Gen. Chemistry, U. Padova. Research, numerous publs. on inorganic, nuclear and structural chemistry. Home: Via A. Medin N2, Padova, Italy.*

CROCCHI, Peter Andrews, Argentine radiologist; b. Buenos Aires, Argentina, Jan. 1, 1901; grad. Med. Sch., Nat. U.; m. Maria Italia Casazza, Dec. 19, 1925; children—Betty Lucia (Mrs. Arturo Hugo Simon). Practician in pub. assistance, 1920; asst. Ramosmejia Hosp., 1921-24; physician, asst. radiology, C. R. Mejia, 1925-30; radiologist Piñero Hosp., 1930-32; chief radiology service Zubizamata Hosp., 1932-36; practice medicine, specializing in radiology, 1936-40; chief service Santojanmi Hosp., 1940-42; chief practical exercises Med. Sch., La Plata, Argentine, 1942-49; asst. prof. radiology Faculty Medicine, La Plata, 1949-51; mem. Nat. AEC, 1951-58; extraordinary chief radiology Santojanni Hosp., Buenos Aires, 1958—. Mem. Hosp. Santojanni Soc. (hon.), Panama Soc. Radiologists, Inter-Am. Coll. Radiologists, Argentine, Internat. socs. radiology, Argentine Med. Assn. Author: Radiology of Pulmonary Tuberculosis, 1950; (with J. A. Aguime) Radiology and Physicotherapy Compendium, 1945; also articles. Research on radioactive isotopes, biol. effects of atomic bombing in populations, effects of drugs on small intestine. Home: 1002 Billinghurst. Office: 2466 Cordota Av., also 63 Rio de Janeiro, Buenos Aires, Argentina.*

CROCKER, Walter James, Am. veterinarian; b. Ada, Minn., Nov. 20, 1885; s. Walter Joseph and Helen (Wiley) C.; B.S.A., Utah Agrl. Coll., 1909; V.M.D., U. Pa., 1911; m. Rosa Binder, Feb. 6, 1915; 1 dau., Helen Marie. Lectr., instr., Vet. Sch., U. Pa., 1911-13, asst. prof. vet. pathology, 1913-14, asst. prof. vet. pathology and bacteriology, 1914-16, prof. vet. pathology, from 1916, asst. dir. Wistar Inst. Anatomy and Biology, U. Pa., 1920-21; dist. mgr. J. Lee Nicholson Inst., 1921; gen. mgr. Globe Labs., 1922-24; clin., pathologist, Phila. Gen. Hosp., from 1925. Mem. Am., Pa., Keystone vet. med. socs., Phila. Pathol. Soc. Internat. Assn. Vet. Med. Museums, Dallas and Ft. Worth Vet. Med. Assn., Tex. Vet. Med. Assn., Pa. Fish and Game Protect. Assn., Sigma Xi, Sigma Alpha, Alpha Psi. Author: Veterinary Post Mortem Technic, 1917. Translator: Mastitls of the Cow and Its Treatment (by Sven Wall), 1918. Contbr. numerous sci. articles on hematology; hemography in diagnosis, prognosis and treatment; hemography in the diagnosis of appendicitis; nonspecific immunotransfusion in treatment septicemia and typhoid fever. Deceased.

CROCKER, William, Am. botanist; b. Medina, O., Jan. 27, 1876; s. Charles David and Catherine C.; grad. Ill. State Normal U., 1898; A.B., U. Ill., 1902, A.M., 1903; Ph.D., U. Chgo., 1906; m. Persis D. Smallwood, Sept. 3, 1910 (dec. July 1948); children—John Smallwood, David Rockwell. Instr. biology No. Ill. State Normal Sch., 1903-04; asso. plant physiology U. Chgo., 1907-09, instr., 1909-11, asst. prof., later asso. prof., 1911-21; mng. dir. Boyce Thompson Inst. for Plant Research, Yonkers, N.Y., from 1921, also pres. Boyce Thompson Research Found.; Walker-Ames visiting prof. U. Wash., winter 1943; trustee Tropical Plant Research Found., 1927-43, acting dir., gen. mgr., 1931-43. Chmn. div. biology and agr. NRC, 1927-28. Fellow A.A.A.S. (vice chmn. Sect. G, 1925-26); mem. Bot. Soc. Am. (pres. 1924-25), Am. Chem. Soc., Am. Phytopathol. Soc., Am. Philos. Soc., Soc. Arts and Sciences (medalist, 1931), Am. Inst. N.Y. (gold medal, 1938). Sigma Xi, Phi Kappa Phi, Gamma Alpha. Author: Growth of Plants—Twenty Years' Research at Boyce Thompson Institute, 1948; (with Lela V. Barton) Twenty Years of Seed Research, 1948; (with Lela V. Barton) Seeds and Germination. Research on delayed germination in seeds, effect of noxious gases on plants, plant hormones, tropisms. Died Feb. 11, 1950.

CROFT, John, Brit. surgeon; b. Pettingoe, Sussex, Eng., Aug. 4, 1833; s. Hugh and Maria C.; after serving a short apprenticeship entered St. Thomas's Hosp., 1850; m. Annie Douglas, 1864; began as house surgeon St. Thomas's Hosp., later full surgeon; lectr. practical surgery and clin. surgery; cons. surgeon to numerous hosps.; fellow Royal Coll. Surgeons; Developed plaster of Paris casts for broken bones; one of 1st to adapt Lister's method of sterile treatment. Died Nov. 21, 1905.

CROFTON, John Wenman, physician; b. Dublin, Ireland, Mar. 27, 1912; s. William Mervyn and Mary (Abbott) C.; M.D., Cambridge U., 1947, M.B., B.Chir., 1937; m. Eileen Chris Mercer, Dec. 15, 1945; children—Richard, Patricia, Pamela, Alison, Ian. Mem. Tb Unit, Med. Research Council, 1947-49; sr. med. asst. Inst. Diseases Chest, Brompton Hosp., London, 1949-50; sr. lectr. medicine Postgrad. Med. Sch., London, 1951; prof. respiratory diseases Edinburgh (Scotland) U., 1952—, dean faculty medicine, 1964-66. Hon. cons. on diseases of chest to Regional Hosp. Bd. Scotland. Fellow Royal Coll. Physicians London, also of Edinburgh; mem. Assn. Physicians Gt. Britain and Ireland, Thoracic Soc., Brit. Tb Soc., Scottish Thoracic Soc., Am. Thoracic Soc. (corr. mem.). Publs. on modern methods of chemotherapy of tuberculosis; syndromes of pulmonary eosinophilia; physiology of bronchial muscle and bronchitis. Home: 7 Pentland Av., Colinton, Edinburgh 13. Office: City Hosp., Greenbank Dr., Edinburgh 10, Scotland.*

CROLL (CROLLIUS), Oswald, German physician, chemist; b. Wetter, Hesse, Germany, 1580; studied at Marburg, Germany, Heidelberg, Germany, Strasbourg (now in France), Geneva, Switzerland; 1st became physician to Prince Christian of Anhalt-Bernberg, later councillor to Emperor Rudolph II; author: Basilica chymica (major 17th century source of Paracelsian and chem. medical preparations), 1609. Described fulminating gold, silver chloride precipitated and fused, mercuric oxide, use of mercurous chloride as a purgative, preparation of calomel by precipitaton, antimonic acid, tin acetate, potassium sulphate, succinic acid; 1st description of calcium acetate. Died 1609.

CROME, Auguste-Frederick-William, German cartographer; b. Sengwarden, Aug. 6, 1753; student theology, Halle, Germany; became tchr. history, geography, Dessau, Germany, 1779; became tchr. polit. economy U. Giessen (Germany), 1787. Author: Handbuch der Statistick des Grosherzogthums Hessen, 1822; Europa's Producte, 1782; Die Staatsverwaltung Toscana's unter Leopold, 1795-97. Pioneer in econ. cartography; produced map Europe, 1782. Died Noedecheim, June 11, 1833.

CROMMELIN, Andrew Claude de la Cherois, Brit. astronomer; b. Cushendun, County Antrim, Feb. 6, 1865; s. Nicholas de la Cherois Crommelin; B.A., D.Sc., Oxford; ed. Marlborough Coll., Trinity Coll., Cambridge; m. Letitia Noble, 1897; 1 son, 1 dau. Asst. master Lancing Coll., 1889; asst. Royal Obs., Greenwich, 1891-1927; mem. eclipse expdns. Brit. Astron. Assn., 1896-1900, 05, expdn. to Brazil, 1919. Fellow Royal Astron. Soc. (council 1906, sec. 1917-

23, pres. 1929-30), Brit. Astron. Assn. (council 1896-1939, pres. 1904-06). Co-author: Comets, 1937; also numerous articles. Investigated (with P. H. Cowell) motion of Halley's Comet from 240 B.C. to A.D. 1910. Died London, Eng., Sept. 20, 1939.

CROMPTON, Alfred Walter, paleontologist; b. Durban, S. Africa, Feb. 21, 1927; s. William Lister and Grace (Bishenden) C.; B.S., U. Stellenbosch, 1947, M.S., 1949, D.Sc., 1951; Ph.D., U. Cambridge (Eng.), 1953; m. Mary Anne Oosthuisen, Feb. 20, 1954; children—Peter, Mary Jane, John. Curator fossil vertebrates Nat. Mus. S.Africa, 1954-56; dir. S.African Mus., 1956-64; dir. Peabody Mus. Natural History, prof. biology, geology Yale, New Haven, 1964—. Fellow London, S.African zool. socs.; mem. Soc. Vertebrate Paleontology, S.African Parliamentary and Sci. Soc. (past chmn.), S.African Assn. Advancement Sci., Sigma Xi. Publs. on problems of origin of mammals and dinosaurs and evolutionary changes during Triassic Period. Home: 202 Prospect St., New Haven, Conn. 06511. Office: Peabody Mus., Yale, New Haven 06520.*

CROMPTON, Rookes Evelyn Bell, Brit. engr.; b. May 31, 1845; s. Joshua S. Crompton; m. Elizabeth Gertrude Clarke, 1871; 2 sons, 3 daus. Founder Crompton and Co., 1878; recipient Faraday medal; fellow Royal Soc. 1933; pres. Instn. Elec. Engrs., Instn. Automobile Engrs.; founder mem. Royal Automobile Club. Author: Reminiscences, 1928. Inventor steam car, circa 1858, introduced it to army in India, 1864; experimented with engine, Eng., 1871, successfully built it in India; built (with Emil Burgin) generator, arc lamp, dynamo with constant tension, direct current dynamo for electric furnaces with 60 volts and 5000 amperes, also used enamel as isolator; built (with Fleming) measuring instruments, a bus powered by accumulators; contbd. to devel. of English tank, 1917; pioneer in electric lighting in Eng. Died Ripon, Eng., Feb. 15, 1940.

CROMPTON, Samuel, English inventor; b. Firwood nr. Bolton, Eng., Dec. 3, 1753; m. dau. of W. I. mcht., Feb. 1780; 1 son, John Crompton; began as mill hand; fiddler Bolton Theatre; invented spinning mule which was 1st machine to reproduce action of left arm, finger and thumb of spinner, 1799; his machine laid basis for muslin manufacture in Eng.; financially unable to patent spinning mule; sold his rights to it. Died Bolton, June 26, 1827.

CROMWELL, Norman Henry, Am. chemist, educator; b. Terre Haute, Ind., Nov. 22, 1913; s. Henry and Ethyl (Harkelroad) C.; B.Ch.E., Rose Poly. Inst., 1935; Ph.D. in Chemistry, U. Minn., 1939; m. Grace Newell Meeker, Jan. 29, 1955; children—Christopher Newell, Richard Earl. Chem. engr. Comml. Solvents Corp., Terre Haute, 1935; teaching asst. U. Minn., 1935-39; research chemist Union Oil Co., Wilmington, Cal., 1938-39; faculty U. Neb., Lincoln, 1939—, prof. chemistry, 1948—, Wilson prof., 1960—, chmn. dept., 1964—; cons. USPHS, Philip Morris, Inc. John Simon Guggenheim Meml. fellow, 1950, 1958; Fulbright Advanced Research scholar, 1950. Mem. Am. Chem. Soc., Chem. Soc. London, Am. Assn. U. Profs., Sigma Xi (nat. lectr. 1964), Phi Lambda Upsilon, Sigma Tau, Alpha Chi Sigma, Gamma Alpha. Contbr. chpt. Heterocyclic Compounds, vol. 6, 1957. Research, publs. on elimination reactions, synthesis of three and four. membered ring heteroegelies, synthesis of chemotherapeutic and carcinogenic compounds. Home: 2417 S. 70th St., Lincoln, Neb. 68505.*

CRONIN, James W., Am. physicist; b. Chgo., Sept. 29, 1931; s. James F. and Dorothy (Watson) C.; A.B., So. Meth. U., 1951; A.M., U. Chgo., 1953, Ph.D., 1955; m. Annette Martin, Sept. 11, 1954; children—Cathryn, Emily. Research asso. Brookhaven Nat. Lab., 1955-58; faculty Princeton, 1958—, prof. physics, 1965—. Mem. Am. Phys. Soc. Contbr. articles to tech. jours. Research on hyperon decays; developed spark chambers; co-discoverer CP-noninvariance. Home: 248 Hartley Av., Princeton, N.J.*

CRONKITE, Eugene Pitcher, Am. med. researcher; b. Los Angeles, Dec. 11, 1914; s. Clarence E. and Anita (Pitcher) C.; A.B., U. Cal. at Los Angeles, 1935; M.D., Stanford, 1940; D.Sc., C.U. Post Coll., 1962; m. Elizabeth E. Kaitschuk, Aug. 17, 1940; 1 dau., Christina. Med. officer USN, 1946-54; head hematology Naval Med. Research Inst., 1954-67; chmn. med. dept. Med. Research Center, Brookhaven Nat. Lab., Upton, L.I., N.Y., 1967—, also head exptl. pathology, asso. med. researcher in atomic bomb field tests, dir. study case of human beings exposed to fallout. Investigation of hematologic effects of ionizing radiations. Home: 5 Ivy Lane, Strong's Neck, Setauket, N.Y. Office: Med. Research Center, Brookhaven Nat. Lab., Upton, L.I., N.Y.

CRONSTEDT, Baron Axel Fredrik, Swedish mineralogist; b. Stroepsta, Södermanland, Sweden, Dec. 23, 1722; studied math. and phys. scis.; metallurgist Bur. Mines. Mineral (cronstedtite) named in his honor. Mem. Acad. Scis., Stockholm, Sweden. Author: Essay on Mineralogy or a Classification of Minerals, 1758; Tables, Collections and Trees of Metallurgy, pub. 1771. Discovered nickel, 1751, and zeolite; developed classification of minerals based on chem. compositions; introduced blowpipe for study of minerals. Died Saters, Stockholm, Aug. 19, 1765.

CROOK, Welton Joseph, metall. engr.; b. Melbourne, Australia, Feb. 1, 1886; s. William Joseph and Grace (Cooper) C.; student Scotch Coll., Melbourne, 1898-1904, Melbourne Sch. Mines, 1906-07; A.B. in Mining and Metallurgy, Stanford, 1911, Met.E., 1919; Dr.Ing., Scoala Poletechnica Regele Carol, Bucarest, Rumania, 1936; m. Maud L'anfipere, May 30, 1911. Came to the United States of America, 1908, naturalized, 1927. Began as chemist, assayer Dept. Mine and Water Supply, Victoria, Australia, 1907-08; chief chemist Pacific Coast Steel Co., 1911-13, 17-19, metallurgist, 1917-21, chief insp., 1919-20; prof. metallurgy Stanford, 1921-51, prof. emeritus, 1951—; prof. metallurgy S.D. State Sch. Mines, 1913-17; cons. metall. engr., 1946—. Cooperating analyst U. S. Bur. Standards, 1920-28, cons. engr., 1922. Mem. Am. Inst. Mining and Metall. Engrs. (Legion of Honor 1963), Am. Soc. Metals, N.Y. Acad. Scis., Sigma Xi, Tau Beta Pi. Author: Tables of Arc Spectrum Lines, 1934; Metallurgical Spectrum Analysis, 1935; Recherches Expérimentales sur la Constitution Mineralogique et sur l'action chemique des Scories de l'acier, 1936; Abacus Arithmetic, 1958; The Vreeland Spectrograph, 1967; also articles. Patentee heat treatment carbon steels, heat treatment cylinders for cannon recoil mechanisms, reforging steel castings for gun carriages, super hard metals for oil well core drills. Home: 675 San Juan St., Stanford, Cal.*

CROOKE, Arthur Carleton, English endocrinologist; b. Frodingham, Eng., May 5, 1905; s. Arthur and Lily (Hudson) C.; ed. Queens' Coll., Cambridge (Eng.) U., 1924-27, London Hosp., 1927-30; m. Nancy Mary Pelham Morter, Mar. 11, 1935; children—Christopher Arthur Pelham, Mary Rose (Mrs. Roger Bowen). Staff, London Hosp., 1930-33; Grocers' Co. Med. Research scholar Bernhard Baron Inst. Pathology, London Hosp., 1933-35, asst. to med. unit, 1935-48; dir. dept. clin. endocrinology United Birmingham (Eng.) Hosps., 1948—; hon. reader medicine Birmingham U., 1948—. Recipient Raymond Horton Smith prize, Cambridge U., 1935, Triennial Liddle prize, 1942. Fellow Royal Soc. Medicine; mem. Soc. for Endocrinology, Soc. for Study Fertility. Contbg. author: Modern Trends in Paediatrics, 1950; Modern Trends in Endocrinology, 1966. Research, numerous publs. on pituitary gland changes in Cushing's syndrome called Crooke's cells and in Addison's disease called Crooke-Russell cells, constn. of human gonadotrophins and their therapeutic use. Home: Cherryholme, Far Forest, nr. Kidderminster, Worcestershire, Eng. Office: Dept. Clin. Endocrinology, Birmingham and Midland Hosp. for Women, Showell Green Lane, Birmingham 11, Eng.*

CROOKES, Sir William, English chemist, physicist; b. June 17, 1832; ed. Royal Coll. chemistry; hon. doctorates, Birmingham, Oxford U., Cambridge U., Sheffield, Durham; m. Ellen Humphrey, 1856; 4 sons, 1 dau. Became lecturer, chemistry Tng. Coll., Chester, 1855; founder, editor Chem. News. Recipient medal Internat. Exhbn., 1862; Gold medal French Acad. Scis., 1880; medal Elec. Exhbn., Paris, 1881; Fergusson Gold medal Soc. Arts, 1885, Albert Gold medal, 1899; medal Exposition Universelle, Paris, 1889; Elliott Cresson Gold medal Franklin Inst., Phila., 1912; Gold medal Soc. Chem. Industry, 1912; Fellow Royal Soc., 1863 (Pres., 1913-15, fgn. sec., 1908-12, recipient Royal, Copley, Davy medals, Bakerian lectr.); pres. Chem. Soc., Brit. Assn., Inst. Elec. Engrs., Soc. Chem. Industry; hon. mem. Royal Philos. Soc. Glasgow, Royal Soc. New S. Wales, Pharm. Soc., Chem. Metall. and Mining Soc. S. Africa, Am. Chem. Soc., Am. Philos. Soc., Royal Soc. Sci. Uppsala, Deutsche Chem. Gesell. Berlin, Psychol. Soc. Paris, Antonio Alzate Sci. Soc. Mexico, Sci. Soc. Bucharest, Reg. Accad. Zelanti; fgn. mem. Acad. Lincei, Rome; Royal Swedish Acad. Sci.; corr. French Acad. Scis.; corr. mem. Bataafsch Genoots., Rotterdam, Soc. for Encouragement Industry, Paris; fgn. asso. Nat. Acad. Scis., Washington. Author: Select Methods in Chemical Analysis, 4th edit., 1905; Manufacture of Beetroot-Sugar in England, 1870; Handbook of Dyeing and Calico-Printing, 1874; Dyeing and Tissue Printing, 1882; (with Ernst Rohrig) Kerl's Treatise on Metallurgy, 1868; Wagner's Chemical Technology; Auerbach's Anthracen and its Derivatives, 2d edit., 1890; Ville's Artificial Manures, 3d edit., 1909; A Solution of the Sewage Question; The Profitable Disposal of Sewage; The Wheat Problem, 1899, 3d edit., 1917; Diamonds, 1909. Discoverer of selenocyanides; discoverer thallium 1861, repulsion resulting from radiation, 1873; inventor radiometer, 1875; Crookes tube; research on illumination of lines of molecular pressure, 1878, radiant matter, 1879, radiant matter spectroscopy, 1881, gadolinite, 1886, genesis of elements, 1887, electricity and wireless telegraphy, 1892; fixation of atmospheric nitrogen, 1898, the spinthariscope, 1903, spectacle glass. Died London, Apr. 4, 1919.

CROOKS, James, Scottish physician; b. Glasgow, Scotland, Mar. 22, 1919; s. William and Christina (Beaton) C.; M.B., Ch.B., U. Glasgow, 1951, M.D. with 1st class honors, 1959; m. Elizabeth Newton MacDonald, Oct. 11, 1940; 1 son, George William. House officer Royal Infirmary and So. Gen. Hosp., Glasgow, 1951; sr. house officer med. unit. So. Gen. Hosp., Glasgow, 1952-54; faculty dept. medicine U. Glasgow, 1954-60; sr. med. registrar, 1955-60; faculty U. Aberdeen (Scotland), 1960—, reader in therapeutics, 1965—. Hon. cons. physician N.E. Regional Hosp. Bd., 1960—. Recipient Bellahouston medal for thesis, 1956. Fellow Royal Faculty Physicians and Surgeons Glasgow, Royal Coll. Physicians Edinburgh, Royal Coll. Physicians Glasgow; mem. Royal Coll. Physicians London, Scottish Soc. for Exptl. Medicine (local sec. 1965—), Scottish Soc. Exptl. Medicine, Med. Research Soc., Assn. Physicians Gt. Britain and Ireland, Soc. for Endocrinology, European Thyroid Assn., Brit. Pharmacological Soc. Research, publs. on devel. a statis. approach to diagnosis of thyroid disease, aspects of iodine metabolism in man and animals and the biol. effects of irradiation on thyroid gland. Home: 162 Broomhill Rd., Aberdeen, AB1, 6HY, Scotland.*

CROOKSHANK, Francis Graham, Brit. physician; b. Wimbledon, Eng., 1873; s. W. H. F. and Alice Harriet (Burton) C.; ed. Univ. Coll. and Hosp., London; M.B., U. London, 1895, M.D., 1896; m. 2d, Hélène Bonhoure, 1925. House physician Univ. Coll. and Brompton Consumption Hosp.; asst. med. supt. Northampton County Asylum; med. supt. Isolation Hosp., also med. officer health, Barnes, Eng.; asst. physician Belgrade Hosp. for Children; physician Hampstead Gen. and N.W. London Hosp., St. Mark's Hosp., French Hosp., London, Prince of Wales' Gen. Hosp., Tottenham; Bradshaw lectr., 1926. Fellow Royal Soc. Medicine, Royal Coll. Physicians (also licentiate); mem. Royal Coll. Surgeons, Med. Soc. Individual Psychology (editor), French Soc. History Medicine, French Soc. Morphology, Paris. Author: Invididual Sexual Problems, 1932; Individual diagnosis, 1930; Epidemiological Essays, 1930; Diagnosis and Spiritual Healing, 1927; Migraine, 1925; Influenza: Essays by Several Authors, 1922; Essays and Clinical Studies, 1911; introductions to Masson-Oursel's Comparative Philosophy, Cumston's History of Medicine, 1927, Blondel's Morbid Conscience, 1928, Adler's Problems of Neurosis, 1929, Wexberg's Individual Psychology and Sex, Prinzhorn's Psychotherapy, 1932; also articles. Died Oct. 27, 1933.

CROONE (or CROUNE), William, Brit. physician; b. London, Eng., Sept. 15, 1633; ed. Merchant Taylors Sch.; fellow Emmanuel Coll., Cambridge (Eng.) U.; M.D., 1662; m. Mary Lorymer. Prof. at Cambridge U.; later became physician in London; apptd. prof. rhetoric Gresham Coll., 1659; founder Croonian Lecture. Fellow Royal Soc. (founding), 1663. Author: De ratione motus musculorum, 1664. Thought muscular motions were produced by Descartes' vital spirits flowing along the nerves and mixing with the spirit in the blood. Died Oct. 12, 1684.

CROS, Charles, French inventor; b. Fabrezan, France 1842; 1 son, Guy Charles; discovered (simultaneously with Ducos du Hauron) trichromatic principle of color photography; inventor phonograph before Edison (shown as paleophone 1877). Author: Le coffret de santal, 1873. Died 1888.

CROSBY, Dixi, Am. surgeon; b. Sandwich, N.H., Feb. 7, 1800; s. Asa and Betsy (Holt) C.; studied medicine under father; grad. Dartmouth, 1824; m. Mary Jane Moody; children—Albert, Alpheus. Practiced medicine (with father), Gilmanson, N.H., 1824-34; apptd. prof. surgery Dartmouth Coll., Hanover, N.H., 1837. Reduced metacarpo-phalangeal dislocation, 1824; 1st to amputate arm, scapula, 3 quarters of clavicle in single operation, 1836; 1st to open hip joint abscess. Died Hanover, 1873.

CROSBY, Elizabeth Caroline, Am. anatomist; b. Petersburg, Mich., Oct. 25, 1888; d. Lewis Frederick and Frances (Kreps) Crosby; B.S., Adrian Coll., 1910, Sc.D., 1939; M.S., U. Chgo., 1912, Ph.D., 1915; Sc.D., Marquette U., 1957; Sc.D., Denison U., 1959; M.D. (hon.), U. Groningen, The Netherlands, 1958; LL.D., Wayne State U., 1958; 1 dau., Kathleen (Mrs. Thomas Palmer) (foster child). Faculty, U. Mich., Ann Arbor, 1920—, prof. anatomy, 1936-59, prof. emeritus, 1959—; cons. neurosurgery U. Mich. Hosp., 1959—; prof. emeritus anatomy U. Ala. Med. Center, Birmingham, 1964—; Henry Russell lectr., Mich., 1946; Mellon lectr. U. Pitts., 1951; Ferris lectr. Yale, 1958; lectr. Montreal Neurol. Inst., 1959, Mayo Clinic, 1963; Distinguished lectr. Tulane U., 1964; Marion Hines lectr. Emory U., 1966. Recipient Achievement award Am. Assn. U. Women, 1950, Distinguished Faculty Achievement award U. Mich., 1956, Galen award for preclin. med. teaching, 1958. Mem. Am. Anat. Assn., Anat. Soc. Gt. Britain and Ireland, Am. Assn. Neuropathologists, Am. Neurol. Assn., Assn. Research Nervous and Mental Diseases, N.Y. Acad. Sci., Sigma Xi, Alpha Omega Alpha, Phi Kappa Phi. Author: (with others) The Comparative Anatomy of the Nervous System, 1936; (with others) The Correlative Anatomy of the Nervous System, 1962; (with others) Correlative Neurosurgery, 1955. Contbr. numerous articles to profl. jours. Studies, publs. on forebrains of reptiles and birds; olfactory system and amygdala in submammals and mammals; correlation between anat. lesions and clin. manifestations in nervous system disorders. Home: Michigan League, Ann Arbor, Mich. 48103; also 222 Poinciana Dr., Birmingham, Ala. 35209.*

CROSBY, William Otis, Am. geologist; b. Decatur, O., Jan. 14, 1850; s. Francis William and Hannah Everett (Ballard) C.; B.S., Mass. Inst. Tech., 1876; m. Alice Ballard, Sept. 4, 1876. Was engaged in mining in N.C. and Colo.; mem. faculty Mass. Inst. Tech., 1883-1907, ret. Cons. geologist, chiefly engaged in original research; spl. geologist, U. S. and N.Y. surveys, U. S. Reclamation Service, U. S. Army Engrs., Met. Water Bd. of Mass., Bd. Water Supply, N.Y.; geologic investigations in connection with engring. projects in U. S., Can., Mexico, Alaska and Spain. Author: Contributions to the Geology of Eastern Massachusetts, 1880; Common Minerals and Rocks, 1881; Guide to Mineralogy, 1886; Tables for the Determination of Common Minerals, 1887; Guide to Dynamical Geology and Petrography, 1892; Geology of the Boston Basin, 1893-94; Geology of Long Island. Died Dec. 31, 1925.

CROSKEY, John Welsh, Am. ophthalmologist; b. Phila., Jan. 26, 1858; s. Henry and Ann (Dunnohew) C.; student Swarthmore Coll., 1886-87; M.D., Medico-Chirurg. Coll., Phila., 1889; certificate of proficiency, Phila. Sch. Anatomy, 1889; m. Elisabeth Estes Browning, Dec. 15, 1880 (dec.); children—Henry B., Elisabeth B. (Mrs. L. E. Bailey, dec.), Marion L., John Welsh Croskey, Jr. (dec.), m. 2d, Marie Lanche Bretschneider, Jan. 21, 1939; stepchildren—Gordon Bretschneider, Louis Lanche Bretschneider. Began as chief asst. to surg. clinic Medico-Chirurgical Coll., 1889, later lectr. on minor and operative surgery; asst. surgeon Wills Hosp., 1891-97, surgeon, 1897-1902; cons. ophthalmic surgeon to George Nugent Home for Baptists, 1899; ophthalmic surgeon Phila. Gen. Hosp., apptd. 1900, cons. surgeon, apptd. 1925; also lectr. tng. Sch. for Nurses; ophthalmic surgeon Samaritan Hosp., Annie M. Warner Hosp., 1902-05; prof. ophthalmology, laryngology and otology, Temple U., 1902-05, etc.; acting asst. surgeon USPHS; ophthalmologist Home of Merciful Saviour for Crippled Children. Formerly editor and owner Internat. Med. Mag.; editor Medico-Chirurg. Jour. Fellow Am. Acad. Ophthalmology and Oto-Laryngology, A.M.A.; mem. Pa., Phila. County (sec. bd. censors) med. socs., W. Phila. Med. Assn. (ex-pres.), Acad. Natural Sci. Phila., Am. Med. Authors Assn., Alumni Assn. Medico-Chirurg. Coll. (ex-pres.). Author: Dictionary of Ophthalmic Terms, 1907; History of Blockley; Anatomy and Physiology of the Eye and Its Appendages. Died July 30, 1951.

CROSS, Charles Frederick, English chemist; b. Brentford, Middlesex, Dec. 11, 1855; s. Charles James and Ella (Mendham) C.; B.Sc., King's College, London, 1878; studied Univ. and Polytechnikum, Zurich, Owens College, Manchester; m. Edith Vernon Stainforth, 1890; 2 sons, 1 dau. Engaged in research, Barrow-in-Furness, 1879-81; with Bevan, Jodrell Lab., Kew; formed Cross & Bevan Research and Consulting Chemists, 1885; formed various companies for development of viscose process, 1893-1900. Awarded medal, Soc. of Chem. Industry, 1916; research medal, Worshipful Co. of Dyers, 1918; Perkin Medal, Soc. of Dyers and Colourists (pres., 1918-20), 1924. Fellow, Royal Soc., 1917. Author: Cellulose (with E. J. Bevan), 1895; Researches on Cellulose (with Bevan and Dorée, 4 vol.), 1895-1921; Textbook of Paper Making, 1918. Devised viscose process of treating cellulose which resulted in production of artificial silk, 1892; work on development of viscose films and cellulose acetate, 1894; research on fibers and paper. Died Hove, England, April 15, 1935.

CROSS, Hardy, Am. civil engr.; b. Nansemond County, Va., Feb. 10, 1885; s. Thomas Hardy and Eleanor Elizabeth (Wright) C.; B.A., Hampden-Sydney Coll., 1902, B.S., 1903, Sc.D. (hon.), 1934; B.S. in Civil Engring., Mass. Inst. of Tech., 1908; M.C.E., Harvard, 1911; M.A. (hon.), Yale, 1937; D.Eng. (hon.), Lehigh U., 1937; m. Edythe Hopwood Fenner, Sept. 5, 1921. Instr. English, Hampden-Sydney Coll., 1902-03; instr. English and math. Norfolk Acad., 1903-06; engr. bridge dept. M.P. Ry., 1908-10; asst. prof. civ. engring., Brown U., 1911-18; gen. practice structural engring., 1918-21; prof. structural engring. U. Ill., 1921-37; prof. civil engring. Yale, 1937, former head dept., now prof. emeritus Strathcoma. Mem. Am. Soc. C.E. (Norman medal, 1933), Am. Ry. Engring. Assn., Am. Concrete Inst. (Wason medal 1936), Western Soc. Engrs., Conn. Soc. Civil Engrs. Am. Soc. Engring. Edn., Royal Soc. Arts, Am. Soc. Cons. Engrs. (Lamme medal, 1944), Am. Acad. Arts and Scis., Sigma Xi, Tau Beta Pi, Sigma Tau, Chi Epsilon, Omicron Delta Kappa. Author: Continuous Frames of Reinforced Concrete (with N. D. Morgan), 1932; also bulls. and tech. articles. Died Feb. 11, 1959.

CROSS, Ira Brown, Am. botanist, economist; b. Decatur, Ill., Dec. 1, 1880; s. Bradford and Orietta (Clemons) C.; A.B., U. Wis., 1905, M.A., 1906, LL.D., 1951; Ph.D., Stanford, 1906; LL.D., U. Cal., 1957; m. Blanche Julia Mobley, July 11, 1911; children—Ira Brown, Carleton Parker (dec.). Faculty, Stanford, 1906-14; sec. Cal. Indsl. Accident Commn., 1913-14; faculty U. Cal. at Berkeley, 1914—, Flood prof. econs., 1919-51, emeritus, 1951—. Mem. Am. (past v.p.) Western (past pres.) econ assns., Nat. Chrysanthemum Soc. (asso. editor bull. 1955—, highest award 1963), Pan Xenia, Phi Beta Kappa, Beta Gamma Sigma, Delta Sigma Rho, Alpha Kappa Psi. Author: Essentials of Socialism, 1912; Foreign Exchange, 1923; Economics, 1931; Money and Banking, 1931; History of the Labor Movement in California, 1934; also numerous articles on econs. and chrysanthemums. Editor: Frank Roney, Irish Rebel and Labor Leader, 1931. Research on hybridization of chrysanthemums. Home: 1454 LeRoy Av., Berkeley, Cal. 94708.*

CROSS, James Cecil, Am. cytologist; b. Sallisaw, Okla., Dec. 27, 1895; s. William Francis and Amanda S. (Foust) C.; A.B., Southwestern U., 1924; M.A., U. Tex., 1928, Ph.D., 1931; m. Opal Lema Smith, Aug. 18, 1934; children—Lowell Merlin, Evelyn Joy. Vis. prof. N. Tex. State Tchrs. Coll., Denton, 1931-32; asso. prof. S.W. Tex. State Tchrs. Coll., 1932-36; prof., head dept. biology Tex. Coll. Arts and Scis., 1936-48; prof., head dept. biology Tex. Technol. Coll., Lubbock, 1948-59, prof, 1959-66, prof. emeritus, 1966—. Fellow A.A.A.S., Tex. Acad. Sci., Acad. Sci. (Agra, India); mem. Am. Genetics Assn., Am. Soc. for Microbiology, Soc. for Study Evolution, Biophys. Soc. (charter), Am. Soc. Zoologists, A.A.A.S., Tex. Acad. Sci., Sigma Xi. Author: An Introduction to Biology, 1941; also numerous articles. Research on virus-like substance produced in vitro, artifically induced adenocarcinoma with glutathione, chromosomes various vertebrates. Home: 2430 20th St., Lubbock, Tex. 79411.*

CROSS, Paul Clifford, Am. phys. chemist; b. Bruin, Pa., July 19, 1907; s. Henry Anderson McLean and Catherine (Smith) C.; B.S., Geneva Coll., 1928, Sc.D., 1963; M.S., U. Wis., 1930, Ph.D., 1932; M.S. ad eundum, Brown U., 1942; Sc.D., Waynesburg Coll., 1963; m. Sara Groves Cross, June 18, 1932; children—Carroll Edward, Robert Henry, Beverly Jane, Elizabeth Anne. Alumni Research Found. fellow U. Wis., 1932-33; NRC fellow Cal. Inst. Tech., 1933-35; Carnegie Found. fellow Stanford, 1935-36, asst. prof. chemistry, 1936-38; asso. prof. chemistry Brown U., Providence, 1938-42, prof., 1942-49, dir. Metcalf research lab., 1946-49, chmn. dept. chemistry, 1947-49; exec. officer, dept. chemistry U. Washington Seattle, 1949-61; pres., chief exec. officer, trustee Mellon Inst., Pitts., 1961—. Dep. research dir. Underwater Research lab. Office Sci. Research and Devel., Woods Hole (Mass.) Oceanographic Instn., 1943-44, research dir., 1944-46; adj. prof. chemistry U. Pitts., 1962—; cons. various govt. agys. Dir. Oakland Corp., Pitts. Recipient Life G Alumni award Geneva Coll., 1958. Fellow Am. Phys. Soc., N.Y. Acad. Sci., Am. Acad. Arts and Sci.; mem. Am. Chem. Soc., A.A.A.S., Spectroscopy Soc. Pitts., Pa. Soc., Western Pa. Conservancy, Newcomen Soc. N.Am., Phi Beta Kappa, Sigma Xi, Phi Lambda Upsilon, Gamma Alpha, Alpha Chi Sigma. Clubs: University, Chemists, Duquesne, Rolling Rock (Pitts.); Chemists (N.Y.C.). Author: (with E. B. Wilson, Jr. and J. C. Decius) Molecular Vibrations, 1955; (with H. C. Allen, Jr.) Molecular Vib-Rotors, 1963; numerous sci. publs. Study of thermodynamics, spectroscopy and structure of molecules and free radicals; photochemistry; theory of infrared and raman vibrational spectra; molecular vib-rotors; explosion phenomena. Home: 1067 Blackridge Rd., Pitts. 15235. Office: Mellon Inst., 4400 5th Av., Pitts. 15213.

CROSS, Roy, Am. chemist; b. Ellis, Kan., Jan. 13, 1884; s. George Washington and Ada (Pendleton) C.; A.B., U. Kans., 1905; M.D., U. Med. Coll., Kansas City, 1908; m. Mary Forbes, Oct. 1, 1917. Teaching fellow chemistry U. Kan., 1905-06; tchr. chemistry U. Med. Coll., Kansas City, 1906-12; mgr. or pres. Kansas City Testing Lab., 1908-47; v.p., cons. Gasoline Products Co., 1922-26; pres. Silica Products Co., 1924-35; v.p. Cross Devel. Co., 1924-35; pres. Cross Engring. Co., Cross Devel. Corp., Cross Labs.; cons. chemist. Trustee, Midwest Research Inst., U. Kan. Research Found. Fellow A.A.A.S.; mem. Am. Inst. Chem. Engring., Am. Chem. Soc., Am. Concrete Inst., Am. Petroleum Inst., Am. Soc. for Testing Materials, Kansas City Engrs., Am. Forestry Assn., Phi Beta Kappa, Sigma Xi. Co-inventor of Cross cracking process and designer of approximately 200 refining plants for gasoline. Holder of about 100 U. S. patents. Author: Handbook of Petroleum, Asphalt and Natural Gas, 1931; Handbook of Bentonite, 1935; Random Recollections of a Chemist, 1941; From A Chemist's Diary, 1943; also bulletins on mineral waters, air conditioning, etc. Died Mar. 21, 1947.

CROSS, (Charles) Whitman, Am. geologist; b. Amherst, Mass., Sept. 1, 1854; s. Moses Kimball and Maria (Mason) C.; B.S., Amherst Coll., 1875, D.Sc., 1925; Ph.D., U. Leipzig, 1880; m. Virginia Stevens, Nov. 7, 1895; 1 son, Richard Stevens. Asst. geologist U. S. Geol. Survey, 1880-88, geologist, 1888-1925, chief sect. petrology, 1903-06. Mem. Nat. Acad. Scis. (treas. 1911-19), Geol. Soc. Am. (pres. 1918), Washington Acad. Scis., NRC (vice chmn. div. geology and geography 1918), Am. Philos. Soc.; fgn. mem. Geol. Soc. London. Author: (with Pirsson, Iddings, Washington) Quantitative Classification of Igneous Rocks, 1903; also numerous reports, maps and papers. Research in petrology, mineralogy, systematic petrography, geology of Colo.; mineral crossite named for him. Died Washington, D.C., Apr. 20, 1949.

CROSSAN, Donald Franklin, Am. plant pathologist; b. Wilmington, Del., Apr. 8, 1926; s. Samuel David and A. Bertha (Spinken) C.; B.S. with distinction in Plant Pathology, U. Del., 1950; M.S., N.C. State U., 1952, Ph.D., 1954; m. Ruth Hilda Swanson, June 11, 1948; children—Connie Christine, Donna Christine, Eric Richard. Faculty, U. Del., Newark, 1954—, asso. prof. plant pathology, 1961-66, asst. dean Coll. Agrl. Scis., 1965—, asst. dir. Del. Agrl. Expt. Sta., 1965—. Recipient Lindbach award for excellence in teaching U. Del., 1961. Mem. Am. Phytopath. Soc., A.A.A.S., Am. Inst. Biol. Scis., Sigma Xi, Phi Kappa Phi. Research, publs. on life cycles and control maj. fungal and bacterial pathogens of vegetable crops in Del., physiology of fungicidal action. Home: 67 Mercer Dr., Newark 19711.*

CROSSEN, Harry Sturgeon, Am. gynecologist; b. Centerville, Ia., Feb. 2, 1869; s. James and Sarah Affinity (Sturgeon) C.; M.D., Washington U., 1892; m. Mary Frances Wright, Mar. 28, 1895; children—Theodore Wright, Ruth Victoria (Mrs. Henry Spence Brookes, Jr.), Robert James, Virginia Mabel (Mrs. Charles McEwen Avery, Jr.), David Frederic. Intern, later asst. supt. St. Louis City Hosp., 1892-95; supt. and surgeon in charge St. Louis Female Hosp., 1895-99; began pvt. practice, St. Louis, 1899; instr. in gynecology, Washington U., 1901-21, prof. clin. gynecology, 1921-35, prof. emeritus, 1935-51; gynecologist, Barnes, St. Luke's, St. Louis Maternity hosps. Fellow A.M.A.; mem. Mo. Med. Assn. Am., St. Louis gynecol. socs. Author: Diseases of Women, 1907; Operative Gynecology, 1915; Gynecology for Nurses, 1927; Synopsis of Gynecology, 1931; Foreign Bodies Left in Abdomen, 1940. Died Mar. 10, 1951.

CROTEAU, John Tougas, Am. economist; b. Holbrook, Mass., Mar. 10, 1910; s. Narcisse L. and Mary (Tougas) C.; A.B., Holy Cross Coll., 1931; M.A., Clark U., 1932, Ph.D., 1935; LL.D., U. Moncton (Can.), 1956; m. Gertrude D. Gallant, June 2, 1936 (dec. Mar. 1961). Carnegie chair econs. and sociology Prince of Wales Coll. and St. Dunstan's Coll., Charlottetown, P.E.I., 1933-45, also mng. dir. Co-op. Union P.E.I.; faculty Xavier U., 1946-47, Cath. U. Am., 1947-53; faculty U. Notre Dame (Ind.), 1953—, prof. econs., 1956—. Cons. Bur. Fed. Credit Unions, 1956, Credit Union Nat. Assn., 1960-61, Nat. Credit Union Mgmt. Assn., 1957—; contbr. ways and means com. study on broadening tax base fed. internal revenue system, 1959; project dir. study credit by low income groups Social Security Adminstrn., 1962-63. Fellow Royal Econ. Soc.; mem. Am. Econ. Assn., Am. Finance Assn., Agrl. History Soc., Canadian Polit. Sci. Assn. (dir. 1946-47), Am. Assn. U. Profs. Author: Cradled in the Waves: The Story of a People's Cooperative Achievement, 1951; The Federal Credit Union: Policy and Practice, 1956; The Economics of the Credit Union, 1963; (with H. B. Chandler) A Regional Library and Its Readers, 1940. Devel. and testing of an econ. theory of co-op. credit. Home: 124 N. Eddy St., South Bend, Ind. 46617.*

CROUCH, James E(nsign), Am. ornithologist, zoologist; b. Urbana, Ill., Jan. 28, 1908; s. Harry Ensign and Mary Jane (Pierce) C.; B.S., Cornell U., 1930, M.S., 1931; Ph.D., U. So. Cal., 1939; m. Mary Vrooman Page, Nov. 28, 1931; children—Jeanette Elnor (Mrs. Alex Rigopoulos), James Page. Mem. faculty San Diego State Coll., 1932—, prof. zoology, 1940—, chmn. div. life scis., 1962—. Fellow San Diego Zool. Soc., San Diego Mus. Natural History; mem. Nat. Audubon Soc., Sigma Xi, Phi Sigma. Author: Anatomy of the Lower Chordates, 1960; Introduction to Human Anatomy, 1958; Functional Human Anatomy, 1965; Atlas of Cat Anatomy, 1968. Research, publs. on bird behavior, particularly of the roadrunner, cedar waxwing, and phainopepla. Home: 10430 Russell Rd., La Mesa, Cal. 92041. Office: San Diego State Coll., San Diego 92115.*

CROUCH, Marshall Fox, Am. physicist; b. St. Louis, Nov. 22, 1920; s. Marshall Choate and Edna (Fox) C.; B.S., U. Mich., 1941; postgrad. Washington U., St. Louis, 1946-47, Ph.D., 1950; postgrad. Harvard, 1947-48; m. Katherine Francis Carmickle, Jan. 30, 1949; children—Thomas, Michael, Kenneth, Katherine. Staff mem. Radiation Lab., Mass. Inst. Tech., 1941-43, Los Alamos Sci. Lab., 1943; prof. physics dept. Case Inst. Tech., Cleve., 1950—; vis. prof. U. Tokyo, 1956-57. Dep. sci. attache U. S. Embassy, Tokyo, 1959-61. Mem. Am. Phys. Soc. Research, publs. on muons and neutrinos in cosmic radiation, electron-neutron interaction, neutron-proton capture cross section. Home: 35800 Chardon Rd., Willoughby Hills, O. 44094. Office: 10900 Euclid Av., Cleve. 44106.*

CROUNSE, Robert Griffith, Am. physician; b. Albany, N.Y., Mar. 23, 1931; s. Kenneth Eugene and Hilda (Leroy) C.; B.S., Yale, 1951, M.D., 1955; m. Marion Conrad, Feb. 11, 1955; children—Richard Conrad, Donald Albert. Clin. fellow dermatology Yale, 1956; practice medicine, specializing in dermatology, Miami, Fla., 1959-64, Balt., 1964—; clin. asso. dermatology Nat. Cancer Inst., NIH, 1957-58, spl. research fellow, 1959-60; asst. prof. dermatology Sch. Medicine U. Miami, 1961-64, asso. prof., 1964; asso. prof., head div. dermatology Sch. Medicine Johns Hopkins, 1964-67; prof. dermatology and of biochemistry, research director of department dermatology Medical College of Georgia, Augusta, 1967—. Consultant to Clin. Center NIH, 1965—. Recipient Research Career Devel. award Nat. Inst. Arthritis and Metabolic Diseases, NIH, U. Miami, 1962-64. Candidate Markle scholar U. Miami, 1963. Mem.

Soc. Investigative Dermatology, Am. Fedn. Clin. Research, Am. Acad. Dermatology. Asso. editor: Jour. Investigative Dermatology, 1966——. Research, publs. on cutaneous pharmacology of antifungal and antibacterial antibiotics, mechanisms of hair growth, biochemistry of unique fibrous ectodermal protein, keratin.*

CROUSAZ, Jean-Pierre de, see De Crousaz, Jean-Pierre.

CROUZON, Octave, French neurologist; b. France, 1874; student Faculty Medicine, Paris, France; apptd. chief clinic and lab. Hôtel de Dieu, Paris, 1906; became physician Sch. Nursing, Salpetrière Hosp., 1908; apptd. to chair med.-social welfare Paris Faculty, 1937. Mem. French Soc. Neurology (pres.), Acad. Medicine, Research and numerous publs. on hereditary nervous diseases; described craniofacial dysostosis (Crouzon's diseases), 1912. Died Sept. 16, 1938.

CROVA, André Prosper, French physicist; b. Perpignan, France, Dec. 3, 1833; aggregation in sci. Coll. Perpignan, 1859; prof. physics École d'agriculture; prof. physics Faculté des Sciences, Montpellier, France, 1870-94; dir. Inst. Physics, Montpellier. Mem. French Acad. Scis., 1886. Research on optics, electricity, solar radiation, hygrometry; built precise hygrometer with circulation of surrounding air. Died June, 1907.

CROW, James Franklin, Am. geneticist; b. Phoenixville, Pa., Jan. 18, 1916; s. H. Ernest and Lena (Whitaker) C.; B.A., Friends U., Wichita, Kan., 1937; Ph.D., U. Tex., 1941; m. Ann Crockett, Aug. 9, 1941; children—Franklin, Laura, Catherine. Faculty, Dartmouth, 1941-48; faculty U. Wis., Madison, 1948——, acting dean Sch. Medicine, 1963-65, prof., chmn. genetics, 1965——. Mem. Nat. Com. Radiation Protection. Mem. Genetics Soc. Am. (past pres.), Am. Soc. Human Genetics (past pres.), Nat. Acad. Scis. Author: Genetics Notes, 1950-66. Research, numerous publs. primarily in field of math. theory of population genetics. Home: 24 Glenway St., Madison, Wis. 53705.*

CROW, William Bernard, English biologist; b. Stratford, London, Eng., Sept. 11, 1895; s. William and Ellen (Moule) C.; student U. London, 1912-15, Ph.D., 1923; D.Sc., 1929; M.Sc., U. Wales, 1926; m. Alice Maude Whalley Welsh, Nov. 7, 1934; 1 step-dau., Pamela Margaret (Mrs. Alan Botterill); 1 dau., Angela Rosemary (Mrs. John LaThangue). Chemist, Ministry Munitions, 1915-18; lab. staff Royal Naval Exptl. Sta., Stratford, 1918; asst. lectr. botany U. Coll., Cardiff, Wales, 1919-23, lectr., 1923-28; head dept. biology Huddersfield Tech. Coll., U. Leeds (Eng.), 1928-38; sr. lectr. biology S.W. Essex Tech. Coll., 1938-44, Leicester (Eng.) Coll. Tech., 1945-60; head dept. biology Davies, Laing, & Dick Ltd., London, 1960——; examiner for degrees in sci. and medicine U. Wales, 1923-28; hon. prof. biology London Coll. Physiology, 1930; lectr. U. Leicester, 1957-59. Fellow Linnean Soc. London, Zool. Soc. London, Royal Soc. Medicine; mem. Soc. for Study Evolution. Author: Principles of Morphology, 1929; Mysteries of Ancients, 18 vols., 1941-45; A Synopsis of Biology, 1960. Research, numerous articles on description of new species and genera of microscopic organisms, evolution and early history mankind. Home: 78 Broadmead Rd., Woodford Green, Essex, Eng. Office: 10 and 11 Pembridge Sq., London W.2, Eng.*

CROWE, Samuel James, Am. physician; b. Washington, Va., Apr. 16, 1883; s. Walter Andrew and Flora (Thompson) A.; A.B., U. Ga., 1904; M.D., Johns Hopkins, 1908; postgrad. Freiburg, Berlin, Vienna; m. Susie Childs Barrow, June 1908; children—Samuel James, David Francis. Asst. surgeon in charge Hunterian Lab., 1908-09; asst. research surgeon Johns Hopkins Hosp., 1909-11, otolaryngologist in charge, 1919-52, founder research lab. for study deafness; adj. prof. otolaryngology, med. dept. Johns Hopkins U., emeritus prof., 1952——. Fellow. A.C.S., A.M.A., Assn. Pathology and Bacteriology, Laryngology Assn.; pres. Otolaryn. Soc., 1935. Contbr. to med. jours. Research on cause and treatment of impaired hearing, radium treatment of nasopharynx and aerotitus; 1st (with H. W. Cushing and J. Homans) to demonstrate relationship between pituitary gland and reproductive system, 1910.

CROWELL, Albert Dary, Am. physicist; b. Dover, N.H., Feb. 12, 1925; s. Milton Frederick and Esther (Dary) C.; Sc.B. summa cum laude, Brown U., 1946, Ph.D., 1950; M.S., Harvard, 1947; m. Janet Louise Wright, June 21, 1947; children—Judith Ann, Susan Wright, Cynthia Dary. Faculty, Amherst (Mass.) Coll., 1950-55, asst. prof., 1953-55; faculty U. Vt., Burlington, 1955——, prof., chmn. dept. physics, 1961——. Mem. Am. Phys. Soc., Am. Assn. Physics Tchrs., Sigma Xi. Author: (with D. M. Young) Physical Absorption of Gases, 1962; also articles. Research on interaction gas molecules with solid surfaces. Home: 30 Warner Av., Essex Junction, Vt. 05452. Office: Williams Sci. Hall, U. Vt., Burlington, Vt. 05401.*

CROWELL, John C(hambers), Am. geologist; b. State College, Pa., May 12, 1917; s. James W. and Helen H. (Chambers) C.; B.S. in Geology, U. Tex., 1939; M.A. in Meteorology, U. Cal. at Los Angeles, 1946, Ph.D. in Geology, 1947; Sc.D. honoris causa, U-Louvain, Belgium, 1966; m. Betty Marie Bruner, Nov. 22, 1946; 1 dau., Martha Lynn. Jr. geologist Shell Oil Co., Inc., 1941-42; mem. faculty U. Cal. at Los Angeles, 1947-68, prof. geology, 1960-68, chmn. dept., 1957-60, 63-64; faculty U. Cal. at Santa Barbara, 1968——. Fellow Geol. Soc. Am.; mem. Am. Assn. Petroleum Geologists, Am. Geophys. Union, A.A.A.S., Am. Inst. Profl. Geologists. Spl. research structural geology, tectonics, interpretation sedimentary rocks, studies San Andreas fault system, tectonics Cal.*

CROWELL, Luther Childs, Am. inventor; b. West Dennis, Cape Cod, Mass., Sept. 7, 1840; s. Francis B. Crowell; ed. West Dennis, Pine Grove Sem., Harwich, Mass., 1844-56, Pierce's Acad., Middlebury, Mass., 1856-57; m. Mrs. Margaret D. Howard, Aug. 18, 1863. Invented an aerial machine, 1860; invented and patented metallic tie paper bag, 1867; square bottomed grocers' paper bag, also machine for making same, 1872; also (placed in Boston Herald, 1873) first mechanism for associating webs of paper in printing, by which the multiple newspaper is produced; also the supplement newspaper press, and the double and quadruple presses, pamphlet printing and combined wire binding machines; received 280 patents from U. S. for printing machinery alone; engaged with R. Hoe & Co., 1879-1903. Died 1903.

CROWELL, Prince Sears, Jr., Am. biologist; b. Natick, Mass., May 2, 1909; s. Prince Sears and Ethel Iona (Moody) C.; A.B., Bowdoin Coll., 1930; M.A., Harvard, 1931, Ph.D., 1935; m. Villa Elizabeth Bailey, July 2, 1938; children—Persis Ann (Mrs. James A. Gessaman), Polly Foster (Mrs. Fred C. Feitler), Prince Sears III. Instr., Bklyn. Coll., 1935-36; instr., then asso. prof. Miami U., Oxford, O., 1936-48; mem. faculty Ind. U., 1948——, prof. zoology, 1962——; instr. invertebrate zoology Marine Biol. Lab., Woods Hole, Mass., 1936-41, 53, 64, trustee, 1958-66, 67——, mem. exec. com., 1962-65, 67——; dir. Ind. Sci. Talent Search, 1962, 63. Fellow A.A.A.S.; mem. Am. Soc. Zoologists (program officer 1957-59, mng. editor Jour. Am. Zoologist 1961-65), Am. Inst. Biol. Scis., Am. Ornithol. Union, Soc. Developmental Biology, Internat. Inst. Embryology, Sigma Xi. Research, publs. in exptl. morphology; regeneration, growth, aging, especially morphogenesis of colonial hydroids. Home: 1717 Ruby Lane, Bloomington, Ind. 47401.*

CROWELL, Thomas Irving, Am. chemist; b. Glen Ridge, N.J., July 9, 1921; s. Thomas Irving and Pauline Patten (Whittlesey) C.; B.S., Harvard, 1943; A.M., Columbia, 1947, Ph.D., 1948; m. Mary Miller Wheat, Sept. 16, 1950; children—Leslie Adams, Mary Allison. Research asst. Manhattan Project, 1943-45; instr. math. Bard Coll., 1946-47; mem. faculty U. Va., 1948——, prof. chemistry, chmn. dept., 1957-62; research chemist E. I. du Pont de Nemours & Co., Inc., summer 1951. Bd. dirs. Camp Robin Hood, Sargentville, Me. Mem. Am. Chem. Soc., Va. Acad. Sci., Sigma Xi. Research in organic reaction kinetics. Home: 1884 Field Rd., Charlottesville. Va.

CROWTHER, Henry, Brit. naturalist; b. Leeds, Eng. 1848; s. John Crowther; M.S. M.Sc.; m. Martha Jane Clarke; 3 daus. Asst. curator, curator Leeds Philos. and Lit. Soc.; lectr. botany Leeds Mech. Inst.; lectr. geology and micros. sci. Stalybridge Mech. Inst.; curator mus., lectr. chemistry, botany, animal physiology, mineralogy Royal Inst. Cornwall, Truro; lectr. geology and mineralogy Mining Sch., Camborne; lectr. metal and coal mining Mining Sch., Chacewater; lectr. agr. Cornwall County Council; curator Mus. City of Leeds, also Abbey House Mus., Kirkstall. Fellow Zool. Soc., Royal Micros. Soc.; founder Leeds Nature Study Soc.; pres., a founder, hon. mem. Conchological Soc. Gt. Brit. and Ireland; asst. sec. Yorkshire Geol. and Poly. Soc. Author: Synopsis of British Butterflies; Pozo Stone Weather Letters, Truro Museum; General Guide, Leeds Museum; A Natural History of Leeds; Wild Nature of Countryside; Section Microscipical Investigations; Record of First Series of British Coal Dust Experiments. Died Nov. 29, 1937.

CROWTHER, James Arnold, Brit. physicist; b. Sheffield, Eng., Aug. 28, 1883; s. J. W. Crowther; ed. (scholar) St. John's Coll., Cambridge; M.A., Sc.D.; m. Florence Maud Billinger, 1912; 2 sons. Mackinnon student Royal Soc.; fellow St. John's Coll., Cambridge, 1909-12, demonstrator, lectr. physics Cavendish Lab., 1912-24, lectr. physics applied to med. radiology, 1921-24; prof. physics U. Reading, 1924-46, prof. emeritus, 1946-50. Fellow Inst. Physics (hon. sec. 1932-46, v.p. 1946-49); mem. Faculty Radiologists (hon.), Brit. Inst. Radiology (pres. 1936). Author: Ions, Electrons and Ionizing Radiations; Molecular Physics; Manual of Physics; Practical Physics; Principles of Radiology; Life of Michael Faraday. Editor: Handbook of Industrial Radiography, also numerous articles. Died Mar. 25, 1950.

CROZIER, Herbert William, Am. cons. engr.; b. San Francisco, June 28, 1875; s. William J. and Elizabeth (Mackeon) C.; student Cogswell Poly. Engring. Sch., San Francisco, 1891-92, Mechanics Inst. Schs., San Francisco, 1892-95; B.S., U. Cal., 1899; m. Elizabeth Hyde, June 9, 1904; children—Elizabeth (Mrs. Milton G. Mauer), Hallett; m. 2d, Mary E. Sevison, Feb. 22, 1925. Engaged in constrn. work, Pacific Electric Ry., between Vallejo and Napa, Stockton and Sacramento; with Ore. Electric Ry., 1902-1905; hydro-electric installation on Stanislaus River, Cal., 1905-09; hydro-electric exploration on Klamath River, Cal., 1912-15, same and devel. work, Hoh, Queets, Cowlitz, North and Columbia rivers, Wash., 1922; cons. engr., State of Nev. (mem. Ariz.-Nev. Engrs. Conf., Ariz., Nev. and Cal. Engrs. Conf., Boulder Canyon project), 1923-26; appraisal commr. Islais Creek Reclamation Dist., San Francisco, and cons. engr. Nev.-Colo. River Commn., 1927-28; v.p. Cal. Desert Products Co. Rep. Monterey Breakwater before Bd. Engrs. for Rivers and Harbors, Washington, 1929; West Coast Naval Airship Base, Sunnyvale, Cal., before House Naval Affairs Committee, 1930; Redwood Harbor project before U. S. dist. engrs., 1931. Cons. engr., Central Valley project, 1935; project mgr. Boulder-Pioche project, 1936-37. Died Apr. 14, 1939.

CRUCHET, Jean René, French physician, neurologist; b. Bordeaux, France, Mar. 21, 1875; student medicine Bordeaux, 1893, also several German univs.; M.D., 1902; m. Adelie Feytit; 5 children. Named prof. medicine Bordeaux, 1920; head physician to 288th regt. during World War I; vis. prof., La. where he lectured on medicine and physicians in French lit.; founder, Center for Med. Examination for Pilots, 1920, Center for Med. Examination for Autoists, 1930. Decorated Officer in Legion of Honor. Mem. Soc. Anatomy and Physiology, Soc. Medicine and Surgery Bordeaux, Acad. Beaux-Arts Bordeaux (became pres. 1937), Royal Soc. Medicine London. Author: Torticolis spasmoiques, 1907; la Tiquose, 1909; Traitement de la diphterie, 1928; l'Encéphalite épidémique ou Cruchet's disease, 1928; En Louisiane, légendes et realités, 1937; le Syndrome hysterique, 1951. First description of epidemic or lethargic encephalitis (Cruchet's disease), 1927. Died Bordeaux, Apr. 14, 1959.

CRUICKSHANK, Durward William John, Brit. chemist; b. London, Eng., Mar. 7, 1924; s. William Durward and Margaret (Meek) C.; B.Sc. in Engring., London, 1944; B.A. (wrangler in math. tripos), Cambridge (Eng.) U., 1949, M.A., 1953, Sc.D., 1961; Ph.D., U. Leeds (Eng.), 1952; m. Marjorie A. Travis, June 27, 1953; children—Helen Margaret, John Durward. Faculty math. chemistry U. Leeds, 1950-62, reader, 1957-62; prof. theoretical chemistry U. Glasgow (Scotland), 1962-67, U. Manchester (Eng.) Inst. Sci. and Tech., 1967——; fellow St. John's Coll., Cambridge U., 1953-56. Mem. Internat. Union Crystallography (chmn. computing commn. 1965——, treas. 1966——), Chem. Soc. (mem. council 1966-67). Research, numerous publs. on theory of accuracy and computing in crystal structure determination, vibrations in crystals, theory chem. bonding by 2d-row elements; determination of crystal structure of compounds of 2d row elements, electron diffraction by gases. Office: Chemistry Dept., U. Manchester Inst. Sci. and Tech., Sackville St., Manchester 1, Eng.*

CRUICKSHANK, Ian Alfred Murray, plant pathologist; b. Rangiora, New Zealand, Aug. 5, 1924; s. Oliver Desmond and Myrtle May (Byles) C.; M.Sc. with honors, U. New Zealand, 1948; D.Sc., U. Canterbury (New Zealand), 1964; m. Jean Margaret Morrison, Apr. 22, 1950; children—Joanna Robyn, Guy Murray, Louise Margaret. Field mycologist plant diseases div. Dept. Sci. and Indsl. Research, Lincoln, New Zealand, 1948-56; research scientist div. plant industry Commonwealth Council for Sci. and Indsl. Research Orgn., Canberra, Australia, 1956——, sr. prin. research scientist, 1966——. Nuffield Travelling fellow, 1953. Mem. Australian Inst. Agrl. Sci. Research, publs. on field crop diseases, isolation of pisatin and phaseollin, physiol. aspects of host-pathogen relationships, sporulation of Peronospora tabacina; devel. Phytoalexin theory of disease resistance in plants. Office: P.O. Box 109, Canberra, A.C.T., Australia.*

CRUIKSHANK, Ernest Wiliam Henderson, Brit. biochemist, physiologist; b. Edinburgh, 1888; s. George Hunter and Sarah (Henderson) C.; M.B., Ch.B., Marischal Coll. Aberdeen U., 1910, M.D., 1920; D.Sc., Univ. Coll., London U., 1926; Ph.D., King's Coll., Cambridge U., 1926; m. Bertha Christina Stevenson, 1930; 2 daus. Carnegie research fellow Univ. Coll., London, 1912-14; asso. in physiology Washington U. Sch. Medicine, St. Louis, 1919-20; prof. Physiology Peking (China) Union Med. Coll., 1920-24; Rockefeller Found. Travelling Research fellow, 1924-26; prof. physiology and biochemistry Prince of Wales Med. Coll., Patna, India, 12 1926-28; Sukhr Raj Ray reader natural sci. U. Patna, 1927; prof. physiology Dalhousie U., Halifax, N.S., Can., 1929-35; Regius prof. physiology Marischal Coll., U. Aberdeen, (Scotland), from 1935. Mem. sci. adv. com. Dept. Health, Scotland, 1944; mem. Nutrition sub-Com., 1940-48, chmn., 1944-48; temporary cons. WHO, Geneva, 1952, 53, 55. Fellow Royal Soc. Edinburgh; mem. Royal Coll. Physicians, London. Author: Practical Biochemistry for Students, 1928; The Value of Scientific Thought in the Advance of Modern Medicine, 1928; Food and Physical Fitness, 1938; Food and Nutrition, 1946, 2d edit., 1951; also papers. Address: Bramblehurst, 113 Anderson Dr., Aberdeen, Scotland.

CRUIKSHANK, William Cumberland, surgeon, chemist; b. Edinburgh, Scotland, 1745; ed. Edinburgh U.; M.A., Glasgow (Scotland) U., 1767, M.D. (hon.). 1 dau., Mrs. Honoratus Leigh Thomas. Became asst. to Dr. William Hunter, 1771; partner (with Hunter) Windmill St. Sch. Fellow Royal Soc., 1797. Author: The Anatomy of the Absorbing Vessels of the Human Body, 1786; also articles, tracts. Research on absorbent system; proved effluence of carbonic acid from skin,

1795; observed liberation of oxygen and hydrogen at poles of metals in water, also acid and ammonia formation from decomposing solutions of salts; analyzed alcohol and ether by exploding them with oxygen; produced muriatic acid by mixing hydrogen and chlorine gas; showed passage of fertilized rabbit ovum through uterine tube to uterus, 1797; traced lymph vessels from periosteum to cortex of bone, 1786. Died June 27, 1800.

CRULL, Harry Edward, Am. astronomer, mathematician; b. Chgo., Feb. 7, 1909; s. Eddy Roy and Janette (Ostrom) C.; A.B., U. Ill., 1930, M.A., 1931, Ph.D., 1933; m. Edna Hale, Sept. 3, 1932; children—Janet (Mrs. Harold Smashey), Royale (Mrs. Donald Boger), Harry Edward. Lectr., Adler Planetarium, Chgo., 1933-34; prof. math., astronomy Park Coll., Parkeville, Mo., 1934-47; head dept., prof. math., astronomy Butler U., Indpls., 1947-65, dir. U. Coll., 1948-54; J. I. Holcomb Obs., 1954-65; prof. astronomy Dudley Obs., State U. N.Y., Albany, 1965—. Study of fourth order geometrical surfaces.

CRUM, Walter, Scottish chemist; b. Glasgow, Scotland, 1796; s. Alexander C.; studied under Thomas Thomson; owner calico-printing works, Thornliebank nr. Glasgow; pres. Anderson's U., Glasgow. Fellow Royal Soc., 1844; mem. Philos. Soc. Glasgow. Research and publs. on colors; discovered lead dioxide test for manganese; copper peroxide; prepared pure nitric oxide; tested bleaching methods; prepared pure indigo blue; used castor oil in sulfated oil for dyeing; determined composition of gun cotton; supported theory that dye adheres to a porous fabric by capillary action. Died Ronken nr. Glasgow, May 5, 1867.

CRUMEYROLLE, Albert, French mathematician; b. Curemonte, France, Dec. 10, 1919; s. Jean and Marguerite (Tronche) C.; ed. Faculty of Sciences of Lyon, Paris, Clermont-Ferrand; m. Marie Pelissier, Mar. 4, 1945; children: Annette, Jean-Francois, Bernard, Pierre, Philippe. Prof. of secondary education; student of Prof. Lichnerowicz, Collège de France; taught modern algebra; now teaching differential geometry, algebra. Officer of Palmes académiques. Author: Notions fondamentalles d'algèbre moderne, 1967; pub. articles. Investigation of theory of relativity, differential geometry, groups and linear algebra. Home: 17 Lauragais, Toulouse, France. Office: 118 route de Narbonne, Toulouse 10, France.*

CRUMPTON, Charles Whitmarsh, Am. physician; b. Minden, La., Oct. 8, 1918; s. Joseph Rufus and Gladys (Whitmarsh) C.; B.S., Tulane U., 1939, M.D., 1942; m. Frances Mixon McInnis, June 12, 1942; children—Charles Whitmarsh, Susan Irene, Marc McInnis. Faculty, U. Pa., Phila., 1948-51; faculty U. Wis., Madison, 1951—, prof. medicine, 1962—. Mem. coms. NIH; cons. cardiology VA Hosp., Madison; gov. Am. Coll. Cardiology, 1966—; pres. Wis. Heart Assn., 1963-64. Mem. Am. Soc. Pharmacology and Exptl. Therapeutics, Am. Soc. Clin. Investigation, Am. Physiol. Soc., A.C.P. Research, numerous publs. on evaluation and application of drugs used in therapy of hypertension, coronary blood flow and myocardial metabolism including effects of drugs, cerebral blood flow and metabolism in hypertension, hemodynamics of circulation in health and disease. Home: 9 Cambridge Rd., Madison, Wis. 53704.*

CRUSCHMANN, Heinrich Jakob Wilhelm, German internist; b. Giessen, Germany, June 28, 1846; s. Johann Heinrich and Anna Maria (Wilhelm) C.; studied medicine under R. Leuckart, K. Eckhard, E. Seitz at Giessen U., 1863-68; M.D., 1868; m. Margarethe Lohde, 1872; 2 sons, Fritz, Hans; 1 dau. Asst. physician Rochusspital, Mainz, Germany, for 3 years; went to Berlin, Germany, 1871; joined faculty, 1875; directing physician Moabit Hosp., Berlin; named med. dir. municipal hosps. in Hamburg, Germany, 1879; founder model hosp. Eppendorfer Allgemeines Krankenhaus; prof. internal medicine Leipzig, Germany, 1888-1910. Publs. on discovery of spirals in expectorations of asthmatics (Cruschmann's spirals), 1882; described icing liver; invented troilcart for puncturing skin edema. Died Leipzig, May 6, 1910.

CRUSELL, Gustaf Samuel, physician; b. Sweden, 1810; 1st to use galvanism successfully in therapy; treated urethral stricture, circa 1841. Died 1858.

CRUSIO, Carlo, Italian physician; b. 1628; 1st to distinguish scleroderma from leprosy, ichthyosis, and other diseases; recorded 1754. Died 1700s.

CRUVEILHIER, Jean, French pathologist; b. Limoges, France, Feb. 9, 1791; s. Leonard and Anne (Reix) C.; studied Paris, (France; docteur en médecine, 1816; aggregation, 1823; m. Jenny Grellet des Prades; a son, Edouard. Became prof. at Paris, 1825; received chair path. anatomy founded by Dupuytren in 1835; physician, anatomist. Mem. French Acad. Medicine. Author: Essai sur l'anatomie pathologique, 1816; Anatomie pathologique, 1816; Anatomie pathologique du corps humain, 1828-42; Discours sur les devoirs et la moralité du médecin, 1836; Anatomie du système nerveux de l'homme, 1845; Traité d'anatomie pathologique générale, 1849-68. Described pyloric atenosis caused by hypertrophy of pyloric muscle, 1829-35, several cases of gastric ulcer (la maladie de Cruveilhier), 1830; 1st to describe disseminated sclerosis, also progressive muscular atrophy (sometimes called Cruveilhier's atrophy or paralysis). Died 1874.

CRUZ, Oswaldo, Brazilian physician; b. Sao Luiz do Parahitinga, Brazil, 1872; worked at Pasteur Inst., Paris, France, for two years; established pub. hygiene service under the Brazilian Govt.; began a mosquito-control program in 1900; cleared Rio de Janeiro and other Brazilian cities of yellow fever, smallpox, and bubonic plague; developed Manguinhos Inst. (now named after him) in Rio into important biol. research center. Studied paludism. Died 1917.

CRUZEN, Richard Harold, Am. engr.; b. Kansas City, Mo., Apr. 28, 1897; s. Nathaniel Greene and Mary Edna (Gearhart) C.; student Va. Mil. Inst., 1914-15, U. S. Naval Acad., 1916-19; grad. U. S. Naval War Coll., 1942; m. Margaret Beaty, May 16, 1940; 1 dau. (by previous marriage), Susan. Entered U. S. Navy, 1916, and advanced through the grades of rear adm.; now comdr. U. S. Naval Forces, P.I. Decorated Legion of Merit, Antarctic Expdn. medal. A foremost polar authority; with Adm. Byrd's expdn., 1947, discovered 2 "bases," one a region of ice-free lakes. Home: Sangley Point, Luzon, P.I.

CRYER, Matthew Henry, oral surgeon; b. Manchester, Eng., July 21, 1840; s. Henry and Elizabeth (Cookson) C.; grad. Phila. Dental Coll., 1876; M.D., U. Pa., 1877; m. Martha Gates Phillips, June 17, 1889. Came to U. S., 1851. Practiced in Phila., 1877-1921; prof. oral surgery, U. Pa., 1898-1921; vis. surgeon, Phila. Gen. Hosp.; cons. dental surgeon, Univ. Hosp. Author: Regional Anatomy, 1886; Studied of Internal Anatomy of the Face, 1901; Imperial Stereoscopic Anatomy of Head (with D. J. Cunningham and David Waterston), 1909. Wrote chapter on Extraction of Teeth, in American Text-Book of Operative Dentistry, and one on General Hygiene of Mouth, in Musser and Keller's Handbook of Practical Treatment. Died Aug. 12, 1921.

CSABA, György, Hungarian histologist; b. Törökszentmiklós, Hungary, May 31, 1929; s. József and Rozália (Pollák) C.; M.D., Med. U., Budapest, Hungary, 1953, C.Sc.M., 1957; m. Klára Hegyi, Dec. 5, 1954; children—Anikó, Klára. Sci. worker Hungarian Acad. Sci., 1956; asst. dept. histology Med. U., Budapest, 1956-59, 1st asst., 1959-63, asst. prof. dept. histology and embryology, 1963—. Mem. Hungarian Biol. Soc., Korányi Sándor Soc. Author: (with I. Törö) The Normal and Pathological Development of the Man, 1964; (with I. Törö, T. Acs) The Philosophical Problems of Biology and Medicine, 1964; also articles. Contbg. author: Handbuch der plastischen Chirurgie, 1965. Editor: The Biological Paradox of Modern Man, 1967. Research on role of mast cells and polysaccharides in def. against tumors, origin and significance of mast cells, role of thymus on polysaccharide metabolism and mast cell genesis, immunological role of thymus, endocrine correlations between thymus and other endocrine glands; discovered new anti-tumor drug, Zitofenton. Home: 22 Paprika, Budapest X., Hungary.*

CSAKY, Tihamer Zoltan, physician; b. Maramarossziget, Hungary, Aug. 12, 1915; s. Tihamer L. G. and Olga (Rudolf) C.; student U. Halle (Germany), 1936-37; M.D., U. Budapest, 1939; m. Susan Dischka, June 18, 1953; children—Catharina M., Karl G. Came to U. S., 1949, naturalized, 1954. Instr., then asst. prof. physiology U. Budapest, 1938-46; research adj. Hungarian Biol. Research Inst., 1946-47; research fellow Biochem. Inst., Helsinki, Finland, 1947-48, Microbiol. Inst., Uppsala-Ultuna, Sweden, 1948-49; research asso. Duke U. Med. Sch., 1949-51; Practiced medicine, specializing in pharmacology, Chapel Hill, N.C., 1951-61, Lexington, Ky., 1961—; faculty Sch. Medicine U. N.C., 1951-61; prof., chmn. dept. pharmacology Coll. Medicine U. Ky., Lexington, 1961—. Guggenheim fellow U. Copenhagen, 1958-59. Mem. A. Chem. Soc., Am. Soc. Pharmacology and Exptl. Therapeutics, Soc. Exptl. Biology and Medicine, Am. Physiol. Soc., Soc. Gen. Physiologists, Am. Coll. Clin. Pharmacology and Chemotherapy, Sigma Xi. Research, numerous publs. on functions of cell membranes, transport of nutrients across them, and effect of drugs on process; pioneered studies of intestinal absorption of sugars by using synthetic, unnatural compounds to establish relationship between chem. structure and absorption rate. Home: 1032 The Lane, Lexington, Ky. 40504. Office: Coll. Medicine, U. Kentucky, Lexington.*

CSÜRÖS, Zoltán, Hungarian chemist; b. Budapest, Hungary, Feb. 6, 1901; s. Zoltán and Verszávia (Muntyán) C.; grad. Poly. U., Budapest, 1924, Ph.D., 1929; m. Valéria Bruckner, Aug. 30, 1930; children—Éva (Mrs. Attila Naszlady), Karola (Mrs. Ádám Horváth), Zoltán. Staff, Poly. U., Budapest, 1924—, prof. organic chemistry tech., head Organic Chemistry Tech. Inst., 1938—, dean Chem. Faculty, 1943-44, rector univ., 1946-49, 58-61. Recipient Kossuth prize, 1953, state decorations, 1956, 58, 61. Mem. Hungarian Acad. Scis. Author: Müanyagok, 1956; (with István Rusznák) Textilkémia, 1964; also numerous articles. Research on glucoside syntheses, correlation between structure and activity of catalysators in hydrogenation and oxidation, reactions catalysed in non-aqueous media by strong acids.*

CTESIBIUS, flourished 2d century, B.C., Alexandria, Egypt; founded engring. tradition of Alexandria; reportedly discovered expansive power of air and applied it as motive force; invented a constant-flow water clock (clepsydra), force pump, hydraulic organ, fire engine, and a bent syphon.

CTSEIAS, Greek physician; b. Cnidus, Carina, Asia Minor; flourished 410 B.C.; historian, physician in Ct. of Artaxerxes Mnemon, King of Persia; Persian envoy to Evagoras and Conon, 398 B.C.; later returned to Cnidus. Author: Persika (a history of Persia in 24 books); Indika (book on India). Wrote on med. use of hellebore and improved its dosage.

CUARON, A. Santisteban, Mexican biochemist; b. Mexico City, Mexico, June 29, 1933; s. Jose and Elisa (Santisteban) C.; B.S., Nat. U. Mexico, 1950; M.D. cum laude, U. Mexico, 1957; postgrad. exptl. radiopathology research unit, med. research council Hammersmith Hosp., London, Eng., 1959-60; fgn. visitor endocrine dept. U. Hosp. Leyden (Holland), 1960-61; m. Cristina Orozco, Sept. 12, 1959; children—Alfredo David, Alfonso, Cristina Elisa. Physician, endocrine dept. Nat. Health Ministry Mexico, 1957-59, radioisotope unit Hosp. 20 de Noviembre Mexico City, 1962; head nuclear med. unit Gen. Hosp. Nat. Med. Centre, Mexico City, 1963—. Lectr., U. Mexico Sch. Medicine, 1963; cons. Nat. Commn. Nuclear Energy Mexico, 1962; tchr. clin. radioisotope techniques Nat. Mexican Inst. Social Security. Recipient Squibb award, 1957. Mem. Royal Soc. Medicine, Soc. Nuclear Medicine (U. S. A.), Sociedad Espanola de Endocrinologia (Spain), Assn. Radiation Research, Sociedad Mexicana de Radiologia. Author: Manual the Medicina Nuclear para Medicos Generales. Research, publs. in biochemistry on effects of maternal hypothyroidism on devel. of foetal brain and studies on biliary excretion of thyroid hormones; in nuclear medicine improved clin. procedures with radioisotopes mainly on interactions between thyroid hormones and serum proteins. Home: 21 Tepeji, Mexico D.F. 7. Office: Av. Cuaumtemoc 330, Mexico 7 D.F., Mexico.*

CUATRECASAS, José, botanist; b. Camprodón, Spain, Mar. 19, 1903; s. José Genís and Carmen (Arumí) C.; Lic. Ph., U. Barcelona, Spain, 1923; Dr. Ph., U. Madrid, 1928; m. Martha Maria Nowack, July 27, 1933; children—Teresa (Mrs. William Rivera), Gil, Pedro. Came to U. S., 1947, naturalized, 1953. Asst. prof. botany U. Barcelona, Spain, 1924-30; prof. systematic botany U. Madrid, Spain, 1931-39; curator Tropical Flora Madrid Bot. Garden, 1932-39, dir. Bot. Garden, Madrid, 1936-39; prof. Instituto Botánico U. Nacional Bogota, Colombia, 1939-42; dir. tropical agrl. sch. Cali, Colombia, 1942-43, dir. Misión Botánica del Valle, prof. faculty agronomy, 1943-47; curator Colombian Botany, Chgo. Natural History Mus., 1947-50; investigator NSF, 1952—; research asso. dept. botany Smithsonian Instn., Washington, 1955—; sci. dir. Flora Neotropica. Guggenheim fellow, 1951-52; recipient Cross of Boyacá, Colombia, 1959; Henry Allan Gleason award, 1963. Mem. Soc. Geog. Colombia (hon.), Acad. Columbiana Ciencia, Mus. History Nat. Paris (hon. asso.), A.A.A.S., Am. Soc. Plant Taxonomists, Soc. for Study Evolution, Ecol. Soc. Am., Soc. Econ. Botany, Assn. Tropical Biology, Am. Inst. Biol. Scis., Internat. Assn. Plant Taxonomy, Internat. Soc. Tropical Ecology, Soc. Botany France, Sigma Xi. Author: Vegetación y. Flora del macizo de Magina, 1928; Observaciones Geobotánicas en Colombia, 1934; Prima Flora Colombiana I-II, 1957-58; Aspectos Vegetacion natural de Colombia, 1958; Taxonomic revision of the Humiriaceae, 1961; Cacao and its Allies, 1964. Research, numerous publs. contbg. to knowledge of flora and plant ecology of Spain and Tropical Am., especially of Colombia, through extensive explorations and collections, taxonomic studies and publ. of results; advance of systematic biology in field of phanerogams through monographs of several groups. Home: 3707 34th St., Washington 20008. Office: Dept. Botany, U. S. Nat. Mus., Smithsonian Instn., Washington 20560.*

CUBA, Johannes von, see von Cuba.

CUBICCIOTTI, Daniel, Am. chemist; b. Phila., June 28, 1921; s. Daniel D. and Ida (Orecchia) C.; B.S., U. Cal. at Berkeley, 1942, Ph.D., 1946; m. Lois de Roos, Dec. 25, 1948; children—Daniel 3d, Roger, Kelly Ann. Faculty, N.Y. U., 1946-47, U. Cal. at Berkeley, 1947-48, Ill. Inst. Tech., 1948-51; research chemist Atomics Internat. div. N.Am. Aviation Co., 1951-55; sr. chemist, sci. fellow Stanford Research Inst., Menlo Park, Cal., 1955—. Vis. prof. chemistry U. Kan., 1962; vis. lectr. Stanford, 1963-67. Mem. Am. Chem. Soc., Electro-chem. Soc., Phi Beta Kappa, Sigma Xi. Research, numerous publs. on high temperature chemistry, thermodynamics, kinetics of metal oxidations, fused salt systems, evaporation of materials. Home: 1125 Las Flores Ct., Los Altos, Cal. 94022. Office: Stanford Research Inst., Menlo Park, Cal. 94025.*

CUBITT, Sir William, Brit. civil engr.; b. Dilham, Eng., 1785; s. Joseph and (Lubbock) C.; ed. village sch., Dilham; 1 son, Joseph. miller, then cabinet-maker, 1800-04; agrl. machine maker, until 1807; millwright, 1807-12; became chief engr. Ramsome's works, Ipswich, 1812, partner, 1821-26; became engr. in charge bldg. Bute Docks, Cardiff, South-Eastern Ry., London, 1823; Berlin Waterworks, Oxford Canal, Liverpool Junction Canal, Middlesborough Docks, coal drops on Tees, Black Sluice drainage; used atmospheric system on Croydon Ry.; cons. engr. Boulogne and Amiens Ry.; ret., 1858. Fellow Royal Soc., 1830, Royal Irish Acad., Instn. Civil Engrs. (mem. council 1831, v.p., 1836, pres. 1850-51). Invented self-regulating windmill sails, 1807, treadmill for grinding corn but was used as punishment for criminals, 1818; during bldg. of South-Eastern Ry. blew up face of Round Down Cliff, and built ry. along beach with tunnel under Shakespeare Cliff; works include 2 landing-stages, Liverpool, Eng., bridge carrying London turnpike across Medway at Rochester. Died Clapham Common, Eng., Oct. 13, 1861.

CUCKLER, Ashton Clinton, Am. parasitologist; b. Wilsonville, Neb., Mar. 16, 1910; s. Edward D. and Jessie (Walker) C.; B.A., U. Neb., 1935; M.A., 1936; Ph.D., U. Minn., 1941; m. Kathleen A. Jackson, Oct. 9, 1941; children—John M., Anne C. Faculty, U. Neb., 1936-38, U. Minn., 1938-40; asst. parasitologist U. Hawaii, 1941-42; asst. prof. U. Minn., 1942-47; with Merck Inst. for Therapeutic Research, Rahway, N.J., 1947—, sr. dir., 1966—. Dir. lab. Italian Med. Mission, UNRRA, 1945; adj. asso. prof. parasitology and adminstrv. medicine Columbia, 1964. Markle Found. fellow, 1944. Fellow N.Y. Acad. Sci.; World Assn. Vet. Parasitologists, mem. Am. Soc. Parasitology, Am. Soc. Tropical Medicine and Hygiene, N.Y. Soc. Tropical Medicine. Studies, numerous publs. on life cycles, physiology, disease prodn. and chemotherapy of parasites of domestic animals and poultry; research on discovery and evaluation of therapeutic agts. for infectious diseases. Home: 31 Hawtorn Dr., Westfield, N.J. 07090. Office: Merck Inst. for Therapeutic Research, Rahway, N.J. 07065.*

CUDWORTH, Ralph, Brit. philosopher; b. Aller, Eng., 1617; s. Ralph Cudworth; M.A., Emmanuel Coll., Cambridge, Eng., 1639, D.D., Cambridge, 1651; m. 1654; children include dau., Damaris (Mrs. Francis Masham). Apptd. master Clare Hall, 1644; Regius prof. Hebrew, Cambridge, 1645-88; became master Christ's Coll., 1654; became rector Ashwell, Eng., 1650; named prebendary, Gloucester, Eng., 1678. Cons. com. on proposed revision of translation of Bible, Ho. of Commons, 1657. Author: Discourse concerning the true notion of the Lord's Supper, 1642; The Union of Christ and the Church, 1642; The True Intellectual System of the Universe (criticism of 2 forms of atheism, atomic of Democritus and Hobbes, and hylozoic of Staato and Spinoza), 1678; Treatise Concerning Eternal and Immutable Morality. Known as chief of Cambridge Platonists; originated theory of plastic nature as opposed to chance and constant divine interference; believed life may be either conscious (spirit) or an unconscious plastic power seen in purposeful but unconscious animal instincts; accepted atomic theory. Died Cambridge, June 26, 1688.

CUEILLERON, Jean Ferdinand Marie, French chemist; b. Paris, France, Oct. 28, 1915; s. Victor Lucien and Blanche (Liardeaux) C.; Licencié ès Sciences, U. Paris, 1939, Docteur-ès-Sciences, 1944; m. Paule Flamens, July 25, 1953; children—Jacqueline, Henri. Asst., Sorbonne, Paris, 1944-46; prof. l'École de Chimie, Mulhouse, France, 1946-50; prof. Faculté des Sciences, Lyon, France, 1951—. Research, numerous publs. on chemistry of boron. Home: 26 Jeanne Hachette, Lyon 3 69, France. Office: 43 11 novembre 1918, Velleurbanne 69, France.*

CUÉNOT, Lucien Claude, French zoologist, geneticist; b. Paris, France, Oct. 21, 1866; s. Charles Auguste and Berthe Marie (Merlet) C.; Sc.B., Municipal Coll., Paris, 1883; Lic. Sc., U. Paris, 1885, Sc.D., 1887; Hon. Dr., U. Louvain, Belgium; m. Geneviève Marguerite de Maupassant, July 31, 1900; 6 children. Became asst. anatomy and physiology Faculty Scis., Paris, 1888; named lectr. zoology Faculty Scis., Nancy, France, 1890, prof., 1898, dir. Zool. Mus. Mem. French Acad. Scis., 1918, Royal Acad. Belgium (asso.), Inst. France, Royal Acad. Denmark, Pontifical Acad. Scis., Nat. Acad. Scis. Mexico, Acad. Stanislas Nancy, London Linnean Soc. (fgn.), London Zool. Soc. (corr.); hon. mem. Roumanian Acad. Scis., Nat. Acad. Metz, Royal Soc. Zoology Brussels, Paris Biol. Soc. Author: La genèse des espèces animales, 1911; Théorie de la préadaptation, 1914; l'adaptation, 1925; L'invention et la finalité en biologie, 1941; Hasard ou finalité, 1946; (with A. Tetry), L'évolution biologique, 1951. Proved selection of species results from pre-adaption (animals in new environments are prepared physiologically by having come from analogous conditions); showed Mendel's laws were applicable to animal world using white rats; in producing yellowish mice with red eyes, discovered lethal genes, 1905. Died Nancy, Jan. 7, 1951.

CUESTA DUTARI, Norberto, Spanish mathematician; b. Salamanca, Nov. 13, 1907; s. Antonio and Juana C. D.; ed. univs. Salamanca, Saragoose, Madrid; Ph.D. in math.; doctorate U. Madrid, 1943. Prof., Inst. Middle Edn., U. Salamanca. Recipient prize Acad. Scis. of Madrid, 1958. Mem. Math. Soc. Research and publs. on math. Home: Calle Pinzones 13. Office: Faculty of Sciences, Salamanca, Spain.

CUGNOT, Nicolas Joseph, French engr.; b. Poid, Meuse, France, Sept. 25, 1725; s. Claude and Marie (Bourget) C.; engr. in army of Maria Theresa; granted pension by Louis XV, 1772; when this was withdrawn during French Revolution, he retired to Brussels. Author: Elements de l'art militaire ancien et moderne, 1766; La Fortification de campagne, théorique et practique, 1769; Théorie de la fortification, 1778. Invented 3-wheeled steam-driven carriage, 1770, 1st true automobile, 1769; improved mil. weapons and machines. Died Paris, France, Oct. 2, 1804.

CUISINIER, Jeanne A(dèle) L(ucie), French sociologist; b. Neuilly-sur-Seine, France, Oct. 30, 1890; diploma Ecole des Langues orientales vivantes, Paris, France, 1929, Institut de Phonetique Malais, 1929; doctorat es lettres, U. Paris, 1944. Prof. social geography, social anthropology U. Jogjakarta (Indonesia), 1952, U. Jakarta (Indonesia), 1953-54. Author: Les Mu'o'ong—Geographie Humaine et sociologie, 1946; La Danse Sacree en Indochine et en Indonesie, 1951; Le theatre d'ombres a Kelantan, 1957. Mem. sci. research missions sponsored by French Nat. Center in Malaya, 1932-33, N. Vietnam, 1937-38, Java, Sumatra, Sumba, Timor, 1952-55, Java, Bali, 1960-61; researches in magical beliefs and ritualistic observances touching on aspects of parapsychology.

CULBERSON, William Louis, Am. botanist; b. Indpls., Apr. 5, 1929; s. Louis Henry and Lucy (Hellman) C.; B.S., U. Cin., 1951; Diplome d'Etude Superieures, Sorbonne, 1952; Ph.D., U. Wis., 1954; m. Chicita F. Forman, Aug. 29, 1953. Faculty, Duke, Durham, N.C., 1955—, asso. prof. botany, 1964—. Mem. Bot. Soc. Am., Am. Bryological Soc., Sigma Xi, Phi Beta Kappa. Editor-in-chief Bryologist, 1962—. Research in taxonomy of lichens, relationships of anat. and chem. variations in lichens, classification and evolution of cryptogamic plants. Home: Villa Pinca, Route 7, King Rd., Durham, N.C. 27707.*

CULBERTSON, Clyde Gray, Am. pathologist; b. Vevay, Ind., July 27, 1906; s. Carl Scott and Anna Mary (Gray) C.; B.S., Ind. U., 1928, M.D., 1931; m. Margaret O'Neil, June 25, 1931. Practice medicine, specializing in pathology, Indpls., 1931-46; faculty Sch. Medicine Ind. U., 1931—, prof., chmn. clin. pathology, 1943-63; prof. pathology, 1963—, dir. clin. lab., 1931-46; dir. Ind. Bd. Health Lab. Hygiene, 1933-46; dir. Biol. Research div. Lilly Research Labs., 1949-63, research adviser, 1963—. Recipient Distinguished Service award Jr. C. of C., 1936; Ward Burdick award, 1960. Diplomate Am. Bd. Pathology. Mem. Am. Assn. Pathologists and Bacteriologists, Am. Soc. Clin. Pathologists, Assn. Clin. Pathologists Eng. (hon.), Coll. Am. Pathologists, Internat. Acad. Pathology, A.M.A., N.Y. Acad. Scis., A.A.A.S., Am. Assn. Immunologists, Am. Inst. Biol. Scis., Am. Soc. Microbiology, Soc. Protozoologists, Tissue Culture Assn., Sigma Xi, Alpha Omega Alpha. Research, publs. on intestinal obstruction, gen. pathology, viral vaccines, antibiotics, pathogenic freeliving amebas. Home: 6060 Park Av., Indpls. 46220. Office: Lilly Research Labs., Indpls. 46206.*

CULLEN, Glenn E(rnest), Am. biochemist; b. Isle Saint George, O., Apr. 1, 1890; s. Charles and Emma (Gould) C.; A.B., U. Mich., 1912, B.Chem. Engring., 1913; Ph.D., Columbia, 1917; m. Marie Wherry, June 22, 1917; children—Mary Alice, Donna Jean, Glenn Wherry. Research chemist Rockefeller Inst. for Med. Research, N.Y., 1913-22; asso. prof. research, medicine U. Pa., 1922-24, prof. biochemistry Vanderbilt U., 1924-31; traveling fellow Rockefeller Found., 1924-25; prof. biochemistry Grad. Sch., and prof. research pediatrics Coll. Medicine, U. Cincinnati, from 1931; dir. labs. Children's Hosp. Research Found., from 1931. Mem. Am. Soc. Biol. Chemists (pres. 1937-39). Writer on enzymes, antiseptics, chemistry of blood in health and disease, diseases of children. Died Apr. 11, 1940.

CULLEN, Stuart Chester, Am. physician; b. Milton Junction, Wis., Jan. 31, 1909; s. Archie Hodge and Myrtle (Killam) C.; B.S., U. Wis., 1931, M.D., 1933; m. Caroline Wells Swannell, June 25, 1932; children—Carol Lynn (Mrs. Dale Edward Walter) Bruce Frederick. Prof. anesthesia, chmn. div. anesthesia State U. Ia., 1938-58; prof. anesthesia, chmn. dept. anesthesia U. Cal. at San Francisco, 1958—, asso. dean Sch. Medicine, 1963—66, dean, 1966—. Cons. numerous hosps.; contbr. services WHO. Recipient Distinguished Achievement award Modern Medicine, 1958; certificate of Recognition, U. Wis. Med. Alumni Assn., 1961. Diplomate Am. Bd. Anesthesiology (v.p. 1960, pres. 1961). Fellow A.A.A.S.; mem. Am. Soc. Anesthesiologists (Distinguished Service award 1964), Am. Soc. Pharmacology and Exptl. Therapeutics, Am. Assn. U. Profs., Assn. U. Anesthetists (pres. 1959), N.Y. Acad. Scis., Internat. Anesthesiology Research Soc., Soc. Exptl. Biology and Medicine, numerous fgn socs., others, Alpha Omega Alpha, Sigma Xi. Author: (with E. G. Gross) Manual of Medical Emergencies, 1958; Anesthesia: A Manual for Students and Physicians 1957. Mem. editorial bd. Anesthesiology, 1941-63; anesthesia sect. editor Year Book of General Surgery, 1950-62; specialty editor Audio-Visual Postgraduate Course, 1958—; editor: Year Book of Anesthesia, 1962—. Publs. on aspects of anesthesiology—use of curare, muscular relaxants, preanesthetic medication; 1st use of xenon for anesthesia in humans. Home 73 Westshore Dr., Belvedere, Cal. 94920.*

CULLEN, Thomas Stephen, gynecologist; b. Bridgewater, Ont., Can., Nov. 20, 1868; s. Thomas and Mary (Greene) C.; ed. Collegiate Inst., Toronto; M.B., U. Toronto, 1890, also LL.D.; hon. D.Sc., Temple U. Specialist in abdominal surgery; prof. gynecology, Johns Hopkins U., vis. gynecologist, Johns Hopkins Hosp. Hon. fellow Edinburgh Obstet. Soc.; mem. A.M.A. (trustee 1929-41), So. Surg. and Gynecol. Assn. (pres. 1916), Phi Beta Kappa; hon. or corr. mem. German and Italian profl. socs. Author: Cancer of the Uterus, 1900; Adenomyoma des Uterus, Verlag von August Hirschwald, 1903; Adenomyoma of the Uterus, 1908; Myomata of the Uterus (with Howard A. Kelly), 1909; Embryology, Anatomy and Diseases of the Umbilicus Together with Diseases of the Urachus, 1916; Henry Mills Hurd, 1920; Early Medicine in Maryland, 1927; also Accessory Lobes of the Liver for Archives of Surgery and articles in med. jours. Editor of 2 vols. on gynecology, Lewis System of Surgery, 1928. First to give complete description of hyperplasia of endometrium, 1900; described bluish discoloration of skin around umbilicus as sign of ruptured ectopic pregnancy (Cullen's sign), 1919. Died Mar. 4, 1953.

CULLERIER, Michel, French physician; b. Angers, France, June 8, 1758; ed. Nantes Sch. Surgery; Docteur es médecine, Paris, France, 1803; became master surgeon Bicetre Hosp., 1787, S. Jacques Hosp. for Venereal Patients, 1792. Mem. Acad. Medicine (became pres. 1826). Asso. with invention of guillotine; research on treatment of syphilitic patients. Died Soulin, France, Jan. 3, 1827.

CULLINGWORTH, Charles James, Brit. gynecologist, obstetrician; b. Leeds, Eng., June 3, 1841; s. Griffith and Sarah (Gledhill) C.; ed. Wesley Coll., Sheffield, 1953-57, Leeds Sch. Medicine, 1861-65; M.D., Durham, 1881, D.C.L., 1893; LL.D., Aberdeen, 1904; m. Emily Mary Freeman, 1882; 1 dau. Gen. practitioner, Leeds, 1864; asst. resident physician Manchester Royal Infirmary, 1867, later resident med. officer; pvt. practice, Manchester, 1869; police surgeon, Manchester, 1872-82; lectr. med. jurisprudence Owens Coll., 1879, appt. chair obstetrics and gynecology, 1885; sec. to bd. studies in medicine Victoria U., 1883; obstetric physician St. Thomas' Hosp., London, 1888-1904; examiner U. London, 1890-95, Cambridge U., 1896-1900; Ingleby lectr. U. Birmingham, 1904. Mem. Royal Coll. Surgeons, Royal Coll. Physicians (Bradshaw lectr. 1902), Manchester Med. Soc., Obstet. Soc. London (chmn. bd. exam. midwives 1895-96, pres. 1897-98). Author: Puerperal Fever a Preventable Disease; Clinical Illustrations of the Diseases of the Fallopian Tubes and of Tubal Gestation, 1895; A Manual of Nursing, Medical and Surgical, 1883; A Short Manual for Monthly Nurses, 1884. Pioneer in gynecology; research on cause of pelvic peritonitis (one of 1st in Eng. to believe it was secondary to other conditions and not a primary disease; prominent in movement to secure legal registration of midwives (passed in 1902). Died May 11, 1908.

CULLIS, Charles Fowler, English chemist; b. London, Eng., Aug. 31, 1922; s. Charles Gilbert and Winifred (Fowler) C.; B.A. with 1st class honors, Trinity Coll., Oxford (Eng.) U., 1944, B.Sc., 1945, M.A., 1948, D.Phil., 1948, D.Sc., 1960; m. Marjorie Elizabeth Anderson, Sept. 3, 1958; children—Jonathan Oliver, Jane Madeleine, Eleanor Judith, Philip James. Demonstrator phys. chemistry Oxford U., 1945-50, tutor Brasenose Coll., Merton Coll., Christ Church, 1945-50, Imperial Chem. Industries Research fellow, 1947-50; lectr. phys chemistry Imperial Coll. London, 1950-59, sr. lectr. chem. engring. and chem. tech., 1959-64; reader combustion chemistry U. London, 1964-66; vis. prof. Coll. Chemistry, U. Cal. at Berkeley, 1966; prof. phys. chemistry City U., London, 1967—. Cons., Laporte Chems. Ltd., 1959—; mem. Brit. Com. Combustion Inst., 1963—. Fellow Royal Inst. Chemistry. Asst. sci. editor Internat. Union Pure and Applied Chemistry, 1964—. Research, numerous publs. on elucidation of mechanism of wide variety of organic and inorganic reactions especially gaseous oxidation of organic compounds, exothermic decomposition yielding solid products, interaction of gases with solids at high temperatures, reactions in stationary flames, liquid-phase oxidation processes, Home: 11 Roedean Crescent, Roehampton, London S.W.15., Eng.*

CULLUMBINE, Harry, physician; b. Rotherham, Eng., Dec. 29, 1912; s. Henry and Mary (Knowles) C.; B.Sc., U. Sheffield, Eng., 1933, M.Sc., 1934, M.B., 1937, M.D., 1946; m. Mary Catherine Fox, Jan. 17, 1959; 1 dau., Diana Pauline. Came to U. S., 1958, naturalized, 1966. Prof., head dept. physiology, pharmacology U. Ceylon, 1947-51; chief med. officer research Chem. Def. Exptl. Establishment, Eng., 1952-56; prof., head dept. pharmacology U. Toronto, 1956-58; exec. v.p., med. dir. Air-Shields, Inc., Hatboro, Pa., 1958-60, pres., 1960-66, also dir.; corp. med. dir., v.p. National Aero. Corp. (name now changed to Narco Sci. Industries), Fort Washington, Pa., 1966—, also dir.; U. Pa. lectr. Grad. Sch. Med., 1958—. Cons. Army Council, Britain, 1948-57, Def. Research Bd. Can., 1956—; Mem. Am. Soc. Pharmacology, Canadian Physiol. Soc., Brit., Canadian Pharmacol. soc., Physiol. Soc. London. Research, numerous publs. on belladonna alkaloids, toxicity of drugs, respiratory physiology. Home: 541 W. Moreland Av., Phila. 19118. Office: Narco Sci. Industries, Commerce Dr., Ft. Washington, Pa. 19034.*

CULMANN, Karl, German engr., mathematician; b. Bergzabern, Germany, July 10, 1821; s. Carl Wilhelm and Caroline Emilie (Boell) C.; studied engring. Polytech. Sch., Karlsruhe, Germany; m. Emilie Mathilde Küss, 1856; 2 sons, 1 dau. Began bldg. mountain r.r. bridge for State of Bavaria, 1841; traveled through Eng. and N.Am. observing bridges, 1849-51; became prof. Eidgenössische Polytechnikum, Zurich, Switzerland, 1855. Author: Reisebericht; Graphische Statik, (developed graphic statics as solution in civil engring. problems), 1866. Presented graphical calculus as a symmetrical whole; used polar theory of reciprocal figures to express relation between force and funicular polygons; tchr. of M. Koechlin. Died Zurich, Dec. 9, 1881.

CULP, Ormond Skinner, Am. physician; b. Toronto, O., Nov. 18, 1910; s. Frank and Dora (Skinner) C.; A.B., Ohio Wesleyan U., 1931, D.Sc., 1963; M.D., Johns Hopkins, 1935; m. Helen M. Ericson, Aug. 19, 1938; 1 dau., Pamela (Mrs. Michel Antoine LaFond). Faculty, Johns Hopkins Sch. Medicine, 1938-42; asso.-surgeon-in-charge, div. urology Henry Ford Hosp., Detroit, 1942-50; cons. urology Mayo Clinic, Rochester, Minn., 1950—, head sect., 1962—; asso. prof. urology Mayo Grad. Sch. Medicine, 1952-62, prof., 1962—; Cons. VA, 1946-52, USAF, 1965—. Mem. A.C.S. (past gov.), Clin. Soc. Genito-Urinary Surgeons (past sec.-treas., pres. 1966), A.M.A., Am. Assn. Genito-Urinary Surgeons, Urol. Forum for Clin. Investigation, Internat., Am. (1st prize for original research 1948, 53; pres. N. Central sect. 1965) urol. assns., Central Surg. Soc., Pan-Pacific Surg. Assn., Henry Ford Hosp. Med. Assn. (pres. 1965—), Phi Beta Kappa, Sigma Xi, Omicron Delta Kappa, Pi Delta Epsilon, Delta Sigma Rho, Nu Sigma Nu, Sigma Chi; hon. mem. Mexican Urol. Soc., Pan-Am. Med. Assn., St. Paul, Bay surg. socs. Contbr. numerous articles to tech. jours. Originated plastic types operations on kidney and male genitalia. Home: 1032 Plummer Circle, Rochester, Minn. 55901.*

CULPEPER, Nicholas, English physician, botanist, astrologer; b. London, Eng., Oct. 18, 1616; s. Nicholas C.; ed. Cambridge, 1634; m.; 7 children. Apprentice to apothecary; settled in Spitalfields and began practice as astrologer and physician, 1640; fought on parliamentary side in civil war. Author: A Translation of the London Dispensatory, 1649; Semeiotica Uranica, or an Astronomicall Judgment of Diseases, 1651; Directory for Midwives, 1651; Galen's Art of Physic, 1652; Idea Universalis Medica Practica, 1652; The English Physician, 1652; The English Physician Enlarged, 1653; Anatomy, 1654; A New Method of Physic, 1654; Culpeper's Last Legacy, 1655; Culpeper's Astrologicall Judgment of Diseases, 1655; Chemistry Made Easy and Useful (tranl. from Sennertus), 1662; The Chirurgeon's Guide, 1677. Opposed College of Physicians' monopoly on medical knowledge; translated medical works and wrote his own in English; published for 1st time many medicines of English herbs. Died Spitalfields, Jan. 10, 1654.

CULPIN, Millais, Brit. surgeon, psychologist; b. Ware, Eng., 1874; s. Millice Culpin; ed. Grocers' Co. Sch., also London Hosp.; M.D.; m. Ethel Maude Bennett, 1913; 1 dau. Receiving room officer, also house surgeon, ophthalmic house surgeon, surg. registrar London Hosp., lectr. on psychoneurosis London Hosp. Med. Coll.; prof. med.-indsl. psychology U. London; surg. specialist, 1914-47; neurol. specialist, 1917-19. Fellow Royal Coll. Surgeons. Author: Spiritualism and the new Psychology; Psychoneuroses of Peace and War; The Nervous Patient; Medicine and the Man; Recent Advances in the Study of the Psychoneuroses; Mental Abnormality; Facts and Theories. Died Sept. 14, 1952.

CULVER, James Fox, Am. physician; b. Macon, Ga., June 10, 1921; s. Peter J. and Mary (Fox) C.; student Va. Mil. Inst., 1939-41, Mercer U., 1942-43; M.D., U. Ga., 1945; postgrad. Northwestern U. Med. Sch.; m. Jean E. Williams, May 4, 1947. Practice medicine, specializing in ophthalmology, Watsonville, Cal., 1952-59; commd. maj. USAF, 1959, advanced through grades to col., 1966; chief ophthalmology br. USAF Sch. Aerospace Medicine, Brooks AFB, Tex., 1959-66, asst. dep. research, devel., 1966—. Mem. exec. council Armed Forces-NRC Com. on Vision NRC, Nat. Acad. Scis.; mem. vision com. Aerospace Med. Panel Adv. Group for Aerospace Research, Devel. NATO. Diplomate Am. Bd. Ophthalmology. Fellow Am. Acad. Ophthalmology; mem. A.M.A., Aerospace Med. Assn. (Arnold D. Tuttle award 1966). Research in area of ocular effects of high intensity radiation and visual problems of aerospace. Home: 123 Vinsant St., San Antonio 78235. Office: Aerospace Med. Div., Brooks AFB, Tex. 78235.*

CUMINGS, Hugh, Brit. naturalist; b. West Alvington, Kingsbridge, Devonshire, Eng., Feb. 14, 1791; apprentice to sail-maker 1819; began collection of shells in Valparaiso, Chile; made expdns. to Pacific Islands and in 1835 to P.I.; returned to Eng. with collection of hundred thirty thousand plant specimens; also numerous zool. collections. Cuminigia (genus of bivalved shells) named in his honor, 1833. Died London, Eng., Aug. 10, 1865.

CUMINGS, John Nathaniel, Brit. physician, chem. pathologist; b. London, Oct. 4, 1905; s. Arthur Nathaniel and Kathleen (Aylott) C.; M.B., B.S., King's Coll., London, 1928, M.S., 1931, grad. King's Coll. Hosp., 1924; m. Mary Phyllis Parish, 1940; children—Rosemary, David. Pathologist, King Edward Meml. Hosp., Ealing, Eng., 1932-45, Wembley Hosp., 1932-58, Nat. Hosp., London, 1945-58; prof. chem. pathology Inst. Neurology, U. London, Nat. Hosp., 1958—, bd. govs. Hosp., 1961—, com. mgmt. Inst. Neurology, 1961—, dep. dean, 1965—. Chmn. med. research com. Muscular Dystrophy Group Gt. Britain. Fellow Royal Coll. Physicians Eng., Coll. Pathologists London; mem. Assn. Brit. Neurologists, Assn. Clin. Pathologists, Royal Soc. Medicine, Biochem. Soc., Internat. Acad. Pathology, Path. Soc. Gt. Britain and Ireland, World Fedn. Neurology, Sci. Counsellors and Commn. Neurochemistry (mem. bd.). Author: Heavy Metals and the Brain, 1959; also numerous articles. Editor: (with L. van Bogaert, A. Lowenthal) Cerebral Lipidoses, 1957; (with Michael Kremer) Biochemical Aspects of Neurological Disorders, 1st series, 1959, 2d series, 1965; Modern Scientific Aspects of Neurology, 1960; (with J. M. Walshe) Wilson's Disease; Some Current Concepts, 1961. Research on diagnosis and treatment of various neurol. disorders, especially hepatolenticular degeneration, muscle diseases, chemistry of cerebral lipids, especially degenerative diseases of brain, heavy metals, migraine, cerebral edema. Home: 17, Elm Grove Rd., Ealing, London, W.5. Office: Dept. Chem. Pathology, Nat. Hosp., Queen Sq., London W.C.1., Eng.*

CUMMING, James, Brit. chemist; b. Eng., Oct. 24, 1777; B.A., Trinity Coll., Cambridge (Eng.) U., 1801; became fellow Cambridge U., 1803, prof. chemistry, 1815-60. Fellow Royal Soc., 1816; mem. Cambridge Philos. Soc. (pres.). Author: A Manual of Electro-Dynamics, 1827; also articles. Independently discovered thermo-electricity; described tangent galvanometer (galvanoscope); 1st to show the electromotive force of a thermocouple can change sign with rise of temperature. Died Norfolk, Eng., Nov. 19, 1861.

CUMMING, James Burton, Am. nuclear chemist; b. N.Y.C., June 6, 1928; s. Samuel B. and Minnie (Widmer) C.; B.S., Yale, 1949; M.A., Columbia 1951, Ph.D., 1954; m. Janet Smith, June 5, 1953; children—James S., Jonathan R. Staff Brookhaven Nat. Lab., Upton, N.Y., 1954—, chemist, 1959—. Mem. Am. Chem. Soc., Am. Phys. Soc. Contbg. author: Annual Reviews of Nuclear Sci., vol. 13, 1963; also articles. Research on nuclear reactions between high-energy particles and complex nuclei using radiochem. and solid state detector techniques. Home: 18 Gen. McLean Dr., Bellport, N.Y. 11713. Office: Dept. Chemistry, Brookhaven Nat. Lab., Upton, N.Y. 11973.*

CUMMING, William, Brit. physician; b. Eng., 1822; pioneered modern ophthalmology; research and publs. on retinal reflex, entrance of optic nerve; 1st to demonstrate that fundus of eye could be made visible. Died Limehouse, Eng., 1855.

CUMMINGS, Martin Marc, Am. physician; b. Camden, N.J., Sept. 7, 1920; s. Samuel and Cecelia (Silverman) C.; B.S., Bucknell U., 1941; M.D., Duke, 1944; m. Arlene Sally Avrutine, Sept. 27, 1942; children—Marc Steven, Lee Bernard, Stuart Lewis. Commd. asst. surgeon USPHS, 1946, med. dir., 1966; cons. office of dir. NIH, 1960-61, chief office internat. research, 1961-63, asso. dir. research grants, 1963-64, dir. Nat. Library Medicine, USPHS, 1964—; with dept. medicine and surgery VA, 1949-59, cons., 1959—; faculty Emory U., 1948-53, George Washington U., 1953-59; prof., chmn. dept. microbiology U. Okla. Sch. Medicine, 1959-61, asso. prof. medicine, 1959-61. Chmn. com. med. research Nat. Tb Assn., 1958-59; chmn. panel on sarcoidosis NRC-Nat. Acad. Scis., 1958—. Recipient Exceptional Service award VA, 1959. Diplomat Am. Bd. Microbiology. Fellow N.Y. Acad. Sci.; mem. Am. Fedn. Clin. Research (sr.), Am. Soc. for Clin. Investigation, Am. Thoracic Soc., Assn. Am. Med. Colls., Friends of U. S. of Latin Am., Med. Library Assn., Nat. Assn. on Standard Med. Vocabulary. Author: (with H. S. Willis) Diagnostic and Experimental Methods in Tuberculosis, 1952. Research, publs. on lab. techniques for diagnosis, evaluation of Tb and infections, epidemiology of sarcoidosis. Home: 11317 Rolling House Rd., Rockville, Md. 20582. Office: 8600 Rockville Pike, Bethesda, Md. 20852.*

CUMMINS, Harold, Am. anatomist; b. Markleville, Ind., May 28, 1893; s. William Herbert and Margaret (McCallister) C.; A.B., U. Mich., 1916; Ph.D., Tulane U., 1925; m. Elizabeth Clay Van Buskirk, Aug. 28, 1918; children—Harold, Robert, Irving. Faculty, Vanderbilt U. Sch. Medicine, Nashville, 1916-19; faculty Tulane Sch. Medicine, New Orleans, 1919-64, prof. anatomy, 1932-64, prof. emeritus, 1964—, head div. microscopic anatomy, 1934-45, chmn. dept. anatomy, 1945-59, asst. dean, 1949-64. Mem. A.A.A.S., Am. Assn. Anatomists (past pres.), Internat. Confs. on Dermatoglyphics (hon. pres. 1961—). Author: (with Charles Midlo) Palmar and Plantar Dermatoglyphics in Primates, 1942, Finger Prints, Palms and Soles, 1943, 61; also numerous articles. Research in dermatoglyphics, morphology, embryology, methodology, personal identification, racial differences, distinctions in mongolism and other constl. groups. Home: 310 Audubon St., New Orleans, 70118.*

CUMMINS, John Edward, Australian chemist; b. Perth, Australia, Oct. 21, 1902; s. Ambrose Michael and Elizabeth (Hamilton) C.; B.Sc., U. Western Australia, 1923; M.Sc., U. Wis., 1929; m. Elizabeth Margaret Lamborne, Jan. 4, 1927; 1 dau., Helen Carol (Mrs. Noble). Staff chem. investigations W.A. Forests Dept., 1924-27; with Commonwealth Sci. and Indsl. Research Orgn., 1927-62, chief sci. liaison officer, London, 1948-54, Washington, 1955-58, dir. IAEA, Vienna, Austria, 1958-60; sec.-exec. officer Ian Clunies Ross Meml. Found., E. Melbourne, Victoria, Australia, 1962—. Exec. council Commonwealth Agrl. Bur., 1948-54, chmn., 1950-52. Fellow Royal Australian Chem. Inst. (past mem. council), Royal Inst. Chemistry (past mem. council). Research, numerous publs. on wood preservation in Australia, tropic proofing service material in New Guinea; invented boric acid process for prevention borer attack on Australian timbers. Home: 10 Nortimer St., Kew, Victoria. Office: 314 Albert St., East Melbourne, Victoria, Australia.*

CUNAUS, physicist; from Leyden, Netherlands; flourished mid-18th century; asst. to Pieter van Musscenbroek in discovering principles of Leyden jar.

CUNEO, Bernard, French surgeon; b. Toulon, France, 1873; intern Paris, France hosps., 1901; aggregate prof. surgery, 1901; surgeon, from 1902; dir. Hôtel-Dieu Clinic, prof. surg. anatomy, 1920-30; author: (with Poirier and Charpy) Traité d'anatomie; research on anatomy of cranial nerves. Died Dec. 11, 1944.

CUNHA, Tony Joseph, Am. nutritionist; b. Los Banos, Cal., Aug. 22, 1916; s. Anthony A. and Maria (Cunha) C.; student Cal. Poly., 1936-39; B.S., Utah State Coll., 1940, M.S., 1941; Ph.D., U. Wis., 1944; m. E. Gwen Smith, Sept. 1, 1941; children—Becky Jane (Mrs. A. T. Mallory), Sharon Marie (Mrs. C. J. Lewis), Susan Ann. Faculty, Washington State U., Pullman, 1944-48; faculty U. Fla., Gainesville, 1948—, prof., chmn. dept. animal sci., 1950—. Fellow A.A.A.S.; mem. Am. Soc. Animal Sci., Nat. Acad. Sci., Am. Inst. Nutrition, Am. Assn. U. Profs., Soc. Exptl. Biology and Medicine, Am. Feed Mfrs. Assn., Fla. Acad. Sci., Sigma Xi, Alpha Zeta, Gamma Alpha, Gamma Sigma Delta, Phi Sigma. Author: Swine Feeding and Nutrition, 1957. Co-editor, Cross-breeding Beef Cattle, 1963; Factors Affecting Calf Crop, 1967. Research, publs. in sci. and profl. jours. on nutrition of animals. Home: 1641 N.W. 10th Av., Gainesville, Fla. 32601.*

CUNINGHAME, John Garry, English nuclear chemist; b. Exmouth, Eng., Apr. 22, 1920; s. John Percy and May (Roberts) C.; B.Sc., Imperial Coll. Sci. and Tech., London, 1949; m. Dorothy Mary Stapleton, Apr. 17, 1948; children—Alastair Ross, Christopher James. Staff, Atomic Energy Research Establishment, Harwell, Eng., 1949—, prin. sci. officer. Fellow Royal Inst. Chemistry; mem. Royal Coll. Surgeons (asso.). Author: Introduction to the Atomic Nucleus, 1964; also articles. Research in radiochemistry and physics of nuclear fission, especially mechanism of fission reaction; nuclear counting techniques, especially low-background beta counting; techniques of using accelerator-produced neutrons in nuclear chemistry. Home: 33 Appleford Dr., Abingdon, Eng. Office: Chemistry Div., Atomic Energy Research Establishment, Harwell, Didcot, Berkshire, Eng.*

CUNNINGHAM, Allan, botanist; b. Wimbledon, Surrey, Eng., July 13, 1791; s. Allan Cunningham; ed. sch. at Putney. With Lincoln's Inn after leaving sch.; clk. to W. J. Aiton; became bot. collector Royal Gardens, Kew, Eng.; became mem. bot. expdn., S.Am., 1815, Australia, 1817, later in Tasmania; colonial botanist, New South Wales, Australia, 1835; resigned, 1836. Coniferous genus Cunninghamia named in his honor; monument erected in his honor Bot. Gardens. Numerous publs. on description of many new species; collected living and dried plants from S.Am. Died June 27, 1839.

CUNNINGHAM, Burris Bell, Am. chemist; b. Springer, N.M., Feb. 16, 1912; s. Charles Chapman and Lora (Beall) C.; B.Sc., U. Cal. at Berkeley, 1935, Ph.D., 1940; m. Irene C. Metcalf, Aug. 1, 1936 (div. June 1964); children—Susan Mae (Mrs. Robert T. Moore), Bruce Jon, Joseph Martin; m. 2d, Juliana B. Weaver, July 8, 1964. Research asst. Shell Devel. Co., Emeryville, Cal., 1936; research asso. U. Cal. at Berkeley, 1941, faculty, 1946—, prof. chemistry, 1954—; staff plutonium project div. Manhattan Project, Chgo., 1942-45. Cons., Los Alamos Sci. Labs., 1960—, Oak Ridge Nat. Lab., 1966—. Guggenheim fellow, 1955. Mem. Am. Chem. Soc., N.Y. Acad. Scis., A.A.A.S., Am. Assn. U. Profs. Research, numerous publs. on chem. properties of transuranium elements; 1st isolation of plutonium, americium, berkelium, californium, einsteinium. Home: 733 Balra Dr., El Cerrito, Cal. 94530. Office: Lawrence Radiation Lab., Berkeley, Cal. 94720.*

CUNNINGHAM, Daniel John, Brit. anthropologist; b. Crieff, Apr. 15, 1850; s. John Cunningham; M.B., U. Edinburgh, 1874, M.D., 1876; M.D., D.Sc., U. Dublin; LL.D., St. Andrews U., Glasgow U.; D.C.L., Oxford U., 1892; m. Elizabeth Cumming Brown; 3 sons, 2 daus. Demonstrator anatomy U. Edinburgh, 1876-82, prof. anatomy, from 1903, also dean Med. Faculty; prof. anatomy Royal Coll. Surgeons, Ireland, 1882, Dublin U., 1883-1903; examiner on anatomy univs. Edinburgh, London, Cambridge, Oxford, Victoria. Fellow Royal Soc., 1891; mem. Royal Zool. Soc. Ireland (past pres.), Royal Dublin Soc. (past v.p.).

Author: (monographs) Monograph on the Marsupials; The Microcephalic Brain; (with E. H. Bennett) The Anatomy of Hernia, many others; Manual of Practical Anatomy, 2 vols., 1893-94; The Anatomy of Congenital Caecal Hernia, 1888. Acting editor: Jour. Anatomy and Physiology. Research and numerous publs. on human and comparative anatomy; largely responsible for inaugurating med. dept. of territorial army in Scotland. Died June 23, 1909.

CUNNINGHAM, David Douglas, surgeon; b. 1843; s. W. B. Cunningham; M.B., C.M., Edinburgh (Scotland) U., 1867; hon. surgeon to Viceroy of India; prof. physiology Med. Coll.; fellow Calcutta U.; hon. physician to King. Fellow Royal Soc., 1889, Linnean Soc., Zool. Soc. Author: Some Indian Friends and Acquaintances; Plagues and Pleasures of Life in Bengal. Noted parasitic organisms (probably Leishman-Donovan bodies) in Delhi boil, recorded 1885. Died Dec. 31, 1914.

CUNNINGHAM, George John, Brit. pathologist; b. Belfast, North Ireland, Sept. 7, 1906; s. George Samuel and Blanche (Harvey) C.; M.B., B.S., St. Bartholomew's Hosp. Med. Sch., 1933, M.D., 1937; m. Patricia Champion, June 8, 1957. Sr. lectr. pathology St. Bartholomew's Hosp., London, Eng., 1946-55; prof. pathology Royal Coll. Surgeons of Eng., U. London, 1955—. Fellow Coll. Pathologists; mem. Internat. Acad. Pathology (past pres. Brit. div.), Assn. Clin Pathologists (pres.), Quekett Micros. Club (pres.). Contbr. numerous articles to sci. jours., contbr. to Cancer (R. W. Raven). Research on gen. pathology of tumors, lung diseases, changes and mechanisms of prodn. of lung cancer, detection of liver damage by histochem. methods. Home: 97 Maze Hill, Greenwich, London S.E. 10. Office: Royal College Surgeons of Eng., London W.C.2, Eng.*

CUNNINGHAM, Hugh Meredith, Canadian animal nutritionist; b. Brandon, Man., Can., Dec. 28, 1927; s. James Russell and Bertha (Banting) C.; B.S., U. Man., 1949, M.S., 1950; Ph.D., Cornell U., 1953; m. Betty Lou McKillop, June 16, 1951; children—Carol, Gail, Nancy, Craig. Research scientist research br. Can. Dept. Agr., Ottawa, Ont., 1953-59, head div. animal, poultry sci., Nappan, N.S., 1959—; tchr. biochemistry Mt. Allison U., Sackville, N.B., Can., 1959-61. Mem. Am., Canadian (past dir.) socs. animal sci., Nutrition Soc. Can., Profl. Inst. Pub. Service Can., Agrl. Inst. Can., N.S. Inst. Agrologists. Sci. editor: Canadian Jour. Animal Sci., 1965—. Research, numerous publs. on physiology of digestion and ways to improve carcass quality in pigs, new surg. techniques for studying digestion of food in pigs and use of nicotine to make pigs leaner; invented machine for finely mincing carcasses of pigs for chem. analysis; originated techniques for studying digestion in cattle. Address: Nappan, N.S., Can.*

CUNNINGHAM, James, botanist, surgeon; b. Scotland; surgeon to factory East India Co., Emouï, China, 1698; driven from Banjar-Massin by native uprising, 1707; became chief of Banjar under East India Co., 1707. Fellow Royal Soc., 1699. Contbr. numerous articles on meteorology and geography to tech. jours. including Trans. Royal Soc. First Englishman to collect bot. specimens in China; discovered many new plants; observed change of inclination in China; species of madder tribe named in his honor. Died circa 1709.

CUNNINGHAM, Joseph Thomas, Brit. zoologist, biologist; b. London, Eng., 1859; s. W. H. Cunningham; M.A. (scholar 1878-81), Balliol Coll., Oxford, (fellow) Univ. Coll., Oxford, 1882-89; m. Sophie Ingo Crossfield. Asst. to prof. natural history U. Edinburgh, 1883; naturalist Marine Biol. Assn., 1887-97; lectr. Tech. Instrn. Com., Cornwall, 1897-1902; lectr. zoology East London Coll., U. London, 1917-26; on expdn. (with D. M. Reid) to Island of Marajo (at mouth of Amazon) to make expts. on function of vascular filaments of pelvic limbs of male of Am. lungfish Lepidosiren, 1930-31. Fellow Zool. Soc. Author: Treatise on the Common Sole, 1890; Marketable Marine Fishes of British Isles, 1896; Sexual Dimorphism in Animal Kingdom, 1900; Hormones and Heredity, 1921; Modern Biology, 1928; also numerous articles. Editor vol. on Reptiles, Amphibia, Fishes in Animal Life, 1912. Died June 5, 1935.

CUNNINGHAM, Robert Morton, Am. meteorologist; b. Boston, July 1, 1919; s. William Hayes and Mildred (Pilpel) C.; S.B., Mass. Inst. Tech., 1942, Sc.D., 1952; m. Claire Steinhardt, June 3, 1945; children—Peter, James, William. Instr., Mass. Inst. Tech., 1942-43; staff, 1943-52, research asso., 1946-52; cloud physicist Air Force Cambridge Research Labs., Bedford, Mass., 1952—, chief cloud physics br., 1960—. Mem. Am. Meteorol. Soc. (chmn. local br. 1965—), Am. Geophys. Union, Royal Meteorol. Soc. Contbr. articles to tech. jours., chpts. in books. Devised explanation for radar bright band, a layer melting snow which gives strong return on radar sets; showed how in quantitative fashion precipitation grows in various parts large cyclones and importance of lowest layers in total rainfall; importance of cumulus clouds in producing errors in precise radio tracking systems for missile operations. Home: R.F.D. 2, Rockwood Lane, Lincoln, Mass. 01773. Office: Air Force Cambridge Research Lab., L. G. Hanscom Field Bedford, Mass.*

CUNNINGHAM, Walter Jack, Am. elec. engr.; b. Comanche, Tex., Aug. 21, 1917; s. Walter Jack and Percy Adele (Moore) C.; A.B., U. Tex., 1937, A.M., 1938; Ph.D., Harvard, 1947; m. Barbara Virginia Lynch, Feb. 26, 1944; children—Lawrence Bradford, John Hartwell. Instr. physics and communication engring. Harvard, 1939-46; part-time research OSRD, in acoustics and electric circuits. 1939-46; asst. prof. elec. engring. Yale, 1946-50. asso. prof., 1950-56, prof., 1956—. Fellow of Trumbull Coll. Coll. in Yale. Mem. Acoustical Soc. Am., I.E.E.E., Am. Soc. Engring. Edn., Sigma Xi. Author: Introduction to Nonlinear Analysis, 1958; tech. papers. Bd. editors Am. Scientist, 1955—. Jour. Franklin Inst., 1962—. Application of mathematics to engineering problems, particularly to those problems where nonlinear effects are important. Home: 200 Dessa Dr., Hamden, Conn. 06517. Office: Dunham Lab., Yale U., New Haven 06520.

CUNNINGHAM, William, Brit. mathematician, physician; b. Norfolk, Eng., 1531; became pensioner Corpus Christi Coll., Cambridge (Eng.) U., 1548, M.B., 1557; M.D., U. Heidelberg (Germany), circa 1559; practiced medicine; became pub. lectr. Surgeons' Hall, 1563. Author: The Cosmographicall Glasse, 1559; Commentaria in Hippocratem; Organographia. Astron. observations and calculations; surveying and mapping; designed math. instruments. Died 1586.

CUNNINGHAM, William Francis, Am. pathologist; b. Baraboo, Wis., May 18, 1885; A.B., Notre Dame, 1907; Ph.D., Catholic U., 1912. Asso. with dept. exptl. pathology Coll. Phys. and Surg., Columbia U., N.Y.C.; early investigator Hodgkin's disease. Died Nov. 19, 1940.

CUREAU DE LA CHAMBRE, Marin, French biologist, physicist; b. Le Mans, France, 1594; Jean d'Asse and Anne (Molet); m. Marie Duchesne. Physician, Paris, France, from 1620, to Louis XIV, 1650-69. Mem. French Acad., French Acad. Scis. Author: Nouvelles pensées sur les causes de la lumière, du desbordement du Nil et de l'amour d'inclination 1635; Conjectures sur la digestion, 1636; Charactères des passions, vol. I (exposing function of soul from med. and physiol. viewpoint), 1640; Des Passions Courageuses; De la connoissance des betes; De la haine et de la douleur; Des larmes, de la crainte, du désespoir; Traité de la connoissance des animaux où tout ce qui a esté dict pour et contre le raissonnement des bestes est examiné, 1647; Discours de l'amitié et de la haine qui se trouvent entre les animaux, 1667. Believed he proved that animals could reason, but from particular propositions, whereas man could reason universally; proposed that best way to acquire knowledge of man was to study his physiognomy; wrote more about sci. in French than any other man before Descartes. Died 1669.

CURETON, Thomas Kirk, Jr., Am. educator; b. Fernandina, Fla., Aug. 4, 1901; s. Thomas Kirk and Annie (Jeffreys) C.; B.S., Yale, 1925; B.P.E., Springfield Coll., 1929, M.P.E. with highest honors, 1930; M.A., Columbia, 1937, Ph.D., 1939; m. Portia Brownell Miller, June 26, 1941; children—Kirk Jeffreys, Richard Dozier. Faculty, Suffield (Conn.) Acad., 1925-29, Springfield (Mass.) Coll., 1929-41; faculty U. Ill., 1941—, prof., dir. phys. fitness research lab., 1945—. Recipient Roberts-Gulick Honor award YMCA, 1943, Presdl. citation Nat. Jr. C. of C. Fellow A.A.H.P.E.R. (Honor award), A.A.A.S.; mem. Am. Physiol. Soc., Am. Psych. Soc., Am. Sch. Health Assn., Am. Coll. Sports Medicine, Royal Soc. Health (Eng.), Fedn. Internat. d'Edn. Physique, Am. Assn. U. Profs., Phi Kappa Phi, Phi Delta Kappa, Phi Epsilon Kappa, Phi Delta Theta, Sigma Delta Psi. Author numerous books, latest include: (with S. C. Staley) Exercise and Fitness; (with others) Health and Fitness in the Modern World; (with W. Raab) Measurement and Evaluation in Health, Physical Education and Recreation, 1950; (with Alan J. Barry) Improving the Physical Fitness of Youth, 1965; Physical Fitness and Dynamic Health, 1965. Research, publs. on phys. fitness, aquatics and health, kinesiology, applied physiology. Home: 501 E. Washington St., Urbana, Ill. 61801.*

CUREUS, Joachim, physician, scientist; b. Freystadt, Silesia, Oct. 23, 1532; s. Gregor Scheer and Margaretha (Jung) C.; student of Melanchthon, Wittenberg, Germany, 1550; Magister, 1554; studied medicine Padua and Bologna, Italy, 1557-58; M.D., 1558; m. Anna, after 1557; children—at least 2 sons, including Adam. Became physician, 1559, Glogau, Silesia. Libellus physicus continens doctrinam de natura . . . , 1567; Exegesis perspicua et ferme integra contoversiae de Sacra Coena . . . , 1574; Spongia exigua et mollis, comparata ad eluendos colores quos illerit controversiae de sacra coena Paulus Eberus . . . , 1575; In febri quartana et epilepsia pro quadam puellula quatuor annoram. studies, writings on theol., med., bot., hist., polit. subjects. Died Glogau, Jan. 21, 1573.

CURIALTI, Pietro, see Da Tossignano, Pietro.

CURIE, Daniel, French physicist; b. Paris, July 27, 1927; s. Maurice and Raymonde (Simonin) C.; Licence ès sciences mathématiques, 1947, Doctorat ès sciences physiques, 1951; m. Germaine Buchader, Oct. 21, 1952; With Faculté des Sciences, Paris, 1947-55, 58—, prof., 1960—; research staff Centre National de la Recherche Scientifique, 1955-58. Mem. Société Française de Physique (Prix Ancel), Société de Chimie-Physique, Electrochem. Soc. Author: (with Maurice Curie) Questions Actuelles en Luminescence Cristalline, 1956; Luminescence Cristalline, 1960; also articles. Research on theories of crystal luminescence and electroluminescence. Home: 10 rue Cuif, Saint-Maurice, France 94. Office: 9 Quai Saint Bernard, Paris, France 75.*

CURIE, Marie (Maria Sklodowska), chemist, physicist; b. Warsaw, Poland, Nov. 7, 1867; d. Vladislav Sklodowski; tutored by her father, also student Warsaw high sch.; student math. and physics Sorbonne, Paris (France), Licence ès sciences physiques, 1893, Licence ès sciences mathématiques, 1894, D. phys. sci., 1903; m. Pierre Curie, July 26, 1895; children—Irene (Mrs. Frederick Joliot), Eve. Governess in Poland, 1885-91; engaged in independent research in labs. of municipal sch. of physics and chemistry, circa 1894; tchr. École Normale for Women, Sèvres, from 1900; succeeded her husband as prof. physics Sorbonne, Paris, 1906, first woman titular prof. at Sorbonne, dir. Radium Inst., 1914-34. Founder Radium Inst., Warsaw, 1913; provided radiological and X-ray services to hosps. during World War II; visited U. S., and was presented with gram of radium-salt by Pres. Harding, 1921. Recipient (with husband and Henri Becquerel) Nobel prize for physics, 1903, (with husband) Davy medal Royal Soc. London, 1903, Nobel prize for chemistry (only person to receive prize twice), 1911. Nominated for French Acad. Scis. (but rejected by 1 vote because of her sex), 1911; mem. Académie de médecine, 1922. Author: Recherches sur les substances radioactives (doctoral thesis), 1903; Traité de radioactivité, 1910; La radiologie et la guerre, 1921; Radioactivité et phénomènes connexes, 1923; L'Isotopie et les éléments isotopes, 1924; Radioactivité, pub. 1935. Famous for work in radioactivity; the curie (quantity of radon in radioactive equilibrium with 1 gram of radium) named in honor of the Curies; studied and measured radiations given off by uranium; showed that amount of radioactivity in uranium compounds is in proportion to amount of uranium they contain; showed that heavy element, thorium is radioactive, 1898; tested all elements for radioactivity; discovered (with husband) polonium in pitchblend, and isolated it; also detected as trace impurity a still more radioactive substance, which they named radium, 1898; isolated radium, 1902; isolated (with Debierne) 1 gram of pure radium metal, 1910; found (with husband) that beta rays carry negative charge, also investigated excited or induced radioactivity; did pioneer work in use of radioactivity in medicine. Died of leukemia caused by overexposure to radioactive substances, Haute Savoie, France, July 4, 1934.

CURIE, Paul-Jacques, French physicist; b. Paris, France, 1855; s. Eugène and Claire (Depoully) C.; lectr. mineralogy at U. Montpellier (France), 1883; studied geology of N. Africa, 1890; prof. Paris Faculty Scis., 1904; recipient le Prix Planté, 1895. Author: Étude succinctes sur les roches éruptives de l'Algérie; Recherches sur le pouvoir inducteur specifique et sur la conductibilité des corps cristallisés, 1888. Discovered (with bro. Pierre) piezoelectricity and invented an improved quadrant electrometer; research on crystalized substances. Died 1941.

CURIE, Pierre, French physicist, chemist; b. Paris, France, May 15, 1859; s. Eugène Curie; doctorate, Sorbonne, Paris (France), 1895; m. Maria Sklodowska, July 26, 1895; 2 daus. Asst. tchr. at phys. lab., Sorbonne, from 1878; dir. labs. municipal sch. physics and chemistry, from 1882, prof., 1895-1904; prof. physics (chair created for him) Sorbonne, 1904-06. Recipient (with wife and Henri Becquerel) Nobel prize for physics, 1903, also other prizes with his wife for their joint work. Mem. Acad. Scis. Author: Les propriétés magnétiques des corps à diverses températures, 1895; Oeuvres de Pierre Curie (collected works, with preface by Marie Curie), 1908. Research on crystallography; discovered phenomenon of piezoelectricity; investigated effect of heat on magnetism, and showed that there is critical temperature (Curie point) above which magnetic properties disappear, 1895; collaborated with his wife on study of radioactivity, also in discovery of polonium and radium. Died Paris, Apr. 19, 1906.

CURL, Herbert Charles, Jr., Am. oceanographer; b. N.Y.C., Feb. 28, 1928; s. Herbert Charles and Erna (Locke) C.; B.S., Wagner Coll., 1950; M.S., Ohio State U., 1951; Ph.D., Fla. State U., 1956; m. Elizabeth Louise Simpson, June 28, 1952; children—Michael C., Sophia Louise. Research asso. Woods Hole Oceanographic Instn., Woods Hole, Mass., 1956-61; asso. prof. dept. oceanography Ore. State U., Corvallis, 1961—. Cons., USPHS, 1963-66. Mem. Am. Soc. Limnology and Oceanography (sec. 1964-66), Am. Inst. Biol. Scis., Phycological Soc. Am., A.A.A.S., Ecol. Soc. Am., Internat. Limnology Assn. Research, publs. on physiol. ecology marine phyto-plankton algae, ecology snow algae.

CURLING, Thomas Blizard, Brit. surgeon; b. London, Eng., 1811; ed. Manor House, Cheswick, Eng.; asst. surgeon London Hosp., 1832-49, surgeon, 1849-69; began practice in Brighton, Eng., 1871; examiner surgery U. London. Fellow Royal Soc., 1850; mem. Coll. Surgeons (pres.). Author: Essay on Tetanus; A Practical Treatise on the Diseases of the Testis, and of the Spermatic Cord, and Scrotum. Described acute ulceration of duodenum caused by burns of the skin and superficial tissues; reported acute perforating duodenal ulcer caused by burns (Curling's ulcer), 1866; 1st to describe cretinism and suggested thyroid deficiency as its cause. Died Cannes, France, 1888.

CURME, George Oliver, Jr., Am. chemist; b. Mount Vernon, Ia., Dec. 24, 1888; s. George Oliver and Caroline Chenoweth (Smith) C.; B.S., Northwestern University, 1909, D.Sc. (honorary), 1933; grad. student Harvard University, 1909-10; Ph.D., University of Chicago, 1913, D.Sc. (hon.), 1954; student, U. Berlin, 1913-14; m. Lillian Hale, June 29, 1916; children—Katharine Hale (Mrs. Randolph C. Neely), George Oliver III, Mary Ellen (Mrs. Charles P. Cooper, Jr.), Florence Louise (Mrs. Richard C. Horton), John Henry. Industrial research chemist for the Mellon Institute, 1914, conducting research on organic synthesis based on natural hydrocarbons, on behalf of cos. which later became units of Union Carbide and Carbon Corp.; became chief chemist, Carbide and Carbon Chemicals Corp., Inc. to commercialize results of preceeding research, with plants near Charleston, W.Va., 1920, vice pres. and dir. of research, 1929-44, vice president 1944-51; vice president Union Carbide and Carbon Research Lab., Inc., 1938-52; chmn. bd., 1952-54; vice president and director Carbide and Carbon Chemicals Ltd., 1944-55; v.p. research Union Carbide Corp. (formerly Union Carbide & Carbon Corporation), 1951-55, member of the board of dirs., 1952-61. Chandler medalist, 1933, Perkin medalist, 1935, Elliott Cresson medalist, 1936, Willard Gibbs medalist, 1944; recipient Nat. Modern Pioneer Award, 1940. Member American Chem. Soc., Am. Inst. Chem. Engrs., Soc. Chem. Industry, A.A.A.S., Nat., N.Y. acads. sci., Phi Beta Kappa, Sigma Xi. Study of organic chemical reactions; plastics; hydrocarbon gases. Home: 135 E. 54th St., N.Y. City 22. Office: 270 Park Av., N.Y.C. 17, N.Y.

CURNEN, Edward Charles, Jr., Am. physician; b. Yonkers, N.Y., Jan 5, 1909; s. Edward Charles and Florence (Mayer) C.; grad. Hill Sch., 1927; A.B., Yale, 1931; M.D., Harvard, 1935; m. Marion Clement, Apr. 18, 1942; children—Sheila, Edward C. III, Constance Avery. Harvard research fellow Thorndike Meml. Lab., Boston City Hosp., 1938; asst. pediatrics Harvard Med. Sch., 1938-39; asst. Rockefeller Inst. Med. Research, 1939-46, asst. resident physicians hosp. of Inst., 1939-46; asst. prof. preventive medicine Yale Sch. Medicine, 1946-47, mem. Poliomyelitis study unit, 1946-52, asso. prof. pediatrics and preventive medicine, 1948-52. fellow Berkeley Coll., Yale, 1947-52; successively asso. physician, asso. pediatrician and head bacteriology labs., Grace-New Haven Community Hosp., 1946-52; prof. pediatrics, chmn. dept. U. N.C. Sch. Medicine, 1952-60; chief pediatrics service N.C. Meml. Hosp., 1952-60, chief staff, 1959-60, also cons. pediatrician N.C. State Bd. Health; Carpentier prof. pediatrics, chmn. dept. Columbia Coll. Phys. and Surg., 1960—; dir. pediatrics service Presbyn., Babies hosps., 1960—. Asso. commn. influenza Armed Forces Epidemiological Bd., 1947—; planning com. studies cardiovascular effects influenza Nat. Insts. Health, 1957, bd. sci. counselors div. biologics standards, 1959-63; infectious diseases and tropical medicine tng. grant com. Nat. Inst. Allergy and Infectious Diseases, 1960-63. Chmn. allergy and infectious diseases panel Health Research Council City of N.Y., 1962—. Recipient Presdl. award Internat. Poliomyelitis Congress. Fellow Am. Acad. Microbiology, A.A.A.S., Am. Pub. Health Assn., N.Y. Acad. Sci.; mem. Am. Acad. Pediatrics (chmn. com. control infectious diseases 1957-59, cons. 1959—), Am. Assn. History Medicine, Am. Assn. Immunologists, Am. Fed. Clin. Research, A.M.A., Am. Pediatric Soc. (council 1956-61, chmn. 1960-61), Am. Soc. Clin. Investigation, Assn. Am. Med. Colls., Beaumont Med. Club, Boyleston, Yale (pres. 1951-52) med. socs., Harvey Soc., Interurbam Clin. Club, Soc. Exptl. Biology and Medicine, Soc. Pediatric Research. Am. Soc. Microbiology, Aesculapian Club. Sigma Xi, Alpha Omega Alpha. Contbr. articles profl. jours. Editorial bd. Medicine, 1955—. Isolated Coxsackie virus from poliomyelitis patients, 1949; further work in infectious diseases, bacteriology, virology, hemagglutination phenomena. Home: 110 East End Av., N.Y.C. 10028. Office: 3975 Broadway N.Y.C. 10032.

CURRAN, Samuel Crowe, physicist; b. Ballymena, North Ireland, May 23, 1912; s. John Hamilton and Sarah (Crowe) C.; M.A., Glasgow (Scotland) U., 1933, B.Sc., 1934, Ph.D., 1937, D.Sc., 1950; Ph.D., Cambridge (Eng.) U., 1940; m. Joan Elizabeth Strothers, Nov. 7, 1940; children—Sheena, John, Charles, James. Minister aircraft prodn. Ministry of Supply, 1939-44; faculty Manhatten U. Cal., 1944-45, Glasgow U., 1945-55; sr. lectr. U.K. Atomic Energy Authority, 1955-58; chief scientist A.W.R.E., Aldermaston, Berkshire, Eng., 1958-59; prin. Royal Coll. Sci. and Tech., 1960-64; prin., vice chancellor U. Strathclyde, Glasgow, Scotland, 1964—. Mem. Adv. Council on Tech., 1964—, chmn. U. Relations Bd., 1966—; mem. Sci. Research Council, 1964—, Scottish Econ. Planning Council, 1965—; chmn. adv. Com. on Med. Research, 1962—, Fellow Royal Soc. Edinburgh, Royal Soc., Phys. Soc., Royal Coll. Physicians and Surgeons. Author: Counting Tubes, 1949; Scintillation Counter, 1953; also numerous articles. Research on nuclear transmutations, proximity fuse, centimetre radar, spectrum long-lived radioelements and work on ultra-sensitive radiation detections devices; discovered scinitillation counter modern gas-filled proportional counter and spectrometer. Home: 6 Sinclair Dr., Helensburgh, Dunbartonshire, Scotland. Office: George St., Glasgow, Lanarkshire, Scotland.*

CURRI, Sergio Bertini, Italian molecular biologist; b. Fiume, Italy, Apr. 25, 1927; s. Dante and Tea (Bolf) C.; grad. in medicine and surgery U. Padova (Italy), 1951; m. Edvige Panin, Aug. 7, 1954 (dec. July 1965). Faculty, U. Padova, 1951—, prof. path. anatomy, 1958—, prof. histochemistry 1962—; dir. Istituto di Biologia Sperimentale, Milano, Italy, 1954-59, Istituto di Biologia Sperimentale Abano Terme, Padova, 1959-66, Center Molecular Biology, Padova, 1966—. Hon. mem. French Soc. Angiology; mem. Italian Soc. Histo-chemistry, Italian Soc. Path. Anatomy. Author: (A. Cavallini, M. Maioli) La malattia varicosa degli arti inferiori. Fistole arterovenose congenite e anastomosi arteovenose, 1963; Angiopatie periferiche e anastomosi arterovenose, 1964; also numerous articles. Research on relationship between morphology and histochemistry especially molecular biology, biol. and therapeutic properties of hyaluronic acid, complexes between phospholipids, proteins and RNA from central nervous system of human subjects and animals; first to isolate phospholipidic fractions from hypothalamus. Home: via Lister n. 3. Office: Centro di Biologia Molecolare, via Lister n. 3, Padova, Italy.*

CURRIE, James, physician; b. Kirkpatrick, Fleming, Dumfriesshire, Scotland, May 31, 1756; s. James Currie; student medicine and metaphysics U. Edinburgh (Scotland), 1776-79; grad. U. Glasgow (Scotland), 1780; m. dau. of William Wallace, Jan. 1783; 1 son, William Wallace. Trader in Va., 1771; came to London, Eng., 1777; physician at Liverpool, Eng., from 1780; advocated abolition of slave trade in 1787. Fellow Royal Soc., 1792. Author: Medical Reports on the Effects of Water, Cold and Warm as a Remedy in Fever and Febrile Diseases, 1797. First systematic English clin. thermometrical observations; 1st to use cold bathing for fevers. Died Eng., Aug. 31, 1805.

CURRIE, Malcolm Roderick, Am. physicist, elec. engr.; b. Spokane, Wash., Mar. 13, 1927; s. Erwin Casper and Genevieve (Hauenstein) C.; A.B., U. Cal. at Berkeley, 1949, M.S., 1951, Ph.D., 1954; m. Sunya Lofsky, June 24, 1951; children—Deborah, David, Diana. With Hughes Aircraft Co., Culver City, Cal., 1954—, asso. dir. research labs., 1961-65, v.p. research labs., 1965-66, v.p. and mgr. research and devel. div. Aerospace Group, 1966—; lectr. U. Cal., Los Angeles, 1955-57. Named Nations Outstanding Young Elec. Engr. Eta Kappa Nu, 1958; One of five Outstanding Young Men of Cal., Cal. Jr. C. of C., 1960. Fellow I.E.E.E., Inst. Aeros. and Astronautics; mem. Am. Phys. Soc., A.A.A.S., Phi Beta Kappa, Sigma Xi. Invented microwave tubes; research, publs., patents on noise in electron beams and devel. of lowest noise tunes yet achieved, on parametric amplifiers, on ion propulsion, plasma physics, millimeter waves, ion beam neutralization, gaseous lasers. Home: 1022 Maroney Lane, Pacific Palisades, Cal. 90272. Office: Hughes Aircraft Co., Culver City, Cal.*

CURRY, Alan Stewart, Brit. forensic scientist; b. Blackpool, Eng., Oct. 31, 1925; s. Richard C. and Margaret (Booth) C.; B.A., Trinity Coll., Cambridge, 1946, M.A., 1948, Ph.D., 1952; m. Miriam Mawdsley, July 26, 1947; 1 son, Christopher C. S. Staff home office Forensic Sci. Labs., Wakefield, Eng., 1952-54, Harrogate, Eng., 1954-64, dir., Nottingham, Eng., 1964-66; dir. home office Central Research Establishment, Reading, Eng., 1966—. Fellow Royal Inst. Chemistry; mem. Coll. Pathologists, Am. Acad. Forensic Scis. Author: Poison Detection in Human Organs, 1963; Methods of Forensic Science, vol. 3, 1964, vol. 4, 1965; (with Stewart, Stolman) Toxicology, vol. I, 1960, vol. II, 1961; also articles. Research on methods of detecting and estimating poisons administered to humans, especially criminal poisoning, use of science in detection of crime. Office: Home Office, Central Research Establishment, Aldermaston, Reading, Eng.*

CURRY, Haskell B(rooks), Am. math. logician; b. Millis, Mass., Sept. 12, 1900; s. Samuel Silas and Anna (Baright) C.; A.B., Harvard, 1920, A.M., 1924; postgrad. Mass. Inst. Tech., 1920-22; Ph.D., U. Göttingen, Germany, 1930; m. Mary Virginia Wheatley, July 3, 1928; children—Anne Wright (Mrs. Richard Shaner Piper), Robert Wheatley. Faculty, Harvard, 1926-27, Princeton, 1928-29; faculty Pa. State Coll. (now Pa. State U.), University Park, 1929-66, Evan Pugh Research prof., 1960-66; prof., dir. Inst. for Foundational Research and Philosophy of Exact Scis., U. Amsterdam, Netherlands, 1966—. Mathematician, Frankford Arsenal, 1942-44, applied physics lab. Johns Hopkins, 1944-45, Ballistic Research Labs., Aberdeen Proving Ground, Md., 1945-46; cons. computing methods U. S. Naval Ordnance Lab., 1948-49. Hon. vis. prof. U. Louvain, 1951. Fellow A.A.A.S.; mem. Academie Internationale de Philosophie des Sciences Brussels (assesseur, 1961-65, v.p. 1966—), Am. Math. Soc. (council 1945-47), Am. Ornithologists Union, Am. Philos. Assn., Assn. Computing Machinery, Assn. Symbolic Logic (v.p. 1935-36, pres. 1937-39), Math. Assn. Am., Wiskundig Genootschap te Amsterdam, Sigma Xi, Pi Mu Epsilon, Sigma Pi Sigma. Author: A Theory of Formal Reducibility, 1950; Outlines of a Formalist Philosophy of Mathematics, 1951; Leçons de logique mathematique, 1952; The Foundations of Mathematical Logic, 1963; (with Robert Feys) Combinatory Logic, vol. 1, 1958. Research in math. logic and founds, math.; theory of formal systems and formal processes; system of combinatory logic; formulations of logical calculus using inferential rules. Home: 99 hs, Amsterdam-Z, Netherlands.*

CURRY-LINDAHL, Kai, Swedish zoologist; b. Stockholm, Sweden, May 10, 1917; s. Kossuth and Margit (Törnblom) C.-L.; student Sigtuna (Sweden) Coll., Lycee Chateaubriand, Rome; m. Anne van der Voordt, Sept. 3, 1947; children—Brigitte, Edithe, Robin. Sci. editor Natur och Kultur pub. house, 1937-44, editor Sveriges Natur, Swedish Soc. Conservation Nature, dir. Svensk Natur pub. house, 1945-53, dir. zoöl. dept. Nordiska Mus. and Skansen, 1953—, (all Stockholm); leader Lund U. and Swedish Congo expdns., 1958-59. Conservation adviser several govts., especially in Africa. Recipient Order of Crown, Belgium, 1961; Order of Leopold, Belgium, 1964; Geoffrey St. Hilaire gold medal, France, 1963; Silver medal Swedish Acad. Scis., 1962; others. Mem., officer, numerous internat. and nat. ecol. and zool. assns. in Am., Europe, Africa, Asia. Author: Ecological Studies on Mammals, Birds, Reptiles and Amphibians in the Eastern Belgian Congo, vols. 1-2, 1956-60; Sarek, 1960; Contribution à l'étude des vertébrés terrestres en Afrique tropicale, vol. 1, 1961; Flyttfaglarnas tropiska vinterhem, 1961; Skogar och djur, 1961; Arktis och tropik, 1963; Natur i Lappland, vols. 1-2, 1963; Nordens djurvärld, 1963; Vara Faglar i Norden, vols. 1-4, 1959-63; Europe—A Natural History, 1964; Fiskarna i färg, 1964; Djuren i färgdäggdjur, kräldjur, groddjur, 1965; numerous others; also numerous articles. Research on ecology, biology, population dynamics of vertebrates, behavior of herons, physiology of migratory birds, conservation of nature Home: Skansen, Stockholm. Office: Nordiska Mus. and Skansen, Stockholm, Sweden.*

CURSCHMANN, Hans, German neurologist; b. Berlin, Germany, Aug. 14, 1875; prof. in Rostock, Germany. Author: (with F. Kramer) Lehrbuch der Nervenheilkunde, 1909; Nervenkrankheiten, 1924; Pathogenese und Therapie der Arteriosklerose, 1920; Endokrine Krankheiten, 1927; Lehrbuch der speziellen Prognostik, 1943. Died Rostock, Mar. 10, 1950.

CURTAIN, Cyril Curtis, Australian biochemist; b. Melbourne, Australia, Jan. 29, 1928; s. Charles Henry and Florence (Seabridge) C.; B.Sc., Melbourne U., 1949, M.Sc., 1951, Ph.D., 1953, D.Sc., 1965; m. Betty Wallace, Dec. 15, 1956; children—Roger Charles, Diana Margery. Research asst. Walter and Eliza Hall Inst. Melbourne, 1951-53; postdoctoral fellow dept. biochemistry Oxford (Eng.) U., 1954-55; biochemist Baker Med. Research Inst., Melbourne, 1955-66; prin. research scientist CSIRO Div. Animal Health, Melbourne, 1967—. Fellow Royal Australian Chem. Inst.; mem. Soc. for Med. and Biol. Electronics (past pres.), Australian Biochem. Soc. Research, publs. on tropical hypergammaglobulinaemia in relation to regulation of antibody synthesis, genetics of Melanesian Peoples. Home: 12 Esplanade, Williamstown, Victoria 3016. Office: CSIRO Div. Animal Health, pvt. bag 1, Parkville, Victoria 3052, Australia.*

CURTIN, Charles Byron, Am. biologist; b. Cohoes, N.Y., Aug. 2, 1917; s. Charles David and Margaret (Byron) C.; B.S., George Washington U., 1945; M.S., Cath. U., 1947; Ph.D., U. Pitts., 1956; m. Antoinette Lisz, Mar. 17, 1945; children—David Anthony, Michele Jean. Sr. fingerprint analyst U. S. Dept. Justice, FBI, Washington, 1941-46; prof. Mt. St. Mary's Coll., Emmitsburg, Md., 1947-57; staff editor McGraw-Hill Ency. Sci. and Tech., Charlottesville, Va., 1957-62; asso. prof. biology Creighton U., Omaha, 1962—. Cons., Campbell Soup Co., 1965—. Mem. A.A.A.S., Ecology Soc., Pa. Acad. Sci., Soc. Systematic Zoology, Am. Microscope Soc., Neb. Acad. Sci., Am. Soc. Zoologists, Am. Mus. Natural History. Research, numerous publs. on invertebrate zoology including taxonomy, ecology, evolution free-living and parasitic forms. Home: 6218 Florence Blvd., Omaha 68110.*

CURTIN, David Yarrow, Am. chemist; b. Phila., Aug. 22, 1920; s. Ellsworth Ferris and Margaretta (Cope) C.; A.B., Swarthmore Coll., 1943; Ph.D., U. Ill., 1945; m. Constance O'Hara, July 1, 1950; children—Susan, David F., Jane. Research asst. Harvard, 1945-46; faculty Columbia, 1946-51; faculty U. Ill., Urbana, 1951—, prof. chemistry, 1954—. Mem. Am., Brit., Swiss chem. socs., Nat. Acad. Scis. Research, numerous publs. stereochemistry, exploratory organic chemistry, mechanisms organic reactions. Home: 3 Montclair Rd. Office: 354 E. Chemistry, U. of Ill., 61801.*

CURTIS, Adam Sebastian, Brit. biologist; b. London, Eng., Jan. 3, 1934; s. H. Lewis and Nora (Stevens) C.; B.A., U. Cambridge (Eng.), 1955, M.A., 1958; Ph.D. Edinburgh (Scotland) U., 1957; m. Ann Park, May 3, 1958; children—Penelope, Susanna. Research worker dept. anatomy U. Coll. London, 1957-62, lectr. zoology, 1962-67; prof. cell biology U.

Glasgow (Scotland), 1967—. Dir. Co. Biologists. Author: The Cell Surface—its Molecular Role in Morphogenesis, 1967; also articles. Editor: Med. and Biol. Illustration, 1962-65; sec. editorial bd. Jour. Embryology and Exptl. Morphology, 1964—. Research on embryology, cellular basis of malignancy. Home: 50 Hillhead St., Glasgow W. 2. Office: Dept. Cell Biology, U. Glasgow, Glasgow W. 2, Scotland.*

CURTIS, Arthur Covel, Am. physician; b. Binghamton, N.Y., June 30, 1898; s. Anson Bartie and Mae (Christie) C.; B.S., U. Mich., 1923, M.D., 1925; m. Affa Myrtice Leek, Aug. 18, 1928; 1 son, James Christie. Faculty, U. Mich., Ann Arbor, 1928—, prof. dermatology, 1942—, chmn. dept., 1946—; mem. staffs Univ., St. Joseph's Mercy, Wayne County hosps. Cons. USPHS, Surgeon Gen. U. S. Army; examiner Am. Bd. Dermatology, 1949-57. Diplomate Am. Bd. Internal Medicine, Am. Bd. Dermatology and Syphilology (pres. 1955-56). Fellow A.C.P., A.A.A.S., Soc. Investigative Dermatology (pres. 1953-54), Am. Soc. for Clin. Investigation, Am. Acad. Dermatology and Syphilology (pres. 1956), Am. Dermatol. Assn.; mem. Assn. Profs. Dermatology (pres. 1963), Am. Venereal Disease Assn. (pres. 1963, 66—), A.M.A. (chmn. dermatology sect. 1961), Am. Social Health Assn. (trustee), Central Soc. for Clin. Research, Detroit (pres. 1947-48), Chgo. dermatol. assns., Sigma Xi, Alpha Omega Alpha, Phi Sigma, Alpha Kappa Kappa, Gamma Alpha. Asso. editor Jour. Investigative Dermatology, 1948-60; dermatology editor Am. Lecture Series, 1949-66. Contbr. articles to med. jours. Home: 2027 Medford St., Ann Arbor, Mich. 48104.*

CURTIS, Charles Gordon, Am. inventor; b. Boston, Apr. 20, 1860; s. George Ticknor and Louise A., C.; C.E., Columbia, 1881, M.S., 1907; LL.B., N.Y. Law Sch., 1883. Patent lawyer, 8 yrs.; organized C. & C. Electric Motor Co., the first to make electric motors and electric fans; organizer, pres. Curtis Electric Mfg. Co.; introduced steam turbine of own design into British, Japanese, German and U. S. Navies. Recipient Count Rumford Gold and Silver medals Am. Soc. of Arts and Sciences. Invented Curtis Scavenging System for 2-cycle engines. Died Central Islip, N.Y., Mar. 10, 1953.

CURTIS, Heber Doust, Am. astronomer; b. Muskegon, Mich., June 27, 1872; s. Orson B. and Sarah E. (Doust) C.; A.B., U. Mich., 1892, A.M., 1893, Ph.D., U. Va., 1902; m. Mary D. Rapier, July 12, 1895; children—Margret Evelyn (Mrs. Alexander Walters), Rowen Doust, Alan Blair, Baldwin Rapier. Prof. Latin, Napa Coll., 1894-97; prof. math. and astronomy U. of Pacific, 1897-1900; fellow in astronomy Leander McCormick Obs., U. Va., 1900-02; asst. Lick Obs., 1902-04, asst. astronomer, 1904-06; acting astronomer in charge of D. O. Mills Expdn. to Southern Hemisphere, 1906-09; astronomer Lick Obs., 1909-20; dir. Allegheny Obs., 1920-30; dir. obs. U. Mich., from 1930. Mem. Nat. Acad. Scis. Spl. editor for astrophysics, 2d edit. Webster's New Internat. Dictionary. Observed 11 total solar eclipses—Georgia, 1900, Sumatra, 1901, Labrador, 1905, Russia, 1914, Washington, 1918, Mexico, 1923, New Haven, 1925, Sumatra, 1926, 29, Nevada, 1930, Maine, 1932; investigated the spiral nebulae. Died Ann Arbor, Mich., Jan. 8, 1942.

CURTIS, Howard James, Am. biologist; b. Lansing, Mich., Dec. 11, 1906; s. Harve L. and Anna (Puffer) C.; B.S., U. Mich., 1928; M.A., Swarthmore Coll., 1929; Ph.D., Yale, 1932; m. Dorothy Albert, Aug. 27, 1932; children—Brian A., Richard H., Barbara Bavaro (Mrs. Joseph Bavaro). Biophysicist, Biol. Lab., Cold Spring Harbor, N.Y., 1932-35; faculty Columbia, 1936-47; fellow Rockefeller Found., Johns Hopkins, 1938-40; head biology div. Clinton Labs., Oak Ridge, 1943-46; prof. physiology, head dept. Vanderbilt U. Med. Sch., 1947-50; sr. biologist, chmn. dept. Brookhaven Nat. Lab., Upton, N.Y., 1950—. Mem. various coms. NRC; chmn. radiation study sect. USPHS, 1956-61. Mem. Am. Physiol. Soc., Radiation Research Soc. (past pres.), Biophys. Soc., Soc. Gen. Physiologists. Author: (with Philip Bard), 1940; Biological Mechanisms of Aging, 1965; also numerous articles. Research on structure and function living cells by electric impedance measurements, mechanism conduction in nerve, biol. effects radiations, biol. mechanisms aging. Home: Fitzgerald Rd., Shoreham, N.Y. 11786. Office: Brookhaven Nat. Lab., Upton, N.Y. 11973.*

CURTIS, James Wylie, Am. psychologist; b. Madison, Ind., July 3, 1913; s. Wylie Ralph and Gertrude (Allison) C.; A.B., U. Ky., 1937, M.S., 1938; postgrad. Princeton, Universidad de Panama; m. Mildred Louise Fisher, Apr. 29, 1942; children—James W., Carol Ann. Asst. instr. U. Ky., Lexington, 1937-38; research psychologist U. S. Forest Service, Winchester, Ky., 1938-39; acting head dept. psychology Pikeville (Ky.) Coll., 1939-41; supervising psychologist Ill. Div. Vocational Rehab., Springfield, 1947—; staff psychologist Meml. Hosp. Sch. Nursing, Springfield, 1951—. Personnel cons. to banks; testing cons. St. John's Sch. Nursing, 1955—. Named col., aide-de-camp Gov. Ky., 1938. Fellow A.A.A.S.; mem. Midwest, Ill. psychol. assns. Contbr. articles to tech. jours. Developed psychol. tests, techniques in adapting standard ability tests for handicapped individuals; pioneered study relationship between hypnotic susceptibility and intelligence. Home: 2117 Noble Av., Springfield 62704. Office: 227 S. 7th St., Springfield, Ill. 62706.*

CURTIS, Morton Landers, Am. mathematician; b. Port Lavaca, Tex., Nov. 11, 1921; s. David Morton and Rosena (Montier) C.; B.S., Tex. A. and I. Coll., 1943; Ph.D., U. Mich., 1951; m. Eleanor Thomas, Aug. 12, 1944; children—Dana Allan, Jacqueline Ann. Instr., Northwestern U., 1950-51, asst. prof., 1953-56; mem. Inst. Advanced Study, Princeton, 1951-53; prof. math. U. Ga., 1956-59, Fla. State U., 1959-64; prof. math., chmn. dept. Rice U., 1964—. Served to lt. USNR, 1943-46. Mem. Am. Math. Soc. (trustee 1966—), Math. Assn. Am., Sigma Xi. Contbr. profl. jours. Research in topology. Home: 4040 San Felipe St., Houston 77027.

CURTIS, Moses Ashley, Am. botanist; b. Stockbridge, Mass., May 11, 1808; s. Jared and Thankful (Ashley) C.; grad. William Coll., 1827; m. Mary de Rosset, Dec. 3, 1834. Ordained to ministry Episcopal Ch., 1835; missionary, N.C., 1835-37; tchr. Episcopal Sch., Raleigh, N.C., 1837-39; pastor in Hillsboro, N.C., 1841-47, 56-72, Society Hill, S.C., 1847-56; studied vegetation of N.C. during his many missionary travels, specialized in study of fungi. Author: Natural History Survey of North America, Part III, Botany; Containing a Catalogue of the Plants of the State, with Descriptions of History of the Trees, Shrubs, and Woody Vines, 1860. Died Hillsboro, Apr. 10, 1872.

CURTIS, Otis Freeman, Am. plant physiologist; b. Sendai, Japan, Feb. 12, 1888 (parents Am. citizens); s. William Willis and Lydia Virginia (Cone) C.; A.B., Oberlin Coll., 1911; Ph.D., Cornell U., 1916; m. Lucy Marguerite Weeks, Aug. 27, 1913; chidren—Otis Freeman, William Edgar, Margaret Ann. Tutor Oberlin Acad., 1911-12; with Cornell U., 1913-49, instr., later asst. prof. botany until 1922, prof. botany, plant physiologist, Expt. Sta., 1922-49; vis. prof. U. Leeds, Eng., 1926-27, Ohio State U., 1930-31. Mem. A.A.A.S., Bot. Soc. Am., Am. Assn. Naturalists. Am. Soc. for Hort. Sci., Am. Soc. Plant Physiologists (v.p. 1936-37, pres. 1937-38), Phi Kappa Phi, Sigma Xi, Gamma Alpha. Author: Translocation of Solutes in Plants; Introduction to Plant Physiology, 1949. Research on translocation of foods in green plants, vegetative propagation and water and temperature relations of plants. Died July 4, 1949.

CURTIS, William, Brit. botanist; b. Alton, Eng., 1746; apothecary licentiate. Demonstrator practical botany at med. schs.; established bot. garden, Lambeth Marsh, Eng., also at Brampton, Eng.; founder Bot. Mag., 1787. Fellow Linnean Soc. Author: Flora Londinensis, 1777; Lectures on Botany, 3 vols., 1813-14; British Grasses; Silpha Grisca; Curculio Lapathi. Combined study of insect life with botany. Died Brampton, July 7, 1799.

CURTISS, Charles Francis, Am. chemist; b. Chgo., Apr. 4, 1921; s. Ralph Charles and Camille (Guthormsen) C.; B.S., U. Wis., 1942, Ph.D., 1948; m. Lois Pauline Hruska, Mar. 23, 1946; children—Larry, Glenn, Ned. Faculty, U. Wis., Madison, 1949—, prof., 1960—. Fellow Am. Phys. Soc., A.A.A.S.; mem. Am. Chem. Soc., Soc. Engring. Sci. Author: (with J. O. Hirschfelder, R. B. Bird) Molecular Theory of Gases and Liquids, 1954. Research, numerous publs. on theory of transport phenomena in gases; on theory of molecular collisions. Home: 5760 Forsythia Pl., Madison 53705. Office: 1101 University Av., Madison, Wis. 53706.*

CURTISS, Glenn Hammond, Am. inventor, aviator; b. Hammondsport, N.Y., May 21, 1878; s. Frank R. and Lua (Andrews) C.; ed. pub. schs.; m. Lena P. Neff, 1898; 1 son, Glenn Hammond. Former pres. G. H. Curtiss Mfg. Co., Curtiss Aeroplane Co., Curtiss Motor Co., Curtiss Engring. Co., Curtiss Aeroplane & Motor Corp.; dir. Curtiss Aeroplane & Motor Corp., Curtiss Flying Service. Early began expts. with motor vehicles, establishing motorcycle factory at Hammondsport, 1902. Set speed records for motorcycle, riding his own machines, 1905; at Ormond Beach, Florida, 1907, made record for mile of 46 2/5 seconds with 110 pound motorcycle. Designed aeronautical motors for dirigibles with Capt. T. S. Baldwin, 1907-09, building for U. S. A., Dirigible No. 1; dir. expts. for Aerial Expt. Assn., 1907-09, and supervised constrn. of and piloted "June Bug," July 4, 1908, for 1st pub. flight of mile in U. S., winning Sci. Am. trophy; experimented with the "Loon," an aeroplane fitted with pontoons, 1908; won Gordon Bennett cup and prix de la Vitesse at Rheims, France, Internat. Aviation Meet as rep. Aero Club America, Aug. 1909, with aeroplane and motor of Curtiss design; won N.Y. World prize of $10,000 in flight from Albany to New York in 2 hours, 51 minutes, May 29, 1910; made pub. demonstration of hydroaeroplane, Jan. 1911, with which had been experimenting for number of years, following demonstration of this invention with that of flying boat (awarded prize by Aero Club Am. 1912). Established flying schs. at Hammondsport, San Diego, Buffalo, Newport News, Miami, Atlantic City, 1909-19; introduced flying boat to Brazil, Russia, Austria, Italy and Germany, 1913-14; designed and built for Rodman Wanamaker, the "America," first multimotored flying boat, and 1st heavier-than-air flying craft designed for transatlantic flight, 1914. With J. N. Willys expanded Curtiss factories to meet war demands of Great Britain, Russia, and U. S., 1917. Developed "Wasp" (holder of world's records for speed, climb, and altitude) and other types of aeroplanes, flying boat types, and with U. S. Navy, Navy-Curtiss flying boats 1, 2, 3 and 4, the latter of which made the first Atlantic crossing, May 16-27, 1919. Hon. mem. Nat. Aero. Assn., 1924. Author: (with Augustus Post) Curtiss Aviation Book. Died Buffalo, N.Y., July 23, 1930.

CURTISS, John Hamilton, Am. mathematician, educator; b. Evanston, Ill., Dec. 23, 1909; s. David Raymond and Sigrid Sofia (Eckman) C.; A.B., Northwestern U., 1930; S.M., State U. Ia., 1931; Ph.D., Harvard, 1935. Faculty, Johns Hopkins, 1935-36, Cornell U., 1936-43; asst. to dir. Nat. Bur. Standards, 1946-47, chief applied math. labs., 1947-53; sr. scientist, adj. prof. math. Inst. Math. Sci., N.Y.U., 1953-54; exec. dir. Am. Math. Soc., 1954-59; prof. math. U. Miami, Coral Gables, Fla., 1959—, chmn. dept., 1959-61. Vis. lectr. Harvard, 1953. Recipient medal Dept. Commerce. Fellow A.A.A.S., Inst. Math. Statistics, Am. Statis. Assn.; mem. Am. Math. Soc., Math. Assn. Am., Econometric Soc., Soc. Indsl. and Applied Math., Phi Beta Kappa, Sigma Xi. Founder, editor Nat. Bur. Standards Applied Math. Series, 1946-53; editor various Nat. Bur. Standards, Am. Math. Soc. symposia vols. Research, publs. in methods of solving boundary value problems; probability theory as applied in statistics and computer scis. Home: 8120 S.W. 54th Av., Miami, Fla. 33143. Office: Dept. Math., U. Miami, Coral Gables, Fla. 33124.*

CURTISS, L(eon) F(rancis), Am. physicist; b. Adrian, Mich., Jan. 5, 1895; s. Arthur Truman and Rosa M. (Dibble) C.; A.B., Cornell U., 1917; Ph.D., 1922; m. Chloe L. James, June 27, 1917; children—John A., Robert E. Instr., Cornell U., 1918-22; with Nat. Bur. Standards, 1924-26, physicist 1926-61. Chmn. com. on Nuclear Sci., NRC, 1946-61. NRC fellow, Cambridge, Eng., 1922-24. Mem. Am. Phys. Soc., Washington Acad. Sci., Washington Acad. Medicine, Sigma Xi. Author: Introduction to Neutron Physics, 1958. Investigation of beta-ray spectra; elec. conductivity of thin metal films in magnetic field; detectors of nuclear radiation; neutron physics; cosmic rays; point and tube electron counters; photography of alpha-ray tracks. Address: 1690 Bayshore Dr., Englewood, Fla.*

CURTISS, Ralph Hamilton, Am. astronomer; b. Derby, Conn., Feb. 8, 1880; s. Hamilton Burton and Emily Wheeler (Curtiss) C.; B.S., U. Cal., 1901, Ph.D., 1905; m. Mary Louise Welton, June 17, 1920. Asst. Astron. Obs., U. Cal., 1900; mem. Lick Obs. Eclipse Expdn. to Sumatra, 1901; fellow at Lick Obs., 1901-04; Carnegie asst., same, 1904-05; astronomer, Allegheny Obs., 1905-07; asst. prof. astronomy, 1907-11, asso. prof., 1911-18, prof., 1918—, U. Mich., and asst. dir. Detroit Obs., dir., 1927—. Prin. line of investigation, the properties of stars having bright line spectra in Classes B to A of the Draper Classification. Died Dec. 25, 1929.

CURTIUS, Ernst, German archeologist; b. Lübeck, Germany, Sept. 2, 1814; s. Carl George and Dorothea (Plessing) C.; studied classics, classical civilization in Bonn, Germany, Göttingen, Germany; doctorate Halle, Germany, 1841; m. Auguste Besser, 1850; a son Friedrich; m. 2d, Clara Reichhelm, 1853; a dau., Dora (Mrs. Richard Lepsius). Employed by A. Böckh, Berlin, Germany, 1835; went to Athens, Greece as tutor, 1837; traveled in Greece with other classicists; returned to Berlin, 1841; joined faculty Berlin U., 1843; named tutor to future Kaiser Friedrich, also apptd. asso. prof. at univ., 1844; prof. classical philology, Göttingen, 1855-67; named prof. classic archeology, Berlin, also dir. Altes Mus., 1868; apptd. dir. Antiquarium, 1872; a founder Deutsches Archäologisches Institut; placed in charge of excavation at Olympia. Mem. Berlin Akademie. Author: Commentatio de portubus Athenarum, 1841; (with K. O. Müller) Anecdota Delphica, 1843; Peloponnesos, eine historischgeographische Beschreibung der Halbinsel, 1851-52; Griechische Geschichte I, 1857, II, 1861, III, 1867; (with I. A. Kaupert) Atlas von Athen, 1878; Stadtgeschichte von Athen im Altertum, 1891. Research on Greek inscriptions; wrote 1st Greek history in German; expdn. in Olympia discovered Praxiteles' Hermes and the sculptures of the Zeus temple. Died Berlin, July 11, 1896.

CURTIUS, Friedrich Wilhelm, German chem. industrialist; b. Goch/Niederrhein, Germany, Apr. 21, 1782; s. Caspar Ludwig and Johanna (Fabritius) C.; m. Theodora Pilgrim, 1815; 4 sons and 7 daus., including Julius, Friedrich, Auguste, Emma. Founder drugs and dyes wholesale house, 1815; began organic chem. manufacture, circa 1824; founder sulfuric acid plant, Duisburg, Germany, 1824; founded (with son-in-law) Matthes & Weber (1st W. German soda factory), 1838, later founded a chloride of lime br.; founder Friedrich Curtius & Co. (alum and alumina sulfate factory), 1840, Julius Curtius (ultramarine factory), 1849; ret., 1857; joined A. von Humboldt's expdn. Advocate of syndicates for price stablzn. in German chem. industry. Died Duisburg, Feb. 12, 1862.

CURTIUS, Ludwig Michael, archeologist; b. Augsburg, Germany, Dec. 13, 1874; s. Ferdinand and Therese (Göhl) C.; studied law, then archeology, Munich, Germany; student of H. von Brunn, A. Furtwängler; doctorate, 1903; m. Editha Wyneken, 1921; 2 daus. Traveled in S. Europe; participated in excavations in Aigina, later Boghazköj in Middle E.; named lectr. Munich, 1907; became lectr. Erlangen, Germany, 1908, prof., 1912; went to Freiburg, Germany, after World War I; became prof. Heidelberg, Germany, 1920; named dir. German Archeol. Inst., Rome, Italy, 1928-38. Author: Geschichte der Altorientalische Kunst I, Gilgamisch und Haebani, 1912; Antike Kunst, 1913; Wandmalerei Pompejis, 1929; Zeus und Hermes, Studien zur Geschichte ihres Ideals und seiner Überlieferung, 1931; Die klassische Kunst Griechenlands, 1938; Das antike Rom, 1944; Interpretationen von sech griechischen Bilderwerken, 1947; Deutsche und antike Welt, Lebenserrinerungen (autobiography), 1950; also numerous articles. Pioneer in archeol. research Nr. East; specialized in hermae. Died Rome, Apr. 10, 1954.

CURTIUS, Theodor, German organic chemist; b. Duisberg, Germany, May 27, 1857; studied under Kolbe; became prof., Kiel, Germany, 1889, Bonn, Germany, 1897; named prof. chemistry, Heidelberg, Germany (successor to Victor Meyer), 1898; editor Jour. für praktische Chemie. Research and numerous publs. on hydrazides, azides of organic acids; research on polypeptides, beginning 1882; discovered 2 gen. reactions named after him (conversion of an acid to amine or aldehyde and ammonia with intermediary compounds of azide and urethane, also conversion of an azide into an isocyanate); discovered hydrazine, 1887, hydrazoic acid, 1890, organic bonds. Died Heidelberg, Feb. 8, 1928.

CUSHING, Harvey, Am. neurologist; b. Apr. 8, 1869; s. Henry Kirke and Betsey M. (Williams) C.; A.B., Yale, 1891, hon. A.M., 1912, Sc.D., 1919; A.M. and M.D., Harvard, 1895, Sc.D., 1931; M.D., hon. causa, Belfast, 1918, Strasbourg and Brussels, 1930, Budapest and Bern, 1931, Paris, 1933; Sc.D., Washington U., 1915; LL.D., Western Res. U., 1919, Cambridge, 1920. Edinburgh and Glasgow univs., 1927; Litt.D., Dartmouth, 1929; m. Katharine Stone Crowell, June 10, 1902; children—Mary Benedict, Betsey, Henry Kirke, Barbara. Engaged in practice surgery, 1895-1933; asso. prof. surgery Johns Hopkins, 1902-12; prof. surgery, Harvard, and surgeon-in-chief, Peter Bent Brigham Hosp., 1912-32; Sterling prof. neurology, Yale, from 1933. Decorated D.S.M. (U. S.); Companion of Bath (Eng.); Officer Légion d'Honneur (France); recipient Cameron prize U. Edinburgh, 1924; Lister medal (London), 1930. Fellow Royal Soc. 1933; hon. fellow Royal Coll. Surgeons (Eng., Ireland and Edinburgh); mem. Nat. Acad. Scis., Am. Neurol. Assn. (pres. 1923), A.C.S. Author: The Pituitary Body and Its Disorders, 1912; Tumors of the Nervus Acusticus, 1917; The Life of Sir William Osler, 1925 (Pulitzer prize); A Classification of the Gliomata (with P. Bailey), 1925; Consecration Medici and other Essays, 1928; Intracranial Tumours, 1932; Pituitary Body and Hypothalamus, 1932. Renowned for brain operations, particularly with local anesthesia; performed 1st successful operation for intracranial hemorrhage in the newborn, 1905; demonstrated (with S. J. Crowe, J. Homans) that excision of pituitary gland results in atrophy of genital organs, 1910; suggested term, third circulation, to designate cerebrospinal fluid system, 1926; demonstrated evolution of different histological types of intracranial and intraspinal neoplasms; formulated Cushing's law (increase of intercranial tension causes increase of blood pressure to a point above pressure exerted against medulla). Died Oct. 8, 1939.

CUSHMAN, Allerton Seward, Am. chemist; b. (U. S. Consulate) Rome, June 2, 1867; s. Edwin and Emma (Crow) C.; B.S., Worcester Poly. Inst., 1888; Freiberg, and Heidelberg, 1889-90; A.M., Harvard, 1896, John Harvard fellow, Ph.D., 1897; m. Sarah Dunn Hoppin, June 20, 1901 (dec. 1921); children—Charles Van Brunt, Agnes Hoppin (dec.). Asso. prof. chemistry, Bryn Mawr Coll., 1900-01; asst. dir. Office Pub. Roads, U. S. Dept. Agr., and chemist in charge investigations, 1902-10; founder and dir. Inst. Indsl. Research, Washington, 1910-24. Prin. researches: extraction of potash from feldspathic rocks; use of ground rock as fertilizer; properties of road materials; cause and prevention of the rusting of iron and steel. Franklin medal, 1906. Author: The Corrosion and Preservation of Iron and Steel, 1910; Chemistry and Civilization, 1920. 2d edit., 1925. Died May 1, 1930.

CUSHMAN, Joseph Augustine, Am. micropaleontologist; b. Bridgewater, Mass., Jan. 31, 1881; s. Darius and Jane Frances (Fuller) C.; S.B., Harvard, 1903, Ph.D., 1909, hon. Sc.D., 1937; m. Alice Edna Wilson, Oct. 7, 1903 (dec. Jan. 1912); children—Robert Wilson, Alice Eleanor, Ruth Allerton; m. 2d, Frieda G. Billings, Sept. 3, 1913. Museum dir. Boston Soc. Natural History, 1913-23; dir. Cushman Lab. for Foraminiferal Research from 1923. Lectr. in micropalaeontology, from 1940. Cons. geologist U. S. Geol. Survey; mem. Carnegie Instn. Expdn. to Jamaica, 1912. Chmn. com. on micropaleontology NRC, from 1930. Recipient Hayden Meml. Gold medal, 1945. Fellow Am. Acad. of Arts and Scis.; mem. Geol. Soc. Am. (v.p. 1938). Washington Acad. Scis., Paleontol. Soc. (pres. 1937); Am. Geog. Soc., N.E. Bot. Club, Am. Assn. Petroleum Geologists, A.A.A.S., Soc. Econ. Paleontology and Mineralogy (pres. 1930-31), Sigma Xi; hon. fellow Royal Micros. London. Author: Monograph of Foraminifera of North Pacific Ocean (Smithsonian Instn.),

Parts I-VI, 1910-17; Foraminifera of Atlantic Ocean, Parts I-VIII, 1918-31; also various papers on fossil and living Foraminifera. Editor Jour. of Paleontology, 1927-30. Developed accurate methods of classifying foraminifera, enabling him to determine (from examination of fossil shells in oil drilling samples) the possibilities of oil deposits in the area. Died Apr. 16, 1949.

CUSHNY, Arthur Robertson, Scottish pharmacologist; b. Speymouth, Morayshire, Scotland, Mar. 6, 1866; s. Rev. John and Catherine (Ogilvie Brown) C.; M.A., 1886, M.B., C.M., 1889, Aberdeen Univ.; studied at Berne and Strasbourg, M.D., LL.D.; hon. deg., Michigan, Aberdeen; m. Sarah Firbank, 1896; 1 dau. Asst. to prof. of pharmacology, Strasbourg, 1892-3; prof. of pharmacology, Univ. Michigan, 1893-1905; prof. pharmacology and materia medica, Univ. London, 1905-18; prof. of materia medica and pharmacology, Univ. Edinburgh, from 1918. Mem., Royal Comm. on Whiskey and other Potable Spirits, 1908; Dohme lectr., Johns Hopkins, 1925. Fellow, Royal Soc., 1907. Author: Textbook of Pharmacology and Therapeutics, 1899; The Secretion of the Urine, 1917; The Action and Uses in Medicine of Digitalis and its Allies, 1924; Biological Relations of Optically Isomeric Substances, 1926. Research on action of digitalis glucosides, involving investigation of physiology of mammalian heart; with Edmunds, suggested that delirium cordis might be identical with auricular fibrillation; studied functions of kidneys and action of diuretics; pharmacological actions of optical isomers; proved that two substances, identical in chemical composition and structure, apart from optical activity, may differ widely in pharmacological activity. Died Feb. 25, 1926.

CUTBUSH, James, Am. chemist; b. Pa., 1788; s. Edward and Anne (Marriot) C. Authored 15 article series "Application of Chemistry to the Arts and Manufactures," Phila. Aurora, beginning 1808; contbr. article about mercury fulminate to Med. Museum, 1808, article describing method of purifying ether and production of ethylene, 1809, article about value of hop to brewers, 1811; contbr. article "Subjects and Importance of Chemistry" to Freemason's Mag., 1811; founder, 1st pres. Columbian Chemistry Soc., 1811; v.p. Linnaean Soc.; mem. Soc. for Promotion Rational System of Edn.; prof. chemistry, mineralogy and natural philosophy St. John's Coll., Phila.; apptd. asst. apothecary gen. U. S. Army, 1814; chief med. officer U. S. Mil. Acad., 1820, acting prof. chemistry and mineralogy; article on improvement Voltaic electric lamp to Am. Jour. of Science, 1820. Author: A Useful Cabinet, 1808; An Oration on Education, 1812; Philosophy of Experimental Chemistry, 1813; A Synopsis of Chemistry, 1821; A System of Phyrotechny, 1825. Died West Point, N.Y., Dec. 15, 1823; buried West Point.

CUTHBERSON, John, inventor; flourished 2d half 18th century; mechanic, Amsterdam, Netherlands, later London, Eng.; built elec. machine which had doubled glass plates and gave sparks 60 centimeters long, circa 1780.

CUTLER, Condict Walker, Jr., Am. surgeon; b. Morristown, N.J., Aug. 9, 1888; s. Condict Walker and Cora (Carpenter) C.; B.S., Columbia, 1910, M.D., 1912; unmarried. Asst. surg. Roosevelt Hosp., 1927-32, asso. surg., 1932-38, attending surgeon, 1948-53, cons. surgeon, 1953—; sec. med. bd., 1948-53; attending surg. Lincoln Hosp., 1929-31; dir. surgery Goldwater Meml. Hosp., 1939-53; pres. med. bd. 1949-51, cons. surgeon, 1953—; dir. surgery Morristown (N.J.) Meml. Hosp., 1952-54, cons. surgeon, 1954—; cons. surgeon and trustee N.Y. Dispensary, 1932-48; Rockland State Hosp., 1943-58; instr. surgery, Columbia, 1920-28, asso. prof. clin. surgery 1947, prof. clin. surgery, 1947-54; pvt. practice, N.Y.C., 1916-58. Decorated Congl. Medal for Meritorious Service, 1946, Columbia Medal for Excellence 1944, Alumni medal, 1946. Trustee, N.Y. Acad. Medicine, 1941-58 (chmn. com. on med. edn. 1945-47, chmn. sect. surgery, 1935-36). Fellow A.C.S.; mem. Am. Surg. Assn., Internat. Surg. Assn., N.Y. Acad. Medicine, Soc. Consultants to Armed Forces, Am. Assn. Surg. of Hand (pres. 1950), A.M.A., Med. Soc. County N.Y. (treas., 1947-52). Author: The Hand: Its Diseases and Disabilities, 1941. Co-author: History of Roosevelt Hospital, 1956. Contbr. numerous monographs and articles to med. publs. Died July 6, 1958.

CUTLER, Elliott Carr, Am. surgeon; b. Bangor, Me., July 30, 1888; s. George Chalmers and Mary Franklin (Wilson) C.; A.B., Harvard, 1909, M.D., 1913; hon. doctorate U. Strasbourg, 1938; D.Sc., U. Vt., 1941, U. Rochester, 1946; m. Caroline Parker, May 24, 1919; children—Elliott Carr, Thomas Pollard, David, Marjorie Parker (dec.), Tarrant. Resident surgeon, Harvard Unit, Am. Ambulance Hosp., Paris, 1915, Mass. Gen. Hosp., 1915-16; alumni asst. in surgery, Harvard, 1915-16; voluntary asst. Rockefeller Inst., N.Y.C., 1916-17; resident surgeon Peter Bent Brigham Hosp., 1919-21, asso. in surgery, 1921-24; prof. surgery, Western Res. U. Sch. Medicine, 1924-32; dir. surg. service Lakeside Hosp., Cleve., 1924-32; consulting surgeon, Peabody Home for Crippled Children New Eng. from 1932, Children's Hosp., Boston, from 1945; Moseley prof. surgery Harvard, from 1932; surgeon in chief, Peter Bent Brigham Hosp., from 1932; chief cons. to profl. services div. VA, 1945; civilian

consultant to Sec. of War from 1946; acting asst. med. dir. VA, from 1947, chief surg. cons. E.T.O., U. S. Army, 1942-45. Decorated D.S.M., Legion of Merit, Croix de Guerre with Palm, Order of British Empire. Diplomate Am. Bd. Surgery. Fellow A.C.S., Internat. Soc. Surgery (chmn. Am. com. 1929-47); mem. A.A.A.S., Am. Geog. Soc., Am. Acad. Surgery, Am. Bur. for Med. Aid to China, Inc., Am. Soc. for Clinical Investigation, Am. Heart Assn., A.M.A., Am. Soc. for Exptl. Pathology, Am. Surg. Assn. (pres. 1947), Am.-Soviet Med. Soc. (regional v.p. 1945-47), Am. Com. for Protection of Medical Research, (chmn. 1926-42), Assn. Mil. Surgeons of the U. S.; Boylston Med. Soc. of Harvard U., L'Europe Medicale (hon. scientific mem., patronage com.), Fedn. Am. Socs. for Exptl. Biology, Soc. Clin. Surgery (pres. 1941-46), U. S. Med. Consultants of World War II (pres. 1946), Alpha Omega Alpha, Sigma Xi. Editor: America Clinica (adv. bd.), American Heart Journal, (adv. bd.), Macmillan Co. Surgical Monograph Series, (editor), Journal of Clinical Investigation (editorial com.), Surgery (adv. council), Am. Jour. of Surgery, (asso. editor), Brit. Jour. Surgery, (editorial bd. and exec. com.), Washington Inst. of Medicine, (editorial and consulting bd. in surgery). Author: Atlas of Surgical Operations (with R. Zollinger), 1939. Specialized in thoracic and neurol. surgery; 1st Am. surgeon to operate on cases of mitral stenosis. Died Aug. 16, 1947.

CUTLER, Hugh Carson, Am. botanist; b. Milw., Sept. 8, 1912; s. Manuel and Mary (Williams) C.; A.B., U. Wis., 1935, A.M., 1936; Ph.D., Washington U., St. Louis, 1939; m. Marian W. Cornell, Aug. 26, 1940; 1 son, William Cornell. Research asso. Washington U., St. Louis, 1939-40, asso. prof. botany, 1953—; asso. dir. Mo. Bot. Garden, 1954-56, acting dir., 1957-58, exec. dir., 1958-64, curator useful plants, 1965—. Guggenheim fellow, 1942-43, 46-47. Mem. Am. Soc. Plant Taxonomists, Soc. for Econ. Botany, Assn. for Tropical Biology, Internat. Assn. for Plant Taxonomy, Soc. for Am. Archaeology. Research, numerous publs. on evolution and taxonomy New World cultivated plants and their wild relatives. Home: 5 Shaw Pl., St. Louis 63110. Office: Mo. Bot. Gardens, St. Louis 63110.*

CUTLER, Manasseh, Am. botanist; b. Killingly, Windham County, Conn., May 13, 1742; s. Hezekiah and Susanna (Clark) C.; grad. Yale, 1765, A.M., 1768, LL.D. (hon.), 1789; m. Mary Balch, Sept. 7, 1766. Admitted to Mass. bar, 1767; ordained to ministry Congregational Ch., 1771, pastor, Ipswich Hamlet (now Hamilton), Mass., 1771-1823; an organizer Ohio Co., colonizers Ohio River Valley, 1786; mem. U. S. Ho. of Reps. (Federalist) from Mass., 7th-8th congresses, 1801-05. Mem. Am. Philos. Soc., Am. Acad. Arts and Scis., Am. Antiquariran Soc. New Eng. Linnaean Soc. Systematized and catalogued flora of New Eng. by Linnean system. Died Hamilton, Mass., July 28, 1823.

CUTLER, Max, surgeon; b. Jitomir, Russia, May 9, 1899; s. Sam and Esther (Tchudnowsky) C.; brought to U. S. in 1907, naturalized, 1914; B.S., U. of Ga., 1918; M.D., Johns Hopkins Med. Sch., 1922; grad. study, Curie Inst., Paris and Radiumhemmet, Stockholm; m. Bertie Burger, Apr. 12, 1946; children—Nina, Nancy, Susie. Surg. intern Johns Hopkins Hosp., 1922-23; cons. in cancer, dir. cancer research, Edward Hines Vets. Hosp. and U. S. Vets Adminstrn., 1931-46; asso. in surg., Northwestern U. Med. Sch., 1935—; vis. prof. surgery Peking Union Med. Coll., China, 1936-37; now member surgical staffs Cedars of Lebanon, St. John's hospitals, Los Angeles, California; founder of Chicago Tumor Inst., past pres., dir. First pres. Am. Assn. Study of Neoplastic Diseases, 1933-34. Mem. Nat. Advisory Cancer Council, 1939-42. Served in United States Army, World War I. Mem. N.Y. Acad. of Medicine, Chicago Inst. of Medicine, Am. Radium Soc., Am. Assn. Cancer Research, A.M.A., Chicago Med. Soc., Internat. Coll. Surgeons; hon. mem. Cuban Radiol. Soc., Radiol. Soc. Chile, Phi Epsilon Pi, Phi Delta Epsilon, Phi Beta Kappa, Alpha Omega Alpha. Jewish religion. Mason. Club: Tamarisk. Author: (with Sir George Lenthal Cheatle) Tumors of the Breast, 1931; Cancer, Its Diagnosis and Treatment, 1938; Tumors of the Breast: Their Pathology, Symptoms, Diagnosis and Treatment, 1961. Contbr. about 100 articles on all phases of cancer. Home: 38550 Florence St., Beaumont, Cal. Office: 436 N. Roxbury Dr., Beverly Hills, Cal.*

CUTLER, Richard Loyd, Am. psychologist; b. Nottawa, Mich., Sept. 6, 1925; s. William Loyd and Louise (Beerstecher) C.; B.S. summa cum laude, Western Mich. U., 1949; M.A., U. Mich., 1951, Ph.D., 1953; m. Loismary Burmeister, June 19, 1948; children—Patrice, Scott. Asst. prof. U. Cal. at Berkeley, 1953-55; faculty U. Mich., Ann Arbor, 1955—, prof. psychology, 1963—, v.p., 1964—, dir. grad. program in edn. and psychology, 1962-64. Mem. Mental Health Commn., State of Mich., 1961-63. Mem. A.A.A.S., Am. Assn. U. Profs., Mich. Psychol. Assn. (pres. 1966—), Phi Beta Kappa, Sigma Xi. Research, publs. on personality theory, child devel.; discovered relationships between parental attitudes and problem devel. in children; devel new theory and methods for practical mgmt. childhood behavior problems in sch. and home: Home: 2105 Greenview St., Ann Arbor, Mich. 48103.*

CUTTER, Ephraim, Am. physician, food expert; b. Woburn, Mass., Sept. 1, 1832; s. Benjamin and Mary (Whittemore) C.; B.A., Yale, 1852, M.A., 1855; M.D., Harvard, 1856; M.D., U. Pa., 1857; LL.D., Ia. Coll., 1887; spl. lab. work in Sheffield Sci. Sch. and at Harvard under Oliver Wendell Holmes and J. P. Cooke; m. Rebecca Smith Sullivan, 1856 (dec. 1899); m. 2d, Mrs. Anna L. Davidson, 1901. Practiced at Woburn, Boston, N.Y., 1875-1917. Inventor of many surg. and gynecol. instruments, and procedures in relation to same; successfully with George B. Harrimon, D.D.S., of Boston, in 1876, used in microphotography of blood and sputum the higher and highest power lenses extant then. Mem. 9th and 10th Internat. Med. Congresses. Author: Versions and Flexions and Food in Motherhood; Fatty Ills and Their Masquerades (with John A. Cutter); Food—Its Relation to Health and Disease (with same), 1907. A pioneer of Am. laryngology; has studied the morphology of raw beef, 1854—; proved that the galvanic currents penetrate the human body, 1871; discovered Tb cattle test, 1894; etc. Died Apr. 24, 1917.

CUTTING, Hiram Adolphus, Am. physician; b. Concord, Vt., Dec. 23, 1832; s. Stephen Church and Eliza Reed (Darling) C.; A.M. (hon.), Norwich (Vt.) U., 1868, Ph.D. (hon.), 1870; M.D. (hon.), Dartmouth, 1870; m. Maranda E. Haskell, Feb. 3, 1856. An extremely precocious child, headed a sch. in Guildhall, Vt., by age 16; mainly self-educated; surveyor for a time; partner (with an uncle) in dry goods store, Lunenburg, Vt., 1854-79, sole owner, 1879-92; examining surgeon U. S. War Dept., 1861-65; prof. gen. science Norwich U., various times before and after Civil War; lectr. in med. coll. Dartmouth, 1870; became state geologist Vt., 1870; also practiced medicine during this period, later (1885-92) specialized solely in practice of medicine; became mem. Vt. Bd. of Agr., 1880; chmn. Vt. Fish Commn., for a time; Vt. del. to Internat. Forestry Congress, 1885; credited with devising improved camera lens; did research on capability of various building stones; also interested in agrl. improvements; contbd. many articles to mags. and newspapers. Author works including: Mining in Vermont, 1872; Microscopic Revelations, 1878; Scientific Lectures, 1884. Died Lunenburg, Apr. 18, 1892.

CUTTING, James Ambrose, Am. inventor; b. Hanover, N.H., 1814; s. Abijah Cutting. Invented a bee hive, patented 1844; patented photog. process ambrotype, 1850, photolithographic process, 1858; committed to asylum, Worcester, Mass., 1862. Died Worcester, Aug. 6, 1867.

CUTTING, Windsor Cooper, Am. physician; b. Campbell, Cal., July 30, 1907; s. Theodore Abijah and Mary (Cooper) C.; A.B., Stanford, 1928, M.D., 1932; m. Mary Estelle Weaver, May 3, 1935; children—Cecil Cooper, John Weaver, David Windsor, Ann Ely, April Bourne (Mrs. Michael Waldman), Susan Mary. Fellow NRC, Courtauld Inst. Biochemistry, Middlesex Hosp., London, Eng., 1935-36; fellow Johns Hopkins, 1936-38; faculty Stanford Med. Sch., 1938-57, prof. exptl. therapeutics, 1957-64, dean, 1953-57; prof. pharmacology U. Hawaii, 1964—, dir. Pacific Biomed. Research Center, 1964-66, dean Sch. Medicine, chmn. exec. com. Coll. Health Scis., 1965—. Mem. Am. Assn. Cancer Research, Cal. Acad. Medicine, A.M.A., Am. Soc. Pharmacology and Exptl. Therapeutics, Hawaii Acad. Sci., Phi Beta Kappa, Sigma Xi, Alpha Omega Alpha. Author: Manual of Clinical Therapeutics, 1943, revised 1948; Actions and Uses of Drugs, 1946; Handbook of Pharmacology, 1962, rev., 1964, 66; also numerous articles. Research in endocrinology and obesity, pharmacology of sulfonamides and antibiotics, virus and cancer chemotherapy, antifertility agts. from natural products. Home: 769 Sunset Av., Honolulu 96816.*

CUVIER, Baron Georges (Jean-Léopold-Nicolas-Frédéric-Dagobert), French naturalist; b. Montbéliard, Doubs, France, Aug. 23, 1769; s. Jean Georges and Clémence (Chatel) C.; student law Acad. Stuttgart (Germany), 1784-88; m. Anne-Marie Coquet de Trazaille Duvaucel, Feb. 2, 1804; 4 children. Tutor to family of comte d'Héricy, nr. Caen, France, 1788; at Jardin des Plantes, Mus. Natural History, Paris, from 1795, apptd. prof. comparative anatomy, 1802; lectr. École Centrale du Panthéon, 1796; prof. natural history Coll. de France, 1799; held numerous positions in gov. and edn., including: insp. edn., 1802, mem. council Imperial U., 1808, councilor of state, 1814, chancellor U. Paris, 1821-27; pres. com. interior, 1819-32. Mem. la section d'anatomie et zoologie de la première classe de l'Académie des sciences, 1795 (sec. for phys. scis., 1800, permanent sec., 1803), Académie des Inscriptions, Académie des belles-lettres, Académie de médecine. Author: Mémoires sur les espèces d'éléphants vivants et fossiles, 1796; Tableau élémentaire de l'histoire naturelle des animaux, 1798; Leçons d'anatomie comparée, 1805; Rapport historique sur les progrès des sciences naturelles depuis 1789 et sur l'état actuel, 1810; Recherches sur les ossements fossiles de quadrupèdes, 1812; Règne animale distribué d'après son organisation, 1817; Mémoires pour servir à l'histoire et à l'anatomie des mollusques, 1817; (with A. Brongmart) Description géologique des environs de Paris, 1825; Discours sur les révolutions de la surface du globe, 1825; Discours sur la théorie de la terre, 1825; (with Valenciennes) Histoire naturelle des poissons (Cuvier died after publ. vol. 8), 22 vols., 1828-49. Founder comparative anatomy; investigated comparative anatomy of fishes, osteology of mammals; used principles of anatomy to extend, perfect classificatory system of Linnaeus; grouped related Linnaean classes into still broader groups called phyla, distinguished 4 distinct phyla in animal kingdom; pioneer in palaeontology, and influential in making it separate sci.; extended taxonomy to fossils; 1st to identify and name pterodactyl; recognized anat. similarities between extinct and living animals; argued that fossil species had disappeared in sudden catastrophes, and was firm anti-evolutionist; adapted Bonnet's theory of catastrophism to account for gradations of fossils with time. Died Paris, May 13, 1832.

CUVILLIER, Jean, French geologist; b. Ambleteuse, Feb. 16, 1899; s. Léonce and Félicie (Blériot) C.; Ph.D. in Sci., Sorbonne. Prof., French Sch., Cairo, U.A.R.; lecture master U. Cairo; prof. Faculty Scis. of Paris; dir. Micropateontologic Lab. Mem. Geol. Soc. France (past pres.), Acad. of Turin. Research and numerous publs. on geology of Egypt, Aquitaine. Home: 39 av. de Saxe, Paris 7. Office: 191 rue Saint-Jacques, Paris, France.

CUYLER, Robert Hamilton, Am. geologist; b. Austin, Tex., May 28, 1908; A.B., U. Tex., 1926, A.M., 1927, Ph.D., 1931, instr. geol. 1927-37, asst. prof., 1937-39, asso. prof., from 1939. Mem. Paleontol. Soc., Geol. Soc. Am., Assn. Petroleum Geologists. Research in stratigraphic and subsurface geology, paleontol. and subsurface correlations, micropaleontology, microlithology, ecology. Died 1944.

CVIJIC, Jovan, Serbian geographer; b. Loznica, Serbia, Sept. 29, 1865; Ph.D., Vienna, Austria, 1892. Prof., Belgrade, Yugoslavia, from 1893; founder, head Serbian Geog. Soc. Author: Das Karstphänomen, 1893; Osnov e za geografiju i geologiju Makedonije i stare Srbije, 3 vols., 1906-11; Geographie humaine, 1918; Geomorfologija, 2 vols., 1924-26; La Péninsule balkanique. Investigated geog. and anthrop. background of Balkans; research in hydrology. Died Belgrade, Jan. 9, 1927.

CYERT, Richard Michael, Am. economist; b. Winona, Minn., July 22, 1921; s. Walter Michael and Anne Fostine (Brown) C.; B.S. in Econs., U. Minn., 1943; Ph.D., Columbia, 1948; m. Margaret Shadick, Sept. 8, 1946; children—Lynn Anne, Lucinda Carol, Martha Sue. Instr., U. Minn., 1946, 48, City Coll. N.Y., 1948; asst. prof. econs. and indsl. adminstrn. Carnegie Inst. Tech., 1949-55, asst. prof. econs. and indsl. adminstr., head indsl. mgmt. dept., 1955-60, prof. econs. and indsl. adminstrn., 1960-62, dean Grad Sch. Indsl. Adminstrn. Mem. adv. com. ednl. TV sta. WQEX, Pittsburgh; director Screw & Bolt Corp. of America. Chmn. adv. com. Nat. Planning Assn. Served USNR, 1943-46. Ford fellow, 1959-60; Guggenheim fellow, 1967-68. Mem. Am. Econ. Assn., Am. Statis. Assn., Econometric Soc., Inst. Mgmt. Scis. (v.p. educational and research), Phi Beta Kappa, Beta Gamma Sigma frat. Co-author: Sampling Techniques in Accounting, 1957; Sampling for Accounting Information, 1962; A Behavioral Theory of the Firm, 1963; Theory of the Firm: Resource Allocation in a Market Economy, 1965. Adv. bd. Cal. Mgmt. Rev.; bd. editors Behavioral Sci. Developed theory of statistical sampling methods applied to accounting and auditing; developed organizational theory of the firm which is integrated with the classical theory of the "firm" of economic theory. Home: 35 Chapel Ridge Rd., Pitts. 15238.

CYSAT, Johann (Cystatus, Jean-Baptiste), Swiss astronomer; b. Lucerne, Switzerland, 1586; s. Renward and Elisabeth (Bosshart) C.; student theology, Ingolstadt, Germany. Entered Jesuit order, Luzern, 1604, later became priest; prof. math. Jesuit Coll., Ingolstadt, also rector. Author: Mathematica astronomica de loco, motu, magnitudine et causis cometae, 1619. Used telescope to survey sky from 1611; discovered and studied spots on sun, 1611-23; discovered Orion nebula, 1618; studied satellites of Saturn and Jupiter; made 1st observations of comets with telescope; observed movements of Mercury. Died Lucerne, Mar 3, 1657.

CYSTATUS, Jean-Baptiste, see Cysat, Johann.

CYVIN, Sven Josef, chem. physicist; b. Louny, Czechoslovakia, Feb. 25, 1931; s. Karle Josef and Stepanka (Brozek) C.; Chem. Engr., Tech. U. Norway, 1956; m. Bjorg Nygaard, June 29, 1957; children—Mette Berit, Einar Dag. Staff, Tech. U. Norway, Trondheim, 1957—, docent in phys. chemistry, 1960—; research asso. in chemistry Ore. State U., Corvallis, 1965. Recipient award for sci. research A/s Norsk Varekrigsforsikrings Fond, 1963. Mem. Royal Norwegian Sci. Soc. Author: Mean Amplitudes and Related Problems of Molecular Vibrations; also numerous articles. Theoretical studies on molecular vibrations, calculations of mean amplitudes of vibration from spectroscopic data; introduced mean-square amplitude matrix. Home: Solhogdvegen 16, Trondheim, Norway.*

CZAPEK, Friedrich Johann Franz, botanist; b. Prague, Czechoslovakia, May 16, 1868; s. Friedrich and Maria Rosina (Blechinger) C.; studied medicine; became student botany under W. Pfeffer, Leipzig, Germany; Ph.D., 1894; m. Irene Margarethe Antoine Lambel; 1 son, 1 dau. Became asst. Plant Physiol. Inst., Vienna, Austria, 1894, lectr., 1895; apptd. asso. prof. botany German Technische Hochschule, Prague, 1896, prof., 1902; named prof. U. Czernowitz (now Chernovtsy, USSR), 1906; made tropical expdns., 1907-08; named prof. German U., Prague, 1909; mil. doctor, 1915-18; apptd. prof. U. Leipzig, 1921. Author: Die Bakterien in ihrer Beziehung zur belebten Natur, 1899; Biochemie der Pflanzen, 2 vols., 1905, 3 vols., 1913-21; also numerous articles. Research on physiol. biology, plant taxonomy, floristics. Died Leipzig, July 31, 1921.

CZAPLEWSKI, (Alexander Emil Hermann) Eugen, German bacteriologist; b. Königsberg (now Kaliningrad, USSR), Nov. 17, 1865; s. Carl and Clara (Seydler) C.; doctorate, Königsberg, 1899; m. Lisa Berding, 1897; 2 daus.; joined faculty U. Königsberg, 1894; apptd. dir. bacteriological lab. City of Cologne, Germany, 1908; prof., dir. Mus. for Pub. Health, Cologne, 1908-31. Author: Kurzes Lehrbuch der Desinfektion, 1909; also numerous articles on epidemic control, pub. health edn., determination of causes of disease. Pioneer in microbiology. Died Cologne, Nov. 15, 1945.

CZAPLINSKI, Bogdan, Polish parasitologist; b. Sycyna, Poland, July 18, 1923; s. Ludwik and Izabella (Dobiecka) C.; Vet. Faculty, U. Warsaw, 1950, D.Biol. Scis., 1956, Docent of Zoology, 1962; m. Danuta Laganowska, May 19, 1945; children—Tomas, Marek. Asst. dept. parasitology Vet. Faculty, U. Warsaw, 1947-53; adj. dept. parasitology Polish Acad. Scis., 1953-63; dir. biol. dept. Med. Faculty, Med. Acad., Warsaw, Poland, 1963—. Mem. Polish Parasitological Soc., Polish Acad. Scis. (sec. parasitological com. 1962—, mem. sci. council dept. parasitology 1962—). Author: Robaczyce Drobiu (Helminthoses of Domestic Fowl), 1960. Research, numerous publs. on parasites of birds, especially cestoda (Hymenolepididae), their morphology, systematics, devel. and specifity. Home: 10a Francuska. Office: 5 Chalubinski, Warsaw, Poland.*

CZECZUGA, Bazyli, Polish biologist, hydrobiologist; b. Plutycze, Poland, Nov. 30, 1930; s. Jerzy and Anna (Poplawska) C.; M.Sc., U. Minsk (USSR), 1956; Doctor Sci., Higher Sch. Agr., Olsztyn, Poland, 1960, Degree of Asst. Prof., 1963; m. Ada Matusewicz, July 15, 1956; 1 dau., Ewa. Staff, Bialystok (Poland) Med. Acad., 1956—, head chair biology, 1963—. Decorated for Excellent Work in Health, 1965, Thousand Years Poland, 1966. Mem. Internat. Chlorella Union, Polish Bot. Soc., Polish Biochem. Soc., Polish Hydrobiol. Soc., Polish Physiol. Soc., Polish Zool. Soc. Author: (monograph) Ecological-Physiological Aspects of the Distribution of Some Species of Tendipedidae (Diptera) larvae in Water Reservoir, 1962; also articles. Research on dyes in plants and invertebrates and discovered they are present in several species; studied climatic changes in Northeastern part of Poland during post ice-age period; 1st to discover that the Jatvings tribe (Baltic tribe living in Northeastern part of Poland in 1-4th century) cultivated rye. Home: 35a Szpitalna, Bialystok, Poland.*

CZERMAK, Johann Nepomuk, physiologist; b. Prague, Czechoslovakia, June 18, 1828; s. Conrad and Josefine C.; studied philosophy, Prague, 2 years; studied medicine at Vienna, Austria, 1845, Breslau, Germany (now Wroclaw, Poland), 1847, Würzburg, Germany, 1849-50; m. Marie v. Lamel, 1843; 2 sons, 1 dau. Named asst., Prague, 1850; named prof. zoology at Graz, Austria, 1855; apptd. prof. physiology Cracow, Poland, 1856, Budapest, Hungary, 1858, Jena, Germany, 1865, Leipzig, Germany, 1869; built lab. and auditorium for exptl. physiology demonstrations. Author: Der Kehlkopfspiegel, 1860; Geschichtlicher Schriften, 2 vols., 1879; also numerous articles. Pioneer in rhinoscopy; disputed with Ludwig Türck for priority of discovery of laryngoscope; made 1st clin. application of laryngoscope, 1858; improved laryngoscope; invented mirror for inspecting nasopharynx, 1858; discovered that pressure exerted on a point in inferior carotid triangle of neck causes heart beat to slow, 1865. Died Leipzig, Sept. 16, 1873.

CZERNIAK, Pinchas, physician; b. Horodec, Poland, Sept. 16, 1909; s. Arie and Szyfra (Jarmuk) C.; student Faculté de Medicine et École d'Hygiene, Montpellier, France, 1930-36; specialization radiotherapy and nuclear medicine Institut Fournier, Saclay, Radiumhemmet, Hadassa U.; m. Gitel Feldstein, Dec. 30, 1941; children—Irith (Mrs. Beni Weiss), Szyfra Cyla, Rami, Jehudith. Asst., adj. U. Hosp., Wroclaw, Poland, 1947-50; asst. Hadassa U. Hosp., Jerusalem, Israel, 1950-54; head dept. radiotherapy and isotopes Tel-Hashomer Govt. Hosp., Tel-Aviv (Israel) U. Med. Sch., 1954—. Adviser, Ministry Health, Jerusalem, 1950-56; med. adviser AEC, Rehovoth, Israel, 1953—. Recipient Prix Dubreuil, 1935, Prix Meyer, 1959; named hon. prof. Central U. Quito, 1966. Mem. Israel Radiol. Soc., Soc. Nuclear Medicine, Société de Radioprotection et Hygiene Atomique, Sociedad Argentina de Medicina Nuclear, Internat. Nuclear Hematology. Author: Ten Years Experience in Nuclear Medicine, 1964. Research, numerous publs. on radiobiology problems, including chem. protection against radiation; radiomicrobiology; nuclear hematology; nuclear medicine, including subacute thryoiditis, spermatogenic activity test, radiol. curetage, liver studies. Home: 3 Shulamith, Tel-Aviv, Israel. Office: Tel-Hashomer Govt. Hosp., Tel-Hashomer, Israel.*

CZERNY, Vincenz, surgeon; b. Trutnoy, Czechoslovakia, Nov. 19, 1842. Asst. to Theodor Billroth; with U. Freiburg (Germany) 1871; became prof. Heidelberg, Germany, 1877, established Inst. for Exptl. Cancer Research, 1906. Mem. French Acad. Scis., 1900. Author: Über die Beziehungen der Chirurgie zu den Naturwissenschaften, 1872; Beiträge zur operativen Chirurgie 1887; Über die Entwicklung der Chirurgie während des 19 Jahrhunderts und ihr Beziehung Zumunterricht, 1903. Research on exptl. and clin. surgery including treatment of bone and joint Tb., operation for inguinal hernia, hysterectomy by vagina, enucleation of subperitoneal uterine fibroids, resection of intestine, plastic surgery; successfully performed pyelolithotomy, 1880; introduced term and concept of osteitis deformans, 1873. Died Heidelberg, Oct. 3, 1916.

CZETSCH LINDENWALD, Hermann von, pharmacist; b. Judenburg, Feb. 28, 1900; s. Otto and Maria (Grögl) C. L.; Ph.D. in pharmacy; m. Maria von Kundratitz; children—Peter, Klaus, Hermann. Asst. pharmacist indsl. co., 1929-45; instr. U. Innsbruck, U. Fribourg, 1932; founder, co-propr. pharm. manf. co. Pan Chemie, Wolfsberg, 1945-56; full prof. tech. pharmacology Faculty Pharmacy, Alexandria, Egypt, 1956. Author: Salben, Puder, Externa, 3d edit.; Pharm. Technologie, 1948; Suppositorier, Planzenextracte, 2d edit.; Hilstoffe d. Pharmazie, Arzneikapseln, 2d edit., also numerous articles. Home: 5 rue Pharaons. Office: Faculty of Pharmacy, Alexandria, Egypt.

CZUBER, Emanuel, mathematician; b. Prague, Czechoslovakia, Jan. 19, 1851; s. Karl and Karoline (Libora) C.; studied engring. German Polytechnikum, Prague; dr. (hon.) Technische Hochschule, Munich, Germany, 1918; m. Adalberta Willigk, 1877; 2 sons; 2 daus. including Berta; asst. to geodesist K. Koristka; became lectr., Prague, 1876; named prof. Brno, Czechoslovakia, 1886; prof. Technische Hochschule, Vienna, Austria, 1891-1919. Author numerous books, including: Geometrische Wahrscheinlichkeitsrechnung und ihre Andwendung auf Fehlerausgleichung, Statistik und Lebensversicherung, 1902-03; Die Kollektivmasslehre, 1908; Einführung in die höhere Mathematik, 1909; Die statistischen Forschungsmethoden, 1921; Die philosophischen Grundlagen der Wahrscheinlichkeitsrechnung, 1923; Mathematische Bevölkerungstheorie, 1923; also studies on probability, ins. math. Died Gnigl, nr. Salzburg, Germany, Aug. 22, 1920.

CZYZYK, Artur Stanislaw, Polish physician; b. Bodzentyn, Poland, Feb. 6, 1927; s. Kazimierz, and Franciszka (Pirog) C.; M.B., Med. Acad., Cracow, Poland, 1951, M.D., 1951, Candidate Med. Scis. 1955; m. Teresa Kasperska, Aug. 17, 1961; 1 dau., Maria-Franciszka, Aug. 5, 1962. Staff dept. pharmacology Med. Acad., Cracow, 1948-50; with III dept. medicine Warsaw (Poland) Med. Acad., 1951-65, adj. 1955-60, asso. prof. medicine, 1960-65; head III dept. medicine and diabetes Praski Hosp., Warsaw, 1965—; head Diabetes Research Centre, Kolobrzeg, Poland, 1964—. Con. physician diabetes Health Resort Kolobrzeg, 1962—. Mem. Polish Med. Assn. Polish Assn. Internal Medicine, Polish Soc. Endocrinology, Deutsche Gesellschaft für innere Medizin, European Assn. for Study Diabetes (mem. council 1966—), Am. Diabetes Assn. Author: Badania czynnosciowe w klinice chorob wewnetrznych, 1961 (with others); also articles. Research on diabetes, oral antidiabetic drugs, amino acid metabolism, frequency of diabetes in Poland. Home: 14 Koszykowa. Office: 67, Al.Gen.Swierczewskiego, Praski Hosp., Warszawa, Poland.*

D

D'ABANO, Pietro (or Pietro d'Abano, Peter of Abano, Petrus Aponensis), Italian physician; b. Abano, nr. Padua, Italy, 1250; traveled in Greece and Nr. East; studied in Paris, France, at least 10 years; studied Greek manuscripts, Constantinople; taught medicine at Padua, circa 1300; tried as magician by Inquisition but died during 2d trial. Author: Concilliator (dealt mainly with med. and astrological issues, tried to synthesize Greek and Arabic med. thought, included only reference in contemporary sci. work to Marco Polo, whom D'Abano may have met); Lucidator astronomiae, (most important astron. work written in Italy during the period), 1310; Tractatus de venenis eorumque remediis (treatise on poisons); Liber compilationis physiognomiae; supplement (on remedies for cardiac and digestive problems) to Grabadin (Mesuë). Translator: Methodus medendi ad Glauconem, Galeni Prognostica de decubitu ex mathematica scientia (both by Galen). A great rep. of med. astrology; held that nerves had their source in the brain and blood vessels in the heart; made nearly accurate estimate of length of year; believed that air had weight. Died circa 1316.

D'ABBADIE, Antoine Thomas, explorer, astronomer; b. Dublin, Ireland, Jan. 3, 1810; s. Michel d'Abbadie;

bachelier Toulouse, France, 1827; studied law, Paris, France, from 1829; m. Virginie de Saint-Bonnet, Feb. 21, 1859. Went to France, 1820; sent to Brazil by French Acad. Scis., 1836; in Ethiopia (with bro. Arnaud), 1838-48. Decorated cross Legion of Honor, 1850; recipient Gold medal Geog. Soc. Mem. Bur. Longitudes, French Acad. Scis. Author: Catalogue raisonné de manuscrits éthiopiens, 1859; Éthiopie (map in 10 sects.), 1862-69; (with Radau) Géodésie de la Haute Éthiopie, 1873; Préparation des voyageurs aux observations astronomiques et géodésiques, 1880; Dictionnaire de la langue amarrinna, 1881; Géographie de l'Éthiopie, 1890. Originated expeditious geodesy (method for taking bearings); invented new theodolite (aba) for measuring horizontal and vertical angles. Died Mar. 19, 1897.

D'ABBANS, Claude-François-Dorothée Jouffroy, French engr., shipbuilder; b. Roches-sur-Rognon, France, 1751; 1 son, Achille-François-Eléanor; page to Dauphin; soldier Bourbon inf., 1871-72; exiled to Provence, France, 1872; later inventor Paris, France. Author: Les Bateaux à vapeur, 1816. Worked on using steam power for sea travel, replaced rudder with paddle wheel, tested on Saone River with 46 meter long ship, 1783. Died 1832.

DABBS, John Wilson Thomas, Am. physicist; b. Nashville, Dec. 11, 1921; s. John Wilson Thomas and Ruth (Fuqua) D.; B.S. in Mech. Engring., U. Tenn., 1944; Ph.D. in Physics, 1955; m. Betty Jane Hicks, Sept. 16, 1945; children—Carol Jane, John Richard, David Frederick. Research asst. in physics, metall. lab. U. Chgo., 1944-45; physicist, applied physics lab. John Hopkins, Balt., 1945; physicist Oak Ridge Nat. Lab., 1946—. Exec. v.p., dir. Oak Ridge Devel. Corp., 1961—. Fulbright lectr., Argentina, 1961. Fellow Am. Phys. Soc.; mem. Research Soc. Am., Delta Tau Delta. Editor: Knox County Government, 1958; Semiconductor Nuclear Particle Detectors, 1961. Research in delayed neutrons, nuclear orientation, neutron gravity, nuclear fission. Home: 106 Osage Rd., Oak Ridge 37830. Office: P.O. Box X, Oak Ridge 37831.*

DABNEY, Samuel Gordon, Am. physician; b. nr. Charlottesville, Va., Aug. 6, 1860; s. William S. and Susan Fitzhugh (Gordon) D.; M.D., U. Va., 1882, U. Louisville, 1883; studied in Zurich and Vienna; m. Louisa Higgins Allen, Dec. 21, 1887; children—Mary Allen (Mrs. George Ezra Woodruff), William Cecil. Practiced at Louisville since 1885; prof. diseases of ear, nose and throat and clin. prof. diseases of eye U. Louisville; mem. staff Louisville City Hosp., Norton Meml. Infirmary. Fellow A.C.S.; mem. A.M.A., Ky. State Med. Assn., Am. Acad. Ophthalmology and Otolaryngology. Died Dec. 14, 1935.

DABOLL, Nathan, Am. mathematician, educator; b. Groton, Conn., Apr. 24, 1750; s. Nathan and Anna (Lynn) D.; m. Elizabeth Daboll; m. 2d, Elizabeth Brown. Discovered errors in almanac prepared by Clark Elliott and pub. by Timothy Green, 1770, employed to revise the calculations; responsible for New Eng. Almanack by Nathan Daboll Philomath (pub. by Timothy Green), 1773, 74, 75; prof. math. and astronomy Plainfield (Conn.) Acad., 1783-88; taught navigation aboard frigate President, 1811. Author: Daboll's Complete Schoolmaster's Assistant, 1799; Daboll's Practical Navigation, pub. 1820. Died Mar. 9, 1818.

DABROWSKA-POPLAWSKA, Polish neuropsychologist; b. Ostrowiec, Poland, Jan. 10, 1929; s. Ludwik and Stanislawa (Stypulska) Dabrowski; M.Sc., U. Lodz (Poland), 1952; Ph.D., Nencki Inst. Exptl. Biology, Warsaw, Poland, 1960; m. Ryszard Poplawski, July 5, 1952; 1 dau., Magdalena. Staff, Nencki Inst. Exptl. Biology, Polish Acad. Scis., to present, now sr. research asso. Mem. Polish Physiol. Soc. Research, publs. on reversal learning, learning set in rats, functions of frontal cortex and hippocampus in dog's behavior. Home: 8 Ladyslawa. Office: 3 Pasteur, Warsaw, Poland.*

DA CARPI, Giacomo Berengario, see Berengario da Carpi.

DA CASTRO, see Rodrigues de Castro, Estevan.

DACK, Simon, Am. physician; b. N.Y.C., Apr. 19, 1908; s. Isidore and Rebecca (Baitch) D.; B.S., Coll. City N.Y., 1928; M.D., N.Y. Med. Coll., 1932; m. Jacqueline Rosett, Jan. 23, 1949; children—Jerilyn, Leonard. With Mt. Sinai Hosp., 1932—, chief prenatal clinic, 1953—, asso. physician for cardiology, 1958—, asso. clin. prof. medicine Mt. Sinai Sch. Medicine, 1966—; lectr. medicine, cardiology Columbia, 1935—; with N.Y. Med. Coll., 1955—, asso. clin. prof. medicine 1959—; attending physician Flower Fifth Av. Hosp., 1966—; attending physician Met. Hosp.; asso. vis. physician Bird S. Coler Meml. Hosp.; cons. cardiologist Prospect Heights Hosp., St. Joseph's Hosp., 1963—; editor-in-chief Am. Jour. of Cardiology, 1958—. Fellow Am. Coll. Cardiology (pres. 1956-57), N.Y. Acad. Medicine, A.C.P., Am. Coll. Chest Physicians; mem. N.Y. Heart Assn., Am. Fedn. Clin. Research, Am. Heart Assn., Alpha Omega Alpha. Contbr. numerous articles to sci. jours. Clin. research in field electro-cardiography, exercise tests of heart, coronary heart disease, heart block, implanted heart pace makers, treatment heart

failure, other forms clin. heart disease. Home: 85 East End Av., N.Y.C. 10028. Office: 1111 Park Av., N.Y.C. 10028.*

DACOS, Fernand, Belgian electronics engr.; b. Liège, Belgium, Mar. 19, 1892; ed. U. Liège. Asst. Montefiore Inst., later head works; prof. Inst. Physics; prof. elec. theory and electronics U. Liège. Author books, textbooks on electricity and electronics, also numerous articles. Address: 33, rue Saint-Gilles, Liège, Belgium.

DA COSTA, Antonio Plácido, Portuguese ophthalmologist; b. Covilha, Portugal, 1848; s. Rafael da Costa; M.D., Escolo do Pôrto, 1879. Prof. Escolo Médico-Cirúrgica do Pôrto. Author: Fisiologia do punctum coecum da retina humana, 1883; O cristalocone polar anterior; Novo instruments para a investigaçao rápida e completa das irregularidades de curvatura de córnea o astigmatoscópio. Invented keratoscope (Placido's disk). Died 1916.

DA COSTA, (John) Chalmers, Am. surgeon; b. Phila., Nov. 15, 1863; s. George T. and Margaretta (Beasley) D.; grad. U. Pa., 1882; M.D., Jefferson Med. Coll., 1885; m. May R. Brick. Resident phys. Phila. Hosp., 1885-86, asst. phys. insane dept., 1886-87; surgeon, 1895—; became asst. demonstrator anatomy Jefferson Med. Coll., 1887, demonstrator of surgery, 1891, clin. prof. surgery, 1898, prof. surgery, 1900; Samuel D. Gross prof. of surgery, 1901—; asst. surgeon Jefferson Hosp., 1887; surgeon St. Joseph's Hosp., 1896—. Fellow Am. Surg. Assn., Coll. Physicians of Phila. Author: A Manual of Modern Surgery, 1895-1925 (10 edits.). Editor English edit. Zuckerkandi's Operative Surgery, 1899, also new Am. edit. of Gray's Anatomy, 1905. Died May 16, 1933.

DA COSTA, Jacob Mendez, physician; b. St. Thomas, W.I., Feb. 7, 1833; grad. Jefferson Med. Coll., Phila., 1852; postgrad. in Paris, Prague, and Vienna; LL.D., U. Pa., Harvard; established practice, Phila., 1854; became lectr. clin. medicine, Jefferson Med. Coll., 1864; prof. theory and practice of medicine, 1872-91; emeritus prof., 1891; a founder of Path. Soc. Phila. (1st sec.), 1857; pres. Coll. Phys. of Phila., 1884-86 and 1895-98; original mem. Assn. Am. Physicians (pres. 1897). Author: Medical Diagnosis, 1864; Harvey and His Discovery, 1879; Strain and Over-Action of the Heart. Influenced the systematization of med. diagnosis; described irritable heart (neurocirculatory asthenia) in Civil War soldiers, known as Da Costa's syndrome; investigated respiratory percussion and typhoid fever; reformed clinical methods. Died Villanova, Pa., Sept. 11, 1900.

DA COSTA, Newton Carneiro Affonso, Brazilian mathematician; b. Curitiba, Paraná, Brazil, Sept. 16, 1929; s. Dimas Affonso and Sylvia (Carneiro) da C.; Civil Engr., U. Paraná, 1952, B.Math., 1955, D.Math., 1961; m. Neusa Feitosa, Jan. 15, 1955; children—Newton Carneiro Affonso, Sylvia Lucia F. A. Faculty, U. Paraná, Curitiba, 1955-59, 1961—, prof. math., 1965—; asst. prof. Instituto Tecnológico de Aeronáutica, Sao Paulo, Brazil, 1960, mem. Inst. Maths., 1965-66; mem. Inst. Maths., U. Paraná, 1960—. Mem. Am. Math. Soc., Assn. for Symbolic Logic, Sociedade Paranaense de Matemática. Author: Espaços Topológicos e Funçoes Continuas, 1959; Introduçao Aos Fundamentos da Matemática, 1962; Sistemas Formais Inconsistentes, 1963; Algebras de Curry, 1966; also articles. Created theory of inconsistent formal math. systems; treated systematically theory of non monotone operations in lattices; research on founds. of set theory and category theory. Home: 97 Jesuino Lopes, Curitiba, Paraná, Brazil.*

DACQUÉ, Edgar Viktor August, German paleontologist, geologist; b. Neustadt/Weinstrasse, Germany, July 8, 1878; s. Eugen and Martha (Andreae) D.; studied paleontology U. Munich (Germany); doctorate, 1903; m. Louise Kölsch, 1903; 1 son, 4 daus. Became asst. U. Munich, 1904, named lectr. paleontol. and hist. geology, 1912; became asso. prof., curator paleontol. collection of State of Bavaria, 1915. Author: Paläontologie, Systematik und Deszendenzlehre, 1911; Grundlagen und Methoden der Paläogeographie, 1915; Vergleich biologische Formenkunde der fossilen niederen Tiere, 1921; Sage, Urwelt and Menschkeit, 1924; Paläogeographie, 1925; Vom Sinn der Erkenntnis, Eine Bergwanderung, 1931; Natur und Erlösung, 1933; Organische Morphologie und Paläontologie, 1935; Das verlorene Paradies, 1938; Bildnis Gottes, 1939; Die Urgestalt, 1951. Research on paleogeography, geology of Alps and paleontology, especially paleogeography and paleobiology of non-vertebrates; faunistic-stratigraphic studies on Regensburger chalk; tried to synthesize paleontology, metaphysics and the evolution of man. Died Solln nr. Munich, Sept. 14, 1945.

DA CREMONA, Jacope, Italian mathematician; b. Cremona, Italy, flourished, circa 1449; tchr. at Mantua and Rome, Italy; wrote complete Latin version of Archimedes works, pub. by Venatorius, 1544.

DA CUNHA, Antonio Brito, Brazilian geneticist; b. Sao Paulo, Brazil, June 17, 1925; s. Antonio Paulo and Olga de Almeida (Brito) da C.; Ph.D., Faculdade de Filosofia Ciencias e Letras Universidade de Sao Paulo, 1948, Livre-Docencia, 1955; m. Lygia Freire Gaspar, Sept. 3, 1949; children—Eduardo Brito,

Alfredo Brito. Faculty, departamento de biologia Faculdade de Filosofia, Ciencias e Letras, Universidade de Sao Paulo, 1945-—, asso. prof. genetics, 1960-—, editorial bd. Univ. Press, 1965-—. Dir. University Collection Companhia Editora Nacional. Mem. Soc. for Study Evolution (past v.p.), Am. Soc. Naturalists, Am. Genetical Assn., Sociedade Brasileira de Genetica., Brazilian Acad. Scis. Author: (with C. Pavan) Elementos de Genetica, 2d edit., 1966). Asso. editor Evolution, 1952-54, 1959-61; editorial bd. Mutation Research, 1964-—, Caryolopia, 1967-—. Publs. on relations between chromosomal variability and adaptation in fruit fly, effects radiation and lethals within its natural population. Home: 55 rua das Perdizes, Sao Paulo, Brazil.*

DA CUNHA, Jose-Anastacio, Portuguese mathematician; b. Lisbon, Portugal, 1744; tchr. math. U. Coimbra, 1773; author: Avoz da razao; Principles of Mathematics, 1782. Died Lisbon, 1787.

DADANT, Camille Pierre, apiarist; b. Langres, France, Apr. 6, 1851; s. Charles and Gabrielle (Parisot) D.; came to U. S., 1863; ed. in France and pub. schs., Hamilton, Ill.; m. Mary Marinelli, Nov. 1, 1875; children—Louisa G. (Mrs. L. G. Saugier), Valentine, Louis C., Henry C., Maurice G., Clemence S., Harriette G. (Mrs. F. A. Bush). Asso. with father, as Dadant & Son, in bee culture, and publisher of book, The Hive and Honey Bee, 1874-1902; with 3 sons and 2 daughters as Dadant & Sons, 1902-—; began mfr. of bee comb foundations, 1879. Mem. Nat. Bee Keepers' Assn. (sec., v.p., pres., treas.). Decorated Order Crown for services rendered Belgian bee-keepers during World War. Author: First Lessons in Bee-Keeping, 1915; Bee Primer: The Dadant System of Bee-Keeping. Editor: The Hive and Honey Bee (L. L. Langstroth), various edits., 1888-—. Editor Am. Bee Jour., 1912-—. Died Feb. 25, 1938.

DADD, Reginald Hugh, zoologist; b. London, Eng., Jan. 3, 1925; s. Stephen William and Winifred (Fathers) D.; B.Sc., Imperial Coll. U. London, 1951, Ph.D., 1954; m. Edith Mary McEwan, Apr. 4, 1956; children—Christopher Julian, Sarah Nicola and Andrew John (twins). With Anti-Locust Research Centre, London, 1954-59, sr. sci. officer, 1956-59; research fellow Entomology Research Inst., NRC Can., Belleville, Ont., 1960-61; asst. research entomologist NIH, NSF grants, dept. entomology U. Cal., Berkeley, 1961-—. Mem. Entomol. Soc. Am. Contbr. articles to sci. jours. Investigated relationship between digestive enzyme secretion in certain beetles, to metamorphosis on one hand and to feeding activity on other; specializing in insect nutrition, devising synthetic diets for locusts, waxmoth, aphids and manipulating composition of these so as to determine in growth expts. specific nutritional requirements of these insects; 1st demonstrated that plant-eating insects may have a requirement for ascorbic acid. Home: 1350 Scenic Av., Berkeley, Cal. 94708.*

DADOURIAN, Haroutune Mugurdich, physicist, educator; b. Everek, Turkish Armenia, Dec. 5, 1878; s. Mugurdich and Ezgule (Kalajian) D.; came to U. S., 1900, naturalized, 1908; Ph.B., Yale, 1903, M.A., 1905, Ph.D., 1906; m. Ruth McIntire, Dec. 28, 1918. Instr. physics Sheffield Sci. Sch., lectr. Yale Grad. Sch., 1906-17; aero. engr. U. S. Govt., 1917-19; asso. prof. physics Trinity Coll., Hartford, Conn., 1919-23, Seabury prof. math. and natural philosophy, 1923-49, prof. emeritus, 1949-—. Bd. dirs. Hartford Civil Liberties Union, Com. for Sane Nuclear Policy. Fellow Am. Phys. Soc.; mem. Am. Math. Soc., Math. Assn. Am., A.A.A.S., Am. Assn. U. Profs., Sigma Xi. Author: Graphic Statics, 1919; Analytical Mechanics, 3d edit., 1931; Introduction to Analytical Geometry and Calculus, 1947; How to Study, How to Solve, 1949; Plane Trigonometry, 1950. Contbr. articles to Am., Brit., French, German sci. jours. Research on radioactivity, x-rays, sound ranging, relativity, mechanics. Address: 177 N. Main St., West Hartford, Conn. 06107.*

DADYKIN, Vsevolod Petrovich, Russian agrobotanist; b. Vilnius 1910; grad. Timiryazev Agrl. Acad., Moscow, 1931; postgrad. All-Union Inst. Phytoculture, 1935-37; D.Biol. Sci. Pioneer farming experimenter in Okhotsk-Kolyma region, 1932-34; sr. asso., learned sec. Obruchev Inst. Permafrost Studies, USSR Acad. Sci., 1940-—. Mem. USSR Acad. Sci. (chmn. Presidium Karelian br.). Research and publs. on northern farming and market gardening. Address: Obruchev Inst. Permafrost Studies, USSR Acad. Sci., Cherkassky p. 2-10, Moscow, USSR.

DA FANO, Corrado Donato, physician; b. Urbino, Italy, June 1, 1879; s. Alessandro and Adele (Maroni) Da F.; ed. Golgi's Inst. Histology and Gen. Pathology, U. Pavia, 1902-05, M.D., 1905, L.D. in Morbid Anatomy, 1912; m. Dorothea Landau, 1915. Resident physician Cernusco dept. Hosp. Milan, 1908, Sangalli traveling scholar, 1908; hon. asst. Ziehen's Neurol. Clinic, U. Berlin, 1908, Imperial Cancer Research Fund, London, 1909; 1st asst., prosector Path. Inst., U. Groningen, 1910; vice dir. Path. Inst. of Milan, 1912-15; lectr. histology U. London, 1918, reader in histology King's Coll., 1922-27. Recipient Fossati prize Istituto Lombardo di Scienze e Lettere, 1906. Fellow Linnaean Soc., Royal Micros. Soc. (council); mem. Real Sociedad Española de Historia Natural, Bologna Med.-Surg. Soc. (corr.). Co-editor: Physiol. Abstracts. Research and numerous publs. on histological, histopath. and histophysiol. subjects. Died Mar. 14, 1927.

DA FOLIGNO, Gentile, Italian physician; b. Foligno, nr. Perugia, latter part 13th century; s. Gentile da Foligno; studied under Alderott at Bologna, before 1303, under Peter of Abano in Padua, before 1315; taught at U. Perugia; lectr. U. Padua, 1337-45; personal physician to Ubertino da Carrara; made several public dissections and autopsies; possibly made discovery of gallstone, Padua, 1341; wrote some 90 Consilia on gen. subjects (fevers, various ailments, other med. questions), containing many practical remarks on diseases and methods of treatment; his Consilium of 1348 was 1 of earliest plague treatises; also wrote many commentaries on works of Galen, Mesuë, Ibn Sina, Giles of Corbeil and others. Died 1348.

D'AGELET, Joseph le Paute (LePaute d'Agelet, Joseph), French astronomer; b. Thoune-le-long, France, Nov. 25, 1751; student under Lalande. Prof. math. Ecole Militaire. Mem. French Acad. Scis. Took part in expdns. with Kerguélen-Trémarec in Australian waters and La Pérouse. Died Isle of Vanikoro, 1788.

DAGNALL, Roy Maurice, English chemist; b. London, Eng., Aug. 19, 1938; s. Maurice and Violet (Stafford) D.; B.Sc. with honors, U. Birmingham (Eng.), 1960, M.Sc., 1961, Ph.D., 1963; m. Jacqueline Rose, July 15, 1961; 1 son, Mark St. John. Faculty, Imperial Coll., U. London, 1963-—, lectr. chemistry, 1965-—. Mem. Soc. for Analytical Chemistry (mem. com.). Research, publs. on devel. instrumental methods of analysis for trace amounts of metal ions in solution and in atomic state (atomic fluorescence spectroscopy); devel. high intensity spectral sources. Home: 23 Cotswold Close, Chalvey, Slough, Eng. Office: Imperial Coll., Chemistry Dept., London, S.W.7, Eng.*

DAGOMARI, Paolo, Italian mathematician, astrologer; b. Prato, Tuscany, circa 1281. Lived in France; notable arithmetician and astronomer; also known as Paolo Dell'Abaco, Paolo Astologico, Pagolo Astrologo, Paoli il Geometra, Paolo Geometra, and Paolo Arismetra. Author: Trattato d'Abbaco, d'Astronomia, edi segreti naturali e medioinali, 1532. Invented various math. instruments; 1st to conceive of composing almanacs with predictions. Died Florence, Italy, 1372.

DAGRON, René, French chemist; b. Beauvoir, France, 1819; inventor microphotography on collodion film, circa 1965, process used with carrier pigeons to carry film to Paris, France, under siege, 1870-71; inventor various inks, also range-finder. Died Paris, 1900.

DAGUERRE, Louis Jacques Mande, French inventor, photographer; b. Cormeilles, France, Nov. 18, 1789; employed in Orleans and elsewhere; utilized optical knowledge for creation of more realistic backdrops Paris Opera; opened (with Bouton) a diorama, Paris, 1822, later opened similar establishment, Regent's Park, London, Eng.; began study of possibility of permanent photography; formed (with Joseph Niepce) Daguerre-Niepce firm (became Daguerre firm 1835). Mem. Am. Art Acad., Société libre des Beaux-Arts. Author 2 treatises on photography. A founder of photog. process; discovered process for making daguerrotypes by using chem. action of light on copper or silver plate coated with silver salt to produce permanent images, 1839. Died Bry-sur-Marne, nr. Paris, July 10, 1851.

D'AGUESSEAU, Henri François, French sci. patron, magistrate; b. Limoges, France, Nov. 27, 1668; s. Henri D'Aguesseau; m. Anne d'Ormesson, 1694; children—4 sons, including Henri Cardin. Lawyer, Parlement de Paris, France, 1690-1700, gen. procurator, from 1700; chancellor France, 1717-1751. Mem. French Acad. (pres. 1729-39). Contbd. to reform of English calendar. Died Feb. 9, 1751.

DAGUIN, Pierre Adolphe, French physicist; b. Poitiers, France, Aug. 4, 1814; s. Pierre and Désirée (Delamotte) D.; ed. École normale supérieure, 1835; agrégé in phys. scis., 1841. Prof. physics Faculty of Toulouse (France), 1847-69; named dir. Toulouse Obs., 1866. Author: Traité de physique avec les applications à la météorologie et aux arts industriels, 1855-59; Cours de physique élémentaire, 1863. Applied physics to meteorology; research on acoustics, decomposition of sound, undulatory systems; invented a new acoustical device, the acoustèle. Died Toulouse, Nov. 20, 1884.

DAHL, Adrian Hilman, Am. biophysicist; b. Mott, N.D., Dec. 6, 1919; s. Frederick A. and Tonetta (Fortney) D.; A.B., St. Olaf Coll., 1941; Ph.D., U. Rochester, 1953; m. Ellen Virginia Lieneman, Feb. 4, 1942; children—David A., Sonje A. Chief radiation physics sect. U. Rochester Med. Center, 1950-58; Fulbright lectr. health physics to Argentina, 1958; prin. scientist Oak Ridge Inst. Nuclear Studies, 1959-61; Internat. Atomic Energy Agy. expert health physics to Indonesia, 1961; dir. Radiation Health Specialist Tng. Program Colo. State U., Ft. Collins, 1961-—. Mem. Am. Phys. Soc., Am. Assn. Physics Tchrs., I.E.E.E., Health Physics Soc., Sigma Xi, Sigma Pi Sigma. Research on radiation measurement techniques, nuclear fallout levels, uptake of fallout by plants and animals, radiation exposures in mining operations, whole body counting techniques. Home: 4785 Venturi Lane, Route 1, Ft. Collins, Colo. 80521.*

DAHL, Andreas, Swedish botanist; b. 1751; student of Linneaus at Uppsala, Sweden, University; genus Dahlia named for him, 1791. Died 1789.

DAHLBERG, Gunnar, Swedish biologist; b. Lofta, Sweden, 1893; s. Henning and Gertrud (Jaede) D.; M.D., U. Uppsala (Sweden), 1926; LL.D., U. Aberdeen (Scotland), 1938; m. Stina Westberg, 1919; 1 dau. Asst. prof. U. Uppsala, 1926-36; head Swedish State Inst. for Human Genetics and Race Biology, 1936-—. Recorder sect. human genetics 7th Internat. Congress Genetics, Edinburgh, Scotland, 1939. Author: Twin Births and Twins from a Hereditary Point of View, 1926; Ovulation Mechanism, 1930; Statistical Methods for Medical and Biological Students, 1939; Race, Reason and Rubbish, 1942; also articles.

DAHLER, John Spillers, Am. theoretical chemist; b. Wichita, Kan., May 7, 1930; s. Raymond Edward and Agnes (Spillers) D.; B.S., U. Wichita, 1951, M.S., 1952; Ph.D., U. Wis., 1953; m. Lanaya Dorothy Williams, June 30, 1954; children—Kurt Williams, Gwendolyn Kay. NSF fellow U. Amsterdam (Netherlands), 1955-56; faculty U. Minn., Mpls., 1959-65, 66-—, prof. depts. chem. engring., chemistry, 1963-—, Alfred P. Sloan fellow, 1963-66; NSF sr. fellow U. Cal. at Berkeley, 1965-66. Mem. Am. Phys. Soc. Research, publs. on theory of thermodynamical properties and equation of state of liquids, kinetic theory of fluids composed of polyatomic molecules, simple, gas-phase chem. reactions. Home: 3820 Chowen S. St., Mpls. 55410.*

DAHLGREN, George, Am. chemist; b. Chgo., Apr. 12, 1929; s. George Axel and Helen (Galloway) D.; B.S., Ill. Wesleyan U., 1951; M.S., U. Wyo., 1956, Ph.D., 1958; m. Mary Basler, Sept. 1, 1951; children—Sarah Jane, Kirsten Anderson. Postdoctoral fellow Cornell U., 1957-59; faculty U. Alaska, 1959-66, asso. prof. chemistry, 1962-66, head dept. chemistry, 1964-66; prof., asst. head dept. U. Cin., 1966-—. Fellow A.A.A.S. (exec. sec. Alaska div. 1960-65); mem. Am. Chem. Soc., Am. Assn. U. Profs., Sigma Xi. Contbr. articles on periodate oxidations, hydrogen bonding, analysis to sci. jours. Editor Sci. in Alaska, Proc. of Alaskan Sci. Confs., vols. 11-15, 1961-65. Home: 3767 Middleton Av., Cin. 45220.*

DAHLGREN, Karl Vilhelm Ossian, Swedish botanist; b. Arboga, Sweden, July 19, 1888; s. Otto Wilhelm and Alma (Tour) D.; Cand.Phil., U. Uppsala (Sweden), 1910, Mag.Phil., 1911, Lic.Phil., 1916, D.Phil., 1916; m. Greta Estelle, Sept. 10, 1921; children—Göran A. W., Östen W., E. Barbro M. (Mrs. Olle Martell), O. Sune W. Amanuensis bot. lab. U. Uppsala, 1912-15, acad. docent in botany, 1916-53, prof. h.c. Inst. Systematic Botany, 1947-—. Spl. tchr. Ultuna Agrl. Coll., Uppsala, 1915-31; lectr. Tng. Sem. Sch., Uppsala, 1921-63. Named knight Order of Pole Star. Fellow Royal Soc. Sci. Uppsala; hon. mem. Swedish Bot. Soc. Research, publs. on phanerogamous morphology especially embryology, genetics, Linnaeus, biography and letters. Home: 18 Geijersgatan, Uppsala 2, Sweden.*

DAHLGREN, Ulric, Am. biologist; b. Bklyn., Dec. 27, 1870; s. Charles Bunker and Augusta (Smith) D.; A.B., Princeton, 1894, M.S., 1896; m. Emillie Elizabeth Kuprion, Sept. 3, 1896. Instr. Princeton, 1896, prof. biology, 1911-39, emeritus since 1939. Asst. dir. Marine Biol. Lab., Woods Hole, Mass., 1899; trustee Harpswell (Me.) Biol. Lab., 1912-16; dir. Mt. Desert Island Biol. Lab., Bar Harbor, Me., 1921, pres. Corp. since 1937. Fellow A.A.A.S., Phila. Acad. Scis.; mem. Am. Soc. Zoologists, Am. Soc. Naturalists, Am. Philos. Soc. Author: Principles of Animal Histology (with W. A. Kepner), 1908. Contbr. series Production of Light by Organisms (Jour. Franklin Inst.), 1915; also zool. memoirs in German and Am. jours., mostly on prodn. of light and electricity by animals. Died May 30, 1946.

DAHL-IVERSEN, Erling, Danish surgeon; b. Nov. 30, 1892; s. Anders and Cathrine (Pedersen) D.-I.; Ph.D. in Medicine, U. Copenhagen; m. Inga Margrethe Thortsen; children—Torsten, Flemming. Research in hosps. of Copenhagen and fgn. countries, 1919-34; prof. clin. surgery U. Copenhagen, chief surgeon state hosp. of Copenhagen, 1935-63. Hon. mem. Soc. Surgery of Lyons, Soc. Brit. and Irish Surgeons, Internat. Soc. Surgery; mem. Belgian Soc. Surgery; asso. Acad. Medicine of Paris. Author: Studies in Surgery. Address: Tranegaardsvej, 24, Copenhagen, Denmark.

DAHLQVIST, Arne Lennart, Swedish physician, biochemist; b. Hälsingborg, Sweden, Sept. 26, 1929; s. Edvard and Ebba (Jönsson) D.; M.D., U. Lund (Sweden), 1957; m. Gun-Britt, Astrid Petrea Olsson, Dec. 31, 1952; children—Marie-Louise, Patrick, Ursula. Asst. prof. physiol. chemistry U. Lund, 1960-—; research fellow Swedish Med. Research Council, Stockholm, 1961-67, asso. prof., 1967-—; vis. prof. biochemistry Chgo. Med. Sch., 1963. Cons. physician dept. pediatrics Clin. Chem. Lab. Lund, 1964-—. Mem. Swedish Physicians Soc., Swedish Biochem. Soc. Research, numerous publs. on carbohydrate metabolism and enzymology, properties of small intestinal disaccharidases, screening procedures for galactosuria, enzyme assay methods. Home: Skolbänksvägen 3, Lund, Sweden.

DAHN, Hans, chemist; b. Kassel, Germany, Jan. 2, 1919; s. Richard and Flora (Kaufmann) D.; student chemistry U. Basel (Switzerland), 1937-44, Dr. Phil., 1944; m. Alice Winkler, Mar. 23, 1959; 1 son, Michael. Asst. in organic chemistry U. Basel, 1944-50, privatdocent, 1950-52, asst. prof., 1952-54, asso. prof., 1954-60; prof., head, dept. organic chemistry U. Lausanne (Switzerland), 1960——. Mem. Swiss Chem. Soc. (v.p. 1966-67). Research and publs. in synthetic organic chemistry, kinetics of hydrolysis of diazo compounds, application of oxygen isotopes in organic chemistry, especially chem. shifts of oxygen-17 in nuclear magnetic resonance of organic compounds. Home: Le Mont-sur-Lausanne. Office: Instit. Chimie Org., Place du Chateau 3, Lausanne, Switzerland.*

DAIGAARD, Jorgen Brems, Danish physician; b. Viborg, June 11, 1918; s. Johannes and Laura (Brems) D.; M.D., U. Copenhagen; m. Ena Bach; children—Lena, Lone, Lise. Asst. prof. anatomy and pathology U. Aarhus (Denmark), 1945-53, prof. legal medicine, 1959, dean Faculty Medicine, 1963-64; prof. U. Bergen (Norway), 1954-58, Mayo Clinic, Rochester, Minn., Howard U., Washington, 1959. Mem. Med.-Legal Soc. (London), Internat. Acad. Legal Medicine and Social Medicine (Liège). Author: Experimental Investigations on the Cause of the Increase in Serum Phosphatase in Obstructive Jaundice. Research and numerous articles on legal medicine and pathology, death by carbon monoxide (suicides, accidents, homicides). Home: Fr. Nansensvej 4. Office: Institute of Forensic Medicine, University of Aarhus, Finsensgade 15, Aarhus, Denmark.

D'AILLY, Marie Joseph Louis d'Albert (Duc de Chaulnes), chemist; b. Paris, France, Nov. 18, 1741. Fellow Royal Soc. Author: Mémoire et Expériences sur l'Air fixe . . . , 1780. Demonstrated that gas evolved in beer fermentation is fixed air, has acid reaction and forms potassium bicarbonate with potassium carbonate; placed density of fixed air at twice that of air. Died after 1789.

D'AILLY, Pierre, French natural philosopher; b. Compiègne, France, 1350; D. Theology, Coll. of Nevarre, U. Paris; 1380; became master Coll. of Nevarre, 1384; named chancellor U. Paris, also confessor of Charles VI, 1389; became bishop of Puy, 1395, of Cambrai, 1596; cardinal, from 1412; participated in Council of Constance, 1414-18; contbd. to efforts to end. Gt. Schism; named pontificial légat, Avignon, by Martin V. Author 170 works, including Imago Mundi. Supported of Ockhamistic nominalism; submitted his reforms of the Julian calendar to Council of Constance; predicted coming of Anti-Christ in 1789; contbd. to discovery of New World by suggesting in Imago Mundi (read by Columbus) that India could be reached by sailing west; ascertained (before Copernicus) rotation of earth on its axis; participated in trial of John Huss. Died Avignon, France, 1420.

DAIMLER, Gottlieb Wilhelm, German inventor, automobile industrialist; b. Schorndorf/Württemberg, Germany, Mar. 17, 1834; s. Johannes and Wilhelmine Friederika (Fensterer) D.; studied at Polytech. Sch., Stuttgart, Germany, 1857-59; m. Emma Kurz, 1867; 3 sons including Paul, Adolf; 2 daus.; m. 2d, Lina Schwend, 1893; 1 step-dau.; 1 son, 1 dau. Made study trips to Paris, France, Eng., 1860; joined Metallwarenfabrik Straub, Geislingen, 1862; dir. engine factory Bruderhaus, Reutlingen, 1865; named head all factories Maschinenbaugesellschaft Karlsruhe (Germany), 1869; named director Gasmotorenfabrik Deutz, 1872; a founder (with Wilhelm Maybach) factory Cannstadt, Germany, 1887 (became Daimler-Motoren-Gesellschaft in 1890; merged with Benz 1926). One of the inventors of the automobile; patented light high speed gasoline engine to replace developed method of gasifying liquid fuel, 1885; heavy stationary gas motor, 1883; mounted motor on bicycle, 1885, built it into a motor coach, 1885-86; improved cooling system, carburetor; developed 1st Mercedes automobile, 1900-01. Died Stuttgart-Cannstatt, Mar. 6, 1900.

DAINS, Frank Burnett, Am. chemist; b. Gouverneur, N.Y., Jan. 15, 1869; s. George G. and Celestia Stone (Burnett) D.; Ph.B., Wesleyan U., Conn., 1890, M.S., 1891, D.Sc., 1940; Ph.D., U. Chgo., 1898; postgrad. Freiburg and Berlin, 1901-02; m. Alice Haight, Sept. 24, 1898. Asst. in chemistry, Wesleyan U., 1891-93; asst. prof., U. Kan., 1893-94, prof. chemistry, 1911-42, acting dean grad. sch., 1926-27; fellow U. Chgo., 1894-95; asst. prof. chemistry Northwestern U. Schs. Medicine and Pharmacy, 1895-1901; prof. chemistry Washburn Coll., 1902-11. Fellow A.A.A.S.; mem. Kan. Acad. Science, Am. Chem. Soc., History of Science Soc., Phi Beta Kappa, Sigma Xi. Died Jan. 5, 1948.

DAINTY, Jack, English biophysicist; b. Mexborough, Eng., May 7, 1919; s. Jack and Evelyn (Vickers) D.; M.A., U. Cambridge (Eng.), 1940; D.Sc., U. Edinburgh (Scotland), 1958; m. Mary Elizabeth Elbeck, July 16, 1941; children—Anton, Christopher, Patrick. Research scientist nuclear physics Cavendish Lab., Cambridge, 1940-46, Canadian Atomic Energy Lab., Chalk River, Ont., Can., 1946-49; lectr. physics U. Edinburgh, 1949-52, sr. lectr., reader biophysics, 1952-63; prof. biology U. East Anglia, Norwich, Eng., 1963——. Mem. numerous sci. socs. Research, publs. on nuclear physics including fast neutron measurements, biophysics, especially transport of ions and water across membranes of plant cells including giant algal cells. Address: Sch. Biol. Scis., U. East Anglia, Norwich, Eng.*

DAJOZ, Roger, French biologist; b. Paris, France, Aug. 22, 1929; s. Georges and Jeanne Dajoz; student Ecole Normale Supérieure de St. Cloud, 1949-54; aggregation, 1954; Docteur ès Scis., 1966; m. Aline Langevin, Aug. 13, 1954; children—Helène, Isabelle. Asst., lab. comparative anatomy Paris Faculty Scis. Mem. Entomol. Soc. France (librarian 1966——), Zool. Soc. France (past mem. bd.), Faunistic Soc. (past dir.), Internat. Assn. Earth Science. Author: Les insecticides, 1958; also papers. Research on morphology, taxonomy, ecology of xylophagous Coleoptera; morphology, taxonomy and anatomy of Coleoptera (Elateridae and Latridiiae), ecology of Coleoptera (Caribidae). Home: 4, rue Herschel, Paris 6, France.*

DAKE, Charles Laurence, Am. geologist; b. Chaseburg, Wis., Apr. 2, 1883; s. George E. and Mary A. D.; A.B., U. Wis., 1911, A.M., 1912; Ph.D., Columbia, 1922; m. Ella Hildagarde Falkenstern, 1912; children—Laurence Falkenstern, Helen Elizabeth, Emilie Louise. Asst. in geology Williams Coll., 1912-13; asst. prof. geology Sch. Mines and Metallurgy (U. Mo.), 1913-18, asso. prof., 1918-21, prof., 1921-—. Asst. Wis. Geol. and Natural History Survey, summer 1910; geologist Mo. Bur. Geology and Mines, 16 summers, Wyo. Geol. Survey, summer 1916. Author: (with others) Field Methods in Petroleum Geology, 1921; Interpretation of Topographic and Geologic Maps, 1925. Died 1934.

DAKIN, Henry Drysdale, chemist; b. London, Eng., Mar. 12, 1880; s. Thomas Burns and Sophia (Stevens) D.; B.Sc., Victoria U., Manchester, Eng., 1901; D.Sc., U. Leeds, 1907, LL.D., 1936; Ph.D. (hon.), U. Heidelberg; Sc.D., Yale; m. Susan Dows Herter, July 1916. Demonstrator in chemistry U. Leeds, 1901-02; research worker Lister Inst. Preventive Medicine, successively at London and Heidelberg, 1901-05, Herter Lab., N.Y.C., 1905-20; sci. adviser Merck Inst. Therapeutic Research; dir. Merck & Co. Decorated chevalier Legion of Honor; recipient Philip A. Conne medal Chemists' Club. Fellow Royal Soc., 1917 (Davy medal 1941); mem. Soc. Exptl. Biology, Inst. Chemistry Gt. Britain and Ireland, London Chem. Soc., others. Author: Oxidation and Reductions in the Animal Body, 1912; (with E. K. Dunham) Handbook of Chemical Antiseptics, 1917. Editor Jour. Biol. Chemistry, 1911-31. Synthesized adrenaline, 1905; developed antiseptic (Dakin's solution) used during W.W. I; discovered antiseptics chloramine T and dichloramine T; (with R. West) isolated substance from liver used in treatment of pernicious anemia. Died Scarborough-on-Hudson, N.Y., Feb. 10, 1952.

DAKIN, William John, Brit. zoologist; b. Liverpool, Eng., 1883; s. William Dakin; student Victoria U., Eng.; B.Sc., U. Liverpool, M.Sc., 1907; studied in Kiel, Germany, also in Italy; D.Sc., 1910; m. C. M. G. Lewis, 1913. Asst. lectr. U. Belfast (Ireland), 1909, Liverpool, 1910; sr. asst., dept. zoology and comparative anatomy Univ. Coll., U. London, 1912; prof. biology U. Western Australia, 1913-20; became Derby prof. zoology U. Liverpool, 1920; prof. zoology Sydney (Australia) U., from 1929; tech. dir. camouflage Commonwealth Govt., 1941-45. Pres. Royal Soc. Western Australia, 1913-15. Author: Whalemen Adventurers (history of whaling), rev. edit., 1938; Osmotic Pressure and Blood of Fishes; textbooks on zoology, also papers. Died Apr. 2, 1950.

DAKSHINAMURTI, Chirravuri, Indian physicist; b. Nadupur, India, July 17, 1914; s. Chirravuri Vedadri and Kameswari (Nerella) D.; B.Sc., Andhra U., 1933; M.Sc., Banaras Hindu U., 1936, D.Sc., 1943; Ph.D., U. London, 1948; m. Raghunath Nerella, June 10, 1952; children—Dikshith, Visalakshi, Jagannath, Subbalakshmi, Viswanath. Sir. C. V. Raman Research scholar Andhra U., 1936-37; lectr. physics Banaras Hindu U., 1937-44; lectr. math. physics Andhra U., 1944-46; Govt. Overseas scholar in soil physics, Rothamstad Exptl. Sta., Eng., 1946-48; exptl. physicist Indian Agrl. Research Inst., New Delhi, India, 1948-54, sr. physicist, 1954-62, head div. agrl. physicist, 1962——, radiol. safety officer, since 1960. Mem. panel research coms. FAO/IAEA, 1964——, mem. research reviewing team div. atomic energy in agr., 1966. Fellow Inst. Physics; mem. Faraday Soc. London, Brit. Soil Sci. Soc. Research, publs. on nuclear physics and soil physics, measurement of ionization potentials of atoms by atomic collisions in positive ray beams, ionic diffusion and elec. conductivity measurements in soil systems, soil structure and plant growth, migration of nutrient ions through soils using active and inactive isotopes, isoconductivity value, new characteristic of soil or clay for determination of cation exchange capacities. Home: B-5 Indian Agrl. Research Inst., New Delhi-12, India.*

D'ALBI, Pierre Gilles, zoologist, naturalist; b. Albi, 1490; traveled in Orient for sci. purposes; one of 1st to successfully cultivate natural history in France; dissected an elephant. Author: Graeco-Latin Lexicon, 1532; On the Nature of Animals (in Latin), 1533; Treatise on the Antiquities of Constantinople, 1561; other works. Died Rome, 1555.

DALBY, W. Ernest, Brit. engr.; M.A., B.Sc.; prof. engring. U. London. Fellow Royal Soc., 1913; mem., v.p. Instn. Civil Engr.; mem. Instn. Mech. Engrs.; hon. v.p. Inst. Naval Architects. Author: Power and the Internal Combustion Engine; The Balancing of Engines; Valves and Valve Gear Mechanisms; Steam Power; Strength and Structure of Steel and other Metals; also articles in Ency. Brit., profl. jours. Died June 25, 1936.

DALE, Sir Henry Hallett, English physiologist; b. London, Eng., 1875; s. C. J. Dale; Scholar, Leys Sch., Coutts-Trotter Student Trinity Coll., Cambridge; St. Bartholomew's Hosp.; George Henry Lewes student, Sharpey Scholar, U. Coll., M.A., M.D., D.Sc., LL.D.; hon. degrees Cambridge, Princeton, Oxford, St. Andrews, other univs.; m. 1904; one son (dec.), 2 daus. Dir. Wellcome Physiol. Research Lab., 1904-14; mem. Gen. Med. Council, 1927-37; dir. Nat. Inst. for Med. Research, Hampstead, 1928-42; Crown nominee to Ct. of London U., 1939-50; Fullerian prof., dir. Davy-Faraday Lab., Royal Instn., 1942-46; Croonian lectr. Royal Soc., 1919, Herter Lectures, Balt., 1919, Harvey lectures, N.Y.C., 1919, 37, Croonian lectr. Royal Coll. Physicians, 1929; Dohme lectures, Balt., 1933, Welch lectures, N.Y.C., 1937, Pilgrim Trust lecture, Phila., 1946; chmn. Wellcome Trust, 1938-60. Knighted, 1932. Decorated Medal of Freedom with Silver Palm (U. S. A.); Orden Pour le mérite (Fed. Republic Germany); Order of Merit, Grand Cross Order Brit. Empire; recipient Nobel prize in physiology and medicine (with O. Loewi), 1936. Fellow Royal Soc., 1914 (sec. 1925-35, pres. 1940-45); fellow Royal College Physicians; pres. Brit. Assn. for Advancement of Sci., 1947; fgn. asso. Nat. Acad. Sci. U. S. A. Author: Adventures in Physiology, 1953; (article) Some Recent Extensions of the Chemical Transmission of the Effects of Nerve Impulses, 1936. Research on chem. transmission of nerve impulses; isolated histamine and ergotoxine. Died Cambridge, Eng., July 24, 1968.

DALE, Thomas Nelson, Am. geologist; b. N.Y.C., Nov. 25, 1845; s. Thomas Nelson and Sarah Patten (Monson) D.; gen. edn. Europe and Williston Sem., Mass.; geol. tng. under Zittel and Pumpelly; m. Margaret Brown, Dec. 22, 1874; children—Sarah, Norman Brown, Nelson Clark, Oswald, Margaret, Arthur. With U. S. Geol. Survey, 1885-1920, geologist, 1892-1920. Instr. geology and botany, Williams Coll., 1893-1901. Author: The Marbles of Western Vermont, 1912; Slate in the United States, 1914; The Commercial Granites of New England, 1923. Died Nov. 16, 1937.

DALECHAMPS, Jacques, French physician, botanist; b. Bayeux, France, 1513; Docteur en medicine, 1547; practiced at Lyons, France, from 1552; dir. Roville Bot. Gardens. Author: Chirurgie françoise, 1570; Historia generalis planatrum, (attempt to unite all previous works on bot. sci., after he worked 30 years on book completion was left to Desmoulins) 1587; also pub. edits. of Pliny, Athenaeus, several med. treatises. Died Lyons, Mar. 1, 1588.

D'ALELIO, Gaetano Francis, Am. chemist; b. Charlestown, Mass., Dec. 26, 1909; s. Severino and Francesca (Polcari) D'A.; A.B., Boston Coll., 1931; Ph.D., Johns Hopkins, 1935; m. Josephine Marie McCarthy, Sept. 5, 1932; children—Denny, Ellen, Jane, William. Head dept. chemistry U. Notre Dame (Ind.), 1955-60, research prof., 1960——; mem. research adv. com. on materials NASA, Washington, 1961-63; mem. labs. com. Franklin Inst., Phila., 1963——; mem. com. on radiation preservation food Nat. Acad. Scis., 1965——; cons. to industry. Mem. Am. Chem. Soc., A.A.A.S., Am. Inst. Chemists, Internat. Mark Twain Soc., Hist. Sci. Soc., Am. Ordnance Soc., Am. Inst. Chem. Engrs., Armed Forces Chem. Assn., Phi Beta Kappa, Sigma Xi, Phi Lambda Upsilon. Author: (with R. L. Guile) Laboratory Manual of Synthetic Plastics and Resinous Materials, 1942; Laboratory Manual of Plastics and Synthetic Resins, 1943; Experimental Plastics and Synthetic Resins, 1946; (with Manual Marin) Substancias plasticas experimentales y resinas sinteticas, 1948; (with Carl Hanser) Kunststoff Praktikum, 1952; Fundamental Principles of Polymerization, 1952. Contbr. numerous articles to profl. jours. Research, publs., numerous inventions including radar insulation; rocket launchers; ion exchange resins; acrylic resins for purification antibiotics; basis for acrylic fibers; devel. polystyrene foam portable refrigeration units in U. S.; heat resistent plastics. Home: 2011 E. Cedar St., South Bend, Ind. 46617. Office: Chemistry Dept., U. Notre Dame, Notre Dame, Ind. 46556.*

D'ALEMBERT, Jean le Rond, French mathematician, physicist, philosopher; b. Paris, Nov. 16, 1717; s. Gen. Chevalier Destouches and Mme. de Tencin; raised by poor family named Rousseau; bachelor's degree, Mazarin Coll., 1735; prepared for bar (admitted as advocate 1738 but did not practice); studied medicine; self-taught in math. Admitted to French Acad. Scis., 1741, became asst. dir., 1768; elected to

French Acad., 1754, permanent sec., 1772-83; invited by Frederick II to become pres. Berlin Acad., refused during visit to Berlin, 1763; declined invitation to tutor son of Catherine II of Russia; granted pension by Louis XV of France. Author: Mémoire sur le calcul intégral, 1739; Sur la réfraction des corps solides, 1741; Traité de dynamique, 1743; On the General Theory of the Winds, 1747; Le précession des equinoxes, 1749; Recherches sur differents points importants du système du monde, 1754-56; Éléments de philosophie, 1759; Opuscules mathématiques, 8 vols., 1761-80; various other writings. Contbr. Discours prélimineaire, also article on air to Diderot's Encyclopédie, 1751. Wrote on theory of differential equations, on calculus and its applications; applied his calculus to motion of vibrating cords, 1749; applied his principles to motion of any body of a given figure, 1749; solved (independently of A. C. Clairaut) dynamic problem of 3 bodies; elaborated Wallis' proposal to extend geometry to embrace study of n dimensional space by calling mechanics a 4-dimensional form of geometry, with time as the 4th dimension; worked on gravitational theory; gave complete solution to problem of precession of equinoxes, 1754; added to Newton's results on motion of heavenly bodies; defined elementary air as homogenous and the fundamental ingredient of air of atmosphere, while a mixture with other substances (impurities) forms common air; stated that elementary air is permanently elastic. Died Paris, Oct. 29, 1783.

DALÉN, Nils Gustaf, Swedish engr.; b. Stenstorp, Sweden, Nov. 30, 1869; mech. engr. Chalmer Tech. Sch., Göteborg, Sweden, 1896; student Fed. Poly., Zurich, Switzerland, 1896-97. With de Laval Steam Turbine Co., Stockholm, 1897-1900; mem. engring. firm Dalén and Celsing, 1900-05; works chief Swedish Carbide and Acetylene Co., 1901-03; chief engr. Gas Accumulator Co., 1906-09; mng. dir. Swedish Gas Accumulator Co., from 1909; blinded in test explosion, 1912. Recipient Nobel prize in physics, 1912. Research on tech. turbines and gases especially acetylene; many inventions on use of acetylene especially non-explosive gas accumulator and light-sensitive valve; constructed Laval steam turbine. Died Stockholm, Sweden, Dec. 10, 1937.

DALENÉE, French physicist; flourished mid 17th century; author: Traité des baromètres, et thermomètres, 1688; devised graduations between two fixed points on thermometer.

DALESME, André, physician; b. 1647; practice medicine Brest, France, 1680-1711; mem. French Acad. Scis.; studied hernia; perfected screw-jack; designed kiln. Died 1727.

DALGARNO, Alexander, physicist; b. London, Eng., Jan. 5, 1928; s. William and Margaret (Murray) D.; B.Sc. with 1st class honors, U. Coll., London, 1947, Ph.D., 1951; m. Barbara Wilma Fletcher Kane, Oct. 31, 1957; children—Penelope Murray, Rebecca Susan, Piers Alexander, Fergus John. Faculty, Queen's U., Belfast, 1951-67, prof. quantum mechanics, 1961-65; prof. math. physics, 1965-67, dir. computing lab., 1961-65; prof. astronomy Harvard, 1967—; staff mem. Smithsonian Astrophys. Obs., 1967—; chief scientist GCA Corp., Bedford, Mass., 1962-63. Cons. U.K. Atomic Energy Research Establishment, 1961-66, Goddard Space Flight Center, NASA, 1963-66. Mem. Phys. Soc. London, Am. Geophys. Union. Editorial bd. Planetary and Space Sci.; Jour. Atmospheric Sci., 1963—, Internat. Jour. Quantum Chemistry, 1966—. Research, numerous publs. on application of quantum mechanics to atomic and molecular physics, physics of planetary atmospheres. Home: 5 Fernald Dr., Cambridge, Mass. 02138.*

DALIBARD, Thomas-François, French physicist, botanist; b. Crannes, France, Nov. 5, 1703; s. Thomas and Marie (Gareau) D.; student natural scis., Angers, France; m. Françoise Thérèse-Aumerle de Saint-Phalier; mfr. porcelain and candles; pub. sec. of Metz. Author: Florae parisiensis prodromus, 1740; Théorie abrégée de l'électricité, 1752. Translator: Experiences et observations sur l'électricité de B. Franklin, 1752. Research on physics including electricity and the weight of bodies immersed in liquids. 1st French author to adopt system of Linnaeus. Died Paris, France, 1779.

DALITZ, Richard Henry, physicist; b. Dimboola, Australia, Feb. 28, 1925; s. Frederick William and Hazel (Drummond) D.; B.A. with honors, Melbourne U., 1944, B.Sc., 1945; Ph.D., Cambridge U., 1950; m. Valda Suiter, Aug. 1946; children—Rodric W., Katrine B., Heather K., Ellyn J. Lectr., reader math. physics Birmingham U., 1950-56; research asso. Newman Lab. Nuclear Studies, Cornell U., 1953-55; mem. Inst. for Advanced Study, Princeton, 1955; prof. physics Enrico Fermi Inst. Nuclear Studies, U. Chgo., 1956-66; Royal Soc. research prof. Oxford (Eng.) U., 1963—. Cons. Brookhaven Nat. Lab., Upton, N.Y., 1958—, Argonne (Ill.) Nat. Lab., 1961—, Rutherford High-Energy Lab., Chilton, Eng., 1963—. Recipient Maxwell medal Inst. Physics, Phys. Soc., 1966; Mem. Phys. Soc. (London), Royal Soc., Am. Phys. Soc., Sigma Xi. Author: Strange Particles and Strong Interactions, 1962; Nuclear Interactions of the Hyperons, 1965. Research, numerous publs. on interactions governing decay and reaction properties of mesons and baryons, nuclear interactions of lambda hyperon, models for elementary particles. Home: 28 Jack Straw's Lane. Office: Clarendon Lab., 12 Parks Rd., Oxford, Eng.*

DALL, William Healey, Am. naturalist; b. Boston, Aug. 21, 1845; s. Charles Henry Appleton and Caroline (Healey) D.; studied natural scis. under Louis Agassiz; spl. courses anatomy and medicine; A.M. (hon.), Wesleyan, 1888; D.Sc., U. Pa., 1904; LL.D., George Washington U., 1915; m. Annette Whitney, Mar. 3, 1880; children—Charles Whitney, Marcus Hele, Marion, William Austin. Lt. in Internat. Telegraph Expdn. to Alaska, 1865-68; in U. S. Coast Survey, Alaska, 1871-84; paleontologist U. S. Geol. Survey, 1884-1925; prof. invertebrate paleontology Wagner Inst. Sci., Phila., Feb. 1893—. Fellow Am. Acad. of Arts and Scis.; mem. Nat. Acad. Scis. Author: Tribes of the Extreme Northwest; Scientific Results of the Exploration of Alaska; Reports on the Mollusca of the Blake Expedition; Alaska and Its Resources, 1870; Coast Pilot of Alaska; Biography of Spencer Fullerton Baird, 1915. Authority on Pacific Coast mollusks. Died Washington, D.C., Mar. 27, 1927.

DALLDORF, Gilbert, Am. pathologist; b. Davenport, Ia., Mar. 12, 1900; s. Julius and Hulda (Leisner) D.; B.S., State U. Ia., 1921; M.D., N.Y. U., 1924; D.Sc., Bowdoin Coll., 1953; D., U. Freiburg, 1957; m. Frances Elizabeth Barnhart, Apr. 6, 1926; children—Elizabeth (Mrs. Robert Martin), Frederic. Fellow pathology Pathologisches Inst., Freiburg, Germany, 1925-26; pathologist N.Y. Hosp., N.Y.C., 1926-29; instr. pathologic anatomy Cornell Med. Coll., 1926-32; pathologist Grasslands Hosp., Valhalla, N.Y., 1929-43; dir. labs. and research Westchester County, 1943-45; dir. div. labs. and research N.Y. Dept. Health, Albany, 1945-57; dir. research Nat. Found., 1958-59; mem. Sloan-Kettering Inst., Walker Lab., 1959—; prof. pathology Sloan Kettering div. Cornell U. Med. Coll., 1960—. Recipient Lasker Award, 1959. Mem. A.A.A.S., Am. Assn. Immunologists, A.M.A., Am. Soc. Exptl. Pathology, Assn. Am. Physicians, Nat. Acad. Sci., N.Y. Acad. Medicine, N.Y. Assn. Pub. Health Labs., N.Y. Med. Soc., Soc. Exptl. Biology and Medicine, Am. Epidemiological Soc. Isolated Coxsackie virus from stools of children (with Grace Mary Sickles), 1948; research in virus diseases, poliomyelitis, choriomeningitis. Home: Oxford, Md. 21654.*

DALLERY, Thomas Charles Auguste, French inventor, engr.; b. Amiens, France, Sept. 4, 1754; s. Charles Dallery; became watchmaker, designed smallest watches of the time, 1793; invented device to produce half tones on the harp; built steam wagon which traveled on roads, 1788; designed or built horizontal mill-wheel; 1st patentee use of the screw with steam ships, 1803. Died Jouy-en-Josas, France, June 1, 1835.

DALLINGER, William Henry, Brit. biologist; b. Devonport, Eng., July 5, 1842; s. Joseph Stephen Dallinger; ed. privately; LL.D., U. Toronto (Ont., Can.), 1884; D.Sc., U. Dublin (Ireland), 1892; D.C.L., U. Durham (Eng.), 1896; m. Emma Goldsmith; 1 son. Qualified for Wesleyan ministry, 1861; lectr. Gilchrist Ednl. Trust; apptd. gov., pres. Wesley Coll., Sheffield, Eng., 1880. Fellow Royal Soc., 1880; pres., Royal Micros. Soc., 1884-87, Quekett Club, 1890-92. Rewrote The Microscope and Its Revelations (Carpenter); pub. many articles in Monthly Micros. Jour., 1873-76. Micros. researches; observed one-celled organisms and discovered the formation of heat resistant spores. Died Nov. 7, 1909.

DALLMEYER, John Henry, optician; b. Loxten, Germany, Sept. 6, 1830; s. William and Catherine Wilhelmina (Meyer) D.; ed., apprenticed in Osnabrück, Germany; m. Hanna Ross; m. 2d, Elizabeth Mary Williams; 5 children, including Thomas R. Went to London, Eng., 1851; employee, then sci. adviser, firm of Andrew Ross, later in charge telescope-making sect., 1859; ret., 1880. Decorated cross Legion of Honour. Fellow Royal Astron. Soc. (council); mem. Royal Photog. Soc. (council). Research and publs. on optical instruments; improved telescope and microscope lenses, condensers for optical lanterns, celestial photography; built photoheliographs for Wilna Obs., 1863, Harvard, 1864, Brit. Govt., 1893. Died aboard ship off New Zealand, Dec. 30, 1883.

DALLOS, Peter John, biophysicist; b. Budapest, Hungary, Nov. 26, 1934; s. Ernest and Marie (Klein) D.; student Technol. U. Budapest, 1953-56; B.S., Ill. Inst. Tech., 1958; M.S., Northwestern U., 1959, Ph.D., 1962; m. Cirla Joan Hammerman, Sept. 9, 1961. Came to U. S., 1956, naturalized, 1961. Research engr. Am. Machine & Foundry Co., Niles, Ill., 1959, cons. engr., 1959-60; asst. prof. Northwestern U., Evanston, Ill., 1962-65, asso. prof., 1965—. Mem. I.E.E.E., Acoustical Soc., Biophys. Soc., N.Y. Acad. Scis., Sigma Xi, Tau Beta Pi, Eta Kappa Nu. Studies on eye fixation and acoustic reflex control systems, psychoacoustic phenomena such as intensity DL's and short-tone thresholds, electrophysiol. investigations of nonlinear distortion in peripheral auditory system, specifically first thorough study of auditory subharmonics.*

DALLY, Eugene, physician, anthropologist; b. Brussels, Belgium, 1833; s. N. Dally; studied medicine in Paris; doctorate, 1859; prof. anthropology and ethnology Sch. Anthropology; mem. Soc. Anthropology (pres.). French translator: De la place de l'homme dans la nature (Huxley), 1868. Research and publs. on muscular pathology, sch. and child hygiene, hydrotherapy, effects of consanguineous marriages. Died L'Etang-la-Ville, France, Dec. 30, 1887.

DALMA, Giovanni, psychiatrist, sociologist; b. Fiume, Yugoslavia, June 18, 1895; s. Desiderius and Ada (Kastel) D.; student medicine U. Vienna (Austria), 1913-14, U. Budapest (Hungary), 1914-17; M.D., U. Padua (Italy), 1920; m. Maria Urso, May 27, 1949. Organizer, chief Psychiat. Hosp., Fiume, 1928-39; staff psychiat. problems Ministry Pub. Health, Rome, Italy, 1944-48; founder Med. Sch., Nacional U. Tucumán (Argentina), 1949-51, prof. neurology, psychiatry, psychology, legal medicine, history of medicine, 1954—. Expert mental health WHO, 1963—. Mem. Nat. Acad. Medicine Buenos Aires (corr.), psychiat., sociol., med. history socs. of Argentina and Italy. Author: The Doctrine of Psychological Density, 1956; The Genetics of Mental Diseases, 1955; The Arabians and Medicine, 1963; also numerous articles. Discovery of new reflexes in sacrogluteal region; determination of action of hyperthermia in Trypanosoma Cruzi, pyretotherapy in Chagas disease, dis-psychobiosis, psychol. census of population, psychol. studies on Leonardo, new interpretation of equestrian monument of Colleoni. Home: 428 Pasaje Sorol, Tucumán, Argentina.*

DAL MONTE, Guidubaldo, Italian astronomer, mathematician; b. Jan. 11, 1545. Mil. engr. of Tuscany; studied perspective, and helped establish it on firm geometric basis; claimed to have invented proportional compass (actually, stolen from Galileo); posed question of composition of forces which became quest of early sci. of mechanics. Died Pesaro, Jan. 6, 1607.

DAL PALU', Cesare, Italian physician; b. Venice, Italy, Jan. 20, 1923; s. Giuseppe and Maria (Guastalla) D. P.; grad. U. Padua (Italy), 1946, Cardiology Libera Docenza, 1950, Internal Medicine, 1955; m. Giuliana Darin, Sept. 20, 1951; 1 dau., Nicoletta. Asst. dept. medicine U. Padua, 1946-60, now asso. prof. path. medicine; staff Inst. Patologia Medica, 1950-60; 1st asst. Clinica Medica, 1961. Mem. Italian Soc. Internal Medicine, Italian Soc. Cardiology (mem. directory), Italian Soc. Gastroenterology. Author: Errori e limiti della diagnostica electrocardiografica, 1961; also numerous articles. Research on electrocardiography, including electrocardiographic alteration due to cranial trauma, 1953-55, physiopathologic studies on portal circulation of man in normal subjects and portal hypertension, 1954-65; demonstrated influence of distal electrode on precordial leads, 1949. Home: 25 G. Barbarigo, Padova, Italia.*

DALQUEST, Walter W., Am. zoologist; b. Seattle, Sept. 11, 1917; s. Neils Walter and Florence (Woelber) D.; B.S., U. Wash., 1941, M.S., 1942; Ph.D., La. State U., 1951; m. Peggy Burgner, Aug. 8, 1940; 1 dau., Linda Lee. Research asso. U. Kan., 1945-49; teaching fellow La. State U., 1949-52; prof. biology Midwestern U., Wichita Falls, Tex., 1953—. Recipient Hardin Prof. award, 1962. Mem. Am. Soc. Mammalogists, Soc. Vert. Paleontology, Paleontol. Soc., Biol. Soc. Washington, Soc. Ichtheologists and Herpetologists. Author: Mammals of Washington, 1948; Mammals of the Mexican State of San Luis Potosi, 1952. Studies, publs. on geog. distbn., ecology, taxonomy of living and fossil mammals of U. S., Mexico, Africa. Home: Route 3, Box 248A, Wichita Falls, Tex. 76308.*

DALRYMPLE, William Haddock, veterinarian; b. Scotland, Apr. 23, 1856; s. Thomas and Mary Eleanor (Haddock) D.; M.R.C.V.S., Glasgow Vet. Coll., 1886; m. Mary Isabel Umpleby, Aug. 27, 1891. Mem. vet. staff Irish Privy Council, Dublin, 1888; came to U. S., 1889; prof. comparative medicine La. State U., also veterinarian La. Expt. Stas., 1889-93, 97-1919, also dean Coll. Agr., vice dir. La. Agrl. Expt. Stas. Mem. agrl. adv. com. U. S. Food Adminstrn. for La., 1917; mem. Internat. Congress on Tb, Washington, 1908, 15th Internat. Congress on Hygiene and Demography (Washington), 1912; reporter on anthrax for the U. S., 10th Internat. Vet. Congress, London, 1914. Fellow Glasgow Vet. Med. Soc., A.M.A., A.A.A.S.; mem. Nat. Live Stock Assn. (exec. com.), La. State San. Assn. (v.p.), U. S. Live Stock San. Assn. (pres. 1908-09), Am. Vet. Med. Assn. (pres. 1907-08), U. S. Expt. Sta. Vet. Med. Assn. (pres. 1901-02), La. State Agrl. Soc. (sec.), La. Stockbreeders' Assn. (sec.). Author: Veterinary Obstetrics and Livestock Sanitation. Editor livestock dept. New Orleans Picayune, Jour. Am. Vet. Med. Assn.; staff collaborators Am. Vet. Rev. Died Sept. 17, 1925.

DALRYMPLE-HAY, Sir Harley Hugh, Brit. civil engr.; b. Oct. 7, 1861; s. George James and Amelia Emily (Maitland) D.-H.; m. Agnes Yelland Waters, 1891; 1 dau. Began career as pupil on S. Wales lines Midland Ry., served as asst., divisional office, Brecon; engring. staff L. & S.W. Ry.; resident engr. Waterloo and City Tube Ry., 1894; worked on design and constrn. of Bakerloo, Hampstead, Piccadilly tubes; cons. engr. Underground Electric Rys. Co., 1902; responsible for new Piccadilly Circus Sta., other extension works on London underground; pvt. practice as civil engr., from 1907; cons. civil engr. Post Office (London) Tube Ry. (opened 1928); designed works for Brit. corps. in connection with power stas.; engr. for tunnel under Hooghly River at Calcutta; worked on extensions of tube rys. for London Passenger Transport Bd. Mem. council Instn. Civil Engrs. Contbr. papers to Instn. Civil Engrs. Inventor hooded shield and clay pocket system of tunnelling, also spl. apparatus for controlling accurate movement of shields. Died Dec. 17, 1940.

DALTON, H(oward) Clark, Am. biologist; b. Bklyn., Aug. 7, 1915; s. Howard A. and H. Elizabeth (Clark) D.; B.A. (Woods Hole scholar), Wesleyan U., Conn., 1936, M.A., 1937; Ph.D. (Royall Victor fellow), Stanford, 1940; m. Eleanora Alice Keene, June 19, 1948. Faculty, U. Rochester, 1940-41, Brown U., 1946-47, Bates Coll., Lewiston, Me., 1947-48; research fellow Carnegie Inst. Wash., Cold Spring Harbor, N.Y., 1948-50; faculty N.Y. U., 1950-67, prof., dept. chmn., 1961-67; prof. Pa. State Coll., University Park, 1967—. Fellow N.Y. Acad. Scis.; mem. A.A.A.S., Am. Assn. Anatomists, Am. Inst. Biol. Sci. (past bd. govs.), Am. Soc. Naturalists (past sec.), Am. Soc. Zoologists, Soc. Study Growth and Devel. (past treas.), Phi Beta Kappa, Sigma Xi. Asso. editor: Journ. Morphology, 1948-61. Demonstrated facets of genetic controlling mechanisms in devel. of pigment patterns in salamanders. Home: 316 Hubler Rd., State College, Pa. 16801. Office: Life Sci. Bldg., University Park, Pa. 16802.*

DALTON, John, English chemist; b. Eaglesfield, Eng., Sept. 6, 1766; s. Joseph and Deborah (Greenup) D.; ed. Quaker's sch., Eaglesfield, tchr., 1778-80; farm worker, also studied math., 1780-81; D.C.L., Oxford U., 1832; LL.D., U. Edinburgh; asst., then joint mgr., sch. in Kendal, 1781-93; pub. lectr. on natural philosophy, Kendal, 1787; tchr. math. and natural philosophy, New Coll., Manchester, 1793-99; pub. and pvt. tchr. math. and chemistry; lectr. on chemistry, mechanics and physics, Royal Instn., London, 1803, 09. Fellow Royal Soc., 1822 (Gold medal, 1826); corr. mem. French Acad. Scis., 1816, fgn. assoc., 1830; mem. Royal Soc. Edinburgh; Berlin, Munich acads. scis., Natural History Soc., Moscow; Lit. and Philos. Soc., Manchester (sec. 1800-08; v.p. 1808-17; pres. 1817-44); co-founder Brit. Assn. for Advancement of Sci., 1831. Author: Meteorological Observations and Essays, 1793; Elements of English Grammar, 1801; New System of Chemical Philosophy, (vol. 1) 1808; also many papers. Developed Dalton's law of partial pressures (each component of a mixture of gases exerts same pressure as it would if it alone occupied whole volume of the mixture at same temperature), 1801; discovered law of expansion of gases by heat, 1801; discovered diffusion of gases, 1801; developed laws of definite and multiple proportions, 1803; formulated atomic theory that all elements are composed of tiny indivisible, indestructible particles called atoms, which preserved their individuality in chemical changes; all atoms of same element identical in all respects including their weight while atoms of other elements have different weights; chemical combination of different elements occurs in simple numerical ratios by weight, 1803 (1st published by Thomas Thomson, 1807); arranged table of relative atomic weights, 1803; developed system of chem. symbols; thought of atoms arranged in gaseous form as a pile of shot; rejected Gay-Lussac's views on simple combining ratios of volumes of gases; 1787-1844 determined composition of ether and found its correct formula; discovered propylene; showed constancy of atmosphere to 15,000 feet, 1837; held that aurora borealis is elec. in origin, 1793; gave 1st detailed description of color-blindness (from which he suffered), 1794. Died Manchester, Eng., July 27, 1844.

DALTON, John Call, Am. physiologist; b. Chelmsford, Mass., Feb. 2, 1825; s. John Call and Julia (Spalding) D.; grad. Harvard, 1844, M.D., 1847; studied physiology under Claude Bernard, Paris, France. First U. S. physician to devote life to exptl. physiology and related scis.; prof. physiology U. Buffalo (N.Y.), 1851-54, U. Vt., 1854-56; prof. physiology Coll. Physicians and Surgeons, N.Y.C., 1855-83, pres., 1884-89; with L.I. Coll. Hosp., 1859-61; served as surgeon M.C., 7th N.Y. Regt., 1861-64. Recipient annual prize A.M.A. for essay on Corpus Luteum, 1851. Mem. Nat. Acad. Scis. Author: Treatise on Human Physiology, 1859; A Treatise on Physiology and Hygiene, 1868; Experimentation on Animals as a Means of Knowledge in Physiology, Pathology, and Practical Medicine, 1875; The Experimental Method in Medical Science, 1882; Doctrines of the Circulation, 1884; Topographical Anatomy of the Brain, 1885; Sugar Formation in the Liver (article), 1871. Introduced Bernard's methods of vivisection and exptl. physiology into U. S.; 1st in America to illustrate lectures on physiology with vivisection expts. Died N.Y.C., Feb. 12, 1889.

D'ALTON, Josef Edouard, naturalist, archeologist; b. Aquileja, Aug. 11, 1772; lived in Weimar and Jena, Germany; apptd. prof. archeology and history of art U. Bonn, circa 1821-40; corr. of Goethe on topics of comparative anatomy and osteology. Author: Natural History of the Horse, 1810; Comparative Osteology, 1821-28. Died Bonn, May 11, 1840.

DALTON, Norman, Brit. physician; b. s. E. T. E. Dalton; ed. Queen's Coll., Brit. Guiana, Christ's Coll., Finchley, King's Coll., London; M.D. Fellow, prof. path. anatomy, King's Coll., London, also lectr. medicine, med. sch., sr. physician, hosp.; examiner medicine U. London; physician Nat. Provident Instn.; med. officer London br. Scottish Union and Nat. Ins. Co. Fellow Royal Coll. Physicians. Research and publs. on influenza, use of subnitrate of bismuth in X-raying stomach. Died Mar. 9, 1923.

DALY, Ivan de Burgh, Brit. physiologist; b. Leamington, Warwickshire, Eng., Apr. 14, 1893; s. James Thomas and Amy (Pritchard) D.; B.A., Gonville and Caius Coll., Cambridge, Eng., 1914, M.A., 1923; M.D., U. Cambridge, 1922; m. Beatrice Mary Daly, Nov. 1, 1920; children—Michael de Burgh, Peter de Burgh (dec.). Beit Meml. research fellow, 1920-24; lectr. physiology U. Wales, Cardiff, 1923-27; 1923; prof. physiology U. Birmingham (Eng.), 1927-33; prof. physiology U. Edinburgh (Scotland), 1933-48; 1st dir. Agr. Research Council Inst. Animal Physiology, Babraham, Cambridge, 1948-58; Sr. Wellcome Trust Research fellow U. Oxford (Eng.), 1958-61; vis. research worker, 1958—. Fellow Royal Soc., 1943; Royal Coll. Physicians (Baly medal 1959); mem. Physiol. Soc., Path. Soc., Biochem. Soc., Thoracic Soc. (past pres.). Named Comdr. Brit. Empire. Author: (with Catherine Hebb) The Bronchial and Pulmonary Circulations, 1966; also numerous articles. Research on relation between pulmonary circulation and ventilation; demonstrated active control of lung blood vessels by nervous system. Home: 25 High St., Lond Crendon, Bucks, Eng. Office: U. Lab. Physiology, Oxford, Eng.*

DALY, Joseph Michael, Am. biologist; b. Hoboken, N.J., Apr. 9, 1922; s. Michael and Julia (Yarwood) D.; B.S., U. R.I., 1944; M.S., U. Minn., 1947, Ph.D., 1952; m. Cecilia Rieger, Sept. 1, 1951; children—Katherine, Stephen, Timothy, Martha, Cecilia, Anne, Constance, Melissa. Instr., U. Minn., 1949; asst. prof. biology U. Notre Dame, 1952-55; faculty U. Neb., Lincoln, 1955—, prof., 1958—, regent's prof., 1966—. Mem. Am. Soc. Plant Physiologists, Am. Phytopath. Soc., A.A.A.S. Studies, publs. on biochemistry and physiology of plant diseases designed to understand nature of disease, disease resistance with expectation for improved control of disease either through natural resistance or applied chems. Home: 1326 N. 38th St., Lincoln, Neb. 68503.

DALY, Reginald Aldworth, geologist; b. Napanec, Ont., Can., May 19, 1871; s. Edward and Jane Maria (Jeffers) D.; A.B., Victoria U., Toronto, Ont., 1891; A.M., Harvard, 1893, Ph.D., 1896, Sc.D. (hon.), 1942; postgrad., Heidelberg 1897-98, Paris, France, 1898; Sc.D. (hon.) U. Toronto, 1923, U. Chgo., 1941; m. Louise P. Haskell, June 3, 1903; 1 son, Reginald Aldworth. Geologist for Can., Internat. boundary surveys, 1901-07; prof. phys. geology Mass. Inst. Tech., 1907-12; Sturgis-Hooper prof. geology Harvard, 1912-42, emeritus, 1942—. Fellow Am. Acad. Arts and Scis., Geol. Soc. Am. (pres. 1932), Royal Soc. Edinburgh (hon.); mem. Am. Philos. Soc., Phila. Acad. Natural Sci., Nat. Acad. Scis., Seismol. Soc. Am., Am. Geophys. Union, Geol. Soc. S. Africa; hon. mem. Norwegian, Russian, Swedish acads., Mineral Soc. of Leningrad, Glasgow, Edinburgh, Stockholm, Belgium geol. socs.; fgn. mem. Geol. Soc. London; fgn. corr. French Acad. Scis. Author: Geology of the North American Cordillera at the 49th Parallel of Latitude, 3 vols.; Igneous Rocks and Their Origin, 1914; Our Mobile Earth, 1926; Igneous Rocks and the Depths of the Earth, 1933; The Changing World of the Ice Age, 1934; Architecture of the Earth, 1938; Strength and Structure of the Earth, 1940; The Floor of the Ocean, 1942. Asso. editor Am. Jour. Sci. Research on shape, emplacement, chem. composition, evolution of igneous rock, also origin of coral reefs and submarine canyons in Pleistocene epoch, isostasy and interior of earth. Died Sept. 19, 1957.

DALZELL, Nicol Alexander, Scottish botanist; b. Edinburgh, Scotland, Apr. 21, 1817; ed. Edinburgh High Sch.; M.A., Edinburgh U., 1837; m.; 6 children. Became asst. commr. custom, Bombay, India, 1841; later forest ranger, Scinde, India; conservator forests, Bombay; ret. on pension, 1870. Author: (with Gibson) The Bombay Flora, 1861; also articles. Named over 200 plants scientifically and described them for 1st time.

DAM, (Carl Peter) Henrik, Danish biochemist; b. Copenhagen, Denmark, Feb. 21, 1895; s. Emil and Emilie (Peterson) D.; M.S. in Chemistry, Poly. Inst., 1920; D.Sc. in Biochemistry, U. Copenhagen, 1934; m. Inger Olsen, July 15, 1924. Instr. in chemistry, Sch. of Agr. and Vet. Medicine, Copenhagen, 1920-23; instr. in biochemistry, U. Copenhagen, 1923, asst. prof., 1928, asso. prof., 1929-41; prof. biochemistry, Poly. Inst., Copenhagen, 1941 (appointment in absentia); sr. research asso. in biochemistry, U. Rochester (N.Y.), 1942-45; asso. mem. Rockefeller Inst., 1945-48; head biol. div. Danish Fat Research Inst., 1956-62. Recipient Christian Bohr award in physiology (for discovery of vitamin K), 1939, Nobel prize in physiology and medicine (with E. A. Doisy), 1943. Hon. fellow Royal Society (Edinburgh); fellow Am. Inst. Nutrition; mem. Soc. Exptl. Biology and Medicine, Am. Soc. Biol. Chemistry, Am. Bot. Soc., Chem. Soc. and Biol. Soc. (Copenhagen), Swiss Chem. Soc., Royal Danish Acad. Scis. and Letters, Danish Acad. Tech. Socs., Danish Nutrition Soc. (pres.), Internat. Union Nutritional Scis. (hon. pres.). Lectr. U. S. and Can., under auspices of Am.-Scandanavian Found., 1940-41, 49; in Copenhagen since 1946. Contbr. many papers relating to vitamins, lipids, nutrition, blood coagulation, to sci. jours. of U. S. and Europe. Discovered Vitamin K, 1934; research in Vitamins K and E, fat, cholesterol; nutritional studies in relation to gall-stone formation. Address: Danmarks Tekniske Hojskole, Afdeling for Biokemi og Ernaering, Ostervoldgade 10 III, Copenhagen 1350, Denmark.

DAMASCHINO, François, French physician; b. Paris, France, 1840; docteur en médicine, Paris, 1867; became head Monneret Clinic, 1867; became aggregate prof. medicine, 1874; named prof. pathology, 1883; physician Laënnec Hosp. Decorated Legion of Honor. Mem. Acad. Medicine. Author: Maladies des voies digestives, 1880. Research on infantile syphilis, infantile paralysis, thrush of esophagus. Died Paris, 1887.

DAMASCIUS OF DAMASCUS, mathematician; b. circa 458; left Greece, 529; took refuge in Persia, 533; last head of Academia of Plato, circa 510-529; wrote commentary on Hippocrates' Aphorisms.

DAMASK, Arthur Constantine, Am. physicist, educator; b. Woodstown, N.J., July 28, 1924; s. Gustave Nicholas and Pauline (McDowell) D.; B.S., Muhlenberg Coll., 1949; M.S., Ia. State U., 1954, Ph.D., 1964; m. Mary Virginia Northrop, Oct. 29, 1965. Frankford Arsenal guest scientist Brookhaven Nat. Lab., Upton, N.Y., 1954-65, vis. scientist, 1965—; prof. physics Queens Coll., City U. N.Y., Flushing, 1965—. Cons. Los Alamos Sci. Lab., 1957—, Lawrence Radiation Lab., 1964—. Mem. solid state scis. adv. panel NRC-Nat. Acad. Scis., 1964—. Fellow Am. Phys. Soc., A.A.A.S., Am. Assn. U. Profs., Sigma Xi. Author: (with G. J. Dienes) Point Defects in Metals, 1963. Research, numerous publs. on quantitative exptl. and theoretical effects of nuclear radiation on rate processes in solids, developer kinetic theory of annealing of point defects in solids and of nucleation of precipitation processes in solids. Home: 39 Thornhedge Rd., Bellport, N.Y. 11713. Office: Brookhaven Nat. Lab., Upton, N.Y. 11973.*

D'AMATO, Francesco, Italian geneticist; b. Grumo Appula, Italy, Oct. 10, 1916; s. Massimo and Caterina (Tateo) D'A.; Dr.Nat.Sci., U. Pisa (Italy), 1939; m. Maria Grazia Avanzi, June 9, 1952; children—Maria Caterina, Massimo, Giacomo. Prof. botany, dir. Inst. Botany, U. Cagliari, 1957-59; prof. genetics, dir. Inst. Genetics, U. Pisa (Italy), 1959—. Cons., Italian Com. for Nuclear Energy, 1959—. Mem. Accademia Nazionale dei Lincei (corr.), Accademia dei Georgofili (corr.), Italian Soc. Agrl. Genetics (pres. 1963—), Italian Genetic Assn. (pres. 1966-67). Recipient Nat. prize Feltrinelli, 1956. Research, numerous publs. on chem. mutagenesis, radiation biology, genetics histological differentiation in higher plants. Home: 20 Via di Parigi, Pisa, Italy.*

DAMBOURNEY, Louis Alexandre, French chemist, botanist; b. Rouen, France, May 11, 1722; adminstr. Rouen Garden of Plants; lived and did his research in Oissel, France. Mem. Rouen Acad. Scis. (sec.), French Acad. Scis., Rouen Soc. Agr. Author: Instruction sur la culture de la garance, 1788; Recueil de procédés et d'expériences sur les teintures solides que nos végétaux indigènes communiquent aux laiges et aux lainages, 1786; Histoire de plantes qui servent à la teinture, 1792. Adapted madderwort to France and used its roots for dyeing flax; research on dyes from native French plants; introduced the potato into Normandy. Died Rouen, June 2, 1795.

DAMEROW, Heinrich Philipp August, German psychiatrist; b. Stettin, Germany, Dec. 28, 1798; s. Henriette (Willett) D.; studied medicine in Berlin, Germany, 1817-21; M.D., 1821; state exam., 1826; m. Franziska; 1 son. Traveled, 1821-22; became lectr. at U. Berlin; held ofcl. state contract to visit mental instns.; named dir. mental instn. Halle, Germany, 1836; became ministerial asst. for psychiatry, 1839; reorganized various instns.; founder mental instn. in Nietleben, Germany, 1844. Author: Die Elemente der nächsten Zukunft der Medizin . . . , 1829; Über die relative Verbindung von Irrenheil-und-pflegeanstalten, 1840; Über die Grundlagen der Mimik und Physiognomik als freier Beitrag zur Anthropologie und Psychiatrie, 1850; Zur Kretinen- und Idiotenfrage, 1858; Irrengesetze und Verordnung in Preussen, 1863; also numerous articles. Pub., Allgemeine Zeitschrift für Psychiatrie, 1844-5. . Developed system for classification and isolation of mentally ill still used today. Died Nietleben nr. Halle, Sept. 22, 1866.

DAMES, Wilhelm Barnim, paleontologist, geologist; b. Stolp, Pomerania, Germany, June 9, 1843;

s. Louis and Elisabeth (Haugk) D.; studied in Breslau (now Wroclaw, Poland), Berlin (Germany); doctorate Breslau; m. Mathilde Wilhelmine Emilie von Toll, 1877. Became asst. Mus. for Geology and Paleontology, U. Berlin, 1871; named lectr. in geology and paleontology in 1874, asso. prof. in 1878, and prof. in 1891. Research and publs. on sea-urchins, paleozoic fossils, mammal and reptile fossils, archaeopteryx siemensi (prehistoric bird); charted Harz foothills, N. German diluvium and its detrital deposits. Died Berlin, Dec. 22, 1898.

DAMIRON, Nicolas, French physician; b. Belleville, France, Oct. 1, 1785; served with French armed forces during Napoleonic wars; asst. physician Val-de-Grâce Hosp., Paris, from 1819, tchr. hygiene. Named officer Legion of Honor. Author: Dissertation sur la sensibilité, 1805. Noted for work on smallpox. Died Sept. 27, 1833.

DAMMIN, Gustave John, Am. physician; b. N.Y.C., Sept. 17, 1911; s. Gustave Frank and Anna Barbara (Anselm) D.; A.B., Cornell U., 1934, M.D., 1938; certificate in trop. medicine, U. Havana, 1937; M.A. (hon.), Harvard, 1953; m. Anita Coffin, July 19, 1941; children—Susan, Tristram, Abigail. House staff mem. Johns Hopkins Hosp., 1939-40; house staff Peter Bent Brigham Hosp., Boston, 1940-41, pathologist in chief, 1952—; instr. Columbia Coll. Phys. and Surg., 1941; prof. pathology, chmn. dept. Wash. U. Med. Sch., 1946-52; prof. pathology, med. sch. Harvard, 1952-62, Elsie T. Friedman professor pathology, 1962—, Niles lectr. Cornell Med. Coll., 1953; Phi Delta Epsilon lecturer Yale University School of Medicine, 1956; I. W. Held lectr. Beth Israel Hosp., N.Y.C., 1963; cons. to surgeon gen. Dept. of Army, USPHS; nat. cons. global preventive medicine and epidemiology to surgeon general U. S. Air Force; laboratory cons. OCDM, 1950-60; president Armed Forces Epidemiol. Bd., 1960—; sci. adv. bd. Armed Forces Inst. Pathology, 1961—; WHO expert adv. panel on enteric diseases, chairman of committee, 1963—. Served from 1st lt. to lt. col. M.C., AUS, 1941-46; dir. labs. div. Office Surgeon Gen., 1945-46; now col. Res. Decorated Legion of Merit. Diplomate Am. Bd. Pathology, Nat. Bd. Med. Examiners. Mem. N.Y. Acad. Scis., Am. Soc. Clin. Investigation, Am. Assn. Pathologists and Bacteriologists, Internat. Acad. Pathology, Am. Soc. Tropical Medicine and Hygiene, Am. Soc. Exptl. Pathology, Assn. Am. Physicians, NIH (tropical medicine and parasitology study sect.), Society of Medical Consultants to the Armed Forces (pres. 1963), Central Society Clin. Research, Res. Officers Assn., Assn. U. S. Army, Japanese-Am. Soc. Pathologists, Assn. Mexican Pathologists, Korean Med. Assn., 38th Parallel Med. Soc. Korea, Sigma Xi, Alpha Omega Alpha. Editorial com. Ann. Rev. Medicine, 1957-60. Research on the definition of the pathogenesis of acute diarrheal disease, including cholera and bacillary dysentery; characterization of the rejection of the kidney as a transplanted organ. Home: 721 Huntington Av., Boston 15.*

DAMON, Albert, Am. physician, anthropologist; b. Boston, July 7, 1918; s. Henry H. and Bessie (Ansell) D.; A.B., Harvard, 1938, M.D., 1951; certificate Oxford U., 1939; Ph.D., U. Chgo., 1946; m. Selma O. Thomsen, Sept. 17, 1952; children—Elsa, Maria, Kristina. Faculty, Harvard Sch. Pub. Health, Cambridge, Mass., 1957-65, lectr. anthropology Harvard, 1965—. Cons. industry, U. S. Armed Forces, USPHS, VA, 1960—; established investigator Am. Heart Assn., 1962-67. Mem. Am. Assn. Phys. Anthropologists, Am. Pub. Health Assn., Am. Eugenics Soc. Author: (with Hertzberg, Churchill, Dupertuis and White) Anthropological Survey of Turkey, Greece, and Italy, 1963; (with Stoudt, McFarland) Human Body in Equipment Design, 1966. Demonstrated constl. factors in several diseases, including cancer and heart disease; showed secular increase in body size of Am. populations; established principles for applying anthropometry to equipment design; related culture and biology, including disease, in primitive societies. Home: 98 Homer St., Newton Center, Mass. 02159. Office: care Peabody Mus., Cambridge, Mass. 02138.*

DAMON, George Alfred, Am. engr.; b. Chesaning, Mich., Apr. 7, 1871; s. Brazil Monroe and Martha Angeline (Gould) D.; B.S., U. Mich., 1895; m. Harriet Diller, June 8, 1904; children—George Alfred, Harriet Antha. On staff Elec. Industries, World's Fair, Chgo., 1893; in shops of Fisher Electric Works, Detroit, 1894; draftsman and engr. Bion J. Arnold, Chgo., 1895-1900; mng. engr. Arnold Co., in charge design and constrn. railroad shops, elec. rys., hydroelectric plants and indsl. work, 1900-08; asso. with Mr. Arnold in reports on subway operation and constrn., N.Y., transp. problems of Pitts., Los Angeles, and surrounding dists., 1911; dean engring. Throop Coll. Tech. (Cal. Inst. Tech.), 1911-17. Cons. engr. Bd. Pub. Utilities, Los Angeles, 1912-13; tech. dir. City Planning Com. Pasadena, 1915-17; cons. engr. City Plan, San José, 1917; cons. City Plan, Long Beach, Calif., 1918-19; co-ordinator Los Angeles Union depot Damon-Judd plan, 1933. Fellow Am. Inst. E.E.; chmn. Joint Tech. Socs., Los Angeles, 1918-19, Am. City Planning Inst.; charter mem. City Planning Assn., Los Angeles (pres. 1918-19, sec. 1922, 25). Author: (papers) Inter and Intra Urban Transit and Traffic as a Regional Planning Problem, 1923, The Influence of the Automobile on Regional Transportation Problems, 1925. Died June 23, 1934.

DAMOUR, Augustine-Alexis, French mineralogist; b. Paris, France, July 19, 1808; began voyage to Antilles Islands, 1860; under-dir. Ministry Fgn. Affairs. Decorated Legion of Honor. Mem. French Acad. Scis. Chem. analysis of rare minerals including volcanic rocks of Antilles, herzolite, beryl, feldspar, periclasite; analyzed water from petrifactive springs, gold-bearing, diamond-bearing, and platinum-bearing sands. Died Sept. 22, 1902, Paris.

DAMPIER, William, Brit. explorer; b. East Coker, Somerset, England, circa June 1652; educated at grammar school near East Coker; served in the Dutch War, 1673; became under-manager of a Jamaican plantation, 1674; local commerce in Gulf of Mexico; long voyages of exploration, privateering and pirating Africa, Spanish Main, W. coast of S.Am., S. Pacific, China, E. Indies, 1678-91; commd. naval officer; comdr. exploratory expdn. to Australia and S. Pacific, 1699-1701; comdr. privateering expdn., 1703-07; pilot Woodes Rogers voyage around the world, 1708-11. Author: New Voyage Around the World, 1697; Voyages and Descriptions, 1699; Voyage to New Holland, 1703, 09. Accurate descriptions in navigation and hydrography; discovered and named New Britain Island, also numerous other geog. discoveries; made accurate meteorological and hydrographic observations. Died London, Mar. 1715.

DAMPIER, (formerly Whetham) Sir William Cecil, English scientist; b. London, Eng., Dec. 27, 1867; s. Charles Langley and Mary Ann (Dampier) Whetham; B.A., Trinity Coll., Cambridge (Eng.) U., 1889, M.A., 1892, Sc.D., 1931; Couts Trotter student, 1889, Clerk Maxwell scholar, 1893; m. Catherine Durning Holt, Dec. 10, 1892; 1 son, 5 daus. Fellow Trinity Coll., 1891, lectr. physics, 1895-1922, tutor, 1907-17, sr. tutor, 1913-17; asst. registrar Cambridge U., 1922-25; fellow Winchester Coll., 1917-47; sec. Agrl. Research Council, 1931-35, mem., 1935-45; a devel. commnr., 1933-51; chmn. Home Office com. factory lighting and machinery com. of Ministry Agr.; mem. Central Agr. Wages Bd., 1925-42; acting chmn. Rural Industries Bur., 1939-46; dir. Cambridge Water Co., Western Enterprises Ltd. Fellow Royal Soc., 1901. Mem. Royal Agr. Soc. (v.p. 1948; Gold medal 1936). Author: Solution and Electrolysis, 1895; Theory of Solution, 1902; The Recent Development of Physical Science, 1904-24; The Theory of Experimental Electricity, 1905; The Foundations of Science, 1912; Matter and Change, an Introduction to Physical and Chemical Science, 1924; Politics and the Land, 1927; A History of Science and its Relations with Philosophy and Religion, 1929-48; A Shorter History of Science, 1944; (with Mrs. Dampier) Studies in Nature and Country Life, 1903; Heredity and Society; An Introduction to Eugenics; (with dau.) Cambridge Readings in the Literature of Science, 1924; also articles. Died Dec. 11, 1952.

DAN, Katsuma, Japanese devel. biologist; b. Tokyo, Japan, Oct. 16, 1904; s. Takuma and Yoshiko (Kaneko) D.; B.A., Tokyo Imperial U., 1929; Ph.D., Pa. U., 1934; m. Jean McNair Clark, July 4, 1936; children—Teruki, Mika D. (Mrs. Makoto Takamiya), Marina D. (Mrs. Tohru Sohkawa), Harumi, Amy. Staff, Misaki Marine Biol. Sta., 1934-43; lectr. Tokyo Imperial U., 1943-49; prof. Tokyo Met. U., 1949—, pres., 1965—. Mem. Japan Zool. Soc. (pres. 1967), N.Y. Acad. Sci. Editor: (with M. Kume) Invertebrate Embryology, 1957; (with T. Yamada) Studies in Developmental Biology, 1958. Research, publs. on cell div., mitotic apparatus, SH-protein fractions of cytoplasm, mechanism of early morphological changes in sea urchin larva, spawning of a Crinoid. Home: 30-27 2-chome, Jingu-mai, Shibuya-ku, Tokyo, Japan.*

DANA, Charles Loomis, Am. physician; b. Woodstock, Vt., Mar. 25, 1852; s. Charles and Charitie Scott (Loomis) D.; A.B., Dartmouth, 1872, A.M., 1875, LL.D., 1905; grad. Georgetown Med. Coll.; M.D., Nat. Med. Coll., 1876, Columbian U., 1876, Columbia, 1877; LL.D., Edinburgh U., 1927; m. Lillian Farlee; children—Marjorie F. (Mrs. William T. Barlow), Elizabeth, Charles Loomis. Became pvt. sec. to U. S. senator from Vt., 1872, also to Spencer F. Baird; asst. surgeon U. S. Marine Hosp. Corps., also prof. physiology Women's Med. Coll., N.Y. Infirmary, 1879-88; prof. diseases of mind and nervous system N.Y. Post-Grad. Med. Sch. and Hosp., 1886-98; prof. nervous diseases Cornell U. Med. Coll., N.Y., from 1902; staff Bellevue Hosp., 1876—. Mem. N.Y. Neurol. Soc. (pres.), Am. Neurol. Assn. (pres.), N.Y. Acad. Medicine (pres.). Author: Textbook of Nervous Diseases and Psychiatry, 1892; Poetry and the Doctors: A Catalogue of Poetical Works Written by Physicians, 1916; also articles. First to describe adequately a condition of combined myotonia and muscular atrophy, 1888. Died Dec. 12, 1935.

DANA, Edward Salisbury, Am. mineralogist; b. New Haven, Conn., Nov. 16, 1849; s. James Dwight and Henrietta F. (Silliman) D.; A.B., Yale, 1870, A.M., 1874, Ph.D., 1876; studied Heidelberg and Vienna; m. Caroline Bristol, Oct. 2, 1883; children—Mary Bristol (Mrs. Alexander C. Brown), James Dwight, William Bristol. Curator mineral collection, Yale, 1874-1922, asst. prof. natural philosophy, 1879-90, prof. physics, 1890-1917, and prof. emeritus, from 1917. Trustee of Peabody Mus., 1885-1929; editor Am. Jour. of Science from 1875. Mem. Nat. Acad. Scis., Am. Philos. Soc.; fellow Am. Acad. of Arts and Scis.; corr. mem. Vienna Acad. Scis. Author: Text-book of Mineralogy, 1877, new edit., 1898; Text-book of Elementary Mechanics, 1881; Dana's System of Mineralogy, 6th edit., 1892; Minerals and How to Study Them, 1895; also many papers on mineral and other scientific subjects. Died June 16, 1935.

DANA, James Dwight, Am. geologist, educator; b. Utica, N.Y., Feb. 12, 1813; s. James and Harriet (Dwight) D.; grad. Yale, 1833; Ph.D. (hon.), U. Munich (Germany), 1872; LL.D., Harvard, 1886, U. Edinburgh (Scotland), 1890; m. Henrietta Silliman, June 5, 1844, 4 children. Math. instr. USN, 1833-36; asst. to Prof. Silliman, chem. lab. Yale, 1936-37; geologist, mineralogist U. S. expdn. to South Seas, 1837-42; prepared reports on expdn., 1842-55; Silliman prof. natural history and geology Yale, 1850-92. Recipient Woolaston medal Geol. Soc. London (Eng.), 1872; Walker prize Boston Soc. Natural History, 1892. Fellow Royal Soc., 1884 (Copley medal 1877), A.A.A.S. (pres. 1854), Geol. Soc. Am. (pres. 1890). Author: System of Mineralogy, 1837; Zoophytes, 1846; Manual of Mineralogy, 1848; Geology, 1849; Crustacea, 1852-55; Manual of Geology, 1862; Textbook of Geology, 1864; Corals and Coral Islands, 1872; Characteristics of Volcanoes, 1890. Editor Am. Jour. Sci., 1840. Provided valuable observations concerning formation of mountains, volcanoes, continents, major features of earth's crust; many of his works are still standards in the field. Died New Haven, Apr. 14, 1895.

DANA, Samuel Luther, Am. chemist; b. Amherst, N.H., July 11, 1795; s. Luther and Lucy (Giddings) D.; grad. Harvard, 1813, M.D., 1818; m. Ann Willard, June 5, 1820; m. 2d, Augusta Willard; 4 children. Served in War of 1812; discovered system of bleaching cotton known as Am. system of bleaching; devised improvements in printing calicoes. Author: Outlines of Mineralogy and Geology of Boston and its Vicinity, 1818; A Muck Manual for Farmers (one of 1st sci. works on agr. written and pub. in U. S.), 1842. Died Lowell, Mass., Mar. 11, 1868.

DANCE, Jean-Baptiste-Hippolyte, French physician; b. St. Pol, France, 1797; M.D., Paris, 1818; intern, 1820; aggregate physician, also became head Cochin Clinic, Paris, 1830. Author: Guide pour l'étude de clinique médicale, 1829. Research on uterine phlebitis, cholera. Died of cholera, 1832.

DANCEL, Christian, inventor; b. Kassel, Germany, Feb. 14, 1847; 2 children. Learned machinist trade in Germany; came to N.Y.C., circa 1865; machinist in various N.Y. shops, 1865-67; devised shoe-sewing machine (bought by Charles Goodyear, Jr. for his shoe-machine factory), became supt. Goodyear's factory; began making machinery for stitching outsoles and sewing shoe-welts, circa 1870; invented machine to sew both welts and turns, 1874 (still used with minor improvements); opened own machine shop and patented some small machines, 1876; called by Goodyear Co. to perfect machine to sew upper and outer sole of shoe, finished it (machine with curved needle sewing a lock-stitch), 1885; patented straight-needle machine, 1891; organized Dancel Machine Co., Bklyn., circa 1895; invented many other shoe-making devices; co-patentee machines for making leather buttonholes, barbed-wire fence, rubbing type. Died Bklyn., Oct. 13, 1898.

DANCIS, Joseph, Am. physician; b. N.Y.C., Mar. 19, 1916; s. Ab and Sarah (Dancis) Goldberg; A.B., Columbia U., 1934; M.D., St. Louis U., 1938; m. Bernice Schrier, July 4, 1948; children—Andrew, Dale. Practice medicine, specializing in pediatrics, N.Y.C., 1947-51; faculty Sch. Medicine N.Y. U., 1951—, prof. pediatrics, 1960—; career investigator NIH, 1962—, mem. reproduction tng. rev. com., 1967—. Recipient Borden award, 1962. Markle scholar in med. sci., 1956-60. Mem. Acad. Pediatrics, Am. Pediatric Soc., N.Y. Acad. Scis., Harvey Soc. Contbr. numerous articles to profl. jours. Research in immunology in the newborn, placental function, inherited metabolic diseases. Home: 217 Melbourne Rd., Great Neck, N.Y. 11021. Office: 550 First Av., N.Y.C. 10016.*

DANCK (or DANCKO), see John of Saxony.

D'ANCONA, Umberto, Italian zoologist; b. Fiume, Italy, Sept. 5, 1896; s. Antonio and Anna (Klas) D'A.; ed. U. Budapest, U. Rome; Ph.D. in natural scis.; D. honoris causa U. Aix-Marseille; m. Luisa Volterra; 1 dau., Silvia. Prof. zoology U. Sienna, Pisa, Padua. Mem. Acad. dei Lincei, Acad. Scis. of Paris (corr.), numerous others. Research and numerous publs. on zoology, biology, especially Lotta per l'esistenza. Home: via Scalcerle 9. Office: via Loredan 10, Padua, Italy.

DANCY, Alexander Brown, Am. ophthalmologist, otologist; b. Holly Springs, Miss., Dec. 23, 1877; s. Clifton and Sarah Torian (Brown) D.; student Southwestern Bapt. U., Jackson, Tenn., 1894-95, 1896-97; M.D., U. Louisville, 1900, Vanderbilt U., 1902; certificate New York Ophthalmic and Aural Inst., 1903; Royal London (Eng.) Ophthalmic Hosp., 1903-04; spl.

certificate Am. Bd for Ophthalmic Examinations, 1916, also Am. Bd. Otolaryngology; postgrad. Harvard, 1917; m. Mary Eloise Happel, June 6, 1911; children—Mary Happel, Alexander Brown. Intern N.Y. Polyclinic, 1902-03; practiced medicine, Jackson from 1904; ophthalmologist and otolaryngologist Crook Sanatorium, also Meml. Hosp. Fellow Am. Acad. Ophthalmology and Otolaryngology, A.C.S. Died June 10, 1933.

DANDELIN, Germinal Pierre, mathematician; b. Bourget, Apr. 12, 1794; s. Anne (Botteman) D.; ed. École polytechnique, Paris, France, 1813. Wounded during the defense of the gate of Clichy (France); became naturalized Dutch citizen, 1817; entered Engr. Corps, 1819, returned as col. after the revolution of 1830; became prof. mechanics U. Liège (Belgium), 1825. Decorated Cross of the Legion of Honor. Author: Mémoire sur quelques propriétés remarquables de la focale parabolique, 1822; Mémoire sur l'hyperboloide de révolution et sur les hexagones de Pascal et de Bounchon, 1824; Memoire sur l'emploi des projection stéréometriques en geometrie, 1825; Lecons sur la mécanique, 1827. Developed theorem on conic sections which was named after him; research on parabolic focus, use of stereometric projections in geometry; studied Pascal's math. and commented on hexagons. Died Brussels, Belgium, 1847.

DANDOLO, Count Vincenzo, Italian chemist, economist; b. Venice, Italy, Oct. 6, 1758. Mem. Gt. Council Cisalpine Republic; lived in Paris, France; became gov. of Dalmatia under Napoleon, 1804-09; ret. to estate at Varese nr. Como. Mem. Instituto Italiano, Società Italiana. Author: Fundamenti della Scienza Chimico-fisica, 1795; Trattato Elementare di Chimica . . . ; Les Hommes nouveaux, ou moyen d'operer une régénération nouvelle, 1799. Wrote valuable treatises on rural economy, regarding prodn. wine, wool, silk, sheep breeding; 1st important Italian follower of Lavoisier; translated Lavoisier's Traité; pub. work on discoveries in chemistry made in last fourth 18th century, 1796. Died Varese, Italy, Dec. 12, 1819.

DANDY, Walter E(dward), Am. surgeon; b. Sedalia, Mo., Apr. 6, 1886; s. John and Rachel D.; A.B., U. Mo., 1907, LL.D., 1928; M.D., Johns Hopkins, 1910, A.M., 1911; m. Sadie Martin, Oct. 1, 1924; children—Walter E., Mary Ellen, Kathleen Louise, Margaret Martin. Began practice medicine, Balt.; prof. neurol. surgery, in charge surgery of nervous system Johns Hopkins. Mem. Am., So. surg. assns., Am. Neurol. Assn., Phi Beta Kappa, Sigma Xi. Author textbook on neurological surgery, several books on various lesions of brain. Contbr. numerous articles to surg. and neurological jours. Introduced new operative procedures for tumors and aneurysms of brain, for hydrocepholus, neuralgias and other disturbances of cranial nerves; introduced ventriculography, ventricular estimation and cerebral pneumography, for diagnosis and localizing of tumors of brain and intracranial lesions; discovered ruptured intervertebral disks which cause low backaches and sciaticas, and introduced surg. procedure for their cure. Died Apr. 19, 1946.

DANES, Zdenko Frankenberger, geophysicist; b. Prague, Czechoslovakia, Aug. 25, 1920; s. Zdenek and Eleanora Rebensteiger (v. Blenkenfeld) D.; Ph.D. in Math—Physics, Charles U., Prague, 1949; m. Marie Hankova, Jan. 20, 1945; children—Peter, Ellen. Came to U. S., 1952, naturalized, 1957. Research asst., asst. prof. Charles U., 1946-50; geophysicist Gulf Research & Devel. Co., Pitts., 1952-59; research engr. Boeing Co., Seattle, 1959-62; vis. prof. U. Minn., Mpls., 1962; asso. prof. U. Puget Sound., Tacoma, 1962—. Cons. to aero-space and oil industry, astrogeologic br. U. S. Geol. Survey. Mem. Am. Geophys. Union, Soc. Exploration Geophysicists, N.Y. Acad. Scis., A.A.A.S. Research, publs. on gravity field of earth, internal constn. of earth, math. theory deformation of rocks, theory of geophys. methods. Home: 4206 N. 13th St., Tacoma, 98406.*

DANESFORT, Baron (John George Butcher), physicist; b. Killarney, Ireland, Nov. 15, 1853; s. Samuel Butcher; ed. Trinity Coll., Cambridge (Eng.) U.; m. Alice Mary Brandreth, 1898. Named fellow Trinity Coll., Cambridge, 1875; barrister Chancery bar, 1878; mem. Parliament, City of York, 1892-1906, 10-23. Author: Viscous Fluids in Motion; Quaternion Forms of General Propositions in Fluid Motion. Died June 30, 1935.

DANFORTH, C(harles) H(askell), Am. anatomist, anthropologist; b. Oxford, Me., Nov. 30, 1883; s. James and Mary (Haskell) D.; A.B., Tufts Coll., 1908, A.M., 1910, D.Sc. (hon.), 1941; Ph.D., Washington U., St. Louis, 1912; m. Florence Wenonah Garrison, June 24, 1914; children—Charles Garrison, Alan Haskell, Donald Reed. Faculty, Tufts Coll.; teaching fellow Harvard Med. Sch., 1910-11, lectr. anatomy, 1929-30; faculty Washington U. Sch. Medicine, 1908-22; faculty Stanford (Cal.) Med. Sch., 1922—, emeritus prof. anatomy, 1949—. Mem. Am. Assn. Anatomists (2d v.p. 1936-38, editorial bd. 1938-48), A.A.A.S. (pres. Pacific div. 1950-51), Soc. for Exptl. Biology and Medicine (v.p. 1941-43, editorial bd. 1936-42), Western Soc. Naturalists (pres. 1942-44), Nat. Acad. Scis., Am. Philos. Soc., Cal. Acad. Scis. (council 1943-53), Am. Soc. Human Genetics (v.p. 1952). Author: Hair, with Special Reference to Hypertrichosis, 1925. Editor: (with E. Allen and E. A. Doisy) Sex and Internal Secretions, 1939. Editorial bd. Growth, 1942-44, Excerpta Medica Sect. 1, 1946-65, Am. Jour. Phys. Anthropology, 1927-42. Studies, publs. on genetic features in components of hereditary traits; fgn. transplants in birds; persistence of hereditary deviants in man. Contbr. pioneering papers on relation of genetic and endocrine factors in skin transplantation. Home: 607 Cabrillo St., Stanford, Cal. 94305.*

DANFORTH, David Newton, Am. physician, educator; b. Evanston, Ill., Aug. 25, 1912; s. William Clark and Gertrude (MacLean) D.; B.S., Northwestern U., 1934, M.S., 1936, M.B., 1938, Ph.D., 1938, M.D., 1939; m. Gladys Blaine, June 28, 1938; 1 son, David Newton. Faculty, Northwestern U., Evanston, Ill., 1947—, prof. obstetrics and gynecology, 1959—, chmn. dept., 1965—; chief obstetrics and gynecology, Evanston Hosp., 1947-65; chmn. obstetrics and gynecology Chgo. Wesley Meml. Hosp., 1965—. Dir. Am. Bd. Obstetrics and Gynecology, 1966—. Recipient Capps prize for med. research, 1940, Barren Found. medal for research, 1966, Merit award Northwestern U., 1966. Mem. Am. Gynecol. Soc., Am. Assn. Obstetricians and Gynecologists, Am. Coll. Obstetricians and Gynecologists, A.C.S., Soc. for Gynecologic Investigation, Soc. for Exptl. Biology and Medicine, A.M.A., A.A.A.S. Editor: Textbook of Obstetrics and Gynecology, 1966. Research and numerous articles on anatomy and physiology of cervix uteri, physiology and conduct of pregnancy and labor, obstet. and gynecol. pathology. Home: 300 Warwick Rd., Kenilworth, Ill. 60034. Office: 2511 E. Superior St., Chgo. 60611.*

DANFORTH, William Edgar, Am. physicist; b. Buffalo, June 22, 1905; s. William Edgar and Edith (Ovens) D.; B.S. in Physics, Union Coll., 1927; M.A., Harvard, 1929, Ph.D., 1931; m. Mary Helen Swartzel, Aug. 9, 1931; children—Richard William, Carolyn Edith (Mrs. Michel d'Obrenovic). Research asst. Gen. Electric Co., 1927, 29, Harvard, 1931-32; fellow Bartol Research Found., Swarthmore, Pa., 1932-46, asst. dir., 1946—. Ordained minister Episcopalian Ch., 1955. Mem. Em. Phys. Soc., Am. Forestry Assn., Sigma Xi; Papers in fields cosmic radiation, phys. electronics, surface physics. Measured primary ionizing events of cosmic ray particle passing through gas; developed improved electron tube cathodes; determined mechanism by which thermion emission of metal is increased by layer of foreign atoms. Home: R.D., Monroeton, Pa. 18832; also, 524 Pine St., Phila. 19106. Office: Bartol Found., Whittier Pl., Swarthmore, Pa.*

DANGEARD, Pierre-Augustin-Clément, French mycologist; b. Segrie, France, Nov. 23, 1862; licence ès sciences naturelles Faculty Caen (France); doctorate ès sciences, 1886; children—Pierre, Louis. Head biol. research Faculty Caen; became lectr. Poitier, France, 1894; professor of botany Sorbonne, Paris, 1924-34. Member of the French Academy of Sciences, 1906. (Grand prize physical science for work with mushrooms 1906, became v.p. 1934, and pres. in 1935). Author: Recherches histologiques sur les urédinées; Reproduction sexuelle des ustilaginées. Research on plant botany and lower animals; developed theory on the distinctive characteristics of plants and animals in which he thought animal and vegetable kingdoms were merging toward a simple plasmatic form; research on reprodn. in mushrooms; invented new method of examination based on the phenomenon of parasitism (enucleophagy) in the field of karyology. Died Segrie, Nov. 10, 1947.

D'ANGHIERA, Pietro Martire (or Petrus Martyr Anglerius), Italian historian; diplomatic rep., ct. of Ferdinand and Isabella of Spain, from 1487; dean Cathedral Granada (Spain), from 1505. Author: De rebus oceanicis et novo orbe (1st account discovery Am.), 1516; De orbe novo decades octo, 1530; Opus epistolarum, 1530. Died 1526.

DANHOF, Ivan Edward, Am. physiologist; b. Grand Haven, Mich., June 24, 1928; s. Benjamin John and Lois (Vandenberg) D.; B.A., M.A., N. Tex. State U., 1949; M.S., U. Ill., 1951, Ph.D., 1953; M.D., U. Tex., Dallas, 1962; m. Martha Aye Crouch, Jan. 22, 1950; children—Janie Marie, Mark Edward, Martha Lynn, Gary Philip. Practice medicine, specializing in physiology, Dallas, 1953—; faculty physiology U. Tex. Southwestern Med. Sch., 1954-56, 1954—, asso. prof., 1967—; vis. asst. prof. physiology So. Meth. U., 1965; dir. Gastrointestinal Research Found. S.W. Fellow Tex. Acad. Sci.; mem. A.A.A.S., Soc. Exptl. Biology and Medicine, Am. Physiol. Soc., A.M.A., Sigma Xi. Research, publs. on effect of various gastrointestinal hormones on secretory and motor function of gastrointestinal tract, isolation and purification of pyloric antral inhibitory hormone, location of secretin, mechanism of action of cholecystokinin, clin. pharmacology. Home: 2322 Ingleside Dr., Grand Prairie, Tex. 75050. Office: 5323 Harry Hines Blvd., Dallas 75235.*

DANIEL, Glyn Edmund, Brit. archaeologist; b. Pembrokeshire, Wales, Apr. 23, 1914; s. John and Mary Jane (Edmunds) D.; ed. St. John's Coll., Cambridge; M.A., Ph.D., Litt.D.; m. Ruth Langhorne. With St. John's Coll., 1938—; reader in archaeology Cambridge U., 1946—. Mem. Soc. Antiquaries (London), German Archeol. Inst. (corr.), Italian Inst. Prehistory and Protohistory (hon.). Author: The Prehistoric Chamber Tombs of England and Wales; A Hundred Years of Archaeology; The Prehistoric Chamber Tombs of France; The Idea of Prehistory; The Megalith Builders of Western Europe. Editor: Antiquity, 1957—. Home: The Merry Boys, St. John's St. Office: St. John's College, Cambridge, Eng.

DANIEL, Herbert Gustav Karl, German physicist; b. Treptow on Rega, Germany, Mar. 30, 1926; s. Herbert Trungott and Marie (Rottschalk) D.; M.A., U. Heidelberg (Germany), 1951, Ph.D., 1954; postgrad. Max Planck Inst., Heidelberg. Staff, Max Planck Inst., 1954-58, 59—, div. leader, 1964—; Research asso. Ames (Ia.) lab. U S, AEC, 1958-59; vis. scientist European Orgn. for Nuclear Research, Geneva, Switzerland, 1966—; privatdozent/tchr. U. Heidelberg, 1961—. Editor: Kolloquium über B-Zerfall und Schwache Wechsel wirkung, 1965. Research, publs. on radioactivity, particularly B decay, in solid state and atomic physics with B decay as a tool, and on pionic and muonic atoms; devel. new types of magnetic particle spectrometers, theoretical work in electron optics. Home: Rte. de Peney 8b, Vernier, Switzerland. Office: CERN, Geneva 23, Switzerland.*

DANIEL, J(ohn) Frank(lin), Am. sociologist; b. O'Fallon, Mo., July 31, 1873; s. John Franklin and Martha Short (Henry) D.; student So. Ill. Normal U., 1901; with dept. of edn., P.I., 1901-05; S.B., U. Chgo., 1906; Adam T. Bruce fellow from Johns Hopkins, Pasteur Inst., Lille, France, 1908-09; Ph.D., Johns Hopkins, 1909; m. Menetta White Brooks, Feb. 16, 1909. Instr. zoölogy U. Mich., 1910-11; faculty U. Cal., 1911—, prof. zoology, 1919—. U. S. del. 12th Internat. Congress of Zoölogy, Lisbon, 1935. Fellow A.A.A.S., Am. Acad. of Arts and Scis., mem. Am. Zoöl. Soc., Western Soc. of Naturalists, Soc. Exptl. Biology and Medicine, Cal. Acad. Scis., Am. Genetic Assn., Société Zoölogique de France, Assn. Anatomistes. Author: Animal Life of Malaysia, 1905; The Elasmobranch Fishes, 1922; also papers on breeding of mice for sci. purposes, exptl. studies on alcohol, morphogenesis. Chmn. U. Cal. Publs. in Zoölogy; collaborator Internat. Jour. of Cytology (Japan). Died Nov. 2, 1942.

DANIEL, John, Am. physicist; b. Perry County, Ala., July 6, 1862; s. John and Susan Lee (Winfield) D.; A.B., U. Ala., 1884, A.M., 1885, LL.D., 1914; postgrad. Johns Hopkins, 1886-88, U. Berlin (Germany), 1892; m. Grace Olive Knight, Sept. 2, 1896; children—Landon Garland, Ray Knight, John Harben Winfield, Robert Bradley (dec.), Grace Olive. Asst. prof. physics U. Ala., 1884-86; fellow Vanderbilt U., Nashville, 1888, instr. physics, 1889, adj. prof., 1890-93, prof., 1894-1939, prof. emeritus physics, 1939-50. Fellow A.A.A.S.; mem. Tenn. Acad. Sci. (charter), Phi Beta Kappa, Sigma Xi. Designed and installed 1st electric dynamo at Vanderbilt U. (before comml. lighting); discovered depilatory and burning effect of X-ray (described in Science, May 1896). Contbr. papers to profl. jours. Died Mar. 2, 1950.

DANIEL, Joseph Carl, Jr., Am. biologist; b. Murphysboro, Ill., Aug. 21, 1927; s. Joseph Carl and Alice (G'Sell) D.; student Rensselaer Poly. Inst., 1945-46; B.S., St. Louis U., 1949; M.S., U. Mich., 1950; Ph.D., U. Colo., 1956; m. Mary Patricia Hurley, June 30, 1951; children—Katherine, Joseph, Mark, Judith, Edward, Alice. Instr., Danver U., 1950-51; faculty Adams State Coll., Alamosa, Colo., 1952-60, asso. prof., 1959-60; vis. lectr. Loretto Heights Coll., Denver, 1963-64; faculty U. Colo., Boulder, 1962—, asso. prof., 1965—, acting chmn. dept. biology, summer 1963, dir. undergrad. ind. research program in biology, 1964-65. Fellow A.A.A.S.; mem. Am. Soc. Zoologists, Am. Genetics Assn., Am. Soc. Cell Biology, Colo.-Wyo. Acad. Sci., Soc. for Study Devel. and Growth, Sigma Xi, Phi Sigma. Research in design of methods and media for growing mammalian embryos in artificial environment, study of effects of various nutrients, hormones, drugs on developing mammal, effects of phys. agts. (laser beams) on embryonic tissue. Home: 1525 Sunset Blvd., Boulder, Colo. 80302.*

DANIEL, Louise Jane, Am. biochemist; b. Phila., Oct. 28, 1912; d. Frank W. and Mabel (Brensinger) Daniel; B.S., U. Pa., 1935; M.S., Pa. State U., 1936; Ph.D., Cornell U., 1945. Tchr., Penn Hall Jr. Coll., 1936-42; staff Cornell U., Ithaca, N.Y., 1942—, prof. biochemistry, 1958—. Mem. Am. Inst. Nutrition, Am. Soc. Biol. Chemists, Am. Chem. Soc., A.A.A.S., Soc. Exptl. Biology and Medicine, N.Y. Acad. Scis., Sigma Xi, Phi Kappa Phi. Author: (with A. L. Neal) Laboratory Experiments in Biochemistry, 1967; also articles. Research on metabolic function of folic acid and vitamin B12; interrelationships of trace minerals. Home: 210 Highgate Rd., Ithaca, N.Y. 14850.*

DANIEL, Lucien Louis, French botanist; b. La Dorée, France, Nov. 1, 1856; ed. École normale primaire, Laval, Paris Faculty of Scis.; doctorate in sci., 1890; m. Marie Robin; 1 dau., Jean. Became prof. Rennes (France) Lycée, 1895; prof. botany Faculty of Scis., Rennes. Mem. Breton Soc. Botany (founder), French Acad. Scis., 1930. Author: Théorie des capacités fonctionnelles, 1902; La question phylloxérique, le greffage et la crise viticole, 1908; Sur des variations produites par des équilibres de nutrition, 1910; Le greffage, sa théorie et ses applications rationelles, 1922; Les plantes médicinales de Bretagne, 1924; Etudes sur la greffe, 4 vol., 1927-34; Les mystères de l'hérédité symbiotique, 1940; La science qui tue. L'infernale chimie de l'alimentation, 1940. Developed several new hybrids including new fodder cabbage, several varieties of stringbeans; grafted eggplant to the tomato; helped ameliorate viticultural crisis, 1900; opposed use of chem. products in foodstuffs, 1940. Died Dec. 26, 1940.

DANIEL, Peter Maxwell, English pathologist; b. London, Eng., Nov. 14, 1910; s. Peter Lewis and Laetitia (Herskind) D.; M.A., M.B., B.Ch., U. Cambridge (Eng.), 1942; M.A., D.M., U. Oxford (Eng.), 1949; D.Sc., U. London (Eng.), 1961; m. Sarah Shelford (div.); 5 children, m. 2d, Frances Dawn Bosanquet; 1 son. Sr. research officer pathology U. Oxford, 1949-56; hon. cons. Radcliffe Infirmary, Oxford, 1949-56; prof. neuropathology Inst. Psychiatry, U. London, 1957—; hon. cons. neuropathology Bethlem Royal, Maudsley hosps., London, 1957—, Brit. Army, 1951—. Mem. Physiol. Soc., Anat. Soc., Path. Soc., Brit. Neuropath. Soc. Research, numerous publs. on physiology, anatomy and pathology of brain, endocrine organs, liver, kidney, muscles and visual system, exptl. cancer research. Office: Dept. Neuropathology, Inst. Psychiatry, Maudsley Hosp., Denmark Hill, London S.E. 5, Eng.*

DANIELL, John Frederic, English scientist, inventor; b. London, Eng., Mar. 12, 1790; hon. D.C.L., Oxford U.; m.; 2 sons, 5 daus. Worked in relative's sugar refining company and introduced improvements there; worked in chemistry and meteorology; prof., chemistry, King's College, London, 1831-1845. Fellow Royal Soc., London, 1813 (foreign sec., 1839-45; Rumford medal, 1830; Copley medal, 1836; Royal medal, 1842); received silver medal of Horticultural Soc., 1824. Author: Meteorological Essays, 1823; Introduction to the Study of Chemical Philosophy, 1839; articles. Invented hygrometer, 1820; described new pyrometer, 1830; invented electric battery (Daniell cell); 1st to attempt to explain main phenomena of atmosphere by physical laws; revolutionized hothouse management by stressing importance of moisture; constructed a water barometer, 1830. Died London, Eng., Mar. 13, 1845.

DANIELLI, James Frederic, biologist; b. Wembley, Eng., Nov. 13, 1911; s. James Frederic and Helena (Hollins) D.; Ph.D., London (Eng.) U., 1933, Cambridge (Eng.) U., 1942; D.Sc., London U., 1938, Gent (Belgium) U., 1956; m. Mary Guy, Jan. 4, 1937; children—Richard, Corinne. Fellow, Princeton, 1933-35, St. John's Coll., Cambridge U., 1942-45; physiologist to Marine Biol. Assn., 1946; reader cell physiology Royal Cancer Hosp., 1946-49; prof. zoology, chmn. dept. King's Coll., London, 1949-61; prof. medicinal chemistry and biochem. pharmacology State U. N.Y. at Buffalo, 1962-65, chmn. dept. biochem. pharmacology, 1962-64, prof. theoretical biology, dir. Center for Theoretical Biology, 1964—, provost Faculty of Science and Mathematics. Cons. to various indsl. firms, pubs., govt. orgns. Fellow Royal Soc.; mem. Inst. Biology, Biochem. Soc., Am. Physiol. Soc., Soc. for Biology, Am. Soc. Exptl. Biology and Medicine, Am. Inst. Biol. Scis. Author: Permeability of Natural Membranes, 1942; Cell Physiology and Pharmacology, 1952; Cytochemistry, 1953; also numerous articles. Editor, Internat. Rev. Cytology, 1951—, Jour. Theoretical Biology, 1960—, Gen. Cytochem. Methods, 1958—. Research on surface chemistry of steroids and proteins; structure, permeability of cell membrane; capillaries; wound healing; cytochemistry of enzymes and proteins; drugs for cancer chemotherapy theoretical biology. Office: Center for Theoretical Biology, 4248 Ridge Lea Rd., Buffalo.*

DANIELLI, Stephano, Italian physician; b. Butrio, Italy, 1656; prof. U. Bologne, Italy; opponent of ideas of Malpighi.

DANIELLS, William Willard, Am. chemist; b. Oakland County, Mich., 1840; s. Nathaniel I. and Lucinda (Reed) D.; B.S., Mich. Agrl. Coll., 1864, M.S., 1867, D.Sc., 1897; Lawrence Sci. Sch. (Harvard), 1867; postgrad. Halle, Berlin, 1881; m. Hontas Augusta Peabody, June 21, 1871. Asst. in chemistry Mich. Agrl. Coll., 1864-68; prof. chemistry U. Wis., 1868-68, prof. agrl. and analytical chemistry, 1869-70, prof. chemistry, 1879-1907. Chemist Wis. Geol. Survey, 1872-76; asst. U. S. Geol. Survey, 1882-83; mem. Wis. Bd. Health, 1885-89. Died Oct. 12, 1912.

DANIELS, Charles Wilberforce, Brit. physician; s. T. Daniels; ed. Trinity Coll., Cambridge, London Hosp.; M.B., Cambridge U.; with Med. Services, Fiji, Brit. Guiana; dir. Inst. Med. Research, Kuala Lumpur, Federated Malay States; dir. lectr. London Sch. Tropical Medicine; cons. physician Albert Dock Hosp.; lectr. tropical diseases London Hosp., London Sch. Medicine for Women, St. George's Hosp.; mem. Royal Soc. Malaria Commn., India and Central Africa. Fellow Royal Coll. Physicians. Author: Laboratory Studies in Tropical Medicine; Tropical Medicine and Hygiene; also reports. Died Aug. 6, 1927.

DANIELS, Edward William, Am. biologist; b. Tracy, Minn., Jan. 19, 1917; s. Azro A. and Nellie (Bundy) D.; B.A., Cornell Coll., 1941; M.S., U. Ill., 1947, Ph.D., 1950; m. Harriet Catherine Zimmerman, Dec. 23, 1943; children—Edward B., Paul G., Thomas F., Lynell K. Faculty, U. Chgo., 1950-54, asst. prof., 1953-54; asso. biologist dept. biol. and med. research Argonne (Ill.) Nat. Lab., 1954—. Cons. biosatellite program USAF and Gen. Electric Co., 1962—. Fellow A.A.A.S.; mem. Soc. Protozoologists (exec. council 1961-64), Am. Inst. Biol. Sci., Am. Soc. Cell Biology, Am. Soc. Zoologists, Am. Assn. Cancer Research, Electron Microscope Soc. Am., Radiation Research Soc., Nat. Soc. Med. Research. Research and numerous publs. on use of electron microscope to determine origin and fate of nuclear envelope during mitosis in amoebae, origin of Golgi apparatus in amoebae, ultrastructural changes within amoeba mitochondria during mitosis and starvation, probable site of nuclear envelope synthesis in leukemia cells, radiation damage and recovery at cellular level.*

DANIELS, Farrington, Am. chemist; b. Mpls., Mar. 8, 1889; s. Franc Birchard and Florence (Farrington) D.; B.S., U. Minn., 1910; Ph.D., Harvard, 1914; D.Sc., U. R.I., 1953, U. 1958, U. Dakar (W. Africa), 1961, U. Louisville, 1964, U. Wis., 1966; m. Olive Miriam Bell, Sept. 15, 1917; children—Farrington, Florence Mary (Mrs. James W. Drury), Miriam Olive (Mrs. Martin Ludwig), Dorin Slater. Faculty, Worcester (Mass.) Poly. Inst., 1914-17; electrochemist U. S. Dept. Soils, 1919-20; faculty U. Wis., Madison, 1920—, professor, 1928-59, chairman of the chemistry department, 1952-59, professor emeritus, 1959—; staff Metall. Laboratory, U. Chgo., 1944-46, dir., 1945-46; vis. prof., Stanford, 1930; Baker vis. prof. Cornell U., 1935. Recipient Willard Gibbs medal Am. Chem. Soc., 1954, Priestley medal, 1956, Norris award, 1957. Mem. Geochem. Soc. (past pres.), Am. Chem. Soc. (past pres.), Solar Energy Soc. (pres. 1964-67), Am. Acad. Arts and Scis., Am. Philos. Soc., Nat. Acad. Scis. (past v.p.), Sigma Xi (pres. 1964-66). Author: Outlines of Physical Chemistry, 1931; (with R. A. Alberty) Physical Chemistry, 1955; (with others) Experimental Physical Chemistry, 1929; Mathematical Preparation of Physical Chemistry, 1928; Chemical Kinetics, 1937; (with J. A. Duffie) Solar Energy Research, 1955; Direct Use of the Sun's energy, 1964; (with T. M. Smith) Challenge of Our Times, 1952; (with L. J. Heidt, R. M. Livingston, E. Rabinovitch) Photochemistry Solid and Liquid State, 1961; also numerous articles. Research in phys. chemistry, chem. kinetics gas reactions, photochemistry, nitrogen fixation air by high temperature and quick chilling, thermoluminescence crystals, geochemistry, direct use sun's energy, electrochemistry. Home: 1129 Waban Hill, Madison, Wis. 53711.*

DANIELS, Fred Harris, Am. mech. engr.; b. Hanover Centre, N.H., June 16, 1853; s. William Pomeroy and H. Ann (Stark) D.; M.E., Worcester Poly. Inst., 1873; spl. study in chemistry, under Dr. Thomas M. Drown, of Lafayette Coll.; m. Sarah Lydia White, May 17, 1883. Entered employ Washburn & Moen Mfg. Co., Worcester, 1873, became gen. supt. and chief engr.; visited Europe, and made spl. studies of advanced methods in mfr. of iron, steel, etc.; apptd. chief engr. Am. Steel & Wire Co. when latter acquired business of Washburn & Moen Co., 1899, and became dir. in co., 1902; apptd. chmn. bd. engrs. U. S. Steel Corp. when latter acquired interests of Am. Steel & Wire Co., 1901; pres. Washburn & Moen Co., Worcester Wire Co., 1900—; also mem. bd. engrs. Ind. Steel Co., Gary, and Minn. Steel Co., Duluth. Recipient grand prize and Gold medal Paris Expn., 1900, for achievements in devel. of wire industry. Died Aug. 30, 1913.

DANIELS, Ralph, Am. chemist; b. N.Y.C., May 2, 1921; s. Sidney M. and Helen (Finkle) D.; A.B. magna cum laude, Bklyn. Coll., 1944; M.S., Harvard, 1949, Ph.D., 1950; m. Shirley Wolhandler, July 1, 1944; children—Lesley Diane, Ethan Howard Michael, Brian Fredrick. Research chemist Givaudan-Delawana Research Inst., N.Y.C., 1944-46; postdoctoral fellow U. Wis., Madison, 1950-51; instr. Purdue U., Lafayette, Ind., 1951-52; faculty U. Ill., Chgo., 1952—, prof. chemistry, 1963—; vis. scientist U. London (Eng.) Chester Beatty Research Inst., 1961-62. Recipient Am. Inst. Chemists award, 1944; U. S. Rubber Co. fellow, 1949. Fellow Chem. Soc. London; mem. Am. Chem. Soc., Am. Assn. U. profs., Sigma Xi, Rho Chi (hon.). Author: (with L. Bauer) Problems in Organic Chemistry, 1964; also articles. Patentee in field. Research on synthesis biologically active compounds in fields radiation protective compounds, local anesthetics, hypnotics, analgesics, antimalarial drugs, course reactions organic compounds, spectra organic compounds. Home: 3838 White Cloud Dr., Skokie, Ill. 60076, Office: 833 S. Wood St., Chgo. 60612.*

DANIELS, Robert S., Am. psychiatrist; b. Indpls., Aug. 12, 1927; s. Harry H. and Mary (Bassett) D.; B.S., U. Cin., 1948, M.D., 1951; m. June Gibson, July 1, 1950; children—Stephen, Allen, Lynn, Judith. Faculty, U. Chgo., 1957—, asso. prof., acting chmn. psychiatry dept., 1963—. Mem. Ill. (sec.), Am. psychiat. assns., Ill. Psychotherapy Soc. (pres.), Am. Group Psychotherapy Assn., A.M.A., Phi Beta Kappa, Sigma Xi, Alpha Omega Alpha. Research, numerous publs. on field effect of interpersonal interactions on treatment of psychiatric patients and on learning process. Home: 6742 Constance Av., Chgo. 60649.*

DANIELSON, Gordon Charles, Am. physicist, educator; b. Dover, Ida., Oct. 28, 1912; s. Gust and Olga (Olson) D.; B.A., U. B.C., 1933, M.A., 1935; Ph.D., Purdue U., 1940; m. Dorothy Edna Thompson, June 24, 1939; children—Ellen Kathleen (Mrs. Karl Richard Fox), Lee Robert, Keith Gordon, Neil David. Research physicist U. S. Rubber Co., Detroit, 1940-41; asst. prof. physics U. Ida., 1941-42; asso. group leader Beacons Radiation Lab., Mass. Inst. Tech., 1942-46; mem. tech. staff Bell Telephone Labs., Murray Hill, N.J., 1946-48; asso. prof. physics Ia. State U., Ames, 1948-53, prof. physics, sr. physicist Inst. for Atomic Research, 1953—, chmn. physics grad. adv. com., 1951-53, mem. metallurgy curriculum com., 1949-54, chmn. physics colloquium com., 1959-60, distinguished prof. scis. and humanities, 1964—. Mem. metallurgy and solid state rev. com. Argonne Nat. Lab., 1960-63, chmn. 1962-63; mem. solid state panel Nat. Acad. Scis.-NRC, 1962; cons. NSF, 1963; civilian cons. U. S. Air Force, Eng., 1944-45; chmn. com. on thermoelectric conversion Office Naval Research, 1958. Recipient Army-Navy certificate appreciation, 1948. Guggenheim fellow U. Cambridge (Eng.), 1958-59. Fellow Am. Phys. Soc. (exec. com. div. solid state physics 1964—), Ia. Acad. Sci.; mem. Am. Inst. Physics (vis. scientist 1960), UN Assn. (pres. Ames chpt. 1964-66), Sigma Xi. Unitarian (past pres., treas. fellowship). Contbr. articles to profl. jours., books. Research on practical Fourier analysis, domain orientation in barium titanate, counting diamonds, thermal diffusivity, tungsten bronzes, semi-conducting compounds. Home: 2007 Country Club Blvd., Ames, Ia. 50012.*

DANIELSSEN, Daniel Cornelius, physician; b. Norway, 1815; considered founder of modern scientific leorology; gave (with C. W. Boeck) 1st modern description of leprosy, 1847. Died 1894.

DANIELSSON, Ingvar, Finnish phys. chemist; b. Inga, Finland, May 24, 1922; s. Paul and Martha (Sevón) D.; Cand.Phil., Abo Academy (Finland), Swedish U., 1950, Phil.D., 1957; m. Raili Lehtonen, 1951; children—Eskil, Jöns, Joneta. Faculty Abo Akademy, 1956—, prof. phys. chemistry, 1966—, also head Inst. Phys. Chemistry. Mem. Finnish Chem. Soc. Author: The Association of Long-Chain Dipotassium alpha-omega Alanedioates in Aqueous Solution, 1956; also articles. Potentiometric, calorimetric and osmometric investigation in field of asso. colloids and lower fatty acid salts. Home: 4 C Stenhuggareg, Abo, Finland.*

DANILOV, Stepan Nikolaevich, Russian organic chemist; born January 6, 1889; graduated Petrograd University, 1914; D.Chem. Science Teacher, later professor Petrograd (now Leningrad) University, 1915-30; prof. Leningrad Tech. Inst., 1930—; dir. lab. Inst. High-Molecular Compounds, USSR Acad. Sci., Leningrad. Mem. USSR Acad. Sci. (corr.). Author: The Theory and Law of Chemical Structure of Butlerov (1861-1961), 1962, also others. Research and publs. on high-molecular compounds, oxyaldehydes and oxyketones; made discoveries which aided in clarifying nature of biochemical processes; introduced new methods for obtaining ethers of cellulose. Address: Leningrad Tech. Inst., Zagorodny prospect 49, Leningrad, USSR.

DANJON, Andre, French astronomer; b. Caen, France, Apr. 16, 1890; prof. U. Strasbourg (France), became dir. obs., 1930; named dir. obs. Paris and Meudon, 1945, then of Astrophysics Inst.; mem. French Acad. Scis., 1948, also Bur. Longitudes; pres. Internat. Astron. Union, 1955-58. Author: Description du ciel, 1926; (with A. Couder) Lunettes et télescopes, 1935; Tables des fonctions trigonométriques, 1949; Pierre-Simon Laplace, 1951; Astronomie générale, 1953; Astronomie populaire, 1955. Contbr. Le ciel et la terre to Encyclopedie Française, Vol. III. Made photometric studies of Venus and Mercury; discovered influence of solar activity on aspects of the moon during eclipse; inventor astrolabe instruments and other instruments for precision measurement of irregularities of earth and measurement of time.

DANKMEIJER, Johan, Dutch biologist; b. Amsterdam, Netherlands, 1907; prof. anatomy, embryology and phys. anthropology Leyden (Netherlands) U., 1932, later prosector, lab. anatomy and embryology; asst., lab. anatomy U. Amsterdam, 1933-35; asst., lab. anatomy U. Utrecht (Netherlands), 1935-36; asst., dept. obstetrics, 1936-46; prof., docent anatomy U. Zurich (Switzerland), 1946-47. Mem. Internat. Inst. Embryology; pres. Assn. Anatomists. Research and publs. on human anatomy, embryology, comparative anatomy, phys. anthropology, history of medicine.

DANNENBERG, Arthur Milton, Jr., Am. physician, educator; b. Phila., Oct. 17, 1923; s. Arthur Mansbach and Marion (Loeb) D.; A.B., Swarthmore Coll., 1944; M.D., Harvard, 1947; M.A., U. Pa., 1951, Ph.D., 1952; m. Aileen Rose Hart, Mar. 30, 1948; children—Arlene Jane, Andrew Loeb, Audrey Ann. Asst. prof. exptl. pathology and microbiology U. Pa. Sch. Medicine, Phila., 1956-64; asso. prof. radiobiology and pathology Johns Hopkins Sch. Hygiene and Sch. Medicine, Balt., 1964—. Diplomate Am. Bd. Microbiology, Nat. Bd. Med. Examiners. Mem. Am. Assn. Immunologists, Am. Soc. Exptl. Pathology, Soc. Exptl. Biology and Medicine, Histochem. Soc., Am. Chem. Soc., Am. Thoracic Soc., Reticuloendothelial Soc., Coll. Physicians Phila. Research, publs. on enzymes involved in pathogenesis of infectious disease, pathogenesis of Tb and melioidosis, histochemistry of phagocytosis, biochemistry of macrophages. Home: 12 Lake Manor Ct., Balt. 21210.*

DANOISEAU DE MONTFORT, Marie-Charles Théodore, French astronomer; b. Besançon, Issy, France, Apr. 9, 1768; mem. arty. regt. of La Fère for 20 years; left France after revolution of 1789; served in armies of Condé; dir. obs., asso. Acad. Scis. Lisbon, Portugal; returned to France with Junot's army, 1807; attached to arty. of Bastia at Antibes; ret. as lt. col., 1817; dir. obs. mil. sch. Mem. Bur. Longitudes, French Acad. Scis. Author: Tables de la lune, 1824; Tables écliptiques des satellites de Jupiter, 1836; Éphémerides nautiques; also numerous articles. Research on return of comet of 1795, small planets. Died Aug. 6, 1846.

DANON, David, research physician; b. Tatar-Pazardjik, Bulgaria, Oct. 17, 1921; s. Moshe H. and Regina (Bassan) D.; grad. Geneva (Switzerland) Med. Sch., 1952; m. Mathilda Galland, 1948 (div. 1958); 1 dau., Daphna; m. 2d, Ariette Behar-Mair, Dec. 14, 1959; children—Ilan-Moshe, Noga. Staff, Inst. Physics, Lab. Biophysics, Geneva, 1951-53; staff Weizmann Inst. Sci., 1956—, asso. prof., 1962-66, Patrick Gorman chair biology, 1966, prof. biology, 1967—; head sect. biol. ultrastructure, 1964—; asso. prof. Hebrew U. Jerusalem (Israel), 1962—; vis. staff Columbia Med. Center, N.Y.C., 1962-63. Mem. European Molecular Biology Orgn., Internat. Soc. Haematology, Internat. Soc. for Electron Microscopy and Biochemistry through Israeli socs., Israeli Soc. for Electron Microscopy (pres. 1964—). Research, numerous publs. on techniques in electron microscopy, of macromolecules of biol. interest, biophys. aspects of red cell aging, biochem. and biophys. studies on red cell differentiation and maturation; invented an ultramicrotome, apparatus for control shadow casting under vacuum, instrument for automatic rec. of red cell osmotic fragility, kit for determination of density distbn. of cells.*

DA NOVARA, Dominico Maria, Italian astronomer; b. Ferrara, Italy, 1464; Dr. arts and medicine. Prof. astrology, Bologna, Italy, 1483-1504; tchr. of Copernicus. Supported Ptolemy's theories; made ann. astrological predictions; estimated obliquity of ecliptic to be 23 degrees and 29 minutes; believed altitude of pole had altered since Ptolemy's time. Died Bologna, 1514.

DANSEREAU, Pierre, ecologist; b. Montreal, Que., Can., Oct. 5, 1911; s. Lucien and Marie (Archambault) D.; B.A., U. Montreal, 1931, B.Sc. in Agr., 1936; D.Sc., U. Geneva (Switzerland), 1939; LL.D. (hon.) U. Sask. (Can.), 1959; D.Sc. (hon.), U. New Brunswick, 1959; m. Françoise Masson, Aug. 29, 1935. Botanist, Montreal Bot. Garden, 1939-42; dir. Service de Biogéographie, 1942-50; asso. prof. dept. botany U. Mich., 1950-55; dean faculty sci. U. Montreal, 1955-61; asst. dir. N.Y. Bot. Garden, Bronx, 1961-66, sr. curator ecology, 1966—; adj. prof. botany Columbia, 1961—. Thomas Alva Edison vis. scholar at Cranbrook Inst. Science, 1967-68. Recipient Fermi medal Toulouse (France) Acad. Sci., 1960, Pariseau medal Assn. Canadienne-Française pour l'Avancement des Sciences 1965; Cons. de Honor, NRC Madrid, Spain, 1960. Fellow Royal Soc. Can.; mem. Ecol. Soc. Am., World Acad. Arts and Scis., Royal Soc. New Zealand. Author: Biogeography, an Ecological Perspective, 1957; Contradictions and Biculture, 1964; also numerous articles. Research on dynamics vegetation compared in tropical, temperature and arctic regions. Office: N.Y. Bot. Garden, Bronx, N.Y. 10458.*

DANSGAARD, Willi, Danish geophysicist; b. Copenhagen, Denmark, Aug. 30, 1922; s. Poul and Oda (Nielsen) D.; Cand.Mag., U. Copenhagen, 1947, Dr.-Phil., 1961; m. Inge Magnus Thomsen, June 28, 1947; children—Finn, Birgitte, Kirsten. Sci. asst. Meteorol. Inst., 1947-50; staff Geomagnetic Obs., Godhavn, Greenland, 1947-48; staff Biophys. Lab., U. Copenhagen, 1950-63, Phys. Lab. II, 1963—, prof., 1961—; head glaciological expdns. to Norway, 1962, Greenland, 1964, 67. Sci. cons. Danish AEC, 1960-65; mem. Commn. on Atmospheric Chemistry and Radioactivity, 1960—; mem. sub.-com. on ice core studies SCAR, 1967. Author: The Isotopic Composition of Natural Waters, 1961; also articles. Applications of environmental isotopes in meteorol., hydrological and glaciological problems, including turnover of water, past climatic conditions and flow patterns of glaciers especially Greenland ice cap. Home: 55, Trorodvej, Vedbaek 2950, Denmark. Office: H. C. Orsted Inst., Universitetspark 5, Copenhagen, Denmark.*

DANTI, Ignazio, Italian mathematician; b. Perugia, Italy, 1536; entered Dominican Order; tchr. sci. Florence, Italy; Pope Gregory XIII apptd. him to reform calendar; made bishop of Alatri, 1583. Author: La prospettiva di Euclide; La prospettiva de eliodor larisseo, 1573. Translated Euclid and Heliodorus into Italian. First to make a gnomon to determine equinoxes and solstices. Died 1586.

D'ANTONI, Alexander-Victor Pahagino, Italian educator, arty. investigator; b. 18th century; officer, prof. Arty. and Engr. Sch., Turin, Italy. Investigated role of percussion primer and of packed powder; in experiments on occurring pressure in a closed area, found a value which doubled.

D'ANTONIO, William Vincent, Am. sociologist; b. New Haven, Feb. 7, 1926; s. Albert and Marie (Nuzzo) D'A.; B.A., Yale, 1948; M.A., U. Wis., 1953; Ph.D., Mich. State U., 1958; m. A. Lorraine Giorgio, June 15, 1950; children—Jo Anne, Albert, Nancy, Carla, Raissa, Laura. Faculty, U. Notre Dame (Ind.), 1959—, now prof., chmn. dept., chmn. edn. com. Catholic Inter-Am. Cooperation Program. Social Sci. Research Council grantee, 1962, 64. Mem. Am. Sociol. Assn., Am. Cath., Ohio Valley (pres. elect) sociol. socs., Am. Assn. U. Profs., Assn. Latin Am. Studies (past pres. Midwest council). Author: (with W. H. Form) Influentials in Two Border Cities, 1965. Editor: (with Howard J. Ehrlich) Power and Democracy in America, 1961; (with F. B. Pike) Religion, Revolution and Reform: New Forces for Change in Latin America, 1964. Studies on community power structures, utility and limitations of various techniques for studying community power structures, importance of longitudinal studies, gradual changing nature of decision-making over time. Home: 1444 S. Bend Av., South Bend, Ind. 46617. Office: O'Shaughnessy Hall, Notre Dame, Ind. 46556.*

D'ANVILLE, Jean-Baptiste Bourguignon, French cartographer; b. Paris, July 11, 1697; s. Hubert and Charlotte (Vaugon) d'A.; student under Longuerue, Paris. Named royal geographer, 1719. Mem. Acad. Inscriptions and Belles-Lettres. Author: Atlas général (most comprehensive geog. work of mid 18th century), 1737-80; Nouvel atlas de Chine, 1737; Atlas antiquus, 1768. Applied critical method to his work; collected ancient authorities and modern traveller's reports; wrote on ancient meteorology; redrew maps of continents (which provided starting point for exploration of inland Africa in 19th century). Died Paris, Jan. 28, 1782.

DANYAU, Antoine Constant, French physician; b. Paris, France, 1803; s. Alexis Constant Danyau; became intern, 1825; docteur en médecine, 1829; m. Roux. Began practice medicine, Paris, 1830; head clinic Charité Hosp., Paris, 1830-34; became surgeon for Bur. central, 1834; taught courses on childbirth Maternité Hosp., Hôtel-Dieu; taught surgery at Paris Faculty of Medicine. Mem. French Acad. Medicine. Author: Des abcès à la marge de l'anus, 1823. Translated publs. of Karl Waegek. Research on gangrenous metritis; wrote on anal abscesses. Died Paris, 1871.

DANYLEVS'KYJ, Vasyl', Russian physiologist; b. Charkov, Ukraine, U.S.S.R., 1852; prof. Charkov; leader Inst. Endocrinology and Organotherapy. Died 1939.

DANYSZ, Andrzej Witold, Polish physician; b. Warsaw, Feb. 22, 1924; s. Michal and Jadwiga (Strzalkowska) D.; M.D., Jagellonian U., 1949; m. Wieslawa Dec, Nov. 29, 1950; children—Beata Anna, Wojciech. Asst., Jagellonian U., Cracow, 1945-46, 47-50; lectr. Mil. Center Higher Edn., Lódz, 1950-57; asst. in gen. pathology Med. Acad., Lódz, 1951-57; faculty Med. Acad., Biatystok, 1957—, head dept. pharmacology, 1961—, dir. Radiobiol. Research Center, 1966—. Mem. med. physics com. Polish Acad. Sci., 1967—, pres. radiobiol. commn., 1966—. Recipient awards Polish Acad. Sci., Polish Ministry Health. Mem. Polish Physiol. Soc., Polish Pharmacological Soc., Polish Physicians Soc., European Soc. Radiol. Biology. Author: Pathology (with Andrzej Gtuszcz), 1966; Pharmacology for Students, 1967; also numerous articles. Research on radiobiology, chem. radioprotection, pharmacology of cell membrane; psychopharmacology. Home: 35 Szpitalna, Biatystok. Office: 1 Kilinskiego, Biatystok, Poland.*

DANYSZ, Jean, Polish pathologist; b. Poland, 1860; worked in France; isolated Salmonella typhimurium (causative agt. of mouse typhoid and of food poisoning in man), 1900; 1st to use radium in treatment of malignant disease, 1903; described Danysz phenomenon that toxin added to antitoxin in an equal amount at once, produces a nontoxic mixture, but when added at intervals in fraction, results in a generally toxic mixture. Died 1928.

DANZ, Ferdinand Georg (Friedrich), German physician; b. Gedern, Germany, Apr. 29, 1770; s. Friedrich Georg and Amalia Susanna Friderica (Graf) D.; studied at Marburg, Jena univs. (both Germany); M.D., U. Giessen (Germany), 1790. Became prosector anat. theater U. Giessen, 1790, became asso. prof., 1791. Author: Versuch einer allgemeinen Geschichte des Keuchhustens (1st monograph on whooping cough), 1791; Von Menschen ohne Haare und Zähne (1st publ. on congenital ectodermal dysplasia), 1792; also chpts. on obstetrics, anatomy and physiology. Died Giessen, Mar. 1, 1793.

DAPPER, Olfert (or Olivier), Dutch geographer, physician; b. 1636; pioneer in functional anthropology; contbd. to understanding of position of customs in the setting of their socs. Author: Description of Africa; A New Description of the African Countries, 1668; A Description of Amsterdam, 1663; The Memorable Expedition of the Netherlanders to the Coasts and into the Empire of Taising or of China, 1670; A Description of America and the Land of the South, 1676; A New Description of Asia, 1680. Died 1689.

D'APRÉS DE MANNEVILLETTE, Jean-Baptiste-Nicolas-Denis, French hydrographer, navigator; b. Le Havre, France, Feb. 11, 1707; s. Jean-Baptiste-Claude and Françoise (Marion) d'A.; m. Lorient Marie-Madeleine-Jaquett de Binard, 1732. 4th officer Maréchal d'Estrées, 1726; comdr. Le Fier, 1730-32; with the Galatée, 1732-35; 2d lt. Prince de Conti, 1736; comdr. Chevalier Marin, 1749, Glorieux, 13 Canons, le Duc de Bourgogne, 1758; director of depository of maps and papers of India Company. Member l'Académie de Marine. Author: Le nouveau quartier anglais, 1739; Routier des côtes des Indes, 1745; Directions for Navigating from the Channel to East Indies, 1768; Instructions pour la navigation des Indes orientales, 1775. Editor: Le neptune oriental, (collection of maps authoritative for 100 years), 1745. 1st to utilize octant on a voyage; determined longitudes by distances of sun and moon, 1749 voyage. Died Lorient, France, Mar. 1, 1780.

DARAN, Jacques, French physician; b. Saint-Frajou, France, Mar. 6, 1701; physician in Army of King of Sardinia; practiced medicine, Marseilles, France; became surgeon to king, 1755. Author: Recueil d'observations chirurgicales sur les maladies de l'urètre, 1745. Research on urinary tract diseases, gonorrhea; originated method of contracting the urethra using dilating candles. Died Paris, France, 1784.

D'ARBELA, Felice, physician; b. Jerusalem, June 30, 1894; s. Gregory and Malvina (Schwarz) D'A.; Ph.D. in medicine and surgery; m. Marcella Maria Ass. Mei Gentlucci; children—Valeria, Serena. Prof. med. semeiology U. Florence; head physician Civil Hosp. of Venice (Italy). Mem. Italian Soc. Internal Medicine, Italian Soc. Cardiology, Italian Soc. Biol. Experimenters, Accad. Medico Fisica (Florence), Soc. Triveneta Med. Chir. Research and numerous publs. on pathology. Home: S. Silvestro 1176. Office: S. Polo 1176, Venice, Italy.

DARBOUX, Jean Gaston, French mathematician; b. Nimes, France, Aug. 14, 1842; ed. Ecole normal, Paris, France; hon. degrees from many fgn. univs. Teaching post at Ecole Normale; asst. to Bertrand, in chair math. physics Collège de France, 1866-67; prof. math. Lycée Louis le Grand, 1867-72; maître de conferences Ecole Normale, 1872-73; asst. to prof. rational mechanics Sorbonne, Paris, 1873-80, prof. higher geometry, 1880-89, dean Faculty Sci., 1889-90. Recipient of the Sylvester medal, 1916. Fellow Royal Society, 1902; member of the French Academy of Sciences, 1884. Author: Leçons sur la théorie générale des surfaces et les applications géométriques du calcul infinitésimal, 1887-96; Leçons sur les systèmes orthogonaux et les coordonnées corvilignes, 1898. Co-founder, Bull. des scis. mathématiques et astronomiques, 1870. Research on orthogonal surfaces, deformation of surfaces, infinitesimal deformation, spherical representation of surfaces, use imaginary geometrical elements, use moving coordinate axes, calculus of variations, partial differential equations of 2d order, theory of integral. discontinuous functions, approximations to functions of large numbers; used imaginary geometrical elements; defined a general integral; perfected Riemann's integral. Died Paris, Feb. 23, 1917.

DARBY, Henry Clifford, Brit. geographer; b. Resolven, Ruanda-Urundi, July 2, 1909; s. Evan and Jennie (Thomas) D.; M.A., Ph.D., Litt.D., Cambridge U.; m. Eva C. Thomson; children—Jennifer Elisabeth, Sarah Caroline. Reader in geography Cambridge U., 1931-45, mem. teaching staff King's Coll., 1932-45; prof. geography U. Liverpool, 1945-49, U. London, 1949—. Recipient Daly medal Am. Geog. Soc. Mem. Royal Geog. Soc. (Victoria medal), Inst. Brit. Geographers, Geog. Assn. Editor, collaborator: An Historical Geography of England; The Draining of Fens; The Medieval Fenland; The Domesday Geography of England, 4 vols. Home: Ivy Gates, Cross Oak Rd., Berhamsted, Herts. Office: University College, Gower St., London W.C.1, Eng.

DARBY, John, see Garretson, James Edmund.

DARBY, John Eaton, Am. physician; b. South Williamstown, Mass., Aug. 20, 1835; s. William and Electa (Edwards) D.; A.B., Williams Coll., 1858; M.D., Western Res. Med. Coll., 1861; m. Julia Frances Wright, Apr. 11, 1861 (dec. 1867); 2d, Emma Maybell Cox, May 1, 1872. Prof. Latin and Greek, Cleve. Inst., 1858-60; demonstrator anatomy Western Res. Med. Coll., 1861-2, prof. materia medica and therapeutics, 1867-1906; surgeon in Civil War, 1862-66; surgeon Cleve. & Pitts. R.R., 1867-87; vis. physician Lakeside Hosp., 1867-87; mem. med. bd. City Hosp., 1895-1908. Died Jan. 4, 1918.

DARCET, Jean, French physician; b. Douazit, France, Sept. 7, 1725; Tutor to son of Montesquieu, Paris, 1742; literary assistant to Montesquieu, 1742-55; prof. Collège de France; dir. Sèvres Porcelain Factory, also Gobelins Tapestry Works; assayer of mint; insp. gen. Assay Office. Mem. French Acad. Scis., 1784. Author several treatises. Improved manufacture porcelain, soap, salt, extraction gelatin from bones; demonstrated diamond is combustible; invented metallic alloy (similar to Wood's metal) with low fusing-point; studied effect of fire on various kinds of earths; investigated geology of Pyrenees; a fusible alloy of bismuth, lead, and tin is named for him. Died Paris, France, Feb. 13, 1801.

DARCET, Jean Pierre Joseph, French chemist; b. Paris, France, Aug. 31, 1777; s. Jean Darcet; apptd. assayer of mint, 1801; became commissaire général des monnaies; chemist. Mem. Acad. Medicine, Soc. Agr., French Acad. Scis., 1823. Author books and treatises. Improved and manufactured various chem. products including soda, alum, and soap; invented method of prodn. of copper alloys, bicarbonate of soda, artificial scales, glue, sulphuric acid, method for extracting gelatin from bones; developed chemical industry; application of chemistry to sanitation. Died Paris, August 2, 1844.

DARCY, Henri-Philibert-Gaspard, French railroad engr.; b. Dijon, France; 1803; engr. French road adminstrn.; hydraulics engr.; investigated movement of water in conduits and pipes, 1857. Died Paris, 1858.

D'ARCY, Patrick Comte, French mathematician, astronomer; b. Galway, Ireland, Sept. 27, 1725; s. Jean and Jeanne (Linch) D'A.; student Claude Clairaut, Paris, 1739; m. a niece, 1777. Joined King's Army and fought in Germany and Flanders; returned to France, 1747; became asst. mechanic, 1749; served in French Army during Seven Years War; became camp marshal, 1770. Mem. Royal Acad. Scis. Author: Essay on Artillery, 1760; Memoir on the Duration of the Sensation of Sight, 1765; Memoir on Hydraulic Machines, 1754. Research on weapons with ballistic pendulum; invented measurement instrument for recoil energy; devel. new solution to problem of the curve of equal pressure in a resisting medium; proposed (with Jean Baptiste LeRoy) construction of repulsion electrometer, 1749; discovered independently the principle of preservation of rotary motion. Died Paris, Oct. 18, 1779.

DARDEN, Sperry Eugene, Am. physicist; b. Chgo., Aug. 16, 1928; s. Sperry E. and Catherine (Mahoney) D.; B.S., Ia. State U., 1950; M.S., U. Wis., 1951, Ph.D., 1955; m. Marcia Brienen, June 26, 1954; children—Timothy, Kristin, Ruth, Stephen. Research asst. U. Basel, (Switzerland), 1955-56, guest asso. prof., 1965-66; faculty U. Wis., 1956-57, 61-62; faculty U. Notre Dame (Ind.), 1957-61, 62-65, 66—, prof. 1965—. Fellow Am. Phys. Soc. Research, publs. on interactions of neutrons with nuclei, spin polarization phenomena in nuclear interactions of neutrons and deuterons. Home: 1929 Beverly Pl., South Bend, Ind. 46616. Office: Physics Dept., U. Notre Dame, Notre Dame, Ind. 46556.*

DA REGGIO, Niccolo, Italian translator; b. circa 1280; student medicine, Greek, Reggio, Italy, also student Salerno, Italy. Tchr., Salerno, Naples, Italy; translator regius; commd. to render Greek works into Latin (patrons were Charles II of Anjou and Robert, King of Naples) Translator: De Usu Partium (Galen), 1322 (treatise on parts of human body). Died circa 1350.

DARESTE, Gabriel Madeleine Camille, French naturalist; b. Paris, 1822. Prof. natural history in provinces; dir. lab. École des hautes études at Paris. Mem. Soc. Anthropology. Author: Recherches sur la production artificielle des monstruositès ou essais de tératogénie experimental, 1877. Research on teratology. Died Paris, 1899.

D'AREZZO, Ristoro, writer, scientist; flourished circa 1282 at Arezzo, Tuscany, Italy; compiled Della composizione del mondo colle sue cagioni (an encylopedic treatise on the composition of the world), 1282.

D'ARGELATA, Pietro, Italian physician; student of Guy de Chauliac, grad. in medicine, Bologna, 1391; tchr. Bologna, Italy; practice medicine; considered greatest Italian surgeon of his period; skilled in operating for stones, fistula, hernia, and dental problems; employed compressing bandages in chronic ulcers; used cautery in varicose veins. Died 1423.

DARIER, Jean F., French dermatologist; b. France, 1856; described Darier's disease, known as keratosis follicularis, characterized by papules containing crusts which can be squeezed out, 1889. Died 1938.

DARKEN, Lawrence S(tamper), Am. phys. chemist; b. Bklyn., Sept. 18, 1909; s. William Henry and Gertrude Ann (Stamper) D.; A.B., Hamilton Coll., 1930; Ph.D., Yale, 1933; m. Margaret Elizabeth FitzGerald, Sept. 6, 1939; children—Joanne Savage, Mary Cummins, Lawrence Stamper, William Henry II, Marjorie Beth, Edward Reynolds. With U. S. Steel Corp., 1935—, dir. fundamental research, 1962—; instr., then adj. prof. Poly. Inst. Bklyn., 1942-53. Recipient Francis J. Clamer medal Franklin Inst., 1966. Howe meml. lectr. Am. Inst. Mining Engrs., 1961. Fellow N.Y. Acad. Scis., A.A.A.S., Metall. Soc.; mem. Nat. Acad. Scis., Am. Inst. Mining, Metall. and Petroleum Engrs., Am. Chem. Soc., Faraday Soc., Am. Soc. Metals (Campbell meml. lectr., 1961), Am. Iron and Steel Inst., Phi Beta Kappa, Sigma Xi. Author: (with R. W. Gurry) Physical Chemistry of Metals, 1953; also numerous articles on basic phenomenology of diffusion and thermodynamics of binary and multicomponent systems, especially in metall. systems at elevated temperature. Home: R.D. 7, Irwin, Pa. 15642. Office: U. S. Steel Corp. Research Center, Monroeville, Pa. 15146.*

DARKSHEVICH, Liveri O., Russian neurologist; b. Russia, 1858; described nerve fibers which run from optic tract to habenular ganglion, 1904. Died 1925.

DARLING, Samuel Taylor, Am. pathologist; b. Harrison, N.J., 1872; s. Edmund Adams and Sarah Ann (Patterson) D.; M.D., Coll. Physicians and Surgeons, Balt., 1903; m. Nannyrle Llewellyn, Feb. 18, 1905. Chief of Lab., Panama Canal, 1906-15; with Gen. Gorgas on sanitary mission to Rand mines and Rhodesia, South Africa, 1913-14; chmn. bd., under Rockefeller Found., to investigate causes of anemia among people of Malaya, Java and Fiji, 1915-17; prof. hygiene and dir. labs of hygiene Faculdade de Medicina e Cirurgia de Sao Paulo (Brazil), 1917-20; fellow Sch. Hygiene and Pub. Health, Johns Hopkins U., 1921-22; dir. field lab. for research in malaria Internat. Health Bd. of Rockefeller Found., Leesburg, Ga., 1922-25. Hon. fellow Royal Soc. Tropical Medicine and Hygiene (London). Research and publs. on bionomics of hookworm, malaria; his sanitation studies helped make Panama Canal possible; identified agt. of trypanosomiasis among horses and checked an epidemic; described Darling's histoplasmosis, the agt. and symptoms. Died in motor accident, Beirut, Lebanon, May 21, 1925.

DARLING, Stephen Foster, Am. chemist; b. De Smet, S.D., May 1, 1901; s. Andrew Delos and Harriet (Sturgeon) D.; B.S., U. Minn., 1922, M.S., 1924; A.M., Harvard, 1926, Ph.D., 1928; Sheldon Traveling fellow, U. Vienna, 1928-29; m. Delphine Deziel, Aug. 20, 1930; children—Stephen D., Charlotte E. (Mrs. Robert Ehrhart), Anne M. (Mrs. Gary Appel), Andrew D. Faculty, Lawrence U., Appleton, Wis., 1929—, prof. chemistry, 1937-66; research asso. Inst. Paper Chemistry, 1930—. Mem. Gov's. Food Law Adv. Com. Fellow Am. Inst. Chemists; mem. Am. Chem. Soc., Wis. Acad. Arts, Scis., Letters (past pres.). Research, publs. primarily in field of cyclopropanes, glucosides. Home: 617 E. Alice St., Appleton, Wis. 54914.*

DARLINGTON, Cyril Dean, English biologist; b. Chorley, Eng., Dec. 19, 1903; s. William H. and Ellen (Frankland) D.; B.Sc., Wye Coll., Kent; m. Margaret Upcott; children—Oliver, Andrew, Clare, Deborah, Rachel; m. 2d, Gwendolen Adshead. Dir., John Innes Hort. Instn., Merton and Bayfordbury, 1939-53; Sherardian professor botany University of Oxford, 1953—. Fellow Royal Society, 1941 (Royal medal 1946); member of the Genetical Society (hon.), Acad. Lincei (Rome), Royal Danish Acad. (foreign mem.). Author: Chromosomes and Plant Breeding, 1932; Recent Advances in Cytology, 1932; Evolution of Genetic Systems, 1958; Genetics and Man, 1964; Chromosome Botany, 1963; Cytology, 1965; Darwin's Place in History, 1959; Conflict of Science and Society, 1948; (with K. Mather) Elements of Genetics, 1949; Genes, Plants and People, 1950; (with L. F. LaCour) Handling of Chromosomes (4th edit.), 1962; (with A. P. Wylie) Chromosome Atlas, 1963; (with A. D. Bradshaw) Teaching Genetics, 1963. Co-editor, co-founder Heredity, 1947—. Investigated evolution of heredity and breeding in plants, animals and man. Home: Woodside, Frilford and Heath, Abingdon. Office: Botany Sch., U. Oxford, Eng.*

DARLINGTON, Thomas, Am. physician; b. Bklyn.; Sept. 24, 1858; s. Thomas and Hannah Anne (Goodliffe) D.; C.E., Ph.B., N.Y. U.; Litt.D. from Juniata Coll., 1924; M.D., Coll., Phys. and Surg. (Columbia), 1880; m. Josephine A. Sargeant, June 9, 1886; children—Clinton Pelbam, Dorothea. Practiced Newark, 1880-82, N.Y., 1882-88, Bisbee, Ariz., 1888-90, N.Y., since 1891. Commr. and pres. N.Y. Bd. Health, 1904-10; mem. New York State Workmen's Compensation Commn., 1914-15; cons. physician N.Y. Founding Hosp.; lectr. indsl. hygiene, Stevens Inst. Tech., Fordham U.; san. engr. N.Y.C. Dept. Health until 1934. Asst. to pres. Am. Iron and Steel Institute. Fellow A.A.A.S., N.Y. Acad. Sci.; mem. Soc. Med. Jurisprudence, A.M.A., Am. Institute, Am. Climatol. and Clin. Assn., Am. Assn. for Promoting Hygiene and Pub. Baths (v.p.), Nat. Inst. Social Sci. (v.p.), Harvey Soc., Internat. Sunshine Soc. (care of blind children), Med. Soc. of County of N.Y., N.Y. Acad. Medicine, Greater N.Y. Med. Assn. (v.p.), Physicians Mut. Aid Assn., Thomas Hunter Assn. (pres. 1923-24, 29-37) N.Y. Soc. Mil. and Naval Officers World War; one of 14 exec. mems. Congress of Physicians and Surgeons, 1907-39. Writer on med. and climatol. subjects, sanitation. Died Aug. 23, 1945.

DARLUC, Michel, French physician, naturalist; b. Grimaud, France, 1717; ed. Oratory Sch. Marseille (France); prof. botany U. Aix (France); founder bot. garden of Aix; practice medicine, Aix. Mem. Acad. Medicine. Author: Histoire naturelle de la Provence, 1782-86. Research on mineral waters especially at Gréoulx; publs. on natural history of Provence region of France including econ. history, diet and hygiene of the people. Died Aix, 1783.

DARMOIS, Émile Eugène, French physicist; b. Èply, France, Apr. 10, 1884; s. Emile and Virginie (Dardar) D.; ed. École normale supérieure, Paris, France, Faculty Scis., Paris; agrégé in physics and chemistry; docteur ès scis.; 8 children. Lectr. physics Faculty Scis., U. Rennes (France), 1910; engr. Westinghouse Cooper Hewitt Co., 1911-14; prof. physics Faculty Scis., U. Nancy (France), 1918-26; apptd. prof. Sorbonne, Paris, 1926, also at École supérieure d'électricitè. Mem. French Acad. Scis., 1951, French Soc. Electricians (pres.), French Soc. Physics (gen. sec.). Research on electromotive forces, electrolytes and rotatory polarization. Died Nov. 4, 1958.

DARMOIS, Georges, French physicist; b. Éply, France, June 24, 1888; s. Emile and Virginie (Dardar) D.; ed. École normale supérieure; agrégé in math., 1909; docteur ès scis. math., 1921. Prof., Faculty Scis., U. Nancy (France), 1919-33; apptd. prof. statistics U. Paris (France), 1925, prof. math. physics, 1949-60; became dir. Paris Inst. of Statistics, 1934; sent to London, 1940; returned to Paris, 1944. Mem. Sci. Com. of Nat. Def. Decorated Officer Legion of Honor, Croix de Guerre. Mem. French Acad. of Scis., 1955, Bur. Longitudes Math. Soc. France (pres.), Meteorol. Soc. France, Statis. Soc. Paris, Internat. Inst. Statistics (pres. 1953-60). Author: Sur les courbes algébriques à torsion constante, 1921; Les Équations de la gravitation einsteinienne de la gravitation, 1934; (with others) Les Mathématiques del la psychologie, 1940; Cours de calcul des probabilités et cours de statistiques, 1945; La Radiesthésie, 1956. Research on differential geometry, gen. relativity and theory of gravitation, theoretical statistics, calculus of probability. Died Jan. 3, 1960.

DARMSTAEDTER, Ludwig, German chemist; b. Mannheim, Germany, Aug. 9, 1846; s. Jonas and Eleonora (Dinkelspiel) D.; studied chemistry under Bunsen and Erlenmeyer, Heidelberg (Germany), beginning in 1865; Ph.D., 1868; also studied under Kolbe at Leipzig (Germany). Mem. staff C. H. Wichelhaus' lab., Berlin; worked with Benno Jaffé in indsl. chemistry, beginning in 1872; founder Vereinigte Chemische Werke AG, 1900. Author: (with R. Du Bois-Reymond) 4000 Jahre Pionierarbeit in den exakten Wissenschaften, 1904; (with Du Bois-Reymond, C. Schaefer) Handbach zur Geschichte der Naturwissenschaft und der Technik, 1908; Nauturforscher und Erfinder, Biographische Miniaturen, 1926; also articles on chemistry and history of chemistry. Collected original manuscripts of important scientists and left them to the Prussian State Library, Berlin, 1907; research on isolation of glycerine and effect of gas liquor on ammonium salts; began the manufacture of lanoline, 1890; (with J. Liftschütz) research on the composition of wool fat. Died Aug. 9, 1846.

DARNALL, Carl Rogers, Am. physician; b. Weston, Tex., Dec. 25, 1867; s. Joseph Rogers and Mary Ellen (Thomas) D.; student Carlton Coll., Transylvania U.; M.D., Jefferson Med. Coll., Phila., 1890; grad. Army Med. Sch., Washington, 1897; m. Annie Estella Major, Apr. 27, 1892; children—Joseph Rogers, William Major, Carl Robert. Commd. 1st lt. and asst. surgeon U. S. Army, 1896, advanced through grades to brig. gen., 1929; comdr. Army Med. Center, Washington, 1929-31; served in Spanish-Am. War, Philippine Insurrection, Boxer Rebellion, World War I. Decorated D.S.M. Fellow A.C.S. Developed Darnall's method for purifying water for troops in the field by use of sodium hypochlorite, 1908. Died Jan. 18, 1941.

DARNALL, William Edgar, Am. surgeon and gynecologist; b. Pearisburgh, Va., 1869; s. Henry T. and Margaret Poague (Johnstone) D.; A.B., Washington and Lee U., 1892; M.D., U. Va., 1895; m. Elizabeth Nesbitt, Feb. 27, 1907; children—William Edgar, Jean Mauzy. Began practice, Covington, Va., 1895; moved to Atlantic City, N.J., 1896; gynecologist and pres. staff Atlantic City Hosp.; surgeon Bamburger Home, Longport, N.J.; cons. surgeon to the Home for Incurables (Longport), N. Am. Sanitarium for treatment of surg. Tb; cons. gynecologist Atlantic City Municipal Hosp. Pres. Atlantic Co. Mosquito Extermination Commn.; pres. N.J. Bd. Med. Examiners, 1933. Diplomate Nat. Bd. Obstetrics and Gynecology. Fellow A.C.S., A.A.A.S., A.M.A. (v.p. 1914). Died Dec. 27, 1937.

DARNELL, James Edwin, Jr., Am. biochemist, virologist; b. Columbus, Miss., Sept. 9, 1930; s. J. E. and Helen (Hopkins) D.; B.A., U. Miss., 1951; M.D., Washington U., St. Louis, 1955; m. Jane Roller, 1957; children—Christopher, Robert, Jonathan. Research virologist NIH, Bethesda, Md., 1956-60; fellow Pasteur Inst., Paris, France, 1960-61; faculty Mass. Inst. Tech., 1961-64, asso. prof., 1962-64; prof. biochemistry and cell biology Albert Einstein Coll. Medicine, N.Y.C., 1964—. Mem. Am. Soc. Microbiology, Am. Assn. Immunologists, Am. Soc. Biol. Chemists. Research, numerous articles on biochem. mechanisms in virus synthesis and control of molecular events in growth of animal cells. Home: 16 Sherwood Dr., Larchmont, N.Y. Office: Dept. Biochemistry, Albert Einstein Coll. Medicine, N.Y.C.*

DA ROCHA-LIMA, Henrique, French physician; b. 1879; worked in Germany; isolated causative agt. of typhus and named it Rickettsia prowazeki (in honor of Howard Taylor Ricketts and Stanislas Josef Mathias Von Prowazek), 1916.

DA ROCHA SERPA PINTO, Alexandre Alberto, explorer; b. Sinfaes, Portugal, Apr. 20, 1846; with mil.

campaign in Mozambique; assigned to study Congo basin, 1877; apptd. gov. Mozambique, 1889; mem. French Acad. Scis. Made African expdns. which helped confirm Portugal's claims to territory; opened comml. route between Lake Nyassa and Mozambique, 1885; gathered data on Zambezi River. Died Lisbon, Portugal, Dec. 28, 1900.

DARONDEAU, Benoit-Henri, French engr.; b. Paris, France, Apr. 3, 1805; s. Henri Benoit François and Lucie (Arkwright) D.; ed. Paris École polytechnique, 1824; Engr., 1835; began 2 year voyage around the world, 1835; apptd. prof. Sch. Naval Constrn., 1850; became dir. expdn. to map Italian coast, 1853. Decorated Cross of Legion of Honor. Mem. Bur. Longitudes. Author: Livret des phares des mers du globe. Translator: Indian Oceans (Horsburgh). First to develop way of using compass on iron ships in France; research on terrestrial magnetism; publs. on lighthouses. Died Paris, Mar. 1, 1869.

DARQUIER DE PELLEPOIX, Augustin, French astronomer; b. Toulouse, France, Nov. 23, 1718; installed obs. in Toulouse; Chevalier de Malte; mem. French Acad. Scis., 1757. Author: Uranographie ou contemplation du ciel, à la portée de tout le monde, 1771. Translated from English, Elements of Geometry (Thomas Simpson); from Spanish, The Eclipse of 1778 (Ulloa). Died Toulouse, Jan. 18, 1802.

DARRACH, William, Am. surgeon; b. Germantown, Phila., Mar. 12, 1876; s. William and Edith Romeyn (Aertsen) D.; grad. Hill Sch., 1893; A.B., Yale, 1897, A.M., 1920; A.M., M.D., Columbia, 1901, D.Sc., 1929; LL.D., St. Andrews U., 1928; D.Sc., Jefferson, 1930; m. Florence Borden, May 22, 1907; children—Edith, William, Effie Brooks, Judith. Intern Presbyn. Hosp., 1901-03, asso. attending surgeon, 1913-16, attending physician, to 1946, cons., 1946—; demonstrator anatomy Coll. Phys. and Surg. (Columbia), 1903-09, instr. surgery, 1903-16, prof. clin. surgery since 1916, dean med. faculty, 1919-30, dean emeritus since 1930; attending surgeon Vanderbilt Clinic, 1903-06; 2d asst. attending surgeon Roosevelt Hosp., 1906-08, 1st asst., 1908-10, jr. surgeon, 1910-13; dir. 1st surg. div., Bellevue Hosp., 1916-19; cons. surg. Neurol. Inst., N.Y. Orthopedic Dispensary and Hosp., Greenwich, Beekman St., Babies', Sloane, Willard Parker, Morristown Meml. N. Westchester hosps., Neuro-Psychiatric Inst. of Hartford (Conn.) Retreat. Civilian cons. to Surg. Gen., and Vets. Adminstrn. Fellow A.C.S. (gov.), Chgo. Surg. Soc. (hon.), Am. Assn. Surgery of Trauma (hon.); mem. Am. Acad. Orthopedic Surg., A.M.A., Assn. Am. Anatomists, Am. Surg. Assn. (pres. 1944), N.Y. Surg. Soc., Soc. Clin. Surgery (pres. 1929-31), N.Y. State, N.Y. County med. socs., Société de Chirurgiens de Paris, N.Y. Acad. Medicine, Société Internationale de Chirurgie, NRC (sub-com. on orthopedics), A.A.A.S., Am. Philos. Soc., Phi Beta Kappa, Sigma Xi. Contbr. articles on surgery, anatomy and med. edn. Died May 24, 1948.

DARRAS, Raymond Lucien, French chemist; b. Persan, Val d'Oise, France, Sept. 9, 1927; student U. Grenoble (France), 1946-49; Docteur-Ingénieur, U. Paris (France), 1963. Staff, metallurgy dept. Commissariat à l'Energie Atomique, CEN Saclay, Gif-sur-Yvette, France, 1952—. Mem. Société francaise de Métallurgie (prix Rist 1963), Am. Nuclear Soc., Soc. des Hautes Températures et des Réfractaires. Research, numerous publs. on corrosion by hot gases and liquid metals; new zirconium alloys for high temperature service in carbon dioxide. Office: BP N°2 91 Gif-sur-Yvette, France.*

D'ARREST, Heinrich Louis, German astronomer; b. Berlin, Germany, Aug. 13, 1822; s. Louis C. and Sophie (d'Espagne) d'A.; ed. Berlin; m. Auguste Emilie Möbius, Nov. 4, 1851. Named observer, 1848; asst. Obs., Berlin; apptd. asso. prof. U. Leipzig (Germany), 1852, prof., from 1858; dir. U. Obs., Copenhagen, Denmark. Gold medal Royal Astron. Soc., 1875. Author: Über das System der kleinen Planeten, 1851; Resultate aus Beobachtungen der Nebelflecke und Sternhaufen, 1856; Siderum nebelosorum observationes Hafniensis, 1867; Untersuchungen über die nebulosen Sterne, 1872. Discovered over 200 new comets, nebulae and star clusters. Died Copenhagen, June 14, 1875.

DARRIEUS, Georges-Jean-Marie-Eugène, French engr.; b. Toulon, France, Sept. 24, 1888; mem. French Acad. Scis., 1946; created (with Fallon) frequency regulation between various power stations by overlapping several stages.

DARROW, Chester W., Am. psychophysiologist; b. Ft. Plaine, N.Y., Nov. 7, 1893; s. William E. and Harriet A. (Mills) D.; M.A., Oglethorpe U., 1922; Ph.D., U. Chgo., 1924; m. Alice Hale; 1 dau., Mrs. Robin Oggins. With Inst. Juvenile Research, Chgo., 1926—, now chief div. psychophysiology; asso. prof. physiology U. Ill. Med. Sch., Chgo., 1944—. Mem. Am., Midwest psychol. socs., Am. Physiol. Soc., Am. Electroencephalographic Soc., Central Assn. EEG, Am. Psychophysiol. Soc. A.A.A.S., Soc. Biol. Psychiatry, Sigma Xi. Publs. on devel. of behavioral study techniques; devel. improved use of electroencephalogram. Home: 5802 S. Blackstone Av., Chgo. 60637. Office: 907 S. Wolcott Av., Chgo.*

DARROW, Robert Arthur, Am. plant physiologist; b. Saratoga Springs, N.Y., Dec. 15, 1911; s. Arthur Elliott and Edith (Foote) D.; B.S. in Forestry, N.Y. State Coll., 1932; postgrad. U. Ida., 1932-33; M.S., U. Ariz., 1935; Ph.D., U. Chgo., 1937; m. Bertha Mathilda Schweitzer, Oct. 23, 1936; children—Gordon Roger, Janet Marjorie. Faculty, U. Ariz., 1936-48, prof. dept. botany, 1947-48; faculty Tex. A. and M. U., College Station, 1948-62, prof. dept. range and forestry, 1950-62; chief chem. br. crops div. U. S. Army, Ft. Detrick, Frederick, Md., 1962—. Fellow A.A.A.S.; mem. Am. Bryological Soc., Am. Soc. Plant Taxonomists, Am. Soc. Range Mgmt., Ecol. Soc. Am., Weed Soc. Am. (mem. terminology com. 1960—), Am. Soc. Plant Physiologists, Research Soc. Am., So. Weed Conf. (past pres.), Phi Kappa Phi. Author (with L. Benson) Trees and Shrubs of Southwestern Deserts, 1958; also articles. Editor: Plant Ecology, Biol. Abstracts, 1949-53, Jour. Range Mgmt., 1954-56. Research on ecology and floristics Southwestern desert and grassland, lichen flora Ariz., range mgmt. and ecology in Ariz. and Tex., woody plant control with herbicides, aerial application herbicides, screening and evaluation defoliants, desiccants and herbicides for mil. use. Home: Route 5, Office: Crops Div., Fort Detrick, Frederick, Md. 21701.*

D'ARSONVAL, Jacques Arsène, French physicist; b. Borie, France, June 8, 1851; ed. Collège de Brive, Lycée de Limoges, Sch. Medicine at Limoges; became dir. lab. biol. physics Collège de France, 1882; named prof. exptl. medicine, 1894, held chair of Brown-Sequard; founder Sch. and Soc. of Electricity. Mem. French Acad. Medicine, French Acad. Scis. (pres. 1917), French Acad. Surgery. Invented magnetotelephone, repeat gas burner, D'Arsonval instrument for measuring direct electric current, aperiodic D'Arsonval galvanometer; used high-frequency current to treat diseases of skin and mucous membranes (D'Arsonvalization); introduced term diathermy. Died Borie, Dec. 31, 1940.

DART, Raymond Arthur, anatomist; b. Toowong, Brisbane, Australia, Feb. 4, 1893; s. Samuel and Eliza Anne (Brimblecombe) D.; B.Sc., U. Queensland, 1913, M.Sc., 1915; M.B., Ch.M., Sydney U., 1917, M.D., 1927; hon. doctorate, Natal, 1956; m. Marjorie Gordon Frew, 1936; 1 son, 1 dau. Demonstrator anatomy, acting prin. St. Andrew's Coll., Sydney, 1917; house surgeon Royal Prince Alfred Hosp., Sydney, 1917-18; sr. demonstrator anatomy, Univ. Coll., London, 1919-20, 21-22, lectr. histology, 1921-22; fellow Rockefeller Found., 1920-21; prof. anatomy U. Witwatersrand, Johannesburg, S. Africa, 1923-58, dean faculty medicine, 1925-43, emeritus prof., from 1959. Mem. Internat. Commn. on Fossil Man, from 1929. Recipient Sr. Capt. Scott Meml. medal S. African Biol. Soc., 1955, Viking medal and award for phys. anthropology Wenner-Gren Found. of N.Y., 1957; Fellow Royal Soc. S. Africa (past v.p.); mem. S. African Assn. for Advancement of Sci. (past v.p.), S. African Archaeol. Soc. (pres. 1951), S. African Museums Assn. (pres. 1961-62), S. African Soc. Physiotherapy (pres. 1961), other profl. assns. Author: Cultural Status of the South African Man-Apes, 1956; The Osteodontokeratic Culture of Australopithecus promethus, 1957; (with Denis Craig) Adventures with the Missing Link, 1959; Africa's Place in the Emergence of Civilisation, 1960. Editor: Africa's Place in the Human Story, 1954. Discovered fossil brain of man-ape child (Australopithecus africanus) at Taungs, S. Africa, 1924. Address: Walmer, 26 Park St., Oaklands, Johannesburg, S. Africa.

DARTON, Nelson Horatio, Am. geologist; b. Bklyn., Dec. 17, 1865; s. William and Caroline M. (Thayer) D.; hon. D.Sc., U. Ariz., 1922; m. Alice Weldon Wasserbach, 1903; children—Arthur B., Horace L., Annunciata (Mrs. William J. Kerlin). Chemist in N.Y., 1880-86; geologist, U. S. Geol. Survey, 1886-1910 and 1913-36 (ret.); geologist Bur. Mines, 1910-13. Inventor of a sugar process; researches in tannic acid and water analysis, etc. Lectured at various colls. Fellow Geol. Soc. Am. (ex-v.p.), A.A.A.S.; mem. Washington Acad. Scis. (former v.p.), Soc. Econ. Geologists, Mining and Metall. Soc., Am. Inst. Mining and Metall. Engrs., Soc. Linn de Lyons, Inst. Français Pi Gamma Mu, Soc. Géol. de France, Assn. Am. Geographers, Internat. Geol. Congress, Soc. Fine Arts, Am. Geophys. Union, Archaeol. Soc., Fed. Bd. of Surveys and Maps, English-Speaking Union, Italy-America Soc., Alliance Française (v.p.), Spanish Atheneum (ex-pres.), Instituto de las Españas; mem. sub-com. Nat. Council. Awarded Daly gold medal by Am. Geog. Soc., 1930, and Penrose gold medal by Geol. Soc. of Am., 1940; Legion of Honor, Am. Inst. Mining and Metall. Engrs., 1944. Author: The Story of the Grand Canyon; Geologic Guide to Santa Fe R.R.; Geologic Guide to Southern Pacific R.R.; Geology of Great Plains, Black Hills, Bighorn Mts., Owl Creek Mts.; Geology Dist. Columbia region; Geologic maps Great Plains, Grand Canyon, S.D., Neb., Ariz., N.M., Lower Calif. and Tex.; many folios U. S. Geol. Survey, etc. Contbr. geol. subjects; geol. maps and reports on many dists., topog. maps states of Ariz., N.M., Tex., Nebr., N.D., S.D.; many articles in Ency. Americana and other publs. Explored ruins of the temple of Cuicuilco, Mexico, for Nat. Geog. Soc. and oil geology of Lower Cal., Santo Domingo, Eastern Cuba and Central Venezuela; made pioneer studies in hydrography, esp. in Dakotas; geologically and topographically mapped most of U. S. Great Plains and Southwest; established stratigraphic units in Western U. S., which have formed basis for all subsequent stratigraphic work there. Died Feb. 28, 1948.

DARWENT, Basil de Baskerville, chemist, educator; b. Trinidad, W.I., May 20, 1913; s. Edgar Nicholas and Mary (Henderson) D.; B.Sc., McGill U., 1941, Ph.D., 1943; m. Jocelyn Margaret Taitt, Apr. 20, 1938; 1 son, John Nicholas de Baskerville. Postdoctoral research McGill U., 1943-44, Oxford U., 1948-49; with NRC, Ottawa, Ont., Can., 1944-48, asso. research scientist, 1949-52; mgr. phys. chemistry research Olin Mathieson Chem. Corp., New Haven, 1952-55; research prof. chemistry Cath. U. Am., Washington, 1955-57, prof., 1957—, head dept. chemistry, 1961-67. Fellow Royal Soc. Can.; mem. Am. Chem. Soc., Chem. Soc. (London), Faraday Soc., Washington Acad. Sci. Research, publs. on factors controlling rates of chem. reactions, elementary reactions of atoms and free radicals, photochemistry. Home: 1736 Q St. N.W., Washington 20009.*

DARWIN, Sir Charles Galton, Brit. physicist; b. Cambridge, Eng., 1887; s. George Howard Darwin; ed. Trinity Coll., Cambridge (Eng.) U.; M.C., M.A., Sc.D.; several hon. degrees; m. Katharine Pember, 1925; 4 sons, 1 dau. Lectr. physics Manchester (Eng.) U., 1910-14; lectr. math. Christ's Coll., Cambridge (Eng.) U., 1919-22, master, 1936-38, also hon. fellow; Tait prof. natural philosophy Edinburgh (Scotland) U., 1923-36; dir. Nat. Phys. Lab., 1938-49; hon. fellow Trinity Coll., Cambridge U. Fellow Royal Soc., 1922 (Royal medal 1935); mem. Am. Philos. Soc. Author: The New Conceptions of Matter, 1931; The Next Million Years, 1952; also papers in theoretical physics. Gave an early statement of theory of X-ray diffraction, 1913; research in atomic physics. Died Dec. 31, 1962.

DARWIN, Charles Robert, English naturalist; b. Shrewsbury, Shropshire, Eng., Feb. 12, 1809; s. Robert Waring and Susannah (Wedgwood) D.; ed. Shrewsbury Sch.; student medicine U. Edinburgh (Scotland), 1825-27; ed. for ministry Christ's Coll., Cambridge U. (Eng.), 1828-31, degree 1831; m. Emma Wedgwood, Jan. 29, 1839; 5 sons, 2 daus. Influenced by botanist John Stevens Henslow and geologist Adam Sedgwick, Cambridge; embarked on 5 year sci. expdn. around world as naturalist on H.M.S. Beagle, 1831; studied flora, fauna, and geology of so. islands, S. Am. coasts, and Australasia; made special studies of fossils and of various species of finches and tortoises on islands of Galapagos Archipelago; returned to Eng., 1836, worked in Cambridge for short time, then moved to London; sec. Geol. Soc. London, 1838-41; left London because of ill health, 1842, and spent remainder of life in Down, Kent, Eng. Fellow Royal Soc., London, 1839 (Royal medal, 1853; Copley medal, 1864); corr. botany sect. French Acad. Scis., 1878. Author: Origin of Species by Means of Natural Selection (most famous work), 1859; Voyage of the Beagle, 1839; Structure and Distribution of Coral Reefs, 1842; Volcanic Islands, 1844; Geological Observations, 1844-46; On the Various Contrivances by which British and Foreign Orchids are Fertilized by Insects, 1862; The Fertilization of Orchids, 1862; The Variation of Plants and Animals under Domestication, 1868; The Descent of Man, 1871; The Expression of the Emotions in Man and Animals, 1872; The Effects of Cross and Self Fertilization in the Vegetable Kingdom, 1876; The Power of Movement in Plants, 1880; The Formation of Vegetable Mould through the Action of Worms, 1881. Father of modern evolutionary theory (though many of same ideas were developed independently by Alfred Russel Wallace); compiled vast amount of information to show evolution had in fact occurred; held that species of plants and animals develop slowly and continuously by process (natural selection) which depends entirely on factors of random variation; later presented evidence that man evolved from subhuman forms, 1871; formulated accurate theory about origin of coral reefs; though Darwinism created enormous controversy for considerable time, it gained rapid acceptance in sci. community. Died Apr. 19, 1882, Down, Kent.

DARWIN, Erasmus, English physician; b. Elton, Newark, Eng., Dec. 12, 1731; s. Robert D.; B.A., Cambridge (Eng.) U., 1754, M.B., 1755; student medicine, U. Edinburgh (Scotland), 1754; m. Mary Howard, Dec. 1757 (dec. 1770); children—Charles, Erasmus, Robert Waring; m. 2d, Mrs. Chandos-Pole, 1781; 4 sons, 3 daus. Physician, Nottingham, Eng., 1756, Litchfield, 1757-81, Derby, Eng., beginning in 1781. Philos. Soc. Derby (founder 1784). Author: The Loves of Plants, 1789; The Botanic Garden (embodied botanical system of Linnaeus), 1791; Phytologia, 1800; Zoonomia, 1794-96; The Temple of Nature, 1803; also writings supporting Linnaean biol. classification system. Anticipated laws of organic life of Lamarckian theory of evolution; cultivated 8-acre botanical garden. Died Breadsall Priory near Derby, Apr. 18, 1802.

DARWIN, Francis, English botanist; b. Down, Eng., Aug. 16, 1848; s. Charles R. and Emma (Wedgwood) D.; B.S., Trinity Coll., Cambridge, Eng., 1870; M.B., St. George's Hosp., London, Eng., 1875; postgrad. under Julius von Sachs, Würzburg, Germany, 1876; m. Amy Ruck, 1874 (dec. 1876); m. 2d, Ellen Wordsworth Crofts, 1883 (dec. 1903); m. 3d, Florence Fisher Maitland, 1913 (dec. 1920). Sec. and asst. to father, 1875-82; named univ. lectr. in botany, Cambridge, 1884; became fellow Christ's Coll., 1886; hon. fellow, 1908; reader in botany, 1888-1904; dep. to Prof. Charles C. Babington, 1892-95. Created knight, 1913. Fellow Royal Soc., 1882; pres. Brit. Assn., 1908. Author: Life and Letters of Charles Darwin, 1887; (with E. H. Acton) The Practical Physiology of Plants, 1894; The Elements of Botany, 1895; (with A. C. Seward) More Letters of Charles Darwin, 1903; Darwin's Work on the Movement of Plants, 1909; Rustic Sounds and Other Studies in Literature and Natural History, 1917; Springtime and Other Essays, 1920. Editor: Naturalist's Calendar (Leonard Blomefield 1820-31); The Foundations of the Origins of Species, 1909. Investigated vegetable physiology, especially plant movements and response to stimuli and transpiration through stomata. Died Cambridge, Sept. 19, 1925.

DARWIN, George Howard, English astronomer; b. Down, Kent, England, July 9, 1845; s. Charles Robert and Emma (Wedgwood) D.; educated (fellow) Trinity College, Cambridge (Eng.) Univ., 1873-83; married Maud duPuy, 1884; 2 sons, 2 daughters. Became Plumian professor astronomy and experimental philosophy, Cambridge Univ., 1883. Fellow Royal Soc., 1879 (Copley medal, 1911). Member of the Royal Astronomical Society (elected president 1899), British Assn. (elected president 1905). Author: The Tides, 1898; The Scientific Papers of G. H. Darwin, 5 vols., 1907-16. Applied math. techniques to astronomy and geology; developed gen. cosmogony of earth-moon system and solar system using research on tides and tidal friction; 1st to base a cosmogony on math. and phys. principles; developed theory that moon was originally thrown off from earth. Died Cambridge, Eng., Dec. 7, 1912.

DARWIN, Leonard, Brit. eugenist; b. Jan. 15, 1850; s. Charles and Emma (Wedgwood) D.; ed. Woolwich, Eng.; Sc.D. (hon.), Cambridge (Eng.) U.; m. Elizabeth Frazer, 1882; m. 2d, Charlotte Mildred Langdon, 1900. Instr. Sch. Mil. Engring., Chatham, 1877-82; staff intelligence dept. War Office, 1885-90; mem. several sci. expdns., including transit of Venus, 1874-82; mem. Parliament, Litchfield div., Staffs., 1892-95; chmn. Bedford Coll. for Women, U. London, 1913-20. Pres. Royal Geog. Soc., 1908-11, Eugenics Edn. Soc., 1911-28. Author: Bimetallism, 1893; Municipal Trade, 1903; The Need for Eugenic Reform, 1926. Died Mar. 26, 1943.

DAS, Bhupendra Chandra, Indian physiologist; b. Kheora, Bengal, India, Sept. 1, 1924; s. Mahim Chandra and Kumud (Pal) D.; B.S., Cornell U., 1949; M.S., U. Ill., 1951, Ph.D., 1955; m. Rhea Stagner, June 9, 1952; children—Linda, Raul. Dir. biometry research unit Indian Statis. Inst., Calcutta, 1955-60, 62-—, asso. prof., 1955-60, prof., 1962-—; research asso. Wayne State U. Coll. Medicine, Detroit, 1960-61; vis. asso. prof. Case Inst. Tech., Cleve., 1961-62. Researcher various instns., Chgo., Eng., Switzerland. Mem. Indian Sci. Congress, Am. Fisheries Soc., N.Y. Acad. Scis., Inst. Mgmt. Scis., Pacific Sci. Assn., Sigma Xi. First to observe differences between avian and mammalian gonadotrophins, and to induce follicle formation in immature avians by pituitary injections, also to regenerate prothrombin activity. Address: 203 Barrackpore Trunk Rd., Calcutta 35, India.*

DAS, N. B., Indian biochemist; b. Engal, India, June 7, 1908; s. D. B. and S. (Choudhury) D.; B.Sc., Calcutta U., 1932; Ph.D., Inst. for Med. Chemistry, U. Szedqed (Hungary), 1937; m. I. Sen, Oct. 21, 1941; 3 children. Research scholar P. C. Roy Lab., Calcutta, India, 1933-35, Biochem. Inst., Stockholm, Sweden, 1935-36; amanuens U. Stockholm, 1937-39; sr. research fellow Bose Inst., Calcutta, 1939-41; biochemist Indian Vet. Research Inst., Muktesway, 1941-48, Indian Agrl. Research Inst., New Delhi, 1948-61; supernumerary biochemist I.A.R.I., Delhi, 1961-—. Mem. tea research liaison com. Indian Tea Bd., Indian Standard Instn., Delhi. Mem. Soc. Biol. Chemists, India Pusa Agrl. Research Soc. Research, numerous publs. on enzyme chemistry, bacterial metabolism, nutrition and cereal chemistry. Home: B3, Indian Agrl. Research Inst., Delhi 12.*

DAS, Satya Ranjan, Indian physicist; b. Bengal, India, Oct. 25, 1926; s. J. C. and Suhasini (Sinha) D.; B.Sc. with honors, Bankura Christian Coll. and Vidyasagar Coll., 1947; M.Sc., Calcutta U. Coll. Sci., 1949, D.Phil., 1960; m. Sadhana Ghosh, June 2, 1955; children—Sourav, Kumar, Sumita. Faculty, Calcutta U., 1950-62; research staff B.T.H. Research Lab., Rugby, Eng., Nat. Phys. Lab., Teddington, Eng., 1956-58; sr. sci. officer in optics Nat. Phys. Lab. of India, New Delhi, 1962-—, head div. optics, 1962-—; tchr. optics, I.I.T., Delhi, 1965. Recipient Gold medal U. Calcutta. Sr. Fulbright Travel grantee, 1965-66. Research, publs. on fundamental mechanisms of color vision and time behavior of visual receptors; work on photometry and colorimetry; electrical discharge in gases; interior lighting design; inventor Rec. Skyscanner for automatic rec. of luminance of daylight. Home: Paikpara (Birbhum), West Bengal, India. Office: Nat. Phys. Lab., New Delhi, India.*

DAS-GUPTA, Niraj Nath, Indian biophysicist; b. Chittagong, India, Feb. 1, 1909; s. Ganga Charan and Indu Lekha (Ray) D.; ed Presidency Coll., Calcutta, India; M.Sc. Coll. Sci., Calcutta U., 1931; Ph.D., King's Coll., London U., 1938; m. Alaka Das Gupta, Nov. 30, 1941. Lectr., Calcutta U., 1942-45; research asso. Stanford, 1945-46; reader Calcutta U., 1946-52; prof., head of biophysics, Saha Inst. Nuclear Physics, Calcutta, 1953-—. Cons. biophysicist Chittaranjan Nat. Cancer Research Centre, 1951-—; sec. biophysics research com. Council Sci. and Indsl. Research, Govt. of India, 1955-58. Recipient Gold medal and prize Calcutta U., 1931; Mahendralal Sarcar Research medal, Indian Assn. for Cultivation Sci., 1938. Exec. Com., Internat. Fedn. Socs. Electron Microscope; Fellow Indian Phys. Soc., Nat. Inst. Scis. India; mem. Biophys. Soc. Am., Nat. Com. Biophysics India, Am. Phys. Soc. Author: (with S. Ghosh) A Report on Wilson Cloud Chamber and its Applications in Physics, 1946. Installed 1st electron microscope in India, 1948; research on electron microscopy of microorganisms causing kala-azar, leprosy, also structure of hemoglobin and DNA molecules; developed sch. of biophysics in Calcutta. Home: 44. Hazra Rd., Calcutta 19. Office: Biophysics Laboratory, 37 Belgachia Rd., Calcutta 37, India.*

DASMANN, Raymond Fredric, Am. ecologist; b. San Francisco, May 27, 1919; s. William H. and Mary (McDonnell) D.; A.B., U. Cal. at Berkeley, 1948, M.A., 1951, Ph.D., 1954; m. Elizabeth Sheldon, May 30, 1944; children—Sandra, Marlene, Lauren. Faculty, Humboldt State Coll., 1954-59, 62-66, prof. ecology, 1964-66, chmn. natural resources, 1965-66; research biologist Nat. Museums Rhodesia, 1959-61; lectr. zoology U. Cal. at Berkeley 1961-62; ecologist Conservation Found., Washington, 1966-—. Mem. Ecol. Soc. Am., Am. Soc. Mammalogists, Wildlife Soc., Am. Geog. Soc., Cal. Acad. Scis., Assn. Tropical Biology, Faunal Preservation Soc. Author: Pacific Coastal Wildlife, 1957; Environmental Conservation, 1959; African Game Ranching, 1963; Last Horizon, 1963; Wildlife Biology, 1964; Destruction of California, 1965; also numerous articles. Research on ecology, populations, social behavior deer, elk, African ungulates, econ. value African wildlife. Home: 3816 Jennifer St., Washington 20015. Office: 1250 Connecticutt Av., Washington 20036.*

DASTRE, Albert-Jules-Franck, French physiologist; b. Paris, France, Nov. 7, 1844; licence ès scis. mathématiques École normale supérieure, 1866; aggregation, 1869; docteur en médecine, 1876. Served in med. corps Franco-Prussian War; became lectr. comparative anatomy and zoology École normale, 1879; apptd. professor of general physiology Paris Faculty Sciences, 1886. Member of the French Academy Scis., 1904, Acad. Medicine. Author: La vie et la mort, 1803; also others. Research on embryology and composition of egg, nutrition, liver, nervous system, circulation of blood, anesthetics, vasomotors. Died Oct. 23, 1917.

DASYPODIUS, Konrad (Cunradus), mathematician; b. Frauenfeld, Switzerland, 1532; ed. at U. Strasbourg (France); s. Peter Dasypodius; Prof. math. U. Strasbourg (France); canon St. Thomas's Ch., Strasbourg. Author: Analysis geometrica sex librorum Euclidis, 1566; Dictionarium mathematicum, 1573; Heron mechanicus, 1580; Institutionum mathematicarum . . . , 2 vols., 1593; Volvmen primum: mathematicum disciplinarum principia, 2 vols., 1567-70; Brevis doctrina de cometis (on astrology), 1578. Edited many Greek math. works. Constructed an astron. clock. Died Strasbourg, Apr. 26, 1600.

DA TOSSIGNANO, Pietro (Curialti, Pietro, or Pietro da Tossignano), Italian physician; flourished 1st half of 14th century; s. Zeto (Gepto or Alberghetto) dei Curialti; student medicine-surgery, Padua, Italy; m. Caterina Ruffini, 1372; children—Antonio, Alberto, Lippa (or Filippa). Became prof. medicine, Bologna, Italy, 1364; joined faculty medicine, Padua, 1390; returned to Bologna, 1396. Author: Receptae super nonum almansoris (A Textbook of Therapeutics and Materia Medica); Consilium pro peste evitanda, 1398; De regimine sanitatis; Pietro's liber de balneis burmi. Died Bologna, 1407.

DATTA, M. N., zoologist; b. Nov. 22, 1901; s. A. C. and Nanibala (Bose) D.; B.Sc., St. John's Coll., 1919; M.Sc., Agra (India) Coll., 1924; m. Syeenati Susama, July 8, 1924; children—Piu (Mrs. Vaskar Sen), Asis, Jharna (Mrs. K. D. Miller), Malay, Samir, Dipak, Jolly (Mrs. Satya Guha), Rini, Baby. Empress Victoria reader Allhabad U., 1925-29; lectr. St. John's Coll., Agra, 1924-29; staff Zool. Survey India, 1929-56; lectr. Calcutta (India) U., 1956-—. Recipient King George V Coronation Jubilee medal. Fellow Zool. Soc. India (life), Acad. Zoology, Helminthological Soc. India. Research, publs. on parasitic infection of helminths of birds and fishes; created nine new genera, two orders and 17 new species. Home: 45 B Central Rd., Jadalpu, Calcutta-32, West Bengal, India.*

DATTNER, Bernhard, neuropsychiatrist; b. Ustron, Silesia, Austria, July 7, 1887; s. Adolf and Anna (Hechter) D.; Jur. D., U. Vienna, 1911; M.D., 1919; m. Margaret Friedrich, Jan. 14, 1939. Came to U. S., 1938, naturalized, 1943. Cons. WHO, since 1950; with N.Y. State Dept. of Health, Albany, since 1949; cons. in neurology USPHS, Marine Hosp., Staten Island and Ellis Island since 1948; spl. cons. USPHS, Washington, since 1946; asso. clin. prof. neurology N.Y. U., 1943-47; asst. clin. prof. neurology Coll. Phys. and Surg., Columbia, 1945-47; acting attending neurologist Goldwater Meml. Hosp., N.Y.C., since 1946, Bronx Hosp. since 1942; attending neurologist, Montefiore Hosp. since 1945; asso. vis. neuropsychiatrist Bellevue Hosp. since 1943. Diplomate Am. Bd. Psychiatry and Neurology. Fellow N.Y. Acad. Medicine, Am. Neurol. Soc., N.Y. Neurol. Soc., Mexican Soc. Neurology and Psychiatry (hon.), Vienna Med. Soc. (corr.), Vienna Neuro-psychiatric Soc. (corr.). Author: Moderne Therapie der Neurosyphilis, 1933; Management of Neurosyphilis, 1944. Died Aug. 11, 1952.

DATZ, Sheldon, Am. chemist; b. N.Y.C., July 21, 1927; s. Jacob and Clara (Green) D.; B.S., Columbia, 1950, M.A., 1951; Ph.D., U. Tenn., 1960; m. Roslyn Gordon, Aug. 25, 1948; children—William Lawrence, Joan Ellen. Technician, Columbia, 1943-44, 46-50; chemist chemistry div. Oak Ridge Nat. Lab., 1951-—, leader molecular beam group, 1960-—; guest scientist Found. for Fundamental Research on Matter, Inst. for Atomic and Molecular Physics, Amsterdam, Netherlands, 1962-63. Cons. atomic beam lab., gen. atomic div. Gen. Dynamics Corp., San Diego, 1960-62, Plasma Propulsion Lab., Republic Aviation Corp., L.I., N.Y., 1961-62. Mem. Am. Chem. Soc., Am. Phys. Soc., A.A.A.S., Research Soc. Am., Sigma Xi. Research, publs. demonstrated applicability of method of crossed molecular beams to study chemically reactive collisions; demonstrated single atom collisions can be studied by interaction of ion beams with metal surfaces; helped develop method for determining atomic potentials in crystals. Home: 109 Ditman Lane, Oak Ridge 37830. Office: P.O. Box, Oak Ridge 37831.*

DATZEFF, Assène Borissoff, Bulgarian physicist; b. Kamenar, Bulgaria, Feb. 14, 1911; s. Boris Datzeff and Gina (Nedelcheva) Diakoff; grad. dept. physics and math. Sofia (Bulgaria) U., 1933; Docteur ès Scis., Sorbonne, Paris, France, 1938; m. 1950 (div.); children—Elina, Uliana. Staff, Sofia U., 1939-—, prof., head theoretical physics dept., 1950-—, dean faculty physics and math., 1950-55. Adviser, Bulgarian Embassy Moscow, USSR, 1946-47; mem. Bulgarian commn. UNESCO. Decorated comdr. Order Ninth of Sept., 1944, Order for Chivalry, 1945, Order Narodna republica Dimitrov prize, 1951. Mem. Bulgarian Acad. Scis. (sec. dept. for math. and physics), IUPAP (corr. mem. commn. on thermodynamics and statis. mechanics 1966-—). Author: Sur le porbléme de la propagation de la chaleur dans les corps solides, 1963; also articles. Developed method solving problem of heat propagation at gen. initial and boundary conditions in solid phases, consisting of 2 layers at 1, 2 and 3 dimensions, also at 2 phases (solid and liquid) having variable boundary; proposed causal interpretation of quantum mechanics; deduced probability equation which generalizes Schrödinger's equation. Home: 9 Vassil Kolarov, Sofia, Bulgaria.*

DAUBENMIRE, Rexford, Am. plant ecologist; b. Coldwater, O., Dec. 12, 1909; s. George Franklin and Ethel R. (Friedline) D.; B.S. magna cum laude, Butler U., 1930; M.S., U. Colo., 1932; Ph.D., U. Minn., 1935; m. Evelyn Jean Boomer, May 20, 1938; 1 dau., Janet. Asst. prof. U. Tenn., 1935-36, U. Ida., 1936-46; faculty Wash. State U., Pullman, 1946-—, prof. plant ecology. Cons. Nat. Park Service, Canadian Dept. Agr., Sierra Club. Mem. Ecol. Soc. Am. (pres. 1966-67), Brit. Ecol. Soc., Pacific Sci. Assn., Sigma Xi, Phi Kappa Phi. Author: Plants and Environment, 1947; Plant Communities, 1968; also numerous articles. Research on effects environment on plants, relation vegetation types to climate, soil and fire, community theory. Home: Route 2, Box 89, Pullman, Wash. 99163.*

DAUBENTON, Louis Jean Marie, French physician, naturalist; b. Montbard, France, May 29, 1716; s. Jean D.; M.D., U. Reims (France), 1741; m. Marguerite Boucheron, 1749; 1 dau. Apptd. keeper, demonstrator, asst. to Buffon, Jardin des Plantes (now Mus. Natural History), Paris, France, 1744; became lectr. Coll. Medicine, 1775; prof. zoology, Collège de France, 1778; lectr. rural economy Alfort Sch., 1783; prof. mineralogy Mus. Natural History, 1793; prof. natural history, Ecole normale, 1793; apptd. mem. Senate, 1799. Fellow Royal Soc., 1755; mem. French Acad. Scis., 1760. Began writing anat. descriptions for Buffon for his book l'Histoire naturelle (1794-1804), 1742; provided anat. descriptions of quadrupeds; also zool. descriptions and dissections (including Daubenton's line, Daubenton's angle, Daubenton's plane); research on comparative anatomy of recent and fossil mammals, vegetable physiology, mineralogy, agrl. expts., introduced Merino sheep into France. Died Paris, Jan. 1, 1800.

DAUBENY, Charles Giles Bridle, English chemist, botanist; b. Stratton, Eng., Feb. 11, 1795; s. James Daubeny; ed. at Winchester and Magdalen Coll., Oxford (Eng.) U., M.D.; B.A., 1814; studied medicine at U. Edinburgh (Scotland), 1815-18. Toured, studied Auvergne (France) volcanic region, 1819; prof. chemistry Oxford U., 1822-55, prof. botany, from 1834, prof. rural economy, from 1840. Fellow Royal Soc., 1822. Author: A Description of Active and Extinct Volcanoes, 1826; Introduction to the Atomic Theory, 1831; Lectures on Agriculture, 1841; Lectures on Roman Husbandry, 1857; Climate: . . . , 1863; Essay on the Trees and Shrubs of the Ancients, 1865; Miscellanies, 2 vols., 1867. Described and attempted to explain European volcanoes; research on physiology and ecology of plants, photosynthesis, chemistry of mineral water, geol. and chem. history of the earth. Died Oxford, Dec. 13, 1867.

DAUBERT, Bernard Forbes, Am. chemist; b. Martins Ferry, O., May 1, 1905; s. Lewis Henry and Elizabeth (Davis) D.; Ph.G., U. Pitts., 1925, B.S., 1930, Ph.D., 1939; m. Dorothy Samuels, June 8, 1925; children—Margaret Helen (Mrs. John J. Jenkins), Sara Jean (Mrs. Allan M. Baum). Faculty, U. Pitts., 1925-51, asso. prof. chemistry, 1930-42, research adminstr. chemistry, 1948-51, acting head dept. chemistry, 1950-51; asst. mgr. lab. sect. Koppers Co., Inc., Pitts., 1951-54; tech. dir. chem. research Gen. Food Corp., White Plains, N.Y., 1954-60, dir. sci. research, 1960-61, dir. nutrition, 1961—. Chmn. adminstrn. liaison com. Inst. Food Technologists/FDA, 1963—. Mem. Am. Chem. Soc., Am. Soc. Biol. Chemists, Inst. Food Technologists, Am. Oil Chemists Soc. Research, numerous publs. on chemistry of fats, fatty acids and derivatives, autoxidation and antioxidation, physiology of fats and fatty acids. Home: 30 N. Broadway, White Plains 10601. Office: 250 North St., White Plains, N.Y. 10602.*

D'AUBIGNE, Robert Merle, French physician; b. Neuilly, France, July 23, 1900; s. Charles Merle and Lucy (Maury) d'A.; grad. Lycee Pasteur, Neuilly, France, 1917, Paris Faculty of Medicine, 1927; m. Anna de Gunzburg, July 16, 1932; children—Catherine (Mrs. Michael Ebersolt), Jean. Chief med. services French Forces of Interior, 1944; aggregate prof. Faculty Medicine, Paris, 1945-49, prof. orthopedic surgery, 1949—; chief surgeon Mil. Center of Reparatory Surgery, Hosp. Foch, Paris, France, 1944-49; surgeon Cochin Hosp., Paris, 1949—; head orthopedic dept. U. Paris, 1949—. Decorated comdr. Legion of Honor, Croix de Guerre, Broze Star medal. Mem. Nat. Acad. Scis., A.C.S., Royal Soc. Medicine, French Acad. Medicine, French Acad. Surgery. Author: Chirurgie Reparatrice, 1945, 50; Traumatologie, 1955; Traumatismes Anciens, 1958; also numerous articles. Research, publs. on traumatic, non-traumatic lesions of hip, treatment of lesions of peripheral nerves, fractures, reparative osteogenesis, bone tumors and new means of treating them. Home: 3 rue Fleurus, Paris. Office: Hôp. Cochin, Faub. St. Jacques 27, Paris, France.*

DAUBLEBSKY VON STERNECK, Robert, geodesist; b. Prague, Czechoslovakia, Feb. 7, 1839; s. Jakob and Maria (Kalina) D. von S.; student Technische Hochschule, Prague, 2 years; Dr. (hon.), Göttingen, Germany. Became Austrian officer, 1859, advanced to col. by 1894; assigned to Mil. Geog. Inst., Vienna, 1862; became dir. inst. obs., 1880; became dir. Astronomic-Geodetic Group, 1894-1906; named Austrian commissar for European degree measurement, 1882. Mem. Vienna Acad. (corr.), Bohemian Sci. Soc. (Prague), Accademia dei Lincei (Rome), Leopoldina (Halle). Research and numerous publs. on force of gravity in earth's center; invented pendulum for measurement of relative gravity (opened new era of gravimetric survey); built tide gauge for measurement of central waters of Adriatic. Died Vienna, Nov. 2, 1910.

DAUBNEY, Robert, English veterinarian; b. 1891; s. Robert Daubney; ed. Liverpool, George Washington, Cambridge univs., Royal Vet. Coll., London; M.Sc.; m. Jean Ethel Winifred Ker, 1919. Became helminthologist Ministry Agrl., 1920; in colonial service, 1925; dir vet. services, Kenya, 1937-47, E. African Central Vet. Research Inst., 1939-47; vet. adviser Egyptian Govt., 1951; vet. cons. FAO UN, from 1951. Mem. Royal Coll. Vet. Surgeons. Research, publs. on helminthology and bacteriology; 1st to describe (with J. R. Hudson) enzootic hepatitis, virus disease of sheep, cattle and man, in Kenya (called Rift Valley Fever), 1931. Address: c/o Glyn Mills & Co. (Holt's Branch), Whitehall, S.W.1, London, Eng.

DAUBRÉE, Gabriel-Auguste, French geologist; b. Metz, France, June 25, 1814; studied at École polytechnique, Paris, France; became qualified mining engr., 1834; in charge of mines at Bas-Rhin, Alsace, 1838; later became prof. mineralogy and geology Strasbourg, France; became engr.-in-chief of mines, 1859; named prof. geology Mus. Natural History in Paris, 1861; became prof. mineralogy École des Mines, 1862, named dir., 1872. Recipient Wollaston medal Geol. Soc. London (Eng.), 1880. Fellow Royal Soc., 1881; mem. French Acad. Scis. (became v.p., 1878 and pres., 1879). Author: Études et expériences synthétiques sur le métamorphisme et sur la formation des roches crystalline, 1860; Ètudes synthétiques de géologie expérimentale, 1879; les Eaux souterraines à l'époque actuelle, 2 vols., 1887; les Eaux souterraines aux époques anciennes, 1887. Research on artificial prodn. of minerals and rocks, the origin of minerals especially bogiron ore, thermal waters, water permeability of rocks and its relation to volcanic phenomena, metamorphism in the earth's crust, earthquakes, composition and classification of meteorites. Died Paris, Apr. 29, 1896.

D'AUBUISSON DE VOISINS, Jean François, French geologist; b. Toulouse, France, Apr. 16, 1769; student mining U. Freiberg, Germany, 1797-1802. Adjunct conservator, collections of École des Mines de Paris, from 1802; engr., Corps des Mines, from 1807; municipal counsellor, Toulouse, 1817-28. Mem. French Acad. Scis. Author: Mines de Freiberg en Saxe et leur exploitation, 1802; Basalts of Saxony, 1803; Géognosie, 1804; Traité de géognosie (at the time most complete book on geology in France), 1819; Histoire des fontaines de Toulouse, 1828; Traité d'hydraulique, 1834; Traité du mouvement de l'eau dans les tuyaux de conduite, 1836. Died Toulouse, Aug. 21, 1841.

DAUDIN, François-Marie, French naturalist; b. Paris, France, 1774; author: Histoire naturelle des reptiles, 1802-03; Histoire naturelle des rainettes, des grenouilles, et des crapauds, 1803. Writer beautifully illustrated, popular editions dealing mainly with mollusks, birds, frogs and toads; made more precise classification of Lacépède's reptiles; classified numerous neglected species. Died Paris, 1804.

DAUGHADAY, William Hamilton, Am. physician; b. Chgo., Feb. 12, 1918; s. Carlos Colton and Marion (Sharpe) D.; A.B., Harvard, 1940, M.D., 1943; m. Hazel Judkins, Jan. 22, 1945; children—Elizabeth Colton, John Freer. Faculty, Washington U., St. Louis, 1947—, prof. medicine, 1963—, cons. in chemistry Barnes Hosp., 1951—. Cons. endocrinology U. S. Army, USAF; bd. sci. counselors Nat. Inst. Arthritis and Metabolic Diseases, 1963-67. Mem. Central Soc. Clin. Research, Am. Fedn. Clin. Research, Am. Soc. Clin. Investigation, Assn. Am. Physicians, Endocrine Soc., Am. Inst. Nutrition, Am. Soc. Clin. Nutrition, Phi Beta Kappa, Alpha Omega Alpha, Sigma Xi. Editorial bd. Nutrition Revs., 1954-58, Jour. Clin. Endocrinology; editorial bd. Jour. Lab. and Clin. Medicine, 1958-60, editor, 1961-66. Research, publs. on conticosteroid binding globulin, radioimmunoassays of pituitary growth hormone, mechanism of inositol excretion in diabetes mellitus; clin. studies of diabetic acidosis. Home: 1414 W. Adams, Kirkwood, Mo. 63122. Office: Barnes Hosp. Plaza, St. Louis 63110.*

DAUGHERTY, J(oseph) Dwight, Am. mathematician; Balt., Jan. 16, 1900; s. Joseph and Rebecca (Prowell) D.; B.A., Lebanon Valley Coll., 1922; M.A., U. Penn., 1926; Ed.D., N.Y. U., 1954; m. Sara Lackey, Sept. 4, 1949. Staff, Ridley Park (Penn.) High Sch., 1922-26, Central High Sch., Paterson, N.J., 1926-60; head math. dept. Eastside High Sch., Paterson, 1926-60; asst. prof. math. Paterson State Coll., 1942-44; asst. prof. math. Fairleigh-Dickinson U., Rutherford, N.J., 1954-60; prof. math. Kutztown (Penn.) State Coll., 1960—. Mem. N.J. Math. Assn. (pres. 1944-45), Math. Assn. Am. (vice chmn. bd. govs. N.J. 1959-60), N.Y. Sch.-masters Club (admission com. 1963-64), Nat. Council Tchrs. Math., Am. Math. Soc., Nat., N.J., Penn. edn. assns., Penn. Council Tchrs. S.A.R., Am. Legion, Kappa Delta Pi, Phi Delta Kappa, Kappa Mu Epsilon. Home: P.O. Box 46, College Hill, Kutztown, Penn., 19530.*

DAUGHERTY, Lewis Sylvester, Am. zoölogist; b. Belmont County, O., Aug. 10, 1857; s. Samuel and Rachel Ann (Mechem) D.; student, Ill. State Normal U., Normal, 1881-82; B.S., U. Ill., 1889, M.S., 1893; postgrad. U. Chgo., 1894-96; Ph.D., Ill. Wesleyan U., 1901; spl. study, various summer schs. and in univs. of Germany, 1907; m. Millie Crum, July 8, 1885. Prof. science, Twp. High Sch., Ottawa, Ill., 1889-92, prof. biology, 1892-94; prof. natural science State Normal Sch., Kirksville, Mo., 1897-1900, prof. zoology, 1900-13; prof. zoölogy and chemistry Mo. Wesleyan Coll., 1913—. Author: (with wife) Principles of Economic Zoölogy, 1912; Field and Laboratory Guide, 1912. Died Feb. 18, 1919.

DAUGHERTY, Richard Deo, Am. anthropologist; b. Aberdeen, Wash., Mar. 31, 1922; s. Charles D. and Audrey (Ross) D.; B.A., U. Wash., 1946, Ph.D., 1954; m. Phyllis J. McCullough, Mar. 2, 1944; children—Melinda, Carol, Richard Deo. Faculty, Wash. State U., Pullman, Wash., 1950—, prof. anthropology, 1963—; Mem. adv. bd. Wash. State U. Radiocarbon Dating Lab., 1963—; mem. Pres.'s Adv. Council on Historic Preservation, 1967—. Mem. Am. Anthrop. Assn., Soc. for Am. Archaeology, N.W. Anthrop. Conf. (past chmn.), A.A.A.S., Gt. Basin Archaeol. Conf. (past chmn.), Am. Assn. U. Profs. (past chpt. pres.), Western Canadian Archaeol. Council, Sigma Xi, Alpha Kappa Delta, Gamma Theta Upsilon, Phi Kappa Phi. Research, publs. on prehistory of Pacific N.W., N. Africa, western Europe; worked on salvage of materials to be submerged by Aswan Dam in Egypt and Sudan. Home: 309 Spring St., Pullman, Wash. 99163.*

DAUGHLISH (OR DAUGLISH), John, Brit. physician; b. London, Eng., Feb. 10, 1824; s. William and Caroline D.; M.D., U. Edinburgh (Scotland), 1855. Went to London, 1855; practiced medicine; founded bakery, Islington, 1859. Recipient Silver medal Soc. Arts, 1860. Patentee improved method of making bread, 1856. Inventor aerated bread (process eliminated fermenting process, also reduced personal contact of workman with produce materials). Died Jan. 14, 1866.

DAUMONT, Arnulphe, French physician; b. Grenoble, France, 1720; prof. Faculty Medicine, Valence, France; contbr. numerous articles for d'Alembert's Encyclopédie. Died 1800.

D'AUNOY, Rigney, Am. pathologist; b. New Orleans, La., Aug. 8, 1890; s. Joseph and Zelina (Chretien) D'A.; B.S., Tulane U., 1910, M.D., 1914. Asst. pathologist New Orleans Dispensary, 1916-17; asst. pathologist Charity Hosp. of La., 1919-24, pathologist and dir. labs., 1928—; instr. pathology and bacteriology Tulane U., 1919-24; prof. pathology and bacteriology La. State U. Sch. Medicine, 1931—, dean, 1937—. Died Sept. 17, 1941.

DAUNT, John Gilbert, physicist, educator; born Killiney, Ireland, June 30, 1913; s. George and Constance C. (Burns) D.; B.A. (Stapledon scholar, Abbott scholar), U. Oxford (Eng.), 1935, M.A., Ph.D. (Scott scholar), 1937. Demonstrator in physics U. Oxford, 1937-39, lectr., 1940-46; faculty physics Ohio State U., Columbus, 1946-65, prof., 1950-65; prof. physics and elec. engring. Stevens Inst. Tech., Hoboken, N.J., 1965—, dir. cyrogenics center, 1967—. Vis. scientist Oak Ridge Nat. Lab., 1949, cons. solid state and radiation damage lab., 1950-53; cons., vis. scientist cryogenics lab. Los Alamos Sci. Lab., 1958-65. Mem. adv. panel for physics NSF, 1960-63; vis. prof. U. Sao Paulo (Brazil), 1961; vis. prof. City Coll. N.Y., 1964, Columbia, 1965. Medalist for physics Free U. Bruxelles, 1955; Guggenheim Meml. fellow U. Amsterdam, 1953, Harvard, 1954, 58. Fellow Am. Phys. Soc., Phys. Soc. London (Duddell medal 1956), Ohio Acad. Scis.; mem. Internat. Inst. Refrigeration, Internat. Union Pure and Applied Physics (pres. commn. for low temperatures 1960-65, U. S. nat. com. 1960—), Phi Beta Kappa (hon.), Sigma Xi. Author: Electrons in Action, 1946; (with C. J. Gorter) Progress in Low Temperature Physics, 1955; Helium Three, 1960. Adv. editor Physics of Condensed Matter, 1962—, Physics Letters, 1962—. Contbr. articles on low temperature and solid state physics, radar tube devel., infra-red detectors to profl. publs.*

DAUPHIN, Joseph, French chemist; b. Clermont-Ferrand, France, Oct. 14, 1908; s. Antoine and Madeleine (Bouyon) D.; diploma chem. engring. Ecole Normale Supérieure de Chimie, Clermont, 1929; Licence ès scis., diploma higher edn., aggregation phys. scis. Faculty Scis. Clermont; doctorate in pharmacy, 1954; aggregation in pharmacy, 1955; m. Simone Chaboissier, July 10, 1930; children—Monique Bon, Gerard. Asst., Clermont Faculty Scis., 1929-33; aggregate prof. lycées, 1933-55; lectr. Faculty Medicine and Pharmacy Clermont, 1957—; mil. engr. in charge powder reserves; prof. chem. engring. Ecole Normale Supérieure de Clermont-Ferrand. Mem. Chem. Soc. France, Soc. Phys. Chemistry. Author: L'extraction par solvants, 1954. Research in chem. engring, chromatography, polarography, radiocrystallography; studies of solutions, mineral chemistry, metallic complexes. Home: 5, rue Ledru, Clermont-Ferrand, 63 France.*

DAUSSE, (Marie-François) Benjamin, French engr.; b. Grenoble, France, Jan. 28, 1801; s. Joseph Henri Christophe and Marie Thèrese (Courturier) D.; ed. École polytechnique, Paris, France, 1817; became chief engr. Bridges and Hwys. Dept., 1840; placed in charge of studying floods in Italy, 1858. Mem. French Acad. Scis. Author: Statistique des variations du niveau de la Seine à Paris, 1831; Note sur l'endiguement de l'Isère, 1850; Sommaire des lois et résultats touchant l'hydraulique fluviale, 1878. Originated the theory that the amount of ann. rainfall increases with the altitude to a certain limit (Dausse's law); developed statistics for variations in water-level of the Seine River; research on the floods of the Isère River; presented projects for the embankment of the Tiber River. Died Grenoble, Jan. 16, 1890.

DAUSSY, Pierre, French hydrographic engr.; b. Paris, France, Oct. 8, 1792; became hydrographic engr. in charge of the Navy, 1806; chief dir. Map Depository, Paris, Mem. Soc. Geography (pres.), French Acad. Scis., 1855, Bur. Longitudes. Author: Tables des. positions géographiques, 1847; also publs. on chronometers, the tides of N. coast of France. Died Paris, Sept. 5, 1860.

DAUTREBANDE, Lucien Jean-Baptiste, Belgian physiologist; b. Gougnies, Belgium, Jan. 25, 1894; s. Emile and Elise Renelle (Poulet) D.; M.D., U. Louvain (Belgium), 1919; Ph.D., U. Edinburgh (Scotland) 1925; Docteur Special en Sciences Pharmacodynamiques et Thérapeutiques, U. Liège (Belgium), 1926; Dr. honoris causa, U. Bordeaux (France), 1933; m. Fernande Fatton, Aug. 5, 1930. Mem. Fondation de Recherches Médicales Reine Elisabeth, 1921-31; prof. pharmacodynamics U. Liège, 1931-64; exchange prof. various univs., Bordeaux, Geneva, Switzerland, Paris, Saragossa, Crakow, Warsaw, Vilna, Yale, Stanford, Buenos Aires, Montreal. Cons. various sci. Instns., U. S. A., Can., France, Germany, Poland. Recipient sci. awards Académie de Médecine, Belgium, France, Académie des Sciences Paris, Institut de France; decorated grand officer l'Ordre de Léopold; grand officer l'Ordre de la Couronne; Légion d'Honneur. Author 12 books including Studies on Aerosols, 1958, Microaerosols, 1962; also numerous articles. Research on microaerosols. Home: 64, av. Emile Duray, Brussels 5, Belgium.*

DAUVILLIER, Alexandre, French astronomer; b. St.-Lubin-des-Joncherets, 1892; s. Joseph and Clarisse (Chalus-Leneveo) D.; license Lyceum St. Louis, Paris, Chem.E., Faculty Scis., U. Paris, 1914, Sc.D., 1920. Became chief M. de Broglie Lab. on Research in X-rays, 1920; lectr. Superior Sch. Electricity, 1925—; lectr. Faculty Medicine, Paris, 1925—; became dir. research Nat. Research Center, 1942; prof. cosmic physics Coll. France, 1944—. Sec. Comité Français des unités radiologiques de dosage, 1928; chief of research Nat. Center for Sci. Research, 1931-—; chief of lab. cosmic physics Obs. at Meudon, 1935. Author: Les premiers ages de la terre, 1939; Le genèse de la vie, 1942; Génétique et origine des rayons cosmiques, 1952; Le magnétisme des corps célestes, 1954; Cosmologie et chimie, 1955; L'origine des planétes, 1956; Quel est pour un laïc le sens de la vie, 1957; L'origine photochimique de la vie, 1958.

D'AUXIRON, Claude François Joseph, French inventor; b. Besancon, France, 1731; capt. of arty.; later ret. in Paris to work on problems of applied science; studied steamship travel; built ship on the Seine, 1772-74, brought it to Meudon, where it sank (sabotaged by a seaman); studied methods of providing pure water for Paris. Died Paris, 1778.

DAVAINE, Casimir Joseph, French biologist; b. Sainte-Amand-les-Eaux, France, Mar. 19, 1812; student of Rayer; M.D., 1837; practiced medicine in Paris; mem. Société de biologie, Acad. Médicine. Author: De la paralysie générale et partielle des deux nerfs de la septième paire, 1852; Recherches sur la génération des huitres, 1853; Recherches sur l'anguillule du blé niellé, 1857; Mémoire sur les anomalies de l'oeuf, 1861; Étude sur la genèse et la propagation du charbon, 1870. Research in exptl. physiology, parasites of animals and vegetation; one of the earliest to study bacteria; discovered charbon bacteria, 1850, anthrax bacillus, 1850-65; made 1st observation of Cercomonas intestinalis, 1857; one of 1st to prove germ theory of disease. Died Garches, France, Oct. 13, 1882.

DAVAINE, Napoléon Emmanuel, French engr.; b. Saint Amand, France, Dec. 30, 1804; studied at Valenciennes, also École polytechnique, Paris, France, 1824; became engr. in 1831 and chief engr. in 1843. Mem. Acad. Arras. Author: Nivellement des routes royales et departementales, 1845. Worked on the constrn. of the railroad Lille-Dunkerque, the Roubaix canal, revolving bridge which was designed by la Lawe; developed new method of building Archimedes' screw. Died Arras, France, Mar. 2, 1864.

DAVANNE, Louis Alphonse, French chemist, photographer; b. Paris, France, Apr. 12, 1824; studied under Pelouze; mem. French Soc. Photography (became pres. 1876). Author: Chimie photographique (with Barreswill), 1854; le Progrès de la photographie, 1877; la Photographie appliquée aux sciences, 1881. Studied chemistry and physics of photography; developed better cameras; originated photog. exhibit at Universal Expn. of 1900; originated method for determining the focal distance of photog. lenses. Died Saint-Cloud, France, Sept. 19, 1912.

DA VARIGNANA, Bartolommeo, Italian physician; b. Bologna, Italy; s. Giovanni da Varignana; studied medicine under Taddeo Alderotti; a son, Guglielmo. Taught medicine, Bologna; held municipal office, Bologna, from 1303; banished from Bologna, 1306; with Henry VII of Luxembourg in Italy, 1311-13; author of Commentaries on the Works of Hippocrates, Galen, and Ibn Sina. Performed (assisted by physician and 3 surgeons) 1st formally recorded post-mortem (on suspected poisoning case), Bologna, 1302. Died Bologna, 1318.

DAVE, Jitendra Vishvanath, atmospheric physicist; b. Ahmedabad, India, May 4, 1926; s. Vishvanath Shivshankar and Motiben Dave; B.Sc., Wilson Coll., Bombay, 1948; M.S., Phys. Research Labs., Ahmedabad, 1952, Ph.D., 1957; m. Kunjbala, Feb. 1949; 1 dau., Sheela Jitendra. Vis. asst. prof., research physicist meteorology U. Cal. at Los Angeles, 1958-60; research asst. in meteorology Imperial Coll., London, Eng., 1961; program scientist Nat. Center for Atmospheric Research, Boulder, Colo., 1962-—; affiliate asso. prof. meteorology Fla. State U., Tallahassee, 1965-—. Fellow Royal Meteorol. Soc.; mem. Optical Soc. Am., Am. Meteorol. Soc., Am. Geophys. Union. Author: (with others) Tables Related to Radiation Emerging from a Planetary Atmosphere, 1960; also articles. Research in solution of problem of transfer of solar radiation in planetary atmospheres, multiple scattering in homogeneous and nonhomogeneous plane-parallel as well as spherical atmospheres. Home: 884 Ithaca Dr. Office: P.O. Box 1470, Boulder, Colo. 80302.*

DAVENPORT, Bennett Franklin, Am. physician; b. Cambridge, Mass., May 28, 1845; s. Charles and Joan Fullerton (Hagar) D.; A.B., Harvard, 1867, A.M., 1870, M.D., 1871; M.D., Coll. Phys. and Surg. (Columbia), 1871; studied U. of Tübingen 1 yr.; m. Annie Emeline Coolidge, July 23, 1873; children—Grace Coolidge (Mrs. Henry J. Winslow), John C., Anna C. (Mrs. Clifford M. Holland), Benita C. Began gen. practice of medicine at Boston; apptd. coroner Suffolk County, Mass., 1875, thereafter devoted mostly to medico-legal specialties, including expert examination of documents to determine authorship of handwriting; expert upon many notable court trials; expert for State Record Commn. on ofcl. record ink, 1895-—. Prof. chemistry, Mass. Coll. Pharmacy, 1879-86; dairy insp., City of Boston, 1882-85; chemist Mass. Bd. Health, 1882-92, Dairy Bur., 1892-1921; chmn. Watertown health, park and water bds. Delegate for the revision of U. S. Pharmacopoeia, 1880, 90, 1900. Died June 2, 1927.

DAVENPORT, Charles Benedict, Am. biologist; b. Stamford, Conn., June 1, 1866; s. Amzl B. and Jane Joralemon (Dimon) D.; B.S., Poly. Inst., Bklyn., 1886; A.B., Harvard, 1889, A.M., Ph.D., 1892; m. Gertrude Crotty, June 23, 1894; children—Mrs. Millia Davenport Harkavy, Mrs. James A. deTomasi, Charles Benedict. Engr. survey of Duluth, S. Shore & Atlantic Ry., 1886-87; asst. zoology Harvard, 1888-90, instr., 1891-99; asst. prof. zoology and embryology U. Chgo., 1899-1901, asso. prof. and curator Zool. Mus., 1901-04; dir. dept. of genetics Carnegie Instn., comprising Sta. for Exptl. Evolution, 1904-34, and Eugenics Record Office, 1910-34, Cold Spring Harbor, N.Y.; dir. biol. lab. Brooklyn Inst. Arts and Scis., 1898-1923. Fellow Am. Acad. Arts and Scis., A.A.A.S. (vice-pres. 1900-01, 1925-26), N.Y. Zoöl. Soc.; mem. Nat. Acad. Scis., Am. Philos. Soc., Am. Soc. Zoologists (pres. 1902-03, 29-30), Am. Genetic Assn., Am. Soc. Naturalists (sec. 1899-1903, v.p. 1906), Soc. Exptl. Biology and Medicine, Eugenics Research Assn. (hon. pres. 1937), N.Y. Acad. Medicine (asso.), Galton Soc. (pres. 1918-30), Nat. Inst. of Social Scis. (Gold medal, 1923), Internat. Fedn. of Eugenic Orgns., (pres. 1927-32), Anthropologische Gesellschaft in Wien, Berliner Gesellschaft für Anthropologie, Ethnologie und Urgeschichte. Author: Graduate Courses, 1893; Experimental Morphology, Part 1, 1897, Part 2, 1899; Statistical Methods in Biological Variation, 4th edit., 1936; Introduction to Zoology (with G. C. Davenport), 1900; Elements of Zoology, 1911; Inheritance in Poultry, 1906; Inheritance of Characteristics of Fowl, 1909; Eugenics, 1910; Heredity in Relation to Eugenics, 1911; Heredity of Skin Color in Negro-White Crosses, 1913; The Feebly-inhibited—Nomadism and Temperament, 1915; Naval Officers—Their Development and Heredity (with M. Scudder), 1919; Physical Examination of First Million Draft Recruits, 1919; Defects Found in Drafted Men, 1920; Army Anthropology, 1921 (the last 3 with A. G. Love); Body Build and Its Inheritance, 1923; Race-Crossing in Jamaica (with M. Steggerda), 1929; Genetical Factor in Endemic Goiter, 1932; How We Came by Our Bodies, 1936. Asso. editor Jour. Exptl. Zoology, from 1898, Genetics, from 1916, Jour. Phys. Anthropology, from 1918. Contbr. to biol. jours. Research in eugenics and heredity; genetic basis of human skin pigmentation; anthropometric studies of Am. soldiers during W.W. I. Died Cold Spring Harbor, N.Y., Feb. 18, 1944.

DAVENPORT, Demorest, Am. biologist; b. Utica, N.Y., Sept. 26, 1911; s. William R. and Alice (Demorest) D.; grad. St. George's Sch., Newport, R.I., 1929; A.B., Harvard, 1933; M.A., Colo. Coll., 1934; Ph.D., Harvard, 1937; m. Winnifred Smith Bailey, June 10, 1941; children—Mary Stuart, Evelyn Curtis. Faculty, Reed Coll., 1937-42, asst. prof., 1938-42; faculty U. Cal., Santa Barbara, 1946-—, prof. biology, 1959-—, chmn. dept. biol. scis., 1963-66; vis. prof. zoology Friday Harbor Labs., U. Wash., 1959-60. Vis. investigator Marine Biol. Lab., Plymouth, Eng., 1953, Lab. Arago, U. Paris, France, Gatty Lab., St. Andrews, Musée Oceanog., Monaco, 1961; vis. research asso. Harvard, 1964. John Simon Guggenheim Meml. Found. fellow, 1952, 61. Fellow A.A.A.S.; mem. Ecol. Soc. Am., Am. Soc. Zoologists, Marine Biology Assn. U.K., Brit. Soc. Study Animal Behavior, Sigma Xi. Research, publs. on behavior in invertebrates. Home: 1045 Winther Way, Santa Barbara, Cal. 93105.*

DAVENPORT, Eugene, Am. agriculturist; b. Woodland, Mich., June 20, 1856; s. George Martin and Esther (Sutton) D.; B.S., Mich. Agrl. Coll., 1878, M.S., 1884, M.Agr., 1895, LL.D., 1907; D.Sc., Ia. State Coll., 1920; LL.D., U. Ky., 1913, U. Ill., 1931; m. Emma Jane Coats, Nov. 2, 1881; children—Dorothy, Margaret (Mrs. H. B. Tukey). Asst. botanist expt. sta. Mich. Agrl. Coll., 1888-89, prof. practical agr. and supt. of farm, 1889-91; pres. Collegio Agronomica, Sao Paulo, Brazil, 1891-92; dean Coll. of Agr., U. Ill., 1895-1922; dir. Agrl. Expt. Station, prof. thremmatology U. Ill., 1896-1922. Author: Principles of Breeding, 1907; Education for Efficiency, 1909; Domesticated Animals and Plants, 1910; Vacation on the Trail, 1923; The Farm, 1927. Leader in devel. of crop rotation; research on soil fertility; developed Illinois agriculturally. Died Mar. 31, 1941.

DAVENPORT, Fred Marshall, Am. physician, virologist; b. Scranton, Pa., Nov. 30, 1914; s. Fred Marshall and Laura May (Church) D.; B.A., Columbia, 1936, M.D., 1940, Sc.Med.D., 1945; m. Clara Josephine Dommerich, June 14, 1941; children—Laura May, Steven Marshall, Clara Josephine. Asst. instr. Columbia, 1946-47; fellow Rockefeller Inst., 1947-49; faculty U. Mich., Ann Arbor, 1949-—, prof. epidemiology and internal medicine, 1958-—. Dir. Comm. on Influenza, Dept. Def. Armed Forces Epidemiol. Bd., 1955-—; cons. on infectious diseases VA Hosp., Ann Arbor, 1953-—. Mem. Assn. Am. Physicians, Am. Assn. Immunologists, Am. Pub. Health Assn., Central Soc. Clin. Research, Am. Acad. Microbiology, Am. Soc. Clin. Investigation, N.Y. Acad. Sci., Am. Epidemiol. Soc., Robert Koch Inst. (hon.). Bd. editors Jour. Lab. and Clin. Medicine, 1964-—. Research, publs. on epidemiology, pathogenesis and prevention of influenza A and B with spl. emphasis on defining by serologic surveys in humans the periods of former pandemic prevalences of type A strains and relation between animal, avian, human infections; devel. and testing purified extracted hemagglutinin vaccines. Home: 1038 Martin Pl., Ann Arbor, Mich. 48104.*

DAVENPORT, George Edward, Am. botanist; b. Boston, Aug. 3, 1833; ed. Boston schs.; resided in Medford from 1875; devoted much attention to botany research; best known for his work on ferns; a founder Middlesex Field Club (became Middlesex Inst.); life mem., Mass. Hort. Soc. (donated his collection of ferns now known as Davenport Herbarium; recipient Appleton Gold medal). Contbr. monographs and papers on ferns. Died 1907.

DAVENPORT, Harold, English mathematician; b. Accrington, Eng., Oct. 30, 1907; s. Percy and Nancy (Barnes) D.; B.Sc., Manchester (Eng.) U., 1927; M.A., Cambridge U., 1932, Sc.D., 1938; m. Anne Lofthouse, Mar. 25, 1944; children—James Harold, Richard Barnes. Asst. lectr. Manchester U., 1937-41; prof. U. Coll. N. Wales, 1941-45; Astor prof. math. U. Coll., London, 1945-58; Rouse Ball prof. math. Cambridge U., 1958-—; Vis. prof. Stanford, 1947-48, U. Mich., 1962, 66; Gauss prof. Akad. Wiss. Göttingen, Germany, 1966. Recipient Adams prize Cambridge U., 1941; Berwick prize London Math. Soc. Fellow Royal Soc. (Sylvester medal 1967); mem. Acad. Scis. Uppsala. Author: The Higher Arithmetic, 1952; Analytic Methods for Diophantine Equations and Inequalities, 1963; Multiplicative Number Theory, 1967; also numerous articles. Research on theory numbers, diophantine equations and inequalities. Home: 8 Cranmer Rd., Cambridge, Eng.*

DAVENPORT, Horace Willard, Am. physiologist; b. Phila., Oct. 20, 1912; s. Horace W. and Elizabeth (Langendorf) D.; B.S., Cal. Inst. Tech., 1935, Ph.D., 1939; B.A., Oxford U., 1937, B.S., 1938, M.A., 1960, D.Sc., 1961; m. Virginia Dickerson, Feb. 1, 1945; children—Thomas L., Robertson D. Lilly fellow pathology U. Rochester Med. Sch.; 1939-40; Sterling fellow physiol. chemistry Yale, 1940-41; faculty U. Pa., 1941-43, Harvard, 1943-45; prof., head dept. physiology U. Utah Med. Sch., 1945-56; prof., chmn. U. Mich. Med. Sch., Ann Arbor, 1956-—. Mem. adv. com. biology, medicine NSF, 1960-62; vis. prof. Mayo Found., 1962-63; cons. govt. agys. Mem. Am. Physiol. Soc. (editorial bd.), Soc. Exptl. Biology and Medicine (editorial bd.). Author: The ABC of Acid-Base Chemistry, 1947; Physiology of the Digestive Tract, 1961. Research, numerous publs. primarily in field of physiology of gastric mucosa. Home: 1918 Day St., Ann Arbor, Mich. 48104.*

DAVENPORT, Thomas, Am. inventor; b. Williamstown, Vt., July 9, 1802; s. Daniel and Hannah (Rice) D.; m. Emily Goss, Feb. 14, 1827. Blacksmith; publisher tech. jour. in N.Y.C., 1839-43. Experimented with electromagnets, circa 1831; constructed his 1st motor, 1834; built small circular ry. (1st electric ry. on record), 1835; invented early model of electric train motor, 1836, patented, 1837, unable to improve and market because of financial difficulties; patented 2d model, 1837; devised 1st electric printing press, 1839. Died Salisbury, Vt., July 6, 1851.

DAVENPORT, William Hunt, Am. anthropologist; b. Upland, Cal., May 26, 1922; s. Milton William and Nelle (Hunt) D.; B.A., U. Hawaii, 1952; Ph.D., Yale, 1956; m. Pearl Yim, Apr. 12, 1952. Instr. anthropology Yale, 1955-58, asst. prof., 1960-63; asso. anthropology Bishop Mus., Honolulu, 1958-60; asso. prof. anthropology U. Mus., U. Pa., Phila., 1963-67, prof., 1967-—. Contbr. monographs and articles on Pacific peoples, Caribbean peoples and social structure to tech. jours. Home: 4411-2A Walnut St., Phila. 19104.*

DAVEY, Harold William, Am. economist; b. Syracuse, N.Y., Sept. 8, 1915; s. William R. P. and Elizabeth (Cleasby) D.; B.A. summa cum laude, Syracuse U., 1936; M.A., Harvard, 1938, Ph.D., 1939; m. Maririta Lauterjung, May 15, 1942; 1 son, William R. Instr., Ill. Inst. Tech., 1939-42; placement, hearing officer, panel chmn. Nat. War Labor Bd., Washington, Chgo., N.Y.C., 1942-45; asst. prof., dir. Inst. Econ. Affairs, N.Y. U., 1945-48; asso. prof. Ia. State U., 1948-50, prof. econs., 1950-—; vis. prof. U. Minn., 1959; vis. Fulbright prof. U. Bologna, Italy, 1959-60. Profl. arbitrator labor disputes, 1944-—. Mem. Am., Midwest (pres. 1964-65), econs. assns., Nat. Acad. Arbitrators (bd. govs. 1958-61), Indsl. Relations Research Assn., Phi Beta Kappa, Phi Kappa Kappa, Phi Kappa Phi. Author: Contemporary Collective Bargaining, 1951, 2d edit., 1959; also articles on collective bargaining process and on arbitration of labor disputes. Home: 416 Oliver Av., Ames. Ia. 50010.*

DAVEY, Henry, Brit. inventor; b. 1843; s. Jonathan and Mary Davey; ed. Tavistock, Eng.; m. Elizabeth Barbenson Le Ber; 1 son. Founder firm Hathorn, Davey & Co., Hydraulic Engrs., Leeds, Eng., 1871; became cons. engr., London, 1887. Mem. Instn. Civil Engrs. (Watt medal, Telford premium). Author: The Principles and Construction of Pumping Machinery (textbook); also papers. Inventor differential pumping engine, also improvements in hydraulic machinery. Died Apr. 11, 1928.

DAVEY, John, agriculturist; b. Somersetshire, Eng., June 6, 1846; s. Samuel and Ann (Shopland) D.; m. Bertha A. Reeves, Sept. 21, 1879. Learned floriculture and landscape architecture, Torquay, Eng., 1866-72;

came to U.S. and settled at Warren, O., 1873; moved to Kent, O., 1881; pres. Davey Tree Expert Co. and Davey Inst. Tree Surgery (school), 1908——. Author: The Tree Doctor, 1901; A New Era in Tree Growing, 1905; Davey's Primer on Trees and Birds, 1905; Instruction Books on Tree Surgery and Fruit Growing, Nos. 1-23, 1914. Introduced tree surgery, 1890. Died Nov. 8, 1923.

DAVEY, Wheeler P(edlar), Am. physicist, chemist; b. Cleve., Mar. 19, 1886; s. Thomas George and Myra Eliza (Christian) D.; A.B., Western Res. U., 1906; M.S., Pa. State Coll., 1911; grad. study U. Chgo.; Huntingdon fellow, Cornell U., 1912-13, Ph.D., 1914; m. Laura L. Gunn, Aug. 28, 1912; children—Myra Ellen, George Thomas, Ruth Barton, Mary Louise, Tchr. physics and chemistry, Central Inst., Cleve., 1906-08; tchr. high sch., Mansfield, O., 1908-09; instr. in physics, Pa. State Coll., 1909-11; asst. in physics, Cornell U., 1911-12, instr. in physics, 1913-14; research physicist research lab. Gen. Elec. Co., Schenectady, 1914-25; prof. phys. chemistry and prof. indsl. research, Pa. State Coll., 1926-31; research prof. physics and chemistry, 1931-49. research prof. emeritus 1949-59. Chmn. Schenectady Civil Service Commn. 1916-17. Lectr. on X-rays and crystal structure, Union U., 1920-26; lectr. in physics dept., Grad. Sch., Pa. State Coll., summers, 1922, 23, 24; lectr. on crystal structure, U. Mich., summer, 1925; Thurston lectr., Am. Soc. M.E., 1928. Mem. elec. insulation com. NRC, 1928-41, chmn. physics sub-com., 1935-40. Mem. optics sub-com., Century of Progress Expn.; mem. Am. Inst. Physics Council of Applied Physics, 1935-38. Mem. editorial bd. Jour. of Chem. Physics, 1933. Award of merit, Am. Soc. Testing Materials, 1952. Fellow Am. phys. Soc., Inst. of Physics (London); mem. gov. bd. Am. Inst. of Physics, 1931-33 and 1937-41; Am. Soc. Testing Materials, Am. Soc. for Metals, Soc. Rheology (pres. 1930-33; asso. editor, 1933-36, editor, 1936-41) Sigma Xi, Phi Lambda Upsilon, Sigma Pi Sigma (chmn. placement bd., 1934-45), Alpha Pi Mu. Mem. joint com. (A.S.T.M., I.P. and Am. Chrystallographic Soc. on chem. analysis by X-ray diffraction methods, 1940-59, chmn. 1940-56. Contbr. to Fairbanks' Laboratory Investigation of Ores, 1928. Author: A Study of Crystal Structure, 1934; also articles on X-rays, crystal structure, automatic X-ray diffraction apparatus, criterions for rating of physics depts., planning new physics bldgs. Died Oct. 12, 1959.

D'AVEZAC DE CASTERA MACAYA, Marie Armand Pascal, French geographer; b. Bagneres de Bigorre, France, Apr. 18, 1800; licienté Faculty Law, Paris; m. Françoise Dumas de Boucheron; 1 son, 1 dau. (Mrs. Charles Defremery). Custodian archives Minister Marine, 1843-60; dir. 2d cabinet bur. of ministry, from 1861. Mem. Société Ethnographique (founder), Société de géographie (sec.-gen.). Author: Essais historiques sur la Biggore, circa 1823; Etudes de géographie critique sur une partie de l'Afrique septentrionale, 1836; Equisse générale de l'Afrique et de l'Afrique ancienne, 1844; Notice sur le pays et le peuple des Yebous, en Afrique, 1845; Notice sur les découvertes faites au Moyen Age dans l'océan Atlantique, 1845; Iles de l'Afrique, 1848; Coup d'oeil historique sur la projection des cartes de géographie, 1863; Le livre de Fernand Colomb, 1873. Editor works on voyages John of Plan-Carpin and Jacques Cartier. Collaborator, Encyclopédie moderne, Revue des deux mondes, Encyclopédie des gens du monde, Nouvelles annales des voyages. Studied ancient and medieval geography, discoveries of 15th and 16th centuries, portolanos, mappemondae and terrestrial globes of Middle Ages. Died Paris, Jan. 14, 1875.

DAVID, Armand, French naturalist; b. Espelette, France, September 7, 1826; ed. Lazarist Sch., 1848; became mem. Congregation of Mission, 1848; ordained in Catholic Ch., 1862; prof. in Italy, 1851-62; sent as missionary priest to Mongolia, 1866, then to Peking, China; developed sci. mus., Peking; Mem. French Acad. Scis. Author: les Oiseaux de la Chine, 1877; Plantae Davidianae (catologue of his bot. collections), 1890; Voyage en Mongolie, 1875; Journal de mon troisième voyage d'exploration dans l'Empire chinois, 2 vols., 1875. Sent rare specimens of animals and plants from China to France; identified 200 species of wild animals in China (including 63 new species, and 800 bird species, 65 of them new; collected reptiles, fish and insects; discovered 52 new species of rhododendron, also many new species of gentian. Died Paris, France, Nov. 10, 1900.

DAVID, Cornel, physician, Rumanian gerontologist; b. Orastie, Rumania, Aug. 13, 1912; s. Cornel and Aurora (Balta) D.; D. in Medicine and Surgery, Faculty Medicine, Cluj, Rumania, 1935; m. Sanda Caterina Constantinescu, Sept. 19, 1955. Humbold Stiftung scholar for specializing in internal medicine and cardiology at clinics in Munich, Frankfurt/Main, Berlin, Bad-Nauheim, Germany, 1936-38; hon. asst. cardiology Clinic Paris, France, 1938-39; asst. prof. Med. Clinic, Bucharest Coltea Hosp., 1947-52; 1st asst., chef de traveaux et medicin des Hopitaux, Bucharest (Rumania) Coltea Hosp., 1947-52; div. dir., chief physician clin. sect. Inst. Geriatrics, Bucharest, 1952——; dir. Polyclinic Ministry of Agr., Bucharest, 1949-52; asst. dir. Hosp. Ministry Constrn., Bucharest. Decorated Star of Rumanian Socialist Republic, Liberation from Fascism medal; recipient Distinction in med. san. work award, 1961. Mem. Rumanian Gerontological Soc., Internat. Assn. Gerontology, French Cardiology Soc. World Union Prophylactic Medicine and Social Hygiene (Austria), German Assn. Stomatology and Biology. Author: (with B. Theodorescu) Rumanian Treatise on Cardiology, 1960; also numerous articles. Research on valuation of biologic age, biologic study of longevity and premature ageing, articular and bone diseases of the aged, teatment of old age by biotrophic substances, role of nervous sytem, hypothalamus in mechanism of ageing, antagonistic actions of thyroid hormone and vitamin A, regulation of venous circulation, factors influencing circulation time; introduced new methods for investigating psychic activity of aged and hypothalamus activity. Home: 3 Bd. Dacia. Office: 9 Str. Minastiea Caldarusani, Bucharest, Rumania.*

DAVID, Sir (Tannatt William) Edgeworth, Brit. geologist; b. St. Fagans, nr. Cardiff, Wales, Jan. 28, 1858; s. William David; ed. New Coll., Oxford; D.Sc. (hon.), 1911; hon. doctorates Cambridge U., 1926, Wales, 1921, St. Andrews, 1926; m. Caroline M. Mallet, 1886; 1 son, 2 daus. Became asst. geol. surveyor to govt. of New S. Wales, Australia, 1882; prof. geology Sydney (Australia) U., 1891-1924. Leader expdn. to Ellice Islands, 1897; sci. officer Shackleton Antarctic Expdn., 1907-09; leader party which reached S. Magnetic Pole, Jan. 16, 1909; mem. Geol. Survey, New S. Wales; hon. fellow New Coll., Oxford, 1926. Recipient Bigsby medal, 1899; Mueller medal Australian Assn. for Advancement Sci., 1909; Conrad Malte-Brun prize Geog. Soc. France, 1915; Wollaston medal Geol. Soc. London, 1915; Clark medal Royal Soc. New S. Wales, 1917; Patron's medal Royal Geog. Soc., 1926. Fellow Royal Soc., 1900. Author: Geological Notes British Antarctic Expedition, 1907-09; The Geology of Australia, 1932; also articles. Contbg. author: The Heart of Antarctica (Shackleton). Leader in investigation of permo-carboniferous glacial phenomena in Australia; mapped Sydney-Newcastle basin; studied origins of coral atolls in Funafuti. Died Aug. 28, 1934.

DAVID, Heinz Werner, German pathologist; b. Tilsit, Germany, Dec. 5, 1931; s. Werner and Edith (Lütke) D.; Dr.Med., Humboldt U., Berlin, 1955, Dr.-Med.Habil., 1960; m. Sabine Richter, Jan. 4, 1964; 1 dau., Christine. Asst. physician Charité, Humboldt U., Berlin, 1955-59, pathologist, 1959——, head dept. electron microscopy, 1959——, asso. chief physician, 1961——, prof., 1965——. Recipient Rudolf Virchow prize, 1960; Mem. German Soc. Pathology, German Soc. Electron Microscopy, Soc. Morphology. Author: Die Leber bei Nahrungsmangel und Mangelernährung, 1961; Submicroscopic Ortho- and Pathomorphology of the Liver, 1964; Elektronenmikroskopische Organpathologie, 1967. Research on pathology and electron microscopy of liver and other organs and of the cell, molecular biology. Address: 35 Kottmeierstrasse, 116 Berlin, East Germany.*

DAVID, Herbert Aron, statistician; born Berlin, Germany, Dec. 19, 1925; s. Max and Betty (Goldmann) D.; B.Sc., Sydney U., Australia, 1947; Ph.D., London U., 1953; m. Vera Reiss, May 13, 1950; 1 son, Alexander John. Came to U. S., 1957, naturalized 1964. Sr. lectr. U. Melbourne, Australia, 1955-57; prof. statis, Va. Poly. Inst., Blacksburg, 1957-64; prof. biostatistics, U. N.C., Chapel Hill, 1964——. Recipient Horsley award Va. Acad. Sci., 1963. Fellow, Am. Statis. Assn., Inst. Math. Statistics, A.A.A.S. Asso. editor Jour. Am. Statist. Assn., 1964-66; editor Biometrics, 1967——. Research, publs. on order statistics. Home: 20 Rolling Rd., Chapel Hill, N.C. 27514.*

DAVID, Jean-Pierre, French physicist, physician; b. Gex, France, Feb. 17, 1737; s. Claude-François; studied medicine Lyons, France; docteur en médecine, Paris, France; m. Miss Lecat, 1766; 8 children. Master surgeon, Paris; prof. chemistry and anatomy U. Rouen (France); lithotomist for Province of Normandy, France; head surgeon Hôtel-Dieu, Rouen. Honored (for study on mechanism of respiration) by Rouen Acad., 1765. Mem. Acad. Scis. Rouen. Author: Dissertation sur la cause de la pesanteur, 1767; Dissertation sur la figure de la terre, 1769; Réflexions sur la force centripète et sur la force centrifuge, 1782; Idées sur la cause du mouvement, 1782. Research on medicine, mechanics, astronomy, geo-phyiscs; described vertebral Tb with bumps and paralysis. Died Aug. 21, 1784.

DAVID, Norman Austin, Am. physician; b. San Francisco, Oct. 22, 1902; s. Louis Nathaniel and Ida Grace (Holiday) D.; A.B., U. Cal. at Berkeley, 1925; M.D., U. Cal., 1931; m. Muriel Stott, July 11, 1931; children—Diane Margaret, Carol Elinor, Jeanne Louise (dec.). Eli Lilly Research fellow, dept. pharmacology, U. Cal. Med. Sch., San Francisco, 1930-32; asst. prof. pharmacology U. W.Va. Sch. Medicine, Morgantown, 1932-35; asso. prof. Coll. Medicine, U. Cin., clinician N. Permanente Hosp., Vancouver, Can., 1944-45; asso. prof. pharmacology, physiology Dental Sch., U. of O., 1943-46; prof. pharmacology U. Ore. Med. Sch., chmn. dept., dir. student health service; cons. toxicologist. Mem. A.M.A., Amer. Soc. Pharmacology, Portland Acad. Medicine, Sigma Xi, Phi Beta Pi. Contbr. articles profl. jours. Work on pharmacology of barbital drugs, chronic effects of opium drugs, synthetic analgesics. Home: 4206 N. E. Glisan St., Portland, Ore.

DAVID, Serge Michel Joseph, French chemist; b. Grenoble, France, Nov. 6, 1921; s. Joseph J. and Valerie (Seaume) D.; ed. Ecole Normale Superieure, Paris; D.Phil., Magdalen Coll., Oxford, Eng.; Dr. ès Sci., Paris; m. Georgette Potier, Sept. 7, 1949; children—Bruno, Dominique, Claire. Lectr., Faculté des Sciences, Nancy, 1949-59, prof. organic chemistry, 1959-63; prof. organic chemistry Faculté des Sciences, Orsay, France, 1963——. Mem. Am. Chem. Soc., Société Chimique de France. Research, publs. on organic chemistry of natural products, biosynthesis of ribose, deoxyribose and thiamine; preparation of semisynthetic antigens for immunochemistry expts. Home: 29 Rue Maginot, Orsay 91, France.*

DAVID DE POMIS, see de Pomis, David.

DAVID KHAN DAVDIANTZ, physician; b. Armenia, 1786; with Calcutta (India) Mil. Sch.; physician in Brit. Army in India; court physician to the Persian king Feth-Ali. Created Khan. First to study a cholera epidemic, 1816. Died Teheran, Iran, 1851.

DAVIDÉNKOV, Sergéi Nikoláevich, Russian neuropathologist; b. Sept. 6, 1880; ed. Moscow U.; became prof. nervous and mental diseases Kiev Women's Med. Inst., 1912; with Institut Usovershenstvovania Vrachei, Leningrad Grad. Sch. Medicine, 1925——. Mem. USSR Acad. Med. Scis. Author: Materials on the Study of Aphasia, 1915; Defense Reflexes, 1918; Evolutionary and Genetic Problems of Neuropathology, 1947; Hereditary Diseases of the Nervous System, 1932; The Problem of Polymorphism in Hereditary Diseases of the Nervous System, 1934; Clinical Lectures on the Nervous Diseases, Numbers 1-3, 1952-57. Described many diagnostic neuropath. symptoms, during 1920's; described two-wave meningoencephalitis virus, 1949.

DAVIDON, William Cooper, Am. physicist; b. Ft. Lauderdale, Fla., Mar. 18, 1927; s. Jack and Ruth (Simon) D.; student Purdue U., 1943-44; B.S., U. Chgo., 1947, M.S., 1950, Ph.D., 1954; m. Phillis Wise, June 8, 1947 (div. June 1953); 1 son, Alan; m. 2d, Ann Morrissett, Dec. 6, 1963; children—Ruth, Sarah. Began career as research director of Nuclear Chicago, 1948-54; research associate at Fermi Inst., U. Chgo., 1954-56; asso. physicist Argonne (Ill.) Nat. Lab., 1956-61; chmn. physics dept. Haverford (Pa.) Coll., 1961-64, 65-66; Fulbright research fellow Aarhos U., Denmark, 1966-67. vis. asso. prof. U. Wash., Seattle, 1958. Named one of Ten Outstanding Young Men of Year for nuclear sci. Chgo. Jr. C. of C., 1960. Mem. Fedn. Am. Scientists (past v.p.), Soc. for Social Responsibility in Sci. (pres. 1965——), Am. Phys. Soc., Am. Math. Soc., Am. Assn. Physics Tchrs., Am. Assn. U. Profs., Sigma Xi. Contbr. articles to tech jours. Developed formulation quantum electrodynamics, theory interactions between electrons and electromagnetic waves; research on pi meson, nucleon scattering, fundamentals relativistic quantum mechanics. Home: 7 College Lane, Haverford, Pa. 19041.*

DAVIDS, Anthony, Am. psychologist; b. Providence, Aug. 28, 1923; s. William J. and Louise (Nahigan) D.; A.B. magna cum laude, Brown U., 1949; A.M., Harvard, 1951, Ph.D., 1954; m. Martha J. St. Germain, Sept. 17, 1949. Research asso., lectr. Harvard, 1953-55; faculty Brown U., Providence, 1955——, prof. psychology, 1964——; chief psychologist Emma Pendleton Bradley Hosp., Riverside, R.I., 1955-64, dir. psychology, 1965——. USPHS Spl. Research fellow Inst. Personality Assessment and Research, U. Cal. at Berkeley, 1963-64. Fellow Am. Psychol. Assn., A.A.-A.S., Am. Orthopsychiat. Assn., Soc. for Projective Techniques; mem. Soc. for Research in Child Devel., Am. Psychosomatic Soc., Psychonomic Soc., Phi Beta Kappa, Sigma Xi. Adv. editor Jour. Cons. Psychology, 1964——, also of Contemporary Psychology, 1968——. Research and publications on personality assessment of normal and abnormal children and adults, intellectual and personality factors related to acad. achievement, psychol. factors in pregnancy, childbirth and early mother-child relations, cognitive functioning in emotionally disturbed children in residential treatment and normal children in pub. schs. Home: 218 Burgess Av., East Providence, R.I. Office: Psychology Dept., Brown U., Providence.*

DAVIDSOHN, Israel, pathologist; born Tarnopol, Austria, Apr. 20, 1895; s. Jacob and Rachel (Halpern) D.; M.D., U. Vienna, 1921, postgrad.; postgrad. U. Berlin; Sc.D.(hon.), Chgo. Med. Sch., 1964; m. Clara Freud, Oct. 10, 1923; children—Samuel James, Ellen Doris (Mrs. Langdon E. Morris, Jr.). Came to U. S., 1923, naturalized, 1930. Fellow Research Inst. Cutaneous Medicine, Phila., 1926-30; dir. dept. pathology Mt. Sinai Hosp., Chgo., 1930-65, dir. dept. exptl. pathology, 1965——; faculty Rush Med. Coll., 1934-41, U. Ill. Coll. Medicine, 1941-47; prof. pathology, chmn. dept. Chgo. Med. Sch., 1947——. Cons. Armed Forces Inst. Pathology, Washington, 1947——. Diplomate Am. Bd. Pathology (trustee 1952——, past pres.). Mem. Coll. Am. Pathologists (past gov.), Am. Soc. Clin. Pathologists (past pres. Gold medal 1943, Ward Burdick award 1954), Am. Assn. Blood Banks (past pres.). Co-editor: A Curriculum for Schools Medical Technology, 1964; Clinical Diagnosis by Laboratory Methods (Todd-Sanford), 13th edit., 1962; editor Am. Jour. Clin. Pathology, 1943-46, asso. editor, 1947——. Research, numerous publs. on Forssman antigen and antibody, differential test for infectious mononucleosis, blood groups, Rh Factor, bacteriogenic hemagglutination, fetal erythroblastosis, exptl. hemolytic anemia in rabbits, mice and guinea pigs, exptl. thrombocytopenic purpura, test for lupus erythematosus cell, effect prolonged immunization, cancer and blood groups. Home: 6927 Oglesby Av., Chgo. 60649.*

DAVIDSON, Anstruther, physician, botanist; b. Watten, Scotland, Feb. 19, 1860; s. George and Ann (Macadam) D.; M.D., C.M., U. Glasgow, 1881, M.D., 1887; came to U. S., 1889; in med. practice, Los Angeles, 1889-1932; m. Alice Merritt, June 24, 1897; children—Ronald A., Merritt T. Practice medicine, Los Angeles, from 1889; asso. prof. dermatology U. So. Cal., from 1909. Fellow So. Cal. Acad. Scis. (pres. 1893-94, editor Bull. from 1902). Author: Plants of Los Angeles County, 1892; Flora of Southern California, 1923. Research in botany and entomology; described several new plant species. Died April 3, 1932.

DAVIDSON, Ben, Am. meteorologist; b. Walden, N.Y., Sept. 6, 1916; s. Maurice and Elizabeth (Klevin) D.; B.A., N.Y. U., 1947, M.S., 1949, Ph.D., 1959; m. Silvia Kimmel, Apr. 13, 1943; children—Elizabeth, Irene, Sarah. Meteorologist, Air Force Cambridge Research Center, 1949-55; research scientist to prof. meteorology N.Y. U., Bronx, 1955—; tng. grants com. USPHS; mem. rev. com. div. radiol. physics Argonne Nat. Labs. Mem. Am. Meteorol. Soc., Am. Geophys. Union, Air Pollution Control Assn., Am. Pub. Health Assn. Author: (with H. Lettau) Exploring the Atmosphere's First Mile, 1957. Research in structure of atmospheric turbulence, theory and observation of valley winds, diffusion of particulate matter, diffusion of radioactive debris, urban meteorology and diffusion of pollutants. Home: 2802 Webb Av., Bronx, N.Y. 10468.*

DAVIDSON, Charles Findlay, Scottish geologist; b. Monifieth, Scotland, July 16, 1911; s. John and Elizabeth (Findlay) D.; B.Sc. with honors, U. St. Andrews, 1933, D.Sc., 1942; m. Helen MacLean Wallace, June 1, 1938; children—John Findlay, Martin MacLean, Charles Stewart, Alistair Graham. With geol. survey of Gt. Britain, also Geol. Mus., London, 1934-55, chief geologist to Brit. atomic energy orgns., 1941-55; prof. geology U. St. Andrews, Fife, Scotland, 1955—, U. Ct., 1961-66. Distinguished vis. lectr. Am. Geol. Inst., 1963. Decorated officer Order of Brit. Empire, 1953; Lyell medal Geol. Soc. London, 1965. Mem. Soc. Econ. Geologists (v.p. for Europe 1955-57), Royal Soc. Edinburgh (v.p. 1964-66), geol. socs. of London, Am., Belgium, Norway, mineral. socs. of Gt. Britain, Am., Can., Soc. Applied Geology, Am. Geophys. Union, Inst. Mining and Metallurgy, Geochem. Soc., Internat. Assn. on Genesis of Ore Deposits. Author: A Prospector's Guide to Radioactive Mineral Deposits, 1949; also numerous articles. Editor, Econ. Geology, 1954—, Applied Earth Sci., 1965—. Research to discover, interpret ore deposits. Home: Gowan Park, Cupar, Fife, Scotland.*

DAVIDSON, Charles S(precher), Am. physician; b. Berkeley, Cal., Dec. 7, 1910; s. Charles Sprecher and Mary (Blossom) D.; A.B., U. Cal. at Berkeley, 1934; M.D., C.M., McGill U., 1939; M.A. (hon.), Harvard, 1953. Intern, house officer medicine San Francisco Hosp., 1939-41; research fellow medicine Harvard Med. Sch. and asst. resident physician Thorndike Meml. Lab., Boston City Hosp., 1941-42; various appointments, 1942-44; associate director of II and IV Harvard Medical Services, Boston City Hospital and asso. physician Thorndike Meml. Lab., 1948—, associate director, 1964—, vis. physician Boston City Hosp., 1965—; asso. prof. medicine Harvard Medical School, 1953—; consultant Veterans Administration Hospital, Boston, Lemuel Shattuck Hospital, Boston, Cambridge City Hospital. Mem. food and nutrition bd. Nat. Research Council; mem. com. metabolism, subcom. on liver Office Surgeon Gen., U. S. Army; sci. adv. com. United Health Founds., Inc. Fellow A.C.P.; mem. Am. Gastroenterological Assn., A.M.A. (council foods and nutrition), Am. Soc. Clin. Investigation, Am. Fedn. Clin. Research, Assn. Am. Physicians. Mem. editorial bd. Archives of Internal Medicine, A.M.A.; Nutrition Rev. Contbr. profl. jours. Research on liver diseases and metabolism in man; nutrition. Home: 100 Memorial Dr., Cambridge 42, Mass. Office: Thorndike Meml. Lab., Boston City Hosp., Boston 18.

DAVIDSON, Eugene A., Am. chemist; b. N.Y.C., May 27, 1930; s. Jack and Sophie (Deutsch) D.; B.S., U. Cal. at Los Angeles, 1950; Ph.D., Columbia, 1955; m. Alice Howell, Jan. 27, 1952; children—Mark, Robin, Steven, Ellen. Research asso. U. Mich., 1955-58; faculty biol. chemistry Duke, 1958-67, prof., 1965-67; prof., chmn. dept. biol. chemistry Milton S. Hershey Med. Center, Pa. State University College of Medicine, Hershey, 1967—. Consultant National Cancer Inst., 1959-62, USPHS, 1963—. Guggenheim fellow, 1965-66. Mem. Am. Assn. U. Profs., Am. Chem. Soc., Am. Rheumatism Assn., Am. Soc. Biol. Chemists, Biochem. Soc. (London), Sigma Xi. Author: Contemporary Carbohydrate Chemistry, 1967. Studies on structure of connective tissue; metabolism of complex structural polysaccharides and their components and relation to biol. role. Home: 198 Governor Rd., Hershey, Pa. 17033.*

DAVIDSON, George, geodesist, astronomer; b. Nottingham, Eng., May 9, 1825; s. Thomas and Janet (Drummond) D.; came to U. S., 1832; A.B., Central High Sch., Phila., 1845, A.M., 1850; PhD., Santa Clara Coll., 1876; Sc.D., U. Pa., 1889; LL.D., U. Cal., 1910; m. Ellinor Fauntleroy, Oct. 5, 1858. Sec. to A. D. Bache, supt. Coast Survey, 1845-46; mem. U. S. Coast and Geod. Survey, 1845-95; in geod. field and astron. work in Eastern states, 1845-50; in coast survey work of Cal., Ore., Wash., Alaska, 1850-95, in charge Pacific Coast work, 1868-95; hon. prof. geodesy and astronomy U. Cal., from 1870, regent, 1877-84, prof. geography, from 1898. Expert U. S. mints, Phila., San Francisco, 1872, 85, 86; mem. U. S. Irrigation Commn. Cal., 1873-74, India, Egypt, 1875; mem. U. S. Adv. Bd. Harbor Improvement, San Francisco, 1873-76; in charge Transit of Venus Expdn. to Japan, 1874, to N.M., 1882; mem. Miss. River Commn., 1888-90; spl. agt. U. S. 9th Internat. Geod. Congress, Paris, 1889. Recipient medal Paris Expn., 1878; Daly gold medal Am. Geog. Soc., 1908; decorated Order of St. Olaf, Norway, 1907. Fellow Am. Acad. Arts and Scis., A.A.A.S.; mem. French Acad. Scis., 1901. Author: Directory for the Pacific Coast of the United States, 1858; Coast Pilot of Alaska, 1869; California, 1887; The Discovery of San Francisco Bay, 1907; Francis Drake on the Northwest Coast of America, 1908; The Origin and the Meaning of the Name California, 1910. Made astron. observations on variations of latitudes; established own observatory in San Francisco (1st in California), 1879; studied total eclipses of sun, transit of Mercury; land reclamation in China, India, Egypt, and Europe. Died Dec. 2, 1911.

DAVIDSON, George, Brit. entomologist; b. Glasgow, Scotland, Aug. 28, 1917; s. Alfred and Elizabeth (Frazer) D.; B.Sc. with honors in Zoology, Liverpool, Eng., 1939, D.Sc., 1966; m. Edna Hall, Nov. 8, 1950. Staff, Ross Inst. Tropical Hygiene, London (Eng.) Sch. Hygiene and Tropical Medicine, 1946—, reader, 1964—. Cons. WHO, 1955, 57, 60, 61, 64, mem. sci. group on vector genetics and insecticide resistance, 1963. Fellow Royal Soc. Tropical Medicine and Hygiene, Royal Entomol. Soc. London, Genetical Soc. Research, publs. on anopheline mosquitoes including their behavior, resistance to insecticides, their role as malaria carriers. Office: Keppel St, Gower St., London W.C. 1., Eng.*

DAVIDSON, J. Brownlee, Am. agrl. engr.; b. Douglas, Neb., Feb. 15, 1880; s. James H. and Margaret Jane (Dickson) D.; B.S., M.E., U. Neb., 1904, A.E., 1914, D.Eng., 1931; m. Jennie Baldridge, June 14, 1906; children—Margaret Elizabeth, Ethel Brownlee, James Vincent (dec.), Helen Mary (dec.). Instr. farm mechanics, U. Neb., 1904-05; asst. prof. to prof. agrl. engring., Ia. State Coll. Agr. and Mech. Arts, 1905-15; prof. agrl. engring., U. Cal., 1915-19, Ia. State Coll. Agr. and Mech. Arts, 1919-56. Dir. Survey of Research in Mech. Farm Equipment for U. S. Dept. Agr. (on leave), 1926; mem. Am. Com. on Colonization, Russia (on leave), 1929; cons. farm equipment mfrs.; Hon. mem. Am. Soc. Agrl. Engrs. (chmn. com. History); life mem. Am. Soc. Engring. Edn., Ia. Engring. Soc.; "Estranger" mem. Swedish Royal Agrl. Soc.; Sigma Xi, Phi Kappa Phi, Gamma Sigma Delta, Sigma Tau (nat. pres.), Tau Beta Pi, Alpha Zeta; Assn. of College Honor Socs. (mem. council). Author: Farm Machinery and Farm Motors, 1908; Agricultural Engineering, 1913; Agricultural Machinery, 1931; (with others) A Study of Extension Service, 1933; Report of an Inquiry into Changes in Quality Values of Farm Machines, 1933; farm equipment cons. WPB, 1943, UNRRA, 1944. Chmn. Com. on agrl. engring., Ministry of Agr. and Forestry, China (on leave, 1946-48). Recipient of Cyrus Hall McCormick medal for achievement in engring. of agr., 1933. Died May 8, 1957.

DAVIDSON, James Norman, Scottish biochemist; b. Edinburgh, Scotland, Mar. 5, 1911; s. James and Wilhelmina (Foote) D.; B.Sc., Edinburgh (Scotland) U., 1934, M.B., Ch.B., 1937, M.D., 1939, D.Sc., 1945; m. Morag McLeod, July 27, 1938; children—Rona McLeod (Mrs. Euan MacKie), Ailsa Morag (Mrs. Tom Campbell). Carnegie Research fellow Kaiser Wilhelm Institut für Zellphysiologie Berlin-Dahlem, Germany, 1937-38; lectr. biochemistry U. St. Andrews (Scotland), 1938-40, U. Aberdeen (Scotland), 1940-45; mem. staff Nat. Inst. Med. Research, Hampstead, London, Eng., 1945-46; prof. biochemistry U. London, St. Thomas' Hosp. Med. Sch., 1946-48; Gardiner prof. biochemistry, head dept. biochemistry U. Glasgow (Scotland), 1948—. Fellow Royal Soc. (London), Royal Inst. Chemistry, Royal Coll. Physicians Edinburgh, Royal Soc. Edinburgh (pres. 1964-67); mem. Assn. Clin. Biochemists (past pres.), Biochem. Soc. (past chmn.). Author: The Biochemistry of the Nucleic Acids, 1965; (with G. H. Bell, H. Scarborough) Textbook of Physiology and Biochemistry, 1965; also numerous articles. Editor: (with W. E. Cohn) Progress in Nucleic Acid Research, vols. 1-7, 1963-67. Research on biochemistry of nucleic acids; 1 of first in U.K. to use isotopes and tissue culture in biochem. investigations. Home: 6 Ledcameroch Rd., Bearsden, Glasgow, Scotland.*

DAVIDSON, Thomas, paleontologist; b. Edinburgh, Scotland, May 17, 1817; studied at U. Edinburgh, 1835-36; also univs. in France, Italy and Switzerland; LL.D., U. St. Andrews (Scotland), 1882. Fellow Royal Soc., 1857. Wollaston medal Geol. Soc. London, 1865, Royal medal Royal Soc., 1870. Author: Monograph of British Fossil Brachiopoda, 6 vols., 1850-86. Research, collections, and classification on brachiopods. Died London, Eng., Oct. 14, 1885.

DAVIDSON, William, see Davison, William.

DAVIDSON, William L(ee), Am. physicist; b. Jonesville, Va., July 14, 1915; s. William Lee and Zelma (Albert) D.; B.S., Coll. of William and Mary, 1936; Ph.D., Yale, 1940; m. Miriam Alloway, Dec. 20, 1941. Research physicist B. F. Goodrich Co., Akron, O., 1940-46, Brecksville, O., 1947-48, dir. phys. research, 1948-52; trainee Monsanto Chem. Co., Oak Ridge, Tenn., 1946-47; became dir. Office Indsl. Development, AEC, Washington, 1952; director Central Research Lab., Chem. divs. Food Machinery & Chem. Corp., 1958-59, coordinating mgr., propellants, 1959-60; chmn. bd., chief exec. officer Trans-Sil Corp., 1960-61; asst. to v.p. research FMC Corp., 1961-64, mgr. central devel. chem. divs., 1964—. Vice chmn. nuclear energy com. N.A.M. Recipient Distinguished Service award, Ohio Jr. C. of C., 1949. Fellow Am. Inst. Chemists, Am. Phys. Soc.; mem. Armed Forces Chem. Assn., American Chem. Soc. Phi Beta Kappa, Sigma Xi, Sigma Pi Sigma, Phi Kappa Phi. Author: Applied Nuclear Physics (with E. Pollard), 1942. Contbr.: Physical Methods in Chemical Analysis (Berl), 1950; also articles in chem. and phys. jours. Mem. Pollard's group which constructed cyclotron of moderate size, 1940, for investigation of energy levels in nuclei through study of energy of particles emitted in transmission reactions (including bombardment of boron, carbon and aluminum by deuterons); worked on X-ray diffraction, radio trace research, self-sealing fueling tanks for war use. Home: 125 Prospect St., Stamford, Conn. Office: 633 3d Av., N.Y.C. 17.

DAVIEL, Jacques, French ophthalmologist; b. La Baree, Normandy, France, Aug. 11, 1696; studied surgery U. Rouen, Paris (both France); practiced medicine specializing in ophthalmology, Paris; named oculist to king, 1749. Mem. French Royal Acad. Surgery. Author: Lettres sur les malades des yeux, 1748; Mémoire sur une nouvelle méthode pour guérir la cataracte par extraction. Developed modern method of treating cataract by removing the lens. Died Geneva, Switzerland, Sept. 30, 1762.

DAVIES, David Allen Lewis, Brit. biochemist; b. Newport Pem, Wales, Mar. 18, 1923; s. David and Francis Mary (Lewis) D.; M.A. (scholar) Cambridge (Eng.) U., 1948; Ph.D., London (Eng), 1953, D.Sc., 1958; m. Ilse Muller, Sept. 11, 1949; children—Detlef, Peter, Juliet. Microbiol. research staff Govt. Service, 1949-64, sr. prin. sci. officer, 1958-64; staff Med. Research Council, Queen Victoria Hosp., East Grinstead, Sussex, Eng., 1964-66; chief immunology div. Searle Research Labs., High Wycombe, Eng., 1966—. Mem. expert adv. panel on immunology WHO, Geneva, Switzerland, 1963—. Recipient Merit award Sci. Civil Service, 1959. Mem. Biochem. Soc., Soc. for Gen. Microbiology, Geol. Assn., Brit. Mycological Soc., Brit. Soc. for Immunology, N.Y. Acad. Scis. Research, publs. on bacterial vaccines, substances responsible for tissue graft rejection, identification of immunogenic substances distinguishing cancer cells from normal cells. Home: 1 Earls Rd., Tunbridge Wells, Eng. Office: Lane End Rd., High Wycombe, Eng.*

DAVIES, David Arthur, meteorologist; b. Barry, Eng., Nov. 11, 1913; s. Garfield Brynmor and Mary Jane (Michael) D.; B.Sc., U. Wales, 1933, 1st class honors Pure and Applied Math., 1934, Physics, 1935, M.Sc., 1940; m. Mary Shapland, Mar. 5, 1938; children—Michael, Rosalind, Margaret. Tech. officer Meteorol. Office, London, 1936-39, prin. sci. officer, 1947-49; dir. East Africa Meteorol. Dept., Nairobi, 1949-55; sec.-gen. World Meteorol. Orgn., Geneva, Switzerland, 1955—, pres. Regional Assn. for Africa, 1951-55. Fellow Inst. Physics, London; mem. Royal Inst. Pub. Adminstrn. Publs. on meteorol. studies of tropical rainfall, artificial stimulation of rain in East Africa. Home: 34 Avenue Krieg. Office: 41 Av. Giuseppe Motta, Geneva, Switzerland.*

DAVIES, Douglas Mackenzie, Canadian entomologist; b. Toronto, Ont., Can., May 11, 1919; s. Gordon Albert and Doris (Stark) D.; B.A. with honors in Sci., U. Toronto, 1942, Ph.D. (Ont. Research Commn. scholar), 1949; m. Sheila Margaret Whittemore, Sept. 11, 1948; children—Ian Whittemore, Kenneth Gordon. Meteorologist, Can. Dept. Transport, 1942-45; research fellow dept. parasitology Ont. Research Found., 1945-51; faculty McMaster U., Hamilton, Ont., 1951—, prof. biology, 1963—. Sr. Research fellow NRC Can., 1963-64. Mem. Entomol. Soc. Ont. (pres. 1960, editor Proc. 1964-66), Entomol. Soc. Am., Entomol. Soc. Can., Royal Entomol. Soc. London, Canadian Soc. Zoologists. Research, publs. on weather factors influencing flying and biting of blood-sucking black-flies, reflected colors and carbon dioxide affecting host selection, egg prodn. in house-fly, enzymes in blood digestion of black flies, parasites and predators of black flies; described new species black flies. Home: 11 Dromore Cres, Hamilton, Ont., Can.*

DAVIES, George Neville, dentist; b. Christchurch, New Zealand, Sept. 20, 1921; s. Christopher A. R. and Lillian (Hawkes) D.; B.D.S., U. Otago, 1943, D.D.S., 1953; D.D.S., Northwestern U., 1947; m. Valerie Todd, Nov. 9, 1948; children—Christopher, John, Rosemary. Sr. lectr., asso. prof. U. Otago, New Zealand, 1948-63; research fellow Eastman Dental Center, Rochester, N.Y., 1954; prof. dentistry, dean faculty dentistry U. Queensland, Brisbane, Australia, 1964—. Mem. expert adv. panel on dental health WHO, 1957-66; mem. Med. Research Council of New Zealand, 1961-63; mem. dental health com. Nat. Health and Med. Research Council Australia, 1964—. Fellow Am. Coll. Dentists, Australian Coll. Dental Surgeons; mem. Royal Soc. Queensland, Australian Dental Assn., Internat. Assn. Dental Research, Internat. Dental Fedn. Author: (with R. M. King) Dentistry for Pre-School Children, 1961. Research and publs. on epidemiology of dental disease, devel. standardized methods for recording and reporting disease. Home: Roedean St., Brisbane, Queensland, Australia.*

DAVIES, Herbert, Brit. physician; b. London, Sept. 30, 1818; s. Thomas Davies; ed. Gonville and Caius Coll., Cambridge; B.A., Queen's Coll., 1842, M.B., 1843, M.D., 1848; student Paris, Vienna; m. Miss Wyatt, Aug. 24, 1850; 7 children. Became asst. physician London Hosp., 1845, physician, from 1854; examiner for med. degrees Cambridge; physician to Bank of Eng. Fellow Queen's Coll., 1844. Fellow Coll. Physicians. Publs. on observations of relative magnitude of areas of 4 orifices of heart; advocated use of blisters on joints in acute rheumatism. Died Hampstead, Eng., Jan. 4, 1885.

DAVIES, J., flourished 17th century; contributed to devel. of math. instruments; introduced Davies' quadrant to enable navigator to sight on a shadow cast by sun's glare, 1595.

DAVIES, Jack Neville Phillips, pathologist; b. Devizes, Eng., July 2, 1915; s. David Osborne and Mabel (Mapham) D.; M.B., Ch.B., U. Bristol, Eng., 1939, M.D., 1948, D.Sc. (hon.), U. East Africa, 1967; m. Daphne Margaret Perry, Jan. 8, 1944 (dec. Jan. 1961); 1 son, Antony; m. 2d, Valerie Elizabeth Davis, Aug. 23, 1961; step-children—Nigel, Rupert. Asst. lectr. physiology, biochemistry, demonstrator pharmacology. U. Bristol, 1941-44; med. officer, pathologist Uganda Med. Service, 1944-49; Commonwealth Fund fellow Duke, 1949-50; prof. pathology Makerere Coll., Uganda, 1950-61; reader morbid anatomy U. London Med. Sch., 1961-63; prof. pathology Albany (N.Y.) Med. Coll., 1963—; mem. East African Council Med. Research. Mem. Pathol. Soc. Gt. Britain and Ireland, Assn. Clin. Pathology, Am. Assn. Path. Bacteriology, Internat. Acad. Pathology, N.Y. Acad. Scis., Internat. Soc. Geog. Pathology. Author (With H. C. Trowell, R. F. A. Dean) Kwashiorkor, 1954. Research, numerous publs. on path. basis of kwashiorkor and protein deficiency states, African cardiovascular diseases; recognition and description of new diseases, as endomyocardial fibrosis in Africans; established 1st cancer incidence studies in tropics, importance of cancers in Africa, especially lymphomas of African children. Home: R.D. 2, Selkirk, N.Y. 12158. Office: Albany Med. Coll., Albany, N.Y. 12208.*

DAVIES, James C., Am. polit. scientist; b. Wauwatosa, Wis., May 6, 1918; s. Howell David and Julia (Merrell) D.; A.B., Oberlin Coll., 1939; postgrad. U. Chgo. Law Sch., U. Tex. Law Sch.; Ph.D., U. Cal. at Berkeley, 1952; m. Eleanor Johnstone Getze, Jan. 10, 1943; 1 dau., Sarah Louise. Supr. prodn. scheduling Consol. Vultee Aircraft Corp., San Diego, Ft. Worth, 1942-45; research tng. fellow Social Sci. Research Council, Cambridge, Mass., 1950-51; Carnegie fellow polit. sci. U. Mich. Survey Research Center, 1951-53; faculty Cal. Inst. Tech., 1953-59, 60-63; vis. asso. prof. U. Cal., Berkeley, 1959-60; prof. U. Ore., Eugene, 1963—, head, dept. polit. sci., 1964-67. Cons. U. S. Office Edn., Dept. Health, Edn. and Welfare, 1966—. Social Sci. Research Council fellow, 1961-62; Rockefeller Found. fellow, 1962-63. Mem. Internat., Am. (dir. 1965-67) polit. sci. assns., A.A.A.S., Phi Beta Kappa. Author: Human Nature in Politics, 1963, also articles. Developed theory of polit. revolution on basis of initial satisfaction and subsequent frustration of phys. and mental needs, 1962; posited systematic relationships between individual and social psychology and gen. polit. theory. Home: 1560 Prospect Dr., Eugene, Ore. 97403.

DAVIES, John Vipond, civil engr.; b. Swansea, South Wales, Oct. 13, 1862; s. Andrew and Emily (Vipond) D.; Wesleyan Coll., Taunton, Eng.; U. of London; m. Ruth Ramsey, Apr. 16, 1895 (died 1931); children—Margaret, Muriel, John Vipond. Engaged in coal mining, steel mfr. and other engring. work to 1889, when came to N.Y. and employed as engr. with Austin Corbin; chief asst. engr. tunnel under East River for East River Gas Co.; pres. Jacobs & Davies, Inc., cons. engrs., N.Y.; v.p., chief engr. Hudson & Manhattan R.R.; chief engr. in charge design and constrn., W.Va. Short Line R.R. and Kanwaha & Pocahontas R.R. in W.Va.; chief engr. Atlantic Av. improvement, L.I. R.R. Bklyn.; cons. engr. Bklyn. Rapid Transit Co.; cons. engr. City of Detroit for water supply tunnel under Detroit River; engr. in charge for contractors of terminal improvement of N.Y.C. R.R., N.Y.C.; designed and built 4 tunnels under Hudson River and under New York, Jersey City and Hoboken, for Hudson & Manhattan R.R.; constrn. engr. for 26 aqueduct tunnels in Mexico; one of bd. of 3 engrs. on constrn. of Moffat Tunnel in Rocky Mountains nr. Denver; firm prepared original studies for Pa. R.R. tunnels under North and East Rivers, N.Y.; engr. in charge Astoria Tunnel, Consol. Gas. Co. of N.Y.; engr. in charge constrn. Hales Bar Dam across Tenn. River at Chattanooga; engr. in charge constrn. intake and discharge tunnels of N.Y. Edison Co.; on constrn. bridge or tunnel crossing of Miss. River at New Orleans; mem. bd. cons. engrs. N.Y. State Bridge and Tunnel Commn. and N.J. interstate Bridge and Tunnel Com. (vehicular tunnel under Hudson River, 1912-22); one of two engrs. on tunnel and bridge crossing of San Francisco Bay. Recipient Telford gold medal Am. Inst. C.E., 1914, Norman gold medal, 1913, Thomas Fitch Rowland prize, 1917; Fowler professorial award, 1930. Died Flushing, N.Y., Oct. 4, 1939.

DAVIES, Kenneth, physicist; b. Merthyr Tydfil, Wales, Jan. 28, 1928; s. William R. and Hannah E. (Broad) D.; B.Sc. summa cum laude in Physics, U. Coll. Wales, Swansea, 1949, Ph.D., 1953; m. Joyce A. DeMerchant, Feb. 20, 1958; children—Russell, Elizabeth, Kenneth. Came to U. S., 1956, naturalized, 1961. Mem. staff Def. Research Telecommunications Establishment, Ottawa, Can., 1952-56; faculty Brown U., 1956-58; with Central Radio Propagation Lab., Boulder, Colo., 1958-60, chief, ionosphere research sect., 1960—; exchange scientist Radio Research Sta., Dept. Sci. and Indsl. Research, Slough, Eng., 1962-63; prof. adjoint astrophysics and geophysics U. Colo., Boulder, 1960—, elec. engring., 1963—. Cons., mem. numerous internat. coms. Mem. I.E.E.E., Am. Geophys. Union, Research Soc. Am., Sigma Xi. Author: Ionospheric Radio Propagation, 1965. Asso. editor Radio Sci., 1963—. Research in propagation of radio waves in the ionosphere and ionospheric effects of solar flares; pioneer (with J. M. Watts) in HF Doppler technique for studying variations in ionospheric phase paths. Home: 4600 Osage Dr., Boulder, Colo. 80302.

DAVIES, Rhisiart Morgan, physicist; b. Feb. 4, 1903; s. Rhys Davies; ed. Univ. Coll. Wales, Aberystwyth, Trinity Coll., Cambridge, Eng.; D.Sc., Wales, Ph.D., Cambridge; m. Elizabeth Florence Davies, 1928; 1 son. Asst. lectr. in physics Univ. Coll. Wales, 1925, lectr., 1928, sr. lectr., 1939-46, prof. physics, from 1946, vice prin., 1954-56; Leverhulme research fellow, Cambridge, 1939-41; researcher for Admiralty, also for Ministry Home Security, Ministry Supply, at Cavendish and univ. engring. labs., Cambridge, 1941-45. Vis. prof. Cal. Inst. Tech., Rensselaer Poly. Inst., 1956-57. Co-editor, contbr.: Surveys in Mechanics, 1956. Contbr. papers on detonation waves in gases and stress waves in liquids and solids, to sci. jours. Died Feb. 18, 1958.

DAVIES, Robert Ernest, biochemist; b. Lancashire, Eng., Aug. 17, 1919; s. William Owen and C. Stella (Spencer) D.; B.Sc., U. Manchester (Eng.), 1941, M.Sc., 1942, D.Sc., 1952; Ph.D., U. Sheffield (Eng.), 1949; M.A., Oxford U. (Eng.), 1956; m. Helen Jean Rogoff, Sept. 8, 1961; step-children—Daniel, Richard. Came to U. S., 1955. Faculty, U. Manchester, 1941, U. Sheffield, 1942-45, 48-54; sci. staff Med. Research Council, Eng., 1945-56; guest prof. U. Heidelberg, Germany, 1954; faculty medicine Oxford U., 1955-59; faculty U. Pa., Phila., 1955—, chmn. dept. animal biology Sch. Vet. Medicine, 1962—, chmn. grad. group com. on molecular biology, 1962—, prof. biochemistry Grad. Sch. Medicine, 1962—. Fellow Royal Soc. (London); mem. Chem. Soc., Biochem. Soc., Physiol. Soc., Soc. Exptl. Biology (all Eng.), Biochemists Club Phila. (pres. 1959), Physiol. Soc. Phila., Biophysics Soc., Soc. Gen. Physiologists, Am. Soc. Biol. Chemists, Am. Chem. Soc., A.A.A.S., John Morgan Soc., N.Y. Acad. Scis., Am. Physiol. Soc., Am. Inst. Biol. Scis., Am. Soc. Cell Biology, Am. Soc. Vet. Physiologists and Pharmacologists, Sigma Xi, Phi Zeta. Mem. editorial bd. Biochem. Jour., 1951-56. Research, numerous publs. on mechanism of hydrochloric acid secretion in stomach, role of enzyme carbonic anhydrase, energy metabolism and active transport of ions in cells and mitochondria from kidney, role of adenosine-triphosphate in muscular contraction, devel. of theory of muscle contraction. Home: 7053 McCallum St., Phila. 19119.*

DAVIES, Rodney Deane, radio astronomer; b. Balaklava, S. Australia, Jan. 8, 1930; s. Holbin James and Rena (March) D.; B.Sc. with Honors, U. Adelaide (Australia), 1951, M.Sc., 1953; Ph.D., U. Manchester (Eng.), 1956; m. Valda Beth Treasure, Jan. 3, 1953; children—Warwick Brian, Rosalyn Anne, Margaret Claire, Stewart John. Research officer radiophysics div. Commonwealth Sci. and Indsl. Research Orgn., Sydney, Australia, 1951-53; faculty Nuffield Radio Astronomy Labs., Jodrell Bank, U. Manchester, (Eng.), 1953—, reader radio astronomy, 1967—. Fellow Royal Astron. Soc. Author: (with H. P. Palmer) Radio Studies of the Universe, 1959; (with H. P. Palmer, M. I. Large) Radio Astronomy Today, 1963; also articles. Research on enhanced radio emission from sun, radio emission from neutral hydrogen and hydroxyl molecule which originated in interstellar space in Milky Way and nearby galaxies, structure of magnetic field in galaxy. Office: Nuffield Radioastronomy Labs., Jodrell Bank, Cheshire, Gt. Britain.*

DAVIES, Thomas Stephens, Brit. mathematician; b. Eng., 1795; math. master Royal Mil. Acad., Woolwich, Eng., 1834; fellow Royal Soc. Edinburgh, Soc. Antiquarians. Contbr. sci. articles to sci. and popular jours., mags. Developed new system of spherical geometry; research on properties of trapezium, Pascal's hexagramme mystique, Brianchon's theorem, symmetrical properties of plane triangles, geometry of 3 dimensions. Died Kent, Eng., Jan. 6, 1851.

DA VIGO, Giovanni, Italian physician; b. 1460; s. Battista di Rappallo; surgeon to Pope Julius II; Author: Practica copiosa in arte chirurgica, 1514; Advised that gunshot wounds be cured by use of boiling elder oil or cautery; described trephine which he invented, also number of other new instruments. Died 1525.

DAVIS, Alexander Jackson, Am. architect; b. N.Y.C., July 24, 1803; s. Cornelius and Julia (Jackson) D.; m. Margaret Beale, July 14, 1853. With firm Town & Davis, 1829-43; self employed as architect, 1843-80; prin. works include: N.Y. Customs House, 1832; Ind. (1832-35), N.C. (1831), Ill. (1837), Ohio (1839) state capitols; U. S. Patent Office, Washington, 1832; Va. Mil. Inst., 1852, 59; Assembly Hall, U. N.C., Alumni Hall, Yale; Gilmer House in Balt. (1832) and U. Mich. (1838); a founder Llewellyn Park, West Orange, N.J.; exponent of classic and gothic styles; an early experimenter with structural iron. Died West Orange, Jan. 14, 1892.

DAVIS, Arthur Powell, Am. civil engr.; b. Decatur, Ill., Feb. 9, 1861; s. John and Martha P. D.; grad. State Normal Sch., Emporia, Kan.; B.S., Columbian (now George Washington) U., 1888, Sc.D., 1917; D.Eng., Ia. State Coll., 1920; m. Elizabeth Brown, June 20, 1888 (dec. Apr. 1917); children—Mrs. Rena Peck, Mrs. Florence Eslin, Mrs. Dorothy Smith, Mrs. Elizabeth Smith; m. 2d, Marie MacNaughton, June 19, 1920. Topographer U. S. Geol. Survey, 1884-94, conducting surveys and explorations in Ariz., N.M., Cal.; hydrographer in charge of all govt. stream measurements, 1895-97; hydrographer in charge hydrographic exam. of Nicaragua and Panama canal routes, 1898-1901; chief engr. U. S. Reclamation Service, 1906-14, dir., 1914-23. Cons., Panama Canal, 1909; examined and reported on irrigation in P.R., 1909, in Turkestan, 1911; mem. bd. engrs. reporting on flood control in China, 1914; cons. engr. on many high dams; mem. joint conf. on standard specifications for Portland cement, 5 yrs.; tech. adviser to U. S. on Pecuniary Claims Arbitration, London, 1923; chief engr. and gen. mgr. E. Bay Municipal Utility Dist., Oakland, Calif., 1923-29; built large reservoir on Mokelumne River, aqueduct and tunnels, 95 miles long, to deliver mountain water to Oakland, San Francisco and 8 other cities around San Francisco Bay; chief cons. engr. for irrigation projects in Turkestan and Transcaucasia 2 yrs. Fellow Am. Acad. Arts and Scis., Washington Soc. Engrs. (pres. 1907; hon. mem. 1923). Author: Irrigation Works Constructed by the U. S. Government, 1917. Died Aug. 7, 1933.

DAVIS, Audrey Kennon, Am. physiologist; b. Tenn., Aug. 7, 1920; d. Earl Kennon and Jewell (Jones), D.; B.S., Memphis State U., 1947; Ph.D., U. Tenn., 1951; m. Hortense Louckes, Aug. 12, 1961; children—Richard Earl, Audrey Irene, Virginia Lee. Chemist, E. I. DuPont, Memphis, 1942-44; research asso. U. Tenn. Coll. Medicine, Memphis, 1946-49, instr., 1951-52; head biophysics Naval Radiol. Def. Lab., San Francisco, 1952-61; dir. exptl. radiobiology div. radiol. health USPHS, Washington, 1961-67; chief long-term effects lab., 1967—; professorial lectr. physiology George Washington U., 1964-67; prof. physiology Colo. State U., 1967—. Mem. Am. Physiol. Soc., Radiation Research Soc., Am. Soc. Health Physics, Am. Soc. Exptl. Biology and Medicine. Research and publs. on thermal injury from nuclear weapons, anemia caused by thermal burns, acute lesions of skin after beta radiation, skin tumors caused by beta radiation, comparative radiosensitivity of embryonic, infantile and adult tissues to X-rays; devel. indices of radiation damage relating long term effects of ionizing radiation to morphological, functional or biochem. changes in organism; functional kidney damage after irradiation. Address: CRHL, Colo. State U., Ft. Collins, Colo. 80521.*

DAVIS, Bergen, Am. physicist; b. White House, N.J., Mar. 31, 1869; s. John and Catherine Marie (Dilts) D.; B.S., Rutgers, 1896, Sc.D. (hon.), 1929; A.M., Columbia, 1900, Ph.D., 1901, Sc.D. (hon.) 1929; postgrad. U. Gottingen 1901-02, Cambridge, 1902-03; m. Matie Pearl Clark, 1922. Instr. physics Columbia, 1903-09, adj. prof., 1909-13, asso. prof., 1913-1918, prof. physics, 1918-39, prof. emeritus, 1939-58. Awarded medal and prize of Research Corp., 1929. Fellow A.A.A.S. (v.p. sect. B 1932), Am. Phys. Soc., Am. Optical Soc.; mem. Nat. Acad. Scis., Sigma Xi. Contbr. articles and papers. Died June 30, 1958.

DAVIS, Bernard D(avid), Am. microbiologist; b. Franklin, Mass., Jan. 7, 1916; s. Harry and Tillie (Shain) D.; A.B., Harvard U., 1936, M.D., 1940; m. Elizabeth Menzel, June 19, 1955; children—Franklin A., Jonathan H., Katherine J. Intern, Fellow Johns Hopkins Hosp., 1940-41; commd. officer USPHS, 1942-54, successively assigned Nat. Inst. Health, Columbia, Pub. ealth Research Inst. of N.Y., Rockefeller Inst., and charge USPHS Tb Research Lab. at Cornell U., 1947-54; prof. pharmacology, chmn. dept. N.Y.U., 1954-57; prof. bact., chmn. dept.,

Harvard, 1957-—, Adele Lehman prof. bacteriology and immunology, 1963-—; divisional com. for biology, medicine Nat. Sci. Found., 1954-57; mem. medical advisory board Hebrew University, 1956-—. Recipient Waksman medal Soc. Am. Bacteriologists, 1953. Mem. A.A.A.S., Am. Soc. Biol. Chemists, National Academy of Sciences, American Academy of Arts and Science, American Society Microbiology, Am. Soc. Cell Biology, Soc. Gen. Physiology (pres. 1964-65), Phi Beta Kappa, Sigma Xi, Alpha Omega Alpha. Study of chemotherapy; genetics and microbial metabolism. Home: 23 Clairemont Rd., Belmont, Mass.

DAVIS, Bradley Moore, Am. botanist; b. Chgo., Nov. 19, 1871; s. Charles Wilder and Emma Frances (Moore) D.; A.B., Stanford, 1892; A.B., Harvard, 1893, A.M., 1894, Ph.D., 1895; research at Bonn, 1898, Naples, 1904; m. Annie Elizabeth Paret, Sept. 22, 1908; 1 dau., Margery French (Mrs. Allen M. Boyden). Faculty U. Chgo., 1895-1906, asst. prof. plant morphology, 1902-06; head dept. botany Marine Biol. Lab., Woods Hole, 1897-1906, in charge of bot. sec., biol. survey Woods Hole, Bur. of Fisheries, 1903-09; asst. prof. botany U. Pa., 1911-14, prof., 1914-19; prof. botany U. Mich., 1919-42, prof. emeritus, 1942-57. Editor statis. div. U. S. Food Adminstrn., 1918. Fellow Am. Acad. Arts and Scis.; mem. Am. Philos. Soc. (sec. 1918-19). Bot. Soc. Am., Am. Soc. Naturalists (sec., 1913-19, pres. 1921), Am. Genetic Assn., A.A.A.S., N.E. Bot. Club. Co-author: Principles of Botany, 1906; Laboratory and Field Manual of Botany (with Joseph Y. Bergen), 1907. Asso. editor Genetics. Contbr. many papers on plant cytology and plant genetics to jours. Died Mar. 13, 1957.

DAVIS, Chandler, mathematician; born Ithaca, N.Y., Aug. 12, 1926; s. Horace B. and Marian (Rubins) D.; B.S., Harvard, 1945, M.A., 1947, Ph.D., 1950; m. Natalie Zemon, Aug. 16, 1948; children—Aaron Bancroft, Hannah Penrose, Simone Weil. Faculty, University of Michigan, 1950-54; associate editor of Mathematical Reviews, Providence, R.I., 1958-62; faculty Univ. of Toronto, 1962-—, now prof. math. Research, publs. on operators, including behaviour of eigenvalues of matrices, topics in analysis and geometry. Home: 85 Crescent Rd., Toronto 5, Ont., Can.*

DAVIS, Charles Albert, Am. geologist; b. Portsmouth, N.H., Sept. 29, 1861; s. Lewis Gilman and Cyrena Frances (Peirce) D.; A.B., Bowdoin, 1886, A.M., 1889; postgrad. Cornell Sch. Forestry, 1900-01; Ph.D., U. Mich., 1905; m. Frances Margaret Humphreys, Aug. 26, 1886. Tchr. natural science Hyde Park (Ill.) High Sch., 1886-87; prof. natural science Alma (Mich.) Coll., 1887-96, prof. biology and geology, 1896-1900; instr. forestry U. Mich., 1900-05, curator Herbarium, 1905-08; peat expert U. S. Geol. Survey, 1907-10, Bur. Mines, Washington, 1910-12; fuel technologist, 1912-14, geologist, 1914-—. Field agt. Mich. Geol. Survey 1896-1907; instr. geology U. Mich. Summer School, 1900, 01; field asst. U. S. Geol. Survey, 1904. Fellow Geol. Soc. Am., A.A.A.S. Editor Jour. Am. Peat Soc., 1907-—. Died Apr. 9, 1916.

DAVIS, Charles Carroll, Am. limnologist, ecologist; b. Azusa, Cal., Nov. 24, 1911; s. William A. and Maude (Snyder) D.; A.B., Oberlin Coll., 1933; M.S., U. Wash., 1935, Ph.D., 1940; m. Sally M. Jacobsen, June 11, 1936; children—Peter T., Betsy A. Research asst. Scripps Inst. Oceanography, La Jolla, Cal., 1938-40; asst. prof. zoology U. Miami (Fla.), 1947-48; faculty Western Res. U., Cleve., 1948-—, prof. biology, 1963-—. Adv. editor Eng. translation Doklady Acad. Sci. USSR, biol. sci. sect. 1959-64; zoology consultant. New World Dictionary. Fellow A.A.A.S., Ohio Acad. Sci.; mem. Ecol. Soc. Am., Am. Soc. Limnology and Oceanography, Internationale Verein der Limnologie, Am. Soc. Zoologists, Phycological Soc. Am., Internat. Phycological Soc., Plankton Soc. Japan, Marine (U.K.), Freshwater (U.K.) biol. assns. Author: The Pelagic Copepoda of the Northeastern Pacific Ocean, 1949; The Marine and Freshwater Plankton, 1955; also numerous articles. Ecol. studies marine and fresh-water plankton, effects water pollution on Lake Erie plankton, study hatching mechanisms in eggs aquatic invertebrates, analysis problems prodn. and productivity in ecology. Home: 3363 Dellwood Rd., Cleveland Heights, O., 44118. Office: Dept. Biology, Case Western Res. U., Cleve. 44106.*

DAVIS, Charles Henry, Am. naval officer, scientist; born Boston, January 16, 1807; the son of Daniel and Lois (Freeman) Davis; married Harriet Mills, 1842, 6 children including Charles Henry, Anna Cabot, Evelyn. Commissioned midshipman in United States Navy, 1824, lt., 1827, comdr., 1854; served as chief of staff, capt. of fleet in expdn. under Dupont which captured Port Royal, S.C., 1861; exec. head Bur. of Detail for selecting, assigning officers during Civil War; assumed command Upper Mississippi Gunboat Flotilla above Ft. Pillow, 1862; commd. chief Bur. of Navigation, 1862; commd. rear adm., 1863; a founder Nat. Acad. Scis., 1863; supt. Naval Observatory, 1864-66, 1870-73; commanded Brazilian Squadron, 1867-69, Norfolk (Va.) Navy Yard, 1870-73. Author: The Coast Survey of the United States, 1849; Narrative of the North Pole Expedition of the U.S.S. Polaris, 1876. Hydrographic studies, especially ocean currents, ocean tides, dangerous waters around Nantucket, Mass. Died Washington, D.C., Feb. 18, 1877.

DAVIS, David Edward, Am. ecologist; b. Chgo., July 18, 1918; s. David John and Almira (Jones) D.; B.S., Swarthmore Coll., 1935; M.S., Harvard, 1936, Ph.D., 1939; m. Emily Rodgers, Oct. 8, 1942; children—Susan, Alice, Jean. Postdoctoral fellow, U. Chgo., 1940; zoologist Rockefeller Found., Rio de Janerio Brazil, 1941-43; scientist USPHS, San Antonio, Tex., 1943-45; faculty John Hopkins Sch. Pub. Health, 1945-59; prof. zoology Pa. State U., University Park, 1959-67, North Carolina State University, Raleigh, 1967-—. Cons., USPHS, 1956-—. Mem. Ecol. Soc., Am. Ornithol. Union, Am. Soc. Mammalogists, Animal Behavior Soc., Wildlife Disease Assn. (pres. 1962-64). Author: Principles in Mammalogy (with Frank Golley), 1963; Integral Animal Behavior, 1966. Research and publs. on regulation animal populations, social behavior birds, role hormones in behavior. Home: 2911 Fairview Rd., Raleigh, N.C. 27608.*

DAVIS, David John, Am. pathologist; b. Racine, Wis., Aug. 9, 1875; s. David W. and Catharine (Jones) D.; B.S., U. Wis., 1898; M.D., Rush Med. Coll., 1904; Ph.D., U. Chgo., 1905; studied univs. of Vienna and Freiburg; m. Myra H. Jones, July 27, 1908; children—Dorland Jones, Edward David. Practiced in Chgo., 1904-—; prof. pathology U. Ill., 1914-43, dean of med. sch., 1925-43, emeritus, 1943-—. Pres. bd. dirs. Chgo. Municipal Tb Sanatorium, 1946; dir. Chgo. Inst. Medicine. Mem. A.M.A., Ill. Med. Soc. (permanent historian 1946-—), Soc. Am. Bacteriologists, Am. Assn. Pathologists and Bacteriologists, Am. Assn. Pathologists and Bacteriologists, Chgo. Pathol. Soc., Sigma Xi, Phi Beta Kappa. Contbr. papers, chiefly on infection with influenza bacilli, streptococci, sporotricha and med. edn. Died Dec. 20, 1954.

DAVIS, David Roy, Am. mathematician, educator; b. Clarkshill, Ind., Apr. 30, 1893; s. David Moss and Carrie L. (Whipple) D.; A.B., Ind. U., 1917, A.M., 1923; Ph.D., U. Chgo,. 1927; m. Vera M. Brooke, May 29, 1920. Asst. prof. math. U. Ore., 1926-31; prof. Montclair (N.J.) State Coll., 1931-57, head dept., 1954-57; prof., head dept. math. East Carolina Coll., Greenville, N.C., 1957-64, prof. emeritus, 1964-—. Prof. math. War Dept., Shrivenham (Eng.) Am. U., U. Vienna (Austria), 1945-46. Mem. Am. Math. Soc., Math. Assn. Am. (v.p., program chmn. N.J. 1955), N.J. Council Tchrs. Math. (pres. 1943-44), Phi Beta Kappa, Sigma Xi, Pi Mu Epsilon. Author: Modern College Geometry, 1949; The Teaching of Mathematics, 1951; (with W. E. Milne) Introductory College Mathematics, 3d edit., 1962. Contbr. numerous articles to profl. jours. on mathematics of annuities and pension funds. Address: 1024 North Shore Dr., St. Petersburg, Fla. 33701.*

DAVIS, Donald Echard, Am. plant physiologist; b. Charleston, Ill., Jan. 12, 1916; s. Leonard Earnest and Jessie (Echard) D.; B.Ed., Eastern Ill. U., 1938; M.S. Ohio State U., 1940, Ph.D., 1947; m. Dorothy Dale Richey, June 5, 1940; children—Leonard Richey, Dorothy Ann. Nat. Research Council fellow Ohio State U., 1945-46, instr. botany, 1946-47; asst. prof. botany Auburn U., 1947-48, asso. prof., 1948-50; U. Tenn. AEC Research fellow, Oak Ridge, 1951-52; with Auburn U., 1952-—, prof. botany, 1955-—, pres. faculty, faculty council, 1956-57. Mem. So. Weed Conf. (pres. 1966-67), Am. Soc. Plant Physiologists, Weed Soc. Am., Am. Inst. Biol. Sci., A.A.A.S., Ala. Acad. Sci., Sigma Xi, Gamma Sigma Delta, Kappa Delta Pi, Phi Epsilon Phi, Gamma Alpha. Author (with N. D. Davis) Alabama Trees, 1963, Guide and Key to Alabama Trees, 1965. Research and publs. on established factors responsible for variability in toxicity of a dinitro herbicide to cotton; determined causes of variability in susceptibility of crop plants to triazine herbicides; determined the ultimate fate of triazine herbicide in plants and soils; partially determined the exact way that triazine herbicides kill plants. Home: 512 Auburn Dr., Auburn, Ala. 36830.*

DAVIS, Dorland J., Am. physician; b. Chgo., July 2, 1911; s. David J. and Almira (Jones) D.; B.S., U. Ill., 1933; M.D., Johns Hopkins U., Dr.P.H., 1940; m. Caroline B. Baker, July 15, 1938; children—David H., Constance E. With NIH, 1939-—, dir. Nat. Inst. Allergy and Infectious Diseases, Bethesda, Md., 1964-—; bd. dirs. Gorgas Meml. Inst. of Tropical and Preventive Medicine. Recipient Edward Rhode Stitt award, 1955. Fellow Am. Pub. Health Assn., Am. Acad. Microbiology, Am. Coll. Preventive Medicine; mem. Am. Assn. Immunologists, Am. Soc. Tropical Medicine and Hygiene. Contbr. numerous articles to sci. jours. Identified cause of epidemic conjunctivitis as Haemophilus aegyptius and described treatment with antibiotics; contbr. to knowledge of transmission of poliomyelitis, hepatitis, influenza. Home: 4841 Broad Brook Dr. Office: NIH, Bethesda, Md. 20014.*

DAVIS, Edward Wilson, Am. metall. engr.; b. Cambridge City, Ind., May 8, 1888; s. Walter Clarance and Della Mendenhall (Wilson) D.; B.S., Purdue U., 1911, E.E., 1918; m. Jessie Mary Campbell, June 4, 1914; children—Jane, Martha, Ruth. Elec. engr. Westinghouse Electric & Mfg. Co., 1911, Gen. Electric Co., 1912; instr. math. U. Minn., 1913-15; engr. Mesaba Iron Co., 1915-18; with U. Minn., 1918-55, prof. and supt. Mines Expt. Sta., 1925-33; dir., 1938-55; metall. cons. to Reserve Mining Co., 1955-—. Mem. Am. Mining Congress, Am. Inst. Mining and Metall. Engrs., A.A.A.S., Eastern States Blast Furnace and Coke Oven Assn. Author Mines Expt. Sta. bulls. on taconite. Encouraged indsl. use of taconite (ore high in iron content) by perfecting a magnetic separation process, 1915; developed a pelletizer; numerous patents on ore dressing machinery, similar devices. Home: 1 Kent Lane. Office: Reserve Mining Co., Silver Bay, Minn.*

DAVIS, Edwin Hamilton, Am. physician, archeologist; b. Hillsboro, O., Jan. 22, 1811; s. Henry and Avis (Slocum) D.; grad. Kenyon Coll., 1833, Cin. Med. Coll., 1838; m. Lucy Woodbridge, 1841, 9 children including John Woodbridge. Prof. materia medica and therapeutics N.Y. Med. Coll., 1850-60; surveyed and described (with E. G. Squire) 100 of more important earthworks of Mound Builders, So. Ohio; compiled findings in Ancient Monuments of the Mississippi Valley (1st work pub. by Smithsonian Instn. and still a standard work), 1847; noted for his collections of cultural objects of Mound Builders. Died N.Y.C., May 15, 1888.

DAVIS, Eli, physician; b. Birzai, Lithuania, June 1, 1908; s. Haim and Rosa (Bergel) D.; B.Sc. in Anatomy and Physiology, U. Manchester (Eng.), 1928, Physiol. with 1st class honors, 1929, M.Sc., 1930, M.B., Ch.B., 1933, M.D., 1935; m. Elise Sara Rosenbloom, Jan. 20, 1939; children—Naomi, Haim. Physiol. research scholar Manchester U., 1929-33; house physician; Manchester Royal Infirmary, 1933-36; asst. med. officer, Salford, Eng., 1936; physician St. Stephen's Hosp., London, Eng., 1937-41, St. Alfege's Hosp., 1941-42; chief medicine St. Andrew's Hosp., London, 1942-46; med. dept. dir. Hadassah Med. Orgn., Jerusalem, Israel, 1946-48, dir., 1948-51; chief med. outpatient dept., chief rheumatic fever clinic, chief capillary research lab. Hadassah U. Hosp., Jerusalem, 1951-—; asso. prof. medicine Hebrew U. Hadassah Med. Sch., Jerusalem, 1965-—. Fellow Royal Soc. Medicine, Internat. Coll. Angiology; mem. European Microcirculation Soc. (exec. com. 1960-—), Royal Coll. Physicians London. Author: (with J. Landau, M. Ivry) Clinical Capillary Microscopy, 1966; also articles. Research on clin., epidemiological and genetic factors in rheumatic fever in Israel, small blood vessels in common maj. med. diseases, familial benign bleedings. Home: 9 Shlomo Molcho St., Office: Hadassah-U. Hosp., Jerusalem, Israel.*

DAVIS, Frank Wilbur, Am. mech. engr.; b. Charleston, W.Va., Dec. 6, 1914; s. Madison T. and Julia (Staunton) D.; B.S. in Mech. Engring., Cal. Inst. Tech., 1936; Sc.D., W.Va. U., 1960; m. Frances Washington Pfeiffer, Mar. 15, 1941; children—Caroline (Mrs. O. D. Calvert, Jr.), Frank Wilbur, William Brewster. Asst. to v.p. engring. Gen Dynamics Corp. (formerly Convair), San Diego, 1940-54; pres. Ft. Worth div., 1954-—. Mem. adv. council Grad. Research Center, Dallas, 1964-—; mem. hon. adv. bd. for history aviation U. Tex., Austin, 1965. Named Engr. of Year, Ft. Worth sect. Tex. Soc. Profl. Engrs., 1957, Golden Deeds award Exchange Club Ft. Worth, 1965. Registered profl. engr. Cal., Tex. Fellow Soc. Exptl. Test Pilots (hon.); mem. Nat. Acad. Scis., Nat. Soc. Profl. Engrs., Air Force Assn., Soc. Automotive Engrs., Inst. Aerospace Scis. Research in engring. aircraft and guided missiles. Home: 6328 Curzon St., Ft. Worth 76116. Office: Gen Dynamics Corp., Ft. Worth div., Box 748, Ft. Worth 76116.*

DAVIS, George H., Am. engr.; b. Oswego, N.Y., 1863; s. Samuel A. and Esther T. (Parks) D.; grad. Oswego State Tchrs. Coll., 1885; M.E., Cornell U., 1892; m. Katherine McGrath, 1898; children—Philip McGrath, Putnam. With design constrn. and mgmt. mills and pub. utilities, Balt., also N.Y., 1892-95; partner and dir. Ford, Bacon & Davis, Inc., New Orleans, N.Y., San Francisco, 1895-1941; engaged with partners in design, constrn. and mgmt. various pub. utilities, railways and indsl. plants, New Orleans, San Francisco, other southern and western cities, 1895-1907, including resurveys and reconstrn. of San Francisco after earthquake and fire of 1906; v.p., mgr. Am. Cities Ry. & Light Co., 1907-11; pres. Am. Cities Co., 1911-13; engaged in design, constrn. and reconstruction various terminals, railways, warehouses, harbor structures, New Orleans, Mobile and Galveston, 1914-18; prepared report to sec. war on strategic seclusion and mil. strength of New Orleans area for airplane, army and naval bases and operating terminals, 1918; supr. prodn. Platt Iron Works, Dayton, O., of army tanks for A.E.F., 1918; supr. constrn. New Orleans Army Supply Base, 1918-19; dir. Atlantic Aircraft Corp., 1925-27, Fokker Aircraft Corp., 1927-30; active in devel. and manufacture of airplane engines; gen. engring. practice and capital mgmt., since 1921. Mem. Am. Soc. M.E., Am. Soc. C.E., La. Engring. Soc. Died May, 1957.

DAVIS, George Kelso, Am. nutritionist, biochemist; b. Pitts., July 2, 1910; s. Ross Irwin and Jennie L. (Kelso) D.; B.S., Pa. State U., 1932; Ph.D., Cornell U., 1937; m. Ruthanna Wood, Jan. 25, 1936; children—Dorothy (Mrs. Arthur C. Aikin, Jr.), Ruthanna M., Mary E. (Mrs. W. E. Benedict), Virginia K. (Mrs. John M. Edison), Robert W., George W. R. Asst. in nutrition Cornell U., 1933-37; asst. prof. Mich. State U., East Lansing, 1937-42; prof. nutrition U. Fla., Gainesville, 1942—, dir. nuclear sci., 1960-65, dir. biol. sci., 1965—. Cons. govts. of Costa Rica, Peru, Chile, Argentina, Brazil, Panama, 1953—, FAO, WHO, 1958—. Named hon. prof. U. Chile. Mem. Am. Chem. Soc. (Fla. award), Am. Inst. Biol. Scis., A.A.-A.S., Am. Soc. Biochemistry, Am. Inst. Nutrition (Borden award), Am. Nuclear Soc., Am. Soc. Animal Sci., Am. Dairy Sci. Assn. Research and publs. on relations nutrition to disease, trace element nutrition, radioactive isotopes as tracers, sources radiation. Home: 429 N.W. 24th St., Gainesville, Fla. 32601.*

DAVIS, Gwilym George, Am. surgeon; b. Altoona, Pa., July 20, 1857; s. Thomas Rees and Catherine (Fosselman) D.; A.B., Central High Sch., Phila., 1876, A.M., 1881; M.D., U. Pa., 1879, U. Göttingen, 1881; LL.D., Lafayette Coll., 1911. Resident physician Pa. Hosp., 1881-82; surgeon St. Joseph's, Episcopal, German, Orthopedic hosps. many years; asso. prof. applied anatomy, U. Pa., 1900-11, prof. orthopedic surgery, 1911—; orthopedic surgeon Phila. Gen. Hosp., 1902-14; chief surgeon Widener Sch. for Crippled Children. Fellow Am. Surg. Assn., Phila. Acad. Surgery, Coll. Physicians Phila.; mem. Royal Coll. Surgeons Eng. Author: The Principles and Practice of Bandaging, 1891; Applied Anatomy, 1910. Died June 16, 1918.

DAVIS, Harmer E(lmer), Am. civil engr.; b. Rochester, N.Y., July 11, 1905; s. Elmer and Charlotte A. (Harmer) D.; B.S., U. Cal., 1928, M.S., 1930; m. Clare Housel Melbin. Research asst., U. Cal., 1928-30, research engr. engring. materials lab., 1930-48, dir. inst. transportation and traffic engring. since 1948, instr. civil engring., 1930-36, asst. prof., 1936-39, asso. prof., 1939-48, prof. since 1948, chmn. div. transportation, 1949-54, chmn. dept. civil engring., 1954-59. Chmn. exec. com. Hwy. Research Bd. Recipient Roy Crum Distinguished Service award, Hwy. Research Bd., 1959; James Laurie prize Am. Soc. Civil Engrs., 1967. Registered engr., Cal. Fellow Am. Soc. C.E. (chmn. exec. com. hwy. div. 1956-57, pres. San Francisco sect. 1959, nat. dir. 1960-63); mem. Nat. Research Council (mem. Div. Engineering and Indsl. Research 1960-63, 63—, mem. exec. com. of div. 1961-64), Am. Soc. Testing Materials, National Academy of Engineering, Am. Concrete Institute, Highway Research Board, Am. (director 1964—), Cal. (dir. 1956—) automobile assns., Am. Pub. Works Assn. (hon. 1961), Sigma Xi, Phi Beta Kappa, Tau Beta Pi, Chi Epsilon. Author: Making and Testing of Plain Concrete, 1938; Testing and Inspection of Engineering Materials, 1941; Composition and Properties of Concrete (with G. E. Troxell), 1956. Cons. editor of McGraw-Hill Rook Company, 1955—. Clubs: Cosmos (Washington); Faculty (U. Cal., Berkeley); Commonwealth, Bohemian (San Francisco). Contbr. articles, tech. publs. Research on soil mechanics; bituminous materials; cement concrete; transportation engineering. Home: 200 Yale Av., Berkeley 8, Cal.

DAVIS, Harold Thayer, Am. mathematician; b. Beatrice, Neb., Oct. 5, 1892; s. Harry Watson and Hellen (Moulton) D.; A.B., Colo. Coll., 1915, LL.D., 1949; A.M., Harvard, 1919; Ph.D., U. Wis., 1926; m. Agnes Marie Holm, Sept. 3, 1921; children—Hellen Dagmar (Mrs. Leon Little), Donald Holm (dec.), Harold Moulton. Instr., U. Wis., 1920-23; faculty Ind. U., 1923-37; faculty, prof. Colo. Coll., econometrics Colorado College, also director Cowles Commn. for Research in Econs., 1937; prof. math. Northwestern U., 1937-60, chmn. dept., 1942-55; prof. Trinity U., San Antonio, 1960—; asso. S.W. Found. Research and Edn., 1960-65. Cons. math. physics Lawrence Radiation Lab., Livermore, Cal., 1958-59. Mem. Math. Assn. Am., Econometric Soc., Phi Beta Kappa, Sigma Xi, Delta Epsilon Tau Kappa Alpha. Author numerous books including: Alexandria, the Golden City, 2 vols., 1955; (with others) Studies in Differential Equations, 1956; The Adventures of an Ultra-Crepidarian, 1962; The Summation of Series, 1962; Introduction to Nonlinear Differential and Integral Equations, 1962; (with Vera Fisher), Tables of Mathematical Functions Vol. 3, 1962; also numerous articles. Computation of math. tables, solution linear and non-linear functional equations, application math. methods to philosophy, the social scis. and theory cancer especially skin cancer. Home: 131 Park Hill Dr., San Antonio 78212.*

DAVIS, Harry Aaron, surgeon; b. London, Eng., Sept. 27, 1905; s. Joseph and Sarah Anna (Kaufman) D.; L.R.C.S., L.R.C.P., U. Edinburgh, 1928; M.D., McGill U., 1931; Ph.D., U. Chgo., 1958; m. Elfrieda Miriam Druhe, Dec. 26, 1942. Practice medicine, specializing in surgery, Los Angeles; faculty Loma Linda U., 1946—, clin. prof. surgery, 1949-58, emeritus clin. prof. surgery, 1958—, dir. div. grad. surgery, 1946-57, dir. Hunterian Lab. Surg. Research, 1946-57; dir. surg. research lab. Los Angeles County Gen. Hosp., 1955-58. Recipient Davidson prize, 1937. Diplomate Am. Bd. Surgery. Fellow A.A.A.S., A.C.S.; mem. Internat. Soc. Surgery, Soc. Exptl. Pathology, Am. Assn. Cancer Research, Soc. Exptl. Biology and Medicine, A.M.A. Author: Shock and Allied Forms of Failure of the Circulation, 1949; Principles of Surgical Physiology, 1957; Blood Volume Dynamics: Studies in Surgical Disease, 1962. Contbr. numerous articles to profl. jours. Studies on shock, water and electrolyte metabolism, dehydration, ascitic fluid transfusion, blood substitutes, devel. blood volume technics and cardiac output technics, lung transplantation, exptl. pulmonary artery occlusion, pulmonary embolism, blood volume changes in disease, basal blood volume, exptl. bowel strangulation, exptl. appendicitis, cardiac surg. technics and instruments. Office: 2010 Wilshire Blvd., Los Angeles 90057.*

DAVIS, Harvey Nathaniel, Am. mech. engr.; b. Providence, R.I., June 6, 1881; s. Nathaniel French and Lydia Martin (Bellows) D.; A.B., Brown U., 1901, A.M., 1902, Sc.D., 1928; A.M., Harvard, 1903, Ph.D., 1906; Sc.D., Northeastern Univ., 1938, Columbia, 1940; LL.D., Rutgers, 1928; E.D., Stevens Inst. Tech., 1948; D.Eng., N.Y. U., 1936, Rose Poly. Inst., 1938, Rensselaer Poly. Inst., 1949; m. Suzanne C. Haskell, June 28, 1911 (dec. Jan. 1919); children—Suzanne, Louisa Frederika; m. 2d, Alice M. Rohde, Sept. 20, 1920 (dec. Aug. 1933); children—Marian, Nathaniel; m. 3d, Helen Clarkson Miller, Feb. 8, 1935. Instr. math. Brown Univ., 1901-02; inst. physics Harvard 1904-10, asst. prof., 1910-19, prof. mech. engring., 1919-28; became pres. Stevens Inst. Tech., 1928, ret., 1951. Engr. in turbine dept. Gen. Electric Co., 1917-18; cons. engr. Franklin Ry. Supply Co., 1920-27, U. S. Bur. Mines, 1921-25, Air Reduction Co. 1922-25. Fellow Am. Acad. Arts and Scis., Am. Phys. Soc., A.A.A.S., Am. Soc. M.E. (pres. 1937-38); mem. Am. Math. Soc. (life), Franklin Inst. (hon.), Instn. of Mech. Engrs., London, Am. Philos. Soc., Newcomen Soc., Washington Acad. Scis., Phi Beta Kappa, Sigma Xi, Tau Beta Pi, Delta Phi. Conglist. Author: (with L. S. Marks) Steam Tables and Diagrams, 1908; (with N. Henry Black) Practical Physics for High Schools, 1913; Elementary Practical Physics, 1938. Participant helium research of Army, Navy, Bur. Mines, during World War I; research in thermodynamics, steam tables, tabulated data on properties of steam; patentee steam turbine, 1922, also processes and apparatus for liquefaction and rectifaction of steam. Died Dec. 3, 1952.

DAVIS, Hassoldt, Am. explorer, writer; b. Boston, July 3, 1907; s. Albert Milton and Lucille (Hassoldt) D., student Harvard 2 1/2 years (class of 1929); div. Visited South Pacific Islands, 1929-30; subsequently worked with Andre Roosevelt on moving picture of Bali named "Goona-Goona"; returned to South Seas for 2 more yrs.; writer and photographer with Denis-Roosevelt Asiatic Expdn., 1939, driving up Burma Rd. into China and later visiting Kingdom of Nepal; made collection of ethnol. data on Nepal and comprehensive photographs of country. During 1947-48, led expdn. sponsored by UNESCO and French Govt. to explore Tumac Humac Mountains of French Guiana and to film nomadic Indians; led ethnol. expdn. to Ivory Coast of Africa, 1949-50. Joined Free French Forces, 1941, serving in Tchad, Lybia, Tunisia, Italy, France and Indo-China campaigns, 1941-46. Awarded Croix de Guerre (twice), Award of Legion of Honor (France). Mem. Royal Geog. Soc. Australia, Cercle de la France d'Outre Mer. Author: Islands Under the Wind, 1933; Save Me the Sun, 1939; Land of the Eye, 1940; Nepal, Land of Mystery, 1942; The Fighting Family Paux, 1943; Half Past When, 1945; Feu d'Afrique, 1946; The Jungle and the Damned, 1952; Sorcerers' Village (film). 1955 (book) 1955; World Without a Roof, 1957; Bonjour Hangover, Captain Billy's Magic, 1959. Died Sept. 10, 1959.

DAVIS, Henry Gassett, Am. surgeon; b. Trenton, Me., Nov. 4, 1807; s. Isaac and Polly (Rice) D.; M.D., Yale, 1839; m. Ellen W. Deering, 1856; 3 children. Practiced in Worcester, Mass., also Milbury, Mass., 1838-54; went to N.Y.C., 1855. Author: Conservative Surgery, as Exhibited in Remedying some of the Mechanical Causes that Operate Injuriously both in Health and Diseases, 1867 (1st significant textbook in history of Am. orthopedic surgery). Founder traction sch. of orthopedic surgery; his theories concerning the nature and treatment of club foot, congenital dislocation of the hip, chronic diseases of the joints and poliomyelitis form the basis for the modern approach to these problems; unique treatment of abscesses anticipated the Cassel-Dahin therapy; 1st to devise a splint for traction and the protection of the hip joint. Died Everett, Mass., Nov. 18, 1896.

DAVIS, Herman S(terns), Am. astronomer; b. Milford, Del., Aug. 6, 1868; s. Thomas Josiah and Mary Jane (Potter) D.; ed. Phillips Acad., Andover, Mass.; A.B., cum laude, Princeton, 1892, grad. study, 1892-93, A.M., 1912; A.M. Columbia, 1894, Ph.D., 1895; m. Coreita Register Hoffecker, May 24, 1894; 1 son, Herman Stearns. Asst. astronomer, U. S. eclipse expdn. to W. Africa, 1889-90; tchr. astronomy and geodesy Columbia, 1895-99; lectr. Bd. Edn., N.Y.C., 1896-99, 1905-07; asst. U. S. Coast and Geodetic Survey, 1900; dir. Internat. Latitude Obs., Gaithersburg, Md., 1900-05; cons. engr. and auditor, N.Y.C., Pitts., 1905-10; sec. to pres. Gulf Refining Co., Pitts., 1910-20; sec.-treas. Indian River Fruit & Vegetable Co., Indian River Grove & Farming Co., Dupont Land Co., Indian River Corp., Matson Oil Co. Astronomer, Carnegie Instn., Nat. Acad. Scis. Am. editor Astronomischer Jahresbericht, 1900-14; dir. New Reduction of Piazzi's Star Catalog, 1895—. Author: Glossary to Homer's Iliad, 1888; Parallax of Eta Cassiopeiae, 1895; Catalogue of 62 Stars about Eta Cassiopeiae, 1895; Computation Forms for the Use of Classes in Practical Astronomy, 1897; An Abbreviated Form for Least Square Solutions, 1898; Private Cipher-Book, 1911; Dictionary of Telegraphic Code-Words, 1912. Died May 23, 1933.

DAVIS, Jared James, Am. ecologist; b. Custer, Wash., July 9, 1920; s. Alvin Jared and Myrtle (Brown) D.; B.S., U. Wash., 1942; M.S., Wash. State U., 1948; m. Audrey Jean Buchanan, Aug. 17, 1945; children—Cheryl Jean, Ronald Jared, Craig James. Aquatic biologist Hanford Labs. Gen. Electric Co., Richland, Wash., 1948-58, mgr. radioecology sect., 1958-62; v.p. Hazleton Nuclear Sci. Corp., Palo Alto, Cal., 1963; ecologist div. biology and medicine U. S. AEC, Washington, 1964-65; tech. asst. ecology Pres.'s Office of Sci. and Tech., Washington, 1965—. Fellow A.A.A.S.; mem. Entomol. Soc. Am., Am. Inst. Biol. Sci., Ecol. Soc. Am., Am. Soc. Limnology and Oceanography, Polar Soc. Am., Sigma Xi, Phi Sigma. Contbr. numerous articles to profl. jours. Home: 13116 Chestnut Oak Dr., Rt. 3, Gaithersburg, Md. 20760. Office: Office of Sci. and Tech., Exec. Office of Pres., 17th and Pennsylvania Sts. S.E., Washington 20506.*

DAVIS, John, Am. zoologist; b. Woodmere, N.Y., Dec. 1, 1916; s. George P. and Bertha (Straus) D.; B.A., Yale U., 1937; Ph.D., U. Cal. at Berkeley, 1950; m. Betty Ruth Schuck, Dec. 20, 1947; children—John Steven, Carol Ann. Asst. prof. biology Occidental Coll., 1950-53; jr. research zoologist Frances Simes Hastings Natural History Reservation U. Cal., Carmel Valley, 1953-56, asst. research zoologist, 1956-62, asso. research zoologist, 1962—. Guggenheim fellowship, 1959. Fellow Am. Ornithologists Union; mem. Ecol. Soc. Am., Cooper Ornithol. Soc. (hon.), Soc. for Study of Evolution, Am. Soc. Ichthyologists and Herpetologists, Sigma Xi. Contbr. articles in field to sci. jours. Studies on the systematics, ecology, and behavior of birds and reptiles. Home and Office: Hastings Reservation, Jamesburg Route, Carmel Valley, Cal. 93924.*

DAVIS, John Dwelle, Am. psychologist; b. Poughkeepsie, N.Y., Apr. 7, 1928; s. Philip H. and Helen (Dwelle) D.; A.B., Brown U., 1954; M.A., U. Ill., 1956, Ph.D., 1962; m. Jane E. Peterson, June 19, 1954; children—Philip, John Dwelle, Ward, Andrew. Asst. prof. Am. U. Beirut (Lebanon), 1958-61, Yale, 1961-65; asso. prof. U. Ill., Chgo., 1965—. Mem. Am. Psychol. Assn., A.A.A.S., Psychonomic Soc., Sigma Xi. Research, publs. on role of blood chems. in hunger control; 1st demonstration that a chem. appearing in blood after a meal inhibits further food intake. Home: 1833 Ashbury Av., Evanston, Ill. 60201. Office: Dept. Psychology, U. Ill., Chgo., 60680.*

DAVIS, John Emerson, Am. pharmacologist; b. Detroit, Jan. 1, 1907; s. Emerson and Marion (Biegler) D.; A.B., Oberlin Coll., 1930; M.S., U. Mich., 1931; Ph.D., U. Chgo., 1936; m. Unni Dorothea Haerem, Sept. 2, 1935; 1 dau., Barbara Jean (Mrs. Peter Nordahl Nustad). Faculty, Sch. Medicine U. Ark., 1942-51, prof. pharmacology, 1945-51, sec. hosp. med. staff, 1950-51; prof. pharmacology U. Tex., Austin, 1951—; mem. pharmacology, endocrinology fellowship review panel NIH, 1960-64. Fellow Am. Coll. Cardiology, Am. Coll. Angiology; mem. Am. Physiol. Soc., Am. Soc. Pharmacology and Exptl. Therapeutics, Soc. Exptl. Biology and Medicine, A.A.-A.S. Contbr. numerous articles to profl. jours. Research, publs. on red blood cell prodn. by adminstrn. cobalt salts to dogs, ducks and humans, injection of epinephrine, pituitrin and similar drugs in animals, phys. tng. in dogs, acetylcholinesterase in rats; produced hyperchromic anemia by injections of acetylcholine in dogs; found high concentrations of acetylcholine in serum of pernicious anemia patients, which is reduced to normal by effective therapy. Home: 1413 Larkwood Dr., Austin, Tex. 78723.*

DAVIS, John Frederick, Am. soil scientist; b. Sandusky, Mich., Sept. 11, 1908; s. John Gould and Lydia (McGregor) D.; B.S., Mich. State U., 1933, M.S., 1939, Ph.D., 1943; m. Helen Elizabeth Morony, June 18, 1938; children—Catherine Lydia, Miriam Elizabeth. Faculty U. Del., 1943-44, Cornell U., 1944-46; faculty Mich. State U., East Lansing, 1934-43, 1946—, prof. soil sci., 1953—. Mem. Soil Sci. Soc. Am., Soc. Sugar Beet Technologists, Soil Conservation Soc., Internat. Soil Sci. Soc., Sigma Xi, Phi Kappa Phi, Phi Sigma, Alpha Zeta. Research, publs. on temperature relationship to set of pods of white pea beans; established relationship of zinc to yield of white pea beans; extensive research on soil fertility problems of organic soils; foliar application of micronutrients and secondary plant nutrients; control of magnesium deficiency in celery by foliar spray. Home: 226 Kenberry Dr., East Lansing, Mich. 48823.*

DAVIS, John Henry, Jr., Am. botanist; b. Cumberland County, Va., July 16, 1901; s. John H. and Susan (Morton) D.; B.S., M.A., Davidson Coll., 1924; Ph.D., U. Chgo., 1929; m. Emma C. Adcock, June 2, 1932; children—E. Virginia (Mrs. Robert M. Jeffers), Susan M. (Mrs. James D. Wiltshire). Prof., Davidson Coll., 1924-31, Southwestern Coll., Memphis, 1931-41; prof. botany U. Fla., Gainesville, 1946—; vis. prof. Auckland (New Zealand) U.,

1950, U. Mandalay, (Burma), 1958-60, Nat. Taiwan (Formosa) U., 1964-65; research asso. Carnegie Instn., summers 1935-41, Fla. Geol. Survey, 1941-46. Mem. Ecol. Soc. Am. Author: Ecology and Geologic Role of Mangroves in Florida, 1940; The Natural Features of Southern Florida, 1943; The Peat Deposits of Florida, 1946; The Forests of Burma, 1961. Described Everglades and other regions so. Fla., mangrove swamp forests coasts Fla., forest Burma. Home: 1729 N.W. 8th Av., Gainesville, Fla. 32601.*

DAVIS, John Staige, Am. plastic surgeon; b. Norfolk, Va., Jan. 15, 1872; s. William Blackford and Mary Jane (Kentie) Howland; Ph.B., Yale U., 1895, A.M., 1925; M.D., Johns Hopkins, 1899; m. Kathleen Gordon Bowdoin, Oct. 26, 1907; children—Kathleen Staige (Mrs. Charles E. Scharlett, Jr.), William Bowdoin, Howland Staige. Surgeon, Balt., since 1899; asso. prof. surgery, Johns Hopkins U., since 1923; visiting plastic surgeon, Johns Hopkins Hosp., Union Memorial Hosp., Hosp. for Women of Md., Children's Hosp. Sch. Chmn. Am. Bd. Plastic Surgery. Fellow A.C.S.; mem. Am. Surg. Assn. (vice pres. 1939), Am. Bd. Surgery, So. Surg. Assn. (pres. 1940), Am. Assn. for Surgery of Trauma, Interurban Surg. Soc., Surg. Research Soc., Am. Assn. Plastic Surgeons (pres. 1945). Died Dec. 23, 1946.

DAVIS, John Williams, Am. engr.; b. Petersburg, Va., Nov. 21, 1887; s. Richard Beale and Annie Warwick (Hall) D.; student Randolph Macon Coll., 1904-06; M.E., Cornell U., 1910; M.S., U. Ill., 1917; m. Elizabeth Grimes Walker, Oct. 22, 1921; children—John Williams, Elizabeth Walker, Timothy Pickering. Instructor in elec. engring. Harvard U., 1910-11; Vanderbilt, 1912-13, Stanford, 1913-14, U. Ill., 1914-17; also with various engring. firms for short periods, 1910-17; research in helium gas, U. S. Bur. Mines (inventor process for separation of helium from natural gas), 1919-25; head develop. div., later head tech. dept. Atmospheric Nitrogen Corp., Syracuse, N.Y., Hopewell, Va., 1925-29; cons. engr., Atmospheric Nitrogen Corp., Solvay Process Co., 1929— (inventor of improvements in processes for nitrogen fixation). Mem. commn. to Toronto, Can., on establishment of schs. of mil. aeronautics in U. S., 1917; asst. in establishment of sch. of mil. aeronautics at U. Ill., 1917; adj. Flying Dept. U. S. Army Kelly Field, San Antonio, Tex., later asst. to exec. officer U. S. AS, Washington, later in charge helium work for AS, World War. Died Oct. 4, 1938.

DAVIS, John Woodbridge, Am. civil engr.; b. N.Y. Aug. 19, 1854; s. Edwin Hamilton and Lucy (Woodbridge) D.; C.E., Columbia Sch. Mines, 1878 (Ph.D., 1880); began practice as civil engr., N.Y.; devised plan for sending life lines ashore from ships by means of kites; with cooperation of U. S. Govt., sent out a stout life line, dragged by a large steerable kite, from Brenton Reef lightship to Brenton's Point, a tongue of land 1 3/4 miles distant, 1893. Author: Dynamics of the Sun, 1891. Died 1902.

DAVIS, Joseph Barnard, Brit. craniologist; b. 1801; M.D., St. Andrews, Scotland, 1862; became surgeon on a whaler, 1821; qualified for Apothecaries's Hall, 1823; practiced medicine at Sheldon, Eng. until 1881. Fellow Royal Soc. Author: (with Thurnom) Crania Britannica, 1856-65, 75; Thesaurus craniorum, 1836; Popular Manual of the Art of Preserving Health. Editor Jour. Anthropology, also Anthropologia. Collected, described, classified skulls and skeletons of various races, especially skulls of Brit. Isles. Died May 19, 1881.

DAVIS, Kingsley, Am. sociologist, demographer; b. Tuxedo, Tex., Aug. 20, 1908; s. Joseph Dier and Winifred (Kingsley) D.; A.B., U. Tex., 1930, M.A., 1932; M.A., Harvard, 1933, Ph.D., 1936; m. Judith Blake, Nov. 4, 1954; 1 dau., Laura Isabelle; children—(by previous marriage) Jo Ann (Mrs. Charles Daily), Jefferson Kingsley. Asso. prof., chmn. div. sociology Pa. State U., College Park, 1937-42; research asso., office of population research Princeton (N.J.) U., 1942-48, asso. prof. sociology and anthropology 1944-48; prof. sociology Columbia, 1948-55; prof. sociology U. Cal. at Berkeley, 1955—, chmn. dept., 1961-62, dir. internat. population and research, 1957—. U. S. rep. to Population Commn., UN, 1954-61; chmn. behavioral scis. div. NRC, 1964. Trustee Population Reference Bur., 1952—. Postdoctoral fellow Social Sci. Research Council, 1940-41; fellow Center for Advanced Study in Behavioral Scis., 1956-57; sr. post-doctoral fellow NSF, 1964-65. Mem. Am. Sociol. Assn. (pres. 1959), Population Assn. Am. (pres. 1962-63), A.A.A.S. (v.p. 1963), Am. Philos. Soc., Am. Acad. Arts and Scis., Nat. Acad. Scis. Editor: A Crowding Hemisphere, 1958. Author: Human Society, 1949; The Population of India and Pakistan, 1951; (with others) The World's Metropolitan Areas, 1959. Home: 199 Hillcrest Rd., Berkeley, Cal. 94705.*

DAVIS, Leverett, Jr., Am. astrophysicist; b. Elgin, Ill., Mar. 3, 1914; B.S., Ore. State Coll., 1936; M.S., Cal. Inst. Tech., 1938, Ph.D., 1941; m. 1943. Fellow Rockefeller Inst. 1940-41; faculty Cal. Inst. Tech., Pasadena, 1941—, asso. prof. physics, 1950—. Mem. Am. Phys. Soc., Astron. Soc. Research on stellar electromagnetic fields, polarization of starlight, rocket ballistics, theory of conduction in nerves, electron theory of metals. Address: Cal. Inst. Tech., Pasadena 4, Cal.

DAVIS, Loyal, Am. surgeon; b. Galesburg, Ill., Jan. 17, 1896; s. Albert Clark and Laura (Hensler) D.; student Knox Coll., 1912-14; M.D., Northwestern U., 1918; M.S., 1921, Ph.D., 1923; D.Sc., Knox Coll., 1933, Temple U., 1961; m. Edith Luckett, May 21, 1929; children—Nancy (Mrs. Ronald Reagan), Richard A. Mem. faculty, Northwestern U. Med. Sch., 1924—, prof. surgery, chmn. dept. surgery, 1932-63, prof. emeritus, 1964—, dir. Lab. Surg. Research, 1925-32; chief surgery Passavant Meml. Hosp., Chgo., 1932-63. Cons. neurol. surgeon U. S. VA Hosp., Hines, Ill., 1927-53; cons. surgeon U. S. VA Research Hosp., Chgo., 1953-63. Diplomate Am. Bd. Neurol. Surgery (founding), Am. Bd. Surgery (founding). Decorated Legion of Merit; Recipient Northwestern U. Alumni medal, 1945; NRC fellow, 1922-24. Fellow Royal Coll. Surgeons (Eng.) (hon.), Royal Coll. Surgeons Edinburgh (hon.), A.C.S. (past chmn. bd. regents, past pres.); hon. mem. Soc. Brit. Neurol. Surgeons, Académie de Chirurgie (France), Los Angeles, Mont Reid surg. socs., Omaha Mid-W. Clin. Soc., Argentine Med. Soc., Argentine Surg. Soc., Société de Neurologie de Paris, Western Neurosurg. Soc.; mem. A.M.A., Am. (past pres.), Western, Central, surg. assns., Am. Neurol. Assn. (past v.p.), Soc. Clin. Surgery, Society of Neurological Surgery (past president), Alpha Omega Alpha. Author numerous books including: The Principles of Neurological Surgery, 1936; Go in Peace, 1954; Fellowship of Surgeons, 1960; (with Richard A. Davis) Principles of Neurological Surgery, 1963. Editor: Surgery, Gynecology and Obstetrics, 1939—, Christopher's Textbook of Surgery, 1955—. Research on tracts, tumors, trauma, pain transmission of central nervous system, Parkinsonism, history of surgery, ednl. methods in surgery and neurosurgery. Home: 535 N. Michigan Ave., Chgo. Office: 700 N. Michigan Av., Chgo. 60611.*

DAVIS, M. Edward, Am. physician; b. Cheyenne, Wyo., Oct. 27, 1899; s. Max and Dora (Flaxman) D.; student U. of Colo., 1916-19; B.S., U. of Chicago, 1920; M.D., Rush Med. Coll., 1922; post grad. study Berlin and Vienna, 1927-28; fellow Carnegie Instn. of Embryo, 1933-34; m. Jeannette Sanger, July 12, 1927; children—Barbara Adele, M. Edward. Intern Los Angeles Gen. Hosp., 1922-23; resident obstetrician Chicago Lying-in Hospital, 1925-27; U. of Chicago Clinics. Instr. obstetrics Northwestern, 1929-30; asst. prof. obstet. and gynecol. U. of Chicago, 1930-34, asso. prof., 1934-42, prof. since 1942; Joseph Bolivar DeLee prof. obstet. and gynecol., 1947—; chairman dept. obstetrics and gynecology, 1954—, and chief of service Chgo.-Lying-in Hospital. Recipient Gold Medal Award of A.M.A. for science work on ergot, 1935; Rubin award, Am. Soc. Study of Sterility, 1960, First Gold Medal award of Barren Inc., 1960. Hon. mem. Sociedade de Obstetrica e Ginecologia de Brasil, Sociedad Cubana de Obstet. y Ginecol., Sociedad de Obstet. y Ginecol. de Venezuela, A.F.A. King Obstet. Soc. of George Washington U., Honolulu. Obstet. and Gynecol. Soc.; corresponding mem. Sociedad Lationo Americana para el Estudio de la Esterilidad, 1950. Diplomate Am. Bd. Obstet. and Gynecol. Hon. fellow History of Medicine Soc., Tulane, 1962. Fellow A.C.S. Am., Chgo. gynec. socs. Am. Coll. Obstetricians and Gynecologists, American Soc. Study Sterility; mem. Soc. Gynecologic Investigation; Am. Assn. for Maternal and Infant Health (president 1958-61), Endocrine Soc., A.M.A., Soc. Experimental Biology and Medicine, Am. Society Pharm. and Exptl. Therapeutics, Inc., Central Soc. Clin. Research, Internat. Corrs. Soc. Obstetricians and Gynecologists. Author: DeLee's Obstetrics for Nurses, 1944, 17th Edit.; Have Your Baby Keep Your Figure, 1963. Editor: Cancer of the Uterus (English editor), 1963. Editor of Journal Fertility and Sterility since 1950. Editorial bd. Geriatrics, Acta Cytologica, Psychosomatics, Excerpta Medica, Current Therapy. Contbr. profl. jours. Fundamental studies of the hormones controlling the reproductive function in women; investigation of menopause, its enlightened management, and the retardation of aging. Home: 5760 Blackstone Av. Office: 5841 Maryland Av., Chgo. 60637.*

DAVIS, Martin Arnold, Canadian chemist; b. Montreal, Que., Can., Apr. 24, 1930; s. Louis H. and Eva (Lauterman) D.; B.Sc., with honors in chemistry, McGill U., Montreal, 1951; Ph.D., London (Eng.) U., 1955; m. Margaret Elizabeth Calvert, Oct. 28, 1956; children—B. Mark, Joseph S. Research chemist Ayerst Labs., Montreal, 1955-64; group leader medicinal chemistry, 1964—. Fellow Chem. Soc. (London); mem. Am. Chem. Soc., Canadian Inst. Chemistry, Patentee in field. Research and publs. on synthesis of new drugs affecting central and autonomic nervous systems, antibacterial agts., mechanisms of oxidation processes, novel heterocyclic compounds as potential medicinals. Home: 4758 Meridian Av., Montreal 29. Office: Ayerst Labs., P.O. Box 6115, Montreal, Que., Can.*

DAVIS, Martin David, Am. mathematician; b. N.Y.C., Mar. 8, 1928; s. Harry and Helen (Gottlieb) D.; B.S., Coll. City N.Y., 1948; M.A., Princeton, 1949, Ph.D., 1950; m. Virginia W. Palmer, Sept. 21, 1951; children—Harold, Nathan. Instr., U. Ill., 1950-52; research scientist Inst. for Advanced Study, 1952-54; asst. prof. U. Cal., Davis, 1954-55; asst. prof. Ohio State U., Columbus, 1955-56; faculty Hartford Grad. Center, Rensselaer Poly. Inst., 1956-59, vis. asso. prof. N.Y. U., 1959-60; prof. math., 1965—, faculty Yeshiva U. Grad. Sch. Sci., N.Y.C., 1960-65. Author: Compatability and Unsolvability, 1958. Editorial bd. Jour. Assn. for Computing Machinery, 1960—, Jour. Symbolic Logic, 1954, Transactions Am. Math. Soc., 1960—. Research and publications in logic, undecidability, and computer techniques for proving theorems. Home: 326 W. 85th St., N.Y.C. 10024.*

DAVIS, Nathan Smith, Am. physician; b. Greene, N.Y., Jan. 9, 1817; s. Dow and Eleanor Smith D.; grad. Coll. Physicians and Surgeons, Fairfield, N.Y., 1837; A.M. (hon.), Northwestern; LL.D., Ill. Wesleyan; m. Anna Maria Parker, Mar. 1838. Practiced medicine at Vienna and Binghamton, N.Y., and 1847-49, at N.Y.; in Chgo., from 1849; lectr. Coll. Phys. & Surg., N.Y., 1848; prof. Rush Med. Coll., Chgo., 1849-59; a founder Chgo. Med. Coll. (now med. dept. Northwestern U.), 1859, prof. for 30 yrs., dean faculty until 1898; a founder Mercy Hosp., physician over 40 yrs.; a founder and trustee Northwestern U., Union Coll. of Law (prof. med. jurisprudence), Washingtonian Home for Reformation of Inebriates. Author: Principles and Practice of Medicine; Medical Education and Reform. Editor Annalist, Chgo. Med. Jour., Chgo. Med. Examiner, also Jour. A.M.A., 6 years. Died 1904.

DAVIS, Nathan Smith, Am. physician; b. Chgo., Sept. 5, 1858; s. Nathan Smith and Anna Maria (Parker) D.; A.B., Northwestern U., 1880, A.M., 1883; M.D., Chgo. Med. Coll., 1883; post-grad. Heidelberg, and Vienna, 1885; m. Jessie B. Hopkins, Apr. 16, 1884. Asso. prof. pathology Northwestern U.; 1884-86, prof. principles and practice of medicine and of clin. medicine, 1886—, also dean; practiced medicine, Chgo.; physician Mercy Hosp., 1883—, Wesley Hosp., 1899—, St. Luke's Hosp., 1909—. Mem. 9th Internat. Med. Congress, Pan-Am. Med. Congress; v.p. U. S. Pharmacopoeia Conv., 1910; chmn. sect. of therapeutics and pharmacology and sec. of sect. of medicine A.M.A.; chmn. sect. of medicine, Ill. State Med. Soc. Author: Consumption, How to Prevent It and How to Live With It; Diseases of the Lungs, Heart and Kidneys; Diet in Health and Disease. Died Dec. 21, 1920.

DAVIS, Nathan Smith, Am. physician; b. Chgo., June 25, 1889; s. Nathan Smith and Jessie Bradley (Hopkins) D.; A.B., Harvard, 1910; M.D., Rush Med. Coll., 1913; m. Cordelia Fairbank Carpenter, July 6, 1923; children—Nathan Smith, Graham, Stephen Fairbank, Alden Carpenter. Asst., asso. medicine Rush Med. Coll., 1915-20; asso. medicine Northwestern U. Med. Sch., 1921-28, asst. prof. medicine, 1928-53, asso. prof., 1953-55, emeritus, 1955—. Treas., trustee Nat. Physicians Com. for Extension Med. Service, 1939-41. Fellow Acad. Internat. Medicine (pres. 1945-46, sec. 1947—), A.C.P.; mem. Am. Med. Writers Assn., Am. Soc. Study Arteriosclerosis, Am. Gerontological Soc., Am. Geriatric Soc., Am. Heart Assn., Central Soc. Clin. Research, A.M.A., Miss. Valley Med. Soc., Am. Therapeutic Soc., A.A.-A.S., Sigma Nu. Died Apr. 20, 1956.

DAVIS, Robert Benjamin, Am. mathematician; b. Fall River, Mass., June 23, 1926; s. Benjamin Franklin and Ethel (Reed) D.; S.B., Mass. Inst. Tech., 1946, S.M., 1949, Ph.D., 1951; m. Rose Margaret Garcia, Sept. 14, 1958; children—Alexandrea Claire, Paul Geoffrey. Instr., Mass. Inst. Tech. 1946-51; asst. prof. U. N.H., Durham, 1951-56; faculty Syracuse (N.Y.) U., 1956—, now prof. math., edn.; dir. Madison Project, 1957—; sci. computing IBM, 1956; vis. lectr. Yale, 1959-60; vis. prof. Webster Coll., St. Louis, 1961—. Mem. steering com. African Math. Program, Ednl. Services, Inc., Watertown, Mass., 1961—. Author: Discovery in Mathematics—A Text for Teachers, 1964. Home: 444 Bradford Av., St. Louis 63119.*

DAVIS, Sir Robert Henry, English inventor; b. London, Eng., June 6, 1870; s. Robert Davis; D.Sc. (hon.) Birmingham, 1951; m. Margaret Tyrrell, 1900 (dec. 1952); 4 s., 2 daus. With Siebe, Gorman and Co., Ltd., 1882-1960. Knighted, 1932. Author: Deep Diving and Submarine Operations; Diving Manual; Breathing in Irrespirable Atmospheres, and at High Altitudes, and Resuscitation. Inventor Davis submersible decompression chamber for divers, 1930-35; helped refine oxygen lung for submarine escape, naval reconnaisance, underwater demolition. Died Mar. 29, 1965.

DAVIS, Robert Houser, Am. physicist; b. N.Y.C., Mar. 20, 1926; s. Robert William and Ilene (Houser) D.; B.S., U. Neb., 1949; M.S., U. Wis., 1950, Ph.D., 1955; m. Patricia B. Barth, Aug. 4, 1958; children—Pamela Jean, John Oliver. Research asso. physics Rice U., 1955-57; faculty Fla. State U., Tallahassee, 1957—, prof. physics, 1964—, prin. scientist Tandem Accelerator Program, 1963—; chmn. bd. Recon, Inc. Mem. Am. Phys. Soc., Phi Beta Kappa, Sigma Xi, Phi Mu Epsilon. Co-developer evapor-ion pump, negative helium ion beam injection into tandem accelerators; co-discoverer deformation of doubly closed shell nuclei. Home: 726 Ivanhoe St., Tallahassee, Fla. 32303.*

DAVIS, Robert Wilson, Am. vet. anatomist; b. Grinnell, Ia., Oct. 20, 1910; s. Phillip Franklin and Bertha (Wilson) D.; student Bakersfield Jr. Coll., 1928-

30; D.V.M., Colo. State U., 1935; postgrad. U. Colo., 1949-50; M.S., Colo. State U., 1952; m. Donna Bailey, Aug. 20, 1938; children—Emily Ann, Phillip Howard, Lee Edward, Jeffrey Wilson. Practice vet. medicine, Great Falls, Mont., 1935-37; staff U. S. Bur. Animal Industry, 1935-37; faculty Colo. State U., Ft. Collins, 1937—, prof. anatomy, 1946—, head dept. anatomy, 1948—. Recipient Harris T. Guard Distinguished Faculty award Colo. State U., 1960, Outstanding Profl. Ednl. Exhibit, Nat. Am. Vet. Med. Assn. Conv., 1949. Fellow N.Y. Acad. Scis.; mem. Am. Assn. Anatomists, Am. Assn. Vet. Anatomists, Am., Colo. vet. med. assns., No. Colo. Vet. Medicine Assn., A.A.A.S., Am. Inst. Biol. Scis., Am. Assn. Zoologists, World Assn. Vet. Anatomists, Mountain States Soc. Electron Microscopists, Wildlife Disease Assn. (exec. bd. 1959—), Sigma Xi, Alpha Psi, Phi Zeta, Phi Kappa Phi. Research and publs. on anatomy, pathology of domestic and game animals. Home: 1728 W. Vine Dr., Ft. Collins, Colo. 80521.*

DAVIS, Russell Edmund, Am. biochemist; b. Leesburg, O., Jan. 7, 1903; s. Isaac Ellsworth and Mary (Link) D.; B.S., Wilmington Coll., 1924; B.A., Ohio State U., 1925, Ph.D., 1928; m. Mary Louise Roads, June 5, 1928; children—Mary Katherine (Mrs. E. R. Penn), Martha Ellen (Mrs. J. M. Aisquith), William Russell. Instr. Wilmington Coll., 1928-29; chemist Spreckles Sugar Corp., Yonkers, N.Y., 1929-31; biochemist USDA, Beltsville, Md., 1931-45, nutritionist, 1945-55, research chemist, 1945-63, asst. chief beef cattle research br., 1963—. Fellow A.A.A.S., Am. Soc. Animal Sci.; mem. Am. Chem. Soc., Optical Soc. Am., Sigma Xi. Research, studies, publs. on metabolism of proteins; fermentation of sugars by molds; oxidation of alcohols and aldehydes; spectro-chem. analysis; effects of vitamins on reprodn.; ruminant nutrition. Home: 6304 Tecumseh Pl., College Park, Md. 20740. Office: Agrl. Research Center, Beltsville, Md. 20705.*

DAVIS, Tenney Lombard, Am. chemist; b. Somerville, Mass., Jan. 7, 1890; s. Thomas Lombard and Martha W. (Tenney) D.; student, Dartmouth, 1907-08; B.S., Mass. Inst. Tech., 1913; M.S., Harvard, 1914, Ph.D., 1917; postgrad. U. Cal., 1916-17; m. Dorothy Theresa Münch, Aug. 28, 1923. Austin teaching fellow Harvard, 1913-16; instr. organic chemistry Mass. Inst. Tech., 1919-20, asst. prof., 1920-26, asso. prof., 1926-38, prof., 1938-42, prof. emeritus, 1942-49. Summer lectr. Western Res. U., 1931, 38; sect. chmn. NDRC, 1940-41; dir. sci. research and devel. Nat. Fireworks, Inc., from 1942. Mem. Am. Chem. Soc. (chmn. history chem. div. 1932-39). History of Sci. Soc. (v.p. 1941), A.A.A.S., Am. Acad. Arts and Scis. (corr. sec. 1930-37, rec. sec. 1937-38), corr. mem. Royal Soc. Bohemia (Prague, Czechoslovakia). Newcomen Soc. Author: The Chemistry of Powders and Explosives. Asso. editor Jour. Chem. Edn. Editor-in-chief Chymia. Isis, Tech. Rev. Contbr. various articles on chem. subjects, history of chemistry, and Chinese alchemy. Died Norwell, Mass., Jan. 25, 1949.

DAVIS, Thomas W(ilders), Am. chemist. b. Upper Nyack, N.Y., Aug. 1, 1905; s. George F. W. and Elizabeth (Guenther) D.; B.S., N.Y. U., 1925, M.S., 1926; Ph.D., 1928; postgrad. Columbia, 1930; m. Ruth Emily George, June 2, 1942; children—George P. Jahn (stepson), William F. W., James J. D. Chemist, Combustion Utilities Corp., Linden, N.J., 1928-29; faculty N.Y. U., N.Y.C., 1929—, prof. chemistry, 1951—, chmn. dept., 1956-64; research chemist Metall. Lab., U. Chgo., 1942, Oak Ridge Nat. Lab., 1946-47, U. S. Naval Ordnance Test Sta., China Lake, Cal., 1952, Argonne (Ill.) Nat. Lab., 1955-56; vis. prof., U. Leeds (Eng.), 1964-65. Mem. Am. Chem. Soc., Am. Inst. Chemists, Fedn. Am. Scientists, Radiation Research Soc., Sigma Xi, Phi Beta Kappa, Phi Lambda Upsilon. Asso. editor: Encyclopedia of Chemical Reactions, Vol. 1-8, 1946-59. Research and publs. in rates and mechanisms chem. reactions including those occurring after absorption of light or exposure to ionizing radiations, equilibria in homogeneous and in heterogeneous systems and chem. uses radioactivity. Home: 42 LeFurgy Av., Dobbs Ferry, N.Y. 10522. Office: Dept. Chemistry, N.Y. U., University Heights, N.Y.C. 10453.*

DAVIS, Trupapur Antony, Indian biologist; b. Nagercoil, Madras, India, Feb. 9, 1923; s. Trupapur Anthony Sebastian and Maria Siluvai D.; B.A., St. Joseph's Coll., Trichinopally, India, 1944; postgrad. Madras Agr. Coll., Coimbatore, India, 1944-47; m. Eunice Thomas, Aug. 29, 1949; children—Bernard Paul, Basil Sebastian, Felix Jerome. Botany asst. Sugarcane Breeding Inst., Coimbatore, 1947-48; supt. Sugarcane Research Sta., Karnal, Punjab, India, 1948-52; plant physiologist Coconut Research Sta., Kayangulam, Kerala, India, 1952-60; prof., head crop sci. unit Indian Statist. Inst., Calcutta, 1960—. Recipient awards Invention Promotion Bd., New Delhi, India, 1962, 66. Fellow or mem. Am. Soc. Plant Physiology, Ecol. Soc. Am., Palm Soc., Animal Behavior Soc., Linnean Soc. London, Royal Microscopical Soc., Royal Entomol. Soc., Royal Hort. Soc., Scandinavian Soc. for Plant Physiology, Internat. Soc. Plant Morphologists, numerous others. Author numerous articles. Research on productive advantages of certain left spiraling coconut palms; designed instruments for agrl. research. Home: 78 Ramanputhoor, Nagercoil, Madras, India. Office: 203 B.T. Rd., Calcutta, West Bengal, India.*

DAVIS, Watson, Am. information scientist; b. Washington, Apr. 29, 1896; s. Allan and Maud (Watson) D.; B.S. in C.E., George Washington University, 1918; Civil Engr., 1920, D.Sc., 1959; m. Helen Augusta Miles, Dec. 6, 1919 (dec.); children—Charlotte, Miles; m. 2d, Marion Shaw Mooney, Nov. 21, 1958. Asst. engr. and physicist U. S. Bur. of Standards, 1917-21; science editor Washington Herald, 1920-22; mng. editor Science Service 1921—, dir., 1933—; editor Science News Letter, 1922—, THINGS of science, 1940—, Chemistry (mag.), 1944-62; Columbia Broadcasting System radio program, 1930-59; pres. Am. Documentation Inst., 1937-47. William L. Honnald lectr. Knox Coll., 1939. Chmn. U. S. delegation World Congress of Documentation, 1937; mem. Nat. Inventors Council, 1940—; dir. Science Clubs Am., 1941—, Nat. Sci. Fair Internat., 1949—; emeritus mem. exec. bd. Nat. Child Research Center; chmn. Science Clubs Com., UNESCO, 1949, Popularization Sci. Conf., Madrid, 1955; mem. Sec. Navy's Adv. Bd. Ednl. Requirements, 1959-61; mem., chmn. Sec. Commerce's Patent Office Adv. Com., 1960-62; mem. Nat. Adv. Dental Research Council, 1949-53. Recipient Syracuse U. Journalism medal, 1944; Westinghouse Sci. Writing award, 1946; War-Navy Cert. Appreciation, 1946; Phila. Sci. Council Award, 1951; Thomas Alva Edison Found. award, 1955, 56; Pioneer medal Nat. Microfilm Assn., 1959; James T. Grady medal Am. Chem. Soc., 1960. Registered profl. engr., D.C. Fellow Am. Inst., A.A.A.S.; mem. Overseas Writers, Congl. Press Gallery, White House Corr. Assn., Am. Soc. for Testing Materials, Am. Eugenics Soc., Am. Polar Soc., Am. Concrete Inst., Nat. Assn. Sci. Writers (founder mem.), Aviation Writer's Assn., Acad. of Medicine Washington (pres. 1956-58), Population Soc. Am., Brit. Assn. Advancement Sci., Assn. francaise pour l'avancement des scis., Hist. Sci. Soc., Newcomen Soc., Seismol. Soc. Am., Sigma Xi, Pi Delta Epsilon, Sigma Delta Chi. Editor: Sci. Today, 1931; New World of Science Series, 1931; The Advance of Science, 1934; Atomic Bombing, 1950. Author: The Story of Copper, 1924; Science Picture Parade, 1940; From Now On, 1950; The Century of Science, 1963. Contbr. to mags. and engring. jours. Participated in sci. expdn. of U. S. Naval Obs., 1925, Sci. Service, 1932; contbns. in interpretation and popularization of sci., history of sci., documentation, microfilm duplication techniques, aux. publs., also in concrete and cement. Died 1967.

DAVIS, William Elias Brownlee, Am. surgeon; b. Trussville, Ala., Nov. 25, 1863; s. Elias and Rhoda (Latham) D.; ed. U. Ala.; M.D., Bellevue Med. Coll., N.Y.C., 1884; also studied at hosps. in London, Eng.; Vienna, Austria; Berlin, Germany; m. Gertrude Mustin, Aug. 12, 1897; 2 daus. Practiced at Birmingham, Ala.; founded pvt. infirmary and sanitarium Birmingham; prof. gynecology and abdominal surgery Birmingham Med. Coll. Mem. Ala. State Bd. Med. Examiners; Hon. pres. sect. on abdominal surgery 1st Pan-Am. Med. Congress. Mem. Ala. Surg. and Gynecol. Assn. (founder), So. Surg. Assn. (sec. until 1900, pres. until 1902), Am. Assn. Obstetricians and Gynecologists (became pres. 1901), A.M.A. (pres. sect. diseases of women and obstetrics 1900), Tri-State Med. Soc. (became pres. 1889). Hon. fellow N.Y. State Med. Soc., La. State Med. Assn. Research and publs. on gynecology and surgery of the biliary tract; developed treatment of liver by drainage for removal of gall-stones; performed surgery on liver and bile duct. Died 1902.

DAVISON, Arlen D., Am. plant pathologist; b. Hastings, Neb., Oct. 27, 1932; s. Lannie D. and Esther (Petersen) D.; B.S., U. Wyo., 1955, M.S., 1956; Ph.D., Ore. State U., 1962; m. Nancy Elizabeth Bane, Aug. 8, 1954; children—Timothy R., Robert A., Sarah C. Survey entomologist, plant pathologist U. Wyo., 1958-59, plant pathologist, 1959-61; research asst. Ore. State U., 1961-63; extension plant pathologist U. Ariz., 1963-67; extension plant pathologist Wash. State U., Puyallup, 1967—. Mem. Am. Phytopath. Soc., A.A.A.S., Am. Inst. Biol. Sci., Sigma Xi, Gamma Sigma Delta. Contbr. numerous articles to sci. jours. Research and extension activities on diseases of vegetables, small fruits, field crops; revision of system of identification of races of Uromyces phaseoli variety phaseoli.*

DAVISON, Charles, Am. surgeon; b. Lake County, Ill., Jan. 13, 1858; s. Peter and Martha Maria (Whedon) D.; M.D., Northwestern U., 1883, A.M., 1917; m. Mary Lavinia Kidd, Oct. 20, 1887; 1 son, Charles Marshall. Intern Cook County Hosp., Chgo., 1883-84, attending surgeon, 1894-1926; in practice at Chgo., 1884—; asst. surgeon Ill. Charitable Eye and Ear Infirmary, 1887-92; attending surgeon West Side Hosp., 1896-1907; surgeon-in-chief Univ. Hosp., 1918—; surgeon Research and Ednl. Hosp., 1925-26; prof. surgery Chicago Clin. Sch., 1896-1907; prof. surg. anatomy U. Ill., 1899-1900, adj. prof. surgery, 1900-05, prof., 1905-26, head of dept. of surgery, 1917-26; a founder W. Side Hosp., Univ. Hosp. Fellow Founder's Group Am. Bd. Surgery; founder, fellow A.C.S. Author: Autoplastic Bone Surgery, 1916. Died Jan. 18, 1942.

DAVISON, Peter Fitzgerald, biophysicist; b. London, Eng., Nov. 12, 1927; s. Stanley F. and Edith (Taylor) D.; B.Sc. with 1st class honors, U. Coll., London, 1949; Ph.D., Inst. Cancer Research, London, 1954; m. Bjorg Viseth, June 5, 1954; children—Gail Berit, Peer Neil. Came to U. S., 1957. Mem. staff Inst. Cancer Research, 1951-56, Indsl. Cellulose Research Ltd., Can., 1956-57; research asso. Mass. Inst. Tech., Cambridge, 1957-61, staff div. sponsored research, 1962—. Research and publs. on fractionation and analysis of histones; phys. chemistry of high molecular weight DNA; structure and phys. chemistry of collagen; fractionation and characterization of nerve proteins. Home: 193 Cedar St., Lexington, Mass. 02173. Office: Dept. Biology, Mass. Inst. Tech., Cambridge, Mass. 02139.*

DAVISON, Thomas Callahan, Am. surgeon; b. Woodville, Ga., Nov. 13, 1883; s. Charles C. and Elizabeth (Callahan) D.; M.D., Emory U., Atlanta, 1906; m. Lucile Goodwin, Apr. 15, 1931; children—Betty, Margaret. Intern Ga. Bapt. Hosp., 1906-07; practiced medicine, Atlanta, 1907—; chief surg. service Grady Hosp., 1928-40, chief staff, cons., 1940-53; pres., founder Sheffield Cancer Clinic, 1934, attending, cons. surgeon, 1937—; chief staff, chief surg. service, Ga. Bapt. Hosp., 1940—; asso. prof. clin. surgery Emory U. Med. Sch. Diplomate Am. Bd. Surgeons (mem. founders group), Am. Bd. Thoracic Surgery, Fellow A.C.S., Internat. Coll. Surgeons; mem. Fulton Co. Med. Soc. (pres. 1931), Med. Assn. Ga., So. Surgeons Assn., Southeastern Surgeons Congress (founder 1929; pres. 1939), A.M.A., Am. Goitre Assn. (pres. 1950), Am. Assn. Thoracic Surgeons. Contbr. articles to med. jours. Died Sept. 17, 1953.

DAVISON, Wilburt Cornell, Am. pediatrician; b. Grand Rapids, Mich., Apr. 28, 1892; s. William L. and Mattie E. (Cornell) D.; A.B., Princeton, 1913. Rhodes scholar, 1913-16; Sr. Demy Magdalen Coll., Oxford, 1915-17; B.A., Oxford (Eng.) U., 1915, B.Sc., 1916, M.A., 1919; M.D., Johns Hopkins, 1917; D.Sc., Wake Forest Coll., 1932; LL.D., U. N.C., 1944, Duke U., 1961; m. Atala Thayer Scudder, June 2, 1917; children—William Townsend, Atala Jane Scudder Levinthal, Alexander Thayer, Instr., asso. prof., acting head dept. pediatrics, asst. dean, Johns Hopkins U. Med. Sch., 1919-27; asso. pediatrician, acting pediatrician in charge, editor of The Bulletin of Johns Hopkins Hosp., 1919-27; dean and James B. Duke prof. pediatrics, Duke U. Sch. Medicine, 1927-61, chmn. com. on hosps. and child care, trustee Duke Endowment, 1961—; v.p., dir. Doris Duke Found. Former mem. div. med. scis. NRC, vice chmn., 1941-43; former cons. office Surgeon Gen., U. S. Army; adv. group Armed Forces Med. Library; past mem. com. on vets. med. problems; mem. com. atomic casualties, NRC, med. adv. panel, Oak Ridge Inst. Nuclear Studies; former mem. council, chief cons. V.A.; dir. Platex Park Research Inst.; past mem. med. adv. com. Research Found., nat. adv. com. Chronic Dis. and Health of Aged; trustee Ednl. Council Fgn. Med. Grad.; mem. Civilian Health and Med. Adv. Council. Awarded Alvarenga Prize, 1917. Fellow A.C.P., Am. Acad. Pediatrics, Am. Coll. Clin. Adminstrn. (hon.); mem. Am. Pediatric Soc., Soc. for Pediatric Research, Am. Soc. Clin. Investigation, Am. Acad. Gen. Practice (hon.), Phi Beta Kappa, Sigma Xi, Omicron Delta Kappa, Alpha Omega Alpha (pres.). Author: Pediatric Notes, 1925; Enzymes (with S. A. Waksman), 1926; The Compleat Pediatrician, 1934, 38, 40, 44, 46, 49, 57, 61; numerous articles in med. and scientific jours. Home: Roaring Gap, N.C. 28668. Office: Duke Endowment, North Carolina National Bank, Charlotte, N.C. 28202.

DAVISON, William (Davidson, Davisson, D'Avissone, D'Avissonus, Davidsoune), physician, alchemist, botanist; b. Aberdeenshire, Scotland, circa 1593; M.D. degree. Went to Paris, France, circa 1620; councillor and physician to Louis XIV; apptd. 1st prof. chemistry Jardin des Plantes; keeper Royal Bot. Garden, Paris, 1648; moved to Poland, 1650, to become physician to John Casimir, King of Poland. Author: Philosophia Pyrotechnica, 1633-35 (major chemical text book, which went through 7 edits. to 1675); Oblatio Salis . . . , 1641; Observations sur l'antimoine . . . , 1651; Commentaria in Ideam Medicinae Philosophicae Petri Severini Dani, 1660; Collectanea Chimica Medico-Philosophica, 1698. Enthusiastic partisan of Paracelsian chemical medicine; work contains notable early treatment of crystallography. Died Poland, circa 1669.

DAVISON, William Henry Thomas, English chemist; b. London, Eng., June 12, 1916; s. William Phillip and Grace (Ayton) D.; B.A., Jesus Coll., Cambridge (Eng.) U., 1938, M.A., 1947; m. Joyce Evelyn Coe, Dec. 25, 1938; children—Clare Nicolette, Lesley Marguerite. Research chemist Dunlop Research Centre, Birmingham, Eng., 1946-50, group leader, 1951-55; group leader T.I. Research Labs., nr. Cambridge, 1955-61, asst. dir., 1961—. Fellow Royal Inst. Chemistry, Brit. Interplanetary Soc. Research and publs. on applications of infra-red spectroscopy to analysis, structure and configuration of organic compounds and polymers; indsl. uses of radioisotopes and radioactivation; radiation chemistry and devel. indsl. radiation processing of polymers and coatings. Home: 18 Landscape View, Saffron Walden, Essex. Office: T.I. Research Labs., Hinxton Hall, nr. Saffron Walden, Essex, Eng.*

DAVISSON, Clinton Joseph, Am. physicist; b. Bloomington, Ill., Oct. 22, 1881; s. Joseph and Mary

(Calvert) D.; B.S., U. Chgo., 1908; Ph.D., Princeton, 1911, Sc.D. (hon.), 1938; D.Sc. (hon.), Purdue U., 1937, Colby, 1940; Dr. (hon.), Lyon, 1939; m. Charlotte Sara Richardson Aug. 4, 1911; children—Clinton Owen Calvert, James Willans, Elizabeth Mary Dixon, Richard Joseph. Instr. physics, Carnegie Inst. Tech., 1911-17; mem. tech. staff Bell Telephone labs. (formerly engring. dept. Western Electric Co.), 1917-46; vis. prof. physics, U. Va., 1947-49; Mem. NRC. Fellow A.A.A.S. (chmn. sect. B 1933), Am. Phys. Soc., Optical Soc. Am.; mem. Nat. Acad. Scis., Am. Philos. Soc., Am. Acad. Arts and Scis., Franklin Inst., Am. Inst., N.Y. Acad. Scis. (hon. life), Sigma Xi, Phi Beta Kappa. Awarded Comstock prize Nat. Acad. Scis., 1928; Elliot Cresson medal, 1931; Hughes medal, Royal Soc. London, 1935; Nobel prize for physics, 1937; Alumni medal U. Chgo., 1941. Editorial bd. Phys. Rev. Contbr. articles to sci. jours. Discoverer (with L. H. Germer) of diffraction of electrons by crystals (1st empirical proof of De Broglie's wave theory of electrons), 1927; investigated thermionics, electron diffraction, radiation, electron devices and optics, crystal physics. Died Charlottesville, Va., Feb. 1, 1958.

DAVITASHVILI, Leo Shiovich, Russian paleontologist, geologist; b. 1895; grad. Moscow U., 1925. Prof., Gubkin Petroleum Inst., Moscow, 1930-48, Moscow U., 1948-50; prof., 1936—; founder, dir. Inst. Paleobiology, Georgian Acad. Sci., 1950—. Recipient Stalin prize, 1949. Mem. Georgian Acad. Sci. Author: Paleontology, 1933, 36; The Development of Concepts and Methods in Paleontology after Darwin, 1940; History of Evolutionary Paleontology from Darwin to the Present, 1948; Course of Paleontology, 1949; Outline History of Theory on Evolutionary Progress, 1956; The Theory of Sexual Selection, 1961. Research on stratigraphy of upper tertiary deposits in southern USSR, theory and history of paleontology; compiler (with A. D. Arkhangelsky) summary of prin. fossils in oilbearing regions of Crimea and Caucasus. Address: Georgian Acad. Sci., Tbilisi, Georg, SRR, USSR.

DAVY, Edmund William, Brit. chemist; b. Penzance, Eng., 1785; s. William Davy; asst. Royal Instn., 1804-12; prof. chemistry Royal Cork (Ireland) Instn., 1813-26; prof. chemistry Royal Dublin (Ireland) Soc. until 1851. Fellow Royal Soc., 1826; mem. Chem. Soc. London, Société Francaise Statistique Universelle (hon.). Author: An Essay on the Use of Peat or Turf, 1850. Discovered acetylene, nitrosyl chloride; developed method for detecting certain metallic poisons; preparation of numerous platinum compounds; invented a lactometer. Died Dublin, Nov. 5, 1851.

DAVY, Edward, physician; b. Ottery, Eng., June 16, 1806; s. Thomas and Elizabeth (Boutflower) D.; ed. sch. of Rev. Richard Houlditch; apprenticed to Charles Wheeler, surgeon of St. Bartholomew's Hosp.; m. Mary Ann Bryant; m. 2d, a son, George Boutflower; joined St. Bartholomew's Hosp., 1825; passed Apothecaries' Hall, 1828; practiced as chemist under name of Davy & Co.; sailed as med. supt. on emigrant ship to Australia, 1839; placed in charge Govt. Assay Office, Adelaide, Australia, 1852, Melbourne, Australia, 1853-54; practiced surgery, Malmesbury, Australia. Mem. Soc. Telegraph Engrs. and Electricians (hon.), Royal Coll. Surgeons. Author: An Experimental Guide to Chemistry, 1836; Outline of a New Plan of Telegraphic Communication, 1836. Invented and patented cement (Davy's diamond cement), 1835; invented needle telegraph, 1837. Died Jan. 27, 1885.

DAVY, Sir Humphrey, English chemist; b. Penzance, Cornwall, Eng., Dec. 17, 1778; s. Robert and Grace (Millet) D.; student grammar schs., Penzance and Truro; apprenticed to apothecary, 1795-98; self-taught in chemistry; LL.D. (hon.), Trinity Coll., Dublin, 1811; m. Lady Jane Kerr, Apr. 11, 1812. Apptd. supt. lab. Beddoes Pnematic Inst., Bristol, 1798-1801; asst. lectr. chemistry Royal Instn., London, from 1801, apptd. prof., also asst. editor Jours. of Royal Instn., 1802, resigned 1812; investigated agrl. chemistry at request of Bd. Agrl., 1802; lectr. electro-chemistry by invitation of Dublin Soc., 1810, 11; made Continental tour, 1813-15; examined papyri of Herculaneum by commn. of Brit. govt. at Neapolitan Mus., Italy, 1818-20; a promoter of London Zool. Soc., 1826. Won prize instituted by Napoleon for best work of year in electricity. Fellow Royal Soc., 1803 Copley, 1805, Rumford, 1816, Royal, 1827, medals; (sec., 1807-12, pres., 1820-27); corr. la section de la première Classe de l'Institut national, 1813, fgn. asso., 1819. Author: Researches, Chemical and Philosophical; chiefly concerning Nitrous Oxide, 1800; A Discourse Introductory to a Course of Lectures on Chemistry, delivered in the Theatre of the Royal Institution, 1802; Outlines of a Course on Chemical Philosophy, 1804; Elements of Chemical Philosophy, 1812; Elements of Agricultural Chemistry, 1813; On the Safety Lamp for Coal Miners, with some Researches on Flame, 1818; On the Safety Lamp for Preventing Explosions in Mines, Houses Lighted by Gas, Spirit Warehouses, or Magazines in Ships, etc., with some Researches on Flame, 1825; Six Discourses delivered before the Royal Society at their Anniversary Meeting, 1827; Salmonia, or Days of Fly-Fishing, 1828; Consolations in Travel; or the Last Days of a Philosopher, 1830. Famous for his study of effects of voltaic electricity on chem. compounds; developed theory of galvanic decomposition; discovered exhilarating and anesthetic effects of nitrous oxide (laughing gas), 1799, also experimented with inhalation of other gases; disproved old caloric theory and postulated that heat is form of motion, 1799; proved that thermal radiations can be transmitted through vacuum; research on tanning, 1801, agrl. chemistry, 1802; experimented (with Wedgewood) in photography, 1802; isolated sodium, potassium, boron, calcium, magnesium, strontium, and barium, 1807-08; proved that chlorine and flourine are elements but was unable to isolate the latter; showed that iodine is similar to chlorine and that it is mem. of group of electronegative elements, 1813; demonstrated that diamond is carbon and that rare earths are oxides of elements; advanced elect. theory of chem. affinity, 1807; demonstrated that some acids are free from oxygen, and showed that base is compound of oxygen and metal, also suggested that hydrogen is responsible for the acidity of acids, 1810-15; investigated fire-damp, and invented Davy safety lamp (still used in mines), 1815. Died Geneva, Switzerland, May 29, 1829.

DAVY, John, Brit. physician, chemist; b. Penzance, Eng., May 24, 1790; s. Robert and Grace (Millett) D.; M.D., U. Edinburgh (Scotland), 1814; m. Margaret Fletcher, 1830, assisted bro. Humphry with research; army surgeon; insp. hosps.; lived in W.I. for 3 years before 1854; apptd. insp. gen. Army Hosps., 1862. Fellow Royal Soc., 1814. Author: An Account of the Interior of Ceylon, 1821; Researches, Physiological and Anatomical, 1839; Notes and Observations on the Ionean Islands . . . , 2 vols., 1842; Lectures on Chemistry, 1849; also numerous articles. Editor: Memoirs of Sir Humphry Davy, 1836. Discovered carbonyl chloride (which he called phosgene), also was 1st to synthesize urea from it, 1811; research on hydrofluoric acid, corrosive sublimate, glow of phosphorus, combinations of metals with chlorine. Died Lesketh-how nr. Ambleside, Eng., Jan. 4, 1868.

DAVY, Richard, Brit. surgeon; b. 1838; s. John Croate and Elisabeth (Sweet) D.; M.B., U. Edinburgh, 1862; m. Edith Cutcliffe; 3 daus. Practiced medicine, specializing in surgery, London, 30 years; cons. surgeon Westminster Hosp. Fellow Royal Coll. Surgeons Eng., Royal Soc. Edinburgh. Author: Surgical Lectures, 1880. Died Sept. 25, 1920.

DAVYDOV, Aleksandr Sergeevich, Russian physicist; b. Yevpatoriya (now Ukraine), 1912; grad. Moscow U., 1939. Sr. asso. Inst. Physics, Ukrainian Acad. Sci., 1934-50, dep. dir., 1950-53; prof. Moscow U.; sr. asso. Phys. Inst., USSR Acad Sci., 1953—. Mem. Ukrainian Acad. Sci. Author: The Theory of the Atomic Nucleus; Electrons on Thin Crystals, 1963. Research on theory of atomic nucleus, absorption and dispersion of light in molecular crystals, theory of luminescence. Address: Moscow University, Leninskie gory, Moscow, USSR.

DAVYDOVSKY, Ippolit Vasilevich, Russian pathoanatomist; born August 1, 1887; graduated from Medical Faculty, Moscow University, 1910; D.Med. Sci., 1921. Assistant dept. pathological anatomy Pirogov 2d Moscow Med. Inst., 1912-21, prosector, later lectr., 1921-30, head chair path. anatomy, 1930—; prosector, later asst. Moscow Yauza Hosp., 1913-21; chief pathoanatomist USSR ministry Health, 1941-45; dir. Lab. Pathology of Old Age, Inst. Human Morphology, USSR Acad. Med. Sci.; prof., 1930—. Decorated Order of Lenin (2). Mem. USSR Acad. Med. Sci. (Presidium mem., v.p. 1953-60), All-Union (chmn., hon. bd. mem.), Moscow (hon. chmn.) socs. pathoanatomists, All-Union Soc. Oncologists (chmn., hon. bd. mem.). Author over 260 works including The Pathological Anatomy and Pathogenesis of the Major Diseases of Man, 1938, 3d edit., 1956; Gunshot Wounds in Man, Vols. 1-2, 1950-54; The Theory of Infections, 1956; General Pathology of Man, 1961; Causality in Medicine, 1962; Gerontology, 1966. Editor: Archives of Pathology; mem. editorial bd. Large Med. Ency., 2d edit.; mem. editorial council Clin. Medicine. Research on path. anatomy, pathogenesis of diseases and war injuries; introduced teaching of anatomy in terms of nosological principles. Address: 2d Moscow Med. Inst., Malaya Pirogovskaya 1, Moscow, USSR.

DAWE, Albert Rolke, Am. physiologist; b. Milw., June 1, 1916; s. Albert James and Hattie (Rolke) D.; B.A., Yale, 1938; M.A., Harvard, 1951; Ph.D., U. Wis., 1953; m. Eleanor Thorn, Sept. 22, 1942; children—Albert Timothy, Constance Ann, Cynthia Lee. Faculty, U. Wis., Milw., 1946-50, asst. prof. Med. Sch., 1953-56; biologist Office Naval Research Chgo., 1956-57, chief scientist, 1958—; research investigator U. Strasbourg (France), 1965-66; vis. asso. prof. Stritch Med. Sch., Loyola U., Chgo., 1964—. Chmn. biology and medicine panel World Book Ency., 1963—; co-dir. Hibernation Information Exchange, 1960—. Mem. Am. Physiol. Soc., Am. Zool. Soc., Am. Soc. for Engring. Edn. (mem. information retrieval com. 1964—.) Author: (with Schopp) Laboratory Manual of Human Physiology, 1957; also articles. Editor: (with Lyman) Mammalian Hibernation, 1960. Research on natural mammalian hibernation. Home: 1050 Knollwood Rd., Deerfield, Ill. 60015. Office: Office Naval Research, 219 S. Dearborn St., Chgo. 60604.*

DAWES, Edwin Alfred, English biochemist; b. Goole, Yorkshire, Eng., July 6, 1925; s. Harold and Maude (Barker) D.; B.Sc., U. Leeds (Eng.), 1946, Ph.D., 1948, D.Sc., 1961; m. Amy Rogerson, Dec. 19, 1950; children—Michael Edwin, Adrian Stewart. Asst. lectr., lectr. biochemistry U. Leeds, 1947-50; lectr., sr. lectr. biochemistry U. Glasgow (Scotland), 1951-63; Reckitt prof. biochemistry U. Hull (Eng.), 1963—. Vis. lectr. Meml. U., St. Johns and Dalhousie U., 1959, U. Brazil, Rio de Janeiro, 1960. Fellow Royal Inst. Chemistry, Inst. Biology; mem. Biochem. Soc. (com. 1964-65), Soc. Gen. Microbiology (com. microbiol. teaching group 1965-66), Am. Soc. Microbiology. Author: Quantitative Problems in Biochemistry, 1956, 3d edit., 1965. Mem. editorial bd. Biochem. Jour., 1958-65. Research, publs. on growth, energetics, metabolism of micro-organisms, enzymology, quantitative evaluation of pathways of glucose metabolism, regulation of metabolism, storage compounds and survival of bacteria. Home: Dane Hill, 393 Beverley Rd., Anlaby, E. Yorkshire, Eng. Office: Dept. Biochemistry, Hull U., Eng.*

DAWES, William Rutter, English astronomer; b. Christ's Hospital, Eng., Mar. 19, 1799; ed. Charterhouse; studied medicine at St. Bartholomew's Hosp.; practiced medicine, Haddenham, Buckinghamshire; astron. studies at Ormskirk, circa 1830-33; in charge South Villa Regent's Park, 1839-44. Gold medal Astron. Soc., 1855. Fellow Royal Soc., 1865; mem. Royal Astron. Soc. Author: Astronomical Observations at South Villa, 1852; Catalogue of Micrometrical Measures of Double Stars. Observed and measured double stars; invented wedge photometer, 1865; established non-atmospheric nature of redness of Mars, 1865; denied Nasmyth's discovery of solar willow-leaves. Died Feb. 15, 1868.

DAWKINS, Sir William Boyd, Brit. geologist; b. Buttington, Welshpool, Dec. 26, 1837; s. Richard Dawkins; ed. Jesus Coll., Oxford, M.A., D.Sc.; D.Sc. (hon.), Manchester; m. Frances Evans, 1866 (dec. 1921); 1 dau.; m. 2d, Mary Congreve, 1922. Geologist, Geol. Survey Gt. Britain, 1861-69; curator Manchester Mus., 1870; became lectr. Owens Coll., 1860, prof., 1872; hon. prof. geology and paleontology Victoria U., Manchester; cons. geologist in mining and civil engring., from 1870; travelled in N.Am., Australia, 1874-90; Lowell lectr., Boston, 1880; hon. fellow Jesus Coll., Oxford. Fellow Royal Soc., Geol. Soc., Soc. Antiquaries; mem. Instn. Mech. Engrs. Author: Cave Hunting, 1874; Early Man in Britain, 1880; British Pleistocene Mammalia, 1866-87; Research, excavations on fossil mammals and early man, role of fossil mammals in dating Paleolithic and Neolithic periods, division of Tertiary into stages; discovered 1st Brit. caveman's art; research on location of coal deposits, surveyed Brit. and French coasts for tunnel. Died Jan. 15, 1929.

DAWSON, Alden Benjamin, biologist; Tryon, P.E.I., Can., Oct. 12, 1892; s. Bruce Edward and Maria (Leard) D.; A.B., Acadia U., 1915, D.Sc., 1938; Ph.D., Harvard, 1918; m. Iva Evelyn Coldwell, Sept. 17, 1918; 1 dau., Enid Barbara (Mrs. Richard William Renaud). Came to U. S., 1915, naturalized, 1932. Prin., Tryon Consol. Sch., 1910-12; prof. biology Mt. Allison U., Sackville, N.B., 1918-19; faculty Loyola U. Med. Sch., Chgo., 1919-25, asso. prof. anatomy, 1923-25; faculty N.Y. U., 1925-29, prof., 1928-29; faculty Harvard, 1929—, prof. zoology, 1938-59, prof. emeritus, 1959—, dir. Biol. Labs., 1935-40, chmn. dept. biology, 1940-45. Cited in zoology U. Buffalo, 1960. Fellow Am. Acad. Arts and Scis., A.A.A.S.; mem. Am. Soc. Naturalists, Am. Soc. Zoologists (past v.p.), Am. Assn. Anatomists, Soc. for Devel. and Growth, Am. Micros. Soc. (past pres.), Internat. Soc. Cell Biology. Research and publs. on processes secretion in glands with or without ducts, growth and differentiation rat skeleton, response red blood cells to lead poisoning in lower vertebrates, reproduction in cat, changes in ovary, uterus and pituitary gland, neurosecretion, devel. spl. methods for staining tissues for microscopic study. Home: 12 Scott St., Cambridge, Mass. 02138.*

DAWSON, Bertrand (Dawson of Penn), English physician; s. Henry D.; ed. U. College, London Hospital, M.D., B.Sc.; Sc.D., U. Pennsylvania, 1925; hon. D.C.L., Oxford, 1926; hon. LL.D., Edinburgh, 1927; m. Ethel Yarrow, 1900; 3 daus. Asst. physician, London Hospital, 1896; physician, 1906; physician in ordinary to Prince of Wales, 1923-36; mem., Medical Research Council, 1931-35; physician in ordinary to George VI, Queen Mary, Edward VIII, George V, Edward VII; consulting physician, Edward VII Sanatorium, Midhurst; tchr. and examiner, London Hospital and Royal College of Physicians. Made 1st Viscount of Penn, 1936; baronet, 1920; other knightly orders. Fellow Royal College Physicians (pres. 1931-38); Fellow Royal Soc. Medicine; pres., Brit. Medical Assn., 1943. Author numerous articles. Research on influenza, diabetes, rheumatoid arthritis, physical exam. of stomach and intestine, diseases of lymphatic vessels, haematemesis, gastric ulcer, use of röntgen rays, diseases of stomach and intestine. Died London, Mar. 7, 1945.

DAWSON, Charles, English paleontologist; b. 1864. Practiced law at Lewes, Sussex, England; collected fossils as hobby. Found cranial fragments with part of jawbone, Piltdown, 1912; this discovery accepted as Lower Pleistocene human skull until testing in 1953 demonstrated that jaw was that of an ape treated to show age; cranial fragments of Piltdown skull believed to have been put on site by Dawson. Died 1916.

DAWSON, Charles R., Am. chemist; b. Peterboro, N.H., Apr. 9, 1911; s. John C. and Eva B. (Trueman) D.; B.S., U. N.H., 1933, M.S., 1935, D.Sc. (hon.) 1953; fellow, Columbia, 1937-38, Ph.D., 1938; Cutting traveling fellow Cambridge (Eng.) U., 1938-39; m. Dorothea A. Lockard, Aug. 21, 1937; children—Patricia Louise, Sarah Mae, John Harold. Instr. chemistry U. N.H., 1933-35; asst. organic chemistry Columbia, 1935-37, instr., 1939-42, asst. prof., 1942-46, asso. prof. organic chemistry, 1946-52, now prof. chemistry; asst. to dean Columbia Coll., 1944-55. Bd. directors New York Botanical Gardens, 1962—; pres. Bd. Edn. Leonia, N.J., 1958-60. Mem. American Chemical Soc., American Association Advancement of Science, N.Y. Acad. Sci., Am. Soc. Biol. Chemists, Sigma Xi, Phi Lambda Upsilon, Phi Kappa Phi, Alpha Chi Sigma. Republican. Contbr. articles profl. jours. Research in the fields of copper protein enzymes and the naturally occurring alkenyl phenols; such as are found in poison ivy and related plants; study of the chem. nature and mechanism of function of ascorbic acid oxidase and the pigmentation and respiratory enzyme, tyrosinase. Home: 177 Lakeview Av., Leonia, N.J. Office: Columbia U., N.Y.C. 27.*

DAWSON, Elmer Yale, Am. botanist; b. Creston, Ia., Mar. 31, 1918; s. Elmer C. and Mabelle (Campbell) D.; A.B., U. Cal. at Berkeley, 1940, Ph.D., 1942; m. Caroline Maxine Christianson, Mar. 7, 1942; children—Dawn Carol, Renée Nicolet. Oceanographer, Scripps Inst., La Jolla, Cal., 1943-45; faculty U. So. Cal., Los Angeles, 1945-55, 62-63; vis. asso. prof. botany U. Hawaii, Honolulu, 1955-56; research asso. Los Angeles County Mus., 1956-60, curator botany, 1957; research dir. Beaudette Found. Biol. Research, Santa Ynez, Cal., 1958-62; dir. San Diego Mus. Natural History, 1964; curator div. cryptograms Smithsonian Instn., Washington, 1965—; asso. dir. Inst. Tropical Biology, San Jose, Costa Rica, 1963. Cons., Cal. Water Pollution Control Bd., 1957-59. Guggenheim fellow, 1945-46; recipient Darbaker prize in phycology Am. Bot. Soc., 1963. Fellow So. Cal. Acad. Sci., San Diego Soc. Natural History, San Diego Zool. Soc., Cactus and Succulent Soc. Am.; mem. Internat. Charles Darwin Found. (sec. for Ams. 1964—), Western Soc. Naturalists (past pres.). Author: Marine Algae of the Gulf of California, 1944; How to Know the Seaweeds, 1956; How to Know The Cacti, 1963; Introduction to Marine Botany, 1966; Cacti of California, 1966; also numerous articles, other books. Research on Marine floras Pacific Mexico, Gulf Cal., El Savador, Peru, Marshall Islands, Palmyra Atoll, Viet Nam, cacti Galapagos Islands; marine algal herbaria. Home: 6377 Dockser Terrace, Falls Church, Va. 22041. Office: Smithsonian Instn. Constitution Av. at 10th St., Washington 20560.*

DAWSON, George Mercer, Canadian geologist; b. Pictou, Nova Scotia, Canada, August 2, 1849; son of John William Dawson; ed. McGill University, Royal School Mines, London, Eng.; LL.D. Geologist, naturalist Her Majesty's N.Am. Boundary Commn., 1873-75; became mem. Geol. Survey Can., 1875, asst. dir., 1883, dir., 1895. Mem. Behring Sea Commn., 1891; commr. under Behring Sea Joint Commn. Agreement, 1892. Fellow Royal Soc., 1891; pres. Royal Soc. Can., 1894. Author: (with W. F. Tolmie) Comparative Vocabularies of the Indian Tribes of British Columbia, 1884; also, numerous sci. and tech. reports, papers. Explored Yukon Valley; pioneer in geology and mapping of Canadian N.W. Territory; Dawson, capital of Yukon Territory, named for him. Died Ottawa, Ont., Can., Mar. 2, 1901.

DAWSON, Harry Medforth, Brit. chemist; b. Nov. 11, 1875; s. Stephen and Emma (Medforth) D.; ed. Yorkshire (Eng.) Coll., univs. Berlin, Leipzig, Giessen (all Germany); D.Sc., Ph.D.; m. Philis Mary Barr, 1907; 3 sons, 2 daus. Asst. lectr. chemistry Yorkshire Coll., 1899-1904, lectr., 1904-19; prof. phys. chemistry U. Leeds (Eng.), from 1919. Fellow Royal Soc. Contbr. papers on mechanism of chem. change in solution to sci. jours. Died Mar. 11, 1939.

DAWSON, John Keith, Brit. chemist; b. Hednesford, Eng., Dec. 13, 1923; s. John and Bernice (Dando) D.; B.Sc. with honors, Birmingham (Eng.) U., 1944, Ph.D., 1946, D.Sc., 1966; m. Dorothy Mary Grice, Aug. 7, 1948; children—Patricia Mary and Nicholas John (twins). Sci. officer Ministry Supply, Chalk River, Ont., Can., 1946-47, at Atomic Energy Research Establishment, Harwell, Eng., 1947-54; staff U.K. Atomic Energy Authority, Atomic Energy Research Establishment, Harwell, 1954—, sr. prin. scientist, 1957—. Fellow Royal Soc., 1933; Royal Inst. Chemistry; mem. Chem. Soc., Brit. Nuclear Energy Soc. Author: (with G. Long) Chemistry of Nuclear Power, 1959; (with R. G. Sowden) Chemical Aspects of Nuclear Reactors, 3 vols., 1963; also articles. Research on chemistry of actinide elements especially magnetic properties of compounds of elements from thorium to americum; chem. properties of oxides and fluorides of uranium and plutonium; chem. aspects of nuclear reactors; design and basic data for chemonuclear reactors; corrosion problems of water-cooled reactors. Home: 124 Oxford Rd., Abingdon, Berkshire. Office: Atomic Energy Research Establishment, Harwell, Berkshire, Eng.*

DAWSON, John Myrick, Am. physicist; b. Champaign, Ill., Sept. 30, 1930; s. Walker Myrick and Emily (Stephan) D.; B.S., U. Md., 1952, M.S., 1954, Ph.D., 1957; m. Nancy Louise Wildes, Dec. 28, 1957; children—Arthur Walker, Margaret Louise. Research physicist Project Matterhorn, Princeton, 1956-62, plasma physics lab., 1962-64, asso. head theoretical group, 1964-66, head theoretical group, 1966—; Fulbright fellow Inst. Plasma Physics, Nagoya, Japan, 1964-65; cons. RCA (N.J.), 1962-63, Boeing Co., Seattle, 1964, 66—. Fellow Am. Phys. Soc.; mem. N.J. Acad. Scis., Sigma Xi, Phi Kappa Phi, Sigma Phi Sigma. Research, publs. in plasma physics, theory of plasma oscillations, kinetic theory of plasmas, numerical models of plasma. Home: R.D. 2, Cherry Hill Rd., Princeton, N.J. 08540.*

DAWSON, John William, Canadian geologist; b. Pictou, N.S., Can., Oct. 30, 1820; s. James D.; ed. at Pictou Coll.; grad. Edinburgh (Scotland) U., 1842; m. Margaret A. Y. Mercer. Supt. edn., N.S., 1850-55; prof. geology, prin. McGill U., 1855-93; field studies of Nova Scotia, 1842, 52; field studies for Canadian Geological Survey, 1871-73. Fellow Royal Soc., 1862; member Royal Soc. Can. (1st pres.), A.A.A.S. (pres.); Knighted, 1884. Author: Acadian Geology, 1855; Air-breathers of the Coal Period, 1863; The Story of the Earth and Man, 1873; The Dawn of Life, 1875; Fossil Men and their Modern Representatives, 1880; Geological History of Plants, 1888; The Canadian Ice Age, 1894; Fifty Years of Work in Canada, 1901. Research on Canadian geology; pioneered in paleobotany; 1st description of Eozoon; studied fossil plants especially fossil forest of carboniferous period, early reptiles; opposed Darwin's ideas on origin of man. Died Montreal, Que., Can., Mar. 2, 1899.

DAWSON, Ray F(ields), Am. biochemist; b. Muncie, Ind., Feb. 13, 1911; s. Emmett Hamilton and Elsie (Fields) D.; A.B., DePauw U., 1935; Ph.D., Yale, 1938; m. Helen Dunham, Aug. 11, 1942. Instr., DePauw U., Greencastle, Ind., 1939-40; asst. prof., Princeton, N.J., 1942-45; Univ. of Mo., Columbia, 1940-42; faculty Columbia U., N.Y.C., 1945-66; prof. plant biology, dir. internat. programs at Coll. Agr., Rutgers University, 1966—. Industrial and agrl. cons., 1943-66; treas., dir. Lancaster Labs. and Lancaster Products, Inc. (Pa.), 1960—. Recipient 2d Ann. Research award Cigar Mfrs. and Cigar Inst. Am., 1957. Mem. Am. Chem. Soc., Assn. for Tropical Biology, Sigma Xi. Research and publs. on nicotine, chemistry tobacco and tobacco smoke, adaptation Chichona to cultivation as source quinine, domestication Dioscorea species. Home: 152 Westcott Rd., Princeton 08540. Office: Coll. of Agr., Rutgers U., New Brunswick, N.J. 08903.*

DAWSON, William Leon, Am. ornithologist; b. Leon, Ia., Feb. 20, 1873; s. William Edwy and Ada Eliza Sarah (Adams) D.; student Washington U., 1887-90; A.B., Oberlin, 1897, A.M., 1903; B.D., Oberlin Theol. Sem., 1899; m. Frances Etta Ackerman, May 1, 1895; children—William Oberlin, Giles Edwin, Barbara Dorothy. Ordained Congl. ministry, 1899; pastor North Ch., Columbus, O., 1900-02; organizer Wheaton Pub. Co., Columbus, 1902, Occidental Pub. Co., Seattle, Wash., 1905. Birds of Calif. Pub. Co., 1911, Birds of Ohio Pub. Co., 1926, Birds of Fla. Pub. Co., 1927. Dir. Internat. Mus. Comparative Oölogy, Santa Barbara, Cal. Author: The Birds of Ohio, 1903; The Birds of Washington, 2 vols., 1909; The Birds of California, 4 vols., 1923. Died Apr. 30, 1928.

DAY, Alan Charles Lynn, English economist; b. Chesterfield, Eng., Oct. 25, 1924; s. Henry C. and Ruth (Lynn) D.; student Queen's Coll., Cambridge (Eng.) U., 1942-43, 1947-49; m. Diana Hope Barry, Sept. 21, 1962. Lectr., London Sch. Econs., U. London (Eng.), 1949-56, reader, 1956-63, prof. econs., 1963—; econ. adviser U.K. Treasury, 1954-56. Mem. bd. Brit. Airports Authority. Author: The Future of Sterling, 1954; Outline of Monetary Economics, 1956; Money and Income (with S. T. Beza), 1959; also numerous articles. Research on internat. monetary econs. and econs. urban planning. Home: 11 Christchurch Hill, London N.W.3., Eng.*

DAY, Albert M., Am. biologist; b. Humboldt, Neb., April 2, 1897; s. John Breese and Laura (Thayer) D.; B.S., U. Wyo., 1921; m. Gertrude Wichmann, Aug. 1923; m. 2d, Eva Kendall, Sept. 2, 1944. children—Doris Jeane, Richard Thayer, John Kendall. Field biologist, U. S. Biol. Survey, 1918; biol. survey in Wyo. for control of predatory animals and rodents to protect livestock and crops, 1920-30; asst. chief (supervising throughout the U. S.) div. Predator and Rodent Control, Washington, 1930-35, chief 1935-38; chief div. Fed. Aid in Wildlife Restoration, promoting improved game mgmt. in cooperation with all states, P.R., Virgin Islands, 1938-42; liaison officer between Fish and Wildlife Service, and War, Navy and other govt. agencies, Washington, 1942-43; asst. dir. Fish and Wildlife Service (govt. wildlife conservation agy.), Dept. Interior, Chgo., 1943-46; dir. Fish and Wildlife Service, 1946-53, asst. to dir., 1953-55; dir. wildlife research Arctic Inst. N.A., 1955-58; dir. Ore. Fish Commn., 1958-60; exec. dir. Pa. Fish Commn., 1960-64; staff cons. Dept. Interior, 1965—; adv. UN Conf. on law of the sea; adv. N. Pacific Fisheries Commn.; apptd. adminstr. Def. Fisheries Adminstrn., 1950; mem. Pacific Marine Fisheries Commn., internat. Pacific Salmon Fisheries Commn., 1947-54; Pa. Water Bd., Pa. Water and Power Resources Bd. Mem. Wildlife Soc., Am. Forestry Assn., Nat. Audubon Soc., Am. Wildlife Inst., Am. Soc. Mammalogists, Am. Fisheries Soc., Izaak Walton League (nat. dir.), A.A.A.S., Internat. Assn. Game, Fish and Conservation Commn. Sigma Xi. Author: North American Waterfowl, 1949. Contbr. articles to mags. A conservationist; made device to provide selectivity in traps set for predatory animals. Home: 1810 Pine St., Camp Hill, Pa. 17011.*

DAY, David Talbot, Am. geologist; b. East Rockport (Lakewood), O., Sept. 10, 1859; s. Willard Gibson and Caroline (Cathcart) D.; A.B., Johns Hopkins, 1881, Ph.D., 1884; m. Elizabeth Eliot Keeler, Mar. 17, 1886; children—Mrs. Stanley G. Breneizer, David Eliot. Demonstrator chemistry University of Maryland, 1884-85; special agent United States Geological Survey, 1883-85, chief, mining and mineral resources division, 1886-1907, expert in charge petroleum investigations, 1907-14; cons. chemist U. S. Bur. Mines, Washington, 1914-20. Exhibitor Centennial Exhbn., 1870; in charge petroleum exhibits Chgo. Expn., 1893; dir. mining Cotton States and Internat. Expn., Atlanta, 1896; sec. Jury of Awards, Tenn. Centennial, 1897; dir. mining Trans-Miss. Expn., 1898; in charge petroleum Phila. Centennial, 1899, Paris Expn., 1900; in charge mining Buffalo Expn., 1901; hon. chief dept. of mines and metallurgy, St. Louis Expn., 1904; hon. commr. mining Lewis and Clark Expn., Portland, Ore., 1905, Jamestown Expn., 1907; U. S. commr. Internat. Commn. for Petroleum Tests, 1907-09; pres. fuel sect. Internat. Congress Applied Chemistry, 1912. Fellow Geol. Soc. Am.; mem. Am. Chem. Soc., Am. Inst. Mining Engrs., Nat. Statis. Soc., Geog. Soc. Switzerland, Nat. Geog. Soc. Compiler: Mineral Resources of the United States, 1885-1904. Author: Day's Handbook of the Petroleum Industry. Helped develop complete statistics for mineral prodn. in U. S.; investigated oil shale distillation, cracking process for deriving gasoline from heavier oils. Died Apr. 15, 1925.

DAY, Edgar A., Am. food chemist; b. Romney, W. Va., Sept. 11, 1928; s. James Edgar and Francis (Louthan) D.; B.S., U. Md., 1953; M.S., Pa. State U., 1955, Ph.D., 1957; m. Esther Virginia Doverspike, Apr. 15, 1955; children—James E., Mary E., George L., Katherine A., Scott A. Asst. prof. U. Md., 1956-58; with Ore. State U., 1958-66, prof., 1964-66; v.p. flavor research Internat. Flavors & Fragrances, Inc., N.Y.C., 1966—. Recipient Inst. of Food Technologists Research award, 1964, Am. Chem. Soc. Borden award, 1965, Basic Research award Ore. State U., 1965. Mem. Am. Chem. Soc., Am. Oil Chemists Soc., Am. Dairy Sci. Assn., Inst. Food Technologists, A.A.-A.S. Author (with H. W. Schultz, R. O. Sinnhuber): Symposium on Foods: Lipids and Their Oxidation, 1962, (with H. W. Schultz, L. M. Libbey): Symposium on Foods: Chemistry and Physiology of Flavors, 1966. Contbr. numerous articles to profl. jours. on dairy chemistry, lipid oxidation, and protein-lipid reactions. Home: 153 Bingham Av., Rumson, N.J. 07760. Office: 521 W. 57th St., N.Y.C. 10019.*

DAY, Elbert J., Am. nutritionist; b. Cullman, Ala., Mar. 11, 1925; s. Albert and Edna (Milligan) D.; B.S., Auburn U., 1952, M.S., 1953, Ph.D., 1956; m. Marie Agnes Davis, May 13, 1948; children—Brenda Joyce, Elbert J. II, Gary Lynn. Asst. animal nutritionist Auburn U., 1955-56; faculty Miss. State U., State College, 1956—, prof., nutritionist, 1961—. Mem. Assn. So. Agrl. Workers, Poultry Sci. Assn., Am. Inst. Nutrition, Sigma Xi. Research, numerous publs. on carotenoids, fat soluble vitamins A, D, K; mineral interrelationships. Home: Rt. 2, Starkville, Miss. 39759.*

DAY, Emerson, Am. physician; b. Hanover, N.H., May 2, 1913; s. Edmund Ezra and Emily (Emerson) D.; B.S., Dartmouth Coll., 1934; M.D., Harvard, 1938; m. Ruth Fairfield, Aug. 7, 1937; children—Edmund P., Robert F., Nancy E., Martha M., Sheryl J. Practice medicine, specializing in preventive medicine N.Y.C., 1947—; mem. staff Meml. Hosp., 1950-63, dir. Strang Clinic, Inc., also pres. Preventive Medicine, 1963—; vis. physician James Ewing Hosp., 1950-63; mem. chief div. preventive medicine Sloan-Kettering Inst., Sr. mem. PMX Med. Group, 1956—; faculty Cornell Med. Coll., 1947-64; adj. prof. biology N.Y. U., 1965—. Recipient Distinguished Achievement award Dartmouth Club, N.Y., 1955, Bronze medal Am. Cancer Soc., 1956. Fellow N.Y. Acad. Scis. (past pres.), A.C.P., Am. Soc. Cytology (past pres.); mem. Am. Cancer Soc. (past pres. N.Y.C. div.). Author: (with Walter O'Donnell and Louis Venet) Early Detection and Diagnosis of Cancer, 1962. Pioneer in fields of preventive medicine and cancer detection. Home: 91 Greenacres Av., Scarsdale, N.Y. 10583. Office: PMI-Strang Clinic, 55 E. 34th St., N.Y.C. 10016.*

DAY, Harry Gilbert, Am. biochemist; b. Lovilia, Ia., Oct. 8, 1906; s. John Freeman and Minta E. (Spencer) D.; A.B., Cornell College, Iowa, 1930, Sc.D. (honorary), 1967; Sc.D., Johns Hopkins University, Balt., 1933; m. Marie Miller, July 10, 1933;

children—Margaret Louise (Mrs. Michael Craig), Barbara Jean (Mrs. Robin P. Baumann), Robert Miller. NRC fellow Johns Hopkins U., 1933-34; Gen. Edn. Bd. fellow Yale, 1934-36; asso. in biochemistry Johns Hopkins U., 1936-40; faculty Ind. U., Bloomington, 1940—, prof. biochemistry, 1950—, chmn. dept. chemistry, 1952-62, asso. dean research and advanced studies, 1967—. Fellow Ind. Acad. Sci. (pres. 1962), Am. Inst. chemists; mem. Am. Chem. Soc. (vice chmn. examination com. 1963—), Am. Soc. Biol. Chemists, Am. Inst. Nutrition, Soc. Exptl. Biology and Medicine, Phi Beta Kappa, Sigma Xi. Author: (with E. V. McCollum and E. Orent-Keiles) Newer Knowledge of Nutrition, 5th edition, 1939; also numerous articles. Research in biochem. significance of trace inorganic elements including zinc, fluorine and boron, conversion of carotene to vitamin A, stannous fluoride and dentifrices in prevention of dental cavities. Home: 916 E. University St., Bloomington, Ind. 47401.*

DAY, Jeremiah B., Am. natural philosopher; b. New Preston, Conn., Aug. 3, 1773; s. Rev. Jeremiah Osborn and Abigail (Noble) D.; grad. Yale, 1795; LL.D. (hon.), Williams Coll., Middlebury Coll., 1817; D.D. (hon.), Union Coll., 1818, Harvard, 1831; m. Martha Sherman, Jan. 14, 1805; m. 2d, Olivia Jones, Sept. 24, 1811. Licensed to preach, 1800, ordained to ministry Congl. ch., 1817; prof. math. and natural philosophy Yale, 1803-17, pres., 1817-47. Author: Introduction to Algebra, 1814; An Inquiry Respecting the Self-Determining Power of Will, 1838; An Examination of President Edward's Inquiry on the Freedom of the Will, 1841. Died New Haven, Aug. 22, 1867.

DAY, Mahlon Marsh, Am. mathematician; b. Rockford, Ill., Nov. 24, 1913; s. Mahlon Harlow and Mary (Marsh) D.; student U. Ore., 1930-32; B.S., Ore. State Coll., 1935; Sc.M., Brown U., 1937, Ph.D., 1939; m. Elizabeth G. Coone, July 15, 1939 (dec. Feb. 1951); children—(Mahlon) Michael, Susan E., George H.; m. 2d., Frances Morfoot Mautner, July 21, 1952; children—Jean Mautner (stepdau.), Donald M., Dorothy F. Mem. staff Inst. for Advanced Study, Princeton, N.J., 1939-40, 1948-49; faculty U. Ill., Urbana, 1940—, prof., 1948—, head dept., 1958-65. NSF fellow U. Wash., 1956-57. Author: Normed Linear Space, 1958. Address: Dept. Math., U. Ill., Urbana, Ill. 61801.*

DAY, Paul Louis, Am. chemist; b. Grants Pass, Ore., Dec. 26, 1899; s. Marcus C. H. and Adella (Scott) D.; A.B., Willamette U., 1921; M.A., Columbia, 1923, Ph.D., 1927; LL.D., U. Ark., 1960; m. Mildred Garrett, June 22, 1922; children—Peggy (Mrs. Peter K. Leppmann), Dorothy Ann (Mrs. James Ciarlo). Instr. sci. Columbia Jr. Coll., Milton, Ore., 1921-22; prof. chemistry Mont. Wesleyan Coll., Helena, 1923-25; prof., head dept. biochemistry Sch. Medicine, U. Ark., Little Rock, 1927-58, asst. dean Grad. Sch., 1956-58; sci. dir. FDA, U. S. Dept. Health, Edn., and Welfare, Washington, 1959-61; scientist adminstr. Nat. Heart Inst., NIH, Bethesda, Md., 1961—. Mem. Am. Inst. Nutrition (past pres., editorial bd. Jour. Nutrition 1948-52, Mead Johnson award 1948), Am. Soc. Biol. Chemists, Am. Chem. Soc. (Mid-W. award 1948, S.W. Regional award 1952, prime organizer, chmn. Central Ark. sect. 1952), Soc. for Exptl. Biology and Medicine, (editorial bd. proc. 1950-52), A.M.A. (spl. affiliate), So. Soc. for Clin. Investigation (founder mem.), Sigma Xi (past pres. Ark. chpt.), Alpha Omega Alpha (hon.). Contbr. numerous articles to tech. jours. Research on exptl. cataract in rats from riboflavin deficiency and tryptophane deficiency, folic acid deficiency, nutrition rhesus monkey. Home: 5405 W. Cedar Lane. Office: NIH, Bethesda, Md. 20014.

DAY, William Cathcart, Am. chemist; b. Urbana, O., May 30, 1857; s. Willard G. and Caroline Cathcart Day; grad. Johns Hopkins (Ph.D.); spl. studies in chemistry and physics; m. Jane Leamy, Dec. 27, 1884. Prof. chemistry Swarthmore (Pa.) Coll. Author series of tech. reports for U. S. Geol. Survey. Research papers on chrome iron ore, brom. cymene, asphalt prodn., action of carbon dioxide on sodium aluminate. Deceased.

DAYAL, Bisheswar, Indian physicist; b. Muzaffarnagar, India, July 9, 1908; s. Gauri and Prasanna (Devi) Shankar; B.Sc., Banaras Hindu U., India, 1928, M.Sc., 1930, D.Sc., 1946; A.I.I.Sc., Indian Inst. Sci., Bangalore, Ind., 1945; m. Prakashi Wati Devi, June 1, 1931; 1 son, Shiva Prasad. Mem. staff Banaras Hindu U., Varanasi India, 1932—, prof., head dept. physics, 1965—; research scholar Indian Inst. Sci., 1940-42; also dir. summers insts. AID. Recipient Wattumull award, 1962. Fellow Nat. Inst., Inst. Physics London, Nat. Acad. Scis. India. Author: (with Dayal, Verma, Pandey) Heat and Thermodynamics, 1966. Research, publs. primarily on laltice dynamics, theoretical computation of vibration spectra, elastic constants, specific heats, thermal expansion and equation of state of solids; exptl. work on X-ray diffuse scattering, thermal expansion measurement and crystal structure determination. Address: New G-1 Quarters, Banaras Hindu U., Varanasi-5, India.*

DAYKIN, David Edward, Brit. mathematician; b. Aldershot, U.K., June 24, 1932; s. David and Maria (Bowden) D.; student Southall Tech. Coll.; B.Sc., Reading U., 1956, M.Sc., 1958, Ph.D., 1961; m. Kathleen V. A. Mills, Jan. 30, 1954; children—Kathleen Susan, Jacqueline Wendy. Engr. trainee AEC Ltd., 1949-53; with Reading U., 1953-59, lectr., 1961-65; mathematician Plessey Co., Romsey, Hants, 1959-61; prof. pure math., dir. computer centre U. Malaya, 1965—. Research, publs. on pure math. and computing. Home: 3, Armour Rd., Reading, Berks, U.K. Office: Math. Dept., U. Malaya, Kurla, Lumpur, Malaysia.*

DAYRIT, Conrado, Philippine pharmacologist, physician; b. Manila, Philippines, May 31, 1919; s. Conrado V. and Eufronia (Singian) D.; A.A., Ateneo de Manila, 1938; M.D., U. Philippines, 1943; postgrad. U. Mich. Sch. Medicine, Cornell U. Med. Coll.; m. Milagros A. Millar, June 1, 1949; children—Manuel, Conrado III, Fabian, Eduardo, Rafael, Ignacio, Regina, Francis. Faculty, U. Philippines Coll. Medicine, 1943—, prof. pharmacology, 1958—, chmn. dept. pharmacology, 1958-66; staff Philippines Gen. Hosp., 1954—, asso. attending physician, 1963—, chief sect. clin. pharmacology, 1965—; dir. med. research United Labs., Inc., 1966—. Hon. cons. internat. panel for humid tropics research program UNESCO, 1961; cons. cardiology Philippine Am. Life Ins. Co., 1961—; hon. adviser Far E. Med. Jour., 1965—; mem. NRC Philippines. Recipient Republic of Philippines Cultural Heritage award; several other prizes for research. Fellow Philippines Coll. Physicians; mem. Internat. (past councilor), Asian Pacific (mem. research com. 1960—) socs. cardiolgoy, Am. Coll. Clin. Pharmacology and Chemotherapeutics (charter mem., fellow), Philippine Heart Assn. (past pres.). Editor-in-chief Physicians Drug Index, 1965. Research, publs. on digitalis and quinidine compounds in man, antihypertensives, actions of drugs on pulmonary circulation, electrolyte content of human hearts, ECG tests for coronary insufficiency, atherosclerosis and cardiovascular problems in Philippines, Philippine medicinal plants, cholinergic transmission in sympathetic nerves, anticholinesterase drugs, newer amebicides, anthelmintics. Home: 680 Gen. Miguel Malvar, Manila. Office: 82 Mayflower, Mandaluyong, Rizal, Philippines; also U. Philippines Coll. Medicine, Herran St., Manila, Philippines.*

D'AZAMBUJA, Lucien Henri, French astronomer; b. Paris, France, Jan. 28, 1884; s. Antonio and Blanche (Gagniot) D'A.; D.Sc.; m. Marguerite Roumens, July 4, 1935. Joined Observatoire de Meudon, 1899, named titular astronomer, 1938; titular astronomer astrophysics sect. Observatoire de Paris. Lauréat de l'Institut, 1915, 27, 35, 43, Société Astronomique de France, 1932, 48. Mem. Société Astronomique de France (pres. 1949-51), Conseil des Observatoires Astronomique, Fédération des Sociétés Françaises de Physique, Comité Français de Radio-Electricité Scientific, Bur. Longitudes (corr.). Research and publs. on structure of solar chromosphere and evolution of protuberances. Home: Laclotte, Quartier Arribourdès, Salies-de-Béarn (B.-P.), France.

DEACON, George Edward Raven, English oceanographer; born Leicester, England, March 21, 1906; son George Raven and Emma (Drinkwater) D.; B.Sc., King's Coll., London, 1926, M.Sc., 1936, D.Sc., 1938; D.Sc. (hon.), U. Liverpool, 1961; m. Margaret Elsa Jeffries, May 11, 1940; 1 dau., Margaret Brenda. Voyages to Antarctic with Discovery Investigations, 1927-39; mem. staff various Admiralty research establishments, 1939-49; dir. U.K. Nat. Inst. Oceanography, Wormley, Surrey, Eng., 1949—. Comdr., Order Brit. Empire; Polar medal, 1942; Alexander Agassiz gold medal, 1962. Fellow Royal Soc., 1944, Royal Soc. Edinburgh, Inst. Navigation (pres. 1961-64); mem. Swedish Royal Acad. Scis. (fgn. mem.), Royal Soc. New Zealand (hon. mem.), Internat. Assn. Phys. Oceanography (pres. 1960-63), U.K. Antarctic Club (pres. 1962-63), U.K. Royal Geog. Soc. (v.p. 1965-66). Chmn., U.K. Nat. Com. for Geodesy and Geophysics, 1955-60, Nat. Com. for Oceanic Research, 1960—. Editor: Oceans, 1962. Research and numerous publs. on gen. circulation of water in Antarctic Ocean and gen. theoretical and practical approaches to problems of phys. oceanography. Home: Flitwick House, Milford, Surrey, Eng. Office: Nat. Inst. Oceanography, Wormley, Surrey, Eng.*

DEACON, George Frederick, English engr.; b. Bridgewater, Eng., July 26, 1843; s. Frederick and Katharine (Charlton) D.; ed. Heversham and Glasgow U.; LL.D. (hon.) Glasgow U., 1902; apprenticed to R. Napier & Sons, Glasgow; m. Emily Zoe Thomson, 1 son, 3 daus.; m. 2d, Ada Emma Pearce. Asst. to C. F. Varley sci. adviser to company on Atlantic Telegraph expdn., 1895; cons. engr., Liverpool, 1865-71; borough and water engr., Liverpool, 1871-79; water engr. until 1890; cons. practice, Westminster, 1890-1909. Recipient Telford medal and premium, 1875, G. Stephenson medal and Telford premium, 1892, Watt medal and Telford premium, 1879. Fellow Royal Meterot. Soc.; mem. Inst. Civil Engrs. (council 1900-09), Inst. Mech. Engrs., Brit. Assn. (pres. mech. sect. 1897), Assn. Municipal and County Eng. (pres. 1878), Royal Inst. Pub. Health (pres. engring. sect.), San. Inst. (pres. reng. sect. 1894). Author on waste supply in 9th and 10th edit. Ency. Brittanica; also many papers and addresses. Recognized authority Mersey estuary; laid inner-circle trampway rails Liverpool, 1877, on own system; introduced wood pavement into Liverpool, improved method of setting paving; invented waste water meter, 1873; devised (with T. Hawksley) and supervised constrn. of system which supplied Liverpool with water from Vyrnwy river in N. Wales, 1876-92; constructed waterworks for Kendal, Methyr Tydfil, Todmorden, Rigglewade, Milton (Kent) and other places, from 1890; at time of death he was working on plans for works at Birkenhead and Ebbw Vale; investigated the utilization of Niagara Falls, 1890, also studied water supply system of London, 1897 and Coolgardie water supply scheme, 1897. Died Westminster, June 17, 1909.

DEACON, Henry, English indsl. chemist; b. London, England, July 30, 1822; apprentice engring. firm at age 14; joined foundry of Nasmyth and Gaskell, Patricroft, Eng.; joined glass firm Pilkington, St. Helens, 1841, went to Widnes with Gossage where he worked on ammonia-soda process, 1851, became partner, 1853; became partner with Gaskell, 1855. Patentee manufacture of heavy chems. Invented process for manufacture of chlorine by oxidation of hydrochloric acid using air and cupric chloride as catalyst. Died Widnes, July 23, 1876.

DE ACOSTA, José, Spanish naturalist; b. Medina del Campo, Spain, 1539; mem. Soc. of Jesus; tchr. theology at Ocana; sent as missionary to Peru, 1570; returned to Spain, 1588; became rector U. Salamanca, Spain. Author: De procuranda salute Indorum, 1589; Historia natural y moral de las Indias (descriptions of plants and animals of Peru, Mexico), 1590. First to describe symptoms of mountain sickness (Acosta's disease), 1590; described simplified modification of amalgamation process used in silver mines, Potosí, Bolivia; worked on plant and animal classification; known as Pliny of New World. Died Salamanca, Feb. 15, 1600.

DEADERICK, William Heiskell, Am. physician; b. Knoxville, Feb. 7, 1876; Thomas Oakley and Josephine (Heiskell) D.; student Southwestern Presbyn. U., 1891-95, U. Louisville, 1895-97; M.D., Vanderbilt U., 1898; m. Ava Van Leer Lusby, Jan. 8, 1921; children—Elizabeth, Margaret, William H. Practiced at Clarksville, Tenn., 1898-99, Marianna, Ark., 1899-12, Hot Springs, Ark., since 1912. Mem. med. adv. bd. Hot Springs Clinic, USPHS; mem. advisory com. div. of venereal diseases USPHS. Fellow A.C.P. (gov.); hon. mem. Nat. Malaria Com., Am. Congress on Internal Medicine. Author: A Practical Study of Malaria, 1909; The Epidemic Diseases of the Southern States (with L. O. Thompson), 1915. Editor Am. Jour. Syphilis, 1920-30. Deceased.

DE ALFARO, Vittorio, Italian physicist; b. Palermo, Dec. 14, 1933; s. Ugo and Bianca (Baschiera) de A.; Doctor's degree, U. Turin (Italy), 1955; Libera Docenza, 1962; m. Brunilde Quassiati, 1957; children—Marco, Luca. Research asso. Istituto Nazionale di Fisica Nucleare, 1957-61; asso. prof. U. Turin, 1961—. Mem. Italian Phys. Soc. Author: (with T. Regge) Potential Scattering, 1965; also articles. Research on elementary particle theoretical physics, including potential scattering, dispersion relations, current algebra. Home: 37/3 Strada Delnobile, Turin, Italy.*

DE ALVAREZ, Russell Ramon, Am. physician; b. N.Y.C., June 20, 1909; s. Juan Enrique and Isidra de Torres (y Sanchez) de A.; B.S., U. Mich., 1931, M.D., 1935, M.S., 1940; m. Betty Jane Casey, Sept. 10, 1943; children—Ann, Russell Ramón. Instr. obstetrics and gynecology U. Mich. Hosp. and Med. Sch., 1938-44; asst. prof. obstetrics and gynecology, dir. gynecology Cancer Clinic, U. Ore. Sch. Medicine, 1946-48; 1st exec. officer, prof. obstetrics and gynecology U. Wash. Sch. Medicine, Seattle, 1948-64; prof. obstetrics and gynecology, chmn. dept. Temple U. Sch. Medicine, Phila., 1964—; obstetrician, gynecologist-in-chief Temple U. Health Scis. Center, 1964—. Mem. sub-com. on chemotherapy for gynecologic cancer Nat. Cancer Inst., ovarian malignancy sect., 1960—. Diplomate Am. Bd. Obstetrics and Gynecology. Fellow A.C.S., Am. Coll. Obstetricians and Gynecologists (past asst. sec.); mem. Am. Gynecol. Soc., Am., central assns. obstetricians and gynecologists, A.A.-A.S., Soc. for Gynecologic Investigation (past pres.), Am. Fedn. for Clin. Research, A.M.A., Norman F. Miller Gynecologic Soc. (past pres.), Am. Soc. Human Genetics, Pacific Coast Obstet. and Gynecol. Soc., Pacific N.W. Obstet. and Gynecol. Assn. (past pres.), Am. Nephrology Soc., Sigma Xi, Alpha Omega Alpha, Nu Sigma Nu, others. Editor; Quar. Rev. Surgery, Obstetrics and Gynecology, 1952—; editorial bd. Western Jour. Surgery, Obstetics and Gynecology, 1949—; editorial com. Am. Jour. Obstetrics and Gynecology, 1952—. Research, numerous publs. on cancer in women, lipid and water metabolism. Home: 810 Waverly Rd., Bryn Mawr, Pa. 19010. Office: Temple U. Health Scis. Center, Phila., 19140.*

DEAN, Alastair Campbell Ross, Brit. biophysicist; b. Glasgow, Scotland, Apr. 3, 1919; s. Robert Forbes and Elizabeth (Ross) D.; B.Sc. with 1st class honors, Glasgow U., 1941; D.Phil., Oxford (Eng.) U., 1951, M.A., 1956, D.Sc., 1964; m. Nesta Jones, Feb. 28, 1948. Staff research dept. Imperial Chem. Industries Ltd., 1941-45, Brit. Drug Houses, Ltd., 1945-49; Carnegie Sr. scholar Oxford U., 1949-51, Imperial Chems. Industries fellow, 1951-55, sr. research officer phys. chemistry, 1955—; fellow St. Cross Coll. Oxford, 1965—. Fellow Royal Inst. Chemistry, Pharm. Soc.

Gt. Britain; mem. Faraday Soc., Biochem. Soc., Brit. Biophys. Soc., Soc. Gen. Microbiology. Author: (with Cyril Hinshelwood) Growth, Function and Regulation in Bacterial Cells, 1966; also articles. Application of laws of physl. chemistry to bacterial growth and to problems asso. with drug resistance, adaptation to new substrates, cell div. and continuous culture of cells. Home: 21 Templar Rd., Oxford, Eng.*

DEAN, Arthur Lyman, Am. chemist; b. Southwick, Mass., Oct. 1, 1878; s. William Kendrick and Nellie May (Rogers) D.; A.B., Harvard, 1900; Ph.D., Yale, 1902; m. Leora Elvena Parmlee, Aug. 11, 1904; children—Sylvia, Lyman Arnold, Pierson Goddard. Asst. instr., Sheffield Sci. Sch. (Yale), 1902-03; instr. plant physiology, Yale, 1903-07, instr. indsl. chemistry, 1908-09, asst. prof., 1909-14; with A.D. Little, chemist and engr., Boston, 1907-08; pres. U. Hawaii, 1914-27; dir. expt. sta. Assn. of Hawaiian Pineapple Canners; pres. Alexander and Baldwin Co. Research asst. Carnegie Instn., 1904-05; chief, sec. of wood chemistry U. S. Forestry Service, 1905-07. Mem. A.A.A.S., Am. Chem. Soc., Sigma Xi, Kappa Gamma Chi, Phi Sigma Kappa. Developed process for refinement chaulmoogra oil for use in treatment leprosy. Died June 1952.

DEAN, Bashford, Am. zoologist, medieval armor expert; born New York, October 28, 1867; A.B., College City of New York, 1886; A.M., Columbia University, 1889, Ph.D., 1890; married Mary Alice Dyckman, 1893. Tutor natural history, Coll. City N.Y., 1886-90; instr. biology, 1891-96, adj. prof. zoology, 1896-1904, prof. vertebrate zoology 1904-27, hon. prof., 1927-28, Columbia. Asst. N.Y. State Fish Commn., 1886-88; asst., 1889-92, biologist, 1900-01, spl. investigator U. S. Fish Commn.; dir. Biol. Lab., Cold Spring Harbor, N.Y., 1890; mem. Adv. Bd., N.Y. Aquarium, 1902-28; curator of herpetology and ichthyology, 1903-26, hon. curator of ichthyology, 1926-28, Am. Mus. Natural History; curator arms and armor, Met. Mus. Art, 1903-28; prof. Fine Arts, N.Y. U., 1925; pres. Dyckman Inst., curator Dyckman House Mus. Trustee N.Y. Mus. Chevalier Legion of Honor. Adviser on armor U. S. War Dept.; maj. of Ordnance U. S. A.; mem. Mission to France, Belgium, Eng., 1917. Author: Fishes, Living and Fossil, 1895; Bibliography of Fishes, 3 vols., 1916-23; numerous works on palaeichthyology and embryology of fishes (myxinoid, chimaeroid and ganoid), and of bibliography of fishes (50,000 titles). Many publs. on armor and arms, including Handbook of Arms and Armor, European and Oriental . . . , 1915; Notes on Arms and Armor, 1916; Helmets and Body Armor in Modern Warfare, 1920. Died Battle Creek, Mich., Dec. 6, 1928.

DEAN, Burton Victor, Am. mathematician; born Chicago, Illinois, June 3, 1924; the son of Samuel and Dorothy (Eisner) Dean; B.S., Northwestern University, 1947; M.S., Columbia Univ., 1948; Ph.D., U. Ill., 1952; m. Barbara Louise Arnoff, Nov. 26, 1958; children—Howard David, Paul Evan, Heather Diana. Instr. math. Columbia, 1947-49, Hunter Coll., 1949-50; research fellow math. U. Ill., 1950-52; mathematician Nat. Security Agy., Washington, 1952-55; research mathematician Operations Research Inc., Silver Spring, Md., 1955-57; asso. prof. operations research, Case Inst. Tech., Cleve., 1957-65, prof. organizational scis., chmn. operations research group, 1965—; vis. prof. Israel Inst. Tech., Haifa, 1962-63. Cons. U. S. industry, govt., 1957—, TAHAL Water Planning for Israel, 1962-64. Fulbright prof., 1962-63. Fellow A.A.A.S. (sec. sect. 1965—); mem. Operations Research Soc. (mem. edn. com., 1957-59, Lancaster prize screener 1958—), Inst. Mgmt. Scis. (council mem. 1966-67, chmn. No. Ohio chpt. 1964-65), Am. Math. Soc. Author: Operations Research in Research and Development, 1963; (with Maurice S. Sasieni, Shiv K. Gupta) Mathematics of Modern Management, 1963. Editor: Mgmt. Sci., 1962—. Home: 2920 Broxton Rd., Cleve. 44120. Office: University Circle, Cleve. 44106.*

DEAN, George, bacteriologist; born Balquhain, N.B., Can., 1863; s. John Dean; ed. univs. Aberdeen (Scotland), Berlin (Germany), Vienna (Austria); M.A., 1885; C.M., M.B., 1889; m. Laura Hope Geddie; 1 son, 1 dau. Univ. asst. to prof. pathology Aberdeen U., Regius prof. pathology, from 1908; asst. pathologist Royal Infirmary; pathologist Royal Hosp. Sick Children, 1891-97; became bacteriologist in charge serum dept. Brit. (later Jenner and Lister) Inst. Preventive Medicine, 1897; became chief bacteriologist Lister Inst., London, 1906; lectr. on bacteriology London U. Mem. War Office Commn. on Typhoid Inoculation. Mem. several sci. socs. Research and publs. on diptheria immunization, antitrypsin, phagocytosis, Tb, rat leprosy. Died May 30, 1914.

DEAN, George Adam, Am. entomologist; b. Topeka, Kan., Apr. 19, 1873; s. Thomas Jackson and Harriet (Reese) D.; tchrs.' certificate, Kan. State Teachers Coll., 1898; B.S., Kan. State Coll., 1895, M.S., 1905; D.Sc. Southwestern Coll., Kan., 1943; m. Minerva Blachly, August 30, 1903; children—Helen Elizabeth, George Thomas, Loua Marjorie, Paul McConnell, Dorothy. With Kan. State Agrl. Coll. from 1902, prof. entomology, also Expt. Sta. entomologist, state entomologist of Kan. from 1912; sr. entomologist, U. S. Dept. Agr., 1923-25. Mem. Mediterranean Fruit Fly com. Fed. Fruit Fly Board. Fellow A.A.A.S., Entomol. Soc. Am. (pres. 1925); mem. Am. Assn. Econ. Entomologists (pres. 1921), Sigma Xi, Phi Kappa Phi, Gamma Sigma Delta. Contbd. extensively to knowledge of control measures for mill insects and to studies of field crop insects; developed heat method for control of injurious insects, and poison bait method for control of grasshoppers, cut worms and army worms. Died 1956.

DEAN, H(enry) Trendley, Am. epidemiologist; born at Winstanley Park (now East St. Louis), Ill., Aug. 25, 1893; s. William Ware and Rosalie Harriet (Trendley) D.; D.D.S., St. Louis U., 1916; grad. Officers Sch., USPHS, 1931; m. Ruth Martha McEvoy, Sept. 14, 1921; children—Ruth Celestine, Dorothea Virginia, Mary Harriet. Engaged in pvt. practice, Wood River, Ill., 1916, 20; with USPHS, 1921-53, dental dir., 1945-53, ret. 1953; professorial lectr. epidemiology U. Chgo., 1953—. Dir. Nat. Inst. Dental Research, 1948-53; duty in div. infectious diseases, div. physiology Exptl. Biology and Medicine Inst., 1931-53; sec. Council Dental Research, Am. Dental Assn. since 1953. Recipient Gorgas award, 1949, John M. Goodell prize, 1950, Jarvie medal, 1951. Lasker award, 1952; Georgetown U. Sch. Dentistry award of Merit, 1953; Distinguished Service award Am. Assn. Pub. Health Dentists, 1953. Fellow A.A.A.S., Am. Coll. Dentists (chmn. Washington sect. 1941; asso. editor Jour. Am. Coll. Dentists 1946-48), Am. Pub. Health Assn.; mem. Am. Dental Assn., Internat. Assn. Dental Research (pres. 1944; trustee since 1948), Washington Acad. Scis., Assn. Mil. Surgeons, Am. Epidemiol. Soc., Nat. Research Council, Fedn. Dentaire Internationale (v.p. from U. S. 1947-52), Am. Water Works Assn. (Com. Policy in Re-fluoridation of Pub. Water Supplies). Contbr. chpts. to sci. books, also numerous articles in field. Research on periodontal diseases, flouridation of pub. water, Vincent's infection, mandibular fractures, radium dial painters' poisoning. Died 1962.

DEAN, Lee Wallace, Am. otolaryngologist; b. Muscatine, Ia., Mar. 28, 1873; s. Henry Munson and Emma (Johnson) D.; B.S., State U. Ia., 1894, M.S., 1896, M.D., 1896; studied in Vienna, 1896-97; m. Ella May Bailey, Dec. 29, 1904; 1 son, Lee Wallace. Prof. and head of otolaryngology and oral surgery, State U. Ia. until July 1927, also dean Coll. of Medicine, 1912-27; prof. otolaryngology Washington U. Sch. Medicine, since 1927; mem. staff Barnes, St. Louis Children's and Jewish hosps.; otolaryngologist in chief McMillan Eye, Ear, Nose & Throat Hosp., Oscar Johnson Research Inst., St. Louis (emeritus 1943). Recipient de Roaldes prize, 1937. Diplomate Am. Bd. Otolaryngology. Fellow A.C.S.; mem. Am. Laryngol. Assn. (pres.), Am. Laryngol., Rhinol. and Otol. Soc. (pres.), Am. Otol. Soc. (pres.), Am. Peroral Endoscopists, Mo. State Med. Soc., Am. Acad. of Ophthalmology and Otolaryngology (pres.), La Societe de Laryngologie des Hopitaux de Paris. Editor Annals of Otology, Rhinology, and Laryngology. Died Feb. 9, 1944.

DEAN, Richard Doggett, Am. dentist; b. Nesbitt, Miss., Sept. 10, 1884; s. Thomas Jefferson and Eliza Francis (Doggett) D.; student Randall U. Sch., Hernando, Miss., 1900-04; B.S., Miss. State, 1908, postgrad. in elec. engring., 1909; D.D.S., U. Tenn., 1922, M.D., 1931; m. Marguerite Gladys Taylor, Sept. 5, 1914. Insp., Municipal Elec. Testing Labs., Seattle, 1909-12; dir., 1912-18; active practice dentistry, Memphis, 1922-24, student instr. U. Tenn. Coll. Dentistry, 1922, prof. applied dental physics, metall. and materials, 1922-24, prof. surgery and pathol., chief div. oral medicine and surgery from 1924, dean Coll. of Dentistry from 1941. Honored by colleagues who founded the Richard Doggett Dean and Marguerite Taylor Dean hon. Odontol. Soc., 1948. Mem. Am., Tenn. dental assns., Internat. Assn. Dental Research, Ninth Dist. Dental Soc., Omicron Kappa Epsilon, Alpha Omega Alpha, Delta Sigma Delta. Research and investigation (with wife) in physical properties of dental materials, bacteriol., serol. and immunogenic studies on Vincent's Infection, bacteriophage as a therapeutic measure in treatment of dental pulps, etc. Contbr. articles to profl., sci. jours. Died Aug. 29, 1950.

DEAN, William Thornton, Brit. geologist; b. Whitby, Yorkshire, Eng., July 25, 1926; s. William Thornton and Kathleen (Hammond) D.; B.Sc., U. Leeds, 1952, D.Sc., 1966; Ph.D., U. Bristol, 1955; m. Janette Fleming Simpson, May 14, 1955; children—Joanne Julia, Tracey Helen. Geologist, H.M. Geol. Survey, London, Eng., 1955-56; sci. officer Brit. Mus. (Natural History), London, 1956-62, prin. sci. officer, 1962-66, dep keeper dept. paleontology, 1966—. Recipient Liverpool Geol. Soc. Silver medal, 1966. Fellow Geol. Soc. London) Lyell Fund award 1961); mem. Geologists Assn., Paleontographical Soc., Paleontol. Assn.

DE ANDRADA E SILVA, José Bonifacio (or Bonifacio, José), Brazilian geologist; b. Santos, Brazil, 1765; studied U. Coimbra, Portugal; returned to Brazil, 1819; instigated Brazil's independence from Portugal, 1822; prime minister of Brazil, 1822-23; apptd. guardian of emperor's minor children, 1831; wrote several sci. treatises. Died 1838.

DEANE, Helen Wendler, Am. anatomist; b. Franklin, N.C., Mar. 16, 1917; d. Julian S. and Bertha (Wendler) Deane; B.A., Wellesley Coll., 1938; M.A., Brown U., 1940, Ph.D., 1943; m. George F. Markham, June 14, 1947. Lectr. in zoology McGill U., Montreal, Que., Can., 1943-44; faculty Harvard Med. Sch., Boston, 1944-54, research asso. Harvard U. Biology Labs., Cambridge, 1955-57; faculty Albert Einstein Coll. Medicine, Yeshiva U., Bronx, N.Y., 1957—, prof. anatomy, research asso. prof. pathology, 1962—, dep. chmn. dept. anatomy, 1963—. Mem. Histochem. Soc. (past treas., councilor 1965—), Am. Assn. Anatomy, A.A.A.S., Am. Inst. Biol. Sci., Am. Soc. Cell Biologists, Am. Soc. Zoologists, Electron Microscope Soc. Am., Endocrine Soc., Marine Biol Lab., Soc. Developmental Biology, Soc. Endocrinologists (Gt. Britain), Soc. for Study Fertility (Gt. Britain), Sigma Xi. Subeditor: Handbuch der experimentellen Pharmakologie, Ergänzungswerk XIV/1, 1962, Ergänzungswerk XIV/2, 1964. Asso. editor Histochemie, 1958—, Jour. Histochemistry and Cytochemistry, 1961—. Research and publs. on histo- and cytophysiology mammalian liver, intestine, adrenal cortex, ovary, uterus, placenta, testis, seminal vesicle; on histochemistry glycogen, lipids, proteins, nucleoproteins, antigens, ascorbic acid, phosphatases, hydroxysteroid dehydrogenases. Home: 541 Pelham Rd., New Rochelle, N.Y. 10805. Office: 1300 Morris Park Av., Bronx, N.Y. 10461.*

DEARBORN, Earl Hamilton, Am. pharmacologist; b. Manhattan, Kan., June 10, 1915; s. Edgar Hamilton and Gladys (Nichols) D.; A.B., U. Kan., 1938; M.A., U. Chgo., 1940, Ph.D., 1942; M.D., Johns Hopkins, 1949; m. Margaret Ann Kuchta, Dec. 24, 1943; children—Margaret Kathleen, Barbara Ann, Earl Hamilton II, Patricia Lee. Faculty, U. Kan., 1937-40, U. Chgo., 1940-43, Johns Hopkins Med. Sch., 1943-52; prof. pharmacology, chmn. dept. Boston U. Sch. Medicine, 1952-56; head dept. pharmacology research Lederle Labs., Pearl River, N.Y., 1956-60, asst. dir. exptl. therapeutics, 1960-63, dir. exptl. therapeutics, 1963-65, asst. dir. research, 1965—. Mem. Am. Soc. Pharmacology and Exptl. Therapeutics, Soc. Exptl. Biology and Medicine, Am. Chem. Soc., Soc. Toxicology, European Soc. Drug Toxicity, N.Y. Acad. Scis., Phi Beta Kappa, Sigma Xi, Alpha Omega Alpha, Phi Sigma. Research and publs. on pharmacology of antimalarials, reserpine, renal enzymes, antibiotics, drug toxicity. Home: 4 Garden Lane, Montvale, N.J. 07645. Office: Lederle Labs., Pearl River, N.Y. 10965.*

DEARBORN, George Van Ness, Am. psychiatrist; b. Nashua, N.H., Aug. 15, 1869; s. Cornelius Van Ness and Louisa Frances (Eaton) D.; Litt.B., Dartmouth, 1890; M.D., Coll. Phys. and Surg. (Columbia), 1893; A.M., Harvard, 1896; Ph.D., Columbia, 1899; m. Blanche V. S. Brown, June 18, 1893; 1 dau., Lucia Eaton (Mrs. Seabury B. Hough). Asst. in philosophy Harvard, 1896, asst. in physiology, 1899; prof. and dir. lab. of physiology Tufts Coll., 1900-16; prof. psychology and edn. Sargent Normal Sch., Cambridge, 1906-21; instr. psychology Sch. Eugenics, Boston, 1912-15; cons. physiologist and psychologist Forsyth Dental Infirmary for Children, Boston, 1913—; asst. physician for nervous diseases Boston City Hosp., 1919-21; physician Augusta (Me.) State Hosp., June-Nov. 1921; surgeon (R) USPHS, neuropsychiatric sect., 1921—; med. officer expert, U. S. Vets.' Bureau, 1924. Fellow Boston Soc. Natural History, Am. Psychiat. Assn. Author: The Emotion of Joy, 1899; A Textbook of Human Physiology, 1908; Moto-Sensory Development, 1910; Relations of Mind and Body, 1914; The Physiology of Exercise, 1918; The Influence of Joy, 1916; How to Learn Easily, 1916; The Psychology of Clothing, 1918; Physiology and Hygiene. Editor: Our Senses Series, 1916. Died Dec. 12, 1938.

DEARBORN, Henry M., Am. physician, b. Epsom, N.H., Nov. 1846; studied Harvard Med. Coll.; grad. Bowdoin Med. Coll., 1869; m. Sadie Smith, Jan. 1873. Located in N.Y. 1880; vis. physician Met. Hosp., from 1883; prof. principles and practice medicine, N.Y. Med. Coll. and Hosp. for Women, 1883-1903; prof. dermatology N.Y. Homeo. Med. Coll. and Hosp., from 1893; asso. editor N. A. Jour. Homoeopathy, 1883-91; cons. dermatologist Flower Hosp., and cons. physician several hosps., N.Y., Bklyn. Author: Diseases of the Skin, 1903. Died 1904.

DEARING, William Hill, Am. physician; b. Memphis, Dec. 3, 1908; s. William Hill and Theresa (Trenham) D.; B.A., U. Pa., 1930, M.D., M.A., 1934; Ph.D., U. Minn., 1941; m. Edith Wintersteen, Aug. 29, 1936; children—Jane C., John C. (dec.), Carl B. Asso. medicine Geisinger Meml. Hosp., Danville, Pa., 1935-36; with Mayo Clinic, Rochester, Minn., 1936—, head sect. medicine, 1955—, bd. govs., 1955-60; prof. medicine Mayo Grad. Sch. Medicine, U. Minn., 1962—. Diplomate Am. Bd. Internal Medicine, Bd. Gastroenterology. Mem. A.C.P., Am. Fedn. Clin. Research, A.A.A.S., Central Soc. Clin. Research, Central Clin. Research Club, Am. Gastroent. Assn. Contbr. numerous articles to med. jours. Research in gastroenterology. Home: Sunny Slopes. Office: Mayo Clinic, Rochester, Minn. 55901.*

DEATHERAGE, Fred E., Am. biochemist; b. Waverly, Ill., Dec. 30, 1913; s. Fred E. and Marian (Sevier) D.; A.B., Ill. Coll., 1935; A.M., U. Ill., 1936; Ph.D., State U. Ia. 1938; D.Sc., Ill. Coll., 1960; m. Nellie Lou Carothers, Jan. 3, 1942; children—Fred Sevier, Catherine Margaret, Marilyn Nan. Instr., State U. Ia., 1938-40; research fellow Ohio State U., Columbus, 1940-42, faculty, 1946—, prof. biochemistry,

1951—, chmn. dept., 1951-64; agrl. biochemist Ohio Agrl. Expt. Sta., 1949—. Prof. food sci., tech. AID, U. Sao Paulo, Escola Superior de Agricultura, Luiz de Queiroz, Piracicaba, Brazil, 1964—; cons. to food industry, 1952—. Mem. Am. Chem. Soc., Am. Soc. Biol. Chemists, Am. Inst. Nutrition, Inst. Food Technologists, Am. Chem. Soc., Am. Soc. for Animal Prodn., A.A.A.S. Patentee in field. Research, numerous publs. on biochem. nature food quality and processing, spoilage, use antibiotics in meat. Home: Am. Consulate Gen., Sao Paulo, APO N.Y.C. 09676. Office: 2121 Fyffe Rd., Columbus, O. 43210.*

DEAVER, John Blair, Am. surgeon; b. 1855; M.D., U. Pa. 1878; Sc.D., Franklin and Marshall; LL.D., Villa Nova. Prof. surgery, U. Pa.; surgeon in chief Lankenau Hosp. Fellow A.C.S., Am. Surg. Assn. Author: Surgical Anatomy (3 vols.); Appendicitis, Its History, Pathology, Treatment, 4th edit., 1905; Enlargement of the Prostate (with A. P. C. Ashhurst), 1895; Surgery of the Upper Abdomen (with same), Vol. 1, 1909; Diseases of the Breast. Pioneer in surgery of appendix and peritoneal cavity; also contbd. to urology, breast surgery. Died Sept. 25, 1931.

DE AYLLON, Lucas Vasquez, Spanish explorer; b. Toledo, Spain, circa 1475. Came to Hispaniola (Santo Domingo) with Gov. Nicholas de Ovando, 1502; judge of Supreme Ct. of Hispaniola; had 400 Indians to use to work his land; entered partnership with clk. of Audiencia of Hispaniola, sent a ship under Francisco Gordillo (capt.) and Alonzo Sotil (pilot) to sail north until reaching land, 1520 (reached Am. coast in approximate area of N.C. and took possession, June 30, 1521); went to Spain and reported discovery, took along a native of the country who reported on great riches there; granted right to explore Am. coast in effort to find western route to Indies by Emperor Charles V, who also granted him and his descendants title of adelantado and gov.; sent 2 ships to explore new lands, 1525 (they returned same year with gold, silver and pearls); sailed from La Plata, Hispaniola, with 3 ships carrying from 500 to 600 settlers, landed at what is now either North or South Carolina, 1526; possibly went as far south as Santee River (S.C.), began settlement of San Miguel de Gualdape (colony abandoned after fever outbreak which left only about 150 survivors who returned to Hispaniola). Died San Miguel de Gualdape, Oct. 18, 1526.

DE AZARA, Félix, naturalist; b. Barbunales, Spain, May 18, 1746; mil. engr.; commd. lt. by 1775, brig. gen. by 1781; sent to settle border dispute Spanish and Portuguese Commns., S.Am., 1781, remained as naturalist and collector geog. data 20 years. Author: 3 studies of nat. history Paraguay, 1801-02; Voyage dans l'Amérique méridionale, 4 vols., 1809. Probably 1st and most thorough of early S.Am. naturalists in animal life histories. Died Aragon, Spain, 1811.

DEB, C(handicharan), Indian physiologist; b. Uttarpara, India, Nov. 1, 1928; s. Amullya Nath and Sovabati (Mitra) D.; B.Sc., M.Sc., Presidency Coll., Calcutta, India, 1948; D.Phil., Calcutta U., 1953; m. Anjali Mitra, May 31, 1953; children—Soumitra, Suparna. Research scholar Govt. W. Bengal, India, 1950-52; research fellow Nat. Inst. Scis., India, 1952-54, 56; postdoctorate research fellow NRC Can., McGill U., Montreal, Que., 1955-56; reader physiology, Calcutta U., 1960—, hon. lectr. biochemistry; research asso. surg. research Sinai Hosp., Balt., 1965-67; fellow surgery John Hopkins, Balt., 1966-67. Recipient Griffith Meml. prize Calcutta U., 1956. Mem. Physiol. Soc. India, Indian Sci. Congress. Research, publs. on vitamin hormone interrelationship, hibernation in toad, histometry, applied histochemistry of liver and kidney, comparative endocrinology; devel. new histochem. method for light and electron microscopy. Home: 11 Banerjee Para St., Uttarpara, Hooghly, India. Office: Dept. Physiology 92, Acharya P.C. Rd., Calcutta, India.*

DE BAILLOU, Guillaume, French physician; b. Paris, France, 1538; 1st physician to Dauphin Louis XIII; prof., head physician at Paris; author: Epidemiorum libri duo, 1640; Liber de rheumatismo, 1642; De virginum et mulierum liber, 1643; Definitionum medicinarum liber, 1639. Founder of modern epidemiology; distinguished between small pox and measles; introduced the term rheumatism as used in the modern sense; 1st description of whooping cough which he called tussis quintana, 1578. Died Paris, 1616.

DE BAKEY, Michael Ellis, Am. surgeon; b. Lake Charles, La., Sept. 7, 1908; s. Shaker Morris and Raheiga (Zerba) DeB.; B.S., Tulane U., 1930, M.D., 1932; M.S., 1935; Dr. honoris causa, U. Lyon (France), 1961, U. Brussels (Belgium), 1962, U. Ghent (Belgium), 1964, U. Athens (Greece), 1964; m. Diana Cooper, Oct. 15, 1936; children—Michael Maurice, Ernest Ochsner, Barry Edward, Denis Alton. Resident U. Strasbourg, 1935-36, U. Heidelberg, 1936; faculty dept. surgery Tulane U., 1937-48, asso. prof., 1946-48; prof. surgery, chmn. dept. Baylor U. Coll. Medicine, Houston, 1948—, v.p. for med. affairs, 1968—; surgeon-in-chief Ben Taub Gen. Hosp., Houston, 1949—; sr. attending surgeon Meth. Hosp., Houston, 1950—; cons. surgery VA Hosp., Houston, 1948—; clin. prof. surgery U. Tex. Dental Br., Houston; cons. Tex. Inst. for Rehab. and Research, also numerous hosps.; surg. cons. U. S. Army Surgeon Gen.; area cons. thoracic surgery VA. Mem. Nat. Adv. Heart council NIH, 1957-61, Nat. Adv. Health council, 1961-65, also various coms.; chmn. President's Commn. on Heart Disease, Cancer and Stroke, 1964. Bd. regents Nat. Library Medicine, 1957-60, chmn., 1959. Recipient Rudolph Matas award in vascular surgery, 1954, Modern Medicine award, 1957, Roswell Park medal, 1959, Great Medallion, U. Ghent, 1961, Grand Cross Order Leopold, Belgium, 1962, St. Jude Man of Year award, 1963, Albert Lasker award for clin. research, 1963. Diplomate Am. Bd. Surgery, Am. Bd. Thoracic Surgery. Fellow A.C.S., Am. Coll. Cardiology (hon.); mem. A.A.A.S., Am. Assn. for Thoracic Surgery (pres. 1959), m. Heart Assn., A.M.A. (Distinguished Service award 1959, Hektoen Gold medal 1954), So. Med. Assn., Am., Soc., Western surg. assns., Am. Thoracic Soc., Assn. Am. Med. Colls., Halsted Soc., Internat. Cardiovascular Soc. (pres. 1959, pres. N.Am. chpt. 1964), Internat. Soc. Surgery (Distinguished Service award 1958, Leriche award 1959), Nat. Assn. on Standard Med. Vocabulary (pres. 1964), Royal Soc. Medicine, Sociedad Nacional de Cirurgia (Cuba, hon.), Soc. Clin. Surgery, Soc. for Exptl. Biology and Medicine, Soc. Med. Cons. to Armed Forces, Soc. for Vascular Surgery (pres. 1954), Soc. U. Surgeons, So. Soc. for Clin. Research, Southwestern Surg. Congress (pres. 1952), Tex. Acad. Sci., Am. Geriatrics Soc., Mexican Acad. Surgery (hon.), Sigma Xi, Alpha Omega Alpha, Phi Beta Pi, Alpha Pi Alpha. Author: (with Robert A. Kilduffe) Blood Transfusion, 1942; (with G. W. Beebe) Battle Casualties, Incidence, Mortality, and Logistic Considerations, 1952; (with A. Ochsner) Christopher's Minor Surgery, 7th edit., 1955, 8th edit., 1959; (with T. F. Whayne) Cold Injury, Ground Type, 1958. Editor: Year Book of General Surgery, 1957—; General Surgery, Vol. II, History of World War II, 1954. Mem. editorial bd. numerous med. jours. Contbr. numerous articles to profl. jours. Devised roller-type blood pump (later used in heart-lung machines), 1934; performed 1st successful carotid endarterectomy, 1953; pioneer in devel. dacron tubing replacements for blood vessels; inventor numerous sur. instruments; developed surgical techniques for removing aneurysms from weakened aortas, 1966; research on devel. artificial human heart. Home: 5323 Cherokee St., Houston 77005.*

DE BALL, Leon Anton Carl, astronomer; b. Lobberich, Nov. 23, 1853; s. Victor W. Joseph and Josephine Petronella Clara (Roeffs) de B.; student in Berlin (Germany), Bonn (Germany); worked at obs. Gotha and Bothkamp; with U. Luttich, 1883-91; apptd. dir. Kuffner's Obs., Vienna, Austria, 1891. Author: Eigenbewegung des Sonnensystems, 1871; Sphärische Astronomie, 1912. Discovered planetoid Athamantis; research on spherical astronomy and atmospheric refraction; computed a refraction table (1906) which is still in use. Died Vienna, Dec. 12, 1916.

DE BARTHEZ, Antoine, French physician; b. Narbonne/Aude, France, Aug. 6, 1811; s. Antoine-Michel François B.; M.D., Paris, France, 1839. Apptd. chief clinic at Chome, also doctor to adjoining prison, 1844. Mem. Acad. Medicine. Author: Treatise on the Diseases of Children, 1843. With Rilliet did comprehensive study on child health for French govt., also research on infantile paralysis and tuberculous meningitis. Died Dec. 11, 1891.

DE BARY, Heinrich Anton, botanist; b. Frankfort/Main, Germany, Jan. 26, 1831; student medicine Heidelberg, Marburg, Berlin (all Germany), 1849-53; M.D., Berlin, 1853; 2 sons, 1 dau. Surgeon in Frankfort, 1853; became private docent biology Tübingen, 1954; prof. botany Freiburg, 1855-66; prof. botany U. Halle (all Germany), 1867-72; 1st rector U. Strasbourg (France), 1872-88; founded in Freiburg 1st bot. lab. in Germany, also founder bot. labs., Halle, Strasbourg. Fellow Royal Soc., 1884. Editor Botanische Zeitung for 21 years. Author: Vergleichende Anatomie der Vegetationsorgane der Phanerogamen und Farne, 1877; Vergleichende Morphologie und Biologie der Pilze, Mycetozoen, und Bacterien, 1884; and many others. Founder of modern mycology (study of fungi); described life history of many varieties of fungi; helped establish difference between parasite and saprophyte, also to systemize bacteriology; research on infectious fungi, sexuality of fungi, related epidemic diseases especially potato diseases and wheat rust, also symbiosis and lichens, ferns and phanerograms. Died Strasburg, France, Jan. 19, 1888.

DEBAUVE, Alphonse Alexis, French engr.; b. Mureaux, France, Aug. 10, 1845; ed. École des Ponset chaussées; named engr. first class, 1886; lived at Beauvais; insp. gen. of roads and bridges. Author: Guide des conducteurs, 1881; Distributions d'eau, égouts, 1897; publs. on safety in the mines. Research on defects in the then newly installed Paris subway system, 1900; studied ways of purifying drinking water. Died 1906.

DEBAY, Auguste, French physician; b. Clermont-Ferrand, France, 1802; s. Jean Baptiste Joseph D.; surgeon mil. corps Spanish campaign; went to Algeria, 1845. Author: Hypnologie: du sommeil et des songes, 1843; Histoire des monstruosités sexuelles, 1845; Hygiène des mains, des piedes, de la poitrine, 1851. Wrote quasi sci. lit. in popular editions on sleep and dreams, sexual monstrosities, hygiene (of voice, chest, feet), and cosmetics (promoting their sale). Died Colombes, 1890.

DE BEAUCHAMP, Pierre-Joseph, French astronomer; b. Vesoul, France, June 29, 1752; s. Xavier de B.; ed. Coll. Bernardins, Coll. France; began voyage to Bagdad, 1781, later went to Middle East; came to Paris, 1790; visited Alexandria, Egypt, 1795; consul to Arabia; founder of astronomical observatory, Bagdad. Member of French Academy of Sciences, 1785. Author: Mémoire sur Trébizonde, 1813. Collected Arabian manuscripts; made drawings of Mideastern monuments and map of Babylon; numerous observations of planet Mercury, including passage of Mercury across sun, 1786, Bagdad. Died Nice, France, Nov. 19, 1801.

DE BEAUGRAND, Jean, mathematician; born circa 1600; sec. to chancellor Séquier, France; pub. works of Viète; supporter Descartes, detractor Desargues. Died 1640.

DE BEAUMONT, see Élie de Beaumont.

DE BEAUNE, Florimond, French mathematician; b. Blois, France, Oct. 7, 1601; s. Florimond de Beaune; student of law. Counsellor to pres. city of Blois (position held for most of his life); founded an obs. and optical workshop. Author: Florimundi de Beaume in Cartesii geometriam notae breves de aequationum constructione et limitibus, 1659; Held math. dicussions and corresponded with Decartes; considered a leading commentator on Descartes in 17th century and developed his geometrical method; among 1st to give sci. treatment to upper and lower limits of roots of numerical equations; studied deducing nature of curves from properties of their tangents (called Beaune's problem, he was 1st to use this approach, which led to inverse methods of tangents); paved way for development of integral calculus; constructed astron. glasses and invented telescopes. Died Aug. 18, 1652.

DE BEAUREGARD, Claude Guillermet, see de Berigard, Claude Guillermet.

DE BEAUSOLEIL, Martine de Bertereau du Chatelet (baroness); see du Châtelet, Martine de Bertereau.

DE BEAUVAIS, Vincent, sci. writer; b. France, circa 1190; mem. Order of St. Dominic. Author: Speculum majus (most complete sci. ency. of 13th century), printed 1473. Died circa 1264.

DE BEAUVAL, Firmin, French astrologer. Author: De mutatione aeris (influence of planets on meteorol. phenomena), 1338; treatise on reform of calendar (Julian), 1345.

DE BEER, Sir Gavin Rylands, English biologist; b. Malden, England, Nov. 1, 1899; s. Herbert Chaplin and Mabel (Rylands) de B.; M.A., Magdalen Coll., Oxford (Eng.) U., 1924, D.Sc., 1932; hon. degrees include D.-es-L., U. Lausanne (Switzerland), 1950, Sc.D., Cambridge (Eng.) U., 1958, D. de l'U. Bordeaux (France), 1961; m. Cicely Glyn Medlycott, Mar. 20, 1925. Fellow, Merton Coll., Oxford, 1923-38; lectr. Oxford, 1921-38; prof. embryology U. Coll., London, Eng., 1939-50; dir. Brit. Mus. Natural History, 1950-60. Wilkins lectr. Royal Soc., 1961; Rede lectr. U. Cambridge, 1965; guest lectr. Sorbonne, Paris, France, 1955. Recipient Royal Soc. Darwin medal, 1958, Linnean Soc. Gold medal, 1958, Société d'Acclimatation, Geoffroy-St. Hilaire Gold medal, 1954, Royal Soc. Arts Duke of Edinburgh medal, 1958; named Chevalier Légion d'honneur, 1953; created knight, 1954. Fellow Royal Soc., 1940, Linnean Soc. (past president; member British Association Advancement Science (past pres. zoology sect.); corr. mem. French Acad. Scis., 1952, Zool. Soc. India, Soc. Neuchateloise d'Hist. et d'Archeol.; hon. mem. Soc. zool. de France, Soc. Royal Zoology de Belgique. Author: numerous books including: Archaeoptery, 1954; Embryos and Ancestors, 1962; Charles Darwin, 1963, Altas of Evolution, 1964; also numerous articles. Established theory of paedomorphosis in opposition to theory of recapitulation, and established theory of clandestine and mosaic evolution in opposition to germ-layer theory. Research in comparative developmental anatomy; paleontology and history of sci. Address: La Colline, Bex, Switzerland.*

DE BEGUELIN, Nicolas, philosopher; b. Courlori, France, 1714; ed. U. Basle (Switzerland); LL.D.; mem. Prussian legation in Dresden, Germany until 1744; held chair Coll. Joachimstal; apptd. tchr. to Frederick William nephew of Frederick the Gt. Mem. Acad. Berlin (dir.). Author: Observations météorologiques; also publs. on color, lights and numbers, poems. Died Berlin, Germany, 1789.

DE BELDAMANDI, Prosdocimo, Italian mathematician; b. Padua, Italy, circa 1370-80; ed. U. Padua; prof. U. Padua, 1422-28. Author: Algorithmus de integris, 1410 (1st printed 1483). Discovered rule for summing geometric series; research and publs. on arithmetic, music, and astronomy. Died Padua, 1428.

DE BELIDOR, Bernard Forest, French engr.; b. Catalonia, Spain, circa 1693; prof. arty. sch. of La-Fère; served in army as a.d.c. of lt. gen. de Ségur, later under Duke of Harcourt, then Prince of Conti; camp marshal; insp. arty. Mem. French Acad. Scis., Acad. Berlin. Author: Sommaire d'un cours d'architecture militaire . . . , 1720; Cours de mathematiques . . . , 1725; la Science des ingenieurs . . . , 1729; le Bombardier français, 1731; Traité des fortifications, 1735; Architecture hydraulique, 4 vols., 1737-53; Ecole de la fortification permanente, 1769. Research and publs. on fortification methods, arty., hydraulic architecture; describes Pater Noster works, water wheels with curved paddles; extended meridian line at Paris to N. coast. Died Paris, France, 1761.

DE BELLEUAL, Pierre Richer, French biologist; b. Chalons-sur-Marne, France, 1564; student U. Montpellier (France); med. degree Avignon (France); one of 1st botany tchrs. in France. Author: Onomatologia (nomenclature of plants in Royal Garden of Montpellier), 1598; Recherche des plantes du Languedoc, 1605. Died Montpellier, 1632.

DE BENEDETTI, Sergio, physicist; born Florence, Italy, August 17, 1912; the son of Guido and Amelia (Passigli) De Benedetti; Ph.D., University Florence, 1933; m. Emma Falco, Apr. 10, 1944; children—Lydia, Vera, Gilbert. Came to U. S., 1940, naturalized, 1946. Asst. prof. U. Padua (Italy), 1933-38; U. Naples (Italy) fellow Curie Lab., Paris, France, 1935, French Caisse Nationale des Recherches fellow, 1938-40; research asso. Bartol Found., Swarthmore, Pa., 1940-42; asso. prof. Kenyon Coll., Gambier, O., 1943-44; sr. physicist Monsanto Chem. Co., Dayton, O., 1944-45; prin. physicist Oak Ridge Nat. Lab., 1946-48; asso. prof. Washington U., St. Louis, 1948-49; prof. physics Carnegie Inst. Tech., Pitts., 1949——. Staff Manhattan Project, cons. AEC, 1941-45; vis. prof., lectr. various instns. Fulbright fellow U. Turin (Italy), 1956-57. Fellow Am. Phys. Soc.; mem. Fedn. Am. Scientists (nat. council 1961). Author: Nuclear Interactions, 1964. Editorial bd. Rev. Modern Physics, 1962-64, Nuclear Instruments and Methods, 1960-63. Contbr. articles to profl. jours. in U. S., France, Italy. Home: 122 Hastings St., Pitts. 15206.*

DE BENEDICTY, Mario Gustavo, mathematician; b. Trieste, Italy, July 16, 1922; s. Gustavo Carlo and Emma (Serti) de B.; D. in Math. Sci., U. Rome, 1946, Libero Docente in Geometria, 1951; m. Alfonsina Mucciante, July 16, 1947; children—Gustavo, Franca. Came to U. S., 1958. Asst., then asso. prof. U. Rome, 1947-58; mem. faculty U. Pitts., 1958——, prof. math., 1960——, chmn. dept., 1962——; vis. prof. U. Pitts., 1957, U. B.C., 1960-62. Mem. Unione Matematica Italiana, Circolo Matematico di Palermo, N.Y. Acad. Scis., Canadian Math. Congress, Math. Assn. Am., Am. Math. Soc., A.A.A.S., Sigma Xi. Author: (with F. Conforto) Introduzione alla Topologia, 1960. Co-editor: Introduzione alla Geometria Algebrica, 1948; Lezioni sulle Funzioni Analitiche di piu' Variabili Complesse, 1958. Contbr., editor math.: Dizionario Enciclopedico dell'Enciclopedia Italiana, 1952-58, Enciclopedia Italiana Appendix III. Contbr. articles on algebraic geometry, theory of quasi-Abelian functions. Home: 314 Hoodridge Dr., Pitts. 15234.*

DE BERIGARD (or DE BEAUREGARD), Claude Guillermet, philosopher; b. Moulins, France, 1578 or 1591; studied the philosophy of medicine, Aix, France; called to Pisa, Italy by grand duke of Florence, 1628; called to teach natural philosophy at Padua, Italy by the Venetian Govt., 1640. Author: Dubitationes in dialogum Galilei pro terra immobilitate, 1632; Circulus Pisanus Claudii Berigardi . . . , 1643. Proposed that atoms are different sized spheres, that light is made of particles; that weight is caused by the mut. action of bodies; described an air thermoscope; discussed the nature of elements in compounds. Died Padua, circa 1664.

DE BESSY, Bernhard Frenicle, mathematician; b. circa 1602; mem. French Acad. Scis. Discovered number of magic squares increased with the order; formulated rule of writing down magic squares of even order; research on Pythagorean numbers (numbers forming sides of right angle triangles). Died 1675.

DE BÉTANCOURT Y MOLINA, Augustín, engr., inventor; b. Tenerife Island; ed. Mil. Sch. Madrid (Spain); began career as Spanish engr.; entered Russian service, 1808; founded engr. corps and high sch. for exact scis. in Russia. Mem. French Acad. Scis., 1809. Author: Mémoire sur la force expansive de la vapeur d'eau, 1790; Mémoire sur un nouveau systéme de navigation intérieure, 1807; (with Lanay) Essai sur la composition des machines, 1808; Plans du pont de bateaux de Sainte-Isaac, sur la grande Neva, 1820. Studied elasticity of saturated water vapor; 1st to experiment with elasticity of other vapors such as alcohol vapor; calculated steam pressure curve of water from 0 to 135 degrees Celsius; (with Bréguet) invented a telegraph; built earth windlass (capstan) whose cable was wound several times around 2 drums equipped with screw-shaped round timbers. Died St. Petersburg, Russia, July 26, 1824.

DE BÉTHENCOURT (or DE BETTENCOURT), Jacques, French physician; flourished 1520; practiced at Rouen, France; author: Nova poenitentialis quadragesima et purgatorium in morbum gallicum sive venereum, 1527. Originated the term, venereal; named syphilis, maladie vénérienne.

DE BETHUNE, Pierre Felix, Belgian geologist; b. Ixelles, Belgium, Mar. 23, 1909; s. Gaston S. and Marthe (Terlinden) de B.; Ingenieur Civil des Mines, U. Louvain (Belgium), 1931; M.Sc., U. Wis., 1933; postgrad. Stanford, 1934, Columbia, 1935; m. Marguerite Calmeyn, June 15, 1942; children—Andre, Philippe, Stanislas, Marie-Pierre, Xavier. Prof. geology Louvain U., 1936——; vis. prof. U. Cal. at Los Angeles, 1967 Mem. Geol. Council Belgium. CRB Advanced fellow, 1947. Research, numerous publs. on devel. land forms in regions of folded structure (Appalachian Valley mount of Pa., Condroz of Belgium); petrography of Carbonatites; zonal growth of garnet in metamorphic rocks. Home: 83 Meunierstraat, Louvain, Belgium.*

DE BEURMANN, Charles Lucien, French dermatologist; b. 1851; (with Henri Gougerot) described sporotrichosis (chronic infection characterized by formation of nodular lesions which have tendency to break down and form ulcers), caused by Sporotrichum schenckii, also known as de Beurmann-Gougerot disease, 1906. Died 1923.

DEBIASI, Ettore, Italian gynecologist; b. Ala de Trento, May 30, 1900; s. Valentino and Margherita (De Gresti) D.; ed. in obstetrics and gynecology; m. Zorta Baljac; children—Valentina, Danillo. Dir., Clinic Obstetrics and Gynecology, U. Genes. Mem. Italian Cancer Soc., Italian Soc. Obstetrics and Gynecology. Author 130 monographs. Home: via Montallegro 28. Office: viale Benedetto XV 10, Genoa, Italy.

DEBIERNE, André Louis, French chemist; b. Paris, France, 1874; studied at l'Ecole de physique et chimie; prof., dir. l'Ecole de physique et chimie; prof. Sorbonne, Paris; succeeded Marie Curie as director Inst. Radium. Discovered actinium in pitchblende (with Giesel) and showed its emanation formed helium; (with Marie Curie) isolated pure radium; research on effect of nuclear explosions on meteorol. phenomena. Died Paris, Aug. 1949.

DE BIRAN, Marie-François-Pierre Gonthier (or Maine de Biran), French philosopher; b. Bergerac, France, Nov. 29, 1766; studied medicine, Perigueux; became adminstr. Dept. of Dordogne, 1795; mem. Council of 500, 1797; dep. and quester of the Chambre; apptd. conseiller d'Etat, 1816; founder med. and lit. soc., Bergerac. Author: De l'influence de l'habitude sur la faculté de penser, 1803; La décomposition de la pensée, 1805; L'apperception immédiate, 1807; Rapports du physique et du moral, 1814; Examen des leçons de philosophie de Laromiguière. Nouvelles considérations sur le sommeil, les songes, et le somnambulisme; Observations sur le système du Docteur Gall. Studied sleep, dreams, sleep-walking, effect of habit on thought; criticized ideas of Gall and Laromiguière; considered outstanding defender of spiritualism in France. Died July 16, 1824.

DE BLAINVILLE, Henri Marie Ducrotay, French biologist, physician; b. Arques, France, Sept. 12, 1777; s. Pierre and Catherine (Pauger) de B.; student Mil. Sch. Beaumont; studied medicine, comparative anatomy under Cuvier; Docteur es Médicine, École de Médecine, 1808. Prof. zoology Athenaeum, from 1811; prof. Mus. Natural History, from 1830, titulary prof., 1832. Mem. French Acad. Scis. Author: De l'organisation des animaux (new classification of animals based on external rather than internal organs), 1822; Manuel de malacologie, 1825; Description iconographique des mammifères fossils, 4 vols., 1839-64; also article in Dictionnaire d'histoire naturelle on silk worms pointing out previously unnoticed similarities between certain types. First to describe congenital assymmetry of two ears (known as Blainville's ears), 1814; showed that details of physiognomy were arranged in perfect harmony according to providential design; classified animals into 3 main groups: amorphozois, actionozoids and zigomorphozoids; collaborated with Cuvier on comparative anatomy work. Died May 1, 1850.

DE BLECOURT, Jan Johannes, Dutch physician; b. Wyk by Duurstede, Holland, Oct. 24, 1916; s. Lucas and Johanna (Eist) de B.; M.D., Med. Sch., Leyden, Netherlands, 1941; postgrad. internal medicine and rheumatology U. Leyden, 1940-45. Dep. chief Mil. Blood Transfusion Service, 1945-48; chief, reader rheumatism unit U. Groningen, Netherlands, 1948——. Med. dir. Rheumatism Service Groningen, 1948——; med. adviser Netherlands Rheumatic Assn., 1950——. Hon. mem. Dutch, French, Belgian and Argentine socs. against rheumatism. Author: (with J. J. Bode) books on rheumatology. Publs. on clin. trials with steroids, antimalarials; genetic, population studies in rheumatism; liver function in rheumatoid arthritis. Home: Quintuslaan, Groningen, Netherlands.*

DE BLÉGNY, Nicolas, French physician; b. Chaumont-en-Bassigny, France, circa 1646; s. Etienne and Marie (Villars) de B.; studied medicine in Paris; m. Charlotte Gallois; s., Marc-Antoine. Became apothecary, 1674; surgeon, Hosp. Saint Côme; apptd. surgeon to the King, 1678. Founder of an iatrochem. acad. in Paris. Author: L'art de guérir les maladies vénériennes, 1673; L'art de guérir les hernies, 1676. Published 1st med. jour. in Paris, 1679. Invented glasses for strabismus, also truss for hernia; started factory for manufacturing elastic supports for hernia. Died Avignon, France, 1722.

DE BOCK, Alphonse, Belgian physicist; b. Antwerp, Belgium, Mar. 24, 1920; s. Henry T. and Helen (Hazen) De B.; D. in Physics, U. Louvain (Belgium), 1945; m. Emily Swinnen, Feb. 1, 1944; children—Walter, Hilde. Asst., U. Louvain, 1943; research fellow Nat. Found. Sci. Research, 1944-46; asst. lectr., 1947-52, prof., 1952——, also dean faculty sci. Recipient prize Acad. Sci. Mem. I.E.E.E. (sr.). Contbr. articles to jours. Added to knowledge of properties of metallic films and properties of matter by means of ultrasonic power. Home: 47 Prinses Lydialaan. Office: 28 Kapeldreef, Heverlee, Belgium.*

DE BOER, Abraham Adolf, Dutch tech. economist; b. Deventer, Holland, Mar. 30, 1928; s. A. J. and T. H. (Crom) deB.; candidate degree Utrecht (Holland) U., 1949, doct.ex., 1955, Ph.D., 1962; postgrad. (scholar) U. Oslo (Norway), 1950; m. Lisbet Heier, July 2, 1953; children—Morten, Erik, Kristien. Chemist, Joint Establishment for Nuclear Energy Research, Norway, 1951-54; tech. economist Reactor Centrum Nederland, 1956-58. Euratom, Brussels, Belgium, 1958-61; adviser to Dutch mem. EURATOM-Commn., Brussels, 1961——. Sec., Commn. for Standardization in field Nuclear Energy, 1956-58. Mem. Koninklyk Instituut van Ingenieru. Author: Kernenergie, inleiding tot de reactor-kunde, 1955; Inleiding tot het denken van Keynes, 1962; Economische aspecten van de ontwikkeling der kernenergie, 1962; Weten en regeren: noodzaak van een nieuwe politiek, 1967; also articles. Research on econs. of nuclear energy, electricity, grid planning energy planning; math. model studies about introduction nuclear central stas. in electricity prodn. grid. Home: 41, Av. des Eperviers, Brussels 15, Office: 51, Rue Belliard, Brussels, Belgium.*

DE BOER, Jan Hendrik, chemist; b. Ruinen, Netherlands, Mar. 19, 1899; s. J. and J. B. (Somer) deB.; Ph.D. in Natural Scis., U. Groningen; Dr. honoris causa U. Hanover; m. E. A. Malcolm Swanson. Asst. at Groningen, 1922-23; chemist Naamloze Vennootschap Philips' Gloeilampenfabriek, Eindhoven, 1923, later head chem. research; head lab. Ministry of Supply, London, 1940-45; dir. Lab. at Unilever Soc., Gt. Britain, 1946-50; sci. cons. mines at Limbourg, 1950-62; asso. prof. Tech. U. of Delft, 1946——. Pres., Sci. Council for Nuclear Energy, 1962——, Central Council for Nuclear Energy, 1963——. Mem. Royal Acad. Sci., Faraday Soc., Brit. Assn. for Advancement Sci., Netherlands Assn. Natural Scis.; hon. mem. Royal Netherlands, Flemish assns. chemistry. Author: (with A. E. van Arkel) Chemische Binding; Electron Emission and Adsorption Phenomena; The Dynamical Character of Adsorption; others. Home: Zorgvliet, Alexander Gogelweg 47. Office: Duinweg 35 Postbus 5086, 's-Gravenhage, Netherlands.

DE BOER, Thymen Jan, Dutch chemist; b. Grootegast, Netherlands, Feb. 24, 1924; s. Souwe and Antje (v.d. Weg) de B.; Ph.D., Rijks U., Groningen, Netherlands, 1953; m. Eva M. M. Rink, Apr. 17, 1954; children—Judith E., Claire A. Faculty U. Amsterdam (Netherlands), 1957——, prof. organic chemistry, 1960——, dir. lab. organic chemistry, head chemistry dept., 1967——. Cons. Chem.Fabr. Naarden. Ramsay fellow, 1952-53. Research, publs. on organic reaction mechanisms, including small ring chemistry, photochem. nitrozation, alkaline olefin isomerization; organic spectrometry, including conformation analyses by ultraviolet and NMR, mass spectral fragmentation patterns of labelled compounds. Home: 36 Gunterstein, Amsterdam, Netherlands.*

DE BOISBAUDRAN, Paul Émile Lecoq, French chemist; b. Cognac, Charente, France, 1838; studied with Kirchhoff, Bunsen and Crookes; research in own lab. on agrl. and phys. chemistry; discovered rare earths samaria, dysprosia, gadolinia, gallium; discovered relationship between spectroscopic lines of members of same family of metals and their atomic weights; developed use of spark spectra. Died 1912.

DE BONNARD, Augustin-Henri, French geologist; b. Paris, France, Oct. 8, 1781; ed. École polytechnique, 1799, École des mines; head engr. mines, 1810; became gen. insp. mines, 1837. Mem. French Acad. Scis., 1837. Studied tin deposits in Cornouailles, France, coal regions in Nord region of France. Died Jan. 5, 1857.

DE BOODT, Anselmus Boëtius, Belgian physician; b. Bruges, Belgium, circa 1550; became physician, lapidary to Emperor Rudolf II of Bohemia, 1604. Author: Historia gemmarum et lapidum, 1609; also works on botany. Distinguished 5 degrees of hardness in stones; research on precious stones; discussed atomic theory; accepted Aristotle's four elements and the three principles of Paracelsus. Died 1632.

DE BOOY, Theodoor, archeologist, explorer; b. Hellevoetsluis, The Netherlands, Dec. 5, 1882; s. C. J.

G. and Mary (Hobson) de B.; ed. Royal Inst.; m. Elizabeth Hamilton Smith, Mar. 29, 1909. Came to U. S., 1906; naturalized, 1916; in charge West Indian archeol. work of Mus. of Am. Indian, New York, from 1911. Explored previously unknown regions of Santo Domingo and Venezuela; conducted archeol. investigations in Bahamas, Cuba, Jamaica, Hayti, Santo Domingo, Turks and Caicos Islands, Margarita, Trinidad, Martinique, Venezuela and Virgin Islands of U. S. Author: The Newly Acquired Virgin Islands of the U. S. and the British Virgin Islands, 1918. Died Feb. 18, 1919.

DE BORDA (or BORDA D'ORO), Jacques François, French naturalist; b. Dax, France, May 25, 1718; studied in Paris (met d'Alembert there) practiced law; ret. at early age to his estate at Oro to do sci. research. Mem. French Acad. Scis., 1753, Acad. Bordeaux. Author: Mémoire sur les fossiles des environs de Dax. Studied Gascony and its use in heating; analyzed mineral water of Tercis; worked to perfect techniques for manufacture tiles, ceramics, pottery; discovered carved pieces of silica which attracted attention of prehistorians. Died Jan. 4, 1804.

DE BORDA, Jean Charles, French mathematician, nautical astronomer; b. Dax, France, May 4, 1733; s. Jean-Antoine and Marie-Thérèse (de la Croix) de B.; ed. Jesuit Coll., La Flèche, France. Commd. in cav.; after battle of Hastembeck, joined navy; visited Azores and Canary Islands (which he mapped); captured by Brit., 1782; became naval constrn. insp., 1783. Mem. French Acad. Scis., 1756. Author: Mémoire sur le mouvement des projectiles, 1756; Voyage en diverses parties de l'Europe et de l'Amérique, 2 vols., 1778; Descriptions et usage du cercle de réflexion, 1787; Tables trigonométriques décimales (augmented by Delambre 1804); also numerous memoirs presented to Acad. Scis. Investigated hydrodynamics; studied relative motion, especially in connection with theory of water wheels; improved reflecting and repeating circles used in nautical astronomy; helped introduce metric system in France; measured (with J. B. J. Delambre and P. F. A. Méchain) an arc of meridian; invented instruments for measuring arc. Died Paris, France, Feb. 20, 1799.

DE BORDEU, Théophile, French physician; b. Izeste, France, Feb. 22, 1722; s. Antoine and Adrienne (de Tovya) de B.; docteur en médecine Montpellier, France, 1743; studied anatomy under J. L. Petit, Paris, France, 1746. Prof. anatomy, Montpellier; named demonstrator anatomy U. Pau (France), 1749; became doctor Paris Med. Faculty, 1754; insp. mineral water springs; physician to Mme. du Barry. attended dying Louis XV. Author: Lettres sur les eaux minérales du Béarn, 1744; La position de glandes et leur action, 1751; Traité des glandes, 1752; Chilificationis historia, 1757; Les eaux thermales du Béarn; Tous les organs du corps participent—is à la digestion?; La Chasse est-elle plus salubre que les autres exercises?; Recherches sur le pouls; Recherches sur l'inoculation, 1764; Recherches sur le tissu muqueux, 1768; Recherches anatomiques . . . ; Les maladies chroniques, 1775; also memoirs on scrofula (Tb of lymphatic glands), Collaborator, Encyclopédie. Founder modern hydrology; proponent of vitalism; recorded earliest concept of internal secretion, 1775; foresaw role of conjunctive tissues, also coenesthesis, functional synergy, cerebral localizations. Died Dec. 23, 1776.

DE BORGO, Luca (Paccioli, Paciolus), Italian mathematician; b. Borgo-San Lorenzo, Italy, 1445; Franciscan monk; taught math. in various Italian cities. Author: Summa de arithmetica, geometria, proportioni et proportionalita . . . , 1494; Libellus in tres partiales tractatus divisus quorum cumque corporum reularum . . . , 1508; De divina proportione, 1509. Originated rules of false singular and double position (rules of Elkatham); developed numerous algebraic and geometric solutions. Died Florence, Italy, 1570.

DE BORHEGYI, Stephan Francis, anthropologist; b. Budapest, Hungary, Oct. 21, 1921; s. Ernest Francis and Hildegard Maria, (Geiger de Karácsonymező) de B.; Ph.D., Royal Péter Pázmány U., Budapest, 1946; postgrad. U. Ariz., Yale; m. Suzanne Catharine Sims, July 5, 1949; children—Ilona Maria, Stephan Ernest, Carl Robert, Cristopher Francis. Came to U. S., 1948, naturalized, 1955. Instr., Péter Pázmány U., 1946-48; curatorial asst. Hungarian Nat. Mus., 1946-48; asso. prof. anthropology San Carlos U. Guatemala, 1949-51; asst. prof. U. Mo., 1952; asst. prof., mus. dir. U. Okla., 1954-59; asst. prof. anthropology U. Wis., Milw., 1959—; mus. dir. Milw. Pub. Mus., 1959—. Recipient diploma merit Govt. Guatemala, 1951. Mem. Am. Assn. Museums (past councilor), Midwest Museums Assn. (v.p. 1963—), Soc. for Underwater Archaeology, Internat. Council Museums (mem. com. 1964—), Wis. Archeol. Survey (past pres.), Am. Anthrop. Assn., Central States Anthrop. Soc. (past pres.), Archaeol. Soc. Am., Royal Anthrop. Soc. Gt. Britain and Ireland, Archaeol. Inst. Am. Author: Prehistoric Settlement Patterns in the New World, 1956; (with others) Essays in Pre-Columbian Art and Archaeology, 1961; (with Elba Dodson, Irene Hanson) Bibliography of Museums and Museum Works, 1900-1961, vol. 1, 1960, vol. 2, 1961; Handbook of Middle American Indians vol. 2, (with others) Archaeology of Southern Mesoamerica, 1965. Pioneered research in underwater archaeology in Central Am.; research pre-historic settlement-pattern devels. in Mexico, Central Am., Am. S.W., Maya religion and prehistory, hallucinogenic mushroom cult pre-columbian C.Am.; pioneered research in testing visitor reactions to mus. exhibits. Home: 2515 N. Terrace Av., Milw. 52311.*

DE BORT, Teisserenc, French physicist; b. 1855; 1st to use balloons for exploring upper atmosphere; with others, began technique of radio probing atmosphere by means of transmitter on balloon (used in meteorology). Died 1913.

DE BORY, Gabriel, navigation scientist; b. Paris, France, Mar. 1720; gov. of Santo Domingo (now Dominican Republic). Mem. Marine Acad., French Acad. Scis., 1765. Described octant invented by Hadley; determined exact positions of Azores Islands; observed passage of Mercury over sun and eclipse of sun on Oct. 26, 1753. Died Oct. 8, 1801.

DE BOUELLES, Charles, French mathematician; b. Saucourt, Picardy, France, circa 1470; studied math. under Lefévre d'Étaples; canon 1st at Saint-Quentum; later canon, prof. theology, Noyon, France; traveled to Germany in 1505, to Rome, 1507, also Spain. Author: Metaphysicum introductorium, 1503; Le Livre de l'art et science de géométrie (1st work on geometry in French), 1511; Prover biorum vulgarium libri tres, 1531; Liber de differentia vulgarium linguarum et gallici sermonis varietate, 1533. First to study cycloid; publs. on perfect numbers. Died Noyon, circa 1553.

DE BOUGAINVILLE, Louis Antoine, French navigator; b. Paris, Nov. 11, 1729; s. Pierre-Yves and Marie (d'Arbouville) de B.; Practiced law; a.d.c. to Montcalm in Can.; established colony, Falkland Islands, 1763-65; commd. by govt. to make exploratory voyage around world, accompanied by naturalists and astronomers, 1767-69, visited Tahiti, Samoan group, New Hebrides, rediscovered Solomon Islands; apptd. sec. to king, 1772; vice adm. French Navy. Mem. French Acad. Scis., Bur. Longitudes, Royal Soc. London, Marine Acad. Author: Traité du calcul integral, 2 vols., 1754-56; Voyage autour du monde (contbd. to diffusion of theory of goodness of man in state of nature), 1771. Bougainville (largest of Solomon Islands) named in his honor, also 2 straits in South Pacific, and S. Am. climbing plant Bougainvillea. Died Paris, Aug. 31, 1811.

DEBOVE, Maurice Georges, French physician; b. Paris, France, 1845; M.D., 1878; prof. pathology, later clin. medicine, Paris. Pres., Consultative Com. Hygiene France. Mem. Acad. Medicine (perpetual sec. 1913). Author: (with Rémond) Traité des maladies de l'estomac; (with Gaillard) Précis de pathologie interne. Described basement membrane (noncellular layer under epithelium of bronchial, tracheal mucosa, alimentary tract), essential splenomegaly. Died Paris, 1920.

DE BRAGELOGNE, Christophe-Bernard, French mathematician; b. Paris, France, 1688; s. Christophe-François and Charlotte (de Charmois) De B.; studied under Malobranche; took religious vows; ecclesiatical dean St. Julien of Brioude Ch. Mem. French Acad. Scis., 1711. Author: Traité des lignes du quatrième ordre; Examen des lignes du quatrième ordre; also publs. on quadrature of curves. Died Feb. 20, 1744.

DE BRAHM, William Girard, geographer; b. 1717; m. 2d, Mary (Drayton) Fenwick, Feb. 18, 1776; at least 1 child. Came to Am., 1751; founded Town of Bethany, Ga., 1751; surveyor of Ga., 1754-64, planned towns of Ebenezer, 1757, Ft. George, 1761; supervised constn. of fortifications at Charleston, S.C., 1755, Savannah, Ga., 1762; drew 1st map of Ga. and S.C., 1757; surveyor gen. for So. Dist., 1764-70; commr. to mark No. boundary line of N.J., 1765; drew map of Atlantic Ocean, 1772. Author: The Atlantic Pilot, 1772; The Levelling Balance and Counter-Balance, 1774; DeBrahm's Zonical Tables for the Twenty-Five Northern and Southern Climates, 1774; Time and Apparition of Eternity, 1791; Apocalyptic Gnomon Points Out Eternity's Divisibility, 1795. Sounded and Mapped inlets of Fla. coasts, platted land along rivers flowing into them. Died 1799.

DEBRAY, Henri Jules, French chemist; b. Amiens, France, July 26, 1827; studied at École normale, 1847-50; prof. Lycée Charlemagne; asst. École normale, also École polytechnique; assayer to the mint; became prof. at Sorbonne, Paris, France, 1881. Mem. French Acad. Scis., 1877. Research on the salts of beryllium, molybdenum, tungsten, and cerium, also on alkaline earths, decomposition of oxides, carbonates of copper; (with Deville) studied separation of rhodium, acids of ruthenium, metals of platinum group. Died Paris, France, July 19, 1888.

DE BRÉAUTÉ, Éléonore-Nell-Suzanne, French astronomer; b. Rouen, France, June 29, 1794; dau. Jean and Marie Letellier; Bréauté Penninsula named in her honor, 1823. Mem. French Acad. Scis. Financed voyage of Blosseville to Greenland, 1833; first to observe comet of 1823; instrumental in constrn. of map of Normandy showing different points of elevation. Died Feb. 3, 1855.

DEBREYNE, Pierre-Jean-Corneille, French physician; b. Quaëdypre, Nov. 7, 1786; s. Pierre Cornil Francois, and Victoire (Blaevoet) D.; docteur en médecine, Paris Faculty Medicine, 1814. Practiced medicine in Lille, France; became Trappist monk, 1817; ordained priest, 1820; opened clinic at Trappist monastery for people of the surrounding villages. Author: Pensées d'un catholique croyant sur le materialism modern, 1839; Essai sur la théologie morale dans ses rapports avec la physiologie, 1842; Physiologie catholique et philosophie, 1863. Publs. on modern materialism; research on physiology as related to theology. Died Trappe, France, 1867.

DEBRUIJN, Nicolaas Govert, Dutch mathematician; b. Hague, Netherlands, July 9, 1918; s. Cornelis Pieter and Jorina (Van der Boom) DeB.; student U. Leyden (Netherlands); Dr.'s Degree, Free U., Amsterdam, Netherlands, 1943; m. Elizabeth De Groot, Aug. 30, 1944; children—Jorina, Frans, Elisabeth, Judith. Asst., Technol. U., Delft, Netherlands, 1939-44, prof., 1946-52; collaborator Philips' Labs., Eindhoven, Netherlands, 1944-46; prof. U. Amsterdam, 1952-60; prof. Technol. U., Eindhoven, 1960—. Mem. Royal Netherlands Acad. Scis. Author: Asymptotic Methods in Analysis, 1958; also articles. Research in pure and applied math., including analytic number theory, asymptotics and combinatorial analysis. Home: 2 Eikenlaan, Nuenen, Netherlands. Office: Technol. U., Eindhoven, Netherlands.*

DE BRUX, Jean André, French pathologist; b. Soissons, Aisne, France, Apr. 22, 1911; s. Gaston and Julia (Pages) de B.; M.D., Faculté de Medecin de Toulouse (France), 1939; m. Lou Elizabeth Bailey, Mar. 20, 1951; children—Jean-Louis, Elie-Pierre. Prof. agrege anatomie pathologique Faculté de Ecola Medecine d'Angers, 1952-63; faculté U. Paris, (France), 1963—, biologiste des hopitaux, 1966—; dir. Centre de Depistage Broca-Boucicaut, 1956—. Mem. Am. Assn. Pathologists and Bacteriologists, Am. Assn. Clin. Pathology, Internat. Acad. Pathology, Internat. Acad. Cytology, Am. Soc. Cytology, European Soc. Pathology. Research, numerous publs. on gynecol. pathology and cytology, electron-microscopy. Home: 11 Bis Dufrenoy, Paris 16. Office: Dept. Pathology, Hosp. Boucicaut, 78 Rue de la Convention, Paris 15, France.*

DEBRUYN, Peter Paul Henry, anatomist; b. Amsterdam, Holland, Jul. 28, 1910; s. Henry and Marianne (van den Nieuwenhuysen) DeB.; M.D. (Stokvis Fund Travel fellow) U. Amsterdam, 1938; m. Jeannette Meershoek, Sept. 6, 1931; children—Anneke (Mrs. Oliver Overseth), Yolande (Mrs. Ray Markel). Came to U. S., 1941, naturalized, 1946. Asst., U. Amsterdam, 1936-39; faculty U. Chgo., 1941—, prof. anatomy, 1952—, chmn. dept. anatomy, 1952-61. Editor., Scientist's Library, U. Chgo. Press., 1952—; departmental editor, advisor Ency. Britannica, 1955—. Mem. Am. Assn. Anatomists, Am. Soc. Naturalists, Internat. Soc. Cell Biology, Radiation Soc. Research and publs. on bone repair, formation of blood cells, fine structure of blood-forming organs. Home: 1605 E. 50th St., Chgo. 60615.*

DEBUS, Heinrich, chemist; b. Wolfshagen, Hesse, Germany, July 13, 1834; asst. to Bunsen in Marburg, Germany; tchr. at Queenswood Coll., Stockbridge, Hampshire, Eng.; later prof. Naval Acad., Greenwich, Eng.; prof. Guy's Hosp., London, Eng. Author of biography of Robert Bunsen. Research on thionic acids, affinity; discovered hexathionic acid, glyoxylic acid, glyoxal, glyoxaline; reduced hydrocyanic acid to methylamine. Died Kassel, Germany, Oct. 9, 1916.

DEBUS, Kurt Heinrich, engr.; b. Frankfort Main, Germany, Nov. 29, 1908; s. Heinrich P. J. and Melly (Graulich) D.; M.S. in Elec. Engring., Darmstadt (Germany) Tech. U., 1936, Ph.D. in Elec. Engring., 1939; m. Irmgard Helene Brueckmann, June 30, 1937; children—Ute, Sigrid. Came to United States, 1945, naturalized, 1959. Assistant professor elec. engineering Darmstadt Tech. U., 1939-42; test engr., later flight test dir. Peenemuende Rocket Center, 1942-45; dep. dir. guidance and control div., later staff asst. to Wernher von Braun rocket research and devel. div., U. S. Army Ordnance, 1945-52; dir. missile firing lab. Army Ballistic Missile Agy., 1952-60; dir. Launch Operations Center, NASA, 1960-63, John F. Kennedy Space Center, NASA, 1963—, member of management council Office Manned Space Flight, 1962—, mem. sr. council Office Space Sci., 1962. Mem. univ. adv. bd. Brevard Engring. Coll.; mem. Fla. Bd. Control and Edn., GENESYS. Recipient Exceptional Civilian Service award, 1959; Frank A. Scott Gold Medal award, Am. Ordnance Assn., 1964; NASA Outstanding Leadership award, 1964; Pioneer Windrose award, Order Diamond, 1965. Fellow Am. Inst. Aeros. and Astronautics; mem. Am. Ordnance Assn. (life), Nat. Geog. Soc., Hermann Oberth Gesellschaft (hon.). Directed launches of 1st U. S. earth satellite, Explorer 1; 1st U. S. space probe to orbit sun, Pioneer IV; 1st flight of primates in Jupiter rocket; 1st Mercury program primate launch; 1st manned suborbital flight of Freedom 7; and 10 successful launches of NASA's Saturn 1 vehicles. Home: 100 N. Riverside Dr., Patrick AFB, Fla. Office: Kennedy Space Center, Fla.*

DE BUSSY, Louis (Marie-Anne-), French naval engr.; b. Nantes, France, Mar. 22, 1822; ed. École polytechnique, 1841. Engr. naval constrns., Lorient, France, head constrns., 1875; inspector gen., naval constrn., from 1885. Recipient Prix Plumey, 1886. Mem. French Acad. Scis., 1888. Important in bldg. models for French war fleet; introduced use of steel in naval constrn.; made several improvements in form of armours, and important work on use of steel with nickel in protection of ships; built armored cruiser, Dupuy-de-Lôme, also battleships of Redoutable type and dispatchboats of Forbin and Condor types were constructed according to his designs; also investigated ships rolling in swells. Died Paris, France, Apr. 24, 1903.

DEBUYS, Laurence Richard, Am. pediatrician; b. New Orleans, Nov. 12, 1878; s. James and Stella (Rathbone) DeB.; B.S., Tulane U., 1899, M.D., 1904; postgrad. in pediatrics Harvard, 1907, 08; clinics in Germany, Austria. Eng., France; m. Miriam Duggan, June 14, 1904; children—Laurence Richard, William Eno. Herbert Fowler, John Forester, Henry Duggan. Practiced medicine, Houma, La., 1904-07, New Orleans, from 1907, specializing in pediatrics, from 1910; chief of clinic, dept. gynecology and obstetrics and clin. asst., dept. pediatrics Tulane U., 1907-08. asso. prof. pediatrics, 1912-17, clin. prof., 1917-19, asst. prof. pediatrics post grad. sch. medicine, 1912-17, prof. of pediatrics, 1919-29; chief pediatric staff New Orleans Presbyn. Hosp., 1910-11; chief of staff in pediatrics, Charity Hosp. La., 1919-22; mem. staff Touro Infirmary, from 1910, chief of pediatric dept., 1919-39, cons. pediatrician, from 1939; prof. pediatrics nurses Tng. Sch., 1924-34; physician in charge Jewish Childrens Home, 1925-36, Isadore Newman Sch., 1925-36. Mem. White House Conf. on Child Health, 1929; mem. follow-up com., 1929-31. Diplomate Am. Bd. Pediatrics. Fellow A.C.P. (bd. govs. 1925-26), Am. Acad. Pediatrics (emeritus); mem. La. Pediatric Soc. (organizer and pres. 1924-28), So. Med. Assn. (hon.) chmn. pediatrics sect. 1925), A.M.A. (hon., chmn. sect. diseases of children 1917-18), Assn. Am. Med. Colls., Archivos Americanos de Medicina (corr. editor), Pan. Am. Med. Assn. (v.p. pediatrics sect. 1933), Am. Assn. Med Milk Commrs. (pres. 1919-20), Assn. for Study of Internal Secretions (councillor 1916-31), Am. Child Health Assn., Am. Pediatric Soc. (v.p. 1930-31), Assn. Study Internal Secretions (life); Soc., Tri-State Dist. Med. Assn., Alpha Omega Alpha, Phi Chi (charter). Del., 2d Internat. Pediatric Congress, Stockholm, 1930, Rome, 1936; Internat. Hygiene Congress, Germany, 1930. Contbr. chpts. Abt's Pediatrics, Feer's Pediatrics. Member editorial board American Journal Diseases of Children, 1926-39; mem. cons. staff Archives of Pediatrics, 1919-26; collaborator Am. Jour. of Syphilis, 1916-23. Contbr. numerour articles to med. jours. Pioneer in use of motion pictures in medicine, demonstrating peristaltic waves by motion pictures, 1913. Died June 20, 1957.

DEBYE, Peter Joseph William, phys. chemist; b. Maastricht, Netherlands, Mar. 24, 1884; s. Wilhelmus and Maria (Reumkens) D.; E.E., Engring. Sch., Aachen; Ph.D., U. Munich; hon. degrees Harvard, Bklyn. Polytech., St. Lawrence U., Colgate U., Oxford, Brussels, Liège, Sofia Eidgenossiche Technische Hochschule, Boston College; m. Mathilde Alberer Apr. 10, 1913; children—Peter Paul Ruprecht, Mathilde Maria Gabriele. Came to U. S., 1940. Prof. theoretical physics U. Zurich, 1911; prof. univs. Utrecht, Goettingen, Leipzig, Berlin; prof. chemistry and chmn. dept. Cornell U., 1940-50, emeritus, 1950—. Recipient Lorentz, Faraday, Rumford, Franklin medals; Nobel prize in chemistry, 1936; Willard Gibbs medal, 1949; Max Planck medal, 1950; Kommandeur des Ordens Leopold II; Kendall award, colloid chemistry, Am. Chem. Soc., 1957; Nichols medal N.Y. sect. Am. Chem. Soc., 1961, Priestley medal, 1963; Am. Phys. Soc. High-Polymer prize, Ford Motor Co., 1965; Madison Marshall award N.Ala. sect. Am. Chem. Soc., 1965; Nat. Medal Sci., 1965. Mem. Royal Soc. Amsterdam, Pontifical, Royal Irish, Royal Danish, Berlin, Göttingen, Munich, Brussels, Liège, Nat. (U.S.A.), N.Y., Am. (Boston), Indian (Bangalore), acads., Franklin Inst. Phila., Royal Instn. Gr. Brit., USSR, Hungarian, Argentina acads. scis., Real Sociedad Espanola de Física y Química Madrid, Am. Philos. Soc., Fellow Royal Soc., 1933. Author: Collected Papers of Peter Debye, 1954. Research on polar molecules, dipole moments and molecular structure; discovered structure analysis of powdery crystalline substances by means of X-rays; formulated theory of behavior of strong electrolytes; pioneer in study of polymers. Died Nov. 2, 1966.

DECAISNE, Joseph, French botanist; b. Brussels, Belgium, Mar. 7, 1807; s. Victor and Marie (Maës) D. Made anat. sketches and charts, Pavis; then Worked with Colin in Jardin des Plantes; worked in herbarium and fruit room Natural History Mus., became prof. 1845; prof. agrl. statistics Coll. de France. Fellow Royal Soc., 1877; mem. French Acad. Scis. (elected pres. 1865), Bot. Soc. France (a founder). Author: Le jardin fruitier du Muséum, 9 vols., 1858-75. Research on anatomy and physiology of madder plant, sugar beet, China grass, the potatoe and its diseases, studied pollination, structure of mistletoe, sexuality of fucoids. Died Feb. 8, 1882.

DE CANISTRIS, Opicinus (or Opicinus de Papia), Italian cartographer; b. Lomello, Italy, Dec. 24, 1296; priest; became curate of Santa Maria Capella of Pavia, Italy, 1323; exiled to Avignon, France, 1329; apptd. clerk Penitentiary Tribunal of the Curia by Pope John XXII. Author: Codex palatinus latinus (a collections of drawings including maps and schemas with text). His maps were best of time. Died 1352.

DE CARCAVI, Pierre, French mathematician geometer; b. Lyons, France, 1600; apptd. by Colbert to head his library, 1661; chief adminstr. King's library, 1663-83; friend and corr. of Fernat, Pascal, Huygens, other scientists; early mem. French Acad. Scis., 1666. Died Paris, Apr. 1684.

DE CARLI, Luigi, Italian geneticist; b. Trento, Italy, Apr. 25, 1930; s. Giulio and Erminia (Lona) De C.; doctorate U. Pavia, 1953; m. Fiorella Nuzzo, Feb. 12, 1961; children—Giulio, Giovanni, Paola. Research worker in microbiology Istituto Sieroterapico Milanese, Milan, 1953-58; asst. in dept. genetics U. Pavia, 1958—, tchr. gen. biology, 1964—; head tissue culture group Internat. Lab. of Genetics and Biophysics, Pavia, 1962—. Mem. Biometric Soc., Am. Eugenics Soc. Studies, publs. on genetics of mammalian cells cultured in vitro. Home: 8 Viale Gorizia. Office: 14 Via S. Epifanio, 27100 Pavia, Italy.*

DE CARVALHO, Jair Correa, nematologist; b. Tres Pontas, Brazil, Mar. 31, 1903; s. Joao Correa and Ana (Custódia) de C.; student Escola Superior de Agricultura de Lavras, 1921-22, Escola de Agronomia e Veterinaria de Belo-Horizonte, 1923-24; m. Eunice Salles, Oct. 19, 1925 (div.); 1 son, Maurício Salles Correa. Staff Inst. Biológico, Sao Paulo, 1937-40; insect control, 1941-48, nematologist, 1949-66; head Secao de Parasitologia Vegetal, Sao Paulo, 1966—. Mem. Soc. Nematologists. Research and publs. on description new species. Home: 228 Rua Gen. Jardim. Office: 1252 Av. Rodrigues Alves, Sao Paulo.*

DE CASTRO, see Rodrigues de Castro, Estevan.

DE CAUS (or DE CAUX), Salomon, engr., architect; b. Dieppe, France, 1576; Became tutor to Henry, Prince of Wales, London, Eng., 1612; laid out gardens at Heidelberg (Germany) Castle, 1613; served as engr. to elector of Palatine, Heidelberg, 1614-20; became royal architect and engr., 1621; returned to France, 1623. Author several books including: Institution harmonique, 1615; les Raisons des forces mouvantes avec diverses machines tant utiles que plaissantes ausquelles sont adjoints plusieurs desseins de grotes et fontaines, 1615. Pioneered theory of steam engine; studied ways of raising water; maintained movements of the lever accorded with that of balance, and that power increased by wheels and gears; invented whistling or drinking mech. birds, saw mill, fire engine, lathe, music boxes run by water power, organs; suggested running clocks by water power instead of rewinding them. Died 1626.

DE CEFFONA, Pierre, see Ceffons, Pierre.

DE CESPEDES, Andrés García, Spanish mathematician; flourished 16th, 17th centuries; Author: Regimiento de navegacion; Hidrographia y theoria de planetas; Libro de instrumentos nuevos de geometria muy necesarios para distancias y alturas, 1606; manuscripts including atlases.

DE CHABERT, Joseph Bernard (Marquis), see Bernard, Joseph.

DE CHALES, Claude François Milliet, see Des Chales, Claude François Milliet.

DECHAMBRE, Amédée, French physician; b. Sens, France, Jan. 12, 1812; ed. Paris, France, 1831; M.D., U. Strasbourg (France), 1844; Bachelier ès sciences, Sens, 1830; editor, prin. staff mem. Gazette médicale, 1944-53; founder Gazette hebdomadaire; founder (with Auguste Mercier) l'Examinateur médical. Decorated Legion of Honor. Mem. Acad. Medicine. Author: (with Duval, Lereboullet) Dictionnaire usuel des sciences médicales, 1844; le Médecin, 1833; also published (with Raigé-Delorme) Dictionnaire encyclopédique des sciences médicales, 100 vols., 1864-89. Died Paris, Jan. 3, 1885.

DE CHANCOURTOIS, Alexandre Émile Béguyer, French geologist; b. Paris, France, Jan. 2, 1819; ed. l'École polytechnique, Paris, 1838, l'École des mines, 1840. Became prof. descriptive geometry, 1848; named prof. geology l'École des Mines, 1852; insp. gen. mines; geol. explorations in France, Asia Minor, Iceland, Greenland. One of 1st to try to establish a periodicity of chem. elements; he noted that if the elements were mounted spirally on a cylinder, vertical lines covered elements which were related. Died Paris, Nov. 14, 1886.

DE CHARLES, Guy, French surgeon; b. Auvergne, France, circa 1300; ed. at Montpellier, France; Bologna, Italy; Paris, France. Took orders in the church; canon, commensular chaplain, Lyons, France; physician to Pope Clement VI, Pope Innocent VI, and Pope Urban V; practiced medicine in Lyon and Avignon, France. Author: Cyrurgia magna, 1363. Described amputations, eye diseases, various injuries, ulcers, inflammations, çataracts, treatment of fractures using lead weight extensions. Died 1368.

DE CHARMS, Richard, Am. psychologist; b. Wilkes Barre, Pa, Dec. 13, 1927; s. Richard and Carita (Pendleton) de C.; B.A., Swarthmore Coll., 1952; M.A., Wesleyan U., 1954; Ph.D., U. N.C., 1956; m. Hana Gach, Mar. 1, 1964; children—Richard Christopher. Faculty, Washington U., St. Louis, 1957—, prof. psychology, 1965—; prin. investigator Office Naval Research, 1958-65; research asso. dept. social relations Harvard, 1965-66. Carnegie Corp. grantee, 1966—. Mem. Am. Psychol. Assn., A.A.A.S., Sigma Xi. Author: Personal Causation, 1968; also articles. Demonstrated incidence of achievement and affiliation motivation in culture of U. S. from 1800 to 1950; studies in human motivation, especially individual freedom of choice in group process and ednl. settings. Home: 702 Radcliffe St., St. Louis 63130.

DE CHARPENTIER, Jean, see Charpentier, Johann H.

DE CHAULIAC, Guy, French physician, surgeon; b. Chauliac, France, circa 1300; studied medicine, Toulouse, France, under Raymond of Molières at U. Montpellier (France), U. Bologna (Italy). Practiced medicine in Lyons, France; also canon and prévôt in Lyons; apptd. canon of Reims, France, 1353, of Mende, France 1367; papal physician to Clement VI, Innocent VI, and Urban V in Avignon, France, 1342-70; sr. officer of the Chapt. of Saint-Just. Author: La grande chirurgie (1st summary of surgery); Inventorium sive collectorium in parte chirurgiciali medicine, finished in 1363. The father of French surgery; earliest known description of femoral hernia in 1361; described plague in Europe, 1348; invented several surg. instruments. Died Lyons, July 23, 1368.

DE CHAZELLES, Jean Matthieu, French mathematician, engr., hydrographer; b. Lyons, France, July 16, 1657; studied with the Jesuits; apptd. prof. hydrography at Marseille, France, 1685; visited Greece, Turkey, and Egypt. Mem. Acad. Scis. Engaged to pub. 2d vol. of Neptune Française which includes the hydrography of the Mediterranean, also his maps and plans of the port of Ponant, France. Measured the pyramids to check the invariability of the meridians; determined the meridian of Alexandria; made various coastal surveys in France. Died Paris, Jan. 16, 1710.

DECHEN, Ernst Heinrich Karl von, see von Dechen, Ernst Heinrich Karl.

DE CHÉZY, Antoine, French mathematician; b. Chalons-sur-Marne, France, Sept. 1, 1718; s. Jacques and Marie (Bernard) de C.; ed. Châlons Oratory Sch. Engr., Paris (France) Sch. Bridges and Hwys., 1751; m. Pollin. Dir., Sch. Bridges and Hwys.; became chief engr., Chalons, 1763. Author: Méthode pour la construction des équations indéterminées relatives aux sections coniques. Invented air-bubble level for use in mountains; developed theory on uniform motion of water; built bridges of Nantes, Mantes, and Trèport. Died Paris, Oct. 5, 1798.

DECKER, Wayne Leroy, Am. meteorologist; b. Patterson, Ia., Jan. 24, 1922; s. Albert Henry and Effie (Holmes) D.; B.S., Central Coll., Pella, Ia., 1940-43; postgrad. U. Cal. at Los Angeles, 1943-44; M.S., Ia. State U., 1947, Ph.D., 1955; m. Martha Jane Livingston, Dec. 29, 1943; 1 dau., Susan Jane. Meteorologist U. S. Weather Bur., Washington, Des Moines, 1947-49; faculty U. Mo., Columbia, 1948—, prof. meteorology, 1958— chmn. grad. program in atmospheric sci., 1960—. Mem. Am. Meteorol. Soc. (vis. scientist 1959—), Am. Geophys. Union, Am. Agronomy Soc., A.A.A.S., Sigma Xi, Gamma Sigma. Delta. Research on radiative and convective exchange heat between earth's surface and atmosphere, use energy by evaporation from free water and soil surfaces also by transpiration by plants. Home: 1228 Ridge Rd., Columbia, Mo. 65201.*

DE CLAPIÉS, Jean, French engr.; b. Montpellier, France, Aug. 28, 1670; s. Pierre and Suzanne (de Loys) de C.; mem. cadet co. in army; commd. 2d lt. for Picardie regt.; lt. for Zanterre regt.; with Neerwinden; apptd. dir. bridges of Rhone, 1712; named prof. math. U. Montpellier, 1718; apptd. dir. gen. pub. works at Languedoc also of Pont S.-Esprit, 1711. Decorated Chevalier de S. Michel, 1726. Mem. French Acad. Scis., 1702. First to apply rectilinear trigonometry to graphic constrn. sun dials; observed solar eclipse, 1706. Died Montpellier, Feb. 19, 1740.

DÉCLAT, Gilbert, French physician; b. Saint-Martin-d'Estréaux, France, 1827; studied medicine, Paris, France; practiced infantile surgery. Author: Nouvelles applications de l'acide phénique en médecine, 1865; Traité de l'acide phénique, 1874; manual de médecine antiseptic, 1890. Editor med. mag. la Médecine des ferments. Discovered antiseptic properties of carbolic acid before Lister, 1861, and used it to treat wounds; recognized presence of microorganisms in infections before Pasteur. Died Nice, France, Nov. 26, 1896.

DE CLAVASIO, Dominicus, Italian mathematician; b. Chivasso nr. Turin, Italy; flourished 1346; mem. Faculty of Arts, Paris, France, 1349-50; mem. Med. Faculty, Paris, 1356-57; became tchr. philosophy Coll. Constantinople, Paris, 1349; named ct. astrologer to John the Good, before 1368. Author: Practica geometriae, 1346. Realized that 3 1/7 is only an approximate value of pi; studied measurements of triangles, rectangles and other surfaces. Died circa 1357-62.

DE CLÉRAMBAULT, Gaétan-Henri-Alfred-Édouard-Léon-Marie Gatian, French psychiatrist; b. Bourges, France, July 2, 1872; Bachelier, Stanislas Coll., Paris, 1881. Intern, charity hosps., Seine; thesis, 1899; named asst. physician, spl. infirmary for insane Prefecture de Police, 1905, head physician, 1920. Author: Les toxiques; l'Automatisme mental; Oeuvre psychiatrique, 1942. Investigated symptomatology of deliria caused by chloral, hashish, ether; formed concepts of chronic hallucinatory psychoses; described condition of patient who believes mind is under control of another person or outside influence (Clérambault-Kandinsky complex), circa 1905; studied costumes of various tribes. Died (suicide) Nov. 17, 1934.

DE CLERCK, Jean, Belgian chemist; b. Brussels, Belgium, Dec. 24, 1902; s. Louis and Jeanne (Vandenberge) de Cl.; Diplom Humanites Anciennes, Coll. St. Michel, Brussels, 1920; Ingenieur Chimiste Agricole, U. Louvain (Belgium), Ingenieur Brasseur, 1924; Dr. Honoris Causa, Technische Hochschule Munchen (Germany), 1965; m. Marie Rosalie Debecker, May 23, 1928; children—Louis, Anne-Marie, Jean-Paul, Pierre, Etienne, Agnes (Mrs. Francois Paternotte), Bernadette (Mrs. Ivan Znamensky), Michel, Therese (Mrs. Leo Van Laer), Cecile, Philippe. Asst., Brewing Sch., U. Louvain, 1925-26, prof. brewing tech. and chemistry, 1939—; chemist Brasserie de Haecht (Belgium), 1926-33; brewmaster Brasserie Roelants, Brussels, 1929-30; chief chemist Brasserie Artois, Louvain, 1933-39; dir. research and analysis lab. for brewing U. Louvain, 1939—. Pres. sci. direction com. Research Centrum Belgian Breweries, 1962—. Decorated Chevalier de l'ordre de Leopold, Commandeur de l'ordre de la Couronne, Grand officier de l'ordre de Leopold, II, Laureat du Travail de Belgique. Hon. mem. Deutsche Braumeister and Malzmeister Bund, Versuchs and Lehranstalt fur Brauerei in Berlin; mem. European Brewery Conv. (mem. council 1949—, pres., 1963—). Author: Cours de Brasserie, 2 vols., 1948; Lehrbuch der Brauerei, 1950; A Textbook of Brewing, 2 vols., 1958; Compléments au Cours de Brasserie, 1965; also numerous articles. Research on oxido reduction potential of beer and its influence on quality, respiration during germination of barley, hop analysis and bittering of beer, influence of water composition on beer quality, rational bldg. of micro- and pilot plants for brewing industry, brewing tech. and analytical methods. Home: P.O. Box 31, Louvain, Belgium.*

DE COMBASLE, Mathieu, French agronomist; b. Nancy, France, Feb. 26, 1777; s. Jean-Antoine-Mathieu and Marie Le Febvre (de Montique) de D.; ed. Nancy école centrale. Founder beet sugar refinery, 1810; founder model farm, Roville, France, 1822; manufactured farm equipment. Member of the French Academy of Sciences, 1825. Author: Calendrier du bon cultivateur, 1821; Instruction sur la fabrication du sucre de betteraves, 1839; Traité d'agriculture, 5 vols., 1861; Annales agricoles de Roville, 9 vols. Developed triennial rotation for cultivation of cereals and vegetables, maceration process of refining beet sugar. Died Nancy, Dec. 27, 1843.

DÉCOMBE, Jean Léon, French chemist; b. Paris, France, June 26, 1900; s. Louis and Berthe (Wallet) D.; Ingenieur Chimiste, Ecole Nationale Supérieure de Chimie de Paris, 1924; Docteur ès Scis. Physiques, Faculté des Scis. de Paris, 1932; m. Jadwiga Rosinska, Mar. 23, 1949. Asst. in chemistry Faculté des Scis. de Paris, 1926-41; prof. chemistry Faculté des Scis. de Dijon (France), 1941—. Mem. Chem. Soc. France. Contbr. papers to jours. Research on esters, formula of organo-magnesium compounds, action of organo-magnesiums on polynitrites, synthesis of new quinolines. Home: 10, rue Mets, Dijon, France.*

DE COMBEROUSSE, Charles-Jules-Felix, French mathematician; b. Paris, France, 1826; s. Alexis de Comberousse; student L'Ecole central; engr. ry. of Saint-Germain and of l'Est; later prof. applied mechanics l'Ecole centrale; prof. rural engring. Conservatory Arts and Crafts; prof. math. Chapta Coll., 1854-88. Mem. French Acad. Scis., 1847, Soc. Civile Engrs. France (elected pres. 1885). Author: Etude des résistances au mouvement des trains sur les chemins de fer, 1853; Cours complet de mathématiques, 1860; Cours de cinématique, 1865; (with Eugene Rouche) Traité de géométrie élémentaire, 1865; Histoire de l'Ecole centrale des arts et métiers depuis sa fondation jusqu'à nos jours, 1879; (with Serret) Traité d'arithmétique, 1882. Died Paris, 1897.

DE CONDORCET, Marie Jean Antoine, see Condorcet, Marie-Jean-Antoine-Nicolas Caritat, Marquis de.

DE COPPET, Louis Casimir, chemist; b. N.Y., July 21, 1841; Ph.D., U. Heidelberg (Germany), 1866; worked with Frankland in London, Eng.; research and publs. on confirmation of law of proportionality; calculated an atomic depression in freezing point measurements and used it to classify salts; calculated amounts of hypothetical hydrates of a salt in solutions. Died Nice, France, Aug. 1911.

DE CORIOLIS, Gaspard Gustave, French physicist; b. Paris, France, May 21, 1792; s. Jean-Baptist-Elzéar and Marie-Sophie (de Maillet) C.; studied engring. in Paris. Began as tchr. of math. and physics at L'Ecole Polytechnique, apptd. dir. studies, 1838; head engr. Dept. Civil Engring.; prof. analytical geometry and gen. mechanics Ecole centrale des arts et manufactures. Mem. French Acad. Scis., 1836. Author: Traité de la mécanique des corps solides et du calcul de l'effet des machines, 1829; Théorie mathematique des effets du jeu de billard, 1835; also wrote on the constrn. of highways, mechanics, force and motion. Edited publs. on mechanics, powers, and motions. Research on the form of the whirlpool in 1836; calculated piston swing lever and friction; calculated compound centrifugal force which affected moving bodies in a rotating system; 1st modern definitions of kinetic energy and work; developed theorem of relative motion. Died Paris, Sept. 19, 1843.

DECOSTA, Edwin J., Am. gynecologist; b. Chgo., Mar. 25, 1906; s. Lewis M. and Grace (Myers) Dec.; B.S., U. Chgo., 1926; M.D., Rush Med. Coll., 1929; m. Mari H. Bachrach, Jan. 5, 1935; children—Mari Jane (Mrs. David M. Terman), Catherine (Mrs. Stuart S. Burstein), Louise (Mrs. Burton V. Wides), John Lewis. Pvt. practice medicine, specializing in obstetrics, gynecology, surgery, Chgo., 1934—; attending obstetrician, gynecologist Passavant Meml. Hosp., Chg.; attending gynecologist Cook County Hosp., Chgo.; prof. Northwestern U. Med. Sch. Mem. A.M.A., A.C.S., Am. Coll. Obstetrics and Gynecology, Am. Gynecol. Soc., Am. Assn. Obstetrics and Gynecology, Central Assn. Obstetrics and Gynecology (pres., 1961), Chgo. Gynecol. Soc. (pres., 1960), Phi Beta Kappa, Sigma Xi, Alpha Omega Alpha. Author: (with John I. Brewer) Textbook of Gynecology, 1967. Contbr. numerous articles to med. jours. on gynecologic endocrinology. Home: 821 Highland Pl., Highland Park, Ill. 60035. Office: 707 N. Fairbanks Ct., Chgo. 60611.*

DE COURSEY, Russell Myles, Am. zoologist, entomologist; b. Indpls., Jan. 17, 1900; s. Arthur I. and Sarah E. (Sims) Dec.; A.B. (Edward Rector scholar), DePauw U., 1923; M.A., U. Ill., 1925, Ph.D., 1927; m. Mary E. Tucker, Aug. 20, 1930; children—Lowell A., Marilyn J. (Mrs. Donald B. Richardson). Faculty, La. State U., 1927-29; faculty U. Conn., Storrs, 1929—, prof. zoology and entomology, 1935—. Vis. prof. U. Miami (Fla.), 1936; research asso. U. Cal., Berkeley, 1955. Fellow A.A.A.S.; mem. Entomol. Soc. Am., Soc. Systematic Zoologists, Am. Soc. Zoologists, Am. Inst. Biol. Scis., Conn. Entomol. Soc., Am. Assn. U. Profs., Sigma Xi. Author: The Human Organism, 1961, 3d edit., 1968; (with F. Dolyak) Laboratory Manual of Human Anatomy and Physiology, 1963; El Organismo Humano, 1966. Research on order Hemiptera (Insecta) immature or larval forms of family Pentatomidae, Keys, life histories. Home: 24 Storrs Heights Rd., Storrs, Conn. 06268.

DÉCOURT, Luiz Venere, Brazilian cardiologist; b. Campinas, Sao Paulo, Brazil, Dec. 7, 1911; s. Paulo L. and Alzira (Venere) D.; Med. Dr., Sao Paulo U. Sch. Medicine, 1935; m. Guiomar Correa Sampaio, Dec. 7, 1938. Faculty, Sao Paulo U. Sch. Medicine, 1940-42, 47—, prof. clin. medicine, 1950—; prof. clin. medicine Sao Paulo U. Sch. Nursing, 1944-49; head sect. cardiology Hospital Clinicas, 1953—. Mem. Sao Paulo Soc. Medicine (past pres. sect. medicine), Brazilian (past pres.), Argentine, French, Interam., Internat. socs. cardiology, Am. Coll. Cardiology, Brazilian Soc. Nephrology. Author: Lessons on Disorders of the Heart and Circulation, 1945; The Q-T Interval of the Electrocardiogram, 1950; Biopsy of the Human Heart, 1964 (all published in Portuguese). Mem. editorial bd. Revista Paulista Medicina, 1949-50, Arquivos Brasileiros Cardiologia, 1948-58, Archives of Interam. Rheumatology, 1958-64, Am. Heart Jour., 1966—. Research and publs. on humoral, electrocardiographic and histopathol. aspects of rheumatic fever, relations between elastic pattern of pulmonary artery and degree of pulmonary hypertension, electrocardiographic evolution in patients with mitral and/or aortic valvular prostheses, method for human percutaneous biopsy of heart for early diagnosis of endocardial and myocardial diseases, endocardial fibroelastosis. Home: 3 Rua José Freitas Guimaraes. Office: 601 Rua Itacolomi; also Hospital das Clinicas, Caixa Postal 8091, Sao Paulo, Brazil.*

DE CRÈVECOEUR, Hector Saint-John, (Crèvecouer, Hector Saint-John de), French agronomist; b. Caen, France, Jan. 31, 1735; went to U. S. and operated large farm nr. N.Y.; returned to France, 1784; mem. French Acad. Scis., 1783, Soc. Agr.; author: Lettres d'un cultivateur américain, 1784; Voyage dans la haute Pensylvanie et dans l'État de New-York, 1801. Introduced potato culture to Normandy, France. Died Sarcelles, France, Nov. 12, 1813.

DECROLY, Ovide, Belgian psychologist; b. Renaix, Belgium, July 23, 1871; founder sch. pedagogical psychology, Uccle; prof. at Brussels (Belgium); Author: Faits de psychologie individuelle et de psychologie experimentale, 1908; Fonction de globalisation, 1923; L'intelligence et sa mesure, 1921; Les intérets chez les enfants, 1925; Evolution de l'affectivité, 1927; Developpement du langage, 1930. Proposed children should study their natural and social surroundings. Died Renaix, Sept. 12, 1932.

DE CROUSAZ, Jean-Pierre, mathematician; b. Lausanne, Switzerland, Apr. 13, 1663. Prof. math. and philosophy U. Lausanne, 1700-24, 1735-48; prof., U. Groningen (Holland). Author: Logique ou Système de reflexions qui peuvent conduire à la netteté de nos connaissances, 1712; Examen de pyrrhonisme ancien et moderne, 1733; Observations critiques sur l'abrégé de la logique de Wolf, 1744; Traité sur l'education des enfants, 1722. Advocated sci. as basis of edn.; critic of Leibnitz and Wolf. Died Lausanne, Mar. 22, 1750.

DE CUBIERES, Simon-Louis-Pierre, French naturalist; b. Roquemaure, France, Oct. 12, 1747; studies in natural sci. encouraged by Buffon. Visited Italy, 1775; made a voyage to Eng. and brought back rare plants in 1780; conservator of monuments Palace of Versailles (France). Mem. French Acad. Scis., 1810. Author: Histoire abrégée des coquillages de mer, 1800; Mémoire sur les abeilles, 1800; Histoire du tulipier, 1800; Mémoire sur le genévrier rouge de Virginie, 1805; Mémoire sur le frene du Canada, 1804. Research on lighter-than-air balloons, the tulip tree, Canadian oak, Virginian red juniper tree, auriculate magnolia tree and other rare plants. Died Paris, Aug. 10, 1821.

DE CYON, Elie, Russian physiologist; b. Russia, 1843; prof. St. Petersburg (Russia) Acad. Medicine, from 1870; with Karl Friedrich W. Ludwig, German physiologist, discovered branch of vagus nerve (in rabbits) the stimulation of which causes a lowering of blood pressure, 1866; studied and reported vasomotor reflexes. Died 1912.

DEDEKIND, Julius Wilhelm Richard, German mathematician; b. Brunswick, Germany, Oct. 6, 1831; s. Julius and Caroline (Emperius) D.; studied under C. F. Gauss, U. Göttingen (Germany, Ph.D., 1852; hon. degrees, Oslo, Norway, Zurich, Switzerland, Brunswick. Docent U. Göttingen, 1854-58; faculty Fed. Inst. Technology, Zurich, 1858-62; prof. Technische Hochschule, Brunswick, 1862-1912. Mem. French Acad. Scis., 1900, acads. Berlin (Germany), Rome (Italy). Author: Stetigkeit und irrationale Zahlen, 1872; Über die Theorie der ganzen algebraischen Zahlen, 1879; Was sind und was sollen die Zahlen? 1888; Gesammelte mathematische Werke, 2 vols., 1930-31; preface to collected works of Riemann, also appendix in later edits.; papers on vibrations of liquid ellipsoid, binomial equations, theory of modular and Abelian functions. Editor Dirichlet's work on theory of numbers, 1894. Prominent 19th century contbr. to theory of algebraic functions and number theory; using Dedekind cuts, originated theory of irrational numbers, as early as 1858. Died Brunswick, Feb. 12, 1916.

D'EDELCRANZ, Abraham Nicholas, inventor, physicist; b. Aabo, Sweden, 1754; research in physics, mechanics, econ. and indsl. agr.; invented a telegraph, mercury lamp, machine for mfg. of canvas. Died Stockholm, 1821.

DE DELLINGHAUSEN, Nicolas, see von Dellinghausen, Nicolas.

DE DIETRICH, Phillippe-Frédéric, French mineralogist; b. Strassbourg, France, Nov. 14, 1748; s. Jean de Dietrich; insp. mines and factories. Mem. French Acad. Scis., 1780. Author: Descriptions des gites de mineral des Pyrénées; Descriptions des gites de mineral et des bouches à feu en France, 1786-1800. Research on mineral deposits in the Pyrenees. Died Dec. 29, 1793.

DE DION, Albert (Marquis), French inventor; b. Nantes, France, 1856; pioneer automobile inventor; mem. Société de Dion-Bouton (co-founder with Georges Bouton); 1st worked on constrn. steam vehicles; won races Paris to Versailles, 1887, Paris to Rouen, 1894 (tractor pulled coach); built radial engine with rotary cylinders, 1889, gasoline powered 3-wheeler with 1 cylinder with cooling ribs, battery ignition, surface carburetor, device for advanced ignition, 1896, vehicle with tubular chassis and drive shaft transmission, 1899. Died 1946.

DE DION, Henri, French engr.; b. Montfort, France, 1828; studied at École central, Paris, France, 1848; began work at Bayeux, 1853; made voyage to Guadeloupe, 1860, and built sugar refinery, 1862; (without use of tie-beams), prof. civil constrns. Conservatory of Arts and Crafts, from 1865; head engr. Santander Co.; became head of metallic constrns. 1878 Universal Exposition, 1875. Author: Cathédrale de Bayeux, 1861; De la production et du commerce des sucres, 1864. Built exhibition halls (without use of tie-beams), 1878 Exposition; saved cathedral of Bayreux by supporting vaults and tower with iron framework. Died l'Amaury, France, 1878.

DE DIVINI, Eustachius, Italian optician; b. San Severino, 1620; made optical instruments, Rome; produced eyeglasses with 2 eyepieces and 2 objectives (object lens); built a microscope; noted for his telescopes; rival of Campani.

DE DOLOMIEU, Dieudonné-Sylvain-Guytancrède Gratet, French geologist, mineralogist, seismologist; b. La Tour-du-Pin, France, June 23, 1750; s. François, marquis de Dolomieu and Mirie-Françoise de Berenger; ed. Order of Malta Mil. Sch., Paris; pursued mil. career until 1786; became prof. mineralogy Ecole centrale des mines, 1794; apptd. insp. mines, 1795; prof. geology and mineralogy Mus. Natural History, from 1796; mem. Napoleon's expdn. to Egypt, 1798, captured and imprisoned at Messina during return. Mem. French Acad. Scis., 1795, Inst. Egypt,

Royal Acad. Scis. Author: Mémoire sur les tremblemens de terre de la Calabre pendant l'année, 1783; Memoire sur la constitution physique de l'Egypte, 1794; Mémoire sur les iles Ponces et les produits volcaniques de l'Etna, 1788; Dernier voyage dans les Alpes, 1802; Traité de philosophie minéralogique, Memoire sur l'espèce minerale (both written with improvised materials during term in prison), 1801. Discovered volcanic formations in Lisbon, 1778; considered father of Alpine geology (with Saussure); explored Pyrenees, Sicily, the Alps, 1782; founder obs., Malta; discovered dolomite, 1791; established relationship between earthquakes and volcanic phenomena. Died Nov. 16, 1801.

DE DOMINIS, Marcus Antonius, physicist; b. Arbe, Dalmatia, 1566; prof. philosophy and math., Padua, Italy; then became archbishop of Spalato (now Split, Yugoslavia), primate of Dalmatia, circa 1611; went to Eng., circa 1615; became Protestant and was apptd. dean of Windsor by James I; pub. anti-papal articles in On the Ecclesiastical Republic, 1617, but abjured doctrines after return to Italy, 1622. Author: De radiis visus et lucis in perspectivis et iride, 1611; Euripis seu de fluxu et refluxu maris sententia Marci Antonii de Dominis Archiepiscopi Spalantensis ad illustrissimum principem franciscum barberiunum S.R.E. Card. Amplissimum, 1614. Used principles of refraction to explain rainbow; considered light a substantial element that produced colors of bodies without being an inherent part of them; attributed tides to influence of sun and moon. Imprisoned on suspicion he might recant again and died in Rome, 1624.

DE' DONDI, Giovanni, Italian physician astronomer; b. Chioggia, Italy, 1318; s. Giacomo de' Dondi; chosen personal physician of Charles IV, 1349; joined faculty astronomy Padua, Italy, 1350; lectr. medicine Florence, Italy, circa 1367-70; with U. Pavia (Italy), 1379-88. Author: De fontibus calidis agri Patavini in which he discussed origin of hot mineral springs and their analysis. Constructed elaborate clock and orrery. Died Genoa, Italy, 1389.

DE DUVE, Christian Rene, biochemist; b. Thames-Ditton, Eng., Oct. 2, 1917; s. Alphonse and Madeleine (Pungs) de D.; M.D., B.Sc., U. Louvain, 1946; m. Janine Herman, Sept. 30, 1943; children—Thierry, Anne, Francoise, Alain. Lectr., U. Louvain, Belgium, 1947-51, prof. biochemistry, 1951—; prof. biochemistry Rockefeller U., N.Y.C., 1962—. Decorated officer Order of Leopold; Comdr., Order of Merit, Chile; Pfizer award, 1957; Francqui award, 1960. Mem. Royal Acad. Medicine, Royal Acad. Sci. of Belgium, Belgium, French, Brit. biochem. socs., Am. Soc. Cell Biology, others. Author: Glucose, Insuline et Diabete, Paris, 1945; also numerous articles. Research on insulin and glucagon; devel. of cell fractionation methods; discovery of lysosomes and peroxisomes and their roles. Home: 80 Central Park W., N.Y.C. 10023.*

DEE, Arthur, alchemist; b. Mortlake, Eng., July 13, 1579; s. John and Jane (Fromond) D.; ed. Westminster Sch., Oxford (Eng.) U.; m. Isabella Prestwyck; 7 sons, 6 daus. Travelled in Germany, Poland, and Bohemia; apptd. physician to czar and remained in Russia 14 years; lived in London after returning to Eng.; later started med. practice in Norwich, Eng.; physician in ordinary to King Charles I. Author: Faciculus chemicus, 1631, collection of alchemical extracts written in Moscow. Died Norwich, Sept. 1651.

DEE, John, English mathematician, astrologer, alchemist; b. London, Eng., July 13, 1527; possibly s. Rowland and Johanna (Wild) D.; B.A., St. John's Coll., Cambridge, 1545, fellow 1545-46; one of original fellows Trinity Coll., Cambridge; M.A., Cambridge, 1548; postgrad. Louvain, Belgium, 1548-50; (probably received doctorate in law); m. 2d, Jane Fromond, Feb. 5, 1577 or 78; 11 children, including Arthur. Became acquainted with Continental mathematicians and their work through travels in Europe, 1547-50; lectr. on Euclid, Paris, France, 1550; astrologer, sci. and med. adviser English court, circa 1550-80; rector, Upton-upon-Severn, Eng., 1553; accused but found innocent by Star-chamber of practicing sorcery against queen's life, 1555; suggested to Queen Mary that royal library of ancient manuscripts be formed, 1556; instr. nav. at various times; visited Venice, Italy, also St. Helena, Hungary; cons. on Queen's health and sent to Germany to consult with other physicians, 1578; practiced (with Edward Kelly and others) alchemy and crystallomancy, Europe, circa 1580-89; warden Manchester (Eng.) Coll., 1595-1604; unsuccessfully petitioned James I to be cleared of charges of being magician, 1604. Author many works (some unpublished) on nav., hydrographics, ancient learning, various occult subjects; established as intellectual leader of applied math. by his math. preface to Billingsley's transl. of Euclid (1st complete English transl.), 1570. Helped revive math. learning in Eng., especially applied math., geometry, trigonometry; influenced by Archimedes in math. thought; in communication with leading scientists of the time; made hydrographical and geog. charts of newly discovered lands, 1580; worked on calculations to facilitate adoption of Gregorian calendar in Eng.; collected large sci. and math. library. Died in poverty, Mortlake, Eng., Dec. 1608.

DE EGAS MONIZ, Antônio Caetano de Abreau Freire, Portuguese physician; b. Avança, Portugal, Nov. 29, 1874; ed. univs. Coimbra (Portugal), Bordeaux (France). Began career as mem. med. faculty Coimbra, prof., 1902; dir. med. faculty Lisbon, Portugal, 1911-44. Fgn. minister Portugal, 1917; head Portugese delegation to Paris Peace Conf. Mem. numerous European sci. socs. Author: Altera coes anátamo—patalogicas nadifteria, 1900; A vida sexual (fisiologia), 1901; A vida sexual (patalogia), 1901; Cursode neurolgia, 1912; A Neurologia na guerra Lisboa, 1917; Clinica neurológica, 1925; O Padre Faria na historia de hipnotismo, 1925; Diagnostic des tumeurs cerebrales et epreuve de l'encéphalographie artérielle, 1931; L'angiographie cérébrale . . . , 1934; Tentatives opérations dans le traitement de certaines psychoses, 1936; Cerebrale Arteriographie und Phlebographie, 1940; Trombosses y otra obstructiones de las carótidas, 1941; also numerous other works. Co-recipient Nobel prize in medicine and physiology (with W. R. Hess) for discovery of therapeutic value of prefrontal leucotomy in certain psychoses, 1949; 1st to relieve mental disorder by surgery; introduced arteriography; introduced cerebral angiography for diagnosis of cerebral tumors by method of visualizing blood vessels of brain, 1927; inventor arterial encephalograph, 1927; successfully localized (with others) intra-cranial tumors, 1929; performed (with Almeida Lima) prefrontal lobotomy by cutting communication between frontal lobes and thalamus for relief of psychoses, 1935; inventor leucotome for use in surgery of frontal lobes. Died Lisbon, Dec. 13, 1955.

DE ENCISO, Don Martin F. (Enciso, Don M.F. de), navigator; b. Seville, Spain; b. circa 1450; author: Suma de geografia, que trata de todas las partidas del munda, 1519; founded city of Santa-Maria, St. Domingo; at isthmus of Panama noted difference in levels of Atlantic and Pacific oceans, 1519. Died 1519.

DEER, William Alexander, Brit. geologist, mineralogist; b. Oct. 26, 1910; s. William and Cunningham (Davina) D.; M.Sc., U. Manchester; Ph.D., Cambridge U.; m. Margaret Kidd; children—David, Diana, Stephen. Lectr. at U. Manchester, 1937-38; instr. St. Johns Coll., Cambridge, 1938-40, lectr., 1946-50; prof. geology U. Manchester. 1950-61; prof. mineralogy and petroleum chemistry Cambridge U., 1961—. Recipient medal Royal Soc. of Bruce of Edinburgh, Murchison Fund Geol. Soc. Mem. Royal Soc. of London, Geol. Soc. of Gt. Britain, Mineral. Soc. (U. S.). Author: (with L. R. Wager) Petrology of the Skaergaard Intrusion; (with R. A. Howie and J. Zussman) Rock Forming Minerals, 5 vols. Home: 154 Hinton Way. Office: Downing Pl., Cambridge, England.

DEERE, Don Uel, Am. civil engineer; b. Corning, Ia., Mar. 17, 1922; s. Ora Uel and Wilma (Hanna) D.; student N.W. Mo. State Coll., 1939-41; B.S., Ia. State Coll., 1943; M.S., Colo. U., 1948-49; Ph.D., U. Ill, 1955; m. Carmen Pilar Garcia, Aug. 15, 1944; children—Diana, Don William. Jr. mining engr. Phelps Dodge Corp., 1943-44; exploration engr. Potash Co. Am., 1941-46; asst. prof., then asso. prof. civil engring. U. P.R., 1946-51, head dept., 1949-51; mem. faculty U. Ill. at Urbana, 1955—, now prof. civil engring. and geology; cons. to industry, govt.; spl. research rock constrn. Mem. Am. Soc. C.E., Geol. Soc. Am., Am. Geophys. Union, Am. Inst. Profl. Geologists, Assn. Engring. Geologists, Am. Inst. Mining Engrs., Am. Soc. Testing Materials, Nat. Acad. Engring., Sigma Xi, Phi Kappa Phi, Tau Beta Pi. Study of the stability of natural slopes; regional susidence caused by withdrawal of petroleum, mined products, or groundwater. Home: 307 E. Pennsylvania Av., Urbana, Ill. 61801.

DEERE, John, Am. inventor; b. Rutland, Vt., Feb. 7, 1804; s. William Rinold and Sarah (Yates) D.; m. Damaris Lamb, Jan. 28, 1827; m. 2d, Lucinda Lamb, 1867. First mfr. plow steel in U. S., organizer various mfg. firms; incorporated, became pres. firm Deere & Co., 1868. Mayor, Moline, Ill.; pres. 1st Nat. Bank Moline. Originated, developed idea that successful self-scouring of a steel moldboard depended upon its shape; devised new plow with steel blade; manufactured 1st cast plow steel in U. S. and other farm implements. Died May 17, 1886.

DE ESPINOSA, Fray Antonio Vazquez, Spanish missionary; b. Juarez, Spain, 1570. Joined Carmelite order Roman Catholic Ch., travelled in Mexico and Peru, returning to Spain, 1622. Author: Viaje y navegación del año de 1622 que hizo la flota de Nueva España y Honduras, 1623; Circumstancias tratos y contratos de las Indias del Perú y Nueva para los tratos y contratos de las Indias del Perú y Nueva España, 1624; Compendium and Description of the West Indies, 1628. Described preparation of mercury from cinnabar and the amalgamation process. Died Seville, Spain, 1630.

DEESZ, Louis A(spell) Am. elec. engr.; b. Denver, May 22, 1888; s. Louis Phillip and Lucy (Soper) D.; student Colo. Coll., 1907-11; B.S., Carnegie Inst. Tech., 1922, E.E., 1936; m. Henrietta Davis, Aug. 13, 1913; children—Lucy Ann (Mrs. Duncan Heubner); m. 2d, Myrtle May Robbins, May 28, 1939. Various engring. positions Colo. Light & Power Co., Cripple Creek, 1912, Fed. Light & Power Co., Trinidad, Colo., 1913, Fed. Light & Power, Deming, N.M., 1914, Intermountain Ry. Light & Power Co., Colo. Springs, Colo., 1915-16, Colo. Fuel & Iron Co., Pueblo, 1917-19; gen. engr. Westinghouse Elec. Corp., 1919-11; engr. of tests Colo. Fuel & Iron Co., 1922-30; cons. engr. Frevn Engring. Co., Chgo., assigned to "Energocenter" and "Stalproect" Moscow and Siberia, USSR, 1930-33; lectr. on engring. U. Moscow, Russia, 1931; chief dist. combustion engr. Republic Steel Corp., Youngstown dist., 1933-42; prof. elec. engring. Youngstown (Ohio) Coll., from 1939, dean Wm. Raven Sch. of Engring., from 1942. Decorated Udarnick of the USSR, 1932. Registered engr., Colo. Mem. Nat. Soc. Profl. and Registered Engrs., Am. Soc Engring. Edn., Am. Inst. E.E., Kappa Sigma, Mu Pi Epsilon. Contbr. articles on electrical precipitation, combustion engring., and rotating elec. machinery. Died Apr. 19, 1950.

DEETZ, Charles Henry, Am. cartographic engr.; b. Sellersville, Pa., Apr. 10, 1864; s. Thomas Berger and Caroline (Nase) D.; Phillips Exeter Acad., 1885; studied civil engring. (geodetic course), Mass. Inst. Tech., 1886-88; m. Clarissa Hannah Wilson, Dec. 7, 1892. Field worker with U. S. Coast and Geod. Survey in Ala. and Fla., 1888, also served in Eastern States; assigned to cartographic work in preparation nautical charts; specialized in map projections. Mem. Philos. Soc. Washington, Am. Numis. Assn. Author: Lambert Conformal Conic Projection, 1918; Lambert Projection Tables for France, with Conversion Tables, for the Use of the Army, 1918; (with Oscar S. Adams) Elements of Map Projection, with applications to map and chart construction. 1921, revised edit., 1944; Cartography, a review and guide for constrn. and use of maps and charts, 1936; revised edit., 1943. Contbr. cartographic articles to sci. and tech. publs. Died Mar. 1946.

DE FAGNANO, Giulio Carlo, see Fagnano.

DEFANT, Albert Joseph Maria, Austrian meteorologist, oceanographer; b. Trient, Tyrol, Austria, July 12, 1884; s. Joseph O. and Maria (Veith) D.; degree in physics and math, U. Innsbruck (Austria), 1906; Ph.D., U. Berlin (Germany), 1910, Nat. Sc.D. honoris causa; m. Mimi Krepper, 1909; m. 2d, Maria Schletterer, 1952. Head Central Inst. Meteorology and Gen. Dynamics, Vienna, Austria, to 1919; asst. U. Innsbruck from 1907; extraordinary prof., ordinary prof. from 1924, dir. Inst. Meteorology and Geophysics from 1945, prof. emeritus from 1955; prof. Inst. Oceanography U. Cal. at La Jolla, 1949-50, prof. (hon.) U. Berlin, 1927-45, U. Hamburg (Germany), 1951-54. Recipient Vega-Agasiz and Galatea medal, E. Wiechert medal, Ring of Honor U.N. Mem. acads. scis. cities Austria, Germany, Sweden, Norway, Finland, N.Y.C., others. Author: Lufthülle und Klima, 1923; Wissenschaftliche Ergebnisse der Deutschen Atlantischen Expedition Meteur 1924-26; Wetter und Wettervorhersage, 1914; Gezeitenproblemedes Meeres in Landnähe, 1925; Dynamische Ozenographie, 1929; Physikynamik der Atmosphäre, 1958; Physical Oceanography, 2 vols., 1960. Research, publs. on circulation of atmosphere, theoretical meteorology, marine meteorology, interaction of ocean and atmosphere. Address: Sternwartestrasse 38, Innsbruck, Tyrol, Austria.

DEFANT, Friedrich Richard Wolfgang, meteorologist, geophysicist; b. Vienna, Austria, Apr. 14, 1914; s. Albert and Maria (Krepper) D.; Ph.D. in Meteorology, U. Berlin, 1940; m. Christine Schmoll von Eisenwerth; children—Beate, Maria, Andreas, Martin, Wolfgang. In armed forces, 1940-45; Prof. meteorology and geophysics U. Innsbruck, 51-53; vis. prof. univs. Chgo., Calif. at Los Angeles, Stockholm; prof. meteorology U. Kiel, 1961—. Author: (manual) Physik, Dynamik der Atmosphäre. Research on gen. circulation of atmosphere. Home: Wehdenweg 65. Office: Klaus Grothplatz 2, Kiel, Germany.

DEFAY, Raymond, Belgian engr., phys. chemist; b. Anderlecht, Belgium, Feb. 1, 1897; s. Ernest and Clémentine (Schoutens) D.; Ingénieur civil, U. Libre de Bruxelles, 1921, Docteur Spécial en Chimie Physique, 1932; m. Suzanne van Marcke de Lummen, Oct. 18, 1921; children—Nicole, Jacques, Nadine (Mrs. Janson). Faculty, U. Libre de Bruxelles, 1941—, prof., 1956—, sec. Faculty Applied Scis., 1957-60, v.p., 1960-62, 65-66, pres., 1962-65; engr. Ernest Defay et fils, 1922-46. Author: Etude thermodynamique de la tension superficielle; Thermodynamique chimique conformèment aux méthodes de Gibbs et De Donder; Chemical Thermodynamics; Thermodynamics of Clouds; Surface Tension and Adsorption; also articles. Research in chem. thermodynamics, electrochemistry and surface chemistry. Home: 23, avenue de l'Orée, Bruxelles 5, Belgium.*

DE FERRANTI, Sebastian Ziani, English inventor; b. Liverpool, Eng., Apr. 9, 1864; s. César Ziani and Juliana (Scott) de F.; ed. St. Augustine's Coll., Ramsgate, Eng.; D.Sc. (hon.) Manchester, Eng., 1912; m. Gertrude Ruth Ince, 1888; 3 sons, 4 daus. Founded (with Francis Ince) bus. for manufacture elec. apparatus, London, Eng., 1883; engr. Grosvenor Galley Electric Supply Corp., 1886; chief electrician London Electric Supply Corp., 1887-92; contracted to light Portsmouth, Eng., 1894; founded Ferranti, Ltd., Lancashire, Eng., 1896. Fellow Royal Soc., 1927; mem. Instn. Elec. Engrs. (pres. 1910-11), Instn. Civil and Mech. Engrs. Patentee 176 devices including mercury meter, a gas-turbine procedure, textile yarn prodn. Constructed (with Siemens) induction furnace; (with A. Thompson) built alternating current generator Thompson-Ferranti, 1882; invented electrolytic meter, 1883, dynamo with coreless disk, 1884, improved transformer, 1885, use of oil under pressure in prodn. of cables, 1888, cable for high frequency alternating current, differential protection against mistakes in insulation; originated long distance transmission of high power elec. current; manufactured steam turbines with spl. steel blades, after 1900. Died Jan., 1930.

DE FILIPPI, Cav. Filippo, surgeon; b. Turin, Italy, Apr. 6, 1869; s. Giuseppe and Olimpia (Sella) De F., Sch., Turin U.; m. Caroline Fitzgerald, 1901. Former asst. surg. clinics U. Bologna, U. Genoa; reader operative surgery U. Bologna; mem. Alaskan Exptn. of Duke of Abruzzi, climbed Mt. St. Elias, 1897, also Expdn. of 1909 to Western Himalayas and Baltoro Glacier in Karakoram; organizer, leader sci. expdn. to Karakoram and Central Asia under auspices of Italian Govt., Govt. of India, various scis. bodies, 1913-14. Recipient Gold medal Royal Italian Geog. Soc., Am. Geog. Soc. Fellow Royal Geo. Soc. (Gold medal); mem. Accademia dei Lincei, Pontificia Accademia delle Scienze (Rome); Reale Societá Geografica (Rome), Gesellschaft für Erdkunde (Berlin), Société de Géographie (Gold medal) (Paris), Royal Rumanian Geog. Soc. Author: The Ascent of Mount St. Elias, 1900; Ruwenzori, 1909; Karakoram and Western Himalaya, 2 vols., 1912; Himalaya, Karakoram and Eastern Turkestan, 1932; An Account of Tibet: The Travels of Impolito Desideri of Pistoia, S.J. (1712-1727), 1932-37, also articles on surgery and biol. chemistry. Died Sept. 23, 1938.

DE FILIPPI, Filippo, Italian naturalist; b. Milan, Lombardy, 1814. Prof. natural history, Paris (France) Mus.; prof. zoology, U. Turin (Italy). Author: Reproductive Functions in Animals, 1850; Terrestrial Creation, 1854; Research, publs. on reproductive functions animals; recognized parasitic origin of oyster pearls, 1852. Died Hong Kong, China, 1867.

DE FLEUR, Melvin Lawrence, Am. sociologist; b. Portland, Ore., Apr. 27, 1923; s. Robert H. and Dorothy (Foster) DeF.; B.S. cum laude, St. Louis U., 1949; M.S., U. Wash., 1952, Ph.D., 1954; m. Lois Begitske, Dec. 5, 1961. Faculty, Ind. U., 1954-63; prof. U. Ky., 1963-67; prof. Wash. State U., Pullman, 1967-—. Fellow Am. Sociol. Assn.; mem. Ohio Valley Sociol. Assn., So., Argentine sociol. socs., Internat. Institut de Sociologie. Author (with Otto N. Larsen) The Flow of Information, 1958; Techicas y Metodos de Investigacion Social, 1965; Theories of Mass Communication, 1966. Contbr. to theory of attitudes and their relationship to overt behavior, analysis of media of mass communication in Am. soc., formalization of theory in field of sociology. Home: 703 Skyline Dr., Pullman, Wash. 99163.*

DE FLEURIEU, Charles Pierre Claret, French hydrographer, explorer; b. Lyons, France, July 2, 1738; joined French Navy, 1754; made voyage to Canary Islands, 1768-69; became gen. dir. of ports and arsenals, 1776. Decorated Grand Officer Legion of Honor; A bay in Tasmania which was discovered by Baudin in 1802 was named in his honor. Mem. Bur. Longitudes, Marine Acad., French Acad. Scis. Author: Mémoire sur la construction des navires, 1763; Histoire générale des navigations de tous les peuples; Découvertes des Français en 1768. Worked on frigate constrn., 1763; invented nau. watch (with Berthoud) which kept time with reference to the time of departure and thus solved the problem of determining longitude at sea. Died Paris, France, Aug. 18, 1810.

DE FLEURY, Maurice, French physician; b. Bordeaux, France, 1860; s. Armand de F.; intern, Bordeaux; practiced at Paris. Author: Recherches cliniques sur l'épilepsie, 1900; les Grands symptômes neurasthéniques, 1901; Manuel pour l'étude des maladies du système nerveux, 1904; Brévaire de l'arithritique, 1912; L'angoisse humaine, 1925; les Fous, les pauvres fous, et la sagesse qui'ils enseignent, 1929. Studied joints and arthritis, nervous disorders, including neurasthenic symptoms, epilepsy. Died Paris, 1931.

DE FONSECA BENEVIDES, Francisco, Portuguese physicist; b. Lisbon, Portugal, 1835; joined Portuguese navy, 1851; became prof. physics Lisbon Indsl. Inst., 1854; mem. Lisbon Acad. Author: Cours d'artillerie, 1858; Traité élémentaire de l'électricité et du magnétisme, 1868; Principles d'optique, 1868; Éléments de balistique, 1872; Physique moderne, 1880. Research on electricity, magnetism, optics, ballistics. Died 1911.

DE FONTANEY, Father Jean, French astronomer; b. Leon, France, Feb. 17, 1643; Jesuit missionary to China during rule of Kuang-Hsi (Manchu Dynasty); sent on sci. and religious mission to China, 1685; returned to France, 1699. Mem. French Acad. Scis., 1684. Observed total solar eclipse from Siam, 1685. Died La Flèche, France, Jan. 16, 1710.

DE FONTELLE, Jean Sebastien Eugène Julia, see Julia Fontelle, Jean Sebastien Eugène.

DEFORD, Donald Dale, Am. chemist; b. Alton, Kan., Dec. 28, 1918; s. H. Dale and Helena (Hadley) DeF.; A.B., U. Kan., 1940, M.S., 1947, Ph.D., 1948; m. Leora Miriam Adams, June 21, 1942; children—Ruth Irene, David Lynn. Asst. instr. U. Kan., 1946-48; faculty Northwestern U., 1948-—, prof. chemistry, 1960-—, chmn. dept. chemistry, 1962-—. Cons. research and devel. dept. Phillips Petroelum Co., 1956-—. Mem. Am. Chem. Soc., A.A.A.S., Instrument Soc. Am., Am. Assn. U. Profs. Research and numerous publs. on coulometric titrations, electrochem. measurement and instrumentation, gas chromatography. Home: 619 Juniper Rd., Glenview, Ill. 60025. Office: Dept. Chemistry, Northwestern U., Evanston, Ill. 60201.*

DE FOREST, Lee, Am. inventor; b. Council Bluffs, Ia., Aug. 26, 1873; s. Henry Swift and Anna Margaret (Robbins) D.; B.S., Sheffield Sci. Sch. (Yale), 1896; Ph.D., Yale, 1899, D.Sc., 1926; D.Sc., Syracuse, 1919, D.Eng., 1937; LL.D., Talladega Coll., 1951, Beloit Coll., 1951; D.Sc., College Osteopathic Surgery, 1951; m. Nora Stanton Blatch, 1908; m. 2d, Mary Mayo, 1912; m. 3d, Marie Mosquini, 1930; children—Harriet S., Eleanor Peck, Marilyn Swanke. Pioneer in devel. of wireless telegraphy in Am.; called father of radio; started radio broadcasting cos., 1902; officer numerous pioneering broadcasting orgns.; inventor triode, audion amplifier, 4 electrode valve; introduced grid into wireless valve; 1st to use alternating current generator and transmitter; patentee 300 inventions in wireless telegraphy, radio, wire telephone, sound-on-film, high speed facsimile and picture transmission and TV, numerous smaller items in field; broadcast voice of Caruso, by radio, 1910, 1st radio news broadcast, 1916; established broadcast sta., 1916; showed sound-on-film program Rivoli Theatre, N.Y.C., Apr. 1923. Recipient Gold medal World's Fair, St. Louis, 1904, Panama Pacific Expn., San Francisco, 1915; medal of honor Inst. Radio Engrs.; Elliot Cresson medal Franklin Inst.; John Scott medal City of Phila.; Prix La Tour. Inst. of France; Edison medal, 1946. Asso. several profl. socs. Author: Television Today and Tomorrow; (autobiography) Father of Radio; Conqueror of Space; also various sci. papers. Died Hollywood, Cal., June 30, 1961.

DE FOUCHY, Jean Paul Grandjean, French astronomer; b. Paris, France, Mar. 10, 1707; s. Philippe Grandjean de Fouchy; studied with Clairaut; mem. French Acad. Scis. (became perpetual sec. 1743). Originated an easily readable form of astron. tables, 1731; developed new method for observing the transit of Mercury, 1737; invented various astron. instruments. Died Paris, Apr. 16, 1788.

DE FOURCROY, Antoine François, see Fourcroy, Antoine François, Comte de.

DE FOURCROY, Charles Renè de Ramecourt, French physicist, engr.; b. Paris, France, Jan. 19, 1715; dir. mil. fortifications; mem. French Acad. Scis. Author: Mémoires sur la fortification perpendiculaire, 1786; l'Art du tuilier-briquetier. Publs. on mil. fortifications especially perpendicular fortifications; brick and tile manufacture; studied improvement of the limeburner. Died Paris, Jan. 12, 1791.

DE FREYCINET, Charles-Louis de Saulses, French engr.; b. Foix, France, Nov. 14, 1828; ed. Écoel Polytechnique, 1846; became engr. Dept. Mines, 1852; in charge civil constrn. Midi region, 1856-62; named chief engr., 1892; minister pub. works, 1877. Mem. French Acad. Scis., French Acad. Author: Traité de mécanique rationelle, 2 vols., 1858; Essai sur l'analyse infinitésimale, 1858; Sur un choix de nouvelles unités mètriques, 1887. Research and publs. on mechanics on metaphys. aspects of infintesmal analysis, san. conditions in factories; proposed new metric system; devised program for devel. communications (Freycinet plan). Died May 14, 1923.

DE FREYCINET, Louis de Saulses, French earth scientist; b. Montélimar, France, Aug. 8, 1779; mem. (with Duperry) expdn. to study terrestrial magnetism Rio de Janeiro, Cape of Good Hope, 1817. Decorated Chevalier de Saint-Louis, Comdr. of Legion of Honor; portion of So. Australian coast named in his honor. Mem. French Acad. Scis., Bur. Longitudes, Soc. Geography (a founder). Author: Voyage autour du monde, 13 vols, 1820; Voyage de découvertes aux terres australes, 1807-1816. Drew maps of Flinders Land, the Champagny, New Holland; discovered Rosa Island (named after his wife); research on geography, ethnography, history of Papua; observations on gravity and magnetism in Sidney, Australia. Died Freycinet, France, Aug. 18, 1842.

DE GARENGEOT, René-Jacques Croissant, French surgeon; b. Vitré, France, July 30, 1688; ed. Angers, France, also École de Médecine, Paris. Worked in navy hosps., then with barber-surgeon, Paris, 1711, under masters Winslow Méry and J. L. Petit; mem. Community Surgeons Paris, 1725; Royal demonstrator in med. matters, 1728, then in operations of sch. surgery; apptd. surgeon maj. King's Inf. Regt., 1742. Mem. Acad. Surgeons. Author: Traité des opérations de chirurgie, 1720; Nouveau traité des instruments de chirurgie les plus utiles et de plasieurs nouvelles machines propres pur les maladies des os, 1723; Myotomie humaine et canine, 1724; Splanchnologie, 1728; l'Operation de la taille par l'appareil latéral, 1730. Made important modifications in surg. techniques and instrumentation; perfected dental key (Garengeot's Key), rotary extracting instrument; described lumbar hernia (Petit's Hernia), 1731. Died Cologne, Dec. 10, 1759.

DE GASPARIN, Adrien-Étienne-Pierre, French agronomist, veterinarian; b. Orange, France, June 29, 1783; s. Thomas de Gasparin; children—Agenor, Paul-Joseph. Minister of Interior, 1836-37, 39; dir. Versailles Agronomic Inst., 1848-52; prefect of La Loire, then Isere, later Rhone, after 1830. Mem. French Acad. Scis., 1829, Central Agrl. Soc. Author: Manuel d'Art vétérinaire, 1817; Memoires d'agriculture et d'économie rurale sur la gourme des chevaux; Course of Agriculture, 5 vols., 1843-49. Contbd. to application of phys. and chem. scis. to agronomy; revived vet. medicine in France and assisted in its application to agr.; contbns. in agrology, treatment of soil through fertilizers, classification and composition of soil; also active in prison reform. Died Orange, Sept. 7, 1862.

DEGAULLE, Jean Baptiste, French engr., cartographer; b. Attigny, France, July 6, 1732; s. Leonard and Marie (Lentremauze) D.; cabin boy on voyage to Antilles Islands, 1751-55; in Am. during French and Indian War; later went to Que., Can.; became cartographer, Le Havre, France, 1765; prof. hydrography, Le Havre, 1771-91; became hydrographic engr., 1777. Mem. French Acad. Scis., Marine Acad. Author: Construction et usage d'un nouveau compas azimutal à réflection, 1779. Designed and installed guide lights in Le Havre and Honfleur ports; made precise map of English Channel and the Seine estuary; invented nautical speed indicator; designed reflex compass. Died Honfleur, Apr. 18, 1810.

DE GEER, Carl, Swedish entomologist, biologist; b. Finspang, Sweden, 1720; ed. classical learning Utrecht, Netherlands; studied under Linnaeus, Upsala, Sweden; mem. Acad. Stockholm, French Acad. Scis., 1748. Author: Mémoires pour servir à l'histoire des Insectes, 1752-78. Died Stockholm, Sweden, Mar. 8, 1778.

DE GEER, Gerard Jakob, geologist; b. Oct. 2, 1858; prof. geology U. Stockholm (Sweden), also founder geochronological inst.; authority on glacial age. Died Aug. 1943.

DE GENNES, Pierre Gilles, French physicist; b. Paris, 1932; s. Robert and Yvonne (Morin-Pons) de G.; student Ecole Normale Supérieure, 1951-55; m. Anne Marie Rouet, June 3, 1954; children—Christian, Dominique, Christine. Research engr. CEN, Saclay, France, 1955-59; postdoctoral fellow U. Cal. at Berkeley, 1959; maitre de conférences Faculté des Sciences, Orsay, France, 1961-65; prof. titulaire a titre personnel, 1965-—. Cons. Energie Atomique, 1961-—. Recipient Prix Ancel, Société F. de Physique, 1959; Médaille D'Argent, CNRS, 1965. Author: Superconductivity of Metals and Alloys, 1966; also articles. Research on magnetic materials (neutron scattering, nuclear resonance, nature of interactions), superconductors (surface superconductivity, neutron scattering and tunneling in presence of vortices, proximity effects).*

DE GENSSANE, Antoine, mineralogist; b. France; mining engr.; dir. Languedoc Mining Operations. Mem. French Acad. Scis., 1757. Author: Traité de la fonte des mines, 1770; Géométrie souterraine, 1776. Studied geol. history of Languedoc region. Died 1780.

DEGERING, Edward Franklind, Am. radiation chemist; b. Dodge City, Kan., May 17, 1898; s. Irving Harrison and Talitha (Cowgill) D.; student Walla Walla Coll., 1917-19; Canadian Junior Coll., Alta, Can., 1919-20; B.A., Union Coll., Lincoln, Neb., 1924; M.S., U. Neb., 1929, Ph.D., 1930; postgrad. Cornell U., 1931; m. Clara Mae Ogden, Aug. 12, 1921; 1 son, John Edward (dec.). Prof. math. Canadian Jr. Coll., 1919-20; prin. Oriens High Sch., Seattle, 1921-23; grad. asst. and fellow, U. Neb., 1924-30; faculty Purdue U., 1930-49, prof., 1942-49; asst. chmn. chemistry and chem. engring. Armour Research Found., 1949-50; research cons. Miner Labs., 1950-51; research mgr. Buckman Labs., Inc., 1951-53; chief chems. and plastics OQMG, Washington, 1953-54; chief phys. chemistry sect. pioneering research div. Q.M. Research and Devel. Command, U. S. Army, 1954-59; head Radiation Chem. Lab., 1959-—. Mem. Ind. Acad. Sci. (pres. 1945-46), Ind. Chem. Soc. (pres. 1940), Am. Chem. Soc. (chmn. med. div., 1945-46) A.A.A.S. (sec. div. chemistry 1946-—), Am. Inst. Chemists, Am. Science Teachers Assn., Alpha Chi Sigma, Phi Lambda Upsilon, (editor The Register, 1936-38), Sigma Xi. Holder of numerous patents on organic syntheses. Author: An Outline of Organic Chemistry, 1937; The Quadri-Service Manual of Organic Chemistry, 1938; The Workbook of Fundamental Organic Chemistry, 1941; Fundamental Organic Chemistry,

1942; An Outline of the Chemistry of the Carbohydrates, 1943; An Outline of Organic Nitrogen Compounds, 1945. Research and publs. on synthesis of isopropenyl acetate; 1st statis. evaluation of effects of exptl. variables; polymerization of vinyl monomers. Home: Lakeview Garden Apts., 15 Kansas St. Office: Radiation Chemistry Lab., Pioneering Research Div., U. S. Army Natick Labs., Kansas St., Natick, Mass. 01762.

DEGERLACHE DE GOMERY, explorer; b. Hasselt, Belgium, Aug. 2, 1866; ed. Ecole Polytechnique, Brussels: D. honoris causa, U. Louvain. Pres., Nat. Geog. Com., Brussels; leader 1st modern Antarctic sci. exppdn.; 1st to winter with ship Belgica in Antarctic pack-ice; discovered Gerlache's Strait (No. Graham Land), 1897-99, farthest north with ship in Greenland Sea; leader sci. expdn. to Persian Gulf, 1901; went on 3 expdns. with Duc d'Orleans on board Belgica; hon. dir. gen. Belgian Marine Adminstrn. Mem. Inst. France (corr.), Royal Geog. Soc. (London) hon. corr.), Geog. Soc. France, Gesellschaft für Erdkunde (Berlin); hon. mem. geog. soc. Antwerp, Brussels, Copenhagen, Edinburgh, Geneva, Neufchatel, Phila., Rio de Janeiro, Rome, Rouen. Author: Quinze mois dans l'Antarctique; Relation succincte du Voyage de la Belgica dans la mer du Gröntland en 1905; Journal de bord, 1907; Carte bathymétrique de la mer du grönland. Died Dec. 4, 1934.

DE GIMBERNAT, Antonio, Spanish surgeon, anatomist; b. Spain, 1734; originated spinal method for operation of crural hernia; demonstrated lacunar ligament, known as Gimbernat's ligament, 1768; published report on findings, 1779. Died 1816.

DE GIRARD, Philippe, French inventor; b. Lourmarin, France, 1775; at least 1 son, Joseph. Employed as painter, Mahon, Minorca; mfr., Leghorn, Italy; founder factory chem. products, Marseille, France, after Thermidor 9 (July 27, 1794); tchr. natural history, Nice, France; tchr. chemistry Coll. Marseille; established in Paris, France, 1795-1815; went to Hirtenberg; then opened spinning mill and factory, Vienna, Austria; went to Poland, 1825, opened spinning mill and factory, Warsaw, became engr.-in-chief of Poland; returned to France, 1844. Invented an achromatic telescope, 1806; improved methods of preparing linen yarn, 1810; invented flax spinning machine, 1812, rotating steam engine and steam canon, 1813, also beet press, water wheel, cylindrical boiler, hackle and wire drawing machine, machine for mech. machining of rifle butts, brick press, warming device for blast furnace gases, chrono-thermometer, tremolophon (mus. instrument). Died 1845.

DEGLI ARMATI, Salvino, Florentine inventor; b. Florence, Italy, flourished circa 1300; considered inventor of spectacles, 1299. Died 1317.

DEGLI STABILI, Francesco, see Cecco d'Ascoli.

DE GOBINEAU, Count Joseph Arthur, French social philosopher; b. Ville d'Avray, France, July 14, 1816. Served in French Diplomatic Corps, Europe and Persia, 1855-58, Athens, Greece, 1864-68, Rio de Janeiro, Brazil, also Stockholm, Sweden. Author: Essai sur l'inégalité des humaines, 1854; Trois ans en Asie, 1855-58; Les religions et les philosophies dans l'Asie centrale, 1865; Histoire des Perses d'après les auteurs orientaux grecs et latins, 1869. Advanced theory (known as Gobinism) which held that blond Aryan or Teuton (in this case English, Belgian, and French from northern France) constituted superior race of humans, and maintained inferiority of Negro and Semitic races; considered problem of develop. and decay of socs., with racial interpretation; 1st of selectionists. Died Turin, Italy, Oct. 13, 1882.

DE GOLYER, E(verette) L(ee), Am. geologist; b. Greensburg, Kan., Oct. 9, 1886; s. John William and N. Kagy (Huddle) De G.; A.B., U. Okla., 1911; D.Sc., Colo. Sch. Mines, 1925, So. Meth. U., 1945, Tulane U., 1954; LL.D., Trinity Coll., 1947, Princeton, 1949, U. Mexico, 1951, Washington U., 1952; m. Nell Virginia Goodrich, June 10, 1910; children—Nell Virginia, Dorothy Margaret, Cecelia Jeanne, Everette Lee. With U. S. Geol. Survey, 1906-09; geologist and chief geologist Mexican Eagle Oil Co., 1909-14; cons. practice, 1914-19; v.p. and gen. mgr. Amerada Corp., also Amerada Petroleum Corp., Amerada Refining Corp., 1919-26, pres. and gen. mgr. 1926-29, chmn. bd., 1929-32; v.p. and gen. mgr. Geophys. Research Corp.; pres. Atlatl Royalty Corp., from 1932, Felmont Corp., 1934-39; sr. mem. DeGolyer & McNaughton, from 1936. Asst. dep. petroleum adminstr. for war, Washington, 1941-43; head Petroleum Adminstrn. for War mission to Mexico, 1942; head Dept. Interior Petroleum Reserves Corp. mission to Middle East, 1943-44; tech. adviser N.R.A. oil code, 1933; mem. Nat. Petroleum Council, from 1946, U. S. Mil. Petroleum Adv. Bd., from 1947; mem. adv. com. on raw materials U. S. Atomic Energy Commn., from 1947; Aldred lectr. Mass. Inst. Tech., 1929; Cyrus Fogg Brackett lectr. Princeton, 1929, Lewis Clark Vanuxem lectr. 1941. Anthony F. Lucas medalist, 1941, John Fritz medalist, 1942; Sidney Powers gold medal Am. Assn. Petroleum Geologists, 1950. Fellow Geol. Soc. Am., A.A.A.S., N.Y. Acad. Scis., Brit. Inst. of Petroleum, Am. Geog. Soc.; mem. Am. Assn. Petroleum Geologists (pres. 1925; hon. mem. 1945), Am. Inst. Mining and Metall. Engrs. (pres. 1927), Nat. Acad. Scis., Am. Petroleum Inst. (dir.), Am. Geophys. Union, Instituto Sudamericano del Petroleo, Soc. Econ. Geologists, Soc. Exploration Geologists, Soc. Exploration Geophysicists, Pan Am. Inst. Mining Engring. and Geology, Phi Beta Kappa, Sigma Xi, Tau Beta Pi, Pi Epsilon, Sigma Gamma Epsilon, Pi Gamma Mu. Chmn. editorial bd. Sat. Rev. Lit., from 1948. Asso. editor New Colophon, Southwest Rev. Often regarded as father of Am. geophysics; developed geophys. techniques which enable geologists to determine position, form and structure of concealed rock units; assembled outstanding library of sci. and hist. books, gave 12,000 vols. to U. Okla. Died Dec. 14, 1956.

DE GORDON, Bernard, French physician; b. Gordon, Rouergue, France, flourished circa 1283-1308; studied at Salerno, Italy; taught at U. Montpellier (France); practiced medicine in Montpellier. Author: Lilium medicinae, 1303. His book is 1st to mention spectacles; 1st description of a truss, also petit mal or short epileptic seizures.

DEGOS, Robert, French physician; b. Mugron (Landes), France, Nov. 8, 1904; s. Louis and Marguerite (Tisset) D.; ed. Lycee Buffon; U. Paris (France); M.D., 1927; m. Monique Lortat-Jacob, Aug. 8, 1936; children—Jean-Denis, Claude, Laurent, Bernadette. Internship, Hosp. Paris, from 1936, then physician, 1946; apptd. chief physician St. Louis Hosp., 1946; aggregate prof. Paris Faculty Medicine, prof. clinical medicine, specialist in cutaneous and syphlitic diseases, from 1951. Mem. com. Pub. Hygiene France; cons. on treponematosis WHO, 1948; pres. Internat. Com. Dermatology, 1962, Internat. League Nat. Socs. Dermatlogy, 1962. Mem. principle dematol. socs. of world. Author: Dermatology (with ann. supplements), 2 vols., 3rd. edit.; also numerous articles. Research on eyrthema and dermo-hypodermic streptococci, malignant atrophied papules (Degos' disease), malignant cutaneous reticulosis, mastyocytosis, treatment myocardial weakness and syphilis, pseudo-peladic states, dermatomyositis. Home: 20 rue de Penthièvre, Paris 8. Office: 2 Place du Dr Alfred Fournier, Paris 10, France.*

DEGOUSÉE, Joseph Marie Anne, engr.; b. Rennes, France, July 8, 1795; studied at Brussels (Belgium) Lycée; active in politics including July Revolution. Decorated Cross of July. Author: Guide du sondeur, 1847. Developed method of prospecting for natural resources by sounding the earth; discovered coal and oil deposits in France; invented steam drills; created several artesian wells; dug well in Sahar, 1856. Died 1862.

DE GRAAF, Regnier, Dutch anatomist, biologist; b. Schoonhoven, Holland, July 30, 1641; ed. univs. Louvain, Utrecht, Leyden; M.D., U. Angers, 1665; practiced medicine, Delft, from 1667. Author: De natura et usu succi pancreatici, 1663; De mulierum organis generationi inservientibus, 1672. Conducted studies of generative organs and reproductive system; described fine structure of testicles, 1668, of ovary, 1673; discoverer ovarian follicles (Graffian follicles). Died Delft, Aug. 17, 1673.

DE GRAMONT, (Antoine-Alfred) Arnaud-Xavier-Louis, French physicist; b. Paris, Apr. 21, 1861; mem. French Acad. Scis., 1913. Author: Analyse spectrale, 1923. Research on spectral series, relation of mineral chemistry to spectral analysis. Died Oct. 31, 1923.

DEGRÉ, Gerard, cultural sociologist; b. Havana, Cuba, Jan. 21, 1915; s. Lorand and Emelina (Andreu) DeG.; came to U. S., 1916, naturalized, 1933; B. Social Scis. Coll. City N.Y., 1937; M.A., Columbia, 1939, Ph.D., 1943; m. Muriel Evelyn Harris, Dec. 13, 1933; 1 dau., Erica (Mrs. Guy Ducornet). Instr., Butler U., 1941-42, St. Lawrence U., 1942-43; faculty Bard Coll., Annandale-on-Hudson, N.Y., 1946—, chmn. dept. sociology, anthropology and social philosophy, 1952—, prof. sociology and social philosophy, 1956—; Fulbright prof. sociology Am. U. at Cairo, Egypt, 1954-55, U. Chile, Santiago, 1961, U. San Marcos and Lima, Peru, 1966. Sr. research cons. fgn. area studies div. Am. U., Washington, 1961. Fellow Am. Sociol. Assn. Author: Science as a Social Institution, 1957; Society and Ideology, 1943; also articles. Research in comparative cultures, sociology of knowledge and culture, phenomenological social theory, European social philosophy. Home: Annandale-on-Hudson, N.Y. 12504.*

DE GRIBEAUVAL, Jean-Baptiste Vaquette, French inventor; b. Amiens, France, Sept. 15, 1715; in service of Maria Theresa during Seven Years' War; became lt. gen., 1765. Author: Tables des constructions des principaux attirails de l'artillerie, 1764; also articles. Responsible for best movable arty. in Europe at time of French Revolution; introduced open sights; standardized basic implements of arty. and vehicle trains, made them interchangeable. Died Paris, May 9, 1789.

DE GROOT, Jeanne Lampl, Dutch psychiatrist; b. Schiedam, Netherlands, Oct. 16, 1895; d. M. C. M. and Henriette (Dupont) De Groot; ed. U. Leiden; M.D., U. Amsterdam; Dr., Psychiatrist, U. Vienna; m. Hans Lampl, Apr. 7, 1925 (dec.); children—Henriette, Edith (Mrs. R. N. P. Berkovits). Practice as psychoanalyst, Berlin, 1925-33, Vienna, 1933-38, Amsterdam, 1938—. Tchr. psychoanalytic insts. Berlin, Vienna, Amsterdam, Frankfurt. Mem. Internat. Psychoanalytical Assn. (hon. v.p. 1949), Netherlands Soc. Psychiatry and Neurology. Author: The Development of the Mind, 1965. also articles. Research and publs. on theoretical and practical problems of psychoanalysis, and of the interaction with other brs. of sci. Home: 39 Haringvlietstraat, Amsterdam, Netherlands.*

DE GROOT, Sybren Ruurds, Dutch physicist; b. Amsterdam, Netherlands, Apr. 18, 1916; s. Ruurd and B. (Ketelaar) de G.; Ph.D., U. Amsterdam, 1945; Ph.D. honoris causa, U. Strasbourg (France), 1956; m. Silvia W. Rosbergen, Oct. 29, 1942. Chef de service Commissariat à l'Energie Atomique, Paris, France, 1947-48; prof. theoretical physics U. Utrecht (Netherlands), 1948-53, U. Leyden (Netherlands), 1953-64, U. Amsterdam, 1964—; vis. prof. U. Md., 1951, Varenna Sch. (Italy), 1955, U. Montreal (Que., Can.), U. Strasbourg, 1963, Rockefeller U., N.Y.C., 1965. Mem. Royal Netherlands Acad. Scis., Netherlands Phys. Soc., Società Italiana di Fisica. Author: L'effet Soret, 1945; Thermodynamics of Irreversible Processes, 1951; (with P. Mazur) Non-Equilibrium Thermodynamics, 1962; also numerous articles. Research on statis. mechanics and thermodynamics of irreversible phenomena; beta- and gamma-radioactivity especially of oriented nuclei; relativistic theory of statis. electromagnetic phenomena. Home: 65 Valckenierstraat, Amsterdam, Netherlands.*

DE GROUCHY, Jean William, geneticist; b. Deventer, Holland, Aug. 10, 1926; s. Armand and Meta (van den Broek d'OBrenan) de G.; M.D., Faculty of Paris, 1952; grad. Faculty Scis., Paris, 1955. Rockefeller Found. fellow, U. S., 1956-57; biol. asst. in hosps., 1960—; dir. research CNRS, 1963—; head cytogenetics lab. Hôpital des Enfants Malades, Paris, 1960—. Recipient Prix du Cancer, 1964; named Laureat de L'Académie des Sciences, 1966. Mem. Societ de Génétique, Paris. Author: Hérédité Moléculaire Conditions Normales et Pathologique, 1958; Le Message Héréditaire, 1965; L'Homme et l'Hérédité, 1967; also numerous articles. Discovered new chromosome anomalies and clin. syndromes; studied chromosomal theory of cancer. Home: 31 R. de Tournon, Paris. Office: Hopital des Enfants Malades, Paris, France.*

DE GUA, Jean Paul, French mathematician; b. Malves, France, c. 1713; prof. philosophy Collége de France from 1743; Mem. French Acad. Scis., 1741, Royal Soc. London. Author: Usage de l'analyse de Descartes, 1740. Demonstrated Descartes' rule of signs; investigated powers of analytic geometry; discussed the idea of an Encyclopedie before Diderot. Died Paris, June 2, 1786.

DEHAAN, Robert Lawrence, Am. biologist; b. Chgo., Nov. 18, 1930; s. Oliver and Annette (Dubkin) DeH.; A.A., Pasadena City Coll., 1950; B.A., U. Cal. at Los Angeles, 1952, M.A., 1954, Ph.d., 1956; postgrad. U. Amsterdam (Holland), 1952-53; m. Virginia Selmanoff, Feb. 21, 1957; children—Tracy Ann, Benjamin Jeremy. Jr. research physiologist dept. physiology U. Cal. at Los Angeles, 1954-56; research embryologist Carnegie Instn. Washington, Balt., 1956—; asso. prof. embryology Johns Hopkins, 1964—; vis. asst. prof. Haverford Coll., 1961; vis. investigator Inst. Zoology, U. Zurich (Switzerland), 1962-63. Mem. A.A.A.S., Am. Inst. Biol. Sci., Am. Soc. Zoologists, Soc. Developmental Biologists, Md. Acad. Scis. (v.p. 1965—). Co-editor: Organogenesis, 1965. Research and publs. on embryonic devel. vertebrate heart, spontaneous activity beating heart cells isolated in tissue culture. Address: 115 W. University Pkwy., Balt. 21210.*

DE HAEN, Anton, physician; b. 1704; student Boerhaave at Leyden (Netherlands); taught at Vienna, Austria. Author: Ratio medendi in nosocomio practico, 18 vols., 1758 (considered most important work on therapeutics of its time). Developed clin. methods; one of 1st to use thermometer. Died 1776.

DE HALDAT DU LYS, Charles, French physicist; b. Bourmont, France, Dec. 24, 1770; mil. physician during Revolution; prof. physics, Nancy, France; univ. insp., 1824-31; became dir. Nancy Secondary Sch. Medicine, 1843; mem. French, Nacy (sec.) acads. scis. Author: Recherches chimiques sur l'encre, 1803; Recherches sur la cause du magnétism par rotation, 1841; Histoire du magnétisme. Inventor hydrostatic device named after him; research on magnetism, optics, propagation of sound; made chem. studies of ink. Died Nancy, Nov. 26, 1852.

DE HARO, Andrés, Spanish biologist; b. Mazarrón, Spain, Dec. 27, 1925; s. José and Josefa (Vera) de H.; D.Biology, U. Barcelona (Spain), 1952; m. Josefa Ollé, July 2, 1960; children—Josefina, Juan Jose. Asst. prof. zoology U. Barcelona, 1958-66; prof. zoology U. Salamanca, Spain, 1966—; sec. Instituto de Biologia Aplicada, Barcelona, 1963-66. Mem. Real Sociedad Espanola de Historia Natural, A.A.A.S. Author: El Mundo Organico, 1963; A Textbook of Invertebrate Zoology, 1965; also articles. Research on anatomy, phylogeny and ethology of invertebrates especially pycnogonida, brachiopoda and hexapoda. Home: 35, Paseo de Canalejas, Salamanca, Spain.*

DE HAUTEFEUILLE, Jean, French physicist; b. Orléans, France, Mar. 20, 1647; patronized by duchess of Bouillon; ordained monk. Research and publs. on watch mechanisms, acoustics, optics, tidal phenomena; invented spiral spring moderating movements of a watch, micrometermicroscope, measuring instrument. Died Orléans, Oct. 18, 1724.

DEHOFFMANN, Frederic, physicist; b. Vienna, Austria, July 8, 1924; s. Otto and Marianne (Halphen) deH.; student U. London; B.S., Harvard U., 1945, M.S., 1947, Ph.D., 1948; m. Patricia Lynn Stewart, June 10, 1953. Staff mem. Los Alamos Sci. Lab., 1944-46, 1948-55, alternate asst. dir., 1950-51; cons. AEC, 1947-48, com. sr. responsible reviewers, 1947-51, cons. joint congl. com. on atomic energy, 1954; v.p. Gen. Dynamics Corp., San Diego, Cal., 1955-67, also gen. mgr. Gen. Atomic div., 1955-59, pres., 1959-67; pres. Gen. Atomic Europe, Zurich, Switzerland, 1960-—; v.p. Gulf Oil Corp., 1967-—, pres. Gulf Gen. Atomic div., 1967-—, pres. Gulf Gen. Atomic Europe, 1967-—; sci. sec. UN Internat. Conf. Peaceful Uses Atomic Energy, 1955; mem. U. S. tech. del. 2d, 3d UN Internat. Conf. Peaceful Uses Atomic Energy, 1958, 1964; dir. Atomic Indsl. Forum, 1961-62, 64-—. Hon. prof. theoretical physics University of Vienna (Austria), 1968. Fellow American Physical Society, American Nuclear Society (dir.), Atomic Indsl. Forum (dir.), Sigma Xi. Author: The Science and Engineering of Nuclear Power, Vol. II, 1949, (with K. M. Case, George Placzek) Introduction to Neutron Diffusion Theory, Vol. I, 1953, (with H. A. Bethe, S. S. Schweber) Mesons and Fields, Vol. I, 1955, Vol. II, 1955. Research and publs. field exptl. reactor physics, termination of delayed neutron constants; methodology of measuring absolute reactivity; theoret. reactor physics; fluxuations of neutron chance in nuclear reactors; plasma physics theory of thermonuclear reactions; treatment of collison of interstellar gas masses; high energy physics; theory of scattering reactions; nuclear power plants: design and economics of high temperature gas cooled reactors. Home: 9736 La Jolla Farms Rd., La Jolla, Cal. 92137. Office: P.O. Box 608, San Diego, Cal. 92112.*

DE HONESTIS, Christophorus, Italian physician; b. Italy, early 14th century; M.D., Bologna, Italy, 1367; tchr. logic, philosophy, medicine Bologna, possibly Perugia, Italy, about 1380, Florence, Italy, 1385-89. Author: Commentary on Antidotarium of Mesuë the Younger, circa 1386. Died Bologna, 1392.

DEIBNER, Leonce (Leonty), chemist; b. St. Petersburg, Russia, Aug. 21, 1900; s. Nicolas and Tatiana (Sokolov) D.; B. degree, U. Paris, 1929, M.Sc., 1929; postgrad. High Nat. Sup. Sch. Petroleum, Strasbourg, High Nat. Sup. Sch. Chem., Montpellier, Doct.-Eng., 1949; m. Louise Squarcioni, Aug. 21, 1941; 1 son, Georges. Engr.-chemist petroleum industry, 1931-45; master research central sta. tech. vegetable products Nat. Inst. Agronomy Research, Narbonne (Aude), France, 1945-65; expert mem. subcommn. methods of analysis and of appreciation of wines Office Internat. de la Vigne et du Vin, Paris. Research, publs. on analytical and phys. oenochemistry, plant physiology (vine), oenometry; contbrn. to sci. devel. of vine and wine. Address: Narbonne (Aude), France.*

DEICHER, Helmuth Richard Georg, German physician; b. Berlin, July 31, 1929; s. Heinrich and Martha (Doden) D.; M.D., U. Göttingen, 1955; m. Gertrud Heisler, June 1, 1957; children—Susanne, Christiane, Robert. Research asst. Rockefeller Inst., N.Y.C., 1957-59, U. Marburg (Germany), 1959-65; dozent internal medicine U. Marburg, 1964-65, Hanover Med. Sch., 1965-—. Mem. German Soc. Arts and Sci., German Soc. Internal Medicine. Co-author: Klinik der Gegenwart, 1962; also articles. Research on clin. immunology; collagen diseases, especially systemic lupus erythematosus and rheumatoid arthritis; serum groups. Home: 7 Haeckelstrasse, 3000 Hanover, Germany.*

DEICHMANN, William Bernhard, pharmacologist; b. Kiel, Germany, Sept. 2, 1902; s. J. F. Wilhelm and Mathilde (Bollenhagen) D.; brought to U. S., 1924, naturalized, 1930; A.B., Western Res. U., 1932, M.S., 1934; Ph.D., U. Cin., 1939; m. Hedy Gruebler, Aug. 4, 1928; children—Herbert William, Herta (Mrs. John Holly). Asst. biochemistry Western Res. U., 1927-34; pharmacologist Dupont Haskell Lab. Indsl. Toxicology, Wilmington, Del., 1934-37; with Kettering Lab. Applied Physiology and Pharmacology U. Cin., 1937-47; asso. prof., head pharmacology Union U., 1947-50; prof. pharmacology Albany Med. Coll., 1950-53; prof., chmn. dept. pharmacology U. Miami (Fla.) Sch. Medicine, 1953-—; cons. NASA, 1962-—; mem. Endrin com. Food and Drug Adminstrn. Nat. Acad. Scis., 1966. Fellow Ohio, N.Y. acad. sci., A.A.A.S., Royal Soc. Medicine (Eng.), Am. Coll. Clin. Pharmacology and Chemotherapy; mem. Soc. Toxicology (founder); Am. Soc. Pharmacology and Exptl. Therapeutics, Deutsche Pharmakologische Gesellschaft, Am. Therapeutics Soc., Am. Indsl. Hygiene Assn., N.Y. State Soc. Med. Research. Author: (with others) The Toxicity of DDT, 1950; (with Frank T. Kurzweg) First Aid Manual, 1960; Signs, Symptoms and Treatment of Certain Acute Intoxications, 1955; Symtomethology and Therapy of Toxicological Emergencies, 1964. Research field toxicology. Home: 1931 S. Bayshore Dr., Miami, Fla. 33133.*

DEIDIER, Antoine, French physician; b. 1746; docteur en médecin, 1691; became prof. chemistry U. Montepellier (France), 1697; mem. Royal Soc. London. Author: Lettre sur la maladie de Marseille, 1720; Dissertatio de tumoribus, 1714. Worked in Marseilles, France, during plague of 1720; advocated use of arsenic for treatment of cancer; proposed that blood and lymph were different only in densities. Died Marseilles, Apr. 30, 1746.

DEIMAN, Johann Rudolf, chemist, physician; b. Hagan, Ostfriesland, 1743; grad. U. Halle (Germany), 1776; physician to King Louis Bonaparte, circa 1806. Author: Dissertatio de indicatione vitali generatin, 1770; Observations sur l'electricité, 1779; Investigations physiques et chimiques, 1793; Pharmacopea Batava, 1805; De Geesten Stekking der kritische Wysbegeerte, 1805; Treatise on Medical Electricity. Codiscoverer of olefiant gases and Holland's fluid; he was center of reunion known as the Dutch Chemists. Died 1808.

DEINOSTRATOS, See Dinostratos.

DE ISLA, Rodrigo Ruiz Díaz, Spanish physician; b. 1462; physician Hosp. of All Saints, Lisbon, Portugal; later practiced in Barcelona, Spain; treated syphilitic sailors of Columbus returning from Am. Author: Tratado llamado de totos los santos contra el mal serpentino (probably written between 1504-06), pub. 1542. Died 1542.

DEISS, Charles F(rederick) Am. geologist; b. Covington, Ky., Mar. 18, 1903; s. Charles Fred and Anna Dorothea (Reinhart) D.; A.B., Miami Univ., 1925; Ph.D., U. Mich., 1928; m. Minnette Blanche Davison, Jan. 22, 1929. Asst. prof. geology Mont. State Univ., Missoula, 1928-30, asso. prof., 1930-36, prof. 1936-42, dir. library 1937-40; cons. Mont. Power Co., 1940-41; asst. geologist U. S. Geol. Survey, hdqrs., Missoula, Mont., 1940-41, geologist since 1942, in charge exploration for dolomite in western U. S., 1942-45, and for phosphate in Ida., 1944-45; prof., chmn. dept. geology Ind. U., also state geologist, Ind., from 1945. Mem. Geol. Soc. Am., Paleontological Soc., A.A.A.S., Am. Inst. Mining, Metall. and Petroleum Engrs., Am. Assn. Petroleum Geologists. Soc. Econ. Geologists, Ind. Acad. Sci., Assn. Am. State Geologists (pres. 1954), Sigma Xi. Author sci. bulls. Contbr. articles on geologic subjects to various jours. Died June 13, 1959.

DÉJARDIN, Georges Louis, French physicist; b. Roye, France, June 30, 1893; s. Louis and Berthe (Mouton) D.; student U. Paris, 1912-14, Agrégé des Sciences Physiques, 1919, D.Sc., 1924; student U. Clermont (France), 1915-16; m. Suzanne Dubreuil, July 12, 1922. Asst. physics U. Paris, 1919-20; prof. high schs., Paris, 1920-25; asso. prof. U. Lyon, France, 1925-30, prof., 1930-35, head dept. physics, 1935-63, emeritus, 1963-—; vis. prof., lectr. various Canadian, Am., Dutch univs., U. St. Joseph, Beirut, Lebanon, since 1929-—. Mem. Académie des Sciences (corr.), Am. Phys. Soc., Phys. Soc. London, I.E.E.E. (sr.), Illuminating Engrs. Soc., Soc. France de Physique, Soc. France des Elec. Author: Les Quanta, 1929; also numerous articles. Spectroscopic studies; discovered spectra NeII and NeIII, continuum of heavy rare gases in visible region; spectrophotometric measurements; fluorescence studies, especially sodium salicylate; discovered constancy of quantum yield; photoelectric layers; thermionic emission, especially oxide cathodes and MgO cathodes; astrophysics and atmospheric physics, especially ozone altitude, light of night sky, contribution to discovery of atmospheric sodium. Home: 32 Rue du Commandant Fuzier, 69-Lyon (3), France.*

DE JAUCOURT, Chevalier Louis, French scholar; b. Paris, France, 1704; student U. Geneva (Switzerland), Cambridge (Eng.) U., Leiden (Netherlands) U.; returned to Paris, 1736; editor of articles on physiology, chemistry, botany, pathology, polit. sci. and history in Encyclopédie. Author: Histoire de la vie et des oeuvres de Leibniz, 1734; also worked on Bibliothèque raisonnée des savants de l'Europe, 1728-40; Description du musée de Séba, 1734-65. Died Compiègne, France, 1779.

DEJEAN, Pierre François Marie Auguste, French entomologist; b. Amiens, France, Aug. 10, 1780; s. Jean François Aimé Dejean; soldier serving under Napoleon; became gen. of brigade, 1810, of div., 1813; exiled 1815-19; created peer of France, 1824. Author numerous publs., including: Catalogue des Coléoptères de la collection de M. le Comte Dejean, 1802, 33, 36; Histoire générale des Coléoptères, 7 vols., 1825-39; Observations sur l'ordnance de 1829, 1838; (with Latreille) Iconographie des Coléoptères d'Europe, 1822; (with Boisduval, Aubé) Histoire naturelle et iconographie des Coléoptères, 1929. First great amateur entomol. collector; specialized in Carabidae of world (described at least 40 species from Cal.). Died Paris, France, Mar. 18, 1845.

DEJERINE, Joseph Jules, neurologist; b. Geneva, Switzerland, Aug. 3, 1849; studied medicine at Paris; studied under Vulpian; Docteur en médecine, 1879; aggregate prof., 1886; m. Augusta Dejerine-Klumpke, 1890. Named prof. clin. medicine Charité Hosp., 1896; apptd. prof. med. history Faculty Medicine, Paris, 1901; became prof. internal pathology, 1907; received chair of nervous disorders at Salpetrière Hosp., 1910. Mem. Acad. Medicine. Author: L'hérédite dans les maladies du système nerveux, 1886; De l'agraphie, 1891; Anatomie des centres nerveux, 1895; (with Thomas) Traité des maladies de la moelle épinière, 1902; Les manifestations fonctionnelles des psycholonevroses, 1911; Sémiologie des affections du système nerveux, 1914. Described several diseases including Déjerine's disease (a form of infantile neuritis), Déjerine-Klumpke paralysis, Déjerine-Roussy syndrome of the thalamus, Déjerine-Sottas disease (a form of neuropathy). Died Feb. 26, 1917.

DEJERINE-KLUMPKE, Augusta, French neurologist; b. France, 1859; grad. Paris, 1889; m. Joseph Jules Dejerine, 1890. Patroness of pupils of Joseph Jules Dejerine until her death. Author: (with husband) Anatomie des centres nerveux, 1895-1901. Collaborated with husband in research on anatomy of nervous centers; 1st description of paralysis and atrophy of hand muscles caused by lesion in brachial plexus (Klumpke's paralysis). Died 1927.

DE JOINVILLE, Jean, French historian; b. Joinville, Champagne, France, circa 1224; s. Simon and Beatrix (Etienne; de J.; m. Alais de Gran-Pré; children include Jean; m. 2d, Alix de Resnel; children include Ancel. Seneschal of Champagne; accompanied St. Louis to Egypt. Author: Credo (articles on his religious belief), rev. 1287; Histoire de Saint Louis (remains best source on period of Louis IX). First to record description of scurvy, 1250. Died circa 1317.

DE JONG, Gerben, Dutch geographer; b. Jan. 23, 1904, Scharnegoutum, Netherlands; s. Johannes and Stilma (Wybrig) de J.; Ph.D. in Geography, U. Amsterdam; m. J. G. H. Robbers. Asst., Internat. Inst. Social History, Amsterdam; sci. ofcl. U. Leyden Library; lectr. econ. geography; prof. Free U. Amsterdam. Mem. Royal Netherlands Assn. Geography, Netherlands Assn. Econ. and Social Geographers. Author: De duitse Landbeschrijving in de 18e eeuw, 1947; Het Karakter van de Geographische Totaliteit, 1955. Home: Morgenstond 14, Amstelveen. Office: Koningslaan 31-33, Amsterdam, Netherlands.

DEJONG, Russell Nelson, Am. neurologist; b. Orange City, Ia., Mar. 12, 1907; s. Conrad and Cynthia (Bursma) DeJ.; A.B., U. Mich., 1929, M.D., 1932, M.S. in neurology, 1936; m. Madge Anna Brook, Apr. 23, 1938; children—Mary Cynthia, Constance Jacqueline, Russell Nelson II. Instr. neurology U. Mich. Med. Sch., 1934-37, asst. prof., 1937-41, asso. prof., 1941-50, prof., chmn. dept., 1950-—; cons. neurology VA Hosp., Battle Creek, and Ann Arbor, Mich., Ypsilanti State Hosp., Wayne County Gen. Hosp.; cons. Nat. Inst. Neurol. Diseases and Blindness NIH, 1953-—; mem. residency review com. for psychiatry and neurology, 1961-—. Fulbright fellow Inst. Neurology U. London, 1954-55. Mem. Am. Epilepsy Soc. (pres. 1955-56), Am. Bd. Psychiatry and Neurology, Am. Neurol. Assn. (pres. 1964-65), Am. Acad. Neurology (v.p. 1961-63), Pan-Am. Congress Neurology (v.p. 1963), Nat. Multiple Sclerosis Soc. (adv. bd.), Am. Psychiatric Assn., Assn. for Research in Nervous and Mental Disease, Central Neuropsych. Assn., Am. Med. Writers Assn., Am. Assn. History of Medicine. Author: The Neurologic Examination, 1950, 2d edn. 1958. Research, publs. on clin. neurology, tchg. of neurology. Home: 1526 Harding Rd., Ann Arbor, Mich. 48104.*

DE JONGH, Samuel Elzevier, Dutch pharmacologist; b. 1898; ed. univs. Utrecht, Amsterdam; M.D.; became prof. pharmacology U. Leyden (Netherlands), 1935, dismissed by Germans during war; adviser pharm. firm Organon Ltd.; supr. med. sect. Nat. Def. Orgn.; hon. mem. Flemish Soc. for Advancement of Medicine (hon.), Royal Acad. Scis., Royal Flemish Acad. Medicine. Author: (with others) Het autonome zenuwstelsel, 1934, Aanwinsten op diagn. en therap. gebied, 1935; (with J. H. Gaarenstroom) Contribution to the Knowledge of the Influences of Gonadotropins and Sex Hormones on the Gonads of Rats, 1946; (with E. Laqueur, M. Tausk) Hormonologie, 1948; Inleiding Tot de Algemene Farmacologie, 1959. Identified (with Laqueur, Dingemanse, Hart) estrogenic activity in urine of males, 1927. Address: Boerhaavelaan 33, Leyden, The Netherlands.

DEJORGE, Francisco Bastos, Brazilian biochemist; b. Sumidouro, R.J., Brazil, Dec. 9, 1918; s. Francisco Assis and Carminda (Bastos) DeJ.; B. and licentiate in chemistry Faculdade Nacional de Filosofia, U. Brasil, Rio de Janeiro, 1948-51; M.D., Faculty Medicine, U. Minas Gerais, Belo Horizonte, Brazil, 1951-57; m. Daisy Nogueira, July 25, 1951; children—Francisco Nogueira, Angela Maria, Eliane. Head lab. biochemistry dept. medicine Sao Paulo (Brazil) U., 1958-—, asst. doctor Hosp. das Clinicas, 1964-—. Mem. Associaçao Medica Brasileira, Associaçao Paulista de Medicina. Author: Chemical Constitution of the Brazilian Foods, 1961; also articles, monograph. Research on biochemistry of eye and ear, nutritional content of various diets, analysis of trace elements in normal pregnancy, chem. states of various diseases, biochemistry tropical invertebrates. Home: 1667 Oscar Freire, Sao Paulo City, Sao Paulo, Brazil.*

DE JUSSIEU, see Jussieu.

DE KAY, James Ellsworth, naturalist; b. Lisbon, Portugal, Oct. 12, 1792; s. George and Catherine (Colman) De K.; attended Yale, 1810-12; M.D., U. Edinburgh (Scotland), 1819; m. Janet Eckford, July

31, 1821. Librarian, editor transactions Lyceum of Natural History, N.Y.C., 1819-30; asso. in literary life with Joseph Rodman Drake, James Fenimore Cooper; visited Turkey (study of Asiatic cholera familiarized him with first epidemics later appearing in U. S.), 1831-32; commissioned to prepare zool. sect. Natural History Survey of State of N.Y., 1836-44. Author: Sketches of Turkey by an American, 1833; Zoology of New York, 5 vols., 1842-44. Died Oyster Bay, L.I., N.Y., Nov. 21, 1851.

DEKEYSER, Willy Clement, Belgian physicist; b. Ostend, Belgium, Feb. 16, 1910; s. Georges Julien and Eugenie (Vanderputte) D.; Dr.Physics, U. Gent (Belgium) D.; m. Marie-Madeleine, Feb. 16, 1943; 1 dau., Nica. Asst. U. Gent, 1939-44, prof. crystallography, 1944—, dir. lab. crystallography and study solids, 1948—. Mem. bd. Institut de Recherches Scientifiques dans l'Industrie et l'Agriculture. Mem. Belgian Phys. Soc. (pres. 1965—), Société d'Etudes et de Recherches appliqués à l'Industrie (mem. bd.), Société Belge d'Optique et d'Instruments de précision (mem. bd.), Am. Phys. Soc., Phys. Soc. (U.K.), Nederlandse Natuurkundige Vereiniging, Société francaise de Cristallographie. Author: (with S. Amelinckx) The Structure and Properties of Grain Boundaries, 1959, Les dislocations et la croissance des cristaux, 1955; also numerous articles. Research on growth of crystals at low supersaturation, visualization and properties of dislocations in ionic crystals. Home: 57 Rijsenbergstraat, Gent, Belgium.*

DEKKER, Adrianus Jacobus, Dutch physicist; b. The Hague, Holland, Nov. 8, 1918; s. Johannes and Dirkje (Bongers) D.; Ph.D. in Physics, U. Amsterdam (Holland), 1945; m. Albertina van der Sluis, Aug. 8, 1947; children—Anita Joan, Beatrix Edith. Physicist, Philips Research Labs., Eindloven, Holland, 1943-47; asso. prof. U. B.C., Vancouver, Can., 1947-52; faculty U. Minn., Mpls., 1952-60, prof., 1956-60; prof. U. Groningen (Holland), 1960—. Cons. Honeywell Research Center, Mpls., 1954-59; adviser Philips Research Lab., Eindhoven, 1960—. Guggenheim fellow, 1958-59. Fellow Am. Phys. Soc.; mem. Dutch Phys. Soc. Author: Solid State Physics, 1957; Electrical Engineering Materials, 1959; also articles. Research on various aspects of solid state physics. Home: 18 Esserlaan, Groningen, Holland.*

DEKKER, Jacob Christoph Edmond, mathematician; b. Hilversum, The Netherlands, Sept. 6, 1921; s. Anton and Maria (Zeegers) D.; M.A., Syracuse U., 1949, Ph.D., 1950; m. Hedwig C. H. Schemering Reelfs, Dec. 19, 1950. Came to U. S., 1947, naturalized, 1958. Faculty U. Chgo., 1951-52, 55-56, Northwestern U., Evanston, Ill., 1952-55; mem. Inst. for Advanced Study, Princeton, N.J., 1956-58; faculty U. Kan., Lawrence, 1958-59; faculty Rutgers U., New Brunswick, N.J., 1959—, prof. math., 1961—. Cons. IBM. Mem. Am. Math. Soc., Math. Assn. Am., Assn. for Symbolic Logic. Author: (with J. Myhill) Recursive Equivalence Types, 1960. Research and publs. in theory of recursive functions and applications to set theory. Address: Dept. Math., Rutgers U., New Brunswick, N.J. 08903.*

DEKKERS, Frederick, Dutch physician; b. S'Hertogenbosch, Holland, 1648; studied medicine Leiden, Netherlands; grad., 1668; became prof. medicine, Leiden, 1694. Author: Exercitationes practicae circa methodum medendi (1st reported detection of albumin in urine using heat and acetic acid), 1673. Died 1720.

DE KLEINE, William, Am. physician; b. Jamestown, Mich., Nov. 28, 1877; s. Hilbert and Alice (Kremers) DeK.; A.B., Hope Coll., 1902; hon. D.Sc., 1937; M.D., Northwestern Univ., 1906; M.Sc., U. Mich. Sch. of Pub. Health, 1915; postgrad. Mass. Inst. Tech., 1924; m. Lottie Maria Hoyt, June 28, 1906; 1 son, Edwin Hoyt (M.D.). Practiced medicine, Grand Haven, Mich., 1906-14; dir. Mich. Tb Survey Campaign, State Bd. of Health, 1915-17; health officer, Flint, Mich., 1917-22, Saginaw, Mich., 1922-25; dir. child health demonstrations, Mansfield, O. (conducted by Am. Child Health Assn.), Fargo, N.D., and Salem, Ore. (conducted by Commonwealth Fund of New York), and organized full-time health dept. in each city, 1925-28; became asso. with A.R.C. during Miss. flood, 1927, med. dir. A.R.C., 1928-42; participated in all major disasters as organizer med. relief activities; in pvt. practice of internal medicine (spl. interest in nutritional therapy). 1942-43; state commr. health, Mich., 1944-47; engaged in private practice of medicine, 1947-56. Pres., Mich. Tb Assn., Mich. Pub. Health Assn.; dir. Nat. Tb Assn. Fellow Am. Pub. Health Assn., A.M.A.; mem. So. Med. Assn. Contbr. to med. and health jours. Pioneer in field of pub. health and traveling Tb clinics and nutrition in clinical medicine, also in combating pellegra in South. Died Sept. 20, 1957.

DE KLERK, John, physicist; b. Dordrecht, South Africa, Oct. 16, 1917; s. William Adrian and Muriel (Norval) deK.; B.S., U. Cape Town, 1942, M.S., 1946; Ph.D., U. London, 1954; m. Ann Margaret Howlett, June 19, 1954; children—Susan Margaret, Peter John. Asst. prof. Brown U., 1955-59; fellow physicist Westinghouse Research and Devel. Center, Pitts., 1960—. Mem. I.E.E.E., Am. Phys. Soc., A.A.A.S., Sigma Xi. Contbr. chpt. to Physical Acoustics, 1966. Research on mech. properties of solids, particularly dielectric materials, using microwave acoustic techniques; developed method growing thin piezoelectric film transducers on single crystals of material being studied. Office: 1310 Beulah Rd., Pitts. 15235.

DE KONINCK, Laurent-Guillaume, Belgian paleontologist; b. Louvain, Belgium, May 3, 1809; studied medicine U. Louvain, chemistry in Paris, France, Berlin, Giessen (both German); tchr. chemistry, Ghent, also Liège, Belgium; named prof. paleontology and organic chemistry U. Liège. Recipient Wollaston medal Geol. Soc. London, 1875. Author: Eléments de chimie inorganique, 1839; Description des animaux fossiles qui se trouvent dans le terrain carbonifère de la Belgique, 1842-51; Monographie des genres produtus et chonetes, 1847; (with le Hon) Recherches sur les crinoides du terrain carbonifère de la Belgique, 1854. Research on Paleozoic Era; described brachiopods, crustacea, mollusca, crinoids of Carboniferous limestone in Belgium, discovered phoridzin; Died Liège, July 15, 1887.

DE KRUIF, Paul, Am. bacteriologist, sci. writer; b. Zeeland, Mich., Mar. 2, 1890; s. Hendrik and Hendrika J. (Kremer) de K.; B.S., U. Mich., 1912, Ph.D., 1916; m. Rhea Barbarin, Dec. 11, 1922 (dec. July 1957); m. 2d, Eleanor Lappage, Sept. 1, 1959. Bacteriologist, U. Mich., 1912-17, Rockefeller Inst., 1920-22; reporter for Curtis Pub. Co., 1925—; cons. Chgo. Bd. Health, Mich. State Health Dept. Labs. Mem. Mich. Med. Soc. (hon.). Collaborator with Sinclair Lewis, on Arrowsmith, 1925. Author: Our Medicine Men, 1922; Microbe Hunters, 1926; Hunger Fighters, 1928; Seven Iron Men, 1929; Men Against Death, 1932; Why Keep Them Alive?, 1936; The Fight for Life, 1938; Health Is Wealth, 1940; Kaiser Wakes the Doctors, 1943; The Male Hormone, 1945; Life Among the Doctors, 1949; A Man Against Insanity, 1957; The Sweeping Wind, 1962. Contbr. to Readers Digest. Wartime research on poison and antitoxin of bacillus gangrene; made 1st prophylactic injections of gas gangrene serum. Home: Holland, Mich. Address: care Harcourt Brace & Co., 750 3d Av., N.Y.C.

DE KUPFFER, Adolph Theodore, Russian scholar; b. St. Petersburg, Russia, Jan. 6, 1799; studied under Haüy, Paris, France. Prof. chemistry, physics, and mineralogy U. Kasan (Russia), 1824-28; employed by govt. to explore Urals, Russia, 1828; prof. Sch. Bridges and Hwys, also Pedagogical Sch.; later dir. Central Obs. Physics. Founder, Bur. Meteorology, 1843, also dir. Recipient Berlin Acad. prize, 1826. Mem. Acad. St. Petersburg. Author: Handbuch der rech. Krystallonomie, 1831; Sur l'élasticité des métaux, 1859. Research on crystallography and terrestrial magnetism, influence of heat on elasticity of metals. Died St. Petersburg, May 23, 1865.

DELABARRE, Christophe-François, French dentist; b. Lisieux, France, 1784; studied medicine, Paris, France; docteur en médecine, 1806; a son, Antoine François Adolphe. Became prof. dental hygiene Central Hosp. Adminstrn., Paris, 1817. Author: Odontologie ou observations sur les dents humaines, 1815; Traité de la seconde dentition, 1819; Traité de l'art chirurgien-dentiste, 1820. Invented Delabarre syrup (used for infant dentition in France); research and publs. on dental surgery techniques, 2d cutting of teeth. Died 1862.

DELABARRE, Edmund Burke, Am. psychologist; b. Dover, Me., Sept. 25, 1863; s. Edward and Maria (Hassell) D.; student Brown U., 1882-83; A.B., Amherst, 1886; A.M. Harvard, 1889; Ph.D., U. Freiburg, 1891; m. Dorothea Esther Cotton. Mar. 14, 1907; children—Maria Elizabeth, Edmund Burke, Barbara Melville, Dorcas Hope. Asso. prof. psychology, Brown U., Providence, 1891-96, prof., 1896-1932. Dir. Psychol. Lab., Harvard, 1896-97. Fellow A.A.A.S., Am. Acad. of Arts and Scis.; mem. Am. Psychol. Assn., Phi Beta Kappa, Sigma Xi; corr. mem. Geog. Soc. Phila., Geog. Soc. Lisbon, Archaeol. Soc. Portugal. Author: Ueber Bewegungsempfindungen, 1891; Report of Brown-Harvard Expedition to Nachvak, Labrador, in 1900, 1902; Dighton Rock History, 1916, 17, 19; Inscribed Rocks of Narragansett Bay, 1919-23; Dighton Rock—A Study of the Written Rocks of New England, 1928. Contbr. to psychol., archaeol., hist. publs. Died Providence, R. I., Mar. 16, 1944.

DE LA BECHE, Sir Henry Thomas, English geologist; b. London, Eng., Feb. 10, 1796; student Gt. Marlow Coll., circa 1810-15; studied geology in France, Switzerland, Jamaica. Mem. Ordnance Survey; apptd. 1st dir. Geol. Survey Gt. Britain, 1835, developed geol. mus. and ednl. facilities. Wollaston medal Geol. Soc. Lond., ca. 1850. Fellow Royal Soc., 1819; mem. French Acad. Scis., 1853, Geo. Soc. London (pres. 1848-49). Author: A Geological Manual, 1831; Researches in Theoretical Geology, 1834; How to Observe Geology, 1835, rev. as The Geological Observer, 1851. Collected geol. materials during extensive trips through Eng., France, Switzerland; began geol. map of Eng.; research on geology of Eng. especially rocks of Cornwall and Devon; 1st description of Jamaican rocks, 1824; organized study of geology. Died London, Apr. 13, 1855.

DE LÁBERGERIE, Jean-Baptiste Rougier, (baron), French agronomist; b. Beaulieu, France, 1757; mem. Commune of Paris, 1789; later dep. Legislative Assembly from Yonne; prefect of Yonne, 1800-11. Author: Traité d'agriculture pratique, 1795; Rapport général sur les étangs de la Republique, 1795; Histoire de l'agriculture française, 1815; Les forets de la France, 1817; Historie de l'agriculture des Gaulois, 1829; Histoire de l'agriculture ancienne des Romains, 1834. Died Paris, 1836.

DE LA BILLARDIÈRE, Jacques-Julien Houtou, see La Billardiere, Jacques-Julien Houtou de.

DE LA BLANCHÈRE, Pierre-René-Marie-Henri Moulin du Coudray, see La Blanchère, Pierre-René-Marie-Henri Moulin du Coudray.

DE LA BOË, see Sylvius de la Boë, Franciscus.

DE LABOULAYE, Charles-Pierre Lefebvre, French inventor; b. Paris, France, 1813; studied at Poly. Sch., later at l'Ecole d'Application de Metz; gave up mil. career to go into industry; established a foundry. Author: Organisation du travail, 1848; Traité de cinématique, 1849; Essai sur l'art industriel, 1856; Essai sur l'équivalent mécanique de la chaleur, 1859; Dictionnaire des arts et manufactures, Economie des machines et des manufactures d'après l'ouvrage anglais de Charles Babbage, 1879. Invented various machines; analyzed composition of some alloys. Died Paris, 1886.

DE LABOULAYE-MARILLAC, Pierre-Charles-Madeleine, see Laboulaye-Marillac, Pierre-Charles-Madeleine de.

DE LA BROSSE, Guy, physician, botanist; b. Rouen, France; physician to Louis XIII; founder (with Jean Hérvard) Jardin du Roi (later called Jardin Royale des Plantes Medicinales), Paris, 1626, became 1st dir. of garden. Author: Traité de la Peste, 1623; Dessin du Jardin Royal pour la Culture des Plantes Médicinales à Paris . . . , 1626. Collected plants from all over the world, catalogued 2,000 specimens, 1636; left more than 400 engraved plates at death (only 50 extant). Died 1641.

DE LACAILLE, Nicolas Louis, see Lacaille, Nicolas Louis de.

DE LACAZE-DUTHIERS, (Felix) Henri, see Lacaze-Duthiers, (Felix) Henri de.

DE LACÉPÈDE, Comte Bernard Germaine Étienne de la Ville, French naturalist; b. Agen, France, Dec. 26, 1756; studied under Buffon; mem. Royal Cabinet; after Revolution chair of reptiles and fish created for him at Mus. Natural History; mem. legislature during Revolution; became senator, 1799, pres. senate, 1801; named minister of state, 1804; grand maître de l'université during 100 Days. Decorated grand chancellor Legion of Honor. Author: Histoire général et particulière des quadrupèdes ovipares et des serpents; Histoire naturelle des reptiles, 1789; Histoire naturelle des poissons, 1798-1803; Histoire naturelle des cétacés, 1804; Oeuvres complètes d'histoire naturelle, 1826; also wrote on electricity. Died Oct. 6, 1825.

DE LA CHAMBRE, Marin Cureau, French physician, natural scientist; b. Maine, France, Nov. 29, 1596; at least 1 son, Pierre; royal physician to Louis XIII; instr. botany and anatomy Royal Bot. Gardens; mem. French Acad. Scis., 1666. Author: Nouvelles pensées sur les causes de la lumière du debordement du Nil et de l'amour d'inclination, 1634; Chiromacie, 1653. Wrote some of earliest sci. works in French; opponent of scholasticism; studied flooding of Nile; theorized on cause of light; wrote on analyzing the appearance of the hand. Died Paris, Nov. 29, 1669.

DELACOUR, Alfred-Charlemagne Lartigue, French physician; b. Bordeaux, 1815; M.D., Paris, 1841; wrote med. tracts, Paris; gave up medicine to write plays, after 1847. Author: Encyclopédie médicale, 9 vols.; L'angine de poitrine, 1846. Studied inflammation of chest; invented and sold Lartigue pills for gout; wrote popular home medicine guides for gout and rheumatism sufferers. Died Paris, 1883.

DELACOUR, Jean Theodore, naturalist; b. Paris, France, Sept. 26, 1890; s. Theodore and Marguerite (Rousseau) D.; ed. Jesuit Sch., Rue de Madrid, Paris; Lic.S., U. of Lille, 1914; unmarried. Naturalized citizen of U. S., 1946—. Engaged in work as naturalist since 1908; maintained home, Chateau de Clères, Normandy, as site of gardens and buildings for living collections of rare animals and birds obtained on own expeditions and by special collectors throughout world, 1919-39; the collections were practically destroyed with records and library during World War II, now restored; ex-dir. dept. history, sci., arts Los Angeles County. Research asso. Am. Mus. Natural History, N.Y.C., 1942-60; collaborator Fish and Wildlife Service, Dept. Interior; president Ligue Française pour la Protection des Oisenux; v.p. Société Nationale d'Acclimatation de France, Avicultural Soc. London; pres. Avicultural Soc. Am.; mem. com. Am. Com. for Internat. Wild Life Protection; council Zool. Soc. London (hon.), Société Ornithologique et Mammalogique de France (editor 1920-40), Société Zoologique de France; v.p. Council Internat. de la Chasse (v.p.), Académie des Sciences, Arts et Belles-Lettres de Rouen, N.Y. and Phila. zool. socs., Brit. Ornithologists Union, Ornithol. Soc. Japan. Editor: L'Oiseau et la Revue Française d'Ornithologie (mag.), 1920-40. Author: numerous scientific books and articles relating to ornithology and mammalogy published in learned journals of France, Eng., U. S. and Germany. Home: 538 S. Flower St., Los Angeles 17.

DE LACROIX DILLON, Jacques-Vincent-Marie, engr.; b. 1760; m. Henriette de Meulan; 1 dau., Andrée-Élisa. Became capt. in Corps Hydraulic Engrs., 1795; named insp. metric system, 1797; prof. metric measurement École pratique; built 1st iron bridge in France (Pont des arts) Paris, 1798. Died 1807.

DE LA FAILLE, Charles, Flemish mathematician; b. Antwerp, Belgium, 1597; prof. to John of Austria; author: Theses mechanicae Theoremata de centro gravitatis, 1632; did research on gravity. Died Barcelona, Spain, 1652.

DE LA FAYE, Georges, French surgeon; b. Paris, France, 1701; demonstrator Royal Acad. Surgery; author of Principes de Chirurgie, 1739. Removed crystalline lens of eye by pressure after incision; modified le Dran's operation for disarticulation of shoulder. Died 1781.

DE LAFAYE, Jean-Elie Leriget, engr.; b. Vienna, Austria, Apr. 15, 1671; soldier who gave leisure time to study of sci. Died Apr. 20, 1718.

DELAFIELD, Edward, Am. physician; b. N.Y.C., May 7, 1794; s. John and Ann (Hallett) D.; A.B., Yale, 1812; M.D., Coll. Phys. and Surg., N.Y.C., 1816; m. Elina Elwyn, Oct. 1821; m. 2d, Julia Floyd, Jan. 1840; 5 children. Practiced medicine, specializing in ophthalmology, surgery; founder Eye Infirmary, 1818; prof. obstetrics and diseases of women and children Coll. Phys. and Surg., 1825-38, pres., 1858-75; physician N.Y. Hops., 1834-38; founder, 1st pres. Soc. Relief of Widows and Orphans of Med. Men, 1842; a founder, 1st pres. Am. Ophthalmol. Soc., 1864. Died N.Y.C., Feb. 13, 1875.

DELAFIELD, Francis, Am. physician; b. New York, Aug. 3, 1841; s. Dr. Edward and Julia (Floyd) D.; A.B., Yale, 1860; M.D., Coll. Phys. and Surg. (Columbia), 1863; studied London, Berlin and Paris; (LL.D., Yale, 1890, Columbia, 1904); m. Katherine Van Rennsselaer, Jan. 17, 1870. Surgeon New York Eye and Ear Infirmary, 1871; pathologist, Roosevelt Hosp., 1871; physician Bellevue Hosp., 1874, consulting phys., 1885—; adj. prof., 1875-82, prof. pathology and practice of medicine, prof. emeritus, 1901—, Coll. Phys. and Surg. Author: Studies in Pathological Anatomy (2 vols.), 1882; Hand Book of Pathological Anatomy and Histology (with Dr. T. Mitchell Prudden, 1885, 5th edit., 1897. Investigations of nephritis and colon diseases; 1st to show difference between bronchial and lobar pneumonia. Died Norton, Conn., July 17, 1915.

DE LA FOLLIE, Louis Guillaume, French chemist; b. Rouen, France, 1739; insp. mfg. Author: Le philosophe sans prétention ou l'homme rare, 1775. Invented technique for fixing indigenous colors in wool fabrics (la Follie's finishing); discovered permanent color: Indian red; research on application of chemistry to industry especially for wool mfg. Died 1780.

DELAFOND, (Mamert) Onésime, French veterinarian; b. Saint-Amand, France, Feb. 13, 1805; ed. Alfort Sch., 1823; studied under Dupuy; received chair therapeutics Alfort Sch., 1833, became dir., 1860. Author: Instruction sur la pleuro-pneumonie, 1840; Traité de therapeutique générale vétérinaire, 1843-44; Traité sur la maladie de sang des betes bovines, 1848; Traité de pathologie generale comparée des animaux domestiques, 1855; Typhus de l'espèce bovine. Research on intestinal villus, composition of animal chyle, internal parasites of domestic animals, contagious diseases; discovered and isolated bacterial anthrax which he considered to be cryptogams, 1850. Died Nov. 15, 1861.

DELAFONTAINE, Marc, chemist; b. 1837; student under J. C. A. de Mariguac; tchr. U. Geneva, Switzerland; came to U. S., 1870; tchr. high schs. Chgo.; analytical chemist, expert Chgo. Police Dept.; research on spectrum analysis, showed that spectrum of didymia varied according to its source. Died 1911.

DELAFORGE, Louis, French philosopher; flourished 17th century. Author: De homine. Attempted to reconcile ideas of Descartes (his friend) with those of St. Augustine; foreshadows Malebranches' work.

DELAFOSSE, Gabriel, French crystallographer; b. San Quentin, France, Feb. 24, 1796; ed. École normale, Paris, France; held chair in mineralogy Faculty Scis., Paris, also at École normale; named conservator natural history collection Paris Faculty Scis., 1822; apptd. lectr. natural history École normale, 1826; apptd. prof. mineralogy Paris Faculty Scis., 1840, and later at Natural History Mus. Mem. French Acad. Scis. Author: Traité de cristallographie, 1821; Traité des minéralogie, 1822; Rapport sur les progrès de la minéralogie, 1867. Founder of crystallography; 1st to show relationship between direction of rotating power of crystals and orientation of their facets. Died Paris, Oct. 13, 1878.

DE LA FRATTA E MONTALBANI, Marco Antonio, Italian mineralogist; b. Bologna, Italy, 1630; student mineralogy Germany, Hungary, Poland, Adriatic coast. Author: Practica Minerale . . . , 1670; Catascopia Minerale . . . , 1678; Dell' Acque Minerali Del Regno Di Ungheria, 1687; described amalgamation process for extracting silver; separation of gold and silver by aqua fortis, separation of silver from copper by lead, crystallization of vitriol, assay and cupellation furnaces, weights and others. Died Apr. 30, 1695.

DELAGE, Yves, French zoologist; b. Avignon, France, May 13, 1854; ed. medicine and zoology at various colleges; m.; 1 son, 1 dau. Asst. lectr., zoology, 1878, lectr., Sorbonne, 1880; asst. lectr., Faculty of Sci., U. Paris, 1881, asst. prof., 1882; lectr., Caen, and dir., zoological station, Luc-sur-mer, 1883; prof., 1884; lectr., U. Paris, 1885, prof., 1886; asst. dir., laboratory for research on experimental zoology, 1889; dir., Roscoff Zoological Station, 1901; founded L'Année biologique. Received Darwin medal, Royal Soc., London, 1916. Mem., French Acad. of Sci., 1901. Author: L'Appareil circulatoire ces crustacés édriophhalmes, 1881; Evolution de la sacculine, 1884; Fonction des canaux demicirculaires et otocystes des invertébrés, 1887; Embryogénie des éponges, 1892; L'Hérédité et les grands problèmes de la biologie générale, 1895; Traité de zoologie concrète (6 vols.), 1896-1903; Les théories de l'évolution, 1909; La Parthénogénèse naturelle et expérimentale, 1913; numerous articles. Research on heredity, reproduction, evolution, hybridism; authority on sponge culture; dissected non-segmented eggs of seaurchins, 1899; fundamental research on parthenogenesis; studied commercial utilization of seaweeds. Died Sceaux, France, Oct. 7, 1920.

DELAGENIÈR, Henry-Yves, French surgeon; b. Paris, France, 1856; studied at Med. Sch. of Angers; docteur en médecine, Paris, 1890. Founded one of 1st med. clinics outside of Paris, at Le Mans, France, 1890. Mem. Acad. of Medicine. Publs. on several operations including total and partial removal of the stomach, 1908, crural hernia, pleurotomy. Died Royan, France, 1930.

DE LAGNY, Thomas Fantet, French mathematician; b. Lyons, France, Nov. 7, 1660; ed. Acad. Scis., Paris, France; became prof. hydrography, Rochefort, France, 1697; named dir. bank of law, Paris, 1716; conservator royal library. Mem. French Acad. Scis., 1696. Author: Méthodes nouvelles et abrégées pour l'extraction et l'approximation des racines, 1691; Nouveaux éléments d'arithmétique et d'algèbre, 1697; Cubature de la sphère, 1702; Arithmétique nouvelle, 1703; Analyse générale, 1733. Gave algebraic expressions yielding approximate cube and 5th roots; 1st to derive gen. formulae and discuss periodicity of multiple-angle trigonometric functions; cited advantages of unit hypotenuse in trigonometric functions. Died Paris, Apr. 12, 1734.

DE LAGUNA, Andrés, anatomist; b. 1499; physician to Charles V. Author: Anatomica methodus, 1535; Methodus cognoscendi extirpandique nascentes in vesicae collo caruncula, 1594. Gave 1st description of ileo-cecal valvule. Died 1560.

DE LAGUNA, Frederica A.; Am. anthropologist; b. Ann Arbor, Mich., Oct. 3, 1906; d. Theodore and Grace (Andrus) de Laguna; B.A., Bryn Mawr Coll., 1927; Ph.D. (U. fellow), Columbia, 1933. Asst. Am. sect. U. Pa. Mus., Phila., 1931-33; asso. soil conservationist U. S. Dept. Agr., Albuquerque, 1935-36; faculty Bryn Mawr (Pa.) Coll., 1938-42, 46—, prof. anthropology, 1955—. Vis. lectr. U. Pa., 1947-49; vis. prof. U. Cal., Berkeley, 1959-60. Bryn Mawr Coll. European fellow, 1928-29, NRC fellow, 1936-37, Rockefeller fellow, 1945-46, Viking Fund fellow, 1949-50, Social Sci. Research Council fellow, 1962-63. Fellow Arctic Inst. N.Am., Am. Anthrop. Assn. (president 1966-67); member of the Society for American Archeology (served as first vice pres. 1949-50), Am. Ethnol. Soc., Sigma Xi (pres. Bryn Mawr 1964-66). Author: The Archaeology of Cook Inlet, Alaska, 1934; Chugach Prehistory, 1956; The Story of a Tlingit Community, 1960. Editor: Selected Papers from the American Anthropologist, 1960. Archeol., ethnol. studies and publs. on various tribes of Alaskan Indians. Home: 221 Roberts Rd., Bryn Mawr, Pa. 19010.*

DELAHAY, Paul, chemist; b. Sas van Gent, Netherlands, Apr. 6, 1921; s. Jules and Helene (Flahou) D.; B.S., U. Brussels, 1941, M.S., 1945; M.S., U. Liege, 1944; Ph.D., U. Ore., 1948; m. Yvonne Courroye, Dec. 1, 1962. Came to U. S., 1946, naturalized, 1955. Instr., U. Brussels, 1945-46; research asso. U. Ore., 1948-49; mem. faculty La. State U., Baton Rouge, 1949-65, Boyd prof. chemistry, 1956-65; prof. chemistry N.Y. U., N.Y.C., 1965—; Guggenheim fellow Cambridge U., 1955; Fulbright prof. U. Paris, 1962-63. Recipient Univ. medal U. Brussels, 1963; Heyrovsky medal Czechoslovak Acad. Sci., 1965. Mem. Electrochemical Society (Turner prize 1951, Palladium medal award 1967), American Chemical Soc. (award in pure chemistry 1955, Southwest award 1959), A.A.A.S., Sigma Xi. Author: New Instrumental Methods in Electrochemistry, 1954; Instrumental Analysis, 1957; Double Layer and Electrode Kinetics, 1965; also numerous articles. Home: 1 Washington Sq. Village, N.Y.C. 10012.*

DE LALANDE, Joseph Jérôme Le François, see Lalande, Joseph Jérôme Le Français de.

DE LALANDE, Michel-Jean-Jérôme Le François, French astronomer; b. Courcy, France, Apr. 21, 1766; m. Marie Harlay, 1788; substitute prof. at Coll. France. Mem. French Acad. Scis., 1801, Bur. Longitudes. Described sky visible from France; established orbit of Mars; collaborated with Delambre in triangulation research. Died Paris, France, Apr. 8, 1839.

DELALANDE, Pierre Antoine, French naturalist; b. Versailles, France, 1787; author: Précis d'un voyage entrepris au cap de Bonne-Espérance, 1822; brought back important mineral collections from voyages to South Africa. Died Paris, France, 1823.

DE LALOUBÈRE, Antoine, see Laloubère, Antoine de.

DELAMAINE THE ELDER, Richard, English mathematician; b. Eng.; flourished 1631; studied under W. Oughtred; 10 children including Richard. Tutor math. to Charles I; tutor, engr. for govt. Author: Grammelogia or the Mathematicall Ring, 1631; The Making Description and Use of a small portable Instrument called a Horizontall Quadrant, 1631. Described circular slide rule and other math. instruments.

DE LAMANON, Robert de Paul (Lamanon, Robert de Paul de), French Geologist; b. Salon-en-Provence, France, Dec. 6, 1752; mem. French Acad. Scis., 1783; participated in La Perouse expdn.; research on geology and paleontology of south and midi region of France. Died Dec. 11, 1787.

DELAMARRE-DEBOUTTEVILLE, Edouard, French engr.; b. Rouen, France, 1856; inventor cotton reaming machine, 1879, (with Malandin) 1st gasoline powered automobile in world to be successfully driven, 1883. Died Fontaine-le-Bourg, France, 1901.

DE LAMARTINIÈRE, Germain-Pichaut, (Lamartinière, Germain-Pichaut de), French surgeon; b. France, 1696; ed. Coll. Saint-Côme; Aggregation; apptd. royal surgeon to Louis XV, 1747; pres. Acad. Surgery; author Memoires presentes au roi. Important organizer of French medicine; placed surgeons independent of Faculty Medicine. Died 1783.

DELAMATER, Cornelius Henry, Am. mech. engr.; b. Rhinebeck, N.Y., Aug. 30, 1821; s. William and Eliza (Douglass) D.; m. Ruth O. Caller, 6 children. Built iron boats; built 1st steam fire engines used in U.S.; built engines for Monitor; partner (with Peter Hogg, ret. 1856) in iron works, 1850-56, pres. Delameter Iron Works, 1856-89, constructed 30 gunboats for Spanish govt., 1869; noted for propellors, air compressors, for constrn. 1st successful submarine torpedo boat, 1881; an original mem. Am. Soc. M.E. Died N.Y.C., Feb. 7, 1889.

DE LAMATER, Edward Doane, Am. microbiologist; b. Plainfield, N.J., Jan. 24, 1912; s. Van Ness and Jacqueline (Newton) DeL.; student Oberlin Coll., 1930-32; M.A., Johns Hopkins, 1937; Ph.D., Columbia, 1941, M.D., 1942; m. Jean Edgar, Jan. 31, 1943; children—Gretchen, Peter, David; m. 2d Margaret Henderson Turner, Feb. 25, 1961; children—Margaret Murray Maynadier, Anna Van Ness. Practice medicine, specializing in microbiology, Phila., 1948-63, N.Y.C., 1963-66; faculty Sch. Medicine U. Pa., 1948-63, research prof. dermatology, microbiology, 1951-63, dir. sect. on cytology, genetics, 1953-63; prof., chmn. dept. microbiology N.Y. Med. Coll., 1963-66; dean Coll. of Sci., Fla. Atlantic U., Boca Raton, 1966 68, Distinguished professor of Science, 1960—; sci. adviser Mikroskopie, 1955—; cons. to hosps., govt. agys., indsl. firms, Guggenheim Meml. Found. fellow, 1953. Fellow A.C.P., N.Y. Acad. Scis.; mem. Royal Soc. Medicine, Nat. Multiple Sclerosis Soc., Swedish, Venezuelan dermatol. socs., Am. Acad. Microbiology, A.A.A.S., Am. Assn. Cancer Research, Am. Genetic Assn., A.M.A., Am. Soc. Cell Biology, Am. Soc. Human Genetics, Am. Soc. Microbiology, Am. Soc. Zoologists, Bot. Soc. Am., Electron Microscope Soc. Am., Genetics Soc. Am., Genetics Soc. Can., Mycol. Soc. Am., Soc. Gen. Microbiology, Torrey Bot. Club, Optical Soc. Am., N.Y. Acad. Medicine, Sigma Xi. Research and publs. on agts. of fungal, syphilitic diseases; cytologic, chromosomal studies of various bacteria. Home: 888 Oleander Av., Boca Raton, Fla. 33432.*

DE LAMBALLE, Antoine Jobert, French surgeon; b. Lamballe, France, 1799; student Lamballe, Paris, France; surgeon to emperor France, from 1854; author: Traité de Chirurgie plastique, 1849; discovered surg. process for cure of fistula vesico-vaginale, known as elitroplastie. Died 1867.

DE LAMBERTYE, Léonce (Lambertye, Léonce de), French botanist; b. Montluçon, France, 1810; author Catalogue des plantes vasculaires de la Marne, 1877; made definitive study of vascular structure of plants in Marne region, also momography on culture of strawberry. Died 1877.

DELAMBRE, Jean-Baptiste, French astronomer; b. Amiens, France, Sept. 19, 1749; student of Jacques Delille, Amiens coll., also studied at U. Paris; later student of Lalande. Tutor to son of M. d'Assay, receiver-gen. of finances; used astron obs. installed for him by d'Arcy, from 1788; insp.-gen. of studies; succeeded Lalande as prof. astronomy Coll. de France,

1807; treas. imperial univ., 1808-15. Mem. French Acad. Scis., 1792, perpetual sec. math. sect., 1803; mem. Bur. Longitudes. Author: Tables écliptiques des satellites de Jupiter, 1792, 1817; Methodes analytiques pour la détermination d'un arc du meridien, 1799; Tables du soleil, 1806; Base du système metrique, 3 vols., 1806, 07, 10; also works on history of astronomy. Noted for astron. tables of Uranus, also prepared tables for Jupiter and Saturn; precisely measured arc of meridian from Dunkirk to Barcelona (as basis for determination of length of meter), 1792-99, also provided confirmation of oblateness of earth at the poles; discovered Dalambre's analogies (in spherical geometry). Died Paris, Aug. 19, 1822.

DE LAMÉTHERIE, Jean-Claude, French naturalist; b. La Clayette, 1743; began practice of medicine, Paris, 1780; named asst. prof. natural scis. Coll. France, 1812. Author: Vues physiologiques sur l'organisation végétale, 18 1781. Editor Jour. de physique. Research in mineralogy and geology; studied physiol. orgn. of plants; formulated a theory that matter originated by crystallization and is always in movement; described phosphorescence of some diamonds and certain minerals, 1802; discovered aluminum silicate in France (named it macle). Died 1817.

DE LA METTRIE, Julien Offroy, see La Mettrie, Julien Offroy de.

DE LANA (or LANA-TERZI), Francesco, Italian physicist, inventor; b. Brescia, Italy, Dec. 10, 1631; ed. Roman Coll.; asso. of Kircher in Rome, Italy, 1652; mem. Soc. of Jesus; tchr. in numerous Italian villages, also in Terni, Italy; prof. math., Ferrara, Italy, 1677-79; prof. physics, Brescia, Italy; founder Academia Philexotilorum Naturae et Artis. Author: Podromo overo saggio di alcune invenzioni nuove promesso all'Arte Maestra, 1670; Magisterium naturae et artis, Vol I, 1684, Vol. II, 1686, Vol III, 1692. Experimented on ballistics, also on barometer; invented clocks, also projector of airship raised by copper globes; described condensation of moisture on cold vessel; wrote on concentrating alcohol by passing it in vapor form through pig's bladder; conceived plans for lighter- and heavier-than-air aircraft, also methods of writing for the blind, speaking from long distances, building a sowing machine. Died Brescia, Feb. 22, 1687.

DE LANGE, Cornelia Catharina, Dutch physician; b. 1871; grad. U. Amsterdam, Netherlands, 1897; prof. pediatrics U. Amsterdam (1st woman to become full prof. on med. faculty at this univ.), 1927-38. Died 1950.

DELANGE, Hubert Marie, French mathematician; b. Angers, France, Mar. 1, 1914; s. Gaston Alexandre and Jeanne (Godard) D.; Agrégation de Mathématiques, Ecole Normale Supérieure, 1935; postgrad. Fondation Thiers, 1937-39; Doctorat ès Sciences, Paris, 1940; m. Madeleine Desarméniens, Mar. 10, 1945; 1 dau., Anne-Béatrice. Tchr., Lycée Malherbe, Caen, France, 1941; prof. U. Clermont-Ferrand (France), 1942-58; prof. math. U. Paris, 1958—; dean Faculty Sci., Clermont-Ferrand, 1955-58. Mem. Société Mathématique de France, Am. Math. Soc., Société Royale des Sciences de Liège (corr.). Research, numerous publs. on sequences of polynomials or entire functions, Laplace integrals, Tauberian theorems, arithmetic functions. Home: 22 Allée des Troënes, Bures-sur-Yvette (Essonne). Office: Faculté des Sciences Bat+425, d'Orsay, Orsay (Essonne) 91, France.*

DELANNE, Gabriel, engr.; b. Mar. 23, 1857; s. Alexandre D.; grad. École centrale d'ingénieurs; engr. for Cie Co., 1877-92; editor Revue Scientifique et morale du spiritisme. Founder-mem. French Assn. for Study Physic Phenomena (apptd. pres. 1899). Institut Metapychique Internat. Author: le Spiritisme devant la Science, 1885; Le phenomene spirite, 1896; l'Evolution animique, 1897; Recherches sur la médiumnité, 1898; l'Ame est immortelle, 1899; les Apparitions materialisées des vivants et des morts, Vol. I, 1900, Vol. II, 1911; Documents pour servir à l'étude de la réincarnation, 1927. Research on spiritualist mediums, 1900-1920's. Died Feb. 15, 1926.

DE LA NOE, Gaston-Ovide, (La Noë, Gaston-Ovide de), French geographer, scholar; b. Limoux, France, 1836; ed. École polytechnique, 1857; general in French army; became dir. Mus. of Army, 1895; founder Geog. Bur. of army; author Les Formes du terrain, 1888. Died Paris, 1902.

DELANY, Patrick Bernard, electrician; b. Kings County, Ireland, Jan. 28, 1845; s. James and Margaret D.; ed. pvt. and parochial schs., Ireland and U. S.; m. Annie M. Ovenshine, Mar. 31, 1869. Learned telegraphy in Hartford, Conn., and advanced from office boy to supt. of lines; expert operator, newspaper corr., editor, writer. Recipient Gold medal and diploma Internat. Inventions Exhbn., London, Eng., 1885; Elliott Cresson Gold medal, twice, also John Scott Legacy medal Franklin Inst.; Gold medal Buffalo Expn., 1901, St. Louis Expn., 1904. Holder 150 patents, covering anti-induction cables, synchronous multiplex telegraphy; developed automatic systems for ocean cables; rapid machine telegraphy for land lines; vox Humana talking machines; improved system of automatic telegraphy for transmitting and plainly recording 3,000 words per minute over single wire; patented method for locating submerged metallic bodies. Died Oct. 19, 1924.

DE LA PÉROUSE, Jean François, see La Pérouse, Jean François, Comte de Galaup.

DE LAPERSONNE, Felix, (Lapersonne, Felix de), French ophthalmologist; b. Toulouse, France, 1853; Aggregation, 1886; chief Paris Ophthalmology Clinic, 1883-89; mem. Acad. Med. Research on occular syphilis, hysteric and tabetic eye, occular paralysis, purulent conjunctiuitis. Died Paris, 1937.

DE LA PEYRONIE, François Gigot, see La Peyronie, François Gigot de.

DE LAPOUGE, George Vacher, French anthropologist; b. Neuville, France, Dec. 12, 1854; lawyer; librarian, tchr. race and culture at various French univs. Author: les Sélections sociales, 1896; L'aryen, son rôle social, 1899. Supported theory of racial significance in cultural devel. Died Poitiers, France, Feb. 20, 1936.

DE LAPPARENT, Albert-Auguste Cochon, French geologist; b. Bourges, France, Dec. 30, 1839; ed. École Polytechnique, École des Mines; at least 1 son, Jacques. Apptd. sec. of com. which drew up plans for submarine tunnel between France and Eng. (never built), 1874; prof. geology and mineralogy Paris Catholic Inst. Mem. French Acad. Scis., 1897; elected perpetual sec. Soc. Agr., 1907. Author: Traité de géologie, 1882; La géologie en chemin de fer, 1888; Cours de minéralogie, 1884; La géologie en chemin de fer, 1888; Lecons de géographie physique, 1896; Le globe terrestre, 3 vols., 1899; Science et apologétique, 1905; La philosophie minérale, 1910. Collaborated on great geol. map of France; made detailed study of geology of LaBray region in France; studied formation of sediment in plateau regions; authority on stratigraphy; made submarine study of rock formations of Pas-de-Calais. Died Paris, May 4, 1908.

DE LAPPARENT, Marie-Jacques Cochon (Lapparent, Marie-Jacques Cochon de), French petrographer; b. Paris, Apr. 5, 1883; s. Albert Auguste; became prof. mineralogie Strasburg Faculty Scis., 1918; lectr. geology, Lille; mem. French Acad. Scis., 1936. Author: Leçons de petrographie, 1923; Les Calcaires à globicérines du crétacé supérieur, 1925. Research on structure and classification of rocks; discovered certain minerals; specialist in phenomenon of alteration of surface. Died Paris May 18, 1948.

DE LA QUINTINE, Jean, French agronomist; b. Chabanais, France, 1626; practiced law, Poitiers; became farmer. Author: Instructions pour les jardins fruitiers, 1690. Wrote on transplantation, trimming of fruit trees; introduced spl. trellis for fruit trees. Died Versailles, 1688.

DE LA RAMÉE, Pierre, see Ramus, Petrus.

DE LA RIVE, Charles Gaspard, see La Rive, Charles Gaspard de.

DE LA RIVE, Lucien, physicist; b. Geneva, Switzerland, 1834; with Sarasin used a Hertz resonator to prove stationary electric waves occurred along insulated wires under influence of oscillating discharge, 1890; determined that wave length in air, where it was reflected in a mirror was same as in a wire.

DE LA RIVIÈRE, Roch Le Baillif, Sieur, French physician; b. Falaise; practiced medicine, Paris, also Rennes, France; apptd. 1st physician to Henry IV, 1594. Author: Bains curans la lepre, 1577; Premier traité de l'homme et son essentielle anatomie, 1580; Conformité de l'ancienne et moderne médecine, d'Hippocrate à Paracelse, 1592. Studied leprosy, hydropsy, paralysis, ulcers, pleurisy, anthrax. Died 1605.

DE LA RUE, Sir Ernest, English inventor; b. 1852; s. Warren de la Rue; ed. King's Coll.; m. Florence Gay Octavia Williams, 1852. Knighted, 1921. Designed and decorated Albert Hall; made 1st model of electric starting gate (used by Jockey Club); inventor new method of molding papier-maché surg. splints and boots; organized and financed Dullingham Br. for prodn. splints and boots. Died Aug. 18, 1929.

DE LA RUE, Warren, English astronomer, inventor; b. Guernsey, Channel Islands, Jan. 15, 1815; s. Thomas and Jane (Warren) de la R.; ed. Paris; D.C.L., Oxford (Eng.) U.; m. Georgiana Bowles, 1840; children—4 s., 1 dau. Entered father's printing firm; founded obs., Canonbury, Eng., circa 1850, which was moved to Cranform, Eng., 1857; dir. expdn. to observe solar eclipse, Avabellosa, Spain, 1860. Fellow Royal Soc. (gold medal 1864); mem. Royal Astron. Soc. (gold medal 1862, pres. 1864-66), Chem. Soc. (charter, pres. 1867-69, 1879-80) French Acad. Scis. Author: Researches in Solar Physics, 1865-68; also articles. Invented 1st envelope-making machine, 1851, photo-heliograph for photographing sun daily at Kew Obs., 1858, (with H. Müller) silver chloride battery; proved prominences (or flames observed during solar eclipses belong to sun, not the moon, 1860; research (with Hugo Müller) on Rangoon tar, 1859, glyceric acid, 1859, terephthalic acid, 1861, electric discharge through gases, 1868-83; observed (with Balfour, Steart, Benjamin Loewy) sun spots, 1862. Died London, Apr. 19, 1889.

DE LA RUEL, Johannes Jean, French physician, botanist; b. 1479; physician to Francis I of France; dean Paris (France) Faculty of Medicine; author: De Natura Stirpium, 1536. Died 1537.

DE LA TOUR, Charles Cagniard, see Cagniard-Latour, Charles.

DE LA TOURRETTE, Marc-Antoine-Louis Claret (La Tourrette, Marc) French botanist; b. Lyons, France, Aug. 11, 1729; worked with Jussieu, Linnaeus and Haller; mem. French Acad. Scis., 1772, Lyons Acad. (perpetual sec.); author Chloris lugdunesis, 1785; gave detailed description of mosses and fungi of Lyons region. Died 1793.

DE LAUBENFELS, Max W., Am. zoologist; b. Mt. Pleasant, Ia., May 9, 1894; s. Harry J. and Hattie M. (Walker) de L.; A.B., Oberlin Coll., 1916; A.M., Stanford, 1926, Ph.D., 1929; also student U. Cal., Art Inst. Chgo., several European univs.; m. Beth Jones, Aug. 10, 1921; children—Peter Max, Leroy Arthur, David John, Allan Neal, Marilyn Beth. Business in Chgo., 1916-21; instr. Oberlin Coll., 1927, Pasadena City Coll., 1928-47; prof. zoology U. Hawaii, 1947-50, Ore. State Coll., 1950—; research for various univs., also Bur. Fish, Office Naval Research, NRC, State Fla., various biol. stas., museums and pvt. corps., others. Fellow A.A.A.S.; mem. many sci. socs. Author sci. articles, including 7 monographs on Porifera; also textbooks. Died Feb. 4, 1958.

DELAUNAY, Albert, French biologist; b. Mortagne /Sèvres, France, Oct. 17, 1910; s. Augustin and Angèle (Ernoul) DeL.; student Sch. Medicine, Nantes, France, 1929-35; M.D., U. Paris, 1938; m. Marcelle, Ramon, July 24, 1941; children—Jean, Pascale, Philippe. Asst., Sch. Medicine, Nantes, 1933-36, Pasteur Hosp., Paris, 1936-38; Rockefeller scholar, Boston, 1938-39; staff Pasteur Inst. Paris, 1936—, head lab., Garches, 1943-56, head dept. exptl. pathology, 1956—; prof. microbiology Ecole Normale Supérieure de Fontenay-aux-Roses, 1950—. Named Laureate, Faculty Medicine in Paris, (3 times) Acad. Medicine, Acad. Scis., French Acad. Mem. French Microbiology Soc., French Biochem. Soc., Tissue Culture Club, French Connective Tissue Club (head 1963—). Author: (with A. Boivin), Phagocytose et Infection, 1947, L'organisme en luttre contre les Microbes, 1947; Pasteur et la Microbiologie, 1951; Jean Rostand, 1956; Journal d'un Biologiste, 1959; Histoire de l'Institut Pasteur, 1961; also numerous articles. Research on phagocytosis in vitro by polymorphonuclear leukocytes and macrophages, influence of non specific opsonins, diapedesis of leukocytes during early phase of inflammation, influence of scurvy, intoxication by bacterial lipopolysaccharides, mechanism of action of bacterial lipopolysaccharides in resistance of animals to infections, non specific resistance to infections, role of nutritional internal antibiotics, metabolic factors, role of intermolecular interactions in stability and functions of connective tissue. Home: 8 rue de Suresnes, Vaucresson, Hauts de Seine, France. Office: Institut Pasteur, Garches, Hauts de Seine, France.*

DELAUNAY, Charles-Eugène, French astronomer; b. Lusigny, Aube, France, Apr. 9, 1816; s. Jacques-Hubert and Catherine (Choiselat) D.; student École Polytechnique, Paris, France, 1834; licensed in physics and math.; entered Sch. Mines, 1836. Became instr. mech. physics and descriptive geometry Sch. Mines, 1844; prof. spl. math. Ste.-Barbe, 1845-51; placed in charge of mech. physics and exptl. physics U. Paris, 1849; apptd. prof. mechanics l'École polytechnique, 1854, also U. Paris; became dir. French Obs., 1870. Mem. French Inst., Bur. Longitudes. Author: Théorie complète du mouvement de la lune autour de la terre; Mémoire sur la théorie des marées, 1844; Nouvelle theorie analytique du mouvement de la lune, 1846-67; Cours élémentoir de mécanique théorique et pratique, 1851; Traité de mécanique rationelle, 2d edit., 1857; Rapport sur les progrès de l'astronomie, 1867; Les etoiles et les comètes, 1878. Founder, Bulletin mensuel de l'observatoire, also Annuare météorologique. Research on mechanics of solar system, phys. constn. of universe (especially precise measurement of equinoxes), perturbations of Uranus, inequality of lunar periods, distance from sun to earth, phys. constn. of moon, ocean tides. Died nr. Cherbourg, France, Aug. 5, 1872.

DE LAUNAY, Louis Alphonse Auguste (Launay, Louis de); French geologist; b. Paris, July 19, 1860; gen. inspector of mines; became prof. applied geology Paris École des Mines, 1889; mem. French Acad. Scis., 1912 (pres. 1931), Acad. Agr. Author: L'Argent, géologie, metallurgie, rôle economique, 1896; Formation des gites métalligeres ou Métallogenie, 1905; Géologie de la France, 1921; La vie des Montagnes, 1926. Known for work dealing with origin of metalliferous deposits; made geol. studies of Bulgaria, France, Asia, Quebec. Died Paris, June 30, 1938.

DELAUP, Sidney Phillip, Am. surgeon; b. New Orleans, June 5, 1863; s. Alfred and Josephine (Gastinel); B.Sc., Tulane, 1883, M.D., 1890; m. Gabrielle Roux, Jan. 8, 1900. Practiced at New Orleans, 1890-1923; prof. genito-urinary and rectal surgery New Orleans Polyclinic (Tulane U.). Pioneer and advocate of spinal analgesia in U. S. Died Oct. 29, 1923.

DELAVAL, Edward Hussey, English chemist; b. 1729; M.A., Cambridge, Eng.; studied chemistry and exptl. philosophy; named fellow Pembroke Hall, Cambridge, 1759; apptd. (with Benjamin Franklin and others) to report to Royal Soc. on ways of protecting St. Paul's Cathedral from lightning. Recipient Gold medal Royal Soc.; Gold medal Manchester Lit. and Philos. Soc. Fellow Royal Soc., 1759 (Copley medal, 1766); mem. Royal Soc. Göttingen, Royal Soc. Uppsala, Inst. Bologna. Author: The Cause of Changes in Opaque and Coloured Bodies, 1777; also articles. Studied effects of lightning; advocated use of blunt lightning conductors for ordinary houses; studied specific gravities of several metals and their colors when bonded with glass; manufactured artificial gems, most complete set of musical glasses then known in Eng.; developed method of abstracting fluor from glass. Died Westminster, Aug. 14, 1814.

DE LAVAL, Carl Gustaf Patrik, Swedish inventor; b. Orsa, Sweden, May 9, 1845; B.S., Uppsala (Sweden) U., Ph.D., 1872; engr. in steel works of Klosters-Bruck; del., senator Swedish Reichstag. inventor centrifugal cream separator, 1878, steam turbine, 1893; used rotation principle to produce glass bottles; inventions in iron and zinc industries. Died Stockholm, Sweden, Feb. 2, 1913.

DE LA VALLÉE-POUSSIN, Charles-Jean-Gustave-Nicolas, mathematician; b. Louvain, Belgium, Aug. 14, 1866; D.Sc., U. Louvain; prof. U. Louvain; tchr. Sorbonne, also Coll. France, during World War I; became cons. French Inst., 1916. Mem. French Acad. Scis., 1945. Author: Cours d'analyse infinitésimale, 1926-28; Intégrales de Lebesgue, Fonctions d'ensemble, Classe de Baire, 1926. Research on Riemann zeta function, also on integration of functions of Lebesgue; established theorem relative to number of prime numbers lower than 4.

DE LA VALLÉE-POUSSIN, Charles Louis Joseph Xavier, Belgian mineralogist; b. Namur, Belgium, 1827; ed. Coll. Notre Dame de la Paix, Belgium, also Paris; became prof. geology and mineralogy, Louvain, Belgium, 1863; v.p. council for preparation of geol. map of Belgium, 1903; mem. Royal Acad. Belgium. Died Brussels, Belgium, 1903.

DELAVEAU, Pierre Georges, French biochemist; b. Charenton le Pont, France, June 4, 1921; s. Georges Louis and Suzanne Gabirielle (Neuville) D.; Bachelier Ensignement secondaire, 1939; pharmacist, 1944; Licencie es Scis., 1944; Docteur en Pharmacie, 1953; Docteur en Medecine, 1954; Docteur es Scis., 1967; m. Christiane Gautier, May 18, 1949; children—Philippe, Jean Francoise, Louis. Monitor, Paris Faculty Medicine, 1944-46, asst. in pharmacology; researcher Nat. Inst. Agronomic Research, 1948-52; dir. central lab. Foch Med.-Surg. Center, 1952—; temporary lectr. Nat. Sch. Medicine and Pharmacy, 1956-58; aggregate lectr. Nat. Sch. Medicine, and Pharmacy, Rouen, 1958-60; prof. Faculty Pharmacy, Paris, 1960—, asst. dean, 1966—. Author various surveys. Research and publs. on repartition, biogenesis and metabolism of sulphuric heterosides and of flavonoids of plants, also on biochemistry of polyphenolic substances, quinons, coumarines, desiccation and stabilization of medicinal plants. Home: 13, rue Soufflot, Paris 5, France.*

DE LAVERGNE, E., French virologist; b. Nancy, France, Feb. 7, 1926; student Faculty Medicine, Nancy; m. Simone Galmiche, Aug. 19, 1955; 1 dau., Sylvie. Head research Nancy Faculty Medicine, aggregate prof.; certified biologist Nancy Hosps.; dir. Laboratorie Central de Microbiologie de Hopitaux, Nancy. Mem. French Soc. Microbiology, French Soc. Med. Biology. Author: Les Bactéries, Coll. Que saisje?; also numerous articles. Research on effect of hormones including cortison ACTH on antibodies, modifications of intestinal flora under effect of antibiotics, entertoxic power of staphylococci, bacterial antagonism, sensitivity of anaerobes to antibiotics, allergy and pasteurelloses, work on virology-enterovirus, reovirus, respiratory virus (parainfluenza virus). Home: 8 473 LaFayette, Nancy, France.*

DELAVIGNETTE, Pierre Oscar, Belgian physicist; b. Monceau-sur-Sambre, July 31, 1931; s. Rodolphe and Gabrielle (Castiaux) D.; student U. Libre de Bruxelles, Brussels, Belgium, 1950-55, Licencié en Sciences Physiques, 1955, Doctor's grade with greatest distinction, 1961; m. Anne Haillot, July 18, 1959; children—Evelyne, Yves, Marc. With Centre d'Etudes de l'Energie Nucléaire, Mol, Belgium, 1957—, in charge electron microscopy sect., 1959—. Mem. Belgian, French phys. socs., Belgian, French electron microscopist socs. Research and publs. on dislocation interactions and dislocation fine structure (dissociation of dislocations into partials) using electron microscope, stacking faults in crystals, measurement of their energy by direct methods, contrast of crystalline defects in electron microscope, radiation damage in metals and graphite, crystal growth, effect of growth stacking defects on electron diffraction pattern. Home: 22 place Jean Vander Elst, Bruxelles 18, Belgium. Office: C.E.N., 200 Boeretang, Mol, Belgium.*

DELBET, Pierre, French surgeon; b. La Ferté-Gaucher, France, Mar. 15, 1861; s. Ernst Pierre Julien Delbet; ed. Paris (France) Faculty Medicine, 1892; aggregate prof., 1893; became prof. clin. surgery Necker Hosp., 1909, Cochin Hosp., 1919-32. Mem. French Soc. for Study Cancer (pres.), Acad. Medicine, Acad. Surgery. Author: Traité de chirurgie, 1893; Asepsie opératoire, 1901; (with Mendaro) Cancer du sein, 1926; Sels halogénes de magnésium et cancers, 1928; l'Agriculture et la santé, 1946. Research on pelvic suppuration, breast tumors, varicose veins; perfected propidon (anti pyrogenic vaccine); maintained magnesium helps prevent formation of precancerous lesions in humans and campaigned against excessive refining of salt especially in a manner which lowered magnesium content. Died La Ferté-Gaucher, France, July 17, 1957.

DELBOEUF, Joseph-Remy-Leopold, Belgian philosopher; b. Liege, Belgium, 1831. Author: Prolégomènes philosophiques de la géométrie, 1860; Essai de logique scientifique, 1865; De la psychologie comme science naturelle, 1870; Logique algorithmique, 1877; Eléments de psycholophysique générale et speciale, 1883; Questions de philosophie et de la science, 1883; la Matière brute la Matière vivante, 1887; De l'origine des effets curatifs de l'hypnotisme, 1887. Noted for work in use of suggestion in hypnosis. Died Bonn, Germany, 1896.

DELBRUCK, Max, biologist; b. Berlin, Germany, Sept. 4, 1906; s. Hans and Lina (Thiersch) D.; Ph.D., U. Göttingen 1930; m. Mary Bruce, Aug. 2, 1941; children—Jonathan, Nicola. Came to U. S., 1937, naturalized, 1945. Rockefeller Found. fellow physics Copenhagen, also Zurich, 1931; asst. Kaiser Wilhelm Inst. fur Chem., 1932-37; Rockefeller Found fellow biology Cal. Inst. Tech., 1937-39; instr. physics Vanderbilt U., 1940-45, asst. prof., 1945-46, asso. prof., 1946-47; prof. biology Cal. Inst. Tech., 1947—. Mem. Nat. Acad. Scis. Research on genetic processes of bacteriophage and fruit fly, quantum theory of chem. bond, nuclear physics theory, mutations in Drosophila, bacterial viruses, sensory physiology. Home: 1510 Oakdale St. Pasadena, Cal. 91106.

DELBRÜCK, Max Emil Julius, German inventor; b. Bergen auf Rügen, Germany, June 16, 1850; s. Berthold and Laura (v. Henning) D.; studied chemistry in Berlin and Greifswald (both Germany); hon. dr.engring. Technische Hochschule Munich (Germany); m. Marie Spuhn, 1882; children—Berthold, Konrad, Joachim, Hans. Asst. to M. Maercker in Halle, Germany; founded exptl. inst. of Verein der Spiritusfabrikanten in Deutschland, (assn. spirit mfrs. in Germany), 1874; founder Institut für Gärungsgewerbe, Landwirtschaftliche Hochschule, Berlin; became prof., 1882 and rector, 1898-1900; founder Zeitschrift für Spiritus-Industrie, also Wochenschrift für Brauerei. Mem. Vorstand der Deutschen Chemischen Gesellschaft, Verein Deutscher Chemiker, Kaiser-Wilhelm Gesellschaft. Nat. Acad. Sci. fgn. asso. Author: (with M. Maercker) Handbuch der Spiritusfabrikation, 1880; (with G. Foth) Anleitung zum Brennereibetrieb, 1898; (with F. Schönfeld) System der natürlichen Hefe-Reinzucht, 1903; (with E. Struve) Beiträge zur Geschichte des Bieres under der Brauerei, 1903; (with A. Schrohe) Hefe, Gärung und Fäulnis, 1904; (with F. Hayduck) Die Gärungsführung in Brauerei, Brennerei und Presshefefabrik, 1911. Founder fermentation tech.; research on potato and hop cultures, natural pure cultivation of micro-organisms, effect of motion on fermentation and yeast, influence of enzymes and proteins on yeast. Died Berlin, May 4, 1919.

DEL BUONO, Candido, Italian physicist; b. Florence, Italy, 1618; mem. Accademia del Cimento; inventor water gauge for collecting gases and steams rising out of watercourses, apparatus for measuring pressure of gases on quicksilver and for comparison of their density as well as for measurement of compressibility of air and water. Died 1676.

DEL CANO, Juan Sebastián, Spanish navigator; b. Guetaria, Spain, circa 1460; sailed with Magellan and, after Magellan's death, brought ship home to Spain after circumnavigating globe (1st to do so), 1519-22. Died Pacific Ocean, Aug. 4, 1526.

DEL CARPINE, Giovanni del Pian, Italian missionary, explorer; b. Perugia, Italy, circa 1182; Franciscan monk; companion of St. Francis of Assisi; 1st European explorer of Mongol empire; named by Pope Innocent IV as head of diplomatic corps to Mongols, 1245-47; archbishop of Antivario, Italy. Author: Historia Mongolorum quos nos Tartaros Appellamus (or Liber Tartarorum), one of 2 most important books of its kind before Marco Polo. Died 1252.

DEL CHIARO, Adolfo, Italian mathematician; b. Pisa, Feb. 27, 1910; s. Luigi and Assunta (Carli); Ph.D. in math.; m. Liliana Bertini. Dir., head service Central Inst. Statistics; instr. U. Bari; instr. financial math., prof. statistics Faculty Econ. and Comml. Scis. of Cagliari; prof. statistics of ins. Faculty Statis., Demographic and Actuarial Scis., U. Rome. Mem. Italian Inst. Actuaries, Internat. Congress Actuaries of Brussels (permanent com.), Soc. Statistics of Paris, Soc. Statistics, Am. Eugenic Soc., Econometric Soc., Italian Soc. Econ. and Statis. Demography, Italian Assn. Sociol. Scis., Italian Math. Union. Author: Tavale di Eliminazione; Analisi delle Distribuzioni Statistiche. Research and publs. on pure math., calculus of probabilities, financial math., demography, theoretical and applied statistics. Home: via Aristide Leonori 113. Office: Institute of Actuarial Sciences, Faculty of Statistical, Demographic and Actuarial Science, University of Rome, Rome, Italy.

DELCROIX, Jean-Loup, French physicist; b. Voiron, France, July 15, 1924; s. Pierre and Simone (Recoura) D.; Agregation de physics, Ecole Normale Supéieure, 1948, Ph.D., 1953; m. Mariette Dreyfus, Nov. 9, 1957; children—Christine, Bernard, Francoise. Staff, Laboratoire de physique l'Ecole Normale Supérieure, 1948-60, under-dir. lab., 1956-60; faculty Faculté des Sciences d'Orsay, France, 1960—, prof., 1963—. Mem. French Physics Soc., Am. Phys. Soc. Author: Introduction à la theorie des gaz ionisès, 1959; (with J. F. Denisse) Ondes dans les plasmas, 1961; Physique des plasmas, 1963; also numerous articles. Research in kinetic theory of weakly and fully ionized gases; elec. properties of weakly ionized gases; propagation of waves in plasmas; interaction of two waves in a plasma. Home: 37 rue des Longs Pres 92, Boulogne, France. Office: Faculté des Sciences, 91, Orsay, France.*

DELEAU, Nicolas, French physician; b. Vézelise, France, Apr. 21, 1797; studied at Brussels (Belgium) Hosp., 1814; a son, Léon. Surgeon to 4th Cav., Brussels, 1814-18; became physician Orphans' Hosp., Paris, France, 1829; began practice medicine at Larchant, 1840. Recipient Monthyon prize, 1825. Author: Aperfu sur l'abus du vomissement provequée, 1820; Mémoire sur la perforation de la membrane du tympan, 1822; Honoré Trézel, 1825; Recherches pratiques sur les maladies de l'oreille, 1834. Research on deafness, ear drum, case of Trézel, who learned to speak although deaf-mute from birth; opposed induction of vomiting as treatment in certain diseases. Died Paris, Nov. 30, 1862.

DE L'ECLUSE, Charles, see Clusius, Carolus.

DE LEE, Joseph Bolivar, Am. obstetrician; b. Cold Springs, N.Y., Oct. 28, 1869; s. Morris and Dora (Tobias) D.; student Coll. City N.Y., 1 year; M.D., Chgo. Med. Coll. (now Northwestern U. Med. Sch.), 1891; postgrad. univs. Vienna and Berlin. 1893-94, Paris, 1894; A.M. (hon.), Northwestern, 1906. Demonstrator anatomy, Chgo. Med. Coll., 1892-93; lectr. in physiology Dental Sch., 1892-93; became demonstrator obstetrics Northwestern U. Med. Sch., 1894, lectr. obstetrics 1895; took chair of obstetrics, 1896, given title prof. obstetrics, 1897; prof. emeritus obstetrics and gynecology U. Chgo.; founder Chgo. Lying-in Hosp. and Dispensary, 1895, opened hosp. in connection with them, 1899, later cons. in obstetrics; founder Chicago Maternity Center, 1932, cons. Fellow Edinburgh Obstet. Soc. (hon.), Am. Gynecol. Soc. (v.p. 1929), A.C.S.; A.M.A., Chgo. (councillor 1902), Ill. (sec. 1899) med. socs., Chgo. Gynecol. Soc. (pres. 1908), Miss. Valley Med. Assn., Chicago Hist. Soc. Author: Obstetrics for Nurses, 1904, 12th edit., 1941, Notes on Obstetrics, 1904; Yearbook of Obstetrics, 1904-41, The Principles and Practice of Obstetrics, 1913, 7th edit., 1938. Died Apr. 2, 1942.

DE LEHOCZKY, Tibor, Hungarian physician; b. Eperjes, Hungary, May 14, 1897; s. Alexander and Rose (de Dokus) de L.; M.D., Med. U. Budapest (Hungary), 1921; m. Margit Halasy, June 23, 1951. Became asso. with Inst. Neuropathology, 1922, later with Neurol. Clinic of Budapest; chief dir. neurol. dept. Istvan Hosp., Budapest; extraordinary prof. of Univ., named hon. prof., 1963. Mem. Hungarian Neurol. Soc. (past pres.), Neuropath. Congress (v.p. Zurich, 1965), Am. Neurol. Acad., Am. Neuropath. Soc., Neurol. Soc. Paris, German Neuropath. Soc. Author: Die neuroallergischen Beziehungen in der Histopathologie der Multiplen Sklerose; (with J. Soc, Margit Halasy) Animal Experiments on the Aetiology of Myelopathy; also articles. Research on multiple sclerosis, organic neurol. diseases, Parkinsonism, tumors of brain and spinal cord, glycogenic myopathy, pain, corticosteroid therapy of central nervous system, haloperidol. Home: Nepköztarsasag-ut, Budapest VI. Office: Istvan Hosp., Neurological Dept., Budapest, Hungary.*

DELÉPINE, Marcel Stéphane, French chemist; b. Etocquigny, France, Sept. 19, 1871; s. Léopold Delépine; agrégé in pharmacology, Paris Sch. Pharmacy; Ph.D. in Sci., Paris Sch. Scis.; m. Marguerite Dorveaux, Oct. 4, 1904; children—Madeleine (Mme.

Tard), Marie. Pharmacy intern, 1892-97; pharmacist Paris hosps., 1902-27; prof. Ecole superieure de pharmacie, 1913-30, Collège de France, 1930-41. Recipient Cannizzaro prize, del Lincei Acad., 1923. Mem. French Acad. Scis. (laureate 4 times, Lavoisier medal), Nat. Acad. Medicine, French Acad. Pharmacy (hon. resident), Royal Acad. Belgium, Royal Acad. Pharmacy Madrid. Author: (with Armand Gautier) Traité de Chimie organique; Vies et Oeuvres de Le Bel; Armand Gautier et Fourneua; numerous biographies. Research on rhodium, organic sulphur compounds, terpenes, pyridines. Died 1966.

DELEPINE, Sheridan, pathologist; b. Jan. 1, 1855; s. Antoine and Henriette (Mennet) D.; ed. College Charlemagne, Paris, Acad. Lausanne, U. Geneva, Edinburgh U.; M.B., C.M. with 1st class honours, M.Sc.; B.Sc. with distinction, Lausanne, 1872; med. studies, 1877-82; m. Florence Rose; 1 dau. Lectr. physiology and pathology St. George's Hosp. Med. Sch., London; demonstrator anatomy and pathology, 1881-82; prof. pub. health and bacteriology, former prof. pathology U. Manchester, 1891-1921, dir. Pub. Health Labs., 1901-21. Fellow Inc. Soc. Med. Officers of Health (hon.). Research and publs. on anatomy, physiology and pathology. Died Nov. 13, 1921.

DE LERMA, Baldassarre, Italian biologist; b. San Giogio a Cremano, Aug. 25, 1908; s. Eduardo and Teresa Anna (Duranti) L.; Ph.D.; m. Barbaro Brunella, June 30, 1958; children—Andrea, Lorenzo. Asst. U. Naples (Italy), 1932-38, chair of zoology, later titular prof. zoology until 1952, chair gen. biology, dir. gen. giol. and genetics inst., 1960—; titular prof. U. Bari (Italy), 1952-60. Mem. Nat. Acad. Entomology. Author many works on gen. bio. biophysics, zoology, physiogenetics; collarborator Handbuch der Histochemie. Home: via Santo Strato a Posillipo 25, Naple. Office: Istituto di Biologia e Genetica, via Mezzocannone 8, Naples, Italy.

DELESSE, Achille-Ernest-Oscar-Joseph, French geologist; b. Metz, France, Feb. 3, 1817; ed. Ecole Polytechnique, Paris, France, École des mines; apptd. prof. mineralogy and geology, Besançon, France, 1845, prof. geology Sorbonne, Paris, 1850; named prof. agr. École des mines, 1864; apptd. insp.-gen. of mines, 1878; pub. (with A. Laugel, A. de Lapparent) Revue de progrès de géologie 1860-80. Mem. French Acad. Scis., 1879. Author: De l'azote et des matières organiques dans l'écorce terrestre, 1861; Recherches sur l'origine des roches, 1865; Études sur le métamorphisme des roches, 1868; Lithologie des mers, 1872. Studied metamorphism; described rocks, including melaphyre, syenite, igneous rocks of Vosges, Alps, Corsica; prepared geol. and hydrographic maps of Paris, 1858, agronomic map of Sein-et-Marne; described lithology of undersea deposits. Died Paris, Mar. 24, 1881.

DELEUZE, Joseph Philippe François, French naturalist; b. Sisteron, France, Mar. 1753; ed. Mézières sch., Paris, France, 1772; family moved to Paris, 1772; became asst. naturalist at Mus. Natural History, 1795. Author: Histoire critique du magnétisme animale, 2 vols., 1813. Pub. French translation of The Loves of Plants (Erasmus Darwin), 1799. Supported Mesmer's theory of animal magnetism and suggested its study by Acad. Medicine. Died Paris, Oct. 31, 1835.

DE LEY, Jozef, Belgian microbiologist; b. Gent, Belgium, July 1, 1924; s. Leo and Josephine (Van de Velde) De L.; D.Sc., U. Gent, 1949, Agregate, 1959; m. Briers Lili, Nov. 27, 1947; children—Annemarie, Hilde. Faculty, U. Gent, 1946—, prof. microbiology, 1959—. Guest prof. bacteriology U. Cal., 1957; vis. prof. microbiology U. Ill., 1961. Rockefeller Found. fellow, 1955-56. Mem. Royal Acad. Scis. Belgium (corr., J. B. Van Helmont award 1948, Alumni award 1959), others. Research and numerous publs. on carbohydrate metabolism of bacteria, concept of comparative biochemistry for understanding bacterial evolution, renewal of bacterial classification by biochem. and molecular-biol. methods; discoverer new carbohydrates produced by bacteria. Home: 12 Molendreef, Deurle (O.Vl.), Belgium. Office: 21 Casinoplein, Gent, Belgium.*

DELEZENNE, Camille, French physiologist; b. Genech, France, 1868; ed. Armentières Coll., Faculty Medicine, Lille, France; doctor, 1892; prof. at Pasteur Inst.; prof. Faculty Medicine, U. Montpellier (France). Mem. Acad. Medicine. Author: Le zinc constituant celluaire de l'organisme animal, 1919. Research on coagulation of blood, physiology of pancreas and duodenum, venoms, zinc constituent of cells of living organisms, defense mechanisms of injured tissues, anticoagulation function of liver. Died July 1932.

DELEZENNE, Charles E., French physicist; b. Lille, France, Oct. 4, 1776; baccalaureat in scis., 1813; named premier consul by Lacroix, 1800; at College, S. Germain, until 1803; became prof. in lycée, Paris, France, 1803; math. tchr. College Lille, 1805-36; ret. 1936; taught math. at Lille until 1848. Mem. French Acad. Scis., Author: Notions élémentaires sur les phénomènes d'induction, 1845; Expériences et observations sur les cordes des instruments à archet, 1853; Considératons sur l'acoustique musicale, 1855; Note sur le ton des orchestres et des orgues; Sur la constitution et la suspension des nuages, 1856; Table de logarithmes acoustiques, 1857; Les pigeons voyageurs, 1862; also studies on musical acoustics. Invented siphon barometer, aerometer, magnetic hoop to measure primary currents, polariscope to study sun, dry cells. Died Aug. 26, 1866.

DEL FERRO, Scipione (Ferri, Ferreo), Italian mathematician; b. Bologna, Italy, 1465; prof. pure math., Bologna, 1496-1526. Studied solution of equations of 3d and 4th degrees, constructions depending on single opening of compasses; discovered method of solving a spl. case of cubic equation. Died Bologna, 1526.

DELFINO, Frederic, Italian physician; b. Padua, Italy, 1477; practice medicine Venice, Italy; prof. anatomy Padua; author: De fluxu et refluxu aquae maris, 1559; studied tidal patterns. Died 1547.

DELGADO, José Manuel Rodriguez, neurophysiologist; b. Ronda, Spain, Aug. 8, 1915; s. Rafael Rodriguez and Amada (Orozco) D.; M.D., Madrid U., 1940, D.Sc., 1942; m. Caroline Stoddard, May 26, 1956; children—Jose Carlos, Linda Amada. Practice medicine specializing in physiology, New Haven, 1950—; prof. physiology Yale Med. Sch., 1950—. Recipient Countess of Maudes prize, 1944, Roel prize, 1945, Ramon y Cajal prize, 1952. Guggenheim fellow, 1963. Fellow N.Y. Acad. Scis., Am. Coll. Neuropychopharmacology; mem. Am. Physiol. Soc., Eastern EEG Assn. Asso. editor: Internat. Review of Neurobiology, 1964—; mem. editorial bd. Psychosomatic Medicine, 1962—. Research and publs. on autonomic and somatic functions, individual and social behavior, emotional and mental reactions may be evoked, maintained, modified or inhibited, both in animals and man by elec. stimulation of specific cerebral structures. Address: 333 Cedar St., New Haven 06510.*

DEL GARBO, Dino, Italian physician; b. Florence, Italy; b. during last part 13th Century; s. Buono del Garbo; doctorate Bologna (Italy), circa 1300; 1 son, Tommaso. Tchr., Bologna, Siena, Padua, Florence (all Italy); physician to John XXII; commd. by Robert of Naples to translate Arab writings into Latin. Author: Treatise on the Qanun of Ibn Sina, written between 1311-19; commentaries on two of Hippocrates' aphorisms; Commentary on Galen's De malitia complexionis diversae; also treatises on weights and measures, plasters and unguents. Tried to prove that germs of hereditary disease are in suborgans of heart by using astrology. Died Florence, 1327.

DEL GARBO, Tommaso, Italian physician; s. Dino del Garbo; influenced by teachings of his father, Dino, and ideas of Taddeo Alderotti. Tchr., Perugia; later succeeded his father as prof. medicine, Bologna; prof. U. Florence, 1364 (all Italy). Author: Summa medicinalis (gen. pathology giving summary of med. knowledge and controversies of time); also wrote treatise on plague of 1348, suggesting best ways to survive epidemic, advocating evacuation of stricken areas, fumigation, and cleaning of houses. Died Aug. 8, 1370.

DEL GRECO, Francesco, physician; b. Lanciano, Italy, Aug. 23, 1923; s. Gaetano and Gilda (Borga) del G.; M.D., Rome (Italy) U., 1946; m. Geraldine Flynn, June 21, 1956; 1 son, Paul. Came to U. S., 1951, naturalized, 1959. Research fellow Cleve. Clinic, 1951-54, staff, 1955-57; research fellow Postgrad. Med. Sch. and St. Thomas Hosp., U. London (Eng.), 1954-55; faculty Northwestern U., Evanston, Ill., 1960—, asso. prof. medicine, 1964—, dir. clin. research center, Med. Sch., 1961—; adj. staff Passavant Meml. Hosp., Chgo., 1960-64, staff, 1964—, head artificial kidney unit, 1960—; staff VA Research Hosp., Chgo, 1960—; Mem. Am. Physiol. Soc., Soc. Exptl. Biology and Medicine, Central Soc. Clin. Research, Am. Soc. Artificial Internal Organs, Am. Heart Assn. (mem. med. adv. bd. council on circulation 1962—), Am. Fedn. Clin. Research, A.A.A.S., A.M.A., others. Research and publs. on mechanism of urine concentration, renal pressor system in exptl. and human hypertension, metabolism of uremia, dialysis. Address: 303 E. Superior St., Chgo. 60611.*

DELHAYE, Jean Robert Emile, French astronomer; b. Lourches (North), Feb. 27, 1921; ed. U. Rennes, U. Paris; Ph.D. in sci.; m. Jeanne Guézel; children—Jean-Loic, Geneviève, Anne, François. With obs. of Leyden, Netherlands, Stockholm, Sweden, Obs. of Paris, 1944-57; prof. Faculty Sci., U. Besancon, 1957—; dir. Obs. of Besançon. Research and publs. on astronomy and stellar astronomy. Home: 43, av. de l'Observatoire. Office: 34, av. de l'Observatoire, Besançon, France.

DELHAYE, Michel, spectroscopist; b. Fresnes sur Escaut, Mar. 10, 1929; s. Henri and Rachel (Jansone) D.; Doctorat d'Etat, Faculté des Scis. de Lille (France), 1960; m. Marie-Berthe Buisset, Aug. 6, 1951; children—Colette, Anne, Florence. With Centre Nat. de la Recherche Scientifique, France, 1949-50, 53-62; asst. Faculté des Scis. de Lille, 1950-52, maitre de confs., 1962-64, prof. without chair, 1965—. Sci. adviser Soc. of Conversion of Energy, 1965. Recipient Prix Bardet, 1960. Mem. Chem. Soc. France, French Soc. Physics. Contbr. papers to jours. Research on structure of small molecules, fast recording techniques of Raman spectra, study of quickly evolving chem. reactions by photoelectric recording of Raman spectra in less than one millisecond, excitation of Raman effect by cw and pulsed lasers. Home: 25 Sq. du Portugal, Lille 59, France.*

DE L'HOPITAL, Guillaume François Antoine (or de l'Hospital, Marquis de Saint-Mesme, Count d'Autremont), French mathematician; b. Paris, 1661; studied under Johann and Jacob Bernoulli; served with army, then began work in math.; mem. French Acad. Scis., 1693. Author: L'Analyse des infiniment petits pour l'intelligence des lignes courbes, 1696; Traité analytique des sections coniques, 1707. Devised L'Hopital's rule; showed (with Maclaurin) that parabolas of higher order have entirely different shapes; among those who found that a body moves in shortest possible time along cycloid; participated in challenges issued by Leibniz, the Bernoullis and other mathematicians; gave solution to brachistochrone (important in calculus of variations), also investigated form of solid of least resistance (result stated in Newton's Principia). Died Paris, Feb. 2, 1704.

D'ELHUYAR Y DE SUVISA, Faust (Elhuyar y de Suvisa, F.), Spanish chemist, mineralogist; b. Lorgrono, Spain, 1755; ed. Paris, France, Freiberg, Germany. prof., Vergara; charge of Mexican mines, from 1788; wrote on theory of amalgamation of metals; succeeded in isolating tungsten (wolfram) discovered by Scheele. Died Madrid, 1833.

DELIGNÉ, Pierre, French physician, anesthesiologist; b. Fontaines, Vendée, France, June 24, 1926; s. Auguste and Madeleine (Bonnaud) D.; ed. Lycée Fontanes, Niart, Collège François Viète, Fontenay-le-Comete, Lycée Henri IV, Poitiers; baccalauréat lettres-mathématiques, 1943, 44; M.D., Faculté de Medecine de Paris, 1953; m. Germaine Harscöet, Aug. 25, 1954; children—Pascal, Francois, Daniel. Anesthesiologist neuro-surg. center St. Anne's Hosp., Paris, 1953-59; anesthesiology asst. Hosp. Paris, from 1959; work under Dr. M. David neuro-surg. clinic Pitié Hosp. Paris, 1960-66; aggregate prof. Paris Faculty Medicine, from 1966; certified anesthesiologist, 1966. Mem. Soc. Neuro-Surgery in French-speaking Countries; Assn. Anesthesiologists France. Contbr. numerous articles to jours, also chpts. to books. Editor jours: Anesthésie-Analgésie Réanimation, 1954-60, Agressologie, 1960-62, Annales de l'Anésthesiologie Francaise, 1960-66. Research on anesthesia and intensive care in neuro-surgery and in functional sterestaxic neuro-surgery, hypothermia and artificial hibernation, also artificial respiration in neuro-surgery, neurolept analgesia and vigil anesthesia; also studies on cerebral circulation and brain-swelling, intravenous-nutrition, pharmacological and clinical studies of new drugs. Home: 72 Avenue du Général Leclerc, Paris 15, France.*

DELILLE, Arthur, French physician; b. 1876; described Renon-Delille syndrome (with Louis Renon), condition marked by hypotension, oliguria, tachycardia, insomnia, hyperhidrosis and other manifestations. Died 1950.

DELIMARSKY, Yuri Konstantinovich, Ukrainian chemist; b. 1904; grad. Kiev U. 1928. Instr. Kiev Poly. Inst., 1928-34; asso. inst. chemistry, Ukrainian Acad. Sci., 1934-41; head lab. electrochem. fused salts inst. Gen. and Inorganic Chemistry, 1945—; prof. Kiev U., 1950-60. Academician Ukrainian Acad. Sci., 1957—; dir. inst. Gen. and Inorganic Chemistry, 1960—. Research and publs. on phys.-chem., electrochemistry, polarography fused salts. Address: Gen. and Inorganic Chemistry, Ukrainian Acad. Sci., Novo-Belichanskaya 32-34, Kiev, Ukraine, USSR.*

DELISLE, Guillaume, French cartographer, geographer; b. Paris, France, Feb. 28, 1675; s. Claude and Nicole (de la Croyère) D.; student of G. D. Cassini; apptd. 1st geographer of King Louis XV, 1718; pub. maps of Europe, Asia, and Africa, 1700; by rejecting Ptolemy's statements on longitude and comparing them with new data prepared most accurate world map of his time, 1720. French Acad. Scis., 1702. One of the founders of modern geography and modern cartography, made world globe; wrote treatises on mensuration and ancient geography; instrumental in reforming map making in France by introducing a prime meridian (meridian 20° W. of Paris); used positions of many places fixed by astron. observations at Paris obs.; reduced gross errors in longitude found in earlier maps. Died Paris, Jan. 25, 1726.

DELISLE, Jean Baptiste Rome, crystallographer; b. 1736; measured a number of angles of crystals with contact goniometer, made over 500 drawings;

showed clearly a tabular crystal of monoclinic sulfur. Died 1770.

DELISLE, Joseph Nicolas, French astronomer; b. Paris, France, Apr. 4, 1688; s. Claude Delisle; made astron. observations at obs. at Luxembourg, 1710 to circa 1724; visited Eng., circa 1724; founder, dir. obs. St. Petersburg, Russia, 1725-47; apptd. geog. astronomer to French Navy, 1747; joined obs. at Hôtel Cluny, Paris, circa 1747; became prof. math. Collège de France. Mem. Paris Acad. Scis., Royal Soc. London. Author: Mémoires pour servi à l'histoire et au progrès l'astronomie, de la géographie et de la physique, 1738; also others. Originated theory that sun's corona was caused by refraction, 1715; developed method for observing transits of Venus and Mercury across sun, 1743, 1st method for determining heliocentric co-ordinates of sun spots; invented Delisle's thermometer, 1736; proposed measuring longitude from gun powder explosion signals. Died Paris, Sept. 11, 1768.

DELIUS, Heinrich Friedrich von, see von Delius, Heinrich Friedrich.

DELL, Richard Kenneth, biologist; b. Auckland, New Zealand, July 11, 1920; s. Francis and Alice (Deery) D.; student Auckland U., 1937-40; B.A., Victoria U., Wellington, New Zealand, 1940, B.Sc., 1948, M.Sc., 1950, D.Sc., 1957; m. Miriam Patricia Matthews, Aug. 3, 1946; children—Margaret Miriam Elaine, Sharon Elizabeth, Judith Anne Mary, Robin Rosemary. Sch. tchr., 1937-41, 46; conchologist Dominion Mus., Wellington, New Zealand, 1947-61, asst. dir., 1961—, dir., 1966—. Nuffield Travelling fellow, 1959-61. Fellow Royal Soc. New Zealand (mem. council 1965—, editor, 1964—, Hamilton prize 1955, Hector medal 1965), Mus. Assn. New Zealand; mem. Zool. Soc. London, Malacological Soc. London, Art Galleries and Museums Assn. New Zealand. Research and numerous publs. on systematics, distbn. and relationships of shellfish, and crabs of New Zealand especially from deep water, shellfish of Antarctica. Home: 204 Waiwhetu Rd., Lower Hutt, New Zealand. Office: Dominion Mus., Wellington, New Zealand.*

DELLA BELLA, Davide, Italian pharmacologist; b. Busto Arsizio, Italy, Jan. 1, 1925; s. Mario and Anna (Colombo) D. B.; degree in Medicine and Surgery, Università degli Studi, Milan, Italy, 1949; Libera Docenza in Pharmacology, 1956; m. Gabellini Claudia, Oct. 14, 1953; children—Paolo, Chiara, Silvia. Asst., Pharmacological Inst., U. Milan, 1949-56; grantee Institut de Pathologie et Thérapeutique Générales, also Institut de Physiologie L. Fredericq, U. Liège (Belgium), 1952-53; dir. pharmacological research dept. Laboratorio Bioterapico SELVI, Milan, 1956-62; dir. pharmacological research dept. ZAMBON S.P.A., Bresso-Milan, 1962—. Mem. N.Y. Acad. Scis., Accademia Medica Lombarda, Società Italiana di Farmacologia. Research, numerous publs. on physiopharmacology of peripheral autonomic nervous system, pharmacological properties of sulfonium compounds, new drugs; patentee beta-receptors blocking, antiinflammatory, analgesic, antidepressant. Home: 15, via Lazzaro Papi, Milan, Italy. Office: 12, via L. del Duca, Bresso-Milan, Italy.*

DELLA CROCE, Norberto, Italian marine biologist; b. La Spezia, Italy, Mar. 24, 1926; s. Azzeglio and Rina Ines (Picchiotti) Della Croce; Dr. Natural Scis. cum laude, U. Genoa, 1951, Libero docente in zoology, 1962; m. Frieda Den Haan, Oct. 1, 1960. Research fellow, asst. Istitut Italiano di Idrobiologia, Pallanza, 1952-55; research fellow U. Wis., Madison, 1956, Zool. Sta., Naples, 1957, U. Genoa Inst. Zoology, 1958—, Columbia U. Sandy Hook Marine Lab., 1962; planktonologist Italian Program in Oceanography, Internat. Geophys. Year, 1956-57; biologist oceanographic expdns., also with U. S. Biol. Program in Internat. Indian Ocean Expedition, 1964; prof. charge oceanography U. Genoa, 1964—. Research and publs. on microdistbn. of zooplankton in interfacies between atmosphere and ocean; seasonal cycles of zooplankton and its geog. distbn.; biology and ecology of marine cladoceran Penilia avirostris; deep scattering layer in Mediterranean Sea; also translator from German and English to Italian of books on marine scis. Home: 248/6 XXV Aprile, Pieve Ligure, Genoa, Italy.*

DELLA FRANCESCA, Piero (or Pietro di Benedetto dei Franceschi), Italian painter; b. Borgo San Sepolcro, Italy, circa 1410-20; asst. to Domenico Veneziano, 1439; worked as painter in San Sepolcro, Florence, Urbino, Ferrara, Rome, Arezzo. Author: De perspectiva pingendi (treatise on perspective, especially that of Brunelleschi and Uccello), 1478; De quinque corporibus regularibus (on 5 regular solids), 1482. Advanced theory of perspective; paintings reflect his study of geometry, math., perspective. Died Borgo, Oct. 12, 1492.

DELLA PENNA, Giovanni (Johannes de Pegna or de Pinna), Italian physician; mem. faculty medicine U. Naples, 1344-87. Author: Concilium Magistri Johannes della Penna contra pestem; Tractatus de peste compositus a Magistro Johanne de Penna excellentiore aliis (both on Black Death). Refuted most of Gentile da Foligno's theories on cause, prevention and cure of pestilence; recognized nosological unity of Black Death epidemic of 1348. Died circa 1387.

DELLA TORRE, Giovanni Maria, Italian naturalist; b. Rome, Italy, 1712. Prof. philosophy and math., Venice, also other Italian cities; librarian to King of Sicily; corr. mem. Royal Soc., London, French Acad. Scis., 1750. Author: Course of Physics (in Latin and Italian); The History and Phenomena of Vesuvius Explained, 1755. Made improvements in microscope. Died Naples, Italy, Mar. 7, 1782.

DELLINGER, John Howard, Am. physicist; b. Cleve., July 3, 1886; s. John Pfohle and Catherine (Clark) D.; student Western Res. U., 1903-07; A.B., George Washington U., 1908; Ph.D., Princeton, 1913; D.Sc., George Washington U., 1932; m. Carol Van Benschoten, Oct. 11, 1909. Instr. physics, Western Reserve U., 1907-08; physicist, Nat. Bureau of Standards, 1907-48, chief of Radio Sect., 1918-46; chief Central Radio Propagation Lab., 1946-48, chief engr. Fed. Radio Commn., 1928-29; chief of radio sect. of research div., Aeronautics br., Dept. of Commerce, 1926-34; chief Interservice Radio Propagation Lab., 1942-46; chmn. Radio Tech. Com. for Aero, 1941-57 (Collier award for 1948); chmn. Radio Tech. Commn. for Marine Services, 1947-56; chmn. study group of radio propagation Internat. Radio Consultative Com., 1949-57. Chmn. gen. arrangements com. for 1957 Gen. Assembly Internat. Sci. Radio Union; del. Conf. Inter-allied Tech. Com. on Radio Communication, Paris, 1921, Internat. Tech. Cons. Com. on Radio Communications, The Hague, 1929, Copenhagen, 1931, and chmn. U. S. delegation, Lisbon, 1934, Bucharest, 1937; U. S. rep. at Internat. Electrotechnical Commn., Italy, 1927, and Scandinavia, 1930; tech. adviser Internat. Telecommunication Conf., Madrid, 1932; chmn. U. S. com. on radio frequency allocations for scientific research Nat. Acad. Sci.-NRC; mem. U. S. delegation Five-Power Telecommunications Conf., Moscow, 1946; U. S. del. Internat. Radio Consultive Com., Stockholm, 1948, Geneva, Switzerland, 1951, London, 1953, Los Angeles, 1959, Internat. Scientific Radio Union, Zurich, 1950, Sydney, Australia, 1952, The Hague, 1954, Boulder, Colo., 1957, London, Eng., 1960. Recipient Pioneer award I.R.E., 1960. Fellow I.R.E. (v.p. 1924; pres. 1925); chmn. U. S. Govt. Interdepartment Radio Adv. Com., 1941-43; mem. Inst. of Navigation (v.p. 1949-52), Am. Geophys. Union, Internat. Sci. Radio Union (hon. pres. 1952—), Associazone Italiana di Acrotecnica (hon.), Phi Beta Kappa, Eta Kappa Nu. Radio editor: Webster's Dictionary. Author: (with others) The Principles Underlying Radio Communication (govt. publ.), 1918; Radio Instruments and Measurements (govt. publ.), 1918; Radio Handbook (with L. E. Whittemore), 1922; also many articles and treatises on radio and elec. topics. Research in math., elec. properties of copper and elec. insulating materials, electric units, devel. and applications of radio, particularly in aviation. Address: 3900 Connecticut Av., Washington 8.

DELLINGHAUSEN, Baron Nicolas von, see von Dellinghausen.

DELLON, Gabriel, French physician; b. 1649; with East Indian Co., made voyage to south seas, imprisoned by Portuguese Inquisition, 1669. Author: Traité des maladies particulières aux pays orientales, 1685. Physician to Prince of Conti; author works on tropical diseases of south seas.

DE L'OBEL (or LOBIUS), Matthias, naturalist; b. Lille, France, 1538; studied at Montpellier, France, Rondelet; went to Eng. after death of his patient, Prince of Orange, and remained there most of his life; botanist, physician to James I. Plant genus, Lobelia, named in his honor. Author: Stirpium adversaria nova; Plantarum seu stirpium historia, 1576; Plantarum stirpium icones, 1581. Described flora and ecology of area around Montpellier and the Cévennes. Died Highgate, Eng., Mar. 2, 1616.

DELONE, Boris Nikolaevich, Russian mathematician; b. Mar. 15, 1890; grad. Kiev U., 1913; D.Phys.-Math. Scis., 1934. Prof., Leningrad U., 1922-35, Kiev U., 1926, Moscow U., 1935—; with Inst. Math., USSR Acad. Scis., 1932—. Mem. USSR Acad. Scis. (corr.). Author numerous works, including: (with A. D. Alexander) Mathematical Foundation for the Structural Analysis of Crystals, 1934; (with D. K. Faddeev) Theory on Irrationality of the Third Power, 1949; St. Petersburg School for the Theory of Numbers, 1947. Research and publs. on number theory, solution by integers of 3d degree equations with 2 unknowns, geometry of numbers, math. crystallography, geometrizing Galois' theory. Home: Ryatnitskaya 12. Office: Steklov Mathematics Institute, USSR Acad. Scis., 1-y Akademicheskii Proyezd 28, Moscow, USSR.

DELONG, Dwight Moore, Am. entomologist; b. Corning, O., Apr. 6, 1892; s. George Washington and Addie (Moore) DeL.; B.S., Ohio Wesleyan U., 1914, D.Sc., 1941; M.S., Ohio State U., 1916, Ph.D., 1922; m. Fanny Merchant, Dec. 22, 1917; children—Joan Elizabeth (Mrs. Robert L. Snouffer), Eleanor Jane (Mrs. David A. Wiedie), George Wesley. Grad. asst., asst. zoology, entomology Ohio State U., 1914-17, instr., 1918, asst. prof., 1921-23, prof., 1923-62, prof. emeritus, 1962—, dir. Franz Theodore Stone Lab., 1936-37; entomologist Pa. Dept. Agriculture, 1918-21; sci. expdns. to Mexico, 1939, 1941, 1945, 1954; entomologist Nat. Sci. Found., Europe, 1960, Alaska, 1964, Panama, 1967. Recipient Distinguished Tchg. award Ohio State U., 1962. Mem. A.A.A.S., Ohio Acad. Sci. (v.p., 1930, 1932, pres. 1959), Entomol. Soc. Am. (chmn. North Central br. 1960, Founder's Meml. award, 1964.), Washington Entomol. Soc., Soc. Systematic Zoology, Sigma Xi. Author: (with D. J. Borror) An Introduction to the Study of Insects, 1958. Contbr. numerous articles in field to sci. jours. Formulation, publs. recommended controls for several field and household economic insect pests; described several genera and several hundred species of leafhoppers new to sci. Home: 1967 Collingswood Rd., Columbus, Ohio 43221. Office: 1735 Neil Av., Columbus, Ohio 43210.*

DELORME, Edmond, French physician; b. Lunéville, France, Aug. 2, 1847; s. Georges and Rosine (Flouparl) D.; ed. Strasbourg (France) Sch. Mil. Medicine; docteur en médecine, 1871; m. Leonie Antoni; 2 daus. including Marguerite. Army physician at Givet Hosp. during Franco-Prussian War of 1870; made voyage to Algeria, 1872; became prof. clin. surgery and ophthalmoscopy Val de Grâce Hosp., Paris, France, 1877-87; dir. Versailles (France) Hosp.; sent to organize Casablanca Hosp. in Morocco, 1907; became gen. med. insp. of army, 1908; insp. san. condition in Army, 1914-16. Anatomo-path. research on traumas caused by bullet wounds; developed new method for ligature of arteries of hand and foot; discovered new way of opening chest (Delorme's opening). Died Paris, 1929.

DELORME (or DE LORME), Jean, French physician; b. Moulins, France, 1547; M.D., Montpellier, France, 1577; 1 son, Charles. First physician to Duke Charles of Lorraine; physician to Marie of Medicis; apptd. 1st physician to Louis XIII, 1622; corresponded with all scholars of his time including Balzac, Patin, Scaliger. Died Moulins, Jan. 14, 1637.

DELORY, George Edward, biochemist; b. London, Eng., June 10, 1909; s. Peter William and Susan (Jackson) D.; B.Sc., Poly. Inst. London, 1932, Ph.D., 1948; m. Florence J. Marklew, June 9, 1935; children—Colin, Beryl (Mrs. G. Frederick), Susan (Mrs. T. Drew), Margaret. Research asst. Courtauld Inst. Biochemistry, London, 1927-35; lectr. Postgrad. Med. Sch., London, 1935-41; biochemist Royal Infirmary Preston, Lancashire, Eng., 1941-48; asso. prof. biochemistry U. Man. Med. Sch., Winnipeg, Can., 1948-64. Mem. Biochem. Soc., Am. Assn. Biol. Chemists; Author: Photoelectric Biochemical Analysis, 1948; (with L. F. D. White) Practical Course in Biochemistry, 1952; Photoelectric Colorimetry in Clinical Biochemistry, 1966; also numerous articles. Developer convenient methods for chem. analysis biol. fluids, method for assessing fat absorption in infants; research on conditions in which enzymes appear in blood in abnormal amount. Home: 1228 Hector Bay, Winnipeg, Man., Can.*

DELOST, Paul, French physiologist; b. Paris, France, June 21, 1922; s. Gaston and Elisa (Alquier) D.; M.D., U. Toulouse (France), 1948; Sc.D., U. Paris, 1954; m. Helen Vincent, Aug. 11, 1945; Asst., Sorbonne, Paris, 1956-58; master research C.N.R.S., Paris, 1958-59; lectr. Faculty Scis., Clermont-Ferrand, France, 1959—, prof. physiology, 1961—, dir. dept. physiology, 1959—. Recipient Acad. decoration, 1964, Acad. Scis. prize, 1955. Mem. Soc. Biology, European Soc. for Comparative Endocrinology, Soc. Endocrinology. Research and numerous publs. on endocrine and reprodn. in mammals, adrenal and thyroid gland, endocrine seasonal cycles, congenital malformations induced by hormones and antibiotics and vitamin deficiencies effects on reproduction. Home: 44 Av. Libération Ceyrat 63, France. Office: I Av. Vercingétorix Clermont-Fd 63, France.*

DE LOUVILLE, J. E. d'Allonville, French astronomer; b. Louville, France, July 14, 1671; studied astronomy, Orleans, France; col. of dragoons of Rhine; mem. French Acad. Scis. First to use Picard's micrometer for measuring quadrant of circle; studied earth's curvature and obliquity of eliptic and set its rate of diminuation at 60 seconds a century. Died Carré, France, Sept. 10, 1732.

DE LOZ, Albert Auguste, Belgian physiotherapist; b. Apr. 1, 1915; s. Felix De Loz; M.D., Free U. Brussels; m. Rosa Biard; 1 son, Réginald. Former head service of physiotherapy med.-surg. dispensary Nat. Assn. War Disabled; prof. Higher Inst. Kinestherapy, Brussels. Mem. Belgian Group Med. Specialists, Belgian Med. Fedn., Com. Sic. Investigation of Paranormal. Author: Chroniques Médicales, 4 vols.; Traité de radioélectro-physiothérapie; Physicodiagnostic et physiothérapie; Electrothérapie; Histoire de d'Electricité Médicale. Home: 1, av. de l'Horizon, Ohain. Office: 194a, av. de Tervueren, Brussels 15, Belgium.

DELPECH, Jacques Mathieu, French orthopedic surgeon; b. Toulouse, France, Oct. 12, 1777; s. Claude Mathieu Delpech; began internship at la Graue Hosp., 1789; studied at Toulouse Sch. Surgery, beginning in 1791; began studies for doctorate at St. Jacques Hosp., 1801. Practiced medicine at Toulouse, later prof. anatomy and surgery; chair clin. surgery U. Montpellier (France), 1812; founded orthopedic clinic in Paris. Mem. French Acad. Scis., 1814, Acad. Medicine. Proved tubercular origin of Patt's disease, 1816; research on animal grafting, autoplasty, rhinoplasty, scar tissue used to heal deformities such as trichiasis. Died Montpellier, Oct. 29, 1832.

DELPORTE, Augustin, astronomer; b. Tournai, Belgium, 1844. Author: Astronomie et Cartographie pratiques à l'usage des explorateurs de l'Afrique, 1889; Exploration du Congo, 1890. Early sci. pioneer of Congo; studied astronomy and topography. Died Manyanoa, Congo, 1891.

DEL POZO, Efrén Carlos, Mexican physiologist; b. San Luis Potosi, Mexico, Sept. 11, 1907; s. Francisco C. and Romana (Rangel) del P.; B.Sc., U. San Luis Potosi, 1929; M.D., U. Nacional de México, 1936. Curator, Natural History Lab., U. San Luis Potosi, 1929-30; asst. prof. botany San Luis Coll., 1928-30; prof. physiology Nat. Sch. Biol. Scis., 1936-46, dean, 1943-44; head physiology lab. Inst. Estud. Méd. Biol., Mexico City, Mexico, 1943—; prof. physiology Nat. U. Sch. Medicine, 1943-60, sec. gen., 1953-61. Sec. gen. Union Latinoam. Univs., 1960—. Research fellow Harvard Med. Sch., 1940-43, Nat. Inst. for Med. Research, London, Eng., 1947; Guggenheim fellow., 1941-43. Mem. Am. Physiol. Soc., A.A.A.S., Academia Nacional de Medicina (past pres.), Soc. Mex. de Ciencias Fisiologias (past pres.). Editor: Libellus de Medicinalibus Indorum Herbis. Méx., 1964. Research and numerous pubs. on physiology of muscular fatigue, activity of cerebral cortex, physiol. actions of scorpion venom, pharmacology of indigenous plants. Home: 620 Colegio, México 20, D.F., Mexico.*

DEL REGATO, Juan Angel, physician, therapeutic radiologist; b. Camagüey, Cuba, Mar. 1, 1909; s. Juan and Damiana (Manzano) del R.; came to U. S., 1938, naturalized, 1941; student U. Havana, Cuba, 1926-30; student U. Paris, France, 1930-34, diploma D.Medicine, 1937; postgrad. (research fellow) Curie Found., Paris; m. Inez Gertrude Johnson, May 1, 1939; children—Ann (Mrs. Don T. Jaeger), Juanita I., John Carl. Asst. roentgentherapist Radium Inst. Paris, 1936-37, Chgo. Tumor Inst., 1938; radiotherapist Warwick Cancer Clinic, Washington, 1939-40, Tumor Clinic Inst., Balt., 1941-42, Ellis Fischel Cancer Hosp., Columbia, Mo., 1943-49; dir. Penrose Cancer Hosp., Colorado Springs, 1949—; prof. clin. radiology U. Colo., Colorado Springs, 1965—. Cons. to govt. agys., hosps. Recipient Laureat, Faculty Medicine, U. Paris, 1937, Nat. Acad. Medicine France, 1948, Diploma of Honor, Liga Contra el Cancer Cuba, 1950. Diplomate Am. Bd. Radiology, Fellow A.M.A., Am. Coll. Radiology; mem. Radiol. Soc. N.Am. (past v.p., Gold medal 1966), Am. Roentgen Ray Soc., Am. Radium Soc. (treas.), Am. Soc. Therapeutic Radiologists (sec.), Assn. Am. Med. Colls., Colegio Interam. de Radiologia, Internat. Club Radiotherapists, numerous others. Author: (with L. V. Ackermann) Cancer, 1947; also numerous articles. Research on radiotherapy cancer; designer lighting device used for localization and outlining fields in clin. radiotherapy. Home: 2 E. Columbia St. Office: 2215 N. Cascade Av., Colorado Springs, Colo. 80907.*

DEL RIO, Andrés Manuel, mineralogist; b. Madrid, Spain, Nov. 10, 1764; ed. in theology San Isidro; studied mining, Freiberg, Germany, chemistry in Paris, France. Fled to Eng. because of his assn. with Lavoisier; became prof. mineralogy at Sch. Mines, Mexico, 1793. Discovered vanadium which he named erthronium (Sefström now usually given credit for discovery of vanadium). Died Mexico City, Mexico, Mar. 23, 1849.

DELUC, Jean André, geologist, meteorologist; b. Geneva, Switzerland, Feb. 8, 1727; s. François Deluc; ed. by father. Became mem. of Council of Two Hundred, 1770; settled in Eng., 1773; reader to Queen Charlotte Sophia; named hon. prof. geology U. Göttingen (Germany), 1798. Fellow Royal Soc., 1773; mem. French Acad. Scis., 1803, Royal Soc. Dublin. Author numerous books including: Recherches sur le modifications de l'atmosphere, 1772; Lettres physiques et morales sur l'histoire physique de la terre, 1798; Lettres à Blumenbach sur l'histoire physique de la terre, 1798; Bacon, 1802; Traité élémentaire de géologie, 1809; Letters on the Origin and Formation of the Earth; Geological Travels, 1803; Introduction à la physique terrestre, 1803; also numerous articles. Research on latent heat, effects of heat and pressure on mercury barometer; discovered water had maximum density at 4° C.; developed theory on quantity of water vapor in given space and its independence of air density; built hygrometer; invented dry pile; 1st accurate rules of measuring mountain height with barometer; tried to reconcile sci. with Bible by interpreting biblical days as epochs. Died Windsor, Berks, Eng., Nov. 7, 1817.

DE LUCA, Ferdinand, mathematician; b. Serracapriola, Italy, 1785; named prof. math. mil. sch., 1810, (lost position at return of old regime), renamed to it, 1848; elected dep., sec. Parliament, Naples; became mem. Parliament, 1848. Author: Traité d'analyse des coordonnées, 1812; Analyse géometrique des anciens; Trigonometrie analytique; Nouveau système d'études géometriques; Instituzioni elementari di agrimensura. Studied earthquakes and volcanoes. Died 1869.

DELUCA, Hector F., Am. biochemist; b. Pueblo, Colo., Apr. 5, 1930; s. Louie S. and Mary (Longo) DeL.; B.A., U. Colo., 1951; M.S., U. Wis., 1953, Ph.D., 1955; m. Emily Bishop Swan, July 31, 1954; children—Camille Renee, James Louis, Deborah Barcliffe, Thomas Henry. Research asst. to H. Stecmbock, 1951-55; postdoctorate fellow U. Wis., Madison, 1955-57, faculty, 1957—, Harry Stecmbock research prof., 1965—. Vis. scientist Strangeways Research Lab., Cambridge, Eng. Mem. Am. Chem. Soc., Am. Soc. Biol. Chemists, Am. Inst. Nutrition (Mead Johnson award 1968), Biochem. Soc., Soc. Exptl. Biology and Medicine, Soc. Gen. Physiology, Phi Beta Kappa, Phi Lambda Upsilon. Research, numerous publs. on understanding of how Vitamin D and parathyroid hormone control calcium and phosphorus levels of blood, how Vitamin D is metabolized in body, how Vitamin D controls membrane permeability to calcium. Home: 5130 Minocqua Ct., Madison, Wis. 53705.*

DE LUZZI (or DE LUICCI), Mondino, Italian anatomist; b. Bologna, circa 1275; student U. Bologna; Doctor's Degree, 1300. Tchr., also practice medicine, Bologna, 1300-26. Author: Anatomia mundini (1st book entirely on anatomy), 1316. Called restorer of anatomy; first since Erasistratos and Herophilos to perform human dissections at pub. lectures. Died 1326.

DEL VECCIO, Ettore, Italian mathematician; b. Ancône, Apr. 7, 1891; s. Michele and Artemisia (Duranti) Del V.; Ph.D. in math. Full prof. financial math. U. Genes. Mem. Italian Inst. Actuaries. Author monographs on math. analysis, financial math., actuarial scis. Address: via Caboto 44, Turin, Italy.

DELWICHE, Edmond Joseph, agronomist; b. Orbais, Belgium, Mar. 25, 1874; s. Désiré Joseph and Marie Joseph (Dethy) D.; brought to U. S., 1879; student Dixon (Ill.) Coll., Interstate Sch. of Correspondence (Northwestern U.); B.S.A., U. Wis., 1906, M.S., 1909; m. Alice Josephine Collin, 1899; children—Mary A. (Mrs. W. E. Hansen), Anthony J., Edmond D., Joseph J., Francis R., Richard O., Eugene A., Constant C. Tchr. country schs. until 1903; with U. Wis. since 1904, successively as field asst. Expt. Sta., supt. branch stations, asst. prof. agronomy, 1910-12, asso. prof., 1913-19, prof. agronomy, 1920-45 (emeritus since 1945), also supt. br. expt. stas. Fellow A.A.A.S.; mem. Am. Soc. Agronomy, Am. Genetics Assn., Wis., Acad. Art Letters and Scis. Originator 2 varieties of corn, also varieties of wheat, disease resistant peas, oats, soybeans. Author numerous expt. sta. publs. Died Jan. 19, 1950.

DELWICHE, Eugene Albert, Am. microbiologist; b. Green Bay, Wis., Nov. 26, 1917; s. Edmond Joseph and Alice (Collin) D.; B.S., U. Wis., 1941; Ph.D., Cornell U., 1948; m. Constance Nott, July 9, 1949; children—Christine Jean, Michael Joseph, Anne Teresa, Stephen Richard. Faculty, Cornell U., Ithaca, N.Y., 1948—, prof. bacteriology, 1955-65, prof. microbiology, 1965—; cons. biology div. Oak Ridge Nat. Lab., 1951-59. Guggenheim fellow Karalinska Institutet, Stockholm, Sweden, 1964. Mem. Canadian Soc. Microbiologists, Am. Soc. Biol. Chemists, Am. Soc. Microbiology (past pres.), Am. Acad. Microbiology. Mem. editorial bd. Jour. of Bacteriology, 1954-58. Research and publs. on contbn. to knowledge of basic physiology of living cells. Home: 117 Pine Tree Rd., Ithaca, N.Y. 14850.*

DEMACHY, Jacques-François, French chemist; b. Paris, Aug. 30, 1728; ed. Beauvais Coll.; studied under Rouelle, Paris Bot. Garden; m. Élizabeth Gigot; 1 dau., Marie-Élisabeth-Adélaide; worked as pharmacist under Gillet; named 1st dir. Central Pharmacy of Hosps., 1756; apptd. demonstrator natural history Coll. Pharmacy, 1777, taught there 25 years; became head pharmacist St. Denis Hosp., 1793, Hôtel Dieu, 1794. Mem. Acad. Berlin. Author: Économie rustique, 1769; L'art du distillateur des eaux-fortes, 1775; Manuel du pharmacien; Dialogues de morts; Histoires et contes; Fables; also pamphlets. Research on chemistry of distillation, prodn. of vinegar and various liqueurs; analyzed mineral waters of S. Corneille and Passy; investigated cocoa butter and oil of sweet almonds. Died Paris, July 7, 1803.

DE MAGELLAN, Jean Hyacinthe (or Joao Jacinto de Magalhaes), sci. investigator; b. probably Talavera, Spain, 1723; descendant Portugese navigator Ferdinand Magellan; ed. Lisbon, Portugal; studied chemistry, mineralogy. Joined Augustinian Order; left monastic life for sci. research, 1763; went to England circa 1764; tutor; supt. constrn. astron., meteorol. instruments Court of Madrid, Spain. Fellow Royal Soc., 1774; corr. mem. Paris (France), Madrid, St. Petersburg (now Leningrad, USSR) acads. scis. Author: Collection de Différens Traités sur des Instrumens d'Astronomie, 1775-80; Description des Octants et Sextants Anglois, 1775; Description of a Glass Apparatus for Making Mineral Waters, 1777; Description et Usages des Nouveaux Baromètres pour Mesurer la Hauteur des Montagnes et la Profondeur des Mines, 1779; Essai sur la Nouvelle Théorie du Feu Élémentaire, et de la Chaleur des Corps, 1780; An Essay towards a System of Mineralogy, 1788. Research, publs. on constrn. instruments sci. observation; pub. one of most comprehensive books of period on English reflecting instruments, 1775. Died nr. London, Feb. 7, 1790.

DE MAILLET, Benoit (Benoist, Bernard), French natural philosopher; b. Saint-Mihiel, France, 1656. Consul-gen. of France in Egypt, circa 1692-1702. Author: Telliamed, printed 1735, released 1748; A Description of Egypt, 1735. Precursor of ideas of evolution, transformism, extinction of species; maintained that all land plants and animals had origin in marine forms that survived the flood, also that transformations were sudden and direct (such as birds from flying fish, lions from sea lions, man from mermaid); suggests modification of organisms by environment and hereditary transmissions of these modifications; maintained right of scientists to examine nature directly. Died 1738.

DEMAINBRAY, Stephen Charles Triboudet, Brit. astronomer; b. Eng., 1710; s. Stephen Triboudet; ed. Westminster Sch., also Leiden (Netherlands); LL.D., U. Edinburgh (Scotland); 1 son, Stephen George Francis Triboudet. Became tchr. Prestonpans, 1745; apptd. tutor to George III, then Prince of Wales, 1754; astronomer Royal Obs., Kew, Eng., 1768-82. Mem. French Acad. Scis. Discovered that electricity stimulated growth of plants. Died Feb. 20, 1782.

DE MAIRAN, Jean Baptiste Dortous, French physicist, mathematician; b. Béziers, France, 1678; ed. at Toulouse, France, Paris, France; settled in Paris, 1717; founder (with W. J. Bouillet, A. Portalon) of acad. to spread interest in sci. in So. France, 1723; mem. French Acad. Scis., 1718 (became perpetual sec. 1740), French Acad. Author: Dissertation sur les variations du baromètre, 1715; Dissertation sur la glace, 1715; Traité physique et historique de l'aurore boréale, 1733. Research on geometry, astronomy, physics, natural history. Died 1771.

DEMAL, Jean, Belgian zoologist; b. Gembloux, Belgium, July 9, 1923; s. Arthur and Delina (Robert) D.; Bachelier en Philosophie, U. Louvain (Belgium), 1946, Docteur en Scis., 1952. Prof., U. Louvain, 1960—. Mem. Belgian Soc. Zoology, Biology Soc. Strasbourg. Author: Zoologie, 1963; Anatomie, 1964; also articles. Research on in vitro cultures of insect organs, invertebrate histology and histogenesis. Home: Kardinal Mercierlaan, 74, Heverlee, Belgium. Office: Naamsestraat, 59, Louvain, Belgium.*

DE MALEBRANCHE, Nicolas, see Malebranche, Nicolas de.

DEMANGEON, Albert, French geographer; b. Gaillon (Eure), France, June 13, 1872; ed. École Normale Supérieure, agrégé in history and geography, 1895. Prof. at U. of Lille and at Sorbonne from 1911. Author: La Picardie et les régions voisines, 1905; Les sources de la géographie de la France aux Archives Nationales, 1905; Le declin de l'Europe, 1920; l'Empire Britannique, 1923; Géographie universelle, 1927; Paris, la ville et sa banlieu, 1933; Problèmes de géographie humaine, 1942; others. Editor of Annales de Geographie. Known especially for his contribution to field of regional and human geography; insisted on influence of problems of circulation; proposed classification of rural dwelling in France, studied geography of rural occupancy using natural, social and demographic conditions; created method for study of human geography based on solid historical research. Died Paris, June 25, 1940.

DEMARCAY, Eugène, French chemist; b. Paris, France, Jan. 1, 1852; studied under Cahours at École polytechnique; also student of Wurtz, Dumas and Deville. Asst. to Cahours; research in his own pvt. lab.; prof. École polytechnique. Author: Spectres électriques (on separation of rare earths), 1895. Research on essences and ethers of unsaturated acids, 1878, volatility of metals at low temperatures and pressures; discovered tetroxide acid, europium, 1896; prepared pure salts of neodymium, praseodymium; built machine for compressing gases which produced low temperatures on expansion; developed method for fractional crystallization of rare earths in aqueous solution; gave spectroscopic proof of radium. Died Paris, 1903.

DE MARCO, Carlo, Italian biochemist; b. Castrovillari, Italy, Apr. 11, 1929; s. Mario and Carmelina (Gargano) de M.; Laurea in Medicine, U. Rome, 1951, Libera Docenza in Biochemistry, 1958, Applied Biochemistry, 1960; m. Maria Teresa Balmas, Aug. 18, 1951; children—Maria Laura, Carmelina, Gabriella, Mario. Asst., Inst. Biochemistry, U. Rome, 1953-67, asso. prof. biochemistry Faculty Pharmacy, 1959-62; asso. prof. Faculty Scis., U. Modena (Italy), 1963; asso. prof. applied biochemistry Faculty Pharmacy, U. Camerino (Italy), 1964; asso. prof. biochemistry Faculty Medicine, U. Cagliari (Italy), 1965-67, prof., 1967—. Decorated Ufficiale al Merito della Repubblica Italiana. Mem. Società Italiana di Biochimica, Fedn. European Biochem. Socs. Research, numerous publs. on amino acid composition of human adult and fetal hemoglobins, some pathways of cisteine and cysteamine metabolism, synthesis of hypotaurine and thiotaurine, transulfurations of organic sulfinates, biogenesis of thiosulfates. Home: 10/B Mogadiscio, 00199, Rome, Italy; also 13 Trentino, 09100, Cagliari. Office: 4 via Porcell, 09100 Cagliari, Italy.*

DE MARGERIE, Emmanuel-Marie-Pierre-Martin Joaquin, French geologist; b. Paris, France, Nov. 11,

1862; s. Eugène and Charlotte (Demion) de M.; ed. U. Paris; hon. D.Sc. U. Lausanne, U. Toronto; m. Renée Ferrère, 1903; 1 dau. Hon. prof., Strasbourg U., 1933; premio Pio XII: Officer of Legion of Honor, Crown of Belgium; Commander of Nile; Commander of Order of Leopold II. Chairman, French Geological Soc., 1899-1919; mem., French Acad. Scis., 1939; Fellow, Royal Soc., 1931; fgn. mem., Am. Acad. of Arts and Scis., Am. Philosophical Soc.; hon. mem., Royal Geog. Soc., London, (Lyell medal, 1921; Victoria medal, 1930). (with A. Heim) Author: La Dislocation de l'écorce terrestre, 1888; les Formes du terrain, 1888; Catalogue des Bibliographies géologiques, 1896; la Face de la terre (3 vol.), 1897-1918; Oeuvre de Sven Hedin, 1928; Description tectonique du Jura français, 1936; Critique de Géologie (vol. 1), 1943; reports on maps and surveys. Made geological map of Pyrenees; studied Jura, Ardenne, geology of Alsace and the Saar; added new concepts and vocabulary to geomorphology; with Gaston de la Noë, showed existing slopes were formed by atmospheric influences; with Heim, 1st named and classified in German, French, English, various accidents that happen to landscape. Died Paris, Dec. 21, 1953.

DE MARIA, Anthony John, physicist; b. Santa Crece, Italy, Oct. 30, 1931; s. J. A. and N. (Daddona) DeM.; came to U. S., 1935; B.S., U. Conn., 1956, Ph.D., 1965; M.S., Rensselaer Poly. Inst., 1960; m. Katherine M. Waybright, Aug. 29, 1953; 1 dau., Karla Kay. Research engr. Anderson Labs., West Hartford, Conn., 1956-57, Hamilton Standard div. United Aircraft Corp., Windsor Locks, Conn., 1957-58; prin. scientist United Aircraft Research Labs., East Hartford, 1958-—. Mem. I.E.E.E., Optical Soc. Am., Am. Phys. Soc., Research Soc. Am., Sigma Pi Sigma. Research and publs. on utilization of optical-acoustical interactions for controlling output of lasers; generated ultrashort optical pulses. Home: 19 Garfield St., West Hartford, Conn. 06107. Office: Silver Lane, East Hartford, Conn.*

DE MARIGNAC, Jean Charles Galissard, Swiss chemist; b. Geneva, Switzerland, Apr. 24, 1817; s. Jacob; student under Liebig at Giessen, then Polythech. Sch., Paris, 1835, later Sch. Mines. Apptd. prof. chemistry, Geneva, 1841, Zurich, Switzerland, 1878. Fellow Royal Soc. London, 1881 (Davy medal 1886); mem. French Acad. Scis., 1866. Studied atomic weights; suggested that deviations from Prout's hypothesis might be due to elements being made up of mixtures of atoms of differing masses (now known as isotopes), 1865; discovered silicotungstic acid, 1862; worked on rare earths, from 1840, theromochemistry of solutions, 1870; discovered and isolated ytterbium, 1878, gadolinum, 1880; isolated samaria, also proved tantalum and niobum are not identical; demonstrated formula for silicon dioxide and greatly contbd. to study of complex silicates; studied isomorphism of salts of nobium, tin, tungsten, also of fluostannates, fluosilicates, fluozirconates, circa 1858-60; discovered true nature of ozone. Died Geneva, Apr. 16, 1894.

DE MARINONI, Jean-Jacques, mathematician, engr., astronomer; b. Udine, Italy, 1676; completed studies in Vienna, Austria; designed plan of Vienna, 1706; executed design of cadastral (crematory) of Milan, Italy, 1719-22; went to Vienna, 1730 and constructed obs. Author: De astronomica specula domestica et organo apparatu astronomico libri duo, 1746; De re ichnometrica, 1751. Invented planimetric balance which measured surfaces. Died Vienna, 1755.

DEMARQUAY, Jean-Nicolas, French surgeon; b. Longueval, France, Dec. 14, 1814; studied in Paris, France; Docteur en Médecine, 1847; head Blandin Clinic; anat. asst. Paris Faculty Medicine. Mem. Acad. Medicine, Soc. Surgery. Author: Traité des tumeurs de l'orbite, 1860; Traité clinique des maladies de l'utérus, 1876. Research and publs. on gas and its role in physiology and therapeutics. Died Longueval, France, June 21, 1875.

DE MARSIGLI, Comte Louis-Ferdinand, Italian naturalist, oceanographer; b. Bologna, Italy, 1658; joined service of Emperor Leopold; became gen. in army, circa 1684; demoted in 1703; traveled through Switzerland and France as a naturalist; commanded troops of Pope Clement XI; founder Inst. Arts and Scis., Bologna. Author: Breve Ristretto del saggio fisico intorno alla storia del mare, 1711. Founder oceanography; made observations on sea in Provence. Died 1730.

DE MARTEL, Thierry, French surgeon; b. Maxéville, France, 1876; sr. dir., clinic of prof. Segond. Author works on gynecology, surgery of nervous system; 1st to perform nerve surgeries; inventor trepanation instrument. Died Paris, France, 1940.

DE MARTONNE, Emmanuel, French geographer; b. Chabris, France, Apr. 1, 1873; ed. École normale supérieure; aggregation in history and geography; Docteur ès scis.; from instr. geography U. Rennes (France), 1899, to prof.; mem. faculty, univs. Lyons (France), Paris, from 1909; became dir. gen. geography lab. École Pratique de Hautes Études, 1926; named dir. Inst. Geography, 1928. Decorated officer Legion of Honor. Mem. French Acad. Scis. Author: Traité de géographie physique, 1909; Europe Centrale, 2 vols., 1931; La France, 1942. Contbg. author: Géographie universelle (Vidal). Became a dir. Annales de géographie, 1920. Made geog. studies of newly established countries in Europe, 1918. Died Sceaux, France, 1955.

DE MAYERNE, Théodore Turquet, see Turquet de Mayerne, Théodore.

DE MAYO, Paul, chemist; b. London, Eng., Aug. 8, 1924; m. Mary DeMayo; children—Ann, Philip. Asst. lectr. dept. chemistry Birkbeck Coll., London, 1954-55; lectr. dept. chemistry U. Glasgow, Scotland, 1955-57; lectr. dept. chemistry Imperial Coll., London, 1957-59; prof. dept. chemistry U. Western Ont., London, Can., 1959-—. Recipient Merck, Sharp & Dohme Lectr. award, 1966. Research, numerous publs. on organic chemistry and photochemistry. Address: U. Western Ont., London, Ont., Can.*

DEMBER, William Norton, Am. psychologist; b. Waterbury, Conn., Aug. 8, 1928; s. David and Henrietta (Siegel) D.; A.B., Yale, 1950; M.A., U. Mich., 1951, Ph.D., 1955; m. Cynthia Fox., Dec. 21, 1958; children—Joanna, Laura, Gregory. Instr., dept. psychology U. Mich., Ann Arbor, 1954-56; asst. prof. Yale, 1956-59; faculty U. Cin., 1959-—, prof. psychology, asst. dean, graduate school 1965-—. Mem. Am. Psychol. Assn., A.A.A.S. Author: Psychology of Perception, 1960; Visual Perception, 1964; also articles. Developed and tested theory motivation applying to behavior human beings and animals. Home: 8378 Mockingbird Lane, Cin. 45231.*

DE MENT, Jack Andrew, Am. chemist; b. Portland, Ore., Feb. 6, 1920; s. Andrew Thomas and Bernadine (Michaels) De Ment; student Reed Coll., 1938-41; D.Sc. (hon.), Western States Coll., 1955. Pres., De Ment Labs., Portland, 1941-—; research asst. U. Ore. Dental Sch., 1948-50; co-investigator NIH, 1953-58, research cons. in biophysics and pharmacology, 1961-—; pres. Polyphoton Corp., Portland, 1965-—. Del. White House Conf., 1957; sci. cons. Sec. War (attached to UN), Bikini Tests, 1946. Recipient Gravity Research Found. prize, 1951; named to Exec. and Profl. Hall of Fame, 1966. Diplomate Am. Bd. Bioanalysts. Fellow Am. Coll. Med. Technologists; mem. Sigma Xi, Chi Beta Phi. Author numerous books including: Fluorochemistry, 1949; Rarer Metals, 1946; Uranium and Atomic Power, 1941; Biophysical Theory of Cancer, 1954; Handbook Fluorescent Gems and Minerals, 1949; Handbook Uranium Minerals, 1948. Research and numerous articles on luminescence, dental sci., mineralogy, uranium; invented radioactive photographs, chemically pumped lasers, new radiol. weapons and fallout countermeasures, smoke-forming munitions, explosively-pumped lasers, basic laser generators. Home: 4847 S.E. Division St., Portland 97206. Office: 1717 N.E. 19th Av., Portland, Ore. 97212.*

DEMENY, M. G., inventor; b. circa 1850; worker (with Marey) in color photography; inventor phonoscope, 1892, device for taking motion pictures for house of Gaumont, 1895. Died circa 1920.

DEMEREC, Millislav, geneticist; b. Kostajnica, Yugoslavia, Jan. 11, 1895; s. Ljudevit and Ljubica (Dumbovic) D.; B.Sc., Coll. of Agr., Krizevci, Yugoslavia, 1916; student Grignon, France, 1919; PhD., Cornell U., 1923; LL.D., Hofstra Coll.; Dr. hon. causa, U. Zagreb, 1960; D.Sc., Long Island University, 1961; married Mary Alexander Ziegler, Aug. 24, 1921; children—Zlata Elizabeth, Vera Radoslava. Came to U. S., 1919, naturalized, 1931. Adjunct Krizevci Experimental Sta., Yugoslavia, 1916-19; asst. in plant breeding, Cornell U., 1921-23; resident investigator dept. genetics, Carnegie Instn. Washington, 1923-35, asst. dir., 1936-41, acting dir., 1942, dir., 1943-60; faculty medicine U. Chile, dir. Biol. Lab., 1941-60; sr. geneticist Brookhaven Nat. Lab., 1960-—; vis. prof. Rockefeller Inst. Med. Research, N.Y.C.; chmn. section zoology and anatomy NRC, 1958-61; asso. genetics Columbia University, 1943-—. Member of the permanent internat. com. Genetics Congress, 1939-53; council 6th Internat. Congress, Genetics, Ithaca, N.Y., 1932; v.p. 7th Congress, Edinburgh, 1939, mem. organizer Cold Spring Harbor Symposia on Quantitative Biology, 1941-60. Recipient Order of St. Sava by Yugoslav Govt.; Kimber Genetics award National Acad. Sciences, 1962. Fellow A.A.A.S., N.Y. Acad. Sciences, Am. Acad. Arts and Scis.; mem. Acad. Scis. Yugoslavia (hon.), Genetics Soc. Japan, Soc. Biol. of Santiago, Royal Danish Academy Scis. and Letters, American Philosophical Society, American Society Naturalists (treasurer, vice pres., pres.), Radiation Research Society (mem. council), National Acad. Science, Soc. Am. Bacteriologists, Soc. for Study of Evolution, Genetics Society of America (sec.-treas., v.p., pres.), American Genetic Assn., Bot. Soc. Am., Am. Soc. Zoölogists, Sigma Xi. Contbr. many articles sci. jours. Research on genetics of maize, delphinium, Drosophila, bacteria, viruses; unstable genes; nature of genes; spontaneous and induced mutability. Died Apr. 12, 1966.

DEMETRIOS PEPAGONENOS, Byzantine physician; flourished 1260; physician to Greek emperor Michael VIII Palaeologos, 1259-82; author book on dogs, important book on govt., treatise on feeding and nursing of hawks; believed gout to be a diathesis caused by defective elimination of excreta.

DE MEYRONNES, François, French natural philosopher; licentia docendi Paris, France, 1323; joined Order of St. Francis; preached in Avignon, France, from 1324. Author: Commentary on Aristotelian Logics and Physics (discussion on rotation of earth and plurality of worlds). Strongly influenced by ideas of Plato, St. Augustine, Dionysios Areopagita, and Duns Scotus. Died 1325.

DE MEZIRIAC, Bachet, French mathematician; b. Bourg-en-Bresse, France, Oct. 9, 1581; mem. French Acad.; author: Problèmes plaisaints et déleciables qui se font par les nombres. Re-discovered solution in integers of linear indeterminate equations; gave rule for writing down magic squares of odd order; proposed theorum that every integer is equal to sum of four or less number of squares. Died Feb. 25, 1638.

DEMIANOV, Nikolai Jakovlevich, Russian chemist; b. Russia, Mar. 15, 1861; s. Jakob; prof. Moscow (Russia) Acad.; wrote on agrochemistry; discovered a method to separate glycols. Died Mar. 19, 1938.

DEMIKHOV, Vladimir P(etrovich), Russian surgeon; b. Moscow, 1916; ed. First Medical Inst.; m., 2 children. Exptl. work Vishnevsky Surg. Inst., USSR Acad. Med. Scis., after World War II; developed technique for replacement hearts and lungs of dogs with hearts and lungs of other dogs, replacement heart of a dog with mechanical instrument; established lab. First Med. Inst., Moscow; successfully grafted head of one dog to neck of another, 1954; demonstrated transplantation of dog's heart Leipzig U., 1958; demonstrated head grafting of dogs, resulting in two-headed dogs, Inst. Reconstructive Plastic Surgery, N.Y. U.-Bellevue Med. Center, 1959; performed successful human hand graft. Recipient Academician Burdenko prize for giving dogs a second or auxiliary heart, 1951. Address: Organ Transplantation Laboratory, First Medical Inst., Moscow, USSR.*

DE MIRBEL, Charles François Brisseau, French biologist; b. Paris, France, Mar. 27, 1776; prof. botany Paris Mus. Natural History; mem. French Acad. Scis., 1807 (pres. 1829), Soc. Agr. Author: Traité d'anatomie et de physiologie végétale, 2 vols., 1802; Eléments de physiologie végétale et de botanique, 3 vols., 1815 (world famous for these books). One of founders of cytology and plant physiology; known for work on cell and in embryology. Died Champerret, France, Sept. 12, 1854.

DEMIS, D. Joseph, Am. physician; b. N.Y.C., Apr. 19, 1929; s. Joseph R. and Mary (Connolly) D.; B.S., Union Coll., Schenectady, 1950; Ph.D., U. Rochester, 1953; M.D., Yale, 1957. Chief dept. dermatology Walter Reed Army Inst. Research, Washington, 1960-63; dir. div. dermatology, asso. prof. medicine Washington U. Sch. Medicine, St. Louis, 1963-67; dermatologist-in-chief Barnes Hosp., St. Louis, 1963-67; prof. dermatology, head sub-dept. dermatology Albany (N.Y.), Med. Coll., 1967-—. Cons. Scott AFB, 1964-67, Jewish Hosp., St. Louis, 1964-67, VA Hosp., Albany, 1967-—. Mem. Am. Soc. Clin. Investigation, Soc. Investigative Dermatology, A.C.P. Research and publs. on skin disease, effects of drugs on skin disease. Home: Tamarac Farm, 3547 Indian Fields Rd., Feura Bush, N.Y. Office: 47 New Scotland Av., Albany, N.Y.*

DEMMER, Fritz, Austrian surgeon; b. Vienna, Austria, Apr. 6, 1884; s. Edouard and Bertha (Fritz) D.; ed. U. Vienna; m. Auguste Schmidsfelden; children—Heinrich, Walter, Elisabeth. Demonstrator in anatomy; 1st asst., instr., prof. surgery U. Clinic of Prof. von Hochenegg; head surgeon hosp. of Bros. of Misericorde; with François-Joseph Polyclinic, Vienna; cons. surgeon of Vienna. Mem. German Soc. Surgeons, Soc. Surgeons (Vienna), Soc. Doctors (Vienna). Contbr. numerous articles to profl. publs. Address: Grubthal 14, Wilhelmsburg N.Ö., Austria.

DEMOCEDES, Greek physician; b. Crotona, Magna Graecia, circa 500 B.C.; s. Calliphon; probably ed. by his physician father; m. dau. of Milo of Crotona, after return from Persia. Left Crotona, practiced medicine, Aegina; later Athens, then Samos; captured in Sardis by Persians, taken to Court of Darius 1, Susa, practiced medicine there and gained great favor; circa 521-485 B.C.; sent with Persian nobles to Greece on spying mission, he arranged to have them seized in Tarentum, while he escaped to Crotona; settled and practiced medicine, Crotona. Author work on medicine according to Suidas; mentioned by (with Aelian, John Tzetzes; regarded by Dion Cassius (with Hippocrates) as most celebrated physicians of antiquity; cured Darius and Queen Atossa of their ills.

DEMOCRITOS, Greek natural philosopher, b. Abdera, Thrace, Greece, circa 470-460 B.C.; s. Hegesistratus; apparently studied oriental lore from Magi as child; later studied under Leucippus and possibly Anaxagoras; spent several years traveling in Asia

Minor; lived in Egypt, and studied math. and phys. systems of ancient sch., circa 7 years. Among most successful Greek natural philosophers; covered almost every then-known field of study in his theorizing; best known for his cosmological and atomic theories; held that universe consists of void space filled with tiny, invisible, unchangeble, indestructible particles, which he named atoms; maintained that atoms are solid and that they fill entirely the space they occupy; also that they are homogeneous in substance, but differ in size and shape, which accounts for differing properties of various substances; one of first mechanists; believed behavior of atoms is governed by definite and unbreakable natural laws, and that they form aggregates upon contact with one another by kind of hook and eye mechanism; believed that atoms, void and motion are uncaused, and that the void is infinite in extent; concluded that there are an infinite number of worlds in various stages of formation and decay; had vague notion of concepts now known as conservation of matter and energy; maintained that Milky Way was vast conglomeration of tiny stars; rejected notion of deity taking part in creation or government of universe, but admitted existence of class of super-human beings; investigated structure of human body; regarded soul as material, but made distinction between soul and body; studied physiology of perception and made one of first attempts to explain color; compiled one of first true collections of ethical concepts; believed true pleasure consists in possession of good humor and of taking advantage of what one has. Died circa 370 B.C.

DEMOCRITOS (pseudo Democritos, or Bolos Democritos of Mendes), Greek natural scientist; flourished, circa 200 B.C.; lived in Alexandria. Author: Physica et Mystica; De Arte Sacra Magna. Probably earliest Greek writer on alchemy; compiled writings from tech. and natural scis., including lit. on Persian magic.

DE MOFRAS, Duflot, French naturalist, botanist; b. Toulouse, France, 1810; French attache to Madrid, Spain, 1828-40; insp. for France, U. S. Pacific coast, 1840; polit. writer France, 1842-84; noted fatal disease among Indians in Cal. (possibly cholera), 1841. Died 1884.

DE MOIVRE, Abraham, mathematician; b. Vitry, Champagne, France, May 26, 1667; son of a surgeon at Vitry; student Greek, U. Sedan, France, 1678-82; studied logic at Namur, Belgium, 1682-83, physics at College d'Harcourt, Paris, 1684; studied math. under Ozenam, 1684-85; was secluded in Priory of St. Martin's, Paris, after revocation of Edict of Nantes, 1685; fled to Eng., 1688; settled in London; obtained various teaching positions; commr. apptd. by Royal Soc. to arbitrate on claims of Newton and Leibniz to invention of infinitesimal calculus, 1712. Fellow Royal Soc. (recommended by Newton and Halley), 1697; member Berlin Academy Sciences, French Academy Sciences, 1754 (fgn. asso.). Author: Methods of Fluxions, 1695; De mensura sortis, 1711; Doctrine of Chances, 1718; Annuities upon Lives, 1725; Miscellanea analytica, 1730; contbg. author: Cheynaei Tractatum de fluxionum methodo inversa (D. Georgii), 1704. Founder of sci. of life-contingencies; invented De Moivre's theorem which created an imaginary trigonometry using recurring series; generalized Cotes's theorem on property of the circle; research on laws of probability. Died London, Nov. 27, 1754.

DE MOLIÈRES, Abbé Joseph Privat, French physicist; b. Les Baux, France, 1676. Oratorian; tchr. philosophy Coll. Angers (France); prof. philosophy Coll. France, Paris, 1723. Mem. French Acad. Scis., 1729. Author: Leçons de physique contenant les éléments de la physique determinés par les seules lois des mécaniques, 1733-39. Tried to reconcile systems of Descartes and Newton. Died Paris, 1742.

DEMOLON, Albert-Omer, French biologist; b. Lille, France, Apr. 30, 1881; ed. Agronomic Inst., 1901; docteur ès sciences, 1926. Taught chemistry and physics Chesnay Sch. Agr.; dir. Aisne Bacteriological Lab., 1909-27; named insp. gen. for Ministry Agr., 1927; teacher soil sciences Ecole supérieure de génie rural, 1945-50. Member of the French Academy of Sciences, 1946, Acad. Agr., Internat. Soc. Soil Sci. (pres.). Author: Contribution à l'étude des produits volatils dans la fermentation alcoolique, 1907; la Dynamique du sol, 1932; le Fumier artificial, 1935. Research on artificial fertilizers, effect of acid water and bacteria on soil, bacteria, radicicola, which grows on alfalfa; explained role of iron in treatment of chlorosis of grapevine. Died Paris, France, Oct. 23, 1954.

DE MONDEVILLE, Henri, French physician, surgeon; b. Normandy, France, 1260; studied medicine, Paris, Montpellier (both France), Bologna, Italy; pupil of Theodoric; apptd. surgeon to king of France, circa 1301; went to Montpellier, 1301, lectr. on anatomy, 1303. Author: Cyrurgia (1st French book of surgery, advised use of sutures, advocated methods of Theodoric), 1320, printed 1892. Died 1320.

DE MONTALEMBERT, Marc René, Marquis, French engr.; b. Angouleme, France, July 16, 1714; fought in French Army, 1732-42; founder arsenal, Ruelle; promoted to marechal de camp, 1761; returned to France after Revolution; became gen. of a div., 1792; mem. French Acad. Scis., 1747. Author: La Fortification perpendicularie, 1776. Precursor of modern mil. fortifications; abandoned Vauban's system. Died Paris, Mar. 29, 1800.

DE MONTESSUS DE BALLORE, Fernand, French seismologist; b. 1851. Author: les Tremblements de Terre, Geographie Séismologique, 1906; la Science Séismologique, 1907; la Géologie Séismologique, 1924. Compiled notable earthquake records; prepared catalog that demonstrated epicenters of maj. earthquakes lie mainly in 2 belts (1 around Pacific Ocean, other through Mediterranean region eastward through Asia); described destructive shock at Bauispe, Sonora, Mexico (May 3, 1887) as accompanied by surface faulting; summarized destructive seismic sea waves following earthquakes of Africa (1868) and Iquique (1877).

DE MONTMORT, Pierre Rémond, French mathematician; b. Paris, Oct. 27. 1678; traveled in Eng., Germany; studied philosophy and math. under Malebranche in France, 1699; made canon in ch. of Notre Dame, Paris; corresponded with Newton, Leibniz, De Moivre. Fellow Royal Soc., 1715; mem. French Acad. Scis., 1716. Author: Essai d'analyse sur les jeux du hasards, 1703; contbr. essay on infinite series to Philos. Trans. Royal Soc. Editor: On the Quadrature of Curves (Newton); The Application of Algrebra to Geometry (Guisnee). Research on theory of probability; applied calculus of finite differences to probability theory; gave 1st gen. solution of problem of points; a solver of De Moivre's problem. Died Paris, Oct. 7, 1719.

DE MOOR, Pieter, Belgian physician; b. Borgerhout, Belgium, June 12, 1921; s. Albert and Marie Therese (Van Sinja) De M; M.D., U. Leuven (Belgium), 1945, D. Hygiene, 1946; m. Marie Therese Hoet, Aug. 6, 1949; children—Clara, Anna, Stefaan, Kristin, Lucia, Cecilia. Faculty U. Leuven, 1955—, prof. endocrinology, 1959—, head sect. nutrition Sch. Pub. Health, 1963—, head lab. for exptl. medicine, 1958—. Grad. fellow Belgian-Am. Edn. Found., 1949-50, advanced fellow, 1964; Research fellow Belgian N.F.-W.O., 1947-49. Mem. Flemish Acad. Medicine (corr.), Belgian Soc. Internal Medicine, Belgian, French socs. endocrinology, Am. Diabetes Assn. Author: Masson, Paris on Alloxan Diabetes, 1953; (with A. Hendrix) Clinical Dietetics, 1966; also numerous articles. Research on prodn. exptl. diabetes with alloxan, action growth hormone, cortisol metabolism, protein binding of hormones, chromatography of steroids in plasma and urine. Home: 254 Heideberg, Kessel-Lo, Belgium. Office: Rega Inst., Leuven, Belgium.*

DE MORGAN, Augustus, mathematician, logician; b. Madura, India, June 27, 1806; s. John De Morgan and John Dodsons' dau.; entered Trinity Coll., Cambridge, Eng., 1823; m. Sophia Elizabeth Frend, 1837; children—Elizabeth Alice, William Frend, George Campbell, Edward Lindsey. Prof. math. Univ. Coll., London, Eng., 1828-31, 36-66. Mem. Royal Astron. Soc., Math. Soc. (co-founder, 1st pres. 1865). Author: Elements of Arithmetic, 1830; The Elements of Algebra, 1835; The Elements of Trigonometry and Trigonometrical Analysis, Preliminary to the Differential Calculus, 1837; Essay on Probabilities and on Their Application to Life Contingencies and Insurance Offices, 1838; Formal Logic, or the Calculus of Inference, Necessary and Probable, 1847; Trigonometry and Double Algebra (forerunner of quaternions, contained complete geometrical interpretation sq. root of minus 1 equals i), 1849; Syllabus of a Proposed System of Logic (in which he developed a new logic of relations), 1860; Budget of Paradoxes, 1872; also numerous articles. Independently discovered principle of quantification of predicate; developed new logic of relations, also new system of nomenclature for logical expression, 1850-63; eponym for De Morgan's theorem; advocate of decimal coinage; contbd. commutation columns to calculation of ins.; in mathematics worked on foundations of algebra and arithmetic, probability theory and calculus. Died London, Eng., Mar. 18, 1871.

DE MOROGUES, Sebastien, French ballistician; b. Brest, France, Apr. 5, 1705. Capt. French navy. Mem. French Acad. Scis., 1735, Marine Acad. Author: Essai sur l'application de la theorie des forces centrales aux effets de la poudre a canon, 1737. Research on gun powder, ballistics, ventilation of ships. Died Orleanais, France, Aug. 26, 1781.

DE MOSS, Ralph Dean, Am. microbiologist; b. Danville, Ill., Dec. 29, 1922; s. Guy and Ruby (Walker) DeM.; A.B., Ind. U., 1948, Ph.D., 1951; student Ciemson Coll., 1943, St. Louis U., 1943-44; m. Patricia H. Day, June 2, 1946; children—Susan L., G. Newton, Guy R., Kurt S. AEC postdoctoral fellow Brookhaven Nat. Lab., 1951-52; asst. prof. McCollum-Pratt Inst., Johns Hopkins, 1952-56; asso. prof. microbiology U. Ill., Urbana, 1956-59, prof. microbiology, 1959—. Served with AUS, 1942-46; ETO. Mem. Am. Soc. Microbiology, Am. Acad. Microbiology, Am. Soc. Biol. Chemists, A.A.A.S., Soc. Gen. Microbiology, Sigma Xi. Research and publs. in field of microbial biochemistry and physiology. Editor: Jour. of Bacteriology, 1965—. Investigation of intercellular regulatory mechanisms; carbohydrate dissimilation; alternate pathways. Home: 307 S. Garfield St., Champaign, Ill. 61820. Office: Dept. of Microbiology, U. Ill., Urbana, Ill. 61801.

DEMOSTHENES, Philalethes, Greek physician; flourished 1st century A.D.; pupil of Alexander Philalethes; belonged to sch. medicine founded by Jerophilys; oculist. Author: Diseases of the Eyes; also wrote on the pulse.

DEMOURS, Pierre, French surgeon; b. Marseilles, France, 1702; oculist of Louis XV; physician on Faculty of Paris; keeper of cabinet Jardin du Roi. Mem. French Acad. Scis., 1769. Author: Tables of the History of the French Royal Academy of Sciences, vols. V-IX. Discovered aqueous humor membrane of eye; proved cornea is not continuation of sclera (white fibrous membrane covering eyeball). Died 1795.

DEMPSEY, Edward Wheeler, Am. anatomist; b. Buxton, Ia., May 15, 1911; s. Aid S. and Julia (Wheeler) D.; A.B., Marietta Coll., 1932, D.Sc., 1954; M.S., Brown U., 1934, Ph.D., 1937; A.M., Harvard, 1946; m. Betsey Mills Beach, June 13, 1936; children—Charles Gates, Julia Wheeler (Mrs. S. B. Webb, Jr.), Richard Clinton. Faculty, Harvard Med. Sch., 1938-50; prof., head dept. anatomy, Sch. Medicine Washington U., St. Louis, 1950-66, asst. to dean, 1956-58, dean, 1958-64; spl. asst. to sec. health, med. affairs Dept. Health, Edn. and Welfare, 1964-65; prof., chmn. anatomy Coll. Phys. and Surg., Columbia, 1966—; vis. prof. L.I. Coll. Medicine, 1943, Stanford, 1945; cons. Nat. Cancer Inst., Armed Forces Inst. Pathology. Mem. President's Commn. on Heart Disease, Cancer and Stroke, 1963-64, Nat. Adv. Health Council, 1960-64, Nat. Adv. Gen. Med. Scis. Council, 1965—. Fellow A.A.A.S., N.Y. Acad. Scis.; mem. Am. Assn. Anatomists (past pres.), Am. Soc. Physiologists, Anat. Soc. Gr. Brit. and Ireland, Am. Acad. Neurology, Am. Acad. Arts and Scis., Soc. Study Internal Secretions. Contbr. numerous articles to profl. jours. Research and publs. on reproduction, neurology, endocrinology, electron microscopy, chem. cytology, med. edn. Home: 75 E. End Av., N.Y.C. 10028.*

DEMPSEY, James Raymon, Am. aerospace engr.; b. Red Bay, Ala., Oct. 4, 1921; s. Newman W. and Maude (Berry) D; student U. Ala., 1937-39; B.S., U. S. Mil. Acad., 1943; M.S. in Aero. Engring., U. Mich., 1947, D.Engring., 1964; m. Dolores Barnes, Jan. 1943; children—Susan, David Barnes, Anne. Commd. 2d lt. U. S. Air Force, 1943, advanced through grades to lt. col., 1950; chief guided missile projects research and devel. Air Staff, 1948-49; exec. dep. chief of staff-devel., 1950-51; range operations officer Atlantic Missile Range, Cape Canaveral, 1952-53; resigned, 1953; asst. to v.p. planning Convair, 1953; program dir. Atlas, 1954-57; div. mgr. Convair-Astronautics, San Diego, 1957-61; v.p. Gen. Dynamics, pres. astronautics div., San Diego, 1961-65; pres. Convair div., 1965; dep. group v.p. missiles, space, electronics group Avco Corp., Wilmington, Mass., 1966—. Fellow Am. Inst. Aeros. and Astronautics, Am. Astronaut. Soc Prin. developer Atlas (1st intercontinental ballistic missile); provided booster for Project Mercury; led devel. Centaur (1st liquid hydrogen powered rocket stage). Home: Hickory Hill Rd., Manchester, Mass. 01944. Office: 201 Lowell St., Wilmington, Mass. 01887.*

DEMPSTER, Arthur Jeffrey, physicist; b. Toronto, Ont., Can., Aug. 14, 1886; s. James and Emily (Cheney) D.; A.B., U. Toronto, 1909, A.M., 1910, Sc.D., 1937; postgrad. univs. Göttingen, 1911-12, Munich, 1912, Wurzburg, 1912-14; Ph.D., U. Chgo., 1916; m. Came to U. S., 1914, naturalized, 1918. Asst. prof. physics, U. Chgo., 1919-23, asso. prof., 1923-27; prof. from 1927. Mem. Nat. Acad. Sci., Am. Philos. Soc. Research in positive ray analysis of chem. elements, excitation of light and elec. discharges in gases; discovered uranium-235, 1935. Died Stuart, Fla., Mar. 11, 1950.

DE MURIS, Jean (or Normanus), astronomer, mathematician; b. Normandy, France, circa 1290; prof., Sorbonne. Author: De musica practica; De musica speculativa; Summa musice; Ars novae musicae, 1319. Wrote on medieval ideas of numerical perfection based on multiples of 3; used math. reasoning to defend notational improvements; espoused changes in musical style and notation introduced by Philippe de Vitry. Died Paris, circa 1351.

DEMYER, William Erl, Am. neurologist; b. South Charleston, W.Va., Aug. 7, 1924; s. Solomon and Dalvah (Hammonds) DeM.; B.S. in Anatomy and physiology Ind. U., 1952; m. Marian Kendall, Oct. 14, 1952; children—David, Lawrence. With Ind. U. Sch. Medicine, Indpls., 1957—, now prof. neurology, dir. neuroanatomy lab. dept. neurology, 1954—; cons. neurology Marion County Gen., VA hosps. (both Indpls.). Diplomate Am. Bd. Psychiatry and Neurology. Fellow Am. Acad. Neurology, World Fedn. Neurology, Am. Assn. U. Profs., Alpha Omega Alpha. Author: Neurohistology for Clinical Medicine, 1958, 2d edit. 1961. Contbr. articles to sci. jours. Home: 4000 Cooper Rd., Indpls. 46208.*

DENAEYER, Marcel E., Belgian mineralogist; b. Brussels, Belgium, Jan. 29, 1893; s. Alphonse and Eugenie (Van Loey) D.; ed. U. Brussels, U. Geneva; Ph.D. in sci.; m. Maria del Pilar Fernandez-Aguilar; children—Jose-Luis, Rolando, Maria del Sol, Graciela, Lucila. Asst., 1922; head of works, 1922-24; instr., 1926; full prof. mineralogy and petroleum chemistry U. Brussels, 1928. Mem. Royal Acad. Overseas Scis. (asso.), Nat. Center Volcanology, Internat. Assn. for Study Zones of Earth's Crust. Author: Esquisse Géologique, Notice et Bibliographie de l'Afrique Équat. Française, du Cameroun et Régions Voisines, 1928-33; Tableaux de Pétrographie, 1951;

Travaux sur les Structures Cone-in-Cone, 1938-52; Les Syénites Métasomatiques du Massif du Kirumba, 1959; Carte Volcanologique du Congo Belge et du Ruanda-Urundi et Notice (Atlas Général du Congo), 1961. Home: 4, Square de Biarritz, Brussels 5. Office: University of Brussels, 50, av. Fr. Roosevelt, Brussels 5, Belgium.

DENBOW, Carl Herbert, Am. mathematician; b. Zanesville, O., Dec. 13, 1911; s. Edward C. and Edna (Gigax) D.; S.B., U. Chgo., 1932, S.M., 1934, Ph.D., 1937; m. Stefania A. Bjornson, June 5, 1939; children—Carl Jon, Stefania Augustine, Elise Signe. Instr. to asso. prof., Ohio U., 1936-46; asso. prof. U. S. Naval Postgrad. Sch., 1946-50; prof. math. Ohio U., Athens, 1950—, chmn. math. dept., 1954-55, 66-67. Ford Faculty fellow Harvard University, 1955-56. Member of the American Mathematical Society, Assn. Am., Nat. Council Tchrs. Math., Phi Beta Kappa, Sigma Xi. Author: (with V. Goedicke) Foundations of Mathematics, 1959; also papers in profl. jours. Contbns. to calculus of variations and foundations of math.; pioneered new techniques of math. instrn. Home: 61 Columbia Av., Athens, O. 45701.*

DENCKMANN, (Heinrich Wilhelm Martin) August, German geologist; b. Salzgitter, Germany, May 6, 1860; s. Ludolf and Emilie (Meyenberg) D.; ed. U. Göttingen (Germany); m. Klara Funke, 1901; 1 son. Became asst. geologist Preussische Geologische Landesanstalt (Prussian Geol. Inst.), 1883; named local geologist, 1898; named state geologist, 1901; became prof. Bergakademie, Berlin, Germany, 1906; became mem. mining bd., 1915. Author: Stratigraphie des Oberdevons im Kellerwald, 1894; Übersichtskarte des Kellerwaldes 1:100,000, 1900; Übersicht über Tektonik und Stratigraphie des Kellerwald-Horstes, 1902; Gliederung der Siegener Schichten, 1909; Neue Beobachtungen über die tektonische Natur der Siegener Spatusinsteingänge, 1909, 12-18; Stratigraphie des tiefen Underdevons im nördlichen Siegerlände, 1911. Research on paleozoic mountains of Western Germany, Triassic and Jurassic alps especially Kellerwald; practical geology in mining industry. Died Siegburg, Germany, Mar. 7, 1925.

DENEUX, Louis-Charles, French obstetrician; b. Heilly, France, Aug. 25, 1767; docteur en médecine, Paris, France, 1804. Became master surgeon at Amiens (France) Sch. Medicine, 1790, prof. anatomy, 1795; named head master surgeon 24th div. French Army, 1791; established practice, Paris, 1810; gave courses on childbirth, 1814-16; became obstetrician to the Duchess of Berry, 1816. Mem. Acad. Medicine. Author: Observations sur la terminaison des grossesses extrautérines, 1819; Sur la sortie du cordon umbilical, 1820; Recherches sur les causes de l'accouchememnt spontané après la mort, 1823. Research on cutting of the umbilical cord, spontaneous childbirth after death, extra-uterine pregnancy. Died Nogent-le-Rotrou, France, 1846.

DEN HERTOG, Herman Johannes, Dutch chemist; b. Amsterdam, Feb. 12, 1902; s. Herman J. and Gertruida (Delfgou) den H.; Ph.D. in Chemistry and Natural Scis., Communal U. Amsterdam; m. Gertje W. Berlage; children—Herman Johannes, Henriette Geertruida. Prof., V.H.M.O., 1931-42; instr. U. Amsterdam, 1942; curator, 1941-45; reader Communal U. Amsterdam, 1945-49; prof., dir. Lab. Organic Chemistry, Higher Sch. Agr., 1949—. Mem. Royal Netherlands Assn. Chemists, Chemists Soc. (Gt. Britain), Am. Chem. Soc. Author: De Bromering van Pyridine, 1931; Discours d'Ouverture, 1950; Stadia der Organische Chemie, also numerous publs. Home: Heidepark 51. Office: De Dreyen 5, Wageningen, Netherlands.

DENICOLA, Pietro, Italian physician; b. Bologna, Italy, Aug. 12, 1919; s. Alfonso and Cristina (Prelli) DeN.; M.D., U. Bologna, 1943, m. Gianna Gnagnatti, Sept. 16, 1953; children—Caterina, Alfonso, Francesca, Paolo. Faculty, U. Pavia (Italy) Med. Sch., 1943—, asso. prof. Clinica Medica, 1956—, prof. gerontology, geriatrics, 1963—, libero docente in med. pathology, 1950—, occupational medicine, 1960—; research fellow, Milan, Italy, 1947, Zurich, Switzerland, 1950, Detroit and Boston, 1952. Mem. cons. com. for study thrombosis and vascular diseases, WHO, 1963; mem. Internat. Com. for Nomenclature of Blood Coagulation Factors, 1954—. Named prof. ad honorem Med. Faculty, U. Montevideo (Uruguay), 1964; recipient award for best thesis U. Bologna, 1943, for best lecture Argentine Assn. for Internal Medicine, 1952. Mem. Nat. Acad. Medicine Buenos Aires, Argentine Med. Assn., German Soc. for Blood Coagulation Research, Internat., European Swiss, Argentine socs. hematology. Research, numerous publs. on hemorrhagic diseases, physiopathology of blood coagulation, thromboembolic diseases and anticoagulant therapy, gerontology and geriatrics. occupational diseases, plasma proteins, lung and liver diseases. Author: (with A. Baserga) Le malattie emorragiche, 1950; The Laboratory Diagnosis of Coagulation Defects, 1954; (with P. Introzzi) La terapia dei difetti di coagulazione, 1955; Thrombelastography, 1957; Patologia professionale della coagulazione, 1961. Co-editor Italian Textbook Internal Medicine (24 vols.), 1959—. Address: 3, Lungoticino Visconti, Pavia, Italy.*

DENIKER, Joseph (Iosif Egorovich), anthropologist; b. Astrakhan, Russia, Mar. 8, 1852 (of French parents); ed. Moscow and Technol. Inst., St. Petersburg; Licence ès Scis., Faculty Scis., U. Paris, 1882; D.Sc., 1886; LL.D., Aberdeen; m.; 3 sons, 1 dau. Named librarian Natural History Mus., Paris, 1888. Mem. Soc. Anthropology of France (past pres.), Soc. Geography. Author: Anatomie et Embryologie des singes anthropides, 1886; (with Hyades) Anthropologie et Ethnographie de la Mission du Cap Horn, 1892; Les Races de l'Europe, 1896-1908; Bibliographie des Societes savantes de la France, 1889-95; The Races of Man, 1900; Les six races composant la population actuelle de l'Europe, 1904; Les Races de l'Europe, 1908; Les Races et les Peuples de la Terre, 2 edit., 1916. Co-editor: Dictionnaire de géographie universelle. Made ethnol. classification of Europeans according to stature, cranial index and color of hair (basis of common modern classification); maintained no single characteristic could serve as basis of sound classification; studied ethnography and anthropology of Asian peoples; showed structural differences between men and apes are not as great as those between apes and monkeys; characteristics used in his classification of human races still used today. Died Paris, Mar. 18, 1918.

DENIKER, Pierre G., French psychiatrist, psychopharmacologist; b. Paris, France, Feb. 16, 1917; s. Georges J. and Marguerite (Delgobe) D.; M.D., Faculté des Sciences, Faculté de Médecine, Paris, 1945; m. Nadine Vincent, Dec. 6, 1941. Head clinic Faculté de Médecine, Paris, 1949-61, aggregate prof. neurology and psychiatry, 1961—; physician in chief Hôpitaux Psychiatriques, since 1949—; physician Hôpitaux de Paris, since 1962—. Mem. Comité Fonctions et Maladies du Cerveau de la Recherche Scientifique, 1961—, Commision de Neurologie et Psychiatrie de l'INSERM, 1965—. Decorated chevalier la Légion d'Honneur; recipient Albert Lasker award, 1957. Fellow Am. Psychiat. Assn. (corr.); mem. Order Psysicians Seine (mem. council), Med.-Psychol., Therapeutic and Pharmacodynamic Soc. Author: (with Jean Delay) Les Méthodes chimiothérapiques en Psychiatrie, 1961; Psychopharmacologie, 1966. Numerous publs. on pioneer work in psychiat. chemotherapies, psychopharmacology. Home: 21 avenue Pierre 1er de Serbie-Paris 16°. Office: 1 rue Cabanis-Paris 14°, France.*

DENIS, Jean Baptiste, French physician; b. 1643; s. Claude Denis; studied medicine at Montpellier, France; established himself in Paris, 1665; physician to Louis XIV. Author: Discours sur les comètes, 1665; Lettres (on blood transfusion), 1667-68. Performed 1st recorded transfusion using blood from lamb for human being, 1667; believed transfusion could cure insanity. Died 1704.

DENIS, Pierre Maurice, physicist; b. Geneva, Switzerland, Jan. 19, 1923; s. Charles Antoine and Hélène (Miville) D.; Docteur ès Sciences, U. Genève, 1951; m. Paulette Cottier, June 15, 1965; 1 son, Luc. With U. Genève, 1954-56, head exptl. physics research, 1956; head nuclear dept. IVNIC, Caracas, Venezuela, 1956-57; dir. nuclear research lab. Institut de Physique de Genève, 1958-60; head lab. new methods for ray measurements Centre d'Etudes Nucléaires de Grenoble, France, 1954; studies in transport theory Istabanbul (Turkey) Teknik Universitesi-Nükleer Enerji Enstitüsü. Mem. French Physics Soc., Am. Phys. Soc. Author: (with others) La résonance paramagnétique, 1955, Research and publs. on theory of nuclear relaxation. Home: 4 Place Paul Mistral. Office: B.P. 269, Grenoble 38, France.*

DENISON, Charles Simeon, Am. engr.; b. Gambier, O., July 12, 1849; s. George and Janett Balloch (Ralston) D.; student Norwich U., 1 yr.; C.E., U. Vt., 1871 (Sc.D., 1907). Asst. engr. in constrn. Milw. & No. R.R., 1871-72; instr. engring. U. Mich., from 1872, prof. stereotomy, mechanism and drawing from 1885; U. S. astronomer and surveyor, locating boundary line between Wash. and Ida., 1873, 74. Died July 30, 1913.

DENISON, Robert Howland, Am. vertebrate paleontologist; b. Somerville, Mass., Nov. 9, 1911; s. William Kendall and Florence (Howland) D.; A.B., Harvard, 1933; M.A., Columbia U., 1934, Ph.D., 1938; m. Marion Swift, June 29, 1940 (div.); children—John H., David O.; m. 2d, Mary S. Maynard, Aug. 3, 1965. Asst. curator Dartmouth Coll. Mus., Hanover, N.H., 1937-47; faculty Dartmouth, 1938-47, asst. prof., 1943-47; paleontologist U. Cal. African Expdn., 1947-48; curator fossil fishes Field Mus. of Natural History, Chicago, Illinois, 1948—; lecturer on evolutionary biology University of Chicago, 1965—. Recipient A. Cressy Morrison prize N.Y. Acad. Scis., 1937. Guggenheim fellow, Europe, 1953-54. Fellow Geol. Soc. Am.; mem. Soc. Vertebrate Paleontology (past pres.), Am. Soc. Zoologists. Research and articles on fossil mammals, early fossil vertebrates with emphasis on morphology, histology, taxonomy, ecology. Home: 717 Park Av., Winnetka, Ill. 60093.*

DENIZOT, Alfred, physicist; b. Poznan, Poland, Oct. 21, 1873; s. August and Nowacka Denizot; Phil.D.Dipl., U. Berlin, 1897. Asst., Tech. Highsch., Aachen, Germany, until 1900; became pvt. lectr. Lvov (USSR) U., 1807, asso. prof., 1908, prof., from 1909. Mem. Acad. Sci., Nuovi Lincei, Rome, Berlin Math. Scoc., German Phys. Soc., French, Polish socs. physics, Soc. Polish Naturalists Kopernik, Soc. of Friends of Sci. in Poznan, Math. Circle of Palermo. Address: Kolejova 29, Poznan, Poland.

DENK, Wolfgang Karl Josef, Austrian surgeon; b. Linz, Austria, Mar. 21, 1882; s. Karl and Elizabeth (Wisgrill) D.; M.D., U. Vienna (Austria), 1907; Hon. D.Medicine, U. Graz (Austria), 1962; m. Ilse Walzel, Jan. 6, 1916; children—Liselotte (Mrs. Pius Prutscher), Gerhart. Prof. surgery U. Vienna, 1931-53, U. Graz., 1928-31; prof. surgery Rektor U., Vienna, 1948-49; chief Austrian Cancer Research Inst., Vienna, 1953-66; chief Billroths Surg. Klinik, Wien, Austria, 1931-53. Pres. Oberster Sanitartsrat, 1946-58. Decorated Ehrenzeich für Wissenschaft und Kunst. Hon. mem. Am. Surg. Assn., Am. Assn. for Thoracic Surgery, Austrian Acad. Sci., Soc. Thoracic Surgeons Gt. Britain and Ireland. Research and numerous publs. on thoracic surgery, treatment of lung cancer using surgery and chemotherapy, prevention of cancer in mice. Home: 26 Wickenburggasse A 1080, Vienna, Austria.*

DENMAN, Ira O., Am. physician, surgeon; b. Lenna, Kan., June 9, 1872; s. Francis M. and Lydia (Harding) D.; Allen County Normal Sch., Iola, Kan.; M.D., Hahnemann ed. Med. Coll., 1897; postgrad. Post-Grad. Med. Coll., N.Y. Eye and Ear Infirmary, Harvard Med. Coll., univs. Vienna and Freiburg, Morfields Hosp., London, Eng.; m. Sabra Blair, Sept. 14, 1893; children—Loraine, Ira O., Patti. Practiced gen. medicine and surgery, also specialized in eye, ear, nose and throat, Charleston, Ill., until close of 1907; moved to Toledo, 1908; practice limited to eye, ear, nose and throat; chief of staff Toledo Hosp., from 1913; v.p. Vocaphone Co.; oculist Pa. R.R., Detroit & Toledo Shore Line R.R., Nickel Plate Ry.; chmn. Bd. Health, Charleston, 1903-07. Fellow A.C.S.; mem. Am. Bd. Oto-Laryngology, Am. Coll. Phys. Therapy. Editorial staff Archives of Phys. Therapy, X-Ray and Radium. Originated technique and designed chair for tonsilectomy under nitrous oxide and oxygen gas anesthesia; co-inventor vocaphone and artificial larynx for use in talking after laryngectomy. Died Sept. 28, 1933.

DENMAN THE ELDER, Thomas, English physician; b. Bakewell, Derbyshire, Eng., June 27, 1733; s. John Denman; studied medicine at St. George's Hosp., beginning in 1753; M.D., U. Aberdeen (Scotland), 1764; m. Elizabeth Brodie; children—Thomas, Mrs. Richard Croft, Mrs. Matthew Baillie. Surgeon in navy, 1757-63; physician, obstetrician Middlesex Hosp., 1769-83. Licentiate in midwifery Coll. Physicians. Author: A Letter to Dr. Richard Huck on the Construction and Method of Using Vapour Baths, 1768; Essays on the Puerperal Fever and on Puerperal Convulsions, 1768; An Introduction to the Practice of Midwifery, 1782; Aphorisms on the Application and Use of the Forceps and Vectis on Preternatural Labours, on Labours attended with Hemorrhage and with Convulsions, 1783; also articles. Instituted practice of inducing premature labour in cases of narrow pelvis; 1st accurate description of nasal and laryngeal catarrh of congenital infantile syphilis. Died Nov. 26, 1815.

DENNETT, John, inventor; b. 1790; apptd. custodian Carisbrooke Castle; mem. Brit. Archeol. Assn. Invented Dennett rocket apparatus to rescue shipwrecked crew ashore, 1832. Died July 10, 1852.

DENNING, William Frederick, Brit. astronomer; b. nr. Radstock, Somerset, Nov. 25, 1848; s. Isaac Poyntz Denning; ed. various schs., Bristol; M.S. (hon.), Bristol U. Began study of astronomy; 1865. Recipient bronze comet medals Astron. Soc. of Pacific, 1890, 92, 94, Valz prize Acad. Scis. Paris, 1895. Fellow Royal Astron. Soc. (Gold medal 1898), Royal Astron. Soc. Can.; mem. Liverpool Astron. Soc. (pres. 1887-88). Author: Telescopic Work for Starlight Evenings; The Great Meteoric Shower of November; General Catalogue of Radiant Points of Meteoric Showers: The Planets Mercury and Venus. Discovered 5 comets, new nebulae, new star of June 8, 1918, also new star in Cygnus, Aug. 20, 1920; research and numerous publs. on surface markings, rotation periods and phenomena of Mars, Jupiter and Saturn; calculated heights and velocities of 1220 fireballs and shooting stars leading to deductions as to heat of atmosphere at great altitudes, concluding that temperature above 50 miles is higher than formerly believed. Died June 9, 1931.

DENNIS, Foster Leroy, Am. mathematician; b. West Milton, Pa., Jan. 1, 1910; s. Jacob Myron and Blanche (Huntington) D.; B.S., Ursinus Coll., 1931; M.A., Cornell U., 1932; Ph.D., U. Ill., 1938; m. Elizabeth Ann McCadskey, June 1938 (div. Dec. 1963); children—John Robert, Pamela Blanche, Lawrence Alfred. Tchr. high sch., Milton, Pa., 1933-34; instr. U. Ill., 1935-38; prof. math. Ursinus Coll., Collegeville, Pa., 1938—. Cons. Hankins Marketing Service, Phila., 1948-53, Philco Corp., 1953-62. Mem. Am. Math. Assn. Mason. Home: 95 W. 5th Av., Collegeville, Pa. 19426.*

DENNIS, Frederic Shepard, Am. surgeon; b. Newark, Apr. 17, 1850; s. Alfred Lewis and Eliza (Shepard) D.; A.B., Yale, 1872; M.D., Bellevue Hosp. Med. Coll. (N.Y. U.), 1874, Royal Coll. Surgeons, Eng., 1877; m. Fannie Rockwell, Feb. 5, 1880. Vis. surgeon, St. Vincent's Hosp., 1882-1934; prof. surgery Bellevue Hosp. Med. Coll., 1883-98; prof. clin. surgery, Cornell U. Med. Coll., 1898-1910 (emeritus); cons. surgeon Montefiore Home, N.Y., from 1888, also Bellevue and St. Vincent's hosps. Fellow Royal Coll. Surgeons, A.C.S., Am. Surg. Assn. (pres. 1894). Died Mar. 8, 1934.

DENNIS, Louis Munroe, Am. chemist; b. Chgo., May 26, 1863; s. Joseph S. and Faustina (Munroe) D.; Ph.B., U. Mich., 1885, B.S., 1886; D.Sc., Colgate, 1923, U. Mich., 1926; postgrad. U. Munich, Polytechnikum of Dresden and of Aix-la-Chapelle, pvt. lab. of Fresenius, Wiesbaden; m. Minnie Clark, Aug. 25, 1887; children—Faustine, Clark M., Frank S. Instr. chemistry Cornell U., 1887-89, asst. prof., 1891-93, asso. prof. inorganic and analytical chemistry, 1893-1900, prof. inorganic chemistry from 1900, head dept. chemistry, 1903-1933. Mem. com. on design of labs. of chemistry NRC. Fellow A.A.A.S. Author: Elementary Chemistry (with Frank W. Clarke), 1902, Laboratory Manual of Elementary Chemistry, 1902; Manual of Qualitative Analysis (with Theodore Whittlesey), 1902; Gas Analysis, 1913; Gas Analysis (with M. L. Nichols), 1929; Laboratory Manual and Problems (with A. W. Laubengayer); The Baker Laboratory of Chemistry at Cornell. Died Dec. 9, 1936.

DENNIS, Olive Wetzel, Am. engr.; b. Thurlow, Pa., Nov. 20, 1885; d. Charles Edwin and Annie (Wetzel) Dennis; B.A.; Goucher Coll., 1908; M.A. in Math., Columbia, 1909; C.E., Cornell U., 1920. Tchr. math. Washington, U. Wis., from 1909; service engr. B. & O. R.R., from 1920. Mem. Am. Cryptogram Assn., Nat. Puzzlers' League, Phi Beta Kappa. 1st service engr. in U. S., position created to find technol. means for improving service. Died 1957.

DENNIS, Wayne, Am. psychologist; b. Sitka, O., Sept. 1, 1905; s. Samuel R. and Mary (Fox) D.; B.A., Marietta Coll., 1926; M.A., Clark U., 1928, Ph.D., 1930; m. Marsena Galbreath, Mar. 17, 1928 (dec. July 1966); children—Mary (Mrs. Geoffrey Ravenhall), Anne, Gill. Faculty, Mich. State U., 1929, U. Va., 1929-42; prof., chmn. psychology La. State U., 1942-46; prof., chmn. psychology U. Pitts., 1946-51; prof. psychology Bklyn. Coll., 1951-—, chmn., 1951-61; vis. prof. Clark U., 1937-38, Am. U. of Beirut, 1955-56, 58-59; exec. officer doctoral program psychology City U. N.Y., 1962-64. Fellow Am. Psychol. Assn., Soc. Research Child Devel., Gerontol. Soc.; mem. Phi Beta Kappa, Sigma Xi, Psi Chi (past nat. pres.). Author: Hopi Child, 1940; Readings in History of Psychology, 1948; Readings in Child Psychology, 1951, rev., 1963; Group Values Through Children's Drawings, 1966. Research and numerous articles primarily in field of cross-cultural differences in child rearing, child devel. Home: 565A 3d St., Bklyn. 11215.*

DENNIS, William B., Am. mining engr.; b. Cin., Dec. 8, 1865; s. Mendenhall John and Sophia D.; A.B., Central U. Ky., 1884; m. Queen H. Littlefield, June 1900. Entered newspaper work at Dayton, O., as spl. corr. Dayton Daily Jour. and other papers; pub. Farmers' Home, 1885-90; editor and mgr. Port Townsend (Wash.) Leader (daily and weekly), 1890-92; was several yrs. pres. and gen. mgr. Eureka-Pacific Consol. Mining Co. Ida.; pres. and gen. mgr. Black Butte Quicksilver Mine Ore.; v.p., mgr. Carlton & Coast R.R. Co.; v.p., mgr. Carlton Consol. Lumber Co. Chmn. Ore. Bur Mines and Geology, 1917-25; mem. Ore. Bd. Engring. Examiners, 1919-31; rep. from Yamhill Co. in Ore. Ho. of Reps., 1919-20. Inventor Dennis Roasting Furnace. Died Jan. 7, 1937.

DENNISON, David Mathias, Am. theoretical physicist; b. Oberlin, O., Apr. 26, 1900; s. Walter and Anna (Green) D.; A.B., Swarthmore Coll., 1921, D.Sc. (hon.), 1950; Ph.D., U. of Mich., 1924; m. Helen Lenette Johnson, Aug. 28, 1924; children—Edwin Walter, David Severin. Gen. Edn. Bd. fellow, Inst. for Theoretical Physics, Copenhagen, Denmark, 1924-26; Guggenheim fellow, 1940; prof. of physics, U. Mich., 1935-66. Harrison M. Randall prof. physics, 1966-—, chairman of dept., 1955-65, Henry Russell, lectr., 1952; cons. OSRD, 1942-45. Recipient Distinguished Faculty Achievement award, 1963. Fellow Am. Phys. Soc., Am. Optical Soc.; mem. Nat. Acad. Scis., Phi Beta Kappa, Sigma Xi, Phi Kappa Phi. Solved the problem of the specific heat of hydrogen and proved the existence of the spin of the proton, 1927; research in the field of molecular structure; the theory of infrared spectra; the theory of orbits in the large accelerators; the optical properties of thin films. Home: 2511 Hawthorn Rd., Ann Arbor, Mich.*

DENNISON, Henry Strugis, Am. management scientist; b. Boston, Mar. 4, 1877; s. Henry B. and Emma J. (Stanley) D.; A.B., Harvard, 1899; Sc.D., U. Pa., 1927; D.B.A., U. Mich., 1929; m. Mary Tyler Thurber, Feb. 12, 1901 (died Mar. 31, 1936); m. 2d, Gertrude B. Petri, Oct. 11, 1944. Pres. Dennison Mfg. Co., Framingham, Mass., 1917-—; dep. chmn. Fed. Res. and dir. Federal Reserve Bank of Boston, 1937-45. Asst. dir. Central Bureau of Planning and Statistics, Washington, World War I; ex-dir. service relations U. S. Post Office Dept.; mem. Pres. Wilson's Indsl. Conf., 1919, Pres. Harding's Unemployment Conf., 1921; mem. Bus. Adv. and Planning Council of U. S. Dept. of Commerce, 1933; apptd. mem. Nat. Labor Bd., 1934; chmn. Indsl. Adv. Bd. under NRA, 1934; mem. Nat. Resources Planning Board, 1935-43, Nat. Manpower Council, 1951-52. Mem. Am. Acad. Arts and Sciences. Trustee, The Twentieth Century Fund, N.Y.C. Author: Organization Engineering, 1931; (with others) Profit Sharing and Stock Ownership for Employees, 1926; Toward Full Employment, 1938; Modern Competition and Business Policy, 1938. Pioneered employee profit-sharing plan; unemployment insurance; executive development program; formed management research association. Died Framingham, Mass., Mar. 29, 1952.

DENNISTON, Rollin Henry, II, Am. psychologist; b. Chgo., Dec. 16, 1914; s. Rollin H. and Helen (Dobson) D.; B.A., U. Wis., 1936, M.A., 1937; Ph.D., U. Chgo., 1941; m. Katherine D. Beverstock, July 26, 1941; children—Rollin Henry III, Thomas A., Nancy L. Instr., U. Ariz., 1941-43; faculty U. Wyo., Laramie, 1943-46, 48, 53-60, prof., 1953-60, dir. research devel., 1965-—; Ford Found. fellow psychology Yale, 1952-53; spl. sr. research fellow Nat. Inst. Mental Health, NIH, Bethesda, Md., 1960-61. Fellow American Association for Advancement Science, Am. Soc. Exptl. Biology; mem. Animal Behavior Soc. (founding), Am. Physiol. Soc., Am. Psychol. Assn., Am. Soc. Mammalogists, Am. Soc. Zoologists, Am. Ecol. Soc., Colo.-Wyo. (past sect. chmn.), N.Y. acads. scis., Nat. Commn. U. Research Adminstrs., Nat. Council Adminstrs. Research, Sigma Xi, Alpha Epsilon Delta. Author: Comparative Psychology, 1964; Sex Inversion, 1965; contbr. to Physiologie de L'Hippcampe, 1961; also articles. Synthesized information from endocrinology, neurophysiology and behavior in teaching, research, and research direction.

DENNY, Floyd Wolfe, Jr., American pediatrician; born Hartsville, South Carolina, Oct. 22, 1923; the son of Floyd Wolfe and Marion (Porter) D.; B.S., Wofford Coll., 1944; M.D., Vanderbilt U., 1946; m. Barbara Horsefield, Apr. 27, 1946; children—Rebecca, Mark, Timothy. Asst. prof. pediatrics U. Minn. Sch. Medicine, 1952-53; asst. prof. pediatrics Vanderbilt U. Sch. Medicine, 1953-55; with Western Res. U. Sch. Medicine, 1955-60, asso. prof. preventive medicine, asst. prof. pediatrics, 1960; prof., chmn. dept. pediatrics U. N.C. Sch. Medicine, 1960-—; cons. Surgeon Gen. U. S. Army; cons. Surgeon Gen. USPHS; mem. editorial bd. Jour. Pediatrics; mem. commn. on streptococcal diseases Armed Forces Epidemiol. Bd., 1954-—, dep. dir., 1959-63, mem. commn. on acute respiratory diseases, 1955-—, dep. dir., 1962-67, director, 1967-—. Diplomate Am. Bd. Pediatrics; mem. Soc. Exptl. Biology and Medicine, Soc. Pediatric Research, Am. Fedn. Clin. Research, Am. Assn. Immunologists, Am. Soc. Clin. Investigation, Am. Pediatric Soc., So. Soc. Pediatric Research, So. Soc. Clin. Research, Am. Heart Assn., Phi Beta Kappa, Alpha Omega Alpha. Research and publs. on epidemiology, pathogenesis, prevention and treatment of streptococcal infections; rheumatic fever; upper respiratory tract infections and pneumonia. Home: 512 Redbud Rd., Chapel Hill, N.C. 27514.

DENNY, M(aurice) Ray, Am. psychologist; b. Terre Haute, Ind., Nov. 5, 1918; s. Maurice Ray and Marie (Williams) D.; B.S., U. Mich., 1942, M.A., 1943; Ph.D., U. Ia., 1945; m. Audrey M. Deeks, Aug. 22, 1942 (div. Aug. 1963); children—Michael, Richard, Douglas; m. 2d, Ruth E. Wehner, June 12, 1964. Faculty, U. Okla., 1945-46; faculty Mich. State U., East Lansing, 1946-—, prof. psychology, 1957-—; vis. prof. U. Mich., 1959-—, Ind. U., 1961-—, U. Wis., 1961-—. Research con. VA Hosp., Battle Creek, Mich., 1952-—; study panelist NIH, 1964-—. Mem. Am., Mich. (past pres.), Midwestern psychol. assns., Psychonomic Soc., Animal Behavior Soc., A.A.A.S., Sigma Xi, Phi Sigma, Psi Chi (Tchr. of Year 1964). Author: Comparative Psychology: Research in Animal Behavior, 1964. Research and numerous publs. in learning, inhibition, application motivation learning principles to mentally retarded and schizophrenics; author elicitation theory. Home: 4565 Hawthorn Lane, Okemos, Mich. 48864. Office: Olds Hall, Psychology Dept., Mich. State U., East Lansing, Mich. 48823.*

DENON, Baron Dominique-Vivant, French archaeologist; b. Cholon-sur-Saône, France, Jan. 4, 1747; studied at Paris, France; accompanied Napoleon to Egypt in 1798; sec. French embassy, Naples, Italy, 1780-87; apptd. dir. gen. of museums, 1804. Mem. Royal Acad. Author: Voyage dans le basse et la haute Egypte, 1802; Monuments des arts du dessin, 1829; also numerous drawings of ancient Egyptian ruins and artifacts; instrumental in bringing fgn. masterpieces to Louvre. Died Paris, Apr. 27, 1825.

DENONVILLIERS, Charles Pierre, French surgeon; b. France, Feb. 4, 1808; docteur en médecine Paris, France, 1840; m. Cordier; named surgeon to Paris Central Bur., 1840, received chair of anatomy, 1842, later chair external pathology; head surgeon St. Louis Hosp.; became gen. insp. med. instrn., 1858. Mem. Acad. Medicine (named pres. 1870). Author: Description des pièces pathologiques déposée au musée Dupuytren: (with Berard) Compendium de chirurgie pratique; Traité théoriques et pratique des maladies des yeux, 1855; Rapport sur le progrés de la chirurgie, 1867. Described rectovesical fascia between rectum and prostate (Denonvillier's fascia). Died July 5, 1872.

DE NOTARIS, Giuseppe, Italian botanist; b. Milan, Italy, 1805. Author: Repertorium florae ligusticae, 1844. Research on Italian mosses, mushrooms, lichens. Died Rome, 1877.

DENSLOW, John Stedman, Am. osteo. physician; b. Hartford, Conn., Dec. 19, 1906; s. George Henry and Maud (Stedman) D.; D.O., Kirksville Coll. Osteopathy and Surgery, 1929; D.Sc., Chgo. Coll. Osteopathy, 1941; m. Mary Jane Laughlin, Aug. 22, 1934; children—Martha Stedman (Mrs. Calvin H. Van O'Linda), Michael George Taylor, Peter Ross. Asst. dir., dir. clinics Chgo. Coll. Osteopathy, 1930-38; gen. practice osteo. medicine, Chgo., 1932-38; prof. Kirksville (Mo.) Coll. Osteopathy and Surgery, 1938-45, dir. Still Meml. Research Trust, 1940-—, dir. research affairs, 1945-65, v.p., 1965-—. Mem. Pub. Health Service surgeon gen.'s rev. com. on constrn. med., osteo. colls. Recipient Ann. Distinguished Service award Sigma Sigma Phi, 1941. Mem. Am. Osteo. Assn., A.A.-A.S., Am. Physiol. Soc., N.Y. Acad. Scis., Mo. Assn. Osteo. Physicians and Surgeons, Mo. Assn. for Mental Health. Research and publs. on reflex activity of spinal cord, neuromuscular physiology, reflex and postural muscle contraction. Home: Thousand Hills Farm, Kirksville, Mo. 63501.*

DENT, Charles, Brit. physician; b. Burgos, Spain, Aug. 25, 1911; s. Frank and Carmen (Colsa) D.; ed. Imperial Coll. Sci., London, Univ. Coll. Hosp., London; Ph.D., M.D.; m. Margaret Ruth Coad; children—Ann, Susan, Thomas, Sarah, Christine, Margaret. Research chemist Imperial Chem. Industries, Ltd.; asst. in research med. unit Univ. Coll. Hosp.; reader in medicine U. London, also prof. human metabolism; physician Univ. Coll. Hosp. Mem. Med. Research Soc., Biochem. Soc., Brit. Pediatric Assn., Assn. Physicians, Royal Inst. Chemistry, Royal Coll. Physicians, Royal Soc. Research and numerous publs. on amino acid, calcium and phosphorus metabolism, metabolic illnesses of bone. Home: Eaton Rise 77, Ealing, London. Office: University College Hospital, Gower St., London W.C.1, Eng.

DENT, James Norman, Am. zoologist; b. Martin, Tenn., May 10, 1916; s. James Rolandus and Alta (Norman) D.; A.B., U. Tenn., 1938; Ph.D., Johns Hopkins, 1941; m. Val Nielsen, Dec. 27, 1945; children—Julie Anne, Martha Elizabeth. Asst. prof. Marquette U., Milw., 1945-46; asst. prof. biol. scis. U. Pitts., 1946-49; faculty U. Va., Charlottesville, 1949-—, prof. biology, 1958-—. Cons., biology div. Oak Ridge Nat. Lab, 1954-—. Guggenheim fellow St. Andrews (Scotland) U., 1959-60. Fellow A.A.A.S.; mem. Am. Assn. Anatomists, Am. Soc. Zoologists, Radiation Research Soc., Am. Microscop. Soc., Va. Acad. Sci., Assn. Southeastern Biologists. Research on function thyroid and pituitary glands in amphibians, regeneration limbs in amphibians. Home: 1940 Thomson Rd., Charlottesville, Va. 22903.*

DENTON, Derek Ashworth, Australian physician; b. Launceston, Tasmania, Australia, May 27, 1924; s. Arthur and Catherine (Edwards) D.; student U. Tasmania, 1940; M.B., B.S., U. Melbourne (Australia), 1947; m. Margaret Scott, Mar. 13, 1953; children—Matthew, Angus. Nat. Health and Med. Research Council Australia fellow U. Cambridge (Eng.), 1952-53; sr. research fellow Nat. Health Med. Research Council Australia, Melbourne U., 1956-64, prin. research fellow, 1965-—. Mem. Am. Endocrine Soc., Animal Behavior Soc. (Eng.), Brit. Physiol. Soc. Research and numerous publs. on control orgn. of secretion of adrenal gland, instinctive behavior especially salt appetite. Home: 270 Orrong Rd., 3142 Victoria, Australia. Office: Howard Florey Labs., Melbourne U., Melbourne, Australia.*

DENUCÉ, (Jean-Louis)-Paul, French surgeon; b. Ambarès, France, Jan. 21, 1824; studied medicine at Paris, France; 1 son, Maurice. Became asst. prof. clin. surgery Bordeaux (France) Sch. Medicine, 1855, apptd. prof. in 1858 and dean in 1878. Mem. French Acad. Medicine. Author: Traité clinique de l'inversion utérine, 1883; Mémoire sur les luxation du coude 1854; contbg. author: Nouveau dictionnaire de médecine, 1882. Died Bordeaux, 1889.

DENUES, A(rthur R(ussell) T(aylor), Am. cancer researcher, consulting bio-engineer; b. York, Pennsylvania, August 16, 1914; son John and Constance Haviland-(Taylor) D; B.Engring., Johns Hopkins, 1935, M.Gas Engring., 1937; Ph.D., U. Md., 1939; student biology U. Pitts., nights 1939-42; Am. Cancer Soc.-NRC fellow biology Mass. Inst. Tech., 1947-48; m. Florence Mildred Smith, June 27, 1942; children—John II, Anne, Ginna, Arthur II, Katharine, Elizabeth. Part-time research asst. gas engring. Johns Hopkins, 1935-37; part-time lab. asst. Dept. Agr., 1935-36; coop. investigator U. S. Bur. Mines, also Chile Exploration Co., 1937-38; asst. chem. engr. Bur. Mines, 1938-39, chem. engr., 1939-47; research asso. Detroit Inst. Cancer Research, 1948-50; asst. Sloan-Kettering Inst., N.Y.C., 1950-53, asso., 1953-56, dep. dir., 1956-59, acting dir., 1959-60, v.p., 1960; instr. biology Sloan-Kettering Div. Cornell Med. Soc., 1952-54, asst. prof., 1954-59, asso. prof., 1959-61; grad. faculty Cornell U. Med. Sch., 1958-61; mem. dessemination and field testing com. adv. council, Chem. Corps U. S. Army, 1959-60, 61-—; mem. cancer research tng. com., cons. Surg. Gen., USPHS, 1960-64; president, program dir. CANCIRCO, Inc. (non-profit), Cancer Internat. Research Co-op., 1962-—; cons. BERMCO, 1963-—; lab. instr. biology Norwalk (Conn.) Community Coll., 1963-—. Founding trustee N.Y. Center for Biomathematical Research, 1960. Commd. 2d lt. C.E., U. S. Army, 1935; to maj. chem. warfare service, AUS, 1942-46; lt. col. Residence. Decorated Legion Merit. A. P. Sloan, Jr. scholar cancer research, 1950-53. Fellow N.Y. Acad. Scis., A.A.A.S., American Assn. Cancer Research; mem. Am. Pub. Health Assn., Am. Soc. Cell Biology, Am. Chem. Soc., Electron Miscoscope Soc. Am. (program chmn. 1953, dir. 1955-57), Friends of Kresge-Hooker Sci. Library (life), N.Y. Acad. Scis., N.Y. Microscopical Soc. (bd. mgrs. 1953), N.Y. State Soc. Med. Research (asso.), N.Y. Soc. Electron Microscopists (pres. 1952), Soc. Study Development and Growth, Operations Research Soc. Am. (asso.), Sigma Xi. Phi Sigma. Episcopalian (layreader). Clubs: Chemists, Johns Hopkins, Mass. Inst. Tech. of N.Y. Cons. to editor Biology, Yearbook of Cancer, 1964-—. Research on cancer chemotherapy; viral etiology of human cancer; fine structure of chromosomes; isolated chromosomes; electron microscopy; biochemistry and biophysics in carcinogenesis and invasion and metastasis of tumors; fuels and combustion; munitions and weapon development; explosives and explosions. Home: 12 Manitou Ct. Address: Box 302, Saugatuck, Conn. 06882.

DE NUYSEMENT, Jacques, French alchemist; flourished France, 1621; receiver-gen., Ligny-en-Barrois. Author: Poem philosophic de la verite de la phisique mineralle, 1620; Traittez de l'harmonie et constitution generalle du vray sel (alchem. work), 1621.

DENZA, Francesco, Italian astronomer; b. Naples, Italy, June 7, 1834; studied theology, meteorology and astronomy, Rome, Italy; joined order of Barnabites, 1850; mem. faculty of Barnabite coll., Moncalieri, Italy, 1856-90; built obs., Moncalieri; founder Bolletino mensile di meteorologia, 1859; named dir. Vatican Obs., 1890. Became dir., chmn. awards of sci. sect., nat. exposition, Turin, Italy, 1883; represented pope at internat. sci. congresses. Mem. Italian Meteorol. Soc. (founder 1881), Accademia dei Nuovi Lincei (became pres. 1890). Research and publs. on terrestrial magnetism; began work on mapping of skies using 18 obs. Died Rome, Dec. 14, 1894.

DE OLIVEIRA, Lejeune Pacheco Henriques, Brazilian biologist; b. Suassui, Brazil, Nov. 16, 1915; s. Aristides Henriques and Noami (Pacheco) de O.; Dr., Faculdade Nacional de Medicina, 1938; student Instituto Oswaldo Cruz, 1936; m. Maria Helena Queiros Santos, Dec. 28, 1943 (div. May 1951); 1 son, Jose Carlos; m. 2d, Luiza Krau de Oliveira, July 14, 1953; 1 dau., Luiza Cristina Krau. Asst. aux. Sta. of Hydrobiology, 1939-50, in charge, 1950-63, head sector Hydrobiology, 1963-—; prof. hydrobiology Instituto Oswaldo Cruz, Rio, Brazil; survey lagoons Lab. Hydraulics, Escritorio Saturnino de Brito. Recipient D. Joao VI medal for forest preservation, protection waters and nature. Mem. Sociedade Biology Rio de Janeiro, Academia Brasileira de Ciencias, Internat. Assn. Limnology. Research, publs. on systematics of crustaceans, especially genus Uca, carcinology, ethology; survey of Bentos of Guanabara Bay, crustacea Amphipoda gammaridea of Rio, lagoons with brackish water of Rio; studies of shore pollution of Guanabara Bay. Home: 385 Ubiraci, St. Guanabara, ZC 24. Office: Instituto Oswaldo Cruz, Brazil Avenida, Caixa Postal 926, Rio, Brazil.*

DEORAS, P. J., Indian entomologist; b. Bilaspur, H.P., India, July 20, 1909; s. J. S. and Main (Harpal) D.; B.Sc., Sci. Coll. U. Nagpur, 1931, M.Sc., 1933, LL.B., 1935; Ph.D., U. Durham (U.K.), 1940; m. Vatsala Ghate, May 21, 1936; children—Mukunda, Maya (Mrs. S. G. Surange), Chayya, Vijaya, Dinesh, Sandhya. Asst. to entomologist, New Delhi, 1941-43; prof. zoology Furgussan Coll., Poona, India, 1943-45; post-grad. tchr. U. Bombay and Poona, 1945-—; entomologist Geigy Insecticides Bombay; head entomology dept. Maharashtra Assn. for Cultivation of Science, Poona; head dept. entomology, asst. dir. Haffkine Inst. Bombay, 1951-—. Recipient WHO Expert Panel plaque. Fellow Entomology Soc. India; Mem. N.Y. Acad. Scis., Pan Am. Med. Assn., Internat. Soc. Toxinology. Author: Snakes of India, 1966; Handling of Lab. Animals, Snakes, 1966; Venomous and Poisonous Animals . . . Studies on Bombay Snakes, 1963. Pioneered studies on poisonous snakes, scorpions, farm breeding, behavior, venom, antigenecity, devel. of methods of venom extraction and standardization. Home: Tilaknagar, Bilaspur, M.P., India. Office: Haffkine Inst., Bombay—No. 12, Maharashtra, India.*

DE OVIEDO Y VALDÉS, Gonzalo Fernandez, Spanish naturalist; b. Madrid, Spain, 1478; received govt. appointment on Island of Hispaniola. Author: Quinquagenas; Historia general y natural de las Indias Occidentales (1st natural history of Am.), 1548. Described many mammals, edible and medicinal plants and birds. Died Valladolid, Spain, 1557.

DEPAGE, Antoine, Belgian surgeon; b. Boitsfort, Brussels, Belgium, Dec. 23, 1862; prof. external pathology, clin. surgery La Haye Faculty of Medicine; organizer, dir. La Panne Hosp. during World War I; prof. chem. surgery Brussels, Belgium. Mem. Belgian Soc. Surgery (founder 1893), French Acad. Scis., 1916. Pioneered gynecol. surgery; reorganized Red Cross in Belgium. Died La Haye, Belgium, June 10, 1925.

DEPARCIEUX, Antoine, French mathematician; b. d'Uzès, France, Oct. 28, 1703; s. Jean-Antoine and Jeanne (Donzel) D.; ed. Alès Jesuit Coll.; became secretary to de Moncarville, Paris, France, 1730; appointed royal censor, 1765; substitute prof. at Collège de France. Member of the French Acad. of Sciences, 1746. Author: Traité de trigonométrie rectiligne et sphèrique, 1738; Traité complet de gnomonique, 1740; Essai sur les probabilités de la vie humaine, 1746, supplement, 1760. Invented machine for pumping water, tobacco press; built sundials; 1st to set up mortality table for ins. cos. Died Paris, Sept. 1, 1768.

DEPAUL, Jean Anne Henri, French obstetrician; b. Morlaas, France, June 26, 1811; s. Bernard and Miss Claverie; studied at Paris, France; docteur en médecine, 1839; aggregate prof., 1847. Named prof. obstetrics Paris Faculty of Medicine, 1862; practised at Paris Maternity Hosp., infant mortality decreased sharply while he was on staff; called to U. S. to assist famous patients, titular dir. Vaccination Service, 1864-73. Mem. Acad. Medicine (became pres. 1873). Author: Traité théorique de l'auscultation obstétricale, 1847; Convulsions des femmes enceintes, 1854; Sur la véritable origin du virus-vaccin, 1864; De la vaccine, 1879. Promoted smallpox vaccinations; research on origins of viruses, sickness of pregnant women. Died Morlaas, Oct. 4, 1883.

DE PEIRESC, Nicolas Claude Fabri, French astronomer, archeologist; b. Beaugensier, Provence, France, Dec. 1, 1580; legal degree Aix, France, 1604; became senator for Provence; judge highest ct. in Provence, 1605; made abbot Monastery of Guistres. With Gassendi made 1st map of moon; 1st to observe Mercury in broad daylight; used numismatics for hist. research; verified Harvey's discovery of blood circulation; introduced Angora cat to Europe, exotic plants to France. Died Aix, June 24, 1637.

DEPERET, Charles, French paleontologist, geologist; born Perpignan, France, June 25, 1854; entered School Military Medical Corps, 1873; studied paleontology and geology under Hébert and de Gaudry; commissioned medical major, 1883; placed in charge geology Faculty Sciences, Marseilles, France, 1886; left Med. Corps, 1888; became prof. Lyons, France, 1889, apptd. dean Faculty Sci., 1896. Mem. Assn. Human Paleontology and Prehistory, (founder 1922), French Acad. Scis., 1898. Author: Rechérches sur la succession des faunes de vertébrés miocènes de la vallée du Rhône, 1887; Notes stratigraphiques sur le bassin tertiare de Marseille, 1889; Études stratigraphiques et paléontologiques pour servir à l'hist. de la période tertiare dans le bassin du Rhône; les Terrains tertiares marins de la côte de Provence 2 vols., 1889-92; la Faune de mammifères miocènes de la Grives-Alban, 1892; (with Delafond) Les Terrains tertiares de la Bresse et leur gites de lignite et de minerais de fer, 2 vols., 1893; les Animaux pliocènes du Roussilon, 1897; Bassins tertiares du Rhône, 1900; (with F. Roman) Monographie des pectinidés néogènes de l'Europe, 1902-05; (with Douxami) les Vertébrés oligocènes de Pyrimont-Challonges, 1902. Studied some stages of devel. in Europe including Spain, Danube, Russia, also Greece; studied quaternary fossils in Nice, France, and basin of Mediterranean after 1903, tertiary fauna especially of Miocene; excavation work at Mont d'Or lyonnais, S.-Albon, La Grive, Isère, Villefranche. Died Lyons, May 18, 1929.

DE PETIT, François Pourfour, French physician, physiologist; b. Paris, June 24, 1664; M.D., Montpellier, France; physician Royal Army; mem. French Acad. Scis., 1722. Author: Mémoire dans lequel il est démontré que les nerfs intercostaux—portent des esprits dans les yeux, 1727; Sur le cristallin de l'oeil de l'homme, 1730. Described spaces around capsule or periphery of crystalline lens, 1715, vasomotor nerves, 1727; produced in animals syndrome similar to Horner's syndrome in humans, 1727; disproved belief that intercostal nerves originate in cerebrum, 1727. Died Paris, June 18, 1741.

DE PIGNANO, Francisco (da Pignano, Francisco, Pignano, Francisco de), Italian physicist; flourished 1320. Joined Order of St. Francis; lectr. on Sentences, Paris, France; believed that virtus motiva of projectile is caused by initial impulsion rather than airy medium, yet felt that medium might somehow assist in maintaining motion; extended theory to celestial bodies which move because of initial impulsation from God.

DEPORT, Joseph-Albert, French engr.; b. Saint-Loup-sur-Semouse, France, 1846; French arty. officer; dir. factory, Puteaux, France; engr. Compagnie de Chatillon, Commentry et Neuves-Maisons. Inventor gun-sighting mechanism; worked on 7.5 cm cannon with long barrel recoil and hydropneumatic brake (later perfected by others); built long range cannon for U. S., Italy, also early arty. drawing engine. Died 1929.

DE POURTALES, Louis François, naturalist; b. Neuchatel, Switzerland, Mar. 4, 1823; m. Elise Bachmann, 1 child. Accompanied Jean Agassiz on expdns. to study glaciers of Alps, 1840; came to U. S., 1846; with U. S. Coast Survey, 1848-73, in charge of tidal div., 1864-73; keeper Mus. Comparative Zoology, Harvard, 1873-80; collected and studied animal life at great depths; engaged in explorations carried on by Coast Survey steamer Bibb in waters of So. Fla., No. Cuba, Western Bahamas, located Pourtales Plateau off Southeastern Fla.; accompanied Agassiz on voyage in Hassler around Cape Horn to San Francisco, in charge of dredging and other deep sea work, 1871; did his most important work Deep Sea corals, pub. 1871; collected sea-urchin off So. Fla., named Pourtalesia. Died Beverly Farms, Mass., July 18, 1880.

DEPREZ, Marcel, French physicist; b. Aillant-sur-Milleron, France, Dec. 29, 1843; studied at Ecole superieure mines; served as cannoner during the siege of Paris, France, in 1871; sec. to dir. Ecole Superieure Mines; prof. Conservatoire des arts et métiers, 1890-1918. Recipient Fourneyron prize, 1884, Montyon prize mechanic, 1876, Fremont prize for ballistics, 1878. Mem. French Acad. Scis., 1886. Author: Traité d'électricité industrielle théorique et pratique, 3 vols., 1896-1900; Lois fondamentales de l'electrotechnique, pub. 1919; also articles. Invented absolute electrodynamometer, (with d'Arsonval) the aperiodic galvanometer; wattmeter, and ammeter; 1st solution to problem of transporting energy; (with C. Herz) transmitted electrical power over 35-mile distance from Miesach to Munich, Germany, 1882; research on regulation of electric motors, role of iron core in armatures, friction, equivalence of heat and work; discovered the compound principle; perfected use of spark gap and tuning fork to record ballistic phenomena; (with Carpentier) perfected dynamo. Died Vincennes, France, Oct. 16, 1918.

DE PRIMA, Charles Raymond, Am. mathematician; b. Paterson, N.J., July 10, 1918; s. Mario and Louise (Ruggiero) DeP.; A.B., Washington Sq. Coll., 1940; Ph.D., N.Y. U., 1943; m. Annemarie Boerschmann, June 14, 1952. Lectr., N.Y. U., 1942-46, vis. prof., 1962-63; with Cal. Inst. Tech., Pasadena, prof. math., 1956-—. Mem. applied math panel OSRD, 1942-46; head math. div. Office Naval Research, 1951-52; mem. Com. on Undergrad. Programs in Math., 1961-—. Mem. Am. Math. Soc., Math. Assn. Am., Soc. Indsl. and Applied Math. (mem. council 1961-—), Phi Beta Kappa. Research and publs. on functional analysis, partial differential equations, gas dynamics, fluid mechanics. Home: 3791 Hampstead Rd., Pasadena, Cal. 91103.*

DE PRONY, Baron Gaspard-Clair-François-Marie Riche, French mathematician, engr.; b. Chamelet, France, July 22, 1755; ed. École des Ponts et Chaussées. Became engr.-in-chief bridges and causeways, 1791, dir. project to draw up Tables du Cadastre, 1792, inspector gen., 1798; dir. École des Ponts et Chaussées, 1798; prof. math. scis. École Polytechnique, 1794; supt. engring. operations to regulate course of Po at Ferrara, also to drain Pontine marshes, and to regulate course of Rhone. Mem. Bur. Longitudes. Fellow Royal Soc., 1818; mem. French Acad. Scis., 1795. Author: Memoire sur la poussée des voutes, 1783; Architecture hydraulique, 1790-96; Researches sur la poussée des terres, 1802; Researches physico-mécaniques sur la théorie des eaux courantes, 1804; Cours de mécanique concernaent les solides, 1815; Novelle méthode de nivellement trigonométrique, 1822; Description hydrographique et historique des marais Pontins, 2 vols., 1822-23. Invented dynamometric absorption brake (Prony Brake); drew up gen. formula on theory of fluid motion; built Pont de la Concorde, Paris; restored harbor at Dunkerque; directed drawing up of new trigonometric tables extended to 25 decimal places due to establishment of metric system; 1st to derive equation of curve (Watt's Curve). Died Asnières, France, July 29, 1839.

DE PROSPO, Nicholas Dominick, Am. biologist; b. Bronx, N.Y., July 16, 1923; s. Domenick and Lucy (Riccio) De P.; B.A., N.Y.U., 1946, M.A., 1947, Ph.D., 1957; m. Margaret Joyce Bing, Dec. 31, 1960; 1 son, Douglas Francis. Mem. faculty Seton Hall U., 1947-—, chmn. dept. biology, 1958-—, prof., 1959-—. Recipient Founders Day award N.Y.U., 1958. Mem. A.A.A.S., Am. Soc. Zoologists, Am. Inst. Biol. Scis, N.J. Acad. Sci., Am. Assn U. Profs., Sigma Xi, Alpha Epsilon Delta (hon.) Author: (with Gallo) Guide to Chordate Anatomy, 1962, textbook on Comparative Vertebrate Anatomy. Research, publs. on role of adrenals in quiniline dye carcinogenesis effects of Ehrlich's ascites carcinoma in pregnancy, interaction of certain pox virus and Ehrlich's tumor; age and susceptibility to Kreb's ascites tumor. Home: 757 Mitchell Av., Union, N.J. 07083. Office: Seton Hall U. South Orange, N.J. 07079.*

DE QUATREFAGES DE BREAU, Jean Louis Armand, French naturalist, ethnologist b. Berthezène, Gard, Feb. 10, 1810; M.D., D.Sc., U. Strassbourg, 1829. Practiced medicine, Toulouse, 1833-38, prof., 1838, resigned to continue researches, 1839; prof. natural history Lycée Napoleon, 1850; prof. anthropology and ethnography Musee d'histoire naturelle, 1855. Mem. Acad. Sci., 1852, hon. mem. Royal Soc., London, 1879; mem. Inst. and Acad. Medicine. Author: Sur les aerolithes, 1830; Considérations sur les caractères zoologiues des rongeurs, 1840; Souvenirs d'un naturaliste, 2 vols., 1854; Physiologie comparée, metamorphoses de l'homme et des animaux, 1862; Les Polynésiens et leurs migrations, 1866; Histoire naturelle des annelés marins et de l'eau douce, 2 vols., 1866; Rapport sur les progrès de l'anthropologie, 1867; Darwin et ses précurseurs francais, 1870; La race prussiene, 1871; (with Hamy) Crania Ethnica, 2 vols., 1875-82; L'Espèce humaine, 1877; Nouvelles Etudes sur la distribution geographique des negritos, 1882; Hommes fossiles et hommes sauvages, 1884; Histoire générale des races humaines, 2 vols., 1886-89. Made accurate observations and descriptions of zool. phenomena; described parietal angle (Quatrefages's angle); refuted view that pygmies were retrograde negro type; extensive work in human anthropology. Died Paris, France, Jan. 12, 1892.

DE QUERVAIN, Marcel Roland, Swiss meteorologist; b. Zurich, May 17, 1915; s. Alfred and Elisabeth (Nil) de Q.; Ph.D. in Natural Sci., Fed. Poly. of Zurich; m. Rita Wismer; children—Ines-Anne, Elisabeth-Lucie, Martin-Alfred. Sci. collaborator Fed. Inst. for Study Snow and Avalanches, Weissfluhjoch/Davos, Switzerland. Mem. Internat. Glaciological Expdn. to Greenland, 1959. Mem. Assn. Nat. and Internat. Sci. Research and publs. on crystallography, mechanics, hydrology of snow. Home: Arvenhugel, Davos-Dorf. Office: Federal Institute for the Study of Snow and Avalanches, Weissfluhjoch/Davos, Switzerland.

DERANIYAGALA, Paulus Edward Pieris, Sinhalese anthropologist, zoologist; b. Colombo, Ceylon, May 8, 1900; s. Paulus E. Pieris and Hilda (Obeyesekere) D.; edn. Trinity Coll., Cambridge (Eng.) U., M.A., 1923; A.M., Harvard, 1924; D.Sc., Vidyodaya U., 1960, Ceylon U., 1963; m. Prini Molamure, 1934; children—Arjun, Ranil, Siran, Isanth. Dir. nat. museums Govt. of Ceylon, Colombo, 1939-63, now emeritus, acting dir. fisheries, 1939, acting archaeol. commr., 1956-57, 60; faculty Faculty Arts, Vidyodaya U., Ceylon, 1959-64, prof. anthropology, 1961-64, dean Faculty Arts, 1961-64; mem. U. Cal. African Expdn., 1947-48. Recipient U. S. Govt. Leader Exchange award, 1956. Fellow Am. Soc. Vertebrate Paleontologists, Internat. Union Prehistoric and Protohistoric Scis. (com. mem.), Indian Paleontol. Assn. (com. mem.). Author numerous books, articles on zoology, geology, anthropology, archaeology. Home: 26 Guildford Crescent, Colombo 7, Ceylon.*

DE RATTE, Estienne-Hyacinthe, French astronomer; b. Montpellier, France, Sept. 1, 1722. Perpetual sec. Royal Soc. Sci., Montpellier; mem. French Acad. Scis., 1796. Author: Researche sur la Pesanteur; also contbd. articles, especially on physics to Encyclopedie. Observed passage of Venus in front of sun, 1761; research in stem growth. Died Montpellier, Aug. 15, 1805.

DERBES, Vincent Joseph, Am. physician; b. New Orleans, May 8, 1912; s. Albert Joseph and Hazel (de Generes) D.; M.D., Tulane U., 1934; postgrad. U. Ill., 1954-57, McGill U., 1959, Georgetown U., 1960; m. Christine Smith Musgrove, Apr. 2, 1936; children—Nancy Anne, David Raoul. Practice medicine specializing in dermatology and allergy, New Orleans, 1941—; instr. medicine, med officer Hutchinson Meml. Clinic, Tulane U. Sch. Medicine, New Orleans, 1948-54, prof. dermatology, chief sect. allergy and dermatology, 1954—. Mem. Am. Acad. Dermatology and Syphilology, Am. Math. Assn., Sigma Xi. Author: (with H. T. Engelhardt) The Treatment of Bronchial Asthma, 1946; (with T. E. Weiss) Untoward Reactions of Cortisone and ACTH, 1952; (with Andrew Kerr, Jr.) Cough Syncope, 1955. Numerous publs. on etiological discoveries in various mycologic, allergic, and dermatologic diseases. Home: 1001 Falcon Rd., Metairie, La. Office: 1430 Tulane Ave., New Orleans, La. 70112.*

DERBY, George Strong, Am. ophthalmologist; b. Boston, May 29, 1875; s. Hasket and Sarah (Mason) D.; A.B., Harvard, 1896; M.D., Harvard Med. Sch., 1900; surg. interne, Mass. Gen. Hosp., 1900-01; postgrad. in pathology and ophthalmology, Austria, Germany, Holland, France and Eng., 1901-02; m. Mary Brewster Brown, Aug. 5, 1901. Ophthalmic chief, Mass. Charitable Eye and Ear Infirmary; prof. ophthalmology Harvard Med. Sch., mem. adv. com. Mass. Commn. for Blind. Died Dec. 12, 1931.

DERCUM, Francis X., Am. neurologist; b. Phila., Aug. 10, 1856; s. Ernest and Susanna (Erhart) D.; grad. Central High Sch., A.M., 1878; M.D., U. Pa., 1877, Ph.D., 1877; Sc.D., Jefferson Med. Coll., 1927; m. Elizabeth Comly, Aug. 5, 1891; children—Elizabeth Comly (Mrs. Samuel Wright Mifflin), Ernest Comly, Mary DeHaven. Began practice at Phila., 1877; prof. nervous and mental diseases Jefferson Med. Coll., 1892-1925; cons. neurologist Phila. Gen. Hosp. Fellow Coll. Physicians of Phila. Am. (pres. 1896), Phila. (founding mem.) neurol. socs., Soc. Physicians Vienna, Royal Soc. Medicine, Am. Philos. Soc. (pres. 1925-31), Author: Rest, Suggestion and Other Therapeutic Measures in Nervous and Mental Diseases; A Clinical Manual of Mental Diseases; Hysteria and Accident Compensation, 1916; The Biology of the Internal Secretions, 1924; The Physiology of Mind, 1925; also papers, Editor: Text book of Nervous Diseases by American Authors, 1895. Described adiposis dolorosa (Dercum's disease), 1892; 1st (with Muybridge) to photograph persons in convulsions; photographed men with abnormal path. gaits; produced psychogenic convulsions by means of hypnotism. Died Apr. 23, 1931.

DE RENZI, Giuseppe, Italian pathologist; b. Pisa, Italy, Jan. 30, 1928; s. Alfredo and Marcella (Bartoli) de R.; M.D., U. Pisa., 1957, postgrad.; Ph.D., U. Rome (Italy), 1965; m. Teresa Galati, Sept. 25, 1965. Coroner, City of Livorno, Italy, 1957—; practice medicine specializing in pathology, Livorno, 1957—; med. cons. DOW Chem. Co., Livorno, 1961-65; asst. prof. legal medicine U. Pisa., 1957-61, asst. prof. pathology, 1961—; prof. radiobiology Inst. Biochemistry, U. Genoa, 1965—; chief path. sect. CAMEN, 1958-65. Mem. Italian Soc. Pathology, Italian Soc. Biology and Nuclear Sci., Med. Commn. for Radiation Protection. Author: Argomenti di Radiobiologia, vol. I. and II. (with Lamarche), 1963, vol. III (with Salvetti), 1963. Research and publs. in diagnosis and treatment of cancer with radioactive substances. Home: 18, via Forte Cavalleggeri, Livorno. Office: 18, Piazza Guerrazzi, Livorno, Italy.*

DE RENZO, Edward Clarence, Am. biochemist; b. Passaic, N.J., Sept. 29, 1925; s. Francis and Antonetta (Paterno) DeR.; B.S., Fordham U., 1945, M.S., 1947, Ph.D., 1950; m. Dorothy Jones, Oct. 22, 1950; children—Anne, Susan, Lisa, Linda. Research scientist Lederle Labs., Pearl River, N.Y., 1951-54, sr. research biochemist, 1955-67, head dept. metabolic chemotherapy, 1967—. Member of the Am. Soc. Biol. Chemists. Research and publs. on importance of vitamin B6 in metabolism of sulfur containing amino acids, role for molybdenum in normal animal nutrition and as component of certain enzymes, mode of action of streptokinase activation of human plasminogen in formation of plasmin (fibrinolysin), enzyme which dissolves blood clots. Home: 43 Buff Lane, Hillsdale, N.J. 07642. Office: Metabolic Chemotherapy Dept., Lederly Labs., Pearl River, N.Y. 10965.*

DE REYNIER, Jean Pierre, Swiss otologist; b. Leysin, Switzerland, Jan. 16, 1914; s. Leopold and Hélène (Suchard) R.; Diploma of Medicine, U. Basel (Switzerland), 1939, M.D., 1940; m. Gilberte de Watteville, July 4, 1942; children—Micheline, Jean-Francois, Olivier. Asst., Children's Hosp., Lausanne, Switzerland, 1940, U. Pathology Inst., Zürich, Switzerland, 1941, Policlinic Medicine, Basel, 1942, ORL Clinic, U. Zürich, U. Bern, 1943-47; practice medicine specializing in otolaryngology, Lausanne, 1947—; otologist Hôpital de l'Enfance, Lausanne, 1947-53, Deaf Children's Inst., Moudon, Switzerland, 1963—; founder Centre Audiology for Children. Mem. Swiss Assn. for Help Deafmute (chmn.), Swiss, French socs. ORL, Internat. Soc. for Rehab. Disabled, Internat. Soc. Logopedy, Internat. Bur. d'audio-phonologie (treas.). Author: (with F. R. Nager) Monograph on Ear Bay Congenital Head Malformation, 1959; Deafmuteness in Switzerland, 1953; also articles. First application of hearing aids to very young children; 1st audiology center for young deaf children. Home: 29 Avenue de Cour. Office: 4 Place de la Gare, Lausanne, Switzerland.*

DERGE, Gerhard Julius, Am. metall. engr.; b. Lincoln, Neb., Feb. 11, 1909; s. Matt L. and Wilma (Gesell) D.; A.B., Amherst Coll., 1930, Ph.D., Princeton U., 1934; m. Katharine McKechnie, Oct. 26, 1937; children—Portia, Jeffrey. With Carnegie-Mellon U., 1934—, prof., 1949—, Jones and Laughlin professorship metall. engring., 1951-65, assigned to Manhattan Engring. Dist., U. Chgo., Oak Ridge, 1943-45, Indian Inst. Tech., Kanpur, U.P., 1962-63. Recipient Robert W. Hunt medal. Mem. Am. Ceramic Soc., Am. Soc. Metals, Am. Inst. Mining, Metall. and Petroleum Engrs. (Howe Meml. lectr. 1967), Iron and Steel Inst., Am. Chem. Soc., Am. Soc. Testing Materials. Contbr. numerous articles to sci., tech. jours. Home: 8 Longfellow Rd., Pitts. 15215.*

DERHAM, William, Brit. physicist, biologist; b. Stoulton, Eng., Nov. 26, 1657; B.A., Trinity Coll., Oxford, Eng., 1679; D.D., Oxford U., 1730; m. Anne (Scott); children including William. Became chaplain to dowager Lady Grey of Werke, circa 1679; ordained deacon, 1681, priest, 1682; became vicar of Wargrave, 1682, Upminster, Eng., 1689; chaplain to Prince of Wales (George II), upon accession of George I; named canon of Windsor, 1716; fellow, pres. St. John's Coll., Oxford. Fellow Royal Soc. 1702. Author: The Artificial Clockmaker, 1696; Physico-Theology, or a Demonstration of the Works and Attributes of God from His Works of Creation (Boyle lectures 1711, 12), 1713; Astro-theology, or a Demonstration of the Being and Attributes of God from a Survey of the Heavens, 1715; also notes for Eleazar Albin's histories of birds and insects, 1724-31; rev. edit. Miscellanea Curiosia (collection of some of the greatest curiosities of nature accounted for by greatest philosophers of time), 1726; papers on weather, barometer, great storm of 1703, migration of birds, will o' the wisp, habits of deathwatch and of wasps. Editor: Synopsis of Birds and Fishes (Ray), 1713; Philosophical Letters (Ray and Willoughby), 1718. Studied motion of pendulum in vacuum; showed influence of wind speed on speed of sound; invented portable barometer, also instrument for determining meridian; introduced rain gauge for wagons. Died Apr. 5, 1735.

DE RHEITA, Anton Maria Schyrlaus, see von Rheita.

DERIAGIN, Boris Vladimirovich, Russian chemist, physicist; b. Aug. 4, 1902; ed. Moscow (USSR) U.; staff, inst. phys. chemistry USSR Acad. Scis., 1935—, also corr. mem. acad. Author: (with N. A. Krotova) Electrical Theory of Adhesion, Adhesion, Studies in Adhesion and Adhesive Action. Research and publs. in phys. chemistry and molecular physics; developed molecular theory of surface friction of solid bodies, theory of coagulation of dispersed systems by electrolytes; research on spl. properties of thin layers of fluids, technique of lubricating by thin coatings.

DERICKSON, Samuel Hoffman, Am. biologist; b. Perry County, Pa., Apr. 9, 1879; s. Henry Benner and Lizzie Naomi (Hoffman) D.; B.S., Lebanon Valley Coll., 1902, M.S., D.Sc. (hon.), 1925; postgrad. Johns Hopkins, 1903, 10, biol. labs., Cold Spring Harbor, N.Y. and Bermuda Islands; m. Jennie Vallerchamp, June 28, 1905; children—George Vallerchamp, Mary Elizabeth. Acting prof. biology Lebanon Valley Coll., 1903, prof. 1907-50, emeritus, from 1950, acting pres., 1912. Fellow A.A.A.S.; mem. Bot. Soc. Am., Am. Fern Soc., Pa. Acad. Science (pres.), Am. Soc. Zoölogists (asso.), Torrey Bot. Club. Died Nov. 27, 1951.

DE RIEMER, Pieter, Scandinavian anatomist; b. Scandinavia, 1760; reputed to have been 1st to prepare frozen anat. sects., 1818. Died 1831.

DERKACH, Vasiliy Stepanovich, Russian microbiologist, immunologist; b. Korsun (now Donetsk Oblast), 1894; grad. Med. Faculty, Kharkov U., 1917; D. Med. Sci., 1939. Asst., later lectr. dept. microbiology Kharkov Med. Inst., 1918-31, head chair microbiology, 1932—; prof., 1932—; head dept. microbiology Mechnikov Inst. Vaccines and Sera, Kharkov. Decorated Order of Lenin. Mem. USSR Acad. Med. Sci. (corr.). Author over 80 works including Study of the Antibiotic Properties of Pyocyanine, 1946; Antibiotics against Malignant Neoplasms, 1956. Co-editor Microbiology sect. Large Med. Ency., 2d edit.; mem. editorial council Antibiotics. Research on nature of typhoid toxin and antitoxin, 1930-36, antiobiotic properties of aniline dyes and compounds of phanazine series (pyocyanine), 1939; devised method of bacteriophagia in mixed cultures, 1943, developer method of obtaining pyocyanine, 1946, introduced biol. and immunological methods of combatting paratyphus of bees and sugarbeet pests. Address: Kharkov Med. Inst., ulitsa Sumskaya 1, Kharkov, Ukraine SSR, USSR.

DERMAN, Cyrus, Am. math. statistician; b. Phila., July 16, 1925; s. Samuel and Bessie (Segal) D.; A.B., U. Pa., 1948, A.M., 1949; Ph.D., Columbia U., 1954; m. Martha Winn, Feb. 24, 1961; 1 son, Adam Jason Winn. Instr. Syracuse U., 1954-55; faculty Columbia, 1955—, prof. indsl. engring. and operations research, 1965—; vis. prof. Israel Inst. Tech., Haifa, 1961-62, Stanford, 1965-66. Fellow Inst. Math. Statistics, Am. Statis. Assn.; mem. Internat. Assn. Statistics in Phys. Scis. Author: (with Morton Klein) Probability and Statistical Inference for Engineers, 1959. Research and publs. on theory of Markou chains, Brownian motion, statis. inference, mgmt. sci. and operations research. Home: 15 Pond Hill Rd., Chappaqua, N.Y. 10514. Office: Mudd Bldg., Columbia U., N.Y.C. 10027.*

DERMEN, Haig, botanist; b. Kindjilar, Constantinople, Oct. 22, 1895; s. Dikran and Elizabeth (Shamlian) Deyirmendjian; came to U. S., 1920, naturalized, 1928; B.S., U. Conn., 1925; M.S., U. Me., 1927; M.S., Harvard, 1931, Ph.D., 1933; m. Margaret Schneider, Aug. 29, 1927; children—Armen, Diran, Susan. Research cytologist Hort. Crops Research br. Agrl. Research Service, U. S. Dept. Agr., Beltsville, Md., 1937-65, collaborator in research, 1965—. Mem. Bot. Soc. Am., Genetics Soc. Am., Genetics Association, Am. Soc. Hort. Science, Washington Academy of Science, Society Am. Naturalists, A.A.-A.S., Sigma Xi. Research and numerous articles on plant breeding, cytology and basic bot. problems; developed principles in inducing polyploidy in plants with use of drug colchicine; studies of plant shoot tips toward understanding of changes brought about by use of colchicine. Home: 5915 Pontiac St., College Park, Md. 20740. Office: Fruit Research, U. S. Plant Industry, Beltsville, Md. 20705.*

DE ROALDÈS, Arthur Washington, Am. surgeon; b. Opelousas, St. Landry Parish, La., Jan. 25, 1849; s. Abel (M.D.) and Coralie Testas de Folmont D.; ed. by Jesuits in France; grad. in letters, 1865, in science, 1866, U. of France; M.D., U. La., 1869, U. Paris, 1870. m. Laura Pandely, 1873 (died 1874); m. 2d, Annie E. Miller, La., 1885. In charge of Charity Hosp. of La., 1880-83; founded 1889, Eye, Ear, Nose and Throat Hosp. of New Orleans of which became trustee and surgeon-in-chief; emeritus prof. diseases ear, nose and throat, Post-Grad. Dept. Tulane U. of La. Mem. Internat. Med. congresses, Berlin, 1890, Rome, 1894, Paris, 1900, Internat. congresses Otology, Florence, 1895, London, 1899 (mem. Am. Com. of Organization 7th Otol. Congress, Bordeaux, 1904); v.p. La. State Med. Soc., 1892; pres. New Orleans Diphtheria Antitoxin Commn., 1894; fellow Am. Laryngol. Assn. (pres. 1906-07). Officer Legion d'Honneur of France, 1903, in recognition of services rendered indigent sick and med. edn. by his hosp. found. Died June 12, 1918.

DE ROCHON, Alexis Marie, French astronomer; b. Brest, France, Feb. 21, 1741. Guardian of French king's Cabient of Optics and Physics, 1774; astronomer optician for French Navy, from 1787; dir. Brest Obs., from 1802. Mem. French Acad. Scis., 1795, Marine Acad. Author: Opuscules Mathematique, 1768; Recueil de Memoirs sur la Mécanique et la Physique, 1783; Noveau Voyage à la Mer du Sud, 1783; Voyages à Madagascar, 1783; Voyage aux Indes Orientales et en Afrique, avec une Dissertation sur les Iles de Salomon, 1807. Made voyage to French E. Indies (determined best route around reefs), 1768; inventor new nav. lighting system for ships; developed spyglass, 1777. Died Paris, France, Apr. 5, 1817.

DE ROMAS, Jacques, French physicist; b. Nérac, France, Oct. 13, 1713; ed. to become magistrate. Advisor to Nerac Presidial; became interested in scis. Invented brontomètre which gave protection against lightning, 1750; used elect. kite to observe atmospheric electricity (apparatus rose to 180 metres and its metal wire produced impressive elect. discharge), 1752; conducted expts. without knowledge of Franklin's discoveries. Died Nérac, Jan. 21, 1776.

DEROSNE, (Louis) Charles, French chemist; b. Paris, France, Jan. 23, 1780; became asso. with his bro., François in pharm. business in Paris, 1806; founder factory building locomotives, 1839. Author: De la fabrication du sucre aux colonies, 1812. Translated Archard's treatise on European sugar beet into French. Perfected manufacture of animal black and used it to decolor sugar sap; studied prodn. beet sugar; made refined beet sugar, 1811. Died Paris, Sept. 21, 1846.

DE ROSSET, Moses John, Am. physician; b. Wilmington, N.C., July 4, 1838; s. Armand John and Eliza Jane (Lord) De R.; attended U. Cologne (Germany), 1856-57; studied medicine under Dr. Gunning Bedford, N.Y.C.; M.D., U. City N.Y., 1860; m. Adelaide Savage Mears, Oct. 13, 1863, 7 children. Resident physician Bellevue Hosp., N.Y.C., 1860-61; commd. asst. surgeon Confederate Army, 1861, promoted surgeon 1863; in charge of Gen. Hosp. Number 4, Richmond, 1863, later became insp. hosps. for Dept. of Henrico; prof. chemistry med. dept. U. Md. Dental Coll., 1865-73; practiced medicine specializing in eye and ear diseases, Wilmington, Del., 1873-78, N.Y.C., 1878-81; wrote many articles. Translator: Annual Abstract of Therapeutics Materia Medica, Pharmacy and Toxicology for 1867 by A. Bouchardet. Died May 1, 1881.

DEROY, Henri, see Duroy, Henri.

DERRA, Enrst, German surgeon; b. Passau, Mar. 6, 1901; s. Ernst and Magdalena (Osterkorn) D.; ed. univs. Munich, Heidelberg, Vienna; m. Hanna Laura Siebel, 1937; intern, med. clinics univs. Innsbruck, Leipzig; staff Univ. Surg. Clinic Bonn (Germany), 1929-45, prof. surgery, 1943; head surgeon Marien Hosp., Bonn, 1945-46; prof. surgery Dusseldorf, Germany, from 1946. Collaborator: Handwörterbuch für gerichtliche Medizin; Operations-Lehr; Handbuch für Unfallheilkunde und Verischerungsmedizin; Lehrbuch für Chirurgie.

DERRICK, Edward Holbrook, Australian pathologist; b. Blackwood, Victoria, Australia, Sept. 20, 1898; s. Clement Herbert and Elizabeth (Sweetman) D.; M.B., B.S., Melbourne (Australia) U., 1920, M.D., 1922; m. Margaret Gina Quadrio, Mar. 11, 1930; children—Noel Edward, Graham Holbrook. Grice Cancer Research scholar U. Melbourne, 1921-22; pathological asst. London Hosp., 1922-23; dir. Lab. Microbiology and Pathology, Queensland Health Dept., 1934-47; dep. dir. Queensland Inst. Med. Research, Brisbane, Australia, 1947-61, dir., 1961-66; spl. lectr. Faculty Medicine, U. Queensland, Brisbane, 1939-65. Recipient Joint Cilento medal, 1939, Britannica Australia award for medicine, 1965. Fellow Australia New Zealand Assn. for Advancement Sci., Royal Soc. Tropical Medicine and Hygiene, Royal Australasian Coll. Physicians, Australian Acad. Sci.; mem. Australian Med. Assn., Coll. Pathologists Australia, Australian Soc. Allergists. Research and numerous publs. on fevers of obscure nature in Queensland including 1st identification of Q fever and pomona type of leptospirosis. Home: Mt. Nebo via Brisbane, 4520 Queensland, Office: Queensland Inst. Med. Research, Herston Rd., Herston, 4006 Brisbane, Queensland, Australia.*

DE RUBEIS, see Rossi, Girolamo.

DERWIDUÉ, Léon, Belgian mathematician; b. Fontaine-l'Eveque, Belgium, June 7, 1914; s. Jules and Léonie (Beugnies) D.; licencé in math. scis. U. Liège (Belgium), 1937; D.Sc., 1945; agrégé, 1947; m. Jeanne Vanbockrijck, June 26, 1945; 1 son, Jules. With Nat. Found. Sci. Research (FNRS), 1939-53, asso., 1949-53; asst. U. Liège, 1947-48, in charge of works, 1948-53; faculty Poly. Faculty Mans (Belgium), 1953-—, prof., 1955-—; prof. Univ. Center Mans, 1965-—. Mem. Royal Socs. Scis. Liège, Belgian Center Math. Research. Author: Introduction à l'algèbre supérieure et calcul numérique algebrique, 1957; Compléments d'analyse numérique et mathématiques pour ingéneirues et physiciens, 3 vols., 1967-68; 5 other books, also many articles. Research in algebraic geometry encompassing fundamental elements of irrational transformations, gen. irrational transformations, recomposition of singularities of algebraic varieties, recomposition of gen. irrational transformations in products of elementary transformation, studies of exceptional varieties and cyclic involutions carrying an algebraic variety; work in numerical analysis including devel. of original methods of numerical calculus of solutions of algebraic equations, of values and vectors of matrices, of periodic solutions of differential systems; research in nonlinears analysis on existence of periodic solutions. Home: 134, rue Grande, Nimy (Hinaut), Belgium. Office: Faculté Polytechnique, rue de Houdain, Mans, Belgium.*

DERYAGIN, Boris Vladimirovich, Russian phys. chemist, born August 4, 1902; graduated Moscow University, 1922; D.Sc. (hon.), Clarkson Coll. With Inst. Phys. Chemistry, USSR Academy of Sciences, 1920-32, dir. dept. surface phenomena, 1935-—. Del., Internat. Congress on Surface Activity, London, 1956. Corr. mem. USSR Acad. Scis. (Lomonosov prize 1958). Co-author: Adhesion, 1949; Film Coating Theory, 1964, others. Editor: Research in Surface Forces, 3 vols., 1962, 66, 68. Research on spl. properties of thin layers of liquids, discovered disjoining pressure of thin layers, boundary phases; proposed theory of adhesion of solid bodies, 1933-34; developer electric theory of adhesion (with N. A. Krotova); proposed theory of coagulation of dispersed systems by electrolytes; measured molecular attraction of solid bodies, 1954-58. Address: USSR Acad. Scis., Leninsky prospect 14, Moscow, USSR.

DERZHAVIN, Aleksandr Nikolaevich, Russian ichthyologist; b. Dec. 5, 1878; ed. Kazan (USSR) U.; organizer, head Baku Ichthyologic Lab., 1912-26; prof. Baku U., 1919-23; dir., dep. dir. Pacific Fisheries Inst., 1927-32; dir. sect. water animals, inst. zoology Azerbaijan SSR Acad. Scis., also mem. acad. Research and publs. on water fauna of Caspian and the Pacific.

DESAGULIERS, John Théophile, natural philosopher, mathematician; b. La Rochelle, France, Mar. 13, 1683; s. Jean Desaguliers; brought to Eng., 1685; B.A., Christ Ch., Oxford, Eng., 1710, M.A., 1712; LL.B., Oxford, 1718; m. Joanna Pudsey, 1712; children—Thomas, John Theophilus, Alexander. Ordained as deacon to ministry, 1710, served at Whitechurch, Eng., from 1714, also chaplain to Duke of Chandos and Frederick, Prince of Wales; lectr. exptl. philosophy Hart Hall, Oxford, 1710. Recipient prize for best article on electricity Acad. Bordeaux (France), 1752. Fellow Royal Soc. (demonstrator, curator 1714, Copley Gold medal 1742); mem. French Acad. Scis. Author: Treatise of Fortifications, 1711; Fires Improved; being a new Method of Building Chimneys, so as to prevent their Smoaking, 1716; Physics: Mechanical Lectures, 1717; A Course of Mechanical and Experimental Philosophy, 1724; an Experimental Course of Astronomy, 1725; A Course of Experimental Philosophy (influenced Benjamin Franklin), 2 vols., 1734; also various papers. Translator: The Motion of Water and other Fluids, 1718. Played important part in propagation of Newtonian philosophy in Europe; inventor planetarium; carried on Gray's work in electricity and magnetism; coined term conductor; introduced concept of conductors and non-conductors into teaching of electricity, 1742; measured flow resistance of water in tubes, also flow resistance of freely falling body in various media; tried to prove experimentally existence of ratio between kinetic energy and speed of a body; determined friction of cylinders; calculated air speed of new centrifugal ventilators; investigated pyrometric and moisture measurements, also causes of cloud formations and rain patterns; studied human muscle power. Died London, Eng., Feb. 29, 1744.

DESAINS, Edouard, French physicist; b. Saint-Quentin, France, 1812; prof. Metz, France, Lycée of San Luis, Paris; measured (with La Provostaye) specific heat of ice, 1845; studied melting of phosphorus, polarisation on glass surface areas; confirmed theory of capillary phenomenon through accurate measurement, 1854. Died 1865.

DESAINS, Paul Quentin, French physicist; b. St-Quentin, France, July 12, 1817; studied at St. Quentin and in Paris, France; tchr. sci. in Caen, France and Paris; held chair physics Sorbonne, Paris, 1853-85. Research on terrestrial magnetism, law cooling, reflection heat rays on metals; measured Newtonian rings; proved (with La Provostaye) radiant heat is disturbance in air which spreads by waves. Died Paris, May 3, 1885.

DE SAINT-COME, Jean, French surgeon, monk; b. France, 1703; improved operation for removing bladder stones and invented lithotome caché instrument for this purpose, 1747. Died 1781.

DE SAINT-VENANT, Adhémar-Jean-Claude Barré (comte), French mathematician, physicist; b. Billiers-en-Bière, France, Aug. 23, 1797; ed. Polytechnic Sch.; with explosives service; became asso. with service of bridges and roads, 1823; dep., then asso. prof. applied mechanics Sch. Bridges and Roads, 1837-42; named prof. Sch. Agronomy, Versailles, France, 1850; ret. with rank of chief engr., 1852; mem. French Acad. Scis., 1868. Author: Table of Formulas on the Theory of Curves in Space, 1844; Roll on the Swelling Sea, 1871; On the Diverse Manners of Presenting the Theory of Light Waves, 1872. Experimented (with Wantzel) on law of effusion of gases, 1839, set up a corresponding law; gave solution for light deformations of elastic curves, 1844; extended theory of elasticity in research on torsion and flexure, 1855-64, also in his amplification of rev. math. treatment due to Clebsch; formulated slide theory (St.-Venant's problem), 1855-56; also investigated infinitesimal geometry, hydrodynamics. Died St.-Ouen, France, Jan. 6, 1886.

DE SAINT-VINCENT, Grégoire, Flemish mathematician; b. Bruges, Flanders, Sept. 8, 1584; studied under Clavius, Rome Italy. Became Jesuit; prof. at Prague for 2 years; tutor of Don Juan of Austria; librarian, prof. city of Ghent, Flanders. Author: These de cometis, 1619; Theoremata mathematica scientiae staticae, 1624; Opus geometrium quadrature circuli et sectionum coni (sect. on squaring circle attacked by Descartes, Mersenne, and Roberval, but contains several basic discoveries on conic sects.) 1647; Opus geometricum ad mesolabium per rationum proportionalitatumque novas proprietates, 1668. Worked on problem of squaring circle; applied pre-calculus methods to various quadrature problems; 1st to use geometric series on paradoxes of Zeno; employed infinitesimals when dealing with conics, surfaces and solids; used method of transformation of 1 conic into another (per subtendas); deduced property of hyperbola; described constrn. of certain quartic curves, called virtual parabolas of St. Vincent. Died Ghent, Jan. 27, 1667.

DE SAINTE-ANDRÉ, Nathaniel, physician; b. Switzerland, 1680; m. Lady Elizabeth (Capel) Molyneux, May 17, 1730; probably unqualified practitioner (no Barber-Surgeons apprenticeship recorded); anatomist in household of George I of Eng., from 1723; lost court favor after supporting and publishing claim of Mary Tofts of Godalming that she gave birth to rabbits (depicted in Hogarth's engraving Cuniculari, or the Wise Men of Godleman in Consultation), also suspected (without proof) of poisoning his wife's 1st husband; gave pub. lectures on anatomy; pub. Garengeot's Treatise of Chirurgical Operations, 1723; affirmed that Paracelsus' 3 principles could be reduced to alkaline and acid salts. Died Southampton, Eng., Mar. 1776.

DE SAINTE-HILAIRE, Augustin François César Prouvencal, French naturalist; b. Orleans, France, Oct. 4, 1799; mem. expdns. to Brazil, 1816-22; prepared large bot. and zool. collections described in his works; mem. French Acad. Scis. Author: (with A. de Jussieu and Cambessedes) Tropical Flora of Brazil, or a History and Description of All the Plants which Grow in the Different Provinces of Brazil, 3 vols., 1824-25; Voyage in the Provinces of Rio-Janerio Minas-Geraes, 1830; Voyage in the Diamond District and on the Seaboard of Brazil, 1833; Voyage to the San Francisco River, 1847-48; Agriculture and the Raising of Cattle in the Campos-Geras, 1849. Died Orleans, Sept. 30, 1953.

DE SAINTE-JACQUES DE SILVABELLE, Guillaume, French astronomer, mathematician; b. Marseille, France, Jan. 18, 1722; became dir. Marseille obs., 1764; mem. French Acad. Scis. Author: On the Procession of the Equinoxoes and, in General, on All the Movements of the Earth and on the Variations of the Orbital Planes of All the Planets. Research in astronomy, hydrodynamics, clockmaking; formulated law of flow of fluids through a hole in a container. Died Marseille, Feb. 10, 1801.

DESALVA, Salvatore Joseph, Am. pharmacologist; b. N.Y.C., Jan. 14, 1924; s. Nicolo C. and Frances (Caldarella) DeS.; B.S., Marquette U., 1947; M.S., U. Ill., 1949; Ph.D., Stritch Sch. Medicine, Loyola U., Chgo., 1958; m. Elaine Mae Radloff, June 14, 1948; children—Salaine, Christopher, Stephanie, Steven, Gregory, Peter, Philip, Deirdre. With Marquette U., 1947-49, Milwaukee County Gen. Hosp., 1950, U. Ill. Med. Center, 1951-52, Chgo. Coll. Optometry, 1951-53, Armour Lab., 1953-59, Loyola U., 1957-59; head pharmacology Colgate-Palmolive Research Center, Piscataway, N.J., 1959-—. Lectr., Hahnemann Med. Sch., 1962-—. Mem. Am. Soc. Pharmacology and Exptl. Therapeutics, Soc. Exptl. Biology and Medicine, A.A.-A.S., N.Y. Acad. Scis. Editor: Biomedical Electronic Instrumentation, 1965. Research and numerous publs. on exptl. neurology, evolution of new and old mammalian brain, pharmacology of centrally acting muscle relaxant drugs, interdependency of endocrine and nervous system, electronic instrumentation; discoverer styramate and inhibitory spinal reflexes; inventor nasal probe for measuring nasal decongestion in humans and animals. Home: 83 DeMott Lane, Somerset, N.J. 08873. Office: 909 River Rd., Piscataway, N.J. 08854.*

DE SANCTA SOPHIA, Giovanni (Giovanni da Santa Sofia), Italian physician, philosopher; flourished middle 14th century; s. Niccolo (XIV-1); med. degree, 1353. Tchr. medicine, Padua, 1353, then Bologna, 1388; returned to Padua, 1389. Author: Practica Medicine (180 chapters); Consilium ad Pestilentiam (tract on plague, giving advice considered more sensible and rational than most of his contemporaries); also commentaries on Hippocrates, Galen and Ibn Sina. Became popular and famous as physician and tchr. of philosophy and medicine. Died Padua, 1389.

DE SAPORTA, (Louis-Charles-Joseph) Gaston, Marquis, French paleontologist; b. Ste.-Zacharie, France, July 28, 1823; mem. French Acad. Scis. Author: Prodrome d'un flore fossile des travertins anciens de Sezanne, 1868; Les recherches sur les végétaux fossils de Meximieux, 1875; Les phanérogames, 1885; Origine paléontologique des arbres cultivés, 1888. First in France to study fossil dicotyledons. Died Aix-en-Provence, France, Jan. 17, 1895.

DESARGUES, Gérard, French mathematician; b. Lyons, France, Mar. 2, 1593; s. Jane (Croppet) Desargues; engr., architect; army officer engring. sect. at siege of LaRochelle (1627-28); lectr. math. Paris, France, 1620-30. Writings include: Traité de la section perspective, 1636; Brouillon project, 1639. Investigations in geometry; studied conic sections, developed theory of involution and transversals, theorems known by his name; with B. Pascal formulated basic rules of synthetic projective geometry, introducing theory of perspectives; his writings lost until 1864. Died Lyons, Oct. 8, 1662.

DESAULT, Pierre Joseph, French anatomist; b. Feb. 6, 1744, Magny Vernois, France; asst. to barber surgeon at Magny Vernois; worked in mil. hosp. at Belfort; practiced medicine under doctors at Belfort; went to Paris, France, to study under Antoine Petit, 1762; founded sch. anatomy in Paris, 1766; became chief surgeon Charity Hosp., 1782; surgeon at Hôtel Dieu, 1788; later founded surgical school there. Author: Traité des maladies des voies urinaires, 1802; Oeuvres chirurgicales de Desault, 2 vols., 1798. Invented surg. technique (Desault's apparatus); developed new surgical instruments; introduced clinical method of instruction in France; observed patient to study diseases; popularized topographical conception of anatomy. Died Paris, June 1, 1795.

DE SAUSSURE, Ferdinand, Swiss linguist; b. Geneva, Switzerland, 1857; s. Henri; student U. Geneva; studied linguistics U. Leipzig (Germany), 1877-78, U. Berlin (Germany), 1878-79; doctorate U. Leipzig, 1880. Received chair comparative grammar Sch. Higher Learning, Paris, 1891; later became prof. U. Geneva. Author: Mémoire sur le systéme primtif des voyelles dans les langues Indo-Européenes, 1879; Cours de linguistique genérale, 1916; Recueil, 1922. Worked on theory of Indo-European langs. Died 1913.

DE SAUSSURE, Henri, Swiss entomologist; b. Geneva, Switzerland, 1829; mem. expdn. to C. Am.; research and publs. on orthoptera and hymenoptera (classes of insects). Died Geneva, 1905.

DE SAUSSURE, Horace Bénédict, Swiss naturalist, physicist; b. Geneva, Switzerland, Feb. 17, 1740; at least 1 son, Nicolas-Théodore; prof. physics and exptl. philosophy, Geneva, 1762-86; founder Soc. for Advancement of Arts, Geneva, 1772; Fellow Royal Soc., 1788; mem. French Acad. Scis. Author: De praecipuis errorum nostrorum causis ex mentis facultatibus oriundis, 1762; De electricitate, 1766; De aqua, 1771; Voyages dans les Alpes, 1779-96; Sur l'hygrométrie, 1783. Made 1st ascension of Mt. Blanc (except for guides), thus considered founder of modern mountaineering; introduced word geology into sci. use; studied terrestrial structure of globe and high mountains, effects of currents on mountains; developed theory on accumulation of water reservoirs on high mountains, also early theory on geol. strata, their distbn. and deformation; 1st to take a sci. attitude toward meteorology; gave 1st report on ann. variation in fair weather field; discovered 15 varieties of minerals; eponym of mineral saussurite; inventor filiment hygrometer, 1783, cyanometer, 1791, dicephanometer for comparing color of sky and transparence at different altitudes, 2 kinds of electrometers, an aneurometer. Died Geneva, Jan. 22, 1799.

DE SAUVAGES, François Boissier, French physician; b. Alais, France, 1706; M.D., Montpellier, became prof., circa 1740; fellow Royal Soc.; mem. royal socs. Stockholm, Berlin. Author: Nouvelles classes de maladies, 1732; Nosology methodica, 5 vols., 1732, 1763. Compiled a classification of diseases, 1732. Died 1767.

DE SAVITSCH, Eugene, surgeon, pathologist; b. Petrograd, Russia, Aug. 19, 1903; A.B., Denver U., 1932; postgrad. Pasteur Inst., Paris, France; M.D., U. Chgo., 1935; m., 1952. Asst. immunologist N.J. Hosp., Denver, 1927-31; Dennison Found. fellow, 1928-29; research staff Albert Merritt Billings Hosp., Chgo., 1935; asso. surgeon Laennec Hosp., Paris, 1936-38; practics medicine, specializing in surgery, 1939—; staff Home for Incurables, 1940—, Doctor's Hosp., Washington. Mem. Congo expdn., 1936. Dennison Found. fellow, 1928-29; Belgian-Am. Edn. Found. fellow, Brussels, 1936-37. Fellow, Internat. Coll. Surgeons, A.M.A., Coll. Chest Surgeons; mem. Soc. Exptl. Biology, Soc. Exptl. Pathology. Research and publs. on vitamins and antibiotics in Tb, sympathetic nervous system, radon in indolent ulcers.

DESBOIS DE ROCHEFORT, Louis, French physician; b. Paris, France, 1750; s. Louis-René and Marie (Syraud) D. de R.; ed. at St. Barbara Coll., Paris; studied medicine Faculty of Paris; m. Marie LeRoi, 1782; 2 daus. Became physician at Charity Hosp., 1780, also taught clin. medicine. Author: Cours élémentaire de matière médicale, 1789. One of 1st to teach clin. medicine. Died Jan. 1786.

DESCARTES, René (in Latin: Cartesius; also known as Du Perron, after a small estate destined for his inheritance), French mathematician; b. La Haye (now La Haye-Descartes), near Poitiers, Touraine, France, Mar. 31, 1596; s. Olympe-Joachim Ier, conseiller in parlement of Rennes, in Brittany, and Jeanne (Brochard) D.; ed. in La Haye by his maternal grandmother and a nurse, 1597-1606; studied at Jesuit Collège Royale de La Flèche, Anjou, 1606-14 (possibly 1605-13); studied medicine (possibly) and law at Univ. of Poitiers, 1614-16; bachelier and licencie en droit (canon and civil law), 1616; never married; one dau., Francine D. (b. 1635, d. 1640) by Hélène Jans. Went to Holland, served as a volunteer in army of Maurice of Nassau, autumn 1618; met Isaac Beeckman, with whom he discussed problems of mathematics, philosophy, music, Nov. 1618; left Holland for Denmark, then Germany, where he volunteered for service with the Catholic forces of Duke Maximilian of Bavaria; experienced famous night of enthusiasm, had three dreams which inspired discovery of foundations of an admirable science, Nov. 10-11, 1619; renounced military life, 1620; traveled through Germany and Holland to France, arriving 1622; sold estate in Poitou, invested proceeds so that he might have adequate annual income to allow pursuit of his studies; visited Italy, possibly accomplishing pilgrimage to Loretto, autumn 1623-spring 1625; in France, mostly Paris, summer 1625-autumn 1628; Cardinal de Bérulle told Descartes study of philosophy an obligation of conscience for him, autumn 1627; in Holland, Oct. 1628; returned to Paris, winter 1628-29; permanent residence in Holland, 1629-49, with many changes of address (at least 24 residences in 13 cities and towns, including Franekar, Amsterdam, Leyden, Deventer, Utrecht, Santpoort, etc.); made only three short return trips to France: May-Nov. 1644, to settle family affairs after father's death; June-Nov. 1647, to receive pension from Cardinal Mazarin; met Pascal; May-Aug. 1648, to receive royal honors; traveled to England, 1630; to Denmark, 1634; involved in philosophical-theological polemics, Utrecht, 1641-45, Leyden, 1647; left Amsterdam for Stockholm, Sept. 1649, at invitation of young Queen Christina (she requested philosophical instruction at 5:00 A.M.). Author: Compendium musicae, 1618; Regulae ad directionem ingenii, compiled 1628, pub. 1701; Le monde, finished 1633, pub. 1664; Discours de la méthode, 1637; La Dioptrique, 1637; Les Météores, 1637; La géometrie, 1637; Meditationes de prima philosophia, 1641; Principia philosophiae, 1644; Les passions de l'ame, 1649; others. Founded analytic geometry by showing how geometric forms may be systematically studied by analytic (i.e. algebraic) menas; introduced several improvements in mathematical notation; wrote exponents in Arabic numerals; used small letters throughout equations; used letters at beginning of alphabet to denote known quantities; introduced symbols for equality, square and cube roots; announced Descartes rule of signs, which sets a limit to number of roots in equation (in any equation there can be as many positive roots as there are changes of signs, as many false roots as there are successive pairs of plus or minus signs); described various kinds of ovals (Descartes' ovals) and investigated their properties; described method of drawing tangent to curve; made many other mathematical advances; sought to elaborate a physics to replace at once moribund scholastic Aristotelianism and animistic Renaissance philosophies of nature; made radical separation between mind and matter; identified matter with space (extension); hence argued there can be no void; sought mechanical explanations for all phenomena, i.e. to explain qualitative phenomena solely in terms of matter and motion; introduced vortices to account for planetary motions, gravity and magnetism; formulated laws of nature, including law of inertia; studied (not very successfully) impact phenomena; investigated optics; discovered law of refraction; explained rainbows; observed parhelia (false suns) and undertook study of meteorology (atmospheric phenomena, including clouds, winds, rain, etc.); made numerous anatomical dissections, and investigated anatomy of eye and mechanism of vision; also well known for philosophic thought; his influence has been profound. Died Stockholm, Sweden, Feb. 11, 1650.

DESCEMET, Jean, French surgeon; b. Paris, France, Apr. 20, 1732; ed. Faculty Medicine, Paris; practice surgery, Paris. Author: Observations sur la choroïde, 1768. Research on structure of eye, skin; treatment of gout and measles; described membrane between substantia propria and endothelial layer of cornea (Descemet's membrane). Died St. Denis, France, Oct. 17, 1810.

DESCH, Cecil Henry, Brit. metallurgist; b. 1874; s. Henry Thomas Desch; ed. Finsbury Tech. Coll.; Ph.D., Würzburg (Germany) U.; D.Sc., Univ. Coll., London; LL.D., U. Glasgow; hon. doctorate, Leoben, S. Austria; m. Elison Ann Macadam, 1909; 2 daus. With metall. dept. King's Coll., London, 1902-07; lectr. metall. chemistry U. Glasgow, 1909-18; prof. metallurgy Royal Tech. Coll., Glasgow, 1918-20; prof. metallurgy U. Sheffield (Eng.), 1920-21; supt. metallurgy dept. Nat. Phys. Lab., 1932-39. Fellow Royal Soc., 1923; mem. Faraday Soc. (pres. 1923-28), Inst. Metals (pres. 1938-40), Iron and Steel Inst. (pres. 1946-48), French Acad. Scis. (corr.). Author: Metallography) Chemistry of Cement and Concrete; Intermetallic Compounds; Chemistry of Solids; also articles on sociology. Died June 19, 1958.

DE SCHAEPDRYVER, André-Felix-Georges-Marie-Ghislain, Belgian physician; b. Gent, Belgium, Apr. 6, 1926; s. Paul and Marie (Verbrugge) De S.; M.D., Gent U., 1951, Ph.D., 1959; m. Thérèse Caenepeel, Aug. 19, 1954; children—Jan, Luc, Patricia, Marc. Prof., head dept. pharmacology U. Gent, 1962—. Fellow Swedish Inst., Karolinska Inst., 1951-52, Mayo Found., Mayo Clinic, Rochester, Minn., 1952; Fulbright Research fellow, Riker fellow in pharmacology NIH, Bethesda, Md., also Duke, Durham, N.C., 1959-60; NATO Research fellow, Gt. Britain, 1965-66; laureate Royal Flemish Acad. Medicine Belgium, 1958, Royal Acad. Medicine Belgium, 1960; recipient R.I.T. prize Royal Flemish Acad. Medicine Belgium, 1960-62. Mem. Belgian Soc. Physiology and Pharmacology, Belgian Soc. Internal Medicine, Belgian Soc. Clin. Investigation, Belgian Soc. Pharmacotherapy, Belgian Soc. Endocrinology, Royal Flemish Acad. Medicine (corr.), Argentina Soc. Pharmacology (hon.), N.Y. Acad. Scis., Royal Soc. Medicine Eng. (affiliate). Author: (with C. Heymans, A. Simonart, G. De Vleeschhouwer) Beginselen der Farmakoterapie, 1959. Editor: (with M. Hebbelinck) Doping, 1965. Research on pharmacology of cardiovascular and autonomic nervous systems, pharmacodynamics of anti-hypertensive drugs, diagnosis of amine producing tumors. Home: 3 R. Soenenslaan St. Denijs-Westrem, Belgium. Office: 3 A. Baertsoenkaai, Gent, Belgium.*

DES CHALES, Claude François Milliet, French mathematician, physicist; b. Chambery, France, 1621; mem. Soc. of Jesus; missionary in Turkey; taught in various Jesuit schs.; prof. math. and philosophy, Lyons, France; rector at Chambéry. Author: Euclidis elementorum libr., VIII, 1660; Cursus seu mundus mathematicus, mathesin tribus tomis complectens, 3 vols., 1674; Principes généreau de la géographie mathématique, 1676; les Élémens d'Euclide expliqués d'une manière nouvelle et trés facile, 1677; l'Art de naviguer, 1677; l'Art de fortifier, 1677; Mundus mathematicus, pub. 1690. Showed air resistance increased with descent; demonstrated the falsity of Descartes' theory on thrust. Died Turin, Italy, Mar. 28, 1678.

DESCHAMPS, Joseph-François-Louis, French surgeon; b. Chartres, France, 1740; ed. Paris Sch. Surgery; 1 son, Joseph-Louis. Apptd. dir. Charity Hosp., 1788; became cons. surgeon to Emperor Napoleon, 1801. Decorated Cross of Legion of Honor, 1816. Mem. French Acad. Scis., 1811, Acad. Medicine. Author: Traité historique et dogmatique de l'opération de la taille, 1796-97. First in France to perform operations on waist using Hunter's method. Died Paris, Dec. 8, 1824.

DESCHIENS, Robert Édouard André, French physician, biologist; b. Ile St.-Denis, France, Mar. 12, 1898; s. Victor and Pauline (Astruc) D.; Docteur en Médecine, Faculty Medicine Paris, 1921; m. Andrée Joubert, Nov. 29, 1932. Head lab. Hosps. of Paris, 1923-25; head lab. Mil. Hosp. Fes, Morocco, 1926-27; asst. Pasteur Inst., Paris, 1927-36, head lab., 1936-41, head dept., prof., 1941—. Cons. WHO, 1951—; mem. Higher Council Pub. Hygiene, 1951—. Recipient Colonial medal, 1926. Mem. Soc. Exotic Pathology (sec. gen. 1932—), Nat. Acad. Medicine France, Acad. Overseas Scis., Soc. Biology, (all French socs.), Royal Soc. Tropical Medicine and Hygiene, Eng., Belgian Soc. Tropical Medicine, Soc. Tropical Medicine Netherlands. Author: L'ambiasis, 1965; La coprologie en pratique médicale, 1930; also articles. Research in epidemiology, parasitology and immunology, including amibiasis (role of associated bacterial flora), bilharziosis (prophylaxis by chem. molluscacides), verminour and hypereosinophilic toxic substances, anthelmintic medications (piperazine, immunology of parasitic diseases). Home: 9, Av. Henri Martin, Paris XVI. Office: 25, rue Docteur Roux, Paris XV, France.*

DESCHISAUX (or DESCHIZAUX), Pierre, botanist; b. Macon, France, Mar. 31, 1690; s. Jean and Claudine (Foillard) D.; student medicine Caen, France. Went to Russia, 1724; designer bot. garden, St. Petersburg, Russia. Author: Voyago de Moscovie, 1727; wrote on natural history of Russian plants. Died 1730.

DE SCHWEINITZ, George Edmund, Am. physician; b. Phila., Oct. 26, 1858; s. Edmund and Lydia de S.; A.B., A.M., Moravian, 1876; M.D., U. Pa., 1881, LL.D., 1914; L.H.D., Moravian Coll; D.Sc., U. Mich., 1922, Harvard, 1927. Prof. ophthalmology, U. of Pa. Grad. Sch. Medicine, 1902-24, emeritus, 1924—; cons. ophthalmologist Phila. Hosp.; Bowman lectr., London, 1923. V.p. Pa. Inst. for Instrn. of Blind. Recipient plaque French Soc. Ophthalmology, 1924; Howe prize in ophthalmology, 1927; Leslie Dana prize for prevention of blindness, 1930. Mem. A.M.A. (pres. 1922-23), Am. Ophthal. Soc. (pres. 1916). Author: Diseases of the Eye, 1924; Diseases of the Eye, Ear, Nose and Throat (with Dr. Randall), 1899; Toxic Amblyopias, 1896 (Alvarenga prize essay). Am. editor Haab's Ophthalmoscopy and External Diseases of Eye and Operative Ophthalmology; Pulsating Exophthalmos (with Dr. Holloway); Ophthalmic Year Book (with Dr. Jackson), 1905-09; editorial bd. for med. and surg. history of World War I. Contbr. numerous articles and monographs on ophthal. and neurol. subjects. Died Aug. 22, 1938.

DES CLOIZEAUX, Alfred-Louis-Olivier Legrand, French mineralogist; b. Beauvais, France, Oct. 17, 1817. Lectr., École Normale Supérieure; prof. Mus. Natural History. Mem. French Acad. Scis., 1869 (v.p. 1888, pres. 1889). Author: Manuel de minéralogie, 1862; also over 100 memoirs on mineralogy. Research on pseudomorphism, and optical properties of natural and artificial crystals; examined optical properties of 468 minerals and salts; discovered rotatory polarization of cinnabar and strychnine sulphate; perfected microscopes of Amici and Nörrenberg. Died Paris, May 6, 1897.

DES COUDRES, Theodor, German physicist; b. Veckerhagen/Weser, Germany, Mar. 13, 1862; s. Julius and Anna Henriette (Rosenstock) Des C.; studied in Geneva, Switzerland, Leipzig, Munich, Berlin (all Germany); doctorate, Berlin, 1887. Became asst. to G. H. Wiedemann, Leipzig, 1889; joined faculty in Leipzig, 1891, apptd. prof., 1903; became prof. applied electricity U. Göttingen (Germany), 1895; named prof. theoretical physics Würzburg (Germany), 1901. Developed method of measuring velocity of cathode rays which was 1st executed by Emil Wiechert; measured velocity and specific charge of alpha rays. Died Leipzig, Oct. 8, 1926.

DESCROIZILLES, François-Antoine-Henri, French chemist; born Dieppe, France, June 11, 1751; s. François Descoizilles, studied chemistry at Paris under Rouelle, pharmacy at Rouen, France; 1 son, Paul. Royal demonstrator pharmacy, Rouen; founder factory for bleaching canvas at Rouen, 1787; dir. mine inspections; became gen. counselor for mfg., 1806; manufactured hardware and housewares; set up factories for mfg. salts of alkali and chlorine. Mem. Acad. Rouen. Author: Instruction sur l'aérométrique, 1806; Notice sur la fermentation vineuse, 1822; Description du berthollimeter, 1823; numerous treatises. Research on phys. properties of air, fermentation of wine, refining of salt-peter; contributed to volumetric analysis with apparatuses he derived, alkalimeter, acetimeter, and Berthollimeter studied technical chemistry; devised method of conserving stored grains. Died Paris, France, Apr. 14, 1825.

DE SELYS-LONECHAMPS, Michel Edmond, naturalist; b. Paris, May 25, 1813; ed. U. Liège (Belgium); mem. senate of Waremme, pres., 1880. Author: Études de micromammalogie, 1939; Tableau de dibellubdées d'Europe, 1840; Belgian Fauna, Vol. I, 1842. Described birds, reptiles, fish of Belgium. Died 1900.

DE SENARMONT, Henri Hureau, French physicist; b. Broué, France, Sept. 6, 1808; student at a poly. sch.; prof. minerology Ecole supérieure des mines; prof. physics Ecole polytechnique; head engr. Dept. Mines, 1848. Mem. Franch Acad. Scis. Author: Traité des modifications sur la lumière polarisée par les reflections aux surfaces cristallines, 1840; Essai de description géologique des départements de Seine-et-Oise et de Seine-et-Marne, 1844. Research on optical and caloric properties of crystals, isomorphic bodies, reflection of polarized light on a plain metal, petrification and the prodn. of minerals in veins of ore; proposed theory of double refraction. Died Paris, June 30, 1862.

DESER, Stanley, physicist; b. Rowne, Poland, Mar. 19, 1931; s. Norman and Miriam (Melamed) D.; B.S. summa cum laude, Bklyn Coll., 1949; M.A., Harvard, 1950, Ph.D., 1953; m. Elsbeth Klein, Aug. 25, 1956; children—Toni, Eva, Clara, Abigail. Mem. Inst. for Advanced Study, Princeton, N.J., 1953-55; instr. for theoretical physics, Copenhagen, Denmark, 1955-57; lectr. Harvard, 1957-58; faculty Brandeis, U., 1958-—, prof. physics, 1965-—; vis. scientist Centre Europeende Recherche Nucleare, 1962-63; vis. prof. Sorbonne, Paris, 1966-67. Mem. Am. Phys. Soc. Research and publs. on quantum field theory, elementary particles, gen. relativity. Home: 45 Whitney Rd., Newtonville, Mass. 02160. Office: Physics Dept., Brandeis U., Waltham, Mass. 02154.*

DE SERRES, Olivier, French agronomist; b. Villneuve-de-Berg, France, 1539; s. Jean de Serres; m. Marguerite d'Harcaus, 1559. Became deacon Protestant ch. at Berg; established model farm, Pradel, France, called to Paris by king. Author: Cueillette de la soye par la nourriture des vers qui la font, 1599; Theatre d'agriculture des champs, 1600; la Seconde richesse du murier blanc, 1603. Founder French agrarian sci.; 1st to practice systematic crop rotation; imported madder, hops, maize and mulberry. Died Pradel, July 2, 1619.

DES ESSARTZ, Jean-Louis-Charles, French physician; b. Bragelogne, France, Oct. 29, 1729; ed. Beauvais Coll., Paris, France; docteur en médecine, U. Rheims (France); prof. pharmacy and surgery Paris Faculty Medicine, 1770-75, also dean. Mem. French Acad. Scis., 1795. Author: Traité de l'éducation corporelle des enfants, 1760; Reflexions sur la musique . . . comme moyen curatif, 1803. Research on smallpox and vaccination, electric treatment of nervous diseases, childhood edn., therapeutic value of music. Died Paris, Apr. 12, 1811.

DESFONTAINES, René Louiche, French botanist; b. Tremblay, France, Feb. 14, 1750; s. René-Jean and Yvonne (Boulmerson) D.; docteur-ès-sciences, U. Rheims (France); studied at Paris, France beginning in 1782; m. Angèlique Perrosset, 1814; 1 dau., Marie. Made voyage to Tunisia and Barbary States, 1783; returned to France, 1785; became prof. Jardin du Roi, 1786; received chair of botany Museum of Natural History, Paris, 1793. Member of the Academy of Medicine, Soc. Agr., French Academy Sciences, 1783. Author: Flora atlantica, 3 vols., 1798; Tableau de l'École de botanique du Muséum, 1804; Voyages dans les régences de Tunis, 1830; Expériences sur la fécondation artificielle des plantes; Histoire des arbres et arbrisseaux qui peuvent etre cultives en pleine terre sur le sol de France. Discovered Algerian hardgrained wheat (Triticum durum), six-rowed barley (Hordeum hexastychum), a variety of oak with edible acorns (Quercus ballota); 1st to study date tree. Died Paris, Nov. 16, 1833.

DESGENETTES (Des Genettes), Nicolas-René Dufriche, French physician; b. Alençon, France, May 23, 1762; ed. Coll. St. Barbara, Paris; made voyage to Eng. and Italy, 1782-85; tchr. med. physics Val-de-Grâce Hosp., 1795; became chief physician of army for Egyptian campaign; named chief physician Val-de-Grâce, 1825; became prof. hygiene Paris Faculty Medicine, 1819. Mem. Acad. Medicine, French Acad. Scis., 1832. Author: Sur la topography physique et médicale de l'Egypte, 1799; Histoire médicale de l'armeé d'Orient, 1799. Research on physiology of lymphatic vessels, med. topography of Egypt; worked with plague in Napoleon's forces at Jaffa. Died Paris, Feb. 3, 1837.

DESGREZ, Alexandre, French chemist; b. Bannes, France, July 15, 1863; docteur en médicine, Paris (France) Faculty Pharmacy, 1895; became prof. chemistry, Paris Faculty Medicine, 1912; apptd. dir. research lab. Ministry War, 1914; became dir. Hydrological Inst., 1924. Decorated Commdr. Legion of Honor. Mem. French Acad. Scis., Acad. Medicine. Developed methods of gas protection for soldiers; research on transformation of ketones, hydration of acetylene hydrocarbons, application of chemistry to medicine; pioneered chem. research on causes of disease; suggested diabetics use a low sugar diet. Died Mennecy, France, Jan. 20, 1940.

DESHALIT, Amos, Israeli physicist; b. Jerusalem, Israel, Sept. 29, 1926; s. Moshe and Ada (Rapoport) De S.; M.Sc., Hebrew U., Israel, 1949; Dr.Sc.Nat., Eidg. Tech. Hochsch., Zurich, Switzerland, 1951; m. Nechama Maisus, Sept. 25, 1950; children—Arie-Ehud, Avner. Research asst. Princeton, 1952; research asso. Mass. Inst. Tech., 1952-53, vis. prof., 1960-61; head nuclear physics dept. Weizmann Inst. Sci., Rehovoth, Israel, 1954-64, sci. dir., 1961-63, dir. gen., 1966-—; vis. prof. Hebrew U., Jerusalem, 1956-61; vis. prof., Stanford, 1960. Mem. Israel AEC, 1956-58; chmn. Com. for Promotion of Teaching of Natural Scis. in Secondary Schs., 1964-—; sci. cons. Ministry Def., 1956-58. Recipient Israel prize, 1965. Mem. Israeli Acad. Scis., Israel Phys. Soc. (past pres.). Author (with Igal Talmi) Nuclear Shell Theory, 1963. Editor (with H. Feshbach, L. Van Hove) Preludes in Theoretical Physics, 1966. Research and numerous publs. on nuclear structure, nuclear radiations and nuclear reactions. Home: Neve Weizmann, Rehovoth, Israel.*

DESHAYES, Gérard Paul, French geologist; b. Nancy, France, May 13, 1795; student medicine, Strasbourg, France; also studied at Paris Mus. Prof. conchology Paris Mus. Natural History. Author: Ossements et coquilles fossiles, 2 vols., 1844; Catalogue of the Conchifera, 1853; Conchyliologie de l'ile de la Réunion, 1863; Traité élémentaire de conchyliologie, 1834. Research on fossil shells and mollusks of Paris basin. Died Boran, Oise, France, June 9, 1875.

DESIDERALISSIMUS, Dr., see Gilbert the Englishman.

DE SIEBOLD, Charles Gaspard, German surgeon; b. Nideck, Germany, 1737; became dir. Sch. Midwives, Wurtzbourg, Germany, 1774. First in Germany to perform sectioning of pubic symphysis. Died Wurtzbourg, 1807.

DE SIGUENZA Y GONGORA, Don Carlos, Mexican astronomer, mathematician; b. Mexico City, Mexico, 1645; ed. U. in Mexico City. Jesuit priest; prof. math. and astronomy U. Mexico. Author of publs. including: Infortunios de Alonso Ramirez; Relación de lo sucedido a la Armada de Barlovento; Alboroto y Motin de México del 8 de Junio de 1692. Studied comet of 1680. Died Aug. 22, 1700.

DESIO, Ardito, Italian geologist; b. Palmanova (Udine), Apr. 18, 1897; s. Antonio and Caterina (Zorzella) D.; Ph.D. in Geology, U. Florence; m. Aurelia Bevilacqua; children—Gian-Luca, Maria-Emanuela. Dir., Inst. Geology, U. Milan (Italy), also prof. geology. Mem. Soc. Geology and Geography. Contbr. over 285 articles to profl. publs. Mem. 16 overseas sci. expdns. Home: viale Maino 14. Office: Piazzale Gorini 15, Milan, Italy.

DESKINS, Wilbur Eugene, Am. mathematician; b. Morgantown, W. Va., Feb. 20, 1927; s. Wilbur Lawrence and Avis (Creasy) D.; B.S., U. Ky., 1949; M.S., U. Wis., 1950, Ph.D., 1953; m. Barbara Brown, Apr. 18, 1953; children—Lucinda Eugenie, Samantha Eugenie. Teaching assistant at the University of Wisconsin, 1949-51, fellow 1951-52, teaching asst., 1952-53, instr., 1953; instr. Ohio State U., 1953-55, asst. prof., 1955-56; asst. prof. Mich. State U., East Lansing, 1956-59, asso. prof., 1959-63, prof., 1963-—. Author: Abstract Algebra, 1964. Research and articles on algebra and group theory. Home: 2634 Roseland Av., East Lansing, Mich.*

DESLANDES, Pierre de Launay, French inventor; b. Avranches, France, May 21, 1722; ed. Ovatian Coll. at Soissons, École des ponts et chaussées. Became dir. St. Gobain Royal Mirror Mfg., 1758. Decorated chevalier Order of St. Michel, 1773. Member of the French Academy of Sciences, 1774. Improved methods of manufacturing mirrors by eliminating glass blowing, thus succeeded in mfg. mirrors of large size. Died Dec. 10, 1803.

DESLANDRES, Henri Alexandre, French astrophysicist; b. Paris, France, July 24, 1853; engr. École Polytechnique, Paris, 1874; became dir. Meudon Obs., 1907; apptd. dir. Paris Obs., 1927. Gold medal Royal Astron. Soc., 1913. Fellow Royal Soc., 1921; mem. French Acad. Scis., 1902, (became pres. 1920), Bur. Longitudes. Author: Spectres et bandes ultraviolets des métalloides avec une faible dispersion, 1881; Histoire des idées et des recherches sur le soleil, 1906; Recherches sur l'atmosphèresolaire, 1910. Invented velocity recorder, instrument related to the spectroheliograph; discovered independently of G. E. Hale, reversal of calcium H and K lines on solar disc; proved structural identity of various bands emitted by a molecule under given conditions, 1886-91; showed wave numbers of individual band heads are like those of individual lines in a band nearly in arithmetical progression; derived gen. formula for wave numbers; inferred presence of harmonic oscillations. Died Paris, Jan. 15, 1948.

DE SLUSE, René François Walter, Belgian mathematician; b. Visé-sur-Meuse, July 7, 1622; studied under van Schooten; student in Italy; became canon, Liège, Belgium. Author: Über die Construction der Gleichungen, 1659; Mesolabum s. duae mediae proportionales inter extremas datas per circulum et per infinitas hyperbolas vel ellipses, 1688; also numerous publs. on spiral points of inflection, finding of geometric means. Discovered method for drawing tangents to all geometrical curves, 1672; studied theory of primary numbers and celestial mechanics; demonstrated method of constructing equation of a power using circle and conic sections; pearls of Sluse, a family of curves, named after him. Died Liège, Mar. 19, 1685.

DES MARAIS, Andre, Canadian biologist; b. Pierreville, Que., Can., Apr. 23, 1919; s. Edmond and Antoinette (Shooner) DesM.; B.A., U. Montreal, 1940, B.Sc., 1943; Ph.D., Laval U., 1948; m. Simone Descheneaux, Aug. 15, 1944; children—Gilles, Pierre, Louis, Denise. Lectr. U. Montreal, 1943-45; asst. prof. U. Laval, 1945-55; asso. prof. U. Ottawa (Ont., Can.), 1956-58, prof., 1958-—, chmn. dept. biology, 1962-—, sec. Faculty Sci., 1963-67. Mem. biology selection com. Nat. Research Council, Ottawa, 1965-68. Fellow Rockefeller Found., 1950-51; Royal Soc. Can. fellow, 1962. Mem. A.A.A.S., Am. Physiol. Soc., Canadian Physiol. Soc., Assn. des Physiologistes (Paris). Research and numerous publs. on cold-acclimation, secretory activity of thyroid gland; excretion rates and metabolism of thyroid hormones; mode of action of inhibitors of calorigenic activity of thyroxine; relationship with cathecolamines; effects of ascorbic acid adminstrn.; relative importance of thyroidal and adrenocortical hormones. Home: 92 Drouin St., Ottawa 7. Office: 550 Cumberland St., Ottawa, Ont., Can.*

DESMAREST, Anselme-Gaétan, French zoologist; b. Paris, France, Mar. 6, 1784; s. Nicolas Desmarest; ed. École centrale, Paris; als studied under Cuvier; became prof. Alfort (France) Vet. Sch., 1814. Mem. Philomatic Soc., Acad. Medicine, French Acad. Scis., 1825, Royal Acad. Medicine, Entomol. Soc. France (hon.). Author: Manuel de minéralogie, 1827; Histoire naturelle des crustacés, 1828; Mammalogie, 1820-22; also numerous articles on fossils, mineralogy, crustaceans and French fauna. Described new mammal species; studied new species of rats. Died Alfort, June 4, 1838.

DESMAREST, Nicolas, French geologist; B. Aube, France, Sept. 16, 1725; ed. at Oratorium colls. in Troyce and Paris; insp. gen., dir. manufactures for France, 1788-92. Mem. French Acad. Scis., French Acad. (became dir. 1786). Author: Conjectures physico-méchaniques sur la propagation des secousses dans les tremblements de terre, 1756; Atlas encyclopédique; Dictionnaire de géographie physique. First clear proofs that valleys have been eroded by their streams; identified deposits of active and extinct volcanoes; distinguished between synclines and anticlines; demonstrated volcanic nature of basalt; in charge of inventory of natural resources of France; made detailed volcanic map of Auvergne. Died Paris, Sept. 28, 1815.

DESMARRES, Louis Auguste, French ophthalmologist; b. Évreux, France, Sept. 1810; docteur en médecine, 1839; opened eye clinic, Paris. Author: Mémoire sur une nouvelle méthode d'employer le nitrate d'argent dans quelques ophtalmies, 1842; Traité Thèorique et pratique des maladies des yeux, 1847. A founder of ocular surgery; introduced iridectomy for treatment of glaucoma in France; described type of dacryolith consisting of masses of Nocardia foersteri (Desmarres' dacryolith). Died Aug. 22, 1882.

DESMAZIÈRES, Jean-Baptiste-Henri-Joseph, French botanist; b. Lille, France, July 10, 1786; author: Agrostographie des départment du nord de la France, 1812; Plantes cryptogames du nord de la France, 1823; Plantes cryptogames de la France, 1853-60. Classified plants of N. France and Belgium; research on rhizomes of grasses, also cryptograms (mosses, ferns, algae) of N. Europe; made special library of bot. scis., also valuable collection of plants (willed to city of Lille). Died July 25, 1862.

DESMOULINS, Antoine, physiologist; b. 1796. Author: (With Magendie) Anatomie des systémes nerveux des animaux vertébrés, 1825; attributed senility to atrophy of brain from discovery that brains of elderly are lighter than those of younger adults. Died 1828.

DESNOS, Charles-Jules-Pierre, French engr.; b. Paris, 1832; ed. Paris École centrale; editor mag. L'Invention, 1852-70; joined Patent Office, 1860, became dir., 1877. Author: L'Annuaire des inventeurs, pub. annually, 1858-79; Du Régime de l'invention, 1862. Publs. on inventions patented in France, 1850-1880. Died 1882.

DESNOYERS, Jules-Pierre-François Stanislas, French earth scientist; b. Nogent-le-Rotrou, France, Oct. 9, 1800; student law, from 1820, geology, 1822-30; children—Edouard, Mrs. Milne-Edwards; named naturalist aide in geology Mus. d'histoire naturelle, Paris, France, 1833, became librarian, 1834. Founding mem. Commn. on Archives, 1941. Recipient prize l'Academic Inscriptions and Belles-Lettres, 1838. Mem. Société de l'histoire de France (became sec. 1833), French Geol. Soc. (sec. 1831-32). Author: Tracer l'histoire des différentes incursions des Arabes d'Asie et d'Afrique tant sur le continent de l'Italie que dans les iles qui en dépendent; Mémoire sur la craie et sur les terrains tertiaires du Cotentin; Observations sur un ensemble de depôts marins plus récents que les terrains tertiaires du bassin de la Seine, et constituant une formation géologique distincte, précédeés d'un aperçu de la non-simultanéité des bassins tertiares, 1829; Recherches géologiques et historiques sur les cavernes à ossements, 4 vols., 1845; Recherches sur la coutume d'exorciser et d'excommunier les insectes et autres animaux nuisibles à l'agriculture, 1853. Collected 6,000 vols. on history and archeology of French provinces. Died Nogent-le-Rotrou, France, 1887.

DE SOLO, Gerald, French physician; prof. med. faculty Montpellier, France; author: Introductorium Juvenum Sive De Regimine Corporis Humani In Morbis (med. primer for students of Montpellier). Died circa 1360.

DESOR, Édouard, geologist; b. Friedrichsdorf, 1811. Author: Synopsis des echinides fossiles, 1858; Palafittes du lac de Neuchâtel; also writings on geology, prehistoric Switzerland. Died Nice, France, 1882.

DESORMEAUX, Antonin, French urologist; b. 1815; s. Marie-Alexandre Desormeaux; docteur en médecine, U. Paris (France), 1844; became surgeon in Paris hosps., 1849. Author: De l'endoscope et de ses applications, 1865. Invented endoscope and used it to perform 1st removal of papilloma from urethra. Died 1882.

DESORMEAUX, Marie-Alexandre, French obstetrician; b. Paris, France, May 5, 1778; s. Jean-Saturnin Desormeaux; ed. Harcourt Coll., also Paris Faculty Medicine. Served in army of Rhine; returned to Paris, 1802; made preceptor for life; became head dept. obstetrics Faculty of Paris; named physician in chief Maternity Hosp., 1828. Mem. French Acad. Medicine (treas.), French Acad. Surgery. Recipient 1st prize École pratique, 1798. Translated Morgagni's works on causes of diseases into French; publs. on childbirth and abortion. Died Apr. 24, 1830.

DESORMES, Charles Bernard, French chemist; b. Dijon, France, June 3, 1777; ed. École Polytechnique; at least 1 dau., Mrs. N. Clement. Asst. to Guyton in chemistry until 1804; asst. lectr. École Polytechnique; founder (with Montgolfier and Clement) chem. mfg. plant, Verberie, France. Research (with Clement) on iodine, catalysts, especially nitrogen oxides. Died Verberie, Aug. 30, 1862.

DE SOTO, Hernando, Spanish explorer; b. Barcarrota, Spain, 1500; grad. U. Salamanca (Spain); m. Isabel Davila; at least 1 child (dau., by Inca mistress Curicuillar). Served as capt. Spanish Army, under Gov. Pedrarias Davila in Central Am., 1519-32; served in conquest of Peru, 1532-33; led group which captured Inca chieftain Atahualpa, 1532; received contract from King Charles V of Spain to conquer Fla., 1537; gov. Cuba, 1537-42; landed with army of 100 men in Fla., 1539; wandered through what is now Southern U. S. in search of gold and Indian treasures similar to those found in Mexico and Peru, 1539-42; conquered those Indian tribes he encountered enroute; suffered severe defeat by Mauvilian Indians in Ala., fall 1540; discovered and crossed Mississippi River nr. what is now Memphis, Tenn., 1541; wintered nr. what is now Ft. Smith, Ark., 1541-42; turned back to Fla., Apr. 1542. Died on bank of Mississippi River nr. what is now Arkansas City, Ark., May 21, 1542.

D'ESPAGNAT, Bernard Georges, French physicist; b. Fourmagnac, France, Aug. 22, 1921; s. Georges and Marguerite (de Ginestet) d'E.; grad. Ecole Polytechnique; Doctorate d'état-ès-sciences physiques, U. Paris, 1950; m. May de Schoutheete de Tervarent, Dec. 27, 1950; children—Isabelle, Anne. Staff, Centre National de la Recherche Scientifique, 1947-58, maitre de recherches, 1956-58; research asso. Inst. for Nuclear Study, U. Chgo., 1951-52; research fellow Inst. for Theoretical Physics, Copenhagen, Denmark, 1953-54; physicist European Orgn. for Nuclear Research, 1954-59; faculty U. Paris, 1959—, prof. physics, 1964—. Author: Conceptions de la Physique Contemporaine-Les Interprétations de la Mécanique Quantique et de la Mesure, 1965; also articles. Research on relationship between pion and beta decay; introduced (with J. Prentki) group theoretical methods into theory of strange particles and of hyperchange; contbns. to spurion scheme, octet behaviour of non-leptonic weak interactions; intermediate boson model. Home: 7 rue Fustel de Coulanges, 75, Paris 5e, France. Office: Physique Théorique, Bat. 211, Faculté des Sciences, 91 Orsay, France.*

D'ESPAGNET, Jean, French alchemist; flourished 1600-50. Pres. parliament of Bordeaux, France. Author: Arcanum Hermeticae Philosophiae Opus; Enchiridion Physicae Resitutae, 1623. Stressed light as natural force; affirmed influence of stars, but doubted possibility of successful astrological prediction; admitted possibility of heliocentric view; wrote that earth, air, and water are not pure elements but are compounds, also that air is essential to life, that vegetation is nourished by air as well as by earth and water; wrote treatise which was in opposition to Aristotle's physics; also wrote manual for those in search of philosopher's stone.

DE SPARRE, Magnus-Louis-Marie, mathematician; b. Turgouie, Switzerland, May 12, 1849; Docteur ès scis. mathématiques; prof. Catholic U. Lyons (France); mem. French Acad. Scis. Author: Mouvement des projectiles autour de leur centre de gravité, 1927. Demonstrated that form for type of polyhedron as conceived by Poinsot was erroneous; research in ballistics; studied Foucault's pendulum. Died Lyons, Feb. 27, 1933.

DESPARS (or DESPARTS), Jacques, French physician; b. Tournay, France, 1380; M.D., U. Paris (France), 1409; also studied at U. Montpellier (France). Canon of Tournay; chancellor Ch. of Paris; dept. Univ. of Council of Constance, 1410; physician to Philippe de Bourgogne, also King Charles VII. Founded, endowed Rue de la Bucherie, sch. medicine, Faculty Paris. Author: Commentaire sur Avicenne, 18 vols. Died Jan. 3, 1458.

DESPEYROUS, Théodore, French mathematician; b. Beaumont-de-Lomagne, France, May 15, 1815; studied at Toulouse (France), also Lectoure; prof. Faculty of Dijon (France); mem. Faculty of Toulouse; dir. Obs. Toulouse. Author several publs. including: Sur les fonctions elliptiques; Sur la théorie des permutation; (with notes by Darboux) Cours de mécanique, 1884. Research on attraction of ellipsoids, effect of earth's rotation on tree growth, gen. motion of solid bodies. Died Toulouse, 1883.

D'ESPINE, Adolphe, physician; b. Geneva, Switzerland, 1846; M.D., Paris, 1873; intern Paris hosps., 1870; became prof. internal pathology U. Geneva, 1876. Mem. acads. medicine Paris, St. Petersburg. Author: (with Picot) Traité des maladies des enfants, 1899. Research on infantile diseases, bacteriology, cardiography. Died 1930.

DESPLACES, Philippe, French astronomer; b. Paris, 1659; produced calendars under title État du ciel, 1721-35; continued publ. of Beauleiu's Éphémérides des mouvements célestes, 3 vols., 1716-34. Died Paris, 1736.

DESPOIS, Jean Jacques, French geographer; b. Paris, France, Jan. 19, 1901; s. Andre and Louise (Kortz) D.; D.Litt., Faculté des lettres U. Strasbourg (France), 1935; m. Pauline Vacherot, Dec. 29, 1924; children—Raymond, Odile, Hélène, Nicole. Prof. Sakiki Coll., later Lycée of Tunis, Tunisia, 1924-37; prof. Algiers, Algeria, 1937-57; prof. geography Faculté des lettres U. Paris, 1957—. Sec. gen. Nat. Geography Com. Recipient Bodin prize Acad. Scis., 1949, Pelliot prize Univ. Presses France, 1949, Rockefeller scholar, 1932-33. Mem. Paris Soc. Geography (6 prizes, v.p.), Assn. French Geographers (v.p.), Acad. Sci. Overseas. Author: la Colonisation italienne en Libye; le Djebel Nefousa; Sahel et Basse steppe; l'Afrique du Nord; le Djebel Amour; le Hodna; la Tunisie. Known for studies of geography of Africa. Home: 95 Blvd. Romain-Rolland, Montrouge, Seine, France. Office: 191 rue Saint-Jacques, Paris 5, France.

DESPRES, Armand, French surgeon; b. Paris, France, Apr. 13, 1834; s. Charles-Denis Despres; studied medicine under Velpeau in Paris; M.D., 1861. Surgon of hosps.; prof. of faculty; became town councillor of Paris, 1884; elected dep. in 1889. Author: Traité de l'érysipèle, 1862; Dictionnaire de thérapeutique médicale et chirurgicale, 1866; Du début de l'infection syphilitique, 1869; Traité inconographique de l'ulcération et des ulcères du col de l'utérus, 1870; Traité théorique et pratique de la syphilis, 1873; La chirurgie journalière, 1877; La prostitution en France, 1883. Research on erysipelas, venereal diseases; opposed religious hosps.; opposed antisepsis and asepsis and was known for his technique of "dirty surgery". Died Paris, July 29, 1896.

DESPRETZ, Cézar Mansuète, physicist; b. Lessines, Belgium, May 11, 1798; studied at Paris, France; taught in Bruges, Belgium; mem. faculty l'École polytechnique; taught physics at Collège Henri IV; prof. physics at Sorbonne, Paris, France, 1837; became naturalized citizen of France, 1838. Fellow Royal Soc., 1862; mem. French Acad. Scis., 1841. Research and publications on determination of specific heats and heat conductivities of metals, 1817, pressure curves of steams with density and latent vaporization heat, 1821-23, compressibility of gases, 1827; research on elasticity of vapors, animal heat, volatilization of solids, action of galvanic pile, limits of sound, mercury thermometer, carbon product used for polishing gems; observed fluctuation of freezing points and determined greatest density of water; research on expansion of salt solutions and sulphur, 1837-45. Died Paris, May 11, 1863.

DESPREZ, Florimond, French agronomist; b. 1820; contbr. numerous articles to Journal d'agriculture pratique; renowned for 50 years work on devel. of seeds of weed and forage plants, also indsl. prodn. perfected cereal seeds. Died Capelle, France, 1900.

DESSAIGNES, Victor, French chemist; b. Vendôme, France, Dec. 30, 1800; s. Jean-Philibert Dessaignes; docteur en médecine, Paris, France, 1835; m. Renou Dessaignes, 1837. Prof. anatomy, physiology Collége Vendôme; tax collector, Vendôme. Recipient Jecker prize (with Berthelot), 1860. Mem. Chem. Soc., French Acad. Scis., Royal Soc. London. Discovered malonic, tartronic, tricarballylic acids; studied tartaric acid, succinic fermentation, quercite, methyluramine; 1st to synthesize hippuric acid from benzoyl chloride and zinc glycocollide, 1853. Died Paris, Jan. 5, 1885.

DESSAUER, Friedrich J., biophysicist; b. Aschaffenburg, Switzerland, July 19, 1881; s. Philipp Dessauer; ed. Coll. of Aschaffenburg, insts. of tech. at Munich and Darmstadt, U. Frankfort; Ph.D. in Natural Philosophy, 1916; m. Elisabeth Elshorst, Apr. 17, 1909; 4 children. Prof. biophysics, dir. biophysics dept. Goethe U., 1924-33; became prof. radiology and biophysics, Istanbul, 1934; head prof. physics U. Fribourg, from 1937. Hon. mem. Inst. Tech., Vienna, Kaiser Wilhelm Inst., Frankfort. Author: Statement of Fundamental Lawsin Depth Therapy with X-Rays, 1905; Introduction of Quantum Theory in Biophysics, 1922; Wissen und Bekenntis, 1944; Philosophy der Tecknik. Contbd. to devel. of X-ray aparati; investigated biol. effects of ionized air; developed an optimistic philosophy of technology. Address: Atressemann-Allée 36, Frankfurt/Main, Germany.

DESSAUER, Herbert Clay, Am. biochemist; b. New Orleans, Dec. 30, 1921; s. Herbert Andrew and Shirley (Patin) D.; student U. N.M., 1943; certificate in meteorology Cal. Inst. Tech., 1944; B.S., La. State U., 1949, Ph.D., 1952; m. Frances Jane Moffatt, Dec. 10, 1949; children—Dan Winston, Rebecca Lynn, Bryan Clay. Faculty, Sch. Medicine La. State U., New Orleans, 1951—, prof., 1963—. Cons. VA Hosp., 1960—; panel mem. NSF Advanced Scis. Ednl. Program, 1965—. Mem. Am. Physiol. Soc., Soc. Exptl. Biology and Medicine, Am. Soc. Icthyologists and Herpetologists, Herpetologists League, A.A.A.S., Sigma Xi, Phi Kappa Phi. Contbr. numerous articles to profl. jours. Research and publs. on relationships and evolution of amphibians and reptiles; discovered seasonal differences in reproductive and somatic structures both in morphology and chemistry. Home: 7100 Dorian St., New Orleans 70126. Office: 1542 Tulane Av., New Orleans 70112.*

DESSAUER, John Hans, chemist; b. Aschaffenburg, Germany, May 13, 1905; s. Dr. Hans and Bertha (Thywissen) D.; student Inst. Tech., Munich, Germany, 1924-26; D.Eng., Inst. Tech., Aachen, Germany, 1929; m. Margaret B. Lee, June 29, 1935; children—John Philip, Margot, Thomas David. Came to U. S., 1929, naturalized, 1936. Chemist, Ansco, Binghamton, N.Y., 1929-35; chemist Haloid Co., Rochester, N.Y., 1935-38, research dir., 1938-46, v.p. charge research and product devel., dir., 1946-59 (name changed to Haloid Xerox, Inc., 1958), executive vice president research and advanced engineering division Xerox Corporation, 1959—, vice chmn. bd., dir.; dir. Xerox Internat., Rank Xerox, Ltd., Electro-Optical Systems. Mem. the advisory board of regents of St. John Fisher Coll.; mem. Rochester bd. regents

LeMoyne Coll. Fellow Photog. Soc. Am.; mem. Am. Chem. Soc., Optical Society of America, National Academy of Engineering, Soc. of Motion Picture Engineers, C. of C. Co-author: Xerography and Related Processes. Research in organic chemistry; study of xerography; photo research. Home: Twin Oaks, Pittsford, N.Y.; also Hillsboro Beach, Fla. Office: Xerox Corp., P.O. Box 1540, Rochester 3, N.Y.

DESSLER, Alexander Jack, Am. space scientist; b. San Francisco, Oct. 21. 1928; s. David Alexander and Julia (Shapiro) D.; B.S., Cal. Inst. Tech., 1952; Ph.D., Duke, 1956; m. Lorraine Hudek, Apr. 18, 1952; children—Pauline Karen, David Alexander, Valerie Jan, Andrew Emory. Sect. head Lockheed Missiles and Space Co., 1956-62; prof. Grad. Research Center, Dallas, 1962-63; prof., chmn. space sci. dept. Rice U., 1963—. Served with U. S. Navy, 1956-58. Recipient Outstanding Young Scientist award Tex. wing Air Force Assn., 1963. Fellow Am. Geophys. Union (Macelwane award 1963); mem. Am. Phys. Soc., Am. Astronomical Soc., Internat. Sci. Radio Union, Coeditor Jour. Geophys. Research; asso. editor Revs. of Geophysics; adv. bd. Planetary and Space Sci. Research in plasma physics; interplanetary physics; low temperature physics; hydromagnetism; geomagnetism; theory of geomagnetic storms. Home: 5126 Loch Lomand Dr., Houston, 77035.

DESSOIR, Max, German psychologist, philosopher; b. Berlin, Germany, Feb. 8, 1867; s. Ludwig (Leopold) and Auguste (Grünemeyer) D.; Ph.D., U. Berlin, 1889; M.D., U. Würzburg (Germany), 1892; m. Susanne Triepel, 1899. Became lectr. in philosophy, 1892; named asso. prof. Berlin, 1897; prof. emeritus, 1934-47. Mem. Gesellschaft für Ästhetik und allgemeine Kunstwissenschaft (founder 1909, set up internat. congresses 1913-31). Author: Bibliographie der modernen Hypnotismus, 1888; K.Ph. Moritz als Ästhetiker, 1889; Das Doppel-Ich, 1890; Geschichte der neueren deutsche Psychologie, 1894; Ästhetik und allgemeine Kunstwissenschaft, 1906; Geschichte der Psychologie, 1911; Vom Jenseits der Seele, 1917; Abriss einer Geschichte der Psychologie, 1911; Vom Diesseits der Seele, 1923; Der Okkultismus, 3 vols., 1925; Beiträge zur allgemeinen Kunstwissenschaft, 1929; Einleitung in die Philosophie, 1936; Die Rede als Kunst, 1940; Buch der Erinnerung, 1946. Research on physiol. psychology, occult phenomena; founder parapsychology. Died Königstein/Taunus, July 19, 1947.

DE STEIN, Walter, German mathematician, physicist; b. Altona, Aug. 20, 1904; s. Hermann and Margaretha (Stahl) De S.; ed. univs. of Hamburg, Kiel, Tübingen; m. Hildegard Grubbert (dec.); m. 2d, Renate Gümpel; children—Frauke, Elke, Hans Konrad. Prof. agrege Sch. Navigation of Leer, 1932-36; with Sch. Navigation of Stettin, 1936-42, dir., 1942-46; dir. Sch. Navigation of Bremen, 1946—. Mem. Olbers Soc. of Bremen, Vereinigung der Sternfreunde, Astron. Soc., Soc. Meteorology. Author: Von Bremer Astronomen und Sternfreunden; Das kleine Sternebuch; Von Wind und Wetter; Wetter- und Meereskunde für Seefahrer, Navigation Leicht Gemacht. Home: Werderstrasse 39/41-11. Office: Werderstrasse 73, Bremen, Germany.

DESUA, Frank C., Am. mathematician; b. Monessen, Pa., Oct. 26, 1921; s. Joseph and Rose (Frasinelli) DeS.; B.S. cum laude, U. Pa., 1944, Ph.D., 1956; m. Elizabeth Laverne Daniels, June 3, 1945; 1 son, Joseph Paul. Instr. U. Pa., 1944-54, Ohio U., Athens, 1954-56; asst. prof. U. Pitts., 1956-57; mem. tech. staff Bell Telephone Labs., Inc., N.J., 1957-58; asso. prof. Coll. William and Mary, Williamsburg, Va., 1958-60; prof., chmn. math. dept. Simmons Coll., Boston, 1960—. Address: Simmons Coll., Boston 02115.*

DESVAUX, Auguste-Nicaise, French botanist, mineralogist; b. 1784; dir. Bot. Garden, Angers, France; directed preparation herbarium in Mus. Paris (France); prof. botany Angers. Author: Traité général de botanique, 1838; Notice sur un nouveau genre de plantes, 1808; also constructed table of minerals, wrote on beekeeping. Research on geol. formations of Loire region; discovered new genus of plants in Cypericae group. Died Angers, 1856.

DE SWART, Johan Jacob, Dutch physicist; b. Dordrecht, Netherlands, Jan. 31, 1931; s. J. J. and M. S. (Hoefnagel) de S.; Natuurkundig Ingenieur, Technische Hogeschool, Delft, Netherlands, 1953; Ph.D., U. Rochester (N.Y.), 1959; m. Betty Jean Miessen, June 5, 1959; children—Robert J., Suzanne M., Johan F., Annemieke E. Research asso., asst. prof. (p.t.) U. Rochester, 1959-60; research asso. Fermi Inst., U. Chgo., 1960-62; prof. theoretical physics U. Nijmegen, (Netherlands), 1962—; research asso. CERN, Geneva, Switzerland, 1962-63; NSF sr. fgn. vis. scientist U. Pitts., 1966-67. Mem. Am. Phys. Soc., Nederlandse Natuurkundige Vereniging. Research, publs. on nuclear physics and high energy physics.

DE SY, Albert Leon, Belgian metallurgist; b. Oost-Eeklo, Flanders, Aug. 16, 1909; s. Arthur and (De Smet) De Sy; ed. in civil engring. U. Ghent; m. Lea Vols; children—Walter-Arthur, Rita-Albert. With steel mills; instr. U. Ghent, full prof., dir. Lab. Metallurgy and Metallography; Francqui prof. U. Brussels, 1960-61. Recipient prize of honor Internat. Com. of Foundry Tech. Assns., 1958. Mem. Flemish Acad. Sci., Letters and Fine Arts, Am. Soc. Metals, Am. Inst. Mining, Metall. and Petroleum Engrs., Am. Foundrymen's Soc., others. Author: Leerboek der Algemene Metallurgie, Sider-Electrosideren Metallografie, 1946; Algemene Metallurgie, 2 vols; (with J. Van Eeghem) Recherches sur les Fontes Grises à Graphite Lamellaire. Etude de la Corrélation: Propriétés Physiques, Massiveté; (with J. Vidts) Theoretische en toegepaste Metaalkunde, 1961, Traité de Métallurgie Structurale Théorique et Appliquée, 1962. Home: 293, Chaus. de Courtrai. Office: 41, rue Neuve Saint-Pierre, Ghent, Belgium.

DE TAKATS, Geza, surgeon; b. Budapest, Hungary, Dec. 9, 1892; s. Emile de Grosz and Margaret De Takats; M.D., U. Budapest, 1915, M.Surgery, 1918; m. Carol Beeler, Oct. 22, 1924. Traveling fellow Rockefeller Found., 1923-24; asst. prof. surgery U. Budapest, 1925; Elisabeth J. Ward fellow in surgery Northwestern U. Med. Sch., 1926-28, head vascular clinic, 1925-34; head Vascular Clinic, U. Ill., Chgo., 1934-61, clin. prof. surgery Coll. Medicine, 1952-61, clin. prof. emeritus, 1961—; attending surgeon Presbyn-St. Luke's Hosp., Chgo., 1959-61, cons. surgeon, 1961—. Fellow A.C.S.; mem. NRC (past mem. vascular subcom.), Am., Chgo. (past pres.) heart assns, Soc. for Vascular Surgery (past pres.), Internat. Cardiovascular Soc. (pres. 1965-67, Cardiovascular Surgeons Club, Central Soc. for Clin. Research, Am. Acad. Nat. Scis., Inst. Medicine Chgo., Am. Hungarian Studies Found. (adv. com. 1961—), Internat., Ill., Chgo. (past pres.) surg. socs.; hon. mem. Minn., St. Paul, New Orleans, Lyon surg. socs., Internat. Soc. Anesthesia, Acad. Surgy Detroit. Author: Local Anesthesia, 1928; Thromboembolism, 1955; Vascular Surgery, 1959; numerous articles on varicose veins, thromboembolism, peripheral arterial disease, hypertension aneurysms, calcareous pancreatitis, causalgia, Raynaud's phenomenon, microcirculation. Home: 1326 Greenwood Blvd., Evanston, Ill. 60201. Office: 9701 N. Kenton, Skokie, Ill. 60076.*

D'ETAPLES, Jacques Lefévre, philosopher; b. Étaples, France, circa 1450; studied Paris, France; M.A., 1492; tchr. Charles de Bouelles; lived nr. Paris, 1507-20; condemned for heresy in religious and critical writings; fled to Strasbourg, France, 1525. Wrote commentaries on Aristotle. Pub. Arithmetic (Nemorarius), 1496; Arithmetic (Boethius), 1503; Elements (Euclid), 1516. Editor: Sphaera (Sacrobosco), 1499; Works (Nicholas of Cusa), 1514. Studied theory of the Three Worlds (intelligence, celestial bodies, matter); used gnosticism as a basis of his theories on natural magic, planetary harmonies, number mysticism. Died circa 1537.

DE TARANTA, Valescus (Valescus de Taranta), physician; b. 1382; leading tchr., practitioner U. Montpellier, France. Author: Philonium, pub. 1490; tract on pests, 1473-74. Explored doctrine of signatures or application of remedy because of some imagined relation in shape or color to disease (as a red rag for smallpox). Died 1417.

DETENGOF, Fedor Fedorovich, Russian psychiatrist; b. Moscow, USSR, 1898; grad. Med. Faculty, 1st Moscow U., 1920, Marxist-Leninist U., Moscow, 1940; D.Med. Sci., 1938. Intern, 1st City Psychiat. Hosp., Moscow, 1928-32; former asso. Gannushkin Psychiat. Research Inst., Moscow; sr. asso. psychoneurol. dept. Central Disabled Labor Research Inst., Moscow, 1932-33; asst. dept. psychiatry Pediatric Faculty, 2d Moscow Med. Inst., head clin. dept. Solovev Nervous and Psychiat. Hosp., 1933-37; asst. psychiat. clinic 1st Moscow Med. Inst., 1937-40; prof., head chair psychiatry Tashkent Med. Inst., 1940—; chief psychiatrist Uzbek Ministry Health, 1940—. Chmn., Uzbek Forensic Psychiat. Consultative Bd.; cons. Tashkent Mil. Hosp. Author: Treatment of Neuroses and Psychoses. Mem. editorial council S. S. Korsakov Jour. Neuropathology and Psychiatry; co-editor Neuropathology and Psychiatry sects. Large Med. Ency., 2d edit. Research and numerous publs. on neuropathology and psychiatry. Address: Tashkent Med. Inst., ulitsa Karla Marksa 103, Tashkent, Uzbekistan SSR, USSR.

DETERLING, Ralph Alden, Jr., Am. surgeon; b. Williamsport, Pa., Apr. 29, 1917; s. Ralph Alden and Edith Pauline (Ritter) D.; A.B., Stanford, 1938, M.D., 1942; M.S. in Surgery, U. Minn., 1946, Ph.D., 1947; m. Mary Ann Gilson, June 21, 1947; children—Ralph III, William R., John S., Paul A. Fellow surgery Mayo Found., 1943-47; asst. attending surgeon Presbyn. Hosp., N.Y.C., 1948-50, asso. attending surgeon, 1950—; surg. cons. U. S. Naval Hosp., St. Albans, Manhattan VA Hosp., N.Y.C., Paterson (N.J.) Gen. Hosp.; asso. surgeon attending Francis Delafield., N.Y.C.; asst. prof. surgery Coll. Phys. and Surg., Columbia, 1948-50, asso. prof., 1950-54, asso. clin. prof. surgery, 1954-59, dir. surg. research labs.; prof., chmn. dept. surgery Tufts U. Sch. Medicine, Boston, 1959—; surgeon-in-chief New Eng. Med. Center hosps.; dir. 1st surg. service Boston City Hosp.; cons. Boston VA, Mt. Auburn, St. Elizabeth's, Lemuel Shattuck, Newton-Wellesley hosps. Recipient Malmo (Sweden) Sur. Found. award, 1963; Outstanding Achievement award U. Minn., 1964. Diplomate Nat. Bd. Med. Examiners, Am. Bd. Surgery, Bd. Thoracic Surgery. Fellow A.C.S., Am. Coll. Chest Physicians; mem. Am., N.Y. thoracic socs., Harvey Soc., Colombian Coll. Surgeons (hon.), Med. Strollers (pres. 1957), Soc. Vascular Surgery, Internat. Cardiovascular Soc. N.Am. chpt. pres., 1966—, A.A.A.S., Am. Soc. Artificial Internal Organs, Angiology Soc. Argentina (hon.), Soc. Univ. Surgeons, N.Y. Acad. Scis., Internat. Soc. Surgery, Am. Assn. Thoracic Surgery, Am., Mass. heart assns., Am. Philatelic Soc., Am. Acad. Arts and Scis., Am. Surg. Assn., New Eng., Boston surg. socs., New Eng. Cardiovascular Soc., Nat. Surg. Soc. Cuba (hon.), Phi Beta Kappa, Sigma Xi (chpt. sec.), Alpha Omega Alpha, Alpha Kappa Kappa. Contbr. articles to med., sci. jours. Research in blood vessel prostheses; hypotheimia; extracorporeal circulation; laser; organ transplantation. Office: 171 Harrison Av., Boston; also 818 Harrison Av., Boston.*

DETHIER, Vincent Gaston, Am. insect physiologist; b. Boston, Feb. 10, 1915; s. Jean Vincent and Marguerite Frances (Lally) D.; A.B., Harvard, 1936, A.M., 1937, Ph.D., 1939; Sc.D. (honorary), Providence College, 1964; m. to Lois E. Check; children—Jehan Vincent, Paul. Harvard fellow Atkins Inst. of Arnold Arburetum, Soledad, Cuba, 1939-40; instr. biology John Carroll U., Cleve., 1939-41, asst. prof., 1941-42; prof. zoology Ohio State U., 1946-47; asso. prof. biology Johns Hopkins, 1947-51, prof., 1952-58; professor zoology U. Pa., 1959-67, prof. psychology, 1959-67, associate Neurological Inst., 1959-67; professor of biology Princeton Univ., 1967—; Hixon lectr. Cal. Inst. Tech., 1949; speaker Internat. Congress Entomology, Amsterdam, 1951; Belgian-Am. Ednl. Found. fellow Belgian Congo, 1952; sr. Fulbright research scholar London Sch. Hygiene and Tropical Medicine, 1954; Guggenheim fellow, Netherlands, 1964-65; cons. Canadian Defense Bd., Office Surgeon Gen., U. S. Served as maj. AUS, World War II. Fellow Entomol. Soc. Am., Nat. Acad. Sci., Am. Academy Arts and Sciences; mem. A.A.-A.S., Am. Soc. Zoologists, Soc. Gen. Physiologists, Am. Physiol. Soc., Am. Soc. Naturalists, Assn. Study Animal Behavior, American Association University Profs. Author: Chemical Insect Attractants and Repellents, 1951; (with Stellar) Animal Behavior; Physiology of Insect Senses; To Know a Fly. Editor Journal Insect Physiology; editorial bd.; Am. Rev. Entomology, Jour. Comparative Physiol. Psychology. Contbr.: Insect. Physiology (editor K. D. Roeder), 1953; also various periodicals. Studies of the host/plant relationships of plant feeding insects and the chemical senses of insects; mechanisms of hunger and satiation in insects. Home Stony Ford, Pretty Brook Rd., Princeton, N.Y.*

DE TILLY, Joseph Marie, Belgian mathematician; b. Ypres, Belgium, 1837; s. Charles Joseph Marie De Tilly; grad. L'Ecole Militaire, 1853. Prof. L'Ecole Militaire, 1868-79, head (rank lt. gen.), from 1889; dir. workshop, Antwerp, Belgium, 1879-89. Mem. Royal Acad. Author: Note sur Deux Traités Récents de Balistique et sur L'État Actuel de Cette Science, 1874; Essai sur les principes Fondamentaux de la Geometrie et de la Mecanique, 1878; L'Integration des Equations Lineaires du Deuxième Ordre; L'Essai de Geometrie Analytique Generale. Studies on non-Euclidean geometry; established relation between 10 distances of any 5 points, as related to Lobachenskian, Niemannian, Euclidean geometries. Died Schaerbeck, Belgium, 1906.

DETMOLD, William Ludwig, surgeon; b. Hanover, Germany, Dec. 27, 1808; M.D., U. Göttingen, 1830; surgeon Royal Hanoverian Grenadier Guard, 1830-37; came to U. S., 1837; prof. mil. surgery and hygiene Columbia, 1862-66, prof. emeritus, from 1866. Author: Opening an Abscess in the Brain, 1850. Pioneer orthopedic surgery U. S.; credited with 1st operation in which lateral ventricle was opened for drainage of a brain abscess, 1850; vol. surgeon U. S. Civil War, as such introduced knife and fork for one-handed men known as Detmold's knife. Died 1894.

DE TONI, Giovanni Battista, Italian botanist; b. Venice, Italy, Jan. 2, 1864. Prof. botany U. Modena (Italy). Author: Flora Algologica della Venezia, 6 vols., 1885-1924; Sylloge Algarum, 5 vols., 1889-1907; Sylloge Funorum, 6 vols., (with Saccardo vols 7, 8). Editor Revue Internationale d'Algologie. Research on cryptogam plants, gen. botany, systematic botany, history bot. sci.; specialist algae. Died Modena, July 31, 1924.

DE TOURNEFORT, Joseph Pitton, see Tournefort, Joseph Pitton de.

DE TRESSAN, Count Louis Elisabeth de la Vergne, French amateur scientist; b. LeMans, France, Nov. 4, 1705; 1 son, L'Abbé de Tressan. Received rank of Mestre de Camp in Italy, mil. career in War of Polish Succession, 1733-38, marechal de Camp, Flanders, 1741, aide-de-camp to Louis XV, Fontenoy, served to lt. gen., 1747; gov. Lorraine, France, from 1750, ret. to Franconville, France. Grand marechal at ct. of King Stanislas, 1752. Mem. French Acad. Scis., 1749, Académie francaise. Author: Le Petit Jehan de Saintre; Traité sur L'Electricité, 1749. Don Ursino de Navarrin; 3 vol. extract of series Amadis de Gaulle, also Roman de la Rose. Translator Aristotle. Died Paris, France, Oct. 31, 1783.

DETTWEILER, Peter, German physician; b. Winterheim, Germany, Aug. 4, 1837; mil. surgeon serving in

Danish-German, Austrian-German, Franco-German wars; founder Tb. sanatorium of Falkenstein in the Taunus, became dir. 1876. Author: Die Behandlung der Lungenschwindsucht in geschlossenen Heilanstalten, 1880; Die hygienische-diatetische Anstaltsbehandlung der Lungentuberkulose, 1899; also contributed to Handbuch der Ernährungstherapie (E. von Leiden), 1898. Introduced rest cures in the open air, also portable receptacles for spitting; contbd. to establishment of sanatoriums in Germany, especially to those for working class. Died Kronberg, Germany, Jan. 12, 1904.

DE TUBIÈRES, Anne Claude Philippe, see Caylus, Anne Claude Philippe.

DETWILER, Samuel Randall, Am. anatomist; b. Ironbridge, Pa., Feb. 17, 1890; s. Isaiah H. and Mary (Hallman) D.; student Ursinus Coll., 1910-12; Ph.B., Yale U., 1914, A.M., 1916, Ph.D., 1918, hon. M.S., 1931; m. Gladys I. Hood, July, 1942; children (previous marriage) Samuel Randall (dec.), Ross Harrison. Asst. instr. biology, Yale, 1914-17, instr. in anatomy, 1917-20; asso., Peking (China) Union Med. Coll., 1920-23; asst. prof. zoölogy, Harvard, 1923-26, asso. prof., 1926-27; prof. anatomy and exec. officer of dept. Columbia, 1927—. Mem. Am. Assn. Anatomists (pres. 1954-56), Am. Physicians Art Assn., N.Y. Acad. Scis., Am. Naturalists, Am. Neurol. Assn., Soc. Exptl. Biology and Medicine Am. Acad. Arts and Scis., Nat. Acad. Scis., Harvey Soc., Am. Philos. Soc., Sigma Xi, Gamma Alpha, Nu Sigma Nu, Omicron Kappa Upsilon. Author: Neuroembryology, 1936; Vertebrate Photoreceptor, 1943; also articles in profl. jours. Mem. editorial bds. Exptl. Biology, monograph series; Jour. Exptl. Zoology; Columbia Biol. series. Mem. adv. bd. Human Biology. Studies in exptl. neuro-embryology and related fields of exptl. embryology, morphogenesis of nervous system, physiomorphology of eye; showed compensatory regulation of nervous ability to modify in response to conditions imposed upon it, whether by transplantation of limbs or by transposition of segment of spinal cord; demonstrated completely the malleable nature of nervous system; did careful analysis of form and function of the eye; gathered an eye collection that was technically finest and most complete in world. Died May 2, 1957.

DEUCHAR, E. M., Brit. embryologist; b. Eng., 1927; s. George Lindsay and Margaret (Giddings) D.; B.A. in Zoology, St. Hugh's Coll., Oxford, 1948, M.A., 1953; Ph.D., U. Edinburgh (Scotland), 1951. Research asst. to a prof., 1951-53; faculty anatomy dept. U. Coll., London, 1953-66, reader embryology, 1963—. Office: Anatomy Dept., U. Coll., London, Eng.*

DE ULHOA CINTRA, Antonio B., Brazilian physician; b. Sao Paulo, Brazil, Sept. 13, 1907; s. Arnaldo P. and Antonia A. (Barros) de U. C.; M.D., U. Sao Paulo, 1930; postgrad. Harvard, Vanderbilt univs.; m. Grace de Ulhoa Cintra, Feb. 23, 1932; children—Lelia (Mrs. Luiz Almeida Prado Galvao), Regina (Mrs. Francisco Escobar). Rockefeller research fellow 1941-42, 48; faculty U. Sao Paulo Med. Sch., prof., head dept. internal medicine, 1941—, and other depts.; dean U. Sao Paulo, 1960—. Recipient awards for outstanding papers. Mem. A.A.A.S., Am. Diabetes Assn., Am. Endocrine Soc., several Brazilian med. and sci. assns. Author: Metabolic Bone Disease, 1951; also numerous articles. Address: 57 Marina Cintra, Sao Paulo, Brazil.*

DE ULLOA, Antonio, mathematician; b. Seville, Spain, Jan. 12, 1716; joined navy, 1733; with expdn. of French Acad. to measure degree of meridian at equator in Peru, 1735-45, captured by Brit. on return trip, imprisoned in Eng.; mem. several sci. commns. upon return to Spain; gov. of La., 1766-68; apptd. lt. gen. of navy, 1779; a founder obs., Cadiz, Spain, also 1st mus. of natural history, 1st metall. lab. in Spain. Fellow Royal Soc.; mem. French Acad. Scis., 1748. Author: Relación histórica del viaje a la America meridional, 1748. Died nr. Cadiz, July 5, 1795.

DEUSING, Anton, German physician; b. Meurs, Westphalia, Germany, 1612; M.D., 1637; practiced medicine, tchr. math., Meurs; named prof. physics and math Harderwick, 1639 and prof. medicine, 1642; named 1st prof. medicine U. Groningen (Germany), 1646, apptd. rector, 1649; archiâtre of the Count of Nassau, 1652. Author numerous works including: Cosmographia catholica et astronomia secundum hypotheses Ptolemaei, 1642; Do vero systemate mundi dissertatio mathematica qua Copernici systema mundi reformatur sublatis interim infinitus pene orbibus quibus in systemate Ptolemaico humana mens distrahitur, 1643; Naturae theatrum universale, 1645; Hexaemeron rerognitum, sive de creatione meditationes, 1645; Synopsis medicinae universalis, 1649; Fasciculus dissertationum selectarum, 1660. Translated into Latin: Institutiones de médecine (Avicenna); Aphorismes (Mesué). Research on astronomy, elasticity of air; attacked use of weapon-salve in medicine. Died 1666.

DEUTSCH, Felix, German elec. industrialist; b. Breslau, Germany (now Wroclaw, Poland), May 16, 1858; s. Moritz and Amalie (Ausch) D.; numerous hon. degrees including Dr.rer.pol., U. Cologne (Germany); hon.dr.engring. Technische Hochschule Karlsruhe, Germany; m. Lili Kahn; children—George, Gertrud (Mrs. Gustav Brecher). Joined Deutsche Edison-Gesellschaft für angwandte Elektricität (later became Allgemeine Elektrizitäts-Gesellschaft), became head of firm, also chmn. bd. responsible for fgn. bus. connections, electrotech. projects both domestic and fgn., internal constrn. and adminstrn. of factories. Died Berlin, Germany, May 19, 1928.

DEUTSCH, Martin, physicist; b. Vienna, Austria, Jan. 29, 1917; s. Felix and Helena (Rosenbach) D.; S.B., Mass. Inst. Tech., 1937, Ph.D., 1941; Docteur honoris causa, U. Algiers, 1959; married Suzanne Zeitlin, June 6, 1939; children—L. Peter, Nicholas A., Came to U. S., 1935, naturalized, 1941. Instr. Mass. Inst. Tech., 1941-44, asst. prof., 1945-49, asso. prof., 1949-53, prof., 1953—; mem Los Alamos Sci. Lab., 1944-46; Guggenheim fellow, 1953-54, 60-61; cons. Mem. Am. Phys. Soc., A.A.A.S., Am. Acad. Arts and Scis., Soc. Francaise de Physique, Fedn. Am. Scientists (v.p. 1956), Nat. Acad. Sci. Research in nuclear spectroscopy; elementary particle physics; study of radioactive radiations; study of fission process. Home: 43 Reservoir St., Cambridge 38, Mass.

DEUTSCHMANN, Richard Heinrich, German ophthalmologist; b. Liegnitz, Germany, Nov. 17, 1852; studied under eye specialist, Theodor Leber; m. Hedwig Albertine A. Scheneyg. Joined Faculty of Göttingen (Germany), 1877, became asso. prof., 1883; became head physician eye polyclinic Heinesches Hosp., Hamburg, Germany, from 1887. Recipient Graefe prize for monograph, 1889. Author: Über die Ophthalmia migratoria, 1889; also articles in med. and ophthal. jours. Established (with Leber) migration theory of sympathetic ophthalmia; developed irritant therapy (Deutschmann-Serum); research on origin of detached retina which he treated by separating the tearing tissues in vitreous humour. Died Hamburg, Nov. 13, 1935.

DEV, Sukh, Indian chemist; b. Chakwal, India, June 17, 1923; s. Hari Chand Lala and Mya Vanti; B.Sc. with honors, D.A.V. Coll., Lahore, India, 1944, M.Sc., 1945; Ph.D., Indian Inst. Science, Bangalore, India, 1950, D.Sc., 1960; m. Sashi Prabha, Dec. 7, 1951; children—Indu Bala, Poornima, Deepak Kumar. Sr. fellow Nat. Inst. Scis., 1951-53; lectr. Indian Inst. Sci., Bangalore, 1953-57, 58-59; research asso. U. Ill., 1957-58; head div. organic chemistry Nat. Chem. Lab., Poona, India., 1960—. Recipient Sudborough medal Indian Inst. Sci., 1949; Guha Research medal, 1958; Shanti Swarup Bhatnagar Meml. award chemistry, 1964. Mem. Am. Chem. Soc. Research, publs. on structure determination and synthesis in sesquiterpene and diterpene field. Home: Kalyan Kuteer, 136C-Civil Lines, Bareilly (U.P.), India. Office: Nat. Chem. Lab., Poona 8, Maharashtra, India.*

DE VALLAMBERT, Simon, pediatrician; b. France, 1512; mem. French Acad. Scis.; 1st to publish observation on syphilis in children, also credited with 1st book on pediatrics in French, 1565. Died 1578.

DE VALOIS, Russell L., Am. physiol. psychologist; b. Ames, Ia., Dec. 15, 1926; s. John James and Henriette (Hofland) de V.; B.A., Oberlin Coll., 1946, M.A., 1948; Ph.D., U. Mich., 1952; m. Rikki Herrick, Apr. 7, 1951; children—Geoffrey H., Gregory N., Gordon A. Research asso., lectr. U. Mich., 1953-59; faculty Ind. U., Bloomington, 1959—, prof. physiol. psychology, 1964—. Mem. Am. Psychol. Assn., Sigma Xi. Contbg. author Psychology, 1958. Research and publs. on characteristics of color vision in monkeys, analyzing and encoding color and brightness information in primate visual nervous system. Home: 1900 Viva Dr., Bloomington, Ind. 47405.*

DE VARGAS, Perez, author; flourished late 16th century. Author: De re metallica en el qual se tratan muchos y diversos secretos del conocimento de toda suerte de minerales . . . , 1569; also wrote on steel tempering, engraving on metals with aqua fortis, gilding metals with gold amalgam.

DE VAUBAN, Sébastien le Prestre, French mil. engr.; b. Saint-Léger-de-Foucherest, France, May 1633; ed. by Carmelite prior of Sémur; m. Jeanne D'Aulnay, circa 1659. Joined Condé in war of Fronde, at age of 17; captured by Royal troops and converted to King's side; became lt. at siege of St. Ménéhould; besieged and took Clermont; made royal engr., 1655; given company in Picaroy Regt., 1663; served under King in War of Devolution; made lt. in Royal Guards, 1667, gov. of Lille Citadel, 1668; became commissary gen. of fortifications, 1667; began to direct successful sieges when Dutch War broke out in 1672; became inf. brigadier, 1674, promoted again in 1676, lt. gen., 1688, marshal of France, 1702. Mem. French Acad. Scis., 1699, pres., 1701, 05. Author: Mémoire pour servir a l'instruction dans la conduite des sièges, 1740; Instructions pour la défense, 1740; Traité de l'attaque des placès, 1704; Traité des fortifications de campagne; De la défense des places; Projet d'une dixme royal, 1707; Mes oisivets; Le directeur général des fortifications, 1683. Invented socket bayonet; 1st to use ricochet fire as principle methods of breaking down defense; developed method of parallels in attacking fortified positions; rearmed French inf. with flint-lock muskets; conducted numerous sieges and built or rebuilt numerous fortifications. Died Paris, Mar. 30, 1707.

DE VAUCANSON, Jacques, French inventor; b. Grenoble, France, Feb. 24, 1709; student anatomy, music, mechanics, Paris, France, from 1735; became insp. silk mfg., 1741. Mem. French Acad. Scis., 1748. Developed mech. weaver's loom; perfected silk mfg. machines; made hydraulic pump for city of Lyons, France. Died Paris, Nov. 21, 1782.

DEVAUX, Henri, French botanist; b. Saintonge, France, July 6, 1862; ed. La Rochelle, France; Dr. ès scis., 1889; studied medicine and pharmacy Bordeaux (France) Faculty Scis., 1884-87; studied under Bonnier, Paris. Became lectr. U. Bordeaux, 1891, prof. physiology, 1916-32. Author: Du mécanisme des échanges gazeux chez les plantes aquatiques, 1889. Investigated gaseous exchange in aquatic plants, function of gases in tissue of some fruits, absorption and its importance in mineral nutrition of plants. Died 1956.

DÉVAY, Jozsef, Hungarian chemist; b. Szombathely, Hungary, Feb. 26, 1926; s. Jozsef and Ilona (Soos) D.; Dipl. Chem., Eötvös L. U., Budapest, Hungary, 1950, Candidate Chem. Scis., 1959, Doctor Rer. Nat., 1962, D.Chem. Scis., 1963; m. Etelka, Oct. 14, 1950; children—D. Laszlo, D. Piroska. Faculty sci. dept. phys. chemistry Eötvös L. U., 1950-64, sr. lectr., 1956-64; prof., head dept. phys. chemistry U. Chem. Industries, Veszprém, Hungary, 1964—; head dept. for electrochem. research Hungarian Acad. Scis., 1966—. Research and publs. on electrochem. kinetics especially effect of alternating current on electrode reactions and electrolytic corrosion. Home: 3 Bimbo Budapest, Hungary. Office: 12. Schönherz, Veszprém, Hungary.*

DEVELEY, E., Swiss astronomer, mathematician; b. La Bretonniere, Vaud, Switzerland, May 27, 1764; prof. Acad. Lausanne, Switzerland. Author: Arithmétique d'Émile; Mémoires sur les puissances des nombres et sur leurs racines, 1799; Physique d'Emile, 1802; Algèbre d'Emile, 1805; Cours élémentaire d'astronomie, 1835. Died Lausanne, May 22, 1839.

DE VERDIER, Carl Henric Anders Olof, Swedish clin. chemist; b. Stockholm, Sweden, June 30, 1924; s. Anders and Hedvig (Steuch) de V.; Med.Kand., U. Uppsala (Sweden), 1946, Med.Lic., 1957, Med.Dr., 1957; m. Karin M. Ruback, Sept. 6, 1952; children—Kerstin, Britt-Marie, Ulla, Katarina. Docent, U. Uppsala, 1955, 61, asst. dir. dept. clin. chemistry, since 1965—. USPHS research fellow McArdle Lab. for Cancer Research, Madison, Wis., 1958-59; research fellow Swedish Med. Research Council, 1961-64. Research and publs. on isolation of phosphothreonine from casein, studies of turnover of phosphorus in phosphoproteins from different origins, nucleoside phosphorylases, regulation of glycolysis in human erythrocytes, galactose oxidase for galactose tolerance tests. Home: Backvagen 5 B, Uppsala 9, Sweden.*

DEVEREUX, George, ethnologist; b. Lugos, Hungary, Sept. 13, 1908; student Sch. Oriental Langs., Paris; licencé ès lettres U. Paris, 1932; Ph.D. in Anthropology, U. Cal., 1935. Research sociologist Worcester (Mass.) State Hosp., 1939-40; asst. prof. sociology Middlesex U., Waltham, Mass., 1940-41; instr. sociology U. Wyo., 1941-43; lectr. anthropology Columbia, 1944; vis. lectr. Wellesley (Mass.) Coll., 1945; vis. prof. U. Haiti, summer 1945; dir. research, staff ethnologist Winter Vets. Hosp.; faculty Menninger Sch. Psychiatry; lectr. Topeka Inst. for Psychoanalysis, 1946-54; dir. research Devereux Found. and Schs., Devon, Pa., 1954-56; now prof. research in ethnopsychiatry Temple U. Sch. Medicine, Phila. Mem. Fellow Am. Anthrop. Assn.; mem. Am. N.Y. State psychol. assns., Am. Ethnol. Soc., N.Y. Soc. Clin. Psychologists, Phila. Assn. for Psychoanalysis (affiliate). Editor, contbg. author: Psychoanalysis and the Occult, 1953. Editorial bd. Archives of Criminal Psychodynamics, Jour. Clin. and Exptl. Psychopathology, Ann. Survey Psychoanalysis, Psychoanalysis and the Social Scis. Research on beliefs concerning the parapsychol. Address: Sch. of Medicine, Temple U., Phila.

DEVERGIE, Marie Guillaume Alphonse, French dermatologist; b. Paris, France, 1798; became chief of clinic Hôtel-Dieu, 1823; prof. chemistry, 1825-33, prof. agrégé, from 1833; physician with central bur., 1829-34, médecin titulaure de hôpitaux, from 1834. Mem. French Acad. Medicine. Author: Médecine légale théorique et pratique, 1835-36; Mémoire sur les plaies d'armes à feu, 1849; Traité des maladies de la peau, 1854; Ou finit la raison ou commence la folie, 1859. Described Devergie's disease (lichen ruber acuminatus), pityriasis rubra pilaris, chronic skin disease marked by patches of small follicular papules and pinkish scaling patches, 1856. Died Paris, Oct. 2, 1879.

DE VERNEUIL, Philippe Edouard Poullettier, French geologist, paleontologist; b. Paris, France, Feb. 13, 1805; student law; attended lectures on geology by Elie de Beu Beaumont. Travelled through Europe for several years; expdns. to Russia, U. S. Recipient Wollaston medal Geol. Soc. London, 1853. Mem. French Acad. Scis., 1854, Royal Soc. (fgn.), 1860. Author: The Geology of Russia and the Ural Mountains, 2 vols., 1846; (with E. Collomb) Carte Geologique de L'Espagne et du Portugal, 1864. Geol. investigations of Crimea, Devonian rocks, fossils of Bas-Boulonnais, (with Sedgwick and Murchison) older paleozoic rocks of Rhenish provinces and Belgium, 1839, (with Murchison) Russian empire, paleozoic rocks in U. S., Spain. Died Paris, May 29, 1873.

DEVÈZE, Jean, French physician; b. Rabastens, France, 1753; ed. France; founder, headmaster, maison de santé, Cap-Français, Haiti, 1778-93; head Bush Hill Hosp., Phila., during yellow fever epidemic; practice medicine Fontainbleau, France, from 1798; physician-in-ordinary to French king. Author: Traité de la fièvre jaune, 1820; Recherches et observations sur les causes et les effets de la maladie épidémique qui a ravagé Philadelphie en 1793, 1793; Mémoire au roi or Protestation contre le travail de la commission sanitaire centrale du royaume, 1821. Noted for theory that yellow fever is not contagious although propagated by infection; exposed himself repeatedly to this disease out of his conviction. Died 1820's.

DE VICK, Heinrich, German clockmaker; flourished circa 1380; clockmaker for Palais de Justice, Paris, France, 1362-70; made 1st mech. clock of which there is complete description, 1379.

DE VICO, Francesco, astronomer; b. Macerata, Italy, May 19, 1805; studied under P. Dumouchel, Obs., Jesuit Coll., Rome, 1835-39. Joined Soc. Jesus; tchr. math. and astronomy, dir. obs. Coll. Rome, 1839; became dir. obs. Georgetown U., Washington, 1848. Recipient Lalande medal French Acad. Scis. Mem. Italian Soc. Sci. Research and publs. on rotation of Venus; discovered 6 new comets; set length of day at 23 hours, 21 minutes, 22 seconds. Died London, Nov. 15, 1848.

DE VIGENÉRE, Blaise, French chemist, cryptologist; b. France, 1546; credited with 1st description of benzoic acid, 1608; one of 1st to describe simple square table known as Vigenére cipher although he did not claim to have invented it. Died 1619.

DE VIGHNE, Harry Carlos, Am. physician; b. Santa Barbara, Cal., 1876; ed. Hahnemann Med. Coll. of Pacific, 1904. Commr. health Alaska Ty., Juneau, from 1907; chief surgeon Alaska Juneau Gold Mining Co. Sec. Alaska Bd. Med. Examiners. Mem. Alaska Territorial Med. Assn. (sec.), A.M.A. (ho. of dels. 1926), Am. Acad. Gen. Practice. Author: Pale Star; The Time of My Life; History of the Santa Barbara County Medical Society. Died Santa Barbara, Aug. 7, 1957.

DE VILLA DEI, Alexandre, see de Villedieu, Alexandre.

DE VILLAFRANCA, George Warren, Am. zoologist; b. Meriden, Conn., Nov. 21, 1923; s. Jose F. and Ruth E. (Pease) de V.; B.S., Yale, 1948, Ph.D., 1953; m. Suzanne E. Crane, July 26, 1947 (dec. 1960); children—Suzanne Christina, Ruth Place, George Warren; m. 2d, Erica Satzinger, July 13, 1965. Faculty, Smith Coll., Northampton, Mass., 1951-—, prof. zoology, 1963-—, asst. to pres., 1961-66; Lalor Found. fellow Marine Biol. Lab., Woods Hole, Mass., summers, 1953, 56. Am. Cancer Soc. fellow Inst. for Muscle Research, Woods Hole, 1954-55. Mem. A.A.A.S., Biophys. Soc., Soc. Gen. Physiologists, Am. Soc. Zoologists. Contbr. articles to tech. jours. Research on comparative biochem. mechanisms muscular contraction; formulated theory to provide uniform explanation of protein and muscle contraction. Home: 88 N. Elm St., Northampton, Mass. 01060.*

DE VILLEDIEU, Alexandre (or de Villa Dei, Alexandre), mathematician; b. Villedieu, Normandy, France; studied at Paris, France; mem. Franciscan Order; canon Ch. of St. André, Auranches until his death; taught at Paris. Author: Carmen de Algorismo, circa 1240; Doctrinale puerorum; also didactic poems on arithmetic, works on grammar. Helped spread use of Hindu numerals; 1st Latin text in which zero is considered one of numerals and number of operations is given. Died circa 1240.

DEVINATZ, Allen, Am. mathematician; b. Chgo., July 22, 1922; s. Victor and Kate (Bass) D.; B.S., Ill. Inst. Tech., 1944; A.M., Harvard, 1947, Ph.D., 1950; m. Pearl Moskowitz, Sept. 16, 1956; children—Victor Gary, Ethan Sander. Instr., Ill. Inst. Tech., 1950-52; NSF Postdoctoral fellow, 1952-53; fellow Inst. for Advanced Study, 1953-54; asst. prof. U. Conn., 1954-55; faculty Washington U., St. Louis, 1955-67, prof., 1961-67; prof. Northwestern U., Evanston, Ill., 1967-—. Sr. NSF Postdoctoral fellow, 1960-61. Mem. Am. Math. Soc., Sigma Xi, Tau Beta Pi. Research and publs. in theory and applications of Hilbert space. Home: 626 La Vergne Av., Wilmette, Ill. 60091. Office: Northwestern U., Evanston, Ill.*

DE VISIANI, Roberto, Italian botanist; b. Sebenico, Italy, Apr. 9, 1800; s. Giovanne Battista and Maddalena (Drassich) De V.; student, sem. Spalato, Italy; grad. U. Padua, Italy, 1822. Asst. to prof. botany U. Padua, until 1826, prof. botany, from 1835, dir., bot. gardens, 1837-78. Author: Introduzione allo studio dei vegetabili di Nicolo Giuseppe de Jacquin, 1824; Illustrazione di alcun piante della Grecia e dell' Asia Minore, 1843; Proposta di una nuova distribuzione delle labiate Europee; Dell' origine ed anzianità dell' orto botanico di Padova, 1839; Considerazioni intorno al genere ed alla specie in botanica. Collected bot. specimens in Dalmatia. Died May 4, 1878.

DEVLIN, Thomas Francis, Am. physician; b. Phila., Jan. 20, 1869; s. Thomas and Helen (Sanford) D.; A.B., LaSall Coll., Phila., 1887; student Georgetown U., 1887-88; M.D., U. Pa., 1891, post grad., 1891-92; m. Stella Hill, May 29, 1905; children—Thomas Francis, John Joseph. Physician, Phila., 1892-—; asso. with St. Mary's Hosp.; mem. staff Misericordia Hosp., Archbishop Ryan Meml. for Deaf-mutes; pioneer in endocrinology as applied to children. Dir. and partner with wife, Marydell Sch. for physically and mentally retarded children. Died June 30, 1952.

DE VOGÜÉ, Charles Jean Melchoir, Marquis, French archeologist; b. Paris, France, 1829; s. Leonce, Marquis de Vogüé; student religious history, oriental art, archaeology; 3 sons, 3 daus. Explored Palestine, Syria, 1853-54; ambassador to Constantinople, Turkey, from 1871, to Vienna, Austria, from 1875. Pres. French Red Cross. Decorated Legion Honor. Mem. Academie des Inscriptions, French Acad. Scis., 1901, Société des Agriculteurs de France (pres.), Author: Les Eglises de la Terre-Saint, 1859; Les Evènements de Syrie, 1860; Le Temple de Jerusalem, 1864-65; Mélanges D'Archéologie Orientale, 1869; L'Architecture Civile et Religieuse du Ier au VIIe Siecle, dans la Syrie Centrale, 1865-77; Inscriptions Semitiques, 1869-77; Villars D'Apres sa Correspondance et des Documents inedits, 1888; Le Duc de Bourgogne et Le Duc De Beuavilliers, 1900. Editor: Memoires du Marechal de Villars, 1884. Studies antiquities Palestine, Syria, Christian archeology; research on discovery Venus de Milo. Died Paris, 1914.

DEVONS, Samuel, physicist; b. Bangor, N.Wales, U.K., Sept. 30, 1914; s. David Isaac and Edith (Edlestein) D.; B.A., Trinity Coll., Cambridge (Eng.) U., 1935, M.A., Ph.D. (Exhbn. 1851 scholar 1939), 1939; M.Sc. (hon.), Manchester (Eng.) U., 1959; m. Celia Ruth Toubkin, Sept. 7, 1938; children—Susan Danielle, Judith Rosalind, Amanda Jane, Cathryn Ann Julie. Came to U. S., 1959. Sr. sci. officer Air Ministry, Ministry Supply, U.K., 1939-45; fellow, dir. studies, lectr. physics Trinity Coll., 1946-49; prof., physics Imperial Coll., London, Eng., 1950-55; Langworthy prof. physics, dir. phys. labs. U. Manchester, 1955-60; prof. physics Columbia, 1960-—, chmn. dept., 1963-—. Mem. Tech. Assistance-UNESCO Team of UN to S. Am., 1957. Fellow Phys. Soc. London (past v.p.), Royal Soc. London; mem. Am. Phys. Soc. Author: Excited States of Nuclei, 1949; Angular Correlations, 1957. Devel. anti-aircraft barrages, magnetrons, microwave and radar techniques; investigations of nuclear resonance processes, angular correlations, properties of excited states of nuclei; properties of pions, muons, and other elementary particles; spectroscopy of muonic x-rays. Home: Lewis Rd., Irvington-on-Hudson, N.Y. Office: Dept. Physics, Columbia Univ., N.Y.C. 10027.*

DE VRIES, Adolf Eduard, Dutch chemist; b. Groningen, Holland, July 12, 1924; Ph.D., U. Amsterdam, 1956; m. Tine Elisabeth Hamerling, June 10, 1953; children—Olyne, Maryke. With Inst. for Atomic and Molecular Physics, Amsterdam, Holland, 1951-—, head dept. chemistry and molecular physics, 1957-—. Postdoctoral fellow Brookhaven Nat. Lab., 1956-57; Weizmann fellow Weizmann Inst., Rehovoth, Israel, 1961-63. Research and publs. on transport properties, thermal diffusion, phys. differences between isotopic species, rotational and vibrational energy transfer in collisions, all research in gaseous state. Home: 9, Avogadrostraat. Office: 407, Kruislaan, Amsterdam, Holland.*

DE VRIES, Daniel Alexander, Dutch physicist; b. Leeuwarden, Netherlands, Nov. 25, 1915; s. Izak and Betje (Israëls) de V.; doctoral physics State U., Leiden, Netherlands, 1939, Ph.D., 1952; m. Berendina Johanna Wisselink, Aug. 20, 1947; children—Johannes, Bettine, Izak David, Micha Chaim. Physicist Lab. Physics and Meteorology, Agrl. U., Wageningen, Netherlands, 1939-41, 48-55; meteorologist Royal Netherlands Air Force, 1945-48; physicist Commonwealth Sci. and Indsl. Research Orgn., Deniliquin, Australia, 1955-58; prof. physics Technol. U., Eindhoven, Netherlands, 1958-—. Cons. research dept. Koninkyke Nederlandsche Heidemaatschappij, 1961-—. Co-recipient Robert E. Horton award Am. Geophys. Union, 1958. Mem. Nederlandse Naturkundige Vereniging, Royal Meteorol. Soc. (fgn.), Am. Geophys. Union, Koninklijk Instituut van Ingenieurs. Author: (with W. R. van Wijk) Weer en Klimaat, 1952. Research and publs. on heat and moisture transfer in porous media, energy balance of earth's surface, theory of influence of irrigation on climate, evaporation under irrigation conditions; developed method for calculating thermal conductivity of granular materials from composition. Home: 10 t Hof, Lieshout, Netherlands. Office: 2 Insulindelaan, Eindhoven, Netherlands.*

DE VRIES, Hugo Marie, Dutch botanist; b. Haarlem, Netherlands, Feb. 16, 1848; student Leiden (Netherlands), Würzburg (pupil of Julius Sachs), Heidelberg (both Germany) univs.; M.D., 1870. Holder chair Realschule, Amsterdam, Netherlands, from 1871; named prof. extraordinary of botany U. Amsterdam, 1877, prof. ordinary, 1880-1918; established exptl. garden, Hilversum, nr. Amsterdam. Faculty U. Würzburg, 1897. Recipient Darwin medal 1906, Linnean Gold medal Linnean Soc. London, 1928. Corr. mem. botany sect. Acad. Scis. Author: Eine Methode zur Analyse der Turgorkraft, 1884; Intrazellulare Pangenesis, 1889; Die Mutations theorie, 1901-03. Study of discontinuous variations, especially on Oenothera (evening primrose) led him to rediscover works of Gregor Mendel (independently rediscovered by K. Correns, E. von Tschermak) and Mendelian laws of inheritance, 1900; developed theory mutations which states that new species arise through sudden change in offspring of normal parents; maintained every quality subject to change is represented by single phys. unit of heredity which he called pangen; work filled gap in Darwinian theory (also modified theories of Weismann by showing that germ plasm can be altered); research on osmosis, plasmolysis. Died Luntern, nr. Amsterdam, May 21, 1935.

DE VRIES, Simon Izak, Dutch hematologist; b. Hoorn, The Netherlands, May 1, 1902; s. Bernard S. and Anna (Elzas) De V.; student U. Amsterdam (The Netherlands), 1919-26, D. Medicine, 1937; m. Ruth Gelber, Dec. 17, 1945; children—Nicoline, Linda Simone. Practice gen. medicine, Amsterdam, 1928-38; asst. dept. hematology Brit. Post-grad. Sch., London, 1938-40; clin. hematologist U. Hosp. Amsterdam, 1940-—, lectr. hematology U. Amsterdam, 1963-—; head Bloodbank of the Whilhelmina Gasthuis; cons. The Universital Hosp., Army, Navy. Mem. Dutch Soc. Hematology (pres.), Internat. Soc. Hematology (v.p.), Swiss Soc. Hematology (corr.). Author: Textbook of Hematology, 1947; Hematological Laboratory Investigations, 1955. Research and publs. on treatment of malignancies; iron metabolism in man; human blood-coagulation. Home: Albrecht Durerstraat 17, Amsterdam. Office: 2e Helmersstraat 104, Amsterdam, The Netherlands.*

DE VRY, Herman Adolf, inventor; b. Mecklenburg, Schwerin, Germany, Nov. 26, 1877; s. William and Kunegunde (Dirnberger) DeV.; came to U. S., 1886; ed. pub. schs.; D.Sc. (posthumous), Lincoln Meml. U., 1941; m. Ida Schoelkopf, Apr. 1, 1916; children—Emma Wilhelmina (Mrs. Herbert Carlson), Edward Bernard, William Charles. Asso. with Lee DeForest; founder, pres. DeVry Corp. Founder, Nat. Conf. on Visual Edn. and Film Exhbn. (DeVry Found.). Named to Honor roll of Internat. Soc. Motion Picture Engrs., 1941. Patentee numerous motion picture machines. Pioneered use of motion pictures for ednl. purposes; developed 1st portable motion picture projector, 1912, one of 1st projectors with both sound and picture, 1934, 1st sound picture for use in home. Died Chgo., Mar. 23, 1941.

DEVYATKOV, Nikolay Dmitrievich, Russian electronics specialist; b. Apr. 11, 1907; grad. Leningrad Poly. Inst., 1931. Asso., Physicotech. Inst., USSR Acad. Scis., 1925-—, head dept. superhigh-frequency electronics, dir. Inst. Radio Engring. and Electronics, 1954-—; instr. Moscow Power Engring. Inst., 1944-—. Mem. USSR Acad. Scis. (corr.). Co-author: Electrovacuum Production Technology, 1962. Research and publs. on gas discharge and construction of gas-discharge equipment for protection of communication lines against overvoltage and acoustic shocks. Address: Inst. Radio Engring. and Electronics, Mokhovaya 11, Moscow K-9, USSR.

DE WAARD, Dirk, geologist, b. Hilversum, Holland, Feb. 5, 1919; s. Dirk and Christina C. (Spoel) de W.; B.Sc., Utrecht U., 1940, M.Sc., 1943, Sc.D. cum laude, 1947; m. Maire E. Aavasalo, Dec. 14, 1950. Came to U. S., 1958. Sci. asso. Utrecht U., 1947-52, privat-docent, 1949-52; prof. petrology U. Indonesia, 1952-58; vis. lectr. U. Cal. at Berkeley, 1958-59; asso. prof. geology Syracuse (N.Y.) U, 1959-64, prof., 1964-—. Guggenheim Found. fellow, 1964. Fellow Geol. Soc. Am., Mineral. Soc. Am., A.A.A.S., mem. geol. socs. India, Finland, Norwegian Geol. Soc., Mineral. Assn. Can., Royal Netherlands Acad. Scis. and Letters (corr.), Royal Geol. Mining Soc. Netherlands, Am. Geophys. Union. Research and numerous publs. on rocks and minerals of crystalline regions, mineral assns. in rocks, and reactionary textures revealing pressure and temperature conditions of rock formation, time and place of origin of metamorphic and magmatic rocks in mountain belts. Home: 868 Westmoreland Av., Syracuse, N.Y. 13210.*

DE WAARD, Hendrik, Dutch physicist; b. Groningen, Netherlands, Apr. 20, 1922; s. Sytze Klaas and Maria (Radys) DeW.; Candidate U. Groningen, 1946, Doctorandus, 1949, Doctorate, 1954; m. Paula Dekking, Sept. 16, 1950; children—Marietta, Karin, Anita. Stipendiate, Nobel Inst., Stockholm, Sweden, U. Uppsala (Sweden), 1954-55; with U. Groningen, 1955-—, prof. physics, 1958-—, dir. lab. physics, 1965-—; vis. prof. U. Ill., Urbana, 1962-63. Cons. Baird-Atomic Europe, 1959-—. Mem. Dutch, Am. phys. socs. Author: Electronica, 1959; (with D. Lazarus) Modern Electronics, 1966; also articles. Research on nuclear spectroscopy including properties of nuclear decay, short lived nuclear states, Mössbauer effect including solid state and nuclear effects, especially in isotope iodine-129; nuclear electronics; invented method for stabilizing scintillation spectrometers. Home: 8 Werfstraat, Groningen, Netherlands.*

DEWAR, Sir James, Brit. chemist; b. Kincardine-on-Forth, Scotland, Sept. 20, 1842; s. Thomas D. and Ann (Eadie) D.; ed. univs. Edinburgh, Ghent; m. Helen Rose Banks, 1891. Worked with Kekulé at Ghent, 1867; lectr. chemistry Royal Vet. Coll., Edinburgh, 1869, later prof.; Jacksonian prof. natural exptl. philosophy Cambridge U., 1875-1923; Fullerian prof. chemistry Royal Instn., London, 1877-1923. Mem. Balfour Commn. on London Water Supply, 1893-94; mem. Com. on Explosives, 1888-91. Recipient Rumford medal, 1894; 1st Hodgkins Gold medal Smithsonian Instn., 1899; Lavoisier medal French Acad. Scis., 1904; Matteucci medal Italian Soc. Scis., 1906; Albert medal Soc. Arts, 1908. Fellow Royal Soc., 1877; mem. French Acad. Scis., 1920; Author: Collected Papers, 1927. Research on organic chemistry, liquefaction of so-called permanent gases, temperatures approaching absolute zero; obtained temperatures as low as 14°K built device for producing liquid oxygen in quantity, 1891; demonstrated that liquid oxygen and liquid ozone are attracted by a magnet, 1891; discovered (with Frederick Augustus Abel) cordite, 1891; produced (with Moissan) liquefied flourine, 1897; inventor Dwar flask for storing liquid gases, 1892; obtained liquid hydrogen, 1898, solid hydrogen, 1899; determined melting and boiling points of liquid hydrogen and refractive index of solid hydrogen; discovered that absorbent power of charcoal for gases is increased by cold, 1902; experiment on phosphorescence at very low temperatures; devised model of carbon atom; studied physiol. action of light on eye; investigated vapor densities of potassium and sodium, specific heat of hydrogen; measured rate of prodn. of helium from pure radium salt, 1910. Died Mar. 27, 1923.

DEWAR, Michael James Steuart, chemist; b. Ahmednagar, India, Sept. 24, 1918; s. Francis and Nan (Keith) D.; B.A., Oxford U., 1940, D.Philosophy, 1942, M.A., 1943; m. Mary Williamson, June 3, 1944; children—Robert Berriedale Keith, Charles Edward Steuart. Came to U. S., 1959. Researcher Oxford U., 1942-45, Imperial Chem. Industries fellow, 1945; phys. chemist Courtaulds, Ltd., 1945-51; prof. chemistry, head dept. Queen Mary Coll., U. London (Eng.), 1951-59; prof. chemistry U. Chgo., 1959-63; Robert A. Welch prof. chemistry U. Tex., Austin, 1963—; Reilly lectr. Notre Dame U., 1951; vis. prof. Yale, 1957; Falk-Plaut lectr. Columbia U., 1963; William Pyle Phillips vistor Haverford Coll., 1964; Arthur D. Little vis. prof. Mass. Inst. Tech., 1966. Cons. Monsanto Co., 1959—. Fellow Royal Soc., 1960, Am. Acad. Arts and Scis.; mem. Chem. Soc. (Tilden lectr. 1954), Am. Chem. Soc. (Harrison Howe award 1961), Sigma Xi. Author: The Electronic Theory of Organic Chemistry, 1949; Hyperconjugation, 1962; An Introduction to Modern Chemistry, 1965. Research and numerous publs. on phys. and chem. behavior organic compounds in terms of molecular structure, synthesis potential pharms. Home: 6808 Mesa Dr., Austin, Tex. 78731.*

DE WECK, Alain Ladislas, Swiss immunologist; b. Montreux, Switzerland, July 26, 1928; s. Ladislas and Blanche (Bachmann) De W.; M.D., U. Geneva, 1953; P.D., U. Bern (Switzerland), 1966; postgrad. U. Lausanne, Geneva, Paris, France; m. Christine Blasel, Apr. 27, 1965. Fgn. USPHS post-doctoral research fellow Washington U., St. Louis, 1958-61; head div. allergy and clin. immunology, asso. prof. immunology, U. Bern, 1961—, also cons. Mem. Collegium Internationale Alergologicum, Brit., French, Swiss socs. allergy and immunology. Author: Immunotolerance to Simple Chemicals, 1966. Research, publs. on immunochem. mechanisms of penicillin allergy, devel. test procedures to detect penicillin allergy; mechanisms of contact dermatitis; exptl. prodn. of immunological tolerance and desensitization to chem. allergens; study of immune response to chem. allergens in man. Home: 12 Ch. St. Marc, 1700 Fribourg, Switzerland. Office: Allergy Div., Inselspital, 3000 Bern, Switzerland.*

DE WECKER, Louis, ophthalmologist; b. Frankfort/Main, Germany, 1832; Author: Traité des maladies des yeaux, 1863; (with Jäger) Atlas des maladies du fond de l'oeil, 1870; (with Landolt) Traité d'ophthalmologie, 1880-89. Popularized sclerotomy, iridectomy; improved cataract operation. Died Paris, 1904.

DEWEES, William Potts, Am. pediatrician, obstetrician; b. nr. Pottstown, Pa., May 5, 1768; studied medicine under 2 neighboring doctors; M.D., U. Pa.; m. Martha Rogers; m. 2d, Mary Lorrain, 1802. Practiced medicine, Abington, Pa.; moved to Phila., 1793; specialized in obstetrics at time when delivering babies was considered beneath dignity of doctor; his health failed for time, returned to practice, 1817; adj. prof. obstetrics U. Pa., 1825-34, prof., 1834-35; forced to retire by failing health. Author: (chief work) A Compendious System of Midwifery; credited with authorship 1st pediatrics textbook in Am., 1825. Died Phila., May 20, 1841.

DEWEESE, James Arville, Am. surgeon; b. Kent, O., Apr. 5, 1925; s. Arville and Vergie (Jenkins) DeW.; student Harvard, 1942-43; Kent State U., 1943-44; M.D., U. Rochester, 1949; m. Margaret Brown, June 20, 1950 (dec. 1960); children—James Arville, Margaret Ann, Elizabeth Lynn, Joanne Spencer; m. 2d, Patricia Bidwell, May 5, 1962; children—Robert Bidwell, Jamie Susan. Faculty, U. Rochester, 1956—, asso. prof. surgery, 1963—; staff Strong Meml. Hosp., Rochester, 1956—, asso. surgeon, 1958—. Cons., Rochester Gen. Hosp., 1959—, Bath and Batavia VA Hosp., Rochester, 1963—. Fellow A.C.S.; mem. A.M.A., N.Y., Monroe County med. socs., Rochester Acad. Medicine, Internat. Cardiovascular Soc., Am. Coll. Angiology, Soc. for Vascula Surgery, Soc. U. Surgeons. Research, numerous publs. on diseases of heart and blood vessels, use x-rays of veins and arteries for diagnosis and evaluation of treatment; demonstrated use of patient's own veins for replacement of diseased arteries and veins. Home: 121 Newcastle Rd., Rochester, N.Y. 14610.*

DEWET, Johannes Martenis Jacob, botanist; b. Vredefort, South Africa, July 4, 1927; s. Johannes M. J. and Martha (Wessels) DeW.; B.Sc. U. Pretoria (South Africa), 1949; Ph.D., U. Cal. at Berkeley, 1952; m. Alessa Eleonore Hudec, Sept. 3, 1950; children—Martin Christian, Giselle Isabelle. Came to U. S., 1960, naturalized, 1966. Plant breeder South African Dept. Agr., 1952-54; biosystematist Nat. Herbarium, Pretoria, 1954-59; faculty Okla. State U., Stillwater, 1960-67, prof. botany, 1966-67; professor cyrogenetics U. Ill., Urbana, 1967—. South African Govt. fellow, 1949-50; Canadian NRC fellow, 1959-60. Mem. Am. Bot. Soc., Am. Soc. Plant Taxonomists, Assn. pour l'Etude Tax., Flor d'Afr. Trop. South Africa. Research and publs. on phylogenetic relationships between species, genera and tribes of various plant groups, how methods of reprodn. and types of polyploid effect speciation, how man is contbg. to speciation by changing environment to fit his own needs. Home: 1501 Hillcrest, Urbana, Ill. 61801.*

DEWEY, Bradley, Jr., Am. chem. engr.; b. Pitts., Apr. 10, 1916; s. Bradley and Marguerite (Mellon) D.; B.S., Harvard, 1937; Sc.D., Mass. Inst. Tech., 1940; m. Jane Holcombe, Aug. 10, 1940; children—Margot, Bradley III, John Holcombe, Carolyn, Joan. With Dewey & Almy Chem. Co., Cambridge, Mass., 1940-51, v.p., 1951-56, dir., 1952-56; pres. The Cryovac div. W. R. Grace & Co., Duncan, S.C., 1956-64; sr. v.p. W. R. Grace & Co., 1964—. Mem. Am. Chem. Soc., Am. Inst. Chem. Engrs., Soc. Chem. Industry, Am. Inst. Chemists, Sigma Xi, Alpha Chi Sigma. Improved packaging through manufacture of sealing compounds for tin, glass and steel containers; research in colloid chemistry, rubber latex and plastics, chelating agts. Home: Arden Way, Spartanburg, S.C. Office: The Cryovac Div., W. R. Grace & Co., Duncan, S.C.

DEWEY, Frederic Perkins, Am. chemist; b. Hartford, Conn., Oct. 4, 1855; s. Daniel S. and Elizabeth (Perkins) D.; Ph.B., Sheffield Sci. Sch. (Yale), 1876; m. Charlotte Esther Candee, Apr. 12, 1877. Asst. analytical chemistry Lafayette Coll., 1876-77; chemist with iron and steel mfrs. until 1881; (with Dr. George W. Hawes) investigated bldg. stones of U. S. for 10th census, 1881; curator metallurgy, U. S. Nat. Mus., 1882-89; propr. comml. lab., 1890-1903; assayer Mint Bur., U. S. Treasury; acting dir. Mint Bur., 1913. Author: Descriptive Catalogue Collections in Economic Geology and Metallurgy (bull. 42), 1891. Deceased.

DEWEY, John, Am. philosopher, psychologist; b. Burlington, Vt., Oct. 20, 1859; s. Archibald S. and Lucina A. (Rich) D.; A.B., U. Vt., 1879, LL.D., 1910; Ph.D., Johns Hopkins, 1884; LL.D., U. Wis., 1904, Peking Nat. U., 1920, U. Paris, 1930; D.Sc., U. Pa., 1946; Ph.D., Oslo U., 1946; m. Alice Chipman, July 28, 1886; children—Fred'k A., Evelyn, Morris, Lucy A., Gordon, Jane U., Sabino L. (adopted); m. 2d, Mrs. Roberta Grant, Dec. 11, 1946; children (adopted)—John, Adienne. Instr., asst. prof. philosophy U. Mich., 1884-88, prof., 1889-94; prof. philosophy U. Minn., 1888-89; prof. and head dept. philosophy U. Chgo., 1894-1904, dir. Sch. Edn., 1902-04; prof. philosophy Columbia from 1904. Mem. Nat. Acad. Scis., Am. Psychol. Assn. (pres. 1899-1900), Am. Philos. Soc. (pres. 1905-06) Inst. France (corr.). Author: Psychology, 1886; Leibnitz, 1888; Critical Theory of Ethics, 1894; Study 1894; School and Society, 1899; Studies in Logical Theory, 1903; How We Think, 1909; Influence of Darwin on Philosophy, and Other Essays, 1910; German Philosophy and Politics, 1915, rev. edit., 1942; Democracy and Education, 1916; Reconstruction in Philosophy, 1920; Human Nature and Conduct, 1922; Experience and Nature, 1925; The Public and Its Problems, 1927; The Quest for Certainty, 1929; Art as Experience, 1934; A Common Faith, 1934; Liberalism and Social Action, 1935; Logic: The Theory of Inquiry, 1938; Culture and Freedom, 1939; Education Today; Problems of Man, 1946; Knowing and the Known (with Arthur Bently), 1949. His work served as organizing principle of Chgo. "school" of functional psychology; opposed elementistic psychologies of his time; anticipated Gestalt psychology in his stress on total coordinations; contbd. to devel. of dynamic psychology. Died June 2, 1952.

DEWEY, Lyster Hoxie, Am. botanist; b. Cambridge, Mich., Mar. 14, 1865; s. Francis Asbury and Harriet (Smith) D.; B.S., Mich. Agrl. Coll., 1888; m. Etta Conkling, Aug. 22, 1889; children—Grace Marguerite, Mary Genevieve. Instr. botany Mich. Agrl. Coll., 1888-90; asst. botanist U. S. Dept. Agr., 1890-1902, botanist in charge of fiber investigations since 1902; conducted investigations on grasses and troublesome weeds; U. S. rep. to Internat. Fiber Congress, Soerabaia, Java, 1911. Fellow A.A.A.S.; mem. Washington Acad. Scis., Bot. Soc. Washington, Biol. Soc. Washington. Died Nov. 1944.

DEWEY, Richard (Smith), Am. psychiatrist, neurologist; b. Forestville, N.Y., Dec. 6, 1845; s. Elijah and Sophia (Smith) D.; M.D., U. Mich., 1869, A.M., 1900; postgrad. under Virchow, Berlin, Germany, 1871; m. Lillian Dwight, Jan. 2, 1873 (dec. 1880); children—Richard Dwight, Ethel Lillian, Robert Strong; m. 2d, Mary E. Brown, June 22, 1886; children—Ellinor Maria, Donald Mack. Intern Bklyn. City Hosp., 1870; vol. asst. surgeon, Franco-Prussian War, with 7th corps, field hosp., Pont à Mousson, France, and Res. Hosp., Hesse-Cassel, Germany, 1870-71; asst. physician, State Hosp. for Insane, Elgin, Ill., 1872-99; med. supt. State Hosp. for Insane, Kankakee, Ill. (1st example in U. S. of detached ward or cottage plan of bldgs. for care of mental diseases), 1879-93; prof. mental and nervous diseases Chgo. Post-Grad. Med. Sch., 1893-1909; in charge Milw. Sanitarium, 1895-1921. Pres. Am. Medico-Psychol. Soc. (now Am. Psychiat. Assn.), 1896. Died Aug. 4, 1933.

DEWEY, William Cornet, Am. radiation biologist; B. Omaha, Nov. 4, 1929; s. Carroll Wright and Julia (Stephens) D.; B.S., U. Wash., 1951; Ph.D., U. Rochester, 1958; m. Helen Mesman, June 14, 1951; children—Marc, Lisa, Sara, Jana. Asst. prof. radiophysics U. Tex. Postgrad. Sch. Medicine, 1958-60, asso. physicist, asso. prof. biophysics, 1960-65; prof. radiation biology Colo. State U., Ft. Collins, 1965—. Mem. Radiation Research Soc., Sigma Xi. Research, and numerous publs. contbg. to advances in scanning techniques for localizing radioisotopes in human beings and in understanding mechanisms of ionizing radiation effects of mammalian cells. Home: 1920 Seminole St., Ft. Collins, Colo. 80521.*

DEWHURST, Christopher John, English obstetrician, gynecologist; b. Garstang, Eng., July 2, 1920; s. John and Agnes (Kirkham) D.; M.B., Ch.B., Manchester U., 1943; m. Hazel Mary Atkin, Sept. 3, 1952; children—Charles, Alan, Mary. House physician Dept. Child Health, Manchester, 1946; former registrar, sr. registrar St. Mary's Hosps., Manchester; lectr. obstetrics and gynecology Sheffield U., 1951-56, sr. lectr., 1956-60, reader, 1960-66; hon. cons. obstetrician and gynecologist United Sheffield Hosps., also Sheffield Regional Hosp. Bd., 1955-67; prof. obstetrics and gynecology Inst. Obstetrics and Gynaecology, Queen Charlotte's Hosp. and Chelsea Hospital for Women, London, 1967—. Fellow Royal Coll. Surgeons; mem. North Eng., Birmingham, Midland obstetrics and gynecol. socs., Royal Coll. Obstetricians and Gynecologists. Author: Students Guide to Obstetrics and Gynecology, 2d edit., 1966; Gynaecological Disorders of Infants and Children, 1963. Research and publs. on major complications and accidents of childbirth; chromosomal abnormalities in gynecology; diseases of female genital tract in infants. Home: Glendale House, Old Slade Lane, Iver, Bucks., Eng. Office: Queen Charlotte's Hosp., Goldshawk Rd., London W. 6, Eng.*

DEWHURST, David John, Australian biophysicist; b. Melbourne, Australia, Jan. 8, 1919; s. John Heyliger and Winifred (Tatchell) D.; B.A. with honors, U. Melbourne, 1940, M.Sc., 1952, Ph.D., 1958; m. Hilda Marjorie Wilmot, Dec. 13, 1947; children—Penelope Anne, Peter, John, Timothy David. Faculty, U. Melbourne, 1949—, reader biophysics, 1964—. Hon. cons. in biophysics Royal Melbourne Hosp., 1963—. Mem. Faraday Soc., Australian Physiol. Soc., Australian Fedn. for Med. and Biol. Engring. Author: Physical Instrumentation in Medicine and Biology, 1966. Research and publs. on mechanisms of fine control of human muscular movement, mechanisms connecting distbn. of ions across a living cell membrane and elec. potentials arising from this distbn. Home: 37 Culwell Av., Mitcham, Victoria. Office: Physiology Dept., U. Melbourne, Parkville, N.2., Victoria, Australia.*

DEWIEST, Roger J. M., civil engr., earth scientist; b. Lebbeke, Belgium, Feb. 6, 1925; s. Louis O. and Maria (Van Nuffel) DeW.; came to U. S., 1954, naturalized, 1963; M.Sc. Electr. Engring., Ghent (Belgium) U., 1949; M.Sc. in Civil Engring., Cal. Inst. Tech., 1955; Ph.D., Stanford, 1957-59; m. Odette M. Gewillig, Oct. 4, 1952; children—Daniel, Maria, Denise, Roger. Engr., Belgian Dept. Labor, 1950-54, SOFINA, Brussels, Belgium, 1956-57; faculty Princeton, 1959—, prof. civil and geol. engring., 1965—. Cons. engring. and hydrology to industry, state and fed. agys. Freeman fellow Am. Soc. C.E., 1963-64. Mem. Am. Soc. C.E., Am. Geophys. Union, Soc. Royale Belge des Tujeniers et des Industriels, Sigma Xi. Author: (with S. N. Davis) Geohydrology, 1965; also articles. Translator: Theory of Ground Water Motion, 1962. Applied math. techniques to solve problems in ground water flow; analysis basic properties flow in porous media both theoretical and exptl. Home: 631 Lake Dr., Princeton, N.J.*

DE WIJN, Harold W., Dutch physicist; b. Naarden, Netherlands, June 2, 1936; s. C. J. and A. (de Kievit) de W.; D.Physics cum laude, U. Amsterdam, 1963. Sci. worker Zeeman Lab., U. Amsterdam (Netherlands), until 1963, Natuurkundig Laboratorium, 1963—, faculty microwave spectroscopy and magnetic resonance, 1964—. Research, publs. on microwave spectroscopy of gases, magnetic resonance of solid state. Home: Van der Helstlaan 52, Naarden, Netherlands. Office: Valckenierstraat 65, Amsterdam, Netherlands.*

DE WITT, Jan (Johan), Dutch mathematician; b. Dort, Netherlands, Sept. 24, 1625; s. Jacob D. de Witt; ed. at Leiden, Netherlands; m. Wendela Bicker, 1655; 2 sons, 3 daus. Visited France, Italy, Switzerland, Eng. with his bro. Cornelius, 1645; on his return became advocate at The Hague, Netherlands; apptd. pensionary at Dort, 1650; apptd. grand pensionary of Holland, 1653, re-elected in 1663, 1668, resigned 1672; restored the finances of Holland and extended its comml. supremacy in East Indies. Author: Elementa curvarum linearum, (gives number of typical equations and geometric characteristics for curves), 1659. Developed new method of generating conics using Cartesian analysis. Died The Hague, Aug. 20, 1672.

DE WOLF, Frank Walbridge, Am. geologist; b. Vail, Ia., Mar. 22, 1881; s. John Horton and Carrie M. (Tempest) D.; S.B., U. Chgo., 1903, post-grad., 1903-04; m. Fanny Davis, Dec. 26, 1904; children—John Walbridge, Eleanor, Robert Williams, Frank Tempest. Geologic aid and asst. geologist U. S. Geol. Survey, Washington, 1904-08; asst. state geologist, Ill., 1908-09; acting dir. Ill. Geol. Survey, 1909-11, dir. 1911-23. Asst. dir. U. S. Bur. of Mines, 1917-18; mem. com. on geology and paleontology and chmn. sub-com. on geology of cantonments NRC, 1917, div. of states relations, 1919-23; chief geologist Humphreys Corp., 1923-27; v.p., gen. mgr. La. Land and Exploration Co., oil producers, 1927-31; head dept. geology U. Ill., 1931-46, emeritus, from 1946. cons. on petroleum geology. Fellow Geol. Soc. Am. Soc. Econ. Geologists; mem. Am. Inst. Mining Engrs., Am. Assn. Petroleum Geologists (v.p.), Phi Delta Theta, Sigma Xi. Explored and mapped parts of Tex., Okla. and La. Gulf Coast; established highest discovery rate in petroleum prospecting of the time (late 1920's); pioneered use of seismograph in petroleum exploration, design and use of swamp-buggies for exploration in swampy areas. Died Sept. 16, 1957.

DEWOLFE, Barbara Blanchard, Am. zoologist; b. San Francisco, May 14, 1912; d. Marion Sargeant and Elizabeth (Dewing) Blanchard; A.B., U. Cal. at Berkeley, 1933, Ph.D., 1938; m. Nels Oakeson, Sept. 2, 1950 (div. Sept. 1954); m. 2d, Robert Hill DeWolfe, May 28, 1960. Instr. life sci. Placer Jr. Coll., Auburn, Cal., 1939-42; instr. zoology U. Cal., Davis, 1942-43; instr. zoology Smith Coll., 1943-45; faculty zoology. U. Cal., Santa Barbara, 1946—, prof., 1957—, asso. dean Coll. Letters and Sci., 1961-64, acting dean, 1965. Mem. Am. Ornithol. Union, Cooper Ornithol. Soc., Ecol. Soc. Am., Am. Soc. Zoologists, Sigma Xi. Author: (with T. T. McCabe) Three Species of Peromyscus, 1950. Research, publs. on behavior and physiol. cycles in whitecrowned sparrows; ann. cycles in native vertebrates in relation to environment under natural conditions. Home: 1361 Holiday Hill Rd., Goleta, Cal. 93017. Office: Biol. Sci. Dept. U. Cal., Santa Barbara, Cal. 93106.*

DE WREDE, Fabian Jacob, Swedish physicist; b. 1802; research in theoretical physics and phys. chemistry. Died 1893.

DEXLER, Hermann, veterinarian; b. Teesdorf, Austria, May 10, 1866; s. Josef and Mathilde (Fischer) D.; studied vet. medicine in Vienna, Austria, 1884-88; doctorate, 1888; hon. dr. vet. medicine from vet. schs. of Vienna and Budapest, Hungary, 1928; m. Maria von Perko. Began practice vet. medicine in Leoben, Steiermark, 1888; became asst. Tierärztliche Hochschule, Vienna, 1893; research staff brain research inst. of Heinrich Obersteiner; became chief asst., 1895; named lectr., 1897; became asso. prof. on epidemics, chmn. vet. inst. of med. faculty German U. in Prague (Czechoslovakia), 1898; named prof., 1925. Author: Die Nervenkrankheiten des Pferdes (1st vet.-neurol. work), 1899; also other monographs, articles. Research on neuropathology of house pets, morphology of dugongs in Australia, binocular sight of horses, pyramidal paths of sheep and goat herds; specialized in psychology and psychotic diseases of animals. Died Prague, May 9, 1931.

DEXTER, David Lawrence, Am. physicist; b. Ashland, Wis., July 2, 1924; s. Stephen T. and Mildred (Carlstrom) D.; B.S. in Math., Mich. State U., 1947; M.S., U. Wis., 1948, Ph.D. in Physics, 1951; m. Doris Grigsby, Sept. 4, 1948; children—Cecily, Diane, Philip (dec.), Jefferson, Pamina, Melanie. Research asso. U. Wis., 1950-51, U. Ill., 1951-52; cons. U. S. Naval Research Lab., 1952—; prof. Inst. Optics and Physics, U. Rochester (N.Y.), 1952—; cons. to numerous cos. and govt. agys. Home: 184 Penfield Rd., Rochester, N.Y. 14610.*

DEXTER, Lewis, Am. physician; b. Concord, Mass., Mar. 1, 1910; s. Smith Owen and Helen (Denison) D.; A.B., Harvard, 1932, M.D., 1936; m. E. Cassandra Kinsman, Dec. 12, 1941; children—Lewis Dexter, Smith Owen III, Cassandra Kinsman. Fellow in medicine Harvard, 1938-40, staff, 1941—, clin. prof. medicine, 1958—, tutor Med. Sch., 1948—; practice medicine specializing in cardiology, Boston, 1942—; staff Peter Bent Brigham Hosp., Boston, 1939-40, 41—, physician, 1952—. A.C.P. fellow Instituto Physiology, Buenos Aires, Argentina, 1940-41. Diplomate Am. Bd. Internal Medicine. Mem. A.M.A., Am., New Eng. heart assns., Mass. Med. Soc., Am. Soc. for Clin. Investigation, A.C.P., Am. Fedn. for Clin. Research, Assn. Am. Physicians, Am. Clin. and Climatol. Assn., Am. Physiol. Soc., Am. Acad. Arts and Scis., Sociedad Cardiologia de Argentina, Interurban Clin. Club. Author: (with L. Dexter, S. Weiss) Pre-Eclamptic and Eclamptic Toxemia of Pregnancy, 1941; translator Renal Hypertension, 1946. Editorial bd. Circulation, 1968—. Research, numerous publs. on measurement of circulatory changes in various heart diseases, correlating them with manifestations of patients' disease. Home: 108 Upland Rd., Brookline, Mass. 02146. Office: Peter Bent Brigham Hosp., 721 Huntington Av., Boston 02115.*

DEXTER, Ralph Warren, Am. ecologist; b. Gloucester, Mass., Apr. 7, 1912; s. Brant Mess and Bessie (Clark) D.; B.S., U. Mass., 1934; Ph.D., U. Ill., 1938; m. Jean Westwater, June 13, 1938; children—Carol Jean (Mrs. Louis A. Hovancsek), Diane Christine (Mrs. Barry P. Jones). Mem. faculty Kent (O.) State U., 1937—, prof. biol. scis., 1948—. Named Outstanding Faculty Mem., Kent State U., 1957. Fellow A.A.A.S.; mem. Am. Malcol. Union (pres. 1965-66), Ecol. Soc. Am., Am. Soc. Limnology and Oceanography, Am. Ornithol. Union, Wilson Ornithol. Soc., Inland Bird Banding Assn., Soc. Systematic Zoology, History of Sci. Soc. Research and numerous publs. on ecology of marine communities, mollusks and crustaceans, life history of chimney swift, history of Am. naturalists. Home: 1228 Fairview Dr., Kent, O. 44240.*

DEXTER, Richard N., Am. physicist; b. Ashland, Wis., Nov. 22, 1927; s. Stephen Torrey and Mildred (Carlstrom) D.; B.S., Mich. State Coll., 1949; M.S., U. Wis., 1949, Ph.D., 1955; m. Gladys R. Eastman, Dec. 20, 1958; children—Carol E., Beth C., Andrea E. With Lincoln Lab., Mass Inst. Tech. Cambridge, 1952-55; faculty U. Wis., Madison, 1955—. Cons. IBM Research, Alfred P. Sloan Found. fellow, 1955-59. Fellow Am. Phys. Soc.; mem. A.A.A.S., Sigma Xi. Editorial bd. Revs. Sci. Instruments, 1960-63. Contbr. numerous publs. on research to profl. jours. Home: 925 Waban Hill, Madison, Wis.*

DEYEUX, Nicolas, French pharmacist; b. Paris, France, Mar. 21, 1745; ed. Mazarin Coll.; began career in uncle's pharm. house; pharmacist to Napoleon; prof. of medical chemistry and pharmacy at Sch. Medicine, Paris. Mem. French Acad. Scis., 1797 (became titulary mem. com. on pub. welfare 1797). Author: Précis d'expériences et d'observations sur les différences espèces de lait, 1800; Considérations chimiques et médicales sur le sang des ictériques, 1804. Manufactured sugar from beets during English blockade. Died Paris, Apr. 27, 1837.

D'EYRINIS, Eyrinis, physician; b. Russia, flourished circa 1712; tchr. Greek, Neuchatel, Switzerland. Author: Description des lois des mines, 1721; Avis sur l'usage des asphaltes. Discovered rock-asphalt, Travers, Switzerland, made asphalt mastic by blending powdered rock-asphalt with hot pitch, 1712.

DE' ZANCARI, Alberto, Italian physician; b. Bologna, Italy, 1280; s. Galvano de' Zancari; Doctor Physicae, 1310; final med. degree, 1326; physician Ravenna, Italy, from 1314; advocated use of human cadavers in demonstrating anat. principles. Author: De Cautelis Medicorum. Died after 1348.

DEZEIMERIS, Jean-Eugene, French physician; b. Villefranche-de-Longchapt, France, 1799. Author: Dictionnaire historique de la médecine ancienne et moderne, 1838-39; studied history of medicine; wrote dissertations on Hippocratic doctrines. Died 1851.

DEZOTEUX, François, French physician; b. Boulogne-sur-Mer, France, 1724; chief sch. surgery founded by Louis XVI, France. Author: Pièces justificatives concernant l'inoculation, 1765. Advocated innoculation for smallpox. Died Versailles, France, 1803.

DHAR, N. R., chemist; b. Jessore, Bengal, E. Pakistan, Jan. 2, 1892; s. P. K. and Nirod Mohini (Ghosh) D.; B.Sc., Calcutta (India) U., 1911, M.Sc., 1913; D.Sc., London (Eng.) U., 1917; Docteur es Sciences, Sorbonne, Paris, France, 1919; m. Mira Chattopadhyaya, Oct. 14, 1950. Staff, Indian Edn. Service, 1919-46; faculty Allahabad (India) U., until 1953, head chemistry dept.; hon. dir. Sheila Dhar Inst. Soil Sci., 1950—; founder Research Centre in Chemistry and Soil Sci., India; lectr. U. Cambridge (Eng.), U. Edinburgh (Scotland), U. London, U. Paris, U. Göttingen (Germany), U. Wageingen (Germany), U. Giessen (Germany), Rothamsted Exptl. Sta., Royal Coll. Agr., Sweden, Uppsala. Mem. Nobel Com. for Selecting Nobel Prize in Chemistry, 1938-52. Named 1st Indian Scientist; Gold medalist Calcutta U. Fellow Royal Inst. Chemistry London, Chem. Soc. (London); mem. Internat. Congress Soil Sci. (past pres.), Nat. Acad. Sci. India (past pres.), Indian Soc. Soil Sci. (pres.), Soc. Biol. Chemists India (pres.), Agr. Soc. India (pres.), Indian Chem. Soc. (past pres.), French Acad. Agr. (fgn.), French Acad. Sci. (corr.). Author: Chemical Action of Light; New Conceptions in Biochemistry; Influence of Light on Some Biochemical Processes; Amader Khadya; Jamir Urbarata Briddhi; also numerous articles. Established that sunlight and phosphates augment nitrogen fixation in soils by slow oxidation of organic matter created by photosynthesis, air in nitrate formation in soils. Address: 2-D Beli Rd., Allahabad, U.P., India.*

D'HÉNOUVILLE, Baron Hyacinthe-Théodore, French chemist; b. Paris, June 17, 1715; s. Hyacinthe Théodore d'H; studied under Rouelle, Jardin du roi; demonstrator Jardin du roi; became asst. chemist Faculty Medicine, Paris, 1752. Author: Expériences pour servir à l'analyse du borax, 1750. Editor: Cours de chymie (Lemery), 1756. Made 1st determination of the composition of borax (combination of soda with boric acid), 1747, independently demonstrated by Scheele, 1768. Died Paris, Mar. 10, 1768.

DHÉRAIN, Pierre-Paul, French agronomist; b. Paris, France, Apr. 19, 1830; orphan raised by Decaisne; studied in Fremy's Lab., 1850; License ès sciences physiques; m. M. Page, 1866; 1 son, Henri; m. 2d, Hulleu François, 1873; became prof. plant physiology Mus. Naval History, also prof. natural history Guignon (France) Sch., 1880; set up exptl. agrl. stas. at Guignon. Mem. French Acad. Scis., 1887, Soc. Agr., Soc. Rural Economy. Author: les Engrais, les ferments de la terre, 1895; Traité de chemie agricole. Founder, Annales agronomiques, 1875; publs. on practical agronomy and agrl. chemistry; Research on effect of phosphates on growth, devel. of rye, barley, wheat, and sugar beet, use of fertilizers; demonstrated that ultra-violet light had harmful effects on plants; proved that absorption is regulated by consumption in plant nutrition. Died Paris, Dec. 7, 1902.

DHERMY, Pierre, French ophthalmologist, pathologist; b. Paris, June 25, 1922; s. Hippolyte and Accia (Guilbaut) D.; M.D., U. Paris, 1949; m. Andree Vallee, Mar. 21, 1946; children—Didier, Agnes. Chief, Ophthalmic Pathology Lab., Paris, 1954—. Mem. Paris Soc. Ophthalmology, European Ophthalmic Pathology Soc. Contbr. numerous articles to profl. jours. Home: 70 rue M. Thorez, Nanterre, France. Office: Hotel-Dieu, Pl. Parvis Notre Dame, Paris, France.*

D'HONT, Maurice Docile Emile, Belgian chemist; b. Bruges, Belgium, Dec. 2, 1922; s. Prosper and Rachel (Loof) D'h; bacch. phil., 1944; dr. sc. U. Louvain (Belgium), 1948; postgrad. U. Cambridge (U.K.), 1948, Argonne (Ill.) Internat. Sch., 1952; m. Ma ie-Thérèse, Aug. 30, 1949; children—Myriam, Godelieve, Ann, Marc. Research asso. Masst. Inst. Tech., also Harvard, 1941-51; research scientist Métallurgie Hoboken, Olen, Belgium, 1951; head chemistry and metallurgy dept. CEN, Nat. Nuclear Research Center, Mol, Belgium, 1952-59, sci. mgr., 1959-63, sci. adviser to pres., 1963—; faculty U. Louvain, 1951—, prof. extr., 1963—, head dept. radiol. protection, 1964—; pres. Controle-Radioprotection, 1966—. Mem. tech. com. Eurochemic, 1957—; dep. dir. Euratom and Cuming, Europe, NV, 1964—. Mem. Am. Nuclear Soc., Royal Metall. Soc. (London), Royal Flemisch Acad. Sci. (corr.). Mem. adv. bd. Jour. Nuclear Materials, 1962—. Research and publs. on chemistry, metallurgy, reprocessing, irradiation effects, waste treatment of nuclear program. Home: 265 Boeretang. Office: 200 Boeretang, Mol, Belgium.*

DIAMOND, Louis Klein, Am. physician; b. N.Y.C., May 11, 1902; s. Lazer and Lena (Klein) D.; A.B., Harvard, 1923, M.D., 1927; m. Flora Kaplan, July 2, 1929; children—Jared Mason, Susan Judith. Successively intern, asst. resident, chief resident pediatrics Children's Hosp., Boston, 1927-31, asso. med. chief, 1955—, chief hematology div., 1951—; fellow in hematology, 1927-28, 31-33; dir. Blood Grouping Lab., Boston, 1947—; prof. pediatrics Harvard Med. Sch., 1963—. Med. dir. blood program A.R.C, 1948-51, bd. dirs. Met. Boston chpt., 1964—; mem. study sects. NIH, USPHS, NRC, 1949—. Recipient Karl Landsteiner award Am. Assn. Blood Banks, 1963, Distinguished Pub. Service in Sci. medal Theodore Roosevelt, 1965, 4 awards sci. research Joseph P. Kennedy, Jr. Meml. Found., 1966; Carlos J. Finlay gold medal Cuban Govt., 1951; award of merit Netherland Red Cross, 1959; George R. Minot lectr. A.M.A., 1965; Rachford Meml. lectr. Children's Hosp., Cin., 1966. Mem. Am. Pediatrics Soc., Am. Acad. Pediatrics (Mead Johnson award 1946), Soc. Clin. Investigation, Internat. Hematology Soc., Soc. Pediatric Research, Mass. Med. Soc., Am. Acad. Arts and Scis., Am. Hematology Soc., Internat. Transfusion Soc. Author: Erythroblastosis Fetalis, 1958; Atlas of Blood in Children, 1944; also numerous articles. Research on children's diseases, especially diseases of the blood. Home: 29 Lowell Rd., Brookline, Mass. 02146. Office: 300 Longwood Av., Boston 02115.

DIAMOND, Richard Martin, Am. nuclear chemist; b. Los Angeles, Jan. 7, 1924; s. William and Bess (Weizenhaus) D.; B.A., U. Cal. at Los Angeles, 1947; Ph.D. in Nuclear Chemistry, U. Cal. at Berkeley, 1951; m. Marian Cleeves, Dec. 18, 1950; children—Catherine T., Richard C., Jeffrey B., Ann E. Faculty, Harvard, 1951-54, Cornell U., 1954-58; sr. staff nuclear chemistry div. Lawrence Radiation Lab., Berkeley, 1958—. Guggenheim fellow Niels Bohr Inst., U. Copenhagen, 1966-67. Mem. Am. Chem. Soc., Am. Phys. Soc., Phi Beta Kappa, Sigma Xi, Phi Lambda Upsilon. Co-editor: Proc. 3d Internat.

Conf. on Reactions between Complex Nuclei, 1963. Research and publs. on theory and practice of ion-exchange resin processes and of solvent extraction systems, coulomb excitation, nuclear spectroscopy. Home: 574 Santa Clara Av., Berkeley 94707. Office: Lawrence Radiation Lab., Berkeley, Cal. 94720.*

DIANIN, Aleksander Pavlovich, Russian chemist; b. Russia, Apr. 18, 1851; s. Pavel; prof. medicine Peterburg War Acad.; discovered reaction of condensed ketons with phenols (reaction of Dianin). Died Nov. 23, 1918.

DIANNELIDIS, Themistokles, Greek botanist; b. Portaria, Apr. 27, 1910; s. Diannelos and Barbara (Tragari) D.; Ph.D. in Natural Scis., U. Athens; m. Ilse Passler; children—Barbara, Alexander. Asst., Inst. Botany, U. Athens; prof. of gymnasium; asso. prof. U. Salonica, 1948, full prof. botany, dir. Inst. Botany, 1951. Mem. German Soc. Botany, Internat. Commn. for Sci. Exploration of Mediterranean Sea. Author: Beitrag zur Elektrophysiologie Pflanzlicher Drüsen; Cytologische Untersuchungen au Spontanen Diploiden Allium Carinatum; Über des Elektrische Potential und über den Errengungsvorgang bei Myxomyceten; Contribution à la Connaissance des Algues Marine des Sporades du Nord; Über die Elaeoplasten der Rotalge. Home: Palaion Patron Germanou 2. Office: Institute of Botany, University of Salonica, Thessalonica, Greece.

DIAZ, Francisco, Spanish surgeon; b. Spain, 1550; physician to Phillip II; tchr. philosophy Alcalá de Henares; went to P.I., 1632, later to China. Author: Tratado de todas las enfermidades de los riñones, vexigo y carnosidades de la verga y vexiga de la orina, 1588. Founder urology; publ. on diseases of kidney, bladder and urethra. Died 1646.

DIBDIN, William Joseph, Brit. chemist; b. London, 1850; s. Thomas Colman Dibdin; m. Marian Aglio, 1878; 3 sons, 5 daus. Chief, chem. and gas dept. Met. Bd. Works and London County Council, 1882-97; pvt. practice, from 1897. Fellow Inst. Chemistry, Chem. Soc., Royal San. Inst., Royal Micros. Soc.; pres. Inst. San. Engring., Assn. Mgrs. of Sewage Disposal Works; v.p. Soc. Pub. Analysts. Author: Practical Photometry; Purification of Sewage and Water; Lime, Mortar, and Cement; Public Lighting; Composition and Strength of Mortars; also papers. Editor, co-author: Churchill's Chemical Technology, Vol. IV. Research on use of micro-organisms for purification of sewage in contact and slate beds, 1897; inventor Bd. of Trade standard pentane argand, radial photometer, Dibdin's hand photometer. Died June 9, 1925.

DI BIRAGO, Carlo, Italian engr.; b. Cascina-d'-Olmo, 1792. Author (in German): Essay on a System of Bridge Construction, 1839. Constructed type of light-weight mil. bridges which bear his name; also built fortifications of Linz. Died 1845.

DI BRUNO, Faa, Italian mathematician; b. Alessandria, Italy, Mar. 7, 1825; student Paris, France, Turin, Italy. Capt. army, Sardinia; ordained to priesthood; tchr. math. Turin, till 1888. Wrote on theory of elliptic functions, binary forms. Died Turin, Mar. 26, 1888.

DICARLO, Frederick Joseph, Am. biochemist; b. N.Y.C., Nov. 24, 1918; s. Amilcare and Anna (Manieri) DiC.; B.S., Fordham U., 1939, M.S., 1941; Ph.D., N.Y. U., 1946; m. Nancy R. Cucco, Oct. 16, 1943; children—Nancyann, Frederick, Paul. Instr., N.Y.U., N.Y.C., 1943-45; scientist Squibb Inst. for Med. Research, New Brunswick, N.J., 1945-46; head biochemistry div. Fleischmann Labs., Stamford, Conn., 1946-60; sr. research asso. Warner-Lambert Research Inst., Morris Plains, N.J., 1960—. Mem. Reticuloendothelial Soc. (sec.-treas. 1965—), Am. Chem. Soc. (sec. biochem. discussion group 1966—), Am. Soc. Biol. Chemists, A.A.A.S., Soc. for Exptl. Bioltoy and Medicine. Research and numerous publs. on penicillin, enzymes and nucleic acids, metabolism of sulfa and other drugs, mechanism by which body protects itself from infection, cardiovascular drugs. Home: 341 Boulevard St., Mountain Lakes, N.J., 07946. Office: Warner-Lambert Research Inst., Morris Plains, N.J. 07950.*

DICE, Lee Raymond, Am. biologist; b. Savannah, Ga., July 15, 1887; s. Theodore Franklin and Eleanora A. (Spangenberg) D.; student Wash. State Coll., 1908-09, U. Chgo., 1909-10; A.B., Stanford, 1911; postgrad. U. Mont., 1913; M.S., U. Cal. at Berkeley, 1914; Ph.D. in Paleontology and Zoology, 1915; m. Dora Sibyl Lemon, June 24, 1918; children—Elizabeth Jane, John Raymond, Dorothy Anne (Mrs. John Allen Foster). With Alaska Fisheries Service, 1911-13; instr. Kan. State Agr. Coll., 1916-17; asst. prof. U. Mont., 1917-18; field asst. U. S. Biol. Survey, 1918; curator mammals Mus. Zoology, U. Mich., Ann Arbor, 1919-38, faculty, 1919—, prof. biology, 1942-57, prof. emeritus, 1957—, dir. Lab. Vertebrate Biology, 1934-50, dir. Inst. Human Biology, 1950-56; research asso. Carnegie Instn. Washington, 1924-39. Mem. A.A.A.S., Am. Eugenics Soc., Am. Soc. Human Genetics, Brit. Ecol. Soc., Ecol. Soc. Am. (Eminent ecologist 1964), Am. Soc. Mammalogists, Animal Behavior Soc., Mich. Acad. Sci., Arts, and Letters, Nature Conservancy, Population Reference Bur., Soc. Study Evolution, Soc. Systematic Zoology, Soc. Vertebrate Paleontology. Author: Biotic Provinces of North American, 1943; Natural Communities, 1952; Man's Nature and Nature's Man: the Ecology of Human Communities, 1955. Research and numerous articles on classification ecologic communities, selective advantage protective coloration, analysis variation in behavior in natural populations animals, ecology human communities, differential fertility within human populations; organized heredity clinics. Home: 1120 Brooks St., Ann Arbor, Mich. 48103.*

DICK, George Frederick, Am. physician, bacteriologist; b. Ft. Wayne, Ind., July 21, 1881; M.D., Rush Med. Coll., 1905; D.Sc. (hon.), U. Cin., 1925, Northwestern U., 1927; m. Instr. pathology U. Chgo., 1910-11; staff McCormick Inst. Infectious Diseases, Chgo., 1911-14; pathologist St. Luke's Hosp., Chgo., 1916-18; prof. clin. medicine Rush Med. Coll., Chgo., 1918-45, chmn. dept. medicine, 1933-45, emeritus, from 1945. Recipient Mickel prize, Cameron prize, 1933. Mem. Assn. Physicians, Assn. Pathologists and Bacteriologists, A.M.A., Soc. Exptl. Biology, others. In work (with wife Gladys Dick) on scarlet fever, devised skin test for determining susceptibility; isolated streptococci, developed effective antitoxin, 1924. Home: 1665 Middlefield Rd., Palo Alto, Cal.

DICK, Robert, Brit. geologist, botanist; b. Tullibody, Clackmannanshire, Eng., Jan. 1811; apprenticed to baker; served as journeyman in Leith, Glasgow, Greenock (all Scotland); went to Thurso in Caithness in 1830; asst. to Sir. Roderick Murchison and Hugh Miller in research. Made collection Brit. flora; also collected fossils; re-discovered northern holy-grass. Died Dec. 24, 1866.

DICK-READ, Grantly, physician; b. Beccles, Suffolk, Eng., Jan. 26, 1890; s. Robert John and Fanny Maria (Sayer) D.; student Bishop's Stortford Coll., Hertfordshire; B.A. with honors, St. John's Coll., Cambridge, 1911, M.B., B.Ch., 1916, M.A., 1916, M.D., 1920; m. Dorothea Cannon, Apr. 7, 1920; m. 2d, Jessica Bennett, Apr. 10, 1952. House physician London Hosp., 1911-14, resident accoucheur, 1918-22; demonstrator pathology Cambridge U., 1918-20; pvt. practice of medicine, Eastbourne, Eng., 1922-23, Woking and Harley St., 1926-48, Johannesburg, S. Africa, 1949-53; ret. from active practice, 1953; lectr., writer, 1953—; lectr. Maternity Center Assn. N.Y., French Acad. Medicine. Paris, 1947, Ireland, Scotland, Wales, Am., English univs., South Africa, 1948, German Swiss, French Italian univs., 1957. Mem. Royal Coll. Surgeons, Royal Coll. Physicians. Author: Natural Childbirth, 1933; Childbirth Without Fear, 1944; Birth of a Child, 1947; Introduction to Motherhood, 1951; No Time for Fear, 1955; Antenatal Illustrated, 1955, Contbr. sci. articles to profl. publs. Died June, 1959.

DICKE, Robert (Henry), Am. physicist; b. St. Louis, May 6, 1916; s. Oscar H. and Flora (Peterson) D.; A.B., Princeton, 1939; Ph.D., U. Rochester, 1941; m. Annie Currie, June 6, 1942; children—Nancy Jean, John Robert, James Howard. Micro wave radar development Radiation Lab. Mass. Inst. Tech., 1941-46; physics faculty Princeton, 1946—, Cyrus Fogg Brackett professor physics, 1957—, chairman physics department, 1967—; industrial consultant. Recipient Rumford medal American Academy of Arts and Sciences, 1967. Member National Academy of Sciences, American Geophysical Union, American Physical Society, also member Am. Astron. Soc., Am. Academy Arts and Scis., Phi Beta Kappa, Sigma Xi. Author: (with Montgomery and Purcell) Principles of Micro-wave Circuits, 1948; (with J. P. Wittke) An Introduction to Quantum Mechanics, 1960; The Theoretical Significance of Experimental Relativity, 1964. Study of coherent radiation processes; microwave radiometer; microwave atomic spectroscopy; relativity; gravitation; astrophysics. Home: 321 Prospect Av., Princeton, N.J.*

DICKEN, Samuel Newton, Am. geographer; b. nr. Colfax, Ky., Jan. 26, 1901; s. William Francis and Sarah (Harry) D.; A.B., Marietta Coll., 1924, Sc.D., 1964; Ph.D., U. Cal. at Berkeley, 1931; m. Emily Fry Puehler, Jan. 23, 1929; 1 son, Charles Francis. Faculty, U. Minn., 1929-47; prof. geography U. Ore., Eugene, 1947—, head dept. geography and geology, 1947-57, head, dept. geography, 1957-63. Cons., U. S. Air Force, 1941-43, OSS, 1943; mem. commn. geography NRC, 1957-60. Fellow Am. Geog. Soc.; mem. Am., Pacific Coast (pres. 1951-52) assns. geographers, Am. Assn. U. Profs., Phi Beta Kappa, Sigma Xi. Author: Regional Economic Geography, 1949; The Pacific Northwest, 1958; Oregon Geography, 1965; (with F. R. Pitts): Introduction to Human Geography, 1963. Applied Walter Penck's principles of slope formation and demonstrated unique factors in soil erosion in Ky. Karst; developed spl. course in geography for U. S. Air Force; studied effect of land expropriation on Mexican hacienda for 30 year period; measured effects of man's activities on phys. changes on Ore. coast. Home: 2385 Madrona Dr., Eugene, Ore. 97403.*

DICKENS, Frank, English biochemist; b. Northampton, Eng., Dec. 15, 1899; s. John and Elizabeth Ann (Pebody) D.; student (scholar) Magdalene Coll., Cambridge, Eng., 1919-21; M.A., Cambridge U., 1924; Ph.D., U. London (Eng.), 1923, D.Sc., 1936; m. Molly Jelleyman, Feb. 12, 1925; children—Pamela Jane, Diana Elizabeth. Research staff Imperial Coll. Sci., 1921-23; lectr. biochemistry Middlesex Hosp. Med. Sch., 1923-29; sci. staff Med. Research Council, 1931; dir. North of Eng. Council Brit. Empire Cancer Campaign, 1933-46; Philip Hill prof. exptl. biochemistry Middlesex Hosp. Med. Sch., London, 1946-67; dir. Tobacco Research Council Labs., Harrogate, Eng., 1967—. Chmn., Brit. Nat. Com. for Biochemistry, 1962-67. Fellow Royal Soc., 1946. Author: (with E. C. Dodds) Chemical and Physiological Properties of the Internal Secretion, 1925. Editor Biochem. Jour., 1937-47. Research and numerous articles on aspects of metabolism of tissues both normal and cancerous, nature of cancer induction by chem. agts. Home: 48 Rutland close. Office: Tobacco Research Council Labs., Otley Rd., Harlow Hill, Harrogate, Yorkshire, Eng.*

DICKERSON, Roy Ernest, Am. geologist; b. Parkersburg, W.Va., 1878; student Denison U.; B.S., U. N.M., 1901; Ph.D., 1903; D.Sc., Columbia U., 1929; hon. degrees from univs. of Grenoble, Nancy, Montpellier (all France), Denison U.; Tchr., geologist U. S. Geol. Survey in S.W., 1897-1905; named instr. Mass. Inst. Tech., 1903, asst. prof., 1905-07; chief fgn. geologist Atlantic Refining Co.; tech. sect. chief petroleum div. Fgn. Econs. Adminstrn.; asst. prof. physiography Harvard, 1906-12; became asso. prof. physiography Columbia U., 1912, named prof., 1919, named exec. officer geol. dept., 1937. Mem. Geol. Soc. Am. Author: Shore Processes and Shoreline Development, 1919; The New England-Acadian Shoreline, 1925; Paysages Americaines et problémes geographique, 1927; Stream Culture on the Atlantic Slope, 1931; The Origins of Submarine Canyons, 1939. Mem. com. of geologists who disproved theory that Atlantic seaboard was sinking, 1926-29; discovered and traced fault in the ocean floor from Bay of Fundy to Isle of Shoals, 1922. Died 1944.

DICKEY, Donald Ryder, Am. zoologist; b. Dubuque, Ia., Mar. 31, 1887; s. Ernest M. and Anna (Ryder) D.; B.A., Yale, 1910; hon. M.A. Occidental Coll., Los Angeles, 1925; m. Florence Van Vechten Murphy, June 15, 1921; 1 son, Donald R. Writer, lectr., field naturalist; specialized in mammals and birds of N. and C.Am.; research asso. in vertebrate zoology, Cal. Inst. Tech. 1926-32; collected over 50,000 specimens of birds and animals, regarded as largest such pvt. collection in U. S. Trustee S.W. Mus., 1920-28. Died Apr. 15, 1932.

DICKEY, Herbert Spencer, Am. physician, explorer; b. Highland Falls, N.Y., Feb. 24, 1876; s. Charles Henry and Marie (Brosseau) D.; prep. edn., Phillips Exeter Acad., Exeter, N.H.; student New York U. Med. Sch., 1895-98; M.D., Boston, 1899; m. Elizabeth Staley, Oct. 6, 1925 (div. 1933). Served as surgeon the Tolima Mining Company, Colombia, S.A., 1900-06; resident physician Peruvian Amazon Co., 1907-08, 11-12, Antunes Rubber Estates, Remate de Males, Brazil, 1908-10, La Romana Sugar Estates, Dominican Republic, 1914-16; chief surgeon Guayaquil & Quito R.R., 1923-25. Associate in S. Am. research Southwest Mus., Los Angeles. Mem. Am. Ethnol. Soc. Author: The Misadventures of a Tropical Medico (with Daniel Hawthorne), 1929; My Jungle Book, 1932. Contributor to the New York Times. Believed to be 1st white man to descend Caqueta River from Colombia to its mouth; explorations include 5 trips on foot over Ecuadorian Andes, exploring affluents of Amazon; explored River Tomo, affluent to Orinoco; located source of Orinoco, July 14, 1931; organized and led 1st "Dude" expdn. over Andes and down Amazon, 1932; accompanied Sir Roger Casement on his trip to Amazon, 1911; discovered and removed from Ecuador nearly 500 archaeol. specimens for Southwest Mus., 1936; originated process for extracting quinine from lowgrade cinchona bark in Ecuador, 1941. Died Oct. 28, 1948.

DICKINSON, Edmund (Dickenson), English alchemist, physician; b. Appleton Berkshire, Eng., Sept. 26, 1624; s. William and Mary (Colepepper) D.; B.A., Merton Coll., Oxford (Eng.) U., 1647, M.A., 1649, M.D., 1656. Became acquainted with Theodore Mandanus, French alchemist, 1656, who prompted him to devote attention to alchemy; practiced as physician, Oxford, from 1656; superior reader Linacre's lectures U. Oxford; move to London to practice, 1684; apptd. physician in ordinary to Charles II (who built him lab. and showed interest in his chem. expts.), also physician to household; physician to James II, 1685-88; ret. to study and write, 1688. Fellow Coll. Physicians, 1677, St. John's Coll., Oxford U. Author: Diatriba de Moae in Italiam Adventu, 1655; Delphi Phoenicizantes, 1665; Epistola ad T. Mandanum de Quintessentia Philosophorum, 1686; Physica vetus et vera, 4 vols., 1702. Believed in alchemy; maintained he had witnessed transmutation and knew secret of philosopher's stone; tried to show that all sci. is to be found in writings of Moses, also taught that Moses has disolved the golden calf by chem. means; denied that metals grow from seeds, also denied spontaneous generation; defended atomic theory; extracted metal regulses from human feces, maintaining that metal was taken in in food we eat. Died Apr. 3, 1707.

DICKINSON, Hobart Cutler, Am. physicist; b. Bangor, Me., Oct. 11, 1875; s. George Lyman and Emma T. (Cutler) D.; A.B., Williams Coll., 1900, A.M., 1902; studied Clark U., 1902-03, Ph.D., 1910; m. Elizabeth Wells, 1903 (dec. 1921); children—David, Bradley Wells (adopted); m. 2d, Mabel V. Kitson, 1923; 1 dau., Anne Katherine. Became connected with Bur. of Standards, Washington, 1903, asst. physicist, 1906-10, asso., 1910-16; physicist, 1916-21; research mgr. Soc. Automotive Engrs., 1921-23; chief Div. of Heat and Power, Bur. of Standards, 1923-45. Fellow A.A.A.S., Am. Phys. Soc.; mem. Am. Soc. Testing Materials, Am. Soc. Refrigerating Engrs., Washington Acad. Scis., Washington Philos. Soc. Died Nov. 27, 1949.

DICKINSON, Joshua Clifton, Jr., Am. ornithologist; b. Tampa, Fla., Apr. 28, 1916; s. Joshua C. and Mary (Martin) D.; student U. Fla., 1934-36, U. Va., 1936-39, Cornell U., summer 1938; B.S., U. Fla., 1940, M.S., 1946, Ph.D., 1950; postgrad. U. Va., summer 1940; m. Lucy Freeman Jackson, Apr. 13, 1936; children—Joshua Clifton III, Martin Freeman, Susan Ellissa. Faculty, U. Fla., Gainesville, 1939—, asso. prof. zoology, 1956—, curator ornithology, chmn. dept. natural scis. Fla. State Mus., 1951-61, dir., 1961—; research fellow Harvard, 1951, Woods Hole Oceanographic Instn., 1951. Chmn., Fla. Antiquities Commn, 1965—. Mem. Assn. Southeastern Biologists (past sec.), Fla. Acad. Scis. (past mem. council editor Quar. Jour. 1952-62), Am. Assn. Museums (mem. council 1964—), Am. Ornithologists' Union, Am. Soc. Naturalists, Wilson Ornithol. Soc., Am. Soc. Zoologists, Sigma Xi, Phi Sigma. Research and publs. on systematics and faunistics of birds. Home: 1804 S.W. 35th Pl., Gainesville, Fla. 32601.*

DICKINSON, Robert Eric, English geographer; b. Manchester, Eng., Sept. 2, 1905; s. William and Mary (Jones) D.; B.A., M.A., U. Leeds (Eng.), 1926; Ph.D., U. London (Eng.), 1932; m. Mary Winwood, Nov. 6, 1941. Asst. lectr. U. Coll., Exeter, Eng., 1926-28; faculty U. Coll., London, 1928-47, reader, 1941-47; prof. Syracuse (N.Y.) U., 1947-58; prof. U. Leeds (Eng.), 1958-65, research prof., 1965—, vis. prof. U. Ariz., 1967—. Mem. Royal Geog. Soc., Am. Geog. Soc., Inst. Brit. Geographers, Assn. Am. Geographers. Author: (with O. J. R. Howarth) Making of Geography, 1932; German Lebensraum, 1943; Regions of Germany, 1944; City, Region and Regionalism 1947; West European City, 1951; Germany, 1953; Population Problem of Southern Italy, 1955; City and Region, 1964; Makers of Geography, 1967; City Region in Western Europe, 1967. Home: 4 Dunstarn Gardens, Leeds 2, Eng.

DICKINSON, Robert Latou, Am. gynecologist; b. Jersey City, Feb. 21, 1861; s. Horace and Jeannette (Latou) D.; ed. Poly. Inst. Bklyn., and in Switzerland and Germany; M.D., L.I. Coll. Hosp., 1882; m. Sarah Truslow, May 7, 1890 (dec.); children—Margaret (dec.), Dorothy (Mrs. George B. Barbour), Jean (Mrs. Truman Squire Potter. Gynecologist and obstetrician Bklyn. Hosp.; prof., L.I. Coll. Hosp.; asst. chief, med. sec. Nat. Council Defense, Washington, 1917; med. adviser and mem. Gen. Staff, Washington, 1918-19; on mission to China for USPHS, 1919, Near East, 1926. Fellow of A.C.S. (dir.); Mem. Am. Assn. Marriage Counsellors, Am. Assn. for Study Sterility, Am. Gynecol. Soc. (pres.), N.Y. Acad. Medicine, Brit. Gynecol. Soc., Am. Gynecol. Club (pres.) Am. Geog. Soc.; sec. Nat. Com. on Maternal Health, 1923-37 (hon. chmn. from 1937); sr. v.p. Planned Parenthood Fedn., from 1939; pres. Euthanasia Soc. from 1946. Recipient A. and M. Lasker award for research, 1946. Author: (booklet) Palisades Guide, 1921; New York Walk Book (with others), 1923, 39, 50; A Thousand Marriages, 1931; Control of Conception, 1931-38; Atlas of Human Sex Anatomy, 1933; The Single Woman, Her Sex Education, 1933; co-author Sex Variants, 1941; Birth Atlas, 1941; Techniques of Conception Control, 1941; Human Sterilization, 1950; also 200 researches and reports on obstetrics, diseases of women, hosp. orgn. and sex problems. Coeditor of American Text Book of Obstetrics, 1895. Illustrator of own writings. Sculpture (with A. Belskie) of Birth Series for N.Y. World's Fair, 1939; pelvic teaching models, 1941-47. Died Nov. 30, 1950.

DICKINSON, William Howship, physician; b. Brighton, 1832; ed. St. George's Hosp.; M.D., Cambridge U.; m. Laura Wilson; 1 son, 4 daus. Cons. physician St. George's Hosp., Hosp. for Sick Children; demonstrator in anatomy St. George's Hosp.; med. registrar, curator mus., asst. physician, physician, cons. physician, lectr. materia medica, pathology, medicine, examiner in medicine Coll. Physicians, Coll. Surgeons, univs. Cambridge, London, Durham. Fellow Royal Coll. Physicians (Croonian, Lumleian, Harveian lectr., censor); mem. Royal Med. and Chirurg. Soc. (past pres.), Path. Soc. (past pres.). Contbr. articles to med. publs. Proved that when nerve is cut that proximal end eventually atrophies, 1869; gave early account of paroxysmal hemoglobinuria, 1865. Died Jan. 9, 1913.

DICKSON, Henry Newton, meteorologist; b. Edinburgh, Scotland, June 24, 1866; student Edinburgh U.; M.A., D.Sc., Oxford U.; m. Margaret Stephenson, 1891; 1 son, 1 dau. Prof. geography Univ. Coll., Reading, 1906-20; head. geog. sect. Intelligence Dept., Admiralty, 1915-19. Fellow Royal Soc. Edinburgh; pres. Royal Meteorol. Soc., 1911-12, sect. E, Brit. Assn., 1913. Author: Elementary Meteorology, 1893; Climate and Weather, 1912; Maps and Map Reading, 1912; also papers. Asst. editor Ency. Brit. Died Apr. 2, 1922.

DICKSON, James Douglas Hamilton, Brit. mathematician; b. Glasgow, Scotland, May 1, 1849; s. J. R. Dickson; ed. univs. Glasgow, Cambridge, Edinburg; M.A.; m. Isobel Catharine Banks; 1 son, 1 dau. Asst. to Sir William Thomson (Lord Kelvin), 1867-69; asst. for laying of 1869 Atlantic cable, Societe du Cable Transatlantique Francais, electrician-incharge, Brest, until 1870; became scholar, tutor Peterhouse, Cambridge (Eng.) U., fellow, 1874, sr. fellow, from 1907; Eglinton fellow Glasgow U. Fellow Royal Soc. Edinburgh, Cambridge Philos. Soc.; mem. Royal Inst., London Math. Soc. Coadjutors of Lady Dewar in publ. of collected papers of Sir James Dewar. Contbr. papers on thermodynamics and thermoelectricity to jours.; also contbr. to Dictionary Nat. Biography. Constructed (with W. F. King, C. Cuttriss) apparatus and made preliminary expts. for Lord Kelvin's determination of v (ration of electromagentic to electrostatic unit of electricity); perfected (with W. F. King, Theophilus Varley) Lord Kelvin's siphon recorder, 1868-69. Died Feb. 6, 1931.

DICKSON, Leonard Eugene, Am. mathematician; b. Independence, Ia., Jan. 22, 1874; s. Campbell and Lucy (Tracy) D.; B.S., U. Tex., 1893, M.A., 1894; Ph.D., U. Chgo., 1896; postgrad. U. Leipzig (Germany), 1896, U. Paris (France), 1897; hon. D.Sc., Harvard, 1936, Princeton, 1941; m. Susan Davis, Dec. 30, 1902; children—Campbell, Eleanor (Mrs. Harlow Higinbotham). Instr., U. Cal., 1897; prof. math. U. Chgo., editor, Am. Math. Monthly, 1902-08. Trans. Am. Math. Soc., 1910-16. Corr. de l'Académie de l'Institut de France. Mem. Nat. Acad. of Science, Phi Beta Kappa, Sigma Chi. Author works including: Linear Groups; Exposition of the Galois Field Theory, 1901; Linear Algebras, 1914; History of the Theory of Numbers, 3 vols., 1919-23; Studies in the Theory of Numbers, 1930. A leading specialist in groups; worked with groups of linear homogeneous substitutions on "n" variables; constructed division algebras in "n" fundamental units with coefficients in any field F, 1914; began the congruencial theory of forms, 1907; studied relationship between theory of invariants and theory of numbers; investigated the Galois group of the equation of the 27 lines of cubic surface. Died Jan. 17, 1954.

DICKSON, William Elliot Carnegie, pathologist; b. Edinburgh, 1878; s. George Dickson; B.Sc. with spl. distinction in anatomy and anthropology, Edinburgh U., 1898, M.B., Ch.B. with 1st class honours, 1901, M.D. with 1st class honours, 1905; m. Frances Edith Greenfield; 2 sons, 2 daus. Baxter scholar in natural scis., U. Edinburgh, U. London, 1898; Stark scholar in clin. medicine, 1901; Chrichton research scholar in pathology, 1902-05; ret. cons. pathologist and bacteriologist; cons. dir. pathology dept. West End Hosp. for Nervous Diseases and Pathology, Grosvenor Hosp. for Women, London; lectr. bacteriology U. Edinburgh, 1907-14; asst. pathologist Edinburgh Royal Infirmary; pathologist Royal Hosp. for Sick Children, Edinburgh, 1906-14; dir. pathology dept. Royal Chest Hosp., London. Fellow Royal Coll. Physicians Edinburgh, Royal Med. Soc. Edinburgh (past pres.); mem. Path. Soc. Gt. Britain and Ireland, Assn. Clin. Pathology, Assn. Brit. Neurologists, Royal Soc. Medicine, Brit. Med. Assn., Royal Coll. Physicians. Author: Bacteriology, The People's Book Series, No. 6, 2d edit., 1919; The Cytology of the Bone Marrow in Health and Disease, 1908. Editor, co-author: Textbook of Pathology (Beattie and Dickson), 5th edit., 1948, also numerous articles. Died Nov. 25, 1954.

DICQUEMARE, Jacques-François, French biologist, physicist; b. Havre, France, Mar. 7, 1733; studied sci. under Abbé Nollet, took religious vows; prof. exptl. physics at Havre. Mem. Acad. of Rouen, French Acad. Scis., 1786, Maritime Acad. Author: Idée générale de l'astronomie, 1769; Description du cosmoplane, 1769; Index géographique, 1769; An Essay toward elucidating the History of Sea Anemonies, 1774; (with Maneville) Neptune oriental. Invented cosmoplane; research on nautical astronomy; prepared naval maps; research on jellyfish, molluscs especially oysters, octopus; worked to improve Havre port, 1780-84. Died Havre, Mar. 29, 1789.

DICUIL, Irish geographer, astronomer; b. Ireland, circa 757; flourished 814-25; studied under Suibenus; mem. Irish monastery. Author of treastise on astronomy and computus (without title) in 4 books, c. 814-16; Epistola de Quaestionibus decem artis grammaticae (lost); Liber de mensura orbis terrae (earliest account of Iceland, refers to ancient canal connecting Nile and Red Sea); also treatise on astronomy, 825.

DIDEROT, Denis, French natural philosopher; b. Langres, France, Oct. 5, 1713; ed. Harcourt Coll., Paris, France; m. Antoinette Champion, 1743; 2 daus., including Angélique (Mme. de Vandeal). Tchr. math.; writer, man of letters; imprisoned for seditious article, 1749; gained patronage of Catherine the Great, 1765, in appreciation for her purchase of his library, visited Russia, 1773-74. Author: Mémoires sur différents sujets de mathématiques, 1748; Pensées philosophiques (defended deism and supported philos. rationalism), 1746; Les bijoux indiscrets, 1748; Lettre sur les aveugles, 1749; Lettre sur les sourds et muets (1st truly sci. study of deaf and dumb), 1751; Pensées sur l'interpretation de la nature (foresaw era of great discovery in biol. scis., advanced theories of transformism and natural selection), 1753; Discours sur la poésie dramatique, 1758; La neveu de Rameau, 1762; La religieuse (presents characters and situation with penetrating psychol. realism), 1796; Éléments de physiologie, 1774-78; Le reve de d'Alembert (written 1769), 1830; Correspondance de Grimm et Diderot, 1829; began transl. (with d'Alembert, also Voltaire, Rousseau, Buffon, Montesquieu, others, who later withdrew because of ofcl. opposition) Ephraim Chambers' Cyclo., expanded it to include summary of sci. knowledge of the times, pub. as Encyclopédie, 34 vols., 1751-77. First to build philos. system based on sci. facts and discoveries in physics, anatomy, physiology, rather than on logic and reason; developed philosophy of sci. materialism; wrote on doctrines of evolution and cellular composition of living organisms; developed methods later followed by Lamark, Claude Bernard, Taine. Died Paris, July 31, 1784.

DIDIO, Liberato Joao Affonso, anatomist; b. Sao Paulo, Brazil, May 7, 1920; s. Pascoal and Lydia (Cacace) DiD.; B.S., U. Sao Paulo, 1939, M.D. summa cum laude, 1945, M.S. summa cum laude, 1949, Ph.D. summa cum laude, 1951; m. Lydia Leite Silva, Mar. 12, 1960; children—Lydia Nivalda, Arthur. Faculty, U. Sao Paulo, 1942-53, Ciencias Medicas, Belo Horizonte, Minas Gerais, Brazil, 1953-54, U. Minas Gerais, 1954-63, Northwestern U., Chgo., 1963-67; prof., chmn. dept. anatomy Toledo State Coll. Medicine, 1967—; vis. prof., cons. numerous instns. Rockefeller fellow, 1960-61. Recipient Rockefeller Found. award, 1945, Alvarenga prize Brazilian Acad. Medicine, 1956; Order of Merit, Brazil, 1957; Medal Sci. Merit, State of Gerais, 1957; Great Medal of Inconfidencia, 1966; Medal for Cultural Merit, Republic of Italy, 1963. Fellow A.A.A.S.; mem. Am. Assn. Cell Biology, Am. Assn. Anatomists (ofcl. del. to numerous internat. congresses), Anatom. Soc. Gt. Britain, and Ireland, Assn. Am. Med. Colls., Brazilian Anat. Coll., Brazilian Assn. Med. Schs., Brazilian Soc. Anatomy, Electron Microscopy Soc., Am., French Soc. Anatomists (life), German, Brazilian, Italian, Japanese socs. anatomy, Internat. Coll. Surgeons, German Soc. Anthropology, Nat. Council Research Brazil. N.Y. Acad. Scis., PanAm. Assn. Anatomy (founder, U.S.A. del.), Societe d'Anthropologie de Paris, Sigma Xi, others. Research and publs. on anat. background for varicose veins; anatomy, physiology and radiology of terminal ileum; variations of promontory; congenital and acquired structures in left renal vein; lymphatic system in animals; ultrastructure of prostate, heart and pineal body. Home: 2805 Oatis Av., Toledo 43606.*

DIDION, Isidore, French mil. scientist; b. Thionville, France, Mar. 22, 1798; s. Jean-Baptiste and Catherine (Simonet) D.; ed. École polytechnique, 1817, Metz Mil. Sch., 1825. Named prof. arty. Metz Sch., 1837; apptd. dir. armaments manufacture, Paris, France, 1848; became brig. gen. French Army, 1858. Author: Traité de balistique, 1848; Lois de la résistance de l'air sur les projectiles, 1857; Calcul des probabilités appliqué au tir des projectiles, 1858. Invented instrument for testing centering of cannons (Didion scale), 1826; (with Piobert and Morin) research on air resistance of projectiles, 1834-39; introduced corrections for projectile motion in a vacuum; developed idea of probability of error. Died Nancy, France, July 4, 1878.

DIDOMENICO, Mauro, Jr., Am. physicist; b. N.Y.C., Jan. 12, 1937; s. Mauro and Elizabeth (Zittola) DiD.; B.S., Stanford, 1958; M.S., 1959, Ph.D., 1963; m. Angela M. Carracino, Aug. 29, 1964. Mem. tech. staff Bell Telephone Labs., Murray Hill, N.J., 1962—. Mem. Am. Phys. Soc., I.E.E.E., Sigma Xi, Tau Beta Pi. Research and publs. on quantum electronics especially optical communications, high-speed electro-optic light modulators, high-speed solid-state photodetectors, single-frequency laser configurations and laser modulation techniques for generating repetitive optical pulse trains, ferroelectricity in areas of transport and optical properties. Home: 59 Addison Dr., Basking Ridge, N.J. 07920. Office: Bell Telephone Labs., Murray Hill, N.J. 07974.*

DIDOT, François-Ambroise, French inventor; b. Paris, France, 1730; book dealer; inventor velin paper, type of printing press, other printing machines, stereo type printing; builder fine paper mills; contbr. improvements in type founding; 1st to attempt vellum printing, 1780. Died 1804.

DIECKERHOFF, (Friedrich Julius Heinrich) Wilhelm, German veterinarian; b. Lichtendorf, Germany, Oct. 18, 1835; s. Johann Friedrich Wilhelm and Henriette (Vogt) D.; student vet. medicine Tierarzneischule, Berlin, Germany, 1853-57; hon. M.D., U. Greifswald, 1888. Practice vet. medicine, Bochum, for 12 years; came to Tierärztliche Hochschule (vet. sch.), Berlin, 1870; became prof., dir. largest horse clinic in Germany, 1877; apptd. dir. internal dept., 1884. Author: Die Geschichte der Rinderpest und ihrer Literature, 1890; Gerichliche Tierarzneikunde, 1902;

Lehrbuch der speziellen Pathologie und Therapie, 1903; Die Pferdestaupe, 1882. Developed subcutaneous and intravenous methods in vet. medicine. Died Berlin, Dec. 14, 1903.

DIECKHOFF, Josef Hubert, German pediatrician; b. Cologne, Germany, Mar. 21, 1907; s. H. W. and Sybille (Fischer) D.; Med. Staatsexamen, U. Cologne, 1933, Dr.med.habil, 1939, also dozent, 1946; m. Margarete Hilden, Apr. 15, 1950. Mem. staff U. Cologne, 1933-46, dozent, 1946-49; dozent, prof. Acad. Dusseldorf (Germany), 1949-50; head, dept. pediatrics, dir. children's clinic U. Halle, 1950-58, U. Leipzig, 1958-60, Berlin-Humboldt, Univs. Charite, 1960—. Recipient Nat. prize Verdienter Arzt des Volkes. Mem. Polish, Bulgarian socs. pediatrics, Purkinje-Soc.-Prag, Deutsche Ges.F.Kinderheilkunde. Author: Kurzgefasstes Lenrbuch f. Kinderneilkunde, 1963, 2d edit., 1966; Padiatrie und Grenzgebite, 2 vols. 1965. Research and publs. on ekto and endotoxin and ektotoxin shock; enzymes in fetal tissues of men and animals; cortisone and biogenesis of histamine in allergy; histamine metabolism in healthy and allergic children. Home: 4 Offland-Sheet, Berlin. Office: Kinderklnik Humboldt U. Charite, 22 Schumann, Berlin GDR, Germany.*

DIEFFENBACH, Ernest, German naturalist; b. Giessen, Germany, 1811; prof. geography Giessen. Author: Voyages in New Zealand. Explorer (with Brit.) New Zealand. Died Giessen, 1855.

DIEFFENBACH, Johann Friedrich, German surgeon; b. Königsberg, Germany, Feb. 1, 1792; s. Conrad and Sophie (Buddik) D.; student theology in Rostock, and Greifswald; student medicine Königsberg, 1916-20; M.D., Würzburg (Germany), 1822; state exam. in Berlin, 1824; m. Johanna Tillheim, 1824 (div. 1833); 1 dau.; m. 2d, Emilie Friederike Wilhelmine Heydecker, 1833; 1 son, 1 dau. Attending physician Prince Protashow in Paris and Montpellier, France; practice medicine and surgery, Berlin; asst. to J. N. Rust at surg. clinic Charité Hosp., Berlin, named dir. surgery, 1829; apptd. prof., 1832; joined Berlin Med. Faculty, 1840. Recipient Monthyon prize (twice) Institut de France. Author: Die operative Chirurgie, 2 vols., 1844, 49; Die Durchschneidung der Sehnen und Muskeln, 1841; Der Aether gegen den Schmerz, 1847; Chirurgische Erfahrungen besonders über die Wiederherstellung, 3 vols, 1829-34. Editorial bd. Hamburger Zeitung für die gesamte Medezin. Research on plastic surgery and transplants of facial features, urinary tract, perineum; amputation at thigh; ovariotomies; treated cholera by infusions and transfusions; used tenotomies for correcting club foot; first successful performance of strabotomy. Died Berlin, Nov. 11, 1847.

DIEHL, Harold Sheely, Am. physician; b. Nittany, Pa., Aug. 4, 1891; s. William Kleinfelter and Annie Belle (Sheely) D.; B.A., Gettysburg (Pa.) Coll.,1912, Sc.D., 1935; postgrad. Syracuse U.; M.D., U. Minn., 1918, M.A., 1921; m. Julia Louise Mills, Sept. 7, 1921; children—Annabelle Louise (Mrs. Robert P. Bush), Antoni Mills. Faculty U. Minn., 1920—, prof., 1929-58, dean med. scis., 1935-38, emeritus prof., dean, 1958—; sr. v.p. research, med. affairs, dep. exec. v.p. Am. Cancer Soc., Inc., N.Y.C., 1957—. Mem. adv. com. cancer control program USPHS, 1959-65; vice chmn. Nat. Interagency Council on Smoking and Health, 1964—; mem. Nat. Adv. Food and Drug Council, 1964—. Recipient 1st Edward Hitchcock award Am. Coll. Health Assn., 1961, 1st Distinguished Service certificate Gettysburg Coll. Alumni Assn., 1962. Diplomate Am. Bd. Preventive Medicine. Fellow A.M.A. (past council mem.), Am. Pub. Health Assn.; mem. Am. Coll. Chest Physicians, Am. Coll. Health Assn., A.A.A.S., Phi Beta Kappa, Sigma Xi, Alpha Omega Alpha. Author: Textbook of Healthful Living, 1935, 7th edit., 1964; (with Ruth E. Boynton) Healthful Living for Nurses, 1944; Personal Health and Community Hygiene, 1951; (with Anita Laton) Health and Safety For You, 1954, 2d edit., 1960. Research and numerous publs. on bacterial enzymes, blood pressure variability; treatment and prevention of colds, developing a codeine-papaveration combination, med. edn. Home: 11 Riverside Dr., N.Y.C. 10023. Office: Am. Cancer Soc., Inc., 219 E. 42d St., N.Y.C. 10017.*

DIEHL, Harvey, Am. chemist; b. Detroit, Nov. 2, 1910; s. Harvey C. and Ella F. (Straass) D.; B.S., U. Mich., 1932, Ph.D., 1936; m. Helen Louise Clark, June 21, 1936; children—Margaret Sue (Mrs. Gavin Scott), Byron Clark, Rosemary, Barbara Lyon, Harvey R., Elizabeth Ann. Instr., Cornell U., 1936-37, Purdue U., 1937-39; faculty Ia. State U., Ames, 1939—, prof. chemistry, 1945-65, distinguished prof., 1965—. Recipient Wilkinson award for meritorious teaching Iowa State University, 1965, Anachem award Association of Analytical Chemists, Detroit, 1966. Member of the Iowa Academy of Science, American Chem. Soc. (Fisher award 1956, Gold medal Ia. sect. 1961), Soc. for Analytical Chemistry, Sigma Xi, Phi Lambda Upsilon. Author: (with H. H. Willard) Advanced Quantitative Analysis, 1942; (with G. Frederick Smith) Analysis, 1952; Research and numerous articles on metal derivatives of organic compounds. Home: 3310 Oakland St., Ames, Ia. 50012.*

DIEHL, J(ohannes) Friedrich, German chemist; b. Ilbesheim, Germany, June 18, 1929; s. Heinrich and Gertrud (Linn) D.; M.Sc., U. Ky., 1953; diploma in chemistry, Heidelberg U., 1955, Dr.rer.nat., 1957; m. Eva M. Muller, Mar. 14, 1955; children—Margaret, Annette, Jennifer, Susan. Research fellow German Wool Research Inst., Aachen, 1957; faculty U. Ark. Med. Sch., 1957-65, asso. prof., 1963-65; dir., prof. Inst. Radiation Tech., Fed. Research Center for Food Preservation, Karlsruhe, Germany, 1965—. Guest scientist Max Planck Inst. for Med. Research, Heidelberg, 1963-64. Mem. Am. Inst. Nutrition, Soc. for Exptl. Biology and Medicine, Am., German chem. socs., Sigma Xi. Research, publs. on biol. function of vitamin E, protein metabolism in muscular dystrophy, amino acid and peptide transport through biol. membranes, chem. synthesis of peptides, radiation effects on vitamins. Home: Basler Torstrasse 71, Karlsruhe-Durlach. Office: Engesserstrasse 20, Karlsruhe, Germany.*

DIEL, (August Friedrich) Adrian, German physician, pomologist; b. Gladenbach, Germany, Feb. 4, 1756; s. Kaspar Ludwig and Marianne Christine (Zimmermann) D.; m. Dorothea Scriba, 1787; 7 children. Summer practice as balneologist in Bad Ems, winters as ofcl. physician in Diez/Lahn. Hon. mem. Assn. for Advancement Horticulture in Prussia. Author: System der in Deutschland vorkommenden Kernobstarten, 9 vols., 1799-1819; Über den Gebrauch der Thermalbäder in Ems, 1825; Über den innerlichen Gebrauch der Thermalbäder in Ems, 1832; Anzeigen und Gegenanzeigen über die Zulässigkeit des Gebrauches einer Heilquelle. First to analyze mineral waters with modern methods; cultivated Diels Butterbirne (butter pear). Died Bad Ems, Apr. 20, 1839.

DIELS, (Friedrich) Ludwig (Emil), German botanist; b. Hamburg, Germany, Sept. 24, 1874; s. Hermann Alexander and Bertha (Dübell) D.; student botany U. Berlin (Germany); doctorate 1896; hon. degrees include hon. dr. Cambridge (Eng.) U.; m. Gertrud Biesenthal, 1908; 1 son, 2 daus. Became lectr. U. Berlin, 1900; apptd. asso. prof. U. Marburg (Germany), 1906; became 2d dir. Bot. Garden and Mus., Berlin-Dahlem, Germany, 1914; apptd. 1st dir., prof. 1921; many expdns. including Australia, 1901-02. Mem. Prussian Acad.; hon. or corr. mem. numerous fgn. and German assns. Editor. Syllabus der Pflanzenfamilien; Die Botanische Jahrbücher für Systematik, Pflanzengeschichte und Pflanzengeographie; Das Pflanzenreich. Research and numerous publs. on plant taxonomy, morphology and geography. Died Berlin-Dahlem, Nov. 30, 1945.

DIELS, Otto Paul Hermann, German chemist; b. Hamburg, Germany, Jan. 23, 1876; s. Hermann Alex and Bertha (Dübell) D.; student of E. Fischer, U. Berlin (Germany); doctorate 1899; hon. M.D.; m. Paula Geyer, 1909; 3 sons, 2 daus. Apptd. prof. chemistry U. Berlin Chem. Inst., 1906; became asso. prof. U. Berlin, 1914; became prof. U. Kiel (Germany), 1916, also dir. Chem. Inst. of univ. Recipient (with Kurt Alder) Nobel prize in chemistry for diene synthesis, 1950. Mem. academies of Halle, Göttingen, Munich. Author: Einführung in die organische Chemie, 1907; Einführung in die anorganische Experimental chemie, 1924. Discovered carbon suboxide, method of dehydrating cyclical hydrocarbons using selenium, structure of steroids, (with Kurt Alder) diene synthesis (Diels-Alder reaction) 1927-28, which led to improved methods of analyzing organic compounds and synthesizing them. Died Kiel, Mar. 7, 1954.

DIEMER, Hugo, Am. industrial engr., management scientist; b. Cincinnati, O., Nov. 18, 1870; s. Theodore and Bertha L. (Huene) D.; M.E. in Electrical Engring., Ohio State U., 1896; student U. of Chicago, 1900; B.A. in history and polit. science, Pa. State Coll., 1913; m. Mabel N. Hudson, June 26, 1901; children—Theodore Hudson, Natalie Elizabeth, Dorothy Arnold, Mary Louise. With Addyston Pipe & Steel Co., Cincinnati, 1888-92; Bullock Electric Mfg. Co. and Westinghouse Electric & Mfg. Co., 1896-1900; asst. prof. mech. engring., Mich. Agrl. Coll., 1900-01; asso. prof. mech. engring., U. of Kan., 1901-04; cons. engr., Indianapolis and Chicago, 1904-07; prof. mech. engring., in charge of dept., 1907-09, prof. industrial engring., 1909-19, Pa. State Coll. Commd. maj. ordnance dept., 1917, then lieut. col.; in charge at U. S. Cartridge Co., Lowell, Mass., 1917, Bethlehem Steel Co., 1918; personnel supt. Winchester Repeating Arms Co., 1919-20; dir. management courses and personnel, La Salle Extension U., 1920—. Summer lecturer on organization and management, University of Chicago, 1915; lecturer on industrial orgn., dept. of univ. extension Mass. State Bd. of Edn., 1917. Fellow and dir. Inst. of Management (div. of Am. Management Assn.), 1927—. Author: Factory Organization and Administration, 1910, 5th edit., 1935; Modern Foremanship and Production Methods (with Meyer Bloomfield), 1921; Personnel Administration (with Daniel Bloomfield), 1921; Foremanship Training, 1927; Production Control, 1930. Wrote classic text on factory management; pioneer in development of management education in U. S.; early proponent of training for foremen. Died Chicago, Ill., Mar. 3, 1939.

DIEMERBROECK, Isbrand van, see van Diemerbroeck.

DIENA, Benito Baruch, bacteriologist; b. Rhodes, Greece, Apr. 17, 1926; s. Giorgio and Ida (Ventura) D.; D.V.M., U. Pazma, Italy, 1950; M.Sc., McGill U., 1954, Ph.D., 1956; m. Sara Costi, Feb. 19, 1950; children—Daniel, Emanuel, Tamar. Lectr., Inst. Path. Anatomy, Parma, Italy, 1950; with Lab. Hygiene, Ottawa, Ont., Can., 1956—, head research div. Biol. Control Labs., 1966—. Fellow Am. Acad. Microbiology; mem. Canadian Soc. Microbiology, Biol. and Biochem. Soc. Ottawa, Royal Soc. Medicine. Contbr. articles to sci. jours. Devel. exptl. bacterial vaccines; role of L forms in vaccination and disease; devel. new serol. methods for diagnosis of Tb. and other diseases; studies with mycobacteria: new media. Home: 1440 Lepage Av. Office: Tunney's Pasture, Ottawa, Ont., Can.*

DIENER, Carl, Austrian paleontologist; b. Vienna, Austria, Dec. 11, 1862; s. Karl and Marie (Wechtl) D.; student geography in Vienna; doctorate 1883; m. Marie Glanz, 1884; 2 sons. Mem. expdns. to Syria, Lebanon, 1885, Auvergne, Pyrenees, 1886, Himalayas, 1892, Spitzbergen, 1893, Urals, Caucasus, 1897, N.Am., 1901, Mexico, 1906, Japan, Hawaii, Can., 1913; changed field to geology in 1893; became asso. prof. geology, 1897 and asso. prof. paleontology, 1903; apptd. prof. U. Vienna, 1906. Mem. and hon. mem. sci. acads. and assns. in Austria, Germany, Eng., Russia, U.S.A. Author: Die marinen Reiche der Triasperiode; Grundzüge der Biostratigraphie. Co-author: Fossilium Catalogus. Research on geography, geology, stratigraphy and biology especially in Asiatic areas and eastern Alps, also on ammonites. Died Vienna, Jan. 6, 1928.

DIENER, Urban Lowell, Am. phytopathologist; b. Lima, Ohio, May 26, 1921; s. Urban Edward and Ethel (Hoverman) D.; A.B., Miami U., 1943; M.A., Harvard, 1945, student Columbia, 1945-46; Ph.D., N.C. State U., 1953; m. Mary Jacqulyn Maund, Aug. 11, 1956. Mycologist, Sindar Corp., N.Y.C., 1945-47; asst. plant pathologist Clemson U.,1947-48; grad. research asst. N.C. State U., 1949-51; with Auburn U., 1952—, prof. plant pathology, 1963—. Mem. Am. Phytopathol. Soc., A.A.A.S., Am. Inst. Biol. Sci., Sgma Xi. Research related to knowledge of fungus toxins in feeds, seeds and foodcrops, especially peanuts; fungicidal control of diseases of fruits, vegetables, pecans; knowledge of deterioration of stored peanuts by fungi. Home: 750 Sherwood Dr., Auburn, Ala. 36830.*

DIENES, G(eorge) J(ulian), physicist; b. Budapest, Hungary, Apr. 28, 1918; s. Kalman and Aranka Fekete (de Galantha) D.; B.S., Carnegie Inst. Tech., 1940, M.S., 1942, D.Sc. in Phys. Chemistry and Physics, 1947; M.S. In Math., Columbia, 1946; m. Margaret T. Dienes, Sept. 23, 1940; 1 dau., Claire (Mrs. L. Hill). Came to U. S., 1936, naturalized, 1945. Instr. chemistry Washington and Jefferson Coll., 1940-41; teaching asst. Carnegie Inst. Tech., 1941-43; research chemist Ridbo Labs., 1943-44; group leader physics div. Bakelite Corp., 1944-49; research specialist atomic energy research dept. N.Am. Aviation, 1949-51; sr. physicist Brookhaven Nat. Lab., Upton, L.I., N.Y., 1951—. Fellow Am. Phys. Soc.; mem. Soc. Rheology (sec.-treas. 1949-53), Am. Inst. Physics (gov. 1952-53), Sigma Xi. Author: (With D. H. Gurinsky) Nuclear Fuels, 1956; (with G. H. Vineyard) Radiation Effects in Solids, 1957; (with A. C. Damask) Point Defects in Metals, 1963. Mem. editorial bd. Phys. Rev., 1955-58; N.Am. editor Jour. Physics and Chemistry of Solids, 1956—. Research on theory of diffusion in crystals, physics of high polymers, solid state physics, imperfections in crystals, radiation effects in solids. Home: Harbor View Rd., Stony Brook, L.I. Office: Brookhaven Nat. Lab., Upton, L.I., N.Y.*

DIENSTBIER, Zdenek, Czechoslovakian physician; b. Chrudim, Czechoslovakia, May 30, 1926; s. Josef and Anna (Fenclova) D.; MUDr., Charles U., Prague, Czechoslovakia, 1950, C.Sc., 1955, Dr.Sc., 1964; m. Vera Volkova, Oct. 17, 1950; children—Jan, Helena. Staff. Faculty Hosp., Prague, 1950-57; staff Biophys. Inst., Charles U., 1957—, dir. Biophys. Inst., vice dean Med. Faculty, 1960-63, vice rector Charles U., 1964—; staff Botkin's Hosp., Moscow, USSR, 1957, Middlesex Hosp., London, Eng., 1960. Mem. European Soc. Radiation Biology (council), Czechoslovakian Physicians Soc., Czechoslovakian Soc. Nuclear Medicine (chmn.). Author: (with M. Arient and V. Kofranek); also numerous articles. Editor: (with J. Brousil) Isotopes Diagnostics, 1965; (with M. Arient) Experimental Postirradiation Syndrome, 1966. Research in nuclear medicine and biology, biophysics. Home: 560 Kladenska, Prague 6, Czechoslovakia.*

DIERBACH, Johann Heinrich, German botanist; b. Heidelberg, Germany, Mar. 23, 1788; s. Friedrich and Anna Margaretha (Bastian) D.; student botany U. Heidelberg; M.D., 1816; state exam, 1816; m. Elisabeth Münch, 1820; 5 daus. Became lectr. U. Heidelberg, 1817; apptd. asso. prof. medicine, 1820. Recipient gold medal U. Heidelberg, 1815. Author numerous books including: Grundriss der Recepturkunst, 1818; Handbuch der medizinisch-pharmaceutischen Botanik . . . , 1819; Anleitung zum Studium der Botanik, 1820; Abhandlung über die Arzneikräfte der Pflanzen verglichen mit ihrer Struktur und ihren chemischen Bestandteilen, 1831; also articles in numerous bot., med. and pharm. jours. Research on pharmacognosy. Died Heidelberg, May 11, 1845.

DIERNHOFER, Karl Ludwig, Austrian veterinarian; b. Schwertberg, May 12, 1895; s. Karl and Mayr Diernhofer; D.M.V., Vet. U. of Vienna; D. honoris causa, U. Budapest, 1962; m. Elisabeth Hruschka. Asst. prof. of univ.; full prof. for illnesses of ruminants and pigs; rector Vet. U. of Vienna, 1952. Mem. Vet. Med. Specialists (Vienna), Gesellschaft Deutscher Naturforscher und Arzte, Austrian Soc. for Microbiology and Hygiene, Austrian Soc. for Study Sterility and Fertility (founder), Berlin Sci. Soc. fur Veterinarians (corr.), Italian Soc. for Progress Zootechnics (fgn.), Italian Soc. Vet. Sci. (hon.). Author: (with Wirth) Lehrbuch der Inneren Krankheiten der Haustiere. Address: Linke Bahngasse 11, Vienna III, Austria.

DIESEL, Rudolf Christian Karl, inventor, engr.; b. Paris, France, Mar. 18, 1858; s. Theodor and Elisabeth (Strobel) D.; studied mech. engr. Technische Hochschule, Munich, Germany, 1875-79; passed exam with highest mark in its history; m. Martha Flasche, 1883; 2 sons including Eugen; 1 dau. Began work as manager in ice factory of Carl Linde, Paris, 1880-90; then worked in Berlin to 1893; later became dir; with Maschinenfabrik Augsburg; visited America, 1904, 1912; lectured during 1912 trip. Author: Theorie und Konstruktion eines rationellen Wärmemotors zum Ersatz der Dampfmaschinen und der heute bekannten Verbrunnungsmotoren, 1893; Die Entstehung des Diesel-Motors, 1913; Solidarismus, natürliche wirtschaftliche Erlösung der Menschen, 1903. Invented diesel engine, 1897; achieved ignition by compression. Drowned en route to Eng., Sept. 29, 1930.

DIETERICI, (Karl Friedrich) Wilhelm, German statistician; b. Berlin, Germany, Aug. 23, 1790; s. Wilhelm and Eva Sophia (Haacke) D.; studied polit. sci. U. Königsberg, U. Berlin (both Germany); m. Wilhelmine v. Wedell, 1819; 4 sons including Friedrich, 4 daus. Engr.-geographer; entered pub. service in Potsdam, Germany, 1816; apptd. asst. Kultusministerium (ministry pub. edn.), 1820; apptd. prof. polit. sci., Berlin, 1834; became dir. Bur. Statistics, 1844; elected mem. 1st chamber, 1848. Mem. Prussian Acad. Scis. Research, numerous publs. on statistics related to state adminstrn. Died Berlin, July 29, 1859.

DIETL, Józef, Polish physician; b. Poland, 1804; contbd. ideas to sch. of therapeutic nihilism; described attacks resulting from partial twisting of kidney pedicle, marked by chills, nephralgia, nausea and other symptoms, 1864. Died 1878.

DIETRICH, Adam, German botanist; b. Ziegenhain, Germay, Nov. 4, 1711; s. Alomo and Anna (Röders) D.; m. Anna Katherine Huber, 1734; 3 sons, 2 daus. Farmer in Ziegenhain; corresponded with Linnaeus; called "the Botanist of Ziegenhain." Author: (with A. Haller) Flora Jenesis, 1745. Died Ziegenhain, July 10, 1782.

DIETRICH, (Konkordie) Amalie Nelle, German botanist; b. Stebelehn nr. Meissen, Germany, May 26, 1821; d. Gottlob Leberecht and Johanna Regina (Bormann) Nelle; m. Wilhelm August Salomo Dietrich, 1846; 1 dau. Collected and sold bot. specimens in Germany and Europe; sent to Australia by I. C. Godefroy & Sohn, Hamburg, Germany; became mem. expdn. to Tonga Island, 1872; returned to Germany in 1873; curator Hamburg (Germany) Bot. Mus. Mem. Entomologische Gesellschaft in Stettin. Many species named in her honor including Nortonia amaliae, Odynerus dietrichianus (wasps), Endotrichella dietrichiae (moss), Amansia dietrichiana, Sargassum amaliae (algae). Collected numerous bot., zool. and ethnographic objects in Australia and Tonga Islands. Died Rendsburg, Germany, Mar. 9, 1891.

DIETRICH, (Johann Christian) Gottlieb, German botanist; b. Ziegenhain, Germany, Mar. 9, 1765; s. Johann Adam and Maria Barbara (Böttcher); ed. at Jena (Germany); received Ph.D.; travelling companion, adviser to Goethe; gardener for Duke Karl August of Weiner, Germany, 1792-1801; became court gardener, 1794; dir. ducal bot. gardens in Wilhelmsthal, Germany, 1801-45; prof. botany. Plant species Dietrichia named in his honor. Mem. Botanische Gesellschaft zu Regensburg, Gesellschaft naturforschender Freunde zu Berlin, Leipziger ökanomische Societät. Author: Oekonomische-botanisches Garten-Jour., 6 vols., 1795-1804; Die Weimarische Flora, 1800; Vollständiges Lexikon der Gärtnerei und Botanik, mit Nachträgen, 30 vols., 1902-40; Handbuch der botanischen Lustgärtnerei, 2 parts, 1826, 28; Handlexikon der Gartnerei und Botanik, 2 vols., 1829, 30. Died Eisenach, Germany, Jan. 2, 1850.

DIETRICH, L(aRoy) S(eibert), Am. biochemist; b. Gettysburg, Pa., Feb. 17, 1926; s. LaRoy S. and Margaret (Horner) D.; B.S. in Chemistry, Wagner Coll., 1948; M.S., U. Wis., 1950, Ph.D., 1952; m. Elisabeth L. Hintermeister, June 12, 1948; children—Richard L., David C., Mark E. Research asst. biochemistry Columbia, 1952-53, research asso., 1953-56, asst. prof., 1956-58; faculty U. Miami (Fla.), 1958—, prof. biochemistry, 1965—. Sr. research fellow USPHS, 1959-63, career devel. awardee, 1963—. Mem. Am. Soc. Biol. Chemists, A.A.A.S., Harvey Soc., N.Y. Acad. Scis., Am. Assn. Cancer Research, Am. Assn. U. Profs., Am. Chem. Soc., Am. Soc. Cell Biologists, Am. Inst. Biol. Scis., Soc. Exptl. Biology and Medicine, Sigma Xi. Contbr. numerous articles to tech. jours. Elucidated metabolic pathways used in synthesis vitamin-containing coenzymes and control pathways by hormones and other substances; research on drug action at molecular level.*

DIETRICH, Nathan Franz David, German botanist; b. Ziegenhain, Germany, 1800; scholar U. Jena (Germany); became bot. gardner of univ, 1828; curator univ. herbarium, Jena. Author: (with C. Zenker) Musci Thuringici, 1821-23; Synopsis plantarum, 5 vols., 1838-52; Flora universalis (which included 4,760 color illustrations), 1831-61; Deutschlands Flora, 1833-64, 5 vols. Research on mosses. Died Jena, Oct. 23, 1888.

DIETRICH, Baron Philippe Frédéric, French chemist, mineralogist; b. Strasbourg, France, 1748; held several civil offices; elected mayor of Strasbourg, 1790; left after French Revolution 10th of Aug.; mem. French Acad. Scis. Author: Description of the Deposits of Ore and the Forges of France, 3 vols., 1786-1800; also a French transl. of Scheele's chem. treatise on air and fire. Tried upon return to France and guillotined in Paris by Jacobins, 1793.

DIETRICH OF FREIBERG (or Theodoricus Teutonicus de Vriberg), optician, physicist; b. Freiberg, Saxony; master theology, Paris, France, probably 1297; Dominican of German province; elected provincial Strasbourg (now in France) 1293; prior Wurzburg (Germany) monastery; apptd. vicarius of his province, 1310. Author: De luce et eius originie; De coloribus; de iride et radialibus impressionibus (on optical meteorology, especially rainbows). Trying to find middle path between explanations given by Averroes and by Avicenna, Dietrich explained colors by theory of inner total reflection in transparent bodies; held that even most transparent bodies can act as mirrors because of refraction and inner reflection; made simple expts. on dispersion of light by crystals 362 years before Newton, but was hampered by lack of exact law of refraction and dispersion of light; concluded that astron. phenomena cannot be accounted for without theory of eccentrics and epicycles, but accepted supremacy of exptl. method. Died 1311 or soon after.

DIETTRICH, Sigismond, geographer; b. Budapest, Hungary, Feb. 7, 1906; s. M. Anton deR. and Julia (Papp) D.; B.A., U. Budapest, 1927, M.A., 1928; Ph.D., Clark U., 1931; postgrad. U. Chgo., D.Sc., Royal Hungarian Palatine-Joseph U., Budapest, 1936; m. Iren Dokupil, July 27, 1932; 1 dau., Rosemary (Mrs. Charles Edward Leedham). Faculty, U. Fla., Gainesville, 1931-59, prof. geography, 1948-59, chmn. div. geography and geology, 1945-48; vis. prof., acting dir. Geographic Inst., Royal Hungarian Palatine-Joseph U., 1935-36; regional geographer U. S. Bd. Geog. Names, U. S. Dept. Interior, 1944, geographer, asst. to chief div. geography and cartography, 1944-45; educationist U. S. Dept. State, 1945-53; sr. Fulbright lectr. U. Dacca (Pakistan), 1959-61; chmn. social scis. Inter-Am. U. P.R., San Juan, 1962-65, chairman of the department of geography, 1965—. Recipient Distinguished Service award U. Fla., 1957; U. S. Ednl. Found. grantee, 1960; Asia Found. grantee, 1960; Mem. Am. Assn. Geographers (past dir., past pres. Southeastern sect), Fla. Acad. Scis. (past pres.), Nat. Council Geog. Edn., Hungarian, Am. geog. socs., Am. Acad. Polit. and Social Scis., Soil and Crop Sci. Soc. Fla. Author: Miami, 1960; The Philippines, 1961. Research and numerous articles on human and econ. growth of Fla.; field studies of P.I., Burma, Ceylon, India, Pakistan, Scotland, West Germany, P.R. with spl. studies on role of govt. planning in econ. devel. Home: 637 El Monte Hostos 165, Hato Rey 00918. Office: P.O. Box 1293 Hato Rey, P.R. 00919.*

DIETZ, Johann Simon Jeremias von, see von Dietz, Johann Simon Jeremias.

DIETZ, Johannes, German physician; b. Halle/Saale, Germany, Dec. 18, 1665; s. Johann and Maria Magdalene (Hitzsche) D.; student barber-surgeon G. Schober, Halle, 1681-84; m. Elisabeth Watzlau, Dec. 3, 1694; 1 dau.; m. 2d, Maria Magdalene Müller, June 5, 1727; 3 daus. Worked for barbers in Berlin; apptd. asst. med. officer 2d Turkish war of Leopold I in Hungary, 1686; with Andreas Horch, surgeon, Berlin; then med. officer with Danish troops; doctor whaling ship for 2 trips to Greenland; again became Danish mil. doctor; opened free barber shop, Halle, 1694; court barber. Author: Meister Johannes Dietz, der Grossen Kurfürsten Feldscher und königliche Hofbarbier, written circa 1730, pub. 1915. His book is important as source of med. history. Died Halle/Saale, Mar. 4, 1738.

DIETZEL, Adolf Hugo, German chemist; b. Pforzheim, Feb. 3, 1902; s. Adolf Karl and Emilie Dietzel; ed. Technische Hochschule Karlsruhe; Ph.D. in engring.; m. Margarethe Madlener. Instr. silicates Technische Hochschule Karlsruhe; head div. research of silicates Kaiser-Wilhelm Inst., Berlin-Dahlem; prof. Technische Hochschule Berlin; dir. brs. Max-Planck Inst. for Silicate Research; dir. silicate research MPI, Würzburg; hon. prof. U. Würzburg. Mem. Am. (hon.), German (Plaquette Seger) caramic socs., German Glass Tech. Soc. (Gehlhoff gold ring), Soc. German Chemists, Verein Deutscher Emailfachleute (Plaquette Louis-Vielhaber). Author: Auflärung der Hafmechamismus zwischen Glas (Email) und Metallen; Deutung der Feldstärken von Kationen in Mehrstoffsystemen; Elektrochemische Untersuchungen an Gläsern; Texturuntersuchungen an Keramischen Körpern mit Mikrowellen. Home: Wörthstrasse 22. Office: Neunerplatz 2, Würzburg, Germany.

DIEUDONNÉ, Adolf, German physician; b. Stuttgart, Germany, Aug. 29, 1864; s. Eduard and Marie (Krauss) D.; student medicine in Tübingen, Berlin, Munich, and Würzburg (all Germany). Joined Bavarian army as physician, 1888; built bacteriological lab., Würzburg, 1891; assigned to Imperial Health Office, Berlin, 1894; went to Bombay, India to study plague, 1897; later garrison doctor in Würzburg for 6 years; joined univ. faculty, 1898; became tchr. at Mil. Acad., Munich, 1904 also at univ.; named hon. prof. Munich, 1906; med. adviser Bavarian Ministry of Interior; ret. from mil., 1919; dir. pub. health in Bavaria, 1919-29. Author: Schutzimpfung und Serum-therapie, 1896; Taschenbuch des Feldarztes Übertragbare Krankheiten, 1914. Research on epidemics including plague, leprosy, cholera, smallpox, malaria (in India), typhus, dysentery, diphtheria, meningitis, scarlet fever, also on nutrition. Died Bad Heilbrunn, Germany, Oct. 26, 1944.

DIEUDONNÉ, Jean Alexandre, French mathematician; b. Lille, France, July 1, 1906; s. Ernest and Leontine (Lebrun) D.; student École normale supérieure, Paris, 1924-27; agrégé de mathématiques, 1927; Docteur ès Scis. mathématiques, Paris, 1931; m. Marie Odette Clavel, July 22, 1935; children—Jean-Pierre, Francoise (Mrs. Roger Dournaux). Maitre de confs. U. Rennes (France), 1933-37; maitre de confs., then prof., Nancy, France, 1937, 1937-46, 48-52; prof. U. Sao Paulo (Brazil), 1946-47, U. Mich., 1952-53, Northwestern U., 1953-59, Institut des Hautes Études Scientifiques (Bures), 1959-64; prof. U. Nice (France), 1964—, dean faculty scis., 1965—. Mem. Math. Soc. France, Am. Math. Soc. Author: La géométrie des groupes classiques, 1955; Foundations of Modern Analysis, 1960; Algèbre linéaire et géométrie élémentaire, 1965; also articles. Research in abstract algebra, theory of Lie groups, topology, functional analysis. Home: Villa Nancago, Corniche Fleurie, 06-Nice, France. Office: Faculté des Sciences, Parc Valrose, 06-Nice, France.*

DIEULAFOY, Georges, French physician; b. Toulouse, France, Nov. 18, 1839; med. degree Paris, France, 1869; apptd. prof. internal pathology Paris Faculty Medicine, 1884; named prof. clin. medicine, chief med. service Hôtel-Dieu, 1896. Mem. French Acad. Medicine (became pres. 1910). Author: Manuel de pathologie interne, 1890. Research on Bright's disease, appendicitis, Tb, typhoid fever and sudden death caused by it; invented suction pump for evacuating path. fluids from body. Died Paris, Aug. 16, 1911.

DIGBY, Sir Kenelm, English natural philosopher; b. Gayhurst, Buckinghamshire Eng., July 11, 1603; s. Everard and Mary (Mulsho) D.; student Worcester Coll., Oxford, 1618-20, Gresham Coll., 1633-35; m. Venitia Stanley, 1625; 4 sons, 1 dau. Travelled to Paris and Angers, 1620, Spain, 1623; returned to Eng., 1623; privateer, 1627-29; deeply involved with cause of English Catholics throughout life; lectr. plant vegetation Gresham Coll., 1660-61. Knighted, 1623. One of the founders of Royal Soc., 1663. Author of Two Treatises: Of Bodies, and of the Immortality of Mans Soul, 1644; Commercium epistolicum, 1658; Discours fait en une célèbre assemble, 1658; a Discourse concerning the vegetation of plants, 1661; Choice and Experimented Receipts, 1688; The Closet of Sir Kenelm Digby, 1669. Described expts. on blood circulation and supported Harvey; noted that plants receive nourishment not only from soil but also from air; credited as 1st to use saltpeter as fertilizer; observed magnetic and elec. attractions and acoustic resonance; advocate of "powder of sympathy" (said to have healed wounds at a distance); performed original experiments in embryology; supported epigenesis. Died June 11, 1665.

DIGGES, Leonard, English mathematician; b. Digges Court, Kent, Eng.; s. James and Philippa (Engham) D.; studied at U. Coll., Oxford (Eng.) U.; m. Bridget Wilford; a son, Thomas. Mathematician, land surveyor, architect. Author: A Prognostication Everlasting (rules to judge weather using sun, moon, stars), 1553; A Booke named Tectonicon, (showed measurements of lands, squares, timber, stones, steeples), 1556; A Geometricall Practise, named Pantometria (a geometry of regular and Platonical bodies), 3 books, 1571; 1553; (with T. Digges) An Arithmeticall Militare Treatise, named Stratioticos, 1572. His expts. with combinations of concave and convex lenses anticipated invention of telescope, 1571; invented modern theodolite used for measuring angles. Died circa 1571.

DIGGES, Thomas, English mathematician; b. Kent, Eng., circa 1543; s. Leonard and Bridget (Wilford) D.; B.A., Queen's Coll., Cambridge (Eng.) U., 1550/51, M.A., 1557; m. Agnes St. Leger; children—Dudley, Leonard, Margaret Ursula, William, Mary. M.P. for Wallingford, Eng., from 1572, for Southampton, Eng., from 1585; muster-master-gen. English forces in Netherlands, 1586; commd. (with others) to equip expdn. for exploration Cathay, Antarctic Seas, 1590. Author: A Geometrical Practise, named Pantometria . . . , 1571; epistle to the reader, Parallacticae Commentationis Prax . . . Nucleus quidam (John Dee), 1573; Alae seu Scalae Mathematical, quibus visebilium remotissima Theatra Caelorum conscendi, et Planetarum omnium itinera novis et inauditis Methodis explorari . . . , 1573; A Prognostication . . . contayning . . . rules to judge theWeather by the Sunne, Moone, Stars . . . with a briefe judgement for ever, of Plenty, Lacke, Sickenes, Daerth, Warres, etc., opening also many natural causes worthy to be knowen (pub. by Leonard Digges, augmented, corrected by Thomas Digges), 1578; An Arithmeticall Militare Treatise, named Stratioticos (begun by father), 1579; England's Defence: A Treatise concerning Invasion, 1586; Instructio exercitus apud Belgas, 1586; Perfect description of the celestial orbs, according to the most antient doctrine of the Pythagoreans, 1592. Continued pub. father's jour. Prognostication Everlasting. Pub. some of father's works; wrote numerous works on applied math.; considered (with John Dee, Thomas Hariot) one of 3 gt. mathematicians of period; as Copernican, attacked Ptolemaic system; gave summary of Copernicus' planetary system; denied existence of sphere of fixed stars (stated that orb in which fixed stars are set extends infinitely upwards); 1st to declare that stellar universe is infinite; one of 1st to destroy classical objection that if earth were moved, stone dropped from tower would be left behind. Died London, Aug. 24, 1595.

DIJKGRAAF, Sven, Dutch biologist; b. The Hague, Netherlands, Apr. 17, 1908; s. Pieter Cornelis and Priska (Exner) D.; student natural scis. U. Vienna (Austria), 1927-31; Dr.phil., U. Munich (Germany), 1933; m. Elisabeth Kunz, July 17, 1948; children—Pieter Christiaan, Sven, Claudia. With U. Groningen (Netherlands), 1935-48, lectr. comparative physiology, 1946-48; prof. comparative physiology, head lab. comparative physiology, 1948——, dean Faculty Natural Scis., 1966-68. Mem. Dutch Zool. Soc., Dutch Soc. Physiologists. Research, publs. on comparative animal sense physiology and orientation, including lateral line function, hearing, pressure sense in fishes, optomotor reactions, statocyst functions in crustacea and molluscs, electroreception in fishes especially elasmobranchs; discovered independently from Griffin, echolocation in bats.

DIKUSHIN, Vladimir Ivanovich, Russian mech. engr.; b. 1902; grad. Bauman Higher Tech. Sch., Moscow, 1928. With Exptl. Research Inst. of Metal-Cutting Machine Tools, Moscow, 1933-39, chief designer 1939——; prof. Moscow Higher Tech. Sch., 1932——; bur. mem. dept. tech. sci. USSR Acad. Scis. Decorated Order of Lenin; recipient Stalin prize, 1941, 51. Mem. USSR Acad. Scis. Author: Mechanical Engineering, 1949; Problems of Process Automation in Mechanical Engineering, 1956. Research and publs. on assembly machines and automatic tooling lines, basic problems of machine tool manufacture, developer theory and methodology of designing metal-cutting machine tools. Address: Exptl. Research Inst. of Metal-Cutting Machine Tools, 5-y Donskoy pr. 21b, Moscow, USSR.

DILGER, (Johann) Friedrich, German clockmaker; b. Urach nr. Donaueschingen, Germany, Feb. 27, 1712; s. Simon and Anna (Rissler) D.; m. Agnes Gfell, 1735. Went as clock salesman to Paris where he investigated French clock industry for year (used information in own factory on returning to Germany); reawakened clock making industry in Black Forest; later studied in Switzerland. Manufactured precision tools; invented graduated plate making, movable figures; built clocks with month and year; introduced Swiss metal bells as substitute for glass bells. Died Urach, Sept. 13, 1773.

DILICH, Wilhelm, German topographer, engr.; b. Wabern, Germany, circa 1571; s. Heinrich Scheffer; student Wittenberg, Marburg (both Germany) m. Anna Stubenrauch, circa 1600; a son, Johannes Wilhelm. Topographer, historiographer Landgrave Moritz, Hessen-Kassel, Germany; companion to a Hessian prince, Netherlands; became engr., circa, 1607; under arrest by Landgrave Moritz, 1621-24; escaped and became overseer all fortifications of Saxony, Wittenburg, Germany; designed (with son) fortifications for Frankfurt/Main, beginning 1627. Wrote topography of Saxony, descriptions of cities and countries, including Hessia and Bremen; also engravings of cities (used as models by Matthacus Merian the Elder). Died Dresden, Germany, Apr. 1655.

DILL, David Bruce, Am. physiologist; b. Eskridge, Kan., Apr. 22, 1891; s. David White and Lydia (Dunn) D.; B.S., Occidental Coll., 1913, Sc.D., 1959; M.A., Stanford, 1914, Ph.D., 1925; m. Olive Lillian Cassel, June 10, 1915; children—Elizabeth Cassel (Mrs. Steven M. Horvath), David Bruce; m. 2d, Chloris Luella Fuller Gillis, Jan. 3, 1946. Asst. prof. to prof. Harvard, dir. Fatigue Lab., 1927-47; dep. dir. med. research U. S. Army Chem. Research & Devel. Labs., 1947-61; research scholar Ind. U., Bloomington, 1961-66; research prof. Nev. So. U., Las Vegas, 1966——. Leader physiol. exploratory expdns. to Andes, C.Z., Colo. desert; vis. lectr. Harvard, 1950-61; mem. exec. com. div. biology, agr. Nat. Acad. Scis.-NRC, 1955-58. Fellow Am. Acad. Arts and Scis., A.A.A.S., Am. Coll. Cardiology, Am. Coll. Sports Medicine (past pres.), Am. Acad. Phys. Edn.; mem. Am. Soc. Biol. Chemists, Fedn. Am. Soc. Exptl. Biology (past chmn. bd.), Am. Physiol. Soc. (past pres.), Aerospace Med. Assn., Am. Chem. Soc., Phi Beta Kappa, Sigma Xi, Phi Delta Kappa, Phi Lambda Upsilon. Author: (with A. V. Bock) Physiology of Muscular Exercise, 1931; Life, Heat and Altitude, 1938. Editor: Adaptation to the Environment, 1960-63. Research and numerous publs. on comparative and environmental physiology; physiology of aging. Home: 303 Wyoming St., Boulder City, Nev. 89005.*

DILLENIUS, Johann Jakob, botanist; b. Darmstadt, Germany, Dec. 22, 1684; s. Justus Friedrich and Anna Elisabeth (Finck) D.; Lic.med., U. Giessen (Germany), 1710; M.D., 1719; M.D., Oxford (Eng.) U., 1735; botanist, London, Eng., from 1721, also James Sherard's Bot. Garden in Eltham; 1st prof. botany Oxford U. (chair founded by William Sherard), 1728-47. Fellow Royal Soc., 1724. Author: Historia muscorum, 1741; Catalogus Plantarum circa Gissam sponte nascentium, 1718; Hortus Elthamensis, 2 vols., 1731. Described numerous new species; 1st to use morphological characteristics of fruit for classification; Linnaeus adopted this classification system for fungus, without change; 1st to describe species Bryum, Hypnum, Sphagnum, Amanita, Boletus, Morchella; began sci. of mosses; Linnaeus named genus Dillenia (tropical trees) after him. Died Oxford, Apr. 13, 1747.

DILLER, William F., Am. zoologist; b. Lancaster, Pa., July 26, 1902; s. William F. and Lida (Schofield) D.; A.B., Franklin and Marshall Coll., 1923; Ph.D., U. Pa., 1928; m. Irene Corey, June 18, 1938. Instr. zoology U. Pa., 1923-25, Franklin and Marshall Coll., 1925-27, U. Pa., 1927-30; asst. prof. zoology Dartmouth, 1931-39; faculty zoology U. Pa., 1939-42, 46——, prof., 1958——; vis. scientist Johns Hopkins Zool. Lab., 1939. Sterling Research fellow Yale, 1930-31; fellow by courtesy Laboratoire Arago, Sorbonne, Banyuls, 1939; USPHS fellow Central Coll., Bangalore, India, 1955. Mem. Am. Soc. Zoologists, Am. Soc. Protozoologists, A.A.A.S., Am. Soc. Naturalists, Am. Micros. Soc., Phi Beta Kappa, Sigma Xi. Research, numerous publs. on nuclear behavior and genetics of one-celled animals, including finding that these organisms may develop either bisexually or by self-fertilization. Home: 2417 Fairhill Av., Glenside, Pa. 19038. Office: Leidy Lab., U. Pa., 38th and Woodland Av., Phila. 19104.*

DILLING, Walter James, pharmacologist; b. Aberdeen, May 15, 1886; s. William Dilling; ed. Robert Gordon's Coll., Aberdeen; M.B., Ch.B. with honours, U. Aberdeen; m. Vida J. B. Ducat, 1914; 2 daus. Second asst. in physiology Aberdeen U., 1907, lectr. pharmacology, 1910-14; Carnegie research scholar in physiology, 1907-09; Carnegie research fellow in pharmacology, 1909-10; 1st asst. prof. pharmacology and physiol. chemistry Rostock U., 1909-10; Dr. Robert Pollok lectr. materia medica and pharmacology Glasgow U., 1914-15, 19-20; lectr. pharmacology and gen. therapeutics Liverpool U., 1920-26, dean Med. Faculty, 1923-33, 39-45, mem. council, 1923-29, 40-41, prof. pharmacology and gen. therapeutics, 1930-50; examiner in materia medica, pharmacology, others univs. St. Andrews, Oxford, Cambridge, Leeds, Belfast, Bristol, Sheffield, London, Cardiff, Birmingham. Mem. Gen. Med. Council. Mem. Pharm. Soc. (council). Author: Atlas of Crystals and Spectra of the Haemochromogens, 1910; Charts of Blood Spectra, 1911; Pharmacology and Therapeutics of Materia Mecia, 19th edit., 1950; (with S. Hallam) Dental Materia Medica, Pharmacology and Therapeutics, 3d edit., 1946, also articles. Died Aug. 19, 1950.

DILLON, John Henry, Am. physicist; b. Ripon, Wis., July 10, 1905; s. Frank George and Hattie (Barnes) D.; B.A., Ripon Coll., 1927, Sc.D. (hon.), 1950; M.A., U. Wis., 1928, Ph.D., 1931; M.Sc. (hon.), Lowell Technol. Inst., 1951; m. Bernice Olmsted, June 18, 1935 (dec. Aug. 1960); m. 2d, Rena Quinn Perkson, Apr. 4, 1963; stepchildren—Howard N. Perkson, Jr., Pamela Dare Perkson. Research asst. physics U. Wis., 1927-31; physics research group Firestone Tire & Rubber Co., Akron, O., 1931-37, head physics div., 1937-45, asst. dir. research, 1945-46; dir. research Textile Found. and Textile Research Inst., Princeton, N.J., 1946-51, dir. Textile Research Inst., 1951-59, pres., 1959——. Vis. lectr. chemistry Princeton, 1947-52, prof., 1952——; chmn. Gordon Research Conf. on Textiles, 1949; chmn. physicists group Nat. Acad. Scis.-NRC adv. panels to Nat. Bur. Standards, 1961-64. Recipient Harold DeWitt Smith medal Am. Soc. for Testing and Materials, 1955. Fellow Am. Phys. Soc. (chmn. div. high polymer physics 1944), Textile Inst. Gt. Britain (mem. Am. panel); mem. Am. Inst. Physics (gov. bd. 1957-59), Am. Assn. Textile Chemists and Colorists, Am. Assn. for Textile Tech. (adv. council 1963——), Am. Chem. Soc., Fiber Soc. (v.p. 1960-61, pres. 1961-62), Soc. Rheology (v.p. 1939-41, 53-57, pres. 1957-59), Nat. Council for Textile Edn. (pres. 1958-59), Soc. Chem. Industry (Gt. Britain), Sigma Xi, Phi Beta Kappa, Phi Psi (hon.). Editorial bd. Jour. Applied Polymer Sci., 1959——. Research and publs. on photoelectric properties single crystals, rheology of rubbers, mech. behavior rubbers and fibers, triboelectric properties of fibers, applications of naturally radioactive metals in spark breakdown devices. Patentee U. S. and Gt. Britain. Home: 237 Elm Rd. Office: Textile Research Inst., Princeton, N.J. 08540.*

DILLON, Lawrence Samuel, Am. biologist; b. Reading, Pa., Apr. 6, 1910; s. LeRoy Victor and Emma (Culp) D.; B.S., U. Pitts., 1933; M.S., Tex. A. and M. U., 1951, Ph.D., 1954; m. Elizabeth Jane Schatz, Jan. 4, 1932; 1 dau., Patricia Jane (Mrs. Henry Albert Brown). Research and devel. chemist Glidden Co., Reading, Pa., 1934-37; curator zoology, asst. dir. Reading Pub. Mus., 1937-48; faculty Tex. A. and M. U., College Station, 1948——, prof. biology, 1961——. NSF Faculty fellow U. Queensland, Brisbane, Australia, 1959-60. Fellow A.A.A.S., Tex. Acad. Sci., Instituto Americano, Acad. Zoology; mem. Am. Assn. U. Profs., Am. Soc. Zoologists, Ecol. Soc. Am., N.Y. Acad. Scis., Soc. for Study Evolution, Soc. Systematic Zoology. Author: (with Elizabeth S. Dillon) Manual of the Common Beetles of North America, 1961; The Science of Life, 1964; Principles of Life Science, 1964; Principles of Animal Biology, 1965; Research and publs. on evolution of life and origins of phyllum Chordata. Home: 1700 Jersey, College Station, Tex. 77840.*

DILLON, Thomas, Irish chemist; b. Inniscrone, Jan. 15, 1884; s. John B. and Elisabeth (Sullivan) D.; ed. Queen Coll., Cork, Ireland; m. Geraldine Plunkett; 5 children. Asst. chemistry dept. Univ. Coll., Dublin, Ireland, 1909-19; prof. chemistry Univ. Coll. of Balway, 1919-54. Mem. R.I.A., I.C.I. Research and publs. on chemistry. Address: 13, Marlborough Rd., Donnybrook, Dublin 4, Ireland.

DILTHEY, Wilhelm, German philosopher; b. Biebrich, Germany, Nov. 19, 1833; studied for ministry U. Heidelberg (Germany), also U. Berlin (Germany). Prof. philosophy U. Basel (Switzerland), 1867-68, U. Kiel (Germany), 1868-71, U. Breslau (now Wroclaw, Poland), 1871-82, U. Berlin, 1882-1911. Author numerous works including: Einleitung in der Geisteswissenschaften; Ideen über eine beschreibende und Zergliedende Psychologie; Der Aufbau der geschichtlichen Welt in der Geisteswissenschaften; Das Erlebnis und die Dichtung (works available in Gesammelte Schriften). Distinguished between cultural, natural scis.; held man can understand what humans have done, created, thought; meaningfulness depends upon present-day and subjective values, but can also be discerned in hist. record through sympathetic understanding, re-experiencing of past by analogy with present (thus meaning in history is not fixed but changes with time and culture of historian); rejected positivism, as cultural scis., having own aim, method, need not be subservient to natural scis., also rejected metaphysics; attempted to unfold cultural history of modern world in series of important works. Died Seis, nr. Posen (now Poland), Oct. 1, 1911.

DILUZIO, Nicholas Robert, Am. physiologist; b. Hazleton, Pa., May 4, 1926; s. Nicholas and Carmela (Searfella) D.; B.S. with honors, U. Scranton, 1950; Ph.D., U. Tenn., 1954; m. Gertrude Alma Dezagattis, June 10, 1948; children—Nicholas Mark, Tamara Ann, Daniel Val. Investigator, Dorn Lab. for Med. Research, Bradford, Pa., 1954, Oak Ridge Nat. Lab., 1956, U. S. Naval Radiol. Def. Lab., 1958; faculty U. Tenn. Med. Units, Memphis, 1955——, prof., 1962——, chmn. physiology and biophysics dept., 1965——. Mem. Tenn. Adv. Com. on Atomic Energy, 1958; mem. sci. adv. bd. Nat. Council on Alcoholism, 1963——. Fellow USPHS, 1952-54; recipient Lederle Med. Faculty award, 1958-61. Mem. A.A.A.S., Am. Physiol. Soc., Am. Heart Assn., Radiation Research, Transplantation Soc., Soc. for Exptl. Biology and Medicine, Reticuloendothelial Soc. Contbr. numerous articles to tech. jours. Research on elucidation role reticuloendothelial system, host def. mechanism, chem. agts. altering reticuloendothelial activity; developed new concept chem. induced liver injury. Home: 5953 Brierdale, Memphis 38117.*

DILWORTH, Robert P., Am. mathematician; b. Hemet, Cal.; s. John Norman and Myrtle (Palmer) D.; B.S., Ph.D., Cal. Inst. Tech.; m. Miriam White, Dec. 23, 1940; children—Robert P., Gregory L. Sterling research fellow Yale, 1939-40, instr. math., 1940-43; faculty math. Cal. Inst. Tech., Pasadena, 1943——, prof. math., 1951——. Chmn. com. examiners Coll. Bd., 1958-61; mem. adv. bd. Inst. for Def. Analyses, 1961——; mem. math. edn. steering com. African Edn. Program, Ednl. Services, Inc., 1962——. Mem. Math. Assn. Am. (chmn. publs. com. 1961——). Research, publs. on abstract algebra and mathematical statistics, lattice theory. Home: 3121 Doyne Rd., Pasadena Cal. 91107.*

DI MACCO, Gennaro, Italian physician; b. Syracuse, Sept. 9, 1895; s. Giuseppe and Bianca (Leboffe) Di M.; M.D.; m. Maria Di Macco; children—Giuseppe, Bianca. Full prof. gen. pathology; dir. Gen. Pathology Inst., U. Rome. Author: Patologia Generale, 2 vols., 1951; Patologia del Metabolismo, 1950; Malattia e Disposizione, 1958. Founder numerous jours., including Medicina Sperimentale, Ormonologia, Neoplasie. Home: via Nibby 5. Office: University of Rome, Rome, Italy.

DIMLER, Robert Julius, Am. chemist; b. Pekin, Ill., Sept. 28, 1914; s. Paul John and Amelia (Dietrich) D.; B.S., Bradley U., 1936; M.S. (Alumnae Research Found. scholar and fellow) U. Wis., 1938, Ph.D., 1940; m. Elizabeth Ruth Dregne, June 23, 1941; children—Bruce G., Paul T., Steven R. With No. utilization research and devel. div. Agrl. Research

Service, U. S. Dept. Agr., Peoria, Ill., 1941—, in charge starch structure group, 1948-60, chief cereal properties lab., 1960-64, dir. div., 1964—. Instr., carbohydrate chemistry Bradley U., Peoria, 1947-63. Recipient Superior Service awards U. S. Dept. Agr., 1962, 63, Distinguished Service award, 1965; Distinguished Alumni award Bradley U., 1965. Mem. Am. Chem. Soc. (mem. at large exec. com. carbohydrate div. 1966—), Am. Assn. Cereal Chemists, Agrl. Research Inst., Sigma Xi, Phi Kappa Phi. Patentee in field. Research and numerous publs. on structure of starch and dextran, reaction of dextrose, qualitative and quantitative determination of sugars, cereal grain proteins, utilization of cereal grains, soybeans, flaxseed Home: 2532 N. Rockwood Dr. Office: 1815 N. University St., Peoria, Ill. 61604.*

DIMOCK, William Wallace, Am. veterinarian; b. Tolland, Conn., Feb. 20, 1880; s. Henry Eugene and Ellen M. (Clark) D.; B.Agr., Conn. Agrl. Coll., 1901; D.V.M., N.Y. State Veterinary Coll. (Cornell U.), 1905, U. Habana (Cuba), 1908; m. Ruth Attwill Mudge, Nov. 27, 1909; children—Phoebe, Betty Anne, Shubael Eugene, Gladys Eusebia, Ruth Mudge. Began practice as veterinarian in Conn., 1905; asst. chief animal husbandry Cuban Expt. Sta., Santiago de los Vegus, Cuba, 1906-08; chief veterinarian Nat. Bd. Health, Cuba, 1908-09; prof. pathology and bacteriology State Coll. of Ia., 1909-19; prof. vet. science U. Ky., since 1919, head dept. animal pathology Ky. Agrl. Expt. Sta. Mem. Am. (pres. 1942-43), Ky., Ia. vet. med. assns., A.A.A.S., Ky. Acad. Sci., U. S. Live Stock San. Assn., Sigma Xi. Author or joint author many publs. on animal diseases. Died Oct. 1953.

DIMOND, Edmunds Grey, Am. physician; b. St. Louis, Dec. 8, 1918; s. Edmunds G. and Gertrude Ruth (Smith) D.; student Purdue U., 1938-41; B.S., Ind. U., 1942, M.D., 1944; m. Audrey Stone, Apr. 1, 1945; children—Lark Grey, Lea Grey, Sherri Grey. Professor of medicine, chairman of the department of University Kansas Med. Sch., 1953-60; dir. Inst. for Cardiopulmonary Diseases, Scripps Clinic and Research Found., La Jolla, Cal., 1960—; research asso. Physiology Research Lab., Scripps Inst. Oceanography, La Jolla, 1963—; prof. medicine U. Cal. at San Diego, 1967—. Fulbright prof., Netherlands, 1956; mem. heart disease control adv. com. USPHS, 1966—, mem. nat. adv. com. to regional med. center program, 1966—; scholar-in-residence Nat. Library Medicine, 1967. Mem. Am. Coll. Cardiology (trustee, past pres.), Am. Fedn. Clin. Research, Central Western socs. clin. research, A.A.A.S., Am. Heart Assn., Harveian Soc. of London. Author: Electrocardiography, 1952, 4th edit., 1967. Research on cholesterol metabolism, electrocardiography, clin. medicine. Home: 1600 Ludington Lane. Office: 476 Prospect St., La Jolla, Cal. 92038.*

DIMOPOULLOS, George Takis, Am. microbiologist; b. N.Y.C., Nov. 24, 1923; s. Takis and Anna (Sepides) D.; B.S., Pa. State U., 1949, M.S., 1950; Ph.D., Mich. State U., 1952; m. Annie S. Wainwright, July 28, 1945; 1 son, James Cameron. Research asso. U. Wis., 1952-53; bacteriologist U. S. Dept. Agr., Greenport, N.Y., 1953-57; prof. microbiology La. State U., Baton Rouge, 1957—. Diplomate Am. Bd. Microbiology. Fellow Am. Acad. Microbiology, A.A.A.S.; mem. Soc. Exptl. Biology and Medicine, Am. Soc. Microbiology, Conf. Research Workers in Animals Diseases. Research and numerous publs. on characterization of pathologic changes in animals diseases and infectious organisms at molecular level, immunochemistry, plasma proteins, lipids, Home: 1245 Stephens Av., Baton Rouge 70808.*

DIMOTAKIS, Paul Nicholas, Greek chemist; b. Athens, Greece, Feb. 19, 1928; s. Nicholas and Aikaterini Dimotakis; B.Sc., U. Athens, 1953; Ph.D., Cambridge (Eng.) U., 1964. Staff, Wine Inst., Athens, 1954-56; staff Greek Atomic Energy Commn., Democritus, Greece, 1956—; staff Pa. State U., also Argonne (Ill.) Nat. Lab., 1956-57; dir. chem. div. Democritus Nuclear Research Center, 1960-62; head nuclear chemistry lab. Democritus N.R.C., 1964—; dir. programming and coordination office Greek Atomic Energy Commn., 1965—, sci. dir., 1967—. Mem. Greek Chemists' Assn., Am. Nuclear Soc. Author: How To Prospect for Uranium, 1956. Publs. on discovery that isothermal kinetics of annealing in neutron irradiated solid chem. compounds constitutes curve with many maxima and minima, new mechanism of annealing alpha particles by secondary transport of energy over large distances in solids. Home: 74, Nichopoleos Athens—218, Greece. Office: Nuclear Chemistry Lab., Democritus, NRC, Athens, Greece.*

DIMROTH, Otto, German chemist; b. Bayreuth, Germany, Mar. 28, 1872; s. Karl and Adele (Geys) D.; student U. Munich (Germany), U. Strasbourg (France); doctorate 1895; m. Aloysia Bayer, 1900; 3 sons including Herman, Karl; 1 dau., Gertrud; m. 2d, Vera Julia Wilhelmine Schütt, 1919; 3 sons, 1 dau. Spent two years in industry; apptd. lectr. U. Tübingen (Germany), 1900; prof. organic dept. state lab. U. Munich, 1905-13; prof., dir. chem. inst. Greifswald, Germany, 1913-18; became prof. U. Würzburg (Germany), 1918. Research and publs. on border zone between organic and phys. chemistry, mercurization of aromatic compounds, triazol, tetrazol and triazene, desmotropic compounds, relationship between solubility and arrangement of tautomeric equilibrium, insect dyes, use of lead tetracetate as oxidation medium. Died Aschaffenburg, Germany, May 16, 1940.

DIMSDALE, Thomas, English physician; b. Essex, Eng., May 6, 1712; s. Sir John and Susan (Bowyer) D.; ed. St. Thomas's Hosp.; M.D., 1761; m. dau. of Nathaniel Brassey (dec. 1744); m. 2d, Anne Iles, 1746. Became vol. under Duke of Cumberland, 1745; inoculated Empress Catherine, various Russian princes and Hawaiian prince Omai for smallpox; made baron and councillor of state in Russia; became M.P. from Hertford, Eng. in 1780, 84; visited Joseph II in Vienna, 1784; banker in London, Eng.; founder, supr. inoculating house, Hertford. Fellow Royal Soc., 1769. Author: The Present Method of Inoculation for the Small Pox, 1767; Tracts on Inoculation, 1768, 81; Thoughts on General and Partial Inoculation, 1776; Observations on the Plan of a Dispensary and General Inoculation, 1780. Died Essendon in Hertfordshire, Eng., Dec. 30, 1800.

DI NATALE, Luigi, Italian surgeon; b. Alanno (Pescara), Nov. 1, 1896; s. Natale and Anna Concetta (Pignatari) Di N.; D. Medicine and Surgery, U. Rome; m. Alessandra Peretti. Specialist in spl. pathology of demonstrative surgery and operative medicine; 1st surgeon Fatabenefratelli-Fatabenesorelle Ciceci Agnesi Hosp., Milan, Italy. Mem. Italian, Piedmont, Toscoumbra socs. surgeons, Italian Soc. Thoracic Surgeons, Internat. Coll. Surgeons. Author over 120 works and monographs, including: Il Cancro dello Stomaco, 1939; La Gastrectomia Totale, 1950; La Evoluzione della Chirurgia Gastria Nell' Ultimo Decennio, 1956. Home: Piazza della Repubblica 9. Office: Corso Porta Nuove 23, Milan, Italy.

D'INCARVILLE, Pierre-Noël Le Chéron, French botanist; b. Louviers, France, Aug. 21, 1706; became Jesuit missionary, Que., Can., 1730-39, Peking, China, 1740-57. Mem. French Acad. Scis., 1750. Research and publs. on flora of China; sent Chinese plant drawings to Jussieu; introduced many Chinese trees into France. Died Peking, June 17, 1757.

DI NEGRO, Andalo, Italian astronomer, mathematician; b. Genoa, Italy, 1260; tchr. astronomy to Boccaccio; sent by Signoria of Genoa as ambassador to Alexios II Commenos, emperor of Trebizond, 1314. Author: **Tractatus sphaerae; Theorica planetarum;** Theorica distantiarum omnium sperarum et planetarum a terra et magnitudinem eorum; Canones super Almanach Profatii, circa 1323; Opus preclarissimum astrolabi; Practica astrolabii; De operationibus scale quandrantis in astrolabio scripte; De compositione astrolabii; Tractatus quadrantis; De infusione spermatis; Ratio diversitatis partis. Wrote on theoretical and practical astronomy; claimed to have improved knowledge of dimensions of celestial spheres as transmitted by al-Farghani, al-Battani, Ibn Rusta; opposed theory that land sphere and water sphere (in the earth) have different centers; discussed and published tables on astrolabe; applied astrology to physiology. Died circa 1340.

DINELLI, Dino Ferdinando Gaspero, Italian chemist; b. Rome, Italy, June 27, 1909; s. Ferruccio and Corinna (Catola); m. Cordelia Catola; children—Fiamma, Serena. Prof. organic chemistry; dir. United Labs. Study and Research, S.N.A.M.; councillor of adminstrn ANIC-Gela. Mem. Italian, Am. chem. socs., Nat. Union Chemists. Contbr. articles to profl. publs. Home: via Giuseppe Gatteschi 23, Rome, Italy. Office: United Labs., S. Donato Milanese, Milan, Italy.

DINES, William Henry, English meteorologist; b. Pimlico, London, Aug. 5, 1855; s. George and Louisa Sara (Cokes) D.; B.S., Corpus Christi Coll., Cambridge (Eng.) U., 1881; m. Catharine Emma Tugwell, 1882; 2 sons. Pvt. corr. tutor in math.; became dir. work on upper air for Brit. Meteorol. Office, 1905. Fellow Royal Soc., 1905, also Inst. Physics, Royal Aero. Soc. (hon.), Royal Meteorol. Soc. (pres. 1901-02, Symons medal 1914, Buchan prize 1924). Author: The Characteristics of the Free Atmosphere, 1919; Collected Papers, 1931. Leading exponent of exptl. meteorology in Eng.; research on anemometry, upper air, wind force, solar and terrestrial radiation, pressure of water vapor in air, wind force; used kites and balloons to investigate upper air; developed an anemometer. Died Benson, Oxfordshire, Dec. 24, 1927.

DINGLE, A(lbert) Nelson, Am. meteorologist; b. Bismarck, N.D., May 22, 1916; s. Victor Stanley and Nanna B. (Nelson) D.; B.S., U. Minn., 1939; M.S., Ia. State Coll., 1940; S.M., Mass. Inst. Tech., 1945, Sc.D., 1947; m. Eleanor A. Nelson, Nov. 20, 1941; children—Karen L., Timothy N. Head physics dept. Hampton Inst., Hampton, Va., 1941-43; research asso. dept. meteorology Mass. Inst. Tech., 1943-47; asst. prof. dept. physics Ohio State U., Columbus, 1947-54; asso. research meteorologist U. Mich., Ann Arbor, 1954-56, faculty, 1956—, prof. meteorology, 1963—. Cons., Pres.'s Adv. Com. on Weather Control, 1954-56. Mem. Am. Meterol. Soc., Am. Assn. U. Profs., Am. Geophys. Union, A.A.A.S., Internat. Soc. Biometeorology. Research and numerous publs. on air transport allergenic materials, cleansing action rain on atmosphere; developed instrument to observe sizes raindrops. Home: 8040 W. Huron River Dr., Dexter, Mich. 48130. Office: E. Engring. Bldg., U. Mich., Ann Arbor, Mich. 48104.*

DINGLE, Herbert, English astronomer, physicist; b. London, Eng., Aug. 2, 1890; s. James Henry and Emily Jane (Gorddard) D.; B.Sc., Imperial Coll. Sci. and Tech., London, 1918, diploma, 1920. D.Sc., 1930; m. Alice Westacott, Dec. 7, 1918; 1 son. Demonstrator, lectr., reader, prof. natural philosophy Imperial Coll. Sci. and Tech., 1918-46; Lowell lectr. Boston, 1936; mem. eclipse expdns. Brit. Govt., 1927, 32, 40; prof. history, philosophy of sci. London U. Coll., 1946-55, prof. emeritus from 1955. Mem. Royal Astron. Soc. (hon. sec. 1929-32, v.p. 1938-39, 42-44, 48-50, 53-54, pres. 1951-53), Internat. Astron. Union, Inst. Coimbra, Royal Coll. Sci. (asso.), Internat. Union History Sci. (v.p. 1953-56), Brit. Soc. Hist. Sci. (pres. 1955-57). Author: Relativity for All, 1922; Modern Astrophysics, 1924; Science and Human Experience, 1931; Through Science to Philosophy, 1937; The Special Theory of Relativity, 1940; Mechanical Physics, 1941; Subatomic Physics, 1942; Science and Literary Criticism, 1949; Practical Applications of Spectrum Analysis, 1950; The Scientific Adventure, 1952; The Sources of Eddington's Philosophy, 1954; (with Viscount Samuel) A Threefold Cord, 1961; (with others) Splendour of the Heavens, 1923, Life and Work of Sir Norman Lockyer, 1929, The New World Order, 1932, The New Learning, 1933, Science Today, 1934. Research, publs. on astrophysics, relativity, spectrum analysis. Address: 104 Downs Court Rd., Purley, Surrey, Eng.

DINGLE, John H(olmes), Am. physician; b. Cooperstown, N.D., Nov. 24, 1908; s. John Geech and Harriet (Holmes) D.; Ph.C., B.S., U. of Wash., 1930, M.S., 1931; Sc.D., Johns Hopkins, 1933; M.D., Harvard (James Jackson Cabot fellow, 1936-39, Francis Weld Peabody fellow in medicine, 1940-42), 1939; married to Doris V. Brown, January 18, 1946; children—Eva M., David R. Assistant, McDermott Foundation University of Washington, 1929-31; assistant bacteriologist State Department Health, Md., 1933; bacteriologist The Upjohn Co., Kalamazoo, Mich., 1933-35; house officer in medicine Infants and Children's Hosp., Boston, 1939-40; asst. depts. of medicine, bacteriology and immunology Harvard, 1940-41, instr. bacteriology and immunology, 1940-42, instr. dept. medicine, 1941-42, asso. medicine, 1942-46, asst. physician Boston City Hosp., 1941-46; prof. preventive medicine Sch. Medicine, Western Res. U., 1946—, asso. prof. medicine, 1946-65, professor of medicine, 1965—, associate physician Univ. Hosps., Cleveland, since 1946. Consultant to Sec. of War on epidemic diseases, 1941-44; mem. Commn. on Acute Respiratory Diseases, Armed Forces Epidemiol. Bd., 1942-55, director, 1942-55, associate member, 1955—; mem. Armed Forces Epidemiol. Bd., 1951—, pres., 1955-57; mem. bd. cons. med. and pub. health Rockefeller Found., 1952-55; mem. Cleve. Health Council, 1946—. Served from maj. to lt. col. M.C., AUS, 1944-46. Decorated Legion of Merit; Albert Lasker award, 1959; James D. Bruce Meml. award. Mem. A.A.A.S., Am. Assn. of Immunologists (v.p. 1956, pres. 1957), Am. Soc. Microbiology, Association Teachers Preventive Medicine, American Federation Clinical Research, Am. Soc. Clin. Investigation, Soc. Exptl. Biol. and Medicine, Association American Physicians, Central Society for Clinical Research (vice pres. 1958, president 1959), American Epidemiology Soc., (pres. 1958), Am. Clinical and Climatol. Assn., Harvey Soc., Nat. Acad. Scis., Phi Beta Kappa, Sigma Xi, Alpha Omega Alpha. Club: Cosmos. Research on infectious diseases; immunology. Home: 2344 Roxboro Rd., Cleve Heights, O. 44106. Office: 2064 Abington Rd., Cleve. 44106.

DINGLER, Hermann, German botanist; b. Zweibrücken, Germany, May 23, 1846; s. Johann Gottfried and Elisabeth (Lindemann) D.; student medicine U. Zurich (Switzerland), U. Erlangen (Germany); U. Munich (Germany); M.D., 1870; student U. Vienna (Austria); state exam Munich, 1872, student botany, 1875; doctorate U. Leipzig (Germany), 1882; m. Marie Erlenmeyer, 1880; 1 son, Hugo; 2 daus. Made bot. expdns. to Palestine and Asia Minor; mil. doctor in Turkey; became lectr. Munich, 1883; went to Forestry Sch. Aschaffenburg, Germany where he was prof. botany until 1910, after retirement built up its natural sci. collection; made expdns. to Asia Minor, 1892, Ceylon, 1909, Sicily, 1912, Caucasus, 1914. Mem. Naturwissenschaftlicher Verein (chmn.), Local Com. for Preservation Nature (founder 1907). Author: Die Bewegung der pflanzlichen Flugorgane, 1889. Research on movement of plant flight organs, apical growth of gynosperms, 1882, flat shoots of phanerograms, 1885, constrn. of grape vine, 1885, cause of defoliation, 1902; forced elongation of bamboo shoot, 1896-97. Died Aschaffenburg, Dec. 30, 1935.

DINGLER, Hugo Albert Emil Hermann, German mathematician; b. Munich, Germany, July 7, 1881; s. Hermann D. and Marie (Erlenmeyer) D.; ed. univ. Erlangen, Göttingen, also Munich Tech. Inst.; m. Marie Stach von Goltzheim. Asst. for higher math. and descriptive geometry Tech. High Sch. Munich, 1907-12; mem. faculty U. Munich from 1912, extraordinary prof. from 1920; prof. philosophy, pedagody and psychology Tech. High Sch. Darmstadt, 1932-34, later prof. emeritus; founder, head Inst. für Methodozogische Forschung. Author: Die Grundlagen der Physik, 1919; Die Philosophie Ernst Machs, 1923; Der Zusammenbruch der Wissenschaft und der Primat der Philosophie, 1926; Metaphysik als Wissenschaft von Leyzten, 1929; Geschichte der Naturphilosophie, 1932; Die Methode der Physik, 1938; Grundriss der methodischen Philosophie, Die Lösungen der philosophischen Hauptprobleme, 1949; Das physikalische Weltbild, 1952; Über die Geschichte und das Wesen des Experimentes, 1952. Investigations in math., philosophy exact science, philosophy of math. Died 1954.

DINGMAN, Joseph Francis, Am. physician; b. N.Y.C., Aug. 20, 1921; s. Dominic and Frances (Augenti) S.; M.D., U. Buffalo, 1950; m. Caroline G. McKenzie, Feb. 10, 1945; children—Joseph Francis, Kathleen J., Kenneth J., Brian M., Gary R. Asst., Harvard Med. Sch., 1951-55, instr., 1955-56; faculty Tulane Med. Sch., 1956-61, asso. prof., dir. endocrinology, 1959-61; dir. med. research Lahey Clinic Found., Boston, 1961-66; staff Peter Bent Brigham Hosp., 1961—, sr. asso. medicine, 1965—; lectr. medicine Harvard Med. Sch., 1961—. Recipient Bacelli Research award U. Buffalo, 1950. Fellow A.C.P.; mem. Endocrine Soc., Am. Physiol. Soc., Am. Diabetes Assn., Am. Fedn. Clin. Research, A.A.A.S., N.Y. Acad. Scis. Author: (with G. W. Thorn) Diseases of Neurohypophysis, Principles of Internal Medicine, 1954, 4th edit., 1966. Research and numerous publs. on hormone secretion in pituitary disorders, methods for measuring posterior pituitary hormones in blood, devel. glass paper chromatography for steroid and peptide hormones, brain regulation of pituitary hormone secretion, hormonal regulation excretion of water and salt by kidneys, methods of measuring protein hormones with radioactive labeling. Home: 123 Benvenue St., Wellesley, Mass. 02181. Office: Peter Bent Brigham Hosp., Boston, 02115.*

DINGWALL-FORDYCE, Alexander, Brit. physician; b. Mar. 8, 1875; s. James and Penelope Gordon (Miller) D.-F.; ed. Edinburgh U.; M.D.; m. Bessie Ianthe Warren; 1 son, 1 dau. Physician, lectr. Royal Edinburgh Hosp. for Sick Children; lectr. on diseases of children U. Liverpool (Eng.); physician Royal Liverpool Children's Hosp. Author: Diseases of Children, 1916, 2d edit., 1921; many other publs. on children's diseases. Died Jan. 7, 1946.

DINI, Ulisse, Italian mathematician; b. Pisa, Italy, Nov. 14, 1845; s. Pietro and Teresa (Marchionneschi) D.; baccalaureate in sci. U. Pisa; studied under G. Bertrand and C. Hermite at Paris (France), 1864-65; joined faculty in advanced algebra and theoretical geometry U. Pisa, 1866, named prof., 1867, received chair as prof. math. and physics, 1871, prof. analysis and infinities until 1918; rector of Atheneum of Pisa, 1888-90; dir. Normal High Sch. Pisa, 1908-18; dir. Sch. Applied Engring. Pisa; mem. Council Higher Edn. Italy, 1893-1917; mem. Govt. of Commerce of Pisa, 1871-73, 74-87, 89-95; elected to Nat. Parliament from Pisa, 1882, 86, 90; became mem. House of Depts., 1881, served on many legislative coms. A square in Pisa is named in his honor. Author: Fondamenti per la teorica della funzioni di variabili reali, 1878; Analisi infinitestimali, 1878; also articles. Research on infinitesimal geometry and analytical math.; derived and proved formulae of differential equations. Died Oct. 28, 1918.

DINITZ, Simon, Am. sociologist; b. N.Y.C., Oct. 29, 1926; s. Morris and Dinah (Schulman) D.; student Coll. City N.Y., 1943-44; B.A., Vanderbilt U., 1947; M.A., U. Wis., 1949, Ph.D. (K. K. Knapp fellow), 1951; m. Mildred Harriet Stern, Aug. 20, 1949; children—Jeffrey H., Thea I., Risa M. Faculty, Ohio State U., Columbus, 1951—, research asso. psychiatry, 1956—, prof. sociology, 1963—. Mem. Am. Sociol. Assn., Am. Psychopath. Assn., Am. Soc. Criminology (editor Criminologica 1966—), Soc. Study Social Problems. Author: (with R. R. Dynes, A. C. Clarke, I. Ishino) Social Problems: Dissensus and Deviation in an Industrial Society, 1964; (with B. Pasamanick and F. Scarpitti) Schizophrenics in the Community, 1967 (Hofheimer prize Am. Psychiat. Assn. 1967). Research and publs. in causes of delinquency, prediction and prevention; studies of mental hosp. as social system, outcome and prevention of hospitalization in mental hosp. Home: 298 N. Cassady St., Columbus, O. 43209.*

DINMAN, Bertram David, Am. physician; b. Phila., Aug. 9, 1925; s. Myer and M. (Kaufman) D.; M.D., Temple U., 1951; Sc.D., U. Cin., 1957; m. Gabrielle Stamm, June 11, 1950; children—Stefanie, Jonathan D., Emily, Joshua. Faculty, Ohio State U. Coll. Medicine, 1957-65, prof., 1962-65; prof. indsl. health U. Mich. Sch. Pub. Health, Ann Arbor, 1965—, research asso. Inst. Indsl. Health, 1965—. Mem. commn. on environmental hygiene Armed Forces Epidemiology Bd., 1965—; cons. to surgeon Air Force Logistics Command, 1964—; mem. adv. com. toxicology NAS-NRC-NAE, 1966—. Fellow Am. Acad. Occupational Medicine (sec. 1966—); mem. Am. Coll. Preventive Medicine (v.p. 1965-66), Am. Indsl. Hygiene Assn. (dir. 1965—), Ramazzini Soc., Soc. Toxicology, Sigma Xi. Research and publs. on environmental agts. effects on human health and productivity especially effects of toxic chems. on biol. systems. Home: 1132 Aberdeen Dr., Ann Arbor, Mich. 48104.*

DINNING, James Smith, Am. biochemist; b. Franklin, Ky., Sept. 28, 1922; s. James Starks and Fanny (Smith) D.; student U. Okla., 1943-44, U. Tenn., 1944-46; B.S., U. Ky., 1946; M.S., Okla. State U., 1947, Ph.D., 1948; m. Sally Sue Hensley, Oct. 28, 1944; children—Kay (Mrs. Lenord Jordon), James M., Robin Joann, Randal S. Asst. prof. U. Ark., 1948-52, asso. prof., 1953-59, asst. dean Sch. Medicine, 1957-59, prof. biochemistry, 1959-63, asst. dean Grad. Sch., 1959-63; asst. prof. U. Pitts., 1952-53; staff mem. Rockefeller Found., N.Y.C., 1963-65, asst. dir. med. and natural scis., 1965—. Spl. cons. NIH, 1957-61. Recipient Lederle Med. Faculty award, 1955. Mem. Soc. Exptl. Biology and Medicine, Am. Inst. Nutrition (Mead-Johnson award 1964), Am. Soc. Biol. Chemists, Sigma Xi. Asso. editor Nutrition Revs., 1953-58; editorial bd. Jour. Nutrition, 1955-59. Research, numerous publs. on mechanism of action of vitamins folic acid B12, C, B6; E; prodn. of vitamin E deficiency in monkey and demonstration of accompanying anemia, demonstration that anemia of children with protein-calorie malnutrition would respond to Vitamin E. Home: G.P.O. Box 2453, Bangkok, Thailand. Office: 111 W. 50th St., N.Y.C. 10020.*

DINOSTRATOS (or DEINOSTRATOS), Greek mathematician; flourished 4th century; bro. of Menaechmos; pupil of Plato; collaborated on Theudios' textbook of geometry; contbd. (with bro.) to devel. of geometry; applied Hippias' quadratix to squaring of the circle (quadratix of Dinostratus); found area of circle with circumference and radius known; mentioned by Pappus in his large compendium; work on area of circle expanded later by Archimedes.

DINSMOOR, James Arthur, Am. psychologist; b. Woburn, Mass., Oct. 4, 1921; s. Daniel Stark and Jean (Masson) D.; A.B., Dartmouth, 1943; A.M., Columbia U., 1945, Ph.D., 1949; m. Marise Kay Sawyer, Jan. 1, 1956; children—Daniel Stark, II, Mara Jean, Robert Scott. Instr., Newark Colls., Rutgers U., 1945-46; lectr. Columbia U., 1946-51; faculty Ind. U., Bloomington, 1951—, prof. psychology, 1963—. Mem. Soc. for Exptl. Analysis Behavior (sec.-treas. 1964—, editorial bd. Jour. 1961-67), Am. Psychol. Association (member council representatives 1967—), Behavioral Pharmacology Soc., Psychonomic Soc. Editorial bd. Psychol. Reports, 1955-61. Research and publs. on theoretical analysis suppression of punished behavior by incompatible responses avoiding punishment, empirical studies of how originally ineffective stimuli become rewarding and can be used to strengthen behavior, control of behavior by stimuli that indicate future reward or punishment of this behavior. Home: 1511 Maxwell Lane, Bloomington, Ind. 47401.*

DINTENFASS, Leopold, colloid chemist; biorheologist; b. Tarnow, Poland, Apr. 29, 1921; s. Isser and Anna (Katzner) D.; Dipl. Ing. in Chemistry, Poly. Lvov (Poland), 1946; M.Sc., U. New S. Wales, 1958, Ph.D., 1961; m. Irene Kurzer, Sept. 26, 1959. Research chemist BALM Paints div. ICIANZ, Cabarita, Australia, 1950-56; sr. research chemist Taubmans Pty., Ltd., St. Peters, Australia, 1956-59; doctoral fellow U. New S. Wales, 1959-61; research biorheologist Sydney (Australia) Hosp., 1961-62, hon. cons. biorheologist, 1963—; hon. cons. rheologist children's Med. Research Found., Crown St. Women's Hosp., Royal Alexandra Hosp. for children, 1967—; sr. research officer Nat. Health and Med. Research Council dept. medicine U. Sydney, 1962—, sr. research fellow Nat. Heart Found. Australia, 1964—. Fellow Royal Australian Chem. Inst., Chem. Soc. (London), Internat. Coll. Angiology; mem. Am. Fedn. for Clin. Research, Am. Geriatrics Soc., Brit. Soc. Rheology (pres. 1961-63), Hematology Soc. Australia, Instn. Engrs. Australia (asso.). Research, publs. on rheology of suspension and polymeric solutions, rheology and coagulation of blood in normal and path. conditions, devel. of viscometers. Home: 74 Gilgandra Rd., North Bondi, N.S.W., Australia. Office: Dept. Medicine, U. Sydney, Sydney, Australia.*

DINTER, Zvonimir, virologist; b. Osijek, Yugoslavia, Oct. 24, 1914; s. Emil and Liana (Niderlender) D.; student Faculty Vet. Medicine, Zagreb, Yugoslavia, 1933-38; Vet.Med.Dr., Royal High Sch. Vet. Medicine, Stockholm, Sweden, 1958; m. Elsie Maria Bucht, Aug. 22, 1955; children—Peter, Anna, Lena, Richard. Asst. prof. virology Faculty Vet. Medicine, Zagreb, 1943-49; asst. dept. bacteriology Nat. Vet. Inst., Stockholm, 1949-53, head, 1953-66; prof. virology Royal High Sch. Vet. Medicine, Stockholm, 1966—; staff Inst. Virology, U. Uppsala (Sweden), 1955-66. Cons. Royal Vet. Bd. Sweden, 1965—. Decorated Knight of Royal No. Star. Mem. Swedish Microbiol. Soc. (mem. bd.). Research, numerous publs. on classification of avian and bovine viruses, relation of viruses to animal diseases, behavior of foot-and-mouth disease virus in cell cultures; detected new influenza A virus subtype in chickens. Home: 25 Orrvagen Nasbypark, Sweden. Office: Inst. Virology, U. Uppsala, Sweden.*

D'INVILLIERS, Edward Vincent, Am. geologist; b. Germantown, Pa., Aug. 2, 1857; s. Camille S. and Ann S. (Maitland) d'I.; B.S., U. Pa., 1878, D.Sc., 1913; spl. studies in geology and mining engring.; m. Ann Maitland, June 6, 1894. Asst. geologist, 2d Geol. Survey of Pa., 1875-85; geologist and cons. engr. 1885-1919. Fellow Geol. Soc. Am. Died 1928.

DIOCLES OF CARYSTOS, Greek physician; flourished 300 B.C.; s. Archidamus; mem. sect of dogmatics; practiced in Athens; ranked by Pliny next to Hippocrates; 1st physician to write in Attic; influenced by Sicilian sch., also by Empedocles, Hippocrates, Aristotle, Polybus, but was independent in details of sci. research; adherent of Empedocles' 4 humors theory, idea of pneuma, concept of central role of heart, and of Hippocrates' total organism view of body; gave 1st recorded description of early human fetus; believed that both sexes contribute to formation of embryo; described cotyledonous placenta; saw fever as symptom of other disorders; author at least 16 books, including On Anatomy, others on physiology, symptomatology, prognostics, aetiology, dietetics, botany, animal anatomy.

DIOGENES, Greek natural philosopher, physician; b. Apollonia (probably Phrygian, not Cretan, Apollonia); flourished 435 B.C.; s. Apollothemis; contemporary of Anaxagoras; eclectic philosopher in Ionian tradition; also influenced by Anaxagoras and Leucippus; tchr. philosophy, Athens. Author: On Nature (quoted extensively by Simplicius); book on heart (best anat. work of Hippocratic corpus); other works no longer extant include Against the Sophists, Meteorology, The Nature of Man. Revived Anaximenes' teaching that primary substance is air; believed that air was principle of soul and intelligence in living creatures and was changed into substance by infinite transformations of rarefaction and condensation; agreed with Anaxagoras that Nous (a principle of intelligence) governed and diffused air; held pneumatic theory of medicine; introduced physiological theories of generation and respiration; knew about pulse; noted blood vessels pass to left ventricle, also described vena cava and its branches (his account of veins was preserved by Aristotle); interested in sensation; held novel but incorrect views on vision; his cosmology included a flat round earth, but still resembled that of Anaxagoras; described heavenly bodies as pumice stones filled with fire; although Diogenes revived older teachings, he had sci. interest in details characteristic of Hippocratic writers of his age.

DIOGENES LAERTIUS, historian; b. probably Laerte, Cilicia, early 3d century; probably lived during reigns of Severus or Caracalla (211-235); settled in Athens. Author: Lives, Doctrines, and Maxims of Famous Philosophers (history of Greek schs. of philosophy to that time, contains much biog. information not found elsewhere), 10 vols.

DIOLES, mathematician, physicist; flourished circa 2d century B.C.; invented cissoid curve and applied it to duplication of cube; wrote on burning-glasses (possible invented parabolic burning-glass); used conics to solve Archimedes' problem of division of sphere into 2 spherical segments with volumes of given ratio.

DION, Roger Pierre, French geographer; b. Argenton-sur-Creuse, France, Oct. 28, 1896; s. Albert and Apolline (Geoffrey) D.; ed. Ecole normale supérieure; Ph.D. in Geography; m. Marie Thévenin; children—Pierre, Francoise (Mme. Jacquin), Denis, Remi, Anne-Marie; m. 2d, Hermite, 1952. Sec., Ecole normale supérieure, 1925; became asst. in geography Faculté des lettres, Paris, 1927, prof., 1945; named prof. Faculté des lettres, Lille, France, 1934; apptd. prof. geography U. Sao Paulo (Brazil), 1947; became prof. College de France, 1948. Recipient prize French Acad., 1960. Mem. Assn. French Geographers, Nat. Com. Geography (pres. commn. hist. geography). Author: Le Val de Loire, 1934; Essai sur la formation du paysage rural francais, 1934; Le frontières de la France, 1947; Le site et la croissance de Paris, 1951; Les routes de l'etain, 1952; Histoire de la vigne et du vin en France, des origines au XIXe siecle, 1959. Home: 10, rue Benouville, Paris 16. Office: Place Marcelin-Berthelot, Paris 5, France.

DIONIS, Pierre, French surgeon; b. Paris, France, 1643; studied surgery at Confraternity of S. Côme. 1st surgeon to Queen Maria Theresa; apptd. prof. anatomy, surgery, dissection Jardin des plantes, 1672. Author: l'Antomie de l'homme suivant la circulation du sang et les dernières découvertes, 1690; Dissertation sur la mort subite, 1709; Traité général des accouchements, 1718. Died 1718.

DIONIS DU SÉJOUR, Achille Pierre, French astronomer, mathematician; b. Paris, France, Jan. 11, 1734; studied math. at Lycée de Louis-le-Grand, Paris; student of Isaac Newton. Mem. French Acad. Scis., 1765. Author: Essai sur les comètes en général . . . , 1775; Essai sur les phénomènes relatifs aux disparitions périodiques de l'anneau de Saturne, 1776; Traité analytique des mouvements apparents des corps célestes, 1787-98. Research on eclipses, comets, Saturn's ring; analytical study of higher degree curves. Died Angerville, France, Aug. 22, 1794.

DIONYSODOROS OF AMISOS, Greek mathematician; b. probably Amisos, Pontos; flourished 2d century B.C.; preceded Diocles; author book on tores, (quoted by Heron); credited by Vitruvius with invention of cylindric form of sundial (unknown whether concave or convex); applied intersection of hyperbola and parabola to solve problem of cutting a sphere so that its segments are in a given ratio (found in Archimedes' sphere and cylinder).

DIOPHANTOS, mathematician; flourished 250 (possibly as early as 75); m. at age 34; 1 son; lived in Alexandria, Egypt. Author: Arithmetica, 13 books (6 extant); De multiangulis numeris; Porismata. First Greek to write important algebraic treatises; treated number arithmetically (instead of geometrically); introduced symbols for minus, also for unknown quantities and for exponents; solved in rational numbers indeterminate equations of 2d, 3d, 4th, 6th degrees; concentrated on indeterminate or semi-determinate equations, both single and simultaneous, of 2d degree; sought solutions in positive numbers; failed to recognize negative numbers or negative roots of equations standing alone; developed no gen. method for solutions of equations, solved each math. equation separately; eponym for Diophantine equations (indeterminate equations with rational coefficients which require a rational solution). Died at age 84.

DIOSCORIDES, Pedanius, Greek physician; b. Anazarbus, nr. Adana (now Turkey), flourished 40-68 A.D., mil. surgeon in Nero's army. Author of De materia medica, 5 books (1st systematic pharmacopeia describing about 600 plants and almost 1,000 drugs). One of earliest botanists; studied use of plants as drug source; use of mandrake to put patients to sleep before operation (1st mention of anesthesia); books translated into Arabic; highly regarded by Moslems; until dawn of modern medicine in 17th century, Dioscorides highly esteemed in medicine and botany.

DI PALMA, Joseph Rupert, Am. pharmacologist, educator; b. N.Y.C., Mar. 21, 1916; s. Frank and Anna (Attanasio) Di P.; B.S., Columbia, 1936; M.D., State U. N.Y., Bklyn., 1941; m. Mary Solowey, June 26, 1948; children—Maria, Dorothea, Joan, Yvonne, Mary-Jo. Asst. prof. medicine State U. N.Y. Downstate Med. Center, Bklyn., 1944-50; prof., chmn. dept. pharmacology Hahnemann Med. Coll., Phila., 1951—. Recipient Alumni medallion for distinguished service to medicine State U. N.Y. Downstate Med. Center, 1965. Mem. Am. Soc. for Pharmacology and Exptl. Therapeutics, Am. Soc. for Clin. Investigation, A.M.A., Am. Coll. Clin. Pharmacology, Pa. Med. Soc. Editor: Drill's Pharmacology in Medicine, 3d edit., 1965. Research and numerous publs. on physiology and pharmacology of human skin reactions, pharmacology of drugs for heart irregularities, antidotes for war gases, plant pigments and physiology of carnivorous plants. Home: 100 Pembroke Av., Wayne, Pa. 19087. Office: 235 N. 15th St., Phila. 19102.*

DI PIEDIMONTE, Francesco, (or Francescus de Pedemontium), Italian physician; b. Piedimonte, nr. Alife, 2d half 13th century; ed. Salerno; prof. U. Naples; physician at courts Robert of Naples and his son the Duke of Calabria. Author: Complementum (one of most complete medieval compendia of practical medicine, most thorough medieval account of gynecology and midwifery, scholastic in tone). Died 1319.

DIPPEL, Johann Conrad, alchemist, theologian; b. Schloss Frankenstein, Germany, Aug. 10, 1673; s. Johann Philip and Anna Eleonora (Münchmeyer) D.; student theology Giessen, Germany; M.D., Leiden, Netherlands, 1711. Settled in Giessen; practiced alchemy; began practice medicine, Berlin, 1704, later Amsterdam, Netherlands; went to Altona (now annexed Hamburg), 1714; went to Christianstadt, Sweden; personal physician to King Friedrich I; lived in exile with other radical pietists, Berleburg, Prussia, 1729-34; temporary positions as chemist, Liebenburg, Prussia. Author theol. books. Discovered an oil (oleum animale Dippeli) distilled from animal bones and excrement which was used for medicinal purposes (later pyridine was discovered in it); pigment, Prussian Blue, was discovered by G. E. Stahl in an alkali which he used for oil purification. Claimed he would live until 1808. Died Schloss Wittgenstein nr. Berleburg, Apr. 25, 1734.

DIPPEL, Leopold, German botanist; b. Lauterecken, Germany, Aug. 4, 1827; s. Carl and Susanna (Purpus) D.; student forestry Aschaffenburg, Munich, Karlsruhe (all Germany), 1845-48; student physiol. Inst. Jena (Germany), until 1850; hon. dr. Bonn (Germany) U., 1865; m. Sophia Fries, 1851; 1 son. Became tchr., 1850; became tchr. at Realschule, Idar, 1856; dir. bot. garden Darmstadt (Germany), also prof. botany Technische Hochschule, 1869-96. Recipient Prix Bourdin, Paris Acad., 1863, various prizes Niederländische Gesellschaft für experimentelle Naturwissenschaft, 1864-65. Mem. Royal Micros. Soc. London (hon.), Deutsch Dendrologische Gesellschaft (a founder; v.p. 1892-96). Author: Das Mikroskop und seine Anwendung, 1867-72; Handbuch der Laubholzkunde, 1889-93. Research on microscopy, plant histology, intracellular substance, 1867, fine structure of cell sheath, 1878, the diatoms of Rhein-Main plane, 1905. Died Darmstadt, Mar. 4, 1914.

DIRAC, Paul Adrien Maurice, English physicist; b. Bristol, Eng., Aug. 8, 1902; s. Charles Adrien Ladislas and Florence Hannah (Holten) D., B.Sc., Bristol U., 1921; Ph.D., Cambridge U., 1926; m. Margit Wigner. Fellow St. John's Coll., Cambridge, from 1927. Lucasian prof. math. Cambridge U., Eng., 1932—; vis. lectr., U. Wis., 1929, U. of Mich., summer, 1929. Princeton U., 1931; mem. Inst. for Advanced Study, Princeton, N.J., 1934-35, Sept.-Dec., 1946, 1947-48, 58-59. Fellow Royal Soc., 1930 (Royal medal 1939, Copley medal 1952). Awarded Nobel Prize for Physics (with E. Schrodinger), 1933. Author: Principles of Quantum Mechanics, 1930. Formulated math theory to describe relativistic electron, 1928; his subsequent theory of negative-energy holes predicted the existence of positron; his theories marked beginning of investigation of antimatter. Home: 7 Cavendish Av. Office: St. John's College, Cambridge, England.

DIRICHLET, Peter Gustav Lejeune, mathematician; b. Düren, Germany, Feb. 13, 1805; s. Arnold L. and Anna Elisabeth (Linteren) D.; ed. Cologne, Göttingen (both Germany), Paris (France) U.; m. Rebecca Mendelsohn-Bartholdy, 1832; 3 sons, 1 dau. Became docent Breslau (now Wroclaw, Poland) U., 1828; tchr. at mil., then constrn. sch., Berlin, Germany; became asso. prof. U. Berlin, 1831, prof., 1839; succeeded Gauss at U. Göttingen, 1855. Fellow Royal Soc., 1855; mem. Berlin Acad., French Acad. Scis. Author: Vorlesungen über Zahlentheorie, 1839; Bestimmte Integrale, 1871; research on complex numbers pub. in Berichte der Berliner Akademie (edited by Dedekind 1863), 1841, 42, 46; memoirs on series and hydrodynamics in Crelle's Jour.; also writings on theory of potential, 5th degree equations, definite integrals, Gauss' work; collected works edited by Kronecker, 1889-97. First lectr. on theory of numbers in Germany; work on potentials influenced theory of mechanics; worked on binary, prime numbers, infinite series, application of analytical functions, definite integrals; advanced study of number theory by application of higher analysis, proved Fermat's theorem for the case $n=5$. Died Göttingen, May 5, 1859.

DIRNAGL, Karl, German biologist; b. Munich, Germany, Oct. 31, 1917; s. Karl and Fransziska (Amtmann) D.; ed. Technische Hochschule, Munich, U. Munich; m. Anne-Marie Mittermayr; children—Stefanie, Johanna, Ulrich. With Med.-Bioclimatic Inst., Riederau, Balneologic Inst., div. medicine and climatology U. Munich. Mem. Internat. Soc. Biometeorology, Fachausschuss für Aerosolfragen der Fraunhofer Gesellschaft. Author: Aerosoltherapie, 1957. Research and numerous publs. on aerosols, meteorobiology, balneology. Home: Hackenstrasse 4, Munich 2. Office: Ziemssenstrasse 1, Munich 15, Germany.

DI SANT'AGNESE, Paul Artom, physician; b. Rome, Italy, Apr. 23, 1914; s. Valerio Artom and Rosita (Sinigaglia) di S'A.; M.D., U. Rome, 1939; D.Med. Sci., Columbia, 1948; M.D. (hon.), Justus Liebig U., 1962; m. Elizabeth Boryzewski, Feb. 14, 1943; children—Paul Anthony, Valerie Ann. Came to U. S., 1939, naturalized, 1945. Faculty dept. pediatrics Columbia, 1944-59; chief cystic fibrosis and celiac clinic Presbyn. Hosp., N.Y.C., 1951-59; lectr. Johns Hopkins, 1960-63; chief pediatric metabolism br. Nat. Inst. Arthritis and Metabolic Disease, NIH, Bethesda, Md., 1960—; clin. prof. pediatrics Georgetown U. Med. Sch., 1960—; dir. cystic fibrosis care, research and teaching center Children's Hosp. of D.C., 1960—; acad. staff Children's Hosp. of D.C., 1960—. Chmn. med. council Nat. Cystic Fibrosis Research Found., 1964—; chmn. med.-sci.-adv. council Internat. Cystic Fibrosis (Mucoviscidosis) Assn., 1965—. Diplomate Am. Bd. Pediatrics. Mem. Am. Pediatric Soc., Soc. for Pediatric Research, Am. Acad. Pediatrics, Harvey Soc., A.M.A., Am. Inst. Nutrition, N.Y. Acad. Medicine, N.Y. Acad. Scis., N.Y. County Med. Soc., Med. Soc. of D.C., Am. Pub. Health Assn. Editor: Pathogenesis of Cystic Fibrosis of the Pancreas, 1966. Research and numerous publs. on immunization in new-born infants, glycogen storage disease, malabsorption, cystic fibrosis of pancreas, developer sweat test. Home: 5310 Falmouth Rd., Washington 20016. Office: NIH, Bethesda, Md. 20014.*

DISERENS, Paul, Am. engr.; b. Cin., Jan. 9, 1882; s. Albert Day and Alice (Jefferies) D.; B.S., Purdue U., 1904, M.E., 1906; postgrad. U. Ill. 1906-08. Research asst. with Dr. W. F. M. Goss., 1904-08; in charge locomotive tests in study of superheated steam locomotive service Carnegie Inst of Washington, 1905-1906; research asst., studying Ill. coal U. Ill., 1907-09; engr. of test in charge research. Laidlaw Dunn Gordon Co., Cin., 1909-19; asst. chief engr. Worthington Pump and Machine Corp., N.Y.C., 1919-28, chief cons. engr., 1928—, dir. research, 1944-45, dir. research and devel. for corp. and subsidiaries, Worthington-Gamon Meter Co., Ransome Machinery Co., Electric Machinery Co. 1945-54; tech. advisor Compressed Air and Gas Inst. 1954—, Cons. Nat. Def. Research Com., 1941-44. Fellow Am. Soc. M.E. (mem. power test codes com.); mem. Am. Soc. Refrigerating Engrs. U. S. Nat. Com. Internat. Electrotech. Com. (dir. secretariat, chmn. tech. com. on internal combustion engines). Contbr. to jours. Inventor expander engines for refrigeration in gasoline industry, valves for compressor, hot oil pumps for oil refineries; holder U. S. and fgn. patents. Died Oct. 6, 1958.

DISERTORI, Beppino Giuseppe, Italian physician; b. Trento, June 19, 1907; s. Marcello and Maria (Nicolini) D.; M.D.; m. Elena Rosita Banfi; children—Donatella, Marcello. Prof. social psychiatry and psychopathology U. Trento. Mem. Italian Soc. Psychiatry, Neurology, Philosophy and Parapsychology, Italian Assn. Metaphysics, Italian Assn. Med. Writers, Agiati Rovereto Acad. Author 9 books, numerous articles. Home: via Grazioli 19. Office: via Petrarca 32, Trento, Italy.

DISLERE, Paul, French engr.; b. Douai, France, 1840; ed. Paris (France) Ecole polytechnique, 1859-61; sec. French Maritime Council, 1871-82, under sec. in colonial affairs, from 1882. Author: Les croiseurs, la guerre de course, 1874; reorganized French arsenal at Saigon, Indo-China; studied constrn. battleships. Died Paris, Apr. 17, 1928.

DISMAS, Friedrich Carl Joseph (Count von Stadion), chemist; b. Mentz, Germany, Sept. 1, 1774; canon of Bamberg, Germany; ambassador to Stockholm, London, St. Petersburg; minister fgn. affairs, 1806-09, reapptd., 1813; signed peace of Paris, 1814. Discovered perchloric acid; adopted Davy's theory of chlorine. Died Naples, Italy, 1821.

DISSELHORST, Rudolf, German physician, veterinarian; b. Rinteln/Weser, Germany, Jan. 4, 1854; s. Adam Heinrich and Louise (König) D.; m. Auguste Kuhlmann, 1888. Asst., Landwirtschaftliches Institut of U. Halle (Germany), 1881-86; apptd. prosector Anatomisches Institut, Berlin (Germany) Tierärztliche Hochschule, 1887; physician until 1898; named prosector, lectr. Anatomisches Inst., Tübingen, Germany; became prof. in Halle, dir. vet. clinic, 1898; emeritus in 1924. Mem. Leopoldina. Author: Vergleichende Anatomie und Physiologie der Haussäuger, 1923; Die Tierseuchen, 1920. Research on anatomy of sexual organs. Died Halle/Saale, Jan. 28, 1930.

DISTANT, William Lucas, English naturalist; b. Nov. 12, 1845; s. Alexander Distant; m. Edith Blanche de Rubien, 1874; 3 s., 3 ds. Dir., hon. sec. Anthrop. Inst., 1878-81; sec. Entomol. Soc., 1878-80, v.p., 1881, 1900; visited and made natural history collections in Transvaal and Malay Peninsula. Author: Rhopalocera Malayana; Monograph of the Oriental Cicadidae; A Naturalist in the Transvaal; Rhynchota (Vol. I of Godman and Salvin's Biologia Centrali Americana); Insecta Transvaaliensia; also many publs. on Coleoptera, Lepidoptera, Rhynchota, as well as some on anthropology. Died Feb. 4, 1922.

DITCHBURN, Robert William, English physicist; b. Waterloo, Lancashire, Eng., Jan. 14, 1903; s. William and Martha Kathleen (Hutchison) D.; B.Sc., Liverpool (Eng.) U., 1922; M.A., Ph.D. (Isaac Newton student) Trinity Coll., Cambridge (Eng.) U., 1928; m. Doreen May Barrett, Jan. 14, 1929; children—Barbara, Robert, Moira (Mrs. Stephen Holt), Clodagh. Fellow, Trinity Coll., Dublin, Ireland, 1928; prof. natural and exptl. philosophy U. Dublin, 1929-45; prof. physics Reading (Eng.) U., 1946-68. Fellow Royal Soc., 1962, Inst. Physics; mem. Royal Irish Acad. Author: Light, 1952. Research and numerous publs. on spectroscopy of vacuum ultra violet especially measurement probabilities photo-ionization processes, solid state physics especially physics of diamond, physiol. human visual process. Home: 14 Betchworth Av., Earley, Reading, Berks, Eng.*

DITMARS, Raymond Lee, Am. naturalist; b. Newark, N.J., June 20, 1876; grad. Barnard Mil. Acad., 1891; Litt.D. degree. Asst. curator entomology Am. Museum Natural History, 5 yrs.; court reporter N.Y. Times, 1898-99; entered N.Y. Zoöl. Park as curator of reptiles, 1899 and in charge dept. mammals from 1910. Fellow N.Y. Zool. Soc., Am. Inst.; corr. mem. Zoöl. Soc., London. Author: The Reptile Book, 1907; Reptiles of the World, 1909; Snakes of the World, 1931; Strange Animals I Have Known, 1931; Thrills of a Naturalist's Quest, 1932; Forest of Adventure, 1933; Confessions of a Scientist, 1934; The Book of Zoögraphy, 1934; The Book of Prehistoric Animals, 1935; The Book of Living Reptiles, 1936; The Making of a Scientist, 1937; The Book of Insect Oddities, 1938; The Fight to Live. Contbr. on entomology and herpetology. Erected studio at Scarsdale, N.Y., for prodn. of ednl. motion pictures, 1913. Pioneered antivenom research which led to devel. of serums for treatment snake-bite. Died May 12, 1942.

DITTE, Alfred, French chemist; b. Rennes, France, Oct. 20, 1843; studied at École normale supérieure; became prof. physics in Caen, France, 1873, apptd. prof. chemistry, 1879; named prof. in Sorbonne, Paris, France, 1888. Mem. French Acad. Scis. Author: la Constitution de la matière, 1876; Exposé de quelques propriétés générales des corps, 1881; l'Uranium et ses composées, 1884; Étude générale des sels, 2 vols., 1906; Research and numerous publs. on vanadium compounds, iodic acid, chem. equilibrium, crystalline substances, borates, uranates, apatities, wagnerites, vanadates, uranium and its compounds. Died Paris, Nov. 1908.

DITTEL, Leopold (Ritter) von, see von Dittel, Leopold (Ritter).

DITTMARSCH, Alfred Ludwig, German mining engr.; b. Dresden, Germany, Apr. 4, 1836; s. Albert Ludwig and Caroline Emilie (Ranft) D.; ed. Dresden Polytechnikum, Bergakademie, Freiburg, Germany; m. Marie Flocon, 1864; 1 son, 6 daus. Worked in France, 1859-70; went to Colo. as geol.-mining expert; dir. copper works in Vigsnä, Norway, 1872-75; became tech. dir. Lugau-Niederwürschnitzer Steinkohlenbauverein, 1877; apptd. dir. mining sch., Zwickau, Germany, 1881-1906; founder, curator Richter mineral collection in König-Albert-Mus., Zwickau. Mem. Leopoldina (elected sec. 1875). Author: Leitfaden der Bergbaukunde, 1894. Research on mining engring. Died Dresden, May 7, 1926.

DITTMER, Howard James, Am. botanist; b. Pekin, Ill., Jan. 29, 1910; s. William J. and Anna (Lehman) D.; A.B., U. N.M., 1933, M.A., 1934; Ph.D., State U. Ia., 1938; m. Lois Andersen, June 28, 1941; 1 dau., Deborah Ann. Faculty, Chgo. Tchrs. Coll., 1938-43; faculty U. N.M., Albuquerque, 1943—, prof. biology, 1953—, asst. dean Coll. Arts and Scis., 1955—. Mem. A.A.A.S. (pres. elect Southwestern and Rocky Mountain div. 1966—), Bot. Soc. Am., Sigma Xi, Phi Kappa Phi. Author: (with M. C. Coulter) The Story of the Plant Kingdom, 1959, rev., 1964; Phylogeny and Form in the Plant Kingdom, 1964. Research and numerous publs. on root systems of plants, particularly desert plants. Home: 600 Vassar St. N.E., Albuquerque 87106.*

DITTON, Humphry, English mathematician; b. Salisbury, Eng., May 29, 1675; educated by a clergyman, Dr. Olive; m. Miss Ball. Became master math. sch. at Christ's Hosp., 1706. Author: The General Laws of Nature and Motion, 1705; An Institution of Fluxions, containing the first Principles, Operations and Applications of that Admirable Method as Invented by Sir Isaac Newton, 1706; A Treatise of Perspective, Demonstrative and Practical, 1712; A Discourse Concerning the Resurrection of Jesus Christ, 1714; The New Law of Fluids, or a Discourse Concerning the Ascent of Liquids, in Exact Geometrical Figures Between Two Nearly Contiguous Surfaces, 1714; also other math. publs. First to try to explain capillarity mathematically; invented (with William Whiston) method of determining longitude. Died London, Eng., Oct. 15, 1715.

DITTRICH, Franz, physician; b. Germany, 1815; studied in Prague, Czechoslovakia, under Hyrtly, 1841, then in Vienna, Austria and Prague under Jaksch and Kiwisch; became prof., Prague, 1848, Erlangen, Germany, 1850. Author: Ueber den Laennecischen Lungeninfarkt, 1850. Described Dittrich's plugs, or masses of whitish or yellowish color in sputum of patients suffering from pulmonary abscess, bronchiectasis, bronchitis, or other lung diseases, 1850. Died 1859.

DITTRICH, (Georg Paul) Max, German chemist; b. Görlitz, Germany, Sept. 9, 1864; s. Oskar and Liddy (Schelle) D.; student natural scis. in Leipzig, Munich, Göttingen (all Germany); doctorate Heidelberg, 1890, later studied food chemistry; certification, 1895; m. Helen Schultz, 1902; 1 son, 2 daus. Asst. univ. chem. lab., Heidelberg, then at Badische Geologische Landes anstalt, Heidelberg, 1897; named prof., 1903; in charge A. Bernthsen's Lab. in Brunnengasse. Author: Anleitung zur Gesteinsanalyse, 1905; Chemisches Praktikum: Qualitative Analyse, 1906; Qualitative Analyse, 1908; Chemische Experimentierübungen für Studierende und Lehrer, 1911; also numerous articles. Quantitative separations using persulphates; analysis of spring waters and rocks around them; research on absorption phenomena in decomposed rock. Died Heidelberg, June 5, 1913.

DIU, Bernard, French physicist; b. Pau (France), Oct. 11, 1935; s. Paul and Marie (Caubisens) D.; Doctorat Spécialité, École Normale Supérieure, 1960; m. Marie-Ange Nadal, July 21, 1956; children—Isabelle, Anita. Faculty, Sci. Faculty, U. Paris, researcher Lab. Theoretical Physics, Orsay, France, 1961-64, 1966—; research asso. European Center Nuclear Research, 1964-66. Mem. French Soc. Physics. Author: Qu'est-ce qu'une particule élémentaire, 1965; also articles. Research on group theoretic classification of particles, on S-matrix theory. Home: 17 Adjudant Petit, Dammarie, Seine-et-Marne 77. Office: Lab. Theoretical Physics, Bat. 211, Faculty of Sciences, 91 Orsay, France.*

DIVDATI, Theodore, physician; B. Geneva, Switzerland, circa 1574; M.D., Leyden Netherlands, 1615; brought up in Eng.; became physician; attended Prince Henry and Princess Elizabeth; practised in parish of St. Bartholomew the Less; licentiate Coll. Physicians, 1617. Died 1651.

DIVERS, Edward, Brit. chemist; b. London, Eng., Nov. 27, 1837; ed. City of London Sch., Royal Coll. Chemistry, Queen's Coll., Galway, Ireland, 1860; M.D., D.Sc.; m. Margaret Theresa Fitzgerald, 1865; 2 daus. Lectr. med. jurisprudence Middlesex Hosp. Med. Sch., 1870; prof. chemistry Imperial Coll. Eng., Japan, 1873; prin. imperial Coll. Engring., Japan, 1882. Fellow Royal Soc., 1885; mem. Chem. Soc. (v.p. 1900-02), Brit. Assn. (pres. chem. sect. 1902), Inst. Chemistry (v.p. 1905), Soc. Chem. Industry (pres. 1905). Research and publs. on sulfur and nitrogen compounds, investigated ammonium carbonates; discovered hyponitrites. Died London, Apr. 8, 1912.

DI VIGO, Giovanni, Italian physician; b. Genoa, Italy, 1460; practice medicine Rome, Italy; physician to Pope Julius II. Author: Practica, 1514. Taught that wounds were poisoned burns and should be treated without a 1st dressing of boiling oil. Died 1520.

DIVISCH, Procopius, natural philosopher; b. Senftenberg, Bohemia, Aug. 1, 1696; ed. Premonstratensian cloister sch., Bruck, Styria; D.Theol., Salzburg, 1733. Ordained Premonstratensian monk, at Bruck Abbey, 1726 (changed name from Wenceslaus to Procopius); prof. philosophy abbey school soon after; abbot at Bruck Abbey; became priest, Prenditz, Moravia, 1736. Author: Tractatus de Dei unitate sub inscriptione Alpha et Omega, 1733; Theoretischer Tractat über die längst verlangte Theorie von der metrologischen Electricität, 1765. Studied hydraulics and electricity; carried out experiments at court in Vienna; applied electricity to medicine and chemistry; inventor of lightning rods, 1759; proposed wide use of lightning rods; invented organ played with hands and feet (Golden Denis). Died Prenditz, Moravia, Dec. 21, 1765.

DIWISH, Procope, see Divisch, Procopius.

DIX, Dorothea Lynde, Am. mental health reformer; b. Hampden, Me., Apr. 4, 1802; d. Joseph and Mary (Bigelow) Dix. Conducted a girl's sch., Boston, circa 1817-35, put primary emphasis on devel. of moral character, gave up sch. because of poor health; began Sunday sch. class East Cambridge (Mass.) House of Correction, 1841; visits to the jails made her aware of inhuman treatment of insane as criminals; made a thorough investigation of the treatment of the insane in Mass., 1841-43; demanded intelligent and humane treatment of insane by keepers of almshouses and jails, later realized that state-supported asylums with intelligent personnel were necessary; responsible for re-founding or enlarging of state mental hosps. in Mass., R.I., N.J., Pa., and Toronto, Ont., Can., 1841-45; convinced state legislatures to found state hosps. in Ky., Ill., Ind., Tenn., Mo., Miss., La., Ala., S.C., N.C., Md., 1845-52; a 12 million dollar bill for land to be set aside for taxation to support the care of insane vetoed by Pres. Pierce, 1854; travelled in Europe, 1854-57; due to her efforts on this trip hosps. were founded on Isle of Jersey and Rome, and Queen Victoria began a royal commn. to investigate condition of insane in Scotland; apptd. supt. Women Nurses, 1861; continued acivities on behalf of insane after Civil War. Died Trenton, N.J., July 17, 1887.

DIX, John Homer, Am. ophthalmologist; b. Boston, Sept. 30, 1811; s. John and Sarah Taffrey (Eddy) D.; grad. Harvard, 1833; M.D., Jefferson Med. Coll., 1836; m. Helen Perhan Curtis, June 9, 1859. Became mem. Mass. Med. Soc., 1837; practiced medicine, specializing in eye and ear diseases, Boston; became interested in ophthalmology, one of 1st to import ophthalmoscope developed by Helmholtz. Mem. Am. Ophthal. Soc. Author: Treatise on Strabismus, or Squinting, and the New Mode of Treatment, 1841; also wrote papers on rare diseases of eyes. Died Aug. 25, 1884.

DIXEY, Frederick Augustus, Brit. zoologist; b. London, Eng., Dec. 9, 1855; s. A. W. Dixey; M.A. Univ. Coll., London; M.D., Oxford U.; m. Isabel Atkins, 1892 (dec. 1916); 2 sons, 1 dau. Demonstrator physiology Univ. Coll., London, 1880-83; demonstrator physiology Oxford U., 1883-91, hon. fellow, lectr. Wadham Coll., former curator Hope collections. Fellow Royal Soc., 1910; mem. Assn. Brit. Zoologists (vice chmn.), Entomol. Soc. London (past pres.), Brit. Assn. (past pres. zool. sect.). Author: Necessity of Pain, 1888; Epidemic Influenza, 1892, also numerous articles. Died Jan. 16, 1935.

DIXON, Alfred Cardew, Brit. mathematician; b. 1865; s. G. T. Dixon; ed. Trinity Coll., Cambridge; M.A., London; Sc.D., Cambridge; D.Sc., Queen's U., Belfast, Ireland 1932; m. G. L. Smallpage, 1895. Fellow Trinity Coll., Cambridge; prof. math. Queen's Coll., Galway, Ireland, 1893-1901; prof. math. Queen's U., Belfast, 1901-30. Fellow Royal Soc., 1904; pres. London Math. Soc., 1931-33. Author: Elementary Properties of Elliptic Functions; also papers. Died May 4, 1936.

DIXON, Andrew Derart, anatomist; b. Belfast, No. Ireland, Oct. 27, 1925; s. Andrew and Martha (Stewart) D.; Licentiate in Dental Surgery, Queens U., Belfast, 1948, B.Dental Surgery, 1949, M.Dental Surgery, 1953, B.S. (Nuffield Found. dental fellow), 1954, D.Sc., 1965; Ph.D., U. Manchester, 1958; m. Mary Elizabeth Henderson, Oct. 14, 1948; children—Penelope Jane, Melinda Sara, Alison Mary. Asst. lectr. anatomy U. Manchester, 1954-56, lectr., 1956-62, sr. lectr., 1962-63; vis. asso. prof. anatomy U. Ia., 1959-61; prof. dental sci. U. N.C., Chapel Hill, 1963-65, prof. dental sci., prof. anatomy, 1965—, asst. dean, coordinator research Sch. Dentistry, 1966—. Fulbright Sr. Travel award, 1959-61, Commonwealth Fund Travel fellow, 1961. Mem. Anat. Soc. Gt. Britain and Ireland, Am. Assn. Anatomists, Brit. Dental Assn., Internat. Assn. Dental Research, Electron Microscopy Soc. Am., A.A.A.S., Am. Soc. Cell Biology, Internat. Soc. Craniofacial Biology, Sigma Xi, Psi Omega. Author: (with J. H. Scott) Anatomy for Students of Dentistry, 1966. Contbr. numerous articles to profl. jours. Studies on early devel. and growth of the jaws, sex chromatin in oral smears as a diagnostic tool, nerve supply to oral mucous membrane, facial tissues and temporomandibular joint, facial skeletal growth, trigeminal pathway, including trigeminal ganglion, using histological, histochem. and electron microscopy methods. Home: 1514 Cumberland Rd., Chapel Hill, N.C. 27514.*

DIXON, Andrew Francis, Brit. anatomist; s. George and Ro Rebecca (Yeates) D.; ed. Trinity Coll., Dublin, also Leipzig, Germany; M.B., Sc.D.; m. Margaret Kerr Johnston; 1 son, 1 dau. Prof. anatomy Univ. Coll., Cardiff, 1897-1903; Univ. prof. anatomy Trinity Coll., Dublin, from 1903, also dean faculty of physic. Pres. Royal Zool. Soc., Ireland, 1927-e 31, Anat. Soc. Gt. Britain and Ireland, 1934. Author: Manual Human Osteology; sect. in Cunningham's Text-book of Anatomy; also papers. Died Jan. 15, 1936.

DIXON, Arthur Lee, Brit. mathematician; b. Nov. 27, 1867; s. G. T. Dixon; ed. Worcester Coll., Oxford; M.A.; m. Hélène Rieder; 1 dau. Fellow, tutor Merton Coll., Oxford; Waynflete prof. pure math.; fellow Magdalen Coll., Oxford. Fellow Royal Soc., 1912. Contbr. to sci. publs. Died Feb. 20, 1955.

DIXON, Charles, English naturalist; b. London, Eng., July 20, 1858; s. Charles Thomas Dixon; ed. in London, 1889-91, Devonshire, 1891-1900; m. Mary Knight, 1886. Devoted life to studying natural history; asso. of Henry Seebohm, 5 years. Author numerous books, including: Rural Bird Life, 1880; Evolution without Natural Selection, 1885; Our Rarer Birds, 1888; Annals of Bird Life, 1890; The Migration of Birds, 1892 (Russian edit., 1895); The Game Birds and Wild Fowl of the British Islands, 1893, 1899; The Nests and Eggs of British Birds, 1893; The Story of the Birds, 1900; Birds' Nests: An Introduction to the Science of Caliology, 1902; The Migration of British Birds, 1904. Research on migration of birds, geog. distbn. of species. Died June 17, 1926.

DIXON, Frank James, Jr., Am. physician; b. St. Paul, Minn., Mar. 9, 1920; s. Frank James and Rose (Kuhfeld) D.; M.D., U. Minn., 1943; m. Marion Edwards, Mar. 14, 1946; children—Janet, Frank, Michael. Research asst. in pathology Harvard Med. Sch., Boston, 1946-48; instr. pathology Washington U., St. Louis, 1948-50, asst. prof., 1950-51; prof., chmn. dept. pathology U. Pitts. Sch. Medicine, 1951-61; head div. exptl. pathology Scripps Clinic and Research Found., La Jolla, Cal., 1961—; vis. prof. Duke, 1961; prof. in residence, dept. biology U. Cal. at San Diego, 1965. Cons. dept. biophysics and nuclear medicine U. Cal. at Los Angeles, 1962; cons. Francis I. Proctor Found. for Research in Ophthalmology U. Cal. San Francisco Med. Center, 1965—. Recipient award for distinguished achievement Modern Medicine, 1961. Diplomate Am. Bd. Pathology. Fellow Am. Acad. Allergists; mem. A.A.A.S. (Theobald Smith award 1952), Am. Soc. for Exptl. Pathology (Parke-Davis award 1957, pres. 1966), Am. Assn. Pathologists and Bacteriologists, Am. Assn. Immunologists, Am. Assn. Cancer Research, N.Y. Acad. Sci., Soc. for Exptl. Biology and Medicine, Internat. Acad. Pathology, Interurban Pathology Soc., Am. Soc. Clin. Investigation, Am. Coll. Allergists, Western Assn. Physicians, Western Soc. for Clin. Research, Harvey Soc. (hon.), Sigma Xi. Contbr. numerous articles to profl. jours. Research in immuno-pathology, tissue injury induced by antigen-antibody reactions. Home: 2355 Avenida de La Playa. Office: 476 Prospect St., La Jolla, Cal. 92037.*

DIXON, Harold Baily, English chemist; b. London, Eng., Aug. 11, 1852; William Hepworth Dixon; ed. Christ Ch., Oxford (Eng.) U., 1st class honors in natural sci., 1875; D.Sc., Ph.D., M.A.; m. Olive Beechey Hopkins, 1885; 1 son, 1 dau; m. 2d, Muriel Kinch, 1918; 1 dau. Millard lectr. Trinity Coll. Oxford U., 1879-86; Bedford lectr. Balliol Coll., 1881-86; prof., Owens Coll., 1891-1922; mem. Royal Commn. Explosions Coal Dust in Mines, 1891-94, Royal Commn. Coal Supplies, 1902-05, home office Exec. Com. on Explosions in Mines, 1911-14, Alcohol and Fuel Com., 1918; dep. inspector High Explosives Manchester Area, from 1915; chrmn. Royal Tech. Coll., Salford, from 1916, Salford Higher Edn. Com., from 1919, Selective Com. Northwest Dist. Ministry Labor, 1922; supr. research ignition gases Safety in Mines Research Bd., 1927. Fellow Royal Soc., 1886; pres. Manchester Lit. and Philos. Soc., 1907-09, 23-25, Chem. Soc., 1909-11. Author: Conditions of Chemical Change in Gases, 1884; Rate of Explosion in Gases, 1893; Movements of Flame in Explosions of Gases, 1902; Atomic Weight of Chlorine, 1905; The Ignition-Temperatures of Gases, 1909; The Firing of Gases by Compression, 1914; The Velocity of Sound in Gases at High Temperature, 1921; The Propagation of the Explosion-Wave, 1923; The Velocity of Sound in Gases and Vapours, 1924; The Ignition of Carbon Disulphide, 1925; The Explosion-Wave in Cyanogen Mixtures, 1926; The Burning of Gases in Nitrous Oxide, 1927. Research on gaseous explosions, especially important to mining; studied effect of moisture on gas explosions showing that dry mixtures were less likely to explode, but that explosive wave travels faster in dry gas once explosion occurs (led to important work on role of water in chem. reactions); showed that carbon monoxide reduces steam at high temperatures, also that N203 is not completely dissociated in gaseous state; investigated explosion of carbon monoxide and hydrogen mixtures in conditions of low oxygen availability. Died Sept. 18, 1930.

DIXON, Henry Horatio, Irish botanist; b. Dublin, Ireland, 1869; s. George and Rebecca (Yeates) D.; Sc.D., Trinity Coll., Dublin; also ed. U. Bonn (Germany); m. Dorothea Mary Franks, 1907; 3 sons. Asst. to prof. botany Dublin U., 1892-1904, univ. prof. botany, 1904-50; prof. plant biology Trinity Coll., Dublin, from 1922, dir. bot. gardens, 1906-51, keeper herbarium, 1910-51. Trustee Imperial Library Ireland, 1914; commr. Irish Lights, 1924; vis. prof. U. Cal., 1927; hon. chmn. 6th Internat. Bot. Congress, Amsterdam, Netherlands, 1935; hon. pres. Internat. Bot. Congress, Stockholm, Sweden, 1950. Recipient Boyle medal, 1917. Fellow Royal Soc. (Croonian lectr.), 1908; mem. Internat. Inst. Agr. (council, chmn. com. on biochemistry Rome 1927), Royal Dublin Soc. (council 1908, v.p. 1930, pres. 1945-49), Am. Soc. Plant Physiologists (corr.), Brit. Assn. Advancement Sci. (pres. bot. sect. 1922). Author: Transpiration and the Ascent of Sap in Plants, 1914; Practical Plant Biology, 1922; The Transpiration Stream, 1924; On the Physics of the Transpiration Current; First Mitosis of the Pollen Mother Cells of Lilium, others. Research on plant transpiration. Died Dec. 20, 1953.

DIXON, Joseph, Am. inventor; b. Marblehead, Mass., Jan. 18, 1799; s. Joseph and Elizabeth (Reed) D.; m. Hannah Martin, July 28, 1822. Inventor machine for cutting files in his youth; took up printing, made wood type, became skilled in wood-engraving and lithography; later invented matrix for casting metal type; recognized value of mineral graphite, helped open up markets that used this substance (such as pencils and stove polish); opened mfg. plant, Salem, Mass., 1827; invented photolithographic process and process for producing colored inks to prevent counterfeiting, 1832; patentee for anti-friction bearing metal, 1845; relocated his mfg. plant, Jersey City, 1847; received patents on processing graphite crucible, 1850, also received other patents on mfg. improvements; organizer, head Joseph Dixon Crucible Co., 1867-69; made improvements in lens grinding. Died June 15, 1869.

DIXON, Samuel Gibson, Am. bacteriologist; b. Phila., Pa., Mar. 23, 1851; s. Isaac and Ann (Gibson) D.; grad. Mercantile Coll.; studied law, admitted to bar, 1877; M.D., with honors, U. Pa., 1886, LL.D., 1909; postgrad. dept. of bacteriology, King's Coll., London, State Coll. of Medicine, London, and Pettenkofer's Lab. of Hygiene, Munich; m. Fannie Gilbert. Prof. hygiene in med. and scientific depts. and dean aux. dept. of medicine U. Pa., 1888-1910; prof. bacteriology and micros. technology Acad. Natural Scis., Phila., from 1890, curator, 1891-92, exec. curator, from 1892, pres., from 1896. Mem. Bd. Pub. Edn., Phila., 1898; Commr. of Health, Pa., from 1905. Trustee U. Pa., Wistar Inst. of Anatomy. Author: Physiological Notes, 1886. Died Feb. 26, 1918.

DIXON, Walter Ernest, English pharmacologist; b. Darlington, Eng., June 2, 1870; s. Robert Bland and Mary Ann Whitecomb (Parr) D.; B.Sc., St. Thomas's Hosp., 1891, M.D., 1898; LL.D. (hon.), U. Man., Can., 1930; M.A. honoris causa, Cambridge (Eng.) U., also D.P.H.; M.D., S.B., B.Sc., London; m. Hope Glen-Allen, 1907. Asst. to Downing prof. medicine, Cambridge, 1899, then lectr. Cambridge, 1909 and 1st reader in pharmacology, 1919-31; concurrently chair materia medica King's Coll., London, 1919; dir. pharm. lab. Cambridge, also assessor to Regius prof. physics Cambridge U.; examiner in pharmacology univs. Cambridge, Oxford and London; mem. Govt. Com. on Food Preservatives, Drug Addiction and Ethyl Petrol.; mem. League of Nations Expert Com. on Drugs of Addiction; Dohme lectr. Johns Hopkins; William Withering lectr. U. Birmingham. Fellow Royal Soc., 1911, Royal Coll. Physicians, 1930; mem. Royal Soc. Medicine (pres. for therapeutics), Brit. Assn. (pres. physiology), Brit. Med. Assn. (pres. therapeutics); hon. mem. Pharm. Soc. Great Britain, Acad. Onorario, Reale Acad., Medica di Roma. Author: A Manuel of Pharmacology, 1905; Practical Pharmacology, 1907; contbr. Albutt's System of Medicine, also Heffner's Pharmacology. Played maj. role in establishment pharmacology as an exptl. sci., also as the sci. basis of therapeutics; investigated physiology and pharmacology of bronchial muscles, pulmonary circulation and cerebro-spinal fluid; leading expert on problems of drug addiction, studying use of tobacco, alcohol, cocaine, morphine and hashish; noted similarity in effect of certain drugs and of nerve stimulation on muscle tissue and he proposed (1907) that the effect of nerve excitation may be due to liberation of chem. substance (it remained for later researchers to verify this suggestion).

DIZÉ, Michel-Jean-Jérôme, French chemist; b. Aire, France, Sept. 29, 1764; s. Michel Dizé; studied under Parcet in Paris; m. Adelaïde Barthélémy, 1797; 2 sons; m. 2d, Agathe Dumont; m. 3d, L. C. F. Mondet; 2 daus. Became pharmacist St. Denis Hosp., 1793; prof. pharmacy Paris École de pharmacie, 1796-98; founder factory at St. Denis, France, 1791; factory for refining silver bullion by sulfuric acid, Brussels, Belgium, 1823. Mem. Acad. Medicine. Inventor an indelible ink, process of preparing sodium carbonate, process for preservation of meat through dehydration. Died Paris, France, Aug. 1, 1852.

DJAKOV, Emil Stefanov, Bulgarian physicist; b. Svishtov, Bulgaria, Mar. 2, 1908; s. Stefan Ivanov and Olga (Mandikova) D.; student Sofia (Bulgaria) U., 1927-31; m. Elena Nikolova Monkova, July 1, 1938; children—Boian, Assen. Faculty, Sofia U., 1932—, prof. electronics, 1942—; vice dir. Inst. Physics, Bulgarian Acad. Scis., Sofia, 1955-63, dir. Inst. Electronics, 1963—; vice dir. Joint Inst. for Nuclear Research, Dubna, USSR, 1959-61, mem. sci. council, 1956—. Decorated Order for Civil Merit, 1946; Order Cyril and Methodius I degree, 1958; Order of Red Flag of Labour, 1959; Dimitrov Prize II degree, 1952. Mem. Bulgarian Acad. Sci. (mem. presidum 1963). Author: Physics, Elementary Course, 1946; Bases of Electrotechnics II, 1964; Bases of Radiotechnics III, 1965; also articles. Research on phys. electronics, vacuum measurement in split anode magnetrons, saturated current dependencies in diodes, low and high frequency oscillations in cesium thermoelectronic converters, microwaves, electron oscillations in diodes and magnetrons, measurement of complex resistances, delay line of a new type, applied electronics, integrater magnitude depending on fluctuating parameter, delay line with high equilibrium states. Home: 17 Venelin, Sofia, Bulgaria.*

DJEHANGUIRE, see Jahangir.

DJERASSI, Carl, chemist; b. Vienna, Austria, Oct. 29, 1923; s. Samuel and Alice (Friedmann) D.; came to U. S., 1939, naturalized, 1945; A.B. summa cum laude, Kenyon Coll., 1942, D.Sc., 1958; Ph.D., U. Wis., 1945; D.Sc., U. Mexico, 1953; m. Norma Lundholm; children—Pamela, Dale. Research chemist CIBA, Summit, New Jersey, 1942-43, 45-49; associate director of chemical research Syntex, S.A., Mexico City, 1949-52, vice president in charge research, 1957-60; faculty Wayne State U., 1952-59; prof. chemistry Stanford, 1959—. Dir. Syntex Corp. Recipient Am. Chem. Soc. award in pure chemistry, 1958, Baekeland medal, 1959, Fritzsche medal, 1960; Fellow Brit. Chem. Soc. (hon., Centenary lectr. 1964); member National Academy Scis., Brazilian Acad. Sci., Mexican Acad. Sci. Investigation, Am., German, Swiss, French chem. socs., Am. Soc. Biol. Chemists, Soc. Chem. Industry, Phi Beta Kappa, Sigma Xi, Phi Lambda Upsilon. Author: Optical Rotatory Dispersion, 1960; (with others) Interpretation of Mass Spectra of Organic Compounds, 1964, Structure Elucidation of Natural Products by Mass Spectrometry, 1964; Mass Spectrometry of Organic Compounds, 1967. Editor: Steroid Reactions, 1963. Contbr. numerous articles to profl. jours. Research on chemistry of steroids, notably synthesis of female sex hormones and adrenal hormones, chemistry of natural products, notably antibiotics, alkaloids and terpenoids, application of phys. methods, notably optical rotatory dispersion, circular dichroism and mass spectrometry, organic and medicinal chemistry; discoverer several oral contraceptives, corticosteroid anti-inflammatory agts., antihistamines. Office: Dept. Chemistry, Stanford U., Stanford, Cal. 94305.*

DJORDJEVIC, Slobodan, Yugoslavian otorhinolaryngologist; b. Valjevo, Yugoslavia, Apr. 24, 1921; s. Petar and Leposava (Siskovic) D.; M.D., U. Belgrad, 1950, specialist degree in oto-rhino-laryngology, 1954. Mem. Med. Faculty, U. Belgrade, 1950—, prof. oto-rhino-laryngology, 1966—, head of ward, dept. ear, nose and throat, 1957—. Editor in chief Nature and Sci., 1948-60, Serbian Archives of Medicine, 1950—, Acta Historica Medicinae, 1960-65. Mem. Collegium Otorhinolaryngologicum, Yugoslav Soc. Ear, Nose, Throat, Soc. Alergology, Soc. Plastic Surgery, Soc. Cancer, Soc. Med. History, numerous other Yugoslav, French med. socs. Author: Textbook of Oto-rhino-laryngology for Students, 1967; also numerous articles. Research on hearing and vestibular apparatus in population of strumogenic regions; extraction of fgn. bodies from respiratory and digestive tract; treatment of neck cancer matastasis; history of medicine. Home: 1 Loznicka, Belgrade, Yugoslavia.*

DLHOS, Ernest, Russian physician; b. Zvolen, USSR, Sept. 5, 1910; s. Lúdevít and Jana (Kohnová) D.; M.D., Komensky U., Bratislava, USSR; m. Viera Dlhos, July, 13, 1940; children—Viera (Mrs. Tregerova), Luba, Peter. Mem. staff gynecol. clinics U. Bratislava, also Prague, Czechoslovakia; pvt. docent U. Bratislava, 1949-64, prof., 1964-66, doctor med. sci., 1966—, also mem. sci. council; chief clinic obstetrics and gynecology Inst. Postgrad. Edn. Physicians and Pharmacologists, also mem. sci. council. Author ten books. Research and publns. on female sterility, disorders of menstrual cycle and hemorrhage in 3d stage of labor. Home: 20 Hviezdoslavova. Office: 45 Sovietskej, Trencín, USSR.*

DMOCHOWSKI, Leon Ludomir, physician, virologist; b. Tarnapol, Poland, July 1, 1909; s. Roman and Ludmilla (Lazurkiewicz) D.; M.B., Ch.B., U. Lwow (Poland), 1933; M.D., U. Warsaw (Poland), 1937; Ph.D., U. Leeds (Eng.), 1949; m. Sheila Bessie Jurdon, Jan. 14, 1950; 1 son, Roger Roman. Came to U. S., 1953, naturalized, 1959. Lectr., research prof. cancer research U. Leeds, 1946-53; practice medicine, specializing in microbiology, Houston, 1954—; prof. anatomy Coll. Medicine Baylor U., 1954-55, clin. prof. microbiology, 1955—; head virology and electron microscopy sect. U. Tex. M.D. Anderson Hosp. and Tumor Inst., 1955—, prof. virology, chmn. dept., 1965—; prof. exptl. pathology Postgrad. Sch. Medicine U. Tex., 1959—, prof. virology Grad. Sch. Biomed. Scis., 1965—. Diplomate Am. Bd. Microbiology. Fellow Am. Acad. Microbiology, Royal Soc. Medicine, N.Y. Acad. Scis., Royal Soc. Tropical Medicine and Hygiene; mem. Brit. Med. Assn., Am. Assn. Cancer Research (past pres. S.W. br.), A.A.A.S., Electron Microscope Soc. Am., Soc. Exptl. Biology and Medicine, Pan-Am. Cancer Cytology Soc., Soc. Am. Bacteriologists. Contbr. chpts. to Advances in Cancer Research, 1953; Cancer, 1957; Diseases of Poultry, 1959, 1965; The Book of Health, 1962; Progress in Experimental Tumor Research, 1963. Contbr. numerous articles to profl. jours. Studies on relationship of viruses to origin of cancer in animals and man by means of tissue culture, light microscopy, electron microscopy, biophys. and biochem. techniques and tests in animals. Home: 3502 Durness St. Office: M.D. Anderson Hosp. and Tumor Inst., Houston 77025.*

DOAN, Charles Austin, Am. physician; b. Nelsonville, O., June 5, 1896; s. Robert Austin and Lelia M. (Welch) D.; B.Sc., Hiram Coll., 1918; M.D., Johns Hopkins, 1923; Sc.D., Ohio State U., 1964; m. Margaret Dixon Riggs, May 28, 1926; children—Elizabeth (Mrs. Robert Bushell), Ellen Virginia (Mrs. William Haagen). Instr. dept. anatomy Johns Hopkins, 1923-24; instr., research asso. Thorndike Meml. Lab., Harvard Med. Sch., 1924-25; asso. Rockefeller Inst. for Med. Research, 1925-30; prof., chmn. dept. med. and surg. research Ohio State U. Med. Sch., Columbus, 1930-44, dir. med. research, 1936-44, dean Coll. Medicine, 1944-61, dean emeritus, 1961—, dir. div. hematology, 1961-65. Cons. to govt. agys. Recipient Distinguished Service gold medal, citation A.M.A., 1960. Master A.C.P. (chmn. bd. govs. and regents 1951-54); mem. Ohio Acad. Scis. Central Soc. Clin. Research (pres. 1939-40), Am. Soc. Hematology (past pres.), Internatl Reticulo-Endothelial Soc. (pres. 1954-57), Assn. Am. Physicians, numerous others. Contbr. numerous articles to tech. jours., monographs, chpts. in books. Research on origin blood cells, role of R-E system in health and disease, Tb., hypersplenic syndromes, chemotherapeutic research on leukemias and lymphoma, autoimmune hemolytic anemias, thrombocytopenic purpuras, pernicious and hypoplastic anemias. Home: 4935 Olentangy Blvd., Columbus, O. 43214.*

DOAN, Richard Lloyd, Am. physicist; b. Lapel, Ind., Sept. 7, 1898; s. Arthur W. and Dora B. (Lloyd) D.; A.B., Ind. U., 1922, A.M., 1923; Ph.D., U. Chgo., 1926; D.Sc. (hon.) Coll. Ida., 1958; m. Melba Pyle, Dec. 15, 1926; children—Robert L., Virginia (Mrs. David Fanger). NRC fellow U. Chgo., 1926, research asso., 1942-43, dir. Metall. Lab. (name changed to Argonne Nat. Lab); staff, Western Elec. Co., 1927-33; asso. dir. Phillips Petroleum Co., Bartlesville, Okla., 1936-42, dir. research, 1950, mgr. atomic energy div., 1951-63; dir. reactor licensing AEC, Washington, 1964—; dir. research Clinton Lab. (later Oak Ridge Nat. Lab.), 1944-45. Mem. AEC Adv. Com. on Reactor Safeguards, 1951-61. Cited by AEC, 1963. Fellow Am. Phys. Soc., A.A.A.S., Am. Nuclear Soc.; mem. Am. Chem. Soc. First to get X-ray spectra from ruled metallic grating. Home: 6601 Millwood Rd., Bethesda, Md. Office: U. S. AEC, Washington 25.*

DOBBERSTEIN, Johannes Christian Albert, German veterinarian; b. Graudenz, Sept. 19, 1895; s. Eugen and Marie (Mooslehner) D.; D.V.M., U. Humboldt, Berlin; D.V.M. honoris causa univs. Leipzig, Stockholm, Budapest; m. Frieda Leipacher. Dir., Inst. Comparative Pathology, Acad. Scis., Berlin. Mem. Acad. Scis. of Berlin, Acad. Rural Econs. (Berlin), Soc. Comparative Pathology (Paris), Italian Soc. Vet. Sci. Author: Lehrbuch der Allgemeinen Pathologie, 5th edit., 1963; Handbuch der Pathologischen Anatomie der Haustiere; Neubearbeitung des Joest'schen Handbuches, vol. 2, ed edit., 1937, 3d edit., 1963; Gerichlichte Tierheilkunde, 11th edit., 1955; Vergleichende Anatomie der Haustiere, 1st edit., 1953-58, 2d edit., 1961; Richtlinien für die Sektion der Haustiere, 8th edit., 1957. Home: Am Birkenwerder 18, Berlin-Kaulsdorf. Office: Wilhelmstrasse 4, Berlin-Friedrichsfelde, Germany.

DOBBIE, Sir James Johnston, Brit. chemist; b. Glasgow, Scotland, Aug. 4, 1852; s. Alexander Dobbie; LL.D. (Clark scholar), U. Glasgow; also ed. U. Edinburgh (Scotland), U. Leipzig (Germany); M.A., D.Sc.; m. Violet Chilton, 1887; 1 son, 2 daus. Asst. to prof. chemistry U. Glasgow, 1881-84; prof. chemistry U. Coll. N. Wales, Bangor, 1884-1903; dir. Royal Scottish Mus., Edinburgh, 1903-09; prin. Govt. Lab., London, Eng., 1909-20. Mem. Royal Commn. on Awards to Inventors, Univ. Grants Com. Fellow Royal Soc., 1904; mem. Inst. Chemistry (pres. 1915-18), Chem. Soc. (pres. 1919-21). Research, publs. on chem. constn. alkaloids; studied relation between chem. const. and absorption spectra of organic compounds, also absorption spectra of vapors of elements. Died Fairlie, Ayrshire, Scotland, June 19, 1924.

DOBBING, John, English physiologist; b. Sheffield, Eng., Aug. 14, 1922; s. Alfred Herbert and May Gwendoline (Cattell) D.; B.Sc. with 1st class honors, U. London, 1950, M.B., B.S., 1953, M.C. Path., 1964; m. Rachel Lamb, Dec. 6, 1946; children—Christopher J., Katherine J., Mary C. House physician St. Mary's Hosp., 1953; obstetric house surgeon Perivale Maternity Hosp., 1954; lectr. pathology, Gull scholar Guy's Hosp., 1954-60; sr. lectr. physiology London Hosp. Med. Coll., 1961-65; sr. lectr. Inst. of Child Health,

1966—— (all London). Hon. cons. Hosp. for Sick Children, London. Fellow Royal Soc. Medicine; mem. Physiol. Soc., Neonatal Soc., Internat. Soc. for Neurochemistry, Coll. Pathologists. Author: (with A. N. Davison) Applied Neurochemistry. Research and publs. on nature of blood brain barrier; metabolism of the myelin sheath; effects of malnutrition on the brain. Home: 38 Kidbrooke Gardens, London S.E. 3. Office: 30 Guilford St., London W.C. 1, Eng.*

DOBELL, Clifford, English protistologist; b. Feb. 22, 1886; s. William Blount and Agnes (Thurnely) D.; exhibitioner Trinity Coll., Cambridge (Eng.) U., 1905, B.A. (major scholar), 1906, M.A. with 1st class nat. sci. tripos, 1910, fellow, 1908-14; student Zool. Inst., Munich, Germany, 1907-08, Zool. Sta., Naples, Italy, 1908, Zool. Sta., Ceylon, 1909; m. Monica Baker, 1927. Lectr., then asst. prof. protistology and cytology Imperial Coll. Sci., 1910-19; protistologist to Med. Research Council, Nat. Inst. Med. Research. Walsingham medalist Cambridge U., 1908; Rolleston prizeman Oxford and Cambridge univs., 1908; Balfour student Cambridge U., 1908-09. Fellow Royal Soc., 1918. Author: The Amoebae Living in Man, 1919; (with F. W. O'Connor) The Intestinal Protozoa of Man, 1921; Leeuwenhoek and his Little Animals, 1932; also articles. Research on protozoa, bacteria and related organisms, especially related to human health; research in amoebic dysentery. Died Dec. 23, 1949.

DÖBEREINER, Johann Wolfgang, German chemist; b. Hof, Bavaria, Dec. 15, 1780; s. Johann Adam and Johanna Susanna (Göring) D.; self-educated; D.Phil., U. Jena (Germany), after appointment on faculty; m. Clara Knab, 1803; 5 sons, 4 daus. Asst. to various apothecaries, Münchberg, Dillenburg, Karlsruhe, Strasbourg; became pharmacist, Hof, 1802; named asso. prof. chemistry, pharmacy, tech. U. Jena, 1810, prof., 1819-49. Friend and teacher of Goethe. Author: Lehrbuch der allgemeinen Chemie, 3 vols., 1811-12; Grundriss der allgemeinen Chemie, 1816, supplement, 1837; Elemente der pharmaceutischen Chemie . . . , 1816; Neueste stöchiometrische Untersuchungen und chemische Entdeckungen, 1816; Anleitung zur Bereitung verschiedener Essige, 1816; Über die chemische Constitution der Mineralwässer . . . , 1821; Zur mikrochemischen Experimentierkunst, 1821; Zur Gährungschemie, 1822; Darstellung der Zeichen und Verhältnisszahlen der irdischen Elements zu chemischen Verbindungen, Part I, 1823; Beiträge zu physikalischen Chemie, 1835. Fist to note relationship between atomic weights of barium, strontium, calcium, 1817; discovered catalytic effect of platinum (illus. in Döbereiner's lamp, which he invented), 1823; by his classification of elements into groups of three (Döbereiner's triads) with related atomic weights and other properties (1829) aided devel. of periodic law and Mendelejeff's periodic table; prepared furfural, also aldehyde ammonia, 1831; investigations of relation between acetic acid and alcohol, 1832 (led to quick vinegar process); conducted practical classes in analytical chemistry before Liebig in Giessen. Died Jena, Mar. 24, 1849.

DOBO, Marie-Antoine, mechanic; co-inventor worsted spinning machine; inventor process of rubbing silver to make it cohere (method of roving), 1815. Died circa 1825.

DOBREANU-ENESCU, Viorica, Rumanian physician; b. Porcesti, Rumania, Sept. 9, 1924; d. Mihail G. and Maria (Dimitriu) Dobreanu; M.D., Bucharest Sch. Medicine, 1949; m. Nicolaie Enescu, Mar. 5, 1948. Asst. prof. Cardiol. Clinic of Sch. Medicine, Bucharest, 1949-51, asst. prof. Clinic of Internal Medicine, 1951-57; chief research fellow Inst. Internal Medicine Rumanian Acad., Bucharest, 1957-58, chief electrocardiographic dept., 1958——; research fellow cardiology Inst. for Med. Research, Los Angeles, Cal., 1965-66. Fellow Am. Coll. Cardiology; mem. European Soc. Cardiology. Author (with Iuliu Popescu) The Myocardium, Physiopathology and Clinical Treatment, 1957; also numerous articles. Research on essential hypertension, pathogenesis of the disease; hemodynamic changes in various stages and types of disease; long term treatment with Guanethidine; electrocardiograph. Home: 21 Galati St., Bucharest 13, Rumania. Office: Inst. Med. Interna, 19-21 Stefan Cel Mare, Bucharest 10, Rumania.*

DOBRIANSKY, Lev Eugene, Am. economist; b. N.Y.C., Nov. 9, 1918; s. John I. and Eugenia (Greshchuk) D.; student Fordham U., 1937-43; B.S., N.Y. U., 1941, M.A., 1943, Ph.D., 1951; LL.D., U. Munich, 1952; m. Julia Kusy, June 29, 1946; children —Larisa Eugenia, Paula Jon. Faculty, N.Y. U., 1942-48, Nat. War Coll., Washington, 1957-58; faculty Georgetown U., Washington, 1948——, prof. econs., 1960——. Researcher, Am. Bankers Assn., 1944; cons. to govt. and industry. Recipient award Freedom's Found., 1961; tribute for authorship of Captive Nations Week resolution U. S. Congress, 1959; medal Shevchenko Sci. Soc., 1964; Freedom award, 1965; others. Mem. Am. Econ. Assn., Economists' Nat. Commn. on Monetary Policy, Am. Assn. U. Profs., Inst. Am. Strategy, Am. Security Council. Author: The Free Trade Ideal, 1954, Veblenism, A New Critique, 1957; Nations, Peoples & Countries in the USSR, 1964; (with A. Jacobs, et al) Decisions for a Better America, 1960; The Vulnerable Russians, 1967. Editor: Europe's Freedom Fighter: Taras Shevchenko, 1960; Captive Nations Week, 1966. Collaborator: Peace and Freedom Through Cold War Victory, 1964. Research in instl. econs., politico-econ. thought on the USSR, studies in problems of non-Russian nations in USSR. Home: 4520 Kling Dr., Alexandria, Va. 22312.*

DOBRIN, Milton B., Am. geophysicist; b. Vancouver, B.C., Can., Apr. 7, 1915 (parents Am. citizens); s. Harry and Lillian (Zinn) D.; B.S., Mass. Inst. Tech., 1936; M.S., U. Pitts., 1941; Ph.D., Columbia, 1950; m. Stefanie Zink, Sept. 1, 1948; children—Susan, David, Daniel, Bruce, Barbara. Geophysicist, Gulf Research & Devel. Co., Pitts., 1937-42; physicist Naval Ordnance Lab., Washington, 1942-49; research geophysicist Magnolia Petroleum Co., Dallas, 1949-55; geophys. supr. Triad Oil Co. Ltd., Calgary, Alta., Can., 1956-61; chief geophysicist United Geophys. Corp., Pasadena, Cal., 1961——, v.p., 1967——. Mem. Soc. Exploration Geophysicists (v.p. 1961-62, Best Paper award 1966), Am. Geophys. Union, Am. Assn. Petroleum Geologists, European Assn. Exploration Geophysicists. Author: Introduction to Geophysical Prospecting, 1952, 2d edit., 1960. Research, publs. on rec., analysis and identification of noise made underwater by fish, determination of subsurface geol. structure of Bikini Atoll, analysis and identification of surface waves from small explosions, application of optical filtering techniques to enhancement of seismic oil exploration data. Home: 2080 Midlothian Dr., Altadena, Cal. 91001. Office: 2650 E. Foothill Dr., Pasadena, Cal. 91109.*

DOBRINER, Konrad, research physician; b. Elberfeld, Germany, Oct. 14, 1902; s. Paul and Laura (Drey) D.; student Gymnasium Cologne, Gymnasium Lennep, 1909-21, U. Freiburg, 1921-25; M.D., D.M.S., U. Munich, 1927; m. Shirley FitzGerald, June 28, 1945; children—Madeleine Joan, Mark George. Came to U. S., 1934, naturalized, 1940. Intern II Medizinische Abteilung Krankenhaus Munich-Schwabing, 1927-28, assistenzarzt, 1928-33; specialist for internal med., Munich, 1932; research fellow U. Rochester Med. Sch., 1934-36, Hosp. of Rockefeller Inst., 1936-39; head dept. research chemistry Meml. Hosp., N.Y.C., 1939-47; mem. Sloan-Kettering Inst., N.Y.C., from 1947. Mem. Am. Soc. Biol. Chemists, Assn. Study Internal Secretions, Am. Assn. Cancer Research, Am. Soc. Clin. Investigation, Soc. Exptl. Biol. and Med., Harvey Soc., N.Y. Acad. Scis. Research in metabolism of steroid hormones; cancer research; prophyrin metabolism, metabolism of carcinogens, infra-red spectrometry. Died Mar. 10, 1952.

DOBRYNIN, Nikolay Fedorovich, Russian psychologist; b. 1890; grad. History and Philology Faculty, Moscow U., 1915; postgrad. Moscow Psychol. Inst., 1922-25; D.Psychol. Sci., 1937. Instr. psychology 2d Moscow U. (now Lenin Pedagogical Inst.), 1922——, prof., head chair psychology, 1960——; prof. Liebknecht Indsl. Pedagogical Inst., Moscow, until 1940; head chair psychology Potemkin Pedagogical Inst., Moscow, 1942-60. Chmn. learned commn. on psychology RSFSR Ministry Edn.; chmn. psychologists sect. House of Scientists, USSR Acad. Sci. Mem. All-Union Znanie Soc. (sci. methods council pedagogical sect.). Author: An Introduction to Psychology, 1929; Interest and Attention, 1941; The Dynamics of Attention, 1947; The Basic Problems of the Psychology of Attention, 1959. Past mem. editorial bd. Problems of Psychology. Address: Lenin Pedagogical Inst., M. Pirogovskaya ulitsa 1, Moscow, USSR.

DOBSON, Ernest Lowry, Am. physiologist; b. Peking, China, Apr. 25, 1914 (parents Am. citizens); s. Robert J. and Mabel (Lowry) D.; B.S., U. Cal. at Berkeley, 1937, Ph.D., 1950; m. Martha Anne Taylor, Aug. 12, 1939; children—Kathleen Elizabeth, Margaret Anne. Tchr. high sch., Dunsmuir, Cal., 1939-45; instr. Napa Jr. Coll., 1945-46; with Lawrence Radiation Lab., U. Cal. at Berkeley, 1946——, physiologist, 1950——. Mem. Soc. Exptl. Biology and Medicine (pres. Pacific Coast sect. 1964-65), Am. Physiol. Soc., A.A.A.S., Reticulo-endothelial Soc. (v.p. 1962-63). Contbr. articles to sci. jours. Research on kinetics of phagocytosis, salt and water regulation, regulation of volume of body fluids, liver blood flow. Home: 15 Canyon Rd., Berkeley, Cal. 94704.*

DOBSON, George Edward, zoologist; b. Edgeworthstown, Longford, Eire, Sept. 4, 1848; s. Parke Dobson; ed. Royal Sch. Enniskillen, Trinity Coll. Dublin (Ireland); B.A., 1866, M.B., M.Ch., 1867, M.A., 1875; entered army med. dept., 1868; served in India; ret. as surgeon-maj., 1888; became curator Royal Victoria Mus., Netley, Eng., circa 1878. Fellow Linnaean Soc., Royal Soc., 1883, Zool. Soc.; corr. mem. Acad. Natural Scis. Phila., Biol. Soc. Washington. Author: Catalogue of the Chiroptera in Collection of British Museum, 1878; Monograph of the Asiatic Chiroptera, 1876; Monograph of the Insectivora, Systematic and Anatomical, 1882; also numerous articles. Research on structure and classification of chiroptera and insectivora. Died Nov. 26, 1895.

DOBSON, Richard Lawrence, Am. physician, educator; b. Boston, Apr. 12, 1928; s. Joseph W. and Celia (Beatrice) D.; student U. N.H., 1945-46, 47-49; M.D., U. Chgo., 1953; m. Marie C. Mollomo, Aug. 19, 1950; children—Richard Lawrence, Pamela B., Lisa M. Faculty U. N.C. Med. Sch., 1957-51; asso. prof. dermatology U. Ore. Med. Sch., Portland, 1961-64, prof., 1964——. Mem. gen. medicine study sect. NIH, 1965——. Diplomate Am. Bd. Dermatology. Fellow A.A.A.S.; mem. Soc. Investigative Dermatology (dir.), Am. Acad. Dermatology (dir. physiology and biochemistry), Am. Physiol. Soc., Am. Assn. Cancer Research, Histochem. Soc., Am. Fedn. Clin. Research, Western Soc. Clin. Research. Asst. chief editor Archives of Dermatology. Contbr. numerous articles to sci. jours. Research on secretion of salt and water by eccrine sweat gland, abnormality in sweating in cystic fibrosis, sequential devel. of exptl. skin cancer. Home: 505 S.W. Glenn Rd., Portland 97219. Office: 3181 S.W. Sam Jackson Park Rd., Portland, Ore. 97201.*

DOBY, John Thomas, Am. social psychologist; b. Gray, Ky., May 29, 1920; s. Daniel W. and Minnie (Farris) D.; A.B. cum laude, Union Coll., Ky., 1946; M.S., U. Wis., 1950, Ph.D., 1956; m. Rose C. Hopper, Dec. 21, 1942; children—Mary Catherine, Nancy Hopper. Faculty, Wofford (S.C.) Coll., 1950-58; faculty Emory U., Atlanta, 1958——, prof. sociology and anthropology, 1963——, chmn. dept., 1961——. Mem. fellowship panel NSF, Washington, 1965-66; chmn. Sci. and Tech. Adv. Com. on Mental Retardation, Ga., 1965-66. Fellow Am. Sociol. Assn.; mem. So. Sociol. Soc. (1st v.p. 1965-66), Am. Council Learned Socs. (sci. faculty fellowship panel 1965-66), So. Soc. Psychology and Philosophy, A.A.A.S., Alpha Kappa Delta, Pi Gamma Mu. Author: Introduction to Social Psychology, 1966; (with Francis, Suchman, Dean, McKinney) Introduction to Social Research, 1954, coauthor, editor 2d edit., 1967. Research on effects of previous learning upon subsequent learning; demonstrated a negative linear relationship between rates of learning and levels of difficulty of materials to be learned, interference of attitude laden previous learning with subsequent learning of related but contrary materials. Home: 1897 Breckenridge Dr., Atlanta 30329.*

DOBYNS, Brown McIlvaine, Am. surgeon; b. Jacksonville, Ill., May 14, 1913; s. Harry D. and Leah (McIlvaine) D.; B.A., Ill. Coll., 1935; M.D., Johns Hopkins, 1939; M.S. in Surgery, U. Minn., 1944, Ph.D., 1946; m. Mary Meredith Davis, Sept. 21, 1940; children—Mary Meredith (Mrs. Donald B. Steffa), Courtney Sara, Brown McIlvaine. Faculty surgery Harvard Med. Sch., 1946-51; faculty Western Reserve U., Cleve., 1951——, prof. surgery, 1958——; mem. staff Cleve. Met. Gen. Hosp., 1951——, associate dir. surgery, 1967——; mem. staff of Univ. Hosps., 1951——. Recipient Van Meter prize, Am. Goiter Assn., 1946, Award of Merit, 1954; Outstanding Achievement award Mayo Found., 1964. Diplomate Am. Bd. Surgery. Mem. Am., Central surg. assns., Soc. U. Surgeons, A.C.S. Cleve. Surg. Soc. (pres. 1966-67), Am. Soc. Clin. Investigation, Am. Thyroid Assn. (pres. 1956-57), Endocrine Soc., A.A.A.S., Sigma Xi. Contbr. articles to med. jours. Research in surgery, thyroid physiology, neoplasms. Home: 2904 Huntington Rd., Shaker Heights, O. 44120.*

DOBZHANSKY, Theodosius, geneticist; b. Nemirov, Russia, Jan. 25, 1900; s. Gregory and Sophia (Voinarsky) D.; ed. U. Kiev (Russia); D.Sc. (hon.), U. Sao Paulo (Brazil), U. Munster (Germany), U. Montreal (Que., Can.), U. Oxford (Eng.), U. Louvain (Belgium), U. Wooster, U. Chgo., U. Columbia, Clark Coll.; m. Natalie Siverzev, Aug. 8, 1924; 1 dau., Sophia (Mrs. M. D. Coe). Came to U. S., 1927, naturalized, 1936. Instr., U. Kiev, 1921-24, U. Leningrad (Russia), 1924-27; fellow Internat. Edn. Bd., Rockefeller Found., 1927-28; faculty Cal. Inst. Tech., Pasadena, 1929-40, prof. genetics, 1936-40; prof. zoology Columbia U., N.Y.C., 1940-58, Da Costa prof., 1958-62; prof. Rockefeller U., N.Y.C., 1962——. Recipient Nat. Medal of Sci., 1964. Mem. Am. Soc. Geneticists (past pres.), Am. Soc. Naturalists (past pres.), Soc. for Study Evolution (past pres.), Am. Soc. Zoologists (past pres.), Nat. Acad. Scis. (Elliott medal 1946, Kimber Genetics award 1958), Am. Philos. Soc., Am. Acad. Arts and Scis., Royal Soc. (London, Eng.), Academia Leopoldina (Germany), Accademia dei Lincei (Italy), Royal Swedish, Royal Danish, Brazilian acads. scis. Author: Genetics and the Origin of Species, 1937; Evolution, Genetics and Man, 1956; Biological Basis of Human Freedom, 1957; Mankind Evolving, 1963; Heredity and the Nature of Man, 1964; also numerous articles. Research on heredity and cultural evolution, studies of Drosophila. Home: 425 E. 63d St., N.Y.C. 10021.

D'OCAGNE, Philibert Maurice, French geometer; b. Paris, Mar. 25, 1862; became prof. Sch. Civil Engrng., 1894, École polytechnique, Paris, from 1912; named gen. insp. bridges and roads, 1920; mem. French Acad. Scis., 1922. Author: Traité de monographie, 1899. Applied geometry to calculus, modifying it with graphic representations; founder monography; made sci. studies of calculators, classified them according to constrn. Died Le Havre, France, Sept. 23, 1938.

DOCHEZ, Alphonse Raymond, Am. physician; b. San Francisco, Apr. 21, 1882; s. Louis and Josephine (Dietrich) D.; A.B., Johns Hopkins, 1903, M.D., 1907; Sc.D., N.Y. U., 1925, Yale U., 1926, Western Res. U., 1931, Columbia U., 1954. Formerly member staff Rockefeller Inst. for Med. Research; asso. prof. medicine, Johns Hopkins Med. Sch., 1919-21; asso. prof. medicine, Coll. Phys. and Surg. (Columbia), 1921-25, prof., 1925-37, John E. Borne prof. of med. and surg.

research, 1939-49, emeritus prof. med. and surg. research, 1949——; vis. physician Presbyn. Hosp. Former trustee Rockefeller Inst. for Med. Research, Mem. Assn. Am. Physicians, Am. Soc. for Clin. Investigation, Am. Soc. for Exptl. Pathology, Soc. for Exptl. Biology and Medicine, Harvey Soc., Nat. Acad. Science, Alpha Delta Phi. Conducted investigations infectious disease; prepared specific antiscarlatinal serum; research on pneumonia and common cold. Home: 1 W. 54th St. Office: 620 W. 168th St., N.Y.C.

DOCK, William, Am. physician; b. Ann Arbor, Mich., Nov. 1, 1898; s. George and Laura (McLemore) D.; B.S., Wash. U., 1920; M.D., U. Chgo., 1923; m. Marie S. Malard, Oct. 2, 1934; children—George, Christopher. With Stanford Med. Sch., 1925-41, prof. pathology, 1936-41; prof. pathology Cornell Med. Sch., 1941-44; prof. medicine L.I. Coll. Medicine, 1944——; prof. medicine State U. N.Y., 1944——; chief med. service VA Hosp., Bklyn., 1964——. Mem. Am. Soc. Clin. Investigation (pres. 1941-42), Assn. Am. Physicians, Am. Heart Assn., A.C.P. Author: (with H. and R. Mandelbaum) Ballistocardiography, 1953, Russian edit., 1956. Research, publs. on blood flow in organs during hypertension, localization of tb in apex of lung, coronary arteriosclerosis. Home: 145 E. 16th St., N.Y.C. 10003. Office: VA Hosp., Poly Pl., Bklyn. 11209.*

DOCKERTY, Malcolm Birt, pathologist; b. Cardigan, P.E.I., Can., Sept. 9, 1909; s. Robert Alexander and Addie May (Birt) D.; student Prince of Wales Coll., Charlottetown, P.E.I., 1925-28; M.D., Dalhousie U., 1934; M.S., U. Minn., 1937; m. Marjorie Olive Stoddart, Dec. 29, 1937; 1 son, John M. (foster child). Practice medicine, specializing in pathology, Rochester, Minn.; faculty U. Minn. Mayo Grad. Sch., 1938——, prof. pathology, 1952——; head surg. pathology Mayo Clinic, Mayo Found., 1958——. Recipient Anderson Gold medal Prince of Wales Coll., 1928; Gold medal Dalhousie U., 1934. Mem. Am. Soc. Clin. Pathologists, Sigma Xi. Research, numerous publs. on tumors of uterus, ovaries, stomach, small intestine, breast, salivary glands, spinal cord, brain. Home: 1344 2d St. N.W., Rochester 55901. Office: Mayo Clinic, 200 1st St. S.W., Rochester, Minn. 55901.*

DOD, Daniel, Am. inventor, steam engine builder; b. Va., Sept. 28, 1788; s. Lebbens and Mary (Baldwin) D.; ed. Queen's College (now Rutgers); m. Nancy Squier, 1801; 8 children, including Albert Baldwin. Granted U. S. patents on steam engines, including boilers and condensers, for use in steamboats and mills, 1811, manufactured ferryboats, put 1st product into service, 1813; greatest contbn. was machinery for Savannah (1st steamboat to cross Atlantic Ocean 1819). Died in boiler explosion in steamboat test on East River, N.Y.C., May 9, 1823.

DODART, Denis, French physician, botanist; b. Paris, France, 1634; s. Jean Dodart; M.D., U. Paris, 1660; 1 son, Claude Jean. Physician to Princess of Conti, Louis XIV of France; head physician Convent of St. Cyr, France; prof. pharmacy, from 1666. Mem. French Acad. Scis., 1673. Author: De febribus balneum, 1660; Non ergo carnes quovis alio cibo salubriores, 1677; De cancro hydraugyro, 1682; Médecine des pauvres, 1692; Ergo febribus acutis e carnibus juscula, 1700; An omnis morbus a coagulatione, 1730; co-author: Mémoires pour servir à l'histoire des plantes, 1676. Noted for treating indigent; champion of smallpox vaccination; interested in phenomena of voice functioning and problems of perspiration; 1st to use incineration as means of analysing chem. composition of plants. Died 1707.

DODD, Alvin Earl, Am. management scientist; b. Hudson, N.Y., Mar. 11, 1883; s. Alvin Harvey and Edith (Merrill) D.; B.S., Armour Inst. Tech., 1905; LL.D., Temple U., 1948; m. Catherine Filene, Dec. 10, 1921 (div. 1930); 1 dau., Joan; m. 2d, Henrietta F. Coster, Aug. 31, 1941. Asst. prin. Fifth Ward Manual Tng. Sch. Allegheny, Pa., 1905-06; head manual arts dept. Mass. Normal Sch., North Adams, 1906-07; pres. Eastern Arts Assn., 1907-08; prin. North Bennett Indsl. Sch., Boston, 1908-12; dir. Nat. Soc. for Promotion of Indsl. Edn., 1912-16; mem. Com. on Classification of Personnel, Gen. Staff U. S. Army, 1917; head War Service Com., Retail Dry Goods Industry, 1918; dir. Retail Research Assn. and Asso. Merchandising Corp., 1917-21; mgr. distbn. dept. U. S. C. of C., 1921-27; mem. bd. dirs. Fgn. Policy Assn., 1927-28; dir. gen. Wholesale Dry Goods Inst., 1927-29; lectr. on trade and indsl. problems Northwestern U., U. Chgo., U. Wash., Stanford U., 1927-29; asst. to pres. Sears, Roebuck & Co., 1929-30; v.p. Kroger Grocery and Baking Co., 1930-33; exec. v.p. Am. Mgmt. Assn., 1934-35, pres. 1936-48, hon. pres., 1949-51. Cons. mem. staff William L. Batt, Chief of ECA mission to United Kingdom, 1950. Recipient Henry L. Gantt gold medal, 1947. Past mng. dir. U. S. council Internat. C. of C., exec. vice chmn., 1949. Mem. Am. Econ. Assn., Nat. Inst. Social Sciences, Am. Acad. Polit. and Social Sci., Fgn. Policy Assn., Am. Trade Assn., N.Y. Trade Assn. Execs. Led in development of management acceptance of social responsibility. Died New York, N.Y., June 2, 1951.

DODD, Charles G(ardner), Am. physical chemist; b. St. Louis, Jan. 26, 1915; s. Harry Gardner and Ruth Esther (Hauskins) D.; B.S. with distinction, Rice U., 1940; M.S., U. Mich., 1945, Ph.D., 1948; m. Edel Marie Bovbjerg, June 10, 1943; children—Sally Little, Karen Elise, Mary Bartlett, Frederick Porter. Chem. engr. Freeport Sulphur Co., New Orleans, 1940-42; instr. chemistry, research investigator war research project U. Mich., 1942-47; asst. prof. dept. ceramics Pa. State Coll. Sch. Mineral Industries, 1947-48; sr. phys. chemist Fed. Bur. Mines, Bartlesville, Okla., 1948-52; research asso. Continental Oil Co., Ponca City, Okla., 1952-55; asso. prof. dept. chemistry Lehigh U., 1955-56; Erle P. Halliburton prof. petroleum engineering University of Oklahoma 1956-62; chief advanced materials research section of Owens-Illinois Technical Center, Toledo, 1962——. Chairman 8th National Clay Conference, University of Oklahoma, 1959. Mem. Clay Minerals Society, American Ceramic Soc., N.Y. Acad. Scis.; Ohio Acad. Sci., Am. Chem. Soc. (past mem. exec. com. div. colloid chemistry), A.A.A.S., Am. Crystallographic Assn., Mineral. Soc., Phi Beta Kappa, Sigma Xi, Phi Lambda Upsilon, Tau Beta Pi, Phi Kappa Phi. Author sci., tech. articles. Investigation of chem. bonding and structure of glass and glass-ceramic materials as determined by x-ray spectroscopy; theory of chem. analysis by differential absorptiometry —x-ray absorption spectroscopy; identification and characterization of clay minerals; new interfacial visiometer. Home: 4507 Indian Ridge Rd., Sylvania, O. Office: 1700 N. Westwood, Toledo 43607.*

DODD, Henry Work, surgeon; b. Victoria, B.C., Can.; s. Charles Dodd; ed. St. Bartholomew's Hosp.; m. Agnes Shuter; 1 dau., 2 sons. Clin. lectr. ophthalmic medicine and surgery, ophthalmic surgeon Royal Free Hosp.; surgeon Royal Westminster Ophthalmic Hosp.; ophthalmic surgeon Hosp. for Nervous Diseases, London. Licentiate Royal Coll. Physicians, Soc. Apothecaries. Fellow Royal Coll. Surgeons. Contbr. papers to jours. Died June 28, 1921.

DODD, Matthew Charles, Am. biologist; b. Circleville, O., Mar. 24, 1910; s. John A. and Mary Rose (Smith) D.; A.B., Ohio State U., 1933; Ph.D., U. Mich., 1942; m. Josephine Hayes, Sept. 3, 1933; children—Marilyn Ann (Mrs. Robert Reid Haley), Sarah Virginia, Matthew Charles. Research bacteriologist Parke Davis & Co., Detroit, 1933-37; research bacteriologist, asst. instr. U. Mich., Ann Arbor, 1937-41; research bacteriologist, head research in bacteriology and pharmacology Eaton Labs., Norwich, N.Y., 1941-46; faculty Ohio State U., Columbus, 1946——, prof. bacteriology and immunology, 1954——, prof. microbiology, immunology and pathology, 1963——. Fellow Am. Acad. Microbiology, Ohio Acad. Sci., A.A.A.S.; mem. Am. Soc. Microbiology, Am. Assn. Immunologists, Fedn. Biol. Socs., Am. Fedn. for Clin. Research, Soc. Exptl. Biology and Medicine, Tissue Culture Assn., Reticular Soc., Sigma Xi, Alpha Epsilon Delta. Research, numerous publs. on early antisyphilitic drugs such as Mapharsen, anti-bacterial nitrofuran compounds, auto-immune diseases, serology of tumor antigens, immunology of cancer. Home: 1411 W. 2d Av., Columbus, O. 43212.*

DODD, Stuart Carter, behavioral scientist; b. Talas, Turkey, Oct. 3, 1900 (parents Am. citizens); s. William S. and Mary (Carter) D.; B.S., Princeton, 1922, M.A., 1924, Ph.D., 1926; m. Elizabeth M. Cairns, July 28, 1928; children—Peter Carter, Bruce Cairns. Prof. sociology, dir. social sci. research sect. Am. U., Beirut, Lebanon, 1927-47; research prof. sociology U. Wash., Seattle, 1947——, dir. Washington Pub. Opinion, Lab., 1947-61. Decorated Gold Order of Cedar, Republic of Lebanon. Mem. Am., Pacific (past pres.) sociol. assns., World, N.W. (past pres.) assns. for pub. opinion research, Soc. for Gen. Systems Research, Am. Psychol. Assn., Am. Statis. Assn. Author: Social Relations in Middle East, 1931; a Controlled Experiment on Rural Hygiene in Syria, 1934; Dimensions of Society, 1942; Polling in Syria and Sicily, 1943; Systematic Social Science, 1947; Probable Acts of Men, 1963; also numerous tech. articles. Research on reliable measurement attitudes, opinions, behavior, value systems, changing culture whole communities, on item diffusion and consensus-forming, dimensional systematizing human soc. and cosmos, linguistic and semantic instruments for communication in world and sci. behavior of scientists, polling that can represent all people in world on all behaviors and in any culture. Home: 1140 38th Av., Seattle 98122.*

DODDS, Sir Edward Charles, English physician, biochemist; b. Darlington, Eng., 1899; s. Ralph Edward and Jane D.; ed. Middlesex (Eng.) Hosp. Med. Sch.; D.Sc., Ph.D., M.D.; U. London (Eng.); numerous hon. degrees; m. Constance Elizabeth Jordan, 1924; 1 son. Courtauld prof. biochemistry U. London, dir. Courtauld Inst. Biochemistry, Middlesex Hosp. Med. Sch., 1927——. Simms traveling prof. Royal Coll. Surgeons, 1952; mem. numerous coms. on foods, poisons, contaminants, fisheries, others; holder 16 distinguished lectureships, 1934-60. Recipient Meyer prize, Walker prize Royal Coll. Surgeons, 1946, Gold medal W. London Med. Soc., 1938, Cameron prize U. Edinburgh (Scotland), 1940, Garton prize and medal Brit. Empire Cancer Campaign, 1948, medals Ghent, Brussels (both Belgium) univs., Berzelius medal Swedish Med. Soc., Pasteur medal Société de Chimie biologique Francaise, Gold medal Soc. Apothecaries, 1951; Gold medal Soc. Chem. Industry, Herben medal Royal Inst. Pub. Health, 1952. Fellow Royal Soc. (council, v.p. 1957-59), 1942, Royal Inst. Chemists, Royal Soc. Edinburgh, Royal Coll. Physicians (pres. 1962——), hon. numerous socs.; mem. numerous fgn., domestic profl. socs. Author: (with Beaumont) Recent Advances in Medicine; (with Dickens) Chemical and Physiological Properties of Internal Secretions; (with Whitby) The Laboratory in Surgical Practice. (With others) isolated hexestrol (synthetic estrogen), 1938; introduced dienestrol, 1938; discovered stilbestrol, 1938; greatly contbd. to furtherance of med. research, also to application of sci. knowledge to problems of world soc.

DÖDERLEIN, Albert Siegmund Gustav, German obstetrician; gynecologist; b. Augsburg, Germany, July 5, 1860; s. Gustav and Natalie (Casella) D.; studied medicine in Erlangen, and Munich (both Germany); M.D., 1884; hon. dr. artis obstetriciae, Groningen, Germany, 1914; m. Anna Deichert, 1885; 1 son, Gustav; 4 daus.; m. 2d, Helen v. Zwehl, 1918. Asst. to P. Zweifel; became lectr., 1887 and asso. prof. in 1893; became prof. Groningen and Tübingen, Germany; given chair at Munich, 1907; emeritus, 1934——. Author: Leitfaden für den geburtshilflichen Operationskurs, 1893; (with J. Veit) Handbuch der Gynäkologie, 1897; (with O. Küstner) Kurzes Lehrbuch der Gynäkologie, 1901; Operative Gynäkologie, 1905; Handbuch der Geburtshilfe (with F. v. Winckel), 1906; (with F. Penzoldt, R. Stintzing) Handbuch der gesamten Therapie, 1911; (with P. Zweifel, E. Payr) Klinik der bösartigen Geschwülste, 1927. Editor: Handbuch der Geburtshilfe, 3 vols., 1915-21. A founder of gynecol. bacteriology; research on radiation treatment in uterine cancer; introduced rubber glove in obstetrics, operation for narrow pelvis, hebotomy, the extraperitoneal cervical cesarean sect.; discovered Döderlein's bacillus in vaginal secretion. Died Munich, Dec. 10, 1941.

DÖDERLEIN, Ludwig Heinrich Philipp, zoologist, paleontologist; b. Bergzabern, Germany, Mar. 3, 1855; s. Wilhelm and Marie Clothilde (Zwicky) D.; student univs. Erlangen and Munich (both Germany); doctorate U. Strasbourg (France), 1877; m. Auguste Schoen, 1883; 3 sons, 2 daus. Began as asst. tchr. secondary sch., Mulhouse; prof. descriptive natural scis. U. Tokyo (Japan), 1879-81; apptd. curator, dir. zool. collection in Strasbourg, 1882; became lectr. U. Strasbourg in 1883, asso. prof. zoology, 1891; expelled as German scholar from Strasbourg, 1919; became hon. prof. taxonomic zoology, Munich, 1921; dir. state zool. collection, 1923-27. Author: Wegweiser für Pilzfreunde, 1918; Elemente der Paläontologie (Vertebrata); Bestimmungsbuch für deutsche Land- und Süsswassertiere, 1931. Research on fauna of Alsace-Lorraine and Japan, vertebrate fossils; synthesis of zoology and paleontology through phylogenesis and paleobiology. Died Munich, Mar. 23, 1936.

DODGE, Barnett Fred, Am. chem. engr.; b. Akron, O., Nov. 29, 1895; s. Fred Bradley and Charlotte (Barnett) D.; S.B., Mass. Inst. Tech., 1917; D.Sc., Harvard, 1925; D.Sc., Worcester Poly. Inst., 1956; Diplôme de Docteur, U. Toulouse (France), 1961; m. Constance Caroline Woodbury, June 5, 1918; children—Richard, Phyllis (Mrs. Robert J. Putney). Faculty, Yale, New Haven, Conn., 1926-64, head chem. engring. dept., 1931-64, dean engring., 1960-61, prof. emeritus, 1964——; vis. prof. chem. engring. various instns. in France, Spain, Venezuela, Argentina, Uruguay, 1951-66; cons. to bus. firms and govtl. orgns. Mem. Am. Inst. Chem. Engrs. (pres. 1955, Walker award, 1950, Founder's award 1962, W. K. Lewis award 1963), Am. Chem Soc., A.A.A.S., Am. Soc. Engring. Edn., Sigma Xi. Author: Chemical Engineering Thermodynamics, 1944; also articles. Patentee in field. Study of thermodynamic properties at high pressure; rate of absorption and heat transfer; chem. equilibria; treatment of saline water and industrial waste water; catalytic gas reactions. Home: 108 Middle Rd., Hamden, Conn. 06517. Office: 225 Prospect St., New Haven, Conn. 06520.*

DODGE, Charles Richards, Am. entomologist; b. Miss., July 17, 1847; s. Jacob Richards and Frances Gove (Buxton) D.; 2 yrs. spl. course Sheffield Scientific Sch. (Yale); m. Mira Reab, Jan. 23, 1868. With Dept. Agr., 1867, asst. entomologist, and had charge Agrl. Mus. 10 yrs.; began study of fibres, 1870; spl. agt. in charge fibre investigation, Dept. Agriculture, from 1890. Dir. agr. U. S. Commn., Paris Expn., 1900; mem. jury awards expdns., Paris, 1889, Chicago, 1893, Atlanta, 1895, Nashville, 1897, Omaha, 1898, Paris, 1900, Buffalo, 1901, St. Louis, 1904, Jamestown Expn., 1907. Chevalier du Mérite Agricole France; Chevalier Légion d'Honneur. Author: Dictionary of the Fibre Plants of the World, 1897; also numerous spl. reports on fibers and fiber industries. Described Arctia Williamsii, 1871. Deceased.

DODGE, Homer Levi, Am. physicist, educator; b. Ogdensburg, N.Y., Oct. 21, 1887; s. Orange Wood and Isabella (Donaghue) D.; A.B., Colgate U., 1910, Sc.D., 1932; M.S., State U. Ia., 1912, Ph.D., 1914; Sc.D., U. Vt., 1945; LL.D., Middlebury Coll., 1945; m. Margaret Mary Wing, Sept. 5, 1917; children—Alice Isabella (Mrs. Stewart R. Wallace), Norton Townshend. Instr., asst. prof. State U. Ia., 1910-19; prof. physics U. Okla., Norman, 1919-44, head dept., 1919-42, dir. Sch. Engring. Physics, 1924-42, dean Grad. Sch., 1926-44, founder, dir. Research Inst., 1941-44; pres. Norwich U., Northfield, Vt., 1944-50, pres. emeritus, 1950——. Dir., cons. various govtl. burs.; mem. Nat. Adv. Com. for Engring., Sci. and Def. Tng., 1941-45, acad. adv. bd. U. S. Mcht. Marine Acad., 1947-52, Engring. Edn. Mission to Japan, 1952. Fellow A.A A.S., Am. Phys. Soc. (council 1936); mem. Am. Assn. Physics Tchrs. (1st pres. 1930-33, Oersted medal 1944), Am. Inst. Physics (governing bd. 1932-39), Am. Soc. for Engring. Edn. (council 1942-47), Am. Assn. U. Profs., Newcomen Soc., Phi Beta Kappa, Sigma Xi, Sigma Pi Sigma (nat. pres. 1947-50, nat. lectr. on Soviet edn. 1957-58). Author: Problems in Physics Derived from Military Situations and Experience, 1919; (with D. E. Roller) Laboratory Manual of Physics, 1926. Contbr. articles on advances in physics, ednl. systems to profl. publs. Research on elec. measurements, elasticity. Address: 409 S. Union St., Burlington, Vt. 05401.*

DODGE, Raymond, Am. psychologist; b. Woburn, Mass., Feb. 20, 1871; s. George S. and Anna (Pickering) D.; A.B., Williams Coll., 1893, D.Sc., 1918; Ph.D., U. of Halle, 1896; D.Sc., Wesleyan U., 1931; m. Henrietta C. Cutler, Aug. 18, 1897. Asst. librarian, Williams Coll., 1893-94; asst. to Benno Erdmann, U. of Halle, 1896-97; prof. philosophy Ursinus Coll., 1897-98; instr. in psychology Wesleyan U., 1898-99, asso. prof., 1899-1902, prof., 1902-24; prof. Inst. of Psychology, Yale, 1924-29, prof. psychology, Inst. Human Relations, Yale, 1929-36, prof. emeritus from 1936. Exptl. psychologist of nutrition lab., Carnegie Instn., 1913-14; lectr. psychology, Columbia, 1916-17, E. K. Adams research fellow, 1916-18. Mem. psychology com., chmn. div. anthropology and psychology NRC; chmn. program com. IX Internat. Congress of Psychology, 1929. Mem. Am. Psychol. Assn. (pres. 1916-17), Am. Philos. Assn., Am. Acad. Arts Scis., Nat. Acad. Scis. Author: Die Motorischen Wortvorstellungen, 1896; Psychologische Untersuchungen über das Lesen, 1898; Experimental Study of Visual Fixation, 1907; Psychological Effects of Alcohol (with F. G. Benedict), 1915; Elementary Conditions of Human Variability, 1927; Sensorimotor Consequences of Passive Oscillation (with R. C. Travis), 1928; Autobiography—A History of Psychology in Autobiography (Vol. 1), 1930; Conditions and Consequences of Human Variability, 1931; The Craving for Superiority (with Dr. Eugen Kahn), 1931; also numerous scientific articles. Asso. editor Psychol. Bull., 1904-10, Psychol. Rev., 1910-15, Psychobiology, 1917-20, Jour. Exptl. Psychology, from 1916, Jour. Comparative Psychology, from 1921. Discovered fundamental law of visual perception in reading; devised specialized (Erdmann-Dodge) tachistoscope, pendulum tachistoscope, transparent (Dodge) mirror tachistoscope; developed photographic method of eye-movement recording, falling plate camera, eye testing apparatus; research on inhibition modification of reflex response. Died Tryon, N.C., Apr. 8, 1942.

DODGE, Richard Elwood, Am. geographer; b. Wenham, Mass., Mar. 30, 1868; s. Robert Francis and Sarah Elizabeth (Wood) D.; A.B., Harvard, 1890, A.M., 1894; m. Stella Pomeroy Dalton, Aug. 19, 1896; children—Stanley Dalton, Margaret Belden, Philip Elwood. Instr., Harvard, 1894-95; instr. geology and geography Columbia Tchrs. Coll., 1895-96, asso. prof. natural science, 1896-97, prof. geography, 1897-1916; county agent leader, extension service, Connecticut, 1918-20; dean 2 yr. course, Conn. State Coll., 1920-30; prof. geog., 1926-38, prof. emeritus, U. Conn., from 1938. Asst. U. S. Geol. Survey, Northeastern, Southern Appalachians, summers 1890-95. Recipient Distinguished award Nat. Council Geog. Teachers, 1946. Fellow N.Y. Acad. Scis., Geol. Soc. Am., A.A.A.S., Assn. Am. Geog. (pres. 1915), Sigma Xi; hon. corr. mem. Geog. Soc. of Australasia. Author: Reader in Physical Geography for Beginners, 1900; Dodge's Geographies, 1903; Dodge's Geographical Note Books, 1912; Dodge-Lackey Geographies, 1927; also numerous articles on the teaching of geography; co-author Dodge and Kirchwey's Teaching of Geography in Elementary Schs.; Bowman and Dodge's English Edition of Brunhes' La Géographie Humaine; Foundations of Geography (with son Stanley Dalton Dodge), 1937; Economic Geography (with W. Harrison Carter, Jr.), 1939. Inaugurated Jour. Sch. Geography (Jour. Geography), 1897. Died Apr. 2, 1952.

DODGEN, Harold Warren, Am. chemist; b. Blue Eye, Mo., Aug. 31, 1921; s. James Monroe and Lora (Myers) D.; student Long Beach Jr. Coll., 1939-41; B.S., U. Cal. at Berkeley, 1943, Ph.D., 1946; m. Harriet Keddie Ralston, Jan. 20, 1945; children—Cynthia Jeanne, Gilbert Keddie, Stephen La Rele. Postdoctorate fellow Inst. for Nuclear Studies, U. Chgo., 1946-48; faculty Wash. State U., Pullman, 1948—, prof. chemistry, physics, 1963—, dir. nuclear reactor project, 1955—, chmn. chem. physics program, 1966—. Mem. Am. Chem. Soc., Am. Nuclear Soc., Am. Assn. U. Profs., A.A.A.S., Phi Beta Kappa, Sigma Xi, Alpha Chi Sigma. Research, publs. on radioactive and stable isotopes used to study mechanisms of chem. reactions, motion of molecules in crystals, nature of chem. bonds in crystals, speed of very fast chem. reactions in solutions. Home: 1607 Fisk St., Pullman, Wash. 99163.*

DODGSON, Charles Lutwidge (pseudonym Lewis Carroll), English mathematician; b. Daresbury, Eng., Jan. 27, 1832; s. Charles and Francis Jane (Lutwidge) D.; B.A., Christ Ch. Coll., Oxford Eng., 1854, M.A., 1857. Lectr. in math., Christ Ch. Coll., 1855-81; ordained deacon, 1861. Author: A Syllabus of Plane Algebraical Geometry, 1860; The Formulae of Plane Trigonometry, 1861; A Guide to the Mathematical Student in Reading, Reviewing, and Writing Examples, 1864; Alice's Adventures in Wonderland, 1865; An Elementary Treatise on Determinants, 1867; Phantasmagoria and other Poems, 1869; Songs from Alice's Adventures in Wonderland, 1870; Through the Looking-Glass and what Alice Found There, 1871; Euclid, Book V., Proved Algebraically, 1874; The Hunting of the Snark: an Agony in Eight Fits, 1876; Euclid and His Modern Rivals, 1879; Doublets: a Word Puzzle, 1879; Euclid, Books I and II, 1882; Rhyme? and Reason? 1883; A Tangled Tale, 1885; Alice's Adventures Underground, 1886; The Game of Logic, 1887; Curiosa Mathematica, Part L.—A New Theory of Parallels, 1888; Symbolic Logic, 1896. While he made no original contbns. to math., he posed problems in logic of interest to later mathematicians. Died Jan. 14, 1898.

DODINVAL, Pol Alfred, Belgian physician; b. Louveigne, Belgium, Nov. 26, 1931; s. Camille Jules Joseph and Jeanne (Malempre) D.; M.D., U. Liège (Belgium), 1956; m. Jeanne Versie, Apr. 23, 1960; 1 son, Luke. Asst., lab. of blood groups U. Liège, 1956, research leader depts. forensic medicine and human genetics, 1958—, dir. Human Genetics Center, 1964—; asst., lab. path. anatomy U. Brussels, Belgium, 1957-58. Asso. geneticist U. Hawaii, 1967. Recipient Count of Launoit award U. Liège. Mem. Forensic Medicine Soc., Soc. Biol. and Clin. Research (both Belgium), Human Genetics Soc. (The Netherlands), Internat. Acad. Legal and Social Medicine, World Fedn. Neurology (problem commn. neurogenetics). Research, publs. on med. and population genetics, primarily in heritable disorders of urinary excretion of aminoacids; treatment of cystine stone disease, detection of gene carriers and physiopathology of kidney in phenylketonuria. Home: Parc de Méhagne, chenee (Liege). Office: 40, q. Godefroid Kurth, Liège, Belgium.*

DODOENS, Rembert, botanist; b. Malines, Belgium, June 29, 1517; med. degree from Louvain, Belgium; also studied at Italian and German univs. Prof. in Leiden, Netherlands; physician to Maximilian II, also Rudolf II; fled to Holland during Protestant upheavals in Belgium. Author: Herbal, 1554; Pemptades, 1583. Died Leiden, Mar. 10, 1585.

DODSON, Calaway H., Am. botanist; b. Selma, Cal., Dec. 17, 1928; s. Homer C. and Leona (Jones) D.; A.B., Fresno State Coll., 1954; M.A., Claremont Coll., 1956, Ph.D., 1959; m. Piedad Marmol, Jan. 15, 1960; children—Debra Ann, David C., Thomas A. Dir. Inst. Botanico, Universidad de Guayaguil, Ecuador, 1959-60; research asso. Mo. Bot. Garden, St. Louis, 1960-61, taxonomist, curator living plants, 1961-64; asst. prof. botany, curator Herbarium U. Miami, 1964-65, asso. prof. biology, curator Herbarium, curator orchids Fairchild Trop. Gardens, 1965—; Fulbright lectr. U. Nac. Amazonia, Peru, 1964-65. Mem. Am. Inst. Biol. Sci., A.A.A.S., Soc. for Evolution, Am. Soc. Plant Taxonomists, Internat. Assn. for Plant Taxonomy. Author: Agentes de Polinizacion y su Influencia Sobre la Evolucion en la Familia Orquidacea, 1964; (with L. van der Pijl) The Orchid Flower, Its Pollination and Evolution, 1966; (with R. J. Gillespie) The Biology of the Orchids, 1967. Research, numerous publs. on classification of orchid family and selected orchid genera, effect of attraction of specific pollinators on evolution in orchids, chem. nature of fragrances which attract specific pollinators, discovery of nature of Euglossine pollination and certain types of floral mimicry such as pseudocopulation, pseudoparasitism, pseudoantagonism. Home: 10601 S.W. 74th Av., Miami, Fla. 33156. Office: U. Miami Biology Dept., Coral Gables, Fla. 33124.*

DODSON, Edward O(ttway), Am. biologist; b. Fargo, N.D., Apr. 26, 1916; s. Samuel Ottway and Fanny (Markley) D.; B.A., Carleton Coll., 1939; Ph.D., U. Cal. at Berkeley, 1946; m. Mary Katherine Street, Aug. 10, 1940; children—Thomas Stephen, Peter, Marie Louise, Kathleen Ann, Monica, David Charles. Instr., Dominican Coll., San Rafael, Cal., 1946-47; faculty U. Notre Dame, 1947-57, asso. prof., 1952-57; faculty U. Ottawa (Ont., Can.), 1957—, prof. biology, 1959—; vis. prof. biology U. Montreal (Que., Can.), 1961; vis. research prof. Roswell Park Meml. Inst., Buffalo, 1964-65. Mem. A.A.A.S., Soc. Study Evolution, Genetics Soc. Am., Genetics Soc. Can., French-Canadian Assn. for Advancement Sci., Albertus Magnus Guild, Assn. Profs. U. Ottawa (past pres.). Author: Textbook of Evolution, 1952; Genetics, the Modern Science of Heredity, 1956; Evolution: Process and Product, 1960; also numerous articles. Research on chromosomes, genetics especially mutation and evolution. Home: 15 Laval Rd., Aylmer E., Quebec, Que., Can. Office: 30 Somerset St., Ottawa 2, Ont., Can.*

DODSON, James, mathematician; flourished circa 1742; children—James, Thomas, Elizabeth. Tchr. math., master Royal Math. Sch., Christ's Hosp. mem. Fellow Royal Soc., 1755. Author: The Anti-Logarithmic Canon, 1742; The Calculator, 1747; The Mathematical Miscellany, 1747; Accountant, or a Method of Bookkeeping, 1750. Editor: Wingate's Arithmetic; An Account of the Methods Used to Describe Lines on Dr. Halley's Chart of the Terraqueous Globe, 1758. Died Nov. 23, 1757.

DODSON, John Milton, Am. physician; b. Berlin, Wis., Feb. 17, 1859; s. Nathan Monroe and Elizabeth Osborn (Abbot) D.; A.B., U. Wis., 1880, A.M., 1888; Sc.D., 1925; M.D., Rush Med. Coll., Chgo. 1882, Jefferson Med. Coll., Phila., 1883; postgrad., Berlin, 1896; m. Maie Van Slyke, July 1, 1884 (died 1887); 2d, Jessie Palmer Kasson, Nov. 12, 1890 (died 1914); children—Kasson M. (dec.), Elizabeth Palmer (Mrs. Lester J. Michael); m. 3d, Mary Hyde Webb, Jan. 17, 1923. Practiced in Wis. and Chgo., from 1882; lectr. and demonstrator anatomy, Rush Med. Coll., 1889-92, prof. physiology and demonstrator anatomy, 1892-98, prof. pediatrics, from 1899, dean students, 1901-24; professorial lectr. on medicine, and dean med. courses, U. Chgo., 1901-24; dir. Bur. Health and Pub. Instrn., A.M.A.; prof. pediatrics, Northwestern U. Woman's Med. Coll., 1894-97. Died Aug. 15, 1933.

DODT, Eberhard, German physiologist; b. Bielefeld, Germany, Feb. 22, 1923; s. Martin and Else (Pfeiffer) D.; student U. Freiburg (Germany), 1943-48, U. Erlangen (Germany), 1944-45, U. Marburg (Germany), 1945-46; m. Anne Marie Nauck, Feb. 5, 1948; children—Hans-Ulrich, Tobias. Faculty, U. Giessen (Germany), 1954—, prof. physiology, 1960—; head dept. dept. exptl. ophthalmology physiology dept. W. G. Kerckboff Inst., Max-Planck Soc., Bad Nauheim, Germany, 1966—. Mem. Max-Planck Soc. Research and numerous publs. on cone electroretinography by flicker, mode of action of warm receptors, detection of color abnormalities by electroretinography, light sensitivity of pineal organ of lower vertebrates. Home: Park Str. 1, 6350 Bad Nauheim, Germany.*

DOEBNER, Oskar Gustav, German chemist; b. Meiningen, Germany, Nov. 20, 1850; s. August Wilhelm and Pauline (Schmid) D.; student botany Jena, Germany, 1869, botany, chemistry Munich, Germany, 1869-70, chemistry Leipzig, Germany, from 1871; Ph.D., Tübingen, Germany, 1873. Asst. to Robert Otto, Braunschweig, Germany, 1874-75, to A. W. Hofmann, Berlin, 1875-79; privat dozent Chemisches Institut, Berlin, 1879-84; asso. prof. U. Halle (Germany), 1884-99, prof. chemistry, pharmacy, 1899-1907. Contbr. numerous articles to Berichte der Deutschen Chemischen Gesellschaft, others. Discovered malachite green; (with Wilhelm von Miller) synthesized quinaldine, 2.5 and 2.8 dimethylquinoline, several quinaldine carboxylic acids; research on synthesis and decarboxylizing double unsaturated acids. Died Marseilles, France, Mar. 28, 1907.

DOEHLE, Paul, German pathologist; b. Mühlhausen, Germany, June 6, 1855; s. Gustav and Amalie Rebecca (Mehler) D.; student medicine, Tübingen, Leipzig, Kiel (all Germany), Strasbourg, France; state exam Kiel, 1881. Asst., Kiel Path. Inst., 1883; became lectr. gen. pathology and path. anatomy, 1889, named chmn. dept. and hon. prof., 1908; named full prof., 1921. Research and publs. on syphilitic aorta disease (later known as Heller-Doehle aortitis); discovered lumps in protoplasm of neutrophilic leukocytes (which he considered indicative of scarlet fever). Died Kiel, Dec. 7, 1928.

DOELL, Richard Rayman, Am. geophysicist; b. Oakland, Cal., June 28, 1923; s. Raymond Arthur and Mable (Frost) D.; student U. Cal. at Los Angeles, 1940-43; A.B., U. Cal. at Berkeley, 1952, Ph.D., 1955; m. Ruth Gertrude Jones, Dec. 13, 1950; children—Patricia Kerstin, Shirley Kathleen. Lectr. U. Toronto (Ont., Can.), 1955-56; asst. prof. geophysics Mass. Inst. Tech., 1956-58; geophysicist U. S. Geol. Survey, Menlo Park, Cal., 1958—. Fellow Royal Astron. Soc., Am. Geophys. Union, Geol. Soc. Am. Research and publs. on behavior of geomagnetic field during past several million years by studies of remnant magnetization in rocks; discovered that field has repeatedly reversed its polarity. Home: 6 Patricia Pl. Office: 345 Middlefield Rd., Menlo Park, Cal. 94025.*

DOELLINGER, Johann Ignaz J., German physician; b. Bamberg, Germany, 1770; apptd. prof. physiology, Bamberg, 1794; named prof. anatomy Wurtzbourg, 1803, Landshut, 1823, Munich, 1826; dir. Anat. Mus., Munich. Mem. Munich Acad. Scis. (pres.), Bavarian Acad. Discovered a form of thickening of Descemet's membrane around the cornea, 1817; founder with C. F. Wolff and C. H. Pander of modern embryology. Died 1841.

DOELTER (Y CISTERICH), Cornelio August, mineralogist; b. Arroyo, P.R., Sept. 16, 1850; s. Karl August and Maria (Cisterich de la Torre) D.; ed. Freiburg, Heidelburg (both Germany); doctorate, 1872; m. Eleonore (div.); 1 son, 1 dau.; m. 2d, Maria Schilgerius, 1919. Joined Verband der Geologischen Reichsanstalt, Vienna, Austria, 1872; became lectr. U. Vienna; named asso. prof. U. Graz (Austria); became prof. mineralogy and petrography, 1883; succeeded G. Tschermak at U. Vienna, 1907-21. Author: Chemische Mineralogie, 1890; Edelsteinkunde, 1893; Physikalischchemische Mineralogie, 1905; Petrogenesis, 1906; Das Radium und die Farben, 1910; Die Farben der Mineralien, 1915; also numerous articles. Editor: Handbuch der Mineralchemie, 4 vols. last appearing in 1931. Founder of phys.-chem. mineralogy, also exptl. silicate chemistry; research on influence of x-rays on mineral colors, expert on vulcanology. Died Vienna, Aug. 8, 1930.

DOERFEL, Georges-Samuel, German astronomer; b. Plauen, Germany, 1643; studied comets, 1st astronomer to support the parabolic movement of stars and to recognize sun as common source of trajectories of all comets. Died Weida, Germany, 1688.

DOERR, Arthur Harry, Am. geographer; b. Johnston City, Ill., Aug. 28, 1924; s. Arthur H. and Nettie (Felts) D.; student U. Cal., Berkeley, 1943-44; B.A. with High Honors (Valedictory scholar), So. Ill. U., 1947; M.A., Ind. U., 1948; Ph.D., Northwestern U., 1951; m. Dale A. Lantrip, Aug. 15, 1947; 1 son, Marc M. Field team chief Rural Land Classification Program, P.R., 1950; faculty U. Okla., Norman, 1951—, prof. geography, 1960—, dean Grad. Coll., 1961-66; prof. geography U. Pa., 1966-67. Field team chief devel. project for Pahlavi U., Shiraz, Iran, 1966-67; intelligence expert Dept. Army, 1956; cons. ednl. panels Dept. Health, Edn. and Welfare, 1965—. Fellow Okla. Acad. Sci.; mem. A.A.A.S., Assn. Am. Geographers, Southwestern Social Sci. Assn., Nat. Council Geog. Edn., Sigma Xi, Gamma Theta Upsilon, Pi Kappa Sigma. Author: Oklahoma Coal Mining and Its Landscape Modifications, 1960; (with Lee Guernsey) Principles of Geography, 1959, Principles of Physical Geography, 1964. Contbr. articles, revs. to books, newspapers. Systematic study and analysis of climatic change, particularly in Okla.; work in area of land use and its effect on phys. and biotic environment and its econ. ramifactions. Home: 1827 Chenystone, Norman, Okla. 73069.*

DOERR, Georg Wilhelm, German physician; b. Langen (Hessen), Aug. 25, 1914; s. Erich and Marie (Hofmann) D.; ed. Gymnasium of Darmstadt, U. Heidelberg, U. Marbourg; M.D.; m. Eva Neuroth; children—Monika, Hans-Wilhelm, Renate. Prof. agrege, 1942; prof. U. Heidelberg, 1948, prof., 1963; asso. prof. U. Berlin, 1953, U. Kiel, 1956-63; dir. Inst. Pathology of Heidelberg. Mem. German Soc. Natural Sci. and Medicine (adminstrv. council), Soc. Medicine of Berlin, German Soc. Pathologists, Internat. Acad. Pathologists. Co-editor: Berichte d. Pathologie; Zwanglose Abh. aus der Normal und Patholog. Anatomie; Das Herz des Menschen, 2 vols., 1963. Research and numerous publs. on failure of system of heart vessels, forms of evolution of metabolic illnesses. Home: Am Wingertsberg 13, Heidelberg-Ziegelhausen. Office: Vossstrasse 2, Heidelberg, Germany.

DOERR, Robert, bacteriologist; b. Tescö, Hungary (now Tiacovo, USSR), Nov. 1, 1871; s. Moritz and Mina (Dujardin) D.; student medicine Vienna, Austria; state exam, 1897; m. Bertha Maria Herzog, 1924; 1 son, 2 daus. Mil. physician; apptd. lectr. gen. and exptl. pathology, Vienna, 1908, asso. prof., 1912; named prof. hygiene and bacteriology, dir. Hygiene Inst., U. Basel (Switzerland), 1919, emeritus, 1943—. Author: Die Immunitätsforschung—Ergebnisse und Probleme in Einzeldarstellung, 1947. Editor: (with C. Hallauer) Handbuch der Virusforschung, 2 vols., 1938-50; editor: Archiv für die gesamte Virusforschung, from 1939. Demonstrated pappataci fever is a viral disease, transmitted by the sandfly, phlebotomus; pioneered research on viruses and immunization; research on epidemic diseases including dengue fever, dysentery, also toxins, allergies and anaphylaxis. Died Basel, Switzerland, June 6, 1952.

DOFLEIN, Franz Theodor, zoologist; b. Paris, France, Apr. 5, 1873; s. Franz and Jane (Cooper) D.; studied zoology and medicine Munich, Germany, from 1893, doctorate, 1897; student Strasbourg, France, 1895-96; m. Leonie Rössle, 1899; 1 son, Erich; 2 daus. Asst., Munich Biol. Inst. for Research on Fish Diseases, 1897-98; became asst. State Zool. Collection, Munich, curator, 1901; apptd. lectr. U. Munich, 1903, asso. prof. zool. taxonomy and biology, 1907; appdt. 2d dir. State Zool. Collection, 1910; apptd. prof. Freiburg, Germany, 1912; became prof. Breslau, Germany, 1918; ret. 1923. Author: Lehrbuch der Protozoenkunde, 1906; (with R. Hesse) Tierbau und Tierleben, 1914; Der Ameisenlöwe, 1916; Studien zur Naturgeschichte der Protozoen, 10 parts, 1897-1918; Ostasienfahrt, 1906; Mazedonien, 1921. Research on ecology of animals, origin of species, genetics, termites and ants, taxonomy of decapods, protozoa especially as diseases carriers. Died Obernigk, nr. Breslau, Aug. 24, 1924.

DOGGETT, Wesley Osborne, Am. physicist; b. Brown Summit, N.C., Jan. 24, 1931; s. Banks Chandler and Elizabeth (Dobbs) D.; B.Nuclear Engring., N.C. State U., 1952, B.E.E., 1953; M.A. in Physics, U. Cal. at Berkeley, 1954, Ph.D. (NSF fellow), 1956; m. Leonor Pinzon, June 13, 1953; children—Kevin W., Marc G., Norman A., Eric L., Valerie G., Nydia L., Steven N. Physics research asso. U. Cal. Radiation Lab., Berkeley, 1954-56; faculty N.C. State U., Raleigh, 1958—, prof. physics, 1962—, asst. dean Sch. Phys. Scis. and Applied Math., 1964—. Dir. Troxler Electronics, Inc., Raleigh. Cons. Research Triangle Inst., 1962—, Office Civil Def., 1963—. Recipient N.C. Acad. Sci. Poteat award, 1961, 62. Mem. Am. Inst. Physics, A.A.A.S., Am. Phys. Soc., Am. Nuclear Soc., Am. Assn. Physics Tchrs., Am. Soc. for Engring. Edn., Sigma Xi (Research award N.C. State U. chpt. 1962), Phi Kappa Phi. Research, publs. on devel. and analysis of statis. methods for investigating half-lives of daughter radioactivities that are in secular equilibrium with parent activities; theoretical analysis of gamma-ray penetration in matter; discovered radioactive isomer rubidium 81-m. Home: 2452 Oxford Rd., Raleigh, N.C. 27608.*

DOGIGLI, Hans, German engr.; b. Traunstein, Bavaria, Oct. 24, 1916; s. Anton and Franziska (Biberger) D.; ed. Poly. and Higher Tech. Sch., Munich; m. Gertrude Frahm; children—Johannes, Michael. Mem. Technico-Lit. Soc. Author: Drahtloses Jahrhundert; Magie der Strahlen; Entfesselte Atomkraft; Entfesselte Naturkraft; Strahlende Materie; Von 1000 Lebenswundern; Erste Ultraviolett-Raumentkeimung in Operationssälen in Europa Eingerichtet. Made 1st calculator of antennas of world. Home: Gebelestrasse 12, Munich 27. Office: Theresienstrasse 69, Itab-Verlag, Munich, Germany.

DOHERTY, Henry Latham, Am. inventor; b. Columbus, O., May 15, 1870; s. Frank and Anna (McIlvaine) D.; ed. pub. schs.; hon. D.Eng., Lehigh U., 1931; m. Mrs. P. F. Eames, 1929. With Columbus Gas Co. circa 1882-1890; engr. or mgr. pub. utility cos., Madison, Wis., St. Paul, San Antonio, Denver, and 25 other cities until 1905; organizer, mgr. Henry L. Doherty & Co., bankers and operators of pub. utility corps., 1905—; organizer, pres. Cities Service Co., (holding co. for more than 190 pub. utility and petroleum properties), 1910—. Mem. orgn. bd. World's Congress on Electricity, St. Louis, 1904. Leader in Am. in gas and electric arts and industries; leader in movement for oil conservation by means of unit operation of pools under fed. control; patentee many combustion processes and apparatus and originator of many standard practices. Recipient 1st Beall gold medal, Am. Gas Light Assn., for paper Gas for Fuel 1898; Walton Clark medal, Franklin Inst. in consideration of outstanding and valuable work in devel. of the manufactured gas industry, 1930. Died Dec. 26, 1939.

DOHERTY, Robert Ernest, Am. elec. engr.; b. Clay City, Ill., Jan. 22, 1885; s. Anthony and Clara (Sauther) D.; B.S., U. Ill., 1909; M.S. Union Coll., Schenectady, N.Y., 1921; hon. M.A., Yale, 1931; LL.D., Tufts Coll., U. Pitts., 1936; D.Sc., Waynesburg Coll., 1948; m. Pearl Edna Mills, June 20, 1911; children—Robert Ernest, Vera Maud, James Anthony. With Gen. Electric Co., Schenectady, 1909-31, test engr., 1909-10, designing engr., 1910-18; asst. to C. P. Steinmetz, 1918-23, cons. engr., 1923-31; prof. elec. engring. Yale, 1931-33, dean Sch. of Engring., 1933-36; pres. Carnegie Inst. Tech., Pitts., 1936-50; chmn. Engrs. Council for Profl. Devel., 1941-43; mem. NACA, 1940-41; mem. Adv. Com. for Engring. Sci. and Mgmt. War Tng., 1940-46, adv. com. Army Specialized Tng. Div., 1943-46; Civil Adv. Council. Office Chief of Ordnance, 1942-45. Recipient Lamme medals: for engring., Am. Inst. E.E., 1937; for edn., Soc. for Promotion Engring. Edn., 1945. Mem. Am. Inst. E.E., Social Science Research Council, Am. Soc. for Engring. Edn. (pres. 1943-44). Theta Tau, Sigma Xi, Tau Beta Pi, Eta Kappa Nu. Omicron Delta Kappa. Author: Mathematics of Modern Engineering (with E. G. Keller), 1936. Helped develop high tension elec. system of U. S.; extended theory of alternating current in machinery. Contbr. tech. and ednl. articles. Died Oct. 19, 1950.

DOHRN, Anton Felix, German zoologist; b. Stettin, Germany (now Szczecin, Poland), Dec. 29, 1840; s. Carl August and Adelheid (Dietrich) D.; studied at Königsberg, Bonn, Berlin, Jena (all Germany); doctorate Breslau, Germany; m. Maria von Baranowska, 1874; 4 sons including Wolfgang, Reinhard. Lectr. zoology in Jena, 1867-71; mem. sci. expdns. to Scotland and Messina; founder Zool. Sta., Naples, Italy, 1874, dir. until 1909. Named Hon. Citizen of Naples. Fellow Royal Soc., 1899. Author: Fauna und Flora des Golfes von Neapel; Der Ursprung der Wirbeltiere und das Prinzip des Funktionswechsels, 1875; Studien zur Urgeschichte des Wirbeltierkörpers, 1881-1907. Research on origin and evolution of vertebrates (which he believed could be traced to annelids), crustaceans and other marine animals of Mediterranean and English coasts. Died Munich, Germany, Sept. 26, 1909.

DOHRN, Carl August, entomologist; b. Stettin, Poland, June 27, 1806; s. Heinrich and Johanna (Hüttern) D.; m. Adelheid Dietrich, 1837; hon. Ph.D., U. Königsberg (Germany), 1862; children—Heinrich, Anton, Wilhelm, Anna (Mrs. Gustav) Wendt. Educated as businessman; first dir. Pommersche Prov. Zuchersiederei until 1872; mem. Entomologischer Verein (founder 1837, pres.). Author: Catalogue Coleopterorum Europae, 1855; also numerous articles. Editor: Linnaea entomologica, 16 vols., 1846-66. Editor, Entomologische Zeitung, 1843-87. Research on entomology; his well known beetle collection given to Stettin Mus. Died Stettin, May 10, 1892.

DOHRN, Max, German pharm. and physiol. chemist; b. Farnroda, Thuringia, Germany, Aug. 30, 1874; s. Wilhelm and Marie (Jungnickel) D.; student at Berlin, Leipzig, Heidelberg (all Germany); doctorate, 1899; m. Luise Berta Antoinie Harlacher, 1910; 1 dau. Asst., U. Marburg (Germany), U. Berlin; joined sch. lab. Chemische Fabrik (formerly E. Schering), Berlin, 1902; founder physiol. lab., 1904. Discovered phenyl-quinoline-carboxylic acids, drugs against bacterial infections; research on hormones, function of cow's pancreas and its significance against diabetes. Died Rothaus, Germany, June 17, 1943.

DOIG, Peter, Brit. astronomer, marine engr.; b. Jan. 26, 1882; s. Peter and Martha (Rodger) D.; ed. Glasgow, Scotland; m. Margaret Paterson Scott, 1908. Ship draughtsman, trainee John Brown and Co., Clydebank, Scotland; engaged in shipbldg. in Scotland, Ireland, U. S., Shanghai, China; gen. sec. Assn. Engring. and Shipbldg. Draughtsmen, 1918-45. Author: An Outline of Stellar Astronomy; A Concise History of Astronomy; also papers on marine propulsion and astron. topics. Editor Brit. Astron. Assn. Jour., 1930-37, also from 1948. Specialist on screw propeller design. Died Oct. 13, 1952.

DOISY, Edward Adelbert, Am. biochemist; b. Hume, Ill., Nov. 13, 1893; s. Edward Perez and Ada (Alley) D.; A.B., U. Ill., 1914, M.S., 1916, Sc.D., 1960; Ph.D., Harvard, 1920; Sc.D., Washington U., 1940, Yale, 1940, U. Chgo., 1941, Central Coll., 1942, U. Paris, 1945; LL.D., St. Louis U., 1955; Sc.D., Gustavus Adolphus Coll., 1963; m. Alice Ackert, July 20, 1918 (dec. Aug. 1964); children—Edward Adelbert, Robert A., Philip P., Richard J.; m. 2d, Margaret McCormick, Apr. 19, 1965. Prof., Sch. Medicine St. Louis U., 1923—, Distinguished Service prof., 1951-65, emeritus prof., 1965—. Mem. Nat. Adv. Cancer Council, 1941-42, 46-51; mem. com. biology, medicine AEC, 1952-58. Recipient Gold medal St. Louis Med. Soc., 1935; Philip A. Conne medal Chemists Club N.Y., 1935; St. Louis award, 1939; Willard Gibbs award, 1941; Nobel prize in physiology and medicine (with Henrik Dam), 1943; Am. Pharm. Mfg. Assn. award, 1942; Squibb award, 1944; St. Louis U. Fleur de Lis, 1951; Comml. Solvent award, 1952; Illini Achievement award U. Ill., 1958. Mem. Am. Soc. Biol. Chemists, Am. Chem. Soc., Soc. Exptl. Biology and Medicine, Endocrine Soc., Am. Cancer Soc., A.A.A.S., Am. Acad. Arts and Scis., Nat. Acad. Scis., Am. Philos. Soc., Pontifical Acad. Scis. Author: Sex Hormones, 1936; (with Edgar Allen, C. H. Danforth) Sex and Internal Secretions, 1939; Female Sex Hormones, 1941. Contbr. numerous articles to profl. jours. Publs. on isolation of first pure crystalline female sex hormone, theelin (1929), dihydrotheelin (1936), used in treatment of female disorders, and first crystalline vitamin K (1939), used in treatment of impaired coagulation; studied relationship between menstrual flow and estrogen activity; metabolism, blood buffers; also insulin. Home: 4B, Colonial Village Ct., Webster Groves, Mo. 63119. Office: 1402 S. Grand Blvd., St. Louis 63104.*

DOKHMAN, Genrietta Isaakovna, Russian phytocenologist, geobotanist; b. Kremenchug, 1897; grad. Physico-Math. Faculty, Moscow U.; postgrad. Moscow U., 1927-30; Cand. Biol. Sci., 1937; D.Biol. Sci., 1942. Instr. botany Krupskaya Acad. Communist Edn., 1926-30; geobotanist Inst. Agropedology, 1930-31; with dept. territorial orgn., Territorial Orgn. Research Inst., USSR Peoples Commissariat of Agr., 1931-32; instr. Moscow U., 1932-38, lectr., 1938-46, prof. chair geobotany Biol. Faculty, 1946—. Author: The Social Life of Plants, 1927; The Classification of Steppes, 1937; A History of the Vegetation of the USSR, 1938; The Plant and Its Life, 1940; The Study of the Vegetation of Steppe Reservations, 1949; The Beginnings of Experimental Phytocenology, 1959. Address: Moscow University, Leninskie gory, Moscow, USSR.

DOKUMOV, Stoyan Ivanov, Bulgarian physician, gynecologist, endocrinologist; b. Pazardjik, Bulgaria, Sept. 14, 1918; s. Ivan and Maritza (Koprivshka) D.; M.D., U. Sofia, 1944. Community doctor, 1944-47; mem. Clinic Obstetrics and Gynecology, Postgrad. Med. Inst., Sofia, Bulgaria, 1947-57, mem. Clinic Endocrinology and Metabolic Diseases, 1957—, sr. research officer, 1961—, tchr. endocrine gynecology, 1957—. Mem. Bulgarian Endocrine Soc., Bulgarian Soc. Obstetrics and Gynecology. Co-author: Diagnosis of the Endocrine and Metabolic Diseases, 1962; Laboratory and X-ray Diagnostics in Obstetrics and Gynecology, 1962. Research, publs. on human intersexuality, classification; of human hermaphroditism, vaginal smears, normal and path. ovaries, pregnancy and newborn. Home: 10 Assen Zlatoarov-str., Sofia 4. Office: 6, Gen.Chr. Mihailov-str., Sofia 3, Bulgaria.

DOLAMORE, William Henry, Brit. dental surgeon; cons. dental surgeon, dean Royal Dental Hosp.; cons. dental surgeon St. Mary's Hosp.; external examiner in dental surgery, univs. Liverpool, Leeds; mem. Gen. Med. Council; treas. Dental Bd. U.K. Licentiate in dental surgery. Fellow Royal Coll. Surgeons (bd. dental examiners); pres. odontological sect. Royal Soc. Medicine; pres. Brit. Dental Assn., 1915-18. Died Apr. 19, 1938.

DOLBEAR, Amos Emerson, Am. inventor; b. Norwich, Conn., Nov. 10, 1837; grad. Ohio Wesleyan U., 1866; A.M., M.E., Ph.D., all U. Mich.; LL.D., Tufts Coll., 1902. Prof. physics, Tufts Coll., from 1874. Recipient bronze medal for acoustic apparatus, Centennial Expn., Phila., 1876, and silver medal, Paris, 1881, and gold medal, London, 1882, for static telephone. Author: Chemical Tables; Art of Projecting, 1876; The Speaking Telephone; Matter, Ether and Motion, 1892; Modes of Motion; Natural Philosophy; Machinery of the Universe, 1897. Invented writing telegraph, 1864; magneto telephone, 1876; static telephone, 1879; spring balance ammeter, 1889; air space telegraph cable, 1882; discovered convertibility of sound into electricity, 1873; telegraphing without wires, 1881; photographing with electric waves, 1893. Died Feb. 23, 1910.

DOLBEAR, Samuel Hood, Am. mining engr.; b. Somerville, Mass., Dec. 6, 1886; s. Amos Emerson and Alice Jeannette (Hood) D.; spl. and summer courses Tufts Coll., 1901, 04-06; student Clark University, Worcester, Massachusetts, 1904, M.Sc. (honorary), 1939, D.Sc. (honorary), 1962; m. Inza Snowden Jordan, May 31, 1941. In charge constrn. soda-ash plant Cal. Trona Co. (now Am. Potash & Chemical Co.), 1906-09; pres. Am. Steel Bar Mfg. Co., 1913-14; Pacific Coast cons. Am. Refractories Co., 1914-20; pres. Piedra Magnesite Co., Cal., 1917-18; operated chrome, magnesite and manganese mines in Cal., Ore., Ariz., 1916-20; pres. Selective Treatment Co., Ltd., 1922-31; engaged in research and development beneficiation of asbestos ore in Can.; exam. mines in Korea, 1936; pres. Wright Dolbear Co., Inc., engrs. and geologist, N.Y., Toronto and San Francisco, 1937-43; pres. Behre Dolbear and Company, Incorporated, mining engineers and geologists, 1944-61, consultant, 1961—, engaged in mineral exams., geol. and econ. studies in U. S., Can., Mexico, China, Philippines, Burma, Turkey, Cuba, Haiti, Central and S.A. Adv. Com. and cons. U. S. Bur. of Mines, 1937-40; mining engr. cons. Cal. State Div. of Mines, 1944-45, making statewide study of postwar outlook for mine operations and employment in state mineral industries; mineral advisor Provisional Gov. Republic of Korea, 1945; tech. advisor Chinese Nationalist Gov. study reserves, development and mechanization of tin mines, Yunan, natimony mines, Hunan and Tungsten mines, Kiangsi Provinces, 1946-47; tech. advisor on minerals Burma Gov., 1948, Gov. of Haiti, 1949; cons. Govt. of Denmark, 1958; lectr. U. So. Cal., 1943. Trustee Clark U., 1940-61, vice chmn. trustees, 1959-60. Hal Williams Hardinge award, outstanding achievement, Am. Inst. Mining and Metall. Engrs., 1960. Mem. Am. Inst. Mining and Metall. Engrs. (1st chmn. indsl. minerals div. 1935), Soc. Econ. Geologists, Mining and Metall. Soc. America, Am. Mining Congress. Author: Economic Mineral Resources of California, 1945; also 75 tech. papers and bulls. Editor Ind. Minerals and Rocks, 1937-47. Research on steel reinforcement of concrete curbs; production of barium chloride; extraction of asbestos from ore; solution mining of soluble minerals with abrasives. Patentee 35 U. S. and fgn. metall. processes and equipment. Home: 1000 Park Av. Office: 11 Broadway, N.Y.C. 4.*

DOLBEAU, Henri Ferdinand, French surgeon; b. Paris, France, Aug. 3, 1830; became surgeon in hosps. of Paris; named prof. external pathology Paris Faculty Medicine, 1877; named surgeon of Bur. Central, 1858. Mem. Acad. Medicine. Author: Recherches sur les vaisseaux du bassin, 1855; Traité de la pierre dans la vessie, 1864; De la lithotritie périnéale, 1872. First to perform perineal lithotrity (Dolbeau's operation); research on bladder stones, cartilaginous tumors of jaw and pelvis. Died Mar. 10, 1877.

DOLBEAULT, Pierre, French mathematician; b. France, Oct. 10, 1924; s. Marcel and Cécile (Rablat) D.; agrégation, Ecole normale supérieure, Paris, 1947; postgrad. Princeton, 1949-50; Sci.Dr., U. Paris, 1955; m. Simone Lemoine, Dec. 24, 1955; 1 son, Jean. Attaché, Centre National de la Recherche Scientifique, Paris, 1947-53; with U. Bordeaux (France), 1954-60, lectr., 1956-60; prof. U. Poitiers (France), 1960—. Mem. Société mathématique de France, Am. Math. Soc. Research, publs. on several complex variables including global properties of complex manifolds by methods of algebraic topology, differential forms, differential forms with singularities, divisors. Home: La Fenetre 86, Biard, France. Office: route de Chauvigny 86, Poitiers, France.*

DOLE, Hollis Mathews, Am. geologist; b. Paonia, Colo., Sept. 4, 1914; s. Edwin Enyart and Mary (Mathews) D.; B.S., Ore. State U., 1940, M.S., 1942; postgrad. U. Cal. at Los Angeles, 1941; Ph.D., U. Utah, 1955; m. Ruth Josephine Mitchell, Sept. 29, 1942; children—Michael Hollis, Stephen Eric. Mining engr. U. S. Bur. Mines, Scappoose, Ore., 1942-43; geologist U. S. Geol. Survey, Tucson, 1946-47; field geologist Ore. Dept. Geology and Mining Industries, Grants Pass, Ore., 1947-49, staff geologist, Portland, 1949-54, asst. dir., 1954-55, state geologist, 1955—; prof. Ore. Bd. Higher Edn., Portland State Coll., 1948-49. Mem. Am. Mining Congress (mem. interstate oil compact commn. 1955—), Am. Inst. Mining Engrs., Am. Assn. Petroleum Geologists, Am. Assn. State Geologists, Sigma Xi. Contbr. articles to tech. jours. Research on mechanics chromite emplacement in S.W. Ore., stratigraphic succession Upper Jurassic-Lower Cretaceous strata in S.W. Ore. Home: 2612 N.E. 23d St., Portland 97212. Office: State Office Bldg., 1500 S.W. 5th St., Portland, Ore. 97201.*

DOLECEK, Richard Leroy, Am. physicist; b. Wilson, Kan., Dec. 28, 1911; s. Vitus E. and Sophia (Macek) D.; A.B., U. Kan., 1933, M.A., 1934, Ph.D., 1937; m. Avis E. Preusch, June 7, 1937; children—Quentin E., Gayle R., Elwyn H. Asso. prof. physics S.D. State Coll., Brookings, 1937-42; prof. physics Tex. Technol. Coll., Lubbock, 1942-46; supt. solid state div. U. S. Naval Research Lab., Washington, 1946-63, asso. dir. for materials, 1963—. Fellow Am. Phys. Soc., A.A.A.S. Research in low temperature physics, solid state physics; magnetism and electricity; cryomagnetics; cryogenics. Home: 102 Rolph Dr., Washington 20021. Office: Code 6000, U. S. Naval Research Lab., Washington 20390.

DOLEZAL, Eduard, geodesist; b. Mähr, Budwitz, Mar. 2, 1862; s. Franz and Eleonore D.; ed. Tech. U., U. Vienna (Austria); numerous hon. doctor degrees. Began as asst. practical geometry Technische Hochschule Vienna, 1887; became prof. Tech. High sch., Sarajevo (now in Yugoslavia), 1889; returned to Technische Hochschule Vienna, 1896; first to lectr. on photogrammetry; apptd. prof. descriptive and practical geometry, 1899; later became prof. practical geometry and mining surveying Mining U., Leoben; became prof. lower geodesy Tech. U., Vienna, 1905. Mem., pres. Austrian Commn. for Internat. Geodesy, 1913-37. Recipient Goethe medal for art and sci., 1942. Mem. Leopoldina, Spanish, Austrian acads. sci., Internationale Gesellschaft für Photogrammetrie (founder). Author: Anwednung der photographie in der praktischen Messkunst, 1896; Theoretische und praktische Anleitung zum Nivellieren, 1902; Hand- und Lehrbuch der Niederen Geodäsie, Neuarbeitung des Hartnerschen Lehrbuches, 3 vols., 1903-04; also numerous articles. Founder, editor Internat. Archive for Photogrammetry, 6 vols., 1908-23. Reformed mining research; centralized state surveying systems; created fed. bur. weights and measures. Died Baden nr. Vienna, July 7, 1955.

DOLEZALEK, Friedrich, physico-chemist; b. Szigeth, Hungary, Feb. 5, 1873; s. Carl Borromäus and Adelheid (Frankenberger) D.; student Tech. U. Hannover (Germany), 1893-95; doctorate Göttingen, Germany, 1898; m. Helene Samwer, 1900; m. Paula Bomhoff, 1910; 2 sons, 1 dau. Asst., Inst. Phys. Chemistry, Göttingen; joined State Phys.-Tech. Inst., Berlin-Charlottenburg, Germany, 1900; with Siemens & Halske, 1901; became lectr. on chemistry and iron forging Tech. U. Berlin, 1902; named asso. prof. Tech. U. Danzig (Poland), 1904; became dir. Inst. for Phys. Chemistry, Göttingen, 1905; apptd. prof. physics Technische Hochschule Berlin, 1907, prof. Physical chemistry and electrochemistry in 1913. Author: Die Theorie des Bleiakkumulators, 1901; Theorie der binaren Gemische und Konzentrierten Lösungen, 1908; also articles. Research in electrotechnics, phys. chemistry, thermodynamic theory homogeneous mixtures, chem. reactions in lead accumulator, solubility and conductivity of solutions, theory of binary mixtures and concentrated solutions; improved long distance telephone; invented quadrant electrometer. Died Berlin-Charlottenburg, Dec. 10, 1920.

DOLGINOV, Shamaia Shlemovich, Russian geophysicist; b. 1917; grad. physics dept. Leningrad (USSR) U., 1942; candidate physics-math. scis., 1945. Head magnetic lab. Inst. Earth Magnetism, Ionosphere and Distbn. of Radiowaves, 1946—. Recipient Lenin prize for part in discovery and study of outer radiational belt of earth and moon, 1960.

DOLGOPLOSK, Boris Aleksandrovich, Russian organic chemist; b. Nov. 12, 1905; grad. Moscow U., 1931. With synthetic rubber plants, 1932-46; instr. Yaroslavl Tech. Inst., 1944-45, prof., 1945-46; asso. All-Union Research Inst. Synthetic Rubber, 1946—, Inst. High-Molecular Compounds, USSR Acad. Sci., 1948—. Recipient Stalin prize, 1941, 49. Mem. USSR Acad. Sci. Research and publs. on chemistry and practical uses of polymerization processes, connection between structure and properties of rubber, carboxylate rubber synthesis and rubber extraction from it with properties resembling natural rubber. Address: All-Union Research Inst. of Synthetic Rubber, Gapsalskaya ulitsa 18, Leningrad. USSR.

DOLGUSHIN, Donat Aleksandrovich, Russian agrobiologist, selectionist; b. Penza, 1903; grad. Tiflis Poly. Inst., 1927; D.Biol. Sci., 1936. With Azerbaijan Central Agrl. and Selection Exptl. Sta., Gandzha (now Kirovabad), All-Union Selection and Genetics Inst., Odessa, until 1944; head selection dept. exptl. base All-Union Lenin Acad. Agrl. Sci., Moscow, 1944-51; dep. sci. dir. Lysenko All-Union Selection and Genetics Inst. Decorated Order of Lenin; recipient Stalin prize, 1941. Mem. All-Union Lenin Acad. Agrl. Sci. Author: Seed-Growing of Cereals, 1941; Darwinism and Plant Selection, 1960. Research and publs. on selection of ramified and other high yield strains of winter and spring wheat, developer new wheat strains, theory of stage devel. of plants and its use in selection and growing of field crops. Address: Lysenko All-Union Selection and Genetics Inst., Ovidiopolskaya doroga 5, Odessa, Kurainian SSR, USSR.

DOLL, William Richard Shaboe, English epidemiologist; b. Hampton, Eng., Oct. 28, 1912; s. Henry William and Amy Kathleen (Shaboe) D.; M.B., B.S., St. Thomas Hosp. Med. Sch., London, Eng., 1937; M.D., U. London, 1945, D.Sc., 1958; m. Joan Mary Faulkner (Blatchford) Oct. 4, 1949; children—Nicholas, Catherine. House physician St. Thomas's Hosp., 1937, Brit. Postgrad. Med. Sch., 1939; research asst. Central Middlesex Hosp., 1946; mem. statist. research unit Med. Research Council, London, 1948-61, dir., 1961—, dep. dir. designate Clin. Research Centre, 1966—; hon. asso. physician Central Middlesex Hosp., 1949—; hon. lectr. epidemiology U. Coll. Hosp., Med. Sch., London, 1962—. Mem. adv. com. on med. research WHO, 1963-66, mem. sci. council Internat. Agy. for Cancer Research, 1966—. William Julius Mickle fellow U. London, 1955. Decorated Order Brit. Empire. Recipient David Anderson Berry prize Royal Soc. Edinburgh, 1958; Bisset Hawkins medal Royal Coll. Physicians, 1962; UN award for cancer research, 1962. Research and numerous publs on relationship of cigarette smoking to cancer and other diseases, X-rays and the prodn. of leukemia, efficacy of gastric ulcer treatment. Home: 24 Lansdowne Rd., London W.11. Office: Med. Research Council Statis. Research Unit, 115 Gower St., London W.C. 1, Eng.*

DOLLAR, Alexander Melville, food chemist, biochemist; b. Vancouver, B.C., Can., Apr. 7, 1921; (parents Am. citizens); s. Alexander M. and Alice Mae (Allen) D.; B.S. in Tech. Criminology, U. Cal. at Berkeley, 1948; M.S., U. Cal. at Berkeley, 1949; postgrad. U. So. Cal., 1953-55; Ph.D., U. Reading (Eng.), 1958; m. Jean Pitchford Rosen, Apr. 15, 1944; 1 dau., Cathlin. Microchemist, Don Baxter, Inc., Glendale, Cal., 1949-52; staff Griffin-Hasson Labs., Los Angeles, 1952-54; cons. on food products and metal processing chem. problems, Los Angeles, 1952-55; research staff U. Cal. at Berkeley, 1958-59; prof. food sci. Coll. Fisheries, U. Wash., Seattle, 1959-67; supr. Hawaii Devel. Irradiator, Hawaii Dept. Agr., Honolulu, 1967—. Cons. on trout and other fish feeds, 1960—. Mem. Am. Chem. Soc., A.A.A.S., Inst. Food Tech., Biochem.Soc. (London), Nutrition Soc. (London), Sigma Xi. Research and publs. on biochemistry of foods, radiation processing of foods and seeds, mechanism of deterioration in food products, devel. animal and fish feed formulations, comparative nutrition of animals. Home: 1615 Wilder St., Honolulu. 96822.*

DOLLARD, John, Am. psychologist; b. Menasha, Wis., Aug. 29, 1900; s. James E. and Ellen (Brady) D.; A.B., U. Wis., 1922; A.M., U. Chgo., 1930, Ph.D., 1931; M.A. (hon.), Yale, 1952; m. Victorine Day, Nov. 28, 1930 (div. 1959); children—Julie (Mrs. R. R. Bradford), John Day, Victorine, Peter Day; m. second, Joan Ganis Palance, Dec. 23, 1961. Secretary Memorial Union Building Com., U. Wis., 1923-26; asst. to the pres. U. Chgo., 1926-29; Social Sci. Research Council fellow in social psychology, Germany, 1931-32; research work Inst. Human Relations, Yale, 1932-52, prof. psychology, 1952—, fellow Berkeley Coll.; expert cons. to Sec. of War, 1942-45. Pres. Conn. Mental Health, 1949-52. Served as pvt. U. S. Army, 1918. Certified profl. psychology, 1947. Fellow Am. Psychol. Assn. (diplomate in clin. psychology), A.A.A.S., Am. Acad. Arts and Scis.; mem. Phi Beta Kappa, Sigma Xi, Alpha Tau Omega. Author of ten books including: Caste and Class in a Southern Town, pub. 1937; Social Learning and Imitation (with N. E. Miller), 1941; Personality and Psychotherapy (with N. E. Miller), 1950; Steps in Psychotherapy (with Frank Auld, Jr., and Alice M. White), 1953; Scoring Human Motives (with Frank Auld, Jr.), 1959. Home: 102 York Square, New Haven. Office: 333 Cedar St., New Haven.*

DÖLLINGER, Ignaz Christoph, German anatomist, physiologist; b. Bamberg, Germany, May 24, 1770; s. Ignaz and Magdalene (Flöckher) D.; student medicine Bamberg, Würzburg (both Germany), Vienna, Austria, Pavia, Italy; doctorate Bamberg, 1794; m. Theresia Schuster, 1797; 5 sons including Ignaz; 3 daus. Physician to poor, Bamberg; became prof. medicine Bamberg U., 1796; apptd. prof. physiology and normal and path. anatomy, Würzburg, 1803; became curator Acad. Scis., Munich, Germany, 1823; later became prof. anatomy and physiology, dir. Anat. Mus., U. Munich, 1826; sec. math.-phys. class Acad. Scis., 1827-38. Author: Grundriss der Naturlehre des menschlichen Organismus, 1805; Beiträge zur Entwicklungsgeschichte des menschlichen Gehirns, 1814; Über den Wert und der Bedeutung der vergleichenden Anatomie, 1814; Illustratio iconographica fabricae oculi humani, 1817; Was ist Absonderung und wie geschieht sie?, 1819; Lehrbuch der Physiologie, 1835-36; Grundzüge der Physiology der Entwicklung des Zell-, Knocken- und Blutsystems, 1842. Research on embryology, use of microscope in medicine, blood and blood circulation, spleen, portal veins, liver, glandular secretion, eye, comparative anatomy. Died Munich, Jan. 14, 1841.

DOLLOND, George, English optician; b. London, Eng., Jan. 25, 1774; sent to learn trade of math. instrument-making in Fairborne's factory; began apprenticeship to his uncle, 1788; became partner with uncle, Peter Dollond, 1805. Fellow Royal Soc., 1819, Royal Geog. Soc. (charter); mem. Astron. Soc. (founding). Contbr. articles to tech. jours. Invented an improved altazimuth, 1821, double altitude instrument, 1823, atmospheric recorder, micrometer of quartz, achromatic magnification lenses, instrument for measuring horizontal and vertical angles, a kilometer counter; constructed a heliometer. Died 1852.

DOLLOND, John, English optician; b. Spitalfields, London, Eng., June 10, 1706; trained as silk weaver; self-educated in sci.; 2 sons, including Peter; 3 daus., including Mrs. Jesse Ramsden; began as silk weaver; joined son Peter as optician, 1752; named optician to king, 1761. Recipient Copley medal, 1758. Fellow Royal Soc., 1761. Contbr. articles. to profl. jours. Invented triple objectives, 1757-58, achromatic telescope (independently invented by Chester Moore Hall), 1758, modern heliometer, 1754; showed Newton's idea that chromatic aberration could not be avoided was wrong. Died London, Nov. 30, 1761.

DOLLOND, Peter, English optician; b. London, Eng., 1730; s. John D.; 2 daus.; worked as silk weaver, Spitalfield, several years; optician, London, 1750-1819. Mem. Am. Philos. Soc. Author: Some Account of the Discovery Made by the Late Mr. John Dollond, F.R.S., which Led to the Grand Improvement of Refracting Telescopes, 1789. Built triple achromatic lens, 1765, eirometer, 1811, heliometer, goniometer, patent binnacle compass; improved achromatic telescope, 1800, also refracting telescope; observed from Greenwich, transit of Venus, 1769. Died Kennington, Eng., July 2, 1820.

D'OLMALIUS D'HALLOY, Jean-Baptiste Julien, geologist; author: Carte Minéralogique de l'Empire Française, 1825; contbr. theory of metamorphism; defined the cretaceous system of mesozoic group, 1822.

DOLMAN, Claude Ernest, microbiologist; b. Porthleven, Cornwall, Eng., May 23, 1906; s. John Ernest and Peternal (Holloway) D.; M.B., B.S., U. London, 1930, D.P.H., 1931, Ph.D., 1935; m. Clarisse Leonore Askanazy, Dec. 6, 1955; children—John Frederick, Jennifer Elizabeth, Peter Julian. Dir. div. labs. Dept. Health and Welfare B.C., 1935-56; prof., head dept. bacteriology, preventive medicine U. B.C., Vancouver, 1936-54, prof., head dept. bacteriology, immunology, 1954-65, research prof. dept. microbiology, 1965—; cons. bacteriologist Vancouver Gen. Hosp., 1936—. Fellow Am. Pub. Health Assn., N.Y. Acad. Sci., Royal Soc. Can. (past pres. sci. sect.), Royal Coll. Physicians London, Royal Coll. Physicians Can.; mem. Am., Canadian socs. microbiologists, Am. Assn. Immunologists, Canadian Pub. Health Assn. (past pres. lab. sect.), Canadian Assn. Med. Bacteriologists (past pres.). Contbr. numerous articles to profl. jours. Research in various fields of med. microbiology, food-borne infections and toxemias, staphylococcal food poisoning and botulism. Home: 1611 Cedar Crescent, Vancouver 9, B.C., Can.*

DOLOMIEU, Dieudonné-Sylvain-Guytancrède Gratet de, see de Dolomieu, Dieudonné-Sylvain-Guytancrède Gratet.

DOLUKHANOV, Armen Georgievich, Russian geobotanist, phytocenologist; b. Tbilisi, 1900; grad. Forest Mgmt. Faculty, Tbilisi Timber Inst., 1931. Asso., Kirovakan Exptl. Forestry Sta., 1931-33; sr. asso. Inst. Botany, Georgian Acad. Sci., 1933—. Author: The Vegetation of the Lagodekhi National Park, 1942; The Forests of Zangezur, 1949; The Dendroflora of the Caucasus, 1960. Research and publs. on geobotany of forests, typology of mountain forests in Transcaucasia. Address: Inst. of Botany, Georgian Acad. Sci., Tbilisi, Gruz, SSR, USSR.

DOMAGK, Gerhard, chemist, pathologist; b. Lagow, Brandenburg, Germany, Oct. 30, 1895; s. Paul and Martha (Reimer) D.; M.D., U. Kiel, 1921. Privatdozent in medicine, U. Greifswald, 1924-25; apptd. to Pathological Institute, Münster, 1925; extraordinary prof. of general pathology and pathological anatomy, U. Münster, 1928; dir. of research, I.G. Farbenindustie, Elberfeld (now Wuppertal), 1927. Awarded Nobel Prize, Medicine, 1939 (arrested by Nazis and forced to decline prize; in 1947 received certificate and Nobel gold medal). Author: Chemotherapie bakterieller Infektionen, 1940; Chemotherapie der Tuberkulose mit Thiosemikarbazonen, 1950. Discovered antibacterial properties of prontosil; revealed extraordinary powers of sulfonamide drugs (ushered in age of chemotherapy); studied anti-tubercular drugs and cancer. Died Burberg, Baden-Wurttenberg, Apr. 24, 1964.

DOMANIG, Erwin, Austrian surgeon; b. Vienna, Austria, Mar. 21, 1898; s. Karl and Irmgard (Müller) D.; M.D., Faculty Medicine, Vienna; m. Maria Schwan; children—Maria, Helene, Erwin, Englebert, Karl. Intern, 1923-24; surgeon in Vienna, Graz, Clinic of Surgery, Dr. Wolfgang Denk, Vienna; head physician Sankt-Johannspital, Salzburg, Austria, 1934, dir., 1950. Mem. Roman (hon.), German Internat. socs. surgeons, German Soc. Blood Transfusion, Order Doctors of Vienna (corr.). Research and publs. on chemistry, anesthesia, blood transfusion, osteomyelitis, thoracic surgery. Home: Schwarzstrasse 32. Office: Sankt-Johannspital, Salzburg, Austria.

DOMANUS, Józef Czeslaw, Polish engr.; b. Bielsko, Poland, July 31, 1919; s. Henryk and Adela (Mrozinska) D.; M.S. in Elec. Engring., Politechnika Warszawska, 1945; m. Stefania Adler, Apr. 15, 1941; 1 son, Jerzy. Sr. asst. Politechnika Warszawska, Warsaw, Poland, 1945-46, adjunct, 1949-51; head x-ray servicing sta. High Voltage Equipment Factory, Warsaw, 1947-49; head indsl. radiology dept. Electrotech. Inst., Warsaw, 1947—, asst. prof., 1960—, mem. sci. council, 1960—; mem. sci. secret., 1967—. Vice chmn. Nat. Com. Radiation Protection, 1958—; mem. presidium commn. on non-destructive testing Polish Acad. Sci., 1955—; chmn. nuclear energy commn. Polish Standards Com., 1957—. Recipient award State Council for Peaceful Applications Nuclear Energy, 1959, 64; Golden Cross of Merit, 1958; Order Polonia Restituta, 1964. Mem. Com. on Non-destructive Testing, Council for Sci. and Tech. Author: X-ray Technique, 1948; X-ray Darkroom, 1951; New Trends in X-ray Technique, 1952; Radiology, 1956; Technical Problems of Application of Radioisotopes, 1957; also articles. Research on applications of X-ray and radioisotope technique, especially for non-destructive testing of materials, properties of X-ray films used in radiography and radiation protection; devel. standard methods of testing for packaging of radioisotopes. Home: 44 Polna, Warsaw 10, Poland.*

DOMAR, Evsey David, economist; b. Lodz, Poland, Apr. 16, 1914; s. David O. and Sarah (Slonimsky) Domashevitsky; came to U. S., 1936, naturalized, 1942; student State Faculty of Law, Harbin, Manchuria, 1930-31; B.A., U. Cal. at Los Angeles, 1939; M.A., U. Mich., 1941; M.A., Harvard, 1943, Ph.D., 1947; m. Carola Rosenthal, Apr. 16, 1946; children—Erica C., Alice D. Economist, Bd. Govs. of Fed. Res. System, 1943-46; asst. prof. econs. Carnegie Inst. Tech., 1946-47; asst. prof., research asso. Cowles Commn., U. Chgo., 1947-48; asso. prof. Johns Hopkins, 1948-55, prof., 1955-58; prof. econs. Mass. Inst. Tech., 1958—; lectr. George Washington U., 1944, U. Mich., 1946; vis. prof. U. Buffalo, 1949, Stanford, 1957, Harvard, 1958, 62, Oxford (Eng.), 1952-53, Columbia, 1951-55, Centro de Estudios Monetarios Latinoamer, 1954, U. de Los Andes, Bogota, Colombia, 1965. Cons. econs. div. RAND Corp., Santa Monica, Cal., 1951—; cons., govt. agys., pvt. orgns.; fellow Center for Advanced Study in Behavioral Scis., Stanford, 1962-63; research asso. Harvard Russian Research Center, 1958—. Recipient John R. Commons award Omicron Delta Epsilon, 1965. Fellow Am. Acad. Arts and Scis.; mem. Am. Econ. Assn. (past mem. exec. com.), Am. Council Learned Soc. (past chmn. com. on slavic grants), Royal Econ. Soc., Econometric Soc., Am. Assn. U. Prof., Assn. Study Soviet-Type Econs., Nat. Conf. Research in Income and Wealth (exec. com. 1966—), Phi Beta Kappa, Pi Gamma Mu, Omicron Delta Epsilon (trustee). Author: Essays in the Theory of Economic Growth, 1957; also articles, revs. Editorial bd. Am. Econ. Rev., 1957-59, Am. Economist, 1963—. Pioneered theory of econ. growth and devel. which has revolutionized modern econs. Home: 27 Heath's Bridge Rd., Concord, Mass. 01742. Office: Dept. Econs., Mass. Inst. Tech., Cambridge, Mass. 02139.*

DOMBEY, Joseph, French botanist; b. Mâcon, France, Feb. 22, 1742; s. Marie (Carra) D.; doctorate in natural history, Montpellier, France; studied plants under Gowan and Cusson; began mission (with Ruiz and Pavon) to Chile, Lima, Peru, S. Am., to discover species of plants which could be cultivated in France, 1776; returned to France, 1784, received royal pension; sent by Com. of Pub. Health to buy grain in U. S., 1793; seized by Spaniards and died in prison. Mem. Acad. Medicine Madrid, French Acad. Scis. Introduced Araucanian pine wood from Chile for naval constrn. (now named in his honor); collected numerous new plants in Latin Am., including 60 new genera; collected rare butterflies. Died Antilles, Feb. 18, 1794.

DOMBROVSKAYA, Yuliya Fominchna, Russian pediatrist; b. 1890; grad. women's med. inst., 1913; D.Med. Sci., 1936. With children's clinic Med. Faculty, Moscow U. (now Sechenov 1st Moscow Med. Inst.), 1916-36, prof., 1936—, dir. clinic, head chair children's diseases, 1950—; head Moscow City Dept. Mother and Child Care, 1918-21. Del., 9th Internat. Congress Pediatrists, Montreal, 1959. Mem. USSR Acad. Med. Aci., All-Union (dep. chmn.), All-Russian (hon., chmn. 1955-57, bd. mem.), Moscow (chmn.) socs. pediatrists. Author: Vitamins in Pediatrics, 1948; Infantile Pneumonia, 1955; Diseases of the Respiratory Organs in Children, 1957; The Clinical Aspects and Pathogenesis of Hypoxemia of the Growing Body, 1961; co-author: Propedeutics of Children's Diseases, 1953. Dep. editor Pediatrics sect. Large Med. Ency., 2d edit.; mem. editorial bd. Pediatrics. Research and numerous publs. on child pathology, respiratory disorders, dystrophy and avitaminosis. Address: Sechenov 1st Moscow Med. Inst., B. Pirogovskaya ulitsa 2-6, Moscow, USSR.

DOMBROVSKY, Bronislav Aleksandrovich, Russian zoologist; b. Yelisavetgrad (now Kirovograd), Ukraine, Jan. 18, 1885; grad. Kiev U., 1912. With Kiev U., 1913-24, Kiev Zool. and Vet. Inst., 1924-29, Alma-Ata Zool. and Vet. Inst., 1929—; head chair Kazakhstan U., Alma-Ata, 1954—. Mem. Kazakhstan Acad. Sci. Author: Functional Morphology of the Vascular and Nervous Systems of Vertebrates, 1963. Research and publs. on biomorphological trend in comparative anatomy of vertebrates, devel. artificial methods of establishing correlation, integral anatomy, morphogenetic role of trophic processes. Address: Kazakhstan University, ulitsa Kirova 136, Alma-Ata, Kazakhstan SSR, USSR.

DOMEYKO, Ignacio (or Ignacy), mineralogist, chemist; b. Niezdwiadka, Poland, 1802; ed. at Wilna. Forced to leave Poland after participating in Polish insurrection; went to Chile; founder sch. chemistry and mineralogy at Coquimo; became prof. U. Santiago (Chile), 1839, rector, 1867. Author: Aspects of Mineralogy; Introduction to the Study of Natural Sciences; Treatise on Chilean Mineralogy; Treatise on Chilean Geology; The Araucanians; Philarètes et les Philomathes, 1872. Discovered mineral desposits in Andes; helped develop Chilean resources. Died Santiago, 1889.

DOMINICUS DE CLAVASIO, see de Clavasio, Dominicus.

DOMINIK, Tadeusz, Polish botanist, ecologist; b. Wloclawek, Poland, Nov. 7, 1909; s. Peter and Eveline (Petren) D.; D.Sc., Faculty Scis., U. Poznan, 1936, Ing., Faculty Forestry, 1945; m. Eva Dobrowolska, July 15, 1950; children—Ingeborga, Georg. Asst., Faculty Scis., U. Poznan, 1930-39, 45-46, dir. plant protections, 1946-49; forester, Poznan, 1939-45; prof. U. Wroclaw, 1949-53; dir. dept. forest microbiology Forestry Inst., Varsav, 1954-62; dir. lab. of plant protection Coll. Agr. in Szezecin, 1957—. Mem. Mycol. Soc. Am., Ecol. Soc. Am., Mycol. Soc. France. Author: Polyporaceae, 1957; also numerous articles. Research on mycorrhizae and classification of ectotrophic mycorrhizae of trees; ecology of mycorrhizae; ecology and systematics of keratinolytic fungi of soils; systems of fungi; plant anatomy and pathology; discovered American mycorrhizae of Douglas fir in Poland. Home: 7 Kusnierska, Szcsecin, Poland.*

DOMINO, Edward Felix, Am. neuropsychopharmacologist; b. Chgo., Nov. 20, 1924; s. James I. and Mary (Dolerzek) D.; B.S., U. Ill., 1948, B.S. in Medicine, 1949, M.S., 1951, M.D., 1951; m. Antoinette Kaczorowski, Nov. 20, 1948; children—Karen Barbara, Laurence Edward, Debra Ann, Kenneth Edward, Steven Edward. Faculty, U. Ill., 1951-53; faculty U. Mich., Ann Arbor, 1953—, prof. pharmacology, 1962—; vis. asso. prof. pharmacology U. Cal., 1961; vis. dir. clin. neurophysiology dept. neurosurgery St. Barnabas Hosp., N.Y.C., 1963; cons., dir. neuropsy chopharmacology lab. Lafayette Clinic, Detroit, 1959—; vis. prof. pharmacology Wayne State U., 1965—; director of Michigan Neuropharmacology Research and Training Program, 1966—; member study section pharmacology and chemistry National Institute of Mental Health, 1965-68. Recipient Sigma Xi prize in medicine U. Ill., 1951; Research award Mich. Soc. Neurology and Psychiatry, 1955; First prize Am. Soc. Anesthesiologists, 1963; certificate of merit A.M.A., 1964. Fellow A.A.A.S.; mem. Central EEG Soc., Am. Soc. Pharmacology and Exptl. Therapeutics, N.Y. Acad. Sci., Internat. Brain Research Orgn., Soc. Psychophysiol. Research, Sigma Xi, Alpha Omega Alpha. Editor: Atlas of the Canine Brain, 1964. Mem. editorial bd. Jour. Pharmacology and Exptl. Therapeutics, 1958-65, Jour. Neuropharmacology, 1962—, U. Mich. Med. Jour., 1964—; adv. bd. Psychopharmacologia, 1967—. Contbr. numerous articles to profl. jours. Research and publs. on effect of drugs on the brain and behavior; chem. communication between neurons controlling wakefulness, sleep, states of consciousness. Home: 3071 Exmoor Dr., Ann Arbor, Mich. 48104.*

DOMNINOS OF LARISSA, mathematician; b. probably Larissa, Syria, circa 430 B.C.; disciple of Proclos; studied at Plato's Acad. under Syrianos, Athens, head Acad. after Syrianos, before Proclos; author summary of theory of numbers, marking reaction against Nicomachos and return to Euclidean tradition, also probably 2d treatise on theory of numbers, and a tract on Plato's opinions; applied neo-Platonic ideas to math.

DOMOKOS, Gabor, physicist; b. Budapest, Hungary, Mar. 5, 1933; s. Laszlo and Aranka (Szekely) D.; Dipl. Phys. with highest honors Eötvös Lorand U., Budapest, 1956; D.Phys. and Math. Sci., Joint Inst. for Nuclear Research, Dubna, USSR, 1963. Research asso. Central Research Inst. for Physics, Budapest, 1956-60; research physicist Joint Inst. for Nuclear Research, Dubna, 1960-63; sr. research physicist Central Research Inst., Budapest, 1964-65, 67—; lectr. Johns Hopkins, Balt., 1965-66; research physicist U. Cal. at Berkeley, 1966-67. Mem. Eötvös Lorand Phys. Soc., Am. Phys. Soc. Research and publs. on theory of interactions of elementary particles at very high energies, interactions of particles of cosmic radiation. Address: Central Research Inst. for Physics, Budapest, Hungary.*

DONAHUE, Roy Luther, Am. soil scientist; b. Ringgold, Tex., Nov. 3, 1908; s. Peter Jerome and Nellie Grace (Boicourt) D.; B.S., Mich. State U., 1932; Ph.D., Cornell U., 1939; m. Lola Purtee, Nov. 4, 1928; children—Roger Purtee, Julian Purtee, Jane Lee (Mrs. Douglas Bridges). Faculty, Mich. State U., East Lansing, 1934-35, 66—, now prof. soil sci.; faculty Miss. State U., 1935-37, Cornell U., 1937-39, 44, Tex. A. and M. U., 1939-43, 45-52, U. N.H., 1952-56; sr. forester Rubber Devel. Corp., 1943; agronomy cons. Greek Feasibility Fertilizer Survey, 1955; prof. agronomy Kan. State U.-U.S. AID-India Team, Hyderabad, Andhra Pradesh, India, 1956-61; cons. on fertilizers Ford Found., New Delhi, India, 1961-66; agronomist African mechanization study Mich. State U., 1967—. Fellow Tex. Acad. Sci., mem. Soil Sci. Soc. Am.,

Indian Soc. Agronomy, Soc. Am. Foresters, Internat. Soil Sci. Soc., Indian Society Soil Science, Soil Conservation Soc. India, Ecol. Society America, Sigma Xi, Phi Kappa Phi, Alpha Zeta, Xi Sigma Pi. Author: (with E. F. Evans) Our South, Its Resources and Their Use, 1949, Exploring Agriculture, 1962; (with others) The Range and Pasture Book, 1956; Our Soils and Their Management, 1968; (with others) Soil Management in India, 1962; (with others) Agriculture in India, 1963; (with others) Soils-Their Chemistry and Fertility in Tropical Asia, 1964; Soils—An Introduction to Soils and Plant Growth, 1965; (with others) Soil Testing in India, 1965. Contbr. numerous articles to profl. jours. Home: 133 Kenberry Dr., East Lansing, Mich. 48823.*

DONALDSON, Coleman duPont, Am. aero. engr.; b. Phila., Sept. 22, 1922; s. John Wilcox and Renee (duPont) D.; B.Aero. Engring., Rensselaer Poly Inst., 1943; M.A., Princeton, 1954, Ph.D., 1957; m. Barbara Goldsmith, Jan. 17, 1945; children—B. Bierne, Coleman duPont, Evan F., Alexander M., William F. Staff, NACA, Langley Field, Va., 1943-44; aero. engr. Bell Aircraft Corp., Niagara Falls, N.Y., 1946; head aerophysics sect. NACA, 1946-52; pres., sr. cons. Aero. Research asso. Princeton, Inc. (N.J.), 1954—. Cons. to pvt. cos. Asso. fellow Am. Inst. Aeros and Astronautics; mem. Am. Phys. Soc., Inst. Navigation, Acad. Applied Sci., N.Y. Acad. Scis., Sigma Xi. Contbr. numerous articles to tech. jours. Pioneered research on air inlets for supersonic air-breathing aircraft; research on theory turbulent skin friction and heat transfer at very high Mach numbers. Home: 162 Library Pl., Office: 50 Washington Rd., Princeton, N.J. 08540.*

DONALDSON, Henry Herbert, Am. neurologist; b. Yonkers, N.Y., May 12, 1857; s. John J. and Louisa Goddard (McGowan) D.; A.B., Yale, 1879; Sheffield Sci. Sch. (Yale), 1880, D.Sc., 1906; Coll. Phys. and Surg. (Columbia), 1881; Ph.D., Johns Hopkins, 1885; m. Emma Brace, Apr. 6, 1907. Instr. biology Johns Hopkins, 1883-84, asso. prof. psychology, 1887-88; asst. prof. neurology Clark U., 1889-92; prof. and head of dept. neurology, 1892-1906, dean Ogden (Grad.) Sch. Sci., U. Chgo., 1892-98; prof. neurology, Wistar Inst. Anatomy and Biology, Phila., from 1906. Author: The Growth of the Brain, 1895; The Physiology of the Central Nervous System in An American Text-Book of Physiology, 1896; The Rat, 1924. Research in neurology and growth of central nervous system; carried out one of most thorough studies ever done of a single human brain, 1890-91. Died Phila., Jan. 23, 1938.

DONALDSON, Robert, pathologist; b. Dalkeith, Midlothian; s. James and Annie (Macdonald) D.; ed. U. Edinburgh, U. Marburg; M.A., M.D. with honours, Ch.B., D.P.H., D.T.M.; m. Annie Owen; 2 daus. Sir William Dunn prof. pathology U. London, lectr. pathology and forensic medicine St. George's Hosp. Med. Sch., also asso. examiner in pathology; dir. path. dept. Guy's Hosp. Med. Sch.; demonstrator pathology U. Sheffield, U. Bristol; pathologist Royal Berks Hosp., Reading, Ministry Health Venereal Diseases Clinic, Reading, St. Mary Abbott's Hosp., Princess Louise Children's Hosp., Kensington, Royal Waterloo Hosp. for Women and Children, London; pathologist, curator Path. Mus., St. George's Hosp., London; bacteriologist Royal Borough Kensington; hon. pathologist Guy's Hosp.; cons. pathologist Hosp. St. John and St. Elizabeth; hon. asst. pathologist Royal Infirmary, Bristol. Fellow Royal Coll. Surgeons Edinburgh. Author: Practical Morbid Histology: A Handbook for the Use of Students and Practitioners, also articles. Died Jan. 3, 1933.

DONATH, Fred A(rthur), Am. geologist; b. St. Cloud, Minn., July 11, 1931; s. Arnold C. and Elizabeth (Crary) D.; B.A., U. Minn., 1954; M.S. Stanford, 1956, Ph.D., 1958; m. Mavis Eleanor Hagen, July 19, 1952; children—Robert William, Deborah Ann. Faculty, San Jose (Cal.) State Coll., 1957-58; faculty Columbia, N.Y.C., 1958-67, prof. geology, 1966-67, prof.; head dept. U. Ill., Urbana, 1967—. Vis. lectr. Am. Geol. Inst., 1965—. Recipient Semicentennial medallion Rice U., 1962. Fellow Geol. Soc. Am., A.A.A.S.; mem. Am. Geophys. Union (sec. tectonophysics sect. 1964—), Am. Assn. Petroleum Geologists (lectr. continuing edn. program), Phi Beta Kappa, Sigma Xi. Asso. editor Geol. Soc. Am., 1963—, Tectonophysics, 1963—. Contbr. articles to profl. lit. Fundamental research in high pressure geophysics; stress-strain analysis in the earth's crust; design of high pressure equipment. Home: 2021 Cureton Dr., Urbana, Ill. 61801.*

DONATI, Giovanni Battista, Italian astronomer; b. Pisa, Italy, Dec. 16, 1826; ed. at Pisa; apptd. asst. obs., Florence, Italy, 1852; became dir., 1864; research on stellar spectra; 1st to observe spectrum of a comet; discovered comet (named after him), 1858, also 5 others, 1854-58; built spectroscope with 25 prisms; discovered gaseous compositions of comets; instrumental in founding observatory at Arcetri, nr. Florence. Died Florence, Sept. 20, 1873.

DONCASTER, Leonard, English zoologist; b. Sheffield, Eng., Dec. 31, 1877; s. Samuel Doncaster; ed. King's Coll., Cambridge, (Eng.) U.; Mackinnon student Royal Soc., 1904-06; M.A., Sc.D.; m. Dora Priestman, 1908; 1 son, 2 daus. Lectr. zoology Birmingham (Eng.) U., 1906-10; spl. lectr. on variation and heredity, 1909; prof. zoology U. Liverpool (Eng.); became fellow King's Coll., 1910. Fellow Royal Soc., 1915. Author: Heredity in the Light of Recent Research, 1910; The Determination of Sex, 1914; also papers on zoology. Died May 28, 1920.

DONDERS, Franciscus Cornelis, Dutch ophthalmologist; b. Tilburg, Netherlands, May 27, 1818; became prof. anatomy and physiology, mil. sch., Utrecht, Netherlands, 1842; named prof. physiology U. Utrecht, 1852; founder eye clinic, Utrecht. Fellow Royal Soc., 1866; mem. French Acad. Scis. Author: De Leer der Slofwisseling, 1845; Physiologie des Menschen, 1850; Anomalies of Accommodation and Refraction, 1866; also classic report on physiology of speech, 1870. Founder (with von Gräfe) Archiv für Ophthalmologie. Research on physiology and pathology of eye; formulated Donder's law that rotation of eye around line of sight is involuntary, 1847; defined condition of hyperopia (hypermetropia), 1864; probably 1st to measure reaction time involved in a mental process, 1868; pioneer in study of astigmatism, also of function of nerves; explained chem. passage of breath (dissimilation); introduced use of cylindrical and prismatic lenses in eyeglasses. Died Utrecht, Mar. 24, 1889.

DONDI, Giacomo, Italian physician, astronomer; b. Padua, Italy, 1298; elected municipal physician of Chiogga, Italy, 1313; tchr. medicine, Padua, by 1342; author: Aggregatio-medicamentorum, or Promptuarium medicine (a collection of recipes from Greek and Arabic sources); Planetarium (astron. tables based on those of Alphonsine); Tractatus de causa salsedinis squarum et modo conficiendi salis ex eis; De fluxu et refluxu maris. Pioneered the sci. of balneology; 1st to recommend the extraction of salts from mineral waters for medicinal purposes; believed the movement of tides was influenced by Venus or Jupiter when the planets were near enough. Died 1359.

DONHAM, C(harles) R(umpel), Am. veterinarian; b. Rockport, Ind., Aug. 1, 1898; s. Lewis Singleton and Amelia Rebecca (Rumple) D.; D.V.M., Ia. State Coll., 1921; M.S., Ore. State Coll. 1927; student, sch. of med., Washington U., also U. Minn., 1929-35; m. Margaret Hyde Lysinger, June 18, 1921; children—Marion Margaret (Mrs. Joseph F. Jamison), James Charles. Pvt. practice in vet. medicine, 1921-22; instr., asst. prof. vet. medicine Ore. State Agr. Coll., 1922-29; asst. prof., asso. prof. vet. medicine U. Minn., 1929-35; prof. vet. medicine Ohio State U., 1935-40; prof. vet. sci. Purdue U., 1940—. Mem. com. on brucellosis U. S. Live Stock San. Assn.; adv. com. Bur. Animal Industry, U. S. Dept. Agr.; mem. Nat. Com. on Brucellosis. Mem. Conf. of Ofcl. Research Workers in Animal Diseases in N. Am., Ind., Am. vet. med. assns., Sigma Xi. Died Apr. 24, 1956.

DONHOFFER, Szilárd, Hungarian physician; b. Budapest, Hungary, July 3, 1902; s. Géza and Hilda (Trautmann) D.; M.D., U. Budapest, 1926; m. Margit Mittag, May 6, 1929; children—Agnes (Mrs. George Illei), Hilda (Mrs. Tibor Heim). Staff, U. Clinic Internal Medicine, Pécs, Hungary, 1926-30; staff dept. physiology U. Aberdeen (Scotland), 1931; med. faculty U. Pécs, 1932—, head dept. pathophysiology, 1944—, dep. rector Med. Sch., 1961-64, rector, chancellor, 1964-67; vis. prof. McGill U., Montreal, Que., Can., 1964, dept. physiology U. Cal. at Los Angeles, 1964. Recipient State prize for achievement in sci., 1961. Mem. Hungarian Physiol. Soc., Hungarian Acad. Sci. Author: Pathophysiology, 1957, 61; also numerous articles. Research on regulation of body temperature, and energy metabolism in homeotherm animal, regulation of food intake and food selection. Home: 3 János utca, Pécs, Hungary.*

DONINI, Piero, Italian chemist; b. Fermignano, Pesaro, Italy, Oct. 9, 1910; s. Gino and Giuseppina (Falasconi) D.; ed. Bologna (Italy) U., Chemistry Sch., 1928-33; m. Alessandra Brilli-Cattarini, Sept. 1, 1938; children—Ilaria, Silvia. Chemist, research labs. Istituto Farmacologico Serono, Rome, Italy, 1935-52, dir. research labs., 1952—. Decorated Maltese Cross. Mem. Am. Chem. Soc., Royal Soc. Medicine. Contbg. author Ovulation. Research and publs. on urinary gonadotropins, purification and separation of follicle-stimulating hormone and luteinizing hormone; use of these hormones for therapeutic purposes. Home: 95 Corso Trieste, Rome. Office: 125 Via Casilina, Rome, Italy.*

DONITA, Nikolae, ecologist; b. Kisinew, U.S.S.R., Nov. 20, 1929; s. Boris and Emilia (Renner) D.; forestry engr. Brasov Forestry Coll., 1955; m. Doina Ivan, Nov. 27, 1965. Research work at Forestry Group, 1955-58, at dept. geobotany and ecology Inst. Biology, Acad. Scis., Bucharest, 1958—. Recipient prize Rumanian Acad. Scis., 1961. Author: (with S. Pascovschi) Wood Vegetation in Wood-Steppe of Rumania, 1967; also articles. Research on distbn. and evolution of wood vegetation of wood-steppe, zonation of vegetation, phytogeog. div. terrs., geobot. map of Rumania. Home: 3 Dacia, Bucharest, Rumania.*

DONITZ, Friedrich Karl Wilhelm, German bacteriologist; b. Germany, 1838; ed. Berlin, Germany. Lectr. anatomy Tokyo, Japan; asso. Koch Sch., Germany, from 1886, acting dir.; staff mem. Ehrlich's Inst. für Serumforschung, Steglitz, Germany, from 1896. Organized anat. teaching Japan; worked on standardization of diphtheria anti-toxin and tuberculin. Died 1912.

DONKIN, Bryan, Brit. engr.; b. Sandree, Northumberland, Eng., Mar. 22, 1768; apprenticed to Mr. Hall of Dartford; m. Mary; a son, John. Constructed model of 1st paper-making machine, 1801-02; established factory for canning meats and vegetables, Bermondsey, 1812; civil engr., 1815-55. Fellow Royal Soc., 1838; mem. Soc. Arts (gold medal), Royal Astron. Soc., Instn. Civil Engrs. (v.p.). Invented polygonal printing machine, 1813, also composition printing-roller; invented process of canning meat and vegetables, 1812; built paper-making machine, Frogmore, Kent, 1804. Died Feb. 27, 1855.

DONKIN, Sir (Horatio) Bryan, physician; b. Blackheath, Kent, Feb. 1, 1845; s. Bryan Donkin; ed. Queen's Coll., Oxford, M.D.; m. Auguste Margaretha Elizabeth de Langhi, 1888; m. 2d, Marie Louise Reston, 1923. Cons. physician Westminster Hosp., East London Hosp. for Children, King George Hosp., 1915; lectr. medicine London Sch. Medicine for Women. Mem. adv. com. to home sec. His Majesty's Preventive Detention Prison; mem. Com. Prisons; med. adv. to Prison Com.; examiner in medicine Royal Coll. Physicians; mem. Royal Commn. on Control Feeble-minded, 1904-08. Fellow Royal Coll. Physicians (Harveian orator on inheritance mental characteristics 1910); mem. Royal Medico-Psychol. Assn. (hon.). Author: Diseases of Childhood. Research and numerous publs. on mental pathology, crime, psychical research, med. prevention venereal disease, fallacies of psychoanalysis. Died July 26, 1927.

DONNAN, Frederick George, Brit. chemist; b. Sept. 6, 1870; s. William Donnan; ed. Queen's U., Belfast, univs. Leipzig, Berlin, London; M.A., Ph.D., S.Sc., LL.D.; studied under William Ramsay, Ostwald, van't Hoff, Letts; hon. degrees U. St. Andrews, Belfast, Liverpool, Edinburgh, Durham, Balt., Princeton, Coimbra, Athens, Oberlin, Nat. U. Ireland. Jr. fellow, examiner Royal U. Ireland, 1898-1901; asst. prof. Univ. Coll., London, 1902, prof. chemistry, 1913-37, dir. chem. lab., 1928-37, also fellow; lectr. chemistry Royal Coll. Soc., Dublin, 1903-04; prof., phys. chemistry, dir. Muspratt Lab. Phys. and Electro-Chemistry, U. Liverpool, 1904-13. Recipient Longstaff medal Chem. Soc., London, 1924. Fellow Royal Soc. (Davy medal 1928), Royal Inst. Chemistry, Royal Soc. Edinburgh, Nat. Inst. Scis., India, Indian Acad. Scis.; hon. mem. Am., Dutch, Austrian chem. socs., German Bunsen Soc., Instn. Chem. Engrs., French Soc. Indsl. Chemistry, Soc. Philomathique, Paris. Contbr. papers to sci. publs. Gave theory of Hall and Thomson effects for solutions of an electrolyte; devised drop pipette for expts. in oil dispersion; proved adsorption detergent action of soap; arrived at theory of negative surface tension (supposedly reason for colloids); research in chem. genetics; produced theory of membrane equilibria which involved osmotic pressure and elec. considerations. Died Canterbury, Eng., Dec. 16, 1956.

DONNÉ, Alfred, French bacteriologist; b. Noyon, France, 1801; M.D., Paris, France, 1831; named head of clinic Charité Hosp., Paris, 1829; sub-librarian Faculty of Medicine, Paris; insp. gen. U. Paris. Believed to have been 1st to describe blood platelets as 3d element of blood, 1842; described Trichomonas vaginalis which he believed was causative agent of gonorrhea, 1836. Died Paris, France, 1878.

DONNER, Henry F(rederick), Am. astronomer, geologist; b. Wilson, N.Y., Sept. 1, 1902; s. Henry and Louise M. (Schultz) D.; B.S. in Elec. Engring., U. Mich., 1925, M.S. in Astronomy, 1927, Sc.D. in Geology, 1936; m. Florence L. Mudge, July 30, 1927. Test man Gen. Elec. Co., Schenectady, 1925-26; research astronomer Lamont-Hussey Obs. U. Mich., Bloemfontein, S.Africa, 1927-33; faculty Western Res. U., Cleve., 1936—, prof. astronomy, 1946-61, prof. geology, 1946—. Fellow A.A.A.S., Geol. Soc. Am., Ohio Acad. Sci. (v.p. geology sect. 1953-54), mem. Am. Assn. Petroleum Geologists, Am. Astron. Soc., Am. Geophys. Union, Am. Soc. Photogrammetry, No. Ohio Geol. Soc. (treas. 1963-65), Sigma Xi. Discovery and study of double stars, especially So. Hemisphere; study and interpretation of geologic structure and history of N.W. Colo. Home: 3094 E. Overlook Rd., Cleveand Heights, O. 44118.

DONNER, Joakim Jalmar, Finnish geologist; b. Helsinki, Finland, Dec. 19, 1926; s. Kai and Greta (von Bonsdorff) D.; M.A., U. Helsinki, 1951; Ph.D., 1951; Ph.D., U. Cambridge (Eng.), 1956; m. Ruth Hilda Goldstein, Aug. 10, 1956; children—Gabriel Kim, Julia Ruth, Sara Ann. Faculty, U. Helsinki, 1957—, prof. geology and paleontology, 1965—; postdoctoral research fellow Yale, 1959. Mem. Societas Scientiarum Fennica. Research, publs. on geol. and vegetational history of last 15,000 years especially in Finland, Scotland, S. France. Home: 12 Pohjoisranta, Helsinki 17, Finland.*

DONNER, Martin Walter, physician; b. Leipzig, Germany, Sept. 5, 1920; s. Walter Theodore and Elsa (Ruehl) D.; M.D., U. Leipzig, 1945; m. Adelheid Ilse Wimmer, Mar. 17, 1951; children—Cornelia, Stephanie, Thomas. Came to U. S., 1954, naturalized, 1963. With Johns Hopkins U. and Hosp., Balt., 1957-

—, prof. radiology, 1966—, asso. prof. radiologic scis., 1964—; vis. investigator Carnegie Instn. Washington, 1960—; vis. prof. Free U. of Berlin (Germany), 1964; cons. radiology VA Hosp., Balt., Rosewood State Hosp., Owings Mills, Md.; mem. editorial staff Am. Jour. Med. Scis. Contbr. numerous articles in field to sci. jours. Cineradiographic analysis of normal and disturbed physiology; studies of the effect of drugs on intestinal motility; radiographic test to diagnose esophagitis with acid barium; radiographic evaluation of blood flow through the placenta of rhesus monkeys in the various stages of pregnancy; radiography of diabetic complications and diseases involving the hands; radiotherapy and blood coagulation. Home: 215 N. Tyrone Rd., Balt. 21212. Office: The Johns Hopkins Hosp., B-27, Balt. 21205.*

DONOVAN, Bernard Thomas, Brit. physiologist; b. London, Eng., Nov. 8, 1927; s. Thomas and Rose (May) D.; student Chelsea Poly. Inst., 1948-49; B.Sc., U. Coll. London, 1952; Ph.D., U. London, 1954, D.Sc., 1967; m. Heather Isabel Jackson, July 1, 1950; children—Susan Heather, Iain Hugh. Beit Meml. research fellow Inst. Psychiatry, U. London, 1954-58, lectr. physiology, 1958-61; sr. lectr., 1961-66, reader neuroendocrinology, 1966—; vis. prof. Inst. Physiology, U. Lund, Sweden, 1959; vis. prof. dept. anatomy U. Cal., San Francisco, 1963-64. Recipient CIBA Found. prize for Basic Research to Problems of Aging. Mem. Physiol. Soc., Soc. for Endocrinology (council 1967—), Endocrine Soc., Soc. for Study of Fertility, Royal Soc. Medicine, A.A.A.S. Author: (with J. J. van der Werff ten Bosch) Physiology of Puberty, 1965; (with G. W. Harris) The Pituitary Gland, 1966. Research, numerous publs. field physiology of reprodn., especially interaction between nervous and endocrine systems. Home: 48 Kelsey Lane, Beckenham, Kent, Eng. Office: Dept. Neuroendocrinology, Inst. Psychiatry, de Crespigny Park, London, S.E.5, Eng.*

DONOVAN, Desmond Thomas, English geologist; b. Cheam, Surrey, Eng., June 16, 1921; s. Thomas Bartholomew and Marie Augusta (Benker) D.; B.Sc., U. Bristol, 1942, Ph.D., 1951, D.Sc., 1960; m. Shirley Louise Saward, July 31, 1959; children—Thomas Hector, Tessa May, Daniel Seumus. Asst. lectr. geology U. Bristol, 1947-50, lectr., 1950-62; prof. geology U. Hull, 1962-66; Yates-Goldsmid prof. geology U. Coll. of London, 1966—. Mem. Geol. Soc. London (Murchison Fund award 1959), Palaeontographical Soc. (v.p. 1965—). Author: Stratigraphy: An Introduction to Principles, 1966. Research on paleontology of ammonites with application to correlation of Jurassic rocks; evolution and classification of cephalopods; mesozoic geology of East Greenland; geology of sea floor around Britain. Office: University Coll., Gower St., London W.C. 1, Eng.*

DONOVAN, Michael D., Irish physician; b. Ireland, 1809. Author: Observations and Experiments Concerning Mr. Davy's Hypothesis of Electrochemical Affinity, 1811; A Treatise on Chemistry, 1832; On the Extemporaneous Preparation of Hydrocyanic Acid From Cyanide of Potassium. Inventor Donovan's solution (solution of arsenic and mercuric iodides used in skin diseases and formerly as tonic), 1839. Died 1876.

DONY, l'Abbé Jean-Jacques-Daniel, Belgian chemist; b. Liège, Belgium, 1759; dir. Vielle-Montagne Mines, Belgium; separated zinc from its ore, 1805; began metall. prodn. of zinc, 1809. Died Liège, 1819.

DOOB, Joseph Leo, Am. mathematician; b. Cin., Feb. 27, 1910; s. Leo and Mollie (Doerfler) D.; B.A., Harvard, 1930, M.A., 1931, Ph.D., 1932; m. Elsie Haviland Field, June 26, 1931; children—Stephen, Peter, Deborah. Faculty U. Ill., 1935—, successively asso., asst. prof., asso. prof., 1935-45, prof. math., 1945—. Mem. Nat. Acad. Scis., Am. Acad. Arts and Scis. Author: Stochastic Processes, 1952. Work with various aspects of probability theory to obtain meth. results and to demonstrate necessary role of measure theory. Home: 208 W. High St., Urbana, Ill.*

DOOB, Leonard William, Am. social psychologist; b. N.Y.C., Mar. 3, 1909; s. William and Florence (Doob) D.; B.A., Dartmouth, 1929; M.A., Duke, 1930; postgrad. Frankfurt (Germany) U., 1930-32; Ph.D., Harvard, 1933-34; m. Eveline Bates, Mar. 21, 1936; children—Christopher, Anthony, Nicholas. Faculty, Yale, 1934—, prof. social psychology; staff Office Coordinator Inter-Am. Affairs, Mil. Intelligence, War Dept., Overseas Br., Office War Information, 1940-45. Fellow Am. Acad. Arts and Scis.; mem. African Studies Assn. Author: Propaganda, 1935; (with M. A. May) Competition and Cooperation, 1937; (with others) Frustration and Aggression, 1939; The Plans of Man, 1940; Public Opinion and Propaganda, 1948; Social Psychology, 1952; Becoming More Civilized, 1960; Communication in Africa, 1961; Patriotism and Nationalism, 1964; Ants Will Not Eat Your Fingers, 1966; also articles. Analyses of pub. opinion, propaganda, communication, acculturation. Home: Clark Rd., Woodbridge, Conn. 06525. Office: 333 Cedar St., New Haven 06510.*

DOODSON, Arthur Thomas, English oceanographer; b. Worsley, Eng., Mar. 31, 1890; s. Thomas and Eleanor (Pendlebury) D.; B.Sc., Liverpool (Eng.) U., 1911, B.Sc. with honours, 1912, M.Sc., 1914, D.Sc., 1919; m. Margaret Galloway, Apr. 15, 1919; (dec. Mar. 1931); children—Joan, Thomas; m. 2d, Elsie Mary Carey, Dec. 2, 1933. Asst., Karl Pearson Statics, 1916-18; dir. Ballistic Computations, 1918-19; sec. Tidal Inst., 1919-29, asso. dir., 1929-45; dir. Liverpool Obs. and Tidal Inst., 1945-60. Mem. Hydraulics Research Bd., 1947-60. Decorated comdr. Order Brit. Empire. Fellow Royal Soc., 1933, Royal Soc. Edinburgh (hon.), Brit. Assn. (past sect. com. on tides), Oceanographic Research Council. Author: (with H. D. Warburg) Admiralty Manual of Tides, 1942; also numerous articles, chpt. in book. Statis. studies of relation of mean, median and mode theory of tides; computation of Riccati-Bessel functions and asso. Legendre functions; study of storm surges. Home: 10 Ingetre Ct., Ingetre Rd., Birkenhead, Eng.*

DOOLEY, James Creswell, Australian geophysicist; b. Melbourne, Australia, Jan. 30, 1919; s. Norval Henry and Olive (Haynes) D.; M.Sc., Melbourne (Australia) U., 1941; m. Nanette Eleanor Norris-Smith, Feb. 19, 1949; children—Anthony Haynes, Nicholas Morphett, Gillian Mary Adele, Michael Creswell, Philip Maxwell. Research asst. Mt. Stromio Obs., Canberra, Australia, 1941-44; geophysicist Bur. Mineral Resources, Melbourne, Canberra, 1944—, asst. chief geophysicist in charge geophys. obs., regional surveys and engring. geophysics, 1959—. Fellow Australian Inst. Physics (vice chmn. geophysics group 1965, sec., 1966—), Royal Astron. Soc. (London, Eng.); mem. Inst. Physics (London asso.), Soc. Expln. Geophysics, Am. Geophys. Union. Research and publs. on geophys. exploration for oil and other minerals in Australia; established standard reference network for gravity measurements in Australia; math. studies in interpretation of gravity anomalies, and application of seismology in dam site testing and investigations of structure of earth's crust. Home: 28 Ulverstone St., Lyons, A.C.T. 2606. Office: P.O. Box 378, Canberra City, A.C.T. 2601, Australia.*

DOOLEY, M(arion) S(ylester), Am. physician; b. Cedar Grove, Mo., Dec. 23, 1879; s. Thomas Jefferson and Elizabeth Caroline (Howell) D.; A.B., U. Mo., 1907; med. student U. Mo., Harvard, 2 yrs.; M.D., Syracuse U., 1914; m. Mary Elizabeth Jadwin, Sept. 1, 1908; children—Elizabeth (Mrs. Frederick D. Becker), Alice Ann (Mrs. David Radford Serpell); m. 2d, Constance Howell, Mar. 1, 1943. Successively instr., asst. prof., asso. prof., prof. physiology and pharmacology Syracuse U. Coll. Medicine, 1907-17, prof. pharmacology, 1917-45, emeritus prof., 1945—; drugs cons. Univ. Hosp. staff, 1922-47, Bur. Hosp. Standards and Supplies, Inc., 1935—, WPB, 1943-45. Mem. U. S. Pharmacopoeia Revision Com., 1920-50. Fellow Internat. Coll. Anesthesia, A.M.A., Syracuse Acad. Medicine; mem. Internat. Anesthesia Research Soc. (hon. pres.), Am. Soc. Pharmacology and Therapeutics, Soc. Exptl. Biol. and Medicine, Sigma Xi. Author: Pharmacology and Therapeutics in Nursing, 1948. Chmn. editorial com. Practitioners and Interns Handbook, 4 edits., 1928-49; co-chmn. editorial com. Drug Manual, 1949. Died Dec. 13, 1958.

DOOLEY, Thomas Anthony III, Am. physician; b. St. Louis, Jan. 17, 1927; s. Thomas Anthony and Agnes (Wise) D.; student U. Notre Dame; M.D., St. Louis U., 1953; postgrad. U. Paris. Served with USNR, 1944-46, 53-56; served at Naval Hosp., Camp Pendleton, Cal., 1 year, trans. to Naval Hosp., Yokosuka, Japan; volunteered for duty in USS Montague, transporting no. Vietnamese refugees to Saigon, 1954; later French interpreter, med. officer Navy preventive medicine unit, Haiphong, 1954; duty at Naval Hosp., Yokosuka, 1955; made lecture tour of U. S. under auspices U. S. Navy, 1956; organized pvt. med. mobile unit to work in Laos, 1956; an organizer Med. Internat. Corp. (MEDICO). Named 1 of ten outstanding men of 1956, Look mag.; decorated Legion of Merit; officer Ordre National de Vietnam. Author: Deliver Us from Evil, 1955; The Edge of Tomorrow, 1958; The Night They Burned the Mountain, 1960. Died New York, N.Y., Jan. 18, 1961.

DOOLITTLE, Arthur K(ing), Am. chem. engr.; b. Oberlin, O., Nov. 15, 1896; s. Frederick Giraud and Maud (Tucker) D.; A.B., Columbia, 1919, B.S., 1920, Chem. E., 1923; m. Dortha Bailey, Aug. 8, 1923; children—Robert Frederick II, Elizabeth May (Mrs. Donald Charles Peckham). Research engr. The Dorr Co., Westport, Conn., 1923-25; plant engr. Sherwin-Williams Co., Chgo., 1925-29, chief lacquer div., Newark, 1929-31; development engr. spray drying Bowen Research Corp., N.Y.C., 1931; dir. lacquer research Bradley Vrooman Co., Chgo., 1931-32; tech. head coatings research Carbide and Carbon Chemicals Co., South Charleston, W.Va., 1932-44, asst. dir. research, 1944-45, sr. scientist, 1955-61; pres. Arcadia Inst. for Sci. Research, Inc., Charleston, W.Va., 1959—; consultant, 1959—; partner of Dorr Consultants, New York City, 1959-61; professor chemistry, Drexel Inst. Tech., Phila., 1961-64. Mem. adv. bd. chem. engring. dept. Princeton, 1955-58. Served from pvt. to 2d lt. A.S., AUS, 1917-19; acceptance test pilot, 1919. Registered professional engineer, W.Va., N.Y., N.J., Del. Mem. Am. Inst. Chem. Engrs. (dir. 1951-54, v.p., 1955, chmn. Charleston sect. 1943-44), A.A.A.S. (chmn. mgmt. com. Gordon Confs., 1955-56, adv. bd. 1950-58), Am. Chem. Soc. (chmn. paint plastics and printing ink div. 1952-53, division councilor 1952-56, council com. on Nat. meetings and divisional activities 1952-56, adv. bd. indsl. and engring. chemistry 1954-56). Phi Beta Kappa, Tau Beta Pi, Chi Beta Phi, Alpha Chi Rho, Sigma Xi. Author: The Technology of Solvents and Plasticizers, 1954; also articles in sci. jours. Contbr. to Ency. of Chem. Tech., 1954. Patentee in field. Research on mechanistic theory of solutions; free-space viscosity equation; internal force equation; directed research that created vinyl coatings technology that launched plasticizers as a major petrochemical industry. Home: 406 Osborne Lane, Wallingford, Pa. 19086.*

DOOLITTLE, Charles Leander, Am. astronomer; b. Ontario, Ind., Nov. 12, 1843; s. Charles and Celia D.; C.E., U. Mich., 1874, Sc.D. (hon.), 1897; LL.D., Lehigh U., 1912; m. Martha Cloyes Farrand, Sept. 18, 1866, m. 2d, Helen Eugenia Wolle, May 11, 1882; 1 son, Eric D. On U. S. Boundary Survey, 1873-75; prof. math. and astronomy, Lehigh U., 1875-95; prof. astronomy U. Pa., also dir. Flower Astron. Obs., 1895-1912. Author: Practical Astronomy as Applied to Geodesy and Navigation; Results of Observation with Zenith Telescope, Sayre Obs., 1876-95; Results of Observation with Zenith Telescope, Flower Obs., 1894-1911. Research on variation of latitude; concluded that observations under a variety of instrumental conditions is needed to isolate sources of systematic error in cases of latitude variation; used reflex Zenith tube in observations. Died Mar. 3, 1919.

DOOLITTLE, Eric, Am. astronomer; b. Ontario, Ind., July 26, 1869; s. Charles Leander (q.v.) and Martha Cloyes (Farrand) D.; ed. prep. sch. Lehigh U., 1883-87; C.E., Lehigh U., 1891; postgrad. in astronomy, 1894-96; m. Sara Bitler Halliwell, Mar. 31, 1902. Instr. astronomy Lehigh U., 1891-92, State U. Ia., 1892-93, U. Pa., 1896-1904; asst. prof. astronomy U. Pa., 1904-12; prof. astronomy and dir. Flower Astron. Obs., from 1912. Author: Measures of 900 Double and Multiple Stars, 1901; Measures of 1066 Double and Multiple Stars, 1905; Catalogue and Remeasurement of the 648 Hough Double Stars, 1907; Measures of 1954 Double Stars, 1914. Computed secular perturbations of elements of orbits of four inner planets; measured double stars. Died Phila., Sept. 21, 1920.

DOOLITTLE, Howard Daniel, Am. physicist; b. Willimantic, Conn., Feb. 18, 1910; s. Sherwood B. and Ida (Steeves) D.; B.S., Trinity Coll., Conn., 1931; Ph.D., U. Chgo., 1936; m. Phyllis Elizabeth Johnson, July 1, 1933. Staff mem. Radiation Lab. Mass. Inst. Tech., 1940-45; asst. dir. engring. Machlett Labs., Inc., Stamford, Conn., 1945-60, dir. tech. Machlett Labs., Inc. div. Raytheon Co., 1960—. Fellow Am. Phys. Soc.; mem. N.Y. Acad. Scis., Electrochem. Soc., Phi Beta Kappa. Research in nuclear physics, magnetism, phys. electronics, pulse modulators for radar and x-ray applications, power transmitting tubes, microwave triode tubes, gas tubes and pulse forming network circuits, high voltage insulation in vacuum. Home: 42 Pembroke Dr., Stamford 06903. Office: 1063 Hope St., Stamford, Conn. 06907.*

DOOLITTLE, Thomas Benjamin, Am. engr., inventor; b. Woodbury, Conn., June 30, 1839; s. Benjamin and Betsey C. (More) D.; ed. Woodbury Acad.; (Sc.D., Dartmouth, 1909); m. Mary Louise Bradley, Dec. 24, 1866. In early life was a mfr. of brass articles at Bridgeport, in which he made many inventions in connection with mfr. of barbed wire; was originator of buffer platform and coupler, of which modified types came into general use on passenger cars; became connected with Bell Telephone Co., originated 1st telephone switchboard, the hard drawn copper and telephone call bell, related devices; originated and placed in use a fare registering device on street cars; retired from Am. Tel. & Tel. Co., 1909. Recipient Edward Longstreth medal Franklin Inst. of Phila., 1898, for origination of process of producing hard drawn copper wire. Died Apr. 4, 1921.

DOORENBOS, Norman John, Am. chemist; b. Flint, Mich., May 13, 1928; s. Garrett Jake and Victoria (Manary) D.; B.S., U. Mich., 1950, M.S., 1951, Ph.D., 1953; m. Fumiko Ikemori, Nov. 10, 1951; children—Beverly Jean, Phyllis Ann, Donna Louise, Alice Joanne, Gail Susan, David Isamu, Martha Lillian. Research chemist Ansco, Binghamton, N.Y., 1953-56; asst. prof. U. Md., 1956-58, asso. prof., 1958-63, prof., 1963-65; prof. pharm. chemistry U. Miss., 1965—. NSF vis. scientist, 1963—; cons. Mallinckrodt Chem. Works, 1963—, NSF Sci. Curriculum Project, U. Ill., 1964, 1965; Merck Sharp and Dohme lectr. W. Va. U., 1964. Mem. Md. Acad. Scis., Am. Chem. Soc., Am. Pharm. Assn., Miss. Acad. Scis., A.A.A.S., Am. Assn. U. Profs., Am. Assn. Colls. Pharmacy, Sigma Xi, Rho Chi, Phi Lambda Upsilon Contbr. numerous articles to sci. jours. Research on stblzn. of photog. materials, new sensitizing dyes for color film, synthetic methods for the preparation of certain 7, 8 and 9 membered rings; new methods for the synthesis of heterocyclic steroids, synthesis of a large variety of heterocyclic steroids; discovery of a new class of highly active bacteriocidal and fungicidal agents which are derivatives of azasteroids. Home: Box 455, University, Miss. 38677.*

DOPPELMAYR, Johann Gabriel, mathematician, physicist; b. Nuremberg, Germany; christened Sept. 30, 1677; s. Johann Siegmund Doppelmayr; studied law, Altdorf from 1696, math. and physics, Halle (both

Germany), astronomy and lens grinding in Holland and Eng.; m. Maria Susanna Kellner, Feb. 18, 1716; 4 children. Returned to Nuremberg in 1702; prof. math. Egidisches Gymnasium, 1704-50. Fellow Royal Soc., 1733; Leopoldina, acads. of Berlin, St. Petersburg. Author: Kurze Erklärung des Copernicanischen Systems, 1707; Neue vermehrte Welpersche Gnomica, 1708; Neue und grünliche Anweisung . . . grosse Sonnenuhren . . . richtig zu verzeichnen . . . , 1719; Atlas novus coelestis, 1742; Neue entdeclete Phaenomena . . . welche bei der fast allen Cörpernzukommenden Electrischen Krafft . . . hervorgebracht werden (on newly discovered phenomenon of electricity), 1744. Significant work in spreading sci. knowledge through his writings; pub. several earth and sky globes. Died Nuremberg, Dec. 1, 1750.

DOPPING-HEPENSTAL, Lambert John, Brit. inventor; b. 1859; s. R. A. and Diana (Hepenstal) D.-H.; m. Amy Maude Tottenham, 1920; 1 son, 1 dau. Commd. Royal Engr., 1879, ret. list as maj., 1907; high sheriff County Wicklow, 1909, Longford, 1910. Inventor direct engine-room telegraph for steamers, modified ry. points. Died Dec. 5, 1928.

DOPPLER, Christian Johann, Austrian physicist, mathematician; b. Salzburg, Austria, Nov. 29, 1803; s. Johann and Therese (Seelentner) D.; ed. Polytechnisches Institut, Vienna, Austria, Lyceum, Salzburg; m. Mathilde Sturm, 1836; 5 children. Tchr., math. high sch., Prague, Czechoslovakia, 1835; lectr. Tech. Inst., Prague; prof. math., geometry, 1841-47; prof. math. physics and mechanics Mining Acad., Chemnitz, Germany, 1847-49; prof. geometry Vienna Tech. U., 1849-50; prof. exptl. physics U. Vienna, 1850, also dir. Phys. Inst. Author: Versuch einer analytischen Behandlung von Flächen und Korper . . . , 1839; Über das Farbige Licht der Doppelsterne, 1842. Noted for discovery of principle in physics (named for him) which explains how apparent frequency of sound waves is related to the relative motion of sound source and hearer, also predicted that similar effect would hold for light waves, 1842; investigated colors of double stars, 1843-52, effect of rotation of the medium on properties of light and sound waves, 1845; indicated method of determining distance and absolute diameter of fixed stars (method perfected by Fizeau, 1848). Died Venice, Italy, Mar. 17, 1853.

DORABIALSKA, Alice Dominica, Polish chemist; b. Sosnowiec, Poland, Oct. 14, 1897; d. Thomas S. and Helen (Kaminska) D.; student U. Moscow (USSR), 1915-18; doct.dis., U. Warsaw (Poland), 1922; postgrad. Curie Inst. Radium, Paris, France, 1925-26, 29; Habilitation, Tech. U. Warsaw, 1927. With Tech. U. Warsaw, 1918-34, 40-44, prof. phys. chemistry Tech. U. Lvov (Poland), 1934-39; prof. phys. chemistry Tech. U. Lodz (Poland), 1945—. Recipient Sci. award Lodz, 1958; Polonia Restituta Commandery, 1966. Mem. Polish Chem. Soc. (mem. governing body 1922—, hon. mem.), Sci. Soc. Lodz. Author: Natural Radioactivity of Elements, 1952; also numerous articles. Editor sec. chemistry Zeszyty Nauk, Pol. Lodz, 1954—. Research on energy of nuclear transformations by means of originally developed apparatus; applications of radioactive elements to physico-chem. processes at surfaces and to measurements of rates of chem. reactions. ul. Zwirki 36, Lodz, Poland.*

DORAN, Alban Henry Griffiths, Brit. surgeon, gynecologist; b. Londo, London, 1849; s. Dr. Doran; ed. St. Bartholomew's Hosp.; staff (under Sir William Flower) Mus., Royal Coll. Surgeons, 8 years; cons. surgeon Samaritan Free Hosp., 1877-1909; practiced medicine, specializing in operative surgery of abdomen. Fellow Royal Coll. Surgeons Eng., Obstet. and Gynecol. Soc. Bordeaux (hon.), Gynecol. Soc. Bordeaux (hon.), Buenos Aires, Am. gynecol. socs., Royal Soc. Medicine; pres. Obstet. Soc. London, 1899-1900. Author: Tumours of the Ovary, 1884; Handbook of Gynaecological Operations, 1887. Compiler (with James Paget, James Goodhart) Catalogues of the Pathological Series in the Museum, 2d edit.; prepared descriptive catalogue of surg. instruments, 1912-25. Died Aug. 21, 1927.

D'ORBIGNY, Alcide Dessalines, French naturalist; b. Couëron, France, 1802; travelled in Bolivia and Patagonia, 1826-34; became prof. paleontology Muséum National d'Histoire Naturelle, 1854. Mem. Geol. Soc. France (pres.). Author: L'Homme Américain, 1840; Paleontologie francaise, 1840-60; Cours élémentaire de paléontogolie et de géologie stratigraphique, 1849; Prodromede paléontogolie stratigraphique et universelle des animaux mollusque et rayonnes, 1850-52; also 1st comprehensive map of S.Am. continent, 1842. A founder stratigraphical paleontology; studied Paleozoic fossils in Andes and Cretaceous fossils of Chile, 1826-34; 1st to propose subdivision of geol. formations into stages of deposit; divided fossil-bearing rocks into 6 periods (giving 1st 5 periods 27 stages); believed each stage represented an independent fauna made by a spl. act of creation; studied foraminifers. Died Pierrefitte, France, 1857.

DORCUS, Roy Melvin, Am. psychologist; b. Woodsboro, Md., Feb. 9, 1901; s. Charles W. and Emma (Feiser) D.; A.B., Johns Hopkins, 1922, M.A., 1924, Ph.D., 1925; m. Mildred Elizabeth Day, Sept. 8, 1925. Mem. faculty Johns Hopkins, 1925-37; faculty U. Cal. at Los Angeles, 1937—, dean, div. life scis., 1949—, dir. Vets. Vocational Counselling Unit, 1935-40. Researcher numerous activities NRC; cons. to industry and govt. Bd. dirs. Los Angeles Psychiat. Service. Mem. Soc. Advancement Mgmt. (exec. com.), Am. (dir.), Western, So. Cal. (pres. 1945), Cal. (past pres.) psychol. assns., So. Soc. Philosophy and Psychology (pres. 1934), Soc. Clin. and Exptl. Hypnosis (pres. 1958-59), Sigma Xi. Author: (with G. W. Shaffer) Textbook of Abnormal Psychology, 1934; (with Margaret Jones) Handbook of Employee Selection, 1950, also articles; (with Ethel DaSilva) Science in Betting: The Players and the Horses, 1961. Editor, contbr. Hypnosis and Its Therapeutic Applications, 1957. Mng. editor Jour. Comparative Psychology, 1936-45, Comparative Psychology Monographs, 1936-45; bd. editors Jour. Psychology, 1935, Am. Jour. Exptl. and Clin. Hypnosis, 1956—; cons. editor Brit. Jour. Med. Hypnotism, 1955—. Home: 1242 A-B Berkeley St., Santa Monica, Cal. 90404.*

DÖRELL, Georg Ludwig Wilhelm, German mining engr.; b. Clausthal, Germany, Dec. 17, 1793; s. Carl August and Auguste Eleonore (Kurtz) D.; studied at Bergschule mining sch., Clausthal, 1812-13, Bergakademie, Freiburg, Germany, 1815; m. Caroline Charlotte Louice Kathenbein, 1824. Engaged in mining industry, Katzhütte from 1822, then Zellerfeld, Germany, later Clausthal, from 1833; became high civil ofcl., 1841; mining master, Zellerfeld, 1848. Invented mech. movable ladder to haul miners in and out of shafts. Died Zellerfeld, Oct. 30, 1854.

DOREMUS, Charles Avery, Am. chemist; b. N.Y.C., Sept. 6, 1851; s. Robert Ogden and Estelle Emma (Skidmore) D.; A.B., Coll. City N.Y., 1870; A.M., Ph.D., U. Heidelberg, 1873; postgrad. U. Leipzig, 1873; hon. M.D., U. Buffalo, 1879; m. Elizabeth Johnson Ward, Aug. 4, 1880. Reporter on photography for U. S. govt., Vienna Expn., 1873; chemist U. S. Dairy Co., 1873; asst. chemistry, toxicology and med. jurisprudence, Bellevue Hosp. Med. Coll., 1874-79, adj. prof., 1879-97; prof. chemistry and toxicology, med. dept., U. Buffalo, 1879-82; prof. chemistry Am. Veterinary Coll., 1882-92, emeritus prof., 1892-98; asst. prof. chemistry and physics Coll. City N.Y., 1897-1901, asst. prof. chemistry, 1901-03, acting prof., 1903-04. Wrote sect. on Gaseous Poisons, Text-Book of Legal Medicine and Toxicology, 1903. Died Dec. 2, 1925.

DOREMUS, Robert Ogden, Am. chemist; b. N.Y.C., Jan. 11, 1824; grad. N.Y. U., 1842; med. dept., 1850, LL.D., 1872; studied chemistry in Paris, 1847-48. Established lab. in N.Y., 1848; prof. chemistry and physics, U. City of N.Y., 1852-1903; prof. chemistry and toxicology, Bellevue Hosp. Med. Coll., from 1864. In Paris, 1862-64, developing the use of compressed granulated gunpowder, which was adopted by the French govt.; patented methods of extinguishing fires and other chem. processes; an authority on toxicology. Died N.Y.C., Mar. 22, 1906.

DORF, Erling, Am. geologist; b. Nysted, Neb., July 19, 1905; s. Alfred Thorkil and Thyra (Axelsen-Dreier) D.; B.S., U. Chgo., 1925, Ph.D., 1930; m. Ruth Kemmerer, Apr. 3, 1934; children—Thomas Alfred (dec.), Norman Kemmerer, Robert Erling, Martha Dreier. Asst. instr. U. Chgo., 1926-27, fellow geology, 1928-30; faculty Princeton, 1926—, prof. geology, 1946—, curator paleobotany, 1932—; research asst. Carnegie Inst. Washington, 1926-45; research curator Phila. Acad. Sci., 1936-46; vis. prof. geology Wagner Free Inst. Sci., Phila., 1946—, Villanova (Pa.) U., 1963—. Sci. collaborator Nat. Park Service, 1956-63; cons. to pvt. cos., museums. Named hon. mem. Princeton Class of 1933, 1965—. Fellow Geol. Soc. Am., Paleontol. Soc. Am. (past v.p.), A.A.-A.S. (past mem. council); mem. Nat. Assn. Geology Tchrs. (Neil Miner Teaching award 1963, past pres. Eastern sect.), Yellowstone Bighorn Research Assn. (pres. 1964-66), Paleobot. Soc. India (fgn. hon.), Royal Danish Acad. Scis. and Letters (fgn.), Am. Geol. Inst., Am. Inst. Biol. Scis., Bot. Soc. Am., Geol. Soc. N.J., Internat. Assn. Plant Taxonomists, Internat Paleobot. Assn., Am. Assn. U. Profs., Atlantic Coastal Plain Soc., Nat. Geog. Soc., Sigma Xi, Kappa Epsilon Pi, Alpha Tau Omega. Author: Pliocene Floras of California, 1933; Upper Cretaceous Floras of the Rocky Mountain Region, 1942; also articles. Discovery of oldest land plants in U. S.; research on cretaneous and tertiary fossil forests of U. S.; geologic climatic changes. Home: 283 Mercer Rd., Princeton, N.J. 08540.*

DÖRFFEL, Georg Samuel, German astronomer; b. Plauen, Saxony, Germany, Oct. 21, 1643; s. Friedrich and Maria (Tröger) D.; studied theology U. Leipzig (Germany), from 1658, philosophy and natural scis., Jena, Germany, from 1662; student of astronomer Hevel; m. Maria Elisabeth Gebhardt, Nov. 12, 1668; m. 2d, Maria Salome Gottsmann, Sept. 17, 1679; m. 3d, Judith Francke, Aug. 26, 1687; 10 children. Became substitute pastor (for his father), Plauen, 1667, then pastor, 1672; apptd. supt., Weida, Thuringen, Germany, 1684. Moon mountain range named in his honor, 1791. Author: Cometographia, 1668; Astronomische Betrachtung der Grossen Cometen . . . , 1681; Eilfertige Nachricht von dem itzund am Himmel stehenden neuen Cometen . . . , 1682; Methodus nova phaenomenorum coelestium intervalla a terris facillime determinandi . . . , 1685. Discovered comet orbit was parabola around sun, 1680; discovered new comet later known as Halley's Comet, 1682. Died Weida, Aug. 16, 1688.

DORFMAN, Albert, Am. pediatrician; b. Chgo., July 6, 1916; s. Aaron and Anna (Schwartzman) D.; S.B., U. Chgo., 1936, Ph.D., in Biochemistry, 1939, M.D., 1944; m. Ethel Steinman, Sept. 1, 1940; children—Abby, Julie. Mem. faculty U. Chgo., 1958—, prof. pediatrics, 1957—, chmn. dept., 1962—, Richart T. Crane prof. pediatrics, 1965—, prof. biochemistry, 1961—; dir. research LaRabida Jackson Park San., Chgo., 1950-57, dir. LaRabida-U. Chgo. Inst., 1957—; cons. to surgeon gen. USPHS, 1952-57, 58-59, 61—. Mem. research com. Arthritis and Rheumatism Found., 1960—; bd. govs. Chgo. Heart Assn. Recipient E. Mead Johnson award, 1958, Phi Sigma Epsilon award, 1959. Mem. Am. Chem. Soc., Am. Soc. Biol. Chemists, Soc. Exptl. Biology and Medicine, Am. Pediatric Soc., Soc. for Pediatric Research, Midwest Soc. Pediatric Research (pres. 1963—), Am. Acad. Arts and Scis., Am. Heart Assn., Am. Rheumatism Soc., Am. Soc. Cell Biology, Am. Acad. Pediatrics, Brazilian Pediatric Soc. (hon.), Phi Beta Kappa, Sigma Xi, Alpha Omega Alpha, others. Editor: Circulation, 1959—, Excerpta Medica, 1961—, Rheumatism and Arthritis, 1964—. Research and publs. on growth-promoting substances; aspects of connective tissue diseases, including hyaluronidase, effects of ACTH, mucopolysaccharides, pathology. Home: 2231 E. 67th St., Chgo. 60649.*

DORFMAN, Ralph Isadore, Am. biochemist; b. Chgo., June 30, 1911; s. Aron and Anna (Schwartzman) D.; B.S. in Chemistry, U. Ill., 1932; Ph.D. in Physiol. Chemistry and Pharmacology, U. Chgo., 1934; m. Margaret Patricia Cameron, Feb. 19, 1965; children—Gerald Allen, Ronald Arthur. Instr., La. State U., 1935-36; faculty Yale, 1936-41, asst. prof., 1939-41; faculty Western Res. U., 1941-51, asso. prof., 1950-51; asso. dir. labs. Worcester Found. for Exptl. Biology, Shrewsbury, Mass., 1951-56, dir. labs., 1956-64; affiliate prof. chemistry Clark U., Worcester, Mass., 1956-64; research prof. biochemistry Boston U., 1951-67; dir. inst. hormone biology Syntex Research Center, Palo Alto, Cal., 1964—, sr. v.p., 1964—. Mem. Endocrine Soc., Am. Chem. Soc., A.A.A.S., Am. Acad. Arts and Scis., Soc. Exptl. Biology and Medicine, Am. Assn. Cancer Research, Am. Statist. Soc., Sigma Xi, Phi Lambda Epsilon. Author: (with Soffer, Gabrilove) Androgens, 1956; Human Adrenal Gland, 1961; (with Castro) Pituitary-Ovarian Endocrinology, 1963; (with F. Ungar), Metabolism of Steroid Hormones, 1965; also numerous articles. Editor: Methods in Hormone Research, Vol. I and II, 1962, Vol. III, 1964, Vol. IV, 1965, Vol. V, 1966. Research on house dust allergens for asthma, normal and abnormal aspects biosynthesis and catabolism steroid hormones; devel. corticoids, anabolic agts., progestational agts., estrogens, anti-cancer substances for various diseases and control fertility in humans and animals. Home: 10465 Berkshire Dr., Los Altos Hills, Cal. 94022. Office: 3401 Hillview St., Palo Alto, Cal. 94304.*

DORFMAN, Robert, Am. economist, educator; b. N.Y.C., Oct. 27, 1916; s. Samuel M. and Mina (Gordon) D.; B.A., Columbia 1936, M.A., 1937; Ph.D., U. Cal., 1950; m. Nancy Schelling, Nov. 6, 1949; children—Peter J., Ann Elizabeth. With Bur. Labor Statistics, 1939-41, OPA, 1941-43; faculty U. Cal., 1950-55; faculty Harvard, 1955—, prof. econs., 1957—. Mem. President's Committee to Appraise Employment and Unemployment Statistics, 1962. Served as operations analyst USAAF, World War II. Ford Found. faculty research fellow. Fellow A.A.A.S., Econ. Soc.; mem. Am. Econ. Assn., Operations Research Soc. Am. (council 1959-62), Econometric Society (member council 1961—), National Research Council, American Statis. Assn., Inst. Mgmt. Scis. (pres. 1965). Author: Application of the Linear Programming to the Theory of the Firm, 1951; (with P. O. Steiner) Economic Status of the Aged, 1956; (with P. A. Samuelson, R. Solow) Linear Programming and Economic Analysis, 1958; The Price System, 1964. Application of linear programming, operations research techniques to econ. theory, public investment decisions. Home: 81 Kilburn Rd., Belmont, Mass.*

D'ORGEVAL-DUBOUCHET, Bernard Marie Jules, French mathematician; b. Caluire et Cuire, France, Aug. 2, 1909; s. Francois Ludovic and Jeanne (Billet) d'Orgeval; Agrégé des Sciences Mathematiques, 1932, Docteur ès Sciences, 1943, Docteur en Droit, 1947; m. Albine Aimée Yoillemot, July 21, 1946; children—Brigette, Michel, Isabelle, Pierre, Régis, Benedicte, Agnes, Blandine, Fabienne. Prof., École Normale Supérieure, Teheran, Iran, 1932-34, Lycée, Beauvais, France, 1934-35; postgrad. U. Rome, 1936-38; prof. Lycée, Orléans, France, 1938-39; research asso. Centre National de la Recherche Scientifique, 1945-48; lectr. Faculty Scis., Grenoble, France, 1946-52; prof. Algiers (Algeria) Faculty Scis., 1951-55, Dijon (France) Faculty Scis., 1955—. Mem. French, Austrian math. socs., Italian Math. Union, History of Law Soc. Research and publs. on geometry of algebraic surfaces. Home: 9 boulevard St. Jacques, 21 Beaune, France. Office: Faculté des Sciences, Boulevard Mamart, 21, Dijon, France.*

DORN, Friedrich Ernst, German chemist; b. Guttstadt, East Prussia, Germany, 1848; ed. Königsberg,

Germany; taught physics at Darmstadt, Halle (both Germany); discovered emanation from radium known as niton (now named radon or radium emanation), 1900. Died Halle, Germany, June 13, 1916.

DORN, Gerard, German chemist, physician; flourished 2d half 16th cent.; lived in Frankfurt, Germany. Author and editor of numerous works and collected editions relating to chem. medicine, including Clavis totius philosophiae chymisticae, 1567; Chymisticum artificium naturae, Theoricum and Practicum, 1568-69; Congeries Paracelsicae Chemiae de Transmutationibus Metallorum, 1581. Edited several works of Paracelsus and added a commentary to the Archidoxes. Defended Paracelsus in controversies with other authors. Distilled urea; described methods for distilling medicinal water and oils; described and studied sulphuric acid.

DORNBLÜTH, Otto Wilhelm Albert Julius, German psychiatrist; b. Rostock, Germany, Mar. 19, 1860; s. Friedrich and Katherine Elisabeth Sophie (Wetzel) D.; student medicine, Rostock, Tübingen, Munich (all Germany); m. Lillie Seeler, 1884 (div.); 1 son; m. 2d, Hedwig v. Klingspor, 1906. Asst. physician Rostock Med. Clinic, also Munich Psychiat. Clinic; became psychiatrist in Silesia, 1886; apptd. dir. insane asylum, Freiburg, Germany, 1892; neurologist in Rostock, from 1895, Frankfurt/Main, Germany, 1900; dir. pvt. clinic, Frankfurt/Main, 1902-08, later in Wiesbaden, Germany. Author: Compendium der Inneren Medizin, 1892; Klinisches Wörterbuch, Die Kunstausdrücke der Medizin, 1894; Compendium der Psychiatrie, 1894; (with K. v. Noorden, Hedwig Dornblüth) Diätetisches Kochbuch, 1897; Arzneimittel der heutigen Medizin, 1898; Moderne Therapie, 1906; Pathologie und Therapie der nervösen Erkrankungen, 1908; Psychoneurosen, 1911; Die Schlaflosigkeit und ihre Behandlung, 1912; Gesunde Nerven, 1916. Popular med. writer in rational-moral style of time; popularized sci. of psychology. Died Wiesbaden, Dec. 29, 1922.

DORNO, Carl Wilhelm Max, bioclimatologist; b. Königsberg, Germany, Aug. 3, 1865; s. Carl and Emma (Lehnhard) D.; began studies of chemistry, physics, econs. and law, in 1899; doctorate in chemistry, Königsberg, 1904; m. Erna Hundt, 1892; 1 dau. Took over family bus. in Königsberg, 1891; founder Physikalische-Meteorologische Observatorium, Davos, Switzerland, 1907, dir. until 1926; dir. 1st internat. climate congress, Davos, 1925. Recipient Leibniz medal, 1919, Niels-Finsen medal Comité Internat. de la Lumière, 1937; named Hon. Citizen, Davos, 1924. Author: Studien über Licht und Luft des Hochgebirges, 1911; Physik der Sonnen- und Himmelsstrahlung, 1919; Klimatologie im Dienste der Medizin, 1920; Ausstattung moderner Strahlungsobservatorien, 1927; Grundzüge des Klimas von Muottas-Muraigl, 1927; (with F. Lahmeyer) Assuan, eine meteorologisch-physikalisch-physiologische Studie 1932; Das Klima von Agra, eine 3 und letzte meteorologisch-physiologische Studie, 1934; also numerous articles. Founder modern bioclimatology and radiation climatology; invented instruments for measurement effect climate especially high altitude climate on man; research on therapeutic effects of natural radiation. Died Davos-Platz, Apr. 22, 1942.

DORODNITSYN, Anatoliy Alekseevich, Russian geophysicist; b. Dec. 2, 1910; grad. Grozny Petroleum Inst., 1931. Instr. higher edn. and research establishments, Moscow, Leningrad, 1936-41; with Central Aerodynamics Inst., Moscow, 1941-55; with Math. Inst., USSR Acad. Sci., 1945-55, dir. Computing Center, 1955——; prof. Moscow U., 1947——. Del. 44th session Assn. Sci. Congress, Calcutta, 1957, Congress on Electronic Computers, Paris, 1959. Recipient Stalin prize, 1943, 44, 46, 47, 51. Mem. USSR Acad. Sci. Author: The Boundary Layer in Compressible Gas, 1942; The Solution of Mathematical and Logical Problems on High Speed Electronic Computers. Research and publs. on dynamic meteorology, aerodynamics, applied math., effects of irregularities in earth's surface on air currents. Address: Computing Center, USSR Acad. Sci., Akedemichesky pr. 28, Moscow, USSR.

DOROZHKIN, Nikolay Afanasevich, Russian phytopathologist, mycologist; b. 1905; grad. Byelorussian Agrl. Acad., 1927; D.Agrl. Sci., 1934. Asst. dept. phytopathology and microbiology Byelorussian Agrl. Acad., 1927-29; dir. Borisovo Agrl. Technicum, 1929-30, Minsk Plant Protection Sta., 1930-31; lectr. Horticulture and Truck Gardening Inst., 1931-32; head lab. phytopathology Byelorussian Acad. Sci., 1932-34, dep. dir., 1934-36, dir. Inst. Biology, 1936-38, sr. asso. Inst. Socialist Agr., 1956; prof., head chair plant systematics Byelorussian U., 1934-41, prof., 1948——; dir. Research Inst. Fruit, Vegetable and Potato Growing. Editor: The Flora of the Belorussian SSR. Vol. 1-3, 1949-50. Research and publs. on methods of obtaining high potato yields, studied potato cancer and counter-measures, 1944-68. Address: Research Inst. of Fruit, Vegetable and Potato Growing, Samokhvalovichi, Minsk, Byelorussian SSR, USSR.

DORP, Willem Anne van, Dutch chemist; b. Rotterdam, Netherlands, 1847; Ph.D., U. Heidelberg (Germany), 1871; (with Behr) obtained anthracene, pyridine, anthraquinone; (with Hoogewerff) determined constn. of isoquinoline.

DÖRPFELD, Wilhelm, German archeologist; b. Barmen, Germany Dec. 26, 1853; s. Wilhelm Friedrich and Christine (Keller) D.; student Acad. Architecture, Berlin, 1873-76; m. Anna Adler, 1886; 1 son, 2 daus. Assigned to Olympia excavations, 1877-81; succeeded Schliemann in excavations at Troy, 1893, worked chiefly along outer border of site, revealed ruins of Middle and Late Bronze ages; excavated at Pergame, Athens, Attica, Thermos, Etoly, Thebes, Corfre; prof. U. Jena (Germany), 1921-24. Mem. Prussian Acad. Scis., Vienna Acad. (corr.), Greek Archeol. Soc. (sec.). Author: Troie et Ilion, 1902; Le retour d'Ulysse, 1924; Ithaque ancienne, 1927; Olympic ancienne, 1935; Athenes ancienne, 1937-39. First to study constrn. of ancient Greek theaters. Died Apr. 26, 1940.

DORR, John A., Jr., Am. geologist; b. Grosse Pointe Park, Mich., Oct. 25, 1922; s. John A. and Velma (Read) D.; B.S., U. Mich., 1947, M.S., 1949, Ph.D., 1951; m. Ruth Muriel Pritchett, Nov. 4, 1943; children—John A., III, James, Robin. Curator vertebrate paleontology Carnegie Mus., Pitts., 1951-52; prof. U. Mich., 1952——, chmn. geology dept., 1966——, research asso. Mus. Paleontology, 1952——, dir. Geol. Expdns., 1965——. Fellow Geol. Soc. Am.; mem. Soc. Vertebrate Paleontology (pres. 1966-67), Soc. for Study of Evolution, Mich. Acad. Sci., Sigma Xi. Contbr. articles to sci. jours. Research and publs. on geol. and paleontol. history of Middle Rocky Mountains, Colo., Wyo., Mont. Home: 291 Gralake St., Ann Arbor, Mich. 48103.*

DORSET, Marion, Am. chemist; b. Columbia, Tenn., Dec. 14, 1872; s. Walter Clagett and Jane (Mayes) D.; B.S., U. Tenn., 1893; M.D., Columbian (now George Washingon) U., 1896; also studied at U. Pa.; m. Emily K. Jackson, Oct. 10, 1900; children—Walter Clagett (dec.), Jane Mayes (dec.), Virgil Jackson. Taught bacteriology and pathology; asst. chemist biochemic lab., Dept. Agr., Washington, 1894-1903; chief biochemic div., from 1904. Research on chemistry and biology of tubercle bacillus, chemistry and bacteriology of meats; co-discoverer hog-cholera filtrable virus, 1903, and later hog-cholera serum. Died Washington, D.C., July 15, 1935.

DORSEY, George Amos, Am. anthropologist; b. Hebron, O., Feb. 6, 1868; s. Edwin Jackson and Mary Emma (Grove) D.; A.B., Denison, 1888, LL.D., 1909; A.B., Harvard, 1890, Ph.D., 1894; m. Ida Chadsey, Dec. 8, 1892; children—Dorothy Ann, George Chadsey; m. 2d, Sue McLellan. Conducted anthrop. investigations in S. America for Chgo. Expdn., 1891-92; supt. archaeology, dept. anthropology, same, 1892-93; asst. in anthropology Harvard, 1894-95, instr., 1895-96; asst. curator anthropology Field Mus. Natural History, Chgo., 1896-98, curator, 1898-1915; prof. comparative anatomy, Northwestern U. Dental Sch., 1898-1913; asst. prof. anthropology U. Chgo., 1905-08, asso. prof., 1908-15. Lectr. anthropology, New School for Social Research, N.Y.C., from 1925. Visited Europe, Egypt, India, Ceylon, Java, Australia, Bismarck Arch., New Guinea, P.I., China and Japan, for Field Mus., 1908; mem. edit. staff and fgn. commr. Chicago Tribune, 1909-12, investigating sources of emigration in Italy, Austria, Hungary, Rumania, Serbia and Bulgaria, and studying polit. conditions in India, China, Japan, Australia and South Africa. U. S. del. Internat. Congress Anthropology and Prehistoric Archaeology, Paris, 1900. Author: Why We Behave Like Human Beings, 1925; The Nature of Man, 1927; The Evolution of Charles Darwin, 1927; Hows and Whys of Human Behavior, 1929. Considered one of most thorough students and research workers in his field. Died N.Y.C., Mar. 29, 1931.

DORSEY, Herbert Grove, Am. oceanographer, physicist; b. Kirkersville, O., Apr. 24, 1876; s. Edwin Jackson and Mary Elma (Grove) D.; B.S., Denison U., 1897, M.S., 1898, Sc.D., 1938; Ph.D., Cornell Univ., 1908; m. Virginia Rowlett, June 21, 1906; children—Herbert Grove, William Rowlett. Instr. physics U. Me., 1898-1900; asst. prof. U. Fla., 1901-03; instr. physics and electricity Mechanics Inst., Rochester, 1903-04; asst. instr. physics Cornell U., 1904-05, instr., 1905-10; engr. research br., Western Electric Co., 1910-12; research engr. Nat. Cash Register Co., 1912-16, Hammond Radio Research Lab., 1916-22, Submarine Signal Co., 1922-26; sr. elec. engr. U. S. Coast and Geodetic Survey, 1926-28, prin. elec. engr. and chief research sect., 1923-46; lectr. in physics, George Washington Univ., 1947; instr. physics Capitol Radio Engring. Inst. since 1948. Invented Dorsey Phonelescope, dynamic loudspeaker, Fathometer, Sono Radio Buoy. and improved acoustics contrivances in telephone and radio fields; Sonar, the fathometer used horizontally has been employed extensively to locate enemy craft and submarines. Recipient 1st annual award, Washington Soc. Engrs., 1941. Fellow A.A.-A.S., Am. Phys. Soc., Acoustical Soc. Am., I.R.E. (chmn. Washington sect., 1933), Am. Inst. E.E. (chmn. Washington sect., 1934); mem. Am. Optical Soc., Am. Assn. Physics Tchrs., Internat. Com. on Radio, Am. Geophys. Union, Am. Radio Relay League, Washington Acad. Sci., Philos. Soc. of Washington, Washington Soc. Engrs., Sigma Xi, Phi Kappa Phi, Beta Theta Pi. Home: 3708 33d Pl., Washington 8.

DORSEY, James Owen, Am. ethnologist; b. Balt., Oct. 31, 1848; s. Thomas Anderson and Mary Sweetser (Hance) D.; attended Protestant Episcopal Theol. Sem. Ordained deacon Episcopal Ch., 1871; missionary to Pawnee Indians, Dakota Ty., 1871-80; worked with Omaha Indians as mem. Bur. Am. Ethnology; mem. Anthrop. Soc. Washington, A.A.A.S. Author: The Cegiha Language, 1890; contbr. to Annual Reports U. S. Bur. Am. Ethnology. Studied Omaha Indians; made linguistic and ethnological studies on Western Indians. Died Feb. 4, 1895.

DORST, Jean (Pierre), French biologist; b. Mulhouse, France, Aug. 7, 1924; s. Joseph Victor and Gabrielle (Rusch) D.; Doctorat es Sciences, Faculte des Sciences, Sorbonne, Paris, France, 1949; m. Eliane Gherardi, Sept. 16, 1966. With Nat. Mus. Natural History, Paris, 1947—— prof., 1954——, dir. lab. zoology of mammiferes and birds. Vice pres. Internat. Council for Bird Preservation, 1955; dir. Revue Mammalia 1960; pres. Charles Darwin Found. for Gallagos Islands, 1964. Recipient Gold medal Acad. Agr. France. Corr. fellow Am., Brit. ornithologists unions, others; mem. Souek Ornithologifre de France (pres.), Société Zoologists France (sec.-gen.), others. Author: The Migrations of Birds, 1956, 62; Les Oiseant, 1957; Les animaux voyageurs, 1965; Avant que nature meme, 1965. Research, publs. on ecology of tropical birds and high Andes, anatomy and evolution of several groups of birds; conservation of nature. Home: 114 Ter Avenue de Versailles, 75 Paris 16. Office: 55 rue de Buffon, 75 Paris 5, France.*

DORTET DE TESSAN, Louis Urbain, French engr.; b. Vigan, Aug. 25, 1804; ed. École polytechnique, Paris, France, 1822. Hydrographic engr. under Beautemps-Beaupré, 1831-33; mem. expdn. to map coastline of Algeria, 1836-39; meteorologist Dupetit-Thouars mission to Cape of Good Hope. Mem. French Acad. Scis. Author: Description nautique des côtes de l'Algeria (an atlas of the Algerian coast), 1837. After trip of 1839, he described the temperature of the waters, wave motion, the flight of birds, currents, and the aurora borealis of the region; hydrographic exploration of the Strait of Gibralter, 1848; research on use of muddy sand dredged from Brit. Channel for fertilizer. Died Paris, Sept. 30, 1879.

DORWART, Harold Laird, Am. mathematician; b. Greenville, Pa., Aug. 27, 1902; s. George Wilson and Clara (Laird) D.; A.B., Washington and Jefferson Coll., 1924; Ph.D. in Math., Yale, 1931; m. Carolyn Frances Yeisley, Jan. 2, 1933; 1 son, Roger Wilson. Asst., then instr. math. Yale, 1924-28; instr. Williams Coll., 1928-30, 31-35; from asst. prof. to prof. math. Washington and Jefferson Coll., 1935-49; Seabury prof. math., chmn. dept. Trinity Coll., Hartford, Conn., 1949-67, dean of the coll., 1967-68. Member of American Mathematical Society, also Math. Assn. Am. (bd. govs. 1948-51), Phi Beta Kappa, Sigma Xi. Author: The Geometry of Incidence, 1966. Contbr. numerous articles profl. jours. Study of number theory; algebra; criteria for the irreducibility of polynomials. Home: 125 Vernon St., Hartford, Conn. 06106.

DOSITHEOS OF PELUSIUM, Greek geometer; flourished circa 230 B.C.; pupil of astronomer Conon. Continued the relationship between the Alexandrian astronomers and Archimedes which had resulted from Archimedes' studies in Alexandria (several of Archimedes' books dedicated to him); made observations on times of appearance of fixed stars (some at points farther north than Alexandria, and on weather-signs (recorded in writings of Geminos and others); authored work containing discussion of Aratos' Phaenomena and Eudoxos' researches, also a work on the calendar.

DOSS, (Karl) Bruno, German geologist; b. Auerbach, Germany, Nov. 1, 1861; s. Karl Friedrich and Christiana Friederike (Piering) D.; student mineralogy-petrography Munich, Germany, 1882-84; doctorate Leipzig, Germany, 1886. Asst., Tech. U., Dresden, Germany; became lectr., 1889; apptd. prof. geology Polytechnikum, Riga, Latvia, 1889; exiled to Orel, Russia as German, 1914; expelled from Russia, 1915; worked in Saxony, Germany; became leader group geologists on Eastern front, 1917. Research and publs. on geology of Baltic countries especially diluvial and alluvial geologic phenomena; discovered recent colloidal iron bisulfide; research on mineral deposits, petrography and crystallography, Kurland meteorite. Died Dresden, May 28, 1919.

DOS SANTOS, Reynaldo, Portuguese urologist; b. Vila France de Xira, Portugal, Dec. 3, 1880; s. Clemente and Maria Amelia Dos Santos; M.D., U. Lisbon; m. Susana Cid, Aug. 18, 1906; 1 child. Prof. surg. pathology, dir. sch. medicine U. Lisbon. Recipient Gold medal Internat. Soc. Urology, 1935. Mem. numerous European sci. and med. socs.; pres. Soc. Med. Scis., 1932-33. Contbr. to med. jours., texts. Introduced (with J. Caldas) thorium dioxide or thorotrast in arteriography, circa 1931.

DOST, Friedrich Hartmut, German physician; b. Dresden, July 11, 1910; s. Georg D. and Else (Dähne) D.; ed. univs. Rostock, Fribourg, Innsbruck, Leipzig; M.D.; m. Felicitas Webel; children—Frank, Bettina, Eva, Stephan. Titular prof. U. Leipzig, 1940, instr.,

1943; full prof., dir. Clinic of Charity; with U. Berlin, 1953; full prof., dir. Univ. Clinic for Children, Giessen, 1959. Mem. German Acad. Naturalists (Leopoldina). Author numerous monographs and articles in profl. publs. Home: Gutenbergstrasse 24. Office: Klinikstrasse 28, Giessen, Germany.

DOSTAL, Rudolph, Czech. botanist; b. Proruba, Czechoslovakia, Mar. 28, 1885; s. Frantisek and Anna (Fiedlerová) D.; Ph.D., Charles U., Prague, Czechoslovakia, 1908; habil., U. Brno (Czechoslovakia), 1919; m. Marie Vánová, Nov. 8, 1911; children—Vera, Kaderábková, Olga. Tchr., gymnasium, 1919-33; prof. Vet. Coll., 1928-33; prof. Agrl. Coll., Brno, 1933-60, mem. sci. council. Decorated Order Labor. Mem. Acad. Agrl. Scis., Acad. Scis. (corr.), Bot. Soc. Prague. Author: Experimental Morphology, 1930; Vegetable Food Stuffs, 1929; On the plant Integration, 1959; Physiology of Growth, Development, and Movements, 1962; also numerous articles. Research on growth correlations of germinating plants, organs and reprodn. Calulerpa, marine algae. Home: 51 Vackova, Brno, Czechoslovakia.*

DOTEN, Samuel Bradford, Am. entomologist; b. Gold Hill, Storey County, Nev., Dec. 14, 1875; s. Alfred and Mary Calista (Stoddard) Doten; B.A., U. Nev., 1898, M.A., 1912; D.Sc., 1950; m. Laura Katherine Schweis, June 16, 1915. Instr. history and math. U. Nev., 1898-1900, instr. math. and entomology, 1900-02, asst. prof., 1902-03, asst. prof. entomology, meteorology and math. 1903-05, prof. entomology, since 1906; also entomologist and dir. Nev. Expt. Sta., 1913-46. Fellow A.A.A.S. Author numerous bulls. and articles on entomol. subjects. Died May 9, 1955.

DOTT, Robert H(enry), Am. geologist; b. Sioux City, Ia., Jan. 8, 1896; s. Richard M. and Delia (Rood) D.; B.S.F., U. Mich., 1917, A.M., 1920; m. Esther Reed; children—Esther (Mrs. S. F. Bird), Robert Henry. Began career as a geologist with Empire Gas & Fuel Co., 1917-18, 19, Standard Oil Co. N.J., 1920-22, Carter Oil Co., 1922-26, Mid-Continent Petroleum Corp., 1926-29, Sunray Oil Co., 1929-31; cons. geologist, 1931-35; dir. Okla. Geol. Survey, Norman, 1935-52; exec. dir. American Association of Petroleum Geologists, 1952-63, editorial consultant, 1963—. Served as pvt. U.S.A.A.F., 1918. Mem. Geol. Soc. Am., Am. Assn. Petroleum Geologists (hon.), Sigma Xi. Contbr. articles profl. jours. Investigation of Pennsylvanian strata in northeastern Oklahoma; structural history of Arbuckle Mountains, Oklahoma; analysis of hypotheses on origin, migration, and accumulation of oil. Home: 2550 E. 24th St., Tulsa 74114. Office: P.O. Box 979, Tulsa 74101.*

DOTTER, Charles Theodore, Am. physician, educator; b. Boston, June 14, 1920; s. John Maury and Rosalind (Allin) D.; A.B., Duke, 1941; M.D., Cornell, 1944; m. Pamela Beattie, Sept. 30, 1944; children—Barbara, Jeffrey, Jane. Faculty Cornell, 1948-52, asst. prof. radiology, 1951-52; asst. attending radiologist N.Y. Hosp., 1950-52; prof. radiology, head dept. U. Ore. Med. Sch., 1952—. Mem. heart tng. com. Nat. Heart Inst., 1966—. Diplomate Am. Bd. Radiology. Mem. Inst. Radio Engrs., Am. Coll. Radiology, Am. Roentgen Ray Soc. (Silver medal 1965), Am. Heart Assn., Czechoslovakia Med. Soc. (hon.), Western Soc. Clin. Research, Pacific N.W. Radiology Soc., Radiol. Soc. N.Am., Assn. U. Radiologists, Internat. Cardiovascular Soc., Ore. Radiol. Soc., Ore., Multnomah County med. socs. Author: (with I. Steinberg) Angiocardiography, 1951. Research and publs. include descriptions of several new techniques in cardiovascular radiology, especially the non-surg. catheter treatment of arteriosclerotic obstruction. Home: 4004 S.W. Greenleaf Dr., Portland, Ore. 97221.*

DOTTERWEICH, (August Adolf) Heinz, German zoologist; b. Dresden, Germany, Sept. 25, 1904; s. August and Helen (Schmorl) D.; began studies natural scis. and philosophy U. Kiel (Germany), 1924; doctorate, 1927; m. Thea van Lindt, 1930; 1 dau. Became asst. zool. collection Tech. U., Dresden, 1929; apptd. lectr., 1931; became asso. prof. Zool. Inst., 1938, named dir., 1939; founder Zentraluntersuchungsstelle für Angorawolle (Central Research Sta. for Angora Wool), Resse, Germany, 1946. Author: Das biologische Gleichgewicht und seine Bedeutung für die Hauptprobleme der Biologie, 1940; also numerous articles. Research on physiology including breathing of birds, heredity analysis, fur and hair especially angora wool. Died Resse, July 12, 1949.

DOTY, Delbert Malcolm, Am. chemist; b. nr. College Corner, Ind., Apr. 27, 1908; s. Delbert Vernon and Martha (Spenny) D.; B.S., Purdue U., 1929, M.S., 1932, Ph.D, 1941; m. Ruth Sheets, Sept. 3, 1930; children—Donna Margaret (Mrs. Arthur L. Kunz), Dale Anson. Faculty, Purdue U., 1929-48, asso. prof., 1946-48, asst. chief dept. agrl. chemistry, 1943-48; staff Am. Meat Inst. Found., Chgo., 1948-64, dir. research and edn., 1961-64; tech. dir. Fats and Proteins Research Found., Inc., Des Plaines, Ill., 1964—. NRC adv. coms. 1962—. Recipient Reciprocal Meat Conf. Signal Service award, 1963; award for patriotic service Dept. Army, 1966. Mem. Nat. Acad. Sci., Am. Chem. Soc., Am. Oil Chem. Soc., Am. Leather Chem. Soc., Animal Nutrition Council, Inst. Food Technologists, Am. Meat Sci. Assn., Research and Devel. Assos., Sigma Xi, Alpha Zeta, Phi Lambda Upsilon. Author: (with others) The Science of Meat and Meat Products, 1960; also articles. Research on influence of heredity and environment on composition and nutritive value of plants, enzymes and their role in food quality, better utilization of animal fats and proteins, evaluation of factors influencing quality of meat and meat products. Home: 21W237 Gove St., Itasca, Ill. 60143. Office: 3150 Des Plaines Av., Des Plaines, Ill. 60018.*

DOTY, Paul Mead, Jr., Am. chemist; b. Charleston, W.Va., June 1, 1920; s. Paul Mead and Maud Katherine (Stewart) D.; B.S., Pa. State Coll., 1941; M.A., Columbia, 1943, Ph.D., 1944; m. Margaretta Elenor Grevatt, Oct. 31, 1941; 1 son, Gordon Sutherland; m. 2d, Helga Boedtker, Feb. 27, 1954; children—Marcia, Rebecca, Katherine. Engaged as instructor and research associate Polytechnic Institute, Bklyn., 1943-45, asst. prof. chemistry, 1945-46; asst. prof. chemistry, U. Notre Dame, 1946-48; asst. prof. Harvard, 1948-50, associate professor chemistry, 1950-56, prof. chemistry, 1956—; Harvey lecturer, 1959-60. Rockefeller fellow Cambridge U., Eng., 1946-47; Guggenheim fellowship, 1958; recipient Am. Chem. Soc. pure chemistry award, 1956. Fellow Nat. Acad. Scis.; mem. Fedn. Am. Scientists, American Acad. Arts and Science, American Chemical Society, Biophysical Society, American Society Biological Chemists. Author research papers. Editor Jour. Polymer Sci., 1945-61; editorial bd. Jour. Molecular Biology. Research in molecular biology; functions, properties, and structure of proteins and nucleic acids. Home: 4 Kirkland Pl., Cambridge, Mass.

DOUB, Howard Philip, Am. physician; b. Hagerstown, Md., Sept. 30, 1890; s. Louis P. and Ella (Newcomer) D.; A.B., Western Md. Coll., 1913; M.D., Johns Hopkins, 1917, postgrad., 1917-18; m. Helen Ringrose, June 21, 1919; 1 son, Gerald P. Practice medicine specializing in radiology, Detroit, 1918-23; radiologist-in-chief Henry Ford Hosp., Detroit, 1923-55, cons. radiology, 1955—. Pres., Mich. Cancer Found., 1960. Diplomate Am. Bd. Radiology. Fellow Am. Coll. Radiology (past pres. Gold medal 1962); mem. Radiol. Soc. N.Am. (past pres., Gold medal 1950), Am. Cancer Soc. (past pres. S.E. Mich.), A.M.A. (Gold medal 1927), Mich. State, Wayne County med. socs., Detroit X-ray and Radium Soc., Am. Radium Soc., Am. Roentgen Ray Soc. Contbr. numerous articles to tech. jours. Research on exptl. nephritis produced by irradiation, heart lesions produced by X-rays, X-ray diagnosis, treatment cancer by radiation. Home: 700 Seward St. Office: Henry Ford Hosp., Detroit 48202.*

DOUBLE, François-Joseph, French physician; b. Verdun-sur-Garonne, France, Mar. 11, 1776; mem. French Acad. Scis., 1832; a founder (with Portal) Acad. Medicine; author memoir on croup which won 1st honorable mention in competition founded by Napoleon in 1811. Died June 12, 1842.

DOUBLEDAY, Henry, English naturalist; b. Epping, Essex, Eng., July 1, 1808; s. Benjamin Doubleday; made many collecting expdns. to Eastern countries until 1848. Mem. Entomol. Soc. London. Author: Nomenclature of British Birds, 1838; Synonymic List of British Lepidoptera, 1847; also articles on habits of mammals, birds and insects. Attempted to establish uniform system of entomol. nomenclature (Doubleday's list); introduced practice of catching moths at sallow blossoms and sugaring. Died June 29, 1875.

DOUCAS, Christophe, Greek dermatologist; b. Myteline, Greece, Dec. 25, 1890; s. Aristides and Hadji Theologou (Paleologo) D.; student univs. Athens, Berlin, Paris; M.D.; m. Lydia Pouliezos, Aug. 27, 1949; Asst., Syggros Hosp., 1914-18; chief clinic, 1923-29; dir. dermatol. sect. Evangelismos Hosp., Athens, 1935-36, cons. hosp.; became asso. prof. dermatology and venerology U. Athens. Expert on venereal diseases and trepanosomatoses for WHO. Mem. French, Greek, Israeli, Austrian, Yugoslavian dermatol. socs. Contbr. numerous articles to profl. jours. Address: Odos Sekeri 4, Athens, Greece.

DOUCET, Jean, French parasitologist; b. Paris, Mar. 4, 1919; s. Leon and Louisa (Renaud) D.; Docteur en Médecine, U. Paris, 1947, Docteur es Sciences, 1965, Agrégé des Facultés de Médecine, 1965; m. Andrée Labuzan, Sept. 20, 1958; children—(by previous marriage) Cathérine, Béatrice. With Office de la Recherche Scientifique et Technique d'Outre-mer, 1948—, dir. research, 1962—; prof. lectures U. d'Abidjan; biologist Abidjan Hosps., 1966—; chief lab. med. entomology Inst. Recherche Sci. Madagascar, 1948-51; chief lab. parasitology, medicine Inst. d'Enseign Recherche Trop. Abidjan, 1951-65; chief lab. Centre Hospitalier. Mem. Soc. entomol. Fr., ARBO-virus Information Exchange, Assn. Experts Coop. Tech. Internat., Soc. Fr. Parasitol., Internat. Filariasis Assn. Author: Les Anophèles de Madagascar, 1950; Les Serpents de Côte d'Ivorie, 1964; Contribution à l'Étude des Pentastomes, 1965; also articles. Biol. and systematic studies of 58 mosquito-species including 18 new species and a new sub-genus, Madagascar, breeding activity of mosquitoes in trees using steel tower, Ivory Coast, W. Africa; distbn. in Ivory Coast of Anopheles, Culicine, horse flies, snakes; anat. study of Pentastomida (parasities of snakes and man). Home: B.P. 20.632 Ecole Médecine, Abidjan—Rép. Côte d'Ivoire, Afrique.*

DOUDOROFF, Michael, microbiologist; b. Petrograd, USSR, Nov. 14, 1911; s. Boris and Natalie (Shulgin) D.; came to U. S., 1923, naturalized, 1928; A.B., Stanford, 1933, M.A., 1934, Ph.D., 1939; m. Mary Gottlund, July 15, 1934 (div. 1944); 1 son, Michael John; m. 2d, Rita Whelton, Oct. 10, 1944 (dec. 1951); m. 3d, Olga Lott, Aug. 15, 1952. Mem. faculty U. Cal. at Berkeley, 1940—, prof. bacteriology, 1952—, research prof. Miller Inst., 1960-62, prof. molecular biology, 1964—; spl. research microbial physiology and metabolism, enzymology. Recipient Sugar Found. award, 1947; Guggenheim fellow, 1949-50; spl. fellow NIH, 1963. Mem. Nat. Acad. Scis., Am. Soc. Biol. Chemists, Soc. Am. Microbiologists, Soc. Gen. Physiologists. Home: 611 San Luis Rd., Berkeley, Cal.*

DOUGHERTY, Raymond Philip, Am. Assyriologist; b. Lebanon, Pa., Aug. 5, 1877; s. Joseph Brant and Mary Elizabeth (Shaeffer) D.; A.B., Lebanon Valley Coll., 1897, A.M., 1903; B.D., Bonebrake Theol. Sem., 1910; Ph.D., Yale, 1918; m. Lulu E. Landis, Oct. 4, 1910. Prin. normal dept. Leander Clark Coll., Toledo, Ia., 1900-02; prin. Albert Acad., Freetown, Sierra Leone, West Africa, 1904-14; Am. vice consul at Sierra Leone, 1905-06, 1912-13; prof. Bibl. lit. Goucher Coll., Balt., 1918-26; William M. Laffan prof. Assyriology and Babylonian lit., curator Babylonian Collection, Yale, 1926—. Annual prof. Am. Schs. Oriental Research, Jerusalem and Bagdad, 1925-26. Attended Internat. Archaeol. Congress, Syria and Palestine, 1926; conducted archaeol. survey in so. Babylonia, 1926; mem. com. on Mediterranean antiquities Am. Council Learned Socs. Author: Records from Erech, Time of Nabonidus, 1920; The Shirkutu of Babylonian Deities, 1923; Archives from Erech, Time of Nebuchadnezzar and Nabonidus, 1923; Nabonidus and Belshazzar, 1929. Died July 13, 1933.

DOUGHERTY, Robert Watson, Am. veterinarian; b. Newcomerstown, O., Feb. 5, 1904; s. Robert Watson and Amie (Dinsmore) D.; student Muskingum Coll., 1922-24; B.S., Ia. State Coll., 1927; D.V.M., Ohio State U., 1936; M.S., Ore. State Coll., 1941; m. Ruth Eileen McKinley, Jan. 1, 1948; children—Michael Robert, Susan Eileen. Faculty, Ore. State U., 1936-42, asst. prof., 1937-42; asso. prof., chmn. div. vet. sci. Wash. State Coll., Pullman, 1946-48; prof. physiology Cornell U., 1948-61; leader physiolpath. investigation Nat. Animal Disease Lab., Ames, Ia., 1961—. Fulbright scholar, New Zealand, 1956-57; Distinguished Alumnus award Ohio State U., 1965. Mem. Am. Vet. Med. Assn. (Borden award 1963), Am. Soc. Vet. Physiologists and Pharmacologists, Am. Physiol. Soc., Soc. Exptl. Biology and Medicine, Am. Assn. Vet. Nutritionists. Contbg. author: Seventh International Grassland Congress Proceedings, 1957; Forages, 1962. Editor in chief: Physiology of Digestion in the Ruminant, 1965. Research and publs. on reproductive physiology in ruminants, bovine spermatozoa and vaginal pH of female, formation and absorption of toxins from bovine rumen, hematological changes in certain path. conditions in ruminants, physiol. studies of acute indigestion in ruminants, isotopic studies physiologic disposition of inhaled eructated gas in sheep and goats. Home: Rural Route 2, Ames, Ia. 50010.*

DOUGHERTY, Thomas F(rancis), Am. histologist; b. Forman, N.D., Mar. 27, 1915; s. Thomas Francis and Mary (Brandenburg) D.; B.S., U. of Minn., 1936, A.M., 1937, Ph.D., 1942; fellow, Donner Foundn., 1942-43; m. Jean Ann Hay, Apr. 5, 1941; children—Michael Bruce, Ann Marie. Instr. in anatomy Yale Sch. of Medicine, 1943-47; prof. anatomy (histology) U. Utah, 1947, now chmn. dept. anatomy, director radiobiology lab., 1954—; consultant to Surgeon Gen. of U. S. Army. Member of Reticuloendothelial Society (president 1957), American Assn. for Cancer Research, Am. Assn. for Study of Internal Secretions, Am. Assn. of Anatomists, Internat. Soc. of Hematology, Soc. for Exptl. Biology and Medicine, Phi Sigma Kappa. Investigation of estrogen-induced leukemia; endocrine control of release of antibodies and growth and physiology of lymphoid tissue; site of antibody formation; adrenal cortex; allergic phenomenon; inflammation; relation of morphology to function of lymphocytes. Home: 1097 Bonneville Dr., Salt Lake City.

DOUGLAS, Alexander Edgar, Canadian physicist; b. Melfort, Sask., Can., Apr. 12, 1916; s. Donald and Jessie (Carwardine) D.; B.A., U. Sask., 1939, M.A., 1940; Ph.D., Pa. State U., 1949; m. Phyllis Helene Wright, July 26, 1946; children—Nancy Phyllis, Donald James, Andrew A. With Acoustics Lab., Nat. Research Council, Ottawa, Ont., Can., 1941-47, Spectroscopy Lab., 1949—, asso. dir. div. pure physics 1967—. Mem. Canadian Assn. Physicists (dir. 1966-67), Royal Soc. Can. (treas. 1965-68), Am. Phys. Soc. Research and numerous publs. on structure and spectra of simple molecules and optical instruments. Home: 150 Blenheim Dr. Office: Nat. Research Council, Ottawa, Ont., Can.*

DOUGLAS, Bodie Eugene, Am. chemist; b. New Orleans, Dec. 31, 1924; s. Harvey Bodie and Lennie (Cochran) D.; B.S., Tulane U., 1944, M.S., 1947; Ph. D., U. Ill., 1949; m. Gladys Elenora Backstrom, Nov. 1, 1945; children—Judy Ann, Stephen Bruce, Sharon Lynn, Janice Lee. Faculty, Pa. State U., 1949-52; faculty U. Pitts., 1952—, prof., 1963—. Fulbright lectr. U. Leeds (Eng.), 1954-55. Mem. Am. Chem. Soc., A.A.A.S., Sigma Xi, Alpha Chi Sigma, Phi Lambda Upsilon. Author: (with Darl H. McDaniel) Concepts and Models of Inorganic Chemistry, 1965. Contbr. numerous articles to profl. jours. Studies of metal compounds related to those which occur in nature. Home: 1209 Sherbrook Dr., Pitts. 15241.*

DOUGLAS, Carstairs Cumming, physician; b. Kirkcaldy, Fifeshire, Oct. 1866; ed. George Watson's Coll. and U., Edinburgh, U. Berlin; M.B., C.M., Edinburgh U., 1890, M.D., 1896; D.Sc. in Pub. Health, 1906; m. Anita Helena Lockhart, 1896; 1 son, 1 dau. Prof. med. jurisprudence and pub. health Anderson Coll. Medicine, Glasgow; dir. West of Scotland Clin. Research Lab. Fellow Royal Faculty Physicians and Surgeons Glasgow, Royal Soc. Edinburgh. Author: Chemical and Microscopical Aids to Clinical Diagnosis, 1899; The Laws of Health; Manual of School Hygiene, 1907, also numerous articles. Died Sept. 28, 1940.

DOUGLAS, Claude Gordon, Brit. physiologist; b. Feb. 26, 1882; s. Claude Douglas; ed. Wellington Coll., Magdelen Coll., Oxford, Guy's Hosp.; M.C., B.Sc., D.M. Hon. Fellow, fellow, tutor natural sci. St. John's Coll., Oxford; prof. gen. metabolism Oxford U.; Oliver-Sharpey lectr. Royal Coll. Physicians, 1927. Recipient Radcliffe prize, 1911, Osler Meml. Medal, 1945. Fellow Royal Soc., 1922. Research and publs. on respiration and other physiol. subjects. Died Mar. 23, 1963.

DOUGLAS, Clifford Hugh, cons. engr., economist; b. Jan. 20, 1879; s. Hugh and Louisa Arderne (Hordern) D.; ed. Pembroke Coll., Cambridge; m. Constance Mary Phillips; m. 2d, Edith Mary Dale; 1 dau. Chief engr., mgr. in India, Brit. Westinghouse Co., Ltd.; asst. chief elec. engr. Buenos Aires and Pacific Ry.; asst. supt. Royal Aircraft Factory; chief reconstrn. adviser Govt. of Alta., 1935. Author: Economic Democracy; Credit Power and Democracy; Social Credit; The Monopoly of Credit; The Alberta Experiment. Died Sept. 29, 1952.

DOUGLAS, David, Scottish botanist; b. Scone, Scotland, 1798; s. John Douglas; apprentice in gardens of Earl of Mansfield; became undergardener to Sir Robert Preston, Valleyfield, 1817; later with Bot. Garden, Glasgow, Scotland; collected specimens in U. S. for Royal Hort. Soc., 1823; made other expdns. to Am. and Pacific, especially Columbia River area and southward toward California, 1824-25; crossed Rocky Mts. to travel to Hudson Bay, 1827; began last voyage, 1829; in Cal., on Fraser River, 1832-34, Fellow Linnean Soc., Geol. Soc., Zool. Soc. Eponym for Douglas squirrel; genus Douglasia named in his honor by Lindley, 1828. Contbr. articles to tech. jours. Introduced ribes and other new world plants to Europe; discovered Douglas fir, 1825, also several species of pine, Cal. vulture, Cal. sheep. Died Hawaii, July 12, 1834.

DOUGLAS, Donald Wills, Am. aero. engr.; b. Bklyn., Apr. 6, 1892; s. William Edward and Dorothy (Locker) D.; student U. S. Naval Acad., 1909-12; B.S., Mass. Inst. Tech., 1914; m. 2d, Marguerite Tucker, Mar. 6, 1954; children (by previous marriage)—Donald Wills II, William Edward, Barbara Jean, James Sholto, Malcolm Angus. Chief engr. Glenn L. Martin Co., aircraft mfrs., Los Angeles, Calif., 1915-16; chief civilian aero. engr., U. S. Signal Corps, 1916-17; again chief engr. Glenn L. Martin Co., Cleve., 1917-20; pres. Douglas Co., Santa Monica, Cal., 1920-28; pres. Douglas Aircraft Co., 1928-57, chmn. bd., chief exec. officer, 1957—. Recipient USAF Exceptional Service Award, French Legion of Honor, Comdr. Order House of Orange and Nassau (Netherlands). Mem. Nat. Acad. Scis. Designed and built 1st airplane to carry load equal to its own weight; developed B-19 exptl. bomber, DC models, other aircraft; assisted in devel. of early guided missiles; work with satellites and spacecraft. Office: Douglas Aircraft Co., Santa Monica, Cal.

DOUGLAS, Frederic Huntington, Am. anthropologist; b. Evergreen, Colo., Oct. 29, 1897; s. Charles Winfred and Mary Josepha (Williams) D.; A.B., U. Colo., 1921, D.Sc., 1948; post-grad. U. Mich., 1921-22; student Pa. Acad. Fine Arts, 1922-26; m. Freda Bendix Gillespie, May 21, 1926; children—Ann Pauline and Eve (twins), David. Painter and wood-carver, 1926-29; pres. sch. bd. Evergreen, Colo., 1929-34; curator, dept. Indian Arts, Denver Art Museum, 1929-47, curator, dept. Native Arts, since 1947; dir. Denver Art. Mus., 1940-42; asst. prof. anthropology U. Denver, from 1934; lectr. anthropology U. Colo., from 1946; research fellow in ethnology Harvard, 1952; dir. edn. Fed. Indian Exhibit, San Francisco Fair, 1938-39; co-dir. N. Am. Indian Art, Museum Modern Art, N.Y.C., 1940-41; mem. Anglo-Am. group inspecting Swedish and Finnish museums, 1946; commr. Fed. Indian Arts and Crafts Bd., from 1946. Fellow A.A.-A.S. (v.p. Southwest Div., 1942-47, pres., 1947-48), Royal Anthrop. Soc. of Gt. Britain; mem. Am. Anthrop. Assn., Soc. for Am. Archeology, Am. Folklore Soc., Societe des Americanistes de Paris. Author and editor, Denver Art Mus. publs. in Indian art; Indian Leaflet Series, since 1930, Indian Design Series since 1938. Author (with Rene d'Harnoncourt), Indian Art of the United States, 1941; The Inner Light (verses, 4 vols.), 1946-53. Contbr. articles to newspapers and jours. Research on design, styles and techniques of Indian tribes north of Mexico since circa 1800. Died Apr. 23, 1956.

DOUGLAS, George John, see Argyll, Duke of.

DOUGLAS, James, British physician; b. Scotland, 1675; M.D., Rheims, France; settled in London, Eng., circa 1700; practiced midwifery; physician to Queen Caroline; prof. at London. Fellow Royal Soc., 1706, Coll. Physicians (hon.). Author: Myographiae comparatae specimen, 1707; Description of the Periotoneum and of the Membrana Cellularis which is on its Outside; Lilium Sarniense or a Description of the Guernsey Lily, 1725; History of the Lateral Operation for the Stone, 1726; Arbor gemensis fructum café ferens, 1727; also numerous articles. First to compile a systematic bibliography of med. lit., 1715; Described crescentic line which marks end of posterior sheath of rectus muscle, 1707, fold in peritoneum (cul-de-sac of Douglas), circa 1730. Died London, Aug. 2, 1742.

DOUGLAS, James, metallurgist; b. Que., Can., 1837; s. James Douglas; ed. U. Edinburgh (Scotland); grad. Queen's U., Ont., Can.; tchr. chemistry Morrin Coll., Que., 3 years; visited mining centers of U. S., 1875; in charge of copper plant at Phoenixville, Pa., 1875; developed Clifton and Bisbee copper deposits in Ariz., also Copper Queen Mine, Aris.; head Phelps, Dodge & Co., pres. several railroads and mining corps. in southwestern U. S. and northern Mexico. Recipient John Fritz medal; gold medal Instn. Mining and Metallurgy. Mem. Am. Inst. Mining Engrs. (twice pres.). Research and numerous publs. on hydrometallurgy of coppers; developed process (with T. Sherry Hunt) for wet extraction of effected reforms in copper industry. Died N.Y.C., June 25, 1918.

DOUGLAS, James Nathaniel, Am. astronomer; b. Dallas, Aug. 14, 1935; s. Loyd and Nell (Curtis) D.; B.S., Yale, 1956, M.S., 1958, Ph.D., 1961; m. Charlotte Cummings, Aug. 30, 1956; children—Neva Jean, James Loyd, Alan Nevins. Instr. astronomy Yale, 1960-61, asst. prof., 1961-65; asso. prof. astronomy U. Tex., Austin, 1965—, dir. Radio Astronomy Lab., 1965—. Mem. Am. Astron. Soc., Am. Geophys. Union, A.A.A.S., Internat. Astron. Union, Union Radio Sci. Internat., Phi Beta Kappa, Sigma Xi. Research, publs. on 20-MHz radio emission from Jupiter, observations of solar wind, observations of discrete radio sources. Home: 2300 Leon St., Austin, Tex. 78705.

DOUGLAS, Jesse, Am. mathematician; b. N.Y.C., July 3, 1897; s. Louis Douglas and Sarah (Kommel) D. B.S., Coll. City N.Y., 1916; Ph.D., Columbia, 1920; m. Jessie Nayer, June 30, 1940 (dec. 1955); 1 son, Lewis Philip. Instr. math. Columbia, 1920-25, lectr. math., 1951-55; NRC fellow Princeton, 1926-27, Harvard, 1927, Chgo., 1928, Paris, 1928-30; asst. prof. math. Mass. Inst. Tech., 1930-34, asso. prof., 1934-37; research Inst. Advanced Study, Princeton, 1934-35, 38-39; Guggenheim fellow, 1940-42; asst. prof. math. Bklyn. Coll., 1942-46; prof. math. Coll. City N.Y., 1955—, City U. N.Y., 1963—. Recipient Fields medal Internat. Congress Mathematicians, Oslo, Norway, 1936; Townsend Harris medal Assn. Alumni Coll. City N.Y., 1939. Fellow Am. Acad. Arts and Scis.; mem. Am. Math. Soc. (council mem. 1935-38, Bocher prize 1943), Math. Assn. Am., Nat. Acad. Scis., Sigma Xi, Phi Beta Kappa. Asso. editor Transactions Am. Math. Soc., 1930-32, 36-40. Contbr. articles to various Am., European math. jours. Gave complete math. solution of problem of least surface area made by soap bubble on wire model, 1931. Home: Butler Hall, 88 Morningside Dr., N.Y.C. 10027.

DOUGLAS, Silas Hamilton, Am. chemist; b. Fredonia, N.Y., Oct. 16, 1816; s. Benjamin and Lucy (Townsend) D.; M.D., Coll. Surgeons and Physicians, Balt. Practiced medicine, Dearborn, Mich., for a time; instr. chemistry U. Mich., Ann Arbor, 1844-45, lectr. in chemistry and geology, 1845-46, prof. chemistry, geology, mineralogy, 1846-51, an organizer dept. medicine, prof. chemistry and pharmacy, dept. medicine, 1855-70, prof. chemistry, 1870-75, prof. mineralogy, 1875-77. Author: Tables for Qualitative Chemical Analysis, 1864; co-author: Qualitative Chemical Analysis: A Guide in the Practical Study of Chemistry, 3d edit., 1880. Died Ann Arbor, Mich., Aug. 26, 1890.

DOUGLAS, Stewart Ranken, Brit. bacteriologist; b. Feb. 22, 1871; s. J. A. Douglas; ed. Haileybury Coll., St. Bartholomew's Hosp.; m. Frances Miriam Clare Dayrell, 1920. Served with Plague Commn., 1899; mem. China Expdn., 1900; asst. inoculation dept. St. Mary's Hosp., London, 1907-14, lectr. bacteriology, 1912-14; dep. dir., dir. dept. exptl. pathology Nat. Inst. Med. Research, Hampstead, Eng., from 1921. Fellow Royal Soc., 1922; licentiate Royal Coll. Physicians; mem. Royal Coll. Surgeons. Contbr. papers on bacteriology and immunity. Research on wounds and dysentery; demonstrated (with A. E. Wright) presence of opsonins in normal and immune serums, 1903. Died Jan. 20, 1936.

DOUGLAS, Thomas Basil, Am. physicist; b. Elkins, W.Va., May 15, 1909; s. Henry Thomas and Cora (Skidmore) D.; B.S. in Chemistry, U. N.C., 1931, M.S., 1932; Ph.D., Ohio State U., 1938; m. Susan Harry Pettengill, Nov. 28, 1949. Comley fellow in surg. research Ohio State U., Columbus, 1939, research asso. Manhattan project, 1943; faculty Mont. State U., Bozeman, 1939-46, asst. prof. chemistry, 1942-46; asst. prof. Western Res. U., 1946-47; staff Nat. Bur. Standards, Washington, 1947—, supervisory physicist, 1957—, asso. coordinator thermodynamic research program for chem. propulsion, 1958—. Fellow Washington Acad. Scis.; mem. Am. Phys. Soc., Am. Chem. Soc., A.A.A.S., Phi Beta Kappa, Sigma Xi, Alpha Chi Sigma. Contbg. author: Experimental Thermodynamics, Vol. I, 1968. Research and publs. on high accuracy high-temperature calorimetry, vaporization equilibrium, and other chem. thermodynamics; devel. numerous detailed correlations of interrelated thermodynamic properties; contbd. to semi-empirical theory especially of lattice defects in solid solutions. Home: 3031 Sedgwick St. N.W., Washington 20008. Office: Nat. Bur. Standards, Washington 20234.*

DOUGLASS, Andrew Ellicott, Am. astronomer; b. Windsor, Vt., July 5, 1867; s. Malcolm and Sarah E. (Hale) D.; A.B., Trinity Coll., Conn., 1889, hon, D.Sc., 1908; D.Sc., U. Ariz., 1938; m. Ida E. Whittington, Aug. 3, 1905. Asst. Harvard Coll. Obs., 1889-94; 1st asst. astronomer Lowell Obs., Flagstaff, Ariz., 1894-1901; probate judge Coeonino County, Ariz., 1903-06; instr. Northern Ariz. Normal, 1905-06; prof. physics and astronomy U. Ariz., 1906-18, acting pres., 1910-11, dean Coll. of Letters, Arts and Scis., 1915-18, dir. Steward Obs., 1918-38, prof. astronomy from 1918; prof. dendrochronology from 1936; dir. Lab. of Tree Ring Research from 1937. Research asso. Carnegie Instn. of Washington, 1925-38. Recipient award for studies in tree rings and chronology Research Corp., N.Y., 1931. Fellow Royal Astron. Soc., A.A.A.S. (pres. southwest div., 1921); mem. Nat. Geog. Soc. (hon. life), Am. Philos. Soc., Am. Meteorol. Soc. (v.p., 1924-25), So. Cal. Acad. Sci., Ariz. Archeol. and Hist. Soc. (pres. 1929-30), Am. Astron. Soc., Astron. Soc. Pacific, Phi Beta Kappa, Phi Kappa Phi, Sigma Xi. Author: Annals of Lowell Observatory, Vol. I, pt. II, and Vol. II; Climatic Cycles and Tree Growth, Vols. I, II and III (Carnegie Institution), 1919, 1928, 1936. Made photographs of shadow bands, zodiacal light, Mars; research in dating prehistoric ruins by tree rings (dendrochronology); invented the cycloscope for climatological and cycle studies of same; researched history of sun by examining ann. rings of pines and sequoias in Ariz., estimated that the time of the sun-spot cycle remained much the same in period covered. Died Tucson, Ariz., Mar. 20, 1962.

DOUGLASS, Earl, Am. geologist, paleontologist; b. Medford, Minn., Oct. 28, 1862; s. Fernando and Abigail Louisa (Carpenter) D.; ed. U. S.D., 1888, S.D. Agrl. Coll., 1889, 92; B.S., Ia. State Coll., 1893; M.S., U. Mont., 1900; fellow, Princeton, 1900-02; m. Pearl C. Goetschius, Oct. 20, 1905; 1 son, Gawin Earl. Taught sch. with interruptions, 1883-1900; prin. schs., Virginia City, Mont., 1897-98; taught geology, phys. geography and physics, U. Mont., 1899-1900; asst. under Prof. William Trelease in Mo. Bot. Gardens, St. Louis, 1890-91; engaged in research work in dept. of vertebrate paleontology, Carnegie Mus., Pitts., 1902-24; in charge obtaining U. Utah collection of skeletons from Dinosaur Nat. Monument, 1923, 24. Studied geology; made collections especially of fossil vertebrates, 1894-1900; with Princeton sci. expdn. in Mont., summer 1901, collecting Cretaceous dinosaurs and marine reptiles, also made collections from Tertiary and other formations of Mont. and N.D.; discovered, 1909, an immense deposit of Comanchean dinosaur skeletons near Jensen and Vernal, Utah (now Dinosaur Nat. Monument); investigated oil problems, including origin of oilshales, asphalts. Died Jan. 13, 1931.

DOUGLASS, Frederick Melvin, Am. surgeon; b. Kalida, O., June 26, 1890; s. Curry Frederick and Kathrine Willoughby (Melvin) D.; M.D., U. Toledo, 1911; postgrad., Boston, London, Eng., Mayo Clinic, Chgo., 1914, 15; m. Ruth Jacobson, Nov. 2, 1913 (dec.); children—Kathryn Jane (Mrs. Nelson Montgomery Loud), Frederick Melvin (M.D.); m. 2d, Beatrice Ossege, Nov. 21, 1926. Intern St. Vincent's Hosp., Toledo, 1911-12; dir. surgery, 1926—; resident physician Lucas County Hosp., Toledo, 1912-13; practicing gen. surgeon, with J. H. Jacobson, 1913-18, with C. W. McNamara and Richard Hotz, 1918—. Diplomate Am. Bd. Surgery (founders group). Fellow U. S. chpt. Internat. Coll. Surgeons (past chmn. bd. govs.), A.C.S. (life mem.), Am. Assn. Obstetricians, Gynecologists, and Abdominal Surgeons; mem. Acad. Medicine Toledo and Lucas County (pres. 1937-38, trustee 1938—), Northwestern Ohio Med. Soc. (pres. 1928-29), Am., Ohio med. assns. Presbyn. Mason (32°), Contbr. sci. and med. articles. Died July 4, 1950.

DOUGLASS, Raymond Donald, Am. mathematician; b. Gorham, Me., Dec. 29, 1894; s. Edward K. and Josephine (Chick) D.; A.B., U. Me., 1915, A.M., 1916, Sc.D., 1941; Ph.D., Mass. Inst. Tech., 1931; m. Ollave Elizabeth Norton, Feb. 2, 1918; children—Charlotte Ollave (Mrs. Willard Mott), Eleanor Norton, Marjorie Ann (Mrs. Charles Humphreys). Prof., Mass. Inst. Tech., 1919-65, prof. emeritus, 1965—; chmn. com. war tng. programs Greater Boston, state coordinator Mass., 1940-45. Recipient Charles F. Park medal, 1965. Mem. Am. Math. Soc., Math. Assn. Am., Am. Engring. Soc., Phi Beta Kappa, Sigma Xi. Author: Nomographic Charts, 1950; Analytic Geometry, 1951; Calculus and its Applications, 1952. Contbr. sects. on nomography to Plant Engrs. Handbook, 1955, Ency. Sci. and Tech., 1957. Research and publs. primarily in field of graphical math., math. series, nomography, definite integrals and differential equations. Home and office: 18 Oak Av., Belmont, Mass. 02178.*

DOUGLASS, William, physician; b. Gifford, Scotland, circa 1691; studied medicine, Paris, Leyden and Edinburgh. Arrived in Boston, 1718; strongly opposed smallpox innoculation, wrote 4 pamphlets on subject. Author: The Practical History of a New Epidemical Eruptive Military Fever . . . , 1736; A Summary, Historical and Political of the First Planting, Progressive Improvements and Present State of the British Settlements in North America, 2 vols., 1749-51. Credited with 1st listing of scarlet fever cases in New Eng., 1736. Died Oct. 21, 1752.

DOUNCE, Alexander Latham, Am. biochemist; b. Syracuse, N.Y., Dec. 7, 1909; s. George Alexander and Cornelia L. (Esty) D.; student Hamilton Coll., 1926-28; A.B., Cornell U., 1930, Ph.D., 1935; m. Anna-Elizabeth Reddick, Aug. 20, 1937; children—Helen Anne (Mrs. José Sanchez), Eric Alexander, George Harry. Chemist, Continental Can Co., Syracuse, 1935; instr. biochemistry Cornell U., Ithaca, N.Y., 1936-41; faculty U. Rochester Sch. Medicine and Dentistry, 1941-43, 49—, prof., 1965—, cancer research 1946-49, biochemist Manhattan Project Med. Sch., 1943-46. Mem. Am. Soc. Biol. Chemists, Am. Chem. Soc., A.A.A.S., Soc. for Exptl. Biology and Medicine, Am. Assn. for Cancer Research, Am. Soc. Cell Biology, N.Y. Acad. Scis., Rochester Acad. Sci., Sigma Xi. Contbg. author: Pharmacology and Toxicology of Uranium Compounds, 1949; The Enzymes, 1950; The Nucleic Acids, vol. II, 1955; The Nucleohistones, 1964. Research, numerous publs. in enzymology, biochemistry and physiology of uranium poisoning, chemistry of cell nucleus, enzymes of isolated cell nuclei, nucleic acids, nucleoprotein, histones of cell nuclei, DNA nucleoprotein; 1st crystallization (with J. B. Sumner) enzyme catalase from beef, horse and lamb liver; developed gen. mechanism for action of uranium on kidney; developed methods for isolating cell nuclei on chem. scale. Home: 615 Edgewood Av., Rochester, N.Y. 14618.*

DOUSMANIS, George Christos, physicist; b. Aigion, Greece, Jan. 13, 1929; s. Christos and Photinee (Papageorge) D.; came to U. S., 1948, naturalized, 1962; B.A., Columbia, 1951, M.A., 1953, Ph.D. in Physics, 1956; m. Marilyn Sarah Derby, June 16, 1956; children—Alexandra Photinee, Christos, Thanos. Research asst. Columbia U. Radation Lab., 1951-56; research physicist RCA, David Saranof Research Center, Princeton, N.J., 1956-66. Mem. Am. Phys. Soc., I.E.E.E., Phi Beta Kappa. Research and publs. on exptl. and theoretical study effect of rotation of a molecule on coupling between its spinning electrons and nucleus, detection of electronis in crystals that behave as if they had negative mass, optimization of performance of injection lasers. Died Dec. 9, 1966.

DOUSTE-BLAZY, Louis Léonce Stanislas, French biochemist; b. Mirande, France, Feb. 11, 1921; s. Vincent and Francoise (Canton-Bacara) D.-B.; ed. Faculté de Médecine, 1947; Faculté de Sciences, 1955, Faculté de Pharmacie, 1956; m. Geneviève Beguere, Apr. 21, 1949; children—Bernard, Philippe. Agrégé de Biochimie médicale, 1955; named lectr., Toulouse, France, 1955; became head biochem. services Centre Anticancéreaux, 1958; became titular prof. biochemistry Toulouse Faculty Medicine and Pharmacy, 1963. Mem. Biol. Chemistry Soc., Soc. Clin. Biology, Am. Chem. Soc. Research and numerous publs. on plasmatic and globular lipids with evidence of lysolecithines in counterclockwise distbn., 1954; purification of A phospholipases of spleen and their specificity of action while fractionating them into A-1 and A-2 phosphalipases. Home: 24 rue d'Aubuisson, Toulouse 31, France.*

DOUTHIT, Thomas D. Nathan, Am. geophysicist; b. Lubbock, Tex., July 9, 1918; s. Thomas Levi and Mercy (West) D.; B.S. in Geology with honors, Tex. Tech. Coll., 1951; M.S. in Geophysics, St. Louis U., 1959; M.B.A. in Research and Devel. Mgmt., U. Chgo., 1962; m. Billie Sue Brown, Mar. 21, 1948; children—Leslie Nathan, Debra Sue. Commd. 2d lt. USAF, 1944, advanced through ranks to lt. col., 1966; staff Terrestrial Scis. Lab., Air Force Cambridge Research Lab., Bedford, Mass., 1956-61, project officer for Arctic terrain research, 1959-60, project officer project Vela Uniform, 1960-61; mem. staff, chmn. Air Force Cambridge Research Labs. working group for Internat. Year of Quiet Sun, Commdr.'s staff, 1962-64; chief Space Physics Lab., Air Force Cambridge Research Lab., 1964—. Mem. Soc. Expln. Geophysicists, Am. Geophys. Union, Seismol. Soc. Am., Sigma Xi. Research on earthquakes, meteorology, aeronomy, geomagnetism, aurora, airglow, ionospheric physics, cosmic rays, solar physics, outer space environment, physics outer space. Home: 24 Raymond Rd., Chelmsford, Mass. 01824. Office: Air Force Cambridge Research Lab., L. G. Hanscom Field, Bedford, Mass. 01730.*

DOUTRELEPONT, (Louis Guillaume) Joseph, dermatologist; b. Malmédy, Belgium, June 3, 1834; s. Joseph and Catherine D.; student medicine Berlin, Germany, 1858; student of surgeon Wilhelm Busch, Bonn, Germany; m. Marie Bettendorf, 1861; 3 sons, 1 dau. Joined faculty U. Bonn., 1863; instigated founding of ind. skin clinic U. Bonn, 1882; apptd. prof. 1869, hon. prof., 1894; emeritus, 1910. Author: Die Ätiologie des Lupus vulgaris, 1884; Über die Bedeutung der Hauttuberkulose, 1912; also numerous articles. Research in skin and venereal diseases especially etiology and therapy of lupus vulgaris, therapy of skin tuberculosis; discovered cause of tertiary syphilitic skin changes; proved tubercular bacilli could be bred from lupus cells. Died Bonn, Apr. 30, 1918.

DOVE, Heinrich Wilhelm, German physicist, meteorologist; b. Liegnitz, Germany (now Leignica, Poland), Oct. 6, 1803; s. Wilhelm Benjamin and Susanne (Brückner) D.; began studies natural scis. in Breslau, Berlin, (both Germany), 1821; doctorate, 1826; m. Luise Etzel; 6 sons including Richard, Alfred, Heinrich; 4 daus. Became lectr. U. Königsberg (Germany), 1826; apptd. asso. prof. physics, 1828; went to Berlin, 1829; tchr. secondary schs. arty. sch., mil. sch. and trades inst. also at univ.; became prof. physics, 1845; became adviser Prussian Meteorol. Inst., 1849; created observation network in Prussia. Recipient Copley medal, 1853, Order pour le mérite (v.p. of order 1867). Mem. Berlin Acad. Scis. Author: Meteorologische Untersuchungen, 1837; Temperaturtafeln, 1848; Temperaturcurven, 1852; Optische Studien, 1853; Darstellung der Wärmeersteinungen durch fünftägige Mittel, 1856; Das Gesetz der Stürme, 1857; also numerous articles. Founder comparative climatology and its application to agr.; introduced monthly isotherms and isanomalia; formulated meteorol. law gyration; developed theory of polar and equatorial currents; introduced use of telegraph in weather service; research on induced electricity, circularly polarized light. Died Berlin, Apr. 4, 1879.

DOVE, Karl Wilhelm, German geographer; b. Tübingen, Germany, Nov. 12, 1863; s. Richard and Caroline (Nobiling) D.; student geography, physics and econs., Göttingen, Freiburg (both Germany), 1883-88; doctorate Göttingen, 1888; m. Frieda Glese, 1911; 1 son. Became lectr. on geography and climatology U. Berlin (Germany), 1890; on contract from German Colonial Soc., German S.W. Africa, 1892-94; asso. prof. U. Jena (Germany), 1899-1907; hon. lecturing prof. Freiburg. Author: Deutsch-Südwest-Afrika, 1896; Landeskunde von Deutsch-Südwest-Afrika, 1902; Die deutschen Kolonien, 4 vols., 1909-13; Die angelsächsische Riesenreich, I, Das britische Weltreich, II, Die Vereinigten Staaten von Nord-Amerika, 1906, 07; (with F. Frankenhäuser) Deutsche Klimatik, 1910; Wirtschaftsgeographie von Afrika, 1917; Allgemeine politische Geographie, 1920; Allgemeine Wirtschaftsgeographie, 1921; Allgemeine Verkehrsgeographie, 1921. Research on econ. geography of Africa, hygienic meteorology, application of climatology to balneology and medicine; founder med. geography. Died Jena, July 30, 1922.

DOVE, W(illiam) Franklin, Am. biologist; b. Marion, Ia., Apr. 11, 1897; s. William Franklin and Edith (Gregory) D.; B.S., Ia. State Coll., 1922; M.S., U. Wis., 1923, Ph.D., 1927; m. Ruth Rebecca Stone, Sept. 5, 1933; children—E. Felicia, William F(ranklin), Ellen R., Christopher S., John G. Biologist, head biol. lab. Me. Agrl. Expt. Sta., U. Me., 1931-43; biologist, chief Food Acceptance Research Br., Subsistence Research and Devel. Lab., 1944-46, Q.M. Food and Container Inst. of Armed Forces, 1946-48; biologist, nutrition br. USPHS, 1950; biologist Dept. Pub. Health, Coll. Medicine, U. Ill., Chgo., 1950-66. Adv. bd. U. S. Soil Plant Nutrition Lab., 1939-42. Contbr. numerous articles to profl. jours. Research on transplantation of embryonic tissues, linkage of climate-soil-plant-animal-man, devel. of food acceptance research. Home: 339 N. Grove Av., Oak Park, Ill. 60302.*

DOVGYALLO, Georgiy Khrisanfovich, Russian internist, hematologist; b. Vitebsk, Belorussia, 1902; grad. Med. Faculty, Belorussian U., 1928; Cand. Med. Sci., 1939; D.Med. Sci., 1955; Head Rural Dist. Hosp., Kokhanovo, Vitebsk Oblast, head therapeutic dept. Polotsk R.R. Hosp., asst. dept. hosp. therapy Vitebsk Med. Inst., 1928-41; lectr. dept. hosp. therapy Minsk Med. Inst., 1946-53, prof., head chair hosp. therapy, 1955—, also chmn. circuit therapy commn.; chief therapeutics Belorussian Ministry Health, 1953-59, now presidium mem. learned med. council. Mem. Belorussian Republic Med. Soc. Therapeutists (chmn. 1953—), All-Union Soc. Therapeutists (bd. mem. 1953—). Author: Red Cell Treatment in Pernicious Anemia, 1955; Pernicious Anemia, 2 edits. Research and publs. on internal medicine and hematology, correlation between liver and intestine functions in carbohydrate metabolism. Address: Minsk Med. Inst., Leninsky prosp. 6, Minsk, Belorussia SSR, USSR.

DOVGYALLO, Nikolay Dmitrievich, Russian anatomist; b. 1899; grad. Odessa Med. Inst., 1925, post-grad., 1928-30; D.Med. Sci. Asst. dept. normal anatomy Odessa Med. Inst., 1928-30; head chair normal anatomy Kazakhstan Med. Inst., 1941-43; founder dept. anatomy Stalino Med. Inst., 1930, past dep. dir.; head chair normal anatomy Donetsk (Stalino) Med. Inst., 1930—. Mem. All-Union Soc. Anatomists, Histologists and Embryologists (bd. Mem.), Znanie Soc. Author: A General Study of the Form of Joint Surfaces and Movement in Joints, 1932; Certain Aspects of Arthrology, 1958; Cybernetics and Certain Aspects of Morphology, 1964; co-author: A Practical Anatomy Course, 1939. Research and publs. on anatomy of nervous and articular systems. Address: Donetsk Med. Inst., Makeevskoe sh., Donetsk, Ukraine SSR, USSR.

DOVRING, Folke, economist; b. Rystad, Sweden, Dec. 6, 1916; s. Karl Gustav and Naëmi (Arntan) Ossiannilsson; Ph.D., Lund U., (Sweden), 1947; m. Karin Dovring, May 30, 1943. Asso. prof. agrarian history Lund U., 1947-53; economist, statistician FAO, Rome, Italy, 1954-60; prof. agrl. econs. U. Ill., 1960—. Cons. Econ. Commn. for Europe (UN), Geneva, Switzerland, 1953; OECD, Paris, France, 1963-64; guest lectr. various European, Am. univs. Rockefeller fellow, 1953-54. Mem. Am. Assn. for Advancement Slavic Studies. Author: (with K. Dovring) Land and Labor in Europe, 1956, rev., 1965; History as a Social Science, 1960; several books in Swedish; also articles. Editor: Inleidinge (Grotius), 1952. Invented method drawing maps of medieval villages; discovered unknown manuscript by Hugo Grotius; explored European land systems; formulated gen. theory of differential growth in transition from agrarian to indsl. economies; research on theory and analysis of productivity; postulated theory of acceleration as pervasive feature of world history. Home: 613 W. Vermont Av., Urbana, Ill. 61801.*

DOW, Charles, Brit. veterinarian; b. Bute, Scotland, Aug. 15, 1930; s. Charles and Rae (Stevenson) D.; B.V.M.S., Glasgow (Scotland) U., 1954, Ph.D., 1958; m. Mary Paterson, Aug. 17, 1956; children—Carol Ann, Rosemary Catriona. Agrl. Research Council fellow, 1954-56; lectr. pathology Glasgow U. Vet. Sch., 1956-60; pathologist Vet. Research Labs., Ministry Agr. for No. Ireland, 1960-66, dir., 1966—; prof. comparative pathology Queen's U., Belfast, No. Ireland, 1966—. Mem. Royal Coll. Vet. Surgeons, Brit. Vet. Assn., Path. Soc. Gt. Britain. Research, publs. on vet. pathology, including devel. pyometra syndrome in dogs and cats (1st exptl. prodn. of disease in both), path. and virol. studies of encephalitides in domestic animals especially pseudorabies. Home: 11 Glencraig Park, Craigavad, County Down, No. Ireland. Office: Vet. Research Labs., Stormont, Belfast, No. Ireland.*

DOW, Herbert Henry, chemist; b. Belleville, Ont., Feb. 26, 1866; s. Joseph H. and Sarah J. (Bunnell) D.; grad. Case Sch. Applied Sci., 1888, hon. D. Eng., 1924; D. Engring., U. Mich., 1929; m. Grace A. Ball, Nov. 16, 1892; children—Willard, Osborn, Alden, Helen, Ruth, Margaret, Dorothy. Prof. chemistry and toxicology Huron Street Hosp. Coll., Cleveland, 1888-89; mfr. chemicals, from 1889, as partner, or officer Midland Chem. Co., Inc. (Mich.), Dow Process Co., and Dow Chem. Co. (founded 1897). Recipient Perkin medal Soc. of Chem. Industry, 1930. Mem. adv. com. Council of Nat. Defense, The Chem. Alliance, Inc.; trustee Case Sch. Applied Sci. Research on lithium and bromine in brines; developed process for extracting bromine by oxidation without evaporating the brine, 1888; replaced bleaching powder with chlorine as oxidizing agt.; developed process for making phenol from chlorbenzol, also new process for aniline. Died Rochester, Minn., Oct. 15, 1930.

DOW, Robert Stone, Am. physician; b. Wray, Colo., Jan. 4, 1908; s. Simon Stone and Edna (Sisson) D.; B.S., Linfield Coll., 1929, D.Sc. (hon.), 1964; M.A., U. Ore., 1934, M.D., 1934, Ph.D., 1935; m. Margaret Willetta Leever, July 9, 1934; children—Margretta, Barbara. Practice medicine, specializing in neurology, Portland, Ore., 1946—; mem. staff Good Samaritan Hosp., Portland, 1943, chmn. dept. neurology, 1960—, pres. staff, 1965; faculty U. Ore. Med. Sch., Portland, 1939—, asso. prof. neurology, 1952-66. Cons. neurology VA Hosp., Vancouver, Wash., 1950—, Portland 1950—; mem. neurology study sect. NIH, 1961-64. NRC fellow, 1936-37; Belgian Am. Edn. Found. fellow, 1937-38; Rockefeller Inst. fellow, 1938-39; Fulbright Research scholar, 1953-54. Fellow Am. Acad. Neurology, A.C.P.; hon. fellow Mexican Acad. Medicine; mem. A.M.A., Am. Assn. Anatomists, Am. Physiol. Soc., Am. Neurol. Assn., Am. Electroencephalographic Soc. (pres. 1958), Harvey Cushing Soc., Am. Epilepsy Soc. (pres. 1964), Western Soc. Electroencephalography (pres. 1948), N. Pacific Soc. Neurology and Psychiatry (pres. 1961), numerous others. Author: (with Giuseppe Moruzzi) The Physiology and Pathology of the Cerebellum, 1958. Research on anatomy, physiology, and pathology of cerebellum; electroencephalography as a diagnostic tool. Home: 2716 N.W. Monte Vista Terrace, Office: 2525 N.W. Lovejoy St., Portland, Ore. 97210.*

DOW, Sterling, Am. archeologist; b. Portland, Me., Nov. 19, 1903; s. Sterling Tucker and Alice (Verrill) D.; A.B. cum laude, Harvard, 1925, A.M., 1928, Ph.D., 1936; postgrad. Trinity Coll., Cambridge, Eng., 1925-26, Am. Sch. Classical Studies, Athens, Greece, 1931-32, 35-36; LL.D., U. Cal. at Berkeley, 1965; m. Elizabeth Sanderson Flagg, June 5, 1931; children—Elizabeth (Mrs. Robert George Lown), Sterling, III. Faculty, Harvard, 1941—, prof. history and Greek, 1946-49, Hudson prof. archaeology, 1949—, war archivist, 1944-45; with OSS, Washington, Cairo, Egypt, 1942-44. A founder Am. Research Center,

Cairo. Ann. prof., Guggenheim fellow Am. Sch. Classical Studies, Athens, 1966-67. Mem. Archaeol. Inst. Am. (past pres.). Author: Prytaneis, 1937; Fifty Years of Sathers, 1965; (with R. F. Healey) A Sacred Calendar of Eleusis, 1965; also numerous articles. Hist. studies based on inscriptions and archaeology from Bronze Age to late Roman Empire; anticipated Ventris' discovery that Lines B tablets of Knossos and Pylos are Greek; worked to perfect technique of studying classical inscriptions; deciphered religious texts; discovered classical allotment machine in Athenian democracy. Home: 159 Brattle St., Cambridge, Mass. 02138.*

DOW, Willard Henry, Am. chemist; b. Midland, Mich., Jan. 4, 1897; s. Herbert Henry and Grace Anna (Ball) D.; B.S., Mich., 1919, D.E., 1941; D.Sc., Mich. Coll. Mining and Tech., 1939; D. Eng., Ill. Inst. Tech., 1944; m. Martha L. Pratt, Sept. 3, 1921; children—Helen Dow Whiting, Herbert Henry II. Began as chem. engr. Dow Chem. Co., 1919, dir., from 1922, asst. gen. mgr., 1926-30, pres. and gen. mgr., from 1930, chmn. bd., from 1941; pres. Ethyl-Dow Chem. Co., 1933-46, Midland Ammonia Co., 1937-45, Dow of Can., Ltd., 1942-46, Io-Dow Chem. Co., 1936-39, Dowell Inc., 1932-39, Cliffs Dow Chem. Co., 1935-39; dir. Dowell, Inc., Dow Magnesium Corp., Dow Corning Corp., Midland Ammonia Co., Dow Chem. of Can., Ltd., Saran Yarns Co., Ethyl-Dow Chem. Co., Cliffs Dow Chem. Co. Mem. Am. Chem. Soc., Am. Inst. Chem. Engrs., Am. Inst. Chemists, Soc. Chem. Industry, A.A.A.S., Engring. Soc. of Midland. Recipient Chandler medal Columbia, 1943; Gold medal. Am. Inst. Chemists, 1944; Chem. Industry medal Am. sect. Soc. of Chem. Industry, 1946; Medal for Advancement of Research, Am. Soc. for Metals, 1948. Mem. Am. Chem. Soc., Am. Inst. Chem. Engrs., Am. Inst. Chemists, Soc. Chem. Industry, A.A.A.S., Engring. Soc. Midland. Directed Dow Chem. Co. work in extraction of bromine and magnesium from sea water, also in prodn. of synthetic rubbers and plastics. Died Mar. 31, 1949.

DOWBEN, Robert Morris, Am. biologist; b. Phila., Apr. 6, 1927; s. Morris and Zena (Brown) D.; A.B., Haverford Coll., 1946; M.S., U. Chgo., 1947, M.D., 1949; m. Jane Carla Lurie, June 20, 1950; children—Peter, Jonathan, Susan. Faculty, U. Pa., 1952-55, Northwestern U., 1956-63; asso. prof. biology Mass. Inst. Tech., Cambridge, 1963—; lectr. Med. Sch. Harvard. Mem. A.A.A.S., Am. Chem. Soc., Am. Physiol. Soc., Am. Soc. Clin. Investigation, Biophys. Soc. Contbr. numerous articles to profl. jours. Developed metal filled microelectrode for neurophysiol. studies, biosynthesis of estrogens, effects of steroid hormones on membrane function, modification of course of muscular dystrophy by hormones, study of membrane structure. Home: Wilson Rd., Concord, Mass. 01742.*

DOWDELL, Ralph Lewis, Am. metall. engr.; b. St. Paul, Sept. 27, 1896; s. George Dillon and Emma (Brasemer) D.; Metall. Engr., U. Minn., 1918, M.S., 1921, Ph.D., 1926; m. Katherine Norton Canfield, Aug. 10, 1922; 1 son, Robert Alan. Asst. metall. engr. U. S. Bur. Mines, 1918; faculty U. Minn., 1918-29, 30-55, prof. metallurgy, 1930-55, head dept. metallurgy, 1945-55; sr. metallurgist U. S. Bur. Standard, Washington, 1929-30; cons. metall. engr., St. Paul, 1955—. Registered profl. engr., Minn. Mem. Am. Soc. Metals (Henry Marion Howe Gold medal 1928), Engring. Club Mpls., Sigma Xi, Phi Lambda Upsilon. Author: (with Henry S. Jerabet, Arthur C. Forsyth) General Metallography, 1943; also numerous articles. Research on mechanism hardening steel, non-toxic lead shot for waterfowl, permanent magnet steels, microstructure steels. Address: 1432 W. California Av., St. Paul, 55108.*

DOWDY, Andrew Hunter, Am. physician; b. Longwood, Mo., Nov. 24, 1904; s. Nathaniel M. and Emma (Patterson) D.; A.B., Central Coll., Fayette, Mo., 1929; D.Sc., 1957; M.D., Washington U., St. Louis, 1931; m. Helen M. Brandes, Aug. 27, 1930; children—Andrew Hunter, Robert Alan. Faculty, Sch. Medicine and Dentistry, U. Rochester, 1937-48, prof., dir. atomic energy project, 1946-47, chmn. dept. radiology, chief radiologist, 1947, Manhattan project, 1943-46; prof., chmn. dept. radiology Sch. Medicine, U. Cal. at Los Angeles, 1948-66. Cons. on radiology, isotopes, cancer control, nuclear energy to govtl. agys., pvt. orgns., 1953—; mem. com. on radiology NRC, 1948—. Recipient certificate of appreciation, U. S. Army-Navy, 1948, Am. Coll. Radiology, 1963. Diplomate Am. Bd. Radiology, Pan Am. Med. Assn.; Fellow A.A.A.S., Am. Coll. Radiology. Research and publs. on modern treatment of gas gangrene and use of antitoxin and vaccines; role of free oxygen in extent of tissue damage from irradiation; treatment of cancer. Home: 1984 Stradella Rd., Los Angeles 90024.*

DOWELL, Walter Charles Thomas, Australian physicist; b. Melbourne, Australia, May 22, 1925; s. Walter Luxmore and Leola (Ellis) D.; M.S., Melbourne U., 1951; Dr.Rer.Nat., Berlin U., 1951; m. Lieselotte Hildegard, June 21, 1963; stepchildren—Torsten, Valborg. With div. chem. physics Commonwealth Sci. and Indsl. Research Orgn., Melbourne, 1953-58, 63—; staff Institut für Elektronenmikroskopie, Fritz-Haber-Institut, Berlin, 1959-63. Research, publs. on electron optics of electron microscope, thin crystals, theory of moire pattern formation in electron micrographs of overlapping crystals; 1st to apply techniques of tilted illumination to imaging of lattice planes in single crystals achieving 1st resolution of planes spaced in region of atomic diameter. Home: 18 Hartlands Rd., East Ivanhoe, Melbourne, Victoria. Office: P.O. Box 160, Clayton, Victoria, Australia.*

DOWLING, Geoffrey Barrow, physician; b. Cape Town, S. Africa, Aug. 9, 1891; s. Thomas Barron and Minna (Grant) D.; ed. Dulwich Coll.; M.D., Guy's Hosp.; hon. M.D., univs. Utrecht and Pretoria; m. Mary Elizabeth Kelly, 1923; 2 sons, 2 daus. Watson Smith lectr.; cons. physician to dept. skin diseases St. Thomas's Hosp.; cons. dermatology to RAF; dir. Inst. Dermatology. Fellow Royal Coll. Physicians, 1955; hon. fellow Royal Soc. Medicine (past pres. sect. dermatology); hon. mem. Brit. Assn. Dermatology (pres. 1956), French Dermatol. Soc., others. Author papers on med. subjects. Introduced use of calciferol in treatment of Lupus (with E. W. P. Thomas), 1945. Address: 52 Ravenscourt Gardens, London W. 6, Eng.

DOWLING, Harry Filmore, Am. physician; b. Washington, D.C., Nov. 11, 1904; s. William Alexander and Mae Clara (Krause) D.; A.B., Franklin and Marshall Coll., 1927; M.D., George Washington U., 1953; D.Sc., Franklin and Marshall Coll., 1953; m. Edith Laine, June 27, 1931; children—Harry Filmore, William Laine, John Nelson. Faculty George Washington U., 1934-50; prof., head dept. preventive medicine U. Ill., 1950-51, prof., head dept. internal medicine Coll. Medicine, 1951—. Mem. revision com. U. S. Pharmacopeia, 1956—; mem. Med. adv. bd. FDA, 1965—. Recipient Merit prize in medicine for research in pneumonia Washington Med. and Surg. Soc., Alumni Achievement award George Washington U. Diplomate Am. Bd. Internal Medicine. Fellow A.C.P.; mem. A.M.A. (chmn. council on drugs, 1963-66), Infectious Diseases Soc. Am. (pres. 1964-65), Am. Assn. History Medicine, Am. Soc. Clin. Investigation, Am. Fedn. Clin. Research, Assn. Am. Physicians, Central Soc. Clin. Research, Phi Beta Kappa, Alpha Omega Alpha. Author: The Acute Bacterial Diseases: Their Diagnosis and Treatment, 1948; That the Patient May Know, 1959; also numerous articles, monographs. Research on infectious diseases, methods action serums, sulfonamides, and antibiotics, pathogenesis respiratory infections. Home: 208 Bliss Lane, Great Falls, Va. 22066. Office: 840 S. Wood St., Chgo. 60612.*

DOWLING, Herndon G(lenn), Am. zoologist; b. Cullman, Ala., Apr. 2, 1921; s. Herndon Glenn and Ada (Camp) D.; B.S., U. Ala., 1942; M.S., U. Fla., 1948; Ph.D., U. Mich., 1951; postgrad. Mexico City (Mexico) Coll., summer 1948. Instr. biology Haverford (Pa.) Coll., 1951-52; faculty U. Ark., Fayetteville, 1952-59, asso. prof., 1956-59; asso. curator reptiles N.Y. Zool. Park, Bronx, 1959-60, curator reptiles, 1960-67; research asso. dept. herpetology Am. Mus. Natural History, N.Y.C., 1957—; U. Fla. Postdoctoral Research fellow, 1956-57; adj. prof. zoology U. R.I., Kingston, 1964—; adj. prof. biology N.Y. University, 1965—. Recipient NIH grant, 1964-67; recipient NSF grants, 1957, 58, 68-70. Fellow N.Y. Zool. Soc., A.A.A.S. Am. Assn. Zool. Parks and Aquariums, Herpetologists League (past pres.); mem. Am. Soc. Ichthyologists and Herpetologists (prize paper award, 1947, 49), Acad. Zoology (India), Am. Assn. U. Profs., Am. Inst. Biol. Scis., Am. Soc. Zoologists, Brit. Herpetol. Soc., Internat. Soc. on Toxinology (asso.), Soc. Study Evolution, Soc. Systematic Zoology, Soc. Vertebrate Paleontology, Southwestern Assn. Naturalists (past sec-treas., past gov.), Sigma Xi, Phi Sigma. Research and publs. on taxonomic relationships of ratsnakes, evolution of higher categories of snakes, geog. distbn. reptiles, temperature relations of reptiles, venomous snakes and treatment of snakebite, conservation of plants and animals of Galapagos Islands. Home: 127 W. 79th St. Office: Am. Mus. Natural History, N.Y.C. 10024.*

DOWN, John Langdon Haydon, Brit. physician; b. London, Eng., 1828; M.B., London, 1858, M.D., 1859; lectr. materia medica; supt. Earlswood Asylum, 1858-68; fellow Royal Coll. Physicians; author treatise on degeneration of race as result of marriages of consanguinity, 1866; described clin. picture of mongolism, also made observations on ethnic classification of idiots, 1866. Died 1896.

DOWNES, Sir Arthur Henry, Brit. physician; b. Oct. 11, 1851; s. T.R.C. Downes ed. Univ. Coll., London, Aberdeen U.; M.B., C.M. with highest honours, 1873; M.D., Aberdeen U., 1875; D.P.H., Cambridge U., 1877; m. Evelyn Downes, 1875; 1 son, 1 dau.; m. 2d, Florence Chapman, 1902; 1 dau. Dep. med. officer health Shropshire Combined Dists., med. officer health Essex Combined Dists., 1876-89; med. insp. Local Govt. Bd., 1889-1919. Research and publs. on effect of light on bacteria, influence of light on protoplasm (revealed destructive action of light on organisms of putrefication and disease). Died Mar. 11, 1938.

DOWNEY, John Charles, Am. zoologist; b. Eureka, Utah, Apr. 12, 1926; s. John Charles and Cleone (Owens) D.; B.S., U. Utah, 1949, M.S., 1950; Ph.D., U. Cal. at Davis, 1957; m. Norine Margaret Simpson, June 25, 1949; children—John Charles III, Michael, Mary Ann, Dennis James, Patrick Joseph. Faculty, U. Utah, 1947-52, instr. biology, 1951-52; faculty U. Cal. at Davis, 1952-56; asst. prof. zoology So. Ill. U., Carbondale, 1956-61, asso. prof., 1961-67, prof., 1967—. Recipient research grants Am. Philos. Soc., 1959, NSF, 1959-61, 62, 64-66, 66-68. Member of Society Study Evolution, Lepidopterists Foundation, Pacific Coast Entomologists Society, Lepidopterists Society (secretary 1964—), Soc. Systematic Zoologists (charter mem.), Entomologists Soc. Am., So. Cal., Ill. acads. scis., Sigma Xi, Phi Sigma, Phi Kappa Phi. Publs. on variation and evolution using insects, especially butterflies as the tool organism; studies of population dynamics showing how host-plants, parasites and symbiotic associations can influence evolution; studies on mimicry; sound prodn. in immature stages of butterflies. Home: 43 Hillcrest Dr., Carbondale, Ill. 62901.*

DOWNIE, Allan Watt, Brit. microbiologist; b. Rosehearty, Scotland, Sept. 5, 1901; s. William and Margaret (Watt) D.; M.B., Chirurg. B., 1923, M.D., 1929, D.Sc., 1938, LL.D., 1956; m. Annie McHardy, Sept. 2, 1936; children—Alison May (Mrs. Christopher Hewetson), William Alan, Susan Margaret. Lectr., Aberdeen (Scotland) U., 1923-26, Manchester (Eng.) U., 1927-34; asst. Rockefeller Inst. Med. Research, 1934-35; sr. freedom research fellow London (Eng.) Hosp. Med. Sch., 1935-39; mem. sci. staff Nat. Inst. Med. Research, London, 1939-43; prof. bacteriology Liverpool (Eng.) U., 1943—. Smallpox cons. Ministry Health, 1947—; hon. cons. Liverpool United Hosps., 1943—. Fellow Royal Soc.; mem. Path. Soc. Gt. Britain and Ireland (hon. sec. 1954-63), Soc. Immunology, Soc. Gen. Microbiology, Royal Soc. Medicine, Harvey Soc. N.Y. Author: (with S. P. Bedson, F. O. MacCallum, C. H. Stuart Harris) Virus and Rickettsial Diseases, 1950; also numerous articles. Research on viral, bacterial infections and immunity. Home: 26 Blundell Dr., Southport, Eng. Office: Bacteriology Dept., Univ., Liverpool, Liverpool, Eng.*

DOWNIE, Norville Morgan, Am. psychologist; b. Troy, N.Y., Dec. 13, 1910; s. James N. and Elizabeth (Morgan) D.; B.S., St. Lawrence U., 1932, M.A. in Edn., 1933; postgrad. Cornell U., 1933,34,39; Ph.D., Syracuse U., 1948; m. Johanna Munson, Apr. 2, 1942; children—Elizabeth M., Johanna C., James N., Jasper Paul. Faculty Robert Coll., Istanbul, Turkey, 1936-39; tchr. sci. high sch., Elbridge, N.Y., 1939-42; faculty Wash. State U., 1948-51; prof. psychology Purdue U., Lafayette, Ind., 1951—. Cons., research psychologist VA, Marion, Ind., Indpls., 1951—. Fellow A.A.A.S., Am. Psychol. Assn.; mem. Am. Ednl. Research Assn., Sigma Xi, Phi Beta Kappa. Author: Fundamentals of Measurement, 1967, Types of Test Scores, 1967; (with R. Heath) Basic Statistical Methods, 1965; (with W. Cottle) Procedures and Preparation for Counseling, 1959. Research includes application of psychol. principles to problems of human adjustment, especially as related to mental illness and alcoholism; investigations in vocational interests of employees near bottom of vocational scale. Home: 505 Lingle Terrace, Lafayette, Ind. 47901.*

DOWNING, Elliot Rowland, Am. zoölogist; b. Boston, Nov. 21, 1868; s. Orrien Elliot and Mary Jane (Rowland) D.; B.S., Albion Coll., 1889; M.S., 1894; Ph.D., U. Chgo., 1901; postgrad. Columbia, 1907-08; U. of Würzburg, Naples Aquarium, 1908; m. Grace Emma Manning, June 24, 1902; children—George Elliot, Mary Elizabeth, Lucia Grace. Instr. science Ft. Payne (Ala.) Acad., 1890-91, Beloit (Wis.) Coll. Acad., 1891-96; supt. Bklyn. Tng. Sch. for Boys, 1896-98; sec. Bklyn. Children's Aid Soc., 1898-99; instr. embryology, summer sessions U. Chgo., 1900, 01, asst. prof. natural science, 1911-13, asso. prof., 1913-34, emeritus asso. prof. from 1934, asst. dean Sch. Edn., 1913-16; prof. biology No. State Normal Sch., Marquette, Mich., 1901-11. Pres. Mich. State Non-Game Bird Commn., 1907-11. Fellow A.A.A.S.; mem. Am. Eugenics Assn., Wis. Acad. Sci., Am. Nature Study Soc., Nat. Assn. Research in Science Teaching (pres. 1930-32), Sigma Xi. Author: Elementary Eugenics; A Naturalist in the Great Lakes Region; Our Living World; Our Physical World; Teaching Science in the Schools; Science in the Service of Health; Introduction to the Teaching of Science; Living Things and You. Editor Nature Study Rev., 1911-17. Died Sept. 10, 1944.

DOWNS, Cornelia Mitchell, Am. microbiologist, b. Kansas City, Kan., Dec. 20, 1892; d. Henry Mitchell and Lily (Campbell) Downs; A.B., U. Kan., 1915, A.M., 1920, Ph.D., 1924. Prof. bacteriology U.Kan., Lawrence, 1935-61, Solon Summerfield distinguished prof. microbiology, 1961-63, emeritus, 1963—. NIH spl. fellow, Oxford, Eng., 1959-60. Contbr. over 100 articles to sci. jours. Research on typhoid fever, Q fever, tularemia, fluorescent antibody method. Home: 1646 Alabama St., Lawrence, Kan. 66044.*

DOWNS, Wilbur George, Am. epidemiologist; b. Perth Amboy, N.J., Aug. 7, 1913; s. James Cloyd and Mabel (Lehman) D.; A.B., Cornell U., 1935, M.D., 1938; M.P.H., Johns Hopkins, 1941; m. Helen Hartley Geer, Sept. 20, 1940; children—Helen (Mrs. Christian J. Haller, III), Anne, W. Montague, Isabel, Martha. Asst. biologist N.Y. State Biol. Survey, 1934-36; staff Rockfeller Found., N.Y.C., 1941—, asso. dir., 1962—; prof. epidemiology Yale, 1963—; dir. WHO World Reference Center for Arboviruses, also Yale Arbovirus Research Unit, 1965—. Chmn. commn. for malaria Armed Forces Epidemiol. Bd., 1965—. Mem. Am. Soc. Trop. Medicine and Hygiene, Am. Pub. Health Assn., Royal Soc. Trop. Medicine and Hygiene, Am. Soc. Parasitologists. Research and publs. on epidemiology of malaria, use of naturalistic methods and insecticides in malaria control, epidemiology of insect transmitted virus diseases. Home: 10 Halstead Lane, Branford, Conn. 06405. Office: Yale Med. Sch., 333 Cedar St., New Haven, Conn. 06410. Office: Rockefeller Found., 111 W. 50th St., N.Y.C. 10020.*

DOWSON, Joseph Emerson, English inventor; b. London, Sept. 26, 1844; ed. Dulwich (Eng.) Coll. Recipient Silver medal Soc. Arts, Watt Gold medal, also Telford premium Instn. Civil Engrs., premium Instn. Elec. Engrs. Author book on producer gas, article on gas plants for engines in Ency. Brit., several publs. on metric measures and weights. First to use producer gas in working gas engines; eponym for Dowson gas; inventor Dowson gas producing apparatus. Died Jan. 3, 1940.

DOYEN, Eugène-Louis, French surgeon; b. Reims, France, Dec. 16, 1859; lived in Paris, France. Author: Traité de la chirurgie de l'estomac, 1895; Technique chirurgicale, 1897. Traitement de cancer, 1904; Le malade et le médecin, 1906; Traité de thérapeutique chirurgicale et de technique opératoire, 1907; Le cancer, 1909. Innovator in surgery; introduced cinematography in teaching of surgery; claimed to have discovered microbe causing cancer. Died Nov. 22, 1916.

DOYÈRE, Louis-Michel-François, French naturalist; b. Saint-Jean-des-Essartiers, France, 1811; prof. natural history Lycée Henri IV; prof. zoology applied to agr. Agronomic Inst., Versailles, France; prof. École centrale des arts et manufactures; research on anatomy and physiology of man and animals. Died Corsica, 1863.

DOYLE, William Thomas, Am. physicist; b. New Britain, Conn., Dec. 5, 1925; s. Thomas W. and Kathleen (McConn) D.; student Colby Coll., 1947-49; M.Sc. in Physics, Brown U., 1951; M.Sc., Yale, 1952, Ph.D., 1955; m. Barbara May Grant, June 16, 1951; children—Peter, Jeffrey. Faculty, Dartmouth, Hanover, N.H., 1955-—, prof., physics, 1964-—. NSF fellow, 1953-54, 54-55, 58-59. Mem. Am. Phys. Soc., Am. Inst. Physics, Am. Assn. Physics Tchrs. Research, numerous publs. on photoconductivity, optical absorption on electron paramagnetic resonance of defects in solids, solid state physics. Home: 6 Tyler Rd., Hanover, N.H. 03755.*

DOZY, Jean Jacques, Dutch geologist; b. Rotterdam, June 18, 1908; s. Eduard and C. E. (Van Doesburgh) D.; Ph.D. in Natural Scis., U. Leyden; m. Margo Schröter; children—Catherine, Pauline, Eduard. Geologist, Royal Dutch Shell, New Guinea, Guatemala, Equador, Iran, Venezuela, Indonesia; adminstr. Bataafse Petroleum Mij, Sumatra; prof. geology Delft Tech. U., 1968-—. Mem. Royal Netherlands Mineral and Geologic Assn., Royal Netherlands Assn. Geography, Am. Assn. Petroleum Geologists, Mining Council Netherlands. Contbr. articles to profl. publs. First to climb Carstensz mountains. Home: Pompstationsweg 21, The Hague, Netherlands.*

DRACH, Jules, French mathematician; b. Saint-Marie-aux-Mines, France, Mar. 13, 1871; lectr., Clermont, also Lille, France; prof. at Poitiers, France, Toulouse, France; named prof. Sorbonne, Paris, 1913; prof. mech. analysis, later prof. higher analysis Faculty Scis., Paris. Mem. French Acad. Scis. Research on group theory (in 1890's) led to structural theory which defines concepts of reducibility and group of rationality for any system of ordinary differential equations; investigated Galois's theory, essentials of rational groups; introduced idea of irreducibility in algebra; studied applications of mechanics and math. physics. Died Cavalaire-sur-Mer, France, Mar. 7, 1949.

DRAGANIC, Ivan, Yugoslavian chemist; b. Belgrade, Yugoslavia, Sept. 10, 1924; s. Gedeon and Sofija (Nikolic) D.; student U. Beograd, 1946-50; D.Sc., U. Paris (France), 1958; m. Zorica Kamperelic, Feb. 25, 1950; children—Sofija, Milan. Research chemist Boris Kidric Inst. Nuclear Scis., Beograd, 1950-—; vis. scientist in nuclear research labs. in France and Denmark; dept. dir. Atomic Energy Commn. Yugoslavia, 1961-62. Author: Introduction to Radiochemistry, 1957; also articles. Editor: Radiochemical Procedures, 1959; Radioactive Isotopes and Radiations, vol. I-III, 1963. Research on kinetics of radiation induced chem. reactions in aqueous solutions; use of chem. changes for measuring absorbed doses especially in nuclear reactors. Home: 16 Jevremova. Office: Boris Kidric Inst. Nuclear Scis., POB 522, Belgrade, Yugoslavia.*

DRAGENDORFF, (Johann) Georg (Noël), German pharmacist; b. Rostock, Germany, Apr. 20, 1836; s. Ludwig and Ernestine (Klingel) D.; pharmacy apprentice, journeyman; student Rostock; exam., 1958; Ph.D., U. Rostock, 1861; hon. dr. med. Dorpat; m. Sophie Spohn, 1868; 4 sons including Ernst, Hans. Pharmacist, Heidelberg, Germany, 1858-60; asst. chem. inst. U. Rostock, 1860-61; went to St. Petersburg as editor pharm. jour. for Russia, 1862; became prof. pharmacy Dorpat, Estonia, 1864. Author: Die Heilpflanzen der verschiedenen Völker und Zeiten, 1898. Research on foods and nutrition especially on mineral waters; chem. plant research; toxicologial-foensic chem. research. Died Rostock, Apr. 7, 1898.

DRAGO, Russell Stephen, Am. chemist; b. Turners Falls, Mass., Nov. 5, 1928; s. Stephen R. and Lillian (Pucci) D.; B.S., U. Mass., 1950; Ph.D., Ohio State U., 1954; m. Ruth Ann Burrill, Dec. 30, 1950; children—Patricia, Stephen, Paul. Faculty, U. Ill., Urbana, 1955-—, prof. chemistry, 1965-—. Cons., Am. Cyanamid, 1957-—. Mem. Am. Chem. Soc., Chem. Soc. (London). Author: (with T. L. Brown) Experiments in General Chemistry, 1958; Physical Methods in Inorganic Chemistry, 1965; Prerequisites for College Chemistry, 1966; also numerous articles. Quantitative correlations of molecular structures with reactivity, formulation of coordination model for nonaqueous solvents; measurement of covalency in metal-liquid bonds; correlation of result from molecular orbital calculations with reactivity and spectroscopic properties of molecules. Home: 1704 Henry St., Champaign, Ill. 61820. Office: Noyes Lab., U. Ill., Urbana, Ill. 61820.*

DRAGSTEDT, Carl Albert, Am. pharmacologist; b. Anaconda, Mont., Oct. 26, 1895; s. John Albert and Carrie (Selene) D.; S.B., U. Chgo., 1916; M.S., 1917, Ph.D., 1923; M.D., Rush Med. Coll., 1921; m. Ethel Johnson, Mar. 3, 1919; children—Carl Albert, Jane (Mrs. Donald Boyes); m. 2d, Naomi Alfred, Oct. 8, 1960. Instr., U. Ia., 1917-18; faculty Northwestern U., Evanston, Ill., 1926-64, prof. pharmacology emeritus, 1964-—. Chmn. pharmacology study sect. NIH, 1952-53. Mem. Soc. Exptl. Biology and Medicine (pres. 1952-53), Am. Soc. Pharmacology and Exptl. Therapeutics (pres.), Am. Physiol. Soc., A.M.A., Am. Soc. Pharmacology, Inst. Medicine, A.A.A.S., Phi Beta Kappa, Sigma Xi, others. Author: Personal Health Record, 1956; also numerous articles. Established role of histamines in anaphylactic and allergic reactions. Address: 909 S. Knight St., Park Ridge, Ill. 60068.*

DRAGSTEDT, Lester Reynold, Am. physician; b. Anaconda, Mont., Oct. 2, 1893; s. John Albert and Caroline (Seline) D.; M.S., U. Chgo., 1916, Ph.D., 1921; M.D., Rush Med. Coll., 1921; postgrad. U. Vienna, 1925-26; Dr. honoris causa, U. Guadalajara (Mexico), 1956, U. Lyon (France), 1961; m. Gladys Shoesmith, Sept. 4, 1922; children—Charlotte (Mrs. Thomas E. Jeffery), Carol (Mrs. Robert Stauffer), Lester Reynold II, John Albert. Asst. prof. U. Chgo., 1920-23, Distinguished Service prof. surgery, chmn. dept. surgery, 1925-59; prof. physiology and pharmacology Northwestern U. Med. Sch., 1923-25; research prof. surgery U. Fla., Gainesville, 1959-—. Recipient Distinguished Service award and medal A.M.A., 1963, Henry Jacob Bigelow medal Boston Surg. Soc., 1964, Julius Friedenwald medal Am. Gastroenterol. Assn., 1964. Fellow Royal Coll. Physicians and Surgeons Can. (hon.), Royal Coll. Surgeons Eng. (hon.); mem. Nat. Soc. for Med. Research (pres. 1949-60), Chgo. Surg. Soc. (past pres.), Nat. Acad. Scis., Am. Surg. Assn., Am. Gastroenterol. Assn., Am. Physiol. Soc. Contbr. numerous articles to tech. jours. Discovered lipocaic; that animals can be kept alive indefinitely after removal parathyroid glands; demonstrated cause duodenal ulcers; introduced vagotomy and drainage operation for treatment duodenal and gastric ulcers. Home: 2224 N.W. 11th Av., Gainesville, Fla. 32601. Office: J. Hillis Miller Health Center, U. Fla., Gainsville, Fla. 32603.*

DRAIN, Leslie Ernest, English physicist; b. London, Eng., Jan. 29, 1927; s. Leonard Richard and May (Clark) D.; B.A., Oxford U., 1947, D.Phil., 1950, M.A., 1952. Staff div. pure chemistry Nat. Research Council Can., Ottawa, 1950-52, crystallography sect chemistry dept. U. Coll., London U., 1952-54; staff solid state physics div. Atomic Energy Research Establishment, Harwell, Berkshire, Eng., 1955-—. Fellow Phys. Soc. Research and publs. on thermodynamic properties of gases absorbed on surfaces of fine powders, nuclear magnetic resonance applied to molecular motion in solids, crystallographic problems and electronic structure of metals and alloys. Home: Maisonette 1, Icknield Rd., Goring-on-Thames, Oxon, Eng. Office: Solid State Physics Div., Atomic Energy Research Establishment, Harwell, Berkshire, Eng.*

DRAIS VON SAUERBRONN, Karl Friedrich Christian Ludwig, German inventor; b. Karlsruhe, Germany, Apr. 29, 1785; s. Karl Friedrich and Margarete Ernestine (von Kaltenthal) D. von S.; student Privat-Fortschule, Karlsruhe, 1800-03; later studied physics, architecture and agr. in Heidelberg (Germany). Named chief forester Gengenbach, 1810; ret. from forestry with title prof. mechanics, 1818; with G. H. v. Langsdorff in Brazil, 1825-29. Patentee 4 wheeled wagon propelled by passengers; foot propelled 2 wheeler. Invented original form of bicycle, periscopic mirror-system, meat chopping machine, typewriter. Died Karlsruhe, Dec. 10, 1851.

DRAKE, Daniel, Am. physician; b. Plainfield, N.J., Oct. 20, 1785; s. Isaac and Elizabeth (Shotwell) D.; attended Med. Coll. of U. Pa., 1805-06, M.D., 1815; m. Harriet Sisson, 1807. Studied with Dr. William Goforth at Ft. Washington (Cin.), 1800-05; practiced medicine, Mays Lick, O., 1806; taught materia medica Transylvania U., Lexington, Ky., 1817-18, became dean, 1825-27; founder Ohio Med. Coll. (now Med. Coll. of U. Cin.), 1819, Commercial Hospital and Lunatic Asylum, 1821, Eye Infirmary, 1827, Ohio Sch. for the Blind, 1837; tchr. Jefferson Med. Coll., 1828; taught in various med. schs., Ohio, Pa., in 20 year period. Hon. mem. Phila. Acad. Natural Scis., Am. Philos. Soc., Wernerian Acad. Natural Scis. of Edinburgh. Contbr. to many med. jours. Author: A Systematic Treatise, Historical, Etiological and Practical, on the Principal Diseases of the Interior Valley of North America; As They Appear in the Caucasian, African, Indian and Eskimoux Varieties of Its Population, 1850-54. Studied disease as related to geography. Died Nov. 6, 1852.

DRAKE, Edwin Laurentine, Am. petroleum engineer; b. Greenville, N.Y., Mar. 29, 1819; married, 1845; m. 2d, Laura Dow, 1857; 4 children. Night clk. on steamboat between Buffalo and Detroit; engaged in farming 1 year; employed as hotel clk., 2 years, dry goods clk.; express agt. Boston & Albany R.R. Springfield, Mass., 1845-50; conductor N.Y. & New Haven R.R., 1850-57; 1st to tap oil at its source, Titusville, Pa., 1859; became impoverished, granted $1500 annual stipend by Pa. Legislature, 1873. Credited with 1st proof of oil reservoirs within earth; 1st to drive pipe to bedrock to prevent drill hole from filling in. Died Bethlehem, Pa., Nov. 8, 1880.

DRAKE, Sir Francis, English explorer, navigator; b. nr. Tavistock, Devonshire, Eng., circa 1540; married twice; m. 2d, 1585. Capt. English ship Judith, San Juan de Ulloa in Gulf of Mexico, circa 1568; became privateer by commn. of Queen Elizabeth, 1570; plundered Nombre de Dios on Isthmus of Panama, burned Portobeio, circa 1572; planned to sail into South Seas through Straits of Magellan, started on his journey, 1577, reached Brazil, then entered Straits, 1578; continued up coast of Chile and Peru and America, trying to discover passage to Atlantic; sailed possibly to 48° latitude (Wash. State), repaired his ship (The Golden Hind) at San Francisco Bay, 1579; named new country New Albion in name of Queen Elizabeth; sailed West to Molucca Islands, the Celebes, Java, 1580, then set sail for home; arrived in Eng., 1580; created knight by Queen Elizabeth; sailed against Spanish Indies, 1585, plundered St. Augustine settlement, Fla.; rescued Roanoke Island (N.C.) colony; vice-adm. of fleet that defeated Spanish Armada, 1858; made unsuccessful expdn. against Spanish West Indies, 1595. 1st Englishman to sail around world (2 years, 10 months). Died off Portobelo, Jan. 28, 1596.

DRAKE, Frank Donald, Am. astronomer; b. Chgo., May 28, 1930; s. Richard Carvel and Winifred (Thompson) D.; B.Engring. Physics, Cornell U., 1952; M.A., Harvard, 1956, Ph.D., 1958; m. Elizabeth Buckner Bell, Mar. 7, 1953; children—Stephen D., Richard P., Paul R. Scientist, Nat. Radio Astronomy Obs., Green Bank, W.Va., 1958-63, head telescope operations sci. services divs., 1961-63; chief lunar and planetary scis. sect. Jet Propulsion Lab., Cal. Inst. Tech., Pasadena, 1963-64; asso. prof. astronomy, 1964-65, prof., 1966-—, asso. dir. Center for Radiophysics and Space Research, Cornell U., Ithaca, N.Y., 1964-—. Mem. Am. Astron. Soc., Internat. Union Radio Sci., Internat. Astron. Union, I.E.E.E., A.A.A.S. Author: Intelligent Life in Space, 1962; also articles. High Resolution radio studies of Galactic Center; discovery and radio study of Jupiter radiation belt; radio studies of moon and planets; theoretical and observational studies of possible radio signals from other civilizations. Home: 121 Pine Tree Rd., Ithaca, N.Y. 14850.*

DRAKE, Noah Fields, Am. geologist; b. Summers, Ark., Jan. 30, 1864; s. Wesley and Martha (Kellam) D.; C.E., U. Ark., 1888; A.B., 1894, A.M., 1895, Ph.D., 1897, Stanford U.; m. Mary Elenor Shockley, July 30, 1904 (dec. Dec. 1926); m. 2d, Lota West Fairchild, Dec. 23, 1932. Geologic work, Ark. Geol. Survey, 1887, Tex. Geol. Survey, 1889-93, U. S. Geol. Survey, 1897; prof. geology and mining Pei Yang U., Tientsin, China, 1898-1900, 05-11; asso. prof. econ. geology, Leland Stanford Jr. U., 1911-12; prof. geology and mining, U. Ark., 1912-20; cons. geologist from 1920. Engr. Pub. Works Dept., Tientsin, China, 1900-01; cons. geologist Am. China Devel. Co., 1902-04; chmn. bd. Tientsin Land & Investment Co., Ltd., 1904-11; v.p. Am. Machinery & Export Co., Tientsin, 1910. Fellow Geol. Soc. Am.; mem. Am. Inst. Mining Engrs., Seismol. Soc. Am., China Philos. Soc. Died May 4, 1945.

DRAKE, Robert Mortimer, Jr., Am. engr., educator; b. Eagle Cliff, Ga., Dec. 13, 1920; s. Robert Mortimer and Elizabeth Margaret (Foushee) D.; B.S. in Mech. Engring., U. Ky., 1942; M.S. in Mech. Engring., U. Cal. at Berkeley, 1946, Ph.D., 1950; m. Jane Mardelle Smith, Aug. 19, 1944; children—Dianne Elizabeth, Kevin Robert. Asso. prof. mech. engring. U. Cal. at Berkeley, 1947-55; engine design cons. aircraft gas turbine div. Gen. Electric Co., 1954-56; prof. mech. engring. Princeton, 1956-63, chmn. dept., 1957-63; vis. prof. U. of Ky., Lexington, 1964-65, prof. mechanical engineering, 1965-—, chairman department mechanical engring., 1966-67; dean College of Engineering, 1967-—, director of engineering experiment sta., 1967-—. Cons. Air Preheater Corp.; sr. staff cons. Arthur D. Little, Inc., Cambridge; dir. Intertech Corporation. (N.J.). Mem. Ky. State Bd. Registration Profl. Engrs. Served from 2d lt. to capt. USAAF, 1942-47. Asso. fellow Am. Inst. Aeros. and Astronautics; mem. Am. Soc. M.E., Am. Ordnance Assn., Sigma Xi, Pi Tau Sigma. Author: (with E. R. G. Eckert) Introduction to Transfer of Heat and Mass, 1950, Heat and Mass Transfer, 1959; also numerous articles. Cons. editor: McGraw Hill Book Co. Research in area of connective heat transfer, mechanics and heat transfer in rarefield gases; optical interferometry; invented viewing system for interferometers. Home: 1087 The Lane, Lexington, Ky.*

DRAKE, (John Gibbs) St. Clair, Am. anthropologist; b. Suffolk, Va., Jan. 2, 1911; s. John Gibbs St. Clair and Bessie Lee (Bowles) D.; B.S., Hampton Inst., 1931; postgrad. U. Chgo., 1937-40, 46-47, Pd.D. in Anthropology, 1954; m. Elizabeth Dewey Johns, June 17, 1942; children—Sandra, Karl J. Instr. sociology and anthropology Dillard U., 1935-37, 41-42; Julius Rosenwald fellow U. Chgo., 1937-38, 47-48; asso. dir. Ill. Commn. Condition Urban Colored Pop., 1940-41; statis. med. div. U. S. Maritime Service, N.Y.C., 1943-45; asst. prof. sociology and anthropology Roosevelt Univ., Chgo., 1946-48, asso. prof., 1948-54, prof. sociology, 1954—; Rosenwald fellow for study of race relations in Great Britain, 1947-49; vis. lectr. social anthropology Boston U., 1953; research asso. Twentieth Century Fund's Survey of Tropical Africa, 1953-54; vis. prof. social scis. U. Liberia, Republic of Liberia, W. Africa, 1954, studied impact of press, film and radio on W. Africa, under grant from Ford Found., 1954-55; cons. Ford Found. African fellowship tng. program, 1955-57; prof., chmn. dept. sociology U. of Ghana, 1958-61; vis. prof. sociology Stanford, 1963. Fellow Am. Anthrop. Assn., African Studies Assn. U. S. A.; mem. Internat. Soc. for Study Race Relations, Am. Soc. African Culture, Phi Beta Sigma. Author: (with Horace R. Cayton) Black Metropolis, 1945 (selected for 1946 honor roll Schomburg collection N.Y.C. Pub. Library, as race relations book of the year; also Anisfeld Wolf award as one of two books contbg. most to race relations for year 1945), revised edit., 1962; (with Peter Omari) Social Work in West Africa, 1963; also articles profl. publs. Research in Negro life in Chgo. Office: Roosevelt U., 430 S. Michigan Av., Chgo.

DRANCE, Stephen Michael, ophthalmologist; b. Bielsko, Poland, May 22, 1925; s. George Henry and Ida (Berger) D.; M.B., Ch.B., U. Edinburgh (Scotland), 1948; M.D., U. Sask., Can., 1960; m. Betty Joan Palmer, Jan. 17, 1952; children—Jonathan Stephen, Michael George, Elisabeth Joan. Research asso. U. Oxford (Eng.), 1955-57; asso. prof. U. Sask., Saskatoon, 1957-63; asso. prof. U. B.C., Vancouver, Can.; 1963—, dir. Glaucoma Services, 1963—. Cons. ophthalmologist D.V.A. Hosp., Vancouver, 1963—. Fellow Royal Coll. Surgeons; mem. Brit., Canadian med. assns., Ophthal. Soc. U.K., N.Y. Acad. Scis., Am. Acad. Ophthalmology, Canadian Ophthal. Soc., Assn. Research Ophthalmologists. Translator: Light Coagulation (Meyer-Schwickerath), 1961. Research, publs. in glaucoma medication, diurnal intraocular pressure changes, effects of intraocular pressure on visual function. Home: 1561 Wesbrook Cr., Vancouver, B.C., Can.*

DRANT, Patricia (Hart), Am. dermatologist; b. Grenola, Kan., Jan. 27, 1895; d. James Lafayette and Nora Coombs (Demmitt) Hart; B.S., U. Kan., 1918; M.D., U. Pa., 1920; grad. student dermatology, U. Pa. Grad. Hosp., Phila., 1921-22, St. Louis Hosp., Paris, 1924-25, Vienna, 1925, Budapest, 1925; m. Reginald Drant, Sept. 1, 1920 (div.); m. 2d, William Warren Rhodes, Aug. 18, 1934 (div. 1949); m. 3d, James S. Collins, June 14, 1952. Dermatologist Abington Mem. Hosp., Pa., Methodist Hosp., Phila., Woman's Hosp., Phila. Diplomate, mem. Am. Acad. Dermatology and Syphilology. Fellow A.A.A.S.; mem. A.M.A., Phila. Dermatol. Soc., Phila. County Med. Soc. Died Oct. 3, 1955.

DRAPER, Charles Stark, Am. aero. engr.; b. Windsor, Mo., Oct. 2, 1901; s. Charles Arthur and Martha Washington (Stark) D.; student, U. Mo., 1917-19; A.B., Stanford, 1922, B.S. in elec. chem. engring., Mass. Inst. Tech., 1926, M.S. in aero. engring., 1928, Sc.D. (physics), 1938; m. Ivy Willard, Sept. 7, 1938; children—James, Martha, Michael, John Clayton. Operated lab. to develop infra-red signaling devices for U. S. Navy, 1927; research asst. in aero. engring. Mass. Inst. Tech., 1929-30, research asso., 1930-35, asst. prof., 1935-38, asso. prof., 1938-39, prof., 1939—, head aeros. and astronautics dept., until 1966, dir. Instrumentation Lab.; Wilbur Wright Meml. lectr. Royal Aero. Soc., 1955. Mem. Sci. Adv. Bd.; cons. to USN, USAF, and comml. orgns. in field of aeronautics and control. Recipient Exceptional Civilian Service award, Dept. of Air Force, 1951, Navy Distinguished Pub. Service award, 1956; Airpower award Mass. Wing., Air Force Assn.; Thurlow award Inst. Navigation; Airpower trophy Air Force Assn.; Holley medal Am. Soc. M.E., 1957; Blandy medal Am. Ordnance Assn., 1958; Godfrey L. Cabot award Aero Club New Eng., 1959; Potts medal, 1960; Navy Distinguished Pub. Service award Dept. Navy, 1961; Golden Plate award Acad. Achievement, 1961; Mo. Honor award U. Mo., 1962; Nat. Soc. Profl. Engrs. award, 1962; Louis W. Hill Space Transp. award, 1962; Comdrs. award Ballistic System div. USAF, 1964; Montgomery award Nat. Soc. Aerospace Profls. and Aerospace Mus., 1965; Nat. Sci. medal Presdl. award, 1965. Fellow N.Y. Acad. Scis., Am. Phys. Soc., Inst. Aero. Sci. (hon.), Am. Acad. Arts and Sci., Am. Soc. M.E., I.E.E.E., Am. Inst. Aeros. and Astronautics, A.A.A.S.; mem. Am. Soc. Engring. Edn., Internat. Acad. Astronautics (pres.), Soc. Automotive Engrs., Nat. Inventors Council (chmn.), Nat. Acad. Scis., Am. Ordnance Assn., Am. Inst. of Cons. Engrs., Sigma Xi. Awarded Sylvanus Albert Reed Award, Inst. of Aeronautical Scis., Medal for Merit, Naval Ordnance Development Award, for development of major improvements in antiaircraft fire control equipment. Contbr. articles in field of aero. instruments to tech. jours. Contbns. in advanced weapon technology, especially in navigational and guidance systems using gyroscopes. Home: 62 Bellevue St., Newton, Mass. 02158. Office: Mass. Inst. of Tech., 68 Albany St., Cambridge, Mass. 02139.*

DRAPER, George Otis, Am. inventor; b. Hopedale, Mass., July 14, 1867; s. William Franklin and Lydia Warren (Joy) D.; ed. Mass. Inst. Tech.; m. Lily Duncan, Apr. 28, 1892. Learned details of mfr. of cotton machinery in father's shops; partner, George Draper & Sons, Hopedale, from 1889 to its consolidation with Draper Co., 1897; officer in some 25 corps. connected with textile mfg., quarrying, mining and the devel. of patented inventions. Pres. Draper Realty Co., Draper-Hansen Co., Michener Stowage Co., Sapphire Record & Talking Machine Co., Draper-Latham Magneto Co., Scholz Fireproofing Co., Farrington Co., Phillips Mfg. Co., Hilton Mfg. Co. Author: Searching for Truth, 1902; Still on the Search, 1904; More, 1908. Inventor of many patented mech. devices and improved details of the Northrop loom; expert on patents, and in cotton mfr. Died Feb. 7, 1923.

DRAPER, Henry, Am. astronomer; b. Prince Edward County, Va., Mar. 7, 1837; s. John William and Antonia Coetana de Paiva Pereira (Gardner) D.; M.D., U. City N.Y., 1858; m. Mary Anna Palmer, 1867. Mem. staff Bellevue Hosp., N.Y.C., 1858-60; prof. natural sci. U. City N.Y., 1860-66, prof. physiology, 1866-73, dean of faculty, also prof. analytical chemistry, 1870-82, prof. chemistry, 1882; built obs. at Hastings-on-Hudson, N.Y., 1860; directed photog. work in U. S. govt. expdn. to observe transit of Venus, 1874; led expdn. to Wyo. to observe total eclipse of sun, 1878. Mem. Nat. Acad. Scis., Am. Philos. Soc. Honored by standard Draper Catalogue (meml. completed by Harvard Obs. 1924). Worked in celestial photography; made 1st photograph of a stellar spectrum in which dark lines showed, 1872. Died N.Y.C., Nov. 20, 1882.

DRAPER, James Edward, Am. physicist; b. Kansas City, Mo., Sept. 14, 1924; s. Raymond Edward and Doris (Aiken) D.; B.A., Williams Coll., 1944; Ph.D., Cornell U., 1952; m. H. Caryl Andrews, June 12, 1948; children—Lorna Beth, Jeffrey, John, Ann. Asso. physicist Brookhaven Nat. Lab., Upton, N.Y., 1952-56; faculty Yale, 1956-64, asso. prof., 1959-62, sr. research asso., 1962-64; prof. U. Cal., Davis, 1964—, chmn. dept. physics, 1966—. Fellow Jonathon Edwards Coll., Yale, Am. Phys. Soc., A.A.A.S.; mem. Am. Assn. Physics Tchrs., Phi Beta Kappa, Sigma Xi. Research on resonance neutron reactions, neutron capture gamma rays, photonuclear reactions, nuclear physics instrumentation. Home: 10 Almond Lane, Davis, Cal. 95616.*

DRAPER, John William, chemist; b. St. Helen's, Eng., May 5, 1811; s. Rev. John Christopher Draper; attended London U.; M.D., U. Pa., 1836; m. Antonia Coetana de Paiva Pereira Gardner, circa 1830; children—Henry, John Christopher, Daniel, Virginia, Antonia. Came to Am., circa 1832; operated own sci. lab.; prof. chemistry and natural philosophy Hampden-Sydney (Va.) Coll., 1836-38; prof. chemistry Univ. of City N.Y. (now N.Y. U.), 1838; founder sch. of medicine of Univ. City of N.Y., 1839, prof. chemistry and physiology, 1839-50, pres. med. sch., 1850. Author: A Treatise on the Forces Which Produce the Organization of Plants, 1844; Human Physiology, Statical and Dynamical, 1856; History of the Intellectual Development of Europe, 1863; Thoughts on the Future Civil Policy of America, 1865; History of the American Civil War, 3 vols., 1867-70; History of the Conflict Between Religion and Science, 1874. Pioneer in photography, photochemistry; took 1st complete portrait of person by sunlight, 1839, 1st photograph of moon, 1840; photographed dark lines in solar spectrum, 1843; conducted expts. in phosphorescence; observed Draper effect; proposed Draper's law in regard to gas combinations in producing light; proved experimentally law of reciprocity for photochem. changes. Died Hastings-on-Hudson, N.Y., Jan. 4, 1882.

DRASCHE, Anton von, see von Drasche, Anton.

DRAVNIEKS, Andrew, chemist; b. Petersburg, Russia, Oct. 3, 1912; s. Fricis and Vera (Sokolovskis) D.; Chem.E., U. Latvia, 1938; postgrad. U. Marburg (Germany), 1946; Ph.D., Ill. Inst. Tech., 1949; m. Erika Dorotea Veitbergs, Sept. 10, 1939; children—Sarma Sue (Mrs. Alfred Gort), Dzintar Eric. Came to U. S., 1946, naturalized, 1952. Research asso. Ill. Inst. Tech., 1947-49, sci. adviser, head Olfactronics and Odor Sci. Center, Research Inst., 1960—; sect. head corrosion research Standard Oil Co. Ind., Whiting, 1949-60. Recipient IR 100 award for polar vapor detector, 1964. Mem. A.A.A.S., Am. Chem. Soc., Electrochem. Soc., Nat. Assn. Corrosion Engrs., N.Y. Acad. Sci., Inst. Food Technologists. Author: Friction, Lubrication and Lubricants, 1940; Selection and Use of Lubricants, 1944; (with Bregman) Surface Effects in Detection, 1964. Contbr. numerous articles to profl. jours. Research in devel. of instruments to measure electrically rates of corrosion, clarification of sulfidic corrosion mechanism; contbr. to theories of olfaction; devel. of techniques and instruments for characterization and measurement of odors. Home: 211 Tampa St., Park Forest, Ill. 60466. Office: 10 W. 35th St., Chgo. 60616.*

DRAWIN, Hans Werner, physicist; b. Hamburg, Germany, Jan. 21, 1930; s. Werner Franz and Gerda (Millahn) D.; Diploma in Physics, Kiel (Germany) U., 1954, Ph.D., 1956; m. Marianne Kunow, Dec. 27, 1957; children—Thomas, Angela, Stefan. Sci.-tech. staff Atlas-Werke A.g., Bremen, 1956-60, lab. chief for devel. mass spectrometers and high vacuum equipment, 1957-60; sci. staff E.U.R.A.T.O.M., Fontenay-Aux-Roses, France, 1960—, group leader research group on controlled thermonuclear reactions, 1961—. Mem. Physikalische Gesellschaft. Author: (with P. Felenbok) Data for Plasmas in Local Thermodynamic Equilibrium, 1965; also articles. New constructions of mass analyzers; combination of mass-spectrometric and optical methods for interpreting thermodynamic state of ionized gases; research on radiation from ionized gases submitted to strong magnetic fields. Home: 2, Rue de la Ronce, Ville D'Avray, France. Office: Euratom-CEA, Fontenay-Aux-Roses, France.*

DRAY, Sheldon, Am. immunologist; b. Chgo., Nov. 20, 1920; s. Harry and Sarah (Zaas) D.; B.S., U. Chgo., 1941; M.D., U. Ill., 1946, M.S., 1947; Ph.D., U. Minn., 1954; m. Margaret Berman, Mar. 7, 1953; children—Tevian Gordon, Nancy Laraine. Med. officer USPHS, NIH, Bethesda, Md., 1947-65; prof. microbiology, head dept. U. Ill. Coll. Medicine, 1965—. Cons. immunology sect. WHO, 1964—. Mem. Am. Assn. Immunologists (sec.-treas. 1964—), Am. Soc. Human Genetics, Am. Chem. Soc., A.A.A.S., Alpha Omega Alpha, Phi Lambda Upsilon. Asso. editor Jour. Immunochemistry, 1964—, Jour. Immunology, 1967—. Research, numerous publs. on theory of elec. potentials across ion-selective membranes, allotypes, genetic variation of serum proteins detected by antibodies, genetic control of immunoglobulins, maternal-fetal interactions resulting in allotype suppression. Office: 835 S. Wolcott St., Chgo. 60612.*

DREBBEL, Cornelis, philosopher, inventor; b. Alkmaar, Netherlands, 1572; s. Jacob Drebbel; m. sister of Hubert Goltzius. Worked as engraver; received by James I in Eng., 1604; visited court of Rudolph II; imprisoned by elector Palatine on capture of Prague, Czechoslovakia, 1620; released at James I's intercession; in charge of fireships on Rochelle expdn., 1627. Author: Tractat oder Abhandlung von Natur und Eigenschafft der Elementen, 1608. Supposedly invented perpetual motion machine presented to James I; invented a telescope, self-regulating ovens, incubator for hatching eggs; built submarine, circa 1610; claimed to have invented air thermometer, microscope (both earlier invented by Galileo); suggested way of making sulphuric acid by oxidation of sulphur; discovered method of dyeing cloth scarlet; improved drainage pumps; experimented with problems of light. Died London, Eng., 1634.

DREBENEVA-UKHOVA, Varvara Pavlovna, Russian entomologist; b. 1903; grad. Biology Faculty, 1st Moscow U., 1925; postgrad. Timiryazev Research Inst., 1927-30; D.Biol. Sci., 1945. Asso. ecology dept. Timiryazev Research Inst., 1925-27; asst. dept. gen. biology Ivanovo Med. Inst., 1930-33; sr. asso., dept. head, later head fly extermination lab. Martsinovsky Inst. Med. Parasitology and Tropical Medicine, USSR Ministry Health, 1933-63, head dept. entomology, 1963—. Expert, WHO. Author: Flies and Their Epidemiological Importance, 1952. Research and publs. on gen. exptl. and med. entomology, structure and devel. of female reproductive system of various species of flies, drew up scale of maturation of ova of flies in connection with gen. devel. of ovarian tubes, also ecol. classification of flies; 1st in USSR to use DDT on flies, 1946. Address: Inst. Med. Parasitology and Tropical Medicine, Pirogovskaya ulitsa 20, Moscow, USSR.

DRECHSEL, (Ferdinand Heinrich) Edmund, physiol. chemist; b. Leipzig, Germany, Sept. 3, 1843; s. Carl Ferdinand and Frederike Auguste (Günther) D.; student chemistry at Leipzig and Marburg, Germany; doctorate Leipzig, 1864; m. Alida Esther Hamming, 1875; 3 sons, 1 dau. Worked in lead and silver works of Dumont Bros., Selaigneaux, Belgium, until war of 1870; became asst. to Scheerer at Bergacademie, Freiberg, Germany, 1870, also lectr. chem. tech.; apptd. dir. chem. dept. C. Ludwig's Physiol. Inst. in Leipzig, 1872; joined faculty philosophy 1875; became asso. prof. physiol. chemistry in med. faculty, 1878; named prof. med. and physiol. chemistry and pharmacology U. Bern (Switzerland), 1892. Research and numerous publs. on protein chemistry; discovered lysine; invented alternating current electrosynthesis in organic preparative chemistry, wash bottle; synthesized sodium oxalate; showed protein molecule contains both monoamino and diamino acids. Died Naples, Italy, Sept. 22, 1897.

DRECHSLER, Charles, Am. biologist; b. Butternut, Wis., May 1, 1892; s. Louis and Bertha (Schulz) D.; B.S., U. Wis., 1913, M.S., 1914; Ph.D., Harvard, 1917; m. Mary Florence Morscher, July 30, 1930; children—Charles Edward, Mary Kathryn (Mrs. David Taylor), Robert Norton. With U. S. Dept. Agr., 1917-

18, 19-62, Plant Industry Sta., Beltsville, Md., 1935-62, prin. mycologist, 1958-62. Recipient Distinguished Service award U. S. Dept. Agr., 1958. Mem. A.A.A.S., Am. Inst. Biol. Scis., Bot. Soc. Am., Am. Phytopath. Soc., Am. Mycol. Soc., Soc. Econ. Botanists, Torrey Bot. Club, Am. Micros. Soc., Washington Acad. Sci., Wis. Acad. Arts and Sci., Deutsche Botanische Gesellschaft. Contbr. numerous articles to profl. jours. Research on leaf-spot and stem-rot diseases of grasses; discoverer, describer numerous fungi that capture and destroy nematodes, amoebae, shelled rhizopods, tardigrades and springtails found infesting decaying plant materials in or on ground, also many fungi related to common house-fly fungus. Address: 6915 Oakridge Rd., Hyattsville, Md. 20782.*

DREIKURS, Rudolf, psychiatrist; b. Vienna, Austria, Feb. 8, 1897; s. Sigmund and Fanny (Cohn) D.; M.D., U. Vienna, 1923; m. Sadie J. Ellis, Sept. 12, 1943; children—Eric, Eva (Mrs. Alexander Ferguson). Came to U. S., 1937, naturalized, 1943. Practice medicine, specializing in psychiatry, Chgo., 1939—; prof. psychiatry Chgo. Med. Sch., 1942—; dir. Alfred Adler Inst., 1943—; vis. prof. U. Rio de Janeiro, 1946, Northwestern U., 1947, Ind. U., 1951-54, U. Ore., 1957-63, Roosevelt U., 1954-56, Loyola U., 1959, Barilan U., Israel, 1963-64, Tel-Aviv U., Israel, 1963, So. Ill. U., 1964, Tex. Technol. Coll., 1965. Fellow Am. Psychiat. Assn.; mem. Am. Humanist Assn., Internat. Assn. Individual Psychology, Am. Soc. Group Psychotherapy and Psychodrama (past pres.), Am. Soc. Adlerian Psychology. Author: Seelische Impotenz, 1931; Das Nervose Symptom, 1932; Einfuhrung in Die Individual-Psychology, 1933; The Challenge of Marriage, 1946; The Challenge of Parenthood, 1948; Fundamentals of Adlerian Psychology, 1950; Psychology in the Classroom, 1957; Encouraging Children to Learn, 1963; (with Vicki Solts) Children: The Challenge, 1964. Contbns. to theory and practice of psychotherapy, group psychotherapy, edn. and guidance; originator multiple psychotherapy, methods of tng. for parents and tchrs. Home: 2608 N. Lake View Av., Chgo. 60614. Office: 6 N. Michigan Av., Chgo. 60602.*

DREIZLER, Erich Helmut, German physicist; b. Mannheim, Germany, Mar. 30, 1929; s. Erich Friedrich and Paula (Schmidt) D.; abitur, Gewschister School Gymnasium Dusseldorf, 1949; Dipl. Phys., U. Freburg, 1955, Dr.rer.nat., 1957; m. Marianne Pfannensteil, Oct. 15, 1959; children—Annette, Stefan, Andreas. Research asso. U. Freiburg, 1957-60, asst. Insts. Phys. Chemistry and Physics, 1960-66, habilitation in physics, 1966—, dozent physics. Research, publs. on oetermination of structure hindering potential dipole moment of several molecules with microwave spectroscopy; contbns. to group theoretical analysis of rotational spectra. Home: 28 Schwarzwaldstrasse, Badkrozingen 7812, Germany. Office: 3 Hervnanh Hender Strasse, Freiburg 1 Br 78, Germany.

DRELINCOURT, Charles, physician; b. Paris, France, Feb. 1, 1633; s. Charles and Marguerite (Bolduc) D.; docteur en médecine, Montpellier, France, 1650; m. Suzanne Jacob, 1662; 1st physician for French Armies in Flanders; named physician to king; named prof. anatomy, Leiden, Netherlands, 1668; physician to William of Orange. Author: Experimenta anatomica, 1681; De conceptione adversaria, 1685. Proved children born a month prematurely were not necessarily mentally deficient, also that theory of procreation from eggs dated from ancient Greeks rather than 17th century. Died Leyden, May 31, 1697.

DRELL, Sidney David, Am. physicist; b. Atlantic City, Sept. 13, 1926; s. Tully and Rose (White) D.; A.B., Princeton, 1946; M.A., U. Ill., 1947, Ph.D., 1949; m. Harriet J. Stainback, Mar. 22, 1952; children—Daniel W., Persis S., Joanna H. Research associate at the University of Illinois, 1949-50; instructor Stanford (Cal.) University, 1950-52, asso. prof., 1956-60, prof., 1960—; research asso. Mass. Inst. Tech., 1952-53, asst. prof., 1953-56; Loeb lectr., vis. prof. Harvard, fall 1962. Cons., Office of Sci. And Tech., Exec. Office Pres., 1960, Inst. Def. Analyses, 1960—; member President's Sci. Adv. Committee, 1966—. Guggenheim fellow, 1961-62. Fellow Am. Phys. Soc. Author: (with F. Zachariasen) Electromagnetic Structure of Nucleons, 1961; (with J. D. Bjorken) Relativistic Quantum Mechanics, 1964, Relativistic Quantum Fields, 1965. Editorial bd. Phys. Rev., 1960-63, 66—. Research and publs. on quantum electrodynamics and elementary particle processes. Home: 570 Alvarado Row, Stanford, Cal.*

DRESBACH, Melvin, Am. physiologist; b. Hallsville, O., July 8, 1874; s. Harvey and Amanda (Orr) D.; B.Sc., Ohio State U., 1897, M.Sc., 1900; M.D., Ohio Med. U., 1903; spl. work, Woods Hole, Mass., summer 1903, Würzburg, Germany, 1904, Leyden, 1911; m. Sylvia Reedy, Dec. 28, 1899. Asst. in physiology Ohio State U., 1897-1902, instr., 1902-05; instr. physiology Cornell Med. Coll., 1908-09, asst. prof. pharmacology, 1909-10, asst. prof. physiology Albany Med. Coll. (Union U.), 1918-36; researcher U. Pa. Mem. Am. Physiol. Soc., Soc. Exptl. Biology and Medicine, A.A.A.S., Sigma Xi. Contbr. to med. publs. Published a description of elliptical human red blood cells, 1904. Died Oct. 15, 1946.

DRESCHFELD, Julius, physician, pathologist; b. Bavaria, (now Germany) 1846; s. Samuel and Elizabeth (Giedel) D.; ed. Owens Coll., Manchester, Eng., Manchester Royal Sch. Medicine; M.D., U. Würzburg (now Germany), 1864; m. Selina Gaspari, 1888; m. 2d, Ethel Lilley, 1905; 4 children. Surgeon Bavarian army, 1864; came to Manchester, 1869; mem. staff Manchester Royal Infirmary from 1873; Bradshawe lectr. Victoria U., Manchester, 1887, prof. pathology, 1881-91, prof. pathology and medicine, from 1891. Fellow Royal Coll. Physicians. Research, publs. on field pathology. Almost forestalled Pasteur in research hydrophobia, 1882-83, first to recommend wide use of dye eosin in watery solution as stain for animal tissues; recorded 1st post-mortem case of primary lateral sclerosis; credited with description of creeping pneumonia (now known as influenzal pneumonia), alcoholic paralysis (researched with James Ross), and lung complications of diabetes. Died Manchester, June 13, 1907.

DRESDEN, Arnold, mathematician; b. Amsterdam, Netherlands, Nov. 23, 1882; s. Mark and Anna (Meyerson) D.; student U. Amsterdam, 1901-03; S.M., U. Chgo., 1906, Ph.D., 1909; m. Louise Schwendener, June 12, 1907; 1 son, Mark Kenyon. Came to U. S., 1903, naturalized, 1912. Tchr. math. Univ. High Sch., Chgo., 1906-09; instr. math. U. Wis., 1909-12, asst. prof., 1912-21, asso. prof., 1921-27; prof. math. Swarthmore Coll., 1927—. Fellow A.A.A.S.; mem. Math. Assn. Am. (pres. 1933-35), Am. Math. Soc., Acad. of Macon (France), Math Soc. France. Author: Plane Trigonometry, 1921; Solid Analytical Geometry and Determinants, 1930; An Invitation to Mathematics, 1936; Introduction to the Calculus, 1940. Died Apr. 10, 1954.

DRESDEN, Max, theoretical physicist; b. Amsterdam, The Netherlands, Apr. 23, 1918; s. Abraham and Henriette (Smit) D.; student U. Amsterdam, 1935-37, U. Leyden, 1937-39; Ph.D. in Physics, U. Mich., 1946; m. Bertha Evelena Cummins, Aug. 8, 1948; children—Janna, Danielle. Came to U. S., 1939, naturalized, 1948. Teaching asst. U. Mich., 1939-46; faculty physics U. Kan., 1946-57, prof., 1951-57; prof., chmn. dept. physics Northwestern U., Evanston, Ill., 1957-60; prof. physics U. Ia., 1960-64; prof. physics State U. N.Y., Stony Brook, L.I., 1964—, exec. dir. Inst. Theoretical Physics. Cons., Argonne Nat. Lab.; vis. sci. Am. Inst. Physics. Fellow Am. Phys. Soc. Research in statis. mechanics, behavior of positrons, meth. problems of quantum field theory, transport theory, parastatistics, symmetries. Home: Beacon Hill Dr., Stony Brook, L.I., N.Y. 11790.*

DRESEL, Peter E., pharmacologist; b. Ulm, Germany, Feb. 27, 1925; s. Kurt M. and Mathilde D.; came to U. S., 1938, naturalized, 1944; B.S. in Chemistry and Biology, Antioch Coll., 1948; Ph.D. in Pharmacology, U. Rochester, 1952; m. Anita Pitts, June 22, 1947; children—K. Michael, Patricia L., P. Evan. Faculty U. Cin., 1952, Emory U., 1953-54; sr. research pharmacologist William Merrell Co., Cin., 1954-56; faculty U. Manitoba, Winnipeg, Can., 1956—, prof. pharmacology, 1965—. Mem. Am. Soc. Pharmacology and Exptl. Therapeutics, Canadian Pharmacol. Soc., Cardiac Muscle Soc., A.A.A.S., Sigma Xi. Studies on mechanism of cardiac irregularities, heart failure and action of digitalis drugs. Home: 681 Cordova St., Winnipeg 9, Man., Can.*

DRESLER, Heinrich Wilhelm, German metallurgist; b. Weidenau/Sieg, Germany, Nov. 29, 1849; s. Daniel D. and Sara (Neff) D.; student Bau- und Gewerbeschule, Berlin, Germany; m. Marie Sell; 1 dau.; m. 2d, Marie Menne; 2 sons, 1 dau. Asst. to Fritz W. Lümann at Gorgsmarienhütte in Osnabrück; went to Sweden to study charcoal blast furnaces, 1873; first became dep. dir. Cöln-Müsener Bergwerks- und Hüttenverein, Kreuztal, Germany, then dir. until 1918. Patentee prodn. slag bldg. blocks. Research on forging fine ores; built first Gröndal sinter system with tunnel kiln in Germany; application of oxygen smelting process; devel. blast furnace operation. Died Kreuztal nr. Siegen, Germany, May 24, 1929.

DRESSEL, Francis George, Am. mathematician; b. Hart, Mich., Sept. 22, 1904; s. George W. and Minnie (Savage) D.; B.S. (scholar), Mich. State Coll., 1928, M.S. (fellow), 1929; postgrad. U. Chgo.; Ph.D., Duke, 1933; m. Hazel C. Johnson, June 25, 1932; children—David Leroy, John Michael. Faculty, Duke, Durham, N.C. 1929—, prof., 1952—, dir. undergrad. studies in math., 1949—, acting chmn. dept. math., 1963-64. Sr. sci. adv. Army Research Office, Durham, 1951—; cons. Air Force Operations Analysis Standby Unit, Chapel Hill, 1951—. Editorial bd. Bull. Am. Math. Soc., 1948-50, Duke Math. Jour., 1951-57; rev. staff Math. Rev. 1940-58. Contbns. to study integral equations, fundamental solutions of parabolic equations, harmonic functions, mapping theory, non-linear differential equations, eigenvalue problems for Bessel difference equations. Home: 2502 Francis St., Durham, N.C. 27707.*

DRESSER, John Alexander, Canadian geologist; b. Richmond, Que., Can., June 27, 1866; s. George F. and Alzina M. (Healy) D.; student Harvard 1894-95; B.A., McGill U., 1893, M.A., 1897, LL.D.; m. Florence McLean, 1895 (dec. 1902); m. 2d Elizabeth Lindsay, 1911. Lectr. geology McGill U., from 1907; geologist Canadian Geol. Survey, from 1909; directing geologist Que. Bur. Mines, from 1927. Mem. Geol. Soc. Am., Soc. Econ. Geology, Royal Soc. Can. Noted for petrographic studies; geology of copper-bearing rocks; investigated asbestos and chromite deposits Que., also geology of Lake St. John Dist., petroleum resources of Peace River Dist. Died 1954.

DRESSLER, William, physician; b. Budzanow, Poland, Aug. 9, 1890; s. Jacob M. and Henrietta (Ehrenbrod) D.; M.D. U. Vienna Med. Sch., 1915; m. Margaret Aldor, Nov. 25, 1942; 1 dau., Ruth Harrie (Mrs. Hermann F. Weiss). Came to U. S., 1938, naturalized, 1945. Asso. chief Heart Hosp., Vienna, Austria, 1924-38; chief cardiac clinic, cardiographi lab. Maimonides Hosp., Bklyn., 1939—; emeritu lectr. medicine State U. N.Y.; cons. physician cardiology VA, Bklyn. Hosp. Med. Center. Mem. A.M.A., Am Heart Assn., Cardiol. Soc. Brazil. Author: Clinica Cardiology, 1942, (with Dr. Hugo Roesler) Atlas o Clinical Cardiology, 1948. Described the postmyocardial infarction syndrome, 1956; research and publs. in phys. diagnosis, which stimulated cardiac diagnosi without lab. aid at the bedside. Home: 160 Riverside Dr., N.Y.C. 10024. Office: 1150 Fifth Av., N.Y.C 10028.*

DRETTNER, Börje, Swedish otolaryngologist; b Linköping, Sweden, Dec. 7, 1926; s. Olov and Hjördi (Wiman) D.; M.D., Uppsala U., 1961; m. Gunnel Rosengren, Sept. 13, 1952; children—Agneta, Johan. Asst. Inst. Med. Chemistry, Uppsala, Sweden, 1947-48; physician dept. otolarngology Univ. Hosp., Uppsala, 1953-61, asst. prof. U. Uppsala, 1961-65, asso. prof. otolaryngology, 1966—; cons. hosp., Enköping, Sweden 1960—; guest lectr. socs., univs. Mem. Internat (bd.), European (bd.) rhinologic socs., Internat. Soc Biometeorology, Swedish Otologic Assn. (sec.), Upplands Gen. Med. Assn. (vice-chmn.). Author: Vascular Reactions of the Human Nasal Mucosa on Exposure to cold, 1961. Research, publs. on vascular reactions of nasal mucosa and pathophysiology of para nasal sinuses, hearing defects after shooting, audiometric testing using vascular responses. Home: 1 Tegnérgatan, Uppsala, Sweden.*

DREVERMANN, Friedrich Ernst, German paleontologist, geologist; b. Auhammer, Germany, Feb. 15 1875; s. Ernst and Emilie (Funke) D.; student geolog and paleontology Munich, Berlin, Bonn; doctorate Marburg, 1900; m. Marie Christine Steinbrück, 1905 3 sons. Asst. to Emanuel Kayser, Marburg; joined faculty Marburg, 1903; joined Senckenberg Mus., Frankfort/Main, Germany, 1905, later dir. geol. dept.; became prof., dir. geol. inst. Frankfort U., 1914, rector 1928-29. Mem. Geologische Vereinigung (co-founder) Paläontologische Gesellschaft (pres.). Founder, edito jours. Senckenbergiana, Natur und Museum; edito Paläontologische Zeitschfift. Research on paleontology, stratigraphy, vertebrate fossils of Rhenish Paleozoic; enlarged collection of Senckenberg Mus. Die Frankfort/Main, May 16, 1932.

DREW, Charles Richard, Am. surgeon; b. Washington, June 3, 1904; s. Richard Thomas and Nora R (Burrell) D.; grad. Amherst Coll., 1926; M.D., C.M. Faculty Med., McGill U., 1933; Med.D.Sc., Columbi Coll. Phys. and Surg., 1940; hon. D.Sc., Va. State 1945, Amherst Coll., 1947; m. Minnie Lenore Robbins Sept. 23, 1939; children—Bebe Roberta, Charlene Rosella, Rhea Sylvia, Charles Richard. Instr. in pathology Howard U., Washington, 1935-36, asst. in surgery, 1935-36, prof. surgery, head dept., from 1941; Rockefeller fellow in surgery Coll. Phys. and Surg. Columbia, also resident in surgery Presbyn Hosp., N.Y.C., 1938-40; chief surgeon, chief of staf Freedman's Hosp., Washington, med. dir. 1946-47 surg. cons. ETO (Army), 1949. Dir. first plasma div Blood Transfusion Assn., supplying plasma to Brit. 1940-41; first dir. A.R.C. Blood Bank, supplying plasma to U. S. forces, 1941. Diplomate Am. Bd. Surgery, became examiner, 1942. Recipient Springarn award, 1944, dir. D.C. chpt. Nat. Founds. Polyomyelitis, 1946, D.C. Soc. for Crippled Children. Mem. Am. Soviet Science Com., 1944. Died Apr. 1, 1950.

DREW, Gilman Arthur, Am. biologist; b. Newton Ia., Nov. 15, 1868; s. Orrin Gilman and Mary Emil (Drew) D.; B.S., State U. Ia., 1890; Ph.D., John Hopkins, 1898; m. Lena E. Slawson, Nov. 24, 1892 Acad. tchr., 1890-91; high sch. tchr., Oskaloosa, Ia. 1892-94; fellow 1897-98, Bruce fellow, 1898, asst in zoology, 1898-1900. Johns Hopkins; prof. biology, U. Me., 1900-11. Asst. dir. Marine Biol. Lab. Woods Hole, Mass., 1909-26. Fellow Am. Acad. Art and Scis. Died Oct. 26, 1934.

DREWS, Gerhart, German microbiologist; b. Berlin Germany, May 30, 1925; s. Wilhelm and Hilm (Schumann) D.; Staatsexamen, U. Halle (Germany) 1951, Dr.rer.nat., 1953, Dr.rer.nat. habil., 1960; m Christiane May, June 18, 1960. Sci. asst. dept. botan U. Halle, 1952-53; sci. mem. Inst. for Microbiolog and Exptl. Therapy, Acad. Sci., Berlin, Jena, Germany 1953-60; faculty scis. U. Freiburg, Germany, 1960—, prof. microbiology, 1964—, dir. Inst. Botany 1964—. Mem. Soc. Gen. Microbiology, Deutsch Gesellschaft Hygiene und Microbiologie, Deutsch Gesellschaft Elektronenmikroskopie, Deutsche Bot Gesellschaft. Co-author: Ergebnisse der Biologie, 1960 Handbuch Pflanzenphysiologie, 1962; Fortschritte de Botanik, 1964; Die Zelle, 1966; also articles. Researc on phototaxis of blue green algae, substructure c blue green algae and bacteria, metabolism of polyphos phates in mycobacteria, morphogenesis, fine structur and biosynthesis of bacterial chromatophores, regula tion of biosynthesis of bacteriochlorophyll, cell wall composition; discovered structure of photosynthesis apparatus in blue green algae. Home: 15 Bleiche Strasse, Freiburg i. Br., 78, Germany. Office: 9 Schaenzlestrasse, Freiburg i. Br., 78 Germany.

DREYER, Georges, pathologist; b. Shanghai, July 4, 1873; s. G. H. N. and Dagmar Alvilde (Ovistgaard) D.; M.A., Oxford U.; M.D., Copenhagen (Denmark) U.; m. Margrete Caroline Jörgensen, 1900; Prof. pathology Oxford U., from 1907, also mem. hebdomadal council, curator univ. chest, fellow Lindoln Coll.; mem. Med. Research Council. Fellow Royal Soc., 1921, Royal Danish Acad. Letters and Scis. Research and publs. on immunity, exptl. pathology, biochem. effect of light rays. Died Aug. 17, 1934.

DREYER, John Louis Emil, astronomer; b. Copenhagen, Denmark, Feb. 13, 1852; s. F. Dreyer; ed. Copenhagen U.; Ph.D., M.A. (hon.), Oxford (Eng.) U.; D.Sc., Belfast, Ireland; m. Katherine Hannah Tuthill; 3 sons, 1 dau. Astronomer, Earl of Rosse's Obs., 1874-78; asst. astronomer Dublin U. Obs., 1878-82; dir. Armagh (Ireland) Obs., 1882-1916. Fellow Royal Astron. Soc. (Gold medal 1916, pres. 1923-25); mem. Royal Danish Acad. Scis. Author: (with Prof. Copeland) Copernicus: an International Journal of Astronomy, 1881-84; Second Armagh Catalogue of 3300 Stars, 1886; New General Catalogue of Nebulae and Clusters of Stars, 1888; Supplements, 1895, 1908; Tycho Brahe: a Picture of Scientific Life and Work in the 16th Century, 1890; Planetary Systems from Thales to Kepler, 1906. Editor W. Herschel's Scientific Papers, 1912; Opera Omnia (Tycho Brahe), 15 vols., 1913-29. Studied motion of stars, nebulae. Died Oxford, Eng., Sept. 14, 1926.

DREYFUS, Camille Edouard, chemist; b. Basel, Switzerland, Nov. 11, 1878; s. Abraham and Henreitta (Wahl) D.; M.A., U. Basel, Ph.D., 1902; postgrad. Sorbonne, U. Paris (France); m. Jean Tennyson, Sept. 18, 1931: Founder (with brother Henry) chem. lab., Basel, began mfg. cellulose acetate; incorporated Am. Cellulose and Chem. Mfg. Co., Md., 1918, pres., 1918-45, chmn. bd., from 1945. Recipient Modern Pioneer award Nat. Assn. Mfrs., 1940. Mem. Am. Chem. Soc., A.A.A.S. Perfected (with bro.) cellulose acetate fiber, 1st to manufacture it commercially. Home: Hotel Pierre, Fifth Av. at E. 61st St., N.Y.C. 10021. Office: Celanese Corp. Am., 180 Madison Av., N.Y.C. 10016.

DREYSE, (Johann) Nikolaus von, see von Dreyse, (Johann) Nikolaus.

DRICKAMER, Harry George, Am. chemist; b. Cleve., Nov. 19, 1918; s. George H. and Louise (Strempel) D.; B.S., U. Mich., 1941, M.S., 1942, Ph.D., 1946; m. Mae Elizabeth McFillen, Oct. 28, 1942; children—Lee Charles, Lynn Louise, Lowell Kurt, Margaret Ann, Priscilla. Chem. engr. Pan Am. Refining Corp., Texas City, Tex., 1942-46; faculty U. Ill., Urbana, 1946—, prof. phys. chemistry, chem. engring., 1953—. Recipient Colburn award Am. Inst. Chem.E., 1947, Ipatieff prize Am. Chem. Soc., 1956. Fellow Am. Phys. Soc., (O. E. Buckley Solid State Physics award 1967), Am. Geophys. Union; mem. Nat. Acad. Scis., Am. Chem. Soc., Am. Inst. Chem. Engrs. (Alpha Chi Sigma award 1967), Faraday Soc. Research and publs. primarily in field of use of very high pressure to study electronic structure of solids. Home: 405 W. Washington St., Urbana, Ill. 61801.*

DRIESCH, Hans Adolf Eduard, German biologist; b. Bad Kreuznach, Germany, Oct. 28, 1867; s. Paul and Josephine (Raudenkolb) D.; student zoology in Freiburg, Jena, Munich (all Germany); doctorate Jena, 1889; m. Margarete Reifferscheidt, 1899; children—Kurt, Ingeborg (Mrs. Tétaz). Went to Zool. Sta., Triest, 1891; staff Zool. Sta., Naples, Italy 1891-1900; ind. scholar in Heidelberg Germany, until 1920; Gifford lectr. U. Aberdeen (Scotland), 1907-08; became privatdocent, 1909; became prof. philosophy Heidelberg, 1911; apptd. prof. systematic philosophy, Cologne (Germany), 1919; apptd. prof. Leipzig (Germany), 1921; guest prof. in China, 1922-23, U. S. A., 1926-27, Buenos Aires; became prof. emeritus, 1933. Author: Analytische Theorie des Organischen Entwicklung, 1899; Der Vitalismus als Geschichte und Lehre, 1905; Philosophie des Organischen, 2 vols, 1909; Ordungslehre, 1912; Leib und Seele, 1916; Wirklichkeitslehre, 1917; The Crisis in Psychology, 1925; Parapsychologie, die Wissenschaft von den occulten Erscheinungen, 1932; Alltagsraetsel des Seelenlebens, 1938. Research on sea urchin eggs which he split into parts and allowed to develop; in his philosophy of vitalism postulated non-mechanistic agency, entelechy, which involved theory of autonomy of life and whole organism; psychic research. Died Leipzig, Apr. 17, 1941.

DRIESSENS, Jules-Pierre, French biologist; b. Roubaix, France, Mar. 28, 1906; s. Jules and Helene (Debrander) D.; ed. Lille (France) Faculty Medicine; m. Melle Marcelle Pages, Dec. 29, 1939. Lectr., aggregate prof., titular prof. path. anatomy, prof. gen. cancerology Lille Faculty Medicine; lectr., lab. head, asst. dir., dir. Lille Fight Against Cancer Center; head dept. research, dir. Lille Cancer Research Inst. Contbr. papers. to sci. publs. Research on metastatic diffusion of malignant tumors, cytology of cancerous cells, viruses and cancer, radiobiology. Address: Centre Anticancéreux, Sac Postal 38, Cite Hospitalière, Lille (Nord), France.*

DRIGALSKI, (Karl Rudolf Arnold Artur) Wilhelm von, see von Drigalski, (Karl Rudolf Arnold Artur) Wilhelm.

DRIGO, Angelo Lorenzo, Italian physicist; b. Padua, Italy, May 29, 1907; s. Antonio and Zaira (Bragadin) D.; Ph.D. in Physics, U. Padua; m. Giulia Alocco; children—Laura, Paolo, Antonio, Marialuisa. Titular dir. U. Ferrara, also Inst. Gen. Physics. Mem. Ferrara Acad. Scis. Research and numerous publs. on instructional physics, ferromagnetism. Home: via J. Stellini 3, Padua, Italy. Office: via del Paradiso 12, Ferrara, Italy.

DRINKER, Cecil Kent, Am. physician; b. Phila., Mar. 17, 1887; s. Henry Sturgis and Aimee Ernesta (Beaux) D.; B.S., Haverford Coll., 1908, D.Sc. (hon.), 1933; M.D., U. Pa. Med. School, 1913; Am. (hon.) Harvard, 1942; m. Katherine Livingston Rotan, Sept. 7, 1910; children—Anne, Cecil. Acting head dept. physiology, Harvard Med. Sch., 1917-18, asst. prof. physiology, 1918-19, asso. prof., 1919-23; sec. Com. on Indsl. Hygiene, 1918-22; mng. editor Jour. Indsl. Hygiene, 1919-32; asst. to vis. physicians Boston City Hosp., 1922-24; prof. physiology, Harvard Univ. Sch. Pub. Health, 1923-48, asst. dean, 1924-35, dean, 1935-42; asso. editor Jour. of Indsl. Hygiene and Toxicology, 1932-48; lectr. in physiology, Cornell Med. Coll., 1948-49; cons. physiologist U. S. Navy from 1951. Emeritus mem. American Soc. for Clin. Investigation; mem. Assn. Am. Physicians, Am. Physiol. Soc., Royal Danish Acad. Sci. and Letters, Phi Beta Kappa, Alpha Omega Alpha, Sigma Xi. Died Apr. 14, 1956.

DRIPPS, Robert Dunning, Jr., Am. anesthesiologist; b. Phila., June 19, 1911; s. Robert Dunning and Madge (Heron) D.; A.B., Princeton, 1932; M.D., U. Pa., 1936; m. Diana Rogers, Feb. 11, 1911; 1 son, Robert Dunning. Faculty U. Pa. Sch. Medicine, Phila., 1938—, prof. chmn. dept. anesthesia, 1949—, dir. anesthesia, hosp., 1943—; cons. in anesthesia U. S. Naval Hosp., Phila., 1947—, VA Hosp., Phila., 1953—, Surgeon Gen., U. S. Army, 1953—. Mem. Nat. Adv. Research Resources Com., USPHS, 1963-67; selection committee Merck Foundation, 1965—. Trustee Princeton University. Recipient of the Lindbach award for Distinguished Teaching, U. Pa., 1962, John F. Lewis prize Am. Philos. Soc., 1965. Diplomate Am. Bd. Anesthesiology (pres. 1965). Mem. Am. Soc. Clin. Investigation, Am. Coll. Anesthesiologists, Am. Soc. Anesthesiologists (Distinguished Service award 1965, Distinguished Service award 1967), American Physiological Society, Am. Soc. Pharmacology and Exptl. Therapeutics, A.M.A., Assn. Am. Physicians, Am. Surg. Assn., Halstead Soc., Assn. U. Anesthetists, John Morgan Soc., N.Y. Acad. Scis., Soc. Phi Zeta, Phi Beta Kappa, Sigma Xi, Alpha Omega Alpha, others. Author: (with others) Physiological Basis for Oxygen Therapy, 1953; (with others) Introduction to Anesthesia, 1963. Asso. editor Anesthesiology, 1948-53, Digest of Treatment, 1949—, Jour. of Pharmacology and Exptl. Therapeutics, 1950-55; editorial bd. Modern Medicine, 1954-56, Jour. Surg. Research, 1963-67, Rev. of Surgery, 1964—; cons. editor Survey of Anesthesiology, 1957-60. Research, publs. on effects of anesthetics on respiration and circulation; spinal anesthesia; oxygen toxicity. Home: 526 Avonwood Rd., Haverford, Pa. 19041.*

DRIVER, Harold Edson, Am. anthropologist; b. Berkeley, Cal., Nov. 17, 1907; s. John Rush and Florence (Flook) D.; B.A., U. Cal. at Berkeley, 1930, M.A., 1934, Ph.D., 1936; m. Wilhelmine Schaeffer, Sept. 15, 1950. Research fellow U. Cal. at Berkeley, 1936-37, research asso., 1948-49; fellow Social Sci. Research Council, 1937-38; faculty Ind. U., Bloomington, 1949—, prof. anthropology, 1958—. Expert witness U. S. Dept. Justice in Indian land claims cases, 1953-55. Mem. Am. Anthropol. Assn., Am. Ethnol. Soc., A.A.A.S., Am. Assn. U. Profs., Ind. Acad. Sci. Author: (with William C. Massey) Comparative Studies of North American Indians, 1957; Indians of North America, 1961; (with Wilhelmine Driver) Chichimeca-Jonaz of Northeast Mexico, 1963, Indian Farmers of North America, 1967; The Americas on the Eve of Discovery, 1964; also numerous articles. Comparative studies N.Am. Indians. Home: 1824 E. Hunter St., Bloomington, Ind. 47401.*

DROBISCH, Mortiz Wilhelm, German mathematician; b. Leipzig, Germany, Aug. 16, 1802; s. Karl Wilhelm and Wilhelmine (Klotz) D.; m. Emilie Leichsenring, 1827; 3 sons, 3 daus. Prof. math . U. Leipzig, 1826-40; prof. philosophy there, 1840-96. Author: Beiträge zur Orientierung über Herbarts System der Philosophie, 1834; Neue Darstellung der Logik nach ihren einfachsten Verhälnissen, mit Rücksicht auf Mathematik und Naturwissenschaft, 1836; Grundlehren der Religionsphilosophie, 1840; Empirische Psychologie nach naturwissenschaftlichen Methode, 1842; Erste Grundlinien der mathematischen Psychologie, 1850; Über die Fortbildung der Philosophie durch Herbart, 1876. Research on logic, metaphysics, ethics and religious philosophy; adherent of ideas of Johann F. Herbart; stressed formal character of logic. Died Leipzig, Sept. 30, 1896.

DROBOTKO, Viktor Grigorevich, Russian microbiologist, epidemiologist; b. Degtyary (now Chernigov Oblast), Nov. 23, 1885; grad. Med. Faculty, Kiev U., 1913; D.Med. Sci. Asst., Kiev Bacteriological Inst., 1925-31; sr. sci. asso., head med. microbiology dept. D. K. Zabolotny Inst. Microbiology and Epidemiology (now Inst. Microbiology, Ukrainian Acad. Sci.), Kiev, 1931—, dir., 1944—, also prof. dept. pathogenic micro-organisms. Decorated Order of Lenin (2). Mem. Ukrainian Acad. Sci., Ukrainian Soc. Microbiologists, Epidemiologists and Infectionists (chmn. 1948-58). Author: Microbiology, 1936; Modern Chemical Therapy of Infectious Diseases, 1946, 49. Mem. editorial council Antibiotics; editor Microbiology . Jour. Research and numerous publs. on devel. new vegetable antibiotic Imanin, developer method of differential staining of bacteria and synthetic nutrient media, designer original filter for working with phages. Address: Inst. Microbiology, ulitsa S. Razina 4, Kiev, Ukrainian SSR, USSR.

DROKE, George Wesley, Am. mathematician; b. Morgan County, Ind., Sept. 26, 1854; s. George and Diana (Etter) D.; A.B., U. Ark., 1880, A.M., 1884, LL. D., 1929; postgrad. U. Mich., Johns Hopkins, U. Chgo.; LL.D., Hendrix Coll., 1919; m. Josephine Campbell, Sept. 24, 1879 (dec. 1886); children—George Prentice, Leila Ruth, Marvin Josephine; m. 2d, Inez James, Aug. 18, 1887 (dec. Apr. 1931); children—Albert Hill, Mary Inez, Louise Blanche, James Walling (adopted); m. 3d, Mrs. Belle Clayton Fenner, Feb. 20, 1932. Asst., 1st asst. tchr., prep. dept. U. Ark., 1880-85, with univ., 1887-1929, prof. math., astronomy, 1897-1929, emeritus, from 1929, dean Coll. Arts and Scis., 1915-25; head dept. English, Coronal Inst., San Marcos, Tex., 1885-86; prin. high sch., Bentonville, Ark., 1886-87. Mem. Math. Assn. Am., Phi Beta Kappa (1932). Died Sept. 4, 1936.

DRONKERS, Jo Johannis, Dutch mathematician; b. Poortvliet, The Netherlands, May 24, 1910; s. Willem and Cornelia (van der Slikke) D.; doctorate in math. U. Leiden, 1934; m. Desiree Westenburg, July 12, 1945; children—Hillegonda, Job, Marius, Cornelia. Mathematician, Netherlands Rijkswaterstaat (Pub. Works and Waterways Dept.), The Hague, 1939-50, chief mathematician, 1950-63, head hydraulic dept. of Deltaworks, 1963—. Cons., Netherlands Engring. Consultance in India, Korea, Can., Japan. Recipient Conrad medal Royal Inst. Engrs. of Netherlands, 1965. Author: Tidal Computations in Rivers and Coastal Waters of North Holland, 1964; also articles. Theoretical studies on computation of propagation of tidal waves in rivers and coastal waters, hydraulic studies on bldg. of dams in sea-arms inlets and riverbranches., statis. studies on extreme values. Home: 180 Ruychrocklaan, The Hague. Office: 60 Hogenhoucklaan, The Hague, The Netherlands.*

DROWN, Thomas Messinger, Am. chemist; b. Phila., Mar. 19, 1842; s. William Appleton and Mary E. M. (Pierce) D.; grad. Phila. High School, 1859; M.D., med. Dept., U. Pa., 1862. Practiced medicine a short time; studied chemistry at Yale and Harvard, chemistry and metallurgy at Freiberg, Saxony, School of Mines and at Heidelberg; LL.D., Columbia, 1895; m. Helen Leighton. Analytical chemist, Phila., several years; prof. analytical chemistry, Lafayette Coll., Easton, Pa., 1874-81; prof. chemistry, Mass. Inst. Tech., 1885-95; pres. Lehigh U., Bethlehem, Pa., from 1895. Sec. Am. Inst. Mining Engrs., 1873-83; in charge chem. dept., Mass. State bd. of health, 1887—; pres. Am. Inst. Mining Engrs., 1897-98. Research and publs. on analytical chemistry and san. aspect of city water supplies. Died Nov. 16, 1904.

DROZ, Pierre Jean, engraver, inventor; b. Chaux-de-Fond, Switzerland, 1746; studied at Paris, France; children include Jules Antoine. Mechanic, engraver; settled in Paris, 1766; began work under Watt and Boulton in Eng., 1790; made English coins; chief engraver pub. medals and coins during Empire in France. Invented circular scissors; improved material of coinage die under Louis XVI. Died Mar. 2, 1825.

DRUCE, George Claridge, Brit. naturalist; b. Potters Pury, Eng., 1850; M.A., D.Sc., Oxford U.; LL.D., St. Andrews; Fielding curator Oxford U.; mem. bd. examiners Pharm. Soc. Gt. Britain; pres. Brit. Pharm. Conf., 1900-01. Fellow Royal Soc., Bot. Soc. Edinburgh (hon.); pres. Brit. Assn., Oxford, 1926, Archeol. Soc., Oxfordshire; Ashmolean Natural History Soc., Oxfordshire; mem. Bot. socs. Geneva, Czechoslovakia; hon. mem. Birmingham, Banbury, Chiltern, Northamptonshire natural history socs. Author: The Flora of Northamptonshire, 1879, 30, of Oxfordshire, 1886, 2d edit., 1927, of Berkshire, 1897; An Account of the Herbarium in the University of Oxford; The Flora of West Rossshire; An Account of the Dillenian Herbaria; (with Vines) An Account of the Morisonian Herbarium and Biography of Morison and the two Bobarts, 1914; North African Experiences; British Plant List, 1908; Hayward's Botanist's Pocket Book, 1909; The Dubious Plants of Britain; Flora Zetlandica, Mosses and Hepatics of Oxfordshire; Flora of Buckinghamshire; Comital Flora of the Birtish Isles, 1932; also articles. Editor yearly reports Brit. Bot. Soc. Discoverer of Bromus interruptus, Potamogenton Drucei. Died Feb. 29, 1932.

DRUCKER, Daniel Charles, Am. engineer; b. N.Y.C., June 3, 1918; s. Moses Abraham and Henrietta

(Weinstein) D.; B.S., Columbia U., 1937, C.E., 1938, Ph.D., 1940; m. Ann Bodin, Aug. 19, 1939; children—David, Miriam. Instr., Cornell U., 1940-43; supr. mechs. solids Armour Research Found., 1943-45; asst. prof. Ill. Inst. Tech., 1946-47; faculty Brown U., Providence, prof. engring., 1950-—, chmn. div. engring., 1953-59, chmn. phys. scis. council, 1961-63, L. Herbert Ball Univ. prof., 1963-—. Cons. to indsl. firms, govtl. agys.; mem. Providence Bldg. Ordinance Bd. Rev., 1964-—; chmn. U. S. Nat. Com. on Theoretical and Applied Mechanics, 1967-—; Marburg lectr. Am. Soc. Testing and Materials, 1966. Guggenheim fellow, 1960-61; recipient Illig medal Columbia U., 1938; Lamme award 1967. Fellow Am. Acad. Arts and Scis.; mem. Nat. Acad. Engring., American Society C.E. (von Karman medal 1966), A.A.A.S., Am. Inst. Aeros. and Astronautics, Am. Soc. C.E. (past pres. New Eng. council), Am. Soc. E.E., Am. Soc. M.E., Nat. Soc. Profl. Engrs., Soc. Exptl. Stress Analysis (past pres.), Soc. Rheology, Sigma Xi, Nat. Acad. Scis. Author: Introduction to Mechanics of Deformable Solids, 1967; also numerous articles. Pioneered research and devel. three-dimensional photoelasticity, exptl. photoelastic techniques; research on formulation of gen. theorems of analysis and design in plastic range, basic concept stability of inelastic material; solved early fundamental problems of plastic behavior of shells and three-dimensional continua. Home: 230 Arlington Av., Providence 02906.*

DRUCKREY, Hermann, German pharmacologist; b. Greifswald, Germany, July 27, 1904; s. Otto and Dorothea (Bettermann) D.; student U. Giessen (Germany), 1926-28, U. Heidelberg (Germany), 1928-29; M.D., U. Leipzig (Germany), 1932; m. Annemarie Eick, Aug. 29, 1936; children—Eike, Inge. Guest asst. Inst. Exptl. Pathology, Prague, Czechoslovakia, 1929-33, Inst. Organic Chemistry, Göttingen, Germany, 1933; asst. prof. Pharmacological Inst., Berlin, 1933-43; prof. pharmacology and toxicology U. Berlin, 1942-48; dir. labs. Chirurgical Clinic, Freiburg, Germany, 1948-64; dir. Forschergruppe Praeventiv Medizin, Max Planck Institut für Immunbiologie, Freiburg, 1964-—. Recipient Scheele medal biochemistry, Stockholm, Sweden, 1955. Mem. Royal Soc. Medicine London (affiliate), Italian Cancer Soc., Internat. Commn. on Cancer Research (past chmn.). Author: (with K. Küpfmüller) Dosis und Wirkung, 1949; Fortpflanzung und Wachstum, 1959; also numerous articles. Research in exptl. biochem., and theoretical pharmacology, etiology, prevention, pathogenesis, and chemotherapy of cancer. Home: 18 Reinhard Booz Str., Freiburg Merzhausen 7802, West Germany. Office: 8 Stefan Meier Str., Freiburg 78, W. Germany.*

DRUDE, (Carl George) Oscar, German botanist; b. Brunswick, Germany, June 5, 1852; s. Carl and Therese (Giesecke) D.; began studies natural scis. and chemistry Collegium Carolinum (now Tech. U. Braunschweig), 1870; studied at Göttingen, Germany, from 1871; doctorate 1873; m. Lydia Coester, 1879; 3 sons, 3 daus. Herbarium asst. to F. G. Bartling in Göttingen; became lectr., 1876; apptd. dir. Dresden (Germany) Bot. Garden, also prof. botany Dresden Polytechnikum (now Tech. U.), 1879; emeritus, 1920; dir. botanical gardens there, from 1921. Mem. Isis, Flora, Landesverein Sächsischer Heimatschutz. Author: Atlas der Pflanzenverbreitung, 1887; Handbuch der Pflanzengeographie, 1890; Deutschlands Pflanzengeographie, Vol. I, 1896; Die Ökotogie er Pflanzen, 1913. Editor: (with Ad. Engler) Die Vegetation der Erde, 1902 (author Vol. 6). Research on plant geography especially taxonomy of palms and umbellifers; created bot. garden in Dresden using systematic arrangement of plants; a founder of plant ecology. Died Dresden, Feb. 1, 1933.

DRUDE, Paul Karl Ludwig, German physicist; b. Brunswick, Germany, July 12, 1863; s. Carl and Luise (Schrodt) D.; began studies Göttingen, Germany, 1822; student Freiburg and Berlin, Germany; doctorate Göttingen, 1887; m. Emilie Regelsberger, 1894; 1 son, 3 daus. Became asst. Math.-Phys. Inst., Göttingen, 1887, apptd. lectr., 1889; asso. prof. tech. physics Leipzig, 1894-1900; prof. physics Giessen (Germany), 1900-05; prof. Berlin after 1905. Mem. Berlin Acad. Scis. Author: Lehrbuch der Optik, 1900; Die Physik des Äthers auf elektromagnetischer Grundlage, 1884; Optische Theorie in der Physik, 1895; Lehrbuch der Optik, 1900. Became editor Annalen der Physik, 1900. Research on optics, relationship between dielectricity constants and optical refraction exponents, between conductivity, absorption and refraction elec. waves in liquids; developed theory of electrons, physics of metals based on electronic theory; applied Maxwell's electromagnetic theory of optics. Died Berlin, July 5, 1906.

DRUM, Ryan, Am. biologist; b. Milw., Sept. 25, 1939; s. Warren William and Virginia (Kruger) D.; B.Sc. in Chemistry, Ia. State U., 1961, Ph.D. in Phycology, 1964; m. Jane Isabel Rowley, Mar. 1, 1961; 1 dau., Hillary Jane. NATO fellow U. Bonn (Germany), 1964-65, U. Leeds (Eng.), 1965; asst. prof. U. Ark., 1966; asst. prof. botany U. Mass., Amherst, 1966-—. Mem. Internat. Limnological Soc., Bot. Soc. Am., Phycological Soc. Am., A.A.A.S., Internat. Phycological Soc. Author: (with Pankratz, E. Stoermer) Electron Microscopy of Diatom Cells, Vol. 6, 1966; also articles. Research on electron microscopy of diatom cytoplasm, diatom locomotion using electron microscopy, microcinematographic techniques and drugs; discovered an endosymbiotic blue-green alga in a diatom genus. Home: Box 78-A, Deerfield, Mass. 01342. Office: Botany Dept., U. Mass., Amherst, Mass. 01003.*

DRUMMOND, Sir David, physician; b. Dec. 1852; s. David Drummond; ed. Trinity Coll., Dublin, univs. Prague, Vienna, Strasburg; M.A., M.D., D.C.L., LLD.; m. Margaret Horsley; 6 sons, 2 daus. Former vice chancellor U. Durham, also past pres., emeritus prof. medicine Coll. Medicine; cons. physician Royal Victoria Infirmary, Newcastle-on-Tyne; former physician Sick Children's Hosp., Newcastle; sr. physician Northumberland War Hosp. Mem. Royal Commn. on Lunacy, 1925-26. Fellow Royal Soc. Medicine, Royal Acad. Medicine (Ireland); mem. Brit. Med. Assn. (pres. 1921), Assn. Physicians Gt. Britain and Ireland (pres.). Author: Diseases of the Brain and Spinal Cord; Paralysis from Peripheral Neuritis; Anaesthesia in Functional and Organic Disease of the Nervous System, also numerous articles. Introduced (with James Rutherford Morison) omentopexy to relieve ascites due to cirrhosis of liver (known as Morison-Talma operation), 1896. Died Apr. 28, 1932.

DRUMMOND, Sir Jack Cecil, English biochemist; b. Jan. 12, 1891; s. J. Drummond; ed. Queen Mary's Coll., King's Coll., U. London; D.Sc., London; hon. doctorate U. Paris; m. Anne Wilbraham, 1940; 1 dau. Research asst. King's Coll., London, 1913; became research asst., biochem. dept. Cancer Hosp., Research Inst., London, 1914; dir. biochem. research Cancer Hosp., London, 1918; became reader in physiol. chemistry Univ. Coll., London, 1919, prof. biochemistry, 1922-45; dir. research, dir. Boots Pure Drug Co. Ltd., from 1946. Sci. adviser Ministry Food, 1939-46; Lane lectr. Stanford (Cal.) U., 1933; Harvey lectr., N.Y., 1933; Fullerian prof. physiology Royal Instn., 1942-44; adviser on nutrition Allied Post-War Requirements Bur. and SHAEF, 1944-45; adviser on nutrition Control Commn. for Germany and Austria (Brit. element), 1945-46. Knighted, 1944. Fellow Royal Soc., 1944, Royal Inst. Chemistry; mem. N.Y. Acad. Sci. (hon.). Author: (with Anne Wilbraham) The Englishman's Food. Recognized existence of vitamin C, 1919; suggested vitamine be spelled vitamin (accepted 1920). Died Aug. 4, 1952.

DRUMMOND, Thomas, British engr.; b. Edinburgh, Scotland, Oct. 10, 1797; s. James and Elizabeth (Somers) D.; studied at U. Edinburgh, math. and chemistry at Royal Instn.; m. Miss Kinnaird, Nov. 19, 1835; children—Mary Elizabeth, Emily, Fanny. Became cadet at Woolwich, 1813; entered Royal Engrs., 1815; joined ordnance survey of Col. Thomas Frederick Colby, 1820; head boundary commn. Gt. Reform Bill, 1832-34; became pvt. sec. to Lord Althoys, the chancellor of the Exchequer; named undersec. of Dublin Castle, 1835; gov. of Ireland, 1835-40. Statue erected in his honor in City Hall, Dublin, 1843. Invented a limelight (Drummond's light), utilizing incandescent properties of lime; demonstrated its use for lighthouse; improved the heliostat; helped perfect Colby-Drummond compensation bar. Died Dublin, Ireland, Apr. 15, 1840.

DRURY, Dru, English naturalist; b. London, Eng., Feb. 4, 1725; son of silversmith and Mary (Hesketh); m. Esther Pedley, June 7, 1748; 17 children including Mrs. André. Silversmith in Strand; entomol. collector. Fellow Linnean Soc. Author: Thoughts on the Precious Metals, 1801; Illustrations of Natural History Exhibiting upwards of 240 figures of Exotic Insects, 1770-82; Illustrations of Exotic Entomology, 1837; Directions for Collecting Insects in Foreign Countries, 1800. Research on natural history and entomology especially libelludidae and insects of Sierra Leone. Died Dec. 15, 1803.

DRUYVESTEYN, Mari Johan, Dutch physicist; b. Groningen, Netherlands, Feb. 13, 1901; s. Willem Frederik and S. C. (Mees) D.; student U. Utrecht (Netherlands), 1919-25; doctor's degree U. Groningen, 1928; dr. honoris causa in tech. sci. U. Brussels (Belgium); m. M. A. H. Nievergeld, Nov. 16, 1927; children—Bianca Maria, Willem Frederik. Staff, Philips Lab., 1927-45; prof. Physics, Technol. U. Delft (Netherlands), 1945-66. Author: (with F. M. Penning) The Mechanism of Electrical Discharges in Gases of Low Pressure; also articles. Research on gas discharges, including velocity distbn. of electrons in an electric field, Ne-He bands; solid state physics, including internal oxidation of metals, influence of imperfections on properties of metals.*

DRYANDER (A[E]ICHMANN), Johannes, German anatomist, astronomer; b. Wetter, Germany, June 27, 1500; grad. U. Erfurt, Germany, 1518; student Bourges, Paris, France, 1528-33; M.D., U. Mainz, Germany, 1533; m. Susanne Bruel; 5 children. Lectr. math., astronomy, Paris, 1528-33; physician to Johannes von Metzenhausen, Coblenz, Trier, Germany; prof. medicine, math. U. Marburg, Germany, from 1535, rector, 1548-60; staff, hosps., Haina, Merxhausen, Germany. Author: Ein new Artzney und Practicyr Büchlein . . . , 1527; De Balneis Emsensibus, 1535; Anatomia capitis humani, Novi annuli astronomici nuper anno 29 excogitati, Zubereitung unnd warer verstanndt eines Quadranten, 1536; Annulorum trium diversigeneris instrumentorum astronomicorum, componendi ratio atque usus, 1537; Anatomia Mundini, 1541; Der gantzen Artzeney gemeiner Inhalt . . . , 1542; Vom rechtem christlichen Brauch des Artztes und der heylsamen Artzeney, 1543; Libellus de peste, 1553. One of 1st to conduct dissections in Germany, illustrate anatomy texts with own drawings; inventor tripie sun-dial. Died Marburg, Dec. 20, 1560.

DRYANDER, Jonas, botanist; b. Sweden, 1748; studied at U. Gottenburg (Sweden); grad. Lund (Sweden) U.; studied under Linnaeus; tutor to nobleman; came to Eng., became librarian to Sir Joseph Banks, in 1782; later became librarian to Royal Soc. A founder, 1st librarian, v.p., and chief author of laws of Linnean Soc. Genus Dryandra named in his honor. Author: Catalogus bibliothecae historico-naturalis Josephi Banks, Baronetti, 1796-1800; prin. author of 1st edit. of Acton's Horties Kewensis, 1789. Editor: Plants of the Coromandel Coast (Roxburgh), 1795-1798. Among 1st to explain anat. definitions by figures; invented several astron. instruments. Died London, Eng., Oct. 19, 1810.

DRYDEN, Charles Early, Jr., Am. chem. engr.; b. Hagerstown, Md., Oct. 15, 1917; s. Charles Early and Mathilde (Krieger) D.; B.Ch.E. summa cum laude, Drexel Inst. Tech., 1939; M.Ch.E., Princeton, 1943; Ph.D., Ohio State U., 1951; m. Rosalie Ewing, Dec. 22, 1957; children—Donald, John, Mark, Cynthia. Devel. engr. Silmo Chem. Corp., Vineland, N.J., 1935-41, M. W. Kellogg Co., Jersey City, 1942-43; head process devel. Napco Chem. Corp., Harrison, N.J., 1943-48; asst. div. chief Battelle Meml. Inst., Columbus, O., 1948-54; prof. chem. engring. Ohio State U., Columbus, 1954-67; vis. prof. Indian Inst. Tech., Kanpur, 1963-65. Cons. Wright AFB, Battelle Meml. Inst., Nat. Steel Corp., Gen. Electric Co., E. I. DuPont. Recipient Drexel Inst. Tech. Distinguished Alumni award. Mem. Am. Chem. Soc., Am. Inst. Chem. Engring. (sec. nuclear engring. div. 1956-61), Am. Soc. Engring. Edn. (chmn. nuclear engring. div. 1961-63), Instrument Soc. Am., Am. Nuclear Soc. Author: (with F. C. Vilbrandt) Chemical Engineering Plant Design, 1959; Outlines of Chemical Technology, 1965; Chemical Engineering Cost Handbook, 1966. Contbr. numerous articles to sci. jours. Devel. chem. and pharm. proceses from lab. research data to large-scale plants producing such items as vitamins A, D, B-complex, petrochems., titanium, zirconium. Patentee in field. Died Columbus, Sept. 23, 1967.

DRYDEN, Hugh Latimer, Am. physicist; b. Pocomoke City, Md., July 1, 1898; s. Samuel Isaac and Nova Hill (Culver); D.; A.B., Johns Hopkins, 1916, A.M., 1918, Ph.D., 1919, LL.D., 1953; Sc.D., Poly. Inst. Bklyn., 1949, U. Pa., 1951, Western Md. Coll., 1951; D.Eng., N.Y.U., 1950, Rennsalaer Poly. Inst., 1951, U. Md., 1955; m. Mary Libbie Travers, Jan. 29, 1920; children—Hugh Latimer, Mary Ruth, Nancy Travers. With Nat. Bur. of Standards, 1918-47, asso. physicist, chief of aerodynamics sect., 1920-34, chief of mechanics and sound div., 1934-46, asso. dir., 1946-47; dir. Nat. Adv. Com. for Aeros., 1947-58; dep. adminstr. Nat. Aeros. and Space Adminstrn., 1958-—; U. S. del. adv. group aero. research and development, NATO. Mem. Nat. Inventors Council, Dept. of Commerce, Wilbur Wright lectr., Royal Aero. Soc. Soc., London, 1949. Hon. officer Order Brit. Empire. Recipient Medal of Freedom, 1946; Daniel Guggenheim medal, 1951; Wright Brothers Meml. Trophy, 1955; Nat. Civil Service League Career award, 1958; Rice medal, Am. Ordnance Assn., 1958; Langley medal Am. Philos. Society, 1962; Rockefeller Public Service award, 1962. Hon. fellow Inst. Aero. Scis. (pres. 1943), Royal Aero. Soc. (London); home sec. Nat. Acad. Scis.; mem. Rocket Soc., Internat. Council Aero. Scis., Nat. Geographic Soc. (life mem., board trustees), Am. Soc. M.E., Am. Phys. Soc., Washington Acad. of Sciences, Philos. Soc. (Washington), Phi Beta Kappa, Sigma Xi. Author books; contbr. to tech. jours. Research in aerodynamics, especially boundary layer flow, wind tunnel turbulence, wind pressure on structures. Home: 5606 Overlea Rd., Washington 16. Address: 4th and Maryland Av. S.W., Washington 25.

DRYGALSKI, Erich Dagobert von, see von Drygalski, Erich Dagobert.

DRYSDALE, Thomas Murray, Am. surgeon; b. Phila., Aug. 14, 1831; grad. Pa. Med. Coll., 1852; m. Mary L., d. Dr. Washington Atlee, Oct. 1857. Prof. chemistry, Wagner Inst. Sci., 1855; lectr. microscopy Franklin Inst., 1862; cons. gynecologist, Medico-Chirurg. Hosp.; cons. surgeon Girard Coll. Mem. Am. Gynecol. Soc. (a founder 1876), Internat. Med. Congress. Specialist in surgery and gynecology; discovered and described ovarian cell which exists in ovarian tumors. Died 1904.

DUANE, William, Am. physicist; b. Phila., Feb. 17, 1872; s. Charles Williams and Emma Cushman (Lincoln) D.; A.B., U. Pa., 1892, Sc.D., 1922; A.B., Harvard, 1893, A.M., 1895; postgrad. univs. Berlin and Göttingen, 1895-97, Ph.D., Berlin, 1897; Sc.D., U. Colo., 1923; m. Caroline Elise Ravenel, Dec. 28, 1899; children—William, Arthur Ravenel (dec.), John P., Margaretta C. Asst. in physics, Harvard, 1893-95, Tyndall fellow, 1895-97; prof. physics U. Colo., 1898-1907; research, Curie Radium Lab., U. Paris, 1907-12; asst. prof. physics Harvard, 1913-17, prof. bio-physics, from 1917. Mem. Am. Acad. Arts and Scis. Recipient John Scott medal and premium, 1922; Comstock prize Nat. Acad. Sci., 1922; first Leonard prize, Am. Roentgen Ray Soc., 1923; (all three prizes for researches in radioactivity and X-rays). Discovered Duane Hunt Law (there is a sharp upper limit to frequency of X-rays emitted from a target under electron bombardment). Died Mar. 7, 1935.

DUBARRY, Jean-Jacques, French physician; b. Casteljaloux, France, Sept. 4, 1906; s. Raoul and Armande (Malbec) D.; student Bordeaux (France) U., 1923-34, M.Medicine, 1939; m. Colette Richard, Oct. 21, 1937; children—Annie (Mrs. Michel Roux-Dessarps), Christine (Mrs. Daniel Datcharry), Brigitte (Mrs. Bernard Babin), Bertrand, Francois, Florence, Michel. Chief clin. medicine Bordeaux Hosps., since 1936—, staff, 1942—; prof. hydroclimatology U. Bordeaux, since 1953—; staff Clinique Médicale de l'Appareil Digestif, 1958—. Dir. Bordeaux Savs. Bank. Decorated chevalier Légion Honour; chevalier Italian Merit; officer Pub. Edn. and Social Merit. Mem. Soc. Gastro-Entérology France (past vice chmn.), Soc. Gastro-Enterology Belgium, Soc. Gastroenterology Spain, Latin Med. Union (vice chmn. 1964—), Author: Globular Sedimentation, 1933; Formulary of Gastro-enterology, 1949; Intestinal Tuberculosis, 1951 (with M. D. Roux); also numerous articles. Research on genetic factors, atypical forms, ulceriform states and therapeutics of ulcerous diseases, digestive cancers in the same family, digestive polyposis, ulcero-hemorrhagic recto-colitis, functional biliary states, cirrhosis therapeutics, hydroclimatology and thalassotherapy, thermo-mineral water of Rochefort sur mer, penetration of chem. elements through teguments, atmospheric ionization, thermal and seamads, biologic properties. Home and office: 126 rue de Saint Genés, Bordeaux, Gironde, France.*

DUBINI, Angelo, Italian physician; b. Milan, Italy, 1813; described electric chorea characterized by sudden and violent movements (Dubini's disease), also myoclonic form of epidemic encephalitis, 1846. Died Cassano Magnago, Italy, 1902.

DUBININ, Mikhail Mikhaylovich, Russian phys. chemist; b. Jan. 1, 1901; grad. Moscow Practical Acad. Comml. Sci., 1917, Chemistry Faculty, Moscow Higher Tech. Coll., 1921; D.Chem. Sci., 1936. Instr., asst., sr. asst., lectr. dept. chemistry Moscow Higher Tech. Coll., 1921-32; prof., dept. dir. Mil. Acad. Chem. Def., 1932-49; prof., 1933—; dir. lab. sorption processes Inst. Phys. Chemistry, USSR Acad. Sci., 1946—, acad. sec. dept. chem. sci., 1948-57. Mem. Mendeleev All-Union Chem. Soc. (pres. 1946-50), USSR Acad. Sci. (presidium mem. 1960-63). Recipient Stalin prize, 1942, 50. Author: The Physical and Chemical Principles of Sorption Techniques, 1935; The Sorption and Structure of Active Carbons, 1947; co-author: The Physical and Chemical Principles of Anti-Gas Work, 1939. Research on adsorption phenomena adsorption of gases and vapors, developed method of calculating installations for separation of gaseous vapors, 1932-35, role of ultraporosity of adsorbent in adsorption of vapors from substances with molecules of unequal size, 1936-37, solution of various theoretical and practical problems of anti-chem. defense, classified structural types of adsorbents, 1935-49, porous structure and adsorption properties of active carbons, 1949-67, primary and secondary structure of synthetic zeolites and their adsorption properties, 1961-66, others. Address: Inst. of Physical Chemistry, USSR Acad. Sci., Leninsky prospect 31, Moscow, USSR.

DUBININ, Nikolay Petrovich, Russian biologist; b. Jan. 4, 1907; grad. Moscow U., 1928. With Moscow Zootech. Inst., 1928-32, with Inst. Cytology, Histology and Embryology, USSR Acad. Sci., 1932-48, Inst. Forestry, 1949-55, Inst. Biophysics, 1955-58, dir. Siberian dept. Inst. Cytology and Genetics, USSR Acad. Scis., 1958-60, Presidium mem. Siberian dept., 1958—; prof., 1935—. Recipient Darwin medal, 1959. Mem. USSR Acad. Sci. (corr.). Author: Birds of the Lower Ural Valley Forests, 1953; Problems of the Physical and Chemical Bases of Heredity, 1956; Problems and Tasks of Radiation Genetics, 1956; Birds of the Ural Valley Forests, 1956. Research in genetics, cytogenetics, genetic principles of selectivity, and theory of evolution. Address: Siberian Dept., USSR Acad. Sci., Norosibirsk, RSFSR, USSR.

DUBIS (or DUBOIS), Paul Antoine, French obstetrician; b. Paris, France, Dec. 7, 1795; s. Antoine Dubis; M.D., Paris, 1818; joined Faculty Paris, 1823; prolific writer obstetrics, particularly cesarean sections; known for simplicity of style. Died Paris, Dec., 1871.

DUBISCH, Roy, Am. mathematician; b. Chgo., Feb. 5, 1917; s. Otto and Flora (Gossing) D.; B.S., U. Chgo., 1938, M.S., 1940, Ph.D., 1943; m. Joyce Marie Nielsen, Nov. 18, 1939; children—Jill Susanna, Russell John, Ralph Emiel. Instr. math. Mont. State U., 1942-46; asst. prof. math. Syracuse (N.Y.) U., 1946-48; asso. prof. math. Fresno (Cal.) State Coll., 1948-55, prof., chmn. dept., 1955-61; prof. math. U. Wash., Seattle, 1961—. Mem. Nat. Council Tchrs. Math. (v.p. 1963-64), Math. Assn. Am. (gov. 1964-67), Am. Math. Soc., A.A.A.S., Phi Beta Kappa, Sigma Xi. Author: Nature of Number, 1952; Trigonometry, 1955; Intermediate Algebra, 1961; Teaching of Mathematics, 1963; Lattices to Logic, 1964; Introduction to Abstract Algebra, 1965. Editor: Math. Mag., 1964—. Home: 19248 93d Pl. W., Edmonds, Wash. 98020. Office: U. Seattle, Seattle, Wash.*

DUBLIN, Louis Israel, statistician; b. Kovno, Lithuania, Nov. 1, 1882; s. Max and Sarah (Rosenzweig) D.; brought to U. S., 1886; B.S., Coll. City N.Y., 1901; Ph.D., Columbia, 1904; m. Augusta Salik, Apr. 5, 1908; children—Elizabeth, Mary, Thomas, Amos. Instr. math. Coll. City of N.Y., 1901-08; statistician Met. Life Ins. Co. from 1911, 3d v.p., 1931. Lectr. on vital statistics, Yale. Fellow Am. Pub. Health Assn. (pres. 1932); fellow Am. Statis. Assn. (pres. 1924), Population Assn. Am. (pres. 1935-36), chmn. bd. Am. Museum of Health. Author: Mortality Statistics of Insured Wage Earners and Their Families, 1919; Health and Wealth, 1928. Collaborator (with Lee K. Frankel and Miles M. Dawson) Workingmen's Insurance in Europe, 1910; (with Lee K. Frankel; Principles of Life Insurance, 1911; (with A. J. Lotka) The Money Value of a Man, 1930; To Be or Not To Be—A Study of Suicide (with B. Bunzel), 1933; Length of Life (with A. J. Lotka), 1936; Twenty-five Years of Health Progress (with A. J. Lotka), 1937; A Family of Thirty Million, 1943. Also many monographs, papers and addresses on vital statistics and public health. Editor: Population Problems in the United States and Canada, 1926; The American People—Studies in Population, 1936. Devised system for birth and death registration used throughout U. S.; expert on race and occupational mortality, incidence of mortality from diseases. Home: Westport, Conn. Office: 1 Madison Av., N.Y.C.

DUBOIS, Arthur Brooks, Am. physiologist, physician; b. N.Y.C., Nov. 21, 1923; s. Eugene Floyd and Rebeckah (Rutter) DuB.; student Harvard, 1941-43; M.D., Cornell U., 1946; m. Roberdeau Callery, June 21, 1950; children—Anne Roberdeau, Brooks, James E. F. Life ins. Med. Research fellow physiology, Rochester, N.Y., 1949-51; faculty U. Pa. Grad. Sch. Medicine, Pitts., 1952—, prof. medicine, 1958—, prof. physiology, 1960—. Cons. Phila. Gen., U. S. Naval hosps., Phila., 1958—. Recipient Research Career award NIH, 1963. Mem. Assn. Am. Physicians, Am. Soc. for Clin. Investigation, Am. Physiol. Soc. Author: (with Comroe, Forster, Briscoe, Carlson) The Lung, 1955; also articles. Originated several methods for measuring lung function in health and disease, including body plethysmographic methods for airway resistance, lung vol. and pulmonary capillary bloodflow; diving and space physiology. Home: 505 W. Chestnut Hill Av., Phila. 19118.*

DUBOIS, Charles-Frédérick, naturalist; b. Barmen, Germany, 1804; author of Planches colorées des oiseaux de l'Europe; Catalogue systématique des Lepidoptères de la Belgique. Research on butterflies and birds; pub. colored plates on butterflies and their larvae. Died Brussels, Belgium, 1867.

DU BOIS, Cora, Am. anthropologist; b. N.Y.C., Oct. 26, 1903; d. Jean Jules Philip and Mattie (Schreiber) Du Bois; B.A., Barnard Coll., 1927; M.A., Columbia, 1928; Ph.D., U. Cal. at Berkeley, 1932; Sc.D., Wilson Coll., 1958; LL.D., Mills Coll., 1959; H.H.D. (hon.), Mt. Holyoke Coll., 1960. Research asso. anthropology U. Cal., 1932-35, NRC fellow, 1935-36; tchr. Hunter Coll., 1936-37; Social Sci. Research fellow Columbia, 1937-39; tchr. Sarah Lawrence Coll., 1939-42; br. chief OSS, also Dept. State, 1942-50; social sci. cons. WHO, UN, 1950-51; dir. research Inst. Internat. Edn., 1951-54; Zemurray-Stone prof. anthropology Harvard and Radcliffe Coll., 1954—; vis. prof. U. Cal., 1948, U. Colo., 1954; Carnegie vis. prof. U. Hawaii, 1957. Fellow Center for Advanced Study Behavioral Scis., Palo Alto, Cal., 1958-59; field research in India, 1961-63. Recipient exceptional civilian service award, U. S. Army, 1946; Order Crown of Thailand 3d class, 1949; Achievement award Am. Assn. U. Women, 1961. Mem. Far Eastern Assn. (dir.), Am. Acad. Arts and Scis., Am. Anthropol. Assn. (vice pres. 1949), Phi Beta Kappa, Sigma Xi. Author: People of Alor, 1944; Social Forces in Southeast Asia, 1949; Fgn. Students and Higher Education in the United States, 1956. Editor: Lowie's Selected Papers in Anthropology, 1960. Contbr. profl. jours. Studies on "revitalization" (millenial) religious cults among west coast Indians; ethnography of Cal. Indians, of Indonesia; contbns. to field of personality and culture. Home: 20 Coolidge Hill Rd., Cambridge 38, Mass.

DUBOIS, Emmanuel, French physicist; b. Paris, France, Nov. 19, 1895; s. Emmanuel and Marie Sophie Thérèse (Thiebault) D.; Lic. Sc., Faculty Scis., Paris, 1915; diploma in advanced phys. studies École Normale Supérieur, 1917, Sc.D., 1923; m. Marthe Marie Anne Besnier, 1933; 3 children. Asst., Faculty Scis., Paris, 1918-19; with Central Sch. Art and Manufacture, 1920-24; lectr. Faculty Scis., Caen, France, 1924-31; prof. gen. physics Faculty Scis., Clermont-Ferrand, France, from 1931, dean, from 1934. Mem. French, Am. phys. socs. Author: Recherches sur l'effet Volta, 1930; Recherches sur les ondes issues de sources explosives, 1947. Address: 34 Av. Carnot, Clermont-Ferrand, Puy de Dôme, France.

DUBOIS, Eugene Floyd, Am. physiologist; b. West New Brighton, N.Y., June 4, 1882; s. Eugene and Anna Greenleaf (Brooks) DuB.; A.B., Harvard 1903; M.D., Coll. Phys. and Surg. (Columbia), 1906; Sc.D., U. Rochester (N.Y.), 1948; m. Rebeckah Rutter, June 4, 1910. Instr. applied pharmacology, Cornell U. Med. Coll., N.Y., 1910-17, asso. prof. medicine, 1919-30, prof. medicine, 1930-41, prof. physiology, 1941-50; med. dir. Russell Sage Inst. Pathology, 1913-50; dir., vis. physician Second Med. Div. of Bellevue Hosp., 1919-32; cons. physician, from 1932; physician-in-chief N.Y. Hosp., 1932-41. Chmn. com. on aviation medicine NRC, 1940-45. Decorated Navy Cross. Mem. Assn. Am. Physicians, Am. Assn. Clin. Investigation, Am. Physiol. Soc., Nat. Acad. Scis. Author: Basal Metabolism in Health and Disease, 1924, 27, 36; The Mechanism of Heat Loss and Temperature Regulation. Research in metabolism; established basal metabolism standards shown in tabular form (calories per sq. meter per hour), circa 1924. Died Feb. 1959.

DUBOIS, Georges, French geologist; b. Armentières, France, Sept. 10, 1890; s. George H. A. and Philomène (Lampin) D. ed. U. Lille (France); licence ès scis., 1910, aggregation, 1912; m. Camille Droulez, Aug. 28, 1919; 3 children. Became asst. geology dept. Lille Faculty Scis., 1919, prof. gen. geology, 1928; dir. Commn. for Geol. Map. Mem. French Geol. Soc., Northern Geol. Soc., Soc. Biogeography. Author: Les terrains quaternaires du Nord de la France, 1924; also more than 10 maps. Investigated geol. history of No. France and Alsace-Lorraine; specialized in quaternary formations; precursor of paleontology through studies of fossil pollen. Died Oct. 2, 1953.

DU BOIS, Henri, physicist; b. Velz, Belgium, June 24, 1863; ed. at Delft, Netherland, Glasgow, Scotland; Ph.D., Strasbourg, France, 1887; prof. at Berlin, Germany, Utrecht, Netherlands. Mem. Amsterdam Acad. Sci. Author: Magnetic Circuit, 1896. Research on optics, magnetic balance, half-ring electro-magnet, ironclad galvanometer. Died Utrecht, 1918.

DUBOIS, Jacques (Sylvius), French physician; b. Souvilly, France, 1478; studied under Güther of Andernach, Paris, France; prof. medicine Collège Royal (now Collège de France). Author: Methodus medicamenta . . . , 1541; Pharmacopoeia, 1542; also compiled pharm. works from Galen and classical authors. Used intravascular injections to demonstrate complex structure of arterio-venous network; used human cadavers for anat. studies; supported Galenic system. Died Paris, Jan. 13, 1555.

DUBOIS, Jacques Emile, French chemist; b. Lille, France, Apr. 13, 1920; s. Paul Eugene and Emilienne (Cherrier) D.; Ph.D., U. Grenoble (France), 1947; Ramsey fellow Univ. Coll., London, also sci. adviser to French cultural counsellor in London, 1948-50; m. Bernice Claire Shaaker, May 24, 1952; children—Rhoda Nicole, Alain. Prof. phys. chemistry and petrochemistry, also dir. Chemistry Inst., U. Saar, 1949-57; dean sci. faculty U. Paris (France), 1952-57, prof. phys. organic chemistry, 1957; research fellow Columbia, 1956; sci. adviser to French minister edn., 1962-63, joint dir. higher edn., 1963-65; dir. research Ministry Def., 1965—. Mem. directorate Nat. Research Council France, 1963—; mem. Internat. Com. Electrochem. Thermodynamics and Kinetics, 1966—; bd. dirs. Palais de la Découverte, 1961—, Inst. de Biologie physico-chimique, 1967—. Decorated chevalier Legion of Honor, Médaille de la Résistance, comdr. des Palmes Académiques; recipient Jecker, prize, also Berthelot medal French Acad. Scis., Stas medal Belgian Chem. Soc. Mem. French (council 1965—; Le Bel, also Ancel prize), Soc. Phys. Chemistry (council 1967—), Am. Chem. Soc. Author chpts. in Internat. Chemistry Ency. Research and publn. on kinetics, fast reaction rates, electrochemistry; patentee automatic titrating apparatus in polarovoltry, Titravit; created Chem. Inst. of Sarrebruck and trilingual formation of chem. engrs.; automation applied to chemistry; DARC topological system including coding of organic compounds by use of matrixes and systematic approach thereby to correlations between structure and reactivity. Home: 100 rue de Rennes, Paris 6 75. Office: 1 rue Guy de la Brosse, Paris 5 75, France.*

DUBOIS, Kenneth Patrick, Am. pharmacologist; b. Aberdeen, S.D., Aug. 9, 1917; s. Clarence C. and Mary (Cronin) DuB.; B.S., S.D. State Coll., 1939; M.S., Purdue U., 1940; Ph.D., U. Wis., 1943; m. Jere A. Deroin, Nov. 16, 1957; children—Elizabeth, Kenneth, Thomas. Mem. faculty U. Chgo., 1943—, prof. pharmacology, 1956—, dir. USAF Radiation and Toxicity Labs., 1953—. Cons. USPHS, 1956—, U. S. Army, 1960—; mem. NDRC, 1941-43, NRC-Nat. Acad. Sci. com. 1965—. Mem. A.M.A., Soc. Toxicology (v.p. 1961-62), Am. Soc. Pharmacology and Exptl. Therapeutics, Am. Chem. Soc., Soc. for Exptl. Biology and Medicine, Radiation Research Soc., A.A.A.S., Am. Indsl. Hygiene Assn., N.Y., Ill. acads. scis. Author: Textbook of Toxicology, 1959, also articles. Asso. editor Radiation Research, 1954-56; mng. editor Jour. Toxicology and Applied Pharmacology, 1960-64; field editor Toxicology, 1965—; editorial bd. Jour. Pharmacology and Exptl. Therapeutics, 1953-65. Research in evaluation of health hazards of radiation, drugs in toxic doses, pesticides, indsl. and environmental chems.; explanation on a biochem. basis of the exact manner in which toxic substances produce harmful and adverse effects. Home: 1214 E. 48th St., Chgo. 60615.*

DUBOIS, Marie Eugene François Thomas, Dutch paleontologist, physician; b. Eijsden, Netherlands, Jan. 28, 1858; student medicine, natural history; mil. surgeon Dutch E. Indies, 1887-95; prof. U. Amsterdam (Netherlands), from 1899. After Darwin's theories pub., interested in finding missing link; discovered animal apparently intermediate between man and existing anthropoid apes (Pithecanthropus Erectus or Java Man), Java, subsequent findings substantiated evidence (Dubois later inexplicably changed his attitude and maintained that it was only an advanced ape form). Died Halen, Belgium, Dec. 16, 1940.

DUBOIS, Paul Antoine, see Dubis, Paul Antoine.

DUBOIS, Paul Charles, Swiss neuropathologist; b. Chaux de Fonds, Switzerland, Nov. 28, 1848; s. Charles Ulysse and Marie Louise (Geiser) D.; student In Geneva and Bern, Switzerland; M.D., 1874; state exam, 1876; m. Maria Bertha Dinichert, 1881; 1 son, 1 dau. Lectr. on internal medicine, Bern; apptd. asso. prof. neuropathology; founder Schweizer Archiv für Neurologie und Psychiatrie. Mem. Swiss Soc. Neurology (founder, pres. 1911, 16). Author: De l'influence sur le corps, 1901; Les psychonéuroses et leur traitement moral, 1904; Die Einbildung als Krankheitsursache, 1907; L'éducation de soi-meme, 1908; Pathogenese der neurasthenischen Zustände, 1909. Research on neuroses and their treatment, electrotherapy; designed volta-galvanometer; defined persuasion as distinct from hypnosis and suggestion; created and defined term, psychoneurosis. Died Bern, Nov. 4, 1918.

DUBOIS, Pierre, French chemist; b. Saint Crepin, France, Aug. 3, 1889; Docteur ès sciences physiques, Paris Faculty Scis., 1935; m. Raymonde Chevallier, Aug. 3, 1933; 1 dau., Elsie (Mrs. Deryil). Full prof. plastics Conservatoire National des Arts et Métiers, Paris; dir. Center Research Plastic Materials, 1955-67. Mem. Soc. Phys. Chemistry, Soc. Mech. Chemistry. Author: Généralites technologiques; Manuel des Plastiques, Plastochimie, Plasturgie; also articles. Research on chem., thermal, mech. and elect. properties of plastics. Address: 16 rue Achille, Lachaire, Paris 14, France.*

DUBOIS, Raphael, biologist; b. 1849; studied phys.-chem. mechanism of phosphorescence; on the Pholades showed that prodn. of light results from action of diastasic ferment (luciferiase) on a product of glandular secretion. (luciferine) Died 1929.

DUBOIS D'AMIENS, Frédéric, French physician; b. Amiens, France, 1799; Docteur en médecine, U. Paris (France), 1828; named aggregate prof., 1832. Mem. Acad. Medicine (became perpetual sec. 1847). Author: Des différences de l'hystérie et de l'hypocondrie, 1830. Wrote on hysteria and hypochondria; supported theory of animal magnetism. Died Amiens, 1873.

DUBOIS DE MONTPÉREUX, Frédéric, geologist; b. Môtiers, May 18, 1798; m. Thérèse Dubois; 1 dau. Sent to Berlin, Germany as tutor to a Polish noble; sent on expdn. to Ukraine and Caucasus Mountains by Govt. of Russia, 1831-34; became prof. archeology Acad. Neuchatel (France), 1848. Recipient Grand prize Paris Soc. Geography, 1838, Order of St. Stanislas. Mem. geol. socs. of London, Eng.; Berlin; St. Petersburg, Russia; Paris, France. Author: Voyage autour du Caucase, 1839-43; Antiquités neuchate loises, 1843. Research on geography, geology, archeology of Caucasus Mountains; publs. on hist. monuments of Neuchatel, fossil shells. Died Môtiers, May 5, 1850.

DU BOIS-REYMOND, Emil Heinrich, German physiologist; b. Berlin, Germany, Nov. 7, 1818; s. Felix-Henri and Minette (Henry) D.; studied geology at Bonn (Germany) and under Johannes Müller, Berlin; received degree, 1843; m. Jeanette Claude, 1853; 4 sons, 5 daus. Became asst. lectr., 1846; named lecturing prof., 1854; became prof. physiology Berlin, 1858; rector, 1869-70, 82-83. Fellow Royal Soc., 1877; mem. Berlin Acad. Scis. (became sec. math. div. 1867), Physikalische Gesellschaft Berlin (founder, pres.), Physiol. Soc. (chmn.), Author: Untersuchungen über tierische Elektrizität, 2 vols., 1848-60; Geschichte Abhandlungen zur allgemeinische Muskel un Nervenphysik, 2 vols., 1875-77; Über die Grenzen des Naturerkennens (Ignoramus et ignorabimus), 1872; Die sieben Welträtsel, 1880; Übereine Akademie der deutsche Sprache, 1874; Reden, 2 vols., 1912. Pioneered research in exptl. physiology including animal electricity; invented instruments for research in animal electricity, 1840; used galvanometer to prove an electric impulse in a nerve is an elec. wave of negativity passing along the nerve; showed electrical phenomena occur in muscular activity; stated his law that variation of current density, not its absolute value at a given moment, is stimulus to motor center or muscle; studied theory of polarization of animal tissues, physiology of muscles, metabolic processes; suggested velocity of nerve impulse is measurable; opposed theory of vitalism. Died Berlin, Dec. 26, 1896.

DU BOIS-REYMOND, Paul David Gustave, German mathematician; b. Berlin, Germany, Dec. 2, 1831; s. Felix-Henri and Minette (Henry) D.; studied medicine U. Zurich (Switzerland), math. U. Königsberg (Germany); grad. Berlin, 1859; m. Henriette Massute, 1875; 1 son, Pascal. Became asst. lectr., 1865; named asso. prof. U. Heidelberg (Germany), 1868; apptd. prof. U. Freiburg (Germany), 1868, U. Tübingen (Germany), 1874, Berlin Tech. Coll., Charlottenburg, Germany, 1884. Author: Beiträge zur Interpretation der Partiellen Differentialgleichungen mit drei Variabeln, 1864; Abhandlungen über die Darstellung der Functionen durch trigonometrischen Reihen, 1876; Die Allgemeine functionentheorie, 1882; Grundlagen der Erkenntnis in den Exakten Wissenschaften, 1890. Research on theory of functions, infinite series theory, partial differential equations, math. ideas of consistency, continuum. Died Freiburg, Apr. 7, 1889.

DUBOS, René Jules, microbiologist, pathologist; b. St. Brice, France, Feb. 20, 1901; s. George Alexander and Adeline Madeleine (De Bloedt) D. student Coll. Chaptal, Inst. Nat. Agronomique, Paris, France, 1915-21; Ph.D., Rutgers U., 1927; numerous hon. degrees from univs. and colls. including Harvard, Liege U., U. Paris, U. Rio, Dartmouth, Yeshiva U., U. Alta., U. Pa., L'Academie de Lille; m. Marie Louise Bonnet, Mar. 23, 1934 (dec. 1942); m. 2d, Letha Jean Porter, Oct. 16, 1946. Came to U. S., 1924, naturalized, 1938. Asst. editor Internat. Inst. Agr., Rome, Italy, 1924-27; faculty Rutgers U., 1924-27; prof. dept. environmental medicine Rockefeller U., N.Y.C., 1927—, faculty Harvard, 1942-44. Recipient John Phillips Meml. award A.C.P., 1940; Lasker award Am. Pub. Health Assn., 1948; Trudeau medal Nat. Tb. Assn., 1951; Centennial award Robert Koch Inst., also Passano Found. award, 1960; Phi Beta Kappa award, 1963; A.M.A. award, 1965; Arches of Sci. award, 1966; numerous others. Mem. Soc. Am. Bacteriologists, Nat. Acad. Scis., Am. Philos. Soc., others. Author: Louis Pasteur—Free Lance of Science, 1950; Pasteur and Modern Science, 1960; The Dreams of Reason, 1961; The Torch of Life, 1962; The Unseen World, 1962; Bacterial and Mycotic Infections of Man, 4th edit. 1965; Health and Disease, 1965; Man Adapting, 1965, others. Editor: Jour. of Exptl. Medicine, 1946—. Research and publs. on antibiotics; acquired immunity; tuberculosis; indigenous gastrointestinal bacteria; discovered crystalline form of soil-bacteria agent which destroys "gram-positive" germs; this discovery laid basis for new field of chemotherapy. Home: Garrison, N.Y. 10524. Office: Rockefeller U., N.Y.C. 10021.*

DUBOSC, Jules, French optician; b. Paris, France, 1817; studied under Soleil; optician in Paris; received medal from London Exposition, 1851, medal from N.Y. Universal Expn., 1853. Collaborated with Soleil on his inventions; invented the phenakistikop (a projector), 1850; developed a regulator for arc lamps, 1856, an electric tuning fork for projection of vibrations, 1879; improved stereoscope. Died Paris, 1886.

DUBOSCQ, Octave, French biologist; b. Rouen, France, Oct. 1, 1868; Licence ès scis., U. Caen (France); M.D., Paris; became zoology asst. Caen Faculty Scis., 1900; named prof. zoology Montpellier (France) Faculty Scis., 1904; became prof. maritime biology, Paris, 1923; dir. Arago Lab., 1924-37. Author: Sur la terminaison des nerfs sensitifs des chilopodes, 1897. Specialized in unicellular plants and animals; studied morphology, cytology, reproduction of Protista. Died Nice, France, Feb. 18, 1943.

DUBOWITZ, Victor, physician, histochemist; b. Beaufort West, S. Africa, Aug. 6, 1931; s. Charlie and Olga (Schattel) D.; B.Sc., U. Cape Town (S. Africa), 1950, M.B., Ch.B., 1954, M.D., 1960; Ph.D., U. Sheffield (Eng.), 1965; m. Lilly Sebok, July 11, 1960; children—David Julian, Michael Nathan, Gerald Peter. Demonstrator, U. Cape Town, 1950-52; research fellow Postgrad. Med. Sch., London, 1958-60; lectr. clin. pathology U. London, Nat. Hosp. for Nervous Diseases, 1960-61; faculty U. Sheffield, 1961—, sr. lectr. child health, 1965—, reader in child health and developmental neurology, 1967—; research associate at the Institute for Muscle Disease, New York City, 1965-66; asst. pediatrician New York Hosp., Cornell Med. Coll., 1965-66. Cons., United Sheffield Hosps., 1965—. Recipient Westdene prize U. Cape Town, 1954. Fellow Royal Soc. Medicine; mem. Royal Coll. Physicians, Brit. Med. Assn., Histochem. Soc., Brit. Paediatric Assn., Paediatric Research Soc. Contbg. author Disorders of Voluntary Muscle, 1964; Methods in Experimental Pathology, 1966. Research and publs. on muscle disease, structure and function of muscle, histochem. studies developing human and animal muscle, influence of nervous system on muscle structure and function, neurology of newborn infants, congenital abnormalities, dwarfism, rheumatoid disorders, leukemia, enzymology and genetics; 1st descriptions of enzymic fiber types in human and animal muscle, path. changes in symptom-free carriers muscular dystrophy.*

DUBREIL, Paul, French mathematician; b. Le Mans, France, Mar. 1, 1904; s. Léon and Léonie (Carpentier) D.; Lic.Sc., Superior Normal Sch., 1925, Sc.D., 1930; Rockefeller fellow, univs. Hamburg, Groningen, Frankfort/Main (all Germany), Rome 1929-31; m. M. L. Jacotin, June 28, 1930; 1 dau. Lectr., Faculty Scis., Lille, France, 1931; prof. differential and integral calculus Faculty Scis., U. Nancy (France), 1937-46; lectr. gen. math. Faculty Scis., U. Paris, 1946—. Lectr. geometry Poly. Sch. Nancy, 1943-46. Recipient Prix Francoeur, 1942, Prix Marquet, 1944. Mem. French Math. Soc. Author: Quelques propriétés des variétes algébriques se rattachant aus théories d'algèbre moderne, 1935; Contribution à la théorie des demi-groupes, 1941, Algèbre, Vol. I, Equivalences et opérations, 1946. Home: 11, rue René Bazin, Paris 16, France.

DUBREUIL, Jean, mathematician; b. Paris, France, 1602. Mem. Soc. of Jesus; dir. novices at Dijon, France. Author 3 vol. work on geometrical aspects of perspective, 1642-49; L'art universal des fortifications, 1665. Died 1670.

DUBRIDGE, Lee Alvin, Am. physicist; b. Terre Haute, Ind., Sept. 21, 1901; s. Frederick Alvin and Elizebeth (Browne) DuB.; A.B., Cornell Coll., Mt. Vernon, Ia., 1922, Sc.D., 1940; A.M., U. Wis., 1924, Ph.D., 1926, Sc.D., 1957; numerous hon. degrees including, Sc.D., Washington U., St. Louis, Ind. U., Columbia; LL.D., U. Cal. at Los Angeles, U. So. Cal., Northwestern U.; m. Doris May Koht, Sept. 1, 1925; children—Barbara Lee (Mrs. David MacLeod), Richard Alvin. Mem. faculty U. Wis., 1922-26, Washington U., 1928-34, U. Rochester, 1934-46; pres. Cal. Inst. Tech., Pasadena, 1946—. Mem. panel of com. sci. and astronautics U. S. Ho. of Reps., 1960—; bd. dirs. Nat. Merit Scholarship Corp.; trustee Rockefeller Found., 1956-67. Decorated King's Medal (Great Britain). Recipient Research Corp. award, 1947; U. S. medal for merit, 1948; Golden Key award NEA, 1959; Arthur Noble award City of Pasadena, 1961. Benjamin Franklin fellow Royal Soc. Arts, London, Eng.; fellow A.A.A.S., Am. Phys. Soc. (v.p. 1946, pres. 1947); mem. Am. Philos. Soc., Nat. Acad. Scis., Phi Beta Kappa, Sigma Xi, Tau Kappa Alpha, Tau Beta Pi, Sigma Pi Sigma, Eta Kappa Nu. Author: (with A. L. Hughes) Photoelectric Phenomena, 1932; New Theories of the Photoelectric Effect, 1935; Introduction to Space, 1960. Research and publs. on photoelectric and thermionic emission, nuclear disintegration, radar, direct current amplification. Home: 415 S. Hill Av., Pasadena 91106.*

DUBROV, Yakov Grigorevich, Russian orthopedist, traumatologist; b. Pokrovskoe (now Zaporozhe Oblast), 1903; grad. Kharkov Med. Ist., 1927; postgrad. Kharkov Inst. Orthopedics and Traumatology, 1932-34; Cand. Med. Sci., 1936; D.Med. Sci., 1948; Head traumatology dept. Khrustalnoe Hosp., 1934-39; dep. dir. Ukrainian Research Inst. Orthopedics and Traumatology, Kharkov, 1939-40, dir., 1940-41; sr. asso. Inst. Surgery, USSR Acad. Med. Sci., Central Inst. Prosthetics, USSR Ministry Health, 1945-55; head orthopedics and traumatology clinic Vladimirsky Moscow Oblast Clin. Research Inst., 1955—; chief traumatologist Moscow Oblast. Recipient Pirogov prize. Mem. All-Union (bd. mem.), Moscow (dep. chmn.) socs. traumatologists and orthopedists. Author: First Aid Underground, 1934; Plastic Surgery of Damaged Wrist and Finger Tendons, 1940. Mem. editorial bd. Khirurgiya. Address: Moscow Oblast Clin. Research Inst., 3-ya Meschanskaya ulitsa 61-62, Moscow, USSR.

DUBRUNFAUT, Augustin Pierre, French indsl. chemist; b. Lille, France, Sept. 1, 1797. Prof. tech. chemistry Paris, France, école de commerce until 1833. Publs. on alcohol distilling industry; patentee method of preparing crystallizable sugar using Dutrochet osmometer, 1854. Proposed incorrectly that carbon does not burn in oxygen without an appreciable amount of water; discovered invert sugar contains a d- and l-sugar (named glucose and fructose by Fischer). Died Bercy, Oct. 7, 1881.

DU BUAT, Pierre Louis Georges, French engr.; b. Tortisambert, Normandy, France, Apr. 23, 1734; mil. engr.; mem. French Acad. Scis., 1804. Author: Principes d'hydraulique, 1779. Research on movement of water in rivers, canals, and supply pipes; studied working of sluices, reservoirs, water pressure, ship travel on rivers and canals. Died Vieux-Condé, France, Oct. 17, 1809.

DUBUC, Guillaume, French chemist; b. Sierville, France, 1764; studied under Lavoisier; recipient of Prix Montyon. Author: Mémoire de l'encollage des étoffes, 1821. Research on sugar of fruits, preparation of cider, sizing of cloth; analyzed arable soil. Died 1837.

DUBUISSON, François-René-André, French naturalist; b. Nantes, France, 1763; apptd. dir. Nantes (France) Mus. Natural History, 1810. Author: Essai d'une méthode géologique, 1819; Catalogue de la collection minéralogique de la Loire-Inférieure. Catalogued the rocks of the region around Nantes; 1st well known research on mineral deposits of Brittany. Died Paris, France, circa 1836.

DUBULIER, William, Am. engr., inventor; b. N.Y.C., July 25, 1888; s. Alie and Anna Dubulier; ed. Tech. Inst., N.Y.C., Cooper Union; m. Florence Don, 1923; 2 sons. Author: Wireless Telegraphy Equipment, 1905. Inventor power condensor, med. apparati, nylon window screens, numerous elec. devices; developed 1st wireless communication for airplane, 1912.

DUBYAGO, Aleksandr Dmitrievich, Russian astronomer; b. 1903; s. D. I. Dubyago (astronomer); grad. Kazan U., 1925; D.Physico-Math. Sci., 1941. Prof., 1941—, Kazan U., 1944—. Author: Some Problems of the Motion, Structure and Decay of Comets, 1942; Secular Acceleration of the Motion of Periodic Comets, 1948; The Structure of Comet Nuclei and the Formation of Meteor Showers, 1950. Research and publs. on motion of Brooks' and Daniell's comets, structure and decay of comets, observed variable stars, research on geodesy and gravimetry, author guide for determining orbits of small planets, comets and meteors. Address: Kazan University, ulitza Lenina 18, Kazan, Tatarskaya ASSR, USSR.

DU CAMP, Théodore-Joseph, French surgeon; b. Bordeaux, France, 1792; docteur en médecine, 1815; surgeon at Paris Val-de Grâce Hosp. Mem. Soc. Medicine. Author: Des polypes de la matrice, 1815; Traité des rétentions d'urine, 1822. Research on urinary disorders; developed method of cauterization for injuries to the mid-section, which avoided operation; invented instrument to replace an umbilical cord which protruded prematurely. Died Paris, France, 1823.

DU CHAILLU, Paul Belloni, Am. explorer; b. New Orleans, July 31, 1838; sailed from N.Y. to French settlement at mouth of Gaboon river, W. Africa; at his own expense traveled 8,000 miles, with only native companions; covered much previously unexplored country and added 60 species of birds and 20 species of mammals (including the gorilla) to the known zoölogy of Africa; his accounts of the gorillas and Obongo dwarfs were contradicted by scientists, but were later confirmed; made 2d exploration, 1863-65; discovered many new species of animals and birds; also traveled in Sweden, Norway, Lapland and Finland and other countries. Author: Wild Life Under the Equator; My Apingi Kingdom; The Country of the Dwarfs; Lost in the Jungle; The Land of the Long Night, 1899; The World of the Great Forest, 1900; How Animals, Birds, Reptiles and Insects Talk, Think, Work and Live, 1900. Died 1903.

DUCHARTRE, Pierre, French botanist; b. Béziers, France, Oct. 27, 1811; prof. plant physiology Paris Agronomic Inst., 1848-52; became prof. botany Paris (France) Faculty Scis., 1861. Mem. French Acad. Scis., 1861 (pres. 1890), Bot. Soc. France (a founder), Soc. Agr. Author: Éléments de botanique, 1867; La géographique botanique des environs de Béziers. Died Paris, France, Nov. 5, 1894.

DU CHATELET, Martine de Bertereau (baroness de Beausoleil), mineralogist; m. Jean du Châtelet (baron de Beausoleil), 1610; 1 dau., Anne; explored mines and mineral deposits; traveled with husband in Germany, Sweden, Am., 1610-26. Author: Diorismus verae philosophiae de materia prima lapidis, 1627; Véritable déclaration faite au roi et à nos siegneurs de son conseil, des riches et inestimables trésors, nouvellement descouverts dans le royaume, 1632; La restitution de Pluton a l'Eminentissime cardinal duc de Richelieu . . . , 1640. Imprisoned for witchcraft, Vincennes, France, 1642; died in prison.

DUCHENE-MARULLAZ, Pierre, French physiologist, pharmacodynamist; b. Belfort, France, Jan. 24, 1921; s. Henri and Marie (Hincky) D.-M.; M.D., U. Strasbourg, 1946. Faculty, Clermont-Ferrand (France) Sch. Medicine, 1946——, prof. physiology, 1963——; mng. dir. research Mauvernay's European Center Research, Riom, France, 1959——; mng. dir. Cardiologic Research Inst., Royat, France, 1965——. Recipient Médaille de la reconnaissance Franco-Alliée, 1945, Acad. Palms, 1960. Mem. Physiologists Assn., Soc. Biology, Soc. Therapeutics and Pharmacodynamy, European Assn. for Drug's Toxicity, N.Y. Acad. Scis. Research and numerous publs. on cardiac rhythm, extrinsic innervation, variations of veinous ammonemia during muscular activity, including heart, pharmacol. studies on numerous synthetic drugs. Home: 19 Bd. Bazin, Royat, France (63). Offices Faculty Medicine, Clermont-Ferrand, France (63).*

DUCHENNE, Guillaume Benjamin Amand, French physician, neurologist; b. Boulogne-sur-Mer, France, Sept. 17, 1806; student, Douai, France; M.D., Paris, France, 1831; m. circa 1831; 1 son. Practiced gen. medicine, Boulogne, 1831-33; resumed work in electro-physiology, Paris, 1842. Author: l'Electrisation localisée, 1855; Physiologie des mouvements, 1867; Mécanisme de la physiognomie humaine (with atlas), 1862. Pioneer in neurology; investigated elec. stimulation of muscles; founder electro-therapeutics; pioneered in methods of electroanalysis; gave 1st descriptions of prog. muscular atrophy, 1850, locomotor ataxia; gave 1st classic description of tabes dorsalis (Duchenne's disease), 1858, 59; 1st to recognize prog. bulbar palsy (under name labioglossolaryngeal paralysis), 1861; clarified some forms of lead palsy, also established their elec. reactions, 1872; founded sci. of biopsy by discovery of method for examining muscle from living patients (called Duchenne's histological harpoon by Gowers). Died Paris, Sept. 15, 1875.

DUCHESNE, Antoine-Nicolas, French naturalist; b. Versailles, France, Oct. 7, 1747; s. Antoine Duchesne; prof. natural history École centrale of Sein-et-Oise, Lycée de Versailles; also with Versailles Trianon Bot. Garden. Recipient honorable mention award French Acad. Scis. Mem. Soc. Agr. Paris, Soc. Agr. Versailles. Author: Manuel de botanique, 1764; Histoire naturelle des fraisiers, 1766; Jardinier prévoyant, 11 vols., 1770-81. Research on natural history of strawberry; discovered mutation of monophyllous strawberry, 1763; gave common names to plants around Paris; ideas were original and often ahead of his time. Died Feb. 18, 1827.

DUCHESNE, Edouard-Adolphe, French botanist; b. Paris, France, 1804; Docteur en médecine, 1827. Mem. Paris Com. on Hygiene and Sanitation. Recipient award from Acad. Medicine, 1830. Author: Répertoire des plantes utiles et des plantes vénéreuses du globe; L'Emploi du maïs, 1830; Des danger des papiers colorés, 1854. Research on use of corn as a food for lactating women, Turkish wheat, san. conditions; discovered that chickens fed on certain meats were producing unfit eggs, that enamel workers easily contracted lead poisoning, that certain colored papers used at that time were toxic; listed useful and poisonous plants of the world. Died 1869.

DUCHESNE, Joseph (or Josephus Quercetanus), physician, chemist; b. L'Esture, Gascony, France, circa 1544; student natural scis. Germany; M.D., Basel, Switzerland, circa 1573. Took up residence in Geneva, 1584, and engaged in peace negotiations between Geneva and its neighbors which ended successfully, 1592; physician-in-ordinary to King Henry IV of France, Paris, from 1593. Author: Ad Iacobi Auberti Vindonis de Ortu et Causis Metallorum contra Chymicos Explicationem . . . brevis Responsio, 1575; De Priscorum Philosophorum verae medicinae materia, 1603; Ad Veritatem Hermeticae Medicinae, 1604; other works on medicine, chemistry, poetry and drama. An extensive selection of his works appeared as Quercetanus redivivus, 1648, and various books by him were translated into French, German, English and Italian. Duchesne was one of the most influential of the 17th cent. iatrochemists, he engaged in many controversies on behalf of use of med. remedies; supported by Theodore Turquet de Mayerne in controversy with Galenists at Paris, 1603-09; introduced five element system, 1584; discussed local "circulations" of blood, 1603; introduced many new chemicals to med. use. Died Paris, 1609.

DUCHESNE, Jules Charles Gérard Léon, Belgian chem. physicist, biophysicist; b. Seraing, Belgium, Dec. 1, 1911; s. Jules Louis and Jeanne (Bovy) D.; B.Sc., U. Liège (Belgium), 1937, Ph.D., 1941, D.Sc., 1944; m. Audrey Cripps, Aug. 12, 1948; children—Hélène, Jules, Jean. With U. Liège, 1939——, prof., dir. dept. atomic and molecular physics, 1957——; Francqui prof. U. Brussels (Belgium), 1947-48; hon. research asso. U. Coll. London, 1955-56; asso. prof. U. Paris, 1964. Recipient Gold medal of Francqui Prize; numerous prizes Belgian Royal Acad. Sci. Advanced fellow Belgian Am. Found., 1948. Mem. Belgian Phys. Soc. (vice chmn., 1959-63), Belgian Biophys. Soc. (chmn. 1967), Belgian Biophys. Com. (chmn., 1964), Académie Royale des Sciences de Belgique, Accademia Tiberina Roman, Accademia Teatina per le Scienze Chieti (hon.). Editor: Molecular Structure, 1947; Radiofrequency Spectroscopy, 1961; Structure and Properties of Biomolecules, 1964. Research, numerous publs. in molecular physics, optical and radiofrequency spectroscopy, cosmochemistry and biophysics; discovered meaning of signs of interaction terms in vibrational force fields, 1939, method to orient photochem. reactions, 1948, free radicals in carbonaceous rocks and meteorites, 1961, semi-conductivity in nucleic acids and relations between molecular structure and radio-resistance of matter, 1957, variability of isotopic ratios in carbon dioxide exhaled by living beings, 1967; transfer of free radicals, using photodynamic effects, in biological systems, and correlation with mutations in viruses, 1967. Home: 3 boulevard Frère Orban, Liège, Belgium. Office: University of Liège, Institute of Physics, Sart-Tilman Par, Liège 1, Belgium.*

DUCHESNE, Simon, mathematician; b. Dale, France; flourished 1583. Because of persecution as a Calvinist, went to Delft where he taught math. Author: Quadrature du cercle, où manière de trouver un quarré égal au cercle donné, 1584. Research on squaring the circle.

DUCKWORTH, Henry Edmison, Am. physicist; b. Brandon, Man., Can., Nov. 1, 1915; s. Henry Bruce and Ann (Edmison) D.; B.A., U. Man., 1935, B.Sc., 1936; Ph.D., U. Chgo., 1942; m. Katherine Jane McPherson, Nov. 21, 1942; children—Henry William, Jane Edmison. Lectr. physics United Coll., Winnipeg, Man., 1938-40; with NRC Can., 1942-45, mem. hon. adv. council, 1961——; asst. prof. physics U. Man., 1945-46; asso. prof. physics Wesleyan U., 1948-52; prof. physics McMaster U., Hamilton, Ont., Can., 1951——, chmn. dept. physics, 1956-61, dean faculty grad. studies, 1961——. Sec. commn. on nuclidic masses I.U.P.A.P., 1960——. Recipient Distinguished Alumni award U. Man., 1962. Fellow Royal Soc. Can. (pres. sci. sect. 1964-65), Am. Phys. Soc.; mem. Canadian Assn. Physicists (pres. 1960-61, medal 1964). Author: Mass Spectroscopy, 1958; Electricity and Magnetism, 1960; Little Men in the Unseen World, 1965. Research in mass spectroscopy, determination of atomic masses, loss of energy by atoms in traversing matter. Home: 89 Dalewood Crescent, Hamilton, Ont., Can.*

DUCLAUX, Pierre Émile, French biochemist, bacteriologist; b. Aurillac, France, June 24, 1840; student at L'École normale supérieure; D.Sc., 1865; studied under Pasteur; m. Mary Darmesteter. Became prof. chem. biology Sorbonne, Paris, France, 1885; prof. Faculty Scis., Lyon, France; prof. microbiology at Institut agronomique; apptd. dir. Pasteur Inst., Paris, 1895. Mem. French Acad. Scis., French Acad. Medicine, French Agrl. Soc. Author: Sur l'iodure d'amidon, 1872; Etudes sur la nouvelle maladie de la vigne dans le sud-est de la France, 1873-75; Ferments et maladies, 1882. Collaborated with Pasteur; research and publs. on diseases of plants and microbes; compared surface tension with a stretched elastic membrane, 1872; research on flow in capillaries. Died Paris, May 3, 1904.

DUCLO, Gaston, French alchemist; b. Nivernois, France, circa 1530; studied law; practiced law at Nevers; author of Apologia chrysopoeiae . . . , 1598; De recta . . . , 1592. Probably produced calcium nitrate and gold chloride. Died circa 1600.

DUCLOS (COTTEREAU DU CLOS), Samuel Cottereau, French physician; b. Paris, France; 1st physician ordinary to King Louis XIV. Mem. French Acad. Scis., 1666. Author: Observations sur les eaux minérales, 1675; Dissertation sur les principes des mixtes naturels, 1680. Tried to base pharmacy on chem. knowledge; analyzed mineral water; opposed Boyle's corpuscular theory of chemistry. Died 1715.

DUCOFF, Howard S., Am. physiologist; b. N.Y.C., May 5, 1923; s. Dave and Tillie (Machinist) D.; B.S., City Coll. N.Y., 1942; Ph.D. in Physiology, U. Chgo., 1953; m. Rose Hirsch, Aug. 25, 1946; children—Sandra, Barbara, Paul J., Laura. With Argonne (Ill.) Nat. Lab., 1946-63, asso. scientist, 1951-57, cons., 1957-63; mem. faculty U. Ill. at Urbana, 1957——, prof. physiology and biophysics, 1965——; USPHS spl. fellow dept. zoology U. Cambridge (Eng.), 1964-65. Served with AUS, 1943-46. Mem. Am. Soc. Cell Biology, Am. Soc. Zoologists, Radiation Research Soc., Soc. Gen. Physiologists, Soc. Protozoologists, N.Y. Acad. Scis., Sigma Xi. Editor: (with C. F. Ehret) Mitogenesis, 1959. Author articles. Research on cellular radiobiology; physiology of cell division; insect physiology. Home: 1516 W. Charles St., Champaign, Ill. 61822. Office: Burrill Hall, Urbana, Ill. 61801.

DUCOS, Jean Bernard, French immunologist; b. Toulouse, France, Mar. 4, 1926; s. Louis and Emilie (Morere) D.; M.D., U. Toulouse, 1954; m. Liliane Pecoul, July 19, 1951; children—Philippe, Bernard, Elisabeth. Head lab. procedures U. Toulouse, 1949-51, asst. Blood Transfusion and Hematology Center, 1951-53; asst. Nat. Inst. Hygiene, Paris, 1953; head lab. Blood Transfusion and Hematology Center, Toulouse, 1954-56, head dept., 1956-62, asst. dir., 1962——; asst. Toulouse Faculty Medicine, 1957, lectr., 1958-63, prof., 1963——, prof. gen. immunology, 1965——; dir. immuno-hematology and cytogenetics research unit and Med. Research. Mem. French Soc. Immunology, French Soc. Hematology, Internat., Nat. socs. blood transfusions, Internat. Soc. Transplantation (mem. com. research and study of transplantation). Research, numerous publs. on determination of Rh group of dried blood stains, Rh group of fetus in utero, demonstration of Rh antigens in blood platelets, immunological diagnosis of ABO hemolytic disease of new born, modifications of Gm antigens in leukemias, study of reaction of passive hemagglutination and its applications to study of proteic antigens, study of blastic transformation of lymphocytes in vitro and its immunological applications. Home: 17, rue de l'Obelisque 31, Toulouse, France. Office: C.H.U. Purpan 31, Toulouse, France.*

DUCOS DU HAURON, Louis Arthur, French inventor; b. Langon, Gironde, France, 1837. Research in several fields of photography, patented process similar to motion picture, 1864; inventor (with Cros) trichrome process for reproduction colors, 1869; proposed Tripack in which 3 emulsions separated by filter were laid on top of each other but were not uniformly exposed, 1895. Died Agen, 1910.

DUCREY, Augusto, Italian dermatologist; b. Naples, Italy, 1860; prof., Naples, also Rome; discovered Hemophilus ducreyi (causative agt. of chancroid), circa 1889. Died Rome, 1940.

DUDA, Seweryn Jozef, geophysicist; b. Chorzow, Poland, Apr. 20, 1933; s. Josef and Hedwig (Hammerling) D.; Magister geofizyki, Warsaw (Poland) U., 1955; Ph.D., Uppsala (Sweden) U., 1961, filosofie doktor, docent, 1966; m. Theresia Mrziglod, Dec. 24, 1955; children—Chrysanth-Caesar, Laurent-Claudius. Asst., Inst. Geophysics, Warsaw U., 1955-58; scientist Seismos G.m.b.H., Hanover, Germany, 1958-60; project scientist Seismol. Inst., Uppsala U., 1961-65; vis. prof. dept. geophysics and geophys. engring. St. Louis University, 1966-67, assistant professor, 1967——; served as research fellow Seismol. Lab., Cal. Inst. Tech., Pasadena, Cal., 1964. Mem. Seismol. Soc. Am., Soc. Expln. Geophysicists, European Assn. Expln. Geophysicists, Deutsche Geophysikalische Gesellschaft, Polskie Towarzystwo Geofizyczne, Sigma Xi. Research and publs. on wave propagation in anisotropic elastic media, improvement of seismic signal to noise ratio and seismic data processing, patterns of earthquake occurrence in time and space, rheological properties of the upper part of earth, radiation pattern and energy of earthquakes, scale model expts. related to stress changes in earth with reference to earthquake prediction. Home: 9727 Pauline Pl., St. Louis 63123.*

DUDDELL, William Du Bois, English engineer; b. London, Eng., 1872; ed. Collège Stanislass, Cannes; Central Technical College, London; unmarried. Apprenticed to Davey, Paxman & Co., Colchester, 1890-3; obtained Whitworth Exhibition, 1896, and Whitworth Scholarship, 1897; consulting engineer. Hon. sec., Internat. Electrical Congress, St. Louis, 1904; Internat. Conference on Electrical Units and Standards, 1908; mem., Admiralty board of invention and research; Advisory Council for Industrial Research. Received gold medals, Paris Exhibition, 1900, St. Louis, 1904. Fellow Royal Soc., 1912 (Hughes medal, 1912); pres., Röntgen Soc., pres., Institution of Electrial Engineers; hon. treas., Physical Soc. Author numerous articles on electricity. Research on radio-telegraphy; invented improved oscillograph, 1897. Died London, Eng., Nov. 4, 1917.

DUDEN, Paul, German chemist; b. Soest, Germany, Oct. 30, 1868; s. Konrad and Adeline (Jakob) D.; student chemistry, Marburg and Würzburg, Germany, Geneva, Switzerland; doctorate Jena, Germany; numerous hon. dr. degrees; m. Berta Hebe, 1897; 5 children. Apptd. lectr. Jean, 1896, became asso. prof., 1899; named dir. of a research lab. in firm Hoechst, 1905; in charge incorporating Hoechst into IG-Farbenindustrie, later became mem. bd., dir. operation in middle Rhine; ret. from industry, 1932. Recipient Goethe medal for art and sci. City of Frankfurt/Main, Germany. Mem. bd. Deutsches Mus. Mem. Berufsgenossenschaft der Chemischen Industrie (elected chmn. 1929), Verein Deutscher Chemiker (chmn. 1929-37), Dechema (chmn. 1934-40). Research and publs. on dyes, aliphatic chemistry; introduced catalytic oxidation from acetylene to acetaldehyde and acetic acid. Died Schliersee, Bavaria, Germany, Feb. 7, 1954.

DUDGEON, Robert Ellis, homeopath; b. Leith, Scotland, Mar. 17, 1820; M.D., U. Edinburgh (Scotland), 1841; ed. also at Vienna, Austria, Berlin, Germany, Dublin, Ireland; m. twice; 5 children. Practiced homeopathy London, Eng., from 1845; founder (with others) Hohnemann Hosp. and Sch. Homeopathy, 1850; lectr. theory and practice homeopathy, anatomy. Mem. Brit. Homeopathic Soc. (sec., 1848, v.p. 1874-75, pres. 1878, 90). Author: Organon, 1849; materia medica Pura, 1880; The Human Eye, its Optical Construction popularly explained, 1878; The Sphygmograph . . . , 1882; Lectures on Theory and Practice of Homeopathy, 1854; The Influence of Homeopathy on General Medicine since the death of Hohnemann, 1874. Editor Brit. Jour. Homeopathy, 1846-84. Translator works by Hohnemann, 1849-50. Inventor Dudgeon's sphygmograph for registering pulse, 1878. Died Sept. 8, 1904.

DUDLEY, Benjamin Winslow, Am. surgeon; b. Spotsylvania County, Va., Apr. 12, 1785; s. Ambrose Dudley; M.D., U. Pa., 1806; m. Anna Maria Short, 1821, 3 children. Studied medicine under Dr. Fred Ridgely; began practice of medicine, Lexington, Ky.; traveled, studied in Europe, 4 years; mem. Royal Coll. Surgeons; returned to Lexington, 1814; prof. anatomy and surgery Transylvania U., 1817-50; retired from active medical practice, 1853. Contbr. essays to med. jours. Performed 1st successful cataract operation in West, 1836. Died Fairlawn, nr. Lexington, Jan. 20, 1870.

DUDLEY, Dud, English inventor; b. Worcestershire, Eng., 1599; s. Edward Sutton and Elizabeth (Tomlinson) D.; ed. Balliol Coll., Oxford (Eng.) U.; m. Elinor Heaton, Oct. 12, 1626. Became ironmaster, circa 1630; built and operated blast furnace nr. Bristol, Eng., 1651; col. in army of Charles I; gen. of the ordnance to Prince Maurice. Author: Metallum martis, 1665. Patentee process of making iron into any sort of cast works, 1638. First to smelt iron ore with coal, 1619. Died Oct. 25, 1684, Worcester, Eng.

DUDLEY, Emelius Clark, Am. physician; b. Westfield, Mass., May 29, 1850; s. John Harmon and Marana P. (Mason) D.; A.B., Dartmouth, 1873; M.D., L.I. Coll. Hosp., 1875; (hon. A.M., Ia. Coll., 1892; LL.D., Grinnell Coll., 1917); m. Anna M. Titcomb, June 29, 1882; children—Katharine Dorothy (Mrs. H. B. Harvey), Helen, Prescott, Caroline (Mrs. Daniel J. Reagon). Engaged in practice in Chgo., 1875-1928; prof. gynecology, Northwestern U., 1882-1928. Maj. O.R.C., 1917. Mem. Chgo. Bd. Edn., 1901-06. Fellow Am. Gynecol. Assn. (pres. 1904), Brit. Gynecol. Soc. Author: Principles and Practice of Gynecology, 1908, 6th edit., 1914. Died Dec. 1, 1928.

DUDLEY, Harold Ward, Brit. biochemist; b. Derby, Eng., Oct. 20, 1887; s. Joshua and Florence Dudley; ed. Truro Coll.; M.Sc., U. Leeds, 1910; Ph.D., U. Berlin, 1913; m. Louisa Mary Nettleship, 1921. Asst., Hertar Lab., N.Y., 1912-13; lectr. biochemistry, Leeds, 1914-15; biochemist Med. Research Council's Labs., from 1919. Fellow Royal Soc. 1930. Editor Biochem. Jour., 1924-30. Contbr. papers to jours. Isolated (with John C. Moir) ergometrine, 1935. Died Oct. 3, 1935.

DUDLEY, Paul, Am. botanist; b. Roxbury, Mass., Sept. 3, 1675; s. Gov. Joseph and Rebecca (Tyng) D.; grad. Harvard, 1690; studied law, Temple, London, Eng.; m. Lucy Wainwright, 1703; children—Thomas, Joseph. Admitted to Mass. Bar, 1700; commd. by Queen Anne atty. gen. Province of Mass. Bay, 1702-18; advocate in Vice Admiralty Ct. Boston; apptd. atty. gen. by Mass. gov. and council; mem. Mass. Legislature, Mass. Exec. Council; judge Mass. Superior Ct. of Judicature, 1718-51; chief justice Mass., 1745-51; founder Dudleian Lecture on Religion at Harvard Coll. Author: Observations on Some of the Plants in New England with Remarkable Instances of the Nature and Power-of Vegetation (article in Philos. Trans., described his discovery of possibility of hybridizing maize, formed basis for 2 entries in Miller's Gardener's Dictionary), 1724. Died Roxbury, Jan. 25, 1751.

DUDLEY, Pemberton, Am. physician; b. Torresdale, Phila., Pa., Oct. 17, 1837; grad. Homoeo. Med. Coll., Phila., 1861; prof. chemistry and toxicology there, 1868-69; prof. physiology and microscopic anatomy, 1876-90; LL.D., Rutherford Coll., N.C., 1899; m. Sarah K. Hall, Dec. 25, 1867. Prof. insts. medicine and hygiene, Hahnemann Med. Coll., 1890-1907, dean 1896-1907. Gen. sec. and editor ann. Transactions, Am. Inst. Homoeopathy, 1887-94; editor Hahnemannian Monthly, 1880-88; became State bd. of health, 1885. Died 1907.

DUDLEY, Plimmon Henry, Am. civil and metall. engr.; b. Freedom, O., May 21, 1843; s. Charles and Sarah (Leete) D.; Ph.D., Hiram Coll.; m. Lucy May Bronson, Dec. 12, 1871. Chief engr. Valley Ry., 1872-74; chief engr. City of Akron O., 1866-72; later cons. engr. N.Y.C. lines. Invented dynamometer, 1874; track indicator, 1880, designed, 1883, the first 5-inch steel rail used in U. S., and in 1892 introduced the first 6-inch 100-lb. rails; invented stremmatograph for obtaining and registering strains in rails under moving trains; made first announcement, 1884, that fungi were the cause of decay in wood. Reported for U. S. on the "Nature of the Metal for Rails" to Internat. Ry. Congress, 6th session, Paris, 1900, and on "Rails for Lines with Fast Trains," to 7th session, Washington, 1905. Died N.Y.C., Feb. 25, 1924.

DUDLEY, Sir Sheldon Francis, surgeon; b. Lisbon, Aug. 16, 1884; s. John Dudley; M.D., St. Thomas's Hosp.; D.P.H.; LL.D., U. Edinburgh, 1953; D.T.M.; m. Ethel E. Franklyn, 1913; 1 child. Joined Royal Navy, 1906; prof. pathology Royal Naval Med. Sch., Greenwich; dir. med. studies to Royal Navy, dep. med. dir. gen., 1935-38, med. dir. gen., 1941-45. Fellow Royal Soc., 1941, Royal Coll. Physicians (London), Royal Coll. Surgeons; mem. Royal Soc. Medicine (pres. epidemiological sect. 1935). Author: The Four Pillars of Wisdom, 1950. Research and numerous publs. on spread of infectious diseases, diphtheria, Schick test, spread of droplet infection. Died May 6, 1956.

DUDLEY, William Lofland, Am. chemist; b. Covington, Ky., Apr. 16, 1859; s. George Reed and Emma (Lofland) D.; B.S., U. Cin., 1880; hon. M.D., Miami Med. Coll., 1885. Demonstrator of chemistry Miami Med. Coll., 1879-80, prof. chemistry and toxicology, 1880-86; prof. chemistry Vanderbilt U., from 1886, dean Medical Dept., 1895-1913. Sec., sect. inorganic chemistry, Internat. Congress Arts and Sciences, St. Louis, 1904; U. S. commr. 7th Congress Applied Chemistry, London, 1909; v.p. sect. on law and legislation as affecting chem. industry, 8th Congress Applied Chemistry, 1912. Devised process for working and electro-plating with iridium; did spectrographic studies on tellurium and atomic weight determinations of tellurium; discovered that poisonous quality of tobacco smoke is carbon monoxide. Died Nashville, Tenn., Sept. 8, 1914.

DUDLEY, William Russel, Am. botanist; b. Guilford, Conn., Mar. 1, 1849; s. Samuel William and Lucy (Chittenden) D.; B.S., Cornell, 1874, M.S., 1876; studied natural history Agassiz Sch., Penikese Island, 1874, Harvard Summer Sch., 1876; postgrad. univs. Strassburg and Berlin, 1887-88. Instr. botany Cornell, 1873-76, asst. prof., 1876-83, asst. prof. in charge cryptogamic botany, 1883-92; prof. botany, Leland Stanford Jr. U., 1892-1911. Asso. editor Sierra Club Bull., 1898—. Author: The Cayuga Flora, 1886; Lackawanna and Wyoming Flora, 1887; Manual of Histology (with M. B. Thomas), 1894. Founder Dudley Herbarium at Stanford; helped establish redwood parks in Cal. Died June 4, 1911.

DUE, John Fitzgerald, Am. economist; b. Hayward, Cal., July 11, 1915; s. Jackson Angelo and Emmarene (Hurd) D.; A.B., U. Cal. at Berkeley, 1935, Ph.D., 1939; A.M., George Washington U., 1936; m. Margaret Jean Mann, Aug. 18, 1950; children—Allan, Kevin. Instr., U. Utah, 1939-42, asst. prof., 1945-48; economist Treasury Dept., 1942; mem. faculty U. Ill., 1948—, prof. econs., 1951—, chmn. dept., 1963-67. Mem. Am. Econ. Assn., Nat. Tax Assn., Internat. Inst. Pub. Finance, Phi Beta Kappa. Author: Intermediate Economic Analysis, 5th edit., 1966; Government Finance, 4th edit., 1968; State Sales Tax Administration, 1963; Taxation and Economic Development in Tropical Africa, 1963. Co-author: The Electric Interurban Railway in America, 1960. Analysis of sales taxation, including incidence and burden, distbn., econ. effects, adminstrn., structure. Home: 808 Dodds Dr., Champaign, Ill. Office: Com. W., Univ. Ill., Urbana, Ill.*

DUERDEN, J. E., biologist; ed. Royal Coll. Sci., London, Eng., Johns Hopkins, Balt.; M.Sc., Ph.D.; post in zoology Royal Coll. Sci., Dublin, Ireland, also made fishery investigations; curator mus., Kingston, Jamaica, also studied living corals; positions in several Am. univs.; hon. curator Am. Mus. Natural History, N.Y.C.; head expdn. to study Pacific corals in Hawaiian Islands for Carnegie Instn.; prof. zoology Rhodes Univ. Coll., Grahamstown, S. Africa; dir. wool research Union S. Africa, Grootfontein Sch. Agr.; hon. fellow zoology Leeds (Eng.) U.; researcher Wool Industries Research Assn., Leeds. Research and publs. on Irish marine zoology, corals and Actiniaria of W.I., aboriginal remains, Jamaica, morphology of Madreporaria, ostriches and ostrich farming, S. Africa. Died Sept. 4, 1937.

DU FAY, Charles François de Cisternay, French chemist; b. Paris, France, Sept. 14, 1698; became supt. of gardens to King of France. Mem. French Acad. Scis., 1723. Author: Six memoires sur l'electricite, 1733; Mémoires sur le barometre luminieux, 1723. Research and publs. on principles of elec. induction, 1733; discovered positive and negative electricity and repulsion which exists between them; distinguished between electrics and nonelectrics related to scale of increasing conductivities; research on phosphorescence, caustic lime, magnetic needle and double refraction in crystals. Died Paris, July 16, 1739.

DUFAY, Jean Claude Barthélemy, French astronomer; b. Blois, July 18, 1896; s. Pierre and Marie-Louise (Berteloot); Ph.D. in Sci., Faculty Scis., Paris; agrégé of univ.; m. Helénè Toyes; children—Maurice, Claude. Prof., Lycée of Montpellier, 1921, Charlemagne Lycée, 1925, Saint-Louis Lycée, Paris, 1928; astronomer Obs. of Lyons, 1929, dir. 1933; dir. Obs. of Haute-Provence, 1936; prof. Faculty Scis. of Lyons, 1955. Mem. Inst. France, Acad. Scis., Bur. Longitudes (corr.), Royal Astron. Soc. (asso.), Royal Soc. Scis. of Liege (corr.). Author: Nébuleuses Galactiques et Matière Interstellaire, 1954; Introduction a l'Astrophysique, 1961. Research and publs. on light of night sky, spectres of novae, central region of Milky Way. Died Nov. 6, 1967.

DUFF, G(eorge) Lyman, Canadian pathologist; b. Hamilton, Ont., Can., Jan. 26, 1904; s. Charles and Elizabeth Anne (Ostler) D.; B.A., U. Toronto, 1926, M.A., 1927, M.D., 1929, fellow pathology, 1929-31, Ph.D., 1932; NRC fellow medicine Johns Hopkins, 1931-32; m. Isobel Farrell Griffiths, Oct. 23, 1935; children—Sheila Louise, Graham Lyman, Ian Griffiths, Catharine Isobel. Asst. pathology Johns Hopkins, 1931-33, instr. pathology, 1933-35; asst. pathologist Johns Hopkins Hosp., 1933-35; lectr. pathology U. Toronto, 1935-37, asst. prof. pathology, 1937-39; asst. pathologist Toronto Gen. Hosp., 1935-39; asso. editor Am. Jour. Med. Sci. from 1938; prof. pathology, dir. path. inst. McGill U. from 1939, hon. curator Royal Canadian Army Med. Corps Med. Mus., 1941-48, dean faculty medicine from 1949; cons. pathologist Montreal Gen., Childrens Meml., Alexandra, Reddy Meml. hosps., Jewish Hosp. of Hope, Montreal; examiner pathology Royal Coll. Phys. and Surg. Can. from 1944; hon. cons. pathology Royal Victoria Hosp.; lectr. medicine Royal Coll. Physicians Can., 1947; editorial bd. Am. Heart Jour. from 1950, Lab. Investigation from 1953. Certified specialist in pathology Royal Coll. Phys. and Surg. Can. Recipient Flavelle medal, Royal Soc., Can., 1956. Fellow Royal Society Can., Royal Coll. Physicians Can.; mem. Nat. Cancer Inst., Can. (pres. 1954-55), Path. Soc. Gt. Brit. and Ireland, Am. Assn. Pathologists and Bacteriologists (pres. 1954-55), Canadian Cancer Soc. (nat. dir.), Internat. Assn. Med. Mus. (pres. 1950-51), Montreal Medico-Chirurgical Soc., Que. Association Pathologists (pres. 1946-47), Am. Soc. Study Arteriosclerosis (pres. 1951-52), Soc. Exptl. Biology and Medicine, Canadian Med. Assn., Canadian Assn. Pathologists. Contbr. articles med. jours. Died Nov. 1, 1956.

DUFF, Ivan Francis, Am. physician; b. Pendleton, Ore., July 20, 1915; s. Frank and Nina (Jack) D.; A.B., U. Ore., 1939; M.D., U. Mich., 1940; m. Betty Anne Macduff, Feb. 14, 1942; children—David Bruce, Frank Nelson. Faculty, U. Mich. Med. Sch., Ann Arbor, 1946—, prof. internal medicine, 1960—, in charge Rackham Arthritis Research unit, 1953—. Commonwealth Fund fellow in Far East, 1964-65; Faculty Research grantee in Far East, 1967. Fellow A.C.P.; mem. Am. Fedn. Clin. Research, Central Soc. for Clin. Research, Am. Rheumatism Assn., Mich. Rheumatism Assn. (past pres.), Nat. Soc. Rheumatologists. Diplomate Am. Bd. Internal Medicine. Research, publs. on submarine medicine, anticoagulant therapy, diseases of joints. Home: 4 Ridgeway St., Ann Arbor, Mich. 48104.*

DUFFENDACK, O. Stanley, Am. physicist; b. Napoleon, Mo., May 7, 1890; s. John Henry and Caroline (Suhre) D.; student Central Mo. State Coll., 1906-10; B.S., U. Chgo., 1917; A.M., Princeton, 1921, Ph.D., 1922; m. Cora Harriet Chittenden, June 24, 1914; children—Geil Harriet (Mrs. Guy Henderson Orcutt), Stanley. Prin. sch., Lincoln, Mo., 1911-12; prin. high sch. Blaine, Wash., 1912-14; prof. physics Kendall Coll., U. Tulsa, 1917-20; from instr. to prof. physics, U. Mich., 1922-44; pres., dir. research Philips Labs., Irvington, N.Y., 1944-56; hon. fellow physics U. Wis., Madison, 1962—. Dir. N. Am. Philips Co., Amperex Elec. Corp., Ferrox Cube Corp. Sect. chief NDRC, 1943-46, chmn. panel Research and Devel. Bd., 1946-48. Decorated King's medal for service in the cause of freedom (Britain). Recipient President's Certificate of Merit. Fellow Am. Phys. Soc., A.A.A.S., Optical Soc. Am., I.E.E.E. Conglist. Dissociated hydrogen in a tungsten furnace, developed modern method of spectrochemical analysis. Home: 4110 Council Crest St., Madison, Wis. 53711.*

DUFFIEUX, Pierre Michel, French physicist; b. St. Macaire, France, Feb. 21, 1891; s. Francois and Alice (Maneyrol) D.; agrégé Normal Superior Sch., 1920,

D.Sc., 1925; m. Fernande Carrete, Mar. 29, 1921; 1 dau., Monique (Mrs. Christian Le Naour). Asst., lab. of C. Fabry and H. Buisson, 1920-27; lectr. U. Rennes (France), 1927-45; prof. physics and geology Nat. Sch. Agr. Rennes, 1929-40, Nat. Sch. Agr. Grignon, 1940-45; prof. Faculty Scis. Besancon (France), ret., 1963, now instr. In charge works Superior Sch. Commerce Marseilles (France), 1924-27; pres. Commn. Meterology Bretagne, 1932-40. Mem. French Acad. Scis. (officer), French Soc. Physics, Sci. Soc. Bretagne, League of Mental Hygiene. Author: L'intégrale de Fourier et ses applications à l'optique; also papers. Research on photog. resolvant powder, 1917, glue for polarisers, 1917-19, thermal stability and conduction, polarizing crystals, size of spectral rays, discharges in gaseous currents, rolling action of mercury on glass, constrn. of new Fabry standards, measurement of reflection powers, spectrographs with cylindrical lens, atmospheric sodium, azote oxides, harmonic analysis and defects of fringes of interference, photon and corpuscular optics, harmonic analysis of optical images, Fourier transformation in optics and its numerous applications. Home: 92, rue des Granges, Besancon (Doubs), France.*

DUFFORD, Ray Theodore, Am. physicist; b. Bonfield, Ill., Dec. 2, 1891; s. Theodore and Mary A. (Hitchcock) D.; B.S., Northwestern U., 1918, M.S., 1921; postgrad. U. Chgo.; Ph.D., U. Mo., 1931; m. Mamie Ericson, Aug. 16, 1924; children—Marian (Mrs. Floyd F. Helton), Ellen (Mrs. James M. Logan), Catherine (Mrs. Norman L. Paulu). Asst. in physics Northwestern U., 1916-18, instr., 1918-21; instr. physics U. Mo., 1921-31, asst. prof., 1931-44; prof. physics Missouri Valley Coll., Marshall, Mo., 1944-52, chmn. sci. div., 1944-52; prof. physics Evansville (Ind.) Coll., 1952-67, prof. emeritus, 1967—, chmn. dept., 1952-62. Physicist, Radium Inst., Chgo., summers 1922-23; guest research physicist Nela Labs., Cleve., summers 1927-28, USN Underwater Sound Lab., New London, Conn., summers 1952, 54; vis. prof. physics N.M. Highlands U., summer 1940; lectr. astronomy U. Vincennes (Ind.), 1962-65; Planetarium lectr. Evansville Mus., 1968—; cons. Deaconess Hosp., Evansville, 1959—. Fellow Am. Phys. Soc., A.A.A.S.; mem. Am. Physics Tchrs. Assn., Am. Astron. Soc., I.E.E.E. (sr.), Am. Assn. U. Profs., Phi. Beta Kappa, Sigma Xi, Phi Kappa Phi. Contbr. articles on luminescence, photovoltaic effect, raman effect to profl. jours. Home: 512 S. Weinbach Av., Evansville, Ind. 47714.*

DUFFUS, James Edward, Am. plant pathologist; b. Detroit, Feb. 11, 1929; s. John and Dorothy (Pellow) D.; B.S. with high honor, Mich. State U., 1951; Ph.D., U. Wis., 1955; postgrad. Oak Ridge Inst. Nuclear Studies; m. Rachael B. Anderson, May 17, 1952; children—Mark Craig, John Scott, Lisa Kay. Plant pathologist tobacco and sugar crops research br., sugarbeet investigator U. S. Dept. Agr., Salinas, Cal., 1955—. Mem. Am. Phytopath. Soc., Am. Soc. Sugarbeet Technologists, Sigma Xi, Phi Kappa Phi. Research and publs. on nature of virus strains responsible for sugarbeet diseases, their origin, distbn., importance and purification; describer of new virus diseases causing econ. damage on worldwide basis. Home: 114 Primrose Dr. Office: U. S. Research Station, P.O. Box 5098, Salinas, Cal. 93901.*

DUFFY, Elizabeth, Am. psychologist; b. New Bern, N.C., May 6, 1904; d. Francis and Lida (Patterson) D.; A.B., Woman's Coll. U. N.C., 1925; M.A., Columbia, 1926, postgrad. (NRC fellow); Ph.D. (NRC scholar), Johns Hopkins, 1928; m. John E. Bridgers, Jr., Aug. 27, 1938; 1 dau., Elizabeth Duffy. Faculty, Sarah Lawrence Coll., 1929-37; prof. psychology Woman's Coll. U. N.C., Greensboro, 1937—; part-time vis. prof. N.Y. U., 1931, Bklyn. Coll., summer 1936; vis. scientist NSF various colls., Va., 1959. So. Fellowships Fund grantee, 1959; Research Council U. N.C. grantee-in-aid, 1959-62, 64-66; certificate of Meritorious Service City Greensboro for services to Redevel. Commn., 1964. Mem. Am. (council mem.), Southeastern psychol. assns., Soc. For Research in Psychophysiology, Psychonomic Soc., So. Soc. Philosophy and Psychology, A.A.A.S., Am. Assn. U. Profs., Phi Beta Kappa, Sigma Xi. Author: Activation and Behavior, 1962. Publs. of proposed concepts in descriptive psychology particularly idea of activation. Home: 1412 W. Lake Dr., Greensboro, N.C. 27408.

DUFFY, James Joseph, Am. physician; b. Webster, Mass., 1892; s. James H. and Elizabeth (Curwin) D.; A.B., Holy Cross Coll., 1915; M.D., Harvard, 1918. Contbd. numerous papers in jours. Specialized in radium and x-ray treatment of cancer; noted for judicious use of combined surgery and radiation treatment. Died 1941.

DUFOUR, Didier, Canadian immunologist; b. St. Hilarion, Charlevoix, Que., Can. Feb. 7, 1926; s. Adrien and Lumina (Tremblay) D.; B.A., U. Montreal, 1947, V.M.D., 1952, Ph.D., 1956; m. Ghislaine Simard, Oct. 16, 1954; children—Danielle, Andre, Renee, Jean Didier et Robert. Research asst. Inst. Exptl. Medicine U. Montreal, 1953; research asst. Inst. Human Physiology U. Laval, Que., Can., 1954-56; asst. prof. dept. biology U. Ottawa, 1956; faculty U. Laval, 1957—, asso. prof. immunochemistry, 1964-67, director biochem. center, 1965—, prof. biochemistry (immunochemistry), 1967—. Author: (with others) Precis de Biochimie, 1965. Research and publs. on tumor immunity produced by a specific antigenic factor which appears in the livers of exptl. animals after induction of tumors. Home: 1636 Chemin Ste-Foy, Quebec. Office: Centre de Biomedecine, University of Laval, Quebec 10, Que., Can.*

DUFOUR, Jean Marie Léon, French entomologist; b. Landers, France, Apr. 10, 1782; Docteur en médecine, 1806; became physician Spanish Expeditionary Force, 1823; practiced medicine at Saint-Sever, France, beginning in 1826. Recipient Prix Cuvier Acad. Scis., 1861. Mem. French Acad. Scis., Soc. Agr. Author: Recherches anatomiques et physiologiques sur les hémiptères, 1833; Propriétés des végétaux et leur application à l'alimentation, 1861; Research and publs. on anatomy and physiology of different families of insects which extended work of Straus-Dürchheim; research on plants and their relation to alimentation; anatomy and physical studies of hemiptera insects. Died Apr. 18, 1865.

DUFOUR, Louis, Swiss physicist; b. Lausanne, Switzerland, 1832; hon. doctorate Basel (Switzerland) U.; prof. physics Lausanne U., 1853-77; research on supercooled melted sulphur and phosphorus; determined boiling points, 1861; studied earth streams, polarization in the base of submerged metallic conductors, 1866, diffusion of gas through porous tissues. Died 1892.

DU FOUR, Vital, French natural philosopher; studied under Jacques de Carceto; lectured at U. Montpellier (France), 1295-96; master of theology; joined Franciscan Order; became minister of Aquitanian province of the Franciscan Order, 1307; made cardinal by Clement VI, 1312; and was given various monasteries in commendam. Wrote commentary on the four books of the Sentences which included material on natural phenomena including magnetic attraction, phosphorescence. Author: Pro sanitate, which includes material on water, metals, plant and animal substances. Died 1327.

DUFRÉNOY, Ours Pierre Armand Petit, French geologist; b. Sevran, France, Sept. 5, 1792; s. Adelaide Gillette (Billet) D.; ed. École polytechnique, 1811. Dir. Sch. Mines. Mem. French Acad. Scis., 1840. Author: Mémoires pour servir à une description géologique de la France, 1836-38 (with Beaumont); Traité complet de minéralogie, 1844. Treatise on Mineralogy, 4 vols., 1847; also geol. monographs. Produced (with Elie de Beaumont) geol. map of France; introduced new system of classification based on crystallography; studied volcanic formations. Died Paris, Mar. 20, 1857.

DUFTSCHMID, Caspar Erasmus, Austrian physician, entomologist; b. Gmunden, Austria, Nov. 19, 1767; s. Anton Duftschmid; M.D., U. Vienna (Austria), 1790; m. Theresia Elsasser v. Grünwald, 1792; son, Johann. Began practice medicine in Linz, Austria, 1791; became dist. physician, 1815; became first physician for Austria above the Enns, 1819. Author: Fauna Austriae oder Beschreibung der Österreichischen Insecten für angehende Freunde der Entomologie, 1805-26; Tractatus de scarlatina, 1820. Described many new types of insects; used priority prin. in nomenclature. Died Linz, Dec. 17, 1821.

DUGAN, Raymond Smith, Am. astronomer; b. Montague, Mass., May 30, 1878; s. Jeremiah Welby and Mary Evelyn (Smith) D.; B.A., Amherst, 1899, M.A., 1902; Ph.D., Heidelberg, 1905; m. Annette Rumford Odiorne, July 28, 1909; children—Kenneth Langdon, Hannah Priscilla. Instr. math. and astronomy, acting dir. obs. Syrian Protestant Coll., Beirut, Syria, 1899-1902; asst. Astrophys. Obs., Heidelberg, 1902-04; instr. astronomy Princeton, 1905-08, asst. prof., 1908-20, prof., 1920—; exchange prof. Lowell Obs., 1929; acting dir. Princeton U. Obs., 1929-30. Mem. Lick Obs. eclipse expdn. to Spain, 1905. Research on elipsing variable stars (double stars); noted for accuracy of photometric observations; made numerous measurements of stars. Died Aug. 31, 1940.

DUGÉS, Antoine-Louis, French physician; b. Mezieres, France, 1797; docteur en médecine; aggregation from Paris (France) Faculty of Medicine, 1824; prof. childbirth and external pathology U. Montpellier (France). Mem. Acad. Medicine. Author: Recherches sur les maladies des nouveaux-nés, 1821; Traité de physiology comparée de l'homme et des animaux, 1838; Mémoire sur la conformité organique dans l'échelle animale. Research and publs. on diseases of new-born, comparative physiology of man and animals; tried to integrate the doctrines of Brown, Broussais and Hippocrates; renewed classification of animals, 1832. Died Montpellier, 1838.

DUGGAN, Hector Ewart, Canadian physician; b. Medicine Hat, Alta., Can., July 3, 1915; s. Hector Ossian and Jacqueline (Ewart) D.; M.D., U. Alta., 1938; m. Sadie Jones, Aug. 17, 1940; children—Frances Layne (Mrs. Kenneth Charles Cuthbert), Jacqueline Louise, Ronald Hector. Practice gen. medicine, 1940-41; staff U. Hosp., Edmonton, Alta., 1946-65; dir. dept. radiology Foothills Hosp., Calgary, Alta., 1965—; faculty U. Alta., 1946-65, prof. radiology, 1957-65. Mem. Canadian Med. Assn., Canadian Assn. Radiologists, Am. Coll. Radiology, Pacific N.W. Radiol. Soc., Radiol. Soc. N.Am. Research and publs. on diagnostic and therapeutic radiology, nuclear medicine, measurement of radiation dosage from diagnostic radiography, radioactive isotopes, assessment radiation damage by peripheral cell cultures, mechanism underlying response of polycythemia rubra vera to radiophosphorus. Home: 2032 Uralta Rd. Office: Foothills Hosp., Calgary, Alta., Can.*

DUGGAR, Benjamin Minge, Am. botanist; b. Gallion, Ala., Sept. 1, 1872; s. Reuben Henry and Margaret Louisa (Minge) D.; student U. Ala., 1887-89; B.S., Miss. A. and M. Coll. (1st honors), 1891; M.S., Ala. Poly. Inst., 1892; A.B., Harvard, 1894, A.M., 1895; Ph.D., Cornell, 1898; postgrad. German univs. and at Naples, Paris, Montpelier, 1899-1900, 1905-06; LL.D U. Mo., 1944; D.Sc., Washington U., 1953; m. Marie L. Robertson, Oct. 16, 1901 (dec. 1922); children—Marie Louise, Benjamin Minge, Anna St. Julian Guerard, George Strowan, Emily Westwood; m. 2d, Elsie Rist, June 6, 1927; 1 dau., Gene Lorraine. Asst. dir. Uniontown (Ala.) Agrl. Expt. Sta., 1892-93; asst. botanist Ill. Lab. Natural History, 1895-96; cryptogamic botanist Agrl. Expt. Sta., also instr. plant physiology Cornell U., 1896-1900, asst. prof., 1900-01, prof., 1907-12; physiologist bur. plant industry U. S. Dept. Agr., 1901-02; prof. botany U. Mo., 1902-07; research prof. plant physiology Mo. Bot. Garden and Washington U., 1912-27; prof. physiology and econ. botany U. Wis., 1927-43; cons. mycological research and prodn. Lederle Labs. div. American Cyanamid Co., Pearl River, N.Y., 1944. Acting prof. biology chemistry Washington U. Med. Sch., 1917-19. Recipient Medal of Honor of Pub. Edn., Venezuela, 1951. Fellow A.A.A.S.; mem. Nat. Acad. Scis., Am. Philos. Soc., Phila Acad. Sci., Bot. Soc. Am. (pres. 1923), Am. Soc. Plant Physiology (pres. 1946-47), Am. Phytopathol. Soc., Am. Chem. Society, Soc. Am. Naturalists, Am. Pub. Health Assn., Torrey Bot. Club, NRC (chmn. div. biology and agr. 1925-26), Sigma Xi, Phi Beta Kappa. Editor for physiology Bot. Abstracts, 1917-26, Biol. Abstracts, 1926-33, Biol. Abstracts of Radiation, 1936. Author: Fungous Diseases of Plants, 1909; Plant Physiology, 1911; Mushroom Growing, 1915; A Textbook of General Botany (with G. M. Smith, et al); also research articles. Discovered and introduced aureomycin, 1948. Died Sept. 10, 1956.

DUGGAR, John Frederick, Am. agriculturist; b. Faunsdale, Ala., Aug. 24, 1868; s. Reuben Henry and Margaret Louisa (Minge) D.; ed. So. U., Greensboro, Ala.; B.S., Miss. A. and M. Coll., 1887, M.S., 1888; student Columbian (now George Washington) U., Cornell U., U. Colo.; m. Frances Ambrose Camp, June 17, 1891; children—John Frederick, Frances Camp, Mrs. Margaret McCormick, Ambrose Camp, Llewellyn Goode, Dorothy. Asst. prof. agr. Tex. A. and M. Coll., Bryan, 1887-89; editor So. Live Stock Jour., Starkville, Miss., 1890; asst. dir. Expt. Sta., Clemson Coll., S.C., 1890-92; editor dept. field crops Expt. Sta. Record, U. S. Dept. Agr., 1893-95; prof. agr. Alabama Poly. Inst., 1896-1921, research prof. farm mgmt., spl. investigator, 1922-31, research prof., from 1931; also dir. Ala. Expt. Sta., 1903-21, dir. Ala. Extension Service, 1914-20. Lectr. agronomy, U. Cal. 1922-24. Recipient medal for distinguished service Assn. So. Agrl. Workers, 1939. Mem. Phi Beta Kappa. Author: Agriculture for Southern Schools; Southern Field Crops; Southern Forage Crops; also numerous articles and pamphlets. Research on field and forage crops of South, cotton leguminous plants, corn, oats, animal feeling, plant breeding. Died Dec. 25, 1945.

DÜGGELI, (Franz) Max, Swiss bacteriologist; b. Lucerne, Switzerland, July 29, 1878; s. Karl D. and Martina (Meyer) D.; student agr. Fed. Polytechnikum, Zurich, Switzerland, 1897; agrl. engring. degree, 1900; doctorate U. Zurich, 1902; m. Maria Emma Weber, 1907; 2 son, 2 daus. Asst., Zurich Polytechnikum (now Tech. U.); made study trip to Agrl. U., Berlin; joined faculty agrl. bacteriology Tech. U. Zurich, 1907; named titular prof., 1910; became 1st prof. agrl. bacteriology in Europe; taught until 1946. Mem. Swiss Sci. Nat. Park Commn. Mem. Schweizerische Naturforschenden Gesellschaft. Research and numerous publs. on soil and milk bacteriology especially relationship between higher vegetation and soil microorganisms, water conservation. Died Zurich, Aug. 14, 1946.

DUGUE, Daniel, mathematician; b. St. Louis, Senegal, Sept. 22, 1912; s. Pierre and Berthe (Crepin) D.; student Normal Superior Sch., 1930-33; Univ. Coll., London, 1937-38; D.Sc., 1937; m. Lucie Canaud, Nov. 30, 1941; children—Catherine (Mme. Corbasson), Elisabeth, David, Marc. Mem. Faculty Scis. Algiers, 1942-48, Caen, 1948-53, Paris, 1953—; faculty Polytech. Sch., 1952—; dir. statistics inst. U. Paris, 1960—. Mem. Internat. Inst. Statistics; corr. Bur. Longitudes. Author: Arithmétique des lois de probabilités, 1958; Traité de statistique théorique et appliquée, 1959; also articles. Research on applications of theory of functions to calculation of probabilities, also on arithmetic of laws of probabilities, applied statistics. Home: 24 J-L Sinet, Sceaux 92, France. Office: 9, Quai St. Bernard, Paris 5, France.*

DUGUNDJI, James, Am. mathematician; b. N.Y.C., Aug. 30, 1919; s. Basil and Rosa (Finale) D.; B.A., N.Y. U., 1940; Ph.D., Mass. Inst. Tech., 1948; m. Merope Alban, Dec. 23, 1944. Teaching fellow U. N.C., 1940-42; research asso. Mass. Inst. Tech., Cambridge, 1946-48; faculty U. So. Cal., Los Angeles,

1948-51, 1953—, prof. math., 1958—; mem. Inst. for Advanced Study, Princeton, N.J., 1951-53; guest prof. U. Frankfort/Main (Germany), 1965-66; research mathematician U. Pisa (Italy), 1966; vis. prof. Rice U., 1967-68. Mem. Am. Math. Soc., Phi Beta Kappa, Sigma Xi. Editor: Pacific Jour. Math., 1963—. Contbr. research articles to math. jours. Author: Topology, 1965. Research on information theory, deformation topology. Home: 400 S. Hauser Blvd., Los Angeles, 90036.

DUHAMEL, Bernard Georges, French surgeon; b. Paris, France, May 11, 1917; s. Georges and Blanche (Sistoli) D.; M.D., Faculty Medicine Paris, 1946; m. Liliane Amati, Sept. 26, 1959; children—Pascal, Claude-Alain, Dominique, Jerôme, Isabelle. Became anat. asst. Faculty Medicine Paris, 1942, demonstrator anatomy, 1946, head clinic, 1946; asst. surgeon hosps. Paris, 1948; became head surgeon dept. pediatric surgery Hosp. St. Denis, 1954; aggregate prof., 1955—. Mem. Soc. Pediatrics, Soc. Orthopedics, Soc. Infantile Surgery, Soc. Plastic Surgery (all French), Brit. Assn. Pediatric Surgeons, German, Spanish socs. infantile surgery. Author: Nourrisson, 1953; Technique chirurgicale infantile, 1957; Morphogénèse pathologique, 1966; also papers. Research on surgery of the newborn and of congenital malformations, causes and treatment of malformations, treatment of diaphragmatic hernia and ano-rectal malformations, new operation for Hirschsprings' disease. Home: 31, rue de Liège, Paris 8. Office: Hôpital de Saint-Denis, Seine, France.*

DUHAMEL, Gerard, French physician; b. Paris, France, Jan. 28, 1918; s. Victor and Laure (Bejot) D.; m. Nicole Moussié, July, 1942; children—Veronique, Laurent, Catherine, Nicolas. Dr. in medicine, 1946—; prof. agregé Faculté de Médecine, Paris, France, 1961—; mem. staff Hosp. St. Antoine, Paris, 1964—. Mem. Internat. Soc. Haematology. Author: (with G. Marchal) Atlas of Bone Marrow Biopsy, 1962, Daily Consultations in Haematology, 1966. Research, publs. on cancer propagations in bone marrow; pathology of bone marrow obtained by needle-biopsy; spl. histology of haematological tissues. Office: Hosp. St. Antoine, Paris 12, France.*

DU HAMEL, Jean-Baptiste, French astronomer, physicist; b. Vire, Normandy, France, June 11, 1623; studied at Paris (France) Sch. Oratory, 1643-53. Curé of Neuilly-sur-Marne; almoner to King Louis XIV; prof. physics Bourgogne Coll., Mem. French Acad. Scis., 1666 (1st perpetual sec.). Author: Astronomia physica, 1659; De meteoris et fussilibus, 1659; Histoire de l'Académie des Sciences. Pub. work criticizing Descartes; also wrote of the spheres of Theodosa. Died Paris, Aug. 6, 1706.

DUHAMEL, Jean-Marie-Constant, French mathematician; b. Saint-Malo, France, Feb. 5, 1797; ed. at École polytechnique; licentiate, 1814. Founder, École Sainte-Barbe; prof. Faculty Scis., U. Paris (France), also of École polytechnique. Mem. French Acad. Scis. Proposed a theory of the transmission of heat in crystals; improved equations of elastic equilibrium of Navier; research on vibration of gases in tubes; invented recording apparatus, 1840. Died Paris, Apr. 29, 1872.

DUHAMEL, Joseph Pierre, French physicist; b. Prades, France, July 26, 1920; s. Bernard and Elisabeth (Perebosch) D.; Ingeneer, Ecole Polytechnique, 1940; Dr., Faculty Medicine, Bordeaux, France, 1950; m. Anita Reiss, Nov. 7, 1964; daus., Monique, Hélène. Prof. med. physics Faculté de Médecine, Bordeaux, since 1965. Med. counsellor Office professionnel de Prévention du Batiment et des Travaux Publics. Mem. Institut Elec. Engrs. Author: Traité de Physique, 1964; Précis de Physique Médicale, 1962; also numerous articles. Research on tomography (radiol. method for obtaining body sects.). Home: 4, rue Gabriel Fauré-33-Pessac, France. Office: Laboratoire de Physique, Faculté de Médecine, Rue Leyteire, Bordeaux, France.*

DUHAMEL DU MONCEAU, Henri Louis, French chemist, botanist; b. Paris, 1700; student of Jussieu, Paris; insp.-gen. of navy. Mem. French Acad. Scis. (dir. 1743, 56, 68), French Agrl. Soc., Naval Acad., royal socs. London, Edinburgh, acads. St. Petersburg, Stockholm, Palermo, Padua, Inst. Bologna, Soc. Agr., Leiden; free asso. Royal Soc. Medicine. Author: Treatise on fishing, and history of the fish which they furnish, 1782. Research on plant nutrition and diseases, drugs, animals; 1st to distinguish between potash and soda, 1736. Died Paris, Aug. 23, 1782.

DUHEM, Pierre, French physicist, historian of science; b. Paris, June 10, 1861; ed. Stanislas Coll., Paris, Ecole normale supérieure, Paris, 1881-85, aggrégation, 1885; became lectr. Lille (France) Faculty Scis., 1887; prof. theoretical physics U. Bordeaux (France), 1893-1916. Mem. French Acad. Scis., 1913. Author: Le potential thermodynamique et ses applications à la mécanique chimique et à la théorie des phénomènes électriques, 1886; Commentaires sur le thermodynamique; Traité d'énergetique générale; Leonardo de Vinci, ceux qu'il a lus et ceux qui l'ont lu; Le système du monde (with numerous anonymous collaborators), 8 vols.; La théorie physique; son object, sa structure; others. Introduced notion of thermodynamic potential; tried to construct a general energetics and abstract thermodynamics using axiomatic-deductive approach (had antipathy to all pictorial models; hence vigorous opponent of atomism); spread ideas of Willard Gibbs in France; studied hydrodynamics, theory of elasticity, propagation of waves of impact in fluids; preferred electromagnetic ideas of Helmholtz to those of Maxwell; pioneer in history of medieval science; argued advances in Renaissance mechanics and astronomy had deep medieval roots; work of Galileo, etc., was a continuous devel. of work of medieval scholars, esp. those at U. Paris; in philosophy of science, held for separation of physics from metaphysics; usually ranked among positivists (but in fact his position is very subtly qualified positivism). Died Sept. 14, 1916.

DÜHREN, Eugen, see Bloch, Iwan.

DÜHRING (Karl) Eugen, German economist; b. Berlin, Germany, Jan. 12, 1833; s. Wilhelm Ferdinand and Auguste Friederike Louise (Zeibig) D.; student law, philosophy and econs. U. Berlin, 1856-59; Ph.D., 1861; m. Emilie Gladow, 1862; 2 sons including Ulrich. Became lectr. philosophy U. Berlin, 1863, later lectr. econs., until 1877; ind. scholar and writer afterwards; pub. mag. Personalist und Emanzipator, 1899-1921. Recipient prize Philosophy Faculty, U. Göttingen (Germany). Author: Kritische Geschichte der Philosophie von ihren Anfängen bis zur Gegenwart, 1869; Logik und Wissenschaftstheorie, 1878; Kritische Geschichte der Nationalökonomie und des Sozialismus von ihren Anfängen bis zur Gegenwart, 1871; Kritische Geschichte der allgemeinen Prinzipien der Mechanik, 1873; Neue Grungesetze zu rationellen Physik und Chemie, 1878. Pioneered history of sci. especially econs. and sociology; forwarded philosophy of optimistic positivism, which derived from Auguste Comte; in socioeconomic thought, objected to theories of both Marxists and utopian socialists; proposed modified capitalism. Died Nowawes nr. Potsdam, Germany, Sept. 21, 1921.

DUHRING, Louis Adolphus, Am. dermatologist; b. Phila., Dec. 23, 1845; s. Henry and Caroline (Oberteuffer) D.; M.D., U. Pa., 1867; postgrad. dermatology in hosps. of Paris, London and Vienna. Opened a dispensary for skin diseases in Phila., 1870, of which was physician, 1870-80, cons. phys., 1880—; clin. lectr. U. Pa., 1871-76, prof. diseases of skin, from 1876. Author: Atlas of Skin Diseases, 1876; Practical Treatise on Diseases of the Skin, 1877; Cutaneous Medicine, 1898; also numerous other works on dermatology. Reported form of dermatitus now known as Duhring's disease (dermatitus multiformis or herpetiformis), 1884. Died May 8, 1913.

DÜHRSSEN, Alfred, German gynecologist, obstetrician; b. Heide, Germany, Mar. 23, 1863; s. Jacob and Adele (Dohrn) D.; student medicine U. Marburg (Germany), Kaiser-Wilhelm Akademie, Berlin. Germany; m. Gertrud Passmann, 1888; 3 sons. Mil. physician; later asst. Frauenklinik, Königsberg, Germany; became asst. Obstet. Polyclinic, Berlin, 1886; named midwife instr., lectr., 1888, prof. 1895; founder pvt. clinic for women's diseases, 1892; resigned from faculty U. Berlin, 1913. Author: Geburtshilfliches Vademecum, 1890; Gynäkologisches Vademecum, 1891; Der vaginale Kaiserschnitt, 1896; Die Einschränkung des Bauchschnittes durch die vaginale Laparotomie, 1899; Die neue Geburtshilfe, 1923; also numerous articles. One of founders of modern surg. gynecology; 1st to perform vaginal cesarean sect.; developed other vaginal operation methods; research on abdominal cancer. Died Berlin, Oct. 11, 1933.

DUISBERG, (Friedrich) Carl, German chemist; b. Barmen, Germany, Sept. 29, 1861; s. Carl and Wilhelmine (Weskott) D.; doctorate chemistry U. Jena (Germany), 1882; hon. dr. degrees from univs. of Dresden, Munich, Bonn, Tübingen, Heidelberg, Cologne, Berlin, Marburg (all Germany); m. Johanna Seebohm, 1888; 3 sons including Carl Ludwig, Curt; 1 dau. Joined as chemist Farbenfabriken Bayer, Elberfeld, Germany, 1883, later held adminstrv. positions, founded its prodn. of medicines, set up sci. labs, became mem. bd., 1900; after merger of German chem. industries into IG-Farbenindustrie, became chmn. bd., 1925. Mem. Nat. Assn. German Industry (chmn.). Important in chem. industry of Germany. Died Leverkusen, Germany, Mar. 19, 1935.

DUJARDIN, Félix, French zoologist; b. Tours, France, Apr. 5, 1801; self educated; became librarian and prof. in Tours, Grenoble and Toulouse (all France), 1839; prof. zoology, Rennes, France. Mem. French Acad. Scis., 1859. Author: Histoire naturelle des infusoires, 1840; Manuel de l'observateur au microscope, 1842; Histoire naturelle des helminthes, 1844. Research on geology, botany and zoology, especially cytology; studied Infusoria and bacteria; described protoplasm in unicellular animals, naming it sarcode, Died Rennes, Apr. 8, 1860.

DUKE, William Waddell, Am. physician; b. Lexington, Mo., Oct. 18, 1882; s. Henry Buford and Susie (Waddell) D.; Ph.B., Yale, 1904; M.D., Johns Hopkins, 1908; postgrad. Mass. Gen. Hosp., 1909-10, U. Vienna, 1910-12, U. Berlin; m. Frances Thomas, May 18, 1920; children—Henry Basil, Frances Suzanne. Practiced medicine, specializing in internal medicine, Kansas City, Mo., from 1912; prof. exptl. medicine U. Kan. Sch. of Medicine, Rosedale, 1914-18; vis. physician Christian Ch. Hosp., Kansas City, 1918-24. Fellow A.M.A.; mem. A.C.P. Recipient silver medal, A.M.A. for research in allergy, 1924; annual gold medal Midwest Forum of Allergy for outstanding contbns. in field of allergy., 1941. Author: Oralsepsis in Relationship to Systemic Disease, 1918; Allergy, Asthma, Hay Fever, Urticaria and Allied Manifestations of Reaction, 1925; also chapters in Practitioners' Library of Medicine and Surgery, Cyclo. of Medicine, Modern Home Med. Adviser. Contbr. tech. articles. Discoverer in field of allergy and physical allergy, oral sepsis transfusion and anemia, palm color test, bleeding time, relation between platelets and hemorrhagic disease; co-discoverer physiology of heart beat in relationship to the potassium and calcium content of the blood; also made pollen surveys. Died Apr. 10, 1946.

DUKE-ELDER, Sir William Stewart, Brit. surgeon, ophthalmologist; b. Dundee, Apr. 22, 1898; s. Neil and Isabella (Duke) D.-E.; ed. U. St. Andrews; M.A., Ph.D., D.Sc., LL.D., M.D. Cons. ophthal. surgeon St. George's Hosp, London, Moorsfield Hosp., Brit. army and air force; pres. Inst. Ophthalmology, U. London. Mem. Ophthal. Soc., U.K., U. S., Can., Mexico, France, Switzerland, Netherlands, Belgium, Denmark, Sweden, Italy, Greece, Egypt, India ophthal. socs. Author: Textbook of Ophthalmology; System of Ophthalmology; Diseases of the Eye; The Practice of Refraction, also articles. Home: 28 Elm Tree Rd. Office: 63 Harley St., London W.1, Eng.

DU LAURENS, André, French physician; b. Tarascon, France, 1558; prof. U. Montpellier (France), 1586-98; physician ordinary to Henry IV; 1st physician to Marie de Médicis; chancellor Montpellier U. Author: Apologia pro Galeno, 1593 (translated into French, 1646); Historia anatomica humani corporis (most thorough history of anatomy of the time), 1595; De crisibus libri tres, 1596. Supported Galen; described fetal cardio-pulmonary vessels; publs. on scrofula, gout, old age, syphilis. Died Paris, France, 1609.

DULBECCO, Renato, biologist; b. Catanzaro, Italy, Feb. 22, 1914; s. Leonardo and Maria (Virdia) D.; M.D., U. Torino (Italy), 1936; m. Giuseppina Salvo, June 1, 1940 (div. 1963); children—Peter Leonard, Maria Vittoria; m. 2d, Maureen Muir, July 17, 1963. Came to U. S., 1947, naturalized, 1953. Asst. U. Torino, 1942-47; research asso. Ind. U., 1947-49; sr. research fellow Cal. Inst. Tech., 1949-52, asso. prof., then prof. biology, 1952-63; sr. fellow Salk Inst. Biol. Studies, San Diego, 1963—; vis. prof. Royal Soc. Gt. Britain, 1963-64; Clowes Meml. lectr., Atlantic City, 1961. Mem. Cal. Cancer Adv. Council, 1963. Guggenheim and Fulbright fellow, 1957-58; recipient John Scott award City Phila., 1958; Kimball award Conf. Pub. Health Lab. Dirs., 1959; Albert Lasker Med. Research award, 1965; Ricketts award, 1965; adj. prize to Paul Ehrlich and Ludwig Darmstaedter prize, 1966. Mem. Nat. Acad. Scis., A.A.A.S., Genetics Soc. Am., Assn. Cancer Research. Discovered phenomenon of photoreactivation of phage inactivated by ultraviolet light; invented method for assaying animal virus by plague formation in tissue culture; physiology and genetics of animal viruses; discovered mutants of poliovirus; current research in biology of viruses causing cancer. Home: 7206 Rue de Roark, La Jolla, Cal. 92037. Office: Salk Inst., P.O. Box 1809, San Diego 92112.*

DULCO, Gaston, see Duclo, Gaston.

DULONG, Pierre Louis, French chemist, physicist; France, Feb. 12, 1785; studied medicine École polytechnique, 1801-03. Practiced medicine Paris, France; later worked on botany, then chemistry in Berthollet's Lab.; maitre de conférénces École normale; prof. chemistry Faculty Scis., also École d'Alfort, Paris; became prof. physics École polytechnique, 1820, apptd. dir., 1830. Fellow Royal Soc., 1826. Discovered nitrogen chloride, 1811; nitrogen trichloride (an explosive), 1813; proposed that acids were compounds of hydrogen which could be replaced by a metal, 1815; research (with Petit) on refractive indices and specific heats of gases; formulated Dulong's and Petit's law that specific heats of elements are inversely proportional to their atomic weights and of the constancy of atomic heats, 1819; devised empirical formula for the calculation of the heat value of fuels from their chem. composition (Dulong's formula); studied elasticity of steam at high temperatures. Died Paris, July 18, 1838.

DUMANSKII, Anton Vladimirovich, Russian chemist; b. Apr. 20, 1880; grad. Kiev (USSR) Poly., 1903, remained until 1913; organized lab. for colloidal chemistry, Voronezh, USSR, 1913, reorganized into All-Union Inst. for Colloidal Chemistry, 1932, dir., until 1942; dir. inst. gen. and inorganic chemistry USSR Acad. Scis., Kiev, 1946—, also corr. mem. acad. Author: Liophilicity of Dispersed Systems, 1940; Study of Colloids, 1948. Founder, editor Colloidal Jour. Devised phys. methods for research in colloidal chemistry; introduced collodion membranes in place of animal membranes; introduced powerful centrifuge for use in measuring size of colloidial particles. Address: Inst. Gen. and Inorganic Chemistry, Ulitsa Leontovicha 9, Kiev, Ukranian SSR, USSR.

DUMAS, Charles-Louis, French physician; b. Lyons, France, Apr. 8, 1765; docteur en médecine U. Montpellier (France), 1785; physician Montpellier Hôtel-Dieu; prof. anatomy and physiology Faculty of Montpellier, later rector. Recipient award Royal Soc. Medicine, 1785. Mem. French Acad. Scis., 1800. Author: Essai sur la vie, 1785; Système de classification des muscles du corps humain, 1797; also publs. against the vitalist sch. Classified human muscles according to their attachments in order to simplify the existing nomenclature; studied asphyxiation through chem. changes in the air during respiration; studied chronic diseases and divided them into simple and complex diseases. Died Montpellier, Apr. 3, 1813.

DUMAS, Georges, French physician, psychologist; b. Lédignan, France, 1866; ed. École normale supérieure, docteur ès lettres, docteur en médecine. Became asst. prof., 1902; apptd. titulary prof. at the Sorbonne, Paris, France, 1912; founded several French insts. including those at Buenos Aires, Argentina; Santiago, Chile; Rio de Janeiro, Brazil. Mem. Acad. Medicine, Acad. Moral and Polit. Scis. Author: Philosophie de Léon Tolstoi, 1893; les États intellectuels dans la melancolie, 1894; la Tristesse at la joie, 1900; Troubles mentaux et nerveux de la guerre, 1919; Traité de psychology, 10 vols., 1930. Founder, (with Pierre Janet) Journal de psychologie. Research on relation between war and mental disorders, the psychology of several famous figures such as Tolstoy, A. Comte, Saint-Simon. Died Lédignan, 1946.

DUMAS, Jean Baptiste André, French chemist; b. Alais, France, July 15, 1800; ed. under Theodore de Saussure, De Candolle, Gaspard de la Rive, in Geneva, Switzerland, also in Paris, France. Apprentice to apothecary, Alais; entered pharm. lab. of Le Royer, Geneva, 1816; became lecture asst. to Thenard, École Polytechnique, Paris, 1823, succeeded him as prof., 1835; evening lectr. Athenaeum; co-founder L'École Central des Arts et Manufactures, 1829; prof. organic chemistry École de Medecine; succeeded Gay-Lussac as asst. prof. Sorbonne, Paris, 1832, prof., 1841; dean faculty of sciences; lectr. Coll. France; deputy in French assembly, minister agr. and commerce, 1849-51; minister edn., French senator under the Second Empire. Fellow Royal Soc., 1840; mem. French Acad. Scis., 1832 (perpetual sec. from 1868), Acad. Medicine, French Acad. Author: Traité de chimie appliquée aux arts, 8 vols., 1828-48; Leçons sur la philosophie chimique, 1837; Essai de statique chimique des etres organisés, 1841. also numerous articles. Worked (with Coindet) on use of iodine compounds as goiter cure, also (with Jean Louis Prevost) on elec. muscular phenomena and blood corpuscles, 1818; enumerated some differences between animals and vegetables, 1818; discovered anthracene in coal tar; discovered cymene; obtained oxamide; worked (with Boullay) on ether; discovered (with Peligot) methyl alcohol, 1834; investigated atomic weights, atomic theory, theory of chem. types, fermentation, vapor intensity of elements, hydrogen and amide compounds, med. and physiol. chemistry; studied (with Boussingault) composition of water and atmosphere, (with Stas) composition of carbon monoxide; discovered homologues; discovered combustion method for determination of nitrogen; classified organic compounds according to reactive group; studied theory of substitution in organic compounds; supported unitary view of matter; opposed dualistic theory of chem. structure held by Berzelius and Liebig; put forth theory of replacement. Died Cannes, France, Apr. 11, 1884.

DUMBLE, Edwin Theodore, Am. geologist; b. Madison, Ind., Mar. 28, 1852; s. James F. and Mary A. D.; B.S., Sc.D., Washington and Lee U., Va.; m. Fanny Doswell Gray, June 15, 1876; children—Mrs. Milly Gray Mitchell, Mrs. Rosalie McCoy Davis. State geologist of Tex., 1887-96; cons. geologist and mgr. oil properties So. Pacific Co. (Rio Bravo Oil Co., Tex., East Coast Oil Co., Mexico), 1897-1925; cons. geologist Pacific Oil Co.; cons. practice, Houston. Fellow Geol. Soc. Am., A.A.A.S., Tex. Acad. Sci. Author: Brown Coal and Lignite; Geology of East Texas. Died Jan. 26, 1927.

DUMÉE, Jeanne, French astronomer; b. Paris, France; flourished 1680-85; ed. privately. Author: Entretiens sur l'opinion de Copernic touchant la mobilité de la terre, in which she defended Galileo and pointed out errors in the Bible.

DUMÉRIL, André Marie Constant, French zoologist; b. Amiens, France, Jan. 1, 1774; studied pharmacy, Rouen, France; student Sch. Health, Paris children—Henri Andre, Auguste, others. Asst. in anatomy Rouen Hosp.; became prof. anatomy and pathology Paris, (France) Faculty of Medicine, 1801; prof. zoology and ichthyology Paris Mus. Natural History; installed reptile collection Mus. Natural History; named prof. herpetology Mus. Natural History, Paris, 1803. prof. herpetology Paris Bot. Garden; taught anatomy to Cuvier. Mem. French Acad. Scis., 1816 (became pres. 1831), Acad. Medicine. Author: Traité d'erpétologie, 10 vols.; Ichthyologie analytique, 1856; (with Bibron) Erpétologie générale, 9 vols. and atlas, 1834-54; Pioneered research on reptiles and amphibians, called father of herpetology; made analytic studies in ichthyology and entomology; research on anat. structure of human neck. Died Aug. 2, 1860.

DUMÉRIL, Auguste Henri André, French naturalist; b. Paris, France, Nov. 30, 1812; s. André Marie Constant Duméril; docteur en médecine; docteur en science. Prof., Faculty of Scis., 1844-46; named prof. geology Chantal Coll., 1947; apptd. prof. herpetology Paris (France) Bot. Garden, 1857; prof. Paris Sch. Medicine. Mem. French Acad. Scis. Author: Historie naturelle des poissons, 1865-70; Des odeurs, de leur nature et de leur action physiologique, 1843. Research on fluctuations of body temperature of medicated animals. Died Paris, Nov. 12, 1870.

DU MEZ, Andrew Grover, Am. pharm. chemist; b. Horicon, Wis., Apr. 26, 1885; s. Andrew Alexander and Anna (Meister) Du M.; Ph.G., U. Wis., 1904, B.S., 1907, M.S., 1910, Ph.D., 1917; m. Mary Elizabeth Fields, June 9, 1912. Instr. in pharm. chemistry U. Wis., 1905-10; prof. chemistry Pacific U., Forest Grove, Ore., 1910-11; asst. prof. chemistry Okla. Agrl. and Mech. Coll., 1911-12; dir. Sch. of Pharmacy, U. of Philippines, 1912-16; Hollister fellow, U. Wis., 1916-17; asso. pharmacologist, Hygenic Lab., USPHS, Washington, 1917-26; dean of Sch. of Pharmacy, U. Md., from 1926. Pharmacy consultant to the surgeon general, U. S. Army; vice chmn. Revision Com. of Pharmacopoeia of U. S., 1930-40; chairman sub.-com. on nomenclature from 1920; U. S. Govt. del. to Second Conf. on Unification of Standards for Potent Remedies, Brussels, 1925; sec. Am. Council on Pharmaceutical Edn. from 1932. Fellow A.A.-A.S.; mem. Am. Assn. Colls. Pharmacy (pres. 1929), Am. Chem. Soc., Am. Pharm. Assn. (pres. 1939-40, Remington Medal, 1948), Am. Pub. Health Assn., Wis. Acad. Sci., Arts and Letters, Sigma Xi, Phi Delta Chi. Author: Quantitative Pharmaceutical Chemistry (with Glenn L. Jenkins), 1931. Adv. editor, joint author, American Pharmacy. Editor: Digest of Comments on the Pharmacopoeia of the U. S. and the National Formulary, 1916-22; Yearbook of Am. Pharm. Assn., 1921-34, Pharm. Abstracts, 1935-41. Science editor Jour. Am. Pharm. Assn., 1938-41. Contbr. to profl. jours. Died Sept. 26, 1948.

DU MONCEL, Théodore, French engr.; b. Paris, France, 1821; elec. engr.; inventor anemograph (electric sound recording device); observed change in resistance as opposed to flow of an electric current at a contact between conductor and carbon (origin of microphone), 1857. Died 1884.

DUMOND, Jesse William Monroe, Am. physicist; b. Paris, France (parents Am. citizens), July 11, 1892; s. Frederick Melville and Louise Adèle (Kerr) DuM.; B.S. in Elec. Engring., Cal. Inst. Tech., 1916, Ph.D. in Physics, 1929; M.S. in Elec. Engring, Union Coll., 1918; m. Blanche I. Gaebel; children—Adele (Mrs. W. K. H. Panofsky), Andree (Mrs. Richard Wilson); m. 2d, Louise M. Baillet, 1944. With Gen. Elec. Co., N.Y.C., 1916-17; French Thomson-Houston Co., Paris, 1919-20, Nat. Bur. Standards, 1920-21; faculty Cal. Inst. Tech., 1921-—, prof. physics, 1946-63, prof. emeritus, 1963-—; faculty Stanford, 1931. Mem. com., NRC, 1944-—; mem. joint commn. on nuclidic masses, Internat. Union Pure and Applied Physics, 1960; mem. U. S. Nat. Acad. Sci., 1953-—. Fellow Am. Phys. Soc.; mem. Societe Francaise de Physique. Author: (with Cohen, Crowe) Fundamental Constants of Physics, 1957. Designed and developed precision instruments for research in X-ray, atomic and nuclear physics. Home: 530 S. Greenwood Av., Pasadena, Cal. 91107.*

DUMONT, André-Hubert, Belgian geologist; b. Liège, Belgium, Feb. 15, 1809; docteur ès sciences, U. Liège; 1 son, André; faculty Liège, 1835-57. prof. mineralogy and geology, rector U. Liège. Recipient gold medal Brussels Acad. Scis., 1830, Wollaston prize London, 1840, grand medal of honor Paris Universal Expn. 1855. Author: Carte géologique de la province de Liége, 1830; Sur le terrain ardennais et rhenan, 1848. Worked on geol. map of Belgium, 1849, mineralogy of Belgium; determined epoque and type of metalliferous deposits in Belgium; distinguished upper level of Rhenish terrain from lower anthraciferous level. Died Liège, Feb. 28, 1857.

DUMONT, François-Marcellin-Aristide, French engr.; b. Crest, France, 1819; ed. Ecole polytechnique, Sch. Bridges and High-Ways, 1838; engr. from 1839. Author: Essai sur la canalisation du Rhône, 1842; Pratique des distributions d'eau, 1863; Sur le projet du canal d'irrigation du midi. Engr. on constrn. of irrigation canals in Midi region, France and in Rhone Valley; writings deal with motion and flow of bodies of water.

DUMONT DE COURSET, Georges-Louis-Marie, French agronomist; b. Boulogne, France, Sept. 16, 1746; m. in 1775; mem. Royal Agrl. Soc. Paris, French Acad. Scis., 1798, Agrl. Soc. Arras. Author: la Météorologie des cultivateurs, 1798; le Botaniste cultivateur, 1798-1805; Observations géorgica-méteorologiques. Research on application of meteorology to agr., 1798; described 8,900 plants, and correlated nomenclatures of Jussieu and Linnaeus, 1789-1805; introduced English methods of agr. in the gardens of his chateau, 1784-94. Died 1824.

DUMONT D'URVILLE, Jules Sébastien César, French navigator; b. Condé-sur-Noireau, Calvados, France, May 23, 1790; m. Adélie. Went to sea aboard Aquilon, 1807; worked on hydrographic survey of Mediterranean, 1820; with circumnavigating expn. of Coquille, 1822; apptd. commdr., 1825; sent to discover traces of lost explorer, J. F. La Perouse; commdr. voyage of Astrolabe in South Atlantic, 1826-29, South Polar regions, 1837-41; conveyed exiled Charles X to Eng., 1830; apptd. rear admiral, 1840. Recipient Gold medal Société de Géographie, 1841. Author: Voyage de découverte autour de monde et à la recherche de La Perouse, 1822-34; Enumeratio plantarum quas in insulis Archipelagi aut littoribus Ponti Euxini . . , 1822; Voyage de la corvette "l'Astrolabe", 1826-29, pub. 1830-35; Voyages autour de monde, résumé général des voyages de Magellan . . . , 1833; (with Vincendon-Dumoulin) Voyages au pôle sud et dans l'océan, 1837-40, pub. 1842-54. Recognized Venus de Milo in Greek statue that had recently been unearthed and reported it to French ambassador at Constantinople, thereby securing its preservation, 1820; discovered Joinville Island and Louis Philippe land, 1838, Adélie coast, 1840, while on voyage in Antarctic region, 1837-41; studied botany, entomology, and languages. Died nr. Meudon, France, May 8, 1842.

DUMONTPALLIER, Victor-Alphonse-Amédée, French physician; b. Honfleur, France, 1826; physician Paris (France) hosps., from 1866; recipient Prize, Acad. Scis., Montyon award, 1857. Author: Infection purulente à la suite de l'accouchement, 1857; Traitement local de l'endométrite chronic, 1890. Studied postpartum infections, hypnotism; indicated treatment of chronic endometritis. Died Paris, 1899.

DUMREICHER VON OESTERREICHER, Johann Heinrich Georg, surgeon; b. Triest, Italy, Jan. 13, 1815; s. Johann and Amalia Waldburga (Fechtig von Fechtenberg) D. von O.; student physics U. Verona (Italy); M.D., U. Vienna (Austria), 1838; m. Franziska Sautier, 1844; 7 children including Armand. Joined Surg. Inst., 1839; became asst. Wattmann's Surg. Clinic, 1841; named lectr. on surgery in 1844 and head surgeon in 1846; became adminstrv. asst. Allgemeines Krankenhaus (Gen. Hosp.), 1848; succeeded Wattman as prof., dir. of clinic, 1849-79. Author: Über die Notwendigkeit von Reformen des Unterrichts an den medizinischen Facultäten Österreichs, 1878. Reorganized Austrian mil. medicine; pioneered operative orthopedics. Died nr. Zagreb, Croatia, Nov. 16, 1880.

DUN, Patrick, physician; b. Aberdeen, Scotland, Jan. 1642; s. Charles and Katherine (Burnet) D.; M.D., Dublin (Ireland); m. Mary Jephson, 1694; 1 son. Became M.P. in Irish House of Commons for Killileagh, 1692, for Mullingar, 1695 and 1703; apptd. physician-gen. to army, 1705; obtained new charter for Dublin Coll. Physicians, 1692, left money to found professorship of physic. Mem. Dublin Philos. Soc. (a founder). Created knight, 1696. Wrote paper The Analysis of Mineral Waters, 1683. Died May 24, 1713.

DUNAL, Michel-Felix, French botanist; b. Montpellier, France, 1777; studied botany with Candolle; prof. botany Montpellier; mem. French Acad. Scis., 1819. Author: Monographie des anonacées, 1816. Specialist annonceous plants. Died Montpellier, 1856.

DUNBAR, Carl Owen, Am. geologist; b. Hallowell, Kan., Jan. 1, 1891; s. David and Emma (Thomas) D.; A.B., U. Kan., 1913; Ph.D., Yale, 1917, Dana Research fellow, 1917-18; m. Lora Beamer, Sept. 18, 1915; children—Carl Owen, Lora Louise (Mrs. Sidney M. Johnson). Faculty, U. Minn., 1918-20; faculty Yale, 1920-59, prof. paleontology and stratigraphy, 1930-59, also curator Peabody Mus., 1924-42, dir., 1942-59, emeritus, 1959. Recipient Hayden medal Phila. Acad. Sci., 1959, Distinguished Service citation U. Kan., 1961. Mem. Paleontol. Soc. (pres. 1940), Geol. Soc. Am. (v.p. 1940), Nat. Acad. Scis., Am., Philos. Soc., Am. Acad. Arts and Sci., Geol. Soc. Am., Am. Assn. Petroleum Geologists, Soc. Econ. Paleont., geol. socs. Mexico, London. Author: Historical Geology, 1949, 2d edit., 1960; (with John Rodgers) Principles of Stratigraphy, 1957; The Earth, 1966, others. Asso. editor Am. Jour. Sci., 1939-—, Jour. Paleontology, 1930-38. Research, publs. on devel. of basis for correlation of Pennsylvanian and Permian coal and oil fields; study of geology of late Paleozoic Era; synthesization of geologic history of N. Am. Address: 1615 Santa Barbara Dr., Dunedine, Fla. 33528.*

DUNBAR, (Helen) Flanders, Am. psychiatrist; b. Chgo., 1902; d. Francis William and Edith (Flanders) Dunbar; B.A., Bryn Mawr Coll., 1923; M.A., Columbia, 1924, Ph.D., 1929, Med. Sc.D., 1935; B.S., Union Theol. Sem., 1927; M.D., Yale, 1930; m. George Henry Soule; 1 dau., Marcia Winslow Dunbar-Soule. Hospitant in Gen. and Psychiatric-Neurological Hosp. and Clinic of U. of Vienna, Austria, and asst. at Burghölzli, Zurich, Switzerland, 1929-30; asst. in medicine Columbia U. Coll. of Phys. and Surg., also Presbyn. Hosp., 1930-34; clin. asst. vis. physician Bellevue Hosp., 1935-37; instr. in psychiatry Columbia Univ. Coll. of Phys. and Surg. and asst. attending psychiatrist Presbyn. Hosp. and Vanderbilt Clinic, 1931-36, asso. in psychiatry, asst. physician and asso. attending psychiatrist, 1936-49; in charge psychosomatic research, 1932-49; asso. mem. staff Greenwich (Conn.) Hosp., 1944-49; instr. N.Y. Psychoanalytic Inst., 1941-49. Diplomate Nat. Bd., Am. Bd. Psychiatry and Neurology. Fellow N.Y. Acad. Medicine, Am. Psychiat. Assn. (sec.-treas. 1939-40), Internat.

Psychoanalytic assns., Am. Psychopath. Assn. (v.p. 1942-45), Am. Com. for World Fedn. for Mental Health. Author: Symbolism in Medieval Thought, 1929; Emotions and Bodily Changes, 1935, rev. 1954; Psychosomatic Diagnosis, 1943, rev. 1956; Mind and Body: Psychosomatic Medicine, 1947, rev. 1955; Synopsis of Psychosomatic Diagnosis and Treatment, 1948; Your Child's Mind and Body, 1949; Psychiatry in the Medical Specialties, 1959. Translator: Eugen Kahn's Psychopathic Personalities, 1931. Inaugurator Jour. Psychosomatic Medicine, exptl. and clin. studies, monograph supplements, editor in chief, 1938-47. Collaborating editor Psychoanalytic Quarterly, 1938-40, editor, 1939-40. Editor Acta Psychotherapeutica, Psychosomatica et Orthopaedagogica. Died Aug. 1959.

DUNBAR, Maxwell John, oceanographer; b. Edinburgh, Scotland, Sept. 19, 1914; s. William and Elizabeth (Robertson) D.; B.A., Oxford (Eng.) U., 1937, M.A., 1939; Ph.D., McGill U., 1941; m. Joan Jackson, August 1, 1945; children—Douglas, William; m. 2d, Nancy Wosstroff, December 14, 1960; children—Elisabeth, Andrew. Member faculty of McGill University, Montreal, Quebec, Canada, 1946—, prof., 1959—; Guggenheim fellow, Denmark, 1952-53; dir. Eastern Arctic Investigations, Can., 1947-55. Recipient Bruce medal Royal Soc. Edinburgh, 1950. Fellow Royal Soc. Can., Linnaean Soc. London; mem. Arctic Inst. N.Am. (gov., past chmn.). Author: Eastern Arctic Waters, 1951. Contbr. numerous articles to profl. jours. Research in arctic marine biology and oceanography. Home: 488 Strathcona St., Montreal 6, Que., Can.*

DUNBAR, Paul Brown, Am. chemist; b. Lebanon, Pa., May 29, 1882; s. William H. and Jennie (Chonberlain) D.; B.S., Gettysburg Coll., 1904, M.S., 1906; Ph.D., Johns Hopkins, 1907; Ph.D. (hon.), Phila.; m. Alice Lenore Davison, 1910; children—Lucy (Mrs. John W. Bisselle), Emilie (Mrs. Stanley E. True), Jan (Mrs. Romulo D. Tedeschi). Chemist, bur. chemistry U. S. Dept. Agr., 1907-25, asst. chief, 1925-27, with FDA, 1927-40; asst. commr. FSA, 1940-42, asso. commr., 1942-44, commr., 1944-51. Mem. Am. Chem. Soc., Assn. Food and Drug Ofcls., Assn. Agrl. Chemists. Research in food chemistry; developed policies of FDA. Address: 4711 Cumberland Av., Somerset, Chevy Chase 15, Md.

DUNBAR, William Philipps, hygienist; b. St. Paul, Minn., Oct. 27, 1863; s. Lewis and Johanna Emilie (Naumann) D.; student medicine U. Giessen (Germany); state med. exam, 1892; m. Nelly Pascoe, 1892; 2 sons, 1 dau.; m. 2d, Anita v. Hillern-Flinsch, 1914; 1 dau. Became dir. Staatliches Hygienischen Institut (State Pub. Health Inst.), Hamburg, Germany, 1893. Recipient Pettenkofer Prize, 1903, Grand prix Chgo. World's Fair, 1904. Author: Leitfaden für die Abwasserreinigungsfragen, 1907; also numerous articles. Research on water purification and sewerage, bacteriology of drinking water, evolutionary relationships between microorganisms; discovered pollen caused hay fever and described treatment; discovered healthy people can be carriers of diseases. Died Hamburg, Mar. 19, 1922.

DUNCAN (THE ELDER), Andrew, Scotch physician; b. Pinkerton nr. St. Andrews, Scotland, Oct. 17, 1744; s. Andrew and the dau. of Prof. William Vilant; M.A., St. Andrews, 1762, M.D., 1769; m. Elizabeth Knox, Feb. 1771; 12 children including Andrew, Alexander. Became surgeon on bd. East Indianman, Asia Found. for China, 1768; founder Royal Pub. Dispensary, Edinburgh which was incorporated, 1818; prof. physiology Edinburgh, 1790-1821; named 1st physician to the king in Scotland, 1821. Mem. Royal Med. Soc. (pres.), Edinburgh Medico-Chirurg. Soc. (pres.), Edinburgh Coll. Physicians (pres.), Harveian Soc. (founder). Author: Elements of Therapeutics, 1770; Medical Cases, 1778; Hoffmann's Practice of Medicine, 1783; The New Dispensatory, 1786; Observations on the Distinguishing Symptoms of three different Species of Pulmonary Consumption, 1813. Founder, Med. and Philos. Commentaries, quar. jour., 1773. Proposed a pub. lunatic asylum be built in Edinburgh; it was built at Morningside, 1807. Died July 5, 1828.

DUNCAN, Christopher John, Brit. zoologist; b. Croydon, Eng., Feb. 23, 1932; s. Jack William and Muriel (Kirlew) D.; B.Sc. with spl. honors 1st class, Queen Mary Coll., U. London, 1953, Ph.D., 1956; m. Jennifer Jane Powell, Sept. 6, 1958; children—Stephen Richard, James Michael, Alastair John. Civil Service Research fellow Physiol. Lab., Cambridge, Eng., 1956-58; lectr. U. Liverpool (Eng.), 1958-64; reader animal physiology U. Durham (Eng.), 1964—. Mem. Soc. for Exptl. Biology, Marine Biol. Assn., Assn. for Study Animal Behaviour. Author: The Molecular Properties and Evolution of Excitable Cells, 1966; also articles. Editorial bd. Malacologia, 1962—. Research on molluscan physiology, vertebrate neurophysiology, especially taste, molecular properties of excitable cell membrane. Home: Oaklea The Avenue, Durham, Eng.*

DUNCAN, Donald Pendleton, Am. forester; b. Joliet, Ill., Feb. 24, 1916; s. Kenneth Whitney and Nettie (Pendleton) D.; student North Park Coll., 1934-35, Mich. Technol. U., 1935; B.S.F., U. Mich., 1937, M.S., 1939; Ph.D., U. Minn., 1951; m. Dymer Mercein Benzie, July 6, 1956; children—Kenneth Houlton, Nancy Susan, Debra Mercein. Faculty, Sch. Forestry U. Minn., 1947-65; prof., dir. Sch. Forestry U. Mo., Columbia, 1965—; cons. Minn. Natural Resources Council, 1963-64; vis. scientist NSF, Soc. Am. Foresters, 1964-65. Mem. A.A.A.S., Soc. Am. Foresters (past div. and sect. chmn.), Ecol. Soc. Am., Forest Products Research Soc. Am. Assn. U. Profs., Am. Inst. Biol. Scis., Sigma Xi, Phi Sigma, Gamma Alpha, Xi Sigma Pi. Author: A Laboratory Manual for Farm Forestry in the Upper Midwest, 1951. Studies, publs. on forest tree species ecology, forest mgmt. to meet timber, water and recreational demands, variations in forest tree species as affected by geographic origin. Home: 209 W. Brandon Rd., Columbia, Mo. 65201.*

DUNCAN, James Matthews, Scottish gynecologist; b. Aberdeen, Scotland, Apr. 1826; s. William and Isabella (Matthews) D.; M.A., Marischal Coll., Aberdeen, 1843; began study medicine at Edinburgh (Scotland) U., 1845; M.D., Aberdeen, 1846; m. Jane Hart Hotchkis, 1860; 13 children. Became asst. to James Young Simpson, Edinburgh, 1847; lectr. on midwifery; became physician for diseases of women Edinburgh Royal Infirmary, 1861; named obstet. physician St. Bartholomew's Hosp., London, Eng., 1877. Fellow Royal Coll. Physicians Edinburgh, Royal Soc., 1883, Royal Coll. Physicians London. Author: Fecundity, Fertility and Sterility, 1866; Researcher in Obstetrics, 1868; Treatise on Parametritis and Perimetritis, 1869; The Mortality of Childbed and Maternity Hospitals, 1870; also numerous articles. Research (with Simpson) on anesthetics, 1847; described folds of loose peritoneum covering uterus after expulsion of fetus (Duncan's folds), 1854. Died 1890.

DUNCAN, John, physician; b. Scotland, Oct. 25, 1839; comdr. Bombay Army, 1896-98; reputed to have demonstrated that in chlorosis the hemoglobin content is lessened but number of red blood cells is nearly unchanged, 1867. Died Sept. 5, 1898.

DUNCAN, Louis, elec. engr.; b. Washington, D.C., Mar. 25, 1862; s. Thomas and Maria (Morris) D.; grad. U. S. Naval Acad., 1880; Ph.D., Johns Hopkins U., 1885; m. Edith McKee, 1887. Resigned from navy, 1887; maj. 1st vol. engrs., Spanish War; asso. and asso. prof. applied electricity, Johns Hopkins U., 1887-98; head dept. elec. engring., Mass. Inst. Tech., 1902-04. Fellow Am. Philos. Soc.; hon. mem. Franklin Inst.; pres. Am. Institute of Electrical Engineers, 1895. Author many articles on engineering. Research in electrical traction; supervised electrification of above and underground transportation systems in N.Y. and Washington. Died Pelham Manor, N.Y., Feb. 13, 1916.

DUNCAN, Peter Martin, English geologist; b. Twickenham, Eng., Apr. 20, 1821; s. Peter King and dau. of Capt. R. Martin; M.B., U. London (Eng.), 1846; m. Jane Emily Cook, 1851; 4 sons, 7 daus.; m. 2d, Mary Jane Emily Liddel Whitmarsh, 1869; 1 son. Apprenticed to med. practitioner, London, 1840; practiced at Colchester, Eng., 1848-60, Blackheath, Eng., beginning in 1860; became prof. geology at King's Coll., London, 1870, and at Cooper's Hill Coll., circa 1871. Fellow Geol. Soc. (sec. 1864-70, pres. 1876-78), Zool. Soc., Linnaean Soc., Royal Soc., 1868. Recipient Wollaston medal, 1881. Author: Observations on the Pollen Tube, 1856; Abstract of the Geology of India, 1875; The Sea-shore, 1879; Primer of Physical Geography, 1882; Heroes of Science, 1882; also numerous articles. Editor: Natural History (Cassell), 1876-82. Research on corals and echinids, ophiurids, sponges and protozoa; described fossil corals of Malta, Java, Hindustan, Australia, Tasmania and W.I., also echinids of Sind; used distbn. of species to study ancient phys. geography. Died May 28, 1891.

DUNCAN, Robert Kennedy, chemist; b. Brantford, Ont., Can., Nov. 1, 1868; s. Robert Augustus and Susan (Hawley) D.; B.A., U. Toronto (with 1st class honors in physics and chemistry), 1892; fellow in chemistry, Clark U., 1892-93; postgrad. in chemistry, Columbia, 1897-98; studied abroad, 1900, 03, 04, 07; m. Charlotte M. Foster, Dec. 27, 1899. Instr. physics and chemistry, Auburn (N.Y.) Acad. High Sch., 1893-95, Dr. Julius Sach's Collegiate Inst., N.Y., 1895-98, Hill Sch., Pottstown, Pa., 1898-1901; prof. chemistry Washington and Jefferson Coll., 1901-06; prof. indsl. chemistry, U. Kan., 1906-14; dir. indsl. research, U. Kan. 1910-14; indsl. research and prof. indsl. chemistry, U. Pitts., 1910-14; vis. lectr. Clark U., 1911-14; cons. in chemistry. Author: The New Knowledge, 1905; The Chemistry of Commerce, 1907; Some Chemical Problems of Today, 1911; Editor: New Science Series. Research in industrial chemistry; discoverer and patentee of new process for mfg. phosphorus, of a new low-melting glass, and of processes of decorating glass. Died Feb. 18, 1914.

DUNCAN, William Jolly, Brit. aero. engr.; b. Apr. 26, 1894; s. Robert and Mary (Jolly) D.; ed. Dulwich Coll., Univ. Coll., London; D.Sc.; m. Enid Meyler Baker, 1936; 4 daus. Mem. firm Ross & Duncan, marine engrs., 1919-26; sci. staff, aerodynamics dept. Nat. Phys. Lab., 1926-34; became 1st head. dept. aeros. Univ. Coll., Hull, 1934, Wakefield prof. aeros., 1938; research on armaments and aerodynamics, Air Def. Research Dept., Exeter, Eng., World War II; chief scientist Ministry Aircraft Prodn., 1945; became prof. aerodynamics Coll. Aeros., Cranfield, 1945; Mechan prof. aero. and fluid mechanics U. Glasgow, from 1950. Chmn. Aero. Research Council; fellow Univ. Coll., London. Fellow Royal Soc., 1947, Royal Aero. Soc. (hon.); mem. Instn. Mech. Engrs. Author: The Principles of the Control and Stability of Aircraft, 1952; Physical Similarity and Dimensional Analysis, 1953; also papers, reports. Co-author: Elementary Matrices, 1938; An Elementary Treatise on the Mechanics of Fluids, 1960. Died Dec. 9, 1960.

DUNDEE, John Wharry, Irish physician; b. Larne, No. Ireland, Nov. 8, 1921; s. William Wharry and Matilda (McRoberts) D.; M.B., B.Ch., B.A.O., Queen's U. Belfast (Ireland), 1946, M.D., 1951; Ph.D., U. Liverpool (Eng.), 1957; Diploma in Anesthesiology, Royal Coll. of surgeons, Dublin, Ireland, 1948, London, Eng., 1950; m. Sarah Irwin Houston, Sept. 6, 1949; children—Ellen Elizabeth, William John Wharry, Ann Matilda, Cathryn Houston. Staff, Walton Hosp., Liverpool, 1948-49; sr. registrar Royal Infirmary, Liverpool, 1950-51; faculty U. Liverpool, 1952-58, sr. lectr., 1956-58; research fellow U. Pa., 1955-56; head dept., prof. anesthetics Queen's U., Belfast, 1958—. Fellow Royal Coll. Surgeons, Assn. Anaesthetists Gt. Britain and Ireland, Royal Soc. Medicine; mem. Brit. Med. Assn. Author: Thiopentone and Other Barbiturates, 1956; Relief of Pain, 1965; also numerous articles. Research on intravenous anesthetics and their improvement, methods of pain relief. Home: 12 Cadogan Park, Belfast, No. Ireland.*

DUNÉR, Nils Christofer, Swedish astronomer; b. Billeberga, Malmöhus, Sweden, May 21, 1839; studied at Lund, Sweden; studied geography; then astronomy, from 1864; prof. astronomy at Uppsala, Sweden, dir. obs., 1888-1909. First to discover different regions of sun rotate at different speeds, 1891; research on stellar spectroscopy, variable and double stars. Died Stockholm, Sweden, Nov. 10, 1914.

DUNGEY, James Wynne, physicist; b. Stamford, Eng., Jan. 30, 1923; s. Ernest and Alice (Jeudwine) D.; M.A., Magdalene Coll., Cambridge (Eng.), Ph.D., 1950; m. Christine S. Brown, Sept. 30, 1950; children—Timothy, Catherine. Fellow, U. Sydney (Australia), 1950-53; vis. asst. prof. Pa. State Coll., 1953-54; fellow Cambridge U., 1954-57; lectr. King's Coll., Newcastle, Eng., 1957-59; sr. prin. sci. officer Atomic Weapons Research Establishment, Aldemaston, Eng., 1959-63; fellow Imperial Coll., London, Eng., 1963-65, prof. physics, 1965—. Cons., Ionosphere Research Lab., Pa. State U., 1959—, Goddard Space Flight Center, 1962—. Mem. Am. Geophys. Union, Royal Astron. Soc. Author: Cosmic Electrodynamics, 1958; also articles. Research on plasma theory, theory of magnetosphere, reconnection of lines of force at magnetic neutral points, diffusion of particles in radiation belts. Home: 6 Picton Way, Reading, Berkshire, Eng. Office: Physics Dept., Imperial Coll., London, Eng.*

DUNGLISON, Richard James, Am. physician, author; b. Balt., Nov. 13, 1834; s. Robley and Hariette (Leadam) D.; grad. U. Pa., 1852 (A.M., 1855); Jefferson Med. Coll., 1856; m. Aug., 1877, Mrs. Violette Fisher. Acting asst. surgeon U. S. A., on duty in mil. hosps. at Phila. during Civil war; practiced at Phila., but relinquished practice for literary work. Edited: The College and Clinical Record, 1880-99; one of original editors of The Phila. Med. Times. Author: The Practitioners' Reference Book; A Handbook of Diagnosis, Therapeutics and Dietetics; The Present Treatment of Disease; A New School Physiology and Hygiene; An Elementary Physiology and Hygiene. Died Phila., Mar. 4, 1901.

DUNGLISON, Robley, physician; b. Keswick, Eng., Jan. 4, 1798; s. William and Elizabeth D.; attended Royal Coll. of Surgeons, London, 1818; M.D., U. Erlangen (Germany), 1823; M.D. (hon.), Yale, 1825; m. Harriette Leadham, Oct. 5, 1824, at least one child, Dr. Richard J. Prof. medicine U. Va., 1824-33; prof. U. Md., 1833-36, Jefferson Med. Coll., Phila., 1836-68; prof. U. Md., 1833-36, Jefferson Med. Coll., Phila., 1836-68. Mem. (v.p.) Am. Philos. Soc. Author: Human Physiology, 1832; Dictionary of Medical Science and Literature (1st complete med. dictionary in U. S.), 1833; Elements of Hygiene, 1835; The Practice of Medicine, 1842; A Dictionary of the English Language for the Use of the Blind, 1860. Credited with 1st recognition of chronic hereditary chorea, 1842. Died Phila., Apr. 1, 1869.

DUNHAM, Charles Little, Am. physician, govt. ofcl.; b. Evanston, Ill., Dec. 28, 1906; s. William Huse and Margaret (Little) D.; B.A., Yale, 1929; M.D., U. Chgo., 1933; m. Lucia Elizabeth Jordan, June 22, 1932; children—George Stuart, Carol Jordan, Sara Gale. Intern, internal medicine U. Chgo. Clinics, 1933-34, asst. Sch. Medicine, 1936-42, instr., 1942-46, asst. prof. medicine, 1946-49; asst. resident medicine New Haven Hosp., 1934-35; asst. chief med. br. AEC, Washington, 1949-50, chief med. br., div. biology and medicine, 1950-54, dep. dir. div., 1954, dir., 1955-67; chmn. div. med. Scis. Nat. Acad. Scis.-NRC, 1967—. Mem. com. NRC, 1955—; mem. exec. com. Nat. Com. Radiation Protection, 1955—; expert adv. panel on radiation WHO, 1959—. Recipient Distinguished Service award AEC, 1957. Mem. Soc. Nuclear Medicine, A.M.A., N.Y. Acad. Sci., Am. Nuclear Soc., Health Physics Soc., Radiol. Soc. N.Am., Am. Rheumatism Assn., A.A.A.S., Radiation Research Soc., Sigma Xi. Research on radiation biology, arthritis. Home:

5302 Carvel Rd., Washington 20016. Office: U. S. Atomic Energy Commn., Washington 20545.

DUNHAM, Edward Kellogg, Am. pathologist; b. Newburgh, N.Y., Sept. 1, 1860; s. Carroll and Harriet E. (Kellogg) D.; Ph.B., Columbia, 1881; M.D., Harvard, 1886; m. Mary Dows, June 4, 1893; children—Theodora (Mrs. Herbert Bodman), Edward Kellog. Prof. bacteriology and hygiene Univ. and Bellevue Hosp. Med. Coll. (New York U.), 1898-1900, prof. pathology, 1898-1907. Invented solution (peptone and sodium chloride in distilled water) for use as culture medium for bacteria to determine formation of indole, 1887. Died Apr. 16, 1922.

DUNHAM, Henry Kennon, Am. physician; b. Fairview, O., Mar. 3, 1872; s. William Henry and Mary (McPherson) D.; student U. Cin., 1888-91, M.D., 1894; student Miami Med. Coll., 1891-94; postgrad. Johns Hopkins Hosp. also research work to demonstrate specific Roentgen markings characteristic of pulmonary Tb; postgrad. Great Ormond St. Hosp. and St. George's Hosp., London; m. Amelia Hickenlooper, Mar. 14, 1905; children—Harry, Amelia. Asst. in medicine Miami Med. Coll., 1896-99; prof. electrotherapeutics U. Cin., 1904-40; dir. Tb clinic, 1914-40; head of dept. of tuberculosis, and asso. prof. of medicine, Med. Coll., U. Cin., also dir. Tb service Cin. Gen. Hosp. until 1940. Pres. Cin. Anti-Tb League. Mem. A.M.A., Am. Roentgen Ray Soc., Cin. Acad. Medicine, Am. Coll. Chest Physicians, Am. Coll. Tb Physicians, Am. Clin. and Climatological Soc., Nat. Tb Assn., Ohio State Med. Soc., Ohio Pub. Health Assn. Author: Stereo-Roentgenography-Pulmonary Tuberculosis, 1915; also many tech. articles. Died Apr. 27, 1944.

DUNHAM, Henry Warren, Am. sociologist; b. Omaha, Jan. 24, 1906; s. Henry Warren and Elizabeth (Cowan) D; student U. Omaha, 1924-26; Ph.B., U. Chgo., 1929, M.A., 1935, Ph.D., 1941; m. Vera Sandomirsky, Nov. 1, 1942; 1 dau., Eugenie. Mem. faculty Ill. Sch. Psychiat. Nursing, 1934-36, Vanderbilt U., 1940; prof. sociology Wayne State U., Detroit, 1940—; dir. epidemiology lab. Lafayette Clinic, Detroit, 1959—; cons. state and city agys., instns. Recipient Susan-Colver Rosenberger award U. Chgo., 1941. Fulbright scholar, 1956-57, 66-67. Mem. Am. Assn. U. Profs., Am. Sociol. Assn., Mich. Acad. Sci., Arts and Letters, A.A.A.S., Am. Pub. Health Assn., Ohio Valley Sociol. Soc. (past pres.), Sigma Xi, Alpha Kappa Delta., others. Author numerous books including: (with R. E. L. Faris) Mental Disorders in Urban Areas, 1939; Crucial Issues in the Treatment and Control of Sexual Deviation in the Community, 1951; Community and Schizophrenia, 1965. Editor: The City in Mid-Century, 1957. Research and publs. on investigations of distbn. of schizophrenia; effects of environment on functional mental illness; negative correlation of schizophrenia to criminal behaviour. Home: 446 Fisher Rd., Grosse Pointe, Mich. Office: 951 E. Lafayette St., Detroit 48207.*

DUNHAM, Kingsley Charles, English geologist; b. Sturminster, Eng., Jan. 2, 1910; s. Ernest Pedder and Edith (Humphreys) D.; B.Sc. with class 1 honors, Hatfield Coll., U. Durham (Eng.), 1930, Ph.D., 1933, D.Sc., 1946; S.D. Harvard, 1935, U. Liverpool, 1967; m. Margaret Young, Jan. 2, 1936; 1 son, Ansel C. With Geol. Survey Gt. Britain, 1935-48, chief petrographer, 1948-50; prof. geology U. Durham, 1950-66; sub-warden Durham Coll., 1959-61. Geol. adviser to pvt. cos. 1951-66; dir. Inst. Geol. Scis., London, 1966—. Recipient Consol. Gold Fields Gold medal, 1944, Bigsby medal Geol. Soc. London, 1954, Sorby medal Yorkshire Geol. Soc., 1964. Fellow Royal Soc. (mem. council 1965—), Geol. Soc. London (pres. 1966, Murchison medal 1966), Mineral. Soc. Am.; mem. Instn. Mining and Metallurgy (past pres.), Geol. Soc. Am., Geologische Vereinigung (Germany), Yorkshire Geol. Soc. (past pres.), Mineral. Soc., Geotech. Soc. Author: Geology of the Organ Mountains, New Mexico, 1935; Geology of the Northern Pennine Orefield, 1948; Fluorspar, 1952; (with G. A. L. Johnson) Geology of the Moor House Nature Reserve, 1963; (with F. W. Anderson) Geology of Northern Skye, 1966; also numerous articles. Research on origin of metalliferous mineral deposits especially, lead, zinc, flourine, barium, iron, igneous and metamorphic rocks, stratigraphy of carboniferous sedimentary rocks, mineralogy of salt deposits. Home: Charleycroft, Quarryheads Lane, Durham, Eng., also 29 Bolton Gardens, London, S.W. 5.*

DUNHAM, Theodore, Jr., Am. astronomer; b. N.Y.C., Dec. 17, 1897; A.B., Harvard, 1921; M.D., Cornell U., 1925; A.M., Princeton, 1926, Ph.D., 1927; m., 1926; 1 child. Nat. research fellow physics Mt. Wilson Obs., Carnegie Instn., 1926-38, asst. astronomer, 1928-36, astronomer, 1936-47; asso. prof. astronomy Princeton, 1934-46; research asso. Oxford U., 1937-39; sci. dir. Fund Astrophys. Research, Inc., 1936—. Mem. sect. instruments NDRC, 1940-42, chief sect. optical instruments, 1942-46; asso. Peter Bent Brigham Hosp., Boston, 1946—; Henry E. Warren fellow Harvard Med. Sch., 1946—; research asso., obs., 1947—; with Mass. Inst. Tech., 1947—; mem. vision com. armed forces NRC; mem. Nat. Geog. Soc.-U. S. Nat. Mus. USN eclipse expdn., Canton Island, 1937. Fellow Phys. Soc., Royal Astron. Soc.; mem. Astron. Soc., Optical Soc., Astron. Soc. Pacifis. Discovered (with W. S. Adams) interstellar rays, 1937-41; research on stellar spectrographs and telescopes, photoelectric spectrophotometry, ultra-violet spectrophotometry of cells, planetary atmospheres, material in interstellar space, application of phys. methods to med. research. Address: 545 Hammond St., Chestnut Hill 67, Mass.

DUNHAM, Wolcott Balestier, Am. microbiologist; b. Boston, June 15, 1900; s. Theodore and Josephine (Balestier) D.; student Harvard, 1920-22; A.B., Columbia, 1924; M.D., Columbia U., 1928; m. Isabel Bosworth, Oct. 7, 1940; children—Wolcott Balestier II, Anne Huntington. Research bacteriologist N.Y. Post-Grad. Med. Sch. and Hosp., 1936-46; asso. microbiology Squibb Inst. Med. Research, New Brunswick, N.J., 1942-46; dir. med. research labs. VA Hosp., Memphis, 1946—; asso. prof. microbiology U. Tenn. Coll. Med., 1952—. Diplomate Am. Bd. Microbiology. Fellow A.C.P., Am. Acad. Microbiology; mem. Theobald Smith Soc. (pres.), Am. Soc. Microbiology, N.Y. Acad. Medicine, N.Y. Acad. Sci., Soc. Exptl. Biology and Medicine, Am. Assn. Immunologists, Sigma Xi (pres. med. units chpt. 1964-65), numerous others. Contbr. numerous articles to sci. jours., chpt. Virucidal Agents in Antiseptics, Disinfectants, Fungicides and Sterilization, 1954. Inventor inoculator for virus culture on chick embryo membranes; studies of virucidal action of antiseptics; studies, publs., demonstration that cells present in normal blood are capable of continuous growth in culture. Home: 67 N. Reese St., Memphis, 38111. Office: VA Hosp., 1030 Jefferson Av., Memphis 38104.*

DUNIN, Mikhail Semenovich, Russian phytopathologist; b. 1901; grad. Moscow U., 1925; D.Agrl. Sci., 1935. Dir., Agrl. Lab. (became Central Lab. Mass Experimentation, All-Union Lenin Acad. Agrl. Sci. 1930, later divided into Lab. Seed Study and Lab. of Plant Protection, Inst. Soya and Spl. Crops), 1925-39; asst. dir. for sci. work All-Union Inst. Soya and Spl. Crops, 1931-32; organizer, dir. virus lab. All-Union Inst. Plant Protection, 1932-33, dir. immunobiol. lab., 1932-44; former lectr. phytopathology, plant virus diseases, diseases of agrl. crops Moscow U., also courses and seminars throughout USSR; prof., 1939—; head chair phytopathology Timiryazev Agrl. Acad., Moscow, 1944—. Decorated Order of Lenin. Author: The Classification of Phytopathogenic Viruses, 1938; Plant Immunogenesis and Its Practical Use, 1945; The Immunity of Plants to Diseases, 1948; co-author, editor: Among the Farming Patriarchs of the Near and Far Eastern Peoples, 1960. Research and publs. on toxicity of linseed oil, detection and study of ambary diseases and devel. countermeasures, developer new means of determining contamination of grain by blight and forms of fungus infection. Address: Timiryazev Agrl. Acad., Novoe sh. 51, Moscow, USSR.

DUNITZ, Jack David, chemist; b. Glasgow, Scotland, Mar. 29, 1923; s. William and Mildred (Gossman) D.; B.Sc., U. Glasgow, 1944, Ph.D., 1947; m. Barbara Steuer, Aug. 11, 1953; children—Marguerite, Julia. Research fellow Oxford (Eng.) U., 1946-48, 51-53, Cal. Inst. Tech., Pasadena, Cal., 1948-51, 53-54; vis. scientist NIH, Bethesda, Md., 1954-55; sr. research fellow Royal Instn., London, Eng., 1956-57; prof. chem. crystallography Swiss Fed. Inst. Tech., Zurich, 1957—. Mem. Chem. Soc. (London), Am., Swiss chem. socs., Am. Crystallographic Assn. Editor: (with J. A. Ibers) Perspectives in Structural Chemistry, vol. 1, 1967. Research, publs. on determination of crystal and molecular structures of various substances, especially cyclic molecules, antibiotics, metal complexes by X-ray diffraction; correlation of structure with chem. and phys. properties. Home: Gartenstrasse 640, 8704 Herrliberg (ZH), Switzerland. Office: Organic Chemistry Lab. (ETH), Univeritätstr. 6, 8006, Zürich, Switzerland.*

DUNKER, Erich, German physiologist; b. Hamburg, Germany, May 8, 1911; s. Karl and Agnes (Wulff) D.; student U. Berlin, 1930-35, M.D., 1937; m. Liselotte Heynsen, Feb. 10, 1945. Sci. asst. Pathologisches Institut, Robert Koch Hosp., Berlin, Germany, 1937-38; sci. asst. II. Medizinische Universitätsklinik, Charité Berlin, also research worker in aviation medicine, 1938-45; practice internal medicine, Hamburg, Germany, also co-worker physiologist Gollwitzer-Meier, Hamburg, 1945-49; staff Physiologisches Institut, U. Hamburg, 1950—, prof. physiology, 1957—, chief sect. neurophysiology, 1962—, substitute dir. dept. physiology, 1959-61. Mem. German Soc. Physiology, German Soc. Photography. Research, publs. on heart vibration, regulation of blood circulation, altitude sickness (lung gas exchange), potassium exchange of red blood cells and cerebrospinal fluid, contraction properties of small muscles, interaction of auditory, trigeminal and vagus nuclei in lower brain stem; improvement of volumetric, spectrophotometric and electronic physiol. methods. Home: 6 Am Krähenberg, 2 Hamburg 55, Germany.*

DUNKER, Wilhelm Bernard Rudolph Hadrian, German geologist, paleontologist; b. Eschwege, Germany, Feb. 21, 1809; s. George Leopold and Caroline (Sommer) D.; studied at U. Göttingen (Germany); Ph.D., U. Jena (Germany), 1838; m. Elise Sommer, 1841. Became tchr. mineralogy and geology Höhere Gewerbeschule Kassel, 1837; named prof. mineralogy and geognosy U. Marburg (Germany); dir. geol. survey of Hessia (Kurhessen). Author: Index Molluscorum maris Japonici, 1882; also numerous other publs. Founder (with H. v. Meyer) quar. Paleontographica, which formed the basis of descriptive paleontology. Research on malacology including molluscs of prehistoric Germany; new descriptions of Japanese marine life; set up a mollusc collection. Died Marburg, Germany, Mar. 13, 1885.

DUNKIN, Edwin, astronomer; b. Truro, Aug. 19, 1821; s. William Dunkin; ed. London, Eng., Guines (Calais); m. Maria Hadlow, 1848; Became mem. staff Royal Obs., Greenwich, Eng.; 1838, 1st class asst., 1856, chief asst., 1881, ret., 1884. Fellow Royal Soc. (mem. council 1879-81), Royal Astron. Soc. Author: On the Probable Error of Transit Observations, 1860-64; On the Movement of the Solar System in Space, determined from the Proper Motions of 1167 Stars, 1863; The Midnight Sky; Familiar Notes on the Stars and Planets, 1869; Obituary Notices of Astronomers, 1879; Presidential Addresses, 1885-86, 1890-91; also various astron. papers. Died Dec. 4, 1897.

DUNLAP, Knight, Am. psychologist; b. Diamond Spring, Cal., Nov. 21, 1875; s. Elon and Sarah Calista (Knight) D.; Ph.B., U. Cal., 1899, M.L., 1900; A.M., Harvard, 1902, Ph.D., 1903; L.H.D., Gallaudet Coll., 1931; m. Mary Durand, May 3, 1906; children—Anna Cecelia, Mary Knight, Sarah Calista. Le Conte fellow U. Cal., 1900-01, asst. and instr. psychology, 1902-06, prof. at Los Angeles, 1936-47; instr., asso. and asso. prof. psychology Johns Hopkins, 1906-16, prof. exptl. psychology, 1916-36. Chmn. div. of anthropology and psychology NRC., 1927-29. Fellow A.A.A.S.; mem. Am. Psychol. Assn., Am. Philos. Assn., Phi Beta Kappa, Sigma Xi. Author: A System of Psychology, 1912; An Outline of Psychobiology, 1914; Personal Beauty and Racial Betterment, 1920; Mysticism, Freudianism and Scientific Psychology, 1920; Elements of Scientific Psychology, 1922; Social Psychology, 1925; Old and New Viewpoints in Psychology, 1925; Habits, Their Making and Unmaking, 1932; The Dramatic Personality of Jesus (with R. S. Gill), 1933; Civilized Life, 1935; Elements of Psychology, 1936; Personal Adjustment, 1946; Religion, Its Functions in Human Life, 1946. Died Aug. 14, 1949.

DUNLOP, James, astronomer; b. Dalry, Ayrshire, Eng., Oct. 31, 1793; keeper of Brisbane Obs. at Paramatta, 1823-27; dir. Parmatta Obs., 1829-42. Recipient Gold medal Astron. Soc., 1828. Fellow Royal Astron. Soc., French Acad. Scis. Author: An Account of observations for Brisbane Catalogue of 7,385 southern stars, 1826. Discovered a number of nebulae. Died Bora-Bora, Society Islands, Sept. 23, 1848.

DUNLOP, John Boyd, British inventor; b. Dreghorn, Scotland, Feb. 5, 1840; grad. Irvines Acad., Edinburgh, Scotland, 1859; received vet. diploma; m. 1876; children—John, Jean (Mrs. McClintock). Practice vet. medicine in Edinburgh, 1859-67, Belfast, Ireland, 1867-92; co-owner Pneumatic Tyre, also Booth's Cycle Agy., Dublin, Ireland, 1892-96; engaged in drapery bus. in Dublin, beginning in 1896. Dunlop Rubber Co., Ltd. in Britain named in his honor. Invented pneumatic tire, circa 1887 (1st used on tricycles, later on bicycles and automobiles in 1888). Died Dublin, Oct. 23, 1921.

DUNN, Frederick Lester, Am. epidemiologist, anthropologist; b. Seneca Falls, N.Y., Dec. 24, 1928; s. William Harold and Lora (Lester) D.; A.B., Harvard, 1951, M.D., 1956; Diploma in Tropical Medicine and Hygiene, London Sch. Hygiene and Tropical Medicine, 1960; m. Alice Winthrop Roberts, June 9, 1953; children—Alice Winthrop, Eric Campbell Williams. Asst. chief influenza unit epidemiology br. Communicable Disease Center, USPHS, Atlanta, 1957-58, chief influenza and staphylococcal disease unit, 1958-59, mem. smallpox epidemiol. mission, E. Pakistan, 1958; staff U. Cal. at San Francisco, 1960—, asso. clin. prof., asso. research epidemiologist, 1964-66, asso. prof. epidemiology, 1967—; lectr. epidemiology U. Cal. at Berkeley Sch. Pub. Health, 1965—. Officer-in-charge U. Cal. Internat. Center for Med. Research and Tng. at Inst. for Med. Research, Kuala Lumpur, Malaysia, 1962-64, 68—. Recipient Evans Teaching award U. Cal. at San Francisco Sch. Medicine, 1966. Mem. A.A.A.S., Am. Pub. Health Assn., Am., Royal socs. tropical medicine and hygiene, Am. Anthrop. Assn., Royal Anthrop. Inst., Am. Soc. Parasitologists, Am. Inst. Biol. Scis., Phi Beta Kappa, Sigma Xi. Research, publs. on epidemiology of influenza, malaria, and intestinal parasites; ecol. studies and surveys of parasitism in non-human primates; med.-anthrop. field studies of human hunter-gatherers and tropical forest agriculturalists. Address: Hooper Found., U. Cal. Med. Center, San Francisco 94122.*

DUNN, Gano, Am. elec. engr.; b. New York, N.Y., Oct. 18, 1870; s. N. Gano and Amelia (Sillick) D.; B.S., Coll. City of New York, 1889; M.S., 1897; E.E., Columbia, 1891, hon. M.S., 1914; hon. D.Sc., Columbia, 1938, Rutgers, 1938, N.Y. University, 1941; Doctor of Engineering, Lehigh University, 1942; D.Sc. (honorary), City College, New York, 1947; LL.D. (honorary), Bowdoin College, 1947; married Julia Gardiner Gayley, Aug. 26, 1920. With the Western Union Telegraph Co., 1886-91, also Crocker-Wheeler Electric Mfg. Co., v.p. and chief engr., 1898-1911; v.p. in charge engring. and constrn., J. G.

White & Co., Inc., New York, 1911-13; pres. The J. G. White Engring. Corp. since 1913; trustee Greenwich Savings Bank; dir. Guaranty Trust Co.; trustee and dir. Panhandle Eastern Pipe Line Co., 1936-43; dir. Radio Corp. of Am., Nat. Broadcasting Co., R.C.A. Communications, Inc.; dir. Regional Plan Assn., Inc. President New York Elec. Soc., 1900-02, Am. Inst. Elec. Engrs., 1911-12, United Engring. Soc., 1913-16; chmn. The Engring. Foundation, 1915-16; chairman Nat. Research Council, 1923-28; v.p. Internat. Elec. Congress, Turin, 1911; U. S. delegate and mem. exec. com., World Power Conf., 1936; chmn. Am. Com. World Powers Conf., 1946-52; mem. N.Y. State Committee on Tech. Industrial Development since 1944. Member War Dept. Nitrate Commn., 1916-18; chmn. Visiting Com. Bur. of Standards since 1928; chmn. State Dept. special com. on submarine cables, 1918. Hon. mem. Assn. Iron and Steel Engrs., Am. Institute E.E. (pres. 1911-12). Fellow Inst. Radio Engrs., Royal Micros. Soc., A.A.A.S., N.Y. Acad. Sciences, N.Y. Micros. Soc.; member Pan-American Society (honorary president), Am. Society Mech. Engrs. (hon. mem. 1944), Am. Soc. C.E., British Instn. E.E. (hon. sec. for U. S.), Franklin Inst., Illuminating Engring. Soc., New York Hist. Soc., N.Y. Zoöl. Soc., Pilgrims (chmn. exec. committee, Optical Society of America, Horological Institute of America, Nat. Acad. Sciences, Am. Acad. Arts and Sciences, Am. Philos. Soc. Mem. S.R., N.Y. Chamber Commerce. Mem. visiting com. Harvard Engineering School. War Dept. cons., 1947; mem. N.Y. State Univ. Commn., 1947. Awarded Townsend Harris medal, 1933, Edison medal, 1937; Hoover medal, 1939; Egleston medal, 1939; Modern Pioneer Award, National Association of Manufactures, 1940; Pan-American Soc. Medal, 1947; Petér Cooper medal, 1950. Decorated Order Honor and Merit, Republic of Haiti, 1940. Contributor various papers on elec. and engring. subjects. Died Apr. 10, 1953.

DUNN, Leslie C(larence), Am. geneticist, b. Buffalo, N.Y., Nov. 2, 1893; s. Clarence Leslie and Mary Eliza (Booth) D.; B.Sc., Dartmouth College, 1915. D.Sc. (honorary), 1952; M.Sc. in Zoology, Harvard University, 1917, D.Sc., 1920; married Louise Porter, May 2, 1918; children—Robert Leslie, Stephen Porter. Asst. in zoölogy, Harvard, 1915-17 and 1919; geneticist Conn. (Storrs) Agrl. Expt. Sta., 1920-28, cons. geneticist since 1930; prof. zoölogy, Columbia, 1928-62, emeritus prof., 1962—, research asso. Nevis Biol. Sta., 1962—; dir. Inst. Study Human Variation, 1952-58, exec. officer department of zoology, 1940-46; visiting lectr. in biology Harvard University, 1949-50; research asso. Galton Lab., Univ. Coll., London, 1960-61, 1st lt. infantry, U. S. Army, 1917-19, World War; with A.E.F., 1918-19. Fellow Am. Acad. Arts and Scis.; mem. Am. Soc. Human Genetics (pres. 1961), Academia Patavina, Fedn. Am. Scientists, Am. Soc. Zoologists (sec.-treas. genetics sect. 1925-28), Am. Soc. Naturalists (pres. 1960), Genetics Soc. America (pres. 1932), Norwegian Acad. Scis., Am. Philos. Soc., Nat. Acad. Sci., Phi Beta Kappa. Mem. editorial bd. Genetics, 1935-62, Advances in Genetics; mng. editor: The Am. Naturalist, 1950-60, Genetics, 1936-41. Author: Principles of Genetics (with E. W. Sinnott), 1925, 32, 39, 51, 58 Heredity and Variation, 1932; Heredity, Race and Society (with T. Dobzhansky), 1946; Biology and Race, 1951; Genetics in the 20th Century, 1951; Heredity and Evolution in Human Populations, 1958; A Short History of Genetics, 1966. Research on genetics and gene distribution in human and animal populations; history of genetics; effects of mutations on development in mammals; mice. Home: 635 W. 247th St., N.Y.C. 71.

DUNN, Naughton, orthopedic surgeon; b. Nov. 22, 1884; s. John A. Dunn; M.A., M.B., Ch.B., Aberdeen U., LL.D., 1937; m. Ethel Violet Jackson; 1 son, 3 daus. Lectr. orthopedic surgery Birmingham U.; cons. surgery Derbyshire County Council and Orthopedic Hosp., Hartshill; resident surg. officer Royal So. Hosp., Liverpool, 1910-12; pvt. asst. to Sir Robert Jones, 1912-14; asst. demonstrator anatomy Liverpool U., 1912-13; hon. surgeon Royal Cripples, Warwickshire Orthopedic, Coleshill, Robert Jones and Agnes Hunt Orthopedic hosps.; hon. orthopedic surgery Corbett Hosp., Stourbridge, Birmingham Ednl. Authority. Fellow Surgeons Gt. Britain and Ireland, (asso.); mem. Royal Soc. Medicine (pres. orthopedic sect.), Brit. (pres. 1938-39, council), Am. (corr.) orthopedic assns., Am. Acad. Orthopedic Surgeons (corr.), Internat. Soc. Orthopedic Surgery. Research and publs. on lesions of radial nerve, bone grafting, paralytic deformities of foot, common injuries of knee joint. Died Nov. 19, 1939.

DUNN, Samuel, Brit. mathematician; b. Crediton, Devonshire, Eng.; m. Elizabeth; master of acad. at Ormond House, Chelsea, Eng., 1758-63; math. examiner to East India Co. Mem. Philos. Soc. Phila. Author numerous books including: The Navigator's Guide to the Oriental or Indian Seas, 1775; Astronomy of Fixed Stars, 1792; also astronom. articles. Invented universal planispheres or terrestrial and celestial globes in plano. Died Jan. 1794.

DUNNE, Howard Walter, Am. veterinarian; b. Omaha, Nebraska, March 19, 1913; s. Gilbert F. and Anna (Schlekau) D.; D.V.M., Iowa State University, 1941; Ph.D., Michigan State University, 1951; m. Ida M. Doll, Aug. 18, 1934; children—Margaret Ann (Mrs. Ralph C. Stevenson), James H., John E., Patricia M. Pvt. practice vet. medicine, Maquoketa, Ia. 1941-42; asst. prodn. mgr., prod. mgr. Biol. Lab. sect. Corn States Serum Co., Omaha, 1942-46; AVMA research fellow dept. path. Mich. State U., 1946-48, asst. prof., 1948-50, asso. prof. dept. bacteriology, 1950-52; dep. chief vet. microbiology Biol. Lab. U. S. Army Chem. Corps, Ft. Detrick, Md., 1952-53; prof. vet. sci. Pa. State U., 1953—; cons. USDA, FAO, National Academy Sci., USPHS, U. S. Army. Diplomate Am. Bd. Veterinary Pathologists, American College Vet. Microbiologists. Fellow Am. Vet. Med. Assn., Am. Acad. Vet. Microbiology; mem. U. S. Livestock San. Assn., N.Y. Acad. Sci., Nat. Conf. Vet. Labs. Diagnosticians, Am. Pub. Health Assn., Internat. Assn. Vet. Pathology. Editor: Diseases of Swine, 1964. Research and publs. field pathogenesis of lesions and devel. of diagnostic procedures for hog cholera; serum block in vaccination failure; in vitro cultivation of leukocytes in virus studies; intra-uterine death and infertility from virus infections SMEDI and hog cholera; classification of entero viruses in N.Am. Home: 314 Nimitz Av., State College, Pa. 16801.*

DUNNE, John William, aero. engr.; b. Roscommon, Ireland, 1875; s. John Hart and Julia Elizabeth Dunne; m. Cecily Saye, 1928; 1 son, 1 dau. Began aero. expts., 1900; invented stable, tailless type of aerofoil, 1904, built, flew both monoplanes and biplanes of this type; designed, built 1st Brit. mil. airplane, 1906-07. Fellow Royal Aero. Soc. Author: Sunshine and the Dry-Fly, 1924; An Experiment with Time, 1927; The Serial Universe, 1934; The New Immortality, 1938; Nothing Dies, 1940. Devised theory (called serialism) postulating not only that time is fourth dimension but also that infinite series of dimensions exist within time itself; he regarded "now" as existing partly in past and extending into future. Died Aug. 24, 1949.

DUNNING, Henry Sage, Am. oral surgeon; b. 1881; D.D.S., N.Y. Coll. Dentistry, 1904; M.D., Coll. Phys. and Surg., Columbia, 1911; B.Sc., N.Y. U., 1915. Practiced medicine, N.Y.C., from 1904; prof. oral surgery Columbia U. Sch. Dental and Oral Surgery; cons. oral surgeon to numerous hosps., including Knickerbocker, St. Luke's, Roosevelt, St. Vincent's, Nassau, Minneola, Vassar Brothers', Harlem Eye and Ear Hospital, Manhattan Eye, Ear, Nose and Throat, Sloane, Babies', Mary Immaculate, Jamaica, North Hudson, St. Joseph's, Stamford, Conn., Jersey City Med. Center. Fellow A.C.S.; mem. Am., N.Y. State med. socs., Am. Bd. Plastic Surgery, Am. Laryngol., Rhinol. and Otol. Acad. of Medicine. Died Feb. 10, 1957.

DUNNING, John Ray, Am. physicist, univ. dean; b. Shelby, Neb., Sept. 24, 1907; s. Albert Chester and Josephine (Thelen) D.; A.B. with highest honors, Neb. Wesleyan U., 1929, Sc.D., 1945; Ph.D., Columbia, 1934; LL.D., Adelphi Coll., 1951, Phi'a. Coll. Osteopathy, 1961; Sc.D., Temple U., 1955, Whitman Coll., 1958, Trinity Coll., 1958; D.Sc. in Edn., Coll. Puget Sound, 1957; m. Esther Laura Blevins, Aug. 28, 1930; children—John Ray, Ann Adele. Mem. faculty Columbia, N.Y.C., 1933—, Cutting Traveling fellow, 1935-36, Ernest Kempton Adams fellow, 1940-41, prof. physics, 1946—, dean Faculty Engring. and Applied Sci., 1950—; sci. dir. atomic energy Office Naval Research, 1946—, chmn. sci. and engring. research, 1949—, dir. sci. research, 1950—. Mem. numerous state and nat. sci. adv. coms., chmn. sci. adv. council to N.Y. Legislature, 1963—; mem. coms. NSF, 1958—; Bd. dirs. Oak Ridge Inst. Nuclear Studies, Fund for Peaceful Atomic Devel.; trustee Armstrong Meml. Research Found., Sci. Service. Recipient Presdl. citation 1946, medal for merit, 1946, medal for distinguished service City N.Y., 1948, Stevens award Stevens Inst. Tech., 1958, Pupin medal, 1949, Pegram medal, 1964. Fellow A.A.A.S. (dir.), Am. Nuclear Soc., Am. Phys. Soc. (v.p. 1951), N.Y. Acad. Scis.; mem. Am. Soc. M.E., Am. Assn. Physics Tchres, Am. Inst. Mining, Metall. and Petroleum Engrs., Am. Soc. for Engring. Edn., I.E.E.E., Mining and Metall. Soc. Am., Nat. Acad. Scis., Newcomen Soc. N.Am., Soc. History Tech., Optical Soc., Sigma Xi, Phi Kappa Tau, Phi Kappa Phi, Sigma Pi Sigma, Tau Beta Pi. Author: (with H. W. Farwell) Matter, Energy and Radiation—Syllabus, 1937; (with H. C. Paxton) Matter, Energy and Radiation, 1941. Contbr. articles on atoms, atomic transmutations, neutrons, nuclear physics, uranium mining and refining, U-235 separation plants, atomic power applications to profl. jours.; pioneered 1st neutron expts. in U. S., 1932; demonstrated 1st uranium fission, 1939, 1st fission separated U-235, 1940; invented gas diffusion system for separation U-235, 1939-40. Home: Spring Lake Rd., Sherman, Conn. Office: Sch. Engring. and Applied Sci., Columbia U., N.Y.C. 10027.*

DUNNING, Wilhelmina Frances, Am. biologist; b. Topsham, Me., Sept. 12, 1904; d. Fred Jewel and A. Evelyn (Williams) Dunning; A.B., U. Me., 1926; D.Sc., 1960; M.A., Columbia U., 1928, Ph.D., 1932. Research asso. Columbia U., 1930-41; instr. pathology Wayne U. Coll. Medicine, 1941-46, asst. prof. oncology, 1948-50; research asso. Detroit Inst. for Cancer Research, 1944-50; prof. zoology U. Miami, Coral Gables, Fla., 1950-52, prof. exptl. pathology, dir. Cancer Research Lab., 1952—. Mem. Am. Assn. for Cancer Research, Am. Soc. Zoologists, Soc. Exptl. Biology and Medicine, Soc. for Growth and Devel., N.Y., Fla. acads. scis., A.A.A.S., Sigma Xi. Research and numerous publs. on etiology of experimentally induced cancer in mice and rats; chemotherapy for isologously transplantable neoplasms in rats and mice; devel. steroid responsive neoplasms in rats and mice. Home: 2850 Coconut Av., Miami, Fla. 33133. Office: Cancer Research Lab., U. Miami, Box 8215, Coral Gables, Fla. 33124.*

DUNNINGTON, Frank Glass, Am. physicist; b. Colorado Springs, Colo., May 7, 1903; s. Frank Hobbs and Gertrude (Flagler) D.; B.S. in Elec. Engring., U. Cal. at Berkeley, 1929, Ph.D. in Physics, 1932; m. Georgia Louise Melton, May 12, 1928 (dec. Apr. 1943); children—Frank Melton, Mary Louise (Mrs. John Frank Kohut); m. 2d, Frances L. Tulin, May 12, 1944. NRC fellow Cal. Inst. Tech., 1932-35, research fellow, 1935-37; faculty Rutgers U., New Brunswick, N.J., 1937—, prof., 1946—, chmn.; dept. physics, 1946-52, chmn. com. on radioisotopes, 1946—, project dir. radiologic health tng. program, 1961—, director of the Radiation Science Center, 1966—. staff mem., sect. leader radiation lab. Mass. Inst. Tech., Cambridge, 1941-45. Chmn. N.J. Radiation Protection Commn., 1958—. Fellow Am. Phys. Soc.; mem. Am. Inst. Physics, Am. Assn. Physics Tchrs., Am. Assn. U. Profs., Health Physics Soc., Phi Beta Kappa, Sigma Xi, Tau Beta Pi, Mu Theta Epsilon. Research and publs. on atomic constants and low energy nuclear physics; directed devel. of modulators (transmitters) for land- and airborne radar; research on devel. of tng. program in radiol. health. Home: 27 Salem Rd., East Brunswick, N.J. 08816. Office: Radiation Sci. Center, Rutgers U., New Brunswick, N.J. 08903.*

DU NOUY, Pierre Lecomte, biophysicist; b. Paris, France, Dec. 20, 1883; studied at Sorbonne, Paris; received degrees in philosophy, sci. and law; with research staff Rockefeller Inst., N.Y.C., 1920-27; returned to France and joined Pasteur Inst., Paris, where he developed lab. of molecular biophysics, left in 1936. Author: L'Homme devant la science, 1939; L'Avenir de l'esprit, 1941, Le Temps et la vie; La Dignité humaine. Research on cicatrization of wounds, absorption phenomenon of surface tension, characteristics of plasma, immunity and molecular properties of serum. Died N.Y.C., Sept. 22, 1947.

DUNOYER, Louis, French physicist; b. Versailles, France, Nov. 14, 1880; s. Anatole and Jeanne (Roquet D.; student Superior Normal Sch., 1902-05; Sc.D., Coll. France, 1909; m. Jeanne Emile Picard, June 4, 1907; 2 children. Asst., Coll. France, 1905; research fellow Carnegie-Curie, 1909-12; acting prof. Conservatory of Arts and Handicrafts, 1913; prof. Optical Inst., 1921-41; prof. Sorbonne, U. Paris (France), 1941-45. Recipient Becquerel prize, Danton prize, Walz prize. Mem. Bur. Longitudes, Paris, French Phys. Soc. (v.p., pres. 1939-43). Author: La technique du vide, 1923; also papers. Address: Société SCAD, 135 rue du Theatre, Paris 15, France.

DUNPHY, Edwin B., Am. ophthalmologist; b. Newark, Dec. 26, 1895; s. Alphonse and Hortense (Blakeslee) D.; A.B., Princeton, 1918; M.D., Harvard, 1922; M. Virginia Delano, June 23, 1923; children—Joan (Mrs. Darcy Corwen), Priscilla (Mrs. Charles Burlingham). Chief ophthalmology Mass. Eye and Ear Infirmary, Boston, 1941-62; faculty Harvard Med. Sch., 1925—, Williams prof. Ophthalmology, 1950-62, also ophthalmologist univ. health services. Recipient Howe prize ophthalmology U. Buffalo, 1957. Fellow A.C.S., Royal Soc. Medicine (hon.); mem. A.M.A. (prize in ophthalmology 1962), Am. Ophthalmology Society (president 1962, recipient Howe medal 1963), New England Ophthalmology Society, the American Academy of Ophthalmology and Otolaryngology (pres. 1965), Pan Am. Soc. Ophthalmology, Assn. for Research in Ophthalmology, Nat. Soc. Prevention Blindness (dir.). Research and publs. on use radioactive isotopes for diagnosis intraocular tumors; surgery cataracts. Home: 255 Woodland Rd., Chestnut Hill, Mass. 02167. Office: 75 Mt. Auburn St., Cambridge, Mass. 02138.*

DUNS SCOTUS, Joannes, Scholastic philosopher and polit. thinker; b. Maxton, Roxburgh, Scotland, circa 1266; presumably studied in Paris, France, 1293-96. Entered Franciscan Order, 1278; ordained priest, 1291; returned to Britain after Parisian study to lecture at Oxford U.; taught at Paris, 1302; was banished, 1303, during struggle between King Philip IV and the Pope; taught again in Paris, 1304, and was apptd. regent of theological school at U.; won title, Doctor Subtilis, after brilliantly defending doctrine of Immaculate Conception; sent to Cologne to take side in religious controversy and assist in founding a university, 1308. Author of commentaries on Peter Lombard and Aristotle. Founder of Scotism, one school of scholastic philosophy whose teachings were different from teachings of Thomas Aquinas; taught the "univocity (ultimate abstraction) of being" and use of "formal distinction;" his political thought was far in advance of his time; taught that state arose from consent of people in a kind of social contract; denied that private property was ordained by natural law. Died Cologne, Germany, Nov. 3, 1308.

DUNSTAN, Gilbert Hall, Am. sanitary engr.; b. Santa Ana, Cal., Feb. 20, 1903; s. John and Myrtle (Hall) D.; B.C.E., U. So. Cal., 1926, B.E.E., 1927; M.S., U. Ia., 1929, 1937. Instr. drawing and design

Tulane U., 1928-29; instr. civil, gen. engring. U. So. Cal., 1929-33, asst. prof., 1933-36; asst. prof. mechanics, san. engring. U. Ala., 1938-42, asso. prof. san. engring., 1942-49; san. engr. USPHS, Fairbanks, Alaska, 1949-50; prof. san. engring. U. Ala., 1950-52; prof. san. engring., head san. engring. research Wash. State U., Pullman, 1952-—. Mem. Ala. Water Improvement Adv. Commn., 1948-49, 50-52; mem. Wash. Water Resources Adv. Com., 1959-65. Water Pollution Control Fedn. Fellow Am. Pub. Health Assn., Am. Soc. C.E.; mem. Pacific Northwest Pollution Control Assn. (Albert Sidney Bedell award, pres. 1964), Am. Water Works Assn. (George Warren Fuller award), Am. Soc. Engring. Edn., Am. Indsl. Hygiene Assn., Sigma Xi, Phi Kappa Phi. Contbr. articles in field to sci. jours. Developed information for design of sprinkler irrigation systems for cannery waste disposal, anaerobic-aerobic lagoons for sewage and certain wastes. Home: 1619 Clifford St., P.O. Box 176, Pullman, Wash. 99163.*

DUNSTAN, Sir Wyndham Rowland, Brit. chemist; b. Chester, 1861; s. John Dunstan; M.A. (hon.), Oxford U.; LL.D., Aberdeen U.; m. E. Fordyce Maclean, 1886; m. 2d, Violet Tracy, 1900. Demonstrator in chemistry Univ. Labs., Oxford, 1884, univ. lectr. chemistry as related to medicine, 1885; prof. chemistry Pharm. Soc., 1886, St. Thomas's Hosp., 1892-1900; examiner hon. sch. natural sci. and med. degrees Oxford, natural sci. tripos and med. degrees Cambridge U., U. London; dir. sci. and tech. dept. Imperial Inst., 1896-1903, dir., 1903-24; gov. London Sch. Econs., Bedford Coll., London Sch. Medicine for Women. Pres., Internat. Congress Tropical Agr., 1910, 14; mem. Govt. Com. on Oil-Seeds, 1915-16, on Mineral Resources, 1917, On Empire Cotton Growing, 1917. Fellow Royal Soc., 1893 (council 1905-07); mem. Chem. Soc. (sec. 1893-1903, v.p. 1904-06), Brit. Assn. (pres. chem. and agrl. sect. 1906), Royal Geog. Soc. (council 1916), Internat. Assn. for Tropical Agr. (pres. 1910), Aristotelian Soc. (hon., a founder), Egyptian Inst. (corr.), Agrl. Soc. Trinidad (hon.). Author: British Cotton Cultivation, 1904-08; The Agricultural Resources of Cyprus, 1906; Agriculture in Asia Minor, 1908; Cotton Cultivation of the World, 1910; Editor: Imperial Inst. Handbooks to the Commercial Resources of the Tropics, also numerous articles. Died Apr. 20, 1949.

DUNTHORNE, Richard, Brit. astronomer; b. Ramsey in Huntingdonshire, Eng., 1711; ed. at Free Grammar Sch. at Ramsey; butler at Pembroke Hall, Cambridge; sci. asst. to Dr. Roger Long; supt. of works of Bedford Level Corp. Author: The Practical Astronomy of the Moon, 1739; A Letter Concerning the Moon's Motion, 1747; Letter Concerning Comets, 1751; Elements of New Tables of the Motions of Jupiters Satellites, 1761; also worked on Long's Astronomy, 1770. Assigned the secular rate of 10 to the acceleration of the moon's mean motion; computation on basis of medieval observations. Died 1775.

DUPERREY, Louis Isidore, French naval officer, scientist; b. Paris, France, Oct. 21, 1786; accompanied de Freycinet on the Uranei around the world to study meteorology and terrestrial magnetism, 1817-20; comdr. Coquille to Oceania and S. Am., 1822-25. Mem. French Acad. Scis. Contbr. vols. on hydrography and phys. sci. to Voyage autour du monde éxecuté par ordre du roi sur la corvette la Coquille pendant les années, 1822, 1823, 1824, et 1825, 1826-30. Determined positions of magnetic poles and magnetic equator; discovered several groups of islands including one named in his honor, 1825. Died Paris, Sept. 10, 1865.

DUPETIT-THOUARS, Louis Marie Aubert, French botanist; b. Bournois, France, Nov. 5, 1758; ed. l'Ecole militaire de la flèche; visited Madagascar and near-by islands, 1792-1802. Mem. French Acad. Scis., 1820, French Agrl. Soc. Author: Histoire des Végétaux Recueillis dans les Iles de France de Bourbon et de Madagascar. Proposed new theory for formation of ann. layers in wood and for bud prodn. Died Paris, May 12, 1831.

DUPIN, (François Pierre) Charles (Baron), French mathematician; b. Varzy, France, Oct. 6, 1784; ed. Ecole Polytechnique; a founder, prof. mechanics Conservatoire des Arts et Métiers, Paris; founder Maratime Mus., Toulon, France, 1813; served with Navy Engr. Corps. Mem. French Acad. Scis., 1814. Author: Développments de la géométrie, 1813; Voyage dans la Grande-Bretagne, 1820-24; Applications de géométrie et de mechanique (prophetic in differential geometry), 1822. Introduced concept of conjugate tangents; gave Dupin's theorem; used indicatrix more than had been done before; investigated triply orthogonal families of surfaces; inventor cyclides (significant in work of Klein and Bôcher); wrote on descriptive geometry and theory of curves; denied applicability of theory of probability to matters of morality; perfected Malus' optical work; formulated theory on stability of floating bodies, introducing idea of surface curvature; set down principles for durability of material used in wooden ships. Died Paris, Jan. 18, 1873.

DUPLAY, Simon-Emmanuel, French surgeon; b. Paris, France, 1836; aggregation, 1866; became surgeon Paris hosps., 1867; prof. clin. surgery. Author: Leçons sur les traumatismes cérébraux, 1883; Traitement des fractures transversales, 1887; (with Follin) Traité de pathologie externe, 1892-99. Research on cerebral traumatism; developed method for setting transverse fractures; specialized in external pathology. Died Paris, 1924.

DU PONT, Francis Irénée, Am. chemist; b. Wilmington, Del., Dec. 3, 1873; s. Francis Gurney and Mrs. Ellise Wigfall Simons; studied at U. Pa.; grad. Yale, 1895; m. Marianna Rhett, 1897; children—Emile Francis, Hubert Irénée, Edmond Rhett, Mrs. Earl M. Elrick, Mrs. Powell Glass, Mrs. Taleasin Davies, Jr. Opened a brokerage firm, N.Y.C., 1931; sr. partner firm Francis I. du Pont & Co. and Chrisholm & Chapman. Held over two hundred patents. Invented minerals separation process; worked on explosives; developed processes leading to orgn. of Ball Grain Explosives Co. which made fuses during war. Died N.Y.C. Mar. 16, 1942.

DUPONT, François, French chemist; b. Charvonnex, France, 1847. Author: Guide pour l'achat de betterave, 1887; Traité de la fabrication du sucre, 1891; inventor devices controlling manufacture sugar (differential press, fecula meter); contbd. to progress of sugar beet refining France. Died Paris, France, 1914.

DU PONT DE NEMOURS, Eleuthére Irénée, chemist; b. Paris, France, June 24, 1771; s. Pierre Samuel and Nicole Charlotte Marie Louise (Le Dée) Du Pont de Nemours; m. Sophie Madelaine Palmas, Nov. 26, 1791, at least 2 children including Henry. Trained in French state gunpowder works; studied under Lavoisier; came with family to Newport, R.I., 1801; founded gunpowder firm E. I. Du Pont de Nemours & Co., Wilmington, Del., 1802; fulfilled govt. contracts, during War of 1812, which helped firm to expand. Died Phila., Oct. 31, 1834.

DU PONT DE NEMOURS, Pierre Samuel, polit. economist; b. Paris, Sept. 14, 1739; studied medicine and economics; wrote articles and edited pamphlets which advocated views of Economist sch. or physiocrats; sec., council of pub. instrn., Poland, 1772; financial and econ. adviser to Turgot in France, 1774; ret. to Gatinais to work in agr.; called by C. G. Vergennes (minister fgn. affairs) to negotiate for France, aided in negotiations for recognition of U. S., 1782, commerce treaty with Gt. Britain, 1786; became councillor of state under Calonne, apptd. commissary-gen. of commerce; advocated constitutional monarchy during Revolution, apptd. dep. to States-Gen.; elected pres. Constituient Assembly, 1790; forced out of office because of his conservative polit. views; imprisoned in La Force, 1794; mem. Council of 500; came to U. S., 1799; asked by Thomas Jefferson to negotiate with France for La. Purchase and to prepare scheme for nat. edn.; returned to France, 1802; elected to Institut; became sec. provisional govt. after downfall of Napoleon, 1814; returned to U. S. when Napoleon regained power. Author: Tableau raisonne des principes de l'économie politique, 1775; Philosophie de l'univers, circa 1790; Sur l'education nationale dans les Etats-Unis d'Amerique, 1800. Editor Jour. de l'agr., 1765-66, Ephemerides du citoyen, 1768-72. Died nr. Wilmington, Del., Aug. 6, 1817.

DUPOUY, Gaston-Léopold, French physicist; b. Marmande, France, Aug. 7, 1900; s. Joseph and Jeanne (Duzan) D.; ed. Faculté des sciences, Paris, France; docteur ès sciences; agrégé des sciences physiques. Became lectr. Rennes (France) U.; named prof. Toulouse (France) U., in 1937; apptd. dir. Nat. Center for Sci. Research, 1950; dean Toulouse Faculty Scis., 1950-57. Mem. nat. commn. UNESCO; mem. com. on atomic energy Sci. Com. for Nat. Def. Mem. French Acad. Scis., French Soc. of the Microscope, Soc. Radioelectricians (elected pres. 1958). Author: Élément d'optique éléctronique, 1952; le Microscope éléctronique, 1950. Research on magnetic properties of crystals, magnetoptics, electronic optics, radioactivity; constructed 1st electronic microscope with magnetic lenses in France.

DUPPA, Baldwin Francis, chemist; b. Rouen, France, Feb. 18, 1828; began work with Hofmann in Royal Coll. Chemistry, 1855, later with W. H. Perkin, Sr., in Maidstone; moved to London where he collaborated with Frankland in 1860. Improved (with Frankland) preparation of mercury ethyl, prepared zinc methyl and recognized presence of hydroxyl and carboxyl group in lactic acid; (with Perkin) converted succinic acid to tartaric (racemic) acid; proposed theory of acetoacetic ester structure. Died Budleigh-Salterton, Devonshire, Eng., Nov. 10, 1873.

DUPRÉ, August, chemist; b. Mainz, Germany, Sept. 6, 1835; s. J. F. and J. A. (Schafer) D.; ed. in Darmstadt, Giessen; M.A., Ph.D., U. Heidelberg; m. Florence Marie Robberds, 1876; 4 sons, 1 dau. Chem. adviser explosives dept. of Home Office, 1873-1907; pub. analyst for Westminster; asso. mem. Ordnance Com.; analytical and cons. chemist; came to London, 1855, naturalized, 1866; lectr. chemistry Westminster Hosp. Sch., 1864-97; chem. referee Med. Dept. of Local Govt. Bd., 1871. Fellow Royal Soc., 1867; Royal Inst. Chemists. Co-author: The Nature, Origin and Use of Wine, 1869; Manual of Inorganic Chemistry, 3d edit., 1901, also numerous articles. Died July 15, 1907.

DUPRÉ, Ernest, French physician; b. Marseilles, France, 1862; studied psychiatry under Motet; agrégé en médecine, 1898; became head physician spl. infirmary Police Prefecture, 1905; taught legal psychiat. medicine St. Ann Hosp., beginning in 1918. Research and publs. on spleen infections, salivary infections, 1891; identified several path. conditions including mythomania, puerilism. Died Deauville, France, 1921.

DUPRÉ LA TOUR, François, biophysicist; b. Ecully, Rhône, France, Sept. 18, 1900; s. Laurent and Marguerite (Gindre) D.; Ph.D. in phys. scis.; M.D. Prof. biophysics, French Faculty Medicine and Pharmacy, Beirut, Lebanon, 1934-—, dean, 1942-59. Mem. French Soc. Physics, French Soc. Mineralogy and Crystallography, Sci. Soc. Brussels, Lebanese Council Sci. Research. Author: Importance de la Radio-Activité du Potassium en Biologie; Statistiques Biologiques; Action des Électrons sur les Globues Rouges du Sang. Research and publs. on polymorphism of fatty acids. Address: French Faculty of Medicine, Beirut, Lebanon.

DUPUY (or DUPU DE BORDES), Henri-Sébastien, French mathematician; b. Grenoble, France, 1746; taught math. to Napoleon; prof. math. Valence Sch. Mil. Sci. Author: Nouveaux principes d'artillerie, 1771; Traité de mathématiques, 1774; also articles on mil. fortifications in Diderot's Encyclopédie. Perfected a pianoforte in 1791. Died 1815.

DUPUY, Louis, French mathematician; b. Chassey, France, Nov. 23, 1709; came to Paris, France, 1735; became lectr., librarian. Mem. Acad. Inscriptions (perpetual sec.). Author: Anthémius sur des paradoxes de mécanique, 1777; Observations sur les infiniment petits et les principes métaphysiques de la géometrie, 1778. Studied Archimedes mirror and its effects; publs. on infinity and metaphys. aspects of geometry, ideas of the Greek, Anthémius, on paradoxes of mechanics. Died Apr. 6, 1795.

DUPUY, Paul, French physician; b. Lamonzie-Saint Martin, France, 1827; became prof. internal pathology U. Bordeaux (France), 1878. Author: la Métaphysique et la science, 1864; Essai théorique et critique de philosophie médicale, 1864; De la nécessité des études métaphysiques, 1862. Worked in sci. philosophy; related metaphysics and medicine; criticized the views of Vacherot, 1862. Died Fleix, France, 1917.

DUPUY DE LOME, Stanislas Charles Henri Laurent, French engr., b. Ploemeur, France, Oct. 15, 1816; ed. École polytechnique, 1835; studied use of ironwork in naval constrn., Eng. 1845; designed navy vessels for French Navy, Toulon; named insp. gen. Dept. Naval Constrn., 1859. Mem. French Acad. Scis., 1886. Built 2 propellor dispatch boats (fastest ships of time) the Canton and Ariel, 1844, 49, later the Napoleon (steam cruiser with speed of 13 knots,) and Algesiras (with direct drive propellor); built Gloire, (first armored frigate with sails and steam engine), 1858-60, hand-driven airship, 1870, lighter than air balloon (driven by a screw propeller moved by driver's arms), 1872. Died Paris, France, Feb. 1, 1885.

DUPUYTREN, Guillaume, French surgeon, anatomist; b. Pierre-Buffière, France, Oct. 6, 1777; ed. in Paris, France, Coll. de Magnac-Laval. Prof. clin. surgery U. Paris; prof., surgeon Hôtel-Dieu; chief surgeon to Louis XVIII. Created baron by Louis XVIII. Mem. French Acad. Scis., also Bichat's Soc., Soc. Anatomy (founder with Bayle and Laennec). Author: Leçons orales de clinique chirurgicale, 4 vols., 1830-34; Traité théorique et pratique des blessures par armes de guerre, 2 vols., 1834; also (paper) De la rétraction des doigts, 1831. Invented early form of stomach tube, 1810; discovered and described disorder which deforms fingers by affecting tissue which invests and connects muscles of palm; ligated external iliac artery, 1815; invented cutting forceps or enterotome; gave comprehensive description of fracture of lower end of fibula with rupture of lateral ligament and dislocation (Dupuytren's fracture), 1819. 1st clearly to describe congenital dislocation of femur in hip joint, 1826; formulated early classification of burns, 1832; invented splint to prevent eversion of fracture of lower part of outer leg bone. Died Paris, Feb. 8, 1835.

DURAN, Armando Miranda, Spanish physicist; b. Lugo, Spain, July 10, 1913; s. Angel Cao Duran and Maria Teresa Miranda; M.S. Math., M.S. Physics, U. Madrid, 1934, Ph.D., 1943; m. Maria Antonia Escribano, Aug. 2, 1940; children—Armando, Angel, M. Antonia, Pilar, Ignacio, Javier, Marta, Pablo, Rosario, Carlos. Faculty, Faculty Sci., U. Madrid (Spain), 1934-—, prof. optics, 1945-—; research asso. Nat. Inst. for Physics, 1934-36; research asso. Inst. for Optics, 1939-49, vice dir., 1947-49; dir. Inst. for Sci. Instruments, 1952-58; dir. Inst. for Nuclear Studies, 1956-—. Dir. gen. for tech. edn., 1951-56; dean Faculty Sci., Madrid, 1957-61. Mem. Spanish AEC, 1948-—; counselor Superior Bd. Sci. Research, 1947-—. Recipient Badge with plaque Order Alfonso X el Sabio; Gt. Cross Civil Merit. Mem. Spanish Royal Soc. for Physics and Chemistry (past pres.), Optical Soc. Am. Author: (with J. Montes), Fiscia General, 1965; also articles. Research in physiol. optics, accomodation, geometrical optics, aberrations and nonspherical surfaces; discovered (with J. M. Otero) that vision with low luminosity gives rise to a myopia of around 2 dioptries; design several instrument used by Spanish Navy. Home: 1, Isaac Peral. Madrid (15) España.*

DURAND, Élle Magloire, pharmacist, botanist; b. Mayenne, France, Jan. 25, 1794; s. André Durand; m. Polymnia Rose Ducatel, Nov. 20, 1820; m. 2d, Marie Antoinette Berauld, Oct. 25, 1825; at least 1 son. Apprentice to pharmacist; pharmacist in French army, 1813-14; came to U. S., 1816; in pharmacy partnership with Edme Ducatel, Balt., 1817-24; made trip to France, 1824-25; began drugstore (became profl. and social center of Phila.'s physicians and botanists), Phila., 1825; mem. Phila. Acad. Natural Scis., Am. Philos. Soc.; v.p., Coll. of Pharmacy, 1844; ret., 1851, devoted rest of life to bot. studies; transported his herbarium to Paris, France, 1868, gave it to Jardin des Plantes. Co-translator: Manual of Materia Medica and Pharmacy, 1829. Author: Memoirs of François André Michaux and Thomas Nuitall. Died Aug. 14, 1873.

DURAND, William Frederick, Am. mech. engr.; b. Bethany, Conn., Mar. 5, 1859; s. William L. and Ruth (Coe) D.; grad. U. S. Naval Acad., 1880; Ph.D., Lafayette Coll., 1888; LL.D., U. Cal., 1927; m. Charlotte Kneen, Nov. 23, 1883. Served in Engr. Corps, USN, 1880-87; prof. mech. engring. A. and M. Coll., Mich., 1887-91; prof. marine engring. Cornell U., 1891-1904; prof. mech engring. Leland Stanford Jr. U., 1904-24. Sci. attaché Am. Embassy, Paris, mem. Interallied Commn. on Inventions, 1918-19; mem. President's Aircraft Bd., 1925; adv. bd. engrs. Boulder Dam Project, 1929; chmn. Navy Depts. Spl. Com. in Airship Design and Constrn., 1935. Mem. NACA, 1915-33, 41-45, NRC, 1915-45. Recipient Guggenheim medal award, 1935; John Fritz medal award, 1935; Franklin Inst. medal award, 1938; Presdl. Award of Merit, 1946; Wright Mem. Trophy, 1948. Fellow Am. Acad. Arts and Scis., A.A.A.S., Royal Aero. Soc., Inst. Aero Scis. (hon.); mem. Nat. Acad. Scis. (J. J. Carty medalist 1944), Am Philos. Soc., Soc. Naval Architects and Marine Engrs., Société Technique Maritime, Am. Soc. Naval Engrs. (life mem., Gold medalist), Am. Soc. M.E. (hon. mem., medalist 1945), Assn. Italiana di Aerotecnica (hon.). Author: Fundamental Principles of Mech., 1889; Resistance and Propulsion of Ships, 1898; Practical Marine Engineering, 1901; Motor Boats, 1907; Hydraulics of Pipe Lines, 1921; Biography of Robert Henry Thurston, 1929. Gen. editor Aerodynamic Theory, 6 vols., 1934. Contbr. to engring. jours. Research in hydrodynamics, aero dynamics; conducted studies of marine and air propellers. Died Aug. 9, 1958.

DURAND DE GROS (PHILIPS), Joseph Pierre, French physiologist; b. Gros, France, 1826; expatriated to Eng., 1851; made voyage to Am. under name of Philips, 1855; returned to Paris, 1860. Author: Éléctrodynamisme vital, où relations physiologiques de l'esprit et de la matière, 1855; Cours de braidisme ou hynotisme nerveux, 1860; Questions de philosophie morale et sociale, 1901; Polypsychisme; Les Origines animales de l'homme éclairées par la physiologie et l'anatomie comparatives, 1871. Invented term, braidism (analogous to mesmerism); originated theories on hypnosis; studied electrodynamism; formulated relationships between mind and matter; studied origins of man using comparative anatomy and physiology; his work led to founding of schs. of Salpetrière and Nancy. Died 1901.

DURAND-FARDEL, Charles Louis Maxime, French physician; b. Paris, France, 1815; docteur en médecine, 1840; prof. medicine École pratique; insp. of springs Vichy, France. Author: Traité clinique et pratique des maladies des vieillards, 1854; Dictionnaire des eaux minérales, 1859. Research on chronic diseases especially those found in the aged; analyzed mineral water; studied softening of the brain. Died Paris, 1899.

DURANT, Charles Person, Am. aeronaut; b. N.Y.C., Sept. 19, 1805; s. William and Elizabeth (Woodruff) D.; m. Elizabeth Hamilton Freeland, Nov. 14, 1837. First native Am. balloonist, made about 40 ascensions; printer, lithographer. Author: Algae and Corallines of the Bay and Harbor of New York; Exposition, or a New Theory of Animal Magnetism with a Key to the Mysteries, 1837. Died Mar. 2, 1873.

DURANTE, Castor, Italian physician, botanist; b. Gualdo, Italy; physician to Pope Sixtus V. Author: De bonitate et vitio alimentorum centuria, 1565; De usu radicis et baliorum mechoacanae, 1587; Theatrum Planatarum Animalium, Piscium et Pentarum, 1636; many highly esteemed med. works. Died Viterbo, Italy, 1590.

DURANTE, Francesco, Italian surgeon; b. Letoianni, Italy, 1845; introduced bone grafting into nerve surgery; died Messina, Italy, 1934.

D'URBINO, Baldi, see Baldi, Bernardino.

DURCK, Herman Ludwig Friedrich Franz, German pathologist; b. Munich, Germany, Feb. 11, 1869; s. Friedrich and Maria (Ludorff) D.; student in Munich; doctorate Munich, 1892; m. Maximiliane v. Ritter zu Groenesteyn, 1914. Became lectr. on path. anatomy and bacteriology in 1897 and asso. prof. in 1902; apptd. prof. Path. Inst., U. Jena (Germany), 1909; became dir. Path. Inst., Krankenhaus recht der Isar, Munich, 1911; named hon. prof. U. Munich, 1919. Author: Atlas der Pathologischen Histologie, 3 vols., 1900-03. Research on pathogenesis of infectious and tropical diseases, histopathology of nervous system, path. anatomy of plague, beri-beri and malaria, etiology and histology of pneumonia; discovered new type of fiber in connective tissue and blood vessel walls, granule formations in brains of patients dying of malaria. Died Munich, Jan. 9, 1941.

DURELL, Jack, Am. psychiatrist; b. N.Y.C., July 5, 1928; s. Sam and Helen (Schwartzman) D.; A.B. summa cum laude, Harvard, 1949; M.D. cum laude, Yale, 1953; m. Viviane diGioja, May 19, 1955. Research biochemist Nat. Inst. Mental Health, Bethesda, Md., 1954-57, research psychiatrist, sect. on psychiatry, 1960-63, chief, sect. on psychiatry Lab. Clin. Sci., 1963—; psychiatric resident Yale, 1957-59; clin. asso. Psychiat. Inst., Maudsley Hosp., London, Eng., 1959-60. Faculty Georgetown U. Med. Center, Washington, 1965—; cons. George Washington U. Hosp., Washington, 1966—, asso. clin. prof. psychiatry George Washington U.; clin. dir. Psychiat. Inst. of Washington, 1967—. Recipient 1st prize award Anna-Monika Commemorative Fund. Mem. Psychiat. Research Soc., Am. Psychiat. Assn., Assn. Research in Nervous and Mental Diseases. Contributor to American Handbook Psychiatry. Assistant editor-in-chief Journal of Psychiatriac Research. Research in mechanism of enzyme action, particularly transmethlation; purification of thetin-homocysteine methylpherase; studies on possible biochem. mode of action of acetylcholine in membrane depolorization; studies on possible metabolic factors in etiology of psychoses and studies on group and family therapy of psychiat. patients. Office: care Nat. Inst. Mental Health, Bldg. 10, Bethesda, Md. 20014.*

DÜRER, Albrecht, German artist, mathematician; b. Nuremberg, Germany, May 21, 1471; s. Albrecht and Barbara (Holper) D.; trained in goldsmith's trade by father; apprentice to artist-printer Michael Wolgemut; m. Agnes Frey, July 7, 1494. Lived and worked in Nuremberg; spent 1 year in Netherlands. Author: Unterweissung der Messung mit dem Zirkel und Richtscheid, 1525; Etliche underricht zu Befestigung der Stett Schloss und flecken, 1527; Vier Bücher von menschlicher Proportion, 1528. Sci. contbn. in math. theories; measurements with compass and ruler, fortifications and human proportions; gave 1st description of an epicycloid, 1525; 1st to use magic square in art work; 1st to use cross-hatching to represent shadows and degrees of shading in wood engraving, 1528. Died Nuremberg, Apr. 6, 1528.

DURET, Louis, French physician; b. Bagé-La-Ville, France, 1527; docteur en médecine, 1552; 1 child, Jean. Named prof. medicine Royal Coll., 1568; 1st physician to Henri III. Author: Adversaria sive scholia in Jacobi Hollerii, 1571; In Hippocratis librum de humoribus purgandis, 1631. Stressed importance of clin. observation; pointed out errors in Arabic polypharmacy, astrology and in Galen. Died Paris, France, 1586.

DURGAPRASAD, Nemani, physicist; b. Vijayavada, India, Oct. 11, 1934; s. Venkatramiah and Annapoornamma (Goda) D.; M.Sc. in Physics, Andhra U., India, 1954; Ph.D., U. Bombay, 1964. Came to U. S., 1964. Research physicist Tata Inst. Fundamental Research, Bombay, India, 1955-64; Nat. Acad. Scis.-NASA postdoctoral research asso. Goddard Space Flight Center, Greenbelt, Md., 1964—. Lectr., Atomic Energy Establishment Tng. Sch., Bombay, 1962-64. Mem. Am. Geog. Union, Am. Phys. Soc. Research on primary cosmic radiation, high energy nuclear interactions, particle physics, nuclear emulsion techniques. Home: 7212 Forest Rd., Hyattsville, Md. 20785. Office: Code 611-3, Goddard Space Flight Center, Greenbelt, Md. 20771.

DURHAM, Herbert Edward, bacteriologist; s. Arthur E. Durham; ed. Univ. Coll. Sch., London, Guy's Hosp., London, Hygiene Inst., Vienna. M.B., B.C., Sc.D., Cambridge U.; m. Maud Lowry Harmer. Working mem. tsetse fly disease com. Royal Soc.; in charge expdns. to investigate yellow fever in Brazil for Liverpool Sch. Tropical Medicine, to investigate beri-beri for London Sch. Tropical Medicine. Fellow Royal Coll. Surgeons (Eng.), Herefordshire Assn. Fruit Growers and Horticulturists (past pres.). Research on fermentation; discovered Salmonella typhimurium in patients with food poisoning, 1898. Died Oct. 25, 1945.

DURHAM, John Wyatt, Am. paleontologist; b. Okanogan, Wash., Aug. 22, 1907; s John W. and Sarah E. (Vandiver) D.; B.Sc. in Geology, U. Wash., 1935; M.A., U. Cal. at Berkeley, 1936, Ph.D., 1941; m. Jane Roberts, Aug. 6, 1935; 1 son, John Wyatt. Geologist-paleontologist Netherlandsche Pacific Petroleum Mij., Java-Sumatra, 1936-39; chief paleontologist Tropical Oil Co., Colombia, 1943-46; asso. prof. Cal. Inst. Tech., Pasadena, 1946-47; faculty U. Cal. at Berkeley, 1947—, prof. paleontology, 1953—, chmn. dept., 1956-58. Guggenheim fellow, 1954-55, 65-66. Fellow A.A.A.S., Cal. Acad. Scis. (trustee 1952-54, 55—, pres. 1966—), Geol. Soc. Am. Paleontol. Soc. (pres. 1965-66), Palaeontol. Assn., Palaeontographical Soc. (London, Eng.), Paleontol. Research Inst., Paleontol. Soc. Japan, Soc. Systematic Zoologists (councillor 1963-67). Soc. Econ. Palaeontologists and Mineralogists, Am. Assn. Petroleum Geologists, Sigma Xi, Phi Sigma. Research and publs. on description of early Cambrian, Mesozoic, and Cenozoic fossil invertebrates; co-describer Helicoplacoidea; co-reviser classification Echinoids; research in paleoclimatology, biogeography past geologic epochs. Home: 968 Cragmont St., Berkeley, Cal. 94708.*

DURKHEIM, Émile, French sociologist; b. Epinal, France, 1858; student Ecole Normale Supérieure de Paris, 1879-82; doctorate Sorbonne, 1892; tchr. in various lycées, 1882-87; holder 1st chair in sociology U. Boredeaux, 1887-1902, prof., from 1893; prof. U. Paris, from 1902. Author: De la division du travail social, 1893; Les règles de la méthode sociologique, 1894; Le suicide, 1897; Les formes élémentaires de la vie religieuse, 1912. Founder, editor L'année sociologique, 1897-1912. Studied anthropology, folk psychology, social conditions in Germany; drew heavily on statistics, anthropology and other scis. in evolving his sociol. theories; analyzed group unity in terms of collective conscience; held that collective mind of society, through constraint of the individual, is the wellspring of religion and morality; developed concept of div. of labor; classified causes of suicide as egoistic, altruistic and anomique. Died Paris, Nov. 15, 1917.

DURM, Joseph Wilhelm, German architect; b. Karlsruhe, Germany, Feb. 14, 1837; s. Philipp and Catherine (Singer) D.; studied under Gottfried Semper and Jacob Burckhardt; m. Marie Saal, 1877; a son, Leopold. hon. Ph.D., Heidelberg (Germany); hon. dr. engring. Berlin (Germany). Traveled through Asia Minor, Greece, Italy, So. France and Eng. studying archtl. form, 1860-61; built many bldgs. especially in Mannheim, Heidelberg, and Karlsruhe (all Germany); prof. Tech. U. Karlsruhe; also archtl. writer. Author: Lehrbuch der Baukunst, 1880; Konstruktion und polychrome Details der griechischen Baukunst, 1880; Die Kuppeln von St. Peter in Rom und St. Maria del Fiove in Florenz, 1887; Zur Baugeschichte des Residenzschlosses, 1893; Das Heidelberger Schloss, 1894. Contbr. to Handbuch der Architektur, became co-editor, 1881. His bldgs. were usually in the style of Italian Renaissance and early baroque; worked on archeol. problems. Died Karlsruhe, Apr. 3, 1919.

DURMISHIDZE, Sergey Vasilevich, Russian biochemist; b. 1910; grad. Georgia Agrl. Inst., 1931, postgrad., 1935-37; D.Biol. Sci., 1955. With All-Union Inst. Viticulture and Viniculture, 1932-35, Georgia Agrl. Inst., 1935-43; dir. Inst. Viticulture and Viniculture, Georgian Acad. Sci., 1943-53; dir. Georgian Ministry Agr., 1953-55, acad. sec., 1955—. Mem. Georgian Acad. Sci. Research and publs. on formation of alcohol fermentation by-products, oxidation conversions of phenol compounds in viniculture, chemistry and biochemistry of racemic tannins and dyestuffs. Address: Georgian Acad. Sci., Tbilisi, Gruz. SSR, USSR.

DUROY (or DEROY), Henri, Dutch physician; b. Utrecht, Netherlands, 1598; practiced in Waerden, Holland; became prof. medicine, Utrecht, 1634. Author: Physiologia, 1641; Fundamenta physices, 1647; De arte medica, 1657. Supported Descartes and introduced his ideas to med. practice. Died Utrecht, 1679.

DUROZIEZ, Paul Louis, French physician; b. 1826; 1st to describe congenital mitral stenosis, Duroziez's disease, 1877. Died 1897.

DURRELL, Gerald Malcolm, zoologist; b. Jamshedpur, India, Jan. 7, 1925; s. Lawrence George and Louisa Florence (Dixie) D.; ed. in Europe by pvt. tutors, including Dr. Theodore Stephanides in Greece; m. Jacqueline Sonia Rasen, Feb. 26, 1951. Student/keeper Whipsnade Park, Bedfordshire, Eng., 1945; financed, organized, led animal collecting expdns. to Brit. Cameroons, Brit. Guinea, Argentina, Paraguay, Australia, New Zealand, Malaya and Mexico, which supplied zoos, wildlife instns., also museums with specimens, 1946-59; founder Jersey (Channel Isles) Zoo, 1959, Jersey Wildlife Preservation Trust, 1963; scriptwriter BBC, 1951—; television appearances on own series, also other wildlife programs, 1954—. Fellow Internat. Inst. Arts and Letters, Zool. Soc. London; mem. Am. Soc. Herpetologists, Am. Assn. Zoo Parks and Aquariums, Brit. Ornithol. Union, Fauna Preservation Soc., Australian Mammal Soc., Nigerian Field Soc., Bombay Natural History Soc., Malayan Nature Soc., S. African Wild Life Protection Soc., Avicultural Soc., Univ. Soc., Fedn. Animal Welfare, Brit. Mammal Soc. Author: Overloaded Ark, 1952; 3 Tickets to Adventure (Annual Book award Secondary Edn. Bd. 1956), 1953; Bafut Beagles, 1953; The Drunken Forest, 1954; The New Noah, 1955; My Family and Others Animals, 1956; Zoo in My Luggage, 1958; Encounters with Animals, 1959; Island Zoo, 1961; Look at Zoos, 1961; Whispering Land, 1962; My Favourite Animal Stories, 1963; Menagerie Manor, 1964; Two in the Bush, 1966; (with BBC Television Natural History Unit) Expedition to Sierra Leone, 1965; Rosy is My Relative, 1968; The Donkey Rustlers, 1968. Home: Les Augres Manor, Trinity. Office: Jersey Wildlife Preservation Trust, Trinity, Jersey, Channel Islands.*

DURRELL, Laurence Wood, Am. plant pathologist; b. Lincoln, Neb., Feb. 16, 1888; s. Willis G. and Bellita (von Stubenrauch) D.; B.S., Ohio State U., 1914; M.S., Ia. State U., 1917, Ph.D., 1923; Sc.D., Colo. State U., 1956; m. Nana E. Kenoyer, Aug. 27, 1917; children—Dorothy (Mrs. Armas Laupa), Mary (Mrs. A. D. Kaiser). Field asst. U. S. Dept. Agr., 1917-20; plant pathologist Mo. Agr. Expt. Sta., 1919-21; asst. prof. Ia. State U., 1921-24; prof., head dept. botany and plant pathology Colo. State U., 1924-53, prof. emeritus, 1953—, dean Coll. Sci. and Arts, 1945-53, a founder Colo U. Research Found., 19—. Recipient Merit award Colo. State U. Research Found. Fellow A.A.A.S.; mem. Colo.-Wyo. Acad. Sci. (past pres.), Am. Phytopathology Soc., Am. Mycol. Soc., Phycological Soc. Am., Am. Micros. Soc., Internat. Phycological Soc., Sigma Xi (named Colo. Man of Sci.) Phi Kappa Phi, Omicron Delta Kappa. Research and publs. on plant diseases, soil fungi, soil algae, oat rust, corn root rot control, control wheat smut, control potato ring rot, peach mosaic, plants poisonous to livestock. Home: 1003 Remingston St., Ft. Collins, Colo. 80521.*

DURVILLE, Hector, b. Mousseau, France, Apr. 8, 1849. Founder Ecole Pratique de Magnétisme et de Massage, 1893; mem. Société Magnétique de France (founder 1887). Author: Bibliographie du Magnétisme et des Sciences occultes, 1895; Theories et procédés du magnétisme; Magnétisme personnel, 1905; Le Fantôme des vivants, 1909; Cours supérieur d'influence personnelle. Editor: Journal du Magnétisme. Studied animal magnetism, its phenomena and use in care of sick.

DURYEE, William Rankin, Am. cellular physiologist; b. Saranac Lake, N.Y., Nov. 11, 1905; s. George Van Wagenen and Margaret Van Nest (Smith) D.; B.A., Yale, 1927, Ph.D., 1933; m. Louise Adams Johnson, June 30, 1931; 1 son, Sanford Huntington. Eldridge fellow Yale, 1928-30; instr. comparative anatomy, embryology Northwestern U., 1930-35; research staff U. Bern, Switzerland, U. Munich, Germany, U. Copenhagen, Denmark, 1935-36; asst. prof. biology N.Y. U., 1937-40; profl. asso. NRC Com. on Growth, 1945-46; cytologist, head cell physiology Nat. Cancer Inst., NIH, Bethesda, Md., 1949-55; research prof. physiology George Washington U. Sch. Medicine, Washington, 1955-60; research prof. exptl. pathology, 1960—. Cons. Biochem. Research Found., 1950-62; Voice Am. lectr., 1961—; mem. panel cell physiology com. on growth NRC, 1952-54. Fellow N.Y. Acad. Sci., A.A.-A.S.; mem. Am. Inst. Biol. Scis. (past mem. gov. bd. and exec. com.), Soc. Gen. Physiologists (past mem. council), Internat. Soc. Cell Biology (Belgium), Radiation Research Soc. (charter), Royal Soc. Medicine (Eng.). Research, numerous publs. on cell nucleus and chromosome function; dissected living chromosomes, nucleoli and genetic units in amphibia and fish eggs; made first radioautographs of single cell; pioneered virus and DNA research, demonstration of indirect action radiations and nature of infective DNA in cancer cell transformations. Home: 3241 N. Woodrow St., Arlington, Va. 22207. Office: 2300 K St. N.W., Washington D.C., 20037.*

DU SAULT, Lucille Anne, Am. radiol. physicist; b. Chippewa Falls, Wis.; d. Joseph and Lulu (Miller) Du Sault; A.B., U. Cal. at Berkeley. Staff, Henry Ford Hosp., Detroit, 1954—. Fellow Am. Coll. Radiology (asso. in physics); mem. Am. Radium Soc., Radiol. Soc. N.Am., Am. Assn. Physicists in Medicine, Sigma Xi, others. Contbg. author: Progress in Radiation Therapy, Vol. I, 1958. Research and publs. in oxygen and fractionation effects in radiotherapy. Home: 4219 Cornwall Rd., Berkley, Mich. 48072. Office: Henry Ford Hosp., Detroit 48202.*

DUSCH, (Georg) Theodor von, see von Dusch.

DUSHMAN, Saul, phys. chemist; b. Rostov, Russia, July 12, 1883; s. Samuel and Olga (Hurwitz) D.; came to America, 1891, naturalized, U. S., 1917; B.A., U. Toronto (Ont., Can.), 1904, Ph.D., 1912; hon. D.Sc., Union Coll., Schenectady, 1940; m. Amelia Gurofsky, May 1, 1907 (dec. May 1912); m. 2d, Anna Leff, June 28, 1914; 1 dau., Beulah. Demonstrator electro-chemistry, U. Toronto, 1904-09, lectr., 1909-12; with Research Lab., Gen. Electric Co., Schenectady, from 1912, asst. dir., 1928-48; dir. research div. Edison Lamp Works, Harrison, N.J., 1922-25; research cons, from 1950. Mem. Am. Phys. Soc., Am. Chem. Soc. Author: High Vacuum, 1923; Elements of Quantum Mechanics, 1938; Scientific Foundations of Vacuum Technique, 1949; Fundamentals of Atomic Physics, 1951; also numerous tech. articles. Research in quantum mechanics, determination of electromotive force, atomic structure, electron emission, unimolecular reaction, high vacuum. Died July 7, 1954.

DUSHNIK, Ben, mathematician; b. David-Gorodok, Russia, Aug. 15, 1897; s. Aaron Mordecai and Sarah Kolodny) D.; A.B., U. Mich., 1924, M.S., 1925, Ph.D., 1930; m. Clara Long, June 7, 1930. Came to U. S., 1914, naturalized, 1920. Instr. U. Mich., Ann Arbor, 1925-35, asst. prof., 1935-47, asso. prof., 1947-64, prof. math., 1964—. Contbr. papers to math jours. on point set theory, elementary number theory, transfinite numbers. Home: 111 S. Revena St., Ann Arbor, Mich.*

DUSKA, Leslie Tibor, geophysicist; b. Bartfa, Hungary, Feb. 5, 1912; s. George and Margaret (Simon) D.; B.Sc., Royal Hungarian Mil. Acad, Ludoviceum, 1934; G.S., Gen. Staff Acad., Budapest, Hungary, 1945; m. Yolande Keresztesi-Simon, May 14, 1937; children—Leslie Tibor, Adrienne (Mrs. Lorand Szojka), Emese (Mrs. Gustav Nikli), Eve-Marie (Mrs. Andrew Szakony). Geophysicist, Exploration Consultants, Inc., Calgary, Alta., Can., 1951-65, Geophys. Service, Inc. div. Tex. Instrument, Calgary, 1965—. Recipient Gold medal of honor Regent of Hungary, 1944. Mem. Assn. Profl. Engrs. Alta., Soc. Exploration Geophysicists, Canadian Soc. Exploration Geophysicists. Research and publs. on underground propagation of seismic energy, rapid solutions for complicated cases, differentiating lithology purely from seismic evidence. Home: 710 52d Av. S.W. Office: 640 12th Av. S.W., Calgary, Alta., Can.*

DUSSAUD, Frantz, Swiss physicist; b. Geneva, Switzerland, 1870; became prof. physics Faculty of Scis., U. Geneva, 1891. Author: Action physiologique des parfums, 1897. Improved edn. of deaf and dumb; invented phonograph for deaf which substituted touch for hearing, cinematograph which substituted touch for sight, writing devices for the blind; presented one of 1st electric phonographs to the Sorbonne, Paris; publs. on physiologic effect of odors. Died Paris, 1953.

DUSSER DE BARENNE, Joannes Gregorius, physiologist; b. Brielle, Netherlands, June 6, 1885; s. Elize Marie and Dorothea (Vogelzang) Dusser de B.; M.D., U. Amsterdam, 1909; hon. M.A., Yale, 1930; m. Kate Snellen, Oct. 12, 1911 (dec. 1931); children—Charlotte, Dorothea Rebecca, Elizabeth Maria; m. 2d, Emily Lockwood Greene, Aug. 12, 1935; 1 dau., Marion. Instr. physiology U. Amsterdam, 1909-11; psychiatrist Meerenberg Asylum, 1911-14; lectr. in physiology U. Utrecht, 1919-30; neurologist St. Antonius Hosp., Utrecht, 1919-30; Sterling prof. physiology, Yale, from 1930. Mem. editorial board Am. Jour. Physiology; co-editor, a founder Jour. Neurophysiology. Research on the brain, particularly functional divisions and interrelations of cerebral cortex. Died June 9, 1940.

DUSSIK, Karl Theo, psychiatrist, neurologist; b. Vienna, Austria, Jan. 9, 1908; s. Karl and Mimier Dussik; M.D., U. Vienna; m. Alexandrine Brusvida. With Clinic Psychiatry and Neurology, U. Vienna, 1932-38; dir. dept. nerve diseases Allgemeine Poliklinik, Vienna, 1938-41; head psychiatrist Met. State Hosp., Waltham, Mass. Research and publs. on diagnosis by ultrasonic waves, neurol. psychiatry, ultrasonic waves; developer (with Manfred Sakel) insulin shock therapy for schizophrenia. Address: 76 Concord Av., Lexington, Mass.

DUTCHER, James Dean, Am. biochemist; b. Denver, Oct. 15, 1912; s. O. Dean and Hendrika (Te Roller) D.; B.S., U. Denver, 1933, M.S., 1935; Ph.D., Columbia, 1939. Postdoctoral fellow Cornell U., Ithaca, N.Y., 1939-41; staff Squibb Inst. for Med. Research, New Brunswick, N.J., 1941—, dept. head biochemistry, 1947-62, sr. research asso. biochemistry, 1962—. Mem. Am. Chem. Soc., Am. Soc. Biol. Chemists, A.A.-A.S. Research and publs. on discovery of antibiotics, determination of structure, isolation of pharmacologically active natural products. Patentee in field. Home: 15 Mine St., New Brunswick, 08901. Office: The Squibb Inst., New Brunswick, N.J. 08903.*

DUTCHER, Ray Marvin, Am. microbiologist; b. Jersey City, N.J., Oct. 1, 1926; s. Ray Woodruff and Margaret (Holpp) D.; B.S., Fla. So. Coll., 1949; M.S., u. ky., 1957; Ph.D., U. Mass., 1960; m. Camille Maye Todd, June 9, 1956; children—Margaret Maureen, Christina Camille. Faculty, head virus research lab. dept. bacteriology and pub. health U. Mass., 1957-59; faculty Sch. Vet. Medicine U. Pa., 1960—, research asso. prof. virology, 1964—; head dept. virology Inst. Med. Research (formerly named S. Jersey Med. Research Found.), Camden, N.J., 1964—; vice chmn. med. adv. com. Leukemia Soc., Inc., 1964-67; mem. etiol. sub-group Acute Leukemia Task Force, 1964-67; vice sec. World Com. For Comparative Leukemia Research, 1965-67. Recipient Exchangeite of Year award Cherry Hill Exchange Club, 1966; Key to city New Orleans. Fellow Am. Pub. Health Assn.; mem. Am. Soc. Microbiology, Am. Inst. Biol. Scis., N.Y. Acad. Scis., U. S. Livestock San. Assn., A.A.-A.S., Electron Microscopy Soc. Am., Am. Assn. Cancer Research, Soc. Exptl. Biology and Medicine. Research and publs. on discovery that tissue cultures derived from cattle with leukemia were resistant to other viruses; developed high temperature short-time thermal inactivation method for preparation of virus vaccines. Home: 1210 Wyndmoor Rd., Barclay Farms, Cherry Hill, N.J. 08034. Office: Leukemia Research Center, Sch. Vet. Medicine, U. Pa., New Bolton Center, Kennett Sq., Pa.*

DUTENS, Joseph Michel, French engr.; b. Tours, France, 1765; ed. École des ponts et chaussées, 1787; sent to study interior nav. system of Eng., 1818; named head engr. for dept. of Léman, 1805. Mem. Acad. Moral and Polit. Scis., French Acad. Scis. Author: Mémoires sur les travaux publics de l'Angleterre, 1819; Histoire de la navigation intérieure de la France, 1829; Philosophie de l'économie politique, 1835; Essai comparatif sur la formation et la distribution du revenu de la France en 1815 et en 1835 (best statis. summary of French economy of that time), 1842. Research on inland waterways of France and Eng.; supported econ. ideas of physiocrats; publs. on philosophy of econs. Died Paris, France, 1848.

DU TERTRE, Jean Baptiste, French botanist; b. Calais, France, 1610; joined Dominican Order, Paris, 1635; sent to Antilles as missionary, 1640; returned to France, 1658. Author: Histoire générale des Antilles habitées par les Français, 4 vols., 1667-71. First description of yellow fever in an account of outbreaks in Guadeloupe and St. Kitts in 1635, 40, 48, 1667. Died 1687.

DUTHIE, Herbert Livingston, Brit. surgeon; b. Glasgow, Scotland, Oct. 9, 1929; s. Herbert William and Margaret (Livingston) D.; M.B., Ch.B., U. Glasgow, 1952, Ch.M. with honors, 1959, M.D. with honors, 1962; m. Maureen McCann, Mar. 31, 1959; children—Mark Livingston, James Jardine, Mairi Jean, Andrew Marr. Jr. surg. positions Western Infirmary, Glasgow, 1956-59; Rockefeller Traveling fellow Mayo Clinic, Rochester, Minn., 1959-60; lectr. surgery U. Glasgow, 1958-61; reader U. Leeds, Eng., 1961-64; prof. surgery head dept. surgery U. Sheffield, Eng., 1964—. Fellow Royal Coll. Surgeons Edinburgh, Royal Coll. Surgeons Eng., Surg. Research Soc., Brit. Soc. Gastroenterology, Assn. Surgeons Gt. Britain. Research, publs. on clarification of effects of gastric resection, role duodenum in control of gastric secretion, intestinal absorption, established theory of anal continence. Home: 48 Whirlow Park Rd., Sheffield 11, Eng.*

DUTHIE, William Dwight, Am. meteorologist; b. Pullman, Wash., June 30, 1912; s. Ora Lee and Ellen (Carlson) D.; A.B., U. Wash., 1935, M.S., 1937; Ph.D. (J.S.K. fellow in math.), Princeton, 1940; postgrad. U. S. Naval Acad., 1942-43. Teaching fellow math. U. Wash., 1935-38; instr. math. U. Mich., 1940-46; prof. aerology U. S. Naval Postgrad. Sch., Annapolis, Md., 1946-47, chmn. dept. aerology, Monterey, Cal., 1947-59, chmn. dept. meteorology and oceanography, 1959-64, prof. meteorology, 1959—. Mem. Am. Math. Soc., Am. Meteorol. Soc., Am. Assn. U. Profs., Am. Geophys. Union, Sigma Xi, Pi Mu Epsilon. Research and publs. in mathematics and meteorology, including applications of numerical and dynamical meteorology to computerized weather charts and forecasts; identified the aeropause. Home: Route 1, 62 Mt. Devon Rd., Carmel, Cal. Office: U. S. Naval Postgraduate Sch., Monterey, Cal. 93940.*

DU TOIT, Alexander Logie, geologist; ed. Bishops Coll., Cape Town, S. Africa, Royal Tech. Coll., Glasgow, Scotland, Imperial Coll. Sci., South Kensington, Eng.; D.Sc. Research tour of S.Am. for Carnegie Instn., 1923; leader govt. expdn. to Kalahari, 1925; geologist Union Irrigation Dept.; cons. geologist De Beers Cons. Mines; pvt. practice as con. geologist. Recipient Wollaston Fund, 1919; S. Africa medal Sci. Assn., 1930; Murchison and Draper medals, 1933. Fellow Royal Soc., Geol. Soc. Author: Physical Geography for South African Schools, 1912, 2d edit., 1926; Geology of South Africa, 1926, 2d edit., 1939; Our Wandering Continents, 1937. Co-author: Geology of Cape Colony, 1909. Died Feb. 25, 1948.

DU TOIT, Stefanus Jakobus, South African physicist; b. Aliwal Noord, South Africa, Aug. 31, 1919; s. Charl Wynand M. and Maria (Du Plessis) Du T.; B.Sc., Univ. Coll. for Christian Higher Edn., Potchefstroom, 1940; M.Sc., U. Stellenbosch, 1942, D.Sc., 1948; m. Elma Basson, July 29, 1943; children—Johanna Frederika Huibrecht (Mrs. Frans van der Walt), Charl Wynand, Jan Hendrik. Lectr. physics and applied math. Potchefstroom Univ. Coll., 1943-46; research officer, sr. research officer, prin. research officer, chief research officer Council for Sci. and Indsl. Research, Pretoria, South Africa, 1947-59; chief physicist, dir. physics Atomic Energy Bd., Pelindaba, South Africa, 1959—. Hon. prof. physics Potchefstroom U. for Christian Higher Edn., 1963—. Mem. South African Acad. Arts and Sci. Research, publs. on constrn. and mechanism of Geiger-Müller tubes and asso. electronics, nuclear spectroscopy, photo-disintegration of deuteron, life-times of excited states of nuclei, cosmic ray research, neutron physics research. Home: 314 Clark St., Brooklyn, Pretoria. Office: Private Bag 256, Pretoria, Transvaal, South Africa.*

DUTOUR, Étienne François, French physicist; b. Riom, France, 1711; corr. mem. Acad. Scis.; author treatises on magnetism, electricity, diffraction of light and others; research optics, electricity, magnetism. Died Riom, 1784.

DUTRA DE OLIVEIRA, José Eduardo, Brazilian physician; b. Sao Paulo, Brazil, Dec. 10, 1927; s. José Dutra and Alcina (Meirelles) de O.; M.D., U. Sao Paulo, 1951; m. Maria Helena Arnaldo Silva, Aug. 15, 1952; children—Beatriz, Eloisa, Isabel, Susana, José, Marina. Faculty, Med. Sch. Ribeirao Preto, Sao Paulo, 1955—, prof. nutrition, 1963-65, prof. medicine, 1965—; asst. in nutritional research Vanderbilt U., Nashville, 1953-54. Rockefeller fellow, 1954-55; Pan Am. Health Orgn. fellow, 1961. Mem. Brazilian Med. Assn., Soc. Sigma Xi, Latin Am. Nutrition Soc. Research and publs. on food and nutrition in Brazil especially deficiency diseases in young children, devel. nutritional diets at low cost. Address: Med. Sch., Ribeirao Preto, Sao Paulo, Brazil.*

DU TREMBLAY, Prosper Paul Verdat, inventor; b. Lyons, France, 1810; inventor combined, variously applicable engine with different liquids, 1842; thought to have been initiator of barge shipping between Paris, France and Rouen, France; mfr. extruded copper tubes; inventor (with André Martin) atmospheric hammer, (forerunner of vacuum and air brake) 1860. Died Lisbon, Portugal, 1875.

DUTROCHET, (René Joachim) Henri, French physiologist; b. Chateau de Néon, Poitou, France, Nov. 14, 1776; ed. Paris; M.D., 1806. Army physician in Spanish campaign, 1808; became chief of medicine Burgos (Spain) Hosp. (under Joseph Bonaparte), 1808; lived at Chateau Renault, Touraine, France, 1809-31, then in Paris Mem. French Acad. Scis., 1831. Author: Recherches sur l'accroissement et la reproduction des végétaux 1821; Nouvelles recherches sur l'endosmose et l'exosmose, 1828; Mémoire pour servir à l'histoire anatomique et physiologique des végétaux et des animaux, 2 vols., 1837. Made 1st quantitative expts. on osmosis; found that "exosme" and "endosme" always occur simultaneously; devised "endosmometer" for expts.; confused osmosis with "electroendosme"; thought that Brownian movements he observed were an optical illusion; showed that only plant cells containing green pigment can absorb carbon dioxide in presence of light, 1837; also studied evolution of bird. Died Paris, Feb. 4, 1847.

DUTROULEAU (or DUTROULAY), Auguste-Frédéric, French physician; b. 1808; docteur en médecine, 1847; med. service Dept. Navy, France, 1827-39, head physician, 1839-56; med. insp. baths at Dieppe, France, from 1857. Author: Traité des maladies des Européens dans les pays chauds, 1860. Investigated European susceptibility to tropical diseases. Died Brest, France, 1872.

DUTTA, Nirmal Kumer, Indian physician; b. Bengal, India, Dec. 1, 1913; s. Asu Tosh and Saralabala (Basu) D.; M.B.B.S., U. Calcutta (India), 1937; D. Phil., U. Oxford (Eng.), 1949; D.Sc., U. Oxford, 1964; m. Geeta Basu, May 27, 1939; 1 son, Kalyan. House physician, asst. dept. pathology and bacteriology C.M. Coll., Calcutta, 1937-39; staff Central Drugs Lab., Calcutta, 1940-46, head bio-assay and drug standardization div., 1944-46; research worker dept. biol. standard Nat. Inst. for Med. Research, London, also Oxford (Eng.) U., 1947-49; asst. dir. Haffkine Inst., Bombay, India, 1949-68, dir., 1968—; postgrad. faculty U. Bombay (India), 1954—; guest worker Walter Reed Army Inst. Research, Washington, 1962-63. Cons., WHO, Geneva, Switzerland, 1963—. Recipient Basantidevi Ami Chand prize Indian Council Med. Research, New Delhi, India, 1956, Watumull Meml. award Watumull Found., Honolulu, 1965, Shant: Swarup Bhatnagar Meml. award Council Sci. and Indsl. Research, New Dehli, 1965. Fellow Nat. Inst. Scis. India. Research and publs. in exptl. medicine; discovered model for study human cholera in animals; demonstrated signs and symptoms of clin. cholera could be reproduced in rabbits by products of cholera germs in absence of living organisms. Home: Haffkine Inst., Staff Quarters. Office: Haffkine Inst., Bombay 12, India.*

DUTTON, Clarence Edward, Am. geologist; b. Wallingford, Conn., May 15, 1841; s. Samuel Henry and Emily (Curtis) D.; A.B., Yale, 1860; m. Emeline C. Babcock, Apr. 18, 1864. Served with U. S. Army, advancing through grades to maj., 1862-1901. Joined U. S. Geog. and Geol. Survey of Rocky Mountain Region (Powell Survey), 1875, spent 10 years in plateaus of Utah, Ariz. and N.M. investigating causes of volcanic action, other geol. problems; investigated Charleston earthquake, 1886; studied volcanic regions of Hawaii, Cal. and Ore., 1889. Author: Report on the High Plateau of Utah, 1879-80; Tertiary History of the Grand Canyon District, 1882; Mount Taylor and the Zuni Plateau, 1884; Hawaiian Volcanoes, 1884; The Charleston Earthquake, 1886; The Cascades, 1889; Earthquakes in the Light of the New Seismology, 1904. Discovered method for determining the depth of focus of an earthquake, published important observations on nature and speed of earthquake wave motion; invented term, isostasy, and was pioneer advocate of isostatic theory according to which the floor under continents is composed of heavy basaltic rock and continents are of lighter granitic rock; balance between continent heights and ocean depths is called isostasy; published theory that volcanism is caused by radioactivity, 1906. Died Englewood, N.J., Jan. 4, 1912.

DUTTON, Jack, Brit. physicist; b. Plymouth, Eng., July 18, 1925; s. Ralph and Emma (Pester) D.; B.Sc., U. Coll. Swansea, U. Wales, 1946, Ph.D., 1951; m. Eurwen Mary Roberts, Aug. 20, 1949; children—Helen Margaret, Judith Ann. Faculty, U. Wales, U. Coll., Swansea, Wales, 1951—, prof. physics, 1966—; vis. fellow Joint Inst. for Lab. Astrophysics, Boulder, Colo., 1966-67. Chief examiner physics Joint Matriculation Bd. No. Univs., 1965-66. O.E.C.D. Sr. vis. fellow accelerator research div. Orgn. for European Nuclear Research, CERN, Geneva, Switzerland, 1963. Fellow Inst. Physics. Research, publs. on behaviour of ionized gases through studies of fundamental collision processes of electrons and ions with gas atoms and molecules and with metal surfaces, mechanism of electric spark at high voltage.*

DUTTON, Joseph Everett, English physician, biologist; b. Higher Bebington, Eng., Sept. 9, 1874; s. John and Sarah Ellen (Moore) D.; studied at U. Liverpool (Eng.), 1888-92; M.B., C.M., Victoria U., 1897, elected Holt fellow pathology, 1897; house surgeon to Prof. Rushton Parker, also to Dr. R. Caton, Liverpool Royal Infirmary; joined 3d expdn. of Liverpool Sch. Tropical Medicine to No. Nigeria, 1900; went to Gambia, 1901; mem. expdn. to Congo, 1903-05. Recipient Gold medal in anatomy and physiology U. Liverpool, also medal in materia medica, 1895; medal in pathology Victoria U.; granted Walter Myers fellowship in medicine, 1901. Discovered the flagellate infusorians in human blood which cause sleeping sickness (named it Trypanosoma gambiense), 1901; showed (with J. L. Todd) the organism now called Borrelia duttonii caused relapsing fever in monkeys, 1902; studied method of preventing malaria. Died Kosongo, Congo, Feb. 27, 1905.

DUVAL, Jacques-René, French dental surgeon; b. Argentan, France, 1758; master surgeon, 1786; mem. Acad. Medicine. Author: L'Art du dentiste chez les Anciens, 1791; L'Odontalgie considérée dans ses rapports avec d'autres maladies, 1803; wrote on dentistry in ancient Greece and Rome, dental fistulae, odontology. Died Paris, France, 1854.

DUVAL, Mathias Marie, French anatomist, biologist; b. Grasse, France, Feb. 10, 1844; s. Joseph Duval; M.D., Strasbourg (France) U., 1869. Began teaching anatomy, physiology at Faculty of Paris (France), 1873; prof. anatomy l'École des beaux artes; named dir. lab. of anthropology l'École des hautes études, 1880, became prof. histology, 1887. Mem. Acad. Medicine. Author: Cours de physiologie, 1872; Structure et usages de la rétine, 1873; Manuel de microscope, 1873; Précis de technique microscopique et histologique, 1878; Etude sur la ligne primitive de l'embryon, 1879; (with Charles Morel) Manuel de l'anatomiste, 1883; Dictionnaire usuel des sciences médicales, 1885; le Darwinisme, 1885; l'Anatomie général et son histoire, 1886; Eléments d'histologie, 1896; Leçons sur la physiologie du système nerveux, 1883. Introduced collodion for embedding histologic preparations, 1879; described mass of multipolar ganglion cells in medulla oblongata ventrolateral from hypoglossal nucleus; research in anatomy, histology and embryology. Died Paris, 1907.

DU VAL, Patrick, Brit. mathematician; b. Cheadle Hulme, Eng., Mar. 26, 1903; s. Bartram and Margaret (Shimwell) DuV.; B.Sc., U. London, 1926; Ph.D., U. Cambridge, 1930; m. Isobel Blades, Aug. 2, 1945; children—Nicholas Shimwell, Paula Gülen, Belinda Jane. Fellow Trinity Coll., Cambridge, Eng., 1930-34; mem. Inst. for Advanced Study, Princeton, N.J., 1934-35; asst. lectr. math. U. Manchester, Eng., 1936-41; ordinarius prof. geometry U. Istanbul, Turkey, 1941-49; prof. math. U. Ga., 1949-52; sr. lectr. math. U. Bristol, Eng., 1952-54; reader in geometry U. Coll. London, Eng., 1954—. Mem. London (council 1957-65), Am. math socs., Cambridge Philos. Soc. Author: Analitik Geometri, 1947; Homographies, Quaternions and Rotations, 1964. Research, numerous publs. on classical algebraic geometry, classification of singular points of surfaces, applications of crystallographic methods. Home: Pym Gate, The Broad Walk, Northwood, Middlesex, Eng. Office: U. Coll., Gower St., London W.C.1, Eng.*

DUVAL, Pierre-Alfred, French surgeon; b. Paris, France, 1874; became prof. clin. surgery Faculty of Medicine, Paris, 1919. Developed surg. operations of liver and intestines; perfected pulmonary surgery. Died Paris, 1941.

DUVAL, Vincent, French orthopedic physician; b. Saint-Maclou, France, 1796; docteur en médecine, 1820; founder orthopedic clinic at Chaillot, Paris, France. Author: Traité sur le pied bot, 1839; Aperçu des principales difformités du corps humain, 1939. Performed operation on clubfoot by cutting the achilles tendon, 1835. Died 1876.

DUVAL LE ROY, Nicolas-Claude, French physicist; b. Sainte-Honorine-des-Pertes, France, July 14, 1731; prof. math. Brest, France, 1764-73, Royal Sch. Nav., le Havre, France, 1773-77, Rochefort, France, from 1777; mem. Royal Marine Acad., French Acad. Scis., 1789. Author: Instructions sur les baromètres marins, 1784; Éléments de navigations, 1802. Died Brest, Dec. 6, 1810.

DUVAUCEL, Alfred, French naturalist; b. 1792; studied natural history under Cuvier; became naturalist to his majesty, 1807; went on mission to India, 1818; (with Diard) explored Bengal; author: Sur le Sorex glis. studied Sorex (type of shew found in No. India). Died Madras, India, 1824.

DUVEL, Joseph William Tell, Am. crop technologist; b. Wapakoneta, O., Nov. 16, 1873; s. August and Amanda (Myers) D.; B.Sc., Ohio State U., 1897; D.S., U. Mich., 1902; m. Elva Smith, May 11, 1904; children—Maxine, William August. Asst. botanist Ohio Agrl. Expt. Sta., 1898-99; with bur. plant industry U. S. Dept. of Agr., Washington, 1902-18, crop technologist, in charge of grain standardization investigations, 1910-18; with U. S. Grain Corp., N.Y., 1918-20; grain mcht., Winnipeg, Can., 1920-21; U. S. grain exchange supr., Chgo., 1922-25; chief, Commodity Exchange Adminstrn., Washington, 1925-42. Fellow A.A.A.S.; mem. Washington Bot. Soc., Potomac Grange, Washington Biologists' Field Club, Royal Agrl. Soc. New S. Wales, Australia (hon. life). Died Jan. 8, 1946.

DUVERNEY, Joseph Guichard, French anatomist; b. Feurs, France, Aug. 5, 1648; M.D., U. Avignon (France); lectr. anatomy Paris, France; apptd. prof. anatomy Jardin du roi, 1679; mem. French Acad. Scis., 1676. Author: Traité de l'organe de l'ouïe (1st book on the anatomy, physiology and pathology of ear), 1683. Noted tensor tarsi muscle (later described by Horner), 1689; named nerves of brachial plexus, 1697; discovered ophthalmological ganglion (at approximately the same time as Tiedman and Bartholin) greater vestibular glands which are now known as Duverney's glands; 1st to notice simultaneous extrauterine and intrauterine pregnancy; research on physiology of sight and relation between external and internal ear. Died Paris, Sept. 10, 1730.

DUVERNOY, Georges-Louis, French anatomist, zoologist; b. Montbéliard, France, Aug. 6, 1777; docteur en médecine; began work with Cuvier, 1803; named prof. natural history Faculty of Scis., Paris, France, 1809; became prof. natural history Faculty Scis., Strasbourg, France, 1827; apptd. prof. natural history College of France, 1837. Author: (with Cuvier) Leçons d'anatomie comparée, 1803; Leçons sur l'histoire naturelle, 2 vols., 1839, 42. Studied poisonous snakes, monkeys, crocodiles. Died Paris, Mar. 1, 1855.

DU VIGNEAUD, Vincent, Am. biochemist; b. Chgo., May 18, 1901; s. Alfred Joseph and Mary Theresa (O'Leary) duV.; B.S., U. Ill., 1923, M.S., 1924, Sc.D., 1960; Ph.D., U. Rochester, 1927, Sc.D., 1965; Sc.D., N.Y. U., 1955, Yale, 1955, St. Louis U., 1965, George Washington U., 1968; m. Zella Zon Ford, June 12, 1924; children—Vincent, Marilyn (Mrs. Barry Nicholas Brown). Prof., head dept. biochemistry Cornell U. Med. Coll., N.Y.C., 1938-67, emeritus, 1967—; prof. chemistry Cornell U., Ithaca, N.Y., 1967—. Recipient Nichols medal N.Y. sect. Am. Chem. Soc., 1945, Lasker award, 1948, Osborne and Mendel award, 1953, Passano award, 1955, Nobel prize in chemistry, 1955, Willard Gibbs medal Chgo. sect. Am. Chem. Soc., 1956, A.C.P. award, 1965. Mem. Nat. Acad. Scis., Am. Philos. Soc., Am. Chem. Soc., Am. Soc. Biol. Chemists. Author: A Trail of Research in Sulfur Chemistry and Metabolism and Related Fields, 1952. Research and numerous publs. on metabolism of biologically important sulfur-containing compounds; discovery of transmethylation in mammalian metabolism; isolation and proof of structure of the vitamin Biotin, synthesis of penicillin, isolation and proof of structure by synthesis of oxytocin and vasopressin, hormones of the posterior pituitary gland. Home: 100 Fairview Sq. 6E, Ithaca, N.Y. 14850.*

DUYSENS, Louis N. M., Dutch biophysicist; b. Heerlen, Netherlands, Mar. 15, 1921; s. L. W. W. and H. (Eykeboom) D.; Ph.D., U. Utrecht, 1952; m. W. A. A. Kesler, 1952; children—Frank, Tom, Inpe. Mem. Biophys. Research Group, Utrecht, Netherlands, 1947-56; fellow Carnegie Inst. Washington, Stanford, 1952-53; research asso., photosynthesis project, 1953-54; faculty, head biophys. lab. U. Leiden (Netherlands), 1956—, prof., 1961—. Recipient Kettering award, 1964. Mem. Netherlands Found. Biophysics (sec.), numerous other socs. Contbr. numerous articles to profl. jours. Devel. sensitive absorption and fluorescence difference spectrophotometry and other methods for research in photosynthesis, photochem. components in electron transfer chains of photosynthesis. Home: 2 Ch. de Bourbonlaan, Oegstgeest. Office: Schelpenkade 14A, Leiden, Netherlands.*

DVIGUBSKI, Ivan Alexseevich, physicist, botanist; b. Russia, Dec. 30, 1771; s. Alexsei; prof. Moscow (Russia) U., 1807-33; wrote handbooks on physics and chemistry. Died Jan. 11, 1839.

DVORAK, Jaroslav, Czechoslovakian physician; b. Prague, Czechoslovakia, July 27, 1926; s. Jaroslav and Milada (Janouskova) D.; medicinae universae doctor Charles U., Prague, 1951; Candidatus scientiarum, Czechoslovak Acad. Scis., 1959; m. Alena Kinclova, Sept. 10, 1951; children—Milada, Jaroslava. Asst., Inst. Microbiology, Med. Faculty, Hradec Kralové, Czechoslovakia, 1951-58, chief, 1959-61; chief Central Lab. for Microbiology, Faculty Clinics, Hradec Kralove, 1958-59; chief microbiol. labs. for Dist. Pardubice, 1961—; chief mycoparasitological group Inst. Paras. Czechoslovakian Acad. Scis., Prague, 1963—; chief microbiol. labs. Inst. Research in Pharm. Biochemistry, Prague, 1964—. Mem. Am. Soc. for Mycology, Czechoslovakian Soc. for Parasitology. Author: (with Docekal) Manual of Medical Mycology, 1954; also numerous articles. Research on hyphal fusions occurring between all dermatophytes; new animal hosts of dermatophytes; probable life cycle of Emmonsia crescens in nature; seasonal incidence of T. overrucosum; dynamics of dermatophytes in E. Bohemia; keratinophilic fungi in Cuba and Czechoslovakia; first finding of some dermatophytes and Cryptococcus neoformans in Czechoslovakia. Home: 1094 Ant. Dvorak Hradec Kr., Czechoslovakia. Office: 500 Kyjevska Pardubice, Czechoslovakia.*

DVORYANKIN, Fedor Andianovich, Russian agronomist, selectionist; b. 1904; grad. Timiryazev Agrl. Acad., Moscow, 1938; postgrad. All-Union Lenin Acad. Agrl. Sci., 1938-41. Prof., Moscow U., 1949—. Author: Michurinist Genetics, 1939; Darwinism in the Mendelian Mirror, 1946; The Victory of Michurinist Biological Science, 1948. Sec. jour. Agrobiology. Research and numerous publs. in defense of Michurin's teachings, theoretical works. Address: Moscow University, Leninskie gory, Moscow, USSR.

DWELSHAUVERS-DERY, Victor Auguste Ernest, Belgian engr.; b. Dinant, Belgium, Apr. 25, 1836; prof. civil engring., rector U. Liège (Belgium). Mem. French Acad. Scis., 1900. Calorimetric research on steam engine, 1899; developed theory of regulators. Died Liège, Mar. 15, 1913.

DWIGHT, Arthur Smith, Am. mining and metall. engr.; b. Taunton, Mass., Mar. 18, 1864; s. Benjamin Pierce and Elizabeth Fiske (Dwight) S.; grad. Poly. Inst., Bklyn., 1882; E.M., Sch. Mines (Columbia); 1885, M.Sc., 1914, D.Sc., 1929; m. Jane Earl Reed, June 4, 1895 (dec. Feb. 1929); m. 2d, Mrs. Anne Howard Chapin. Engaged in mining and metall. work, 1885-1906, in charge smelting operations at Pueblo and Leadville, Colo., El Paso, Tex., Argentine, Kan., San Luis Potosi and Cananea, Mexico; cons. practice, dir. bus. of Dwight & Lloyd cos., N.Y., since 1906; pres. Dwight & Lloyd Sintering Co.; pres. Dwight & Lloyd Metall. Co. Mem. Am. Inst. Mining and Metall. Engrs. (life mem., pres., Douglas medallist), Mining and. Metall. Soc. Am., Soc. Am. Mil. Engrs., Instn. Mining and Metallurgy of London (hon.), Soc. Engrs. Louvain (Belgium) U. (hon.), Sigma Xi. Inventor (with R. L. Lloyd) system of ore treatment. Died Apr. 1, 1946.

DWIGHT, Edwin Welles, Am. physician; b. Auburn, N.Y., Aug. 11, 1863; s. Henry Williams and Mary Jane (Winslow) D.; M.D., Harvard, 1891. Engaged in practice in Boston, from 1891; asst. commr. pub. instns., Boston, 1895-96; med. dir. N.E. Mut. Life Ins. Co.; instr. legal medicine and surgery, Harvard Med. Sch.; prof. legal medicine, Tufts Coll. Med. Sch.; asst. vis. surgeon Boston City Hosp. Author: Medical Jurisprudence, 1903; Toxicology, 1904. Died Jan. 14, 1931.

DWIGHT, Jonathan, Am. surgeon, ornithologist; b. N.Y., Dec. 8, 1858; s. Jonathan and Julia Lawrence (Hasbrouck) D.; A.B., Harvard, 1880; M.D., Coll. Phys. and Surg. (Columbia), 1893; m. Georgina Gertrude Rundle, June 12, 1901 (dec. 1903); m. 2d, Ethel Gordon Wishart Adam, Dec. 9, 1914. Mem. 7th Regt. N.G. N.Y., 1889-96; asst. surgeon, dept. laryngology Vanderbilt Clinic, 1894-1904. Fellow Am. Ornithologists' Union (treas., later pres.), N.Y. Acad. Scis. Died Feb. 22, 1929.

DWIGHT, Thomas, Am. anatomist; b. Boston, Oct. 13, 1843; s. Thomas and May Collins (Warren) D.; A.B., Harvard, 1866, M.D., 1867, A.M., 1872; studied abroad 2 yrs.; LL.D., Georgetown, 1889; m. Sarah C. Pasigi, Sept. 18, 1883. Instr. comparative anatomy, Harvard, 1872-73; lectr., prof. anatomy, Bowdoin Coll., 1872-76; instr. in histology Harvard, 1874-83, instr. topog. anatomy, 1880-83, Parkman prof. anatomy, from 1883, Harvard. Editor Boston Med. Journal, 1873-78; gave course of lectures, Lowell Inst., on Mechanism of the Bone and Muscle, 1884. Author: Anatomy of the Head, 1876; Variations of the Bones of the Hand and Foot, 1907; Thoughts of a Catholic Anatomist, 1911. Died Nahant, Mass., Sept. 8, 1911.

DWIGHT, William Buck, Am. geologist; b. Constantinople, Turkey, May 22, 1833; s. Harrison Gray Otis (Am. missionary) and Elizabeth (Barker) D.; came to U. S. permanently, 1849; A.B., Yale, 1854, A.M. 1857, B.S. 1859, Union Theol. Sem., 1857; m. Nov. 17, 1859, Eliza Howe Schneider (dec. 1901). Founder and prin. Englewood (N.J.) Female Inst., 1859-65; in mining explorations Va. and Mo., 1865-67; taught at West Point, N.Y., 1867-70; asso. prin. and prof. natural scis. State Normal Sch., New Britain, Conn., 1870-78; prof. natural history, dept. of geology and mineralogy, and curator mus. Vassar Coll. Apptd. univ. examiner in geology, State of N.Y., 1894. Recipient Bronze medal Paris Expn., 1900. Invented and patented rockslicing machine for scientific section of minerals, 1891. Fellow A.A.A.S.; original fellow Am. Soc. Naturalists and Geol. Soc. Am. Died 1906.

DWYER, Paul Sumner, Am. mathematician; b. Chester, Pa., Dec. 8, 1901; s. Edmund B. and Anna Belle (Tracy) D.; A.B., Allegheny Coll., 1921; M.A., Pa. State Coll., 1923; Ph.D., U. Mich., 1936; m. Florence B. Brown, June 29, 1932; children—John Michael, David James. Instr. math. Pa. State U., University Park, 1921-26; asst. prof. math. Antioch Coll., Yellow Springs, O., 1926-29, asso. prof., 1929-33, prof., 1933-36; faculty math. U. Mich., Ann Arbor, 1937—, prof., 1946—, research asso. in Ednl. Investigations and Statis. Research Lab., 1936-40, cons., 1940—. Research asso. Princeton, 1942; cons. personnel research sect. Dept. Army, 1951-53. Recipient Faculty Award for Distinguished Achievement, U. Mich., 1958. Fellow Am. Statis. Assn., Inst. Math. Statistics (sec.-treas. 1943-49, pres. 1951), A.A.A.S. (mem. com. on agenda and resolutions 1958-60); mem. Psychometric Soc. (mem. council 1954-56), Biometric Soc., Math. Assn. Am. (sec. Mich. 1937-41), Econometric Soc., Phi Beta Kappa, Sigma Xi, Phi Kappa Phi, Sigma Pi Sigma, Alpha Chi Rho. Mem. editorial bd. Psychometrika, 1956—. Author: Linear Computations, 1951. Application of statis. techniques to problems in matrix derivatives; finite sampling; simultaneous equations; least squares, regression, and multivariate analysis; transp. and group assembly. Home: 1343 Ardmoor Av., Ann Arbor, Mich.*

DYAR, Harrison Gray, Am. biologist; b. N.Y., Feb. 14, 1866; s. Harrison Gray and Eleonora Rosella (Hannum) D.; B.S., Mass. Inst. Tech., 1889; A.M., Columbia, 1894, Ph.D., 1895; m. Zella Peabody, Oct. 14, 1889; children—Dorothy, Otis Peabody; m. 2d, Wellesca Pollock Allen, Apr. 26, 1921; children—Roshan Allen, Harrison Golshan, Wallace Joshan. Asst. bacteriology Columbia, 1895-97; custodian of lepidoptera U. S. Nat. Mus., from 1897; entomol. asst., Bur. of Entomology, U. S. Dept. Agr., 1915-17. Co-author: (with L. O. Howard and Frederick Knab) The Mosquitoes of North and Central America and the West Indies (Carnegie Inst., Washington), 4 vols., 1912-17; The Mosquitoes of the Americas, 1928. Editor Jour. N.Y. Entomol. Soc., 1904-07, Procs. of Entomol. Soc., Washington, 1909-12. His large entomol. collection now housed in U. S. Nat. Mus. Died Jan. 19, 1929.

DYBING, Ottar, Norwegian veterinarian; b. Stavanger, Aug. 11, 1907; s. Wilfred and Elisabeth; Ph.D. in Vet. Medicine, Vet. Coll., Copenhagen; m. Ingrid Dybing; children—Sverre, Erik, Ole. Sec., Vet. Coll. of Norway, Oslo, 1936, instr., 1942, prof., 1950. Pres., Com. Pesticide Toxicity, Com. Pharmacopoeia. Mem. Acad. Scis. Author: Zur Kenntnis des Choralosenarkose, 1940; Ether Concentration in Brain, Blood and Muscles, 1945-47; Toxicity of Red Squill, 1952-54; Toxicity of Senecio, 1958-59; Toxicity of Lindane, 1961. Home: Trosteveien 1, Bekkestua. Office: Ullevälsveien 72, Oslo, Norway.

DYBOWSKI, Jean, French agronomist; b. Paris, France, 1856; ed. Nat. Sch. Agr., Grignon, France; lectr. on horticulutre, Grignon; sent on expdn. to South Algeria, 1889; insp. gen. colonial agr. in France, 1892; founder Nogent-sur-Marne Agrl. Sta. author: Traité de culture potagère, 1885; les Jardins d'essai coloniale, 1897; Guide du jardinage, 1899. Improved agr. in French colonies; research on culture of vegetables. Died 1928.

DYCE, Rolf Buchanan, elec. engr.; b. Guelph, Ont., Can., Oct 12, 1929; s. Elton James and Evelyn (Buchanan) D.; B.S., Cornell U., 1951, Ph.D. (Republic Aviation fellow), 1955; m. Sherry Ann Ward, Nov. 28, 1959; children—Eric, Karen. Mem. staff Stanford Research Inst., 1957-64; mem. staff Arecibo (P.R.) Ionospheric Obs., 1964—, asso. dir., 1965—. Mem. I.E.E.E., Am. Geophys. Union, Sigma Xi, Tau Kappa Epsilon. Asso. editor: I.E.E.E. Transactions, 1964—. Exptl. radar echo studies of the aurora; research on radar and radio effects of the ionosphere; co-discoverer of 59-day rotation period of Mercury. Home and office: Box 995, Arecibo, P.R. 00613.*

DYCHE, Howard Edward, Am. elec. engr.; b. Spring Valley, O., Jan. 19, 1884; s. Samuel Edward (M.D.) and Flora Alice (Carey) D.; M.E. in E.E., Ohio State U., 1906; m. Edith Mae Guy, Feb. 2, 1910; 1 son, Howard Edward. Engr., ry. dept. Westinghouse Electric & Mfg. Co., 1906-11; prof. elec. engring. Westinghouse Tech. Sch., Pitts., 1909-11; instr. math. and physics U. Pitts., 1911, instr. elec. engring., 1912-14, asst. prof. elec. engring., 1914-17, asso. prof., 1917-19, prof., head of dept., 1919—, dir. grad. work in industry, 1927-52, acting dean Sch. Engring. and Mining, 1950. Mem. Am. Inst. E.E., Soc. Promotion Engring. Edn., Sigma Xi, Sigma Tau, Eta Kappa Nu. Died Apr. 11, 1954.

DYCHE, Louis Lindsay, Am. zoologist; b. Berkeley Springs, W.Va., Mar. 20, 1857; s. Alexander and Mary (Reilly) D.; B.A., B.S., U. Kan., 1884, A.M., 1886, M.S., 1888. Asst. prof. zoology U. Kan. 1885-86, prof. comparative anatomy, 1886-90, prof. zoology and curator of birds and mammals, 1890-1900, prof. systematic zoology and taxidermy, curator birds and mammals, 1900-15; game and fish warden State of Kan., 1909. Made 23 sci. expdns., securing outstanding collection of large N.Am. mammals for U. Kan. Died Jan. 20, 1915.

DYCK, Walther Franz Anton von, see von Dyck.

DYCKERHOFF, Gustav, German industrialist; b. Mannheim, Germany, Oct. 12, 1838; s. Wilhelm Gustav and Caroline (Eglinger) D.; m. Luise Helmreich, 1867; 4 sons including Karl; 1 dau. Worked for several years in France and Eng.; took over family Portland cement factory Dyekerhoff und Söhne, Amöneburg nr. Biebrich/Rhein, Germany (made tech. Improvements in prodn. and won Netherlands Market, 1869), remained mgr. (with his brother Rudolf) until 1911 when it became corp.; advanced social benefits for his workers; founder health ins. plan for employees, 1864. Died Bibrich nr. Wiesbaden, Germany, Jan. 12, 1923.

DYE, Clair Albert, Am. pharmacist; b. Zeno, O., June 23, 1869; s. Elza A. and Emma O'Neil (Garrett) D.; Ph.G., Ohio State U., 1891, grad. study, 1895-97; Ph.D., U. Berne (Switzerland), 1901; m. Oleta Sinclair, June 30, 1902; m. 2d, Flora Alice Elder, June 19, 1930. Asst. in chemistry and pharmacy Ohio State U., 1890-94, asst. in pharmacy, 1894-98, asst. prof. pharmacy, 1898-1906, asso. prof., 1906-09, prof. since 1909, also acting dean Coll. Pharmacy, 1915-21, dean, 1921-39. Mem. revision com. U. S. Pharmacopoeia. Pres. Am. Assn. Colls. of Pharmacy, 1921-22, chmn. sect. edn. and legislation, 1919-20. Mem. Am., Ohio State pharm. assns., Sigma Xi. Contbr. papers and reports to drug jours., pharm. assns. Died Oct. 10, 1949.

DYE, Henry A., Am. mathematician; b. Dunkirk, N.Y., Feb. 14, 1926; s. Henry A. and Mildred (Morse) D.; M.S., U. Chgo., 1947, Ph.D., 1950; m. Anne Garvey, July 29, 1950; children—Constance, John. Bateman fellow Cal. Inst. Tech., Pasadena, 1950-52, instr. math., 1952-53; mem. Inst. for Advanced Study, Princeton, N.J., 1953-54; asst. prof. math. State U. Ia., Iowa City, 1954, 56, asso. prof., 1959-60; asso. prof. math. U. So. Cal., Los Angeles, 1956-59; prof. math. U. Cal. at Los Angeles, 1960—. Research in functional analysis, especially von Newmann algebras and Ergodic theory. Home: 3624 Mandeville Canyon Rd., Los Angeles 90049.*

DYE, Marie, Am. nutritionist; b. Chgo., Sept. 13, 1891; B.S., U. Chgo., 1914, M.S., 1917, fellow Michael Reese Hosp., 1921-22, Ph.D., 1933. Tchr. Starretts Sch. for Girls, Ill., 1916-17, Tech. Normal Sch., 1917-18, high sch., Chgo., 11 1918-19; research asso. prof. home econs. Mich. State Coll., 1922-30, prof., dean, 1929—. Mem. Chem. Soc., Soc. Biol. Chemistry, Dietetic Assn., Soc. Research Child Devel., Home Econs. Assn. (sec. 1921-33, treas 1942-44, pres. 1948-50). Research in basic problems of nutrition, basal metabolism, relation of vitamin A to chlorophyll content of plants, metabolism in obesity, nutritional requirements of children.

DYE, William David, Brit. physicist; b. Dec. 30, 1887; ed. London U.; D.Sc.; with B.T.H. works, Rugby; mem. staff, later head elec. standards and measurements dept. Nat. Phys. Lab. Fellow Royal Soc., 1928; asso. City and Guilds London Inst. Contbr. sects. on magnetic properties and testing, radio frequency measurements to Dictionary of Applied Physics, also papers to other publs. Died Feb. 18, 1932.

DYER, Eldon, Am. mathematician; b. Corpus Christi, Tex., June 19, 1929; s. S. E. and LeClare (Grace), D.; A.B., U. Tex., 1947, Ph.D., U. Tex., 1952; m. Marian Hall, Aug. 31, 1950; children—Dorothy LeClare, James Eldon. Asst. prof. mathematics U. Ga., 1952-55, Johns Hopkins U., 1955-57; Inst. for Advanced Study, 1956-57; faculty U. Chgo., 1957-64; prof. math., Rice U., Houston, 1964—. NSF postdoctoral fellow, 1956-57; A. P. Sloan Found. fellow, 1960-62. Research and publs. in topology. Asso. editor: Transactions Am. Math. Soc., 1959-64. Editor: Proceedings Am. Math. Soc., 1961—. Home: 3832 Olympia St., Houston 77019.*

DYER, Frank Lewis, Am. inventor; b. Washington, D.C., Aug. 2, 1870; s. George Washington and Kate (Huntress) D.; ed. Columbian (now George Washington) U. Law Sch.; m. Annie Augusta Wadsworth, 1892; children—John Wadsworth, Frank Wadsworth; m. 2d, Isabelle Dawson Archer, 1924; m. 3d. Eliza J. Martin, 1939. Practiced patent law, Washington, 1892-97, N.Y., 1897-1903; gen. counsel Edison interests, 1903-08; exec. officer T. A. Edison's indus. corps., 1908-1912; pres. Gen. Film Co., 1912-14; treas. Condensite Co. Am., 1910-20; officer and dir. numerous corps.; has secured over 100 patents in various arts, including talking books for the blind; practicing from 1914 as mech. and elec. expert. Author: Edison—His Life and Inventions (with T. Commerford Martin), 1910-29. Died June 4, 1941.

DYER, Isadore, Am. dermatologist; b. Galveston, Tex., Nov. 2, 1865; s. Isadore and Amelia Ann (Lewis) D.; Ph.B., Sheffield Sci. Sch. (Yale), 1887; postgrad. U. Va., 1887-88; M.D., Tulane U., 1889; m. Mercedes Louise Percival, July 31, 1905. Intern and grad. N.Y. Skin and Cancer Hosp., 1890-92; lectr. N.Y. Post-Grad. Med. Sch., 1891-92; lectr. diseases of skin, Tulane U., 1892-1905, asso. prof., 1905-08, prof., 1908-20, asso. dean, 1907-08, dean, 1908-20; vis. dermatologist Charity Hosp., cons. dermatologist to Eye, Ear, Nose and Throat Hosp. 1892-1920; prof. diseases skin, New Orleans Poly., 1893-1908, sec.-treas., 1895-1905, cons. Presbyn. Hosp., 1916-20. Founder, 1894, and pres. 1st bd. of control La. Leper Home; cons. leprologist, 1902-20; pres. examining bd. Med. R.C., New Orleans, 1917-18; chmn. med. sect. of State Com. Def., 1917-18; mem. Nat. Bd. Med. Examiners, 1915-20. Author: The Art of Medicine and Other Addresses, 1913. New Orleans Med. and Surg. Jour., from 1896; Am. Jour. Tropical Diseases and Preventive Medicine. Died Oct. 12, 1920.

DYER, Leonard Huntress, Am. inventor; b. Washington, May 13, 1873; s. George Washington and Kate (Huntress) D.; student Corcoran Sci. Sch., Columbian (now George Washington) U., 1893, Georgetown U. Law Sch., 1895, Nat. Law Sch., Washington, 1896; D.Sc., Rollins Coll., 1949; m. Josephine Duncan, July 10, 1905; children—Duncan, Katherine Huntress (Mrs. Elmer Puddington); m. 2d, Jessica Hofstetter, Oct. 14, 1927. Admitted to D.C. bar,

1894; practiced patent law with brother, Frank L. Dyer, until 1897; practiced alone, 1897-1903, with others, N.Y.C., until 1917. Fellow Am. Geog. Soc. Author: The Evolution of the Motor Vehicle as Shown by Patents, 1955. Invented an automobile with a direct drive, sliding transmission, selective gear shift and unit power plant; more than 100,000,000 automobiles have been made embodying this invention; also invented the flying boat, elec. steering gear, and improvements in steam turbines. Died Nov. 16, 1955.

DYER, Robert Allen, S. African botanist; b. Pietermaritzburg, S. Africa, Sept. 21, 1900; s. Robert Macey and Ellen Elizabeth (Legge) D.; B.Sc., Natal U., 1922, M.Sc., 1923, D.Sc., 1937; m. Adeline Beatrice Cooke, Dec. 22, 1926; children—Michael Arthur Allen, Rosemary Allen (Mrs. Anthony Campbell Young), Tristan Allen. Analytical chemist Huletts Sugar Mill, Amatikulu, Natal, 1924; apptd. bot. survey officer Dept. Agr., stationed Grahamstown, Cape Province, 1925, became officer in charge bot. survey, eastern Cape, 1926, S.African botanist Royal Bot. Garden, Kew, 1931-34, returned to Pretoria, 1934, became chief div. botany and plant pathology Dept. Agr., 1944, later chief of Bot. Research Inst., ret., 1963. S. African rep. 2 internat. bot. confs., also Internat. Symposium on Antarctic Research, Paris, 1962. Fellow Cactus and Succulents Soc. Am., Royal Soc. S. Africa; mem. S. African Biol. Soc. (Sr. Capt. Scott medal 1942), Biology Soc. S. Africa (pres. 1948), S. African Bot. Soc. (hon. life), Am. Amaryllis Soc. (corr. mem., Herbert medal 1948), S. African Assn. Advancement Sci. (medal 1951, pres. 1961). Author: The Vegetation of the Divisions of Albany and Bathurst, 1937; (with White and Sloane) The Succulent Euphorbieae of Southern Africa; (with Letty, Codd and Verdoorm) Wild Flowers of the Transvaal; also numerous articles. Contbns. in field of systematic botany, particularly of succulent Euphorbieae of S. Africa. Home: 923 Pretorius. Office: 590 Vermulen, Pretoria, Transvaal, South Africa.

DYER, Rolla Eugen, Am. surgeon b. nr. Galena, O., Nov. 4, 1886; s. Rolla and Nettie (Ryant) D.; A.B., Kenyon Coll., 1907, LL.D., 1932; M.D., U. Tex., 1915; LL.D., Tulane U., 1959; Sc.D., Washington U., St. Louis, 1960; Emory U., 1962; m. Esther Gibney, June 24, 1916; children—Sarah (Mrs. Hugh C. Gracey), Mary (Mrs. David G. Bryce), William Eugene. Commd., asst. surgeon USPHS, 1916, advanced through grades to asst. surgeon gen., 1948; dir. NIH, Bethesda, Md., 1942-50; ret., 1950; dir. research Emory U., Atlanta, 1950-57, clin prof. medicine, 1957—. Mem. Sci. adv. bd. NRC, 1945-50, Armed Forces Epidemic Bd., 1942-50, Chem. Warfare Adv. Bd., 1945-50. Recipient Carlos J. Finlay medal Republic of Cuba, 1944, Typhus Commn. medal, 1944, Lasker award Am. Pub. Health Assn., 1948, Sedgwick Meml. medal, 1950, James D. Bruce medal A.C.P., 1951, Theobald Smith medal Am. Acad. Tropical Medicine, 1953. Fellow A.M.A.; mem. Am. Pub. Health Assn., A.A.A.S., Soc. Tropical Medicine, Am. Epidemiological Soc., Assn. Am. Physicians. Research and publs. on rickettsial diseases; discovered murine typhus transmitted by fleas, presence Rocky Mountain spotted fever in Eastern states, presence Q fever in U. S. Home: 2150 E. Lake Rd., N.E., Atlanta 30307.*

DYHRENFURTH, Gunter Oskar, geologist; b. Breslau (now Wroclaw, Poland), Nov. 12, 1886; s. Oskar and Kate (Bayer) D.; ed. univs. Freiburg, Vienna; M.D., U. Breslau, 1913; m. Hettie Heyman (div. 1911); m. 2d, Irene Klar, 1948. Profl. geology and paleontology U. Breslau, 1919-33; prof. geography Rosenberg Coll., 1939-54. Recipient Olympic Gold medal, 1936. Organized Author: Alpine Geologie, 1931; Himalaya Fahrt, 1942; Zum Dritten Pol, 1952; Das Buch vom Nauga Parbat, 1954; Das Buch vom Kautsch, 1955; (with T. Hagen, Füer, Schneider) Mount Everest, 1959; Der Dritte Pol, 1960. Organized and led internat. Himalayan expdns. in Kaugchendzonga mountains, 1930, Baltoro region, 1934; climbed Jongsang Peak (highest mountain climbed by man until 1930).

DYK, Václav, Czechoslovak parasitologist; b. Strakonice, Czechoslovakia, Feb. 27, 1912; s. Václav and Olga (Vsetecková) D.; vet. doctors degree Vet. Faculty, Brno, Czechoslovakia, 1938, D.Vet. Scis., 1961; m. Silvia Kucerová, Sept. 27, 1938; children—Jana (Mrs. Jirí Sindlár), Iva, Petr. Asst. lectr. dept. biology and parasitology Vet. U., Brno, 1938-40; staff fishery research Vodnany, 1940-46, Opava, Czechoslovakia, 1946-48; Vet. Faculty, Brno, 1948—, prof., 1955—, head dept. parasitology, 1948—; editor chief sci. papers U. Agr., Brno, 1949—. Recipient award from City of Brno for book, 1952, Golden Badge, State Orgn. Fishery, 1964, Diploma, State Game Orgn., 1963, highest award for merits in fishery, 1967. Mem. Czechoslovak Sci. Soc. Parasitology, Czechoslovak Sci. Soc. Zoology, Czechoslovak Sci. Soc. Entomology, Czechoslovak Sci. Soc. Ornitholgy, Czechoslovak Sci. Soc. Museal Mammaliology. Author: Nase ryby, 1944; Ceské perly, 1947; Nemoci nasich ryb, 1952; (with Podubsky, Stedronsky) Nase rybárství, 1948, Základy naseho rybárství, 1956; (with Klimes, Zavadil) Cizopasníci a invasní choroby drubeze, 1957; also numerous articles. Research on natural food of brown trout, physiology of fish reprodn., influence of some chems. and acid water on fishes, influence of vitamins on resistance of young trout, Helminthofauna of Salmonidae, fish reservoirs of disease, Helminthofauna of Ovis musimon. Home: 13 nám.Slov.-nár. povstání, Brno, Czechoslovakia.*

DYKE, Cornelius Gysbert, Am. neurologist; b. Orange City, Ia., 1900; s. Charles Dyke; M.D., U. Ia., 1926; m. Doris Dyke; children—Neil, Gysbert. Dir. Neurol. Inst. N.Y.; asso. prof. radiology Columbia. Mem. Am. Roentgen Ray Soc. Author: (with Leo M. Davidoff). The Normal Encephalogram, The Roentgen Treatment for Disorders of the Nervous System. Specialist in x-ray diagnosis, treatment of diseases of nervous system, skull; research various nervous disorders. Died N.Y.C., Apr. 23, 1943.

DYKSTERHUIS, Edsko Jerry, Am. ecologist; b. Hospers, Ia., Dec. 27, 1908; s. Jerry and Jantina (Brouwer) D.; B.S., Ia. State U., 1932; Ph.D., U. Neb., 1945; m. Margarett A. Cox, Mar. 26, 1933; children—Jantina Kay (Mrs. Archie R. Clegg), Leona Marge (Mrs. Edward F. McCoy), Jerry Edsko. Forester, range examiner U. S. Forest Service, Utah, 1930, Ariz., N.M., 1933-38, Tex., Ark., Mo., Kan., 1938-43; range conservationist, soil conservationist U. S. Soil Conservation Service, Tex., La., Ark., Okla., 1946-49, regional range conservationist Mont., Wyo., N.D., S.D., Neb., 1949-64; prof. range ecology Tex. A. and M. U., 1964—. Recipient Merit, Outstanding Service, Authorship awards U. S. Dept. Agr. Fellow A.A.A.S.; mem. Soc. Am. Foresters, Ecol. Soc. Am. (Mercer award), Soil Conservation Soc. Am., Am. Soc. Range Mgmt. (pres.), Sigma Xi. Contbr. numerous articles to sci. jours., related industry mags. Developed quantatitive ecological approach to inventory and management of rangelands; climatic and edaphic gradients related to gradation of climax vegetation, making it possible to quantify range degeneration and predict potentials. Home: 3807 Oaklawn St., Bryan, Tex., 77801.*

DYMSZA, Henry Adam, Am. nutritional biochemist; b. Newton, N.H., Jan. 14, 1922; s. Alexander and Mary (Lucas) D; B.S., Pa. State U., 1943; M.S., U. Wis., 1950; Ph.D., Pa. State U., 1954; m. Janina Kaminski, June 3, 1956; children—Valerie A., Darlene V., Andrea M. Research nutritionist Gen. Foods Corp., N.Y.C., 1954-59; sr. research asso. Mass. Inst. Tech., Cambridge, 1959-64; head metabolism group, food div. U. S. Army Natick (Mass.) Labs, 1964-66; research affiliate Mass. Inst. Tech., 1964-66; asso. prof., chmn. dept. food and nutrition U. R.I., Kingston, 1966—. Mem. A.A.A.S., Am. Inst. Nutrition, Am. Chem. Soc., Inst. Food Technologists, Animal Care Panel, Assn. for Applied Gnotobiotics, Sigma Xi. Research, publs. on new synthetic food sources, high-energy foods; developer nutrient defined and other specially designed diets for exptl., mil. and space nutrition, technique for artificial feeding of exptl. neonatal rats, relation of food and intestinal micro-organisms to health. Office: Quinn Hall, U. R.I., Kingston, R.I. 02881.

DYR, Josef, Czechoslovak chemist; b. Borontín, Czechoslovakia, Jan. 7, 1904; s. Josef and Bozena (Petrova) D.; M.A., Inst. Agr. and Inst. Chemistry, Prague, Czechoslovakia, 1938, Ph.D., 1937; m. Karla Riedlová, Dec. 31, 1938; children—Zdenka (Mrs. Jirí Kucera), Pavel, Ludmila, Jan. Faculty, Inst. Agr., Prague, 1936-39, asso. prof., 1945-51; staff Brewery Research Inst., Prague, 1939-45; faculty Inst. Chem. Tech., Prague, 1951—, prof., head dept. fermentation and tech., 1954—, pres., 1953-56, dean Inst. Food Faculty, 1952-53. Recipient Nat. prize, 1952. Mem. Czechoslovak Chem. Soc., Czechoslovak Biochem. Soc., Czechoslovak Microbiol. Soc. Author several books, numerous articles. Research on hop resins, lactic acid fermentation, butanol-acetone fermentation, reprodn. of zygomycetes, microbiol. synthesis of fats and carotenoids, microbiol. inhibitors in molasses, recirculation of waste in yeast industry, continuous processes in brewing. Home: 95 Na Pískách, Prague 6, Czechoslovakia.*

DYRSSEN, David Waldemar, chemist; b. Tarrytown, N.Y., June 10, 1922; s. Waldemar and Eugene (Baker) D.; Fil.lic. (Ph.D.), U. Stockholm (Sweden), 1953, Fil.dr. (docent), 1956; m. Margareta Ising, June 3, 1946; children—Agneta, Catharina, Henrik, Cecilia, Jan. Research asst., sect. chief Swedish Research Inst. Nat. Def., 1948-60; docent inorganic chemistry Royal Inst. Tech., Stockholm, 1960-63; prof., head dept. analytical chemistry U. Gothenburg (Sweden), 1963—. Mem. IUPAC Commn. on Equilibrium Constants. Mem. Swedish Chem. Soc. (chmn. Gothenburg br., chmn. bd. analytical chemistry), Swedish Soc. for Oceanographic Research (sec.). Author: (with Daniel Jagner, Fredrik Wengelin) Computer Calculations of Ionic Equilibria and Titration Procedures; also articles. Research in analytical chem. phys., inorganic and radio chemistry; investigations of chelating reagents and equilibria in aqueous solutions, analytical procedures for determinations of constituents in sea water. Home: 5, Lunnatorpsgatan, Göteborg S. Office: 5A, Gibraltargatan, Göteborg S, Sweden.*

DYSON, Sir Frank Watson, English astronomer; b. Measham, Eng., Jan. 8, 1868; s. Watson and Frances (Dodwell) D.; ed. Trinity Coll., Cambridge (Eng.) U., Sc.D. (hon.); LL.D., Edinburgh (Scotland) U., 1910; D.Sc. (hon.), Oxford (Eng.) U.; other hon. degrees English, Canadian, Australian univs.; m. Caroline Bisset Best, 1894; 2 sons, 6 daus. Chief asst. Royal Obs., Greenwich, Eng., 1894-1905, astronomer-royal, 1910-33; astronomer-royal for Scotland, 1906-10. Mem. eclipse expdns. to Portugal, 1900, Sumatra, 1901, Tunisia, 1905. Recipient Bruce Gold medal Astron. Soc. Pacific, 1922. Fellow Royal Soc. (v.p. 1913-14, Royal medal 1921), 1901; mem. French Acad. Scis., 1914, Brit. Horological Inst. (pres., Gold medal 1928), Royal Astron. Soc. (pres. 1911-13, Gold medal 1925), Brit. Astron. Assn. (pres. 1916-18), Internat. Astron. Union (pres. 1928-32). Author: Determination of Wave Length from Spectra Obtained at the Total Solar Eclipses of 1900, 1901, and 1905, 1906; Astronomy, 1910; Observations of Stellar Parallax from Photographs, 1925; (with R. Woolley) Eclipses of the Sun and Moon, 1937; Research, publs. on distbn. and movements of stars and structure of stellar universe; (with W. G. Thackeray) reobserved and derived motions of 4,000 circum polar stars (originally observed by S. Groombridge), 1806-19; proved that phenomenon of star streaming was confirmed by stars of large proper-motion; introduced study of stellar parallaxes; under his direction, distbn. of time signals from Greenwich (through Brit. Broadcasting Co.), was begun, 1924, also radio time signals of world wide range were 1st broadcast from Rugby, Eng.; made studies of spectra of sun's chromosphere and corona; verified Einstein's theory of effect of gravity on light, 1919. Died at sea, May 25, 1939.

DYSON, Freeman John, physicist; b. Crowthorne, Eng., Dec. 15, 1923; s. George and Mildred (Atkey) D.; B.A., U. Cambridge (Eng.), 1945; m. Verena Esther Huber, Aug. 11, 1950; children—Esther, George; m. 2d, Imme Jung, Nov. 21, 1958; children—Dorothy, Emily, Miriam, Rebecca. Came to the United States of America, 1951, naturalized, 1957. Fellow of Trinity College, Cambridge University, 1946-47; commonwealth fellow Cornell and Princeton, 1947-49; prof. physics Cornell U., Ithaca, N.Y., 1951-53; prof. Inst. for Advanced Study, Princeton, N.J., 1953—. Cons. to AEC, NASA. Chmn., Fedn. Am. Scientists, 1962-63. Fellow Royal Soc., 1952; mem. Nat. Acad. Scis. Contbr. to Quantum Electrodynamics, 1949. Research and publs. on math. physics, especially math. form of the theory of electromagnetic radiation. Home: 105 Battle Rd. Circle, Princeton, N.J. 08540.*

DYSON, James, Brit. physicist; b. Bradford, Yorks, U.K., Dec. 10, 1914; s. George and Mary (Bateson) D.; B.A., Christ's Coll., Cambridge (Eng.) U., 1933, M.A., 1960, Sc.D., 1960; m. Marie Florence Chant, June 21, 1948; 1 dau., Gaynor Jacqueline (Mrs. Anthony C. Wagstaffe). With Brit. Thomson-Houston Co. Rugby, U.K., 1936-46, research engr., 1939-46; cons. optics Asso. Elec. Industries Research Lab., Aldermaston, Berks., U.K., 1946-63; supt. div. optical metrology Nat. Phys. Lab., Teddington, U.K., 1963—. Mem. Inst. Physics, Phys. Soc., Brit. Astron. Assn., Royal Micros. Soc., Royal Instn. Gt. Britain. Research and publs. on devel. new optical instruments including interference microscope, image-splitting eyepiece, instruments for alignment and engring. metrology, new types of interferometer; spl. design of optical instruments for immunity to disturbances. Patentee in field. Home: 19 Hansler Grove, East Molesey, Surrey, U.K. Office: Nat. Phys. Lab., Teddington, Middlesex, U.K.*

DYSON, James Lindsay, Am. geologist; b. Lancaster, Pa., May 23, 1912; s. Herbert Pannebecker and Mary Emma (Lindsay) D.; B.S., Lafayette Coll., 1933; M.A., Cornell U., 1935, Ph.D., 1938; m. Lolita Gill Brown, Oct. 10, 1942; children—Dolores Gill, Deborah Anne Lindsay. Faculty, Lafayette Coll., Easton, Pa., 1947—, prof., head dept. geology and geography, 1948-63, Markle prof. geology, 1963—; cooperating geologist Pa. Geol. Survey, 1952; chmn. geology selection com. Fulbright awards Nat. Acad. Scis., NRC, 1961-62; with Earth Sci. Curriculum Project Am. Geol. Inst., Boulder, Colo., 1964-65. Recipient Thomas L. Jones awards Lafayette Coll., 1956, 57, 62, Distinguished Alumnus citation, 1964; Nat. Phi Beta Kappa award, 1962. Fellow Geol. Soc. Am.; mem. Am. Geog. Soc., A.A.A.S., Nat. Assn. Geology Tchrs., Glaciological Soc., Arctic Inst. N.Am., Glacier Natural History Assn., Sigma Xi. Author: The World of Ice, 1962. Research and publs. on activity of mountain glaciers, method of uranium occurrences in Appalachian region. Home: 32 McCartney St., Easton, Pa. 18042.*

DYSON, Robert Harris, Am. archaelogist; b. York, Pa., Aug. 2, 1927; s. Robert H. and Harriet (Duck) D.; A.B., Harvard, 1950, postgrad. (Soc. Fellows jr. fellow), 1951-54, Ph.D., 1966. Mem. faculty Univ. Mus., dept. anthropology, U. Pa., Phila, 1954—, asso. curator, asso. prof., 1962-67, field dir. Tikal Guatemala project, 1962, asst. chmn. dept. anthropology, 1964—, professor anthropology, curator, 1967—; field dir. Iran Expdn., University Museum, also Met. Mus. Art, N.Y.C., 1956—. Named Chevalier des Arts et Lettres, Govt. France 1962. Mem. Council Old World Archaeology (editor for So. Asia, 1956-62, charter mem., trustee 1961—), Am. Anthrop. Soc., Iran-Am. Soc., (dir. 1964—), A.A.A.S., Far Eastern Prehistory Assn., Am. Oriental Soc., Archaeol. Inst. Am., Soc. for Am. Archaeology. Research and publs. on domestic animals ancient Nr. E.; established major sequence cultures for N.W. Iran from 6000B.C. to 400 B.C. Home: 320 S. 2d St., Phila. 19106.*

DZHANELIDZE, Aleksandr Illarionovich, Russian geologist, paleontologist; b. Nov. 5, 1888; grad. U. Paris, 1911. With dept. geology Kazan U., 1917; prof. Tbilisi U., 1924—; dir. Geol. Inst., Georgian Acad. Sci., 1925-56, acad. sec. dept. math. and natural sci., 1941-64. Decorated Order of Lenin. Mem. Georgian Acad. Sci. (Presidium mem. 1941-64, v.p. 1951-55). Author: The Problem of the Georgian Block, 1942; The Birth of Orogeny, 1949; Current Theory of Orogenetic Phases, 1952; Orogenetic Cycles, 1963. Address: Tbilisi University, prospect Chavchavadze 48, Tbilisi, Gruz. SSR, USSR.

DZHAPARIDZE, Levan Ivanovich, Russian anatomist, plant physiologist; b. Tbilisi, Apr. 26, 1905; grad. Tbilisi U., 1927; Cand. Biol. Sci., 1936. With Tbilisi U., 1927-30, Agrl. Inst., Georgia, 1930-31, Tbilisi Forestry Inst., 1930-39, Tbilisi Zool. and Vet. Inst., 1932-34, Tbilisi Pharm. Inst., 1939-43; with Inst. Botany, Georgian Acad. Sci., 1930—, head dept. anatomy and physiology of plants, Bot. Inst., 1948—. Mem. Georgian Acad. Sci. Author: The Structure of Wood, 1934; A Dictionary of Terms used in the Description of Wood, 1936; A Short Practical Manual on Plant Physiology, 1937; Transpiration in Diclinous Plants, 1949; A Manual on the Microscopic Chemistry of Plants, 1953; Sex in Plants, 1963; co-author: The Principles of Microbiology, 1950. Research and publs. on physiology of wood, physiology of stability of plants, physiol. and biochem. distinctions in sexes in plants. Address: Botanical Inst., Georgian Acad. Sci., Tbilisi, Gruz. SSR, USSR.

DZHELEPOV, Boris Sergeevich, Russian physicist; b. Dec. 12, 1910; grad. Leningrad U., 1931. Asso., Physicotech. Inst., USSR Acad. Sci., 1931-43, asso. Radium Inst., 1945—; prof. Leningrad U., 1935—, also head chair nuclear spectroscopy; with All-Union Research Inst. Metrology, 1939-41, 46—. Head Soviet del. Internat. Conf. on Nuclear Reactions, Amsterdam, 1956. Mem. USSR Acad. Sci. (corr.). Author: (with L. N. Syryanova) Influence of the Electric Field of an Atom on Beta Disintegration. Research and publs. on energy spectra of electrons from radioactive nuclei; angular distribution of photons from pair annihilation; designed gamma spectrometer based on pair formation; analyzed data on beta disintegration, mirror nuclei, and isotopic spin. Address: Leningrad University, Universitetskaya n. 7-9, Leningrad, USSR.

DZHORBENADZE, Arkadiy Vissarionovich, Russian pathoanatomist; b. Georgia, 1902; grad. Med. Faculty, Tbilisi U., 1928; D. Med. Sci., 1947. Asst. dept. path. anatomy Med. Faculty, Tbilisi U. (now Tbilisi Med. Inst.), 1928-37, lectr., 1937-48; lectr. dept. path. anatomy Georgian Zootech. and Vet. Inst., 1933-1933-45; prof., head chair path. anatomy and forensic medicine Tbilisi Postgrad. Med. Inst., 1948—. Mem. learned med. council Georgian Ministry Health; chief pathoanatomist Tbilisi City Dept. Health. Decorated Order of Lenin. Mem. Georgian Soc. Pathoanatomists (dep. chmn.), Tbilisi Soc. Forensic Med. Experts and Criminologists (hon.). Author: Data on Typhus. Research and numerous publs. on tumors, pathology of infectious diseases, ulcers, disorders of metabolism, fetal and infant pathology, alloplasty of major vessels, teratology, produced typhus in hamsters and guinea pigs under exptl. scurvy conditions. Address: Tbilisi Postgrad. Med. Inst., Tbilisi, Gruz SSR, USSR.

DZHRBASHYAN, Mkhitar Mkrtychevich, Russian mathematician; b. 1918; grad. Yerevan U., 1941; D.Physico-Math. Sci., 1949. Instr., Yerevan U., 1947-51, prof., 1951—; with math. and mech. sect. Armenian Acad. Sci., 1945-55, asso. Inst. Math. and Mechanics, 1955—, acad. sec. dept. physico-math. sci., 1963—. Mem. Armenian Acad. Sci. Research and publs. on theory of functions. Address: Yerevan University, ulitsa Abovyana 104, Yerevan, Armenia SSR, USSR.

DZIERZON, Johann, German apiculturist; b. Lowkowitz, Upper Silesia, Poland, Jan. 16, 1811; studied theology at Breslau (now Wroclaw, Poland); pastor at Karlsmarkt, 1835-69; after 1869 concentrated solely on apiculture. Author: Theorie und Praxis des Neuen Bienenfreundes, 1848; Rationelle Bienenzucht, 1861; Der Zwillingsstock, 1890. Founder jour. l'Ami des abeilles de Silésie. Discovered parthogenesis in bees; developed new type of beehive with cupboard opening at one end. Died Lowkowitz, Dec. 26, 1906.

DZOTSENIDZE, Georgii Samsonovich, Russian geologist; b. Feb. 23, 1910; grad. Tbilisi U., 1929, staff, 1929—, prof., 1947—; chief lithology sect., geology inst. Georgian SSR Acad. Scis., 1941—, acad. sec., 1949, v.p., 1955—. Recipient Stalin prize, 1950. Author: Domiotsenovy Effuzivny Vulitanism Gruzii, 1948; Lilologia Batskikh Otlozheniy Okribi, 1956; other works. Research in petrography of magmae, lithology, mineralogy; demonstrated genetic connection between nature of vulcanism and geotectogenesis, in studies of Georgia.

E

EABORN, Colin, English chemist; b. Cheshire, Eng., Mar. 15, 1923; s. Tom Stanley and Caroline (Cooper) E.; B.Sc., Ph.D., U. Coll. N. Wales, Bangor, 1941-47; D.Sc., Wales, 1907; m. Joyce Thomas, July 27, 1949. Faculty, U. Leicester (Eng.), 1947-62; prof. chemistry, dean Sch. Molecular Scis., U. Sussex, Brighton, Eng., 1962—; research asso. U. Cal. at Los Angeles, 1960-61; Robert A. Welch vis. scholar Rice U., 1961-62; Erskine vis. fellow U. Canterbury, New Zealand, 1965. Mem. govtl. coms. Mem. Chem. Soc., Am. Chem. Soc. (F. S. Kipping award 1964), Royal Inst. Chemistry, Chem. Soc. (hon. sec. Gt. Britain). Author: Organosilicon Compounds, 1960; also numerous articles. Editor series Reaction Mechanisms in Organic Chemistry, 1963—; regional editor Jour. Organometallic Chemistry, Organometallic Chem. Revs. Research on properties and mechanism of reactions of organosilicon and related compounds, elucidation of effects of substituents in electrophilic aromatic substitutions, reactions of optically active germanium compounds, chemistry of compounds with transition metals bonded to germanium reactions of organametallic free radicals. Office: U. Sussex, Brighton, Sussex, Eng.*

EADES, James Beverly, Jr., Am. aero. engr.; b. Bluefield, W.Va., July 22, 1923; s. James Beverly and Harriet (Smith) E.; student Bluefield Coll., 1940-42; B.S., Va. Poly. Inst., 1944, M.S., 1949, Ph.D., 1958; m. Sara Marshall Porterfield, Dec. 20, 1950; children—Sara Leslie, Beverly Anne, James Christian. Mem. faculty Va. Poly. Inst., Blacksburg, 1946—, prof. aero. engring., 1958—, head aerospace engring., 1961—, dir. Conf. on Lunar Exploration, 1963. Research asso. (Nat. Acad. Sci. fellow) Goddard Space Flight Center, Greenbelt, Md.; cons. to govt. agys., industry. Asso. fellow Am. Inst. Aeros. and Astronautics, mem. Va. Acad. Sci. (chmn. engring. sect. 1960), Am. Astronautical Soc. (st.), Sigma Xi, Tau Beta Pi, Sigma Gamma Tau. Research and publs. in aerodynamics, aerodynamic stability of bridges and structures, investigations of behavior of vehicles, bldgs. to wind loads, theoretical investigations in transonic flight phenomenon, dynamics of space flight, trajectory analysis. Home: 115 McConkey St., Blacksburg, Va. 24060.*

EADIE, W(illiam) Robert, Am. zoologist; b. Manchester, N.H., May 5, 1909; s. James and Maria (Bremner) E.; B.S., U. N.H., 1932, M.S., 1933; Ph.D., Cornell U., 1939; m. Eva Selina Wentzell, Dec. 20, 1933 (dec.); children—William James, Dennis Robert; m. 2d, Laura Coldwell Keenahan, May 19, 1967. Faculty U. N.H., 1933-42, asst. prof., 1939-42; faculty Cornell U., Ithaca, N.Y., 1942—, prof. zoology, 1954—; vis. prof. U. Ore., 1949, Mont. State U., 1964. Cons., United Fruit Co., Guatemala, 1958. Fellow A.A.A.S.; mem. Am. Soc. Mammalogists (dir. 1956-60, 62-64, 65—), Am. Soc. Zoologists, Ecol. Soc. Am., Wildlife Soc., Sigma Xi, Phi Sigma, Phi Kappa Phi. Author: Animal Control in Field, Farm, and Forest, 1954; also numerous articles. Editor, Jour. Mammalogy, 1952-56. Studies reprodn. in shrews, moles, otter, fisher, predation in shrews, skin glands moles and shrews, rodent control methods. Home: 1410 Hanshaw Rd., Ithaca, N.Y. 14850.*

EADS, James Buchanan, Am. engr., inventor; b. Lawrenceburg, Ind., May 23, 1820; s. Thomas C. and Ann (Buchanan) E.; LL.D., U. Mo., 1877; m. Martha Dillion; m. 2d, Eunice Eads, 1857; 5 children. Began career as a clerk in a dry goods house, St. Louis, 1833; worked on a Miss. steamboat, 1839; became partner in steamboat salvaging firm, 1842; called by Pres. Lincoln to Washington, D.C., 1861, recommended means of employing Western rivers for Union war operations, undertook constrn. of fleet of steam-powered armor-plated gunboats which he had proposed, built total of 14 armoured vessels (in record time), featuring his own patented ordnance inventions; constructed Eads Bridge (steel and masonry bridge across Mississippi at St. Louis, best known achievement), incorporated engring. features conquering difficulties considered insuperable by prominent authorities of day, 1867-74; reputation as hydraulic engr. established by river control work completed 1879 at South Pass in Mississippi River, controlled placement of river's sediment so as to keep channel clean; improved harbor facilities at Liverpool (Eng.), Toronto (Ont., Can.), Tampico and Veracruz (Mexico). Recipient Albert medal Brit. Soc. Encouragement Art, Manufacture and Commerce, 1884. Fellow A.A.A.S.; mem. Am. Soc. C.E. (v.p. 1882), Brig. Instn. Civil Engrs., Brit. Assn. Inventor diving bell; designed boats for raising sunken steamers. Died Nassau, Bahama Islands, Mar. 16, 1887.

EAGAN, Charles Joseph, physiologist; b. Tottenham, Ont., Can., Sept. 25, 1921; s. Thomas G. and Theresa M. (Walsh) E.; B.A., U. Western Ont., 1949, M.Sc., 1956; Ph.D., U. Wash., 1961; m. Marlene Joan Varey, Oct. 28, 1952; children—Maureen, Paul, Edmund, Conrad, Mark, Colleen, Monica. Sci. officer Def. Research Bd. Can., 1954-57; research physiologist Arctic Aeromed. Lab., 1959—. Fellow A.A.A S. (past pres. Alaska div., chmn. exec. com. 1965—), Arctic Inst. N.Am.; mem. Am., Canadian physiol. socs., Canadian Physiol. Soc. Biometeorology, Am., Canadian socs. zoology, Soc. Gen. Physiology, Internat. Soc. Biometeorology. Research on temperature regulation, human and animal energetics, cold tolerance and adaption, work physiology, bioclimatology, peripheral circulation, altitude acclimatization. Home: Bldg. 4012-7 Pine St., Ft. Wainwright, Alaska. Office: Arctic Aeromed. Lab., APO Seattle 98731.*

EAGLE, Edward, Am. physiologist, toxicologist; b. Balt., Nov. 27, 1908; s. Louis and Sadie (Kushnoy) E.; A.B., Johns Hopkins U., 1929, postgrad. 1931-34; M.S., U. Va., 1931; Ph.D., U. Chgo., 1940; m. Mary Frances Rowe, Dec. 28, 1942; children—Nancy Louise, Ellen Rowe, Margaret Lillian, Louis Rowe, Faculty U. Chgo., 1936-42, research asst. in physiology, 1938-42; aviation physiologist U. S. Air Force, 1942-46; with Swift & Co., Chgo., 1946—, head div. physiology, toxicology, histology and nutrition, 1961—; mem. adv. bd. div. biol. scis. Air Force Office Sci. Research, 1962—. Mem. Nat. Acad. Scis. (industry com., food protection com.), Am. Physiol. Soc., Soc. Toxicology, Am. Soc. Pharmacology and Exptl. Therapeutics, Soc. Exptl. Biology and Medicine, Council on Arteriosclerosis, Am. Heart Assn., N.Y. Acad. Scis. Research and publs. on adrenal extracts, conditioned reflexes, choline, metabolism, energy output, cottonseed physiology, nutrition, chems. in foods, toxicol. evaluations. Home: 2230 Asbury Av., Evanston, Ill. 60201. Office: Packers and Exchange Avs., Chgo. 60609.*

EAGLE, Harry, Am. microbiologist; b. N.Y.C., July 13, 1905; s. Louis and Sadie (Kushnoy) E.; A.B., Johns Hopkins, 1923, M.D., 1927; M.S., Yale, 1948; D.Sc., Wayne State University, 1965; m. Hope Whaley, August 31; 1 daughter, Mrs. Robert B. Kyle. Instructor Johns Hopkins University, Balt., 1928-32, dir. lab. exptl. research lab. Sch. Hygiene and Pub. Health, also adj. prof. bacteriology, 1946-48; asst. prof. microbiology U. Pa. Med. Sch., 1933-36; dir. Venereal Disease Research Lab., Johns Hopkins Hosp., 1936-46; sci. dir. research br. Nat. Cancer Inst., Bethesda, Md., 1947-49, chief sect. exptl. therapeutics Nat. Microbiol. Inst., NIH, 1949-59, chief lab. cell biology Nat. Inst. Allergy and Infectious Diseases, 1959-61; ret., 1961; prof., chmn. dept. cell biology Albert Einstein Coll. Medicine, Bronx, N.Y., 1961—. Mem. divisional com. for biol. and med. scis. NSF, 1962-67; mem. sci. adv. council Found. for Microbiology, Helen Hay Whitney Found., Sloan-Kettering Inst. Cancer Research. Trustee Cold Spring Harbor (L.I., N.Y.) Lab. Recipient Eli Lilly Co. award in bacteriology, 1936, Alvarenga prize Coll. Physicians Phila., 1936, Presdl. Certificate of Merit, 1948; Borden award Am. Assn. Med. Coll., 1964. NRC fellow, 1932-33. Mem. Nat. Acad. Scis., Am. Acad. Arts and Scis., Am. Soc. for Cell Biology (past mem. council), Soc. Am. Microbiologists (past pres.), Assn. Am. Physicians, Soc. Exptl. Biology and Medicine (past pres.), Am. Assn. Immunologists (past pres.), Am. Soc. Biol. Chemists, Am. Assn. Cancer Research (past dir.). Research and numerous publs. on bacterial physiology; immunochemistry; coagulation; chemotherapy and antibiotics; cell and tissue cultures. Home: 370 Orienta Av., Mamaroneck, N.Y. 10543. Office: Albert Einstein Coll. Medicine, Bronx, N.Y. 10461.*

EAGLESON, Peter Sturges, Am. civil engr.; b. Phila., Feb. 27, 1928; s. William Boal and Helen (Sturges) E.; B.S. in Civil Engring., Lehigh U., 1949, M.S., 1952; Sc.D., Mass. Inst. Tech., 1956; m. Marguerite Anne Partridge, May 28, 1949; children—Helen Marie, Peter Sturges, Jeffrey Partridge. Designer, insp. George B. Mebus, cons. engr., Glenside, Pa., 1950; faculty Mass. Inst. Tech., Cambridge, 1954—, prof. civil engring., 1965—. Mem. Am. (Research prize 1963), Boston (Desmond Fitzgerald medal 1959, Clemens Herschel prize 1965) socs. civil engrs. Am. Geophys. Union, Internat. Assn. Sci. Hydrology, Internat. Assn. Hydraulic Research. Formulation of mechanics of bed sediment movement due to wave action and application of this to study of beach processes; formulation of devel. longshore currents on beaches; demonstrated feedback mechanism in vortex-induced vibration of structural elements; application of theory of communications to analysis of hydrologic systems. Home: 34 Lowell Rd., Wellesley Hills, Mass. 02181. Office: Mass. Inst. Tech., Cambridge, Mass. 02139.*

EAKIN, Richard Marshall, Am. zoologist; b. Florence, Colo., May 5, 1910; s. S. Marshall and Mary (Jack) E.; A.B., U. Cal. at Berkeley, 1931, Ph.D., 1935; m. Mary Mulford, Aug. 8, 1935; children—David Marshall, Dorothy Alice. Faculty U. Cal. at Berkeley, 1936—, prof. zoology, 1949—, chmn. dept. zoology, 1942-48, 52-57, asst. dean Coll. Letters and Sci., 1939-42, research prof., Miller Inst., 1961, chmn. exec. com., 1961-67. Nat. Research fellow Erlangen, Freiburg, (both Germany), 1935-36; Guggenheim fellow, Stanford, 1953; Nat. Sci. sr. postdoctral fellow, Bern, Switzerland, 1957; recipient 1st sr. citation Distinguished Teaching U. Cal. at Berkeley, 1962. Fellow Institut Internat. d'Embryologie Utrecht, Cal. Acad. Scis. (past rec. sec.); mem. A.A.A.S., Am. Soc. Zoologists, Soc. for Exptl. Biology and Medicine, Western Soc. Naturalists (past pres.), Soc. Growth and Devel. Author: Vertebrate Embryology, 1964; also numerous articles. Discovered cone-like cells in 3d eyes lizards, frogs, and lampreys; ascertained time in devel. when rods and cones in amphibian embryo are irreversibly determined; demonstrated dependence intermediate lobe pituitary on brain for devel. fine structure and evolution of photoreceptors. Home: 1627 Spruce St., Berkeley, Cal. 94709.*

EAKLE, Arthur Starr, Am. mineralogist; b. Washington, July 27, 1862; s. Elias H. and Mary Frances (Byington) E.; B.S., Cornell U., 1892; Ph.D., U. Munich, 1896; m. Fannie V. Kinney, Aug. 29, 1899; 1 dau., Alice Frances. Asst. and instr. mineralogy, Cornell U., 1892-94, Harvard, 1897-1900; instr. mineralogy U. Cal. at Berkeley, 1900-03, asst. prof., 1903-12, asso. prof., 1912-19, and prof., 1919-31. Fellow Geol. Soc. Am., Mineral. Soc. Am. (pres. 1925), A.A.-A.S. Author: Mineral Tables for the Determination of Minerals by Their Physical Properties; California Minerals. Died July 5, 1931.

EARDLEY, Armand John, Am. geologist; b. Salt Lake City, Oct. 25, 1901; s. John Alma and Elizabeth (Brown) E.; A.B., U. Utah, 1927; Ph.D., Princeton, 1930; m. Norma Ashton, May 6, 1930; 1 son, Michael John. With U. S. Geol. Survey, 1929-31; faculty U. Mich., 1930-49, prof., 1940-49; prof. geology U. Utah, Salt Lake City, 1949—, head dept. geology, 1949-54, dean Coll. Mines, 1954-65; faculty Army U. in France, 1945; Nat. Sigma Xi lectr., 1955-56. Recipient Distinguished Service award Utah Acad., 1960, James E. Talmage Sci. award Brigham Young U., 1962. Fellow Geol. Soc. Am.; mem. Am. Assn. Petroleum Geologists (past pres. Rocky Mountain sect., Distinguished lectr. 1953-54), Am. Inst. Mining Engrs., Soc. Exploration Geophysicists, Am. Geophys. Union, Am. Inst. Profl. Geologists, Nat. Assn. Geology Tchrs. (past pres.), Am. Geol. Inst. (pres. 1965). Author: Aerial Photographs, 1943; Structural Geology of North America, 1951, 62; General College Geology, 1965; also numerous articles. Regional geology N.Am., structural geology Rocky Mountains, geologic use aerial photographs, Great Salt Lake, Lake Bonneville. Home: 2618 Skyline Dr., Salt Lake City 84103.*

EARL, George Goodell, Am. civil engr.; b. Monmouth Co., N.J., Oct. 9, 1863; s. Holmes and Annie (Taylor) E.; C.E., Lafayette Coll., 1884, D.Sc., 1918; m. Anna L. Riddell, June 1890 (dec. 1911); children—Anna Taylor (dec.), Ralph; m. 2d, Frances H. Fowler, Jan. 1912; 1 son, Thomas Collins. With U. S. Geol. Survey in N.J., 1884-85; r.r. location and constrn. with A.T.&S.F. R.R., in Mo., 1886-87; sewer constrn., Montgomery, Ala., 1888; gen. engring. practice, 1888-92; city engr., Americus, Ga., 1890-91; chief engr. New Orleans Sewerage Co., 1892-99; gen. supt. and chief engr., Sewerage and Water Bd., New Orleans, 1900-30, cons. engr., 1931-40. Has specialized in devel. new methods for regulation, measurement and recording of fluid flows and pressures, especially in proportional flows, and in improvement of liquid meters to record fully on low rates of draft heretofore not recordable. Died Sept. 16, 1940.

EARLE, Clarence Edwards, Am. engr.; b. Bengies, Md., Aug. 27, 1893; s. William George and Annie Rebecca (Edwards) E.; student Balt. Poly. Inst., 1913; B.S. in Chem. Engring., George Washington U., 1923; m. Dorothy B. Stone, April 16, 1919; children—Leslie Marie (Mrs. Richard O. Thomas), Richard Stone. Head aero. gas sect., Bur. Aeros., U. S. Navy, 1920-25; rep. U. S. helium plant, Ft. Worth, 1925-30; head chem. research and devel. sect., 1930-42, chief chem. cons., 1942-43; dir. Earle Research Lab., 1939-53; research dir. Balt. Engring. and Chem. Co.; chief tech. cons. R.R. Engring. Co. Mem. Am. Chem. Soc. Automotive Engrs. Discovered and developed Lithium soap lubricating greases used in aircraft mfg., U. S. and fgn., World War II; originated, developed all-purpose hydraulic oil, chem. polar compounds for thin film preservation of metallic surfaces against corrosion, aircraft carbon monoxide detector; pioneered discovery and devel. of series of chem. compounds known as phenyl-amino salts used as mycotic drug in South Pacific. Died Nov. 25, 1953.

EARLE, David Prince, Am. physician; b. Englewood, N.J., May 23, 1910; s. David Prince and Paula (Benner) E.; A.B., Princeton University, 1933; M.D., Columbia University, New York City, 1937, Med. Sc.D., 1942; m. Elizabeth Temple Ingraham, June 27, 1936; children—David Prince III, Paul W., Kevin C., Charles B. Practice medicine, specializing in internal medicine, Chgo., 1954—; prof. medicine Northwestern U., 1954—, chmn. dept. medicine, 1965—; mem. coms. Nat. Bd. Med. Examiners; mem. sci. adv. bd. Nat. Kidney Found. chmn. urology and renal training committee National Institutes of Health, 1967. Member of the Assn. Am. Physicians, Am. Soc. Clin. Investigation, Central Soc. Clin. Research (past pres.), Am. Physiol. Soc., Am. Clin. and Climatology Assn., Alpha Omega Alpha. Editor: Jour. Chronic Diseases. Research, publs. on clinical and lab. aspects of renal physiology, pathology and immunology, antimalarial drugs in man; discovered new serum albumin in man; pathology and course of nephritis, derangements in epidemic hemorrhagic fever, epidemiology of renal diseases. Home: 1034 Westmoor Rd., Winnetka, Ill. 60093. Office: 303 E. Chicago Av., Chgo. 60611.*

EARLE, Fontaine Richard, Am. chemist; b. Fayetteville, Ark., Sept. 8, 1906; s. Fontaine Pylant and Martha Jane (Moore) E.; B.Chem. Engring., U. Ark., 1927; postgrad. Cornell U., 1927-28; M.S., Okla. A. and M. Coll., 1931; m. Florence Elizabeth Bender, Apr. 20, 1938. Chemist Atlantic Refining Co., Phila., 1928-29, Columbian Rope Co., Auburn, N.Y., 1929-30; tchr. Bentonville (Ark.) High Sch., 1933-34; engr. U. S. Engrs., Memphis, 1934; seafood insp. FDA, U. S. Dept. Agr., 1934-36; chemist Soybean Indsl. Products Lab., Urbana, Ill., 1936-40; chemist No. Regional Research Lab., Peoria, 1940—, investigations head, new crops screening investigations Indsl. Crops Lab., 1956—. Mem. Am. Chem. Soc., Am. Oil Chemists' Soc., Soc. for Econ. Botany, Tau Beta Pi, Phi Lambda Upsilon, Alpha Chi Sigma, Gamma Alpha. Research and publs. on composition of cereal grains, oilseeds and vegetable oils, work on uncultivated plant species to discover new components or species of potential indsl. value. Home: 4611 N. Sheridan Rd., Peoria, 61614. Office: 1815 N. University St., Peoria, Ill. 61604.*

EARLE, Franklin Sumner, Am. botanist, agriculturist; b. Dwight, Ill., Sept. 4, 1856; s. Parker and Melanie (Tracy) E.; student U. Ill.; M.S., Ala. Poly. Inst., 1902; m. Susan B. Skeham, 1886; children—Melanie Tracy (Mrs. William L. Keiser), Ruth Esther (Mrs. David Sturrock); m. 2d, Esther J. Skehan, 1896. Connected with U. Ill., 1886, doing spl. mycol. work, results of which were pub. as The Erysiphaceae of Illinois (joint author with T. J. Burrill); connected with Miss. Agrl. Expt. Sta., 1894-95; joint author (with S. M. Tracy), Mississippi Fungi, 1895; asst. pathologist in charge mycol. herbarium, U. S. Dept. Agr., 1895-96; horticulturist, Ala. (Coll.) Agrl. Expt. Sta., Jan. 1896; prof. biology, Ala. Poly. Inst., 1896-1901; issued Preliminary List of Alabama Fungi (with Dr. L. M. Underwood), 1897; asst. curator in charge mycol. collections, N.Y. Bot. Garden; sent by N.Y. Bot. Garden to Jamaica and Cuba, and by U. S. Dept. Agr. to make sci. investigations in P.R., 1903; dir. Estacion Central Agronómica of Cuba, 1904-06; cons. agriculturist to Cuban-Am. Sugar Co., 1908-11. Sent to P.R. by U. S. Dept. of Agr., to investigate serious sugar cane disease, 1918; expert in sugar cane disease Insular Govt. of P.R., 1919-21; cons. agriculturist Aguirre Sugar Co.; dir. agriculture Gen. Sugar Co., Havana, Cuba, 1923-24; sugar cane technologist Tropical Plant Research Found. (in charge of work with sugar cane varieties in Cuba), 1924-29. Author: Southern Agriculture, 1907. Died Jan. 31, 1929.

EARLE, Sir James, English surgeon; b. London, 1755; studied under Percival Pott; ed. St. Bartholomew's Hosp., London; m. dau. of Percival Pott; at least 3 sons, including Henry. Became asst.-surgeon St. Bartholomew's Hosp., 1770, surgeon, 1784-1815; surgeon extraordinary to George III. Mem. Coll. Surgeons (became pres. 1802). Fellow Royal Soc., 1794; Created Knight, 1802. Author: A Treatise on the Hydrocele, 1791; Practical Observations on the Operation for Stone, 1793; Observations on the Curve of Curved Spine, 1799; On Burns, 1799; A New Method for Operations for Cataract, 1801; Letter on Fractures of the Lower Limbs, 1807; On Haemorrhoidal Excrescences, 1807. Research on deformations of spinal column, lithotomy; developed method of treating hydrocele, cataracts. Died London, 1817.

EARLE, Kenneth Martin, Am. pathologist; b. Jacksonville, Tex., Dec. 29, 1919; s. Allen and Flora (Martin) E.; B.A., Rice U., 1942; M.D., U. Tex., 1945; M.Sc., McGill U., Montreal (Que., Can.) Neurol. Inst., 1951; m. Mary Ellen Sammons, Mar. 12, 1944; children—Mary M., Katherine L., Thomas H. Fellow neuroanatomy and neuropathology Montreal Neurol. Inst., McGill U., 1949-51, univ. fellow pathology, 1951-52; instr. U. Cal. at Los Angeles Sch. Medicine, 1952-53; faculty med. br. U. Tex., Galveston, 1953-62, prof. pathology, 1960-62, asst. dean, 1959-60, dean medicine, 1960-62; chief neuropathology br. Armed Forces Inst. Pathology, Washington, 1962—. Cons., Am. Bd. Pathology, 1963—; lectr. George Washington U. Sch. Medicine, 1962—. Diplomate Am. Bd. Pathology. Mem. Am. Assn. Neuropathologists (pres.-elect 1965—), Coll. Am. Pathologists, Am. Assn. Pathologists and Bacteriologists, Am. Soc. Clin. Pathologists, Internat. Acad. Pathology, Am. Acad. Neurology (asso.), Am. Soc. Exptl. Pathologists, Assn. for Research Nervous and Mental Diseases, Am. Assn. Mil. Surgeons, Am. Crystallographic Assn., Soc. for Applied Spectroscopists, Alpha Omega Alpha, Sigma Xi. Co-translator: The Microscopic Anatomy of Tumors of the Central and Peripheral Nervous System (Pio del Rio-Hortega's). Research and publs. on pathology nervous system, epilepsy, brain tumors, X-ray diffraction, X-ray fluorescent spectroscopy, biol. effects laser radiation, rabies encephalitis. Home: 11402 Charlton Dr., Silver Spring, Md. 20902. Office: Armed Forces Inst. Pathology, Washington 20305.*

EARLE, Pliny, Am. physician; b. Leicester, Mass., Dec. 31, 1809; s. Pliny and Patience (Buffum) E.; M.D., U. Pa., 1837. Supt. Friend's Hosp. for Insane, Frankford, Pa., 1840-44, Bloomingdale (N.Y.) Asylum, 1844-49; lectr. mental diseases Coll. Physicians and Surgeons, N.Y.C., 1853-55; supt. State Lunatic Hosp., Northampton, Mass., 1864-85, mem. Phila. Med. Soc. (1837), N.Y. Med. and Surg. Soc. (1845), A.M.A. (a founder, pres. 1845), Am. Medico-Psychol. Assn. (co-founder, pres. 1884), N.Y. Psychol. Assn. (1st pres. 1874), Mass. Med. Soc. (councillor 1876). Author: A Visit to Thirteen Asylums for the Insane in Europe, 1841; Institutions for the Insane in Prussia, Germany and Austria, 1853; The Curability of Insanity, 1887; Popular Fallacies Concerning the Insane, 1890. Helped introduce humane methods and attitudes into the work of mental asylums. Died Northampton, May 17, 1892.

EARNSHAW, Laurence, inventor; b. Wednescough, Eng., circa 1707; apprentice to father (a weaver), 7 years, to tailor, 4 years, also to clockmaker; inventor machine for spinning and reeling cotton in one operation, 1753, also astron. clock. Died May 1767.

EARNSHAW, Thomas, English inventor; b. Lancashire, Eng., Feb. 4, 1749; apprenticed to watch maker at the age of 14; set up watch making bus., London. Improved and simplified Graham's transit clock, Greenwich Obs.; 1st to make inexpensive chronometers; improved chronometers; invented cylindrical balance spring, detached detent escapement (anticipated by Berthoud in France); one of discoverers of longitude, 1793. Wrote pamphlets on timekeepers. Died Mar. 1, 1829.

EAST, Edward Murray, Am. biologist; b. Du Quoin, Ill., Oct. 4, 1879; s. William Harvey and Sarah Granger (Woodruff) E.; student Case Sch. Applied Sci., 1897-98; B.S., U. Ill., 1900, M.S., 1904, Ph.D., 1907; LL.D., Kenyon Coll., 1926; m. Mary Lawrence Boggs, Sept. 2, 1903; children—Elizabeth Woodruff, Margaret Lawrence, Edward Murray (dec.). Asst. chemist U. Ill. Agrl. Expt. Sta., 1900-03, 1st asst. in plant breeding, 1903-05, agronomist Conn. Agrl. Expt. Sta., 1905-09; asst. prof. exptl. plant morphology Harvard, 1909-14, prof. exptl. plant morphology, 1914-26, prof. genetics, 1926-38. Collaborator tobacco investigations U. S. Dept. Agr., 1908-18; chmn. bot. raw products com., mem. bot. and agr. com. NRC, 1917-18. Harvard lectr. Yale, 1924-25; lectr. U. Chgo., 1911, Grad. Sch. Agr., 1914; De Lamar lectr. Johns Hopkins, 1920; lectr. Cornell U., 1922; Larwill lectr. Kenyon Coll., 1927; lectr. U. Mich., 1930; Harvey lectr. N.Y. U., 1931. Vice pres. 2d Internat. Congress on Eugenics, 1921; hon. v.p. 6th Internat. Bot. Congress, 1935. Fellow Am. Acad. Arts and Scis., A.A.A.S. Author: Heterozygosis in Evolution and Plant Breeding, 1912; Inbreeding and Outbreeding, 1919; Mankind at the Crossroads, 1923; Heredity and Human Affairs, 1927. Editor: Biology in Human Affairs, 1931; editorial bd. Genetics, 1916-38; contbg. bd. Bot. Abstracts, 1918-22. Research on genetics and breeding of tobacco; improved Indian corn for animal nutrition by applying inbred strain breeding; started new method of seed prodn. (with George Shull). Died Nov. 9, 1938.

EAST, William Gordon, English geographer; b. London, Eng., Nov. 10, 1902; s. George Richard and Jemima (Nicoll) E.; B.A. (Open scholar Peterhouse) with honours in History, U. Cambridge (Eng.), 1924, M.A., 1928; m. Dorothea Small, July 28, 1934; children—Penelope Jane (Mrs. Frank Hagen Menuge), Jonathan Robert, Catherine (Mrs. Ram Singh), Richard Gordon. With London Sch. Econs., 1927-41; adminstrv. officer Ministry Econ. Warfare, 1941-44, Fgn. Office, London, 1944-45; reader U. London, 1945-47, prof. geography, Birkbeck Coll., 1947—, chmn. dept. geography, 1941—; vis. prof. U. Minn., 1952, U. Cal. at Los Angeles, 1959-60, U. Mich., Ann Arbor, 1967. Recipient Thirlwall prize in hist. research, 1927. Fellow Royal Geog. Soc.; mem. Inst. Brit. Geographers (past pres.), Royal Geog. Soc. (past mem. council), Royal Inst. Internat. Affairs, Geog. Soc. N.Y., Inst. Brit. Geographers, Econ. History Soc. Author: The Geography Behind History, 1938; The Union of Moldavia and Wallachia, 1859, 1929; An Historical Geography of Europe, 1935; (with S. W. Wooldridge) The Spirit and Purpose of Geography, 1951; The Soviet Union, 1964; Mediterranean Problems, 1940; also articles. Editor: (with O. H. K. Spate) The Changing Map of Asia, 1950; (with A. E. Moodie) The Changing World, 1956; Regions of the British Isles series (Nelson), 1960; Caxton Atlas, 1960. Editor Batsford's hist. geography series. Research on application of hist. and geog. knowledge to present-day polit. problems, effect of geography on history. Home: Wildwood, Danes Way, Oxshott, Surrey, Eng. Office: Birkbeck Coll., Malet St., London, W.C.1., Eng.*

EASTLICK, Herbert Leonard, Am. zoologist; b. Platteville, Wis., Apr. 24, 1908; s. Daniel and Nora (Israel) E.; A.B., U. Mont., 1930; M.S., Washington U., St. Louis, 1932, Ph.D., 1936; m. Margaret L. Gardiner, Aug. 1, 1935. Instr., Stephens Coll., 1936-37; U. Mo., 1937-39; faculty Wash. State U., Pullman, 1940—, prof. zoology, 1964—. NRC fellow U. Chgo., 1939-40. Fellow A.A.A.S.; mem. Am. Micros. Soc., Soc. Devel. and Growth, Soc. for Exptl. Biology and Medicine, Am. Assn. Anatomists, Am. Soc. Zoologists, Am. Soc. Naturalists, Am. Assn. U. Profs., Western Soc. Naturalists, Sigma Xi. Research and publs. on golgi substance in striated muscle vertebrates, origin pigment cells in birds, transplanted embryonic limbs chick, fat bodies in birds, exptl. prodn. tumors in birds. Home: 408 Garfield St., Pullman, Wash. 99163.*

EASTMAN, Charles Rochester, Am. geologist, paleontologist; b. Cedar Rapids, Ia., June 5, 1868; s. Austin V. and Mary (Scoville) E.; A.B., Harvard, 1890, A.M., 1891; Ph.D., Munich, 1894; m. Caroline A., Clark, 1892. Studied natural sci. at Harvard, Johns Hopkins, and abroad; served on U. S., Iowa, N.Y., N.J., and other state geol. surveys; taught geology and palaeontology in Harvard and Radcliffe colls.; curator Carnegie Mus., Pitts., prof., U. Pitts., 1910-13; engaged in sci. research and editor at Am. Mus. Natural History, N.Y.C., 1913-18. Editor Am. Palaeontol. Soc.; translator Von Zittel's Palaeontology, 3 vols., 1900-2. Died Sept. 27, 1918.

EASTMAN, George, Am. inventor, mfr.; b. Waterville, N.Y., July 12, 1854; s. George W. and Maria (Kilbourn) E.; ed. Rochester, N.Y. Became amateur photographer and experimenter and perfected process for making dry plates; patented coating machine, 1879; in partnership with H. A. Strong, began to mfr. dry plates on small scale, 1880; originated kodak and transparent film for use in same; reorganized firm as Eastman Dry Plate and Film Co., 1884; chmn. bd. Eastman Kodak Co. of N.Y., Eastman Kodak Co. of N.J. Patented 1st photog. film (gel spread over paper), 1884, original Kodak camera, 1888; introduced daylight loading film, 1891; developed noncurling film, 1903; color film, 1928. Died Rochester, Mar. 14, 1932.

EASTMAN, John Robie, Am. astronomer; b. Andover, N.H., July 29, 1836; s. Royal Friend and Sophronia (Mayo) E.; M.S., Dartmouth, 1862; Ph.D., 1877; m. Mary J. Ambrose, Dec. 25, 1866. Asst., U. S. Naval Obs., 1861-65; prof. mathematics U.S.N., 1865-1913. Engaged in astron. observations, computations and research, 1862-1913; in charge Meridian Circle work at obs., 1874-91; observed total solar eclipses Aug. 7, 1869, at Des Moines, Dec. 22, 1870, Syracuse, Sicily, July 29, 1878, West Las Animas, Colo., May 28, 1900, Barnesville, Ga. Prepared and edited Second Washington Star Catalogue, which contains results of nearly 80,000 observations made at U. S. Naval Obs., 1866-91. Author: Transit Circle Observations of the Sun, Moon, Planets and Comets, 1903. Died Sept. 26, 1913.

EASTMAN, Joseph Rilus, Am. surgeon; b. Indpls., Apr. 18, 1871; s. Joseph and Mary Katherine (Barker) E.; B.Sc., Wabash Coll., 1891, A.M., 1904, LL.D., 1926; M.D. magna cum laude, U. Berlin, 1897; spl. work Princeton, 1 yr.; m. Frieda Grumpelt, Feb. 1, 1910 (dec.); 1 son, Joseph Rilus; m. 2d, Neva Bonham, 1926. Surgeon, Indpls. City Hosp., 1900-42; prof. surgery Ind. U. Sch. Medicine, Fellow A.C.S. (founder and ex-gov.); mem. Am., Western (pres. 1913-14) surg. assns., A.M.A., Société Internationale de Chirurgie, Sigma Xi, Phi Rho Sigma. Contbr. many articles to Am. and fgn. med. publs. Original worker in surg. pathology; has devised surg. procedures and instruments. Died Nov. 29, 1942.

EASTOE, John Eric, English chemist; b. London, Eng., Nov. 3, 1926; s. Eric Wilfred and Winifred (Trundel) E.; B.Sc., A.R.C.S., Imperial Coll. of Sci. and Tech., London, 1947, Ph.D., D.I.C., 1950, D.Sc., 1965; m. Beryl Musson, Apr. 14, 1951; children—Sally Ann, Richard John. Research officer Brit. Gelatine and Glue Research Assn., London, Eng., 1950-57; research fellow Royal Coll. Surgeons Eng., London, 1957—. Recipient Colgate prize, 1960. Mem. Biochem. Soc., Bone and Tooth Soc., Soc. for Applied Bacteriology, Internat. Assn. for Dental Research, Organisme Européan de Recherche sur la Carie (prize 1965). Author: Practical Analytical Methods for Connective Tissue Proteins, 1963; Practical Chromatographic Techniques, 1964; also numerous articles. Investigated phys. and chem. properties of porous earthenware plant pots and their effects on growth; developed analytical chemistry of collagen and gelatin; studied their interrelationship and variation with species; discovered characteristic proteins in immature dental enamel, their role in devel. Home: 24 Avondale Av., London N.12. Office: Lincoln's Inn Fields, London W.C.2, Eng.*

EASTON, David, educator; b. Toronto, Ont., Can., June 24, 1917; s. Albert and Mary (Nisker) E.; B.A., U. Toronto, 1939, M.A., 1943; Ph.D., Harvard, 1947; m. Sylvia Isobel Victoria Johnstone, Jan. 1, 1942; 1 son, Stephen Talbot. Came to U. S., 1943. Teaching fellow Harvard, 1944-47; asst. prof. U. Chgo., 1947-53, asso. prof., 1953-55, prof., 1955—; chmn. exec. com. information behavioral scis. div. Nat. Acad. Sci.-NRC; fellow Center for Advanced Study in Behavioral Scis., Stanford, 1957-58; cons. Brookings Instn., 1955, Mental Health Research Inst., U. Mich., 1955-56, Royal Commn. on Bilingualism and Biculturalism, Can., 1964-66. Ford prof., 1960-61. Fellow Am. Acad. Arts and Scis.; mem. Am. Polit. Sci. Assn. (council 1964-66), Internat. Com. Social Scis. Documentation (exec. com. 1965—). Author: The Political System, 1953; A Framework for Political Analysis, 1965; A Systems Analysis of Political Life, 1965. Mem. bd. editors Behavioral Sci., 1956—. Editor: Varieties of Political Theory, 1966. Formulated theory at high level of generality—called systems analysis—for use in understanding the functioning of any type of political system, whether small or large, developing or traditional, modern or primitive, industrialized or peasant, democratic, totalitarian, or authoritarian; interpreted nature and meaning of the development of political science as a discipline and science, indicating particularly the crucial role of behavioral theory as against moral (philosophical) theory; contributed to opening up a new field of political science, namely study of nature and source of political attitudes, knowledge, and values as they originate in childhood. Office: Dept. Polit. Sci., Univ. Chicago, Chgo. 60637.*

EASTON, Elmer Charles, Am. engr., b. Newark, Dec. 23, 1909; s. Frank and Helen (Hagny) E.; B.S., Lehigh U., 1931, M.S. (Research fellow), 1933, D.Eng. (hon.), 1965; Sc.D., Harvard, 1942. Mem. faculty Newark Coll. Engring., 1935-42, Grad. Sch. Engring. Harvard, 1942-48; dean Coll. Engring. Rutgers U., 1948—; cons. FOA, Kenya, 1954, ICA, Korea, 1960. Fellow I.E.E.E., A.A.A.S.; mem. Nat. Soc. Profl. Engrs., Am. Soc. Engring. Edn. (pres. 1964-65), Tau Beta Pi, Eta Kappa Nu, Phi Beta Kappa, Sigma Xi, others. Author: (with J. H. Fithian) Calculus Laboratory Manual, 1939; (with C. L. Dawes, R. B. Power) Elements of Electrical Engring, 1944. Research and publs. on conduction of electricity through gases, vapor streams in arcs, time lag of spark breakdown. Home: 4 Orchard Rd., Piscataway, N.J. 08854. Office: Coll. Engring., Rutgers U., New Brunswick, N.J. 08903.*

EASTWOOD, Alice, botanist; b. Toronto, Ont., Can., Jan. 19, 1859; d. Colin Skinner and Eliza Jane (Gowdey) E. Curator, herbarium Cal. Acad. Scis., 1892-49, ret. Fellow A.A.A.S., Cal. Acad. Scis. Author: Popular Flora of Denver, Colo., 1893; Popular Flora and Pacific Coast Edition, Bergen's Botany, 1897; Popular Flora and Rocky Mountain Edition, Bergen's Botany, 1900; Hand-Book of Trees of California, 1905; also many papers on systematic botany and articles for sci. mags. Died Oct. 29, 1953.

EASTWOOD, Thomas Alexander, chemist; b. London, Ont., Can., Nov. 27, 1920; s. Charles Willard and Grace (Carswell) E.; B.A., U. Western Ont., 1942, M.A., 1943; Ph.D., McGill U., 1946; D.Phil., Oxford U., 1951; m. Katharine Margaret Justus, Sept. 24, 1949; children—Diane Elizabeth, Jennifer Ann, (Alexander) Murray, Victoria Katharine. Chemist, Imperial Oil Ltd., Sarnia, Ont., 1943-44; hon. lectr. McGill U., Montreal, Que., 1947; research officer atomic energy project Nat. Research Council Can., Chalk River, Ont., 1947-52; research officer Atomic Energy Can., Ltd., Chalk River, 1952-65, head, research chemistry br., 1965—. Fellow Canadian Inst. Can.; mem. Am. Phys. Soc., Am. Chem. Soc. (sr.). Research in chem. kinetics, nuclear properties of radioactive isotopes, thermal and resonance neutron capture cross sects., nuclear fission. Home: 9 Tweedsmuir Pl., Deep River, Ont. Office: care Atomic Energy Can., Ltd., Chalk River, Ont., Can.*

EATON, Amos, Am. botanist; b. Chatham, N.Y., May 17, 1776; s. Abel and Azuba (Hurd) E.; grad. Williams Coll., 1797; attended Yale, 1815-17; m. Polly Thomas, Oct. 16, 1799; 1 son; m. 2d, Sally Cady, Sept. 16, 1803, 5 sons; m. 3d, Ann Bradley, Oct. 20, 1816, 3 children; m. 4th, Alice Johnson, Aug. 5, 1827, 1 son. Admitted to N.Y. bar, 1802; became lawyer, civil engr., and land agt., Catskill, N.Y., also studied geology, mineralogy, and chemistry; lectured on botany geology throughout N.E.; prof. natural history Med. Sch., Castleton, Vt., 1820-24; prof. Rensselaer Sch. (now Poly. Inst.), 1824-42; under patronage of Stephen van Rensselaer made geol. and agrl. survey of Albany and Rensselaer counties, 1820-21, survey of area along Erie Canal, 1824; wrote papers in all sci. fields. Author: A Manual of Botany for the Northern States, 1817; An Index to the Geology of the Northern States, 1818; A Manual of Botany of North America, 1833; Geological Notebook, 1841; North American Botany, 1840. Died Troy, N.Y., May 10, 1842.

EATON, Daniel Cady, Am. botanist; b. Ft. Gratiot, Mich., Sept. 12, 1834; s. Amos B. and Elizabeth (Seldon) E.; grad. Yale, 1857; postgrad. Harvard, 1857-60; m. Caroline Ketcham, Feb. 13, 1866. Served with army commissary, N.Y.C., during Civil War, 1860-64; prof. botany Yale, 1864-95; specialized in study of ferns, did much field work; wrote bot. definitions for Webster's Internat. Dictionary. Author: The Ferns of North America, 2 vols. Described new Cal. ferns in classical Latin form. Died New Haven, June 29, 1895.

EATON, L(ealdes) McKendree (Lee M.), Am. physician; b. Owaneco, Ill., Feb. 3, 1905; s. Jordan Stewart, Sr., and Margaret (Barrett) E.; student James Millikan U., 1923-25; B.S., U. Chgo., 1927, M.D., 1932; M.S., U. Minn., 1938; m. Mary Louise Long, Apr. 2, 1936; children—Elizabeth Barrett, Lynne St. Pierre, Emily Jordan, Charles McKendree, Thomas Lee. Fellow internal medicine Mayo Found., Rochester, 1933-34, fellow neurology, 1934-36, instr. neurology, 1938-41, asst. prof., 1941-46, asso. prof., 1946-50, prof., 1950-58; cons. neurology Mayo Clinic, 1936-47, head neurologic sect., 1947-54, chmn. neurology sects., 1954—; research myasthenia gravis, polymyositis, other neuromuscular diseases. Med. adv. bd. Myasthenia Gravis Found., Inc., 1954-58. Diplomate Am. Bd. Psychiatry and Neurology. Mem. Central Neuropsychiat. Assn. (pres. 1953), Am. Neurol. Assn., Am. Acad. Neurology, Assn. Research Nervous and Mental Disease (pres. 1958), A.M.A. (chmn. sect. on nervous and mental disease 1956), Sigma Xi, Alpha Omega Alpha. Author articles on neurologic subjects. Died Nov. 17, 1958.

EATON, Lucien, Am. mining engr.; b. St. Louis, July 6, 1879; s. Lucien and Hannah Orr (Noyes) E.; A.B. magna cum laude, Harvard, 1900, M.S., 1902; S.B. in mining and metallurgy, Lawrence Sci. Sch., 1901; m. Eleanor Archibald Stevens, June 15, 1907 (dec. May, 1949); children—Elizabeth Stevens, Eleanor, Lucien; m. 2d, Charlotte Vose, July 8, 1952. On engring. staff Cleveland Cliffs Iron Co., Ishpeming and Ironwood, Mich., 1902-06, supt. Iron Belt & Shores Mines, Iron Belt, Wis., 1906-09; supt. Ishpeming Dist., 1909-29; cons. mining engr. Roan Antelope Copper Co., Rhodesian Selection Trust, Ltd., 1929-33, gold mines of Australia, Western Mining Corp., Bendigo Mines, North Broken Hill, Broken Hill S., Zinc Corp., and other Australian mining cos., 1934-35; cons. engr. Copper Range Co. and Isle Royale Copper Co., 1937-38; gen. mgr. Isle Royale Copper Co., 1938-39; cons. mining engr. with H. A. Brassert & Co., N.Y., 1940-46; cons. engr. Pierce Mgmt., Scranton, Pa., for Turkish Govt. in Turkey, 1947, Copper Range Co., 1947-52, Big Sandy and other mining cos. and with H. A. Brassert & Co. (Mex.), 1946; pres. dir., Inca Mining & Devel. Co., Santo Domingo, Peru, 1925. Mem. Am. Inst. Mining and Metall. Engrs., Mining and Metall. Soc. Am., Am. Mining Congress (chmn. standardization com. 1929). Made various improvements in mining practice and in design of rock drills and mech. equipment for handling ore. Author: Practical Mine Development and Equipment, 1934, also articles. Died Dec. 9, 1952.

EATON, Monroe Davis, Am. microbiologist; b. Stockton, Cal., Dec. 2, 1904; s. Monroe Davis and Ida Virginia (Petty) E.; A.B., A.M., Stanford, 1928; M.D., Harvard, 1930; m. Laura Mitchell, Aug. 9, 1933; children—John Monroe, Lydia, Emily, (Mrs. Leonard Lyons), Katharine (Mrs. Carl Dennis Andrews). Asst., Harvard Med. Sch., 1930-33, asso. prof. bacteriology and immunology, 1947—; asst. prof. bacteriology Washington U. Sch. Medicine, St. Louis, 1936-37; staff Internat. health div. Rockefeller Found., 1937-47; dir. virus lab. Cal. State Dept. Pub. Health, 1939-47; asso. prof. bacteriology and immunology, Harvard Med. Sch., 1947—. Fellow A.A.A.S.; mem. Am. Soc. Bacteriologists, Am. Acad. Arts and Scis., Am. Soc. Microbiology, Soc. Exptl. Biology and Medicine, Am. Assn. Immunologists, Sigma Xi, Phi Beta Kappa. Contbg. author: Medicine, 1940; Handbuch der Virusforschung, 1950; Annual Reviews of Microbiology, 1950, 65; also numerous articles. Asso. editor Am. Jour. Epidemiology, 1960—. Research, publs. on purification of diphtheria toxin, immunology of malaria, devel. influenza vaccination, isolation of causative agt. primary atypical pneumonia; other research on viral pneumonia, chem. inhibition of influenza and other viruses, epidemiology of jaundice asso. with yellow fever vaccines, immunology of murine leukemia and other viral induced tumors, action of viruses on cancer cells, biochemistry of viral growth. Home: 138 High St., Brookline, Mass. 02146. Office: Dept. Bacteriology and Immunology, Harvard Med. Sch., 25 Shattuck St., Boston 02115.*

EATON, Theodore Hildreth, Jr., Am. zoologist; b. Boston, Nov. 16, 1907; s. Theodore Hildreth and Theodora (West) E.; A.B., Cornell U., 1930; Ph.D., U. Cal. at Berkeley, 1933; m. Grace Worthy Janlen, Aug. 25, 1934; children—George Theodore, Lois West (Mrs. William Merwin Bueler), Margaret Rose (Mrs. Theodore Stallknecht). Naturalist, Nat. Park Service, Berkeley, Cal., 1934-36; faculty Union Coll., Schenectady, 1937-40, U. Buffalo, 1942-45, Georgetown Med. Sch., 1945-46, Southwestern Coll., Winfield, Kan., 1947-50; prof. biology East Carolina Coll., Greenville, N.C., 1950-58; faculty U. Kan., Lawrence, 1958—, prof. zoology, curator vertebrate fossils, 1960—; Fulbright prof. zoology U. Saigon, Viet Nam, 1965-66. Mem. Am. Soc. Zoologists, Soc. Vertebrate Paleontology, Soc. Study Evolution, Am. Soc. Ichthyology and Herpetology, others. Author: Comparative Anatomy of the Vertebrates, 1951, 2d edit. 1960. Research and publs. on muscles, skeleton in fishes, amphibians; early fossil fish, amphibians, reptiles; origin of tetrapods; soil fauna. Address: Natural History Mus., U. Kan., Lawrence 66044.*

EAVES, James Clifton, Am. mathematician; b. Hillside, Ky., June 26, 1912; s. John Ridley and Agnes (Williams) E.; A.B., U. Ky., 1935, M.A., 1941; Ph.D., U. N.C., 1949; m. Maona Shinkle, Aug. 20, 1938; children—James C., Mona Jane. Haggin fellow in math. U. Ky., Lexington, 1940-41, grad. asst. math. to instr. dept. math. and astronomy, 1941-43, 46, prof., head dept. math. and astronomy, 1954-63, prof., 1964—, dir. Space Flight Seminar, NASA, 1962; instr., grad. asst. math., U. N.C., Chapel Hill, 1946-49; asst. prof. U. Ala., Tuscaloosa, 1949-50; asso. prof. math. Auburn (Ala.) U., 1950-51, research asso. prof., 1951-52, prof., adminstrv. asst. to head dept. math, 1953-54; prof. math., research asso. Auburn Research Found., 1952-53. Dir. Ford Found. Grant for Study of Large Classes, 1956-58, Ky. Space Flight Program in Space Math. and Astronomy, 1959-63, Inst. Cons. in Math., Statistics and Patent Law, 1956—. Fellow A.A.A.S.; mem. Ky. Assn. Colls. and Secondary Schs. (pres. math. sect. 1957, 64), Math. Assn. Am. (pres. Ky. sect. 1956, 63, lectr. 1955-57, NSF lectr. 1960-62), Ala. Assn. Coll. Math. Tchrs. (pres. 1953-54), Ala. Edn. Assn. (pres. Auburn div. 1954-55), Am. Math. Soc., Soc. Indsl. and Applied Math., Am. Soc. for Engring. Edn., N.E.A., Nat. Assn. Higher Edn., Nat. Council Tchrs. Math., Ky. Acad. Sci., Internat. Congress Mathematicians, Am. Assn. for Computing Machinery, Ky. Soc. for Promotion of Useful Knowledge, Newcomen Soc. N. Am., Sigma Xi, Pi Mu Epsilon (nat. councilor gen. 1960-63), Mu Alpha Theta (internat. pres. 1959-62), Phi Delta Kappa (recipient Service award 1960). Author: Development of Electronic Models, 1945; (with Parker) Matrices, 1960; (with Robinson) An Introduction to Euclidean Geometry, 1956; (with S. E. Pence, others)

College Algebra, 1958; (with Pignani) Computer Programming, 1959; (with Pence) Mathematics Honors Tests, 1962; The Kentucky Program for Large Classes, 1959; Editor: College Algebra and Basic Set Theory, 1962. Patentee dial setting device, buckle device. Inventor spl. submarine tracking plotting bds., automatic clock set, vacuum models, Pythagorean models. Research on delayed commutative, high order and three dimensional matrices, cubic forms, and space trajectories including in-flight corrections for space missiles and three body problem. Home: 557 Culpepper Rd., Lexington, Ky. 40502.*

EBASHI, Setsuro, Japanese biochemist, pharmacologist; b. Tokyo, Japan, Aug. 31, 1922; s. Haruyoshi and Hisaji (Watanabe) E.; grad. Faculty Medicine, U. Tokyo, 1944, M.D., 1944, Ph.D., 1954; m. Fumiko Takeda, May 20, 1956. Research asso. Faculty of Medicine, U. Tokyo, 1946-58, prof. pharmacology, 1959-—, prof. grad. biochemistry, 1963-—. Guest investigator Rockefeller Inst., N.Y.C., 1959; vis. prof. U. Cal. at San Francisco, 1963. Mem. Japanese Pharmacological Soc. (sec. gen.), Japanese Biophys. Soc., Japanese Biochem. Soc. Chief editor, contbg. author Molecular Biology of Muscular Contraction, 1965; also articles. Discovered particulate relaxing factor of muscle, 1955, Ca-binding of sarcoplasmic reticulum and its role in relaxation, 1960, regulatory structural proteins of muscle, native tropomyosin, troponin alpha-actinin. Home: 1-49 7 Chome, Takinogawa, Kitaku, Tokyo, Japan.*

EBBIGHAUSEN, Edwin George, Am. astronomer; b. Crookston, Minn., June 28, 1911; s. George and Rose (Vonderbeck) E.; B.A., U. Minn., 1936; Ph.D., U. Chgo., 1940; m. Ardis Lundgren, Aug. 7, 1937; children—Kristin, Gerard. Instr., Wilson Coll., 1939-41; asst. prof. U. Pitts., 1941-44; research engr. Westinghouse Research Labs., East Pittsburgh, 1944-45; lectr. Buhl Planetarium, Pitts., 1942-46; with U. Ore., Eugene, 1946-—, now prof., dir. N.W. Travelling Sci. Program, 1960-61. Mem. Am., Royal astronomy socs., Astronomy Soc. Pacific, A.A.A.S. Research, articles on spectroscopic and eclipsing double stars. Home: 2750 Emerald St., Eugene, Ore.*

EBBINGHAUS, Hermann, German psychologist; b. Barmen, Germany, 1850; s. Carl and Juliane (Klewitz) E.; Ph.D., U. Bonn (Germany), 1873; also ed. Halle, Berlin univs.; independent study Berlin, France, Eng., 1874-1881; m. Adele Gorlitz, 1884; 2 sons, 2 daus. Became dozent Berlin, 1880, ausseror dentlicher prof., from 1886; founder (with Arthur Konig) Zeitschrift für Psychologie und Physiologie der Sinnesorgane, 1890; received chair philosophy, Breslau (now Wroclaw, Poland), 1894; went to Halle, 1905. Author: Ueber das Gedachtnis, 1885; Grundzuge der Psychologie, 1897; others. Noted for originality of work on memory and for clear, interesting expository style; recognized Fechner's work on sensations, attempted to apply Fechner's method to problem of measurement of memory; postulated that frequency of repetition was essential condition of assn., thus made repetition basis for exptl. measurement of memory; inventor nonsense syllable as an instrument for testing memory (these syllables having no assn. to lang. habits); expts. on brightness contrasts, 1880's; postulated theory of color vision, 1893; devised new method for testing mental capacity of sch. children (Ebbinghaus completion method) still used in many modern intelligence tests, 1897. Died Halle, 1909.

EBEL, Johann Gottfried, physician, natural scientist; b. Züllichau, Germany, Oct. 6, 1764; studied medicine, Frankfort/Oder, beginning in 1784; M.D., 1789; traveled in Switzerland, 1790-93; began practice medicine, Frankfort/Main, 1793; forced to leave Germany, 1796; med. research, Swiss politics in Paris, until 1801; went to Switzerland, 1803. Named hon. citizen, Zürich. Author: Anleitung, auf die nützlichste und genussreichste Art die Schweiz zu bereisen, 2 vols., 1793; Schilderung der Gebirgsvölker in der Schweiz, 2 vols., 1798-1802; Über den Bau der Erde in dem Alpengebirge, 2 vols., 1808; Abriss der politischen Zustände der Schweiz am Ende des Jahres 1813, 1814. Studied geology, geodesy, cretinism, magnetism of earth; developed theory on constrn. of Alps. Died Zurich, Oct. 12, 1830.

EBELING, August, German physicist, engr.; b. Schöpfurth nr. Eberswalde, Aug. 31, 1859; s. August and Bertha (Kühne) E.; studied at Freiburg, Berlin, Germany; Ph.D., 1884; hon.dr.engring. Tech. U., Danzig, Germany, 1922; m. Hedwig Messerschmidt, 1897; 1 son, 2 daus. Joined firm Siemens & Halske, Berlin, 1889, named dir. magnetic work State Phys.-Tech. Inst. Physikalisch-Technische Reichsanstalt, 1892; returned as head of weak current cable dept. Siemens & Halske, 1897; named dir. Siemens-Kabelwerk Berlin, 1920, dir. all cable work, 1921; apptd. mem. bd. Siemens & Halske AG & Siemens-Schuckertwerke GmbH, 1925-29. Mem. German Long Distance Cable Soc. (dir. 1921-35). Research and numerous publs. on efficient long distance telephone cables. Died Berlin-Charlottenburg, Germany, Jan. 16, 1935.

EBELMEN, Jacques Joseph, French chemist; b. Baume-les-Dames, France, July 10, 1814; ed. l'Ecole Polytechnique, l'École des Mines; student math. Henry IV Coll. Engr., Vesoul, France; named asst. prof. l'École des Mines, 1840, prof., 1845; became adminstrv. asst. Manufacture de Sevres, 1845, adminstr., 1847; prof. Conservatory Arts and Scis. Author: Recueil des travaux scientifiques de M. Ebelmen, 1855. Discovered ethyl borate; perfected technique of casting metal; developed process for making imitations of several precious stones, including emeralds, peridots, corundum and spinel; replaced wood with coal in the baking of masonry blocks; studied composition and use of gas in blast-furnace. Died Paris, Mar. 31, 1852.

EBERHARD, Christoph, German geographer; b. Eisleben, Germany, 1675; s. Johann E.; m. Anna Müller, 1719; children—Johann Paul, Johann Peter. Theologian; became chaplain Russian Gen. Staff, 1711; afterwards settled in Germany; traveled to Amsterdam, 1717, The Hague, London, 1718; named v.p. of Altona by Friedrich IV of Denmark; later went to Russia and directed bldg. of ships for exploration of Am. Coast; on the death of Peter I, the project was dissolved; returned to Germany and lived in Halle for the rest of his life. Author: Specimen theoriae magneticae quo ex certis principiis magneticis ostenditur vera et universalis methodus inveniendi longitudinem et latitudinem, 1718, published as Versuch einer magnetischen Theorie, 1720. Invented geog. measuring instruments, including (with C. Semler) instrument for measuring longitude and latitude; math.-geog. expts. Died Halle, July 22, 1750.

EBERHARD, (Paul Alexander Julius) Gustav, German astronomer; b. Gotha, Germany, Aug. 10, 1867; s. Bruno and Emilie (Messmer) E.; doctorate, Munich, 1892; m. Gertrud Müller, 1903; a son, Wolfram. Traveled through Vienna, Gotha, Bamberg, Potsdam, 1898; with Astrophys. Obs., Potsdam, Germany, 1898-1932, named chief observer, prof., 1916. Editor: Handbuch der Astrophysik, 7 vols., 1928-36. Research and publs. on spectroscopy of rare elements, photog. photometry. Died Potsdam, Jan. 3, 1940.

EBERHARD, Johann Peter, German physician, mathematician, physicist; b. Altona, Germany, Dec. 2, 1727; s. Christoph and Anna (Müller) E.; began studies U. Giessen at the age of 14; studied math in Göttingen, Germany, surgery, Helmstedt, Leipzig, Halle (all Germany); M.D., Halle, 1749. Became asso. prof. Philosophy Faculty, U. Halle, 1753, prof. in Med. Faculty, 1756, prof. math., 1766, prof. physics, 1769; named chief overseer Halle Bot. Garden, 1770. Mem. Leopoldina. Author: Erste Gründe der Naturlehre, 1753; Sammlung derer ausgemachten Wahrheiten in der Naturlehre, 1755; Vorschläge zur Verbesserung der Kriegsbaukunst, 1766; Vorschläge zur bequemerem und sicheren Anlegung der Pulvermagazine, 1771; Abhandlungen von physikalischen Aberglauben und der Magie, 1778. Tried to put med. scis. on a math. found.; studied Newton's color theory, sound, storms, aurora borealis, hydrotechnics, mining machinery, optics, mil. sci. Died Halle, Dec. 17, 1779.

EBERHARD, Otto (Eduard Hermann) von, see von Eberhard.

EBERHARD, Wolfram, sinologist; b. Potsdam, Germany, Mar. 17, 1909; s. Gustav B. and Gertrude (Müller) E.; Diploma, Seminar für Orientalische Sprachen, Berlin, Germany, 1929; Ph.D., U. Berlin, 1933; m. Alida Rosmer, May 27, 1934; children—Rainer, Anatol. Came to U. S., 1948, naturalized, 1952. Research asst. Mus. Ethnology, Berlin, 1929-34; lectr. Peking Nat. U., 1934-35; head Asian div. Grassi Mus., Leipzig, Germany, 1936; prof. Chinese Ankara U., Turkey, 1937-48; prof. sociology U. Cal. at Berkeley, 1948-—; guest prof. U. Frankfort, Germany, 1956, U. Punjab, Lahore, Pakistan, 1957-58, U. Heidelberg, Germany, 1964. Adviser, Asia Found., San Francisco, 1956-58, 60; cons. Asia Found., 1961-—. Corr. mem. Deutsche Akademie der Wissenschaften und der literature, Bayerische Academie der Wissens chaften Munchen; mem. Am. Sociol. Assn., Am. Folklore Assn., Am. Oriental Soc., Soc. for Asian Folklore. Author numerous books including: Guilt and Sin in Traditional China, 1967; Settlement and Social Change in Asia, 1967; also numerous articles. Analysis of roots and early devel. Chinese soc. and culture, social and ethical values of Chinese middle and lower class; research on early Chinese astronomy and astrology, systematic analysis of Chinese and Turkish folk tales, quantitative methods to determine social mobility in traditional China, interrelations between Chinese and their no. neighbors down to 1000 A.D. Home: 604 Panoramic Way, Berkeley, Cal. 94704.*

EBERLE, Christoph, German engr.; b. Oberauerbach, Germany, Dec. 7, 1869; s. Michael and Friederike (Lang) E.; studied at Tech. U., Darmstadt, Germany; m. Sophie Hammer, 1896; 2 sons, 2 daus. Became asst. Tech. U., Karlsruhe, Germany; tchr. Maschinenbauschule, Duisburg, Germany, 1895-99; became dir. econ. dept. Bayerischer Dampfkessel-Revisionsverein, Munich, 1899; gen. dir. Steinbeis enterprises, Brannenburg, 1910-17; named prof. Tech. U. Karlsruhe, 1917; became 1st prof. heating techniques and economy Tech. U. Darmstadt, 1921. Author: Üben den Einfluss des Gegendruckes und der Zwischendampfentnahme auf den Dampfverbrauch von Kolben dampfmaschinen, 1909; Wärmewirtschaft in der Textilindustrie, 1923; Der Einfluss des Hochdruckdampfes auf die Entwicklung industrieller Anlagen, 1924; Die Abwärmeverwertung in Orts- und Fernheizwerken, 1925; Neuzeitliche Kraft- und Heizanlagen industrieller Werke und ihre Wirtschaftlichkeit, 1927; Unterschungen über den Wärmerverbrauch der Wohnung, 1931; Die Heizung von Schulgebäuden, 1931. First process for bleeding in compound engine; studied coupling power and heat prodn. in steam engines; originated hypotheses for devel. fuel conservation; studied heat loss by transport of saturated and super-heated steam. Died Darmstadt, Sept. 30, 1929.

EBERLE, Paul Josef, German anthropologist, biologist; b. Eisental, Germany, May 9, 1928; s. Friedrich J. and Theresia (Schemel) E.; Dr.rer.nat., U. Freiburg, 1955; m. Maria Keller, May 22, 1956; children—Susanne M., Markus G. Sci. asst. in botany U. Freiburg, 1955-60; sr. sci. asst. in genetics U. Nijmegen, Netherlands, 1960-62; sci. asst. Inst. Human Genetics, U. Göttingen, Germany, 1962-—, head lab. for chromosal investigation, 1962-—, acad. lectr., 1965-—. Mem. Gesellschaft Deutscher Naturforscher und Arzte., Ges. fur Mediz. Theol. Gemeinschaftsarbeit., German Bot. Soc., Soc. Anthropology and Human Genetics, Gottingen Med. Soc. Publs. on light microscopic investigation of chromosomal structures in cell division; studies of genetic disorders. Home: 15 Ludwig-Beck Strasse, Göttingen, Germany.*

EBERLEIN, William Frederick, Am. mathematician; b. Shawano, Wis., June 25, 1917; s. Michael Gustav and Lora (Rather) E.; A.B., Harvard, 1938, Ph.D., 1942; M.A., U. Wis., 1939; m. Mary Barry, May 29, 1943 (div. 1952); children—Patrick, Kathryn, Michael, Robert; m. 2d Patricia James, June 23, 1956; 1 dau., Kristen. Faculty, Purdue U., spring 1946, U. Mich., 1946-47, U. Wis., 1948-55; vis. prof. Wayne State U., 1955-56; prof. math. U. Rochester, 1957-—. Mem. ergodic theory panel Internat. Congress Mathematicians, 1950, Am. Math. Soc., Math. Assn. Am., Swiss Math. Soc. Contbr. papers in profl. jours. on functional analysis; reviewer, referee various math. jours. Home: 1350 Highland Av., Rochester 14620.*

EBERMAYER, (Wilhelm Ferdinand) Ernst, German agrl. chemist, meteorologist; b. Rehlingen, Germany, Nov. 2, 1829; s. Friedrich and Karoline (Dentler) E.; studied chemistry, natural history, Munich; state exam. in pharmaceutics, 1851; Ph.D., Jena, Germany, 1855; m. Marie Julie Friederike Kiderlin, 1857; 2 sons, 4 daus., 2 step-daus. Began teaching Landwirtschaft-und Gewerbeschule, Nördlingen, 1853, Centralforstlehranstalt, Aschaffenburg, 1858; became prof. agrl. chemistry and soil sci., 1861; named prof. soil sect. forestry U. Munich, 1878-1900, also chmn. Inst. for Forestry Experimentation. Author: Die physikalischen Einwirkungen des Waldes auf Luft und Boden, 1873; Naturgesetzliche Grundlagen des Wald- und Ackerbaues, 1882; Einfluss der Wälder auf die Bodenfeuchtigkeit, 1900. Father of forestry meteorology; founder sci. forest soils; 1st research on metabolism processes in humus; founded network forestry-meteorol. stas., Bavaria, 1866-68; application phys. and chem. research methods to forestry; studied forest climate and effect. Died Hintersee nr. Berchtesgaden, Germany, Aug. 13, 1908.

EBERS, Edith Heirich, German geologist; b. Nuremberg, Dec. 4, 1894; d. Carl and Hermine (Knote) Heirich; ed. U. Munich, U. Heidelberg; state exam. for natural scis.; Ph.D. in geology; m. Hermann Ebers. Named laureate of prize of honor Bavarian Acad. Scis., 1962. Hon. mem. German Quaternary Soc., French Prehistoric Soc., Group for Preservation Nature in Bavaria, Union German Academicians, Soc. Geographers of Munich. Author: Verständlichen Wissenschaft, vol. 66. Research and publs. on geology of glacial period in Alps, preservation of nature. Address: 8121, Haunshofen, Obb., Germany.

EBERS, Georg Moritz, German Egyptologist; b. Berlin, Mar. 1, 1837; s. Meier Moses and Fanny (Levysohn) E.; studied law, Göttingen, Germany; studied hieroglyphics under R. Lepsius; Ph.D., 1862; m. Antonie Beck, 1865; 6 children including Hermann, Marie (Mrs. Heinrich Triepel); 2 step-daus. Became lectr. Jena, Germany, 1865, asso. prof., 1869; named asso. prof. Leipzig, Germany, 1870, prof. Egyptology, 1875. Author: Memnon und die Memnonsage, 1862; Ägypten und die Bücher Mosis, 1868; Durch Gosen zum Sinai, 1872; Complete Works, 32 vols., 1893-97; also numerous novels, including Ägyptische Königstochter, 1864. Editor: Papyrus Ebers, das hermetische Buch über die Arzneimittle die alten Ägypten in hieratischer Schrift, 1875. Discovered, edited and published the Papyrus Ebers from Egypt, the oldest med. manual in the world, 1502 B.C. Died Tutzing, Germany, Aug. 7, 1898.

EBERT, Hermann, German physicist; b. Leipzig, Germany, June 20, 1861; s. Viktor Cäsar and Wilhelmine (Schönenmann) E.; studied astronomy, physics, U. Leipzig; doctorate U. Erlangen (Germany); m. Elsbeth Mitscherlich, 1894; 2 daus.; Asst. to Eilhard Wiedemann at Erlangen, for 7 years; named asso. prof. theoretical physics, Leipzig, 1894, prof. exptl. physics, Kiel, Germany, 1894, prof. exptl. physics Tech. U., Munich, Germany, 1898. Mem. acads. scis. Munich, Göttingen, Leopoldina. Author: Anleitung zum Glasblasen, 1887; Magnetische Kraftfelder, 1897; Lehrbuch der Physik, 1912. Research on devel. electronic and atomic physics, spectroscopy and electric discharging of gases; pioneered work on atmospheric electric phenomena especially ion content, earth's elec. field, radioactive emanations from earth. Died Munich, Feb. 12, 1913.

EBERT, James David, Am. research biologist; b. Bentleyville, Pa., Dec. 11, 1921; s. Alva Charles and Frances (Brundege) E.; A.B., Washington and Jefferson Coll., 1942; Ph.D. (Adam T. Bruce fellow), Johns Hopkins, 1950; m. Alma Christine Goodwin, Apr. 19, 1946; children—Frances Diane, David Brian, Rebecca Susan. Faculty, Mass. Inst. Tech., 1950-51, Ind. U., 1951-55; dir. dept. embryology Carnegie Inst. Washington, 1956—; prof. biology Johns Hopkins, also prof. embryology Sch. Medicine, 1956—; mem. panels NSF, Dept. Health Edn. and Welfare. Trustee Marine Biol. Lab.; bd. dirs. Internat. Inst. Embryology. Fellow Am. Acad. Scis.; mem. Am. Assn. Anatomists, Am. Soc. Cell Biology, Am. Soc. Naturalists, A.A.A.S., Am. Inst. Biol. Scis., Internat. Inst. Embryology, Soc. Gen. Physiologists, Phi Beta Kappa, others. Author: Interacting Systems in Development, 1965. Research, publs. on molecular level of embryonic heart formation, immunology in the embryo; infecting and transforming ability of tumor viruses. Home: 6728 Glenkirk Rd., Balt. 21212.*

EBERT, (Johannes) Ludwig, German chemist; b. Würzburg, Germany, June 19, 1894; s. Emil and Emilie (Förster) E.; studied chemistry, U. Würzburg; doctorate, 1923; m. Ottilie Stock, 1923; 3 sons, 1 dau. Served in World War I; worked with H. Bjerrum, Copenhagen, Denmark, then with W. H. Keesom, Leiden, Holland, 1923-25, with F. Haber, Kaiser-Wilhelm Institut, Berlin-Dahlem, 1926-28; became lectr. U. Berlin, 1928; shortly afterward named asso. prof., Würzburg; named prof. Tech. U., Karlsruhe, Germany, 1934; chmn. Chem. Inst., U. Vienna, 1940-56. Mem. acads. scis. Vienna, N.Y., Bologna. Research and numerous publs. on thermodynamics of liquid mixtures, theory of melting point, electrolytes, including strong electrolytes, conductivity, and transference numbers of liquid electrolytes; showed relationship between dielectric polarization and molecular constrn. Died Vienna, Nov. 2, 1956.

EBERTH, Karl Joseph, German anatomist, pathologist; b. Würzburg, Germany, Sept. 21, 1835; student medicine under H. Müller, A. Kölliker, E. Leydig, F. Rinecker, R. Virchow, U. Würzburg; M.D., 1859; m. Elisabeth Hollensteiner, 1870; 3 daus. Became prosector Kölliker's Zootomical Inst., Würzburg, 1859, also lectr., 1862; named asso. prof. anatomy and pathology U. Zurich (Switzerland), 1865, prof., 1869; also became prof. pathology, histology and evolution Zürich Sch. Vet. Medicine, 1874; prof. comparative anatomy, histology and evolution U. Halle (Germany), 1881-1911, prof. emeritus, 1911-26. Research and publs. on nematodes, cell origin and structure, liver, intestinal diseases, Tb, infection, amyloids, histology of male sex organs, growth of fetal bones, lung epithelium, role of platelets in path. coagulation; discovered typhoid bacillus (Eberth-Gaffky baccilus or Eberthella typhosa), 1880; proved diploccocus is causative agt. of pneumonia; studied anatomy of parasitic worms; bacteriology. Died Berlin, Dec. 2, 1926.

EBLE, John Nelson, Am. pharmacologist; b. St. Louis, May 19, 1927; s. Otto H. and Esta (Nelson) E.; B.S., Mo. U., 1949; M.S., U. Wis., 1951, Ph.D., 1953; m. Jane Brewer, June 15, 1950; children—John Nelson, Mark David, Elizabeth Jane. Asst. prof. Kirksville (Mo.) Coll., 1954-60; asso. scientist Dow Chem. Co., Indpls., 1961—. Mem. Am. Soc. Pharmacology and Exptl. Therapeutics, Am. Physiol. Soc., Soc. for Exptl. Biology and Medicine, Am. Chem. Soc., A.A.A.S. Research and publs. on neurophysiology and action neurohumors, specific receptors for tryptamine and dopamine, visceromotor reflexes, central nervous system stimulants. Home: 5345 Delaware St., Indpls. 46220. Office: Dow Chem. Co., Zionsville, Ind. 46077.*

EBNER VON ROFENSTEIN, (Anton Gilbert) Viktor, Austrian physician, histologist; b. Bregenz, Austria, Feb. 4, 1842; s. Johann Nepomuk and Johanna (Schueller) E. von R.; studied in Innsbruck, Austria, Göttingen, Germany; doctorate, Vienna, 1866; m. Adde Steffan, 1872. Became asst. to Alex. Rollett, U. Graz (Austria), 1870; named lectr. U. Innsbruck; apptd. prof. histology and evolution, Graz, 1873, titular prof., 1879, prof., 1885; became prof., Vienna, 1888; ret., 1912. Author: Die acinösen Drüsen der Zunge und ihre Beziehungen zu dem Geschmacksorganen . . . , 1873; also articles. Editor: Handbuch der Gewebelehre des Menschen, vol. 3, 1899-1903. Described glands in back of tongue and their relation to taste (Ebner's glands); studied constrn. of aorta wall and muscle fibers of heart, hearing nerve endings, growth and change of hair, causes of anisotropy of organic substances, fine structre of bones and teeth. Died Vienna, Mar. 20, 1925.

EBRINGER, Libor, Czech. microbiologist; b. Bratislava, Czechoslovakia, Feb. 3, 1931; s. Ján and Augusta (Bohunská) E.; student Komensky U., Britislava, 1951-56, Dr.degree, 1958; m. Anna Vojtísktvá, Nov. 8, 1958; children—Libor, Susanna. Faculty dept. microbiology Komensky U., 1956—, asso. prof. microbiology, 1965—; lectr. Brasilian univs., 1964. Mem. Czechoslovak Soc. Microbiologists. Author: Ludia a micróby, 1962; also articles. Discovered specific form of antagonism among Actinomycetes, permanent irreversible inhibition of chloroplast's synthesis in some green algae, 1960; discovery of existence of new group of antibiotics with specific activity against protozoa only. Home: 20 Karlova Ves, Bratislava, Czechoslovakia.*

EBSTEIN, Wilhelm, German internist, pathologist, med. historian; b. Jauer, Germany, Nov. 27, 1836; s. Louis and Amalie (Schlesinger) E.; studied medicine, Breslau (now Wroclaw, Poland), Berlin; doctorate, Berlin, 1859; m. Elfriede Nicolaier; children include Erich. Became physician, prosector Breslau Allerheiligen-Hosp., 1861; named lectr., placed in charge of a municipal welfare house, 1869; named dir. U. Göttingen (Germany, later of med. clinic which he built himself. Author: Medizin im Alten Testament, 1901; Dorf-und Stadthygiene, 1902; Rudolf Virchow als Arzt, 1903; Medizin im Neuen Testament und im Talmud, 1903; Die Fettliebigkeit und ihre Behandlung nach physiologischen Grundsätzen, 1905; Über die Lebensweise der Zuckerkranken, 1905; Die Pathologie und Therapie der Lepra, 1909; Pathologie und Therapie der Leukämie, 1909; also numerous articles. Described Disorder of kidneys during diabetes (Ebstein's disease), 1881, lymphadenoma characterized by pyrexia (Pel-Ebstein's disease), 1887; invented and developed palpation; studied metabolism diseases, especially gout and diabetes, medicine in the Bible and Talmud, pathography of various important men, including Luther, Berzelius, Schopenhauer. Died Göttingen, Oct. 22, 1912.

ECCLES, Sir John Carew, physiologist; b. Melbourne, Australia, Jan. 27, 1903; s. William James and Mary (Carew) E.; M.B., B.S., Melbourne U., 1925; M.A., Oxford U., 1929, D.Phil., 1929; LL.D., Melbourne U., 19—; D.Sc. (hon.), U. B.C., 1966, Cambridge U., 1960, U. Tasmania, 1964, Gustavus Adolphus, 1967; m. Irene Frances Miller, July 3, 1928; children—Rosamond Margaret (Mrs. Richard Mason), Peter James, Alice Catherine, William, Mary Rose (Mrs. Brian Mennis), John Mark, Judith Clare, Frances Joan, Richard Aquinas. Research fellow Exeter Coll., Oxford U., 1927-34, tutorial fellow Magdalen Coll., 1934-37; dir. Kanematsu Meml. Inst. Pathology, Sydney (Australia) Hosp., 1937-43; prof. physiology Otago U., Dunedin, New Zealand, 1944-51; prof. physiology Australian Nat. U., Canberra, 1951-66; mem. A.M.A./E.R.F. Inst. Biomed. Research, Chgo., 1966—. Decorated Knight Bachelor, 1958; Royal medal Royal Soc., 1962; Cothenius medal Deutsche Akademie der Naturforscher Leopoldina; Nobel Prize in physiology and medicine (with A. L. Hodgkin and A. F. Huxley), 1963. Fellow Royal Soc., 1941, Australia Acad. Sci. (pres. 1957-61); mem. Am. Philos. Soc. (hon.), Accademia Nazionale dei Lincei (fgn. hon.), Nat. Acad. Sci. (foreign associate), Am. Physiological Soc. (fgn. hon. mem.), Am. Acad. Arts and Scis. (fgn. hon. mem.). Author: (with others) Reflex Activity of the Spinal Cord, 1932; The Neurophysiological Basis of Mind: The Principles of Neurophysiology, 1935; The Physiology of Nerve Cells, 1957; The Physiology of Synapses, 1964; (with Ito, Szentagothai) The Cerebellum as a Neuronal Machine, 1967; The Inhibitory Pathways of the Central Nervous System, 1968. Research, numerous publs. on the nature, physiology of synapse of the nervous system and its chem. transmitters. Home: 1300 N. Lake Shore Dr. Office: 535 N. Dearborn St., Chgo.*

ECCLES, Robert G(ibson), physician; b. Scotland, Jan. 1, 1848; s. David and Isabella E.; M.D., L.I. Coll., 1881; Pharm.D., Scio Coll., 1903; m. Mary Hance, Sept. 1876; 1 son, David Charles. Was chemist U. S. Dept. Indian Affairs, prof. and dean Bklyn. Coll. Pharmacy, editor Merck's Archives, and mem. com. of revision of U. S. Pharmacopoeia. Fellow A.A.A.S., N.Y. Acad. Scis. Author: Food Preservatives, 1905; Darwinism and Diabetes, 1908; Letters from Foreign Lands, 1908; Darwinism and Malaria, 1909; Parasitism and Natural Selection, 1909; Touring the Lands Where Medical Science Evolved, 1909; Darwinism and Anaphylaxis, 1911. Discovered alkaloids calycanthine, glaucosine, calycanthic acid; devised 1900 ofcl. method of assaying pepsin; investigated effects of drugs on peptic digestion.

ECCLES, William Henry, English physicist; b. Ulverston, Eng., Aug. 23, 1875; s. Charles and Annabel (Todd) E.; B.Sc. in Physics, Royal Coll. Sci. London, 1898; D.Sc. in Physics, 1901; m. Nellie Paterson, Jan. 4, 1924. Asst. to Guglielmo Marconi, from 1899; head physics dept. Southwestern Poly. London, from 1902; prof. elec. engring. City and Guilds of London Tech. Coll., 1916; dir. Admiralty Elec. Engring. Lab., London, 1917-18; vice-chmn. Imperial Wireless Commn., 1920-26. Mem. Imperial Wireless Telegraphy Com., 1919. Fellow Royal Soc., 1921; mem. Assn. Sci. Workers (pres.), Phys. Soc. (pres.), Inst. Physics (pres.), Instns. Elec. Engrs. (pres.), Radio Soc. Gt. Britain (pres.), Internat. Sci. Radio Union (hon. pres.). Author: Handbook of Wireless Telegraphy, 1915; Continuous Wave Wireless, 1919; Wireless, 1932. Inventor (with F. W. Jordan) multiple change of poles, 1919. Home: Midway, the Ridings, East Preston, Sussex, Eng., U.K.

ECHAVE-LLANOS, Julian Miguel, Argentine histologist; b. Bragado, Buenos Aires, Argentina, May 18, 1926; s. Julian R. Llanos and Maria de la Luz; student Faculty Med. Scis., La Plata, Argentina, 1944-52; m. Roubini Cavoura, Oct. 1, 1961; children—Carlos Federico, Pablo Gabriel, Pedro Julian, Julian, Alejandro. Faculty, Faculty Med. Scis., La Plata, 1944-49; prof. histology Faculty Med. Scis., U. Cuyo, Mendoza, Argentina, 1956-58; prof. gen. pathology, dir. Inst. Gen. and Exptl. Pathology, U. N. Cuyo, Mendoza, 1958-66; prof. embryology and histology, head dept. U. La Plata Faculty Med. Scis., 1966—. Mem. Sociedad Argentina de Biologia, Soc. for Biol. Rhythms, N.Y. Acad. Scis., Am. Assn. for Cancer Research, A.A.A.S. Research, publs. on effect of circadian rhythms, responsiveness of individual to tissue growth control factors and antitumour agts. Home: 11-716 La Plata, Buenos Aires, Argentina.*

ECK, Heinrich Adolf von, see von Eck.

ECKARDT, Wilhelm Richard Ernst, German meteorologist; b. Hildburghausen, Germany, Oct. 11, 1879; s. Wilhelm and Maria Agnes (Meyer) E.; studied geography, biology, meteorology, Leipzig, Heidelberg, Berlin (all Germany); doctorate, Jena, Germany, 1906; m. Hermine Köhler, 1912. Became tchr., Weimar, also sci. collaborator pub. house B. G. Teubner, Leipzig, 1907; named asst. pub. weather service, Aachen, 1908, then in Weilburg; became dir. weather service, Essen, Germany, 1913; also dir. Meteorol. Obs., beginning in 1919. Author: Das Klimat problem der geologischen Vergangenheit und historischen Gegenwart, 1909; Paläoklimatologie, 1910; Grundzüge einer Physioklimatologie der Festländer, 1922; also articles. Research on climatology and its applications, paleoclimatology, meteorology in Ruhr, Münsterland, and Sauerland. Died Essen, May 11, 1930.

ECKART, Carl (Henry), Am. physicist; b. St. Louis, May 4, 1902; s. William E., and Lily (Hellwig) E.; B.S., Wash. U., 1922, M.S., 1923; Ph.D. (Edison Lamp Works fellow), Princeton, 1925. Nat. Research fellow Cal. Inst. Tech., Pasadena, 1925-27; asst. prof. to asso. prof. U. Chgo., 1928-46; asst. dir. to dir. div. war research University of California at San Diego, 1942-46; professor geophysics Scripps Instn. Oceanography, 1946-65, 67—, vice chancellor Univ. of California, 1965-67. Guggenheim fellow, 1927-28. Fellow Am. Phys. Soc., Acoustical Soc. Am., A.A.A.S. Mem. Nat. Acad. Scis. (Agassiz medal 1965), Am. Acad. Arts Scis. Research and publs. on Quantum theory acoustics, hydrodynamics, irreversible thermodynamics. Home: 375 Via Del Norte, La Jolla, Cal.*

ECKART, Gottfried, German engr.; b. May 23, 1906; s. Johannes and Sophie (Merz) E.; Ph.D. in Engring., Technische Hochschule, Munich; m. Charlotte Rohmer; children—Sabine, Gustav, Wolfgang. Asst., Technische Hochschule, Munich, 1929-34; made aviation tests, 1934-36; with Luftwaffe Test Center, 1936-43; dir. Ohm Inst., Center Research of High Frequencies, Heidelberg, 1943-45; instr. U. Heidelberg, 1943-45; research engr. Nat. Office Aero. Study and Research, Paris, 1948-53; prof. U. Saarbrücken, 1951. Mem. Profl. Assn. Germany and France. Research and numerous publs. on electrotech. theories, radiometeorology, propagation of radio waves and ultrashort waves. Home: Eichendorff 13, Scheidt-Saar. Office: Saarbrücken 15, Germany.

ECKEL, Edwin Clarence, Am. geologist; b. N.Y.C., Mar. 6, 1875; s. August and Helena S. K. (Butt) E.; student Coll. City N.Y.; B.S., N.Y. U., 1895; C.E., 1896; m. Julia Egerton Dibblee, 1902; children—Edwin Butt, Julia Dibblee, Richard Egerton. Asst. geologist N.Y. State Mus., 1899-1901, Fed. Survey, 1901-06; individual practice geology, 1906-33; chief geologist TVA, from 1933; also for C. W. Hayes. Mem. Soc. Chem. Industry, Am. Acad. Polit. and Social Sci. Author: Cements, Limes and Plasters, 1905; Iron Ores, 1912; Coal, Iron and War, 1920. Contbr. articles to sci. jours. Developed alumina cements; authority on iron ores; responsible for soundness of dam founds. in Tenn. Valley area. Died Nov. 22, 1941.

ECKEL, Othmar, Austrian meteorologist; b. June 7, 1906; s. Michael and Therese (Polzlbauer) E.; Ph.D. in Meteorology, U. Vienna; m. Friederike Mittendorfer. Asst., Zentralanstalt für Meteorologie, Vienna, 1936-37, Physico-Meteorol. Obs., Davos, Switzerland, 1937-38; meteorologist for flying weather and for Reichsamt für wetterdienst, Berlin, 1938-45; meteorologist Zentralanstalt für Meteorologie und Geodynamik, Vienna, 1946, dep. dir., 1956. Mem. Sonnblickverein, Meteorolog Gesellschaft, Limnologische Gesellschaft. Research and numerous publs. on radiation of sun, sky, water, phys. characteristics of now, thermal energy of streams and lakes. Address: Hohe-Warte 38, Vienna XIX, Austria.

ECKENER, Hugo, German aero. pioneer; b. Flensburg, Germany, Aug. 10, 1868; s. Johann Christoph and Anna Maria Elisabeth (Lange) E.; student philosophy and econs. univs. Munich, Berlin, Leipzig (all Germany); several hon. dr. degrees; m. Johanna Maass, 1897; 1 son, Knut, 2 daus., including Lotte (Mrs. Simon). Free lance writer, Friedrichshafen, Germany; worked on airships with Count Zeppelin; became head Zeppelin works, 1917-37. Recipient Guggenheim medal, 1937, Gold medal Brit. Aero. Soc., 1938. Author: Im Zeppelin über Länder und Meere, 1949. Pioneered (with Zeppelin) airship constrn. and travel; made over 2000 trips in zeppelin; 1st to cross Atlantic, 1924; flew around world in the Graf Zeppelin, 1929; taught navy airship flyers during World War I. Died Friedrichshafen, Aug. 14, 1915.

ECKENHOFF, James Edward, Am. anesthesiologist; b. Easton, Md., Apr. 2, 1915; s. George Lewis and Ada (Ferguson) E.; student Transylvania Coll., 1933-

36; B.S., U. Ky., 1937; M.D., U. Pa., 1941; m. Bonnie Lee Youngerman, June 4, 1938; children—J. Benjamin, Edward A., Walter L., Roderic G. Harrison fellow in anesthesiology U. Pa., 1945-47, faculty, 1947-65, prof. anesthesiology, 1955-65, asso. dir. dept. anesthesia Hosp. of U. Pa., 1952-65; prof., chmn. dept. anesthesia Northwestern U., Med. Sch., Chgo., 1966-—; chmn. dept. anesthesiology Chgo. Wesley Meml. Hosp., Passavant Meml. Hosp., Chgo., 1966-—. Cons. in anesthesiology WHO, Denmark, 1951; mem. surgery study sect. NIH, 1962-—, mem. anesthesia tng. grants com., 1966-—; cons. to Surgeon Gen., USN, 1964-—. Fellow, Faculty of Anaesthetists, Royal Coll. Surgeons, London, Hunterian Prof., 1965. Diplomate Am. Bd. Anesthesiology (dir. 1965-—). Fellow Am. Coll. Anesthesiologists; mem. Assn. U. Anesthetists, Assn. U. Profs., Am. Physiol. Soc., Royal Soc. Medicine London, A.M.A., Am. Soc. Anesthesiologists. Author: (with Robert D. Dripps; Leroy D. Vandam) Introduction to Anesthesia, 1957, 61; Science and Practice in Anesthesia, 1965; Anesthesia from Colonial Times, 1966. Research and publs. in physiology and pharmacology of coronary circulation, in analeptic agts. and narcotics; first to use narcotic antagonists in man; research in effect of anesthetics and adjuvants on respiration and circulation of man, effect of deliberate hypotension on circulation, deliberate hypotension for surg. procedures. Home: 823 Westwood Lane, Wilmette, Ill. 60091. Office: Northwestern Med. Sch., Chgo. 60611.*

ECKER, Alexander, German anatomist; b. Freiburg im Breisgau, Germany, 1816; prof. anatomy and physiology, Basel, Switzerland, 1844; prof. of comp. anatomy, Freiburg, 1850. Author: Convolutions of the Cerebral Cortex. Described the posterior occipital convolution of the brain known as Ecker's convolution, 1869; much original work on the devel. of cerebral convolutions in the fetus. Died Freiburg, 1887.

ECKER, Günter Heinz, German physicist; b. Siegburg, Germany, June 12, 1924; s. Paul and Margarete (Wirrmann) E.; student U. Bonn (Germany), U. Göttingen (Germany), U. Munich (Germany), 1941-46; Diplom-Examination, U. Bonn, 1946, Dissertation, 1948, Habilitation, 1953; m. Marianne Rosenbaum, Sept. 8, 1956; children—Sandra Sabine, Sara Susanne, Frank Oliver. Asst. theoretical physics U. Bonn., 1949-53, faculty, 1956-59, apl. prof., 1957-59; Royal Soc. London scholar U. Belfast (Ireland), 1953-54; vis. prof. U. Okla., 1955-56; prof. USNPGS, Monterey, Cal., also cons. fusion group U. Cal. at Berkeley, 1959; asso. prof. U. Bonn, 1960-63; prof. U. Bochum, Bonn., 1963-—, dir., 1964-—, dean, 1966-—. Cons. to pvt. cos. Mem. Deutsche Physikalische Gesellschaft, Am. Phys. Soc., Sigma Xi. Research, publs. on quantum mechanics, gaseous electronics, statis. problems, especially plasma physics, theory of polarization of canal rays, theory of arc cathode and anode, theory of plasma column, theory of plasma in equilibrium. Home: 23 Nordstrasse, 4324 Blankenstein, Germany. Office: Inst. für theor. Physik, Ruhr-Universität Bochum, 463 Bochum, Germany.*

ECKER, Jean Alexandre, Czechoslovakian physician; b. Trinitz, Czechoslovakia, Feb. 26, 1766; ed. Prague, Czechoslovakia. Prof. U. Freiburg, Germany, from 1797; obstetrician, surgeon. Credited with introducing vaccine in Bohemia. Died Aug. 5, 1829.

ECKERT, Ernst Rudolf Georg, mech. engr.; b. Prague, Czechoslovakia, Sept. 13, 1904; s. Georg and Margarete (Pfrogner) E.; Dipl.-Ing., German Inst. Tech., Prague, 1927, Dr.-Ing., 1931; Dipl. habil, Inst. Tech., Danzig, 1938; m. Josephine Binder, Jan. 20, 1931; children—Rosemarie Christa (Mrs. Peter Kohler), Elke, Karin, Dieter. Came to U. S., 1945, naturalized, 1955. Asst., German Inst. Tech., Prague, 1928-34, lectr. Inst. Tech. Dansig, 1935-38; sect. chief Aero. Research Inst., Braunschweig, 1938-45; docent Inst. Tech., Braunschweig, 1939-43; prof., dir. Inst. Thermodynamics, 1943-45; cons. USAF, Wright-Patterson AFB, O., 1945-49, NASA, Cleve., 1949-51; prof. mech. engrng. U. Minn., Mpls., 1951-—, dir. thermodynamics and heat transfer div., 1955-—, Regents' prof., 1966-—. Vis. prof. Purdue U., 1958-—; U. S. rep. aerodynamics panel Internat. Com. Flame Radiation 1955-—; chmn. Am. div. Commonwealth and Internat. Library Sci., Engrng. and Liberal Studies 1962-—. Recipient Max Jakob award Am. Soc M.E., 1961; award Inst. Tech., U. Minn., 1965; Western Electric Fund award, 1965. Fellow Am. Inst. Aeros. and Astronautics, N.Y. Acad. Sci.; mem. Am. Soc. M.E., Am. Soc. Engrng. Edn., Wissenschaftliche Gesellschaft fur Luft-und Raumfahrt, Sigma Xi, Pi Tau Sigma, Tau Beta Pi. Author: Technische Strahlungsaustauschrechnungen und ihre Anwendung in der Beleuchtungstechnik und beim Warmeaustausch, 1937, Einfuhrung in den Warme-und Stoffaustausch, 1949, Introduction to Heat and Mass Transfer, 1963; (with R. M. Drake, Jr.) Heat and Mass Transfer, 1950. Chmn. hon. editorial adv. bd. Internat. Jour. Heat and Mass Transfer, 1960-—. Research in field of convective and radiative energy exchange as applied to cooling of parts in gas turbines for stationary applications and for jet propulsion; contbns. to solution of problem to return satellites or space vehicles safely to the ground by research and devel. cooling methods; contbns. to temperature control of satellites in space by research on energy exchange through thermal radiation. Home: 60 W. Wentworth Av. W., St. Paul 55118.*

ECKERT, John Presper, Jr., Am. engr.; b. Phila., Apr. 9, 1919; s. J. Presper and Ethel M. (Hallowell) E.; student William Penn Charter Sch.; B.S., U. Pa., 1941, M.S., 1913; m. Hester Caldwell, Oct. 28, 1944 (dec.); children—J. Presper, III, Christopher. Research asso. Moore Sch. Elec. Engrng., U. Pa., 1941-46, instr. engrng. sci. mgmt. def. tng. courses; designer radar range systems, 1941-42; co-designer, co-inventor electronic numerical integrater and calculator ENIAC (with Dr. J. W. Mauchly), 1942-46; co-designer, co-inventor automatic computers BINAC, 1946-49 and UNIVAC, 1948-51; partner Electronic Control Co., 1946-47; v.p. Eckert-Mauchly Computer Corp., 1947-50; dir. engrng. Eckert-Mauchly div. Remington Rand, Inc., 1950-54; v.p. Remington Rand, Inc., in charge engrng. Eckert-Mauchly div., 1954-55; v.p. Remington Rand UNIVAC div. Sperry Rand Corp., dir. research Phila. operation. Recipient Howard N. Potts medal (with John W. Mauchly), Franklin Inst., 1949. Mem. Inst. Radio Engrs. Patentee in electronic and computer field; designed and constructed large and small scale electronic digital computing devices. Fellow Inst. Radio Engrs. Home: 528 Conshohocken State Rd., Gladwyne, Pa. Office: 1900 W. Allegheny Av., Phila. 29.

ECKERT, Joseph Webster, Am. plant pathologist; b. St. Louis, Mar. 27, 1931; s. Richard Edgar and Louise (Webster) E.; B.S., U. Cal., Los Angeles, 1952; M.S. (Research fellow), Rutgers U., 1953; Ph.D., U. Cal., Davis, 1957; m. Nancy Ann Muenchow, Jan. 19, 1957; children—Patricia Ann, Pamela Jeanne, John Robert. Biologist U. S. Army Chem. Corp., 1953-55; research asst. plant pathology U. Cal., Davis, 1955-57, asst. prof., Riverside, 1957-64, asso. prof., 1964-—; cons. indsl. firms on develop. of antimicrobial agts., 1963-—. Mem. Am. Phytopathol. Soc., Sigma Xi, Alpha Zeta. Research and publs. on develop. of fungicides and bactericides to control postharvest diseases of perishables; discovered the fungistatic properties of 2-aminobutane and the application of this compound to control storage diseases of fruit and flowers; several patents on this process.*

ECKFORD, Henry, marine architect, shipbuilder; b. Irvine, Scotland, Mar. 12, 1775; s. John and Janet (Black) E.; m. Marion Bedell, Apr. 13, 1799, 2 daus., Settled in N.Y.C., 1786; supr. shipbldg. in Lake Ontario during War of 1812; reduced size of stern frame, altered details of rigging; naval constructor Bklyn. Navy Yard, 1817-20; built steamer Robert Fulton (made 1st successful steam voyage from N.Y.C. to New Orleans and Havana, Cuba, 1822); built frigates for So. Am. republics. Died Istanbul, Turkey, Nov. 12, 1832.

ECKHARD, Conrad, German physiologist; b. Hamburg, Germany, Mar. 1, 1822; student anatomy and physiology, Marburg, and Berlin, Germany, 1845-49; M.D., Giessen, 1849. Prosector of anatomy, Marburg, 1848-50; prof. anatomy and physiology, Giessen, 1856-1905. Author: Beiträge zur Anatomie und Physiologie, 1858; Experimentelle Physiologie des Nerven Systems, 1867. Investigated mechanism of penile erection using dogs; influence of nerves on the secretion of milk; salviary nerves of dogs; sympathetic saliva; caudal hearts of the eel; olfactory nerves. Died 1905.

ECKHARDT, Sándor, Hungarian physician; b. Budapest, Hungary, Mar. 27, 1927; s. S. E. and I. M. (Huszar) E.; grad. Med. U., Budapest, 1927; m. Maria Petranyi, June 30, 1951; children—Peter, Gabor. Staff, II Med. Clinic, U. Budapest Cancer Research Lab., 1950-51, Communal Hosp. Uzsoki utca, Budapest, 1951-53; mem. Nat. Cancer Inst. Budapest, 1953-—, asso. prof. hematology, 1964-—. Cons. hematologists to instns., hosps. E. Roosevelt Cancer Research fellow Nat. Cancer Inst., NIH, Bethesda, Md., 1965. Mem. Hungarian Cancer Soc. (sec. gen. 1964-—). Author: (with Kellner, Lapis) die Lymphknotengeschwulste, 1966. Research and publs. on clin. evaluation of cancer chemotherapy drugs; discovered new antileukemic compound. Home: 1 Julia, Budapest, Hungary.*

ECK OF SULZBACH (Sultzbach), Paul, alchemist; flourished, 15th century. Author: Clavis Philosophorum; also probably, De lapide Philosophico, Ludus puerorum Labor Mulierum. Described formation of silver tree (tree of Diana) by action of mercurous nitrate on solution of silver nitrate; possibly discovered increase in weight in metals due to calcination.

ECKSTEIN, Herbert Bernhard, pediatric surgeon; b. Dusseldord, Germany, July 11, 1926; s. Albert and Erna (Schlossmann) E.; M.A., Cambridge Coll., 1953, M.B., B.Ch., 1950, M.D., 1960, M.Chir., 1962; m. Maria Schroeder, July 14, 1955; children—Philip Albert, Susan Jane, Michael Bernard. House surgeon, surg. registrar Addenbrook's Hosp., Cambridge, 1950-57; cons. pediatric surgeon Ankara U., Turkey, 1957-60; resident asst. surgeon Hosp. for Sick Children—, London, 1961-63, cons. surgeon, 1965-—; cons. surgeon Queen Mary's Hosp. for Children, Carshalton, 1963-—, Children's Hosp., Sydenham, 1965-—. Adviser in Spina Bifida to Wessex Regional Hosp. Bd., 19—-—. Fellow Royal Coll. Surgeons; mem. Royal Soc. Medicine, Brit. Assn. Pediatric Surgeons (council), German Soc. Pediatric Surgeons, Soc. Research into Hydrocephalus and Spina Bifida, Soc. Pediatric Urological Surgeons. Research, publs. in pediatric surgery, especially urinary stones in children, hydrocephalus, meconium ileus; colostomies. Address: 4, Appledore Close, Bromley, Kent, Eng.*

ECONOMO, Baron Constantin Alexander (San Serff) von, see von Economo.

ECPHANTOS, Greek astronomer; b. Syracuse, Greece, circa 350; combined Pythagorean with atomistic doctrines; taught earth rotates on its axis, that the universe is spherical and limited, and earth is not necessarily its center.

EDDINGTON, Sir Arthur Stanley, English mathematician, astrophysicist; b. Kendal, Westmoreland, Eng., Dec. 28, 1882; s. Arthur Henry E. and Sarah Ann (Shout) E.; B.Sc., Owens Coll. (now Manchester U.), 1902; ed. Trinity Coll., Cambridge; hon. degrees from numerous univs. Chief asst. Royal Obs., Greenwich, 1906-13; Plumian prof. astronomy and exptl. philosophy Cambridge U., 1913-44, dir. Cambridge U. Obs., 1914-44. Fellow Royal Soc. (Royal medal 1928), 1914; mem. Royal Astron. Soc. (pres. 1921-23), Internat. Phys. Soc. (past pres.), Internat. Astron. Union (pres. 1938-44), also mem. numerous European and Am. acads. sci. Author: Stellar Movements and the Structure of the Universe, 1914; Report on the Relativity Theory of Gravitation, 1918; Space, Time and Gravitation, 1920; The Mathematical theory of Relativity, 1923; The Internal Constitution of the Stars, 1926; Stars and Atoms, 1927; The Nature of the Physical World, 1928; Science and the Unseen World, 1929; The Expanding Universe, 1933; New Pathways in Science, 1935; Relativity Theory of Protons and Electrons, 1936; The Philosophy of Physical Science, 1939; Fundamental Theory, 1946. Famous for his theoretical investigations of internal structure, motion, and evolution of stars; introduced principle that chief agt. in transport of heat from inner to outer regions of star is not convection but radiation; held that stars consist of almost perfectly ionized atoms; calculated life of sun and its interior temperature; enunciated mass-luminosity law, 1924 (relates mass of star to its temperature, radiation pressure and luminosity); calculated theoretical limit for stellar masses; investigated nature of stars known as white dwarfs; one of 1st to appreciate significance of Einstein's theories of relativity and one of observers who verified one of predictions of gen. theory during observations of total solar eclipse, May 29, 1919; worked in linking relativity and quantum theories; gave analysis of nature and propagation of gravitational waves, devised method to include electromagnetic phenomena in scheme; created new type of connection between neighboring portions of space-time which proved significant in devel. of differential geometry; gave interpretation of gen. relativity; discovered and developed certain new principles in physics which together formed coherent system which he named fundamental theory; used principles of fundamental theory of calculate constants of nature. Died Cambridge, Nov. 22, 1944.

EDDY, Henry Turner, Am. mathematician, physicist; b. Stoughton, Mass., June 9, 1844; s. Henry and Sarah Hayward (Torrey) E.; A.B., Yale, 1867; Ph.B., Sheffield Sci. Sch., 1868, A.M., 1879; C.E., Cornell, 1870, Ph.D., 1872; postgrad. U. Berlin, Physikalische Inst., Berlin, 1879, Sorbonne, College de France, Paris, 1880; LL.D., Center Coll., 1892; m. Sebella Elizabeth Taylor, Jan. 4, 1870; 1 son, 4 daus. Instr. in field work Sheffield Sci. Sch., 1867-68; instr. Latin, math., U. Tenn., 1868-69; asst. prof. math., civil engrng., Cornell U., 1869-73; adjunct prof. math Princeton, 1873-74; prof. math. and astronomy, civil engrng., U. Cin., 1874-90, dean acad. faculty, 1874-77, 84-89, acting pres. and pres.-elect, 1890; pres. Rose Poly. Inst., Terre Haute, Ind., 1891-94; prof. engrng. and mechanics, U. Minn., 1894-1907, head prof. math. and mechanics Coll. Engrng., 1907-12, dean grad. sch., 1906-12, ret. 1912. Mem. Internat. Congress Arts and Scis., St. Louis, 1904. Fellow A.A.-A.S. (v.p. for math. and physics 1884); mem. Am. Philos. Soc., Am. Math. Soc., Am. Phys. Soc., Soc. Promotion Engrng. Edn. (pres. 1896), Phi Beta Kappa, Sigma Xi. Author: Analytical Geometry, 1874; Researches in Graphical Statics, 1878; Thermodynamics, 1879; Neue Construction aus der Graphischen Statik, 1880; Maximum Stresses Under Concentrated Loads, 1890; also numerous sci. and tech. papers. Gave graphic solutions of problems on the maximum stresses in bridges under concentrated loads with aid of reaction polygons; studies on properties and stresses in reinforced concrete slabs. Died Dec. 11, 1921.

EDDY, Nathan Browne, Am. pharmacologist; b. Glens Falls, N.Y., Aug. 4, 1890; s. Charles A. and Aletta (Norcross) E.; M.D., Cornell U., 1911; D.Sc., U. Mich., 1963; m. Wilhelmina E. Ahrens, Sept. 7, 1913; 1 son, Charles E. Practice gen. medicine, N.Y.C., 1911-1916; instr. physiology McGill U., 1916-20; asst. prof. physiology and pharmacology, U. Alta., 1920-28, asso. prof., 1928-30; research prof. pharmacology, U. Mich., 1930-39; prin. pharmacologist; NIH, Bethesda, Md., 1939-49, med. officer, 1949-60, cons. on narcotics, 1960-—; profl. asso., exec. sec. com. drug addiction and narcotics Nat. Acad. Scis.-NRC, Washington, 1961-67. Panelist, intermittent cons. addiction producing drugs, WHO, 1947-—; tech. adviser U. S. delegation to U.N. Commn. on Narcotic Drugs, intermittently 1947-—; del. Planning Conf. on Research in Drug Abuse, Tokyo, Japan, 1964. Recipient, 1st Annual Sci. award, Am.

Pharm. Mfrs. Assn., 1939; Pub. Health Service Performance award, Dept. Health Edn. and Welfare, 1960; Snow award, 1967; Hillebrand award, 1968; named Meml. lectr., 1959; Kelynak lectr., 1965. Author: (with Kruger, Sumwalt), Pharmacology of Opium Alkaloids, 1941. Originator of coded information on narcotics, research on the chem. structure and action of potential analgesics. Home: 7055 Wilson Lane, Bethesda, 20034.*

EDDY, Samuel, Am. zoologist; b. Decatur, Ill., Mar. 26, 1897; s. Orlan Thomas and Blanche (Allsop) E.; A.B., James Millikin U., 1924; M.A., U. Ill., 1925, Ph.D., 1929; m. Vera Conel, June 6, 1925; children—Samuel Conel, Richard Lee. Faculty, James Millikin U., 1924-26; aquatic biologist Ill. Natural History Survey, 1926-29; faculty U. Minn., Mpls., 1929-—, prof. zoology, 1944-64, prof., emeritus, 1964-—, curator fishes, 1964-—; head fisheries research Minn. Dept. Conservation, 1937-40. Cons., U. S. Forest Service, 1935-37. Recipient Distinguished Service award Sigma Xi, 1965. Fellow A.A.A.S.; mem. Am. Micros. Soc., Ecol. Soc. Am., Soc. Ichthyologists and Herpetologists, Am. Fisheries Soc., Am. Soc. Limnology and Oceanography, Sigma Xi. Author: several books including: How to Know Freshwater Fishes, 1957; (with A. C. Hodson) Taxonomic Keys to the Common Animals of the North Central States, 1961; (with C. P. Oliver, J. P. Turner) Guide to the Study of the Anatomy of the Shark, 1964, Atlas of Drawings for Vertebrate Anatomy, 1964; Atlas of Drawings for Chordate Anatomy, 1964. Research and publs. on ecol. dynamics lakes, conditions necessary for ideal fish populations. Home: 2100 Dudley Av., St. Paul 55108. Office: Dept. Zoology, U. Minn., Mpls.*

EDDY, Walter Hollis, Am. physiol. chemist; b. Brattleboro, Vt., Aug. 26, 1877; B.S., Amherst Coll.; 1898; A.M., Columbia, 1908; Ph.D., 1909, Asso. in biochemistry Columbia Coll. Phys. and Surg. 1908-13; research chemist N.Y. Hosp., 1913-59; prof. physiol. chemistry Columbia Tchrs. Coll., 1922-41, now emeritus; dir. Bur. Food, Sanitation and Health, Good Housekeeping Mag., 1927-41; sci. dir. Am. Chlorophyll, Inc., Lake Worth, Fla., 1948-52, Am. Chlorophyll div. Strong Cobb Co., 1952; research dir. Mangrove Products, Inc., West Palm Beach, Fla., cons. chem. engr., 1953-59; research cons. U. S. Vitamin and Pharm. Corp. Author: Experimental Physiology and Anatomy, 1906; Vitamin Manual, 1921; Nutrition, 1928; The Avitaminoses, 1937, rev. edit., 1941, 1945. What Are the Vitamins?, 1941; We Need Vitamins, 1941; Vitaminology 1949, also many articles on foods and vitamins. Died Oct. 1959.

EDDY, William Abner, Am. meteorologist; b. N.Y., Jan. 28, 1850; s. H. J. and Amanda (Doubleday) E.; prep. course U. Chgo.; m. Cynthia S. Huggins, Apr. 21, 1887. Began aerial and kite experiments in 1890; took mid-air kite photograph in Western Hemisphere, May 30, 1895; mid-air temperature in world from kites, Feb. 4, 1891. Experimented extensively with atmospheric electricity drawn from kite sustained wire, and with flying machine aeroplanes dismissed from kites' lines in mid-air. Studied air currents and layers by means of kites; Eddy kite (quadrilateral tailless kite) named for him. Died 1909.

EDE, Allan John, English mech. engr.; b. London, Eng., Sept. 4, 1915; s. John Philip and Nellie (Smith) E.; B.Sc., London U., 1937; M.A., Cambridge (Eng.) U., 1947; m. Gertrude Marion King Day, Apr. 1, 1944; children—Martin, Alison. Research asst. Imperial Coll., London, 1937-40; sci. officer Low Temperature Research Sta., Cambridge, 1940-49; dep. to head heat div. Nat. Engring. Lab., East Kilbride, Scotland; 1949-57, head, 1957-66; prof. mech. engring. U. Aston, Birmingham, Eng., 1966-—. Mem. Instn. Mech. Engrs. (chmn. fluid and thermodynamics design data com. 1964-—). Fellow Inst. Physics, Inst. Math. Author: Introduction to Heat Transfer: Principles and Calculations, 1966; also articles. Coeditor Internat. Jour. of Heat and Mass Transfer, 1960-—. Research on heat and mass transfer including freeze-drying, free convection, heat transfer in pipes. Home: 60, Selwyn Rd., Birmingham 16, Eng.*

EDEBOHLS, George Michael, Am. surgeon; b. N.Y., May 8, 1853; s. Henry and Catherine (Brull) E.; A.B., St. Johns Coll., 1871, A.M., 1886. LL.D., 1903; M.D., Columbia Coll. Phys. and Surg., 1875; m. Barbara Leyendecker, Sept. 19, 1882. House physician and surgeon St. Francis Hosp., N.Y., 1875-79, gynecologist, 1887-1903; cons. surg., 1903-08; prof. diseases of women, N.Y. Post-Grad. Med. Sch. and Hosp., 1893-1908; cons. gynecologist St. John's, Nyack hosps. Fellow N.Y., Acad. Medicine, Am. Gynecol. Soc.; mem. A.M.A. Author: The Surgical Treatment of Bright's Disease, 1904. Invented operating table, speculum, leg holders, needle holders; 1st to perform renal decortication for relief of chronic nephritis. Died 1908.

EDELCRANZ, Abraham Nicolas d', see d'Edelcranz, A.N.

EDELEN, Dominic Gardiner Bowling, Am. mathematician; b. Washington, Jan. 3, 1933; s. I. Gardiner and Marybeth (Bowling) E.; B. Engring. Sci., 1954, M. Sci. Engring., 1956, Ph.D., 1965; m. Erica E. von Eschenburg, Dec. 22, 1954; children—Damien S.P., Marybeth B., Mittie-ellen P., Dominique A.N., Clarissa W., John D. B. Jr. instr. Johns Hopkins, Balt., 1954-56; research and devel. scientist Martin Co., Balt., 1956-59; staff engr. Hughes Aircraft Co., Culver City, Cal., 1959-60; mathematician RAND Corp., Santa Monica, Cal., 1960-66, cons., 1966-—; prof. math. Purdue U., Lafayette, Ind., 1966-—. Cons., Douglas Advanced Research Lab., Huntington Beach, Cal., 1966-—. Fellow Royal Astron. Soc.; mem. Tensor Soc., Soc. for Natural Philosophy, Cometé Gravitation et Relativité, Internat. Soc. Engring. Sci. Author: The Structure of Field Space, 1962; also articles. Research on field theories by application math. disciplins to their founds.; analysis rigerous solutions and properties of such solutions and detailed description of them in terms of geometry, systems of equations arising from calculus of variations and group theory. Address: Dept. Math., Purdue U., Lafayette, Ind. 47907.*

EDELMAN, Gerald M., Am. biochemist; b. N.Y.C., July 1, 1929; s. Edward I. and Anne (Friedman) E.; B.S., Ursinus Coll., 1950; M.D., U. Pa., 1954; Ph.D., Rockefeller Inst., 1960; m. Maxime Morrison, June 11, 1950; children—Eric, David, Judith. Asst. physician Hosp. Rockefeller Inst., 1957-60; postdoctoral fellow Rockefeller Inst., 1957-60, asst. prof., 1960-63, asso. prof., asso. dean grad. studies, 1963-—. Recipient Spencer Morris award U. Pa., 1954. Mem. Harvey Soc., Am. Chem. Soc. (Eli Lilly award in biol. chemistry 1965), Sigma Xi, Alpha Omega Alpha. Research, publs. on elucidation structure antibody molecules, fluorescence spectroscopy macromolecules. Home: 144 E. 84th St., N.Y.C. 10021.*

EDELMAN, Jack, English plant physiologist; b. London, Eng., May 8, 1927; s. Samuel and Netta (Smith) E.; B.Sc., Imperial Coll. (London), 1948, A.R.C.S., 1948; Ph.D., Sheffield U., 1950; m. Joyce Mason, Aug. 19, 1958; children—Jane, Simon, Alex, Daniel. Sci. officer Agrl. Research Council, London, 1950-57; faculty U. London, 1957-—, prof. botany, Queen Elizabeth Coll., 1964-—. Rockefeller Found fellow U. Cal. at Berkeley, 1953-55, vis. prof. biochemistry at Davis, 1966. Recipient Huxley medal in natural science Imperial Coll., 1962. Fellow Inst. Biology. Research and publs. on plant physiology and biochemistry especially in fields of devel. and carbohydrate metabolism. Office: Queen Elizabeth Coll., Campden Hill, London W.8, Eng.*

EDELMAN, Nils Holger, Finnish geologist; b. Helsignfors, Finland, Dec. 13, 1918; s. Georg and Sigrid (Osterman) E.; Fil. Kand., U. Helsingfors, 1946, Fil. Dr., 1950; m. Ethel Solveig Mörne, Dec. 21, 1946; children—Chris Burjam, Birgitta. With Geol. Survey Finland, 1946-54; mining geologist Oy Vuoksenniska Ab, 1954-58; prospector, mining geologist Boliden Mining Co., Sweden, 1958-64; prof. geology and mineralogy Abo (Finland) Akademy, 1964-—. Sec. Nordic Geologists Meeting, Helsingfors, 1954; pres. VII Nordic Geologists Winter Meeting, Abo, 1966. Mem. Geol. Soc. Finland, Geogr. Sällsk. Finland, Geol. Fören. Stockholm, Norsk Geol. Fören., Geologische Vereniging, Bergsmannaföreningen. Research on structural evolution of granitic domes and gneiss zones, glacial abrasion, ice movements, postglacial faults in S.W. Finland, also structural evolution, stratigraphy, metamorphism, intrusion of sulphide ores, Kristineberg area, Sweden. Home: 30 C 30 Tavastgatan, Abo, Finland.*

EDELSON, David, Am. chemist; b. Bklyn., Nov. 27, 1927; s. Adolph and Louise (Goldman) E.; B.S., Poly. Inst. Bklyn., 1946; Ph.D., Yale, 1949, Sterling Research fellow, 1949-50; m. Ellin Sue Robinson, Aug. 3, 1962; children—Edward, Kenneth, Ronald, Alan. Mem. tech. staff Bell Telephone Labs., Murray Hill, N.J., 1950-—. Mem. Am. Chem. Soc., Am. Phys. Soc., N.Y. Acad. Sci., Sigma Xi. Research and publs. on polymers, polyelectrolytes, dielectrics, gaseous electronics, ion-molecule interactions. Home: 77 Parkside Rd., Plainfield, N.J. 07060. Office: Bell Telephone Labs., Murray Hill, N.J. 07974.*

EDEN, Richard John, English physicist; b. London, Eng., July 2, 1922; s. James A. and Dora (Clark) E.; B.A., Peterhouse Coll., Cambridge (Eng.) U., 1943, M.A., 1948; Ph.D., 1951; m. Elsie Jane Edwards; children—Christine M., Michael, R., Carolyn. J. Bye fellow Peterhouse, Cambridge, 1949, fellow Clare Coll., 1951-55, 57-66, 66-—; faculty U. Cambridge, 1957-—, reader theoretical physics, 1964-—. Author: Brandeis Lectures, 1961; (with Landshoff, Olive, Polkinghorne) The Analytic S Matrix, 1966; High Energy Collisions of Elementary Particles, 1967; also articles. Research on structure of atomic nuclei, elementary particles. Home: 19 Almoners Av., Cambridge, Eng.*

EDEN, William Gibbs, Am. entomologist; b. Talladega, Ala., May 3, 1918; s. G. G. and Maude (Parnell) E.; B.S., Auburn U., 1940, M.S., 1947; Ph.D., U. Ill., 1950; m. Evelyn Smith, June 1, 1940; children—Brenda (Mrs. W. F. Powell, Jr.), Jane (Mrs. J. M. Buttram). Faculty, Auburn (Ala.) U., 1940-44, 48-65, prof. dept. zoology-entomology, 1940-65; chmn. dept. entomology U. Fla., Gainesville, 1965-—. Mem. Entomol. Soc. Am., Sigma Xi, Phi Kappa Phi, Gamma Sigma Delta. Research and numerous publs. on control of insects of cotton, corn, small grains, legumes, ornamentals, peanuts. Home: 4411 N.W. 17th Pl., Gainesville, Fla. 32601.*

EDER, Joseph Maria Ludwig, Austrian photochemist; b. Krems, Austria, Mar. 16, 1855; s. Joseph and Karoline (Borutzky) E.; student natural scis. U. Vienna (Austria), Tech. U. Vienna; hon. dr. Tech. U. Vienna; m. Anna Valenta, 1885; 1 dau. Became asst. Tech. U. Vienna, 1879, named asso. prof. 1892, prof., 1902; became tchr. chemistry State Indsl. Trades Sch., Vienna, 1880, chemistry prof. at upper level, 1882; organizer Graphische Lehrund Versuchsanstalt, Vienna, 1888, then dir. Mem. Austrian Acad. Scis. Editor: Ausführliches Handbuch der Photographie, 1884; Jahrbuch für Photographie und Reproduktionstechnik, 1887; Rezepte, Tabellen und Arbeitsvorschriften für Photographie und Reproduktionstechnik, 1889. Invented (with G. Pizzighelli) silver chloride gelatin emulsion (basis for less sensitive plates), 1881; invented yellow-green sensitizer for dry plates (orthochromatic plates); studied photometry, sensitometry, spectroanalysis of 3-color printing, history of photography. Died Kitzbühel, Austria, Oct. 18, 1944.

EDERSTROM, Helge Ellis, physiologist; b. Torsas, Sweden, Feb. 28, 1908; s. Herman A. and Alma (Jonsson) E.; came to U. S., 1912, naturalized, 1929. B.S., Beloit Coll., 1937; M.S., Northwestern U., 1939, Ph.D., 1941. Faculty, U. Mo. Med. Sch., 1941-47, asst. prof., 1943-47; faculty St. Louis U. Med. Sch., 1947-52, asso. prof. physiology, 1948-52; prof. physiology U. N.D. Med. Sch., Grand Forks, 1952-—. Mem. Am. Physiol. Soc., Soc. Exptl. Biology and Medicine, Aerospace Med. Soc., A.A.A.S., Am. Assn. U. Profs., Sigma Xi. Research and publs. on devel. circulation homeostasis in newborn animals, regulation of body temperature.*

EDGAR, Arlan Lee, Am. zoologist; b. nr. Alma, Mich., June 3, 1926; s. Sherman J. and Letha (Perdew) E.; B.A., Alma Coll., 1949; M.A., U. Mich., 1950, M.S., 1957, Ph.D., 1960; m. Bonnie Jean Anderson, Mar. 30, 1952; children—Rosemary, Amy, Andrew. Faculty, Alma (Mich.) Coll., 1950-51, 53-—, prof. biology, 1965-—, dir. NSF undergrad. research participation program, 1962-66. Mem. A.A.A.S., Am. Inst. Biol. Sci., Am. Microsc. Soc., Am. Soc. Zoologists, Sigma Xi (pres. Central Mich. 1967-68), Phi Kappa Pi, Beta Beta Beta. Author: The Biology of the Order Phalangida in Michigan, 1966; also articles. Research on dirunal behavior isopods, phalangids, fresh-water clams and terrestial snails, energy transformation in terrestrial arthropods. Home: 602 Woodworth St., Alma, Mich. 48801.*

EDGAR, Edward Charles, English chemist; b. London, 1881; student U. Manchester (Eng.); m. Lectr., Manchester U.; became head chemistry dept. Regent St. Poly., 1923; named prin. Rutherford Tech. Coll., Newcastle on Tyne, Eng., 1926; sr. insp. dept. explosive supplies Ministry of Munitions; chief research asst. Brit. Dyes. Mem. Fabian Soc., Tech. Examinations Council. Became Dalton scholar, 1904. Determined densities of hydrogen and oxygen; synthesized water from pure gases; studied atomic weight of chlorine, burning of chlorine in hydrogen, behavior of hydrogen in contact with pallidium, conditions for activating pallidium. Died Newcastle on Tyne, 1938.

EDGAR, Samuel Allen, Am. pathologist, educator; b. Stafford, Kan., Feb. 6, 1916; s. Joshiah Dodds and Zada (Patton) E.; A.B., Sterling Coll., 1937; student Fresno State Tchrs. Coll., 1934-35; M.S., Kan. State U., 1939; Ph.D., U. Wis., 1944; m. Phyllis Leone Wells, Jan. 30, 1939; children—Philip Allen, Susan Jane. Prof., poultry path. Auburn U., 1947-49, 1950-—; sr. scientist USPHS, Tahiti, 1949-50; mem. edl. bd. Exptl. and Molecular Pathology; research cons. D & M Labs., 1960-—. Fellow A.A.A.S., N.Y. Acad. Sci.; mem Soc. Parasitology and Microscopic Soc., Poultry Sci. Assn., Soc. Trop. Medicine and Hygiene, Soc. Zoology, Entomol. Soc. Am. Mosquito Control Assn., World's Poultry Sci. Assn., Sigma Xi. Research and publs. on diseases, parasites of poultry, emphasis on immunity to parasitic infections, and pathol. response of host to parasite; developed vaccine for immunization poultry against coccidiosis and virus diseases. Home: 612 Cary Dr., Auburn, Ala. 36830.*

EDGERTON, Harold Eugene, Am. elec. engr.; b. Fremont, Neb., Apr. 6, 1903; s. Frank E. and Mary (Coe) E.; B.S., U. Neb., 1925, D.Engring., 1948; D.Sc., Mass. Inst. Tech., 1931; m. Esther May Garrett, Feb. 25, 1928; children—Mary L. Dixon (Mrs. Charles Dixon), Robert. Prof. elec. measurements Mass. Inst. Tech., 1931-—; with Edgerton, Germeshausen and Grier, Boston, 1932-—, v.p., chmn. bd., 1947-65, hon. chmn. bd., 1965-—. Recipient numerous awards including Boston Sea Rovers award, 1961, Gordon Y. Billard award, Mass. Inst. Tech., 1961; named New Eng. Engr. of Year, Engring. Socs. New Eng., 1959, Indsl. Photograhers Assn., 1963. Fellow Am. Inst. Elec. Engrs., I.E.E.E. (Morris E. Leeds award 1965), Royal Photog. Soc. Gt. Britain (Silver Progress medal 1964), Photog. Soc. Am. (Progress medal 1955), Soc. Motion Picture and Television Engrs. (Progress award 1959, E. I. DuPont Gold medal 1962), Eta Kappa Nu. Mem. Acad. Underwater Arts and Scis., Acad. Applied Sci., Am. Acad. Arts and Scis., Photographers Assn. New Eng., Soc. Photog. Engrs., Nat. Acad. Engring., Am. Phys. Soc., Confédération Mondiale des Activités Subaquatiques, Club de Exploraciones y Deportes Acuaticos de Mexico, Internat. Sci. Film Assn., Marine Tech. Soc., Mus. Sci., Nat. Acad. Scis., Sigma Xi,

Sigma Tau. Author: (with J. Killian, Jr.) FLASH, Seeing the Unseen, 1939; 2d edit., 1954; Research and publs. on synchronous machine theory; designed high intensity, short duration stroboscopes, especially for photographic use, e.g., in vitamin B2 balance studies, to photograph capillaries in white of eye without injuring patient; underwater photography; sonar measuring devices. Home: 100 Memorial Dr., Cambridge, Mass. 02142.

EDGREN, Richard A(rthur) E.; Am. biologist; b. Chgo., May 28, 1925; s. Richard A. and Helga (Corydon) E.; B.S., Northwestern U., 1949, M.S., 1950, Ph.D., 1952; m. Margery Kelly, June 7, 1952; children—Susan Ann, Jean Elizabeth. Sr. investigator G. D. Searle and Co., Skokie, Ill., 1952-60; asst. mgr. nutrition and endocrinology sect. Wyeth Labs., Radnor, Pa., 1960-67, mgr. endocrine sect., 1968—. Mem. The Endocrine Soc., Soc. Exptl. Biology and Medicine, Am. Soc. Zoologists, Royal Soc. Medicine, American Institute Biological Sciences, the American Society of Ichthyologists and Herpetologists, Herpetologists' League, Sigma Xi. Editorial bd. Gen. and Comparative Endocrinology, since 1963—. Research and publs. in gen. biology and systematics reptiles and amphibians, biol. effects steroid hormones, mode-of-action oral contraceptives and physiology pregnancy. Home: 2403 Buttonwood Rd., Berwyn, Pa. 19312. Office: Wyeth Labs., Radnor, Pa. 19101.*

EDHOLM, Otto Gustaf, Brit. physiologist; b. London, Eng., June 14, 1909; s. Frans Gustaf and Maria (Kinell) E.; B.Sc. with honors, King's Coll., London, 1930; M.B., B.S., St. George's Hosp., London, 1934; m. Elizabeth Bevan, Dec. 16, 1936; children—Brita, Corinna, Felicity. Lectr. physiology King's Coll., 1935-38; lectr. physiology Queen's U., Belfast, Ireland, 1938-44; prof. physiology Royal Vet. Coll., London, 1944-47; prof. physiology U. Western Ont., Can., 1947-49; head div. human physiology Nat. Inst. for Med. Research, London, 1949—. William Julius Mickle fellow U. London, 1963. Mem. Hampstead Sci. Soc. (pres.), Ergonomics Research Soc. (sec. 1955-61), Physiol. Soc., Am. Physiol. Soc., Med. Research Soc., Nutrition Soc. Author: (with A. Burton) Man in a Cold Environment, 1955. Co-editor: Exploration Medicine, 1965; The Physiology of Human Survival, 1965. Research, numerous publs. on physiology of fainting, with demonstration of dilatation in muscle vessels being responsible for fall of blood pressure; demonstrated that blood flow through skin is largely controlled by vasodilator mechanisms, including role of bradykinin produced by activity of sweat glands; effects of environment on man in extremes of temperature, demonstrating effects of energy expenditure and food intake; relationship between these two variables in temperate conditions and lag that exists between energy expenditure and food intake in man. Home: 14A Wedderburn Rd., London N.W.3. Office: Nat. Inst. for Med. Research, Hampstead Labs., London N.W.3, Eng.*

EDINGER, Lewis Joachim, polit. scientist; b. Frankfort, Germany, Feb. 1, 1922; s. Friedrich E. and Dora E. (Meyer) E.; came to U. S., 1936, naturalized, 1943; A.B., Wabash Coll., 1943; M.A., Columbia, 1947, Ph.D., 1951; m. Hanni Bluemenfeld, Sept. 11, 1950; children—Monica Ruth, Susan Yvonne. Instr., N.Y. U., 1947-49; asst. prof. Sweet Briar Coll., 1950-51; lectr. Vassar Coll., 1951-52; asst. prof. U. N.C., 1952-53; research asso. U. S. Dept. Def., 1953-57; asst. prof. to prof. Mich. State U., 1957-63; prof. Washington U., St. Louis, 1963-67; prof. govt., mem. European Inst., Columbia, N.Y.C., 1967—. Research asso. Yale, 1964-65; vis. research prof. Free U. Berlin, 1959-60, U. Bonn (Germany), 1964-65; research fellow Stanford, 1956-57. Mem. Am. Polit. Sci. Assn. Author: West German Armament, 1955; German Exile Politics, 1956; (With Deutsch), Germany Rejoins the Powers, 1959; Kurt Schumacher, 1964; also other books, articles. Research on individual and group leadership in various polit. systems, especially comparisons in space and time and polit. leadership in modern indsl. societies, post-totalitarian politics particularly in Europe. Home: 83 Lefurgy St., Dobbs Ferry, N.Y. 10522. Office: Columbia, N.Y.C. 10027.*

EDINGER, Ludwig, German neurologist, anatomist; b. Worms, Germany, Apr. 13, 1855; s. Markus and Julie (Hochsbaetter) E.; studied medicine in Heidelberg, Germany, Strasbourg, France; M.D., 1876; m. Anna Goldschmidt, 1886; 1 son, 2 daus. Asst. to Kussmaul; practiced in Frankfort, Germany, 1883-1914; named prof. neurology U. Frankfort, 1914. Author: Vorlesungen über den Baü der nervösen Zentralorgane, 1885. First description of ventral and dorsal spinocerebellar tracts, 1885; described gray nucleus under the aqueduct of Sylvius from which some of the trochlear nerve arise (Edinger's nucleus), 1885; 1st to distinguished between paleocerebrum and neocerebrum, 1885. Died Frankfort, Jan. 26, 1918.

EDINGER, Tilly, vertebrate paleontologist; b. Frankfurt, Germany, Nov. 13, 1897; d. Ludwig E. and Anna (Goldschmidt) Edinger; Dr. phil.nat., U. Frankfurt, 1921, Dr.med. h.c.(hon.), 1964 Sc.D. (hon.), Wellesley Coll., 1950; Dr.rer.nat.(hon.), U. Giessen, 1957; Came to U. S., 1940, naturalized, 1945. Asst. geology-palentology inst. U. Frankfurt, 1921-27; curator fossil vertebrates Naturmus. Senckenberg, Frankfurt, 1927-38; research asso. Mus. Comparative Zoology, Harvard, 1940-48, research paleontologist, 1948-64, hon. asso. paleontology, 1964—; instr. zoology Wellesley (Mass.) Coll., 1944-45. Guggenheim fellow, 1943-44. Mem. palaontologische Gesellschaft (hon.), Senckenberg naturf Gesellschaft (hon.), Soc. Vertebrate Paleontology (past pres.), Soc. for Study Evolution, Am. Soc. Zoologists, Am. Acad. Arts and Scis., Akademie der Wissenschaften und der Literatur (hon. corr.), Sigma Xi. Author: Die fossilen Gehirne, 1929; Evolution of the Horse Brain, 1948; also numerous articles, revs. Research in vertebrate paleontology. Home: 64 Oxford St., Cambridge, Mass. 02138.*

EDINGTON, George Millar, pathologist; b. Glasgow, Scotland, Apr. 18, 1916; s. George and Catherine M. (McDougall) E.; M.B., Ch.B., U. Glasgow, 1939, M.D., 1953; m. Mary J. Hamilton, Jan. 18, 1949. Specialist in pathology Med. Research Inst., Chana, 1950-56; prof. pathology U. Ibadan, Nigeria, 1958—. Hon. life mem. Brit. Red Cross, 1956; named Mem. Brit. Empire, 1957. Fellow Coll. Pathologists (a founder); mem. Royal Coll. Physicians London, Royal Soc. Tropical Medicine and Hygiene, Royal Soc. Medicine, Assn. Physicians of W. Africa, Pathol. Soc. Gt. Britain and N. Ireland, Assn. Clin. Pathologists, Internat. Acad. Pathologists. Research and publs. on pathology of abnormal haemoglobin diseases in W. Africa; co-discoverer of Haemoglobin G and persistence of high foetal haemoglobin; studies on pathology of malaria, schistosomiasis, fungal, renal and cardiovascular disease. Address: Dept. Pathology, U. Ibadan, Ibadan, Nigeria.*

EDISON, Thomas Alva, Am. inventor; b. Milan, O., Feb. 11, 1847; s. Samuel and Nancy Edison; Ph.D., Union Coll., 1878; D.Sc., Princeton, 1915; LL.D., U. State N.Y., 1916; m. Mary G. Stillwell, 1873; children—Marion Estelle, Thomas A., William L; m. 2d, Mina Miller, 1886; children—Madeleine, Charles Theodore. At 12 years of age became newsboy on G. T. Ry; learned telegraphy and worked as operator in U. S. and Can.; pub. Weekly Herald; employee Stock and Gold Co., Boston; founder 1st firm of cons. engrs., 1869; established workshop, Newark; founded 1st indsl. research lab., Menlo Park, N.J., 1876 (moved to West Orange, N.Y. 1887); organizer Edison Electric Co. (became Gen. Electric Co. after merger with Thousou-Houston Co.); designed, built, operated benzol plants, 2 carbolic acid plants, plants for mfg. myrbane aniline oil, aniline salt, paraphenylene diamine, during World War I; worked with Ford and Firestone to produce rubber from domestic plants. Hon. chief cons. engr. St. Louis Expn., 1904; pres. Naval Cons. Bd., from 1915. Decorated comdr. Legion of Honor (France); recipient John Fitz medal, 1908, Rathenau medal (German), Am. Mus. Safety, 1914, Congl. Gold medal, 1928. Inventor electric vote recorder, automatic repeater for telegraph, quadruplex telegraph, printing telegraph, machines for quadruplex and sextuplex telegraphic transmission, the electric pen and mimeograph, the carbon telephone transmitter, 1877-78, the phonograph, 1877, the microphone, the microtasimeter for detection of small changes in temperature, the megaphone, the incandescent lamp and light system, system of wireless telegraphy to and from moving ry. trains, motion pictures, the telescribe, alkaline storage battery, many war inventions for U. S. Govt.; discovered "Edison effect," 1883 (adapted to produce 1st vacuum valve to detect radio waves, 1904); patentee more than 1300 inventions. Died West Orange, N.J., Oct. 18, 1931.

EDKINS, John Sydney, English physiologist; b. July 12, 1863; s. John Edkins; ed. Gonville and Caius Coll., Cambridge (Eng.) U., St. George's Hosp.; M.A., M.B., D.Sc.; m. Mabel Hayman, 1896; m. 2d, Nora Tweedy, 1919; 1 dau. Demonstrator physiology Owens Coll., Manchester, Eng., St. Bartholomew's Hosp.; later lectr. in physiology St. Bartholomew's Hosp., Bedford (Eng.) Coll.; prof. U. London, Bedford Coll., from 1926. Examiner in physiology Manchester, Liverpool, Leeds, Cambridge, Oxford (all Eng.) univs., Royal Coll. Physicians; sec. to Chem. Warfare Med. Com., World War I; mem. com. concerned with grain pests Royal Soc. Co-author several textbooks. Contbr. to Jour. Physiology, Proceedings Royal Soc., Jour. Parasitology, others. Contbns. in parasitology, physiology and grain pests; 1st to describe gastric secretin, 1905. Died Aug. 17, 1940.

EDLBACHER, Siegfried Augustin Johann, physiol. chemist; b. Linz, Austria, Mar. 7, 1886; s. Maximilian and Johann (v. Risch) E.; studied chemistry in Vienna, Giessen, Germany, Jena, Germany; doctorate, Graz, Austria, 1911; M.D. (hon.), Basel, Switzerland, 1946. Asst. med.-chem. insts. in Graz. and Innsbruck, Austria, with chem. industry in Ludwigshafen/Rhein, for a short time; became asst. to A. Kossel, Physiol. Inst., U. Heidelberg (Germany), 1913, lectr., 1919, asso. prof. physiol. chemistry, 1924; named prof. physiol. chemistry U. Basel, 1932. Author: Praktikum der Physiologischen Chemie, 1932; Kurzgefasstes Lehrbuch der Physiologischen Chemie, 1932; Handbuch der Biochemie der Menschen und der Tiere, 1933; Deskriptive Chemie der Zelle, 1933; also numerous articles. Research on intermediate metabolism of the amino acids, arginine and histidine, also metabolism of cancer cells, biochemistry of cell nucleus; discovered ferment histidasis. Died Basel, June 5, 1946.

EDLÉN, Bengt, Swedish physicist; b. Gusum, Sweden, Nov. 2, 1906; s. Gustaf F. and Maria (Rundberg) E.; D.Sc., Uppsala (Sweden) U., 1934; Dr. rer. nat. h.c., Kiel (Germany) U., 1965; m. Elfriede I. Mühlbach, June 30, 1940. Asst. prof. Uppsala U., 1934-43; prof. U. Lund (Sweden), 1944—. Chmn. com. on wavelength standards Internat. Astron. Union, 1952—; mem. Comité consultatif pour la définition du metre, 1952—, Nobel com. for physics, 1961—. Recipient Gold medal Royal Astron. Soc. London, 1945, Howard N. Potts gold medal Franklin Inst., 1946. Mem. Swedish (Arrhenius gold medal 1944), Danish, Norwegian acads. sci., Am. Acad. Arts and Scis., Phys. Soc. (London), Am. Phys. Soc., Optical Soc. Am. (C.E.K. Mees medal 1966). Research, publs. on measurements and term analyses of atomic spectra, especially in extreme ultraviolet, application of results to astrophys. problems, such as interpretation of spectra of hot stars and identification of emission lines in solar corona, implying temperature of 1 to 2 million degrees in extended outermost part of sun's atmosphere. Home: 38 Östervangsvägen, Lund. Office: 14 Sölvegatan, Lund, Sweden.*

EDLING, Nils Peter Gosta, Swedish radiologist; b. Bracke, Oct. 17, 1906; s. Per Gustaf and Anna (Nilsson) E.; ed. Ostersunds laroverk; M.D., U. Uppsala (Sweden), 1934; m. Marta Elisabeth Elofson, Apr. 2, 1938; 1 son, Peter. Became lectr. radiodiagnostics Karolinska Inst., Stockholm, Sweden, 1948, acting prof. radiodiagnostics, 1963-65; asst. radiologist Serafimerlasarettet, also Karolinska Sjukhuset, Stockholm, 1936-49; chief physician, roentgen diagnostic dept. Karolinska Sjukhuset, 1949—. Mem. Svenska Lakarsallskapet, Svensk Forening for Medicinsk Radiology, Svenska Radiologforbundet. Research and publs. on lower urinary tract, contrast medium excretion in kidneys (including autoradiography), nephrographic effect, renal damage by contrast medium examination of renal arteries, ulcerative colitis and risks of malignant degeneration, pneumonoconiosis, classification of bone lesions. Home: 5 St. Martins vag, Solna, Sweden. Office: Karolinska Sjukhuset, Stockholm 60, Sweden.*

EDLUND, Erik, Swedish physicist; b. 1819; prof. Stockholm, Sweden; head Tech. High Sch. Research and publs. on induction, polarization and telegraphy, super-current, relation between Peltier effect and elec. energy, atmospheric electricity. Died 1888.

EDLUND, Milton Carl, Am. physicist; b. Jamestown, N.Y., Dec. 13, 1924; s. Carl John and Ellen (Nyman) E.; B.S., U. Mich., 1948, M.S., 1948, Ph.D., in Nuclear Sci., 1966; postgrad. U. Tenn., 1949-51, Princeton, 1951-52; m. Lois Elkins, June 21, 1945; children—Randall Carl, Susan Ellen. Physicist, Union Carbide Corp., Oak Ridge, 1948-55; mgr. math. and physics dept., 1955-60, mgr. devel. dept., 1960-65; asst. div. mgr. atomic energy div. Babcock & Wilcox, Lynchburg, Va., 1965-66; prof. nuclear engring. U. Mich., Ann Arbor, 1966—. Mem. adv. com. reactor physics AEC, 1958-66. Recipient E. O. Lawrence award, AEC, 1965. Fellow Am. Nuclear Soc. (past dir.); mem. Nat. Assn. Mfg. (mem. research com. 1964—), Sigma Xi. Author: (with Samuel Glasstone) Elements of Nuclear Reactor Theory, 1952; also articles. Directed nuclear design and devel. Indian Point Nuclear power sta. and nuclear ship, Savannah reactors; invented concept of spectral shift control of nuclear reactors.*

EDMONDSON, Don Elton, Am. mathematician; b. Dallas, Sept. 6, 1925; s. Leroy B. and Lena (Upton) E.; B.S. Mech. Engring., 1945, M.S., 1948; postgrad. U. Chgo.; Ph.D., Cal. Inst. Tech., 1954; m. Blance Marian Scheffler, Sept. 8, 1951; children—D. Chris Cathy, David Goodwin, John Everett. Faculty, So. Meth. U., Dallas, 1949-51, 55-60, asso. prof., 1957-60; instr. Tulane U., 1954-55; faculty U. Tex., Austin, 1960—, prof., 1964—. Chmn., Conf. for Advancement Sci. and Math. Teaching, 1964; cons. Tex. Instruments, Tex. Edn. Agy. Mem. Am. Math. Soc., Math. Assn. Am., Tex. Acad. Sci. (pres. elect 1967), Soc. for Indsl. and Applied Math., Nat. Council Tchrs. Math., Am. Assn. U. Profs., Sigma Xi. Research in abstract and topological lattices. Home: 4427 Crestway St., Austin, Tex. 78731.*

EDMONDSON, Hugh Allen, Am. physician, educator; b. Maysville, Ark., Jan. 3, 1906; s. James Turner and Julia Ann (Phillips) E.; A.B., U. Okla., 1926; M.D., U. Chgo., 1931; m. Dorothy E. Mossman, July 14, 1930; children—Hugh Allen, James Paul, Marian Ann (Mrs. Brownell Merrell, Jr.), Marjorie Jean. Faculty dept. pathology U. So. Cal. Med. Sch., Los Angeles, 1934—, prof., 1948—, chmn. dept., 1951—; attending pathologist Los Angeles County Hosp.; cons. pathologist Childrens, Santa Fe hosps. Fellow A.C.P., Coll. Am. Pathologists; mem. A.M.A. (chmn. sect. pathology and physiology 1963-64), Internat. Acad. Pathology (council 1960-63), Am. Soc. Clin. Pathologists, Am. Soc. Pathologists and Bacteriologists, Cal., Los Angeles County med. assns., Cal. Soc. Pathology, Los Angeles Soc. Pathologists, Los Angeles Acad. Medicine. Contbr. numerous articles to profl. jours., chpts. to books. Research on diseases of gallbladder, biliary tract, kidney. Home: 4111 Circle Dr., San Marino, Cal. 91108. Office: 1200 N. State St., Los Angeles 90033.

EDMONDSON, W. Thomas, Am. limnologist; b. Milw., Apr. 24, 1916; s. Clarence Edward and Marie (Kelley) E.; B.S., Yale, 1938, Ph.D., 1942; m. Yvette

Hardman, Sept. 26, 1941. Teaching asst. zoology U. Wis., 1939, biology Yale, 1939-41; asst. phys. oceanographer Naval Ordnance Lab. contract Am. Mus. Natural History, 1942-43; research asso. Bur. Ships contract Oceanographic Inst., Woods Hole, Mass., 1943-46; lectr. biology Harvard, 1946-49; asst. prof. zoology U. Wash., Seattle, 1949-52, asso. prof., 1952-57, prof., 1957-—. Nat. Sci. Found. fellow, 1959-60. Mem. A.A.A.S., Ecol. Soc. Am., Am. Soc. Limnology and Oceanography, Internat. Assn. Theoretical and Applied Limnology. Editor: Ward and Whipple's Fresh Water Biology. Studies of the natural control of biol. productivity in aquatic communities. Office: Dept. Zoology, U. Wash., Seattle 98105.*

EDMONDSTON, Laurence, naturalist; b. Lerwick, Scotland, 1795; s. Laurence E.; studied medicine, Edinburgh; children include Thomas, Rev. Biot, Jessie Margaret (Mrs. Henry L. Saxby), Thomas. Practiced in Unst, Shetland. Mem. Royal Phys. Soc., Wernerian Soc.; hon. mem. Yorkshire Philos. Soc., Manchester Natural History Soc. Author: Research and publs. on natural history; added several birds to list of Brit. birds including snowy owl, Glaucus, Iceland and Ivory gulls. Died 1879.

EDMONSON, Munro Sterling, Am. anthropologist; b. Nogales, Ariz., May 18, 1924; s. Everett Sterling and Lillian (Munro) E.; B.A., Harvard, 1945, M.A., 1948, Ph.D., 1952; m. Barbara Wedemeyer, Aug. 1, 1953; children—Evelyn Mila, Ann Munro, Sallie Ross. Instr., Washington U., St. Louis, 1951; faculty Tulane U., New Orleans, 1951-—, prof. anthropology, 1960-—; vis. prof. U. San Carlos, Quezaltenango, Guatemala, 1960-61, Purdue U., West Lafayette, Ind., 1964, Harvard, 1965-66. Fellow Am. Anthrop. Assn.; mem. Am. Ethnol. Soc. (past pres.), A.A.A.S., Am. Assn. U. Profs. Author: (with John Rohrer) The Eighth Generation, 1959; also articles. Research on cultural values of Spanish-speaking New Mexicans, social life of N.Am. Indians, psychology of New Orleans Negroes, lang. and lit. of Mayan Indians, growth of Neolithic culture, blood groups and race, Am. Indian religious movements, New Orleans carnival, anthrop. theory and folklore. Home: 901 Cherokee St., New Orleans 70118.*

EDMUNDS, George Francis, Jr., Am. entomologist; b. Salt Lake City, Apr. 28, 1920; s. George Francis and Fern (Barratt) E.; B.S., U. Utah, 1943, M.S., 1946; Ph.D., U. Mass., 1952. Faculty, U. Utah, Salt Lake City, 1945-—, prof., 1959-—. Cons. Kaiser Aluminum & Chem. Corp., Kaiser Steel Corp., U. S. Steel Corp., Aluminum Co. Am. Mem. Entomol. Soc. Am., Soc. Study Evolution, Soc. Systematic Zoologists. Research, numerous publs. on classification, zoogeography, evolution and ecology of mayflies of world, ecology and population dynamics of certain forest insects in relationship to air pollution. Home: 3476 Virginia Way, Salt Lake City 84109.*

EDRIDGE-GREEN, Frederick William, Brit. ophthalmologist; b. Eng., 1863; ed. St. Bartholomew's Hosp., Durham U., St. John's Coll., Cambridge; M.B., M.D.; m. Minnie Hicks, 1893. Ophthalmic surgeon London Pensions Bds.; spl. examiner, adviser on vision and color-vision Ministry Transport; chmn. ophthalmic bd. Central London Med. Bds. Nat. Service; Beit Meml. research fellow. Recipient Thomas Gray Meml. prize for color perception lantern, 1936. Fellow Royal Coll. Surgeons Eng. (Hunterian prof., Arris and Gale lectr.); pres. Durham Med. Grads. Assn.; mem. Brit. Assn., Internat. Code of Signals Com. Author: Colour Blindness and Colour Perception, 1909; The Hunterian Lectures on Colour Vision and Colour Blindness, 1911; Colour Vision and Colour Blindness, 1911; Memory and its Cultivation, 1897; The Physiology of vision, 1920; Science and Psuedo-Science, 1933; Card Test for Colour Blindness; also articles in jours, Ency. Brit. Inventor color perception spectrometer, color perception lantern (ofcl. test of Brit. Navy), also bead test (ofcl. test of Brit. Nat. Service). Died Apr. 17, 1953.

EDRISI MOHAMMED, see Idrisi.

EDSALL, David Linn, Am. physician; b. Hamburg, N.J., July 6, 1869; s. Richard E. and Emma E. (Linn) E.; A.B., Princeton, 1890, Sc.D., 1913; M.D., U. Pa., 1893; Sc.D., Harvard U., 1928; m. Margaret Harding Tileston, 1899 (dec. 1912); children—John Tileston, Richard Linn, Geoffrey; m. 2d, Elizabeth Pendleton Kennedy, June 1915; m. 3d, Louisa C Richardson, 1931. Prof. therapeutics and pharmacology, U. Pa., 1907-10, medicine, 1910-11; prof. preventive medicine Washington U., 1911-12; Jackson prof. clin. medicine Harvard Med. Sch., 1912-23, dean, 1918-35, also dean Sch. Pub. Health, 1921-35, dean faculty medicine, Harvard Sch. Pub. Health emeritus, 1935-45; chief med. service Mass. Gen. Hosp., 1912-23. Fellow Am. Acad. Arts and Scis.; mem. Assn. Am. Physicians, Am. Philos. Soc. Contbr. to med. jours. Died Aug. 12, 1945.

EDSALL, Geoffrey, Am. immunologist; b. Phila., Jan. 28, 1908; s. David Linn and Margaret (Tileston) E.; student Harvard, 1925-28, 29-30, M.D., 1934; student U. Cal. at Berkeley, 1928; m. Helen Fogle, Aug. 13, 1935; 1 son, Richard Anthony. Research fellow Harvard Med. Sch., 1936-39; asst. dir. biologic labs. Mass. Dept. Pub. Health, Boston, 1939-42, dir., 1942-49, supt. Inst. Labs., 1960-—; prof. microbiology Boston U. Sch. Medicine, 1949-52; dir. immunology div. Walter Reed Army Inst. Research, 1951-56, communicable disease div., 1956-60; prof. applied microbiology Harvard Sch. Pub. Health, Boston, 1960-—; vis. prof. U. Cal., Berkeley, 1955; with OSRD, 1942-45. Cons., U. S. Army, 1960-—, USPHS, 1963-—; mem. expert panel on immunology WHO, 1963-—. Recipient U. S. Army Meritorious Service award, 1962. Mem. A.M.A., Am. Assn. Immunologists (past pres.), Am. Pub. Health Assn. (chmn. lab. sect. 1950), Am. Epidemiological Soc., Am. Acad. Arts and Scis. Contbg. author: Modern Trends in Immunology, 1963; Bacterial and Mycotic Diseases of Man, 1965. Research, numerous publs. on quantitative relationship between dose of diphtheria or tetanus toxoid and immune response, exptl. typhoid fever in chimpanzees, value of vaccine in preventing disease especially typhoid vaccine. Home: 85 Gate House Rd., Chestnut Hill, Mass. 02167. Office: 375 South St., Boston 02130.*

EDSALL, John Tileston, Am. biochemist; b. Phila., Nov. 3, 1902; s. David Linn and Margaret Harding (Tileston) E.; A.B., Harvard, 1923, M.D., 1928; postgrad. Cambridge U., Eng., 1924-26; m. Margaret Dunham, May 1, 1929; children—Lawrence Dunham, David Tileston, Nicholas Cranford. Tutor in biochem. sci., Harvard, 1928-—, chmn. bd. tutors in biochem. scis., 1931-57, prof. biochem., 1951-—; Fulbright vis. lectr. U. Cambridge, 1952, U. Tokyo, 1964; vis. prof. Coll. de France, Paris, 1955. Pres. 6th Internat. Congress Biochemistry, N.Y.C., 1964. Recipient Passano Found. award, 1966. Mem. Nat. Acad. Scis. Am. Philos. Soc., Am. Acad. Arts and Scis. (v.p. 1956-57), Am. Soc. Biol. Chemists (pres. 1957-58), Am. Chem. Soc., History of Sci. Soc., Deutsche Akademie Naturforscher. Author: (with Jeffries Wyman) Biophysical Chemistry, Vol. I., 1958, (with E. J. Cohn) Proteins, Amino Acids and Peptides, 1943. Editor: Advances in Protein Chemistry, 1944-—; Editor-in-chief Jour. Biol. Chem., 1958-68. Research and publs. on structure, function, size and shape of proteins, including those of muscle (myosin) and blood. Home: 3 Berkeley St., Cambridge, Mass. 02138.*

EDSON, William Alden, Am. electronic engr.; b. Burchard, Neb., Oct. 30, 1912; s. William Henry and Pearl (Montgomery) E.; B.S. in Elec. Engring. (Summerfield scholar), U. Kan., 1934, M.S. in Elec. Engring., 1935; D.Sc. (Gordon McKay scholar) Harvard, 1937; m. Saralou Peterson, Aug. 23, 1942; children—Judith Lynne, Margaret Jane, Carolyn Louise. Mem. tech. staff Bell Telephone Labs. Inc., N.Y.C., 1937-45; mem. faculty Ill. Inst. Tech., Chgo., 1941-42; prof., dir. sch. elec. engring. Ga. Inst. Tech., Atlanta, 1945-52; vis. prof. Stanford, 1952-55; mgr. microwave techniques Gen. Electric Microwave Lab., Palo Alto, Cal., 1955-61; dir. research, a founder Electromagnetic Tech. Corp., Mountain View, Cal., 1961-—, pres. 1962-—, also dir.; cons. to govt. and industry. Registered profl. engr., Cal. Fellow I.E.E.E.; mem. A.A.A.S., Am. Phys. Soc., Tau Beta Pi, Sigma Tau, Phi Kappa Phi, Eta Kappa Nu, Sigma Xi. Author: (with R. I. Sarbacher) Hyper and Ultrahigh Frequency Engineering, 1943; Vacuum Tube Oscillators, 1953. Research and publs. in electronics, especially frequency memory circuits, electronic oscillators, broadband amplifiers, resonators and filters. Home: 25346 La Loma Dr., Los Altos Hills, Cal. 94022. Office: 486 Ellis St., Mountain View, Cal. 94040.*

EDWARD, John Chinnayya, Indian plant pathologist; b. Sholapur, India, Oct. 10, 1920; s. Paul and Priscilla (Francis) E.; B.Sc., U. Madras (India, 1941, B.Sc.Agr., 1945, Ph.D., 1952; postgrad. nematology, Holland; m. Premalata Shanthappa, May 4, 1955; 1 son, Deepak Paul. Sci. Tchr. Mission Sch., Gulbarga, India, 1941-42; dir. agrl. work Meth. Mission, Medak, India, 1945-48; research scholar Madras U., 1948-50, Govt. of India Research scholar, 1950-53, faculty, 1953-64, asso. prof. biology dept., 1959-64; prof., head biology dept. Allahabad (India) Agrl. Inst., 1964-—, coordinator research, 1961-—. Recipient Pearce prize Madras U., 1941; Sir Venkataraman Gold medal, 1943. Rockefeller Found. fellow plant pathology dept. U. Minn., Mpls., 1958-59. Mem. Soc. European Nematologists. Author: Manual on Weeds, 1955; Fungi, 1957; Plant Parasitic Nematodies, 1960; also articles. Editor: Allhabad Farmer, 1955-58. Research on fungi and nematodes causing plant diseases, taxonomy of fungi and nematodes, including discovery and description of some; studied Guava wilt disease and its control using resistant root stocks, plant parasitic nematodes and fungi asso. with rhizospheres of field and fruit crops, fungi antagonistic to pathogenic fungi. Home: 7 down, P.O. Agrl. Inst., Allahabad U.P., India.*

EDWARD, Thomas, Brit. naturalist; b. Gospart, Hampshire, Eng., Dec. 25, 1814; m. Sophia Reid, 1837. Curator Mus. Banff Inst. until 1882; mem. Linnean Soc., Aberdeen and Glasgow Nat. Hist. Soc. Conducted marine zool. research; studied habits, behavior of birds; credited with discovery of 20 new species of sessile-eyed crustacea. Died Apr. 27, 1886.

EDWARDS, Cecile Hoover, Am. nutritionist; b. East St. Louis, Ill., Oct. 20, 1926; d. Ernest Jack and Annie (Jordan) Hoover; B.S., Tuskegee Inst., 1946, M.S. (Swift and Co. Carver Found. fellow), 1947; Ph.D., Ia. State U. Sci. and Tech., 1950; m. Gerald Alonzo, June 29, 1951; children—Gerald Alonzo, Adrienne Annette, Hazel Ruth. Research asso., asst. prof. foods and nutrition, Carver Found., Tuskegee Inst., 1950-56, head dept. foods and nutrition, 1952-56; prof. nutrition and research Agrl. and Tech. Coll. N.C., Greensboro, 1956-—, dir. research projects NIH, 1956-—. Recipient award for contributions to sci. Nat. Council Negro Women, 1963. Diplomate Am. Bd. Nutrition. Mem. Am. Inst. Nutrition, Am. Dietetic Assn., N.C. Acad. Sci., Nat. Inst. Sci., Am. Assn. U. Women, Am. Tchrs. Assn., Sigma Xi, Beta Kappa Chi, Sigma Delta Epsilon, Iota Sigma Pi, Phi Kappa Phi, Alpha Kappa Mu. Research and publs. on amino acid compositions of various foods, pica, utilization of methionine, interrelationships of methionine and vitamin B12, diets of elementary sch. children, pregnant women, low income rural So. families, adolescent youth, utilization of wheat in adult human subjects. Home: 2711 McConnell Rd., Greensboro, N.C. 27401.*

EDWARDS, Corwin D., Am. economist; b. Nevada, Mo., Nov. 1, 1901; s. Granville Dennis and Ida (Moore) E.; B.A., U. Mo., 1920, B.J., 1921; B.Litt., Oxford U. (Eng.), 1924; Ph.D., Cornell U., 1928; m. Janet Ward, Dec. 20, 1924 (div. June 1947); children—Ward Dennis, Corinne Jennifer (Mrs. Gerald Greenwald); m. 2d, Gertrud Greig, Feb. 15, 1948. Tech. dir. Consumers Adv. Bd., NRA, Washington, 1934-35, coordinator trade practice studies, 1935; economist Pres.'s Com. Indsl. Analysis, Washington, 1936; asst. chief economist, FTC, Washington, 1937-39; economist, chmn. policy bd. antitrust div. Dept. Justice, Washington, 1939-44; faculties Northwestern U., 1944-48, Cambridge U. (Eng.), 1953-54, U. Va., 1954-55, U. Chgo., 1955-63; dir. Bur. Indsl. Econs., FTC, Washington, 1948-53; prof. econs. U. Ore., Eugene, 1964-—. Mem. numerous govt. missions; cons. Pres.'s Asst. for Consumer Affairs, 1965-66. Mem. Am. Econ. Assn. (v.p. 1951). Pioneered in comparative internat. analysis of antitrust policies; comprehensive econs. study of Am. price discrimination law. Home: 2355 Van Ness St., Eugene, Ore. 97403.*

EDWARDS, Edward, Welsh marine zoologist; b. Corwen, Wales, Nov. 23, 1803; ed. at Corwen; draper, Bangor, Wales, until 1839; founder foundry and ironworks, Menai Bridge; began study habits and characteristics of fish, 1864. By copying natural conditions under which fishes flourished, improved constrn. aquaria, which included invention darkwater chamber slope-back tank. Died Aug. 13, 1879.

EDWARDS, Edward Allen, Am. physician, anatomist; b. East Lyme, Conn., July 30, 1906; s. Max and Nellie (Gordon) E.; student Tufts U., 1922-24, M.D., 1928; m. Elizabeth Betsy Borwick, May 31, 1931; children—Alicia Betsy, Frederic Thomas. Practice surgery specializing in blood vessels of limb, Boston, 1931-—; faculty Harvard Med. Sch., 1934-—, asso. clin. prof. anatomy, 1955-—; staff Peter Bent Brigham Hosp., Boston, 1950-—, sr. asso. in surgery, 1957-—. Cons. to various hosps.; impartial examiner Mass. Indsl. Accident Bd., 1951-—; mem. adv. bd. Am. Heart Assn., 1938-—. Diplomate Am. Bd. Surgery. Fellow A.C.S.; mem. Soc. for Vascular Surgery, Internat. Cardiovascular Soc., Am. Assn. Anatomists, New Eng., Boston surg. socs. Author: Thrombosis in Arteriosclerosis of the Lower Extremities, 1950; also numerous articles. Research on anatomy and pathology of venous valves, arteries of hand, foot and muscles, collaterals of portal veins, systemic veins and arteries, spectrophotmetry of living skin with effects of testicular and ovarian function, dynamics of arterial closure, application of phys. methods to clin. study vascular disease. Home: 44 Edge Hill Rd., Brookline, Mass. 02146. Office: 110 Francis St., Boston 02215.

EDWARDS, George, English naturalist; b. Stratford, Eng., April 3, 1693; ed. pub. sch., Leytonstone; apprenticed in Fenchurch St.; traveled to Holland, Norway, France; returned to Eng. where he began to make and sell colored drawings of animals; apptd. librarian Royal Coll., Physicians, 1788-64. Recipient Gold medal Royal Soc., 1750. Fellow Royal Soc., 1750, Soc. Antiquaries. Author: Natural History of Birds, 3 vols., 1743-50; Natural History of Uncommon Birds and other Rare and Undescribed Animals, 4 vols., 1743-51; Gleanings of Natural History, 3 vols., 1758-64; Essays of Natural History, 1770; Elements of Fossilology, 1776. Drew nearly 600 birds and animals. Contbd. to knowledge European birds and animals; used Linnean names in his listings. Died London, July 23, 1773.

EDWARDS, Gerald Alonzo, Am. chemist; b. Henderson, N.C., Nov. 22, 1921; s. Gaston Alonzo and Catherine Ruth (Norris) E.; B.S., N.C. Coll., 1941; Ph.D., U. Buffalo, 1951; m. Cecile Annette Hoover, June 29, 1951; children—Gerald Alonzo, Adrienne Annette, Hazel Ruth. Chemist, shift supr. labs. Lake Ont. Ordnance Works, 1942-43; chem. operator Hooker Electrochem. Co., Niagara Falls, N.Y., 1943-45; chemist Bell Aircraft Corp., Niagara Falls, 1945-47; grad. asst. U. Buffalo, 1947-50; instr. dept. chemistry N.C. Coll. at Durham, 1950-51; research asso. Carver Found. Tuskegee Inst., 1951-56, asst. prof. chemistry, 1951-55, asso. prof., 1955-56; prof., chmn. dept.

Agrl. and Tech. Coll. N.C., Greensboro, 1956—. Dir. various NSF insts.; cons. Columbia U. Tchrs. Coll.-AID India Program, 1964-65; cons. Ohio State U.-AID Regional Coll. Edn., Mysore, India, 1966—. Mem. Am. Chem. Soc., Nat. Inst. Sci., Am. Assn. U. Profs., N.C. Acad. Sci., Sigma Xi. Research, publs. on utilization of Methionine by adult rat; chemistry of polymers; transport of amino acids. Home: 2711 McConnell Rd., Greensboro, N.C. 27401.*

EDWARDS, Hardy Malcolm, Jr., Am. nutritional biochemist; b. Ruston, La., Nov. 16, 1929; s. Hardy M. and Lillian (Egan) E.; B.S., Southwestern La. U., 1949; M.S., U. Fla., 1950; Ph.D., Cornell U., 1953; m. Aldies Olafson, Mar. 11, 1954; 1 son, Hardy Malcolm III. Research biochemist Internat. Minerals and Chem. Corp., Chgo., 1954-55, sr. research biochemist, 1955-57; faculty U. Ga., Athens, 1957—, prof. poultry sci., 1966—. Research asso. Oak Ridge Inst. Nuclear Studies, 1958, U. Lund (Sweden), 1964-65; mem. animal nutrition com. NRC-Nat. Acad. Sci. Recipient Gamma Sigma Delta Sci. Research award, 1962; Nutrition Research award Am. Feed Mfrs. Assn., 1962. Mem. Soc. Exptl. Biology and Medicine, Am. Inst. Nutrition, Poultry Sci. Assn., Animal Sci. Assn., Sigma Xi, Phi Kappa Phi, Gamma Sigma Delta. Research and publs. on contbn. of fatty acids of specific chem. structure to well being of animal organism, role of fatty acids in metabolism of phospholipids in liver cells, devel. of chem. methods for studying fatty acid structure, determination of kinetics of cholesterol metabolism in mature fowl, absorption and metabolism of various essential trace elements. Home: Route 1, Winterville, Ga. 30683. Office: Poultry Sci. Dept., U. Ga., Athens, Ga. 30601.*

EDWARDS, Henry, entomologist; b. Herefordshire, Eng., Aug. 27, 1830; ed. London, Eng., came to U. S., 1867; actor attached to Wallach Theatre; collector lepidoptera in Australia, Peru, Panama. Author: Bibliographical Catalogue of the Described Transformations of North American Lepidoptera, 1889; also contbr. articles to sci. jours. Editor: Papilio, 1881-83. Donated collection of 250,000 specimens to Am. Mus. Natural History, N.Y.; pioneered in effects of insects on crops. Died June 9, 1891.

EDWARDS, Howard Dawson, Am. physicist; b. Athens, Ga., Dec. 11, 1923; s. Howard Thomas and Inez (Glenn) E.; B.S., U. Ga., 1944; Ph.D., Duke, 1950; m. Mary Lemon, Mar. 10, 1961; children—David, Jerry, David Bradley (stepson), Kathy. Chief, atmospheric energy br. Air Force Cambridge Research Labs., Bedford, Mass., 1949-55; operations research scientist Lockheed Aircraft Corp., Marietta, Ga., 1956-58; faculty Ga. Inst. Tech., Atlanta, 1959—, prof. aerospace engring., 1965—, dir. Space Scis. Lab. Sch. Aerospace Engring., 1963—. Pres., Space Instruments Research, Inc., 1963—. Mem. Am. Geophys. Union, Am. Inst. Aeros. and Astronautics, Ga. Acad. Sci. (pres. 1966). Research, publs. on physics and chemistry of upper atmosphere and space using sounding rockets and balloons. Home: 365 Amberidge Trail, Atlanta 30328.*

EDWARDS, Jesse Efrem, Am. pathologist; b. Hyde Park, Mass., July 14, 1911; s. Max and Nellie (Gordon) E.; B.S., Tufts Coll., 1932, M.D., 1935; m. Marjorie Helen Brooks, Nov. 12, 1952; children—Ellen Ann, Brooks Sayre. Faculty Boston U. Sch. Medicine, 1938, Tufts Coll. Med. Sch., 1939-40; cons. pathology Mayo Clinic, Rochester, Minn., 1946-60; dir. labs. Charles T. Miller Hosp., St. Paul, 1960—; faculty U. Minn., Mpls., 1946—, prof. pathology Grad. Sch., clin. prof. pathology Med. Sch., 1960—. Cons. to Surgeon Gen., 1947-67. Recipient Modern Medicine award, 1965. Mem. A.M.A., Am. Heart Association (vice pres. 1965-66, pres. 1967-68), International Academy of Pathology (president 1956-57), Am. Assn. Pathologists and Bacteriologists, Am. Soc. Exptl. Pathology, Coll. Am. Pathologists, Am. Acad. Pediatrics. Author: An Atlas of Acquired Diseases of the Heart and Great Vessels, 3 vols., 1961; (with others) Congenital Anomalies of the Heart and Great Vessels, 1948, An Atlas of Congenital Anomalies of the Heart and Great Vessels, 1954; (with R. S. Fontana) Congenital Cardiac Disease. A Review of 357 Cases Studied Pathologically, 1962; (with C. A. Wagenvoort and D. Heath) The Pathology of the Pulmonary Vasculature, 1964; (with J. R. Stewart, O. W. Kincaid) An Atlas of Vascular Rings and Related Malformations of the Aortic Arch System, 1964; (with L. S. Carey, H. N. Neufeld and R. G. Lester) Congenital Heart Disease. Correlation of Pathologic Anatomy and Angiocardiography, 2 vols., 1965. Mem. editorial bd. Circulation, 1956-67, Laboratory Investigation 1950-66, Am. Heart Jour., 1967—, Am. Jour. Cardiology, 1968—. Numerous publs. on cardiovascular anatomy of acquired and congenital diseases and their clin. manifestations; function of pulmonary blood vessels in diseases with septal defects. Home: 1565 Edgecumbe Rd., St. Paul 55116.*

EDWARDS, John Hilton, English human geneticist; b. London, Eng., Mar. 26, 1928; s. Harold Clifford and Ida (Phillips) E.; B.A., Cambridge U., 1949, M.B., B.Chir., 1952, M.R.C.P., 1956; m. Felicity Clare Troussaint, July 18, 1953; children—Vanessa, Conrad, Penelope, Matthew. Lectr., Birmingham (Eng.) U., 1956-58, 61-65, reader in human genetics, 1966—; sci. staff Med. Research Council Unit on Population Genetics, Oxford, Eng., 1958-60; geneticist Children's Hosp. Phila., 1960-61; cons. geneticist Children's Hosp., Birmingham, 1965—. Cons. investigator AEC, Iceland, 1965—. Mem. Genetical Soc., Biometric Soc., Royal Statis. Soc., Royal Soc. Medicine, Brit. Paediatric Assn., Royal Coll. Physicians. Editorial bd. Cytogenetics, 1962—, Brit. Jour. Human Genetics, 1963—. Research and publs. on devel. and application of lab. and computational techniques to study disease especially chromosomal disorders, congenital malformations, and multiple pregnancies and abortions. Home: 82 Oxford Rd., Birmingham 13. Office: Inst. Child Health, Francis Rd., Birmingham 16, Eng.*

EDWARDS, Joshua Leroy, Am. pathologist; b. Jasper, Fla., Aug. 9, 1918; s. Joshua Leroy and Julia (Miller) E.; B.S., U. Fla., 1939; M.D., Tulane U., 1943; m. Jeane Augusta Perrin, July 7, 1953; children—Julia Elisabeth, Jean Augusta, Joshua Leroy III. Gen. practice medicine, Lake City, Fla., 1946-48; teaching fellow Harvard, 1949-51; instr. Duke Med. Sch., 1951-53; asst. prof. Rockefeller Inst., 1953-55; faculty U. Fla. Coll. Medicine, Gainesville, 1955-65; prof. pathology, dir. combined degree program in med. scis. Ind. U., Bloomington, 1965—. Mem. Am. Soc. Exptl. Pathology, Internat. Acad. Pathology, Am. Soc. Cell Biology, A.A.A.S., Tissue Culture Assn., N.Y. Acad. Scis., Fla. Acad. Sci., Fla. Soc. Pathology. Research on cytology and immunology. Home: 1319 E. 10th St., Bloomington, Ind. 47401.*

EDWARDS, Martin Hassall, physicist; b. St. Anne's-on Sea, Eng., Nov. 10, 1927; s. Thomas Maddock and Daisy Evelyn (Harvey) E.; B.A., U. B.C., 1949, M.A., 1951; Ph.D., U. Toronto, 1953; m. Norma Edith Bloom, Sept. 3, 1949; children—David, Barbara. Research asso. U. Toronto (Ont., Can.), 1953-54; asst. prof. physics Royal Mil. Coll., Kingston, Ont., 1954-56, asso. prof., 1956-61, prof., 1961—; research asso. U. Pa., summer, 1959, Stanford U., 1964-65. Dir. Fedn. Ont. Naturalists, 1961—, v.p., 1966-68, pres., 1968—. Fellow Am. Phys. Soc.; mem. Canadian Assn. Physicists, Am. Assn. Physics Tchrs. Research in exptl. low temperature physics, properties of He4 liquid and vapor, expansion coefficient, refractive index and density, critical point. Home: 19 Jane Av., Kingston, Ont., Can.*

EDWARDS, Philip R., Am. bacteriologist; b. Owensboro, Ky., Aug. 30, 1901; s. Elza and Gelishia (Rarick) E.; B.S., U. Ky., 1922, D.Sc. (hon.), 1959; Ph.D., Yale, 1925; m. Katherine Brewer, Sept. 20, 1927; children—Katherine (Mrs. D. W. McClanahan), Dorothy (Mrs. H. E. Ivey). Bacteriologist, Ky. Agrl. Expt. Sta., Lexington, 1925-48; chief, enteric bacteriology unit Communicable Disease Center, Atlanta, 1948-62, chief, bacteriology sect., 1962—. Recipient Distinguished Service award U. S. Dept. Health, Edn. and Welfare, 1955; Kimble Research award, 1956. Mem. Am. Soc. for Microbiology (pres. 1959), Am. Pub. Health Assn., Soc. for Exptl. Biology and Medicine, Research Soc. Am., Sigma Xi, Gamma Alpha. Author: (with W. H. Ewing) Identification of Enterobacteriaceae. Research on intestinal infections in man and animals. Home: 207 Upland Rd., Decatur, Ga. 30030. Office: Communicable Disease Center, Atlanta 30333.*

EDWARDS, Raymond Richard, Am. chemist; b. Ft. Smith, Ark., Dec. 18, 1917; s. David Weems and Alva (Coleman) E.; B.A., U. Ark., 1939; Ph.D., Mass. Inst. Tech., 1948; m. Ruth Skinner, Sept. 7, 1961. Faculty, U. Ark., 1948-60, prof., chmn. dept. chemistry, 1951-56, dir. Grad. Inst. Tech., 1957-60; asst. dir. tech. programs div. Internat. Affairs U. S. AEC, 1956-57; guest research staff Lab. Nuclear Sci. Mass. Inst. Tech., 1961; tech. dir. Nuclear Sci. & Engring. Corp., Pitts., 1961-65; sr. research chemist Carnegie-Mellon U., Pitts., 1965—; chmn. subcom. on nuclear geophysics Nuclear Sci. Com. NRC-Nat. Acad. Sci. Mem. Am. Chem. Soc., Am. Phys. Soc., Am. Nuclear Soc., Am. Geophys. Union, N.Y. Acad. Scis., A.A.A.S., Sigma Xi, Alpha Chi Sigma. Research and publs. on chem. effects of nuclear transformations; inorganic reaction mechanisms; natural fission; isotope ratio variation; radioactivity; application of tracers to biochem. problems; the Mossbauer effect. Home: 220 Colony Ct., Pitts. 15205. Office: Carnegie-Mellon U., Pitts. 15213.*

EDWARDS, Robert Lomas, Am. ecologist; b. Phila., Aug. 24, 1920; s. Robert Davis and Francis (Lomas) E.; student Cornell U., 1939-40; B.S., Colgate U., 1947; A.M. (teaching fellow); Harvard, 1949, Ph.D., 1951; m. Sylvia Bitler Pierce, May 27, 1942; children—Carol, Eric, Susan, Anabel. Instr. Tufts. Coll., 1949-50, Brandeis U., 1950-53; ecol. research Canadian Arctic, 1954; chief research program N.A. Fisheries Investigation; Woods Hole, Mass., 1955-59; asst. dir. Bur. Commn. Fisheries, Biol. Lab., 1959—. Mem. Am. Soc. Mammalogists, Am. Soc. Parasitologists, Am. Fisheries Soc., Wilson Club (life), Soc. Am. Archealogy, Sigma Xi. Research and publs. on evolution bird ectoparasites, ecology vertebrates Canadian Arctic.*

EDWARDS, Ward, Am. psychologist; b. Morristown, N.J., Apr. 5, 1927; s. Corwin D. and Janet (Ward) E.; B.A., Swarthmore Coll., 1947; M.A., Harvard U., 1950, Ph.D., 1952; m. Ruth G. Page, Aug. 28, 1949; children—Tara Anne, Page Corwin. Instr. Bklyn. Coll., 1947-48; lectr. Boston U., 1948-49; teaching fellow Harvard U., 1949-51; instr. Johns Hopkins U., 1951-54; research psychologist Personnel and Tng. Research Center U. S. Air Force, 1954-56, chief intellectual functions sect., 1956-58; faculty U. Mich., 1958—, prof. psychology, 1963—, head engring. psychology lab. Inst. Sci. and Tech., 1963—; cons. N. Am. Aviation, System Devel. Corp., Santa Monica, Cal., GE-TEMPO, Santa Barbara, Cal., Army Personnel Research Office, Human Resources Research Office George Washington U., Consad, Inc. IBM, Dunlap & Asso., RAND Corp. Mem. Eastern, Am., Midwestern psychol. assns., Psychonomic Soc., Sigma Xi. Research and publs. on human information processing, human decision making. Home: 3111 Washtenaw Rd., Ann Arbor, Mich.*

EDWARDS, Willard Eldridge, Am. elec. engr.; b. Chatham, Mass., Dec. 11, 1903; s. Arthur Robbins and Mabel (Eldridge) E.; student Mass. Inst. Tech., 1922-26; B.S., U. Okla., 1929; postgrad. U. So. Cal., 1939-40; M.S., Jackson Coll., 1960, M.A., 1961, Litt.D., 1960; m. Dorothy Shiell, June 13, 1942; children—Willard Eldridge, Annabelle, Arthur, Jo-Anna. With research dept. RCA, 1926-28, engring. dept. Alexander Eaglerock Aircraft Co., Colorado Springs, 1929, telephony and radio-telephony, devel. Am. Tel. & Tel. Co., Denver, Dixon, Cal., 1929-33; with radio broadcast devel., NBC Radio Stas., Los Angeles, 1933-40; elec. designer Lockheed Aircraft Co., Burbank, Cal., 1940-41; electronics design and installation, CAA and U. S. Navy, 1941-53; corrosion control design and cons., Honolulu, since 1960—. Tchr., Jackson Coll., Honolulu, 1959-61. Mem. I.R.E., A.A.A.S., Am. Inst. E.E., Am. Radio Relay League, Nat. Assn. Corrosion Engrs., Soc. Am. Mil. Engrs., Aero. Kokua Club, Hawaiian Astron. Assn. Author: The Perpetual Calendar, 1943; American Samoa, 1949; Research and publs. on devel. electronic devices especially timing devices, safety devices, corrosion control, radar. Home and office: 3038 Oahu Av., Honolulu 96822.*

EDWARDS, William, Am. inventor, tanner; b. Elizabethtown, N.J., Nov. 11, 1770; s. Timothy and Rhoda (Ogden) E.; m. Rebecca Tappan, 1793, 11 children. Built tannery which was for years largest in U. S., 1815; settled in Hunter, N.Y., 1817; opened model tannery; a founder hide and leather industry in U. S.; introduced process used in 11 Am. tanneries by which leather was made in one-fourth time used in European process; invented a leather-rolling machine to save hammering labor; reduced cost of tanning sole leather from 12c to 4c a pound. Died Bklyn., Dec. 29, 1851.

EDWARDS, William Henry, Am. entomologist, naturalist; b. Hunter, N.Y., Mar. 15, 1822; s. William W. and Helen Ann Mann E.; grad. Williams Coll., 1842; admitted to N.Y. bar, 1847; made voyage up the River Amazon, 1846, collecting objects of natural history. m. Catherine Colt Tappan, May 29, 1851. Author: Voyage up the River Amazon, 1847 (describes people, vegetation, climate, animals); The Butterflies of North America, 18 vols., illustrated (3 series), 1879, 1884, 1897. First naturalist from U. S. to travel and write his experiences in S.Am.; collected animals along Amazon. Died April 4, 1909.

EDWARDS, William Sterling, Am. surgeon; b. Birmingham, Ala., July 23, 1920; s. W. Sterling and Elizabeth (Wyman) E.; B.S., Va. Mil. Inst., 1942; M.D., U. Pa., 1945; m. Ann Rorer Dudley, July 13, 1946; children—Bruce, Peter, Catherine, Wyman. Faculty, Med. Coll. Ala., Birmingham, 1952—, prof. surgery, 1962—. Cons. surgeon VA Hosp., Montgomery, also Birmingham, 1956—. Mem. Am., So. surg. assns., Am. Assn. for Thoracic Surgery, Soc. for Vascular Surgery, Internat. Cardiovascular Soc., Author: Plastic Arterial Grafts, 1957; also numerous articles. Invented crimped nylon and teflon synthetic arterial graft; developed new techniques for reconstructing obstructed femorial arteries, palliatiative surgery in congenital heart disease of infants. Home: 3835 S. Cove Dr., Brimingham, Ala.*

EECKHAUT, Zoël Prosper, Belgian chemist; b. Balegem, Belgium, Dec. 14, 1909; s. Polydor and Maria (van Herpe) E.; Chem. Engr. with gt. honors, U. Ghent (Belgium), 1933, Dr.Applied Scis., 1940; m. Emilia Delmulle, July 27, 1937; 1 dau., Liliane (Mrs. Johan Haemers). With U. Ghent, 1934—, prof. analytical chemistry, 1953-57, prof. gen. and inorganic chemistry, 1957—, head lab. analytical chemistry, 1949-57, head lab. gen. and inorganic chemistry, 1957—. Co-worker several research groups C.O. B.E.A., 1945-49, C.N.R.M., 1947-53, I.I.K.W., 1949-61. Decorated officier de Orde van Leopold II, 1952; commandeur de Kroonorde, 1961; recipient Burgerlijkmedaille 1° Klas, 1961, prijs voor chemie Kon. Vl, Akad. voor Wet., 1940. Mem. Natuur-en Geneeskundige Vennootschap, Koninklijke Vlaamse Ingenieursvereniging, Vlaamse Chem. Vereniging. Author 3 books, numerous articles. Analysis and determination of different elements by spectrochem. analysis, polarography, spectrophotometry, activation analysis; equilibrium studies of oxidation; reduction systems; determination of stability constants of metal complexes; structure and chem. bond of halogenated aliphatic compounds and organo-metal compounds. Home: 20, E. Anseelelaan, Gentbrugge, Belgium. Office: 35, K. Ledeganck, Ghent, Belgium.*

EFFLER, Donald Brian, Am. thoracic surgeon, b. N.Y.C., Aug. 15, 1915; s. Louis R. and Marie (Kennelly) E.; A.B., U. Mich., 1937, M.D., 1941; m. Jo-

Anna Stemplinski, Oct. 12, 1943; children—Gretchen, Donald Brian, Brian Donald. Practice medicine, specializing in thoracic surgery, Cleve., 1948—; chief dept. thoracic, cardiovascular surgery Cleve. Clinic Found., 1950—, also bd. govs., trustee. Mem. Bd. Thoracic Surgery. Mem. editorial bd. Jour. of Thoracic and Cardiovascular Surgery, 1964—. Research, publs. and chpts. in books on devel. of modern cardiac surgery employing extracorporeal circulation, surg. revascularization of ischemic myocardium; performed first stopped heart operation. Home: 20001 S. Woodland St., Cleve. 44122. Office: Cleve. Clinic Found., 2020 E. 93d St., Cleve. 44106.*

EFRAIM IBN AL-ZAFFAN, see al-Zaffan.

EGAMI, Nobuo, Japanese biologist; b. Kanazawa City, Japan, Jan. 5, 1925; s. Hideo and Fumiko Egami; student Tokyo (Japan) U., 1944-47, Dr.Sci., 1956; m. Kazuko Egami, Mar. 30, 1952 children—Shigeo, Michiko. Asst., U. Tokyo, 1949-50, lectr. zoolcgy, 1950-61, lectr. Faculty Sci.; chief 1st lab. div. biology Nat. Inst. Radiol. Scis., Chiba, Japan. 1961-64, head div. biology, 1964—. Mem. Zool. Soc. Japan, Japan Radiation Research Soc. Research, numerous publs. on biol. effects of hormonic substances on fish reproductive organs, effects of ionizing radiation on cell population in vertebrate animals from endocrinological, cytological and biochem. view points. Home: 2-2-85, Yawata, Ichikawa, Japan. Office: 4, Anagawa, Chiba, Japan.

EGAS MONIZ, see de Egas Moniz.

EGEBERG, Olav, Norwegian hematologist; b. Trogstad, Norway, May 22, 1916; s. Ludvig and Asta (Aamodt) E.; Degree in Math., Oslo (Norway), 1938, Degree in Economy and Statistics, 1943, M.D., 1952; Degree in State Medicine, Gothenburg U. (Sweden), 1952; Dr.'s degree Oslo U., 1964; m. Signe Wollo, Sept. 3, 1949; children—Gunnar, Kjartan, Tore, Karin. Research asst. Oslo U., 1936-41; sec., cons., statistician Norwegian Dept. Supply and Reconstrn., Oslo, 1945-47; practice medicine, Norway, Sweden, 1952-57; physician Inst. for Thrombosis Research med. dept. A, U. Hosp., Rikshospitalet, Oslo, 1957—. Mem. Norwegian Med. Assn., Norwegian Assn. for Internal Medicine, Internat. Soc. Hematology. Contbg. author: The Hemophilias, 1964. Research, publs. in human blood coagulation and hemostasis; demonstrated various coagulation hyperactivity reactions in conditions asso. with high frequency of thrombosis; showed blood antihemophilic globulin (factor VIII) assay values are correlated with changes in body metabolic rate; discovered blood antithrombin deficiency, which is accompanied by increased tendency to thrombosis. Home: 1 Erich Mogensonsvei, Oslo 5. Office: Rikshospitalet, Oslo, 1, Norway.*

EGELER, Cornelis Geoffrey (Kees), Dutch geologist; b. Jan. 9, 1917; s. Rudolph and Kolff Egeler; Ph.D. in Natural Scis., U. Amsterdam; m. Marie-Louise ter Haar; children—Lars-Adriaan, Fridtjof, Michiel, Marie-Louise, Rudolph-Maarten, Diederik. Asst., curator Geologic Inst.; prof. U. Amsterdam. Mem. Netherlands Expdns. to Andes, 1952, 56 59; head Netherlands expdn. to Himalayas, 1962. Mem. Netherlands Council Alpenverenig, Netherlands Geol. Council Mijnb Gen., Netherlands Council Aandr. Gen. Research and numerous publs. on geology of Indonesia, Corsica, Peru, S.E. Spain. Home: Gravelandseweg 180, Hilversum. Office: Geologic Institute, Nieuwe Prinsengracht 180, Amsterdam, Netherlands.

EGER, Ernest, engr.; b. Kromeriz, Czechoslovakia, Jan. 12, 1892; s. Nathaniel and Rosalie (Weiss) E.; ed. engring. sch., Brno, Czechoslovakia, degree in mech. engring., 1912; m. Betty Hansell, Oct., 1920; children—Herbert, Edgar. Came to U. S., 1913, naturalized, 1920. With rubber industry for 20 years; leading scientist in devel. of self-sealing gas tanks (designed to withstand machine-gun bullets); patentee in rubber (over 50 patents in U. S. and other countries), including rubber-cutting machines; liquid weighing device, Eger valve of vented tubes and rubber valves, stell-grip airplane tires, non-metallic replacements for sheet aluminum.

EGER, Wilhelm, German physician; b. Apr. 26, 1909; s. William end Babetta N. Eger; M.D., U. Göttingen; m. Annemarie Hensel; children—Knut, Wilhelm, Friedericke. Asst. at Breslau, 1936-45; practice medicine, Mainbernheim, 1945-51; asst., chief physician Inst. Pathology, Göttingen, 1951—. Mem. Arbeitsgemeinschaft für Histochemie. Research and over 150 publs. on skeletons, parathyroid, organic tissues, histochemistry. Home: Beuferzweg 9. Office: Goszlerstrasse 10, Göttingen, Germany.

EGERTON, Sir Philip De Malpas Grey, English paleontologist; b. Chesire, Eng., Nov. 13, 1806; s. Philip Grey Egerton; B.A., Eton and Christ Church, Oxford (Eng.) U., 1828; m. Anna Elizabeth Legh, 1832 (dec. 1882); 2 sons, 2 daus. Traveled through Germany, Switzerland, Italy, studying fossil fishes; elected to Parliament for Chester, 1830, for West Chesire, 1868. Trustee Brit. Mus. Recipient 1st Kingley medal Chester Soc. Natural Sci., 1879. Fellow Geol. Soc. (Wolloston medal 1873), Royal Soc., 1831; mem. Royal Coll. Surgeons (trustee). Author: Alphabetical Catalogue of Type Specimens of Fossil Fishes, 1871; also numerous articles. Collected numerous fossil fishes (many later described by Agassiz). Died London, Eng., Apr. 5, 1881.

EGGAN, Fred R(ussell), Am. anthropologist; b. Seattle, Sept. 12, 1906; s. Alfred Julius and Olive M. (Smith) E.; Ph.B., U. Chgo., 1927, A.M., 1928, Doctor of Philosophy, 1933; m. Dorothy Way, Aug. 9, 1938 (dec. 1966). Began his career as a research asso. in Philippine ethnology U. Chgo., 1934-35, instr. anthropology, 1935-40, asst. prof., 1940-42, asso. prof., 1942-48, prof., 1948—, chmn. dept., 1948-52, 61-63, dir. Philippine studies program 1953—. Harold H. Swift Distinguished Service prof., 1963—; Morgan lectr. U. Rochester, 1964. Official U. S. delegate to the 8th Pacific Sci. Congress, Manila, 1953, 9th, Bangkok, 1957, 10th Honolulu, 1961, 11th, Tokyo, Japan, 1966; member Pres.'s Com. on Scientists and Engrs., 1956-57; chmn. bd. Human Relations Area Files, Inc., 1964-67; research asso. Lab. Anthropology, Santa Fe, 1964—; adv. com. social scis. NSF, 1961-62; hon. cons. Bernice P. Bishop Mus., Honolulu, Hawaii; member Commission on Coll. Geography, 1965-67; councillor Smithsonian Instn., 1966—. Served as capt., AUS, 1943; dir. Civil Affairs Tng. Sch. for Far East, U. Chgo., 1943-45. Fulbright research scholar, Philippines 1949-50; Guggenheim fellow, 1953; fellow Center for Advanced Study in Behavioral Scis., Stanford, 1958-59; recipient Viking Fund medal and award, 1956; hon. curator Chgo. Natural History Mus., 1962. Fellow Royal Anthrop. Inst. (hon.), Am. Acad. Arts and Scis.; mem. Am. Philos. Soc., Nat. Acad. Scis. (mem. com. on scis. and pub. policy 1965—, mem. council 1967—), Assn. Asian Studies (dir. 1966—), Am. Anthrop. Assn. (pres. 1953, Memoirs editor 1960-64), Am. Ethnol. Soc., NRC (chmn. com. Asian anthropology 1952-53), Social Sci. Research Council (chmn. bd. 1955-56), Phi Beta Kappa, Sigma Xi, Tau Kappa Epsilon. Author: Social Organization of the Western Pueblos, 1950; The American Indian, 1966. Editor: Social Anthropology of North American Tribes, 1937, enlarged edit., 1955. Supr. Handbook on The Philippines, 4 vols., 1956. Mem. sr. adv. com. Ency. Brit., 1953—. Research on acculturation, ethnology; social anthropology. Home: 1321 E. 56th St., Chgo. 60637. Office: 1126 E. 59th St., Chgo. 60637.

EGGER, Joseph Georg, German physician, geologist, micropaleontologist; b. Kelheim, Germany, Dec. 24, 1824; s. Johann Henomuk and Johanna Maria (Raith) E.; studied medicine and natural scis., Munich, Vienna, Prague; m. Franziska Barbara Kitzinger, 1830; 2 sons; m. Therese Kitzinger, 1862; 1 dau.; m. 3d, Mathilde Keyser, 1869. Began practice medicine, Ortenburg, Germany, 1849, Passau, 1859; named ofcl. dist. physician, Passau, 1871; named govt. adviser to Bayreuth, 1881, to Landshut, 1886; ret. to Munich, 1896. Research and publs. on local geology and paleontology, foraminifers and diatoms, including description of numerous new types; zool. and meteorol. studies in Passau. Died Munich, Mar. 22, 1913.

EGGERT, John, photochemist; b. Berlin, Germany, Aug. 1, 1891; s. Emil Max and Martha (Zwingmann) E.; Dr. Phil., U. Berlin, 1914; hon. doctorates Karlsruhe, also W. Berlin; m. Margarete Ettisch, Dec. 22, 1921; children—Anneliese, (Mrs. Peter Leiser), Toni (Mrs. Kyle Packer). Became mem. faculty U. Berlin, 1914; head. sci. lab. AGFA, 1921-45; prof., Munich, Germany, 1945; prof. photography Fec. Inst. Tech., Zurich, Switzerland, 1946-61, now ret. Recipient Gold medal Photographic Soc. Vienna; Corr. mem. acads. sci. Munich, Heidelberg, Liège, Leipzig, Japan; hon. mem. German Bunsen Soc., photog. socs. London, Paris, Am., Switzerland. Author: (with L. Hockaud, G. M. Schwale) Textbook on Physical Chemistry, 9 German edits., 1926-68, also fgn. edits. Research and publs. on cinetics, photochemistry and photography. Address: 44 Höhe Str., CH 8702, Zollikon, Switzerland.*

EGGERT, (Hermann Paul) Otto, German geodesist; b. Tisit, Germany, Feb. 4, 1874; s. Hermann and Johanna (Bluhm) E.; studied geodesy under C. A. Vogler, Landwirtschaftliche Hochschule, Berlin; later studied math. and astronomy; m. Elsa Schultz, 1903; 1 dau. Worked for several years as surveyor; first asst. then lectr. Landwirtschaftliche Hochschule, Berlin; became prof. geodesy Tech. U., Danzig, Poland, 1904; returned to Berlin as Vogler's successor, 1921; became prof. Tech. U., Berlin, 1925, rector, 1933-34, 36-39; also dir. Potsdam (Germany) Geodetic Inst. Author: Einführung in die Geodäsie, 1907; Logarithmisch-trigonometrische Tafeln für Neue (zentesimale) Teilung mit sechs Dezimalstellen, 1942. Improved tng. of surveyor-engrs., including addition of astronomy and photogrammetry to curriculum. Died Danzig, Jan. 20, 1944.

EGLER, Frank E(dwin), Am. ecologist; b. N.Y.C., Apr. 26, 1911; s. Charles J. and Florence E. (Wilshusen) E.; student N.Y. State Coll. Forestry, Syracuse, 1929-31; B.S., U. Chgo., 1932; M.S., U. Minn., 1934; Ph.D., Yale, 1936, research fellow, 1936-37. Faculty N.Y. State Coll. Forestry, 1937-44; dir. chicle Devel. Co. Expt. Sta., Brit. Honduras, 1941-44; dir., pres. Aton Forest, Norfolk, Conn., 1943—; guest lectr., also vis. prof. various univs.; cons. vegetationist various instns. Chmn., Right of way Resources Am., 1963-66. Fellow A.A.A.S., Am. Geog. Soc.; mem. Am. Assn. U. Profs., Arctic Inst. Am., Ecol. Soc. Am., Nat. Audubon Soc, Nature Conservancy, Wilderness Soc., Wildlife Soc., others. Author several books. Research, publs. on ecology, pesticides, regional vegetation, mgmt. of vegetation, conservation of natural resources. Address: Aton Forest, Norfolk, Conn. 06058.*

EGLESTON, Thomas, Am. mining engr., metallurgist; b. N.Y.C., Dec. 9, 1832; grad. Yale, 1854, A.M., 1857; grad. Ecole des Mines, Paris, E.M., 1860, Ph.D., Princeton, 1874; LL.D., Trinity, 1874; m. Augusti McVickar, May 2, 1865 (dec. 1895). Apptd. supt. mineral and metall. products Smithsonian Instn., 1861-64; founder Sch. of Mines, Columbia, 1864, prof. mineralogy and metallurgy, 1864-97. Egleston medal of Columbia U. Sch. Engring founded in his honor, 1939. Author: Lectures on Mineralogy, 4 vols., Metallurgy of Silver; Metallurgy of Gold; Tables for Determination of Weights, Measures and Coins in the Metric and English Systems, 2 vols., 1887-90. Died N.Y.C., Jan. 15, 1900.

EGLI, (Johann) Jakob, Swiss geographer; b. Laufen-Uhwiesen, Switzerland, May 17, 1825; s. Heinrich and Barbara (Witzig) E.; doctorate in geology under O. Heer and A. Escher von der Linth, 1865; m. Elisabetha Hofmann, 1846; a son, Emil. Began as secondary sch. tchr.; joined faculties in geography, of both Zürich univs., 1866; placed in charge geography edn. in canton sch., 1872; became prof. geography U. Zürich. Author: Nomina geographica, 1872; Der Völkergeist in den geographischen Namen, 1894; Geschichte der geographischen Namenkunde, 1886; Neue Erdkunde, 6th edit., 1881; Neue Handelsgeographie, 3d edit., 1882; Neue Schweizerkunde, 7th edit., 1883. Founder modern sci. geographic nomenclature. Died Zürich, Aug. 24, 1896.

EGLI, Paul Henry, Am. chemist; b. Mt. Vernon, Ind., Apr. 14, 1917; s. William P. and Lula (Schnur) E.; student Evansville U., 1934-35; M.S. in Chem. Engring., Purdue U., 1939; children—Alice L., David W., Kathryn F. With Gen. Petroleum Corp., Los Angeles, 1939-40, Eli Lilly Co., Indpls., 1940-41; head crystal br. U. S. Naval Research Lab., Washington, 1941-65, coordinator energy conv., 1960-64; v.p. PanAura Corp., Pennsauken, N.J., 1965—. Adj. prof. U. Detroit. Fellow Am. Phys. Soc., A.A.A.S.; mem. Washington Acad. Scis., Am. Chem. Soc., Soc. Information Display. Author: Thermoelectricity, 1960. Editor: Internat. Jour. Advanced Energy Conversion, 1961-63. Research, publs. on design of chem. process equipment, especially efficient multiple plate distillation equipment; extensive research in mechanism of crystal growth, pioneered alteration of solutions to suppress nucleation and enhance flawless growth; first successful synthetic quartz process and produced improved quality of many crystals; also substantial work in energy conversion processes, cathode ray displays and biomed. engring. Home: 73 Maple Av., Collingswood, N.J. 08108. Office: 6704 N. Crescent Blvd., Pennsauken, N.J. 08110.*

EGLINTON, Geoffrey, Brit. chemist; b. Cardiff, Wales, Nov. 1, 1927; s. Alfred Edward and Lillian (Blackham) E.; B.Sc., Manchester U., 1948, Ph.D., 1951, D.Sc., 1966; m. Pamela Joan Coupland, Apr. 9, 1955; children—David Geoffrey, Timothy Ian, Fiona Jane. Research fellow Ohio State U., 1951-52; I.C.I. fellow Liverpool U., 1952-54; lectr. chemistry U. Glasgow (Scotland), 1954-64, sr. lectr., 1964-67; reader in chemistry U. Bristol (Eng.), 1967—. Vis. prof. U. Cal. at Berkeley, 1963-64. Mem. Chem. Soc., Phytochem. Soc., Geochem. Soc. Author: (with Brand) Application of Spectroscopy to Organic Chemistry, 1965. Research and publs. on synthesis of acetylenic compounds, structure of natural products, especially leaf waxes, applications of infrared spectroscopy in organic chemistry, especially with reference to hydrogen bond, application of gas liquid chromatography in separation of complex mixtures, application of organic chemistry and phys. methods in study of organic components of terrestrial sediments, fossils and meteorites. Home: 7 Redhouse Lane, Stoke Bishop, Bristol, Eng.*

EGLOFF, Gustav, Am. petroleum researcher; b. N.Y.C., Nov. 10, 1886; s. August and Mary E.; A.B., Cornell U., 1912; A.M., Columbia, 1913, Ph.D., 1915; D.Sc., Poly. Inst. Bklyn., 1938, Armour Inst. Tech., 1940, Phila. Coll. Pharmacy and Sci., 1944; m. Clara Mellor, July 3, 1914. Asst. Chandler Mus., Barnard research fellow, 1914-15; chemist U. S. Bur. Mines, 1915-16, Aetna Chem. Co., Pitts., 1916-17; dir. research Universal Oil Products Co., Chgo., 1917-55. Holder of 250 patents on processing petroleum oil, coal, shale oil and chem. derivatives of hydrocarbons; Lectr. univs. Columbia, Princeton, Chgo., Northwestern, Cal., So. Cal., Stanford, Mo., Mich., N.Y., Johns Hopkins. Recipient gold medal Am. Inst. Chemists, 1940; Octave Chanute medal (1939-40), by Western Soc. Engrs.; named Modern Pioneer N.A.M., 1940; Columbia Medal for Excellence, 1943. Head mission sponsored by Nat. Resources Commn. of China adviser establishment and devel. modern petroleum refining industry in China; Fellow Am. Inst. Chemists (pres. 1942-46), A.A.A.S., Royal Soc. Arts of Gt. Britain, Am. Geog. Soc.; mem. Am. Chem. Soc. (councillor at large, chmn. petroleum div. 1946-47), Am. Inst. Chem. Engrs. (dir.), Am. Inst. Mining and Metall. Engrs., Chmn. Inst. Can., Franklin Inst., Soc. Chem. Industry, Am. Petroleum Inst. (dir.), Nat. Petroleum Assn., Inst. Petroleum Technologists (London), Soc. Am. Mil. Engrs., Western Soc. of Engrs., (v.p.), Soc. Automotive Engrs., Am. Soc. for Testing Materials, Sigma

Xi, Phi Lambda Upsilon. Author: Earth Oil, 1933; The Reactions of Pure Hydrocarbons, 1937; The Physical Constants of Hydrocarbons, Vol. 1, 1939, Vol. 2, 1940; Catalysis, 1940; Emulsions and Foams, 1941; Isomerization of Pure Hydrocarbons, 1942; Physical Constants of Hydrocarbons, Vol III, 1946. Contbr. over 425 articles to tech. and trade journals on petroleum industry, particularly cracking and refining of oil; over 300 patents in petroleum processing. Devised multiple-coil process for cracking crude oil which increases yield of high-octane gasoline from crude oil, made possible 40 percent increased efficiency of planes and automobiles; devised way to make synthetic rubber from butane gas, 1939. Died Chgo., Apr. 29, 1955.

EHLERS, Edward, Danish dermatologist; b. Copenhagen, Denmark, 1863; hon. doctorate U. Paris; prof. clin. medicine, Copenhagen; pres. Danish sect. Alliance Française; research on leprosy; described hyperelasticity of skin (Ehler's-Danlo's syndrome), 1901. Died 1937.

EHLERS, Ernst Heinrich, German zoologist; b. Lüneburg, Germany, Nov. 11, 1835; s. Wilhelm and Caroline (Meyer) E.; began study medicine, Göttingen, 1857; M.D., 1861; Ph.D. (hon.), Göttingen, 1874; m. Marianne Hasse. Became lectr. anatomy, comparative anatomy, zoology, Med. faculty, Göttingen, 1863; prosector under Henle, 1861-69; became prof. zoology, comparative anatomy, vet. medicine, Erlangen, Germany, 1869; prof. zoology and comparative anatomy Philosophy Faculty, Göttingen, 1874-1946; founder mus. Göttingen Zool. Inst., 1873-78. Mem. Leopoldina, Göttingen Soc. Scis., Vienna Acad. Scis. (corr.), Deutsche Zoologische Gesellschaft (a founder 1890). Editor, Zeitschrift für wissenschaftliche Zoologie, 1875-1925. A founder modern academic zoology; research and publs. on lower forms of sea animals, nematodes, sharks, sea mammals, anthropoid apes, armadillos, sponges, siphonophores, bryozoa, tunicates, modern and fossil polychaetes. Died Göttingen, Dec. 31, 1925.

EHRENBERG, Christian Gottfried, German naturalist; b. Delitzsch, Germany, Apr. 19, 1795; s. Johann Gottfried and Christiane Dorothea (Becker) E.; studied theology, then medicine, Leipzig, Germany; M.D., Berlin, 1818; m. Julie Rose, 1831; 1 son, Herman Alexander, 4 daus.; m. 2d, Karoline Friederike Fricius, 1852. Mem. expdn. with W. F. Hemprich to coastlands of Red Sea, 1820-26; became asso. prof. U. Berlin, 1826, prof. medicine, 1839; accompanied A. v. Humboldt on Asiatic Russian jour., 1929. Recipient Pour le mérite, 1842. Mem. Leopoldina, Berlin Acad. Scis. Fellow Royal Soc., 1837. Author: Die Infusionstheirchen als volkommene Organismen, 1838; Die Korallentiere des Roten Meeres, 1834; Mikrogeologie, 1854. Pioneered in microbiology and micropaleontology; described 248 new fungus types for Berlin; collected 34,000 specimens from his Red Sea expdn.; 1st description of Bacillus subtilis, 1838, his work important to subsequent study and classification of microorganisms; his theory that Infusoria (fossil protozoa) organs were multicellular was disproved; 1st to demonstrate Noctiluca cause of much marine phosphorescence. Died Berlin, June 27, 1876.

EHRENBERG, Paul Richard Rudolf, German agrl. chemist; b. Brandenburg, Germany, May 16, 1875; s. Carl and Meta (Lange) E.; studied agr., Jena, Germany; doctorate, 1899; m. Maria Adam, 1911; 2 sons, 3 daus. Asst. to T. Remy, Berlin, T. Pfeiffer, U. Breslau (now Wroclaw, Poland); became lectr., Breslau, 1907; named prof. forestry, Hannover Münden, 1910; became prof. agrl. chemistry, Göttingen, Germany, 1911; joined Agrl., Chem. and Bacteriological Inst. U. Breslau, 1921; dir. Inst. for Agrl. Chemistry, Tech. U., Munich-Weihenstephan, 1945-48. Author: Düngerlehr, 1924; Allgemeine und besondere Bodenkunde, 1947; also textbooks on bacteriology. Studied soil chemistry, botany, animal nutrition, antagonism between various plant foods; applications of agrl. chemistry to practice, including fodder, fertilization, soil; studied prodn. horse fodder from by-products of beet sugar industry. Died Freising, Germany, Jan. 18, 1956.

EHRENBERG, Wilhelm Wolfgang, German chem. physicist; b. Munich, Germany, Jan. 8, 1909; s. Paul and Lilly Ehrenberg; Ph.D. in Chem. Physics, U. Munich. With Kaiser-Wilhelm Inst. Metall. Research, Dusseldorf, Junkers Motorenwerke, Dessau; Argentine Commn. Atomic Energy, Bariloche, Mendoza; sci. cons., Munich, 1953——. Mem. German Soc. Physics, Psychophys. Soc. (founder). Editor: Jour. Psychophysics, PPZ, Erfahrungswissenschaftliche Blätter. Research and numerous publs. on nuclear physics, phys. chemistry, colorings, philosophy, psychology. Address: Office: Wolfratshauserstrasse 26, Munich 25, Germany.

EHRENDORFER, Friedrich, Austrian botanist; b. Vienna, Austria, July 26, 1927; s. Friedrich and Isabella (Kuschel) E.; Ph.D., U. Vienna, 1949; m. Eva Provaznik, Sept. 7, 1959; children—Wolfgang, Johanna, Stefan, Friedrich. Faculty, U. Vienna, Inst. Systematic Botany 1949-64, curator Mus. Natural History, Vienna, 1960-64; prof. botany, dir. Inst. Systematic Botany and Bot. Garden, U. Graz (Austria), 1965——; research asso. U. Cal. at Davis and Berkeley, 1959. Fulbright fellow, 1952; Rockefeller grantee, NSF grantee, 1959; recipient Vienna prize, 1956, prize Körner Found., 1960, 62. Sr. research fellow Australian Acad. Sci., 1966. Author: Check List of Higher Plants in Central Europe, 1967; also numerous articles. Research on mechanisms of evolution and systematics in flowering plants. Home: 23 Pausingergasse, Vienna, Austria A-1140. Office: 6 Holteigasse, Graz A-8010, Austria.*

EHRENFELS, (Maria) Christian (Julius Leopold Karl) von, see von Ehrenfels, (Maria) Christian (Julius Leopold Karl).

EHRENFEST, Paul, nuclear physicist; b. Vienna, Austria, Jan. 18, 1880; prof., Leyden, Netherlands; defined and analyzed ferromagnetic points of Curie, 1933; investigated Loschmidt's paradox; devised method of applying quantum theory to rotating bodies; recognized that Ampères molecular currents are incompatible with classical statistical mechanics; invented term "adiabatic invariants." Died Leyden, Sept. 25, 1933.

EHRENFREUND, David, Am. psychologist; b. N.Y.C., May 24, 1917; s. Leo and Henrietta (Huppert) E.; B.A., State U. Ia., 1943, M.A., 1945, Ph.D., 1947; m. Eleanor Stern, Mar. 6, 1943; 1 dau., Cathy Ann. Faculty, Wash. State U., Pullman, 1947-56, asso. prof., 1952-56; prof., chmn. psychology dept. Adelphi U., Garden City, L.I., N.Y., 1956-62; prof., chmn. psychology dept. So. Ill. U., Carbondale, 1962——. Fellow Am. Psychol. Assn., A.A.A.S.; mem. Eastern (past chmn. program com.), Midwestern (council 1964-67), Wash. State (past pres.) psychol. assns., Psychonomic Soc. (charter), Sigma Xi. Research, publs. on learning and motivation, especially discrimination learning, invention and devel. automated apparatus for maintaining constant hunger levels in lab. animals. Home: 47 Hillcrest Dr., Carbondale, Ill. 62901.*

EHRENHAFT, Felix, physicist; b. Vienna, Apr. 24, 1879; s. Leopold E. and Louise (Egger) E.; ed. U. Vienna; m. Olga Steindler. Became prof. U. Vienna, 1912. Recipient Lieben prize, Haitinger prize Acad. Sci. Vienna, Voigtländer medal. Discovered Brownian motion in gases, 1907, photophorese. Died Mar. 4, 1952.

EHRENREICH, Henry, Am. physicist; b. Frankfort, Germany, May 11, 1928; s. Nathan and Frieda (Rosenstein) E.; came to U. S., 1940, naturalized, 1946; A.B., Cornell U., 1950, Ph.D., 1955; postgrad. Columbia, 1950-51; M.A. (hon.), Harvard, 1963; m. Tema P. Hasnas, Feb. 1, 1953; children—Paul, Beth-Ida, Robert. Theoretical physicist Gen. Elec. Research Lab., Schenectady, 1955-63; vis. lectr. Harvard, 1960-61, Gordon McKay prof. applied physics, 1963——. Fellow Am. Phys. Soc.; mem. Phi Beta Kappa, Sigma Xi. Contbr. profl. jours. Bd. editors Phys. Rev., 1965——. Investigation of theory of metals; semiconductors; solid state physics; optical properties of solids. Home: 80 Douglas Rd., Belmont, Mass. 02178. Office: Pierce Hall, Harvard U., Cambridge, Mass. 02138.*

EHRENSTEIN, Maximilian Richard, chemist; b. Munich, Germany, May 11, 1899; s. Richard and Amanda (Engels) E.; Ph.D., U. Goettingen, 1921; Dr. rer.nat. (hon.), Free U. Berlin, 1965; m. Elsa Meyer, Apr. 5, 1925. Came to U. S., 1934, naturalized, 1944. Asst. in organic chemistry U. Goettingen, 1921-23, Inst. Tech., Breslau, 1923-25; Rockefeller Found. fellow U. Zurich, 1925-26; research asso. U. Munich, 1926-29; asst. in pharm. chemistry U. Berlin, 1929-31, privat dozent, 1931-34; research asso. Med. Sch. U. Va., 1934-37; faculty Sch. Medicine, U. Pa., Phila., 1937——, prof. biochemistry, chief of div. steroid research, 1949——. Recipient Officer's Cross of Order of Merit, Pres. Fed. Republic Germany, 1966; hon. prof. physiol. chemistry U. Hamburg, 1963——; spl. travelling fellow Rockefeller Found., 1952. Mem. Am. Chem. Soc., Am. Soc. Biol. Chemists, Swiss Chem. Soc. Research, patents, numerous publs. in alkaloid and steroid chemistry; discoverer several new constituents in tobacco; synthesis of numerous analogs of naturally occurring hormones, especially those of cortisone and progesterone series; 1st to synthesize 19-norsteroids, led to devel. of therapeutic agts. including contraceptives. Home: 4840 Pine St., Phila. 19143. Office: 806 Maloney Clinic, Hosp. of U. Pa., Phila. 19104.*

EHRET, Charles Frederick, Am. biophysicist; b. N.Y.C., Mar. 9, 1923; s. Henry Richard and Harriet Rose (Adolff) E.; B.S., Coll. City N.Y., 1946; M.S., U. Notre Dame, 1948, Ph.D., 1951; m. Dorothy Mae Armstrong, Aug. 11, 1945; children—William Charles, Henry Richard II, Thomas Joseph, Peter Francis, Louise Elizabeth, Albert Lawrence, Julia Harriet. Asst. scientist Argonne (Ill.) Nat. Lab., 1949-51, asso. scientist, 1951-65, sr. scientist biology div., 1966——; USPHS fellow Institut de Biologie Moleculaire et Biophysique, Geneva, Switzerland, 1960-61; Fridley lectr. U. Minn., 1963; vis. prof. zoology Ind. U., Bloomington, 1963; NASA fellow theoretical biology, 1966. Mem. biology com. Asso. Midwest Univs.-Argonne Nat. Lab., 1965——; cons. NSF, AEC, 1963——. Recipient Centennial award medal in sci. U. Notre Dame, 1965. Fellow A.A.A.S.; mem. Radiation Research Soc., Am. Soc. Zoologists, Am. Soc. Naturalists, Soc. Gen. Physiologists, Sigma Xi. Author: (with Howard S. Ducoff) Mitogenesis, 1959; also articles. Patentee sci. games. Co-builder Argonne Biol. Spectrograph, 1956; discovered circadian rhythms in non-photosynthetic protozoa, action spectrum for resetting biol. clocks, that ultraviolet light resets and is photoreactivable, nucleic acids are involved in cellular clockworks, some mitochondria have tubules, subunits of Paramecium cell surface, and how organelles develop from preorganelles by means electron microscopy; measured free radicals in X-irradiated bacteria; developer autoradiographic techniques for electron microscopy. Home: 122 S. Stough St., Hindsdale, Ill. 60421. Office: Biology div. Argonne Nat. Lab., Argonne, Ill. 60440.*

EHRHARDT, Carl, German surgeon, gynecologist; b. Weyer, Oberlahn, Feb. 24, 1895; s. Karl and Lina (Duill) E.; ed. at Marbourg, Königsberg, Giessen, U. Frankfurt-am-Main; M.D.; m. Elsbeth Hoffman; 1 son, Wolfgang Volker. Med. officer, 1922; asst. Hosp. of City of Offenbach am Main, 1922-25; asst., chief physician Univ. Clinic for Women of Frankfurt-am-Main, 1925-39; dir. Univ. Clinic for Women in Maternity, Graz, Austria, 1939-45; practice medicine specializing in gynecol. and obstet. surgery Dist. of Laufen, 1945-51, Frankfurt-am-Main, 1951——. Mem. German Assn. for Gynecol. Surgery and Obstetrics. Research and over 150 publs. on gynecology, obstetrics, x-rays. Address: Mainzer Landstrasse 123, Frankfurt-am-Main, Germany.

EHRHART, Eugène, French mathematician; b. Guebwiller, France, Apr. 29, 1906; s. Jean and Emma (Eggers) E.; licence, U. Strasbourg (France), 1928, agrégation de math., 1948; Doctorat d'état de mathématique, U. Grenoble (France), 1964; m. Hilda-Anne Sternberger, May 18, 1940; 1 son, Jean. Prof. math. U. Nancy, 1932-34, Toul, France, 1934-36, U. Lille, 1936-39, U. Metz, 1939-45, Strasbourg-Lycee Kleber, 1945——; researcher Centre National de la Recherche Scientifique, 1962-63; lectr. Faculté de Droit et des Sciences Economiques, U. Strasbourg, 1961——. Laureate, Freycinet prize French Acad. Scis., 1959; named officier de l'instruction publique, 1960. Mem. Soc. Math. Profs. in Pub. Edn. Author: Le triangle orienté, 1951; Sur un problème de géométrie diophantienne linéaire, 1966; also numerous articles. Research on polytopes in a system, measurement of their maxima; systems of diophantine equations and inequations which depend linearly on a parameter, their multiple solutions; ovals and ovoids including their properties concerning their center of gravity; inequalities, including new proof of inequalities of Schwarz and Holder and their applications to integral calculus. Home: 11, rue de Bruges, Strasbourg 67, France.*

EHRHART, (Jakob) Friedrich, German botanist; b. Holderbank, Germany, Nov. 4, 1742; s. Johannes and Magdalena (Wild); m. Hedwig Margarete Sonnenberg, 1783. Trained as farmer, then pharmacist; began work for pharmacist Johann G. R. Andreae, Hanover, Germany, 1770; spent several years in Stockholm and Uppsala, Sweden; returned to Hanover, 1776; held ofcl. contract to explore botany in vicinity of Hanover, 1780-83; then became ofcl. botanist of prince-elector, Herrenhausen, Germany; set up herbaria. Editor: Chloris hanoverana (Meyer), 1836; Supplementum systematis vegetabilium, generum et specierum plantarum (K. v. Linné), 1781. Editor, Beiträge zur Naturkunde, 1789-92. Improved many of Linnaeus' designations; described many new plant species, especially sedges, ferns, and mosses; studied thickness growth of plants, pollenization of flowers by insects, flora of Hanover vicinity. Died Herrenhausen, June 26, 1795.

EHRICH, William Ernest Herman Heinrich, pathologist; b. Dahmen, Germany, Nov. 29, 1900; s. Wilhelm Maria and Hedwig (Menzell) E.; student Munich and Freiburg Germany, 1919-24; M.D., U. Rostock, 1924; M.D. (hon.), Freiburg, 1957; m. Marie Louise Goldschmidt, Sept. 7, 1926; children—William G., Anna Louise (Mrs. Peter Batson), Helen M. (Mrs. André Armbruster). Came to U. S., 1926, naturalized, 1942. Pathologist, Rockefeller Inst., N.Y.C., 1926-30; privatdozent, Rostock U., 1930-35; faculty U. Pa., Phila., 1936——, prof. pathology, 1947——, chmn. dept. grad. path., 1948-67; chief dept. anatomic pathology Phila. Gen. Hosp., 1942——. Recipient Gold medal Pathol. Soc. Phila., 1963. Mem. Deutsch. Akademie Naturforscher Leopoldina, Am. Assn. Pathologists and Bacteriologists, Am. Soc. Exptl. Pathology, Soc. Exptl. Biology and Medicine, Deutsch. Pathologische Gesellschaft, A.A.A.S., Harvey Soc. Author: Lymphoid Tissues: Their Morphology and Role in the Immune Response in Immunological Diseases, 1965. Research and publs. on the introduction of experimentation into the study of structure and function of lymphnodes, the discovery that antibodies are formed not by the reticuloendothelium as was believed at that time, but by lymphoid cells (plasma cells), and elucidation of the kinetics and function of the germinel centers of lymphoid tissues; other projects deal with inflamation the pathogenesis of renal diseases and other path. phenomena. Home: 540 W. Hortter St., Phila. 19119.*

EHRICKE, Krafft Arnold, missile engr.; b. Berlin, Germany, Mar. 24, 1917; s. Arnold F. and Ruth (Konietzko) E.; B.S., Tech. U., Berlin, 1942; grad. courses atomic physics and celestial mechanics, U. Berlin, 1941-42; L.H.D., Nat. Tchrs. Coll., Evanston, Ill., 1961; m. Ingeborg Mattull, Jan. 16, 1945; children—Krista, Astrid, Doris. Came to U. S., 1947, naturalized, 1955. Devel. engr. V-2 propulsion system, Peenemuende, Germany, 1942-45; jet propulsion engr. Dept. of Army, Ft. Bliss, Tex., 1947-50; chief gasdynamics sect. Army Ballistic Missile Center, Redstone Arsenal, Ala., 1950-52; preliminary design engr. Bell Aircraft Corp., 1952-54; with Convair, div. Gen. Dynamics Corp., 1954-—, design specialist, 1954-55, chief design and systems analysis, 1956-57, asst. to tech. dir. Convair-Astronautics, 1957-58, program dir. Centaur space vehicle, 1958-—, also dir. advanced studies dept. Recipient 1st Guenther Loeser medal for best paper (The Satelloid) presented during 6th Internat. Astronautical Congress, Internat. Astronautical Fedn., 1956, G. Edward Pendray award Am. Rocket Soc., 1961. Fellow Am. Astron. Soc., Brit. Interplanetary Soc.; mem. Am. Rocket Soc. (Astronautics award 1957; dir. 1957-59), Deutsche Gesselschaft für Weltraumforschg. (pres. 1942-43), Inst. Aero. Scis., Internat. Acad. Astronautics. Author: Space Flight, Vol. I, Environment and Celestial Mechanics, 1959, Vol. II, Dynamics, 1961; Vol. III, Missions, Propulsion and Space Vehicles, 1961; numerous articles. Work on jet propulsion projects, Atlas ICBM (successfully test fired, Cape Kennedy, Fla., 1957). Home: 845 Lamplight Dr., La Jolla, Cal. Office: care Convair-Astronautics, San Diego 92112.

EHRINGHAUS, Arthur Erich, German mineralogist; b. Mettmann, Germany, Nov. 29, 1889; s. Gustav and Caroline (Comberg) E.; studied math., physics, chemistry, mineralogy, geology, philosophy, Göttingen, Germany; teaching degree, 1913; doctorate, 1916. Researcher crystal optics Mineral.-Petrographic Inst., U. Göttingen, 1913-16; became scientist indsl. optical works R. Winkel GmbH, Göttingen, 1916. Author: Das Mikroskop, seine wissenschaftlichen Grundlagen und seine Anwendung, 1921; also numerous articles. Introduced relative dispersion of double refraction in crystal physics; improved polarization microscopy, various optical instruments. Died Göttingen, Jan. 11, 1948.

EHRLICH, Danuta, Am. psychologist; b. Warsaw, Poland, Jan. 3, 1931; s. Seweryn and Felice (Lubelczyk) Ehrlich vel Sluszny; B.A., Queens Coll., New York, 1953; M.A., U. Minn., 1957, Ph.D., 1958. Came to U. S., 1946, naturalized, 1952. Research asst. Lab. for Research in Social Relations, U. Minn., 1953-57; trainee VA, St. Paul, 1957-58; research asso. Psychosomatic and Psychiat. Inst., Michael Reese Hosp., 1958-63; lectr. sociology U. Chgo., 1961-62; psychologist Ill. State Psychiat. Inst., 1963-—; asst. prof. psychology dept. psychiatry U. Ill. Coll. Medicine, Chgo., 1963-—. Mem. Am. Psychol. Assn., Am. Sociol. Assn., Soc. for Psychol. Study Social Issues. Author: (with others) Psychiatric Ideologies and Institutions, 1964; also articles. Research on post-decision exposure to relevant information, impression formation and intolerance of ambiguity, verbal commonality and influencibility, birth order and influencibility, values in psychotherapeutic settings, sociotherapeutically oriented psychiatrists, psychiat. ideologies and psychotherapeutic functions. Home: 2951 S. Parkway, Chgo. 60616. Office: 1601 W. Taylor St., Chgo. 60612.

EHRLICH, Felix, German biochemist; b. Harriehausen/Harz, Germany, June 16, 1877; s. Louis and Luise (Lange) E.; studied organic chemistry under Emil Fischer, Berlin; doctorate, 1900. Mem. staff Inst. for Sugar Industry, Berlin, until 1909; named lectr. chemistry U. Berlin, 1906, Agrl. U., Berlin, 1907; became asso. prof., dir. Inst. for Biochem. and Agrl. Tech., U. Breslau (now Wroclaw, Poland), 1909, prof., 1920; forced to retire by Nazis, 1935. Recipient Ladenburg medal, 1911, Emil Fischer Meml. medal, 1931, Verein Deutscher Chemiker, 1931; named hon. citizen Agricult, U. Berlin. Mem. Leopoldina. Research and numerous publs. on mucous substances, chemistry and biochemistry of pectin, especially its enzymatic decomposition; discovered and explained structure of amino acid, isoleucin, origin of fusel oil as fermentation product of leucin and isoleucin. Died Obernigk nr. Breslau, 1942.

EHRLICH, John, Am. microbiologist; b. N.Y.C., Dec. 13, 1907; B.S., Cornell U., 1928; A.M., Duke, 1928; S.M., Harvard, 1930, Ph.D., 1930. Asst., Harvard, also Radcliffe Coll., 1929-30; Austin teaching fellow Harvard, 1930-31; NRC fellow Imperial Mycologic Inst., Kew, Eng., 1933-34; research grantee Royal Bot. Gardens, 1935; asst. prof. forestry U. Ida., 1935-40, asso. prof., 1940-44; research asso. U. Minn., 1944; div. head, research labs. Parke, Davis & Co., 1944-52, lab. dir., antibiotic research, Detroit, 1952-—. Mem. A.A.A.S., Mycol. Soc., Soc. Microbiology, Soc. Indsl. Microbiology, Acad. Microbiology. Discovered (with Q. R. Bartz, R. M. Smith, D. A. Joslyn, P. R. Burnholder) chloramphenicol; research on beech bark disease, white pine blister rust, armillaria rot, nectriaceae, microbiology of antibiotics, antiviral antibiotics, viomycin, azaserine, elaiomycin, griseoviridin, viridogrisein, actinobolin, paromycin, streptimidone, chalcomycin. Office: Parke, Davis & Co., Detroit 32.

EHRLICH, Paul, German bacteriologist; b. Strehlen, Silesia (now Strezelin, Poland), Mar. 14, 1854; s. Ismar E. and Rosa (Weigert) E.; ed. at Breslau, 1870-71, Strasbourg U., U. Fribourg-en-Brisgau; M.D., U. Leipzig, 1878; hon. doctorates Oxford U., U. Chgo.; m. Hedwig Pincus, 1883; 2 daus. asst. head physician Charité Hosp., Berlin, 1878; began working with Koch on Tb, 1882, contracted light case of Tb, 1886, went to Egypt to recover, 1886-89; asso. prof. U. Berlin, 1890-96; dir. lab. Inst. Infectious Diseases, Berlin, 1890-96; dir. State Inst. Serum Research, Steiglitz, nr. Berlin, 1896-99, Royal Inst. Exptl. Therapeutics, Frankfort, 1899; prof. exptl. therapeutics U. Frankfort, 1914, also 1st rector. Recipient (with Élie Metchnikoff) Nobel prize in physiology and medicine, 1908. Fellow Royal Soc., 1910. Author: Beiträge zur Therapie und Praxis der histologischen Färbung, 1887; Das Sauerstoff bedürfnis des Organismus, 1885; Chemotherapeutische Trypanosomen-Studien, 1907; Farberanalytische Untersuchungen über Histologie und Klinik des Blutes, 1891; Gesammelte Arbeiten zur Immunitätsforschung, 1904; (with Sachs) Über die Beziehungen zwischen Toxin und Antitoxin, 1905; (with Hata) Die experimentelle Chemotherapie der Spirillosen, 1910; Collected Papers (edited by F. Himmelweit, from 1900). Bridged scis. of chemistry, biology and medicine; laid foundation for modern hematology; discovered new methods and materials for staining, discovered stains for tubercle bacillus, living nerves, white blood corpuscles, also other cells; credited with formulating basis of differential blood count technique, 1879; pioneer in immunology and founder of chemotherapy; studied cellular pathology; developed anti-toxin/anti-body-antigen theory of immunization, circa 1900, side-chain theory, 1885 (postulates that organic molecules consist of stable central groups and less stable lateral groups with latter taking part in immunological transformations); undertook researches on cancer; discovered dye (trypan red) which kills trypanosomes (cause of diseases such as sleeping sickness); discovered salvarsan (specific remedy for syphilis), 1907 (announced in 1910); worked out details for diphtheria anti-toxin immunization, circa 1892. Died Homburg, Germany, Aug. 20, 1915.

EHRLICH, Paul R(alph), Am. population biologist; b. Phila., May 29, 1932; s. William and Ruth (Rosenberg) E.; A.B., U. Pa., 1953; M.A., U. Kan., 1955, Ph.D., 1957; m. Anne Fitzhugh Howland, Dec. 18, 1954; 1 dau., Lisa Marie. Field officer No. Insect Survey, summers 1951-52; asso. investigator USAF research project, Alaska and U. Kans., 1956-57; research asso. Chgo. Acad. Scis. and U. Kan. Dept. Entomology, 1957-59; faculty Stanford (Cal.) U., 1959-—, prof. biol. scis., 1966-—. Cons., Behavioral Research Lab., 1964-—; editor, cons. McGraw-Hill Book Co., 1964-—. Sr. postdoctoral fellow NSF, U. Sydney, 1965-66. Fellow Cal. Acad. Scis.; mem. Lepidopterists Soc. (sec. 1957-63), Soc. for Study of Evolution, Am. Soc. Naturalists, Soc. Systematic Zoology, Am. Mus. Natural History (hon. life), Am. Inst. Biol. Sci., American Association of University Professors, Population Crisis Committee, Sigma Xi. Author: (with A. H. Ehrlich) How to Know the Butterflies, 1961; (with R. W. Holm) The Process of Evolution, 1963; also articles. Elucidated comparative morphology and higher classification of butterflies, their comparative reproductive biology and relationship to larval foodplants; helped develop theory of numerical taxonomy; research in population biology, especially of water snakes and butterflies with emphasis on selective changes in natural populations. Home: 936 Valdez Pl., Stanford, Cal. 94305.*

EHRMANN, Charles Henri, French surgeon; b. Strasbourg, France, Sept. 15, 1792; a son, Albert. Became chief anat. work Strasbourg Faculty, 1822; became prof. anatomy Surg. Clinic, 1826; named chief med. accoucheur hospices civils, 1837; founder boarding sch. for students of midwifery; dean Strasbourg Faculty, 1867, also dir. Anat. Mus. Author: Laryngotomie, . . . , 1847; Histoire des polypes du larynx, . . . , 1850; Nouveau recueil de mémoire, d'anatomie pathologiques, . . . , 1862. 1st to successfully remove a laryngeal polyp, 1844. Died Strasbourg, June 19, 1878.

EHRMANN, (Hermann Felix) Paul, German zoologist; b. Leipzig, Germany, Dec. 21, 1868; s. Theodor and Sophie Bertha (Görsch) E.; studied at U. Leipzig, Ph.D. (hon.), 1934; m. Lizzi Spindler; 2 daus. Tchr. at deaf mute sch., Leipzig, for 12 years; became tchr. Gaudischule, Leipzig, 1901; numerous study expdn., especially to the Alps; worked at Zool. Sta., Naples., Italy. Author: Entwicklungsgeschichte, 1914. Coeditor, contbg. author: Die Tierwelt Mitteleuropas, 1933. Studied taxonomy, distbn., life history of molluscs, especially European and Asiatic species, snails and mussels of central Europe. Died Leipzig, Oct. 6, 1937.

EHRMANN, Salomon, dermatologist; b. Ostrowetz, nr. Pisek, Bohemia, Dec. 19, 1854; s. Markus and Fanny (Sachs) E.; studied art history, Praque; M.D., Vienna, 1880; studied dermatology under M. Kaposi; m. Anna Kuhe; a dau., Edith (Mrs. Brünauer). Worked under E. v. Brüke, Physiol. Inst., U. Vienna, became asst. physician, then lectr. on dermatology and syphilis, 1887, titular prof., 1900, asso. prof., 1908, prof., 1917, dir. 2 clinics, 1889; became head physician Wiedner Krankenhaus, 1904, Allgemeines Krankenhaus, 1908-23. Author: Vergleichend-diagnostischer Atlas der Hautkrankheiten und der Syphilide, 1912; Einführung in das mikrokopische Studium der normalen und kranken Haut, 1906; also numerous articles. Studied origin of skin's pigments, syphilis, skin Tb. Died Vienna, Oct. 24, 1926.

EHRMANN, Winston W., Am. sociologist, educator; b. Reno, May, 17, 1912; s. Emil Ernest and Imogene (Bell) E.; student Mass. Inst. Tech., 1930-31; B.S., Yale, 1934, Ph.D., 1943; m. Margaret Clothilde Berlin, Sept. 13, 1934; children—Gretchen (Mrs. J. Bruce Maclachlan), Sally Sherwood (Mrs. David A. Hieronymus). Indsl. engr. U. S. Rubber Co., 1935-36; asst. Yale, also instr. New Haven Coll., 1936-38; mem. faculty U. Fla., 1938-59, prof. sociology, 1945-59; prof. sociology Colo. State U., 1959-62; staff asso. Am. Assn. U. Profs., Washington, 1962-64, asso. sec., 1964-—. Fellow A.A.A.S., Am. Sociol. Soc.; mem. Am. Assn. U. Profs., Nat. Council on Family Relations, Groves Conf. on Family Relations. Author: (with others) The Physical Environment, 1940; (with Williams) Energy and Material Resources, 1941; Premarital Dating Behavior, 1959; also articles. Research in relationship between premarital sexual behavior and love relationship, comparative social class of companion, peer and personal codes of sexual behavior. Address: 1785 Massachusetts Av., Washington 20036.*

EIBNER, Alexander Paul Friedrich, German chemist; b. Munich, Germany, Sept. 11, 1862; s. Friedrich and Luise (Weissenberger) E.; studied in Munich; later asst. to organic chemist W. v. Miller, Tech. U. Munich, doctorate, 1892; m. Martha Rosenberger, 1906; 1 dau. Lectr., research staff aromatic chemistry Tech. U., Munich, 1894-1903; became asst. to G. A. Schulz, Versuchsanstalt und Auskunftsstelle für Maltechnik, 1903, dir., 1908. Author: Malmaterialien als Grundlage der Maltechnik, 1909; Über fette Öle, Leinölersatzmittel und Ölfarben, 1922; Entwicklung und Werkstoffe der Wandmalerei vom Altertum bis zur Neuzeit, 1926; Entwicklung und Werkstoffe der Tafelmalerei, 1928; Öltrocknen, ein Kolloider Vorgang aus chemischen Ursachen, 1930. Studied and improved properties and characteristics of paint materials, especially pigments, binding substances, their influence on painting techniques. Died Munich, May 1, 1935.

EICHELBERGER, Lillian, Am. biochemist; b. Macon, Miss., Mar. 2, 1897; d. Philander W. and Huldah (Richards) Eichelberger; B.S., Miss. State Coll. for Women, 1914; M.S., U. Chgo., 1919, Ph.D., 1921; m. Ralph Hardin Cannon, Aug. 11, 1923. Research instr. dept. chemistry U. Chgo., 1921-24; chem. research Municipal Tuberculosis Sanitorium, Chgo., 1924-27; asst. prof. dept. medicine U. Chgo., 1929-44, asso. prof., 1944-50, prof. biochemistry dept. surgery, 1950-62, prof. emeritus, 1962-—; asso. investigator neurogenic shock OSRD, 1943-44, asso. investigator pharmacology, clin. testing of anti-malarial agts., 1944-47. Recipient Citation, OSRD, 1945. Diplomate Am. Bd. Clin. Chemistry. Mem. Am. Chem. Soc., Am. Soc. Biol. Chemists, Soc. Exptl. Biology and Medicine. Contbr. numerous articles to sci. jours. Studies in Tb.; acid-base balance in the body; distbn. of body water and electrolytes in the soft and hard tissues of body; clin. standardization of many drugs in induced malaria in man; biochem. studies of articular cartilage following disuse by immobilization and by denervation of an extremity; water, electrolyte studies on canine homotransplanted hearts; biochem. studies on hypertrophy of heart muscles. Home: 5849 N. Kostner Av., Chgo. 60646.*

EICHELBERGER, William Snyder, Am. astronomer; b. Balt., Sept. 18, 1865; s. Albert G. and Martha (Snyder) E.; A.B. (fellow), Johns Hopkins, 1888; Ph.D., 1891. Instr. math., asst. astronomy, 1890-96; asst. nautical almanac officer, 1889-90, 96-98; computer U. S. Naval Obs., 1898-1900; became prof. math. U. S. Navy, 1900; head div. Meridan Instruments, 1902-07, in charge various depts. astron. observations, 1907-10; dir. Nautical Almanac, 1910-29; sci. staff mem. Eastman Kodak Co., 1929; mem. U. S. Eclipse Expdn., Pinehurst, N.C., 1900; in charge U. S. eclipse stas., Fort de Kock, Sumatra, 1901, Daroca, Spain, 1905. Fellow A.A.A.S.; mem. Washington Acad. Scis., Am., Royal astron. socs. Author: The Orbit of Hyperion Satellite of Saturn; Positions and Proper Motions of 1504 Standard Stars; The Orbit of Neptunes Satellite and the Pole of Neptunes Equator. Research on positions and proper motions of standard stars. Died 1951.

EICHELBRENNER, Ernest A., physicist; b. Rüstringen, Germany, June 28, 1913; s. Ernst C. and Katharina (Koehn) E.; student univs. Hamburg (Germany). Rome (Italy), 1931-37; Staatsexamen, Hamburg, 1938, Dipl.-Math., 1943; Dr. ès Sc., Paris, France, 1955, Lic. ès Lettres, 1956; m. Emilie M. Martin, Aug. 16, 1939; children—Hans-Michael, Detlef, Ursula, Etienne. Research engr. Deschimag, Bremen, Germany, 1938-45; asst. U. Hamburg, 1946; research engr. Onera, Chatillon-sous-Bagneux, France, 1947-58, cons., 1958-65; maitre de confs. U. Poitiers (France), 1958-60, prof. without chair, 1961-62, titular prof., 1962-65; titular prof. Laval U., Quebec, Can., 1965-—. Cons. BTZ, Brunoy, France, 1958-63. Recipient Henry Bazin prize and medal French Acad. Scis., 1965. Mem. Am. Soc. Mech. Engrs., Math. Soc. France, other profl. socs. Reviewer Zentralblatt für Mathematik, 1958-—. Research in fluid mechanics particularly boundary layer theory; 3-dimensional (laminar and turbulent) boundary layers, characteristics and criteria of separation and reattachment of 3-dimensional boundary layers; problems of heat transfer in 3-dimensional flow. Home: 1586 Côte Ross, Ste.-Foy, Quebec 10, Can.*

EICHENGRÜN, (Ernst) Arthur, German chemist; b. Aachen, Germany, Aug. 13, 1867; s. Julius and Emma (Mayer) E.; doctorate U. Erlangen (Germany), 1890; hon. dr.engring. Tech. U. Hannover, Germany, 1929; hon.dr.rer.nat. Tech. U., Berlin-Charlottenburg, 1948; m. Elizabeth Fechheimer, 1894 (div.); m. 2d, Madeleine Mynssen (div.); 6 children from 1st and 2d marriages; m. 3d, Lucie Henriette Bartsch. Asst. to C. Graebe, U. Geneva; chemist for C. H. Boehringer, Ingelheim, also C. L. Marguardt, Bonn, Germany; named head pharm. lab. Frdr. Bayer & Co., Elberfeld, Germany, 1895, resigned 1908; founded his own works in Berlin which eventually became Cellon Werken Dr. A.E. Berlin-Charlottenburg; confiscated by Nazis, 1938; spent 14 months in concentration camp, Theresienstadt, during World War II. Research and publs. on devel. (with Bayer) new medicines, including aspirin; developed (with T. Becker, H. Guntrum) acetyl cellulose known as cellit (the basis of safety film, acetate artificial silk), 1905; developed nonburning acetyl cellulose celloids (cellon), cast plastics, and injection technique. Died Bad Wiessee, Germany, Dec. 23, 1949.

EICHER, George John, Am. aquatic biologist; b. Bremerton, Wash., Aug. 27, 1916; s. George John and Caroline Agnes (Wolfer) E.; student Wash. State U., 1938; B.S., Ore. State U., 1941; m. Patricia Jane Davies, Feb. 17, 1951; children—George C., Kenneth. Research party leader in Alaska, U. S. Bur. Fisheries, 1939-41; free-lance writer, 1941-43; fish biologist Ariz. Game and Fish Commn., 1943-47; charge salmon research in Western Alaska, U. S. Fish and Wildlife Service, 1947-56; chief aquatic biologist Portland Gen. Elec. Co., 1956—; cons. in field, 1958—. Mem. Gov. Ore. Outdoor Recreation Council, 1960—; sci. bd. cons. demonstration grants water quality and pollution control Dept. Health, Edn. and Welfare, 1963—; adviser Fisheries Nat. Izaak Walton League Am., 1960—. Mem. Am. Fisheries Soc. (pres. 1965), Assn. Power Biologists (pres. 1958-60), Am. Inst. Fishery Research Biologists, Wildlife Soc., Am. Soc. Limnology and Oceanography, Am. Inst. Biol. Scis., Pacific Fishery Biologists (sec. 1950), Sigma Chi. Discovered major cause of trout losses in Ariz. mountains, also red salmon fluctuations in Western Alaska; designed successful major fish passage facilities at North Fork project, Clackamas River, Ore. Home: 60 N.W. 87th Av., Portland 97229. Office: 621 S.W. Alder St., Portland, Ore. 97205.*

EICHHORST, Hermann L., physician; b. Königsberg, Prussia, Germany, Mar. 4, 1849; s. Johann Friedrich and Friederike A. (Bott) E.; ed. U. Königsberg; m. Julie von Ried, 1877; 2 sons, 3 daus. Asst., Berlin, to 1876; assoc. prof., Univ. Jena, 1876; prof., Medical Polyclinic, Göttingen, 1877; dir., Medical Clinic, Zurich, 1884-1921. Author: Handbuch der speziellen Pathologie und Therapie (2 vol.), 1883-84; Lehrbuch der praktischen Medizin innerer Krankheiten, 1899. Noted therapist and enforcer of medical ethics; described neuritis fascians (called Eichhorst's neuritis), 1873; studied pernicious anemia, innervation of the heart, methods of physical exploration; representative of casuistic period in development of medicine. Died Zurich, Switzerland, July 26, 1921.

EICHLER, August Wilhelm, German botanist; b. Neukirchen, Germany, Apr. 22, 1839; s. Johann Adam and Elise (Nöding) E.; doctorate, Marburg, 1861; m. Luise Katherine Dorothea Brill, 1871; 6 children. Asst. to C. F. P. v. Martius, Munich; became lectr., Munich, 1865; joined Tech. U., Graz, Austria, as dir. Bot. Garden there, 1871; became prof., Kiel, Germany, 1872; named prof. bot. taxonomy, dir. herbarium, Berlin, 1878; also dir. Bot. Gardens nr. Schöneberg. Mem. French Acad. Scis. Author: Blüthendiagramme, 2 vols., 1875-78; Syllabus, der Vorlesungen über Phanerogamenkunde, 1876; Beiträge zur Morphologie und Systematik der Marantaceen, 1884; Zur Entwicklungsgeschichte der Palmenblätter, 1885. Worked with v. Martius on Flora Brasiliensis, 1840-1906, and took it over at his death in 1868. Described 22 plant families, especially coniferae and cycadaceae; studied Brazilian flora, taxonomy of higher plants, comparative structure of flowers; flower part symmetry; developed plant classification system (improved by A. Engler), which European botanists widely adopted. Died Berlin, Mar. 2, 1887.

EICHLER, Martin, mathematician; b. Pinnow, Germany, Mar. 29, 1912; s. Max Emil and Katharina (Pirwitz) E.; ed. U. Königsberg (Germany), 1930-31, U. Zürich (Switzerland), 1931-32, U. Halle (Germany), 1932-35; Dr. sc. nat.; m. Erika Paffen, Jan. 3, 1947; children—Ralph Alexander, Norbert Friedrich. Asst., U. Halle, 1935-36; asst. U. Göttingen (Germany), 1936-47, dozent, 1939—, prof., 1947; sci. staff Royal Aircraft Establishment, Farnborough, Eng., 1947-49; apl. prof. U. Münster (Germany), 1949-55; with Tata Inst. Fund. Research, Bombay, India, 1955-56; full prof. U. Marburg (Germany), 1956-58, U. Basel (Switzerland), 1958—. Author: Quadratische Formen und orthogonale Gruppen, 1952; Algebraische Zahlen und Funktionen, 1963, English edit., 1966. Research, publs. on number theory, algebra, algebraic function theory. Home: 14, Magnolienpark, Basel. Office: 21, Rheinsprung, Basel, Switzerland 4000.*

EICHLER, Oskar, German pharmacologist; b. Gilgenburg, Germany, Aug. 20, 1898; s. Carl and Margarete (Neumann) E.; student U. Munich, 1921-22; Dr.med., U. Königsberg; m. Lore Täubert, July 16, 1932; children—Jörg, Gerlind (Mrs. Wolfgang Wiesner), Irmhild (Mrs. Günther Jacobs). Faculty, U. Giessen, 1930-34; prof., dir. Pharm. Inst., U. Breslau, 1934-45; prof. clin. pharmacology U. Heidelberg, 1948—, dir. Pharm. Inst., 1958—. Mem. German Pharm. Soc. Author: Kaffee und Coffein, 1938; Principien des Lebendigen, 1949; Anorganische Anionen, 1950; (with Heinzel) Locale durch Blutungsstörungen, 1961; (with Korkhaus, Sauerwein) Kariesprophylaxe und Sozialhygiene, 1951; also numerous articles. Research on histamine, fluorine and other anorganic anions; bronchial secretion; fluoridated dentifrice. Home: 18 Gutleuthofweg, Heidelberg-Schlierbach. Office: 47-51 Hauptstrasse, Heidelberg, West Germany.*

EICHSTEDT, Karl Ferdinand, German physician; b. Greifswald, Germany, Sept. 17, 1816; s. Johann Philipp and Henriette Christiane Wilhelmine (Geerds) E.; studied medicine, Berlin; doctorate, Greifswald, 1839; m. Hermine Wilhelmine Charlotte Friderike Kettner; 1 son, 4 daus. Worked under F. A. G. Berndt; later worked under surgeon, W. Baum, beginning in 1843; became lectr., Greifswald, 1846, took up gynecology and became asso. prof., head obstet. polyclinic; resigned from univ., 1858; built pvt. polyclinic; pvt. practice medicine. Discovered fungus causing pityriasis versicolor (named Microsporon furfur by Charles P. Robin 1853). Died Greifswald, Dec. 31, 1892.

EICHWALD, Charles-Edward, Russian naturalist; b. Mittau, Russia, 1795; prof. zoology and medicine, Dorpat, Estonia, also St. Petersburg, Russia; pub. work on animal fossils of Russia, 1857, also on antediluvian period in Russia. Died St. Petersburg, 1877.

EICHWALD, Ernest Julius, pathologist; b. Hanover, Germany, Dec. 13, 1913; s. Paul A. and Jenny (Grahn) E.; M.D., U. Freiburg, 1938, U. Utah, 1953; m. Beverly M. Plath, Sept. 9, 1967; children—Jenny, Paul, John. Faculty, U. Utah, 1948-53; practice medicine, specializing in pathology, Great Falls, Mont., now Salt Lake City; dir. labs. Mont. Deaconess Hosp., 1954-65; dir. Lab. Exptl. Medicine, 1954-65; dir. McLaughlin Research Inst., also prof. microbiology Mont. State U., 1966-67; prof. pathology, research prof. surgery U. Utah, Salt Lake City, 1968—. Chmn. transplant com. Nat. Acad. Scis., 1955—. Editor: Transplantation Bull., 1953-63; Transplantation, 1964—. Research, publs. on phenomenon male-specific graft rejection. Home: 3759 Adonis Dr., Salt Lake City.*

EICHWALD, Karl Eduard von, see von Eichwald, Karl Eduard.

EICKEMEYER, Rudolf, inventor; b. Altenbamberg, Bavaria, Germany, Oct. 31, 1831; s. Christian and Katherine (Brehm) E.; m. Mary True Tarbell, July 1856, 6 children. Came to U. S., 1850; devised 1st whip-stitch for hatbands; 1st successful hat stretching machines; patented 1st hatblocking machine, 1865; designed machine to pounce hats, 1869; developed 1st direct-connected ry. motor for use on N.Y. Elevated R.R.; developed differential gear for mowing and reaping, 1870; secured about 150 patents in U. S. and abroad, including 1st symmetrical drum armature, iron clad dynamo; discovered and was 1st employer of Charles P. Steinmetz; bus. consol. with Gen. Electric Co., 1892. Died Washington, Jan. 23, 1895.

EIDE, Carl John, Am. plant pathologist; b. Carrington, N.D., Aug. 20, 1904; s. Mathias and Amelia (Somsen) E.; B.S., U. Minn., 1928, M.S., 1929, Ph.D., 1934; m. Johanna Larsen, Dec. 27, 1930; children—David Alan, Charles Edward. Instr., La. State U., 1929-30; mem. faculty U. Minn., St. Paul, 1930—, prof. plant pathology, 1947—, investigator in charge vegetable disease research Agrl. Expt. Sta., 1938—. Spl. sci. aide Rockefeller Found., 1955; agrl. office UN Food and Agr. Orgn., 1960. Mem. Am. Assn. U. Profs., Am. Phytopathological Soc., Minn. Acad. Sci., Potato Assn. Am., European Assn. Potato Research. Research and numerous publs. in causes and control of vegetable diseases, especially potato, nature of disease resistance, devel. of testing methods for types of disease resistance remaining stable in agrl. practice. Home: 2228 Hillside Av., St. Paul 55108.*

EIDELBERG, Eduardo, physician, educator; b. Lima, Peru, Apr. 30, 1930; s. Oscar and Sarah (Oklander) E.; B.M., Nat. U., Lima, 1954, M.D., 1955; m. Nora Dajes, Feb. 17, 1956; children—Sonia, Elaine. Came to U. S., 1956, naturalized, 1962. Kenny Found. scholar, 1960-62; asso. research prof. dept. anatomy U. Cal. at Los Angeles Med. Sch., 1958-62; chmn. div. neurobiology Barrow Neurol. Inst., Phoenix, 1962—; research prof. neurophysiology Ariz. State U., 1963—. Mem. Am. Physiol. Soc., Am. Electroencephalographic Soc., Biophys. Soc., A.A.A.S., N.Y. Acad. Scis. Author: Textbook of Neurophysiology (pub. in Spanish), 1965. Research and publs. on brain wave prodn. by cerebral cortex; mechanisms of chem. transmission between nerve cells; design of spl. purpose computer for brain wave analysis. Home: 1110 W. Edgemont St., Phoenix 85007. Office: 350 W. Thomas Rd., Phoenix 85013.*

EIDSON, William Whelan, Am. physicist; b. Indpls., July 22, 1935; s. Alonzo Duncan and Gertrude (Whelan) E.; B.S., Tulane U., 1957; M.S., Ind. U., 1959, Ph.D. (NSF fellow), 1961; m. Susan Edwards, June 11, 1960; children—William Benjamin, Duncan McBrayer. Faculty Ind. U., Bloomington, 1961-67, asso. prof. physics, 1966-67; prof., chmn. physics U. Mo., St. Louis, 1967—. Mem. Am. Phys. Soc., Am. Assn. Physics Tchrs., A.A.A.S., I.E.E.E., Sigma Xi, Sigma Pi Sigma. Research in gamma-ray spectroscopy and charged-particle reaction; studies in nuclear energy levels and nuclear reaction mechanisms, nuclear instrumentation, solid state detectors, ion-electron recombination. Home: 66 Bellerive Acres, St. Louis 63121.*

EIDUS, Yacov, Russian chemist; b. Dvinsk, Russia, Mar. 15, 1907; s. Tevel Ch. and Sherel (Mindlin) E.; student Moscow U., 1926-30, Dr.Chem. Scis., 1941; m. Elizabeth M. Terentieva, Feb. 21, 1932; 1 son, Vladimir. Sr. asst. Chem. Faculty, Moscow U., 1932-38; with Inst. Organic Chemistry, Acad. Scis., 1934-37, sr. sci. worker, 1938-45; group leader B.A. Kazansky Lab. of Catalytic Synthesis, Moscow, 1945—. Vice prof. I. M. Gubkin Petroleum Inst., 1944-45; vice chmn. sci. council Inst. Organic Chemistry, Acad. Scis., 1964—. Author: Liquid Fuel and the War, 1943. Contbr. numerous articles to profl. jours. Research on chem. reactions in elec. discharges, Fischer-Tropsch reaction catalysts and mechanisms; discovered initiated heterogeneous catalysis, new reactions initiated by carbon monoxide or oxygen catalytical hydropolymerization of olefins, carboxylation of alcohols, carbalkoxylation of olefins, alcohols by means of carbon monoxide or formic acid. Home: 4/34 D. Ulianova, No. 26. Office: 47 Lenin Prospect, Moscow, USSR.*

EIDUSON, Samuel, Am. biochemist; b. Buffalo, Dec. 15, 1918; s. Aaron and Tillie (Neyman) E.; B.S. in Chemistry, U. Cal. at Los Angeles, 1947, Ph.D., 1952; m. Bernice F. Tabackman, June 21, 1942. Prin. scientist VA, Los Angeles, 1955-60; chief research biochemist Neurophys. Inst., Med. Center, U. Cal. at Los Angeles, 1961—, asso. prof. dept. biochemistry and psychiatry, 1963—. Mem. Am. Soc. Biol. Chemists, Sigma Xi, Phi Lambda Upsilon. Author: (with Edward Geller, Artur Yuwiler, Bernice Elpuson) Biochemistry and Behavior, 1964; also articles. Research on biochemistry of the developing brain and behavior. Home: 941 Stonehill Lane, Los Angeles 90049.*

EIFFEL, Alexandre Gustave, French engr.; b. Dijon, France, Dec. 15, 1832; ed. colls. Dijon, Sainte-Barbe, Central Sch. Arts and Manufacture; grad. 1855. m. Marguerite Gaudelet; 4 children. Constructed iron bridge over Garonne river, Bordeaux, 1858; built bridge over Douro, Portugal, 1876, Viaduct of Garabit, Cantal, 1882, Eiffel tower, Paris, 1889; designed sluices for Panama Canal; founder 1st lab. for aerodynamics, Auteuil, France, 1912. Pres., Civil Engrs. of France Soc.; hon. mem. Mech. Engrs. Instn., London. Author: Resistance to Air, 1913. Research on effects of air currents on airplanes; study of meteorology and aerodynamics; pioneer in use of air caissons in bridge constrn. Died Paris, Dec. 28, 1923.

EIGEN, Manfred, physicist; b. Bochum, Germany, May 9, 1927; s. Ernst E. and Hedwig (Feld) E.; ed. physics and chemistry, U. Göttingen (Germany); hon. degrees, Harvard, U. Chgo., Washington U., St. Louis; m. Elfriede Müller; two children. Sci. asst. Inst. Phys. Chemistry U. Göttingen, 1951-53; mem. staff Max Planck Inst. of Phys. Chemistry, Göttingen, 1953—, now chmn.; vis. lectr. Cornell U. Recipient (with G. W. Norrish and G. Porter) Nobel prize in chemistry, 1967. Mem. Bunsen Soc. Phys. Chemistry (Bodenstein preis), Faraday Soc., Nat. Acad. Scis. Author tech. papers. Studies on rate of hydrogen ion formation through dissociation of water; research on control of enzymes. Address: 19 Stettiner Strasse, Göttingen, Germany.

EIGENMANN, Carl H., zoölogist; b. Flehingen, Germany, 1863; s. Philip and Margaretha (Lieb) E.; A.B., Ind. U., 1886, A.M., 1887, Ph.D., 1889; m. Rosa Smith, Aug. 20, 1887; children—Lucretia Margaretha, Charlotte Elizabeth, Theodore Smith, Adele Rosa E. (Mrs. John O. Eiler), Thora Marie. Prof. zoölogy, Ind. U., 1891-1927, dean Grad. Sch., 1908-27; curator fishes Carnegie Mus., Pitts., 1909-18. Founder and dir. biol. sta., Ind. U., 1895-1920. Explorations in Cal., Ore., Ida., Mont., Dak. and Western Can., 1890-92, Cuba, 1902-04, Brit. Guiana, 1908, Colombia, 1911-12, Peru and Bolivia, 1918, Chile, 1919. Fellow A.A.A.S., Am. Micros. Soc., Washington, Ind. acads. scis., Soc. Am. Naturalists, Western Naturalists, Am. Zool. Soc., Western Zoologists, Am. Philos. Soc., Nat. Acad. Sci., Soc. Geog. Lima, Phi Beta Kappa, Sigma Xi; hon. mem. Sociedad Ciencias Naturales Bogota. Author: Cave Vertebrates of North America, The Archiplata-Archhelenis Theory, The American Characidae, The Fresh-water Fishes of British Guiana, The Fishes of Western South America, and The Doradidae; A Study of Degenerative Evolution, 1909; The Fresh Water Fishes of Chile, 1927. Studied S.Am. catfishes and Cal. fishes, also blind fish of Ind. and Ky. caves. Died Apr. 24, 1927.

EIGSTI, Orie Jacob, Am. geneticist; b. Morton, Ill., Sept. 23, 1908; s. Jacob and Lydia (Gerig) E.; Ph.D., U. Ill., 1935; m. Agnes Weaver, May 22, 1936; children—Karl Jacob, Nicholas Weber. Researcher, Carnegie Inst. Washington, Cold Spring Harbor, L.I.,

N.Y., 1935-36; faculty Triarch Coll., Ripon, Wis., 1937; prof. Greenville Coll., 1937-38, Goshen Coll., 1938, U. Okla., Norman, 1939-45, Northwestern U., 1946-50; bot. researcher Ill. Normal Sch., 1951-52; Fulbright prof., Pakistan, 1952-53; vis. research specialist Med. Sch., Brussels, Belgium, 1953-54, Chgo. Tchrs. Coll., 1955-64; prof. botany Ill. Tchrs. Coll., Chgo., 1965—. Pres., Am. Seedless Watermelon Seed Corp., Goshen, Ind., 1954—. Fellow A.A.A.S.; mem. Bot. Soc., Hort. Soc., Kennicott Linnean Soc., Genetics Soc., Ill., Pakistan acads. sci., Internat. Soc. Phytomorphology, Am. Assn. U. Profs., Internat. Soc. Hort. Sci., Sigma Xi. Author: (with P. Dustin, Belge) Colchicine, 1955; contbr. to Fifty Years of Botany, 1958; also articles. Pioneered use colchicine on plant cell mitosis to induce polyploidy; discovered C-mitosis in plant cells. Home: 1814 W. 79th St., Chgo. 60620.*

EIJKMAN, Christiaan, bacteriologist, physician, hygienist; b. Nijkerk, Netherlands, Aug. 11, 1858; M.D., U. Amsterdam, 1883; studied bacteriology under Koch in Berlin; went to Batavia as army surgeon, 1886; dir. Path. Lab., Batavia (now Djakarta, Indonesia), 1888-96; prof. hygiene, Utrecht, Netherlands, 1898-1928. Recipient (with Sir F. G. Hopkins) Nobel prize in physiology and medicine, 1929. Produced polyneuritis in chickens by feeding them polished rice; proved beriberi is caused by dietary deficiency, thus refuting the universality of germ theory of disease; his experiments led to concept of vitamins and vitamin deficiency. Died Utrecht, Nov. 5, 1930.

EIK-NES, Kristen Borger, physiologist; b. Sparbu, Norway, Sept. 28, 1922; s. Knut and Nina L. (Dahler) Eik-N.; B.S., Coll. Steinkjer, 1942; M.D., U. Oslo, 1951. Faculty dept. biochemistry U. Utah, Salt Lake City, 1949—, asso. prof., 1959—. Tng. program dir. U. S. Pub. Health Tng. Program in Steroid Biochemistry, 1956—; biomed. adviser NASA, 1961—; mem. NIH Study Sect. on Reproductive Biology, 1966. Mem. Endocrine Soc., Am. Physiol. Soc., N.Y. Acad. Scis., Soc. Reprodn. and Fertility, Royal Soc. Medicine (U.K.). Asso. editor Canadian Jour. Biochemistry, 1962—. Contbr. numerous articles to profl. jours. Research, publs. on biochemistry of steroid hormones, regulation of endocrine glands, physiology of reprodn. Home: 1967 Foothill Dr., Salt Lake City 84118. Office: Cancer Research Wing, U. Utah, Salt Lake City 84112.

EILENBERG, Samuel, mathematician; b. Warsaw, Poland, Sept. 30, 1913; M.A., U. Warsaw, 1934. Ph.D., 1936. Came to U. S., 1939, naturalized, 1948. married to Natasha Chterenzon, August 1960. Instr. mathematics U. Mich., 1940-41, asst. prof., 1941-45, asso. prof., 1945-46; prof. mathematics Ind. U., 1946-47, Columbia, 1947—; vis. lectr. Princeton, 1945-46; vis. prof. U. Paris, 1950-51, Tata Inst. Bombay, 1953-54, Hebrew U., Jerusalem, 1954. Fulbright and Guggenheim fellowships, Paris, 1950-51. Member National Academy of Sciences, Am. Mathematical Soc. (councillor 1947-59), Mathematical Association of America. Author: (with N. E. Steenrod) Foundations of Algebraic Topology, 1952; (with H. Cartan) Homological Algebra, 1956. Research on algebra and topology. Address: Columbia U., N.Y.C. 27.

EIMER, Theodor Gustav Heinrich, zoologist; b. Stäfa, Switzerland, Feb. 22, 1843; s. Heinrich and Albertine (Pfenninger) E.; studied medicine and natural scis., Tübingen, Freiburg, Heidelberg, Berlin (all Germany); M.D. under Virchow; state exam, 1868; Ph.D. under v. Kölliker, Würzburg, Germany, 1869; m. Anna Lutteroth, 1870; 2 sons, including Helmuth, 2 daus. Became lectr. zoology, Würzburg, 1870; named insp. Darmstadt Mus., also asso. prof. Polytechnikum, 1874; became prof. zoology, Tübingen, 1875; traveled in Italy, Turkey, Balkans, Egypt, 1878-79. Author: Die Medusen, 1878; Über das Variieren der Mauereidechse, 1881; (with C. Fickert) Die Entstehung der Arten auf Grund von Vererben erworbener Eigenschften, 3 parts, 1888-1901; (with C. Fickert) Die Artbildung und Verwandtschaft bei den Schmetterlingen, 2 parts, 1889-95. Research in morphology, physiology and histology; opposed Darwin's theory of selection; believed variation in organisms occurred according to law, that origin of new characteristics and types stemmed from physiol.-chem. causes. Died Tübingen, May 29, 1898.

EIMMART, George-Christophe (Eimart), German astronomer, engraver; b. Regensburg, Germany, Aug. 22, 1683; dir. Nuremberg (Germany) Obs., until his death; mem. French Acad. Scis., 1699. Author: Iconographia nova contemplationum de sole, 1701. Designed and made plates of birds and plants; invented several astron. instruments, including armillary sphere. Died Nuremburg, Jan. 5, 1705.

EINHOF, Heinrich, German chemist; b. Bahrendorf, Germany; prof. chemistry Thaer's Agrl. Inst., Möglin, Germany; author: Grundriss der Chemie für Landwirthe, 1808; contbr. numerous articles; studied process of vegetation and action of acids and alkalis on it; analyzed foodstuffs; indicated that gluten is not homogeneous; believed that plants grown on soil free lime form lime. Died Möglin, Feb. 28, 1808.

EINHORN, Jerzy, Polish endocrinologist; b. Sosnowiec, Poland, Mar. 17, 1919; s. Oskar and Karolina (Birman) E.; B.Medicine, U. Poznan (Poland)-Med. Acad., 1951, M.D., 1960; m. Jadwiga Piaskowska, Mar. 17, 1946; children—Janusz, Robert, Eva. Vol. asst. Internal Diseases Clinic, Poznan, 1950-51; asst. path. dept. Silesian Med. Acad., Katowice, Poland, 1951-54, faculty, 1954—, sr. lectr. endocrinology, 1959—; researcher Postgrad. Med. Sch., Hammersmith Hosp., London, Eng., 1965; staff Katowice County Endocrinological Out-Patients Clinic, 1951—, dir., 1954—, cons. in endocrinology, 1958—. Decorated Silver Service Cross, 1961. Mem. Polish Endocrinological Assn. (1st prize 1962, exec. com. 1960—), Polish Internal Diseases Assn. Author: (with others) Contemporary Therapy, 1962; also articles. Research in endocrinology, hormone influence on transplanted tissue, influence of emotional factors on the incidence of hyperthyroid activity, complex nature of secretion from simple euthyroid adenomatous goitre, effect of thyroid hormones on calcium metabolism, Einhorn test of thyroid function resulting from research on phenomenon of varied response to thermal stress. Home: 12 Kopernika, Katowice. Office: III Klinika Chorób Wewn. Sl. Akad. Med., Katowice, Poland.*

EIHNORN, Max, physician; b. Grodno, Russia, Jan. 10, 1862; s. Abraham and Sarah (Hoffmann) E.; ed. U. Kiev (Russia), 1879-80, U. Berlin, 1880-84, German Hosp. (now Lennox Hill Inst.), N.Y.C., 1895; M.D.; m. Flora Strauss, Mar. 30, 1892. Prof. medicine N.Y. Post-Grad. Med. Sch. and Hosp., from 1888; vis. physician German Hosp. Mem. N.Y. Acad. Medicine, A.M.A., N.Y. County Med. Soc., Med. Assn. State of N.Y., Med. Soc. Munich (corr). Author: Diseases of the Stomach, 1896; Practical Problems of Diet and Nutrition, 1905; Lectures on Dietetics, 1914; The Duodenal Tube and its Possibilities, 1920. Research on lymphogranulomatosis, leukemia, other diseases; discovered Achylia gastrica, 1892; inventor fermentation saccharometer, 1887, gastrodiaphane, 1889; deglutable stomach electrode, 1891, gastrograph, 1890, gastric spray, 1892, stomach powder blower, 1899, an oesophagoscope, 1901, the radiodiaphane, 1904, radium receptacles for oesophagus, stomach and rectum, 1904, the bead test, 1907, duodenal pump, 1909, cardiodilator, 1909, pyloric dilator, 1910. Died N.Y., Sept. 25, 1953.

EINSTEIN, Albert, theoretical physicist; b. Ulm an der Donau, Germany, Mar. 14, 1879; s. Hermann and Pauline (Koch) E.; ed. Luitpold Gymnasium, Munich, Aarauer Kantonsschule, Aarau, Switzerland, Technische Hochschule, Zurich, 1895-1900; Ph.D., U. Zurich, 1905; D. honoris causa univs. Oxford, Cambridge, Manchester, Princeton, Paris, Madrid, Harvard, London, numerous others; m. Mileva Marec, 1901; children—Albert, Eduard; m. 2d, Elsa Einstein, 1917 (dec. 1936). Tech. asst. Swiss Patent Office, Berne, 1902; privatdozent, Bern, 1908; prof. extraordinary, Zurich, 1909; prof. theoretical physics, Prague, 1911; returned to corr. post, Zurich, 1912; prof. physics U. Leyden, 1912-28; dir. Kaiser Wilhelm Inst. for Physics, prof. physics U. Berlin, 1914-33; lectr. in Europe, U. S., Far East, 1920's; vis. lectr., Eng., U. S., 1930-33; renounced German citizenship, came to U. S., 1933, naturalized, 1940; apptd. for life as mem. Inst. Advanced Studies, Princeton, N.J., 1933; leading figure in world govt. movement after war; offered presidency of Israel, declined; collaborated with Chaim Weizmann in establishing U. Jerusalem. Recipient Nobel prize for physics, 1921; Fellow Royal Soc., 1921 (Copley medal 1925), medal Franklin Inst., 1935. Mem. Prussian, French (fgn.) acads. scis. Author: Meaning of Relativity, 1923; Sidelights on Relativity, 1923; Investigation on the Theory of Browning Movement, 1926; Builders of the Universe, 1932; On the Method of Theoretical Physics, 1933; (with Sigmund Freud) Why War?, 1933; The World as I See It, 1934; (with Leopold Infeld) Evolution of Physics, 1938, also articles. Originator of spl. and gen. theories of relativity, 1905, 16, which deal with systems in uniform, non-accelerated motion, gen. accelerated systems, respectively; assumed for his theory constancy of speed of light in vacuum, particle-like properties of light and relativity of motion and rest; developed law of simultaneity; worked out interrelationship of mass and energy, showed these are different aspects of same phenomenon; developed in math. detail new theory of gravitation of which Newton's classic theory is but spl. case; developed law of photoelectric effect by applying quantum theory to light, did much to establish quantum mechanics, 1905; worked out math. equation to explain Brownian motion, showed it could be used to determine sizes of molecules; deduced influence of gravity on propagation of light; developed gas theory, derivation of Planck's radiation law, concept of duration of excited states, calculation of probability coefficients of emission and absorption, law of photochem. reactions, theory of specific heats; worked on unified field theory in later years. Died Princeton, N.J., Apr. 18, 1955.

EINTHOVEN, Willem, physiologist; b. Samarang, Java, May 22, 1860; M.D., U. Utrecht (Netherlands), 1885; LL.D., Aberdeen, Scotland; M.D., Edinburgh, 1923; m., 1886; 4 children. Became asst. to Donders and Snellen, 1879; prof. physiology U. Leiden (Netherlands), from 1886. Recipient Nobel prize in medicine, 1924. Fellow Royal Soc., 1926. Contbr. papers to jours., chpts. to Handbuch der Laryngologie und Rhinologie (Heyman), Handbuch der normalen und pathologischen Physiologie (Bethe). Editor Onderzoekingen Physiol. Laborat. Inventor forerunner of electrocardiograph by devising string galvanometer for registering elec. changes in human heart, 1903; used electrocardiograms to diagnose heart disorders, 1906. Died Leiden, Netherlands, Sept. 29, 1927.

EISCH, John Joseph, Am. chemist; b. Milw., Nov. 5, 1930; s. Frank Joseph and Gladys (Riordan) E.; B.S. summa cum laude, Marguette U., 1952; Ph.D. (Procter and Gambel fellow 1955, Union Carbide fellow 1956), Ia. State U., 1956; m. Joan Terese Scheuerell, Sept. 5, 1953; children—Margaret, Karla, Joseph, Paula, Amelia. Postdoctoral fellow Max Planck Inst. für Kohlenforschung, Mulheim, Germany, 1956-57; research asso. European Research Assos., Brussels, Belgium, 1957; faculty St. Louis U., 1957-59, U. Mich., 1959-63; faculty Catholic U. Am., Washington, 1963—, prof. chemistry, 1967—, head dept., 1966—; cons. to various business, Ethyl Corp., 1964—. Mem. Am. Chem. Soc., Sigma Xi, Phi Kappa Phi, Phi Lambda Upsilon. Author: The Chemistry of Organometallic Compounds, 1967; (with R. B. King) Organometallic Syntheses, 1965. Research and publs. on the synthesis and properties of organometallic compounds (those with carbon-metal bonds) and heterocycles, with emphasis on the kinetics and stereochemistry of carbon-metal bond and hydrogen-metal bond additions to olefins, acetylenes; radical-anion, halogenation, non-benzenoid aromatic studies. Home: 317 Waterford Rd., Silver Spring, Md. 20901. Office: Maloney Chem. Lab., Catholic U. Am., Washington 20017.*

EISELEY, Loren (Corey), Am. anthropologist; b. Lincoln, Neb., Sept. 3, 1907; s. Clyde Edwin and Daisy (Corey) E.; A.B., U. Neb., 1933; A.M., U. Pa., 1935, Ph.D., 1937; L.H.D., Western Res. U., 1959, N.Y. U., 1960, Washington Coll., 1963, Pace Coll., 1964; Litt. D., U. Neb., 1960, Brown U., 1964, U. Chattanooga, 1968; Sc.D., Franklin and Marshall Coll., 1960, Hahnemann Med. Coll. Phila., 1965; LL.D., Alfred U., 1963; other hon. degrees; m. Mabel Langdon, Aug. 29, 1938. Faculty U. Kan., 1937-44, Oberlin Coll., 1944-47; prof. anthropology, U. Pa., 1947-62, chmn. dept., 1947-59, provost, 1959-61, Benjamen Franklin prof. anthropology and history sci., 1961—, curator early man U. Pa. Mus., Phila., 1961—, chmn. dept. history and philosophy sci. Sch. Arts and Sci., 1961-64. Host TV program Animal Secrets, 1966-68. Bd. dirs. Samuel S. Fells Fund, Phila. Social Sci. Research Council fellow, 1940-41; Wenner-Gren Found. fellow, 1952-53, Center for Advanced Studies in Behavioral Scis. fellow, 1961-62; Guggenheim fellow, 1963-64; recipient Phi Beta Kappa Sci. prize, 1959, Athaeneum Soc. Phila. award, 1959, John Burroughs Meml. Assn., medal 1960, du Nouy Found. award, 1961, Phila. Arts Festival award, 1962; cited by Pa. Dept. Pub. Instruction, 1962. Fellow A.A.A.S. (v.p. elect for 1968), Am. Anthropol. Assn. (past v.p.); N.Y. Acad. Scis., Am. Acad. Arts and Scis., World Acad. Art and Science; member American Philosophical Society, Am. Inst. Human Paleontology (past pres.), Am. Acad. Polit. and Social Scis. (bd. dirs.). Author: The Immense Journey, 1957; Darwin's Century, 1958; The Firmament of Time, 1960; The Mind As Nature, 1962; Francis Bacon and the Modern Dilemma, 1962; also numerous articles, essays and verse. Research in human evolution, early man in Am. and his assn. with extinct Pleistocene fauna. Home: Wyndon Apts., Wynnewood, Pa. 19096. Office: Dept. Anthropology, U. Mus., 33d and Spruce St., Phila. 19104.*

EISELSBERG, Baron Anton von, see von Eiselsberg, Baron Anton.

EISEMAN, Ben, Am. surgeon, b. St. Louis, Nov. 2, 1917; s. Frederick B. and Justine (Godchaux) E.; A.B., Yale, 1939; M.D., Harvard, 1943; m. Mary Harding, Dec. 21, 1946; children—Jane Harding, John Frederick, Lucy Ann, Andrew Thomas. Mem. faculty Washington U. Med. Sch., St. Louis, 1953-61, U. Colo. Sch. Medicine, 1953-61; prof., chmn. dept. surgery U. Ky. Coll. Medicine, 1961-67; prof. surgery, Denver Gen. Hosp., 1967—; cons. to USN. Chmn. com. Nat. Inst. Gen. Med. Scis., NIH, 1965—; mem. trauma com. NRC, 1960—. Diplomate Am. Bd. Surgery, Nat. Bd. Med. Examiners Am. Bd. Thoracic Surgery. Fellow Colombian Coll. Surgeons; mem. A.C.S. Am. Heart Assn. (exec. council 1962—), A.M.A., Am. Surg. Assn., So., Central Surg. Assn., Internat. Soc. Surgery, Soc. U. Surgeons (pres. 1962), Surg. Biology Club, Soc. Vasculary Surgery, Am. Assn. Thoracic Surgery, Am. Gastroent. Soc., Internat. Cardiovascular Soc., Soc. Surgery of Alimentary Tract, Alpha Omega Alpha, others. Research and publs. on the surg. mgmt. of liver failure and the treatment of shock. Home: 3 Village Rd., Denver 80110. Office: Denver Gen. Hosp., W. 6th and Cherokee St., Denver 80204.*

EISEN, Gustavus A(ugustus), biologist, archeologist; b. Stockholm, Sweden, Aug. 2, 1847; s. Frans August and Amalia (Markander) E.; Ph.D., U. Upsala, 1873; came to U. S., 1873, naturalized, 1887; unmarried. Publs. on antique glass, antique beads and antique bronzes; author of The Fig, and The Raisin Industry, articles on biol. nature of caprification, explorations in Lower Cal., Mexico, C. Am. republics (1880-1903); archeol. explorations, especially Christian archeology and antique glass (1903-15), in Spain, Italy, Algiers, Tunis, Morocco and Egypt; studies in prin. museums of Europe, 1910-15. Former docent in zoölogy U. Upsala, collaborator, U. S. Dept. Agr.; late curator Cal. Acad. Scis. Originator of Sequoia Nat. Park in Cal. 1890. Described and classified Gellatly and Freer Mus. collections of antique glass, Smithsonian Inst. Wrote: (monograph) The Great Chalice of Antioch, 1924; Glass—Antique Glass—Its History and Classification (monographs), 2 vols., 1927. Known specially for researches in Oligochaeta (earthworms) of Am., researches in elements of blood of batrachians and man, and amoeba of carcinoma. Died Oct. 29, 1940.

EISEN, Herman N., Am. immunologist; b. N.Y.C., Oct. 15, 1918; s. Joseph Moses and Lena (Karush) E.; A.B., N.Y. U., 1939, M.D., 1943; m. Natalie V. Aronson, Feb. 29, 1948; children—Ellen, Jane, James, Thomas, Matthew. Asst., Columbia U. Coll. Phys. and Surg., 1945-46, N.Y. U. Sch. Medicine, 1947-48, Sloan-Kettering Inst., N.Y.C., 1949; faculty N.Y. U. Sch. Medicine, 1950-55; prof. medicine Washington U. Sch. Medicine, St. Louis, 1955-61, prof. microbiology, head dept., 1961—. Mem. study sect. allergy and immunology NIH, 1957-61, 62—, chmn., 1963—; mem. commn. on immunization Armed Forces Epidemiological Bd., 1961—. Mem. Am. Acad. Arts and Scis., Harvey Soc., Am. Assn. Immunologists, Am. Soc. Microbiology, Am. Soc. Clin. Investigation, Am. Assn. Physicians, Soc. Exptl. Biology and Medicine, Biophys. Soc., Phi Beta Kappa, Sigma Xi, Alpha Omega Alpha. Research and publs. on structure antigens, haptens, antibodies, nature antibody-antigen reaction, immune response to simple substances. Home: 7154 Washington Av., University City, Mo. 63130. Office: 600 S. Kingshighway, St. Louis 63110.*

EISENBERG, Leon, Am. psychiatrist; b. Phila., Aug. 8, 1922; s. Morris and Elizabeth (Sabreen) E.; A.B., U. Pa., 1944, M.D., 1946; m. Ruth Harriett Bleier, June 18, 1948 (div. Aug. 1967); children—Mark Philip, Kathy Bleier; m. 2d, Carola B. Guttmacher, Aug. 31, 1967. Faculty, Johns Hopkins, Balt., 1952-67, prof. child psychiatry, 1961-67; prof. psychiatry Harvard Med. Sch., 1967—; became psychiatrist-in-charge Children's Psychiat. Service, Johns Hopkins Hosp., Balt., 1959, psychiatrist-in-chief Mass. Gen. Hosp., Boston, 1967—. Cons., Balt. City Hosp., Sinai Hosp. Diplomate Am. Bd. Psychiatry. Mem. Md. Psychiat. Soc. (past pres.), Am. Assn. U. Profs., Phi Beta Kappa (pres. 1965—). Research and publs. on infantile autism, psychopharmacologic agt. in child psychiatry, social psychiatry, cognitive devel. in disadvantaged children, reading disability, school phobia, adolescence, devel. intelligence. Address: Mass. Gen. Hosp., Boston 02114.*

EISENBUD, Merril, Am. physicist; b. N.Y.C., Mar. 18, 1915; s. Kalman and Leonora (Kopaloff) E.; B.S. in Elec. Engring., N.Y. U., 1936; Sc.D., Fairleigh Dickenson U., 1960; m. Irma Ruth Onish, Jan. 22, 1939; children—Elliott, Michael, Fredrick. Indsl. hygienist, Liberty Mut. Ins. Co., 1936-47; dir. Health and Safety Lab., AEC, N.Y.C., 1947-57, mgr. N.Y. Operations Office, 1957-59, cons., 1959—; prof. environmental medicine N.Y. U., N.Y.C., 1959—. Expert com. on radiation WHO, 1957—; cons. Edison Co., N.Y.C., 1960—, NRC, 1952—; mem. Nat. Council on Radiation Protection, 1964—. Fellow Am. Nuclear Soc. (dir. 1961-64), A.A.A.S.; mem. Health Physics Soc. (pres. 1965), Am. Indsl. Assn. (dir. 1955-57), Sigma Xi, Eta Kappa Nu. Author: Environmental Radioactivity, 1963. Research in radiol. health, reactor safety, fallout, indsl. toxicology. Home: W. Lake Rd., Tuxedo, N.Y.

EISENHART, Luther Pfahler, Am. mathematician, educator; b. York, Pa., Jan. 13, 1876; s. Charles Augustus and Emma Catherine (Pfahler) E.; A.B., Gettysburg Coll., 1896, D.Sc., 1921, LL.D., 1926; Ph.D., Johns Hopkins, 1900, LL.D., 1953; Sc.D., Columbia, 1931, U. Pa., 1933, Lehigh U., 1935, Princeton, 1952; LL.D., Duke; m. Anna Maria Dandridge Mitchell, Aug. 17, 1908 (dec. Mar. 1913); 1 son, Churchill; m. 2d, Katharine Riely Schmidt, June 1, 1918; children—Anna Small, Katharine Riely. Instr. Gettysburg Coll., 1896-97; faculty Princeton, 1900—, prof., 1909-45, prof. emeritus, 1945—, dean faculty, 1925-33, dean Grad. Sch., 1933-45. Trustee Pa. Coll., 1907-14. Chmn. div. phys. scis. NRC, 1937-46. Recipient Officer Order Crown of Belgium, 1937. Mem. Assn. Am. Colls. (pres. 1930), Am. Math. Soc. (pres. 1931-32), Am. Philos. Soc. (exec. officer 1942-59), A.A.A.S. (past v.p.) Nat. Acad. Sci. (v.p. 1944-48), Acad. Sci. Lima (corr.), Phi Beta Kappa, Phi Kappa Psi. Author: Differential Geometry of Curves and Surfaces, 1909; Transformations of Surfaces, 1923; Riemannian Geometry, 1925; Non-Reimannian Geometry, 1927; Continuous Groups of Transformations, 1933; Co-ordinate Geometry, 1939; An Introduction to Differential Geometry, 1940; The Educational Process, 1945. Editor Trans. Am. Math. Soc., 1917-23. Research on geometrical surfaces; differential geometry and its phys. applications. Home: 25 Alexander St., Princeton, N.J. 08540.

EISENLOHR, August Adolf, German Egyptologist; b. Mannheim, Germany, Oct. 6, 1832; s. Wilhelm and Auguste (Catoir) E.; studied first theology, then chemistry, Heidelberg, Germany; doctorate, 1860; m. Pauline André, 1859; m. 2d, Sofie Schreiber, 1880; children—Friedrich, Ernst. Founder chem. factory, Heidelberg, circa 1860; began study Chinese and Egyptian langs., 1865; became lectr. Egyptology, U. Heidelberg, 1869, asso. prof., 1872, hon. prof., 1885, mem. faculty until his death. Editor: Ein mathematisches Handbuch der alten Agypter (Papyrus Rhin or arithmetic book of Ahmes, written 17th century B.C. which contained most important source of ancient Egyptian math., especially introduction of fractions), 1877. Discovered Gt. Harris Papyrus. Died Heidelberg, Feb. 24, 1902.

EISENLOHR, Wilhelm Friedrich, German physicist; b. Pforzheim, Germany, Jan. 1, 1799; s. Wilhelm and Karoline Heinrike (Sachs) E.; studied pub. finance, math., Heidelberg, Germany, 1817-19; hon.dr., Freiburg, Berlin; m. Gertrud v. Itzstein, 1824; 1 son. Became tchr. physics and math., Lyceum, Mannheim, Germany, 1819, later Gewerbeschule, Mannheim; became prof. physics Polytechnikum, Karlsruhe, 1840-65. Author: Lehrbuch der Physik (1st German physics textbook not based on the French model), 1836; also articles. Research on ultra-violet light, especially that produced by gas discharging, 1854; 1st (with E. Esselbach) to measure wave lengths of ultra-violet. Died Karlsruhe, July 10, 1872.

EISENMAN, George, Am. biophysicist; b. N.Y.C., May 6, 1929; s. William and Nadia (Geffen) E.; A.B., Harvard, 1949, M.D., 1953; m. Nancy Crawford Dunn, Dec. 20, 1952; children—Richard Newton, Frederick Maxwell. Research fellow Harvard, 1950, 53-55, research asso., 1955-56; sr. scientist Eastern Pa. Psychiat. Inst., Phila., 1956-62; asso. prof. Coll. Medicine U. Utah, 1962-65; prof. physiology U. Chgo., 1965—; cons. Corning Glass Works, 1962—. Mem. Am. Physiol. Soc., Biophys. Soc. Editor: Glass Electrodes for Hydrogen and Other Cations: Principles and Practice, 1966. Inventor cation selective glass electrodes; fundamental contbns. to theory of ionic specificity and mechanisms of ionic permeation across membranes. Home: 4860 Kimbark Av., Chgo. 60615.*

EISENSCHMID, Johann Caspar, French geodesist, mathematician; b. Strasbourg, France, Sept. 15, 1656; Ph.D., Strasbourg, 1676; later studied medicine; m. Catherine Elisabeth Boeler, Apr. 30, 1692. Practice medicine, Strasbourg, 1684-96; then became mathematician, geodesist, ancient historian. Mem. French Acad. Scis. (corr.). Author: Diatribe de Figura Telluris elliptico-sphaeroide . . . , 1691; De ponderibus et mensuris veterum romanorum, graecorum, hebraeorum, nec non de valore pecuniae veteris, disquisito, Accesserunt hac editione tabulae Scioppii nummariae et ex variis auctoribus de pecunia romanorum excerpta, 1708; also articles. Re-edited logarithm tables of Kepler-Bartsch, 1700. Studied elongated shape of earth (proof against Newton-Huygens theory of flattened sphere), 1691; publs. on ancient measures, weights, money values. Died Strasbourg, Dec. 4, 1712.

EISENSON, Jon, Am. psychologist; b. N.Y.C., Dec. 17, 1907; s. Abraham Eli and Sarah (Eisenson) E.; B.S., City Coll. N.Y., 1928; M.A., Columbia, 1930, Ph.D., 1935; m. Freda Francke, June 28, 1931; children—Elinore Ruth (Mrs. Laurence Bennett Lurie), Arthur Michael. Prof. speech, dir. Queens Speech and Hearing Center Queens Coll. City U. N.Y., 1946-62; prof. speech pathology and audiology, dir. Inst. For Childhood Aphasia Sch. Medicine Stanford, Palo Alto, Cal., 1962—; cons. VA, 1946—, Nat. Inst. Neurol. Diseases and Blindness, 1965—; guest lectr. numerous univs., 1939-63. Diplomate Am. Bd. Examiners in Profl. Psychology. Fellow Am. Speech and Hearing Assn. (past pres.), Am. Psychol. Assn.; mem. Am. Speech and Hearing Assn. (past pres.), Speech Assn. Am., Phi Beta Kappa. Author: Confirmation and Information in Rewards and Punishments, 1935; The Psychology of Speech, 1938; (with Pintner and Stanton) The Psychology of the Physically Handicapped, 1940; (with Berry) The Defective in Speech, 1942, Speech Disorders, 1956; Examining For Aphasia, 1946, rev., 1954; Basic Speech, 1950; Improvement of Voice and Diction, 1957, rev., 1965; (with Mardel Oglivie) Speech Correction in The Schools, 1957, rev., 1963; (with J. J. Auer and J. Irwin) Psychology of Communication, 1963; (with Paul Boase) Basic Speech, 1964. Contbr. to Handbook of Speech Pathology, 1957. Editor, contbr. Stuttering: A Symposium, 1958. Studies, publs. on organic basis of stuttering; tng. of aphasic children; chronic adult aphasia. Home: 853 Mayfield Av., Stanford, Cal. 94305. Office: Sch. Medicine, Stanford U., 300 Pasteur, Palo Alto, Cal. 94304.*

EISENSTADT, Shmuel Noah, sociologist; b. Warsaw, Poland, Sept. 10, 1923; s. Michael and Rosa (Baruchin) E.; M.A., Hebrew U., Jerusalem, 1944, Ph.D., 1947; m. Shulamith Yarusheusky, Sept. 6, 1948; children—Michael, Erit, Alexander. Mem. faculty Hebrew U., Jerusalem, Israel, 1948—, prof. sociology, 1956—, chmn. dept., 1951—, dean Faculty of Social Scis., 1966-68; fellow Center for Advanced Study in Behavioral Scis., 1955-56; vis. prof. U. Oslo, 1958, U. Chgo., 1960, Harvard, 1966; Carnegie vis. prof. polit. sci. Mass. Isnt. Tech., 1962-63. Recipient McIver award Am. Sociol. Assn., 1964. Mem. Israeli Acad. Scis. and Humanities, Am. Sociol. Soc., Royal Anthrop. Inst., Inst. Differing Civilizations. Author: Absorption of Immigrants, 1954; From Generation to Generation, 1956; Political Systems of Empires, 1963; Essays on Comparative Institutions, 1965; Modernization, Protest and Change, 1966; Israeli Society, 1968; The Protestant Ethic and Modernization, 1968. Research and publs. in sociol. theory and comparative sociology; sociol. analysis of hist. and modern polit. systems; analysis of processes of change and transformation of societies and social systems; social devel. and modernization. Home: 30 Radak, Jerusalem, Israel.*

EISENSTEIN, Ferdinand Gotthold, German mathematician; b. Berlin, Apr. 16, 1823; S. Johann Constantin and Helen (Pollack) E.; ed. Berlin U., 1843; Dr.h.c., Breslau (now Wroclaw, Poland). Private docent, Berlin, 1847-52; later prof. math. Berlin U. Mem. Göttingen Acad. Author: Neue Theoreme der Höheren Arithmatik; Mathematisch Abhandlungen, 1847. Mem. Berlin Acad. Research and publs. on theory numbers and elliptic functions, determinants, theory binary quadratic forms, doubly infinite products using analytical methods; developed theory of complex numbers; extended work of Gauss in ternary quadratic forms from 2 to 3; discovered 1st covariant used in analysis; proved gen. theorem in representation of numbers by sums of squares to be limited to 8 squares and gave solutions for 3 and 5 squares. Died Oct. 11, 1852.

EISENTRAUT, Bruno Martin, German zoologist; b. Oct. 21, 1902; s. Johannes and Anna (Bischoff) E.; Ph.D. in Natural Sci., U. S. Halle; m. Johanna Rissmann; 1 dau., Hannelore. Asst. Zool. Mus., U. Berlin, 1926; asst. State Mus. Natural Sci., Stuttgart, 1926-39, curator, 1939-50, chief curator, 1950-57; dir. Inst. Zool. Research, A. Koenig Mus., Bonn, 1957—. Mem. German Zool. Soc., Soc. Mammiferous Scis., German Ornithol. Soc. Research and publs. on mammal hibernation, regulation of heat, bioloby of cheiropterans; made zool. study trips to Bolivia, Camerouns, Fernando Po. Home: Buschstrasse 45. Office: Koblenzerstrasse 150-164, Bonn, Germany.

EKEBERG, Anders Gustaf, Swedish chemist, mineralogist; b. Stockholm, Sweden, Jan. 16, 1767; grad. U. Uppsala (Sweden), 1788; became prof. chemistry U. Uppsala, 1794; introduced Lavoisier's new chemistry to Sweden; analyzed minerals of Finland and Ytterby, 1802; located and isolated element, tantalum, 1802. Died Uppsala, Feb. 11, 1813.

EKHOLM, Ragnar Erik Gustav, Swedish morphologist; b. Stockholm, Sweden, Apr. 23, 1921; s. Gunnar Erik and Ester (Rewell) E.; M.D., U. Uppsala (Sweden), 1949, Ph.D., 1951; m. Ann-Marie Ryström, Mar. 31, 1951; children—Carl, Catarina. Faculty, U. Gothenburg (Sweden), 1952—, acting prof. anatomy, 1966—; vis. asso. prof. U. Cal. at Los Angeles, 1960, vis. prof., 1967. Decorated knight Order of N. Pole Star. Mem. Swedish Soc. Physicians (mem. bd.), Scandinavian Soc. for Electron Microscopy, European Thyroid Assn. Editorial bd. Jour. Ultrastructure Research, 1965—. Research, publs. on morphology and functions of joints, electron micros. technique, electron micros. studies of exocrine pancreas, electron micros. and biochem. studies of thyroid gland, vital and electron micros. studies on structure and function of blood capillaries. Home: 13 Berggrensgatan, Gothenburg, Sweden.*

EKLUND, Arne Sigvard, physicist; b. Kiruna, Sweden, June 19, 1911; s. Severin and Wilhelmina (Pettersson) E.; Ph.D. in Sci., U. Uppsala; Dr. h.c., U. Graz (Austria); m. Anna-Greta Johannson; children—Kerstin, Anders, Gudrun. Asst., dean Inst. Research for Nat. Def., Stockholm; asst. prof. nuclear physics Royal Inst. Tech., Stockholm, 1946-56; dir. research AB Atomenergi, Stockholm, 1950-56, later dir. reactor devel. div., adminstrv. del. dir. reactor devel., 1950-61; gen. dir. IAEA, Vienna, 1961—. Mem. Royal Swedish Acad. Tech. Scis., Am. Nuclear Soc., Soc. Britannica Nuclear Energy (hon.), others. Author: Studies in Nuclear Physics, 1946. Research and publs. on nuclear physics, instrumentation, atomic energy. Home: Krapfenwaldgasse 48, Vienna XIX. Office: Kärtnerring 2, Vienna 1, Austria.*

EKMAN, (Axel) Gösta, Swedish psychologist; b. Lund, Sweden, June 3, 1920; s. Axel and Elida (Johnsson) E.; B.A., U. Lund, 1944; Ph.D., U. Stockholm, 1947; m. Brita Knutson, July 8, 1944; children—Cecilia, Elisabeth. Psychologist, Inst. Mil. Psychology, 1944-46; asst. prof. psychology U. Stockholm (Sweden), 1947-52, prof., chmn. dept., 1952-67; research prof. psychophysics and scaling methodology Swedish Council for Social Sci. Research, 1967—. Mem. Royal Acad. Letters, History and Antiquities, Swedish Psychol. Assn., Association de Psychologie Scientifique de Langue Francaise, Soc. for Multivariate Exptl. Psychology. Author several textbooks, numerous articles to profl. jours. Research on test theory and psychology of individual differences, psychophysics and scaling methodology, developer methods for measuring perceptual and other subjective events, application of measurement methods to psychol. problems, mainly in perception. (psychol. functions, temporal sensory integration, adaptation and recovery), study of quantitative psychol. mechanisms. Home: Älgstigen 5, Stocksund, Sweden. Office: Psychol. Labs., U. Stockholm, Stockholm VA, Sweden.*

ELBEK, Bent, Danish physicist; b. Uldum, Denmark, Jan. 2, 1929; s. Helge and Lilly (Madsen) E.; mag. scient. et cand.mag., 1953, dr.phil., 1963; m. Eja

Hagen Christiansen, July 18, 1950; children—Birgitte, Jakob, Thomas. Research asst., Niels Bohr Inst., U. Copenhagen (Denmark), 1955-58, lectr., 1958—, sci. dir. Tandem Lab., 1960—; vis. scientist Mass. Inst. Tech., Cambridge, 1955, U. Cal. at Berkeley, 1959-60. Recipient Gold medal U. Copenhagen, 1957. Mem. Danish Phys. Soc. (past chmn.). Research, publs. on properties of heavy nuclei, constn. spectrographs for charged particles. Home: St. Valbyvej, Aagerup, Roskilde, Denmark. Office: Niels Bohr Inst., Risoe, Denmark.*

ELBS, Karl Joseph Xaver, German chemist; b. Breisach, Germany, Sept. 13, 1858; s. Karl and Bertha Katherine (Löw) E.; studied natural scis., Freiburg, Germany; doctorate, 1880; m. Julie Rodeckher v. Rotteck, 1894; 2 sons, 1 dau. Joined faculty U. Freiburg, 1887; named prof., dir. phys.-chem. lab. U. Giessen (Germany), 1894, prof. chemistry, 1913. Mem. German Bunsen Soc. (co-founder). Author: Die synthetische Darstellung der Kohlenstoffverbindungen, 2 vols., 1889-91; Die Anthrachinone, 1900; Übungsbeispiele für die elektrolytische Varstellung chemischer Präparate, 1902; Die Akkumulatoren, 1900. Studied electrochem. reduction of aromatic nitro compounds, anodic oxydation of azo compounds and aliphatic alcohols, chem. processes in lead accumulator; research (with O. Schönherr) on per-sulfuric acid; discovered (with F. Fischer) lead-4-sulfate. Died Giessen, Aug. 24, 1933.

ELDER, Albert Lawrence, Am. chemist; b. Lexington, Ill., June 19, 1901; s. George Clinton and Ann (Rowlands) E.; A.B., U. Ill., 1923, M.S., 1925, Ph.D., 1928, D.Sc., 1962; m. Ruth Dixon, Aug. 31, 1925; children—Louise (Mrs. Marshall Gillispie), John. Research instr. Oberlin Coll., 1928-30; asst. prof., then prof. Syracuse U., 1930-41; dir. research Corn Products Co., 1944-59, co-ordinator research, 1959-61; dir. Corn Products Inst. Nutrition, 1961-66. Head chem. adviser WPB, 1941-44. Mem. Research and Devel. Assos. Food and Container Inst. (pres. 1951-53), Am. Chem. Soc. (pres. 1960), A.A.A.S., Am. Inst. Chemists, Inst. Food Technologists, Sigma Xi, Sigma Pi, Alpha Chi Sigma, Phi Lambda Upsilon. Author: Demonstrations and Experiments in General Chemistry, 1937; Laboratory Manual of General Chemistry, 1941; Textbook of Chemistry, 1941. Research in medicinal chems., proteins, colloid chemistry, also with bacteriophage. Home: 612 S. Stone St., LaGrange, Ill.*

ELDERFIELD, Robert C(ooley), Am. chemist; b. Niagara Falls, N.Y., May 30, 1904; s. Charles James and Nellie (Cooley) E.; A.B., Williams Coll., 1926, D.Sc., 1952; Ph.D., M.I.T., 1930; m. Mary Elizabeth Betts, Aug. 7, 1930; children—Margaret Helen, Anne Elizabeth. Instr. chemistry, Colby Coll., 1930; asst., Rockefeller Inst. Med. Research, 1930-32, asso., 1932-36; asst. prof. chemistry, Columbia Univ., 1936-37, asso. prof. 1937-41, prof., 1941-52; prof. chemistry Univ. Michigan since 1952; instructor chemistry, U. of Ill., summer 1938; sect. mem. Nat. Defense Research Com., 1941-45; exec. sec., Panel on Synthesis, Bd. for Coordination of Malarial Studies, 1943-46; mem. malaria study sect., Nat. Institutes of Health, U. S. Pub. Health Service, 1946-49, consultant 1946-49; sci. cons. Sloan-Kettering Inst., 1952—; visiting scientist Nat. Research Council, Ottawa, Can., 1959-60; Sigma Xi Nat. Lectr., 1959. Received Presidential Certificate of Merit, 1948. Mem. Am. Chem. Soc. (chmn. organic chemistry div. 1952), American Society Biol. Chemists, National Malaria Society, National Academy Sciences, Sigma Xi, Phi Lambda Upsilon, Alpha Chi Sigma, Delta Kappa Epsilon. Author or co-author of about 80 papers in sci. jours. Editor (5 vols.), Treatise on Heterocyclics, 1950; editor, Jour. Organic Chemistry. Research in chemotherapy; synthetic organic chemistry; pharmacology; heterocyclic compounds; alkaloids; cardiac drugs. Home: 1800 Hermitage Rd., Ann Arbor.

ELDJARN, Lorentz, Norwegian physician; b. Masöy, Mar. 23, 1920; s. Lorentz and Astrid (Russanes) E.; M.D., U. Oslo; m. Torunn Weydahl; children—Knut, Astrid, Lorentz-Ivar. Head dept. biochemistry Norsk Hydro's Inst. Research on Cancer, 1950-59; prof., pres. Inst. Clin. Biochemistry, U. Oslo, 1959—. Mem. Norwegian Acad. Sci., Norwegian Med. Assn. Author: The Metabolism of Cystamine, 1955; Mechanisms of Action of Protective and Sensitizing Agents, 1959, also numerous articles. Home: Skogfaret 29, Haslum. Office: Institute of Clinical Biochemistry, Rikshospitalet, Oslo, Norway.

ELDREDGE, Hanford Wentworth, Am. sociologist; b. N.Y.C., Oct. 16, 1909; s. Hanford W. and Elizabeth (Taylor) E.; A.B., Dartmouth, 1931; Ph.D., Yale, 1935; m. Diana Younger, Apr. 22, 1947; children—James, Alan. Faculty, Dartmouth, Hanover, N.H., 1935-42, prof. sociology, 1946—, chmn. sociology dept., 1953-57, 65—, chmn. internat. relations program, 1959-63, chmn. city planning/urban studies, 1960-65; vis. lectr. city planning Harvard, fall 1963; vis. prof. U. Cal. at Berkeley, spring 1967. Dir. Planning and Devel. Collaborative, Washington, 1966—. Mem. Am. Inst. Planners, Am. Sociol. Assn. Author: Culture and Society (with F. E. Merrill), 1952; The Second American Revolution, 1964. Editor: Taming Megalopolis, 1967. Assisted in introducing wholistic planning concepts to urban planning; studied long-range planning concepts in govtl. structuring and function. Home: Tarn House, Elm St., Norwich, Vt. 05055. Office: Dartmouth, Hanover, N.H. 03755.*

ELEY, Daniel Douglas, English phys. chemist; b. Wallasey, Eng., Oct. 1, 1914; s. Daniel and Fanny (Ross) E.; B.Sc., Manchester (Eng.), 1934, M.Sc., 1935, Ph.D., 1937; Ph.D., Cambridge (Eng.) U. 1940, Sc.D., 1954; m. Brenda May Williams, Aug. 31, 1942; 1 son, John Daniel Roderick. Researcher, Ministry of Supply at colloid sci. dept. Cambridge U., 1939-45; faculty Bristol (Eng.) U., 1945-54, reader biophys. chemistry, 1950-54; prof. phys. chemistry Nottingham (Eng.) U., 1954—. Decorated officer Order Brit. Empire. Fellow Royal Soc.; mem. Faraday Soc. (mem. council 1951-54, 59-62, v.p. 1963—), Brit. Biophys. Soc. (meeting sec. 1961-62, hon. sec. 1963-64), Chem. Soc., Faraday Soc., Biochem. Soc., Brit. Biophys. Soc. Editor: Adhesion, 1961. Research, numerous publs. on discovery of electronic semiconduction properties of conjugated organic molecules, demonstration of importance of holes in D-band in heterogeneous catalysis, cationic character of vinyl ether polymerisations, thermodynamic function of aqueous solutions in terms of water structure. Home: 35 Brookland Dr., Chilwell, Notts. Office: Chemistry Dept., Nottingham U., University Park, Nottingham, Eng.*

ELFORD, William Joseph, Brit. bacteriologist; b. Jan. 4, 1900; s. Joseph Elford; B.Sc., Bristol U., 1923, Ph.D., 1925. Colston research fellow, 1923-25; became mem. sci. staff, Med. Research Council, Nat. Inst. for Med. Research, 1925, later head div. phys. chemistry. Fellow Royal Soc., 1950; pres. (Bristol) Univ. Chem. Soc., 1922-23. Research and publs. on study of viruses and bacteriophages by means of phys. chem. methods. Died Feb. 14, 1952.

ELFTMAN, Herbert, Am. anatomist; b. Mpls., Oct. 31, 1902; s. Arthur H. and Ida C. (Magnusson) E.; B.A., U. Cal. at Berkeley, 1923, M.A., 1925; Ph.D., Columbia, 1929; m. Alice Gooding, Jan. 23, 1930; children—Barbara, Eric. Faculty, Columbia, N.Y.C., 1925—, prof. anatomy in charge gross anatomy, 1948—. Cons. to com. on artificial limbs NRC, 1946-57, mem. com. on prosthetics research and devel., 1957—, chmn., 1965—. Mem. Am. Assn. Anatomists, Am. Physiol. Soc., Histochem. Soc., A.A.A.S., N.Y. Acad. Sci., Phi Beta Kappa, Sigma Xi. Contbr. articles to tech. jours. Research on biomechanics, human locomotion, muscular activity in movement and structure and function of foot, evolution upright posture, histochemistry endocrine glands and target organs. Home: 407 Park Av., Leonia, N.J. 07605. Office: 630 W. 168th St., N.Y.C. 10032.*

ELFVING, Erik Gustav, Finnish mathematician; b. Helsinki, Finland, June 25, 1908; s. Fredrik E. W. and Thyra (Ingman) E.; Ph.D., U. Helsinki, 1934; m. Ira Aminoff, Sept. 6, 1936; children—Johan, Jörn, Tord. Lectr., Abo (Finland) Acad., 1933-38; lectr. Inst. Tech., Helsinki, 1938-48; prof. U. Helsinki, 1948—; vis. prof. Cornell U., 1949-51, Columbia, 1955, Stanford, 1960, 66. Mem. Societas Scientiarum Fennica (past treas., chmn. 1967-68). Editor: Mathematica Scandinavica, 1953—; Annals Math. Statistics, 1964-67; asst. editor Zeitschr. für Wahrsch. theorie, 1961—. Research on Riemann surfaces, Markov processes, various problems in math. statistics. Home: 7B Fabriksgatan, Helsinki 14, Finland.*

ELHUYAR Y DE SUVISA, F., see d'Elhuyar y de Suvisa, F.

ELIACHAR, Edouard, French pediatrician; b. Paris, June 6, 1918; s. Maurice David and Renée (Weil) E.; ed. Faculté de Médecine de Paris; m. Nicole Levy, June 29, 1949; children—Laurence, Bernard. Became chief of clinic Faculté de Paris, 1950; apptd. asst. Hôpital des Enfants Malades de Paris, 1952; pediatrician Hôpital d'Aulnay, 1955—. Recipient Medaille des Epidemies. Mem. French Soc. Pediatrics. Co-author: La Coqueluche, 1951; Le diabète infantile, 1954; also papers. Research on path. anatomy and treatment of whooping cough, complications and treatment of diabetes mellitus in childhood, etiology, diagnosis and treatment of acute gastroenteritis in infants. Address: 15, Av. Carnot, Paris 17, France.*

ELIAS, Hans Georg, German chemist; b. Bochum, Germany, Mar. 29, 1928; s. Hermann Ludwig and Elisabeth (Rowlin) E.; Dipl.Chem., Hanover Inst. Tech., 1954, Dr.Rer.Nat., Munich Inst. Tech., 1957; m. Maria Hanke, Mar. 23, 1956; children—Peter, Rainer. Faculty, Fed. Inst. Tech., Zurich, Switzerland, 1960—, asso. prof., 1963—. Cons. several European and Am. cos. Mem. Swiss, German chemists assns., Bunsen Soc. Phys. Chemistry, others. Author: Ultrazentrifugen-Methoden, 1961; also numerous articles. Research on molecular weight determination methods, especially osmosis at non-semipermeable membranes and ultracentrifugation; macromolecular chemistry. Address: 86 Buchzelgstrasse, Zurich, Switzerland.*

ELIAS, (Michael) Hans, biologist; b. Darmstadt, Germany, June 28, 1907; s. Michael and Anna (Oppenheimer) E.; B.A., Realgymnasium, Darmstadt, 1926; M.S., Darmstadt Polytechnicum, 1931; Ph.D., U. Giessen, 1931; m. Anneliese J. Buchthal, Oct. 11, 1936; children—Peter M. S., Thomas D. Research fellow, ETH Zurich, 1935; U. Padua, 1936-37; dept. head, Nat. Research Council, Rome, Italy, 1937-39; prof. embryology and histology, Middlesex Vet. Sch., 1939-45; project Supr., USPHS, 1945-49; prof. anatomy, Chicago Medical Sch., 1949—. Founder and pres., Internat. Soc. for Stereology. Mem., Am. Assoc. Anatomists; Am. Soc. Zoology; A.A.A.S., Anatonische Gesellschaft, Soc. Italiana di Anatomia. Author: (with J. E. Pauly) Human Microanatomy, 1960; approx. 200 articles. Research on basic microanatomy of liver; surgical anatomy of liver; recruitment phenomenon in cancer growth; established stereology as formal science. Home: 2670 Birchwood Lane, Deerfield, Ill. 60015. Office: 710 S. Wolcott Ave., Chicago, 60612.*

ELIASBERG, Wladimir Gottlieb, psychiatrist; b. Wiesbaden, Germany, Dec. 10, 1887; s. Samuel and Rachel Eliasberg; student U. Berlin, 1906-08; M.D., Heidelberg U., 1912; Ph.D., U. Munich, 1924; m. Esther Talbot, Sept. 3, 1952; children—Eva, Hannah Avriel, Miriam Rosenzweig, Suzanne. Practice medicine, specializing in psychiatry, N.Y.C., 1939—; med. dir. Bklyn. Treatment Unit Citizens Com., 1953—, Group Community Guidance Clinics, 1953—; asso. Columbia Seminars, 1962. Diplomate Am. Bd. Psychiatry and Neurology. Fellow A.M.A., Am. Psychiat. Assn., N.Y. Acad. Medicine, Am. Sociol. Assn.; mem. Soc. For History Medicine, Am. Correctional Soc., American Psychological Association, Assn. For Psychiat. Treatment Offenders (past president honorary member), American Mental Health Found. (chmn. profl. bd.), Eastern Psychiat. Research Assn., Am. Soc. Psychoanalytic Physicians (past pres.), Am. Soc. Criminology. Author numerous books including: Psychotherapy and Society, 1959. Contbr. chpts. to books, numerous articles to profl. jours. Research, publs. on abstraction theory, social, indsl. psychology, accident neuroses; criminology, psychoanalysis. Address: 151 Central Park W., N.Y.C. 10023.*

ELIASSEN, Arnt, Norwegian meteorologist; b. Oslo, Norway, Sept. 9, 1915; s. Georg and Helfrid (Strömberg) E.; Dr.Phil., U. Oslo, 1950; m. Ellen Kristine Nome, June 24, 1944; children—Anton, Jörgen. Meteorologist, Norwegian Meteorol. Inst., 1942-52; lectr. U. Oslo, 1953-58, prof. geophysics, 1958—; vis. meteorologist Inst. for Advanced Study, Princeton, N.J., 1948-49, 55, U. Cal. at Los Angeles, 1949, 55, Mass. Inst. Tech., 1962, U. Stockholm, 1951-52. Recipient Meisinger award Am. Meteorol. Soc., 1950; Carl Gustaf Rossby research medal, 1965. Mem. Norwegian Phys. Soc., Norwegian Acad. Scis., Societas Scientiarum Fennica. Research and publs. on theory of almost-balanced motions in atmosphere with applications to fronts, hurricanes, gen. circulation of atmosphere; atmospheric waves generated by mountains, numerical weather prediction. Home: 20 Husebygrenda, Oslo 3, Norway.*

ELIASSEN, Einar, Norwegian physiologist; b. Bergen, Norway; s. Andreas and Helen (Johansen) E.; Mag.Sc., U. Copenhagen (Denmark), 1951; Dr.Phil., U. Bergen, 1960; m. Alis Jorgensen, Dec. 30, 1950; children—Marianne, Bente, Oyvind, Nina-Helene. Prof. dept. physiology U. Bergen. Recipient award Det Norske Videnskap, Akad. Oslo. Research, publs. on circulation, hypothermia, animal physiology. Home: 20 Arstadveien, Bergen, Norway.*

ELIE DE BEAUMONT, (Jean Baptiste Armand Louis) Léonce, French geologist; b. Canon, Caen, France, Sept. 25, 1798; studied at L'École polytechnique, Paris, L'École des mines. Named prof. geology L'École des mines, 1827, Collège de France, 1832; chief engr. mines, from 1824; insp. gen., from 1847. Mem. French Acad. Scis., 1835 (became v.p. 1844, pres. 1845, permanent sec. 1853), Geol. Soc. (became senator 1852, also pres.), Soc. Agr. Author: Notice sur les systèmes des montagnes, 3 vols., 1852; Mémoire pour servir à une description géologique de la France, 4 vols., 1833-38. Pioneered in geology; described age and origin of mountains; prepared (with Dufrénoy) geol. map of France (which was a basis for all geol. work in France), began 1823, published 1841. Died Sept. 21, 1874.

ELIEZER, C. J., math. physicist; b. Ceylon, June 12, 1918; s. Jacob Richard and Ponnamma (Vyrakiam) E.; B.Sc. (Gen.), Ceylon U. Coll., 1937, B.Sc. (Sp.), 1938; M.A. Cambridge U. (Eng.), 1941, Ph.D., 1945, D.Sc., London extension, 1949; m. Jeevaranec Handy, Dec. 29, 1944; children—Thamayanthy, Rahrajothy, Ananthajothy, Renuka. Lectr., U. Ceylon, 1943-46, prof., head dept. math., 1949-59, dean faculty sci., 1954-57; fellow Christ's Coll., Cambridge, Eng., 1946-49; prof., head dept. math. U. Malaya, Kuala Lumpur, 1959—, dep. prin., vice chancellor, 1960-63. Recipient Smith's prize Cambridge U., 1943; Charles Mayer award U. S. Acad. Scis., 1946. Fellow Inst. Math.; Mem. Am. Inst. Physics. Cambridge Philos. Soc., Malayan Math. Soc. Author: Concise Vector Analysis, 1963; Modern Text on Statics, 1964; also articles in field. Research on interaction of elementary particles with radiation; on classical and quantum theories. Home: 36 Jalan University, Petaluy Jaya. Office: U. Malaya, Kuala Lumpur, Malaysia.*

ELIOT, Jared, Am. physician; b. Guilford, Conn., Nov. 7, 1685; s. Joseph and Mary (Wyllys) E.; grad. Yale, 1706; m. Hannah Smithson, 1710, 11 children. Ordained to ministry Congregational Ch.; pastor Killingworth (now Clinton) (Conn.) Ch., 1708-63; trustee Yale Corp., 1730-63; fellow Royal Soc. of London, 1757-63; one of 1st to develop iron ore beds in Northwestern Conn.; helped introduce silk culture into colony; his med. practice wide and successful. Author: Essay on Field Husbandry in New England, 1759. Died Killingworth, Apr. 22, 1763.

ELIOT, Sir John, Brit. meteorologist; b. Lamesley, Durham, Eng., 1839; M.A., St. John's Coll., Cambridge, 1869, fellow, 1869-76; m. Mary Nevill, 1877. Prof. math. Roorkee Engring. Coll., 1869-72, Muir Central Coll., Allahabad, India, 1872-72; prof. physics Presidency Coll., Calcutta; meteorol. reporter Govt. of Bengal, 1874-86; meteorol. reporter of Govt. of India, also dir.-gen. Indian obs., 1886-1903. Fellow Royal Soc., 1895. Author: Handbook of Cyclonic Storms in the Bay of Bengal; also papers. Died Mar. 19, 1908.

ELISEIEV, Nikolai Aleksandrovich, Russian geologist; b. Dec. 19, 1897; grad. Leningrad U., 1924. Prof., Leningrad Mining Inst., 1938-47, Leningrad U., 1947—; mem. staff Pre-Cambrian Geology Lab., USSR Acad. Scis., 1949—. Mem. USSR Acad. Scis. (corr.). Author: Structural Petrology, 1953; Methods of Petrographic Investigations, 1956. Research on methodology of petrographic investigations, petrography of Altai and Kola Peninsula. Address: Lab. of Pre-Cambrian Geology, USSR Acad. Scis., Leningrad, USSR.

EL KHADEM, Hassan Saad, Egyptian chemist; b. Cairo, Egypt, Mar. 24, 1923; s. Saad Saad and Nimet (Zulficar) El K.; B.Sc. with honors, Faculty Sci., Cairo U., 1946; D.Sc., Tech., E.T.H., Zürich, Switzerland, 1949; Ph.D., Imperial Coll. London, 1952, D.Sc., 1967; D.Sc., Alexandria U., 1964; m. Nadia Said, Sept. 5, 1952; children—Samiha, Saad. Faculty, Faculty Sci., Alexandria (Egypt) U., 1952—, prof. organic chemistry, 1964—. Fulbright scholar Ohio State U., 1963-64. Recipient Nat. Sci. award and medal, Cairo, 1963. Mem. Am., Brit., Swiss, Egyptian chem. socs. Research, publs. on synthesis of heterocyclic compounds from sugars, leading to discovery of 2 new reactions of sugar osazones. Home: 19 Marouf El Rassafi, Ramleh, Alexandria, Egypt.*

ELKIN, Daniel Collier, Am. surgeon; b. Louisville, Mar. 26, 1893; s. Robert and Roberta (Collier) E.; A.B., Yale, 1916; M.D., Emory U., 1920; Sc.D., Northwestern U., 1952; D.Sc., Centre Coll., 1956; m. Helen McCarty, Nov. 3, 1923; 1 son, Daniel C. Intern, resident surgeon Peter Bent Brigham Hosp., Boston, 1920-24; asst. in surgery, Harvard, 1924; Whitehead prof. surgery Emory U., 1929-55, now emeritus. Recipient Matas medal for vascular surgery Tulane U., 1940. Fellow A.C.S. (pres.); mem. Soc. Vascular Surgery (pres. 1948), Soc. Med. Consultants (pres. 1949), So. Surg. Assn. (pres. 1946), Am. Surg. Assn. (past pres.), Am. Assn. Thoracic Surgery, Soc. Clin. Surgery (pres. 1947). Author: Medical Reports of John Y. Bassett, 1941. Contbr. numerous papers on surgery of heart and blood vessels to sci. publs. Devised method to suture wounded heart by using deep stay suture to control bleeding until edges of wound are approximated. Died Nov. 3, 1958.

ELKIN, Milton, Am. physician; b. Boston, Feb. 24, 1916; s. Philip and Rose (Dexter) E.; A.B., Harvard, 1937, M.D., 1941; m. Gloria King, Nov. 12, 1943; children—Philip, Karen, Laura. Asso. radiologist Peter Bent Brigham Hosp., Boston, 1951-52; dir. radiology Cambridge (Mass.) City Hosp., asst. radiologist New Eng. Med. Center, Boston, 1952-53; asso. radiologist Cedars of Lebanon Hosp., Los Angeles, 1953-54; prof., chmn. dept. radiology Albert Einstein Coll., Medicine, N.Y.C., 1954—; dir. radiology Bronx Municipal Hosp. Center, N.Y.C., 1954—. Spl. cons. radiology tng. com. Nat. Inst. Gen. Med. Scis., NIH, USPHS, 1966—. Fellow Am. Coll. Radiology; mem. A.M.A., Harvard Med. Soc., Am. Roentgen Ray Soc., Radiol. Soc. N. Am., Assn. U. Radiologists, N.Y. Roentgen Soc. (v.p. 1966—). Research, publs. on demonstration of vascular responses of canine kidney to various drugs, other parameters of renal physiology and disease using radiologic methods. Home: 13 Kingston Rd., Scarsdale, N.Y. 10583. Office: 1300 Morris Park Av., N.Y.C. 10461.*

ELKIN, William Lewis, Am. astronomer; b. New Orleans, Apr. 29, 1855; s. Lewis and Jane (Fitch) E.; C.E., Royal Poly. Sch., Stuttgart, 1876; Ph.D., U. Strassburg, 1880; (hon. M.A., Yale, 1893, Ph.D., Christiania, 1911). Asso. with Sir David Gill, Royal Obs., Cape of Good Hope, investigating parallaxes of southern stars, 1881-83; astronomer, 1884-96, dir., 1896-1910 (emeritus), Yale Obs. Lalande prize, Paris Acad. Scis., 1908. Fellow A.A.A.S.; mem. Nat. Acad. Scis., Astron. and Astrophys. Soc. Am., Astronomische Gesellschaft, Royal Astron. Soc. (London) (fgn. asso.). Measured distance (with David Gill) of 9 first-magnitude stars at Cape Town Obs., 1881-84 and determined distance of sun from earth by observing 3 asteroids; measured stellar parallax and solar parallax from asteroid observations; invented photog. method to observe meteor trails; determined direction of meteor flight and their distance from the earth; measured distances of more than 200 stars at Yale. Died May 30, 1933.

ELKINGTON, Richard, Brit. inventor; b. 19th century; industrialist Birmingham, Eng.; inventor wet gold plating, 1836, electro-gold and silver plating using alkaline baths, 1840.

ELKINTON, J(oseph) Russell, Am. physician; b. Moylan, Pa., Oct. 12, 1910; s. J. Passmore and Mary (Bucknell) E.; A.B., Haverford Coll., 1932; M.D., Harvard, 1937; m. Mary Teresa Sturge, Dec. 14, 1940; children—Mary Gwyneth, Joseph Sturge. Practice medicine, specializing in internal medicine, Phila., 1948—; faculty Sch. Medicine U. Pa., Phila, 1948—, prof. medicine, 1962—, chief chem. sect., 1948-66, ward, staff physician Hosp. U. Pa., 1948—; cons. to surgeon gen. USPHS, NIH study sect., 1954-58. Fellow A.C.P.; mem. Assn. Am. Physicians, Am. Soc. Clin. Investigation, Am. Physiol. Soc., A.M.A. Author: (with T. S. Danowski) The Body Fluids, 1955. Editor: Annals of Internal Medicine, 1960—. Contbr. numerous articles to profl. jours. Studies on physiology of water, electrolytes, acid-base balance in health and disease, especially kidney disease. Home: 441 Oak Lane, Moylan, Pa. 19065. Office: 3400 Spruce St., Phila. 19104; also 4200 Pine St., Phila. 19104.*

ELLENBERGER, Wilhelm, German vet. anatomist; b. Beiseförth nr. Kassel, Germany, Mar. 28, 1848; s. Heinrich and Marie Amalie (Eberhard) E.; studied at Sch. Vet. Medicine, U. Berlin. Served in military, 1870-71; local veterinarian for short time; became prof. Sch. Vet. Medicine, Berlin, 1876; named prof. physiology Sch. Vet. Medicine, Dresden, Germany, 1879; head Dresden Sch. (raised to univ. level 1902), 1896-1923. Author: (with W. Schütz, O. A. Siedamgotzky), Lehrbuch der allgemeinen Therapie der Hausthiere, 1884; Handbuch der vergleichenden Histologie und Physiologie der Haussäugethiere, 1887-92; (with H. Baum) Systematische und topographische Anatomie des Hundes, 1891, Topographische Anatomie des Pferdes, 1893-97. Co-editor: Handbuch der vergleichenden Anatomie der Hausthiere, 1890-1932. Studied comparative vet. anatomy and physiology, including digestive tract, digestion of house animals; discovered food in stomach of house pets is divided not mixed; devel. vet. histology. Died Dresden, May 5, 1929.

ELLENBOG, Ulrich, physician; b. Feldkirch (Vorarlberg), Austria, circa 1435; student Vienna, Austria and Heidelberg, Germany; artium medicinarumque doctor, Pavia, Italy, circa 1459; m. Margareta Weber, before 1470; 8 sons, including Nikolaus, 3 daus. Returned to Feldkirch, 1459; probably settled in Memmingen, Germany, before 1464; physician to bishops, Augsburg, 1470-78; with U. Ingolstadt, 1472; in Biberach, 1478; city physician, Memmingen, 1482 until his death. Author: Tractatulus de balneis, 1460; Tractatus de simplicibus, 1464-68; Consilium de veneis, 1470; Consilium pro maris riatore, 1473; Von den gifftigen Besen Tempffen und Reuchen (1st work on indsl. hygiené, it began sci. research on indsl. diseases), written 1473, pub. 1524; Ordnung . . . wider die gifftigen anrur der prestilentzlichen prechen, 1494. Died Memmingen, Jan. 19, 1499.

ELLER, Eugene Rudolph, Am. paleontologist, geologist; b. Centerville, Pa., May 17, 1904; s. Joseph and Maude (Cook) E.; B.S., Alfred U., 1920; M.S., U. Pitts., 1933; Sc.D., Waynesburgh Coll., 1946; m. Ildra Harris, Aug. 22, 1934; children—Stephen B., Margaret E. Staff, Carnegie Mus., Pitts., 1931—, curator geology and invertebrate paleontology, 1948—; lectr. U. Pitts., 1931-45. Fellow Geol. Soc. Am.; mem. Pa. Acad. Sci., Soc. for Systematic Zoology, Am. Geol. Inst., Pitts. Geol. Soc., Paleontol. Soc., Soc. for Study Evolution, German, Pa. folklife socs., Pewter Soc., Gourd Soc., Paper Weight Soc. Pa. Research, numerous publs. on scolecodonts, fossil annelid jaws resulting in determining age of surface rock outcrops and subsurface geol. horizons, useful and decorative arts: glass and lighting devices. Home: 7543 Rosemary Rd., Pitts. 15221. Office: 4400 Forbes St., Pitts. 15213.*

ELLER, Johann Theodor, physician; b. Plotzkau, Germany, Nov. 29, 1689; ed. in law Quedlinburg and Jena, Germany, in medicine and sci. especially chemistry at Halle, Germany, Leiden, Netherlands, Amsterdam, Paris, London. Became physician to Prince of Anhalt-Bernburg, 1721; became prof. anatomy, dean Med. Coll., Berlin, also physician to army, 1724; named physician to Frederick the Gt., 1735, who made him dir. Collegium Medico-Chirurgicum, 1755, privy-councillor, 1755. Mem. Royal Berlin Acad. Scis. (named dir. 1755). Editor: J. T. Eller's Physikalisch-Chymisch-Medicinische Abhandlungen . . . , 1764. Measured rate of evaporation of water at different temperatures; studied exothermic and endothermic dissolution of salts in water, theories of solution, including interposition of salt particles between water particles; described effect various reagents on color, shape and size of blood corpuscles; 1st to observe Leyden Frost phenomena, 1746. Died Berlin, Sept. 13, 1760.

ELLERBECK, (Ernst) Leopold, German civil engr.; b. Bromberg, Germany, Jan. 14, 1872; s. Ernst and Ida (Erxleben) E.; student Tech. U., Berlin, Germany; certification exam., 1899; d.engring.; m. Mathilde Stieve, 1905; 1 son, 3 daus. Field engr., Münster, Kosel, Potsdam, Germany; dir. reconstrn. Meml. Bridge, Tilsit, USSR; dir. dike maintenance inspection, Meppen, Germany, 8 years; joined Ministry Pub. Works, Berlin, 1916, named ministerial adviser, 1921; expert on bridges, difficult statics problems and materials waterways dept. Reich Traffic Ministry until 1937; dir. bldg. of Niederfinow ship hoist Berlin-Stettin channel, beginning in 1934. Mem. Deutsche Gesellschaft für Bodenmechanik, Deutscher Normenausschuss. Collaborator on an internat. tech. dictionary. Research and publs. on rd. bridges, reinforced concrete, binding agts. for mortar and concrete; reworked cement standards, coating and painting techniques. Died Berlin-Wilmersdorf, May 8, 1945.

ELLERMAN, Ferdinand, Am. astronomer; b. Centralia, Ill., May 13, 1869; s. Mathias and Rosa Augusta (Fleischbein) E.; spl. studies U. Chgo.; A.M. (hon.), Occidental Coll., 1927; m. Hermine Louise Hoenny, May 16, 1895; children—Leola, Louise. Asst., Kenwood Obs., 1892-95, Yerkes Obs., 1895-1901; instr. astrophysics Mt. Wilson Obs. of Carnegie Instn., Washington, 1901-04, asst. astronomer, 1905-15, astronomer, 1915-37. Fellow A.A.A.S., Royal, Am. astron. socs., Astron. Soc. Pacific, Sigma Xi. Research on stellar spectra, spectra of sunspots, chromosphere, solar physics. Died Mar. 20, 1940.

ELLERY, Robert Lewis John, astronomer; b. Granleigh, Eng., July 14, 1827; s. John Ellery; ed. in medicine; m. 1853 (dec. 1858); m. 2d, Margaret Shields, 1858. Went to Melbourne, Australia, 1851; established obs., nr. Melbourne (later moved to Melbourne), 1853, dir. until 1895; dir. Geodetic Survey of Victoria, 1858-74; organized Victorian Torpedo Corps (later named Submarine Mining Engrs.), 1873, comdr., ret., 1889. Fellow Royal Soc., 1873, Royal Soc. Victoria (pres. 23 years). Author works on astronomy and meteorology, also papers. Died Jan. 14, 1908.

ELLET, Charles, Am. civil engr.; b. Penn's Manor, Pa., Jan. 1, 1810; s. Charles and Mary (Israel) E.; attended Ecole Polytechnique, Paris, France; m. Elvira Daniel, 1837, 1 son, Charles Rivers. Chief engr. James River and Kanawha Canal, 1836; designed, built wire suspension bridge across Schuylkill River nr. Phila., 1842; designed, built suspension bridge across Niagara River below falls, 1847; completed Wheeling (W.Va.) Bridge for B. & O. R.R., 1849; engr. Hempfield R.R., 1851-55, Va. Central R.R., 1853-57; considered one of Am.'s great engrs. Author: Physical Geography of the Mississippi Valley with Suggestions as to the Improvement of Navigation of the Ohio and Other Rivers, 1853; Coast and Harbor Defenses, or the Substitution of Steam Battering-rams for Ships of War, 1855. Advanced plans for flood control on the Mississippi. Died Cairo, Ill., June 21, 1862.

ELLETT, Alexander, Am. physicist; b. Chillicothe, Mo., Sept. 5, 1894; s. Andrew Jackson and Bessie (Lane) E.; A.B., U. Colo., 1922; Ph.D., Johns Hopkins, 1923; m. Onabelle Townsend, Jan. 16, 1921; children—Elizabeth Onabelle, Charles Alexander, Norman Townsend, Wilma June. NRC fellow, 1923-25; asst. prof. physics U. Ia., 1925-28, asso. prof., 1928-29, prof., 1929-40; chmn. sect. E. div. A. Nat. Def. Research Com., 1939-41, chief div. 4, 1941-45; dir. research Zenith Radio Corp., Chgo., 1945-59, v.p. charge research, 1949-63, v.p. spl. sci. projects, 1963—. Decorated Medal for Merit, 1948. Fellow Am. Phys. Soc.; mem. I.R.E. Contbr. articles profl. jours. Research in nuclear physics; spectroscopy; effect of weak magnetic fields on the polarization of resonance radiation; electronics. Home: P.O. Box 5695, Carmel, Cal. Office: 6001 Dickens Av., Chgo. 39.

ELLICKSON, Raymond Thorwald, Am. physicist; b. Charlson, N.D., Mar. 5, 1910; s. John and Christine (Quale) E.; B.A., Reed Coll., 1935; M.A., Ore. State U., 1936; Ph.D., U. Chgo., 1938; m. Loene Gibson, June 17, 1938; children—Bryan, Mary, James. Faculty, Poly. Inst. Bklyn., 1938-46, asst. prof., 1942-46; faculty Reed Coll., 1946-48, prof., 1947-48; prof. U. Ore., Eugene, 1948—, head dept. physics, 1948-60, asso. dean Grad. Sch., 1948-55, acting dean, 1955-58; researcher United Aircraft, 1941, RCA, 1943, Signal Corps, U. S. Army, 1943-60; sr. physicist Chgo. Midway Labs., 1952. Mem. Am. Phys. Soc., Am. Assn. Physics Tchrs. (past pres. Ore. sect.), A.A.A.S. (cons., writer sci. writing program for elementary schs. 1963—), Am. Assn. U. Profs. Contbr. numerous articles to encys., jours. Research on phosphors. Home: 2120 Summit St., Eugene, Ore. 97403.*

ELLICOTT, Andrew, Am. surveyor, mathematician; b. Bucks County, Pa., Jan. 14, 1754; s. Joseph and Judith (Bleaker) E.; m. Sarah Brown, 1775; 9 children. Founder Ellicott City, (Md.) 1774; served to maj. during Am. Revolution; publisher U. S. Almanack, from 1782; completed Mason Dixon Line in West Pa., 1784; mem. Md. Legislature, 1786; mem. Pa. commns. for running Western (1785) and No. (1786) state boundaries; surveyed islands in Ohio, Allegheny rivers, 1788; pub. 1st map of Territory of Columbia (now D.C.), 1793; drew up Ellicott Plan (survey of Washington); redrew L'Enfant's plans of Washington under direction of Thomas Jefferson; a commr. to lay out town of Presqu' Isle (Erie), Pa., 1794; commd. to survey frontier between Fla. and U. S., 1796-1800; fixed boundary between Georgia and South Carolina, 1811; prof. mathematics U. S. Mil. Acad., 1813-20. Died West Point, N.Y., Aug. 28, 1820.

ELLINGER, Alexander, German pharmacologist; b. Frankfort/Main, Germany, Apr. 17, 1870; s. Philipp and Mathilde (Ruben) E.; studied chemistry in Berlin, Bonn, 1887-92; Ph.D., Bonn, 1892; later studied medicine and pharmacology, Munich, Strasbourg, France, Königsberg, Germany; M.D., 1898; m. Rosa Simon, 1899; 2 sons, including Friedrich, 2 daus. Became asst. M. Jaffe's Lab. for Med. Chemistry and Exptl. Pharmacology, U. Königsberg, 1897, lectr., 1899, succeeded Jaffe, 1911; named prof. pharmacology U. Frankfort, 1914. Contbg. author: Handbuch der experimentellen Pharmakologie, 1920-24; Handbuch der biologischen Arbeitsmethoden, 1920-25. Co-editor: Handbuch der normalen und pathologischen Physiologie mit Berücksichtigung der experimentellen Pharmokologie, 1925-27. Studied water exchange between tissue and blood, formation of lymph and urine, immunity; research on toxicology, pathology and therapy of gas poison during World War I. Died Frankfort/Main, July 26, 1923.

ELLINGER, Philipp, German biochemist; b. Frankfort, Germany, 1888; studied natural scis. under von Bayer and Roentgen, Munich, Germany; D.Phil. in Chemistry, Greifswald (Germany) U., 1911; D.Med., Heidelberg, Germany, 1913; 2 sons, 1 dau. Served as med. officer German Army, World War I.; successively asst., lectr., dept. dir. Pharmacological Inst., Med. Acad. Düsseldorf, Germany, 1931-33; fled to Britain from Germany for polit. reasons, 1933; research staff Lister Inst., 1933-52; visited Egypt to study pellegra, Med. Research Council, 1937. Discovered (with W. Koschara) chem. formula of riboflavin, 1934; discovered 2 factors in pellegra, latent pellegra from malnutrition, active pellegra from hard phys. work. Died 1952.

ELLIOT, Daniel Giraud, Am. zoologist; b. N.Y., Mar. 7, 1835; s. George Thompson and Rebecca Giraud (Foster) E.; studied zoölogy; Sc.D., Columbia, 1906; m. A. E. Henderson, 1858 (dec. 1905). Traveled in Europe, Africa. Palestine and Asia Minor, 1856-78 later in greater part of U. S., Can., Alaska, S. America; hon. and supervisory curator zoölogy Field Mus. Natural History, Chgo., 1895-1915. Led expdn. into interior of East Africa, 1896, and into recesses of Olympic Mountains, 1898, being 1st naturalist to penetrate that little-known range. Decorated 10 times by European govts. for labors in natural sci. Fellow Royal Soc. Edinburgh; mem. Am. Ornithologists Union (a founder, 2d pres. 1890-91). Elliot medal of Nat. Acad. Scis. founded in his honor. Author: Shore Birds of North America, 1895; Gallinaceous Game Birds; Wild Fowl of the United States and the British Possessions, 1898; Synopsis of the Mammals of North America and the Adjacent Seas, 1901; Land and Sea Mammals of Middle America and West Indies (2 vols.), 1894, and many others. Part author The Deer Family, 1902; Check List Mammals N. American Continent and W. Indies, 1905; Catalogue Mammals in Field Columbian Museum, 1906; Review of the Primates, 3 vols., 1912. His own collection of birds was best pvt. collection extant. Died Dec. 22, 1915.

ELLIOT, William Whitfield, Am. mathematician; b. Prince Edward County, Va., 1898; s. James and Alma (Watson) E.; B.A., Hampden Sidney Coll., 1918; M.A., U. Ky., 1919; Ph.D., Cornell U., 1924. Instr., Ga. Sch. Tech., 1920-21; asst. Cornell U., 1921-23; instr. Yale, 1923-25; asst. prof. Duke, Durham, N.C., 1925-26, prof., 1926-64. Mem. Sigma Xi. Author: First Year College Mathematics; also articles. Home: 4721 Duke Sta., Durham, N.C.*

ELLIOTSON, John, English physician, mesmerist; b. London, Eng., Oct. 29, 1791; M.D., Edinburgh (Scotland) U., Cambridge (Eng.) U.; postdoctoral study St. Thomas Hosp., Guy's Hosp., London. Prof. principles and practices of medicine U. London, 1831-38, resigned because of belief in mesmerism, 1838; founder, lectr., physician U. Coll. Hosp., 1834-38; founder a mesmeric hosp., London, 1849; founder, editor Zoist (mag. devoted to mesmerism). Lumley lectr., 1829; Harveian orator, 1846. Fellow Royal Soc., 1829, Royal Coll. Physicians; mem. Medico-Chirurg. Soc. London (pres. 1837), Phrenological Soc. (founder, pres.). Author: Lectures on Diseases of the Heart, 1830; Principles and Practice of Medicine, 1839; Human Physiology, 1840; Surgical Operations in the Mesmeric State Without Pain, 1843. Used iodine in treatment of goiter, 1819; demonstrated that glanders in the horse is communicable to man, 1830; ascertained that pollen is cause of hay fever, 1831; 1st Brit. physician to use stethoscope; early advocate of advantages of clin. lecturing; experimented with mesmerism as anesthetic and cure for nervous diseases. Died London, July 29, 1868.

ELLIOTT, Alfred Marlyn, Am. zoologist; b. Humbolt, S.D., June 19, 1905; s. William H. and Emily (Board) E.; B.A., Yankton Coll., 1928; M.S., N.Y. U., 1931, Ph.D., 1934; m. Lulu Maynard, July 19, 1928; children—Marilyn E. (Mrs. Sterling Crandall), William D. Prof., chmn. sci. div. State Coll., Bemidji, Minn., 1935-47; faculty U. Mich., Ann Arbor, 1947—, prof. zoology, 1955—, dir. Acad. Year Inst., 1962-65. Fellow A.A.A.S., N.Y. Acad. Sci.; mem. Am. Microscopical Soc. (past pres.), Soc. Protozoology (past pres.), Am. Type Culture Collection (trustee 1960—), Soc. Cell Biology, Am. Soc. Zoologists, Soc. Protozoologists, Soc. for Study Evolution, Sigma Xi. Author: Zoology, 1952; (with C. Ray) Biology, 1960; also numerous articles. Research on cytology, physiology biochemistry, genetics and distbn. protozoa, particularly ciliates. Home: 2204 Needham Rd., Ann Arbor, Mich. 48104.*

ELLIOTT, Edwin Bailey, English mathematician; b. Oxford, Eng., June 1, 1851; s. Edwin Litchfield and Mathilda (Bailey) E.; B.S., Magdalen Coll., Oxford, 1873; M.A.; m. Charlotte Amelia Mawer, 1893. Fellow Queen's Coll., 1874-92, hon. fellow; lectr. Corpus Christi Coll., 1884-93; 1st Waynfelte prof. pure math., fellow Magdalen Coll., 1892-1921; Fellow Royal Soc., 1891, Royal Astron. Soc., pres. London Math. Soc., 1896-98; a founder, sec., pres. Oxford Math. Soc. Author: Algebra of Quantics, 1894-1895, 1913; also papers. Contbd. work on an inequality of importance to theory of integral equations, circa 1928. Died July 21, 1937.

ELLIOTT, Francis Edward, Am. oceanographer; b. Vienna, Austria, July 24, 1909; s. Hugo Engel and Elsa (Pollack) E.; came to U. S., 1941, naturalized, 1943; M.A., U. Va., 1948; postgrad. Yale; Ph.D., Clark U., 1952; m. Esther Dorothy Bender, May 11, 1945; 1 dau., Roberta Bender. Supervisory oceanographer USN Hydrographic Office, Suitland, Md., 1950-56; prof., head dept. geography Butler U., 1956-59; oceanographer Gen. Electric Co., Washington, 1959—; vis. prof. Cornell U., 1962-63. Fellow Royal Geog. Soc. London, Washington Acad. Scis.; mem. Marine Tech. Soc. Author: (with Eric Fischer) A German and English Glossary of Geographical Terms, 1950. Contbr. numerous articles to profl. jours. Co-designer thermistor cable, which measures ocean temperature vs. depth. Home: 7507 Grange Hall Dr., Washington 20022. Office: 777 14th St. N.W., Washington 20005.*

ELLIOTT, Henry Wood, Am. pharmacologist; b. Seattle, Apr. 10, 1920; s. Lionel Henry and Threse (Pariseau) E.; B.S., U. Wash., 1941, M.S., 1943; Ph.D., Leland Stanford Jr. U., 1946; M.D., U. Cal. Sch. Medicine, 1953; m. Donna Dell, Sept. 6, 1947; children—Henry Wood, John Lionel, Kathleen Dell, Edward Robert. With U. Wash., 1941-43, asso. physiology, 1943; asst. chemist Wash. State Planning Council, 1943; research asso. physiology Stanford, 1943-46; instr. physiology Coll. Physicians and Surgeons San Francisco, 1944-46, research fellow pharmacology, 1946-47; lectr., research asso. pharmacology, lectr. dental pharmacology and toxicology U. Cal. Sch. Medicine, 1947-53, faculty pharmacology, 1954—, prof., 1964—; cons. pharmacology, anesthesiology Kaiser Found. Hosp., Oakland, Cal., 1960—. Recipient Lederle Med. Faculty award, 1955-57. Fellow A.A.A.S., Am. Coll. Clin. Pharmacology and Chemotherapy; mem. Am. Fedn. Clin. Research, Am. Soc. Pharmacology and Exptl. Therapeutics, Am. Assn. U. Profs., Am. Chem. Soc., N.Y. Acad. Sci., Soc. Exptl. Biology and Medicine, Am. Soc. Anesthesiologists, A.M.A., Western Pharmacol. Soc., Am. Therapeutic Soc., Sigma Xi, Alpha Omega Alpha, others. Editor: Ann. Rev. Pharmacology, 1965—. Research, publs. on effect of drugs on tissue metabolism, physiology and clin. application of hypothermia, pharmacology of narcotic analgesic drugs, pharmacology of drug addiction.*

ELLIOTT, Kenneth Allan Caldwell, biochemist; b. Kimberley, S. Africa, Aug. 24, 1903; s. Kenneth Caldwell and Venetia (Leppan) E.; B.S., Rhodes U. Coll., Grahamstown, 1923, M.S., 1924; Ph.D., Selwyn Coll., Cambridge, 1930, Sc.D., 1950; m. Frances Howland, Dec. 26, 1936; children—Venetia (Mrs. Milton Crawford), Joan (Mrs. Terence Watt), Kenneth Howland. With Montreal Neurol. Inst. McGill U., 1944-59, neurochemist, dir. Donner Lab. Exptl. Neurochemistry, 1951, faculty, 1944—, prof., chmn. dept. biochemistry, 1959; Norman Bethune exchange prof. Chinese Med. Coll., Peking, 1964. Fellow Royal Soc. Can., A.A.A.S.; mem. Am. Soc. Biol. Chemists, Biochem. Soc. (Brit.), Canadian Biochem. Soc. Editor: Canadian Jour. Biochemistry and Physiology, 1956-59; (with others) Neurochemistry, 1955. Studies, numerous publs. on biol. oxidation-reduction systems, metabolism of normal and cancer tissue, metabolism of brain, on electrolytes in brain, brain swelling and inhibitory agts. and amino acids in brain. Home: 3440 Grey Av., Montreal 28, Que., Can.*

ELLIOTT, Larry Paul, Am. cardiac radiologist; b. Manhattan, Kan., Oct. 16, 1931; s. Leonard Paul and Mary (Myers) E.; B.S., U. Fla., 1954; M.D., U. Tenn., 1957; m. Betty Lou Hawkins, June 13, 1956; children—Laurie Lou, Mary Elizabeth, Larry Paul. Cardiac pathology fellow Charles T. Miller Hosp., St. Paul, 1961-63; diagnostic radiology fellow U. Minn. Hosp., Mpls., 1963-66; asso. prof. Washington U. Sch. Medicine, St. Louis, 1966-67; prof. radiology, head div. cardiac radiology U. Fla. Coll Medicine, Gainesville, 1967—. Mem. Am. Bd. Radiology, Acad. U. Radiologists. Author: (with G. L. Schiebler) X-ray Diagnosis in Congenital Heart Disease, 1968. Research, publs. on correlating and defining the clin.-roentgenologic and pathologic findings in congenital and acquired heart disease; devel. of a multidisciplined tng. program for acad. cardiac radiologists. Home: 1708 N.W. 24th St., Gainesville, Fla. 32601.*

ELLIOTT, Lloyd George, Canadian research physicist; b. Clarence, N.S., Can., Aug. 16, 1919; s. Lorenzo Leander and Eunice (Wotton) E.; B.Sc., Dalhousie U., 1938, M.Sc., 1940; Ph.D., Mass. Inst. Tech., 1943; D.Sc., Carleton U., 1967; m. Margaret Isobel Wilson, July 15, 1944; children—George Arthur, Bruce Edgar, Martin Lloyd. Research officer Can. NRC, 1943-51, asst. dir. physics subdiv., 1951-52; dir. physics div. Atomic Energy of Can. Ltd., Chalk River, Ont., 1952-67, dir. research Chalk River Nuclear Labs., 1967—. Vis. prof. Dalhousie U., 1961. Fellow Royal Soc. Can., Am. Phys. Soc. (council mem. 1960-64); mem. Canadian Assn. Physicists (pres. 1959), A.A.A.S., Sigma Xi. Editor: Canadian Jour. Physics, 1962—. Research, publs. on specific heat of metals, beta and gamma-ray spectrometry, structure of atomic nuclei. Home: 21 Beach Av., Deep River, Ont. Office: Atomic Energy of Can. Ltd., Chalk River, Ont., Can.*

ELLIOTT, Roger James, English physicist; b. Chesterfield, Eng., Dec. 8, 1928; s. James and Gladys (Hill) E.; B.A. in Math., New Coll., Oxford (Eng.) U., 1949, D.Phil., 1952; m. Olga L. Atkinson, July 30, 1952; children—Jane Susan, Rosalind Kira, Martin James. Research asso. U. Cal. at Berkeley, 1952-53; research fellow Atomic Energy Research Establishment, Harwell, Eng., 1953-55; lectr. Reading (Eng.) U., 1955-57; fellow St. Johns Coll., Oxford, 1957—; reader theoretical physics Oxford U., 1963—; vis. prof. U. Cal. at Berkeley, 1961, U. Ill., 1966. Fellow Phys. Soc. London; mem. Am. Phys. Soc. Research, publs. on theory of solids. Home: 11 Crick Rd., Oxford, Eng.*

ELLIOTT, Rush, Am. anatomist; b. New Concord, O., Apr. 21, 1903; s. Fred and Mary (Forsythe) E.; A.B., Ohio U., 1924; A.M., Ohio State U., 1926; Ph.D., U. Mich., 1930; m. Frances J. Gray, Aug. 20, 1927; children—Margaret (Mrs. J. T. Ehrhart), Susanne (Mrs. R. S. Szijarto), Rush E. Faculty, Ohio U., Athens, 1924—, prof. anatomy, 1938—, dir. summer session, 1946-51, dean Univ. Coll., 1951-54, dean Arts and scis., 1954-65, Rush Elliott prof. anatomy, 1965—. Mem. Ohio Acad. Sci. (pres. 1954-55), Am. Assn. Anatomists, Am. Soc. Zoologists, Phi Beta Kappa, Sigma Xi, Omega Delta Kappa, Tau Kappa Alpha. Author: (with Reighard, Jennings, Elliott) Anatomy of the Cat, 1935, Dissection of the Cat, 1935; also articles. Research on distbn., nerve supply, numerical distbn. during life span end organs for taste. Home: 3 Marietta Av., Athens, O. 45701.*

ELLIOTT, Thomas Renton, English physician; b. Durham, Eng., Oct. 11, 1877; s. A. W. Elliott; ed. Trinity Coll., Cambridge; M.A., M.D.; m. Martha M'Cosh, 1918; 3 sons, 2 daus. Fellow Clare Coll.; cons. physician Brit. Armies in France, 1914-19, also Univ. Coll. Hosp.; prof. medicine London U.; mem. Med. Research Council; mem. Inter-Departmental Com. (Goodenough) on Med. Schs., 1942; named hon. fellow Trinity Coll., Cambridge, 1947. Fellow Royal Soc., 1913, Royal Coll. Physicians; Contbr. papers on physiology of supra-renal glands and innervation of viscera, gunshot wounds of chest. Pioneer in theory of chem. mediation of nerve impulses; stated that nerve impulse inhibits adrenin at smooth muscle cell, chemically stimulating the muscle in turn, 1904. Died 1961.

ELLIOTT, William Hueckel, Am. biochemist, educator; b. St. Louis, June 4, 1918; s. William C. and Edna A. (Hueckel) E.; B.S. in Chemistry, St. Louis U., 1939, M.S., 1941, Ph.D., 1944; m. Dorothy E. Singer, Aug. 6, 1949; children—William J., Mary C., Martha A. E., Robert J. Research asst. chemistry Ind. U., 1944; faculty St. Louis U., 1944—, prof. biochemistry, 1959—. Cons. USPHS, Nat. Inst. Gen. Med. Sci., 1963-65. Mem. Am. Chem. Soc. (chmn. St. Louis sect. 1953), Am. Soc. Biol. Chemists, Soc. for Exptl. Biology and Medicine (past chmn. Mo. sect., nat. council), A.A.A.S., Sigma Xi, Pi Mu Epsilon. Contbr. numerous articles to profl. jours. Research on bile acids, cholesterol and related lipids in body. Home: 6300 Tholozan Av., St. Louis 63109.*

ELLIS, Calvin, Am. physician; b. Boston, Aug. 15, 1826; s. Luther and Betsey E.; grad. Harvard, 1846, M.D., 1849. Asst. in pathology Harvard Med. Sch., prof., 1867-83, dean, 1869; admitting physician, pathologist Mass. Gen. Hosp.; leading exponent of diagnosis by elimination of symptoms of disease. Mem. Am. Acad. Arts and Scis. 1859. Author: Obstruction of the Lungs, caused by Pressure on the Primary Bronchus; The Tendency of Disease in One Part to Excite it in Another; also 40 or more med. articles, including Boylston prize essay Tubercle, 1860. Described curved line of dullness which marks limits of exudate in pleuritic effusions (known as Ellis's curve). Died Boston, Dec. 14, 1883.

ELLIS, Carleton, Am. research chemist, author; b. Keene, N.H., Sept. 20, 1876; s. Marcus and Catharine (Goodnow) E.; B.Sc., Mass. Inst. Tech., 1900; m. Birdella M. Wood, Nov. 28, 1901; children—Eleanor Josephine (Mrs. Edward Foerster), Marjorie Olive (Mrs. John Ordway), Carleton, Bertram. Instr., Mass. Inst. Tech., 1900-02; has worked extensively in field of edible oils, fats, waxes, synthetic resins, paints, varnishes, petroleum products, and gasoline mfr.; has taken out about 750 patents; dir. Ellis-Foster Co. Recipient Gold medal for inventions at Jamestown Expn., 1907, Edward Longstreth medal, Franklin Inst., 1916. Author: Hydrogenation of Organic Substances, 3d edit., 1930; Synthetic Resins and Their Plastics, 1923; 2d edit., 1935; Chemistry of Petroleum Derivatives, 1934; Vol. 2, 1937; Chemistry of Printing Ink, 1939. Co-author: The Vital Factors of Foods; Ultra Violet Light; Soilless Growth. Died Jan. 13, 1941.

ELLIS, Florence Hawley, anthropologist; b. Canagnea, Sonora, Mexico, Sept. 17, 1906; d. Fred Graham and Amy (Roach) Hawley; A.B., U. Ariz., 1937, M.A., 1928, Ph.D., 1934; m. Bruce T. Ellis, June 1, 1950; 1 dau., Andrea Ellis (Mrs. Richard Eastin). Instr., research asso. U. Ariz., 1928-33; faculty U. N.M., Santa Fe, 1934-—, prof. anthropology, 1954-—; asso. prof. U. Chgo., 1937-42. Mem. Soc. for Am. Archaeology, Am. Anthrop. Assn., Archaeol. Soc. N.M., Ariz. Archaeol. and Hist. Soc., No. Ariz. Soc. Sci. and Art, Am. Soc. Ethnohistory (pres. elect 1967), Sigma Xi, Phi Kappa Phi. Author: The Significance of the Dated Prehistory of Chetro Ketl, Chaco Canyon, N.M., 1934; Field Manual of Prehistoric Southwestern Pottery Types, 1936; Tree Ring Analysis and Dating in the Mississippi Drainage, 1941; A Reconstruction of the Basic Jemez Pattern of Social Organization, 1964; also numerous articles. Research in patterns of social orgn. of Southwestern pueblos in terms of linguistic groups; dating of Southwestern ruins by tree ring analysis; identification of specific areas in Southwestern archaeology as ancestral to living Pueblos; excavation of much of first Spanish capital of N.M., second oldest in U. S.; excavation of largest pueblo known in Southwest. Home: 1666 Cerro Gordo St., Santa Fe, N.M. 87501.*

ELLIS, Henry Havelock, Brit. psychologist, author; b. Croydon, Eng., Feb. 2, 1859; s. Edward Peppen and Suzanne (Wheatly) E.; studied medicine St. Thomas' Hosp., 1881-89; license from Soc. Apothecaries, 1889; m. Edith M. O. Lees, 1891 (dec. 1916). Tchr. in various sects. of New South Wales, 1875-79; returned to Eng., practiced medicine for short time, then turned his attentions to sci. studies, also worked as writer-editor; editor: Mermaid Series of Old Dramatists, 1887-89, Contemporary Science Series, 1889-1914. Fellow Royal Coll. Physicians. Author numerous works including: The New Spirit, 1890; The Criminal, 1890; Man and Woman: A Study of Human Secondary Sexual Characters, 1894; Studies in Psychology of Sex series: Sexual Inversion, vol. 2, 1897, The Evolution of Modesty, vol. 1, 1899, Eonism and other Supplementary Studies, vols. 2 through 7, 1928; Analysis of the Sexual Impulse, 1903; Sex in Relation to Society, 1910; The World of Dreams, 1910; The Erotic Rights of Women, 1918; The Dance of Life, 1923; Psychology of Sex: A Manual for Students, 1933; Questions of Our Day, 1936. Applied anthrop.-sociol. approach to psychology of sex abnormality; important work pub. in 7 vols. on psychology of sex (1897-1928) resulted in change of public attitudes toward sex. Died July 8, 1939.

ELLIS, Job Bicknell, Am. mycologist; b. Potsdam, N.Y., Jan. 21, 1829; s. Freeman and Sarah E.; grad. Union Coll., 1851; m. Arvilla J. Bacon, Apr. 19, 1856. Tchr. at Germantown, Pa., Albany, N.Y., Poughkeepsie, N.Y., Alexander, Ga., Canton, N.Y., Potsdam; located in Newfield, N.J., 1865; from 1878 devoted entirely to mycol. research and preparation and sale in sets of collections of N. Am. fungi, and 2d edit. under title of Fungi Columbiani. Author: (with B. M. Everhart) North American Pyrenomycetes. Died 1905.

ELLIS, John, naturalist, botanist; b. Ireland, 1710; 8 children, including Martha (Mrs. Alexander Watt). Mcht. in London until 1764; became agt. for W. Fla., 1764, for Dominica, 1770. Fellow Royal Soc., 1754. (Copley medal 1768). Author: An Essay towards the Natural History of the Corallines, 1755; Directions for bringing over Seeds and Plants from the East Indies, 1770; also treatises on coffee, bread-fruit tree, articles. Described coffee tree, mangostan, breadfruit, 1774-75, Dionoea muscipula (now known to be an insectivorous plant), 1770, new genera Halesia, Gardenia, and Gordonia. Linnaeus named group of boraginaceous plants Ellisia in his honor. Died Oct. 15, 1776.

ELLIS, Leslie Lee, Jr., Am. zoologist; b. Norfolk, Va., Sept. 13, 1925; s. Leslie Lee and Mary Teresa (Stickley) E.; B.S., Tulane U., 1948, M.S., 1949; Ph.D., U. Okla., 1952; m. Dorothy Copes Diboll, Dec. 22, 1949; children—Leslie Lee, Christian Hart, Robin Chamblin, Daniel Diboll. With USPHS, 1952; faculty Miss. State U., 1952-68, prof. zoology, 1960-68, chmn. dept., 1962-68; chmn. dept. biol. sci. Fla. Technol. U., 1968-—. spl. research aquatic biology, parasitology Cons. in field, 1962-—; panelist NSF, 1959-—; participant Commn. on Undergrad. Edn. in Biol. Scis. confs. Mem. Entomol. Soc. Am., Am. Inst. Biol. Scis., Sigma Xi. Contbr. profl. jours. Described aquatic insect fauna of S. La. and the mammalian ectoparasites of Wichita Mountains, Okla. Home: Maitland, Fla. 32801. Office: Dept. Biol. Sci., Fla. Technol. U., Orlando, Fla. 32801.*

ELLIS, Norman Richard, Am. psychologist; b. Springville, Ala., Sept. 14, 1924; s. Olin Otis and Willie (Brock) E.; A.B., Howard Coll., 1951; M.A., U. Ala., 1953; Ph.D., La. State U., 1957; m. Mattie Katherine Martin, Aug. 23, 1953; children—David N., Emily K., Janet M., Ben E. Susan E. Faculty, U. Ala., University, 1954-55, prof. psychology, 1964-—; chief psychologist State Colony and Tng. Sch., Pineville, La., 1956-60; asso. prof., prof. Peabody Coll., 1960-64. Mem. Am., Southeastern psychol. assns., Am. Assn. on Mental Deficiency (past v.p., cons. editor jour. 1960-66), Sigma Xi. Author: Handbook of Mental Deficiency, 1963. Editor: International Review of Research in Mental Retardation, 2 vols., 1966. Research, publs. on learning and memory relating to normal and retarded human beings, theory of short-term memory as possibly providing behavioral distinction between normal and retarded humans. Home: 12 Thomas St., Tuscaloosa, Ala. 35401. Office: P.O. Box 6223, University, Ala.*

ELLIS, Stanley, Am. biochemist; b. California, Pa., Sept. 2, 1923; s. Peter and Caroline (Gierko) Eleshevich; B.S., Wayne U., 1946, M.S., 1949, Ph.D., 1951; m. Lucy Drobinsky, Apr. 30, 1943; children—Carol, Jeanne. Research asso. Inst. Exptl. Biology, U. Cal. at Berkeley, 1951-55; asst. prof. biochemistry Emory U., Atlanta, 1955-60; sr. research chemist Cutter Labs., Berkeley, Cal., 1960-62; chief biochem. endocrinology br. Ames Research Center, NASA, Moffett Field, Cal., 1962-—. Recipient Research Career Devel. award USPHS, 1959-60. Mem. Am. Assn. Biol. Chemists, Endocrine Soc., Biochem. Soc., A.A.A.S. Research, publs. on devel. methods for purification anterior pituitary hormones, properties and chemistry of growth hormone, discovery and characterization of proteolytic enzymes in anterior pituitary gland. Home: 2504 Webster St., Palo Alto, Cal. 94301. Office: Ames Research Center, Moffett Field, Cal. 94035.*

ELLIS, William Carlton, Am. metallurgist; b. Dundee, N.Y., Aug. 23, 1902; s. William L. and Agnes (Roberts) E.; Ch.E., Rensselaer Poly. Inst., 1924, M.S., 1925, Ph.D., 1927; m. Beatrice Wright, July 6, 1929; children—John R., Robert W., Elizabeth J. With Bell Tel. Labs., Inc., Murray Hill, N.J., 1927-67, head dept. phys. metallurgy research, 1960-67; faculty Newark Coll. Engring., 1967-—. Russell Sage fellow, 1926-27; Mathewson Gold medalist, 1966. Mem. Am. Soc. Metals, Am. Inst. Mining and Metall. Engring., Sigma Xi. Research, numerous publs. on copper alloys, magnetic materials, semiconductors and growth of crystals, discovery and devel. methods of growing crystals. Address: 323 High St., Newark 07602.

ELLISON, Robert Gordon, Am. surgeon; b. Millen, Ga., Dec. 4, 1916; s. John Gordon and Frederica (Fisher) E.; A.B., Vanderbilt U., 1939; M.D., Med. Coll. Ga., 1943; m. Lois Taylor, Feb. 11, 1945; children—Robert Gordon, Gregory, Mark James, John. Faculty, Med. Coll. Ga., Augusta, 1947-—, prof. surgery, 1959-—, chief div. thoracic surgery, 1955-—. Diplomate Am. Bd. Surgery, Bd. Thoracic Surgery. Fellow Southeastern Surg. Congress, A.C.S., Am. Coll. Cardiology (gov. Ga. 1956-—), Am. Coll. Chest Physicians; mem. Ga. (pres. 1965-—), Am. (councilor Ga. 1963-—) thoracic socs., So. Thoracic Surg. Assn. (past pres.), Soc. Thoracic Surgeons (treas. 1963-—), Tb Assn. (hon. life), Richmond County Med. Soc. (past v.p.), Med. Assn. Ga., A.M.A., Am. Physiol. Soc., Am. Soc. for Artificial Internal Organs, So. Soc. for Clin. Investigation, So. Surg. Soc., Soc. U. Surgeons, Am. Assn. for Thoracic Surgery. Research, numerous publs. on physiol. alterations assoc. with thoracic and cardiovascular surgery. Home: 606 Wellesley Dr., Augusta, Ga. 30904.*

ELLISON, Samuel Porter, Jr., Am. geologist; b. Kansas City, Mo., July 1, 1914; s. Samuel P. and Mary Frances (Edwards) E.; student Jr. Coll. Kansas City, 1930-31; A.B., with honors, U. Kansas City, 1936; M.A., U. Mo., 1938, Ph.D., 1940; m. Dorothy M. Cannady, June 9, 1940; children—Samuel David, John Robert, Stephen Paul. Faculty, U. Mo. Sch. Mines and Metallurgy, 1939-44, asst. prof., 1943-44; geologist Stanolind Oil & Gas Co., 1944-47, dist. geologist, 1947-48; prof. geology U. Tex., Austin, 1948-—, chmn. dept. geology, 1952-62. Named Alumnus of Year, U. Kansas City, 1955. Fellow Geol. Soc. Am. (councilor 1963-66); mem. A.A.A.S., Paleontol. Soc., Am. Assn. Petroleum Geologists, Soc. Econ. Paleontologists and Mineralogists sec.-treas. 1953-58, pres. 1959-60 Soc. Petroleum Engrs., Am. Inst. Profl. Geologists, Nat. Assn. Geology Tchrs. (pres. 1965), Sigma Xi. Research, numerous publs. on geol. distbn. and meaning microfossils in Paleozoic strata U. S., extraction microfossils with new technique by organic acids, use fossils in relative age dating rocks, geol. understanding origin porosity and permeability in oil field reservoir rocks. Home: 5948 Highland Hills Dr., Austin, Tex. 78731.*

ELLSBERG, Edward, Am. marine engr.; b. New Haven, Nov. 21, 1891; s. Joseph and Edna (Lavine) E.; B.S., U. S. Naval Acad., 1914; M.Sc., Mass. Inst. Tech., 1920; D.Engring., U. Colo., 1929; D.Sc., Bowdoin Coll., 1952; L.H.D., U. Me., 1955; m. Lucy Knowlton Buck, June 1, 1918; 1 dau., Mary Phillips (Mrs. Goldwin Smith Pollard). Commd. ensign USN, 1914, advanced through grades to rear adm., 1951; with Line and Constrn. Corps., 1914-26, 41-45; chief engr. Tidewater Oil Co., 1926-35; cons. engr. petroleum and marine, 1936-41, 45-60. Registered profl. engr. N.Y. Mem. N.J. Soc. Profl. Engrs., Soc. Am. Mil. Engrs., P.E.N., Naval Inst., Soc. Am. Historians, Am. Polar Soc. Author numerous books including Under the Red Sea Sun, 1946; No Banners, No Bugles, 1949; Mid Watch, 1954; The Far Shore, 1960; also articles. Pioneered vacuum desalinization sea water for marine use; invented torch for cutting steel underwater, pontoons technique; designed low pressure evaporator system; developed harbor clearance techniques. Home: Windswept, Southwest Harbor, Me. 04679.*

ELLSWORTH, Lincoln, Am. explorer, civil engr.; b. Chgo., May 12, 1880; s. James William and Eva (Butler) E.; student Columbia 2 years; M.S., Yale; LL.D., Kenyon Coll.; m. Mary Louise Ulmer, May 23, 1933. Axman on 1st survey Grand Pacific R.R. surveys of transcontinental route across Can., 1902-07, then resident engr. Prince Rupert Terminal; later resident engr. on constrn. work west of Montreal, C.P. Ry.; prospector for gold, Peace River, 1909; became asst. engr. Kougarock Mining Co., Alaska, 1910; organizer Ellsworth Expdn., sponsored by Johns Hopkins, making geol. cross section of Andes Mountains from Pacific Ocean to headwaters of Amazon, 1924; condr. and navigator N 24 on Amundsen-Ellsworth Polar Flying Expdn., reaching 88° N. Latitude; co-leader of Amundsen-Ellsworth-Nobile Transpolar flight from Kingsbay, Spitzbergen, to Teller, Alaska, May 11-13, 1926; dir. sci. investigation Wilkins-Ellsworth Trans-Arctic Submarine Expdn., 1931; represented Am. Geog. Soc. on Graf Zeppelin Arctic flight, 1931; made airplane flight of 2,300 miles, crossing Antarctic, Nov. 1935; in recognition of his flight and for claiming 30,000 square miles of new land for U. S., awarded spl. gold medal by Congress; in 1939 made flight into interior of Antarctica on Indian Ocean side, south of Australia, claiming 81,000 square miles of ty. for U. S. Fellow Am. Mus. Nat. History, Royal Geog. Soc.; asso. mem. Am. Soc. C. E. Recipient Great King Humbert medal Italian Geog. Soc., 1920; spl. gold medal, by President Hoover from Congress of U. S., 1931; David Livingston Centenary medal. Am. Geog. Soc.; Hubbard gold medal. Nat. Geog. Soc., also Explorer's gold medal; Elisha Kent Kane gold medal. Phila. Geog. Soc.; gold medal Geog. Soc. Chgo., Patron's gold medal, Royal Geog. Soc. Author: The Last Wild Buffalo Hunt, 1915; (with Capt. Roald Amundsen) Our Polar Flight, 1925; First Crossing of the Polar Sea, 1926; Search, 1932; Beyond Horizons, 1938. Died N.Y.C., May 26, 1951.

ELLWOOD, Walter Breckenridge, physicist; b. Columbia, Mo., Feb. 15, 1902; s. Charles Abram and Ida (Breckenridge) E.; B.A., U. Mo., 1924; A.M., Columbia, 1926, Ph.D. in Physics, 1933; m. Glendoris Calbreath, Aug. 18, 1927; children—Joan (Mrs. Lee B. Thomas, Jr.), Elizabeth. Mem. tech. staff Bell Telephone Labs., Inc., N.Y.C., 1930-40, 40-—, supr. devel. group, since 1943-—; sci. cons., scientist in charge group Bur. Ordnance, Navy Dept., Washington, 1940-43. Fellow Am. Phys. Soc.; mem. A.A.A.S. (sr.), I.E.E.E., Am. Soc. for Metals, N.Y. Acad. Scis., Sigma Xi; Am. Inst. Physics. Contbr. articles on magnetic measurements, contact physics to profl. publs. Patentee underwater warfare devices, contact protection, relay amplifier, mfg. methods for reed switch, magnetomotive force gauge; inventor sealed reed switch and relays. Died New York, N.Y., Dec. 9, 1965.

ELMADJIAN, Fred, physiologist; b. Aleppo, Syria, Oct. 5, 1915; s. Yervant Sarkis and Zabel (Yenovkian) E.; came to U. S., 1923, naturalized, 1929; B.S., Mass. Coll. Pharmacy, 1940, M.S. (Pharmacology fellow), 1942; postgrad. Boston U.; M.A., Clark U., 1947; Ph.D., Tufts Coll., 1949; m. Evelina Ivany, Feb. 2, 1952. Mem. staff physiol. lab. Clark U., 1943-44; staff mem. Worcester Found. for Exptl. Biology, Shrewsbury, Mass., 1944-62, sr. scientist, 1956-62, cons., 1962-63; asst. chemist dementia precox research project Worcester (Mass.) State Hosp., 1945-46, dir. labs., 1955-62, neurophysiologist, 1957-62, dir. biol. research, 1958-62; research asso. physiology dept. Tufts Med. Sch., 1950-52; physiologist Nat. Inst. Mental Health Field Sta., Shrewsbury, 1950-53; faculty Mass. Coll. Pharmacy, 1950-53; chief, biol. scis. sect. behavioral scis. tng. br. Nat. Inst. Mental Health, Bethesda, Md., 1962-—. Cons. Johns Hopkins, 1952-61, New Eng. Med. Center, 1959-62; mem. Gen. Com. on Revision U. S. Pharmacopeia, 1955-60. Mem. Am. Pharm. Assn., Am. Physiol. Soc., Endocrine Soc., Am. Inst. Biol. Scis., N.Y. Acad. Scis., A.A.A.S., Mass. Soc. for Research in Psychiatry, Soc. for Biol. Psychiatry, Royal Soc. Medicine. Research in field neuroendocrine mechanisms related to adaptive processes such as biochem. changes related to adrenal-pituitary and sympathico-adrenal systems, biochem. correlates of psychotic and neurotic conditions, emotional disturbances of normal individuals to stressful environment. Home: 11508 Stonewood Lane, Rockville, Md. 20852. Office: Nat. Inst. Mental Health, Chevy Chase, Md. 20203.*

ELMAN, Robert, Am. surgeon; b. Boston, Nov. 9, 1897; s. Samuel and Bessie Marian (Schmidt) E.; B.S., Harvard, 1919; M.D., Johns Hopkins, 1922; m. Mima Kreykenbohn, June 15, 1928. Resident house officer Johns Hopkins, 1922-23; asst. in pathology Rockefeller Inst., 1923-25; mem. teaching staff Washington U. Med. Sch., St. Louis, 1925, 56, prof. clin. surgery; asso. surgeon Barnes Hosp. and St. Louis Children's Hosp.; in pvt. practice surgery, St. Louis, 1928-56; dir. surg. service and chief of staff Homer Philips Hosp. Mem. com. on infected wounds and burns, com. on convalescence and rehab. NRC, 1943-45. Diplomate Am. Bd. Surgery. Fellow A.C.S., A.A.A.S., Internat. Soc. Surgery; mem. Harvey Soc., Am. Surg. Assn., Am. Gastroent. Assn. (president, 1955). Society Exptl. Biology and Med. (pres. Mo. sect. 1948), A.M.A., Phi Beta Kappa, Sigma Xi, Alpha Omega Alpha. Author: (with Warren H. Cole) Textbook of General Surgery, 5 edits. since 1936; Parenteral Alimentation in Surgery, 1946 (awarded the quinquennial Samuel D. Gross award. Phila. Acad. Surgery, 1945); numerous single chapters in surg. texts, many papers on surg. subjects including 1st report on use plasma transfusions in burns and on successful use of amino acid mixtures injected intravenously in human for parenteral protein feeding. Mem. editorial bd. Gastroenterology, Archives of Surgery. Died Dec. 23, 1956.

ELMEN, Gustaf Waldemar, elec. engr.; b. Stockholm, Sweden, Dec. 22, 1876; s. Claes Julius and Josephine (Ericson) E.; B.S., U. Neb., 1902, M.A., 1904, D.Eng., 1932; m. Ruth M. Halvorsen, 1907; children—James Frederick, Richard Spencer, Paul Halvorsen. Came to U. S., 1893, naturalized, 1918. Fellow in physics, U. Neb., 1902-04; elec. engr. for Gen. Electric Co., 1904-06, Western Electric Co., 1906-25, Bell Telephone Labs., 1925-41; magnetic cons. Naval Ordnance Lab., Washington Navy Yard, 1941-57. Recipient John Scott medal City of Phila., 1927, Elliott Cresson medal Franklin Inst., 1928. Mem. Sigma Xi. Research and publs. on magnetic properties of alloys; inventor magnetic materials used in elec. communications. Died Dec. 10, 1957.

ELMING, Niels, Danish chemist; b. Mariager, Denmark, July 9, 1922; s. Jens Christian and Valborg (Jeppesen) E.; chem. engring. degree, Tech. U. Denmark, 1946; m. Lis Poulsen, Oct. 26, 1946; children—Peter, Hanne. Research chemist A/S Sadolin & Holmblad, Copenhagen, Denmark. 1946-49, 50-57; vis. scientist Fed. Inst. Tech., Zurich, Switzerland, 1949-50; dir. research A/S Pharmacia, Copenhagen, 1957-63, dir. devel., 1963——. Mem. Instn. Danish Civil Engrs., Danish Chem. Soc. Contbr. numerous articles to profl. jours. Patentee in field. Organic synthetic research in furan and cevine alkaloid fields; research on structure-activity relations of terpene amines. Home: 4 Vermlandsgade, Copenhagen S. Office: 48 Lindeallé, Copenhagen/Vanlose, Denmark.*

EL NADI, Mohamed Abdel Maksoud, Egyptian nuclear physicist; b. Samanud, Egypt, Jan. 27, 1918; s. Abdel Maksoud and U. (Sief) El N.; B.Sc. with 1st class honors in Physics, Cairo (Egypt) U., 1940, M.Sc., 1945; Ph.D., London (Eng.) U., 1948, D.Sc., 1968; m. Samira Labib, Apr. 2, 1942; children—Adel, Nabil, Nadia, Nazih. Faculty Sci., Cairo U., 1940——, prof. nuclear physics, 1961——; head physics dept. UAR Atomic Energy Establishment, 1961——; research asso. Yale, 1959-60. Recipient UAR State prize, 1960; 66. Mem. Phys. Soc. London, Am. Inst. Physics. Research and numerous publs. on theory of low-energy nuclear reactions especially those in which two or more nucleons are stripped from projectile by target nucleus. Home: 50 Dokki St., Giza, Cairo, UAR.*

El-NESR, Mohamed Saad, Egyptian physicist; b. Assiout, Egypt, May 31, 1934; s. Ahmed Mohamed and Fahima El-Nesr; B.Sc. in Physics with Honor, Cairo (Egypt) U., 1956, M.Sc., 1959, Lecanciate Fil-Lic, Inst. Physics, Uppsala U., Sweden, 1961, Philosophy D., 1962; m. Ekbal Bashandy, Oct. 8, 1959; children—Osama Mohamed, Eman Mohamed. Demonstrator, Atomic Energy Establishment, Cairo, 1956; sr. researcher Atomic Energy Establishment, Cairo, head spectroscopy group; nuclear physicist Assiout U., Middle Eastern Regional-Radio Isotope Centre for Arab Countries in co-operation with IAEA, Vienna; expert nuclear physics Coll. Advanced Tech., Tripoli, Libya, 1966——. Mem. Math. and Phys. Soc. Egypt. Research, publs. on nuclear beta and gamma-ray spectroscopy, internal conversion coefficients measurement of gamma radiations, decay schemes, discoveries in nuclear properties and structure. Home: 16 El-Shiekh Ali Mahmoud, Cairo. Office: Nuclear Physics Dept. Atomic Energy Establishment, Cairo, Egypt.*

ELSASSER, Walter Maurice, geophysicist; b. Mannheim, Germany, Mar. 20, 1904; s. Moritz and Johanna (Masius) E.; Ph.D., U. Goettingen (Germany), 1927; m. Margaret Trahey, July 17, 1937; children—Barbara, William; m. Suzanne Rosenfeld, June 24, 1964. Came to the U. S., 1936, naturalized, 1940. Instr., U. Frankfurt (Germany), 1930-33; research fellow Sorbonne, Paris, 1933-36, Cal. Inst. Tech., 1936-41; war research on radar U. S. Signal Corps, also Columbia, 1941-47; prof. physics U. Pa., 1947-50, U. Utah, 1950-56, U. Cal. at La Jolla, 1956-62; chmn. dept. physics U. N.M., Albuquerque, 1960-61; prof. geophysics, dept. geology Princeton, 1962-68; research prof. Inst. Fluid Dynamics and Applied Math., U. Md., College Park, 1968——. Fellow Am. Phys. Soc., Am. Geophys. Union (Bowie medal 1959); mem. Nat. Acad. Sci. Author: The Physical Foundation of Biology (theory of earth's magnetic field), 1958. Formulated dynamo theory of permanent terrestrial magnetic field, providing explanation of permanence of field and presence of secular variation; writings on relationship between biol. scis. and quantum-mech. concepts of theoretical physics. Home: 6700 Belcrest Rd., Hyattsville, Md. 20782. Office: Inst. Fluid Dynamics and Applied Math., U. Md., College Park, Md. 20742.*

EL SAYED, Mostafa Amr, chemist; b. Ziffa, Egypt, May 8, 1933; s. Amr and Zakia (Ahmed) El S.; came to U. S., 1954, naturalized, 1965; B.Sc., Ein Shams U., Cairo, Egypt, 1953; Ph.D., Fla. State U., 1959; m. Janice Jones, Mar. 13, 1957; children—Lyla, Tarric, James, Doréa. Research asso. Fla. State U., 1958-59; Harvard, 1959-60, Cal. Inst. Tech., Pasadena, 1960-61; faculty U. Cal. at Los Angeles, 1961-——, asso. prof. chemistry, 1964——. Cons. Electro Optics, Pasadena, Cal., 1963——. Recipient U. Cal. at Los Angeles Distinguished Teaching award, 1964; Alfred P. Sloan fellow, 1965-67; Guggenheim fellow, 1965. Mem. Am. Phys. Soc., Am. Chem. Soc., A.A.A.S., Am. Assn. U. Profs., Phi Lambda Upsilon. Research, publs. on mechanisms of molecular origin of phosphorescence, mechanism of energy transfer processes, spectra and structure of molecular crystals, vacuum ultraviolet spectroscopy and photoionization processes, laser spectroscopy. Home: 3325 Colbert Av., Los Angeles 90066.*

ELSBERG, Louis, laryngologist; b. Iserlohn, Prussia, Apr. 2, 1836; s. Nathan and Adelaide E.; M.D., Jefferson Med. Coll., 1857; m. Mary Van Hagen Scoville, 1876. Resident physician Mt. Sinai Hosp., N.Y.C.; an editor N. Am. Med. Reporter, 1859; mem. faculty med. dept. U. City N.Y.; held 1st course of lectures on diseases of throat in U. S., 1862; recipient gold medal A.M.A. for publ. Laryngoscopal Surgery Illustrated in the Treatment of Morbid Growths within the Larynx, 1865; published Regeneration, or the Preservation of the Organic Molecules (his most important contbn. to med. science), 1874; founder, 1st pres. Am. Laryngol. Assn., 1878; founder quarterly Archives of Laryngology, 1880; Neuroses of Sensation, 1882; Structure of Hyaline Cartilage, 1881-82; On Angioma of the Larynx, 1884. Pioneer laryngologist; introduced use of laryngoscope in U. S.; inventor several instruments. Died N.Y.C. Feb. 19, 1885.

ELSCHNIG, Anton Philipp, Austrian ophthalmologist; b. Leibniz, Austria, Aug. 22, 1863; s. Anton and Antonie (Braun) E.; M.D., Graz, Austria, 1886; m. Emma Eichler, 1893; 1 son, 2 daus. Became lectr. U. Graz, 1892; joined U. Vienna, 1895, became prof., 1900; prof. ophthalmology German U., Prague, 1907-33. Author: Die Funktionsprüfung des Auges, 1896; Über die Discision, 1896; Molluscum contagiosum und Conjunctivitis follicularis, 1897; Augenmuskellähmungen durch Geschwulstmetastasen, 1898; Zur Anatomie der Schnervenatrophie bei Erkrankungen des Zentralnervenssystems, 1898; Stereoskopischphotographischer Atlas der pathologischen Anatomie des Auges, 1901; Handbucher der pathologischen Anatomie und Histology, 1928; Die intrakapsuläre Starextraktion, 1932. Research on glaucoma, optic nerves, keratitis parenchymatosa, theory on origin of sympathetic ophthalmia; developed new surg. methods, including improvement of A. v. Hippel's corneal grafting, operations on cataracts; introduced motion pictures in teaching demonstrations for eye surgery. Died Vienna, Nov. 13, 1939.

EL-SHERBINI, Mahmoud Ahmed, physicist; b. Minet El-Gamh, UAR, Jan. 22, 1909; s. Ahmed Mohammed and Nabeha (El-Sherbini) El-S.; B.Sc., Cairo (UAR) U., 1930, M.Sc., 1935; Ph.D., King's Coll., London (Eng.) U., 1935; m. Rawhia Ali El-Sherbini, Nov. 2, 1940; children—Tharwat, Bahgat, Effat. Demonstrator, Faculty Sci., Cairo, 1930-35, lectr., 1935-45; asst. prof. Faculty Sci., Alexandria, UAR, 1945-49, prof., head physics dept., 1949-64, 66——, vice dean, 1950-51, dean, 1957-61; head physics dept. Libya U., Tripoli, 1964-66. Recipient award Math. and Phys. Soc. Egypt. Mem. Phys. Soc. London, Nat. Com. Pure and Applied Physics, Internat. Edn. Com. in Physics (corr.), also Egyptian sci. socs. Author works on physics in Arabic. Research on effect of strong electric fields on atoms, electron reflection at low energy regions, surface structure and secondary emission, reversal of rectification on cooling, devices for measuring small electric currents, cosmic rays in Egypt, fallout in Alexandria. Home: 34 Khalil Motran, Alexandria, UAR.*

ELSHOLTZ, Johann Sigmund, German physician; b. Frankfort/Oder, Germany, Aug. 26, 1623; physician to elector of Brandebourg. Author: Destillatoria curiosa, 1674. Research and publs. on metals, minerals, animals, vegetables; one of 1st to make intravenous injections, 1665. Died Berlin, Feb. 28, 1688.

ELSLAGER, Edward Faith, Am. chemist; b. Decatur, Ill., Oct. 7, 1924; s. Hubert H. and Nira (Faith) E.; B.S., James Millikin U., 1947; M.S., U. Ill., 1948, Ph.D., 1951; m. Faye Louise Pike, Jan. 26, 1946; children—Linda (Mrs. Michael John McGuigan), Cheryle Anne, James Edward. Teaching asst. U. Ill., 1947-49, U. fellow, 1949-51; asso. research chemist Parke, Davis & Co., Detroit, 1951-52, research chemist, 1952-54, research leader, 1954-58, lab. dir. in organic chemistry, 1958-63, group dir. organic chemistry, Ann Arbor, Mich., 1963——. Recipient James Millikin U. Henderson prize in chemistry, 1946, 1947. Mem. Am. Chem. Soc., Am. Soc. Tropical Medicine and Hygiene, Sigma Xi, Alpha Chi Sigma. Author chpt. Antiamebic Agents in Medicinal Chemistry, 2d edit., 1960; Human Antiparasitic Agents in Annual Reports in Medicinal Chemistry, 1965, 66. Editorial bd. Jour. Heterocyclic Chemistry, 1964——. Research, numerous publs. concerned primarily with synthesis of amines, heterocyclic compounds, sulfur compounds, quinones, dyes, chemotherapy of parasitic diseases and leprosy, instrumental in devel. of anthelmintics pyrvinium pamoate and triclofenol piperazine and long-acting antimalarial drugs cycloguanil pamoate and DADDS. Patentee in field. Home: 4081 Thornoaks Dr., Ann Arbor 48104. Office: 2800 Plymouth Rd., Ann Arbor, Mich. 48106.*

EL-STAZLY, K., Egyptian nutritionist; b. Tanta, Egypt, Jan. 14, 1923; s. Abd El Salam Mohamed and Nazly (Rustom) El-S; student Cairo U., 1939-43, Edinburgh U., 1945-48, Rowett Research Inst. Aberdeen, 1948-51; m. Nazly Ali El Dorry, Apr. 24, 1950; children—Mohiel-Din, Ali el-Olin. Demonstrator, Faculty Agr., U. Alexandria, Egypt, 1943-51, lectr., 1951-58, asst. prof., 1958-65, prof., 1965——; with Ohio Agrl. Expt. Sta., 1958-59, bacteriology dept. U. Cal., 1963-64. Mem. Biochem. Soc. Research, publs. on protein degradation in rumen; discovery of branched chain volatile fatty acids as break down products of rumen fermentation; stickland type reaction was discovered and separation of S-amino valeric acid; method for estimation of diamino pimelicacid in rumen micro-organisms and estimation of net growth; starch inhibition of cellulose digestion; role of urea in its relief. Home: 27 El-Geish St., Alexandria, Egypt.*

ELSTER, Julius, German physicist; b. Blankenburg, Germany, Dec. 24, 1854; s. Ludwig Carl and Johanna Elisabeth Clara (Stegmann) E.; student natural scis. Berlin, Germany, 1875; doctorate Heidelberg, Germany, 1879; m. Auguste Friederike Luise Fink. Faculty, Gymnasium, Wolfenbüttel, Germany, 1881-1919; ret. as prof. Mem. Leopoldina. Studied (with Geitel) ion conduction in gases, atmospheric electricity, photoelectric effects, radioactivity; built 1st photoelectric cell, 1st photoelectric photometer, Tesla transformer; 1st to determine charge on falling raindrops, 1899; demonstrated that lead is not radioactive of itself, that radioactive substances in atmosphere, responsible for atmospheric conductivity, easily break down into unstable elements. Died Wolfenbüttel, Apr. 8, 1920.

ELSTON, Jean, French physico-chemist, ceramist; b. Saumur, Maine-et-Loire, France, Mar. 21, 1921; s. Alexandre and Maria (Mouate) E.; Ingenieur-chimiste, Ecole Nationale Superieure de Chimie de Paris, 1945; Dr. es Sciences Physiques, Faculte des Sciences de Paris, 1949; m. Louise Laylle, July 30, 1952. Research engr. Nat. Indsl. Bur. Nitrogen, St. Gobain Co., 1949-56; research engr. French Atomic Energy Commn., Gif-sur-Yvette, France, 1957-59, chief sect. studies on refractory material, 1959——. Tchr., Nat. Instn. for Nuclear Sci. and Tech., 1962-65. Recipient Montyon prize Acad. Scis., 1966. Mem. High Temperatures Soc., Chem. Physics Soc., Metallurgy Soc., French Chem. Soc. Research, numerous publs. on sintering of refractory oxides, irradiation induced damage in ceramics. Home: 26 Parc d'Ardenay, Palaiseau, Essonne. Office: C.E.N.-SACLAY-B.P. No. 2, Gif-sur-Yvette, Essonne, France.*

ELTON, Lewis Richard Benjamin, physicist; b. Frankfort/Main, Germany, Mar. 25, 1923; s. Victor Leopold and Eva (Sommer) Ehrenberg; B.A., Cambridge (Eng.) U., 1945, M.A., 1948; pvt. study, 1945-47; B.Sc., U. London (Eng.), 1947, Ph.D., 1950, D.Sc, 1961; m. Kathleen Mary Foster, Aug. 24, 1950; children—Bridget, Thomas, David, Benjamin. Asst. master St. Bees Sch., Cumberland, Eng., 1944-46; faculty King's Coll. London, 1950-57; vis. research asso. Mass. Inst. Tech., 1955-56; head physics dept. Battersea Coll. Tech., 1957-66; prof., head physics dept. U. Surrey, London, 1966——, dir. research in sci. teaching, 1967——. Vis. research asso. Inst. Theoretical Physics, Copenhagen, Denmark, 1962; vis. research prof. U. Wash., Seattle, 1965. Fellow Inst. Physics, Inst. Math. and Its Applications; mem. Am. Inst. Physics, Assn. for Sci. Edn. (past mem. physics panel). Author: Introductory Nuclear Theory, 1959; Nuclear Sizes, 1961; also articles. Research on nuclear charge and mass distbns. through elastic scattering high energy electrons, nucleons and mesons, radiative correction to electron scattering, inelastic scattering and nuclear reactions, optical model at high energies, single particle wave functions and shell model, hyperfragments. Home: 52 Ravensbourne Park, London S.E. 6., U.K.*

ELVEHJEM, Conrad Arnold, Am. biochemist; b. McFarland, Wis., May 27, 1901; s. Ole Johnson and Christine (Lewis) E.; B.S., U. Wis., 1923, M.S., 1924, Ph.D., 1927; student Cambridge U., 1929-30; D.Sc., Ripon Coll., 1942, Beloit Coll., 1958, Northwestern U., 1959; LL.D., U. Cal., 1959; m. Constance Walts, June 30, 1926; children—Peggy Ann (Mrs. Calvin Henninger), Robert Stuart. Instr. biochemistry U. Wis., Madison, 1925-29, asst. prof., 1930-32, asso. prof., 1932-36, prof. 1936-58, chmn. dept., 1944-58, dean Grad. Sch., 1946-58, pres., 1958-62. Harvey Soc. lectr., 1940, Herter lectr., 1942; Sigma Xi lectr., 1943. Recipient Mead Johnson award for research in Vitamin B Complex 1939, Grocery Mfrs. Am. award, 1942. Willard Gibbs medal, 1943, Nicholas Appert Medal, 1948, Osborne-Mendel award, 1950; Lasker award in med. research, 1952; Charles F. Spencer award Am. Chem. Soc., 1956; Am. Inst. Baking award, 1957. Chmn. Food Nutrition bd. NRC, 1955-58; mem. Nat. Adv. Heart Council, 1948-50, Nat. Sci. Found. Bd., 1960-62, Research Corp. Bd., 1960-62, Sugar Research Found., Inc. Bd., 1959; mem. Council Foods and Nutrition, A.M.A., 1940-59; mem. sci. adv. com. Nutrition Found. 1941-58, Mem. Am. Acad. Arts and Scis., Biochem. Soc. Eng., Am. Chem. Soc., A.A.A.S., Am. Soc. Biol. Chemists, Am. Inst. Nutrition (pres. 1953-54), Soc. Exptl. Biology and Medicine, Inst. Food Technologists, Nat. Acad. Scis., Am. Pub. Health Assn., Am. Philos. Soc., Phi Beta Kappa (hon.), Sigma Xi, Gamma Alpha, Alpha Chi Sigma, Phi Sigma, Alpha Zeta, Delta Theta Sigma, Phi Kappa Phi (hon.) Phi Lambda Upsilon. Author: (with H. A. Waisman) Vitamin Content of Meat, 1941. Editor: Respiratory Enzymes, 1939. Research and publs. on biochemistry and nutrition; discovered that nicotinic acid cured pellagra in dogs, 1937; discovered that copper is essential in formation of hemoglobin; isolated niacin or nicotinic acid (with R. J. Madden, F. N. Strong, D. W. Woolley), 1938; studied respiration in tissues. Died Madison, July 27, 1962.

ELVEY, Christian Thomas, Am. astro-geophysicist; b. Phoenix, Apr. 1, 1899; s. John A. and Lizzie (Miller) E.; A.B., U. Kan., 1921, A.M., 1923; Ph.D., U. Chgo., 1930; m. Marjorie D. Purdy, Sept. 1, 1934; children—Thomas C., Christena V. (Mrs. Lyman B. Jennings). Instr. astronomy U. Kan., 1921-25; fellow in astronomy U. Chgo., 1925-26; instr. astronomy Northwestern U., 1926-28; instr. astrophysics U. Chgo., 1928-32, asst. prof., 1932-42; astronomer, asst. to dir. McDonald Obs., Fort Davis, Tex., 1935-42; physicist Cal. Inst. Tech., Pasadena, 1942-45; head applied research div. Naval Ordnance Test Sta., Inyokern, Cal., 1945-47, head research dept., 1947-49, sr. research sci., 1949-50, head staff, 1951; head dept. geophysics, dir. geophys. inst. U. Alaska, College, 1952-63, v.p. for research and advanced study, 1961-63, univ. research prof., 1963—. Mem. U. S. Nat. Com. for Internat. Geophys. Year, 1957-58, Internat. Com. for Geophysics, 1959-62, Internat. Astronom. Union, 1935—. Research in rotation of stars, galactic light, light of night sky, aurora. Home: P.O. Box 147, College, Alaska 99375.*

ELWES, Henry John, botanist, entomologist; b. Gloucestershire, Eng., May 16, 1846; s. John Henry and Mary (Bromley) E.; ed. Eton; m. Margaret Susan Lowndes-Stone, 1871; 1 son, 1 dau. Entered Scots Guards, 1865; collected plant and insect specimens in Asia Minor, Turkey, India, Tibet, N.Am., Mexico, Chile, Russia, China, Japan, Formosa. Recipient 1st Victoria medal honor Royal Hort. Soc., 1897. Fellow Royal Soc., 1897. Author: Monograph of the Genus Lilium, 1880; (with Augustine Henry) Trees of Great Britain and Ireland, 1906-13; also articles. Discovered many new species; presented his collection of 30,000 specimens of Lepidoptera to Brit. Mus.; grew many rare plants, including several which were named after him (Galanthus elwesii). Died Nov. 26, 1922.

ELWORTHY, Peter Howard, Brit. pharmacist; b. Kingsbridge, Devon, Eng., July 22, 1932; s. Edgar Howard and Winifred (Trenuwan) E.; B.Pharm. with honors, U. London, 1953, Ph.D., 1956, D.Sc., 1965; m. Mary F. Allen, June 9, 1956; children—Andrew, David. Asst. lectr. U. London, 1957-59; faculty U. Strathllyde, Glasgow, Scotland, 1959—, prof. pharm. tech., 1966—; vis. prof. U. So. Cal., 1964. Mem. Royal Inst. Chemistry (asso.), Pharm. Soc., Farraday Soc., Chem. Soc., Sigma Xi. Author: (with Florence, Macfarland) Applications of Solubilisation, 1967; also numerous articles. Research on nonionic detergents in relation to uses as emulsifying, suspending and surface active agts., phys. chemistry of phosphatides, shape of neuromuscular blocking agts. in solution. Home: 59 Beech Av., Newton Mearing, Renfrewshire, Scotland. Office: U. Strathllyde, Glasgow Cl, Scotland.*

ELYUTIN, Vyacheslav Petrovich, Russian metallurgist; b. 1907; ed. Moscow Inst. Steel, Dr.Tech.Sc. Dir., Moscow Inst. Steel, 1945-51; dep. minister of Higher Edn., 1951-54, minister, 1954-59; then minister of Higher and Middle Special Edn.; accompanied Khrushchev to U. S., 1959; dep. chmn. Com. for Lenin Prizes, also Soviet Chinese Friendship Soc. Recipient Stalin prize, 1952, Order of Lenin, 1957. Mem. USSR Acad. Scis. (corr.). Author: Ferrous Alloys Production, 1951; The Viscosity of Fused Titaniumn 1957; Strength of Carbide Base Alloy Bond with NiAl of CoAl Compounds, 1960. Research on phys. and chem. properties of metals and alloys. Address: Ministry of Higher and Middle Special Education, ul. Zhdanova 11, Moscow, Russia.

ELZARAKEEL, see Arzachal.

EMANUEL, Nikolay Markovich, Russian phys. chemist; b. Oct. 1, 1915; grad. Leningrad Poly. Inst., 1938. Asso., Inst. Chem. Physics, USSR Acad. Sci., 1938—; instr. Moscow U., 1944-50, prof., 1950—. Recipient Lenin prize, 1958. Mem. USSR Acad. Sci. (corr., dep. chief learned sec. Presidium). Author: Intermediate Products of Complex Gaseous Reactions, 1946. Research and publs. on chem. kinetics and practical applications, path. processes in biology, chem. manifestations of intramolecular hydrogen bond; discovered formation of intermediary products of free radical type during slow oxidation chain reactions, developer kinetic methods of studying these products, discovered new mechanism of homogeneous catalysis in hydrocarbon oxidation reactions, suggested methods of controlling complex chain reactions during process; developer various methods of stimulating slow chain reactions. Address: Inst. Chemical Physics, USSR Acad. Sci., Vorobyevskoe sh. 2b, Moscow, USSR.

EMBDEN, Gustav Georg, German physiol. chemist; b. Hamburg, Germany, Oct. 10, 1874; s. George Heinrich and Elisabeth Charlotte (Dehn) E.; student physiology Freiburg, Germany, physiol. chemistry, Strasbourg, France; also student Zurich, Switzerland, Frankfort/Main, Germany; m. Hanni Fellner, 1911; 1 son, 3 daus. Named dir. chem. lab. Med. Clinic, Frankfort/Main, 1904; became lectr., Bonn, Germany, 1907, named asso. prof. U. Frankfort/Main, 1909, prof., dir. Inst. for Vegetative Physiology, 1914, rector, 1925-26. Studied physiol. chemistry and chem. constrn. of muscles, organs and organisms; discoveries in metabolic function of liver, diabetes. Died Nassau, Germany, July 25, 1933.

EMBERGER, Louis Marie, French botanist; b. Thann, France, Jan. 23, 1897; s. Emile and Léonie (Hueber) E.; Ph.D. in pharm. scis.; m. Alice Flahault, Apr. 18, 1922; children—Jacques, Odile, Geneviève, Monique, Jean-Marie, Veronique, Marie-Claude. Instr. Faculty Pharmacy, Montpellier, 1921-26; prof. Faculty Scis., Clermont-Ferrand, 1936-37; prof. Faculty Scis., Montpellier, since 1937, also dir. Bot. Inst. Mem. French Acad. Scis. (botany section), 1948, Royal Belgium Acad. Scis. Author: Les végétaux fossiles dans leurs rapports avec les végétaux vivants; Traité de botanique sytématique. Research on vegetal cytôlogy, bot. geography and ecology. Home: 4, rue Provence, Montpellier. Office: Institut Botanique, Montpellier, France.

EMBLETON, Tom William, Am. hort. scientist; b. Guthrie, Okla., Jan. 3, 1918; s. Harry and Katherine (Smith) E.; B.S. with distinction, U. Ariz., 1941; Ph.D., Cornell University, Ithaca, New York, 1949; m. Lorraine Marie Davidson, January 22, 1943; children—Harry Raymond (deceased 1967), Gary Thomas, Wayne Allen, Terry Scott, Paul Henry. Sci. aide to horticulturist U. S. Dept. Agr., Indio, Cal., 1942, 46; asst. horticulturist Irrigation Exptl. Sta., Wash. State Coll., Prosser, 1949-50; staff U. Cal. Citrus Research Center, Riverside, 1950—, horticulturist, 1962—. Recipient Cal. Citrograph Research award, 1965. Fellow A.A.A.S.; mem. Internat., Am. (past sec.-treas., past chmn. Western region) socs. for hort. sci., Am. Soc. Agronomy, Soil Sci. Soc. Am., Western Soc. Soil Sci., Am. Soc. Plant Physiologists, Am. Inst. Biol. Scis., Date Grower's Inst., Cal. Avocado Soc., Fla. State Hort. Soc., Sigma Xi, Alpha Zeta. Research, numerous publs. on improved means of determining nutrient status of trees, devel. techniques of comml. nutritional control, new methods for control of some nutrient deficiencies of trees. Home: 796 W. Spruce St., Riverside, Cal. 92507.*

EMDE, Fritz, German elec. engr.; b. Uschütz, Germany, July 13, 1873; s. Wilhelm and Marie (Grapow) E.; self-educated; hon. dr. engring., Breslau (now Wroclaw, Poland), Zürich Switzerland; m. Anna Maass, 1900, 3 children. Joined Allgemeine Elektrizitäts-Gesellschaft, 1895; later Siemens; joined Bergakademie, Clausthal, Germany, 1911; prof. theoretic electrotechnics, dir. Electrotech. Inst., Tech. U. Stuttgart, Germany, 1912-38. Named hon. citizen Tech. U. Karlsruhe, Germany; recipient Goethe medal, 1943. Mem. Assn. German Elec. Engrs. (hon.). Author: Sinusrelief und Tangensrelief, 1924; Tafeln höherer Funktionen, 1948; Quirlende elektrische Felder, 1949. Studied elec. engines, induction law, transformer field, current displacement, rotary magnets, concepts of motive power of rest and motion, vector analysis; applications of Faraday-Maxwell theory, calculating machines; compiled (with Jahnke) function tables. Died Stuttgart, June 30, 1951.

EMDE, Hermann Karl Christian Maximilian, chemist; b. Opladen, Germany, Dec. 10, 1880; s. Burghard and Fanny (Koch) E.; student pharmacy, chemistry and food chemistry, Braunschweig, Germany; doctorate, Marburg, Germany, 1906; m. Eva Buchler; 1 son, 1 dau. Joined faculty, Braunschweig, 1908; joined Chinin-Fabrik, Buchler & Co., Braunschweig, 1912; officer in World War I; then chief chemist firm Schering, Berlin, Germany; later employed by Hoffman-La Roche, Basel, Switzerland, dir. alkaloid factory, Soekabami, Java for a year; named lectr., Basel, 1928, asso. prof., 1930; became prof. pharm. and nutrition chemistry, Königsberg, Germany, 1931. Author: Beiträge zur Kenntnis und Versuche zur Synthese des Ephedrins und Pseudoephedrin, 1906; Kohlenstoff-Doppelbindung und Kohlenstoff-Stickstoff-Bindung, 1908; also articles. Studied composition and synthesis of natural substances in life process; research on ephedrin including Emde decomposition reaction (quartenary ammonium salts stored with hydrogen which resulted in splitting of molecule). Died Thun, Switzerland, July 19, 1935.

EMDEN, (Jacob) Robert, Swiss astrophysicist, meteorologist; b. St. Gallen, Switzerland, Mar. 4, 1862; s. Moritz Philipp and Emma (Gerstle) E.; student Heidelberg, Berlin, (all Germany); doctorate Strasbourg, France, 1887; m. Klara Schwarzschild, 1907. Became lectr. physics Tech. U. Munich, Germany, 1889, asso. prof. meteorology, aeronautics, 1907, later of theoretical physics; named hon. prof. astrophysics U. Munich, 1924. Mem. Bavarian Acad. Scis., Royal Astron. Soc. (London). Author: Gaskugeln, Anwendungen der mechanischen Wärmetheorie auf kosmologische und meteorologische Probleme, 1907; Grundlagen der Ballonführung, 1910. A founder Zeitschrift für Astrophysik, 1930, editor until 1936. Developed theory of polytropic gas spheres, including application to stellar structure, method of integration of differential equations named after him; studied thermodynamics of stellar bodies, aerodynamics, balloons; 1st studies (with S. Finsterwalder) in photogrammetry from air; developed (with J. H. Lane, A. Ritter) theory of self-gravitating gas sphere into logical survey; applications of Schwarzschild's theory of radiation equilibrium to earth's atmosphere. Died Zurich, Oct. 8, 1940.

EMERSON, Alfred Edwards, Am. zoologist; b. Ithaca, N.Y., Dec. 31, 1896; s. Alfred and Alice (Edwards) E.; B.S., Cornell U., 1918, M.A., 1920, Ph.D., 1925; D.Sc., Mich. State U., 1961; m. Winifred Jelliffe; May 3, 1920 (dec. Sept. 1949); children—Helena Leeming (Mrs. Eugene Wilkening), William Jelliffe; m. 2d, Eleanor Fish, Sept. 3, 1950. Faculty, U. Pitts., 1921-29, asso. prof., 1926-29; faculty U. Chgo., 1929—, prof. zoology, 1934-62, prof. emeritus, 1962—; vis. prof. U. Cal. at Berkeley, 1949; vis. Distinguished Prof. natural sci. Mich. State U., East Lansing, 1960. Research asso. Am. Mus. Natural History, N.Y.C., 1940—. Guggenheim fellow, 1926-27; Belgian Am. Found. fellow, 1948. Mem. Ill. Acad. Sci. (past pres.), Ecol. Soc. Am. (past pres.), Soc. for Study Evolution (past pres.), Soc. Systematic Zoology (past pres.), A.A.A.S. (past v.p.), Nat. Acad. Scis. Author: (with others) Principles of Animal Ecology, 1949; (with Eleanor Fish) Termite City, 1937. Research on evolutionary processes, particularly in termite socs. Home: Huletts Landing, N.Y. 12841. Office: Dept. Zoology, U. Chgo., Chgo. 60637.*

EMERSON, George Albert, Am. pharmacologist; b. San Francisco, Feb. 27, 1912; s. George Waldo and Josephine (Hofmann) E.; B.A., U. Cal. at Berkeley, 1932, M.S., 1933, Ph.D., 1934; m. Opal Bernice Wright, Aug. 31, 1960; children (by previous marriage)—Marjorie Jean (Mrs. Richard Cole), George Frederick, Carolyn Ruth (Mrs. Martin Needleman), Catherine Sue. Research asst. Vanderbilt U. Sch. Medicine, 1934-35; faculty W. Va. U. Sch. Medicine, 1935-44; prof. U. Tex. Med. Br., Galveston, 1943—, chmn. dept. pharmacology, 1943-60, vis. prof. physiology Dental Sch., 1943—; vis. prof. U. Ark. Med. Sch., 1963, U. Baghdad, (Iraq), 1963-65. Mem. Soc. Pharmacology Exptl. Therapy, Soc. Exptl. Biol. Medicine, Sigma Xi. Collaborating editor Tex. Reports Biol. Medicine, 1944-56, Jour. Fac. Medicine, Baghdad, 1963-65, chemotherapy sect. Biol. Abstracts, 1945—, Archives Internat. Pharmacodynamie et de Therapie, 1946—. Research and publs. in anesthesiology, biochem. pharmacology. Home: 1314 Avenue C, Galveston, Tex. 77550.*

EMERSON, Gladys Anderson, Am. biochemist, nutritionist; b. Caldwell, Kan., July 1, 1903; d. Otis and Louise (Williams) Anderson; A.B., B.S., Okla. Coll. for Women, 1925; M.A., Stanford, 1926; Ph.D., U. Cal. (Univ. fellow), 1932; grad. study U. Gottingen, 1932-33. Teaching asst. Okla. Coll. for Women, 1923-25; asst. Stanford, 1925-26; research asso. Inst. Exptl. Biology, U. Cal., 1933-42, vis. lectr. pharmacology med. sch., 1945; research asso. Sloan-Kettering Inst. Cancer Research, 1950-53; head dept., animal nutrition Merck Inst. Therapeutic Research, Rahway, N.J., 1942-46, dir. nutrition, Sharpe & Dohme Research Labs., 1946-57; Marie Curie lectr. Pa. State Coll., 1951; research lectr. Ia. State Coll., 1952; prof., chmn. dept. home econs., U. Cal. at Los Angeles, 1957-61, prof. nutrition and pub. health nutrition, head div. nutritional scis., vice chmn. dept. pub. health, 1962—; vis. lecturer in biochemistry and nutrition, U. Neb., 1958; engaged in research OSRD, 1943-45; mem. liaison and sci. adv. bd. Q.M. Food and Container Inst., 1949-50; food and nutrition research com. NRC, 1952, mem. Food and Nutrition Bd., 1959-64, mem. com. dietary allowances, 1960-64; exec. council Am. Bd. Nutrition, 1959—; panelist Rensselaer Poly. Inst. indsl. council, 1955; organizing com. 5th Internat. Congress Nutrition, S.Cal. sect. WHO, 1963; del. confs. in field; instr. trainees Peace Corps, 1962, 63, 64. Recipient Garvan medal, 1952. Fellow A.A.A.S., N.Y. Acad. Scis.; mem. Am. Chem. Soc. (chmn. women's service com., 1953-58), Am. Inst. Nutrition, (councillor 1952-55, chmn. membership com. 1964), Soc. Exptl. Biology and Medicine, Gordon Research Conf. (chmn. vitamins and metabolism 1952, v. chmn. 1951), Pan Am. Med. Assn. (council 1959-60) Internat. Union Nutrition Scientists (del.; nat. com. 1959-62), Sigma Xi, Delta Omega, Sigma Delta Epsilon, Iota Sigma Pi (nat. v.p. 1945-51, nat. pres. 1951-57. Contbr. articles sci. jours. Asso. editor Jour. Nutrition 1952-56. Research in amino acids, vitamin E, vitamin B Complex, antimetabolites. Home: 319 Amalfi Dr., Sanda Monica, Cal.*

EMERSON, Harrington, Am. management scientist; b. Trenton, N.J., Aug. 2, 1853; s. Edwin and Mary Louise (Ingham) E.; ed. Royal Bavarian Polytechnic,

1872-75; univs. in Italy and Greece, 1875-76; m. June 9, 1879; children—Raffe, Eugene, Eleanor; m. 2d, Mary Crawford Suplee, Feb. 5, 1895; children—Isabel, Margot, Louise. Prof. modern langs., U. of Neb., 1876-82; in banking and land business, 1882-96; pres. Emerson Engineers, efficiency engrs., 1900-23. Author: Efficiency as Basis for Operation and Wages, 1909; The Twelve Principles of Efficiency, 1912; The Scientific Selection of Employees, 1913. As U. S. rep. of British syndicate examined many industrial plants and mines in U. S., Mex. and Can., 1896-98; put into operation some of first long distance mail routes in Alaska and down the Yukon, 1898-1901; reported on all the known coal deposits of N.Am. western coast, and reported on northern submarine cable route to Asia which was largely followed by the War Dept. in laying its Alaskan cables; gained recognition on account of results obtained by efficiency methods installed on the Santa Fé System. Apptd., 1921, mem. Herbert Hoover's Com. for Elimination of Waste in Industry, of Federated Am. Engring. Socs. (assigned to investigate and report on railroads and bituminous coal mining). Died New York, N.Y., May 23, 1931.

EMERSON, Haven, Am. physician; b. N.Y.C., Oct. 19, 1874; s. J. Haven and Susan (Tompkins) E.; A.B., Harvard, 1896; A.M., Columbia, 1899, D.Sc., 1954; M.D., Columbia Coll. Phys. and Surg. 1899; m. Grace Parrish, June 15, 1901; children—Ethel, Robert, J. Haven, Ruth, Ralph. Practiced in N.Y.C., 1899-1957; asso. in physiology and medicine Coll. Phys. and Surg., 1902-14; asst. vis. physician Bellevue Hosp., 1906-14; pres. Bd. Health, and commr. Dept. Health City N.Y., 1915-17; prof. preventive medicine Cornell U.; dir. Cleve. Hosp. and Health pub. Survey, 1919-20; prof. pub. health adminstrn. Columbia, 1922-40, emeritus, 1940-57. Dir. health and hosp. surveys in many cities of U. S., also Athens, Greece, 1921-31; dir. N.Y. Hosp. Survey, 1935-37. Mem. Board of Health, N.Y.C., 1937-57. Recipient Lasker award, 1949, Gold Heart award Am. Heart Assn., 1953. Mem. Am. Pub. Health Assn. (pres. 1933-34, Sedgwick medal 1935), Acad. of Athens (Greece). Author: Alcohol, Its Effects on Man, 1934; reviser of Flint's Manual of Ausculation and Percussion, 1916. Editor: Alcohol and Man, 1932; The Baker Memorial, 1930-39, 1941, Administrative Medicine, 1942, Local Health Units for the Nation, 1945. Died May 21, 1957.

EMERSON, Robert, Am. botanist; b. N.Y.C., Nov. 4, 1903; s. Haven and Grace (Parrish) E.; A.B. cum laude, Harvard, 1925; Ph.D., Friedrich Wilhelm U., 1927; m. Claire Garrison, Feb. 9, 1929; children—Kenneth, Stephen, David, Ruth. Nat. Research fellow, Harvard, 1927-29; asst. prof. biophysics Cal. Inst. Tech., 1929-46; on leave as research asso. Carnegie Inst. Washington, Stanford, 1937-40; research prof. botany U. Ill., 1946-59. Fulbright fellow, 1954. Mem. Nat. Acad. Scis., Am. Bot. Soc., Am. Soc. Plant Physiologists. Author sci. papers. Research on photosynthesis, especially investigation of maximum efficiency. Died Feb. 4, 1959.

EMERSON, Rollins Adams, Am. geneticist; b. Pillar Point, N.Y., May 5, 1873; s. Charles D. and Mary C. (Adams) E.; B.Sc., U. Neb., 1897, LL.D., 1917; Sc.D., Harvard, 1913; m. Harriet Hardin, May 23, 1898; children—Mrs. Thera Kahler, Sterling Howard, Eugene Hardin, Mrs. Myra Ryan. Horticulturist, office of expt. stas., U. S. Dept. Agr., Washington, 1897-98; asst. prof., prof. horticulture, U. Neb., 1899-1914; prof. plant breeding, Cornell U., 1914-42, dean grad. sch., 1925-30. Mem. A.A.A.S., Am. Soc. Naturalists, Nat. Acad. Scis., Am. Genetic Assn. Investigated problems of orchard mgmt., including insect and disease control, winter hardness; effects of environmental factors on seed prodn. in potatoes, heredity studies on common bean, nature of somatic mutations; compiled genetic analysis of corn plant. Died Dec. 8, 1947.

EMERSON, William, English mathematician; b. Hurworth, Eng., May 14, 1701; s. Dudley Emerson; ed. Newcastle, York, Eng.; m. Miss Johnson, 1732; declined fellowship in Royal Soc. Author: Fluxions, 1749; The Projection of the Sphere, 1749; Elements of Trigonometry, 1749; The Doctrine of Proportions, 1763; Elements of Geometry, 1763; The Method in Increments, 1763; Cyclomathesis, 1763; The Laws on Centripetal and Centrifugal Force, 1769; A Short Comment on Sir I. Newton's Principia. Died May 20, 1782.

EMERTON, James H., Am. naturalist; b. Salem, Mass., 1847. Student of Am. spiders. Illustrator of zöol. publs. Illustrations in Packard's Guide to the Study of Insects; Scudder's Butterflies of North America; A. E. Verrill's papers in Reports of U. S. Fish Commn., 1874-84; C. S. Minot's text-book of Embryology. Made models of large octopus and squid in museums at Cambridge, New Haven, N.Y., Washington; anatomical models in med. museums of Harvard, U. Pa., Army Med. Mus., Washington. Author: The Structure and Habits of Spiders, 1878; Common Spiders of the United States, 1902; also The New England Spiders, also articles. Died Dec. 5, 1930.

EMERY, Kenneth Orris, Am. oceanographer; b. Swift Current, Sask., Can., June 6, 1914 (parents Am. citizens); s. Clifford Almon and Agnes (Baird) E.; student N. Tex. Agr. Coll., 1933-35, Scripps Inst. Oceanography, 1938-39; B.S., U. Ill., 1937, Ph.D., 1941; m. Caroline Roberta Alexander, Oct. 3, 1941; children—Barbara Kathryn, (Mrs. Steven Alvarado) Charlet Adelia (Mrs. Paul A. Shave). Mem. staff Ill. State Geol. Survey, Urbana, 1941-43, U. Cal. Div. War Research, San Diego, 1943-45, U. So. Cal., Los Angeles, 1945-62, U. S. Geol. Survey, 1946-60, U. S. Naval Ordnance Test Sta., Pasadena, Cal., 1960-62; staff Woods Hole (Mass.) Oceanographic Instn., 1962—, sr. scientist, 1964—. Guggenheim fellow, Israel, 1959. Mem. Geol. Soc. Am., Am. Assn. Petroleum Geologists, Soc. Econ. Paleontologists and Mineralogists, Phi Beta Kappa, Phi Kappa Phi, Sigma Xi, Sigma Gamma Epsilon. Author: The Sea Off Southern California; A Modern Habitat of Petroleum, 1960. Research and publs. on gen. marine geology, physiography, sediments; regional studies Marshall Islands, Hawaiian Islands, Guam, Japan, Mediterranean, others. Home: 74 Ransom Rd., Woods Hole. Office: Woods Hole Oceanographic Instn., Woods Hole, Mass. 02543.*

EMICH, Friedrich Peter, Austrian chemist; b. Graz, Austria, Sept. 5, 1860; s. Peter and Pauline (Kretz) E.; student chemistry, Tech. U. Graz; Ph.D. (hon.), Graz, 1925; hon. dr. engring. Aachen, Germany, 1925; m. Georgine Baumgartner, 1900; 2 daus. Tchr., Secondary Sch. System, several years; named lectr. chemistry Tech. U. Graz, 1888, asso. prof. pure and analytical chemistry, 1889, prof. gen. chemistry, 1894. Recipient Lieben prize, 1911, Liebig Meml. medal, 1931. Mem. Vienna Acad. Scis. Lepoldina. Author: Lehrbuch der Mikrochemie, 1911; Mikrochemisches Praktikum, 1924; also articles. A founder of inorganic microchemistry, including devel. micromethods; demonstrated that only techniques of handling, observing and measuring must be changed when working with extremely small quantities; built microbalance; developed microspectroscopy, micropolarization. Died Graz, Jan. 22, 1940.

EMLEN, John Thompson, Jr., Am. biologist; ecologist; b. Phila., Dec. 28, 1908; s. John Thompson and Mary C. (Jones) E.; B.S., Haverford Coll., 1931; Ph.D., Cornell U., 1934; m. Virginia S. Merritt, June 25, 1934; children—John Merritt, Stephen Thompson, James Woodroff. Field biologist Fed. Bur. Biol. Survey, 1934-35; faculty U. Cal. at Davis, 1935-43; research asso. Johns Hopkins, 1943-46; faculty U. Wis., Madison, 1946—, prof. zoology, 1950—, chmn. dept., 1952-53, 55; dir. research Jackson Hole Wildlife Station, Moran, Wyo., 1951. Fellow A.A.A.S., Am. Ornithol. Union (v.p. 1960-64). Mem. Wilson (pres. 1957-58), Cooper ornithol. socs., Contbr. numerous articles to prof. jours. Research on population structure, behavior, dynamics in quail, rats, mice, social and reproductive behavior in free-living swallows, blackbirds, gorilla, habitat selection and ecol. distbn. birds in U. S. A. and Africa, origins and devel. behavior patterns in newly hatched birds, distance orientation navigation in swallows, bats, penguins. Home: 2122 Van Hise Av., Madison, Wis. 53705.*

EMMANUEL, David, Rumanian mathematician; b. 1854; hon. mem. Rumanian Acad. Author: Sur les intégrales pseudo-elliptiques, 1904; Lessons on the Theory of Functions, 2 vols., 1924-27. Founder modern math. edn. in Rumania; contbd. to understanding of functions with complex variables, theory of elliptical functions; one of 1st Rumanian writers on higher math. Died 1941.

EMMANUEL, J. Emmanuel, Greek chemist; b. Athens, Greece, Feb. 1, 1886; s. John and Helan (Jordanou) E.; grad. pharmacy U. Athens, 1906; Ph.D., U. Berne (Switzerland), 1912; m. Marie Dambergis, Aug. 1920; 3 children. Instr. pharm. chemistry U. Athens, from 1906, asst. prof. pharm. chemistry, 1918, prof., 1921—, mem. senate; prof. chemistry and explosives Mil. Sch. Navy, 1919-21. Mem. med. council Ministry Health, 1921; prof. Sch. Commerce and Financial Scis., 1930-40, rector, 1937-39. Decorated Cross St. Sepulchre, knight of Order George I. Mem. Acad., Greek Pharm. Soc. (pres.), French, German socs. pharmacy. Author: Chemistry of Foodstuffs, 1916, 22, 25; Analytical Chemistry, 1926; The Art of Dispensing (Pharmacotechnia), 1931; Knowledge in Commercial Articles, 1931; Pharmacognosy, 1933; General Chemistry, 1938; The Story of Pharmacy, 1946. Contbr. monographs on gen., analytic and pharm. chemistry, history galenic pharmacy, hydrology, others. Office: Solonosstreet 104, Chemical Lab. of Univ. Athens, Greece.

EMMET, William LeRoy, Am. elec. engr.; b. New Rochelle, N.Y., July 10, 1859; s. William J. and Julia Colt (Pierson) E.; grad. U. S. Naval Acad., 1881; Sc.D., Union Coll., 1910, Trinity Coll., Hartford, Conn., 1925. D.Eng., Stevens Inst. Tech., 1939; unmarried. Left navy, 1883; elec. engr., 1887-1941; in service Gen. Elec. Co., 1892-1941; most important work has been in steam turbine inventions and devels., invention of mercury vapor power process. Recipient Edison medal, 1919, Elliott Cresson medal, 1920 Gold medal Am. Society M.E., David W. Tayloe medal Am. Soc. Naval Architects and Marine Engrs., 1938. Author: Alternating Current Wiring and Distribution, 1894; The Autobiography of an Engineer, 1931. Died Sept. 26, 1941.

EMMETT, Paul Hugh, Am. chemist; b. Portland, Ore., Sept. 22, 1900; s. John and Vina (Hutchens) E.; B.Chem. Engring., Ore. State Coll., 1922, D.Sc., 1939; Ph.D., Cal. Inst. Tech., 1925; honoris causa, U. Lyon (France), 1964; m. Leila L. Jones, July 14, 1930. Instr., Ore. State Coll., 1925-26; with U. S. Dept. Agr, 1926-37, sr. chemist, 1935-37; chmn. dept. chem. engring. Johns Hopkins, 1937-44, W.R. Grace prof. chemistry, 1955—; staff Manhattan Project, Columbia U., N.Y.C., 1943-44, Mellon Inst., Pitts., 1944-55. Cons., Oak Ridge Nat. Lab., 1955—. Mem. Nat. Acad. Scis., Am. Chem. Soc. (Pittsburgh award 1953, Kendall award, 1958), Consejo Superior Spain (hon.), Am. Inst. Chemists; Editor: Catalysis, 7 vols., 1954-61. Research and numerous publs. on equilibria in gas solid reactions, mechanism of catalytic reactions using radioactive tracers, nature and mode of action of iron catalysts used for ammonia synthesis; developed (with Edward Teller, Stephen Brunauer) method for measuring surface areas of porous and finely divided solids using low temperature gas adsorption (BET method). Home: 5609 Purlington Way, Balt. 21212.*

EMMINGHAUS, Hermann, German psychiatrist; b. Weimar, Germany, May 20, 1845; s. Alexander and Minna Henriette (Winkel) E.; student Göttingen, Germany, Jena, Germany, Vienna, Austria, Leipzig, Germany; doctorate, 1870; m. Mathilde. Became lectr. 1874; named prof. psychiatry U. Dorpat (Estonia), 1880, U. Freiburg (Germany), 1886. Author: Die allgemeine Psychopathologie, 1878; also articles. Proposed a psychology based on natural sci., especially brain physiology. Died Freiburg, Feb. 17, 1904.

EMMONS, Ebenezer, Am. physician, geologist; b. Middlefield, Mass., May 16, 1799; s. Ebenezer and Mary (Mack) E.; grad. Williams Coll., 1818, Rensselaer Inst., 1826; attended Berkshire Med. Sch., 1826-28; m. Maria Cone, 1818. apptd. jr. prof. Rensselaer Inst., 1830; prof. chemistry Albany Med. Sch., 1838, later prof. obstetrics until 1852; apptd. state geologist N.C., 1851. Author: Manual of Mineralogy and Geology (5 papers dealing with investigation of agrl. resources of N.Y. State), 1846-54; Zoology of Mass., 1840; American Geology, 3 vols., 1855. Pioneered work in paleozoic stratigraphy; advanced theory that Taconic rock system is older than Silurian. Died Brunswick County, N.C., Oct. 1, 1863.

EMMONS, Howard Wilson, Am. engr.; b. Morristown, N.J., Aug. 30, 1912; s. Peter Wilson and Margaret (Lang) E.; M.E., Stevens Inst. Tech., 1933, M.S., 1935, D.Engring. (hon.), 1963; Sc.D., Harvard, 1938; m. Dorothy Gertrude Allen, July 9, 1938; children—Beverly Ann Emmons, Scott Wilson, Keith Howard. With Westinghouse Elec. & Mfg. Co., South Phila., 1937-39; asso. prof., U. Pa., 1939-40; prof. div. engring. and applied physics Harvard, 1940—, now Gordon McKay prof. mech. engring. Cons. editor Addison Wesley Co., 1959—; mem. fire research com. Nat. Acad. Scis., 1956—, space tech. panel Pres.'s Sci. Adv. Bd.; cons. aerodynamics, heat transfer, combustion. Fellow Am. Acad. Engring., Am. Soc. M.E. (chmn. applied mechanics div. 1941, hydraulics div. 1960, v.p. 1966—). Research in many fields of fluid mechanics, including discovery of spot process of transition from laminer to turbulent flow and stall propagation in axial flow and centrifugal compressors. Home: 233 Concord Rd., Sudbury, Mass. 01776.*

EMMONS, William Harvey, Am. geologist; b. Mexico, Mo., Feb. 1, 1876; s. St. Clair Peyton and Elizabeth Harvey (Ford) E.; A.B., Central Coll., 1897; Ph.D., U. Chgo., 1904; m. Virginia Cloyd, Sept. 6, 1910; children—Elizabeth, William Cloyd. Geologic aid U. S. Geologic Survey, 1904-06, asst. geologist, 1906-10, geologist, 1910-15; lectr. ore deposits U. Chgo., 1907, on petrology, 1908, asst. prof. petrology and econ. geology, 1908-09, asso. prof. econ. geology, 1909-12; prof., head dept. geology and mineralogy U. Minn., 1911-44, prof. emeritus, 1944-48; also dir. Minn. Geol. Survey. Mem. Geol. Soc. Am. (v.p. 1923), Am. Inst. Mining Engrs., Soc. Econ. Geology (pres. 1928); mem. (hon.) Am. Assn. Petroleum Engrs., geol. socs. France, Belgium, Sigma Xi, Gamma Alpha. Author: Ore Deposits of Maine, 1910; Ore Deposits of Elko, Lander and Eureka Counties, 1916; (with F. C. Calkins) Ore Deposits of Phillipsburg, Mont., 1913; Enrichment of Ore Deposits, 1917; Ore Deposits of Ducktown, Tenn., 1926—(all pub. by U. S. Geol. Survey); also several text books on ore deposits and on petroleum. Died Nov. 5, 1948.

EMMRICH, Hermann Friedrich, German geologist; b. Meiningen, Germany, Feb. 7, 1815; s. George Karl Friedrich and Elisabeth Christiane (Amthor) E.; m. Adolphine Johanna Emilie Müller, 1848; 4 sons, 4 daus. Dir. gymnasium, Meiningen. Research and publs. on geology of Thuringia and Franconia, geodynamics, paleontology, especially trilobites, vertebrates, molluscs; 1st normal profile of S. Tyrolean Triassic, 1844; performed profile on jurassic and chalk layers of no. boundary of Bavarian Alps, 1846. Died Mainingen, Jan. 24, 1879.

EMPEDOCLES OF AGRIGENTUM, Greek philosopher, physician; b. Agrigentum, Sicily, circa 490 B.C.; s. Meton; student, follower of Pythagoreans, Parmenides; asso. with Theophrastus; polit. leader in Akragas; overthrew a tyrant, but refused throne; exiled and wandered through Greece; known as prophet, miracle-worker, poet, physician, philosopher. Author: On Nature, Purifications (both poems). Founder Sicilian sch. of medicine; combined and improved theories of Thales, Anaximenes, Heraclitus of Asia Minor sch.; originated classic doctrine of 4 elements; believed particles of air, water, fire and earth form basis of matter, that love and strife change nature of substances; laid found. of atomism in views criticized and later used by Aristotle; expounded simple but essentially correct theories of vision, sense-perception, magnetism, evolution, adaptation; held earth to be spherical; believed planets move through space, that light moves through space with finite velocity, that atmosphere is corporeal rather than a void; said motion was only sort of change possible; i.e., all other changes could be understood in terms of motion; developed theory of heart as center of blood-vessel system and seat of life; believed blood flows through body; gave earliest description of structure of cochlea of ear, 460 B.C., noted labyrinth, 445 B.C.; drained swamps to prevent malaria; from Pythagoreanism developed belief in metempsychosis. Died Mt. Etna, 430 B.C.

EMPERGER, Friedrich Ignaz von, see von Emperger, Friedrich Ignaz.

EMSCHWILLER, Guy, French chemist; b. Nancy, Jan. 12, 1900; s. Henri and Berthe (Wolff) E.; Ph.D. in Phys. Scis., U. Paris; m. Madeleine Job (dec.); m. 2d, Julia d'Ornano; children—Michel, Daniel, André, Dominique. Asst., work head Nat. Conservatory Arts and Crafts; prof. École Supérieure Physics and Chemistry of City of Paris, 1935—. Named laureate Inst. France. Mem. Soc. Phys. Chemistry (sec. gen.), Internat. Union Pure and Applied Chemistry (sec. div. phys. chemistry). Author: Chimie Physique, 3 vols., 2d edit., 1959-61; Chimie et Thermodynamique, 1962. Research on photochemistry, chemistry of coordination. Home: 3, rue Lagarde, Paris 5. Office: 10, rue Vauquelin, Paris 5, France.

ENATSU, Hiroshi, Japanese physicist; b. Miyakonojo, Japan, Sept. 12, 1922; s. Eizo and Fumi (Kuroiwa) E.; B.Sc., Kyoto U., 1944, Sc.D., 1953; m. Hideko Masui, Apr. 8, 1951; children—Masazumi, Mario. Asst., dept. physics Kyoto U., 1944-57; vis. research asst. Columbia, 1952-53; vis. mem. Inst. Theoretical Physics, Copenhagen, Denmark, 1954-56; asst. prof. physics Ritsumeikan U., Kyoto, 1957, prof., 1957—. Mem. Am. Phys. Soc., Japan Phys. Soc. Research publs. on theory of nuclear forces; quantum theory of fields; structures of elementary particles. Home: 14-2 Yakushido-cho, Shugakuin, Sakyo-ku, Kyoto, Japan.*

ENCELIUS, see Entzelt, Christopher.

ENCISO, Don Martin F. de, see de Enciso, Don Martin F.

ENCKE, Johann Franz, German astronomer; b. Hamburg, Germany, Sept. 23, 1791; s. Johann Michael and Marie (Misler) E.; studied math. and astronomy under K. F. Gauss at U. Göttingen; Ph.D. (hon.), Berlin, Germany, 1825; m. Amalie Becker, 1823; 3 sons, 2 daus. Arty. officer in Napoleonic War, 1813-15; became asst. Seeberg Obs., Gotha, 1816, dir., 1822; named astronomer Berlin Acad., 1825; dir. Berlin obs. (which was rebuilt under his direction, 1832-35), 1825-63; became prof. astronomy U. Berlin, 1844; issued 4 vols. Astronomische Beobachtungen, Berlin Obs., 1840-57. Recipient Gold medal Royal Astron. Soc., 1823, 30. Fellow Royal Soc., 1825; mem. Berlin Acad. Scis. (became sec. 1825), Leopoldina, French Acad. Scis., 1825. Author: Die Entfernung der Sonne von der Erde, aus dem Venusdurchgang von 1761 hergeleitet, 2 vols., 1822-24; Der Venusdurchgang von 1769, 1824; also numerous articles published as Gesamte mathematische und astronomische Abhandlungen, 3 vols., 1888-89. Editor: Berliner Astronomisches Jahrbuch, 37 vols., 1830-66. Computed orbit of comet of 1680; computed orbit and period of comet discovered by Pons in 1818 (Encke's comet); calculated solar parallex from past transits of Venus, 1761, 69; originated new method of determining elliptic orbits from 3 observations, 1849; method of calculating planetary perturbations using rectangular coordinates, 1851; research on smaller planets, including their orbits and perturbations from larger planets. Died Spandau, Aug. 26, 1865.

ENDERS, John Franklin, Am. microbiologist; b. West Hartford, Conn., Feb. 10, 1897; s. John Ostrom and Harriet Goulden (Whitmore) E.; A.B., Yale, 1920, Sc.D. (hon.), 1953; M.A., Harvard, 1922, Ph.D., 1930; Sc.D., 1956; Sc.D., Trinity U., 1955, Northwestern U., 1956; Sc.D., Western Res. U., 1958, Tufts U., 1960; LL.D., Tulane U., 1958; L.H.D., Hartford U., 1960; m. Sarah Frances Bennett, Sept. 17, 1927 (dec.); children—John Ostrom II, Sarah; m. 2d, Carolyn Keane, May 12, 1951; 1 stepson, William Edmund Keane. Asst. dept. bacteriology and immunology, Harvard, 1929-30, instr., 1930-32, faculty instr., 1932-35, asst. prof., 1935-42, asso. prof., 1942-56; prof. Children's Hosp., Harvard Med. Sch., 1956—. Served to lt. (j.g.), Naval Res. Flying Corps, 1917-20. Civilian cons. to Sec. War on epidemic diseases, 1942-46; mem. Commn. on Viral Infections, Armed Forces Epidemiological Bd.; sci. adv. bd. Armed Forces Inst. of Pathology; chief research dept. of infectious diseases, Children's Hosp., Boston; Recipient Passano Found. award for culturing poliomyelitis viruses in living tissues, 1953; Lasker Award, 1954; Nobel Prize (with F. C. Robbins and T. H. Weller) in Medicine and Physiology, 1954; Cameron prize U. Edinburgh, 1960, Howard Taylor Ricketts Meml. award U. Chgo., 1962, Diesel Gold medal, 1962, Robert Koch medal, 1962 (Germany), Sci. achievement award A.M.A., 1963, Presdl. Medal of Freedom, 1963. Fellow Am. Acad. Arts and Scis.; mem. Nat. Acad. Scis., Harvey Soc., Am. Philos. Soc., Soc. Gen. Microbiology (hon.), Soc. Am. Bacteriologists, Am. Assn. Immunologists (pres.), Soc. Exptl. Biology and Medicine, Am. Pub. Health Assn., A.A.A.S., Sigma Xi, Alpha Omega Alpha (hon.); asso. mem. Mass. Med. Soc. Author: (with Hans Zinsser and Leroy D. Fothergill) Immunity: Principles and Application in Medicine and Pub. Health, 1939. Contbr. to Virus and Rickettsial Diseases, 1958, 64. Editor: Jour. of Immunology, 1942-58, Jour. Bacteriology, 1964—. With F. C. Robbins and T. H. Weller grew poliomyelitis virus in test-tube cultures of various tissues, proving that polio virus is not neurotropic; cultivated measles virus, 1954; developed effective measles vaccine, by 1962; also research on mumps. Home: 64 Colbourne Crescent, Brookline, Mass. Office: 300 Longwood, Boston 02115.

ENDLICHER, Stephan Ladislaus, botanist; b. Pressburg, Hungary, June 24, 1804; s. Ignevz and Juliana (Janisch) E.; ed. Pest, Vienna; M.D., 1840; m. Cäcilie Müller; 1 dau. Became asso. with Imperial Ct. Library, Vienna, Austria, 1828; named curator Hofnaturalienkabinet (Royal Natural History Mus.), 1836; apptd. univ. prof. botany, 1839, dir. bot. garden, 1840; mem., co-founder Acad. Scis. Vienna. Author: (with F. Unger) Grundzüge der Botanik, 1843; Atlas von China, 1843; Gesetze des beiligen Stephan, 1849; Genera Plantarum, 1836-50; also articles, other books. Founder Annalen, Royal Natural History Mus. Made a systematic arrangement of herbs. Died Vienna, Mar. 28, 1849.

ENDROCZI, Elemér, Hungarian physiologist; b. Pecs, Hungary, July 31, 1927; s. Dezso Derner and Maria (Kata) E.; M.D., U. Pecs, 1951; m. Magdolna Joody, Oct. 25, 1951; children—Eva, 1952, Gabor, 1953. Staff mem. Inst. Physiology, U. Pecs, 1951—, docent, 1960—. Found. Fund Research in Psychiatry fellow dept. anatomy U. Cal. at Los Angeles, 1960-61, Ford Fellow, 1963-64. Mem. Hungarian Physiol. Soc., Endocrinol. Soc., Internat. Brain Research Orgn. Author: Die Neuroendokrine Adaptationstatigkeit, 1960; Neurohumoral Control of Adaptation, 1965; (with K. Lissak) The Neural and Hormonal Control of Behavior, 1965; also numerous articles. Research on neural and humoral basis of adaptive and reproductive behavior in higher mammals. Home: 35 Kossuth, Pecs, Hungary.

ENDRYS, Jiri, Czechoslovakian physician; b. Jicín, Czechoslovakia, Dec. 25, 1925; s. Josef and Berta (Hronková) E.; M.D., Charles U., Prague, 1950; m. Ladislava Hrdinova, Apr. 24, 1956; children—Ladislava, Jirí. Physician, Charles U., Hradec Králové, Czechoslovakia, 1950-52, lectr., 1952-55, leading cardiologist Cardiosurgical Center, 1959—; lectr. Palacky's U., Olomouc, 1956-59. Recipient Nat. prize Klement Gottwald, 1963. Mem. Czechoslovak Cardiol. Soc. (mem. com.). Author: (with others) Intracardiac Calcification, 1957, Dye Dilution Curves, 1964; also numerous articles. Research on pathophysiology of congenital and acquired heart diseases, left heart catheterization, pharmacologic methods in phonocardiography, bronchial circulation, pathophysiology of pulmonary hypertension, needle biopsy of parietal pleura, quantitation of heart valve insufficiency. Home: 1103 Obráncu míru. Office: Cardiosurgical Center, U. Hosp., Hradec, Králové, Czechoslovakia.*

ENELOW, Allen Jay, Am. physician; b. Pitts., Jan. 15, 1922; s. Isadore M. and Rose (Kasdan) E.; A.B., W.Va. U., 1942; M.D., U. Louisville, 1944; children—David, James, Susan, Margaret, Patience, Abigail. Staff psychiatrist Topeka VA Hosp., 1949-50; staff psychiatrist Topeka State Hosp., 1950-52; pvt. practice medicine, specializing in psychiatry, Beverly Hills, Cal., 1952-58, Pacific Palisades, Cal., 1958-64; instr. Menninger Sch. Psychiatry, 1949-52; clin. instr. U. Cal. Los Angeles Sch. Medicine, 1953-56, asst. clin. prof., 1956-59; faculty U. So. Cal., 1959-67, prof. psychiatry, 1964-67, dir. postgrad. div. dept. psychiatry, 1960-67; dir. Psychosomatic Service Los Angeles County Gen. Hosp., 1964-67; prof., chmn. dept. psychiatry Mich. State U., East Lansing, 1967—; med. dir., med. health unit St. Lawrence Hosp., Lansing, Mich., 1967—. Mem. med. adv. com. State Cal. Div. Indsl. Accidents. Mem. A.M.A., Am. Psychiat. Assn., Am. Psychoanalytic Assn., A.C.P. Author: (with M. Wexler) Psychiatry in the Practice of Medicine, 1966. Research, publs. on teaching and learning processes, psychotherapy. Home: 936 Southland Ave., East Lansing, Mich.

ENGDAHL, Richard Bott, Am. mech. engr.; b. Elgin, Ill., Apr. 16, 1914; s. Walter F. and Emma (Bott) E.; B.S. magna cum laude, Bucknell U., 1936; M.S., U. Ill., 1938; m. Helen Klaas, Nov. 21, 1940; children—Karen, Eric. Research engr. Battelle Meml. Inst., Columbus, O., 1941-45, asst. supr. combustion research, 1945-46, supr. fuels, heat transfer, 1946-53, div. chief fuels, air pollution, 1953-60, staff engr. thermal engring., 1960-65, fellow mech. engring., 1965—. Fellow Am. Soc. M.E., A.A.A.S.; mem. Air Pollution Control Assn., Am. Indsl. Hygiene Assn., Am. Soc. Heating, Refrigerating and Air Conditioning Engrs., Am. Meteorol. Soc., Sigma Xi, Tau Beta Pi, Pi Mu Epsilon. Contbr. chpt. to Air Pollution, 1968. Research, numerous publs. in devel. of heated thermocouple anemometer, residential heating and cooling, combustion of pulverized coal, coal flowmeter, air pollution control for indsl. stokers by overfire air jets, econs. of air pollution control, devel. incinerators, energy utilization and conservation in bldgs. Home: 1165 Glenn Av., Columbus 43212. Office: 505 King Av., Columbus, O. 43201.*

ENGEL, (Christian Lorenz) Ernst, German statistician; b. Dresden, Germany, Mar. 26, 1821; s. George Bernhard and Christiane Rosina (Möbius) E.; m. Johanna Friederike Amalie von Holleuffer, 1848; 2 sons, 2 daus. Made ednl. trips to Eng., France and Belgium; became mem. commn. for investigation indsl. working conditions in Saxony, 1848; organizer German Indsl. Trade Fair, Leipzig, Germany, 1850; dir. Royal Bureau of Statistics, Dresden, 1850-58 became dir. Saxon Mortgage Insurance Co., 1858; dir. Prussian Statis. Bur., 1860-82; Nat.-Liberal mem. Prussian Ho. of Reps., 1867-70. Founder, Central Prussian Statis. Commn., 1860. Mem. Internat. Statis. Congress (a founder), Permanent Statis. Commn., Internat. Statis. Inst., Kommission zur weiteren Ausbildung der Statistik des Zollvereins. An organizer of ofcl. statistics in Germany; made statis. investigations of social conditions of German working class; Engel's law states outlays for food decrease relatively as income increases. Died Radebeaul, nr. Dresden, Germany, Dec. 8, 1896.

ENGEL, Friedrich, German mathematician; b. Lugau, Germany, Dec. 26, 1861; s. Moritz Robert and Marie (Meissner) E.; studied math., Leipzig, Germany, Berlin, Germany, 1879-83; Dr.h.c., Oslo; m. Caroline Ibbeken, 1899. Asst. to Sophus Lie, Christiania (now Oslo, Norway), 1884-86; named asso. prof. Leipzig, 1889, hon. prof., 1899; became prof. math. Greifswald, Germany, 1904, Giessen, Germany, 1913-41. Recipient Golden Lobatchewski medal. Mem. acads. of Saxony, Russia, Norway, Prussia. Author of books including: Theorie der Transformations-gruppen, 3 vols., 1889-90; also numerous articles. Organized W. S. Lie's work and added his own explanations. Died Giessen, Sept. 29, 1941.

ENGEL, George Libman, Am. physician; b. N.Y.C., Dec. 10, 1913; s. Adolph and Esther (Libman) E.; B.A., Dartmouth, 1934; M.D., Johns Hopkins, 1938; m. Evelyn Lipman, Oct. 30, 1938; children—Peter A., Betty L. Faculty, U. Cin., 1942-46; faculty U. Rochester, 1946—, prof. medicine, psychiatry, 1959—. Recipient Career Research award USPHS, 1962. Mem. Am Psychosomatic Soc., Am. Psychiat. Assn., Am. Psychoanalytic Assn., Soc. For Psychosomatic Research (London). Author: Fainting, 1950; Psychological Development in Health and Disease, 1962. Contbr. numerous articles to profl. jours. Research in electroencephalography, neurology, aviation medicine, internal medicine, psychoanalysis psychosomatic medicine, med. edn. Home: 91 San Gabriel Dr., Rochester 14610. Office: 260 Crittenden Blvd., Rochester, N.Y. 14620.*

ENGEL, Hendrik, Dutch zoologist; b. Koog aan de Zaan, Netherlands, Feb. 2, 1898; s. Frederik Hendrik and Antje (Bakker) E.; doctorate U. Amsterdam, 1925; m. Hilda G. Faber, Sept. 2, 1927; 1 dau., Gertrude (Mrs. H. B. Eering); m. 2d, Ottilie Nilant, Oct. 15, 1933; children—Frederik Louis, Ernst Peter, H. W. Boudewyn; m. 3d, Maria S. Ledeboer, Sept. 23, 1946. Faculty, Zool. Mus., U. Amsterdam, 1925—, dir., 1950—, prof. zoology, 1950—. Mem. Ornithol. Soc., Entomol. Soc., Hydrobiol. Soc., Zool. Soc., Soc. History of Sci., others. Author: Viervoeters in beemd en bos, 1933; also numerous articles. Research on taxonomy of opistobranchs, echinodermata, hirudinea; history of zoology. Address: Wëyenberg 12, Almen (Geld.), Netherlands.*

ENGEL, Paul, physician; b. Vienna, Austria, June 7, 1907; s. Julius and Klara (Rosenfeld) E.; M.D., U. Vienna, 1933; Sc.D. (hon.), Free U. Columbia, 1952; m. Josefine Monath, Oct. 22, 1935; children—Ana (Mrs. Peter Rothschild), Teresa (Mrs. Jaime Pienknagura), Juan Jacobo. Asst., U. Montevideo, 1935; asst. surgeon gynecol. clinic U. Vienna, 1935-38; prof. biology Free U. Colombia, 1938-50; prof. pharmacology Nat. U. Colombia, Bogota, 1945-50; prof. biology, gen. pathology, dental faculty Central U. Ecuador, Quito, 1955—, prof. gen. biology, faculty basic sci., 1964—. Mem. Endocrine Soc. Mexico (hon.), Ecuadorian House of Culture, Endocrine Soc., Royal Soc. Medicine, Colombian Soc. Endocrinology (founding mem.), Ecuadorian Soc. Endocrinology, Colombian Soc. Biology. Author: Endocrinologia, 1938; (with J. D. Paltan and J. Homs) Evolucion Filogenetica Emergente, 1958; Vision de la Filosofia del Siglo XX, 1958; (with others) Compendio de Biologia, 1965. Publs. on research on function of pineal gland; establishment of its antagonism to pituitary hormones; studies of inactivation of steroids by the liver; tumors and hormones. Address: Apartado 2213, 125 Carrion, Quito, Ecuador.*

ENGELBERT OF ADMONT, natural philosopher; student grammar, logic, natural history, Prague, Czechoslovakia, 1271-74, logic, philosophy Padua

(Italy) U., 1279-84, theology, Dominican House, Padua, 1284-88. Austrian Benedictine monk; became abbot Benedictine Monastery, Admont, 1297. Author: commentary on the De mundo; commentary on the pseudo-Aristotelian treatise on the rising of the Nile; De fascinatione; De naturis animalium; De quibusdam naturalibus; De causis et signis mutationis aeris et temporum; De causa longaevitatis hominum ante diluvium. Called Austrian Albertus Magnus. Died 1331.

ENGELBRECHT, (Thiess) Hinrich, agrl. geographer; b. Hof Obendeich, Germany, Oct. 6, 1853; s. Johannes and Ida (Lange) E.; student Leipzig, Germany, Strasbourg, France; hon.dr., Breslau, (now Wroclaw, Poland), Kiel, Germany, Agrl. U. Berlin (Germany); m. Margarethe Greve, 1878. Became farmer, Sioux City, Ia., 1880; traveled in U. S., then returned to Germany where he took over family estate; free-conservative mem. Prussian Ho. of Reps., later in Prussian Herrenhaus; held numerous positions in Agrl. Corps., Schleswig-Holstein Chamber Agr., German Agrl. Council, German Agr. Soc. Author: Die Landbauzonen der aussertropischen Lander, 3 vols., 1898-99; Die geographische Verteilung der Getreidepreise in den Vereinigten Staaten von 1862 bis 1900, 1903; Bodenanbau und Viehstand in Schleswig-Holstein, 1905-07; Die geographische Verteilung der Getreidepreise in Indien von 1861 bis 1905, 1908; Die Feldfrüchte Indiens in ihrer geographischen Verbreitung, 1914; Landwirtschaftlicher Atlas des Russischen Reiches in Europa und Asien, 1916; Die Feldfrüchte des Deutschen Reiches, Part I, 1928. Founder agrarian geography; research and publs. on agrarian politics, edn., plant geography, agrl. history; advocated objective sci. research in agr. Died Hof Obendeich, Oct. 18, 1934.

ENGELBRECHT, Mildred Amanda, Am. bacteriologist; b. Marengo, Ill., July 31, 1899; d. Frederick Martin and Sophia (Meier) Engelbrecht; A.B., U. Wis., 1927, M.S., 1930, Ph.D., 1934. Med. technologist Morningside Tb Sanatorium, Madison, Wis., 1930-37; faculty U. Wis., 1930-38; faculty U. Ala., University, 1938—, prof., head dept. bacteriology, 1946—. Fellow A.A.A.S.; mem. Am. Soc. Microbiology, Am. Pub. Health Assn., Royal Soc. Health (London), Sigma Xi, Sigma Delta Epsilon (past nat. pres.), Delta Kappa Gamma. Author: (with W. D. Frost) The Streptococci. Discovered with W. D. Frost two new streptococci, Streptococcus zooepidemicus, Streptococcus equisimilis. Home: 20 Prince Apts., 1105 University Av., Tuscaloosa, Ala. 35401. Office: Lloyd Hall, U. Ala., University, Ala. 35486.*

ENGELGARDT, Vladimir Aleksandrovich, Russian biochemist; b. Moscow, Dec. 3, 1894; grad. Med. Faculty, Moscow U., 1919; D.Biol. Sci., 1935; Hon. Dr., Marseilles U., 1946. Asso., later dept. head Inst. Biochemistry, USSR Peoples Commissariat Health, Moscow, 1921-29; head chair biology Kazan Med. Inst., 1929-33; sr. sci. exper Lab. Biochemistry and Animal Physiology, USSR Acad. Sci., Leningrad, 1933-35, head lab. animal cell biochemistry Pavlov Inst. Physiology, 1944-50, head lab. animal cell biochemistry Bakh Inst., Biochemistry, Moscow, 1935-44, 50—, acting dir., 1941-44, dir. Inst. Molecular Biology, 1959—. head chair biochemistry Moscow U., 1936—; head chair gen. biochemistry Leningrad U., 1939-40; head dept. biochemistry Inst. Exptl. Medicine, USSR Acad. Med. Sci., Leningrad, 1945-52. Decorated Order of Lenin (2); recipient Stalin prize class 1, 1943. Mem. USSR, East German (corr.) acads. sci., USSR Acad. Med. Sci., All-Union Soc. Physiologists and Biochemists (bd. mem. 1937—), All-Union Soc. Natural Sci. Researchers (chmn. chemistry sect. 1953—, council mem. 1954—), Indian Natural Sci. Inst. (hon.). Author over 200 works on biochemistry and molecular biology. Editor: Biokhimia, 1944-65, Molecular Biology, 1966—. Research on regularities in transformation of organic phosphorus compounds in process of cellular metabolism; connection between chem. phenomena in muscle fiber and its function; discovered aerobic resynthesis of adenosine triphosphoric acid; established how myosin obtains energy to function. Address: Acad. Scis. USSR, Mowcow, USSR.*

ENGELHARDT, Victor Josef Karl, electrochemist; b. Vienna, Austria, Oct. 26, 1866; s. Josef and Vittoria (Trouvé) E.; engring. diploma Tech. U. Vienna; student anatomy, zoology, botany U. Vienna; hon. dr. engring. Tech. U. Berlin (Germany), Tech. U. Vienna; m. Karoline Köstlin, 1890; 3 sons, 1 dau. Joined firm Siemens & Halske, Vienna, 1889, with electrochem. plant, Berlin, 1905-32; also became dir. Gesellschaft für Elektrostahlagen mbH, 1907; faculty Tech. U. Breslau (now Wroclaw, Poland), 1910-20, became hon. prof., 1920; lectr. tech. electrochemistry and electrometallurgy Tech., U., Berlin-Charlottenburg. Editor: Monographien über angewandte Electrochemie, 52 vols, 1902-32; Handbuch der technischen Elektrochemie, 3 vols., 1931-35. Developed alkali of gold ore with potassium cyanide and elec. precipitation of gold from it; studied refinement of lead, copper, antimony, zinc, electrolysis of water, electrolytic lead processing, electro steel furnaces, potassium chloride electrolysis; dir. constrn. for 1st plants for prodn. carbides, silicon, copper silicide. Died Berlin-Charlottenburg, Mar. 9, 1944.

ENGELHART, Johann Friedrich, German chemist; b. Wiesbaden, Germany, Feb. 16, 1797; prof. chemistry Tech. Sch., Nürnberg, Germany, from 1829. Author: De Vera Sanguini Purpureum Colorem Impertiensis Natura, 1825. Discoverer (with Berzelius) metaphosphoric acid. Died Nürnberg, June 9, 1837.

ENGELMANN, George, botanist, physician; b. Frankfurt-am-Main, Germany, Feb. 2, 1809; s. George and Julia (May) E.; M.D., U. Wurzburg, 1831; m. Dorothea Horstmann, June 11, 1840, 1 son, George J. Came to U. S., 1832; practiced medicine, St. Louis, 1832-84; an early user of quinine for treatment of malaria; discoverer adaptation of Pronuba moth for accomplishing pollination of yuccas; discovered immunity of Am. grape to phylloxera; organizer St. Louis Acad. Science, 1856. Works collected in Botanical Works of the Late George Englemann Collected for Henry Shaw, 1887. Undertook researches in botany and biology; conducted systematic meteorological observations, 1836-84; eponym for 3 botanical genera and a number of species. Died St. Louis, Feb. 4, 1884.

ENGELMANN, Godefroi, printer; b. 1788; introduced and perfected lithography in France, 1815, water color lithography, 1819; transferred engraved copper proofs on to stone; improved lithography presses, 1830; invented chromolithography, 1837. Died 1839.

ENGELMANN, Wilhelm Theodor, German physiologist; b. Leipzig, Germany, Nov. 14, 1843; student Utrecht, (Netherlands), Jena, Leipzig, Heidelberg, Gottingen, (all Germany); m. Marie Donders, 1869; 1 son, 1 dau.; m. 2d, Emma Brandes, 1874; 2 sons, 2 daus. Became asst. to Donders, Utrecht, 1867. prof. gen. biology and histology, 1871, physiology, 1888; named prof. physiology, dir. Physiol. Inst., Berlin, Germany, 1897. Mem. French Acad. Scis., 1895. Author: Untersuchungen über den Zusammenhang von Nerven und Muskelfaser, 1863; Über die Flimmerbewegung, 1868; Über den Ursprung der Muskelkraft, 1893; Research on cellular physiology and histology; proved chloroplast is most strongly activated by red end of spectrum, then by violet; studied effect polarized light on muscle fibers and microorganisms. Died Berlin, May 20, 1909.

ENGELS, Friedrich, social philosopher, economic theorist; b. Barmen, Prussia, Germany, Nov. 28, 1820; s. Friedrich E.; ed. in economics and commerce, Bremen; m. Mary Burns, 1845 (d. 1863); m. 2nd. Lizzie Burns. Writing under pseudonym, Friedrich Oswald, 1839-43; met Karl Marx, Paris, 1844; traveled from London to Manchester, Barmen, Brussels, Paris, 1844-48; wrote Communist Manifesto with Marx, 1848; employee, partner, dir., father's textile plant, Manchester, Eng., 1849-69; ret. to devote time to writing and politics, 1869. Co-founder of Communist League of Dutch, British, French and German Socialists; with Marx, founded First Socialist International, 1864. Author: Conditions of the Working Class in England, 1845; (with Marx) Communist Manifesto, 1848; Herr Eugen Dühring's Revolution in Science, 1878; The Origin of the Family, Private Property and the State, 1884; The Peasant War in Germany, On Marx's Capital, Dialectics of Nature, pub. 1940; other works; edited Vols. II and III of Marx's Das Kapital. Engel's literary style and skill largely responsible for popularity of Marx's works; stressed economic determinism in history; pointed out necessity of socialistic revolution to alleviate human miseries he believed were caused by capitalism; stressed class warfare; transformed socialism from utopian to scientific stage. Died London, Eng., Aug. 5, 1895.

ENGESSER, Friedrich, German civil engr.; b. Weinheim, Germany, Feb. 12, 1848; s. Joseph and Maria Barbara (Kessler) E.; ed. Polytechnikum Karlsruhe (Germany); hon. dr. engring.; m. Leonie Horn, 1874; 1 son, 1 dau. Became engr. rd. and waterways dept. Baden (Germany) Civil Service, 1870, later State R.R. especially bldg. of Höllental and Schwarzwald R.R.; chmn. Bur. Bridge Constrn.; became prof. statics, bridge bldg., r.r.'s Tech. U. Karlsruhe, 1885. Mem. Preussische Akademie de Bauwesens. Author: Theorie und Berechnung der Bogenfachwerkträger ohne Scheitelgelenk, 1880; Die Knickfestigkeit gerader Stäbe, 1891; Die Zusatzkräfte und Nebenspannungen eiserner Fachwerkbrücken, 2 vols., 1893; Die Berechnung der Rahmenträger, 1913; Die Knickfestigkeit gegliederter Stäbe, 1913. Research on supplementary stresses, secondary tension, buckling, theory of geometric earth pressure, frames, theory iron bridges; used test models. Died Achern, Germany, Aug. 29, 1931.

ENGI, Gadient, Swiss chem. industrialist; b. Chur, Switzerland, Dec. 13, 1881; s. Martin and Margarethe (Färber) E.; student chemistry Fed. Tech. U., Zurich, Switzerland, U. Geneva (Switzerland); Dr.ès sc.; hon.dr.sc.techn. Fed. Tech. U., 1929; Ph.D. (hon.), Basel, Switzerland, 1931; m. Alice Hollenwenger, 1909; 1 son, 1 dau. Became indsl. chemist Gesellschaft für Chemische Industrie, Basel, 1904, dept. head, 1907, dir., 1918, del., 1924, v.p. bd., 1927. Mem. Schweizerische Gesellschaft für chemische Industries (pres. 1927-37, named hon. pres. 1938), Basler Volkswirtschaftsbundes (pres. 1932-38, named hon. pres. 1938—), Verband chemischer Industrieller (pres. 1935-43), Zentralverband Schweizerischer Arbeitgeberorganisationen (mem. bd. 1925-45). Research, numerous publs. and patents in coal tar dyes; numerous discoveries in indigoid vat dyes, including indigo, thioindigo and naphthazarin series. Died Riehen nr. Basel, May 19, 1945.

ENGL, Jo(seph) Benedict, physicist; b. Munich, Germany, Aug. 6, 1893; s. Joseph and Margarethe (Fritzmann) E.; student physics, Göttingen, Germany; doctorate, 1917; m. Frieda Venghaus, 1917 (div.); 1 son 1 dau.; m. 2d, Erika Briesmeister. Became lectr. Tech. U., Berlin, Germany, 1925; with Tri-Ergon AG Zürich, Switzerland, 1918 to circa 1925; founded his own supersonic lab., 1925; research staff Am. film co., 1928-35; later sci. staff Tech. U., Berlin; became cons. physicist, U. S. A., 1939. Author: Der tönende Film, 1927; Raum- und Bauakustik, Ein Leitfaden für Architekten und Ingenieure, 1939. Developed (with J. Massolle, H. Vogt under name of Triergon) complete sound film system, 1918-22, which was 1st demonstrated, Berlin, 1922. Died N.Y., Apr. 8, 1942.

ENGLANDER, Harold Robert, Am. dental researcher; b. N.Y.C., Dec. 11, 1923; s. Samuel Harold and Elsie (Kimless) E.; student City Coll. N.Y., 1941-43, Washington U., St. Louis, 1944; D.D.S., Columbia, 1948, M.P.H., 1951; m. Harriett Beecher, May 8, 1949; 1 son, Mark R. Commd. lt. j.g., USN, 1948, advanced through grades to lt. comdr., 1959, officer-in-charge Dental Research Facility U. S. Naval Tng. Center, Great Lakes, Ill., 1953-58, sr. dental officer U. S. S. Pocono, 1959; asso. prof. dentistry U. Ill., 1959-62; dental dir., chief clin. trials sect. epidemiology and biometry br. Nat. Inst. Dental Research, USPHS, Bethesda, Md., 1962—; cons. to govt. and pvt. orgns. Diplomate Am. Bd. Dental Public Health. Fellow Am. Coll. Dentists, Am. Pub. Health Assn., A.A.A.S.; mem. Am. Dental Assn., Am. Assn. Pub. Health Dentists, Internat. Assn. Dental Researchers, Commd. Officers Assn. USPHS, Omicron Kappa Upsilon, Sigma Xi. Research, publs. and chpt. in book on epidemology of dental caries and periodental disease; effect of fluoride on exptl. dental caries in the hamster; showed that fluoridated water effective in lowering caries in adults; use of mouthpieces for drug application to teeth; field testing possible anticaries agts. Home: 15705 Ancient Oak Dr., Gaithersburg, Md. 20760. Office: Biometry and Field Studies Br. Nat. Inst. Dental Research, NIH, Bethesda, Md. 20014.*

ENGLER, Adolf Gustav Heinrich, German botanist; b. Sagan, Germany, Mar. 25, 1844; s. August and Pauline (Scholtz) E.; Ph.D., U. Breslau (now Wroclaw, Poland), 1866; several hon. doctorates; m. Marie Firle, 1874; 1 son, 2 daus. Custodian herbarium, lectr., Munich, Germany, 1871-78; became prof., Kiel, Germany, 1878, Breslau, 1884; named prof., dir. Bot. Gardens, Berlin, 1889-1921; mem. numerous fgn. expdns. Mem. Prussian, Bavarian acads. scis. Author: Versuch einer Entwicklungsgeschichte der Pfanzenwelt insbesondere der Florengebiete, seit der Tertiäperiode, 2 vols., 1879-82; Syllabus der Volesungen über spezielle und medizinisch-pharmazeatische Botanik, 1892. Editor, contbg. author: (with O. Drude) Die Vegetation der Erde, 15 vols., 1896. Founder, Botanische Jahrbüchr, 1880, editor 1880-1930. Research and numerous publs. on geog. botany, plants of Africa; emphasized importance of geol. history to study of plant geography; 1st to combine phylogenics and geography; developed natural world classification system of plants especially tropical family of Araceae and genus Saxifraga. Died Berlin-Dahlem, Oct. 10, 1930.

ENGLER, Carl Oswald Viktor, German chemist; b. Weisweil, Germany, Jan. 5, 1842; s. Wilhelm and Adelheid (Haufe) E.; student chemistry, Karlsruhe, Germany; doctorate, Freiburg, Germany, 1864; hon. dr.engring., Berlin, Germany, Darmstadt, Germany, 1911, Munich, Germany, 1918; m. Maria Magdalena Bader, 1877; 3 sons, 1 dau. Named lectr., Halle, Germany, 1867, asso. prof., 1872; became prof. chem. tech. Polytechnikum Karlsruhe (became Tech. U. 1885), 1876, prof. pure chemistry, dir. Chem. Inst., 1887; mem. Reichstag, 1887-90; mem. numerous expdns. to oil field of E. Galicia, Caucasus, Egypt, N.Am. German rep. at several internat. petroleum congresses. Mem. acads. of Berlin, Bucharest, St. Petersburg, Turino, Argentina. Author: Handbuch der technischen Chemie, nach Payen's Chimie industrielle, frei bearbeitet, 2 vols., 1872-74; Historischkritische Studien über Ozon, 1879; Der Stein der Weisen, 1889; Vier Jahrzehnte chemische Forschung, 1892; (with H. v. Höfer) Das Erdöl seine Physik, Chemie, Geologie, Technologie und sein Wirtschaftsbetrieb, 6 vols., 1907-25. Founder petroleum sci.; originated theory of formation of petroleum and natural gas from fat of prehistoric animals; built device for determining combustion danger of petroleum (Engler viscosimeter); invented device for discontinuous distillation of petroleum; studied pyridine derivatives, ozone; 1st complete synethesis of indigo. Died Karlsruhe, Feb. 7, 1925.

ENGLISH, Spofford Grady, Am. chemist; b. Mt. Pleasant, Tenn., Nov. 16, 1915; s. Spofford Grady and Ruby May (Warnock) E.; B.S., U. Okla., 1938, M.S., 1940; Ph.D., U. Cal. at Berkeley, 1943; m. Muriel K. Frodin, Sept. 18, 1942; children—Susan P., Helen W., Elizabeth H. Chemist, Okla. Geol. Survey, 1936-40; teaching asst. in chemistry U. Cal. at Berkeley, 1940-42, asst. prof., 1946-47; research asso. Metall. Lab., U. Chgo., 1942-43; sect. chief chem. div. Clinton Labs., Oak Ridge, 1943-46; with U. S. AEC, 1947—, dep. dir. div. research, 1960-61, spl. asst. to gen. mgr. for disarmament, 1959-61, asst. gen. mgr. for research and devel., 1961-—, mem. U. S. delegations UN Disarmament Conf., London, Eng., 1955. Mem. U. S. delegation for drafting Statue IAEA, UN, N.Y.C., 1956, Internat. Conf. on Cessation Weapons Tests, Geneva, Switzerland, 1959. Recipient AEC Outstanding Service award, 1956. Mem. Phi Beta Kappa, Sigma Xi. Research, publs. on phys. and chem. studies heavy elements, methods detection and measurement radiation, tech. methods related to internat. control atomic energy for peaceful purposes, mgmt. research. Home: 8204 Thoreau Dr., Bethesda, Md. 20034. Office: U. S. AEC, Washington 20545.*

ENGMAN, Martin Feeney, Am. dermatologist; b. New Orleans, Aug. 20, 1868; s. Harry A. and Matilda (Feeney) E.; student U. Ky., U. Va.; M.D., N.Y. U., 1891; grad. study U. Heidelberg and in Paris, Berlin, Hamburg; m. Louise Charlott, 1897; children—Martin F., Walter C. Lectr. dermatology and syphilis N.Y. Post Grad. Sch., 1894; began practice, St. Louis, 1897; lectr. dermatology Marion-Sims Coll. Medicine. St. Louis, 1898-99; clin. prof. dermatology Washington U., 1905-53. Fellow A.M.A.; mem. Am. Dermatol. Assn. (ex-pres.); corr. mem. Danish Dermatol. Assn., French Dermatol. Assn. Discoverer of amoebic infection of skin and five other hitherto unrecognized diseases of skin. Died Oct. 12, 1953.

ENGSTROM, Elmer William, Am. elec. engr.; b. Mpls., Aug. 25, 1901; s. Emil and Anna (Nilsson) E.; B.S. in Elec. Engring., U. Minn., 1923; D.Sc., N.Y.U., 1949, Rutgers U., 1963; D.Engring., Drexel Inst. Tech., 1963, Poly. Inst. Bklyn., 1966; others; m. Phoebe Ingeborg Leander, July 28, 1926; 1 son, William Leander. With Gen. Electric Co., Schenectady, 1923-30; with RCA, N.Y.C., 1930-—, exec. v.p. research and engring., sr. exec. v.p., pres., 1961-65, chmn. exec. com., 1966-—, chief exec. officer, 1966-68, also dir.; dir. NBC, RCA Communications, Prudential. Vice chmn. Nat. Commn. Coop. Edn., 1962-67; mem. adv. com. research div. N.Y. U. Coll. Engring., 1963-—; mem. exec. tech. devel. bd. Poly. Inst. Bklyn., 1963-—. Named comdr. Order of Merit, Italian Republic, 1959; Swedish Am. of Year, Vasa Order, Stockholm, 1963; comdr. Royal Order Vasa, Sweden, 1965; recipient Christopher Columbus Civic Inst. of Genoa Internat. award, 1959, Charles Proteus Steinmetz Centennial medal Nat. Acad. Engring., 1965; William Proctor prize for sci. achievement Sci. Research Soc. Am., 1966; others. Fellow I.E.E.E. (Founder's award 1966); mem. Aerospace Elec. Soc. (Merit award 1963), Am. Soc. Metals (Advancement of Research medal 1960), Royal Swedish Acad. Engring. (Silver Plaquette 1949), Nat. Bur. Standards, Nat. Acad. Engring. (charter), (vis. com.), Am. Soc. Swedish Engrs. (hon.), Sigma Xi, others. Research, publs. leading to practical tv, both black and white and color; contbr. to basic and applied research in communications tech. and solid state electronics; a developer of mgmt. techniques for research. Home: 1234 Old Nassau Rd., Jamesburg, N.J. 08831. Office: 30 Rockefeller Plaza, N.Y.C. 10020.*

ENIGK, Karl, German parasitologist; b. Torgau, Germany, Oct. 23, 1906; s. Friedrich Karl and Ida (Reichenau) E.; D.V.M., U. Leipzig, 1931; postgrad. U. Berlin, 1933-37; m. Hildegard Quoss, Dec. 19, 1936. Asst., Vet. Coll. Berlin, 1931-34; sr. asst., parasitology inst. U. Berlin, 1935-41; head parasitology dept. Mil. Vet. Research Labs., Berlin, 1941-44; dir. vet. med. dept. Bernhard-Nocht Inst. for Ship and Tropical Diseases, Hamburg, 1945-53; dir. parasitology inst. Vet. Coll., Hannover, 1953-—. Mem. Exam. Bd. for Govt. Vet. Service, Recipient Bernhard-Nocht medal Institut fur Schiffs und Tropenkrankheiten. Mem. Deutsche Gesellschaften fur Parasitologie, World Assn. for Advancement Vet. Parasitology, Gesellschaft Deutscher Naturforscher und Arzte, Deutsche Gesellschaften fur Hygiene und Mikrobiologie, others. Research, numerous publs. on piroplasmoses of horse, dog, sheep, nematode and fasciola infestation of domestic animals, breeding and biology of ticks, taxonomy of trematodes and nematodes. Home: 7 Rutenbergstrasse, Hannover-Kirchrode, Germany.*

ENKVIST, Terje Ulf Eugen, Finnish chemist; b. Helsinki, Finland, Oct. 3, 1904; s. Ernst and Verna (Aschan) E.; Ph.D., U. Helsinki, 1934; m. Elsa Ulrika Berg, June 10, 1928; children—Elsa, Clarence, Ernst, Karl Ossian. Asst., Central Lab., Helsinki, 1927-35, chief organic chemistry research dept., 1940-45; asst. U. Helsinki, 1926-37, docent, 1937-40, asst. prof., 1940-45, prof. chemistry, 1951-—; laborator of wood chemistry Swedish Forest Prodn. Research Lab., Stockholm, 1945-51. Lectr. Tech. U. Finland, 1943-45, Mem., chmn. numerous Finnish and Swedish profl. orgns. Author: Organic Chemistry, 1965. Research, publs. on constn. of santenone and santenic acid, 1932, organic catalysts, lubricating oil derived from wood tar and tall oil, thiolignin and mechanism of kraft pulpin, organic chems. from kraft black liquors. Home: 30 A Runebergsgatan, Helsinki. Office: 4 S. Hesperiagatan, Helsinki, Finland.*

ENLOW, Donald Hugh, Am. anatomist; b. Mosquero, N.M., Jan. 22, 1927; s. Donald Carter and Martie (Albertson) E.; B.S., U. Houston, 1949, M.S., 1950; Ph.D., Tex. A. and M. U., 1955; m. Martha Ruth McKnight, Sept. 3, 1945; 1 dau., Sharon Lynn (Mrs. Roger Hack). Instr., U. Houston, 1949-52; asst. prof. W. Tex. State U., 1955-56; instr. anatomy Med. Coll. S.C., 1956-57; faculty U. Mich., Ann Arbor, 1957-—, prof. anatomy, 1967-—, dir. cranio-facial program Center for Human Growth and Devel., 1966-—. Lectr., guest speaker numerous instns., profl. socs. and orgns. Mem. Tex. Acad. Sci. (past chpt. pres., Spl. Merit award 1951), Am. Assn. Anatomists, Sigma Xi (Outstanding Mem. award 1955, Outstanding Research award Tex. A. and M. chpt. 1955), Phi Kappa Phi. Author: Principles of Bone Remodeling, 1963; also articles. Research on comparative microscopic structure of fossil and modern bone, remodeling processes in growing human bone, necrosis in bone and mechanisms of regeneration, growth processes in cranio-facial skeleton. Home: 7982 Dexter-Pinckney Rd., Dexter, Mich. 48130.*

ENNEMOSER, Joseph, physician; b. Schönau, Austria, Nov. 15, 1787; student U. Innsbruck (Austria), U. Erlangen (Germany), U. Vienna (Austria); M.D., U. Berlin (Germany), 1816; fought in Lutzow Corps against Napoleon, 1813-14; prof. medicine U. Bonn (Germany), from 1819; practiced medicine, Innsbruck, 1837-41; then went to Munich, Germany. Author: Historisch-Psychologische Untersuchungen über den Ursprung und das Wesen der menschichen Seele, 1824; Anthropologische Ansichten, . . . , 1828; Der Magnetismus im Verhältnis zur Natur und Religion, 1842; Geschichte des thierischen Magnetismus, 1844; Anleitung zur mesmerischen Praxis, 1852. Used hypnotism in treatment of disease; worked on F.A. Mesmer's theory of curative power of "animal magnetism." Died Egern on the Tegernsee, Bavaria, Sept. 19, 1854.

ENNIS, William Duane, Am. engr.; b. in Bergen County, N.J., Jan. 6, 1877; s. William C. and Kate E. (Burroughs) E.; M.E., Stevens Inst. Tech., 1897, Dr. Engring., 1934; m. Margaret Schuyler, Dec. 28, 1898. Mech. engr. various cos. including Am. Linseed Co., 1897-1905; engr. Am. Locomotive Co., 1905-07; prof. mech. engring., Poly. Inst. Bklyn., Columbia, U. S. Naval Acad.; cons. mech. engring. and indsl. mgmt.; Alexander Crombie Humphreys prof. econs. of engring. Stevens Inst. Tech., 1929-44; now prof. emeritus. Fellow A.A.A.S., Royal Econ. Soc.; Am. Soc. M.E. (treas. 1935-44); mem. Am. Econ. Assn., Tau Beta Pi. Author: Linseed Oil, 1909; Applied Thermodynamics, 1910; Vapors for Heat Engines, 1912; Flying Machines Today, 1911; Works Management, 1911; Thermodynamics Abridged, 1920; Business Fundamentals for Engineering Students, 1941. Died Oct. 14, 1947.

ENNO, see Evno.

ENOCHS, Edgar Earle, Am. mathematician; b. McComb, Miss., Sept. 13, 1932; s. Philip Henry and Jean (Lampton) E.; B.S., La. State U., 1958; Ph.D., Notre Dame U., 1958; m. Louise Smith, June 21, 1958; children—Corinne Theresa, Mary Jane, Kathryn Elizabeth, Maureen Margaret, Madelaine Justina, Brigid Anne, John Christian. Faculty, U. Chgo., 1958-60; faculty U. S.C., Columbia, 1960-—, prof., 1965-67; faculty U. Ky., Lexington, 1967-—. Mem. Am., French math. socs. Research in abelian groups; computed homotopy groups of compact abelian groups. Home: 465 W. 3d St., Lexington, Ky. 40508.*

ENRIGHT, John Bautista, Am. virologist, educator; b. San Jose, Cal., Nov. 26, 1909; s. John B. and Maude (Kellogg) E.; Ph.D., Stanford, 1947; m. Zenobia Rogers, Apr. 25, 1931; children—John Roger, Patricia Louise (Mrs. William Liggett). Owner, operator clin. labs., 1934-42; with Cal. Dept. Pub. Health, 1947-50; prof. pub. health U. Cal. at Davis, 1950-—. Cons., USPHS, 1950-—; mem. tropical medicine and parasitology study sect. NIH, 1963-—. Mem. A.A.A.S., Am. Pub. Health Assn., Am. Assn. Immunologists, Am. Soc. for Microbiology. Research, publs. on epidemiology and natural history of viral and rickettsial infections common to man and animals. Home: 1050 W. Capitol Av., West Sacramento, Cal. 95691. Office: University of Cal., Davis, Cal. 95616.*

ENRIQUES, Federigo, Italian mathematician; b. Livorno, Italy, Jan. 5, 1871; s. Giacomo and Matilde E.; hon. doctorate, U. St. Andrews. Prof., U. Bologna, 1894-1922; prof., U. Rome, 1922-49. Received Bordin prize, French Acad. Scis. Pres., Philosophical Soc., Italy, 1905-13; pres., National Assoc. U. Prof., 1912-14; pres., Italian Soc. of Math.; National Assoc. Advancement of Science. Author: Ricerche li geometria sulle superficie algebriche, 1893; Problemi della scienza, 1906; Scienza e razionalismo, 1912; Per la storia della logica, 1925. Studied algebraic systems of curves on surfaces; contributed to development of higher geometry; studied history of science. Died Rome, Italy, June 14, 1949.

ENSKOG, David, Swedish mathematician; b. Amtervik, Sweden, Apr. 22, 1884; s. Nils Olsson and Karolina (Jonasdotter) E.; Ph.D., Uppsala (Sweden) U., 1917; postgrad. U. Göttingen (Germany), 1922, U. Munich (Germany), 1923; m. Anna Jönsson, Dec. 27, 1913; 3 children. Lectr., Skoevde, Sweden 1913-18; lectr. math., physics Gävle, Sweden, 1918-29, Stockholm, Sweden, 1929-30; prof. math. and mechanics Kungl. Tekniska Högskolan, Stockholm, 1930-—, censor matriculation exams., 1931-—; tchr. math. and mechanics Kungl. Sjökrigshögskolan, 1933-39. Recipient Thalens prize, 1923, Wallmarks prize, 1928, Gold medal Royal Swedish Acad. Sci., 1946; Letterstedts travelling fellow, 1921. Mem. Royal Swedish Acad. Engring. Scis., Royal Soc. Sci. Died June 1, 1947.

ENT, Sir George, physician; b. Sandwich, Eng., Nov. 6, 1604; s. Josias Ent; B.A., Sidney Sussex Coll., Cambridge (Eng.) U., 1627, M.A., 1631; M.D., Padua, Italy, 1636; M.D., (inc.) Oxford (Eng.) U., 1638; m. Sarah Meverall, Feb. 10, 1646. Gulstonian lectr. for coll., Oxford, also censor for 22 years, registrar, 1655-70, pres., 1670-75, 1682, 84. Fellow Coll. Physicians (pres.), Royal Soc., 1663. Author: Apologia pro circuitione sanguinis, 1641; Antidiatriba in Malachians thruston de respirationis usu primaris, 1679; also collected works, 1687. Supported Harvey's discovery of blood circulation. Died Oct. 13, 1689.

ENTERLINE, Phillip Ernest, med. statistician; b. Kittanning, Pa., Apr. 25, 1921; s. John Ernest and Charlotte (Bates) E.; B.B.A., Westminster Coll., 1942; M.A., Am. U., 1954, Ph.D., 1960; m. Esther Dickey, Nov. 9, 1945; children—John, Janis, Jeffery. Statistician div. Tb, USPHS, 1946-50, asst. chief program analysis sec. div. chronic disease, 1950-54, chief statistician heart disease control program, 1954-59, chief morbidity analysis br. div. pub. health methods, 1959-61, chief biometrics br. div. occupational health, 1961-64; prof. med. statistics McGill U., Montreal, Que., Can., 1964-—. Professorial lectr. Am. U., 1960-64; mem. health services research study sect. NIH, 1966-—. Fellow Am. Pub. Health Assn., Am. Statis. Assn., Am. Sociol. Soc., Am. Population Assn. Research, publs. on geog. variability in disease, measurement of environmental and social factors affecting disease, factors affecting illness behavior. Home: 1212 Pine St. W., Montreal, Que., Can.*

ENTZELT, Christopher (Encelius), German geologist; b. Saalfeld, Germany, 1517; ed. U. Wittenberg (Germany); pastor. Author: De re metallica, 1551. Studied medicinal uses of metals, stones, gems and minerals, including native and manufactured common salt, soda, saltpetre, alum, pumice, gypsum. Died 1586.

EPHRAIM, Fritz Bruno, chemist; b. Berlin, Germany, Sept. 4, 1876; s. Hermann and Amalie Henriette (Goldstein) E.; ed. Tech. U., Berlin-Charlottenburg; doctorate under C. Liebermann; later student under Emil Fischer; m. Perl Zuckermann, 1905; 1 son, 1 dau. Named asst. to C. Friedheim, U. Bern (Switzerland), 1901, lectr., 1902, asso. prof. analytic chemistry, 1911, prof. organic chemistry, 1932. Author: Lehrbuch für anorganische Chemie, 1922; also numerous articles. Described numerous new compounds, especially double halogens; studied problems of secondary valences, relationship of atomic vols.; tried to prove vol. contraction in formation of compounds. Died Bern, Jan. 17, 1935.

EPHRUSSI, Boris, geneticist; biologist; b. Moscow, USSR, May 9, 1901; s. Samuel and Liuba (Foukelman) E.; M.A., U. Paris (France), 1922, Sc.D., 1932 D.Sc. (hon.), LL.D., U. Glasgow, 1954; U. Brussels, 1960; m. Harriett Taylor, Oct. 5, 1949; 1 dau., Ann. Asso. prof. biology Johns Hopkins, 1941-44; prof. genetics U. Paris, 1946-—; dir. lab. physiol. genetics Centre National de la Rechereche Scientifique, 1946-—; exchange prof. Harvard, 1954, Cal. Inst. Tech., 1959; F. H. Herrick distinguished prof. biology Western Res. U., 1961-—. Decorated Legion of Honour. Mem. Royal Danish Acad.; Nat. Acad. Scis. Author: La Culture des Tissues, 1932; Nucleocytoplasmic relations in micro-organisms, 1953; also articles. Spl. research on cell differentiation, genetic mechanisms, somatic cell variation, cells cultured in vitro. Home: 2127 Cornell Rd., Cleve. 44106.

EPICUROS OF SAMOS, Greek philosopher; b. Samos, Aegean Sea, 341 B.C.; s. Neocles and Charaestrata; student at Samos under the Platonic Pamphilos, at Athens, at Teos under Democritean Nausiphanes; m. Leontion. Founded schs. in Mytilene, Lesbos, and Lampaseus, Asia Minor; in 306 B.C. returned to Athens to found sch. (1st to admit women, began philosophy of Epicureanism). Author: On Nature, 37 books; over 300 treatises, most of which are lost. Philosophy held that pleasure is the only good, but that pleasure must be life of prudence, honor and justice; writings converted Lucretius to Epicureanism; adopted atomism of Democritos as mechanistic explanation of universe (deviated from its deterministic implications by allowing an element of spontaneity; emphasized man's freedom of action. Died Athens, 270 B.C.

EPLEY, John MacNaughton, Am. physician; b. Eugene, Ore., Feb. 8, 1930; s. Malcolm and Jane (Dudley) E.; B.S., U. Ore., 1953, M.D., 1957; m. Norma Marie See, June 27, 1954; children—Cathryn, Cynthia. Med. officer, chief dept. otolaryngology Vandenberg AFB, Cal., 1958-61; practice medicine, specializing in otolaryngology, Portland, Ore., 1965-—; clin. instr. U. Ore. Med. Sch., 1965-—. Diplomate Am. Bd. Otolaryngology. Mem. Phi Beta Pi. Research, publs. on first report on successful multipolar elec. stimulation of auditory nerve in man, demonstrating for first time that stimulation of different fiber groups of auditory nerve will result in perception of different pitch by subject. Home: 5824 S.E. Yamhill St., Portland 97215. Office: 545 N.E. 47th Av., Portland, Ore. 97213.*

EPPING, Joseph, astronomer; b. Bevergern, Germany, Dec. 1, 1835; s. Joseph and Elisabeth E.; student math., Münster, studied theology, 1867-71. Germany; Joined Soc. of Jesus, 1859; ordained priest, 1870; named prof. math. and astronomy, Maria-Laach, Prussia, 1863; tchr. math. Polytechnikum Quito (Ecuador), 1872; tchr. math. and astronomy at sems. in Blijenbeck, also Exaeten, Holland, beginning in 1876. Author: (with J. N. Strassmaier) Astronomisches aus Babylon, 1889; Kreislauf im Kosmos, 1882; also articles. Research and publs. on Assyrian chronology and sci., Babylonian astronomy; attacked materialism of Kant-Laplace nuclear hypothesis. Died Exaeten, Aug. 22, 1894.

EPPINGER, Hans, Jr., physician; b. Prague, Czechoslovakia, Jan. 5, 1879; s. Hans and Anna (Marterer) E.; student path. anatomy and histology under his father; doctorate, 1903; m. Georgine Zetter, 1908; 1 son, 2 daus., including Maria (Mrs. Rühl). Asst., med. clinic, Vienna, Austria; became lectr., 1906; named prof., Vienna, 1914, asso. prof., 1918; joined U. Freiburg (Germany), circa 1918; named prof. U. Cologne (Germany), 1930; prof., Vienna, also I. Univ. Med. Clinic, 1933-45. Author: Zur Pathologie und Therapie des menschlichen Ödems, 1917; (with E. Ranzi) Die hepatolienalen Erkrankungen, 1920; Pathologie und Therspie des Icterus, 1921; (with K. Kloss) Nephritisfragen, 1921; (with F. Kisch) Die Nephritisfrage, 1923; (with L. V. Papp, H. Schwarz) Asthmacardiale, 1924; (with F. Kisch, H. Swarz) Das Versagen des Kreislaufes, 1927; Zur Pathologie der Kreislaufkorrelationen, 1927; (with H. Kaunitz, H. Popper) Die seröse Entzündung, 1935; Die Leberkrankheiten, 1937; Die Permeabilitätspathologie, 1949. Research on endocrinology, especially thyroid and parathyroid, diabetes, function of diaphragm, pathology of liver and spleen; described electrocardiographic changes in bundlebranch block; studied (with C. J. Rothberger) vegetative nervous system, (with L. Hess) theory of physiology balance between 2 branches of autonomic nervous system; introduced (with K. Koss) phenylhydrazine hydrochloride for treatment of polycythemia. Died Vienna, Sept. 25, 1946.

EPSTEIN, M(ichael) Anthony, English physician; b. London, Eng., May 18, 1921; M.A., M.B., B.Ch., Cambridge (Eng.) U., 1949, M.D., 1951; Ph.D., U. London, 1952, D.Sc., 1963. Asst. pathologist Bland Sutton Inst. Pathology, Middlesex Hosp. Med. Sch., 1948-65; reader exptl. pathology U. London, 1965-68; hon. cons. in exptl. virology Middlesex Hosp., 1965-68; prof. pathology U. Bristol (Eng.), 1968—; hon. cons. pathologist United Bristol Hosps., 1968—; vis. investigator Rockefeller Inst., N.Y.C., 1956. Berkeley Travelling fellow, 1952-53; French Govt. Exchange scholar Inst. Pasteur, Paris, France, 1952-53. Fellow Coll. Pathology; mem. Path. Soc. Gt. Britain, Soc. for Gen. Microbiology, Royal Soc. Medicine, Soc. for Exptl. Biology, Brit. Biophys. Soc., Internat. Soc. for Cell Biology. Editor: (with G. W. Richter) International Review of Experimental Pathology, vols. I-VI, 1962-68. publs. on cell structure and function, cell virus interactions with spl. reference to virus neoplasia and possible virus tumors in man. Office: Dept. Pathology, U. Bristol Med. Sch., University Walk, Bristol 8, Eng.*

EPSTEIN, Saul Theodore, Am. physicist; b. Southampton, N.Y., June 14, 1924; s. Joseph Samuel and Jeanette (Friedman) E.; S.B., Mass. Inst. Tech., 1944, Ph.D., 1948; m. Jean Hoopes, Jan. 30, 1948; children—Joanne, Peter, David. Faculty, Columbia, 1948-51, U. Neb., Lincoln, 1954-63; prof. physics U. Wis., Madison, 1963—. Lectr., Stevens Inst. Tech., 1951-52, Boston U., 1952-53; mem. Inst. for Advanced Study, Princeton, N.J., 1947-48. Mem. Am. Phys. Soc., Am. Assn. Physics Tchrs., Sigma Xi. Research in elucidation of certain effects in interaction of radiation with matter, particularly effects concerned with scattering of electrons and structure of electrons; various tech. devels. and applications of approximation methods in quantum mechs. Home: 2325 Kendall Av., Madison, Wis. 53705.*

EPSTEIN, Stephan, physician; b. Nuremberg, Germany, Mar. 14, 1900; s. Ernst and Margaret (Scherbel) E.; student univs. Erlangen, Breslau, Berlin, Heidelberg, Munich; M.D., U. Erlangen, Germany, 1925; m. Elsbeth Lauinger, May 27, 1926; children—Ernst, Wolfgang. Came to U. S., 1936, naturalized, 1941. Dermatologist, Marshfield (Wis.) Clinic, 1936-65; pres. Marshfield Clinic Found. for Med. Research and Edn., 1960-65; clin. asso. prof. dermatology U. Minn., Mpls., 1946—; clin. prof. dermatology U. Wis., Madison, 1965—. Mem. A.M.A., Am. Dermatol. Assn., Am. Acad. Dermatology, Am. Coll. Allergists, Wis. Med. Soc., also others. Editor translator (with Burckhardt) Atlas and Manual of Dermatology and Venerology, 2d edit., 1965. Research, numerous publs. on dermatologic allergy and photosensitivity, distinction between phototoxic and photoallergic reactions; produced exptl. proof of allergy to sunlight. Home: 2921 Harvard Dr., Madison 53705. Office: 2705 Marshall Ct., Madison, Wis. 53705.*

EPSTEIN, William Louis, Am. physician; b. Cleve., Sept. 6, 1925; s. Norman N. and Gertrude (Hirsch) E.; A.B., U. Cal. at Berkeley, 1949, M.D., 1952; m. Joan Goldman, Jan. 29, 1954; children—Wendy Steven. With U. Cal., San Francisco, 1957—, asso. prof. div. dermatology, 1963—, dir. dermatol. research, 1957—, acting chmn. div. dermatology, 1966—, cons. dermatology Outpatient Dept. Cons. various hosps., San Quentin Prison, Cal. Dept. Pub. Health. Mem. A.A.A.S., A.M.A., Am. Acad. Dermatology and Syph., Pacific Dermatologic Assn., Am. Fedn. Clin. Research, Soc. Investigative Dermatology, Am. Dermatol. Assn., Phi Beta Kappa, Sigma Xi, others. Contbr. to immunology in investigative dermatology including initial demonstration passive transfer of allergic contact sensitivity in man; devised a model system for study granulomatous hypersensitivity; investigated immunologic responses of patients with various skin and other diseases; contbr. newer information of biology of epidermal cells in wound healing. Home: 135 San Benito Way, San Francisco 94127. Office: U. Cal., San Francisco, 94122.*

ERASISTRATOS (OF CHIOS), Greek anatomist, physician; b. Iulis, Chios, circa 304 B.C.; s. Cleombrotus; perhaps grandson of Aristotle; studied in Athens under Metrodorus; in close contact with Peripatetic sch.; possibly studied at Cos in Praxagoras' sch.; lived briefly in Antioch, Syria; later lived in Alexandria, founder anatomy sch. there; court physician to king of Syria; continued work of Herophilos; interested in sci. research and experimentation; reputedly performed in vivo expts. and autopsies; particularly noted for studies of brain; 1st to distinguish between cerebrum and cerebellum; held that complexity of convolutions in brain is related to intelligence, opposing Aristotelian view that heart is center of knowledge; recognized difference between sensory and motor nerves; believed that all body functions mech. in nature but believed that nature does nothing in vain; interested in comparative anatomy; held that every organ is connected with rest of body by 3-fold system of vessels, i.e., arteries, veins, nerves; correctly described function of epiglottis and auriculo-ventricular valves; described and named tricuspid valve of heart; named trachea; believed to have introduced term parenchyma; studied growth of body and process of digestion; noted relationship between motion of lungs and pulsation of heart and arteries; recognized changes of body caused by disease; rejected humoral theory of disease, originated theory of pneuma (attributed all illness to plethora or repletion of body through undigested nutrition); 1st physician to distinguish hygiene from therapeutics, attached greater significance to former; emphasized dietetic care and opposed violent cures, excessive use of drugs and practices of phlebotomy and purgation; invented catheter; accepted atomism of Democritos; pub. numerous treatises including 2 anat. works (one on causes of disease). Died Samos, circa 250 B.C.

ERASTUS, Thomas (Thomas Lieber), Swiss physician; b. Baden, Switzerland, Sept. 7, 1524; studied theology, Basel, Switzerland, 1540, philosophy and medicine, Bologna, then Padua, Italy; became physician to count of Henneberg, Saxe-Meiningen, 1553, to Otto Heinrich, electro-palatine, 1558; made privy councillor, mem. ch. consistory, 1559; excommunicated on charge of Socinianism, 1570-75; prof. medicine, Heidelberg, Germany, 1558-80; became prof. ethics, Basel, 1580. Author: Disputationum de medicina nova Philippi Paracelsi, 1572-73; Disputatio de auro potabile . . . , 1578, others. Accepted Zwinglian doctrine; attacked astrology and belief in transmutation of metals; denied man is microcosm; opponent of theories of Paracelsus, especially rejected principles of salt, sulphur, and mercury, while supporting Aristotelian elements of earth, air, fire, and water; doctrine of dominance of state over church called "Erastianism." Died Basel, Dec. 31, 1583.

ERATOSTHENES, Greek astronomer, geographer; b. Cyrene, Libya, circa 275 B.C.; studied under Callimachus and Lysanias in Alexandria, came under influence of Arcesilaus and Ariston in Athens; accepted invitation of Ptolemy III Euergetes to serve as tutor to crown prince, became head of great Alexandrian Library, circa 225 B.C.; most versatile scholar of his time; 1st systematic geographer; dealt with phys., math. and ethnol. geography; made map of known world which was better than any drawn previously; best known for his calculation of circumference of earth as 252,000 stadia which is very accurate (assuming most probable length of stadium); calculated magnitude and distance of sun and moon; measured obliquity of ecliptic with great accuracy; enumerated 675 fixed stars in his astron. catalogue; suggested introduction of extra day into calendar every 4th year; founded sci. chronology; established system of chronology wherein dates were reckoned from conquest of Troy; investigated arithemetical and geometrical problems and dealt with math. definitions; worked out system for determining prime numbers still known as sieve of Eratosthenes; studied methods to find mean proportionals between two given lines; wrote on theory of numbers, lit. criticism, problems of authorship and prodn. of plays, philosophy, mythology, history, prins. of music; author: Geographica; On the Measurement of the Earth; On the Duplication of the Cube; Platonicus; Complete Fragments of Eratosthenes (editor Bernhardy, 1822); died supposedly from voluntary starvation because he was threatened by total blindness. Died Alexandria, circa 195 B.C.

ERB, Ralph Eugene, Am. animal scientist; b. Jerseyville, Ill., May 28, 1917; s. George J. and Anna (Engle) E.; B.S., U. Ill., 1940; M.S., Purdue U., 1942, Ph.D., 1947; m. Rosetta Roodhouse, Aug. 24, 1941; 1 son, George E. Faculty, Wash. State U., 1947-48, 53-62; prof. physiology, asst. head dept. animal sci. Purdue U., Lafayette, Ind., 1962—. Chmn., Creamery Examination Bd. Ind., 1962—. Recipient Borden award, 1961; Master Builder of Men award Nat. Farmhouse Frat., 1962. Fellow A.A.A.S.; mem. Am. Dairy Sci. Assn. (exec. bd. 1963-66, exec. com. 1964-66, v.p.), Am. Soc. Animal Sci. (Physiology and Endocrinology award 1967), Am. Dairy Sci. Assn., Am. Soc. Animal Sci., Endocrine Soc. Research, publs. on physiology of semen, cattle reprodn., milk composition and identification of and methods for measurement of female sex steroids in organs, blood and urine of non-pregnant and pregnant cattle, sheep and swine. Home: 1705 Klondike Rd., West Lafayette, Ind. 47906. Office: Dept. Animal Scis., Purdue U., Lafayette, Ind. 47907.*

ERB, Wilhelm Heinrich, German neurologist; b. Winnweiler, Germany, Nov. 30, 1840; s. Friedrich and Sophie (Hoffmeister) E.; studied medicine, U. Heidelberg (Germany); M.D., 1864; m. Bertha Karoline Hermann; 1 son; m. 2d, Anna Gass, 1876; 2 sons. Became asst. path. anatomy and internal medicine, Heidelberg, 1865, lectr., 1865, asso. prof., 1869; named prof. specialized pathology and therapy, Leipzig, 1880; prof., dir. med. clinic, Heidelberg, 1883-1917. Author: Zur Entwicklungsgeschichte der roten Blutkörperchen, 1865; über die Anwendung der Electrizität in der innern Medizin, 1872; Handbuch der Krankheiten der periopheren cerebrospinalen Nerven, 1874; Handbuch der Elektrotherapie, 1882; Gesamte Abhandlungen, 2 vols., 1910. Research in electrotherapy, including discovery of decomposition reaction in paralysis in tetany (Erb's phenomenon), spastic spinal paralysis, partial paralysis of brachial plexus caused by birth injury, (with C. Westphal) Erb-Westphal syndrome characteristics of tabes dorsalis; studied progressive muscular atrophy (Erb's disease), Thornsen's disease, children's paralysis, brain tumors; connected tabes dorsalis with syphilis; introduced term, tendon reflex, 1875. Died Heidelberg, Dec. 29, 1921.

ERCKER, Lazarus, alchemist, mineralogist, metallurgist; b. Annaberg, Saxony, circa 1530; s. Asmus E., ed. Wittenberg; m. Anna Canitz, 1554, m. 2nd Susanne Thiel, 1564; sons: Joaquim, Hans. Warden mint at Annaberg; assay master to Elector of Saxony, Dresden, 1555, later chief assay master; warden mint to Duke Julius of Brunswick, Goslar, 1558-66, later mine master; assayer, later chief cons. under emperors Maximilian II and Rudolph II, Bohemia, 1568-94. Created baron von Schreckenfels, 1586. Author: Beschreibung Allerfurnemister Mineralischen Ertzt vnd Berckwercks arten, 1574. States that zinc precipitates other metals from solutions, also that speculum metal is made from copper, tin and arsenic, with or without antimony; describes making of brass, making of crucibles in wooden molds in screw press, also technique for sofenting brittle gold, distillation of nitric acid; gave numerous recipes for cement powders; research on chemistry of saltpeter; wrote especially on assaying methods. Died (probably) Prague, Bohemia, 1594.

ERDÉLYI, Arthur, mathematician; b. Budapest, Hungary, Oct. 2, 1908; s. Ignác and Friderike (Roth) E.; cand.Ing., Deutsche Technische Hochschule, Brno, Czechoslovakia, 1928; Dr.rer.nat., Prague (Czechoslovakia) U., 1938; D.Sc., Edinbrugh (Scotland) U., 1940, m. Eva Neuburg, Nov. 4, 1942. Faculty U. Edinburgh, 1941-49, prof. math., 1964—; vis. prof. math. Cal. Inst. Tech., 1947-48, prof., 1949-64; vis. prof. Hebrew U., Jerusalem, Israel, 1956-57. Sci. Cons. Brit. Admiralty, 1942-46. Fellow Royal Soc. Edinburgh; mem. Am., Edinburgh (past co-editor Proc.), London math. socs., Acad. Scis. (fgn. mem. Torinto, Italy). Author: (with others) Higher Transcendental Functions, 1953; Tables of Integral Transforms, 1954; Asymptotic Expansions, 1956; Operational Calculus, 1962. Editor: Mathematical Tables and Other Aids to Computation, 1950-55. Editor, Jour. Indian Math. Soc., 1947—, Jour. Math. and Mechanics, 1955-60, Archive for Rational Mechanics and Analysis, 1957—, Jour. Soc. Indsl. and Applied Math., 1957—, Jour. Math. Physics, 1961-64, Canadian Jour. Math., 1962-65. Research on differential equations, spl. functions, operational calculus. Home: 16 Gilmour Rd., Edinburgh 9, Scotland.*

ERDMANN, Benno, German philosopher; b. Guhrau, Silesia, Prussian, May 30, 1851; s. Karl and Marie (Solf) E.; student Heidelberg, Germany; Ph.D., Berlin, Germany, 1873; m. 2d, Emilie de Gruyter, 1913; 1 son, 2 daus. (by previous marriage). Became lectr. Berlin, 1876; named asso. prof., Kiel, Germany, 1878, prof., 1879; apptd. prof., Breslau (now Wroclaw, Poland), 1884, Halle, Germany, 1890, Bonn, Germany, 1898, Berlin, 1909. Mem. Berlin Acad. Scis. Author: Grundzügen der Reproduktions-psychologie, 1920; also several philos. works, numerous articles. Studied logic, perception, theory, psychology; history of philosophy, especially Kant. Died Berlin, Jan. 7, 1921.

ERDMANN, Hugo Wilhelm Traugott, German chemist; b. Preussisch-Holland, Germany, May 8, 1862; s. Hermann and Rosa Angelica (Hoppe) E.; student chemistry, Halle, Germany, Munich, Germany; doctorate, Strasbourg, France, 1883; m. Marie Bruns, 1897.

Joined faculty U. Halle, 1885, phys. and tech. chemistry under Prof. J. Volhard, became prof., 1894, dir. establishment lab. for applied chemistry, 1899; founder (with Ernst Immanuel Erdmann) pvt. ednl. lab. for chem. tech. work, 1889; named prof., dir. inorganic chemistry lab. Tech. U. Berlin, 1901; made several expdns. to Asia, Siberia, China, Caucasus, Alaska. Mem. Leopoldina. Author: Anleitung zur Darstellung chemischer Präparate, 1891; Lehrbuch der anorganischen Chemie, 1893; (with P. Köthner), Naturkonstanten in alphabetischer Anordung, 1905. Synthesized various organic compounds; practical applications of acetylene; used spectral analysis, conductivity measurement and colorimetry for identification and analytic determination of rarer elements, including rubidium, rare earths. Died Müritzsee, Germany, June 25, 1910.

ERDMANN, Otto Linné, German chemist; b. Dresden, Germany, Apr. 11, 1804; s. Carl Gottfried and Wilhelmine Friederike (Geringemuth) E.; student medicine, Dresden, chemistry, Leipzig, Germany; doctorate, 1824; m. Clara Jungnickel; 3 sons, 1 dau. Became lectr. U. Leipzig, 1825, asso. prof., 1827, prof. tech. chemistry, 1830, rector, 1848-49; dir. nickel works, Hasselrode, Germany, 1 year. Author: Über das Nickel, 1827; Lehrbuch der Chemie, 1828; Grundriss der allgemeinen Waarenkunde, 1833; über das Studium der Chemie, 1861. Founder, editor Jour. für technische und ökonomische Chemie, 1828-33; founder, editor (with others) Jour. für praktische Chemie, 1834-69. Research on nickel extraction, nickel salts, indigo, isatin and its derivatives, illuminating gas; syntheses and discoveries in heterocyclical compounds; determined (with R. F. Marchand) atomic weights of carbon, hydrogen, calcium, copper, mercury, sulphur and iron; discovered pyromelhitic acid. Died Leipzig, Oct. 9, 1869.

ERHARD, Hubert, zoologist; b. Munich, Germany, Jan. 9, 1883; s. Georg and Josepha (Sedlmayr) E.; ed. univs. Munich, Jena, Germany; m. Margaret Berger, 1909. Asst. veterinarian U. Munich, 1912-13; lectr. U. Munster, 1913-14; lectr. U. Giessen, 1914-22, asso. prof., 1922-28; prof. U. Fribourg (Switzerland), since 1928. Mem. Swiss Zool. Soc., Swiss Soc. Natural Researchers. Author: Tierpsychologisches Praktikum, 1936; Geschichte der Zoologie, 1921; Hypnose bei Tieren, 1924; also contbr. numerous chapts. to publs. Research in zoology, cytology, veterinary physiology, zoology of alphs. Home: rue des Ecoles 4, Fribourg. Office: Zoological Inst., U. Fribourg, Fribourg, Switzerland.

ERHARD, Ludwig, engr.; b. Aicha vorm Walde, near Passau, Bavaria, Aug. 25, 1863; s. Alexander and Luise (Lang) E.; student mech. engring. Tech. U., Munich, Germany, 1882-86; hon. dr. engring. Tech. U. Danzig (Poland); m. Ziska Lenk, 1893. Practical tng. Maffei Locomotive Factory, 1 year; joined Bavarian Indsl. Mus., Nuremberg, Germany, 1888; became exec. sec. Vienna (Austria) Technol. Indsl. Mus., also dir. Indsl. Promotion Service; named tech. adviser to re-establishment Tech. Mus. Trade and Industry, Vienna, 1910, dir., 1913-30; became dir. Austrian Research Inst. for Tech. History, 1931. Named hon. citizen Tech. U. Vienna. Research and publs. in tech. mus. work; developed theory of devel. of tech. from organic by way of organic-inorganic to purely inorganic. Died Baden, nr. Vienna, Oct. 28, 1940.

ERHARDT, Albert, German zoologist; b. Rostock, May 7, 1904; s. Franz and Margarethe (Sohm) E.; Ph.D., U. Rostock; m. Martha Mühlenweg; 1 son, Franz Heinrich. Prof., U. Rostock, U. Heidelberg; hon. prof. parasitology and applied zoology U. Munster; specialist in parasitology Asta-Werke A.B., Brakwede, Westphalia. Mem. various zool. and med. socs. Collaborator: Welt-Seuchen-Atlas; Handbücher für Exper. Pharmakologie; Kinderheilkunde; Arbeitsmedizin; Parasitologischer Leitfaden. Research and publs. on parasitology and chem. therapy of helminths. Home: Bielefelderstrasse, Brakwede, Westphalia, Germany.

ERICKSON, Milton Hyland, Am. psychiatrist; b. Aurum, Nev., Dec. 5, 1901; s. Albert and Clara F. (Miner) E.; B.A., U. Wis., 1927, M.A. in Psychology, M.D., 1928; m. Elizabeth E. Moore, June 18, 1936; children—Albert, Lance, Carol Ann (Mrs. V. Deerington), Elizabeth Alice (Mrs. David L. Elliott), Allan H., Robert B., Roxanna L., Kristina K. Practice medicine specializing in psychiatry, Worcester, Mass., 1930-34, Eloise (Mich.) Gen. Hosp. and Infirmary, 1934-48, Phoenix, 1948—; faculty Wayne U. Coll. Medicine, Detroit, 1939-48, prof., 1943-48; sr. mem. teaching faculty seminars on hypnosis Found., Chgo., 1955-61; vis. instr. Roosevelt U., Chgo., 1958-61; sr. mem. teaching faculty Am. Soc. Clin. Hypnosis, Edn. Research Found, Mpls., 1962—. Fellow Am. Soc. Clin. Hypnosis (past pres.), Am. Psychiat. Assn. (life), Am. Psychol. Assn., Am. Psychopath. Assn., Soc. for Clin. and Exptl. Hypnosis, Acad. Psychosomatic Medicine, Pan Am. Med. Assn. (past pres. sect. on clin. hypnosis), La Societe Internat. de Sophrologie et Medicine Psychosomatique (dir. 1963—); mem. Internat. Congress for Hypnosis and Psychosomatic Medicine (mem. internat. hon. com. 1964-65); hon. mem. Brit. Soc. Med. Hypnosis, Nihon Saimin Igaku-Shinri Gakkai (Japan), Sociedad Argentina de Hipnosis Medica e Hipnoanalisis, Sociedad Argentina de Sofrologia, Sociedad de Hipnologia Medica de Cordoba (Argentina), Sociedad Espanola de Sofrologia Medica (Spain), Sociedad de Sofrologia y Medecina Psicosomatica de Nueva Esparta (Venezuela). Author: (with Linn F. Cooper) Time Distortion in Hypnosis, 1954; (with Seymour Hershman, Irving Secter) Practical Application of Medical and Dental Hypnosis, 1961; Advanced Techniques of Hypnosis and Therapy (collected papers edited by J. Haley), 1968; also numerous articles. Editor, Am. Jour. Clin. Hypnosis, 1958-68; asso. editor Diseases of the Nervous System, 1940-55. Research on prins. normal behaviour, therapy of neurotic, psychotic patient, use hypnosis in medicine. Address: 32 W. Cypress St., Phoenix 85003.*

ERICSON, David Barnard, Am. geologist; b. N.Y.C., July 19, 1904; s. David Axel and Susan (Barnard) E.; B.S., Mass. Inst. Tech., 1931; M.S., Cal. Inst. Tech., 1933. Sr. research scientist Lamont Geol. Obs. Columbia, N.Y.C., 1947—. Mem. A.A.A.S., Am. Geophys. Union, Brit. Glaciological Soc., Internat. Assn. For Quaternary Research (past hon. v.p.), Deutsche Quartarvereinigung, Geologische Vereinigung, Geochemical Soc., Sigma Xi. Author: (with Goesta Wollin) The Deep and the Past, 1964; (with G. Wollin) The Ever-Changing Sea, 1967. Research, numerous publs. on sediment cores resulting in first complete record of Pleistocene Epoch during which man evolved; studies of ocean floors and temperatures. Home: 625 S. Broadway, Nyack, N.Y. 10960. Office: Lamont Geol. Obs., Palisades, N.Y. 10964.*

ERICSSON, John, engr., inventor; b. Province of Vermland, Sweden, July 31, 1803; s. Olof ond Brita (Yngstrom) E.; m. Amelia Byam, Oct. 15, 1836. Came to U. S., 1839; built a caloric engine, 1833; developer transmission of power by compressed air, use of centrifugal blowers for boiler forced draft, new types of steam boilers; went to Eng., devised steam locomotive for Ranhill competition, the Novelty, 1826; placed warship engines below water line for protection; constructed r.r. steam locomotive, 1829; introduced screw propellers for boats, 1840, designed much of U.S.S. Princeton (1st screw-propelled vessel of war), 1844; designed and built the Monitor (with a friction recoil mechanism for its guns) for U. S. Navy, 1861; designer, builder 13 inch wrought iron gun for U. S. Govt., 1863; constructed gunboats for Spain. Invented engraving machine, 1820, steam pump, 1828, rotating gun turret, 1855, instruments for measuring sun's rays and reflection capacity of polished metal, 1870-81; made contributions in ordnance, caloric engines, marine engines; made 1st propellor directly connected to engine by drive shaft, 1843; undertook exptl. work in solar physics. Died N.Y.C., Mar. 8, 1889.

ERICSSON, Leif (also spelled Ericson, Erikson), explorer; flourished early 11th Centruy; son of Eric Thorvaldsson (Eric the Red). Made trip from his home in Greenland to Norway, circa 999; on return home, attempted to land at So. tip of Greenland without 1st stopping at Iceland (which was usual practice), probably landed at what is now Labrador, or Newfoundland; named area Vinland because of grapes and grain which grew wild; his description of land makes it possible to believe that he may have landed farther South, perhaps New Eng., although exact area will probably never be determined; some sagas and reports give his wish to spread Christianity as reason for making trip to N.Am. mainland; credited with being 1st European discoverer of N.Am.

ERIC THE RED (Eric Thorvaldsson), explorer; b. Norway; flourished late 10th Century; son of Thorvald; at least 1 son, Leif Ericsson. Nicknamed Eric the Red in youth; accompanied his father into exile, settled (with other families) in Western Iceland; later exiled for outlawry from both Norway and Iceland for 3 year period; decided to spend his exile in exploring area which is now Greenland (known to Icelandic settlers as it could be seen on clear days from extreme N.W. tip of Iceland); sailed 175 miles to Iceland, 981, lived mainly on coast, 982-85, named area Greenland because he thought it better suited to stock raising than Iceland; returned to Iceland to get settlers for Greenland, 986; made 1st settlement, called Brattahild (nr. present town of Julianehaab) on So. coast of Greenland, 986; hero of Icelandic sage Eric the Red.

ERIGENA, Johannes Scotus, philosopher; b. circa 810; invited to take charge of ct. sch. at Paris by Charles the Bald, circa 847; Author: De divisione naturae; De divina praedestinatione. Translated Pseudo-Dionysius and others. His philosophy is a transition from Platonic philosophy to scholasticism; divides nature into 4 categories which begin and end with God. Died circa 877.

ERIKSEN, Charles Walter, Am. psychologist; b. Omaha, Feb. 4, 1923; s. Charles Hans and Louella (Carlson) E.; B.A., U. Omaha, 1943; Ph.D., Stanford, 1950; m. Garnita Tharp, July 22, 1945; children—Michael John, Kathy Ann. Asst. prof. Johns Hopkins, 1949-53, research scientist, 1954-55; lectr. Harvard, 1953-54; faculty U. Ill., Urbana, 1956—, prof., 1959—. Research cons. VA, 1960—; mem. psychobiology panel NSF, 1963; mem. expt. psychology study sect. NIH, 1958-62, 66—. Recipient Stratton award Am. Path. Assn., 1964, Nat. Inst. Mental Health Research Career award, 1964. Mem. Psychonomic Soc., Soc. Exptl. Psychologists, Am., Midwestern psychol. assns., A.A.A.S. Sigma Xi. Author: Behavior and Awareness, 1962; also numerous articles. Asso. editor Jour. Exptl. Research in Personality, 1964—; cons. editor Jour. Exptl. Psychology, 1965—. Research in psychol. def. mechanisms, nature unconscious processes; attentional phenomena in humans; helped distinguish and isolate response processes and effects from gen. perceptual functioning. Home: 1002 W. White St., Champaign, Ill. 61820. Office: 382 Davenport Hall St., Urbana, Ill. 61803.*

ERIKSON, Henry Anton, Am. physicist; b. Mt. Morris, Wis., July 30, 1869; s. Hemming and Elizabeth (Tommeraas) E.; B.E.E., U. Minn., 1896, Ph.D., 1908; student U. Chgo., 1899, Cambridge U., 1908-09; m. Winifred Boynton, of New Lisbon, Wis., June 21, 1899; children—Hemming (dec.), Elizabeth W., Henry B. Instr. physics U. Minn., 1897-1906, asst. prof., 1906-14, asso. prof., 1914-15, prof. 1915-38, also chmn. dept., prof. emeritus, 1938-57. Fellow A.A.-A.S.; mem. Am. Phys. Soc., Sigma Xi, Tau Beta Pi. Author: Elements of Mechanics; Manual of Physical Measurements, 1902. Contbr. on ionization in Phys. Rev., Philos. Mag., others. Died June 22, 1957.

ERIKSSON, Hans, Swedish clin. bacteriologist; b. Stockholm, Sweden, Nov. 11, 1914; s. Ernst A. and Ruth (Schnell) E.; B.Med., Karolinska Sch. Medicine, Stockholm, 1936, M.D., 1941; m. Astri Birgitta Rydin, Oct. 4, 1941; children—Birgitta, Magnus, Margaretha. Asst. bacteriologist State Bacteriologic Lab., 1940-43, 46-48; asst. head bacteriology dept. Karolinska Sjukhuset, Stockholm, 1949-50, head, 1950—; asso. prof. reader bacteriology Karolinska Inst. Sch. Medicine, 1960—. Mem. adv. bd. on antibiotics WHO, 1960—. Decorated Finnish Freedom Cross with red Cross, Order No. Star. Mem. Swedish Med. Soc. (pres. sect. for med. microbiology 1967—). Author: Rational Use of Antibiotics, 1960; also numerous articles. Evolved methods for prodn. vaccine against diphtheria, tetanus; research on hygiene and cross infection in hosps. especially abuse of antibiotics; developed method for determination bacterial sensitivity to antibiotics; introduction of computer technics in clin. bacteriology. Home: 55 Riddaregatan. Stockholm Ö. Office: Dept. Bacteriology, Karolinska Sjukhuset, Stockholm 60, Sweden.*

ERIKSSON, Jakob, Swedish botanist; b. Limbamm, Sweden, Sept. 30, 1848; prof. botany Royal Acad. Stockholm (Sweden); dir. Central Bot. Sta., Sweden. Mem. French Acad. Scis., 1930. Author: Über das vegetative Leben der Getreiderostpilze, 1905; Om Meristemet i Dikotyla växters Rotter (Om the Meristem in Dicotyledons), 1877. Research on systematics and morphology of cereals, parasites of mass-culture plants, especially rusts, meristem in dicotyledons, vegetative life of cereal rusts. Died Apr. 26, 1931.

ERIKSSON, Karl-Erik, physicist; b. S. Rada, Sweden, Nov. 19, 1935; s. Johan and Agnes (Isaksson) E.; m. Kristina Myrberg, Sept. 12, 1958; children—Gunnar, Magdalena, Elin. Research fellow Cern, Geneva, Switzerland, 1959-62; lectr. Inst. Theoretical Physics, U. Uppsala (Sweden), 1962-63; prof. theoretical physics U. Gothenburg (Sweden), 1963—. Research, publs. on calculations of radiative corrections to high energy elementary particle scattering, gauge theories of elementary particle interactions. Home: Ekmansgatan 14, Kungsbacka, Sweden. Office: Inst. Theoretical Physics, Sven Hultins Gata, Goteborg S, Sweden.*

ERINGEN, Ahmed Cemal, Am. aerospace scientist; b. Kayseri, Turkey, Feb. 15, 1921; s. Subru and Meva Eringen; M.S. in Mech. and Aero. Engring., Tech. U., Istanbul, 1943; Ph.D. in Applied Mechanics, Poly. Inst. Bklyn., 1948; m. Jean Dennis, Sept. 8, 1949; children—Meva, Peri, Lisa, Leyla. Head structures dept. aircraft factory Turkish Air League, Ankara, 1946; asst. prof., asso. prof. mechanics Ill. Inst. Tech., Chgo., 1948-53; asso. prof., prof. engring. sci. Purdue U., Lafayette, Ind., 1953-66; prof. aerospace and mech. scis., chmn. solid mechanics program, dir. solid mechanics labs. Princeton (N.J.), 1966—. Cons., Gen. Tech. Corp., Lawrenceville, N.J., 1958—. Recipient certificate of achievement Poly. Inst. Bklyn., 1957, award as outstanding researcher of year Sigma Xi, 1963. Mem. Soc. Engring. Sci. (founder, pres.), Soc. Natural philosophy, Am. Phys. Soc. Author: Nonlinear Theory of Continuous Media, 1962; Mechanics of Continua, 1967. Editor: Proc. of Engring. Sci., 4 vols. Research and numerous publs. on deformable bodies, interaction of electromagnetic waves with deformable media, mech. properties of solid and fluid composites and mixtures, mechanics, thermodynamics and electromagnetic properties of deformable media with microstructure, visco-elastic materials, wave propagations. Home: 129 Braodmead St., Princeton, N.J. 08540.*

ERJAVETS, Fran, naturalist; b. Laibach, Slavonia, 1834; prof. natural scis., Zagreb, Yugoslavia, then Goritz; numerous publs. on natural history; creator sci. terminology of Slovene lang. Died 1887.

ERK, Frank Chris, Am. biologist; b. Evansville, Ind., Dec. 17, 1924; s. Carl Benjamin and Matilda (Schumacher) E.; A.B. magna cum laude, U. Evansville, 1948; Ph.D., Johns Hopkins, 1952; m. Ruth Parker Hobgood, June 12, 1948; children—Susan Patricia,

Elisabeth Carlene, Stephanie Diane. Asso. prof. biology, chmn. dept. Wash. Coll., 1952-57; vis. asso. prof. U. Chgo., 1954-55; prof. biology, chmn. dept. State U. N.Y., Oyster Bay, 1957-61; prof. biol. scis., chmn. dept. State U. N.Y., Stony Brook, 1962—. Vis. investigator Agrl. Research Council's Poultry Research Centre, Edinburgh, Scotland, 1964-65, Istituto di Genetica, Milan, Italy, 1965; cons., writer Biol. Sci. Curriculum Study, 1960—. Mem. Genetics Soc. Am., Genetics Soc. Can., Am. Genetics Assn., Soc. for Study Evolution, Am. Assn. U. Profs. Author: (with C. A. Welch, others) Biological Science: Molecules to Man, 1963; (with N. Abraham, others) Biological Science: Interaction of Experiments and Ideas, 1965. Research, numerous publs. on cytogenetic, evolutionary, nutritional, developmental problems in fruit fly, Drosophila. Home: 33 Yorktown Rd., Setauket, N.Y. 11785. Office: Dept. Biol. Scis. State U. N.Y., Stony Brook, N.Y. 11790.*

ERK, Fritz, German meteorologist; b. Straubing, Germany, Oct. 17, 1857; s. Georg and Magdalene (Jungkunz) E.; student math. and physics, Munich, Germany; doctorate, circa 1883; m. Luise Engelhardt, 1890; children—Sigmund, 1 dau. Became asst. Bavarian Central Meteorol. Sta., 1885; became lectr. U. Munich (Germany), 1887, hon. prof., 1901; named dir. Bavarian Central Meteorol. Sta., 1893. Author: Zwei Studien über die Anwendung einer räumlichgeometrischen Darstellung auf Probleme der Meteorologie, 1887; Das Klima von Oberbayern, 1898. Introduced isopleth method; research on upper atmosphere; founder meteorol. stas. on Wendelstein, Hirschberg, and Zugspitze; made numerous balloon trips. Died Munich, Aug. 31, 1909.

ERKAMA, Jorma, Finnish biochemist; b. Mikkeli, Finland, Jan. 11, 1912; s. Karl August and Lovisa (Paavolainen) E.; M.Sc., U. Helsinki (Finland), 1934, Ph.D., 1947; dipl.engr., Finland Inst. Tech., 1938; m. Tellervo Jortikka, Dec. 28, 1944; children—Juha, Timo, Mikko. Pvt. asst. to Prof. G. Komppa, 1935-36; asst. to Prof. A. I. Virtanen, 1938-40; asst. Biochem. Inst., Helsinki, 1940-48; faculty U. Helsinki, 1948—, prof. biochemistry, 1950—, dir. dept. biochemistry, 1957—. Recipient SLK, 1968. Mem. Suomalaisten Kemistien Seura, Am. Chem. Soc., Scandinavian Soc. for Plant Physiology, A.A.A.S. Research, publs. on trace elements in plant nutrition; aerobic fermentations, nitrogen fixation, nutrition of higher plants. Home: 38, Äyräpääntie, Laajalahti, Finland. Office: 35, Unioninkatu, Helsinki 17, Finland.*

ERLANGER, Bernard Ferdinand, biochemist, educator; b. N.Y.C., July 13, 1923; s. Leo and Frieda (David) E.; B.S., Coll. City N.Y., 1943; M.A., N.Y. U., 1949; Ph.D., Columbia, 1951; m. Rachel Fenichel, June 23, 1946; children—Laura, Louis, Leon. Chemist, U.S. Indsl. Chems. Co., Inc., Newark, 1943-44; tech. adviser Manhattan Project, U. S. Army, Los Alamos, 1944-46; prodn. mgr. Hexagon Labs., Inc., N.Y.C., 1946-48; faculty Columbia, 1951—, prof. microbiology, 1966—. Vis. scientist Instituto Superiore di Sanita, Rome, Italy, 1961-62; cons. to industry; mem. Fulbright-Hays Award Com., 1966—; Fulbright scholar U. Republic, Uruguay, 1967. Mem. Am. Chem. Soc., Am. Soc. Biol. Chemists, Biochem. Soc., Harvey Soc. Studies on mode of action of antibiotics; investigation of mechanisms of enzyme catalysis, immunochemistry of macromolecules concerned with genetics. Office: 630 W. 168th St., N.Y.C. 10032.*

ERLANGER, Joseph, Am. physiologist; b. San Francisco, Jan. 5, 1874; s. Herman and Sarah (Galinger) E.; B.S., U. Cal., 1895, LL.D., 1932; M.D., Johns Hopkins, 1899, LL.D., 1947; Sc.D., U. Wis., 1936, U. Pa., 1936, Washington U., St. Louis, 1946; D. (hon) Free, U. Brussels, 1949; m. Aimée Hirstel, June 21, 1906 (dec. May 22, 1959); children—Margaret Ruth (Mrs. R. H. Swinney), Herman (dec. 1959). Asst. prof. dept. physiology Johns Hopkins, 1900-06; prof., head dept. physiology U. Wis., 1906-10; prof., head dept. physiology Washington U., St. Louis, 1910-48, prof. emeritus, 1948—; Hitchcock lectr. U. Cal., 1930; Johnson lectr. U. Pa., 1936. Recipient Nobel Prize in physiology and medicine (with Herbert S. Gasser), 1944. Mem. A.A.A.S., Am. Physiol. Soc., Am. Philos. Soc., Nat. Acad. Scis., Deutsche Academie Naturforscher, Société Philomathique (Paris), Soc. Exptl. Biology and Medicine, Sigma Xi, Alpha Omega Alpha. Author: (with others) Symposium on the Synapse, 1936; (with Herbert S. Gasser) Electrical Signs of Nervous Activity, 1937; also numerous articles. Research on action potential nerves, heart block, blood pressure; (with Herbert Gasser) devised means for use of cathode ray oscillograph for studying transmission of impulses through single nerve fibers; contbns. to circulatory physiology with devel. of device to record blood pressure; studied influence in field of electrophysiology. Died Dec. 5, 1965.

ERLENMEYER, (Johann Adolph) Albrecht, German psychiatrist; b. Wiesbaden, Germany, July 11, 1822; s. Friedrich Albrecht and Marie Elisabeth (Hanckeroth) E.; student Marburg, Germany, Bonn, Germany; doctorate, Berlin, Germany, circa 1844; m. Emma Tilemann, 1848; 2 sons, including Friedrich Albrecht. Asst. to K. W. M. Jacobi, Siegburg, Germany, 1844-46; visited all larger insane asylum; practiced medicine, Bendorf, Germany; founder eye clinic, asylum for brain and nervous disorders, 1848, added neurol. sect., 1866, agrl. sect., 1867. Mem. Deutsche Gesellschaft für Psychiatrie und Gerichtliche Psychologie (co-founder, editor). Author: Die Gehirnatrophie der Erwachsenen, 1852; Wie sind Seelenstörungen in ihrem Beginn zu behandeln, 1860; Die subcutanen Injektion der Arzneimittel, 1866; Die Embolie der Hirnaterien, 1867; Die luetischen Psychosen, 1876; Die Schrift, 1879; Über stat. Reflexkrämpfe, 1885; Die Prinzipien der Epilepsiebehandlung, 1886; Die Morphiumsucht, 1887; Schriften über Irrenwesen, 1896; Über Entmündigung, 1899; also numerous articles. Research on insanity statistics, psychiatric clinic, care of idiots, and cretins; advocated somatic sch., and close connection of psychiatry with neurology. Died Bendorf, Aug. 9, 1877.

ERLENMEYER, Emil (Richard August Carl), chemist; B. Wehen, Germany, June 28, 1825; s. Friedrich Albrecht and Marie Elisabeth (Hanckeroth) E.; student natural scis., Giessen, Germany, Heidelberg, Germany; grad. 1851; Dr.h.c., Munich, Heidelberg, 1905; m. Auguste Henstenberg, 1850; at least 1 son, 1 dau. Began as pharmacist, Heidelberg, became lectr., 1857, asso. prof., 1863; prof. gen. chemistry Polytechnikum, Munich, Germany, 1868-83. Mem. Bavarian Acad. Sci. Author: (with others) Lehrbuch der organischen Chemie, 3 vols., 1867; Aufgabe des chemischen Unterrichts, 1872. Studied atom, molecule, saturation, constitution of aliphatic compounds; inventor Erlenmeyer flask; introduced terms hydroxyl and structural chemistry; originated law on splitting of secondary and tertiary alcohols; synthesized many compounds and discovered their structural formulas. Died Aschaffenburg, Germany, Jan. 22, 1909.

ERLENMEYER, Friedrich Gustav Carl Emil, German chemist; b. Heidelberg, Germany, July 14, 1864; s. Richard August Carl Emil Erlenmeyer. Became asso. prof., Strasbourg, France, 1896; later mem. staff Kaiser Wilhelm Inst., Dahlem, Germany. Developed theory of partial valencies; studied molecular rearrangements, isohydrobenzoin; synthesized tyrosche, cystine, phenylalanine. Died Heidelberg, Feb. 7, 1921.

ERMAN, Georg Adolf, German physicist; b. Berlin, May 12, 1806; s. Paul and Karoline (Itzig) E.; student natural scis., Berlin, Königsberg, Germany; m. Marie Bessel; at least 1 son, Jean Pierre Adolphe. Mem. expdn. to Siberia, 1828-30; travelled around world making observations on terrestrial magnetism (later used by Gauss); became prof. physics U. Berlin, 1839. Author: Reise um die Erde durch Nordasien und die beiden Oceane, 5 vols., 1833-42; Die Grundlagen der Gaussichen Theorie und der Erscheinungen des Erdmagnetismus, 1874. Determined (with his father) constantly increasing temperature of earth's interior. Died Berlin, July 12, 1877.

ERMAN, Jean Pierre Adolphe (known as Adolf), German Egyptologist; b. Berlin, Oct. 31, 1854; s. Adolphe and Marie (Bessel) E.; ed. Leipzig, Germany; doctorate, Berlin; m. Käte d'Heureuse, 1884; 2 sons, 4 daus. Sci. asst. in library and coin collection Royal Museums, Berlin; became dir. Egyptian dept., Berlin Royal Mus., 1885; named asso. prof. Egyptology, U. Berlin, 1883; prof., 1892. Mem. German and various fgn. acads. Author: Neuägyptische Grammatik, 1880; Ägyptische Grammatik, 1894; Altägyptische Grammatik, 1894; Ägypten und ägyptisches Leben im Altertum, 1885; Deutsche Medailleuve der 16. und 17. Jahrhunderten, 1884; Der Tontafelfund von Tell Amarna, 1888; Die Sprache des Papyrus Westcar, 1889; Gespräch eines Lebensmüden mit seiner Seele, 1896; Bruchstücke koptischer Volksliteratur, 1897; Aus den Papyri der Königlichen Museen, Ein Denkmal memphitischen Theologie, 1899; Zaubersprüche fur Mutter und Kind, 1901; Ägyptische Religion, 1905; Hymnen an den Diadem der Pharaonen, 1911; Die Hieroglyphen, 1912; Die Literatur der Ägypter. 1923; Die ägyptischen Schülerhandschriften, 1925; Mein Werden und Wirken, 1926; Die Welt am Nil, 1936; (with H. Grapow) Wörterbuch der ägyptischen Sprache, 1926-31. Leader of Berlin sch. which laid found. for a generation of Egyptologists; Introduced systematic study of grammatical constrn. of Egyptian lang.; put study of Egyptian lang. on firm scientific basis; author 1st grammar studies of ancient Egyptian. Died Berlin, June 26, 1937.

ERMAN, Paul, German physicist; b. Berlin, Feb. 27, 1764; s. Jean Pierre and Louise (Le Coq) E.; student theology French Séminaire de théologie, Berlin; Ph.D. (hon.), U. Berlin, 1811; m. Karoline Itzig, 1802; 1 son, Adolph, 4 daus. Became tchr., French Gymansium, Berlin, 1781, prof. philosophy, 1791-1820; tchr. Académie des nobles (became Mil. Acad. 1810), 1792-1846; prof. physics U. Berlin, 1810-46. Recipient Galvani prize Paris Institut national, 1806. Fellow Royal Soc., 1827; mem. Berlin Acad. Scis. Research on exptl. physics, conduction of electricity over distances, elec. state of atmosphere; discovered voltage drop in outer circuit of Volta column (preparatory to discovery Ohm's law), principle of conduction mechanism as simultaneous occurance at both poles; proved relationship between friction, electricity and galvanic current; 1st to observe one-sidedness of charged particles; measured (with son) temperature of earth's interior. Died Berlin, Oct. 11, 1851.

ERMENGEM, Émile Pierre Marie van, see van Ermengem, Émile Pierre Marie.

ERMOLENKO, Nikolai Fedorovich, Russian chemist; b. Jan. 29, 1900; grad. Moscow 2d U., 1924. Mem. staff Byelorussian U., Minsk, 1924-31, prof., 1931—; in charge lab. for colloidal chemistry Chemistry Inst., Byelorussian Acad. Scis. Mem. Byelorussian Acad. Scis. Research and numerous publs. on colloidal and inorganic chemistry.

ERMOLIEVA, Zinaida Vissarionovna, Russian microbiologist, bacterio-chemist; b. USSR, Oct. 27, 1898; grad. Rostov/Don, 1921. Dir., lectr., bacteriology dept. Inst. Bacteriology, Rostov U., 1921-25; dir. microbial biochemistry dept. All-Union Inst. Exptl. Medicine, Moscow, USSR, 1925-44, Inst. for Biol. Prevention of Infections, Moscow, 1945-47, dept. exptl. chemotherapy Antibiotics Research Inst., Moscow, 1947—, dept. microbiology Central Postgrad. Med. Inst., Moscow, 1952—. Recipient Order of Lenin, Stalin prize (both 1943). Mem. USSR Acad. Med. Scis. (corr.), Soviet Acad. Scis. (pres. com. antibiotics 1954—). Author: Cholera, 1943; Penicillin, 1946; Antibiotics and their Applications, 1954; Streptomycin Therapy, 1958; Antibiotics, Experimental and Clinical Investigations, vol. I, 1956, vol. II, 1959. Editor: Antibiotiki, 1956—. Investigations into cholera and various antimicrobic agts.; introduced natural immunity factor for use in eye, ear, throat and nose diseases; introduced various new antibiotics of animal origin, forms of penicillin and streptomycin. Office: Dept. Microbiology, Central Postgraduate Medical Inst., 1/2 Vosstanya Sq., Moscow, USSR.

ERNEMANN, (Carl Heinrich) Alexander, German inventor; b. Dresden, Germany, June 3, 1878; s. Heinrich and Therese (Grafe) E.; ed. as engr.; student cinematography, U. S.; Ph.D. (hon.), U. Kiel (Germany); m. Elisabeth Pachtmann, 1910; 1 son, 2 daus. Introduced cinematography to firm Ernemann-Werke, became tech. dir., 1910; became mem. bd. Zeiss ikon AG, Dresden, Germany; after confiscation by Russians, moved to Stuttgart, Germany, 1946; chmn. bd. for reconstrn. Zeiss Ikon AG in W., founder brs. in Kiel, Germany. Named hon. councillor Tech. U. Stuttgart; recipient Messter medal. Mem. Soc. Motion Picture and Television Engrs. (hon.). Built 1st German steel motion picture projector, 1909; manufactured small camera with light intensity of 1:2, 1924, also numerous movie projectors and sound-film installations including spl. accessories such as arc lamps, amplifiers. Died Stuttgart, Oct. 14, 1956.

ERNö, Tyihak, Hungarian chem. engr.; b. Tiszafoldvar, Hungary, Jan. 29, 1933; s. Tyihak and Kerekes (Lidia) E.; dipl. chem. engring., Tech. U. Budapest; m. Gombar Julianna, Aug. 12, 1961. With Sci. Research Inst. for Medicinal Plants, Budapest, 1958—, sci. research worker, 1961—. Mem. Hungarian Chem. Soc. (presidium biochem. sect.), Hungarian Biol. Soc. (presidium bot. sect.). Author: (with others) Thin Layer Chromatography, 1965; New Results of Thin Layer Chromatography, 1966; also articles. Research on isolation of antitumor substances of plant origin, action mechanism of these materials, optically active substances, essential oils. Home: 63 Erzsébet királyné St. Office: 38-42 Daniel St., Budapest, Hungary.*

ERNST, Eugene, Hungarian biophysicist; b. Baja, Hungary, Apr. 16, 1895; s. Lewis and Pauline (Kaufer) E.; grad. U. Budapest (Hungary), 1924. Asst., U. Pécs., 1923-39, prof. Biophys. Inst., 1947—. Chmn. biophys. commn. Hungarian Acad. Sci., 1959—, dir. biophys. research team, 1960—. Decorated Silver medal of war, Order of Labour, Kossuth prize. Mem. Internat. Union Pure and Applied Biophysics (mem. council 1964—), Hungarian Biophys. Soc. (chmn. 1960—), Hungarian Acad. Scis. Author: Muskeltätigkeit, 1958; Biophysics of the Striated Muscle, 1963; Introduction to Biophysics, 1967; also numerous articles. Research on biophysics of striated muscle including vol. constriction and action current denoting excitation, cyrstallization of myosin, bound Kalium and bound water, stretching causing hypertrophy, mechanochem. coupling; biol. concentration work using thermosmosis and thermodiffusion. Address: 80 Rákóczi, Pécs, Hungary.*

ERNST, Harold Clarence, Am. bacteriologist; b. Cin., July 31, 1856; s. Andrew Henry and Sarah H. (Otis) E.; A.B., Harvard, 1876, M.D., 1880, A.M., 1884; studied under Koch; m. Ellen Lunt Frothingham, Sept. 20, 1883. Demonstrator bacteriology, Harvard, 1885-89, instr., 1889-91, asst. prof., 1891-95, prof., 1895-1922; chmn. vis. staff Children's Hosp. Fellow Am. Acad. Arts and Scis. (councilor), Assn. Am. Pathologists and Bacteriologists (sec. 1901-08, 09-22, pres. 1908-09. Author: Infectiousness of Milk, 1896; Infection and Immunity, 1898; Animal Experimentation, 1902; Modern Theories of Bacterial Immunity, 1902. Editor: Jour. Med. Research, 1896-1922. Established 1st diphtheria-antitoxin lab. in Boston. Died Sept. 7, 1922.

ERNST, Otto, German indsl. chemist; b. Wiesbaden, Germany, Oct. 7, 1870; s. Karl and Emma (Eyring) E.; studied chemistry under O. N. Witt, Berlin, Germany, P. Sabatier, Paris, France; Ph.D. Became chemist Meister Lucius und Brüning, Höchst,

Germany, 1894, later dir. catalytic lab.; also artist. Introduced indanthren dyes and self-thickening agt. on a cellulose base for painting; invented cotton azo dyes (basis of Dianil dyes and black wool dye); devel. acetylene chemistry, including design indsl. prodn. of acetylene and its by products, acetal aldehyde, acetic acid, acetone. Died Wiesbaden, Oct. 5, 1936.

ERNST, Theodor Karl Heinrich, German mineralogist; b. Uelzen, Hanover, Jan. 3, 1904; s. Theodor and Luise (Brandes) E.; ed. U. Jena, U. Göttingen; Ph.D. in mineralogy, Ph.D. agrégé in mineralogy; m. Greta Demuth; children—Brigitte, Jürgen, Barbara, Jütta. Asst., U. Göttingen, 1931, prof. agrégé of mineralogy, 1937; sci. cons., Strasbourg, 1942; prof. U. Erlangen, 1943, mandate prof., 1949, titular prof., 1950, dir. Mineral. Inst., 1950, dean, 1954-56; titular prof. Higher Tech. Inst., Munich, 1950. Fellow Mineral. Soc. Am.; mem. Bavarian Acad. Scis., German Soc. Mineralogy, Confedn. of Geology, Geochem. Soc. Research and publs. on crystalline structure, origin of basalt, clay, sediments. Address: Schlossgarden 5a, 852 Erlangen, Germany.

ERSHOFF, Benjamin H., Am. nutritionist; b. Phila., Jan. 31, 1914; s. Harry and Sophia (Belkin) E.; A.B., U. Cal. at Los Angeles, 1937; Ph.D., U. Cal. at Berkeley, 1942; m. Vivian D. Hershberg, June 22, 1941; children—Bonnie Ruth, Harry Daniel. Dir. research Emory W. Thurston Labs., Los Angeles, 1942-53; dir. Western Biol. Lab., Culver City, Cal., 1954-62, Inst. for Biol. Research, Culver City, 1962-64; sci. dir. Inst. for Arteriosclerosis Research, Culver City, 1965—; vis. prof. dept. biochemistry and nutrition, U. So. Cal., 1956-59, adj. prof., 1959-65; research prof. biochemistry Loma Linda U., adj. prof. U. So. Cal. Sch. Dentistry, 1966—. Mem. Am. Inst. Nutrition, Soc. for Exptl. Biology and Medicine, Gerontological Soc., Am. Chem. Soc., Internat. Assn. for Dental Research. Research, numerous publs. on effects unidentified nutritional factors on resistance to stress, effects prolonged exposure to stressor agts. on nutritional requirements, effects of diet on susceptibility to arteriosclerosis and artherosclerosis. Home: 858 Woodacres Rd., Santa Monica, Cal. 90402. Office: 9331 Venic Blvd., Culver City, Cal. 90231.*

ERSPAMER, Vittorio, Italian pharmacologist; b. Malosco, Trent, Italy, July 30, 1909; s. Francesco and Augusta (Gius) E.; Medicine D., U. Pavia, 1935; postgrad. (fellow in pharmacology), U. Berlin (Germany), also U. Bonn (Germany), 1958; m. Falconieri Guiliana, Apr. 15, 1953; children—Francesco, Maria Luisa. Asst., Inst. Comparative Anatomy and Physiology, U. Pavia (Italy), 1935-37, Inst. Pharmacology, U. Rome (Italy), 1938-47; prof. pharmacology U. Bari (Italy), 1947-55, U. Parma (Italy), 1955-67, U. Rome, 1967—. Recipient Feltrinelli prize Accademia dei Lincei, 1955; Golden Merit medal Ministry of Edn., Rome, 1963. Subeditor: Handbook of Experimental Pharmacology, vol. XIX, 1966. Research, numerous publs. on 5-hydroxytryptamine; first description and study of murexine, octopamine, leptodactyline, and several polypeptides, active on vascular and extravascular smooth muscle and on external secretions. Office: Istituto di Farmacologia, Città Universitaria, 00100 Rome, Italy.*

ERUGIN, Nikolai Pavlovich, Russian mathematician; b. May 14, 1907; grad. Leningrad U., 1932; became mem. staff Leningrad U., 1934, prof., since 1943; with Leningrad branch Math. Inst. USSR Acad. Scis., 1939-41, 1951-57; with Byelorussian SSR Acad. Scis., since 1956, then Physics and Math. Inst., 1957; prof. Byelorussian U. Research, publs. in field of differential equations.

ERWALL, Lars Gustaf, Swedish chemist; b. Lund, Sweden, Feb. 21, 1926; s. Gustav and Greta (Mauritzsson) E.; M.Sc.Eng., Royal Inst. Tech., Stockholm, Sweden, 1948, D.Techn., 1958; m. Kerstin Ahlberg, June 23, 1948; children—Barbro, Ann, John, Birgitta. With Royal Inst. Tech., Stockholm, 1948-59, asst. prof. phys. chemistry, 1958-59; head Isotope Techniques Lab., Stockholm, 1960-66, pres. subsidiary AB Isotopteknik, 1966-67, pres. incentive research and devel., 1967—, dir., 1966—. Mem. Am. Nuclear Soc. (mem. exec. com. isotopes and radiation div. 1966—), Swedish Assn. Engrs., Assn. Mems. Parliament and Researchers. Author: (with Hans G. Forsberg, Knut Ljunggren) Industrial Isotope Techniques, 1962; also articles. Introduced radioisotope techniques in Swedish industry; research on application of radioactive tracers to indsl. processes, water pollution by indsl. and municipal wastes. Home: 46 Eskadervägen, Näsbypark, Sweden. Office: 4 Arsenalsgatan, Stockholm C, Sweden.*

ERWIN, Albert R., Am. physicist; b. Charlotte, N.C., May 1, 1931; s. A. R. and Lois (Lee) E.; B.S., Duke U., 1953; Ph.D., Harvard U., 1958; m. Mary Jane Murray, June 12, 1954; dau., Christy Lee. Instr. physics, 1958-59, asst. prof., 1959-62, assoc. prof., 1962-65, prof., U. Wisconsin, 1965—. Research in area of elementary particles of high energy physics; with W. D. Walker, found 1st strong evidence for rho meson in bubble chamber experiment. Home: 5437 Esther Beach Rd., Madison, Wisc. 53713. Office: 3414 Sterling Hall, U. of Wisconsin, Madison 53706.*

ERXLEBEN, Dorothea Christiane, German physician; b. Quedlinburg, Germany, Nov. 13, 1715. M.D. (1st woman to receive this degree), 1754; m. Diakonus Erxleben; 4 children; practiced medicine, Quedlinburg. Died June 13, 1762.

ERXLEBEN, Johann Christien Polycarp, German veterinarian; b. Quedlinburg, Germany, 1744; prof. physics, Göttingen, Germany. Author: Systema regni animalis, 1777. Prominant veterinarian of his time; writings on vet. sci. were widely used. Died Göttingen, 1777.

ESAKI, Leo, physicist; b. Osaka, Japan, Mar. 12, 1925; s. Sohichiro and Niyoko (Itoh) E.; B.S. in Physics, U. Tokyo, 1947, Ph.D., 1959; m. Masako Araki, Nov. 21, 1959; children—Nina, Anna, Eugene. Research mem. Kobe (Japan) Kogyo Corp., 1947-56; chief physicist Sony Corp., Tokyo, Japan, 1956-60; IBM fellow at the T. J. Watson Research Center, IBM, Yorktown Heights, N.Y., 1960—. Recipient Nishina Found. Meml. award, Tokyo, 1959; Asahi Press award, Tokyo, 1960; Toyo Rayon Found. award, Tokyo, 1961; Stuart Ballentain medal Franklin Inst., 1961; Japan Acad. award, 1965. Fellow Am. Phys. Soc., I.E.E.E. (Morris Liebman Meml prize 1961); mem. Phys. Soc. Japan, Inst. Elect. Communication Engrs. Japan. Research in solid state physics and semiconductor electronics, tunneling. Home: 16 Shady Lane, Chappaqua, N.Y. 10514. Office: P.O. Box 218 Yorktown Heights, N.Y. 10598.*

ESAU, Katherine, botanist; b. Ekaterinoslav, Russia, Apr. 3, 1898; d. John J. and Margarethe (Toews) Esau; student Coll. Agr., Moscow, 1916-17, Coll. Agr. Berlin, 1919-22; Ph.D., U. Cal., 1931; D.Sc., Mills College, Oakland, Cal., 1962. Came to U. S., 1923, naturalized 1928. Plant breeder Spreckels Sugar Co., Spreckels, Cal., 1924-28; asst. U. Cal., 1928-31, instr., 1931-37, asst. prof., 1937-43, asso. prof., 1943-49, prof. 1949—, jr. botanist agrl. expt. sta., 1931-37, asst. botanist, 1937-43, asso. botanist, 1943-49, botanist 1949—; John Simon Guggenheim Found. fellow, 1940; John M. Prather lectr. Harvard, 1960. Mem. Nat. Acad. Scis., A.A.A.S., Am. Acad. Arts and Sci., Bot. Soc. (pres. 1951), Am. Philos. Soc., Phi Beta Kappa, Sigma Xi. Author: Plant Anatomy, 1953; Anatomy of Seed Plants, 1960; Plants, Viruses and Insects, 1961; Vascular differentiation in Plants, 1965. Contbr. articles bot. jours. Authority on structure and ontogeny of phloen, food conducting tissue of plants and on plant-host virus relationships. Home: P.O. Box 11428, U. Cal., Santa Barbara, Cal. 93107.

ESBACH, Georges Hubert, French physician; b. 1843; dir. physiol. lab. Paris Faculty of Medicine. Author: Dosage de l'albumine dans l'urine, 1879; les Calculs urinaires et biliaries, 1885. Invented technique for determination of albumin in urine (Esbach tube and Esbach's reagent); made various urinary and biliary calculations. Died 1890.

ESCANDE, Léopold Charles Marie Jean-Baptiste, French physicist; b. Toulouse, France, June 1, 1902; s. Gabriel and Jenny (Desmazures) E.; ed. Inst. Electrotechnique et de Mecanique Appliquee, Toulouse; Ph.D. in Phys. Scis., Toulouse U.; D. honoris causa univs. Lison, Liege, Lima, Vienna, Berlin, Buenos Aires, Rio de Janeiro, others; m. Renée Setzes, Feb. 11, 1924. Prof. hydraulics Toulouse U.; head Ecole Nationale Supérieure d'Electrotechnique et d'Hydraulique; dir. Institut de Mécanique des Fluides, Toulouse; cons. engr. Société Nationale de l'Electricité de France, Schneider works, Enterprises Métropolitaines et Coloniales, Enterprise Industrielle, Empresa Nacional Ribagorzana, Spain; sci. adviser Délégation Générale à la Recherche scientifique. Mem. French Acad. Scis., Acad. Coimbra (Portugal) (corr.). Author: Etude Théorique et Expérimentale sur la Similitude des Fluides Incompressibles Pesants, 1929; Barrages, 1937; Etudes des Veines de Courant, 1940; Hydraulique Générale, 1941-43; Recherches Théoriques et Experimentales sur les Oscillations de l'Eau dans les Chambres d'Equilibre, 1943; Compléments d'Hydraulique, Vol. I, 1947, Vol. II, 1951; Méthodes Nouvelles pour le Calcul des Chambres d'Equilibre, 1949; Nouveaux Compléments d'Hydraulique, Vol. I, 1953, Vol. II, 1955. Authority on fluid mechanics, hydraulics. Home: 2, rue Carmichel, Toulouse, France.

ESCHENBRENNER, Allen Bernard, Am. physician; b. St. Louis, Mo., Apr. 10, 1911; M.D., Washington U., St. Louis, 1935; m. 1940; 3 children. Curator, pathology museum, Washington U., 1934-35; res. and asst. pathologist, Barnes Hospital, 1935-36; asst. 1936-38, med. surgeon, U. S. Public Health Service, 1938-42; staff mem., Nat. Cancer Institute, 1942-53; virus and rickettsia section, 1953-56, hematology and biochemistry section, 1956—, Communicable Disease Center. Mem. Soc. Experimental Biology; Tissue Culture Assn; N.Y. Acad. Scis. Research in virology; radiobiology; tissue culture; histochemistry, histopathology; laboratory research facilities. Office: Communicable Disease Center, Hematology & Biochemistry Section, Box 185, Chamblee, Ga.*

ESCHENHAGEN, (Johann Friedrich August) Max, German physicist; b. Eisleben, Germany, Oct. 22, 1858; s. Gustav and Luise (Böhnert) E.; student math., physics, natural scis. U. Halle (Germany); doctorate, 1880; m. Rosalie Hähnel, 1891; 3 sons. Began as tchr.; assigned to Polar Year, 1883; computed earth magnetic observations German Polar Year Stas., 1882-83; later became asst. Marine Obs., Wilhelmshaven, Germany; named chmn. Magnetic Obs., Prussian Meteorol. Inst., Potsdam, Germany, 1889. Author: Ergebnisse der deutschen Stationen des Internationalen Polarjahres, 1882-83; Ergebnisse der magnetischen Beobachtungen in Potsdam, 1890; Bestimmung der erdmagnetischen Elemente an 40 Stationen in Nordwestdeutschland, 1900. Research on earth magnetism, earth magnetic storms; built new, quick-rec. earth magnetic instruments (led to proof of magnetic elementary waves). Died Potsdam, Nov. 12, 1901.

ESCHERICH, Karl Leopold, German zoologist, entomologist; b. Schwandorf, Germany, Sept. 18, 1871; s. Hermann and Katharine (v. Stengel) E.; student medicine, Munich, Germany, Würzburg, Germany; M.S.; studied zoology under Semper, Boveri, Hertwig, Leuckart; Ph.D., Leipzig, Germany; hon. dr. agr. and forestry sci.; m. Emilie Vinnen, 1901; 2 sons, 1 dau. Asst. to Nüsslin, lectr., Karlsruhe, Germany; worked with O. Bütschli, Heidelberg, Germany, Rostock, Germany, with A. W. Goette, Strasbourg, France; became prof. Forestry Acad. Tharandt, 1907; succeeded Nüsslin, Tech. U., Karlsruhe, 1914; became prof. U. Munich, 1914, rector, 1934-35; made numerous trips to Africa, Nr. E., Ceylon, U. S. A. Mem. Bavarian Acad. Scis. Author: Die angewandte Entomologie in Amerika, 1913; Die Ameise, Schildrerung ihrer Lebensweise, 1906; Die Termiten oder weissen Ameisen, 1909; Termitenleben auf Ceylon, 1910; Leben und Forschen, der Kampf um eine Wissenschaft, 1944; Die Forstinsekten Mitteleuropas, 4 vols., 1913-41. Founder applied entomology in Germany; research on ants, termites, forest insects; discovered yeast fungus in intestinal epithelium of Sitrokepa panicea (laid found. of symbiosis research). Died Kreuth, Germany, Nov. 22, 1951.

ESCHERICH, Theodor, pediatrician; b. Ansbach, Germany, Nov. 29, 1857; s. Ferdinand and Maria Stromer (v. Reichenbach) E. student medicine, Strasbourg, France, Kiel, Germany, Berlin, Germany, Würzburg, Germany; M.D., Munich, Germany, 1881; m. Margaretha v. Hadermur; 1 son, 1 dau. Became asst. to K. Gerhardt, Würzburg, 1882; asst. to Frobenius, also H. v. Ranke, Munich, 1885-90, became lectr., 1886; named asso. prof. children's Clinic, Graz, Austria, 1889, dir., 1890, prof., 1894; named prof. pediatrics U. Vienna, 1902. Author: Die Darmbakterien des Säuglings und ihre Bezichungen zur Physiologie der Verdanung, 1886; Ätiologie und Pathogenese der epidemischen Diphtherie; Der Diphtheriebazillus, 1894; Diphtherie, Croup, Serumtherapie nach Beobachtungen an der Universitäts-Kinderklinik in Graz, 1895; Tetanie der Kinder, 1909. Research on intestinal bacteria in infants, nutritional physiology and pathology of infants, tetanus, children's Tb, diptheria; 1st description of Escherichia coli, 1886; introduced bacteriology to pediatrics. Died Vienna, Feb. 15, 1911.

ESCHER VON DER LINTH, Arnold, Swiss geologist; b. Zürich, Switzerland, June 8, 1807; s. Hans Conrad and Regula (von Orelli) E. von der L.; student Geneva (Switzerland) Acad., Berlin, Germany; study expdns. to S. Italy, Lipari Island, Algeria, Sahara; m. Maria Barbara Ursula v. Latour, 1857. Became lectr. U. Zurich, 1834, prof. geology, 1852; also prof. Fed. Polytechnikum; dir. Zürich Mineral.-Geol. Collection. Mem. Zürich Naturalist Soc. Author: Geologische Übersichtskarte der Schweiz, 1853. A founder Swiss geology; research and publs. on mountain chain of eastern Switzerland, stratigraphic relationships in Central and E. Alps, glacial geology; pub. (with B. Studer) 1st geol. map of Switzerland; discovered stratum reversal in St. Gallen and Glarner Alps, structural stratification of Permian Verrucano on Tertiary Flysch. Died Zürich, July 12, 1872.

ESCHSCHALTZ, Johann Friedrich, Russian zoologist; b. Dorpat, Estonia, Nov. 1, 1793; s. Johann Gottfried and Katherine Hedwig (Ziegler) E.; M.D., U. Dorpat, 1815; m. Christine Friedrike Ledebour, 1819; 2 sons. Ship physician with Kotzebue and Chamisso, 1815-18, 1823-24; named asso. prof., Dorpat, 1819; became dir. Zool. Cabinet, 1822; became prof. anatomy and medicine, Dorpat, 1828. Eschscholtzia Bay on Kotzebue Sound, also genus Eschsholtzia (Cal. poppy) named in his honor. Mem. Natural History Soc. Moscow, Leopold-Caroline Acad. Natural Research, Author: Ideen zur Aneinanderreichung der rüchgratigen thiere auf vergliechen anatomie gegrundet, 1819; Eutoniographien, 1823. Zoologischer Atlas . . . , 1829-33; System der Acalephen, 1829; Collected insects, animals, and flora from Cal., Aleutian Island, S.Am., Pacific Islands, including many new species; studied Buprestidae. Died Dorpat, May 7, 1831.

ESCLANGON, Ernest Benjamin, French astronomer; b. Mison, France, Mar. 17, 1876; student École normale supierure; became astronomer Bordeaux (France) Obs., 1905; named prof. Faculty Scis., Strasbourg, France, also dir. Obs. Strasbourg, 1919; named dir. Obs. Paris (France), 1929. Mem. French Acad. Scis. (became v.p. astron. sect. 1941, pres. 1942). Research on quasi-periodical functions, weights, infra-sonar waves, atmospheric refraction, comets; during war studied synchronization by sound and successfully localized artillery pieces which were bombing Paris; invented speaking clock of Paris Obs., 1932; studied how astronaut can be free of gravity. Died Bordogne, France Jan. 28, 1954.

ESCOBAR, Wladimiro, Colombian physicist; b. Manizales, Colombia, July 9, 1924; s. Luis A. and Anna (Cifuentes) E.; B.Classic Letters, Colegio de S.Cor., 1941; M.Philosophy and Divinity, J. Javeriana, 1950, Electronic Engr., 1964. Subdir. Instituto Geofísico de los Andes Colombianos, 1947-48; prof. physics U. Javeriana, 1954-60, dean electronics faculty, 1964—; dir. Ionospheric Sta. IONBT, Bogotá, Colombia, 1957-60. Mem. Sociedad Colombiana de Fisica, Sociedad Colombiana de Mathemáticas, Asociacion de Profesionales Electronicos. Research, publs. on ionospheric data of Bogota sta., spl. nomogram for resolving quadratic equations, spl. tangents spiral. Home: Calle 69, No. 18-34, Bogota 2, D.E. Office: 40-62 Cra. 7, Bogota D.E. 2, Colombia.*

ESDAILE, James, surgeon; b. Montrose, Eng., Feb. 6, 1808; M.D., Edinburgh (Scotland) U., 1830; m. 1831; m. 2d; m. 3d, 1851. Physician, E. India Co., 1831-35, physician in charge Hooghly Hosp., 1839-46; placed in charge hosp., Calcutta, India, 1846; became Presidency surgeon, Bengal, India, 1848, marine surgeon, 1850. Research and publs. on use of hypnotism for anesthesia, 1845; performed 291 painless operations in India using hypnotism. Died Jan. 10, 1859.

ESHLEMAN, Von Russel, Am. radar astronomer; b. nr. Gettysburg, O., Sept. 17, 1924; s. Earl Ellsworth and L. Mae (Kneisley) E.; student Gen. Motors Inst. Tech., 1942-43, Ohio State U., 1946-47; B.E.E., George Washington U., 1949; M.S., Stanford, 1950, Ph.D., 1952; m. Patricia May Middleton, Mar. 6, 1947; children—Mary Angela, Kathleen Carol, Eric Earl, David Middleton. Faculty, Stanford (Cal.), 1950—, prof. elec. engring., 1961—; co. dir. Center for Radar Astronomy, Stanford and Stanford Research Inst., 1962—. Cons. to govt. agys., pvt. cos. Fellow I.E.E.E.; mem. A.A.A.S., Royal Astron. Soc., Internat. Sci. Radio Union, Am. Geophys. Union, Am. Inst. Aeros. and Astronautics, Internat. Astronautical Congress, Am. Astron. Soc., Sigma Xi, Sigma Tau. Research, publs. on devel. theory radio reflections from trails ionization produced by meteors, application meteor ionization to communications, theory distbn. meteoric dust in space, occulation expt. measuring atmosphere and ionosphere Mars, space-probe expts.; pioneered research on radar studies moon and sun. Home: 576 Gerona Rd., Stanford, Cal. 94305.*

ESHNER, Augustus Adolph, Am. physician; b. Memphis, Nov. 17, 1862; s. James and Jane E.; A.B., Central High Sch., Phila., 1879, A.M., 1884; M.D., Jefferson Med. Coll., 1888; m. Julia Friedberger, 1904; children—Mrs. Annette E. Dalsimer, Mrs. Juliet E. Nathanson. Resident physician Phila. Hosp., 1888-89, registrar neurology dept., 1891-96, physician, 1896-1914; chief clin. asst. outpatient med. dept., Jefferson Med. Coll. Hosp., 1892; prof. clin. medicine, Phila. Polyclinic, 1895-1918; asst. phys. Orthopaedic Hosp. and Infirmary for Nervous Diseases, 1900-17; vis. physician Hosp. for Diseases of Lungs, 1901-02; cons. physician Mercy Hosp., 1910-49. Asst. editor Med. News, 1891-95, Phila. Med. Jour., 1898-99; asso. editor Pa. Med. Jour., 1904-22. Fellow Coll. Physicians of Phila., A.A.A.S.; mem. A.M.A., No. Med. Assn., Sci. League Am., Franklin Inst. Author: Essentials of Medical Diagnosis (with Dr. Solomon Solis-Cohen), 1892, 2d edit., 1900; Hand-Book of Fevers, 1895. Asst. editor: American Text-book of Applied Therapeutics, 1896. Transl. and edited: Atlas and Methods of Clinical Investigation (Dr. Christfried Jakob), 1898; Elements of Clinical Bacteriology (Dr. Ernest Levy and Dr. Felix Klemperer), 1900; A Text-book of the Practice of Medicine (Dr. Hermann Eichhorst), 1900. Collaborator on Annual of the Universal Med. Sciences, 1890-95; Sajous' Annual and Cyclo. of Practical Medicine, 1899; Cyclo. of Practical Medicine and Surgery, 1900; System of Physiologic Therapeutics, 1902-03; Reference Handbook of the Medical Sciences, 1903; A Textbook of Human Physiology (L. Landois), 1904. Has written many med. monographs and articles in med. jours., etc. Died Dec. 20, 1949.

ESMARCH, Johannes Friedrich August von, see von Esmarch, Johannes Friedrich August.

ESMOND, William George, Am. physician; b. Bridgeport, Conn., June 6, 1919; s. William Gregory and Ella (Sanger) E.; B.S., U. Md., 1940, M.D., 1951; m. Ella Marie Weaver, Oct. 6, 1945; children—Jeanne, Mary, David, Robert, Deborah, Catherine, Anne. Practice medicine, specializing in internal medicine, Balt., 1955—; faculty U. Md., 1956—, now asst. prof.; cons. renal disease, dir. Hemodialysis Center. Prin. investigator grant NIH. Mem. A.M.A., A.A.A.S., Am. Soc. Artificial Internal Organs, Am. Soc. Nephrology, Alpha Omega Alpha. Contbr. numerous articles to profl. jours. Invented linear motion ball bearing, lamited fabric-plastic body armour, small compact low cost injection molded artifical kidney in world-wide use in hosp. and home hemodialysis; designer complete integrated gentle low hemolysis artificial heart-lung bypass system for open heart surgery with disposable oxygenators and bubble traps in world-wide use. Home: 537 Stamford Rd., Balt. 21229. Office: 295 Greene St., Balt. 21201.*

ESNAULT-PELTERIE, Robert Albert Charles, French astronautical pioneer; b. Paris, Nov. 8, 1881; ed. Sorbonne, Paris; m. Bernaldo de Quiros; 1 son, Michel. Learned to fly, 1906; engr., Paris; mem. French Acad. Scis., 1936. Author: Astronautics, 1930. Founder (with Louis Andre Louis Hirsch) Pel-Hirsch prize for outstanding astronautical work, 1928. Built 1st monoplane with motor in front, 1906, 1st radial cylinder engine with odd number of cylinders; 1st to study stabilization, 1911; founder astronautics; developed better rocket motors; proposed the study of the upper atmosphere using rockets. Died Nice, France, 1957.

ESPAGNET, see D'Espagnet, Jean.

ESPENSCHIED, Lloyd, Am. elec. engr.; b. St. Louis, Apr. 27, 1889; s. Fred F. and Clara M. E.; grad. Pratt Inst., 1909; m. Ethel L. Lovejoy, Apr. 29, 1912; 2 children. Asst. engr. Telefunken Wireless Telephone Co. Am., 1909-10; mem. engring. staff Am. Tel. & Tel. Co. and Bell Telephone Labs., Inc., 1910-54, research cons., 1937-54, ret., cons., 1954—. Recipient medal of honor I.R.E. Mem. Inst. Aero. Sci., Acoustical Soc. Am., A.A.A.S., Am. Geog. Soc., Phys. Soc., Am. Inst. E.E. Work in devel. of radio telephony, overseas broadcasting telephony, TV, electronic high frequency technology; inventor coaxial cable for wide-band long-distance transmission, electric reflection systems (including radio altimeter), crystal filter, other devices. Home: 99 82d Rd., Kew Gardens, N.Y. 11415.*

ESPER, Eugen Johann Christoph, entomologist; b. Wunsiedel, Bavaria, June 2, 1742; Ph.D., U. Erlangen (Bavaria), 1781. With U. Erlangen, from 1782, appt extraordinary prof., 1783, then prof. natural history, 1799, dir. natural history dept., from 1805. Mem. Berlin Soc. Naturalists, Leopoldina, 1789. Author: Die Schmetterlinge in Abbildunger nach der Natur . . . , Europ Gattungen, 1776-1805; Lehrbuch der Mineralogie, 1810. Studies on Lepidoptera, Coelenterata, and prehistoric caves at Muzzendorf; research on European insects, also collected insects, birds, minerals. Died Erlangen, July 27, 1810.

ESPINE, Adolphe d', see d'Espine, Adolphe.

ESPOSITO, Sergio, Italian physician; b. Saluzzo, Italy, Oct. 7, 1930; s. Alessandro and Maria (Agazzi) E.; degree in medicine and surgery Cairoli U. Coll., Pavia, 1955; m. Rosa Lercara, Dec. 28, 1958; children—Alessandra, Marco, Roberto. Asst. prof. Med. Clinic, Pavia U., 1955—; vice dir. Pavia U. Haematology Service, 1963—, tchr. haematology, 1964—. Sec. provincial dept. Italian League for Research on Cancer. Recipient Giorgio Kropf nat. prize, 1962; Nat. prizes Italian Soc. Internal Medicine, 1964, D. Ganassini, 1965. Mem. Italian Soc. Haematology and Blood Transfusion. Editorial staff Italian Treatise on Internal Medicine, 25 vols., 1958—. Author (with M. Gandolfi and R. Sacco) The Pulmonary Pathology in Blood Diseases, 1959; (with L. Buscarini) Problems of Splenic Physiology and Pathology, 1962; (with G. Marinone and L. Buscarini) The Donation of Blood, 1963; Antineoplastic Treatments and Nucleic Acids, 1965; The Nucleic Acids of Neoplastic Cell, 1966; also articles. Research on transfer of nucleic acids between different cellular populations; induction of antibodies synthesis; modifications of metabolism of leukaemic cells, induction of differentiation of leukaemic cells; modifications of genetic characters of haemoglobins. Home: 99 Liberta, Pavia, Italy.*

ESPY, James Pollard, Am. meteorologist; b. Washington County, Pa., May 9, 1785; s. Josiah and Elizabeth (Patterson) E.; grad. Transylvania U., Lexington, Ky., 1808; m. Margaret Pollard, 1812. Tchr. Xenia, O.; Cumberland, Md.; tchr. math. and classics Franklin Inst., Phila., 1817-35; recipient Magellanic prize Am. Philos. Soc., 1836; apptd. meteorologist U. S. War Dept., 1842; submitted 1st annual weather report, 1843; meteorologist Navy Dept., 1848; formulated convectional theory of storms (his chief contbn.), laid found. of weather forecasting; 1st govt. ofcl. to use telegraph to get weather reports from across nation. Author: Philosophy of Storms, 1841. Died Cin., Jan. 24, 1860.

ESQUIROL, Jean Etienne Dominique, French psychiatrist; b. Toulouse, France, Feb. 4, 1772; ed. St. Lupice sem., Paris; student of Philippe Pinel; intern Salpêtrière, later founder model mental instn., Paris, 1779, visited all asylums in France, 1808, apptd. physician to Salpêtrière, asylum in Paris, 1811; prof., became univ. insp.-gen., 1823; named chief physician, Charenton, 1826. Mem. Acad. Medicine. Author: Des maladies mentales condidérées sous les rapports médical, hygiénique et médico-légal (1st modern textbook of psychiatry), 1838. A founder of modern psychiatry; promoted Pinel's discoveries of origin of mental disorders in path. changes in brain; emphasized role of emotions as source of mental disturbances; supported his views with statistical tabulations; tried to reform mental instns. in France; demanded a more humane treatment of mentally deranged; studied architecture and construction of asylums. Died Paris, Dec. 12, 1840.

ESSEN, Louis, English physicist; b. Nottingham, Eng., Sept. 6, 1908; s. Fred and Ada (Edson) E.; B.Sc. with 1st class honors in Physics, Nottingham U. Coll., 1928; Ph.D., U. London (Eng.), 1941, D.Sc., 1948; m. Joan Margery Greenhalgh, Nov. 6, 1937; children—Judith Katherine (Mrs. David Robinson), Margaret Ann (Mrs. Peter Lawson), Hilary Noel, Lesley Carol. With Nat. Phys. Lab., Teddington, Eng., 1929—, sr. prin. sci. officer, 1956-60, dep. chief sci. officer, 1960—. Decorated Order Brit. Empire; recipient Charles Vernon Boys prize, 1956; Tompion Gold medal, 1957; Wolfe award, 1959; Popov Gold medal, 1959. Fellow Brit. Horological Inst. (hon.), Royal Soc., 1960; mem. Instn. Elec. Engrs., Internat. Sci. Radio Union (internat. chmn. commn.), Internat. Astron. Union. Research, numerous publs. on microwave measurement of velocity of light revealed error in accepted value, time and frequency standards, Essen-ring quartz clock, atomic clocks, radio frequency measurements; provided atomic unit time now internationally accepted. Home: 50 Wensley Dale Rd., Hampton, Middlesex. Office: Nat. Phys. Lab., Teddington, Middlesex, Eng.*

ESSEX, Hiram Eli, Am. physiologist; b. Glasford, Ill., Jan. 28, 1893; s. Hiram and Sarah (Watters) E.; B.S., Knox Coll., 1919; M.S., U. Ill., 1924, Ph.D., 1927; m. Marion Gertrude Murphy, Sept. 11, 1926; children—Sarah Catherine (Mrs. Richard N. Kleareland), Dorothy Jean (Mrs. Charles Cranston Spray, Jr.), Marion Elizabeth (Mrs. William E. Pedersen, Jr.). Faculty, Mayo Grad. Sch. Medicine, Rochester, Minn., 1928—, prof. physiology, 1944-58, emeritus prof., 1958—. Dir. Hormel Inst. Recipient Alumni Achievement award Knox Coll., 1958. Fellow A.A.A.S.; mem. Am. Physiol. Soc. (past pres.), Soc. Pharmacology and Exptl. Therapeutics, Soc. Exptl. Biology and Medicine, Am. Soc. Zoologists, Am. Soc. Parasitologists, Am. Micros. Soc., Nat. Soc. for Med. Research (pres. 1961-65), Am. Assn. U. Profs. Contbr. numerous articles to profl. jours. Research on circulation of heart, liver, behavior of valves of heart and aorta, circulation of spleen, repair of injured peripheral nerves, physiol. effects of venoms of snakes, anaphylasis and anaphytactoic responses, effect of removal of pituitary gland, effects of repeated large doses of epinephrine given to dogs. Home: 711 7th St., Rochester 55901. Office: Plummer Bldg., Rochester, Minn. 55902.*

ESSMAN, Walter Bernard, Am. psychologist; b. N.Y.C., Dec. 25, 1933; s. Louis and Elsie (Eiseman) E.; B.A., N.Y. U., 1954; M.A., U. N.D., 1955, Ph.D., 1957; m. Shirley Glass, June 10, 1962; 1 son, Eric J. Sr. Post-doctoral fellow Albert Einstein Coll. Medicine, 1959-61, research asso. physiology, 1961-63; faculty Queens Coll., N.Y.C., 1962—, asso. prof. psychology, 1965-67, prof., 1967—. research fellow Lab. Neurochemistry, Mt. Sinai Hosp., N.Y.C., 1964—. Mem. Am. Psychol. Assn., Am. Physiol. Soc., Am. Soc. Zoologists, N.Y. Acad. Scis. Research, publs. on role altered body and brain temperature change in behavioral processes learning and memory; significance of changes in brain metabolism for information and fixation of memory in brain. Home: 6 Ashleigh Ct., Glen Cove, N.Y. 11542. Office: Dept. Psychology, Queens Coll., Flushing, N.Y. 11367.*

ESSON, William, English mathematician; b. 1838; M.A., St. John's Coll., Oxford (Eng.) U.; fellow Merton, 1860-97, also bursar; fellow New Coll.; dep. Savilian prof. geometry, Oxford, 1894-97, Savilian prof. geometry, from 1897; curator Univ. Chest. Fellow Royal Soc., 1869. Research and publs. on relation of conditions to amount of chem. change, synthetic geometry, relation of rate of chem. change to temperature variation, plane curves. Died Aug. 25, 1916.

ESTABROOK, Ronald W., Am. biochemist; b. Albany, N.Y., Jan. 3, 1926; s. George Arthur and Lillian (Childs) E.; B.S., Rensselaer Poly. Inst., 1950; Ph.D., U. Rochester, 1954; m. June Templeton, Aug. 23, 1947; children—Linda Ann, Laura Elizabeth, Jill Kathleen, David Edward. With Johnson Research Found., U. Pa., Phila., 1955-68, prof. phys. biochemistry, 1965-68; faculty Southwestern Med. Sch., U. Tex., Dallas, 1968—. Mem. sect. NIH, 1965—; mem. subcom. Nat. Acad. Scis., NRC, 1963—. Mem. Am. Chem. Soc., Am. Soc. Biol. Chemists, Sigma Xi. Editor: (with B. Chance, J. Williamson) Control of Energy Metabolism, 1966; (with B. Chance, T. Yonetani) Hemes and Hemoproteins, 1966; Methods in Enzymology, vol. 10, 1967. Exec. editor Archives of Biochemistry and Biophysics, 1966—. Research, publs on mechanism of biol. oxidation in mammalian tissues to elaborate the process of energy conservation as well as drug and steroid metabolism, purification of enzymes and their role in transport reactions. Address: Southwestern Med. Sch., U. Tex., Dallas.*

ESTERMANN, Immanuel, Am. physicist; b. Berlin, Germany, Mar. 31, 1900; s. Arieh Leo and Rachel (Brenner) E.; D.Sc., U. Hamburg, Germany, 1921; m. Rosa Chwolles, May 18, 1923; children—Hannah (Mrs. John Bergman), Eva Frances. Came to U. S., 1933, naturalized, 1939. Asst. physics U. Rostock, 1921-22; instr. phys. chemistry U. Hamburg, 1922-28, privat dozent, 1928-33, emeritus prof., 1958—; Rockefeller fellow U. Cal., Berkeley, 1931-33; asso. prof. Carnegie Inst. Tech., Pitts., 1933-45, prof. 1945-52; dir. material scis. div. Office Naval Research, Washington, 1951-58, dep. sci. dir., 1955-58, research coordinator, 1958-59, sci. dir., chief scientist, London br., 1959-64; Fulbright lectr. Hebrew U., Jerusalem, Israel, 1963; vis. prof. physics Technion, Haifa, Israel, 1964—. Cons., OSRD, 1940-43, Manhattan dist., 1943-45, Office Naval Research, 1945-50, Nat. Bur. Standards, 1949. Decorated Silver medal (France); recipient Pitts.

Physics award, 1955; Navy Distinguished Civilian Service award, 1964. Fellow A.A.A.S., Am. Phys. Soc., Washington Acad. Sci.; mem. Faraday Soc., Sci. Research Soc. Editor: Methods of Experimental Physics; Advances in Atomic and Molecular Physics, 1959, 65—; Recent Research in Molecular Beams, 1959. Research on molecular beams, solid state physics, low temperatures, rarefied gas dynamics, semiconductors, elucidation of mechanism of crystal growth, basic principles of quantum theory, knowledge of atomic and molecular structure. Home: 18 Hursha St., Haifa, Israel.*

ESTES, William Kaye, Am. psychologist; b. Mpls., June 17, 1919; s. George D. and Mona (Kaye) E.; A.B., U. Minn., 1940, Ph.D., 1943; m. Katherine Walker, Sept. 26, 1942; children—George Edson, Gregory Walker. Faculty, Ind. U., 1946-62, prof., 1955-62; prof. psychology Stanford, 1962—. Fellow, Center for Advanced Study in Behavioral Scis., 1955-56. Mem. Am. Psychol. Assn. (past div. pres., Distinguished Contribution award 1963), Soc. Exptl. Psychologists (Warren medal for psychol. research 1963), Psychometric Soc., Psychonomic Soc., Inst. Math. Statistics, Nat. Acad. Scis. Author: (with others) Modern Learning Theory, 1955; (with R. R. Bush) Studies in Mathematical Learning Theory, 1959; Experimental Study of Punishment, 1944; also numerous articles. Developed math. learning theory. Home: 900 Lathrop Dr., Stanford, Cal. 94305.

ESTEVA, Claudio, Spanish anthropologist; b. Marsella, Nov. 11, 1918; s. Claudio and Mercedes (Fabregat) E.; ed. U. Mexico, U. Madrid, Nat. Sch. Anthropology and History; Ph.D. in history, M.A. in ethnology; m. Saló Juliá; children—Claudio, Mireia, Miguel Angel. Prof., Nat. Sch. Anthropology and History, Mexico, U. Mexico, U. Madrid. Mem. Higher Council Sci. Research; dir. Collection Ciencias Humanas. Mem. Am. Soc. Applied Anthropology, Soc. Anthropology (Washington), Am., Spanish sociol. socs. Author: Desarrollo Social y Planificacion Social; Cultura, Sociedad y Salud Mental; Cultura y Personalidad; Funcion y Funcionalismo en las Ciencias Sociales; El Mestizaje en Iberoamerica. Home: Avenida de America 17, 5-B, Madrid 2. Office: Faculty of Philosophy and Letters, University School, Madrid, Spain.

ESTIENNE, Charles, French anatomist, physician; b. Paris, France, 1504; s. Henry Estienne; M.D., Paris; pupil of Lascaris; children include dau. Olymphe Nicole. Took charge of family printing establishment, Paris, from 1551; printer to king, 1551; bankrupt, 1561. Author: De dissectione partium corporis humani (treatise on human anatomy and dissection giving clear description of cranial nerves), 1548; Thesaurus Ciceronianus, 1557; Dictionarium historicum ac poëticum (1st French encyclopedia), 1553; other works on agriculture and literature. Distinguished sympathetic from pneumogastric nerves; discovered venous valves. Died 1564 (possibly in debtors' prison).

ESTIENNE, Henri, French printer, mathematician; b. Paris, circa 1470; children—Francois, Robert, Charles. Set up printing establishment, Bourges, 1504. Author of a treatise on geometry. Promoted learning by printing over 120 books. Died Paris, 1520.

ESTON, Veronica Rapp, Brazilian radiobiologist; b. Sao Paulo, Brazil, Feb. 21, 1918; d. Robert and Emilie (Ulrich) Rapp; Med.M.D., U. Sao Paulo, 1944; B.Sc., U. Toronto (Ont., Can.), 1948; m. Tede Eston, Sept. 12, 1946; children—Verena Rapp, Marilda Rapp. Fellow U. Ia., Ia. Fedn. Woman's Club, 1946; Mary Putnam Jacobi fellow Women's Med. Assn. N.Y., Toronto, 1947; asst. prof. physiol. chemistry Faculty Medicine, U. Sao Paulo, 1948-58, head div. tng. and biol. research Center Nuclear Medicine, 1959—. Mem. Radiation Research Soc., Reticulo Endothelial Soc. Author several books on tng. program on tracer methodology and tracer applications in biol. research; also articles. Research on def. mechanism of mammalian organism against radiation and certain diseases such as leprosy; pioneered tracer applications in biology and medicine in Latin Am. Home: Rua Agrário de Souza, 150, Sao Paulo, Brazil.*

ETGES, Frank Joseph, Am. zoologist; b. Chgo., June 18, 1924; s. Joseph P. and Anna (Foss) E.; A.B., U. Ill., 1948, M.S., 1949; Ph.D., N.Y. U., 1953; m. Ruth C. Storkan, Sept. 20, 1947; children—Robert, William, Ann, David, Thomas. Mem. faculty U. Ark., 1953-54; faculty U. Cin., 1954—, prof. zoology, 1966—. Research fellow U. S. Army Tropical Med. Research Lab., P.R., 1962-63. Mem. Am. Soc. Parasitologists, Am. Soc. Medicine and Hygiene, Soc. Protozoologists, Am. Soc. Zoologists, Am. Micros. Soc., Sigma Xi (Distinguished Research award 1966), others. Research and publs. on new species of parasitic flatworms and their life histories; investigations of the relationships of the human blood fluke to their snail intermediate hosts, and the several effects of this parasite on the snail; studies on the responses of snails toward molluscicidal chems., and the effects of bacteria on chem. molluscicides. Home: 1297 Sweetwater Dr., Cin. 45215.*

ETHERIDGE, Robert, English paleontologist; b. Ross, Herefordshire, Eng., Dec. 3, 1819; s. Thomas and Hannah (Pardoe) E.; ed. grammar sch., Ross, Eng.; m. 3 times; 1 son, Robert, by 1st wife. Became curator mus. Bristol (Eng.) Philos. Instn., 1850; lectr. physiology and vegetable botany Bristol. Med. Sch., for 5 years; became lectr. Bristol. Mining Sch., 1857; named asst. paleontologist Geol. Survey, 1857, paleontologist, 1863; asst. to T. H. Huxley as lectr. Royal Sch. Mines; asst. keeper geology Brit. Mus. Natural History, 1881-91; named hon. fellow King's Coll. London, 1890. Recipient 1st Bolitho medal Royal Geol. Soc. Cornwall, 1896. Fellow Royal Soc., 1871 (mem. council, v.p., Murchison medal 1880), Geol. Soc. (pres. 1880-82), Paleontol. Soc. (treas. 1880-1903). Author: Geology: its Relation and Bearing upon Mining, 1859; Stratigraphical Geology and Paleontology, 1887; contbg. author: Palaeozoic Species, Vol. 1, 1888. Asst. editor Geol. Mag., 1865-1903. Named invertebrate fossils discovered during geol. survey; listed 18,000 species of fossils, 1888. Died London, Eng., Dec. 18, 1903.

ETO, Hideo, Japanese med. physicist; b. Tokyo, Japan, Oct. 10, 1911; s. Junichi and Shizuko (Saigo) E.; grad. Sch. Sci., Tokyo Imperial U., 1935; Ph.D. in Med. Sci., Tokyo U., 1948; m. Takako Inagaki, Mar. 1, 1942; children—Kuniko, Kazuko. Staff, Tokyo U. Sch. Medicine, 1938-57, asso. prof., 1942-57; head div. radiation hazards Nat. Inst. Radiol. Scis., Chiba, Japan, 1957—; sr. sci. research officer, vice dir., 1965—. Recipient Ofcl. Commendation, Ministry Sci. and Tech., 1961. Mem. Nippon Societas Radiologica, Japanese Soc. Radiology, Japan Radiation Research Soc., Atomic Energy Soc. Japan. Author: Human and Radiation, 1951; Physics for Medicine and Biology, 1952; Medical Application of Radioisotopes, 1953; Fundamentals of X-ray Photography (with Yoshimura, Sato), 1965; (with others) Radiation Protection, 1965. Radiology, 1964; also numerous articles. Research in radiation med. physics, radiotherapy, theoretical analysis and calculation of resolving power of fluorescent screen, camera and film and resultant total resolution in indirect X-ray photography; established theoretical calculation of volume dose; studies in dosimetry and scanning of radioisotopes incorporated in human organs; developed Japan's 1st scintiscanner; dir. research in radiation hazards. Home: 2-4-4, Nishikata, Tokyo, Japan. Office: 4-9-1, Anagawa, Chiba, Japan.*

ETSTEN, Benjamin Edward, Am. physician; b. Lawrence, Mass., May 25, 1908; s. Louis and Bertha (DuFine) E.; B.S., Tufts Coll., 1932; M.D., St. Andrew's Med. Sch., Dundee, Scotland, 1936; m. Jessica Beulah Drooz, Feb. 22, 1938; children—Pamela Susan, (Mrs. Stephen Newman), Edward, Thomas. Faculty, Albany Med. Sch., 1940-48, asso. in physiology and pharmacology, 1943-48, asst. prof., 1942-48; faculty Tufts U., Boston, 1949—, prof. anesthesiology, chmn. dept. anesthesiology, 1952—; anesthetist-in-chief New Eng. Med. Center Hosps., 1949—. Cons., VA Hosps., Boston, West Roxbury, Mass., 1960—. Mem. Assn. U. Anesthetists (charter, pres. 1966—), Mass. Soc. Anesthesiologists (past pres.), Am. Soc. for Pharmacology and Exptl. Therapeutics. Research, numerous publs. on physiol. basis for changes in stages and signs pentothal anesthesia, ventilation and circulation during surgery, effects cyclopropane anesthesia on cardiac output and related hemodynamics in man, ventricular function during various types anesthesia, effects anesthetics upon myocardial catecholamines, mephentermine and methoxamine during spinal anesthesia in man. Home: 37 Gordon Rd., Milton, Mass. Office: 171 Harrison Av., Boston.*

ETTER, Lewis Elmer, Am. radiologist; b. Pitts., Jan. 17, 1901; s. Charles Elmer and Elsie (Gnann) E.; student U. Cal. at Berkeley, 1920-21, U. Pitts., 1921-23; B.S., Sch. Medicine, 1924, M.D., 1927; m. Grace Dorothy Ripple, June 4, 1927; children—Robert Charles. Practice gen. medicine, Warrendale, Pa., 1928-41; faculty U. Pitts. Sch. Medicine, 1946—, prof. radiology, chief radiologic service Falk Clinic, 1957—, Western Psychiat. Inst. and Clinic, 1947—. Cons., C. Howard Marcy State Hosp., Leech Farm, Pitts., 1955—. Fellow Am. Coll. Radiology; mem. Pitts. Roentgen Soc., Pa. Radiol. Soc., Am. Roentgen Ray Soc., Radiol. Soc. N. Am., Am. Soc. Neuroradiology, Pan Pacific Surg. Assn., Pan Am. Med. Assn., Inter Am. Coll. Radiology, Sigma Xi. Author: Atlas of Roentgen Anatomy of the Skull, Paranasal Sinuses and Mastoids, 1955; Glossary of Words and Phrases Used in Radiology and Nuclear Medicine, 1960; Roentgenography and Roentgenology of Middle Ear and Mastoid Processes, 1965; Science of Ionizing Radiation, 1965; also numerous articles. Editor: American Lectures in Roentgen Diagnosis, 1968—; Modern Concepts of Radiology, Nuclear Medicine and Ultra-Sound, 1966—. Historical studies and data on discovery of X-rays; devel. new techniques for, skull paranasal sinuses, temporal bone and roentgen studies; research on radiation dose reduction in dentistry. Home: Pinewood Farm, Warrendale, Pa. 15086. Office: 3601 5th Av., Pitts. 15213.*

ETTINGER, G(eorge) Harold, Canadian physician; b. Kingston, Ont., Can., May 9, 1896; s. John George and Elizabeth Jane (Watts) E.; B.A., Queen's U., Ontario, 1916, M.D., C.M., 1920, LL.D., 1967; post-graduate University of Chicago, Edinburgh (Scotland) University; D.Sc. (hon.), U. of Western Ont., 1958; M.D. (hon.), U. Ottawa (Ont.), 1963; m. Pearl Elizabeth Blyth, Dec. 21, 1920; children—Barbara Joan (Mrs. John E. Hinton). Faculty, Queen's U., 1920, 29-35, prof. physiology, 1935-62, head physiology dept., 1949-62, dean Faculty Medicine, 1949-62; asst. dir. div. med. research NRC, Ottawa, Ont., 1946-58, dir. med. planning Alcoholism and Drug Addiction Research Found., Toronto, Ont., 1962—. Mem. Order Brit. Empire. Fellow Royal Soc. Can.; mem. Canadian, Brit., Am. physiol. socs., Am. Assn. Anatomists, Canadian Soc. for Clin. Investigation (hon.), Research, publs. in field cardiovascular physiology and endocrinology. Home: 219 Union St. W., Kingston, Ont. Office: 344 Bloor St. W., Toronto, Ont., Can.*

ETTINGSHAUSEN, Baron (Johannes) Andreas (Jakob) von, see von Ettingshausen, Baron (Johannes) Andreas (Jakob).

ETTINGSHAUSEN, Constantin von, see von Ettingshausen, Constantin.

ETTLING, Carl Jacob, German mineralogist; b. Russelsheim/Main, Germany, Apr. 15, 1806; became asso. prof. mineralogy, Giessen, Germany, 1849; discovered ethyl carbonate; analyzed creosote; studied salicytic acid; showed (with Liebig) that Dumas' formula for oil of cloves was a mixture of acid of cloves and a terpene. Died Giessen, June 21, 1856.

ETTLINGER, Max Emil, German psychologist; b. Frankfort/Main, Germany, Jan. 31, 1877; s. Emil and Mathilde (Oppenheim) E.; student philosophy, Heidelberg, Germany; Ph.D., Munich, Germany; m. Walburga Seidenschwarz, 1904. Sci. editor mag. Hochland, 1903-07; became lectr. philosophy, Munich, 1914; named prof. philosophy, Münster, Germany, 1917; co-founder, dir. German Inst. for Sci. Pedagogy, 1917. Author: Philosophische Fragen der Gegenwart, 1911; Der Streit um das rechnende Pferd, 1913; Der chrisliche Idealismus des Erzieherberufes, 1921; Beiträge zur Lehre von der Tierseele und ihrer Entwicklund, 1925; Die philosophische Zusammenhänge in der Pädagogik der jüngsten Vergangenheit und Gegenwart, 1925; also articles. Became co-editor: Philosophisches Handbuch, 1920, Vierteljahresschrift für wissenschaftliche Pädagogik, 1926, Handbuch der Erziehungswiessenschaft, 1928. Research on esthetics, psychology, animal psychology, sci. of edn., history philosophy; tried to reconcile Platonic-Aristotelian heritage with modern sci. Died Ebenhausen, Germany, Oct. 12, 1929.

ETTMÜLLER, Michael, German physician; b. Leipzig, Germany, May 26, 1644; M.D., U. Leipzig, 1668; studied in Wittenberg. 2 sons, Ernest, Michael. Mem. staff, Faculty Medicine, Leipzig, from 1676, became prof. botany, prof. surgery and anatomy, 1681. Mem. Academia Naturae Curiosorum Author: Opera Medica Theoretico-Practica, 1708; numerous other works, especially on pharmacy. Critic of Van Helmont's alkahest theory; held that acids engender alkalis; advocate of intravenous injections. Died Leipzig, Mar. 9, 1683.

ETZIONI, Amitai Werner, sociologist; b. Cologne, Germany, Jan. 4, 1929; s. Willi Falk and Gertrude Hannauer (Falk) E.; B.A., Hebrew U., Jerusalem, 1954, M.A., 1956; Ph.D., U. Cal. at Berkeley, 1958; m. Minerva Morales, Sept. 14, 1965; children—Ethan, Oren, Michael. Faculty, Columbia, N.Y.C., 1958—, research asso. Inst. War and Peace Studies, 1961—, prof. sociology, 1967—. Social Sci. Research Council faculty fellow, 1960-61, 67-68; fellow Center for Advanced Study in Behavioral Scis., 1965-66; Guggenheim fellow, 1968. Mem. Am. Sociol. Assn. Author: A Comparative Analysis of Complex Organizations, 1961; Modern Organizations, 1964; Political Unification: A Comparative Study of Leaders and Forces, 1965; Studies in Social Change, 1966; The Active Society, 1968. also numerous articles. Developed organizational analysis, a typology based on means used to control mems.; macrosociol. theory to study socs. able to solve their problems. Home: 450 Riverside Dr., N.Y.C. 10027.*

EUCLID, Greek mathematician; flourished circa 300 B.C.; probably ed. acad. at Athens; lived in Alexandria (under Ptolemy I, 306-283 B.C.), where he taught and probably founded math. sch. Author: (extant works) The Elements, 13 books; Optics, Data; Phaenomena; On Divisions (of figures); Elements of Music; (lost works) Fallacies; Porisms; Surface Loci; Conics. His most famous work, The Elements, is essentially a compilation of math. knowledge rather than an original treatise (incorporates work of Eudoxos and Theaetetos), contains axioms and postulates that virtually defy improvement due to their exceedingly logical arrangement; remained standard text providing almost the sole approach to geometry until 19th century devel. of non-Euclidean geometry; Euclid advanced a severe and accurate method in mathematics; also wrote on ratios and proportions and on theory of numbers; demonstrated that number of primes is infinite; proved that square root of 2 is irrational; wrote treatises on music; also astron. study giving propositions of spherical geometry; developed basic principles of geometrical optics, including laws of reflection and rectilinear propagation of light.

EUCTEMON, Greek astronomer; flourished 5th century B.C.; (with Meton) invented 19 year lunar cycle (basis for Greek calendar), 430 B.C.

EUDEMOS OF RHODES, Greek mathematician, sci. historian; flourished circa 320 B.C.; student of Aristotle; founder sch., Rhodes, Greece. Compiled paraphrase Physics (Aristotle); wrote history of arithmetic, history of astronomy, history of geometry; publs. on angle.

EUDES-DESLONGCHAMPS, Jacques, French naturalist; b. Caen, France; student medicine, Caen, Paris; M.D., 1818; D.Sc., 1826. Dir. Jardin des plantes, Caen, 1825-29; prof. natural scis.; named prof. zoology and animal physiology, 1838; became dean Caen Faculty Scis., 1847. Mem. Linnean Soc. Calvados (founder), French Acad. Scis., 1849. Author: le Poekilopleuron bucklandi, grand fossile sauria, 1835; Essai sur les plicatules fossiles des terrains du Calvados, 1858; Sur la couche à Leptaena, 1859. Research on reptiles, dinosaur fossils of genus, Teleosaurus; called the Cuvier from Normandy. Died 1867.

EUDOXOS, Greek philosopher, mathematician, astronomer, physician; b. Cnidos, 408 B.C.; studied math. under Archytas in Tarentum, medicine under Philistium in Sicily, philosophy, math., rhetoric at Plato's Acad. in Athens, astronomy under priests at Heliopolis, Egypt. Made astron. observations in obs. between Heliopolis and Cercesura, in Egypt, 381-80 B.C.; then established sch. at Cyzicus on sea of Marmora; became tchr. in Athens, circa 365 B.C. Author: The Mirror; Phaenomena; Geography; Octaeteris. Taught that pleasure and joy are the highest good; followed Plato in some philos. views, opposed others; considered by Cicero the greatest astronomer that had yet lived; to save Plato's theory of planetary motion, he suggested a spherical earth at rest in the center of the universe and surrounded by rotating, concentric spheres, or planets, which describe a curve he called horse-fetter; made a map of stars authoritative for centuries; divided sky into degrees of latitude and longitude; probably invented Arachne, or spider (astron. instrument); introduced study of spherics (math. astronomy), also globe into Greece; divided earth into zones, also made new map of known parts; 1st Greek to establish that year is 6 hours longer than 365 days; in geometry, he created gen. theory of proportion (set forth in books, 5, 6 of Euclid's Elements); created method of exhaustion (fundamental for measuring areas and vols.); solved problem of doubling cube; developed theory of golden sect.; showed that vol. of cone and pyramid were respectively 1/3 that of cylinder and prism of same base and height. Died Cnidos, 355 B.C.

EUDOXOS OF CYZICOS, Greek navigator; flourished late 2nd Century B.C.; sent by Ptolemy Euergetes II of Egypt to find sea route to India and to explore the Arabian Sea, after 146 B.C.; later blown off course down coast of E. Africa; set out to circumnavigate Africa to get to India, but failed; sailed later down W. coast of Africa and disappeared.

EULA, Antonio, Italian civil engr.; b. Rome, Italy, May 2, 1897; s. Umberto and Paolina (Gaddi) E.; Ph.D. in civil engring. and aeros.; m. Maria Antonia de Conti. Prof. aerodynamics U. Rome. Mem. Italian Assn. Aerotechnics (pres.), Italian Assn. Rockets (pres.), Italian Aero. Register (pres. tech. com.), Internat. Astronautic Acad., Sci. Soc. for Aviation Research (Germany), Brit. Interplanetary Soc. Research and publs. on aerodynamics, mechanics of flying, astronautics. Home: via Lima 20. Office: via Nazionale 172, Rome, Italy.

EULENBURG, Albert, German physician; b. Berlin, Aug. 10, 1840; s. Michael Moritz and Auguste (Saling) E.; student, Berlin, Bonn, Germany, Greifswald, Germany; m. 2d, Maria Niebauer, 1886. Physician mil. campaigns, 1866, 70-71; later joined Berlin Polyclinic; prof. pharmacology, Greifswald, 1873-82; founder polyclinic for nervous disease, Berlin; asso. prof. neurology U. Berlin, from 1880. Author: Lehrbuch der funktionellen Nervenkrankheiten auf physiologischer Basis, 1871; Die hypodermatische Injection der Arzneimittel, 1875; Die hydroelektrischen Bäder, 1883; Sexuale Neuropathie, 1895; Die Hysterie des Kindes, 1905; Kinder- und Jugendselbstmorde, 1914; Moral und Sexualität, 1916. Editor: Realenzyklopädie der gesamten Heilkunde, after 1893; Deutsche medizinische Wochenschrift. Research in field of neuropathology, sexuality; described paramyotonia congenita which is characterized by spasms of certain neck and facial muscles (Eulenburg's disease). Died Berlin, July 3, 1917.

EULENBURG, Michael Moritz, German orthopedist; b. Wriezen, Prussia, July 15, 1811; s. Sandel Eulenburg; M.D., Berlin, 1832; m. Auguste Saling; children—Albert, Ernst. Began practice medicine, Wriezen, then Berlin, 1840; founder Institut für Orthopädie, Berlin, 1851. Author: Kurzes Handbuch der Akiurgie, 1834; Die schwedische Heilgymnastik, 1853; Mitteilungen aus dem Gebiete der schwedischen Heilgymnastik, 1854; Klinische Mitteilungen aus dem Gebiete der Orthopädie und schwedischen Heilgymnastik, 1860; Die seitlichen Rückgrats-Verkrümmungen, 1876. Pioneered orthopedics; 1st description of congenital abnormal elevation of scapula, 1863; believed cause of scoliosis to be a disturbance of equilibrium in capacity of back muscles. Died Berlin, Dec. 7, 1887.

EULER, Johann Albrecht, Russian mathematician, astronomer; b. St. Petersburg (now Leningrad, USSR), Nov. 27, 1734; s. Leonhard Euler. Dir., Berlin (Germany) Obs., from 1754; sec. St. Petersburg Acad. Scis., 1766-1800, dir. studies, from 1766; prof. physics, St. Petersburg; councillor of state Russia, 1766-1800. Recipient prize (with Bossut) Acad. Paris (France), 1761. Mem. St. Petersburg (prize with Clairaut 1763), French (fgn. 1784) acads. scis., Berlin Acad. Author: Meditationes de motu vertiginis planetarum, 1760; Meditationes de perturbatione motus cometarum ab attractione planetarum orta, 1762; Sur la théorie des comètes, 1763; Sur la théorie de la lune, 1770. Original theories on comets, moon; verified theory of magnetic inclination as well as of kites and objective lenses with water fillings; research on refraction of fluids, mech. drive of ships. Died St. Petersburg, Sept. 6, 1800.

EULER, Leonhard, Swiss mathematician; b. Basel, Switzerland, Apr. 15, 1707; s. Paul E. and Margarete (Brucker) E.; studied under Johann (I) Bernoulli at U. Basel; master's degree, 1723; m. Elisabeth Gsell, 1733; 13 children; m. 2d, Salome Abigail Gsell, 1776/7. Went to St. Petersburg after failing to obtain position at U. Basel, 1727; naval asst., 1727-30; prof. physics Russian Acad. Scis., St. Petersburg, 1730, prof. math., 1733; blinded in one eye, 1735, lost use of other, 1766 (did not check his activity); went to Berlin at invitation of Frederick the Great, 1741; dir. math. Berlin Acad. Scis., 1744; recalled by Catherine II to be dir. St. Petersburg Acad., 1766-83. Fellow Royal Soc., 1746; mem. French Acad. Scis., 1755; recipient numerous prizes and honors; awarded 300 pounds sterling by Brit. Parliament for having furnished theorems useful in navigation, prize of French Acad. Scis., 1727. Author numerous books and about 800 treatises, including Mechanica, sive motus scientia analytice exposita, 1736; Einleitung in die Arithmetik, 1738; Methodus inveniendi lineas curvas, maximi minimive proprietate gaudentes, 1744; Theoria motuum planetarum et cometarum, 1744; Novae et correctae tabulae ad loca lunae computanda, 1746; Tabulae astronomicae solis et lunae; Introductio in analysin infinitorum, 1748; Scientia navalia, seu tractatus de construendis ac dirigendis navibus, 1749; Theoria motus lunae, 1753; Dissertatio de principio minimae actionis, una cum examine objectionum cl. prof. Koenigii, 1753; Institutiones calculi differentialis, cum ejus usu in analysi Infinitorum ac doctrina serierum, 1755; Theoria motus corporum solidorum seu rigidorum, 1765; Institutiones calculi integralis, 1768-70; Lettres à une Princesse d'Allemagne sur quelques sujets de physique et de philosophie, 1768-72; Dioptrica, 1767-71; Theoria motuum lunae nova methodo pertractata, 1772; Opuscula analytica, 1783-85. Most prolific mathematician of all time; worked on algebraic series, 1st to show that such series can be usefully employed only when they are convergent; treated trigonometry as br. of analysis; converted trigonometry from Ptolemaic chord measurements to ratios; introduced short form functional notation f (x), symbols pi, e (Euler's number) for base of modern logarithms, i for imaginary numbers, Sigma for summation symbol; changed logarithms from Napier's geometrical ratios to exponential forms; worked on topology; gave near modern form to spherical trigonometry; gave rules for transformation of coordinates in space; probably one of 1st to work in 3d and 4th quadrants of graph; made numerous improvements in differential and integral calculus (with modification his system still one generally taught in coll. calculus courses); 1st to integrate linear differential equations with constant coefficients; today Eulerian integrals define beta and gamma functions; proved addition theorem for elliptical integrals; introduced partial differential equations; helped found calculus of variations and pub. 1st textbook in field; discussed general equation of 2d degree in 3 variables; showed that conic sections are represented by general equation of 2d degree in 2 dimensions; discovered that cubic equation has 3 roots, showed how these may be determined; defined terms center of mass, center of inertia, moments of inertia, produced famous differential equations which express general motion of solid body about fixed point in terms of forces applied; applied math. to astronomy; worked out nature of several perturbations of planets; in particular worked on lunar theory, made analysis of moon's exact motion; began to replace geometric methods of proof used by Galileo and Newton with algebraic method; worked on tides; investigated mechanics of fluids; advocated wave theory of light, held that color depends on frequency; ennunciated law concerning oscillation of strings and transverse oscillation of bars; did work in optics, acoustics, navigation, also other aspects of practical math.; made important contributions to development of hydrodynamics; developed equations of motion of perfectly compressible fluid; one of 1st to apply calculus to dynamics, 1736; opposed Leibnizo-Wolffian school, especially the monad theory; did much to establish Newtonian thought in Prussia and Russia; Euler's constant, variables, numbers, Euler's line, Euler's equations all named for him. Died St. Petersburg, Sept. 18, 1783.

EULER-CHELPIN, Hans von, see von Euler-Chelpin.

EUPALINOS, Greek architect; b. Mégara, Greece; flourished Samos, Greece, 6th century B.C.; Built water systems, Mégara, circa 530 B.C.; engaged by Polycrates to build aqueduct for Samos, which entailed bldg. tunnel a half mile through a hill.

EURYPHON, Greek physician; flourished in Cnidos, 400 B.C.; ed. Sch. Cnidos, Caria; physician Sch. Cnidos; contemporary of Hippocrates. Studied anatomy; explained pleurisy as lung affliction; used milk and a red hot iron to treat consumption; believed insufficient evacuation of feces caused disease, that hemorrhage could occur from arteries as well as veins; developed obstet. and gynecol. treatments.

EUSTACHIO, Bartolemeo, Italian physician, surgeon, anatomist; b. San Severino, Italy, 1524. Taught at U. Pisa; went to Rome with Cardinal della Rovere; prof. of medicine, Sapienza, Rome, 1555-68; physician to St. Charles Borromeo, Cardinal della Rovere and the Pope; prof. anatomy, Rome, to 1574. Author: Opuscula Anatomica, 1564; Tabulae Anatomicae, completed 1552, pub. 1714. With Vésale and Fallope, one of founders of anatomy; supported Galen and opposed Versalius; described sympathetic nervous system, kidneys, ear, vocal cords, uterus, valves of the heart; 1st to describe adrenal glands; described chorda tympani and identified it as a nerve, 1563; described placenta, 1552; coclea, 1552; semilunar valve, 1563; discovered thoracic duct in animals, 1552; gave 1st accurate description of 1st and 2nd dentitions, 1563; distinguished renal arteries from renal canals; gave classic description of auditory tube, 1563; works included accurate illustrations. Died Fossombrone, August 1574.

EUTOCIOS, Byzantine mathematician; b. Ascalon, Palestine, circa 480; pupil of Isidorus the architect; elder contemporary of Anthemios; author commentaries on 1st 4 books of Apollonios' Conics, also on 3 works of Archimedes, making valuable comments on his Treatise on the Spheres and Cylinder and revealing calculating processes used in sch. of Alexandria, pub. in Bologna, 1556; gave various methods of solving problem of duplication of cube.

EVANARI, Michael (formerly Walter Schwarz), botanist; b. Metz, Germany, Oct. 9, 1904; s. Herman and Karoline (Loewenstein) S.; Ph.D., U. Frankfort (Germany), 1926; m. Liesclotte Wolff, Apr. 6, 1948; 1 son, Eliahu. Asst., U. Frankfurt, 1927-28; asst. German U., Prague, Czechoslovakia, 1928-31; asst. Technische Hochschule, Darmstadt, Germany, 1931-33, privatdozent, 1933; faculty Hebrew U., Jerusalem, Israel, 1937—, prof. botany, 1951—, v.p., 1953-59, dir. dept. botany, 1951—. Fellow Linnean Soc. (London); mem. Am. Soc. Plant Physiologists, Deutsche Botan. Gesellschaft, German Acad. Sci. Am. Bot. Soc. (corr.). Author: General Botany, 1954; also numerous articles. Research on physiology and ecology of germination of seeds, ecology and water balance of desert plants, ancient and modern desert agr. Home: Motza Ilith, Beit Kalonimus, Jerusalem, Israel.*

EVANS, Alfred Spring, Am. physician; b. Buffalo, Aug. 21, 1917; s. John H. and Ellen (Spring) E.; A.B., U. Mich., 1939, M.P.H., 1960; M.D., U. Buffalo, 1943; m. Brigitte Kluge, July 26, 1952; children—John Kluge, Barbara Spring, Christopher Paul. Intern U. Pitts. Hosps., 1943-44; resident Goldwater Hosp., N.Y.C., 1944; USPHS post-doctoral research fellow Yale Med. Sch., 1947-48, from instr. to asst. prof. medicine, 1949-50; resident Buffalo Gen. Hosp., 1948-49; asso. prof. preventive medicine and med. microbiology U. Wis. Sch. Medicine, 1952-59, prof., chmn. dept. preventive medicine, also dir. Wis. State Lab. Hygiene, 1959-66; prof. epidemiology, dir. WHO reference serum bank dept. epidemiology and pub. health Yale U. Sch. Medicine, New Haven, Conn., 1966—. Member microbiology fellowship panel NIH, 1960-64; cons. to Philippine Health Dept., World Health Organization, 1962, 1964. Served to captain, Medical Corps, AUS, 1944-46, 50-52; col. Res. Diplomate American Board Internal Medicine. Fellow Am. Pub. Health Assn.; mem. Assn. of Teachers of Preventive Medicine, American Epidemiol. Soc., Soc. Exptl. Biology and Medicine, Central Soc. Clin. Research, Delta Omega. Author articles infectious diseases. Research studies in epidemiology and etiology of common respiratory diseases, infectious mononucleosis. Home: Dogwood Circle, Woodbridge, Conn. Office: 333 Cedar St., New Haven.*

EVANS, Sir Arthur John, English archeologist; b. Nash Mills, Herfordshire, Eng., July 8, 1851; s. John Evans; grad. Harrow; ed. Brasenose Coll., Oxford (Eng.) U., U. Göttingen (Germany); D.Litt., Oxford U., Dublin (Ireland) U.; M.A., Ph.D. (hon.), U. Berlin (Germany), 1910; LL.D., U. Edinburgh (Scotland); m. Margaret Freeman, 1878. Fgn. travel, 1873-75; keeper Ashmolean Mus., Oxford U., from 1884, supr. reorgn., also fellow Brasenose Coll.; Frazer lectr. in social anthropology Cambridge (Eng.) U., 1930-31. Knighted, 1911. Recipient Royal Gold medal Royal Inst. Brit. Architects, 1909, Petrie medal, 1931. Fellow Royal Soc. (Copley medal 1936), 1901; mem. Soc. Antiquaries (pres. 1914-19), Brit. Assn. (pres. 1916-19). Author: Cretan Pictographs and Prae-Phoenician Script, 1896; Further Discoveries of Cretan and Aegean Script, 1898; Scripta Minoa, 1909; Palace of Minos, Parts I-IV, 1921-36; others. Made archeol. excavations in Crete, 1893-1935; most famous discovery was Minoan civilization of Crete; major excavation Palace at Knossos (structure as large as Buckingham Palace), in which King Minos is said to have lived (information concerning artwork, clothing, architecture, daily life revealed, also hieroglyphic inscriptions of a pre-Phoenician script). Died nr. Oxford, July 11, 1941.

EVANS, Arthur Thompson, Am. botanist; b. Wellington, Ill., May 22, 1888; s. Robert M. and Anna Caldwell (Johnstone) E.; A.B., U. Ill., 1912; studied U. Mich., summers, 1913, 14; M.A., U. Colo., 1915; Ph.D., U. Chgo., 1918; m. Anna Mathilde Hansen, Aug. 22, 1914; children—Margaret Louise, Arthur T., Lewis Hansen, Dorothy Ann. Grad. asst. in botany U. Colo., 1914-15, instr., 1915-17; fellow in botany U. Chgo., 1917-18; in charge cereal disease investigations U. S. Dept. Agr., in Great Pains Region, World War, 1918, corn investigations, 1919; dean coll., prof. botany Huron (S.D.) Coll., 1919-20; asso. prof. agronomy S.D. State Coll., 1920-23, head of dept. and prof. botany and plant pathology, 1923-28; prof. botany Miami U. Fellow A.A.A.S.; mem. Bot. Soc. Am., Sigma Xi, Phi Sigma, Delta Pi. Author: First Course in Botany (with R. J. Pool); Laboratory Manual for First Course in Botany. Research and publs. on morphology and cytology. Died Oct. 5, 1943.

EVANS, Sir Charles Lovatt, English physiologist; b. Birmingham, Eng., July 9, 1884; s. Charles and Alice (Hipkins) E.; B.Sc., U. London (Eng.), 1912, D.Sc., 1913, LL.D., 1957; LL.D., U. Birmingham (Eng.), 1934; m. Laura Stevenson, Apr. 11, 1911; children—Joan Lovatt (Mrs. Peter Prince), Yvonne Lovatt (Mrs. Jorgen Schou). Sharpey scholar U. Coll. London, 1910; prof. physiology U. Leeds, 1918; staff Med. Research Council, 1919-22; prof. physiology St. Bartholomew's Hosp., Med. Coll., 1922-26; Jodrell prof. physiology U. Coll. London, 1926-49; cons. Ministry Def., 1949—. Chmn. council Royal Vet. Coll., 1949-63. Created knight, 1951. Fellow Royal Soc., 1925, Royal Coll. Physicians; mem. Physiol. Soc., Biochem. Soc., Ergonomics Soc. Author: Recent Advances in Physiology, 1925; Principles of Human Physiology, 1930; also numerous articles. Research on circulation and action of drugs. Home: Hedgemoor Cottage, Winterslow, Salisbury, Eng.*

EVANS, Christopher Riche, Brit. psychologist; b. Aberdovey, Wales, May 29, 1931; s. Herbert Riche and Kathleen (Gorst) E.; B.A. with honors, U. Coll., London, 1960; Ph.D., U. Reading, 1963; m. Nancy Jane Fullmer, Aug. 18, 1962; children—Victoria Amelia. Research asst. J. J. Thomson Phys. Lab., U. Reading, 1957-60; with autonomics div. Nat. Phys. Lab., Teddington, Eng., 1963—, prin. research fellow, 1967—. Editor: (with A. D. J. Robertson) Psychology and Physiology of the Brain, 1966. Research, publs. on pattern perception of images held stable on retina, devel. of after-image method of achieving perfect stblzn. superceding contact lens, comparison of dream processes and computer processes. Home: 19 The Barons, St. Margarets, Twickenham, London, Eng. Office: Nat. Phys. Lab., Teddington, Eng.*

EVANS, Clifford, Am. anthropologist, archeologist; b. Dallas, June 13, 1920; s. Clifford and Pearl (Weiss) E.; A.A., San Bernardino Jr. Coll., 1939; A.B., U. So. Cal., 1941; Ph.D., Columbia, 1947; m. Betty J. Meggers, Sept. 13, 1946. Asst. archeology dept. anthropology Columbia, 1946-48; instr. anthropology U. Va.. 1949-51; asso. curator div. archeology U. S. Nat. Mus., Smithsonian Instn., Washington, 1951-62, curator div. archeology, 1962—; archeol. fieldwork, Brazil, Ecuador, Peru, Venezuela, Brit. Guiana, Ponape (Caroline Islands), Dominica, B.W.I. Recipient outstanding achievement award in sci. Washington Acad. Scis., 1954; Gold medal 37th Internat. Congress of Americanists, 1966; Ordem al Merito, Ecuador, 1966. Fellow Am. Anthropol. Assn.; mem. Anthropol. Soc. Washington (treas. 1960-65), Inst. Andean Research (pres. 1960-65, v.p. 1965—), Soc. Am. Archeology. Author: (with W. D. Strong) Cultural Stratigraphy in Viru Valley, Northern Peru, 1952; (with B. J. Meggers) Archeological Investigations at the Mouth of the Amazon, 1957, Archeology of British Guiana, South America, 1960, Archeological Investigations on Rio Napo, Ecuador, 1968; (with Meggers and E. Estrada) Early Formative Period of Coastal Ecuador, 1965. Contbr. articles profl. jours. Reconstructed preColumbian history of man in various parts of world, particularly No. S. Am.; studies in evolution of culture; discovered earliest pottery bearing culture in New World (with Meggers and Estrada). Home: 1227 30th St. N.W., Washington 20007. Office: Mus. Natural History, Smithsonian Instn., Washington 20560.*

EVANS, Cyril David, Am. chemist; b. Anaconda, Mont., May 18, 1909; s. David M. and Jennie (Ulrich) E.; B.S., Mont. State U., 1931; Ph.D., U. Minn., 1938; m. Elizabeth E. Fransham, July 27, 1933; children—Robert O., Barbara E. With Mont. Expt. Sta., Bozeman, 1931-35, U. Minn., 1935-39, Armour Research Found., 1939-41; staff No. Regional Research Lab., Peoria, Ill., 1941—, prin. chemist, 1955—. Recipient Superior Service award U. S. Dept. Agr., 1951. Mem. Am. Chem. Soc., Am. Oil. Chem. Soc., Assn. Food Technologists, A.A.A.S., Sigma Xi. Research, publs., patents on indsl. protein applications, including foods and fibers; edible vegetable oil processing, modifications and evaluation; flavor and odor evaluation of foods and taste panel operation. Home: 5133 Prospect St.; Peoria 61604. Office: 1815 N. University St., Peoria, Ill. 61614.*

EVANS, Daniel Donald, Am. soil physicist; b. Oak Hill, O., Aug. 13, 1920; s. Thomas Herbert and Mary (Jenkins) E.; B.S., Ohio State U., 1947; M.S., Ia. State U., 1949, Ph.D., 1952; m. Frances Louise Merdink, Jan. 16, 1946; children—Robert Keith, Thomas George, Roger Alan, Susan Mary. Faculty, Ia. State U., Ames, 1948-53, Ore. State U., Corvallis, 1953-60, 62-63; soil physicist Ministry Agr., Kenya Govt., Nairobi, 1960-62; prof. U. Ariz., Tucson, 1963—. Mem. Am. Soc. Agronomy, Soil Sci. Soc. Am., Western Soil Sci. Soc., Am. Geophys. Union, Sigma Xi, Gamma Sigma Delta. Research, publs. on soil water flow, energy status soil water, methods evaluating soil water, soil structure characterization, soil-water-plant relationships. Home: 2239 E. Kleindale St., Tucson, 85719.*

EVANS, David Gwynne, English bacteriologist, immunologist; b. Manchester, Eng., Sept. 6, 1909; s. Frederick George and Margaretta (Williams) Evans; B.Sc., U. Manchester, 1933, M.Sc., 1934, Ph.D., 1938, D.Sc., 1948; m. Mary Darby, Aug. 7, 1937; children—John Gwynne, Mary Gwynne. Asst. lectr. dept. bacteriology U. Manchester, 1934-40, reader dept. bacteriology, 1947-55; research staff Nat. Inst. for Med. Research, London, Eng., 1940-47, dir. biol. standards control lab., 1955-58, 61—, mem., 1965—; prof. bacteriology and immunology London Sch. Hygiene and Tropical Medicine, 1961—; mem. governing body Lister Inst. Preventive Medicine, London, 1965—. Mem. adv. panel on biol. standardization WHO, 1956—. Fellow Royal Soc., 1960, Coll. Pathology; mem. Soc. for Gen. Microbiology, Path. Soc., Soc. for Immunology, Internat. Soc. for Microbiol. Standardization. Research, numerous publs. on pathogenesis, prophylaxis of bacterial infections, biol. standardization. Home: Highthwaite, Stony Lane, Great Missenden, Buckinghamshire, Eng. Office: London Sch. Hygiene and Tropical Medicine, Keppel St., London, W.C.1., Eng.*

EVANS, Earl Alison, Jr., Am. biochemist, educator; b. Balt., Mar. 11, 1910; s. Earl Alison and Florence (Lewis) E.; student Balt. Poly. Inst., 1924-28; B.Sc., Johns Hopkins, 1931; Ph.D. (U. fellow) Columbia, 1934-36. Research asst. Johns Hopkins, 1931-32, asst. lab. for endocrine research, 1932-34; research fellow Columbia Coll. Phys. and Surg., 1935-36; faculty U. Chgo., 1937—, prof., chmn. dept., 1942—. Mem. panel on isotopes, com. on growth NRC, 1946-51, div. med. scis., 1962-65; spl. cons. biochemistry and nutrition USPHS, 1947-51, mem. bd. sci. counselors Nat. Inst. Arthritis and Metabolic Diseases, 1960-63; mem. panel on fellowships NSF, 1951-53, div. com. for biology and medicine, 1963-66; cons. Sec. State, 1951-53; adviser Am. Found. for Continuing Edn., 1962—; chmn. postdoctoral fellowship com. div. biology and agr. Nat. Acad. Scis.-NRC, 1963-65. Fellow A.A.A.S.; mem. Am. Chem. Soc. (Eli Lilly medal 1941), Harvey Soc. (hon.), Am. Soc. Biol. Chemists (treas. 1947-51, editorial com. Jour. Biol. Chemistry 1954-57), Biochem. Soc. (Gt. Britain), Soc. Am. Bacteriologists, Asociacion Quimica Argentina (hon. corr.), A.M.A. (Affiliate), Sigma Xi, Tau Beta Pi. Author: Biochemical Studies of Bacterial Viruses, 1952. Editor: Biological Action of the Vitamins, 1942. Research, publs. in chemistry, mechanism of action, metabolism of various organic substances, biochemistry of malarial parasite, bacterial viruses, infectious DNA. Home: 12 E. Scott St., Chgo. 60610.*

EVANS, Elwyn, Am. physician; b. Dodgeville, Wis., Dec. 15, 1901; s. George B. and Kathryn (Jones) E.; B.S., Lewis Inst., 1927; M.D., U. Chgo., 1934; m. Leone Hendrickson, Sept. 6, 1935; children—Sara Jane (Mrs. Dyer) Moss, Elwyn Kim. Practice medicine, specializing in cardiology, Orlando, Fla., 1946—; electrocardiographer Fla. Sanitarium and Hosp., Orlando, 1942—; chmn. dept. cardiology Mercy Med. Center, Orlando, 1965—. Mem. cons. staff to various hosps. Mem. med. adv. bd. Fla. Selective Service, 1951—; mem. cardiac adv. com. Fla. Crippled Childrens Commn., 1963—. Diplomate Am. Bd. Internal Medicine, Am. Bd. Cardiovascular Disease, Pan-Am. Med. Assn. Fellow A.C.P., Am. Med. Writers' Assn., Am. Med. Authors, Council Clin. Cardiology, Am. Heart Assn., Am. Coll. Cardiology (past trustee, past sec. bd. govs.); mem. Am. (dir. 1956-59), Fla. (past pres., dir. 1949—), Orange County (past pres., dir. 1953—) heart assns., Orange County Med. Soc. (past v.p.). Abstractor Excerpta Medica-Amsterdam, Holland, 1952-66. Guest editor Am. Jour. Cardiology, 1961. Research, numerous publs. on electrocardiogram in pneumoperitoneum, hypertension, pericarditis, carotid sinus. Home: 105 Glenridge Way, Winter Park, Fla. 32789. Office: 500 E. Colonial Dr., Orlando, Fla. 32803.*

EVANS, E(rnest) Edward, Jr., Am. microbiologist; b. Parkersburg, W.Va., Dec. 14, 1922; s. Ernest Edward and Vivian (Enoch) E.; A.B. with high honors, Ohio U., 1945; M.S., Ohio State U., 1947; Ph.D., U. So. Cal., 1950; m. Marjorie McConville, Sept. 2, 1947. Instr. bacteriology U. Mich., Ann Arbor, 1950-51, asst. prof., 1951-55; asso. prof. microbiology U. Ala. Med. Center, Birmingham, 1955-61, prof., chmn. dept., 1961—. Fellow A.A.A.S., Am. Acad. Microbiology; mem. Am. Soc. Microbiology (com. on bacteriologic technique 1955-59), Soc. Exptl. Biology and Medicine, Am. Assn. Immunologists, Internat. Soc. Human and Animal Mycology, Sigma Xi. Editorial bd. Jour. Bacteriology, 1961-66. Research, numerous publs. on antigens of fungi, reactions of polyions with microbial cells, evolution of immunity, antibody synthesis, immunological memory; discovered 3 antigenic types of Cryptococcus neoformans. Home: 3525 Crestbrook Rd., Mountain Brook, Ala. 35223.*

EVANS, F., Brit. marine zoologist; b. Croydon, Eng., 1926; s. Alfred and Beatrice (Morgan) E.; B.Sc., U. London, 1952, Ph.D., 1958; m. Rosemary Evans, 1948; children—Susanne Charlotte, Elizabeth Jane. Leader, Petula transatlantic expdn., 1953-54; lectr. zoology U. Durham (Eng.), 1955-58, U. Coll. Ghans, 1958-60, Dove Marine Lab., U. Newcastle-on-Tyne, Eng., 1960-65; dir. marine biology Royal U. Malta, 1966—. Mem. Challenger Soc. Author: Voyage of the Petula, 1957; also articles. Observations on distbn. of planktonic crustacea in tropical Atlantic and N. Sea, on sea currents, formation of sea fog in N. Sea; research on zonation, shore of periwinkles in Europe and W. Africa. Office: The Evans Labs., Royal U., Malta.*

EVANS, Francis Cope, Am. ecologist; b. Phila., Dec. 2, 1914; s. Edward W. and Jacqueline (Morris) E.; B.S., Haverford Coll., 1936; D.Phil. (Rhodes scholar) Oxford U., 1939; m. Rachel Worthington Brooks, June 12, 1942; children—Kenneth R., Katharine C., Edward W. II, Rachel H. Faculty, U. Cal., Davis, 1941-42, Haverford Coll., 1942-48; faculty U. Mich., Ann Arbor, 1948—, prof. zoology 1959—. Recipient H. R. Painton award Cooper Ornithol. Soc., 1963. Guggenheim fellow, 1962-63. Fellow A.A.A.S.; mem. Brit., Am. ecol. socs., A.A.A.S., Am. Inst. Biol. Scis. Editor Ecol. Monographs, 1955-61. Research, publs. on structure and function of natural populations and communities, and on relations between man and his environment. Home: 2019 Day St., Ann Arbor, Mich. 48104.*

EVANS, George William, II, Am. mathematician; b. Houston, June 8, 1920; s. Griffith Conrad and Isabel (John) E.; student U. Colo., 1938-39; A.B., U. Cal. at Berkeley, 1942, M.A., 1942; Ph.D., N.Y. U., 1951; m. Marjorie Louise Woodard, Jan. 30, 1943; children—George William, III, Anne Garvin. Research asso., instr. math. N.Y. U., N.Y.C., 1946-49; research mathematician Argonne Nat. Labs., Lemont, Ill., 1949-53; research mathematician U. Cal. Radiation Lab., Livermore, 1953-54; mgr. math. scis. dept. Stanford Research Inst., Menlo Park, Cal., 1954-66, sr. research mathematician, 1966—. Asso. prof., math. dept. U. Santa Clara (Cal.), 1966—; cons. firm Evans Assos. Mem. Am. Math. Soc., Sigma Xi, Pi Mu Epsilon, Kappa Sigma. Author: Programming and Coding for Automatic Digital Computers, 1961; Stimulation Using Digital Computers, 1967. Contbr. articles on heat conduction, numerical analysis, transp. and prodn. systems, mil. analysis, spectral analysis to profl. jours. Home: 14511 DeBell Dr., Los Altos Hills, Cal. 94022. Office: 333 Ravenswood St., Menlo Park, Cal. 94025.*

EVANS, Griffith, veterinarian; b. Ty Mawr, India, Aug. 7, 1835; s. Evan and Mary Evans; M.D., C.M., McGill U.; student Royal Vet. Coll., London; D.Sc., U. Wales; m. Catherine Mary Jones, 1870; 1 son, 4 daus. With Royal Army Vet. Dept., 1860-90; served in Can. and India. Recipient Mary Kingsley medal Liverpool Sch. Tropical Medicine, John Steel medal Royal Coll. Vet. Surgeons. Contbr. to profl. jours. Named by Govt. of India to study surra (disease fatal to horses and camels), 1880, discovered its cause Trypanosome evansi, demonstrated it can be transferred by subcutaneous injection and by passage into stomach; discovered Filaria Sanguines evansi in blood of adult and embryo camel, 1882. Died Dec. 7, 1935.

EVANS, Griffith Conrad, Am. mathematician; b. Boston, May 11, 1887; s. George William and Mary (Taylor) E.; A.B., Harvard, 1907; A.M., 1908, Ph.D., 1910; student U. Rome, 1910-12; LL.D., U. Cal., 1956; m. Isabel Mary John, June 20, 1917; children—Griffith Conrad, George William, Robert John. Instr. math. Harvard, 1906-07, 09-10; Sheldon fellow in Italy, 1910-12; asst. prof. math. Rice Inst., 1912-16, prof., 1916-34; vis. prof. U. Cal., summers 1921, 28; prof. math., 1934-55, chmn. dept., 1934-49, now emeritus; spl. lectr. colls., univs. Mem. NRC, 1927-31, 40-43, 50-53. Tech. cons., sci. expert ordnance War Dept., 1943-47. Recipient Distinguished Assistance award U. S. Army, 1946; Presdl. Certificate of Merit, 1948. Fellow Econometric Soc.; mem. Am. Acad. Arts. and Scis., Nat. Acad. Scis., Am. Philos. Soc., Am. Math. Soc. (pres. 1938-40), Math. Assn. Am. (v.p. 1932), A.A.A.S. (v.p. 1931-32, 36-37). Catholic. Author: Mathematical Introduction to Economics, 1930; Stabilité et Dynamique de la Production dans l'Economie Politique (Mémorial des Sciences Math. 56), 1932; Functionals and Their Applications, Cambridge Colloquium, Part I, 1918, rev., 1964; The Logarithmic Potential, Vol. 6, 1927; also tech. papers. Contbns. included integral equations with particular types of kernels; applications of elementary and advanced math. to econ. theory, strength of materials, gas dynamics, potential theory; problems of applied math. related to rectifiable curves. Home: 26 Spyglass Hill, Oakland, Cal. 94618. Office: Dept. Math., U. Cal., Berkeley, Cal. 94720.*

EVANS, Harold J., Am. plant physiologist; b. Franklin, Ky., Feb. 19, 1921; s. James H. and Allie (Uhls) E.; B.S., U. Ky., 1946, M.S., 1948; Ph.D., Rutgers U., 1950; postgrad. Johns Hopkins, 1952; m.

Elizabeth M. Dunn, Dec. 14, 1946; children—Heather Mary, Pamela. Faculty, N.C. State Coll., 1950-61, prof. botany, 1956-61; prof. plant physiology Ore. State U., 1961—. Cons., NSF, 1965-66. Co-recipient Hoblitzelle Nat. award, 1960-64; recipient Basic Research award Ore. State U., 1965. Mem. Am. Soc. Plant Physiologists (past. chmn. Western sect.), Am. Soc. Biol. Chemists, Brit. Biochem. Soc., Sigma Xi. Editorial bd. Jour. Plant Physiology, 1965—; editorial com. Ann. Revs. Plant Physiology, 1966—. Research, publs. on requirements and biochem. role minerals; co-discoverer pyridine nucleotide nitrate reductase in microorganisms and higher plants, requirement for cobalt of legumes and nitrogen fixing bacteria. Home: 2939 Mulkey St., Corvallis, Ore. 97330.*

EVANS, Henry John, Brit. geneticist; b. Llanelly, Wales, U.K., Dec. 24, 1930; s. David and Gwladys (Jones) E.; B.Sc. with honors, U. Coll. Wales, Aberstwyth, 1952, Ph.D., 1955; m. Gwenda Rosalind Thomas, Mar. 3, 1956; children—Paul David, William Hugh, John Kelsey, Owen James. Research asst. U. Wales, Aberystwyth, 1952-55; research scientist radiobiol. research unit Med. Research Council, Harwell, U.K., 1955-66; prof. genetics U. Aberdeen (Scotland), 1966—. Mem. Genetics Soc., Assn. for Radiation Research, Soc. for Exptl. Biology, Radiation Research Soc. Author: Radiation-induced Chromosome Aberrations, 1966; also articles. Editorial bd. Radiation Botany, 1961—, Mutation Research, 1964—. Research on chromosome structure and behaviour, especially damage induced by ionizing radiations and chem. agts. Home: 43 Gray St., Aberdeen, Scotland.*

EVANS, Herbert McLean, Am. anatomist, embryologist; b. Modesto, Cal., Sept. 23, 1882; s. C. W. and Bessie (McLean) E.; B.Sc., U. Cal., 1904; M.D., Johns Hopkins, 1908; M.D., honoris causa, Albert Ludwig U. Freiburg, 1930, Univ. Catolica de Chile, 1941; Sc.D., Univ. Mayor de San Marcos de Lima, 1941, U. Birmingham, 1950; Doctor Honoris Causa, U. Paris (France), 1946; m. Anabel Tulloch, Sept. 17, 1905; 1 dau., Marian McLean; m. 2d, Marjorie E. Sadler, June 28, 1932; 1 dau., Gail Bayne; m. 3d, Dorothy F. Atkinson, June 14, 1945. Asst. instr., asso. and asso. prof. anatomy, Johns Hopkins, 1908-15; research asso., Carnegie Instn., Washington, 1913-15; prof. anatomy, U. Cal. at Berkeley, 1915—, Hertzstein prof. biology and dir. Inst. Exptl. Biology, 1930—; hon. prof. Facultad de Biologia y Ciencias Medicas, Universidad de Chile, and Facultad de Ciencias Medicas, Universidad Central del Ecuador, 1941. Faculty research lectr., U. Cal., 1925; Bacon lectr. U. Ill., 1931; Hertzstein lectr. Stanford and U. Cal., 1934; Jackson lectr. U. Minn., 1937; Jones lectr. U. Ore., 1940; Messenger lectr. Cornell U., 1942; Mellon lectr. U. Pitts. 1942; Nat. Sigma Xi lectr., 1942; Guiteras lectr. Am. Urol. Assn., N.Y., 1942; Porter lectr. U. Kan., 1943; Vanuxem lectr. Princeton, 1946; Ludvig Hektoen lectr. Frank Billings Found. of Inst. Medicine Chgo., 1947; William E. Lower lectr. Cleve. Acad. Medicine, 1947; Eastman Meml. lectr. U. Rochester, 1947; William Withering lectr., U. Birmingham, 1950; Macarthur lectr., U. of Edinburgh, 1950. Editor: Am. Anatomical Memoirs, 1918-39; Jour. of Nutrition, 1928-30. Awarded John Scott medal, 1928; gold medal for sci. exhibit, A.M.A. meeting, San Francisco, 1946. Fellow Am. Acad. Arts and Scis.; mem. Am. Physiol. Soc., History Sci. Soc., Nat. Acad. Scis., Am. Assn. Anatomists (pres. 1930-32), Kaiserliche Leopoldinsch-Carolinische Deutsche Akademie der Naturforscher, Buenos Aires, 1942, (hon.) Sociedad Médica de Santiago (Chile), (corr.) Sociedad Argentina de Biología, Phi Beta Kappa, Sigma Xi. Contbr. papers on devel. vascular system from capillaries; action of vital dyes of the benzidine series; physiology of reproduction; relations of fertility and nutrition; endocrine organs, especially the sex glands and hypophysis. Demonstrated origin of body vascular trunks from capillary plexes, 1909; explained physiological behavior of vital stains of benzidine series, 1915; introduced use of certain azo dyes for estimation of blood volume, 1917; charted 48 chromosomes in man, 1918, 1929; with J. A. Long gave first description of oestrous cycle in rat, 1921, of essential value in isolation of female sex hormones; produced gigantism and other specific endocrine effects from anterior-hypophyseal hormones administered parenterally, 1922, and established separation of pituitary growth promoting substance or hormone, 1939, and (with C. H. Li) purified it, 1944; production of permanent diabetes by chronic administration of anterior pituitary extracts, 1932; first detected criterion of vitamin A deficiency in continuous vaginal cornification, 1922; discovered vitamin E, essential for reproduction in higher animals, 1922, and (with D. H. and G. A. Emerson) first purified and determined its impirical constitution, 1935; with C. H. Li first purified anterior hypophyseal adrenocorticotropic hormone, 1942. Home: 511 Coventry Rd., Berkeley, Cal. 94707.*

EVANS, Herbert P(ulse), Am. mathematician; b. Chattanooga, Jan. 5, 1900; s. Oscar Ewel and Effie Gertrude (Pulse) E.; B.S., U. Wis., 1923, M.S., 1927, Ph.D., 1929; student Columbia, 1924-25; m. Rae Elbertine White, Dec. 27, 1929; children—Douglas Sherwood, Gail Kristine. Research engr. Bell Tel. Lab., N.Y.C., 1923-25; instr. elec. engring. U. Wis., 1925-28, research asst. in physics (and Columbia) 1928, mem. dept. math. from 1928, instr., 1928-29, asst. prof., 1929-38, asso. prof., 1938-42, prof., from 1942, charge univ. extension math. dept., from 1945. Fellow A.A.A.S.; mem. Am. Math. Soc., Math. Assn. Am., Inst. Math. Statistics, Am. Statis. Assn., Sigma Xi, Pi Mu Epsilon. Asso. ed. Am. Math. Monthly, 1944-49. Contbr. articles on electric circuit theory, boundary value problems, probability theory, and math. edn. to sci. jours. Died June 2, 1959.

EVANS, John Ellis, Am. physicist; b. Oak Hill, O., Oct. 2, 1914; s. David S. and Sarah (Thomas) E.; student Rio Grande Coll., 1932-34; B.A. and B.S. in Edn., Ohio State U., 1936, M.A., 1937, postgrad., 1941-42; Ph.D., Rice U., 1947; m. Jane Gillespie, Apr. 10, 1948. Tchr. high sch., O., 1937-38; prof. physics and math. Rio Grande (O.) Coll., 1938-41; grad. asst. in physics Ohio State U., 1941-42; mem. staff radiation lab. Mass. Inst. Tech., Cambridge, 1942-45; fellow in physics Rice U., 1945-48; staff Los Alamos (N.M.) Sci. Lab., 1948-52; group leader atomic energy div. Phillips Petroleum Co., Idaho Falls, Ida., 1952-54, sect. head, 1954-56, dir. nuclear physics br., 1956-61; sr. cons. sci., sr. mem. Research Lab., Lockheed Missiles & Space Co., Palo Alto, Cal., 1961—. Mem. U. S. AEC Nuclear Cross Sect. Adv. Group, 1956-61. Fellow Am. Phys. Soc. Research, publs. in fields magnetron operating characteristics, charged particle nuclear physics, homogeneous reactors and delayed neutrons, neutron cross-sections, inelastic scattering of thermal neutrons, auroral physics. Patentee in field. Home: 650 San Antonio Rd., Los Altos, Cal. 94022. Office: 3251 Hanover St., Palo Alto, Cal. 94304.*

EVANS, John Wainwright, Jr., Am. solar astronomer; b. N.Y.C., May 14, 1909; s. John Wainwright and Edith (Clagett) E.; B.A., Swarthmore Coll., 1932; postgrad. U. Pa.; M.A., Harvard, 1936; Ph.D., 1938; m. Elizabeth Fredd Harlan, Aug. 23, 1932; children—Wainwright, Nancy Jane (Mrs. Reinhard Krien), Jeanne Harlan. Instr., U. Minn., 1937-38; asst. prof. Mills Coll., 1938-42; researcher Inst. Optics, U. Rochester, 1942-46; astronomer, asst. dir. High Altitude Obs., Boulder, Colo., 1946-52; dir. Sacramento Peak Obs., Air Force Cambridge Research, Office Aerospace Research, Sunspot, N.M., 1952—. Recipient Dept. Def. Distinguished Civilian Service award, 1965. Fellow Am. Acad. Arts and Scis.; mem. Am. Astron. Soc. A.A.A.S. (co-recipient Cleveland prize 1957), Internat. Astron. Union, Optical Soc. Am., Astron. Soc. Pacific. Editor: The Solar Corona, 1963. Research, numerous publs. on devel. astron. instruments; research on solar atmosphere; discovery inclined bright columnar elements in solar atmosphere. Address: Sacramento Peak Obs., Sunspot, N.M. 88349.*

EVANS, Joseph Patrick, Am. surgeon; b. LaCrosse, Wis., Nov. 29, 1904; s. Edward and Sarah (Thompson) E.; student U. Notre Dame, 1921-23; A.B., Harvard, 1925, M.D. cum laude, 1929; M.Sc., McGill U., 1930, Ph.D., 1937; D.Sc. (hon.), Loyola U., 1964; m. Hermene Eisenman, June 24, 1929; children—Mary F. (Mrs. Francois Bapst), Edward, Frederick N., Caroline (Mrs. Alvaro Villa), Anne W. (Mrs. Claude Lanctot), Hermene W., John Fisher, Thomas More. Asso. prof. surgery in charge neurol. surgery U. Cin., 1937-54; prof. neurol. surgery U. Chgo., 1954—, dir. div. neurol. surgery U. Chgo. Hosps., 1954-67. Fellow A.C.S., American Med. Assn.; member of mem. Acad. Neurol. Surgery, Assn. for Research in Nervous and Mental Diseases, Harvey Cushing Soc., Soc. Neurol. Surgeons, Internat. League Against Epilepsy, others, Sigma Xi, Alpha Omega Alpha. Author: Acute Head Injury, 2d edit., 1960; also numerous articles. Research in head injury, especially cerebral cicatrix; study of neurosurgery and allied fields; trauma. Home: 1160 E. 56th St., Chgo. 60637.

EVANS, Lewis, surveyor, geographer; b. Pennsylvania, circa 1700; in Phila., circa 1736; commd. by Benjamin Franklin. Prepared map of Pa., N.J., N.Y. and 3 Del. counties (used by migrating colonists), 1749. Author: A General Map of the Middle-British Colonies in America, (used by Braddock), 1755; Analysis, (in which he suggested a study of the Ohio Valley), 1755. Died June 12, 1756.

EVANS, Llewellyn Thomas, Am. biologist; b. Elizabeth, Colo., Sept. 15, 1902; s. Llewellyn Price and Sarah Annie (Lewis) E.; A.B., U. Denver, 1925, A.M., 1931; Ph.D., Harvard, 1936; m. Rachel Page Steinhardt, Sept. 4, 1956; stepchildren—Rachel Thorndike Steinhardt, Frederick Page Steinhardt. High sch. tchr., Arcada, Colo., 1925-27; prof. Robert Coll., Istanbul, Turkey, 1927-33; NRC fellow Harvard, 1936-37; asst. prof. Mont. U., 1937-39, Mo. U., 1939-41; research asso. U. Denver, 1941-44; curator Mus. Natural History, Denver, 1944-47; asst. prof. anatomy L.I. Coll. Medicine, N.Y.C., 1948-50; researcher Am. Mus. Natural History, Cold Spring Harbor, L.I. N.Y., 1950-54; prof. histology Albert Einstein Coll. Medicine, Yeshiva, N.Y., 1955-59; prof. comparative psychology Keene (N.H.) State Coll., 1960-65; prof. psychology Research Lab., Troy, N.H., 1959—. Mem. Am. Soc. Zoologists, Am. Soc. Ichthyologists and Herpetologists, Am. Assn. Anatomists, Internat. Soc. Behaviorists, Sigma Xi, Phi Sigma. Research, numerous publs. in devel., structure, hormones and behavior of vertebrates. Home: R.F.D. 1. Office: Research Lab., R.F.D. 1, Troy, N.H. 03465.*

EVANS, Marjorie Woodard (Mrs. George W. Evans, II), Am. chem. physicist; b. Denver, Mar. 15, 1921; d. Raymond George and Mary (Garvin) Woodard; B.A., U. Colo., 1942; Ph.D., U. Cal. at Berkeley, 1945; m. George W. Evans, II, Jan. 30, 1943; children—George W. III, Anne. Research asso. Cal. Research Corp., Richmond, 1945-46; N.Y.U., 1946-49; cons. Princeton, 1950-51; internal cons. Armour Research Found., 1951-53; scientist, dir. Poulter Labs., Stanford Research Inst., Menlo Park, Cal., 1953—. Mem. Am. Chem. Soc., Am. Inst. Physics, Am. Inst. Aeros. and Astronautics, Société de Chimie Physique, Combustion Inst., Phi Beta Kappa, Sigma Xi. Research and publs. on exptl. understanding combustion, explosion, infnition, deflagration, and detonation, critical collections and analysis of theories of deflagration and detonation. Home: 14511 De Bell Dr., Los Altos Hills, Cal. 94002. Office: 300 Ravenswood Av., Menlo Park, Cal. 94025.*

EVANS, Oliver, Am. inventor, steam-engine builder; b. New Castle County, Del., 1755; s. Charles Evans; m. Miss Tomlinson, 1780; 2 children. Perfected machine for wool manufacture that could produce 1500 cards a minute; completed series of improvements in flour-mill machinery operated by means of water-power in 1785, then petitioned legislatures Pa. and Md. for exclusive rights to use his improvements in flour mills and steam carriages in those states (partially granted in Pa., wholly granted in Md.); granted 1st U. S. patent for steam propelled land vehicle, 1789; built high pressure steam engine in Phila., 1801-82; constructed amphibious vehicle for Phila. (dredger boat with drive wheels and paddle wheels, 1803; produced numerous high pressure steam engines for grinding and sawing; improved corn grinding machine with lift, cup chain-conveyor, hopper, slide chute, hauling band; obtained patent for steam engine, 1797; in engine bldg. bus., 1803; established Mars Iron Works, 1807; designed and constructed water works in Phila., 1817; 1st steam engine builder in Am., 50 of his engines in use throughout Atlantic coast states by 1819. Author: The Young Mill-Wright and Miller's Guide, 1795; The Abortion of the Young Engineer's Guide, 1805. Died N.Y.C., Apr. 5, 1819.

EVANS, Richard Isadore, Am. psychologist; b. Chgo., Aug. 29, 1922; s. Louis and Rachel (Selzer) E.; B.S., U. Pitts., 1946, M.S., 1947; Ph.D., Mich. State U., 1950; m. Zena Ann Rubin, May 26, 1952; children—Randolph Warren, Vicki Eve, Dennis Alan, Sharon Renee. Vis. prof. psychology U. Tenn., 1947-48; faculty Mich. State U., 1948-50; prof. psychology, U. Houston, 1950—, Dir. NSF filmed dialogues with notable contbrs. to psychology project, 1963—; cons. U. Tex. Dental Br., 1956—; dir. Nat. Inst. Dental Research Psychology Research Tng. Program, 1963—. NSF, Nat. Inst. Dental Research, U. S. Office Edn., Ford Found., Dept. Health, Edn. and Welfare grantee. Fellow Am. Psychol. Assn.; mem. A.A.-A.S., Tex. Acad. Scis. (v.p. social scis.), Sigma Xi, Psi Chi. Author: Conversations with Carl Jung and Reactions from Ernest Jones, 1964; Dialogue With Erich Fromm, 1966; Dialogue with Erik Erikson, 1967; (with Peter Leppmann) Resistance to Innovation in Higher Education, 1967. Pioneer research in field of ednl. t.v.; devel. innovations in communication in univ. instrn.; basic research on attitude change, juvenile delinquency, polit. extremism, psychology in dentistry. Home: 3624 Linkwood Dr., Houston 77025. Mailing address: Dept. Psychology, U. Houston, Cullen Blvd., Houston 77004.*

EVANS, Richard Joseph, Am. civil engr.; b. Washington, July 14, 1837; s. John and Sarah Zane (Mills) E.; ed. Rittenhouse Acad.; studied architecture and building under his grandfather, Robert Mills; m. Anais D. Lagarde, Feb. 4, 1861. Served several yrs. as aide in U. S. Coast Survey; removed to New Orleans during mil. occupation of that city; apptd. engr. New Orleans, Carrollton & Lake Pontchartrain Ry.; later chief engr. New Orleans, Opelousas & Gt. Western Ry.; afterward in employ of the Morgan Co. (steamship and ry. lines); devised adaptation of stern wheel steamboats for transferring loaded freight cars over Mississippi River; one of 3 engrs. apptd. to pass upon plans for drainage of New Orleans and its protection from overflow; built and superintended Gulf, Western Tex. & Pacific Ry. to Cuero; in Bur. of Steam Engring., Navy Dept., 1875-77; took charge of engring. of terminals, New Orleans Pacific Ry., 1877; later chief engr. Memphis, Selma & Brunswick Ry.; supt. Brunswick & Western Ry. of Ga.; chief engr., 1885, v.p., gen. mgr., 1888 to 1894, Tex., Sabine Valley & N.W. R.R.; in constrn. drainage, sewerage and water system of New Orleans, 1898-1916. Died Dec. 30, 1916.

EVANS, Robert John, Am. biochemist; b. Logan, Utah, Mar. 18, 1909; s. Robert James and Hazel (Stallings) E.; student Brigham Young U., 1927-29; B.S., Utah State U., 1934, M.S., 1936; Ph.D., U. Wis., 1939; m. Alice Pugmire, Aug. 14, 1941; children—Patricia Alice, Robert Pugmire. Prof. biochemistry Mich. State U., East Lansing, 1947—; with Low Temperature Research Sta. U. Cambridge (Eng.), 1963-64. Recipient Research Achievement award Poultry and Egg Nat. Bd., 1958. Mem. Am. Chem. Soc., Am. Inst. Nutrition, Poultry Sci. Assn., Am. Assn. Cereal Chemists, A.A.A.S. Studies, numerous publs. on influence of heat on nutritive value of proteins, effect of fractionated cottonseed fatty acids on eggs from hens fed them; isolated lipoproteins from hens' eggs and studied their structure and protein-lipid binding. Home: 633 N. Dexter Dr., Lansing, Mich. 48910.*

EVANS, Titus Carr, Am. radiobiologist; b. Lorena, Tex., Dec. 9, 1907; s. Charles William and Virginia (Whitsett) E.; B.A., Baylor U., 1929; M.S., State U. Ia., 1931, Ph.D., 1934; m. Mertie Ellen Jahnke, June 1, 1935; children—Titus Carr, Susan Ellen, Lucy Virginia. Asst. prof. Tex. A. and M. U., 1936-38; research asst. prof. U. Ia., Iowa City, 1938-42, research prof. radiology and radiobiology, 1948—, head Radiation Research Lab., 1948—; asst. prof. Columbia Coll. Phys. and Surg., 1942-48. Cons., mem. Policy Adv. bd. Argonne (Ill.) Nat. Lab., 1964—. Mem. Nat. Council Radiation Protection, Soc. Nuclear Medicine (past pres.), Radiation Research Soc., Soc. Nuclear Medicine, Biophysics Soc., Am. Physiol. Soc., Am. Roentgen Ray Soc. Radiol. Soc. N.Am., Cancer Research Soc., Soc. for Exptl. Biology and Medicine, Am. Soc. Zoology, Health Physics Soc., Sigma Xi. Managing editor of Radiation Research, 1952—; associate editor of Nuclear Medicine, 1960—. Research, numerous publs. on effects X-ray on cell div. and embryonic devel. in invertebrates, radio-protective effect anoxia in mammals, relative effectiveness fast neutrons, devel. techniques for radioisotopes in biol. and med. research, effects X-rays on ascites tumor cells. Home: 214 E. Church St., Iowa City, Ia. 52240.*

EVANS, Trevor, mathematician; b. Wolverhampton, Eng., b. Dec. 22, 1925; s. Thomas Richard and Lillian (Landucci) E.; B.A. with 1st class honors, Jesus Coll., Oxford U., Eng., M.A., 1950, D.Sc., Oxford U., 1960; M.Sc., Manchester U., Eng., 1948; m. Ellen Lovell, Dec. 26, 1953; children—Martyn, Judith, Susan, Martha. Faculty, U. Manchester, 1946-50, U. Wis., 1950-51; faculty Emory U., Atlanta, 1951-52, 54—, chmn. math. dept., 1956—, prof., 1958—; mem. Inst. for Advanced Study, Princeton, 1952-53, research asso. U. Chgo., 1953-54; vis. prof. U. Neb., 1959-60. Mem. Am., London math. socs., Math Assn. Am., Assn. for Symbolic Logic, Sigma Xi. Author: Fundamentals of Mathematics, 1959; (with B. K. Youse) Problems in Calculus, 1961. Research, publs. on both solvable and unsolvable decision problems in math., study of structure of gen. algebraic systems. Home: 1304 Bernadette Lane N.E., Atlanta 30329.*

EVANS, Ulick Richardson, English metallurgist; b. Wimbledon, Eng., Mar. 31, 1889; s. Richardson and Amy (Feeney) E.; B.A., U. Cambridge (Eng.), 1910, M.A., 1914, Sc.D., 1932; Sc.D. (hon.), U. Dublin (Ireland), 1964; D.Met. (hon.), U. Sheffield (Eng.), 1961. Asst. dir. research Cambridge U., 1921-45, reader sci. metallic corrosion, 1945-54, reader emeritus, 1954—, hon. fellow King's Coll., 1958—. Recipient Palladium medal Electrochem. Soc. N.Y., 1955; Hothersall medal Inst. Metal Finishing, 1957, Gold medal, 1961. Fellow Royal Soc., 1949, Instn. Metallurgists. Author: Metals and Metallic Compounds, 4 vols., 1923; Corrosion of Metals, 1924; Metallic Corrosion of Metals, 1924; Metallic Corrosion, Passivity and Protection, 1937; Introduction to Metallic Corrosion, 1948, 2d edit., 1963; Corrosion and Oxidation of Metals, 1960; also numerous articles. Research on mechanism of metallic corrosion, properties of thin films on metals, interference colors, prevention of corrosion, protective coatings. Home: 19 Manor Ct., Grange Rd., Cambridge, Eng.*

EVANS, Virginia John, Am. biologist; b. Balt., Mar. 19, 1913; d. John Absalom and Stella (Lewis) E.; A.B., Goucher Coll., 1935; M.Sc., Johns Hopkins, 1940, Sc.D., 1943. Research biologist Nat. Cancer Inst., 1944-64, head tissue culture sect., 1964—. Cons. Navy Tissue Bank, 1955—, Civil Service Rev. Bd. for Biologists, 1962—, Virus Research Resources Panel for Leukemia, 1965—. Nat. Cancer Inst. fellow, 1944-45. Mem. Am. Assn. Cancer Research, Tissue Culture Assn., Soc. for Cell Biology, N.Y. Acad. Scis., Soc. Study Devel. and Growth, Soc. for Exptl. Pathology, Cryobiology Soc., Internat. Soc. Cell Biologists, Am. Inst. Biol. Scis. Contbr. numerous articles to tech. jours. Research on cell physiology and nutrition mammalian cells in vitro; formulated completely chemically designed media for cultivation long term systems cells in vitro, malignant cells and their normal progenitors in tissue culture, cultivatin mass amounts mammalian cells in fluid suspension culture for study malignant transformation phenomenon in vitro. Patentee fluid suspension culture. Home: 5824 Bradley Blvd. Office: 9000 Wisconsin Av., Bethesda, Md. 20014.*

EVANS, William Lloyd, Am. chemist; b. Columbus, O., Dec. 22, 1870; s. William Henry and Anne (Lloyd) E.; B.Sc., Ohio State U., 1892, M.Sc., 1896, LL.D., 1948; Ph.D., U. Chgo., 1905; D.Sc., Capital U., 1949; m. Cora Ruth Roberts, Mar. 9, 1911; children—Lloyd Roberts, Jane Anne (Mrs. Alvin H. Nielsen), William Arthur. Chemist, Am. Encaustic Tile Co., Zanesville, O., 1892-94; U. fellow Ohio State U., 1895-96; asst. dept. ceramics, 1896-98; Univ. fellow and Lowenthal fellow chemistry U. Chgo., 1903-05; lectr. chemistry Starling-Ohio Med. Coll., 1911-15; asst. prof. chemistry Ohio State U., 1905-08, asso. prof., 1908-11, prof., 1911-41, chmn. dept. chemistry, 1928-41, prof. emeritus, 1941-54; cons. chemist Lowe Bros. Co. (Dayton), Columbus Coated Fabrics Corp., Div. Carbohydrates NIH; mem. Nat. Com. on Carbohydrate Research, NRC, 1926-27, Div. Chemistry and Chem. Tech., 1934-40. Recipient William H. Nichols medal, 1929, Gold medal Am. Inst. Chemists, 1942. Fellow A.A.A.S., Am. Acad. Arts and Scis.; mem. Am. Chem. Soc. (chmn. organic div. 1928; mem. exec. com. organic div. 1929; councilor at large 1934-40; pres.-elect 1940; pres., 1941), Phi Beta Kappa (hon.), Sigma Xi, Phi Eta Sigma, Phi Lambda Upsilon, Alpha Chi Sigma, Gamma Alpha. Author: Study and Quiz Outline in Chemistry, 1923. Co-Author: Laboratory Exercises in General Chemistry (with Wm. McPherson and W. E. Henderson), 1928; An Elementary Course in Qualitative Analysis (with J. E. Day and A. B. Garrett), 1938; Semimicro Qualitative Analysis (with A. B. Garrett and L. L. Quill), 1940, rev. edit. (with A. B. Garrett and H. H. Sisler), 1951. Mem. bd. editors Jour. Higher Edn. Contbr. Jour. Am. Chem. Soc. and Jour. Chem. Edn. Known for contbns. to chemistry of carbohydrates, covered oxidation of organic compounds of more simple type, 1st thorough study of exidation of sugars group, his findings concerning their chem. behavior were significant. Died Oct. 18, 1954.

EVANS-PRITCHARD, Edward Evan, English anthropologist; b. 1902; ed. Winchester Coll.; Exeter Coll., Oxford U. (Eng.); M.A., Ph.D.; Sc.D. (hon.), U. Chgo., 1967. With expdns. to central, east and north Africa, 1926-39; prof. sociology Egyptian U. (Cairo), 1930-33; Leverhulme fellow, 1934-35; research lectr. Oxford U., 1935-40, prof. social anthropology, since 1946; reader Cambridge U., 1945-46. Fellow Brit. Acad., Oriental and African Studies London (hon.); hon. mem. Am. Acad. Arts and Scis., Am. Philos. Soc. Author: Witchcraft, Oracles and Magic among the Azande, 1937; The Nuer, 1940; The Sanusi of Cyrenaica, 1949; Kinship and Marriage among the Nuer, 1951; Social Anthropology, 1951; Nuer Religion, 1956; Essays in Social Anthropology, 1962; The Position of Women in Primitive Societies, 1965; Theories of Primitive Religion, 1965; The Zande Trichester, 1967. Address: All Souls Coll., Oxford, Eng.*

EVE, Arthur Steward, physicist; b. Silsoe, Eng., Nov. 11, 1862; s. John Richard and Frederica (Somers) E.; B.A., Pembroke Coll., Cambridge, 1884, M.A., 1887; LL.D., Queen's U., McGill U.; m. Elizabeth Anges Brooks, Apr. 25, 1905; children—Joan (Mrs. Denison Denny), Richard, Cicely. Asst. master Marlborough Coll., Eng., 1886-1902, bursar, 1897; 1902; apptd. lectr. math. McGill U., 1903, asst. prof., 1905, asso. prof., 1909, prof., 1910, dir. dept. physics, 1919-35, dean grad. faculty, 1930-35. Fellow Royal Soc., 1917; pres. Royal Soc. Can., 1919-30. Author: (with D. A. Keys) Applied Geophysics, 1932; Physics, 1934; College Physics; Life and Work of Lord Rutherford, 1939; (with C. H. Creasey) Life and Work of John Tyndall, 1945; also papers on radioactivity and ionization. Died Mar. 14, 1948.

EVE, Joseph, Am. inventor, scientist; b. Phila., May 24, 1760; s. Oswald and Anne (Moore) E.; m. Hannah Singletary, 1800, 1 son Joseph Adams. Inventor machine for separating seed from cotton (early version of cotton gin); manufactured gunpowder and cotton gins, nr. Augusta, Ga., 1810, also experimented with steam; inventor cottonseed huller, 1803, metallic bands for power transmission, 1828, 2 steam engines, 1818, 26. Died Augusta, Nov. 14, 1835.

EVE, Paul Fitzsimons, Am. surgeon; b. Augusta, Ga., June 27, 1806; s. Oswell and Aphra (Pritchard) E.; B.A., Franklin Coll. (now U. Ga.), 1826; M.D., U. Pa., 1828; m. Sarah Twiggs; m. 2d, Sarah Duncan, 1852; 3 children including 2 sons. Practiced medicine in clinics, London, Eng., also Paris, France, 1828-30; participated as physician in Revolution of July 1831, Paris; offered services to Polish Govt., served in hosp., Warsaw; an organizer Med. Coll. Ga., 1832, prof. surgery, 1839-50; prof. surgery U. Louisville, 1850, U. Nashville, 1851-61, 70-77, Nashville Med. Coll., 1877; pres. A.M.A., 1857-58; leading surgeon and tchr. of surgery in South; co-editor So. Med. and Surg. Jour.; asst. editor Nashville Jour. Medicine and Surgery. Author (most noted med. works): A Collection of Remarkable Cases in Surgery, 1857; A Contribution to the History of the Hip-joint Operations Performed During the Late Civil War, 1867. Perfected operation for vesical calculus; 1st Am. surgeon to perform hysterectomy. Died Nashville, Nov. 3, 1877.

EVELETH, Donald Francis, Am. veterinarian; b. Salt Lake City, Nov. 3, 1905; s. George Leslie and Stella (McKay) E.; B.S., U. Cal. at Davis, 1928; M.A., U. Cal. at Berkeley, 1930; Ph.D., Western Res. U., 1932; D.V.M., Ia. State U., 1934; m. Beatrice L. Fugina, Aug. 16, 1952; children—Margaret Frances, George John, Anita Louise. Prof., chmn. dept. vet. sci. N.D. State U., 1943-64; asst. veterinarian in charge Wis. Dept. Agr., Animal Health Lab., Madison, 1965—. Fulbright lectr. U. Cairo, U.A.R., 1961. Mem. Am. Soc. Biol. Chemists, Am. Vet. Med. Assn., U. S. Livestock San. Assn., Am. Assn. Avian Pathologists, World Assn. Advancement Vet. Parasitology. Contbr. numerous articles to profl. jours. Research, publs. on clin. chemistry of physiology and diseases of animals and birds, physiology and anatomy of bovine dwarfs, chems. in control of coccidia and nematodes in animals and birds, metabolism of aluminum compounds in dog. Home: 230 St. Croix Lane. Office: 6101 Mineral Point Rd., Madison, Wis. 53705.*

EVELYN, John, English botanist; b. Wotton, Surrey, Eng., Oct. 31, 1620; s. Richard and Eleanor (Standsfield) E.; became student Middle Temple, 1637, fellow commoner, Balliol Coll., 1637, where he studied dancing and music; left without degree; recipient hon. D.C.L., 1669; m. Mary Browne (circa 12 yrs. old), 1647; 6 sons, 3 daus. Traveled to Holland, 1641; France, 1643-44; Italy 1644-46; France, 1646, 1649-52; lived in England, 1647-49, from 1652; in King's Army, England, 1642-43. Founding Fellow, Royal Soc., 1660 (council 1662; sec., 1672; refused presidency twice, 1682, 1691); Treas., Greenwich Hosp., 1695-1703. Author: Sylva; or A Discourse of Forest Trees and the Propagation of Timber, 1664; Diary, pub. 1818-19; many other works. Remembered for his many services to Royal Soc. and his contemporaries. Died Feb. 27, 1706.

EVEREST, David Anthony, English chemist; b. London, Eng., Sept. 18, 1926; s. George Charles and Ada (Wheddon) E.; B.Sc., U. Coll., London, 1946, Ph.D., 1949; m. Audrey Sheldrick, Mar. 29, 1956; children—Peter Lindsay, Michael David, Richard Martin. Lectr. inorganic chemistry Battersea Coll. Tech., London, 1949-56; head analytical sect. extraction metals group Nat. Chem. Lab., 1956-59, head beryllium sect., 1959-63; head inorganic materials unit Nat. Phys. Lab., Teddington, Middleset, U.K., 1964—. Fellow Royal Inst. Chemistry; mem. Inst. Mining and Metallurgy (asso.). Author: The Chemistry of Beryllium, 1963; also articles. Devel. ion exchange methods for recovery of valuable metals, methods for extraction of beryllium from lean ores; high temperature chemistry especially thermal plasmas; acad. chemistry of germanium, tin. Home: 119 Eastcote Rd., Ruislip, Middlesex. Office: The Nat. Phys. Lab., Teddington, Middlesex, U.K.*

EVEREST, Sir George, mil. engr.; b. Gwernvale, Wales, July 4, 1790; s. Tristram Everest; ed. Royal Mil. Acad., Woolwich, Eng.; m. Emma Wing, Nov. 17, 1846. In mil. service, Bengal (India), Java, 1806-25, became asst. to Col. W. Lombton (founder trigonometrical survey of India), studied short arc measured by V. de Lacaille in S. Africa; became surveyor gen., India, 1830. Mt. Everest named in his honor, 1863. Fellow Royal Soc., 1827, Astron. Soc., Royal Asiatic Soc., Geol. Soc.; mem. Geog. Soc. (v.p.), Asiatic Soc. Bengal (hon.). Author: An Account of the Measurement of the Arc of the Meridian . . . , 1830; An Account of the Measurement of Two Sections of the Meridional Arc of India . . . , 1847; On Instruments and Observations for Longitude for Travellers on Land, 1859. Used his measurements of great meridional arc of India to calculate figure of earth; inventor gridiron system of geodetic triangulation for accurate survey of Indian subcontinent. Died London, Eng., Dec. 1, 1866.

EVERETT, John Wendell, Am. anatomist, educator; b. Ovid, Mich., Mar. 5, 1906; s. Fred Ross and L. Mae (Grimes) E.; A.B., Olivet Coll., 1928; Ph.D., Yale, 1932; m. Marian Della Eggstaff, Sept. 14, 1932; children—Ronald Wilcox, Marian Janice (Mrs. David Rideout). Faculty, Goucher Coll., 1930-31; faculty Duke Sch. Medicine, Durham, N.C., 1932—, prof. anatomy, 1950—. Fellow A.A.A.S., N.Y. Acad. Scis.; mem. Am. Assn. Anatomists, Am. Physiol. Soc., Endocrine Soc., Soc. Exptl. Biology and Medicine, Internat. Brain Research Orgn. Editorial bd. Endocrinology; adv. bd. Neuroendocrinology; asso. editor Anat. Record, 1957-63; sect. editor Biol. Abstracts. Research, numerous articles on brain mechanisms concerned with control of pituitary gland, blockade of ovulation by drugs, effects of ovarian hormones on brain, effects of transplantation of pituitary. Home: 1105 Woodburn Rd., Durham, N.C. 27705.*

EVERETT, Joseph David, English phys. scientist; b. Ipswich, Eng., Sept. 11, 1831; s. Joseph David and Elizabeth (Garwood) E.; ed. Ipswich Mechanics' Instn.; B.A., Glasgow (Scotland) Coll., 1856, M.A., 1857; m. Jesse Fraser, Sept. 3, 1862; 3 daus., 3 sons. Prof. math. King's Coll., Windsor, N.S., Can., 1859-64; worked in Kelvin's lab.; prof. natural philosophy Queens Coll., Belfast, Ireland, 1867-97; became asst. to Dr. Hugh Blackburn, Glasgow, 1884; moved to London, 1898; fellow Royal U. Ireland. Sec., Com. to Determine Selection and Nomenclature Dynamical Units, 1871-78. Fellow Royal Soc., 1879; Royal Soc. Edinburgh; mem. Phys. Soc. London (v.p. 1900-04). Author: Units and Physical Constants, 1875; Elementary Textbook of Physics, 1877; Vibratory Motion and Sound, 1882; Outlines of Natural Philosophy, 1887. Translator: Physics (Deschanel), 1870. Pioneered cycling; invented system of shorthand, 1877, spring hub attachment for spokes of bicycle wheels. Died Ealing, Eng., Aug. 9, 1904.

EVERETT, Mark Reuben, Am. biochemist; b. Slatington, Pa., Nov. 2, 1899; s. Alexander David and Mary (Scheidy) E.; B.S. in Chem. Engring., Bucknell U., 1920, D.Sc., 1948; Ph.D. in Med. Scis., Harvard, 1924; m. Alice Allen, June 21, 1924; children—Mark Allen, Kathleen Elizabeth (Mrs. William F. Upshaw). Chmn., prof. biochemistry and pharmacology U. Okla., Oklahoma City, 1924-35, chmn., prof. biochemistry, 1936-64, dean Sch. Medicine, supt. U.

Hosps., 1947-56, dean Sch. Medicine, dir. Med. Center., 1956-64, dean emeritus, Regents prof. med. scis., cons. prof. biochemistry, 1964—; adminstr. John Archer Hatchett Found., Oklahoma City, 1944—. Mem. Okla. Anat. Bd., 1947-64, Okla. Hosp. Adv. Council, 1947-64; chmn. Okla. Med. Research Commn., 1963-64. Named to Okla. Hall of Fame, Okla. Meml. Assn., 1946, Okla. Med. Scis. Hall of Fame, 1959; named Pa. Ambassador, State of Pa., 1948. Fellow A.A.A.S., Am. Inst. Chemists, Okla. Acad. Sci.; mem. Am. Assn. for Cancer Research, Assn. Am. Med. Colls. (past mem. council), Am. Assn. U. Profs., Am. Chem. Soc., Am. Soc. Biol. Chemists, Soc. for Exptl. Biology and Medicine, Sigma Xi, Phi Beta Pi (hon.), several others. Author: Handbook of Biochemistry, 1935; Medical Biochemistry, 1942; also numerous articles, monographs on metabolism and oxidation of carbohydrates; analytical chemistry. Home 1302 N.W. 21st St., Oklahoma City 73106.*

EVERETT, Newton Bennie, Am. anatomist; b. Dundee, Tex., May 12, 1916; s. Henry B. and Lula Mae (Williams) E.; B.S., N. Tex. State Coll., 1937, M.S., 1938; Ph.D. (teaching fellow 1940-42), U. Mich., 1942; m. Naomi Doris Briggs, Sept. 11, 1940; children—Peter Ben, James Briggs. Instr. anatomy U. Mich., 1942-46; asst. prof. U. Wash., 1946-48, asso. prof., 1948-57, adminstrv. officer, 1955—, prof., 1957—, chmn. dept. anatomy, now prof., chmn. dept. biol. structure. Mem. Am. Assn. Anatomists, Am. Physiol. Soc., Am. Internat. socs. cell biology, Internat. Am. socs. hematology, Royal Soc. Medicine (affiliate), Sigma Xi. Research in fetal physiology; hematology; tracer biology; circulation; radiobiology; cell kinetics; hemocytopoiesis. Home: 16741 37th St. N.E., Seattle 98155.

EVERMANN, Barton Warren, Am. naturalist; b. Albia, Ia., Oct. 24, 1853; s. Andrew and Nety (Gardner) E.; B.S., Ind. U., 1886, A.M., 1888, Ph.D., 1891, LL.D., 1927; LL.D., U. Utah, 1922; m. Meadie Hawkins, Oct. 24, 1875; children—Toxaway Bronté, Edith (Mrs. Wm. E. Humphrey). For 10 yrs. teacher and supt. of schs. in Ind. and Calif.; prof. biology, Ind. State Normal Sch., 1886-91; asst. ichthyologist; 1888-91, ichthyologist, 1891-1914, chief Div. Statistics and Methods of Fisheries, 1902-03, asst. in charge sci. inquiry, 1903-10, chief Alaska Fisheries Service, 1910-14, U. S. Bur. Fisheries, U. S. fur seal commr., 1892; ichthyologist, 1891; chief, div. scientific inquiry, 1903-11; spl. lectr. Stanford, 1893-94, 1926-32; fish and game protection, Cornell U., 1900-03, Yale, 1903-06; chmn. fur seal bd., 1908-14; dir., Cal. Acad. Scis., 1914-32; dir., Steinhart Aquarium, San Francisco, 1922-32. Chmn. com. on zool. investigations of State Council Def.; mem. com. Pacific investigation, NRC, chmn. Com. on Conservation of Marine Life of the Pacific. Author: The Fishes of North and Middle America, 4 vols. (with Dr. David Starr Jordan), 1896, 1900; American Food and Game Fishes, 1902; The Aquatic Resources of the Hawaiian Islands (with David Starr Jordan); The Alaska Salmon Fisheries; Fishes of the Philippines, 1906; The Golden Trout of the Southern High Sierras, 1906; The Fishes of Alaska, 1907; The Fishes of Peru, 1915; A Review of the Giant Mackerel-like Fishes, Tunnies, Spearfishes, and Swordfishes (with David Starr Jordan); Studies of the Pacific Coast Salmon, 1894-97; Fishes of Hawaii, 1905; (with Jordan and H. Walton Clark) Check List of the Fishes . . . of North and Middle America, 1928. Belonged to group of ichthyologists who approached subject from point of view of taxonomy, advocated principles for saving Alaska fur-seal herd from extinction. Died Berkeley, Cal., Sept. 27, 1932.

EVERSHED, John, astronomer; b. Feb. 26, 1864; s. John and Sophia Evershed; m. Mary Acworth Orr, 1906 (dec. 1949); m. 2d, Margaret G. Randall, 1950. Made series of solar observations at pvt. obs., 1890-1906; mem. 6 expdns. to observe total solar eclipse, Norway, 1896, India, 1898, Algeria, 1900, Spain, 1905, Australia, 1922, Yorkshire, 1927; asst. dir. Kodaikanal and Madras Observatories, 1906, later dir.; visited New Zealand to select site for Cawthron Obs., 1914; on astron. expdn. to Kashmir, 1915; built, equipped pvt. obs. at Ewhurst, 1925. Fellow Royal Soc., 1915. Recipient Gold medal Royal Astron. Soc., 1918. Contbr. articles to sci. jours. Discovered radial movement in sunspots, 1909. Died Nov. 17, 1956.

EVES, Howard Whitley, Am. mathematician; b. Paterson, N.J., Jan. 10, 1911; s. Roland Guy and Lydie (Hennion) E.; B.S., U. Va., 1934; M.A., Harvard, 1935; postgrad. Princeton; Ph.D., Ore. State U., 1948; m. Diane Louise Kocher, Oct. 25, 1954; children—Jamie Howard, Cindy Diane, Tammy Susanne, Timothy Jon, Roderick Daniel. Staff, Inst. Math, Bethany Coll., 1938-39; licensed land surveyor, N.J., 1939-40; mem. staff Inst. Math., Allen Acad., 1940-41; mathematician TVA, 1941-42; asst. prof. math. Syracuse (N.Y.) U., 1942-45; asst. prof. math., dept. chmn. Coll. Puget Sound, Tacoma, 1945-46; asst. prof. math Ore. State Coll., Corvallis, 1946-48, asso. prof., 1948-51; prof. math. State U. N.Y., Endicott, 1951-54, U. Me, Orono, 1954—. Contbg. staff Ency. Americana, Chgo., 1958—, Collier's Ency., N.Y.C., 1959—, World Book Ency., Chgo., 1958—; lectr. NSF, summers 1957—. Recipient Carter award Ore. State Coll., 1950. Mem. Math. Assn. Am.; chmn. Northeastern sect. 1956-57, gov. 1958-61, 64-67), Council Tchrs. Math., Am. Math. Soc., Société Mathématiques de France, Phi Kappa Phi, Sigma Mu, Kappa Alpha, Pi Mu Epsilon. Author: (with C. V. Newsom) Introduction to College Mathematics, 1954; Introduction to the Foundations and Fundamental Concepts of Mathematics, 1958, rev., 1965, A Survey Of Geometry, Vol. 1, 1963; Introduction to the History of Mathematics, 1953, rev. 1964; A Survey of Geometry, Vol. 2, 1965; Functions of a Complex Variable, 2 vols., 1966; Elementary Matrix Theory, 1966. Editor: (with E. P. Starke) Otto Dunkel Memorial Problem Book, 1957; problem editor Am. Math. Monthly, 1945—; history of math. editor The Math. Tchr., 1958—. Contbr. articles to tech. jours. on algebraic and differential geometry, theory of dissections. Home: 79 Bennoch Rd., Stillwater, Me. 04489.*

EVNO (or Eyno, Enno), Bavarian meteorologist; flourished Würzburg, Lower Franconia, 1331; author Iudicia de impressionibus quae fiunt in aere (treatise on astrological meteorology and geography). Died 1355.

EVONUK, Eugene, Am. physiologist; b. Springfield, Ore., Oct. 11, 1921; s. Peter and Mary (Racky) E.; B.S., U. Ore., 1952, M.S., 1953; Ph.D., U. Ia., 1960; m. Clarissa M. Kann, Sept., 7, 1946. Faculty, U. Ore., Eugene, 1953-58, asst. prof., 1955-58; research asst. U. Ia., Iowa City, 1958-59, Ia. Heart Assn. fellow, 1959-60; research physiologist Arctic Aeromed. Lab., Fairbanks, Alaska, 1960-63, chief physiology br., 1963-67; dir. Center of Research for Human Performance, U. Ore., Eugene, 1967—. Fellow A.A.A.S., Arctic Inst. N. Am.; mem. Am. Physiol. Soc., Am. Soc. Zoologist, Am. Inst. Biol. Sci, A.A.A.S., Sigma Xi. Research, publs. on thermal dilution technique for measuring cardiac output, norepinephrine induced thermogenesis, cardiovascular responses to surface cooling, cold tolerance. Home: 1060 E. 36th Av., Eugene, Ore. 97405.*

EVVARD, John Marcus, Am. animal nutritionist; b. Saunemin, Ill., Nov. 6, 1884; s. John B. and Mary (Leitel) E.; B.S., U. Ill., 1907; M.S. in Agr., U. Mo., 1909; Ph.D., U. Ariz., 1927; m. Mattie Casey Cooper, Aug. 10, 1911; children—Mary Margaret Batman, John Cooper, Martha Jane Shemer. Asst. to dean and dir. Mo. Agrl. Coll. and Mo. Agrl. Expt. Sta., 1907-10; asst. chief, animal husbandry, Ia. Agrl. Expt. Sta., 1910-14; charge animal husbandry sect. and chief in swine prodn., 1911-30, chief in beef cattle and sheep production, 1919-30, Ia. Agrl. Expt. Sta.; asso. prof. animal husbandry, 1916-18, prof., 1918-30, Ia. State Coll. A. and M. Arts, chief in nutrition, 1914-19; staff Am. Inst. Agr., 1922-23; v.p. McMillen Co., Allied Mills, Ft. Wayne, Ind., 1930-33; research cons. Soya Products, Inc., Chgo., from 1932; adviser to exec. staff Allied Mills, Inc., from 1933; pres. Universal Supply, Inc., Phoenix, 1933-36; prof., head dept. agr. Ariz. State Coll., Tempe, (part time), 1935-37. Contbg. editor Chester White Journal, 1920-30, corr. editor Farm and Fireside, 1924-30. Fellow A.A.A.S.; mem. Am. Soc. Animal Prodn. (past pres.), Sigma Xi, Alpha Zeta, Phi Lambda Upsilon, Phi Kappa Phi, Gamma Sigma Delta, Sigma Delta Chi, Lambda Gamma Delta. Author over 700 publs. covering investigations principally in field of animal feeding and nutrition. Died July 30, 1948.

EWALD, Ernst Julius Richard, German physiologist; b. Berlin, Germany, Feb. 14, 1855; s. Arnold Ferdinand and Auguste (Amelung) E.; student math., physics, medicine, physiology, Strasbourg, France; m. Bertina Schiff, 1884; 2 children. Prof. physiology, Strasbourg, 1900-18. Mem. French Acad. Scis. Author: Physiologische Untersuchungen über das Endorgan des nervus octavus, 1892; also articles. Research on respiration, blood circulation, pancreas, physiology of central nerve system, labyrinth and acoustic receptors of ear; proved (with F. L. Goltz) that a dog may be impregnated after its spinal cord has been severed, 1899. Died Konstanz, Germany, July 22, 1921.

EWALD, Karl (Emil) Anton, German physician; b. Berlin, Germany, Oct. 30, 1845; s. Arnold Ferdinand and Auguste (Amelung) E.; student medicine, Heidelberg, Germany, Bonn, Germany; M.D., Berlin, 1870. Asst. to v. Frerichs, U. Berlin, 1871-76, became lectr., 1874; named dir. Frauensiechenanstalt, Berlin, 1876, asso. prof., 1882; head internal dept. Berlin Augusta-Hospital, 1888-1915, named hon. prof., 1909. Author: Klinik der Verdauungskrankheiten 3 vols., 1879-1902. Research on physiology and pathology of digestive organs, disease of liver and thyroid, diet and nutrition; invented (with Boas) test meal (Boas-Ewald test breakfast), 1885; introduced flexible rubber tube for stomach aspiration, 1875. Died Berlin, Sept. 20, 1915.

EWALD, Paul P., physicist; b. Berlin, Germany, 1888; Ph.D., U. Munich, 1912; D.Sc., Stuttgart U., 1954, U. Paris, 1958, Adelphi U., 1966. Prof. theoretical physics, Poly. U., Stuttgart, 1921-37; prof. math. physics, The Queen's U., Belfast, N. Ireland, 1939-49; head physics dept., Poly. Inst. Bklyn., 1949-57, prof. physics, 1949-59. Fellow Royal Soc., 1958, Phys. Soc. Britain, Am. Phys. Soc. (chmn. solid state div. 1961-62), Am. Acad. Arts and Scis., Acad. Leopoldina (Halle); hon. mem. Deutsche Mineralog. Ges., Société Francaise de Mineralogie et de Cristallographie; corr. mem. Acad. Scis. Göttingen; mem. Internat. Union Crystallography (pres. 1960-63). Author several books and numerous articles on optics and solid state physics. Stated dynamic theory of x-ray interferences in crystals, 1917-24; work in x-ray and light optics, crystal structure analysis, and solid state physics. Address: 19 Fordyce Rd., New Milford, Conn. 06776.*

EWALT, Jack Richard, Am. physician; b. Medicine Lodge, Kan., Jan. 27, 1910; s. Sim and Edith (Crummack) E.; M.D., U. Colo., 1933; A.M., Harvard U., 1958; m. Beatrice Earl, 1931 (div. 1964); children—Ann (Mrs. George Hamilton), Jean (Mrs. John Gelber); m. 2d, Patricia Littlefield, Nov. 18, 1964. Commonwealth Fund fellow U. Colo., 1933-37, asst. prof. psychiatry, 1937-41; prof. psychiatry U. Tex., 1942-51, adminstr. U. Tex. Med. Br. Hosps., 1946-51, dean Post-Grad. Sch. Medicine, 1949-51; commr. Mass. Dept. Mental Health, 1951-58; faculty Harvard, 1951—, Bullard prof. psychiatry, 1961—. Dir. Joint Com. on Mental Illness and Health, 1955-61; psychiat. cons. Mem. Am. Psychiat. Assn. (pres. 1963-64), Am. Bd. Psychiatry (dir. 1960-68, pres. bd. 1964), A.M.A., Am. Acad. Arts and Sci. Author: (with E. A. Streeker, F. G. Ebaugh) Practical University Psychiatry, 8th edit., 1957; (with Dana Farnsworth) Textbook of Psychiatry, 1963; Mental Health Administration, 1957; (with E. A. Parsons, S. L. Warren, S. L. Osborne) Fever Therapy, 1938. Research, publs. on fever therapy of neurosyphilitis and Sydenbams chorea; research on tissue culture of brain tissue, planning of community mental health services, research in treatment of schizophrenia. Home: 39 Colbert Rd. E., West Newton, Mass. 02165. Office: 74 Fenwood Rd., Boston 02115.*

EWAN, George Thomson, physicist; b. Edinburgh, May 6, 1927; s. Alexander Farmer and Jeannie (Taylor) E.; B.Sc. with 1st class honors, Edinburgh U., 1948, Ph.D., 1952; m. Maureen Louise Howard, Aug. 7, 1952; children—Elizabeth Louise, Robert Alexander. Asst. lectr. Edinburgh U., 1950-52; research asso. McGill U., Montreal, Que., Can., 1952-53, NRC Can. fellow, 1953-55; staff Atomic Energy Can. Ltd., Chalk River, Ont., Can., 1955—, sr. research officer, 1962—. Ford Found. fellow Niels Bohr Inst., Copenhagen, Denmark, 1961-62; vis. scientist Lawrence Radiation Laboratory, Berkeley, California, 1966. Recipient Radiation Industry award American Nuclear Society, 1967. Member of American Phys. Soc., Canadian Assn. Physicists. Research, publs. on properties of nuclear energy levels and nuclear decay schemes, high resolution Beta ray spectroscopy, devel. lithium drifted germanium gamma ray spectroscopy and application to low energy nuclear physics research; use of semi-conductor detectors for electron spectroscopy. Home: 15 Frontenac Crescent, Deep River, Ont. Office: Chalk River Nuclear Labs., Chalk, River, Ont., Can.*

EWART, Alfred James, botanist; b. Liverpool, Eng., Feb. 12, 1872; s. Edmund Brown and Martha (Williams) E.; ed. Liverpool Instn., U. Coll., Liverpool; B.Sc., London, Eng., 1893; Ph.D., Leipzig, Germany, 1896; m. Florence Maud Donaldson, 1898; 2 sons; m. 2d, Elizabeth Bilton, 1931. Joined Birmingham (Eng.) U. (formerly Mason Coll.), 1897; prof. botany, plant physiology, also govt. botanist, Melbourne, Australia, 1905-21. Fellow Royal Soc., 1922, Linnean Soc.; mem. Royal Soc. Victoria. Author: The Physics and Physiology of Protoplasmic Streaming in Plants, 1903; A Handbook of Forest Trees for Victorian Foresters, 1925; Flora of Victoria, 1930. Translator: Physiology of Plants (Pfeffer), 1900-06. Research on causes of stock and horse poisoning in Australia, weed identification and control. Died Melbourne, Sept. 12, 1937.

EWART, James Cossar, Scottish zoologist; b. Penicuik, Nov. 26, 1851; s. John Eward; ed. Edinburgh U.; M.D. Demonstrator anatomy Edinburgh U., 1874, Regius prof. natural history, 1882-27; conservator Univ. Coll. Museums, London, 1875-78; visited continental museums, also worked in Strasbourg, 1876; prof. natural history Aberdeen U., 1878-82; began marine sta. (1st marine lab. in Gt. Britain), nr. Aberdeen, 1879; mem. Fishery Bd. for Scotland, 1882; engaged in sci. work in connection with fisheries, 1883-92. Fellow Royal Soc., 1893 (v.p. 1908). Author: (with G. J. Romanes) The Locomotor System of the Echinoderms, 1881; The Natural and Artificial Fertilisation of Herring Ova, 1884; On the Progress of Fish Culture in America, 1884; On Whitebait, 1886; On the Preservation of Fish, 1887; The Electric Organ of the Skate, 1888, 89; The Cranial Nerves and Lateral Sense Organs of Elasmobranchs, 1889-91; The Developement of the Limbs of the Horse, 1894; A Critical Period in the Development of the Horse; The Penicuik Experiments, 1899; Guide to Zebras, Hybrids, etc., 1900; The Multiple Origin of Horses and Ponies, 1904; Horse Skulls from the Roman Fort near Melrose, 1906; On a Prejvalsky Hybrid, 1908; Domestic Sheep and their wild Ancestor, 1913; The Development of the Horse, 1915; Moulting of the King Penguin, 1917; The Nestling Feathers of the Mallard, 1931. Studied fishery investigations in U. S. and Can., 1883, Denmark and Norway, 1884. Studied locomotor system of echinoderms; investigated life history and fertilization of whitebait, herring, and other food fishes; experimented with cross breeding of zebras, quaggas, and horses to disprove theory of telegony; showed that 1st mate of a female does not influence offspring by another mate. Died Penicuik, Dec. 31, 1933.

EWART, Sir Joseph, surgeon; b. 1831; s. Andrew and Catherine (Armstrong) E.; M.D.; m. Madeline Lister, 1856. Entered Honourable E. India Co.'s Service (med.), 1854; prof. medicine, Calcutta, India; dep.-surgeon-gen. Indian Army; ret., 1879; mayor, Brighton, Eng., 1891-94. Author several works on Indian sanitation, pathology, snake poisoning. Prop., editor Indian Annals Med. Sci. Revealed existence of typhoid fever in natives of India. Died Jan. 10, 1906.

EWBANK, Thomas, inventor, govt. ofcl.; b. Durham, Eng., Mar. 11, 1792. Worked in London as tinsmith, 1812-19; came to N.Y.C., 1819; manufactured tin, copper, and lead tubing; developed improved methods of tinning sheet and pipe lead and improved safety valves, patented 1832, improved steam safety valves, patented 1823-32; U. S. commr. patents, 1849-52; founder, active mem. Am. Ethnol. Soc. Author: Descriptive and Historical Account of Hydraulic and Other Machines Ancient and Modern, 1842, 16th edit. 1870; The World a Workshop or the Physical Relation of Men to the Earth, 1855; Thoughts on Matter and Force, 1858; Reminiscences of the Patent Office, 1859. Devised method to increase pressure resistance in building stones. Died N.Y.C., Sept. 16, 1870.

EWERS, John Canfield, Am. anthropologist; b. Cleve., July 21, 1909; s. John Ray and Mary Alice (Canfield) E.; A.B., Dartmouth, 1931; M.A., Yale, 1934; m. Margaret Elizabeth Dumville, Sept. 6, 1934; children—Jane Canfield, Diane Curtis. Field curator Nat. Park Service, Washington, Morristown, N.J., Berkeley, Cal., and Macon, Ga., 1935-40; curator Mus. Plains Indian, Browning, Mont., 1941-44; asso. curator ethnology U. S. Nat. Mus., Smithsonian Instn., Washington, 1946-56, planning officer, 1956-59, asst. dir. Mus. History and Technology, 1959-64, dir., 1964-65, sr. research anthropologist Smithsonian Office Anthropology, 1965—. Mus. planning cons. Bur. Indian Affairs, 1948-49, Mont. Hist. Soc., 1950-54; cons. Am. Heritage, 1959. Recipient 1st Exceptional Service award Smithsonian Instn., 1965. Fellow Am. Anthrop. Assn.; mem. Am. Indian Ethnologist Conf. (pres. 1960-61), Am. Assn. Museums, Anthrop. Soc. Washington. Author: Plains Indian Painting, 1940; The Horse in Blackfoot Indian Culture, 1955; The Blackfeet: Raiders on the Northwestern Plains, 1958; Artists of the Old West, 1965. Editor: Adventures of Zenas Leonard, Fur Trader, 1959; Crow Indian Medicine Bundles, 1960; Five Indian Tribes of the Upper Missouri, 1961. Editor Jour. Washington Acad. Scis., 1955-56. Contbr. articles profl. publs. Research on dynamics and significance of culture change among Plains Indians of N. Am.; refinement of techniques in interpreting anthropology through mus. exhibits. Home: 4432 26th Rd. N., Arlington, Va. 22207. Office: Smithsonian Instn., Washington 20025.*

EWING, George McNaught, Am. mathematician; b. Lexington, Mo., Sept. 30, 1907; s. Joel Harvey and Christie (Vaughan) E.; A.B., U. Mo., 1929, A.M., 1930, Ph.D., 1935; postgrad. Inst. for Advanced Study; m. Mary Alice Jones, Aug. 6, 1937; children—Alicia M., Catharine V., Joel C. Faculty, U. Mo., Columbia, 1930-57, prof. math., 1950-57; instr. Princeton, 1940-41; mathematician research div. Naval Ordnance Lab., Washington, 1944-45, systems evaluation dept. Sandia Corp., Albuquerque, 1951-52, missile div. Ramo-Wooldridge Corp., Los Angeles, 1954; research asso. Research Inst., U. Okla., Norman, 1957-60, prof. math., 1960-63, research prof. math., 1963—. Mem. Am. Math. Soc., Math. Assn. Am., Soc. for Indsl. and Applied Math., A.A.A.S., Am. Assn. U. Profs., Phi Beta Kappa, Sigma Xi. Author: (with H. Betz, P. B. Burcham) Differential Equations with Applications, 1954, 2d edit., 1964. Research in calculus of variations including control theory and design, allocation, testing and predicted effects of weapons. Home: 816 Coll. Av., Norman, Okla. 73069.*

EWING, Henry Ellsworth, Am. entomologist; b. Arcola, Ill., Feb. 11, 1883; s. Joseph Henry and Ann Louisa (McDonald) E.; student Knox Coll., 1902-04; A.B., U. Ill., 1906, A.M., 1908; Ph.D., Cornell U., 1911; m. Bertha May Wood Riley, Aug. 7, 1916; children—Paul McDonald, Lydia Frances. Asst. in zoölogy Ia. State Coll., 1909-10; Schuyler fellow Cornell U., 1910-11; asst. entomologist Ore. Agrl. Expt. Sta., 1911-14; asst. prof. entomology, Ia. State Coll., 1914-16, asso. prof., 1916-19; specialist Bur. of Entomology, U. S. Dept. Agr., 1919-23, asso. entomologist, 1923-29, entomologist, 1929-45, collaborator, from 1947. Fellow A.A.A.S., Entomol. Soc. Am.; mem. Am. Assn. Econ. Entomologists, Soc. Mammalogists, Am. Soc. Parasitologist (pres. 1944), Am. Soc. Ichthyologists and Herpetologists, Ecol. Soc. of Am., Sigma Xi. Author: A Manual of External Parasites, 1929, also articles. Research on arachnids, particularly parasitic external mites and primitive orders of acarology, authority on ectoparasites. Died Jan. 5, 1951.

EWING, J. Franklin, Am. anthropologist; b. N.Y., Oct. 14, 1905; s. Frank Dell and Marie (Schmitz) E.; A.B., Woodstock Coll., 1928, A.M., 1929, S.T.L., 1936; postgrad. Vienna U., 1937-38, Columbia, 1945; Ph.D., Harvard, 1947. Mem. Soc. of Jesus; instr. Ateneo de Manila, 1929-32; tchr. Santa Tomas Internment Camp, 1944-45; asst. dir. excavations at Ksar 'Akil, Lebanon, before World War II, dir. since World War II; dir. Ateneo expdn. to Mindanao, 1941; asst., then asso. prof. anthropology Fordham U., Bronx, N.Y., 1949-67, prof. emeritus, 1967—. Decorated Medaille pour le Merite, Lebanon. Fellow Am. Anthrop. Assn., A.A.A.S.; mem. Am. Assn. Phys. Anthropologists, Sigma Xi. Author: Hyperbrachycephaly as Influenced by Cultural Conditioning, 1950; (with James S. Donnelly) Our Country and Canada, 1960, Living in Eurasia, 1961, Living in Southern Lands, 1962; (with Donnelly and others) Our World Today, 1963; The Ancient Way: Life and Landmarks of the Holy Land, 1964. Editor Proc. Fordham U. Conf. Mission Specialists, 1953, 54, 55, 56, 57, 62. Contbr. numerous articles and book revs. to tech., religious and popular publs. Mem. editorial bd. Rev. of Acad. Religion and Mental Health. Made paleontol. discoveries including "Egbert" (earliest Homo Sapiens in Lebanon) and much older Neanderthaloid specimen. Address: Loyola Coll. and Sem., Shrub Oak, N.Y. 10588.*

EWING, James, Am. pathologist; b. Pitts., Dec. 25, 1866; s. Thomas and Julia R. (Hufnagel) E.; A.B., Amherst Coll., 1888, A.M., 1891; Sc.D., 1923; M.D., Columbia Coll. Phys. and Surg., 1891; Sc.D., U. Pitts., 1911, Amherst, 1923, U. Rochester, 1932, Union U., 1938; LL.D., Kenyon, 1931, Western Res. U., 1931; m. Catherine C. Halsted, June 19, 1900; 1 son, James Halsted. Tutor histology, Coll. Phys. and Surg., 1893-97, Clark fellow, 1896-99, instr. clin. pathology, 1897-98; prof. pathology Cornell U., 1899-1932, prof. oncology, 1932-43. Charles Mickle fellow, Toronto, 1935; John Scott medalist, Phila., 1937; hon. mem. Phila. Pathol. Soc., Acad. Med., Brazil, Swedish Roentgen. Soc., Acad. Med. Budapest; mem. Nat. Acad. Sciences, Assn. Am. Physicians, Am. Roentgen Ray Soc., Am. Med. Museums Soc., Am. Assn. Pathologists and Bacteriologists, Soc. Exptl. Biology and Medicine, Harvey Soc., Am. Assn. for Cancer Research. Author: Clinical Pathology of Blood, 1900-03; articles on Identity, The Signs of Death, and Sudden Death, in Text-Book of Legal Medicine and Toxicology, 1903; on Blood, etc., in Text-Book of Legal Medicine, 1910; Neoplastic Diseases, 1919-27, 40. Described form of bone sarcoma which usually affects shaft of long bones; classified tumors; experimented with cancer cure using x-ray and radium; pioneer in use of radium treatment. Died N.Y.C., May 16, 1943.

EWING, Sir James Alfred, Brit. physicist; b. Dundee, Scotland, Mar. 27, 1855; s. James Ewing; M.A., LL.D., U. Edinburgh (Scotland); LL.D., St. Andrews, Glasgow (both Scotland) univs.; D.Sc. (hon.), Oxford, Durham, Sheffield (all Eng.) univs.; Sc.D. (hon.), Cambridge (Eng.) U.; m. Anne Washington, 1879; m. 2d, Ellen Hopkinson; 2 sons, 1 dau. Engr., until 1878; prof. mech. engring. Imperial U., Tokyo, Japan, 1878-83; prof. engring. U. Coll., Dundee, 1883-90; prof. mechanism, applied mechanics Cambridge U., 1890-1903; dir. Naval Edn., 1903-16; prin., vice-chancellor Edinburgh U., 1916-29. Mem. Explosives Commn., 1903-06, Ordnance Research Bd., 1906-08; head, Admiralty dept. dealing with enemy cipher, 1914-17; chmn. Bridge Stress Com., 1924-28, Com. on Mech. Testing of Timber, 1920-34. Recipient Royal medal for researches in magnetism, 1895, Albert medal, 1929. Fellow Royal Soc., 1887; mem. Brit. Assn. (pres. 1932), Royal Soc. Edinburgh (pres. 1924-29). Author: Treatise on Earthquake Measurement, 1883; Magnetic Induction in Iron and other Metals, 1891; The Steam Engine and other Heat Engines, 1894, 4th edit., 1926; The Strength of Materials, 1899; The Mechanical Production of Cold, 1908; Thermodynamics for Engineers, 1920; An Engineer's Outlook, 1933. Organized system of sci. engring. tng.; modified Weber's theory of induced magnetism, also constructed magnetic model which behaved in accordance with his theory; observed phenomenon which he named hysteresis; investigated magnetic properties of iron, steel, other metals, also crystalline structure of several metals; invented extensometer, an hysteresis tester, permeability bridge, other apparatus for magnetic testing, also several forms of seismographs, 1881; worked on steam and thermal engines; 1 of 1st Europeans to study earthquakes in Japan. Died Jan. 7, 1935.

EWING, John I., Am. geophysicist; b. Lockney, Tex., July 5, 1924; s. Floyd Ford and Hope (Hamilton) E.; B.S., Harvard, 1950; m. Ellen Elizabeth Thomas, June 28, 1948; children—Valerie, Martha, John. Research marine geophysics Columbia Lamont Geol. Obs., 1950—, now sr. research asso. Mem. A.A.A.S., Am. Geophys. Union, Sigma Xi. Contbr. articles to sci. jours. Contbd. to understanding of structure and geologic history of ocean basins. Home: Heyhoe Woods Rds., Palisades, N.Y. 10964. Office: Torrey Cliff, Palisades, N.Y. 10964.*

EWING, W(illiam) Maurice, Am. geophysicist; b. Lockney, Tex., May 12, 1906; s. Floyd Ford and Hope (Hamilton) E.; B.A., Rice Inst., 1926, M.A., 1927, Ph.D., 1931; Sc.D., Washington and Lee U., 1949, U. Denver, 1953, Lehigh U., 1957, U. Utrecht (Holland), 1957, U. R.I., 1960; U. Durham (Eng.), 1963; LL.D., Dalhousie U., 1960; m. Avarilla Hildenbrand, Oct. 31, 1928 (div. 1941); 1 son, William Maurice; m. 2d, Margaret Kidder, Feb. 19, 1944 (div. 1965); children—Jerome H. K., Hope Hamilton, Peter Duryee, Margaret; m. 3d, Harriett Bassett, May 6, 1965. Faculty, Rice Inst., 1929-30, Lehigh U., 1930-44; faculty Columbia, N.Y.C., 1944—, prof. geology, 1947-59, Higgins prof. geology, 1959—; research asso. Woods Hole Oceanographic Instn., 1940-44, asso. geophysics, 1950-60. Com. mem. IGY, 1955-60; com. mem. Nat. Acad. Scis., 1957-64; mem. space scis. working group on lunar explorations NASA, 1959-61, lunar subcom. cons. space scis. steering com., 1960-61. Recipient Agassiz medal Nat. Acad. Scis., 1955, John J. Carty medal, 1963; John Fleming medal Am. Inst. Geonomy and Natural Resources, 1960; Vetlesen prize Columbia, 1960; Joseph Priestly award Dickinson Coll., 1961; medal of honor Rice U., 1962; Vega medal Swedish Soc. Anthropology and Geography, 1965. Guggenheim fellow, 1938-40, 53-55. Mem. Am. Acad. Arts and Scis., A.A.A.S., Am. Assn. Petroleum Geologists, Am. Geog. Soc. (Cullum Geog. medal 1961), Am. Geophys. Union (William Bowie medal 1957), Am. Philos. Soc., Arctic Inst., Argentine Sci. Soc., Geol. Soc. Am. (Arthur L. Day medal 1949), Nat. Geog. Soc., Royal Astron. Soc. (Gold medal 1964), Seismol. Soc. Am., Royal Netherlands Acad. Scis. and Letters (fgn.), Geol. Soc. London (fgn.), Academia Nacional de Ciencias Exactus, Fisicas y Naturales (Argentina, corr.), Phi Beta Kappa, Sigma Xi. Author: (with others) Propagation of Sound in the Ocean, 1948; (with others) Elastic Waves in a Layered Media, 1957; (with others) The Floors of the Oceans, 1959. Research, numerous publs. on earthquake seismology, especially relating to upper mantle of earth; first seismic measurements in open sea (1935); research and devel. instruments for submarine geology and acoustics, oceanography, climatology, geothermal measurements; first deep-sea photographs (1939). Home: Torrey Cliff, Palisades, N.Y. 10964.*

EXELL, Arthur Wallis, English botanist; b. Birmingham, Eng., May 21, 1901; s. William Wallis and Jessie (Holmes) E.; M.A., Emmanuel Coll., Cambridge U.; 1926; Dr.Sc., Coimbra U., Portugal, 1962; m. Mildred Alice Haydon, Aug. 14, 1929. With botany dept. Brit. Mus. (Natural History), 1924-39, 50-62; dep. keeper Fgn. Office, 1940-50; co-editor Flora Zambesiaca, Forestry Dept., Oxford, Eng., since 1962—. Decorated Order of Brit. Empire. Mem. Linnean Soc., Systematics Assn., Assn. for African Studies in U.K. Author: Catalogue of the Vascular Plants of Sao Tomé, 1946; (with others) Conspectus Florae Angolensis, 1937, Flora Zambesiaca, 1960. Research, publs. on flowering plants of islands of Gulf of Guinea and tropical Africa chiefly south of equator; monographic studies of family Combretaceae. Home: Church Gates, Blockley, Moreton-in-Marsh, Glos., Eng. Office: Forestry Dept., S. Parks Rd., Oxford, Eng.*

EXNER, Franz Serafin, Austrian physicist; b. Vienna, Mar. 24, 1849; s. Franz (Seraphim) and Charlotte (Dusensy) E.; student math. and physics, Vienna, Zürich, Switzerland; m. Auguste Bach, 1877; 2 daus.; m. 2d, Friderike Schuh, 1884. Asst. to A. Kundt, Würzburg, Germany, Strasbourg, France; became asst. prof. physics, U. Vienna, 1874, asso. prof., 1879, prof., 1891; dir. Physics Inst. until 1920, Rector, 1908-09. Mem. Vienna Acad. Scis. Author: Die Spektren der Elemente, 3 vols., 1911; Vorlesungen über Elektrizität, 1888. Research in spectroanalysis, colorimetry, single electrode measurements; founder contemporary theoretical research on electricity of atmosphere; proved square of velocity of particles in Brownian movement is approximately proportional to temperature. Died Vienna, Nov. 15, 1926.

EXNER, Karl Franz Joseph, physicist; b. Prague, Czechoslovakia, Mar. 26, 1842; s. Franz Seraphin and Charlotte (Dusensy) E.; teaching certificate in math. and physics U. Vienna (Austria); Ph.D., U. Freiburg (Germany), 1870; m. Henriette Wagner, 1890; 1 son. Became secondary sch. tchr., Troppau, Czechoslovakia, 1871, Vienna, 1878; named lectr. U. Vienna, 1892; became prof. theoretical physics U. Innsbruck (Austria), 1894; returned to Vienna after retirement. Mem. Acad. Scis. Vienna. Author: Vorlesungen über die Wellentheorie des Lichtes, 2 vols., 1881-87; also articles. Research on optics, especially scintillation of stars, theoretically completed problem begun by Newton; studied energy balance of total reflection of light (significant in tunnel effect of wave mechanics). Died Vienna, Dec. 11, 1914.

EXNER, Max Joseph, physician; b. Austria, Mar. 31, 1871; s. Franz and Marie (Zöllner) E.; came to U. S., 1882; M.P.E., YMCA Coll., Springfield, Mass., 1892; B.S., Carleton Coll., 1898; M.D., Univ. Med. Coll., Kansas City, Mo., 1906; m. Elizabeth Wells, May 26, 1899; children—Donald W., Max V., Willard B., Robert M. Phys. dir. Carleton Coll., 1892-98, YMCA, Troy, N.Y., 1898-99, Kansas City, Mo., 1899-1908, China, 1908-11; dir. sex edn., Internat. Com., YMCA, from 1911. A pioneer in sex edn.; started 3 lines of work in China—nat. phys. trng., trng. sch. for native phys. dirs., ednl. propaganda in hygiene and sanitation. Dir. endl. dept. Am. Social Hygiene Assn., 1920-36; epidemiologist Dept. of Health, Newark, 1936-38, dir. venereal div., from 1938. Author: The Rational Sex Life for Men, 1914; The Sexual Side of Marriage, 1932; also many pamphlets. Died Oct. 8, 1943.

EXNER (VON EWARTEN), Sigmund, Austrian physiologist; b. Vienna, Apr. 5, 1846; s. Franz Seraphin and Charlotte (Dusensy) E.; student, Heidelberg, Germany; M.D., Vienna, 1870; hon. doctorate, Leipzig, Germany, Athens, Greece; m. Emilie von Winiwarter (Felicie Ewart), 1875; children—Alfred, Felix, 1 dau. Became lectr. U. Vienna, 1871, asso. prof., 1874, prof., 1891, chmn. Phys. Inst. Mem. Vienna Acad. Sci., Bavarian Acad. Scis., Leopoldina, Wiener Gesellschaft der Arzte (hon. pres.). Author: Untersuchungen über die Lokalisation der Funktionen in der Grosshirnrinde des Menschen, 1881; Die Physiologie des Fliegens und Schwebens in den bildenden Künsten, 1882; Die Innervation des Kehlkopfes, 1884; Die Physiologie der facettierten Augen von Krebsen und Insekten, 1891; Entwurf zu einer physiologischen Erklärung der psychischen Erscheinunge, 1894; also numerous articles. An editor Zeitschrift für psychologie. Research on physiology of senses, brain localization, relation between psychic phenomena and processes in central nervous system, optical illusions, color contrast, insect eyes, bird flight, orientation of carrier pigeons; responsible for creation of 1st record archive, Vienna, for preserving the voices of famous persons, dying dialects. Died Feb. 5, 1926.

EXTERMANN, Richard Charles, physicist; b. Menton, France, Jan. 24, 1911; s. Paul Ernest and Lydia (Schneider) E.; license es sciences, Geneva U., 1933, Ph.D., 1938, hon. prof., 1963; Dr. Honoris Causa, Dijon U., 1957; m. Antoinette Rossiaud, Dec. 22, 1938; children—Charles Ed, Laurent M., Philippe H. Faculty, U. Geneva, 1938-39, 46-63; research physicist Swiss Fed. Sch. Tech., Zurich, 1940-45; head, dept. physics Cooper Union, N.Y.C., 1963—. Head, sci. secretariat, also editor Proc. UNESCO 1st internat. radio isotopes conf., Paris, 1957; cons. Sci. Directorate OECD, Paris, 1960-63. Mem. Am., French, Swiss phys. socs., Am. Assn. Physics Tchrs. Author: (with P. Grivet et al) La Resonance Paramagnetique Nucleaire, 1955; also articles. Contbns. include theory of diffraction of light by ultrasonic waves; nuclear magnetic resonance in low fields. Home: 788 Carroll St., Bklyn. 11215.*

EXTON, William Gustav, Am. pathologist, urologist; b. Savannah, Ga., Feb. 25, 1876; s. Gustav and Rosalie (Unger) E.; M.D., Columbia Coll. Phys. and Surg., 1896; house physician Mt. Sinai Hosp., N.Y.C., 1896-99; postgrad., research Pathol. Inst., Vienna, also in London and Paris, 1906-07; m. Florence Phillips, Sept. 20, 1905; children—William, Manning Mason, John Marshall. Practiced medicine, N.Y.C., 1901-06; spl. practice and research in urology-metabolism, 1907-16; dir. labs., Prudential Ins. Co., Newark, from 1914, planned and directed longevity service, from 1917. Mem. A.M.A., Am. Pub. Health Assn., Am. Urol. Assn., Am. Soc. Clin. Pathologists (pres., Ward Burdick Meml. medal), Assn. Life Ins. Med. Dirs., Optical Soc. Am., A.A.A.S., Am. Chem. Soc. Inventor of gastroscope, 1906; urological table, 1907; immiscible balance, 1915; protein tests, 1918; turbidimeter, 1918; euscope, 1920; spectroscopic method of colorimetry, 1921; scopometer, 1924; junior scopometer, 1927; quantitative microscopy, 1928; photoelectric scopometer, 1929; a new test for sugar tolerance, 1930; new methods for identifying the various sugars that occur in urine, 1932; new method for measuring the number, diameters, volume and hemoglobin content of red blood cells, 1933; the one-hour-two-dose dextrose tolerance test, 1934; incidence of sugars and reducing substances other than glucose in 1000 consecutive cases, 1934; instrument and method for measuring size of sub-microscopic particles photoelectrically by transmitted light and Tyndall beam, 1934; fibrinogen as index of disease and its clinical determination by photoelectric scopometry, 1935; partition of blood fats, 1936; one hour renal condition test, 1937; colorimetric determination of oxygen in blood, 1938; determination of blood water and its distbn. between cells and plasma, 1940; inventor of Photo-panometer: a photoelectric photometer and spectrophometer having scale readings linear with concentration, 1940; clin. significance and measurement of acidosis and alkalosis by colorimetry of carbon dioxide in blood. Author: The Prudential Urinalysis System, 1934. Contbr. on preclinical medicine, longevity, clin. pathology. Died Mar. 12, 1943.

EYDE, Samuel, Norwegian chemist; Arendal, Norway, Oct. 29, 1866; s. Samuel and Elina Kristine Amalie (Stephansen) E.; Founder Norwegian chem. industry; Norwegian minister to Warsaw, Poland, 1920-23. With O. C. Birkeland, developed Birkeland-Eyde process for nitrogen fixation, used for manufacture of fertilizers; 1st to extract nitrogen from air; 1 of 1st industrialists to establish modern housing for indsl. workers. Died June 21, 1940.

EYKHFELD, Iogan Gansovich, Russian botanist, phytoculturist; b. Russia, Jan. 25, 1893; grad. Petrograd. Agrl. Inst., 1923; D.Agrl. Sci., 1935. Charge exptl. agrl. stas., in far north, 1923-31; head polar dept. All-Union Inst. Phytoculture, Leningrad, 1931-40, dir., 1940-46; acad. sec. dept. biol. and agrl. sci. Estonian Acad. Sci., 1946-50, pres., 1950—. Decorated Order of Lenin (2); recipient Stalin prize, 1942. Mem. USSR (corr.), Estonian acads. sci., All-Union Lenin Acad. Agrl. Sci. Research and publs. on polar plant cultivation and agriculture; proved possibility of vegetable and feed base in severe climatic environments of Kola Peninsula and northern Karelian ASSR; instrumental in establishing 1st state farm beyond Arctic Circle, nr. Kirovsk, Murmansk Oblast. Address: Estonian Academy of Sci., Tallinn, Estonia SSR, USSR.

EYNO, see Evno.

EYRIES, Jean Baptiste-Benoit, French geographer; b. Marseille, France, 1767; founder, hon. pres. Soc. Geography of Paris; mem. Asiatic Soc., Library of Acad. of Inscriptions and Belles-Letters; ret., 1844. Author: Voyage de découvertes dans la partie septentrionale de l'ocean Pacifique, 1807; Tableaux de la nature, ou considerations sur les déserts, sur la physionomie des végétaux de l'Amerique, 1808; Voyage en Perse, en Arménie, en Asia Mineure et à Constantinople en 1808 et 1809 by Jacques Morier, 1813. numerous other works, transls. of sci. works from English and German. Died Graville, 1846.

EYRING, Henry, chemist; b. Colonia Juarez, Mex., Feb. 20, 1901; s. Edward Christian and Caroline (Romney) E.; B.S., U. Ariz., 1923, M.S., 1924; Ph.D., U. Cal., 1927; postgrad. U. Berlin, 1929-30; D.Sc., U. Utah, 1952, Northwestern U., 1953, Princeton, 1956, Seoul (Korea) Nat. U., 1963; m. Mildred Bennion, Aug. 25, 1928; children—Edward Marcus, Henry Bennion, Harden Romney. Came to U. S., 1912, naturalized, 1935. Instr. U. Ariz., 1924-25; teaching fellow, U. Cal., 1925-27; instr. U. Wis., 1927-28, research asso., 1928-29; lectr. U. Calif., 1930-31; research asso. Princeton, 1931-36, asso. prof., 1936-38, prof. 1938-46; dean grad. sch. U. Utah 1946—; dir. fundamental research, Textile Found. Textile Research Inst., 1944-46. NRC scholar, 1929-30. Recipient 9th annual prize A.A.A.S., 1932; plaque and $2500 honorarium Research Corp., 1949. Mem. Am. Acad. Arts and Scis., Nat. Acad. Scis., Am. Philos. Soc., Am. Chem. Soc. (dir. 6th dist., 1949-51; pres. 1962—, Peter Debye award 1964), Nat. Sci. Bd., A.A.A.S. (dir., pres.). Author: Theory of Rate Processes (with Glasstone and K. J. Laidler), 1941; Quantum Chemistry (with J. Walter and G. E. Kimball), 1944; (with Frank H. Johnson and Milton J. Polissar) Kinetic Basis of Molecular Biology, 1954; (with Edward M. Eyring) Modern Chemical Kinetics, 1963; (with others) Statistical Mechanics & Dynamics. Editor Ann. Rev. Phys. Chemistr. Author articles sci. publs. Pioneer application of quantum and statistical mechanics to chemistry; developed theory of absolute reaction rates and significant structure theory of liquids, also theories of optical activity, mass spectrography, addition of dipoles and bond lengths in flexible high polymers, and bioluminescence. Home: 1922 E. Ninth South, Salt Lake City.*

EYRING, LeRoy, Am. chemist; b. Pima, Ariz., Dec. 26, 1919; s. Edward Christain and Emma (Romney) E.; B.S., U. Ariz., 1943; Ph.D., U. Cal. at Berkeley, 1949; m. Ruth LaReal Patton, July 21, 1941; children—Michelle, Patricia, Cynthia, Gregory. Faculty, State U. Ia., 1949-61; prof., chmn. dept. chemistry Ariz. State U., Tempe, 1961—. NSF fellow U. Göttingen, Germany, Imperial Coll. and U. Stockholm, Sweden, 1958-59; Guggenheim fellow, Fulbright-Hays Program grantee U. Melbourne (Australia), 1959-60. Mem. Am. Chem. Soc., A.A.A.S., Ariz. Acad. Sci., Phi Beta Kappa, Sigma Xi, Phi Kappa Phi, Phi Lambda Upsilon, Pi Mu Epsilon. Editor: Progress in the Science and Technology of the Rare Earths series, Vol. II, 1966; Advances in High Temperature Chemistry series, Vol. I, 1967; Rare Earth Research III, 1966. Contbr. articles to tech. jours. Research in solid state and high temperature chemistry, existence and nature nonstoichiometry, ordered phases in chem. systems and reactions they undergo with their environment. Home: 6995 E. Jackrabbit Rd., Scottsdale, Ariz. 85251. Office: Dept. Chemistry, Ariz. State U., Tempe, Ariz. 85281.*

EYSENCK, Hans Jurgen, psychologist; b. Berlin, Germany, Mar. 4, 1916; s. Eduard Anton and Ruth (Werner) E.; B.A., U. London, 1938, Ph.D., 1940, D.Sc., 1964; postgrad. U. Dijon (France), Exeter (Eng.) U.; m. Margaret Davies Aug. 1939 (div. Aug. 1950); 1 son, Michael William; m. 2d, Sybil Bianca Guiletta Rostal, Sept. 30, 1950; children—Gary Lionel Laurence, Connie Lilith Tamar, Kevin Russ Mark, Darrin Montague Alan. Research psychologist Millhill Emergency Hosp., 1942-47, Maudsley Hosp., 1947-50; faculty, dir. psychol. labs. Inst. Psychiatry, U. London, 1950—, prof., 1955—; psychologist Maudsley and Bethlem Royal hosps.; vis. prof. U. Pa., Phila., 1949-50, U. Cal. at Berkeley, 1945. Fellow Brit. Psychol. Soc., Am. Psychol. Assn.; mem. Eugenics Soc., Inst. Behaviour Therapy (pres.). Author numerous books, including Dimension of Personality, 1947; Dynamics of Anxiety and Hysteria, 1957; Crime and Personality, 1964; (with S. Rachman) Causes and Cures of Neurosis, 1965; Fact and Fiction in Psychology, 1965; Smoking, Health and Personality, 1965; The Biological Basis of Personality, 1967; also numerous articles. Editor several books, including Experiments in Personality, 1960; Experiments with Drugs, 1963; Experiments in Behaviour Therapy, 1964; Experiments in Motivation, 1964. Devel. behaviour therapy, a new method of treating neuroses based on modern learning theory; devel. and exptl. proof of dimensional personality theory, three factor theory of reminiscence; devel. intensive analysis as method in statis. treatment of personality research data; crime and personality. Home: 10 Dorchester Dr., London S.E. 24. Office: Maudsley Hosp., Denmark Hill, London S.E. 5., Eng.*

EYSTER, William Henry, Am. botanist, agriculturist; b. Sunbury, Pa., July 13, 1889; s. Henry and Alice (Star) E.; A.B. summa cum laude, Bucknell U., 1914, A.M., 1915; Ph.D., Cornell U., 1920; postgrad. Harvard, U. Berlin (Germany), Erlangen (Germany); m. Elmire Snyder, June 18, 1914; children—William Henry, Paul Morris, Helen Elizabeth (Mrs. William Thomas Crisp). Instr., Pa. Forest Acad., 1914-16; head dept. sci. Corland N.Y. State Tchrs.' Coll., 1916-18; instr. Cornell U., 1918-19, asst. genetics., 1919-20; asst. prof. botany U. Mo., 1920-24; prof. botany U. Me., 1924-27; John Simon Guggenheim Meml. Found. fellow, Germany, 1927-38; prof. Bucknell U., 1928-45, Baldwin-Wallace Coll., Berea, O., 1946-47; mgr. edn. Rodale Pub. Schs., 1947-51; spl. researcher on peat for compost John L. Roper Lumber Co., Broadway, N.Y., 1951-54; compost engr., cons., Emmaus, Pa., 1954—; tchr. philos. edn. Moravian Coll., Bethlehem, Pa., 1958. Agr. cons. Zook & Ranck, Inc., Holmes Super-Grow Products, Inc., Millersburg, O., 1962—, Indian Valley Products, Inc., Elgin, Ore. Recipient Alumni award for meritorious achievement Bucknell U., 1962. Fellow A.A.A.S., N.Y. Acad. Scis.; mem. Am., German bot. socs., Am. Genetics Soc., Am. Soc. Plant Physiologists, Am Soc. Naturalists, Torrey Bot. Club, Pa. Acad. Sci., Sigma Xi, Phi Kappa Phi, Phi Sigma. Author: College Botany, 1931; Genetics of Sea Mays, 1934; Bibliogenetica Biographia, 1934; also numerous articles. Discovered chem. plant-fertility hormone; induction of mutations with hormones; created all-natural compost, Eysterite. Address: 235 Harrison St., Emmaus, Pa. 18049.*

EYTELWEIN, Johann Albert, German engr.; b. Frankfort/Main, Jan. 1, 1765; s. Christian Philipp and Anna Elisabeth Katherine (Hung) E.; student hydraulic engring.; m. Dorothea Charlotte Louise Pflaum, 1790; 1 son, 6 daus. Joined Prussian Arty., 1779; discharged as lt.; dike insp. Prussian Hydraulics Adminstrn.; became mem. Bd. Works (dir. regulation numerous E. German rivers, constrn. harbor works), 1794; dir. Bauak-ademie, 1799; named chief dir. constrn., Prussia, 1816; ret., 1830. Mem. French Acad. Scis., 1846. Author: Vergleichungen aller in den Königlichen preussischen Staaten eingefassten Masse und Gewichte, 1798; Praktische Anweisung zum Construction der Faschinenwerke an Flüssen und Strömen, 1800; Handbuch der Mechanik fester Körper, 3 vols., 1808; Handbuch der Perspektive, 2 vols., 1810; Grundlagen der höheren Analysis, 2 vols., 1825; Handbuch der Hydrostatik, 1826; Anweisung zur Auflösung der Höheren numerischen Gleichungen, 1837. Research on hydraulics, hydrostatics, weights and measures system in Prussia. Died Berlin, Germany, Aug. 18, 1849.

EYTH, (Eduard Friedrich) Max(imilian) von, see von Eyth, (Eduard Friedrich) Max(imilian).

EYZAGUIRRE, Carlos Edwards, neurophysiologist; b. Santiago, Chile, Apr. 28, 1923; s. Carlos Gormaz and Ines (Edwards) E.; B.A., U. Chile, 1940, M.D., 1947; m. Elena Fontaine, Aug. 31, 1947; children—Carlos A., Elena M., Rodrigo J. Faculty, Cath. U. Chile, Santiago, 1942-47, 50-52, asso. prof. neurophysiology, 1952-57; fellow Johns Hopkins, 1947-48, Emanuel Libman fellow, 1948-50, fellow Wilmer Inst., John S. Guggenheim Meml. Found. fellow, 1953-55; lectr. physiology Tchrs.'s Coll., U. Chile, 1951-53; asst. research prof. physiology U. Utah Coll. Medicine, Salt Lake City, 1957-59, asso. prof., 1959-62, prof., 1962—, chmn. dept., 1964—. Spl. cons. USPHS neurol. scis. research tng. com., 1966—. USPHS Sr. Research fellow, 1959-64; recipient USPHS Research Career Devel. award, 1964. Mem. Am. Physiol. Soc., Biol. Soc. Santiago, Biol. Soc. Montevideo (hon.). Research, publs. in physiology of nervous system, physiology receptors, and chemoreceptors, initiation of impuses in these areas.*

EZAWA, Hiroshi, Japanese physicist; b. Tokyo, Japan, June 2, 1932; s. Taji and Sue (Yamada) E.; B.Sc., U. Tokyo, 1955, M.Sc., 1957, Ph.D., 1960; m. Yoshiko Kitagawa, Sept. 1, 1963. Research asso. dept. physics U. Tokyo, 1960-67; asso. prof. dept. physics Gakushuin U., Tokyo, 1967—; research asso. U. Md., College Park, 1963-65, U. Ill., Urbana, 1965; vis. asso. prof. U. Wis., Milw., 1965-66; vis. scientist II. Inst. für Theoret. Phys., U. Hamburg (Germany), 1966-67. Fulbright-Hays Travel grantee U. S. Ednl. Commn., 1963. Mem. Phys. Soc. Japan, Am. Phys. Soc. Author: (with H. Umezawa, K. Kawarabayashi) Quantum Mechanics, 1958; Recent Topics in Quantum Field Theory, 1963; also articles. Developed use of impact parameter in analysis of nuclear collisions at extremely high energy; application of field theory methods to statis. mechanics; concepts of elementarity and compositeness of particles in high-energy nuclear physics, canonical formalism of quantum field theory. Home: 25-5, Honkomagome l-chome, Bunkyo-ku, Tokyo, Japan.*

EZEILO, James Okoye, Nigerian mathematician; b. Awka, Nigeria, Jan. 17, 1930; s. Josiah Nwafor and Janet (Ifezue) E.; B.Sc., U. Ibadan, 1953, M.Sc., 1955; Ph.D. in Math., Cambridge U., 1958; m. Phoebe Uzoechina, Dec. 11, 1960; children—Ada Uchechuku, Okechuku Chukunwike, Obasiemeka Nnanedu. Lectr. math. U. Ibadan (Nigeria), 1958-62, sr. lectr., 1962-64, prof., 1964—. Recipient Ayrton prize Council Inst. Elec. Engrs., 1963. Mem. London, Am. math. socs. Research, publs. on oscillation problems in differential equations. Address: U. Ibadan, Ibadan, Nigeria.*

F

FABBRONI, Giovanni Valentino Mattia, Italian chemist; b. Florence, Italy, Feb. 13, 1752; vice-dir. (under Fontana) Archducal Cabinet and Mus., Florence; later dir.; under Napoleon I, apptd. dir. of roads and bridges for subalpine depts.; mem. French Acad. Scis., Soc. Agr. Regarded chem. action as a cause of galvanistic effect, but did not propose chem. theory of galvanism; considered yeast vegetoanimal ferment; research on agriculture and botany; credited with building Col de Genèvre (pass) and initial construction of Corniche road from Genoa to Nice. Died Pisa, Dec. 17, 1822.

FABER, Johannes, botanist; b. Bamberg, Germany, 1574; practiced medicine, Rome, Italy; physician to Pope Urban VIII; helped found 1st scientific soc., Accademia dei Lincei (named for the lynx because of that animal's piercing sight); suggested name "microscope" for recently invented instrument. Died 1629.

FABER, Knud, Danish physician; b. Odense, Denmark, Aug. 29, 1862; M.D., U. Copenhagen, Denmark, 1890; LL.D., Edinburgh; M.D. (hon.), Uppsala, Sweden, 1927, Aarhüs, Denmark, 1946; m. Thyra Petersen, 1891; m. 2d, Ellen Becker, 1910; 2 sons. Prof. internal medicine, head physician Royal State Hosp., Denmark, 1896-1932, retired, 1932, rector, 1916-17. Fellow Royal Soc. Medicine (hon.), Royal Coll. Physicians in England. Hon. mem. numerous European acads. medicine. Author: Tetanus as Infectious Disease, 1890; Pathology of Digestive Organs, 1905; Pernicious Anaemia and Atrophy of Intestines, 1920; Diseases of the Stomach and the Intestines, 1920; Nosography in Modern Internal Medicine, 1923; Tuberculosis in Denmark, 1926; Functional Diseases of the Digestive Organs, 1927; Lectures on Internal Medicine delivered in the United States, 1926; Report on Medical Schools in China, 1931; Gastritis and its Consequences, 1935; Tuberculosis and Nutrition, 1938. Died May 2, 1956.

FABER DU FAUR, Adolf Friedrich von, see von Faber du Faur, Adolf Friedrich.

FABERGÉ, Alexander Cyril, geneticist, cytologist; b. Moscow, Russia, Feb. 26, 1912; s. Alexander J. and Jeanne (Tammermann) F.; B.Sc., U. Reading (Eng.), 1933; Ph.D., U. London (Eng.), 1936, D.Sc., 1945. Came to U. S., 1945, naturalized, 1953. Research worker John Innes Instn., Merton Park, London, Eng., 1935-37; asst. lectr. Galton Lab., U. Coll., London, 1937-45; exptl. officer Army Operational Research Group Brit. Ministry Supply, 1942-45; research asso. botany dept. U. Wis., 1945-47; asso. prof. U. Mo., 1947-55; scientist biology div. Oak Ridge Nat. Lab., 1956-57; research scientist, lectr. zoology dept. U. Tex., Austin, 1957——. Vis. prof. U. Ore., Eugene. Mem. Genetical Soc. Eng., Genetics Soc. Am., Electron Microscopy Soc. Am. Research, numerous publs. on genetics, mechanism chromosome breakage, breakage-fusion bridge cycle in maize, methods for examining chromosomes with electron microscope. Home: Austin, Tex.*

FABIANEK, John, biochemist; b. Perbeta, Slovakia, Dec. 3, 1922; s. John and Joanne (Rosko) F.; Sc.D., Slovak U., 1946; Diploma of Modern French, Alliance Francaise, Paris, 1948; Diploma, Institut des Sciences Politiques, U. Paris, 1950; J.D. in Internat. Law, U. Paris, Sorbonne, 1954, Ph.D. in Biochemistry and Physiology, 1957; Diploma, Institut des Hautes Etudes Internationales, Faculte de Droit, U. Paris, 1955. Came to U. S., 1960, naturalized, 1966. Faculty, Centre National de la Recherche Scientifique, Paris, 1948-60; asso. prof. biochemistry N.Y. Med. Coll., N.Y.C., 1960-65; biochemist U. Cal. Sch. Medicine, Los Angeles, 1965——. Del. French Govt. to Atomic Group of UN, Geneva, 1959. Fellow A.A.A.S., N.Y. Acad. Scis.; mem. Am. Chem. Soc., Am. Inst. Nutrition, Am. Physiol. Soc., Am. Rheumatism Assn., Gerontological Soc., Harvey Soc., Biochem. Soc. London, Inst. Research Danube Basin Problems, Vienna, Société de Chimie biologique Paris, Société Chimique de France, Soc. for Exptl. Biology and Medicine. Author: Relations internationales entre la Tchéco-Slovaquie et l'Union sovietique, 1954. Research, publs. on nutrition, vitamin C, bioflavonoids, mineral metabolism, mucopolysaccharides, collagen, glycoproteins, connective tissue, skin, rheumatoid arthritis, endocrinology. Home: P.O. Box 1907, Long Beach, Cal. 90801.*

FABRE, Jean-Antoine, French hydrographer; b. St.-André, France, Aug. 19, 1748; became head engr. bridges and roads, 1796; dir. nav., Provence, France; mem. French Acad. Scis. Author: Essai sur la manière de construire les machines dydrauliques, 1782; Sur l'irrigation artificielle de la Provence, 1797. Died Jan. 31, 1834.

FABRE, Jean Henri Casimir, French entomologist; b. St. Leons, France, Dec. 21, 1823; ed. Normal Sch. Vancluse, also in Rodez, doctorate, Paris. Tchr. math. and physics, Carpentras, also Ajaccio; tchr., Avignon, 1852, Orange, 1870-79; ret., Serignan (all France). Mem. French Acad. Scis., Royal Acad. Sweden. Author: Histoire de la Buche, 1866; Notions préliminaires de physique, 1867-70; Souvenirs entomologiques, 10 vols., 1879-1907. Studied habits of insects, especially coleoptera, hymenoptera, orthoptera, spiders; his publs. were admired by Dufour and Darwin, yet his observations led him to oppose evolutionary theory. Died Serignan, Oct. 11, 1915.

FABRE, Philippe Achille Joseph, French physician; b. Rivesaltes, Sept. 10, 1892; s. Philippe and Marie (Chastan) F.; ed. Collège of Perpigann, Lycée of Montpellier, Faculty Scis., Montpellier; M.D., Faculty Medicine, Paris. Agrégé in biol. physics; asst. med. physics Faculty in Paris, 1926-28; titular prof. med. physics Faculty Medicine, Lille, 1930-62. Mem. Soc. Biology of Paris (corr.), Acad. Medicine (corr.). Reasearch and publs. on topographical dioptrics, sphygmonanometrical mechanics, excitability of nerves, phase angle of Ranvier's nodes, glottography of phonation and respiration by Hageman factor currents. Home: 19, rue Emile-Zola, Rivesaltes, Pyrenees Orientales, France.

FABRE, Pierre, French surgeon; b. Tarascon, France, 1716; practiced surgery in Paris; mem. Acad. Soc. Surgeons Paris. Author: Traité des maladie vénériennes, 1758; Recherches sur les vrais principes de l'art de guérir, 1790. Research on physiology, pathology, therapeutics, venereal diseases. Died 1791.

FABRE, Pierre Jean, physician; b. circa 1600; M.D., U. Montpellier (France); practiced medicine, Montpellier; physician to Louis XIII of France. Author: Palladium Spagyricum, 1632; Alchymista Christianus, 1632; Hercules Pio-Chymicus 1634; L'Abrege des Secrets chymiques, 1636; others. Voluminous writer whose works were printed in collected form and translated into German. Died 1650.

FABRI (or FABBRI), Honoré, physician; b. Le Bugey, France, 1606; prof. math. and philosophy, Lyons, France. Author: Brevis annotatio in systema saturnium Ch. Hugenii, 1660; Quinquina; Synopsis optica, 1667. Defended use of cinchona bark; opposed ideas of Huygens and of Galileo; falsely claimed to have discovered circulation of blood. Died Rome, Italy, Mar. 9, 1688.

FABRICAND, Burton Paul, Am. physicist; b. N.Y.C., Nov. 22, 1923; s. Irving Kermit and Frances (Sobler) F.; A.B., Columbia, 1947; A.M., 1949, Ph.D., 1953; m. Prudence Diane Montgomery, Sept. 5, 1952; children—Nicole Diane, Lorraine Stewart. Project engr. Philco Corp., Phila., 1952-54; lectr., research asso. U. Pa., 1954-56; sr. research scientist Columbia Hudson Labs., Dobbs Ferry, N.Y., 1957——. Cons., Moore Sch. Elec. Engring., U. Pa., 1954-60, Indsl. Electronic Hardware Corp., N.Y.C., 1960-64. Mem. Am. Phys. Soc., Sigma Xi. Author: Horse Sense: A New and Rigorous Application of Mathematical Methods to Successful Betting at the Track, 1965. Pioneer in determination of many trace elements in the ocean; inventor improved phosphorescent glass dosimeter for measuring gamma ray exposure; furthered understanding of molecular structure and motions of molecules in water solutions; devel. theory of pari-mutual gambling and its applications to sociol. problems. Home: 115 Judson Av. Office: 145 Palisade St., Dobbs Ferry, N.Y. 10522.*

FABRICIUS, David (original surname Goldschmidt), German astronomer; b. Esens, Germany, Mar. 9, 1564; s. Jan Janssen; studied theology under H. Lampadius in Braunschweig, Germany; 3 children, including David, Johannes. Lutheran pastor, Resterhaave, before 1603, Osteel, E. Friesland, Germany, 1603-17. Author: Calendarium historicum . . . , 1590; Kurtzer und Gründtlicher Berichter von Erscheinung und Bedeutung dess grossen neuen Wunder Sterns . . . , 1605; De cometa Anni 1607, 1618; Letters to Kepler. Probably with Tycho Brahe (whose abilities in astron. observations Fabricius equaled) in Prague, Czechoslovakia, 1597-99, 1601; friend of Tycho Brahe and Johannes Kepler (who used Fabricius' observations on Mars); criticized Kepler's finding of Mars' elliptical path; one of 1st to use telescope for observations; built own guadrants and sextants; discovered 1st variable star, Mira Ceti, 1596; observed fixed stars; interested in measuring planets' heights and distances from earth; discovered star Cetus (later named Mira Ceti by J. Hevelius). Murdered by a parishioner, Osteel, Germany, May 7, 1617.

FABRICIUS, Hildanus (or Wilhelm Fabry), German physician, surgeon; b. Hilden, Dusseldorf, Germany, June 26, 1560; apprentice to barber surgeon Master Dümgen, 1576-80; settled in Dusseldorf, 1580; asst. to Kosmas Slot, surgeon of ducal count; worked with prominant surgeons in Geneva, Lausanne (both Switzerland); practiced in Cologne, Germany, Lausanne, Bern, Switzerland; travelled widely. Considered father of German surgery; opposed superstition; made many dissections; 1st to recommend amputation above diseased part in cases of gangrene; made earliest known classification of burns, 1607; credited with 1st amputation of thigh; used improved tourniquet for amputation; credited with 1st removal of gallstone from living patient, 1618; operated for cataract and eye tumors; extracted iron splinter from eye with magnet; credited with 1st recognition of congenital pyloric stenosis; devised new surg. instruments. Died Bern, Feb. 14, 1634.

FABRICIUS, Johan Christian, Danish entomologist; b. Tondern, Denmark, Jan. 7, 1745; studied in Altona, also in Copenhagen, Denmark, under Carolus Linnaeus in Uppsala, Sweden; became prof. natural history, economy and finance U. Kiel (Germany), 1775. Author: Systema entomologiae, 1775, enlarged edit., 1792-94, with supplement, 1798; Genera insectorum, 1776; Philosophia entomologica, 1778. 1 of founders of sci. entomology; introduced new system for classifying insects on basis of mouth structure rather than wings. Died Kiel, Mar. 3, 1808.

FABRICIUS, Johannes, astronomer; b. Resterhave, E. Friesland, Germany, Jan. 8, 1587; s. David Fabricius; studied medicine, Wittenberg, Germany, from 1605; M.D. Author: De manculis in sole observatis, et apparente earum cum sole conversione narratio, 1611. Used (with father) projection method for observation of sun; studied sunspots (anticipated work of Galileo and C. Scheiner); said to have detected sun's rotation on axis by means of sunspot movements; aston. work admired by Kepler. Died 1615.

FABRICIUS, Otto, naturalist; b. Rudkjoebing, Denmark, 1744; ed. Greenland Sem., Copenhagen, Denmark; at least 1 son, Frederik. Missionary, Frederikshaab, 1768-73; became pastor, Drangedal, Norway, 1774, Hobro, 1779, Rise, Aerae Island, 1781, Copenhagen, 1783; became head Greenland Sem., 1783; named pastor, Christianshavn, 1789; became dir. Danish Soc. Natural History, 1789, prof., 1803, dr. theology, 1818. Author: Fauna Groenlandica (descriptions of 473 species, including 130 new to sci.), 1780, also a Greenlandish grammar, 1791, and dictionary, 1801. Made 1st sci. study of fauna in Greenland. Died 1822.

FABRICIUS AB AQUAPENDENTE, Hieronymus, Italian anatomist, surgeon; b. Aquapendente, Padua, Italy, May 20, 1537; pupil of Fallopius at Padua; prof., Padua, 1562-1609; tchr. of William Harvey. Author: De formato foetu, 1600; De venarum ostiolis, 1603. Laid founds. of embryology; studied devel. of foetus in man and animals; gave 1st detailed description of placenta, circa 1600; discovered semi-lunar valves in veins without seeing their significance; made studies in comparative anatomy; gave 1st account of larynx as vocal organ; corrected Vesalius' description of location of lens in eyeball; devised silver stomach tube, also many other surg. instruments. Died Padua, Italy, May 21, 1619.

FABRONI, Giovanni Valentino Mattia, see Fabbroni, Giovanni Valentino Mattia.

FABRY, Charles, French physicist; b. Marseilles, France, June 11, 1867; student Ecole Polytechnique, Paris, France, 1885-87; D.S., 1892; became chief lectr. physics U. Marseilles, 1895, prof. physics, 1904-21; prof. physics U. Paris, 1921-37; 1st dir. Institut d'Optique, Paris. Mem. French Acad. Scis., 1927; Fellow Royal Soc., 1931 (Rumford medal 1918). Author: Physique et astrophysique, 1935; Leçons élémentaires d'acoustique et d'optique, 1898; Les piles électriques, 1899; La lumière monochromatique, sa production et son emploi en optique pratique, 1923; Les applications des interférences lumineuses, 1923; Introduction générale à la photométrie, 1928. Applied interference of multiple waves theory to celestial physics, also to meteorology, spectroscope; inventor (with Perot) interferometer (for study of fine spectral rays); established internat. system. of wave lengths; made 1st direct verification of Doppler-Fizeau principal in optics; discovered ozone in high atmosphere, 1913. Died Paris, Dec. 11, 1945.

FABRY, (Charles) Eugène, French mathematician; b. Marseille, France, Oct. 16, 1856; prof. analysis Faculty Scis. Marseilles (France), then at Montpellier, France; mem. French Acad. Scis. Author: Traité de mathématiques générales, à l'usage des chimistes, physiciens, ingénieurs, 1909; Sur les points singuliers d'une série de Taylor, 1910. Died Oct. 6, 1944.

FABRY, Louis, French astronomer; b. Marseille, France, Apr. 20; 1862; astronomer Marseille obs.; mem. French Acad. Scis., 1919. Died Jan. 26, 1939.

FABRY, Wilhelm, see Fabricius, Hildanus.

FACCHINI, Ugo, Italian nuclear physicist; b. Milan, Nov. 4, 1924; s. Giuseppe and Giuseppina (Villani) F.; D. in Physics, Milan U., 1946; m. Catteneo Eufrasia, Nov. 13. 1947; children—Giuseppina, Francesco. Faculty, Milan U., 1960——, prof. gen. physics, 1957——; dir. nuclear physics lab. C.I.S.E., Milan, 1950——. Research, publs. on neutron physics, nuclear statistics, cross sect. fluctuations, theory of fission, nuclear measurement devices. Home: Via Papa Giovanni 32, Segrate, Italy. Office: Universita di Milano, Milano, Italy.*

FACCIO, Nicolas, mathematician; b. Basel, Switzerland, Feb. 16, 1664; s. Jean Baptiste and Catherine (Basband or Barbaud) F.; ed. Geneva, Switzerland; became citizen of Geneva, 1678; betrayed conspiracy to kidnap Prince of Orange, 1686; went to Eng., 1687; befriended by Newton; disparaged Leibniz, 1699; chief of French prophets; exposed at Charing Cross as imposter; went to Asia to disseminate his theories. Author: Epistola . . . de mari aeneo Salomonis, 1688; Navigation Improv'd, 1728. Demonstrated how to use ship's motion for grinding corn, sawing; improved telescope glasses; discovered how to pierce rubies to receive pivots of balance wheel of watches; developed Cassini's theory of zodiacal light, 1685. Died 1753.

FACCIOLI, Giuseppe, elec. engr.; b. Rome, Italy, Apr. 7, 1877; s. Luigi and Flora (Garbochi) F.; grad.

Royal Poly. Inst., Milan, 1899. Came to U. S., 1902, naturalized, 1914. With Lab. of N.Y. Edison Co., 1902; in employ Interborough Rapid Transit Co., 1903; designer of alternating current machines, Crocker-Wheeler Co., 1904; asst. to William Stanley, Great Barrington, Mass., 1905-07; with ry. dept. Gen. Electric Co., Schenectady, 1907, works engr., Pittsfield, Mass. 1913-27, asso. mgr. and works engr., 1927-30. Fellow Am. Inst. E.E., Am. Phys. Soc., Brit. Instn. E.E., Associazione Elettrotecnica Italiana. Known for investigations in connection with transmission of power at high voltage and for constrn. of high voltage elec. apparatus. Died Jan. 13, 1934.

FAEGRI, Knut, Norwegian botanist; b. Bergen, Norway, July 17, 1909; s. Ole Anton and Gudrun (Stoltz) F.; ed. Bergens Katedralskole, 1920-26, Bergens Mus., U. Oslo (Norway); Cand. Mag., 1930; Ph.D., 1934; m. Nancy Meyer, Oct. 30, 1941; children—Knut II, Alice. Prof. systematic botany and phytogeography, dir. bot. mus. and garden U. Bergen, from 1946, dean Faculty Sci., 1959-65. Mem. Norwegian Council Sci. and Humanities, 1962—. Mem. Det Norske Videnskaps-Akademi (Oslo), Vitenskapelig Avdling Selskapet Til Vitenskapenes Fremme (Bergen), others. Author: (with J. Iversen) Text-book of Modern Pollen Analysis, 1950, 2d edit., 1964; (with L. van der Pijl) Principles of Pollination Ecology, 1966; Krydder pa Kjokkenet og i Verdens Historien, 1966. Editor: Naturen, 1947—. Research, publs. on composition and history of vegetation of western Scandinavia, also role played by man in modifying landscape; theoretical pollen analysis. Home: 37 A, Formansvei, 5000 Bergen. Office: Bot. Mus., P.O. Box 2637, 5010 Bergen, Norway.*

FAELENS, Paul August Pierre, Belgian physicist; b. Bruges, Belgium, Apr. 22, 1923; s. John Bernard and Adriana (Casteleyn) F.; Dr.Phys., U. Ghent (Belgium), 1954; m. Hilda Engelbertha Brusselmans, May 3, 1949; children—Guido, Herman, Veerle, Lydia. Sci. adviser for emulsion tech. and research Gevaert-Agfa Research Labs., Antwerp, Belgium, 1946—. Recipient Silver medal Photographische Gesellschaft in Wien, 1963; De Boelpaepe prize Royal Acad. Scis. Belgium, 1957. Mem. Flemish Chem. Soc., Soc. Photog. Sci. and Engring., Nederlandse Natuurkund Vereniging. Research, publs. on photog. theory especially latent image and chem. sensitization; mechanism of pressure effects on photog. emulsions and mechanism of photog. sensitization using labil S and noble metal salts. Home: 83 Brasschaatse baan, Schilde-Antwerp. Office: Gevaert-Agfa N.V., Mortsel-Antwerp, Belgium.*

FAESSLER, Amand, German physicist; b. Gengenbach, Apr. 26, 1938; s. Amand and Elisabeth (Oschwald) F.; student univs. Freiburg, Munich; diplom, 1961, Ph.D. summa cum laude, 1963; m. Ehrentraud Drechsel, Feb. 20, 1964; 1 son, Amang Georg. Vis. research asso. Fla. State U., Tallahassee, 1963, 65; vis. asst. prof. U. Cal. at Los Angeles, 1965-66; sci. asst. U. Freiburg (Germany), 1964-65, 66—. Contbr. articles to profl. jours. Theoretical studies of nuclei; devel. rotational-vibrational model (with W. Greiner) for heavy deformed nuclei. Address: 4 Lachendaemmle, Freiburg, West Germany.*

FAESSLER, Carl, geologist; b. Steinen, Switzerland, 1895; s. Boniface and Marie Anna (Abegg) F.; Baccalaureate, Classical Coll., Schwyz, Switzerland, 1916; Ph.D., Fribourg (Germany) U., 1923; m. Marie Annen; children—3 sons, 1 dau. Asst. to dir., sch. chemistry Université Laval, Que., Can., 1923-31, titular prof. minerology, 1931-37, Faculté des Sciences, 1937; vis. prof. petrography U. Montreal, Can., 1941-44; lectr. Zurich, Switzerland, 1952. Gold, prospecting cons., 1924, 26; field geologist Que. Bur. Mines, 1927-43. Mem. Geol. Soc. Am., Royal Soc. Can., Swiss Geol. Soc., Swiss Soc. for Minerology and Petrography, Assn. Canadienne-Francaise pour l'Advancement des Sciences. Organized library of geol. publs., cross-indexed maps, illustrations pub. by Geol. Survey and Mines Br., Can.; cross-indexed index to illustrations in publs. Ont. Dept. Mines, 1956, Que. Dept. Mines, 1957; compiled 3000 analyses of Que. minerals, rocks, ores. Died 1957.

FAGAN, Brian Murray, archeologist; b. Birmingham, U.K., Aug. 1, 1936; s. Brian Walter and Gwen (Moir) F.; B.A., Pembroke Coll., Cambridge U., 1959, M.A., 1962, Ph.D., 1963. Keeper prehistory Livingstone Mus., Zambia, 1959-65; dir. Bantu studies project Brit. Inst. Archaeology, E.Africa, 1965-66; vis. asso. prof. U. Ill., Urbana, 1966-67; asso. prof. U. Cal., Santa Barbara, 1967—. Mem. Zambia Nat. Monuments Commn., 1960-65, sec. (hon.), 1961-62. Fellow Royal Geog. Soc., Royal Anthrop. Inst., African Studies Assn., U.K. and U. S.; asso. Current Anthropology. Author: Southern Africa, 1966; Iron Age Cultures in Zambia, I and II, 1967-68. Editor: A Short History of Zambia, 1966. Research, publs. on metal-working peoples of Zambia dating from AD 100 to 1850; excavations on Late Stone Age sites in Central Zambia; investigations of early salt-workers in So. Tanzania, E.Africa; compiler basic source material on African archeology for use in ednl. syllabi in Africa.*

FAGE, Arthur, Brit. aerodynamician; b. Portsmouth, Eng., Mar. 4, 1890; s. William John and Annie (Crook) F.; Diploma (Aero. Research scholar) Imperial Coll. Sci. and Tech., 1914; m. Winifred Eliza Donnelly, Sept. 1, 1920; children—John Donnelly, Christine Mary. (Mrs. Gerald Richards). Sci. staff aerodynamics div. Nat. Phys. Lab., Teddington, Eng., 1914-53, supt. aerodynamics div., 1946-53. Mem. coms. Aero. Research Council, 1925—. Named Comdr. Brit. Empire, 1953. Fellow Royal Soc., Royal Aero. Soc. Author: The Aeroplane, 1916; The Airscrew, 1919. Research, publs. on aerodynamics, fluid motion, aeronautics. Home: 171 Park Rd., Teddington, Middlesex, Eng.*

FAGERSON, Irving Seymour, Am. food chemist; b. Lawrence, Mass., June 7, 1920; s. William and Rose (Zelinsky) F.; S.B., Mass. Inst. Tech., 1942; M.S., U. Mass., 1948, Ph.D., 1950; m. Belle Anita Rost, Jan. 4, 1953; children—Michael L., Judith L. Marketing specialist War Food Adminstrn., 1942-43; Indsl. Research fellow dept. food tech. U. Mass., Amherst, 1947-49, faculty, 1949—, prof. food sci., 1958—. Mem. Inst. Food Technologists, Am. Chem. Soc., Soc. for Applied Spectrometry, Instrument Soc. Am., N.Y. Acad. Scis., A.A.A.S., Sigma Xi, Phi Kappa Phi. Author: (with H. J. Noebels, V. J. Coates) Gas Chromatography, 1958; also articles. Research on flavor chemistry, food analysis.*

FAGET, Guy Henry, Am. physician; b. New Orleans, June 15, 1891; s. Francois and Alice (Beeg) F.; M.D., Tulane U., 1914; m. Isabelle Le Blanc, Feb. 8, 1917; children—Frank Alfred, Max Allen, Elsie Alice, Guy Edmund. Intern Marine Hosp., New Orleans, 1914-16; physician Brit. Colonial Service, Brit. Honduras, 1916-22 apptd. asst. surgeon USPHS, 1922; surgeon, 1934 sr. surgeon, 1942, also med. dir.; with U. S. Marine Hosp., Mobile, 1922-23, Marine Hosp., San Francisco, 1923-24, Seattle, 1924-26, Ft. Stanton, 1926-30, New Orleans, 1930-33, Norfolk, 1933-37, U. S. Marine Hosp., San Francisco, 1937-40; med. officer in charge Nat. Leprosarium, Carville, La. from 1940. Cons. adv. bd. Leonard Wood Meml.; chmn. Pan Am. Conf. on Leprosy, Sect. on Therapeutics, Rio de Janeiro. Recipient Liber Medal in sci. exhibit of Centennial meeting, A.M.A., for exhibit in chemotherapy in leprosy, 1947. Fellow A.C.P., A.M.A.; mem. Internat. Leprosy Assn., Associated Mil. Surgeons, Am. Pub. Health Assn., Am. Physicians Lit. Guild. Contbr. papers to med. jours. Pioneer in sulfone treatment of leprosy. Died July 17, 1947.

FAGET, Jean Charles, Am. physician; b. New Orleans, June 26, 1818; s. Jean Baptiste and Mrs. (Le Mormand) F.; attended Collège Rolin, Paris, France, 1830-37; M.D., Faculté de Paris, 1844; m. Glady Ligeret de Chazet; many children. Practiced medicine, New Orleans, 1845; published articles in La Gazette Médicale, also New Orleans Med. Jour.; published discovery of difference in symptoms between yellow fever and malaria, 1859; apptd. to La. Bd. Health; mem. San. Commn. apptd. by Gen. Nathaniel Banks, 1864. Author: Memoires et Lettres sur la Fièvre Jaune et la Fièvre Pauludeénne, 1864 (named chevalier Legion of Honor by France for this work, 1864). Observed that pulse rate in yellow fever decreases, temperature rises or remains constant (Faget's sign), 1875. Died New Orleans, Dec. 7, 1884.

FAGNANO (or FAGNANI), Giulio Carlo (Marquis of Toschi), Italian mathematician; b. Sinigaglia, Italy, Sept. 26, 1682. Fellow Royal Soc., 1723. His works pub. as Opere mathematiche, 3 vols., 1911, 12. discovered that quadrant of lemniscate can be divided into "n" equal parts by Euclidean constrn.; proved that 2 arcs of any ellipse can be found in an infinity of ways so that a segment of a straight line is their difference (Fagnano's methods suggested those of Euler in proof of addition theorem for elliptic integrals 1761), 1716. Died Fagnano, Italy, Sept. 26, 1766.

FAGON, Guy-Crescent, French physician, botanist; b. Paris, France, May 11, 1638; 1st physician to Louis XIV, France, 1693; dir. Royal Bot. Gardens; prof. botany and chemistry. Author: Hortus regius, 1665; (with Vallet) Qualites de Quinquina, 1703. Mem. French Acad. Scis., 1699. Promoter thesis defending circulation of blood; suggested to Louis XIV that Frenchmen be sent to visit fgn. countries. Died Mar. 11, 1718.

FAHEY, John Leslie, Am. physician; b. Cleve., Sept. 8, 1924; s. Leslie J. and Marguerite (Schardt) F.; M.S., Wayne U., 1949; M.D., Harvard, 1951; m. Jane A. Bishop, June 12, 1954; children—Marguerite Anne, James L., Catharine D. Clin. asso. Nat. Cancer Inst., NIH, Bethesda, Md., 1953-54, sr. investigator, 1954-64, chief immunology br., 1964—, asso. editor Jour. Nat. Cancer Inst., 1960-62. Mem. Am. Soc. for Clin. Investigation, Am. Fedn. for Clin. Research (pres. Washington chpt. 1958-59), Am. Physiol. Soc., Soc. for Exptl. Biology and Medicine, Am. Assn. Immunologists, Sigma Xi. Research on antibodies and immunoglobulins, immunology and cancer, multiple myetoma, macroglobulinemia, diseases of immunity, amino acid and ammonia metabolism, blood coagulation. Office: Bldg. 10, NIH, Bethesda, Md. 20014.*

FAHLBERG, Constantin, chemist; b. Tambow, Russia, Dec. 22, 1850; studied chemistry, Leipzig, Germany; Ph.D., 1873; worked in industry several years; tchr. Johns Hopkins, Balt., 1878; founder (with uncle A. List) small exptl. factory, N.Y., 1884; established Fahlberg, List & Co. (saccharin factory), Salbeke-Westerhusen/Elbe, 1886. Discovered saccharin through synthesis from toluene (on basis of J. Remsen's expts.), 1879. Died Nassau/Lahn, Aug. 15, 1910.

FAHLMAN, Anders Gustaf Viktor, Swedish physicist; b. Skelleftca, Sweden, Apr. 17, 1937; s. Gustaf Adolf and Signe (Wiklund) F.; Fil.Dr., U. Uppsala (Sweden), 1966; m. Eva Britt Kerstin Nilsson, June 30, 1962; 1 son, Nils A. G. Asst. prof. Inst. Physics, U. Uppsala, 1966—. Publs. on measurements of electron binding energies in 50 elements; studies of energy shifts in atomic levels due to chem. effects, structure of valence bond in different elements, Auger electron studies giving evidence of a new atomic excitation and double Auger effect, improved technique for use of photog. detection. Home: 5B Vasagafan, Uppsala, Sweden.*

FAHR, Theodor, German physician; b. Pirmasens, Germany, Oct. 3, 1877; s. Ernst Fahr; studied medicine, Munich, Berlin, Kiel (all Germany); M.D., Giessen, Germany, 1903; m. Martha Maria Drechsler; 2 sons, including Ernst Rudolf. Asst. to Th. Deneke and M. Simmonds, Hamburg, Germany; became prosector Hafenkrankenhaus; became dir. path. inst., Mannheim, Germany, 1909; went to Hamburg-Barmbeck, 1913; named asso. prof., Hamburg, 1919, prof. pathology, 1924. Many publs. on renal pathology and other topics; research (with F. Volhard) on Bright's disease, arteriolar nephrosclerosis, nephrosis; also investigated diabetes mellitus, morphological pathology of circulatory system, rheumatism, polio, scarlet fever, hyperthyroid, gen. path. problems. Died Hamburg, Oct. 29, 1945.

FAHRAEUS, Robin, Swedish pathologist, anatomist; b. Stockholm, Sweden, Oct. 15, 1888; prof. U. Uppsala, Sweden; introduced erythrocyte-sedimentation test, 1918.

FAHRENHEIT, Gabriel Daniel, physicist; b. Danzig, Poland, May 24, 1686; s. Daniel and Concordia (Schumann) F.; studied commerce, Amsterdam, Netherlands, 1802-06; met or corresponded with leading scientists of the time during visits to Berlin, Halle, Leipzig, Dresden (all Germany), Copenhagen, Denmark; established precision mechanics shop (noted for thermometers, barometers, areometers, other astron. instruments. Fellow Royal Soc., 1724. Inventor alcohol thermometer, 1709, mercury thermometer, 1714; devised Fahrenheit temperature scale (now commonly used in U. S. and Eng.); discovered that water can be cooled below its freezing point without becoming solid, also that boiling points of liquids vary with atmospheric pressure. Died Hague, Netherlands, Sept. 16, 1736.

FAHRINGER, Josef, Austrian entomologist; b. Baden, nr. Vienna, Austria, Dec. 21, 1876; s. Karl and Franziska (Schönhuber) F.; Ph.D., U. Vienna, 1904. Secondary sch. tchr., Vienna, 1904-07, Brüx, 1907-10, Brno, Czechoslovakia, 1910-13, Vienna, 1918-36; prin., from 1928; with Fed. Inst. for Alpine Agr., Admont, nr. end of life. Author: Opuscula braconolocica (1st modern monograph on braconids), 4 parts, 1925-37; other works on taxonomy and biology of wasps. Specialist on braconids; studied parasitic hymenopters. Died Vienna, Dec. 18, 1950.

FAHRNEY, Delmer Stater, Am. aero. engr.; b. Grove, Okla., Oct. 23, 1898; s. Albert Franklin and Lillian (Pugh) F.; B.S., U. S. Naval Acad., 1919; M.S. in Aero. Engring., Mass. Inst. Tech., 1930; m. Agnes Whiting Kelly, June 2, 1925; children—Dawn (Mrs. Sanford L. Knotts), Delmer Stater, Carol, Paula (Mrs. André F. Yon). Commd. ensign USN, 1919, advanced through grades to rear adm., 1950; insp. naval aircraft Wright Aero. Crop., 1931-33; main engines officer U.S.S. Lexington, 1933-35; chief insp. Naval Aircraft Factory, Phila., 1936; project officer radio controlled aircraft and guided missiles, 1936-40; head spl. design br. Navy Bur. Aero., 1940-43, dir. pilotless aircraft, 1945-48; comdg. officer naval aircraft modification unit, 1943-44; staff engring. aircraft 7th fleet, 1944-45; commander Naval Air Missile Test Center, Point Mugu, Cal., 1948-50; ret., 1950; cons. White-Rodgers Elec. Co., St. Louis, Coleman Engring. Co. Cal., Los Angeles, 1950-54; historian Naval Bur. Aero., 1954-57; sec. com. sci. and arts Franklin Inst., Phila., 1957—. Decorated Legion of Merit, Gold Star; commended by sec. navy, 1938, 40. Fellow Inst. Aero. Sci. (asso.), Inst. of Aeronautics and Astronautics (associate); mem. Rocket Soc., Am. Inst. Aeros and Astronautics. Author: History of Radio Controlled Aircraft and Guided Missiles, 1958. Pioneered design devel. and testing radio controlled aircraft in U. S., 1st guided missiles in world. Home: Roundelay St., Chadds Ford, Pa. 19317 Office: 20th and Parkway Sts., Phila. 19103.*

FAIR, Gordon Maskew, sanitary engineer; b. Burghersdorp, Union of South Africa, July 27, 1894; s. Charles and Maria (Maskew) F.; grad. Werner Siemens Gymnasium, Berlin, 1913; S.B., Mass. Inst. of Tech., 1916; S.B., Harvard, 1916; hon. M.S., Tufts Coll., 1934; Dr. Ing. Technische Hochschule, Stuttgart, 1951; hon. fellow Imperial Coll. Sci. and Tech., 1951; Dr. honoris causae, Universidad Nacional de Ingenieria, Lima, Peru, 1960; Doctor of Science, Rose Polytechnic Institute, 1963; Doctor of Science, Rutgers, The State Univ., 1965; m. Esther Lansing Mead, December 21, 1918; children—Gordon Maskew, Cornelius Lansing. Began as sanitary engr., 1917; instr. to prof. Harvard Univ., 1918-35, Gordon McKay prof. of sanitary engring. 1935-65, Abbott and James

Lawrence prof. of engring. 1938-65, McKay and Lawrence prof. emeritus, 1965—, dean faculty of engring., 1946-49; master of Dunster House, 1948-61; cons. on san. engring. to govt. agencies, industries; and founds., including National Mil. Establishment 1946-53; internat. health div., Rockefeller Found., 1945-48, 1949-54, Commn. on Environmental Hygiene, Army Epidemiological Bd. 1946-54; chmn. environmental health study sect. NIH, 1952-55; mem. NRC committee on sanitary engineering 1942-64; mem. panel on environmental sanitation WHO. Registered profl. engr., Mass. Served C.E.F., World War I. Fellow Am. Soc. Civil Engrs., Am. Acad. Arts and Scis.; mem. Am. Water Works Assn., Am. Pub. Health Assn., A.A.A.S. (v.p. 1947), Sigma Xi, Nat. Acad. engring., other profl. and scientific socs. and assns. Author books and chpts. on water supply, and waste water disposal. Contbr. scientific articles to jours. Formulated rates of anaerobic decomposition of organic waste; general water purification equation; math. analysis of dissolved oxygen in polluted streams; invented odor-intensity measuring device, ultraviolet disinfecting method. Home: 29 Robinson St., Cambridge 38, Mass.*

FAIRBAIRN, Donald, biologist; b. Ottawa, Ont., Can., Feb. 4, 1916; s. Arthur Edwin and Maria (Spratt) F.; B.A., Queen's U., 1938; Ph.D., U. Rochester, 1942; m. Mary Woodhouse Crawford, June 21, 1944; children—Ian George, Stephanie Ann, Eleanor Mary. Faculty, McGill U., 1946-62; Commonwealth prof., head zoology dept. U. Mass., Amherst, 1962—; vis. prof. Johns Hopkins, 1959-60, U. Mexico, 1965. Cons. NIH, chmn. tropical medicine, parasitology study sect., 1961—. Mem. Am. Chem. Soc., Am. Soc. Biol. Chemists, Am. Soc. Parasitologists, Biochem. Soc. Gt. Britain. Editorial bd. Am. Jour. Epidemiology, 1960—, Jour. Parasitology, 1960—, Exptl. Parasitology, 1958—. Research, numerous publs. on biochem. prins., physiology of animal parasites and host-parasite relations. Home: 301 E. Pleasant St., Amherst, Mass. 01002.*

FAIRBAIRN, Harold Williams, geologist; b. Ottawa, Ont., Can., July 10, 1906; s. Arthur Edwin and Maria (Spratt) F.; B.S., Queen's U., 1929; postgrad. U. Wis.; A.M., Harvard, 1931, Ph.D., 1932; postgrad. U. Innsbruck, U. Göttingen, Tech.-Hochschule Berlin; m. Sheila May Sargent, Apr. 18, 1939; children—Ann, Patrick William, Elspeth, Neil Alastair. Came to U. S., 1937, naturalized, 1949. Instr., Queen's U., Kingston, Ont., 1934-37; with Geol. Survey Can., summers 1926-32, 35-42; faculty Mass. Inst. Tech., Cambridge, 1937—, prof. geology, 1955—; vis. prof. Harvard, 1953. Mem. Geol. Soc. Am., Mineral. Soc. Am., Geochem. Soc., Am. Geophys. Union, Geol. Assn. Can., Am. Acad. Arts and Scis. Author: Structural Petrology of Deformed Rocks, 1949; also numerous articles. Research in areal geol. mapping in Precambrian Shield of Can., petrofabric analysis of deformed rocks by microscopic and X-ray methods, precision and accuracy of analytical methods, geochronologic studies using Rb-Sr isotopic and X-ray methods. Home: 27 Marcia Rd., Watertown, Mass. 02172. Office: 77 Massachusetts Av., Cambridge, Mass. 02139.*

FAIRBAIRN, Sir William, English engineer; b. Kelso, Roxburghshire, Eng., Feb. 19, 1789; s. Andrew and Mrs. (Henderson) F.; hon. LL.D., Edinburgh 1860, Cambridge, 1862; m. Dorothy Mar, circa 1813; seven sons, two daus. Chief fabricator of machinery, Turkish government in England, 1839; with Stephenson, superintended construction of Menai Straits bridge, 1848. Created baron, 1869. Fellow Royal Soc., 1850 (gold medal, 1860); mem., French Acad. Scis., 1852; Brit. Assn. (pres. 1861); Institution of Mechanical Engrs. (pres. 1854); Manchester Literary and Philosophical Soc. (pres. 1855-60). Author numerous papers in scientific pubs. Experimented on properties of iron; introduced riveting machine; invented new wrought iron girders. Died Manchester, Eng., Aug. 18, 1874.

FAIRBANK, Henry Alan, Am. physicist; b. Lewistown, Mont., Nov. 9, 1918; s. Samuel B. and Helen (Martin) F.; B.A., Whitman Coll., 1940; Ph.D., Yale, 1944; M.A. (Guggenheim fellow), Oxford U., 1954; m. Martha E. Edmonds, June 17, 1943; children—Mary, Alan W., Elizabeth. Instr. physics Yale, 1942-44, 45-48, asst. prof., 1948-54, asso. prof., 1954-62; prof., chmn. dept. physics Duke, 1962—; physicist Manhattan Project, Los Alamos, N.M., 1944-45. Cons. Los Alamos Sci. Lab., 1957—; mem. adv. panel Nat. Bur. Standards, 1956—; mem. panel, div. instnl. grants NSF, 1963—. Fellow Am. Phys. Soc.; mem. Phi Beta Kappa, Sigma Xi. Contbr. articles to physics jours; research in low temperature physics, properties of liquid and solid helium, superconductivity, transport properties. Home: 1515 Pinecrest Rd., Durham, N.C. 27705.*

FAIRBANK, William Martin, Am. physicist; b. Mpls., Feb. 24, 1917; s. Samuel Ballantine and Helen Leslie (Martin) F.; A.B., Whitman Coll., Walla Walla, Wash., 1939, D.Sc. (hon.), 1965; post-grad. fellow, U. Wash., 1940-42; M.S., Yale, 1947, Ph.D. (Sheffield fellow), 1948; m. Jane Davenport, Aug. 16, 1941; children—William Martin, Robert Harold, Richard Dana. Mem. staff radiation lab. Mass. Inst. Tech., 1942-45; asst. prof. physics Amherst Coll., 1947-52; asso. prof. Duke, 1952-58, prof., 1958-59; prof. physics Stanford, 1959—; Named Cal. Scientist of Year, Cal. Museum Sci. and Industry, 1961. Fellow Am. Phys. Soc. (Oliver E. Buckley Solid State Physics Prize 1963, Research Corp. award 1965); mem. Nat. Acad. Scis. Research on microwave radar systems, microwave propagation, cryogenics, quantized flux in superconductors, properties liquid helium II, He3, liquid helium bubble chambers, superconducting electron accelerators. Home: 141 E. Floresta Way, Menlo Park, Cal. 94026. Office: Physics Dept., Stanford Univ., Stanford, Cal. 94305.*

FAIRBANKS, Henry, Am. inventor; b. St. Johnsbury, Vt., May 6, 1830; s. Thaddeus and Lucy P. (Barker) F.; A.B., Dartmouth, 1853, A.M., 1856, Ph.D., 1880; grad. Andover Theol. Sem., 1857; m. Annie S. Noyes, Apr. 30, 1862 (dec. 1872); m. 2d, Ruthy Page, May 5, 1874; 1 son, Arthur F. Ordained to Congl. ministry, 1857; in pastorates, 1857-60; prof. natural philosophy Dartmouth, 1859-65, natural history, 1865-68; with E. & T. Fairbanks & Co., mfrs. of scales, 1868—; holder more than 30 patents on various devices. Died June 7, 1918.

FAIRBANKS, Thaddeus, Am. inventor; b. Brimfield, Mass., Jan. 17, 1796; s. Joseph and Phebe (Paddock) F.; m. Lucy Barker, Jan. 17, 1820, 2 children including Henry. In partnership with brother Erastus Fairbanks, 1792-1864; established iron foundry E. & T. Fairbanks Co., St. Johnsbury, Vt., 1823; patented a plow, 1826, flax and hemp dressing machine, 1830; platform scale, 1831; devised parlor stove, cook stove; invented draft mechanism for furnaces, 1843; hot water heater, 1881, also feedwater heater; established St. Johnsbury (Vt.) Acad., 1842. Recipient knightly cross Order St. Joseph from Emperor Austria, Golden medal from King of Siam, token of comdr. Order of Iftikar from Bey of Tunis. Died St. Johnsbury, Apr. 12, 1886.

FAIRBRIDGE, Rhodes Whitmore, geologist; b. Pinjarra, Australia, May 21, 1914; s. Kingsley O. and Ruby E. (Whitmore) F.; B.A., Queens U. (Can.), 1936; B.S., Oxford U. (Eng.), 1940; D.Sc., U. Western Australia, 1944; m. Dolores G. Carrington, June 19, 1943; 1 son, Kingsley. Field geologist Iraq Petroleum Co., 1938-41; asst. prof. geology U. Western Australia, 1946-53; asso. prof. U. Ill., 1953-54; prof. Columbia, N.Y.C., 1955—, leader Nubian Expdn., Sudan, 1961-62. Mem. expdn. Capricorn, Scripps Inst. Oceanography, 1953; founder Internat. Geology Rev., Am. Geol. Inst., 1958; vis. prof. Sorbonne U., Paris, 1962. Mem. Am., London, Swiss, Australian geol. socs., N.Y. Acad. Sci (past pres geology sect)., Am. Assn Petroleum Geologists (past pres. Eastern sect.), Geologische Verein (Bonn), Soc. Geologique France. Author: Australian Stratigraphy, 1953; Encyclopedia of Oceanography, 1966. Research, numerous publs. in world eustatic sea level oscillations, in turn related to evolution of coral reefs; responsible for Fairbridge Curve, first proof of glacial melt-water as cause of climate changes correlated with solar radiation controls; developed concept of world-wide tropical aridity during glacial periods. Home: 420 Riverside Dr., N.Y.C. 10025.*

FAIRBROTHERS, David Earl, Am. biologist, botanist; b. Absecon, N.J., Sept. 24, 1925; s. Earl and Mary (Freas) F.; B.S., Syracuse U., 1950; M.S., Cornell U., 1952, Ph.D, 1954; m. Marguerite Freitag, July 9, 1949; children—Gregg E., Gail A. Instr., Cornell U., 1954; faculty Rutgers-The State U., New Brunswick, N.J., 1954—, prof. botany, 1965—. Lectr. use of serological methods in plant systematics in univs. in Germany, Scotland, Sweden, Czechoslovakia, P.R., U. S.; lectr. in Rumania, England, Jamaica, 1967. Mem. Am. Soc. Plant Taxonomists, Am. Bot. Soc., Internat. Soc. Plant Taxonomists, Torrey Bot. Club, N.J. Acad. Scis., A.A.A.S., Sigma Xi. Contbg. author: Taxonomic Biochemistry and Serology, 1964; Mutational Process, 1966; Modern Methods in Plant Toxonomy, 1968. Research, publs. on introduction and devel. serological and immunochem. methods for bot. research with higher plants, especially in plant systematics. Home: 128 Edgewood Dr., Somerville, N.J. 08876. Office: Botany Dept., Rutgers-The State U., New Brunswick, N.J. 08903.*

FAIRCHILD, David Grandison, Am. agrl. explorer; b. Lansing, Mich., Apr. 7, 1869; s. George Thompson and Charlotte Pearl (Halsted) F.; student Kan. State Agrl. Coll.; Naples Zool. Station, Italy, 1893-95; hon. degrees Oberlin Coll., Fla. State Coll., Kan. State; m. Marian H. Bell, Apr. 25, 1905; children—Graham Bell, Barbara (Mrs. Leonard Mueller), Nancy Bell (Mrs. Marston Bates). Expdsn. in search of plants for introduction to U. S., with Barbour Lathrop, to Dutch E. Indies, 1895, to S. Sea Islands, Siam, Australia, New Zealand, 1896-97, to W. I., S.Am., Egypt, Ceylon and Persian Gulf, 1901-02, Africa, 1903; established garden near Miami, Fla., for introduction and testing of fgn. plants, 1898; later established similar gardens in Cal., Fla., Ga., Md., N.D., and Wash.; organizer, dir., Sect. Fgn. Seed and Plant Introduction, Dept. Agr., 1904-28; with Allison V. Armour expdns. to Morocco, Dutch E. Indies, W. Africa, Carribbean, 1925-27, 32-33; in charge sci. work of Fairchild Garden expdn. to Phillippines, Celebes, Java, Bali and Moluccas, on Mrs. Anne Archbold's Chinese junk, Cheng Ho, 1939-40; collected plants in Colombia, Panama and Guatemala and Yucatan, 1944. Medallist Societe d'Acclimatacion France, Harvard Traveller's Club, Mass. Hort. Soc. (George Robt. White), Nat. Acad. Scis., Garden Club of Am., Men's Garden Clubs (Johnny Appleseed), Nat. Council State Garden Clubs (Gold seal), Fairchild Tropical Garden (Meyer medal, plant introduction); Fairchild Tropical Garden, Coconut Grove, Fla., named in his honor. Fellow Linnean Soc. London; pres. Am. Genetic Assn.; hon. mem. Committee of One Hundred, Miami Beach. Author: Book of Monsters, 1914; Exploring for Plants, 1930; The World Was My Garden, 1938; Garden Islands of the Great East, 1945; The World Grows Round My Door, 1947. Introduced more than 20,000 species of plants from abroad into U. S., including Rhodes grass, Russian durum wheat, Egyptian cotton, and Japanese rice; popularized mangoes, avocados, and other tropical fruits. Died Miami, Fla., Aug. 6, 1954.

FAIRCHILD, Herman Le Roy, Am. geologist; b. Montrose, Pa., Apr. 29, 1850; s. Harmon C. and Mary A. (Bissell) F.; B.S., Cornell, 1874; D.Sc., U. Pitts., 1910; m. Alice Egbert, N.Y., July 25, 1875; children—Katherine (Mrs. Charles T. Lewis), Lillian, Jessie Evelyn (Mrs. Guy Bogart), Le Roy Frink. Tchr. Wyoming Sem., Kingston, Pa., 1875-76; lectr. N.Y. schs., 1876-88, also Cooper Union, 1878-88; prof. geology and natural history U. Rochester, 1888-96, geology, 1896-1920. Mem. N.Y. State Bd. Geog. Names, 1913-23. Fellow A.A.A.S. (gen. sec. 1894, v.p. 1898), Geol. Soc. Am. (sec. 1890-1906, pres. 1912); mem. N.Y. (sec. 1885-88), Rochester (pres.) acads. scis. Author: History of the New York Academy of Sciences, 1887; Revision of Le Conte's Elements of Geology, 1903; Geologic Story of the Genesee Valley and Western New York, 1928; The Geological Society of America (a chapter in Earth Science History), 1932; also monographs and contbns. on geol. and biol. topics. Research in Pleistocene geology; studied effect of most recent ice sheet in N.Y. State. Died Nov. 29, 1943.

FAIREY, Sir Charles Richard, Brit. aero. engr.; b. Eng., May 5, 1887; ed. in engring. and chemistry Finsbury Tech. Coll.; m. 2d, Esther Sarah Whitmey, 1934; 2 sons (one by previous marriage), 1 dau. Mgr., Blair-Atholl Aeroplane Syndicate, 1912-13; chief engr. Short Bros., 1913-15; founder Fairey Aviation Co. Ltd., 1915; chmn., mng. dir. Soc. Brit. Aircraft Constructors, Ltd., 1922-24. Mem. Aero. Research Com., 1923-26; dir.-gen. Brit. Air Commn., Washington, 1942-45; mem. joint aircraft com. Combined Aluminium/Magnesium Com., Washington, 1942-45. Mem. Royal Aero. Soc. (pres. 1930-31, 32-33, Wakefield Gold medal. Contbr. articles on aviation to profl. publs. Pioneer in establishing principles of calculating strength of aircraft; invented and developed wing flap. Died Sept. 30, 1956.

FAIRLEY, James Lafayette, Jr., Am. biochemist; b. Orland, Cal., Oct. 15, 1920; s. James Lafayette and Grace (Tischer) F.; B.A., San Jose State Coll., 1942; Ph.D., Stanford, 1950; m. DeLorez Meyer, Sept. 12, 1948; children—Joel, Laurel. Research asso. Stanford, 1949-51; research biochemist U. Cal. Radiation Lab., Berkeley, 1951-52; faculty Mich. State U., East Lansing, 1952—, prof. biochemistry, 1962—. USPHS fellow biology div. Oak Ridge Nat. Lab., 1962-63. Mem. Am. Chem. Soc., A.A.A.S., Am. Soc. Biol. Chemists, Am. Soc. Plant Physiology, Sigma Xi. Author: (with Gordon Kilgour) Essentials of Biological Chemistry, 1963; also articles. Research on chemistry and structure of nucleic acids, purine and pyrimidine metabolism in mold Neurospora crassa; isolated and studied nucleases from various plants. Home: 4363 Greenwood Dr., Okemos, Mich. 48864. Office: Biochemistry, Dept., Mich. State U., East Lansing, Mich. 48823.*

FAIRLEY, Sir Neil Hamilton, physician; b. Inglewood, Australia, July 15, 1891; s. James and Margaret L. (Fairley) H.-F.; ed. Scottish Coll., U. Melbourne; M.D., D.Sc.; LL.D. honoris causa, U. Melbourne; D.Sc. honoris causa, U. Sydney; M.D. honoris causa, U. Adelaide; m. Mary E. Greaves, 1925; children—James, Gordon. Cons. physician Hosp. Tropical Diseases of London, 1929; Wellcome prof. tropical medicine, London; dir. research on malaria, Cairns, Australia; cons. tropical medicine to Gen. MacArthur. Recipient Manson medal Royal Soc. Medicine and Tropical Hygiene. Fellow Royal Soc., 1942 (Buchanan medal), Royal Coll. Physicians (Moxon medal), Royal Australian Coll. Physicians. Research and numerous publs. on tropical diseases, malaria, snake bite. Died Apr. 19, 1966. Grove, Sonning, Berkshire.

FAIRLIE, Robert Francis, civil engr.; b. Scotland, Mar. 1831; studied locomotive work at Crew & Swindon; m. twice; 5 children. Became supt., gen. mgr. Londonderry & Coleraine Ry., 1853; later with Bombay & Baroda Ry.; became cons. engr., London, circa 1864; left for Venezuela to design system of rys., 1873; became ill and returned to Eng. Recipient gold medal from tsar of Russia. Invented double bogie engine for steep gradients, 1864; introduced narrow gauge lines. Died July 31, 1885.

FAITH, William Lawrence, Am. chem. engr.; b. Hancock, Md., May 12, 1907; s. Eugene Francis and Bernadette (Murray) F.; B.S., U. Md., 1928; M.S., U. Ill., 1929, Ph.D., 1932; m. Mildred Eunice Dixon, Sept. 2, 1932; children—Michael D., Mary E. (Mrs. Russell J. Meskell), Roger L. Spl. research asst. Engring. Expt. Sta., U. Ill., 1931-33; chemist Nat. Aluminate Corp., Chgo., 1933; faculty Kan. State Coll., Manhattan, 1933-42, prof., 1936-42, head dept. chem. engring., 1939-42; prof. chem. engring. State U. Ia., Iowa City, 1942-44; cons. office prodn. research and devel. WPB, Washington, 1942-43, asst. dir., 1944, dept. dir., 1944-45; dir. devel. engring., chem. div. Corn Products Refining Co., Argo, Ill., 1945-48, dir. engring., 1948-54; chief engr., dep. dir. Air Pollution Found., Los Angeles, 1954-57, mng. dir., 1957-61; cons. chem. engr., San Marino, Cal., 1961-—; vis. lectr. Cath. U. Am., 1944; cons. on air pollution to govt., industry. Mem. environmental scis. and engring. study sect. NIH, 1958-62; emergency action com. Los Angeles County Air Pollution Control Dist., 1956-64; mem. adv. com. on reactor safeguards U. S. AEC, 1966-—. Mem. Am. Soc. M.E., Am. Pub. Health Assn., Am. Inst. Chem. Engrs. (dir 1954-56), Air Pollution Control Assn (pres. 1961-62), Am. Chem. Soc., Am. Acad. Environmental Engrs., Sigma Xi. Author: (with D. B. Keyes and R. L. Clark) Industrial Chemicals, 1960, 57, 65; Air Pollution Control, 1959; also numerous articles. Research on nature of air pollution and evaluation of control methods. Home: 2935 Shakespeare Dr. Office: 2540 Huntington Dr., San Marino, Cal. 91108.*

FAJANS, Kasimir, chemist, educator; b. Warszawa, Poland, May 27, 1887; s. Herman and Wanda (Wolberg) F.; student U. Leipzig, (Germany), 1904-07; Dr.Phil.Nat., U. Heidelberg (Germany), 1909; postgrad. U. Zurich (Switzerland), U. Manchester (Eng.); m. Salomea Kaplan, 1910; children—Edgar W., Stefan S. Came to U. S., 1936, naturalized, 1942. Asst., Technische Hochschule, Karlsruhe, Germany, 1911-17, privatdozent phys. chemistry, 1913-17; faculty U. Munich (Germany), 1917-35, prof. phys. chemistry, 1925-35, dir. Inst. for Phys. Chemistry, 1932-35; faculty U. Mich., Ann Arbor, 1936-57, prof. emeritus, 1957-—. Baker lectr. in chemistry Cornell U., 1930; Foster lectr. Buffalo U., 1939; conducted grad. seminar Shell Devel. Co., Emeryville, Cal., 1953, Minn. Mining and Mfg. Co., St. Paul, 1959; sci. collaborator Glass Sci. Research Found., 1944-47; sci. cons. Owens-Ill. Glass Co., Toledo, 1948-55. Recipient Victor Meyer prize U. Heidelberg, 1909, medal U. Liège, 1948; Kasimir Fajans award in chemistry established U. Mich., 1956. Fellow Am. Phys. Soc.; mem. Bunsengesellshaft, Faraday Soc., Am. Chem. Soc. (council 1944, chmn. U. Mich. sect. 1947), A.A.A.S., Am. Crystallographical Assn., Soc. Freedom in Sci., Leningrad, Munich, Cracow acads. sci., Polish Inst. Arts and Scis. in Am., Council Deutsche Chemische Gesellschaft, Sigma Xi; hon. mem. several fgns. socs., Phi Lambda Upsilon. Author: Radioaktivitat und neueste Entwicklung der Lehre von den Chemischen Elementen, 1919, 4th edit., 1922, revised, 1930; (with J. Wuest) Physikalisch-chemisches Praktikum, 1929, 2d edit., 1935; Radioelements and Isotopes, 1931; Chemical Forces and Optical Properties of Substances, 1931; Quanticule Theory of Chemical Binding, 1961; also numerous articles. Editor: Elektrochemie in Handbuch der Experimentalphysik, 1932, (with E. Schwartz), 1933. Co-editor Zeitschrift für Kristallographie, 1924-39; editorial bd. Zeitschrift für Elektrochemie, 1932-33; asso. editor Jour. Phys. and Colloid Chemistry, 1948-49. Established radioactive displacement laws; discovered element 91; adsorption indicators; initiator of concept of heat of hydration of gaseous ions; developed quanticule theory of electronic molecular structure, chem. binding. Home: 1016 Lincoln Av., Ann Arbor, Mich. 48104.*

FAJANS, Stefan S., physician; b. Munich, Germany, Mar. 15, 1918; s. Kasimir and Salomea (Kaplan) F.; came to U. S., 1936, naturalized, 1942; B.S., U. Mich., 1938; M.D., U. Mich., 1942; m. Ruth Stine, Sept. 6, 1947; children—Peter S., John S. Faculty, U. Mich. Med. Sch., 1951-—, prof. internal medicine, 1961;—. Cons. to surgeon gen. USPHS, member of endocrinology study section, 1958-62, member of diabetes and metabolism tng. grants com. NIH, 1966-—. Mem. Am. Diabetes Assn. (dir. 1964-—), Peruvian Soc. Endocrinology (hon.), A.C.P., Am. Fedn. for Clin. Research, Am. Soc. for Clin. Investigation, Central Soc. for Clin. Research, Endocrine Society (member council 1967-—). Member of the editorial board Jour. Clin. Endocrinology and Metabolism, 1963-—; Diabetes, 1962; Metabolism—Clinical and Experimental, 1957-—. Research, numerous publs. on role of amino acids and proteins in physiologic control of insulin secretion in man, natural history of diabetes mellitus, diagnostic tests for recognition of hypoglycemic states, diagnostic tests for early recognition of diabetes mellitus, mechanism of action of drugs used in treatment of diabetes and hypoglycemic states. Home: 2485 Devonshire St., Ann Arbor, Mich. 48104.*

FALAQUERA, Nathan Ben Joel, physician; flourished latter 13th century; may be Nathan of Montpellier, tchr. of unknown author of Sefer ha-Yashar. Author: Zori ha-Guf (collection of opinions of Hippocrates, Galen, Averroes, Avicenna and Maimonides on therapeutics and hygiene).

FALB, Rudolph, astronomer, meteorologist; b. Obdach, Austria, Apr. 13, 1838; studied theology, Graz, Austria, later sci. in Prague, Czechoslovakia, Vienna, Austria, 1869; priest, 2 years; visited N.Am., S.Am., 1877-1880; settled in Berlin, Germany, where he prepared semi-annual weather forecasts until his death. Author: Grundzüge zu einer Theorie der Erdbeden und Vulkan-Ausbrüche, 1870; Gedanken und Studien über den Vulcanismus, 1875; Sterne und Menschen, 1882; Wetterbriefe, 1883; Das Land der Inka in seiner Bedeutung für die Urgeschichte der Sprache und Schrift, 1884; Das Wetter und der Mond, 1887; Von den Umwälzungen im Weltall, 1887; Die Andessprachen in ihrer Beziehung zum semitischen Spachstamme, 1888; Wetterkalender, 1894; Ueber Erdbeden, 1895; Kritische Tage, Sintflut und Eiszeit, 1895. Believed that weather, earthquakes, other phenomena were caused by effect of sun and moon on atmosphere and fluid interior of earth. Died Schoneberg, Berlin, Germany, Oct. 1, 1903.

FALCHI, Giorgio, Italian dermatologist; b. Pavia, Italy, Feb. 12, 1895. Full prof., dir. Inst. Clin. Dermosyphilopathy, U. Pavia. Mem. Italian Soc. Dermatology and Syphilography, Internat. Soc. Microbiology, Med.-Surg. Soc. Sassari, Acad. Fisiocritici, Lombard Med. Soc., Soc. Exptl. Biology, French Soc. Dermatology (corr.). Contbr. numerous articles to profl. publs. Address: Piazza Botta 1, Pavia, Italy.

FALCK, Bengt Olof Torsten, Swedish histologist; b. Malmö, Sweden, Jan. 16, 1927; s. Hans Sigurd T. and Maria (Hagander) F.; M.D., U. Lund (Sweden), 1959; m. Eva Henriette Torp, Nov. 17, 1951; children—Hans Hjalmar, Axel Fredrik, Henriette, Benedikte. Faculty, Inst. Anatomy and Histology, U. Lund, Sweden, 1959-—, asso. prof. histology, 1964-—. Research; publs. on ovarian endocrinology and neuroendocrinology; developed (with Hillarp, Torp) histochem. fluorescence method by which catecholamines and serotonin can be localized at cellular level, 1961. Home: Tirup, Nordana, Sweden. Office: 5 Biskopsgatan, Lund, Sweden.*

FALCK, Carl Philipp, German pharmacologist; b. Marburg/Lahn, Germany, Mar. 1, 1816; s. Johann Christian and Anna Catharine (Schriener) F.; M.D., Marburg, 1843; m. Elisabeth Broeg, 1848; 3 sons, including Ferdinand August, 3 daus.; m. 2d, Emma Schreiner; 1 son, 1 dau. Became mem. faculty U. Marburg, 1845, asso. prof., 1856, prof., 1863; practiced medicine, Marburg; founder pharmacological inst., Marburg, 1867. Author: Handbuch der diätetischen Heilmittellehre, 1850; Handbuch der wissenschaftlichen und praktischen Fleischleunde, 1880. Founder (with son Ferdinand) jour. Beiträge zur Physiologie, Hygiene, Pharmakologie und Toxikologie, 1875. Studied goiter in Hessia and Nassau; made comparative anatomy studies; research on problem of water metabolism; his studies of formation of urine led toward establishment of pharmacology on physiol.-chem. basis. Died Marburg, June 30, 1880.

FALCK, Richard, mycologist; b. Landeck, W. Prussia, Germany, May 7, 1873; s. Julius and Rosa (Baruch) F.; m. Olga Schenkalowski, 1910; 1 dau. Became asst. under Oskar Brefeld, plant physiol. inst., Breslau, Germany, 1900, succeeded Brefeld, 1905; became prof. tech. mycology Forestry Acad., Hannover Munden, Germany, 1910; left Germany, 1933, working in Palestine, Turkmenistan, Russia; settled in Tiberias, 1946, Atlanta, Ga., 1950; Author: Hausschwammforschungen, 1907-37. Worked (with wife) on physiology and ecology of fungi; discovered inhibiting influence of 1 fungus on another and resulting substances (now known as antibiotics, discovered by A. Fleming 1928), 1923; pioneer in applied mycology, especially protection of wood from fungus damage. Died Atlanta.

FALCONER, Douglas Scott, Scottish geneticist; b. Aberdeenshire, Scotland, Mar. 10, 1913; s. Gerald and Lillias (Douglas) F.; B.Sc., U. St. Andrews, 1940; Ph.D., U. Cambridge (Eng.), 1943; m. Margaret Duke, Apr. 6, 1942; children—Andrew Douglas, John Douglas. Temporary lectr. Queen Mary Coll., U. London (Eng.), 1944-45; sci. staff unit animal genetics Agrl. Research Council, 1945-—, dept. dir., 1957-—; faculty dept. animal genetics U. Edinburgh (Scotland), 1947-—. Fellow Inst. Biology; mem. Soc. Exptl. Biology, Genetical Soc. Author: Introduction to Quantitative Genetics, 1960; also numerous articles. Research on genetic control of continuously varying characters, effects of selective breeding and inbreeding on growth and fertility, inheritance of susceptibility to disease in man; genetics of mice. Home: 21 Mansionhouse Rd., Edinburgh 9. Office: Inst. Animal Genetics, West Mains Rd., Edinburgh 9, Scotland.*

FALCONER, Hugh, Scottish paleontologist, botanist; b. Forres, Scotland, Feb. 29, 1808; s. David Falconer; M.A., Aberdeen, Scotland, 1826; M.D., Edinburgh, 1829. Asst. surgeon E. India Co., 1830-32; supt., Botanic Garden, Saharanpur, India, 1832; tea mfr. in India; supt. for exhbn. Indian fossils Brit. Mus., 1844; became supt. Calcutta (India) Botanic Garden, prof. botany Calcutta Med. Coll., 1847; apptd. adviser to govt. on Indian vegetable products, 1848. Recipient Wollaston medal Geol. Soc. London, 1837. Fellow Foyal Soc., 1845. Editor: Fauna Antiqua Sivalensis, 1846-49; Paleontological Memoires and Notes, 1868. Discovered fossil mammals and reptiles in Siwalik Hills, including mastodon elephant, rhinocerous, giant tortoise, and the huge ruminant Sivatherium; 1832; determined tertiary age of Siwalik Hills; made 1st expts. on manufacture of Indian tea; undertook bot. and paleontol. research in Eng. and on Continent. Died London, July 31, 1865.

FALCONER, John, English botanist; b. Eng.; flourished 16th century; traveled in Europe; lived in Ferrara, Italy, 1540 or 1541-47; fellow pupil of William Turner, Bologna, Italy. Author: Maister Falkonner's Boke; probably 1st Englishman to own series of dried plants (method of study used by Luca Ghini of Bologna).

FALCONER, Murray Alexander, neurosurgeon; b. Dunedin, New Zealand, May 15, 1910; s. Alexander R. and Agnes J. (Simpson) F.; M.B.Ch., U. Otago, New Zealand, 1936, M.Ch., 1939; F.R.C.S., 1935, F.R.A.C.S., 1944; m. Valda H. Turley, June 30, 1939; children—Mrs. Barbara H. Martlew, Mrs. Alex M. Hendrick. Held appointment as fellow Mayo Found., Rochester, Minn., 1937-38; Nuffield Dominion fellow in surgery U. Oxford, Eng., 1938-41; asso. prof. neurosurgery U. Otago, 1943-50; cons. in neurosurgery New Zealand Armed Forces, 1943-50; dir. Guy's-Maudsley Neurosurg. unit, London, Eng., also hon. cons. King's Coll. Hosp., 1950-—; tchr. neurosurgery U. London, 1950-—. Vis. neurol. surgeon Johns Hopkins Hosp., 1959-—; vis. prof. surgery U. Cal. at Los Angeles, 1960; Nafzigger vis. prof. neurology U. Cal. at San Francisco, 1966; vis. prof. neurosurgery U. Ore., Portland, 1966. Decorated comdr. Order of Cedars, Lebanon. Mem. Soc. Brit. Neurol. Surgeons; hon., corr. mem. various neurosurg. socs., in N. and S.Am., Europe, Australasia. Research, numerous publs. on surgery of prolapsed intervertebral discs, subarachnoid and intracranial hemorrhage from ruptured intracranial aneurysms, surgery and pathology of epilepsy, surg. treatment of Parkinsonism. Home: 48 Durham Av., Bromley, Kent, Eng. Office: Guy's Maudsley Neurosurgical Unit, London, Eng.*

FALCONER, Randle Wilbraham, physician; b. 1816; s. Thomas; studied medicine, Edinburgh, from 1835; grad. 1839; M.D.; hon. doctorate Queen's U., Ireland, 1879. Went to Tenby, Eng., then moved to Bath, Eng., where he practiced medicine, 1847; elected mayor, Bath, 1857; physician Bath Gen., Mineral Waters hosps. Fellow King and Queen Coll., Dublin. Fellow Medico-Chirurg. Soc., London; mem. Brit. Med. Assn. (became pres. 1878). Author: The Baths and Mineral Waters, 5th edit., 1871; Reports of Cures at the Bath General Hospital, 1860; The Bath Mineral Waters, 1861. Research on curative properties of mineral baths. Died Bath, May 6, 1881.

FALCONER, William, English physician; b. Chester, Eng., Feb. 23, 1744; s. William and Elizabeth (Wilbraham) F.; M.D., Edinburgh, 1766, Leiden, Netherlands, 1767; m. Henrietta Edmunds (dec. 1803); a son, Thomas. Became physician Chester Infirmary, 1767; moved to Bath, Eng., 1770; named physician Bath. Gen. Hosps., 1784-1819. Fellow Royal Soc., 1773, Extralicentiae Coll. Physicians. Author numerous publs. including: An Essay on the Bath Waters . . . , 1770; Observations on Dr. Cadogan's Dissertation on the Gout and all Chronic Diseases, 8 vols., 1772; Observations and Experiments on the Poison of Copper, 8 vols., 1774; Observations respecting the Pulse . . . , 1796; An Essay on the Plague . . . , 8 vols., 1807. Died Bath, Aug. 31, 1824.

FALE, Thomas, mathematician; b. flourished 1604; ed. Caius Coll., Cambridge, 1578; B.A., Corpus Christi Coll., 1582-83, M.A., 1586; B.D., 1597; Author: Horologiographia . . . , 1593. Drew up what was probably the earliest table of sines printed in Eng.

FALK, Isidore Sydney, Am. pub. health economist; b. Bklyn., Sept. 30, 1899; s. Samsin and Rose (Stolzberg) F.; Ph.B., Yale, 1920, Ph.D., 1923; m. Ruth Hill, Mar. 18, 1925; children—Sydney Westervelt, Stephen Ackley. Faculty, U. Chgo., 1923-29; asso. dir. research Com. on Costs Med. Care, Washington, 1929-33; research asso. Milbank Meml. Fund, N.Y.C., 1933-36; dir. Bur. Research and Statistics, U. S. Social Security Adminstrn., Washington, 1940-54; adviser social services World Bank Mission to Malaya, 1954-55; cons. Republie Panama, U. S. C.Z., 1955-58; prof. pub. health, med. care Yale Sch. Medicine, 1961-—. Cons. USPHS, 1933, United Steelworkers Am., 1958-—, U. S. Social Security Adminstrn., 1965-—; mem. numerous govt. coms., 1935-54; adviser social security Govt. Haiti, 1950-54. Recipient Ordre de Honneur et Merite Haiti, 1953; named cabalero Orden de Vasco Nunez de Balboa, Panama, 1956. Fellow A.A.A.S., Am. Pub. Health Assn.; mem. Am. Statis. Assn., Am. Econ. Assn., Am. Hosp. Assn., Am. Pub. Welfare Assn. Author numerous books, including Security Against Sickness, 1936; A Medical Care Program for Steelworkers and Their Families, 1960; (with C. R. Rorem and M. D. Ring) The Costs of Medical Care, 1933; (with K. Stouman) Health Indices . . . An International System, 1936; (with staff) Medical Care Insurance, 1946. Editor (with E. O. Jordan) The Newer Knowledge of Bacteriology and Immunology, 1928; asso. editor Am. Jour. Pub. Health, 1963-—, Med. Care, London, 1963-—. Studies, publs. on statistics in disease; costs, improvement of med. care; nat. health and social security legislation. Home: 472 Whitney Av., New Haven 06511.*

FALK, K. George, Am. chemist; b. N.Y.C., Sept. 8, 1880; s. Arnold and Fannie (Wallach) F.; B.S., Columbia, 1901; postgrad. Johns Hopkins, 1901-03; Ph.D., Strassburg U., 1905; postgrad. Berlin U., 1905-60; m. Dora Lichten, May 31, 1909; m. 2d, Carolyn Rosenstein, Oct. 16, 1935. Asst. in phys. chemistry Columbia, 1906-07, tutor in physics, 1907-09; research asso. in phys. chemistry Mass. Inst. Tech., 1909-11; chemist Harriman Research Lab., Roosevelt Hosp., N.Y., 1911-28, Harriman Research Fund, N.Y. U. Med. Coll., 1929-31; biol. chemist bur. labs. Health Dept., N.Y.C., 1931-39; prof. chem. bacteriology in preventive medicine N.Y. U. Med. Coll., 1933-36; dir. Lab. of Indsl. Hygiene, Inc., from 1936. Mem. Am., London chem. socs., Am. Pub. Health Assn., Am. Soc. Biol. Chemists, A.A.A.S., Soc. Exptl. Biology and Medicine, Harvey Soc., N.Y. Acad. Scis., Am. Assn. for Cancer Research, Am. Inst. Chemists, Sigma Xi. Author: Chemical Reactions, Their Theory and Mechanism, 1920; Chemistry of Enzyme Action, 1921, 2d edit., 1924; Catalytic Action, 1922. Contbr. to sci. jours. Died Nov. 22, 1953.

FALKNER, Thomas, English physician, missionary; b. Manchester, Oct. 6, 1707; s. Thomas F.; ed. Manchester Grammar Sch.; studied medicine under Dr. Richard Mead. Practiced medicine; became surgeon on board slave ship; sailed to Guinea, 1731; fell ill and was treated by Jesuits; converted and became missionary, 1732; spent next 40 years in S.Am.; returned to Eng. after Jesuits were expelled from S.Am., 1768; joined English province Soc. of Jesus, 1771. Author: Volumina duo de anatome corporis humani, quae plurimi sunt pretii apud artis peritos; Botanical, Mineral, and like Observations on the Products of America, 4 vols.; A Treatise on American Distempers cured by American Drugs. Died Plowden Hall, Eng., Jan. 30, 1784.

FALKSON, Geoffrey, South African physician; b. Pretoria, S. Africa, Aug. 9, 1932; s. Isadore and Ethel (Milstein) F.; M.B. Ch.B., Pretoria U., 1955; M.D., Mich. U., Ann Arbor, 1957; M.Med., Pretoria U., 1958, M.D. cum laude, 1965; m. Hendrika Cornelia Van Dyk, Mar. 12, 1956; children—Carla Isadora, Conrad Bif, Lexa Marriane. Clin. asst. in internal medicine Pretoria Gen. Hosp., also U. Pretoria, 1958-62; cons. for cancer chemotherapy, dept. radiotherapy Pretoria Gen. Hosp., 1963, cons., 1964-65, dir. dept. cancer chemotherapy, 1966—. Cancer chemotherapist Primary Liver Cancer unit Chamber of Mines and Nat. Cancer Assn. of S. Africa. Cancer Internat. Coop. fellow; research awards Nat. Cancer Assn. S. Africa, 1962-65; U. S.-S. Africa leader Exchange Program, 1966. Mem. Am. Assn. Cancer Research, Brit. Cancer Research Assn., N.Y. Acad. Scis., European Soc. Biochem. Pharmacology, Am. Gerontol. Assn. Author: Fluorinated Pyrimidines as Potentiators of Ionizing Radiations in the Treatment of Stomach Cancer in Man, 1964; also articles. Pioneered systematic cancer chemotherapy in S. Africa; clin. studies on combining radiation and cancer chemotherapy; demonstrated potentiation of ionizing radiations and fluorinated pyrimidines in treatment of gastro-intestinal cancer; testing of new anticancer drugs, studies on their mechanisms of action; histologic changes in skin caused by cancer chemotherapy; ultraviolet light sensitization by cytostatics. Home: 440 Millers Mile, Lynnwood, Pretoria. Office: 229 Robert Koch Med. Bldg., Pretorius St., Pretoria, S. Africa.*

FALL, Henry Clinton, Am. entomologist; b. Farmington, N.H., Dec. 25, 1862; s. Orrin T. and Annie (Hayes) E.; B.S., Dartmouth, 1884, Sc.D., 1929. Teacher; investigator of Coleoptera, Acmaeodera, Apion, etc., shown in lists, revisions and articles contributed to transactions of entomol. socs. and entomol. mags. Fellow Entomol. Soc. Am. (v.p. and asso. editor Annals), A.A.A.S.; mem. Permanent Com. of Internat. Entomol. Congress. Foremost authority on Coleoptera of So. Cal., named over 1,200 species and varieties from the state. Died Nov. 14, 1939.

FALLERS, Lloyd Ashton, Jr., Am. anthropologist; b. Nebraska City, Neb., Aug. 29, 1925; s. Lloyd A. and Fannie (Lincoln) F.; student U. Utah, 1943-44, Deep Springs Coll., 1944-45; Ph.B., U. Chgo., 1946, M.A., 1949, Ph.D., 1953; postgrad. London Sch. Econs., 1949-50; m. Margaret Elinor Chave, June 18, 1949; children—Winnifred Mary, Beth Laura. Faculty, U. Chgo., 1952-53, 60—, prof. anthropology, 1963—; lectr. Princeton, 1953-54; research fellow E. African Inst. Social Research, Kampala, Uganda, 1950-52, 54-56, dir., 1956-57; faculty U. Cal. at Berkeley, 1957-60, asso. prof., 1959-60. Fellow Center for Advanced Study in Behavioral Scis., 1958-59. Mem. Am. Anthrop. Assn., Assn. Social Anthropologists (Gt. Britain), Internat. African Inst., African Studies Assn., Soc. for Sci. Study Religion. Author: Bantu Bureaucracy, 1956; also articles. Editor: The King's Men, 1964. Asso. editor for anthropology Internat. Ency. Social Scis., 1961—. Research on determinants of marital stability and instability, of social class and social mobility, of polit. devel. in African socs., underlying unities in legal reasoning in modern Western and African customary law. Home: 1361 E. 56th St., Chgo. 60637.*

FALLOPIUS, Gabriel (Gabriello Fallopio), Italian anatomist; b. Modena, Italy, circa 1523; studied medicine at Ferrara. Became canon in Cathedral of Modena; held chair of surgery and anatomy, U. Pisa, 1548-51; held chair of anatomy and surgery (succeeding Vesalius), U. Padua, 1551-62. Author: Observationes anatomicae, 1561; Opera, 1584. Discovered uterine tubes (Fallopian tubes), which he descriptively called tuba (trumpet); discovered tympanum, tympanic membrane of middle ear; described ethmoid and its air cells, cochlea, scala vestibuli, round and oval windows of auditory organ; diagnosed ear diseases with aural speculum; described in detail muscles of eye, tear ducts, facial canal (Fallopian canal), stapes, primary dentition, replacement of primary by secondary teeth; described trigeminal, auditory, glosso-pharyngeal nerves; distinguished between hard and soft palates; first to discern elements of muscle are connective tissue and fiber; introduced terms ovarian tubes, vagina, clitoris, cricoid, placenta, ciliary body, labyrinth (of inner ear); named muscles of forehead, occiput, tongue. Died Padua, Italy, Oct. 9, 1562.

FALLOT, Étienne Louis Arthur, French physician; b. Marseilles, France, 1850; described congenital heart disease characterized by 4 basic defects (teralogy of Fallot), 1888. Died 1911.

FALLOT, Paul, French geologist; b. Strasbourg, France, June 25, 1889; dir. Inst. Applied Geology, Nancy, France, 1923-38; prof. Coll. of France, from 1939; mem. French Acad. Scis., 1948; principal work on Mediterranean geology.

FALLOWS, Fearon, Brit. astronomer; b. Cockermouth, Eng., July 4, 1789; grad. 3d wrangler, St. John's Coll., Cambridge, 1813; M.A., 1816; m. eldest dau. of H. A. Hewey, 1821. Lectr. math. Corpus Christi Coll., for 2 years; then became fellow St. John's, Cambridge, 1820; apptd. dir. astron. obs. Cape of Good Hope, 1820; sailed for Cape of Good Hope, 1821. Fellow Royal Soc., 1820. Author: A Catalogue of nearly all the Principal Fixed Stars between the zenith of Capetown, Cape of Good Hope, and the South Pole . . . , 1824. Catalogued chief so. stars; nearly 4,000 observations. Died Simons Bay, Africa, July 25, 1831.

FALLS, J. Bruce, Can. zoologist; b. Toronto, Ont., Can., Dec. 18, 1923; s. Orville Mervon and Hazel (Ranney) F.; B.A., U. Toronto, 1948, Ph.D., 1953; m. Elizabeth Ann Holmes, Dec. 27, 1952; children—Kathryn Jean, Robert Bruce, Stephen David. Lectr. zoology U. Toronto, 1952-53, 1954-57; post-doctoral overseas fellow Bur. Animal Population Oxford U., Eng., 1953-54; faculty dept. zoology U. Toronto, 1957—, prof., 1966—; vis. scientist wildlife research, Canberra, Australia, 1964; scientific adv. bd. Bishops U., 1966—. Mem. Fedn. Ont. Naturalists (pres. 1962-64), Am. Soc. Mammalogists, Am. Ornithologists Union, Canadian Soc. Zoologists, Can. Soc. Fishery and Wildlife Biologists, Ecol. Soc. Am., Wilson Ornithol. Soc. Studies, publs. on effects of weather on animal activity; role of vocal communication in territorial behavior and species and individual recognition; abundance, local distbn., homing selection of habitat by birds and mammals. Home: 14 Tottenham Rd., Don Mills, Ont., Can. Office: Dept. Zoology, U. Toronto, Toronto 5, Ont., Can.*

FALOON, William Wassell, Am. physician; b. Pitts., July 6, 1920; s. Joseph Coulter and Martha (Wassell) F.; A.B., Allegheny Coll., 1941; M.D., Harvard, 1944; m. Roberta Emery, Sept. 11, 1948; children—Karen Louise, William Wassell, Nancy Elizabeth. Research fellow Thorndike Meml. Lab., Boston City Hosp., 1947-48; faculty Albany Med. Coll., 1948-50, asst. prof. oncology, 1950; faculty medicine Upstate Med. Center, Syracuse, N.Y., 1950—, prof. medicine, 1964—, dir. metabolic unit, 1963-65, program dir. Clin. Research Center, 1965—. Cons., Surgeon Gen., U. S. Army, 1963—. Diplomate Am. Bd. Internal Medicine. Fellow A.C.P.; mem. Endocrine Soc., Am. Fedn. for Clin. Research, Am. Gastroent. Assn., Am. Soc. for Clin. Nutrition, Am. Assn. for Study Liver Disease. Research, publs. in nutrition, liver disease, metabolic and endocrine disease; discovered importance of sodium restriction in controlling fluid accumulation in cirrhosis of liver, effect antibiotics, effects sodium and potassium changes in diet liver disease patients; originated therapy for coma in liver disease using poorly absorbed oral antibiotics; first observed intestinal malabsorption induced by neomycin and related antibiotics; observations on sugar metabolism, water balance and thiamine deficiency occurring in obese patients undergoing fasting. Home: 231 Ambergate Rd., Dewitt, N.Y. 13214. Office: State U. Hosp., Syracuse, N.Y. 13210.*

FALOR, William H., Am. physician; b. Akron, O., June 14, 1911; s. Shelby A. and Elizabeth Ann (Kerr) F.; B.S., U. Akron, 1934; M.D., Western Res. U., 1938; M.Sc., U. Pa., 1940; m. Elizabeth Ann Van Blaricom, Mar. 9, 1940; children—Elizabeth Ann, William H. II, Steven H., Julia A. Practice medicine specializing in surgery, thoracic surgery, Akron, 1946—; mem. surg. staff Akron City, Children's hosps., Edwin Shaw Sanatorium; dir. lymph research lab. Akron City Hosp., 1960—. Recipient certificate of merit A.M.A., 1963; Hektoen Bronze medal, 1964; Alumni Honor award Akron U., 1967. Mem. A.C.S., Am. Assn. for Thoracic Surgery, Am. Soc. for Artificial Internal Organs, Internat. Soc. Lymphology. Author: Proceedings of the First International Symposium on Lymphology, 1966; Deuel Conference on Lipids, 1967. Studies, numerous publs. on identification of cervical reaches of thoracic duct, both right and left ducts with characterization of lipid and protein moiety of chylomicrons and of specific lipid classes in chyle and in lymph. Home: 180 W. Fairlawn Blvd., Akron 44313. Office: 550 E. Market St., Akron, O. 44304.*

FALRET, Jean Pierre, French psychiatrist; b. Marcilhac, France, 1794; M.D., Paris, 1819; practiced medicine, Salpêtrière, from 1831; mem. Acad. Medicine. Author: Traité de l'hypocondrie et de suicide, 1822; Sur le Délire, 1831; Considérations générales sur les maladies mentalues, 1834; Memoire sur la législation relative auxaliénés, 1837; Leçons cliniques sur les maladies mentales, 1854. Known especially for his early description (1854) of what later became known as manic-depressive psychosis. Died 1870.

FALTA, Wilhelm, physician; b. Karlsbad, Czechoslovakia, May 6, 1875; s. Wilhelm and Berta (Seiffert) F.; studied medicine, Strasbourg (now France); M.D., Prague, Czechoslovakia, 1900; m. Maria Herrmann, 1911; 2 sons, 1 dau. Asst. to H. Huppert, Prague, 1900-01; asst. to F. von Muller and W. His, Basel, Switzerland, 1901-06, lectr., 1904-06; became asst. to C. von Noorden and K. F. Wenckebach, Vienna, Austria, 1906; named asso. prof. metabolic and dietetic diseases, 1919; sr. physician Kaiserin-Elisabeth-Spital, Vienna, from 1919; ret., 1947. Author: Die Erkrankungen der Brustdrüsen, 1913; Die Behandlunb innerer Krankheiten mit radioaktiven Substanzen, 1918; Die Mehlfrüchtekur bei Diabetes mellitus, 1920; Die Zuckerkrankheit und d. insulin, 1924; Kostformen bei Diabetes mellitus, 1928; Renaler und insulärer Diabetes, 1929; Die Zuckerkrankheit, 1936; also articles. Specialized in diabetes mellitus and other metabolic disturbances; made 1st attempt to organize glandular diseases with internal secretion into one system. Died Obermarkersdorf, Austria.

FAMINTSYN, Andrei Sergeevich, Russian botanist; b. Sikolnik, Russia, 1835; ed. U. St. Petersburg (Russia); named prof. botany U. St. Petersburg, 1872. Author: Embryologische Studien, 1879. Investigated devel. of embryos of seed plants. Died 1921.

FAN, Hsu Yun, physicist; b. Shanghai, China, June 4, 1912; s. Tsi K. and Ing-Joo (Li) F.; B.S., Harbin Poly. Inst., 1932; M.S., Mass. Inst. Tech., 1934, D.Sc., 1937; m. Li-nien Bien, Feb. 18, 1941; children—David P., Vicky L., Hung, Frances. Came to U. S. 1947, naturalized 1957. Prof., Nat. Tsing Hua U., Kunming-Peiping, China, 1937-48; mem. staff Research Lab. of Electronics, Mass. Inst. Tech., 1948; faculty Purdue U., Lafayette, Ind., 1948—, prof. 1951—, Duncan Distinguished Prof. Physics, 1963. Research, publs. in semiconductor physics, including optical properties, photoelectric effects, damage of crystal lattiaes by high energy particles. Home: 340 Laurel Dr., West Lafayette, Ind.*

FAN, Ky, mathematician; b. Hangchow, China, Sept. 19, 1914; s. Chi Han and Wu-Shien (Fang) F.; B.S., Nat. Peking U., 1936; D.Sci. in Math., U. Paris, 1941; m. Yu-Fen Yen, Apr. 26, 1936. Came to U. S., 1945, naturalized, 1954. French Nat. Sci. fellow, charge de recherches Centre Nat. de la Recherche Scientifique, 1942-45; mem. Inst. for Advanced Study, 1945-47; faculty math. U. Notre Dame, 1947-60, prof., 1952-60; prof. math. Wayne State U., Detroit, 1960-61, Northwestern U., Evanston, Ill., 1961-65, U. Cal. at Santa Barbara, 1965—. Mem. Academia Sinica. Asso. editor: Jour. Math. Analysis and Applications, 1960—. Contbr. articles to math. jours. Research in functional analysis, topology. Home: 1402 Santa Teresita Dr., Santa Barbara, Cal. 93105.*

FANARDZHYAN, Varfolomey Artemevich, Russian roentgenologist; b. 1898; grad. Med. Faculty, Azerbaijan U., 1924; D.Med. Sci., 1936. A founder, developer radio. and oncological service, Armenia, 1926—; asso. Yerevan Med. Inst., 1927—, lectr., 1932—, prof., later dep. dir. for sci. work, 1936—; a founder Research Inst. Roentgenology and Oncology, Armenian Ministry Health, 1946, also mem. learned med. council; founder 1st Armenian sch. for roentgenologists and oncologists. Decorated Order of Lenin. Mem. USSR (corr.), Armenian (Presidium mem. 1956-63) acads. sci., All-Union (bd. mem.), Armenian (chmn.) socs. roentgenologists, radiologists and oncologists. Author: Manual of X-Ray Diagnosis, Parts 1-4, 1934-39; X-Ray Diagnosis of Duodenal Diseases, 1936; X-Ray Diagnosis of Diseases of the Thoracic Viscera, 1957. Co-editor: Large Med. Ency., 2d edit. Address: Inst. Roentgenology and Oncology, Yerevan, Armenia SSR, USSR.

FÄNGE, Ragnar, Swedish zoologist; b. Lund, Sweden, Mar. 27, 1920; s. Petter Brynolf and Gunborg (Andersson) F.; M.B., U. Lund, 1942, D.Sc., 1953; m. Else-Marie Thormodsen, Mar. 2, 1963; children—Thomas, Jens, Cecilia. Research asso. Dept. Zoology, Duke, 1957-58; prof. zoophysiology U. Oslo (Norway), from 1959, U. Göteborg (Sweden), 1963—. Recipient Thunberg medal RNO. Mem. Göteborgs Kungl. Vetenskaps-och Vitterhets-Samhälle. Editor: (with A. Brodal) The Biology of Myxine, 1963. Research, publs.

on comparative animal physiology, especially function of swimbladder in bony fish; investigation of secretion of salt from gland in head of sea gulls, cormorants and other marine birds (with K. Schmidt-Nielsen). Home: Dragspelsgatan 41, Västra Frölunda, Sweden. Office: 4 Langgatan 7B, Göteborg SV, Sweden.*

FANNING, John Thomas, Am. civil engr.; b. Norwich, Conn., Dec. 31, 1837; s. John Howard and Elisabeth (Pridde) F.; acad. and normal sch. edn.; m. M. Louise Bensley, June 14, 1865. Studied architecture and civil engring. until 1861; attained rank of lieutenant colonel in Civil War; after the war resumed engineering practice in Norwich; became prominent in engring., after 1865, particularly in planning and constructing pub. water works and water powers in N.E., and later in the West, notably on the power of Mississippi River at Mpls., of Spokane River in Wash., and Missouri River nr. Helena and at Great Falls, Mont., and on the Weenatahee River in Wash. for the electrification of the G.N. Ry.'s cascade tunnel; was cons. engr. St. Paul, Mpls. & Manitoba R.R. and GN. Ry., and v.p. Mpls. Union Ry.; chief engr., St. Anthony Falls Water Power Co., 1886-1911. Author: A Practical Treatise on Water Supply Engineering (long the standard work on subject), 1877, 16th edit., 1906; Report on a Water Supply for New York and Other Cities of the Hudson Valley, 1881; Homestead and Suburban Sewerage, 1884. Contbr. tech. papers on hydraulic and water-supply engring. to Trans. Am. Soc. C.E. Died St. Anthony Falls, Minn., Feb. 6, 1911.

FANO, Robert Mario, Am. elec. engr., educator; b. Torino, Italy, Nov. 11, 1917; s. Gino and Rosetta (Cassin) F.; student Politecnico di Torino, 1935-39; B.S., Mass. Inst. Tech., 1941, Sc.D., 1947; m. Jacqueline Crandall, Mar. 26, 1949; children—Paola C., Linda C., Carl. Came to U. S., 1939, naturalized, 1947. Teaching asst. elec. engring. dept. Mass. Inst. Tech., 1941-43, instr., 1943-44, staff mem. Radiation Lab., 1944-46, research asso. elec. engring. dept., research lab. electronics, 1946-47, asst. prof. elec. engring. dept., 1947-51, group leader Lincoln Lab., 1950-53, asso. prof. elec. engring dept., 1951-56, prof., 1946-62, Ford prof. engring., 1962—, dir. Project MAC, 1963. Mem. com. for sci. and tech. Library of Congress, Washington, since 19——, Esso Prodn. Research Co., Houston, 1963—. Fellow I.E.E.E., Am. Acad. Arts and Scis.; mem. Assn. for Computing Machinery, Sigma Xi, Eta Kappa Nu. Author: (with R. B. Adler and L. J. Chu) Electromagnetic Fields, Energy and Forces, 1960, Electromagnetic Energy Transmission and Radiation, 1960; Transmission of Information, 1961. Contbg. author: Microwave Transmission Circuits, 1948. Research, publs. in fields of microwave techniques, network synthesis, electromagnetic theory, information theory. Home: 17 Somerset Rd., Lexington, Mass. 02173.*

FANO, Salvador, physician; b. Amsterdam, Holland, 1824; M.D., 1851; aggregate physician, 1857; naturalized French citizen; founder free clinic, Paris. Author: Recherches sur la contusion du cerveau, 1851; Des tumeurs de la voûte palatine et du voile du palais, 1857; Traite pratique des maladies des yeux, 1866; Traité élémentaire de chirurgie, 1868. Studied brain contusion; performed operations on palate tumors. Died Paris, 1898.

FANO, Ugo, physicist; b. Turin, Italy, July 28, 1912; s. Gino and Rosa (Cassin) F.; Sc.D., U. Turin, 1934; m. Camilla V. Lattes, Feb. 8, 1939; children—Mary Rose, Virginia Carol. Came to U. S., 1939, naturalized, 1945. Lectr., U. Rome, 1937-38; physicist X-ray sect. Nat. Bur. Standards, Washington, 1946-49, chief radiation theory sect., 1949-60, sr. research fellow, 1960-66; professor of physics University Chgo., 1966—. Lectr. George Washington U., Washington, 1946-47; vis. prof. U. Cal., Berkeley, summer 1958, Cath. U., Washington, 1963-64. Recipient Rockefeller Pub. Service award, 1956; Exceptional service award Commerce Dept., 1957; Stratton award Nat. Bur. Standards, 1963. Author: (with G. Racah) Irreducible Tensorial Sets, 1959; (with L. Fano) Basic Physics of Atoms and Molecules, 1959. Research, publs. in theoretical physics, especially atomic, nuclear, statis. mechanics, optics, and in radiation genetics and bacterial genetics. Home: 5517 S. Dorchester Av., Chgo. 60637.*

FANTHAM, Harold Benjamin, zoologist; s. B. Fantham; ed. Univ. Coll., London, Royal Coll. Sci., London, Christ's Coll., Cambridge; M.A., Cambridge U., D.Sc., London U.; m. Annie Porter, 1915. Protozoologist to troops in Egypt and Sanonkica; lectr. parasitiology Liverpool Sch. Tropical Medicine; asst. zool. dept. Univ. Coll., London; asst. to Quick prof. biology Cambridge U.; protozoologist Grouse Disease Inquiry; prof. zoology U. Witwatersrand, Johannesburg, S. Africa, 1917-32; Strathcona prof. zoology McGill U., Montreal, Can. Recipient S. Africa medal. Fellow Zool. Soc. London, Cambridge Philos. Soc., Royal Soc. Tropical Medicine, Royal Soc. Medicine, Eugenics Soc. London, Royal Soc. S. Africa (v.p.); pres. S. African Assn. for Advancement of Sci., 1926-27; pres. S. African Biol. Soc., S. African Eugenics Soc.; v.p. Eugenics Research Assn.; corr. mem. Soc. Exotic Pathology, Paris. Author: Some Minute Animal Parasites; Protozoa in Animal Parasites of Man; also papers on parasitic protozoa, soil protozoa, human biology. Died Oct. 27, 1937.

FARABEUF, Louis-Hubert, French anatomist; b. Bannost, France, 1841; studied medicine, Provins, France; aggregation U. Paris, 1876; became asst. in anatomy Paris hosps., 1871; prof. anatomy Paris Sch. Medicine, 1886-1902; mem. French Acad. Medicine. Author: Manuel des amputations, 1881; (with Pinard) Introduction à l'etude des accouchements. Research on amputations, ligatures and resection; described symphysiectomy and ischiopubiotomy. Died Bannost, 1910.

FARADAY, Michael, English physicist, chemist; b. Newington Butts, Surrey, England, Sept. 22, 1791; s. James and Margaret (Hastwell) F.; m. Sarah Barnard, June 12, 1821. Apprenticed to bookbinder, London, 1804, later became master bookbinder and book dealer; read sci. articles, attended lectures on natural philosophy, conducted simple expts.; attended H. Davy's lectures, 1812, became personal sec. and sci. asst. to Davy at Royal Inst., 1813, accompanied Davy on continental tour, 1813-15, apptd. supt. Royal Inst., 1821, dir. lab., 1825, Fullerian prof. chemistry, from 1833; awarded ann. pension of 300 pounds, 1835; sci. adviser to Trinity House, 1836-65; ret. to house in Hampton Court, 1858. Fellow Royal Soc., 1824; mem. French Acad. Scis. (corr. chemistry sect., fgn. mem.), City Philos. Soc. Author: Chemical Manipulation, 1827, 29, 30, 42; Experimental Researches in Electricity, 3 vols., 1839, 44, 55; Experimental Researches in Chemistry and Physics, 1859; The Subject matter of a Course of Six lectures on the Non-Metallic Elements (edited by J. Scoffern), 1853; The Chemical History of a Candle (edited by William Crookes), 1861; Diary being the Various Philosophical Notes of Experimental Investigations made by Michael Faraday during the years 1820-1862 and bequeathed by him to the Royal Institution of Great Britain . . . printed and published . . . under the editorial supervision of Thomas Martin, with a foreward by Sir William H. Bragg, 7 vols., 1932-36; and 158 papers. Was one of greatest experimentalists of all times; best known for work in electricity; discoverer of electromagnetic induction; discovered identity of electricity generated in different ways; introduced concept of magnetic lines of force; demonstrated that magnetism has to be moving or changing in order to generate electric current in neighboring circuits; dealt with properties of dielectrics, diamagnetism and with relations between electricity and light; distinguished paramagnetic and diamagnetic substances; discovered how to produce electromotive force by movement of condr. in magnetic field, thus laid founds. for generator; reduced electrolysis to quantitative terms by announcing what are now known as Faraday's laws of electrolysis; introduced terms electrode, anode, cathode, anion, cation, ion and ionization; developed Volta-electrometer; discovered rotation of plane polarized light by magnetic field, 1845; liquified several gases; discovered two chlorides of carbon; discovered, named and analyzed benzene; produced new kinds of optical glass; quantity of electricity required during electrolysis to liberate one gram atomic weight of element is called faraday, unit of electrostatic capacitance named farad, both in his honor. Died Hampton Court, nr. London, Aug. 25, 1867.

FARAH, Alfred E., pharmacologist; b. Nazareth, Palestine, July 10, 1914; s. Emil S. and Lydia (Rieske) F.; B.A., Am. U. Beirut, 1936, M.D., 1940; m. Mary Stanwood, Apr. 24, 1943. Came to U. S., 1945, naturalized, 1949. Instr. physiology, pharmacology Am. U. Beirut, 1940-43, asst. prof., 1943-45; lectr. dept. pharmacology Harvard Med. Sch., 1945-47; asst. prof. dept. pharmacology U. Wash., 1947-48, asso. prof., 1948-50; asso. prof. pharmacology State U. N.Y., Upstate Med. Center, Syracuse, N.Y., 1950-53, prof., chemn. dept. pharmacology, 1953—. Mem. research com. Am. Heart Assn., 1961-66. Recipient Spl. citation for service to research Am. Heart Assn., 1967. Mem. Am. Soc. Pharmacology and Exptl. Therapeutics, A.A.A.S., Soc. Cell Biology. Editor: Handbook of Experimental Pharmacology, 1953. Research, numerous publs. on cardiac and renal physiology and pharmacology, action of digitalis on heart, mechanism of action of diuretics, mechanism of action of antidiuretic hormone, renal excretion of organic acids and bases, nature of membrane carriers. Home: 309 Buffington Rd., Syracuse, N.Y.*

FARBER, Bernard, Am. sociologist; b. Chgo., Feb. 11, 1922; s. Benjamin and Esther (Axelrod) F.; student Chgo. Jr. Colls., 1939-41; A.B., Roosevelt U., 1943; A.M., U. Chgo., 1949, Ph.D., 1953; m. Annette Ruth Shugan, Dec. 21, 1947; children—Daniel, Michael, Lisa, Jacqueline. Faculty, U. Chgo., 1951-53, Henderson State Tchrs. Coll., Arkadelphia, Ark., 1953-54; faculty Inst. for Research on Exceptional Children and dept. sociology U. Ill., Urbana, 1954—, prof., 1964—. Mem. Am. (council mem. family sect. 1966—), Ill. (pres. 1965-66) sociol. assns. Author: Effects of a Severely Mentally Retarded Child on Family Integration, 1959; Family Organization and Crisis: Families with a Severly Mentally Retarded Child, 1960; Family: Organization and Interaction, 1964; Kinship and Family Organization, 1966; Mental Retardation, 1968; Comparative Kinship Systems, 1968; also articles. Research in areas of sociol. research methodology, family orgn., mental retardation. Home: 711 Sunnycrest St., Urbana, Ill. 61801.*

FARBER, Emmanuel, pathologist, educator; b. Toronto, Ont., Can., Oct. 19, 1918; s. Morris and Mary (Madorsky) F.; M.D., U. Toronto, 1942; Ph.D. in Biochemistry (Am. Cancer Soc. fellow), U. Cal. at Berkeley, 1949; m. Ruth Wilma Diamond, Apr. 16, 1942; 1 dau., Naomi Beth. Came to U. S., 1946, naturalized, 1956. Am. Cancer Soc. fellow Hektoen Inst. Med. Research, Cook County Hosp., Chgo., 1949-50; instr., asst. prof. pathology and biochemistry Tulane U. Med. Sch., New Orleans, 1950-55, asso. prof., 1955-59, Am. Cancer Soc. research prof., 1959-61; prof., chmn. dept. pathology U. Pitts. Sch. Medicine, 1961—. Mem. U. S. Surgeon Gen. Adv. Com. on Smoking and Health, 1961-62; chmn. pathology B study sect. NIH, 1963—; mem. com. on pathology, div. med. scis. Nat. Acad. Scis.-NRC; cons. div. chronic diseases Dept. Health, Edn. and Welfare. Recipient Bertha Goldblatt Teplitz Meml. award 1961. Fellow A.A.A.S.; mem. Am. Assn. for Cancer Research (bd. dirs.), Am. Assn. Pathologists and Bacteriologists, Am. Soc. Exptl. Pathology (Parke Davis award 1958), Am. Chem. Soc., Am. Soc. Biol. Chemists, Biochem. Soc., Pitts. Path. Soc., Pa. Assn. Clin. Pathologists, Soc. for Exptl. Biology and Medicine, Histochem. Soc., N.Y. Acad. Sci., Internat. Acad. Pathology, Sigma Xi. Studies, publs. on biochem. mechanisms underlying disease processes in liver, other organs. Home: 5454 Beacon St., Pitts. 15217.*

FARBER, Erich Alexander, mech. engr.; b. Vienna, Austria, Sept. 7, 1921; s. Ignatz and Hilde (Giebelhauser) F.; B.S. in Mech. Engring., U. Mo., 1943, M.S. 1946; Ph.D., U. Ia., 1949; m. Ellen W. Williams, July 12, 1949; children—Hans F., Webb W. Came to U. S., 1939, naturalized, 1944. Instr., U. Mo., 1943-44, U. Ia., 1946-49; faculty U. Wis., 1949-54, asso. prof., 1953-54; prof. U. Fla., Gainesville, 1954, research prof., 1954—, dir. Solar Energy Lab., 1956—. Cons. pvt. cos., govt. agys.; vis. scientist Fla. Acad. Sci., Nat. Acad. Sci., 1961—. Cited by Air Force, 1962. Mem. Soc. M.E., Am. Soc. Elec. Engrs. (Tech. Paper award 1953), Solar Energy Soc., Fla. Acad. Sci., Am. Radio Relay League, Sigma Xi. Author: (with C. C. Williams) Building an Engineering Career, 1957; contbr. to Solar Energy Utilization, 1964, Mark's Mechanical Engineering Handbook, 1966; also numerous articles. Established complete boiling curve nucleate and film boiling region; discovered surface treatment in heat transfer standardization; pioneered growth high temperature crystals, solar energy utilization. Home: 1218 N.E. 5th St., Gainesville, Fla. 32601.*

FARBER, Eugene Mark, Am. physician; b. Buffalo, July 24, 1917; s. Simon and Matilda (Goldstein) F.; A.B., Oberlin Coll., 1939; M.D., U. Buffalo, 1943; M.S., U. Minn., 1946; m. Ruth Seiffert, Mar. 4, 1944; children—Charlotte, Nancy, Eugene, Donald. Intern Buffalo Gen. Hosp., 1943-44; fellow dermatology and syphilology Mayo Clinic, 1944-48, 1st asst., 1947-48; mem. faculty Stanford Med. Sch., 1948—, dir. div. dermatology, 1950—, prof. dermatology, 1959—, exec. head dept., 1961—; physician in chief Palo Alto Stanford Hospital; consultant physician Presbyn. Med. Center Palo Alto Veterans Administration Hospital; consultant Parks AFB, Travis AFB, 1954-58; attending dermatology Stanford Convalescent Home, 1957—; cons. surgeon gen. USAF, 1957-64; cons. Cal. Bd. Health, 1960—; chmn. sec. cutaneous system NRC, 1962—; chmn. psoriasis symposium Internat. Congress Dermatology, 1962; gen. clin. research center com. NIH, 1965—. Diplomate Am. Bd. Dermatology. Mem. A.M.A., Association of Professors of Dermatology (secretary 1967), American Academy of Dermatology, (dir. 1957-60), Soc. Investigative Dermatology (past dir., pres. 1966), Soc. Clin. Investigation, Am. Fedn. Clin. Research, Cal. Med. Assn., Pacific Dermatol. Assn., Santa Clara County Med. Soc., Soc. Exptl. Biology and Medicine, Am. Dermatol. Assn., Microcirculatory Conf., Am. Assn. U. Profs., A.A.A.S., Sigma Xi; hon. member of Society of Dermatology Austria, Dermatological Society of India, Society Dermatology Yugoslavia, Venezuelan Soc. Dermatology. Author numerous research papers. Editorial bd. World Wide Abstracts of Gen. Medicine. Research on psoriasis; circumileostomy skin; rubefacients; ulcers; other problems in dermatology. Home: 167 Ramoso Dr., Portola Valley, Cal. Office: 300 Pasteur Dr., Palo Alto, Cal.*

FARBER, Maurice Lee, Am. psychologist; b. Jersey City, Mar. 15, 1912; s. Harry and Ethel (Bersin) F.; B.S., N.Y. U., 1932; M.A., Columbia, 1934; Ph.D., U. Ia., 1940; m. Ruth Almond, Oct. 23, 1955. Clin. psychologist Penitentiary City N.Y., 1935-36; research asso. Bur. Applied Social Research, Columbia, 1940-42; faculty U. Conn., Storrs, 1948—, prof. psychology, 1962—; vis. prof. U. London (Eng.), 1955, U. Oslo (Norway), 1961-62. Cons. VA, 1958—, Inst. Polit. Studies, Stanford, 1964. Rockefeller fellow, 1941; Spl. Research fellow Nat. Insts. Mental Health, 1961-62. Mem. Am. Psychol. Assn., Am. Sociol. Assn., Am. Assn. for Pub. Opinion Research. Author: (with others) Authority and Frustration, 1944; also articles. Research on psychology of suicide, relation of polit. attitudes to personality structure, relation future outlook to behavior. Home: R.F.D. 2, Old Stone House Rd., Stafford Springs, Conn. 06076. Office: Dept. Psychology, U. Conn., Storrs, Conn. 06268.*

FARBER, Milton, Am. chemist; b. Los Angeles, Oct. 6, 1916; s. Joshua and Lena (Luner) F.; B.S., U. Cal. at Berkeley, 1938, postgrad.; M.S. in Chem. Engring., U. Minn., 1939; m. Constance Baldwin, Feb. 14, 1942; children—Robert, Richard, Kathleen. Gen. mgr.

Colloidal Products Ltd., Honolulu, 1940-41; area supr. Atlas Powder Co., Paducah, Ky., 1941-42; sr. research engr. Tenn. Eastman, Oak Ridge, 1942-45; pilot plant supr. Carbide & Carbon Corp., Oak Ridge, 1945-46; sr. research engr. Jet Propulsion Lab., Pasadena, Cal., 1946-55; head Astronautics Lab., Aerojet-Gen., Azusa, Cal., 1955-57; head propulsion lab. Hughes Tool Co., Culver City, Cal., 1957-59; v.p. Maremont Corp., also Rocket Power, Inc., Pasadena, 1959——. Cons., Standard Oil Cal., 1948-49, Shell Devel. Co., 1948-50. Mem. Am. Chem. Soc., Am. Phys. Soc., Am. Inst. Aeros. and Astronautics, Sigma Xi. Contbr. articles to tech. jours. Developed new method for determination thermodynamic studies; pioneered research in elec. propulsion in U. S.; originated use heavy particles for elec. propulsion, ion synthesis chem. compounds. Home: 1335 Old Mill Rd., San Marino, Cal. Office: 2275 E. Foothill Blvd., Pasadena, Cal. 91107.*

FARBER, Saul Joseph, Am. physician, educator; b. N.Y.C., Feb. 11, 1918; s. Isidor and Mary (Bunim) F.; A.B., N.Y. U., 1938, M.D., 1942; m. Doris Marcia Balmuth, Mar. 13, 1949; children—Josehua M., Beth Mina. Intern Sinai Hosp., Balt., 1943; research resident Goldwater Hosp., resident Bellevue Hosp., N.Y. U., 1946-48; fellow Sch. Medicine N.Y. U., 1948-49, asst. prof. medicine, 1953-57, asso. prof., 1957-62, prof. medicine, Nathan Friedman prof. cardiovascular renal diseases, 1962-66, Nathan Friedman prof. medicine, chmn. dept. medicine, 1966——, acting dean, dep. dir., 1963-66; dir. 3d and 4th med. divs. Bellevue Hosp., 1966——; dir. medicine Univ. Hosp., 1966——; vis. physician Bellevue, VA Hosp., N.Y.C. Bd. dirs. N.Y.C. Pub. Health Research Inst. Served from lt. (j.g.) to lt., USNR, 1943-46. Mem. Am. Soc. Clin. Investigation, N.Y. Acad. Medicine (chmn. sect. medicine 1963——), Assn. Am. Physicians, Am. Physiol. Soc., Harvey Soc. (v.p. 1967), N.Y. Heart Assn. (dir.). Research, numerous publs. on renal and electrolyte metabolism. Home: 25 Plaza St., Bklyn. Office: 550 1st Av., N.Y.C. 10016.*

FARBER, Seymour Morgan, Am. physician; b. Buffalo, June 3, 1912; s. Simon and Matilda (Goldstein) F.; B.A., U. Buffalo, 1931; M.D., Harvard, 1939; D.H.L., St. Mary's Coll., 1964; m. Lynette True, Dec. 27, 1940; children—Burt, Margaret, Roy. Faculty, U. Cal. at San Francisco., 1942——, prof. clin. medicine, 1961——, asst. dean for continuing edn., medicine and health scis., 1956-63, dir. continuing edn. medicine and health scis., 1963——, dean ednl. services, 1963——, spl. asst. for acad. affairs to pres., 1964——, chief Tb and Chest Service, 1945-65, sr. cons., 1965——. Lectr., Sch. Pub. Health, U. Cal. at Berkeley, 1948——. Nat. cons. for continuing edn. and chest disease to surgeon gen. USAF, 1962——; chmn. Gov.'s Com. on Continuing Med. Edn., 1963——. Fellow Am. Coll. Chest Physician (past pres.), Am. Geriatrics Soc., Am. Coll. Cardiology; mem. A.M.A. (past chmn. chest diseases sect.), Cal., Pan Am. med. assns., Association of American Medical Colleges, American, Cal. Socs. internal medicine, N.Y. Acad. Scis., Am. Fedn. Clin. Research, Internat. Assn. Med. Museums, Am. Trudeau Soc., Am. Cytologic Assn., Internat. Acad. Pathology, A.A.A.S. Author: Cytological Diagnosis of Lung Cancer, 1950; Lung Cancer, 1954; numerous articles. Editor: The Air We Breathe, 1961; Man and Civilization: Control of the Mind, 1961; Man and Civilization: Conflict and Creativity, 1963; The Potential of Woman, 1963; Man Under Stress, 1964; The Family's Search for Survival, 1965; The Challange to Women: The Biologic Avalanche, 1966; Sex Education and the Teen-Ager, 1967; Teen-Age Marriage and Divorce, 1967. Research in lung cancer, air pollution, emphysema, disease of chest, methodology in continuing edn. in medicine and health scis. Office: 1255 Post St., San Francisco 94109.*

FARCOT, Joseph Jean Léon, French engr.; b. Paris, June 23, 1824; s. Marie-Joseph-Denis Farcot; ed. Central Sch. Mfg., 1845; became chief engr. Farcot Works, 1853, dir., 1869. Recipient Plumey prize French Acad. Scis., 1875. Mem. Soc. Civil Engrs. (became pres. 1879). Author: Le servomoteur ou moteur asservi, 1873. Invented isochromal regulator, 1854; improved the servo-motor, centrifugal pump. Died Saint-Ouen, 1908.

FARCOT, Joseph-Marie-Denis, French engr.; b. Paris, Nov. 16, 1798; s. Joseph-Jean-Chrysostome F.; became dir. Scipion Workshops, Paris, 1820; founder Farcot Boiler Works, 1823. Contbd. to devel. of steam engine; invented one of 1st engines in which steam cutoff could be regulated, 1836, continual spray pump, 1829, 1st steam driven kneading machine, 1834, an automatic oil press, 1834. Died St. Ouen, France, Aug. 30, 1875.

FARCOT, Léon, French engr.; b. St. Ouen, France, 1823; s. Joseph-Marie-Denis F.; invented machines leading to improvement of mining, servo-motor, high speed pile drivers, steam hammers, blast engines. Died 1908.

FARDEAU, Michel, French physician; b. Paris, France, Oct. 24, 1929; s. Gaston and Georgette (Laun) F.; M.D., Paris Faculty Medicine, 1959; baccalaureate Le Blanc Coll., India, 1965-66; m. Michèle Gautier, Mar. 25, 1953; children—François, Vincent Christine, Jean-Marie. Hosp. asst., 1961-63; research chief Nat. Center for Sci. Research, from 1963. Mem. French Neurology Soc. (asso.). Author: (with J. Rougerie) Les Cranio-Pharyngiomes, 1962. Research, publs. on clin., histopath., ultrastructural semeiology of diseases of skeletal muscle; various articles on clin. neurology, neuropathology. Home: 24 rue du Dr. Roux, Sceaux 92, France. Office: Salpêtrière, 47 Blvd. Hôpital, Paris 75, France.*

FARDELY, William, telegraph engr.; b. Ripon, Eng., Feb. 16, 1810; s. William and Maria Anna Josepha (Korbach) F. Went to Mannheim, Germany, 1820; observed Cooke-Wheatstone telegraph, London, 1840-42; developed, introduced dial telegraph, to Taunus railroad between Kastel and Wiesbaden, Germany, 1844, Saxon-Silesian and Saxon-Bavarian railroads, 1846-47; dir. laying of telegraph line from Neukirchen to Ludwigshafen, 1851. Author: Die Galvanoplastik, 1842; Die elektrische Telegraphie mit besonderer Berücksichtigung ihrer praktischen Anwendung für den gefahrlosen und zweckmässigen Betrieb der Eisenbahnen, 1844; Der Zeigertelegraph für den Eisenbahndienst, 1856. Pioneer in elec. telegraphy: 1st in continental Europe to use earth as a ground; worked to develop relay transmission, to improve galvanic elements. Died Mannheim, June 26, 1869.

FARDIG, Oliver B., Am. chemist; b. Chgo., Jan. 5, 1918; s. Andrew G. and Esther E. (Andreson) F.; B.A., Carleton Coll., 1940; M.S., Pa. State U., 1942, Ph.D., 1947; m. Helen Elizabeth Archer, Sept. 11, 1943; children—Marsha Jan, Leslie Anne, Elaine Louise, Alison Beth, David Archer. Staff, Pa. State U., 1942-47; rsearch scientist Bristol Labs., Syracuse, N.Y., 1947-50, group leader research, 1950——. Mem. Am. Chem. Soc. (chmn. Syracuse 1967), Sigma Xi. Research publs. on effects food processing on vitamin retention, new nutrients, gen. problems nutrition, antibiotics; isolation pharmacologically active substances from natural sources. Patentee antibiotics. Home: 311 Churchill Lane, Fayetteville, N.Y. 13066. Office: Bristol Labs., Syracuse, N.Y. 13201.

FAREY, John, Brit. geologist, mathematician; b. Woburn, Eng., 1766; sent to sch., Halifax, 1882; studied math.; a son, John. Apptd. agt. for estates by Duke of Bedford, 1792. Author: Survey of the County of Derby, . . . , 8 vols., 1811-13; also articles; contbr. to Encyclopedia (Ree). Collected minerals and rocks; prepared geol. sects. and maps illustrating relative position of Eng.'s strata. Died London, 1826.

FAREY, John, English civil engr.; b. Lambeth, Eng., Mar. 20, 1791; s. John F.; ed. at Woburn; began making drawings for encys., jours., 1805; engaged as civil engr. in constrn. iron-works, Russia, 1819. Mem. Instn. Civil Engrs. Recipient silver medal Soc. Arts, 1813, gold medal, 1813. Author: A Treatise on the Steam Engine, Historical, Practical and Descriptive, 1827. Invented device for making perspective drawings, 1807, machine for drawing ellipses, 1813; introduced steam-engine indicators to Eng. Died Kent, Eng., July 17, 1851.

FARHI, Leon Elie, physiologist; b. Cairo, Egypt, Oct. 9, 1923; s. Elie S. and Victoria (Anzarut) F.; B.Sc., Coll. Internat., Beirut, Lebanon, 1940; M.D., Universite St. Joseph, Beirut, 1947; m. Haya Youlus, July 12, 1949; children—Nitza, Eli R. Came to U. S., 1958, naturalized, 1964. Instr., Hebrew U., Jerusalem, Israel, 1957; asst. prof. U. Buffalo, 1958-62; faculty physiology State U. N.Y., Buffalo, 1962——, prof., 1966——; vis. prof. U.Fribourg (Switzerland), 1965-66. Cons., USPHS, 1963——. NSF Sr. fellow, 1965. Mem. Am. Physiol. Soc., Assn. des Physiologistes, A.A.A.S. Editorial bd. Am. Physiol. Soc., 1964——, Respiration Physiology, 1965——. Research, publs. in pulmonary ventilation, circulation and gas exchange. Home: 158 North Dr., Eggertsville, N.Y. 14226. Office: State U. N.Y., Buffalo 14214.*

FARILL, Juan Solares, Mexican physician; b. Mexico City, Mexico, Feb. 11, 1902; s. Jaime V. and Manuela (Solares) F.; M.D., Nat. U. Mexico, 1926; postgrad. State U. Ia. (Guggenheim fellow), U. Vienna; m. Inés Novelo, Aug. 17, 1927 (dec. June 1966); children—Eugenia (Mrs. Eduardo Gutierrez), Jaime, Lourdes (Mrs. Jaime Pastor). Chief dept. orthopaedic surgery Gen. Hosp., Mexico, 1933-43; chief surgeon Shriners Hosp. for Crippled Children, Mexico City 1945-64, cons.; prof. orthopaedic surgery Nat. U. Mexico, 1942-48; hon. prof. orthopaedic surgery U. Guadalajara, 1943——, U. Zulia State, Venezuela, 1962; gen. dir. Mexican Inst. Rehab., 1960-67; officer council Gen. Pub. Health of Mexico, 1965——. Pres., 1st Pan Am. Congress Rehab., Mexico, 1948. Recipient Lasker Found. award 1954; comdr. Order of Daniel A. Carrion of Lima, Peru, 1965. Mem. Internat. Soc. for Welfare of Cripples (v.p. 1937-42, pres. 1942-48), Sociedad Amigos del Nino Lisiado (founder 1935), Latin Am. Soc. Orthopaedics and Traumatology (founder, pres. 1948-50), Mexican Nat. Acad. Medicine, American, British (corresponding) orthopedic associations, American Academy Orthopaedic Surgeons, Internat. Soc. Orthopaedic Surgery and Traumatology, Internat. Coll. Surgeons. Translator: A Primer of the Prevention of Deformity in Childhood (by Shands and Raney), 1946; Strike Back at Stroke (Internat. Soc. Welfare of Cripples), 1960. Devel., publs. on techniques for treatment club foot and congenital hip dislocation, otho-radiographic measurement of shortening of lower limbs. Address: 854 Av. Morena, Colonia Narvarte, Mexico D.F. 12, Mexico.

FARIS, Robert E. Lee, Am. sociologist; b. Waco, Tex., Feb. 2, 1907; s. Ellsworth and Elizabeth (Homan) F.; Ph.B., U. Chgo., 1928, M.A., 1930, Ph.D., 1931; m. Claire Guignard, Aug. 18, 1931; children—William G., John H., Roger S. Faculty sociology Brown U., 1931-38; asst. prof. McGill U., 1938-40; asso. prof. Bryn Mawr Coll., 1940-43; faculty Syracuse U., 1943-48; prof. U. Wash., 1948——. Mem. Social Sci. Research Council (dir. 1953-59), Pacific (pres. 1954), Am. (pres. 1961) sociol. assns., Sociol. Research Assn. (pres. 1959). Author: (with H. W. Dunham) Mental Disorders in Urban Areas, 1939; Social Disorganization, 1948; Social Psychology, 1952; Chicago Sociology 1920-32, 1967; also articles. Editor Handbook of Modern Sociology, 1964. Research on distbn. of mental disorders in large cities, sociol. causes of high ability; contbns. to sociology of class differentiation. Home: 4318 N. E. 41st St., Seattle 98105.*

FARKAS, Adalbert, chemist; b. Dunaszerdahely, Hungary, July 10, 1906; s. Istvan and Anna (Patzauer) F.; Chem.Eng., Techn. U., Vienna, Austria, 1928, Dr.Eng., 1932; Ph.D., U. Frankfurt-Main (Germany), 1930; m. Deborah Miller, May 9, 1936; 1 son, Edward Joel. Came to U. S., 1941, naturalized, 1944. Instr. phys. chemistry U. Frankfurt, 1930-33; research fellow Cambridge (Eng.) U., 1933-36; instr. Hebrew U., Jerusalem, Israel, 1936-41; research-chemist Union Oil Co., Wilmington, Cal., 1941-46; research supr. Barrett div. Allied Chem., Phila., 1946-56; sect. head Houdry Labs., Marcus Hook, Pa., 1956-59, asst. dir. chem. research, 1960——; dir. Chem. Ednl. Projects, Inc., Phila. Mem. Internat. Congress on Catalysis (past. pres., dir.), Am. Chem. Soc. (past pres. Phila. sect., dir.), Faraday Soc., Chem. Soc. (London), Catalysis Club. Author: Orthohydrogen, Parahydrogen and Heavy Hydrogen, 1935; (with H. W. Melville) Experimental Methods in Gas Reactions, 1939; also numerous articles. Editor: Physical Chemistry of Hydrocarbons, vol. I, 1950, vol. II, 1953. Research, numerous patents on deuterium, hydrocarbon chemistry and catalysis; co-author Bonhoeffer-Farkas mechanism; indsl. prodn. methods of chems., polyurethane foam. Home: 408 Turner Rd., Media, Pa. 19063.*

FARKAS, G. L., Hungarian biologist; b. Budapest, June 15, 1925; s. G. and Ilona (Wirger) F.; B.Sc., Eötvös U., Budapest, 1947, Ph.D., 1949, D.Sc., 1961; m. Lucy Riedel, July 21, 1951; 1 son, Peter. Asst. prof. Budapest U., 1949-52; research officer Inst. Agrl. Research Martonvásár, Hungary, 1952-57; postdoctorate fellow Canadian NRC, 1957-58; sr. research officer Research Inst. for Plant Protection, Budapest, 1958-62; head plant physiology lab. Hungarian Acad. Scis., Budapest, 1962——; vis. prof. dept. biochemistry U. Wis., Madison, 1964-65. Mem. Hungarian Acad. Scis. (corr.). Research, publs. on changes in metabolism in diseased plants, biochem. aspects of disease resistance. Home: 4 Maros, Budapest. Office: 4/A Muzeum KRT, Budapest, Hungary.*

FARKAS, Loránd, Hungarian chemist; b. Budapest, Hungary, Aug. 31, 1914; s. Imre Jenö and Irene (Kovács) F.; Ph.D., U. Szeged (Hungary), 1937; D.Sc., Tech. U. Budapest, 1965. Head, Zemplén Memory Lab., Hungarian Acad. Scis., Budapest, 1940——, asst. prof., 1942——; vis. prof. Faculty Scis., Munich (West Germany) U., 1960, 63, 66; vis. research asso. Fla. State U., Tallahassee, 1965-66; hon. prof. Tech. U. Budapest, 1966——, lectr. natural materials extension, 1960——. Recipient Presdl. prize Presidency Hungarian Acad. Scis., 1959, 62, 64; Chevreul Commemorative medal, (Paris), 1963. Mem. Hungarian, German chemist socs. Research, numerous publs. on chemistry of natural and synthetic dyestuffs especially flavonoids, isoflavonoids and carbohydrates. Patentee chemistry synthetic dyestuffs. Home: 37 Ráday, Budapest IX. Office: Gellért tér 4, Budapest XI, Hungary.*

FARKASS, Imre, physicist; b. Budapest, Hungary, Sept. 26, 1919; s. Imre and Marie (Wagner) F.; Diploma in Elec. and Mech. Engring., Poly. U., Budapest, 1942; m. Violet Romhanyi, May 19, 1951. Came to U. S., 1957, naturalized, 1962. Asst. prof. physics Poly U., Budapest, 1942-46, prof., 1947-49, dept. head vacuum research lab., 1950-57; head dept. math. and physics Agrl. U., Budapest, 1954-57; sr. physicist Nat. Research Corp., Cambridge, Mass., 1957-61; dir. applied physics dept. Ilikon Corp., Natick, Mass., 1961-65; mem. tech. staff Bell Telephone Labs., Whippany, N.J., 1965——. Lectr., Chem. U., Veszprem, Hungary, 1953-56, Mass. Inst. Tech., 1964, Northeastern U., Boston, 1965. Mem Am. Phys. Soc., Am. Vacuum Soc. Author: Theory and Production of Selenium Rectifiers, 1955; also articles. Research and devel. in space physics, ultrahigh vacuum technology, semicondrs., surface physics. Home: 175 Prospect St., East Orange, N.J. 07017. Office: Bell Telephone Labs., Whippany, N.J. 07981.*

FARLEY, Donald Thorn, Jr., Am. physicist; b. N.Y.C., Oct. 26, 1933; s. Donald Thorn and Rebecca (Hamlin) F.; B.Eng., Cornell U., 1956, Ph.D, 1960; m. Jennie Tiffany Towle, June 16, 1956; children—Claire Hamlin, Anne Tiffany, Peter Towle. NATO postdoctoral fellow Cambridge U., 1959-60; docent Chalmers U., Gothenburg, Sweden, 1960-61; physicist Jicamarca Radar Obs., Lima, Peru, 1961-67, dir., 1964-67; faculty Cornell U., Ithaca, N.Y., 1967—. Mem. commns. Internat. Sci. Radio Union, 1963—. Mem. Research Soc. Am., Am. Geophys. Union, Sigma Xi. Asso. editor Revs. of Geophysics, 1963—. Publs. on research in ionospheric physics and plasma physics, coupling between E and F layers of ionosphere, theory of incoherent scattering of radio waves by a plasma, theory of formation of irregularities in E-region of ionosphere by a plasma instability. Home: 710 Henshaw Rd., Ithaca, N.Y. 14850.*

FARLOW, William Gilson, Am. botanist; b. Boston, Dec. 17, 1844; s. John Smith and Nancy White (Blanchard) F.; A.B., Harvard, 1866, A.M., 1869, M.D., 1870; LL.D., 1896; studied botany in Europe several yrs.; LL.D., U. Glasgow, 1901, U. Wis., 1904; Ph.D., Upsala, 1907; m. Lillian Horsford, Jan. 10, 1900. Asst. to Asa Gray, Harvard, asst. prof. botany, 1874-79, prof. cryptogamic botany, from 1879. Fellow A.A.A.S. (v.p. 1887, 1898, pres. 1905), Am. Acad. Arts and Scis.; mem. Nat. Acad. Scis. (pres. 1905); Am. Soc. Naturalists (pres. 1904); French Acad. Scis., 1917. Author: The Gymnosporangia, or Cedar; Apples of the United States, 1880; Marine Algae of New England, 1881; The Potato Rot; A Provisional Host-index of the United States, 3 vols., 1888-91; Bibliographical Index of North American Fungi, 1905. Authority on cryptogamic botany, plant pathology; research on marine algae fungi; diseases of plants; parasitic plants. Died June 3, 1919.

FARMER, Frank Taylor, English physicist; b. Bexleyheath, Eng., Sept. 18, 1912; s. Robert Crosbie and Maud (Sharp) F.; student Eltham Coll., London, Eng., 1921-30; B.Sc. in Engring., London U., 1933; Ph.D., Cambridge (Eng.) U., 1937; m. Marion Bowie Bethune, Sept. 15, 1960. Research engr. Marconi's Wireless Telegraph Co., Gt. Baddow, Eng., 1937-40; physicist Middlesex Hosp., London, 1940-44; chief physicist Newcastle Regional Hosp. Bd., also Royal Victoria Infirmary, Newcastle upon Tyne, Eng., 1945—; hon. prof. med. physics U. Newcastle upon Tyne, 1967—. Fellow Inst. Physics, Inst. Elec. Engrs., Council Engrs. Research, publs. in ionospheric propagation and direction finding, dosimetry of X-rays and gamma rays for treatment of cancer; design high energy accelerators for med. use. Home: 81 Grosvenor Av. Office: Newcastle Gen. Hosp., Newcastle upon Tyne, Eng.*

FARMER, John, Am. psychologist; b. N.Y.C., Sept. 21, 1932; s. Walter and Gertrude (Goadby) F.; B.A., Queens Coll., 1957; M.A., Columbia, 1959, Ph.D., 1962; m. Anne Keller, June 25, 1960; children—John Scott, Michael. Research asso. Columbia, 1962-66; asst. prof. Queens Coll., Flushing, N.Y., 1966—; mem. adv. staff psychology dept. U. Brasilia (Brazil). Reviewer research proposals NSF; co-prin. investigator grant Nat. Inst. Mental Health. Mem. Am., Eastern psychol. assns., A.A.A.S., Sigma Xi. Research in new technique for controlling behavior of organisms, clarification of variables that enhance precision of timing behavior in organisms. Home: 90 Morningside Dr., N.Y.C. 10027.*

FARMER, Moses Gerrish, Am. inventor, electrician; b. Boscawen, N.H., Feb. 9, 1820; s. Col. John and Sally (Gerrish) F.; attended Dartmouth, 1840-43; m. Hannah Shapleigh, Dec. 25, 1844; 1 child. Civil engr., Portsmouth, N.H., 1842; asst. in pvt. sch., Portsmouth, 1843; accepted preceptorship Eliot (Me.) Acad., 1843; prin. Belknap Sch. for Girls, Dover, N.H., 1844; devised machine that printed paper shades for lamps; experimented with electric railroad, 1845, constructed miniature electric train of 2 cars; wire examiner of new electric telegraph between Boston and Worcester, Mass., 1847; telegraph operator between Boston and Newburyport, Mass., 1848; invented electric-striking apparatus for fire alarm service, 1848; supt. 1st electric fire alarm system in Am., installed Boston, 1851; discovered means for duplex and quadruplex telegraph, 1855; became supt. of tobacco-extracting manufactory, Somerville, Mass., 1861; invented an incadescent electric lamp, 1858-59 (20 years before Edison), self exciting dynamo, 1866; electrician U. S. Torpedo Sta., Newport, R.I., 1872-81; cons. electrician U. S. Electric Light Co. of N.Y. Improved the art of torpedo warfare in U. S. Navy; invented an incadescent electric lamp supplied by a wet-cell battery, 1858-59; 1st self-exciting dynamo, 1866. Died Chgo., May 25, 1893.

FARMER, Victor Colin, chemist; b. nr. Galway, Ireland, Dec. 31, 1920; s. Charles Thompson and Sarah (McFadden) F.; B.Sc., Glasgow U., 1943; Ph.D., Aberdeen (Scotland), 1947; m. Jane Wyllie Donald, Apr. 7, 1947; children—Francis, Joan, George. Staff, Macaulay Inst. for Soil Research, Aberdeen, 1946—, sr. research worker, 1957—; vis. prof. Mich. State U., 1964. Fellow Royal Inst. Chemistry, Chem. Soc., Faraday Soc. Research, publs. on application of infrared and ultraviolet absorption spectrometry to biol. problems and to characterizing structure and surface properties of clay minerals. Home: 36 Albury Rd., Aberdeen. Office: Macaulay Inst. for Soil Research, Craigiebuckler, Aberdeen, Scotland.*

FARNER, Donald Sankey, Am. zoologist, educator; b. Waumandee, Wis., May 2, 1915; s. John and Lillian (Sankey) F.; B.S. Hamline U., 1937, D.Sc., 1962; M.A., U. Wis., 1939, Ph.D., 1941; m. Dorothy S. Copps, Dec. 21, 1940; children—Carla M., Donald C. Instr. zoology U. Wis., 1941-43; asst. prof. zoology U. Kan., 1946-47; asst. prof. biology U. Colo., 1947; asso. prof. zoophysiology Wash. State U., 1947-52, prof., 1952-65, dean Grad. Sch., 1960-64; prof. zoophysiology U. Wash., Seattle, 1965—, chmn. dept. zoology, 1966—. President, Internat. Union Biol. Scis. Mem. Am. Soc. Zoologists, Am. Inst. Biol. Scis. (bd. govs.), Am. Physiol. Soc., Am. Chem. Soc., Am. Soc. Naturalists, Am. Ecol. Soc., Am. Ornithologists Union (Brewster medal), Cooper Ornithol. Soc., Soc. Exptl. Biology and Medicine, Western Soc. Naturalists. Research, numerous publs. on characterization of photoperiodic responses in birds, role of hypothalamus in control of anterior pituitary glands in birds. Home: 4533 W. Laurel Dr., Seattle 98105.*

FARNETI, A. Pietro, Italian physician; b. Conselice, Italy, May 24, 1903; s. Antonio and Maria (Tampieri) F.; student U. Ferrara; Laureato in medicina, U. Bologna (Italy), 1926. Asst. med. clinic U. Bologna, 1929-53, prof. hydrological medicine, 1938-62; prof. U. Milano (Italy), 1962—; dir. Inst. Phys. Therapy and Rehab. Hosp. Maggiore Casa Granda, Milan, 1953—; dir. School for Physiotherapists, Milan, 1953—; instr. hydroclimatology and phys. agts. Advanced Istitute of Physical Education Catholic U. of Sacred Heart, Milan, 1966—. Mem. numerous socs. phys. medicine, rehab. and hydroclimatology. Author: Il Massaggio, 1932; Idrologia Medica e Terapia Fisica, I, 1946, II, 1950; Trattato di Terapia Fisica e Riabilitazione, 2 vols., 1966; also numerous articles. Research on phys. therapy, methods of kinesiotherapy, clin. applications of med. hydrology. Home: via Ruggero di Lauria 3 Milano, Italy.*

FARNSWORTH, Dana Lyda, Am. physician; b. Troy, W.Va., Apr. 7, 1905; s. Henry Lyda and Isabell (Waggoner) F.; A.B., U. W.Va., 1927, B.S., 1931; M.D., Harvard, 1933; Sc.D., Salem Coll., 1959, Williams Coll., 1961, U. W.Va., 1965; L.H.D., Lesley Coll., 1962; LL.D., U. Notre Dame, 1964. Instr. chemistry, physics Barrackville (W.Va.) High Sch., 1927-29; asst. dir. health Williams Coll., 1935-41, dir. health, 1945-46; prof., med. dir. Mass. Inst. Tech., 1946-54, acting dean students, 1950-51; Henry K. Oliver prof. hygiene, dir. U. health services Harvard, 1954—. Diplomate Am. Board Psychiatry and Neurology. Fellow Am. Acad. Arts and Scis., Am. Psychiat. Assn.; mem. A.M.A., Am. Coll. Health Assn. (pres. 1953-54), Group for Advancement Psychiatry (pres. 1957-59). Author: Mental Health in College and University, 1957; (with Jack R. Ewalt) Textbook of Psychiatry, 1963; (with Fred Hein) Living, 4th ed., 1965. Editor, contbr. College Health Administration, 1964; Psychiatry, Education and the Young Adult, 1966. Research, publs. on emotional disorders of students; growth and devel. in late adolescence and early adulthood. Home: 52 Old Concord Rd., Belmont, Mass. 02178. Office: 75 Mt. Auburn St., Cambridge, Mass. 02138.*

FARNSWORTH, Harrison Edward, Am. physicist; b. Green Lake, Wis., Mar. 24, 1896; s. Edward H. and Marion (Fortnum) F.; A.B., Ripon Coll., 1918; M.A., U. Wis., 1921, Ph.D., 1922; m. Gertrude Romig, 1925 (dec.); children—Edward Allan, James A. (dec.); m. 2d., Alice Schultze, 1960. Faculty, U. Pitts., 1918-19, U. Me., Orono, 1924-26; faculty Brown U., Providence, 1926—, dir. Barus Lab. Surface Physics, 1946—, Annette L.R. Barstow U. prof., 1963—. Ofcl. investigator NDRC, 1942-43; exec. sec. Panel on Electron Tubes, dir. coordinating group on electron tube reliability, N.Y.C., 1952-53; cons. numerous labs. NRC fellow, U. Wis., 1922-24. Fellow A.A.A.S., Am. Phys. Soc., Am. Acad. Arts and Scis.; mem. Am. Chem. Soc., Sigma Xi, Phi Beta Kappa, Gamma Alpha. Contbr. sci. articles to profl. jours. Research on surface physics and chemistry with metal and semiconductor single crystals; surface catalysis and gas adsorption on crystals. Home: 174 Laurel Av., Providence 02906.

FARQUHAR, Marilyn Gist (Mrs. John W. Farquhar), Am. cell biologist, educator; b. Tulare, Cal., July 11, 1928; d. Brooks D. and Alta (Green) Gist; A.B., U. Cal. at Berkeley, 1949, M.A., 1952, Ph.D., 1955; m. John W. Farquhar, Feb. 4, 1951; children—Bruce Edward, Douglas Gist. Asst. research pathologist Rockefeller Inst., N.Y.C., 1956-58, research asso., 1958-62; asso. research pathologist U. Cal. Sch. Medicine at San Francisco, 1962-64, asso. prof. pathology in residence, 1964—. Mem. Am., Internat. socs. cell biology, Histochem. Soc., Internat. Acad. Pathology, Am. Soc. Exptl. Pathology, Am. Assn. Anatomists, Electron Microscope Soc. Research, publs. in cell biology with electron microscopy and cytochemistry, cellular and capillary permeability and cell secretion, kidney, anterior pituitary, frog skin, bone marrow. Home: 1509 Brandywine Rd., San Mateo, Cal. 94402. Office: Dept. Pathology, U. Cal. Sch. Medicine, San Francisco 94122.*

FARQUHARSON, James, natural philosopher; b. Coull, Scotland, 1781; s. John Farquharson; M.A., King's Coll., Aberdeen, Scotland, 1798, LL.D., 1837; m. Helen Taylor, Oct. 19, 1826; 5 sons, 1 dau. Licenced as preacher; became minister of Alford, Scotland, 1813. Fellow Royal Soc., 1830; hon. mem. French Soc. Universal Statistic. Author: On the Form of the Ark of Noah, 1831; also papers in Philos. Trans.; description of aurora in Edinburgh Philos. Jour., 1823; (paper for Royal Soc.) On the Nature and Localities of Hoar Frost. Traced aurora borealis to devel. of electricity, 1830; made 1 of 1st geometrical measurements of aurora and found its height of less than a mile dependent on altitude of clouds. Died Dec. 3, 1843.

FARR, Clinton Coleridge, physicist; b. Adelaide, Australia, May 22, 1866; s. Julia (Warren) Farr; ed. Univ. Coll., Adelaide; Grad. in Sci., U. Sydney, 1888; D.Sc., Angus Engring. scholar, 1889; m. Maud Ellen Haydon, 1903; 1 son. Lectr. math. and physics St. Paul's Coll., U. Sydney, 1891-95; lectr. elec. engring. U. Adelaide, 1896; made magnetic survey of New Zealand, 1898-1904; established magnetic obs. for new Zealand Govt., Christchurch; lectr. physics and surveying Canterbury Coll., U. New Zealand, 1904-10, prof. physics Canterbury Univ. Coll., 1910-36, later emeritus prof. Fellow Royal Soc., 1928, Inst. Physics. Author: Some continuous Observations on the Rate of dissipation of Electric charges in the Open Air; (with D. B. Macleod) The Visosity of Sulphur; A magnetic Survey of the Dominion of New Zealand and Some of the Outlying Islands; (with N. M. Rogers) Helium in New Zealand. Died Jan. 27, 1943.

FARR, Lee E., Am. physician; b. Albuquerque, Oct. 13, 1907; s. Edward and Mabel (Heyn) F.; B.S., Yale, M.D., 1933; m. Anne Ritter, Dec. 28, 1936; children—Charles Edward, Susan Anne, Frances Alison. Asst. pediatrics Yale Sch. Medicine, 1933-34; from asst. to asso. in medicine Hosp. Rockefeller Inst., 1934-40; dir. research, physician in chief Alfred I. du Pont Inst. Nemours Found., 1940-49; physician-in-chief Brookhaven Nat. Lab., Upton, N.Y., 1948-62; prof. nuclear medicine U. Tex., Houston, 1962-63, prof. nuclear and environmental medicine, 1963—, chief sect. nuclear medicine M. D. Anderson Hosp. and Tumor Inst., 1963—. Exec. com. Nat Acad. Sci.-NRC, 1957—; chmn. adv. com. Atomic Bomb Casualty Com., 1955—; mem. adv. com. on Naval Medical Research. Recipient Mead Johnson award for research in pediatrics; Gold Cross Order of Phoenix, from Greece; Gold Cross Order merit (1st class), Germany. Mem. Am. Acad. Pediatrics (chmn. com. on environmental hazards), A.M.A. (council on post grad. program since, mem. com. on nuclear medicine), Athens Med. Soc. (hon.). Author: Treatment of Nephrotic Syndrome, 1951. Contbr. articles to tech. jours. on research in protein metabolism, renal function, nephroses, nuclear medicine, neutron capture therapy of cancer, design of nuclear reactor facilities for biomed. uses, radioactive isotopes in medicine, environmental contaminants of air, water soil. Home: 5331 Beverly Hill Lane, Houston 77027.*

FARR, Samuel, Brit. physician; b. Taunton, Eng., 1741; ed. Edinburgh; M.D., Leyden, Netherlands, 1765. Physician, Bristol (Eng.) Infirmary, 1767-80; practiced in Bristol, then returned to his native town to practice. Fellow Royal Soc., 1779. Author: An Essay on the Medical Virtues of Acids, 1769; A Philosophical Inquiry into the Nature, Origin, and Extent of Animal Motion, deduced from principals of reason and analogy, 8 vols., 1771; Inquiry into the Propriety of Blood-letting in Consumption, 8 vols., 1775; Elements of Medical Jurisprudence, 1788; also others. Died Mar. 11, 1795.

FARR, Wanda Kirkbride, Am. biochemist; b. New Matamoras, O., Jan. 9, 1895; B.S., Ohio U., 1915; M.S., Columbia, 1918; m. Clifford Harrison Farr, 1918; 1 son, Robert. Instr. botany Kan. State Coll., 1917-18, Tex. A. and M. Coll., 1918-19; research asso. Bernard Skin and Cancer Clinic, St. Louis, 1926-27; instr. Shaw Sch. Botany, 1928; investigator plant physiology Boyce Thompson Inst., 1928-29, dir. Cellulose Lab., Chem. Found., 1936-40; cotton researcher U. S. Dept. Agr., 1929-36; researcher Am. Cyanimid Co., 1940-43; mem. research div. Celanese Corp. Am., 1943-54; research cons., 1954—; asso. prof. botany U. Me., 1957-64, lectr. cytochemistry, 1964—; with Farr Cytochem. Labs., 1956—. Fellow Royal Micros. Soc.; mem. A.A.A.S., Am. Chem. Soc., Am. Bot. Soc., Torrey Bot. Club, Soc. Am. Naturalists, N.Y. Acad. Scis. Discovered origin of cellulose; research on cell physiology in plants, formation and structure of plant cell membranes, root hair growth, microscopic analysis, cotton fibers, x-ray diffraction and microchem. techniques as applied to natural and synthetic materials. Address: 20, Craig Dr., Jericho, L.I., N.Y. 11753.*

FARR, William, Brit. statistician; b. Kenley, Eng., 1807; studied medicine privately; student in Paris, 1829-31, also at Univ. Coll., London; M.D. (hon.), N.Y., 1847; D.C.L., Oxford U., 1857; m. Miss Langford, 1833; m. 2d. M. E. Whittall, 1842; 8 children; Became compiler abstracts, registrar-gen.'s office, 1838; asst. commr., censuses 1851, 61, commr., 1871. Recipient Gold medal Brit. Med. Assn., 1880. Fellow Royal Soc., 1855; mem. Statis. Soc. Author: Vital Statistics, 1885. Laid found. for vital statistics in Eng. Died 1883.

FARRAR, George Elbert, Jr., Am. physician; b. Winter Park, Fla., Mar. 12, 1906; s. George Elbert and Mary (Persinger) F.; B.S., Wesleyan U., 1927; M.D., Johns Hopkins, 1931; m. Elnore Harriet Spangle, Aug. 30, 1933; children—George Elbert III, Alice Taylor (Mrs. M. J. Grinsbergs). Member of the faculty of Univ. of Mich., 1933-35; asso. pharmacologist FDA, Washington, 1935-36; faculty Temple U., Phila., 1936—, clin. prof. medicine, 1964—;

med. dir. Wyeth Lab., Phila., 1949-62, dir. med. services, 1962——. Mem. com. on revision U. S. Pharmacopoeia, 1950-60. Mem. A.M.A., A.C.P., Am. Soc. Pharmacology and Exptl. Therapeutics, Am. Rheumatism Assn., Am. Therapeutic Soc., Coll. Physicians Phila., Pa. Medical Society (president-elect 1968), Philadelphia County Med. Soc. (pres. 1964), Sigma Xi. Author: United States Dispensatory, 1947, 50, 55, 60. Research in clin. internal medicine, including rheumatic diseases, anemia and leukemia, gastrointestinal diseases; devel. of therapeutic drugs such as antibiotics, antacids, enzymes. Home: 528 Scott Rd., Gladwyne, Pa. 19035. Office: P.O. Box 8299, Phila. 19101.*

FARRAR, John, Am. mathematician; b. Lincoln, Mass., July 1, 1779; s. Deacon Samuel and Mary (Hoar) F.; B.A., Harvard, 1803, M.A., 1806; LL.D. (hon.), Brown U., 1833; m. Lucy Buckminster; m. 2d, Eliza Rotch. Tutor of Greek, Harvard, 1805, Hollis prof. math. and natural philosophy, 1807-36; fellow Am. Acad. Arts and Scis., 1808, recording sec., 1811-23, mem. com. on publs., 1828-29, v.p., 1829-31, contbr. articles to its Transactions; wrote various monographs on meteorology and astronomy; helped make European astron. and math. lit. known in America through translations of Euler, Legendre and Bézout. Died Cambridge, Mass., May 8, 1853.

FARRAR, John Laird, Canadian forester; b. Hamilton, Ont., Can., Dec. 31, 1913; s. Robert Watson and Sarah (Laird) F.; B.Sc. in Forestry, U. Toronto (Ont.), 1936; M.F., Yale, 1939, Ph.D., 1955; m. Betty Joan May, Oct. 12, 1946. Forester, Canadian Internat. Paper Co., Trois Rivieres, Que., 1936-37; forest ecologist Can. Dept. Forestry, Petawawa Forest Expt. Sta., 1937-56; Abitibi prof. forest biology U. Toronto, Ont., 1956——. Fellow A.A.A.S.; mem. Ont. Profl. Foresters Assn., Canadian Inst. Forestry, Ecol. Soc. Am., Canadian Soc. Plant Physiologists, T.A.P.P.I. Research, publs. on nature trees and formation wood. Home: 167 Yonge Blvd., Toronto 12, Ont., Can.*

FARRE, Arthur, Brit. obstetrician; b. London, Mar. 6, 1811; s. John Richard F.; ed. Charterhouse Sch., Caius Coll., Cambridge; studied medicine St. Bartholomew's Hosp.; M.B., Cambridge, 1833, M.D., 1841; m. Lectr. comparative anatomy St. Bartholomew's Hosp., 1836-37, lectr. forensic medicine, 1838-40; became prof. obstetric medicine King's Coll., 1841; named physician-accoucheur to King's Coll. Hosp., until 1862; Harvean orator, 1872; examiner in midwifery Royal Coll. Surgeons, 1852-75; attended Princess of Wales and other mems. of royal family; physician extraordinary to Queen. Fellow Royal Coll. Physicians, Royal Soc., 1839; mem. Royal Micros. Soc. (pres. 1851-52), Obstet. Soc. London (elected hon. pres. 1875). Contbr. numerous articles on microscopy to tech. jours. Studied uterus. Died Dec. 17, 1887.

FARRE, Frederic John, English physician; b. London, Dec. 16, 1804; s. John Richard F.; ed. Charterhouse Sch.; 3d wrangler St. John's Coll., Cambridge, 1827; M.A., St. Bartholomew's Hosp., London, 1830, M.D., 1837; m. Julia Lewis, 1848; 2 daus. Became lectr. botany St. Bartholomew's Hosp., 1831, lectr. materia medica, 1854-76, became asst. physician 1836, physician, 1854; physician Royal London Opthalmic Hosp., 1843-86. Fellow Coll. Physicians (became censor, 1841, 42, 54, treas. 1868-83, became v.p. 1885). An editor 1st Brit. Pharmacopoeia, 1864; abridgement of Materia Medica (Pereira), 1865. Studied use of opium for treatment of acute periocarditis. Died Nov. 9, 1886.

FARRE, John Richard, physician; b. Barbados, B.W.I., Jan. 31, 1775; s. Richard John F.; studied medicine under his father; came to Eng. where he studied for a year at Sch. of St. Thomas' and Guy's Hosp.; studied at Edinburgh, for 2 years; M.D., 1806; children—Friederic John, Arthur. Became mem. Corp. Surgeons, 1793; mem. expdn. to France; returned to London, then to Barbados to practice medicine; again returned to Eng., 1800; practiced medicine as physician, beginning in 1806; a founder Royal London Ophthalmic Hosp., physician for 50 years there; ret. 1856. Licentiate Coll. Physicians. Author: The Morbid Anatomy of the Liver, 4 vols., 1812-15; Pathological Researches on Malformations of the Human Heart, 1814. Editor: Arterial and Secondary Haemorrhage (Jones), 1805; On Diseases of the Eye (Saunders), 1811. Editor, Jour. Morbid Anatomy, Ophthalmic Medicine, Pharm. Analysis. Died May 7, 1862.

FARRELL, Gordon Lee, Am. physiologist; b. Helena, Mont., Apr. 16, 1925; s. Richard Henry and Marion Grace (Tintinger) F.; student Carroll Coll., Helena, 1940-43; M.D., St. Louis U., 1947; m. Elizabeth Williams Rauschkolb, June 19, 1959; children—Michael, David, Joseph. Practice medicine, specializing in endocrinology, Cleve., 1952-65, Houston, 1965——; faculty Western Res. U., 1954-65, asso. prof. dept. physiology, 1960-65; asso. prof. depts. biochemistry, psychiatry Coll. Medicine Baylor U., 1965——; prof. mental scis. Grad. Sch. Biomed. Scis. U. Tex., 1967——; chief sect. neuroendocrinology Houston State Psychiat. Inst., 1965-67; head div. mental retardation research Tex. Research Inst. Mental Sci., 1961——; established investigator Am. Heart Assn., 1957-62; sr. visitant in endocrinology VA Hosp., 1957-65. Recipient Ciba award in endocrinology, 1959. Fellow A.A.A.S.; mem. Endocrine Soc. Chila (hon.), Royal Soc. Medicine (London), Am. Heart Assn., Am. Physiol. Soc., Endocrine Soc., N.Y. Acad. Scis, Alpha Omega Alpha. Author: (with Frances) Integrated Anatomy of Physiology, 1957. Research and publs. in endocrinology with emphasis on brainstem and adrenal function. Home: 3614 Montrose St., Houston 77006. Office: 13000 Moursund St., Houston 77025.*

FARRINGTON, Oliver Cummings, Am. geologist; b. Brewer, Me., Oct. 9, 1864; s. Joseph Rider and Ellen (Holyoke) F.; B.S., U. Me., 1881, M.S., 1888; Ph.D., Yale, 1891; m. Clara A. Bradley, Aug. 3, 1896. Sci. tchr. in acads., 1882-87; lab. asst. Yale, 1890-91; asst. U. S. Nat. Mus., 1893; curator geology Field Mus. of Natural History, Chgo., from 1894. Lectr. on mineralogy, U. Chgo., 1894-1904; participated in expdn. to Mexico, 1895; made extensive explorations of caves of Ind. and Ky.; mines and metallurgy, Paris Expn., 1900; mem. Internat. Jury of Awards, St. Louis Expn., 1904; mem. field Mus. Expdn. to Brazil, 1922-23. Pres. Am. Assn. of Museums, 1915-16. Fellow Geol. Soc. Am., A.A.A.S. Author: Gems and Gem Minerals, 1903; Meteorites, 1915. Amassed largest single representative collection of meteorites (almost 700, in Field Mus.). Died Chgo., Nov. 2, 1933.

FASOLI, Angelo, Italian physician; b. Casalpusterlengo, Italy, Oct. 6, 1921; s. Renato and Ilida (Piacentini) F.; M.D., U. Milano (Italy), 1945; m. Anna Tenconi, Jan. 14, 1959; children—Renato, Laura, Ambrogio, Vera. Faculty, U. Milan, 1945——, asst. prof. medicine, 1953——, vice dir. Inst. Gen. Clin. Medicine and Med. Therapy, 1962——, libero docente med. pathology, 1956——, libero docente med. semeiotics, 1965——. Author: Le lipoproteine del plasma sanguigno, 1952; also numerous articles. Research on physiology of serum lipoproteins, especially in atherosclerosis, 1949-66, method for serum lipo-proteins determination by paper electrophoresis, 1952, clin. and lab. studies on protein-losing gastro-enteropathies, 1959——, serum macroglobulins, 1959-63. Home: 2 Mascheroni, Milan, Italy.*

FASSBENDER, Konrad Friedrich Heinrich, German engr.; b. Frankfort/Main, Germany, June 23, 1884; s. Carl Anton and Sophie (Ulrich) F.; ed. Technische Hochschule, Darmstadt, U. Berlin, U. Marburg; Ph.D. in sci. m. Edith Brodhum, July 22, 1916; children—Carlos-Werner, Helga. Prof. electrotechnics U. La Plata (Argentina); chief radiophonic service Service Aerial Transport Research, Berlin; prof. tech. high frequency; dir. Inst. for Study Elec. Oscillations, Heinrich Hertz Inst. Recipient Gauss-Weber medal U. Göttingen. Mem. Inst. Radiophonic Engrs. (U. S.), also others. Author: Hochfrequenztechnik in der Luftfahrt; Einfuhrung in die Messtechnik der Kernstrahlung und die Anwendung der Radiosotope, also numerous publs. Pioneer (with E. Habann) in tech. of carrier frequency. Home: Hardenbergstrasse 34, 1 Berlin 12, Germany.

FASSEL, Velmer Arthur, Am. phys. chemist; b. Frohna, Mo., Apr. 26, 1919; s. Arthur Edward and Alma (Poppitz) F.; B.A., S.E. Mo. State Coll., 1941; Ph.D., Ia. State Coll., 1947; m. Mary Alice Katschke, July 25, 1943. Chemist, Manhattan project Ia. state U., Ames, 1942-47, faculty, 1947——, prof. chemistry, sr. scientist, 1956——. Tech. adviser Atoms for Peace Conf., Geneva, Switzerland, 1958; chmn. panel on analytic methods Nat. Acad. Sci., 1958-61. Recipient Distinguished Alumni award S.E. Mo. State Coll., 1965. Fellow A.A.A.S., Optical Soc. Am., mem. Soc. for Applied Spectroscopy (Ann. medal 1964), Am. Chem. Soc., Am. Inst. Physics, Coblentz Soc., Sigma Xi. Research, numerous publs. on spectroscopy rare earth elements, new spectroscopic technique for determination gaseous elements in metals spectroscopic properties, analytical applications fuel-rich oxyacetylene flames, atomic absorption spectroscopy, spectroscopic analytical applications plasmas. Home: Route 4, Timberland Heights, Ames, Ia. 50010.*

FASTEN, Nathan, zoologist; b. Austria, Dec. 4, 1887; s. Schneier and Jane (Drillman) F.; brought to U. S., 1889, naturalized, 1896; B.S., Coll. City of N.Y., 1910; Ph.D., U. Wis., 1914; m. Frieda Mayer, June 18, 1916; children—Janet Rebecca (Mrs. Leon Benson Levy), Marion (Mrs. Alexander Grinstein), Natalie (Mrs. Gilbert E. Rosenwald, Jr.), Head dept. biology. Marshall College, Huntington, W.Va., 1910-11; asst. instr. zoölogy U. Wis., 1911-14; instr., asst. prof. zoölogy U. Wash., 1914-20; asso. prof. zoölogy, and physiology Ore. State Coll., 1920-21, prof., head dept., 1921-45; chief biologist, chief tech. div. Wash. Pollution Control Commn., 1945-50, ret. chief biologist, cons. biologist, from 1950; summer work as asst. U. S. Bur. Fisheries, Woods Hole, Mass., 1911, investigator Wis. Fish Commn., 1912, 13, 14, Wash. Fish Commn., 1919, Ore. Fish and Game Commn., 1923, in charge invertebrate zoölogy Puget Sound Biol. Sta., 1915, 16, 20, charge animal biology B.C. Summer Session for Tchrs. 1922, 23; spl. investigator Oyster Culture and Pollution Problems, Pacific Spruce Corp., 1927-28; chmn. Ore. Basic Sci. Exam. Com., 1934-45. Fellow A.A.A.S.; mem. Am. Soc. Zoölogists, Am. Soc. Naturalists, Genetics Soc. Am., Western Soc. Naturalists (pres. 1924-25), Sigma Xi, Phi Kappa Phi. Author: Origin Through Evolution, 1929; Principles of Genetics and Eugenics, 1935; Principles of General Zoölogy, 1938; Introduction to General Zoölogy, 1941; General Zoölogy Laboratory Outlines, 1941; also articles in profl. jours. Died Sept. 19, 1953.

FATELEY, William Gene, Am. chem. physicist; b. Franklin, Ind., May 17, 1929; s. Nolan W. and Georgia (Scott) F.; A.B., Franklin Coll., 1951; postgrad. Northwestern U.; Ph.D., Kan. State U., 1955; postgrad. Ind. U.; Sc.D., Franklin Coll., 1965; m. Wanda Lee Glover, Sept. 1, 1953; children—Leslie Kaye, W. Scott, Kevin L., Jonathan H. Research fellow U. Md., summer 1956, U. Minn., 1956-57; head phys. research lab. Dow Chem. Co., Williamsburg, Va., 1957-60; fellow Mellon Inst., Pitts., 1960-65, head sci. relations, 1962-64, asst. to pres., 1963——, sr. fellow ind. research, 1965. Recipient Meml. award for outstanding contbns. to molecular spectroscopy Coblentz Soc., 1965, 1st Outstanding Alumni award Kan. State U. Chemistry Dept., 1964; named Outstanding Young Man of Am., Kiwanis Clubs Am., 1965. Mem. Spectroscopy Soc. Pitts. (past chmn.), Am. Chem. Soc., Sigma Xi, Pi Mu Epsilon, Phi Lambda Upsilon, Alpha Chi Sigma, Sigma Alpha Epsilon. Asso. editor Jour. Applied Spectroscopy, 1965. Research, publs. on structure of matter, far infrared spectroscopy, measurement nonbonding forces in molecules. Home: 3 Alma Dr., Pitts. 15238. Office: 4400 5th Av., Pitts. 15213.*

FATIO DE DUILLIER, Nicolas, mathematician; b. Basel, Switzerland, Feb. 16, 1664; lived in London, Eng., also Geneva, Switzerland, Paris, France, Hague, Netherlands; Fellow Royal Soc., 1688; mem. French Acad. Scis. Author: Lettre à Casini, 1686; Lineae brevissimi descensus investigation geometrica duplex . . . , 1699; Description d'une pièce d'horlogerie très rare et très remarquable, 1704. Also numerous math. works, and writings on nav. and astronomy. Developed hypothesis for explaining form of rings of Saturn, circa 1682; asserted that Leibnitz had taken Newton's ideas in calculus (wrote 1st article in this dispute), 1699; studied dilation and contraction of eye lens, also zodiacal lights; advanced time keeping by method of piercing jewels with minute holes to serve as nearly frictionless and unwearing pivot holes, 1704. Died Maddersfield, Eng., Apr. 28, 1753.

FAUCHARD, Pierre, French dentist; b. Bretagne, France, 1678; practiced in Paris, from circa 1715. Author: Le chirurgien-dentiste ou traité des dents, 1723; Considered father of denistry in France; founder of modern dentistry; inventer prosthetic devices; 1st to oppose retroauricular cauterization as toothache cure; described pyorrhea alveolaris or alveodental periostitis (Fauchard's disease), 1746. Died 1761.

FAUCHER, Joseph A., Am. chemist; b. San Francisco, Oct. 2, 1927; s. Joseph A. and Lavine (Schragg) F.; A.B., Princeton, 1949; Ph.D., Yale, 1953; postgrad. U. Innsbruck (Austria); m. Tonita Jeune Pruitte, Aug. 15, 1959; children—Richard, John. With Union Carbide Chems. div. Union Carbide Corp., South Charleston, W.Va., 1956——, group leader, 1960——, research scientist, 1965——. Mem. Soc. Rheology, Sigma Xi. Contbr. articles to tech. jours. Discovered new method determining molecular weight by ultracentrifuge flotation. Home: Pleasant Hill Rd., Chester, N.J. 07930. Office: Union Carbide Plastics, Bound Brook, N.J. 08805.*

FAUJAS DE SAINT-FOND, Barthélemy, French geologist; b. Montélimart, France, May 17, 1741; ed. Jesuit Coll., Lyons, France; studied law, Grenoble, France; admitted advocate to parliament; became pres. Seneschal's Ct., 1765; invited to Paris by Georges L. L. de Buffon; apptd. asst. naturalist Royal Mus., by Louis XVI; became royal commr. for mines, 1788; 1st prof. geology in France, at Jardin des Plantes, 1793-1818. Author: Les volcans éteints du Vivarais et du Velay 9 (on his theory of origin of volcanoes), 1778; Histoire naturelle de la province de Dauphine, 1781, 82; Description des expériences de la machine aérostatique de MM. Montgolfier . . . , 1783, 84; Mineralogie des volcans, 1784; Voyage en Angleterre, 1797; Essai de géologie, 1803-09; also articles. Studied rock structure and composition in Alps; discovered pozzuolana in Velay, 1775; 1st to point out volcanic nature of basaltic columns in Fingal's Cave, Isle of Statta. Died Saint-Fond, France, July 18, 1819.

FAUL, Henry, nuclear geophysicist; b. Prague, Czechoslovakia, July 17, 1920; S.B., Mass. Inst. Tech., 1941, Ph.D., 1949; M.S., Mich. State U., 1942. Prof. geophysics U. Pa. Research in nuclear geophysics. Address: Dept. Geology, U. Pa., Phila. 19104.

FAULCONER, Albert, Jr., Am. anesthesiologist; b. Arkansas City, Kan., Oct. 24, 1911; s. Albert and Grace (McMillen) F.; B.S., Kan. U. Med. Sch., 1932, M.D., 1936; M.S., U. Minn., 1947; m. Mary Jean White, June 18, 1938; children—Albert III, David White, Barbara. Practice medicine specializing in anesthesiology, Detroit, 1938-41; fellow Mayo Found., Rochester, Minn., 1946-47, cons. sect. anesthesiology Mayo Clinic, 1947——, head sect. anesthesiology, 1953——, chmn. sect., 1966——; prof. anesthesiology Mayo Grad. Sch. Medicine, U. Minn., Rochester, 1961——. Mem. com. on anesthesia NRC, 1950-61; cons. Surgeon Gen., U. S. Army, 1957——. Diplomate Am. Bd. Anesthesiology (pres. 1964, dir. 1956——). Mem. Am. Soc. Pharmacology and Exptl. Therapeutics, Assn. U. Anesthetists (founding, pres. 1955), Sigma Xi, Nu Sigma Nu. Author: (with R. G. Bickford) Electroencephalography in Anesthesiology, 1960; (with T. E. Keys) Foundations of Anesthesiology, 1965; also numerous articles. Research on application instrumentation in physiology and clin. anesthesiology, methods for gas analysis. Home: Mounted Route 72, Rochester, Minn. 55901.*

FAULHABER, Johann, German mathematician; b. Ulm, Germany, May 5, 1580; tchr. math., Ulm. Author: Arithmetischer Wegweiser, 1614; Inginieurs-Schul, erster Theil, darinen durch den Canonem Logarithmicum . . . , (made use of logarithms known in Germany), 1630. Died Ulm, 1635.

FAURE, Jacques Marie Adrien, French physiopathologist, neurophysiologist; b. érigueux, Dordogne, France, Feb. 1, 1912; s. Pierre Jules and Reine (Bordas) F.; Bachelor 1st and 2d degree, Saint Joseph Instn. Périgueux, 1929, 30; M.D., U. Bordeaux, 1941; m. Odette Mathilde Reboul, Apr. 19, 1937. Mem. faculty U. Bordeaux (France), 1942-—, prof., dir. chair physiopathology and neurophysiology, 1961-—, also chief biologist, 1963; chief asso. lab. exptl. neuroendocrinology CNRS, 1965. Decorated Croix de Guerre and Star, 1939, 45; named officer Palmes Académiques, 1961. Mem. French Physiology Soc., EEG and Clin. Neurophysiology Soc., Endocrinology Soc. Paris, Exptl. Psychology Soc., Biology Soc., Gen. Pathology and Clin. Physiology Soc., Pharmacodynamy and Therapy Soc., others. Research, publs. in neurophysiology, psychophysiology, exptl. neuroendocrinology; also studies of brain regulations. Home: 16 rue de Lisleferme, Bordeaux 33, France.*

FAURE, Jean-Louis, French surgeon; b. Ste.-Foyla-Grande, France, Oct. 27, 1863; prof. clin. gynecology Paris (France) Faculty Medicine; mem. French Acad. Scis., 1934, Acad. Medicine, Soc. Surgery (pres.), Congress Surgery (pres. 1926). Author: L'hystérectomie; Le cancer de l'utérus; L'ame du chirurgien; Claude Bernard; (with Siredey) Traité de gynécologie. Died Oct. 26, 1944.

FAURIE, Urbain, botanist; b. Dunières, France, Jan. 1, 1847; ed. Monistozol; became mem. Soc. Fgn. Missions, 1869; ordained, 1873; became tchr., Tokyo, also Niigata, Japan; mission worker, Japan, 1882-94, from 1896. Built herbarium (later transferred to Imperial U.; discovered Fauria japonica; sent speciments to Am. and European museums. Died Formosa, July 4, 1915.

FAUST, Ernest Carroll, Am. med. parasitologist; b. Carthage, Mo., Sept. 7, 1890; s. John Samuel and Elizabeth (Fenner) F.; A.B., Oberlin Coll., 1912; M.A., U. Ill., 1914, Ph.D., 1917; LL.D., Tulane U., 1961; m. Lola Ernesta Swift, Aug. 12, 1919; 1 dau., Jean Louise (Mrs. Kermit Frymire). Asst. parasitologist U. Ill., 1912-14, instr. zoology, 1917-19; faculty Peking Union Med. Coll., 1919-23, 24-28; prof. parasitology Tulane U., New Orleans, 1928-47, William Vincent prof. tropical diseases and hygiene, 1947-58, emeritus prof. parasitology, 1961-—; field coordinator Tulane-Colombia Program Med. Edn., Cali, Colombia, 1956-61. Cons., Armed Forces Inst. Pathology, 1946-56; mem. expert panel on parasitology WHO, 1946-66. Decorated Order San Carlos, Colombia; named hon. and emeritus prof. by many Latin Am. and European univs. Fellow Royal Soc. Tropical Medicine and Hygiene; mem. Am. Soc. Tropical Medicine and Hygiene (past pres.), Am. Soc. Parasitologists (past pres.). Author: (with P. F. Russell), Clinical Parasitology, 1937; (with Beaver, Jung) Animal Agents and Vectors of Disease, 3d edit., 1968; also numerous articles, monographs, chpts. in med. books. Research on parasitic diseases man and animals. Home: 2628 Octavia St., New Orleans 70115.*

FAUST, William Roscoe, Am. physicist; b. Shawnee, Okla., Mar. 9, 1918; s. Hugh Graham and Bertha (Weiman) F.; B.S., Okla. State U., 1939; M.S., Ill. Inst. Tech., 1941; Ph.D., U. Md., 1948; m. Mary Cone Dees, Jan. 23, 1942; children—Hugh H., Margaret Adele, Mary Elizabeth. Radio engr. Naval Research Lab., Washington, 1941-48, physicist, 1948-52, physicist and br. head radiation div., 1952-54, asso. supt. radiation div., 1954-64, supt. applications research div., 1964-—. Instr., U.Md., 1949-50. Recipient Research Soc. award for Applied Sci., Naval. Research Lab., 1956; E. O. Hulbert award, 1961. Fellow Am. Phys. Soc., Washington Acad. Sci.; mem. Philos. Soc. Washington. Research, publs. on plasma physics, radiative transport, and nuclear physics. Home: 6410 Walnut St. S.E., Washington 20031. Office: Naval Research Lab., Washington 20309.*

FAUTH, Philipp Johann Heinrich, German astronomer; b. Bad Durkheim, Germany, Mar. 19, 1867; s. Philipp Friedrich and Johanna (Kinzel) F.; m. Philippine Feil, 1895; 2 sons, 1 dau.; m. 2d, Berta Heinrich, 1923. Began teaching career as sch. adminstr., Kaiserlautern, Germany, 1887; built obs., Kaiserlautern, 1899, transferred it to Landstuhl, 1895, later to Grünwald, nr. Munich, Germany. Moon crater named in his honor, 1932. Author: Astronomische Beobachtungen aus den Jahren 1890 und 1891 . . . , 1893; Astronomische Beobachtungen aus den Jahren 1893 und 1894, 1895; Beobachtungen der Planeten Jupiter und Mars . . . , 1898; (with A. Mang) Wegweiser am Himmel, 1904; Hörbigers Glacial-Kosmogonie, 1913; Fünfzehn astronomische Stereos zur Raumversinnlichung, 1916; 25 Jahre Planetenforschung, 1916; Mondess schicksal, 1925; Unser Mond . . . . mit einer Mondkarte 1:4 Mill., 1936; Mondatlas, 1932; Was Verbirgt die Ruckseite des Mondes? 1937. Work chiefly devoted to map of moon in scale 1:1 million (350 centimeter diameter); his studies made Germany foremost in moon research; also studied planets. Died Grünwald, Jan. 4, 1941.

FAUVEL, Sulpice Antoine, French physician; b. Paris, 1813; M.D., Paris, 1840; became dir. clinic Faculty Medicine, Paris, 1842; apptd. physician central bur., 1845; sent by Inst. San. Medicine to Constantinople, 1847-68; apptd. prof. med. pathologie Imperial Sch. Medicine, Constantinople, 1849; insp.-gen. san. cordons, 1868; returned to France, 1868; mem. staff l'Hotel-Dieu. Mem. Superior Council of Health, Ottoman Empire, 1848. Mem. Soc. Medicine (a founder), Acad. Medicine. Co-founder Gazette Medicale d'-Orient. Contbr. articles to jours. Credited with earliest description of presystolic murmur in mitral stenosis, 1843. Died 1884.

FAVARD, Jean Aimé, French mathematician; b. Peyrat-la-Nonière (Creuse), Aug. 28, 1902; s. Jules and Eugénie (Bonneaud) F.; ed. Ecole normale supérieure, Paris; Ph.D. in sci.; m. Marguerite Holmgren, Dec. 26, 1930; children—Pierre-Jacques, Claude-René, Anne-Karen. Instr., lectr., prof. U. Grenoble, 1927-41; lectr., prof. advanced geometry Faculty Scis., Paris, 1941-—; prof. Ecole polytechnique, Paris, 1954-—. Author: Espace et dimension; Cours de géométrie différentielle locale; Cours d'analyse de l'Ecole polytechnique; Leçons sur les fonctions presque-périodiques. Home: 10, rue de Belgrade, Grenoble. Office: 11, rue Pierre-Curie, Paris, France.

FAVARGER, Claude Philippe Emile, Swiss botanist; b. Neuchatel, Aug. 8, 1913; s. Pierre and Emmy (de Keller) F.; ed. univs. Neuchatel, Lausanne; Ph.D., U. Paris; D. honoris causa, U. Daker, 1963; m. Solange Boutet, Oct. 22, 1938; children—Dominique, Marie-France. Prof., Ecole Suisse de droguerie, 1939-46, Faculty Sci., 1946; dir. Inst. Botany. Author: Flore et Vegetation des Alpes, 2 vols. Research and numerous publs. on systematic botany. Home: 9, rue Chantemerle. Office: 11, rue E.-Argaud, Neuchatel, Switzerland.

FAVARGER, Pierre, Swiss biochemist; b. Geneva, Switzerland, July 5, 1909; s. Samuel and Fanny (Bourgeois) F.; student Coll. Geneva, 1922-28; Ph.D., U. Geneva, 1933; Dr. (hon.), U. Paris; m. May Keller, Mar. 8, 1935; children—Christiane, Pierre Yves. Asst., Inst. Phys. Chemistry, Faculty Scis., Geneva, 1933-34; staff Faculty Medicine, Geneva, 1934-—, prof. biochemistry, dir. Inst. Med. Biochemistry, 1947-—. Mem. Swiss Soc. Biochemistry, Swiss Soc. Physiology, Swiss Soc. Clin. Chemistry, Biochem. Soc. (Eng.), Soc. de chimie biologique (France), Societa di Biologia esperimentale (hon.; Italy), Soc. de Biologie (corr.; France). Research and numerous publs. on metabolism of cholesterol, digestion of fats, metabolism of fats, synthesis of fatty acids. Home: 21B Av. de Miremont, Geneva, Switzerland.*

FAVARO, Giuseppe Alessandro, Italian astronomer; b. Revine (Trevisa), Oct. 22, 1876; Ph.D. in math. scis. Prof. astronomer; regent Internat. Astrogeodesique Sta., Carloforte, 1911-12; adj. astronomer Obs. of Torino, 1912-14; former dir. Astrophys. Obs. of Cantana, Astron. Obs. of Trieste. Mem. various Italian and fgn. sci. orgns. Collaborator: Catalogo astrofotografico Internazionale, 1942. Founder: Ann. of Obs. of Trieste, 1934. Contbr. articles to profl. publs. Address: via Maestra 54, Revine (Treviso), Italy.

FAVÉ, Ildefonse, French engr.; b. Dreux, France, Feb. 28, 1812; became prof. mil. art, Ecole Polytechnique, 1854; comdr. Legion Honor; mem. French Acad. Scis. Author: Nouveau système de défense des places fortes, 1841; Historie de l'artillerie, 1845-47; Histoire des progrès de l'artillerie (written from notes of Napoleon III), 1862; Études sur le passe et l'avenir de l'artillerie, 1862-72. Improved field arty.; built floating batteries (1st armored ships) used in Crimean War. Died Mar. 1, 1894.

FAVE, Louis-Eugène-Napoléon, French hydrographer; b. Paris, France, July 18, 1853; s. Ildephonse Fave; chief hydrographer to French Navy; mem. Bur. Longitudes, French Acad. Scis., 1918; contbr. many hydrographic maps; inventor marégraphe (instrument for measuring rate of flow of tides). Died Paris, July 30, 1922.

FAVELET, Jean-François, Belgian physician; b. Fort de perle, Flanders, Belgium, Apr. 18, 1674; student medicine Louvain, Belgium; prof. medicine, botany, chemistry, Louvain; gen. practice medicine; 1st physician to Marie-Elisabeth of the Low Countries; mem. French Acad. Scis., 1729. Author: Prodromus apologiae fermentations in animabilus, 1721. Worked against process of trituration; studied fermentation. Died Louvain, June 30, 1743.

FAVEZ, Gérard, Swiss chest physician; b. Geneva, Switzerland, Sept. 16, 1916; s. Edouard and Gitla (Bukiet) F.; Baccalauréat es sciences in Math. Collège et Gymnase, Lausanne, Switzerland, 1936; m. Juliette Veillard, Jan. 11, 1939; children—Georges, Francoise. Faculty medicine U. Lausanne, 1936-42; 1st asst. Inst. for Histology, 1943; asst. Phtisiologic Center Leysin (Switzerland), 1944-48, dept. medicine Lausanne Med. Sch., 1949-51; head physician pneumophtisiological dept., chief outpatient chest clinic, U. Lausanne Med. Sch., 1952-—, asso. prof., 1958-—. Recipient Price de Cévenville for thesis Inst. Histology, 1943. Mem. Am. Thoracic Soc., Am. Coll. Chest Physicians, Swiss Soc. Internal Medicine, Swiss Soc. Thoracic Diseases. Author: Radiological Examination of the Lung and Mediastinum with Aid of Josterior Oblique Topography, 1966; also numerous articles. Research on devel. passive co-hemagglutination method for determination of circulating antibodies during active Tb, radiol. examination of chest. Home: Sous le Bouchet, chemin du Pâqueret 1025 Saint-Sulpice, Switzerland. Office: Clinique Sylvana, 1066 Epalinges, by Lausanne, Switzerland.*

FAVILLI, Ranieri, Italian agronomist; b. Pisa, Italy, Feb. 4, 1915; s. Narciso and Micheli Clementina (Assunta) F.; Ph.D. in agrarian sci.; m. Lara Giannoni, Apr. 22, 1940; children—Giuliana, Alberto. Prof. gen. agronomy and herboriculture U. Pisa, also dir. Inst. Gen. Agronomy. Mem. Georgofili Acad. of Florence, Fisiocritici Acad. of Sienna. Research and numerous publs. on gen. agronomy, herboriculture, vegetal biology and horticulture. Home: via Matteotti 1. Office: via S. Michele degli Scalzi 2, Pisa, Italy.

FAVOUR, Cutting Broad, Am. physician; b. Tereva, Ariz., July 19, 1913; s. Richmond and Dorothy (Chapman) F., Jr.; A.B., Hendrix Coll., 1936; M.D., Johns Hopkins, 1940; m. Barbara Hope Griffin, Sept. 6, 1941; children—Joanne Barbara, Emily Howe, Paul Cutting. Successively med. resident, jr. asso. medicine, asso., sr. asso. and cons. bacteriology and immunology Peter Bent Brigham Hosp. 1942-54; also with Harvard Med. Sch., 1941-54, research fellow, asst., instr. asso. medicine; cons. bacteriology Channing Home for Tb, 1948-54; lectr. medicine Simmons Coll., 1947-54; attending medicine Rutland Vets. Hosp., 1949-54; cons. Tb, infectious diseases Boston Lying-In Hosp., 1952-54; cons. medicine VA Hosp., West Roxbury, Mass., 1953-54; pvt. practice, Boston, 1943-54, Palo Alto, Cal., 1954-60; asst. vis. physician Stanford Service, Dept. Pub. Health City and County Hosp. of San Francisco, 1955-60; asst. clin. prof. Stanford, 1955-60; staff Palo Alto Hosp., 1955-60, Palo Alto Med. Clinic, 1955-60; cons.-lectr. internal medicine USN Hosp., Oak Knoll, Cal., 1956-60; dir. dept. immunology Palo Alto Med. Research Found., 1955-60; prof. preventive medicine, chmn. dept., also asso. prof. medicine Georgetown U. Sch. Medicine, 1960-62; chief Georgetown Med. Service, D.C. Gen. Hosp., 1960-62; now chief dept. epidemiology Nat. Jewish Hosp. Denver; vis. investigator Rockefeller Inst. Med. Research, 1946-47. Diplomate Am. Bd. Internal Medicine. Mem. Society Experimental Biology and Medicine, Am. Fedn. Clin. Research, Tissue Culture Assn., A.M.A., A.A.A.S., Soc. Am. Bacteriologists, Am. Rheumatism Assn., Am. Trudeau Soc., Am. Heart Assn., N.Y. Acad. Scis., Am. Acad. Allergy, Soc. Nuclear Medicine, Western Soc. Clin. Research. Studies and publs. on immunity and allergic reactions to infection, particularly Tb; homotransplantation; adaptation to high altitude. Home: 3998 S. Hillcrest Dr. Office: 3800 E. Colfax Av., Denver 80206.

FAVRE, Alexandre Jean Auguste, French physicist; b. Toulon, France, Feb. 23, 1911; s. Auguste Edouard and Annie (Mercure) F.; Baccalaureat Mathématiques, 1927; Engenieur EIM, 1931; Licencié ès scis., 1932; Dr. ès Scis., Paris, 1937; m. Luce Palombe, Dec. 1, 1939; children—Christian, Elyette, Nadine. Mem. Faculty Scis., U. Marseilles (France), 1932-38, 1932-—, prof., 1951-—, dir. Institut de Mecanique Statistique de la Turbulence, 1960-—. Sci. adviser Nat. Office Airspace Studies and Research, 1947, Atomic Energy Commissariat, 1958; mem. Nat. Com. for Sci. Research, 1963. Recipient Prix Marquet, Acad. Scis. Mem. Marseilles Acad. Scis., Letters and Arts, Math. Soc. France, French Soc. Physics, Am. Phys. Soc. Author books, 1938, 53, 62, also articles. Inventor hypersustentation by moving wall, 1934, hyperconvention by moving wall, 1951, centrifugal sub-trans-supersonic compressor, 1940 (used for aircraft jet engines since 1944), apparatus for statis. measurement of correlation coefficient with time delay, 1942, appliance for detection of random noise from periodic signals by autocorrelation, 1952; initiator space-time correlation measurements in turbulent flows, 1942-—; research on devel. of statis. equations for turbulent compressible gas, 1948-—. Home: 122, rue du Commandant Rolland -13 Marseille 3, France.*

FAVRE, (Jean)-Alphonse, Swiss geologist; b. Geneva, Switzerland, Mar. 30, 1815; s. Guillaume Favre; prof. geology U. Geneva; mem. French Acad. Scis., 1879. Author: Description géologique du Canton de Genève, 1879; Carte du phénomène erratique, 1884; described geology of Geneva area, also studied glacial deposits and erratic phenomena in Alps. Died Geneva, July 11, 1890.

FAVRE, Henry George, Swiss civil engr.; b. Geneva, Switzerland, June 10, 1901; s. William Jules and Aimée (Loegel) F.; diploma civil engring. Swiss Fed. Inst. Tech., Zurich, Switzerland, 1923, Dr.ès.-sc.techn., 1929; Dr.h.c., U. Lausanne (Switzerland), U. Poitier (France), U. Liège (Belgium); m. Violette Sautter, Mar. 25, 1926; children—Jean-Pierre (dec.), Nicole (Mrs. Alfred Berchtold), Renaud. Engr., E. Gruner, Basel, Switzerland, also S. A. Züblin & Cie, Strasbourg (France), 1924-27; with Swiss Fed. Inst. Tech.,

1927——, prof. mechanics in French lang., 1938——, dean Faculty Civil Engring., 1944-48, chancellor, 1951-53. Cons. to various coms. in connection with nav. on canals, to various internat. assns. Recipient medal U. Brussels; medal Gustave Trasenster, Liège. Mem. Assn. des Ingénieurs dipl. par l'Universite de Liège (hon. pres.), Groupe Français de Rhéologie. Author: Cours de Mécanique I (Statique), 1946, vol. II (Dynamique), 1947, vol. II (Chapitres choisis), 1949; also articles. Editor: (with F. Stüssi) Oeuvres complètes de L. Euler, 1947. Discovered interferometric method in photoelasticity; studies in hydraulics, theory of elasticity, vibrations, plates and shells and viscoelasticity; history of mechanics. Died May 29, 1966.*

FAVRE, Pierre Antoine, French physicist; b. Lyons, France, Feb. 20, 1813; M.D., 1835; D.Sc. in physics, 1853; asst. prof. Ecole de Medecine, Paris; prof. Marseilles, France, from 1854. Mem. French Acad. Scis., 1863 (Jecker prize 1869, Lacaze prize 1875). Research on reciprocal reactions and their calorimetric results; measured heat in work done by batteries, thermal changes in magnetization, induction under various conditions (thus confirming law of conservation of energy), electromotive forces of solutions of salts of alkali metals; found that chem. and elec. work are unusally unequal and that the reaction determines which is greater. Died Marseilles, Feb. 17, 1880.

FAWAZ, George, pharmacologist; b. Deirmimas, Lebanon, Nov. 22, 1913; s. Amin and Mariam (Hage) F.; B.A., Am. U. Beirut, 1933, M.S., 1935; Ph.D., U. Graz, 1936; student U. Gottingen, 1936-38; M.D., U. Heidelberg, 1955; m. Eva Niemiec, Aug. 26, 1946; children—Katherine Anne, Julia Christine. Came to U. S., 1946, naturalized, 1951. Faculty, Am. U. Beirut, 1939——, prof., chmn. dept. pharmacology, 1953——; Rockefeller fellow in chemistry Harvard, 1946-47; vis. prof. chemistry U. Cal. at Berkeley, 1962-63. Recipient Fed. German Merit Cross, 1964. Mem. Soc. Exptl. Biology and Medicine, German Pharmacol. Soc. (corr. mem.), Alpha Omega Alpha. Research, publs. on diuretics, effect of drugs and hormones on heart, sympathomimetics, antimalarials; phosphorus compounds. Home: Saidi Bldg., Rue Mimari, Ras. Beirut, Beirut, Lebanon.*

FAWCETT, Edward, Brit. anatomist; b. May 18, 1867; s. Thomas Fawcett; B.M., Master of Surgery, M.D., U. Edinburgh, Scotland; m. Edith Dora Wordsworth; children—1 son, 1 dau. Demonstrator anatomy U. Edinburgh, 1887-89, Yorkshire (Eng.) Coll., 1890-93; prof. anatomy U. Coll., Bristol, Eng., 1893-1909, dean, faculty medicine, 1909-34, emeritus, 1934-42. Fellow Royal Soc., 1923. Author several books on eccles. architecture, also wrote on devel. human skeleton and mammalian chrondo-cranium. Died 1942.

FAWCETT, Harold Pascoe, mathematician, educator; b. Upper Sackville, New Brunswick, Can., July 20, 1894; s. George Albert and Agnes (Carey) F.; came to U. S. 1914, naturalized 1919; A.B., Mount Allison U. (N.B., Can.), 1914; tchrs. certificate U. Cal. at Berkeley, 1916; A.M., Columbia, 1924, Ph.D., 1937; m. Frances Muriel Harper, Dec. 30, 1919 (dec. June 1965); children—Dorothy (Mrs. Leon Zechiel), Winifred (Mrs. William Slocum), Helen (Mrs. Charles Cody Hall). Tchr. high sch., Ft. Fairfield, Me., 1914-15, YMCA Sch., San Francisco 1916-17, Home Study div. YMCA Schs., N.Y.C., 1919-24; instr. Columbia, 1924-32; from asst. prof. to prof. edn. Ohio State U., Columbus, 1932-64, asso. dir. U. Sch., 1938-41, chmn. dept. edn., 1948-56. Vis. prof. Northwestern U., 1935-40, Columbia, 1947, U. Va., 1954, U. Mich., 1955, U. Utah, 1956; edn. cons. Coronet Instructional Films, Chgo., 1947——; math. cons. Charles E. Merrill Books, Inc., 1958——. Recipient Ohio State U. Alumni award for Distinguished Teaching, 1961, Alfred J. Wright award, 1964. Mem. Math. Assn. Am., Nat. Soc. for Study Edn., John Dewey Soc., Ohio, Nat. (bd. dirs. Yearbook com. 1952-55, pres. 1958-60) councils tchrs. math. Author: The Nature of Proof, 1938; (with others) Mathematics in General Education, 1940; Advanced High School Mathematics, 1961; Geometry: A Unified Course, 1961; Algebra One, Algebra Two, 1962; also numerous articles. Developed new techniques in teaching math. especially geometry. Address: 4059 Roselea Pl., Columbus, O. 43214.*

FAWCETT, Howard S(amuel), Am. plant pathologist; b. Salem, O., Apr. 12, 1877; s. Thomas F. and Sidney Ann Bonsall) F.; B.S., Ia. State Coll., 1905; M.S., U. Fla., 1908; Ph.D., Johns Hopkins, 1918; m. T. Helen Tostenson, Sept. 15, 1909; 1 dau., Rosamond Annette. Asst. prof. botany and horticulture U. Fla., 1905-06; plant pathologist Fla. Agrl. Expt. Sta., 1906-11, Cal. Commn. Horticulture, 1911-13; asso. prof. plant pathology U. Cal., 1913-18, prof. from 1918; collaborator U. S. Dept. Agr. for study of citrus and date diseases, N. Africa and Mediterranean lands, 1929-30. Fellow A.A.A.S.; mem. Bot. Soc. Am., Phytopathol. Soc. (pres. 1930), Am. Soc. Naturalists, Western Soc. Naturalists, Phi Beta Kappa, Sigma Xi. Author: Citrus Diseases and Their Control (used by the citrus industry as authority), 1926; revised, 1936; also many sci. papers and bulls. Died Dec. 12, 1948.

FAWNES, George, Brit. chemist; b. Eng., May 14, 1815; Ph.D., Giessen, Germany; asst. to prof. Graham, Univ. Coll., until 1840; lectr. chemistry Charing Cross Hosp.; became prof. chemistry Pharm. Soc., 1842, Univ. Coll., 1846. Recipient prize, Royal Agrl. Soc., 1842; Actonian prize; medal Royal Soc. Sec. Chem. Soc. Author gen. text-book of chemistry, 1844; several papers. Discovered benzoline; 1st artificially to produce vegeto-alkali or organic saltbase (furfurine). Died Jan. 31, 1849.

FAWSITT, Charles Edward, chemist; b. Glasgow, Scotland, 1878; s. Charles Albert Fawsitt; ed. U. Birmingham; D.Sc.; U. Edinburgh; Ph.D., U. Leipzig; B.Sc., U. London; postgrad. Technische Hochschule, Aachen, Germany; m. Lena Gertrude Gardner, 1909; 1 dau. Carnegie research fellow, 1903-04; asst. to prof. chemistry Edinburgh U., 1902-04; asst. to prof. chemistry Glasgow U., 1904-07; lectr. metall. chemistry, 1904-08; prof. chemistry U. Sydney (Australia), 1909-46, dean faculty sci., 1923-29. Pres. Royal Soc. New S. Wales, 1919-20. Author: Tables for Students of Quantitative Chemical Analysis, 1902; also papers. Died Nov. 16, 1960.

FAYE, Hervé Auguste Étienne, Albans, French astronomer; b. St. Benoit-du-Sault, France, Oct. 1, 1814; ed. Obs. École Polytechnique, Paris, also in Holland, Paris Prof. geodesy, École Polytechnique, 1848-54; rector, prof. astronomy, Nancy, France, 1855-73; inspector-general of higher edn., 1857-62; apptd. dir., Paris Obs., 1878; became minister pub. edn., 1878. Mem. French Acad. Scis., 1847; Bur. Longitudes (pres. 1876); grand officer, Legion of Honor, 1889. Contbr. to Sci. jours. Discovered comet named after him, 1843; developed theory connecting sunspots with weather, 1872; inventor zenithal collimator; improved methods of astron. measurement; applied photography and electricity to astron. study; added to knowledge of sunspots, comets, meteors. Died Passy, nr. Paris, July 4, 1902.

FAYE, Paul-René, French veterinarian; b. Angoulème, France, Nov. 6, 1914; s. Emile and Renée Faye; Doctorat vétérinaire, Alfort (France) Vet. Sch., 1941; m. Lucette Hinglais, Mar. 28, 1945; children—Catherine, Laurent, Manuele, Pascale. Practice vet. medicine, 1943-48; vet. adviser Nat. Ovine Fedn., also dir. Profl. Center for Experimentation on Diseases of Sheep, 1949-58; chief engr. vet. research Dept. Cattle Pathology Lab., Alfort, 1958——; temporary lectr. Alfort Sch., 1959——. Decorated Chevalier Agrl. Merit. Mem. French Assn. Vet. Microbiologists. Contbr. sci. and popular articles. Research on viruses and intestinal diseases of young ruminants; systematic study of humors in sheep. Home: 22, Servandoni, Paris. Office: Ecole Vétérinaire, Alfort, Seine, France.*

FAYET, Gaston Jules, French astronomer; b. Paris, France, June 5, 1874. With Obs. of Paris, 1889; with Obs. of Nice, 1911, dir., 1917. Mem. French Acad. Scis., Bur. Longitudes. Research on position astronomy, determination of longitudes, measures of double stars, orbits of comets and asteroids; perfected several instruments and methods of observation. Address: 3 villa Hallé, Paris 14, France.

FAYOL, Henri, mgmt. scientist, engr.; b. Constantinople, Turkey, 1841; ed. Lycée, Lyons; School of Mines, St. Etienne, 1860. Engineer, 1860-66; mgr., Commentry pits, 1866-72; genl. mgr., Commentry, Montirco, Berry mines, 1872-88; managing dir., Commentry-Fourchambault, 1888-1918. Author: Administration et industrielle générale, 1916; Enseignement technique superieur, 1918; Incapacité industrielle de l' État, 1921; L'Eveil de l'Esprit Public, 1927. Advanced important seminal ideas in several areas of mgmt. sci., including specialization, unity of direction and command, and responsibility as a corollary of authority. Died Paris, France, 1925.

FAYRER, Sir, Joseph, Brit. surgeon; b. Plymouth, Eng., 1824; s. Robert J. and Agnes (Wilkinson) F.; ed. London, Eng., Edinburgh, Scotland; m. Bethia Spens, 1855; children—4 sons, 1 dau. With Bengal Med. Service, from 1850, served 1st Burmese War, Indian Mutiny, defense of Lucknow garrison; prof. Med. Coll., Calcutta, India; pres. Med. Bd., India Office, 1874-95; physician to Duke of Edinburgh (Saxe-Coburg Gotha); physician extraordinary to king, from 1901; surgeon-gen. Recipient Jubilee medal and clasp, Coronation medal, others. Fellow Royal Soc., 1877. Author: The Thanatophidia of India; Clinical Surgery in India; Clinical and Pathological Observations in India; The Royal Tiger of Bengal; Tropical Diseases; Climate and Fevers of India; Preservation of Health in India. Described venomous snakes of India, 1872. Died May 21, 1907.

FAZIO, Cornelio, Italian physician; b. Garessio, Italy, Mar. 30, 1910; s. Egidio and Giuseppina (Covino) F.; grad. in Medicine and Surgery, U. Turin (Italy) 1935; m. Eleonora Lindner, Aug. 12, 1943; children—Ferruccio, Mara. Staff, Neuropsychiat. Clinic, U. Genoa (Italy) 1937——, prof. neuropsychiatry, 1955——. Recipient Premio Lugaro for research on red softening of brain, 1948. Mem. Collegium Internationale Neuropsychopharmacologicum (bd. dirs.), Soc. Française de Neurologie, Soc. Internationale de Psychopathologie de l'Expression, Tableau d'experts en santé metale, World Fedn. Neurology, Deutsche Gesellschaft für Neurologie. Editorial bd. Sistema Nervoso, 1954——, Rivista Sperimentale di Freniatria, 1958——, Informatore, Medico, 1957——, Archivio Maragliano, 1957——. Research on histology of central nervous system vascularization, especially brain and spinal cord, 1934-42; anatomical and pathophysiol. research on cerebral vascular disorder, 1940——; classification and treatment of depressive states, 1957——; physiology and pharmacology of cerebral circulation, 1958——; psychopharmacology, 1961——. Home: 2 Via S. Giuliano, Genoa, Italy.*

FEA, Carlo, Italian archaeologist; b. Pigna, Piedmont, Feb. 2, 1735; studied ancient monuments in Rome. Sec. to Prince Chigi; Pope Pius VII asked him to continue investigations begun by French, 1814; dir. of excavations, Rome, 1800-34. Author: l'Integrità del Panteon, 1807; Descrizione de Roma e dei Contorni con redute (3 vol.), 1824. A founder of study of Roman topography; works on Roman epigraphy; translated Winckelmann's art history into Italian, 1783-6. Died Rome, Italy, Mar. 18, 1834.

FEARN, Anne Walter (Mrs. John B. Fearn), Am. physician, surgeon; b. Holly Spring, Miss., May 21, 1865; d. Harvey Washington and Martha Fredonia (Brown) Walter; M.D., Women's Med. Coll., 1893; postgrad. N.Y. Polyclinic, U. London, Eng., U. Vienna, Austria; m. John B. Fearn, Apr. 21, 1896 (dec. 1926); 1 dau., Maddie Elizabeth. Practice medicine Soochow, China, from 1893, Shanghai, China, from 1915. Mem. Internat. Council Women, Royal Asiatic Soc., Austro-Am. Inst. Coll. Author: My Days of Strength, 1939. Founder co-ednl. med. sch., Soochow, Fearn Sanitarium, Shanghai, 1915, also China's 1st sun room for Tb patients. Died Berkeley, Cal., Apr. 28, 1939.

FEARON, Robert Earl, Am. physicist; b. Spokane, Wash., Nov. 17, 1912; s. Edwin H. and Edith (Thompson) F.; student Carnegie Inst. Tech., 1929-30, Muskingum Coll., 1930-31; B.S., U. Pitts., 1933, M.S., 1934, postgrad.; m. Ruth B. Strothers, Feb. 23, 1938; children—Lee C., Anne C., (Mrs. A. H. Gill), George C., Edward R., Martha R. (Mrs. James Luther Mims III). Research physicist Engring. Labs., Inc., Tulsa, 1938-39; with Well Surveys, Inc., Tulsa, 1939-40, dir. research, 1944-52; with Stanolind Oil & Gas Co., Tulsa, 1940-44; pres., dir. research, also dir. Electro Chem. Labs. Corp., Tulsa, 1952——. Faculty, U. Pitts., 1933-36, U. Tulsa, 1953-58. Mem. Am. Chem. Soc., Am. Geophys. Union, Am. Inst. Aeros. and Astronautics, Am. Inst. Mining, Metall. and Petroleum Engrs., Am. Inst. Physics, A.A.A.S., Internat. Briquetting Assn., Soc. Exploration Geophysicists, Soc. Am. Mil. Engrs. Research on radiation measurement, leakage testing, chem. analysis, fire starters. Home: 5246 S. 76th East Av., Tulsa 74145. Office: 530 S. Lewis St., Tulsa 74104.*

FEARON, William Robert, biochemist; b. 1892; s. William Fearon; ed. Trinity Coll., Dublin, Sc.D., 1919; B.A., Emmanuel Coll., Cambridge, 1921; M.A., M.B. Research worker Food Ministry and Food Investigation Bd., 1917-19; Mackinnon research student Royal Soc., 1919-21; sr. fellow Trinity Coll., Dublin; prof. biochemistry U. Dublin; physiologist Royal City of Dublin Hosp.; cons. biochemist Rotunda Hosp. Dublin; former mem. Med. Research Council of Ireland. Hon. prof. chemistry Royal Hibernian Acad. Recipient Harvey Research prize Royal Coll. Physicians Ireland, 1918, Carmichael prize Royal Coll. Surgeons Ireland, Buckston Browne prize Harveian Soc., 1935. Fellow Royal Inst. Chemistry, Royal Acad. Medicine (Ireland); mem. Royal Irish Acad. Author: An Introduction to Biochemistry, 3d edit.; 1948; Nutritional Factor in Disease, 1936; Parnell of Avondale, 1937, also articles. Died Dec. 27, 1959.

FEATHERSTONE, Robert Marion, Am. pharmacologist; b. Anderson, Ind., Dec. 24, 1914; s. Marion L. and Adah (Brown) F.; B.A., Ball State U., 1940, LL.D., 1962; M.S., State U. Ia., 1942, Ph.D., 1943; m. Joyce Amanda Byrum, Aug. 31, 1940; children—David, Jean, James, Judith. Instr. biochemistry Med. Coll. S.C., 1943-44; with U. Ia. Coll. Med., 1944-57, prof., 1955-57; prof., chmn. dept. pharmacology U. Cal. Sch. Medicine, San Francisco 1957——; vis. lectr. Middlesex Hosp. Med. Sch., London, 1965-66. Cons. physiology rare gases Linde Co., 1960——; cons. U. S. Dept. Health, Edn., Welfare, 1960——. Commonwealth Fund fellow, 1965-66. Fellow Muncie Acad. Medicine, Royal Soc. Medicine (London); mem. Am. Soc. for Pharmacology and Exptl. Therapeutics, Am. Chem. Soc., A.A.A.S., Soc. Exptl. Biology. Author: (with John Hidalgo) Farmacologia del Sistema Nervibso Autonomo, 1963. Editor: (with A. Simon) Pharmacological Approach to the Study of the Mind, 1959; (with S. C. Cullen) Mechanisms of Anesthesia, 1963. Research, publs. on use of xenon for study of anesthesia, enzyme inducement in vitro and its relation to tumor devel. Home: 15 Corlett Way, Hillsborough, Cal. 94010.*

FEATHERSTONHAUGH, George W., geologist; b. Eng., 1780; m. an American; 2 daus.; came to U. S. and settled in Daunesburg, N.Y.; dir. ry. from Albany to Schenectady; established Monthly Am. Jour. Geology and Natural History, Phila., 1830; returned to Eng., circa 1840; mem. commn. to settle internat. boundary dispute with Can.; Brit. consul, Seine, France. Commd. by U. S. War Dept. to survey geology of Ozark Mountains, 1834-35; surveyed area between Washington, D.C., and Coteau des Prairies, by way of Green Bay and Wis. territory, 1835-36.

Fellow Royal Soc., 1835. First to suggest Am. continent may be older than European; 1st to discuss sediment of Mississippi River as means for determining chronology. Died 1866.

FEBRER CARBO, Joaquin Luis Rafael, Spanish meteorologist; b. Benicarlo, Sept. 26, 1893; s. Joaquin and Julia (Carbo) F.; Ph.D. in sci., U. Barcelona; m. Carolina N. Febrer Carbo, Oct. 24, 1923; children—Maria Nuria, Joaquin. Titular, U. Barcelona, dean Sch. Sci.; dir. astronomy Fabra Obs., Barcelona; asst., meteorologist Meteorol. Service of Catalogna, 1921-32; dir., prof. Febrer Acad. Mem. Spanish and Am. Astronomy Soc. (pres.), Royal Acad. Arts and Scis. of Barcelona (accademician), Internat. Astron. Union, Nat. Commn. Astronomy. Author: Leçons d'Astronomie elementaire; Contribution a l'etude statistique des asteroides. Made pluviometric atlas of Catalogna. Address: calle Guillermo Tell, 49, Barcelona, Spain.

FECHNER, Gustav Theodor, German philosopher, physicist, exptl. psychologist; b. Gross-Sarchen, Saxony, Apr. 19, 1801; ed. Dresden, also U. Leipzig. Prof. physics U. Leipzig, worked chiefly on galvanism, electrochemistry, color, 1834-39, retired; later concentrated on natural philosophy, anthropology, psychology. Author numerous works including: Bewiss, dass der Mond aus Iodine besteht, 1821; (under pseudonym, Dr. Mises) Vergleichende Anatomie der Engel, 1825; Zendavesta, 3 vols., 1851; Elemente der Psychophysik, 2 vols., 1860; Über die Seelenfrage, 1861; Einige Ideen zur Schöpfungsund Entwicklungs- Geschichte, 1873; Vorschule der Ästhetik, 1876; In Sachen der Psychophysik, 1877; Über die physhischen Massprinzipien und das Webersche Gesetz, 1887; wrote poetry and humor under pseudonym, Dr. Mises. Founder of psychophysics, also of exptl. psychology; attempted to make psychology an exact science; known for Weber-Fechner law, which gives math. relationship between intensity of sensation and stimulus; also worked in aesthetics. Died Leipzig, Nov. 18, 1887.

FEDDE, Friedrich, German botanist; b. Breslau, Germany (now Wroclaw, Poland), June 30, 1873; s. Friedrich and Anna (Mittelhaus) F.; doctorate, Breslau, 1896; state certification exam, 1898; m. Käte Woywode, 1902; 2 sons, 1 dau. Univ. tchr., Breslau, Tarnowitz, Berlin (all Germany), until 1924; became colleague Berlin Bot. Mus., from 1901; became prof., 1912. Collaborator Justs Botanischer Jahresbericht, editor from 1903; founder, pub. Repertorium specierum novarum regni vegetabilis, 1905, Repertorium europaeum et mediterraneum, 1913. Contbr. numerous articles to profl. jours. Publs. on nomenclature, 1904-40, on sci. taxonomy, 1928-37; research on anatomy, ecology, taxonomy of corydalis species (family Papaveraceae). Died Berlin—Dahlem, Mar. 14, 1942.

FEDDERSEN, Berend Wilhelm, German physicist; b. Schleswig, Germany, Mar. 26, 1832; s. Berend and Adelheid (Schmidt) F.; student, Göttingen, Germany, 1851-54, Berlin, 3 semesters; doctorate, Kiel, Germany, 1857; m. Dora Feddersen, 1866; m. 2d, Helga Kjaer, 1890. Lived in Switzerland briefly; settled as ind. scholar in Leipzig, Germany. Mem. Saxon Soc. Scis. Contbr. articles to Poggendorff. Editor: (with A. von Oettingen) Poggendorffs biographisch-literarisches Handwörterbuch, 3d vol., 1898. Research on electric waves; proved with expts. that discharging of Leyden flask involves vibration phenomena whose frequence depends on exptl. conditions. Died Leipzig, July 1, 1918.

FEDE, Francesco, Italian pediatrician; b. Petrella, Italy, 1832; ed. U. Naples. Founded journal, La Pediatria, 1893. 1st described sublingual fibroma, which occurs in infants due to abrasion of membrane under tongue (called Fede-Riga disease), 1890. Died 1913.

FEDER, (Johann) Georg Heinrich, philosopher; b. Schornweissbach, nr. Neustadt/Aisch, May 15, 1740; s. Martin Heinrich and Feder; doctorate, Erlangen, Germany, 1765; m. Sophie Häublein, 1767; 3 children; m. 2d, Margarethe Dorothea Best, 1773; children—Karl August Ludwig, Isabella Dorothea Luise, Jeanette Louise Phil. Became pvt. tutor, 1760; named prof. metaphysics, moral and logic Casimirianum, Coburg, Germany, 1765; became prof. philosophy U. Göttingen (Germany), 1767 (resigned); dir. Georgianum, also Royal Library, Hanover, Germany, 1797-1811. Mem. Göttingen Soc. Author: Grundriss der philosophischen Wissenschaft, 1767; Untersuchungen über den menschlichen Willen (1st doctrine on psychol. types), 1779-93; Prolegomena zu einer jeden künftigen Metaphysik, 1783; numerous other philos. works. Held that age of psychology would soon supersede that of philosophy. Died Hanover, May 22, 1821.

FEDER, William Adolph, Am. plant pathologist; b. N.Y.C., Oct. 15, 1920; s. Daniel and Lulu (Wohlauer) F.; A.B., Johns Hopkins, 1941; Ph.D., U. Cal., Berkeley, 1950; m. Leanora M. Siegman, July 29, 1945; children—Eric, Susan, Sandra. Storage disease pathologist U. Hawaii, 1950-51; asso. prof. plant pathology Cornell U., Ithaca, N.Y., 1951-54; research plant pathologist Agrl. Research Service, U. S. Dept. Agr., Orlando, Fla., 1954-66; prof. plant pathology U. Mass., Waltham, 1966—. NSF Sr. Postdoctoral fellow, London, 1958-59; Fulbright Sr. Research fellow Nat. and U. Inst. Agr. Hebrew U., Israel, 1964. Mem. Am. Phytopath. Soc., A.A.A.S., Air Pollution Control Assn., Sigma Xi. Contbr. numerous articles to profl. jours. Developed control methods for foliage and root diseases of crop and ornamental plants, first nematode-resistant citrus rootstocks for use in Fla. groves. Home: 42 Alton Ct., Brookline, Mass. 02146. Office: 240 Beaver St., Waltham, Mass. 02154.*

FEDERER, Walter T(heodore), Am. statistician; b. Cheyenne, Wyo., Aug. 23, 1915; s. John H. and Tony Maria (Brambora) F.; B.S. in Agronomy, Colo. State U., 19—; M.S., in Plant Breeding, Kan. State U., 1941; Ph.D. in Math. Statistics, Ia. State U., 1948; m. Lillian Elizabeth Vasey, Dec. 18, 1945; 1 son, Arthur John. Prof. biol. statistics Cornell U., Ithaca, N.Y., 1948—; asso. geneticist Spl. Guayule Research Project, U. S. Dept. Agr., Salinas, Cal., 1942-44; asso. agrl. statistician Agrl. Marketing Service, U. S. Dept. Agr., collaborator Ia. State U., Ames, 1944-48; head expl. statistics Hawaiian Sugar Planters' Assn. Expt. Sta., cons. Pineapple Research Inst., Honolulu, 1954-55; prof. math. research center U. Wis., Madison, 1962-63. Chmn., exec. sec. Com. Pres.'s Statis. Socs., 1965-67. Fellow Am. Statis. Assn., A.A.A.S., Royal Statis. Soc.; mem. Biometric Soc. (past sec., pres.), Inst. Math. Statistics, Indian Soc. Agrl. Statistics. Author: Experimental Design Theory and Application, 1955; also numerous articles. Statis. procedures for summarizing results of a series of expts., covariance, unequal numbers analyses, variance component estimation, statis. genetics and breeding, constrn. several classes of exptl. designs, treamtent designs. Home: 207 Homestead Terrace, Ithaca, N.Y. 14850.*

FEDERIGHI, Tiziano, Italian physicist; b. Pistola, Italy, Apr. 2, 1927; s. Attilio and Assunta (Fanciullacci) F.; Ph.D. in physics U. Florence, Italy, 1952, L.D., 1960; m. Ebe Ascari; 1 dau. With Instituto Sperimentale Del Metalli Leggeri, Novara, Italy, 1953—, chief metal physics sect., 1960—. Mem. Italian Phys. Soc., Italian Assn. Metallurgy. Contbr. numerous articles to sci. jours. Discovered role of point defects as vacancies in controlling aging behaviour of Al-rich alloys; knowledge of properties of point defects in Al and Al alloys, introducing defects by quenching, neutron-irradiation and cold-working; clarification of aging properties of Al-rich alloys. Home: 11 Via Monte Grappa, Novara, Italy. Office: C.P. 129—I.S.M.L., Novara, Italy.*

FEDERN, Josef (Salomon), physician; b. Prague, Czechoslovakia, Nov. 20, 1831; s. Elias and Ester (Mislap) F.; M.D., U. Vienna (Austria), 1859; m. Ernestine Spitzer, 1867; children—Karl, Paul, Hans Robert, Etta Federn-Kohlhaas. Practiced medicine, Vienna (patients included Rokitansky); contbd. articles to med. jours.; research on measurement and evaluation of blood pressure, physiology and pathology of vascular system, especially viscera; discovered necessity of measuring blood pressure in the sick. Died Vienna, Nov. 9, 1920.

FEDINETS, Aleksandr Vasilevich, Russian surgeon; b. 1897; grad. Med. Faculty, Prague U., 1924; Cand. Med. Sci., 1952. Intern surg. dept. city hosp., Transcarpathia, 1924-28; asst. surg. clinic. Bratislava U., 1928-32; surgeon hosps., Transcarpathia, 1932-45; with Soviet Health Service, 1945-52; lectr., head chair faculty surgery Med. Faculty Uzhgorod U., 1952-61, prof., head chair hosp. and gen. surgery, 1962—. Author 50 articles on surgery. Address: Uzhgorod University, pl. Gorkogo 1-3, Uzhgorod, Ukraine SSR, USSR.

FEDOR, Edward J., Am. physiologist; b. Braddock, Pa., Feb. 15, 1924; s. Michael and Anna (Katus) F.; B.S., Wanesburg Coll., 1947; M.S., W.Va. U., 1950; M.A., Princeton, Ph.D., 1953; Instr., research asso. U. Tenn. Med. Sch., Memphis, 1953-54; asst. research prof. surgery U. Pitts. Med. Sch., 1954-61, asst. dir. Lab. Surg. Research; sr. physiologist Wallace Labs., Cranbury, N.J., 1961-64, Abbott Labs., North Chicago, Ill., 1964—. Proctor fellow Princeton. Mem. Am. Physiol. Soc., Soc. Nuclear Medicine, Soc. for Exptl. Biology and Medicine, A.A.A.S., Am. Assn. U. Profs., N.Y. Acad. Scis., Sigma Xi. Research, publs. on stress, hypertension, organ transplants, fibrinolysis, myocardial infarcts, physiology of hypothermia. Home: 4204 Vista View, West Mifflin, Pa. Office: Sheridan Rd., North Chicago, Ill. 60064.*

FEDORCHENKO, Ivan Mikhaylovich, Ukrainian metallurgist; b. 1909; grad. Kamensk Worker's Evening Inst., 1930; D.Tech. Sci. Engr. various metall. plants, 1930-35; with Research Inst., Ministry Agrl. Machine-Bldg., Moscow, 1935-52; dep. dir. Inst. Powder Metallurgy and Spl. Alloys, Ukrainian Acad. Sci., 1952-57, asso., 1957-61; instr. Kiev Poly. Inst., 1953-55, prof., 1954—. Mem. Ukrainian Acad. Scis. (chief learned sec. 1957-62). Research and publs. on powder metallurgy, properties of spl. steels, heat treatment of steel. Address: Materials Sciences Institute, Ukrainian Acad. Scis., Vladimirskaya ulitsa 54, Kiev, Ukrainian SSR, USSR.

FEDOROFF, Sergey, cell biologist; b. Daugavpils, Latvia, Feb. 20, 1925; s. Paul and Nina (Birokowa) F.; B.A., U. Sask., 1952, M.A., 1955, Ph.D., 1958; m. Muriel Elaine Martin, Aug. 7, 1954; children—Paul, Andrey, Marina, Michael. Faculty, U. Sask. (Can.), Saskatoon, 1955—, prof., head dept. anatomy, 1964—, adminstrv. asst. to dean medicine, 1960-62. Recipient Lederle Med. Faculty award, 1957-60. Mem. Canadian Assn. Anatomists (past pres.), Tissue Culture Assn. (v.p. 1964—), Genetics Soc. Can., Am. Assn. Immunologists, Canadian Soc. Immunologists, N.Y. Acad. Scis., Am. Assn. Anatomists, Am. Assn. Cell Biologists, Canadian Soc. Cell Biologists, Internat. Soc. Cell Biologists, Soc. for Exptl. Biology and Medicine, Transplantation Soc. Research and publs. on toxicity of blood serum to cells of various animal species leading to discovery of heterophile antigen-antibody system in which antigens are present on cell surfaces of several animal species including some bacteria and antibodies which are present in human blood serum. Home: 36 Cantlon Crescent, Saskatoon, Sask., Can.*

FEDOROV, Alexander Konstantinovich, Russian geneticist, plant physiologist; b. Leningrad, USSR, Aug. 25, 1923; s. K. F. and M. I. (Sosova) F.; agronomist in plant breeding Timiryazev Agr. Acad., Moscow, USSR, 1951; cand.biol.sci., Genetic Inst. Acad. Sci. USSR Moscow, 1955. Scientist, Genetic Inst. USSR Acad. Scis., 1955-62, sr. scientist, 1962-65; sr. scientist main bot. garden USSR Acad. Scis., 1965—. Mem. Znanie (methodic council). Author: Peculiar Characters of Development of Wintering Plants, 1959; Biology of Perennial Grasses; also numerous articles in jours. and monographs. Editorial bd. Selskohozjastvennaya Biologie, 1966—. Research on ontogenesis of cereals and perennial grasses, effect of light on devel. of these plants. Home: USSR, Moscow, Efremov str. 18, Kv. 7. Office: USSR, Moscow, Main bot. garden USSR Acad. Scis.*

FEDOROV, Evgraf Evgtafovich, Russian climatologist; b. Nov. 20, 1880; grad. St. Petersburg U., 1909. With Main Phys. Lab., 1910, Magnetic-Meteorol. Obs., Pavlosvk, 1911-34, Inst. Geog., USSR Acad. Sci., 1934-35. Recipient Order of Lenin, Red Star. Author: Climate as an Aggregate of Weather, 1925; The Distribution of Rain Weather and Its Types on the Plains of the European USSR During Six Months of Summer, 1938; Climate in the Weather of the European USSR, 1949. Research on cloud and solar radiation; developed methods for study of climate by means of simultaneous observation of temperature, humidity, etc. Address: Moskva Leninsky Prosp. 14, AN SSSR, USSR.

FEDOROV, Evgraf Stepanovich, Russian mineralogist; b. Orenburg, Dec. 22, 1853; grad. Mil. Sch. Engring., 1872; prof. Moscow Agrl. Inst., 1895-1905; became 1st dir. Mining Inst., St. Petersburg, 1905. Fellow Acad. Scis. (award in crystallography named in his honor 1944). Author: Russkoe telenidiskoe obshchestvo, 1885-93; Symmetry of Regular Systems of Figures (Fedorov 3-dimensional groups described), 1890; Kurs Kristallografii, 1901; Das Krystallreich, 1920; Nutation and Forced Motion of the Earth's Poles from the Data of Latitude Observations, 1963; also papers on crystallography, geometry, petrography, mineralogy, geology. Studied geology of northern Urals, 1885-90; investigated structure of crystals; used new methods in studies in crystallography. Died Petrograd, May 21, 1919.

FEDOROV, Sergey Filippovich, Russian geologist; b. July 13, 1896; grad. Moscow Mining Acad., 1924. Prof., Moscow Petroleum Inst., 1934-54; asso. USSR Acad. Sci., 1934—. Decorated Order of Lenin; recipient Stalin prize, 1950, 52. Mem. USSR Acad. Sci. (corr., Gubkin prize 1952). Research and publs. on geology of oil, mud volcanism, genetic connection between mud volcanoes and petroleum deposits, developer methods for studying mud volcanoes. Address: USSR Acad. Sci., Leninsky prospect 14, Moscow, USSR.

FEDOROV, Yevgeniy Konstantinovich, Russian geophysicist; b. Apr. 10, 1910; grad. Leningrad U., 1932. Magnetologist polar stas., Franz Josef Land, 1932-33, Cape Chelyuskin, 1934-35; geophysicist, astronomer 1st drifting sta. North Pole I, 1937-38; dir. main bd. hydrometeorol. service USSR Council Ministers, 1939-47, now head main meteorol. bd.; with Geophys. Inst., USSR Acad. Sci., 1947-55, dir. Inst. Applied Geophysics, 1955—, chief acad. sec., 1959-62. Decorated Order of Lenin (2); recipient Stalin prize class II, 1946. Mem. USSR Acad. Sci. Author: Astronomical Determination, 1940; Meteorological Instruments and Observations, 1941; The Effects of Atomic Explosions on Meteor Processes, 1956; The Start of Soviet Science, 1963. Research and publs. on magnetology, meteorology, practical astronomy. Address: Inst. Applied Geophysics, Glebovskaya ulitsa 20-b, Moscow, USSR.

FEDOROV, Yevgeniy Pavlovich, Russian astronomer; b. Irkutsk, 1909; grad. Irkutsk U., 1937. With Poltava Gravimetric Obs., Ukrainian Acad. Sci., 1944-59, dir. Main Obs., 1959—. Mem. Commn. for Study Polar Movement, Internat. Astron. Union, 1951—. Mem. Ukrainian Acad. Sci. (corr.). Address: Main Astron. Observatory, Ukrainian Acad. Sci., Vladimirskaya ulitsa 54, Kiev-127, Ukrainian SSR, USSR.

FEDOROVSKY, Aleksey Aleksandrovich, Russian surgeon; b. Russkaya Khalan, Kursk Guberniya, 1897; grad. Kharkov Med. Acad., 1921; D.Med. Sci. Intern, later asst. faculty surgery clinic Kharkov Med. Inst., 1921-35; sci. dir., Kiev Blood Transfusion Inst.,

1935-41; staff Ukrainian Ministry Health, 1946-48; prof. surgery Kiev Med. Inst., 1949-52, head chair surgery, dean Pediatric Faculty, 1952—; chief hematologist Ukrainian Ministry Health, 1948—. Recipient Burdenko prize. Research and numerous publs. on exptl. and clin. surgery, hematology, blood transfusion and blood substitutes. Address: Kiev Med. Inst., b. Tarasa Shevchenko 13, Kiev, Ukraine SSR, USSR.

FEDYNSKY, Vsevolod Vladimirovich, Russian geophysicist; b. May 1, 1908; grad. Moscow U., 1930. With various prospecting establishments of petroleum industry, 1929-56; former mem. expdns. to study force gravity at sea, gas and petroleum prospecting; prof. Moscow U., 1950—; asso. Inst. Geophysics, USSR Acad. Sci., 1957—. Mem. USSR State Geol. Com. Research and publs. on geol. interpretation of gravitational anomalies, application of geophys. methods in studying structure of earth's crust; inventor gravimeteraltimeter, 1944. Address: Inst. of Geophysics, USSR Acad. Sci., B. Gruzinskaya ulitsa 10, Moscow, USSR.

FÉE, Antoine-Laurent-Apollinaire, French botanist; b. Ardentes, France, 1789; prof. botany, Strasbourg, France, 1832-67. Author: la Flore Virgile, 1822; Essai sur les cryptogames des écorces exotiques medicinales, 1824; Commentaires sur la botanique et la matière médicale de Pline, 1833; Mémoire sur la famille des fougeres, 1844-66. Research and publs. on crytogams of rare barks, French ferns, plants mentioned by Virgil and Pliny, especially their med. uses. Died Paris, 1874.

FEENBERG, Eugene, Am. physicist; b. Ft. Smith, Ark., Oct. 19, 1906; s. Louis and Esther (Siegel) F.; B.A., M.A., U. Tex., 1929; Ph.D., Harvard, 1933; m. Hilda Rosenberg, May 30, 1940; children—Andrew L., Daniel Richard. Instr. physics Harvard, 1933-35; lectr. physics U. Wis., 1935-36; fellow Inst. Advanced Study, Princeton, 1936-38; faculty Washington Sq. Coll., N.Y.U., 1938-46; micro-wave engr. Sperry Gyroscope Co., 1941-45; faculty Washington U., St. Louis, 1946—, now Wayman Crow prof. Mem. Am. Phys. Soc. Author: (with G. E. Pake) Notes on the Quantum Theory of Angular Momentum, 1953; Shell Theory of the Nucleus, 1955. Studies in cross sect. theorem, nuclear forces and structure; cosmic radiation; quantum theory. Home: 7128 Kingsbury St., University City, Mo. Office: Washington U., St. Louis.*

FEER, (Walther) Emil, physician; b. Aarau, Switzerland, Mar. 5, 1864; s. Emil and Louise (Grossmann) F.; studied medicine, Munich, Heidelberg (both Germany), Vienna, Austria; M.D., Basel, Switzerland; m. Rosa Sulzer-Grossmann, 1891; 2 sons, 2 daus. Asst., children's Hosp., Basel, 2 years, then began pvt. practice; became mem. faculty in pediatrics U. Basel, 1894; became prof. U. Heidelberg, 1907; prof. U. Zurich (Switzerland), from 1911, founder observation sta. with glass boxes to isolate children with infectious diseases, also reorganized faculty; ret., 1929. Author: Über angeborene spastische Gliederstarve, 1890; Ätiologie und klinische Beitrage zur Diphtherie, 1894; Lehrbuch der Kinderheilkunde (standard work), 1911; Diagnostik der Kinderkrankhei ten, 1921. Developed clin. instrn. with emphasis on observation at sickbed; research on Feer's neurosis (outside of Germany known by names of earlier describers Chardon, Clubbe, Snowball, Selter, Swift). Died Zurich, Oct. 21, 1955.

FEHER, George, physicist; b. Bratislava, Czechoslovakia, May 29, 1924; s. Ferdinand and Sylvia (Schwartz) F.; came to U. S., 1947, naturalized 1953; B.S. U. Cal., Berkeley, 1950, M.S. in Elec. Engring., 1952, Ph.D. in Physics, 1954; m. Elsa R. Rosenvasser, June 1961; children—Laurie Ruth, Shoshanah, Paola. Research physicist Bell Telephone Labs., Murray Hill, N.J., 1954-60; vis. asso. prof. physics Columbia, 1959-60; prof. physics U. Cal., La Jolla, 1960—. Mem. Am. Phys. Soc. (award for inventing and developing ENDOR technique 1960), Sigma Xi. Office: U. Cal., La Jolla, Cal.*

FEHLEISEN, Friedrich, German physician; b. Reutlingen, Germany, 1854; M.D., U. Tübingen (Germany), 1877; asst., surg. clinic, Würzburg, Germany; asst. to von Bergmann, Berlin; gave up acad. career; credited with discovery of Streptococcus erysipelatis, now known as Streptococcus pyogenes, 1882. Died San Francisco, 1924.

FEHLING, Hermann, German gynecologist; b. Stuttgart, Germany, July 14, 1847; s. Hermann Christian and Sophie (von Cless) F.; ed. Tübingen, Leipzig (both Germany); made study trips to Vienna, London, Edinburgh; m. Antonie Muller, 1878; 4 daus., including Margarete, Helene. Asst. physician, head physician Leipzig Women's Clinic, under Carl Crede; became mem. faculty Leipzig, 1876; named dir. State Midwifery Inst., Stuttgart, 1877; became prof. obstetrics and gynecology, Basel, Switzerland, 1887; became prof., Halle, Germany, 1894, prof., Strasbourg (now in France), 1901-18. Author: Lehrbuch der Geburtshilfe für Hebammen (standard textbook for midwives), 1883; Die Physiologie und Pathologie des Wochenbetts, 1890; Die operative Geburtshilfe der Praxis und Klinik, 1908; Entwicklung der Geburtshilfe und Gynäkologie im 19. Jahrhundert, 1925. Studied pelvic form of foetus and infant; devised Fehling's operation. Died Baden-Baden, Germany, Nov. 2, 1925.

FEHLING, Hermann Christian von, see von Fehling, Hermann Christian.

FEHNEL, Edward Adam, Am. chemist, b. Bethlehem, Pa., Apr. 22, 1922; s. Edward Franklin and Marguerite (Paull) F.; B.S., Lehigh U., 1943, M.S., 1944, Ph.D., 1946; M. Dorothy Gary Lynn, Oct. 21, 1944; children—Lynn Susan, Gary Edward. Instr. chemistry Moravian Prep. Sch., Bethlehem, 1943-44; chemist Allied Chem. & Dye Corp., Morristown, N.J., 1944-45; postdoctoral fellow U. Pa., 1946-48; faculty chemistry Swarthmore (Pa.) Coll., 1948—, prof., 1966—. NSF sci. faculty fellow U. Cambridge, Eng., 1962. Mem. Am. Chem. Soc., Phila. Organic Chemists Club, Sigma Xi. Editorial bd., contbr. Organic Electronic Spectral Data, vols. I-IV, 1957-60. Research, publs. on synthesis and properties of organic compounds, especially heterocyclic sulfur and nitrogen compounds, ultra-violet absorption spectroscopy of organic sulfur compounds, new applications of Friedlaender method to synthesis of quinoline derivatives. Home: 600 Elm Av., Swarthmore, Pa. 19081.*

FEHR, Howard Franklin, Am. mathematician, educator; b. Bethlehem, Pa., Dec. 4, 1901; s. Quincy Howard and Minnie (Patterson) F.; B.A., Lehigh U., 1923, M.A., 1928; Ph.D., Columbia, 1940; m. Gladys Emma Thomas, Dec. 25, 1924; children—Patricia Thomas (Mrs. John Hugh McGuigan), Barbara Gladys (Mrs. John P. Savage); m. 2d, Gisela Henle, Sept. 9, 1966. Pub. sch. tchr., Bethlehem, Reading, Pa., Newark, 1923-34; prof. math. Montclair (N.J.) State Coll., 1934-48; prof. math., head, dept. math. edn. Tchrs. Coll. Columbia, N.Y.C., 1948—. Cons., Orgn. Econ. Cooperation and Devel., 1958—; initiator math. reform in Europe, 1958—; U. S. rep. Internat. Commn. Math. Edn., 1959-63; mem. Commn. Math. Coll. Entrance Exam. Bd., 1955-61; organizer internat. confs. on math. edn., France, Yugoslavia, Greece, 1962-65. Mem. Internat., Nat. math. socs. Author numerous math. books, including Mathematics Today, A Guide for Teachers, 1964; Contemporary Mathematics for Elementary Teachers, 1966; Fundamental Concepts of Mathematics, 1966; also articles. Home: 165 W. 66th St., N.Y.C. 10023.*

FEHR, Johann Michael, German physician; b. Kitzingen/Main, Germany, May 9, 1610; s. Michael and Margarete (Martin) F.; studied Medicine, Leipzig, Wittenberg, Jena, Altdorf; M.D., Ph.D., Padua, Italy, 1641; m. Maria Barbara Meister, 1642; 4 sons, including Johann Lorenz, 3 daus.; m. 2d, Anna Maria Otto; 3 sons, including Johann Caspar, 4 daus. Practiced medicine, Schweinfurt; became ofcl. bailiff of Schweinfurt, 1672; named physician to Kaiser, 1686. Co-founder Akademie der Naturforscher (became Leopoldina 1672), 1652, became pres., 1666. Author: Anchora sacra vel Scovzonera . . . , 1666; Hiera picra seu analecta de absynthio, 1667; also essays in bulls. of Leopoldina. Described (with Elias Schmidt) trigeminal neuralgia in eulogy of Dr. Johannes Laurentius Bausch. Died Schweinfurt, Nov. 15, 1688.

FEHRENBACH, Charles, French astrophysicist; b. Strasbourg, France, Apr. 29, 1914; s. Charles and Alma (Holtkemper) F.; ed. Faculty Scis., Strasbourg, Ph.D. in math. scis.; m. Myriam Graff, Aug. 3, 1939; children—Michel, Mireille, François. Became prof. Saint-Charles Lycée, Marseille, 1939; then astronomer Strasbourg Obs., 1941; asst. dir. Haute-Provence, 1943; apptd. dir. Marseilles Obs., 1948; also prof. Faculty Scis., Marseilles. Mem. French Acad. Scis., 1963, Bur. Longitudes (corr.), Royal Soc. Liège, Royal Aero. Soc. (asso). Contbr. numerous articles to jours. Specialist in stellar spectroscopy; devised new instrument for measurement of radial speeds. Address: 2, place Le Verrier, Marseilles, France.

FEIGL, Georg, German mathematician; b. Hamburg, Germany, Oct. 13, 1890; s. Georg and Maria (Pinl) F.; Ph.D., Jena, Germany, 1918; m. Maria Fleischer, 1925. Asst. to Erhard Schmidt, Berlin; became mem. faculty Berlin, 1927, asso. prof., 1933; prof. U. Breslau (Germany), 1935-45. Mem. Nat. Assn. German Math. Socs. and Clubs. Author: Vorlesungen zur Einführung in die höhere Mathematik, edited by H. Rohrbach, 1953. Named editor, ann. on advances in math. Prussian Acad. Scis., 1928. Research on founds. of geometry and topology; studied fixed point theorems for n-dimensional variety; developed modern structural math. thought for use in sch. instrn. Died Wechselburg, Apr. 25, 1945.

FEIGL, Herbert, philosopher of sci.; b. Reichenberg, Austria, Dec. 14, 1902; s. Otto and Camilla (Beck) F.; student U. Munich, 1921-22; Ph.D., U. Vienna, 1927; m. Maria Kasper, June 30, 1931; 1 son, Eric O. Came to U. S., 1930, naturalized, 1937. Faculty, U. Ia., 1931-40; prof. philosophy U. Minn., Mpls., 1941-53, dir. Minn. Center For Philosophy Sci., 1953—; vis. prof. U. Cal., Berkeley, 1946-53, Columbia, 1950; Carnegie vis. prof. U. Hawaii, 1958; vis. prof. Inst. Advanced Studies, Vienna, 1964. Mem. adv. bd. Internat. Ency. Unified Sci., Library of Congress Philosophers. Rockefeller fellow, 1930-31, 40; Guggenheim fellow, 1947; Fulbright fellow Australia, 1965. Mem. Am. Philos. Assn. (past pres. Western div.), Philosophy of Sci. Assn. (gov.), Am. Humanist Assn. (dir.), Academie Internationale de Philosophie, Institute for Unity of Science (president 1967—). Editor: (with W. Sellars) Readings in Philosophical Analysis, 1949, Philosophical Studies, 1949—; (with M. Brodbeck) Readings in the Philosophy of Science, 1953; Minn. Studies in the Philosophy of Science, vols. I, II, III, 1956, 58, 62; (with G. Maxwell) Current Issues in Philosophy of Science, 1959. Research, numerous publs. primarily in field of logic and methodology of physics, psychology, gen. theory knowledge, moral philosophy. Home: 5601 Dupont Av. S., Mpls. 55419.*

FEIN, Johann (or Isidor), Austrian laryngologist; b. Vienna, Austria, Apr. 28, 1864; s. Albert and Franziska (Lowy) F.; M.D., Vienna, 1889. Worked at L. v. Schrotter-Kristelli's clinic U. Vienna, also under O. Chiari at Vienna Polyclinic; dir. ambulatory clinic for ear, nose and throat diseases Wiedner Hosp., Vienna, from 1900; became mem. faculty, 1904, asso. prof., 1914; gen. staff physician, health dir. Montenegro, 1914-18. Author: Rhino-und laryngologische Winke für praktische Ärzte, 1910; numerous monographs. Opposed prevalent notion that tonsils cause obscure internal diseases; discovered that secretion of tonsils is directed toward surface, thus concluded they are not an entry way for microorganisms. Died Vienna, Apr. 26, 1923.

FEIN, (Wilhelm) Emil, elec. engr.; b. Ludwigsburg, Germany, Jan. 16, 1842; s. Karl and Sophie (Weysser) F.; ed. Stuttgart; m. Anna Regina Stuckle, 1869; 5 sons, including Emil, Bertold, Richard, 1 dau. Employee in mech. factories, Karlsruhe, Göttingen, Berlin, (all Germany), also under C. Wheatstone, London; founder shop, Karlsruhe, 1867, moved to Stuttgart, Germany, 1870. Recipient Gold medal for art and sci. State of Württemberg, 1896. Author: Elektrische Apparate, Maschinen und Einrichtungen, 1888; also numerous articles. Built dynamo engine about same time as W. Siemens, 1860's; built 1st electric fire alarm, 1875, 1st portable field telephone, 1875, 1st telephone central system, Barcelona, 1884, 1st telephonic opera transmission, Hoftheater, Stuttgart, 1892, electric single unit power for tools and textile machines, 1892, 1st electric hand drill, 1895. Died Stuttgart, Oct. 6, 1898.

FEINBERG, Gerald, Am. physicist; b. N.Y.C., May 27, 1933; s. Leon and Florence (Weingarten) F.; B.A., Columbia, 1953, M.A., 1954, Ph.D., 1957. Mem. Inst. for Advanced Study, Princeton, N.J., 1956-57; research asso. Brookhaven Nat. Lab., Upton, N.Y., 1957-59; prof. physics dept. Columbia, 1959—. Cons. Brookhaven Nat. Lab., 1960—. Sloan Found. fellow, 1960-64; Overseas fellow Churchill Coll., Cambridge, Eng., 1963-64. Fellow Am. Phys. Soc.; mem. A.A.A.S., Sigma Xi. Research, publs. in theory weak interactions, formulation two neutrino theory; devel. methods for calculating by quantum field theory, higher order corrections to weak interactions. Home: 600 W. 115th St., N.Y.C. 10025.*

FEINBERG, Joseph George, immunochemist; b. Bklyn., Dec. 14, 1914; s. Harry and Annie (Cobden) F.; B.S., Bklyn. Coll., 1933; M.S., Kan. State U., 1936, D.V.M., 1938; m. Patricia Helen Mary Jessel, Nov. 22, 1945; 1 dau., Patricia Ann. Practice vet. medicine specializing in small animals, Norfolk, Va., 1938-41; pharm. chemist Jen-Sal Labs., Kansas City, Mo., 1939-40; veterinarian U. S. Govt., 1941-42; head allergy research unit Beecham Research Labs., Eng., 1954-64; dir. research dept. Miles Labs. Ltd., Stoke Poges, Eng., 1964—; part-time sci. writer, sci. corr. The Observer, Sunday Times, Time-Life, London, 1948-52; lectr. immunology Chelsea Coll., London, 1960—. Recipient Diploma for research contbns. to allergy European Acad. Allergology, 1965. USPHS Postgrad. Research fellow Lister Inst., London, 1952-54. Mem. Royal Soc. Medicine, Am. Assn. Immunologists, Brit. Allergy Soc., Internat. Com. for Microbiol. Standardization. Author: The Story of Atomic Theory and Atomic Energy, 1960; (with Ivor Smith) Chromatography and Electrophoresis on Paper, 1962; also articles. Research, patents on devel. original techniques in immunology; factors producing sensitization to penicillin; invention of silicon emulsions for therapy; discovery of new arsenic and antimony derivaties. Home: 17 Old Church St., Chelsea, London S.W.3, Eng. Office: Research Dept., Miles Labs. Ltd., Stoke Ct., Stoke Poges, Slough, Buckinghamshire, Eng.*

FEINBERG, Samuel Maurice, physician; b. Yurburg, Russia, Mar. 28, 1895; s. George and Anna (Shulman) F.; came to U. S., 1907, naturalized, 1918; B.S., U. Wis., 1917; M.D., Rush Med. Coll., 1919; m. Cecile S. Stern, Mar. 19, 1922; children—Alan Richard, Robert H. and Ruth Ann (twins), Helene (Mrs. Gene Pier). Practice medicine specializing in allergy, Chgo., 1921-61, Winnetka, Ill., 1939—; attending physician Cook County Hosp., 1926-38; staff Passavant Hosp., Chgo., 1942-64, Evanston (Ill.) Hosp., 1964—; faculty dept. medicine Northwestern U. Med. Sch., Evanston, 1921-64, prof., 1950-64; chief regional allergy cons. VA, 1948-61. Bd. dirs., mem. sci. adv. council Ashman and Allergy Found. Greater Chgo., 1964—; mem. tng. grants com. Nat. Inst. Allergy and Infectious Diseases. Decorated Medaille de'Argent de la Ville de Paris (France). Diplomate Am. Bd. Internal Medicine, Am. Bd. Allergy. Hon. fellow Spanish, Scandinavian, Argentinian, Peruvian, Cuban allergy assns.; fellow A.C.P.; mem. Chgo. Soc. Allergy (past pres.), Am. Acad. Allergy (past pres.), Internat. Assn. Allergology (past pres.), A.M.A., Am. Acad. Allergy, Phi Beta Kappa, Sigma Xi. Author: Asthma, Hay Fever and Related Disorders, 1933; Allergy in General Practice, 1934; Allergy in Practice, 1944; Allergy, Facts or Fancies, 1950; (with A. Feinberg, S. Malkiel) The Antihistamines, 1951; Living with Your Allergy, 1958; also numerous articles. Established importance mold spores as airborne cause hayfever and asthma, insect dust as cause respiratory allergy; research on antihistamines, role and mechanism infection in inhibiting allergic phenomena; demonstrated mechanism artificial induction allergy in man. Home: 739 Clarey Rd., Highland Park, Ill.*

FEINBERG, Savelii Moiseevich, Russian nuclear physicist; b. 1910; grad. Azerbaijan Poly., 1932; D. Physico-Math. Scis., 1948. Engr.-planner Azneft, lab. head Azerbaijan Inst. Constn., 1932-40; staff Kurchatov Inst. Atomic Energy USSR Acad. Scis., 1945—, head theoretical sect., 1947—, dir. reactor physics sect., Moscow; prof. Engring. Physics Inst., Moscow. Publs. on burnup of fuel in water-moderated water-cooled power reactors and uranium-water lattice expts. Participated in creation of complex of exptl. aquatic reactors VVR-2, VVR-S, IRT. Address: Nuclear Energy Inst., Acad. Sciences USSR, B. Kaluzhskaya 14, Moscow V-71, USSR.

FEINDEL, William, Canadian neurosurgeon; b. Bridgewater, N.S., Can., July 12, 1918; s. Robert Ronald and Annie (Swanberg) F.; B.A., Acadia U., 1939, D.Sc. (hon.), 1965; M.Sc., Dalhousie U., 1942; M.D., C.M., McGill U., 1945; D.Phil., Oxford U., 1949, F.R.C.S., 1955; m. Faith Lyman, July 28, 1945; children—Christopher, Alexander, Patricia, Janet, Michael, Anna. Demonstrator in biology Acadia U., 1937-39; Rhodes Scholar, Merton Coll., 1939; demonstrator in physiology Dalhousie U., 1940-42, Banting Found. Research grantee, 1941; research asst., demonstrator in anatomy Oxford U., 1946-49; demonstrator in neurosurgery McGill U., 1951-52, lectr., 1952-55, Wm. Cone prof. neurosurgery, 1959—, curator Osler Library, 1962—; research fellow in neurophysiology at the Montreal Neurological Inst., 1951-52, Reford postgrad. fellow, 1953-55, dir. Cone Lab. for Neurosurg. Research, 1959—; prof. neurosurgery U. Sask. and U. Hosp., Saskatoon, Can., 1955-59; neurosurgeon-in-chief Montreal Neurol. Hosp., 1963—; attending neurosurgeon Catherine Booth, Royal Victoria hosps., 1959—; cons. neurosurgeon Sherbrooke Gen. Hosp.; Clarke lectr. U. Cal. at Los Angeles, 1964; Basterfield lectr. U. Sask., 1965; Harvey lectr. Yale, 1966. Fellow A.A.A.S., also A.C.S.; mem. Harvey Cushing Soc., Soc. Neurol. Surgeons, Am. Acad. Neurol. Surgery, Am. Acad. Neurology, Am. Neurol. Assn., Royal Society Medicine, Anat. Soc. Gt. Britain and Ireland, Montreal Neurol. Soc., Canadian Neurosurg. Soc., Alpha Omega Alpha (hon.). Editor: (with A. Dawson) Prospect and Retrospect in Neurology, 1955; Medical Aspects of Traffic Accidents, 1955; Memory, Learning and Language—The Physical Basis of Mind, 1960; Thomas Willis: The Anatomy of the Brain and Nerves, 1965. Research, studies, publs. on anatomy and physiology of pain in central nervous system, localization of epileptic discharge and surg. treatment of epilepsy, surg. treatment of peripheral nerve injuries, localization of intracranial vascular lesions using radioisotopes and brain scanning, cerebral circulation studies. Home: 39 Thornhill Av., Montreal 6. Office: 3801 University St., Montreal 2, Que., Can.*

FEINSTEIN, Alvan Richard, Am. physician; b. Phila., Dec. 4, 1925; s. Joel B. and Bella (Ukasz) F.; student U. Pa., 1942-44; B.S., U. Chgo., 1947, M.S., 1948, M.D., 1952. Research fellow Rockefeller Inst., N.Y.C., 1954-55; asst. med. dir. Irvington House, N.Y.C., 1956-59, med. dir., 1959-62; with Yale Sch. Medicine, 1962—, asso. prof. medicine, 1964—; with West Haven (Conn.) VA Hosp., 1962—, chief clin. biostatistics, 1963—. Fellow A.C.P.; mem. Am. Soc. Clin. Investigation, Am. Fedn. Clin. Research, Am. Statist. Assn., Assn. Computer Machinery, Biometric Soc., A.M.A. Author: Clinical Judgment, 1967. Contbr. numerous articles, abstracts, book revs. to profl. jours. Research in formula diet for treatment of obesity, in clarifications of natural history, therapy and prevention of rheumatic fever; in modes of analyzing prognosis and therapy for cancers of lung, rectum, larynx, other sites, gen. principles of sci. methodology in clin. medicine. Home: 488 Whitney Av. Office: 333 Cedar St., New Haven 06511.*

FEIST, (Eduard Richard) Karl, German pharm. chemist; b. Nordhausen, Germany, May 9, 1876; s. Ferdinand and Ida (Bernigau) F.; student, Marburg, Germany; Ph.D., 1901; pharmacist's diploma, U. S.; m. Dora Schmalz, 1911; 1 son, 2 daus. Pharmacist, Nordhausen, elsewhere in Germany, also in U. S.; became asst. to J. Gadamer, U. Breslau (Germany), 1907; sr. asst., Marburg; became chmn. pharm. chem. dept., chem. inst. U. Geissen (Germany), 1912; navy pharmacist, 1914-18; named prof. U. Göttingen (Germany), 1919, also founder, dir. ind. chem.-pharm. inst., professorship withdrawn by Nazis, 1938, resumed teaching, in food chemistry, 1946. Contbr. numerous articles to profl. jours. Research in alkaloid chemistry and drug synthesis; discovered new constituents in columba root; made qualitative analysis of alkaloids; investigated limonin. Died Göttingen, Feb. 20, 1952.

FEIT, Walter, mathematician; b. Vienna, Austria, Oct. 26, 1930; s. Paul F.; came to U. S., 1947, naturalized, 1953; B.A., M.S., U. Chgo., 1951; Ph.D., U. Mich., 1954; m. Sidnie Marilyn Dresher, Oct. 26, 1957; children—Paul, Alexandra. Faculty, Cornell U., 1953-64, asso. prof., 1960-64; prof. Yale, 1964—. Recipient Cole prize in math., 1965. Mem. Am. Math. Soc., Am. Math. Assn. Contbr. articles to sci. jours. Home: 6 Waite St., Hamden, Conn. Office: Yale U., New Haven.*

FEIT, Wilhelm Friedrich August, German chemist; b. Lippstadt, Germany, Jan. 24, 1867; s. Friedrich and Caroline (Klee) F.; Ph.D., Rostock, Germany, 1888; D. Engring. (hon.), Berlin, Germany, 1932; m. Hedwig Faldix, 1892; 1 son, 1 dau. Became chemist Kaliwerke ((potassium works), Aschersleben, 1887; factor dir. Langelsheim, 7 years, Hercynia (potassium works), Vienenburg, 7 years; gen. dir. Vereinigte Chemische Fabriken (merger with Kaliwerke Aschersleben 1922), 1908-18; lab. researcher, from 1943; independent researcher, Berlin, 1918-43. Founder potassium research inst.; mem. potassium testing sta.; bd. dirs. German potassium syndicate. Recipient Silver Leibniz medal, 1932, Nat. prize 2d class German Dem. Republic, 1950. Hon. mem. Verein Deutscher Chemiker. Discovered 2 new boron minerals; isolated some rare earths in a purity previously unknown; 1st to isolate large quantities of rhenium and develop its comml. prodn.; manufactured gallium, indium, thallium; produced erbium at 99.7 per cent purity. Died Bad Nauheim, Germany, June 19, 1956.

FEJER, Jules Andrew, physicist; b. Budapest, Hungary, Jan. 22, 1914; s. Ernest and Stella (Popper) F.; Diploma in Elec. Engring., Eidgenössische Technische Hochschule, Zurich, Switzerland, 1935; M.Sc., U. Witwatersrand, Johannesburg, South Africa, 1949, D.Sc., 1957; m. Francisca Bongers, Mar. 27, 1943; 1 dau., Stella Hendrika. Research engr. Hungarian Tungsten Lampworks, Budapest, 1936-39; geophysicist Oscar Weiss Cons. Geophysicist, Ltd., Johannesburg, 1939-42; chief sci. officer South African Council for Sci. and Indsl. Research, Johannesburg, 1945-58; def. sci. officer Def. Research Telecommunications Establishment, Ottawa, Ont., Can., 1959-61; tech. specialist Gen. Motors Def. Research Labs., Santa Barbara, Cal., 1961-62; prof. atmospheric and space scis. S.W. Center for Advanced Studies, Dallas, 1962-66; adj. prof. physics So. Meth. U., Dallas, 1966; prof. applied electro physics U. Cal. at San Diego, U. Cal., 1966—. Mem. I.E.E.E., Am. Geophys. Union. Editor: (with C. O. Hines, I. Paghis, T. R. Hartz) Physics of the Earths Upper Atmosphere, 1965. Research, publs. on wave propagation in irregular media, dynamo theory magnetic variations, ionospheric wave interaction, incoherent backscatter radio waves, hydromagnetic wave propagation hydromagnetic helmholtz instability, theory auroral electrojects, impedance of an antenna in a plasma, excitation plasma resonance in ionosphere. Home: 7411 Herschel Av., La Jolla, 92037. Office: U. of Cal., P.O. Box 109, La Jolla, Cal. 92037.*

FEKETE, György, Hungarian physician; b. Budapest, Hungary, Mar. 23, 1926; s. Medard and Irene (Szlavik) F.; grad. Budapest U. Med. Sch., 1951; m. Angela Nagy, Aug. 22, 1958; children—Gábor, Judith. Demonstrator dept. pharmacology Budapest U. Med. Sch., 1947-51; asso. prof. dept. pathophysiology U. Szeged Med. Sch., 1951-55; head dept. pharmacology Chem. Works, G. Richter Ltd., Budapest, 1955—, sci. dir., 1957—; asso. prof. Budapest U. Med. Sch. Mem. Hungarian Acad. Scis. (candidate), Brit., Hungarian (gen. sec.) pharmacological socs., Hungarian Physiol. Soc., Hungarian Endocrinological Soc., Deutsch Pharmakologische Gesellschaft, Internat. Biochem. Pharmacology, European Soc. for Study Drug Toxicity. Author: Hazards of Modern Drug Therapy, 1963; (with P. Braun) Recent Problems of Drug Therapy, Vol. I, 1964; Vol. II, 1967. Editor: Steroids, 1963—; Gyógyszereink, 1963—. Research on pharmacology, physiology, biochemistry of steroidal hormones; described new methods of assay and mechanisms for side effects of synthetic steroidal compounds; patentee new drugs. Home: 5 Mese, Budapest 12. Office: 19-21 Gyömröi, Budapest 10, Hungary.*

FELD, Bernard Taub, Am. physicist; b. N.Y.C., Dec. 21, 1919; s. A. Lewis and Helen (Taub) F.; B.S., Coll. City N.Y., 1939; Ph.D., Columbia, 1945; m. Eliza McCormick, June 2, 1947; children—Elizabeth T., Ellen D. Group leader Manhattan Project, Chgo., Los Alamos, 1942-45; faculty Mass. Inst. Tech., 1945—, prof. physics, 1956—, acting dir. lab. for Nuclear Sci., 1961-62; vis. prof. Rome, Italy, 1953-54; vis. scholar and Ford fellow CERN, Geneva, Switzerland, 1960-61; vis. prof. École Polytechnique, Paris, 1966-67. Cons. Brookhaven Nat. Lab., 1947, AVCO Research Lab., 1956; v.p. Fedn. Am. Scis. 1961-62. Fellow Am. Phys. Soc., Am. Acad. Arts and Scis.; mem. Phi Beta Kappa, Sigma Xi. Author: The Neutron, vol. 2, Experimental Nuclear Physics, 1954. Research in neutron physics; elementary particles and meson physics; atomic and molecular hyperfine structure and nuclear moments. Home: 42 Arlington St., Cambridge, Mass. 02140.

FELD, Joseph M., mathematician, educator; b. Warsaw, Poland, Jan. 21, 1900; s. Israel and Rachel (Lerner) F.; came to U. S., 1907, naturalized, 1917; A.B., Columbia, 1921, Chem.E., 1923, Ph.D. in Math., 1931. Asst. in math. Columbia, 1924-25, instr., 1925-31; instr. Bklyn. Coll., 1931-42; faculty Queens Coll., U. City N.Y., Flushing, 1941—, prof., 1962—. Lectr. math. Columbia Summer Sessions, Sch. Gen. Studies, 1925, 26, 28-31. Mem. Am. Math. Soc., Math. Assn. Am., Am. Assn. U. Profs., Phi Beta Kappa. Research, publs. on contact transformations, expansion analytic functions, differential invariants, non-Euclidean geometry, kinematic geometry. Home: 143-07 Sanford Av., Flushing, N.Y. 11355.*

FELD, Walther, German chemist; b. Neuwied/Rhine, Germany, Nov. 4, 1862; ed. Zurich, Switzerland, also Leipzig, Munich, Berlin (all Germany). Founder Chemische Fabrik Hönningen (barium carbonate factory, formerly Walther Feld & Co. AG), 1890, resigned as dir., 1896; founder Barium-Oxyr-Gubit factory, Hönningen, Germany, 1904; built 1st large installation for preparation of tar constituents from coal gas, Gule-Hoffnungs-Hütte, Sterkrade, Germany, 1914. Developed proress using local raw materials to manufacture white mineral pirment blanc-fixe; worked on purification of local barium sulphate; revolutionized coal industry by his discovery of purification methods and use for by-products of coal gas. Died Linz/Rhine, Germany, Mar. 15, 1914.

FELDBERG, Wilhelm Siegmund, physiologist, pharmacologist; b. Hamburg, Germany, Nov. 19, 1900; s. Emil Daniel and Amalie F.; M.A., Cambridge (Eng.) U.; M.D., Germany; M.D. (hon.), U. Freiburg (Germany), U. Cologne (Germany); m. Katherine Scheffler, 1925. Reader physiology Cambridge (England) University, until 1949; hon. lecturer University of National Institute for Med. Research, London, 1949-65, hon. head div., 1965-66, head lab. of neuro-pharmacology, 1966—; professor emeritus University of Berlin (Germany). Decorated Grand Cross Order of Merit (Germany); recipient Baly medal, 1963; named comdr. Order Brit. Empire. Fellow Royal Soc., 1947; mem. Physiol. Soc., Brit. (hon.), German pharmacol. socs., German Physiol. Soc. Author: (with E. Schilf) Histamin; A Pharmacological Approach to the Brain from its Inner and Outer Surface, 1963; also numerous articles. Research on histamine, acetylcholine, transmission of nerve effects, pharmacology and physiology of central nervous system, roles of monoamines in hypothalamic control of body temperature, action of centrally acting drugs applied to various regions of brain by microinjection. Home: Lavenham, Marsh Lane, Mill Hill. Office: Nat. Inst. for Med. Research, Mill Hill, London, N.W. 7, Eng.*

FELDMAN, David, Am. physicist; b. Bklyn., June 16, 1921; s. Isidore and Ida (Dobsevitch) F.; B.S., City Coll. N.Y., 1940; M.S., N.Y. U., 1946; Ph.D., Harvard, 1949; m. Dorothy M. Appel, Aug. 30, 1946; children—Charles A., Robert E. Mem. Inst. for Advanced Study, Princeton, N.J., 1949-50; asst. prof. physics U. Rochester, 1950-56; asso. prof. physics Brown U., Providence, 1956-59, prof. physics, 1959—; vis. prof. physics U. Paris, France, 1962-63. NSF Sr. Postdoctoral fellow, 1962-63. Fellow Am. Phys. Soc.; mem. Italian Phys. Soc., Phi Beta Kappa, Sigma Xi. Editor Proceedings of the Fifth Annual Eastern Theoretical Physics Conf., 1967. Research, publs. in nuclear physics, quantum field theory, theory of elementary particles. Home: 11 Valentine Dr., Barrington, R.I. 02806.*

FELDMAN, Gerald Lewis, Am. biochemist; b. Buffalo, Jan. 30, 1931; s. Abraham and Bessie (Andelman) F.; A.S., State U. N.Y., 1952; B.S., U. Ga., 1955; M.S., Tex. A. and M. U., 1957, Ph.D, 1959; m. Marjorie Lutrell Slaughter, July 24, 1955; children—Michael Robert, Lois Cecile, Mark Samuel. Chemist, Ga. Expt. Sta., Athens, 1954-55; research asst. Tex. A. and M. U., 1955-57, NIH pre-doctoral fellow, 1957-59; research biochemist Meth. Hosp., Houston, 1959-61, asst. biochemist dept. pathology, 1961—; asst. prof. biochemistry Coll. Medicine Baylor U., Houston, 1961-66, asso. prof., 1966—; sci. adviser Eyes of Tex., Inc., 1961—; cons. biochemist Ben Taub Gen. Hosp., 1961—, VA Hosp., 1961—. Spl. fellow Nat. Inst. Neurol. Diseases and Blindness, 1964—. Mem. Am. Chem. Soc., Am. Oil Chemists Soc., A.M.A., A.A.A.S, Assn. For Research Ophthalmology, Biochem. Soc., N.Y. Acad. Scis., Poultry Sci. Assn. Research, numerous publs. on isolation and structural characterization of lipids from ocular tissues, prodn. exptl. cataracts in chick embryo; devel. of new techniques in analysis of lipids; demonstrated that lipids are an important constituent of ocular tissue and that their concentration is altered by diseases; described lipid composition and metabolism of developing adipose tissue. Home: 7719 Bellerive St., Houston 77036. Office: 1200 Moursund Av., Houston 77025.*

FELDMAN, Harry Alfred, Am. epidemologist; b. Newark, May 30, 1914; s. Joseph and Sara (Pivnick) F.; B.A., George Washington U., 1935, M.D., 1939; m. Lillian Maltz, June 14, 1939; children—Ronald E., Donna E., Jeffrey H., Robert N. Fellow in medicine George Washington U., 1941-42; research fellow in medicine Harvard, 1942; sr. fellow virus diseases NRC, Children's Hosp., Cin., 1946-49; asso. prof. medicine State U. N.Y., Upstate Med. Center, Syracuse, 1949-57, prof., chmn. dept. preventive medicine, 1957—; dir. research Wieting-Johnson Hosp. for Rheumatic Disease, Syracuse, 1949-57. Diplomate

Am. Bd. Internal Medicine, Am. Bd. Microbiology. Mem. Am. Epidemiological Soc. (sec.-treas. 1961-68, pres. 1968—), Commn. on Acute Respiratory Diseases, Armed Forces Epidemiological Bd., Assn. Tchrs. Preventive Medicine, Nat. Bd. Med. Examiners. Contbr. numerous articles to med., sci. jours. Determined that 8-amino-quinolines were causal prophylactics for malaria; co-discoverer toxoplasma dye test system for measuring antibody for that parasite; determined effectiveness of sulfadiazine in treatment and prevention of meningococcal infections; devised hemaglutinating system used in measuring antibodies for mycoplasma pneumoniae; determined system for preventing streptococcal infections in school children; determined effectiveness of killed measles virus vaccine in preventing disease. Home: 704 Crawford Av., Syracuse 12224. Office: 750 E. Adams St., Syracuse, N.Y. 13210.*

FELDMAN, Jacob, Am. mathematician; b. Phila., Jan. 10, 1928; s. Boris and Fannie (Shrager) F.; B.A., U. Penn., 1950, M.A., U. Ill., 1951; Ph.D., U. Chgo., 1954; Mem. Inst. for Advanced Study, Princeton, N.J., 1954-56; mathematician Bell Labs., Murray Hill, N.J., 1956; vis. asst. prof. math. Columbia, N.Y.C., 1956-57; prof. math. U. Cal., Berkeley, 1957-—. Mem. Am. Math. Soc. Research, publs. on operator algebras, especially rings of operators, and probability theory, especially Gaussian processes. Home: 2736 Russell St., Berkeley, Cal.*

FELDMAN, Joseph D., Am. physician; b. Hartford, Conn., Dec. 13, 1916; s. Max and Rebecca (Hurewitz) F.; B.A., Yale, 1937; M.D., L.I. Coll. Medicine, 1941; m. Naomi Granott, Mar. 4, 1949; children—Orna, Danah, Ruth. USPHS Research fellow Yale Sch. Medicine, 1948-50; lectr. Hebrew U., Hadassah Med. Sch., 1950-54; faculty U. Pitts., 1954-61, prof. dept. pathology, 1956-61; mem. Scripps Clinic and Research Found., La Jolla, 1961-—. Cons., NIH, 1961-65; mem. Eli Lilly Award Com., 1964-—. Mem. Am. Assn. Pathology and Bacteriology, Am. Soc. Exptl. Pathology, Am. Assn. Immunologists, Histochem. Soc., Endocrine Soc. Editor: Proc. Soc. for Exptl. Biology and Medicine, 1964-66, Lab. Investigation, 1965-67, Archives Pathology, 1965-—, Jour. Reticulo-Endothelial Society, 1965-—, American Journal of Pathology, 1967-—. Research and numerous publs. on correlation function and structure at intracellular level, endocrine pathophysiology, immunopathology, immunologic renal disease, transplantation immunity, and delayed hypersensitivity. Home: 3654 Fenelon St., San Diego 92106. Office: 476 Prospect St., La Jolla, Cal. 92037.*

FELDMAN, Julian, Am. chemist; b. Bklyn., May 24, 1915; s. Bernard and Fanny (Rossum) F.; B.S., Coll. City N.Y., 1935; M.A., Bklyn. Coll., 1940; Ph.D., U. Pitts., 1950; m. Viola Borden, May 10, 1944; children—Diana, Alan, David. Chief chemist Pro-Medico Labs., Bklyn., 1935-40; asst. chemist U. S. Dept. Agr., Beltsville, Md., 1941; asst. plant mgr. Trubek Labs., East Rutherford, N.J., 1941-42; research asso. Explosives Research Labs., OSRD, Bruceton, Pa., 1942-45; group leader U. S. Bur. Mines, Pitts., 1945-53; sr. research asso. Nat. Distillers & Chem. Corp., Cin., 1953-—. Mem. Am. Chem. Soc., A.A.A.S., Phi Beta Kappa, Sigma Xi, Phi Lambda Upsilon. Patentee in field. Research, publs. on separation of complex organic mixtures by distillation, adsorption and other phys. methods, synthesis of polynuclear compounds, isomerization, hydrogenation, and dimerization by means of catlaysts; inventor processes for making new and useful compounds by these methods. Home: 7511 Sagamore Dr., Amberley, O. 45236. Office: 1275 Section Rd., Cin. 45237.*

FELDMAN, Shaul, neurologist, neurophysiologist; b. Odessa, Ukraine, Aug. 3, 1923; s. Pinhas and Bertha (Drut) F.; student Faculty Scis., Hebrew U., Jerusalem, Israel, 1942-45, M.D., 1952; B.Sc.Med., Sch. Medicine, Geneva, Switzerland, 1946-48; m. Aviva Ben-Arzi, Mar. 23, 1950; children—Nirit, Talya, Ilan. Researcher, U. Cal. at Los Angeles, also Mt. Sinai Hosp., N.Y.C., 1957-59; staff Dept. Nervous Diseases, Jerusalem, 1953-—, chief physician, 1959-—; vis. prof. U. Paris, France, 1965; faculty Hebrew U. Hadassah Med. Sch., Jerusalem, 1957-—, asso. prof. neurology, 1963-—; dir. Lab. Neurophysiology dept. nervous diseases Hadassah U. Hosp., Jerusalem, 1965-—. Mem. panel neurophysiology Internat. Brain Research Orgn., UNESCO, 1965-—. Mem. French Neurol. Assn. (fgn. hon.), Israel Soc. Electroencephalography and Neurophysiology (pres. 1963-—), Internat. Fedn. Socs. for Electroencephalography (mem. council 1963-—), Israel Neuropsychiat. Assn. (chmn. Jerusalem br. 1962-—), Israel Med. Assn. (v.p. sci. council 1965-—), Research, numerous publs. on electrophysiol. investigation of sensory projections to hypothalamus and their modification by other neural structures, role of hypothalamus in regulation of ACTH secretion by pituitary gland, hypothalamic regulation of autonomic function. Home: 16 Iben Ezra, Jerusalem. Office: Hadassah U. Hosp., Jerusalem, Israel.*

FELDMANN, Edward G(eorge), Am. pharm. chemist; b. Chgo., Oct. 13, 1930; s. Edward Louis and Vera (Arneson) F.; foster mother Helen Whitney Feldmann; B.S. in Chemistry, Loyola U., Chgo., 1952; M.S. in Pharmacy (research fellow Am. Found. Pharm. Edn. 1953-55), U. Wis., 1954, Ph.D. in Pharm. Chemistry-Biochemistry, 1955; grad. study Northwestern U., 1956, U. Chgo., 1958; m. Mary J. Evans, Aug. 30, 1952; children—Ann Marie, Edward William, Robert George. Teaching asst. Loyola U., Chgo., 1951-52; research asst. U. Wis., 1952-53; sr. chemist Am. Dental Assn., 1955-58, dir. div. chemistry, 1958-59; asso. dir. sci. div. Am. Pharm. Assn., 1959-60, dir., 1960-—, asso. editor sci. edit. assn. jour., 1959-60, editor, 1960; asso. dir. revision Nat. Formulary, 1959-60, dir. revision, 1960-—, chmn. com., 1960-—; asso. editor Drug Standards, 1959-60, editor, 1960; editor Jour. Pharm. Scis., 1961-—. Mem. adv. panel dental drugs Nat. Formulary, 1955-60; reviewer Internat. Pharmacopeia, WHO, 1958, mem. expert adv. panel, 1963-—; spl. lectr. drug standards George Washington U., 1960-64; del. conf. on fellowships Nat. Health Council, 1960; mem. coordinating com. Nat. Conf. Antimicrobial Agts., Soc. Indsl. Microbiology, 1960; mem. adv. panel pharm. nomenclature A.M.A.- U. S. Pharmacopeia, 1961-—, mem. nomenclature com., 1962-—; adv. panel chems. codex Nat. Acad. Scis.-NRC, 1961-—; mem. lab. com. Am. Pharm. Assn. Found., 1961-—, com. Ebert prize, 1961-—; judge Lunsford - Richardson Pharmacy Awards, 1962-—; cons. Council on Drugs, A.M.A., 1962; vis. scientist Am. Assn. Colls. of Pharmacy, NSF, 1963-65; sec. U. S. Com. on Internat. Drug Standards, 1964-65. Life mem. Am. Pharm. Assn.; mem. Am. Chem. Soc., N.Y. Acad. Scis., Nat. Soc. Med. Research (council 1961-—), Am. Testing Materials, Sigma Xi, Rho Chi, Lambda Chi Sigma. Research and publs. on synthesis of certain therapeutic agts.; methods of analysis and quality control of dental therapeutic agts. Home: 410 Cheryl Dr., Falls Church, Va. Office: 2215 Constitution Av. N.W., Washington 7.

FELICI, Noel Joseph, French physicist; b. Marseilles, France, May 18, 1916; s. Joseph Antoine and Lucie (Ceccaldi) F.; Sc.D., U. Paris, 1940; m. Laure Graux, Oct. 4, 1942; children—Laurent, Vincent, Isabelle, Marianne. Research fellow with P. Neel, Grenoble, 1941-49; faculty Grenoble (France) U., 1949-—, prof. physics, 1953-—. Cons. French AEC, 1955-—; head electrostatic sect., research lab. Nat. Center for Sci. Research, Grenoble. Decorated knight Legion of Honor, 1961. Mem. French Soc. Elec. Engrs. Author: Elektrostatische Hochspannungs-Generatoren, 1957; Accelerateurs de Particules, 1960; also articles. Research on electromagnetic theory of superconductors; inventor types of electrostatic generators, electrostatic spraying and accelerators; conductivity of highly polar liquids, their use as light modulators and electrostatic machines. Home: 13 rue Peguy, Grenoble, France.*

FELIX, Arthur, bacteriologist; b. Andrychow, Poland, Apr. 3, 1887; s. Theodore and Regina Felix; D.Sc., U. Vienna; D.Sc. (hon.), Belfast, Ireland; m. Leah Gluckman, 1923. Bacteriologist on diagnosis typhus fever Austrian Army, 1914-18; researcher U. Prague, 1919-20; dir. bacteriological lab. Hadassah Med. Orgn., Tel Aviv, Palestine, 1921, chief bacteriologist, Jerusalem, 1922-27; chmn. health council Palestine Zionist Exec., Jerusalem, 1924-27; mem. sci. staff Lister Inst. Preventive Medicine, London, 1927-45; past dir. Central Enteric Reference Lab. and Bu., Pub. Health Lab. Service, Med. Research Council, London, specialist Emergency Pub. Health Lab. Service, 1939-45. Chmn. Internat. Com. for Enteric Phage Typing, 1947-56. Fellow Royal Soc., 1943; fgn. corr. mem. Société de Pathologie Exotique (Paris). Contbg. author (with J. A. Arkwright) A System of Bacteriology, 1930. Asso. editor Jour. Hygiene, Excerpta Medica. Research and numerous publs. on bacteriology and immunology; discovered (with Edmund Weil) diagnostic test in typhus (Weil-Felix reaction), 1915; first (with R. Margaret Pitt) to describe Vi antingens, in sheath of bacterium, 1934. Died Jan. 14, 1956.

FELIX, Johannes Paul, German geologist, paleontologist; b. Leipzig, Germany, Sept. 6, 1859; s. Amy Wilhelm and Auguste (Kritz) F.; student, Erlangen, Germany; Ph.D., Leipzig, 1882; m. Anna Limpricht; 1 dau. Became mem. faculty Leipzig U., 1884, asso. prof., 1891-1933; head paleontol. mus., Leipzig, 1914-41; research expdn. (with H. Lenk) to Mexico, 1887-88. Author: (with H. Link) Beiträge zur Geologie und Paläontologie der Republik Mexico, 3 parts, 1889-99; Geologische Reiseskizzen aus Nordamerika, 1895; Leitfossilien aus dem Pflanzen- und Tierreich, 1906; Mammuth von Borna, 1912; Jungtertiäre und guartäre Anthrozoen von Timor und Obi, 1915; Fossile Anthozoen von Borneo, 1921. Research, descriptive work on fossils of coral and woods, territorial diluvial paleontology, Borna Mammut skeleton, prehistory. Died Leipzig, Jan. 25, 1941.

FELIX, Robert Hanna, Am. psychiatrist; b. Downs, Kan., May 29, 1904; s. Tasso Ovid and Neva Lee (Trusdle) F.; student U. Colo., 1923-26, M.D., 1930, Sc.D., 1953; M.P.H., Johns Hopkins, 1942; Sc.D., Boston U., 1953; LL.D., U. Chattanooga, 1957; m. Esther R. Wagner, June 18, 1933; 1 dau., Mary Katherine. Commonwealth Fund fellow in psychiatry, Colo. Psychopathic Hosp., 1931-33; asst. surgeon USPHS, 1933, sr. surgeon, 1948, med. dir., 1950, asst. surgeon gen., 1957; clin. dir. Dept. Justice Med. Center, 1935-36; chief, Psychiatric Service, USPHS Hospital, Lexington, Ky., 1937-38, clin. dir., 1939-41; Rockefeller fellow in pub. health, 1941-42; psychiatrist USCG Acad., 1942-43, sr. med. officer, 1943-44; chief mental hygiene div. USPHS, 1944-49; dir. Nat. Inst. Mental Health, USPHS, 1949-64; dean Sch. Medicine, St. Louis U., 1964-—. mem. expert adv. panel on mental health, WHO. Mem. delegation Internat. Congress on Mental Health, London, 1948; chmn. U. S. delegation Mexico City, 1951; mem. delegation, tech. advisor, 2d World Health Assembly, Rome, 1949; mem. Nat. Com. on Alcoholism. Diplomate, Am. Bd. Psychiatry and Neurology; fellow Am. Public Health Assn., A.C.P., Am. Psychiatric Assn. (mem. council 1947-50); mem. A.M.A., So. Psychiatric Assn. (pres. 1946-47) Am. Orthopsychiatric Assn., Am. Psychol. Assn., Group for Advancement Psychiatry, Royal Medico-Psychol. Assn. (corr.), Nat. Assn. for Mental Health, Alpha Omega Alpha, Phi Chi. Adv. editor The Psychiatric Bulletin; mem. adv. editorial bd. Quarterly Jour. Studies on Alcoholism. Instrumental in obtaining passage of Nat. Mental Health Act, 1946. Home: 550 Twin Fawns Dr., Frontenac, Mo. 63131. Office: 1402 S. Grand Blvd., St. Louis 63104.*

FÉLIZET, Georges-Marie, French surgeon; b. Elbeuf, France, 1844; maj. surgeon during Franco-Prussian War of 1870; anat. research on fractures of skull, infantile inguinal hernia, suture of tendons. Died Paris, 1908.

FELL, George Edward, surgeon; b. Chippewa, Ont., July 10, 1850; s. James Wilkins and Ann Elizabeth (Hoffman) F.; M.D., U. Buffalo, 1882; ad eundum degree Niagara U., 1886; m. Annie Argo Duthie, Oct. 15, 1872; m. 2d, Gertrude Luella Axtell, May 20, 1912. Prof. physiology and microscopy, med. dept. Niagara U., also physician Buffalo Hosp. Sisters of Charity, 1885-95; surgeon Charity, Eye, Ear, Nose and Throat Hosp., Buffalo, 1910-16. Invented 1st successful apparatus to induce artificial mechanical respiration in cases of drowning, asphyxiation, 1887, by which thousands of lives have been saved and thoracic surgery attained. Died July 29, 1918.

FELLENBERG, (Ludwig) Rudolf von, see von Fellenberg, (Ludwig) Rudolf.

FELLER, David Douglas, Am. physiologist; b. San Francisco, July 25, 1922; s. Arthur and Bertha (Graubart) F.; A.B., U. Cal. at Los Angeles, 1944; Ph.D., U. Cal. at Berkeley, 1950; m. Bernice Jean Ross, Feb. 9, 1947; children—Barbette Carol, Richard Ira, Steven Ross. Jr. research chemist Shell Devel. Co., Emoryville, Cal., 1944; instr. Tufts U., Boston, 1950-52; prin. scientist, radioisotope service VA Hosp., research asso. prof. U. Wash., Seattle, 1952-62; asst. chief environmental biology div. Ames Research Center, NASA, Moffett Field, Cal., 1962-—; lectr. Stanford, 1962-—. Pub. Health Service Research fellow Nat. Cancer Inst., 1948-50; Travel fellow Am. Heart Assn., 1958; Nat. Acad. Sci.-NRC fellow, 1959. Mem. Am. Physiol. Soc., Soc. Exptl. Biology and Medicine, Canadian Physiol. Soc., Aerospace Med. Assn., Sigma Xi. Research, publs. in chem. changes in blood and tissues resulting from exposure of rats to high doses of radioactive iodine; showed removal of thyroid gland with radioiodine, defects in carbohydrate metabolism in diabetes with aid of radioactive glucose, metabolism of adipose tissues, alterations in fat metabolism of liver from rats exposed to elevated gravity. Home: 1216 Forest Av., Palo Alto, Cal. 94301. Office: Ames Research Center, NASA, Moffett Field, Cal. 94035.*

FELLER, William, mathematician; b. Zagreb, Yugoslavia, July 7, 1906; s. Eugene V. and Ida (Perc) F.; M.S., Zagreb Univ., 1925; Ph.D., Gottingen Univ., 1926; m. Clara Mary Nielsen, July 27, 1938. Came to U. S., 1939, naturalized, 1944. In charge of applied math. lab. U. Kiel, 1929-33; research asso., U. Stockholm, 1933-39; asso. prof. and exec. editor, Math. Reviews, Brown U., 1939-45; prof. Cornell U. 1945-50; Eugene Higgins prof. math. Princeton, 1950-—; cons. and war work. Hon. fellow Royal Statis. Soc.; fgn. asso. Royal Danish Acad.; mem. Nat. Acad. Scis., Internat. Inst. Statistics Am. Math. Soc., Am. Acad. Arts and Scis., Inst. Math. Statistics (pres. 1949), Yugoslav Acad. Scis., Biometric Soc. Editor Math. Review. Author: An Introduction to Probability Theory and its Applications, 1950. Contbr. articles on probability, statistics, geometry, differential equations, real variables. Largely responsible for making probability theory a math. discipline; devised new framework for nature of chance fluctuations, especially with limit theorems, Markov processes and methodology. Address: Fine Hall, Princeton U., Princeton, N.J.

FELLERS, Carl Raymond, Am. bacteriologist; b. Hastings, N.Y., Oct. 4, 1893; s. Frank and Mary (Baratier) F.; A.B. in Chemistry, Cornell U., 1916; M.S., Rutgers U., 1916; Ph.D., 1918; m. Josephine Sanders, Mar. 28, 1921; children—Francis X., Mary J., Martha L., Anne M., John C., Paul J., David A., Stephen G. Bacteriologist (1st lt.) USPHS, Camps Greene, Bragg and Benning in lab. and field sanitation work, 1918-19; bacteriologist (food law enforcement) U. S. Bur. Chemistry and Soils, Washington, also San Francisco, 1920; research bacteriologist on seafoods and canned foods Nat. Canners Assn., Seattle, also State of Wash., 1921-23; asso. prof., food preservation (tchr. and research in food tech., especially fish and marine products), U. Wash., Seattle, 1924-25; research prof.

food tech. Mass. State Coll. (now U. Mass.) and Expt. Sta., Amherst, 1926, head dept., 1941-57; food and nutrition cons., from 1957. Food research, human and animal nutrition; dir. Blue Channel Corp., Beaufort, S.C. (packing marine products). Recipient Babcock award Am. Inst. Nutrition 1950. Fellow A.A.A.S., Am. Pub. Health Assn.; mem. Am. Chem. Soc., Soc. Am. Bacteriologists, Am. Soc. for Hort. Sci., Am. Fisheries Soc., N.Y. Acad. Sci., Inst. Food Tech. (pres. 1949), Sigma Xi. Contbr. to books, jours. Abstractor, Chem. Abstracts; editor Food Research and Quick Frozen Foods. Inventor methods for pasturizing dried foods, canning Atlantic crabs, ascorbic acid antioxidants and use of chelating agts. in foods. Died Feb. 22, 1960.

FELLINGER, Karl, Austrian physician; b. Linz, Austria, June 19, 1904; s. Karl and Mathilde (Leibetseder) F.; M.D., U. Vienna, 1929; M.D. (hon.), U. Thessaloniki, 1966, U. Athens, 1966; m. Barbara Issakides, Oct. 10, 1953. Faculty, U. Vienna, 1929—, prof., dir. med. clinic, 1946—. Pres., Austrian Supreme Council of Health, 1956—. Recipient decorations from govts. in Europe, Far East. Mem. Austrian Soc. Internal Medicine (chmn.), A.C.P. (corr.), others. Author: Endocrinology, 1938; also numerous articles. Research on thyroid pathology; demonstration of hepatitis-virus importance of ergastoplasma. Home: 7 Garnisongasse, Vienna 1090, Austria.*

FELLNER, William John, economist; b. Budapest, Hungary, May 31, 1905; s. Henry and Margaret (Leipziger) F.; student, U. of Budapest, 1922-23; B.S., Fed. Inst. of Tech., Zurich, 1927; Ph.D., Univ. of Berlin, 1929; m. Valerie Korek, Jan. 4, 1936; 1 dau., Anna Valerie (Mrs. Christopher B. Becker). Partner in a manufacturing firm in Budapest, Hungary, 1929-38; lecturer in economics, U. of Calif., Berkeley, Calif., 1939-40, asst. prof. 1941-42, asso. prof. 1943-47, professor, 1947-52; prof. econs. Yale, 1952—, Sterling prof. econs., 1959—, chairman department of economics, 1962-64; consulting expert U. S. Treas. Dept., 1945, 49-52; consultant Nat. Security Resources Bd., 1948-49; vis. lectr., Harvard, 1950-51; Alfred Marshall lectr. U. Cambridge, 1957. Fellow Am. Acad. Arts and Sciences; member of American Economic Association (member executive committee 1955-58), Econometric Soc., Phi Beta Kappa (hon.). Author: A Treatise on War Inflation, 1942; Monetary Policies and Full Employment, 1946, 2d edit., 1947; Competition Among the Few, 1949; Trends and Cycles in Economic Activity, 1956; Emergence and Content of Modern Economic Analysis, 1960; Probability and Profit, published 1965. Joint author: Survey of Contemporary Economics, 1948; Money, Trade and Economic Growth, 1951; Studies in Income and Wealth, Vol. 16, 1954. Joint editor, Readings in the Theory of Income Distribution, 1946; mem. editorial bd. Am. Econ. Review, 1950-52. Contbr. articles to various economics journals. Interpretation of economic data in terms of theory of "induced bias" of inventions; interpretation of economic behavior, particularly oligopolistic market results, in terms of theory of uncertainty rooted in modern theory of utility and subjective probability. Home: 131 Edgehill Rd., New Haven. Office: Economics Dept., 37 Hillhouse Ave., Yale U., New Haven, Conn.*

FELLOWS, Sir Charles, English archaeologist; b. Nottingham, Eng., Aug. 1799; s. John F.; m. 1st, Eliza Hart, Oct. 25, 1845; m. 2nd, Mrs. Harriet Knight, June 22, 1848. Traveled through Britain; settled in London, 1820; made 13th recorded ascent of Mont Blanc, 1827; lived in Italy and Greece, 1832-42. Knighted, 1845; mem., Brit. Assn. Author: A Narrative of an Ascent to the Summit of Mont Blanc, 1827; A Journal written during an Excursion in Asia Minor, 1839; An Account of Discoveries in Lycia, 1841; The Xanthian Marbles, their Acquisition and Transmission to England, 1843; An Account of the Ionic Trophy Monument Excavated at Xanthus 1848; Travels and Researches in Asia Minor, more particularly in the Province of Lycia, 1852; Coins of Ancient Lycia before the Reign of Alexander, with an Essay on the relative Dates of Lycian Monuments in the British Museum, 1855. Discovered ancient sites in Asia Minor: at Xanthus and numerous other cities. Died London, Eng., Nov. 8, 1860.

FELS, Eberhard Maximilian, mathematician, educator; b. Berlin, Germany, Jan. 13, 1924; s. G. Maximilian and Dora (Lenz) F.; Diplomvolkswirt U. Munich (Germany), 1951, Dr. oec. publ., 1953; postgrad. Harvard; m. Ann Zane Heinrich, Aug. 24, 1957; children—Marc, Ruth. Editorial writer Prisma, Munich, 1947-48, Die Neue Zeitung, 1948-51; research asso. Institut für Wirtschaftsforschung, Munich, 1953; teaching fellow in econs. Harvard, 1955; fellow Center for Advanced Study in Behavioral Scis., Stanford, Cal., 1956; vis. asst. prof. econs. U. Cal. at Berkeley, 1957-58; asso. prof. math. U. Pitts., 1958-64, prof. statistics and social scis., 1962—; prof. statistics U. Munich, 1963—. Contbr. articles to German, Am. tech. jours. Home: Ostmarkstrasse 20, 8 Munich 55. Office: 33 Ludwigstrasse, Munich, Germany.*

FELSEN, Leopold Benno, radiophysicist; b. Munich, Germany, May 7, 1924; s. Markus and Anna (Karfiol) F.; came to U. S., 1940, naturalized, 1943; B.E.E. summa cum laude, Bklyn. Poly. Inst., 1948, M.E.E., 1949, D.E.E., 1952; m. Sima Laks, May 10, 1944; children—Judith, Michael Electronics splst. U. S. Army, 1943-46; with Bklyn. Poly. Inst., 1953—, prof. electrophysics, 1961—; liaison scientist U. S. Office Naval Research, London, 1960-61; cons. radiophysicist to pvt. industry; individual research and supervision research contracts. Fellow I.E.E.E.; mem. Internat. Sci. Radio Union (vice-chmn. U. S. Commn. 6), Am. Assn. U. Profs. Research, publs. in electromagnetic propagation, radiation and diffraction theory, especially in plasma media; devel. modern curriculum in advanced electromagnetic theory. Home: 32 Cary Rd., Great Neck, N.Y. 11021. Office: Bklyn. Poly. Inst., Farmingdale, N.Y. 11735.*

FELSING, William August, Am. chemist; b. Denton, Tex., May 19, 1891; s. William and Anna Judith (Kurner) F.; A.B., U. Tex., A.M., 1916; Ph.D., Mass. Inst. Tech., 1918; m. Stella Elizabeth Scorgie, Sept. 8, 1920; children—Barbara Ann, William August. Successively prof. chemistry, asso. prof., prof. U. Tex., also part-time research scientist, def. research lab.; with Underwater Sound Lab., Harvard, 1943. Mem. Tex. Acad. Scis., Am. Chem. Soc., Sigma Xi, Phi Beta Kappa. Author: Notes on Descriptive Chemistry, 1928; General Chemistry (with E. P. Schoch and G. W. Watt) 1938, rev., 1946. Contbr. research articles to sci. jours. Died Oct. 5, 1952.

FELSON, Benjamin, Am. physician; b. Newport, Ky., Oct. 21, 1913; s. Solomon and Esther (Bissell) F.; M.D., U. Cin., 1935; m. Virginia Raphaelson, Mar. 18, 1936; children—Stephen R., Nancy R., Marcus K., Richard B., Edward J. Asst. dir. radiology Cin. Gen. Hosp., 1948-51, dir., 1951—; with U. Cin., 1936—, prof. radiology, 1951—; practice medicine specializing in radiology, Tulsa, 1941-42; dir. radiology Daniel Drake Meml., Children's, Holmes, Dunham, Longview hosps. Cons. to hosps., govt. agys.; chmn. Commn. on Edn. Fellow Am. Coll. Radiology (chancellor), Am. Coll. Chest Physicians; hon. mem. Nat. Acad. Medicine Colombia, Brazilian Coll. Radiology, Cuban, Colombian, Tex., Rocky Mountain radiol. socs.; mem. Cin. Acad. Medicine, Ohio State Med. Assn., Ohio State, Greater Cin. radiol. socs., Radiol. Soc. N.Am. (past 1st v.p.), Am. Roentgen Ray Soc., Alpha Omega Alpha, Pi Kappa Epsilon. Author: Fundamentals of Chest Roentgenology, 1960; (with A. Weinstein, H. Spitz) Principles of Chest Roentgenology, 1965; (with J. F. Wiot) Case of the Day, 1967. also numerous articles. Editor: Roentgen Techniques in the Experiment Animal, 1968; Seminars in Roentgenology. Asso. editor Diseases of Chest. Research on roentgenology of chest, programmed learning in medicine, clin. radiology. Home: 3994 Rose Hill Av., Cin. 45229.*

FELT, Dorr Eugene, Am. inventor; b. Beloit, Wis., Mar. 18, 1862; s. Eugene Kincaid and Elizabeth (Morris) F.; ed. pub. schs., Beloit; m. Agnes McNulty, Jan. 15, 1891; children—Virginia, Elizabeth, Constance, Dorothea. Inventor 1st key operator calculating machines, also 1st practical adding and listing machines; organized firm Felt & Tarrant, 1886, to manufacture his inventions, and in 1887 inc. Felt & Tarrant Mfg. Co., also pres. Died Aug. 7, 1930.

FELT, Ephraim Porter, Am. entomologist; b. Salem, Mass., Jan. 7, 1868; s. Charles Wilson and Martha Seeth (Ropes) F.; grad. Mass. Agrl. Coll., 1891; S.D., Cornell, 1894; m. Helen Maria Otterson, June 24, 1896; children—Margaret, Ernest Porter, Helen, Elizabeth. Specialist in entomology Gypsy Moth Commn., Mass; tchr. natural sciences, Clinton Liberal Inst., Ft. Plain, 1893-95; asst. to state entomologist, 1895; acting state entomologist, 1898, state entomologist of N.Y., 1898-1928; dir. and chief entomologist Bartlett Tree Research Labs. from 1928. Collaborator U. S. Bur. Entomology, N.Y. State Mus.; chief entomologist Gypsy Moth Bur., N.Y. State Conservation Commn., 1923-24. Fellow Entomol. Soc. Am.; mem. A.A.A.S. (life), Am. Assn. Econ. Entomologists (pres.), N.Y. Entomol. Soc., Entomol. Soc. Washington, Conn. Tree Protective Assn. (pres.), Sigma Xi. Published extended work on park and woodland insects, a manual of tree and shrub insects, key to Am. insect galls, shade trees, plant galls and gall makers, pruning trees and shrubs, shelter trees in war and peace, 25 ofcl. reports, a number of bulls., many articles jours. Hon. editor Jour. of Econ. Entomology. Co-author: (with W. H. Rankin) Insects and Diseases of Ornamental Trees and Shrubs. Died Dec. 14, 1943.

FELT, Vladimír, Czechoslovakian physician; b. Ceská, Lípa, Czechoslovakia, Nov. 25, 1926; s. Alois and Blazena (Nová) F.; M.D., Charles U., Prague, 1950; C.Sc., Czechoslovakian Acad. Sci., Physiol. Inst., 1959; m. Zdenka Boucková, Aug. 28, 1957; 1 son, Vladimír. Staff physician, endocrinologist Zilina and Kladno Hosp., until 1954; research fellow Inst. Cardiovascular Diseases, Prague, 1954-59; chief dept. clin. and exptl. biochemistry Research Inst. Endocrinology, Prague, 1960—, career scientist, 1966—. Mem. Czechoslovakian Soc. Medicine of J. E. Purkyne, European Soc. for Comparative Endocrinology. Contbr. to Blood Lipids and the Clearing Factor, 1956, Metabolismus parietis vasorum, 1962; also numerous articles. Research in mechanism of action of steroid, thyroid and hypophyseal hormones on lipid metabolism in health and disease; expts. in thyroadrenal relations and in devel. and treatment thyrocardiac disease. Home: 525 Jihlavská. Office: 8 Národní, Prague, Czechoslovakia.*

FELTZ, Victor Timothée, French bacteriologist; b. France, 1835; M.D., Strasbourg, France, 1860; prof. anatomy and physiol. pathology U. Nancy, France. Author: Étude clinique et expérimentale des embolies capillaires, 1868; (with Ritter) De l'urémie expérimentale, 1881. pioneer (with L. Coze) in study of germs of diseases, 1866-70. Died 1893.

FENCHEL, Werner, mathematician; b. Berlin, Germany, May 3, 1905; s. Carl and Gertrud (Hirsch) F.; Dr.phil., U. Berlin, 1928; m. Käthe Sperling, Dec. 16, 1933; 1 son, Tom Michael. Asst., Inst. Math., U. Göttingen (Germany), 1928-33; research asso. Inst. Math., U. Copenhagen (Denmark), 1933-42, faculty, 1942—, prof., 1956—. Vis. lectr. U. Lund (Sweden), 1943-45; vis. prof. various univs., U. S., 1950-51, Rockefeller Found. fellow, Rome, Italy, Copenhagen, 1930-31. Mem. Royal Danish Acad. Scis. and Letters, Danish Acad. Tech. Scis., Danish, Am. math. socs. Author: (with T. Bonnesen) Theorie der konvexen Körper, 1934; also articles. Co-editor Mathematica Scandinavica, 1953—. Research on global differential geometry, theory of convex sets and functions, theory of discontinuous groups of isometries. Home: 110 Sönderengen, Söborg 2860, Denmark. Office: 5 Universitetsparken, Copenhagen 0, 2100 Denmark.*

FENG, Paul Yen-Hsiung, nuclear scientist; b. Peiping, China, Aug. 29, 1926; s. Chih-Chung and Pao-Ju (Hu) F.; B.S., Fu-Jen Cath. U., Peiping, 1947; postgrad. Nat. Peking U.; Ph.D., Washington U., St. Louis, 1954; m. Mary Stella Pao-Ching Pai, Oct. 2, 1947; children—Joseph, Dorothy, Alphonso. Came to U. S., 1950, naturalized, 1963. Staff, Washington U., 1950-54; with Manu-Mine Research & Devel. Co., Reading, Pa., 1954-55; tech. dir., 1955; staff Research Inst. Ill. Inst. Tech. (formerly Armour Research Found.), Chgo., 1955—, mgr., 1960-62, sci. adviser, 1962—, lectr. Ill. Inst. Tech., 1956-60; vis. prof. Inst. Nuclear Sci., Formosa, 1958. Mem. Chinese Adv. Com. on Cultural Relations in Am., 1959-60, 65—; Fulbright lectr. Nat. Taiwan U., 1965. Recipient Sci. award Republic of China, 1961. Mem. Am. Nuclear Soc., Am. Chem. Soc., A.A.A.S., Faraday Soc., Radiation Research Soc., Sigma Xi. Research, publs. on method for specific tritiation organic compounds, basic and applied studies in radiation chemistry, high temperature phenomena, colloidal sci., nuclear sci. Patentee method preparing high specific activity cobalt-60, method preparing fluorinated organic compounds by radiation. Home: 88 Robsart Rd., Kenilworth, Ill. 60043. Office: 10 W. 35th St., Chgo. 60616.*

FENN, John Bennett, Am. chemist; b. N.Y.C., June 15, 1917; s. Herbert Bennett and Jeanette (Dingman) F.; A.B., Berea Coll., 1937; Ph.D., Yale, 1940; m. Margaret Elizabeth Wilson, June 6, 1939; children—Margaret Marianne (Mrs. Robert David Steinberg), Barbara Leigh (Mrs. Donald Stewart Leslie), John Bennett. Research chemist Monsanto Chem. Co., 1940-43; Sharples Chems., Inc., 1943-45; research dir., v.p. Expt., Inc., 1945-52; dir. project SQUID, Princeton, 1952-62, prof. aerospace scis., 1959-67; professor applied sci. and chemistry Yale, 1967—. Sci. liaison officer Office Naval Research, London, Eng., 1955; vis. scientist N.Am. Aviation Sci. Center, Thousand Oaks, Cal., 1965-66; cons. to pvt. industry, govt. agys. Recipient Office Naval Research Service award, 1964. Mem. Am. Inst. Aeros. and Astronautics, Am. Rocket Soc., Am. Chem. Soc., A.A.A.S., Sigma Xi. Contbr. articles to tech. jours. Inventor processes for making mercaptans, acetylenes and ethylene by high intensity combustion, propulsion systems; low temperature plasma jet; research on flame ignition phenomena; devel. high intensity and high energy molecular beams for study atomic and molecular collison processes; research on rarefied gas dynamics. Home: 120 Everit St., New Haven.*

FENN, Wallace Osgood, Am. physiologist; b. Lanesboro, Mass., Aug. 27, 1893; s. William Wallace and Faith Huntington (Fisher) F.; A.B., Harvard, 1914, A.M., 1916, Ph.D., 1919; D.Sc., U. Chgo., 1950, U. Paris (France), 1960, U. Brussels (Belgium), 1965, U. Rochester, 1965; Hon. Cathedratico, San Marcos U., Lima, Peru, 1959; m. Clara Bryce Comstock, Sept. 9, 1919; children—William Wallace F., Ruth (Mrs. Lawrence J. Starman), Priscilla (Mrs. John D. Roslansky), David Bryce. Instr., Harvard, 1919-22; Travelling fellow Rockefeller Inst. Med. Research, 1922-24; prof. physiology, chmn. dept. U. Rochester Sch. Medicine and Dentistry, 1924-59, Distinguished U. Prof. physiology, 1959—, dir. Space Sci. Center, 1961-66. Recipient Daniel and Florence Guggenheim award Internat. Acad. Astronautics, 1964; Gold medal award Med. Alumni U. Rochester, 1958; Distinguished Achievement award from Modern Medicine, 1966; Research Achievement award from the American Heart Association, 1967. Honorary member of the Canadian Physiol. Soc., Physiol. Soc. (Britain), Acad. Nazion. dei Lincei Rome (Antonio Feltrinelli Internat. prize for exptl. medicine 1964), Soc. Argentina de Biologia, Italian Soc. for Exptl. Biology; mem. Internat. Union Physiol. Sci. (past sec. gen.), Am. Physiol. Soc. (past pres.), Nat. Acad. Sci., Am. Philos. Soc., Am. Acad. Arts and Sci, Am Inst. Biol. Sci. (past pres.), Soc. Exptl. Biology and Medicine (past pres.). Author: History American Physiological Society, 1937-62, 1963; A Graphical analysis of the Respiratory Gas Exchange: the O2-CO2 Diagram (with H. Rahn),

1955; also numerous articles. Research on heat prodn. muscle during work, force-velocity curve muscle, muscle electrolytes and permeability of tissues to potassium, exchange of K and Na in stimulated muscle, phagocytosis of solid particles, increased oxygen consumption of nerve when stimulated, pressure breathing, pressure-vol. curve of lungs and chest, composition of alveolar air on oxygen-carbon dioxide diagram, oxygen poisoning, nitrogen narcosis. Home: 1394 Highland Av., Rochester, N.Y. 14620.*

FENNAH, Ronald Gordon, English entomologist; b. Ludlow, Shropshire, Eng., Jan. 23, 1910; s. Thomas Edward and Isabel (Norman) F.; B.A., Cambridge (England) University, 1933, M.A., 1937, Sc.D. (honorary), 1967; m. Louise May Florence Lyndon Kerr, Feb. 26, 1949. Asso., Imperial Coll. Tropical Agr., Trinidad, W.I., 1933-34, lectr., 1935-37; entomologist, officer-in-charge Citrus Pests Investigation, Windward and Leeward Islands, W.I., 1937-42, Food-crop Pests Investigation, 1942-48; entomologist Dept. Agr., Trinidad, 1948-51, Regional Research Centre (formerly Cacao Research Scheme), 1951-58; asst. dir. Commonwealth Inst. Entomology, London, 1958-—; sec. Central Plant Quarantine Sta., Imperial Coll. Tropical Agr., 1948-51, sec., officer in charge, 1956-58. John Simon Guggenheim Meml. Found. Research fellow, 1952-—. Mem. Royal Entomol. Soc. London, Biol. Soc. Washington, Entomol. Soc. Washington, Soc. Systematic Zoology, Systematics Assn. (U.K.). Author: The Insect Pests of Food Crops in the Lesser Antilles, 1947; also numerous articles. Research on insect pest of econ. crops in Lesser Antilles and Trinidad, taxonomy of Fulgoroidea, a superfamily of Homopterous insects; established existence of a relationship between outbreaks of cacao-thrips and certain other insect pests and physiol. condition of their host plants. Office: Commonwealth Inst. Entomology, 56 Queen's Gate, London S.W. 7, Eng.

FENNELL, Adolf, instrument designer; b. Kassel, Germany, Mar. 7, 1860; s. Otto and Elisabeth (Zahn) F.; ed. upper indsl. sch., Kassel; D. Eng. (hon.), Stuttgart, Germany; m. Anna Marie Margarete Schmidt, 1887; 1 son, 2 daus. Entered father's firm, 1877, became (with brother) mgr., 1891. Author many publs. Leading designer and mfr. precision surveying instruments; improved accuracy of tachometers, microscopes, dividing machines; holder 12 patents. Died Kassel, Mar. 1, 1953.

FENNEMAN, Nevin Melancthon, Am. geologist; b. Lima, O., Dec. 26, 1865; s. William Henry and Rebecca (Oldfather) F.; A.B., Heidelberg Coll., Tiffin, O., 1883; Ph.D., U. Chgo., 1901; LL.D., U. Cin., 1940; m. Sarah Alice Glisan, Dec. 26, 1893 (dec. 1920). Prof. phys. scis. Colo. State Normal Sch. (now Colo. State Coll. of Edn.), 1892-1900; prof. geology U. Colo., 1902-03, U. Wis., 1903-07, U. Cin., 1907-37, emeritus prof. from 1937; geologist, U. S. Geol. Survey, 1924, Wis. Geol. and Natural History Survey, 1900-02, Ill. State Geol. Survey, 1906-08, Ohio Geol. Survey, 1914-16. Recipient Gold medal, Geog. Soc. Chgo., 1938. Fellow A.A.A.S. (v.p., chmn. sec. E., 1923), Geol. Soc. Am. (pres. 1935); mem. Assn. Am. Geographers (pres. 1918), Am. Soc. Naturalists, Sigma Xi, Phi Beta Kappa (hon.). Mem. NRC, 1917-24, 1932-35, chmn. div. of geology and geography, 1922-23. Author: Physiographic Divisions of the United States; Physiography of Western United States; Physiography of Eastern United States; also numerous govt. bulls., sci. papers. Conducted 1st comprehensive study of physiography of U. S., 1931-38. Died July 4, 1945.

FENNER, Clarence Norman, Am. geologist; b. nr. Paterson, N.J., July 19, 1870; s. William Griff and Elmina Jane (Carpenter) F.; E.M., Sch. Mines (Columbia), 1892, A.M., Columbia, 1909, Ph.D., 1910. Mining and econ. geol. work in U. S., Can., Mexico, Central and S. A., 1892-1907; petrologist Geophys. Lab., Carnegie Instn., Washington, 1910-37, research asso., 1937-38, engaged in researches on application physics and chemistry to geology; mem. N.Y. Acad. Sci. expdn. on geol. reconnaissance of P.R., 1914; researches on optical glass for mil. purposes under War Industries Bd., 1917-1918; geologist Nat. Geog. expdn. to Valley of 10,000 Smokes, Alaska, 1919; leader 2d expdn. sent by Geophys. Lab. to same locality, 1923; geol. investigations in Peru, 1939; rep. Geol. Soc. Am. on NRC, 1925-28, mem. exec. com. div. geology and geography, 1926-28. Fellow Geol. Soc. Am., Am. Phys. Soc., A.A.A.S.; mem. Am. Inst. Mining and Metall. Engrs., Washington, N.Y. acads. scis., Am. Geophys. Union (chmn. sect. of volcanology, 1933-35); fgn. corr. Geol. Soc. London. Writer many sci. papers, especially on geol. and volcanological subjects. Elucidated stability relations of silica minerals. Died 1949.

FENSKE, Merrell Robert, Am. chem. engr.; b. Michigan City, Ind., June 5, 1904; s. William A. and Minna Glassman) F.; A.B. (Rector fellow), DePauw U., 1925, D.Sc., 1946; D.Sc., Mass. Inst. Tech., 1927. Research asso. Mass. Inst. Tech., 1928-29; faculty Pa. State U., University Park, 1929-—, prof. chem. engring., 1936-—, head dept. chem. engring., 1959-—. Chmn. chem. engring. rev. com. Argonne Nat. Labs., 1962-63; mem. adv. panels Dept. Def.; mem. U. S. Nat. Com. World Petroleum Congresses; cons. govt. agys. Recipient certificate of Merit OSRD, 1945; Naval Ordnance Devel. award, 1945. Fellow Inst. Petroleum London (Redwood medal 1964), A.A.A.S., Royal Soc. Arts London; mem. Am. Chem. Soc., Am. Inst. Chem.E., Am. Petroleum Inst., Am. Soc. Testing Materials, Am. Soc. Engring. Edn., Am. Soc. Lubrication Engrs. (Nat. award 1966), Soc. Automotive Engrs., Am. Soc. M.E., Am. Assn. U. Profs., Soc. Chem. Industry, Phi Beta Kappa, Sigma Xi, numerous others. Patentee in field. Research, numerous publs. in chem. engring. and petroleum processes. Home: P.O. Box 202, State College, Pa. 16801. Office: Chem. Engring. Bldg., Pa. State U., University Park, Pa. 16802.*

FENTON, Henry John Horstman, Brit. chemist; b. 1854; s. John Fenton; ed. Magdalen Coll. Sch., Oxford, Eng., King's Coll., London, Eng., Christ's Coll., Cambridge, Eng.; M.A., Sc.D.; m. dau. of George Ferguson, 1892. Hon. fellow Christ's Coll.; univ. lectr. in chemistry. Fellow Royal Soc., 1899. Contbr. papers on organic and phys. chemistry to sci. jours. Died Jan. 13, 1929.

FENTON, Paul Fredric, biochemist; b. Stuttgart, Germany, Nov. 28, 1915; s. Paul and Rose (Spohn) F.; B.S., U. Rochester, 1938; M.S., U. Vt., 1940, Ph.D., 1945; m. Elizabeth Marcy DeForest, Dec. 20, 1941; children—John Paul, Robert Irving. Asst. prof. biochemistry U. Vt. Med. Sch., 1945; research asst. Yale Sch., Medicine, 1945-49; faculty Brown U., Providence, 1949-—, prof., 1953-—, chmn. exec. council div. biol. and med. scis., 1965-—. Mem. com. on growth Nat. Acad. Sci., 1949-52; mem. physiology tng. com. Nat. Inst. Gen. Med. Scis., USPHS, 1961-65. John Simon Guggenheim fellow, 1957-58. Mem. Am. Physiol. Soc., Am. Inst. Nutrition, A.A.A.S., Soc. for Exptl. Biol. Medicine. Contbg. author: Medical Physiology and Biophysics, 1960. Research, numerous publs. on functions of digestive system, quantitative nutritional requirements, regulation of food intake, causes of obesity, metabolic effects of hormones. Home: 55 Townsend St., West Barrington, R.I. Office: Brown U., Providence 02912.*

FENTON, William N(elson), Am. anthropologist; b. New Rochelle, N.Y., 1908; s. John William and Anna Belle (Nourse) F.; A.B., Dartmouth Coll., 1931; Ph.D., Yale, 1937; m. Olive Louise Ortwine, 1936; children—Elizabeth (Mrs. E. Mayo Snyder), John W., Douglas Bruce, Harry (dec.). Community worker, U. S. Indian Service, N.Y. Agency, in charge Tonawanda and Tuscarora Reservations, 1935-37; instr. sociology and anthropology, St. Lawrence U., 1937-38, asst. prof. 1938-39; instr. (summers) Allegany Sch. of Natural History, U. of Buffalo, 1938, St. Lawrence University, 1940; vis. prof. Northwestern U., 1947, U. Mich., 1951, U. Arizona, 1963; lecturer Johns Hopkins, 1949-50, Cath. University of Am., 1950-51; asso. anthropologist, Bureau of Am. Ethnology, Smithsonian Instn., 1939-43, sr. ethnologist, 1943-51, mem. and sec., war com. 1942-44, research asso. Ethnogeographic Board, 1943-45; exec. sec. division anthropology and psychology, National Research Council, 1952-54; dir., asst. commnr. N.Y. State Museum and Sci. Service, 1954-—. U. S. del. IV Internat. Congress Anthrop. and Ethnol. Sci., Vienna, 1952; member American delegation to VII Internat. Congress on Anthropology and Ethnology, Moscow, 1964. Ethnol. field trips to Iroquois Indian Reservations. Mem. com. on Lang. and Areal Implications. Commn. on Implication of Armed Service Ednl. Programs, Am. Council Edn. (1946). Awarded Cornplanter medal for Iroquois Research, 1965. Fellow A.A.A.S., Royal Anthrop. Inst., Am. Folklore Soc. (pres. 1959-60), Am. Anthrop. Assn. (exec. bd. 1963-65), Am. Ethnol. Soc. (president 1959), Am. Indian Ethnohistoric Conf. (president 1962), American Assn. Museums, Anthrop. Society Washington (former sec., vice pres., president), Sigma Xi. Author: Area Studies in American Universities, 1947; Iroquois Eagle Dance, 1953; Indian and White Relations to 1830, 1957, numerous articles. Discovered patterns of sequence governing Iroquois ceremonialism; developed methods of ethnohistory; descriptions of Iroquois cultural activities; factionalism in Am. Indian society. Home: 7 N. Helderberg Pkwy., Slingerlands, N.Y.; (summer) Keene Valley, N.Y. Office: State Education Bldg., Albany 1, N.Y.*

FENYVES, Ervin, Hungarian physicist; b. Budapest, Hungary, Aug. 29, 1924; s. Zoltán and Valeria (Burg) F.; student R. Eötvös U., Budapest, 1942-46, Ph.D., 1950; candidate physics Hungarian Acad. Scis., 1955, dr. physics, 1960; m. Vera Lippai, May 5, 1951; children—Eve, Andrew. Faculty, R. Eötvös U., 1946-—, prof. physics, 1964-—; research fellow Central Research Inst. for Physics, Hungarian Acad. Scis., Budapest, 1951-59, head Lab. for Cosmic Rays, 1959-65, dep. dir., 1965-—. Recipient Nat. Kossuth Price, 1965. Mem. R. Eötvös Phys. Soc. (mem. presidium). Author: Nuclear Radiation Measurements, 1956; (with O. Haiman) The Physical Principles of Nuclear Radiation Measurements, 1965; also articles. Research on nuclear active component of cosmic rays, strong interactions in high energy physics, especially peripheral collisions and resonances. Home: 8/b Gorkij fasor, Budapest, Hungary.*

FERBER, John, mineralogist; b. Karlskrona, Sweden, 1743; visited France, Germany, Eng.; became prof. natural history, Mittau, 1774; mem. Berlin Acad. Contbr. numerous publs. on phys. geography of world to tech. jours. Died Bern, Switzerland, 1790.

FERDMAN, David Lazarevich, Russian biochemist; b. Jan. 7, 1903; grad. Kharkov U., 1925. Asso., Inst. Biochemistry, Ukrainian Acad. Sci., 1928-—; prof. Kiev U., 1944-—. Decorated Order of Lenin. Mem. USSR (corr.), Ukrainian acads. sci. Author: The Biochemistry of Phosphoric Compounds, 1953; The Metabolism of Phosphoric Compounds, 1940; The Biochemistry of Muscular Diseases, 1953; The Chemistry and Biochemistry of Condensed Polyphosphates, 1960. Research and publs. on metabolism of phosphoric compounds, formation and elimination of ammonium in muscles during activity, metabolism of affected muscles; established presence of glutamine in animal tissues. Address: Kiev University, ulitsa Vladimirskaya 64, Kiev, Ukraine SSR, USSR.

FERENCZI, Sandor, Hungarian psychoanalyst; b. Miskole, Hungary, 1873; s. Polish bookstore mgr.; ed. medicine, Vienna, M.D., 1894; m. 1919. Military physician, 1894-95; private practice in neurology, psychiatry; chief neurologist, Elizabeth Poorhouse, 1900; apptd. psychiatric expert to Royal Court of Justice, Budapest, 1905; psychoanalytic practice in Hungary; lectr. before Budapest Med. Assn., 1908; with Freud to Clark U., U. S., 1909; founded Hungarian Psychoanalytic Soc., 1913 (pres. 1913-33); prof. psychoanalysis, U. Budapest, from 1919; lectr., New School for Social Research, N.Y., 1926-27. Author: Thalassa: A Theory of Genitality, 1924; (with Rank) The Development of Psychoanalysis, 1924; Further Contributions to the Theory and Technique of Psychoanalysis; numerous papers. Consolidated and extended basic thought of Freud; many theoretical and clinical ideas on treatment and psychopathology contributed to development of psychoanalysis; investigations of sexual habits, development of sense of reality, active theories in therapy; correlation of biology and psychoanalysis. Died 1933.

FERENTZ, Melvin, Am. physicist; b. N.Y.C., Oct. 14, 1928; s. Sigmond and Rose (Spitz) F.; B.S., Bklyn., 1949; Ph.D., U. Pa., 1953; m. Vivian Gritz, Nov. 24, 1948 (div.); children—Stefan, Adam; m. 2d, Elizabeth Wetherell, July 18, 1964; 1 dau., Ann. Asso. physicist Argonne Nat. Lab., 1952-57; sr. mathematician-analyst IBM, N.Y.C., 1957-59, cons., 1959-—; asso. prof. physics St. John's U., N.Y.C., 1959-65, prof., chmn. dept., 1965-66; prof. physics Bklyn. Coll., 1966-—, dep. chmn., 1967-—; research asso. in music, Columbia, 1964-—; asso. dir. CUNY-SUNY Inst. for Research in Learning, 1967-—. Pres., Ferentz Asso., Inc., N.Y.C., 1962-—; dir. Struthers Realtronics, Inc., Bauer-Mengelberg, Inc., N.Y.C. Mem. Am. Phys. Soc., Assn. for Computing Machinery, Am. Assn. U. Prof., Sigma Xi. Home: 445 E. 86 St., N.Y.C. 10028. Office: Bklyn. Coll. of CUNY, Bklyn. 11210.*

FÉRÉOL, Louis H. F., French physician; b. Paris, France, 1825; s. Louis Second Fereol; M.D., 1859; practice law, from 1847; practice medicine Central Bur., from 1865; mem. Acad. Medicine. Author: Observations et reflexions sure un cas de coloration bronzée de la peau, 1856; De la Perforation de la paroi abdominale antérieure dans les peritonites, 1859; Observations de chromidrose ou chromicrinie, 1885. Described Féréol's nodes, subcutaneous nodosities occurring around the joints in some cases of acute articular rheumatism, 1859. Died 1891.

FERETTI, Bruno, Italian physicist; b. Bologna, Italy, July 1, 1913; s. Mario and Nerina (Baraldini) F.; Ph.D. in physics. Former asst. Inst. Physics, Rome; instr. theoretical physics U. Bologna, 1941, now profl.; asso. prof. theoretical physics U. Milan, 1948; prof. U. Rome, 1950. Mem. dei Lincei Acad. (laureat 1951), Nat. Com. on Nuclear Energy, European Soc. Atomic Energy (work group). Research and numerous publs. on cosmic radiation, quantum electronics. Address: Universita degli Studi, Bologna, Italy.

FERGOLA, Nicolo, Italian geometer; b. Naples, Rome, 1752. Prof. math., Naples; mem. Naples Acad. Scis. Author: Prelezioni sui principii matematici della filosofia naturale del Newton, 1792; L'arte euristica, 1811; Trattato delle sezionoi coniche, 1817. Authority on conic sections; studied math. principles of Newton's natural philosophy. Died Naples, 1824.

FERGUS, Andrew Freeland, Scottish ophthalmologist; b. Glasgow, 1858; s. Andrew and Margaret (Naismith) F.; ed. Glasgow U., also Paris, Utrecht; M.D., LL.D. Ophthalmic surgeon Glasgow Eye Infirmary; prof. ophthalmology Anderson's Coll.; cons. ophthalmic surgeon Corp. Glasgow. Fellow Royal Soc. Edinburgh; pres. Royal Philos. Soc. Glasgow, Med. Chirurg. Soc., Royal Faculty Physicians and Surgeons, Glasgow. Author: Elementary Ophthalmic Optics; also papers on miners' nystagmus, seamen's eyesight, diplopia, bacteriology of conjunctive, light sense. Died Oct. 24, 1932.

FERGUS, Charles Leonard, Am. mycologist; b. Ottawa, Kan., Nov. 11, 1917; s. Hugh Lee and Edna (Ivey) F.; A.B., Ottawa U., 1940; M.A., Kan. U., 1942; Ph.D., Pa. State U., 1948; m. Winifred Ruth Foote, Oct. 31, 1942; children—Charles L., Michael F., Brian S. Instr., Kan. U., 1940-41; faculty Pa. State U., University Park, 1948-—, prof. mycology,

1960—. Mem. Mycol. Soc. Am., Phytopath. Soc. Am., Sigma Xi. Author: Illustrated Genera of Wood Decay Fungi, 1950; Laboratory Guide to Studies in Biology, 1966; also articles. Editor: Phytopathology, 1958-60. Research on biol. activities of disease causing fungi resulting in better control diseases of plants and ornamental plants, physiology of thermophilic molds and actinomycetes. Home: 542 W. Hillcrest Av., State College, Pa. 16801. Office: Buckhout Lab., Pa. State U., University Park, Pa. 16802.*

FERGUSON, Albert Barnett, Am. physician; b. N.Y.C., June 10, 1919; s. Albert Barnett and Vera (McCreary) F.; B.A., Dartmouth, 1941; M.D., Harvard, 1943; m. Louise Enequist, Nov. 20, 1943; children—Sanford, Bruce, Gary, Laurie. Asst. orthopedic surgeon Harvard Med. Coll., 1951-53; Silver prof. Pitts. Med. Coll., 1953—. Author: Orthopedic Surgery Infancy and Childhood, 1957; Metals and Engineering in Surgery (with Laing, Bechtol), 1960; (with Benden) ABC's of Athletic Injuries, 1964; also numerous articles. Research on action ions from metals inserted in living tissue, cause and prevention roundback in children, use radioactive materials to determine bone viability, fundamental pathology degenerative arthritis. Home: 14 James Ross Pl., Pitts. 15215.*

FERGUSON, Denzel E., Am. zoologist; b. La Grande, Ore., Aug. 28, 1929; s. Leonard Denzel and Alta (Harrison) F.; student Eastern Ore. Coll., 1947-49, U. Kan., 1952-53; B.S., Ore. State U., 1951, M.S., 1952, Ph.D., 1956; m. Virginia Louise Ferguson, Mar. 18, 1948; children—Leonard Denzel, Amanda Louise. Instr. biology Eastern Ore. Coll., 1953-54; faculty Miss. State U., 1956—, prof. zoology, 1965—. Recipient Miss. State U. Faculty Achievement award for Research, 1966. Mem. Am. Soc. Ichthyologists and Herpetologists (v.p. Southeastern div.), Am. Fisheries Soc., Wildlife Soc., Animal Behavior Soc., Am. Inst. Biol. Scis., Sigma Xi. Contbr. numerous articles to sci. jours. First demonstration of existence and mechanisms of celestial orientation in frogs and toads; first demonstration and elucidation of pesticide resistance in natural populations of vertebrates, especially fishes. Home: P.O. Box 452, Starkville, Miss. 39759. Office: P.O. Drawer 2, State College, Miss. 39762.*

FERGUSON, George A., Canadian psychologist; b. N.S., Can., July 23, 1914; s. Alexander MacN. and Jean (Dennistoun) F.; B.A., Dalhousie U., 1936; B.Eol., U. Edinburgh, Ph.D., 1940; m. Rowena Sheldon Bellows, May 1955; children—Claudia, Leith. Indsl. psychologist Stevenson & Kellogg, 1945-47; asst. prof. McGill U., Montreal, Que., Can., 1947-49, asso. prof., 1949-52, prof., 1952—, chmn. dept. psychology, 1964—, vice-dean for biol. scis., 1966—. Fellow Royal Soc.; mem. Am., Canadian psychol. assns. Author: Statistical Analysis in Psychology and Education, 2d edit., 1966. Contbr. articles to sci. jours. Research on nature of human intelligence; applications of math. methods to psychol. problems. Home: 3003 Cedar Av., Montreal, Que., Can.*

FERGUSON, Harry George, Brit. industrialist; b. Dromore, Ulster, Ireland, Nov. 4, 1884; s. James and Mary (Bell) F.; ed. privately; D.Sc., Belfast, Louvain, M.A.I., Dublin; married Mary Adelaide Watson, Apr. 21, 1913; 1 dau., Elizabeth. Engaged in automobile distbn. and service; built and flew own airplane in Ireland, 1909; chmn. bd. Harry Ferguson (Motors) Ltd., Belfast, Harry Ferguson Research, Ltd., Coventry, Eng.; governing dir. Harry Ferguson Ltd., Eng., Harry Ferguson Holdings, Ltd., Eng. Joint sponsor with Henry Ford, Nat. Farm Youth Found., Dearborn, Mich. Mem. Instn. Automobile Engrs., Royal Aero. Soc. (both London). Inventor Ferguson system of agrl. prodn., an integrated tractor-implement system, 1936; contbn. to development of auto transmissions, chassis and braking systems; pioneer in automobile and airplane design and development. Died Oct. 25, 1960.

FERGUSON, James, Scottish physicist, astronomer; b. Keith, Scotland, Apr. 25, 1710; s. John and Elspet (Lobban) F.; ed. grammar sch.; m. Isabella Wilson, May 1739; children—James, 2 other sons, Agnes. Showed ability in mechanics at 9 years; sent to work, 1720, shepherd for 4 years; in household of Thomas Grant of Achoynaney, from 1728; patronized by Sir James Dunbar; became portrait painter under patronage of Lady Dipple; lectr. on mathematics and drawing, Edinburgh, 1734-43; lectr. on exptl. philosophy, from 1743. Fellow Royal Soc., 1763. Author: Astronomy explained on Sir Isaac Newton's Principles, 1756; Lectures on Select Subjects in Mechanics, Hydrostatics, Pneumatics and Optics, 1760; Analysis of a Course of Lectures on Pneumatics, 1763; The Young Gentleman's and Lady's Astronomy, 1768; Introduction to Electricity, 1770; The Art of Drawing in Perspective, 1775; numerous other works. Built terrestrial globe from Gordon's Geog. Grammar; used 6 foot reflector to observe transit of Venus; devised astron. rotula; built orrery, 1742; inventor tide-dial, also eclipsareon, 1754, universal dialling cylinder, 1767; presented projection of partial solar eclipse of 1764 to Royal Soc., 1763. Died London, Nov. 16, 1776.

FERGUSON, James Kenneth Wallace, pharmacologist; b. Tamsui, Formosa, Mar. 19, 1907; s. James Young and Harriet (Wallace) F.; B.A., U. Toronto, 1928, M.A., 1929, M.D., 1932; m. Mary Frances Wyndow, Aug. 30, 1933; children—Ian James, Anne Leith (Mrs. Ian Lawrence Russell), Brian Kenneth, Shelagh Mary. Fellow NRC Cambridge, Eng., 1933-34; lectr., asst. prof. physiology U. Western Ont., 1934-36; asst. prof. physiology Ohio State U., 1936-38; faculty U. Toronto, 1938-55; dir. Connaught Med. Research Labs., 1955—. Mem. Pharmacological Soc. Can., Physiol. Soc. Can., Am. Physiol. Soc., Am. Soc. Pharmacology. Author: (with G. H. W. Lucas) Henderson's Materia Medica, 1948, 51, 54. Research, publs. on hemoglobin reduction, respiration at high altitudes, release of oxytocin in parturition, introduction of citrated calcium carbimide for treatment of alcohol.*

FERGUSON, John Howard, physiologist; b. Edinburgh, Scotland, Mar. 1, 1902; s. Howard and Annie (Bowers) F.; B.A., U. Cape Town, 1921, D.Sc., 1957; B.A., U. Oxford, 1925, M.A., 1931; M.D., Harvard, 1928; m. Rosalind Vera Carruthers, July 27, 1927; children—Muriel Joyce (Mrs. Alfred Descloux), Phyllis Mary (Mrs. Harris F. Evans), Margaret Anne (Mrs. James B. Johnston), Helen Forster (Mrs. John A. Rhoades, Jr.), John Carruthers, Colin Carruthers; m. 2d Doris Elizabeth Cummings, June 13, 1955. Came to U. S., 1931, naturalized, 1941. Practice medicine, specializing in physiology, Chapel Hill, N.C., 1943—; prof. physiology University of N.C., 1943—, head department of physiology, 1943-67. Principal investigator annual NIH grants. Fellow A.C.P., A.A.A.S.; mem. Am. Physiol. Soc., Internat. Hematology Soc. Author: Lipoids and Blood Platelets, 1960; Blood and Body Functions, 1965. Research, publs. on clotting and related functions of blood and applications to bleeding, thrombotic disorders. Home: 226 Glandon Dr. Office: Dept. Physiology, Sch. Medicine, U. N.C., Chapel Hill, N.C. 27515.*

FERGUSON, Lloyd Noel, Am. chemist; b. Oakland, Cal., Feb. 9, 1918; s. Noel Swithin and Gwendolyn L. (Johnson) F.; B.S., U. Cal. at Berkeley, 1940, Ph.D., 1943; m. Charlotte O. Welch, Jan. 2, 1944; children—Lloyd Noel, Stephen Bruce, Lisa Annette. Research asst. Nat. Def. Project, U. Cal. at Berkeley, 1941-44; asst. prof. A. and T. Coll., Greensboro, N.C., 1944-45; faculty Howard U., 1945-65, prof. chemistry, 1955-65, head dept., 1958-65; prof. chemistry Cal. State Coll., Los Angeles, 1965—; vis. scientist Am. Chem. Soc., 1956—. Mem. adv. bds. govt. agys. Fellow Chem. Soc. London; mem. Am. Chem. Soc., A.A.AS., Am. Assn. U. Profs., Sigma Xi. Author: Electron Structures of Organic Molecules, 1952; Textbook of Organic Chemistry, 1958; Modern Structural Theory of Organic Chemistry, 1963; also articles. Research on biochem. aspects of sense of taste, correlations of phys. properties with molecular structures. Home: 4221 S. Cloverdale Av., Los Angeles 90008.*

FERGUSON, Robert Earl, Am. phys. chemist; b. Seelyville, Ind., July 28, 1923; s. Earl and Agnes (McNary) F.; student U. Chgo., 1941-42; B.A. in Math., State U. Ia., 1949; Ph.D. in Phys. Chemistry, State U. Ia., 1952; m. Joan Mary Criel, June 14, 1963; stepchildren—Julia C., Mark A., Robyn E., Mary L.; 1 dau., Jennifer McNary. Research chemist Nat. Bur. Standards, Washington, 1952-60, chief elementary processes sect., 1960-64, chief phys. chemistry div., 1965—. Cons. Am. Instrument Co., Silver Spring, Md., 1959—; Mem. evaluation panel NATO Fellowships, 1963—. Recipient Distinguished Service award Nat. Bur. Standards U. S. Dept. Commerce, 1957, 59. Am. Polit. Sci. Assn. Congl. fellow, 1964-65. Mem. Am. Chem. Soc., Washington Acad. Scis., A.A.A.S., Internat. Combution Inst., Sigma Xi. Research, publs. on kinetics and mechanisms high temperature chem. reactions by isotopic tracer techniques; developed retarding-potential techniques for study ion decomposition and ionic collision processes by time-of-flight mass spectrometry. Home: 6307 Tone Dr., Bethesda, Md. 20034. Office: Nat. Bur. Standards, Washington, 20234.*

FERGUSON-SMITH, Malcolm Andrew, Scottish med. geneticist; b. Glasgow, Scotland, Sept. 5, 1931; s. John and Neo (Thorne) F.-S.; M.B., Ch.B., U. Glasgow, 1955, M.C. Path., 1966; m. Marie Eve Gzowska, July 11, 1960; children—Anne Carla, Nicola Mary. Registrar lab. medicine dept. pathology U. and Western Infirmary, Glasgow, 1958-61; fellow medicine Johns Hopkins Sch. Medicine, Balt., 1959-60, instr., 1960-61; faculty dept. genetics U. Glasgow, 1961—, sr. lectr. med. genetics depts. genetics and child health, 1965—. Hon. cons. med. pediatrics Royal Hosp. for Sick Children, Glasgow, 1966—. Recipient R. Thornton Wilson award in genetic and preventive psychiatry, 1961. Mem. Path. Soc. Gt. Britain and Ireland, Am. Soc. Human Genetics, Genetical Soc. Research, publs. on elucidation of incidence and etiology of human sex chromosome aberrations, characterization of normal human chromosomes, identification of chromosomal sites of nucleolus formation. Home: 5 Hamilton Dr., Glasgow W.2., Scotland.*

FERGUSSON, Gordon John, physicist; b. Whangarei, New Zealand, Aug. 2, 1922; s. Andrew Black and Jessie (Spence) F.; B.Sc., Auckland U., 1941; M.Sc., Victoria U., 1945, D.Sc., 1964; m. Barbara Joan Wadey, May 10, 1947; children—Claire Eleanor, Adele Christine, Heather Jane. Physicist, Radio Devel. Lab., Wellington, New Zealand, 1941-45, Ministry of Supply, Chalkriver, Can., Harwell, Eng., 1945-48; chief physicist Inst. Nuclear Sci., Lower Hutt, New Zealand, 1949-59; sci. sec. UN, N.Y.C., 1959-60; faculty U. Cal., Los Angeles, 1960-64; sr. physicist Johnston Labs., Balt., 1964-67; sr. physicist Sci. Research Instruments, Balt., 1967—, also dir. Recipient Mechaelis Meml. prize Otago U., 1954. Fellow Royal Soc. New Zealand (E. R. Cooper Meml. award 1959); mem. A.A.A.S., Am. Geophys. Union, Am. Inst. Physics. Devel. methods for low level radioactivity measurements; use of radiocarbon and tritium for studies of circulation of the ocean, groundwater and the atmosphere. Home: 233 Ridgely Rd., Lutherville, Md. 21093. Office: 6707 Whitestone Rd., Balt. 21207.*

FERGUSSON, Sir William, Scottish surgeon; b. Prestonpans, Scotland, Mar. 20, 1808; s. James Ferguson; ed. Edinburgh (Scotland) Coll. Surgeons, U. Edinburgh, LL.D., 1875; m. Helen Hamilton Ranken, Oct. 10, 1833; children—James Ranken, Charles Hamilton, also 3 daus. Surgeon Edinburgh Royal Infirmary, 1838-40; prof. surgery King's Coll., London, 1840-70; named sgt.-surgeon to Queen Victoria, 1867; clin. prof. surgery, sr. surgeon King's Coll. Hosp. Fellow Royal Soc. Edinburgh, Royal Soc., Eng.; mem. Royal Coll. Surgeons Eng., Path. Soc. (pres. 1859-60), Brit. Med. Assn. (pres. 1873); pres. Coll. Surgeons, 1870. Author: System of Practical Surgery, 1842; Lectures on the Progress of Anatomy and Surgery during the Present Century, 1867; also articles. Noted for skill in operation; 1st to apply term conservative survery to operations for preservation of parts of body, 1852; known for operations of cleft palate and harelip, jaw, excision of joint of hip, knee, elbow and other joints, also for lithotomy, lithotrity, amputations of limbs; devised various bent knives, bulldog forceps, mouth-gag for cleft palate; credited with introducing rack and pinion in structure of lithotrite, circa 1834. Died London, Feb. 10, 1877.

FERLAINO, Frank Ralph, Am. indsl. physician; b. N.Y.C., Feb. 1, 1900; s. John and Angelina (Paola) F.; B.S., Princeton, 1922, M.S., 1923; M.D., Johns Hopkins, 1927; postgrad. Columbia, 1934, 38, 41-43. Intern, resident physician Presbyn. Hosp. Med. Center, N.Y.C., 1927-30; med. asst. Med. Dept. Am. Tel. & Tel. Co., N.Y.C., 1929-38; asst. physician L.I. Med. Coll. and Kings County Hosp., L.I., 1934-37; asst. physician, instr. N.Y. Post Grad. Med. Sch. and Hosp., N.Y.C., 1935-44, asst. attending physician since 1944, co-dir. Ann. Symposium Indsl. Medicine and Surgery, 1943-46; med. examiner N.Y. Life Ins. Co., N.Y.C., 1931-33; physician in charge Compensation Clinic, Gen. Accident Ins. Co., N.Y.C., 1932-34; med. examiner, cons. traumatic diseases and injuries, various ins. cos.; from 1932, med. cons. Stony Wold Sanatorium, Lake Kushaqua, N.Y.; pvt. practice, specializing in heart and chest diseases, indsl. medicine and surgery, traumatic and indsl. med. cons., N.Y.C., since 1930; asso. med. dir. Schenley Labs., N.Y.C., 1944-46; med. dir. Gen. Motors Corp., N.Y.C., from 1946; staff physician Roosevelt Hosp., N.Y.C., since 1945; asst. attending physician U. Hosp., N.Y.C., from 1950; asst. clin. prof. indsl. medicine Inst. Indsl. Medicine, Bellevue Med. Center, N.Y. U., from 1949. Fellow Am. Coll. Chest Physicians; mem. A.M.A., Med. Soc. County Kings, Acad. Medicine Bklyn., N.Y. State Med. Soc., Indsl. Med. Assn., World Med. Assn., Am. Trudeau Soc., Nat. Tb Assn., A.A.A.S., Morgagni Med. Soc., Am. Pub. Health Assn., Am., N.Y. Heart assns., Am. Soc. Bacteriology, Am. Assn. Indsl. Physicians and Surgeons, Research and Edn. Found. for Study Common Cold, Johns Hopkins Med. and Surg. Assn., N.Y. State Soc. Indsl. Medicine, Soc. Med. Jurisprudence, N.Y. Acad. Scis. Asso. editor Indsl. Medicine and Surgery Med. Jour. Contbr. articles on indsl. medicine. Died Sept. 1953.

FERM, Vergil Harkness, Am. embryologist; b. West Haven, Conn., Sept. 13, 1924; s. Vergilius T. A. and Nellie (Nelson) F.; A.B., Coll. Wooster, 1946; M.D., Western Res. U., 1948; M.S., U. Wis., 1950, Ph.D., 1955; m. Ruth Eleanor Rowe, June 5, 1948; children—Daniel W., David V., Judith N., Susan C. Asst. prof. Ind. U., 1955-57; asso. prof. U. Fla., 1957-61; asso. prof. pathology Dartmouth Med. Sch., Hanover, N.H., 1961-66, prof. anatomy and embryology, 1966—. Mem. Am. Assn. Anatomists, Am. Soc. Human Genetics, Teratology Soc. Exptl. Pathology, Phi Beta Kappa, Sigma Xi. Research, publs. on environmental and genetic factors causing birth defects. Home: Dogford Rd., Etna, N.H. 03750. Office: Dartmouth Med. Sch., Hanover, N.H. 03755.*

FERMAT, Pierre de, French mathematician; b. Beaumont-de-Lomagne, France, Aug. 17, 1601; s. Dominique and Claire (de Lons) F.; studied law at Toulouse; m. Louise de Long, June 1, 1631; 3 sons, 2 daus. Commissioner of requests, Toulouse, 1631; King's councillor Parliament of Toulouse, 1648-65; regarded math. as recreation but did much original work in field; greatest writer on theory of numbers since Diophantos; father of modern theory of numbers; founded theory of probability (with B. Pascal); worked out equations for geometric figures shortly before Descartes pub. his Géometrie, worked in 3 dimensions; applied idea of infinitesimals to questions of quadrature and maxima and minima, and to constrn. of tangents, regarded as 1st inventor of calculus by Laplace and Lagrange for this work; defined tangent as secant whose two points of intersection with curve coincide; devised method to construct tangents that is roughly equivalent to that employed in calculus; found area generated by arc of cycloid; wrote on rectification of semicubic parabola; gave equation for cubic curve; solved problem of points by theory of combinations; developed numerous theorems, all of which have since been proved except one known as Fermat's last theorem (most of Fermat's own proofs not extant); studied reflection of light; developed Fermat's principle (1st variational principle of optics), which postulates that light ray traversing different media takes path of least duration; suggested that gravity is reciprocal attraction; his collected works pub. by his son under title Varia Opera Mathematics, 1679. Died Castres, nr. Toulouse, Jan. 12, 1665.

FERMI, Enrico, physicist; b. Rome, Italy, Sept. 29, 1901; s. Alberto and Ida (de Gattis) F.; Doctorate, U. Pisa, 1922; D.S. univs. Heidelberg, 1936, Utrecht, 1936, Columbia, 1946, Washington, 1946, Yale, 1946, Harvard, 1948, Rochester, 1952; LL.D., Rockford Coll., 1947; m. Laura Capon, July 10, 1928; children—Nella, Giullo. Came to U. S., 1939, naturalized, 1945. Lectr. physics U. Florence, 1924-26; prof. theoretical physics U. Rome, 1927-38; prof. physics Columbia, 1939-42; research on atomic bomb U. Chgo., 1942-45, prof. physics Inst. for Nuclear Studies, 1946-54. Recipient Nobel prize, 1938, Hughes medal Royal Soc., 1942, Congl. medal for merit U. S. A., 1946, Barnard medal Columbia, 1950, award of merit Pres. of U. S., AEC, 1954, also numerous others. Fellow Am. Acad. Arts and Scis. (Rumford medal 1953), Royal Soc. Edinburgh; mem. Am. Philos. Soc. (Lewis prize 1946), Nat. Acad. Scis., Franklin Inst. (Franklin medal 1947), Italian Physics Soc., Royal Soc. London, Am. Phys. Soc. (pres. 1953), numerous others. Author: Thermodynamics, 1937; Elementary Particles, 1951, also several books in Italian and articles. Investigated formation of artificial radioactive substances; 1st to bring about nuclear transformations of heavy elements by neutron bombardment, 1934; discovered thermal (or slow) neutrons are more easily absorbed by nuclei than accelerated particles; developed statis. model of atom (Thomas-Fermi model), 1927-28; named neutrino which Pauli had postulated, maintained it has character of real particle, worked out some of math. involved in neutrino emission; responsible for constrn. 1st atomic pile (or nuclear reactor) and 1st self-sustaining chain reaction at U. Chgo., 1942 (marked the beginning of the atomic age); helped develop atom bomb, 1942-45; worked on high energy physics, including pion-nucleon interactions; developed theory of cosmic ray origin; worked on quantum electrodynamics; investigated theory of hyperfine structures of spectrum lines; artificially formed element fermium named in his honor, 1955. Died Chgo., Nov. 28, 1954.

FERMOR, Sir Lewis Leigh, metallurgist; b. Sept. 18, 1880; s. Lewis Fermor; ed. Royal Coll. Sci., Royal Sch. Mines; D.Sc., London; m. Muriel Aileen Taaffe, 1909; 1 son, 1 dau.; m. 2d, Frances Mary Case, 1933. Asso., Royal Sch. Mines, 1901; with Geol. Survey of India, 1902, supt., 1910, officiating dir., 1921, 25, 28, 30-32, dir., 1932-35. Indian Govt. rep. Internat. Geol. Congress, Sweden, 1910, Can., 1913, Spain, 1926, South Africa, 1929; minerals adviser Indian Munitions Bd., 1917, 18; pres. geol. sect. Indian Sci. Congress, Bombay, 1919. Recipient Murchison medal and prize, 1900. Fellow Royal Soc., 1934, Geol. Soc. (Bigsby medal 1921), Indian Assn. for Cultivation Sci. (hon.); mem. Mining and Geol. Inst. India (pres. 1922), Indian Sci. Congress Assn. (pres. 1933), Royal Asiatic Soc. Bengal (pres. 1933-36, P.N. Bose Meml. medal 1944), Nat. Inst. Scis. India (pres. 1935-36), Bristol Naturalists Soc. (pres. 1945-47), Southwestern Naturalists Union (pres. 1948-49), Soc. Econ. Geologists (v.p. 1932), Inst. Mining and Metallurgy (pres. 1951), Royal Sch. Mines (asso.), Mining, Geol. and Metall. Inst. India (hon.), Malayan Chamber Mines (hon.). Asst. editor: Econ. Geology, 1923-37. Research and numerous publs. on geology, petrology, mineralogy, meteorites, ore deposits, mineral statistics. Died May 24, 1954.

FERNALD, Merritt Lyndon, Am. botanist; b. Orono, Me., Oct. 5, 1873; s. Merritt Caldwell and Mary Lovejoy (Heywood) F.; B.S., Harvard, 1897; D.C.L., Acadia U., N.S., 1933; D.Sc., U. Montréal, 1938; m. Margaret Howard Grant, Apr. 15, 1907; children—Katharine, Mary, Henry Grant. Asst. in Gray Herbarium, Harvard, 1891-1902, curator, 1915-36, dir., 1937-—, instr. botany, 1902-05, asst. prof., 1905-15, Fisher prof. natural history 1915-49, prof. emeritus, 1947-—. Hon. pres. Internat. Bot. Congress, Stockholm, 1950. Recipient Gold medal Mass. Hort. Soc., 1944; Marie-Victorin medal Fondation Marie-Victorin, Can., 1950. Fellow Am. Acad. Arts and Scis. Bot. Soc. Am. (pres. 1942), A.A.A.S. (v.p. 1941); mem. Societas pro Fauna et Flora Fennica, Am. Soc. Plant Taxonomists (pres. 1938), Am. Philos Soc., Nat. Acad. Sci., N.E. Bot. Club (pres. 1911-14, asso. editor Rhodora 1899-28), Acad. Natural Scis. Phila. (Leidy Gold medal 1940), Conn. Bot. Soc., Phila. Bot. Club, other Am. and fgn. bot. socs. Author: Edible Wild Plants of Eastern North America (with Alfred C. Kinsey); Gray's Manual of Botany, 8th and Centennial edit., 1950. Editor: (with Benjamin L. Robinson) Gray's Manual of Botany, 7th edit. Contbr. papers to jours. Authority on flora of northeastern U. S. Died Cambridge, Mass., Sept. 22, 1950.

FERNALD, Robert Heywood, Am. engr.; b. Orono, Me., Dec. 19, 1871; s. Merritt Caldwell and Mary Lovejoy (Heywood) F.; B.M.E., Me. State Coll., 1892; postgrad. Mass. Inst. Tech., 1892-93; M.E., Case Sch. Applied Science, 1918; A.M., Columbia, 1901, Ph.D., 1902; Sc.D., U. Pa., 1924; m. Catherine Mason Coupland, June 27, 1905; children—Merritt Caldwell, Frances Mason, Mason. Instr., Case Sch. Applied Science, 1893-96, asst. prof., 1896-1901, dir. dept. mech. engr., 1907-12; dir. dept. mech. engring. Washington U., 1902-07; Whitney prof. dynamical engring., dir. dept. mech. engring. U. Pa., from 1912, dean Towne Sci. Sch., from 1930. Conducted investigations for U. S. Geol. Survey and Bur. of Mines in U. S. and Europe; engr. in charge technologic br. U. S. Geol. Survey, 1904-10; cons. engr. Pub. Service Commn. of Pa., 1913-15; cons. engr., fuel div. Bur. Mines, 1910-20; formulated rules and regulations for gas, heating and water utilities of Pa., 1914; engr. mem. Giant Power Survey Bd. Pa.; mem. sci. adv. com. (mech. engring. dir.), Century of Progress Expn., Chgo., mem. exec. com. Traffic Commn. of Phila., from 1930; engring. mem. adv. com. Phila. agy. Recnstrn. Finance Corp., from 1932; mem. exec. com. 3d World Power Conf., 1935-36. Pres. Cleve. Engring. Soc., 1912. Author: Engineering of Power Plants (with George A. Orrok), 1916, 3d edit., 1927. Died April 24, 1937.

FERNANDES, Abilio, Portuguese botanist; b. Macainhas, Guarda, Portugal, Oct. 19, 1906; s. Jose and Maria (Augusta) F.; student Faculty Scis., U. Coimbra (Portugal), 1923-27, D.Biol. Scis., 1931; m. Rosette M. Saraiva Batarda, Dec. 19, 1942; children—Eduardo Manuel Batarda, José Antonio Batarda. Asst., Bot. Inst., Faculty Scis., Coimbra U., 1927-34, aux. prof., 1934-37, cathedratic prof., botany, 1937-—, dir. Bot. Inst., 1942-—, dir. Agrupamento Científico de Estudos Biológicos junto Faculdade de Ciencias, 1959-—; dir. Bot. Center, Junta de Investigacoes do Ultramar, Lisbon, 1959-—. Recipient William Herbert medal, 1942, Golden medal Am. Daffodil Soc., 1963. Mem. Acad. Scis. Inst. France (corr.), Acad. Scis. Lisbon (corr.), Sociedad Broteriana (chmn. 1942-—). Author: Nouvelles études caryologiques sur le genre Narcissus, 1934; Morfologia e biologia das plantas carnívoras, 1942; Sur la Phylogénie des espèces du genre Narcissus, 1951; also numerous articles. Caryological studies on Narcissus including establishment phylogenetic classification, genetics of heterochromatic chromosomes in Narcissus, detection of genera, species and infraspecific taxa new to sci. particularly in flora of Africa. Home: 21 Filipe Simoes, Coimbra, Portugal.*

FERNANDES, Rosette Mercedes Batarda (Mrs. Abilio Fernandes), Portuguese botanist; b. Redondo, Portugal, Oct. 1, 1916; d. José Ignácio and Berta (Saraiva) B.; grad. in Biol. Scis., U. Lisbon (Portugal), 1941; m. Abilio Fernandes, Dec. 19, 1942; children—Eduardo Manuel, José António. Curator, Herbarium, Bot. Inst., Coimbra (Portugal), U., 1947-59, 65, tech. investigator, 1965-—; investigator Junta de Investigaçoes do Ultramar, Lisbon, Portugal, 1959-64, collaborator Agrupamento Cientifico Estudos Biológicos, 1954-—. Collaborator, Flora Zambesiaca, 1960-—, Flora Europaea, 1964-—; Conspectus Florae Anglensis, 1967-—. Mem. Sociedade Broteriana (sec. treas. 1959-—), Assn. pour l'Étude Taxonomique de la Flore d'Afrique Tropicale, Sociedade Port. Cienc. Nat. Lisboa. Research, numerous publs. on detection of flora of Europe and Africa of genera, species, subspecies and varieties new to sci., discovery of several species and varieties new to flora of Portugal; taxonomic studies on various plant families; caryological studies in sects. Ajax and Hermione of genus Narcissus. Home: 21 Filipe Simoes, Coimbra, Portugal. Instituto Botanico, Arcos do Jardim, Coimbra, Portugal.*

FERNANDEZ-ALONSO, José Ignacio, Spanish chemist, biochemist; b. Ferrol, Sept. 8, 1917; s. José and Casimira (Alonso) F.A.; ed. Inst. Ferrol, U. Santiago, Cal. Inst. Tech., Oxford U. Ph.D. in chemistry; m. Isabel Estelles, July 29, 1946; children—José, Ignacio, Isabel, Marta. Aux. prof. physics, prof. math. U. Saint-Jacques-de-Compostelle; prof. phys. chemistry U. Valencia; research asso. Cal. Inst. Tech. Mem. Royal Spanish Soc. Physics and Chemistry, Dutch Soc. Chemistry. Author: (with L. Carbonell Vila and R. Domingo) Theoretical Study of Nitrogen Heterocyclics. II. Molecular Diagrams and Carciogenci Activities of some Mono- and Dibenzocarbazoles, 1957, also numerous articles, chpts. to books. Home: calle Burriana 18. Office: Faculty of Sciences, Paseo al Mar 13, Valencia, Spain.

FERNANDEZ-GALIANO FERNANDEZ, Dimas, Spanish biologist; b. Barcelona, Spain, Feb. 2, 1921; s. Emilio and Maria Josefa (Fernandez) F.-G.; Ph.D. in Natural Scis., U. Madrid; m. Maria Victoria Ruiz, Oct. 15, 1946; children—Luis Antonio, Dimas, Eladio, Maria Victoria, Maria Teresa, Maria Carmen, Emilio. Prof. natural sci. Nat. Inst. Secondary Teaching Ibanez Martin, Teruel, Goya of Saragossa, Lope de Vega of Madrid; prof. bacteriology and protozoology Sch. Sci., Madrid. Mem. Nat. Royal Acad. Medicine (titular mem.), Royal Spanish Soc. Natural History, Soc. Protozoologists, Assn. Spanish Parasitologists. Author: Observaciones sobre las opalinas; La modernas tecnicas en microscopia electronica aplicada a la biologia; Aspectos actuales de la biologia de protozoos paristos. Research on microbiology, particularly proto-zoology. Home: calle Ministry Ibanez Martin 6, 2A, Madrid 15. Office: Faculty of Sciences, City University, Madrid 3, Spain.

FERNANDEZ-GUARDIOLA, Augusto, physiologist; b. Madrid, Spain, Mar. 24, 1921; s. Augusto and Maria (Guardiola) F.; M.D., Nat. U. Mexico Med. Sch., 1953; IIIeme Cycle Neurophysiology, Faculty Scis., Marseilles, France, 1959; m. Carmen Mas Condes, May 15, 1963; children—Nuria, Isabel, Ana, Rodrigo. Investigator, Inst. Med. and Biol. Studies, Universidad Nacional Autonoma Mexico, Mexico, 1953, prof. psychophysiology, 1956-63, prof. physiology, acting chief dept., 1966-—. Chief EEG research Nat. Inst. Neurology, Mexico, 1966-—. Mem. French EEG Assn., French Physiol. Assn., Nat. Acad. Scis. (Mexico), Mexican Soc. Neurology and Psychiatry, Mexican Soc. Physiol. Sci. (pres. 1967). Author: The Adventure of the Brain (in Spanish), 1963. Contbr. numerous articles to profl. jours. Research on mechanisms of inhibition of convulsive activity, role of pupillary mechanisms in perception and habituation, elec. signs of cerebral anoxia, human reaction time related to EEG changes. Home: 1418 Sanchez Azcona, Mexico City, Mexico 12.*

FERNANDEZ-MORAN, Humberto, biophysicist; b. Maracaibo, Venezuela, Feb. 18, 1924; s. Luis and Elena (Villalobos) F.; M.D., U. Munich (Germany), 1944, U. Caracas (Venezuela), 1945; M.S., U. Stockholm (Sweden), 1951, Ph.D., 1952; m. Anna Browallius, Dec. 30, 1953; children—Brigida Elena, Veronica. Fgn. asst. Neurosurg. Clinic, Stockholm, 1946-48; research fellow Nobel Inst. Physics, Stockholm, 1947-49, Inst. Cell Research and Genetics, Karolinska Institutet, Stockholm, 1948-51; prof., chmn. dept. biophysics U. Caracas, 1951-58; dir. Venezuelan Inst. Neurology and Brain Research, Caracas, 1954-58; faculty Harvard, also asso. biophysicist neurosurg. service Mass. Gen. Hosp., 1958-62; prof. biophysics U. Chgo., 1962-—. Sci. and cultural attache Venezuelan legations, Sweden, Norway, Denmark, 1947-54; head Venezuelan commn. AEC, Geneva, Switzerland, 1955; chmn. Venezuelan commn. 1st Inter-Am. Symposium on Nuclear Energy, Brookhaven, N.Y., 1957; minister edn., Venezuela, 1958; mem. nat. com. UNESCO, 1957. Mem. Academia Ciencias Fisicas y Matematicas, Caracas, Am. Acad. Neurology. Internat. Soc. Cell Biology, Buenos Aires, Santiago, Lima socs. Neurology, Buenos Aires, Santiago, Lima, Porto Alegre socs. surgery, Electron Microscopy Soc. Am., Am. Nuclear Soc., Pan Am. Med. Assn. Mem. editorial bd. Jour. Cell Biology, 1961. Research, publs. on effects of irradiation on various tumors; electron microscopy of cell membranes; electron and neutron diffraction. Home: Apartado 362, Maracaibo, Venezuela. Office: Dept. Biophysics, U. Chgo., 5640 S. Ellis Av., Chgo. 60637.*

FERNBACH, Donald J., Am. physician; b. Bklyn., Apr. 10, 1925; s. Jules J. and Dorothy (Hever) F.; A.B., Tusculum Coll., Greenville, Tenn., 1948; M.D., George Washington U., 1952; m. Freddie Lucille Lieber, Nov. 20, 1954; children—Susan, Judy, Karen, Donald J. Resident pediatric pathology Children's Med. Center, Boston, 1954-55, fellow in hematology, 1956-57; chief resident pediatrics Baylor U., Houston, 1955-56, faculty, 1956-—, associate professor of pediatrics, 1964-—, also director of the Baylor pediatric div. of the Southwest Cancer Chemotherapy Study Group; practice medicine, specializing in pediatrics, Houston; dir. blood transfusion service Tex. Children's Hosp., 1957-—. Fellow Am. Acad. Pediatrics; mem. A.M.A., So. Soc. Pediatric Research, Am. Soc. Hematology. Publs. on studies of cancer chemotherapy in children, cancer-virus relationships. Home: 11714 Winshire Circle, Houston 77024. Office: 6621 Fannin St., Houston 77025.*

FERNEL (or FERNELIUS), Jean François, French physician, astronomer, mathematician; b. Clermont-en-Beauvaisis, France, 1497; M.D., Coll. St.-Barbe, Paris, 1530; practiced medicine, Paris; apptd. prof. medicine, U. Paris, 1534; named 1st physician to Henry II, circa 1547; tchr. of Versalius. Author: Monalosphaerium, 1526; Cosmotheria, 1528; De proportionibus, 1528; De naturali parte medicinae, 1542; De abditis rerum causis, 1548; Universa medicina, 1554; Therapeutica, 1554; Traité sur la composition des médicaments; Therapeutrics universalis, 1571; Febrium curandarum methodus generalis, 1577. Called the modern Galen; corrected erros of Galen; gave earliest description of appendicitis; described peristalsis, 1542, central canal of spinal cord, 1542; gave 1st accurate description of endocarditis, 1554; introduced terms physiology and pathology; computed revolution of wheel; accurately measured degree of latitude between Paris and Amiens, circa 1528; wrote on astronomy and mathematics. Died Paris, Apr. 26, 1558.

FERNER, Bengt, Swedish astronomer; b. Nyeds Prestgard, Sweden, Nov. 10, 1724; prof. astronomy U. Uppsala, Sweden; tutor to royal prince Sweden. Mem. Acad. Stockholm, Sweden, French Acad. Scis., 1769. Contbr. articles Encyclopedie. Meteorol. researcher, observed transits of Venus, 1761, 69; indicated precise method for reclaiming land through reducing sea level. Died Stockholm, Nov. 18, 1802.

FERNER, Helmut Richard, German physician; b. Gratz, Sept. 27, 1912; s. Michael and Mathilde (Wilfer) F.; M.D.; m. Rosel Schumann, Aug. 9, 1941;

children—Ulrike, Regina, Rosi. Physician, German U. Prague; anat. studies U. Leipzig, 1942; prof. U. Hamburg, 1947; full prof. U. Sarre, 1957; full prof. U. Heidelberg, 1961, also dir. anatomy inst. Mem. German Anatomy Assn. Contbr. numerous articles to sci. publs., also monographs. Research in endocrinology, neuroanatomy, embryology. Home: Zähringerstrasse 28. Office: Brunnengasse 1, Heidelberg, Germany.

FERNIE, John Donald, astronomer; b. Pretoria, S. Africa, Nov. 13, 1933; s. John and Nell (Beattie) F.; B.S., U. Cape Town, 1953, B.S., 1954, M.S., 1955; Ph.D., Ind. U., 1958; m. Yvonne Chaney, Dec. 23, 1955; children—Kimberly, Robyn. Lectr. physics, astronomy U. Cape Town, 1958-61; asst. prof. astronomy U. Toronto, 1961-64, asso. prof., 1964—. Fellow Royal Astron. Soc. Gt. Britain; mem. Am. Astron. Soc., Royal Astron. Soc. Can., Astron. Soc. Pacific. Contbr. numerous articles to profl. jours. Research on astron. photoelectric photometry, variable stars and galactic structure. Home: 12 Thornheights Rd., Thornhill, Ont. Office: David Dunlap Obs., Richmond Hill, Ont., Can.*

FERRACINO, Bartolomeo, Italian engineer, inventor; b. Solagna, near Bassano, Italy, 1692. Invented wind-driven saw; hydraulic engine capable of raising water 35 feet, 1749; designed bridge over Brenta at Bassano; St. Mark's clock, Venice. Died Solagna, 1777.

FERRANDO, Raymond, French nutritionist; b. Constantine, France, Mar. 3, 1912; s. Joseph and Etienette (Dessens) F.; B., Vet. Sch. Lyon (France), 1936; D.Vet. Medicine, D.Sc., Faculties Sci. Lyon, 1952. Faculty, Vet. Sch. Alfort (France), 1955—, prof. nutrition, 1955—, dir., 1957-64; hon. prof. U. La Plata (Argentina), 1963. Mem. nutrition com. Institut de la Sante Recherche Med., 1964—; expert WHO, 1962, 67, pres. com. interministerial animal feeding, 1967. Decorated chevalier de la Legion d'Honneur, officier Merite Republique Italienne; named Laureat Acad. Nale Medecine, 1964. Mem. Am. Chem. Soc., Brit. Nutrition Soc., French Soc. Biology, French Therapeutic Soc. (past pres.). Author: Monographies Alimentaires, 1960; Les Bases de L'Alimentation, 1960; Alimentation Equi Libre Biologique, 1962; also numerous articles. Research on antibiotics in animal nutrition, vitamin E and detoxication mechanisms, relationship between vitamine P and unsaturated fatty acids; co-discoverer 1st antivitamin A; research on vegetable proteins and meat by-products. Office: 7 Gal de Gaulle Av., 94 Alfort, France.*

FERRAN Y CLUA, Jaime, Spanish bacteriologist; b. Corbera de Ebro, Spain, 1849; M.D., U. Barcelona, 1873; practiced medicine, Plá de Panadés, then Tortosa; specialized in ophthalmology and electrotherapy; founder bacteriol. inst., Barcelona; studied cholera and immunization; attained notoriety by immunizing against cholera in man; immunized 50,000 in Valencia, 1881-85; advocated similar treatment for other infections. Died 1929.

FERRARI, Carlo, Italian aerodynamicist; b. Voghera, Italy, June 1, 1903; s. Camillo and Emilia (Lorenzola) F.; D.M.E., Politecnico di Torino, Italy, 1926; m. Maria Luisa Soave, Apr. 4, 1934. Lectr. U. Rome, 1929; prof. aerodynamics Politecnico Torino, 1933-48, prof. applied mechanics, 1948—; vis. prof. Brown U., 1961-62; vis. prof. N.Y.U., 1966; dir. Centro Dinamica Del Fluidi Del C.N.R. Torino, 1956—. Recipient Italian Repub. Pres. prize. Mem. Accad. Scienze Torino, Accad. Naz. Lincei Roma, Inst. Lomb. Scienze Lett. Milano, Internat. Acad. Astronomics Paris. Author (with F. G. Tricomi) Transonic Aeronautics, 1960; (with A. Romiti) Meccanica Applicata, 1966. Contbr. numerous articles in field to sci. jours. Research, publs. field aerodynamics of helicopters; application of special techniques for determining pressure on revolution of body in supersonic aeros.; shape of minimum drag, wing-body interference; friction and heat transfer in supersonic aeros. and turbulent flow; shock waves in transonic flow about wing profile. Home: 147 Corso Galileo Ferraris, Torino, Italy. Office: 24 Corso Duca Degli Abruzzi, Politecnico, Torino, Italy.*

FERRARI, Ludovico, Italian mathematician; b. Bologna, Italy, Feb. 2, 1522; doctorate, Bologna; studied under Cardan; became prof. U. Milan (Italy), circa 1543; became prof. U. Bologna, 1565; solved cubic equations; discovered method of resolving equations of 4th degree named after him, circa 1545. Died Oct. 5, 1565.

FERRARI, Pietro, Italian engineer; b. Spoleto, Italy, 1753. Chief engr., Trasimeno District, Italy. Author: (in Italian) On Opening a Canal Connecting the Adriatic Sea with the Mediterranean Ocean, 1825. Made plans to connect Mediterranean with Adriatic by canal; studied problem of drying up of Lake Trasimeno. Died Naples, Italy, 1825.

FERRARI D'OCCHIEPPO, Konradin, astronomer; b. Leibnitz, Dec. 9, 1907; s. Marquard and Emma (von Jagemann) F. d'O.; ed. in astronomy and physics univs. Würzburg, Bonn, Vienna, Leipzig. Asst., Obs. of Bamberg, 1934-35, Obs. of Vienna, 1935-37, 46-55; instr. U. Vienna, 1949, prof. astronomy theory, 1954. Mem. Austrian Acad. Sci., Astronomische Wiener Kathol Akademie, Goerres Gesellschaft (Cologne). Author: Astronomie, 1949; Kunde vom Weitall, 1952; numerous articles. Address: Türkenschanzstrasse 17, A-1180, Vienna Austria.

FERRARIS, Galileo, Italian physicist, engr.; b. Livorno, Italy, Oct. 31, 1847; ed. Livorno, Sch., also 3 years at a univ., 2 years at Scuola d'applicazione, Valentinianum, Italy. Asst., Inst. for Tech. Physics; titular prof. tech. physics U. Turin (Italy). Mem. Acad. Scis., Turin. Author: Le proprieta cardinali degle strumenti diottrici; Sull' intensita delle correnti e delle corrent e delle extracorrenti nel telefono; Sulla teoria matematica della propagazione dell' elettricita nelli solidi onogenei, 1872; Sulla illuminazione electtrica, 1879; Lezioni de elettrotecnia, 1898. His discovery of principle of rotary magnetic field (1885) led to devel. of polyphase motors and hydroelectric industry in Italy; devised alternating current transformers; founded 1st elec. engring. sch. in Italy, 1886-87. Died Turin, Feb. 7, 1897.

FERRARO, John Ralph, Am. chemist; b. Chgo., Jan. 27, 1918; s. Charles and Jennie (Carlotta) F.; B.S., Ill. Inst. Tech., 1941, Ph.D., 1954; M.S., Northwestern U., 1948; m. Mary Leo, June 21, 1947; children—Lawrence, Janice Ann, Victoria. Supr. tnt and tetryl labs Kankakee Ordnance Works, 1941-43; chemist Argonne (Ill.) Nat. Lab., 1948—; permanent staff mem. Canisius Coll. Spectroscopy Inst., Buffalo, 1959—. Mem. Am. Chem. Soc., Research Soc. Am., Coblentz Soc., Soc. for Applied Spectroscopy (pres. 1965). Sigma Xi. Author: (with J. S. Ziomek) Introductory Group Theory and Its Application to Molecular Structure, 1968; also articles. Editor: (with J. S. Ziomek) Developments in Applied Spectroscopy, 1963. Editor Jour. Applied Spectroscopy, 1968—. Research on infrared, raman and high pressure Farinfrared of metal complexes transition metals and rare earths. Home: 568 Saylor St., Elmhurst, Ill. 60126. Office: 9700 S. Cass St., Argonne, Ill. 60439.*

FERREIN, Antoine, French surgeon, anatomist; b. Agen, France, Oct. 28, 1693; studied under Jesuits, Agen; B.A., Montepellier, France, 1716, also med. degree; army surgeon, 1732-34; became prof. anatomy, Montpellier, 1732; began practice medicine, Paris, 1735; named prof. medicine and surgery Coll. Royal Paris, 1842. Mem. French Acad. Scis. Author handbook of practical surgery, also anat. study of lachrymal glands. Credited with originating term vocal cord in a comparision of ligaments of larynx to violin strings and air passing through to the bow. Died Paris, Feb. 28, 1769.

FERREIRA, Alexander Rodrigues, Brazilian naturalist; b. Bahia, 1756; explored Brazil, 1784-93; called Humboldt of Brazil; conducted sci. explorations in Amazonia; author bot. and zool. studies. Died 1815.

FERREIRA, Antonio Jose Fernandes, physician; b. Lisbon, Portugal, July 30, 1923; s. Diogo Antonio and Lydia (Fernandes) F.; M.D., U. Lisbon, 1946; m. Nancy Jane Schumacher, July 21, 1956; children—Carl J., Diane J., Paul J. Came to U. S., 1948, naturalized, 1952. Practice medicine, specializing in psychiatry, San Jose, Cal., 1954—; staff psychiatrist Agnews State Hosp., 1954-55; pvt. practice, 1955—; dir. Adult and Child Guidance Clinic, 1955-56; research asso. Mental Research Inst., Palo Alto, Cal., 1962—; research cons. San Jose State Coll., 1962—. Fellow Am. Psychiat. Assn., Cal. Marriage Counselors Assn.; mem. A.M.A. Research showing pregnant woman's emotions affect the fetus; research on palmar sweat and anxiety; psychoanalytic contbns. to ego-function, empathy and loneliness; research on family decision making and interactional variables in relation to normality or abnormality of families; described split double-bind in delinquency and introduced concept of family myth. Office: 2060 Clarmar Way, San Jose, Cal. 95128.*

FERREL, William, Am. meteorologist; b. Fulton County, Pa., Jan. 29, 1817; s. Benjamin and Miss (Miller) F.; grad. Bethany (W.Va.) Coll., 1844. Taught sch., Liberty, Mo., 1844-46, Todd County, Ky., 1847-54; founded sch., Nashville, 1854; joined staff of Am. Ephemeris and Naut. Almanac, Cambridge, Mass., 1857; mem. Coast and Geodetic Survey, 1867-82; mem. Signal Service, 1882-86, Author: Converging Series Expressing the Ratio between the Diameter and the Circumference of a Circle, 1871; Meteorological Researches, 3 vols., 1877-82; Popular Essays on the Movements of the Atmosphere, 1882; Temperature of the Atmosphere and the Earth's Surface, 1884; Recent Advances in Meteorology, 1886; A Popular Treatise on the Winds, 1889; also notable paper Essay of the Winds and the Currents of the Ocean, in Nashville Jour. Medicine and Surgery, 1856. Inventor maxima and minima tidal predicting machine; formulated Ferrel's law to explain movements of air currents; did research on tides, currents, and storms. Died Maywood, Kan., Sept. 18, 1891.

FERRELL, Richard Allan, Am. physicist; b. Apr. 28, 1926; s. Robert Myers and Elsie (Hopper) F.; B.S., Cal. Inst. Tech., 1948, M.S., 1949; Ph.D., Princeton, 1952; m. Miriam Conover, Jan. 5, 1952; children—Rebecca Ann, Robert Craig. Faculty, U. Md., College Park, 1953. Mem. Am. Phys. Soc., Am. Assn. Physics Tchrs., Wash. Acad. Scis. Author: Short Introduction to Superconductivity Theory, 1967; also numerous articles. Research on theoretical nuclear and solid state physics. Home: 6611 Wells Pkwy., University Park, Md.*

FERRERO, Guglielmo, sociologist; b. Portici, Italy, July 21, 1871; ed. Us. Pisa, Turin. Journalist in Milan; wrote several popular historical works; opponent of fascism, left Italy for Switzerland, 1930; prof., modern history, U. Geneva, until 1942; prof., European Military History Inst., U. des hautes Études Internat. Author: (with C. Lombroso) La donna delinquente, 1893; (with Sighele) Cronachi criminali italiane, 1896; Europa giovine: studi e viaggi nei paesi del Nord, 1897; il Militarismo: diece conferenze, 1898; Grandezza e decadenza di Roma (5 vol.), 1902-07; Fra i due mondi, 1913; Guerra europea, 1915; la Vecchia Europa e la Nuova, 1918; Da Fiume a Roma, Storia di quatro anni, 1919-28; La ruine de la civilisation entique, 1921; la Tragedia della pace, di Versailles alla Ruhr, 1923; Discorsi ai sordi, 1925; Tra civili e barbari, 1930; Aventure: Bonaparte en Italie, 1796-1797, 1935; Reconstruction: Talleyrand at Vienne, 1940; The Principles of Power: Great Political Crises of History, 1942. Studied organization of European societies. Died Geneva, Switz., Aug. 3, 1942.

FERRERS, Norman MacLeod, mathematician; b. Pricknash Park, Eng., Aug. 11, 1829; s. Thomas Bromfield and Lavina (MacLeod) F.; studied under Harvey Goodwin, 1 year; B.A., Caius Coll., Cambridge, Eng., 1851; studied law, London; LL.D., 1883; m. Emily Lamb, Apr. 3, 1866; 4 sons, 1 dau. Called to bar, 1855; became lectr. math. Caius Coll., 1856; became dean Bristol, master Corpus Christi Coll., Cambridge, 1866; master Gonville and Caius Coll., from 1880. Fellow Royal Soc., 1877. Author: Trilinear Coordinates, 1861; Editor: Mathematical Writings of George Green, 1871. Spherical Harmonics, 1877. Contbd. articles to Quar. Jour. Math. Studied quadriplanar co-ordinates, Lagrange's equations, hydrodynamics; found potential at any point of space in zonal harmonics (on basis of Kelvin's work on law of distbn. of electricity in equilibrium on uninfluenced spherical bowl). Died Cambridge, Eng., Jan. 31, 1903.

FERRI, Enrico, Italian criminologist; b. San Bendetto Po, Italy, Feb. 25, 1856; Dr. Laws, U. Bologna, 1877; studied at Pisa, also Paris; student of Limbroso. Became prof. U. Bologna, 1880, also taught at Siena, Pisa, Royal U. Rome, lectr., Paris, Brussels; mem. Chamber of Deps., from 1886; editor Avanti, 1900-05; became head of commn. to revise Italian penal code (his recommendations were rejected in Italy but adopted in Argentina), 1919. Author: La scuola positiva de dirrto criminale, 1883; Criminal Society, 1884; Socialismo e scienza positiva, 1894; Difese penali, studi di giurisprudenza penale, 1898. Studied crime and stressed social and economic factors involved; also preventative treatment of criminals. Died Rome, Apr. 12, 1929.

FERRI, Mario G., Brazilian botanist; b. Sao José dos Campos, Brazil, July 7, 1918; s. Mario and Durvalina de Franca (Guinearaes) F.; grad. Natural Sci., Faculty Philosophy, Sci. and Letters, 1939, dr.'s degree, 1944; m. Ruth Lippi, July 25, 1941. Faculty dept. botany Faculty Philosophy, Sci. and Letters, U. Sao Paulo (Brazil), 1939—, prof., 1955—, dean, 1961—, vice rector, 1963—. Rockefeller Found. fellow Boyce Thompson Inst., Yonkers, N.Y., Columbia, Cal. Inst. Tech., 1944-45. Mem. Brazilian Soc. for Advancement Sci., Brazilian Bot. Soc., Am. Soc. Plant Physiologists, Bot. Soc. Am., Internat. Soc. Tropical Ecology. Author: Ext. Morphology of Plants, 1956; Botany in Sao Paulo since the Foundation of University Anhembi, 1958; also numerous articles. Editorial bd. Rev. Brasil, Biol. Brasil, Rev. Biol. Portugal, Annals of Arid Zone. Research on ecology of cerrado vegetation (savannahs) and castingas; studies in transpiration, phytohormones. Home: 39 c. 37 Leme, Sao Paulo, Sao Paulo, Brazil.*

FERRIÉ, Gustave-Auguste, French inventor; b. St. Michel, France, Nov. 19, 1868; apptd. dir. Mil. Radio Communications, 1914; became head wireless service in France, 1923; mem. French Acad. Scis., 1922. An inventor of wireless telegraphy; research in radio telegraphy from Eiffel tower, 1903; invented electrolyte detector (1st device to detect sound of electric oscillations), 1909; 1st to broadcast meteorol. information, 1911; compared clocks by simultaneously recording their positions, 1922; later used photo cell for clock comparisons. Died Paris, Feb. 16, 1932.

FERRIER, Sir David, Scottish cerebral anatomist, neurologist; b. Woodside, nr. Aberdeen, Scotland, Jan. 13, 1843; s. David and Hannah (Bell) F.; student King's Coll., Aberdeen U.; M.A., 1863, Heidelberg; M.B., Edinburgh, 1868; M.D.; m. Constance Waterlow, 1874; 1 son, 1 dau. Asst. to Dr. Image, Bury St. Edwards, 1868-76; became lectr. physiology Middlesex Hosp. Sch., 1870; prof. medicine King's Coll., London, from 1871; prof. neuropathology, 1889-1908. Knighted, 1911. Croonian lectr., 1874, 75. Recipient Royal medal, 1890; lectureship named in his honor; Ferrier Meml. Library established by Royal Soc. Medicine. Fellow Royal Soc., 1911. Author: The Functions of the Brain, 1876; Cerebral Localisation, 1878. Contbd. to knowledge of physiology of brain, especially in regard to localization of cerebral functions; his work led to removal of brain tumors and to other advances in brain surgery. Died London, Mar. 19, 1928.

FERRIO, Carlo, Italian psychologist, psychiatrist; b. Turin, Italy, Dec. 26, 1898; s. Luigi and Maria (Badariotti) F.; M.D., U. Turin, 1922; m. Candida Schwarz, June 28, 1926; children—Luigi, Vittorio. San. dir. Medico-Paedagogical Sch., Psychopaedological Center of Turin, 1945—; in charge teaching psychology Faculty Letters and Philosophy, U. Turin, 1955—. Mem. Soc. Neurology, Soc. Psychiatry, Soc. Psychology, Soc. Philosophy, Soc. for History Medicine. Author: Dictionary of Medicine, 1961; Treatise of Clinical and Juridical Psychiatry, 1959; History of Mental Medicine, 1948. Research on clin. and juridical psychiatry, history of mental medicine, infantile neuropsychiatry. Address: Via Amedeo Peyron 38, Turin, Italy.*

FERRIS, Deward Olmsted, surgeon; b. Niagara Falls, Ont., Can., Jan. 5, 1907; s. Thomas Elward and Martha (Olmsted) F.; M.D., Queen's U., Kingston, Ont., 1931; M.S., U. Minn., 1943; m. Edythe Culcheth, Mar. 16, 1938; children—Nancy (Mrs. Jeff Robertson), William D. Came to U. S., 1937, naturalized, 1943. Practice medicine, specializing in surgery, Rochester, Minn., 1937—; head surgery sect. Mayo Clinic, 1945—; faculty U. Minn., 1947—, prof. clin. surgery Grad. Sch., 1962—. Diplomate Am. Bd. Surgery. Fellow A.C.S.; mem. A.M.A., Zumbro Valley Med. Soc., Central, Western surg. assns., Internat. Soc. Surgery, Soc. For Surgery Alimentary Tract, Sigma Xi. Contbr. numerous articles to profl. jours. Devised the Ferris bile duct scoop and dilator. Home: 1135 Plummer Circle. Office: 200 1st St. S.W., Rochester, Minn. 55901.*

FERRIS, Eugene B(everly) Jr., Am. physician; b. McNeill, Miss., June 24, 1905; s. Eugene B. and Martha (Reynolds) F.; B.S., Miss. State Coll., 1925; M.D., U. Va., 1930, M.S., 1931; m. Charlotte Gordon Hopkins, June 6, 1936; children—Charlotte Beverly, Ann Gordon, Eugene Beverly. Intern Boston City Hosp., 1931-33; resident U. Mich. Hosp., 1933; asst. resident Thorndike Meml. Lab., also research fellow Harvard Med. Sch., 1933-35; mem. faculty U. Cin. Coll. Medicine, 1935-52, asso. prof. medicine, asst. dir. dept. internal medicine Cin. Gen. Hosp., dir. psychosomatic teaching program, 1947-51, prof. medicine, 1951-52; prof. medicine Emory U., chief of med. services Grady Hosp., from 1952. Cons. in mental health and mem. com. on research of mental hygiene council USPHS, 1946-51, cardiovascular study sect., 1952-57; med. dir. Am. Heart Assn., 1957—; dir. research project on aviation medicine OSRD, 1942-45; mem. sub. com. on decompression sickness NRC, 1942-46. Recipient Horsley prize for research U. Va., 1936. Fellow A.A.A.S., A.C.P. (regent); mem. Am. Soc. for Clin. Investigation (pres., 1950), Assn. Am. Physicians, Am. Psychosomatic Soc. (pres., 1950), A.M.A., Am. Heart Assn. (v.p., bd. govs. 1955-57), Council High Blood Pressure Research (chmn. 1955-56), Ohio Med. Assn., Cin. Acad. Medicine, Central Soc. Clin. Research, Phi Beta Kappa, Sigma Xi. Editor-in-chief, Cin. Jour. Medicine, 1945-47, Journal of Clinical Investigation, 1947-52, mem. editorial bd., 1952—, Annals of Internal Medicine, Am. Jour. Medicine, Jour. Psychosomatic Medicine. Died Sept. 26, 1957.

FERRIS, Harry Burr, Am. anatomist, anthropologist; b. Old Greenwich, Conn., May 21, 1865; s. Samuel Holmes and Mary F. (Clark) F.; B.A. Yale, 1887, M.D., 1890; m. Helen Whiting, June 23, 1892; children—Helen Millington (Mrs. Davenport Hooker), Henry Whiting. Instr. in anatomy, 1891, asst. prof. 1892, prof. of anatomy, 1895-1933, Yale U., prof. emeritus, from 1933. Author: The Indians of Cuczo and the Apurimac, 1916; The Quichua and Machiganga Indians, 1921. Important anthrop. studies of Cuczo, Quichua and Machiganga Indians. Died Oct. 12, 1940.

FERRO-LUZZI, Giovanni, Italian physician; b. Ancone, July 16, 1903; s. Massimiliano and Ersilia (Lanzi) F.-L.; M.D.; m. Sofia Salzman, July 19, 1931; children—Massimiliano, Anna, Giovanna, Michele. Prof. med. pathology, with clinic tropical diseases; head nutrition service Ministry Health. Pres. council nutrition WHO, FAO; founder, dir. Sch. Medicine, Asmara. Research and over 150 publs. on tropical pathology and nutrition. Home: via dei Gozzadini 41. Office: Ministry of Health, Rome, Italy.

FERRONI, Enzo, Italian chemist; b. Florence, Italy, Mar. 25, 1921; s. Guido and Zaira (Ravaglioli) F.; D.Chemistry, U. Florence, 1945, Qualification for univ. teaching, 1954; m. Paola Berchieri, July 25, 1951. Staff, U. Cagliari (Italy), 1947-65, prof. phys. chemistry, 1961-65, extraordinary dir. Inst. Chemistry, 1961-65; ordinary dir. Inst. Phys. Chemistry, U. Florence, 1965—, also mem. faculty. Mem. Societé de Chimie Physique, Societé de Chimie de France, Società Chimica Italiana. Research and numerous publs. on surface tension of liquids in motion, molecular polymorphism, symmetry relations at interfaces in epitaxy, higher order transitions in films of high polymers, X-ray diffraction on thin surface layers, optical antipodes resolution. Home: 11 Piazza Fardella, Florence, Italy.*

FERRY, Clayton Walton, Am. pharm. co. exec.; b. Pingree, N.D., June 9, 1908; s. Rollo H. and Mable (Walton) F.; student Coll. Puget Sound, 1927-29; B.S. in Chemistry U. N.D.); Ph.D. (John Hancock fellow) Johns Hopkins, 1935; m. Dorothy G. Ries, June 16, 1934; children—Margaret (Mrs. John D. Houston), Gardner Walton. Instr., Johns Hopkins, 1934-35; with Burroughs Wellcome and Co., Tuckahoe, N.Y., 1935—, chief devel. chemist, 1941-46, chief control chemist in charge standards control div., 1946—. Mem. com. revision U. S. Pharmacopeia, 1960—. Mem. Am. Chem. Soc., N.Y. Acad. Scis., Pharm. Mfg. Assn., Phi Beta Kappa, Sigma Xi, Phi Lambda Upsilon. Contbg. author: Organic Synthesis; also articles. Synthesis substituted primary amines, ureas, guandidines, substituted mercapto compounds and sulfones, isolation digitalis glycosides. Home: 345 Grand View Blvd., Yonkers, N.Y. 10710. Office: 1 Scardsdale Rd., Tuckahoe, N.Y. 10707.*

FERRY, Ervin Sidney, Am. physicist; b. Croydon, N.H., June 14, 1868; s. Harvey S. and Hattie W. (Eastman) F.; B.S., Cornell, 1889, postgrad., 1891-93, fellow in physics, 1902-03; postgrad., Upsala, Sweden, 1897-98; postgrad., fellow in physics Johns Hopkins, 1893-94; m. Ruth M. White, Aug. 21, 1900; 1 dau., Priscilla Grace. Prof. physics Purdue U., from 1899. Mem. Am. Phys. Soc., Am. Electro-chem. Soc., Astron. and Astrophys. Soc. Am., Société Française de Physique, Sigma Xi. Author: Elementary Dynamics, 1906; Practical Physics (with A. T. Jones), 1907; Pyrometry (with others), 1917; General Physics and Its Application to Industry and Everyday Life, 1921; (with others) Physics Measurements, 2 vols., 1929; Applied Gyrodynamics, 1931. Died Oct. 8, 1956.

FERRY, John Douglass, Am. chemist; b. Dawson, Can., May 4, 1912 (parents Am. citizens); s. Douglass Hewitt and Eudora (Bundy) F.; student U. London, 1932-34; A.B., Stanford, 1932, Ph.D., 1935; m. Barbara Norton Mott, Mar. 25, 1944; children—Phyllis Leigh, John Mott. Instr., fellow Soc. Fellows, research asso. Harvard, 1936-41, 42-45; asso. chemist Woods Hole (Mass.) Oceanographic Instn., 1941-45; faculty U. Wis., Madison, 1946—, prof. chemistry, 1947—, chmn. dept., 1959-67. NSF fellow, Brussels, Belgium, 1959. Fellow Am. Phys. Soc. (High Polymer Physics Prize 1966), Am. Acad. Arts and Scis.; mem. Am. Chem. Soc. (Eli Lilly award in biol. chemistry 1946; Kendall Co. award in colloid chemistry 1960), Am. Soc. Biol. Chemists, Soc. Rheology (Bingham medal 1953, pres. 1961-63), Nat. Acad. Scis., NRC (chmn. com. on macromolecular chemistry 1958-62), Internat. Com. on Rheology (chmn. 1963—), A.A.A.S., Internat. Soc. Hematology, Phi Beta Kappa, Sigma Xi, Alpha Chi Sigma, Phi Lambda Upsilon. Author: Viscoelastic Properties of Polymers, 1961. Co-editor: Fortschritte der Hochpolymeren Forschung, 1958—. Contbr. articles to profl. jours. Research on polymers of high molecular weight; mech. properties of viscoelastic materials. Home: 137 N. Prospect Av., Madison, Wis. 53705.

FERSMANN, Alexander Evgen'evic, Russian mineralogist; b. St. Petersburg, Russia, Nov. 8, 1883. Author: Geochimija, 4 vols., 1933-39; Twenty-five Years of Soviet Natural Science, 1944; The March of Soviet Science, 1945. Research on diamond, pegmatites, geochemistry, chem. elements of earth and cosmos, precious stones of USSR. Died Sochi on the Black Sea, USSR, May 20, 1945.

FERT, Charles, French physicist; b. Carcassonne, France, Sept. 30, 1911; s. Albert and Louise (Cazot) F.; ed. École Normale Supérieure d'Enseignement Technique de Paris; licence ès sciences Faculté des Sciences de Paris; Doctorat ès Sciences; Agrégation de Physique; m. Irmine Signoles, Aug. 10, 1936; children—Albert, André. Prof. agrégé Lycee Bar le Duc; prof. agrégé spl. math. class Lycee de Toulouse, 1937; master confs. Faculty Scis., 1949; prof. Faculty Scis., Toulouse, 1951; subdir. Lab. Electronic Optics Nat. Center for Sci. Research, 1950-63, mem. nat. com.; dir. dept. physics Nat. Inst. Applied Scis., Toulouse. Mem. French Phys. Soc., French Electron Microscopy Soc., French Radioelectricity Soc., French Mineralogy and Cristallography Soc. Research, publs. on electron optics (investigations of magnetic lenses, reflection electron microscopy, electronic secondary emission microscopy, electronic interferometry), solid state physics (secondary emission of monocrystals under ionic bombardment related to channeling phenomena). Home: 8, rue Ozenne, (31) Toulouse. Office: Labouratoire de physique des Solides, Faculté des Sciences, Toulouse, France.*

FERTON, Charles, French entomologist; b. Chierry, France, 1856; ed. École Polytechnique. Concentrated on entomol. research, from 1895. Author: Notes détachées, 1901; La vie des abeilles et guepes, 1923. Specialist in study of hymenopterous insects, particularly bees and wasps. Died Bonifacio, Italy, 1921.

FERY, Charles, French physicist; b. Paris, 1865; prof. Sch. Physics and Chemistry, Paris; inventor a spectrograph, an actinometer, an optical pyrometer, a refractometer; developed chem. theory of lead accumulation; studied high atmosphere; created electric pile of ammonium chloride, electrodes of zinc and carbon, in which air assumes a polarized role. Died Paris, 1935.

FESENKOV, Vasilii Grigorevich, Russian astrophysicist; b. Jan. 13, 1889; grad. Kharkov U., 1911. Prof. Moscow U., 1923—. Mem. Kazakh SSR (presidium 1961—, dir. Astrophys. Inst., 1961—), USSR (chmn. com. on meteorites 1961—) acads. scis., Internat. Astron. Union. Author: On the Origins of the Solar System, 1960; On the Density of Meteoric Material in Interplanetary Space, 1961; On the Nature and Origins of Comets, 1962; On the Optic Characteristics of the Dust Cloud around the Earth, 1964. Research on phys. properties of planets, meteors, physics of stars, evolution of stars, structure of gas and dust nebulae, cosmogony, celestial mechanics, optic of atmosphere; formulated hypothesis of corpuscular photogenesis of stars, criterion of influx stability of celestial bodies, star formation from interstellar gas and dust, dynamic theory of zodiac light. Office: Astrophysics Inst., Shevchenko Ulitsa, 28, Alma-Ata, Kazakh SSR, USSR.

FESHBACH, Norma Deitch, Am. psychologist; b. N.Y.C., Sept. 5, 1926; d. Samuel and Lena (Katz) Deitch; B.S. in Edn., Coll. City N.Y., 1947, M.S., 1949; Ph.D., U. Pa., 1956; m. Seymour Feshbach, Aug. 16, 1947; children—Jonathan Stephen, Laura Elizabeth, Andrew David. Cons. Youth Service, Inc., Phila., 1955-61; lectr. U. Pa., Phila., 1956-57; research asso., 1959-61; vis. asst. prof. Stanford, 1961-62; lectr. U. Cal., Berkeley, U. Cal. at Los Angeles. Cons. Southwestern Regional Lab. Mem. Am., Western psychol. assns., Sigma Xi. Research, publs. on factors influencing projection to others; devel. of empathic behavior in children and its relationship to generosity and altruism; sex differences and social class differences in children's modes of behavior and intellectual development. Home: 743 Hanley Av., Los Angeles 90049.*

FESSARD, Alfred Eugène, French neurophysiologist; b. Paris, France, Apr. 28, 1900; s. Louis and Marie (Erard) F.; Ph.D. in Sci., U. Paris; m. Denise Albe, July 22, 1942; 1 son, Jean. Adj. dir. Ecole pratique des hautes études, 1927-43; dir. Collège de France, 1943-49, prof. gen. neurophysiology, 1949—. Mem. Acad. Sci., Nat. Acad. Medicine, Brazilian Acad. Sci. Author: Les organes électriques, 1958; Corrélations neurophysiologiques de la formation des réflexes conditionnés, 1958. Home: 51, rue Molitor, Paris 16. Office: Collège de France, Paris, France.

FESSENDEN, Reginald Aubrey, physicist, engr.; b. East Bolton, Que., Can., Oct. 6, 1866; s. E. J. and Clementina (Trenholme) G.; ed. Bishop's Coll., Lennoxville, Que.; m. Helen May Trott, 1889; 1 son, Reginald Kenneley. Prin. Whitney Inst., Bermuda, 1885-86; insp. engr. Edison Machine Works, 1886-87; head chemist Edison Lab., 1887-90; electrician Westinghouse Electric & Mfg. Co., 1890-91; prof. elec. engring., Purdue U., 1892-93, Western U. of Pa. (U. Pitts.), 1893-1900; spl. agt. U. S. Weather Bur., 1900-02; gen. mgr. Nat. Elec. Signaling Co., 1902-10; cons. engr. Submarine Signal Co., 1910. Holder of over 300 patents, devised continuous wave principle of wireless transmission and heterodyne system of reception; developed radio telephony, high frequency alternator, a heterodyne, the rotary spark gap, the fathometer, drive for battleships, also in 1902 electrolytic detector. Died Bermuda, July 22, 1932.

FESSENDEN, Richard Warren, Am. chemist; b. Northampton, Mass., Jan. 22, 1934; s. Richard William and Bertha E. (Polley) F.; B.S., U. Mass., 1955; Ph.D., Mass. Inst. Tech., 1958; m. Louise Mae Cooley, June 8, 1957; children—Carol L., Robert G. NSF Postdoctoral fellow Cal. Inst. Tech., 1958-59; fellow Mellon Inst., Pitts., 1959-63, sr. fellow, prof. chemistry Carnegie-Mellon U., 1963—. Mem. Am. Chem. Soc., Am. Phys. Soc., A.A.A.S., Sigma Xi. Research, publs. of reactive or short-lived chem. species produced in irradiated materials using electron spin resonance. Home: 3972 Logans Ferry Rd., Monroeville, Pa. 15146. Office: 4400 5th Av., Pitts. 15213.*

FESTINGER, Leon, Am. psychologist; b. N.Y.C., May 8, 1919; s. Alex and Sarah (Solomon) F.; B.S., Coll. City N.Y., 1939; M.A., State U. Ia., 1940, Ph.D., 1942; m. Mary Olive Ballou, Oct. 23, 1943; children—Catherine, Richard, Kurt. Research asso. State U. Ia., 1941-43; faculty U. Rochester, 1943-45, Mass. Inst. Tech., 1945-48, U. Mich., 1948-51, prof. psychology U. Minn., 1951-55, Stanford, 1955—. Mem. transnat. social psychology com. Social Sci. Research Council. Fellow Am. Psychol. Assn. (pres. div. 8, 1963; Distinguished Scientist award 1959), Am. Acad. Arts and Scis. Author: (with others) Deterrents and Reinforcement: The Psychology of Insufficient Rewards, 1962; Conflict, Decision and Dissonance, 1964; also articles. Research on dissonance theory, visual perception; devised lab. techniques for reproducing under controlled conditions thought processes and motivations regulating prejudice, communication of rumor, and social influence; expts. in tests of validity of psychol. generalization in various cultural settings. Home: 274 Searsville Rd., Stanford, Cal. 94305.*

FESZT, Tiberius, Rumanian physician; b. Cluj, Romania, Sept. 23, 1929; med. diploma U. Tirgu Mures, 1954. Asst. chair histology U. Tirgu Mures, 1954-58; prin. researcher Research Sta., Acad. Socialist Republic of Rumania, Tirgu Mures, 1958—. Research, publs. on normal and pathol. histology, histophysiology, histochemistry; histophysiology of endocrine glands, tissue effects of neuroplegic drugs; histochemistry of normal and experimentally damaged liver tissue; effect of cytostatics on activity of tissue enzymes; extpl. allergic encephalomyelitis. Address: 38, str. Gh. Marinescu, Tirgu Mures, Rumania.*

FETNER, Robert Henry, Am. biologist; b. Savannah, Ga., Feb. 22, 1922; s. William W. and Luciel (Goodrich) F.; B.S., U. Miami, Fla., 1950, M.S., 1952; Ph.D., Emory U., 1955; m. Janet Evans, Jan. 12, 1943; children—Barbara Lynn, Westcott Troy. Research asst. Emory U., Atlanta, 1954; faculty Ga. Inst. Tech., Atlanta, 1955-—, prof. biology, 1963-—, dir. sch. biology, 1964-—. Mem. Ga. Acad. Sci. (editor Bull. 1960-63), Sigma Xi, Phi Sigma Phi. Research, publs. on biol. effects of high energy radiation and some chems. with particular interest in genetic and long-term results. Home: 620 Peach Tree St. N.E., Atlanta 30309.*

FETT, Gilbert Howard, Am. elec. engr.; b. Chgo., June 19, 1909; s. Charles G. and Frieda (Klein) F.; B.S. high honors, U. Ill., 1931, Ph.D. in Engring., 1940; M.S., Ia. State U., 1932; m. Genevieve Weninger, Nov. 26, 1936; children—Miriam (Mrs. Icko Iben, Jr.), Priscilla, (Mrs. Jay P. Mitchell), Catharine (Mrs. Allan Ehler), Faraday Ann. Research engr. Littelfuse Labs., Chgo., 1932-35; prof. elec. engring., U. Ill., Urbana, Ill., 1935-—; research asso. summers Argonne Nat. Lab., Chgo., 1956-61; guest prof. Indian Inst. Tech., and chief of party AID, Kharagpur, 1961-64. Cons. USAF, Washington, 1949-—, Argonne (Ill.) Nat. Lab., AEC, 1955-—; pres. Nat. Electronics Conf., Chgo., 1949, dir., 1945-51. Fellow I.E.E.E., Am. Phys. Soc. Author: (with A. R. Knight) Introduction to Circuit Analysis, 1944; Feedback Control Systems, 1954. Research on electric circuit theory; arc welding; arc discharges; nonlinear systems; servomechanisms. Home: 723 S. Prairie, Champaign, Ill. 61822.

FETTER, Frank Whitson, Am. economist; b. San Francisco, May 22, 1899; s. Frank Albert and Martha (Whitson) F.; B.A., Swarthmore Coll., 1920; A.M. Princeton, 1922, Ph.D., 1926; A.M., Harvard, 1924; m. Elizabeth Garrett Pollard, Jan. 14, 1929; children—Robert Pollard, Thomas Whitson, Ellen Cole (Mrs. John Charles Gille). Faculty, Princeton, 1924-25, 27-34, Haverford Coll., 1934-48; prof. econs. Northwestern U., Evanston, Ill., 1948-67; vis. prof. econs. Dartmouth, 1967-—. With Office of Lend-Lease Adminstrn. and Dept. State, 1943-46; vis. prof. U. Wis., 1951-62. Bd. dirs. Nat. Bur. Econ. Research, 1950-—, v.p., 1963-65, chmn., 1965-67. Mem. Am. (exec. com. 1944-46), Midwest (pres. 1952) econ. assns., Am. Assn. U. Profs., Phi Beta Kappa. Author: Monetary Inflation in Chile, 1931; The Irish Pound, 1955; The Development of British Monetary Orthodoxy, 1965; also articles. Research in Chilean inflation; stress on importance of politics, personality, and compromise in devel. Brit. monetary orthodoxy. Home: 8 Smith Rd., Hanover, N.H. 03755.*

FETTES, Edward Mackay, Am. chemist; b. Bklyn., Jan. 10, 1918; s. Edward Mackay and A. Evelyn (Clarke) F.; B.S., Mass. Inst. Tech., 1940; Ph.D., Poly. Inst. Bklyn., 1957; m. Frances McKean, Mar. 14, 1941; children—Evelyn F. (Mrs. Ernest Tiemann), Carole M. (Mrs. John W. Sinclair), William C., David M.; m. 2d, Lelia Torluemke, February 14, 1953; children—JoAnne M. (Mrs. George Kratas), April R. Began career with the Thiokol Chemical Corp., Trenton, N.J., 1942-60, dir. research, devel., 1958-60; mgr. plastics research Koppers Co., Inc., Pitts., 1960-—; chmn. Gordon Conf. Polymers, 1960. Mem. Am. Chem. Soc., Am. Inst. Chemists, Soc. Rheology, Comml. Chem. Devel. Assn., A.A.A.S., Faraday Soc., Assn. Research Dirs., Sigma Xi, Phi Lambda Upsilon, Delta Upsilon. Editor: Chemical Reactions of Polymers, 1964. Research, publs. on preparation, structure, properties of polysulfide polymers; discovery of disulfide interchange and interactions; polymerization of styrene and ethylene. Home: 850 Hulton Rd., Oakmont, Pa. 15139. Office: 440 College Park Dr., Monroeville, Pa. 15146.*

FEUER, Henry, chemist; b. Stanislau, Austria, Apr. 4, 1912; s. Jacob and Julia (Tindel) F.; M.S., U. Vienna (Austria), 1934, Ph.D., 1936; m. Paula Berger, Jan. 19, 1946. Came to U. S., 1941, naturalized, 1946. Postdoctoral fellow U. Paris (France), 1939; with dept. chemistry Purdue U., Lafayette, Ind., 1943-—, prof. chemistry, 1961-—; vis. prof. Hebrew U., Jerusalem, Israel, 1964. Fellow A.A.A.S.; mem. Am. Chem. Soc., Chem. Soc., Am. Inst. Aeros. and Astronautics, Sigma Xi, Phi Lambda Upsilon. Research, numerous publs. in organic nitrogen compounds; discovered new methods for syntheses nitro compounds, cyclic hydrazides; research on mechanism of these reactions. Home: 726 Princess Dr., West Lafayette, Ind. 47906. Office: Dept. Chemistry, Purdue U., Lafayette, Ind. 47907.*

FEUER, Irving, Am. phys. chemist; b. N.Y.C., Oct. 19, 1922; s. Max and Mina (Engel) F.; B.S. in Chemistry, Coll. City N.Y., 1943; M.S., Bklyn. Poly. Inst., 1955, postgrad.; m. Glenda Silbert, Aug. 3, 1958; children—Mitchell Eric, Douglas Marc (now deceased). Began career as chemist with the Canadian Radium & Uranium Corporation New York City, 1942-44, chem. physicist, 1944-47; dir. research Seederer-Kohlbusch, Inc., Englewood, N.J., 1947-49; cons. U. S. Army Q.M.C., 1951-52; cons. physicist Beth Israel Hosp., N.Y.C., 1950-58; physicist AEC project N.Y. U., N.Y.C., 1954-58; phys. chemist Air Force Cambridge Research Center, Bedford, Mass., 1958-60; dir. applied research and devel. Canrad Precision Industries, 1961-—. Recipient Merit award Air Force Cambridge Research Center, 1960. Mem. Am. Chem. Soc., A.A.A.S., Am. Vacuum Soc., N.Y. Acad. Scis., Sigma Xi. Chem. abstractor Chem. Abstracts, 1953-—. Contbr. numerous articles to tech. jours. Patentee in field; invented radioactive electronic microbalance, electro luminescent devices, promethium and tritium light source structures, magnetic break-seal devices, use Iodine 131 in determination thyroid function, porous electroluminescent detector for vapors; developed radiactive light sources, gas analysis using beta rays, radiation damage to surfaces. Home: 27 Seaview Lane, Port Washington, N.Y. 11050. Office: 43 W. 16th St., N.Y.C. 10011.*

FEUER, Lewis S., Am. sociologist; b. N.Y.C., Dec. 7, 1912; s. Joseph and Fannie (Weidner) F.; B.S. Coll. City N.Y., 1931; A.M., Harvard, 1932, Ph.D., 1935; m. Kathryn Jean Beliveau, Oct. 13, 1946; 1 dau., Robin Kathryn. Asst. in philosophy Harvard, 1935-37; instr. in philosophy Coll. City N.Y., 1939-42; faculty Vassar Coll., 1946-51, U. Vt., 1951-57; prof. philosophy and social sci. U. Cal., Berkeley, 1957-66; prof. sociology U. Toronto, 1966-—. Exchange scholar Inst. Philosophy Soviet Acad. Sci., Moscow, USSR, 1963. Recipient Bowdoin medal, 1935. Mem. Am. Sociol. Assn., Am. Philos. Assn. Author: Psychoanalysis and Ethics, 1955; Spinoza and the Rise of Liberalism, 1958; The Scientific Intellectual, 1963. Research, numerous publs. showing that the hedonistic-libertarian ethic was psychol. basis of origin of modern sci., that Weber-Merton thesis of primacy of Protestant ascetic ethic was invalid; analysis of ambiguities in and emotive appeal of concept of alienation. Home: 29 Roxborough St. E., Toronto 5, Ont., Can.*

FEUERBACH, Karl Wilhelm, German mathematician; b. Jena, Germany, May 30, 1800; s. Paul Johann Anselm von Feuerbach; prof. math. Erlangen (Germany) Gymnasium. Author: Eigenschaften einiger merkwurdiger Punkte des geradlinigen Dreiecks, 1822; Grundriss zu analytischen Untersuchungen der Dreieckigen Pyramid, 1827. Developed 9 point circle known as Feuerbach's circle; proved Feuerbach's theorem; helped introduce geometry of triangle and triangular pyramid, 1827. Died Erlangen, Mar. 12, 1834.

FEUGHELMAN, Max, physicist; b. Czernovitz, Rumania, Dec. 7, 1921; s. Abram and Rosa (Feltstein) F.; B.Sc. with 1st class honours in Math. Physics, U. Sydney (Australia), 1941; m. Dorothy Jean Adams, Jan. 11, 1947; children—Diana, David. Tutor in radio-physics Royal Australian Air Force Sch. Radar, Sydney U., 1941-42; physicist Amalgamated Wireless Valve Co., Sydney, 1942-47; lectr. physics U. New South Wales, 1947-51; physicist Commonwealth Sci. and Indsl. Research Orgn. div. textile physics, Ryde, N.S.W., Australia, 1951-—. Cons. on vacuum and glass tech. to various Sydney firms, 1947-51. Research, publs. on fibrous protein, keratin, relationship between the mech. properties of these fibres under varying thermal and moisture conditions related to molecular structure of protein. Home: 101 Princes St. Office: 338 Blaxland Rd., Ryde, N.S.W., Australia.*

FEUILLET, Louis, botanist; b. Mane, France, 1660; studied math., astronomy; mem. Franciscan order; sent (with Cassini) on royal expdn. to find port cities in Nr. East, 1699; later visited Martinique, Antilles, S.Am.; acquired many bot. specimens on 2d trip to S.Am., 1707-11; named royal mathematician, also awarded obs. at Marseilles, France; pub. sci. observations; observed change of declination and inclination, also northern lights; built hydrostatic hydrometer. Died Marseilles, Apr. 18, 1732.

FEULGEN, Robert, physiol. chemist; b. Essen-Werden, Germany, Sept. 2, 1884; s. Robert and Anna (Volckmar) F.; Ph.D., Kiel, Germany, 1912; m. Frieda Brauns, 1916; 1 dau. Asst. under Hermann Steudel, physiol. inst., Berlin; in Giessen, Germany, from 1919, named asso. prof., 1923, prof. physiol. chemistry, 1927. Author: Chemie und Physiologie der Nucleinstoffe, 1923; numerous articles. Discovered plasmal reaction, in which thymic acid acts as aldehyde and thymonuclein acid also occurs in plants, also nucleal reaction (characteristic reaction for cell nuclei and equivalent structures in bacteria), which led to his discovery of a new class of cell components, acetallipoids. Died Giessen, Oct. 24, 1955.

FEUQUE, Ferdinand Andre, French geologist, petrologist; b. 1828; studied volatile products of volcanic eruptions, formation of volcanic craters; introduced use of microscope for petrographic study into France; reproduced rocks and minerals artificially. Died 1904.

FEUSSNER, Karl, German elec. engr.; b. Rinteln/Weser, Germany, Oct. 17, 1855; s. Heinrich and Marie (Cöster) F.; Ph.D., Marburg, 1882; m. Clara Habicht, 1888; 1 son, Otto. Tchr., 1882-88; became dir. elec. dept. State Phys.-Tech. Inst., Berlin-Charlottenburg, 1888; later worked on electric testing sta. installations. Author: Neue Formen elektrischer Widerstandssätze, 1899; Hochspannungsbatterien, 1899; also articles. Holder numerous patents; created (with S. Lindeck) new resistance alloys manganin, konstantan; developed Poggendorff compensator into precision instrument with capacity for great accuracy in voltage measurement. Died Berlin, Oct. 24, 1915.

FEUSTEL, Irvin Carl, Am. chemist; b. Spokane, Wash., Feb. 3, 1906; s. Otto and Louise (Schafer) F.; B.S., Wash. State U., 1927; M.S., George Washington U., 1929; Ph.D., Am. U., 1934; m. Ruby R. Roberts, Aug. 22, 1928; children—Doris M. (Mrs. Tom C. Voight), Nancy L. (Mrs. Ken A. Jacob). With the United States Department of Agriculture, 1926-65, extension agrl. specialist, Albany, Cal., 1953-65; ret., 1965; Western rep. U. S. Peace Corps, 1962-63; tchr. chemistry Anna Head Sch., Oakland, Cal., 1965-—. Recipient Superior Service award U. S. Dept. Agr. Mem. Inst. Food Technologists, Potato Assn. Am., Sigma Xi. Contbr. numerous articles to sci. jours, books. Research on utilization of fruit and vegetable wastes and devel. process for recovery and purification of rubber from Guayule equivalent to Hevea rubber for critical and strategic purposes. Patentee process for treatment of rubber. Home: 1126 Park Hills Rd., Berkeley, Cal. 94708.*

FEVOLD, Harry Leonard, Am. biol. chemist; b. Badger, Ia., Oct. 21, 1902; s. Even E. and Anne (Thompson) F.; B.A., St. Olaf Coll., 1925; B.S., U. Wis., 1926, Ph.D., 1928; m. Agnes Beatrice Molstad, June 20, 1928; children—Ruth Ann (Mrs. Frederick P. Opem), Harry Richard, Kathryn Ellen. Research asso. U. Wis., 1928-35; faculty Harvard, 1935-41; with U. S. Dept. Agr., 1941-47; chief food research div. Dept. War, Chgo., 1947-51; with Baxter Labs, Morton Grove, Ill., 1951-—, research dir., coordinator, dir. research, devel., 1957-—. Recipient Superior Service award U. S. Dept. Agr., 1950. Fellow N.Y. Acad. Sci., A.A.A.S.; mem. Am. Chem. Soc., Soc. Biol. Chemists, Endocrine Soc., Soc. Exptl. Biology and Medicine, Ill. Acad. Sci., Sigma Xi, others. Research, publs. on chemistry and physiology of ovarian hormones, pituitary hormones, fractionation and characterization of blood proteins and egg proteins, chem. isolation and characterization of antibiotics, lysozyme, subtilin, gramicidin; also food tech. primarily in the elucidation of mechanisms of deterioration and methods for stabilizing food products. Home: 736 Community Dr., LaGrange Park, Ill. 60528. Office: 6301 Lincoln Av., Morton Grove, Ill. 60053.*

FEWKES, J(esse) Walter, Am. ethnologist, geologist; b. Newton, Mass., Nov. 14, 1850; s. Jesse and Susan E. (Jewett) F.; A.B., Harvard, 1875, A.M., Ph.D., 1877; student of Louis and Alexander Agassiz, student zoölogy U. Leipzig (Germany), 1878-1880; LL.D., U. Ariz., 1915; m. Harriett O. Cutler, Apr. 4, 1893. Asst. in Mus. Comparative Zoölogy, Harvard, 1881-89; editor Jour. Ethnology and Archaeology, 1890-94; field dir. Hemenway Southwestern Archaeol. Expdn. field work in Ariz., 1891-94; ethnologist Bur. Am. Ethnology, 1895-1918, chief of bur., 1918-28. Mem. Com. of Overseers to visit Peabody Mus. of Harvard Coll., 30 years. In charge excavation and repair Casa Grande, Ariz., Spruce Tree House, Cliff Palace, Sun Temple, Fire Temple, Far View House, Pipe Shrine House, Mesa Verde Nat. Park, Colo., Wupatki, Elden, Pueblo, Ariz., Weden Island near Tampa, Fla., 1908-26. Fellow Am. Acad. Arts and Scis., A.A.A.S. (v.p. 1911-12), Am. Anthrop. Soc. (pres. 1911, 12), Anthrop. Soc. Washington (pres. 1909, 10), Am. Folk-Lore Soc. (v.p.), Washington Acad. Scis. Author: Snake Ceremonials at Walpi, 1894; Archaeological Expedition to Arizona in 1895; Two Summers' Work in Pueblo Ruins, 1897; Aborigines of Puerto Rico and Neighboring Islands, 1907; Casa Grande, Arizona, 1913. Studied Medusae and formation coral islands, lore and artifacts of Hopi Indians; conceived idea of using Hopi complex as source by which antiquity could be interpreted; his greatest contbn. to Pueblo archeology was preservation of sites for public viewing; studied Pueblo Indians and prehistoric remains in Southwest U. S. and Puerto Rico; 1st to use phonograph for recording Indian music. Died Forest Glen, Md., May 31, 1930.

FEYNMAN, Richard Phillips, Am. physicist; b. N.Y.C., May 11, 1918; s. Melville Arthur and Lucille (Phillips) F.; B.S., Mass. Inst. Tech., 1939; Ph.D., Princeton, 1942. Staff atomic bomb project, Princeton, 1942-43, Los Alamos, 1943-45; asso. prof. theoretical physics, Cornell, 1945-50; prof. theoretical physics Cal. Inst. Tech., 1950-—. Recipient Einstein Award, 1954; Nobel prize in physics, (with J. S. Schwinger and S. Tomonaga) 1965. Mem. Am. Phys. Soc., A.A.A.S., Nat. Acad. Scis. Author: Quantum Electrodynamics, 1961; Theory of Fundamental Processes, 1961; Feynman Lectures on Physics, 3 vols., 1963-64; (with A. R. Hibbs) Quantum Mechanics and Path Integrals, 1966. Developed hypothesis of quantum electrodynamics which resolved inaccuracies of earlier theories dealing with interaction of atoms with radiation fields, 1948; offered math. explanation for behavior of helium; devised Feynman diagram, a means of accounting for possible particle transformation; theory of beta dacay. Address: Physics Dept., California Institute of Technology, Pasadena, Cal. 91109.

FEYRTER, Friedrich, pathologist; b. Vienna, Austria, Feb. 6, 1895; s. Johann and Marie (Artner) F.; Dr.Med., U. Vienna, 1921; m. Josefine Kirchroth Edle von Kirchsfeld, Nov. 6, 1941; 1 dau., Marieluise (Mrs. Fritz Katscher). Prof., dir. Path. Inst., Med. Acad. Danzig, 1936-41; dir. Path. Inst. Graz (Germany), 1941-46; dir. Path. Inst., U. Göttingen, 1951-59, dir. emeritus, 1959—. Recipient Schunk prize U. Giessen, 1961. Mem. German Soc. Pathologists. Research, numerous publs. on neuroma and neurofibromatosis, pathology of vegetative nervous periphery, peripheral endocrine glands. Home: A1040 Vienna IV, Wiedener Hauptstr. 40, Austria.*

FIALA, Anthony, Am. explorer; b. Jersey City Heights, N.J., Sept. 19, 1869; s. Anthony and Annie (Kohout) F.; student Cooper Union and N.A.D., N.Y.C.; m. Mary Clare Puryear, Dec. 6, 1905; children—Anthony, Reid Puryear, Mary Maury, Lenore Fontaine. Began business life as stone artist and designer of lithography; asst. in a phys. and chem. lab., 5 yrs.; newspaper artist and cartoonist, 1890; cartoonist Grit, Williamsport, Pa., 1893-94; studied processes of photo-engraving and photogravure; installed photo-engraving plant for Bklyn. Daily Eagle, 1894, also in charge art and engraving dept., 1894-99; pres. Fiala Outfits, Inc. Photographer Baldwin-Ziegler Polar Expdn., 1901-02; comdg. officer Ziegler Polar Expdn., 1903-05, reaching 82° 4' north, discovered and mapped new islands, also surveyed and mapped accurately greater part of Franz Josef Archipelago, maps and records published by Nat. Geog. Soc., Washington, 1907. Fellow A.A.A.S., Royal Geog. Soc.; mem. Am. Geog. Soc.; hon. mem. Internat. Yukon Polar Inst. Cruising Club of Am. Author: Troop "C" in Service, 1899; Fighting the Polar Ice, 1906. Accompanied Theodore Roosevelt on his trip through Brazilian wilderness, 1913-14; explored Papagaio River and descended Jurnena and Tapajos rivers of Brazil. Died Apr. 8, 1950.

FIALKOW, Aaron David, Am. mathematician; b. N.Y.C., Aug. 9, 1911; s. Abraham and Jennie (Schochet) F.; B.S., Coll. City N.Y., 1931, M.S., 1931; Ph.D., Columbia, 1936; m. Yetta Kleinberg, May 26, 1940; children—Elaine, Lawrence, Diane. Instr., Bklyn Coll., 1937-42, Columbia, 1942-45; asso. prof. Poly. Inst. Bklyn., 1947-51, prof., 1951—; head dept. applied math. Control Instrument Co., Bklyn., 1946-54. Mem. Am. Math. Soc., Phi Beta Kappa, Sigma Xi. Contbr. articles on differential geometry and electric network theory to tech. jours. Research in differential geometry, especially conformal geometry; in electric network theory on properties and synthesis of transformerless networks. Home: 312J Tibbett Av., N.Y. 10463. Office: 333 Jay St., N.Y., 11201.

FIBIGER, Johannes Andreas Grib, Danish pathologist; b. Silkeborg, Denmark, Apr. 23, 1867; studied under Koch and Behring in Berlin, Germany; worked in Copenhagen, Denmark; prof. Inst. Path. Anatomy, Copenhagen; Recipient Nobel prize in medicine and physiology, 1926. 1st to induce cancer experimentally, 1913; infected rats by ingestion of a cockroach carrying Spiroptera neoplastica; recipient Nobel prize in medicine for discovery of Spiroptera carcinoma, 1926. Died Copenhagen, Jan. 30, 1928.

FIBONACCI, Leonardo (Leonardo da Pisa), Italian mathematician; b. Pisa, circa 1170; s. Guglielmo Bonacci; reared in Algeria, tutored by Mohammadan schoolmaster; traveled extensively around Mediterranean, visited Egypt, Syria, Greece, Sicily, southern France; spent considerable time in North Africa; served as minor ofcl. in commune of Pisa, circa 1240. Author: Liber Abaci, 2 versions, 1202, 1228; Practica Geometriae, circa 1220; Liber quadratorum, circa 1225; Flos, circa 1225. Best known for his work Liber Abaci (Book of Abacus), which was instrumental in putting end to old Roman system of numerical notation and gave complete, systematic explanation of use of Hindu and Arabic numerals and value of positional notation; also did work on Lamé series (now called series of Fibonacci); worked on Diophantine analysis; wrote on number theory; had good knowledge of rudimentary algebra and Greek math.; applied algebra to solution of geometric problems; worked on trigonometry; did original work on quadratic equations. Died circa 1230.

FICHERA, Elio Edoardo, Italian astronomer; b. Catania, Jan. 1, 1925; s. Giuseppe and Mimi (Puglisi) F.; Ph.D. in physics; m. Angela Leone, July 24, 1954; children—Maria, Giovanna. Sci. adminstr. of astron. observatories; dir. hydrography cabinet Naval Inst., U. Naples; former dir. Internat. Astron. Sta., Cagliari; Italian dir. sci. delegation IGY, Brussels, 1957-59; head research astron. service artificial satellites Center Spatial Physics, U. Rome; affiliated with RAF, Cambridge Research Center. Recipient Giacomelli prize, 1958. Mem. Italian Astronomy Soc., Internat. Astronomy Union. Research and numerous publs. on problems of variations of latitudes, longitudes, photometry, meridien and geodesic astronomy. Home: via Moiariello 16. Office: Astronomical Observatory of Capodimonte, Naples, Italy.

FICHOT, Eugène, French hydrographer; b. Le Creusot, 1867; dir. hydrographic service French navy; studied tides and their indsl. uses. Died Gironde, Tabanac, 1959.

FICHTEL, Carl E., Am. physicist; b. St. Louis, July 13, 1933; s. Edwin B. and Eleanora (Gutsch) F.; B.S., Wash. U., 1955, Ph.D., 1960. Head gamma ray and nuclear emulsion sect. Goddard Space Flight Center, Greenbelt, Md., 1959—; instr. U. Md., 1963—. Mem. Am. Phys. Soc., Am. Astronom. Soc., Am. Geophys. Union, Sigma Xi. Contbr. articles to sci. jours. Discoverer of heavy nuclei in solar cosmic rays; principal investigator of sounding rocket expts. studying solar cosmic ray composition, sounding rocket expts. on galactic cosmic ray composition, balloon x-ray expts. and a spark chamber for space research. Home: 4330 Hartwick Rd., College Park, Md. 20740. Office: Goddard Space Flight Center, Greenbelt, Md.*

FICHTELIUS, Karl Erik, Swedish physician; b. Stockholm, Sweden, Mar. 29, 1924; s. Erik Hildor and Martha (Ostrom) F.; grad. Med. Sch., U. Uppsala (Sweden), 1952, M.D., docent in histology, 1953. Asst. prof. histology Uppsala U., 1955-60, research asso. prof. histology, 1960-62, prof. histology, 1962—. Am. Heart Assn. vis. prof. Dept. Pediatrics, U. Minn., 1967-68. Recipient Jubilee medal Royal Lymphatic Soc. Uppsala, 1960. Mem. Swiss Soc. Hematology. Author: On the Fate of the Lymphocyte, 1953. Co-author, editor: (with others) The Conditions of Man, 1967. Research, publs. on method for separating lymphocytes with columns of cotton wool, 1951; introduction of isotope labelling in study of homing of lymphocytes, 1952; advocated role of thymus in antibody prodn., 1957; demonstrated negative effect of thymectomy on antibody formation in adolescent guinea pigs, 1961; 1st demonstrated reutilization of H3 thymidine in vivo, 1961; presented theory of communication with dolphins, 1968; demonstrated possible bursa equivalents in bursa-less vertebrates, 1968. Home: 53 Ostra Agatan, Uppsala. Office: 26 Vastra Agatan, Uppsala, Sweden.*

FICHTER, Friedrich, Swiss chemist; b. Basel, Switzerland, July 6, 1869; s. J. Ben. and Maria (Seiler) F.; ed. U. Basel, U. Strasbourg (France); Ph.D., 1894; m. Maria Louise Bernoulli, 1904; 6 children. Prof., Basel. Author: Organische Elektrochemie, 1942; also numerous articles. Editor, Helvetica chimica acta. First to isolate beryllium. Died 1952.

FICINO, Marsilio, Italian philosopher; b. Figline, Tuscany, Oct. 19, 1433; s. Diotifeci d'Agnolo di Giusto and Alessandra (di Nannoccio di Montevarchi) F.; ed. by John Argyropoulos, Florence, 1459; studied medicine at Bologna. Head of Platonic Academy, Florence, 1462, where he lectured on Plato, general philosophical and theological problems; ordained priest, 1473; canon of San Lorenzo, Florence; under patronage of the Medici. Author: Institutiones platonicae, 1556; De Christiana religione, 1474; Theologia platonica, 1482; De vita libri tres, 1489; translator of Hermes Trismegistus, 1463, Plotinus, 1484-86, and others. Contributed to development of Platonic studies in Europe; commissioned by elder Cosimo de'Medici to translate Plato and several neo-platonists into Latin; his translations used for centuries; his philosophical works upheld superiority of Christianity, but saw harmony among all religions as varying forms of neo-platonic universal religion; alchemical works attributed to him. Died Careggi, near Florence, Oct. 1, 1499.

FICINUS, Heinrich David August, physician, natural scientist; b. Dresden, Germany, Sept. 18, 1782; s. David and Johanna Elisabeth (Mücke) F.; entered Berlin Collegium Medico-chirurgicum, 1803; M.D., Wittenberg, Germany, 1806; m. Henriette Seifert, 1809; 1 son; m. 2d, Auguste Ernestine Sausse, 1811, 3 sons, 3 daus. Apprentice in father's pharmacy, circa 1795, then asst. in Kamanz, Prague, Breslau; certified as pharmacist, Dresden, 1804; practiced medicine, Dresden, after 1806; became prof. physics and chemistry Surg.-Med. Acad., Dresden, 1814, also prof. natural history, pharmaceutics, therapy, sch. vet. medicine, from 1817; tchr. chemistry, tech., physics Tech. Ednl. Inst. (now Tech. U. Dreseden), 1828-33; kept father's pharmacy, from 1822. Mem. Leopoldina. Author: Flora der Gegend um Dresden, 1807/08; Anfangsgründe der medizinischen Physik, 1815; Antangsgründe der medizinischen Chemie, 1815; Optik oder Versuch eines folgerechten Umrisses der gesammten Lehre vom Licht, 1828; Physik, allgemeinfasslich dargestellt, 2 vols., 1828; Allgemeine Naturkunde, 1839. One of few contemporaries to recognize value of Goethe's theories of optics and colors; corresponded with Goethe and J. J. Berzelius. Died Dresden, Feb. 16, 1857.

FICK, Adolf Eugen, physiologist; b. Kassel, Hesse, Germany, Sept. 3, 1829; s. Friedrich and Marianne (Spousel) F.; M.D., U. Marburg, 1851; m. Emilie von Cölln, 1862. Asst. to Carl Ludwig, Zurich, 1852; prof. physiology, Zurich, from 1862; prof. U. Würzburg, from 1868. Author: Die medizinische Physik, 1856; Untersuchungen über elektrischen nervenreizung, 1864. Belonged to "mechanistic" school of physiology; made important discoveries in every branch of physiology; proved carbohydrates rather than albumin to be source of muscle energy; constructed 1st pletysmograph, which measured pulse rate; invented myotonograph for measuring and recording muscle tension, circa 1864; developed method to determine cardiac output by gasometry, 1870; discovered law of diffusion in liquids named after him. Died Blankenberghe, Belgium, Aug. 21, 1901.

FICK, Adolf Gaston Eugen, German ophthalmologist; b. Marburg/Lahn, Germany, Feb. 22, 1852; s. Ludwig and Julie Müldner (von Mühlheim) F.; studied medicine, univs. Würzburg, Zurich, Marburg, Freiburg; M.D., 1876; m. Marie Wislicenus, 1884; 2 sons, including Roderich, 5 daus. Asst., physiol. inst., Würzburg, Germany, anatomy inst., Breslau, Germany, eye clinics of Förster and Cohn, Breslau; practiced medicine, specializing in eye disorders, Richmond, 1879-86; became mem. faculty Zurich (Switzerland), 1886/87; founder (with P. von Monakow) pvt. clinic; returned to Germany, 1914; head field hosps. in France, Russia, Turkey, during World War I. Author: Lehrbuch der Augenheilkunde, 1894; also articles. Research on eye anatomy, evolution, light and color sensitivity, exhaustion and recovery of retina, practical optics and surgery; suggested contact lenses for certain refraction faults, 1887. Died Herrsehing/Ammersee, Germany, Feb. 11, 1937.

FICK, Rudolf Armin, anatomist; b. Zurich, Switzerland, Feb. 24, 1866; s. Adolf and Emilie (v. Colln) F.; m. Frieda Prym, 1893; 3 sons, including Wilhelm, 2 daus. Prosector in Würzburg, Germany; became mem. faculty, 1892; prosector, asst. prof. anatomy, Leipzig, Germany; prof., dir. anat. inst. German U. Prague (Czechoslovakia); became asso. with U. Innsbruck (Austria), 1909; succeeded W. von Waldeyer-Hartz, U. Berlin, 1917; ret., 1934. Author: Handbuch der Anatomie und Mechanik der Gelenke, 1904-11. Research on mechanics of muscles and joints, comparative anatomy of anthropoid apes; developed theories of genetics no longer accepted but influential in his time; helped reform anat. instrn. in early 20th century. Died Berlin, May 23, 1939.

FICKEN, Frederick Arthur, Am. mathematician, educator; b. Moore's Hill, Ind., Aug. 13, 1910; s. Richard Oscar and Grace (Fagley) F.; B.A., Oberlin Coll., 1931; M.A., Ohio State U., 1932; B.A. (Rhodes scholar), Exeter Coll., Oxford, Eng., 1934; Ph.D., Princeton, 1938; m. Mary E. Harman, Sept. 11, 1940; 1 son, William H. Instr., Cornell U., 1938-42; prof. math. U. Tenn., Knoxville, 1942-59; research asso. N.Y. U., N.Y.C., 1949-51, prof. math., 1959—, chmn. dept. math. University Heights Campus, 1959-67. Cons., USN, 1944-45, Union Carbide Nuclear Co., Oak Ridge, 1946-64. Mem. Am. Math. Soc., Math. Assn. Am., Soc. Indsl. and Applied Math., Am. Assn. U. Profs., Am. Civil Liberties Union, Phi Beta Kappa, Sigma Xi, Phi Kappa Phi. Author: (monograph) The Simplex Method of Linear Programming, 1961; Linear Transformations and Matrices, 1967. Editor: Am. Math. Monthly, 1962-66. Publs. in various fields of math. Home: 14 Benedict Pl., Pelham, N.Y. 10803. Office: Dept. Math., N.Y. University, N.Y.C. 10453.*

FICKER, Heinrich von, see von Ficker, Heinrich.

FICKER, (Philipp) Martin, bacteriologist; b. Sohland/Spree, Germany, Nov. 17, 1868; s. Julius Gustav and Fanny Auguste (Jacoby) F.; ed. U. Breslau (Germany); m. Lisa Hofmann, 1914; 3 sons. Asst. to F. Hofmann, Hygienic Inst., Leipzig, 1896-1901, became mem. faculty, 1898; named dept. head, hygienic inst. U. Berlin, 1902, asso. prof., 1903, prof., 1908; became asso. with State Bacteriol. Inst., Sao Paulo, Brazil, 1913; returned to Germany, 1917; became dept. head Kaiser-Wilhelm Inst. for Exptl. Therapy, Berlin-Dahlem; founder bacteriol. lab., Sao Paulo, 1923; dir. research sta. for microbiology Kaiser Wilhelm Soc., Sao Paulo, 1926 to World War II. Author: Schulhygiene, 1911; Einfache Hilfsmittel zur Ausführung bakteriologischer Untersuchungen, 1921. Contbg. author handbooks and textbooks. Editor: (with M. Rubner and M. Gruber) Handbuch der Hygiene, 1911-23. Research on bacteriology of air, new coloring and culture methods in bacteriology; developed Ficker's typhus diagnosis; discovered diagnostic flaking reaction for leprosy; worked on burning gas toxins during World War I. Died Sao Paulo, Nov. 22, 1950.

FICKLEN, Joseph Burwell, III, Am. chem. engr.; b. Fredericksburg, Va., Apr. 11, 1902; s. Joseph B. and Ellen Caskie (London) F.; grad. Episcopal High Sch., Alexandria, Va., 1919; student U. Va., 1919-20; B.S. in Chem. Engring., Cal. Inst. Tech., 1928; postgrad. Yale; m. Irene Louise Poole, Dec. 15, 1939; 1 son, Joseph Burwell, IV. Supt., Rappahannock Electric Light & Power Co., 1920-23; chem. engr. Travelers Ins. Co., 1928-29, engr. in charge labs., 1929-41; spl. health officer County of Los Angeles, 1941-42; 44—; cons. City of Los Angeles Health Dept., 1941-42; asst. mgr. U. S. Rubber Co., Charlotte, N.C., 1942-44; owner Joseph B. Ficklen, III, Pasadena, Cal., 1952—; with AEC, 1946-64. Cons. to pvt. cos.; adviser Dominion New Zealand Health Dept., New Zealand Christchurch Tunnel Authority, New South Wales (Australia) Occupational Health Bur., 1964—. Fellow So. Cal. Acad. Scis., Am. Inst. Chemists; mem. Am. Chem. Soc., Am. Inst. Chem. Engrs., Am. Inst. E.E., A.A.A.S., Cal. Soc. Profl. Engrs., Am. Inst. Indsl. Hygienists, Am. Inst. Mil. Engrs., Combustion Inst., Soc. Chem. Industry (Eng.), Nat. Soc. Prof. Engrs., Am. Assn. Engrs., Am. Rocket Soc., Solar Energy Soc. Author: Manual of Industrial Health Hazards, 1940; also articles. Translator: Some Methods for the Detection and Estimation of Noxious Gases and Vapors in Air (A. S. Zithova), 1936. Research on methods for estimation manganese, hydrazine, tin carbon, monoxide and lead in air; co-inventor Cook-Ficklen benzene apparatus, Ficklen-Ott dust camera, dustoscope, Ficklen-Strong thermal precipitators; inventor konisampler, micro-therm precipitator; developed many procedures for handling poisonious material and estimation of explosion damage and protection. Home: 1484 E. Mountain St., Pasadena, Cal. 91104. Office: 30 Lewis St., Hartford, Conn. 06103.*

FIEBER, Franz Xaver, entomologist; b. Prague, Czechoslovakia, Mar. 1, 1807; s. Franz Anton and Maria Anna (Hantsehl) F.; student Poly. Inst. Prague, 1824-28, U. Prague; Ph.D., Jena, Germany, 1848. Early career in finance and tech.; became ofcl., appellate ct., Prague, 1832, later dist. magistrate, Chrudim, Bohemia. Mem. Leopoldina. Author: (monograph) Die europäische Hemiptera, 1860; other monographs, articles in jours. Bot. illustrator, then undertook study of hemiptera, orthoptera; became important amateur entomologist whose work contbd. to devel. of 19th century entomology and formed basis for modern entomology. Died Chrudim, Feb. 22, 1872.

FIEBIGER, Josef, Austrian veterinatian; b. Odrau, Austria, Feb. 7, 1870; s. Josef and Klementine (Hanke) F.; M.D., Vienna, Austria, 1894; degree from sch. vet. medicine, Vienna, 1900; dr. vet. medicine (hon.), Free U. Berlin; m. Margarethe Meyerhoff, 1919; 2 sons, 1 dau. Became lectr. biology and pathology of fish, sch. vet. medicine, 1903; named asso. prof. gen. zoology and parasitology, 1906; became lectr. parasitology, med. sch. U. Vienna, also asso. prof. biology and pathology of fish, histology and embryology, sch. vet. medicine, 1914. Mem. Helminthological Soc., Washington. Author: Die tierischen Parasiten der Haus- und Nutztiere, sowie des Menschen, 1923; (with A. Trautmann) Histologie und vergleichende mikvoskopische Anatomie der Haustiere, 1941-49; many other publs. Research in biology and pathology of fish. Died Vienna, Apr. 9, 1956.

FIEDLER, Carl Ludwig Alfred, German physician; b. Moritzburg, Germany, Aug. 5, 1835; M.D., U. Leipzig, Germany, 1859; asst. prof. City Hosp., Dresden, Germany, 1861-68, chief physician, from 1868. Described Weil's disease, an acute infectious disease marked by fever, jaundice, nephritis and other symptoms, caused by spirochete Leptospira icterohaemorrhagiae, also Fiedler's disease, leptospirosis icterohaemorrhagica. Died 1921.

FIEDLER, Fred Edward, social psychologist; b. Vienna, Austria, July 13, 1922; s. Victor and Hilda (Schallinger) F.; came to U. S., 1938, naturalized, 1943; A.M., U. Chgo., 1947, Ph.D., 1949; m. Judith Miriam Joseph, Apr. 14, 1946; children—Phyllis Elizabeth, Ellen Victoria, Robert Joseph, Carol Ann. Trainee clin. psychology VA, Chgo., 1947-50; faculty U. Chgo., 1949-51; faculty U. Ill., Urbana, 1951—, prof. psychology, 1960—, head div. social and differential psychology, 1962-67, dir. Group Effectiveness Research Lab., 1953—. Cons. to govt. Recipient Outstanding Research award, 1953, Hon. Mention, 1960 Am. Personnel and Guidance Assn; Personal Letter of Commendation for work with Belgian Navy, Chief U. S. Naval Operations, 1964. Fulbright Research scholar U. Amsterdam, 1958-59; Ford Faculty Research fellow U. Louvain, 1963-64. Fellow Am. Psychol. Assn., Soc. Psychol. Study Social Issues; mem. Soc. Exptl. Social Psychologists, Internat. Assn. Applied Psychology, Midwestern Psychol. Assn. Author: Leader Attitudes and Group Effectiveness, 1958; Boards, Management and Company Success, 1959; (with Eleanor Godfrey and D. M. Hall) A Theory of Leadership Effectiveness, 1967. Research in conditions for leadership. Home: 2014 Zuppke Circle, Urbana, Ill. 61801.*

FIEDLER, (Otto) Wilhelm, mathematician; b. Chemnitz, Germany, Apr. 3, 1832; s. Christian Wilhelm and Amalie (Ruppert) F.; ed. indsl. sch., Chemnitz, Mining Acad., Freiberg; Ph.D. (under F. A. Mobius), U. Leipzig, 1858; hon. dr. Tech. U. Vienna; m. Lina Elise Springer, 1860; 7 children, including Ernst, Karl. Tchr. math. and mechanics, new mech. sch., Freiberg, Germany, 1852; became tchr., indsl. sch., Chemnitz, 1853, tchr. math. and descriptive geometry, 1857; named prof. descriptive geometry Tech. U. Prague (Czechoslovakia), 1864; prof. Fed. Polytechnikum, Zurich, from 1867. Recipient Steiner prize Prussian Acad. Scis. Mem. Leopoldina, Bavarian Acad. Scis. Author: Analytische Geometrie der Kegelschnitte, frei nach G. Salmon, 1860; Die Elemente der neueren Geometrie und des Algebra der binaren Formen, 1863; Vorlesungen über die Algebra der linearen Transformationen nach Salmon, 1863; Analytische Geometrie des Raumes nach Salmon, 1863; Die darstellende Geometrie, 3 vols., 1871; Analytische Geometrie der hoheren ebenen Kurven nach Salmon, 1873; Neue elementare Projektionsmethoden? 1879; Zyklographie oder Konstruktion der Aufgaben uber Kreise und Kugeln . . . , 1882. Founder of modern descriptive geometry based on central projection; recognized homogeneous coordinates as cross-ratios invariant in all linear transformations, 1870; pub. works of G. Salmon. Died Zurich, Nov. 19, 1912.

FIELD, Franklyn, Am. meteorologist; b. N.Y.C., Mar. 30, 1923; s. Joseph Judah and Sarah (Berk) F.; B.A., Bklyn. Coll., 1944; B.S., Columbia, 1949; Dr. Optometry, Mass. Coll. Optometry, 1960; m. Joan Sylvia Kaplan, Aug. 10, 1947; children—Elliot David, Pamela Diane, Allison Carol. Meteorologist, U. S. Weather Bur., N.Y.C., 1947-48; gen. mgr. Weather Forecast Co., N.Y.C., 1950-56; pres. Internat. Weather Corp., N.Y.C., 1956-60; dir. meteorology USPHS study of air pollution Med. Health and Research Council of N.Y., 1960; research fellow, instr. Albert Einstein Coll. Medicine, 1961—; dir. research Spray Tech. Corp., 1964—; cons. meteorologist N.Y.C. Dept. Air Pollution Control; meteorologist Sta. WNBC-TV and Radio, 1959—. Mem. Am., N.Y. meteorol. socs., N.Y. Acad. Sci., A.A.A.S., Am. Pub. Health Assn., Am. Inst. Bioclimatology, Air Pollution Control Assn., Am. Geophys. Union, Am. Optometric Assn., Omega Epsilon Chi. Research on bioclimatology; air pollution; meteorology and effects upon health. Home: 111 Fairwater Av., Massapequa, N.Y. 11758. Office: Albert Einstein Coll. Medicine, 1300 Morris Park Av., Bronx, N.Y. 10461; also NBC, 30 Rockefeller Plaza, N.Y.C. 10020.

FIELD, George, English chemist; b. Hertfordshire, Eng., circa 1777; ed. St. Peter's Sch.; went to London at the age of 18; Author: Chromatography . . . , 1835; Rudiments of the Painter's Art . . . , 1850; also others. Invented percolater and drying stove for purification of coloring matter, conical lens for producing continuous rainbow; 1st to grow madder in Eng.; application sci. to improvement artists materials. Died Middlesex, Eng., Sept. 28, 1854.

FIELD, George Brooks, Am. astrophysicist; b. Providence, Oct. 25, 1929; s. Winthrop Brooks and Pauline (Woodworth) F.; B.S., Mass. Inst. Tech., 1951; Ph.D., Princeton, 1955; m. Sylvia Farrior Smith, June 23, 1956; children—Christopher Lyman, Natasha Suzanne. Physicist, U. S. Naval Ordnance Lab., Silver Spring, Md., 1951-52; jr. fellow Harvard Soc. Fellows, 1955-57; asst. prof. astronomy Princeton, 1957-62, asso. prof. 1962-65; prof. astronomy, U. Cal., Berkeley, 1965—; mem. space sci. panel President's Sci. Adv. Com., 1966, Nat. Acad. com. in astronomy adv. to Office Naval Research, 1966—, adv. panel astronomy NSF, 1966—. Mem. Am. Astron. Soc., Astron. Soc. Pacific, Internat. Astron. Union, Am. Geophys. Union, Sigma Xi. Contbr. numerous articles to profl. jours. Research in intergalactic matter, interpretation of interstellar spectral lines in terms of cosmic background radiation, planetary radio emission; study of interstellar gas dynamics and galaxy formation. Home: 48 Senior Av., Berkeley, Cal. 94708.*

FIELD, Henry, Am. anthropologist; b. Chgo., Dec. 15, 1902; B.A., Oxford U., 1925, M.A., 1929, D.Sc., 1937; m. Julie Rand Allen, Feb. 6, 1953; children—Mariana (Mrs. Charles S. Hoppin), Juliana Lathrop. Curator, Field Mus. Natural History, Chgo., 1926-41; govt. research worker, Washington, 1941-45; research fellow in phys. anthropology Peabody Mus., Harvard, 1950—. Trustee Am. Sch. Prehistoric Research, Cambridge, Mass. Mem. A.A.A.S., Am. Assn. Anthropologists, Royal Geog. Soc., Royal Anthropol. Inst. Author numerous books including Anthropological Reconnaissance in West Pakistan, 1959; North Arabian Desert Archaeological Survey, 1925-50, 1959; Bibliography on S.W. Asia, vols. I-VII, 1953-62; studies on Migration and Settlement, 1962. also numerous articles. Developed Hall of Races of Mankind, Field Mus. Nat. History, Chgo.; studied prehistory and phys. anthropology of S. Western Asia. Home: 3551 Main Hwy., Coconut Grove, Miami, Fla. 33133. Office: Peabody Mus., Harvard, Cambridge, Mass. 02138.

FIELD, James Bernard, Am. physician; b. Ft. Wayne, Ind., May 28, 1926; s. Abraham and Clara (Riddner) F.; M.D. cum laude, Harvard, 1951; m. Dorothy Allison Spivey, Sept. 24, 1954; children—Carolyn Rundle, Nancy Terrell, Douglas Andrew, Susan Tillman. Practice medicine, specializing in internal medicine, endocrinology, Bethesda, Md., 1954-62, Pitts., 1962—; sr. asst. surgeon Nat. Inst. Arthritis Metabolic Diseases, NIH, 1954-57, sr. investigator, 1958-62, prof. medicine, 1966—, mem. endocrinology study sect., 1965—; asst. in medicine Diabetic Clinic Kings Coll. Hosp., London, 1957-58; cons. chronic diseases br. USPHS, 1962—. Recipient Eli Lilly award Am. Diabetes Assn., 1958; Van Meter prize Am. Goiter Assn., 1961. Mem. Am. Assn. Physicians, Am. Soc. Clin. Investigators, Am. Physiology Soc., Diabetes Assn., Endocrine Soc., Am. Fedn. Clin. Research. Mem. editorial bd. Metabolism, 1960—, Clinical Research, 1963—. Studies, numerous publs. on mechanism by which thyroid-stimulating hormone regulates thyroid gland function. Home: 109 Hillcrest Rd., Pitts. 15238. Office: Presbyn. Hosp., 230 Lothrop St., Pitts. 15213.*

FIELD, John II, Am. physiologist; b. Phila., Pa., May 11, 1902; s. Thomas Rittenhouse and Rachel Lee (Davidson) F.; Pierre S. DuPont scholar, William Penn Charter Sch., Phila., 1915-19; A.B., Stanford U., 1923, A.M., 1924, Food Research fellow, 1924-25, Ph.D., 1928; m. Sally Miller, Dec. 14, 1929; children—John Austin, Charles Davison, Richard Clark. Mem. faculty, Stanford, 1927-51, prof. physiology, 1942-51 (on leave 1949-51); acting dir. Arctic Research Lab., Office Naval Research, Aug. 1948; head biology br., Office Naval Research, Washington, 1949-51; chmn. dept. physiology, U. Calif. Sch. of Med., Los Angeles, 1951-62, asso. dean School Medicine, 1958—; asst. dir. Nat. Sci. Found., Washington, cons. since 1952; exec. sec. Arctic Research Lab. Adv. Bd., 1949-51; mem. panel on physiology and panel on Arctic environment Research and Development Bd., 1949-51. Member steering committee Alaska Science Conference, Nat. Research Council, 1950. Fellow A.A.A.S.; asso. mem. A.M.A.; mem. Am. Physiol. Soc. (editor-in-chief Handbook of physiology 1955—), Society for Exptl. Biology and Medicine (sec.-treas. Pac. Coast Sect., 1936-38), Western Soc. Naturalists, Western Soc. Clin. Res., Calif. Acad. Medicine, Sigma Xi. Mem. editorial bd. Ann. Rev. Physiology, 1951-57. Contbr. articles. Research on enzyme kinetics; biological oxidation; drug action on cells; cell metabolism; temperature influence on metabolism and enzymes; hypothermia. Home: 521 S. Westgate Av., Los Angeles 49.

FIELD, John Byron, Am. physician; b. N.Y.C., May 15, 1919; s. Michael and Mollie (Light) F.; Pharm. Chemist, B.S., St. John's U., 1940; M.S., U. Wis., 1941, Ph.D., 1944; M.D. with honor, U. Rochester, 1948; m. Sylvia Mandelbaum, Nov. 11, 1944; children—Patricia Laurie, Christopher Adam. Asst. clin. prof. medicine U. So. Cal., 1951-58; asso. clin. prof. medicine Cal. Coll. Medicine, 1963-64; sci. dir. Western Inst. for Cancer and Leukemia Research, Beverly Hills, Cal., 1960-65; dir. Western Foundation for Cancer Research, Culver City, California, 1965—. Consultant to AEC project, 1948-49. Commended by U. S. State Dept. for lecture tour, S.-Am., 1963. Mem. Soc. for Exptl. Biology and Medicine, Am. Chem. Soc., Am. Fedn. for Clin. Research, Am. Soc. Biol. Chemists, Western Soc. for Clin. Research, Am. Assn. for Cancer Research. Author: Cancer: Diagnosis and Treatment, 1959; also numerous articles. Research in blood coagulation, blood biochemistry, and nutrition, clin. and animal studies in cancer and cancer chemotherapy. Home: 607 N. Oakhurst Dr., Beverly Hills, Cal. 90210.*

FIELD, Joseph Herman, Am. chem. engr.; b. Pitts., May 29, 1920; s. Julius and Rebecca (Elinoff) F.; B.S. in Chem. Engring., Carnegie Inst. Tech., 1940; M.S., U. Pitts., 1944; m. Rose Presser, June 22, 1947; children—Robert, Nancy. Jr. engr. U. S. Bur. Mines, Pitts., 1941-44, project to sr. project engr., 1946-58, project coordinator, 1958—. Mem. Am. Inst. Chem. Engrs., Am. Chem. Soc. (Bituminous Coal Research award 1963), Air Pollution Control Assn., Inst. Fuels Gt. Britain, Tau Beta Pi. Contbr. to Organic Synthesis (P. H. Groggins), 1958. Devel. processes for synthesis of gaseous and liquid fuels from coal. Patentee in field. Home: 3177 Shady Av., Pitts. 15217. Office: 4800 Forbes Av., Pitts. 15213.*

FIELD, Joshua, Brit. civil engr.; b. circa 1787. Fellow Royal Soc., 1836; mem. Instn. Civil Engrs. (a founder 1817, v.p., pres. 1848), Soc. Arts. Built (with Maudslay) engines which could propel ships across the Atlantic. Died Surrey, Eng., Aug. 11, 1863.

FIELD, Lamar, Am. chemist; b. Montgomery, Ala., July 19, 1922; s. Samuel Lamar and Nelle (Brock) F.; S.B., Mass. Inst. Tech., 1944, Ph.D., 1949; m. Betty Leyden, Jan. 1, 1948; children—Patricia Leyden, Brock Lamar. Jr. chemist Merck & Co., Rahway, N.J., 1944-46; faculty Vanderbilt U., Nashville, 1949—, prof. chemistry, 1959—, chmn. dept., 1961-67. Cons. Thiokol Chem. Corp., 1955—, NIH, 1965—, Nat. Assn. on Standard Med. Vocabulary; cons. sci. writers. Mem. Am. Chem. Soc. (chmn., councilor Nashville sect., council publs. com. 1963-68, vis. scientist 1966—), Chem. Soc. (London), Tenn. Acad. Sci., Sigma Xi (L. C. Glenn award Vanderbilt University chapter Sigma Xi 1951), Alpha Chi Sigma. Studies and publs. in organic sulfur chemistry, particularly of sulfonic anhydrides, sulfones, thiolsulfonates, disulfides, antiradiation drugs, thiolignin, heterocyclic compounds; new methods of preparation, proof of structure, study of reactions. Home: 610 Lynnbrook Rd., Nashville 37215.*

FIELD, Mark G(eorge), sociologist; b. Lausanne, Switzerland, June 17, 1923; s. Jacques and Mary (Imbert) F.; came to U. S., 1940, naturalized, 1944; A.B. cum laude, Harvard, 1948, A.M., 1950, Ph.D., 1955; m. Anne Bayard Murray, Aug. 28, 1948; children—Alexander James, Michael Bayard, Andrew Murray, Elizabeth Imbert. Asso. dir. hosp. care study Beth Israel Hosp., Boston, 1955-57; research asso. Joint Commn. Mental Health and Illness, Cambridge, Mass., 1957-59; lectr. social relations Harvard, 1957-61, research asso. Russian Research Center, 1959-61, asso. Russian Research Center, 1962—, vis. prof. summer sch., 1961-63, 65-67, research asso. Program on Tech. and Soc., 1967-68; asso. prof. U. Ill., 1961-62; prof. dept. sociology and anthropology Boston U., 1962—, chmn. Soviet and E.Europe Studies Program, 1966—. Project dir. Assos. For Internat. Research, Cambridge, 1959-60; assistant sociologist in the department of psychiatry Massachusetts General Hosp., 1963—; cons. Arthur D. Little, Inc., Cambridge, 1962-63, 66—. Fellow Am. Sociol. Assn., Am. Pub. Health Assn.; mem. Am. Assn. Advancement Slavic Studies, A.A.A.S. Author: Doctor and Patient in Soviet Russia, 1957; (with others) Social Approaches to Mental Patient Care, 1964; Soviet Socialized Medicine: An Introduction, 1967; also numerous articles. Research on med. profession and role of physician under Soviet conditions, mental illness U. S. and USSR, pharm. system U. S. and USSR, social problems, role of women, family in Soviet Union, social environment and its effect on Soviet scientist, nature of Soviet socialized medicine, Soviet social instns., comparative studies. Home: 40 Peacock Farm Rd., Lexington, Mass. 02173. Office: 232 Bay State Rd., Boston 02215.*

FIELD, Stephen Dudley, Am. inventor; b. Stockbridge, Mass., Jan. 31, 1846; s. Jonathan Edwards and Mary Ann (Stuart) F.; ed. Stockbridge, also Poughkeepsie, N.Y.; m. Celestine Butters, Sept 30, 1871; 1 son, 1 dau. Became telegraph operator Cal. State Telegraph Co., 1863; insp. San Francisco Fire Alarm Telegraph Co., until 1872; organized Cal. Elec. Works to develop his elec. improvements, 1872. Invented multiple call dist. telegraph box, 1874, electric elevator, 1878; pioneer of modern trolley ry., 1878-79; made 1st application of dynamo machines to telegraphy, 1879; invented dynamo quadruplex telegraph, 1880, fast stock ticker, 1884; made 1st application of quadruplex telegraph on ocean cable (Key West to Havana), 1909. Died May 18, 1913.

FIELD, Theodore Estes, Am. chemist; b. Auburn, Me., Feb. 19, 1908; s. William K. and Mamie (Estes) F.; B.S., Bates Coll., 1929; Ph.D., Johns Hopkins, 1934; m. Ruth Rivers Graham, July 22, 1939; children—Christopher John, Phyllis Frances. With Corhart Refractories Co., Louisville, 1935—, research dir., 1938-44, tech. dir., 1944-57, tech. cons., 1957—. Fellow A.A.A.S.; mem. Am. Chem. Soc., Am., Brit. ceramic socs., Ky. Acad. Sci. Research, publs., patents on processes and compositions for high temperature refractories cast from fusions. Home: 3017 Sherbrooke Rd., Louisville 40205. Office: 1600 W. Lee St., Louisville 40210.*

FIELDS, John Charles, mathematician; b. Hamilton, Ont., Can., May 14, 1863; B.A., U. Toronto, 1884; Ph.D., Johns Hopkins, 1887; postgrad. univs. Paris, Göttingen, Berlin, 1892-1900. Prof. math. Allegheny Coll., 1889-02; spl. lectr. U. Toronto, 1902-05, asso. prof., 1905-14, research prof. math., mem. senate, from 1914. Fellow Royal Soc., 1913, Royal Soc. Can. (sect. pres.); mem. Russian Acad. Scis. (corr.), Royal Canadian Inst. (pres., hon. life mem.), Internat. Math. Congress (pres. 1924), Internat. Math. Union (hon. pres., 1924), A.A.A.S. (sect. v.p.), Brit. Assn. for Advancement of Sci. (v.p.), Internat. Congress Mathematicians (v.p. 1928), Acad. Scis. of Coimbre (corr.). Author: A Treatise on the Algebraic Functions; also papers on Abelian integrals, other math. subjects. Editor Proc. Internat. Math. Congress, 1924. Died Aug. 9, 1932.

FIELDS, Melvin, Am. chemist; b. Muncie, Ind., June 30, 1920; s. Melvin M. and Clara (Kasten) F.; B.S., Harvard, 1942, M.S., 1943, Ph.D., 1944; m. Zelda Holsinger, June 19, 1940. Research chemist Polaroid Corp., Cambridge, Mass., 1945-48; dir. organic div. Tracer-lab, Inc., Boston, 1948-52; with E. I. DuPont de Nemours & Co., Niagara Falls, N.Y., Wilmington, Del., 1952—, research supr. electrochems. dept., 1955—. Mem. Am., Brit. chem. socs., Sigma Xi, Phi Beta Kappa. Research, publs. on ultraviolet and infrared spectra of organic compounds, organic syntheses, polarography, properties and uses of polymers. Home: 609 Berwick Rd., Edenridge. Office: DuPont Bldg., Wilmington, Del. 19801.*

FIELDS, Paul Robert, Am. chemist; b. Chgo., Feb. 4, 1919; s. Alex and Anna (Green) F.; B.S., U. Chgo., 1941, postgrad.; m. Bernice White, Jan. 3, 1943; children—Marlene, Rita, Donald. Chemist, TVA, 1941-43; Metall. Lab., Chgo., 1943-45, Monsanto Chem. Co., 1945, Standard Oil Co., 1945-46; sr. chemist Argonne (Ill.) Nat. Lab., 1945—. Cons., Webster's Dictionary, 1963—; mem. transplutonium adv. com. AEC, 1964—. Mem. Am. Chem. Soc., Am. Nuclear Soc., Research Soc. Am. Contbr. numerous articles to tech. jours. Codiscovered 3 new elements, new isotopes synthetic elements above uranium; developed new separation techniques for transuranium elements; determined chem. properties and electronic structure of these elements. Home: 9860 S. Calhoun Av., Chgo. 60617. Office: 9700 S. Cass Av., Argonne, Ill. 60440.*

FIELDS, Thomas Henry, Am. physicist; b. Kearny, N.J., Oct. 23, 1930; s. Reuben T. and Anna (Palmer) F.; B.S., Carnegie Inst. Tech., 1951, Ph.D., 1955; m. Karin Rasch, Aug. 25, 1963; children—Timothy, Caroline, Gregory. Asso. prof. physics Carnegie Inst. Tech., 1959-60; prof. physics Northwestern U., Evanston, Ill., 1960—; dir. high energy physics div. Argonne (Ill.) Nat. Lab., 1964—. Mem. Am. Phys. Soc. Research in properties of new subnuclear particles using bubble chamber photographs, properties of hypernuclei, new resonant states in particle interactions at high energy. Home: 5749 Dearborn Pkwy., Downers Grove, Ill. 60515. Office: High Energy Physics Div., Argonne Nat. Lab., Argonne, Ill. 60440.*

FIELDS, William S., Am. neurologist; b. Balt., Aug. 18, 1913; s. Arthur Mortimer and Lenore (Straus) F.; A.B., Harvard, 1934, M.D., 1938; m. Elizabeth May Ritchie, Dec. 18, 1941; children—Susan Katherine, Anne Ritchie. Asso. prof. neurology Baylor U. Coll. Medicine, Houston, 1949-51, prof., 1951-67, chmn. dept. neurology, 1959-65; prof. neurology Southwestern Med. Sch., U. Tex., Dallas, 1967—; cons. neurologist St. Paul, Presbyn., Parkland Meml. hosps., Dallas; area cons. VA, St. Louis, 1951-59. Diplomate Am. Bd. Psychiatry and neurology. Mem. A.M.A., Am. Acad. Neurology, Assn. for Research in Nervous and Mental Diseases, Am. Neurol. Assn., Am. Assn. Neurol. Surgeons, Canadian Neurol. Soc., Sigma Xi. Editor: (with others) Hypothalamic-Hypophysial Interrelationships, 1956, Viral Encephalitis, 1958, Disorders of the Developing Nervous System, 1961, The Biology and Treatment of Intracranial Tumors, 1962, Information Storage and Neural Control, 1963, Neurological Aspects of Auditory and Vestibular Disorders, 1964, Intracranial Aneurysms and Subarachnoid Hemorrhage, 1965; Brain Mechanisms and Drug Action, 1957; Pathogenesis and Treatment of Parkinsonism, 1958; Pathogenesis and Treatment of Cerebrovascular Disease, 1961; Neurological Diagnostic Techniques, 1966, also numerous articles. Research on operation and diseases of brain and their treatment. Home: 13824 Braemar Dr., Dallas 75234. Office: 5323 Harry Hines Blvd., Dallas 75235.*

FIERZ, Hans Eduard, Swiss chemist; b. Zurich, Switzerland, Jan. 5, 1882; s. Eduard and Anna (Wirz) F.; student Fed. Tech. U. Zurich, Royal Coll. Sci., London; Ph.D., U. Zurich, 1905; m. Linda Emma David, 1911; 4 sons. Joined chem. firm. J. R. Giegy, Basel, Switzerland, 1909; prof. organic-tech. chemistry Fed. Tech. U. Zurich, 1917-52. Author: Grundlegende Operationen der Farbenchemie, 1919, later expanded and translated with L. Blangey; Künstliche organische Farbstoffe, 1926; Die Entwicklungsgeschichte der Chemie, 1945; (with E. Merian) Abriss der chemischen Technologie der Textilfasern, 1948. Research on azo, sulphur and indigo dyes, also on products needed for their manufacture, benzol, naphthalin, anthracene derivatives; studied (with W. Jadassohn) relation between chem. make-up and biol. effect of azo dyes. Died Küsnacht, Switzerland, Aug. 25, 1953.

FIERZ, Marlus, Swiss physicist; b. Basel, Switzerland, June 20, 1912; s. Hans E. and Linda (David) F.; ed. univs. Göttingen (Germany), Zurich (Switzerland); Dr.phil., 1936; m. Menga Bibeo, Mar. 1940; children—Lukas, Hans. Asst. to Prof. Pauli, ETH, 1936-40, pvt. docent, 1939, prof., 1960—; prof. U. Basel, 1944-60; dir. theoretical div. CERN, Geneva, 1959-60. Mem. Swiss Phys. Soc. (hon.). Research in quantum theory and relativity, statis. mechanics, history of science. Home: 10 Felscueggstr., Küsnacht 8700, Switzerland. Office: 60 Hochstr., Zurich 8044, Switzerland.*

FIESER, Louis Frederick, Am. chemist; b. Columbus, O., Apr. 7, 1899; s. Louis Frederick and Martha Victoria (Kershaw) F.; A.B., Williams Coll., 1920, hon. D.Sc., 1939; Ph.D., Harvard, 1924; grad. student Frankfort-on-Main, 1924-25, Oxford U., 1925; Dr. honoris causa, U. Paris, 1954; m. Mary A. Peters, June 21, 1932. Asst. and asso. prof. chemistry, Bryn Mawr Coll., 1925-30; asst. prof. and asso. prof. chemistry, Harvard, 1930-37, prof. 1937-39, Sheldon Emery prof. organic chemistry, 1939—. Mem. Surgeon Gen.'s Adv. Com. Smoking and Health, 1963. Recipient Katherine Berkan Judd prize for work on cancer-producing hydrocarbons Meml. Hosp., 1941; award for teaching Mfg. Chemists Assn., 1959; Norris award, 1959; Nichols medal Am. Chem. Soc., 1963. Fellow Nat. Acad. Scis. Author: Experiments in Organic Chemistry, Organic Chemistry, others; also numerous articles in field. Mem. editorial bds. Organic Syntheses, others. Research in cancer and chemotherapeutic studies; synthesized vitamin K-1, 1939; other work on oxidation-reduction, quinones, aromatic chemistry, carcinogenic hydroquinones, steroids, resin acids, napthoquinone antimalarials, cortisone, indsl. products related to pharmaceuticals. Home: 27 Pinehurst Rd., Belmont, Mass.

FIEUZAL, (Marie)-Louis, French ophthalmologist; b. Cahors, France, 1836; head physician Quinze-Vingts Hosp. Author: Clinique ophthalmologique, 1876; Hygiène de la vue dans les Écoles. Founder jour. la Clinique des Quinze-Vingts. Research on ophthalmology; invented yellow-tinted optical glass which is named after him. Died Paris, 1888.

FIFE, Earl Hanson, Jr., Am. microbiologist; b. Elkton, Ky., Apr. 14, 1915; s. Earl Hanson and Mabel (Oldham) F.; B.S., U. Wash., 1948; M.S., U. Md., 1952; m. Mary Mariila Skillings, June 14, 1941; children—Sally Elizabeth (Mrs. Clifton Jordan Johnson), Earl Dennett, Trudy May. Research asst. Swedish Hosp., Seattle, 1947-48; supervisory microbiologist dept. serology Walter Reed Army Inst. Research, Washington, 1948-65, chief dept. serology, 1965—; faculty mem. Walter Reed Army Inst. Research Postgrad. Tng. Program, 1950—. Serology cons. div. biologic standards NIH, 1964. Mem. Soc. for Exptl. Biology and Medicine, N.Y. Acad. Sci., Am. Soc. Tropical Medicine and Hygiene, Am. Soc. for Microbiology, Tropical Medicine Assn. Washington, A.A.A.S. Research on devel. serodiagnostic tests for parasitic and treponemal diseases; isolation and characterization serologically specific antigens, studies on mechanisms immune hemolysis and complement fixation; application immunofluorescence techniques in sero-diagnosis, factors affecting immune response, autoantibodies in health and disease. Home: 6412 Elliott Pl., Hyattsville, Md. 20783. Office: Dept. Serology, Walter Reed Army Inst. Research, Washington 20012.*

FIFKOVLA, Eva, Czechoslovakian neuroanatomist; b. Prague, Czechoslovakia, May 21, 1932; d. Ivan and Marie (Domalípová) F.; student Charles U. Faculty Medicine, Prague, 1951-57, Ph.D., 1963. Asst., Inst. Anatomy, Charles U., 1955-60; staff Inst. Physiology, Czechoslovak Acad. Scis., Prague, 1960—. Author: (with J. Marsala) Stereotaxic Atlas for Cat, Rabbit and Rat Brains, 1960; also articles. Research on mechanism and morphological aspects of spreading depression, reaction of brain tissue to mech., elec. or chem. stimuli; described this reaction in amygdala and thalamus of rat and rabbit brain. Home: 13 Cechova, Prague 7. Office: 1083 Budejovická, Prague 4-Krc, Czechoslovakia.*

FIGARI-BEY, Antonio, naturalist; b. Genoa, Italy, Mar. 16, 1804; s. Lazzaro and Paola (Lanfranco) F.-B.; student univ., 1824; m. Luisa Ucelli, 1827; 6 sons, including Enrico. Apprentice pharmacist, from 1818; dir. pharmacy Alexandria, Egypt, from 1825; prof. botany Mil. Coll. (later Cairo Sch. Medicine), from 1827; chief pharmacist Central Mil. Pharmacy, Cairo, from 1829; insp. pharmacy, mem. Gen. Council on Health, Egypt, from 1839; Author: Studii scientifici sull'Egitto e sue ad iacenze compresa la penisola dell'Arabia Petrea, 2 vols., 1864; Projet pour l'esablissement de colonies agricoles et d'une ferme modele en Egypte, 1867; L'exploration scientifique d l'Egypte sous le regne de Mohamed-Ali, 1895. Travelled in Egypt and Near East collecting fossils, plants and minerals and making observations on landscape (new facts were discovered about natural of life of area); magnificent collection of crystalline rocks gathered by him displayed in Civic Mus. Genoa. Died Genoa, Nov. 8, 1870.

FIGUIER, Louis-Guillaume, sci. writer; b. Montpellier, France, Feb. 15, 1814; aggregate prof. pharmacy. Author: Tableau de la nature; Merveilles de la science. Popularized and spread information on sci. discoveries of 19th century. Died Paris, Nov. 8, 1894.

FIKRY, M(ohammed) Essam, physician; b. Cairo, Egypt, May 1, 1923; s. M. Aziz and Gamal (Sabry) F.; M.B.Ch.B., Alexandria U., 1948, D.M., 1950, M.D., 1953; m. Haifa A. Azab, Apr. 30, 1954; children—Nadia, Hala. Registrar med. dept., Alexandria U., 1951-53, tutor dept. medicine, 1953-60, asst. prof., 1960—; pvt. practice medicine, Alexandria, 1958—; cons. physician in charge med. sect. Med. Ins. Service. Mem. Agyptian Med. Assn., Gastroenterology Soc.' Rheumatism Soc. Research, publs. on gastroenterology and geriatric medicine; discovery normal values for gastric and pancreatic secretions and intestinal absorption of old; discovery of low gastric and pancreatic secretions and reduced intestinal absorption in patients with fibrosis of liver due to schisto-somiasis. Home: 12 Alfred Lyan. Office: 22 Ghorfa Togarria, Alexandria, Egypt.*

FILACHIONE, Edward Mario, Am. chemist; b. Concord, N.H., Oct. 10, 1909; s. Basil and Ida (Scarano) F.; student Morton Jr. Coll., 1927-29; B.S., U. Ill., 1931; Ph.D., Northwestern U., 1935; m. Lynn M. Giusti, May 12, 1945; children—Marie A., Anne M., Jean M., Linda M. Faculty DePaul U., 1935-37, U. Me., 1937-38; DuPont grantee Cornell U., 1938-39; research chemist Pitts. Plate Glass Co., Barberton, O., 1939-41; research chemist U. S. Dept. Agr. Eastern Regional Research Lab., Phila., 1941—. Recipient Alsop award Am. Leather Chemists Assn., 1960. Mem. Am. Chem. Soc., A.A.A.S., Am. Leather Chemists Assn., Delaware Valley Tanners Club, Sigma Xi, Alpha Chi Sigma. Research, numerous publs. in synthetic organic chemistry of aliphatic compounds, monomers and polymers of allyl type esters, lactic acid chemistry, chemistry of callagen and leather. Home: 706 Avondale Rd. Office: 600 E. Mermaid Lane, Phila. 19118.*

FILATOV, Nils Feodorovich, Russian pediatrician; b. Russia, 1847; prof. U. Moscow; founder pediatrics in Russia; described Filatov's disease or infectious mononucleosis (he named it idiopathic adenitis), 1885; described Filatov-Duked disease (a form of German measles with scarlatiniform rash, which he called rubeola scarlatinosa), 1887. Died 1902.

FILCHNER, Wilhelm, explorer; b. Munich, Germany, Sept. 13, 1877; s. Eduard and Rosine (Leistner) F.; studied cartography under M. Schmidt, geography under S. Gunther, Munich; hon. degrees from Königsberg (now Kaliningrad, USSR), Munich Technische Hochschule. Early career with trigonometrical div. Prussia, also cartographic work on Moisel and in Berlin, Germany, work with Potsdam (Germany) Meteorol.-Magnetic Obs., Berlin Earth-Magnetism Inst., Potsdam Astrophys. Inst.; explored in Russia, Balkans, Asia Minor, then Pamir region, 1900, N.E. Tibet (with wife and Albert Tafel), 1903-05, Spitzbergen, 1910; head 2d German antarctic expdn. to Weddell Sea, 1911-12; discovered southwestern extension of Coats Land; mapped region visited in Central Asiatic expdn., 1925-28; visited Tibet, 1935-37; named prof., Berlin, 1938; explored Nepal, 1939; in India during World War II; returned to Europe, 1949; engaged in writing in Zurich, Switzerland. Mem. Leopoldina, Academie Sinica, Nanking, China. Author: Wissenschaftliche Ergebnisse der Expedition Filchner nach China und Tibet, 11 vols., 1907-14; Ein Forscherleben, 1950; Route Mapping and Position Locating in Unexplored Regions, 1957. Established magnetic stations in China and Tibet. Died Zurich, May 7, 1957.

FILEHNE, Wilhelm, pharmacologist; b. Poznan (now in Poland), Feb. 12, 1844; s. Isidor and Rosalie

(Kaufmann) F.; student, Heidelberg; M.D., Berlin, 1866; m. Sophie Deurer. Became asst. to R. Virchow, Berlin, 1868, to L. Traube, after 1871; went to Erlangen, Germany, 1874, became asso. prof., 1876; prof. pharmacology, Breslau, Germany, from 1886. Mem. Paris Acad. Medicine, Med. Soc., Erlangen, Paris Soc. Therapy (hon.). Research in pharmacological testing of synthetic fever drugs; introduced drug antipyrine (discovered by L. Knorv), 1884; recommended drug pyramidon (synthesized by F. Stolz). Died Bensheim/Bergstrusse, Germany, Apr. 29, 1927.

FILHOL, Henri, French paleontologist; b. Toulouse, France, May 11, 1843; s. Edouard Filhol; became lectr. Faculty Scis., Toulouse, 1883; prof. comparative anatomy Mus. Natural History, from 1894; made expdn. to New Zealand, also voyage to Campell Islands to observe passage of Venus across sun. Mem. French Acad. Scis., 1897, Acad. Medicine. Research and works on mammal fossils of phosphorites of Quercy, St.-Gerond-le-Puy, Rongon, Sanson; understood necessity of studying living species in relation to extinct species. Died Paris, Apr. 23, 1902.

FILIPCZYNSKI, Leszek, Polish acoustical scientist; b. Lodz, Poland, Dec. 23, 1923; s. Kazimierz and Apolonia (Caban) F.; M.S., Warsaw (Poland) Tech. U., 1949, D.Sc., 1955; m. Halina Rabcewicz-Zubkowska, June 19, 1950; children—Anna, Ewa. Asst., Warsaw Tech. U., 1948-50; chief Acoustic Lab., Inst. Tech. Physics, Warsaw, 1950-52; chief ultrasonic lab. Inst. Basic Tech. Problems, Warsaw, 1953-65; sci. vice dir. Inst. Basic Tech. Problems, Warsaw Polish Acad. Sci., 1965-—; faculty Warsaw Tech. U., 1955-—, prof., 1962-—. Recipient Polish State award in sci., 1966; Polish and fgn. medals for sci. activity. Mem. Polish Acoustical Soc. (v.p.). Author: Ultrasonic Methods of Testing Materials (with Z. Pawlowski, J. Wehr), 1959; also articles. Ultrasonic propagation in solids, ultrasonic transducers, absolute measurements, nondestructive testing of materials, instrumentation for ultrasonic diagnosis, visualization methods of abdomen and eye, fatigue testing of materials with ultrasonics. Home: 51/7 Lowicka, Warsaw. Office: 21 Swietokrzyska IPPT, Warsaw, Poland.*

FILIPPI, see De Filippi, Filippo.

FILMER, Sir Robert, Brit. polit. writer; b. East Sutton, Kent, circa 1590; s. Sir Edward and Elizabeth (Argall or Argol) F.; matriculated Trinity Coll., Cambridge, 1604; m. Anne Heton; 6 sons, 2 daus.; lived life of rural squire until roused to lit. activity by Puritan revolution; knighted by Charles I at beginning of his reign; ardent supporter of king's cause; strong advocate of divine right of kings; believed God gave Adam patriarchal authority, which he passed to Noah, Noah to his sons, sons passed on their authority, etc.; most effective in his attacks against antimonarchist arguments; attacked social compact doctrines; rejected idea of popular sovereignty; believed majority rule inadequate, that tyranny of many more intolerable than tyranny of one; believed king is free from all human control; author several papers including Patriarcha (most famous), 1680. Died May 26, 1653.

FILON, Louis Napoleon George, mathematician; b. St. Cloud, France, Nov. 22, 1875; s. Pierre M.A. and Marie (Poirel) F.; B.A., Univ. Coll., London, 1896; M.A., King's Coll., Cambridge, 1898; D.Sc., 1902; m. Anne Godet, 1904; 1 son, 2 daus. Naturalized Brit. citizen, 1898. Lectr. pure math. Univ. Coll., London, 1903, Goldsmid prof. applied math. and mechanics, 1912, dir. Univ. Obs., 1929-37; dean faculty sci. U. London, 1926-30, vice chancellor, 1933-35. Fellow Royal Soc. 1910 (council, v.p. 1936), 1910. Author: (with E. G. Coker) A Treatise on Photo-Elasticity; An Introduction to Projective Geometry; A Manual of Photo-Elasticity for Engineers, also numerous articles on pure and applied math. and physics. Research on classical mechanics, mechanics of continuous media, theory of generalized plane stress (showed how average elastic stresses may be determined by 2-dimensional analysis). Died Croydon, Eng., Dec. 29, 1937.

FILSON, John, Am. explorer, historian; b. Chester County, Pa., circa 1747; s. Davidson Filson. Recorded narrative given by Daniel Boone of his expdn. up Chillicothe River, 1782; explored Ohio River area to Lexington, Ky.; acquired 12,000 acres land, Fayette County, Ky.; made 1st map of Ky.; wrote an account of Daniel Boone; a backer of the founding of Cincinnati; Filson Club (hist. soc. with hdqrs., Louisville) named in his honor. Author: Discovery, Settlement, and Present State of Kentucke (1st history of Kentucky), 1st edit., 1784. Killed by an Indian while surveying Cincinnati area, Little Miami, Ky., Oct. 1788.

FINBY, Nathaniel, Am. physician; b. N.Y.C., May 31, 1917; s. Joseph and Celia Finby; A.B., Johns Hopkins, 1938, M.D., 1942; m. Norma Walton, Apr. 5, 1947. Attending radiologist N.Y. Hosp., 1955-61; practice medicine, specializing in radiology, N.Y.C., 1952-—; asso. prof. radiology Cornell Med. Sch., 1959-61; dir. radiology St. Lukes Hosp. Center; asso. clin. prof. radiology Columbia; cons. radiologist Rockefeller U., VA. Fellow Am. Coll. Radiologists, Internat. Coll. Angiology, Am. Geriatric Soc.; mem. Radiol. Soc. N.Am., A.M.A., Am. Roentgen Ray Soc., Internat. Coll. Radiology, Phi Beta Kappa. Contbr. to Surgery Rx Trauma, 1960; Problems in Surgery, 1962; Vasc Roentgenology, 1964. Research, publs. on angiocardiography, contrast media in roentgenology, endocrine radiology, vascular radiology. Home: 535 E. 86th St., N.Y.C. 10028. Office: 421 W. 113th St., N.Y.C. 10025.*

FINCH, Clement Alfred, Am. physician; b. Broadalbin, N.Y., July 4, 1915; s. Percy H. and Marion (Dye) F.; B.A., Union Coll., 1936; M.D., U. Rochester, 1941; m. Eugenia English, Feb. 28, 1966; children—Clifton A., Carin A., Lisa. Pathology fellow Rochester (N.Y.) Hosp., 1938-39; instr. Harvard Med. Sch., 1946-48, asso. in medicine, 1948-49; jr. asso. Peter Bent Brigham Hosp., Boston, 1946-48, asso. in medicine, 1948-49; asso. prof. medicine U. Wash. Sch. Medicine, Seattle, 1949-55, prof. medicine, 1955-—. Mem. Soc. Clin. Investigation (pres. 1963), Am., Western (pres. 1959) assns. physicians, Am. (pres. 1966), European socs. hematology, Sigma Xi, Alpha Omega Alpha. Research, numerous publs. in hematology, erythropoiesis, iron metabolism. Home: 3200 Sierra Dr. S., Seattle 98144. Office: 1959 N.E. Pacific St., Seattle 98105.*

FINCH, Ruy Herbert, Am. geologist, vocanologist; b. Sunbury, O., Aug. 31, 1890; s. Thacker Webb and Ida (Hubbard) F. ed. George Washington U., U. Chgo.; independent studies in meteorology, seismology, volcanology with W. J. Humphreys and T. A. Jaggar, Jr.; m. Margaret Helena Harvey, Oct. 12, 1923; children—Robert H., Harvey E., Amy Jean. With U. S. Weather Bur., 1910, metecrologist, 1919; asst. to Dr. Jagger, dir. Hawaiian Volcano Obs., 1919; asso volcanologist U. S. Geol. Survey, 1924. Established obs. for study of Cascade Volcanoes at Lassen Peak, Cal., 1926, orchardist Watsonville, Cal., 1936-39; volcanologist Nat. Park Service, Hawaii, 1940-47, later U. S. Geol. Survey. Instr. meteorology U S N, forecast at Naval Air Sta., Ireland, during World War I; del. 4th Pacific Sci. Congress, Java, 1929, 7th Congress, New Zealand, 1949. Fellow A.A.A.S., Geophys. Union, Geol. Soc., Am.; mem. Seismol. Soc. Am. Contbr. on sci. topics to various publs. Studied volcanic activity of Halemaumau, Hawaii, 1919-26, Akutan volcano, Alaska, Shishaldin volcano, Aleutian Islands, Vesuvius, Mauna Loa, ash deposits, Kilauea Caldera, also tidal waves, mechanism of phreatic explosions. Died Mar. 25, 1957.

FINCH, Stuart C., Am. physician; b. Broadalbin, N.Y., Aug. 6, 1921; s. Cecil C. and Olga (Lofgren) F.; student Dartmouth, 1938-41; M.D., U. Rochester, 1944; m. Patricia O'Brien, June 15, 1946; children—James, Ellen, Sheldon, Polly. Fellow in hematology Mass. Meml. Hosps., Boston, 1950-53; faculty Yale Sch. Medicine, New Haven, 1953-—, prof., 1967-—. Vis. prof. medicine Hiroshima U. Sch. Medicine, chief medicine Atomic Bomb Casualty Commn., 1960-62. Mem. Am. Soc. for Clin. Investigation, Am. Internat. socs. hematology, Sigma Xi. Research, numerous publs. on iron metabolism, leukemia, leukocyte immunology, leukokinetics, radiation effects. Home: Amity Rd., Woodbridge, Conn. 06525. Office: 333 Cedar St., New Haven 06510.*

FINCKE, Thomas, Danish mathematician; b. Flensborg, Denmark, Jan. 6, 1561; student univs. Strasbourg (France), Basel (Switzerland), Montpellier (France), Padua (Italy), 1577-90; established German library, Padua; physician to Duke Phillipe de Holstein-Gottorp., after 1590; named prof. math. U. Copenhagen (Denmark), 1591, prof. Latin, 1602, prof. medicine, 1603. Author: Geometria rotundi (terms tangent and secant introduced), 1583. Died Copenhagen, Apr. 26, 1656.

FINCKH (or FINLCH), Georg Philipp, cartographer; b. circa 1608; m.; at least 1 son, Georg Philipp; councillor, ct. sec. to prince bishop of Freising; may have been curator at Ottenburg. Author: S. Rom. imperii circuli et electrovatus Bavariae tabula chorographica (new version of Apian's Bavarian maps Bairische Landtafeln, but half their size and without town names Apian included; Finckh's map scaled 1:270,000, included more names of forests and mountains, remained valid a century. Died Jan. 15, 1679.

FINDEISEN, (Theodor Robert) Walter, meteorologist; b. Hamburg, Germany, July 23, 1909; s. Robert and Lina (Hödrich) F.; student, Karlsruhe; Ph.D., Hamburg, 1931; m. Hedwig Schnitzlein (div.); 1 dau.; m. 2d, Ella Steppan, 1942. Became dir. weather air sta., Munich, Germany, 1933; apptd. mem. faculty, 1936; worked with W. Peppler, Friedrichshafen, Germany; dir. cloud research sta. German Bur. Weather Service, Prague, Czechoslovakia, from 1940. Contbr. articles to meteorol. and geophys. jours. Research on cloud formation and precipitation; made 1st analysis of relative effectiveness of processes leading to precipitation; laid found. for study of cloud microphysics; gave impetus to 1st successful attempts to influence rainfall. Died May 9, 1945.

FINDLAY, George William Marshall, Brit. physician; b. 1893; s. George Findlay; ed. U. Edinburgh; M.D., D.Sc.; m. Margaret Williams; 2 daus. Lectr. pathology U. Edinburgh, also asst. pathologist Royal Infirmary, Edinburgh, 1920-23; Alice meml. fellow, asst. Imperial Cancer Research Fund, 1923-28; mem. sci. staff Wellcome Research Instn., 1928-46. Fellow Royal Coll. Physicians; pres. Royal Micros. Soc., 1950-51. Author: Recent Advances in Chemotherapy, 3d edit., 1950; also papers on pathology and bacteriology. Editor Abstracts of World Medicine and Abstracts of World Surgery, Obstetrics and Gynaecology, Brit. Med. Assn., London. Died Mar. 14, 1952.

FINDLAY, John Wilson, physicist; b. Kineton, Eng., Oct. 22, 1915; s. Alexander Wilson and Beatrice (Thornton) F.; B.A., Cambridge U., Eng., 1937, Ph.D., 1950; m. Jean Elizabeth Melvin, Dec. 14, 1953; children—Peter, Stuart, Richard. Came to U. S., 1956, naturalized, 1963. Fellow, physics Queen's Coll., Cambridge; demonstrator in physics Cavendish Lab., Cambridge, 1945-53, Brit. Ministry Supply, London, 1953-56; head electronics div. Nat. Radio Astronomy Obs., Charlottesville, Va., 1956-61, dep. dir., 1961-67, asst. dir., 1967-—, dir. ARECIBO Ionospheric Obs., 1965-66. Decorated Order Brit. Empire. Fellow I.E.E.E. Research, publs. on measurements of intensity of radio astron. sources, design and constrn. of large radio telescopes. Home: Millbank, Route 1, Afton, Va. 22920. Office: Edgemont Rd., Charlottesville, Va. 22901.*

FINDLAY, Robert Artemas, chem. engr.; b. Vancouver, B.C., Can., Jan. 27, 1914; s. James A. and Phoebe (Stevens) F.; B.A., U. B.C., 1934, M.A., 1935; Ph.D., McGill U., 1937; m. Ethel Jean Liddy, Apr. 11, 1940; children—Roslyn Ethel (Mrs. Roland P. Rogers), Marilyn Jean (Mrs. C. F. Bethea, Jr.), Carolyn Elizabeth Ann. Came to U. S., 1943, naturalized, 1949. Research chemist Imperial Oil Ltd., Sarnia, Ont., Can., 1937-39; chief chemist Anglo Canadian Oils Ltd., Brandon, Man., Can., 1939-40; research chemist Trinidad (B.W.I.) Leasholds, Ltd., 1940-43; with Phillips Petroleum Co., Bartlesville, Okla., 1943-—, asst. mgr. devel., 1949-64, dir. chem. and automation devel., 1964-66, dir. mech. and process scis., 1966-—. Mem. Natural Gas Processors (chmn. enthalogy research com. 1961-—), Am. Inst. Chem. Engrs., Am. Petroleum Inst. Contbr. chpts. to Advances in Petroleum Refining and Chemistry, 1958, 59, New Chemical Engineering Separation Techniques, 1962; articles to profl. jours. Patentee in field. Developed continuous crystallization process for purifying chems., processes for mfr. polyolefin plastics and polybutadiene rubber. Home: 1931 Polaris Dr., Bartlesville, 74003. Office: Phillips Petroleum Co., Bartlesville, Okla. 74003.*

FINE, Albert Samuel, Am. biochemist; b. Phila., Oct. 24, 1923; s. Max and Sylvia (Lerner) F.; B.S., Bklyn. Coll., City U., 1950, M.S., 1953; m. Selma Skolnick, Mar. 27, 1959; one son, Martin Jay. Research asst. N.Y. U. Coll. Medicine, 1950-52; research asso. Columbia Coll. Phys. and Surg., 1952-56; biochemist spl. dental research program VA Hosp., Bklyn., 1956-—. Mem. A.A.A.S., Am. Chem. Soc., Am. Inst. Biol. Scis., N.Y. Acad. Scis. Research, publs. on oxidative metabolism, electron transport of cartilage, bone, gingivae, effects poly-ions, sugars, transition metals on enzymes of electron transport system. Office: 800 Poly Pl., Bklyn. 11209.*

FINE, Henry Burchard, Am. mathematician; b. Chambersburg, Pa., Sept. 14, 1858; s. Lambert Suydam and Mary Ely (Burchard) F.; A.B., Princeton, 1880, A.M., 1883; Ph.D., U. Leipzig, 1885; LL.D., Williams, 1909; m. Philena Fobes, Sept. 6, 1888; children—John, Susan Breese Packard, Philena Fobes (Mrs. Bradford B. Locke). Tutor, Princeton, 1881-84, asst. prof., 1885-90, Dod prof. math., from 1891, dean faculty, 1903-12, dean dept. sci., from 1911. Mem. Am. Math. Soc. (pres. 1911), Am. Philos. Soc., Math. Assn. Am. Author: Euclid' Elements, 1891; The Number System of Algebra, 1891; A College Algebra, 1904; Coördinate Geometry (with Henry Dallas Thompson), 1909; Calculus, 1927. Made studies on logic of mathematics; algebra; calculus, research on singularities of curves. Died Princeton, N.J., Dec. 22, 1928.

FINE, Jean Marius, French immunologist; b. Grenoble, France, Apr. 14, 1923; s. Auguste and Aimée (Borrel) F.; Licencié es Sci., Faculté des Sciences de Grenoble, France, 1944; Doctorat en médecine, Faculté de Médecine de Lyon, 1951; Diplômé, Institut Pasteur, 1953; m. Annick Colas, July 8, 1965; 1 dau. (by previous marriage), Anne-Laurence. Asst., Institut Pasteur de Paris Service d'Immunologie des Groupes Sanguins du Pr. Dujarric de la Rivière, 1952-57; named head lab. Centre National de Transfusion Sanguine, 1957, head immunology research on plasma proteins, 1961; chief research Faculté de Médecine, Paris, 1960-—. Recipient Muteau prize French Acad. Scis. Mem. Blood Transfusion Soc. (sec. gen.), Internat. Soc. Blood Transfusion, Biochem. Soc., Hematology Soc., N.Y. Acad. Scis. Author: Les Groupes Sanguins, 1964. Research and numerous publs. on blood groups in man, molecular and genetic variations in haptoglobins and transferrins; immunochemistry of human serum protein, including normal and myeloma immunoglobulins; comparative biochemistry of serum proteins of marine animals; discovered group system based on molecular variation of transferrins in fishes (eel). Home: 124 rue de Javel (15°). Office: 6, rue A. Cabanel (15°), Paris, France.*

FINE, Morris Eugene, metallurgist; b. Jamestown, N.D., Apr. 12, 1918; s. Louis and Sophie (Berrington) F.; B.S., U. Minn., 1940, M.S., 1942, Ph.D., 1943; m. Mildred Eleanor Glazer, Aug. 13, 1950; children—Susan Elaine, Amy Lynn. Faculty, U. Minn., 1942-43; asso. scientist Manhattan Project U. Chgo., Los Alamos, 1944-45; mem. tech. staff Bell Telephone Labs., Murray Hill, N.J., 1946-54; prof. materials sci. Technol. Inst. Northwestern U., Evanston, Ill., 1954—, chmn. materials sci. dept., 1954-60, chmn. Materials Research Center, 1960-64, Walter P. Murphy prof., 1963—. Vis. prof. dept. materials sci. Stanford, 1967-68. Named Chicagoan of Year in Sci., 1960. Fellow Am. Phys. Soc.; mem. Am. Inst. Mining, Metall. Engrs. (chmn. Inst. Metals div. Metall Soc. 1966-67), Am. Soc. Metals (past chpt. chmn.), Am. Ceramic Soc., Nat. Acad. Scis., Sigma Xi, Tau Beta Pi. Author: Introduction to Phase Transformations in Solids, 1964. Contbr. numerous articles to profl. jours. Patentee material for constant frequency vibrating devices; contbr. to theory of precipitation hardening of solids. Home: 3515 Walnut St., Wilmette, Ill. 60091. Office: Dept. Materials Sci., Northwestern U., Evanston, Ill.*

FINE, Oronce (or Orontius Fineus), French mathematician, cartographer; b. Briançon, France, 1494. Prof. math. Royal Coll. (now Coll. de France), 1530. Author: Protomathesis (elementary textbook on pure and applied math.), 1532; De rebus mathematicis hactenus desideratis, pub. 1556; also treatises on Euclidian theory. Drew first map of France printed there, 1525; constructed astron. instruments. Died Paris, 1555.

FINEGOLD, Sydney Martin, Am. physician; b. N.Y.C., Aug. 12, 1921; s. Samuel Joseph and Jennie (Stein) F.; A.B., U. Cal. at Los Angeles, 1943; M.D., U. Tex., 1949; m. Mary Louise Saunders, Feb. 8, 1947; children—Joseph, Patricia, Michael. Research fellow Med. Sch., U. Minn., 1951-52; faculty Med. Center, U. Cal. at Los Angeles, 1955—, asso. prof. medicine, 1962—; chief chest and infectious disease sect. Wadsworth VA Hosp., Los Angeles, 1957-61, chief gen. medicine and infectious disease sect., 1961—. Mem. VA Pulmonary Disease Research Program Commn., 1961-62, Infectious Disease Research Program Commn., 1961-65; mem. Nat. Acad. Scis.-NRC Drug Efficacy Study Group, 1966-67. Diplomate Am. Bd. Internal Medicine. Fellow A.C.P., Am. Pub. Health Assn., Am. Acad. Microbiologists; member Infectious Disease Soc. Am., Am. Soc. Microbiology, A.A.A.S., Am. Thoracic Soc., Western Soc. Clin. Research, Am. Fedn. for Clin. Research, Sigma Xi, Alpha Omega Alpha. Editorial board California Medicine, 1966—. Research in infectious diseases, also on effect of antibacterial agts. on intestinal micro-flora, non-spore-forming anaerobic bacteria, chemotherapeutic compounds, hosp.-acquired infections. Home: 17502 Parthenia St., Northridge, Cal. 91324. Office: Wadsworth VA Hosp., Los Angeles 90073.*

FINERTY, John Charles, Am. anatomist; b. Chgo., Oct. 20, 1914; s. John L. and Hulda (Schulte) F.; B.A., Kalamazoo Coll., 1937; M.S., Kan. State Coll., 1939; Ph.D., U. Wis., 1942; postgrad. (Rackham Found. fellow), U. Mich.; m. Mildred King, Dec. 28, 1940; children—Olivia (Mrs. John L. Moore) Donna (Mrs. James D. Gatewood). Instr., U. Mich., 1943-46; faculty Washington U., St. Louis, 1946-49, asso. prof., 1949; faculty U. Tex. Med. Br., 1949-56, prof. anatomy, 1952-56, asst. dean, 1954-56; prof. anatomy, chmn. dept. U. Miami (Fla.) Sch. Medicine, 1956-66, asst. dean, 1956-62, asso. dean, 1962-66; dean La. State U. Sch. Medicine, New Orleans, 1966—. Chmn., State of Fla. Anatomical Bd., 1956-66. Mem. anatomy test com. Nat Bd. Med. Examiners, 1964-67. Fellow A.A.A.S.; mem. Am. Assn. Anatomists (program sec. 1966—), Soc. for Exptl. Biology, Am. Physiol. Soc., Radiation Research Soc., Tex. Acad. Sci (past pres.), Sigma Xi (past pres.), Phi Kappa Phi, Gamma Alpha, Phi Sigma, Gamma Sigma Delta, Omicron Delta Kappa. Author: (with Cowdry), A Textbook of Histology, 1960; also numerous articles. Research on experimental endocrinology, also pituitary cytophysiology, correlation microscopic structure and function, neurohumoral control respiration, gross human anatomy, parabiosis, protection from X-irradiation, effects fat-deficiency. Home: 6821 Catina St., New Orleans 70124.*

FINESINGER, Jacob Ellis, Am. physician; b. New Castle, Pa., Oct. 28, 1902; s. Hyman Joseph and Fannie M. (Kaplan) F.; A.B., Johns Hopkins, 1923. A.M., 1925, M.D., 1929; studied psychoanalysis in Vienna, 1933-34, conditioned reflex activity in Leningrad. under professor Pavlov, 1934; m. Grace Lubin, June 24, 1932; children—Ruth Joan (Mrs. Sheppard Kellam), and Joe Lubin, intern neurology Boston City Hosp., 1929, resident neurologist, 1930, jr. vis. neurologist, 1930-32, Commonwealth fellow Boston Psychopathic Hosp., 1932-35; asst. in neuropathology Harvard Med. Sch., 1930-32, research fellow in psychiatry, 1932-35, asso. in psychiatry, 1935-36, asst. prof. psychiatry, 1936-49; asst. psychiatrist Mass. Gen. Hosp., 1936-39, psychiatrist, 1939-49; prof. psychiatry and head dept. U. Md. Sch. Medicine from 1949, psychiatrist in chief U. Hosp. from 1949. Prin. investigator on studies on selection of air-craft pilots. USN, NRC, 1941-45; mem. study health sect. for mental health USPHS; mem. com. on psychiatry NRC, mem. dean's subcom. for psychiatry U S. V.A. and research council U. S. Army Chem. Center. Mem. A.M.A., Am. Psychiat. Assn., Am. Neurol. Assn., Am. Soc. for Clin. Investigation (editorial bd. Jour. Clin. Investigation, 1948-54, mem. adv. editorial bd. Human Relations), Am. Psychoanalytic Assn., Assn. for Research in Nervous and Mental Disease. Am. Acad. Neurology, Soc. Biol. Psychiatry, also local and state profl. socs. Tech. adviser series of sound films on therapeutic interviewing. Contbr. to monographs; author papers on neurol., psychiatric topics. Editor in chief Jour. Nervous and Mental Disease, from 1958. Died June 19, 1959.

FINGER, Ernest, dermatologist; b. Prague, Czechoslovakia, July 8, 1856; s. Joseph Theodor and Friedericke (Mudroch) F.; ed. univs. Prague, Vienna; m. Elisabeth Ferdinanda von Stoda, 1888; 2 daus. Named lectr. U. Vienna, 1883, titular prof., 1893, asso. prof., 1901, prof. dermatology and syphiliology; dir. 2d clinic for venereal and skin diseases, from 1904, also founder sero-diagnostic inst. in clinic. Pres. Superior Bd. Health, 1925-31. Author: Die Blennorrhoe der Sexualorgane, 1888; Die Geschlechtskrankheiten als Staatsgefahr, 1924; also articles. Combined traditional sch. of clin. observation with new bacteriological methods in veneral diseases; work in control of outbreak of venereal disease after World War I. Died Vienna, Apr. 17, 1939.

FINGER, Frank W(hitney), Am. psychologist; b. Naples, N.Y., Apr. 16, 1915; s. Jacob and Elizabeth (Lewis) F.; B.A., Syracuse U., 1936, M.A., 1937; Ph.D., Brown U., 1940; m. Eleanor Ford Varn, June 14, 1958; children—Elizabeth Lewis, William Whitney, Thomas McGavock, Eleanor Ford, Louise Anderson. Instr., Brown U., 1940-42; faculty U. Va., Charlottesville, 1942—, prof. psychology, 1955—. Member of American (president division gen. psychology 1960-61, division teaching 1956-57), Eastern, Southeastern, Va. (past pres.) psychol. assns., A.A.A.S. (sec. sect. I 1960—), Va. Acad. Sci. (past chmn. psychology sect.), Phi Beta Kappa, Sigma Xi. Research, publs. on characteristics various indices biol. dr., effect of variables on convulsive behavior in animals, conflict behavior in animals, effect reward on learning, relations between sex practices and beliefs coll. students, visual distortions. Home: 2704 Magnolia Dr., Charlottesville, Va. 22901.*

FINGER, Glenn Charles, Am. chemist; b. Bloomington, Ill., May 3, 1905; s. Charles William and Sophie C. (Henne) F.; student Ill. State U., 1924; B.S., U. Ill., 1927, M.S., 1928, Ph.D., 1938; postgrad. Purdue U.; m. Grace Eleanor Chamberlain, July 26, 1956; 1 stepson, Dale H. Johnson. Chemist, Ill. Geol. Survey, Urbana, 1933—, prin. chemist, head chem. group, 1962—; vis. lectr. U. Heidelberg, Cambridge U., Bristol U., 1955. U. S. del. Internat. Congress Pure and Applied Chemistry, Zurich, 1955, Paris, 1957; plenary lectr. Internat. Fluorine Chem. Symposium, Birmingham, Eng., 1959. Mem. Am., Brit. chem. socs., Sigma Xi, Phi Lambda Upsilon, Alpha Chi Sigma. Contbr. chpt. to Advances in Fluorine Chemistry, 1961. Research in synthetic organic fluorine chems. as new bactericides, fungicides, insecticides, plant growth regulators, weed killers, drugs, research medicinals, spl. chem. intermediates, devel. new indsl. products from Ill. minerals. Home: 1506 S. Race St. Office: Natural Resources Bldg., Urbana, Ill. 61801.*

FINGER, Harold Ben, Am. nuclear engr.; b. N.Y.C., Feb. 18, 1924; s. Beny and Anna (Perlmutter) F.; B.Mech. Engring., Coll. City N.Y., 1944; M.S. in Aero. Engring., Case Inst. Tech., 1950; m. Arlene Karsch, June 11, 1949; children—Barbara Lynn, Elyse Sue, Sandra Ruth. With NASA, 1944—, chief nuclear engine programs, 1958-61, dir. nuclear systems, 1961-65, dir. nuclear systems and space power, 1965-67; dir. space nuclear systems AEC, Washington, 1965-67; mgr. Joint AEC-NASA Space Nuclear Propulsion Office, Washington, 1960-67; asso. adminstr. for orgn. and mgmt. NASA, Washington, 1967—. Co-recipient Manley award Soc. Automotive Engrs., 1957; recipient NASA Outstanding Leadership medal, 1966. Mem. Am. Inst. Aeros. and Astronautics. Research, numerous publs. on operating characteristics of turbomachinery and gas turbine aircraft propulsion and devel. design data and methods; tech. direction research and devel. nuclear propulsion, nuclear electric power, solar and chem. power, electric propulsion systems for space. Home: 6908 Millwood Rd., Bethesda, Md. 20034. Office: NASA, Washington 20546.*

FINGERMAN, Milton, Am. biologist; b. Boston, May 21, 1928; s. Irving and Rose (Goodman) F.; B.S., Boston Coll., 1948; M.S., Northwestern U., 1949, Ph.D., 1952; m. Sue Whitsell, Mar. 31, 1953; children—Stephen Whitsell, David Clay. Faculty, Tulane U., New Orleans, 1954—, prof. biology, 1963—. Mem. Am. Soc. Zoologists, Am. Physiol. Soc., Am. Soc. Naturalists. Author: Chromatophores, 1963; also numerous articles. Research on biol. clock in animals, control mechanisms animal color change; discovered hormones regulating pigment cell activity in crabs and crayfishes. Home: 6120 Magnolia St., New Orleans 70118.*

FINHTENGOLTS, Grigorii Mikhailovich, Russian mathematician; b. June 5, 1888; grad. Novorossiiskii U., Odessa, 1911. Prof., Leningrad U., since 1918. Author: Differential and Integral Calculus, 3 vols., 1947-49. Greatly influenced teaching of math. in Russia.

FINIGUERRA, Maso, Italian goldsmith; flourished in Florence, Italy; b. 1426; said to have studied under Lorenzo Ghiberti; inventor metal engraving (copper- and steel-plate). Died 1464.

FINIKOV, Sergei Pavlovich, Russian mathematician; b. Nov. 15, 1883; grad. U. Moscow, 1906. Prof., U. Moscow, since, 1918. Author: Differential-Projective Geometry, 1937; Cartan's Extrinsic Form Method in Differential Geometry, 1948; Congruence Theory, 1950; Paired Congruence Theory, 1956. Participated in founding modern differential geometry, also founded large sch. of Russian geometrists; demonstrated basic results in classical problems of surface curvature and projective congruence theory. Address: Moskva, Leninskie gory, Gosurdarstvenny universitet, USSR.

FINK, Bruce, Am. botanist; b. Blackberry, Ill., Dec. 22, 1861; s. Reuben and Mary Elizabeth (Day) F.; B.S., U. Ill., 1887, M.S., 1894; A.M., Harvard, 1896; Ph.D., U. Minn., 1899; postgrad. U. Chgo., 1903; m. Ida May Hammond, Jan. 9, 1888; children—Mrs. Lois Honberger, Hugh Willard, Ruth Elizabeth. Prin. of high schs., 1887-92; prof. biology, Upper Ia. U., 1892-1903; prof. botany Grinnell Coll., 1903-06; prof. botany Miami U., from 1906. Mem. Minn. Bot. Survey, 1896-1903; in charge of botany U. Wash. Marine Sta., 1906; mem. Ohio Biol. Survey Bd. Asso. editor Micologia, from 1908. Author: Tobacco, a book on the tobacco problems, 4 rev. edits.; The Lichens of Minnesota, 1910; Laboratory Exercises in Plant Physiology and Ecology, 1911. Made Bot. editor Ohio Jour. Science. Taxonomic studies of lichens. Died July 10, 1927.

FINK, Colin Garfield, Am. electrochemist; b. Hoboken, N.J., Dec. 31, 1881; s. Frederick William and Minnie (Spengeman) F.; B.A., Columbia, 1903; M.A., Ph.D., Leipzig, 1907; Sc.D. (hon.), Oberlin Coll., 1936; m. Charlotte K. Muller, June 6, 1910; children—Frederick William, Ernest Arthur, Harold Kenneth. Asst. in electrochemistry U. Leipzig, 1906-07; research engr. Gen. Electric Co., Schenectady, 1907-10, Edison Lamp Works, Harrison, N.J., 1910-17; head research lab. Chile Exploration Co., N.Y.C., 1917-21; head div. electrochemistry Columbia, 1922-50; also metall. and art research; cons. practice, 1922-50. Chmn. tungsten com. U. S. Munitions Board, 1939. Recipient Edward Goodrich Acheson medal; Perkin medal, 1934; Modern Pioneer award, 1940. Fellow A.A.A.S.; mem. Electrochem. Soc. (pres., sec.), Am. Chem. Soc., NRC, Am. Inst. Mining and Metall. Engrs., Sigma Xi. Editor Electrochemistry, Chem. Abstracts, from 1907; contbg. editor Mineral Industry. Inventor drawn tungsten filament; platinum substitute for tungsten lamps and radio tubes; insoluble anode for copper; process for tin smelting and refining; corrosion resistant metals; comml. chromium and tin plating; aluminum plate on steel; methods for restoring old bronze and other metal objects. Died Sept. 16, 1953.

FINK, Frederick William, Am. engineer; b. Newark, N.J., Mar. 27, 1912; s. Colin G. and Lottie (Muller) F.; E.E., Cornell U., 1935; M.Sc., Cambridge U. (Eng.), 1937; m. Ella Nolen, Jan. 6, 1938; children—Catharina, Carlatta, Colin William. Research engr., 1938-51, asst. chief, 1951-56, chief, 1956-64, fellow, 1964-65, Batelle Memorial Institute, Columbus, O.; lectr., Norwegian Reactor School, 1963; vis. prof., Danish Technical U., 1964. Recipient Frank Newman Speller award, Nat. Assoc. of Corrosion Engrs., 1965. Mem., Electrochemical Soc.; Nat. Assoc. Corrosion Engrs.; A.S.M.: AIME. Author 48 published articles. Work in control of corrosion in water, steam and distillation plants; corrosion control in paper mills, ore and chemical plants; diagnosing causes of failures in metal structures exposed to air, water or soil environments; research on aqueous corrosion, particularly saline water. Home: 2354 Dorset Road, Columbus, Ohio 43221.

FINK, Julius, Austrian geologist, pedologist; b. Vienna, Austria, Apr. 18, 1918; s. Julius and Maria (Kiniger) F.; Ph.D., U. Vienna, 1944; m. Friederike Wieser, Mar. 15, 1943; children—Friedl, Kurt. Staff, U. Vienna, 1944-46, docent soil geography and pedology, 1956—; staff U. Agr., Vienna, 1946—, asso. prof. geology and pedology, 1956—, prof., head dept. geology, 1967—; initiator Austrian Soil-Survey, 1951, sci. head, 1951-67; chmn. Loess-Commn., Internat. Quarternary Orgn., 1961—. Mem. Soil Sci. Soc. Austria (pres. 1962-65, v.p. 1966—). Editor, Bull, S.S.S. Austria, 1959—. Research, publs. on Quaternary especially Paleopedology, loess, geomorphology, problems of soil-mapping and soil-genesis. Home: 11 Alserbachstr., 1090 Vienna, Austria.*

FINK, Kay Ferguson, Am. biochemist; b. State Center, Ia., Feb. 13, 1917; d. Frank H. and Katie (Meyers) Ferguson; B.A., U. Ia., 1938; Ph.D., U. Rochester, 1943; m. Robert Morgan Fink, Jan. 6, 1941; children—Patricia, Suzanne. Research technician Mayo Inst. Exptl. Medicine, Rochester, Minn., 1938-39; NRC fellow U. Rochester, 1939-42, research asso. Manhattan project, 1943-46, research asso. AEC project, 1946-47; with U. Cal., Los Angeles, 1948—, prof. research biophysics and nuclear medicine, 1966—; research biochemist VA Hosp., Long Beach, Cal., 1947-61. Mem. Soc. Exptl. Biology, Am. Soc. Biol. Chemists, Phi Beta Kappa, Sigma Xi. Contbr. articles to sci., med. jours. Research, studies on intestinal secretion, traumatic shock, biol. effects of radiation, thyroid chemistry, amino acids; pyrimidine and purine metabolism, chromatography. Home: 1804 Rial Lane, Los Angeles 90024.*

FINK, Richard Walter, Am. nuclear physicist; chemist; b. Detroit, Jan. 13, 1928; s. Bernard and Ann (Walter) F.; B.S., U. Mich., 1948; M.S., U. Cal. at Berkeley, 1949; Ph.D., U. Rochester, 1953; m. Gunilla Gustafsson, Oct. 4, 1960; children—Kerry Leif, Roger Gunnar. Asso. prof. U. Ark., Fayetteville, 1953-61; vis. prof. Werner Inst. for Nuclear Chemistry, U. Uppsala (Sweden), 1959-60, Inst. for Exptl. Physics, U. Hamburg (Germany), 1963-64; prof. Ga. Inst. Tech. Sch. Chemistry, Atlanta, 1965—; prof. dept. physics Marquette U., Milw., 1961-65; research nuclear chemist Knolls Atomic Power Lab., Schenectady, 1949-50. Cons., Lawrence Radiation Lab., U. Cal., 1963—, Phillips Petroleum Co., Bartlesville, Okla., 1957-65. Fulbright Travel grantee, 1963-64; Fulbright lectr., Europe, 1964. Mem. Am. Phys. Soc., Sigma Xi. Research, publs. on high energy spallation reactions, orbital electron capture, atomic fluorescence X-ray yields and comparison with theory, fast neutron activation cross sects., reactions and radioactive decay schemes for comparison with theory. Office: Nuclear Research Center, Ga. Inst. Tech., Atlanta 30332.*

FINK, Robert Morgan, Am. biochemist; b. Greenville, Ill., Sept. 22, 1915; s. William Harvey and Pearl (Smith) F.; student Kan. State Coll., 1933-35; A.B., U. Ill., 1937; postgrad. Lehigh U., 1937-38; Ph.D., U. Rochester, 1942; m. Kathryn L. Ferguson, Jan. 6, 1941; children—Patricia Kay, Suzanne Joyce. Faculty, U. Cal., Los Angeles, 1947—, prof. biol. chemistry, 1963—; research biochemist VA, 1947-54. Mem. subcom. on internal dose Bur. Standards Nat. Com. Radiation Protection, 1947-49. Mem. Am. Soc. Biol. Chemists. Author: Biological Studies with Polonium, Radium and Plutonium, 1950. Research, publs. on chromatographic and radioactive tracer techniques for thyroid hormone prodn., photosynthesis, and nucleic acid metabolism. Home: 1804 Rial Lane, Los Angeles 90024.*

FINK, William LaVilla, Am. metallurgist; b. Fairmount, Ind., Sept. 14, 1896; s. James Otto and Vella (Lindsey) F.; B.S. Engring., U. Mich., 1921, M.S. Engring., 1923, Ph.D. (Charles A. Coffin fellow), 1926; m. Laura Frances French, July 14, 1926; children—Lucie French (Mrs. Ernest Jackson Felton), Joanna Roslyn (Mrs. Charles E. Manwiller, Jr.). With Aluminum Co. Am., 1925—, chief phys. metallurgy div., New Kensington, Pa., 1943-58, sci. coordinator, 1958-61, cons., 1961—. Instr. phys. metallurgy Carnegie Inst. Tech., 1936-37; mem. adv. bd. Office of Critical Tables. Mem. Am. Inst. Mining, Metall. and Petroleum Engrs. (Matthewson award 1938), Am. Society for Testing and Materials (recipient award of merit 1956, also honorary membership), A.A.A.S., Am. Crystollographic Assn., Am. Chem. Soc., Am. Soc. for Metals, Am. Ordnance Assn., Inst. Metals (Brit.), Soc. for Non destructive Test. Contbg. author: Age Hardening of Metals, 1940; Metals Handbook, 1948; Physical Metallurgy of Aluminum Alloys, 1949; Crystallography and Solid State Physics, 1961; also articles. Patentee in field. Research on aluminum alloys especially hardening, corrosion, non-destructive testing, studies X-ray diffraction. Home: 410 Emerson Av., Pitts. 15215.*

FINKBEINER, Daniel T., Am. mathematician; b. Aspinwall, Pa., Oct. 7, 1919; s. Daniel T. and Elsie (Bair) F.; A.B. summa cum laude, Washington and Jefferson Coll., 1941, M.A., 1943; Ph.D., Cal. Inst. Tech 1949; m. Mary Louise Moffat, Oct. 27, 1945; children—Susan B., Heidi, Ann M., Katherine K. Instr. math. Washington and Jefferson, 1941-43; instr. Yale, 1949-51; asso. prof. math Kenyon Coll., Gambier, O., 1951-56, prof., 1956—, acting dean, 1956-58. NSF faculty fellow Princeton, 1958-59; vis. prof. U. Western Australia, 1964. Mem. Phi Beta Kappa, Sigma Psi. Author: Introduction to Matrices and Linear Transformations, 1960; also articles on lattice theory. Home: Box 405, Gambier, O. 43022.*

FINKE, Leonhard Ludwig, German physician; b. Westerkappeln, Germany, 1747; M.D., U. Halle (Germany), 1772; practiced medicine, Kassel, Germany, specializing in obstetrics, from 1774; apptd. med. officer and instr. to midwives, Tecklenburg County, 1776, Wesphalia, 1802-28 or 29; author 1st med. geography, during 1790s, also Exercitationes physicomedicae. Died 1828 or 29.

FINKEL, Miriam Posner (Mrs. Asher J. Finkel), Am. biologist; b. Chgo., Jan. 22, 1916; d. Jacob I. and Esther (Greenfield) Posner; B.S., U. Chgo., 1938, Ph.D., 1944; m. Asher J. Finkel, Oct. 9, 1943; children—David M., Barry S., Raphael A., Joel R. Scientist, Metall. Lab., U. Chgo., 1944-46; biologist Argonne (Ill.) Nat. Lab., 1946—. Mem. Radiation Research Soc., Am. Assn. Cancer Research, Am. Soc. Zoologists, Am. Soc. Exptl. Pathologists, Radiation Soc. N.Am., Health Physics Soc., A.A.A.S., Soc. for Exptl. Biology and Medicine, Sigma Xi. Research, publs. on toxicity radioactive materials in lab. animals, radiation induced cancer; discovered virus producing bone cancer in mice. Home: 10314 S. Oakley Av., Chgo. 60643. Office: Argonne Nat. Lab., Argonne, Ill. 60440.*

FINKELNBURG, Wolfgang, German physicist; b. Bonn, Germany, June 5, 1905; s. Rudolf and Margot (Zitelmann) F.; Ph.D. in Physics, U. Bonn, 1928; m. Eleonore Schulen, Mar. 6, 1939; 1 son, Wolf-Dieter. Instr., U. Berlin, 1929, Inst. Tech., Karlsruhe, 1931; asso. prof. Inst. Tech., Darmstadt, 1938-42; prof. physics U. Strassburg, 1942-45; dir. reactor devel. Siemens Corp., Erlangen, Germany, 1952—, v.p. reactor devel., 1963—; hon. prof. U. Erlangen, 1955—. Mem. reactor sub-comm. German AEC, mem. Bavarian AEC. Fellow Am. Phys. Soc.; mem. German Phys. Soc. (pres. 1965-67). Author many books including: Structure of Matter, 1964; also numerous articles. Research in spectroscopy, electric arcs, nuclear power. Died Feb. 28, 1967.

FINKELSTEIN, Heinrich, German pediatrician; b. Leipzig, Germany, July 31, 1865; Ph.D., Leipzig, 1888; asst. to Heubner, children's clinic Berlin Charité Hosp., 1894-1900; became lectr. pediatrics, 1899; named nead physician, municipal orphanage, Berlin-Rummelsburg, 1901; apptd. asso. prof., Berlin, 1908; became dir. Kaiser and Kaiserin Friedrich Children's Hosp., 1918; teaching permission withdrawn by Nazis, 1935; cons. physician, Chile, 1931-42. Author: Die durch Geburtstraumen hervogerufenen Krankheiten des Säuglings, 1902 (with L. Ballin) Die Waisensäuglinge Berlins und ihre Verpflegung im städtischen Kinderasyl, 1904; Lehrbuch der Säuglingskrankheiten, 1905; (with Rohr) Die Behandlung der tuberkulösen Bauchfellerkrankungen im (with Galewsky and Halberstädter) Hautkrankheiten und Syphilis im Säuglings- und Kindesalter, 1922. Pioneer in care and nutrition of infants; introduced protein milk; research on skin diseases in children, alimentary fever; stated that most diarrhea in infants is metabolic in nature. Died Santiago de Chile, Jan. 7, 1942.

FINKELSTEIN, Jacob, Am. chemist; b. N.Y.C., Oct. 27, 1910; s. Hyman and Ethel (Singer) F.; B.S., Coll. City N.Y., 1933; M.A., Columbia, 1934, Ph.D., 1939; m. Phyllis Bergman, May 27, 1945; children—Kenneth C., Karen R.A. Research chemist Merck & Co., Inc., Rahway, N.J., 1935-43; sr. research chemist Hoffmann-LaRoche, Inc., Nutley, N.J., 1943—. Mem. Am. Chem. Soc., A.A.A.S., Sigma Xi, Phi Lambda Upsilon. Research in fields of sulfonamides, viral infections, radiopaque agts., hypotensive agts., kidney clearance, mono-amine oxidase inhibitors, mental diseases, fat metabolism, synthesized carbon-phosphorous amino acids, chemotherapeutic investigations in bacterial infections, anti-histamines, cancer, diuretics; determined chem. structure of and synthesized vitamin B-1; isolation, structure and total synthesis of pantothenic acid; synthesized the alkaloid cocluarine. Home: 648 Sunderland Rd., Teaneck, N.J. 07666. Office: Kingsland St., Nultey, N.J. 07110.*

FINKELSTEIN, Richard Alan, Am. microbiologist; b. N.Y.C., Mar. 5, 1930; s. Frank and Sylvia (Lemkin) F.; B.S., U. Okla., 1950; M.A., U. Tex., 1952, Ph.D., 1955; m. Helen Rosenberg, Nov. 30, 1952; children—Sheryl Lyn, Mark Stuart, Laurie Jo. Teaching fellow, research scientist U. Tex., Austin, 1950-55; fellow microbiology. instr. Southwestern Med. Sch., Dallas, 1955-58; chief bioassay sect. Walter Reed Army Inst. Research, Washington, 1958-64; dep. chief dept. bacteriology and mycology U. S. A. med. component SEATO Med. Research Lab., Bangkok, Thailand, 1964-67; vis. asso. prof. microbiology Faculty Grad. Studies, U. Med. Scis., Bangkok, Thailand, 1965-67; asso. prof. dept. microbiology U. Tex. Southwestern Med. Sch., Dallas, 1967—. Mem. nat. com. for coordination of cholera research Ministry Pub. Health, Bangkok, 1965-67. Recipient Outstanding Achievement award U. S. Army Sci. Conf., 1964. Diplomate Am. Bd. Microbiology. Mem. Am. Soc. for Microbiology, Am. Assn. Immunologists, Soc. for Exptl. Biology and Medicine, A.A.A.S., Assn. Microbiologists India, Sig ma Xi. Research, numerous publs. on pathogenesis and immunology of cholera, resistance mechanisms, staphylococci, herpes virus; discovered choleragen. Home: 4218 Mendenhall Dr., Dallas 75234.*

FINKLE, Bernard Joseph, Am. plant biochemist; b. Chgo., Mar. 17, 1921; s. Nathan Ephraim and Lena (Lipa) F.; B.S., U. Chgo., 1942; Ph.D., U. Cal. at Los Angeles, 1950; m. Evelyn Cohen, Dec. 9, 1944; children—Wayne Nathaniel, Claudia. Chemist, Manhattan Project, U. Chgo., also Oak Ridge Nat. Lab., 1943-46; fellow AEC, Molteno Inst., U. Cambridge, Eng., 1950; biochemist U. Cal. at Berkeley, 1951-53; lab. dir. Atomic Research Lab., Los Angeles, 1953-54; biochemist U. Utah, Salt Lake City, 1954-57; chemist Western Regional Research Lab., U. S. Dept. Agr., Albany, Cal., 1957—. Predoctoral fellow NIH, 1948-49; U. S.-Japan Coop. Sci. Program fellow dept. med. chemistry Kyoto (Japan) U., 1966-67. Mem. American Soc. Biol. Chemists, Phytochemical Soc. N.Am. (pres. 1965-66), Am. Soc. Plant Physiology, Am. Chem. Soc., Am. Inst. Biol. Scis. Research in biosynthesis and metabolism of phenolic acids, ascorbic acid, oxalic acid, treatment of fruits to prevent oxidative darkening, properties of thiol groups of papain, chlorophyll biogenesis, cultivation of green algae, properties of radiactive isotopes. Home: 21 Kingston Rd., Berkeley, Cal. 94707. Office: 800 Buchanan St., Albany, Cal. 94710.*

FINLAND, Maxwell, physician; b. Kiev, Russia, Mar. 15, 1902; s. Frank and Rebecca (Povza) F.; came to U. S., 1902, naturalized, 1925; B.S., Harvard, 1922, M.D., 1926; D.Sc. (hon.), Western Res. U., 1964. Faculty, Harvard Med. Sch., Boston, 1928—, George Richards Minot prof. medicine, 1963—; staff Boston City Hosp., 1927—, dir. Thorndike Meml. Lab., 1963—, dir. Harvard Med. services, 1963—; cons. Pondville Hosp., 1939—, Boston VA, Beth Israel, Lemuel Shattuck hosps., U. S. Naval Hosp. at Chelsea, Mass. Mem. numerous coms. govt. agys., founds., orgns.; asso. mem. Commn. on Acute Respiratory Diseases, 1950—; chmn. com. for Lederle Med. Faculty Awards, 1952—; clin. investigators com. VA Dept. Medicine and Surgery, 1955—; vis. prof. Baylor U., 1958, U. N.C., 1960, U. Wis., 1961, U. Tenn., 1965; mem. drug research bd. Nat. Acad. Scis., 1963—. Recipient Dr. Charles Value Chapin award City of Providence, 1960. Mem. Mass. Med. Soc., A.M.A., Am. Soc. Clin. Investigation (councilor 1943-46, v.p. 1947-48;, A.A.A.S., Am. Bd. Internal Medicine, A.C.P., Assn. Am. Physicians, Soc. Exptl. Biology and Medicine, Am. Assn. Immunologists, Am. Acad. Arts and Scis., Am. Soc. Microbiology, N.Y. Acad. Scis., Am. Epidemiol. Soc. (v.p. 1962), Infectious Diseases Soc. Am. (pres. 1964, Bristol award 1966). Editorial bd. New Eng. Jour. Medicine, 1945—, Disease-aMonth, 1954—. Contbr. numerous sci. papers, monographs, clin. articles on infectious diseases and their treatment. Home: 46 Sycamore Rd., Squantum, Quincy, Mass. 02171. Office: 818 Harrison Av., Boston 02118.*

FINLAY, Alexander, Scottish chemist; b. Benholm, Scotland, Sept. 24, 1874; s. William and Catherine (Fyfe) F.; M.A., S.C. D.Sc., U. Aberdeen; Ph.D., U. Leipzig (Germany); m. Alice Mary de Rougemont, 1914; 2 sons. Lectr. organic chemistry U. St. Andrews, 1900; lectr. chemistry, spl. lectr. phys. chemistry U. Birmingham, 1902-11; prof. chemistry Univ. Coll. of Wales, Aberystwyth, 1911—. Lectr. United Free Ch. Coll., Aberdeen, 1915-16. Fellow Inst. Chemistry (council 1915-18); mem. council chem. Soc. Author: The Phase Rule and its Applications (standard work); Physical Chemistry and its Applications in Medical and Biological Science; Practical Physical Chemistry; Osmotic Pressure; Chemistry in the Service of Man; The Treasures of Coal Tar; also papers. Translator: Principles of Inorganic Chemistry (Ostwald). Editor Monographs on Inorganic and Physical Chemistry. Died Feb. 2, 1921.

FINLAY, Carlos Juan, Cuban physician; b. Puerto Principe, Cuba, Dec. 3, 1833; s. Edward and Isabel (de Barrès) F.; ed. Lycée de Rouen, France; M.D., Jefferson Med. Coll., Phila., 1855; postgrad., Paris, France, 1860-61; m. Adele Shine, Oct. 16, 1865. Practiced medicine briefly in Lima, Peru; went to Trinidad for polit. reasons; returned to Cuba, 1870; del. of Cuba to Internat. Sanitary Conf., Washington, 1881; chmn. commn. on infectious diseases, Havana, 1899-1902; chief san. officer of Cuba, 1902-09; del. of Cuba to San. Congress, Washington, 1903; hon. pres. Junta Nacional de Sanidad y Beneficencia, from 1909. Finlay Inst. for Investigation of Tropical Medicine created in his honor by Cuban govt., 1915. Fellow Soc. Tropical Medicine and Hygiene, Eng.; hon. mem. Am. Soc. Tropical Medicine, Société de Médicine Tropicale, Paris. Author numerous works on Tb, beriberi, exophthalmic goiter, filariasis, trichinosis, leprosy, cholera, tetanus. Contbd. to etiology and pathology of yellow fever; founder of doctrine mosquito-borne diseases, which he maintained with respect to yellow fever, from 1881; study of ophthalmology. Died Havana, Aug. 20, 1915.

FINLCH, Georg Philipp, see Finckh, Georg Philipp.

FINLEY, William Lovell, Am. naturalist; b. Santa Clara, Cal., Aug. 9, 1876; s. John Pettis and Nancy Catherine (Rucker) F.; A.B., U. Cal., 1903; D.Sc., Ore. State Coll., 1931; m. Irene Barnhart, Feb. 21, 1906; children—Phoebe Katherine (Mrs. Arthur N. Pack), William Lovell. Writer Reviews of Reviews Co., 1904-05; lectr. Nat. Assn. Audubon Socs., N.Y.C., 1906-25; mem. Bd. of Fish and Game Commrs., Ore., 1911; state game warden, Ore., 1911-15; state biologist, 1915-19; mem. Ore. Game Commn., 1925-27. Mem. adv. bd. U. S. Dept. Agr., Migratory Bird Treaty Act. Vice pres. Izaak Walton League Am., Nat. Wildlife Fedn.; hon. pres. Mem. Am. Ornithologists' Union Outdoor Writers Assn. Am. (dir.), Nat. Parks Assn. (adv. council), Wildlife Soc. (trustee), Sigma Xi; charter mem. Am. Soc. Mammalogists, Cooper Ornithol. Club, Pacific Northwest Bird and Mammal Soc. Author: American Birds, 1907; (with wife) Little Bird Blue, 1915, Wild Animal Pets, 1928; also many sci. papers on bird life and 2 bulls. on Ore. birds. Producer of Finley nature films (motion pictures). Asso. editor The Condor, 1906-09. Editorial staff Nature Mag., 1923-37. Died June 29, 1953.

FINN, John Martin, Jr., Am. chemist; b. Phila., Nov. 16, 1919; s. John Martin and Sara (O'Donnell) F.; A.B., Harvard, 1948; M.S., U. Pa., 1949, Ph.D., 1953; m. Dolores L. Pauloski, Sept. 11, 1954; children—Gregory Gerard, Monica Mary, Brendan Gerard. Project supr. Horizons Inc., Cleve., 1952-55; research chemist Union Carbide Tech. Center, Cleve., 1955—. Mem. Am. Chem. Soc. (dir. Cleve. sect. 1963—), A.A.A.S., Electrochem. Soc., Albertus Magnus Guild, Sigma Xi. Research, publs. in electrolysis alkali metal phosphides in liquid ammonia, functioning graphite anode in chlorine prodn. and anodes for prodn. sodium perchlorate; patentee methods preparing inorganic substances. Home: 7663 Alan Pkwy., Cleve. 44130. Office: Union Carbide Parma Tech. Center, P.O. Box 6116, Cleve. 44101.*

FINNERTY, Frank Ambrose, Jr., Am. physician; b. Montclair, N.J., Nov. 3, 1923; s. Francis A. and Agnes (Fitzsimons) F.; A.B., Georgetown U., 1943, M.D., 1947; m. Catherine Anderson, Dec. 28, 1946; children—Francis, Patty, Peter, Robert, Mary, Shawn, Duffy. Research fellow in medicine Georgetown U. Med. Center, 1949-50, asst. prof. medicine, 1958-63, asso. clin. prof. medicine, 1963—; research fellow Am. Heart Assn., D.C. Gen. Hosp., 1955-57, investigator, 1957-62; chief of medicine Columbia Hosp. for Women, Washington, 1963—; chief cardiovascular research Georgetown U. Med. div. D.C. Gen. Hosp., 1952—. Fellow A.C.P., Am. Coll. Cardiology; mem. Am. Heart Councils of High Blood Pressure Research, Circulation, Epidemiology. Research, numerous publs. to further understanding of heart and blood vessel abnormalities in toxemia of pregnancy which resulted in methods and prevention and treatment of disease; evaluation of mechanisms and clin. trials of drugs and treatment of various types of hypertension. Home: 6 Langley Pl., McLean, Va. 22101. Office: Cardiovascular Research, Georgetown U. Med. div. D.C. Gen. Hosp., Washington 20003.*

FINNEY, Karl F(rederick) Am. chemist; b. Salina, Kan., July 25, 1911; s. Fred Sherman and Mamie (Erickson) F.; A.B., Kan. Wesleyan U., 1935; B.S., Kan. State U., 1936, M.S., 1937, postgrad.; postgrad. Ohio State U.; m. Gertrude Ardean Latham, Mar. 23, 1935; children—Michael F., Patrick L., Kelly R. Asso. chemist Hard Winter Wheat Quality Lab., U. S. Dept. Agr., Kan. State U., Manhattan, 1938-43, research chemist, dir. Hard Winter Wheat Quality Lab., prof. grain sci. and industry, 1946—; chemist Soft Wheat Quality Lab., Ohio Expt. Sta., Wooster, O., 1943-46. Mem. A.A.A.S., Am. Chem. Soc., Am. Assn. Cereal Chemists, Am. Soc. Agronomy, Sigma Xi, Phi Kappa Phi, Gamma Sigma Delta, Alpha Zeta, Alpha Mu. Research, numerous publs. on wheat quality evaluation, micromilling and microbaking, flour fractionating and reconstituting, fractionating and characterizing gluten proteins, increasing protein content and altering protein quality by foliar-spraying wheat plant with urea and substituted ureas, effects of environmental factors, baking techniques and ingredients, fats and lipids, amylases, proteases, varying stages of wheat maturity, and gamma irradiation on phys., chem. and baking properties of wheat varieties. Home: 2315 Bailey Dr., Manhattan, Kan. 66502.*

FINSCH, Otto Friedrich Hermann, zoologist, ethnographer; b. Warmbrunn, Silesia, Aug. 8, 1839; s. Moritz and Mathilee (Leder) F.; Ph.D. (hon.), Bonn, 1868; m. Elisabeth Hoffmann, 1886; 1 dau. Studied birds in Hungary and Balkans, 1858-59; became asst. Netherlands Mus. Natural History, Leiden, 1861; apptd. curator natural history and ethnographic collection Mus. Soc., Bremen, Germany, 1864, dir. collection (then owned by city), 1876-79; traveled in N.Am., 1872; Lapland, 1873, W. Siberia, 1876; made 1st S. Seas expdn., 1879-82; with exploratory expdn. Neuguinea Co., 1884-85; became dept. head Mus., Leiden, 1897, Braunschweig, Germany, 1904. Eponym of various birds and geographic places. Mem. Leopoldina. Author: Die Papageien, 2 vols., 1867; (with G. Hartlaub) Beiträge zur Fauna Central-Plolynesiens, 1867, Die Vögel Ost-Afrikas, 1860; Anthropologische Ergebnisse einer Reise in d. Südsee und d. Mayal. Archipd in den Jahren 1879/82, 1884; Samoafahrten, Reisen in Kaiser-Wilhelms-Land und Englische . . . Neu-Guinea, 1888, revised to 1914; also articles. A leading explorer of 2d half 19th century; made extensive collections from new geog. areas. Died Braunschweig, Jan. 31, 1917.

FINSEN, Niels Ryberg, Danish physician; b. Thorshavn, Faroes, Denmark, Dec. 15, 1860; M.D., U. Copenhagen (Denmark), 1890; established light therapy inst. named after him in Copenhagen, 1896. Recipient Nobel prize in medicine and physiology, 1903. Founder of modern phototherapy; 1st to realize that ultra-violet light has curative effects, especially in reference to lupus; 1896; discovered suppuration in smallpox can be prevented by screening from ultraviolet rays. Died Copenhagen, Sept. 24, 1904.

FINSLER, Paul, mathematician; b. Heilbronn, Germany, Apr. 11, 1894; s. Julius and Elise (Berrer) F.; Ph.D., U. Göttingen. Instr., U. Cologne, 1922; asso. prof. applied math. U. Zurich, 1927, full prof. math., 1944, hon. prof., 1959. Mem. Swiss Soc. Natural Scis. Author: Über Kurven und Flachen in allgemeinen Räumen, 1918, also articles. Research on differential geometry, foundations of math. Address: Zurichbergstrasse 132, Zurich, Switzerland.*

FINSTERER, Hans, Austrian surgeon; b. Weng, nr. Altheim, Austria, June 24, 1877; s. Josef and Fransk Franziska (Brunner) F.; M.D., Vienna, 1902, postgrad., until 1914; m. Emma Fuchs, 1914; 1 son, 3 daus. Became mem. faculty Vienna, 1913; became asso. prof., also head dept. Kaiser-Franz-Josef Ambulatory, head of surgery Hosp. Barmherzigen Brüder, 1919; head of surgery Wiedner Hosp., 1935; head physician I. surg. dept. Vienna Gen. Hosp., 1935-51; prof., from 1948. Recipient Billroth medal. Author: Die Methoden der Lokalanästhesie in der Bauchchirurgie und ihre Erfolge, 1923, English edit., 1924; also papers. Developed local anesthesia for gastrointestinal surgery; modified stomach resection; advocated removal of up to 2 thirds of stomach for stomach ulcers and cancer; developed new surg. methods in duodenal ulcers, rectal cancer. Died Vienna, Apr. 11, 1955.

FINSTERWALDER, Sebastian, German mathematician; b. Rosenheim/Inn, Germany, Oct. 4, 1862; s. Johann Nepomuk and Anna (Amann) F.; student Tech. U. Munich; Ph.D., Tübingen, Germany, 1886; hon. doctorates, Vienna, Fed. Tech. U. Zurich, Innsbruck; m. Franziska Mallepell, 1892; 4 sons, including Richard, 1 dau. Became mem. faculty Tech. U. Munich, 1888, succeeded A. Voss as prof. analytical geometry, differential and integral calculus, 1891, became prof. descriptive geometry, 1911; named mem. Bavarian Commn. for Internat. Geodesy; sci. adviser Bavarian Land Survey; declined offers from univs. Vienna, Potsdam, Berlin. Mem. acads. Munich, Madrid, Swiss Natural Sci. Soc. (hon.). Author: Der Vernagtferner, 1897; Die Geogmetrischen Grundlagen der Photogrammetrie, 1898; Eine Grundaufgabe der Photogrammetrie . . . , 1903. Pioneer in application of aerial photography to topography and cartography; used photogrammetry in glacier studies; inventor instruments for surveying glaciers; founder sch. for study of glacial areas; eponym of geometric flow theory of glaciers. Died Munich, Dec. 4, 1951.

FINSTON, Harmon Leo, Am. chemist; b. Chgo., Feb. 16, 1922; s. Leo M. and Minnie (Fogel) F.; B.S. in Arts and Scis., Ill. Inst. Tech., 1943; Ph.D., Ohio State U., 1950; m. Edythe Heller, Jan. 8, 1950; children—Martin, David, Mira, Leo. Chemist, U. Chgo. Metall. Lab., 1943-45; asso. chemist dept. chemistry Brookhaven Nat. Lab., Upton, N.Y., 1950-51, chemist, head radiochem. analysis sect. nuclear engring. dept., 1951-63; prof. dept. chemistry Bklyn. Coll., City U. N.Y., 1963—. Cons., U. S. Army Nuclear Def. Lab., Edgewood Arsenal, Md., 1964—. Israel AEC fellow, 1965. Mem. Am. Chem. Soc., Sigma Xi, Sigma Phi Sigma, Phi Lambda Upsilon. Research, numerous publs. on application nuclear and radiochem. techniques to chem. analysis, decay scheme studies; developed an incremental polarograph. Home: 3806 Av. H, Bklyn. 11210.*

FINZI, Guido, Italian pathologist; b. Mantona, Aug. 16, 1884; s. Cervo and Domenica (Vecchi) F.; M.D.; D. honoris causa, U. Buenos Aires. Instr., U. Torino, 1913, asso. prof., 1915, full prof., 1919, former dean Vet. Sch.; full prof. spl. pathology and clin. medicine Vet. Sch., U. Milan, former dean; prof. emeritus U. Buenos Aires, U. La Plata. Named laureat of numerous internat. sci. prizes. Mem. numerous Italian and fgn. acads. sci. Research and publs. on bacteriology, immunization, comparative pathology. Address: corso Vittorio Emmanuele 30, Milan, Italy.

FIORAVANTI, Count Leonardo, physician, alchemist; b. Bologna, Italy, circa 1530; studied medicine; practiced medicine, Palermo, 1548-50; went with Spanish fleet to Africa; returned to Naples, 1555; practiced in Rome, Venice, and Bologna. Author: Capprici medicinali, 1561; Il compendio dei secreti rationale intorno alla medicina, cirurgia et alchimia, 1564; Lo Specchio di scienzà universale, 1564; others. Works translated into French, German, and English. Follower of Paracelsus; investigated chem. medicines; introduced balsam for cure of arsenic poisoning; discussed preparation of iron amalgam. Died Sept. 4, 1588.

FIORINI, Ettore, Italian physicist; b. Verona, Italy, Apr. 19, 1933; s. Enoch and Lina (Tosadori) F.; Laurea in Fisica, U. Milan, 1955, Libera Docenza in Fisica Sperimentale, 1962; m. Carla Mangili, Sept. 22, 1958; 1 dau., Beatrice. Faculty, U. Milan, 1955—, prof. physics, 1956—. Research asso., fellow Nat. Acad. Scis., Duke, 1960; research asso. Italian Nat. Inst. for Nuclear Physics, 1957—; chmn. Gargamelle Users Com., European Center for Nuclear Physics. Recipient award Italian Physics Soc., 1966. Research, publs. on cloud and bubble chambers, existence, quantum numbers, prodn. mechanisms of elementary particles. Home: 3 Via Vanvitelli, Milan, Italy.*

FIQUET-FAYARD, Florence Josephe, French phys. chemist; b. Villers-Cotteret, France, Nov. 18, 1929; s. Philippe Désiré and Madeleine (Grenier-Godard) F.-F.; ed. Ecole normale Supérieure; Docteur ès Sciences Physiques; m. Michel Fayard, Oct. 31, 1954; children —Louis, Frédéric. Lectr., Centre National de la Recherche Scientifique, 1963-66; prof. Faculté des Sciences, Orsay, France. Author: Attachement electronique en phase gazeuse, 1965; also articles. Research on isotopic effects of dissociation of small molecules; attachment of slow electrons on conjugated molecules; double-charged ions. Home: 33 Résidence du Val, 91 Palaiseau, France. Office: Laboratoire Collisions Electroniques, 91 Orsay, France.*

FIREMAN, Edward Leonard, Am. physicist; b. Pitts., Mar. 23, 1922; s. Nathan and Anna (Caplan) F.; B.S., Carnegie Inst. Tech., 1943; Ph.D., Princeton, 1948; m. Rita H. Seidman, Sept. 1947; children—Bruce H., Ellen Sue, Gary David. Scientist, Brookhaven Nat. Lab., Upton, L.I., N.Y., 1950-56; guest scientist, 1956—; physicist Smithsonian Astrophys. Obs., Cambridge, Mass., 1956—; lectr. Harvard Obs., 1961—. Pres. commn. meteorites Internat. Astron. AEC fellow, 1948-50. Fellow A.A.A.S.; mem. Phys. Soc., Meteoritical Soc., Am. Geophys. Union. Research, publs. on discovery cosmic ray produced istopes in meteorites, application isotopes to history of solar system. Home: 57 Clifton Rd., Newton, Mass. 02159. Office: 60 Garden St., Cambridge, Mass.*

FIRMICUS MATERNUS, Julius, astronomer; b. Sicily, circa 4th century; of Roman senatorial rank; after conversion to Christianity became ecclesiastic, probably Bishop of Milan (Italy); author De erroribus profanarum religionum (attack on paganism), 346, Matheseos libri viii (infused with urbane Neoplatonic spirit, most comprehensive ancient text on astrology, argues that astrology and free will are compatible, explains horoscope casting), circa 336.

FISCH, Solomon, physician; b. Havana, Cuba, May 30, 1926; s. Abraham U. and Ida (Rosenwald) F.; came to U. S., 1942, naturalized, 1953; B.S., U. Scranton, 1945; Ph.D., U. Rochester, 1949; M.D., N.Y. U., 1953; m. Lila May Mirkin, Dec. 17, 1951; children—Susan B., Melanie Ann, Amy Margo, Steven Lloyd. Chief med. services 1st USAF Hosp., Selfridge AFB, Mich., 1956-58; pvt. practice medicine, N.Y.C., 1958—; research fellow, med. services Bronx VA Hosp., 1958-59, coordinating research fellow, 1959-60, dir. cardiac therapy research unit, 1960—; faculty N.Y. Med. Coll., 1959—, asst. clin. prof. medicine, 1965—; asst. attending physician Flower and Fifth Av., Met., Bird S. Coler hosps. Diplomate Am. Bd. Internal Medicine. Fellow A.C.P.; mem. A.M.A., N.Y. State, N.Y. County med. socs., Pan Am. Med. Assn., Am. Therapeutic Soc., Am. Fedn. Clin. Research, Sigma Xi. Research, publs. on capillary circulation, clin. pharmacology; co-discoverer anti-coagulant; showed nitroglycerin ineffective in treatment of cardiac pain; developed salt-loading approach using cardiacs to study efficacy of diuretic agts. Home: 172 Wood Rd., Englewood Cliffs, N.J. 07632. Office: 1036 Park Av., N.Y.C. 10028.*

FISCHBACH, Frederick Francis, Am. aerospace engr.; b. Chgo., Jan. 31, 1931; s. Glenn Thomas and Ruby (Pope) F.; B.S. with distinction, U. Mich., 1952, M.S., 1958; m. Marylouise Lindquist, Feb. 16, 1952; children—Brett Matthew, Cort Spencer, Kristin Louise, Dirk Frederick. Tech. officer U. S. Naval Guided Missile Service Unit, Seal Beach, Cal., 1953-55; research asso. U. Mich. Research Inst., Ann Arbor, 1956-60, asso. research mathematician, 1960-64, research engr., dept. aerospace engring., 1965—. Mem. Am. Geophys. Union, Phi Beta Kappa, Phi Kappa Phi, Phi Eta Sigma. Contbg. author: Sounding Rockets, 1959; International Astronautics Congress, 1961. Designed, developed Nike-Cajun sounding rocket system, Exos and Strongarm sounding rockets; research on latitude variations in upper air density; headed devel. stellar refraction technique for measurement atmospheric density from a satellite. Home: 2869 Dayton Dr., Ann Arbor, Mich. 48104.*

FISCHBECK, Kurt Hellmuth, German chemist; b. Oldenburg, Aug. 15, 1898; s. Emil and Ella (Logemann) F.; Ph.D. in Chemistry, U. Göttingen; m. Alwine Emme, Aug. 18, 1922; children—Berthild, Helmut, Ilse, Reinhard. Instr., later prof. U. Tübingen; full prof. phys. chemistry and applied physico-chemistry U. Heidelberg. Research and numerous publs. on fresh water from sea, corrosion, computation tables for chemists. Home: Ludolf-Krehlstrasse 29. Office: Institut für angewandte physikalische Chemie, Tiergartenstrasse, Heidelberg, Germany.

FISCHEL, Alfred, embryologist; b. Tschimelitz, Bohemia, Sept. 26, 1868; M.D., Prague, Czechoslovakia, 1894; m. Hedwig Bohm, 1910. Became venia legéndi for anatomy and embryology German U. of Prague, 1898, asso. prof., 1903, dir. new dept. exptl. morphology, 1910; named chmn. embryological inst. U. Vienna, 1916, prof., 1921; ret., 1935. Author: Über die Regeneration der Linse, 1900; Bedeutung der entwicklungsmechanischen Forschung für die Embryologie und die Pathologie des Menschen, 1912; Lehrbuch der Entwicklung des Menschen, 1929; Grundriss der Entwicklung des Menschen, 1931; (with W. Roux) Über vitale Färbung, 1908, Terminologie der Entwicklungsmechanik, 1912. Expanded field of embryology; 1st to use methods of exptl. morphology to study deformities; studied devel. mechanics of organs and organ systems, also their morphological and functional relationships; studied devel. of gonads and liver. Died Vienna, Austria, Jan. 12, 1938.

FISCHEL, Edward Elliot, Am. physician; b. N.Y.C., July 29, 1920; s. Joseph L. and Lisa (Herman) F.; B.A., Columbia, 1941, M.D., 1944, Sc.D. Med., 1948; m. Pauline I. Dunieff, Dec. 26, 1943; children—Robert Elliot, Janet Elizabeth. Fellow, asso. in medicine Columbia, 1947-55; dir. dept. medicine Bronx Hosp. (name later changed to Bronx-Lebanon Hosp. Center), N.Y.C., 1954—; asso. clin. prof. medicine Albert Einstein Coll. Medicine, 1957—. Chmn. research com. Arthritis Found., 1966-67; mem. Health Research Council City N.Y.; mem. research study com. Nat. Insts. Health. Diplomate Am. Bd. Internal Medicine. Fellow A.C.P., N.Y. Acad. Medicine (chmn. sec. microbiology 1962-64), A.A.A.S.; mem. Soc. Clin. Investigation, Am. Council Rheumatic Fever and Congenital Heart Disease, Am. Rheumatism Sect. (pres. 1968—), N.Y. Rheumatism Assn. (pres. 1964), Am. Assn. Immunologists, Am. Heart Assn., Harvey Soc., Soc. Exptl. Biology and Medicine, A.M.A., N.Y. State, Bronx County med. assns., Phi Beta Kappa, Alpha Omega Alpha. Research, publs. on quantitative immunological reactions, Arthus reaction, antibody prodn., effect of cortisone and other drugs on antibodies and hypersensitivity, pathogenesis and mgmt. rheumatic fever, serum complement in disease and inflammation. Home: 337 Engle St., Tenafly, N.J. 07670. Office: 1276 Fulton Av., Bronx, N.Y. 10456.*

FISCHER, Alfred George, geologist, natural scientist; b. Rothenburg, Germany, Dec. 12, 1920; s. George Erwin and Thea (Freise) F.; came to U. S., 1935; student Northwestern Coll., Watertown, Wis., 1935-37; B.A. in Geology, U. Wis., 1939, M.A., 1940; Ph.D., Columbia, 1950; m. Winnifred Varney, Aug. 26, 1939; children—Joseph Fred, George William, Lenore Ruth. Instr., Va. Poly. Inst., 1941-43; petroleum geologist Stanolind Oil & Gas Co., 1943-46; faculty U. Rochester, 1947-48, U. Kan., 1948-51; petroleum geologist Internat. Petroleum Co., 1951-56; faculty Princeton, 1956—, prof. geology, 1963—. Mem. Geol. Soc. Am., Am. Assn. Petroleum Geologists, Paleontol Soc. Am., Soc. Econ. Paleontologists, and Mineralogists, Palaeontologische Gesellschaft, Sigma Xi. Author: (with others) Invertebrate Fossils, 1952; (with others) The Permian Reef Complex, 1953. Research in invertebrate fossils, history of earth and life, nature of ancient marine life communities; stratigraphy, origin of limestones; devel. of mountain belts. Home: 564E Alexander Rd., Princeton, N.J.*

FISCHER, Bernhard, German surgeon; b. Coburg, Germany, 1852; ed. as army, navy surgeon; asst. to Robert Koch, Berlin, Germany; served with Cholera Commn. to Egypt, India; prof. hygiene, Kiel, Germany, from 1899; authority marine bacteria. Died 1915.

FISCHER, Edmond H(enry), biochemist; b. Shanghai, China, Apr. 6, 1920; Maturite Federale, U. Geneva (Switzerland), 1939, License es Sciences Chimiques, 1943, Ingenieur-Chemiste, 1944, Ph.D., 1947; 2 children. Asst. prof. biochemistry U. Geneva, 1950; faculty U. Wash., Seattle, 1953—, prof. biochemistry, 1960—. Mem. biochemistry study sect. NIH, 1959-64; mem. metabolic biology adv. panel NSF, 1965—. Recipient Werner medal for research Swiss Chem. Soc., 1952; Lederle Med. Faculty award, 1956-59; Guggenheim Found. award; Nat. Inst. Gen. Med. Sci. Spl. fellow Pasteur Inst., France, Weizmann Inst., Israel, Lister Inst., Eng. Mem. Am. (asso. editor Biochemistry 1962—, asst. editor 1966—, adv. bd. 1962—), Swiss chem. socs., Am. Soc. Biol. Chemists, A.A.A.S., Brit. Biochem. Soc., Am. Assn. U. Profs., Sigma Xi. Contbr. articles to profl. jours. Home: 3631 43d St. N.E., Seattle 98105.*

FISCHER, (Ludwig) Eduard von, see von Fischer, (Ludwig) Eduard.

FISCHER, Emil, German chemist; b. Cologne, Germany, Oct. 9, 1852; s. Laurenz and Julia (Poensgen) F.; student of Kekulé, Bonn, Germany; grad. (under Baeyer), Strasbourg (now in France), 1874; m. Agnes von Gerlach, 1888; 3 sons. Continued work with Baeyer; became prof. U. Munich (Germany), 1879; named prof. chemistry, Erlangen, Germany, 1882, Würzburg, Germany, 1885, Berlin (as successor to Hofmann), 1892-1919; in charge food and chem. prodn. in Germany, World War I. Recipient Nobel prize for synthetic work in sugar and purine groups of substances, 1902. Fellow Royal Soc., 1899; mem. French Acad. Scis., acads. scis. Berlin, Munich, Vienna. Author: Die Chimie der Kohlenhydrate und ihre Bedeutung fur die Physiologie, 1894; Anleltung zur Darstellung organischer Präparate, 1909; Chemical Research in Its Bearings on National Welfare, 1912; also articles. Outstanding organic chemist of his day; pioneer in research on structure and synthesis of sugars, polypeptides, purines, depsides, synthesis of tannins. Died Berlin, July 15, 1919.

FISCHER, Emil, Swiss physiologist; b. Triengen, Switzerland, Nov. 25, 1868; s. Anton and Katharina (Huber) F.; ed. U. Zurich (Switzerland); mem. leopoldinisch-Karolinische Deutsche Acad. der Naturforscher, Carnegie Inst. for Evolution (U. S.), corr. and hon. mem. numerous entomol. socs. Author: Experimenteller Beweis einer Vererbung erworbener Eigenschaften, 1900; Zur Physiologie der Aberrationen und Var.-Bildung der Schmetterlinge, 1907; Tagfalterpaarung in der Gefangenschaft, 1907; Termperatur-Experimente mit Schmetterlingen, 1908; Fortpflanzungs-fahiger Artbastard, 1924. Expts. on heredity and influence of outside features of organism; introduced new matters for temperature expts. with butterflies, 1894.

FISCHER, Erich Horst, German physicist; b. Allenstein, Germany, July 3, 1910; s. Erich A. and Paula (Blaschy) F.; ed. U. Bonn, U. Munich, U. Berlin; m. Gertrude Krabiell, Dec. 19, 1939; 1 dau., Petra (Mrs. Oskar Meinhardt). Physicist, Max Planck Inst. Physics, 1937-51; faculty U. Berlin, 1942-45, U. Tübingen, 1948-57, U. Ankara, Turkey, 1951-56; sci. dir. Gesellschaft fur Kernenergie, Hamburg, 1957—; faculty U. Hamburg, 1958—. Mem. German Phys. Soc., Soc. for Nuclear Energy. Author: Atomic, Nuclear and Molecular Physics, 1956; also numerous articles. Research on dielectric relaxation and molecular structure, nuclear power research and applications. Home: 2057 Reinbek, Theodor Storm Strasse 5, Germany. Office: 2057 Geesthaiht, Reactor Station, Germany.*

FISCHER, Ernst Georg, Am. mech. engr.; b. Balt., Md., Aug. 6, 1852; s. Georg Ernst and Caroline (Schmidt) F.; ed. Zschogg's Real Schule, Dresden, Germany, 1865-67, Engring. Works of Moritz Kleber, Dresden, 1867-70, subsequently under pvt. tutors; m. Julia Frances Lawson, Apr. 26, 1876. With U. S. Coast and Geod. Survey, 1887-1922, chief Instrument Div., 1889-1922. Made many improvements in instrumental equipment, devised and constructed new apparatus and instruments as follows: improvements in plane table alidade: base bars; spring balance for base tape measurement: tape stretching apparatus; tide gauge and tide indicator; photographic camera; compass declinometer; interferometer for measuring flexure of gravity pendulum supports; electric signal lamp for triangulation; magnetometer; transit micrometer; pendulum apparatus for determination of gravity; plane table; direction theodolite for primary triangulation; astron. transit; designed and constructed, after principles taken from Sir William Thomson and Dr. William Ferrel. U. S. Coast and Geod. Survey tide predicting machine No. 2, the geodetic or precise level, geodetic invar level rod, a new type of pressure sounding tube, a new artificial horizon for sextants. Died Sept. 1935.

FISCHER, Ernst Otto, German chemist; b. Munich, Nov. 10, 1918; s. Karl T. and Valentine (Danzer) F.; Diplom, Munich Inst. Tech., 1949, Dr. rer. nat., 1952, Habilitation, 1954. Asso. prof. inorganic chemistry U. Munich, 1957, prof., 1959; prof., dir. inorganic chemistry inst. Munich Inst. Tech., 1964—. Recipient annual prize, Göttingen Acad. Scis., 1957, Alfred Stock meml. prize, Soc. German Chemists, 1959. Mem. Bavarian Acad. Scis., Soc. German Chemists. Author: (with H. Werner) Metall-pi-Komplexe mit di- und oligoolefischen Liganden, 1963; (transl.) Metal pi-Complexes, Vol. 1, Complexes with di- and oligo-olefinic Ligands, 1966; also numerous articles. Research in organometallic chemistry: metal pi complexes of arenes, olefins, carbene complexes with metals, ferrocene type sandwich compounds, metal carbonyls. Address: 16 Sohnckestrasse, Munich, West Germany.*

FISCHER, Ernst Sigismund, mathematician; b. Vienna, Austria, July 12, 1875; s. Jacob and Emma (Grädener) F.; student, Berlin, Germany; studied under F. Mertens, Vienna, from 1894, Ph.D., 1899; postgrad., Zurich, Switzerland, Göttingen, Germany; m. Elisabeth Strauss, 1917; 1 dau. Became asst. to E. Waelsch, German Tech. U., Brno, Czechoslovakia, 1902, lectr., 1904, asso. prof., 1910; prof. U. Erlangen (Germany), 1911-20; prof. U. Cologne (Germany), 1920-38. Contbr. papers to math. jours. Advanced theory of Fourier's series by developing concept of convergence in center, Fischer-Riess theorem; studied determinants theorems, of Hadamard and Sylvester, Caratheodory problem; foresaw expansion of modern algebra. Died Cologne, Nov. 14, 1954.

FISCHER, (Heinrich August Wilhelm) Ferdinand, German chemist; b. Rödermühle, Germany, May 13, 1842; s. Eduard and Wilhelmine (Schirmer) F.; student, Göttingen, Berlin, Germany; Ph.D., Jena, Germany, 1869; m. Emmi Kolle. Tchr., Realschule, Hanover, Germany, 1871-79; devoted to applied chemistry, from 1880; lectr. chem. tech. U. Göttingen, 1894-98, asso. prof., 1898-1912. Author: Taschenbuch für Feuerungstechniker, 1883; Handbuch der chemischen Technologie, 1889; Die chemische Technologie der Brennstoffe, 2 vols., 1897-1901; Das Studium der technischen Chemie an Universitäten und Technischen Hochschulen Deutschlands und das Chemiker-Examen, 1897; Lehrbuch der chemischen Technologie, 1903; Technologie für Chemiker und Juristen an den preussischen Universitäten, 1903; Die Industrie Deutschlands und seiner Kolonien, 1908. Editor Jahresbericht über die Leistungen der chemischen Technologie . . . , also Zeitschrift für angewandte Chemie. Founder Gesellschaft für angewandte Chemie (became Verein deutscher Chemiker 1896), 1886; helped establish tech. chemistry as academic discipline. Died Homburg, Germany, June 28, 1916.

FISCHER, Frank Ernest Louis, botanist; b. Halberstadt, Russia, 1782. Dir. St. Petersburg Imperial Gardens. Studied botanical taxonomy, physiology of Ficales and serrated leaves; listed Schrenk's plants. Died 1854.

FISCHER, Franz Josef Emil, German chemist; b. Freiburg, Germany, Mar. 19, 1877; s. Emil and Emma (Stenz) F.; ed. Munich, Freiburg, Giessen (all Germany); degree, 1899; postgrad., Paris, Berlin; m. Emma Weuste, 1915; 2 sons, 1 dau. Became lectr. Phys.-chem. Inst., Freiburg, 1903; named prof. electrochemistry Berlin Tech. U., 1911; dir. coal research inst. Kaiser Wilhelm Gesellschaft, 1913-43; named hon. prof. Tech. U. Munich, 1944. Recipient Emil Fischer medal, 1927, Carl Engler medal, 1935, Hofmann Meml. medal, 1937, Goethe medal for Art and Sci., 1939. Contbr. articles to jours. Developed synthetic engine fuel Synthol; discovered (with Hans Tropsch) synthetic benzene (gasoline), 1924; research on origin, structure and use of coal. Died Munich, Dec. 1, 1947.

FISCHER, Fritz, physicist; b. Oberdiessbach, Switzerland, Feb. 9, 1898; s. Ernst and Berta (Stettler) F.; studied elec. engring. Fed. Tech. U., Zurich, Switzerland; Dr. sc. techn., 1924; m. Maud Schaetti, 1936; 2 daus. With Albisrieden telephone works, Zurich; joined central lab. Siemens & Halske Ag, Berlin, German, 1926, dir., from 1928; dir. found. UFA sound film studios, Babelsberg, to 1933; became prof. tech. physics Fed. Tech. U., Zurich, after 1933; founder Gesellschaft sur Forderung der Forschung (GFF), 1936, also firm Contraves for devel, mil. (mostly anti-aircraft) measuring instruments. Contbd. articles to profl. jours. Helped develop remote control ship Zahringen, also remote control plane Ju 52; developed commando or dual system; worked on automatic setting of ship and aircraft arty., also on sound film, phys. problems of color film; developed Eidophor (instrument of transmission color TV); held more than 70 patents. Died Zurich, Dec. 28, 1947.

FISCHER, Hans, German organic chemist; b. Höchst, Germany, July 27, 1881; s. Eugen and Anna (Herdegen) F.; Ph.D., Marburg (Germany) U., 1904; M.D., Munich, Germany, 1908; m. Wilfrud Hanfe, 1935. Chem. researcher Emil Fischer's Inst., Berlin, Germany; succeeded Windaus as prof. med. chemistry, Innsbruck, Austria, 1916-1921; became prof. organic chemistry, Munich, 1921. Recipient Nobel prize for chemistry, 1930. Author: Articles in Liebig's Annalen; Berichte de Deutschen Chemischen Geschichte, und anderen Fachzeitschriften; (with H. Orth) Die Chemie des Pyrrols, Vol. II. Editor various jours. Investigated composition of chlorophyll, also of blood; synthesized hemin, 1928. Died Munich, Mar. 31, 1945.

FISCHER, (Leopold) Heinrich, German mineralogist, zoologist; b. Freiburg im Breisgau, Germany, Dec. 19, 1817; s. Aloys, F.; studied medicine, Freiburg, also Vienna, Austria. Practiced medicine, Freiburg; faculty zoology and mineralogy, Freiburg, from 1845, named asso. prof. mineralogy and geognosie 1845, prof., 1859; became dir. mineral.-geonostic collection (under his dir. it became an early inst. for exptl. work), from 1855. Author: Orthoptera Europaea, 1853; Clavis der Silicate, 1864; Chronologischer Überblick über die allmalige Einführung der Mikroskopie in das Studium der Mineralogie, Petrographie und Paläontologie, 1868; Kritische mikroskopische Studien, 1869, 71-73; Nephrit und Jadeit nach ihren mineralogischen Eigenschaften, sowie nach ihrer urgeschichtlishen und ethnographischen Bedeutung, 1875; also numerous articles. Prominant entomol. taxonomist of his time; significantly advanced mineral. taxonomy by clear classifications of species; made chem. and microscopic studies of minerals; later became interested with prehistoric ethnology and studied stone age instruments to determine origin and distbn. Died Freiburg, Feb. 2, 1886.

FISCHER, Henry George, Am. Egyptologist; b. Phila., May 10, 1923, s. Henry G. and Agnes Beatrice (Hurdman) F.; B.A., Princeton, 1945; Ph.D., U. Pa., 1955; m. Eleanor Armstrong Teel, Dec. 15, 1951; 1 dau., Katherine Fraser. Faculty Am. U. Beirut (Lebanon), 1945-48; asst. Egyptian sect. U. Pa. Mus., 1949-56; mem. univ. expdn. to Mit. Rahineh (Egypt), 1955, 56; asst. prof. Egyptology, Yale Grad. Sch., 1956-58; asst. curator Egyptian art Met. Mus. Art, 1958-63, asso. curator, 1963-64, curator, 1964—; faculty Inst. Fine Arts, N.Y.U., 1962—, adj. prof., 1966—; vis. lectr. art history and archaeology Columbia, 1960-61, Sec.-treas. Am. Com. to Preserve Abu Simbel, 1964—. Trustee Am. Research Center in Egypt, 1965-66. Guggenheim fellow, 1956-57. Mem. Am. Oriental Soc., Am. Inst. Archaeology, Egypt Exploration Soc. (London), Phi Beta Kappa; corr. mem. German Archaeol. Inst. Author: Inscriptions from the Coptite Nome: Dynasties VI-XI, 1964; Ancient Egyptian Turtles, 1968; Dendera in the Third Millennium B.C., 1968. Contbns. include publs. on archaeology, epigraphy, and palaeography of Upper Egypt; occupational titles of the Egyptian Old Kingdom; Egyptian sculpture; protodynastic Egypt. Home: 141 E. 88th St., N.Y.C. 10028. Office: Met. Mus. Art, N.Y.C. 10028.*

FISCHER, (Friedrich Wilhelm) Hermann, German engr.; b. Rödermuhle, Germany, May 2, 1840; s. Eduard and Wilhemine (Schirmer) F.; studied mech. engring., Poly. Sch., Hannover, Germany, 1856-60; hon. dr. engring., Aachen, Germany, 1906; m. Fanny Hörig; 3 sons. Worked in Germany and abroad several years; became civil engr., Hannover, 1867; named lectr. on mech. tech., heating and ventilation installations (successor to K. Karmarsch) Hannover Tech. U., 1876. Re-edited Handbuch der mechanischen Technologie (Karmarsch), 1888-1904; Die Werkzeugmaschinen, 2 vols., from 1900. Noted for success in factory constrn., especially heating and ventilation installations; treated study of machine tools in sci. manner as separate dept. mech. engring. Died Hannover, Feb. 11, 1915.

FISCHER, Johann Karl, German mathematician, physicist; b. Allstedt/Helme, Germany, Dec. 5, 1760; s. Johann Gottfried and Friederike Charlotta Franciska (Hischeback) F.; ed. Jena, Germany; Ph.D., 1788; became asso. prof., Jena, 1792; apptd. prof. math. and physics Archigymnasium, Dortmund, Germany, 1807; prof. math. and astronomy U. Greifswald (Germany), from 1819. Author: Physikalisches Wörterbuch, 7 vols., 3 supplements, 1798-1827; Geschichte der Physik (1st comprehensive and systematically organized history of physics), 8 vols., 1801-08. Died Greifswald, May 22, 1833.

FISCHER, John Lyle, Am. anthropologist; b. Kewanee, Ill., July 9, 1923; s. George Lyle and Ann (Clark) F.; B.A., Harvard, 1946, M.A., 1949, Ph.D., 1955; student New Sch. for Social Research, 1946-47; m. Ann Kindrick, July 9, 1949; children—Madeleine Nikko, Mary Anne. Dist. anthropologist Truk and Ponape, Caroline Islands, 1949-51, asst. dist. adminstr., 1951-53; instr., research asso. Harvard Grad. Sch. Edn., 1955-58; faculty anthropology Tulane U., New Orleans, 1958—, prof., 1963—, chmn., 1967—. NSF grantee for study of Japanese Children, 1961-62; mem. behavioral sci. fellowship rev. com. NIH, 1963-67; mem. anthropol. panel. Behavioral and Social Sci. Survey Com., NRC-Social Sci. Research Council, 1967—. Nat. Inst. Mental Health Spl. fellow Center for Advanced Study in Behavioral Sci., Stanford, 1965-66. Mem. Am., So. anthrop. assns., Am. Ethnol. Soc., Linguistic Soc. Am., Japanese Soc. for Ethnology, Am. Folklore Soc., Am. Sociol. Assn., A.A.A.S. Author: (with A. Fischer) The Eastern Carolines, 1957, The New Englanders of Orchard Town, 1966. Contbr. numerous articles to sci. jours. Research on ethnography of Truk and Ponape, Caroline Islands, relations between social structure and expressive culture, social influences on linguistic change, influences of family structure on personality.*

FISCHER, Karl Tobias, German physicist; b. Nuremberg, Germany, Jan. 18, 1871; s. Johann Peter and Christine Elise (Herzog) E.; student Tech. U. Munich, 1889-93; Ph.D., U. Munich, 1896; m. Valentine Danzer, 1904; 2 sons, including Ernst Otto, 1 dau. Became mem. faculty Tech. U. Munich, 1897, asso. prof., 1903; visited Eng., 1898-99, U. S., 1904; ret., 1952. Named mem. Bavarian Standards Commn., 1907; dir. Bavarian Bur. Weights and Measures, 1923-36, 45-47, built State Bur. Weights and Measures, Munich-Nymphenburg, 1927-30. Hon. mem. German Assn. for Promotion of Math. and Natural Sci. Edn. Author: Der naturwissenschaftliche Unterricht in England, insbesondere in Physik und Chemie, 1901; Neuere Versuche zur Mechanik der festen und flüssigen Körper, 1902; Der naturwissenschaftliche Unterricht, insbesondere in Physik und Chemie bei uns und im Ausland, 1905. Exptl. research in molecular physics, heat theory, elec. and magnetic measuring, cryogenics; worked to develop precision measuring of weights, time, temperature. Died Munich-Solln, Dec. 31, 1953.

FISCHER, Kurt Ernst Paul, German mathematician; b. Magdeburg, Mar. 21, 1910; s. Carl and Agnes Schrader) F.; ed. U. Marburg, U. Göttingen; Ph.D. in math. and natural sci.; m. Alice Hummel, Feb. 13, 1937; children—Renate, Jürgen, Hartmut. Actuary, Gerling-Konzern Lebensversicherungs A.G., Cologne, 1935—, dir., mem. adminstrv. council, 1949—; dir., mem. adminstrv. council Gerling-Konzern Globale Ruckversicherungs A.G., 1952—, Gerling-Konzern Friedrich Wilhelm Lebansversicherungs A.G., 1957—. Mem. German Soc. Actuaries. Address: Beitrage zur Tarifanalyse, 1939; Richttafeln für die Pensionsversicherung, Dr. Heubeck/Dr. Fischer, 1948; (co-author) Die versicherungstrechnischen Rückstellungen im Steuerrecht, 1954; (collaborator) Handwörterbuch der Versicherungswesens, 1958. Study of math. problems as applied in economics, especially in insurance. Home: Kölner Wag 5, Junkersdorf bei Cologne. Office: Gerling-Hochhaus Gereonshof, Cologne, Germany.

FISCHER, Louis, physician; b. Kaschau, Austria-Hungary, Nov. 21, 1864; s. Ignetz and Louisa F.; came to N.Y. in childhood; student Coll. City of N.Y. 2 yrs.; grad. Coll. of Pharamacy, 1882; M.D., Univ. Med. Coll. (New York U.), 1884; also studied in Berlin; m. Clara Robert, Mar. 20, 1895; children—Alfred E., Robert M. Specialist in diseases of children: vis. physician Riverside Hosp., Willard Parker Hosp.; lectr. on diseases of children, N.Y. Post-Grad. Hosp.; physician-in-chief babies wards Suydenham Hosp. Mem. A.M.A. (sec. sect. diseases of children 1900), Acad. Medicine, N.Y. County, Harlem med. socs. Author: Infant Feeding in Health and Disease, 1901; The Health Care of the Baby, 1906; Diseases of Infancy and Childhood, 1907; Health Care of the Growing Child, 1915. Contbr. on diphtheria, infantile diseases and feeding to various mags. Died Apr. 9, 1945.

FISCHER, Nicolaus Wolfgang, German chemist; b. Great Meseritz, Moravia, Jan. 15, 1782; ed. univs. Vienna (Austria), Prague (Czechoslovakia), Breslau (Germany), Berlin; M.D., Erfurt, Germany, 1806. Practiced medicine, Breslau, 1807; became lectr. chemistry U. Breslau, 1808, asst., 1812, asst. prof., 1813, prof. chemistry, 1814. Author: Medicaminum mercurialium praecipua classificatio, 1806; De modia arsenia detegendi, 1812; Ueber die Wirkung des Lichts auf das Hornsilber, 1814; Ueber die chemischen Reagentien, 1816; Chemische Untersuchgen der Heilquellen zu Salzbrunn, 1821; Ueber die Natur der Metallreduction auf Nassem Wege, 1828; Das Verhaltniss der chemischen Verwandschaft zur Galvanischen Elektricitat, in Versuchen Dargestellt, 1830; Systematischer Lehrbegriff der Chemie, in Tabellen Dargestellt, 1838. Research on osmosis, action of light, silver chloride, separation of cobalt and nickel; discovered potassium cobaltnitrite (Fischer's salt), 1830; worked on separating metal from solutions of its salts by another metal and recognized the phenomenon as electrochem. Died Breslau, Aug. 19, 1850.

FISCHER, Otto, German physicist, physiologist; b. Altenburg, Germany, Apr. 26, 1861; s. Friedrich August and Christiane Johanna Marie (Bilger) F.; Ph.D., Leipzig, Germany, 1885; M.D. (hon.), Wurzburg, 1896; m. Anna Schellenberg, circa 1887; 3 children. Became tchr. math. and physics, comml. inst., Leipzig, 1887; became tchr. Petri-Realgymnasium (sci. secondary sch.), Leipzig, 1895, later rector; named to faculty U. Leipzig, 1893, asso. prof. med. sch., from 1896. Mem. Leopoldina, Saxon Soc. Scis. Author: Theoretische Grundlagen für eine Mechanik der lebenden Körper, 1906; Kinematik organischer Gelenka, 1907; also articles in handbooks and jours. Made math.-phys. analyses of movement of human body, mechanics of joints, kinetics of eye muscles, motion of man carrying heavy load compared to that of unburdened man, inertia moment of body, problems of psychophysics. Died Leipzig, Dec. 16, 1916.

FISCHER, Otto Philipp, German chemist; b. Euskirchen, Germany, Nov. 28, 1852; student in Bonn, also Munich (both Germany); Ph.D., Strasbourg (now in France), 1874; prof., dir. chemistry lab., Erlangen, Germany, 1885-1925. Investigated (with cousin Emil) hydrazines; studied dyes; made 1st synthetic alkaloid (kairine). Died Erlangen, Apr. 4, 1932.

FISCHER, Peter-Axel, German neurologist; b. Bremen, Mar. 21, 1929; s. Bertold and Waltraut (Suntheim) F.; M.D., U. Marburg; m. Irminheid Gorn, Aug. 16, 1963. Physician's asst., 1951-53 path. anatomy research at Bremen and Munster, 1953-56; with neurol. and psychiat. sect. Health Center, U. Hamburg-Eppendorf, 1956—, prof., chief physician, 1962—. Mem. various med. socs. Research and publs. on path. anatomy of Tb of lungs, clin. neurology, hypophys. adenoma. Home: Achternfelde 35A, 2 Garstedt. Office: Martinistrasse 52, Pav. 67, Hamburg, Germany.

FISCHER, William August, Am. geologist; b. Litchfield, Ill., Jan. 6, 1919; s. August Ernst and Juliette Maria (Niemeyer) F.; B.S., McKendree Coll., Lebanon, Ill., 1940; postgrad. U. Ill., 1940-41; m. L. Blanche Youngblood, Sept. 8, 1941; children—Judith Lynn (Mrs. Court Soloff), William Jeffrey. Tchr. chemistry, physics Christopher (Ill.) High Sch., 1941; with U. S. Geol. Survey, 1941-44, 46—, chief photogeology sect., 1950-60, charge lunar probe photometric analyses, 1960-62, research geologist charge remote sensing project, 1962-67, coordinator Earth Resources Observation Satellite Program, 1967—; prin. lect. pilot course aerial surveys for geology UN, Tokyo, Japan, 1961. Rep. Dept. Interior orgn. meeting Orgn. Europeenne d'Etudes Photogrammetriques Experimentales, Brussels, Belgium, 1952; chmn. U. S. delegation UN seminar aerial survey methods, Bangkok, Thailand, 1960; chief U. S. rep. orgn. meeting UNESCO conf. integrated surveys, Paris, France; 1963; chief U. S. delegation Internat. Congress Photogrammetry, 1964; mem. com. remote sensing environment Nat. Acad. Scis., 1964—. Recipient Autometric award, 1966; FMA award for Outstanding Contbn. to field of photointerpretation, 1966. Fellow A.A.A.S. (rep. council 1957-58), Am. Soc. Photogrammetry (pres. 1964-65), Dept. Interior Recreation Assn. (pres. 1961). Author numerous papers and maps. Contbns. include establishing geologic usefulness of aerial photographs, infrared images, ultra-violet spectograms and images, and radar images. Home: 228 Noland St., Falls Church, Va., 22046. Office: U. S. Geol. Survey, Washington 20242.*

FISCHER-HJALMARS, Inga Margrete, Swedish physicist; b. Stockholm, Sweden, Jan. 16, 1918; d. Otto and Karen (Wulff) Fischer; Fil.mag., U. Stockholm, 1944, Fil.lic., 1949, Fil.dr., 1952; m. Stig Hjalmars, July 11, 1952; Faculty, U. Stockholm, 1954-59, prof., 1963—; asso. prof. Inst. Tech., Stockholm, 1959-63. Mem. bd. U. Data Machines, Stockholm. Mem. Swedish Chem. Soc., Swedish Nat. Com. Physics, Biophysics Com., Am. Phys. Soc. Publs. on quantum mech. calculations of various properties of small molecules (ab initio calcn.) and of large molecules (semiempirical calcn.), especially molecules of importance in biology. Home: 282 Akerbyvägen, Täby, Sweden. Office: 9 Vanadisvägen, Stockholm, Sweden.*

FISCHER VON WALDHEIM, Gotthelf, entomologist; b. Waldheim, Germany, Oct. 15, 1771; M.D., U. Leipzig, Germany, 1798; prof. natural history, librarian Central Sch. Mayence, 1799-1803; prof., dir. Mus. Natural History, Moscow, Russia, from 1803. Founder Soc. Naturalists Moscow, 1808. Author: l'Os intermaxillaire, 1800; l'Anatomie du Maki, 1804; l'Oryctographie du gouvernement de Moscou, 1830-37; also works on typography, bibliography and transls. Reconstituted collections in Mus. Natural History, Moscow, after fire of 1812. Died Oct. 18, 1853.

FISCHLER, Franz Josef Benedikt, German physician; b. Weisloch, Germany, Mar. 15, 1876; s. Eduard and Emeline (Knauff) F.; studied medicine, Heidelberg, Freiburg, Leipzig, 1895-1900; M.D., 1902; Ph.D. in chemistry, Munich, 1926. Became lectr. U. Heidelberg (Germany), 1906, asso. prof., 1912, chmn. chem. research law. of med. clinic, till 1914; named chmn. sick care dept., municipal food office, Munich, 1917; became colleague pharm. inst. U. Munich, 1921, also German Research Inst. for Food Chemistry. Author: Physiologie und Pathologie der Leber (monograph), 1916; Anleitung zur Harnanalyse, 1943; also articles. Co-author: Anleitung zum Praktikum der analytischen Chemie. Made fundamental studies of liver; research on chemistry and physiology of carbohydrates, degenerative influence of rare earth metals on function of liver, also iodine metabolism, biol. value of fats, food preparation. Died Munich, June 26, 1957.

FISCHTHAL, Jacob Henry, Am. zoologist; b. Bklyn., Apr. 18, 1917; s. Max and Pauline (Denker) F.; B.S., L.I. U., 1937; M.S., State U. Ia., 1938; Ph.D., U. Mich., 1950; m. Lois Clinton, Apr. 18, 1942; children—Barry, Dorene, Glenn, Lynne. Aquatic biologist U. S. Fish and Wildlife Service, N. Atlantic Fishery Investigations, Cambridge, Mass., 1943; aquatic biologist Wis. Conservation Dept. fish mgmt. div., Spooner, 1943-48; prof. biology State U. N.Y. at Binghamton, 1948—, faculty research fellow, 1967, 68. Fulbright lectr. zoology U. Coll. Cape Coast (Ghana), 1965-66; cons. aquatic biology div. water supply N.Y. City Law Dept., 1957—; Fulbright lectr. Haile Selassie I U., Ethiopia, 1968-69. NIH research grantee, 1953-55; Sigma Xi-Research Soc. Am. Research Fund grantee, 1955-56; Office Naval Research grantee, 1956-57. Mem. Am. Soc. Parasitologists, Helminthological Soc. Washington (editorial com. Proc. 1968—). Research, publs. on parasitology of various vertebrates especially taxonomy and ecology of trematodes from U. S., Philippines, N. Borneo, Pacific Islands, Egypt, Ghana. Home: 33 Glann Rd., Apalachin, N.Y. 13732. Office: Dept. Biology, State U. N.Y., Binghamton, N.Y. 13901.*

FISH, Charles John, Am. biol. oceanographer; b. Fall River, Mass., May 13, 1899; s. Charles Frederick and Emily V. (Teale) F.; Ph.B., Brown U., 1921, Sc.M., 1922, Ph.D., 1923; Sc.D., U. R.I., 1966; m. Marie Dennis Poland, Feb. 10, 1923; 1 dau., Marilyn Poland (Mrs. Joseph Barnes Munro, Jr.). Asst. aquatic biologist U. S. Bur. Fisheries, 1922-24, asso. aquatic biologist, 1924-27; dir. Buffalo Mus. Sci., 1927-34; exec. sec., sr. scientist Internat. Passamaquoddy Fisheries Commn., 1931-33; faculty R.I. State Coll., 1934-46, prof. in charge dept. zoology, dir. Narragansett Marine Lab., 1936-46; marine biologist in charge Pacific oceanic biology project Woods Hole Oceanographic Instn., 1946-48; prof. marine biology U. R.I., Kingston, 1948-62, prof. oceanography, 1962-66, emeritus, 1966—, dir. Narragansett Marine Lab., 1948-62, acting dean Grad. Sch. Oceanography, 1961-62; research prof. ichthyology, U. Buffalo, 1929-31; research asso. dept. tropical research N.Y. Zool. Soc., 1925-34; research asso. in marine biology Woods Hole (Mass.) Oceanographic Instn., 1948-64, hon. staff mem., 1964—; mem. corp. Bermuda Biol. Sta. for Research, 1932-35, 48-66. Mem. adv. council Yenching U., Ching, 1934—. Decorated Officier d'Academie (France); Charles J. Fish Oceanographic Lab. named in his honor, U. R.I., 1960; recipient Stamford (Conn.) Mus. award for sci. achievement 1963. Fellow A.A.A.S.; mem. Am. Soc. Limnology and Oceanography, Atlantic Fisheries Biologists, Com. for Sci. Exploration of Atlantic Shelf, Phi Beta Kappa, Sigma Xi, Phi Kappa Phi, Phi Sigma. Research, numerous publs. on ocean plankton populations. Home: 1291 Kingstowne Rd., Kingston, R.I. 02881.*

FISH, Marie Poland, Am. ichthyologist; b. Paterson, N.J., May 22, 1902; B.A., Smith Coll., 1921; postgrad. Coll. Phys. and Surg., until 1923; Sc.D., U. R.I., 1966; m. Charles John Fish, Feb. 1923; 1 dau., Marilyn Poland. Research asst. in cancer problems dept. med. research Carnegie Instn., 1921; field. asst. Bur. Fisheries, 1923-27; asst. dept. tropical research (hon.) N.Y. Zool. Soc., 1925-32; curator ichthyology Buffalo Mus. Sci., 1928-31; research asso. Narragansett Marine Lab., 1937—, U. S. Nat. Mus., 1944-46; biol. oceanographer charge Office Naval Research Biol. Underwater Sound Project, 1946-67. Mem. Soc. Ichthyologists and Herpetologists, Soc. Women Geographers, Buffalo Soc. Natural Sci. Author: Contributions to the Early Life History of Fishes from Lake Erie and Its Tributary Waters. Research on underwater sound of biol. origin, embryology and life histories of marine and freshwater fishes, embryology and devel. of Am. eel, ichthyology of north and south Atlantic and Pacific, Sargasso Sea, Caribbean Sea, Japanese and Indo-Pacific, Great Lakes areas; 1st to discover eggs of eel, 1923. Home: 1291 Kingstowne Rd. Office: Narragansett Marine Lab., Kingston, R.I.

FISHBACK, William Thompson, Am. mathematician, educator; b. Milw., Jan. 28, 1928; s. Richard and Loraine (Thompson) F.; A.B., Oberlin Coll., 1943; M.A., Harvard, 1947, Ph.D., 1952; m. Joan Nelson Landers, Dec. 26, 1960; 1 son, Paul Edward. Staff mem. Mass. Inst. Tech. Radiation Lab., 1943-46; instr. U. Vt., 1950-51, asst. prof., 1951-53; faculty math. Ohio U., Athens, 1953-66, prof., 1964-66; prof. Earlham Coll., Richmond, Ind., 1966—. Mem. Am. Math. Soc., Math. Assn. Am., Phi Beta Kappa, Sigma Xi. Author: Projective and Euclidean Geometry, 1962. Home: 3923 Pinehurst Dr., Richmond, Ind. 47374.*

FISHBEIN, Morris, Am. physician; b. St. Louis, July 22, 1889; s. Benjamin and Fanny (Gluck) F.; B.S., U. Chgo., 1910; M.D., Rush Med. Coll., 1912;

Pharm. D. (hon.), Rutgers U.; LL.D., Fla. So. Coll.; m. Anna Mantel, July 7, 1914; children—Morris, Jr. (dec.), Barbara (Mrs. Morris T. Friedell), Marjorie (Mrs. John Clavey), Justin Mantel. Faculty, U. Chgo., 1942-47; faculty (formerly named Rush Med. Coll.) U. Ill. Coll. Medicine, Chgo., 1941-57, prof. emeritus, 1957; asst. editor Jour. of A.M.A., 1913-24, editor, 1924-50; med. editor Brit. Book of Year, 1938-—; cons. med. editor Doubleday and Col., 1949-—; also editor numerous med. jours. including Hygeia, 1924-50, World-Wide Abstracts Gen. Medicine, 1958-—, Med. World News, 1960-—, others. Named knight comdr. Crown of Italy, 1933, Order of Carlos Finlay of Cuba; comdr. Civil Order of Health (Spain), 1952; recipient Certificate of Merit, Pres. Truman, 1948, Officer's Cross, Order of Orange-Nassau, Netherlands, 1954, others. Fellow Am. Pub. Health Assn., Am. Geriatrics Soc., Royal Soc. Medicine (London, Eng.); mem. Inst. Medicine, A.A.A.S., Alpha Omega, Rho Pi Phi, others. Author: Handbook of Therapy, 1925; Your Diet and Your Health, 1937; The National Nutrition, 1942; Medical Writing: The Technic and the Art, 3d edit. 1957; Popular Medical Ency., rev. 1961; New Advances in Medicine, 1956; Handy Home Medical Adviser and Concise Medical Encyclopedia, 1957, others. Editor: numerous books including Children for the Childless, 1954; Modern Family Health Guide, 1959; Heart Care, 1961; also writer daily health columns, numerous articles in jours., mags. on the communication of medicine and research to the public; sex edn. of children; health information of gen. interest. Address: 5454 S. Shore Dr., Chgo., 60615.*

FISHBERG, Arthur Maurice, Am. physician; b. N.Y.C., June 17, 1898; s. Maurice and Bertha (Cantor) F.; A.B., Columbia, 1919, M.D., 1921; m. Irene Levin, June 16, 1933. Adj., asso. physician Mt. Sinai Hosp., N.Y.C., 1926-46; dir. medicine Beth Israel Hosp., N.Y.C., 1946-—; clin. prof. medicine N.Y. U., 1946-—. Mem. Am. Soc. Clin. Investigation, A.M.A., Am. Heart Assn., Council for Blood Pressure Research, Argentinian Med. Soc., Brazilian Cardiol. Soc. Author: Hypertension and Nephritis, 1930, 5th edit, 1954; Heart Failure, 1937, 2d edit., 1940. Research, publs. on heart disease, kidney disease and high blood pressure; introduced Fishberg Concentration Test; described and named hypertensive encephalopathy; first showed that initial changes in periarteritis nodosa are located in inner layers of artery. Home: 1136 Fifth Av., N.Y.C. 10028. Office: Beth Israel Hosp., 10 Nathan D. Perlman Pl., N.Y.C. 10003.*

FISHER, Albert Kenrick, Am. biologist; b. Sing Sing (now Ossining), N.Y., Mar. 21, 1856; s. Hiram and Susan E. (Townsend) F.; ed. Holbrook's Mil. High Sch., Sing Sing; M.D., Coll. Phys. and Surg. (Columbia U.), 1879; m. Alwilda Merritt; children—Harry Townsend, Walter Kenrick, Mrs. Ethel Merriam White and Mrs. Alberta Merritt Marble (twins). With Death Valley Expdn., 1891, sent out by Dept. Agr., made biol. survey of portions of Cal., Nev., Ariz. and Utah; also made biol. survey in various other western states, 1892-98; mem. Harriman (Alaska) Expdn., 1899; mem. Pinchot South Sea Expdn., 1929; in charge econ. investigations U. S. Biol. Survey, Dept. Agriculture, 1906-31; collaborator U. S. Nat. Mus. One of founders of Am. Ornithologists' Union (pres.); mem. Am. Game Protective Assn. (hon.), Linnaean Soc. N.Y. (corr.), Nuttall Ornithol. Club (corr.), Cooper Ornithol. Club (hon.), Delaware Valley Ornithol. Club (hon.), Internat. Assn. Game and Fish Commrs., Biol. Soc. Washington, Washington Biol. Field Club, Boone and Crockett Club (asso.), Baird Ornithol. Club (pres.). Author: Hawks and Owls of the United States, 1893; Ornithology of the Death Valley Expedition of 1891, 1893; also shorter papers. Died Washington, D.C., June 12, 1948.

FISHER, Cassius Asa, Am. geologist; b. Fremont, Neb., Feb. 15, 1872; s. Marcius Clay and Nellie (Le-Prand) F.; grad. Fremont Normal Sch., 1892; B.A., U. Neb., 1898, M.A., 1900, Sc.D., 1927; post-grad. Yale, 1902-03; m. Evangeline Hazlewood, Aug. 22, 1900; children—Eleonora H., Maurice H., Robert V. Fellow in geology U. Neb., 1898-1902; asst. instr. geology Yale, 1902-03; asst. geologist U. S. Geol. Survey, 1896-1909, geologist, 1909-10, also asst. chief of Fuel Sect.; cons. geologist and engr., splty. fuels, from 1910; mem. Fisher & Lowrie. Cons. mining engr. Bur. Mines, 1911-12; in charge of U. S. Navy fuel expdn. Alaska, 1912; geologist and cons. engr. in devel. of Salt Creek oil field, Wyo., 1910-13. Fellow Geol. Soc. Am. One of 3 originators of former method of valuation by U. S. Govt. of coal lands on pub. domain; co-author Manual for the Oil and Gas Industry, pub. by Treas. Dept. for use in appraising oil properties as basis for taxation; made extensive studies of coal and oil fields in U. S., C.Am., S.Am., Alaska, Can., Europe; del. representing petroleum interests of U. S. at Internat. C. of C. Conv., London, 1921. Died Nov. 4, 1930.

FISHER, Charles, Am. physician, psychoanalyst; b. Los Angeles, Mar. 26, 1908; s. Henry and Anna (Bergs) F.; Ph.B., U. Chgo., 1929; M.D., Northwestern U., 1939, Ph.D., 1934; m. Betty Krowech, Mar. 3, 1930; children—Carla (Mrs. Bennett Gershman), Barbara. Sr. asst. surgeon UPSHS Neuropsychiat. Service, U. S. Marine Hosp., Ellis Island, N.Y., 1942-46; research asso. Mt. Sinai Hosp., N.Y.C., 1954-65, sr. research asso., 1965-—, clin. prof. psychiatry Mt. Sinai Sch. Medicine, 1966-—; practice medicine and psychiatry, N.Y.C.; tng. analyst N.Y. Psychoanalytic Inst., 1956. Recipient Charles F. Menninger award for Meritorious Psychoanalytic Research, 1957; NIH Research grant, 1960-—. Diplomate Am. Bd. Psychiatry and Neurology. Mem. Am. Psychiat. Assn., N.Y. Psychoanalytic Soc. (pres. 1965-67), Am. Psychoanalytic Assn. Author: (with W. R. Ingram, S. W. Ransom) Diabetes Insipidus and the Neuro-Homonal Control of Water Balance, 1938; also articles. Research on functions of hypothalamus especially elucidation of anatomy and pathol. physiology of hypothalamic hypophyseal system in relation to water metabolism and diabetes insipidus; research in subliminary perception, psychophysiology of sleep and dreaming, dream deprivation; discovered nocturnal erection cycle. Home: 141 E. 88th St., N.Y.C. 10028. Office: 65 E. 76th St., N.Y.C. 10021.*

FISHER, (George) Clyde, Am. naturalist, astronomer; b. nr. Sidney, O., May 22, 1878; s. Harrison Jay and Amanda (Rhinehart) F.; student Ohio No. U.; A.B., Miami U., 1905, LL.D., 1926; Ph.D. in Botany, Johns Hopkins, 1913; m. Bessie Wiley, Aug. 29, 1905; children—Ruth Anna, Beth Elinor, Katherine Wiley; m. 2d, Te Ata, Sept. 28, 1933. Tchr. Ohio pub. schs., 6 years; tchr. astronomy, botany and zoology, high sch., Troy, O., 1905-07; prin. Palmer Coll. Academy De Funiak Springs, Fla., 1907-1909; acting pres. Palmer Coll., 1909-1910; in charge courses in ornithology, 3 summers, U. Fla., 1 summer, U. Tenn., courses in nature study Cornell U., summer 1931; mem. sci. staff Am. Mus. Natural History since 1913, later hon. curator astronomy and of Hayden Planetarium. Conductor photog. expdn. to Bermudas, 1924, Arctic Lapland expdn. 1924; visited astron. museums and observatories of Europe 1925, to gather material for proposed Hall of Astronomy at Am. Mus.; mem. of Harvard-Mass. Inst. Tech. Eclipse Expdn. to Siberia, 1936; leader Am. Mus. Eclipse Expdn. to Peru, 1937; mem. Am. Mus. expdn. to volcano Paricutin in Mexico 1943, 44. Group of islands off coast of N. Labrador named in his honor, 1944. Made Hon. Chief, Blood Tribe of Blackfoot Confederacy, 1946. Fellow Royal Astron. Soc., A.A.A.S., N.Y. Acad. Scis., N.Y. Zool. Soc.; mem. Am. Ornithol. Union (asso.), Am. Astron. Soc., Am. Assn. Variable Star Observers, Amateur Astron. Assn. (pres.), Soc. for Research on Meteorities, Torrey Bot. Club, N.Y. Bot. Garden, N.Y. Bird and Tree Club (pres.), Am. Nature Study Soc., Lima (Peru) Geog. Soc. (hon.), Phi Beta Kappa. Author: Nature's Secrets (later edit. titled Nature Encyclopedia), 1921; Exploring the Heavens, 1937; Astronomy (co-author), 1940; The Story of the Moon, 1943 (trans. into Spanish, 1944; printed for the U. S. Armed Forces, 1945); The Life of Audubon, 1949. His views on child edn. formed cornerstone of modern visual edn. Died N.Y.C., Jan. 7, 1949.

FISHER, D. Jerome, Am. mineralogist; b. Canton, N.Y., June 14, 1896; s. Lewis Beals and Fannie A. (Shaw) F.; S.B., U. Chgo., 1917, S.M., 1920, Ph.D., 1922; postgrad. U. Mich., U. Berlin (Germany); m. Dorothy Dorsett, July 27, 1919; children—David L., Donis F. (Mrs. M. Shapiro), Jerome D. (dec.). Faculty, U. Chgo., 1920-—, prof. emeritus, 1961-—; vis. prof. dept. geology Northwestern U., Evanston, Ill., 1965-—; staff U. S. Geol. Survey, 1922-24, 43, Ill. Geol. Survey, 1920-36, S.D. Geol. Survey, 1941-42. Fellow Geol. Soc. Am.; Mineral. Soc. Am. (past pres.); mem. Internat. Mineral. Assn. (past pres.), A.A.A.S., Mineral. Soc. India (hon. fgn.), Mineral. Soc. Gt. Britain, Am. Crystallographic Assn., Sigma Xi. Research, publs. in mineralogy, optical, X-ray, and morphologic crystallography, fuel, areal, and stratigraphic geology, gnomonic and stereographic projections, refractometry, mineral phosphates, pegmatities. Home: 5639 Drexel Av., Chgo. 60637.*

FISHER, Dale John, Am. chemist; b. Omro, Wis., June 4, 1925; B.S., Wis. State U., 1947; Ph.D. (U. fellow), Ind. U., 1951; m. Ruth J. Laird, Apr. 27, 1957. Staff mem. Inst. Paper Chemistry, Appleton, Wis., 1945; chemist city Oshkosh, Wis., summers 1947-49; chemist ionic analyses group Oak Ridge Nat. Lab., 1951-52, group leader analytical instrumentation group, 1952-—. Lectr. before profl. groups. Mem. Am. Chem. Soc., Am. Nuclear Soc., Polarographic Soc., Tenn. Archaeol. Soc., Sigma Xi, Phi Lambda Upsilon. Conceive and direct research programs for devel. of and select and apply physicochem., mech., electronic, and optical principles in design of instrumentation, and for new or improved instrumental methods. Home: 22 Outer Dr. Office: P.O. Box X, Oak Ridge 37830.*

FISHER, David Elimelech, Am. cosmochemist; b. Phila., June 22, 1932; s. Henry R. and Grace (Spicehandler) F.; B.S., Trinity Coll., 1954; Ph.D., U. Fla., 1958; m. Leila Lois Katz, Sept. 4, 1954; children—Lisa, Ronald, Marshall. Postdoctoral fellow in chemistry Brookhaven Nat. Lab., Upton, N.Y., 1958-60; asst. prof. engring. physics Center for Radiophysics and Space Research, Cornell U., Ithaca, N.Y., 1960-66; asso. prof. U. Miami (Fla.) Inst. Marine Sci., 1966-—. Mem. Am. Phys. Soc., Am. Geophys. Union, Geochem. Soc., Meteoritical Soc., A.A.A.S. Research, publs. in nuclear reactions induced by cosmic rays in meteorites, ages of meteorites, origin of meteorites, abundances of elements, geochronology. Home: 9650 Kendale Blvd., Miami, Fla.*

FISHER, Delbert Arthur, Am. physician; b. Placerville, Cal., Aug. 12, 1928; s. Arthur L. and Thelma (Johnson) F.; A.B., U. Cal. at Berkeley, 1950, M.D., U. Cal. at San Francisco, 1953; m. Beverly Jane Carne, Jan. 28, 1951; children—David Arthur, Thomas Martin, Mary Katherine. Practice medicine, specializing in pediatrics, Portland, Ore., 1957-60, Little Rock, 1960-68, Los Angeles, 1968-—; fellow pediatric endocrinology Med. Sch. U. Ore., 1958-60; faculty Med. Sch. U. Ark., 1960-68; prof. pediatrics Univ. of Cal. at Los Angeles, 1968-—. Recipient USPHS Career Program award Nat. Inst. Child Health and Human Devel. U. Ark., 1964. Mem. Am. Acad. Pediatrics, Am. Pediatric Soc., Soc. Pediatric Research, Endocrine Soc., Am. Assn. U. Profs., Phi Beta Kappa, Sigma Xi. Author: (with D. E. Pickering) Fluid and Electrolyte Therapy-A Unified Approach, 1959. Research, numerous publs. on water metabolism and endocrine gland function in newborn infant, kinetics of iodine metabolism, etiologic mechanisms, adrenal corticosteroid drugs. Address: 1000 W. Carson St., Torrance, Cal. 90509.*

FISHER, Edwin Ralph, Am. physician; b. Pitts., Sept. 2, 1923; s. Reuben Fisher and Anna (Miller) F.; B.S., U. Pitts., 1945, M.D., 1947; m. Carole Levy, June 21, 1953; children—Marjorie, Abbe Dava. With dept. pathology and histochemistry NIH, Bethesda, Md., 1952-54; mem. staff, dept. pathology Cleve. Clinic, 1954; chief lab. service VA Hosp., Pitts., 1955-63, dir. research and edn. pathology, 1963-—; prof. pathology U. Pitts., 1958-—. Recipient Parke-Davis award in Exptl. Pathology, 1963. Mem. Am. Assn. Cancer Research, Am. Assn. Pathology and Bacteriology, Internat. Acad. Pathology, Am. Soc. Exptl. Pathology, Soc. Exptl. Medicine and Biology, A.A.A.S., Am. Soc. Clin. Pathology, Soc. Study Arteriosclerosis, Histochem. Soc., Alpha Omega Alpha. Research, numerous publs. in spread of cancer, kidney disease, arteriosclerosis. Home: 5812 Solway St., Pitts. 15217. Office: VA Hosp., Pitts. 15240.*

FISHER, Franklin Marvin, Am. economist; b. N.Y.C., Dec. 13, 1934; s. Mitchell S. and Esther (Oshiver) F.; A.B. summa cum laude, Harvard, 1956, M.A., 1957, Ph.D., 1960; m. Ellen Jo Paradise, June 22, 1958; children—Abraham, Abigail. Asst. prof. econs. U. Chgo., 1959-60; faculty Mass. Inst. Tech., 1960-—, prof., 1965-—. NSF fellow Econometric Inst., Netherlands Sch. Econs., Rotterdam, 1962-63. Fellow Econometric Soc.; mem. Am. Econ. Assn. Author: A Priori Information and Time Series Analysis: Essays in Economic Theory and Measurement, 1962; (with Carl Kaysen) A Study in Econometrics: The Demand for Electricity in the United States, 1962; (with Albert Ando, Herbert A. Simon) Essays on the Structure of Social Science models, 1963; Supply and Costs in the United States Petroleum Industry: Two Econometric Studies, 1964; The Identification Problem in Econometrics, 1966; also articles. Asso. editor Jour. Am. Statis. Assn.; Am. editor Rev. Econ. Studies, 1965-—. Research on econometric methods and studies, econ. theory, cost-benefit analysis. Home: 28 Holden Wood Rd., Concord, Mass. 01742. Office: Dept. Econs., Mass. Inst. Tech., Cambridge, Mass. 02139.*

FISHER, George, English astronomer; b. Sanbury, Eng., July 31, 1794; M.A., St. Catherine's Coll., Cambridge, Eng., 1821. Astronomer to polar expdn., 1818; chaplain, astronomer Parry's N.W. passage expdn., 1821-23; carried out magnetical expts. in Mediterranean, 1827-32; head master Greenwich (Eng.) Hosp. Sch., 1834-60, prin., 1860-63, also erected sch. obs. Fellow Royal Soc., 1825; v.p. Astron. Soc. Contbg. author to appendix for Journal of a Second Voyage for the Discovery of a Northwest Passage (Parry), 1825. Contbr. to jours. various papers, including The Nature and Origin of the Aurora Borealis, 1834. Died May 14, 1873.

FISHER, Granville Chapman, Am. psychologist; b. Nashville, Apr. 11, 1906; s. Henry Gordon and Grace (Dickens) F.; Ph.B., U. Chgo., 1943, M.A., 1946, Ph.D., 1949; B.D., Meadville Theol. Sch., 1944; m. Ijourie Bernice Stocks, Oct. 7, 1940 (div. June 1963); children—Douglas Murray, April (Mrs. Rick Parendes). Psychologist, Cook County Criminal Ct., Chgo., 1945-46; prof. psychology U. Miami, Coral Gables, Fla., 1946-—, chmn. dept. psychology, 1948-60; cons. psychology S. Fla. State Hosp., 1959-—. Mem. S.E., Fla. psychol. assns., Sigma Xi. Contbr. articles to sci. jours. Discovered a pattern of deterioration of intellectual functions in specific brain damage. Home: 3929 Ponce de Leon Blvd., Coral Gables 33134. Office: Dept. Psychology, U. Miami, Coral Gables, Fla. 33146.*

FISHER, Harry Linn, Am. chemist; b. Kington, N.Y., Jan. 19, 1885; s. George Edwin and Emma Adelia (Bray) F.; A.B., Williams Coll., 1909, D.Sc. (hon.), 1953; A.M., Columbia, 1910, Ph.D., 1912; m. Nellie Edna Andrews, June 7, 1910; children—Helen, Ruth (Mrs. Francis B. Rosevear), Robert Andrews. Instr. chemistry (Cornell, 1911-12; instr. organic chemistry Columbia, 1912-19; research chemist B. F. Goodrich Co., Akron. O., 1919-26, U. S. Rubber Co., N.Y.C., Passaic, N.J., 1926-36; dir. organic chem. research Air Reduction Co., 1936-44, U. S. Indsl. Chemicals, Inc., Stamford, Conn., 1936-49, Balt., 1949-50; organizing sec. 12th Internat. Congress Pure and Applied Chemistry, N.Y.C., 1950-51, arranging sec. 16th Conf., N.Y.C., Washington, 1950-51; adminstrv. asst., div. chemistry NRC, Washington,

1950-51; tech. synthetic rubber div. R.F.C. Washington, 1951-52; prof. rubber tech., sch. engring. U. So. Cal., 1953-56; western tech. dir. Ocean Minerals, Inc., 1957——. Tech. cons. Office Rubber Director, 1943-44, Office Rubber Research, 1944-51. Recipient Modern Pioneer award N.A.M., 1940; Charles Goodyear medal, Am. Chem. Soc., 1949. Fellow A.A.A.S., Am. Inst. Chemists (nat. pres. 1940-42), Inst. Rubber Industry Gt. Brit.; mem. Am. Soc. Testing Materials, Am. Chem. Soc. (pres. 1954), Phi Beta Kappa, Sigma Xi, Phi Lambda Upsilon. Author: Laboratory Manual of Organic Chemistry, 1920, rev. edits. 1924, 31, 38; Rubber and its Use, 1941; Chemistry of Natural and Synthetic Rubbers, 1956. Contbr. articles profl. jours. Patentee processes for attaching rubber to metal, non-sulphur vulcanization, synthesis organic chemical compounds, methionine, others. Demonstrated that non-sulfur compounds could produce practical rubber vulcanizates; discovered adhesives for rubber; advanced devel. of series of thermoplastic resinous materials which could be used as substitutes for shellac, balata and gutta-percha. Address: 4116 Santa Tomas Dr., Los Angeles 8.

FISHER, Harvey Irvin, Am. zoologist; b. Edgar, Neb., June 15, 1916; s. Fred Herman and Blanche (Baker) F.; student Kansas City (Mo.) Jr. Coll., 1933-35; B.S., Kan. State U., 1937; Ph.D., U. Cal. at Berkeley, 1942; m. Mildred Leone Hoch, July 11, 1937; children—Frederick Harvey, George Karl, James Rilan. Tech. curator Mus. Vertebrate Zoology, U. Cal., 1942-45, biologist Crocker Radiation Lab., 1942-45; asst. prof. zoology U. Hawaii, 1945-47; exchange prof. U. Nev., 1947-48; asso. prof. U. Ill., 1948-55; prof., chmn. dept. zoology So. Ill. U., Carbondale, 1955——. Mem. Am. Ornithol. Union, Cooper, Wilson ornithol. socs., Soc. Study Evolution, Ill. Acad. Sci., A.A.A.S., Am. Soc. Zoologists, Am. Ecol. Soc., Sigma Xi, Phi Kappa Phi. Author: General Zoology Notes and Study Questions, 1954; (with Donald C. Goodman) The Myology of the Whooping Crane, 1955; (with James B. Kitzmiller) Laboratory Exercises in General Zoology, 1958; (with R. W. Burnett) Zoology. An Introduction to the Animal Kingdom, 1958, Dierenrijk, 1961, Zoologica, 1962, Regne Animal, 1963, Reino Animal, 1964, Il Regno Animale, 1964, Zoologi for Alle, 1965, Fakta om Dyr, 1965; (with Donald C. Goodman) Functional Anatomy of the Feeding Apparatus in Waterfowl, 1962; also numerous articles. Founding editor Pacific Science, 1946-48; editor The Auk, 1948-52, Biol. Monographs, U. Ill., 1951-55, Trans. Ill. Acad. Sci., 1955-59. Home: 33 Hillcrest Dr., Carbondale, Ill.*

FISHER, Irving, Am. polit. economist; b. Saugerties, N.Y., Feb. 27, 1867; s. Rev. George Whitefield and Ella (Wescott) F.; A.B., Yale, 1888, Ph.D., 1891; LL.D., Rollins Coll., 1932, University of Athens and U. of Lausanné, 1937; studied Berlin and Paris, 1893-94; m. Margaret, d. Rowland Hazzard, June 24, 1893 (dec.); children—Margaret (dec.), Caroline (Mrs. Carol Fisher Baumann), Irving Norton. Tutor math., 1890-93, asst. prof., 1893-95, asst. prof. polit. economy, 1895-98, professor political economy, 1898-1935, prof. emeritus, Yale. Hitchcock lecturer, University of California, 1917; lecturer University of London School of Economics and Polit. Science, 1921, Geneva School of Internat. Studies, 1927. Gave lectures on income tax reform, U. of Southern Calif., Feb.-Apr. 1941. President for U. S. of Third Internat. Commn. on Eugenics; mem. Theodore Roosevelt's Nat. Conservation Commn.; chmn. Hygiene Reference Bd. of Life Extension Inst. since 1914; chmn. sub-com. on Alcohol of Council Nat. Defense, 1917-18; pres. Citizens' Com. on War-Time Prohibition, 1917; pres. Com. of 60 on Nat. Prohibition, 1917; chmn. bd. scientific dirs. Eugenics Record Office, 1917; dir. of Cowles Com. for Econ. Research. Vice pres., dir. Gotham Hosp., Gotham Med. Center. Chmn. bd. and dir. Check Master Plan, Inc., Gyrobalance Corp., Automatic Signal Corp.; dir. and mem. exec. com. Remington Rand Inc.; dir. Buffalo Elec. Furnace Corp., Sonotone Corp., Latimer Lab., Life Extension Inst. President American Assn. Labor Legislation, 1915-17, Nat. Institute Social Sciences, 1917, Am. Econ. Assn., 1918, Eugenics Research Assn., 1920, Pro-League Independents, 1920, Econometric Soc., 1931-33, Am. Statis. Assn., 1932; founder and 1st pres. Am. Eugenics Soc., 1923-26; fellow Royal Statis. Soc., A.A.A.S.; rec. sec. New Haven County Anti-Tuberculosis Assn., 1904-14; mem. editorial bd. of Econometrics; mem. Phi Beta Kappa, Royal Econ. Society, Conn. Acad. Arts and Sciences, Am. Acad. Polit. and Social Science, Am. Statis. Assn., Am. Ethnographical Soc., N.E. Free Trade League, Internat. Free Trade League, Nat. Assn. Study and Prevention Tuberculosis, Am. Assn. for Study and Prevention Infant Mortality, Nat. Consumers League, League of Nations Assn., Am. Philos. Soc., Reale Accademia dei Lincei (Rome), Institut Internat. Statisque, Norwegian Acad. of Sci. and Letters, Instituto Lombardo (Milan, Italy), Com. for the Nation. Author: Mathematical Investigations in the Theory of Value and Prices, 1892; Elements of Geometry (with Prof. A. W. Phillips), 1896; Bibliography of Mathematical Economics in (and asst. in translating and editing) Cournot's Mathematical Theory of Wealth, 1897; A Brief Introduction to the Infinitesimal Calculus, 1897; The Nature of Capital and Income, 1906; The Rate of Interest, 1907; National Vitality, 1909; The Purchasing Power of Money, 1911; Elementary Principles of Economics, 1910; Why is the Dollar Shrinking?, 1914; How to Live (with Dr. E. L. Fisk, in collaboration with 93 members of hygiene reference board of Life Extension Institute), 1915; Stabilizing the Dollar, 1920; The Making of Index Numbers, 1922; League or War?, 1923; America's Interest in World Peace, 1924; Prohibition at Its Worst, 1926; The Money Illusion, 1928; Prohibition Still at Its Worst (with H. B. Brougham) 1928; The Theory of Interest, 1930; The Noble Experiment (with H. B. Brougham), 1930; The Stock Market Crash, 1930; Booms and Depressions, 1932; Inflation, 1933; Stamp Scrip, 1933; After Reflation, What?, 1933; Stable Money, a History of the Movement, 1934; 100% Money, 1935; Constructive Income Taxation, 1942; World Maps and Globes (with O. M. Miller), 1944; also numerous articles, monographs, etc. Developed theory of money, equation of exchange, "compensated dollar" theory. Died Apr. 29, 1947.

FISHER, John Crocker, Am. physicist; b. Ithaca, N.Y., Dec. 19, 1919; s. John C. and Ruth (Campbell) F.; A.B., Ohio State U., 1941; Sc.D., Mass. Inst. Tech., 1947; m. Jo Ann Johnson, Oct. 24, 1943; children—Kelly C., Mark E., Holly. Research engr. Battelle Meml. Inst., Columbus, O., 1941-42; instr. Mass. Inst. Tech., Cambridge, Mass., 1942-47; physicist Gen. Electric Research Lab., Schenectady, 1947-64; mgr. information scis. Gen. Electric Tempo, Santa Barbara, Cal., 1964-67, research and devel. mgr. programs and systems Gen. Electric Research and Devel. Center, Schenectady, 1968——. Mem. Am. Phys. Soc. Research, publs. on phys. metallurgy, solid state physics, theoretical physics. Home: 2600 Foothill Rd., Santa Barbara, Cal. 93105. Office: P.O. Box 8, Schenectady 12301.*

FISHER, John Dix, Am. physician; b. Needham, Mass., Mar. 27, 1797; s. Aaron and Lucy (Stedman) F.; grad. Brown U., 1820, Harvard Med. Sch., 1825. One of 1st in Am. to utilize auscultation (an aid to diagnosis); pioneer in use of etherization in childbirth; acting physician Mass. Gen. Hosp.; mem. Mass. Med. Soc.; introduced French methods for educating the blind into America, largely responsible for creation Perkins Instn., Mass. Sch. for Blind, 1829, v.p., dir., 1829-50. Author: Description of the Distinct, Confluent, and Inoculated Smallpox, 1829. Died Mar. 3, 1850.

FISHER, John Gatewood, Am. chemist; b. McLaughlin, S.D., June 9, 1924; s. Glenn and Edna (Gatewood) F.; B.S., Western Ky. U., 1947; Ph.D., Ohio State U., 1951; m. Laura Nell Hendrick, Aug. 3, 1945; children—Sue Ann (Mrs. James Edward Roberts) John Gatewood II, Frances Elizabeth, Glenn Mansfield, Mary Katherine. Research chemist Tenn. Eastman Co., Kingsport, Tenn., 1951-56, sr. research chemist, 1956-66, research asso., 1967——. Mem. Am. Chem. Soc., Am. Assn. Textile Chemists and Colorists, Sigma Xi, Phi Lambda Upsilon. Research on textile dyes and their intermediates, with particular emphasis on disperse, metallized and cationic dyes; heterocyclic compounds; aromatic compounds; quarternary ammonium compounds. Patentee in field. Home: 1157 Watauga St. Office: Tenn. Eastman Co., Bldg. 150, Kingsport, Tenn. 37660.*

FISHER, John Heber, Canadian pathologist; b. Glanworth, Ont., Can., Apr. 18, 1899; s. Maddison Wall and Mary Grace (Barry) F.; M.D., U. Western Ont., 1922, M.Sc., 1924; m. Veda Elizabeth Wigmore, June 10, 1925; children—John Franklin, Jacqueline Miriam (Mrs. Robert A. Lutz). Head, dept. pathology U. Western Ont., London, 1929-65; pathologist in chief Victoria Hosp., London, 1929-65. Regional pathologist Dept. Atty. Gen. Ont., 1937——. Fellow Royal Coll. Physicians and Surgeons Can.; mem. Internat. Acad. Pathologists, Am. Assn. Pathologists and Bacteriologists, Am. Soc. Clin. Pathology, Canadian, Ont. assns. pathologists, Canadian, Ont. med. assns., Alpha Omega Alpha. Research, publs. on pulmonary infections, carcinoma, emphysema, hepatic necrosis, duodenal ulcers in children, postmastectomy lymphangiosarcoma. Home: 383 Huron St., London, Ont., Can.*

FISHER, Leon Harold, physicist; b. Montreal, Que., Can., July 11, 1918; s. Jacob and Rachel (Haimowitz) F.; came to U. S., 1920, naturalized, 1925; B.S., U. Cal. at Berkeley, 1938, M.S., 1940, Ph.D., 1943; m. Phyllis Kahn, Dec. 21, 1941; children—Robert Alan, Lawrence Edgar, Carol Lee, David Bruce. Instr. physics U. Cal., Berkeley, 1943, vis. prof. summers, 1949, 51, 55; instr. physics U. N.M., Albuquerque, 1944; physicist Los Alamos Sci. Lab., 1944-46; asst. prof. physics N.Y. U., N.Y.C., 1946-50, asso. prof., 1950-57, prof. physics, 1957-61; vis. prof. physics U. So. Cal., summer 1948; mgr. plasma physics Lockheed Palo Alto (Cal.) Research Lab., 1961-62, sr. mem. electronics sci. lab., 1963-67, asst. mgr., 1967——. head plasma physics Gen. Telephone & Electronics Labs., Palo Alto, Cal., 1962-63. Cons., Edgerton, Germeshausen & Grier Corp., Boston, 1954-55, Harry Diamond Labs., Washington, 1958-61, Xerox Corp., Rochester, N.Y., 1958-61, Army Research Office, Durham, N.C., 1958——. Chmn. Gaseous Electronics confs., 1948, 67, 68; mem. radiation com. Rome Air Devel. Center, 1959-63; cons. re-entry physics panel Nat. Acad. Scis., 1965-66. Fellow Am. Phys. Soc., A.A.A.S.; mem. Am. Assn. Physics Tchrs. Phi Beta Kappa, Sigma Xi, Pi Mu Epsilon. Asso. editor: Phys. Rev., 1955-58. Research, publs. on elec. currents in gases. Home: 102 Encinal Av., Atherton, Cal. 94025. Office: 3251 Hanover St., Palo Alto, Cal. 94304.*

FISHER, Lloyd Wellington, Am. geologist; b. Feb. 15, 1897; s. George S. and Alice (Mayer) F.; B.A., Lehigh U., 1921; M.S., Pa. State Coll., 1923; Ph.D., Johns Hopkins, 1929; m. Ermelinda McCarthy, Aug. 26, 1938. Instr., Syracuse U., Brown U.; prof. geology Bates Coll., Lewiston, Me.; with U. S. govt., Pa. Survey. Sec. New Eng. Intercollegiate Geol. Excursion. Fellow Geol. Soc. Am.; mem. Mineral. Soc. Am. Contbr. articles to sci. jours. Died 1951.

FISHER, Michael Ellis, math. physicist, chemist; b. Trinidad, W.I., Sept. 3, 1931; s. Harold Wolf and Jeanne Marie (Halter) F.; B.Sc. 1st class honors in Physics, King's Coll., London, Eng., 1951, Ph.D., 1957; m. Sorrel Castillejo, Dec. 12, 1954; children—Caricia J., Daniel S., Martin J., Matthew P. A. Lectr. math. RAF, 1952-53; lectr. theoretical physics King's Coll., London, 1957-62, reader physics, 1962-64; prof. physics U. London, 1964-66; prof. chemistry and maths. Cornell U., Ithaca, N.Y., 1966——; guest investigator Rockefeller Inst., N.Y.C., 1963-64. Mem. Phys. Soc., London, Am. Phys. Soc. Author: (with D. M. MacKay), Analogue Computing at Ultra-High Speed, 1962; also articles. Asso. editor Jour. Math. Physics, 1963-68. Devel. theory and techniques for electronic analog computation at very high speeds, statis. mechanics, theory of polymers, phase transitions, critical phenomena; combinatorial problems. Home: Baker Lab., Cornell U., Ithaca, N.Y. 14850.*

FISHER, Myron Wolf, Am. microbiologist; b. Marlboro, Mass., May 21, 1918; s. Irving and Ida (Bix) F.; B.S., U. Mass., 1939; M.S., Northwestern U., 1950, Ph.D., 1951; m. Eleanor J. Schoenberg, June 1, 1944; children—Steven, Jeanne, Deborah. Instr., research asso. Northwestern U., 1948-51; with Parke, Davis & Co., Detroit, 1951——, lab. dir., 1960-62, dir. research in bacteriology and immunology, 1962——. Mem. Am. Soc. Microbiology, N.Y. Acad. Scis., A.A.-A.S. Research, numerous publs. tuberculosis chemotherapy; exptl. chemotherapy of bacterial infections, evaluation of antibiotics, bacterial immunology, interrelations of immunity and chemotherapy, new bacterial vaccines and therapeutic globulins, immune suppression. Home: 942 S. Shady Hollow St., Bloomfield Hills, Mich. 48013. Office: Parke, Davis & Co., Detroit, Mich. 48232.*

FISHER, Philip C., Am. physicist; b. Rochester, N.Y., Aug. 3, 1926; s. Raymond C. and Alice (Coggins) F.; B.S., U. Rochester, 1947; M.S., U. Ill., 1948, Ph.D., 1953; m. Virginia Ruffner Ball, Aug. 21, 1948; 1 dau. Christine Chapin. Member staff Los Alamos Sci. Lab., 1953-59, cons., 1959-64; cons. scientist Lockheed Missiles & Space Co., Palo Alto, Cal., 1959——. Mem. Am. Phys. Soc., Am. Astron. Soc., Am. Geophys. Union. Research, publs. on stellar X-ray sources, measurement of particle radiations in space and fission gamma spectra, weapons physics. Home: 4 Linaria Way, Menlo Park, Cal. 94025. Office: 3251 Hanover St., Palo Alto, Cal. 94304.*

FISHER, Reginald Brettauer, Brit. biochemist; b. Sheffield, Eng., Feb. 13, 1907; ed. St. John's Coll., Oxford U.; m. 1929; 4 children. Demonstrator in biochemistry, Oxford, 1933; Rockefeller fellow, 1939; with Ministry Nat. Security, 1942, Air Ministry, 1943-45; prof. biochemistry U. Edinburgh. Mem. Biochem. Soc., Physiol. Soc., Royal Soc. Medicine, Royal Soc. Edinburgh. Author: Protein Metabolism, 1954, also articles. Home: Gracemount Farmhouse, Lasswade Rd., Edinburgh 9. Office: Biochemistry Dept., University of Edinburgh Medical School, Edinburgh 8, Scotland, Eng.

FISHER, Robert Joseph, Am. inventor; b. Athens, Tenn., Jan. 23, 1857; s. Richard M. and Ann M. (Gettys) F.; ed. E. Tenn. Wesleyan U.; M. Alice M. Gauche, June 9, 1892. Teller Cleveland (Tenn.) Nat. Bank, 1880-83; organized 1st Nat. Bank, Athens, Tenn., cashier, 1884-96. Recipient John Scott medal City of Phila. on recommendation of Franklin Inst. for meritorious invention, 1899. Inventor Fisher Book Typewriter; holder numerous patents. Died May 1, 1932.

FISHER, Robert Lloyd, Am. marine geologist; b. Alhambra, Cal., Aug. 19, 1925; s. Howard B. and Clara E. (Michalek) F.; B.S. in Geology, Cal. Inst. Tech., 1949; postgrad. Northwestern U.; M.S., U. Cal, 1952, Ph.D. in Oceanography, 1957; m. Shirley Chapman, Aug. 6, 1948; 1 son, Carlos A. With U. S. Geol. Survey, 1949; with Scripps Instn. Oceanography, La Jolla, Cal., 1950——, asso. research geologist, 1965——, dir. Indian Ocean program, 1960-65. Lectr., UNESCO Oceanography Program, Bombay, India, Waltair U., Visaknapatnam, India, 1966. Fellow Geol. Soc. Am.; mem. Seismol. Soc. Am., Am. Geophys. Union Challenger Soc., Sigma Xi. Research, publs. on ocean basin topography, bathymetric and structural studies of islands, seamounts, ridges and trenches; dir. geol.-geophys. expdns. off C.Am., S.Am., S.E. Pacific, Arctic Ocean, Central Indian Ocean and Indonesia, W. and S.W. Pacific Ocean. Home: 1350 Windsor Rd., Cardiff-by-the-Sea, Cal. 92007. Office: Scripps Instn. of Oceanography, La Jolla, Cal. 92037.*

FISHER, Robert Welles, Am. physician; b. Seaford, Del., Oct. 10, 1863; s. Isaac M. and Sarah J.

(Vaughan) F.; grad. Phila. Coll. Pharmacy, 1887; M.D., Jefferson Med. Coll., 1890; m. Margaret Van B. Terry, 1892; children—Vaughan Terry, Anna Bathsheba, Margaret Louise, Robert Welles. Practiced medicine, Salt Lake City from 1890; physician St. Mark's Hosp., 1893-1917; prof. materia medica and pharmacy U. Utah, 1899-1914; sec. State Bd. Med. Examiners, 1901-09; mem. Bd. Health, Salt Lake City, 1892-1916. Died Jan. 16, 1927.

FISHER, Sir Ronald Alymer, Brit. statistician; b. London, Feb. 17, 1890; s. George Fisher; B.A., Gonville and Caius Coll., Cambridge, 1912, M.A., 1920, Sc.D., 1926; m. Ruth Eileen Graltan Guinness, 1917; 8 children. Statistician, Merc. & Gen. Investment Co., 1913-15; pub. sch. tchr., 1915-19; statistician Rothamsted Exptl. Sta., Harpenden, 1919-33; Galton prof. eugenics Francis Galton lab. Univ. Coll., London, 1933-43; Arthur Balfour prof. genetics U. Cambridge, from 1943. Fellow Royal Soc., 1929 (Weldon medal 1929, Royal medal 1938), Royal Statis. Soc. (Guy medal in gold 1946), Am. Statis. Assn. (hon.). Author: Statistical Methods for Research Workers, 1925; The Genetic Theory of Natural Selection, 1930; The Design of Experiments, 1935; (with F. Yates) Statistical Tables, 1938; Theory of Spending, 1949; Contributions to Mathematical Statistic, 1950. Developed statis. techniques for analysis of variance, also for use and validation of small samples; invented term null hypothesis. Died 1962.

FISHER, Russell A(rden), Am. physicist; born Ludington, Mich., Sept. 11, 1904; s. Charles McKinley and Sarah Allen (Judge) F.; B.A., U. Mich., 1927, Ph.D., 1931; m. Helen Foster, Aug. 20, 1929. Instr. physics Northwestern U., 1931-32, asst. prof., 1932-37, asso. prof., 1937-42, prof., 1946—, chmn. dept. physics, 1950-57, acting chmn., 1960-61; tech. cons. govt. agys. Served with AUS, 1942-46. Decorated B.S.M., U. S.; Order British Empire. Mem. Am. Phys. Soc., Optical Soc. of Am., Phi Beta Kappa, Sigma Xi. Contbr. articles to tech. jours. Research in atomic and nuclear spectroscopy; determination and classification of atomic energy levels; hyperfine structure; Zeeman effect. Home: 810 Edgewood Lane, Glenview, Ill. Office: Physics Dept., Northwestern U., Evanston 60201.*

FISHER, Seymour, Am. psychologist; b. Balt., May 13, 1922; s. Sam and Jean (Miller) F.; student Wayne U., 1939-40; M.A., U. Chgo., 1942, Ph.D., 1948; m. Rhoda Feinberg, Mar. 22, 1947; children—Jerid Martin, Eve Phyllis. Pub. health fellow in clin. psychology Ill. Neuropsychiat. Inst., Chgo., 1945-47; chief psychologist Elgin (Ill.) State Hosp., 1949-51; research psychologist VA Hosp., Houston, 1952-56; USPHS career research investigator, asso. prof. Baylor Coll. Medicine, 1957-61; prof. Upstate Med. Center, State U. N.Y., Syracuse, 1961—. Mem. NIH Psychopharmacology Study Sect., 1956-60. Mem. Am., Eastern psychol. assns. Author: (with Sidney Cleveland) Body Image and Personality, 1958. Asso. editor Jour. of Cons. Psychology, 1965—. Research, numerous publs. on developing measures for ascertaining how individual perceives and experiences his own body; using such measures to investigate phenomena related to neurotic and schizophrenic disturbance, severe body disablement, ability to tolerate stress, and orgn. of personality. Home: 214 Brookford Rd., Syracuse 13324. Office: State U. Hosp., 750 E. Adams St., Syracuse, N.Y. 13210.*

FISHER, Seymour, Am. psychologist; b. N.Y.C., Nov. 4, 1925; s. George and Fannie (Hesselson) F.; B.A., N.Y. U., 1948; Ph.D., U. N.C., 1952; m. Carmen Eldridge, June 20, 1959; children—Mark Stephen, Andrew Jonathan. Psychology trainee VA, 1949-52; supervisory research psychologist Walter Reed Army Inst. Research, Washington, 1952-58; research psychologist Nat. Inst. Mental Health, NIH, Bethesda, Md., 1958-60, chief spl. studies unit Psychopharmacology Service Center, 1960-63; research prof. psychology, dir. psychopharmacology lab. Boston U. Sch. Medicine, Boston, 1963—. Cons. Nat. Inst. Mental Health, Office Naval Research, Washington, 1963—. Author: Child Research in Psychopharmacology, 1959; also articles. Research on exptl. hypnosis and persuasibility and social influence; improved methodology of clin. drug evaluations by studying how expectations and personality factors of patients help determine their response to drugs. Home: 78 Leeson Lane, Newton, Mass. 02159. Office: 80 E. Concord St., Boston 02118.*

FISHER, Theodore William, Am. entomologist; b. San Francisco, May 26, 1921; s. Carl M. and Winifred (Holm) F.; A.B., San Jose State Coll., 1943; Ph.D., U. Cal. at Berkeley, 1952; m. Elizabeth M. Mansfield, Sept. 15, 1946; 1 son, Peter Carl. Staff dept. biol. control U. Cal. at Riverside, 1948—, specialist, 1966—. Mem. Entomol. Soc. Am., A.A.A.S., Am. Inst. Biol. Sci., Pacific Coast Entomol. Soc., Ecol. Soc. Am., Sigma Xi. Research, publs. on insect biology, ecology, insectary facilities, biol. control mollusks, taxonomy of Sciomyzidae. Home: 26886 Ironwood Av., Sunnymead, Cal. 92388.*

FISHER, Walter Kenrick, Am. zoologist; b. Ossining, N.Y., Feb. 1, 1878; prof. zoology Stanford, 1925-43; author: Starfishes and Holothurians of Hawaii, 1906; Starfishes of Philippine Waters, 1919; others. Died 1953.

FISHMAN, Alfred Paul, Am. physician; b. N.Y.C., Sept. 24, 1918; s. Isaac and Anne (Tinter) F.; A.B., U. Mich., 1938, M.S., 1939; M.D., U. Louisville, 1943; m. Florence Howitz, Aug. 23, 1948; children—Mark Charles, Jay Alan. Dazian Found. fellow in pathology Mt. Sinai Hosp., N.Y.C., 1943-48; research fellow, established investigator Am. Heart Assn., Chgo. N.Y.C., Boston, 1949-55; asst. prof. medicine Columbia, N.Y.C., 1955-58, asso. prof., 1958-67, dir. Cardio-respiratory Lab., Columbia-Presbyn. Med. Center, 1955-67; prof. medicine U. Chgo., also dir. cardiovascular inst. Michael Reese Med. Center, 1967—. Cons. in medicine St. Mary's Hosp., Med. Sch., London, 1964; Commonwealth Fund fellow Nuffield Inst. for Med. Research, Oxford, Eng., 1964-65; Sir Ernest Finch prof. medicine U. Sheffield, Eng., 1965. Trustee Mt. Desert Island Biol. Lab., Salisbury Cove, Me. Mem. Harvey Soc., Am. Soc. Clin. Investigation, Assn. Am. Physicians, Royal Soc. Medicine, Am. Physiol. Soc., N.Y. Heart Assn. (pres. 1965-67, editor Symposia 1961-65, dir. 1966—). Author: (with D. W. Richards) Circulation of the Blood, 1965; also numerous articles. Editorial bds. Physiol. Revs., Circulation, Circulation Research, Medicina Thoracalis; chmn. editorial bd. Handbooks of Physiology, 1967—. Research in behavior of circulation of lungs in man and animals; analysis of relationships between heart and lungs in normal animals and man and in patients with disorders of heart and lungs. Home: 5825 S. Dorchester Av., Chgo. 60637.*

FISHMAN, Joshua Aaron, Am. social psychologist; b. Phila., July 18, 1926; s. Aaron S. and Sonia (Horwitz) F.; B.S., U. Pa., 1948, M.S., 1948; Ph.D., Columbia, 1953; m. Gella Jeanne Schweid, Dec. 23, 1951; children—M. Manuel, David E., Avrom. Research asso. Coll. Entrance Examination Bd., N.Y.C., 1955-56, dir. research, 1956-58, chmn. com. on research and devel., 1964-67; asso. prof. psychology and human relations U. Pa., Phila., 1958-60; prof. psychology, sociology, dean Grad. Sch. Edn., Yeshiva U., N.Y.C., 1960-66, Distinguished Research prof. social scis., 1966—. Mem. com. on sociolinguistics Social Sci. Research Council, 1964—; mem. research planning com. Yivo Inst. for Jewish Social Research, 1962—. Center for Advanced Study in Behavioral Scis. fellow, 1963-64; sr. specialist E.-W. Center, 1968-69. Mem. Am. Psychol. Assn., Am. Sociol. Assn., Am. Anthrop. Assn., Linguistic Soc. Am., Am. Ednl. Research Assn., Inter-Am. Psychology Soc. Author: Language Loyalty in the United States, 1966; Yiddish in America, 1965; Sociology of Language, 1968. also numerous articles. Research in bilingualism and lang. maintenance, sociology of lang. as a key to social orgn. in developing nations. Home: 3340 Bainbridge Av., Bronx, N.Y. 10463. Office: 55 Fifth Av., N.Y.C. 10003.*

FISHMAN, Robert Allen, Am. neurologist; b. N.Y.C., May 30, 1924; s. Samuel Benjamin and Miriam (Brinkin) F.; A.B., Columbia, 1944; M.D., U. Pa., 1947; m. Margery Ann Satz, Jan. 29, 1956; children—Mary Beth, Alice Ellen, Elizabeth Ann. Faculty, Columbia Coll. Physicians and Surgeons, N.Y.C., 1954-66, asso. prof. neurology, 1962-66; asst. attending neurologist N.Y. State Psychiat. Inst., 1955-66; asst. attending neurologist Neurol. Inst. Presbyn. Hosp., N.Y.C., 1955-61, asso., 1961-66; co-dir. Neurol. Clin. Research Center, Neurol. Inst., Columbia-Presbyn. Med. Center, 1961-66; prof. neurology, chmn. dept. U. Cal. Med. Center, San Francisco, 1966—. Nat. Multiple Sclerosis Soc. fellow, 1956-57; John and Mary R. Markle scholar in med. sci., 1960-65. Mem. Am. Neurol. Assn., Am. Fedn. for Clin. Research, Harvey Soc., Assn. for Research in Nervous and Mental Diseases, Am. Acad. Neurology, N.Y. Neurol. Soc., A.A.A.S., Am. Epilepsy Soc., N.Y. Acad. Scis., A.M.A. (sec. sect. on nervous and mental diseases 1964-67). Research, publs. in metabolic aspects of diseases of nervous system, physiology and biochemistry of cerebrospinal fluid and blood brain barrier in normal and path. states. Home: 26 Southridge E., Tiburon, Cal. 94920. Office: U. Cal. Med. Center, San Francisco 94122.*

FISHMAN, William Harold, oncologist; b. Winnipeg, Man., Can., Mar. 2, 1914; s. Abraham and Goldie (Chmelnitsky) F.; B.S., U. Sask. (Can.), 1935; Ph.D., U. Toronto (Can.), 1939; postgrad. U. Edinburgh (Scotland), 1940, Cornell, 1941; m. Lillian Waterman, Aug. 6, 1939; children—Joel, Nina, Daniel. Came to U. S., 1941, naturalized, 1947. Teaching fellow U. Toronto, 1936-39; Royal Soc. Can. research fellow, 1939-41; faculty Bowman-Gray Sch. Medicine, 1941-45, U Chgo., 1945-48; faculty Tufts U., 1948—, research prof. oncology, 1961—; dir. cancer research New Eng. Med. Center Hosps., Boston, 1958—; Cons. Lemuel Shattuck, VA hosps., NIH; lectr. various instns. Recipient Travel award NSF, 1959; Research Career award NIH, 1962; Fedn. fellow 17th Internat. Physiol. Congress, Oxford, Eng., 1947; Ciba Conf. fellow, 1950. Fellow A.A.A.S.; mem. Am. Soc. Biol. Chemists, Am. Chem. Soc., Soc. Exptl. Biology and Medicine, N.Y. Acad. Scis., Am. Assn. Cancer Research, Am. Fedn. Clin. Investigation, Histochem. Soc., Am. Soc. Exptl. Pathology, Endocrine Soc., Laurentian Hormone Conf., Royal Med. Soc., Am. Soc. Cell Biology, Biochem. Soc., Am. Inst. Biol. Scis., Sigma Xi. Author: Chemistry of Drug Metabolism, 1961. Editor: (with F. Homburger) Physiopathology of Cancer, 1953; regional editor Enzymologia. Research, publs., originator of methods of purification and measurement of beta-glucuronidase and the specific isoenzymes, prostatic acid phosphatase and intestinal alkaline phosphatase; developer enzymorphology as a means of gauging location and degree of activity of a specific enzyme in intact cells and tissues; applied these to problems of diagnosis, pathogenesis and mgmt. of human cancer. Home: 56 Mason Terrace, Brookline, Mass. 02146. Office: 30 Bennet St., Boston 02111.*

FISK, J(ames) B(rown), Am. physicist; b. West Warwick, R.I., Aug. 30, 1910; s. Henry James and Bertha (Brown) F.; B.S., Mass. Inst. of Tech., 1931, Ph.D., 1935; M.A. (hon.), Harvard, 1947; D.Sc., Carnegie Inst. Tech., 1956, Williams Coll., 1958, Newark Coll. Engring., 1959, Columbia, 1960, Colby Coll., 1962, N.Y.U., 1963, Rutgers U., 1967; D.Engring., U. Mich., 1963, U. Akron, 1963; LL.D., Lehigh U., 1967, Ill. Inst. Tech., 1968; m. Cynthia Hoar, June 10, 1938; children—Samuel, Zachary, Charles. Research asst. aero. engring. Mass. Inst. Tech., 1931; Proctor travelling fellow Trinity Coll., Cambridge, Eng., 1932-34; teaching fellow, physics Mass. Inst. of Tech., 1934-36; Soc. of Fellows, Harvard, 1936-38; asso. prof. physics, U. N.C., 1939; electronics research engr., asst. dir., phys. research, Bell Telephone Labs., 1939-47, dir., 1949-51, dir. research, phys. scis., 1952-54, v.p. research, 1954-55, exec. v.p., 1955-58, dir., 1955—, pres. Bell Telephone Labs., 1959—; dir. Am. Cyanamid Co., Equitable Life Assurance Soc., Sandia Corp., Trust Co. Nat. Bank, Cummins Engine Co.; Gordon McKay prof. applied physics Harvard, 1947-49, sr. fellow, Soc. Fellows, 1949; dir. div. research AEC, 1947-48, mem. gen. adv. com., 1952-58; mem. President's Sci. Adv. Com., 1957-60, cons., 1960—. Mem. Mass. Inst. Tech. Corp.; trustee John Simon Guggenheim Meml. Found.; pres. bd. overseers Found. for Advancement Grad. Study in Engring., Newark Coll. Engring.; mem. bd. of overseers Harvard, 1961-67. Recipient Indsl. Research Inst. Medal, 1963. Presidential certificate of Merit, World War II, Fellow Am. Phys. Soc. Am. Acad. Arts and Scis., I.E.E.E.; mem. Nat. Acad. Scis., Nat. Acad. Engring., Am. Philos. Soc., Tau Beta Pi, Sigma Xi. Asso. editor Phys. Rev., 1945-48. Contbr. articles to profl. periodicals. Supervised devel. of microwave magnetron for use in high frequency radar; devised one of 1st complete math. computations for an atomic pile (with Shockley); govt. adviser on defense and disarmament; other work on internal conversion of gamma-rays, scattering problems. Home: Lee's Hill Rd., Basking Ridge, N.J. 07920. Office: Bell Telephone Labs., Murray Hill, N.J. 07974.*

FISKE, Bradley Allen, Am. inventor, naval officer; b. Lyons, N.Y., June 13, 1854; s. William Allen and Susan Matthews (Bradley) F.; grad. U. S. Naval Acad., 1874; LL.D., U. Mich., 1921; m. Josephine Harper, Feb. 15, 1882; 1 dau., Caroline Harper. Commd. ensign USN, 1875, advanced through grades to rear adm., 1911, aid for operations, 1913-15. Mem. U. S. Naval Inst. (Gold medal 1905, pres. 1911-23). Recipient Cresson medal Franklin Inst., 1893; Gold medal Aero Club Am., 1919. Author: Electricity in Theory and Practice, 1883; War Time in Manila, 1911; The Navy as a Fighting Machine, 1917; From Midshipman to Rear Admiral, 1919; The Art of Fighting, 1920; Invention, the Master Key to Progress, 1921. Patentee, gun dir. system and naval telescope sight, 1890; inventor elec. range finder, stadimeter, 1889; radio controller of moving vehicles, 1900; torpedo plane, torpedo capable of being launched from airplane, 1912; reading machine, 1921, others. Died Apr. 7, 1942.

FISKE, Cyrus Hartwell, Am. biochemist; b. St. Paul, Sept. 22, 1890; s. Frederick William and Isabella Tiffany (Hartwell) F.; grad. St. Paul Acad., 1906; A.B., U. Minn., 1910; M.D., Harvard, 1914; m. Josephine K. Bissman, Sept. 22, 1917; children—Catherine (Mrs. James Higgins), Robert Hartwell. Asso. biochemistry Western Res. U. Med. Sch., 1915-17, asst. prof., 1917-18; faculty Harvard Med. Sch., 1918—, prof. biochemistry, 1935-57, prof. emeritus, 1957—. Fellow Acad. Arts and Scis., Am. Inst. Chemists, A.A.A.S.; mem. Am. Soc. Biol. Chemists (past treas.), Am. Chem. Soc., Sigma Xi, Nu Sigma Nu, Alpha Omega Alpha. Contbr. articles to tech. jours. Discovered phophocreatine and adenosine triphosphate, and presence free alpha-glycerophosphate in liver; pioneered treatment pernicious anemia with parenteral liver extract; research on analytical methods and various aspects animal metabolism. Home: 1155 Massachusetts Av., Lexington, Mass. 02173.*

FISKE, Milan Derbyshire, Am. physicist; b. Sharon, Wis., Nov. 15, 1914; s. Rollin H. and Minnie (Derbyshire) F.; B.S., Beloit Coll., 1937; Ph.D., U. Wis., 1941; m. Kathryn H. Barrett, June 29, 1936; children—Robert K., M. Lawrence, Marilyn S. With Gen. Electric Research Lab., Schenedtady, 1941—; trustee, sec. Am. Mus. Electricity, 1962—. Chmn., Internat. Conf. on Electronic Properties of Metals at Low Temperatures, 1958. Recipient Beloit Coll. Distinguished Service award, 1962. Mem. Am. Phys. Soc., I.E.E.E., A.A.A.S., Phi Beta Kappa, Sigma Xi. Research, publs. in microwave radar duplex switches, low temperature physics, especially superconductivity. Patentee in field. Home: 215 Lakehill Rd., Burnt Hills, N.Y. 12027. Office: P. O. Box 8, Schenectady 12301.*

FISKE, Thomas Scott, Am. mathematician; b. N.Y., May 12, 1865; s. Thomas Scott and Clara (Pittman) F.; A.B., Columbia, 1885, A.M., 1886, Ph.D., 1888;

m. Natalie Page, Feb. 1, 1913; 1 dau., Natalie Page. Fellow asst. in math. Columbia, 1885-88, tutor, 1888-91, instr., 1891-94, adj. prof., 1894-97, prof., 1897-1935, emeritus, 1935; acting dean Barnard Coll., 1899; sec. Coll. Entrance Exam. Bd., 1902-36. Fellow A.A.A.S.; mem. Am. Math. Soc. (founder; 1st sec., 1888-95; editor Bull., 1891-99, Trans., 1899-1905; pres. 1902-04), Math. Assn. Am., London Math. Soc., Assn. Tchrs. Math. of Middle States and Md. (pres. 1905-06). Author: Theory of Functions of Complex Variable, 1906; also math. and ednl. papers and articles in periodicals. Died Jan. 10, 1944.

FISON, Alfred Henry, Brit. astronomer; b. Hendon, Middlesex, Eng., 1857; s. Thomas Fison; ed. Royal Sch. Mines; D.Sc., London; m. Alice Maude Williamson; 1888; 1 son, 2 daus. Math. master at 2 pvt. schs., 1879-84; demonstrator physics Univ. Coll., London, 1884-92; lectr. Oxford U. Extension Delegacy, 1892-1912; sec. Gilchrist Ednl. Trust, from 1912; became lectr. on physics Guy's Hosp., 1906, London Hosp., 1910. Author: Recent Advances in Astronomy, 1898; A Text Book of Practical Physics, 1911, 1922; also article on evolution of double stars in Lectures on the Methods of Science, 1905. Died Feb. 4, 1923.

FISTER, George Morgan, Am. physician; b. Logan, Utah, May 22, 1892; s. George and Jennie (Morgan) F.; B.S., Utah State U., 1913, U. Chgo., 1916; M.D., Rush Med. Coll., 1918; LL.D., U. Utah, 1964; m. Ruby Leona Ostler, Sept. 23, 1914; children—Franklin George, Mary F. (Mrs. W. K. Martin). Practice medicine, specializing in urology, Ogden, Utah, 1928—; mem. staffs Thomas D. Dee Meml., St. Benedict's hosps.; faculty U. Utah Coll. Medicine. Civilian cons. Surgeon Gen. USAF, 1961-65. Recipient Pub. Service citation U. Chgo. Alumni, 1941. Mem. Am. (pres. 1961-62), Utah (past pres.), Korean (hon.) med. assns. Research, publs. on cystitis, prostatic cancer, microscopic urine studies, tobacco mosaic virus, socio-econ. problem medicine. Address: 3359 Taylor Av., Ogden, Utah 84403.*

FITCH, Frank Wesley, Am. pathologist; b. Bushnell, Ill., May 30, 1929; s. Harold Wayne and Mary Gladys (Frank) F.; M.D., U. Chgo., 1953, S.M., 1957, Ph.D., 1960; m. Shirley Dobbins, Dec. 23, 1951; children—Mary Margaret, Mark Howard. USPHS postdoctoral research fellow, 1954-55, 57-58; faculty U. Chgo., 1957—, prof. pathology, 1967—; Commonwealth Fund fellow U. Lausanne (Switzerland) Institut de Biochimie, 1965-66. Markle Found. scholar, 1961-66; recipient Borden Undergrad. Research award, 1953; Lederle Med. Faculty award, 1958-61. Mem. Am. Assn. Immunologists, Am. Assn. Pathologists and Bacteriologists, Am. Soc. Exptl. Pathology, Chgo. Pathol. Soc., Radiation Research Soc., Reticuloendothelial Soc., Sigma Xi, Alpha Omega Alpha. Contbr. chpts. to books, numerous articles to profl. jours. Research in immunology and radiation pathology, cellular changes during antibody formation, homeostasis of antibody formation, immunological tolerance, biol. aspects of immune response. Home: 5449 Kenwood Av., Chgo. 60615.*

FITCH, Frederic Brenton, Am. logician, educator; b. Greenwich, Conn., Sept. 9, 1908; s. Ashbel Parmelee and Josephine Hoyt (Smith) F.; grad. The Hotchkiss Sch., 1927; B.A., Yale, 1931, Ph.D., 1934; postgrad. (Dupont Research fellow) U. Va.; m. Marguerite Bailey Rea, Sept. 9, 1933; children—Susan Howell (Mrs. William Bradley Price), Mary Hoyt (Mrs. Elwyn LaVerne Simons). Research asst. in philosophy and psychology Yale, 1936-37, faculty philosophy, 1937—, prof., 1951—, acting dir. grad. studies in philosophy, 1950-51, dir. grad. studies in philosophy, 1951-56. Cons. in logic IBM, Poughkeepsie, N.Y., 1956, Bell Telephone Labs., Murray Hill, N.J., 1957-63. Sterling fellow Yale, 1935-36; Guggenheim fellow, 1945-46; NSF grantee for research on foundations of logic, 1963, on consistency of set theory, 1965-66. Mem. Assn. for Symbolic Logic (pres. 1959-61), Am. Math. Soc., Am. Philos. Assn., Metaphys. Soc. Am. Author: (with others) Mathematico-Deductive Theory of Rote Learning, 1940; Symbolic Logic, 1952. Publs. on constrn. of various consistent systems of logic and math.; log. analysis of action, power, causation, knowledge, desire, value, obligation. Home: 307 Lawrence St., New Haven 06511.*

FITCH, John, Am. inventor; b. Windsor, Conn., Jan. 21, 1743; s. Joseph and Sarah (Shaler) F.; m. Lucy Roberts, Dec. 29, 1767; 1 child. Established brass shop, East Windsor, Conn., 1764; in charge of Trenton Gun Factory during Revolutionary War; surveyed lands along Ohio Valley and N.W. Territory, 1780-85; organized co. to acquire and exploit lands in N.W. Territory, 1782; invented steamboat successfully launched and operated, 1787; launched 60-foot steam paddle propelled boat used to carry passengers from Phila. to Burlington, N.J., 1788; received French and U. S. patents for steamboat, 1791; lost financial support through inefficient handling of financial affairs, even though he had perfected and constructed 4 steamboats. Died Bardstown, Ky., July 2, 1798.

FITCH, John Edgar, Am. marine biologist; b. San Diego, June 27, 1918; s. Edgar G. and Vena (Jorgensen) F.; A.B., San Diego State Coll., 1941; M.A., U. Cal. at Los Angeles, 1963; m. Frances Arline Ley, Aug. 18, 1942; children—John Edgar, Richard A., Janis C. Marine biologist, research dir. Cal. Fisheries Lab., Cal. Dept. Fish and Game, Terminal Island, 1946—. Mem. adv. bd. Sea of Cortez Marine Research Center, Catalina Island Sch. For Boys Marine Lab.; research asso. Los Angeles County Mus. Natural History, 1964—, Scripps Instn. Oceanography, La Jolla, 1966—. Mem. Am. Soc. Ichthyologists and Herpetologists, Am. Fisheries Soc., Am. Inst. Fishery Research Biologists, Am. Malacological Union, So. Cal. Acad. Sci. Author: Offshore Fishes of California, 1958. Editor: Cal. Fish and Game, Fish Bull. series, 1954-66; asso. editor: Transactions Am. Fisheries Soc., 1966—. Contbr. numerous articles to profl. jours. Described four species of fish new to sci. and two new genera; identified fossil fish faunas of Cal. from sagittae found in various Pliocene, Pleistocene deposits. Home: 2657 Averill Av., San Pedro, Cal. 90731. Office: 511 Tuna St., Terminal Island, Cal. 90731.*

FITCH, Val Logsdon, Am. physicist; b. Merriman, Neb., Mar. 10, 1923; s. Fred B. and Frances (Logsdon) F.; B.Engring., McGill U., 1948; Ph.D., Columbia, 1954; m. Elise Cunningham, June 11, 1949; children—John, Alan. Faculty, Princeton (N.J.), 1954—, prof. physics, 1960—. Sloane fellow, 1960-64. Fellow Am. Phys. Soc., Am. Acad. Arts and Scis., A.A.A.S.; mem. Nat. Acad. Scis. Co-discoverer of precise measurement of x-rays from mu-mesonic atoms, leading to precise measurement of nuclear radius; studies of weak interaction through K-meson decay. Home: 292 Hartley Av., Princeton, N.J. 08540.*

FITCH, William Edward, Am. physician; b. Burlington, N.C., May 29, 1867; s. William James and Mary Elizabeth (King) F.; M.D., Coll. Physicians and Surgeons, Balt., 1891; m. Minnie Crump, Oct. 5, 1892; children—Lucille, Elizabeth, William Edward. Specialist in diseases of metabolism, med. hydrology and dietotherapy; lectr. on principles of surgery. Fordham U. Sch. Medicine, 1907-09; attending physician Vanderbilt Clinic; attending gynecologist outpatient dept. Presbyn. Hosp.; asst. in surg. clinic St. Luke's Hosp.; med. dir. and cons. med. hydrologist at French Lick (Ind.) Springs, 1931; cons. med. hydrologist Crazy Hotel and Spa Mineral Wells, Tex. Served as editor Gaillard's Southern Medicine, 1900-09; editor Pediatrics, 1908-17; co-editor and pub. Am. Jour. Electrotherapeutics and Radiology, 1918-19. Mem. med. soc. states of Va., Ind., Am. Med. Editors and Authors' Assn., Med. Assn. Greater N.Y., Nat. Soc. Advancement Gastroenterology, Am. Soc. Balneology, Internat. Soc. Med. Hydrology (London). Author: Fitch's Medical Pocket Formulary, 1914; Dietotherapy, 3 vols., 1918; Mineral Waters of the U. S. and Greater American Spas. 1926; Diseases of Metabolism; The Battle of Alamance; also various writings on early history of N. C. Editor Gaillard's So. Medicine, 1900-09, Pediatrics, 1908-17; co-editor, pub. Am. Jour. Electrotherapeutics and Radiology, 1918-19. Died Sept. 12, 1949.

FITE, Wade Lanford, Am. physicist; b. Apperson, Okla., Oct. 4, 1925; s. Luther Sherman and Floe Ellen (Blue) F.; A.B., U. Kan., 1947; M.A., Harvard, 1949, Ph.D., 1951; m. Ruby Rose Kauffman, Aug. 24, 1947; children—Christopher Sherman, John C., Andrew J., Rebecca Ellen. Research physicist Philco Corp., Phila., 1951-52; instr. physics U. Pa., Phila., 1952-54; NSF post-doctoral fellow Univ. Coll., London, Eng., 1954-55; staff mem. Gen. Atomic div. Gen. Dynamics Corp., San Diego, 1956-63; prof. physics U. Pitts., 1963—. Cons. Nat. Bur. Standards, Washington, 1953, Army Biol. Warfare Lab., Frederick, Md., 1956-58, Inst. Def. Analyses, Washington, 1965—. Fellow Am. Phys. Soc. (chmn. div. electron physics (1965), mem. A.A.A.S., Am. Rocket Soc., Am. Fedn. Musicians, Phi Beta Kappa, Sigma Xi, Pi Mu Epsilon. Research on atomic collision phenomena especially collisions involving free hydrogen atoms; application of atomic processes to instrumentation, controlled thermonuclear research and upper atmosphere physics. Home: 305 Pasadena Dr., Pitts. 15215.*

FITTIG, Rudolph, German chemist; b. Hamburg, Germany, Dec. 6, 1835; s. Johannes Andreas and Anne Catherine Rebecca (Spanhacke) F.; d.Phil. (under Wöhler and Limpricht), Göttingen, Germany, 1858; m. Wilhelmine Dorothea Ehlers, 1864; 3 sons, 3 daus. Became asst. lectr., Göttingen, 1860, asst. prof., 1886; prof., Tübingen, Germany, 1870-76; succeeded Baeyer as prof., Strasbourg (now in France), 1876. Recipient Davy medal Royal Soc. London, 1906. Author: Grundriss der unorganischen Chemie, 1871; also rewrote Wöhler's Grundriss der organischen Chemise, 10th edit., 1877. An editor Annalen der Chemie, 1895-1910. Contbr. numerous papers on synthetic organic chemistry. A founder of organic chemistry; worked on synthesis of lactones; discovered pinacone reaction, 1858; synthesized (with Tollens) toluene, 1864; discovered diphenyl phenanthrene, 1872, coumarone, 1882; introduced Wurtz-Fittig reaction; established the constitution of piperine; investigated structure of napthalene. Died Strasbourg, Nov. 19, 1910.

FITTING, Johannes Theodor Gastav Ernst, German botanist; b. Halle, Germany, Apr. 23, 1877; Dr. Med. Honoris causa, 1950. Became prof. botany U. Strasbourg, 1908, then U. Halle, 1910; apptd. dir. Strasbourg Bot. Inst., 1911, then at Bonn, 1912. Research on rubber-yielding plants, orchids, plant metabolism, also on retention of water by desert plants, assimilation of salts and other substances by living cells.

FITTON, William Henry, geologist; b. Dublin, Jan. 1780; B.A., Trinity Coll., Dublin, 1799; postgrad. U. Edinburgh, 1808; M.D., Cambridge, 1816; m. Miss James, 1820; 5 sons, 3 daus. Went to London to study medicine and chemistry, 1809; began practice medicine, Northampton, Eng., 1812; gave up practice medicine to study geology, 1820. Fellow Royal Soc., 1815; mem. Geol. Soc. (sec. for several years, became pres. 1828, founder Proc.); mem. Linnean Soc., Astron. Soc. Recipient Wollaston medal Geol. Soc., 1852. Determined heights of principal mountains of Ireland, before 1807; listed succession of strata between oolite and chalk, 1824-36. Died London, May 13, 1861.

FITTS, Donald Dennis, Am. chemist; b. Concord, N.H., Sept. 3, 1932; s. Russell P. and Elisabeth (Reille) F.; A.B., Harvard, 1954; Ph.D., Yale, 1957; m. Beverly Hoffman, July 11, 1964; 1 son, Robert K. NSF postdoctoral fellow U. Amsterdam, Netherlands, 1957-58; research fellow Yale, 1958-59; faculty U. Pa., Phila., 1959—, asso. prof. chemistry, 1964—, asst. chmn. dept., 1965—. Cons. Am. Cyanamid Co., 1959-63. Mem. Am. Phys. Soc., Am. Chem. Soc., Assn. Harvard Chemists, Faraday Soc., Yale Chemists Assn., Sigma Xi. Author: Nonequilibrium Thermodynamics, 1962; also articles. Research on theory of optical activity, statis.-mech. theory of transport processes, nonequilibrium thermodynamics, molecular quantum mechanics. Home: 634 Revere Rd., Merion Station, Pa. 19066.*

FITZ, Reginald Heber, Am. physician; b. Chelsea, Mass., May 5, 1843; s. Albert and Eliza Roberts (Nye) F.; A.B., Harvard, 1864, A.M., 1867, M.D., 1868 (LL.D., 1905); m. Elizabeth L. Clarke, June 12, 1879. Instr. pathol. anatomy, 1870-73, asst. prof., 1873-78, prof., 1878-79, Shattuck prof. same, 1879-92, Hersey prof. theory and practice of physic, 1892-1908, emeritus prof., from 1908. Harvard; physician Boston Dispensary, 1871-82, Mass. Gen. Hosp., 1887-1908. Fellow Am. Acad. Arts and Sciences. Author: Perforating Inflammation of the Vermiform Appendix, 1886; (with Dr. Horatio C. Wood) The Practice of Medicine, 1897. Research on appendicitis (named by him), particularly in its relation to peritonitis, also suggested operative treatment of appendicitis; described pancreatitis, 1889; 1st to introduce micros. study of diseased tissue. Died Sept. 30, 1913.

FITZGERALD, George Francis, Irish physicist; b. Dublin, Ireland, Aug. 3, 1851; s. William and Ann Frances (Stoney) FitzG.; B.A., Trinity Coll., Dublin, 1871; m. Harriette Mary Jellett; children—3 sons, 5 daus. Tutor, Trinity Coll., 1877-81, Erasmus Smith prof. natural and exptl. philosophy, 1881-1901; registrar Sch. Engring., Dublin U., from 1886; examiner in exptl. sci. London (Eng.) U., 1888. Fellow Royal Soc., 1883 (recipient medal, 1899); mem. Phys. Soc. London (pres. 1892-93), Inst. Elec. Engrs., Royal Dublin Soc. (hon. sec. 1881-89). Author: On the Rotation of the Plane of Polarisation of Light by Reflection from the Pole of a Magnet, 1876; On the Electromagnetic Theory of the Reflection and Refraction of Light, 1880; On the Possibility of Originating Wave Disturbances in Ether by means of Electric Forces; On the Superficial Tension of Fluids and its Possible Relation to Muscular Contractions; On the Energy Transferred to the Ether by a Variable Current; On an Analogy between Electric and Thermal Phenomena, 1884; On a Model illustrating some Properties of the Ether, 1885; On the Structure of Mechanical Modes illustrating some of the Properties of the Ether, 1885; Note on the Specific Heat of the Ether, 1885; On the Limits to the Velocity of Motion in the Working Parts of Engines, 1886; On the Thermo-dynamic Properties of a Substance whose Intrinsic Equation is a Linear Function of the Pressure and Temperature, 1887. Research electric waves, electrolysis; developed electromagnetic theory of radiation; formulated theory of change of shape of a body due to motion through ether; 1st to propose that head of comet consists of large stones and the tail, small ones. Died Dublin, Feb. 22, 1901.

FITZGERALD, George Patrick, Am. botanist; b. Milw., Sept. 22, 1922; s. Maurice James and Cecelia (Madden) F.; B.S., U. Wis., 1948, M.S., 1949, Ph.D., 1950; m. Charlotte Mae Dickson, May 25, 1943; children—Patricia Mae, Dennis Dickson. With U. Wis., Madison, 1950—, research asso. Engring. Expt. Sta., 1957—. Cons. chem. industry algicide evaluations N.Y. State Bd. Health; cons. chem. industry on nutrition. Mem. Am. Soc. Microbiology, Phycol. Soc. Am., Internat. Phycol. Soc., Weed Soc. Am., Water Pollution Control Fedn. Research, numerous publs. on culture and nutrition of algae, techniques for determining surplus or limited phosphorus supply for algae, relationships between algae and water pollution, evaluations of factors affecting applications of algicides and bactericides. Home: 2018 Dickson Pl., Madison 53713. Office: Water Chemistry Lab., U. Wis., Madison, Wis. 53706.*

FITZGERALD, John Gerald, Canadian physician; b. Drayton, Ont., Can., Dec. 9, 1882; s. William and Alice (Woollatt) F.; M.B., U. Toronto, 1903, M.D., 1920, LL.D., 1925; m. Edna Leonard, 1910; 1 son, 1 dau. Intern Buffalo State Hosp., 1904-05; clin. asst. Sheppard Hosp., Balt., 1906-06; pathologist, clin. dir.

Toronto Hosp. for Insane, also demonstrator in psychiatry U. Toronto, 1907-08; research student Harvard, 1908-09; lectr. bacteriology U. Toronto, 1909-11, prof. hygiene and preventive medicine, dir. Connaught labs. and sch. hygiene, from 1913, dean faculty medicine, 1932-36; research student Pasteur Inst., Paris, also Brussels, 1910, U. Freiburg 1911; asso. prof. bacteriology U. Cal., 1911-13. Fellow Royal Soc. Can. Author: Practice of Preventive Medicine; also papers on bacteriol. topics. Died June 20, 1940.

FITZGERALD, Keane, engr. Fellow Royal Soc., 1756. Attempted to lessen friction in steam engine by changing back and forth movement to rotation using system of safety catches and with addition of fly wheels; inventor metal thermometer. Died 1782.

FITZ GERALD, Leslie Maurice, Am. oral surgeon; b. Cresco, Ia., Aug. 18, 1898; s. Edward A. and Emma (Daly) Fitz G.; D.D.S., U. Ia., 1919; D.Sc., Loyola U., 1954, Temple U., 1957; m. Marcelle Meis, Oct. 8, 1921; children—Shirley Ann (Mrs. F. D. Gilloon, Jr.), Patricia (Mrs. J. A. O'Brien, III), Jacqueline (Mrs. F. Benjamin Merritt II). Instr. U. Ia., 1919-20; resident oral surgery U. Ia. Hosp., 1919-20; pvt. practice of oral surgery, Dubuque, 1920—; chief oral surgery St. Joseph, Finley, Xavier hosps.; cons. oral surgery Central Office V.A., also Surgeon Gen. U.S.N.; pres. Dubuque Thrift Plan Indsl. Bank, 1963—; dir. 1st Nat. Bank, Dubuque Savs. & Loan. Chmn. dental advisory com. Am.-Korean Found., 1954. Recipient Arnold K. Maislen Meml. award, N.Y. U., 1958. Co-founder, and diplomate, sec. Am. Bd. Oral Surgery. Fellow Am. Coll. Dentists (chmn. com. oral surgery), Internat. Coll. Anesthetists, Acad. Internat. Dentistry, Internat. Coll. Dentists; hon. mem. Alaska, Hawaii, Puerto Rico Dental Socs.; mem. Nat. Inter-Assn. Council on Health, 1954; mem. Am. Dental Assn. (pres. 1953-54, mem. editorial bd. Jour. Oral Surgery 1950-56), Am. Soc. Oral Surgeons (pres. 1941-42, 42-43), Fedn. Dentaire Internat., Omicron Kappa Upsilon, Xi Psi Phi. (pres. 1930-31). Contbr. many articles including Oral Surgery for General Practitioner, 1932, Fundamental Principles of Oral Surgery, 1950, to dental and med. publs. Home: Rolling Ridge Farm. Office: Roshek Bldg., Dubuque, Ia.

FITZGERALD, Mark James, Am. educator, economist; b. Olean, N.Y., May 28, 1906; s. Edward William and Helen (York) F.; student St. Bonaventure's, 1925-26; A.B., U. Notre Dame, 1928; M.B.A., Harvard, 1931; postgrad. Holy Cross Coll.; Ph.D., U. Chgo., 1950. Joined Congregation of Holy Cross, 1933, ordained priest Roman Catholic Ch., 1940; faculty U. Notre Dame (Ind.), 1940—, prof. econs., 1956—. Mem. arbitration panels Fed. Mediation and Conciliation Service, 1955—; pres. Holy Cross Ednl. Conf., 1955. Mem. Air Pollution Control Assn., Wilderness Soc., Am. Econ. Assn., Indsl. Relations Research Assn., Cath. Econ. Assn. (pres. 1957), Cath. Assn. Internat. Peace (v.p. 1958-64), Nat. Acad. Arbitrators, Am. Arbitration Assn. Author: Britain Views our Industrial Relations, 1955; The Common Market's Labor Programs, 1966. Research on effect of labor relations on productivity in Gt. Britain and U. S., contbns. of common market to higher working and living standards for labor forces. Home: Corby Hall, U. Notre Dame, Notre Dame, Ind. 46556.*

FITZGERALD, Oliver, Irish physician; b. Waterford, Ireland, Dec. 23, 1910; s. Alexis and Elizabeth (O'Halloran) F.; M.D., U. Coll., Dublin, Ireland, 1941, M.Sc., 1937; postgrad. U. Basel (Switzerland), Cambridge (Eng.) U.; M.R.C.P., London, 1967; m. Cliodna Maiben, Aug. 2, 1944; children—Alexis, Barbara, Cliodna, Nesta, Oliver, Ann. Asst. physician St. Vincents Hosp., Dublin, 1940-49, physician, 1949—; faculty U. Coll., Dublin, 1952—, prof. therapeutics, 1958—. Mem. Med. Research Council Ireland, 1957—; chmn. Irish Nat. Drugs Adv. Bd., 1966—. Mem. Assn. Physicians Gt. Britain and Ireland, Physiol. Soc., Brit., Irish socs. gastroenterology, Royal Acad. Medicine Ireland, Royal Soc. Medicine London. Research, publs. on intestinal absorption, gastric urease function, pancreatic function in health and disease, homocystinuria. Home: 16 Clyde Rd., Dublin, Ireland.*

FITZGERALD, Robert James, Am. microbiologist; b. N.Y.C., Nov. 3, 1918; s. Maurice Edward and Anna (Ledogar) F.; B.S., Fordham U., 1939; M.S., Va. Poly. Inst., 1941; Ph.D., Duke, 1948; m. Dorothea Babbitt, June 20, 1945. Research microbiologist chemotherapy div. Am. Cyanamid Co., Stamford, Conn., 1941-45; research fellow Duke, 1946-48; USPHS officer, scientist dir. Nat. Inst. Dental Research NIH, Bethesda, Md., 1948—. Vis. investigator Karolinska Inst., Stockholm, Sweden, 1962; vis. prof. Microbiology Inst., U. Rio de Janeiro, Brazil, 1966; cons. Pan Am. Health Orgn., 1966—. Recipient Research award Chgo. Dental Soc., 1960; Quinquennial Research award Fedn. Dentaire Internationale, Cologne, Germany, 1962. Mem. Soc. Am. Microbiologists, Am. Acad. Microbiology, Internat. Assn. Dental Research, A.A.A.S. Contbr. numerous articles to profl. jours. Specialist in germfree animal research; demonstrated causative role of specific streptococci in dental caries. Home: 6623 Sulky Lane, Rockville, Md. 20852. Office: Nat. Inst. Dental Research, Bethesda, Md. 20014.*

FITZGERALD, Sir Thomas Naghten, surgeon; b. Tullamore, Ireland, Aug. 1, 1838; s. John Fitzgerald; ed. St. Mary's Coll., Kingston, also Mercer's Hosp., Dublin, Ireland; m. Margaret Robertson, 1870; 3 daus. Commd. in army med. staff; went to Australia, 1858; house surgeon, sr. surgeon Melbourne (Australia) Hosp., 1858-1860, maintained pvt. practice 1860-1862; cons. surgeon St. Vincent Hosp., Melbourne; tchr. Melbourne Med. Sch. Licentiate, fellow Royal Coll. Surgeons Ireland; pres. Med. Soc. Victoria, 1883, 89, Inter-colonial Med. Congress, 1889. Author: Contbd. papers on surg. subjects, including fractured patella, cleft palate, club foot, drilling in bone deformations, to med. jours. Died July 8, 1908.

FITZ-JAMES, Philip Chester, Canadian chemist, microbiologist; b. Vancouver, B.C., Can., Nov. 26, 1920; s. Harold Chester and Gladys (Shaw) Fitz-James; B.S.A., U. B.C., 1943; M.S.A., U. Toronto, 1945; M.D., U. Western Ont., 1949, Ph.D., 1953; m. Monic Allcut, Dec. 17, 1947; children—Meriel, Michael, Bronwyn, Athlyn. Research asst. U. Toronto, 1943-44; research asst. U. Western Ont., 1947-48; faculty U. Western Ont., 1953—, asso. prof. dept. bacteriology, lectr. dept. biochemistry, 1960-67, prof., 1967—. Recipient Royal Soc. Can. Harrison prize, 1963. Mem. Biochem. Soc., Burton Soc. Contbr. articles to sci. jours. Devel. pilot plant for prodn. of penicillin; role of stratification in producing precipitates in gall bladder; study of substances causing foaming of human urine; analysis of phosphorous compounds of bacterial spores; spore structure; elucidation (with Young) of mode of spore formation using electron microscopy and chemistry; observed membranous organelle in bacteria, supplied name "mesosome" and demonstrated their role in membrane synthesis. Home: 17 Northdale St. Office: Dept. Bacteriology Med. Scis. Center, U. Western Ont., London, Ont., Can.*

FITZROY, Robert, English meteorologist; b. Ampton Hall, Suffolk, Eng., July 5, 1805; s. Charles Fitzroy; ed. Royal Naval Coll.; m. Mary Henrietta O'Brien, Dec. 1836; children include Robert. Entered Royal Navy, 1819, advanced through grades to vice adm., 1863; mem. Parliament for Durham, 1842; apptd. gov. New Zealand, 1843; named acting supt. of dockyard, Woolwich, 1848; chief of meteorol. dept. Bd. of Trade, 1854; naval commands included that of Beagle along the Straits of Magellan and coasts of S.Am., with Charles Darwin aboard as naturalist. Fellow Royal Soc., 1851; mem. French Acad. Scis., 1863. Author: Narrative of the Surveying Voyages of H.M. Ships Adventure and Beagle between the years 1826-1836, 3 vols., 1839; Remarks on New Zealand, 1846; Weather Book, 1863. Inventor Fitzroy barometer; set up a storm warning system; simplified wind charts; fixed longitude of many secondary meridians, surveyed Straits of Magellan and S. American coast. Died London, Apr. 30, 1865.

FITZSIMONS, Frederick William, biologist; b. 1875; ed. Natal, Ireland; m.; 2 sons. Dir. Natal Mus., 10 years; creator Snake Park, Port Elizabeth, Union South Africa; dir. Port Elizabeth Mus., 1906-36; later with Fitzsimons' Snake Park and Lab., Durban, South Africa. Author: The Snakes of South Africa: their Venom and the Treatment of Snake Bites; The Natural History of South Africa, (Mammals), 4 vols., (Birds), 2 vols.; The House Fly; A Slayer of Men; The Monkey-folk of South Africa; Pythons and their Ways; Snakes and Treatment of Snake Bites; Ants and Their Ways; Snakes; Opening the Psychic Door; Squire: His Romance. Used detoxicated snake venoms to treat epilepsy. Died May 25, 1951.

FIXMAN, Marshall, Am. chemist; b. St. Louis, Sept. 21, 1930; s. Benjamin and Dorothy (Finkel) F.; A.B., Washington U., St. Louis, 1950; Ph.D., Mass. Inst. Tech., 1954; m. Marian Ruth Beatman, July 5, 1959; children—Laura Beth, Susan Ilene, Andrew Richard. Jewett postdoctoral fellow chemistry Yale, 1953-54; instr. chemistry Harvard, 1956-59; sr. fellow Mellon Inst., Pitts., 1959-61; faculty Inst. Theoretical Sci., U. Ore., 1961-65; prof. chemistry Yale, New Haven, 1965—. Recipient Gov.'s. award Ore. Mus. Sci. and Industry, 1964. Alfred P. Sloan Found. fellow, 1961-63. Mem. Am. Chem. Soc. (award pure chemistry 1964), Am. Phys. Soc., Fedn. Am. Scientists. Asso. editor Jour. Chem. Physics, 1962-64. Studies in statis. mechanics, particularly theory of macromolecules in solution. Address: Sterling Chemistry Lab., Yale, New Haven 06520.*

FIXMILLNER, Placidus, Austrian astronomer; b. Achleuthen, Austria, 1721; studied in Salzburg, Austria; joined Benedictine order, circa 1737; became dir. obs., tchr. canon law, dir. coll. Abbey Kremsmünster (Upper Austria); pub. several astron. works; 1st to compute orbit of Uranus; made observations of Mercury used by Lalande. Died at Abbey Kremsmünster, Aug. 27, 1791.

FIZEAU, Armand Hippolyte Louis, French physicist; b. Paris, Sept. 23, 1819; student Stanislas Coll., studied medicine, then became student at obs. and Coll. de France; m. dau. of Adrien de Jussieu. Worked with Foucault on physical studies, 1845-49; apptd. insp. physics Ecole Polytechnique, 1863; mem. Bur. Longitudes, French Acad. Scis. Decorated officer Legion of Honor; recipient Grand prize French Acad. Scis., 1856. Fellow Royal Soc., 1875 (Rumford medal 1866); mem. French Acad. Scis. (pres. 1878). Experimented (with Foucault) in optics, infra-red portion of the solar spectrum, regularity of light vibrations, identification of radiant heat and light; developed effective induction coil; investigated expansion of crystals; made 1st measurement of velocity of light in air (also in water, later) by expt. confined to earth's surface, 1849; showed how Doppler principle could be used to find star velocity in line of sight; conducted expts. to detect ether drift; studied aspects of photography; devised a method of increasing permanency of dagúerrotypes. Died Nanteuil, France, Sept. 18, 1896.

FJELDSTAD, Jonas Ekman, Norwegian oceanographer; b. Röyken, Norway, Nov. 22, 1894; s. Johan and Emma (Ekman) F.; Cand., U. Oslo (Norway), 1925, Dr.Philos., 1930; m. Ellen Totland, May 24, 1924. Amanuensis, Geophys. Inst., Bergen, Norway, 1922-38; faculty U. Oslo, 1938—, prof. oceanography, 1946-65, prof. emeritus, 1965—, dir. Oceanographical Inst., 1946-65. Decorated knight St. Olav Order; recipient Bergen prize, 1935; Nansen prize, 1945, Nansen prize of Fram Com., 1962; Mem. Norwegian Acad. Sci., Norwegian Math. Soc., Norwegian Geophys. Soc. Author: Contribution to the Dynamics of Free, Progressive Waves, 1929; Interne Wellen, 1933; Wärmeleitung im Meere, 1933; Results of Tidal Observations, 1936; Tidal Waves of Finite Amplitude. Research, publs. on tidal waves and currents, heat conduction in sea, especially internal tidal waves; developed 1st gen. theory of internal tidal waves, 1933; postulated a submarine barrier across the Polar basin, 1936. Home: 41, Carl Kjelsens vei, Oslo 8, Norway.*

FLACHSMEYER, Jürgen, German mathematician; b. Stettin, Germany, Mar. 31, 1935; s. Arnold and Martha (Wolff) F.; Diploma, Greifswald U., 1959; doctor's degree German Acad. Scis., 1960, Dr.habil., 1965; m. Gisela Erdmann, Nov. 6, 1959; children—Andrea, Roland. Asst., Inst. Pure Math., German Acad. Scis., Berlin, 1959-63, head asst., 1963-65, leader sci. investigations, 1965—; faculty U. Berlin, 1965, U. Greifswald, 1964, 66; guest Acad. Scis. USSR, 1962-63. Mem. Math. Soc. German Democratic Republic. Research and publs. on gen. topology, function spaces, extensions of topological spaces, especially relations between Boolean algebras and topology. Home: 87a, Dr.-Wilh.-Külz, Greifswald, DDR, Germany. Office: 6, Rudower Chaussee, Berlin-Adlershof, Inst. Math., German Acad. Scis.*

FLAD, Henry, engr., inventor; b. Rennhoff, Germany, July 30, 1824; s. Jacob and Franziska (Brunn) F.; grad. U. Munich (Germany), 1846; m. Helen Reichard, 1848; m. 2d, Caroline Reichard, Sept. 12, 1855 6; 3 children. Served as capt. co. of army engrs. in Bavaria during German Revolution, 1848; fled to U. S., 1849; ry. constrn. engr., 1849-60; served from pvt. to col. U. S. Army during Civil War; asst. to James B. Eads in constrn. Eads Bridge over Mississippi River; mem. bd. water commrs. St. Louis 1868-76 (during which time city water-works were completed); mem. Am. Soc. C.E., pres., 1866; founder Engrs. Club of St. Louis, pres. 1868-80; pres. St. Louis Bd. Pub. Improvements, 1877-90; mem. Mississippi River Commn., 1890-98; patented filters, water meters, methods of preserving timber and sprinkling streets, systems rapid transit and cable rys.; devised hydrostatic and hydraulic elevator, deep-sea sounding apparatus, pressure gauges, pile driver. Died Pitts., June 20, 1898.

FLAGG, Josiah Foster, Am. dentist, artist; b. Boston, Jan. 10, 1788; s. Josiah Flagg, Jr.; grad. Boston Med. Coll., 1815; m. Mary Wait, Oct. 18, 1818. Engraved plates for Dr. J. C. Warren's book Anatomical Description of the Arteries of the Human Body; designed extracting forceps adaptable to any size tooth; pioneer in use of porcelain for artificial teeth; founder Sch. of Design for Women, Boston. Died Dec. 20, 1853.

FLAHAULT, Charles-Henri-Marie, French botanist; b. Bailleul, France, Oct. 3, 1852; studied plant physiology under Bornet; prof. Faculty Scis., Strasbourg, France, then at Montpellier, France; mem. Acad. Agr., French Acad. Scis., 1904. Author: La distribution géographique des vegetaux dans un coin du Languedoc, 1893; Flore de la Camargue des alluvions du Rhone, 1894; (with Bornet) Révision des nostocacées hétérocystées, 1886-88. Studied plant anatomy and physiology, then specialized in cryptogams, geog. botany; pioneer in phytogeography; studied blue green algae, especially heterocystous nostocacea, also flora of Camargue and Rhône regions in France. Died Montpellier, Feb. 3, 1935.

FLAIANI, Giuseppe, Italian anatomist, physician, surgeon; b. Ascoli, Italy, 1741; doctorates in philosophy and medicine, Rome; founder anat. mus., surgeon, lithotomist Hosp. of Holy Spirit, Sassia, Italy; became head surgeon Holy Ghost Hosps., 1772; apptd. physician to Pope Pius VI, 1775. Author: Collezioni di osservazioni e riflessioni di chirurgia, 1798-1803. Gave 1st description of exophthalmic goiter (Flaiani's disease), 1802. Died 1808.

FLAKS, Joel George, Am. biochemist; b. N.Y.C., Oct. 20, 1927; s. Samuel and Celia (Escher) F.; B.A., Bklyn. Coll., 1950; Ph.D., U. Pa., 1957; m. J. Margot Hoffman, Dec. 24, 1961; children—Sarah E., Judith A. Asst. instr. U. Pa., Phila., 1951-53, faculty, 1957—, asso. prof., 1966—; research asst. Mass. Inst. Tech., 1953-55, instr., 1955-56. Damon Runyon Meml. fund fellow U. Pa., 1956-57; USPHS Career Devel.

award, 1961—; Lalor Found. fellow, 1960. Mem. Am. Soc. Biol. Chemists, Am. Chem. Soc., Am. Soc. for Microbiology, Genetics Soc. Am. Research, publs. on metabolic pathways of purines and pyrimidines, mode action antibiotics and antimetabolites, mechanisms and genetics of resistance to various antimetabolites, mechanism of bacterial virus reprdn. Home: 426 S. 47th St., Phila. 19143.*

FLAMAND, Georges-Barthélemy-Médéric, French geologist; b. Paris, France, 1861; lectr.; prof. phys. geography École supérieure, Algiers, Algeria. Author: De l'Oranie au Gourara, 1898; Recherches géologiques et géographiques sur le haut-oays de l'Oranie et sur le Sahara; Algérie et Territories du Sud, 1911. Research on rivers and waterways in No. Africa, especially geologic and geography of Oran plain, prehistory of Sahara. Died Algiers, 1919.

FLAMEL, Nicolas, alchemist; b. circa 1330. Acted as scrivener, until he began alchemical studies; spent 24 years attempting to make philosopher's stone; claimed expts. were successful, 1382; used wealth to build and restore chs., hosps. Author: Annotata quaedan ex Flamello; le Desir désiré; les Figures hieroglyphiques; Summarium Philosophicum. Died 1418.

FLAMM, Heinz, Austrian hygienist, med. microbiologist; b. Vienna, Austria, July 3, 1929; s. Maximilian and Gretl (Unger) F.; M.D., U. Vienna, 1953; m. Louise Hosnedl, Mar. 16, 1955; children—Heinz, Beatrix, Ulrike. Asst., U. Vienna Inst. Hygiene, 1952-59, asst. prof., 1959-65, asso. prof., 1965-66, prof., dir., 1966—. Mem. Austrian, German socs. microbiology and hygiene, Royal Soc. Tropical Medicine and Hygiene, Soc. Physicians Vienna, N.Y. Acad. Sci., Austrian, German socs. tropical medicine. Author: Die praenatalen Infektionen des Menschen, 1959. Research, publs. on med. microbiology, hygiene, prenatal infections. Home: 150-154 Gersthoferstrasse, 1180 Vienna. Office: 15 Kinderspitalgasse, 1095 Vienna, Austria.*

FLAMM, Oswald, German shipbuilder; b. Dusseldorf, Germany, July 30, 1861; s. Albert and Anna (Arnz) F.; diploma in marine engring. and marine mech. engring. Tech. U. Berlin, 1888, hon. dr. engring.; m. Worker in shipyards of E. Jerninghaus, Duisburg, Germany, Joseph L. Meyer, Papenburg, Germany, Blohme & Voss, Hamburg, Germany; became lectr. Tech. U. Berlin, 1892, asso. prof., 1894, prof. theory and ship design, 1897. Mem. Tech. Shipbldg. Soc. (co-founder), Fleet Assn. (co-founder). Editor Schiffbau, Schiffahrt und Hafenbau. Contbr. to jours. Improved screw propellor and ship stability; worked on submarine; suggested establishment of exptl. shipbldg. inst. on Schleuseninsel, Berlin, 1902. Died Berlin, June 12, 1935.

FLAMMARION, Nicolas Camille, French astronomer; b. Montigny-le-Roi, France, Feb. 26, 1842; ed. in theology; m. Sylvie Pétriaux-Hugo Mathieu, 1874 (dec. 1919); m. 2d, Gabrielle Reuandot, 1919. Apprentice engraver, 1856-58; student astronomer Paris Obs., 1858-62, astronomer, 1876-82; calculator Bur. Longitudes, 1862-76; founder, monthly mag., L'Astronomie, 1882; founder, dir. Juvisy Obs., 1883-1925. Decorated comdr. Legion of Honor. Mem. Soc. for Psychical Research (pres. London 1923), French Astron. Soc. (founder 1887). Author: Cosmogenie universelle; La pluralité des mondes habités, 1862; Voyages en ballon, 1870, 40th edit., 1890; L'atmosphère, 1871, English edit., 1873; Astronomie populaire, 1880, English edit., 1907; Les terres du Ciel, 1877; La planete Mars et ses conditions d'habitation, 1893; L'inconnu et les problèmes psychiques, 1900. Founder mag. L'astronomie, 1882. Made balloon ascents to study upper atmosphere, 1868-73; discovered common proper motion of widely separated stars; noted color changes in crater Pluto and attributed them to vegetation; research on double and multiple stars, color of stars, the moon, topography of planet Mars; attempted to develop psychic research into an exact science. Died Juvisy-sur-Orge, France, June 3, 1925.

FLAMSTEED, John, English astronomer; b. Denby, Eng., Aug. 19, 1646; s. Stephan and Mary (Spateman) F.; Created M.A., Jesus Coll., Cambridge, 1674; m. Margaret Cooke, Oct. 23, 1692. Began systematic observations with Townley's mensurator, 1671; ordained, 1675; named 1st astronomer royal, 1675; Royal Obs. built for him, Greenwich, 1676; Fellow Royal Soc., 1676. Author: Catalogue and Observations of Stars, Vol. 1, 1707; Historia coelestis, 1712; Historia coelestis Britannica (completed by his assistant; 1st accurate star catalogue), 3 vols., 1725. Determined absolute right ascensions through simultaneous observations of sun and a star nr. both equinoxes (laying found. of modern astronomy); provided Newton with results and findings, including data which Newton used for his Principia; invented conical projection in mapmaking; made barometer and thermometer for Charles II and Duke of York, 1674. Died Greenwich, Dec. 31, 1719.

FLANAGAN, Ted Benjamin, Am. chemist; b. Oakland, Cal., July 11, 1929; s. Benjamin J. and Gertrude (Organ) F.; B.S., U. Cal. at Berkeley, 1951; Ph.D., U. Wash., 1955; m. Joyce Ilene Bippes, May 28, 1955; children—Christopher Sean, Jennifer Cara. Asso. physicist Brookhaven Nat. Lab., Upton, N.Y., 1959-61; faculty U. Vt., Burlington, 1961—, asso. prof. chemistry, 1964—. Mem. Am. Chem. Soc., Am. Phys. Soc., Faraday Soc. Investigations of hydrogen absorption by palladium alloys; kinetics of decomposition of solids; radiation damage in solids. Home: 37 Cliff St., Burlington, Vt. 05401.*

FLANDERS, Harley, Am. mathematician; b. Chgo., Sept. 13, 1925; s. Mendel M. and Dorothy (Israel) F.; student Ill. Inst. Tech., 1943-44; S.B., U. Chgo., 1946, S.M., 1947, Ph.D., 1949; m. M. June White, Dec. 24, 1946; children—Daveed D., Zvi B. Bateman fellow Cal. Tech., 1949-51; with U. Cal., Berkeley, 1951-60, asso. prof. math., 1958-60; prof. math. Purdue U., Lafayette, Ind., 1960—. NSF fellow Cambridge U., Hebrew U., 1957-58; research asso. Univ. Coll., London, 1965-66. Mem. Math. Assn. Am. (chmn. No. Cal. sect. 1957, Ind. sect. 1963-64, 2d v.p. 1959-61), Am., London math. socs. Research in algebra, differential geometry. Author: Differential Forms with Applications to the Physical Sciences, 1963. Asso. editor Trans. Am. Math. Soc., 1958-62; asso. editor Am. Math. Monthly, 1966-68, editor-in-chief, 1969—. Office: Math. Dept., Purdue U., Lafayette, Ind.*

FLANDRIN, Pierre, French veterinarian; b. Lyons, France, Sept. 12, 1752; ed. Vet. Sch. Lyons, 1766. Prof. anatomy; dir. Alfort Vet. Sch.; adminstr. for vet. schs. France, from 1786; studied rural economy, sheep raising in Eng., from 1787. Mem. French Acad. Scis., 1792, Soc. Agriculture. Author: Mèmoire sur la possibilité d'améliorer les chevaux en France; Sur l'éducation des betes à laine; Instructions sur les maladies des animaux domestiques, 1782-95. (with Daubenton) introduced merino breed of sheep into France; improved French horse breeding; wrote on diseases of domestic animals. Died Ville-Évrard, France, May 1, 1796.

FLASCHKA, Hermenegild Arved, chemist; b. Cilli, Jugoslavia, June 10, 1915; s. Alois Maximilian and Maria (Bracic) F.; Ph.D., U. Graz, Austria, 1938; m. Johanna F. S. Meralla, Mar. 30, 1944; children—Hermann, Irmgard Monika. Came to U. S., 1957, naturalized 1962. Demonstrator, U. Graz, 1938, asst., 1948-49; sci. co-worker Kaiser-Wilhelm Inst. Physikalische Chemie, Berlin-Dahlem, 1944-45; chemist Fa. Schulz Chem. Works, Berlin, Germany, 1945-47; dir. research A. Zankl., Graz, 1949-53; chemist Allgemeine Chemie, Graz, 1953-55; tchr., examiner analyt. chemistry U. and Tech. U. Graz, 1949-55; head analytical div. Nat. Research Center, Cairo, Egypt, 1955-57; guest prof. U. N.C., 1957-58; with Ga. Inst. Tech., Atlanta, 1958—, Regents prof. Mem. Austrian, Am. chem. socs., Am. Microchem. Soc., Sigma Xi. Author: EDTA Titrations, 1959; (with A. Holasek) Clinical Analysis, 1962; (with G. Schwarzenbach, F. Encke) Komplexometrische Titration, 1965. Research, numerous publs. in analyt. chemistry, especially devel. complexometric titration methods, phytometric titrators; studies on masking reactions. Patentee in field. Home: 2318 Hunting Valley Dr., Decatur, Ga. 30033. Office: Chemistry Dept. Ga. Inst. Tech., Atlanta 30332.*

FLATAU, Theodor Simon, German physician; b. Lyck, E. Prussia, June 4, 1860; s. Theodor Flatau; student Heidelberg, Berlin (both Germany); M.D., 1882; postgrad.; m. Eveline Albu; 2 sons, 1 dau. Began practice medicine, specializing in ear, nose, throat, Berlin, 1885; became tchr. voice physiology Berlin U. Music, 1897, dir. dept. for voice and speech impediments and deafness, I. ear-nose-throat clinic, 1905, asso. prof., 1912; dir. ear ambulatorium Charité, from 1905, dir. (with H. Gutzmann) ambulatory for speech and voice defects, from 1926; mil. physician, France, Balkans, 1914-18. Author: Laryngoskopie und Rhinoskopie, 1890; (with H. Gutzmann) Die Bauchrednerkunst . . . , 1894; Nasen-, Racheno und Kehlkopfrankheiten, 1895; Hygiene des Kehlkopfes und die Stimme, die Stimmstörungen der Sänger, 1898; Intonationsstörung uns Stimmverlust, 1899; Die Hysterie in ihren Beziehungen zu den obeven Luftwegen und zur Ohre, 1899; Die Prophylaxe der Hals- und Nasenkrankheiten, 1900; Das habituelle Tremdieren der Singstimme, 1902; Die funktionelle Stimmschwäche der Sänger, Sprecher und Kommandorufer, 1906; also articles. Co-editor Die Stimme, till 1930. Research on ventriloquism, voice in infancy; developed successful treatment for voice failures in singers and speakers through studies on functional voice defects. Died Berlin, Oct. 29, 1937.

FLATLA, Johannes, Norwegian veterinarian; b. Lunner, Norway, Jan. 11, 1906; s. Lars and Kari (Horgen) F.; Ph.D. in Vet. Scis., Vet. Coll. of Copenhagen; m. Ragna Kirkeby, Dec. 30, 1933; children—Jan, Tore, Lars. Veterinarian, 1930-37; asst. prof. Vet. Sch. of Norway, 1937-47, titular prof. internal vet. medicine, 1947—. Mem. Assn. Veterinarians of Norway, Norwegian Assn. Vet. Physicians, Scandinavian Assn. Agronomical Research. Research on dermatosis of pigs, corynebacterium equi infections, vibriosis of livestock, hypomagnesemis of livestock. Home: Oscarsgate 39. Office: Ullevasveien 72, Oslo, Norway.

FLATT, Adrian Ede, surgeon; b. Frinton, Eng., Aug. 26, 1921; s. Leslie Neeve and Barbara (Allen) F.; B.A., Cambridge U., 1942; M.B., B.Chir., M.A., 1945, M.D., 1951; m. Adele Fulton, June 26, 1955; 1 son, Andrew James. Came to U. S., 1956, naturalized, 1960. Fulbright scholar in surgery of hand Roosevelt Hosp., N.Y.C., 1954-55, New Orleans, Carville Leprosarium, 1955; asso. dept. orthopaedic surgery U. Ia. Hosps., Iowa City, 1956; faculty U. Ia., 1957—, prof. orthopaedic surgery, 1966—; nat. cons. in hand surgery aerospace med. div. U. S. Air Force; cons. VA Hosp., Iowa City. Diplomate Am. Bd. Orthopaedic Surgery. Fellow Royal Coll. Surgeons (Eng.); mem. A.M.A., A.C.S., Brit. Med. Assn., Royal Soc. Medicine, Am. Soc. Plastic and Reconstructive Surgery, Midwestern Assn. Plastic Surgeons (past pres.), Brit. Assn. Plastic Surgery, Am. Soc. for Surgery of Hand, Brit. Hand Club (founder), Am. Rheumatism Assn., Aerospace Med. Assn., Internat. Soc. Orthopaedics and Traumatology, Sigma Xi. Author: The Care of Minor Hand Injuries, 1959; The Care of the Rheumatoid Hand, 1963, also numerous articles. Research on x-rays in treatment of crush injuries, biomechanics of hand function, prosthetic replacement of diseased finger joints, intra-articular alkylating agts., rehab. of rheumatoid hand, pathomechanics of ulnar drift, congenital hand defects. Address: U. Ia., Iowa City 52240.*

FLATZ, Gebhard, physician; b. Graz, Austria, Aug. 14, 1925; s. Emil and Hanna (Baumgartner) F.; student U. Graz, 1947-50; M.D., U. Cologne, 1953. Asst., Inst. Physiol. Chemistry, Cologne, 1954-55; resident Duval Med. Center, Jacksonville, Fla., 1955-58, Children's Hosp. of Mich., Detroit, 1958-60; asst. physician in pediatrics U. Bonn, Germany, 1960-64, dozent in pediatrics and human genetics, 1964—. Research fellow German Research Soc., Thailand, 1962-63. Mem. German Soc. Pediatricians, German Soc. Anthropology and Human Genetics, Siam Soc. Author: Hereditary Erythrocyte Anomalies in Thailand, 1964. Research, publs. on distbn. of hereditary erythrocyte anomalies mainly abnormal hemoglobins, thalassaemia, enzyme deficiencies in S.E. Asia. Address: Dept. Pathology, Chiang Mai U., Chiang Mai, Thailand.*

FLAUBERT, Achille-Cléophas, French physician; b. Mézières, France, 1784; head surgeon Rouen Hosp., for 33 years. Author: Mémoire sur plusieurs cas de luxation, 1827; De la pratique de lithotomie sur les femmes, 1815. First surg. removal of pharyngo-nasal polyps; studied hip dislocations. Died Rouen, 1846.

FLAUGERGUES, (Pierre-Gilles-Antoine-) Honoré, French astronomer; b. Viviers, France, May 16, 1755; became dir. Toulon Obs., 1797; laureate, acads. Lyons, Montpellier, Toulouse, Nimes (all France); mem. French Acad. Scis., 1796. Author: Rétrangibilité des rayons (refrangibility light rays), 1779; Mémoire sur le lieu du noeud de l'anneau de Saturne, 1790; Observations astronomiques faites à Viviers en 1798, 1799. Discovered a famous comet which caused considerable commotion in Europe, 1811; studied ring of Saturn, also satellites of Jupiter, evaporation of water; gave explanation for comet's tail. Died Nov. 26, 1830.

FLAVELL, John Hurley, Am. psychologist; b. Rockland, Mass., Aug. 9, 1928; s. Paul Irving and Anne (O'Brien) F.; A.B., Northeastern U., 1951; M.A., Clark U., 1952, Ph.D., 1955; m. Eleanor Roberts, July 24, 1954; children—Elizabeth, James. Faculty, U. Rochester, 1955-65, asso. prof., 1960-65; prof. psychology U. Minn., Mpls., 1965—. Fellow Am. Psychol. Assn.; mem. Soc. for Research in Child Devel. Author: the Developmental Psychology of Jean Piaget, 1963; also articles. Research on devel. in childhood ability to take other person's role or perspective, and ability to use lang. as instrument thought and communication. Home: 2829 Drew Av. S., Mpls. 55416. Office: Inst. Child Devel., U. Minn., Mpls. 55455.*

FLEAGLE, Robert Guthrie, Am. meteorologist; b. Balt., Aug. 16, 1918; s. Benjamin E. and Frances (Guthrie) F.; A.B., Johns Hopkins, 1940; M.S., N.Y. U., 1944, Ph.D., 1949; m. Marianne Diggs, Dec. 19, 1942; children—Robert Guthrie, John B. Faculty, U. Wash., Seattle, 1948—, prof., 1956—; chmn. dept., 1967—. Staff Exec. Office of Pres., Office Sci. and Tech., Washington, 1963-64, cons., 1964—. Chmn., council mems. U. Corp. for Atmospheric Research, 1966-67; mem. com. atmospheric scis. Nat. Acad. Scis., 1962—. NSF Sr. Postdoctoral fellow 1958-59. Fellow Am. Geophys. Union, Am. Meteorol. Soc. (past mem. council, Meisinger award 1959); mem. Royal Meteorol. Soc. Author: (with J. A. Businger) An Introduction to Atmospheric Physics, 1963; also articles. Research in large scale atmospheric motion and its relation to storm devel., phys. processes energy transfer near earth's surface. Home: 7858 56th Pl. N.E., Seattle 98115.*

FLECHSIG, Paul Emil, German psychiatrist, neurologist; b. Swickau, Saxony, June 29, 1847; s. Emil and Ferdinande (Richter) F.; M.D., Liepzig (Germany), 1870; m. Auguste Hauff; m. 2d, Irene Colditz, 1922. Surgeon, War of 1870-71; asst. to E. Wagner, path. inst. U. Leipzig, then asst. in histology to C. Ludwig, became mem. faculty, 1875, asso. prof. psychiatry, 1877, prof. (succeeded Heinroth), 1882, rector, 1894; also dir. insane asylum, Leipzig, from 1882. Author: Die Lietungsbahnen in Gehirn und Rückenmark des Menschen . . . , 1876; Systemerkrankungen des Rückenmarks, 1878; Plan des menschlichen Gehirns, 1883; Die Localization der geistigen Vorgägne

. . . , 1896; Die Grenzen geistiger Gesundheit und Krankheit, 1896; Anatomie des menschlichen Gehirns und Rückenmarks auf myelogenetischer Grundlage, 1920; also articles. A founder of exptl. psychology; studied physiol. causes of brain disturbances; described fibers of spinal column, also detailed course on conduction in central nervous system which led to his theory of myelogenesis; charted cerebral cortex, dividing it into 50 fields. Died Leipzig, July 22, 1929.

FLECK, Sir Alexander, Brit. chemist; b. Glasgow, Scotland, Nov. 11, 1889; s. Robert and Agnes (Duncan) F.; ed. Glasgow U., LL.D., 1953; D.Sc. (hon.), Oxford (Eng.) U., 1956, London (Eng.) U., 1957; m. Isabelle Kelly, 1917. Faculty, Glasgow U., 1911-13; staff Glasgow Radium Com., 1913-17, Castner Kellner, 1917-26; chmn. Billingham div. Imperial Chem. Industries, 1937-45, dir., 1944-51; chmn. Scottish Agrl. Industries Ltd., 1947-51; dep. chmn. African Explosives and Chem. Industries Ltd., 1953-60. Mem. Council Durham Coll.; chmn. Coal Bd. Orgn. Com., 1953-55; mem. Prime Minister's com. on Windscale accident, 1957-58, adv. com. council Research and Devel., 1958—, Nuclear Safety adv. com., 1960-65. Knighted 1955. Fellow Royal Soc., 1955 (treas., v.p. 1960—); mem. Brit. Assn. Advancement Sci. (pres. 1958), Soc. Chem. Industry (pres. 1960-62), Royal Instn. (pres. 1963). Research and publs. in chemistry of radio elements. Home: Aberleven, Crathorne, Yarm, Yorkshire, Eng.

FLECK, Joseph Amadeus, Jr., Am. physicist; b. Kansas City, Mo., Mar. 10, 1928; s. Joseph Amadeus and Mable (Davidson) F.; A.B., Harvard, 1948; Ph.D., Rice U., 1952; m. Karola Suzanne Becker, Feb. 12, 1961; children—Joseph Amadeus III, Andrea. Asso. physicist Brookhaven Nat. Lab., Upton, N.Y., 1952-57; Fulbright postdoctoral fellow Norway, 1957-58; U. S. AEC cons. European Coop. Reactor Project, Halden, Norway, 1958-59; physicist Lawrence Radiation Lab., Livermore, Cal., 1959—. Cons. Japan Atomic Energy Research Inst., 1959-61; lectr. applied sci. U. Cal. (Davis). Mem. Am. Phys. Soc. Contbr. chpt. to Computational Methods in the Physical Sciences, 1964. Research in theory of nuclear reactors, nuclear weapons and lasers; developed theory of statis. properties of laser radiation. Home: 2409 Chateau Way. Office: Lawrence Radiation Lab., Livermore, Cal. 94550.*

FLECK, Stephen, psychiatrist; b. Frankfort am Main, Germany, Sept. 18, 1912; s. Georg and Anna (Beer) F.; student J. W. Goethe U., Frankfort, 1931-33, U. Amsterdam, 1933-35; M.D., Harvard, 1940; m. Louise Harlan, Oct. 13, 1945; children—Anna Lou, Stephen Harlan, Cara Ruth. Practice medicine, specializing in psychiatry, New Haven; prof. psychiatry, and pub. health Sch. Medicine Yale, 1963—; psychiatrist-in-chief Yale Med. Center, 1964—; cons. VA Hosp., West Haven, Conn., Conn. State Hosp., Middletown, Hamden Vis. Nurse Assn. Mem. Am. Psychiat. Assn., Am. Orthopsychiat. Assn., Am. Psychosomatic Soc., Group for Advancement Psychiatry (chmn. com. on preventive psychiatry 1966), Western New Eng. Psychoanalytic Soc., Am. Pub. Health Assn. Co-author Schizophrenia and the Family, 1966. Co-editor Social Psychiatry. Research, publs. on conditioning, family dynamics of schizophrenics and other patients. Home: 18 Ardmore St., Hamden, Conn. 06517. Office: 333 Cedar St., New Haven 06510.

FLECKENSTEIN, Albrecht, German physiologist; b. Aschaffenburg, Germany, May 5, 1917; s. Anton and Margaretha (Haus) F.; student U. Wurzburg (Germany), U. Vienna (Austria), 1937-42; m. Ilse Brabandt, May 22, 1943; children—Bernhard, Wolfgang, Margaretha. With dept. pharmacology U. Heidelberg (Germany), 1947-51, lectr., 1952-56; asso. prof., Brit. council fellow, dept. pharmacology U. Oxford, 1951-52; prof., head Physiol. Inst., U. Freiburg (Germany), 1956—, dean Med. Faculty, 1961-62. Mem. Deutsch Physiologische Gesellschaft (chmn. 1959-61), Physiol. Soc. (asso. London, Eng.). Author: (with D. Steinkopff) Peripheral Mechanisms of Pain Production and Local Anesthesia, 1950; The Sodium-Potassium Exchange an Energetic Principle in Muscle and Nerve, 1955. Research, publs. on cation movements in active muscle and their significance for excitation and contraction, metabolism of high-energy phosphates in skeletal and heart muscle, determination of intracellular turnover rates, electrophysiology of heart. Home: Reutestrasse 3a, Freiburg, Germany.*

FLEISCH, Alfred, Swiss physician; b. Dietikon, Sept. 29, 1892; s. Johanes and Mina Fleisch; ed. U. Zurich, U. Munich; M.D.; D. honoris causa, U. Nancy; m. Ilse Ullmann, Mar. 18, 1919; children—Alfred, Oscar, Herbert André. Prof. agrégé at Zurich, 1921, titular prof., 1926; full prof. at Tartu (Dorpat); full prof. at Lausanne. Recipient Marcel-Benoit prize, Monthion prize, medal French Acad. Sci. Mem. Acad. Halle, Swiss Acad. Med. Sci., French Nat. Acad. Medicine, Biol. Soc. of Paris, Swiss Assn. Physiology and Pharmacology, Assn. French-Speaking Physiologists, Assn. German Physiologists, Gesellschaft für Nahrungs und Vitalstofforschung (hon.), Schweiz-Gesellschaft für Enährungsforschung (hon.). Author over 300 publs. on physicology, including: L'alimentation et ses erreurs; Ernährungsprobleme in Mangelzeiten; Nouvelles méthodes d'étude des échanges gazeux et de la fonction pulmonaire; Ernähren wir uns richtig?. Address: 6, route du Port, Pully pres Lausanne, Switzerland.

FLEISCH, Herbert André, Swiss physician, biochemist; b. Lausanne, Switzerland, July 22, 1933; s. Alfred and Ilse (Ullmann) F.; Med. degree, U. Lausanne (Switzerland), 1957, M.D., 1960; PD exp. Surgery, U. Basel (Switzerland), 1966; m. Maria Pia Ronchetti, May 18, 1959; children—Marie Gabrielle, Isabelle. Radiation biology AEC, Rochester, N.Y., 1959-60; staff dept. surgery U. Lausanne, 1961-62; head Lab. for Exptl. Surgery, Schweizerisches Forschungsinstitut, Davos, Switzerland, 1963-67; privat dozent exptl. surgery U. Basel, 1966-67; prof. path. physiology U. Berne, 1967—. Author: (with H. J. J. Blackwood, M. Owen) Calcified Tissues, 1965; Springer-Verlag, 1966; also numerous articles. Research on mineral homeostasis; regulatory mechanisms of normal and pathological calcification and bone physiology especially role of inorganic pyrophosphate. Home: Brunnadernbain 3b. Office: Dept. Pathological Physiology, U. Berne, Hügelweg 2, 3000 Berne, Switzerland.*

FLEISCHER, Robert, Am. astronomer; b. Flushing, N.Y., Aug. 20, 1918; s. Leon and Rose (Weissglass) F.; B.S., Harvard, 1940, M.A., 1947, Ph.D., 1949; m. Avis Irene Collins, June 14, 1942; children—Martha C., Warren C., Stephen C. Faculty, Rensselaer Poly. Inst., 1946-62, prof. astronomy charge obs., 1958-62; program dir. solar-terrestrial relations, coordinator Internat. Years of Quiet Sun, NSF, Washington, 1962-66, dep. head Office Internat. Sci. Activities, 1966—; dir. research Dudley Obs., Albany, N.Y., 1956-57; research asso. Nat. Radio Astronomy Obs., Green Bank, W.Va., 1960; cons. Northrop Corp., Boston, 1959-61, Gen. Electric Co., Schenectady, 1960-61. Fellow A.A.A.S., Royal Astronomical Society, London; mem. International Sci. Radio Union, International Astron. Union, Am. Astron. Soc., Geophys. Union, Royal Astron. Soc. Can. Studies on interaction of solar disturbances with earth's atmosphere, nature and distbn. of interstellar particles. Home: 1315 16th St. N.W., Washington 20036. Office: NSF, 1800 G St. N.W., Washington 20550.*

FLEISCHER, Robert Louis, Am. physicist; b. Columbus, O., July 8, 1930; s. Leo H. and Rosalie (Kahn) F.; A.B., Harvard, 1952, A.M., 1953, Ph.D., 1956; m. Barbara L. Simons, June 10, 1954; children—Cathy Ann, Elizabeth Lee. Asst. prof. metallurgy Mass. Inst. Tech., 1956-60; physicist Gen. Elec. Research Lab., Schenectady, 1960—. Sr. research fellow physics Cal. Inst. Tech., 1965-66; adj. prof. physics and astronomy Rensselaer Poly. Inst., 1967-68; cons. U. S. Geol. Survey, 1967—. Recipient awards Indsl. Research, 1964, 65, Spl. award Am. Nuclear Soc., 1964. Mem. Am. Phys. Soc., Am. Geophys. Union, Am. Astron. Soc., A.A.A.S., Sigma Xi. Contbr. articles to profl. lit. Research in charged particle tracks in solids and their use in several fields, including cosmic ray and meteorite sci., geochronology, and nuclear physics; defects in solids and their effects on mech. properties and superconducting properties. Home: 1356 Waverly Pl., Schenectady 12308. Office: care Gen. Elec. Research and Devel. Center, Schenectady 12301.*

FLEISCHHACKER, Hanns, Austrian physician; b. Vienna, Austria, Aug. 8, 1910; s. Anton and M. (Reicher) F.; M.D., U. Vienna; m. Maria Kruljac, 1939. Asst. med. clinic U. Vienna, instr., 1948, prof. internal medicine; chief physician Protestant Hosp.; chief physician dept. clin. medicine Hanuschkrankenhaus. Author: Klinische Haematologie, 1948; Lehrbuch der Blutkrankheiten, 1955; Herkrankheiten, 1958; Almanach für Blutkrankheiten, 1962. Research and publs. on blood diseases, internal medicine. Address: Langegasse 63, Vienna VIII, Austria.

FLEISCHMANN, Albert, German zoologist; b. Nuremberg, Germany, June 28, 1862; s. Wilhelm and Fleischmann; ed. Munich, Heidelberg, Berlin, Strasbourg, Würzburg, Erlangen; Ph.D., 1885; m. Franziska Kiefl, 1902; 1 son, Rudolf. Became asst. Erlangen (Germany) Zool. Inst., 1886, mem. faculty, 1888, asso. prof., 1896, prof. zoology, 1898, ret., 1933; helped found Erlangen bee breeding inst., 1907. Author: Embryologische Untersuchungen, 3 parts, 1889-93; Lehrbuch der Zoologie nach morphogenetischen Gesichtspunkten, 1896-97; Die Deszendenztheorie, 1901; Die Darwinsche Theorie, 1903; Einführung in die Tierkunde, 1928; also articles. Research on head region, urogential system, stomach, lungs, rib cage of vertebrates; made studies of organisms based on research in comparative anatomy, histology, evolution; opponent of Darwinism. Died Erlangen, Nov. 19, 1942.

FLEISCHMANN, Charles Louis, inventor; b. nr. Budapest, Hungary, Nov. 3, 1834; s. Abraham Fleischmann; m. Henrietta Robertson; 3 children including Julius, Max. Formed partnership (with James W. Gaff) to make yeast by Hungarian method, presented huge exhibit of product at Phila. Centennial Expn., 1870, bought Gaff's share of bus. after his death, 1879; patented plow, cotton gin, several distilling improvements and devices, 1866-72; dir. about 25 Cincinnati enterprises; pres. Market Nat. Bank, large vinegar co., cooperage co., newspaper; fire commr., Cin., 1890; mem. Ohio Senate, 1879, 95. Died Dec. 10, 1896.

FLEISHER, Gerard Adalbert, chemist; b. Dresden, Germany, Nov. 10, 1911; s. Siegbert and Else (Prast) F.; student Friedrich Wilhelm U., Berlin, Germany, 1930-34, Verbandsexamen, 1934; Ph.D., Techn. Hochschule Danzig, 1936; m. Gisela Nolte, Jan. 21, 1937; children—Thomas A., Christian A. Came to U. S., 1937, naturalized, 1944. Research chemist Schering Corp., Bloomfield, N.J., 1937-42; research asso. in medicine Cornell U. Med. Coll., N.Y.C., 1942-44; research chemist Gelatin Products Corp., Detroit, 1944-45; staff Mayo Clinic, Rochester, Minn., 1945—; faculty Mayo Grad. Sch. Medicine, U. Minn., Rochester, 1945—, prof. biochemistry, 1964—. Mem. Am. Soc. Biol. Chemists, Am. Chem. Soc., Soc. for Exptl. Biology and Medicine, Am. Assn. Clin. Chemists, Sigma Xi. Fellow A.A.A.S. Research, numerous publs. on structural changes in progesterone group, isolation prolactin and other pituitary hormones, isolation penicillin, partial synthesis in field adrenal hormones, enzyme kinetics, serum enzymes, characterization, effect various disease states, fate enzymes in body fluids, enzymes in human blood cells. Home: 833 11th St. S.W., Rochester, Minn. 55901.*

FLEISHMAN, Bernard Abraham, Am. mathematician; b. N.Y.C., June 16, 1925; s. Saul and Anna (Reingold) F.; B.A. cum laude, Coll. City N.Y., 1944; M.S., N.Y. U., 1948, Ph.D., 1952; m. Ruth Dreizin, June 17, 1950; children—Daniel, Nina, Leo. Faculty, U. Pa., 1943-44; research mathematician Applied Physics Lab. Johns Hopkins, 1952-55; prof. math. Rensselaer Poly. Inst., Troy, N.Y., 1955—. Cons. applied math. Boeing Airplane Co., 1956; engring. specialist Sylvania Electric Products, Inc., 1958; vis. prof. U. S. Army Math. Research Center U. Wis., 1961-62. Mem. Am. Math. Soc., A.A.A.S., Am. Assn. U. Prof. Research in molecular and turbulent diffusion, vibrations in nonlinear systems, automatic control, wave propagation in nonlinear media. Home: Colehamer Av., Troy, N.Y. 12180.*

FLEMING, A(ttie) A(nderson), Am. geneticist; b. Russellville, Ala., Dec. 3, 1921; s. Jesse A. and Mary (Bishop) F.; B.S., Auburn U., 1943, M.S., 1949; Ph.D., U. Minn., 1951. Asst. county agrl. agt. Auburn U. Extension Service, Anniston, Ala., 1943-45; faculty U. Ga., Athens, 1951—, prof. plant genetics, 1962—. Vis. scientist Am. Soc. Agronomy, 1964-66. Named Outstanding Tchr., U. Ga. Coll. Agr., 1955. Mem. So. Corn Improvement Conf., (past chmn.), Am. Genetic Assn., Genetics Soc. Can., A.A.A.S., Am. Soc. Agronomy, Crop Sci. Soc. Am., Ga. Acad. Sci., N.Y. Acad. Scis. (life), Sigma Xi, Gamma Sigma Delta (pres. Ga. chpt. 1963). Research, publs. on cytoplasmic inheritance (female and male), interaction of genes with cytoplasm, effect of male cytoplasm, inbred variation studies in maize (forerunner inbred variation investigations at other research stas.); pioneered work in So. U. S. on field-plot technique with hybrid corn. Home: 140 Hope Av., Athens, Ga. 30601.*

FLEMING, Sir Alexander, Brit. bacteriologist; b. Lochfield, Scotland, Aug. 6, 1881; s. Hugh Fleming; B.S., London (Eng.) U., also M.B., St. Mary's Med. Sch., 1906; hon. doctorates from nearly 20 European and Am. univs.; m. Sarah Marion McElroy, 1915 (d. 1949); 1 son; m. 2d, Amalia Koutsouri-Voureka, 1953. Faculty, St. Mary's Med. Sch., 1908-14, 18-48, prof., 1928-48, emeritus univ. prof. bacteriology, 1948-55; rector U. Edinburgh (Scotland), 1951-54. Hunterian prof., 1919; Arris and Gale lectr., 1929; Cutter lectr. Harvard, 1945. Named William Julius Mickle fellow London U., 1942, Charles Mickle fellow U. Toronto, 1944; Knighted, 1944; recipient John Scott medal City Guild Phila., 1944, Nobel prize in medicine (with E. B. Chain, H. W. Florey), 1945, Cameron prize U. Edinburgh, 1945, Albert Gold medal Royal Soc. Arts, 1946, Gold medal Royal Soc. Medicine, 1947, Medal for Merit, U. S., 1947, Grand Cross Alphonse X the Wise, Spain, 1948. Fellow Royal Soc., 1943, Royal Coll. Surgeons Eng. (Hon. Gold medal 1946), Royal Coll. Physicians London (Moxon medal 1945); mem. French Acad. Scis., 1946, Pontifical Acad. Sci., Soc. for Gen. Microbiology (pres.); hon. mem. numerous fgn. med. and sci. socs. Contbr. many papers on immunology, bacteriology, chemotherapy to profl. jours. Discoverer bacteriolytic substance lysozyme, 1921, penicillin (Nobel prize citation; discovery initiated age of antibiotics), 1928. Died London, Mar. 11, 1955.

FLEMING, Charles Alexander, New Zealand geologist; b. Auckland, New Zealand, Sept. 9, 1916; s. George Herbert and Winifred (Hardy) F.; B.A., U. Auckland, 1940, M.Sc., 1941, D.Sc., 1952; m. Margaret Alison Chambers, Apr. 12, 1941; children—Robin Margaret (Mrs. Peter S. McKinlay), Winifred Mary, Jean Sutherland. Staff, New Zealand Geol. Survey, Lower Hutt, 1940—, sr. paleontologist, 1951-52, chief paleontologist, 1952—. Named officer Order Brit. Empire. Recipient Sir George Gowlds Meml. medal, 1941; Hamilton prize, 1943; Hutton Meml. medal, 1956; New Zealand Research medal, 1951; New Zealand Assn. Scientists, Walter Burfitt prize Royal Soc. New S. Wales, 1965. Fellow Geol. Soc. London, Royal Soc. New Zealand (Hector medal 1963); Am. Ornithologists Union (corr.); mem. Aukland Inst. and Mus., Internat. Soc. for Study Evolution, Soc. for Bibliography Natural History, Malacological Soc. London, Paleontological Soc., Paleontol. Assn. (London, Eng.), Internat. Paleontol. Union. Author: The Geology of Wanganui Subdivision, 1953; The Genus Pectern in New Zealand, 1957; The Jurassic Sequence at Kawhia Harbour (with D. Kear), 1960; The Geology of New Zealand, 1959; Checklist of New Zealand Birds (with others), 1953; also numerous articles. Research on New Zealand birds and molluscs, synthesis of history of life in New Zealand. Home: 42 Wadestown Rd., Wellington, New Zealand. Office: New Zealand Geol. Survey, P.O. Box 30368, Lower Hutt, New Zealand.*

FLEMING, Edward Homer, Jr., Am. chemist; b. University City, Mo., Feb. 27, 1925; s. Edward Homer and Grace (Judlin) F.; student St. Ambrose Coll., 1943-44; A.B. magna cum laude, Wabash Coll., 1949; Ph.D., U. Cal. at Berkeley, 1952; m. Sigrid Holzhuter, Feb. 19, 1960. Research chemist Cal. Research Corp., La Habra, Cal., 1952-55; asst. head chemistry dept. U. Cal. Lawrence Radiation Lab., Livermore, 1955—, tech. dir. project sulky, 1964, chmn. radioactivity working group, 1966—. Mem. A.A.A.S., Sigma Xi, Phi Beta Kappa. Research, publs. on phys. and nuclear chemistry, fallout from atmospheric weapons tests and nuclear excavation expts. Home: 40 Golf Rd., Pleasanton, Cal. 94566. Office: Box 808, Livermore, Cal. 94550.*

FLEMING, John Adam, Am. geophysicist; b. Cin., Jan. 28, 1877; s. Americus V. and Catherine B. (Ritzmann) F.; B.S., U. Cin., 1899, D.Sc., 1933; D.Sc., Dartmouth Coll., 1934; m. Henrietta C. B. Ratjen, June 17, 1903 (dec. Mar. 1912); 1 dau., Margaret Catherine; m. 2d, Carolyn Ratjen, Oct. 30, 1913. Aid, U. S. Coast and Geod. Survey, 1899-1903, asst., 1903, magnetic observer, 1904-10; chief magnetician dept. of terrestrial magnetism Carnegie Instn., Washington, 1904-46, chief obs. div., 1915-18, chief magnetic survey div., 1919-21, asst. dir., 1922-29, acting dir., 1929-34, dir., 1935-46, ret., 1946; adviser in internat. sci. relations, 1946-54. Pres. Assn. Terrestrial Magnetism and Elec. International Union of Geodesy and Geophysics, 1930-48; mem. Internat. Commn. Terrestrial Magnetism and Atmospheric Electricity, 1930-46; mem. Internat. Commn. for Polar Year, 1931-47; mem. exec. council Internat. Council Sci. Unions, since 1937 (pres. 1946-49); mem. NRC, del. to assemblies of Internat. Union Geodesy and Geophysics; acting chmn. Am. sect. Aeroartic, 1929-33, del. from NRC to Stockholm Assembly, 1930, Lisbon Assembly, 1933, Edinburgh Assembly, 1936, Washington Assembly, 1939, of Internat. Union Geodesy and Geophysics. Recipient Charles Chree medal Phys. Soc., London, 1945. Fellow A.A.A.S., Washington Geog. Soc.; mem. Nat. Acad. Scis., Am. Inst. Mining and Metallurgic Engrs., Am. Geophys. Union (gen. sec., 1925-47, hon. pres. 1947—, William Bowie medal 1941), Seismol. Soc. Am., Philos. Soc. Washington (sec. 1913-16, pres. 1925), Sigma Xi; hon. or corr. mem. other sci. socs. Editor and co-author: Scientific Results of the Ziegler Polar Expedition cf 1903-05, 08. Co-author vols. II to VII, Researches of Department of Terrestrial Magnetism, Carnegie Institution, Washington, 1915, 17, 21, 25, 27, 46; Scientific Results of Cruise VII of the Carnegie during, 1928-29, 43-46. Editor of Jour. Terrestrial Magnetism and Atmospheric Electricity, 1927-49. Contbr. numerous articles and revs. on geophysics. Died July 29, 1956.

FLEMING, Sir John Ambrose, English physicist; b. Lancaster, Eng., Nov. 29, 1849; B.Sc., Univ. Coll., London, 1870; postgrad. work. Asst., Cavendish Lab., Cambridge (Eng.) U., 1877-80, demonstrator, 1880-81; prof. physics and math. Univ. Coll., Nottingham, Eng., 1881-82; cons. Edison Electric Light Co., London, 1882-85; prof. elec. tech. Univ. Coll., London, 1885-1926, emeritus, from 1926. Faraday medal Am. Inst. E.E., 1928, Gold medal Inst. Radio Engrs., 1933. Fellow Royal Soc., 1892 Recipient Hughes medal, 1910; pres. TV Soc., 1930-45. Author: The Principles of Electric Wave Telegraphy and Telephony, 1906; Fifty Years of Electricity, 1921; The Thermionic Valve in Radiotelegraphy, 1919. An editor, contbr. articles in Ency. Brit., 10th and 11th edits. Contbd. to devel. of electric lamps and electric lighting, wireless telegraphy, telephony; worked (with Ed Dewar) on elec. resistance at low temperatures; devised 1st electron tube, 1904. Died Sidmouth, Eng., Apr. 18, 1945.

FLEMING, Julian Denver, Jr., Am. chem. engr.; b. Rome, Ga., Jan. 12, 1934; s. Julian Denver and Margaret (Mangham) F.; student U. Pa., 1951-53; B.Chem. Engring., Ga. Inst. Tech., 1955, Ph.D., 1959; J.D., Emory U., 1967; m. Sidney Mack Howell, June 28, 1960. Faculty, Ga. Inst. Tech., Atlanta, 1955—, research engr., prof. chem. engring., 1966—; engring. cons., Decatur, Ga., 1959—. Staff cons. Oak Ridge Nat. Lab., 1962—. Mem. Am. Inst. Chem. Engrs., Am. Nuclear Soc., Am. Inst. Aeros. and Astronautics, Am. Ceramic Soc., Sigma Xi. Author: Fused Silica Manual, 1964; also articles. Developed refractory materials and techniques for their fabrication for use in nuclear reactors and missile systems; research on methods for evaluating thermal stability nuclear reactor fuels; designed and developed systems for determining mech., transport, and chem. properties ceramics at high temperatures, structure and properties composite solids. Home: 2109. Clairmont Rd. Office: P.O. Box 1221, Decatur, Ga. 30033.

FLEMING, Warren Robert, Am. physiologist; b. Enterprise, Ore., Nov. 30, 1922; s. Herbert L. and Josephay (Mackan) F.; B.S., Portland U., 1949, M.A., 1951; Ph.D., U. Ore., 1956; m. Ann Karlen, June 14, 1958; children—Julia, Donald Fredrick. Instr. biology dept. U. Ore., 1953-55; faculty zoology U. Mo., Columbia, 1955—, prof., 1963—. Mem. Am. Physiol. Soc., Soc. Gen. Physiologists, Am. Zool. Soc., Sigma Xi. Field editor biology Life Sciences, 1967—. Research, numerous publs. in hormonal control of electrolyte metabolism in lower vertebrates, comparative physiology and comparative endocrinology. Home: 211 Leslie Lane, Columbia, Mo. 65201.*

FLEMING, Wendell Helms, Am. mathematician; b. Guthrie, Okla., Mar. 7, 1928; s. James Lucian and Helen (Helms) F.; B.S., Purdue, 1948, M.S., 1949; Ph.D., U. Wis., 1951; m. Florence Tatum, Apr. 4, 1948; children—Randall, Daniel, William. Mathematician, RAND Corp., Santa Monica, Cal., 1951-53, 54-55; research asso. U. Wis., 1953-54; asst. prof. Purdue U., 1955-58; faculty Brown U., 1958—, prof., 1963—, chmn. math. dept., 1965—. Mem. Am. Math. Soc., Math. Assn. Am., Soc. Indsl. and Applied Math. Author: Functions of Several Variables, 1965. editorial cons. Math. Revs., 1963-65. Contbr. articles to sci. jours. Basic contbns. to geometric measure theory, differential games, optimal stochastic control theory. Home: 3 Colley Ct., Barrington, R.I. 02806. Office: Brown U., Providence 02912.*

FLEMING, William Adam, Am. aero. engr.; b. Battle Creek, Mich.; s. William and Greta (Wolf) F.; B.S. in Aero. Engring., Purdue U., 1943; m. Evelyn Lasch, Oct. 20, 1945; children—Jack W., Janice Lynne. With NACA, 1943-58, chief engines br., Cleve., 1950-55, asst. chief propulsion research div., 1955-58; with NASA, 1958—, tech. asst. Hdqrs., Washington, 1960-61, dir. program rev. div., 1961—. Mem. Am. Inst. Aeros. and Astronautics. Pioneered turbojet engine research; developed concepts turbojet performance and operating characteristics at altitude flight conditions. Home: 6611 Melody Lane, Bethesda, Md. 20034. Office: NASA, Washington 20546.*

FLEMING, William Leroy, Am. physician; b. Morgantown, W.Va., Aug. 29, 1905; s. Walter Lynwood and Mary (Boyd) F.; A.B., Vanderbilt U., 1925, M.S., 1927, M.D., 1932; m. Beatrice Agnes Heathcote, Apr. 28, 1934; children—Elizabeth Heathcote (Mrs. John Benjamin Hammett), Anne Boyd, Mary Boyd (Mrs. Andrew Davidson), Jane Goodall. Milbank fellow Johns Hopkins Hosp., 1937-39; staff internat. health div. Rockefeller Found., N.Y.C., 1939; research prof. syphilology Sch. Pub. Health, U. N.C., 1939-45, prof. preventive medicine, 1952—, asst. dean Sch. Medicine, 1957—; asso. prof. Boston U. Sch. Medicine, 1946-48, prof. preventive medicine, 1948-52; physician Mass. and Evans Meml. hosps., 1946-52; dir. gen. clinic N.C. Meml. Hosp., 1952-64; vis. prof., cons. preventive medicine Escola Paulista de Medicina, Sao Paulo, Brazil, 1962. Cons., USPHS, 1955—, mem. pub. adv. com. on veneral disease control, 1959—. Fellow Am. Pub. Health Assn.; mem. Am. Venereal Disease Assn. (past pres., dir. 1950—, exec. com. 1958—), Assn. Tchrs. Preventive Medicine (past pres.), Am. Soc. for Clin. Investigation, A.M.A., Am. Coll. Preventive Medicine, Internat. Epidemiological Assn., Elisha Mitchell Sci. Soc., Assn. Am. Med. Colls., Sigma Xi. Research, publs. on bacteriology, syphilology, venereal disease control, epidemiology, med. edn. Home: 406 Morgan Creek Rd., Chapel Hill, N.C. 27514.*

FLEMING, Williamina Paton Stevens, astronomer; b. Dundee, Scotland, May 15, 1857; d. Robert and Mary (Walker) Stevens; ed. Scotland; m. Tchr., Dundee, 1871-76; asst. Harvard Coll. Obs., from 1879; apptd. curator astron. photographs, 1898, in charge of Astrophotog. Bldg.; hon. mem. Royal Astron. Soc. (London); after 1906, hon. asso. in astronomy, Wellesley Coll. Author: A Photographic Study of Variable Stars, 1907; Spectre of Photographic Magnitudes of Stars in Standard Regions, 1911. Discoverer of new stars, variables; found spectrum of meteors on exposed photographic plates, 1897, 1902. Died May 21, 1911.

FLEMMING, Carl Friedrich, physician; b. Jüterbog, Dec. 27, 1799; s. Johann Carl Gottlieb and Johanna Sophie (Baltzer) F.; Ph.D., Berlin, 1820; m. Auguste Winther, 1828; 1 son, Walther, 4 daus. Asst. to E. G. Pienitz, insane asylum on Sonnenstein, nr. Pirna, Germany, 1923-24; became mem. staff insane asylum, Schwerin, Germany, 1825; dir. new asylum, Sachsenberg, nr. Schwerin, from 1830; began pvt. practice, Schwerin, 1854. Author: Beitrag zur Philosophie der Seele, 1830; Die organischen Bedingungen der psychischen Erscheinungen, 1838; Über Nrthwendigkeit, Nutzen und Benützung der Irren-Heilaustalten, 1838; Pathologie und Therapie der Psychosen, 1859; Entwicklungsgang der Psychiatrie, 1859; Zur Kläfung des Begriffs der unbewussten Seelentütigkeit, 1877; also articles. Pioneer in sci. found. of psychiatry, establishment of jours., socs., and instns., promotion of edn.; influenced legislation concerning insane, 1851, also forensic medicine; developed concepts of psychiatry now considered too simple. Died Wiesbaden, Germany, Jan. 27, 1880.

FLEMMING, Kurt Bruno Paul, German radiobiologist, pharmacologist; b. Körlin, Pom., Germany, Feb. 24, 1920; s. Karl Friedrich W. and Emma (Müller) F.; student U. Greifswald, 1946-49; grad. U. Rostock 1951, Doctor, 1953; m. Christa Brandt, Sept. 18, 1948; children—Thomas Flemming, Gabriele. Physician, U. Rostock, 1951-53; sci. asst. Inst. Pharmacology, Rostock, 1953-54, Inst. Pharmacology, Greifswald, 1954-60; lectr. U. Greifswald, 1959-61; faculty U. Freiburg (Germany), 1961—, prof. pharmacology and radiobiology, 1965—; leader dept. radiobiology and pharmacology Heiligenberg-Inst., 1961. Mem. Cometé Internat. de Photobiologie, 1960. Decorated Holthusen-Ring. Mem. Internat. Soc. Research, European Assn. Radiobiology, German Pharmacological Soc., German Roentgen Soc. Research, numerous publs. on pharmacology of radiation effects, formation and destruction of biogenic amines by ultra violet and X-ray irradiation, biol. and chem. radiation protection, effects of ultra violet and X-rays on pituitary-adrenal system and reticulo-endothelial system; irradiation effects on cell permeability. Home: 8 Postplatz, Heiligenberg, Germany 7799 Baden. Office: 23 Albertstr. Freiburg/Br., Germany 78, Baden.*

FLEMMING, Walther, German biologist, anatomist; b. Sachsenberg, Germany, Apr. 21, 1843; s. Carl Friedrich and Auguste (Winther) F.; student medicine, Göttingen, Tübingen, Berlin, Germany, Rostock, Poland; M.D., state exam., Rostock, 1868. Asst. to Thierfelder, Rostock, Semper, Würzburg, Germany; asst. to Kühne, Physiol. Lab., Amsterdam, Netherland, 1869-70; mil. physician during Franco-Prussian War; went to Prague, Czechoslovakia, 1872, became asso. prof. histology, 1873; prof. anatomy, Kiel, Germany, 1876-1901. Author: Text zur Karte des menschlichen Auges, 1877; Beiträge zur Kenntnis der Zelle und ihrer Lebenserscheinungen, 1879-81; Zellsubstanz, Kern und Zellteilung, 1882; also articles. Pioneered in cytology and use of microscope; described (with Strasburger) indirect process of cell division and named it mitosis, 1880; introduced the terms, chromatin, spireme, aster; orginated intra-nuclear research on germ cells with his studies on splitting of chromosomes; discovered (independently of van Beneden) the centrosome, 1876. Died Kiel, Aug. 4, 1915.

FLEMYNG, Malcolm, Brit. physiologist; b. Scotland, early 18th century; studied under Monro at Edinburgh, Scotland, under Boerhave at Leiden, Netherlands; M.D.; practiced surgery, Hull, also Lincolnshire, Eng. Author: The Neuropathia, 1740; The Nature of the Nervous Fluid, or Animal Apirits, 1751; Introduction to Physiology, 1759. Versed in physiol. teaching of the time; made various original expts.; wrote to Haller that motor and sensory nerves may coexist in same bundle but are anatomically distinct (proved experimentally much later). Died 1764.

FLEROV, Georgii Nikolaevich, Russian physicist; b. Mar. 2, 1913; ed. Leningrad Indsl. Inst. Sci. work Leningrad Inst. Physics and Tech., from 1938; later chief, lab. multicharged ions Kurchatov Inst. Atomic Energy, Moscow; dir. Nuclear Reaction Lab.; now with Joint Inst. for Nuclear Research. Mem. USSR Acad. Scis. (corr.). Author: The Absorption of Slow Neutrons by Cadmium and Mercury, 1939; (with K. A. Petrzhak) The Spontaneous Fission of Uranium, 1940; (with L. I. Russinov) Experiments on the Fission of Uranium, 1940, (with others) On the Proton Decay of Radioactive Nuclei, 1964; (with others) Synthesis of Transuranium Elements, 1957, 64. Research on nuclear physics and cosmic rays, energy dependence of cross sects., radiative capture of slow neutrons, emission of secondary neutrons during fission; discovered (with K. A. Petrzhak) spontaneous fission of heavy nuclei, 1940. Office: USSR Acad. Scis., Dept. of Physico-Math. Scis., Pyzhevskii Pereulok, 3, Moscow, USSR.

FLESCH, Peter, physician: b. Budapest, Hungary, Oct. 22, 1915; s. Armin and Marianne (Adler) F.; M.D., Budapest Med. Sch., 1939; M.S., U. Chgo., 1943, Ph.D., 1949; m. Regina Feiner, July 8, 1944. Came to U. S., 1941, naturalized, 1944. Research asso. dept. pharmacology U. Chgo., 1946-49; research asst. Damon Runyon fellow Med. Sch. U. Pa., 1949-50, faculty, 1950—, research prof. dermatology, 1963—. Mem. gen. medicine study sect. USPHS, 1960-64. Recipient 1st prize for best exhibit Am. Acad. Dermatology, 1956; 1st prize Am. Dermatol. Assn. Essay Contest, 1956; Internat. Psoriasis award Taub Found., 1957. Mem. A.A.A.S., N.Y. Acad. Sci., Am. Acad. Dermatologists, Soc. Investigative Dermatology, John Morgan Soc., Sigma Xi. Contbg. author: Physiology and Biochemistry of Skin, 1954; also numerous articles. Research on red iron pigment human hair, chem. effect human sebum, chem. abnormalities in scaling skin diseases, methods for determination sulfhydryl compounds. Home: 4526 Locust St., Phila. 19139.*

FLETCHER, Sir Banister, Brit. architect; s. Banister Fletcher; D.Lit., U. Coll., London, Eng.; student Royal Acad.; m. Alice Maude Mary Bretherton Bamford-Slack, 1914 (died 1932); m. 2d, Mrs. Howard Hazell, 1933 (died 1949). Lectr., asst. prof. King's Coll., London; examiner to city and guilds London Inst.; univ. staff lectr. on architecture London U.; partner firm Banister Fletcher & Sons. Recipient, Medal for Design, Archtl. Assn., 1888; Tite medal for Archtl. Design, 1895; numerous fgn. decorations. Fellow Soc. Architecture, Royal Soc. London; mem. Archtl. Assn. (v.p., pres., hon. sec.). Author: A History of Architecture on the Comparative Method; Andrea Palladio, His Life and Work; Architecture and the Humanities; The Influence of Material on Architecture; The English Home, Architectural Hygiene. Died Aug. 17, 1953.

FLETCHER, Charles Montague, English physician, epidemiologist; b. Cambridge, Eng., June 5, 1911; s. Walter Morley and Maisie (Cropper) F.; M.A., M.D.

Trinity Coll., Cambridge, 1936; M.D., F.R.C.P., St. Bartholomew's Hosp., 1939; m. Louisa Mary Seely, Oct. 24, 1941; children—Mark Walter, Susanna Mary, Caroline Anne. Research fellow Radcliffe Infirmary, Oxford, 1941-43; asst. physician St. Bartholomews Hosp., 1943-44; 1st asst. dept. medicine Postgrad. Med. Sch. London, 1944-45; dir. Med. Research Council Pneumocomosis Research Unit, Cardiff, 1945-52; reader in clin. epidemiology Postgrad. Med. Sch. London, also physician to Hammersmith Hosp., 1952—. TV broadcaster on medicine BBC. Sec., Royal Coll., Physicians Com. on Smoking and Health. Decorated comdr. Order Brit. Empire. Mem. Assn. Physicians Gt. Britain and Ireland. Conducted 1st clin. trials of penicillin, 1945; devel. of classification of radiograph of pneumoconioiis, 1952; studies on diagnosis, classification and epidemiology of chronic bronchitis and emphysema. Home: 20 Drayton Gardens, London S.W. 10. Office: Royal Postgrad. Med. Sch., Ducane Rd., London W. 12, Eng.*

FLETCHER, Edward Abraham, Am. chemist; b. Detroit, July 30, 1924; s. Morris and Lillian (Protes) F.; B.S., Wayne State U., 1948; Ph.D., Purdue U., 1952; m. Roslyn Silber, June 15, 1948; children—Judith, Deborah, Carolyn. With NACA, NASA, 1952-59, head propellant chemistry sect., 1955-59; asso. prof. U. Minn., Mpls., 1959-60, prof. mech. engring., 1960—. Cons., McGraw Hill Book Co.; mem. exchange of sci. personnel with Byellorussian Acad. Scis., 1964; participant adv. group for aero. research and devel. NATO, 1960. Mem. Am. Chem. Soc., A.A.A.S., Am. Inst. Aeros. and Astronautics, Combustion Inst. (papers com.). Publs. on fundamental studies in thermodynamics and combustion ignition, supersonic combustion studies, freons and fluorocarbon thermodynamic and combustion studies, rocket propulsion and combustion. Home: 3909 Beard Av. S., Mpls. 55410.*

FLETCHER, Gilbert Hungerford, Am. radiotherapist; b. Paris, France, Mar. 11, 1911, (parents Am. citizens); s. Walter Scott and Marie (Boudol) F.; B.A., U. Paris, 1929; B.A. in Civil Engring., U. Louvain, 1932; M.S., U. Brussels, 1935, M.D., 1941; m. Mary Walter Critz, June 10, 1943; children—Walter Scott, Thomas. Faculty, U. Tex. M.D. Anderson Hosp. and Tumor Inst., Houston, 1948—, prof. radiology, head dept. radiotherapy, 1965—; prof. U. Tex. Grad. Sch. Biomed. Scis., Houston, 1965—. Chmn. com. for initiation and devel. radiotherapy studies Nat. Cancer Inst., 1963—; cons. Lackland AFB, Tex.; cons. space radiation study panel Nat. Acad. Scis.-Space Sci. Bd., 1964—. Fellow Am. Coll. Radiology; mem. A.M.A., Am. Assn. Cancer Research, Am. Soc. Therapeutic Radiologists, Am. Radium Soc. (past pres.), Am. Roentgen Ray Soc., Am. Soc. Clin. Oncology, James Ewing Soc., Radiol. Soc. N.Am., Soc. Nuclear Medicine, N.Y. Acad. Scis., So., Tex., Harris County med. assns., Tex., Houston radiol. socs. Author: (with W. S. MacComb, R. J. Shalek) Radiation Therapy in the Management of Cancer of the Oral Cavity and Oropharynx, 1962; Textbook of Radiotherapy, 1966; also articles. Mem. adv. bd. Jour. Cancer, 1961—. Research on radiation treatment and care of cancer patient, orthovoltage and megavoltate equipment. Home: 2215 Dorrington St., Houston 77025.*

FLETCHER, Harold Roy, Brit. botanist; b. Glossop, Derbyshire, Apr. 14, 1907; s. James and Leah (Warhurst) F.; B.Sc., U. Manchester; Ph.D., U. Aberdeen; D.Sc., U. Edinburgh; m. Evelyn Betty Veronica Sloan, Dec. 19, 1941; children—Veronica, Andrew. Asst. lectr. in botany U. Aberdeen, 1929-34; botanist Royal Bot. Gardens of Edinburgh, 1934-51; asst. regius keeper, 1954-58, regius keeper, 1958; dir. Royal Hort. Soc. Gardens, Wisley, 1951-54. Recipient Victoria medal of honour in horticulture. Mem. Royal Soc. Edinburgh, Bot. Soc. Edinburgh. Research and numerous articles on flora of Asia. Home: 29 Howard Pl. Office: Royal Botanica Garden, Edinburgh, Scotland.

FLETCHER, Harvey, Am. physicist, engr.; b. Provo, Utah, Sept. 11, 1884; s. Charles E. and Elizabeth (Miller) F.; B.S., Brigham Young U., 1907, D.Sc., 1955; Ph.D., U. Chgo., 1911; D.Sc., Columbia, 1935, Kenyon Coll., 1944, Stevens Inst., 1944, Case Sch. Applied Sci., 1944; m. Lorena Chipman, Sept. 9, 1908; children—Phyllis (Mrs. Kenneth Firmage), Stephen Harvey, James Chipman, Robert Chipman, Harvey, Paul Chipman. Head dept. physics and math. Brigham Young U., Provo, 1911-16, dean coll. phys. and engring. scis., 1952-58, prof. physics, 1958—; with Bell Telephone Labs., N.Y.C., 1916-49, dir. phys. research, 1940-49; prof. acoustical engring. Columbia, N.Y.C., 1949-52. Recipient Presdl. Certificate of Merit, 1945, Naval Ordnance Devel. award Office of Sci. Research and Devel., 1945; named Distinguished Alumni, Brigham Young U., 1937. Fellow Am. Inst. Elec. Engring.; mem. Acoustical Soc. Am. (co-organizer, 1st pres. 1929-31), Franklin Inst. (Gold medal 1924), Soc. Motion Picture Engrs. (Gold medal 1940), Audio Engring. Soc. (hon. mem., Gold medal 1949), A.A.A.S. (v.p. 1937-38), Am. Physics Soc. (pres. 1945), Otol. Soc. (hon.), Phi Beta Kappa, Sigma Xi, Nat. Acad. Scis. Asso. of Dr. Millikan on oil drop expt.; research in determination of elec. charge on electron, determination of NE in gaseous ionization; in charge group which first produced and demonstrated binaural and stereophonic sound; discovered critical bands of frequency in hearing mechanism; research and papers on dynamics of cochlea and middle ear. Home: 1615 N. Willow Lane, Provo, Utah.*

FLETCHER, James, entomologist; b. Ashe, Eng., Mar. 28, 1852; s. Joseph Flitcroft and Mary Ann (Hayward) F.; ed. King's Sch., Rochester, Eng.; LL.D., Queen's U.; m. Eleanor Gertrude Schreiber, 1879; 2 daus. With Bank of Brit. N.Am., 1871-76; with library Parliament, Ottawa, Can., 1876-87; entomologist, botanist Dominion Exptl. Farms, from 1884. Fellow Linnean Soc.; mem. A.A.A.S., Entomol. Soc. Am. Author: Flora Ottawaensis; (with George H. Clark) Farm Weeds of Canada, 1906. Seventeen species of butterflies bear his name. Died Nov. 8, 1908.

FLETCHER, Sir Lazarus, Brit. physicist; b. Salford, Eng., Mar. 3, 1854; s. Stewart Fletcher; M.A., Oxford (Eng.) U.; LL.D., St. Andrews (Scotland) U., Ph.D. (hon.), A.M. (hon.), Berlin (Germany) U.; m. Agnes Ward Holme (dec. 1915); 1 dau.; m. 2d, Edith Holme, 1916. Demonstrator, Clarendon Lab., Oxford, 1875-77; fellow Univ. Coll., Oxford U., 1877-80, Millard lectr. in physics Trinity Coll., 1877-78, pub. examiner, 1880; keeper of minerals Brit. Mus., 1880-1909, dir., natural history depts., 1909-19. Examiner natural scis. tripos, Cambridge (Eng.) U., 1882-83, 89-91, 96-97; mem. bds. electors to professorships mineralogy Oxford, Cambridge univs. Fellow Royal Soc., 1889 (council 1895-97, 1910-12, v.p. 1910-12); mem. Mineral. Soc. (pres. 1885-88, gen. sec. 1888-1909), Geol. Soc. (v.p. 1890-92, Wollaston medal 1912), Phys. Soc. (v.p. 1895-97), Selborne Soc., Hertfordshire Natural History Soc. and Field Club; Ealing Sci. and Micros. Soc., Museums Assn., Brit. Assn. Oxford (pres. geol. sect. 1894); corr. mem. Royal Soc. Göttingen (Germany), Acad. Scis. Munich (Germany), N.Y. Acad. Scis.; hon. mem. Sociedad Cientifica Antonio Alzate (Mexico), Assn. Scientifique et d'Enseignement Med. complementaire. Author: Introduction to the Study of Meteorites, 1881; Introduction to the Study of Minerals, 1884; Introductcatrix, 1892; papers on crystallographical, phys. and mineral. subjects and on meteorites, also articles on meteorites Ency. Britannica. Died Jan. 6, 1921.

FLETCHER, Robert Chipman, Am. physicist; b. N.Y.C., May 27, 1921; s. Harvey and Lorena (Chipman) F.; B.S. in Physics, Mass. Inst. Tech., 1943, Ph.D. in Physics, 1949; m. Rosemary Bennett, Sept. 6, 1945; children—Kathryn, Robert B., Margaret, Christine, David W., Frances Carol, Elizabeth. Staff mem. radiation lab Mass. Inst. Tech., Cambridge, 1942-45, research asst. insulation lab., 1945-49; various positions Bell Telephone Labs., Murray Hill, N.J., 1949-64; v.p. for research Sandia Corp., Sandia Base, Albuquerque, 1964—. Fellow Am. Phys. Soc., mem. I.E.E.E. (sr. mem.). Research on electron dynamics, semicondrs., spin resonance, ferromagnetic relaxation. Home: 7101 Aztec Rd., N.E. Office: Sandia Corp., Sandia Base, Albuquerque.*

FLETCHER, Robert Dawson, meteorologist; b. Lampacitos, Mexico, Feb. 11, 1912; s. Edmond Mc C. and Grace (Dawson) F.; B.S. in M.E., Cal. Inst. Tech., 1933, M.S. in M.E. (Aero), 1934, M.S. in Meteorology, 1935; D.Sc. in Meteorology, Mass. Inst. Tech., 1941; m. Elsie Walser, June 1, 1935; children—Robert Dawson, John E. Meteorologist Am. Airlines, Inc., 1935-39; instr. meteorology U. Cal. at Los Angeles, 1940-42; meteorologist U. S. Weather Bur., 1940-50, supervising forecaster, 1941-46, chief hydrometeorological sect., 1946-50; with USAF Air Weather Service, 1950—, cons., 1950-52, dir. sci. services, 1952-64, director of aerospace sciences, 1964—. Technical cons. OSRD, 1944, USAAF in CBI and Caribbean, 1944-45; U. S. del. World Meteorological Orgn. (UN), 1952—; USAF and NRC del., Manila, 1952, Bangkok, 1957, adv. group aero. research and development NATO Conf. Polar Meteorology, Oslo, Norway, 1956, Australian Conf. Tropical Storms, Brisbane, 1956; mem. meteorology panel U. S. Nat. Com. on Internat. Geophys. Year, 1955-64; liaison rep. com. on high altitude rocket and balloon research Nat. Acad. Scis., 1963—, mem. panel on edn., 1963—. Recipient of USAF decoration for exceptional civilian service, 1962. Mem. Am. Meteorol. Soc. (pres. 1956-57), Am. Geophys. Union, Am. Inst. Aeros. and Astronautics (chmn. tech. com. on atmospheric environment 1964), Royal Meteorol. Soc., Sigma Xi. Contbr. articles profl. jours. Directed establishment 1st electronic-computer weather central, circa 1954; directed establishment 1st operational office predicting solar phenomena, 1962; discovered new features of hurricane and other equatorial and heavy-rainfall phenomena. Home: 135 Roger Dr., Lebanon, Ill. Office: Scott AFB, Ill.*

FLETCHER, Sir Walter Morley, English physiologist; b. Liverpool, Eng., July 21, 1873; s. Alfred Evans and Sarah (Morley) F.; student Trinity Coll., Cambridge, Eng., 1891-95; studied under Michael Foster; hon. degrees from Oxford, Leeds, Birmingham, Eng., Glasgow, Edinburgh, Scotland, Pa.; m. Mary Frances Cropper, 1904. Became fellow Trinity Coll., 1897, tutor, adminstr., 1905-14; became sec. Med. Research Com. (became Council, 1920), 1914. Apptd. chmn. Indian Govt. Com. for Organ. Med. Research, 1928. Mem. Royal Commn. on Univs. Oxford and Cambridge, 1919-22; Croonian lectr., 1915. Fellow Royal Soc., 1915. Research and publs. on respiration of frog's muscle, including discharge of carbon dioxide during successive brief intervals of time; proved main discharge of carbon dioxide occurs during phases of recovery of power in contraction of an isolated muscle; studied (with F. G. Hopkins) cycles of change in lactic acid during and after contraction. Died London, June 7, 1933.

FLETT, Sir John Smith, Brit. geologist; b. Kirkwall, Scotland, 1869; ed. George Watson's Coll., also Edinburgh U.; M.A., M.B., C.M., D.Sc., LL.D.; m. Mary Jane Meason; 2 sons, 2 daus. Asst. to Prof. James Geikie, also lectr. petrology Edinburgh U.; joined Brit. Geol. Survey, 1901, petrographer, 1903, asst. to dir., in charge survey of Scotland, 1911-20; dir. Geol. Survey Gt. Brit., also Mus. Practical Geology, 1920-35. Fellow Royal Soc., 1913. Contbr. to memoirs Geol. Survey. Sent (with Tempest Anderson) by Royal Soc. to report on volcanic eruptions at Soufrière, W.I., 1901. Died Jan. 26, 1947.

FLEURE, Herbert John, geographer, anthropologist; b. Guernsey, Channel Islands, U.K., June 6, 1877; s. John and Marie (LeRougetel) F.; student U. Zurich (Switzerland), 1901; D.Sc., U. Wales, 1904; LL.D., U. Edinburgh (Scotland), 1950, U. Wales, 1954; Sc.D., Bowdoin Coll., 1945; m. Hilda Mary Bishop, Aug. 31, 1910; children—Mary (Mrs. Robert Schlapp), Elizabeth, John Laurence. Faculty, U. Coll. Wales, Aberystwyth, 1904-30, prof., 1910-30, chmn. geography and anthropology, 1919-30; prof. geography U. Manchester (Eng.), 1930-44; vis. prof. Bowdoin Coll., 1944-45; vis. prof. geography Egyptian univs., 1948-50. Recipient Daly medal Am. Geog. Soc., 1942; Research medal Royal Scottish Geog. Soc., Victoria medal Royal Geog. Soc. London, 1946. Fellow Royal Soc., 1936; mem. Brit. Assn. (past pres. sects. H and E), Assn. U. Tchrs. (past pres.), Royal Anthrop. Inst. (Huxley Meml. medal 1936, past pres.), Folklore Soc. (past pres.), Geog. Assn. (past pres., hon. sec.). Author: Human Geography in Western Europe, 1918; (with H. J. E. Peake) The Corridors of Time, 10 vols., 1927-56; A Natural History of Man in Britain, 1951; also articles. Research on interpretation of social evolution, analysis of phys. characters among men. Home: Corner House, 66, West Dr., Cheam, Surrey, Eng.*

FLEURIAN DE BELLEVUE, Louis-Benjamin, French geologist; b. La Rochelle, France, Feb. 23, 1761; mem. French Acad. Scis., 1816; pub. articles on geology, mineralogy, meteorology. Died La Rochelle, Feb. 9, 1852.

FLEURIEU, Charles Pierre Claret de, see de Fleurieu.

FLEXNER, Louis Barkhouse, Am. anatomist; b. Louisville, Jan. 7, 1902; s. Washington and Ida (Barkhouse) F.; B.S., U. Chgo., 1923; M.D., Johns Hopkins, 1927; m. Josefa Barbar, Aug. 23, 1937. Instr., asso. anatomy Johns Hopkins Med. Sch., 1930-39; staff dept. embryology Carnegie Inst. Washington, Balt., 1939-51; prof. anatomy U. Pa., Phila., 1951—, chairman of the department of anatomy, 1951-67, director of the Institute Neurol. Scis., 1953-65; research asso. Carnegie Inst. Washington, 1951-66. Mem. Am. Assn. Anatomists (past sec.-treas.), Nat. Acad. Scis., Am. Physiol. Soc., Am. Soc. Biol. Chemists, Am. Soc. Arts and Scis. Research, publs. in structural and biochem. changes occuring in brain as it develops before birth, chem. events which may account for learning and memory. Home: 4631 Pine St., Phila. 19143.*

FLEXNER, Simon, Am. pathologist, bacteriologist; b. Louisville, Mar. 25, 1863; s. Morris and Esther (Abraham) F., M.D., U. Louisville, 1889; post-grad. Johns Hopkins, univs. Strassburg, Berlin, Prague, Pasteur Inst.; D.Sc., Harvard, 1906, Yale, 1910, Princeton, 1913, LL.D., Johns Hopkins, 1915, Cambridge (Eng.) U., 1920; other hon. degrees; m. Helen Whitall Thomas, 1903; children—William Welch, James Cary Thomas. Asso. prof. pathology, Johns Hopkins, 1895-98, prof. pathol. anatomy, 1898-99; prof. pathology U. Pa., 1899-1903; dir. Ayer Clin. Lab., 1901-03; pathologist Univ. Hosp., Phila. Hosp., 1900-03; dir. labs. Rockefeller Inst., Med. Research, 1903-35, dir. Inst., 1920-35, emeritus from 1935; Eastman prof., Oxford U., 1937-38, fellow Balliol Coll., 1937. Fellow Am. Acad. Arts and Scis. (Boston), Acad. Medicine, A.A.-A.S.; mem. Nat. Acad. Scis., Assn. Am. Physicians, Am. Philos. Soc., Am. Assn. Pathologists and Bacteriologists, A.M.A., Harvey Soc.; corr. or fgn. mem. French Acad. Scis., Royal Soc., 1919; numerous other med. and sci. socs. Author: (with J. T. Flexner) Biography of William Henry Welch, 1941; also many papers and monographs relating to bacterial and pathol. subjects, including pathology of toxalbumin intoxication, biochem. constitution of snake venoms, exptl. pancreatitis and fat necrosis, epidemic cerebrospinal meningitis, poliomyelitis, epidemic encephalitis; exptl. epidemiology; developed Flexner's serum for cerebrospinal meningitis, 1907; directed poliomyelitis research which led to identification of the virus causing the disease; pioneer in microbiology in U. S.; discovered dysentery bacillus (named after him), 1900. Died N.Y.C., May 2, 1946.

FLICHE, Henri-Marie-Thérèse-André (or Paul), French paleobotanist; b. Rambouillet, France, June 8, 1836; student Troyes, France, Nancy (France) Sch. Forestry, 1859. Became gen. guardian of forests, Nancy, 1859; apptd. insp. Nancy Forestry Sch., 1866, asst. prof. natural history, 1880-89, full prof., 1889-1902, prof. natural applied scis., from 1902. Mem. French Acad. Scis., 1904; Soc. Agriculture. Author: Etudes sur la flore fossile de l'Argonne, 1896-97.

Worked with sylviculture and paleobotany; studied fossil flora Aragon, France, demonstrated important influence of chem. composition of soil on physiology of plant. Died Nov. 29, 1908.

FLICK, Lawrence F., Am. physician; b. Carrolltown, Pa., Aug. 10, 1856; s. John and Elizabeth (Schabache) F.; prep. edn., St. Vincent Coll., Pa., LL.D., 1925; M.D., Jefferson Med. Coll., 1879, LL.D., 1929; LL.D., Cath. U. Am., 1915, Villanova Coll., 1916; m. Ella Stone, May 26, 1885; children—Lawrence Francis Patrick, Ella Mary Elizabeth, Seton Mary Mercedes, John Bernard Las Casas, Cecelia Mary Veronica, Thomas Walter George, Joseph Samuel Aloysius. Engaged in med. practice in Phila., from 1879; specialist in the pathology and treatment of Tb; med. dir. Henry Phipps Inst.; founder hosps. and socs. for treatment of Tb, including Pa. Soc. for Prevention of Tb, 1892; a founder Phila. Inst. for Study and Prevention of Nervous and Mental Diseases, 1929, pres. until 1935. Recipient Laetare medal U. Notre Dame, 1920. Author: Consumption, a Curable and Preventable Disease, 1903; The Development of our Knowledge of Tuberculosis, 1925; Tuberculosis—A Book of Practical Knowledge to Guide the General Practitioner of Medicine, 1937. Developed method for control of Tb, 1881-83, advocated isolation in spl. hosps. (leading to controversy among physicians). Died Phila., July 7, 1938.

FLICKINGER, Reed Adams, Am. biologist; b. Council Bluffs, Ia., Apr. 5, 1924; s. Reed Adams and Inslee (Bogart) F.; B.A., Stanford, 1946, Ph.D., 1949; m. Catherine Merola, Dec. 29, 1949; children—Maria Inslee, Reed Adams. Faculty, U. Pa., 1949-52, U. Cal., Los Angeles, 1952-58, U. Ia., 1958-61, U. Cal., Davis, 1961-64; prof. State U. N.Y., Buffalo, 1964—. Mem. Soc. For Developmental Biology, Am. Soc. Cell Biology, Am. Soc. Zoologists. Author: Developmental Biology, 1966. Research, numerous publs. in developmental biology. Home: 265 Sherbrook Av., Williamsville, N.Y. Office: Dept. Biology, State U. N.Y., Buffalo 14214.*

FLIEGEL, (Walter) Gotthard Waldemar, German geologist; b. Nieder-Dammer, Silesia, Dec. 28, 1873; s. Julius and Elise (Elwanger) F.; Ph.D., U. Breslau (now Wroclaw, Poland), 1898; m. Anna Marie Meyer, 1907; 1 son, 3 daus. Entered service Prussian State Geol. Inst., 1902, became dist. geologist, 1911, state geologist, 1920, dept. head, 1923; asso. prof. Berlin Agrl. U., from 1919. Contbr. articles to geol. jours. Research and publs. (with W. Wunstoft) new discoveries on brown coal formations and geology of lower Rhine lowlands; combined basic research and applied sci. in work on geology. Died Klein-Machnow (Berlin), Germany, June 22, 1947.

FLINDERS, Matthew, English hydrographer; b. Lincolnshire, Eng., Mar. 16, 1774; m. Apr. 1801; 1 dau. Began voyage to W.I., 1790; also many voyages to explore Australia. Author: A Voyage to Terra Australis, 1814; also prepared charts and maps of coasts of Australia, Tasmania, 1795-99, 1801-03. One of 1st to notice and correct compass errors in iron ships. Died Eng., July 19, 1814.

FLINN, Edward A., Am. geophysicist; b. Oklahoma City, Okla., Aug. 27, 1931; s. Edward A. and Marion (Prater) F.; B.S., M.I.T.; Ph.D., Calif. Inst. Tech.; m. Jane Margaret Bott, Dec. 29, 1962; dau., Susan Katherin. Fulbright scholar, Australian Nat. U., 1958-60; research seismologist, United Electrodynamics, Inc., 1960-63; chief seismologist; Teledyne Earth Sciences, 1963—. Assoc. editor, Geophysics, 1965-67. Mem., Seismological Soc. of Am.; Am. Geophysical Union; Soc. Exploration Geophysicists, A.A.-A.S., Sigma Xi. Author: (with C. H. Dix, translation and revision) Reflection and Refraction of Progressive Seismic Waves, 1963; numerous articles. Research in methods and techniques for detection of underground nuclear explosions; theoretical seismology; structure of crust and upper mantle of earth.

FLINN, Paul Anthony, Am. engr., educator; b. N.Y.C., Mar. 25, 1926; s. Richard A. and Anna (Weber) F.; A.B., Columbia, 1948, M.A., 1949; Sc.D., Mass. Inst. Tech., 1952; m. Mary Ellen Hoffman, Aug. 20, 1949; children—Juliana, Margaret, Donald, Anthony, Patrick. Asst. prof. Wayne U., 1952-54; staff mem. Westinghouse Research Lab., Pitts., 1954-63; faculty Carnegie Inst. Tech., 1963—, prof. physics and metall. engring., 1964—; cons. metal physics. Fellow Am. Phys. Soc.; mem. A.A.A.S., Am. Inst. Metall. Engring., Phi Beta Kappa, Sigma Xi, Tau Beta Pi. Contbr. articles to sci. jours. Investigation of local atomic arrangements in alloys; contbns. to theory of strengthening mechanisms in alloys; applications of Mossbauer effect to study of alloys. Home: P. O. Box 51, Murrysville, Pa. 15668. Office: Dept. Physics, Carnegie Inst. Tech., Pitts. 15213.*

FLINT, Albert Stowell, Am. astronomer; b. Salem, Mass., Sept. 12, 1853; s. Simeon and Ellen Rebecca (Pollard) F.; A.B., Harvard, 1875; A.M., U. Cin., 1880; m. Helen A. Thomas, Oct. 22, 1884. Computer U. S. Naval Obs., 1881-83, 1888-89; asst. U. S. Transit of Venus Commn., 1883-88; asst. astronomer Washburn Obs., U. Wis., 1889, astron., 1904-19 (emeritus), produced with dir. obs. and others several Obs. publs. Fellow A.A.A.S. Author: Meridian Observations for Stellar Parallax, 1st series (Vol. XI, Publ. Washburn Obs.), 1902; 2d series, Vol. XIII, Part I, 1919. Died Feb. 22, 1923.

FLINT, Austin, Am. physician; b. Petersham, Mass., Oct. 20, 1812; s. Joseph Henshaw Flint; med. degree Harvard, 1833; m. Anne Skillings; 1 son, Austin. Prof. med. theory and practice Rush Med. Coll., Chgo., 1844-45; established Buffalo Med. Jour., 1845; a founder Buffalo Med. Coll., 1847, prof. 1847-61; with U. Louisville, 1852-59; prof. New Orleans Med. Coll., 1859-61; prof. pathology and practical medicine L.I. Coll. Hosp., 1861; a founder, prof. internal medicine Bellevue Hosp. Med. Coll.; pres. N.Y. Acad. Medicine, 1873; del. to Internat. Med. Congress, London, 1881; pres. A.M.A., 1883-84. Author: A Treatise on the Principles and Practice of Medicine, 1866. Credited with introducing terms bronchovesicular respiration, cavernous respiration, circa 1856; described presystolic murmur at apex of heart in aortic regurgitation (Flint's murmur), 1862. Died N.Y.C., Mar. 13, 1886.

FLINT, Charles Louis, Am. agriculturist; b. Middleton, Mass., Mar. 8, 1824; s. Jeremiah and Polly (Howard) F.; grad. Harvard, 1849, attended Harvard Law Sch.; m. Ellen E. Leland, 1857; 2 sons, 1 dau. Admitted to N.Y. bar, circa 1852; sec. Mass. Bd. Agr., Boston, 1853-80; commr. from Mass. to Internat. Exbn., Hamburg, Germany, 1863; a founder Mass. Inst. Tech.; a founder Mass. Agrl. Coll., circa 1862, sec. bd. trustees, 1863-65; pres. New Eng. Mortgage Security Co., circa 1885-89. Author: A Practical Treatise on Grasses and Forage Plants, 1857; Manual of Agriculture for the School, the Farm and the Fireside, 1862. Died Hillman, Ga., Feb. 26, 1889.

FLINT, Richard Foster, geologist; b. Chgo., Mar. 1, 1902; s. Nott W. and Edith (Foster) F.; B.S., U. Chgo., 1922, Ph.D., 1925; M.A., (hon.) Yale, 1945; D.Sc., Trinity Coll., Dublin, Ireland, 1963, U. Wroclaw, Poland, 1966. m. Margaret Cecil Haggott, Dec. 29, 1926; 1 dau., Anne (Mrs. David Ogilvy). Faculty, Yale, 1925—, prof., 1945, dir. Grad. Studies Geology, 1950-57, chmn. dept. geology, 1957-64, Henry Barnard David prof. geology, 1956—. Guggenheim fellow, 1965-66. Fellow Geol. Soc. Am., Glaciological Soc., Am. Acad. Arts and Scis.; hon. fgn. mem. geol. socs. London, Edinburgh, Stockholm, Finland. Author: (with C. R. Longwell, A. Knopf) Physical Geology, 1932; Glacial Geology and the Pleistocene Epoch, 1947; Glacial and Pleistocene Geology, 1957; Introduction to Physical Geology (with C. R. Longwell), 1955; also numerous articles. Research on Pleistocene geology especially galcial geology, synthesis and history glacial ages; application C14 dating to glacial-age events, regional glacial geology. Home: 265 Bradley St., New Haven 06510.*

FLIPPIN, Harrison Fitzgerald, Am. physician; b. Charlottesville, Va., Oct. 26, 1906; s. James C. and Isabel (Anderson) F.; B.S., U. Va., 1930, M.D., 1933; m. Edith Quier, June 15, 1937; children—James C., II, William S., Lucy Lee. Edward Bok fellow medicine U. Pa., Phila., 1935-36, chief sect. infectious diseases, 1948—, also clin. prof. medicine; civilian OSRD, 1942-46. Mem. adv. bd. on pneumonia control State of Pa., 1939-41; active hon. con. Phila. Gen. Hosp.; cons. Lankenau Hosp. Diplomate Am. Bd. Internal Medicine. Fellow A.C.P.; mem. A.M.A., Am. Fedn. for Clin. Research, A.A.A.S., Am. Clin. and Climatol. Assn., Internat. Soc. Internal Medicine, Phila. County Med. Soc., Phi Beta Kappa, Sigma Xi. Author: Medical State Board Questions and Answers, 1950; 9th edit. (with Goepp, R. Max), 1957, 10th edit. (with Eisenberg), 1962; Antimicrobial Therapy in Medical Practice, 1955; also numerous articles. Research on cardiovascular and infectious diseases; chemotherapy. Home: R.D. 2, Douglassville, Pa. Office: Lankenau Med. Bldg., Phila. 19151.*

FLIPSE, Robert Joseph, Am. chemist, educator; b. Topeka, July 18, 1923; s. Joseph C. and Mildred (Hunn) F.; B.S., Kan. State U., 1947; M.S., Mich. State U., 1948, Ph.D., 1950; m. Margaret Estelle Giles, June 24, 1944; children—Robert C., Barbara J., W. Scott, C. Philip, Douglas A. Asst. dairy husbandryman Mich. State U., 1947-49; faculty Pa. State U., 1950—, prof. dairy sci., 1966—. Fellow A.A.A.S.; mem. Am. Inst. Biol. Sci., Am. Dairy Sci. Assn., Am. Soc. Animal Sci. Research, publs. in biochemistry and physiology of gametes, including sources of energy and metabolic products; effects of energy and protein intake on reproductive devel. of bull calves. Home: 832 Hemlock Rd., Boalsburg, Pa. 16827.

FLOCH, Hervé Alexander, French med. biologist; b. Lambézellec, France, Oct. 3, 1908; s. Hervé Marie and Jeanne (Lerouzio) F.; M.D., Faculté de Medecine de Bordeaux, 1932; m. Lucie Henry; children—Thérèse, Hervé Henri, Danièle. Asst. Colonial Hosp., scholar Pasteur Inst., Paris, from 1938; mil. physician, until 1956; dir., founder Pasteur Inst. Cayenne, French Guiana (br. of Pasteur Inst., Paris), 1960-66. Chief Anti-Mosquito Service, French Guiana, 1960-66; malariologist WHO. Mem. French Acad. Medicine (7 prizes), French Acad. Scis. (Prix Muteau), other sci. socs. Research, publs. on leprocology (discovered and used D.D.S. in treatment of leprosy), on malariology, entomology, biology and tropical pathology; studies on alimentation-nutrition. Home: 17 Avenue Ferdinand Buisson, Paris XVI. Office: Institut Pasteur 25 rue du D2Roux, Paris XV, France.*

FLOCK, Eunice Verna, Am. biochemist; b. Kellogg, Ida., Aug. 20, 1904; d. Abraham Lincoln and Florence (Ashby) Flock; B.S., U. Wash., 1926; M.S., U. Chgo., 1930; Ph.D., U. Minn., 1935. Fellow physiol. chemistry Mayo Found., 1933-36, cons. biochemistry, 1936—; faculty Mayo Found., Mayo Grad. Sch. Medicine, Rochester, Minn., 1936—, prof. biochemistry, 1957—. Fellow A.A.A.S.; mem. Am. Chem. Soc., Am. Thyroid Assn., Am. Soc. Biol. Chemists, N.Y., Minn. acads. sci., Alumni Assn. Mayo Found., Sigma Xi, Iota Sigma Pi, Sigma Delta Epsilon. Research, publs. phosphate compounds in liver and muscle, metabolism thyroid hormones and analogues, amino acids brain, utilization Glucose-U14-C by brain. Home: 1012 12th Av. N.E., Rochester, Minn. 55901.*

FLOCKS, Rubin Henry, Am. urologist, educator; b. N.Y.C., May 7, 1906; s. Morris and Rose (Blackman) F.; B.A., Johns Hopkins, 1926, M.D., 1930. Asso. urology U. Ia. Coll. Medicine, Iowa City, 1935-37, prof., head dept. urology, chief urologist U. Ia. Hosps., 1949—. Mem. Nat. Adv. Cancer Council, 1965—. Mem. Am. Acad. Pediatrics, A.A.A.S., Am. Assn. Genito-Urinary Surgeons, Am. Bd. Urology (trustee), A.C.S., Am. Geriatrics Soc., A.M.A., Am. Soc. Study Sterility, Am., Ia. urol. assns., Am. Venereal Disease Assn., Clin. Soc. Genito-Urinary Surgeons, Gerontol. Soc., Ia. Acad. Pediatrics, Am. Inst. Biol. Scis., Ia. Clin. Surg. Soc., Ia. Med. Soc., N.Y. Acad. Scis., Pan-Am. Med. Assn., Pan-Pacific Surg. Assn., Soc. Nuclear Medicine, World Med. Assn. Author: (with David A. Culp) Surgical Urology, 1954. Contbr. numerous articles to med. jours. Pioneer description vasculature of prostate noting its relationship to prostatis surgery; study, description basic mechanisms underlying renal urolithiasis; intensive study prostatic carcinoma, pioneer treatment selected cases interstitial irradiation with radioactive isotopes. Home: 514 Grandview Ct., Iowa City 52241.*

FLOOD, Merrill Meeks, Am. mathematician; b. Seward, Neb., Nov. 28, 1908; s. James Francis and Lydia Jane (Meeks) F.; A.B., U. Neb., 1929, A.M. in Math., 1930; Ph.D. in Math., Princeton, 1935; m. Alice Mae Wikoff, Sept. 20, 1932; children—Susan Rojeane (Mrs. Harold A. Judd), Merrill Meeks, Walter Wikoff, James Francis II, Michael John, Robert Hallam. Instr. math. U. Neb., 1929-31; faculty Princeton, 1931-45, dir. applied math. group, 1944-45, dir. Princeton br. Frankford Arsenal, 1943-44; research supr. N.J. Dept. Instns. and Agys., 1935; owner, dir. Merrill Flood & Assos., photogrammetric engrs., Princeton, N.J., 1942-49; project officer for logistics Rand Corp., Santa Monica, Cal., 1949-52; faculty Columbia, 1953-56, dir. Inst. for Research in Mgmt. Indsl. Prodn., 1954-56; prof. indsl. engring. U. Mich., Ann Arbor, 1956—, head Willow Run Labs., 1956-58, prof., sr. research mathematician Mental Health Research Inst., 1959—, prof. math. biology dept. psychiatry Med. Sch., 1960—. Ofcl. investigator OSRD, 1940-46, field service cons., 1944; spl. cons. guided missiles com. U. S. Joint Chiefs Staff, Washington, 1944; expert cons. Sec. War, 1946-47; asst. dep. dir. research and devel. Dept. Army, Washington, 1947-48; exec. dir. Am. Statis. Assn., Washington, 1948-49; cons. Librarian Congress, 1961-62; cons. numerous indsl. firms, govtl. agys., 1936—; vis. research economist Space Scis. Lab., vis. prof. bus. adminstrn. U. Cal. at Berkeley, 1963-64, vis. prof. mgmt. Mass. Inst. Tech., 1965-66; mem. sci. information council NSF; cons. office Sci. and Tech. of Exec. Office President. Fellow A.A.A.S. (council 1958—), Royal Econ. Soc., N.Y. Acad. Scis.; mem. Am. Inst. Indsl. Engrs. (sr., v.p. 1962—), Inst. Mgmt. Scis. (pres. 1955), Operations Research Soc. Am. (pres. 1961-62), Soc. for Indsl. and Applied Math. (council 1955-58), Am. Math. Soc., Soc. for Gen. Systems Research, Sigma Xi. Author: (with others) Automation and the Library of Congress, 1963. Contbr. numerous articles to profl. jours., books. Pioneer work in operations research, parole prediction, econ. time series forecasting, math. theory games, weapons systems analysis, photogrammetric theory, linear programming, mgmt. sci. and logistics research, exptl. games, stochastic learning theory, computer search methods optimization; introduced new concepts for weapons systems analysis, mgmt. sci., operations research, systems engring.; invented synthetic Wollaston prism, 1941, several components and devices for high precision optical fire control and mapping instruments, 1941-46. Office: Mental Health Research Inst., 205 N. Forest Av., Ann Arbor, Mich. 48104.*

FLOOD, Valentine, Irish anatomist; b. Dublin, Ireland; ed. Trinity Coll., Dublin.; M.D., 1830. Became lectr. anatomy Richmond Hosp. Sch., Dublin, circa 1832; joined staff of a dispensary, Dublin, circa 1835; asso. with a med. sch., London; returned to Ireland, 1846; apptd. by bd. health to fever sheds, Tubrid. Mem. Royal Irish Acad. Author: Anatomy and Physiology of the Nervous System, 1828; The Surgical Anatomy of the Arteries and Descriptive Anatomy of the Heart, 1839. Died Oct. 18, 1847.

FLOREY, Ernst, physiologist; b. Salzburg, Austria, Apr. 3, 1927; s. Gerhard and Hilde (Koch) F.; student U. Salzburg, U. Vienna; Ph.D., U. Graz, 1950; m. Elisabeth Pfanner, Oct. 21, 1952; children—Ellen

Christine, Karen Elisabeth. Postdoctoral fellow U. Goettingen, Germany, Cal. Inst. Tech., U. Graz, U. Wuerzburg, Germany, McGill U.; faculty U. Wash., Seattle, 1956—, prof., 1963—, also chmn. Interdisciplinary Program in Gen. and Comparative Physiology; hon. mem. faculty medicine, biology Universidad Catolica de Chile. Author: An Introduction to General and Comparative Animal Physiology, 1966; also numerous articles. Editor: Nervous Inhibition, 1961. Contbr. Research on nerve and muscle, neuropharmacology. Home: 3530 N.E. 182d St., Seattle 98155.*

FLOREY, Lord Howard Walter, pathologist; b. Adelaide, Australia, Sept. 24, 1898; s. Joseph Florey; ed. St. Peter's Collegiate Sch.; M.D., Adelaide U., Magdalen Coll., Oxford; Rhodes Scholar for S. Australia, 1921; John Lucas Walker student, Cambridge, 1924; Rockefeller Travelling fellow, U. S., 1925; Freedom Research fellow London Hosp., 1926; fellow Gonville and Caius Coll., Cambridge, 1926, Lincoln Coll., Oxford; B. Medicine, B. Surgery, M.A., B.Sc., Ph.D.; m. Mary Ethel Reed, 1926; one son and one dau., Huddersfield lectr. in spl. pathology, Cambridge, 1927; Joseph Hunter prof. of pathology, U. Sheffield, 1931-35; prof. pathology, Oxford, 1935-62; provost Queen's Coll., Oxford, 1962—. Charles Mickle fellow Toronto U., 1944. Recipient Cameron Prize, Edinburgh U., 1945; Lister Medal, R.C.S., Eng., 1945; Berzelius Silver medal Swedish Med. Soc., 1945; Nobel prize for physiology and medicine (with Fleming and Chain), 1945; Harmsworth Meml. award, 1946 Albert Gold Medal, Royal Soc. Arts, 1946; Medal in Therapeutics, Soc. Apothecaries of London, 1946; Gold Medal, Royal Soc. Med., 1947, Royal Medal, 1951; Comdr. Legion d'Honneur, 1948; U. S. medal for Merit, 1948; Addingham Gold medal, 1949; Copley medal Royal Soc., 1957; Gold medal Brit. Med. Assn., 1964; Lomonossor medal USSR Acad. Scis., 1964. Fellow Royal Soc. (pres. 1960), Royal Coll. Physicians. Author: Antibiotics (with others), 2 vols., 1949; papers on physiology and pathology. Editor and part author: General Pathology, 1958. Made systematic investigation (with Chain) of naturally-occurring anti-bacterial substances, 1938, resulting in re-discovery, devel. and prodn. of penicillin, 1939. Address: Queen's Coll., Oxford, Eng.

FLORINSKY, Michael T(imothy), economist, historian; b. Kiev, Russia, Dec. 27, 1894; s. Timothy and Vera (Kremcoff) F.; came to U. S., 1926, naturalized, 1937; student law sch. U. Kiev, 1913-14, 18-19, London (Eng.) Sch. Econs., also King's Coll., London, 1920-23; M.A., Columbia, 1927, Ph.D., 1931; m. Louise Ligott Dear, Sept. 5, 1946. Asso. editor Russian series Econ. and Social History of the World War, 12 vols., 1926-31; faculty Columbia, 1931—, prof. econs., 1956-63, prof. emeritus, 1963—. Mem. Council of Fgn. Relations, Am. Hist. Assn. Author several books including: Towards an Understanding of the USSR, 1939; Russia: a History and an Interpretation, 2 vols., 1953; Integrated Europe?, 1955; Russia: A Short History, 1964; also articles, numerous book revs. Editor: Tariff and Commerical History of Gt. Britain and France, 2 vols., 1939-41; McGraw-Hill Ency. of Russia and the Soviet Union, 1961. Interpretation of the History of countries and polit. and econ. instns., pre-revolutionary and Soviet Russia and selected western European countries. Home: Quai Perdonnet 14, 1800 Vevey, Switzerland.*

FLORY, Paul J(ohn), Am. physical chemist; b. Sterling, Ill., June 19, 1910; s. Ezra and Martha (Brumbaugh) F.; B.S., Manchester Coll., 1931, Sc.D. (hon.), 1950; M.S., Ohio State U., 1931, Ph.D., 1934; m. Emily Catharine Tabor, Mar. 7, 1936; children—Susan, Melinda, Paul J. Engaged in research on synthetic fibers, synthetic rubber and other polymeric substances, DuPont Exptl. Station, Wilmington, Del., 1934-38, U. of Cincinnati, 1938-40, Standard Oil Development Co., Elizabeth, N.J., 1940-43; dir. fundamental research, Goodyear Tire & Rubber Co., Akron, O., 1943-48; prof. of chemistry, Cornell, 1948-57; exec. dir. research Mellon Inst., Pitts., 1956-61; professor chemistry Stanford University, 1961—. Awarded Sullivant medal, Ohio State U., 1945; Baekeland award of Am. Chem. Soc., 1947; George Fisher Baker non-resident lectureship in chemistry, Cornell Univ., 1948. Mem. Am. Chem. Soc., Sigma Xi. Contributor articles in scientific publs. Pioneered research on constitution and properties of substances comprised of giant molecules (rubbers, plastics, fibers, films, proteins, etc.); study of polymerization mechanism; theory of solutions. Home: 210 Golden Oak Dr., Portola Valley, Cal. Office: Stanford U., Stanford, Cal.

FLORY, Walter S., Jr., Am. biologist; b. Bridgewater, Va., Oct. 5, 1907; s. Walter Samuel and Ella (Reherd) F.; B.A., Bridgewater Coll., 1928; M.A., U. Va., 1929, Ph.D., 1931; m. Maude Thomas, Apr. 24, 1930; children—Kathryn Sue (Mrs. John Walter Maier), Walter Samuel III, Thomas Reherd. Prof., head dept. biology Bridgewater Coll., 1934-35; Nat. Research fellow biology Harvard, 1935-36; horticulturist Tex. Agrl. Expt. Sta., 1936-44, Va. Agrl. Expt. Sta., 1944-47; with U. Va., 1947-63, prof. exptl. horticulture, 1947-63, vice-dir., mgr. Blandy Exptl. Farm, 1947-63, curator O. E. White Research Arboretum, 1955-63; Babcock prof. botany, dir. Reynolda Gardens, Wake Forest Coll., Winston Salem, N.C., 1963—; trustee, exec. bd. Nature-Sci. Center of Winston-Salem, 1963—; trustee Highlands Biol. Sta., 1965—. Recipient Horsley award. Fellow A.A.A.S.; mem. Am. Boxwood Soc., Am. Soc. Naturalists, Am. Soc. Hort. Sci., Assn. Southeastern Biol. (pres., 1963), Bot. Soc. Am., Genetics Soc. Am., Va. Acad. Sci. (pres., 1956). Contbr. numerous articles to tech. jours. Research on genetics and cytology of horticultural plants; adaptability, breeding, cytology, and phylogeny of woody ornamentals; cytotaxonomy of Amaryllidiceae. Home: 2025 Colonial Pl., Winston-Salem, N.C. 27104.

FLOSDORF, Earl W., Am. bacteriologist; b. Phila., Jan. 27, 1904; B.S., Wesleyan U., 1925, A.M., 1926; A.M., Princeton, 1928, Ph.D., 1929; (with Stuart Mudd) developed method for dehydrating plasma, sera, etc. by vacuum desiccation in frozen state (lyophilization), 1935. Died 1958.

FLOTTE, C(amille) Thomas, Am. surgeon; b. Phila., June 25, 1922; s. C. Joseph and Agnes (Lavin) F.; B.S., Franklin and Marshall Coll., 1943; M.D., Jefferson Med. Coll., 1946; m. Harriet Shand, Nov. 17, 1945; children—Barbara, Thomas J., John F. Faculty, U. Mich., 1954-60, asst. prof., 1957-60; asso. prof. surgery, surgery staff U. Md., Balt., 1960—; surgery staff Md. Gen. Hosp., Balt., 1960—, co-chief surgery, 1960—. Cons. vascular surgery NIH, 1961—, USPHS, 1961—. Diplomate Am. Bd. Surgery. Fellow A.C.S.; mem. Am. Heart Assn., So. Surg. Assn., A.M.A. Author: Medical Information for Medical Assistants, 1961; also articles. Editor: Md. State Med. Soc. Jour., 1966—; adviser in surgery Med. Digest, 1966—. Research on vascular surgery diseases and vascular surg. operative techniques, lipid metabolism especially in relation to atherosclerosis, thrombophlebitis. Home: 540 Hampton Lane, Towson, Md. 21204. Office: 827 Linden Av., Balt. 21201.*

FLOURENS, Marie Jean Pierre, French physiologist; b. 1794; studied under Cuvieris, Paris; lectr. physiology sensation, Paris; became prof. comparative anatomy Jardin du Roi, 1855; prof. natural history Coll. de France. Fellow Royal Soc., 1835; mem. French Acad. Scis. (named secrétaire perpétuel 1833), French Acad. Author: Recherches expérimentales sur les proprietés et les fonctions du système nerveux dans les animaux vertébrés, 1824; Experiences sur le système nerveux, 1825; De l'instinct et de l'intelligence des animaux, 1841; Examen de la phrénologie, 1842; De la vie et de l'intelligence, 1858; Psychologie comparée, 1864. Discovered anesthetic properties of chloroform, circa 1847; studied physiology of brain; showed relation of semi-circular canals to vertigo, 1828; located respiratory center in medulla oblongata more accurately than had been done before, circa, 1837; demonstrated function of the cerebellum in muscular coordination; studied bone formation. Died 1867.

FLOURESCU, Nicolas Burasa, Rumanian physiologist; b. Craiova, Rumania, 1871; s. Florescu Burada and Irene (Travineanu) F.; Ph.D. in scis. U. Paris. Dir. of Lycee, 1902-07; became prof. U. Cernauti, 1921. Mem. Soc. Scis. Bucharest and Cernauti, French Acad. Scis., Soc. Biology Paris. Author: (with A. Dostre) Fonction marcial; also numerous publs. on digestion, circulation, nervous system.

FLOURNOY, Henri, Swiss psychiatrist; b. Switzerland, Mar. 28, 1886; s. Theodore Flournoy; M.D., U. Geneva, 1911; studied psychiatry with Adolf Meyer, Johns Hopkins, psychoanalysis with Freud, Vienna; m. Elizabeth Richard, 1921; 1 dau., 1 son. Lectr. psychotherapy U. Geneva. Recipient prize Faculty Medicine at Lausanne, Bern, Marburg, Munich. Mem. Swiss Soc. Psychiatry (pres. 1934—), Soc. Phys. and Natural History, Med. Soc., Nat. Inst. Geneva, Swiss Soc. Neurology, Assn. Alienists and Neurologists France, Internat. Psychoanalytical Assn., Med.-Psychol. Soc. Paris (fgn. asso.), Royal Medico-Psychol. Assn. Gt. Britain and Ireland (corr.). Author: L'enseignement psychiatrique d'Adolf Meyer, 1926; Les principales étapes de l'histoire de la psychothérapie, 1929 (English translation by Ruth Putnam); Le caractère scientifique de la psychoanalyse, 1931 (Italian, German, Spanish translations), also articles. Address: 6, rue de Monnetier, Geneva, Switzerland.

FLOWER, Sir William Henry, British zoologist; b. Stratford Upon Avon, Eng., Nov. 30, 1831; s. Edward Fordham Flower; ed. U. Coll., Middlesex Hosp.; M.B., London (Eng.) U., 1851; LL.D., Dublin, Edinburgh; D.C.L., Durham U.; m. Georgiana Rosetta Smyth, 1858; 3 sons, 3 daus. Served in Crimean War; asst. surgeon, lectr. anatomy, curator mus. Middlesex Hosp.; curator Hunterian Mus., Royal Coll. Surgeons, 1861-84, Hunterian prof. anatomy and physiology, 1870-84; dir. Natural History Mus., London, 1884-98. Fellow Royal Soc., 1864 (Royal medal 1882); mem. Brit. Assn. for Advancement Sci., Anthropol. Inst., French Acad. Scis., 1895, Zool. Soc. (pres. 1879-99). Author: Diagrams of the Nerves of the Human Body, 1861; Osteology of the Mammalia, 1870; Catalogue of Specimens . . . , 1880; Fashion in Deformity, 1881; The Horse: A Study in Natural History, 1890; (with Richard Lydekker) An Introduction to the Study of Mammals, 1891; Essays on Museums and Other Subjects, 1898. First to demonstrate that marsupials have a dentition in which only a single tooth is changed throughout life; assisted in consolidation of the group mammalia; studied brain structure of apes and lemurs, classification of carnivora and rhinoceroses, arrangement of the order, edentata, anatomy of whales. Died July 1, 1899, London.

FLOWERS, Alan Estis, Am. engr.; b. St. Louis, Oct. 4, 1876; s. William Pitts and Mary Emma (Cummins) F.; M.E., Cornell U., 1902, M.M.E., 1914, Ph.D., 1915; m. Ida Vandergrift Burns, June 29, 1907; children—George Schluederberg, Nancy Holmes, Priscilla. Engr. apprentice Westinghouse Electric & Mfg. Co., 1902-04; instr. and asso. prof. elec. engring. U. Mo., 1904-12; with Gen. Electric Co., 1912-13; prof. elec. engring., Ohio State U., 1913-18; appraisal engr. Columbus Electric Power & Light Co., 1917-18; test engr. engring. dept. Nat. Aniline & Chem. Co., N.Y., 1919-20; research engr. in charge Research Engring. Sect., 1921-22; mgr. and cons. engr. Chem. Machinery Constrn. Co., 1922-23; engr., in charge devel. De Laval Separator Co., 1923-43, research engr. from 1943; dir. Flakice Corp. from 1943. Mem. Am. Inst. E.E., Am. Soc. M.E. (mem. spl. research com. on lubrication), Am. Soc. for Testing Materials (chmn. sub-com. on sampling and gaging), Am. Phys. Soc., Am. Soc. Metals, Sigma Xi. Contbr. chpt. on centrifuges to Chem. Engrs. Handbook, also papers in jours. Developed apparatus for measuring cylinder friction and lubrication in steam engines, also invented a viscosimeter; developed processes for oil reclamation, improvements in centrifugal separators and ultra high speed test tube centrifuge. Died Dec. 4, 1945.

FLOYD, Edwin E., Am. mathematician, educator; b. Eufaula, Ala., May 8, 1924; s. John Q. and Ludie (James) F.; A.B., U. Ala., 1943; Ph.D., U. Va., 1948; m. Alice Marguerite Stahl, May 11, 1945; children—Judith Lynn, Sally Jean, William James. Instr. math., Princeton, 1948-49; faculty math. U. Va., Charlottesville, 1949—, prof., 1956—, chmn. dept., 1966—. Mem. Inst. for Advanced Study, 1958-59, 63-64. Sloan research fellow 1960-64. Author: (with P. E. Conner) Differentiable Periodic Maps, 1964. Research in topology; open maps; transformation groups. Home: 144 Bennington Rd., Charlottesville, Va. 22901.*

FLOYER, Sir John, Brit. physician; b. Hintess, Eng., 1649; s. Richard Floyer; M.A., Queen's Coll., Oxford (Eng.) U., 1671, M.D., 1680; practiced at Lichfield, Eng. Author: The Touchstone of Medicines; An Enquiry into the Right Use of Baths, 1697; Treatise on Asthma, 1698; The Physician's Pulse Watch, 1707; History of Hot and Cold Bathing, 1709; Medicina gerocomica (1st book on geriatrics), 1724. Studied bathing, asthma; 1st observations on pulse rate using his invention, the pulse watch, 1707; 1st description of changes in the lungs with emphysema, 1726. Died Feb. 1, 1734.

FLUCKIGER, Friedrich August, pharmacist; b. Langenthal, Switzerland, May 15, 1828; s. Friedrich and Anna Maria (Gygax) F.; became student Nobacksche Trade Sch., Berlin, Germany, 1845, pharmacy apprentice, Solothurn, Switzerland, 1847, student univs. U. Bern, Geneva (both Switzerland), then Strasbourg, (France), 1850; Ph.D., U. Heidelberg, 1852; also studied in Paris, France, London, Eng.; M.D. (hon.), Bern, 1884; Ph.D. (hon.), Erlangen, Germany, 1892; m. Esther Louise Frei, 1857; 3 sons, 3 daus. Became owner Grosse Apotheke, Burgdorf, Switzerland, 1853; named Bern State pharmacist, 1860; became lectr. on pharmacognosy, Bern, 1861, asso. prof. pharmacy and pharmacognosy, 1870; dir. pharm. inst. U. Strasbourg, 1872-92; returned to Bern, 1892. Fluckiger Found. established in his honor. Mem. Swiss Pharmacists Assn. (pres. 1857); hon. mem. numerous sci. socs. Author: Lehrbuch der Pharmakognosie des Pflanzenveiches (established pharmacognosy of plants as separate sci. field), 1867; Grundlagen der pharmaceutische Wavenkunde (with 300 of his drawings), 1873, 2d edit. (with A. Tschirch) under title Grundlagen der Pharmakognosie, 1855; (with D. Hanbury) Pharmacographis, a history of the principal drugs of vegetable origin, 1874; also numerous articles. Collaborator: Pharmacopoes Helvetica, 1st edit., 1860, 2d edit., 1872. Realized importance of microchemical methods in pharmacognosy. Died Bern, Dec. 11, 1894.

FLÜCKIGER, Otto, Swiss geographer; b. Bargen, Switzerland, Jan. 19, 1881; s. Gottfried and Marianna (Känel) F.; Ph.D., U. Bern (Switzerland), 1904; m. Frieda Steinmann, 1905. Became tchr. geography, secondary sch., Zurich, 1907; named venia legendi U. Zurich, 1917, asso. prof., 1925, prof., dir. geog. inst., 1940; mem. expdn. to Tanganyika and Kenya, 1932-33. Author: Wanderungen der Berner Bauern, 1920; Pässe und Grenzen, 1928; Glaziale Felsformen, 1934; Der Mensch in der glazialen Landschaft, 1939. Studied effect of man on cultural landscape, also effect of geog. relief on polit. boundaries, glacial morphological problems in Switzerland and Scandinavia. Died Zurich, Jan. 25, 1942.

FLUDD, Robert, English alchemical philosopher, physician; b. Bearsted, Eng., 1574; s. Sir Thomas and Elizabeth (Andros) F.; B.A., 1596, M.A., 1598, St. John's College, Oxford; studied medicine, chemistry and occult sciences in France, Spain, Italy and Germany, 1598-1604; M.B. and M.D., Christ Church, Oxford, 1605; admitted as fellow of Royal College of Physicians after initial opposition because of his opposition to Galen, 1609; served as Censor of the College, 1618, 1627, 1633, 1634; successfully practised medicine in London. Author of many books including: Apologia Compendiaria Fraternitatem de Rosea Cruce Suspicionis et Infamiae Maculis Aspersam, Veritatis quasi Fluctibus abluens et abstergens, 1616; and Tractatus Apologeticus Integritatem Societatis De Rosea Cruce defendens, 1617 (in which he defended the Rosicrucians); The Philosophicall Key, unpublished, circa 1618 (in which he defended himself against charges of witchcraft); Utriusque Cosmi Maioris scilicit et Minoris Metaphysica, Physica atque Technica, 1617 (which formed basis of his cosmological system and was added to in succeeding years); Anatomiae Amphitheatrum effigie triplici, more et conditione varia designatum, 1623 (in which he described both scientific and mystical anatomy of body); Philosophia Moysaica, 1638 (latin), 1659 (English), (which summarized his views on medicine and nature). Attacked Aristotle, Galen, the Universities and all proponents of ancient authority while seeking instead new understanding of nature based on Christian principles; sought new key to nature in (a) Mosaic books of Bible, particularly Creation account in Genesis which he interpreted in terms of divine alchemical process, and (b) Hermetic and neo-Platonic works of later antiquity and the Renaissance which mirrored Christian truths; sought divine truths in macrocosm-microcosm analogy; pictured universe in terms of double "centrality", central earth surrounded by sun, moon and planets whose motions could be explained by mechanical analogies, and central sun because of its position midway between center of earth and God; beyond fixed stars was to be found supernature and region of divinity; relative cosmological distances determined through divine mathematical harmony; rejected four elements of Aristotle and three principles of Paracelsus as truly elementary, preferring instead darkness, light and water which were to be found in Genesis, c.1; all other elementary systems to be considered secondary in nature; Divinity and man linked through nature, more specifically, spirit of life which pervaded atmosphere and emanated from sun, seat of Heavenly Spirit; life possible only through inspiration of this spirit which was identified with an aerial saltpeter and had impressed on it a circular motion because of circular motion of sun; circular motion of this spirit in the body resulted in circulation of blood, described in this mystical fashion by Fludd, 1623; 1st to support Harvey's view of circulation of blood in print, circa 1630, thinking Harvey's experiments confirmed his own views; employed graduated thermoscope in showing effects of heat and cold while seeking universal basis of explanation through doctrine of expansion and contraction; similarly sought truth in concept of sympathy and antipathy, strongly supported sympathetic medicine; engaged in controversy over use of weapon-salve which he supported, here discussing William Gilbert's magnetic experiments as similar examples of action at a distance; refused to reject traditional humoral system of disease, but described it in close connection with astral-atmospheric influences which affected the body; accepted observational and experimental evidence, but felt that eternal truths of Scripture and occult writers of past had more weight than sense evidence. As most detailed author on macrocosm-microcosm universe in early 17th century, his works attracted considerable attention; his many volumes were discussed by Kepler, Mersenne, Gassendi and a host of lesser authors; while his work was proposed as basis for a Christian understanding of universe by John Webster in his plea for reformation of English universities. Died London, Sept. 8, 1637.

FLÜGGE, Carl Georg Friedrich Wilhelm, German bacteriologist; b. Hanover, Germany, Dec. 9, 1847; s. Max Eduard and Marte (von Küster) F.; ed. Göttingen, Germany; hon. doctorate, Aberdeen. Practiced medicine in Franco-Prussian War, then in Neuendorf; became asst. to F. Hoffmann, Leipzig, Germany, 1874; named to faculty U. Berlin, 1878, also employed in pvt. lab.; apptd. asso. prof., Göttingen, 1881, prof., dir. 1st. hygienic inst. in Germany, 1883; became prof., Breslau, Germany, 1887; succeeded M. Rubner in Berlin, 1909. Mem. Reich Health Council. Author: Beitrage zur Hygiene, 1879; Lehrbuch der hygienischen Untersuchungsmethoden, 1881; Die Mikroorganismen, 1886; Grundriss der Hygiene, 1889; Grossstadtwohnung und Kleinhaussiddlung, 1916. Founder, editor (with Koch) Zeitschrift fur Hygiene. Applied bacteriology to social hygiene and epidemiology; studied transmission of Tb and other diseases by coughing and sneezing, also air and room hygiene; proved that complete sterilization changes milk significantly, but that 5 minutes of boiling with quick cooling is sufficient. Died Berlin, Oct. 12, 1923.

FLUHARTY, Rex G., Am. physicist; b. Corvallis, Ore., Nov. 22, 1918; s. Anna George; B.S. in Physics, U. Ida., 1939; Ph.D. in Physics, Mass. Inst. Tech., 1949; m. Muriel Esty, 1943; children—Gail Ann (Mrs. Dickison), Lynn Allen. Staff radiation lab. Mass. Inst. Tech., 1942-45, research asso., 1945-49; physicist Oak Ridge Inst. Nuclear Studies, 1949-52; with atomic energy div. Phillips Petroleum Co., 1952-66, Ida. Nuclear Corp., Idaho Falls, 1966-68, Los Alamos Sci. Lab., 1968—. Fellow Am. Phys. Soc., Am. Nuclear Soc. Research, publs. on med. applications of radioactive tracers; reactor neutron research (particularly nuclear cross sects.); research reactor concepts. Address: Los Alamos Sci. Lab., P.O. Box 1663, Los Alamos 87544.*

FLUKE, Donald John, Am. radiation biophysicist; b. Nankin, O., Feb. 17, 1923; s. Eugene P. and Mary (Swan) F.; B.A., Coll. Wooster, 1947; M.S., Yale, 1948, Ph.D., 1950; m. Margaret Dutcher, Mar. 6, 1954; children—John C., Mary H. Instr. physics Yale, 1950-52; biophysicist Brookhaven Nat. Lab., 1952-58; faculty Duke, Durham, N.C., 1958—, prof. zoology, 1965—. Lectr., U. Cal. at Berkeley, 1956-57; vis. asso. prof., 1958; vis. prof. Fla. State U., Tallahassee, 1964. Mem. Radiation Research Soc., Biophys. Soc., A.A.A.S., Am. Inst. Biol. Scis., Am. Assn. Physics Tchrs. Research, publs. on ultraviolet photobiology, direct action of ionizing radiation, temperature dependence of radiation action. Home: 2703 Sevier St., Durham, N.C. 27705.*

FLURY, Ferdinand, German pharmacologist, toxicologist; b. Würzburg, Germany, May 21, 1877; s. Theobald and Dorothea (Schärfer) F.; diploma in pharmacy and food chemistry, Erlangen, Germany, Ph.D., 1902; M.D., Würzburg, 1910; m. Martha Alzheimer, 1921. Asst. to Otto Fischer, 1902-05; became lectr. pharmacology and toxicology U. Würzburg, 1912, asso. prof., 1915, prof. pharmacology, 1920; named head dept. Kaiser Wilhelm Inst. for Phys. Chemistry and Electrochemistry, Berlin, 1916. Author: (with F. Zernik) Schädliche Gase, 1931; (with K. B. Lehmann) Toxikologie und Hygiene der technischen Lösungsmittle, 1938; (with H. Zangger) Lehrbuch der Toxikologie, 1928; also articles in handbooks and jours. Contbd. to successful measures in insect control through research on toxicology of poisonous gases, steams, fumes, dust. Died Würzburg, Apr. 6, 1947.

FLYNN, Edwin Harold, Am. chemist; b. Dunlap, Iowa, Aug. 16, 1920; B.S. U. Neb., 1944; fellow U. Ill., 1946-49, Ph.D. in biochemistry, 1949; m. 1943; 3 children. Asst. in bacteriology U. Neb., 1942-44; research chemist Merck and Co., 1944-46; with Eli Lilly and Co., 1949-58, research asso., from 1958. Mem. Am. Chem. Soc., Soc. Biol. Chemistry. Research in antibiotics, growth factors, bacterial polysaccharides, infectious disease processes, natural immunity; discovered (with others) erythromycin (or ilotycin), 1952.*

FOA, Joseph Victor, aero. engr.; b. Turin, Italy, July 10, 1909; s. Ettore Ernesto and Lelia (Della-Torre) F.; Dr.Ing. in Mech. Engring., Politecnico di Torino, Turin, 1931; Dr.Ing. in Aero. Engring., U. Rome (Italy), 1933; m. Lucy Yolande Bouvier, June 27, 1942; children—Lelia (Mrs. Robert Stuart Dyer), Sylvana (Mrs. John West Abrashkin), Eugenie, Gay (Mrs. Richard Franklin Woodburn). Came to the United States, 1939, naturalized, 1944. With Piaggio Aircraft Co., Finale Ligure, Italy, 1933-35, project engr., 1937-39; chief engr. Studi Caproni, Reggio Emilia, Italy, 1935-37; project engr. Bellanca Aircraft Co., New Castle, Del., 1939-40; instr. U. Minn., 1940-41; chief engr. Am. Aeromarine Industries, Inc., New Bedford, Mass., 1942; head aero. design research sect. Curtiss-Wright Research Lab., 1943-46; head propulsion br. Cornell Aero. Lab., Buffalo, 1946-52; prof. aero. engring. Rensselaer Poly. Inst., Troy, N.Y., 1952—, chmn. dept. aero. engring. and astronautics, 1958-67. Cons. to aircraft cos., 1952—. Recipient Naval Ordnance Devel. award, 1945. Asso. fellow Am. Inst. Aeros. and Astronautics. Author: Intermittent Jets, 1959; Elements of Flight Propulsion, 1960; also numerous articles. Research on control and utilization energy transformation in steady and nonsteady flows, cryptosteady processes, theory propulsion; invented bladeless propellers and other flow induction devices utilizing cryptosteady energy exchanges, new means high-speed mass transp., nondissipative energy separators for heating, cooling, air-conditioning. Home: 33 Point View Dr., Troy, N.Y. 12180.*

FOA, Piero Pio, physiologist; b. Turin, Italy, Apr. 13, 1911; s. Carlo and Eloisa (Errera) F.; M.D., U. Milano (Italy), 1934, D.Chemistry, 1938; m. Naomi Levin, Apr. 6, 1941; children—Helen S., Richard P. Came to U. S., 1939, naturalized, 1943. Asst. prof. physiology U. Pavia (Italy), 1938-39; Dazian Research fellow Yale, 1939; research fellow U. Mich., 1939-43; faculty Chgo. Med. Sch., 1944-61, prof. physiology and pharmacology, 1951-61, chief endocrine and metabolic clinic, 1944-45; prof. physiology and pharmacology Wayne State U. Coll. Medicine, Detroit, 1961—; chief div. research, staff Sinai Hosp. Detroit, 1961—. Mem. adv. com. Heart Disease Control Program, Chgo. Bd. Health. Recipient Gold medal for one of five best M.D. theses in Italy, 1934; Silver medal for most original work Ill. State Med. Soc., 1953; Morris L. Parker award for meritorious research Chgo. Med. Sch., 1953; award of merit Chgo. Tech. Socs. Council, 1961; Rizzi fellow Italian Soc. for Exptl. Biology and Medicine, 1939. Fellow A.A.A.S.; mem. Am. Physiol. Soc., Soc. for Exptl. Biology and Medicine (past president Illinois, councillor Mich. sect. 1965—, editorial board 1966—), Endocrine Soc., Am. Diabetes Society, Am. Fedn. for Clin. Research, Am. Chem. Soc., Royal Soc. Medicine, Detroit Physiol. Soc., Am. Coll. Clin. Pharmacology and Chemotherapy, Worcester Found. for Exptl. Biology, Sigma Xi, Phi Sigma. Author: (with G. Galansino) Glucagon in Health and Disease, 1962; also numerous articles. Research on renal function in hypertension, hormone glucagon, mode of action and secretion of insulin. Home: 12917 Wales St., Huntington Woods, Mich. 48070. Office: 6767 W. Outer Dr., Detroit 48235.*

FOA, Uriel Gaston, psychologist; b. Parma, Italy, Feb. 25, 1916; s. Enea and Dora (Muggia) F.; Dr. Jur., U. Parma, 1939; Ph.D., Hebrew U., Jerusalem, 1947; m. Edna Ben-Yacob, Aug. 21, 1962; children—Gad, Efraim, Ora Tamar, Hagar Dora. Faculty, Bar Ilan U., Ramat Gan, Israel, 1959-61; UNESCO specialist social research, Bangkok, Thailand, 1962; vis. prof. sociology Stanford, 1962; exec. dir. Israel Inst. Applied Social Research, Jerusalem, 1947—; vis. prof. psychology U. Ill., Urbana, 1965-67; prof. psychology and social research U. Mo., Columbia, 1967—. Chmn. personnel selection com. Israel Civil Service, 1962-65; cons. to industry. Mem. Am., Israel (exec. bd. 1964-65) psychol. assns., Am. Sociol. Assn., Indsl. Relations Research Assn., Am. Acad. Polit. and Social Sci. Contbr. to profl. textbooks. Identified and described personality structures providing criteria of differentiation between cultures as well as between normal and abnormal individuals. Address: Dept. Psychology, U. Mo., Columbia, Mo.*

FOCHEM, Kurt, Austrian physician; b. Pressburg, Czechoslovakia, Aug. 28, 1920; s. Heinrich and Gisella (Pavelka) F.; Dr.Med. univ., U. Vienna (Austria), 1945, dozent, 1958; m. Grete Fritz, May 7, 1945; 1 dau., Brigitte. Staff, U. Klinik, Vienna, 1945-50. Zentralrontgeninstitut, U. Vienna, 1950-52; head of x-ray dept. 1. U. Frauenklinik, Vienna, Sanatorium Hera, 1952—. Recipient Sibernes Ehrenzeichen fur Verdienste, Republic of Austria. Mem. Österreichische Röntgengesellschaft, Deutsche Röntgengesellschaft, Van Swieten Gesellschaft, Gesselschaft der Arzte Wien. Research, publs. contbns. to books on x-ray diagnostics of obstetrics and gynecology; biology of x-rays, treatment of cancer, placentography, gynecography, arteriography, hysterosalpingography, irradiation, syndroma, bone diseases. Home: 74 Langegasse, Vienna 8. Office: 1.Univ. Frauenklinik, Vienna 9, Spitalgasse 23, Austria.*

FOCHLER-HAUKE, Gustav, geographer; b. Katharein, Austria, Aug. 4, 1906; s. Gustav and Anna (Seehof) F.-H.; Ph.D., U. Munich; m. Hildegard Dyck, Dec. 19, 1936; children—Hartmut, Agelinde. Librarian; univ. instr.; dir. German Acad. Munich; prof. U. Munich, U. Tucumán; dir. Tucumán Inst. Geography, 1952. Recipient Carl Ritter silver medal. Mem. Argentine Soc. Geog. Studies (corr.), Nat. Acad. Geography Buenos Aires (corr.), also numerous other profl. orgns. Author: Die Manschurei, 1941; Fischer Weltalmanach, 1960—; Die geteilten Länder, 1967; Das politische Erdbild der Gegenwert, 1968, also over 200 articles on geography. Home: Adelheidstrasse 25c, 8 Munich 13, Germany.

FOCK, Vladimir Alexandrovitch, see Fok, Vladimir Alexandrovitch.

FOCKE, Theodore Moses, Am. mathematician; b. nr. Massillon, O., Jan. 3, 1871; s. Theodore H. and Katherine M. (Brown) F.; B.S. in C.E., Case Sch. Applied Science, 1892, Sc.D., 1944; Ph.D., U. Göttingen (Germany), 1898; m. Anne L. Bosworth, Aug. 7, 1901; children—Helen Metcalf, Theodore Brown, Alfred Bosworth. Instr. math. Case Sch. Applied Sci., Cleve., 1892-93; tutor in physics and chemistry Oberlin Coll., 1893-96; instr. math. and civil engring., 1898-1902; asst. prof. math. Case Inst. Tech., 1902-08, Kerr prof. and head of dept. math., 1908-44, dean faculty, 1918-44, dean emeritus from 1944. Fellow A.A.A.S.; mem. Am. Math. Soc., Math. Assn. Am., Am. Soc. for Engring., Sigma Xi, Phi Kappa Psi. Died Mar. 2, 1949.

FODERA, Michele, Italian physician; b. Girgenti, Sicily, Italy, Apr. 20, 1793; mem. French Acad. Scis., 1823; studied principles of osmosis, 1822. Died Palermo, Italy, Aug. 30, 1848.

FODÉRÉ, François Emmanuel, physician; b. St.-Jean-de-Maurienne, Savoy, Jan. 8, 1764; ed. Provincial Coll., Turin, Savoy; Docteur en medicine; became prof. legal medicine, Strasbourg, France, 1814. Author: Traité du crétinisme, 1789; Essai sur le goitre et le crétinage, 1792; Mémoire sur une affection des gencives, 1795; Les lois éclairées par les sciences physiques, ou traité de médicine légale et hygiène publique (classic work for half a century, earned him title of founder of legal medicine in France), 3 vols., 1796; Leçons sur les épidémies et l'hygiène publique, 4 vols., 1822-24; Essai médico-légale sur la folie, 1832; Recherches toxologiques sur la grande cigue, 1835. Studied Marseilles plague and application of pub. hygiene, also antidotes for hemlock poisoning. Died Strasbourg, Feb. 4, 1835.

FODOR, Gábor Béla, organic chemist; b. Budapest, Hungary, Dec. 5, 1915; s. Domokos Győző and Paula (Bayer) F.; student Tech. High Sch., Graz, Austria, 1932-34; Ph.D., U. Szeged, Hungary, 1937; D.Sc., Acad. Scip., Budapest, 1952; m. Eva Varga, Aug. 1, 1939 (div. May 1964); children—András, Gábor, Judith; m. 2d, Marianne Retfalvi, June 5, 1964; 1 dau., Gabriella Paula. Faculty, U. Szeged, 1935-38, 45-57, prof. organic chemistry, 1949-57; head stereochem. research lab. Acad., Budapest, 1958-64; vis. scientist NRC Can., Ottawa, 1964-65; vis. prof. Laval U., Quebec, Que., Canada, 1965-67, prof., 1967—. Mem. hon. adv. bd. Tetrahedron, 1957—, Tetrahedron Letters, 1959—, Advances in Heterocyclic

Chemistry, 1962—. Overseas fellow Churchill Coll., Cambridge, Eng., 1961; recipient Kossuth award in chemistry, Budapest, 1950, 54, Silver medal U. Helsinki (Finland), 1958. Mem. Chem. Soc. London, Am. Chem. Soc., Chem. Inst. Can., Schweizerische Chemische Gesellschaft, Soc. Chim. France, Hungarian Acad. Scis. Author: Organic Chemistry, 1960, 66; Tropane Alkaloids in Manske, vol. 6, 1960, vol. 9, 1967; Natural Products, vol. 1, 1965; also numerous articles. Patentee in field. Research on synthesis and elucidation of steric structure of natural products, interalia, ephedrines, chloramphenicol, Brainsphingosine, glucosamine, tropane alkaloids, atropine, scopolamine, cocaine, mechanism of acyl migration, Wallach rearrangement, mutarotation of steroid dibromides, selective quaternization of tertiary amines, ring opening of condensed azetidinium salts. Home: 2884 de Vincennes, Ste-Foy, Que., Can.*

FODOR, Josef, Hungarian bacteriologist; b. Hungary, 1843; Prof. Cluj, Rumania, 1872-74, Budapest, Hungary, from 1874; hygienist; pioneer investigator bactericidal action of blood. Died 1901.

FOERSTE, August Frederick, Am. paleontologist; b. Dayton, O., May 7, 1862; A.B., Denison U., 1887, D.Sc., 1927; A.M., Harvard, 1888, Ph.D., 1890; postgrad., Heidelberg and Paris, 1890-92. Asst. U. S. Geol. Survey, 1887-91, 95; with Ohio Geol. Survey, 1892, 1908-10, 17-19, Ind. Geol. Survey, 1896, 97, 99, Ky. Geol. Survey, 1904-12, Geol. Survey of Can., 1911-12; asso. in paleontology U. S. Nat. Mus., from 1932; Marsh fund grantee Nat. Acad., 1928. Rep. of Paleontol. Soc. in NRC, 1932-35. Author reports, contbns. to bulls. and jours. Specialist in Ordovician and Silurian paleontology, with particular reference to Cincinnatian and Silurian faunas of Central States; also in Ordovician and Silurian Cystids and in Ordovician and Silurian Cephalopods of N.Am., especially of Arctic areas.; re-described many invertebrate fossil species; created numerous new genera and species of Cephalopods. Died Apr. 23, 1936.

FOERSTER, Arnold, German entomologist; b. Aachen, Germany, Jan. 20, 1810; s. Caspar Aegidius Arnold and Katherine Theresa Angelika (Duyckers) F.; student, Bonn, 1832-36, Ph.D. (hon.), 1853; m. Maria Barbara Zimmermann, 1842; 3 sons, 3 daus. Tchr., Realschule (sci. secondary sch.), Aachen, 1836-84, headmaster, 1850-84. Mem. Leopoldina. Author: Hymenopterologische Studien I (Formicariae), 1850, II (Chalcididae und Proctotrupii), 1856; Über den Systematischen Wert des Flügelgeäders bei den Insekten und besonders bei den Hautflüglern, 1877. Research in taxonomy of dipterons; described new types of hymenoptera; studied ichneumon fly, chalcidids, proctotrupids, braconids. Died Aachen, Aug. 13, 1884.

FOERSTER, Fritz, German chemist; b. Grünbert, Silesia, Feb. 22, 1866; s. August and Anna (Eichmann) F.; doctorate U. Berlin, 1888; hon. dr. engring., Stuttgart, Germany, 1913; m. Martha Zanke, 1896; 2 sons, 1 dau. Worked (with F. Mylius) at Reich Phys.-Tech. Inst.; mem. faculty Tech. U. Berlin; became mem. faculty Tech. U. Dresden (Germany), 1895, asso. prof., 1898, prof. phys. chemistry and electrochemistry, 1900. Mem. Saxon Acad. Scis., Acad. Scis., Göttingen, Germany. Author: Elektrochemie wässriger Lösungen, 2 vols., 1905. Research on electrolytic purification of copper, processes in alkali chloride electrolysis, Edison accumulator, coal as chem. raw material, sulphur-oxygen compounds, cathodic precipitate forms of metals and related passivity phenomena. Died Dresden, Sept. 14, 1931.

FOERSTER, Max, German engr.; b. Grünberg, Silesia, June 9, 1867; s. August and Anna (Eichmann) F.; studied civil engring. Tech. U. Berlin, until 1890; hon. dr. engring., Darmstadt, Germany, 1919; m. Charlotte Granier; 1 son, 2 daus. With Prussian Waterworks Adminstrn., Kassel, also Münster, Germany, from 1894; became asst. to G. Mehrtens, Tech. U. Dresden (Germany), 1896, lectr. on movable bridges, 1896, asso. prof., 1898, prof., 1900. Mem. Berlin Patent Office. Recipient Schinkel prize. Author: Eisenkonstruktionen des Ingenieurhochbaues, 1899-1902; Lehrbuch der Baumaterialienkunde, 1903-12; Das Material und die statische Berechnung der Eisenbetonbauten, 1907; Balkenbrucken in Eisenbeton, 1908; Grundzuge des Eisenbetonbaues, 1919; Repetitorium fur die Hochbau, 3 vols., from 1919. Research and devel. in iron superstructure, solid constrn., bldg. materials, durability; developed theory and practice of then new field of reinforced concrete constrn., also 1st to lecture on subject in German univ. Died Dresden, June 12, 1930.

FOERSTER, Otfrid, German neuro-surgeon; b. Breslau, Germany (now Wroclaw, Poland), Nov. 9, 1873; s. Richard and Angelika (Lübert) F.; student, Freiburg, Kiel (both Germany); M.D., Breslau, 1897; postgrad. Salpétrière, Paris, 2 years; m. Martha Bauer, 1903; 2 daus. Worked under C. Wernicke, Psychiatric Clinic, 1899-1904; became mem. faculty U. Breslau, 1903, also asso. prof., 1909, prof., 1921-38; mem. staff Wenzel-Hanke Hosp.; began specializing in neurosurgery during World War I; attended Lenin, Moscow, 1922-24. Recipient Hughlings Jackson Meml. medal, 1935; Otfrid-Förster medal established in his honor by German Soc. Neurosurgery, 1953. Author: Kriegsschädigungen der peripheren Nerven, 1921; Die Leitungsbahnen des Schmerzgefuhls, 1927; also numerous articles. Editor: (with O. Bunke) Handbuch der Neurologie, 17 vols., 1935-37. Editor Zeitschrift für die gesamte Neurologie und Psychiatrie. Research on spastic paralysis; introduced (independently of Spiller and Martin) cordotomy, 1913, also (independently of Rosett) hyperventilation test in epilepsy, 1924; developed rhizotomy for treatment of spastic paralysis (Foerster's operation), also Foerster-Penfield operation for traumatic epilepsy; made cytoarchitectonic map of human cerebral cortex, 1936. Died Breslau, June 15, 1941.

FOERSTER, (Carl Friedrich) Richard, ophthalmologist; b. Polish Lissa, Nov. 15, 1825; s. Caroline Polluga; student, Breslau, Heidelberg (both Germany); M.D., Berlin, 1849; m.; 5 children. Became head physician Allerheiligenhospital, Breslau, Germany, 1855; named mem. faculty Breslau, 1857, asso. prof., 1863, prof., 1873, univ. rep. in Prussian Herrenhaus, 1894; ret., 1896. Contbg. author: Handbuch der Ophthalmologie (Gräfe-Sämasch). Research on myopia; inventor perimeter, photometer; pointed out relationship between eye diseases and gen. disease of body; drew attention to water as source of cholera infection before discovery of cholera bacillus. Died Breslau, July 7, 1902.

FOERSTER, Wilhelm Julius, German astronomer; b. Grünberg, Silesia, Dec. 16, 1832; s. Friedrich and Hulda (Seydel) F.; began astron. studies under Enckes, Berlin, 1850, under Argelander, Bonn, 1852; doctorate, 1854; m. Karoline Paschen, 1868; 3 sons, including Friedrich Wilhelm, Karl, 2 daus. Asst., Berlin obs., 1854-63, dep. dir., 1863-65, dir., 1865-1904; became dir. standards commn. N. German Bund, 1868; named dir. German weights and measures bur., 1871; mem. Permanent Commn. Internat. Geodesy. Author: Johann Kepler und die Harmonie der Sphären, 1862; Sammlung von Vortragen und Abhandlungen (articles), 1887, 90. Studien zur Astronomie, 1888; Wissenschaftliche Erkenntnis und sittliche Freiheit, 1896; Kalendar und Uhren am Ende des Jahrhunderts, 1899; Himmelskunde und Weissagung, 1901; Astrometrie, 1905; Lebenserinnerungen and Lebenshoffnungen, 1911. Editor Berliner Astronomisches Jahrbuch, 1867-81. Founder astrophys. obs., Potsdam, Germany, obs., Strasbourg (now in France), Reich Tech.-Phys. Inst.; responsible for Venus expdns., 1874, 82; co-founder Astronomische Gesellschaft, Astron. Calculation Inst.; founder Urania, Berlin, 1888, Assn. Friends of Astronomy and Cosmic Physics, 1891; research on small planets, 1855-65; inventor instruments which contbd. to German tech. precision. Died Bornim (Brandenburg), Germany, Jan. 18, 1921.

FOES, Anutius, physician; b. Metz, France, 1528; ed. Paris; became city physician, Metz, 1556; well-educated in medicine, Greek and Latin langs.; after 40 years of work, pub. Latin transl. of Hippocrates, 1591. Died 1592.

FOFONOFF, Nicholas Paul, Canadian oceanographer; b. Queenstown, Alta., Can., Aug. 18, 1929; s. Paul A. and Anna (Malakoff) F.; B.A., U. B.C., 1950, M.A., 1951; Ph.D., Brown U., 1955; m. Mabel Beryl Deckard, June 16, 1951; children—Paul W., Stephanie A., Timothy W., Nicholas D. Post-doctorate fellow Nat. Inst. Oceanography, Godalming, Eng., 1955-56; asst. scientist Fisheries Research Bd. Can., Pacific Oceanographic Group, Nanaimo, B.C., Can., 1956-57, asso. scientist, 1957-58, sr. scientist, 1958-61, prin. scientist, 1961-62, phys. oceanographer, 1962-63, sr. scientist, 1963—; chmn. dept. phys. oceanography Woods Hole Oceanographic Instn., Woods Hole, Mass., 1967—. Research, publs. on dynamics of ocean circulation, phys. properties and thermodynamics of sea water; application of moored buoy systems to measurement of ocean currents. Home: 6 Greengate Rd., Falmouth, Mass. 02540. Office: Woods Hole Oceanographic Instn., Woods Hole, Mass. 02543.*

FOGEL, Karl Gustav, Finnish physicist; b. Munsala, Finland, June 7, 1927; s. John and Ellen (Smeds) F.; Ph.D., U. Helsinki (Finland), 1949; m. Lisa Margareta Eklund, July 7, 1946; children—Torunn, Bodil, Johan Jakob, Karl Ragnar. Faculty Abo Akad., Turku, Finland, 1948—, prof., 1954—, dean dept. math. and natural scis., 1960-64, vice rector, 1965—. Vis. fellow Cornell U., Ithaca, N.Y., 1957-58; vis. scientist, Saclay, France, 1963; suppleant mem. bd. Nordita, Copenhagen, Denmark, 1960—. Mem. Finnish NRC for Scis., Soc. Scient. Fenn. Research on theory of nuclear matter. Home: 13 Eriksgatan, Turku, Finland.*

FOGG, John Milton, Am. botanist; b. Phila., Nov. 8, 1898; s. John Milton and Grace (Kirby) F.; B.S., U. Pa., 1925; Ph.D., Harvard; Sc.D., LaSalle Coll., 1949; m. Helen Biggs, June 27, 1930; children—Sonia (Mrs. Christopher David), Felicia (Mrs. Joaquin Gonzalez) F. Faculty, U. Pa., Phila., 1925—, prof. botany, 1944—, dean Coll. Arts and Scis., 1941-44, vice provost, 1944-53; dir. Morris Arboretum, Phila., 1953-67; dir. Barnes Aboretum, Merion, Pa., 1967—. Mem. vis. com. Arnold Arboretum, Boston, 1958-66. Mem. A.A.A.S., Bot. Soc. Am., Pa. Acad. Sci., Am. Assn. Bot. Gardens and Arboretums, Phila., N.E., Torrey bot. clubs. Author: Weeds of Lawn and Garden, 1945; also numerous articles. Research on plant life Pa., floristics, plant geography. Home: 6807 Quincy St., Phila. 19119. Office: Barnes Aboretum, Merion, Pa. 19066.*

FOGH, Jorgen Engell, virologist; b. Copenhagen, Denmark, Feb. 6, 1923; s. Knud A. T. A. and Else (Engell) F.; M.D., U. Copenhagen, 1949; m. Helle Yde, Jan. 18, 1947; children—Jens Morten, Peter. Asst. research virologist U. Cal. at Berkeley, 1953-58; asso. med. virologist N.Y. State Dept. Health, 1958-60; asso. dept. microbiology Albany (N.Y.) Med. Coll., Union U., 1959-61; asso. cancer research Roswell Park Meml. Inst., Buffalo, 1960; asso. mem. Sloan-Kettering Inst. for Cancer Research, N.Y.C., 1960—, asso. prof. Sloan-Kettering div. Cornell U. Med. Coll., 1961—. Recipient Lifetime Research Career award Nat. Cancer Inst. NIH, 1964. Mem. Danish Med. Assn., N.Y. Acad. Scis., Internat., Am. socs. for cell biology, Tissue Culture Assn., A.A.A.S., Electron Microscope Soc. Am., Am. Assn. for Cancer Research, Am. Assn. Immunologists, Am. Soc. for Exptl. Pathology, Soc. for Exptl. Biology and Medicine, European Tissue Culture Club, Nordisk Genetiker forening. Contbr. numerous articles to tech. jours. Research on viruses and mammalian cell culture, biol. and morphological characteristics polio and other viruses, cultured human cells, normal and malignant, in vitro transformation of human cells, effects of mycoplasma on cultured mammalian cells. Office: 145 Boston Post Rd., Rye, N.Y. 10580.*

FOGLIA, Virgilio Gerardo, Argentine physiologist, physician; b. Buenos Aires, Argentina, Feb. 13, 1905; s. Virgilio Luis and Adelina (Dufour) F.; M.D., U. Buenos Aires, 1928; m. Maria Angelica Vivanco, July 6, 1950; children—Virgilio Luis, Maria Angelica, Marcelo Daniel. Research and pvt. practice, 1928-35; asst. prof. U. Buenos Aires, 1935-43, prof., dir. physiology dept., 1955—; spl. research fellow NIH, Bethesda, Md., 1940-50. Fellow McGill U., Montreal, Que., Can., 1937-38. Pres. Argentine com. UNESCO, 1962, pres. Venezuela mission, 1964; pres. 7th Latin Am. Congress Physiol. Scis., 1966. Mem. Argentine Soc. Endocrinology (pres. 1950-53), Argentine Soc. Diabetes (pres. 1957-58), Argentine Soc. Biology, Argentine Soc. Physiology, Soc. Exptl. Biology and Medicine, N.Y. Author: La regulación de la secreción interna del pancreas, 1934; (with B. A. Houssay and others) Human Physiology, 1951; Co-editor: Perspectives in Biology, 1963. Research on endocrinology and metabolism, diabetes. Home: 1695 Callao, Buenos Aires. Office: 2155 Paraguay, Buenos Aires, Argentina.*

FOIX, Charles, French neurologist; b. Paris, 1882; intern, 1906; began gen. practice, 1919; author thesis on serum and its leukocytic properties; specialized in vascular diseases; studied albuminocytological dissociations; located (with others) site of lesions in parkinsonism (pub. 1921). Died 1927.

FOK, Vladimir Alexandrovitch, Russian theoretical physicist; b. Dec. 22, 1898; grad. Petrograd U., 1922. With State Inst. Optics, 1919-23, 28-41, Leningrad Inst. Physics and Tech., 1924-36; prof. Petrograd U., 1932; with Inst. Physics, USSR Acad. Scis., 1934-41, 44-53, mem. staff Inst. Phys. Problems, 1954—; faculty Leningrad U., 1961. Recipient Mendeleyev prize, 1936, Stalin prize, 1946, Lenin Prize, 1960. Mem. USSR Acad. Scis. (corr.), Norwegian Royal Soc. (fgn.), Norwegian Acad. Sci., Royal Danish Acad. Author: The Motion of Finite Masses in the General Theory of Relativity, 1939; The Diffraction of Radio Waves around the Earth's Surface, 1946; Some Uses of Lobachevsky's Non-Euclidean Geometry in Physics, 1950; Theory of Space, Time and Gravitation, 1955; (monograph) Works on Quantum Field Theory, 1957; A Poly-electronic Problem of Quantum Mechanics and Atomic Structure, 1957. Research and numerous publs. on quantum electro-dynamics, electromagnetic diffraction and propagation, gen. theory of relativity, math. physics, solution of 2-dimensional static problems in theory of elasticity; proved gauge invariance of wave equation of charged particle in magnetic field, Klien-Fok equation of quantum electrodynamics for zero spin particles, 2d quantization, reinterpreted Einstein's field equations. Address: Physical Institute, University of Leningrad, Leningrad 164, USSR.

FOKKER, Adriaan Daniël, Dutch physicist; b. Buitenzorg, Java, Aug. 17, 1887; s. Anthony Herman Gerard and Suzanna A. (Der Kinderen) F.; Ph.D., U. Leyden (Netherlands), 1913; m. Marg. J. J. Kessler, 1915; m. 2d, Teuntje Van Duk, Dec. 16, 1936; children—Margaretha Suzanna (Mrs. Jan Stoffel), Adriaan Daniel. Grammar sch. tchr., Delft, 1921-23; prof. physics Tech. U. Delft (Netherlands), 1923-27; chief physics dept. and lab. Teylers Mus., Haarlem, Netherlands, also Teylers prof. physics U. Leyden, 1928-55. Mem. Royal Netherlands Acad. Scis., Nederlandse Natuurkundige Vereniging (pres. 1930-36, hon. mem. 1960), Ned. Akoestisch Gen., Brit. Assn., French Soc. Physics, Lid Hollandse Meatschappy der Wetenschappen, Lid Bataafs Genootschap der proefondervindelyke uysbegeete. Author: Relativiteits theorie, 1929; Rekenkundige bespiegeling der muziek, 1945; Just intonation, 1949; Time and Space, 1960; Neue Musik mit 31 Tönen, 1966; also papers. Founder, editor (with others) Nederl. Tydschr. Natuurkunde, 1921-33; founder, editor in charge Physics intern. jour., 1933-59; editor Archives du Musee Teyler, 3d series, 1927-53. Author Fokker-Planck equation in Brownian movement theory; champion of Huygen's refined musical scale of 31 notes in octave; had a 31-key organ built, 1950; contbns. in regard to geodesic precession of free-falling reference systems in rotation-free (parallel) motion, multiple antanairesis, approximation of irrational proportions by integer numbers, body and soul in light of physics, concepts of zero interval and zero mass. Address: 32 Engelander weg., Beekbergen, Netherlands.*

FOKKER, Anthony Hermann Gerard, aero. engr., inventor; b. Kediri, Java, Apr. 6, 1890; s. Anthony G. Fokker; m. Elisabeth von Morgen, 1919 (div. 1923); m. 2d, Viola Lawrence, 1929. Passed pilot exam, also built his 1st single decker airplane, 1911; founder flying sch. (became airplane factory during World War I), Schwerin, Germany, 1914; moved operations to Amsterdam after war; built planes used by KLM, until 1936; came to U. S., 1922, head Fokker Aircraft Corp. Am., several years. Author: Der fliegende Hollander, 1933. Pioneer in design and constrn. of airplanes; built prototype of German fighter planes used in World War I; designed comml. transport planes; 1st to devise synchronous gear for firing machine gun through airplane propeller; developed welded steel fuselage constrn. Died N.Y., Dec. 23, 1939.

FOL, Hermann, physician, zoologist; b. St.-Mandé/Paris, France, July 23, 1845; student of Haeckel; prof. comparative evolution Geneva, Switzerland. Author: Lehrbuch der vergleichenden mikroskopischen Anatomie, 1884. Fundamental research on processes of fertilization; 1st to observe entrance of sperm into starfish egg, 1877; confirmed Hertwig's belief that a single sperm accomplishes fertilization. Died in Atlantic Ocean, 1892.

FOLAND, William Douglas, Am. physicist; b. Knoxville, Tenn., Jan. 15, 1926; s. Fred Douglas and Lou Emma (Boswell) F.; A.B., U. Tenn., 1951, M.S., 1955, Ph.D., 1958; postgrad. U. Innsbruck (Austria); m. Johanna Minna Krüger, June 10, 1955. Faculty, U. Tenn., 1955-58; faculty Univ. of Mass., Amherst, 1958-—, asso. prof. physics, 1964-—. Mem. Am. Phys. Soc., Am. Assn. U. Profs., Sigma Xi, Phi Kappa Phi. Research on explanation fission very heavy nuclides, math. formulation surface energy for charged liquid drop having distorted surface. Home: Central St., Montague, Mass. 01351. Office: Hasbrouck Lab., U. Mass., Amherst, Mass. 01003.*

FOLBORT, Georgii Vladimirovich, Russian physiologist; b. St. Petersburg, Russia, Feb. 16, 1885; grad. St. Petersburg Mil. Med. Acad., 1909. Collaborator, I. P. Pavlov's Physiology Lab.; asst. physiology dept. Mil. Med. Acad., 1912-14, sr. instr., dep. head dept., 1914-24; prof. animal physiology dept. Higher Women's Agrl. Courses, 1916-23; with physiology dept. Leningrad State U., 1923-26; head normal physiology dept. Kharkov Med. Inst., 1926-—; organizer lab. for study physiology of digestion Ukrainian Inst. Exptl. Endocrinology, 1927; founder sect. physiology Ukrainian Inst. Exptl. Biology and Medicine; dir. Lab. Conditioned Reflexes, Ukrainian Psychoneurol. Inst., 1926-—; dir. sect. normal physiology Physiology Inst., Ukrainian Acad. Scis., 1946-—. Mem. Ukrainian Acad. Scis. (Pavlov prize). Research and numerous publs. on study of conditioned reflexes, higher nervous activity, physiology of digestion, endocrinology and other fields of physiology, process of rest and fatigue.

FOLDES, Francis Ferenc, anesthesiologist; b. Budapest, Hungary, June 13, 1910; s. Lipot and Nelli (Friedman) F.; M.D., Pazmany Peter U., Budapest Med. Sch., 1934; m. Edith Ribary, Oct. 9, 1938; children—Eva., Judith, Barbara. Came to U. S., 1941, naturalized, 1947. Practice medicine specializing in anesthesiology, researcher applied pharmacology, Budapest, 1939-41; dir. dept. anesthesiology Mercy Hosp., Pitts., 1947-62; chief div. anesthesiology Montefiore Hosp. and Med. Center, N.Y.C., 1962-—; faculty U. Pitts. Med. Sch., 1948-62; clin. prof. anesthesiology Columbia Coll. Phys. and Surg., 1962-64; prof. anesthesiology Albert Einstein Coll. Medicine, 1964-—; research asso. medicine Mt. Sinai Hosp., N.Y.C., 1962-—. Fellow Am. Coll. Anesthesiologists, N.Y. Acad. Medicine; mem. A.M.A., Internat. Anesthesiology Research Soc., Assn. U. Anesthetists, N.Y. Acad. Scis., Am. Chem. Soc., Am., N.Y. State socs. anesthesiologists. Author: Muscle Relaxants in Anesthesiology, 1957; (with M. Swerdlow, E. S. Siker) Narcotics and Narcotic Antagonists, 1964; also numerous articles. Research on devel. new anesthetic techniques involving use neuromuscular blocking agts., narcotics, narcotic antagonists, local anesthetic agts. Home: 823 Long Hill Rd. W., Briarcliff Manor, N.Y., 10510. Office: Montefiore Hosp. and Med. Center, 111 E. 210th St., Bronx, N.Y. 10467.*

FOLDI, Michael, Hungarian physician; b. Budapest, Hungary, Jan. 10, 1920; s. Michael and Irén (Havas) F.; M.D., Med. U., Szeged, 1944, D.Sc., 1959; m. Ilona Mikuska, June 2, 1941; children—Mary, Katherine. Faculty, 1st Med. Clinic, Budapest, 1945-60; prof. internal medicine Med. U., Szeged, Hungary, 1961-—, dir. 2d Med. Clinic. Mem. Internat. Soc. Lymphology (v.p.), Renal Soc. London, Internat. Soc. Internal Medicine, Hungarian Physiol. Soc. Author: (with Rusznyák and Szabó) Lymphatics and Lymph Circulation, 1960; (with Szabó) The Regulation of Sodium Excretion, 1959; (with Csillik) Histophysics and Histochemistry of Lymphstasis, 1965. Publs. on systematic study of consequences of lymph stasis in various organs leading to description of a new brain disease, lymphogenous encephalopathy, curable with vitamin therapy. Address: 8-10 Koranyi S., Szeged, Hungary.*

FOLEY, Arthur Lee, Am. physicist; b. Hancock County, Ind., Feb. 22, 1867; s. Mansfield Calvin and Clara Alice (Myers) F.; student Central Normal Coll., Danville, 1882-84, Hayward Coll., 1886-87; A.B., Ind. U., 1890, A.M., 1891; postgrad. U. Chgo., 1894; Ph.D., Cornell, 1897; postgrad. European univs., 1908; m. Lorettie Hayworth, Apr. 15, 1885; children—Ernest Lee, Mrs. Eupha May Tugman. Tchr. pub. sch., Vivalia, Ind., 1884-85; prin. pub. sch., Cleveland, Ind., 1885-86, Johnsonville, Ill., 1887-88; instr. physics Ind. U., 1890-91, asso. prof., 1891-97, prof. and head of dept. physics, 1897-1938, emeritus prof. from 1938, Waterman research prof., 1917-25, also chmn. Bur. of Science Service, 1926-35. Dir. of physics research, Affiliated Coll. of Ind., 1917-25; dir. U. S. Radio Sch., Ind. U., 1918; mem. NRC. Fellow A.A.A.S., Ind. Acad. Sci. (pres. 1909; chmn. research com. since 1924), Am. Phys. Soc., Acoustical Soc. Am.; mem. Sigma Xi (pres. 1892), Phi Beta Kappa (pres. 1914). Acoustical engineer, inventor, investigator, contbr. sci. papers, author 4 editions of a physics text. Died Feb. 13, 1945.

FOLEY, George Edward, Am. microbiologist; b. Mechanicville, N.Y., Dec. 1, 1912; s. Edward Joseph and Carolyn M. (Whitman) F.; Sc.D., Universiteit van Amsterdam, 1954; m. Helen A. Gundesen, June 18, 1941; 1 dau., Linda W. Chief, labs. microbiology Children's Cancer Research Found., Boston, 1947-—, bacteriologist, research asso. in pathology Children's Hosp. Med. Center, 1952-—, research asso. pathology, 1952-64; lectr. pathology Harvard Med. Sch., 1964-—. Cons. Nat. Cancer Inst., NIH, USPHS, 1956-—. Recipient Research Career award Nat. Cancer Inst. NIH, 1964. Diplomate Am. Bd. Med. Microbiology. Fellow Am. Pub. Health Assn., A.A.A.S., Am. Acad. Microbiology; mem. Am. Soc. Microbiology (past br. pres.), Am. Soc. Profl. Biologists, Am. Soc. Exptl. Pathology, Am. Assn. Cancer Research, Am. Assn. Pathologists and Bacteriologists, Internat. Assn. Microbiology, N.Y. Acad. Scis., Royal Soc. Health, Sigma Xi. Editorial bd. Antibiotics and Chemotherapy, 1960-62; Cancer Research, 1961-64; regional editor Exptl. Cell Research, 1962-—. Contbr. Research, numerous publs. in diagnostic pediatric microbiology and mycology; cell biology, studies on etiology, immunology, chemotherapy of cancer. Home: 16 Clifton Park, Melrose, Mass. 02176. Office: 35 Binney St., Boston 02115.*

FOLEY, Henry Michael, Am. physicist; b. Palmer, Mass., June 1, 1917; s. Henry M. and Rosemary (O'Neill) F.; B.S., U. Mich., 1938, M.S., 1939, Ph.D., 1942; m. Barbara Mallard, Apr. 19, 1959; children—David, Barbara. With OSRD, Ann Arbor, Mich., 1941-44, Applied Physics Lab., Silver Spring, Md., 1944-45; faculty Columbia, N.Y.C., 1946-—, prof. physics, 1954-—, chmn. dept., 1957-60. Cons. Inst. Def. Analyses, Washington; cons. editor W. H. Freeman Co., San Francisco; mem. advanced physics examination com. Ednl. Testing Service. Recipient Naval Ordnance Devel. award, 1946. Research on radiation theory; infrared spectroscopy; atomic beams and magnetic moments; nuclear quadruple moments; pressure broadening of spectral lines. Home: 460 Riverside Dr., N.Y.C. 10027.

FOLIN, Otto, chemist; b. Sweden, 1867; prof. biochemistry Harvard Med. Sch.; proposed theory of endogenous and exogenous metabolism, 1905; introduced new micro-methods for determination of urea, total nitrogen, ammonia nitrogen, 1912. Died 1934.

FOLINSBEE, Robert Edward, Canadian geologist; b. Edmonton, Alta., Can., Apr. 16, 1917; s. Francis John and Elizabeth (Woolverton) F.; B.Sc., U. Alta., 1938; M.S., U. Minn. 1940, Ph.D., 1942; postgrad. Harvard, U. Cal. at Berkeley; m. Catherine Elizabeth Terwillegar, July 6, 1942; children—Robert Allin, John David, James Terwillegar, Catherine Dee. With Geol. Survey Can., 1936-46; faculty U. Alta., Edmonton, 1946-—, prof., chmn. dept. geology, 1955-—. Fellow Royal Soc. Can., Geol. Assn. Can., Geol. Soc. Am.; mem. Canadian Inst. Mining and Metallurgy (Past Pres.'s medal 1963), Soc. Econ. Geologists, Alta. Soc. Petroleum Geologists, Am. Assn. Petroleum Geologists, Am. Geophys. Union, Edmonton Geol. Soc., Archaeol. Soc. Am. Research, publs. in field applied geochemistry. Home: 11711 Edinboro Rd., Edmonton, Alta., Can.*

FOLKERS, Karl August, American chemist; b. Decatur, Ill., Sept. 1, 1906; s. August William and Laura Susan (Black) F.; B.S., U. Ill., 1928; Ph.D., U. Wis., 1931; post-doctorate research, Yale U., 1931-34; D.Sc., Phila. Coll. Pharmacy and Sci., 1962; m. Selma Leona Johnson, July 30, 1932; children—Cynthia Carol (Mrs. James D. Jamieson), Richard Karl. With Merck & Co., Inc., 1934-—, asst. dir. research, 1938-45, dir. organic and biochemical research, 1945-—, asso. dir. research and devel., 1951-53, dir. organic & biol. chem. research, 1953-56, exec. dir. Fundamental Research, 1956-62, v.p. exploratory research, 1962-63; pres. Stanford Research Inst. Menlo Park, Cal., 1963-68; courtesy prof. chemistry Stanford, 1963-—; prof., dir. Inst. for Biomed. Research, U. Tex., Austin, 1968-—. Baker non-resident lectr. in chemistry Cornell, 1953; regent's lectr. U. Cal. at Los Angeles, 1960; lectr. vitamin chemistry U. Cal. Berkeley, 1963-—; mem. sci. adv. com. Inst. Microbiology, Rutgers, Chmn. adv. council dept. chemistry Princeton, 1958-64. Recipient Am. Chem. Soc. award in pure chemistry, 1941, Spencer award, 1959; Julius Sturmer Lecture award, 1957; Perkin medal Soc. Chem. Industry, 1960. Mem. Nat. Acad. Sci., Am. Chem. Soc. (pres. 1962), Am. Soc. Biol. Chemistry, Am. Inst. Nutrition, Society Exptl. Biology, medicine, N.Y. Acad. Science Am. Soc. Biol. Chemistry, Am. Inst. Nutrition, Soc. Exptl. Biology and Medicine, A.A.A.S., Am. Inst. Chemists, Sigma Xi, Phi Lambda Upsilon, Alpha Chi Sigma. Contbr. sci. jours. on organic chemistry. Author: (with Arthur Wagner) Vitamins and Coenzymes, 1964; also articles. Contbns. in therapeutic agts., particularly vitamins and antibiotics; isolated vitamin B-12, which proved effective in treatment pernicious anemia, 1948; worked on isolation, synthesis and structure of antibiotics in streptomycin series, also on penicillin, neomycin, subtilin, grisein, oxamycin and novobiocin; pioneered in study of Erythrina alkaloid family; discovered mevalonic acid, fundamental in biosynthesis of numerous steroids, carotenoids and terpenes. Address: Inst. for Biomed. Research, U. Tex., Austin, Tex.*

FOLKOW, Björn U., Swedish physiologist; b. Halmstad, Sweden, Oct. 13, 1921; s. Folke U. and Sigrid (Carlson) F.; M.D., U. Lund (Sweden), 1948, Ph.D. in Psychology, 1949; m. Björg U. Hellern, Dec. 28, 1957; children—Viveka, Lars, Helena. Asst. prof. physiology U. Lund, 1949; asso. prof. physiology U. Göteborg (Sweden), from 1950, prof. physiology, 1961-—. Dep. mem. Swedish Med. Research Council; mem. Stouffer prize com., Cleve. Research, publs. concerning integration of local, reflex and central nervous control mechanisms of importance for cardiovascular function; functional orgn. of autonomic nervous system. Home: 38, Gamla Särövägen, Askim, Sweden. Office: 11, Medicinaregatan 11, Göteborg SU, Sweden.*

FOLLETT, Mary Parker, Am. management scientist, political and social philosopher; b. Boston, Mass., 1868; d. Charles Allen and Elizabeth (Curtis) F.; B.A., Radcliffe College, 1898; studied at Newnham College, Cambridge, and in Paris. Engaged in social work, Boston, 1891-16; lectr. in U. S. and England on business management, 1924-28. Mem., Taylor Soc., Nat. Community Centre Assn. Author: The Speaker of the House of Representatives, 1896; The New State, 1918; Creative Experience, 1924; (H. C. Metcalf and L. Urwick, eds.) Dynamic Administration: The Collected Papers of Mary Parker Follett, 1941; (L. Urwick, ed.) Freedom and Co-ordination: Lectures in Business Organization by Mary Parker Follett, 1949. Worked in development of vocational guidance, community centers, civic education; applied social sciences and psychology to management; developed principles of human organization and association in industry based on concept of co-ordination. Died U. S., Dec. 18, 1933.

FOLLEY, Karl Wilmot, mathematician, educator; b. Elgin, Man., Can., Apr. 24, 1905; s. Robert C. and Ada (Sharpe) F.; B.Sc., U. Sask., 1924, M.Sc., 1925; Ph.D., U. Toronto, 1928; m. Helen L. Strange, Mar. 10, 1934; 1 son, Ronald L. Came to U. S., 1929, naturalized, 1932. Asst. prof. Trinity Coll., Hartford, Conn., 1928-29; faculty math Wayne State U., Detroit, 1929-—, prof., 1947-—. Vis. prof. U. S. Naval Postgrad. Sch., Monterey, Cal., 1949-50; sr. mathematician System Devel. Co., Santa Monica, Cal., 1959-60; asso. program dir. NSF, Washington, 1963-64. Am. Assn. U. Profs., Am. Math. Soc., Math. Assn. Am., Sigma Xi. Author: Plane Trigonometry, 1936; Analytic Geometry, 1949; Calculus, 1946; (with Uelson, Coral) Differential Equations, 1964. Research, publs. on point set theory and transfinite number theory. Home: 19230 Gainsborough St., Detroit 48223.*

FOLLI, Cecilio, Italian anatomist; b. Fanano, Italy, 1615; prof. of anatomy, Venice. Described fenestra rotunda and fenestra ovalis of ear, 1645; described "Folian process", arising from anterior aspect of neck of malleus and attached to petrous portion of temporal bone, 1645. Died 1660.

FOLLI, Francesco, Italian physician; b. Poppi in Casentino, Italy, May 31, 1623; s. Domenico di Girolamo and Orsina (Dombosi) F. Author: Stadeva Medici, 1680. Early advocate of blood transfusions; credited with performing transfusion using combination silver tube, bone cannula and animal blood vessel, 1654. Died 1685.

FOLTINY, Stephen, archaeologist; b. Nyiregyhaza, Hungary, July 20, 1918; s. Akos George and Julia (Moroz) F.; Ph.D. sub auspiciis Gubernatoris, Szeged (Hungary) U., 1941; Ph.D., Vienna (Austria) U., 1950; postgrad. U. Berlin (Germany), U. Innsbruck (Austria); m. Ilona Kovacs, Nov. 29, 1949; children—Maria Edith, Stephen Vincent, Peter Joseph, Andrew Akos, Maria Theresa, Miklos Csaba, Tamás Zsolt. Came to the United States of America, in 1951, naturalized, 1957. Research assistant Szeged U., Inst. Archeology, 1940-44; sci. asst. Inst. Prehistory, Vienna U., 1946-50; research asst., librarian Wenner-Gren Found., N.Y.C., 1952-55; research asst. Sch. Hist. Studies, Inst. for Advanced Study, Princeton, N.J., 1955-61, research asso., 1961-—. Wennergren Found. grantee, 1958-59; Am. Philos. Soc. grantee, 1960, 65; grantee Am. Council Learned Socs., 1960,

65, NSF, 1961-64, 67-68; Smithsonian Fgn. Currency Program Archaeology grantee, 1966-70. Fellow Am. Anthropol. Assn.; corr. mem. Deutsches Archaeologisches Institut, Österreichische Arbeitsgemrinshaft für Ur-und Frühgeschichte; mem. Archael. Inst. Am., Anthropologische Gesellschaft in Wien. Author: Das Bronzezeitliche Graberfeld in Szöreg, 1941; Zur Chronologie der Bronzezeit des Karpatenbeckens, 1955; Velemszentvid, 1958; Die Hügelgräberkultur in der Umgebung von Szeged, 1957; The Hungarian Archaeological Collection of the American Museum of Natural History in New York, 1968; also numerous articles, book revs. Research on cultural and econ. history Bronze and Early Iron Age, in Carpathian Basin, cultural interrelations Central Europe and Mediterranean, influence mounted nomads in Bronze and early Iron Age S.E. Europe and N. Italy, Homeric archaeology. Home: 255 Ewing St., Princeton 08540. Office: Inst. for Advanced Study, Princeton, N.J. 08540.*

FOMENKO, Valentine Trofimovich, Russian mathematician; b. nr. Rostov, USSR, July 15, 1938; s. T. Y. and Tatyana (Kolody) F.; ed. Rostov U., 1960, D.Natural Scis., 1963; m. Svetlana Varganova, Apr. 29, 1960; 1 dau., Tatyana. Postgrad. student Rostov U., 1960-62, mem. faculty, 1962—, docent, 1964-65, prof. chair geometry, 1965—. Author numerous articles. Research in prin. results obtained in field of differential geometry as a whole; questions of single-valued definiteness and continuous bendings of convex surfaces with edges; infinitesimal bendings of surfaces of the positive curvature of bushing connections in Euclidean and Reimannian spaces. Home: 122 Djuravleva str. Office: 105 Engels str., Rostov-on-Don, USSR.*

FOMON, Samuel, Am. plastic surgeon; b. Chgo., Jan. 11, 1889; s. Jacob and Rose (Field) F.; M.D., U. Ill., 1906; m. Lenore Davis, Oct. 1944; children—John, Samuel, Robert, Lenore Marie. Practice medicine, specializing in plastic surgery, N.Y.C., 1935—; dir. plastic surgery Manhattan Gen. div. Beth Israel Hosp.; provisional cons. Polyclinic Med. Sch. and Hosp., St. Joseph's Hosp. Recipient R. Clyde Lynch Meml. award, 1953. Mem. A.M.A., Otorhinologic Soc. (hon. pres.), N.Y. Acad. Plastic Surgery (dir.). Author: Medicine and Allied Sciences (3 vols.), 1919; Surgery of Injury and Plastic Repair, 1939; Otolaryngologic Plastic Surgery, 1955; Cosmetic Surgery Principle and Practice, 1960; also numerous articles. Home: 4805 Arlington Av., Bronx, N.Y. 10471. Office: 307 2d Av., N.Y.C. 10003.*

FONDA, Luciano, physicist; b. Pola, Yugoslavia, Dec. 12, 1931; s. Marcello and Lucia (Paliaga) F.; D.Physics, U. Trieste, 1955; m. Thea Arcangeli, Dec. 26, 1957; children—Alessandro, Paola, Gabriella. Asst. prof. Trieste U., 1955-58; research asso. Ind. U., Bloomington, 1958-59; mem. Inst. for Advanced Study, Princeton, N.J., 1958-59; prof. quantum mechanics Palermo (Italy) U., 1961-62, Parma (Italy), 1962-63; prof. Trieste (Italy), 1963—, dir. Istituto di Fisica Teorica, 1966—. Mem. Società Italiana di Fisica, Am. Phys. Soc. Research, publs. on pion-nucleon, nucleon-nucleon, and hyperon-nucleon interaction, hyperfragments, threshold effects, resonances in nuclear (or other) collisions, analytic properties of scattering amplitudes. Home: 1 Ressman, Trieste, Italy.*

FONDARAI, Joseph August, French statistician; b. Marseilles, France, Dec. 22, 1923; s. Raoul Albert and Nella (Pietri) F.; D. Pharmacy, 1948; Licence Sciences, 1949; D.Sc., 1965; m. Sabine Albrand, Mar. 22, 1956; children—Anne, Beatrice. Asst., Faculty Medicine, Marseilles, from 1950; head physics lab., from 1958; instr. physics, statistics, from 1964. Asst. Hotel Dieu de Marseilles. Mem. Société de Biologie; Assn. Information and Operational Research. Research, publs. on peroxide lipids, action of phys. agts. on living tissue, statistics applied to biology (especially field of multivariate analysis and establishment of surveys concerning med. research). Home: 62 Jules Cantini, 13 Marseilles 8°. Office: Blvd. Jean Moulin, 13 Marseilles, France.*

FONG, Jacob, Am. bacteriologist; b. Canton, China, Oct. 29, 1913 (parents Am. citizens); s. Ying King and Sutie (Wong) F.; A.B., U. Cal. at Berkeley, 1934; Ph.D., U. So. Cal., 1944; m. Rose Yee, Aug. 19, 1935; 1 son, Alan Thomas. Bacteriologist, Los Angeles County Med. Assn., 1938-44, Cutter Labs., 1944-45; faculty U. Cal. at Berkeley, 1947—, prof. bacteriology, 1960—, chmn. dept., 1963—. Mem. Am. Assn. Immunologists, Am. Soc. Microbiology, Soc. Exptl. Biology and Medicine, Sigma Xi, Phi Sigma. Research on pathogenesis and immunology of infectious diseases, particularly Tb and influenza; cellular immunity, bacteriophage, replication of myxoviruses, and mammalian cell cultures. Died, Feb. 28, 1967.

FONG, Peter, physicist; b. Tungshang, Chekiang, China, Sept. 3, 1924; s. Pin-Lan and Han-Yin (Mao) F.; B.S., U. Chekiang, 1945; M.S., U. Chgo., 1950, Ph.D., 1953; m. Teresa Wai, Jan. 17, 1959; children—Nora Lillian, Karen Elizabeth, William Peter. Physicist, Inst. Nuclear Studies U. Chgo., 1954; faculty Utica Coll. of Syracuse U., 1954-66, prof. physics, 1962-66; vis. prof. physics Cornell U., 1963-64; vis. prof. Lawrence Radiation Lab. U. Cal., Berkeley, 1965-66; prof. physics Emory U., Atlanta, 1966—. Mem. Am. Phys. Soc., Biophys. Soc., Sigma Xi, Sigma Pi Sigma. Author: Elementary Quantum Mechanics, 1962; Foundations of Thermodynamics, 1963; also articles. Author statis. theory of nuclear fission explaining fission phenomena from first principles; contbr. theory of origin of chem. elements in astrophysics; author of new formulation of thermodynamics as macroscopic theory of equilibrium; author statis. theory of biol. functions of nucleic acids (DNA and RNA) and of origin and evolution of life. Office: Physics Dept., Emory U., Atlanta 30322.*

FONTAINE, Hippolyte, French elec. engr.; b. Dijon, France, 1833; ed. Chalons-sur-Marne Sch.; began as indsl. designer; became head constrns. Cail & Co., 1857; mem. Internat. Soc. Electricians (pres.). Founder jour. Revue industrielle. Discovered reversibility of dynamo and built electric engine using this principle; 1st transmission of elec. energy, Vienne, France, 1873. Died Paris, 1917.

FONTAINE, Maurice Alfred, French marine biologist; b. Savigny-sur-Orge, France, Oct. 28, 1904; s. Emile and Lea (Vadier) F.; grad. Paris (France) Faculty Pharmacy, 1928; Dr. ès sc., Paris Faculty Sci., 1930; m. Yvonne Broca, Aug. 2, 1928; 1 son, Yves-Alain. Asst. Faculty Sci., Paris, 1931, research chief, from 1932; instr. Paris Faculty Pharmacy, 1941-43; instr. comparative physiology Sorbonne, Paris U., 1940-44; prof. Nat. Mus. Natural History, Paris, from 1943, dir., 1966—; dir. lab. Sch. Advanced Studies, from 1946; prof. Paris Oceanography Inst., from 1956, also dir. Mem. Nat. Sci. Research Com., 1946—, council Sci. Inst. for Techniques Ocean Fishery; co-dir. Biol. Year; councillor Physico-Chem. Biology Inst. Recipient Order Agrl. Merit, Johann Schmidt medal, others. Mem. Acad. Sci., Nat. Acad. Medicine, Agrl. Acad., Acad. Pharmacy, N.Y. Acad. Sci., European Comparative Endocrinology Assn. (v.p.), Biochem. Soc. (pres.), Physiol. Soc. (pres.). Author: Physiologie. Research, publs. on fluorescent pigments, especially in riboflavin; biosynthesis vitamin B2 in mushroom Eremoth ashbii; physiol. implications role of osmo-regulator and neuroendocrine systems in fish; intervention of inter-renals and stannius corpuscles; studies on pollution by radioactive elements, phosphorus-calcium metabolism. Home: 25 rue Pierre Nicole. Office: 57 rue Cuvier, Paris 5, France.*

FONTAINE DES BERTINS, Alexis, French mathematician; b. Bourg-Argenta, France, Aug. 13, 1704; studied math. in Paris, France; corresponded with Clairaut; mem. French Acad. Scis., 1733. Author: Mémoires de mathematiques, 1764. Investigated tautochrones, especially those in which resistance is represented by 2d degree trinomial; introduced present system of notation of partial derivatives of functions of several variables. Died Cuiseaux, France, Aug. 21, 1771.

FONTAN, Francis Maurice, French cardiac surgeon; b. Nay, France, July 3, 1929; s. Victor and Jane (Larquir) F.; student Coll. St. Joseph; Baccalauréat, College Immaculée Conception, Pau, France; m. Monique Hoyon, Apr. 30, 1964; children—Marie, Etienne. Surgeon asst., Bordeaux, France, 1960-63; surgeon hosps. of Bordeaux, 1963—; prof. agrégé surgery U. Bordeaux, 1963—. Mem. French Soc. Cardiology, French Soc. Thoracic Surgery, French Coll. Vascular Diseases. Author: Anoxic Cardiac Arrest in Extra-Corporeal Circulation, 1959; also numerous articles. Research in cardiovascular diseases and cardiovascular surgery. Home: 235 Rue Judaique. Office: Hopital du Tondu, Bordeaux, France.*

FONTAN, Jacques, French physicist; b. Brive, France, July 8, 1937; s. Raoul and Paulette (Delbos) F.; D.Sc., U. Toulouse, 1964. Asst., Faculty Scis., U. Toulouse, 1958-61, chief asst., 1961-66, lectr., 1966—. Mem. French socs. Physics, Radioprotection. Research, publs. on atmospheric radioactivity especially natural radioactivity and its application to problems of geophysics. Home: 1 Rue Marceau. Office: 118 Route de Narbonne, Toulouse, France.*

FONTANA, Domenico, Italian architect; b. Mili, Italy, 1543; architect, Rome, Italy; chief architect for Pope Sixtus V. Author: Del Modo Tenuto nel Transportare l'Obelisco Vaticano, 1589. Important works include Egyptian obelisk (in front of St. Peter's Ch., Rome), 1586; obelisks in Piazza del Popolo and San Giovanni Laterano, Rome; Palace of the Lateran, Vatican Library, Quirinal Palace, Rome; Grand Royal Palace, Naples, Italy, 1592. Used phys. properties of cable shortened by wetting in order to install the obelisk in St. Peter's Sq., 1586. Died 1607.

FONTANA, Felice, Italian naturalist, physiologist; b. Pomarolo, Italy, Apr. 15, 1720; ed. Padua, Bologna, Italy; tchr. philosophy, Pisa, Italy; ct. physiologist to Duke Peter Leopold, Florence, Italy; developed mus. at Florence, adding wax anat. models (duplicated for Vienna at request of Joseph II), later became curator. Author: Nuove osservaz. sopra i globetti rossi del sangue, 1766; Dei moti del iride, 1767; Ric. filos. sopra il veleno della vipera, 1767; Traité sur le venin de la vipère . . . , 1781; Choix d'observations physiques et chirurgicales, 1785. Gave 1st description of cell as uniform body with spot in center, 1781; 1st to describe tubules of kidney and nerve sheaths named after Henle; sinus venosus of sclera and space at angle of iris named in his honor: established that hydatids in brain cause sheep disease staggers. Died Florence, Mar. 9, 1805.

FONTANA, Franciscus, Italian astronomer; b. circa 1600; lived in Naples, Italy; pioneer in use of telescope; 1st to see markings on Mars, 1636; observed irregularities on inner edge of crescent of Venus, 1643; probably 1st to substitute convex for concave eye lens in microscope. Author: New Observations on Celestial and Earthly Things, 1646.

FONTANA, Gregorio, Italian mathematician; b. Villa-de-Nogarola, Italy, 1735. Prof. logic and metaphysics, U. Pavia, 1763; prof. math., U. Pavia; apptd. by Napoleon to consulting council on Cisalpine territory, 1800. Author: Analyseos sublimioris opuscula, 1763; Memorie mathematiche, 1796. Inventor of polar co-ordinates, circa 1775; wrote several works on mathematics; translated English works. Died 1803.

FONTANA, Mars Guy, Am. metall. engr., educator; b. Iron Mountain, Mich., Apr. 6, 1910; s. Dominic and Rosalie (Amico) LF.; B.S., U. Mich., 1931; M.S., 1932; Ph.D., 1935; m. Elizabeth Frances Carley, Aug. 21, 1937; children—Martha Jane (Mrs. Gerald W. Worth), Mary Elizabeth, David Carley, Thomas Edward. Metall. engr., supr. engring. dept. Du Pont Co., Wilmington, Del., 1934-45; prof., chmn. dept. metall. engring. Ohio State U., Columbus, 1945—, dir. Corrosion Center, 1946—, supr. metall. research Research Found., 1946—, Regents' prof., 1967—; cons. engr., Columbus, 1945—. Dir. Fontana Aviation, Inc., Iron Mountain; pres., dir. Port au Villa, Inc., Naples, Fla., 1967. Recipient Distinguished Alumnus citation University of Michigan, 1953, Sesquicentennial award U. of Mich., 1967. Mem. Nat. Assn. Corrosion Engrs. (Frank Newman award 1956, past pres.), Am. Soc. Metals, Am. Inst. Mining and Metall. Engrs., Am. Inst. Chem. Engrs., Electrochem. Soc., Am. Soc. Engring. Edn., Sigma Xi, Alpha Chi Sigma, Tau Beta Pi, Phi Lambda Upsilon, Iota Alpha, Phi Eta Sigma Author: Corrosion: A Compilation, 1957; Corrosion Engineering, 1967; also numerous articles. Editor: Jour. Corrosion, 1962—. Research on corrosion; established several mechanisms for different types corrosion, patentee corrosion resistant alloys, corrosion devices, iron ore direct reduction. Home: 2086 Elgin Rd., Columbus, O. 43221.*

FONTANA, Nicola, see Tartaglia, Niccolo.

FONTANEY, Father Jean de, see de Fontanay.

FONTANIEU, Pierre-(Elisabeth), French chemist; b. circa 1730; s. Victor Fontanieu; attendant royal furnishings, France; mem. Acad. Architecture, French Acad. Scis., 1778. Author: L'Art de faire les cristaux colorés imitant les pierres précieuses, 1778; also wrote on art of enamelling and on making crystals which simulate precious stones. Died Paris, France, May 30, 1784.

FONTENELLE, Bernard Le Bovier de, French scientist, man of letters; b. Rouen, France, Feb. 11, 1657; ed. local Jesuit collège; then studied law. Mem., French Acad. Scis. (perpetual sec., 1697-1739); elected to Académie Française, 1691, Académie des Inscriptions, 1701. Author: Nouveaux Dialogues des morts, 1683-84; Entretiens sur la pluralité des mondes, 1686; L'histoire des oracles, 1687; Digression sur les anciens et les modernes, 1688; De l'origine des fables, 1689; Histoire de l'Académie Royale des sciences, 1699-1740; Histoire du Renouvellement de l'Académie, 1708. Great popularizer of philosophical and astronomical issues connected with new Copernican system; served French Acad. of Scis. as perpetual sec. for many years, writing famous obituary notices of prominent European scientists and annual history of Academy's activities. Died Paris, France, Jan. 9, 1757.

FOOSE, Richard Martin, Am. geologist; b. Lancaster, Pa., Oct. 9, 1915; s. Leon K. and Grace (Leinbach) F.; B.S., Franklin and Marshall Coll., 1937; M.S., Northwestern U., 1939; Ph.D., Johns Hopkins, 1942; also M.A. (honorary), Amherst College, 1964; m. Dorothy Jane Kell, Feb. 11, 1943; children—Michele Leslie, Michael Peter, Stephan, Terry. Instr., Northwestern U., 1937-39; prof. and head dept. geology, Franklin and Marshall Coll. 1946-57; sr. geologist Stanford Research Inst., 1957-63, chmn. dept. earth scis.; prof., chmn. dept. geology Amherst (Mass.) Coll., 1963—; asst. geologist Pa. Geol. Survey, 1939-42, asso. geologist, 1942-43, sr. geologist, 1943-46; geologists Pa. Turnpike Commn., 1941; cons. geologist, 1942—; Ford Found. fellow, research asso. Stanford, 1955-56; Nat. Sci. Found. sr. postdoctoral fellow Eide Technische Huchschule, Zurich, Switzerland, 1962-63. Fellow A.A.A.S., Geol. Soc. Am., Am. Geog. Soc., mem. Soc. Econ. Geologists (councillor 1954-57), Am. Inst. Mining Engrs. (chmn. div. indsl. minerals 1950-51, vice chmn. mineral econ. div. 1953-54, del. internat. geol. congress 1952, 56, 60, chmn. com. mineral economy 1962-63; asso. editor Indsl. Minerals and Rocks 1960), Yellowstone Bighorn Research Assn. (v.p. 1955), Pa. Acad. Sci. (pres. 1949-50), Am. Geol. Inst., Assn. Geol. Tchrs. (del. internat. geol. congress 1956). Am. Geophys. Union, Am. Institute of Professional Geologists (charter), Geochem. Soc., Phi Beta Kappa, Sigma Xi. Contbr. articles profl. jours. Research in economics, structural and engineering geology; synthesized structural geology of Beartooth mountains; authority on behavior of ground water and collapse of surface due to ground water withdrawal. Home: 197 South Pleasant, Amherst, Mass., 01002. Office: Dept. of Geology, Amherst Coll., Amherst, Mass., 01002.*

FOOTE, Frank Gale, Am. metallurgist; b. Owosso, Mich., July 21, 1906; s. Grant Ulysses and Nettie (Archer) F.; A.B., Ohio Wesleyan U., 1928; M.A., Ohio State U., 1930; Ph.D. in Metallurgy, Columbia U., 1941; m. Loretta Winifred Fish, July 6, 1935; 1 dau., Ann Gale. Faculty, Cooper Union, N.Y.C., 1938-42; chief metallurgist Metall. Lab., U. Chgo., 1942-46; asso. prof. Sch. Mines, Columbia U., N.Y.C., 1946-48; dir. metallurgy div. Argonne (Ill.) Nat. Lab., 1948-65, sr. metallurgist, 1965——. Fellow Am. Nuclear Soc., A.A.A.S. (mem. council 1962——); mem. Am. Soc. for Metals, Am. Inst. Mining, Metall. and Petroleum Engrs. Contbr. articles to tech. jours. Pioneered in application of X-ray diffraction to study problem in phys. metallurgy, particularly phase diagrams; directed research and devel. of fuel materials for use in nuclear reactors. Home: 4822 Northcott Av., Downers Grove, Ill. 60515. Office: Argonne Nat. Lab., Bldg. 212, Argonne, Ill. 60439.*

FOOTE, Freeman, Am. geologist; b. Orange, N.J., Nov. 8, 1908; s. Will Howe and Helen (Freeman) F.; A.B., Princeton, 1931; postgrad. Columbia; m. Sara Newnham Carlton, July 22, 1939; 1 dau., Nancy Newnham. Faculty, Williams Coll., Williamstown, Mass., 1937——, prof. geology, 1964——, acting chmn. dept., 1952-53, 62-63, chmn., 1964——; asso. prof. Columbia U., summer 1952-53, Wesleyan U., summer 1955. Mem. selection panel NSF, 1964-65. Fellow Geol. Soc. Am.; mem. Nat. Assn. Geology Tchrs. (sec. 1958-60), Am. Geophys. Union, A.A.A.S. Study of structural geology; petrology, Home: Cold Spring Rd., Williamstown, Mass. 01267.

FOOTE, Joe Reeder, Am. mathematician; b. Amarillo, Tex., Aug. 17, 1919; s. Charles Wilson and Ethel (Reeder) F.; B.S., Tex. Tech. Coll., 1940; postgrad. U. Tex.; Ph.D., Mass. Inst. Tech., 1949; m. Elizabeth Ann Young, July 16, 1949; children—Auburn Jane, Craig Warren, Larry Wayne, Marcia Joyce. Faculty, Ia. State U., Ames, 1949-51, U. Okla., Norman, 1953-57, Purdue U., Lafayette, Ind., 1957-58; applied mathematician USAF, Dayton, O., 1951-53; prof. math., dir. Holloman Grad. Center U. N.M., Holloman AFB, 1958-66; prof. applied math. U. Mo., Rolla, 1966——. Cons. Operations Research, Holloman AFB, 1956-65. Mem. Math. Assn. Am., Am. Assn. U. Profs., Am. Soc. Engring. Edn., Am. Inst. Aeros. and Astronautics (chmn. Holloman sect. 1963-64), A.A.A.S. Researcher in fluid mechanics, 1949-53, parachute opening, 1951-57, flight dynamics, 1956——, ICBM defense, 1960——. Home: P.O. Box 851, Rolla, Mo. 65401.*

FOOTE, John A., Am. physician; b. Archbald, Pa., June 9, 1874; s. John (M.D.) and Margaret (McAndrew) F.; Georgetown Coll., 1900-04; M.D., Georgetown U., 1906; post-grad., Berlin, 1913; m. Lois Gibson Dyer, Oct. 12, 1910; children—Mary Virginia, William Dyer. Practiced in Washington, 1906——; asst. prof. therapeutics and materia medica Georgetown U., 1906-08, asst. prof. anatomy, 1908-10, asso. prof. therapeutics and pharmacology, 1911-17, asso. prof. clin. medicine and diseases of children, 1917-20, prof. diseases of children, from 1920, dean faculty medicine, from 1929; pediatrist Providence and Children's hosps.; cons. pediatrist Foundling and Gallinger hosps. Del. Internat. Med. Congress, London, 1913; ofcl. del. from U. S. to Pan Am. Child Conf., Havana, 1927. Author: Essentials of Materia Medica and Therapeutics, 1910; State Board Questions for Nurses, 1916; Diseases of the New-Born, 1925; Diseases of Bones and Joints in Childhood, 1926. Died Apr. 12, 1931.

FOOTE, Paul D(arwin), Am. physicist; b. Andover, O., Mar. 27, 1888; s. Howard Spencer and Abbie Lottie (Tourgee) F.; A.B., Western Res. U., 1909, D.Sc., 1961; A.M., U. Neb., 1911; Ph.D. U. Minn., 1917; D.Sc., Carnegie Inst. Tech., 1953; m. Berenice C. Foote, Feb. 3, 1913 (dec. June 1939); children—Mrs. C. Jane Halliwell, William Spencer; m. 2d, Miriam Sage, June 26, 1940; stepchild—Evan Sage, Jr. Began as asst. physicist, U. S. Bur. of Standards, 1911, sr. physicist, 1924-27; exec. v.p. Gulf Research & Devel. Co., Pitts.; v.p. Gulf Oil Corp., Gulf Refining Co., ret., 1954; asst. sec. def., research and engring., Dept. of Def., 1957-58; chmn. Nat. Acad. Sci. Panels adv. to Bur. Standards, 1960-65. Recipient Achievement medal U. Minn., 1951; Jr. C. of C. science award, 1953; Pitts. award, 1954; Meritorious Civilian Service medal, 1958. Sr. fellow Mellon Inst. Indsl. Research, 1927-29; fellow Am. Phys. Soc. (pres. 1933), A.A.A.S.; mem. Optical Soc. Am., Am. Philos. Soc. (sec. 1956-59), Nat. Acad. Sci., Washington Acad. Scis. (v.p. 1936), Am. Inst. Mining and Metall. Engrs., Am. Geophysical Union, Phi Beta Kappa, Sigma Xi, Sigma Pi Sigma, Sigma Tau, Tau Beta Pi. Author: (with others) Pyrometric Practice, 1921; (with Fred Loomis Mohler) The Origin of Spectra, 1922; (with others) Physics in Industry, 1937. Editor in chief Jour. Optical Soc. Am., Rev. of Scientific Instruments, 1921-32; asso. editor Jour. Franklin Inst. Research in magnetic optics, pyrometry, atomic structure, radiation, petroleum recovery. Home: 5144 Macomb St. N.W., Washington 16.*

FOOTE, Robert Hutchinson, Am. physiologist; b. Gilead, Conn., Aug. 20, 1922; s. Robert E. and Annie (Hutchinson) F.; B.S., U. Conn., 1943; M.S., Cornell U., 1947, Ph.D., 1950; m. Ruth Evelyn Parcells, Jan. 12, 1946; children—Robert Wesley, Dale Hutchinson. Faculty, Cornell, Ithaca, N.Y., 1950——, prof. physiology, 1963——. Fulbright scholar, 1958-59; Cornell U. Traveling fellow, 1956. Fellow A.A.A.S.; mem. Am. Inst. Biol. Sci., Soc. for Cryobiology, Soc. for Study Fertility, Internat. Fertility Assn., Am. Dairy Sci. Assn., Am. Soc. Animal Sci., Sigma Xi, Phi Kappa Phi. Author: (with R. W. Bratton, G. W. Salisbury, C. R. Henderson) Reproduction of Farm Animals, 1954; also numerous articles. Editorial bd. Internat. Jour. Fertility, Jour. Animal Sci., 1960——. Research on fertility and embryonic mortality in relation to aging and stability of DNA in gametes as well as non-genetic maternal factors, nutrient requirements for culturing sperm and eggs, kinetics of spermatogenesis and oogenesis.*

FÖPPL, August Otto, German physicist; b. Gross-Umstadt, Hessia, Jan. 25, 1854; s. Carl August and Christine (Gimbel) F.; studied engring., Darmstadt, Stuttgart, Karlsruhe (all Germany); diploma, 1874; studied physics under G. Wiedemann, 1883; Ph.D., 1886; hon. dr. engring., Darmstadt, Munich, Germany; m. Emilie Schenck, 1878; children—Otto, Ludwig, Gertrud (Mrs. Ludwig Prandtl), Else (Mrs. Hans Thoma). Tchr., constrn. trade sch., Holzminden, Germany; tchr. Indsl. (Trade) Sch., Leipzig, Germany, 1877-92; became asso. prof. agrl. machinery and cultivation U. Leipzig, 1892; named prof. tech. mechanics Tech. U. Munich, also chmn. mech.-tech. lab., 1894; ret., 1922. Mem. Bavarian Acad. Scis. Author: Theorie des Fachwerks, 1880; Theorie der Gewölbe, 1881; Drang und Zwang, 2 vols., Leitfaden und Aufgabensammlung für den Unterricht in der angewandten Mechanik, 1890; Das Fachwerk im Raum, 1892; Einführung in die Maxwellsche Theorie der Elektrizität, 1894; Die Geometrie der Wirbelfelder, 1897; Drang und Zwang, 2 vols., 1920; Grundzüge der Festigkeitslehre, 1923; Vorlesunger über technische Mechanik, 1897-1900, 1951. Influenced devel. of tech. mechanics as a sci. and an acad. field; investigated spacial statics, Laval's turbine wave; introduced Maxwell's theory into Germany. Died Ammerland am Starnberger See, Germany, Aug. 12, 1924.

FORAKER, Alvan Glenn, Am. pathologist; b. Pitts., Mar. 24, 1916; s. Forest Almos, and Jenny (Essex) F.; B.S., U. Pitts., 1936, M.D., 1939. Faculty, Emory U., Atlanta, 1949-53; asso. prof. pathology U. Tex. Postgrad. Med. Sch., Houston, 1953-55; pathologist Bapt. Memorial Hospital, Jacksonville, Florida, 1955——. Member of the Florida Association of Blood Banks (dir. 1962-65), Am. Assn. Pathologists and Bacteriologists, Am. Soc. Clin. Pathologists, Coll. Am. Pathologists, Am. Soc. Cytology, Internat. Acad. Pathology. Research numerous publs. in histochem. studies human tumors, anat. and exptl. studies in pulmonary emphysema. Address: 5545 Stanford Rd., Jacksonville, Fla. 32207.*

FORBES, David, Brit. geologist, philologist; b. Douglas, Isle of Man; Sept. 6, 1828; s. Edward and Jane (Teare) F.; ed. Edinburgh (Scotland) U.; studied at metall. lab. of Dr. Percy, Birmingham, Eng. Supt., Espedal Mining Works, Norway, 10 years; made study trips to Bolivia and Peru to search for nickel and cobalt ores, 1857, 60. Fellow Royal Soc., 1858; mem. Iron and Steel Inst. (fgn. sec.), Geol. Soc. (sec. 1871-76), Ethnol. Soc. Research and publs. on volcanic phenomena of S. Pacific; one of 1st to use the microscope for study of rocks. Died Dec. 12, 1876.

FORBES, Edward, Brit. naturalist; b. Douglas, Isle of Man, Feb. 12, 1815; s. Edward and Jane (Teare) F.; student Edinburgh (Scotland) U.; m. Emily Marianne Ashworth, Aug. 31, 1848; 1 son, 1 dau. Prof. botany King's Coll., London, Eng.; became lectr. Geol. Soc., 1842; named paleontologist Geol. Survey, 1844; lectr. Royal Instn.; named prof. natural history, Edinburgh, 1854; made natural history expdns. to Isle of Man, Norway, France, Switzerland, Germany, Algeria; naturalist to Beacon. Brit. Assn. grantee. Fellow Royal Soc., 1845; mem. Geol. Soc. (became pres. 1853). Author: History of British Starfishes, 1842; (with Hanley) History of British Mollusca, 1848-52; also articles. Collected 3000 plant specimens, Austria, 1838; collected marine animals and studied their relations with plants in the Aegean, 1841; collected molluscs and plants, Lycia, 1842; proved Purbeck beds beloned to oolitic series, 1849; discovered ruins of Termessus; discoveries in littoral zones, deposits formed at various depths in ocean, migration of molluscs. Died Nov. 18, 1854.

FORBES, George, Brit. engr., astronomer; b. 1849; s. James David Forbes; M.A., Cambridge (Eng.) U.; LL.D., St. Andrews (Scotland) U. Prof. natural philosophy Anderson's Coll., Glasgow, Scotland, 1872-80; elec. engr. for 1st works Niagara Falls, N.Y., 1891-95. Spl. corr. London Times, Russo-Turkish war, 1877; Times corr., mem. Jury Awards, Paris Internat. Electric Expn., 1881. Fellow Royal Soc., 1887, Royal Soc. Edinburgh, Royal Astron. Soc.; mem. Instn. Civil Engrs., Instn. Elec. Engrs., Astronomische Gesellschaft, Franklin Inst., Am. Philos. Soc. Author: The Transit of Venus, 1874; The Theory of the Glaciers of Savoy, 1874; Lectures on Electricity, 1888; Alternating and Interrupted Electric Currents, 1895; Elektrische Wechselströme und Unterbochene Ströme, 1896; History of Astronomy, 1909; David Gill, Man and Astronomer, 1916; The Wonder and the Glory of the Stars, 1926; The Earth, the Sun, and the Moon, 1927; The Stars, 1927; Las Estrellas, 1929. Contbd. articles to profl. jours. Crossed Asia from Peking to St. Petersburg by way of Gobi Desert and Siberia, 1875, visited cape, 1895, 1902, 14; surveyed Huka Falls, New Zealand, 1896, also cataracts of Nile, 1897-98; head Brit. Expdn. to observe transit of Venus, Kailua, Hawaii, 1874; perfected infantry range-finder, 1901 (used in S. African war, 1902), research range-finders with Admiralty, 1903-06; inventor gunsight now used in navy, carbon brush for dynamos and motors, meter suitable for alternating electric currents; measured velocity of light, 1876-80; predicted existence of trans-Neptunian planet (Pluto) from evidence relating to families of comets attached to planets, 1880. Died Oct. 22, 1936.

FORBES, George Franklin, Am. mathematician; b. Boston, June 26, 1915; s. Mark Augustus and Leau (Boudreau) F.; B.S., Northeastern U., 1939; postgrad. St. Louis U., Harvard, U. N.H., Mass. Inst. Tech.; m. Mary Eleanor Kimball, Mar. 9, 1943; children—Marie Eleanor (Mrs. Daniel Rector), Dorothy Ann (Mrs. Cleatus Anderson), Carolyn Ruth (Mrs. Francis Davis), Kenneth George, Barbara Joan, Marjorie Jane, Marilyn Lee, Christopher Mark, William Peter, Patricia Sue. Math. analyst Lockheed Aircraft Co., Burbank, Cal., 1952-56; engring. specialist Litton Industries, Woodland Hills, Cal., 1956——. Fellow Brit. Interplanetary Society, American Institute of Aeronautics and Astronautics (associate fellow), member Simulation Council, Assn. Computing Machinery, A.A.A.S., Naval Inst. Author: Digital Differential Analyzers, 1957; System Analyzer, 1961. Research on constrn. and application of computers, especially digital differential analyzers, trajectories of powered rockets. Home: 13745 Eldridge Av., Sylmar, Cal. 91342. Office: 5500 Canoga Av., Woodland Hills, Cal. 91364.*

FORBES, Gilbert Burnett, Am. physician; b. Rochester, N.Y., Nov. 9, 1915; s. Gilbert DeLeverance and Lillian Augusta (Burnett) F.; B.A., U. Rochester, 1936, M.D., 1940; m. Grace Moehlman, July 8, 1939; children—Constance Ann (Mrs. Joseph F. Citro), Susan Young. Practice medicine, specializing in pediatrics, Los Alamos, 1946-47, Rochester, 1954——; faculty Sch. Medicine Washington U., St. Louis, 1943-50, Southwestern Med. Sch., Dallas, 1950-53; asso prof. pediatrics Sch. Medicine U. Rochester, 1953-57, prof., 1957——; cons. Nat. Inst. Child Health and Human Devel.; mem. sci. adv. com. Nutrition Found.; mem. com. infant nutrition, com. dietary allowances NRC, 1960-63. USPHS NIH grantee, 1962——. Mem. Am. Pediatrics Soc. (council), Soc. Pediatric Research (past pres.), A.M.A., Soc. Exptl. Biology and Medicine, A.A.A.S., Am. Acad. Pediatrics (Borden award, 1964), Sigma Xi, Alpha Omega Alpha, Theta Chi. Contbr. articles to profl. jours. Asso. editor: Am. Jour. Diseases Childhood, 1964——; Nutrition Revs., 1961——. Application of isotopic dilution method to study of body composition in man and the changes which occur during growth; application of the whole body scintillation counter to the determination of potassium 40 in children and adults; studies on exchange of isotopes in bone as a function of age. Home: 2021 Westfall Rd., Rochester 14618. Office: 260 Crittenden Blvd., Rochester, N.Y. 14620.*

FORBES, James David, Scotch physicist; b. Edinburgh, Scotland, Apr. 20, 1809; s. Sir William and Williamina (Belches) F.; ed. Edinburgh (Scotland) U.; D.C.L. (hon.), Oxford, 1853; m. Alicia Wauchope, July 4, 1843; children—Edward Balton, George, 3 daus. Became prof. natural philosophy, Edinburgh, 1833-59, dean Faculty Arts, 1837; named prin. St. Andrews, Scotland, 1859. Recipient Keith medal Edinburgh Soc., 3 times. Fellow Royal Soc., 1832 (Rumford medal, Royal medal); mem. French Acad. Scis., 1842, Royal Soc. Edinburgh (sec. 1840-51), Brit. Assn. (co-finder 1831). Author: Travels through the Alps of Savoy and other parts of the Pennine Chain with Observations on the Phenomena of Glaciers, 1843; also numerous articles. Discovered polarization of radiant heat, 1834; 1st to study glacier movements; his claim to be first observer of veined structure of glaciers was contested by Agassiz and Tyndall. Died Dec. 31, 1868.

FORBES, John Campbell, Am. biochemist; b. Melford, N.S., Can., Aug. 27, 1894; s. Allan and Mary (Campbell) F.; B.A., U. Sask., 1920, M.A., 1922; Ph.D., McGill U., 1925; m. Irene Byrd Duval, June 22, 1926; 1 son, Allan Louis. Came to U. S., 1926, naturalized, 1937. Research chemist Biol. Bd. Can., Halifax, N.S., 1925-26; faculty Haverford (Pa.) Coll., 1926-27; faculty Med. Coll. Va., Richmond, 1927-65, prof. emeritus, 1965——. Biochem. supr. child devel. study Med. Coll. Va., NIH, 1959——; mem. research com. McGuire VA Hosp., Richmond, Va., 1959——. Mem. Am. Chem. Soc. (Distinguished Service award Va. 1938-39, chmn. Va. sect. 1938-39), Am. Inst. Chemists, Soc. Exptl. Biology and Medicine, Am. Soc. Biol. Chemists, Am. Assn. U. Profs., Va. Acad. Sci. (Commendation award 1959, pres. 1958-59), Sigma Xi. Initiated grad. degree instrn. at Med. Coll. Va.; dir., in assn. with U. Center Va., vis. scientist program for smaller colls.; research in dental caries, effect of nutritional deficiencies on response of rats to alcohol intoxication, serum lipid fractionation with particular emphasis on factors which may be involved in atherogenesis. Home: Route 2, Montpelier, Va. 23192.*

FORBES, Richard M., Am. nutritionist; b. Wooster, Ohio, Jan. 8, 1916; s. Ernest B. and Lydia (Mather) F.; B.S., Pa. State Coll., 1938, M.S., 1939; Ph.D., Cornell U., 1942; m. Mary Medlicott, Feb. 26, 1944; children—Sally Allen, Anne Mather, Stephen Harding. Research asso. Wayne State Med. Coll., 1942; research asso. Cornell U., 1942-43; asst. prof. animal sci. U. Ky., 1946-49; asso. prof. animal nutrition U. Ill., Urbana, 1949-55, prof. nutritional biochemistry, 1955—. Mem. Am. Chem. Soc., Am. Inst. Nutrition, Am. Soc. Animal Sci., A.A.A.S., Sigma Xi. Research, numerous publs. on protein and mineral metabolism of farm and lab. animals, chem. composition of human body. Home: 2005 S. Vine St., Urbana, Ill. 61801.*

FORBES, Stephen Alfred, Am. naturalist; b. Silver Creek, Ill., May 29, 1844; s. Isaac Sawyer and Agnes (Van Hoesen) F.; ed. Rush Med. Coll.; Ph.D., on exam. and thesis, Ind. U., 1884; LL.D., U. Ill., 1905; m. Clara Shaw Gaston, Dec. 25, 1873; children—Bertha Van Hoesen, Ernest Browning, Winifred, Ethel Clara, Richard Edwin. Curator mus. Ill. State Natural History Soc., 1872-77; tchr. zoölogy Ill. State Normal U., 1875-78; founded, dir. Ill. State Lab. of Natural History, 1877-1917; later chief of natural history survey of Ill.; state entomologist, 1882-1917; prof. zoölogy, U. Ill., 1884-1909, prof. entomology, 1909-21, dean Coll. Sci., 1888-1905. Organizer class medal Internat. Congress Zoölogists, Chicago Expn., 1893; dir. aquarium U. S. Fish Commn., prepared natural history exhibit of Ill. Recipient 1st class medal Soc. d'Acclimatation de France for sci. publs., 1886. Founder, editor, contbr. to Bulls. of Ill. State Lab. of Natural History (now Bull. Ill. State Nat. Hist. Survey), 1877—. Author: Biennial Reports as State Entomologist, 1883-1916; Studies of the Food of Birds, Fishes and Insects; Contagious Diseases of Insects. Final Report on the Fishes of Illinois and Studies on the Biology of the Illinois River (with R. E. Richardson). Died Mar. 13, 1930.

FORBES, Thomas Rogers, Am. anatomist; b. N.Y.C., Jan. 5, 1911; s. James Bruff and Stella (Rogers) F.; B.A., U. Rochester, 1933, Ph.D., in Anatomy, 1937; M.A. (hon.), Yale, 1962; m. Helen Frances Allen, June 19, 1934; children—Thomas Rogers, William M. Fellow in anatomy U. Rochester, 1933-37; asst. in anatomy Johns Hopkins Med. Sch., 1937-38, instr., 1938-45; tech. aide to div. med. scis. NRC, com. on med. research OSRD, 1942-45; faculty Yale U. Sch. Medicine, 1945—, prof. anatomy, 1962—, asst. dean, 1948-60, asso. dean, 1960—, fellow Branford Coll., 1951—. Guggenheim fellow, 1942. Recipient certificate appreciation War and Navy Depts., 1947. Mem. Am. Assn. Anatomists, Soc. for Exptl. Biology and Medicine, Endocrine Soc., A.A.A.S., Am. Assn. History Medicine, Am. Soc. Zoologists, Royal Soc. Medicine (affiliate), Faculty History of Pharmacy and Medicine, Worshipful Soc. Apothecaries (London), Phi Beta Kappa, Sigma Xi. Author: The Midwife and the Witch, 1966; also numerous articles. Editor: Jour. History of Med., 1962-63. Research on human reproductive embryology and endocrinology, reprodn. in reptiles, measurement and action progestins, use steroid hormones in pellet form, occurrence hormones; co-developer assay for minute amounts progestin. Home: 86 Ford St., Hamden, Conn. 06517. Office: 333 Cedar St., New Haven, Conn. 06510.*

FORBES, William Alexander, English zoologist; b. Cheltenham, Eng., June 24, 1855; student Edinburgh U., 1873-75, U. Coll., London, 1875-76, St. John's Coll., Cambridge, 1867. Named prosector to Zool. Soc. London, 1879; lectr. comparative anatomy Charing Cross Hosp. Med. Sch.; made short trip to Pernambuco, 1880; began trip to study fauna of eastern tropical Africa, starting at mouth of Niger, 1882. Research and publs. on muscular structure, voice organs of birds. Died Africa, Jan. 14, 1883.

FORBUSH, Edward Howe, Am. ornithologist; b. Quincy, Mass., Apr. 24, 1858; s. Leander Pomeroy and Ruth Hudson (Carr) F.; m. Etta L. Hill, June 28, 1882; children—Myrtice Elizabeth, Erwin Hill, Lewis Edward, Etta Lorenda. Dir. div. ornithology Mass. Dept. Agr., 1893-1908, 1921; state ornithologist, from 1908. Life mem. Worcester Natural History Soc.; fellow Am. Ornithologists' Union; founder Mass. Audubon Soc. Author: Birds of Massachusetts and other New England States, Vol. 1; Water Birds, Marsh Birds and Shore Birds, 1925; Vol. 2, Land Birds, from Bob-White to Grackeis, 1927. Died Mar. 7, 1929.

FORBUSH, Scott E., math. physicist; b. Apr. 10, 1904; B.S., Case Inst. Tech., 1925; postgrad. Johns Hopkins, 1931-32, George Washington U. Teaching asst. Ohio State U., 1925-26; jr. physicist Bur. Standards, 1926-27; observer dept. terrestrial magnetism Carnegie Instn., Washington, 1927-42, math. physicist, 1942—. Civilian with Naval Ordnance Lab., 1940-44, OSRD, 1944-45, Operations Research Office, Johns Hopkins, 1951-52. Mem. Am. Philos. Soc. (pres. 1953), Am. Geophys. Union, Operations Research Soc. Research in terrestrial magnetism, cosmic radiation statistics, operations research. Home: 7208 Brennon Lane, Chevy Chase, Md. Office: Carnegie Instn. of Washington, 5241 Broad Branch Rd. N.W., Washington 15.*

FORCADEL, Pierre, mathematician; b. Béziers, France; lived in Italy; prof. math. Coll. de France, 1560-73. Translator part of Euclid's Elements (widely used in France). Died 1574.

FORCART, Lothar, Swiss zoologist; b. Basel, Switzerland, Dec. 10, 1902; s. Rudolph and Anita (Bachofen) F.; ed. univs. Basel, Berlin, Cambridge; Ph.D.; m. Anne Müller, Dec. 8, 1937. Head zoology dept. Mus. Natural History, Basel. Mem. Unitas Malacologica Europaea, Agra of Acad. Zool. Studies (India) (hon.). Research and publs. on malacology and herpetology. Home: Zurchestrasse 9. Office: Museum of Natural History, Augustinergasse 2, Basel, Switzerland.

FORCH, Carl Friedrich Otto Hugo, German cinematic engr.; b. Bechtheim, nr. Worms, Germany, Dec. 12, 1870; s. Hugo and Barbara Elisabeth (Neubeck) F.; student, Giessen, Germany, Strasbourg (now in France); Ph.D., Munich, 1895; m. Eleonore Breuer, 1899; 1 son, 2 daus. Asst. to E. Ketteler, Munich, 1894-96, to K. Schering, Darmstadt, Germany, 1889-1906; became mem. faculty Darmstadt, 1899; entered service imperial patent office, Berlin, 1906; became dir. Reich patent office, 1931; ret., 1936; a founder testing and exptl. inst. for cinematics Tech. U. Berlin, 1921, also lectr. on cinematics. Mem. German Cinematic Soc. (a founder 1920, Oskar-Messter medal 1929). Author: Der Kinematograph und das sich bewegende Bild . . . (1st standard cinematic work), 1913; Das Leuchtgas, seine Herstellung und Verwendung, 1914; also articles. Research in physics and phys. chemistry of sea water, other solutions and dissolved substances; contbd. to establishment and devel. of cinematics. Died Koblenz-Lützel, Germany, Apr. 15, 1955.

FORCHHAMMER, Johan Georg, Danish geologist, chemist; b. Husum, Denmark, July 26, 1794; named lectr. U. Copenhagen (Denmark), 1823, prof. chemistry and mineralogy, 1835. Author: Geological Conditions in Denmark, 1835; On the Components of Sea Water, 1859; also numerous articles on geology and chemistry. Father Danish geology; 1st description of coal formation of Bornholm; studied manganate and permanganate salts. Died Copenhagen, Dec. 14, 1865.

FORCHHEIMER, Frederick, Am. physician; b. N.Y., 1853; M.D., Columbia, 1873; prof. chemistry Med. Coll. Ohio, from 1877, prof. diseases of pediatrics, from 1879, prof. physiology, 1882-93, prof. theory and practice of medicine, clin. medicine, 1893-1913, dean coll., from 1905. Author: Diseases of the Mouth in Children, 1892; The Prophylaxis and Treatment of Internal Diseases, 1908. Editor: Therapeutics of Internal Diseases, 1913. Established clinic for diseases of children Med. Coll. Ohio, 1875; described Forchheimer's sign for German measles, 1908. Died 1913.

FORCHHEIMER, Philipp, hydraulic engr.; b. Vienna, Austria, Aug. 7, 1852; s. Eduard and Henriette (Landauer) F.; engring. diploma Tech. U. Zurich (Switzerland), 1873; doctorate, Tübingen, Germany; hon. dr. engring. Tech. U. Vienna, 1930; m. Hilde Charlotte Grohmann, 1891; 1 son. Worked several years on hydraulic problems for various constrn. firms; mem. faculty Tech. U. Aachen (Germany); invited to Ottoman Engring. U., Constantinople, 1891; returned to Aachen as prof., 1892; became prof. hydraulics Tech. U. Graz (Austria); invited to reorganize Engring. U., Constantinople, 1914. Mem. Vienna Acad. Scis. Author: Hydraulik, 1914. Sought to place empirical hydraulics on more sound sci. basis; made theoretical studies of springs, treatment of ground water under foundations, instationary movement. Died Vienna, Oct. 2, 1933.

FORCHIELLI, Enrico Henry, Am. biochemist; b. West Boylston, Mass., Jan. 26, 1918; s. Mariano and Teresa (Bevilacqua) F.; B.S., Mass. Coll. Pharmacy, 1940; B.A., Clark Coll., 1951; M.A., Boston U., 1953, Ph.D., 1956; m. Ruth Doris Harvey, June 21, 1947; children—Lisa H., Serina D., Cynthia V. Sr. scientist Worcester Found., Shrewsbury, Mass., 1947-64; head depts. steroid metabolism, analytical biochemistry Syntex Research Center, Palo Alto, Cal., 1964—. Mem. N.Y. Acad. Sci., Am. Chem. Soc., Am. Soc. Biol. Chemistry, Endocrine Soc. Studies, publs. on analysis, in vitro metabolism of steroids. Home: 4 Alma Ct., Los Altos, Cal. 94022. Office: 3401 Hillview St., Palo Alto, Cal. 94304.*

FORD, Donald Herbert, Am. neuroanatomist; b. Kansas City, Mo., Aug. 18, 1921; s. Horace G. and Gladys (Newell) F.; B.A., Wesleyan U., Conn., 1947, M.A., 1949; Ph.D., U. Kan., 1952; m. Dorothy Ann Glander, Aug. 5, 1944; 1 dau., Linda Ann. Faculty, State U. N.Y., Bklyn., 1952—, asso. prof. neuroanatomy, 1959—. Cons. neuroanatomy Bklyn. VA Hosp., 1955—, L.I. Coll. Hosp., Bklyn., 1956—. Mem. Am. Assn. Anatomists, Endocrine Soc., Am. Physiol. Soc., Sigma Xi, Phi Sigma. Author: (with J. P. Schade) Basic Neurology, 1965, Atlas of the Human Brain, 1965; (with H. Kaplan) Brain Vascular System, 1966; also articles. Research on influence thyroid hormone on metabolic function brain. Home: 224 Verbena Av., Floral Park, N.Y. 11001. Office: 450 Clarkson Av., Bklyn. 11203.*

FORD, Edmund Brisco, Brit. geneticist; b. Papcastle, Cumberland, Eng., Apr. 23, 1901; s. Harold Dodsworth and Gertrude Emma (Bennett) F.; B.A., Oxford (Eng.) U., 1924, M.A., 1927, D.Sc., 1943; D.Sc., Liverpool (Eng.) U., 1964. Research worker dept. zoology Oxford U., 1924-27, faculty, 1927—, prof. ecol. genetics, 1963—, fellow All Souls Coll., 1958—. Recipient Weldon Meml. prize Oxford U., 1959. Fellow Royal Soc. (Darwin medal 1954); mem. Nature Conservancy (founding), Genetical Soc. Gt. Britain (past pres.). Author: Mendelism and Evolution, 1931; (with D. H. Carpenter) Mimicry, 1933; The Study of Heredity, 1938; Genetics for Medical Students, 1942; Butterflies, 1945; Moths, 1955; Ecological Genetics, 1964; Genetic Polymorphism, 1965; also numerous articles. Research on devel. ecol. genetics, evolution of dominance in wild material, research (with Julian Hurley) on genetic controllers growth; defined genetic polymorphism; predicted assn. human blood groups with disease; defined and developed sci. of ecol. genetics. Genetic Labs., Zoology Dept., Parks Oxford, Eng.*

FORD, Harry W., Am. horticulturist; b. Maumee, O., June 21, 1922; s. Wilbur H. and Josephine (Wogaman) F.; B.Sc., Ohio State U., 1946, M.S., 1947, Ph.D., 1950; m. Margaret Gladys Zimola, Sept. 4, 1948; children—William Harry, Donald Wilbur. With U. Fla. Citrus Expt. Sta., Lake Alfred, Fla., 1950—, horticulturist, 1962—. Recipient Fla. Fruit and Vegetable Assn. award, 1965. Mem. Am. Soc. Hort. Sci., Am. Inst. Biol. Sci., Fla. State Hort. Soc. Sigma Xi. Contbr. numerous articles to sci. jours. Established basic root distbn. pattern for citrus in Fla.; developed rapid screening methods for nematode resistant rootstocks; released three rootstocks resistant to burrowing nematodes; determined microbiol. cause of sludge and other deposits in and nr. drain lines; determined that sulfides are principle agt. killing flooded citrus roots in Fla. Home: P.O. Box 1122. Office: P.O. Box 1088, Lake Alfred, Fla. 33850.*

FORD, Henry, Am. inventor; b. Wayne County, Mich., July 30, 1863; s. William and Mary (Litogot) F.; ed. in dist. sch., Dearborn, Mich.; D.Eng., U. Mich., 1926; LL.D., Colgate U., 1935; m. Clara J. Bryant, Apr. 11, 1888; 1 son, Edsel Bryant. Learned machinist's trade; in Detroit since 1887; chief engr. Edison Illuminating Co.; organizer, Ford Motor Co. (largest mfr. of automobiles in the world, employing over 100,000 persons), 1903, also pres.; announced profit sharing plan, 1914; Mem. Soc. Automotive Engrs. Author: My Life and Work, 1925; Today and Tomorrow, 1926; Moving Forward, 1931. Built his 1st automobile, 1893; introduced mass prodn. into automobile manufacture, through standardization of parts, div. of labor, specialization. Died Apr. 7, 1947.

FORD, Hugh, Brit. engr.; b. Northamptonshire, July 16, 1913; s. Arthur and Wearing Constance Ford; D.Sc., Ph.D., Imperial Coll., U. London; m. Wynyard Scholfield, July 4, 1942; children—Clare Victoria, Vanessa. Research engr. Imperial Chem. Industries; chief engr. Brit. Iron and Steel Fedn.; prof. applied mechanics, head dept. mech. engring. U. London. Fellow Instn. Metallurgists, Instn. Mech. Engrs.; mem. Inst. Metals (pres. 1963-64), Instn. Civil Engrs. Author: Advanced Mechanics of Materials, also numerous articles. Home: Shrewsbury House, Cheyne Walk, London. Office: Imperial College, Exhibition Rd., London, Eng.

FORD, Joseph Brandon, Am. sociologist; b. Los Angeles, Jan. 20, 1918; s. William Joseph and Cecilia (Chambers) F.; B.A. with highest honors, U. Cal. at Los Angeles, 1937; M.A., U. So. Cal., 1941; M.A., Harvard, 1947; Ph.D., U. Cal. at Berkeley, 1951; m. Marjorie Henshaw, Dec. 11, 1948; children—Anabel Deirdre, Cecelia Elinore, Stephen Joseph Jerome. Social worker, later social service supr. and supr. Cal. Bur. Personnel Audits, 1939-40; vocational counsellor VA, 1946; spl. agt. War Assets Adminstrn., 1946-48; real estate broker, later v.p., dir. Mortgage Credit Corp., 1948-50; instr. Los Angeles City Coll., 1948-49; asst. prof. to prof. Los Angeles State Coll., 1950-58; prof. sociology, 1958—, chmn. dept. sociology and anthropology San Fernando Valley State Coll., Northridge, Cal. 1958-64; vis. prof. U. Cal. at Berkeley, 1951, U. Cal. at Santa Barbara, 1955, U. Rome, 1955-56, U. Vienna, 1958-59, Nat. U. Mexico, 1960, Am. U., Beirut, 1965-66; Fulbright fellow, vis. prof. U. Madrid, 1961-62; cons. Inacasa Housing program, Italy, 1956; commr. city planning City of Los Angeles, 1961-63; sociol. adv. com. Cal. Dept. Edn., 1963—. Fulbright fellow, 1955-56, 58-59, 65-66; Ford Found and Inst. Internat. Edn. grantee, Poland, summer 1959. Fellow Am. Sociol. Soc., Institut Internat. de Sociologie (councillor, mem. governing bur. 1963—, treas. 1967—), Internat. Sociol. Assn., Pacific Sociol. Soc., Instituto de Estudies Politicos U. Madrid, Phi Beta Kappa, Alpha Kappa Delta. Author: Auguste Comte Philosophy, 1968; (with others) Contemporary Sociology, 1958; (with others) Contemporary American Sociology, 1961; (with C. C. Zimmerman) Sociology of Change, 1964 (also Turkish, Spanish langs.), Philosophy and Sociology of Sorokin; (with Allen, Toynbee, others) Sorokin in Review, 1963. Regional editor for N.Am.: Sociologia Internationalis. Contbr. sci. articles profl. jours. Research in fields of urban sociology, urban and regional planning, propaganda and mass communication, hist. sociology, and social change; chief contbns. in field of social theory, philosophic problems of social scis. and history of sci. Home: 18247 Ludlow St., Northridge, Cal. 91324.*

FORD, Kenneth William, Am. physicist, educator; b. West Palm Beach, Fla., May 1, 1926; s. Paul H. and Edith (Timblin) F.; student John Carroll U., 1945, U. Mich., 1945-46; A.B., Harvard, 1948; Ph.D., Princeton, 1953; m. Joanne Baumunk, June 9, 1962; children—Paul T., Sarah E. (by previous marriage); Caroline A., Adam B., Jason L. Research assistant at the Los Alamos (New Mexico) Sci. Laboratory, 1950-51; research associate Princeton, 1951-52; from research asso. to asso. prof. physics Ind. U., Bloomington, 1953-58; asso. prof. physics Brandeis U., Waltham, Mass., 1958-61, prof., 1961-64; prof. physics U. Cal. at Irvine, 1964—, chmn. dept., 1964-68. Fulbright fellow Max Planck Institut, Göttingen, Germany, 1955-56; NSF fellow Imperial Coll., London, Mass. Inst. Tech., 1961-62. Fellow Am. Phys. Soc. Author: The World of Elementary Particles, 1963; Basic Physics, 1968. Mem. bd. editors Phys. Rev., 1960-62. Contbns. to theory of nuclear structure, scattering theory, and analysis of high energy expts.; popularization of modern physics. Home: 2647 Basswood St., Newport Beach, Cal. 92660. Office: Physics Dept., U. Cal., Irvine, Cal. 92664.*

FORD, Thomas Robert, Am. sociologist; b. Lake Charles, La., June 24, 1923; s. Gervais W. and Alma (Weil) F.; student McNeese Jr. Coll., 1940-42, U. of South, 1942-43; B.S., La. State U., 1946, M.A., 1948; Ph.D., Vanderbilt U., 1951; m. Harriet Lowrey, Aug. 13, 1949; children—Margaret Erin, Janet Patricia, Mark Lowrey, Charlotte Elizabeth. Manpower analyst USAF Personnel and Tng. Research Center, Maxwell AFB, Ala., 1953-56; faculty U. Ky., Lexington, 1956—, prof. sociology, behavioral sci., 1960—, chmn. dept. sociology, 1966—. Research dir. So. Appalachian Studies, 1957-60; mem. Nat. Adv. Commn. on Rural Poverty, 1966-67. Recipient U. Ky. Alumni award for distinguished research, 1965. Guggenheim fellow, 1963. Mem. A.A.A.S., Am. Sociol. Assn., Population Assn. Am., So., Rural sociol. socs., Soc. For Internat. Devel. Author: Man and Land in Peru, 1955; Health and Demography in Kentucky, 1964. Editor: The Southern Appalachian Region-A Survey, 1962. Research in social and cultural factors in transition of agrarian economies; demonstrated relationships between population characteristics and social instns. in rural socs. Home: 1107 Eldemere Rd., Lexington, Ky. 40502.*

FORD, William Ebenezer, Am. mineralogist; b. Westville, Conn., Feb. 18, 1878; s. William Elbert and Caroline Aby (Bishop) F.; Ph.B., Yale, 1899, Ph.D., 1903; m. Mary Treat Jennings, 1920. Asst. in mineralogy Sheffield Sci. Sch., Yale, 1899-1903, instr., 1903-06, asst. prof., 1906-20, prof., 1920. Fellow Am. Acad. Arts and Scis. Editor: Second and Third Appendices to Dana's System of Mineralogy, 1909 and 1915; New Edition of Dana's Manual of Mineralogy, 1912, 29; Dana's Text Book of Mineralogy (new edits.), 1921, 32. Research in mineralogy and crystallography; studied influence of chem. composition on optical properties of isomorphous groups of minerals. Died Mar. 23, 1939.

FORD, William Webber, Am. bacteriologist; b. Norwalk, Ohio, Dec. 15, 1871; s. James B. and Cornelia (Cook) F.; A.B., Western Res. U., 1893; M.D., Johns Hopkins, 1898; fellow, McGill U., 1899-1901, D.P.H., 1902; fellow Rockefeller Inst., N.Y., 1901-02. Instr. bacteriology Johns Hopkins, 1903-05, asso., 1905-06, asso. prof. hygiene and bacteriology, lectr. legal medicine, 1906-16, prof. bacteriology, Sch. Hygiene and Pub. Health, also lectr. hygiene, Med. Sch., 1917-37. Research on poisonous fungi in relation to pub. health, also on early anesthesias. Died Feb. 10, 1941.

FORDOS, Mathurin Joseph, French pharmacist, bacteriologist; b. Sérent, France, Nov. 3, 1816; chief pharmacist Hosp. St. Antoine, then Charité, Paris, France. Isolated pyocyanini from B. pyocyaneus culture; discovered (with Gelis) sodium tetrathionate, also gave correct formula for sulphur nitride. Died Paris, July 1, 1878.

FORDYCE, George, Brit. physician; b. Aberdeen, Scotland, Nov. 18, 1736; s. George F.; M.A., U. Aberdeen, 1750; M.D., U. Edinburgh, 1758; studied anattomy under Albinus; m. 1762; 2 sons, 2 daus. Began lecturing on chemistry, medicine, 1764, and continued for almost 30 years; named physician St. Thomas' Hosp., 1780. Licentiate Coll. Physicians. Fellow Royal Soc., 1776, Coll. Physicians. Author: Elements of Agriculture and Vegetation, 8 vols., 1765; Treatise on the Digestion of Food, 1791; Dissertation on Simple Fever, 1794; also numerous articles. Confirmed Priestley's and Lavoisier's views against Phlogiston theory; thought digestion was a purely physiol. process; demonstrated that humans maintain constant temperature even in heated environments. Died Strand, Eng., May 25, 1802.

FOREL, August Henri, Swiss entomologist, psychologist; b. Morges, Switzerland, Sept. 1, 1848; s. Victor and Pauline (Morin) F.; ed. Zurich (Switzerland) U., 1866-71, U. Vienna (Austria), 1871-72; M.D., Lausanne (Switzerland), 1872; m. Emma Steinheil, 1882; 2 sons, 4 daus. Asst. to von Gudden at Ludwig Maximilian's univ., Munich, Germany, from 1873; privatdozent, from 1877; dir. Burghoizli Asylum, prof. psychiatry U. Zurich, 1879-98, ret., 1898. Author: Der Hypnotismus, 1889; Gehirn und Seele, 1894; Die psychischen Fähigkeiten der Ameisen, 1901; Die sexuelle Frage, 1905; Sexuelle Ethik, 1906; Das Sinnesleben der Insekten, 1910. Known for work on anatomy of brain and nerves; described brain and nerve structures previously unknown; (with others) devised 1st usable brain microtome; independently formulated concept of cellular and functional units (later named neurons by Waldeyer); authority on psychology of ants; pioneer in sex hygiene; known for work in hypnotism and forensic psychiatry; described 3500 new species of Hymenoptera; one of 1st in field to think in terms of social psychology. Died Yvorne, Switzerland, July 27, 1931.

FOREL, François-Alphonse, Swiss physician, naturalist; b. Morges Vaud, Switzerland, Feb. 2, 1841; prof. physiology, also gen. anatomy U. Lausanne (Switzerland), 1869-1895; helped found Seismol. Inst. Switzerland. Author: Expériences sur la temperature du corps humain dans l'acte de l'ascension sur les montagnes, 1872-74; Le Phylloxera Vastatrix dans la Suisse Occidentale, 1875; La faune profonde des lacs suisses, 1885; Le léman (study of Lake Geneva), 3 vols., 1892, 96, 1905; Handbook of Limnology (long the authoritative work on limnology), 1901; also numerous articles on periodic variations and structure of glacier-grain of glaciers in Switzerland, seiches of Swiss lakes, Lake Geneva. Founder of limnology; made limnological studies of Lake Geneva and Constance; studied earthquakes in Switzerland; independently (later with Michele DeRossi) suggested single 10 point scale for earthquakes on both sides of Alps (became most widely used scale, also modified for use throughout Europe and Am.); helped eliminate Phylloxera in Swiss vineyards. Died Morges Vaud, Aug. 7, 1912.

FOREMAN, Kenneth Martin, Am. aero. engr.; b. N.Y.C., July 30, 1925; s. Louis and Zena (Gordon) F.; B.Aero. Engring., N.Y. U., 1950, M.Aero. Engring., 1953; postgrad. Poly. Inst. Bklyn.; m. Shirley Teiger, June 24, 1954; children—Elissa, Michael. Research engr. Bendix Corp., Teterboro, N.J., 1951-52; project engr. Curtiss-Wright Corp., Wood-Ridge, N.J., 1952-56; research engr. Fairchild Engring. div. Fairchild Engine & Airplane Corp., Deer Park, N.Y., 1956-59; specialist sci. research engr. Republic Aviation Corp., Farmingdale, N.Y., 1959-65; chief project engr. Edo Corp., College Point, N.Y., 1965-66; research engr. Grumman Aircraft Engring. Corp., Bethpage, N.Y., 1966—. Staff abstractor Fire Research Abstracts and Revs., NRC, 1960—. Mem. Am. Soc. M.E. (chmn. propulsion com. aviation and space div. 1963-67, systems engring. com., 1967—), Am. Inst. Aeros. and Astronautics, Combustion Inst., Am. Ordnance Assn., Tau Beta Pi. Research, publs. on Mars' atmosphere and surface phenomena, Earth satellite expts., rarefied atmospheres, interactions with spacecraft. Hypersonic gasdynamics and gaseous detonations, interaction low pressure combustible gases with high speed plasmas, advanced aerospace propulsion systems; defined and summarized detonation characteristic for widest extent pressure, temperature and mixture conditions; orginated concept electromagnetic detonation wave propulsion device. Home: 32 Stratford Ct., North Bellmore, N.Y. 11710. Office: Plant 25, Bethpage, N.Y. 11714.*

FOREST, Fernand, French engr.; b. Clermont-Ferrand, France, 1851; pioneer in devel. automobile; patentee gasoline engine, 1880, first 4 cycle, 4 cylinder gasoline engine, 1891; inventor ignition magneto, 1897. Died Monaco, 1914.

FOREST, Petrus, see van Foreest, Pieter.

FORESTIER, Hubert, chemist, metallurgist; b. Tunis, Tunisia, Feb. 13, 1903; s. Gabrielle Michaud; ed. Faculty Scis. of Paris, Coll. of France; Dr. honoris causa, Free U. Brussels (Belgium); m. Magdeleine Canard, Apr. 1, 1914; children—Francois-Hubert, Isabelle. Lectr., Faculty of Scis., 1934-43; dir. Inst. Chemistry, Faculty of Scis., U. Strasbourg (France), 1943, titular prof. chair gen. chemistry Faculty of Scis., 1945; created, 1948, since dir. Superior Sch., Strasbourg. Mem. Council of Advanced Instrn., 1950; mem. Permanent Commn. Nat. Superior Schs., 1967; v.p. Indsl. Bur. Studies of Univ.-Indsl. Relationships. Decorated officer Legion of Honor; officer Order Crown of Oak. Mem. Grand-Duke Acad. Luxembourg, Acad. Scis. Mayence, Chem. Soc. France, Metallurgy Soc. France. Author: Traité de Chimie minerale de Pascal. Discovery and study of magnetic properties of ferrites (10 new ferromagnetic bodies), also magnetic properties of rare earth ferrites (11 new); study influence of adsorbed gas on reactivity between solids and on phys. and mech. properties solid surfaces; discovered maximum of chem. reactivity at the Curie point of ferromagnetic bodies and ferromagnetic alloys; study of alloys on non-ferrous and super refractory alloys. Address: 1 rue Blaise Pascal, Strasbourg 67, France.*

FORFAIT, Pierre-Alexandre-Laurent, French nautical engr.; b. Rouen, France, Apr. 2, 1752; nautical engr. attached to Brest (France) Port; dir. Maritime Service, Le Havre, France, 1803-05; cons. Italian fleet, Genoa, Italy, from 1805. Decorated comdr. Legion of Honor. Mem. Royal Marine Acad., French Acad. Scis, 1789. Author: Traité élémentaire de la mäture des vaisseaux, 1788; Sur le système de construction des mats, 1815. Author treatise on masts and spars; perfected constrn. trans-Atlantic passenger vessels; responsible for large mil. arsenal Anvers, France. Died Rouen, Nov. 8, 1807.

FORGAYS, Donald Gabriel, Am. psychologist; b. Lowell, Mass., Sept. 8, 1926; s. Raymond Gilbert and Stanis (Rogers) F.; B.A., Dartmouth Coll., 1946; M.A., McGill U., 1948, Ph.D., 1950; m. Janet Ruth Wakefield, Dec. 18, 1948; children—Janice Anne, Gabrielle, Ian. Instr., Sir George Williams Coll., Montreal, Que., Can., 1948-50; asst. prof. psychology Western Mich. U., 1950-51; research psychologist Air Force Personnel and Tng. Research Lab., Tex., 1951-55; asst. prof. psychology Cornell U., 1955-57; asso. prof. psychology Rutgers U., 1957-64; prof., chmn. dept. psychology U. Vt., Burlington, 1964—. Mem. Am., Eastern, Midwestern, Vt. psychol. assns., A.A.A.S. Author: Critical Thinking, 1963. Research, numerous publs. on influence of early experience upon later functioning especially intellectual functioning; nature of reinforcement, devel. of visual perception, influence of isolation upon psychol. and physiol. functioning. Home: 28 DeForest Heights, Burlington, Vt. 05401.*

FORGUE, Emile, French surgeon; b. Briançon, France, Dec. 29, 1860; prof. clin. surgery Faculty of Medicine, Montpellier, France. Mem. French Acad. Scis., 1926, Acad. Medicine. Author: Traité de therapeutique chirurgicale, 1892; Précis de pathologie externe, 1902; Au seuil de la chirurgie; wrote on surg. therapeutics and external pathology. Paramed. researcher. Died Grave, France, Feb. 1, 1943.

FORGUE, Stanley Vincent, Am. engring. physicist; b. Cleve., Oct. 6, 1916; s. John N. and Mary (Smith) F.; B.S. in Physics, Ohio State U., 1939, B.E.E., 1940, M.S., 1940, postgrad.; m. Dorothy Jeanne Huber, Feb. 14, 1942; children—Mary Jeanne, Stanley Thomas, Wesley Vincent. Research engr. Ohio Engring. Exptl. Sta., Columbus, 1940-41; research fellow Ohio State U. Research Found., 1941-42; tech. staff RCA Labs., Princeton, N.J., 1942—, research projects head, 1952—. Recipient RCA Lab. awards for outstanding work in research. Fellow I.E.E.E.; mem. Am. Phys. Soc., A.A.A.S., Sigma Xi, Sigma Pi Sigma, Pi Mu Epsilon, Tau Beta Pi. Author: sects. in. Television IV, 1947, Television VI, 1949-50; also tech. articles, numerous govt. reports. Patentee tv camera tubes, color kinescopes, infrared sensitive tubes, storage tubes, photoconductive cells. Home: Grovers Mill Rd., R.D. 1, Cranbury, N.J. 08512. Office: RCA Labs., Princeton, N.J. 08540.*

FORGUS, Ronald Henry, psychologist; b. Cape Town, South Africa, May 18, 1928; s. William A. and Marie (Kleinhaus) F.; B.Sc., McGill U., 1950, M.Sc., 1951; Ph.D., Cornell U., 1953; m. Valerie Lamont, Aug. 9, 1965; children—Michael, Sandra. Came to U. S., 1951, naturalized, 1960. Asst. prof. U. Pa., 1953-58; faculty Lake Forest (Ill.) Coll., 1958—, prof. psychology, 1963—, chmn. dept., 1958—. Vis. prof. Harvard, 1965-66, summer 1967; dir. Nat. Def. Edn. Act. Inst., Wesleyan U., summer 1966. Mem. Am. Psychol. Assn., Sigma Xi. Author: Perception, 1966; also articles. Research on adaptive personality devel. and psychotherapy, perceptual devel. and cognitive style. Home: 659 Green Briar St., Lake Forest, Ill. 60045.*

FORKNER, Claude Ellis, Am. physician; b. Stevensville, Mont., Aug. 14, 1900; s. Allen F. and Lucy A. (Irvine) F.; A.B. cum laude, U. Cal. at Berkeley, 1922, M.A., 1923; M.D., Harvard, 1926; m. Marion Sturges DuBois, Sept. 3, 1927; children—Claude Ellis, Helen (Mrs. David Farley), Lucy (Mrs. Richard Spizzirri). Asst., Rockefeller Inst. Med. Research, N.Y.C., 1927-29; NRC fellow, Freiburg, Germany, also London, Eng., 1929-30; Francis Weld Peabody fellow Harvard, 1930-32; mem. Rockefeller Found. faculty Peking (China) Union Med. Coll., 1932-36; faculty Cornell U. Med. Coll., 1937—, clin. prof. medicine, 1953—; mem. staff Roosevelt Hosp., N.Y. Hosp.; cons. Bronx VA Hosp. Hon. prof. medicine Cheelow U. Med. Coll., Nat. Shanghai Med. Coll., Chengtu, China, also dir., prof. medicine China Med. Bd., 1943-45; mem. panel advisers N.Y. State Com. Med. Edn., 1962—; exec. dir. Med. Passport Found., 1958—, trustee, 1963—. Recipient Medal of Honored Merit, Republic of China, 1942; Gold Medal, Harvard Med. Alumni Assn., 1965; named companion Royal Order of Homayun, Shahinshah of Iran, 1957. Diplomate Am. Bd. Internal Medicine. Fellow N.Y. Acad. Scis.; mem. A.M.A., Am. Soc. Clin. Investigation (emeritus), A.A.A.S., Assn. Am. Physicians, Internat. Soc. Hematology, Nat. Hematological Soc. (pres. Blood Club 1954), PanAm. Med. Assn. (pres. sect. 1962), Research, publs. on blood—the origin of its components, and disease; lymph node studies; tuberculosis, synovial fluid, Kala-azar. Home: Butler's Island, Durien, Conn. Office: 35 E. 69th St., N.Y.C. 10021.*

FORLANINI, Carlo, Italian physician; b. Milan, Italy, June 11, 1847; prof. medicine, Pavia, Italy; began research in 1882 which led to introduction of artificial pneumothorax to treat pulmonary Tb by collapsing affected lung, 1894, thus introducing a new era in treatment. Died Nervi, Italy, May 26, 1918.

FORNASERI, Mario, Italian geochemist; b. S. Massimo all'Adige, Nov. 1, 1913; s. Angelo Vittorio and Matilde (Siviero) F.; Ph.D. in chemistry; m. Elvira Soleri, May 6, 1940; children—Angelo, Vittorio, Marcello. Prof. geochemistry, dir. Inst. Geochemistry, U. Rome. Mem. Acad. Sci. of Torino (corr.), Italian Soc. Mineralogy, Mineral. Soc. Am., Brit. Mineral Soc. Research and publs. on mineralogy, crystallography, geochemistry. Home: via di Villa Ada 10. Office: University of Rome, Rome, Italy.

FORRER, Gordon Randolph, Am. psychiatrist; b. Balt., Apr. 1, 1922; s. William Gordon and Blanche (Shules) F.; student State Tchrs. Coll., Towson, Md., 1939-41, Johns Hopkins, 1941-42; B.A., U. Md., 1945, M.D., 1947; m. Carol Lucille Hanke, May 26, 1951; children—Jane Elizabeth, Susan Ellen, John Gerritt. Clin. dir. Northville (Mich.) State Hosp., 1954-60; practice psychiatry, Detroit, 1960—; chief psychiatry Mt. Carmel Mercy Hosp.; mem. staff St. Mary Hosp., Livonia, Mich., Detroit Rehab. Inst.; clin. asst. prof. Wayne U. Med. Sch. Introduced atropine coma therapy for treatment psychoses, 1950; devel. psychoanalytic theory hallucination, psychoanalytic theory placebo. Diplomate Am. Bd. Psychiatry and Neurology. Mem. Mich. Soc. Psychiatry and Neurology (Research award 1953), Am., Pan-Am. med. assns., Am. Psychiat. Assn. Trustee Schoolcraft Coll., Livonia, 1963—. Contbr. articles in field. Demonstrated hallucinatory experiences to be universal and pathological only when not subject to repression. Home: 46995 W. Main St., Northville, Mich. 48167. Office: 20141 James Couzens Hwy., Detroit 48235.*

FORREST, Irene Stephanie Neuberg, Am. biochemist; b. Charlottenburg, Germany, Aug. 20, 1908; d. Carl and Hela (Lewinski) Neuberg; Ph.D., Berlin U., 1932; m. Fred M. Forrest, Dec. 30, 1949; 1 dau., Joan H. Roberts. Came to U. S., 1936, naturalized, 1944. Research staff Pasteur Inst., Hôtel Dieu, 1932-34; research staff, asst. tchr. dept. chemistry Istanbul (Turkey) U., 1935-36; biochemist N.Y. U., part-time, 1941-43; research staff Moraine Products div. Gen. Motors Corp., Dayton, O., and translations for War Dept., 1944-47; research staff N.Y. U., 1948, research dept. St. Clare's Hosp., N.Y.C., 1949-50; translator, chem. cons. to indsl. concerns, 1951-52; research staff Poly. Inst. Bklyn., also N.Y. Med. Coll., 1953-56; research psychopharmacology, chief biochem. research lab. VA Hosp., Brockton, Mass., 1957-61, Palo Alto, Cal., 1961—; research asso. in psychiatry Stanford Sch. Medicine, 1961—. Mem. com. on pharmacology and chemistry Nat. Inst. Mental Health, NIH, Bethesda, Md., 1962-66. Mem. Am. Chem. Soc., A.A.A.S., Soc. for Exptl. Biology and Medicine, Soc. Biol. Psychiatry, Western Pharmacology Soc., Am. Assn. Clin. Chemists, Am. Soc. Pharmacological and Exptl. Therapeutics. Research, publs. on chemistry and metabolism of phenothiazine tranquilizers, especially chlorpromazine; devised (with Fred M. Forrest) series of urine color tests for tranquilizers. Home: 436 Falk Ct., Menlo Park, Cal. 94025. Office: Biochem. Research Lab., Menlo Park Div., VA Hosp., Palo Alto, Cal. 94304.*

FORREST, Waldie William, Australian chemist; b. Adelaide, Australia, July 20, 1923; s. William Waldie and Isabel (Cameron) F.; B.Sc., Adelaide U., 1947, B.Sc. with honors, 1948; Ph.D., Cambridge (Eng.) U., 1953; m. Patricia Lesley Johnston, Sept. 29, 1949; children—John, Catherine. Lectr., Waite Agrl. Research Inst., Adelaide U., 1949; post-doctoral fellow Yale, 1953-56; research scientist div. nutritional biochemistry Commonwealth Sci. and Indsl. Research Orgn., Adelaide, 1956—; C.N.R.S. Laboratoire de Chimie Bacterienne Marseilles, France, 1965. Mem. Australian Biochem. Soc., Australian Soc. Biomed. Engring. Author: (with others) Biochemical Microcalorimetry, 1967; also articles. Research on quantitative and thermodynamic studies of bacterial energetics including energy of maintenance, design of calorimetric apparatus. Home: Woods Hill Rd., Ashton, S.A. Office: Div. Nutritional Biochemistry, Commonwealth Sci. and Indsl. Research Orgn., Kintore Av., Adelaide, S.A., Australia.*

FORRESTER, A(lvin) Theodore, Am. physicist; b. Bklyn., Apr. 13, 1918; s. Joseph D. and Rose (Kissen) F.; A.B., Cornell U., 1938, A.M., 1939, Ph.D., 1942; m. June Doris Berg, Oct. 5, 1956; children—Bruce H., David A., Cheri J., William C., Susan J. Research asso. U. Cal. at Berkeley, 1942-45; physicist RCA Labs., Princeton, N.J., 1945-46; asst. prof. physics U. So. Cal., Los Angeles, 1946-51, asso. prof., 1951-54; vis. asso. prof. physics U. Pitts., 1954-55; physicist Westinghouse Research Labs., Pitts., 1955-58; nuclear specialist Atomics Internat., Los Angeles, 1948-59; dept. mgr. Electro-Optical Systems, Pasadena, Cal., 1959-65; prof. U. Cal. at Irvine, 1965-67, at Los Angeles, 1967—. Fellow Am. Phys. Soc., I.E.E.E.; mem. Am. Rocket Soc. (Research award 1962, chmn. electrostatic propulsion panel 1960-61), Am. Inst. Aeros. and Astronautics (electric propulsion com. 1964-66), A.A.A.S., Engrs., Am. Assn. Physics Tchrs., Am. Optical Soc., Sigma Xi, Phi Kappa Phi. Research in photoelectric mixing of light, ion propulsion, isotope separation, superconductivity. Home: 1912 N. Edgemont St., Los Angeles, Cal. 90027.*

FORRESTER, James Donald, Am. geologist, mining engr.; b. Salt Lake City, Apr. 6, 1906; s. James Gillon and Diana (George) F.; B.S. in Geol. Engring., U. Utah, 1928, Geol. Engr., 1956; M.S. in Geology, Cornell U., 1929, Ph.D., 1935; m. Lisle Keele, Sept. 27, 1929; 1 dau., Lisle Jean (Mrs. Stanley Francis Johnsen). Geologist, Anaconda Co., Butte, Mont., 1929-32; 35-39; instr. Cornell U., 1934-35; prof. head dept. geology U. Ida., Moscow, 1939-44; prof., head dept. mining engring. Mo. Sch. Mines, Rolla, 1944-54; dean Coll. Mines, dir. Ida. Bur. Mines and Geology, U. Ida., 1954-56; dean Coll. Mines, dir. Ariz. Bur. Mines, U. Ariz., Tucson, 1956—; geologist U. S. Geol. Survey, 1942-45. Cons. on nature and value various mineral commodities. Eleanor Tatum Long scholar in structural geology Cornell U., Ithaca, N.Y., 1933-34. Fellow Geol. Soc. Am.; mem. Engrs. Council for Profl. Devel. (dir. 1965—), Am. Inst. Mining, Metall. and Petroleum Engrs., Soc. Econ. Geologists, Mining and Metall. Soc. Am., Am. Assn. Engring. Edn., Ariz. Acad., Sigma Xi, Tau Beta Pi, Theta Tau, Phi Eta Sigma. Author: Field and Mining Geology, 1946; Laboratory Manual for Physical Geology, 1946; also numerous articles. Research on nature and occurrence mineral deposits, application mil. explosives for mining and quarrying. Home: 5719 E. 8th St., Tucson 85711.*

FORRESTER, Jay Wright Am. elec. engr.; b. Anselmo, Neb., July 14, 1918; s. Marmaduke M. and Ethel Pearl (Wright) F.; B.Sc., U. Neb., 1939, D. Engring., 1954; M.Sc., Mass. Inst. Tech., 1945; m. Susan Swett, July 27, 1946; children—Judith, Nathan Blair, Ned Cromwell. Tchr., x-ray equipment research Mass. Inst. Tech., 1939-40, co-founder servomechanisms lab., 1940, development electric and hydraulic servomechanisms for gun mounts and radar, 1940-44, asso. dir. servomechanisms lab., also supr. Whirlwind I digital computer devel., 1944-51, founder Digital Computer Lab., dir., 1951-56, div. head Lincoln Lab. for Air Def., 1951-56; prof. indsl. mgmt., sch. indsl. mgmt. Mass. Inst. Tech., Cambridge, 1956—; partner Forrester Cattle Ranch, Anselmo, Neb. Fellow I.E.E.E.; mem. Inst. Mgmt. Scis., Am. Phys. Soc. Assn. Computing Machinery, Eta Kappa Nu, Sigma Xi, Sigma Tau. Author: Industrial Dynamics, 1961; also lectures and articles. Invented random-access magnetic core memory for digital computers (led to information-storage method basic to most digital computers); developed indsl. dynamics as means of analyzing behavior of social systems; patentee in servomechanisms, digital information storage, indsl. control. Home: 11 Holden Wood Rd., Concord, Mass. Office: Mass. Inst. Tech., Cambridge 39, Mass.

FORSIUS, Henrik Runar, Finnish ophthalmologist; b. Helsinki, Finland, Aug. 24, 1921; s. Runar and Saima (Granit) F.; Cand. Med. U. Helsinki, 1945; Lic.Med., 1948; m. Harriet Brunila, Mar. 15, 1948; children—Irmelin (adopted), John, Martin, Kaj. Asst. tchr. U. Eye Hosp. Helsinki, 1956-63; faculty U. Oulu (Finland), 1963—, prof. ophthalmology, head ophthalmology, 1965—; sr. researcher Sci. Research Inst. Minerva, Helsinki, 1963; Folkhälsan Inst. for Genetics, Helsinki, 1963—. Mem. Finisch Ophthal. Soc. (past pres.). Author: Arcus senilis corneae, 1948; also numerous articles. Research on hereditary diseases, especially of the eye, anthropology of isolated populations. Office: Univ. Eye Hosp., Uusikatu 50, Oulu, Finland.*

FORSKAL, Peter, naturalist; b. Kalmar, Sweden, Jan. 11, 1736; became pupil of Linnaeus; became prof. natural history and natural sci., Copenhagen, Denmark, also apptd. naturalist by Frederick V of Denmark to sci. expdn. to Egypt and Arabia, 1761. Genus Forskalia of family urticees named in his honor by Linnaeus. Author: Flora aegyptiaco-arabica, 1775; Descriptiones animalium orientalium, 1775; Icones rerum naturalium, 1776. Described many new plants and insects; his collection of more than 500 species included about 200 new. Died Yerim, Arabia, July 11, 1763.

FORSSELL, Gösta, Swedish radiologist; b. Mar. 2, 1876; s. C. A. Forssell and Maria T. (Trahn) F.; med.-lic., Royal Caroline Inst., Stockholm, Sweden. 1903, M.D., 1913; m. Esther Gottlieb, Dec. 19, 1906; 4 children. Radiologist, 1899; chief X-ray dept. Seraphimer Hosp., Stockholm, 1906-40, Caroline Hosp., 1940-41; chief Radiumhemmet, 1910-26, prof. med. radiology, 1916-36; prof. roentgen diagnoses Royal Caroline Hosp. and Inst., 1917-41; prof. X-ray diagnosis, Caroline Inst., 1936-41. Became pres. 2d Internat. Congress Radiology, 1928. Recipient Gold medal Radiol. Soc. N.Am., 1922. Mem. Swedish Soc. Med. Radiology (pres. 1919-33). Author books on anatomy, roentgen diagnosis, radiotherapy, biography, hosp. orgn. Organizer, editor Acta Radiolgica, until 1950. Died Stockholm, Nov. 13, 1950.

FORSSMANN, Werner, German surgeon, urologist; b. Berlin, Germany, Aug. 29, 1904; s. Julius and Emmy (Hindenberg) F.; grad. Askanisches Gymnasium, Berlin; student Friedrich-Wilhelm U., Berlin; m. Elsbet Engel, Dec. 7, 1933; children—Klaus, Knut, Jörg, Wolf, Bernd, Renate. Formerly asst. physician Auguste-Viktoria Hosp., Eberswalde, Germany, surg. clinic Charite Hosp., Berlin, surg. clinic Municipal Hosp., Mainz, Germany, chief resident physician urol. clinic Municipal Rudolf-Virchow Hosp., Berlin, surg. clinic Municipal Hosp., Dresden-Friedrichstadt, Germany, 3d Surg. Univ. Clinic, Robert Koch Hosp., Berlin; captured by U. S. Army during World War II; confined in a prison camp; gen. practitioner, 1945-50; head physician dept. urology Diakonie Insts., Bad Kreuznach, Germany, 1950-57; head physical dept. surgery Evangel. Hosp., Dusseldorf, Germany, 1957—; prof. medicine Johann-Gutenberg U., Mainz, Germany; hon. prof. surgery Universitad Nacional-Cordoba (Argentinien), 1961, medizinische Akademie Düsseldorf, Germany, 1964. Recipient Leibniz medal Berlin Acad. Scis., 1954; Nobel prize in medicine and physiology (with D. W. Richards and A. F. Cournand), 1956. Mem. German Soc. Surgery, German Soc. Urology, Am. Coll. Chest Physicians, Svensk Förenning for Cadiologi (hon.). First to develop technique for catheterization of heart. Home: Holzbüttgen üb. Neub, Feldstrasse 11. Office: Evangelisches Kran-Kenhaus, Dusseldorf Furstenwall 91, Federal Republic of Germany.

FÖRSTER, Arnold, see Foerster, Arnold.

FORSTER, Benjamin Meggot, Brit. natural scientist; b. Walbrook, London, Eng., Jan. 16, 1764; s. Edward (the elder) and Susanna (Furney) F.; ed. Walthamstow. Early mem. anti-slave trade com., 1788; framed Child Stealing Act; studied botany, electricity; made drawings of fungi, also communicated species to Sowerby; invented sliding portfolio, atmospherical electroscope. Author: An Introduction to the Knowledge of Fungusses, 1820; numerous articles. Died Mar. 8, 1829.

FORSTER, Edward, Brit. botanist; b. Walthamstow, Eng., Oct. 12, 1765; s. Edward and Susanna (Furney) F.; comml. studies in Holland; studied botany, Epping Forest, from 1780; m. Mary Jane Greenwood, 1796 (dec. 1846). Cultivated (with 2 bros) all herbaceous plants then grown in father's garden. Fellow Royal Soc., 1821; Linnean Soc. (became treas. 1816, v.p. 1828). Author: Catalogus avium in insulis Britannicis habitantium cura et studio Eduardi Forsteri jun, 8 vols., 1817; also articles; drew up (with bros.) county list of plants in Gough's Camden, 1789. Described Brit. plant species. Died Feb. 23, 1849.

FORSTER, Francis Michael, Am. physician, educator; b. Cin., Feb. 14, 1912; s. Michael Joseph and Louise Barbara (Schmid) F.; student Xavier U., Cin., 1930-32, LL.D., 1955; B.S., U. Cin., 1935, B.M., 1936, M.D., 1937; m. Helen Dorothy Kiley, June 15, 1937; children—Denis, Susan, Kathleen, Mark, Gabrielle. Rotating intern Good Samaritan Hosp., Cin., 1936-37; house officer neurology and neurosurgery Boston City Hosp., 1937-38, resident neurology, 1939-40; fellow psychiatry Pa. Hosp., Phila., 1938-39; asst. neurology Harvard Med. Sch., 1939-40; Rockefeller Found. research fellow physiology Yale Sch. Medicine, 1940-41; instr. neurology Boston U. Sch. Medicine, 1941-43; asst. prof. neurology Jefferson Med. Sch., 1943-47, asso. prof. neurology, 1947-50; prof. neurology, dir. dept. Georgetown U. Sch. Medicine, 1950-58, dean Sch. Medicine, 1953-58; professor, chmn. dept. neurology, University of Wisconsin School of Medicine, 1958—; consultant neurology. Diplomate Am. Bd. Psychiatry and Neurology (dir.). Mem. A.M.A. (chmn. nervous and mental diseases sect. 1952-53), D.C. Med. Soc. (chmn. sect. neurology and psychiatry 1955-56), Am. Acad. Neurology (chmn. survey com. 1948-51; pres. 1957-59), Am. Neurological Assn. (chmn. com. internat. collaboration 1954-55), Am. Epilepsy League, Assn. Research Nervous and Mental Diseases, Am. Physiol. Soc. Am. Psychiat. Assn., Am. Assn. Electroencephalographers, A.A.A.S., Am. Assn. U. Profs., Mass. Med. Soc., State Med. Soc. Pa., N.Y. Acad. Scis., Acad. Medicine of Washington, Sigma Xi. Author: Synopsis of Neurology, 1962. Editor: Modern Therapy in Neurology, 1957; Evaluation of Drug Therapy, 1961. Evaluated drugs used to treat epilepsy; developed use of conditioned reflex treatment in certain kinds of epilepsy. Home: Route 1, Waunakee, Wis. Office: U. Hosp., U. Wis., Madison, Wis.*

FORSTER, Johann Georg Adam, naturalist; b. Nassenhuben, nr. Danzig (now Gdansk, Poland) Nov. 27, 1754; s. Johann Reinhold Forster; ed. St. Petersburg, Russia; m. Therese Heyne, 1783; 1 dau. Went (with father) to Eng., 1766, and assisted him as naturalist on Capt. Cook's 2d voyage; prof. natural history Kassel, Germany, 1778-84, later at Vilna (now in Lithuania); librarian, Mainz, Germany. Fellow Royal Soc. Author: A Voyage around the World (early example of sci. travel book), 1777; Dissertatio botanico-medica de plantis esculentis insularum oceani Australis, 1785; Essais sur la géographie morale et naturelle, l'histoire naturelle et la philosophie usuelle, 1789-97; Views of the Lower Rhine, Brabant, Flanders . . . , 3 vols., 1791-94; collected letters, 2 vols., 1828; Oeuvres completes (pub. by dau., with biography by Gervinus), 1843-44. Described flora of S. Seas. Died Paris, France, Jan. 10, 1794.

FORSTER, Johann Reinhold, German naturalist; b. Dirschau, E. Prussia, Oct. 22, 1729; M.D., Ph.D., U. Halle (Germany); D. in Civil Law, Oxford (Eng.) U.; 1 son, Johann Georg Adam. Pastor, Nassenhuben, nr. Danzig, Germany, from 1753; supt. of new settlements on Volga River, Russia, from 1765; migrated to Eng.; tchr. French, German, Dissenter's Sch., Warrington, Eng.; tchr. French, Episcopal Sch.; writer, translator, London, Eng.; from 1770; naturalist with Cook's 2d Expdn., 1772-75; created councillor by Frederick II of Prussia; chair natural history U. Halle, 1781-99, head bot. garden. Fellow Royal Soc., 1772. Author: (with J. Georg Forster) Characteres Generum Plantarum quas in Itinere ad Insulas Maris Australis collegerunt, descripserunt, delinearunt, annis MDCCXXII-MDCCLXXV, 1775; De Bysso Antiquorum; Observations made during a Voyage round the World on Physical Geography, Natural History, and Ethic Philosophy, 1778; Zoologia Indica, 1781. Translator into English: History of Russia (Lomonosov). Drew up code for Russian govt. based on German civil code; distinguished 18th century naturalist; described 75 new genera of plants, 1775. Died Halle, Dec. 9, 1798.

FORSTER, John Cooper, Brit. surgeon; b. Lambeth, Eng., Nov. 13, 1824; ed. King's Coll. Sch., Guy's Hosp.; M.B., London, 1847. Surgeon Guy's Hosp., 1870-80 (resigned as sr. surgeon in protest to adminstrv. policy); Fellow Royal Coll. Surgeons (pres. 1884-85, Gold medal). Author: The Surgical Diseases of Children, 1860; also various papers. Recorded 1st operation of gastrostomy in Eng., 1858. Died Mar. 2, 1886.

FORSTER, Sir Martin Onslow, Brit. chemist; b. Nov. 8, 1872; s. Martin Forster; ed. Finsbury Tech. Coll., Central Tech. Coll.; D.Sc., London U.; Ph.D., Würzburg (Germany) U.; m. Elena Haynes Parodi, 1925. Demonstrator Royal Coll. Sci., South Kensington, Eng., 1895-1902, asst. prof. chemistry, 1902-13; dir. Brit. Dyes, Ltd., 1915-18; 1st dir. Salter's Inst. Indsl. Chemistry, 1918-22; dir. Indian Inst. Sci., Bangalore, 1922-33. Fellow Royal Soc., 1905; Inst. Chemistry. (v.p.), City and Guilds London Inst.; mem. Chem. Soc. (Longstaff medal 1915), Worshipful Co. Dyers (prime warden 1919), Brit. Assn. (pres. chemistry sect. Edinburgh 1921), Indian Sci. Congress (pres. 1925). Contbr. to Jour. Chem. Soc. Research on camphor and its derivatives. Died May 23, 1945.

FÖRSTER, Otfrid, see Foerster, Otfrid.

FORSTER, Roy Philip, Am. physiologist; b. Milw., Wisc., Sept. 28, 1911; s. Frank M. and Adele M. (Schatz) F.; B.S., Marquette U., 1932; Ph.M., U. Wis., 1935, Ph.D., 1938; M.A. (hon.), Dartmouth, 1948; m. Dorothy F. Seegers, Aug. 31, 1935; 1 dau., Peggy (Mrs. Charles L. Ffolliott). Faculty mem. Dartmouth Coll., Hanover, N.H., 1938—, prof., 1948-64, Ira Allen Eastman prof., 1964—, chmn. zoology dept., 1950-56, chmn. sci. div., 1947-51, lectr. physiology Med. Sch., 1964—; dir. regulatory biology program Nat. Sci. Found., 1959-60; director Mt. Desert Island Biological Laboratory, summers 1940-47, vice president, 1960-63, president, 1963—; dir. John Simon Guggenheim Meml. Found. fellow, Cambridge U. and various European biol. labs., 1949. member Conference on Renal Function of Josiah Macy, Jr., Found., 1949-54; John Simon Guggenheim Meml. Found. fellow for travel and study abroad, 1955-56. Trustee Mt. Desert Island Biological Laboratory, 1940—, Corp. Bermuda Biol. Sta. for Research; mem. sci. rev. com. Health Research Facilities NIH, 1964-66, med. biology rev. com., 1966—. Mem. Am. Heart Assn., Am. Soc. Zoologists, Am. Physiol. Soc., A.A.A.S., N.H. Academy Sci. (exec. com. 1945-48), Society of General Physiologists, Sigma Xi. Author articles profl. jours. Editor kidney sect. of Biological Abstracts, 1947—; asso. editor Jour. Gen. Physiology, 1960—. Mem. editorial bd. Am. Jour. Physiology, Jour. Applied Physiology, 1960-66. Research on many aspects of renal function; hemodynamics; active tubular transport and patterns of nitrogen excretion in vertebrates; developed techniques for evaluation of kidney function; studies of enzymology and cellular physiology of kidney. Home: Hemlock Rd., Hanover, N.H.*

FORSTER, Thomas Furly, English botanist; b. Walbrook, Eng., Sept. 5, 1761; s. Edward the Elder and Susanna (Furly) F.; m. Susanna Williams, 1788; 2 sons, 3 daus. Drew plants, 1775-82; lived at Clapton, 1796-1823; moved to Walthamstow, Eng., 1823. Fellow Linnean Soc.; mem. numerous sci. socs. Author: Flora tonbrigensis, 8 vols., 1816; printed list of additions to Plantae woodfordienses (Warner); drew up (with bros.) county lists of plans in Camden (Gough), 1789. Collected algae, flowering plants, fossils. Died Walthamstow, Oct. 28, 1825.

FORSTER, Thomas Ignatius Maris, naturalist, astronomer; b. London, Eng., Nov. 9, 1789; s. Thomas Furly and Susanna (Williams) F.; 1st studied law, then medicine; M.B., Cambridge, Eng., 1819; m. Julia Beaufoy, 1817; 1 dau., Selena. Fellow Linnean Soc., Royal Astron. Soc. Author: Researches about Atmospheric Phenomena, 1812; Liber rerum naturalium, 1805; Perpetual Calendar, 1824; Recueil de ma vie, 1835; Observations sur l'influence des comêtes, 1836; Sati, 1843; Epistolarium Forsterianum; also sketch Phenological system; published original letters of Locke, Shaftesbury, and Algernon Sydney to his ancestor Benjamin Furley, 1830. Discovered new comet, 1819; studied effect of atmosphere on disease, especially cholera. Died Brussels, Belgium, Feb. 2, 1960.

FORSTER, Werner, pharmacologist; b. Brüx, Czechoslovakia, Aug. 20, 1919; s. Josef and Wilhelmine (Liebig) F.; student German U. in Prague, 1937-39, 42-44; M.D., U. Mainz, 1948; m. Liese Herbrich, July 1, 1944; children—Walter, Gunter, Reinhard, Ulrike. Asst., Pharmacol. Inst., Mainz, 1948-52; dir. pharmacol. dept. Inst. for Indsl. Hygiene, Badische Anilin and Sodafabrik, Ludwigshafen, 1952-57; sr. physician Pharmacol. Inst., Jena, 1957-59; lectr., prof. Pharmacol. Inst., Med. Acad., Magdeburg, 1959-66; prof., dir. Pharmacol. Inst., Martin Luther U. of Halle/Wittenberg, 1966—. Recipient Rudolf Virchow prize East Germany, 1964. Mem. Deutsche Pharmakologische Gesellschaft, East German Pharmakologische Gesellschaft. Author: Rationale Arzneitherapie, 1967; also numerous articles. Studies on structure-effect relationships of heart glycosides, new glycosides, introduced convallatoxol. Home: 27 Schleiermacherstrasse, Halle/Salle, East Germany.*

FORSYTH, Alexander John, inventor; b. Scotland, Dec. 28, 1769; s. James and Isabella (Syme) F.; grad. King's Coll., Aberdeen, Scotland, 1786; LL.D., Glasgow, Scotland. Clergyman. Invented percussion cap for ignition of gunpowder, 1807; Napoleon offered to buy secret, but was refused; Forsyth later pensioned by Brit. govt. Died June 11, 1843.

FORSYTH, Andrew Russell, Brit. mathematician; b. Glasgow, Scotland, June 18, 1858; s. John and Christina (Glen) F.; Became fellow Trinity, 1881. M.A., Sc.D., Cambridge, Eng.; hon. degrees from 8 univs.; m. Marion Amelia Pollock Boys, 1910. With Cambridge, 1895-1910; prof. Imperial Coll. Sci. and Tech., South Kensington, Eng., 1913-23. Fellow Royal Soc., 1886 (Royal medal 1897); mem. London Math. Assn. (became pres. 1903, 04, 36). Author: Treatise on Differential Equations, 1885; Theory of Differential Equations, Part I, 1890, Part II, 1900, Part III, 1902, Part IV, 1906; Treatise on the Theory of Functions, 1893; Lectures of Differential Geometry, 1912; Lectures on Functions of Two or More Complex Variables, 1913; Lectures Introductory to the Theory of Functions of Two Complex Variables, 1927; Calculus of Variations, 1927; Geometry of Four Dimensions, 1930; Intrinsic Geometry of Ideal Space, 1935. Editor, Quar. Jour. Math., 1885-95. Editor: Cayley's Collected Mathematical Papers, vols. 8-13, 1895-98; Burnside's Theory of Probability, 1928. Discovered formulae expressing relations and identities of structural forms invariant under transformations; theory of double theta functions; studied differential geometry. Died London, June 2, 1942.

FORTES, Meyer, social anthropologist; b. Britstown, S.Africa, Apr. 25, 1906; s. Nathan and Bertha (Kerbel) F.; B.A. with distinction, U. Cape Town (S. Africa), 1925, M.A. with distinction, 1926; Ph.D., London (Eng.) U., 1930; M.A., Oxford (Eng.) U., 1946; M.A., Cambridge (Eng.) U., 1950; m. Sonia Donen, May 1, 1928 (dec. 1956); 1 dau., Natalie (Mrs. Trevor William Marshall); m. 2d, Doris Mayer, June 16, 1960; stepchildren—Gail (Mrs. Frederic Petersen), Karl David Mayer. Research fellow Internat. African Inst., London, 1932-37; lectr. London Sch. Econs., 1938-39; asst. lectr. Oxford (Eng.) U., 1939-41; staff Nat. Service, W. Africa, 1941-42; head dept. sociology W. African Inst., Accra, Ghana, 1944-46; reader social anthropology Oxford U., 1946-50; William Wyse prof., social anthropology Cambridge U., 1950—, fellow King's Coll., 1950—. Mem. various adv. coms. to English govt. agys. Fellow Brit. Acad.; Mem. Am. Acad. Arts and Scis. (fgn., hon.) Brit. Assn. for Advancement Sci. (past pres.), Royal Anthrop. Inst. (Welcome medal 1937, Rivers medal 1946, Myers medal 1960; past pres.), Am. Anthropol. Assn., Internat. African Inst., Brit. Sociol. Assn., Assn. Social Anthropologists, African Studies Assn. Author: Dynamics of Clanship among the Tallensi, 1945; Web of Kinship among the Tallensi, 1949; Oedipus and Job in West African Religion, 1959; co-editor, author: (with E. E. Evans-Pritchard) African Political Systems; 1940; also articles. Originated with (Evans-Pritchard) the modern theory of primitive polit. systems; initiated modern ethnographical research in Ghana; research on theory of kinship orgn. in primitive soc., demographical method in preliterate soc., study of ancestor worship, gen. theory of primitive social structure. Home: 113 Granchester Meadows, Cambridge, U.K.*

FORTET, Robert Mary, French mathematician; b. Perigueux, France, May 1, 1912; s. Gabriel Pierre and Madeleine (Queyroulet) F.; Agrégation Math. Scis., École Normale Supérieure, 1934, Doctor ès Sciences, 1939. Prof. mechanics Faculty Scis., Caen, France, 1942-52; prof. calculus probability Faculty Scis. Paris, 1952—. Cons., Laboratoire Central des Télécommunications, (br. Internat. Tel. and Tel.), 1948—. Fellow Inst. Math. Statistics. Author: Elements de la Théorie des Probabilites, 1966; also numerous articles. Research on theory of probability and operation research. Home: 120 Avenue de Suffren, Paris 15, France. Office: 11 rue Pierre-et-Marie Curie, Paris 5, France.*

FORTEZA BOVER, Geronimo, Spanish hematologist, cytogeneticist; b. Alcoy, Alicante, Dec. 11, 1911; s. Geronimo and Isabel (Bover) Forteza; Licenciado en Medicina y Cirujia, U. Valencia (Spain), 1935, Doctor graduado, 1941; m. Elena Vila Ochando, Apr. 18, 1940; children—Jeronimo, Jose Luis. Became physician, provincial hosp., Valencia, 1949; apptd. dir. Inst. Arnaldo Vilanova, 1964; dir. dept. cytogenetics, U. Valencia Sch. Medicine, 1963; dir. Lab. of Cytogenetics, 1964; staff Lab. of Hematology, Valencia. Recipient prizes Inst. Medicine of Valencia, 1945, Royal Acad. Medicine, 1945, 47, Provincial Com. Scis., Valencia, 1963. Mem. Acad. Sci. and Medicine, Barcelona, Valencia Inst. Medicine, Royal Acad. Medicine Valencia, Spanish Assn. Hematology (pres. 1964). Author: El diagnóstico por la punción esternal, 1945; El diagnóstico por la punción ganglionar, 1947; (with J. Garcia-Blanco) Nuevos métodos de coloración en hamatologis, 1948; Atlas de hematología, 1952; Atlas de histopatología de los organos hemocitopoyéticos, 1955; (with R. Baguena Dandela) Atlas de citología sanguinea, 1963; (with E. de Salamanca, M. Valdes Ruiz) Enfermedades de la sangre y de los organos hemocitopoyéticos (vol. 4 of Tratado de patologia medica). Contbr. to devel. of hematic cytology in Spain, in particular to study of leukemia. Home: Ave. Jacinto Benavento 16, Valencia. Office: Jorge Juan 15, Valencia, Spain.*

FORTIER, Claude, Canadian physiologist; b. Montreal, Que., Can., June 11, 1921; s. Carolus and Flore-Edith (Lanctôt) F.; B.A., U. Montreal, 1941, M.D., 1948, Ph.D., 1952; m. Elise Gouin, Sept. 8, 1953; children—Anne, Michele, Nicole, Nathalie. Research asst. U. Montreal Inst. Exptl. Medicine and Surgery, 1948-51; research cons. U. Lausanne, Switzerland, 1952-53; research asso. dept. neuroendocrinology U. London, Inst. Psychiatry, Eng., 1953-55; asso. prof. physiology, dir. neuroendocrine research lab. Baylor U. Coll. Medicine, 1955-60; dir. endocrine lab., prof. exptl. physiology Laval U., Quebec, Que., Can., 1960—, chmn. dept. physiology, 1964—. Vice-chmn. Med. Research Council Can., 1965-67, chmn. com. on computers and their applications, 1968—; vice chmn. med. research coordinating, adv. coms. Def. Research Bd. Can. Fellow Royal Soc. Can., Royal Coll. Physicians of Can.; mem. Canadian (pres. 1966-67), Am. physiol. socs., Endocrine Soc., Am. Thyroid Soc., A.A.-A.S., N.Y. Acad. Sci., Soc. for Exptl. Biology and medicine, Research, numerous publs. on neurohumoral control of adenohypophys. Functions, pituitary-thyroid-adrenocortical interactions, biostatistics, biocontrol systems, role of corticosteroid-binding by plasma proteins in adjustment of pituitary adrenocortical system to thyroid and gonadal activities. Home: 1014 DeGrenoble St., Quebec 10. Office: Dept. Physiology, Laval U., Quebec 10, Que., Can.*

FORTIN, Gerald Adelard, Canadian sociologist; b. Quebec, Que., Can., Dec. 9, 1929; s. Adelard and Bertha (Senechal) F.; M.S.Sc., U. Laval, Que, 1954; Ph.D. in Sociology, Cornell U., 1956; m. Thérèse Bergeron, May 24, 1952; children—Andree, Anne, Marie-Jose. Faculty, Laval U., 1956—, prof., 1960—, secretaire Centre de recherche sociale, 1958-59, dir., 1960-61, directeur dept. de sociologie et d'anthropologie, 1966—. Exec. mem. Canadian Council Regional Planning, 1965—. Mem. Am. Sociol. Assn., Assn. Internationale de sociologie (dir. 1965), Assn. internationale des sociologie de langue francaise (dir. 1966—). Author: (with M. A. Tremblay) Les Comportements economiques des familles salaries du Quebec, 1964; also articles. Research in rural milieu and its transformation due to industrilization, indsl. sociology, econ. sociology, sociology econ. devel. and social change. Home: 653-5679 De Vincennes St., Quebec 10, Que. Office: 656-2226 Cité Universitaire, Ste. Fay, Que., Can.*

FORTIN, Jean, French physicist, engr.; b. Mouchy-la-Ville, France, Aug. 9, 1750; mechanic in Paris, France; mem. Bur. Longitudes; pub. Atlas celeste (John Flamsteed), 1776; invented balance named after him; also a cup barometer; built portable mercury barometer, also clocks, vacuum pumps, numerous other phys. instruments. Died Paris, 1831.

FORTUNATO OF BRESCIA, biologist; b. Brescia, Italy, 1701; became friar minor of Reform, 1718, sec. gen. of order; studied microscopic anatomy, histology and embryology of human organs, central nervous system; developed systematic microscope technique; 1st to distinguish between tissues and organs; completed a morphological classification prior to Bichat's; tried to blend scholasticism and phys. scis. Died Madrid, Spain, 1754.

FORTUNE, Robert, botanist; b. Edrom Parish, Eng., 1813; ed. parish schs.; apprentice in local gardens; visited China, 1842, 48, Formosa, Japan, 1853. Genus Fortunella (including kumquat) named in his honor. Author: Three Years' Wanderings in the Northern Provinces of China . . . , 1847; Report on the Tea Plantations in the North-west Provinces, 1851; A Journey to the Tea Countries of China, 1853; Two Visits to the Tea Countries in China and the British Plantation in the Himalyaas, 1953; A Residence among the Chinese, 1857; Yeddo and Peking, 1863. Introduced double yellow rose from China to Eng., also Japanese anemone; other plants; founded cultivation of tea plant in India. Died 1880.

FORWALD, Haakon Gabriel, elec. engr., inventor; b. Mandal, Norway, Aug. 21, 1897; student Tech. Coll., Oslo, 1915-17; B.A., Technische Hochschule, Danzig, 1925; postgrad. Swiss Fed. Inst. Tech., Zurich, Royal Inst. Tech., Stockholm; m. Birgit Steffenburg, 1932; 2 daus., 1 son. Elec. engr. Brown, Boveri Co., Oslo, Baden, Switzerland, 1918-30, Ateliers de Construction Oerlikon, Zurich, 1930-33; cons. engr. Allmanna Svenska Elektriska Aktiebolaget, Ludvika, Sweden, 1935—; research fellow Duke, 1957, U. Pitts., 1959. Recipient William McDougall award for distinguished work in parapsychology for article on confirmation of psychokinesis placement effect, 1959. Mem. Parapsychol. Assn. (charter). Research on psychokinesis, tests conducted with cubes or dice; developed method for dice by psychokinesis; over 350 patents in Europe and estimating energy involved in lateral displacement of U. S., mainly in circuit-breaker field.

FORWARD, Robert Lull, Am. physicist; b. Geneva, N.Y., Aug. 15, 1932; s. Robert Torrey and Mildred (Lull) F.; B.S., U. Md., 1954, Ph.D., in Physics, 1965; M.S., U. Cal. at Los Angeles, 1958; m. Martha Neil Dodson, Aug. 29, 1954; children—Robert Dodson, Mary Lois, Julie Elizabeth. With Hughes Research Labs. div. Hughes Aircraft Co., Malibu, Cal., 1956-—, sr. staff physicist, 1965-—, asso. mgr. theoretical studies dept., 1966-67, mgr. exploratory studies dept., 1967-—. Gravity Research Found. award, 1962, 63, 64, 65. Mem. A.A.A.S., Am. Astronautical Soc., Am. Phys. Soc., I.E.E.E., Am. Inst. Aeros. and Astronautics, Research Soc. Am., Sigma Pi Sigma. Exptl. research on dynamic gravitational fields, gravitational gradient sensors, tests of gen. theory of relativity and gravitational radiation; radar, masers and lasers. Inventor gravitational mass sensor, laser modulators; co-inventor gravitational radiation antenna. Home: 34 Carriage Sq., Oxnard, Cal. 93030. Office: 3011 Malibu Canyon Rd., Malibu, Cal. 90265.*

FOSBERG, Francis Raymond, Am. botanist, ecologist, geographer; b. Spokane, Wash., May 20, 1908; s. Axel Peter and Helen Rose (Wirtner) F.; B.A., Pomona Coll., 1930; M.S., U. Hawaii, 1935; Ph.D. (Morris fellow), U. Pa., 1939; m. Violet Mildred Oliveira, June 26, 1935; children—Helen Charlotte (Mrs. Everett Kyle), Ilima Jane, Carol Evelyn, Hildegarde Frances. Botanist, U. S. Dept. Agr., 1939-42, 45-46, U. S. Fgn. Econ. Adminstrn., 1942-45, U. S. Comml. Co., Micronesia, 1946-47; Guggenheim fellow, 1947-48; vis. prof. U. Hawaii, 1948; prof. lectr botany George Washington U., 1948-49; research asso. Cath. U. Am., 1949-50; botanist U. S. Geol. Survey, Washington, 1951-65; spl. adviser on tropical biology Smithsonian Instn., Washington, 1966-—. Mem. internat. adv. com. humid tropical research vis. com. tropical herbaria, 1956-64, UNESCO, 1960-—; mem. Pacific Sci. Bd., 1957-62, NRC, 1958-—. Mem. Nature Conservancy (past v.p.), Toulouse Acad. (corr., Fermat medal 1960), Bot. Soc. Japan (corr.), Ecol. Soc. Am., A.A.A.S., Bot. soc. Am., Am. Geog. Soc., Internat. Assn. Plant Taxonomists, Internat. Soc. Tropical Ecology (v.p.), New Eng., So. Appalachian Bot. clubs, Cal. Bot. Soc. Author: (with M. H. Sachet) Island Bibliographies, 1955, Manual for Tropical Herbaria, 1965; Military Geography of the Northern Marshalls, 1956; also numerous articles. Editor: Climate, Vegetation and Rational Land Utilization in the Humid Tropics, 1959; Man's Place in the Island Ecosystem, 1963; founder, editor Atoll Research Bull., 1951-—. Research on systematics various plant groups, biosystematics, devel. vegetational concepts, devel. ecosystem concept in ecology, application entropy trends in ecosystems, coral atoll ecostystem, coral islands and reefs. Home: 3077 Holmes Run Rd., Falls Church, Va. 22042. Office: Smithsonian Instn., Washington 20560.*

FOSCUE, Edwin Jay, Am. geographer; b. Camden, Ark., Aug. 26, 1899; s. Augustus William and Ida May (Bell) F.; student U. Tex., 1918-20; B.A., So. Meth. U., 1922; M.S., U. Chgo., 1925; Ph.D., Clark U., 1931; m. Fannie Louise Knight, June 2, 1925. Faculty, So. Meth. U., Dallas, 1923-—, prof. geography, 1938-65, prof. emeritus, 1965-—. Asst. dir. U. S. Bd. on Geog. Names, 1943-44; dir. Southwestern Internat. Devel. Inst., 1954-55. Recipient Faculty Achievement award Alumni Assn. So. Meth. U., 1960. Mem. Southwestern Social Sci. Assn. (past pres.), Assn. Am. Geographers, Phi Beta Kappa, Sigma Xi. Author: (with C. Langdon White) Regional Geography of Anglo-America, 1943; also articles. Research on econ. geography of regions of Tex., geog. significance of maj. bridges of world. Home: 3225 Hanover St., Dallas 75225.*

FOSHAG, William Frederick, Am. geologist; b. Sag Harbor, N.Y., Mar. 17, 1894; s. William Frederick and Joanna Eva (Riegler) F.; A.B., U. Cal., 1919, Ph.D., 1923; m. Merle Crisler, Sept. 5, 1923; 1 son, William Frederick. Chemist Riverside Portland Cement Co. (Cal.), 1917-18; asst. curator U. S. Nat. Mus., 1919-29, curator, 1929-48, head curator, dept. geology, from 1948. Fellow Mineral Soc. Am. (pres. 1940, Roebling medalist 1953), Geol. Soc. Am. (v.p. 1941), Geophys. Union (v.p. 1947-48, pres. volcanology sect. 1953-55) Soc. Econ. Geologists, Soc. for Research on Meteorites (v.p. 1933-34), Washington Acad. Sci. Geol. Soc. Mexico (hon.), Sigma Xi; corr. mem. Sociédad Cientifica Antonio Alzate de Mexico. Author: (with George P. Merrill) Minerals from Earth and Sky, 1929; Birth and Development of Paricutin Volcano, 1956. Pre-Columbian Art (with others), 1957. Made systematic study of minerals of Mexico. Died May 21, 1956.

FOSSE, Richard-Jules, French chemist; b. Castro, France, July 16, 1870; prof. organic chemistry, plant physics Lille (France) U., Paris Museum; mem. French Acad. Scis., 1931; showed presence of urea in plant tissues. Died Paris, France, Dec. 18, 1949.

FOSSOMBRONI, Vittorio, Italian engineer, mathematician; b. Arezzo, Tuscany, Nov. 15, 1754. Minister Fgn. Affairs, Tuscany, 1796; mem., Tuscany Commission of Finances; pres., Legislative Council. Corr. mem., French Acad. Scis., 1824. Author: Memorie idraulico-storiche sulla Val de Chiana, 1789; treatises on math., hydraulics. Work in hydraulics applied to agriculture; helped develop drainage systems in N. Italy; studied principle of virtual velocity, 1796; draining of Pontine Marshes, 1805. Died Florence, Italy, Apr. 13, 1844.

FOSTER, Adriance S., Am. botanist; b. Poughkeepsie, N.Y., Aug. 6, 1901; s. Raymond and Alice (Adriance) F.; B.S., Cornell, 1923; S.M., Harvard, 1925, Sc.D., 1926; m. Helen N. Vincent, July 29, 1930; 1 son, Richard V. Asst. prof. botany, U. Oklahoma, 1928-34; prof. botany U. Cal. at Berkeley, 1934-—; also chmn. dept. botany. Mem. A.A.A.S., Bot. Soc. Am., Am. Acad. Arts and Sci., Sigma Xi. Author: Practical Plant Anatomy, 2d edit.; (with E. M. Gifford, Jr.) Comparative Morphology of Vascular Plants, 1959. Contbr. sci. jours. Studies on development and anatomy of leaves; investigations on shoot apex of Ginkgo and several genera of cycads. Home: 57 Poplar St., Berkeley, Cal. Office: Dept. of Botany, U. California, Berkeley, Calif.*

FOSTER, Alfred Leon, Am. mathematician, educator; b. N.Y.C., July 13, 1904; s. Henry and Josephine (Sonnen) F.; B.S., Cal. Inst. Tech., 1926, M.S., 1927; Ph.D. (J.S.K. fellow), Princeton, 1931; m. Else Wagner, June 9, 1930; children—Wilfred, Erwin, Elsbeth (Mrs. Jose Ramos), Toni-Ann. Internat. research fellow, Gottingen, Germany, 1931-33; faculty math. U. Cal. at Berkeley, 1933-—, prof., 1950-—. Mem. Am. Math. Soc., Assn. for Symbolic Logic. Author articles on abstract algebra, ring theory, universal algebras, algebraic structure theory, founds. of math. Home: 954 Keeler Av., Berkeley, Cal. 94708.*

FOSTER, Denis Henry, Australian chemist; b. Hobart, Australia, Jan 27, 1924; B.Sc., U. Tasmania (Australia), 1944, M.Sc., 1947; Research officer div. forest products Commonwealth Sci. and Indsl. Research Orgn., Melbourne, Australia, 1947-52; chief chemist Sugar Research Inst., Mackay, Queensland, Australia, 1952-—. Fellow Royal Australian Chem. Inst.; mem. Australian Pulp and Paper Industry Tech. Assn., Am. Chem. Soc. Research, publs. on wood chemistry, cellulose and hemicellulose, tech. and sci. related to raw sugar manufacture. Home: 44 Paulette, Mackay. Office: Nebo Rd., Mackay, Queensland, Australia.*

FOSTER, Donald DeLacy, physicist, govt. ofcl.; b. Torbrook, N.S., Can., Mar. 19, 1900; s. DeLacy Evans and Mary Sophia (Vroom) F.; B.S., Acadia U., 1920; Ph.D. (Loomis fellow), Yale, 1924; m. Gerda Arminel Holman, July 26, 1924; children—James P., Mary (Mrs. Peter Evdokimoff), Ann (Mrs. Ralph Eames Partridge). Came to U. S., 1924, naturalized, 1942. Research physicist Eastman Kodak Co., Rochester, N.Y., 1924-25, Bell Telephone Labs., 1925-36, Westinghouse Electric & Mfg. Co., 1936-38; asst. prof. math. Stevens Inst. Tech., 1939-45, asso. prof., 1945; chief spl. studies lab. Cambridge Research Center, USAF, Bedford, Mass., 1946-51, physicist, 1958-—; research asso. Yale, 1951-54; sr. scientist RCA, Waltham, Mass., 1955-57. Spl. adviser NDRC, 1942; research asso. Harvard, 1943-45. Fellow Am. Phys. Soc. Research, publs. on magnetism, spectroscopy, acoustics, antennas, applied math. Home: R.F.D., Acton, Mass. 01720. Office: Cambridge Research Labs., U. S. Air Force, Bedford, Mass. 01731.*

FOSTER, George Carey, Brit. physicist; b. Lancashire, Eng., 1835; s. George Foster; ed. Univ. Coll., London, U. Ghent, Paris, Heidelberg; B.A., LL.D., D.Sc.; m. Mary Ann Muir, 1868; 4 sons, 4 daus. Became prof. physics Anderson's Coll., Glasgow, 1862, Univ. Coll., London, 1865-1904. Fellow Royal Soc.; Chem. Soc.; pres., Phys. Soc., 1876-78; Soc. Telegraphic Engr. (now Instn. Elec. Engrs.), 1881; gen. treas., Brit. Assn. Advancement Sci., 1898-1904. Gave a formula for acetic acid. Died Feb. 9, 1919.

FOSTER, George McClelland, Jr., Am. anthropologist; b. Sioux Falls, S.D., Oct. 9, 1913; s. George M. and Mary (Slutz) F.; B.S., Northwestern, 1935; Ph.D., U. Cal. at Berkeley, 1941; m. Mary Frazer LeCron, Jan. 6, 1938; children—Jeremy, Melissa. Instr., Syracuse U., 1941-42; lectr. U. Cal. at Los Angeles, 1942-43; anthropologist Inst. Social Anthropology, Smithsonian Instn., Washington, 1943-52, dir., 1946-52; vis. prof. U. Cal. at Berkeley, 1953-55, prof., 1955-—, acting dir. Mus. Anthropology, 1955-57, chmn. dept. anthropology, 1958-61, lectr. pub. health, 1955-64. Cons., AID and predecessors, 1951-52, 55, 57, 61, 62, 65. Guggenheim fellow, 1949. Fellow Am. Anthropol. Assn. (exec. Bd. 1956-58). Author: Empire's Children, 1948; Culture and Conquest, 1960; Traditional Cultures and the Impact of Technological Change, 1962; Tzintzuntzan: Mexican Peasants in a Changing World, 1967; also numerous articles. Research in Cal. Indians, Mexico, Spanish-Am., Spain, processes of social and cultural change among contemporary rural peoples in newly developing countries, social and cultural aspects pub. health, community devel. Home: 790 San Luis Rd., Berkeley, Cal. 94707.*

FOSTER, Henry, Brit. navigator; b. Lancashire, Eng., Aug. 1796; s. Henry Foster; studied under Mr. Saul, Green Row, Eng. Surveyed mouth of Columbia and N. shore of La Plata, 1819; assisted on Basil Hall's voyage to S.Am., 1920; astronomer Parry's polar expdns., 1924-25, 27; given command of govt. sloop, Chanticleer, to determine specific ellipticity of earth, 1829. Recipient Copley medal, 1926. Fellow Royal Soc., 1824. Research and publs. on pendulum in S. Seas, 1828-29; measured difference of longitude across isthmus of Panama by rockets, 1830-31. Died Feb. 5, 1831.

FOSTER, John Stuart, Jr., Am. physicist; b. New Haven, Sept. 18, 1922; s. John Stuart and Flora (Curtis) F.; B.Sc. with honours, McGill U., 1948; Ph.D., U. Cal. at Berkeley, 1952; m. Barbara Anne Wickes, May 23, 1946; children—Susan, Bruce, Scott, John. Civilian air force cons., MTO, 1942-45; with NRC, Chalk River, Ont., Can., summers 1946, 47; mem. staff Lawrence Radiation Lab., Berkeley, 1948-52, asso. dir., 1961-—; mem. staff Lawrence Radiation Lab., Livermore, 1952-—, asso. dir., 1958-61, dir., 1961-—. Mem. Air Force Sci. Adv. Bd., 1956-—, Army Sci. Adv. Panel, 1958-—, Panel Cons. President Sci. Adv. Com., 1959-—; dir. def. research and engring. U. S. Dept. Def., 1965-—. Recipient Ernest O. Lawrence award AEC, 1960. Mem. Sigma Xi. Work in application of nuclear explosives to mil. and peaceful uses; important contbns. in devel. tactical weapons, also in several engring. projects employing nuclear explosives. Home: 6382 Lakeview Dr., Falls Church, Va. 22041. Office: Dept. Def., Washington 20301.*

FOSTER, John Wells, Am. geologist; b. Petersham, Mass., Mar. 4, 1815; s. Festus Foster; grad. Wesleyan U., Middletown, Conn., 1833; studied law, Zanesville, O.; m. Lydia Conerse, Oct. 24, 1838. Practiced law for time, Zanesville; asst. Ohio Geol. Survey, 1837; asso. U. S. geologist, 1847-52; a founder Native Am. Party in Mass., 1852; an organizer Republican Party in Mass., 1855; land commr. I.C. R.R., 1858-63; lectr. geology U. Chgo., 1863-65; pres. A.A.A.S. Author: The Mississippi Valley: Its Physical Geography, Including Sketches of the Topography, Botany, Climate, Geology and Mineral Resources; And of the Progress of Development in Population and Material Wealth, 1869; Mineral Wealth and Railroad Development, 1872; Prehistoric Races of the United States of America, 1873. Research on mineral resources of Lake Superior region, also on paleontology and ethnology of Mississippi Valley; studied central coal bed of Ohio and Carboniferous limestone deposits nr. Columbus, O. Died Chgo., June 28, 1873.

FOSTER, Joseph Franklin, Am. biochemist; b. Marion, Ind., May 17, 1918; s. DeWitt L. and Grace (Cameron) F.; B.S., Ia. State U., 1940, Ph.D., 1943; M. Ruth E. Hobson, June 8, 1940; children—Ann E., Gregory H., Michael C. Postdoctoral fellow Med. Sch. Harvard, 1943-45; research chemist Am. Maize Products Co., Roby, Ind., 1945-46, cons., 1948-59; faculty Ia. State U., 1946-54; faculty Purdue U., Lafayette, Ind., 1954-—, prof., 1957-—, head of the department of chemistry, 1967-—. Cons. Central Research Labs. Gen. Foods Corp., 1960-63, NIH, 1962-—, Army Q.M.C., 1963-64. Mem. Am. Chem. Soc., Am. Soc. Biol. Chemistry, Biophys. Soc., A.A.-A.S., Am. Assn. U. Profs., Sigma Xi. Author: (with S. W. Fox) Introduction to Protein Chemistry, 1957. Research, numerous publs. primarily in field of phys.-chem. technique for studying reversible conformation changes in proteins. Home: 328 Laurel Dr., West Lafayette, Ind. 47906. Office: Dept. of Chemistry, Purdue U., Lafayette, Ind. 47907.*

FOSTER, Sir Michael, Brit. physiologist; b. Marlborough, Eng., 1836; s. Michael and Mercy (Cooper) F.; B.A., London, 1854; M.B., 1856; M.D., 1895; m. Georgiana Edmonds, 1863 (dec. 1869); children—George Michael, Mercy (Mrs. J. Tetley Rowe); m. 2d, Margaret Rust, 1872. Fellow Royal Soc., 1872. Author: Text-Book of Physiology, 1873; Lectures on the History of Physiology; Elements of Embryology, 1874. Co-editor: Scientific Memoirs (Huxley), 1898-1902. Founder, editor Jour. Physiology. Devised (with Huxley) method of practical lab. work; hybridized several plants. Died 1907.

FOSTER, Michael George, Brit. physician; b. Huntingdon, Eng., Dec. 13, 1864; s. Michael and Georgina (Edmonds) F.; ed. Trinity Coll., Cambridge (Eng.) U.; Univ. Coll. Hosp.; M.D.; m. Charlotte Shipley; 3 sons, 1 dau.; m. 2d, Margaret Manning Russell. Cons. physician BEF. Fellow Royal Coll. Physicians. Publs. on cerebro-spinal fever, climate, baths and medicinal waters of Britain and Europe, Hale White's system of therapeutics. Died June 16, 1934.

FOSTER, Richard Bergeron, Am. systems analyst; b. Springdale, Wash., Nov. 21, 1916; s. Raymond Eden and Molly (Bergeron) F.; A.B., U. Cal., 1938; m. Penny Darling, Jan. 28, 1955; children—Molly Eden, Amy Elisabeth. Dir. Strategic Studies Center Stanford Research Inst., Washington, 1954-67, spl. asst. for nat. security research office of Exec. Vice Pres., 1967-—; course dir. Nat. Strategy Seminar, Asilomar, Cal., 1960; lectr. univs., govt. agys. Recipient Patriotic Civilian Service award for pioneering interdisciplinary research methods Dept. Army, 1959. Research fellow Washington Center Fgn. Policy Research. Fellow A.A.A.S.; mem. Operations Research Soc. Am. Phi Beta Kappa. Research, numerous publs. on dynamic program treatment for cost effectiveness, dynamic interaction threat models, devel. of offense, def. cost exchange ratio analytic technique to determine relative rate of change of offense def. tech.; methods of explicit treatment of strategic technol. and operational uncertainties in assessing weapon systems effectiveness, methods of forming and directing interdisciplinary teams to solve complex nat.

security problems. Home: 1500 S. 23d St., Arlington 22202. Office: 1611 N. Kent St., Arlington, Va. 22209.*

FOSTER, Samuel, English mathematician; b. Northamptonshire, Eng.; M.A., Emmanuel Coll., Cambridge, Eng., 1623; Gresham prof. astronomy, 1636, 41-52. Author: The Use of the Quadrant, 1624; The Art of Dialling, 1638; other works pub. posthumously. Mem. group which met in London to discuss new philosophy and from which Royal Soc. evolved; observed eclipses of sun and moon; invented and improved planetary instruments. Died 1652.

FOSTER, William, Am. chemist; b. Hartford, Ky., May 15, 1869; s. William and Sarah Jane (Carson) F.; A.B., Hartford Coll., 1892; B.S., Vanderbilt U., 1893; A.M., Princeton, 1896, Ph.D., 1899; m. Helen Dunham Stewart, Sept. 3, 1902; children—Katharine Sutliff, Helen Stewart, Wilhelmina. Prof. chemistry Central U. of Ky., 1899-1900; instr. chemistry Princeton, 1900-05, asst. prof., 1905-10, prof., from 1910. Fellow A.A.A.S. Author: A Laboratory Manual in General Chemistry, 1905; Introduction to General Chemistry, 1922, 31; The Elements of Chemistry, 1925, 32; Experiments in General Chemistry (with H. W. Heath), 1925; The Romance of Chemistry, 1927, German edit. of same as Welt und Wunder der Chemie, 1931; Inorganic Chemistry for Colleges, 1929; A Laboratory Course of General Chemistry, 1930. Died May 24, 1937.

FOTER, Milton John, Am. microbiologist; b. N.Y.C., Mar. 4, 1909; s. Emil and Barbara (Breina) F.; B.S., Cornell U., 1931, M.S., 1932, Ph.D., 1935; m. Harriet B. Montgomery, June 22, 1935 (div. May 1958); children—Sandra E. (Mrs. Richard P. Deeds), Harriet S. Instr., Cornell U., Ithaca, N.Y., 1931-35; asst. prof. U. Conn., Storrs, 1935-42; chief bacteriology Pet Milk Co., Greenville, Ill., 1942-44, Merrell Co., Cin., 1944-51; chief aerobiology USPHS, Cin., 1951-55, dep. chief milk and food research, 1955-65, asst. program officer environmental health, Washington, 1965-67, dep. chief pesticides program NCDC, Atlanta, 1967——; acting asso. prof. U. Cin., 1959-60. USPHS tech. liaison officer U. S. Senate Subcom. Chem. Hazards in Environment, 1963-65. Diplomate Am. Bd. Microbiology. Fellow Am. Acad. Microbiologists; mem. Am. Soc. Microbiologists (chmn. placement com. 1955——), Am. Inst. Biol. Scis. (vis. lectr. biology 1959-64), Ohio Soc. Microbiologists (past pres.), Nat. Assn. Sanitarians (presdl. citation 1962), Inst. Food Technologists, Royal Soc. Health, Internat. Assn. Milk and Food Tech., Am. Pub. Health Assn. Contbg. author: Microbiological Quality of Foods, 1963; Institutional Sanitation, 1965; also numerous articles. Research on bacterial physiology, bioluminescence in bacteria, dairy and food bacteriology, oral immunization, antibiotics, microbiol. assay methods, surface active agts., disinfection, antisepsis and preservation, oral microbiology, algicides, gaseous sterilization bedding materials, aerobiology, space biology, arctic health, pesticides. Home: 3200 Lenox Rd. N.E., Atlanta 30324. Office: 1600 Clifton Rd. N.E., Atlanta 30333.*

FOTHERGILL, Anthony, Brit. physician; b. Yorkshire, Eng., circa 1732; M.D., U. Edinburgh, 1763; later studied at Leiden, Netherlands, Paris. Practiced medicine, Northampton, Eng.; became physician Northampton Infirmary, 1774; began practice medicine, London, 1780, Bath, Eng., 1784; ret. to Phila., 1803; recalled to Eng. because of impending war, 1812. Recipient Gold medal Royal Humane Soc., 1794. Fellow Royal Soc., 1788. Author numerous publs. including: Hints for Restoring Animation, and for Preserving Mankind against Noxious Vapours, 1783; On the Nature of Disease produced by Bite of a Mad Dog, 1799. Studied restoration of drowning victims. Died London, May 11, 1813.

FOTHERGILL, John, English physician; b. Carr End, Wensleydale, Eng., Mar. 8, 1712; s. John Fothergill; apprenticed to Benjamin Bartlett, apothecary; M.D., Edinburgh U., 1736. Began practice medicine, London, 1740; maintained bot. garden, Upton, Eng.; drew up (with Benjamin Franklin) plan of reconciliation with Am. colonies, 1774. Licentiate Royal Coll. Physicians. Fellow Royal Soc., 1763 Royal Soc. Medicine, Paris. Author: Works (ed. by J. C. Lettsom), 3 vols., 1783-84. Contbr. Account of the Sore Throat (1st recognition of symptoms of diphtheria in Eng.), 1748, also other papers to jours. Described trigeminal neuralgia (Fothergill's disease), 1776; credited with 1st accurate description of migraine (under name sick headache), 1778. Died London Dec. 26, 1780.

FOTIADI, Epaminond Epaminondovich, Russian geologist, geophysicist; b. Jan. 23, 1907; grad. Leningrad U., 1933. With Emba Oil and Gas Trust, 1927-39, Inst. Applied Geophysics, 1946-51, All-Union Research Inst. Geophys. Methods of Exploration, 1951-58; asso. Inst. Geology and Geophysics, Siberian Dept., USSR Acad. Sci., 1958——, now dep. dir. Decorated Order of Lenin. Mem. USSR Acad. Sci. (corr.). Co-author: Course of Gravitational Prospecting, 1941. Exec. editor Problems on Prospecting Geophysics, 1960. Research and publs. on geophys. methods of petroleum prospecting, methods of geol. interpretation of gravitational and magnetic regional anomalies, gravimetric and topogeodetic work. Address: Inst.

Geology and Geophysics, Siberian Dept., USSR Acad. Sci., Novosibirsk, RSFSR, USSR.

FÖTTINGER, Hermann, engr.; b. Nuremberg, Germany, Feb. 9, 1877; s. Carl and Marie (Barthelmess) F.; doctorate Tech. U. Munich, 1904. Mech. engr. Stettin Vulcan-Werft (shipyards); became prof. Tech. U. Danzig (Poland), 1909; prof. hydrodynamics and ship turbines Tech. U. Berlin, from 1924. Recipient Geothe medal for art and sci.; Silver and Gold medals Schiffbautechnische Gesellschaft; named hon. citizen Tech. U. Danzig. Contbr. articles to profl. jours. Inventor torsion indicator for measuring efficiency of ship engines, Föttinger transformer (hydraulic transmission of turbine torque onto propellor). Died Berlin, Apr. 28, 1945.

FOUCAULT, Jean Bernard Léon, French physicist; b. Paris, Sept. 18, 1819; studied medicine and exptl. physics; scientific editor of Jour. des Debâts, from 1845; became physicist, Paris Obs., 1855; mem. French Acad. Scis., Bur. for Linear Measurements. Fellow Royal Soc., 1864. Inventor accurate method of testing lenses for chromatic and spherical aberration; worked (in part with Flezeau) to determine velocity of light by lab. expt.; determined absolute velocity of light more accurately than ever before; proved that light travels faster in air than in water, 1850; inventor pendulum which clearly showed earth's rotation upon its axis, 1851, gyroscope, 1852; constructed Foucault prism, 1857, an improved mirror for reflecting telescopes, 1858; inventor mercury interrupter used with induction coils, an engine gov., an automatic regulator for electric arc lamp; studied binocular vision; eponym of Foucault (or eddy) currents, discovered by generating electricity in a metal plate rotated between poles of a magnet. Died Paris, Feb. 11, 1868.

FOUCHY, Jean Paul Grandjean, see de Fouchy, Jean Paul Grandjean.

FOUCQUET, Father Jean-François, astronomer, historian; b. Autun, France, Mar. 12, 1663; mem. Soc. of Jesus; became prof. math. La Flèche Sch., 1680; sent as missionary to China, 1699; returned to Europe, 1720. Author: Tabula chronologica historiae sinicae, 1729. Developed theory of Chinese lang. which considered the characters hieroglyphics; an inventor of figurism. Died 1736.

FOUGEROUX DE BONDAROY, Auguste-Denis, French naturalist; b. Paris, 1732; mem. French Acad. Scis. Author: Art de tirer des carrières la pierre d'ardoise, 1758. de la fendre et de la tailler, 1762; Art du tonnelier, 1763; Mémoire sur la formation des os, 1763; Recherches sur les ruines d'Herculanum, 1769; also articles on stone-cutting in Encylopèdie (d'Alembert). Research on formation of bones; mem. expdn. to study Herculanum in Italy. Died 1789.

FOUILLEE, Alfred Jules Emile, French philosopher, sociologist; b. La Poueze, Oct. 18, 1838; self-educated; m. Madame Guyot. Tchr. in provincial schs., 1864-68; apptd. to Bordeaux Lycee, 1869; became prof. philosophy Ecole Normale Superieure, Paris, 1872, retired 1879; pres. Institut International de Sociologie. Author: La Science sociale contemporaine, 1880; Critique des systemes de morale comtemporaine, 1893; l'Evolutionnisme des idees-forces, 1890; La Psychologie des idees-forces, 1893; Les elements sociologiques de la morale, 1906. Pursued to reconcile idealism and naturalism, using motor theory of consciousness (idees-forces), in which ideas are inseparable from action; emphasized interdependence of individual and society; believed society to be contractual organism; his theories connected psychol. and sociol. phenomena. Died Lyons, July 16, 1912.

FOUNTAIN, Claude Russell, Am. physicist; b. Ashland, Ore., Nov. 2, 1879; s. James Davis and Grace (Russell) F.; A.B., U. Ore., 1901; univ. scholar in math. Columbia, 1901-02, Ph.D., 1908; m. Lucy E. Landru, June 16, 1914; children—Betty Grace Edwards, Margaret Louise Pinkston. Asst. in physics Columbia, 1902-05, also asst. Columbia Summer Sch. of Practical Astronomy and Geodesy; asso. prof. physics U. Ida., 1905-06; instr. physics Williams Coll., 1906-09; asst. prof. physics Kenyon Coll., Gambier, O., 1909-13; adj. prof. physics U. Ga., 1913-18; prof. physics and astronomy Mercer U., 1918-26; prof. math. and physics George Peabody Coll. for Tchrs., 1926-27, prof. teaching of physics, 1927-40, also prof. physics, summers 1915-19, 23-26; prof. aeros. (teaching all courses at Ground Sch. Civil Aeros. Authority), Southwestern Coll., Memphis, summer, 1940; prof. applied sci. and math. Hume-Fogg Tech. High Sch., 1940-41; vis. prof. physics Amherst Coll., 1941-42; vis. prof. physics Mass. Inst. Tech., 1942-43; sr. physicist U. S. Signal Corps, ground signal service, Camp Evans, Belmar, N.J., 1943-46; research physicist Naval Research Lab. from 1946. Fellow A.A.A.S.; mem. Am. Phys. Soc., Am. Meteorol. Soc., Tenn. Acad. Scis. (pres. 1936). Author: Laboratory Manual of Practical Physics, 1925. Inventor of simplified types of lab. apparatus and devices in automechanics, heat and radio; developed lab. course in which student is led to discover fundamental laws instead of trying to verify them. Died Nov. 28, 1947.

FOUQUÉ, Ferdinand André, French geologist; b. Mortain, France, June 21, 1828; ed. Ecole St.-Cyr, also Ecole normale supérieure; made voyage to Italy, 1860; saw Mt. Etna erupt, 1868; became prof. inorganic substances Coll. France, 1877. Mem. French Acad. Scis., 1881 (pres. 1901). Author: Santorin et ses éruptions, 1879; (with Michel Levy) Introduction à l'étude des roches éruptives francaises, 1879. Studied formation of volcanic craters, volatile products of volcanic eruptions; artificially reproduced rocks and minerals; introduced into France microscope for petrographic study. Died Paris, France, Mar. 7, 1904.

FOUQUET, Henri, French physician; b. Montpellier, France, July 31, 1727; studied under Théophile de Bordeu; grad., Montpellier, 1759; apptd. prof. clin. medicine Faculty Montpellier, circa 1792; med. examiner French Army, 1793-94; mem. French Acad. Scis., 1804. Author: Traitement de la petite vérole, 1772; De Fibrae natural iscribus et morbis in corpore animali, 1759; Essai sur le pouls, considéré par raport aux affections, des principaux organes, 1768; Discours sur la clinique, 1803. Studied heartbeat and pulse, also small pox and its effect on children. Died Oct. 10, 1806.

FOUQUET, Father Jean-Françoise, see Foucquet, Father Jean-Françoise.

FOURCROY, Antoine François, Comte de, French chemist; b. Paris, France, June 15, 1755; grad. in medicine, 1780. Prof. chemistry Jardin du Roi, 1784; prof. Mus. Natural History; minister edn. under Napoleon. Created Count of Empire. Mem., French Acad. Scis., 1784, French Soc. Agr. Author: Système des connaissances chimiques, 1801; Leçons élémentaires d'histoire naturelle et de chimie, 1782; Principes de chimie, 1787; Philosophie chimique, 1792; Tableaux synoptiques de chimie, 1800. A founder Annales de Chimie. Developed (with Lavoisier, Berthollet and Morveau) systematic nomenclature for chemistry; research in physiol. chemistry and pub. health; analysis of mineral waters; believed electricity and galvanism to be different; as member of National Convention used his influence to save other scientists from death. Died Paris, Dec. 16, 1809.

FOURIER, François Marie Charles, French social theorist; b. Besançon, France, Apr. 7, 1772; ed. Besançon. Served in army, 1794-96; worked in mcht.'s office, Lyon, France later as broker; went to Paris, 1826, became clk. in Am. firm; gradually attracted polit. adherents who set up newspaper, 1832. Author: Théorie des quatre mouvements et des destinées générales, 1808; Traité de l'asociation domestique et agricole, 1822; Le Nouveau Monde industriel et sociétaire, 1829. Utopian socialist, who advocated establishment of agriculturally oriented communities (phalanxes) to achieve social harmony; influenced various Ams. including Albert Brisbane, Horace Greeley, leaders of Brook Farm and Red Bank (N.J.) phalanx. Died Paris, Oct. 10, 1837.

FOURIER, Baron Jean Baptiste Joseph, French mathematician; b. Auxerre, France, Mar. 21, 1768; ed. mil. sch., Auxerre; prof. analysis École Polytechnique, shortly after 1794-98; succeeded Laplace as pres. council, 1827; governor of Lower Egypt under Napoleon, 1798-1802; prefect of Isère, 1802-15; sec. Institut du Claire, 3 years; mem. French Acad. Scis. (prize 1812, named perpetual sec. 1822). Author: Théorie analytique de la chaleur, 1822; Analyse des équations determinées, 1831; also papers. Evolved math. series known by his name, which is important in harmonic analysis; thus devised method of representing discontinuous functions of variables by trigonometric series, thus providing source of all modern methods in math. physics involving integration of partial differential equations in problems where boundary values are fixed; originated Fourier's theorem on vibratory motion. Died Paris, May 16, 1830.

FOURMAN, Paul Lucien, Brit. physician; b. London, Dec. 14, 1918; s. Maximillian and Lucie (Cashman) F.; student Guy's Hosp., 1936-44; M.B., B.S., with Honors, U. London, 1941, M.D., 1944, D.Sc., 1960; m. Dec. 4, 1946; children—Michael Paul, Julia Margaret, David Mark, Clive Morel. Med. tutor Guy's Hosp., 1942-44; asst. Nuffield Dept. Medicine, Oxford, 1946-48; Rockefeller fellow Mass. Gen. Hosp., 1948-49; med. tutor Oxford U., 1949-52; staff dept. exptl. medicine Med. Research Council, Cambridge, 1952-54; lectr., reader medicine Welsh Nat. Sch. Medicine, 1954-63; prof. clin. investigation Leeds (Eng.) U. Med. Sch., 1963——. Fellow Royal Coll. Physicians; mem. Med. Research Soc. London, Soc. for Endocrinology, Assn. Physicians Gt. Britain and Ireland, Renal Assn. Author: Calcium Metabolism and the Bone, 1960; also articles. Introduced test for intestinal absorption using xylose, 1948; application of analysis of variance to metabolic studies in man, 1950; prodn. exptl. deficits of potassium and magnesium in man, 1950-56; clin. studies in metabolism and renal disease. Home: 11 Allerton Park, Leeds 7, Yorkshire, Gt. Britain. Office: Clin. Investigation Unit, Leeds Gen. Infirmary, Leeds 1., Gt. Britain.*

FOURNEAU, Ernest François Auguste, French chemist, pharmacologist; b. Biarritz, France, Oct. 4, 1872; s. Jean and Pauline (Bordes) F.; B.A., Lyceum of Boyonne, 1818; m. Veuf De Claudie Seond, May 6,

1906; 3 sons. Intern in pharmacy Hosp. Beaujin, Paris, France, 1892-96; dir. research lab. Establissements Poulenc Frères (pharm. products), Paris, 1900-12; chief of service, lab. therapeutical chemistry, Inst. Pasteur, Paris, 1912-44; mem. faculties Pharmacy, Scis., Paris. Decorated Legion of Honor, also various other decorations; laureate French Inst., 1919, 26, 31, 41, London, Eng., 1930, Madrid, Spain, 1934; recipient Addington medal, Leeds, Eng., 1939. Mem. Chem. Soc. France (sec.-gen. 1911-23), Acad. Medicine, Paris, Royal Spanish Acad. Medicine, Acad. Pharmacy, Madrid, Soc. Austrian Chemists; Soc. Pharmacy (pres. 1931), German Acad. Leopold Caroline, Swiss Chem. Soc., London Royal Soc. Pharmacy (hon.), Pharm. Soc. Great Britain; Soc. Indsl. Chemistry, London. Author: Préparation des Médicaments organiques, 1921; chpt. on aminoalcohols in Traité de chimie organique de Grignard; also pub. studies on local anaesthetics, arsenic, other subjects. Introduced drug moranyl to treat trypanosomiasis, circa 1923; discovered stovarsol, trypanocids, also formula for combatting puerperal fever and meningitis. Died 1949.

FOURNEL, Marie-Jérôme-Henri, French mining engr.; b. Paris, 1799; studied l'École polytechnique, Paris; named gen. insp. mines, 1859. Author: Études des gites houillers et métallifères du Bocage vendéen, 1836; Richesse minérale de l'Algérie, 1850. Studied and mapped coal and metal deposits of Vendée region in France, mineral deposits of Algeria. Died Blois, France, 1876.

FOURNET, Gerard Lucien, French physicist; b. Paris, France, May 2, 1923; s. Paul Yves and Germaine (Tournaud) F.; Engr., Ecole Supérieure de Physique et Chimie, Paris, 1944; Doctorat ès Sciences, Paris U., 1950; m. Noele Tournier, Mar. 13, 1947; children—Jérôme Denis, Christian Francois. With Prof. A. Guinier's Lab., 1946-52; engr., 1952-57; chief engr. Alcatel, 1958-62; asst. Ecole Supérieure de Physique et Chimie, Paris, 1950-59, prof. electronics, 1959—; prof. Ecole Supérieure des Télécommunications, Paris, 1958-62; lectr. Faculté des Scis., Paris, 1961-63, prof. electromagnetism, 1963—. Mem. Société Française de Physique (Prix Louis Ancel 1960); Société des Electriciens, Soc. Française de Mineralogie et de Cristallographie, Am. Phys. Soc. Author: (with A. Guinier) Small Angle Scattering of X Rays, 1954; Physique Electronique des Solides, 1962; also numerous articles; contbg. author: Ency. Physica, 1957. Research on diffuse scattering of X-rays, theory of order-disorder transformations, superconductivity. Home: 41 Av. Le Notre 92 Vaucresson, France. Office: L.C.I.E. 33 Av. du Général Leclerc 92, Fontenay-Aux-Roses, France.*

FOURNET, (Joseph-Jean-Baptiste-) Xavier, French mineralogist, geologist; b. Strasbourg, France, May 15, 1801; student l'École des Mines, Paris, France, 1822-25; Docteur ès scis., 1832; became prof. geology Faculty Scis. Lyons (France), 1834; in charge Pontgibaud Mines, France. Mem. French Acad. Scis., 1853, Royal Soc. London. Author: Sur les sulfres métalliques et leur traitment metallurgique, 1833; De l'extension des terrains houillers, 1855; Du mineur, son rôle et son influence sur les progrès de la civilisation, 1862. Studied metalliferous deposits, eruptive rocks; developed new theory on distbn. coal fields in France; showed that coal formation is continuous, as in Belgium and Eng.; established sulfurability of metals (Fournet's law), 1833. Died Jan. 8, 1869.

FOURNEYRON, Benoit, engr., inventor; b. Bornin St.-France, 1801; ed. l'École des Mineurs, St.-Etienne, France; entered service of Pourtales family (owners of Franche-Comte Iron Work), 1821; Recipient 1st prize Soc. Encouragement for Nat. Industry, 1834, prize for turbine French Acad. Scis., 1836. Built and improved many turbines; inventor hydraulic turbine; used water power in tin rolling mill. Died St.-Etienne, July 8, 1867.

FOURNIER, André, French psychiatrist; b. Caudry, France, June 2, 1926; s. Fernand and Marie (Duvillier) F.; M.D., Sch. Mil. Health Service, Lyons, France, 1951; m. Paule Meresse, July 11, 1950; children—Christian, Isabelle. Asst., army hosps.; chief neuropsychiatry service, mil. hosp. of instrn., Bordeaux, France. Mem. French Neuro-Psychiatry Congress. Research on problems of groups such as parachutists, and problems, of adaptation, including causes of suicide; research on treatment of depression. Home: 60 Chemin Lafitte (33) Talence, Gironde, France. Office: Military Hosp. of Instruction Robert-Picque, Bordeaux, France.*

FOURNIER, (François) Ernest, French nautical engr.; b. Toulouse, France, May 23, 1842; served to vice adm. French Navy; mem. Bur. Longitudes, French Acad. Scis., 1912, Marine Acad.; inventor instruments to regulate nautical compass. Died Neuilly, France, Nov. 6, 1934.

FOURNIER, Georges, geologist; b. Caen, France, 1595; mem. Soc. of Jesus; spiritual coadjutor and prof. math., Tournay, Belgium. Author: Commentaires géographiques, 1642; L'hydrographie, 1643; Euclidis . . . , 1644; Géographica . . . , 1648; Introductio ad cosmographiam; Coriospiti . . . Traité des fortifications, 1649; Asiae nova descripto (new description of Asia), 1656. Died La Flèche, France, Apr. 13, 1652.

FOURNIER, Jean Alfred, French dermatologist; b. Paris, 1832; became prof. Faculty Medicine, Paris, also physician of hosps., 1863; chief of services Hosp. Lourcine, from 1868; prof. Clinic for Cutaneous and Syphilitic Diseases. Author: La syphilis du cerveau, 1879; L'ataxie locomotrice d'origine syphilitique, 1882; La période pré-ataxique du tabes, 1885; La syphilis héréditaire tardiue, 1886. Described severe gangrene of genitals, or Fournier's disease; proved existence of 2 chancres, the chancroid or soft type (nonsyphilitic venereal sore), and the indurated chancre (constitutional venereal sore followed by true syphilis); demonstrated by statistics etiologic relationship of syphilis to tabes dorsalis and dementia paralytica, 1894. Died 1914.

FOURNIER D'ALBE, Edmund Edward, Brit. physicist; b. London, Eng., 1868; s. E. H. Fournier d'Albe; ed. Royal Coll. Sci. London; D.Sc., U. London, U. Birmingham (Eng.); m. Asst. phys. labs. Royal Coll. Sci. Ireland, Trinity Coll., Dublin, Ireland; asst. lectr. in physics U. Birmingham, 1910-14; spl. lectr. in physics Punjab U., Lahore, India (now W. Pakistan), 1914-15. Fellow Inst. Physics; asso. Royal Coll. Sci.; mem. Radio Assn. (v.p.), Belfast Natural History and Philos. Soc. (hon.). Author: An English-Irish Dictionary, 1903; The Electron Theory, 1906; Two New Worlds, 1908; New Light on Immortality, 1909; Wonders of Physical Science, 1911; Contemporary Chemistry, 1912; Life of Sir William Crookes, 1923; The Moon-Element, 1924; Quo Vadimus?, 1925; Hephaestusm or the Soul of the Machine, 1925. Editor: Celtia, 1901-06. Contbr. weekly articles on contemporary elec. sci. The Electrician, 1896-1906, article on phys. sci. Britannica Year-book, 1913, sci. papers mainly on properties of selenium to profl. jours. Organized Pan-Celtic congresses of Dublin, 1901, Carnarvon, 1904, Edinburgh, 1907; inventor optophone (instrument enabling totally blind persons to recognize and locate light by means of the ear), 1912, improved instrument to enable blind to read ordinary print, 1914; transmitted 1st wireless picture broadcast from London, 1923; inventor system of wireless telewriting and telephotography based on acoustic resonance, 1925. Died July 1933.

FOUST, Alan Shivers, Am. chem. engr.; b. Dublin, Tex., June 26, 1908; s. Charles George and Carrie (Lattimore) F.; B.S. in Chem. Engring., U. Tex., 1928, M.S., 1930; Ph.D., U. Mich., 1938; m. Helen Elizabeth Aigler, Nov. 29, 1939; children—H. Patricia (Mrs. William L. Hoppe), Alan Shivers, Carolyn Elizabeth, C. William. With Magnolia Petroleum Co., 1930-32, Tex. Pacific Coal & Oil Co., 1932; faculty U. Tex., 1932-35, Tex. Coll. Mines, 1935-36; faculty U. Mich., 1937-52, prof., 1948-52; head chem. engring. dept. Lehigh U., Bethlehem, Pa., 1952-62, dean engring., 1962-65, McCann Prof. Chem. engring., 1965—. Dir. Bowen Engring. Co., Northbranch, N.J. Cons., U. S. Army Chem. Corps., 1958-65. Mem. Am. Inst. Chem. Engring. (program chmn., div. chmn. heat transfer div. 1962), Nat. Soc. Profl. Engrs., Am. Chem. Soc., Sigma Xi, Tau Beta Pi. Author: Unit Operations, 1950; (with others) Principles of Unit Operations 1960; also articles. Research on heat transfer to two-phase fluids in flow. Home: 917 Prospect Av., Bethlehem, Pa. 18018.*

FOVILLE, Archille-Louis, French physician; b. Pontoise, France, 1799; head physician Rouen (France) Mental Hosp., Charenton (France) Mental Hosp. Author: Traité complet de l'anatomie, de la physiologie et de la pathologie du système nerveux, 1844. Studied anatomy, physiology and pathology of nervous system. Died Toulouse, France, 1878.

FOWDEN, Leslie, English plant biochemist; b. Rochdale, Eng., Oct. 13, 1925; s. Herbert and Amy (Rabbich) F.; B.Sc., U. Coll. London, 1945, Ph.D., 1948; m. Margaret Oakes, July 9, 1949; children—Abigail L., Jeremy S. G. Sci. staff, med. research council Human Nutrition Research unit, 1947-50; faculty U. Coll. London, 1950—, prof. plant chemistry, 1964—. Rockefeller Found. fellow Cornell U., 1955; vis. prof. U. Cal. at Davis, 1963, U. Hong Kong, 1967. Fellow Royal Soc.; mem. Biochem. Soc., Soc. Exptl. Biology, Phytochem. Group. Publs. on contbns. to chemistry, metabolism and function of amino acids and proteins of higher plants. Home: 36A Douglas Rd., London, N. 1, Eng.*

FOWLER, Alfred, Brit. astronomer; b. 1868; Sc.D. (hon.), Cambridge, Bristol, Durham, Leeds (all Eng.) univs.; m.; 1 son, 1 dau. Yarrow Research prof. Royal Soc., 1923-34; emeritus prof. astro-physics Imperial Coll., South Kensington, Eng., Mem. Brit. govt. eclipse expdns., 1893, 96, 98, 1900, 05, 14; mem. adv. council Dept. Sci. and Indsl. Research. Recipient Catherine Wolfe Bruce Gold medal Astron. Soc. Pacific, 1934. Fellow Royal Soc., 1910 (Bakerian lectr. 1914, 24, Royal medal 1918), 1910; mem. Paris Acad. Scis. (corr., Valz prize for contbns. to spectroscopy 1913), Instituto de Coimbra (Portugal), Royal Astron. Soc. (pres., Gold medal 1915), Nat. Acad. Sci. U. S. (fgn. asso., Henry Draper Gold medal 1920), Brit. Assn. (pres. sect. A, Oxford meeting 1926), Inst. Physics (pres. 1935-37). Publs. on solar and stellar spectra, spectra of comets, structure of spectra. After 1901, worked independently and identified previously unknown celestial spectra as being same as spectra obtained previously in lab., especially important was lab. prodn. of spectrum then attributed to cosmic hydrogen (now to ionized helium), proved valuable in establishing Bohr theory of spectra. Died June 24, 1940.

FOWLER, Charles Arman, Jr., Am. physicist; b. Salt Lake City, Apr. 23, 1912; s. Charles A. and Beatrice (Buckle) F.; A.B., U. Utah, 1933, M.S., 1934; Ph.D., U. Cal. at Berkeley, 1940; m. Inez Hanson, Aug. 4, 1934; children—Scott Wellington, Craig Huntington. Faculty, U. Cal. at Berkeley, 1940-46; faculty, chmn. dept. physics Pomona Coll., Claremont, Cal., 1947—, prof., 1950—. Prin. investigator magnetism research Office Naval Research, 1949—; mem. com. on physics faculties in colls. Am. Inst. Physics, 1962-65. NSF Sr. Postdoctoral fellow U. Grenoble, France, 1960-61; R. J. Wig Distinguished Prof. award Pomona Coll., 1960. Mem. Am. Phys. Soc., A.A.A.S., Am. Assn. Physics Tchrs., Phi Beta Kappa, Sigma Xi. Contbr. articles to profl. jours., chpts. to books. Discovered vibrational systems in molecular spectra of diatomic alkaline earth fluorides, tin chloride, cadmium fluoride; co-discoverer of magneto-optic techniques for photographing magnetic domains utilizing Kerr effect, 1952, Faraday effect, 1956; 1st observed domains in thin magnetic films; domain studies in ferrites. Home: 366 Blaisdell Dr., Claremont, Cal. 91711.*

FOWLER, Earle Cabell, Am. physicist; b. Bowling Green, Ky., June 10, 1921; s. William Earle and Reba (Brownfield) F.; B.S. in Chemistry, U. Ky., 1942; A.M. in Physics, Harvard, 1947, Ph.D., 1949; m. Marjorie Jane Land, Oct. 25, 1950; children—Marjorie Anne, Walter Earle, Thomas Land. Asso. physicist Brookhaven Nat. Lab., Upton, N.Y., 1949-52, cons., 1952—; acad. staff Yale, 1952-62; prof. physics Duke, Durham, N.C., 1962—. Cons., Oak Ridge Inst. Nuclear Studies, 1962—. Sr. Fulbright lectureship to U.K., 1958-59. Fellow Am. Phys. Soc.; mem. Phi Beta Kappa, Sigma Xi, Sigma Pi Sigma, Delta Tau Delta. Author: (with Robert K. Adair) Strange Particles, 1963. Research in high energy physics, in cosmic radiation and with large accelerators; helped to develop 1st high pressure diffusion cloud chambers. Home: 1821 Lake Shore Dr., Chapel Hill, N.C. Office: Duke U., Durham, N.C. 27706.*

FOWLER, John, English inventor; b. Melksham, Eng., July 11, 1826; m. Elizabeth Lucy Pease, July 30, 1857; 5 children. Began career in corn trade; joined works Gilke, Wilson & Co., 1847, Kitson & Hewitson, 1860; founder mfg. works at Hunslet, Leeds, Eng. Recipient prize Royal Agr. Soc., 1858. Invented mech. system for draining wastelands, 1849, drain plow, 1850, steam plough, 1856, double engine tackle, 1860; numerous patents on mech. devices including seed drills, horse shoes, slide valves, reaping machines, 1850-64. Died Ackerworth, Eng., Dec. 4, 1864.

FOWLER, Sir John, English engr.; b. Sheffield, Eng., July 15, 1817; s. John and Elizabeth (Swann) F.; ed. pvt. sch., Whitley Hall; studied under J. T. Leather; LL.D., Edinburgh, 1890; m. Elizabeth Broadbent, July 2, 1850; 4 sons, including John Arthur. Ry. work under John Rastnick; joined Stockton & Hartlepool R.R. Line, 1842, successively engr., gen. mgr., locomotive supt.; cons. engr., London, beginning in 1844; sent to Norway to study rys., 1870. Recipient Prix Poncelet, Inst. France, 1890. Mem. Instn. Civil Engrs. (became mem. council 1849, pres. 1866). Knighted, 1885. Designed Pimlico Ry. Bridge, 1860; designed and built (with Benjamin Baker) Forth Bridge (1st maj. steel structure), 1882-90; designed a large part of underground railway system, London. Died Bournemouth, Eng., Nov. 20, 1898.

FOWLER, Joseph Lee, Am. physicist; b. Springfield, O., Nov. 19, 1913; s. Ernest C. and Ruth (Reese) F.; A.B. in Physics, M.S., U. Tenn., 1938; Ph.D., Princeton, 1943; m. Ruth Gray Bowling, Aug. 17, 1945; children—James Reese and Robert Bowling. Leader cyclotron group Los Alamos Sci. Lab., 1946-50; with Oak Ridge Nat. Lab., 1950—, dir. high voltage lab., physics div., 1951-54, asso. dir. physics div., 1954-57, dir. physics div. 1957—. Chmn. nuclear cross sect. adv. group AEC, 1959-61; mem. physics vis. com. Brookhaven Nat. Lab., Upton, N.Y., 1962-66, Carnegie Inst. Tech., Pitts., 1962-64; mem. NRC, 1963—. Fellow Am. Phys. Soc. (chmn. southeastern sec. 1960-61); mem. Tenn. Acad. Sci., A.A.A.S., Tau Beta Pi, Sigma Xi, Phi Kappa Phi. Author-editor: (with others) Fast Neutron Physics, Part I 1960, Part II 1963. Research on deuteron and proton interaction with nuclei; high energy neutron; fragment mass and energy distributions. Home: 116 Wendover Circle, Oak Ridge 37830. Office: P.O. Box X, Oak Ridge 37831.

FOWLER, Noble Owen, Am. physician; b. Vicksburg, Miss., July 14, 1919; s. Noble Owen and Annie Lou (Robertson) F.; student Memphis State U., 1936-38; M.D., U. Tenn., 1941; m. Charlotte Ruth Walters, June 13, 1942; children—Joann (Mrs. Lloyd Buttram), Michael Owen, Anne Stewart. Fellow cardiology U. Cin., 1948-52, U. Cin., 1950-52, asso. prof., 1957-64, prof. medicine, dir. cardiac research, 1964—; asst. prof. medicine State U. N.Y., 1954-57; faculty Emory U., 1954-57, asso. prof., 1957; dir. cardiology Christ Hosp., Cin., 1962—; cons. Cin. VA Hosp., 1961—. Diplomate Am. Bd. Internal Medicine, Am. Bd. Cardiovascular Disease. Fellow A.C.P.; mem. Am. Physiologic Soc., Am. Clin. and Climatological Assn.,

Central Soc. Clin. Research, Assn. U. Cardiologists, Sigma Xi, Alpha Omega Alpha. Author: Physical Diagnosis of Heart Disease, 1962; also numerous articles. Research on oxygen lack, drugs and exercise effects on blood flow through lungs, effect drugs on heart muscle and coronary arteries, effect anemia, changes blood viscosity on blood circulation, effect breathing on circulation blood by normal and diseased heart. Home: 3533 Deep Woods Lane, Cin. 45208. Office: Cardiac Research Lab., Cin. Gen. Hosp., Cin. 45229.*

FOWLER, Sir Ralph Howard, Brit. mathematician; b. Jan. 17, 1889; s. Howard Fowler; ed. Winchester Coll., Trinity Coll., Cambridge; M.A.; m. Eileen Mary Rutherford, 1921; 2 sons, 2 daus. Fellow Winchester Coll.; fellow Trinity Coll., Cambridge, from 1914; asst. dir. anti-aircraft exptl. sect. Munitions Inventions Dept.; Stokes lectr. U. Cambridge, Plummer prof. applied math., from 1932. Recipient Rayleigh prize, 1912, Adams prize, 1925. Fellow Royal Soc., 1925 (Royal medal 1936). Author: Differential Geometry of Plane Curves; Statistical Mechanics; (with E. A. Guggenheim) Statistical Thermodynamics; also papers on external ballistics and theoretical physics. Died July 18, 1944.

FOWLER, Richard, English physician; b. London, Nov. 28, 1765; studied medicine, Edinburgh; M.D., 1793. Began practice medicine, Salisbury, Eng., 1796; physician Salisbury Infirmary, 1796-1847. Licientiate Coll. Physicians London. Fellow Royal Soc., 1802. Author: Experiments and Observations on the Influence lately discovered by M. Galvani and commonly called Animal Electricity, 1793; Observations on the Mental State of the Blind and Deaf and Dumb, 1843; The Physiological Processes of Thinking, especially in Persons whose Organs of Sense are Defective, 1849. Studied Galvani's discoveries; research on effect of opium on nerves and muscles. Died Salisbury, Apr. 13, 1863.

FOWLER, Richard Gildart, Am. physicist; b. Albion, Mich., June 13, 1916; s. Rufus Alexander and Ethel (Gildart) F.; A.B. in Math., Chem., Albion Coll., 1936; M.S., U. Mich., 1939, Ph.D. in Physics, 1942; m. Frances Miriam Holmes, Aug. 26, 1939; children—Lynne Carol (Mrs. Barney Lee Capehart), Nancy Barbara, Patricia Ann, Richard Gerald. Research asst. Dow Chem. Co., 1936-38; research asso. Kimberly Clark Corp., summer 1940; research asst. U. Mich., 1938-41, research physicist, 1941-42, 43-46; instr. physics N.C. State Coll., 1942-43; faculty physics, U. Okla., Norman, 1946—, prof., 1956-61, research prof. physics, 1961—, v.p. Okla. Research Inst., 1962, 63. Sci. cons. Space Tech. Labs., 1958-61, Tech. Research Group, 1959-64, Aerospace Labs., 1960-65, Tex. Instruments, 1960-62. J. S. Guggenheim fellow, 1952-53; Fulbright fellow, 1963. Fellow Am. Phys. Soc., Okla. Acad. Sci., Phys. Soc. London; mem. Am. Assn. U. Profs., A.A.A.S., Okla. Edn. Assn., Phi Beta Kappa, Sigma Xi, Sigma Tau, Sigma Pi Sigma, Delta Tau Delta, Gamma Alpha Author: Infrared Determination of Organic Structure, 1949; Introduction to Electric Theory, 1953; Physics for Engineers and Scientists, 1958. Contbg. author: Handbuch der Physik, 1956; Adv. in Electronics, 1965. Patentee in field. Innovator of electric shock tubes. Home: 1309 Avondale St., Norman, Okla. 73069.*

FOWLER, Samuel, Am. physician; b. Newburgh, N.Y., Oct. 30, 1779; attended Montgomery Acad., Pa. Med. Coll. at Phila. Began practice of medicine, Hamburg, N.J., 1800; moved to Franklin, N.J.; mem. N.J. Council, 1827; mem. U. S. Ho. of Reps. (Jackson Democrat) from N.J., 23d-24th congresses, 1833-37; discovered rare mineral Fowlerite (named in his honor), also franklinite (named by him); owner, developer zinc mines, Franklin, N.J.; owner Franklin Furnace Iron Works. Contbr. to many scientific publs. Died Franklin, Feb. 20, 1844.

FOWLER, Thomas, English logician; b. Burton-Stather, Eng., Sept. 1; s. Henry and Mary Anne (Welch) F.; ed. Merton Coll., Oxford; D.D., LL.D.; became fellow and tutor Lincoln Coll., 1855, hon. fellow, 1901; proctor, 1862-63; select preacher, 1872-73; prof. logic, 1873-88; pres. Corpus Christi Coll., from 1881; vice chancellor U. Oxford, 1899-1901. Author: The Elements of Deductive Logic, 1867; Elements of Inductive Logic, 1870; Bacon's Novum Organum, 2d edit., 1889; Locke, 1880; Locke's Conduct of the Understanding, 3d edit.; 1890; Francis Bacon, 1881; Shaftesbury, 1882; Hutcheson, 1882; Progressive Morality, an Essay in Ethics, 2d edit., 1895; The History of Corpus Christi College Oxford, with Lists of its Members, 1893; (with J. M. Wilson) Principles of Morals, 1884, Part II, 1887, both parts rev., 1894; Popular History of Corpus Christi College, 1898. Died Nov. 20, 1904.

FOWLER, Thomas, English physician; b. York, Eng., 1736; practiced medicine, London, Eng.; author Medical Reports of the Effects of Arsenic, 1786; (with apothecary named Hughes) introduced solution of potassium arsenite (Fowler's solution) to treat remitting fevers, periodic headaches, 1786. Died 1801.

FOWLER, William Alfred, Am. nuclear astrophysicist; b. Pitts., Aug. 9, 1911; s. John McLeod and Jennie (Watson) F.; B.Engring. Physics, Ohio State U., 1933; Ph.D., Cal. Inst. Tech., 1936; m. Ardiane Foy Olmsted, Aug. 24, 1940; children—Mary Emily, Martha Summers. Research fellow nuclear physics Cal. Inst. Tech., Pasadena, 1936-39, faculty, 1939—, prof. physics, 1946—. Guggenheim fellow, Fulbright lectr. Cavendish Lab., U. Cambridge (Eng.), 1954-55; Guggenheim fellow St. John's Coll., also dept. applied math. and theoretical physics U. Cambridge, 1961-62; recipient Medal for Merit, 1948; Lamme medal, Ohio State Univ., 1952; Liège medal U. de Liège (Belgium), 1955; Barnard medal for meritorious service to sci. Columbia U., 1965; named Cal. Co-Scientist of Year, Cal. Mus. Sci. and Industry, 1958. Fellow Am. Acad. Arts and Scis., Am. Phys. Soc.; mem. Nat. Acad. Sci., A.A.A.S., Am. Astron. Soc., Am. Assn. U. Profs., Am. Philos. Soc., Soc. Royal des Scis. Liège (corr.), Am. Geophys. Union, Sigma Xi, Tau Beta Pi, Tau Kappa Epsilon. Research, numerous publs. on measurement of nuclear reaction rates in lab. for application to energy generation and prodn. of elements heavier than hydrogen in sun, stars, and supernovae; gamma radiation. Home: 1565 San Pasqual St., Pasadena, Cal. 91106.*

FOWLER, William Weekes, Brit. entomologist; b. Jan. 1849; s. Hugh Fowler; ed. Jesus Coll., Oxford; D.Sc., M.A.; m. Anne Frances Bonnor, 1875; 2 sons, 2 daus. Asst., house master Repton, 1873-80; head master Lincoln Sch., 1880-1901; rector Rotherfield Peppard, Oxford, 1901-04; vicar of Earley, Reading, from 1904; prebendary, canon Lincoln Cathedral, 1887. Pres. Entomol. Soc. London, 1901-02; v.p. Linnean Soc., 1907-07; mem. sci. com. Royal Hort. Soc. Author: The Coleoptera of the British Islands, 5 vols.; The Membracidae and other groups of Honoptera of Central America (Biologia Centrali Americana); Notes on New Species of Ianguriidae . . . ; Fauna of British India; Coleoptera, General Introduction: Cicindelidae and Paussidae. Died June 3, 1923.

FOWLER-BILLINGS, Katharine, Am. geologist; b. June 12, 1902; d. William and Susan Farnham (Smith) Plumer; A.B. in Geology and Biology, Bryn Mawr Coll., 1925; M.A., U. Wis., 1926; Ph.D., Columbia, 1930; m. James W. Lunn, 1929; m. 2d, Marland P. Billings, 1938; children—George Bartlett, Betty Jean. Geologist, Sierra Leone, West Africa, 1930; with Maroc Gold Co., 1931-32; instr. geology Wellesley Coll., 1935-38, Erskine Jr. Coll., Boston, 1941, Tufts Coll., 1942-43; geologist N.H. Planning and Devel. Commn., 1943-44; asso. geologist New Eng. Mus. Natural History, 1940-47. Fellow Geol. Soc. Am.; mem. Soc. Women Geographers. Research on anorthosites of Laramie Mountains, Wyo., iron ores and molybedenite of Sierra Leone; is mapping Cardigan Quadrangle, N.H., Monadnock and Mt. Washington regions. Address: North Hampton, N.H. 03862.*

FOWLES, Gerald Wilfred Albert, English chemist; b. Uley, U.K., Oct. 9, 1925; s. Albert Edward and Winefred (Bloodworth) F.; B.Sc., U. Bristol (Eng.), 1949, Ph.D., 1952, D.Sc., 1964; m. Joyce Ellen Tocknell, Dec. 30, 1950; children—Andrew Mark, Kathryn Sarah. With R. A. Lister & Co., Dursley, Eng., 1943-44, 45-46; staff Southampton U., 1952-65, reader, 1963-65; prof. chemistry Reading (Eng.) U., 1966—. External examiner B.Sc. degrees Brighton and Lanchester colls. tech., 1965—. Mem. Royal Inst. Chemistry, Chem. Soc., Am. Chem. Soc. Author: (with E. Cartmell) Valency and Molecular Structure, 1956; also numerous articles. Research on chemistry of titanium, vanadium, chromium sub-group elements, non-aqueous solvents, organometallic compounds; devel. new polymers and catalysts, new analytical methods, fungicides and synthetic methods. Home: 61, Church Rd., Woodley, Reading, Berkshire, U. K.*

FOWNES, George, English chemist; b. London, May 14, 1815; student of Liebig, D.Phil., Giessen, Germany; prof. practical chemistry Birkbeck lab. Univ. Coll., London; became prof. chemistry Pharm. Soc., 1842. Fellow Royal Soc., 1845. Author: Manual of Chemistry (popular textbook), 1844, also later edits.; Chemistry as exemplifying the Wisdom and Beneficence of God (Actonian prize essay), 1844; Chemical Tables: chiefly for the use of Junior Students, 1846. Determined equivalent weight of carbon; discovered benzoline (Laurent's hydrobenaamide); obtained furfurol and a derivative; research of fermentation, putrefaction, artificial yeast, preparation of ether, analysis of nitrogenous substances, equivalent and combining vols. of solid bodies; prepared carbon monoxide; discovered phosphates in igneous rocks (thus helping to explain presence in soil for plant assimilation. Died London, Jan. 31, 1849.

FOX, Allen Sander, Am. geneticist; b. Chgo., Mar. 5, 1921; s. Morris E. and Mary (Weinstein) F.; B.S., U. Chgo., 1941, Ph.D., 1948; m. Rima Feinstein, Dec. 31, 1941; children—Natalie Ann, David Gregory. Asst. prof. zoology Ohio State U., 1948-53; faculty Mich. State U., 1954-63, prof. biochemistry, 1959-63; prof. genetics U. Wis., Madison, 1963—; Fulbright research prof., Italy, 1953-54, Australia, 1961. Mem. Genetics Soc. Am., Am. Soc. Zoologists, Am. Soc. Naturalists, Soc. for Exptl. Biology and Medicine, A.A.A.S. Research, numerous publs. on immunogenetics, biochem. genetics, tissue culture, protein synthesis, nucleic acids. Home: 2425 Fontaine Circle, Madison, Wis. 53713.*

FOX, Sir Charles, English engr.; b. Derby, Eng., Mar. 11, 1810; s. Francis F.; m. Mary Brookhouse, 1830; 3 sons, 1 dau. Became constrn. engr. London and Birmingham R.R., 1829; practiced as civil engr., London, from 1857. Fellow Royal Asiatic, Royal Geog. Soc.; mem. Instn. Civil Engrs., Instn. Mech. Engrs. (mem. council), Brit. Assn. (charter, life). Principal works include Crystal Palace of Gt. Exhbn., Eng., 1851, London & Birmingham R.R., Eng., Berlin (Germany) Waterworks, many other railroad stas. and bridges in Ireland, Denmark, France, Canada, India, Cape Colony, Queensland, Australia. Developed new method for tunnel constrn. using a type of skew arch (which aided in constrn. of met. r.r.'s); replaced sliding rail with switch in ry.; improved iron structures. Died Blackheath, Eng., June 14, 1874.

FOX, Denis Llewellyn, Am. marine biologist; b. Udimore, Sussex, Eng., Dec. 22, 1901; s. John James and Florence M. (Fox) F.; came to U. S., 1905, naturalized, 1906; A.B., U. Cal. at Berkeley, 1925; Ph.D., Stanford, 1931; postgrad. (Rockefeller Found. Research fellow), U. Cambridge (Eng.); m. Miriam Perdew, Aug. 31, 1932; children—Ronald L., Stephen J. P. (dec.), Kathleen M. (Mrs. Allan H. Quist), Alan D. Chemist, Standard Oil Co. Cal., 1925-29; faculty Scripps Instn. Oceanography, U. Cal. at La Jolla, 1931—, prof. marine biochemistry, 1948—, vice chmn. dept. marine biology, 1965—. Cons. on fouling by marine animal growths, undersea structures and sea water conduits to pvt. cos., city and state agys. Guggenheim Meml. Found. Research fellow, 1945-46. Fellow A.A.A.S., San Diego Soc. Natural History, San Diego Zool. Soc.; mem. Soc. Gen. Physiologists, Am. Soc. Zoologists, Am. Soc. Limnology and Oceanography, Western Soc. Naturalists. Author: Animal Biochromes and Structural Colours, 1953; also numerous articles. Research on kinds animal pigments, their chem. nature, distbn. in animal phyla and various tissues, their metabolic fractionation and physiol. bearing, marine and comparative biochemistry, growth, nutrition and metabolism marine organisms on comparative basis, biochem. processes and cycles in sea, control marine fouling animal growth. Home: 2459 Ardath Rd., La Jolla, Cal. 92037.*

FOX, George Henry, Am. dermatologist; b. Ballston Spa, N.Y., Oct. 8, 1846; s. Norman and Jane (Freeman) F.; A.B., U. of Rochester, 1867, A.M., 1870; M.D., U. Pa., 1869; postgrad. univs. Berlin, Paris, Vienna and London, 1870-73; m. Harriet Gibbs, Aug. 29, 1872; children—Howard, Adaline (Mrs. Henry R. Russell), Alanson Gibbs, Helen (Mrs. Mason Trowbridge). Surgeon N.Y. Dispensary, 1873-75; clin. prof. diseases of skin Woman's Med. Coll. of N.Y. Infirmary, 1875-79; clin. prof. dermatology Starling Med. Coll., Columbus, O., 1879; clin. prof. diseases of skin Coll. Phys. and Surg. (Columbia), 1881-1904, prof. dermatology, 1904-07; prof. skin diseases. Post-Grad. Med. Sch., 1890-95; cons. dermatologist N.Y. Bd. Health. Author: Photographic Atlas of Diseases of the Skin, 1900; Reminiscences, 1926; The Lineage of One Hundred American Physicians Named Fox, 1927. Died May 3, 1937.

FOX, Herbert, Am. pathologist; b. Atlantic City, N.J., June 3, 1880; s. Samuel Tucker and Hannah Ray (Freas) F.; A.B., Central High Sch., Phila., 1897; M.D., U. Pa., 1901; postgrad. U. Vienna; m. Louise Carr Gaskill, Nov. 9, 1904 (dec. Nov. 1933); children—Margaret, John Freas, Samuel Tucker; m. 2d, Mary Harlan Rhoads, Dec. 3, 1938. Vol. asso. William Pepper Clin. Lab. 1903-06; pathologist Rush Hosp. since 1904, Phila. Zoöl. Soc., since 1906; chief of labs. Pa. Dept. Health, 1906-11; dir. William Pepper Lab. Clin. Medicine, Hosp. U. Pa., from 1911; pathologist Children's Hosp., 1915-26; prof. comparative pathology U. Pa. since 1927. Fellow A.A.A.S.; mem. Coll. Physicians, Am. Philos. Soc., Acad. of Natural Scis., County Med. Soc., Pathol. Soc. (all of Phila.), A.M.A., Am. Assn. Pathologists and Bacteriologists. Author: Elementary Bacteriology and Protozoölogy, 1912, 5th edit., 1931; Text Book of Pathology (with Alfred Stengel), 8th edit., 1927; Disease in Captive Wild Mammals and Birds, 1923. Died Feb. 27, 1942.

FOX, Howard, dermatologist; b. London, Eng., July 4, 1873; s. George Henry and Harriet Lovisa (Gibbs) F.; brought to Am. in infancy; A.B., Yale, 1894; M.D., Coll. Phys. and Surg., Columbia, 1898; post-grad. univs. Berlin and Vienna; Sc.D., Rollins Coll., 1931. Practiced, N.Y.C., from 1903; prof. dermatology. Dartmouth Med. Coll., 1913; dermatol. asst. Vanderbilt Clinic, 1903-13; attending dermatologist USPHS, Dist. 2; prof. dermatology N.Y. Polyclinic Med. Sch., 1923-24; prof. dermatology and syphilology, New York U.; hon. Cons. dermatologist and syphilologist Bellevue Hosp.; cons. dermatologist various hosps. Pres. Am. Bd. Dermatology and Syphilology 1932-45; pres. Med. Soc. County of N.Y., 1938; 1st pres. Am. Acad. Dermatology and Syphilology, 1938; pres. N.Y. Soc. Tropical Medicine, 1939; 1st pres. Assn. Dermato-Syphilologists of Greater N.Y.; mem. Am. Dermatol. Assn. (pres. 1924-25), N.Y. Acad. Medicine, Am. Med. Assn., N.Y. State Med. Soc.; corr. or hon. mem. numerous Am. and fgn. med. socs. Contbr. to Personal Hygiene (Pyle), 1909; Mod. Treatment (Hare), 1910; Skin Diseases in Infancy and Childhood, 1929; Eliot's Quiz. 1938; Clinical Tropical Medicine (Z. T. Bercovitz), 1944. Editor-in-chief Archives of Dermatology and Syphilology, 1937-47. Died Oct. 1954.

FOX, Irving, Am. entomologist; b. Hartford, Conn., Dec. 19, 1912; s. Solomon and Esther (Akst) F.; A.B. with distinction, George Washington U., 1937, A.M., 1938; Ph.D., Ia. State U., 1940; m. Ina Moll, Sept. 13, 1941; 1 son, Robert Irving. Grad. asst. entomology Ia. State U., 1937-40; faculty U. P.R., 1941-42, 1946—, prof. med. entomology Sch. Tropical Medicine, San Juan, 1962—. Program dir. USPHS Grad. Research Tng. Grant in Tropical Medicine, Parasitology and Med. Entomology, 1957—. Fellow A.A.A.S.; mem. Am. Soc. Parasitology, Entomol. Soc. Am., Am. Soc. Tropical Medicine and Hygiene, Entomol. Soc. Washington, Am. Mosquito Control Assn., Sigma Xi. Author: Fleas of Eastern United States, 1940. Studies, numerous publs. on insects and related forms involving descriptions of fleas, mites, ticks, spiders, flies, mosquitoes, toxic chems. against insects and snails, biol. control of fleas, effect of climate on insects, prevention and control of tropical diseases such as malaria, plague, murine typhus fever, insect allergy, myiasis and schistosomiasis. Home: 1960 Cacique, Santurce, P.R. 00911. Office: Sch. Tropical Medicine Bldg., San Juan, P.R. 00905.*

FOX, Jack Jay, Am. chemist; b. N.Y.C., Dec. 21, 1916; s. Samuel and Celia (Stern) F.; A.B., U. Colo., 1939, Ph.D., 1950; m. Ruth C. Inabu, June 13, 1939; children—Dolores M. (Mrs. F. Emspak), John Reed. With Sloan-Kettering Inst. for Cancer Research, N.Y.C., 1952—, mem., sect. head; prof. Cornell U. Grad. Sch. Med. Scis., 1958—. Recipient Alfred P. Sloan award in cancer research, 1956. NRC fellow, 1950-52; Damon Runyon Meml. Fund fellow, 1952-54. Mem. Am., Westchester chem. socs., Am. Soc. Biol. Chemists, Sigma Xi. Research, numerous publs. on synthesis and structural elucidation of compounds of biochem. importance, specific syntheses of compounds related to nucleic acid components, carbohydrate and heterocyclic chemistry. Home: 424 S. Lexington Av., White Plains, N.Y. 10606. Office: 145 Boston Post Rd., Rye, N.Y. 10580.*

FOX, John Perrigo, Am. physician; b. Chgo., Nov. 10, 1908; s. John S. and Myrtle (Perrigo) F.; B.S., Haverford Coll., 1929; M.D., Ph.D., U. Chgo., 1936; M.P.H., Columbia, 1948; m. Helen Duffell, July 14, 1934; children—Judith M., John D., Haigh P., Joanne M. With Northern Trust Co., Chgo., 1929-31; fellow, later asst. dept. pathology U. Chgo., 1933-36; intern Evanston (Ill.) Hosp., 1937-38; staff internat. health div. Rockefeller Found., 1938-49; prof. epidemiology Sch. Medicine Tulane U., 1949-58, Wm. Hamilton Watkins prof. epidemiology, dir. div. grad. pub. health, 1958-60, vis. prof. epidemiology, 1960—; chief dept. epidemiology, mem. Pub. Health Research Inst. City of N.Y., Inc., 1960-65; adj. prof. epidemiology Columbia Sch. Pub. Health, 1960-65, N.Y.U. Sch. Medicine, 1960-65; prof. preventive med. Med. Sch., U. Wash., 1965—. Consultant Communicable Disease Center, USPHS, mem. rickettsial disease commn. Armed Forces Epidemiological Bd., 1955—; chmn. virus and rickettsial study sect. NIH, 1962-64, chmn. bd. sci. counselors div. biologic standards, 1961-63, mem. com. epidemiology and biometry tng. grants, 1965—; cons. expert com. rabies, WHO, 1958—. Recipient Howard Taylor Ricketts prize U. Chgo., 1936. Diplomate Am. Bd. Preventive Medicine and Pub. Health. Fellow Am. Pub. Health Assn. (governing council, 1958-62), N.Y. Acad. Sci., A.A.A.S., Royal Belgian Soc. Tropical Medicine (hon.); mem. Am. Soc. Tropical Medicine and Hygiene, American Society Pathologists and Bacteriologists, Am. Soc. Bacteriologists, Harvey Soc., Am. Coll. Preventive Medicine, Assn. Tchrs. Preventive Medicine (exec. com. 1957-60), Soc. Exptl. Biology and Medicine, Am. Assn. Immunologists, Am. Epidemiological Soc., Am. Acad. Microbiology. Asso. editor Am. Jour. Epidemiology, 1964—. Contbr. articles to sci., med. jours. Study of enterovirus infections; viral tissue culture with yellow fever; active immunization versus yellow fever, poliomyelitis, typhus, and rabies; chemotherapy in typhus; serologic work in yellow fever, poliomyelitis, and typhus. Home: 16529 41st Av. N.E., Seattle 98155.

FOX, Mattie Rae Spivey, Am. nutritionist, biochemist; b. Joy, Tex., Feb. 23, 1923; d. George Lafayette and Almeta (Houston) Spivey; B.S., Tex. Women's U., 1943; M.S., Ia. State U., 1947; Ph.D., George Washington U., 1953; m. Robert B. Fox, June 5, 1954. Chemist, Humble Oil & Refining Co., Ingleside, Tex., 1943-45; nutrition analyst U. S. Dept. Agr., Washington, 1947-49; research biochemist NIH, Bethesda, Md., 1951-62; with FDA, Washington, 1962—, chief micronutrient research br., div. nutrition, 1966—. Fellow A.A.A.S.; mem. Am. Inst. Nutrition, Soc. for Exptl. Biology and Medicine, Am. Chem. Soc., N.Y. Acad Scis., Sigma Xi. Research on metabolic function and requirement of nutrients in exptl. animals, primarily vitamin B12 folic acid, pantothenic acid, vitamin E, essential fatty acids, amino acids, inorganic elements and zinc, factors that affect stability of exptl. diets. Home: 6115 Wiscassett Rd., Washington 20016. Office: 200 C St., Washington 20204.*

FOX, Philip, Am. astronomer; b. Manhattan, Kan., Mar. 7, 1878; s. Simeon M. and Esther (Butler) F.; B.S., Kan. State Agrl., Coll., 1897, M.S., 1901; B.S., Dartmouth Coll., 1902; postgrad. U. Berlin, 1905-06; grad. Army Gen. Staff Coll., Langres, France, 1918; LL.D., Drake U., 1929; D.Sc., Kans. State, 1931; m. Ethel L. Snow, Aug. 28, 1905; children—Stephen Snow, Bertrand, Gertrude, Robert Temple. Comdt., tchr. math. St. John's Mil. Sch., Salina, Kan., 1899-1901; asst. in physics Dartmouth 1902-03; Carnegie research asst. Yerkes Obs., U. Chgo., 1903-05; instr., astro-physics, U. Chgo., 1907-09; prof. astronomy and dir. Dearborn Obs., Northwestern U., 1909-29; dir. Adler Planetarium and Astron. Mus., Chgo., 1929-37; dir. Mus. of Sci. and Industry, Chgo., 1937-40. Fellow Am. Acad. Arts and Scis., A.A.A.S., Royal Astron. Soc.; mem. Am. Astron. Soc. (v.p.), Société Astronomique de France, Astronomische Gesellschaft, Phi Beta Kappa, Sigma Xi. Author: Annals of the Dearborn Observatory (vols. I, II and III), also sci. brochures and contbns. to astron. jours., principally on double stars, stellar parallax and solar physics. Died July 21, 1944.

FOX, R. Fortescue, Brit. physician; s. Joseph John Fox; ed. London Hosp. Med. Coll.; M.D., London; m. Katharine Stewart McDougall, 1885; 1 son, 2 daus. Med. dir. Enham Village Centre for Disabled Ex-Service Men; hon. physician Red Cross Clinic for Disabled Officers, London; hon. pres. Internat. League against Rheumatism. Fellow Royal Coll. Physicians, Royal Meteorol. Soc., Royal Soc. Medicine (Hyde lectr.); pres., founder Internat. Soc. Med. Hydrology, also editor archives; pres., treas. Brit. Balneological and Climatol. Soc.; pres., orator Hunterian Soc. Author: Principles and Practice of Medical Hydrology, 1914; Physical Remedies for Disabled Soldiers, 1917; (with Dr. D Van Breemen) The Causation and Treatment of Chronic Rheumatism. 1934; Arthritis in Women, 1936; The Evolution of Chronic Rheumatism, 1937; The Breakdowns of Middle Life; Principles in the Treatment of Chronic Disease; Modern Physicial Treatment on relation to Rheumatism; Baths as Artificial Climates; also articles in Latham's System of Treatment, The Encyclopedia Medica, Sir. R. Jones' Orthopaedic Surgery and Injuries, Oxford Index of Therapeutics. Editor: Handbook of British Health Resorts. Died June 15, 1940.

FOX, Robert Were, Brit. geologist; b. Falmouth, Eng., Apr. 26, 1789; s. Robert Were and Elizabeth (Tregelles) F.; privately ed.; m. Maria Barclay, 1814; children—Caroline, Barclay, Anna Maria, Charles. Began research with Joel Lean, 1812. Fellow Royal Soc., 1848; mem. Royal Cornwall Poly. Soc. (a founder, v.p.). Research and publs. on elasticity of high pressure steam, internal temperature of earth proving heat increased with depth in a diminishing ratio; built dipping needle used by Sir James Clark Ross in voyage to Antarctic Ocean, 1837, and by Capt. Nares in expdn. to N. Pole, 1875-77; studied magnetic phenomena, especially earth's magnetism. Died July 25, 1877.

FOX, Russell Elwell, Am. physicist; b. Richmond, Va., Dec. 28, 1916; s. Isaac Dudley and Cornelia (Russell) F.; B.S., Hampden-Sydney Coll., 1934-38; M.S., U. Va., 1939, Ph.D., 1942; m. Thelma Todd, June 20, 1942; children—Leslie R., Margaret Louise, Robert Dudley. With Westinghouse Research Labs., Pitts., 1942—, sect. mgr. vacuum physics, 1957-60, mgr. physics dept., 1960-64, dir. atomic and molecular scis. research and devel, 1964—. Fellow Am. Phys. Soc.; mem. Am. Vacuum Soc., A.A.A.S., Am. Soc. for Testing Materials (chmn. com. E-14 1962-64), Phi Beta Kappa, Sigma Xi, Lambda Chi Alpha. Author: Saturday Science, 1960; Science of Science, 1963. Research, publs. on mass spectrometry, and applications to ionization processes; study of ion formation, develop. of ion source and retarding potential difference method for studying electron collisions in gases. Home: 215 Woodside Rd. Pitts. 15221. Office: Westinghouse R. & D. Center, Pitts. 15235.*

FOX, Ruth, Am. psychiatrist; b. N.Y.C., June 21, 1895; d. James Braden and Adelaide (Gomer) Fox; Ph.B., U. Chgo., 1919; M.D., Rush Med. Coll., 1926; m. McAlister Coleman, Mar. 1, 1931; children—Ann (Mrs. Douglass Allen), McAlister. Dir labs. Neurol. Inst., N.Y.C., 1934-38; practice medicine, specializing in psychiatry, N.Y.C., 1938—; medical director of National Council on Alcoholism, Incorporated, 1959—. Member of adv. council on alcoholism N.Y. Dept. Mental Hygiene. Recipient Citation of Merit award Malvern Inst. Psychiat. and Alcoholic Studies, 1963. Fellow Am. Psychiat. and Alcoholic Studies, 1963. Fellow Am. Psychiat. Assn., N.Y. Acad. Medicine, Assn. Advancement Psychotherapy, Am. Acad. Psychoanalysis, Am. Pub. Health Assn.; mem. A.M.A., Pan Am. Med. Assn., Am. Med. Womens Assn., Royal Soc. Health (Eng.), Soc. Study Addiction to Alcohol and Other Drugs (Eng.), Am. Soc. Group Psychotherapy and Psychodrama (award N.Y. chpt. 1963), Am. Group Psychotherapy Assn., N.Y. Acad. Scis., Assn. Med. Group Psychoanalysts, Am. Orthopsychiat. Assn., Am. Soc. Clin. Hypnosis, Soc. Clin. and Exptl. Hypnosis, Am. Acad. Psychoanalysis (trustee). Author: (with Peter Lyon) Alcoholism, Its Scope, Cause and Treatment, 1955; Alcoholism: Behavioral Research and Therapeutic Approaches, 1967. Research, publs. into causes, methods of treating alcoholism; edn. profl. and lay population on alcoholism. Home and office: 150 E. 52d St., N.Y.C. 10022.*

FOX, Sidney W., Am. educator; b. Los Angeles, Mar. 24, 1912; s. Jacob and Louise (Burmon) F.; B.A., U. Cal., Los Angeles, 1933; Ph.D., Cal. Inst. Tech., 1940; m. Raia Joffe, Sept. 14, 1937; children—Jack Lawrence, Ronald Forrest, Thomas Oren. Research chemist Cutter Labs., Berkeley, Cal., 1940-41; research fellow U. Mich., 1941-42; research chemist F. E. Booth Co., Emeryville, Cal., 1942-43; faculty Ia. State Coll., 1943-55, prof., 1947-55; prof. chemistry, dir. Inst. Space Bioscis., Fla. State U., 1955-64; prof. biochemistry, dir. Inst. Molecular Evolution, U. Miami, Fla., 1964—; cons. bioscis. subcom. NASA, 1960—. Sec., U. S. Nat. Center for Biochemistry, Internat. Union Biochemistry, 1956-59. Recipient U. S. Sci. Exhibit medal, 1962. Fellow A.A.A.S.; mem. Am. Soc. Biol. Chemists, Am. Soc. Cell Biologists, Soc. Study Evolution, Am. Chem. Soc., Geochem. Soc. Author: (with Joseph F. Foster) Introduction to Protein Chemistry, 1957. Editor: The Origins of Prebiological Systems, 1965. Research, publs. structure, evolution, spontaneous synthesis of proteins and implications in their initiation of organic life. Home: 7721 S.W. 50th Ct., Miami, Fla. 33143.*

FOX, Thomas G., Am. chemist; b. Union Deposit, Pa., Feb. 19, 1921; s. Thomas G. and Kathryn (Fasnacht) F.; B.S., Lebanon Valley Coll., 1940; Ph.D., Columbia, 1943; m. Joyce Arlene Cake, June 21, 1941; children—Joyce Melinda, Margaret Leigh, Mary Louise. Instr. chemistry Columbia, 1943-44; research chemist Goodyear Tire & Rubber Co., Akron, O., 1946-48; research asso. Cornell U., 1948-50; lab. head research dept. Rohm & Haas Co., Phila., 1950-57; asst. dir. research Mellon Inst., Pitts., 1957-61, dir. research, 1961-62, mem. exec. com., 1961, staff fellow, mem. adv. com. to pres., 1962—. Mem. NASA Research Adv. Com. on Materials, 1963—; sci. adviser to Gov. Pa., 1965—; chmn. Pa. Sci. and Engring. Foundation. Member American Chem. Soc. (chmn. div. polymer chemistry 1956-57), Am. Phys. Soc. (chmn. div. high polymer chemistry), Soc. Rheology, Sigma Xi, Phi Lambda Upsilon. Editor: Jour. of Polymer Sci. Research, publs. of chemistry of preparation, molecular structure, phys. properties of high polymers including free radical and ionic stereospecific polymerizations, intrinsic viscosity, glass temperature, melting temperature and melt viscosity. Home: 715 Old Mill Rd., Pitts. 15238. Office: 4400 5th Av., Pitts. 15213.*

FOX, William Thornton Rickert, Am. polit. scientist; b. Chgo., Jan. 12, 1912; s. John Sharpless and Myrtle (Perrigo) F.; B.S., Haverford Coll., 1932; M.A., U. Chgo., 1934, Ph.D., 1940; m. Annette Baker, Sept. 3, 1935; children—Carol Perrigo (Mrs. Morton Foelak), Merritt Baker. Instr., Temple U., 1936-41; instr., conf. dir. Princeton Sch. Pub. and Internat. Affairs, 1941-43; research asso. Yale Center Internat. Studies, 1943-51, asso. prof. polit. sci., 1946-50; lectr. Harvard, 1947; prof. internat. relations Columbia, N.Y.C., 1950—, dir. Inst. War and Peace Studies, 1951—. Cons., U. S. Dept. State, 1944-53, 62—, Rand Corp., 1950—, Arms Control and Disarmament Agy., 1963-64; Fulbright lectr. Inst. Rio Branco and Pontifical Cath. U. Rio de Janeiro, Brazil, 1966; vis. prof. El Colegio de México, 1967. Fellow Am. Acad. Arts and Scis., Hudson Inst.; mem. NRC, Social Sci. Research Council (chmn. com. on nat. security policy research 1952-63), Am. Polit. Sci. Assn. (v.p. 1965-66), Am. Assn. U. Profs., Council on Fgn. Relations. Author: The Super-Powers, 1944; The Struggle for Atomic Control, 1947; The American Study of International Relations, 1968; also articles. Co-author: Technology and International Relations, 1949; Diplomacy in a Changing World, 1959; The Role of Theory in International Relations, 1964; The Representation of United States Abroad, 1965. Mng. editor: World Politics, 1948-53; editor, contbr.: Theoretical Aspects of International Relations, 1959. Home: 18 Lake Dr., Riverside, Conn. 06878. Office: Columbia U., N.Y.C. 10027.*

FOX, William Tilbury, dermatologist; b. 1836; s. Luther Owen F.; M.B., 1857, M.D., 1858, U. College, London. General practicioner, Bayswater; specialized in midwifery; apptd. physician to Farringdon General Dispensary; traveled in East, 1864; practice in Piccadilly; physician, skin dept., Charing Cross Hospital, 1866; physician, skin dept., U. College Hospital; mem. ed. staff, Lancet. Author: Skin diseases of Parasitic Origin, 1863; Skin Diseases, their Description, Pathology, Diagnosis, and Treatment, 1864; The Classification of Skin Diseases, 1864; Cholera Prospects, 1865; The Action of Fungi in the Production of Disease, 1866; Leprosy, Ancient and Modern, 1866; Eczema, its Nature and Treatment, 1870; (with T. Farquhar) Scheme for Obtaining a better knowledge of Endemic Skin Diseases of India, 1872; Key to Skin Diseases, 1875; Atlas of Skin Diseases, 1875-77; (with T. Farquhar) On certain Endemic Skin and other Diseases of India and Hot Climates generally, 1876; (with T. Colcott Fox) Epitome of Skin Diseases, 1877; On Ringworm and its Management, 1878; numerous papers to medical socs. and journals. Gave 1st descriptions of: pompholyx, 1873, dermatitis herpetiformis, 1880, impetigo contagiosa, 1864, erysipeloid, 1873, epiderolysis bullosa, 1879; investigated problems in obstetrics, bacterial, parasitic, endemic skin diseases. Died Paris, France, June 14, 1879.

FOXON, George Eric Howard, Brit. biologist; b. London, Eng., Sept. 2, 1908; s. George and Edith (Lewis) F.; M.A., Queens' Coll., Cambridge; M.Sc., U. Wales; m. Joan Burlinson, 1932; children—Ann, Anthony. Instr. zoology U. Glasgow, 1932-37; inst., lectr. biology Univ. Coll. of cardiff, 1937-48; head dept. biology Guy's Hosp. Med. Sch., U. London, 1948—, lectr. biology, 1948-55 prof. biology U. London, 1955—. Fellow Linnean Soc. of London, Zool. Soc. of London, Inst. Biology. Research and numerous publs. on anatomy and physiology of heart, circulation in vertebrate animals. Home: Thorpe Cloud, Woodfield Lane, Ashtead, Surrey. Office: Guy's Hospital Medical School, London S.E.1, Eng.

FOXWELL, Arthur, Brit. physician; b. Shepton Mallet, Eng., July 13, 1853; s. Thomas Somerton and Jane (Handock) F.; student Queen's Coll., Taunton; B.A., St. John's Coll., Cambridge, 1877; M.B., 1881, M.A., M.D., 1891; B.A., London, 1873; postgrad. in throat and ear diseases, Vienna, 1887-88; m. Lisette Hollins, 1889; 1 dau. Sr. hon. physician Queen's Hosp., Birmingham, Eng. Mem. Royal Coll. Surgeons London, Royal Coll. Physicians, London. Author: The Causation of Functional Murmurs, 1899. Clin. and path. research on heart murmurs and their causes. Died Aug. 1, 1909.

FOY, Arkadiy Mikhaylovich, Russian obstetrician, gynecologist; b. 1902; grad. Med. Faculty, Baku U., 1926; D.Med. Sci., 1939. Asst. obstetrics and gynecology clinic 1st Leningrad Med. Inst., 1931-37, asst. lectr., 1937-39, lectr., 1939-42, prof. chair obstetrics and gynecology, 1945-52; prof., 1943—; head chair obstetrics and gynecology Krasnodar Med. Inst., 1942-45, Med. Faculty, Saratov Med. Inst., 1952—; chief obstetrician and gynecologist Saratov Oblast Dept. Health, 1953—. Mem. All-Union (bd. mem.), All-Russian (bd. mem.), Saratov (chmn.) socs. obstetricians and gynecologists. Author over 100 works including Some Features of Protracted Labor, 1960. Developed method of exciting labor with androgens, 1965. Address: Saratov Med. Inst., ulitsa 20-letiya VLKSM, 112, Saratov, RSFSR, USSR.

FOY, François, French pharmacologist; b. Fontaine-sous-Mont Aiguillon, France, 1793; prof. pharmacology, Paris; head pharmacist St. Louis Hosp. Author: Cours de pharmacologie, 1830; De choléra en Pologne; Histoire médicale du choréramorbus à Paris, 1832; Manuel théoriques et pratique du pharmacien, 1838; Traité de matière médicale et de thérapeutique, 1843. Research on drug preparation, cholera plagues in Poland, and Paris. Died Paris, 1867.

FOYE, Wilbur Garland, Am. geologist; b. Brockton, Mass., Feb. 8, 1886; s. Josiah Williams and Helen (Howard) F.; A.B., Colby Coll., 1909; A.M., Harvard Univ., 1912, Ph.D., 1915; M.A., Wesleyan U., Conn., 1925; m. Evelyn Louise Ryder, Aug. 9, 1916; children—Howard Ryder, William Dean. Austin teaching fellow Harvard, 1912-15, Sheldon traveling fellow, 1915-16; asst. prof. geology Middlebury Coll., 1916-18; asso. prof. geology. Wesleyan U., Conn., 1918-24, prof., from 1924. Fellow Geol. Soc. Am., Am. Geog. Soc., A.A.A.S. Author: (with William N. Rice) The Geology of Middletown, Conn. (Conn. Geol. and Natural History survey), 1927. Died Jan. 9, 1935.

FOYE, William Owen, Am. chemist; b. Athol, Mass., June 26, 1923; s. Owen Henry and Mildred (Walton) F.; A.B., Dartmouth, 1943; M.A., Ind. U., 1944, Ph.D., 1948. Faculty, U. Wis., 1949-55; faculty Mass. Coll. Pharmacy, Boston, 1955—, prof. chemistry, 1963—; vis. lectr. State U. Ia., 1953. Sec., Bds. and Colls. Pharmacy dist. 1, 1958-63. Mem. Am. Chem. Soc., Am. Pharm. Assn., A.A.A.S., Sigma Xi, Phi Lambda Upsilon, Rho Chi. Research, numerous publs. on correlation of drug activity with other phys. properties, synthesis of drug molecules, radiation protection found in trithiocarbonate derivatives. Home: 178 Waltham St., Lexington, Mass. 02173. Office: 179 Longwood Av., Boston 02115.*

FRACASTORO, Girolamo (or Hieronymus Fracastorius), Italian physician; b. Verona, Italy, 1483; studied math., geology, astronomy, Padua, Italy. Apptd. lectr., Padua, 1502; prof. at Friuli; practiced medicine in Verona; physician to Pope Paul III, who apptd. him physician to Council of Trent. Author: Syphilis sive morbus gallicus (poem in which syphilis received its name, included most knowledge of the time on the disease and recognized its venereal cause), 1530; Homocentricorum seu de stellis liber unus (possibility of telescope with 2 superposed lenses mentioned), 1538; De contagione et contagionsis morbis (study of typhus and plague, suggested germ theory of infection), 1546; De sympathia et antipathia rerum, 1546; others. Collected works pub. 1555. Introduced term fomites for clothing and articles of a sick person that may transmit disease; attributed spread of epidemics to tiny particles or spores, which could transmit infection by direct or indirect contact; 1st to recognize typhus fever; sometimes called father of scientific epidemiology; interests covered astronomy, cosmography, geography, mathematics, physics, biology. Died Casi, nr. Verona, Aug. 6, 1553.

FRAENKEL, Albert, German physician; b. Frankfurt/Oder, Germany, Mar. 10, 1848; studied at Berlin under R. Virchow and L. Traube; prof., Berlin, from 1880; dir. urban hosps., from 1890. Author: Pathologie und Therapie der Krankheiten des Respirationsapparates, 1890-1904. Research on bacteriology; physiol. chemistry; exptl. pathology, leukemia; Tb; showed Diplococcus pneumoniae causes lobar pneumonia, 1884. Died Berlin, July 6, 1916.

FRAENKEL, Benjamin S., physicist; b. Marburg a.d. Lahn, Germany, May 13, 1923; s. Abraham A. and Malka Wilhelmina (Prins) F.; M.Sc., Hebrew U., Jerusalem, Israel, 1951, Ph.D., 1955; fellow U. Liverpool (Eng.), 1955-56; m. Judith Cahn, July 12, 1954; children—Achasah, Ofrah, Yehudah, Yedidah. Faculty, Hebrew U., Jerusalem, 1959—, sr. lectr., 1965—. Mem. Israel Crystellographic Soc. Research, publs. on spectroscopy of highly ionized atoms; first (with others) to identify main lines of highly ionized metal in solar corona; found (with others) distbn. of differently ionized atoms in a vacuum spark. Home: 26 Hapulmach St., Jerusalem, Israel.*

FRAENKEL, Bernhard, physician; b. Germany, 1836; recognized ozena (disease of nose) as clin. entity, 1876; gave 1st description of fungus infection in throat; credited with 1st successful removal of cancer from throat (pub. 1887). Died 1911.

FRAENKEL, Eugen, German pathologist, anatomist; b. Neustadt, Silesia, Sept. 28, 1853; s. Wilhelm and Johanna (Haase) F.; M.D., Breslau, Germany, 1875; postgrad., Vienna; m. Marie Deutsch, 1880; 2 sons, 1 dau. Practiced medicine, Hamburg, Germany. became prospector path. inst. St. George Hosp., Hamburg, 1877; named dir. path. inst. and bacteriol. dept. State Hosp., Eppendorf, Germany, 1889; became 1st prof. pathology U. Hamburg, 1919. Author numerous sci. publs. Research on enteric fever, cholera, pneumonia, typhus fever, influenza, gas gangrene. Died Hamburg, Dec. 20, 1925.

FRAENKEL, George Kessler, Am. physicist, chemist; b. Deal, N.J., July 27, 1921; s. Osmond Kessler and Helene (Esberg) F.; B.A., Harvard, 1942; Ph.D., Cornell U., 1949; m. Johanna-Maria Herzog, June 30, 1951 (div. Aug. 1965); m. 2d, Elizabeth Ruskay R., 1968. Research group leader Nat. Def. Research Com., 1943-46; faculty Columbia, 1949—, prof., 1961—, chmn. dept. chemistry, 1966—, dean of the graduate faculties, 1968—. Member arts coll. adv. council Cornell U., 1964—; chmn. Gordon Research Conf. on Magnetic Resonance, 1967. Fellow Am. Phys. Soc.; mem. Am. Chem. Soc., A.A.A.S. Asso. editor Jour. Chem. Physics, 1962-64; adv. editorial bd. Chem. Physics Letters, 1966—. Research, publs. in electron-spin resonance spectroscopy of organic free radicals with a view to determination of molecular structure and electronic wave functions, relation between the latter and experimentally observed spectra, theories of irreversible statis. mechanics are being employed to interpret variations in spectral linewidths in terms of dynamical inter- and intramolecular interactions. Home: 460 E. 79th St., N.Y.C. 10021.*

FRAENKEL, Gottfried Samuel, biologist; b. Munich, Germany, Apr. 23, 1901; s. Emil and Flora (Weil) F.; came to U. S., 1948, naturalized 1953; Ph.D. in Zoology, U. Munich, 1925; m. Rachel Sobol, Dec. 15, 1928; children—Gideon, Dan. Asst., Hebrew U., Jerusalem, 1928-30; privatdozent U. Frankfurt (Germany), 1931-33; lectr. Imperial Coll., London, Eng., 1935-48; prof. entomology U. Ill., Urbana, 1948—. Mem. Nat. Acad. Scis. Author: (with D. L. Gunn) The Orientation of Animals, 1940, 61. Research on insect physiology, nutrition, cuticle; discovered carnitine vitamin function (1948-52) and insect hormones: ecdyson (1935), busicon (1962). Home: 606 W. Oregon St., Urbana, Ill. 61801.*

FRAENKEL-CONRAT, Heinz, biochemist; b. Breslau, Germany, July 29, 1910; s. Ludwig and Lili (Conrat) Fraenkel; M.D., U. Breslau 1933; Ph.D., U. Edinburgh, 1936; m. Jane Operman, July 15, 1938; children—Richard L., Charles E.; m. 2d, Beatrice Brandon Singer, June 1, 1964. Came to U. S., 1936, naturalized, 1942. With virus lab. dept. virology, dept. molecular biology U. Cal., Berkeley, 1952—, now prof. molecular biology. Recipient Lasker award, Cal. Scientist of Year, 1958. Mem. Am. Soc. Biol. Chemists, Am. Chem. Soc., Biochem. Soc., A.A.A.S. Author: Design and Function at the Threshold of Life, 1962. Editor: The Molecular Basis of Virology, 1968. Studies, numerous publs. on structure and mechanism of biol. function of proteins, nucleic acids, reconstn. of active tobacco mosaic virus, chem. modification of proteins and nucleic acids, mutagenesis. Home: 1230 Brewster Dr., El Cerrito, Cal. 94530.*

FRAENKEN, Karl, see Fränkel, Karl.

FRAENZ, Kurt Otto Max, German physicist, engr.; b. Berlin, Germany, Feb. 2, 1912; s. Georg and Alwine (Stiebert) F.; Dr.phil., U. Berlin, 1935; Dr.-Ing.-Habil, Technische Hochschule, Charlottenburg, Germany, 1943; m. Ilse Gotthold, Dec. 30, 1937. Staff, Heinrich Hertz-Institut, 1933-36; physicist, engr. Telfunken Co., 1936-48; prof. Buenos Aires (Argentina) U., 1949-56; staff Fabricas Militares and Atomic Energy Commn., Buenos Aires, 1948-56; staff Telefunken Research Inst., 1956—, dir. Telefunken Research Inst., 1964—; hon. prof. Technische Hochschule, München, Germany, 1961—. Mem. German Atomic Energy Commn. and Space Research Commn., 1958—. Fellow I.E.E.E.; mem. Nachrichtentechnische Gesellschaft, Deutsche Physikalische Gesellschaft, Phys. Soc. London. Author: (with H. Lassen) Antennas and Propagation, 1956; also numerous articles. Research on ionosphere, antennas, circuits, noise in circuits, radar, nuclear instrumentation, information theory. Home: 90 Weinbergweg 79 ULM/Do. Fr-Germany. Office: 3 Elisabethenstrasse, 79 ULM/Do., FR-Germany.*

FRAHN, W(ilhelm) E(berhard), physicist; b. Duisburg, Germany, Oct. 5, 1926; s. Gerhard Wilhelm and Johanna (van Thiell) F.; Diplom-Physiker, Aachen Inst. Tech., 1951, Dr.rer.nat., 1953; m. Katharine Elisabeth Trumpfheller, Aug. 22, 1953; children—Vera Angela, Carina Valerie, Diana Rita. Staff, Council for Sci. and Indsl. Research, Pretoria, S. Africa, 1955-60; prin. research officer, 1959-60; prof. physics U. Stellenbosch, 1960-63; prof. theoretical physics U. Cape Town, 1964—. Mem. Am., German phys. socs., S. African Inst. Physics. Research, publs. on theory of scattering and reactions between atomic nuclei and fundamental particles (nonlocal interaction, strong absorption model). Home: April Cottage, Pearl Rise, Somerset West, S. Africa. Office: Physics Dept., U. Cape Town, Rondebosch, S. Africa.*

FRAISSE, Paul, French psychologist; b. Saint-Etienne, France, Mar. 20, 1911; s. Jean Pierre and Clotilde Marie (Lyard) F.; Licence de Philosophie, U. Grenoble (France), 1935; Ph.D., U. Louvain (Belgium), 1945; Docteur es Lettres, Paris (France) U., 1956; m. Simone Berthe Bitry, Aug. 28, 1943; children—Jean-Marie, Claire Renée, Geneviève Marcelle, Agnès Thérèse. Attaché, C.N.E.S. (Nat. Center for Sci. Research), 1937-39; asst. dir. lab. exptl. and comparative psychology École Pratique des Hautes Études, 1943-52, dir., 1952—; prof. Sorbonne, Paris, 1957—, dir. Inst. Psychology, 1961—, Co-dir. L'Anée psychologique, 1947-65, dir., 1965—. Decorated Officer des palmes académiques, Prix Paul Pelliot. Mem. French (past sec., pres.), Spanish, Swiss socs. psychology, Internat. Union Psych. Sci. (pres.). Author: Les structures rythmiques, 1956; Psychologie du temps, 1957; Manuel pratique de psychologi expérimentale, 1956. Editor: (with Jean Piaget) Traité de psychologie expérimentale, 1963. Research, numerous publs. on human motor rhythms, perception time, criteria of estimation of duration in children, duration of perceptual processes, threshold of perpetual recognition in function of age and nature of stimuli, verbal reaction time; demonstrated existence of laws of structuration of human motor rhythms.; analysis of aptitudes for rhythm. Home: 19 rue d'Antony Chatenay Malabry 92, France. Office: 28 Rue Serpente, Paris 75, France.*

FRAISSÉ, Roland, French mathematician; b. Bressuire, France, Mar. 12, 1920; s. Ernest-Emil and Jeanne (Gay) F.; licencé Faculté des Scis. de Toulouse (France), 1937; agrégation U. Paris, 1951, doctorat d'État, 1953; m. Christiane Claparede, Oct. 31, 1964; children—Michele (from previous marriage), Alain, Henri. Aero. researcher ONERA, Toulouse, also Paris, 1944-47; asst. Faculté des Scis., Algiers, Algeria, 1947-48, maitre de confs., 1956-60, prof., 1960-62; researcher CNRS, Algiers, 1949-56; tchr. agrégé Lycée Bageand, Algiers, 1951-52; prof. Faculté des Scis., Lyons, France, 1962-64, Marseilles, France, 1964—; chargé de cours Faculté des Scis. Paris, 1962—. Mem. Assn. for Symbolic Logic, U. S., Math. Soc. France. Author: Cours de logique mathématique, 1967; also articles. Research on math. logic, logical algebra, method of local isomorphisms on relations and multirelations. Home: Chateau Sec Bat. A Joseph Aiguier, Marseille 9, France.*

FRALICK, Francis Bruce, Am. physician; b. Northport, Mich., Apr. 8, 1903; s. Francis John and Elizabeth (Bruce) F.; M.D., U. Mich., 1927; M.S. in Ophthalmology, U. Mich., 1935; postgrad., U. Chgo., 1931-32; m. Mary Ellen Appleton, June 8, 1929; children—Margaret (Mrs. Richard Lee Moehl), Elizabeth (Mrs. Geza Leslie Gyorey), Marion (Mrs. Marion Fralick Benz), Martha (Mrs. James Michael Cleary). Instr., U. Hosp., Ann Arbor, Mich., 1929-31, U. Chgo., 1921-32; faculty U. Mich. Med. Sch., Ann Arbor, 1932—, prof., chmn. dept. ophthalmology, 1938—. Cons. to hosps. Diplomate Am. Bd. Ophthalmology (past chmn., cons. 1960-67). Mem. Am. Acad. Ophthalmology and Otolaryngology, A.C.S., Am., Mich. ophthal. socs., A.M.A., Am. Assn. Ophthalmologists, Mich. State Med. Soc. Research and publs. on clin. applications of advances in ophthalmology, anatomy, physiology and pathology. Home: 11332 Algonquin, Pinckney, Mich. 48169. Office: U. Mich. Med. Center, Outpatient Bldg., Ann Arbor, Mich. 48104.

FRAME, J(ames) Sutherland, Am. mathematician; b. N.Y. City, Dec. 24, 1907; s. James Everett and Jean Herring (Loomis) F.; student Ecole Nouvelle, Lausanne, 1921-22, Horace Mann Sch., N.Y. City, 1922-23, Loomis Sch., Windsor, Conn., 1923-25; A.B. summa cum laude, Harvard, 1929, A.M., 1930, Ph.D., 1933; Rogers traveling fellow from Harvard, univs. of Göttingen and Zurich, 1933-34; m. Emily Bogert Boyce, June 25, 1938; children—Barbara Boyce, Paul Sutherland, Roger Everett, Lawrence Henry. Instr. in mathematics Harvard, 1930-33; instr. Brown U., 1934-38, asst. prof., 1938-42, adviser to freshmen, 1936-37, mem. bd. counselors, 1937-42; asso. prof. and head math. dept. Allegheny Coll., 1942-43; prof. math. Mich. State U., 1943-63, prof. math. and engrg. research, 1963—, head of dept., 1943-60; mem. Inst. Advanced Study, Princeton, 1950-51; project dir. Conf. Bd. of Mathematical Sci., 1961-62. Mem. E. Lansing Bd. Edn., 1948-52. Fellow A.A.A.S.; mem. Mich. Edn. Assn., Nat. Council Tchrs. Mathematics, Am. Math. Soc., Math. Assn. of Am. (governor 1950-

52, 58-60; vis. lecturer 1960), American Association University Professors (mem. nat. council 1948-51), Michigan Acad. Sci. Arts and Letters (pres. 1958-59), Phi Beta Kappa Associates, Phi Beta Kappa, Sigma Xi. Author: General Mathematics (with C. H. Currier and E. E. Watson), 1939; Solid Geometry, 1948; Buildings and Facilities for the Mathematical Sciences, 1963. Asso. editor Am. Math. Monthly, 1942-46, Pi Mu Epsilon Jour., 1949-57. Abstractor, Math. Reviews 1940—. Contributor articles to sci. jours. Researcher in theory of representations of finite groups, continued fractions, approximate computations, matrix theory in systems analysis. Inventor trimetric ruler for space drawings, 1943. Home: 136 Oakland Dr., East Lansing, Mich. 48823. Office: Math. Dept., Mich. State U., E. Lansing, Mich., 48823.*

FRAMPTON, Vernon Lachenous, Am. chemist; b. Springer, N.M., Jan. 29, 1906; s. George Vernon and Kathlene Johana (Deacy) F.; A.B., U. Cal. at Los Angeles, 1931; Ph.D., U. Minn., 1936; m. Martha Elizabeth Railsback, June 25, 1930; children—Robert Vernon, Diana Lee, Jerry Lyn. Chemist, Destruxol Corp., Los Angeles, 1930-31; jr. chemist U. S. Dept. Agr., Riverside, Cal., 1931-33, head, protein products investigations So. Regional Research Lab., New Orleans, 1954—; faculty Cornell U., Ithaca, N.Y., 1936-44; exchange prof. chemistry Nat. U. Colombia, Bogota, 1942; chemist Nat. Cotton Council, Dallas, 1944-45; dir. basic cotton research lab. U. Tex., Austin, 1945-54. Cons., Indian Council for Sci. and Indsl. Research, UNICEF, 1957—. NRC fellow, 1936. Mem. Research Soc. Am., Am. Chem. Soc., A.A.A.S., Am. Oil Chemists Soc., Am. Assn. Cereal Chemists, Am. Soc. Plant Physiologists, Sigma Xi, Gamma Alpha, Phi Lambda Upsilon. Contributed to ecology of spread of insect transmitted plant virus diseases; chem. nature of weathering of cellulose by ultra-violet rays of sun; discovery of chemotactic responses in leaf sucking insects, hemostasis activity in peanuts, chem. nature of impairment of nutritive quality of proteins by heat; devel. quantitative analytical method for avilable lysine. Home: 247 Bonnabel Blvd., Metairie, La. 70005. Office: 1100 Robert E. Lee Blvd., New Orleans 70002.*

FRANCESCONI, Daniele, Italian physicist; b. Belvedere di Cardignano, Italy, Mar. 1, 1761; became prof. geometry and physics Collegio San Marco, Padua, Italy, 1793; named prof. U. Padua, 1800. Author: Sopra la corrispondenza degli angoli d'incidenza e di refractione della luce, 1786; Sopra la questione della conservazione della quantità del moto nell' urto, 1807; Sulla teoria della impressioni, ossia resistenze dei solidi, 1808; Prodromo di un teoria della resistenza dei corpi molli, 1809; Sulla velocità della luce, 1814; Sulla velocità degla elastri, 1832. Made physico-math. studies of falling bodies, resistance of elastic bodies, velocity of light, and other subjects. Died Venice, Italy, Nov. 27, 1835.

FRANCHESCHETTI, Adolphe, Swiss ophthalmologist; b. Zurich, Switzerland, Nov. 10, 1896; s. Adolfo and Bertha (Spitzer) F.; student U. Zurich; D. honoris causa, U. Ghent (Belgium), 1963, U. Toulouse (France), 1965, U. Heidelberg (Germany), 1966; m. Antonia Saggiani, July 7, 1937; children—Marguerite (Mrs. Luc Tissot), Albert Th. Asst. ophthalmology, Zurich, 1921-25; chief asst. Ophthal. Clinic, Basle, Switzerland, 1925-33; prof. ophthalmology U. Geneva (Switzerland), 1933-66, hon. prof., 1966—. Hon. fellow Am. Acad. Ophthalmology and Otolaryngology; hon. mem. Swiss, Italian, Austrian, German ophthal. socs., Swiss Acad. Medicine (pres. 1964—), Internat. Assn. for Prevention Blindness (pres. 1954—). Author: (with J. Francois, J. Babel) Les Hérédo-dégénérescences chorio-rétiniennes, 2 vols., 1963; also numerous articles. Research on corneal grafting, clin. relations between eye affections and gen. diseases, heredity in ophthalmology. Home: Le Grand Mancy, Vésenaz (Geneva) 1222. Office: 3, Av. de Miremont, Geneva 1206, Switzerland.*

FRANCHIMONT, Antoine Paul, chemist; b. Leyden, Netherlands, May 10, 1844; became prof. organic chemistry U. Leyden, 1874; discovered (with Kekulé) anthvaquinone and triphenylmethane. Died Leyden, July 2, 1919.

FRANCIS, David Wesson, Am. avian physiologist; b. N.Y.C., Aug. 17, 1918; s. William Winterbottom and Marjorie (Wesson) F.; B.S., Rutgers U., 1941; M.S., U. Del., 1952; Ph.D., U. Md., 1955; m. Marian Louise Reid, Aug. 31, 1950; children—Robert William, Laura Catherine, David Wesson. Grad. asst. U. Md., 1953-55; asso. prof. N.M. State U., Las Cruces 1955-58, prof., head dept. poultry sci., 1958—. Mem. Poultry Sci. Assn., World's Poultry Sci. Assn., U. S. Livestock San. Assn., Wildlife Disease Assn., Am. Assn. Avian Pathologists, A.A.A.S., Am. Assn. U. Profs., Sigma Xi. Contbr. numerous articles to profl. jours. Research in physiology, pathology and mgmt. Home: 2105 Gladys Dr., Las Cruces, N.M. 88001.*

FRANCIS, Edward, Am. bacteriologist; b. Shandon, O., Mar. 27, 1872; s. Abner and Martha Ann (Vaughan) Francis; B.S., Ohio State U., 1894; hon. D.Sc., 1933; M.D., U. Cin., 1897; LL.D., Miami U., 1929. With USPHS, from 1900, surgeon, 1913-30, med. dir. from 1930. Mem. A.M.A., Assn. Am. Physicians, Assn. Mil. Surgeons, Phi Delta Theta, Sigma Xi. Recipient Gold medal Am. Med. Assn. for contbn. to knowledge of tularaemia, 1928. Author: Tularaemia Francis, 1921, a New Disease of Man, (bull.), 1922; also bulls. and papers on yellow fever, pellagra, tetanus, filariasis, rat-bite fever, undulant fever, relapsing fever, athlete's foot and tularemia. Died 1957.

FRANCIS, Frederick John, food scientist; b. Ottawa, Ont., Can., Oct. 9, 1921; s. Frederick Roland and Mary (Dyble) F.; B.A., U. Toronto, 1946, M.A., 1948; Ph.D., U. Mass., 1954; m. Jean Dalton Burrows, Mar. 15, 1952; children—Margaret Ann, John Burrows, Laurie Jean. Came to U. S., 1954, naturalized, 1964. Instr. food sci. U. Toronto, 1946-50; lectr. hort. U. Guelph, Can., 1950-54; asst. prof. U. Mass., 1954-58, asso. prof., 1958-62, prof., 1962-64, Nicolas Appert prof., 1964—. Mem. Inst. Food Technologists, Am. Inst. Biol. Scis., Am. Chem. Soc., U. S. Army Research and Devel. Assn., Inter Soc. Color Council, A.A.A.S., Sigma Xi. Studies and numerous publs. on pigment biochemistry, color measurement, high temperature short-time thermal processing, general physiology and biochemistry. Home: 123 Pine St., North Amherst, Mass. 01059. Office: U. Mass., Amherst, Mass. 01003.*

FRANCIS, James Bicheno, hydraulic engr.; b. Southleigh, Eng., May 18, 1815; s. John and Eliza (Bicheno) F.; m. Sarah Brownell, July 12, 1837; 6 children. Arrived in N.Y.C., 1833; chief engr. group known as Proprs. of Locks & Canals on the Merrimack River, 1837-1840; chief engr., gen. mgr. devel. waterpower facilities, Lowell, Mass., 1845; began constrn. of No. Canal, 1846; following trip to Eng., 1849, built machines for timber preservation; devised water supply for fire protection, Lowell, 1850's; designed and constructed hydraulic lifts for guard gates of Pawtucket Canal, 1870; original mem. Boston Soc. Civil Engrs., pres. 1874, 80; pres. Stonybrook R.R., 20 years; dir. Lowell Gas Light Co., 43 years. Author: Lowell Hydraulic Experiments, 1855, 68, 83; Strength of Cast iron Columns, 1865. Died Boston, Sept. 18, 1892.

FRANCIS, John Wakefield, Am. physician; b. N.Y.C., Nov. 17, 1789; s. Melchior Francis; grad. Columbia, 1809, LL.D., 1860; M.D., Coll. Phys. and Surg. (1st grad. of coll.), N.Y.C., 1811; LL.D., Trinity Coll., Hartford, Conn., 1850; m. Mary Eliza Cutler, Nov. 16, 1829; 1 son, Samuel Ward. Apptd. lectr. in medicine and materia medica Coll. Phys. and Surg.; prof. medicine and materia medica, med. dept. Columbia, prof. forensic medicine, 1817, prof. obstetrics, 1819-26; prof. obstetrics Rutgers Med. Sch., 1826-28; founder N.Y. Acad. Medicine, 1846, pres., 1847-48; financially responsible for establishment Woman's Hosp. Author: Use of Mercury, 1811; Introduction to the Practice of Medicine, 1821; Denmans' Practice of Midwifery, 1825. Editor (with Hosack) Am. Med. and Philos. Register, 1810-14. Died N.Y.C., Feb. 8, 1861.

FRANCIS, Joseph, Am. inventor, mfr.; b. Boston, Mar. 12, 1801; s. Thomas and Margaret F.; m. Ellen Creamer. Produced a wooden boat that withstood severest tests; constructed life boats for U. S. vessels Santee and Alabama, 1829; all U. S. Govt. ships equipped with boats of his invention by 1841; contracted with Novelty Iron Works of N.Y. to manufacture the boats; granted patent for corrugated metal boat, 1845; constructed fleet of light-draft corrugated iron steamers for Russian govt., sometime between 1855-63. Recipient medal Franklin Inst., 1854, Gold medal King Ferdinand III of Sicily, gold snuffbox from Napoleon III, 1856, Congressional medal presented by Pres. Harrison, 1890; named to Royal Order Knighthood St. Stanislaus by Tsar of Russia. Died Cooperstown, N.Y., May 10, 1893.

FRANCIS, Mark, Am. veterinarian; b. Shandon, O., Mar. 19, 1863; s. Abner and Martha Ann (Vaughan) F.; D.V.M., Ohio State U., 1887; 1 yr. at Am. Vet. Coll., N.Y.; postgrad. U. Mich.; postgrad. Berlin and Munich, 1904; LL.D., Miami U., 1929; m. Anna J. Scott, Sept. 10, 1890; children—Andrew Jones, William Bebb. Prof. vet. sci. A. and M. Coll. of Tex., from 1888, dean Sch. Vet. Medicine, 1916—. Veterinarian to Tex. Expt. Sta. Introduced methods of producing immunity to Tex. fever by subcutaneous injections with infected cattle blood. Died June 28, 1936.

FRANCIS, Samuel Ward, Am. physician, inventor; b. N.Y.C., Dec. 26, 1835; s. John W. and Maria (Cutler) F.; B.A., Columbia, 1857; M.D., U. City N.Y., 1860; m. Harriet McAllister, June 16, 1859. Patented printing machine which anticipated typewriter, 1860; invented heating and ventilating device for railroad cars, 1868; developed a sewing machine, 1875; devised signal for telephone and telegraph lines, 1879; active in San. Protection Assn. (founded 1878). Author: Report of Valentine Motts' Surgical Cliniques in the University of New York, 1859-60, pub. N.Y., 1860; Inside and Out, 1862; Biographical Sketches of Distinguished Living New York Surgeons, 1866; Life and Death, 1871. Died Newport, R.I., Mar. 25, 1886.

FRANCIS, Thomas, Jr., Am. epidemiologist, physician; b. Gas City, Ind., July 15, 1900; s. Thomas and Elizabeth Ann (Cadogan) F.; B.S., Allegheny Coll., 1921; M.D., Yale, 1925, M.S. (hon.), 1941; Sc.D. (hon.), Allegheny Coll., 1941; m. Dorothy Packard Otton, June 29, 1933; children—Mary Jane, Thomas, 3d. Intern in medicine New Haven (Conn.) Hosp., 1925-26, prof. bacteriology, dir. bacteriology labs., coll. medicine N.Y. U., vis. physician Third med. div. Bellevue Hosp., 1938-41; vis. physician Willard Parker Hosp., 1940-41; Henry Sewall U. prof. epidemiology and chmn. dept. epidemiology, sch. pub. health also prof. epidemiology, med. school U. Mich. 1941—. Mem. Armed Forces Epidemiology Bd. 1955— (pres., 1958-60), dir. Influenza Commn. 1941-55; cons. Sec. Def., USPHS, Mich. State Dept. Health. Mem. Sci. adv. council Am. Cancer Soc.; dir. Poliomyelitis Vaccine Evaluation Program; bd. sci. advisers Jane Coffin Childs Meml. Fund for Med. Research. Mem. Lobund Advisory Bd., 1958-65. Served with S.A.T.C., 1918. Awarded Medal of Freedom, AUS, 1946; Lasker award Am. Public Health Assn., 1947; Howard Taylor Ricketts award and medal, U. Chgo., 1952; James D. Bruce Meml. medal, A.C.P., 1953. Fellow A.A.A.S., Am. Pub. Health Assn. (mem. governing council), Am. Acad. Arts and Scis.; mem. Nat. Acad. Scis. (mem. governing council 1958-61), Soc. Am. Bacteriologists (pres. 1947; hon. mem.), Am. Soc. Clin. Investigation (pres. 1945-46), Assn. Am. Physicians, A.M.A., Harvey Soc. (sec. 1938-40), N.Y. Acad. Medicine, History Sci. Soc., Am. Soc. Immunologists, Am. Epidemiol. Soc. (pres. 1954-55), Am. Philos. Soc., Constantinian Soc., N.Y. Acad. Sci., Soc. Exptl. Pathology, Soc. Exptl. Biology and Medicine, Nu Sigma Nu, Sigma Xi, Alpha Omega Alpha, Delta Omega, Phi Kappa Phi. Studies in infectious disease, especially pneumonia, poliomyelitis, influenza; isolated diverse influenza viruses; demonstrated effectiveness of Salk vaccine in nation-wide tests. Home: 1178 Heather Way, Office: School of Public Health, University of Michigan, Ann Arbor.

FRANCK, James, physicist; b. Hamburg, Germany, Aug. 26, 1882; s. Jacob and Rebecca (Drucker) F.; student U. Heidelberg, 1901-02; Dr. of Phil., U. Berlin, 1906; LL.D., U. Cal., Berkeley, 1928, Sc.D., Technion, Haifa, Israel, 1954; Dr. rer. nat., U. Heidelberg, 1957; Sc.D., Humboldt U., Berlin, 1960, U. Kiel (Germany), 1960, U. Giessen (Germany), 1962, Gustavus Adolphus Coll., 1963; m. Ingrid Josephson, Dec. 1906 (dec. 1942); children—Dagmar (Mrs. Arthur Von Hippel), Elisabeth (Mrs. Herman Lisco); m. Hertha Sponer, June 29, 1946. Came to U. S., 1935. Asst. in phys. lab. U. Berlin, 1906, pvt. docent 1911-16, asso. prof. physics, 1916-18; head physics division Kaiser Wilhelm Inst. Phys. Chemistry, 1918-20; prof., dir. Phys. Inst., U. Göttingen, 1920-33, now prof. emeritus; guest prof., U. Copenhagen, 1934; prof. physics, Johns Hopkins U., 1935-38; prof. physical chemistry U. Chgo., 1938-47; later prof. emeritus; Hitchcock U. Cal., 1941; during World War II worked on devel. of atom bomb in U. S. Recipient (with Gustav Hertz) Nobel prize in physics (for discovery of laws governing the impact of an electron upon an atom), 1925; awarded Max Planck medal German Phys. Soc., 1953; Rumford medal Am. Acad. Arts and Scis., 1955. Mem. Research Insts. U. Chgo., Nat. Acad. Scis., Am. Philos. Soc., Acad. of Arts and Scis. in Boston, Washington Acad. of Scis., and several European academies and other learned societies. Author: Anregung von Quantensprüngen, 1926; also scientific articles on molecular physics and its application to chemistry. Confirmed Bohr's theory of the atom's energy stages in mercury vapor (with Hertz); discovered energy transmission in atom systems in fluorescence; established principle of constancy of atom distances by electron jumps; studied photochem. processes in chlorophyll molecule. Died Göttingen, May 21, 1964.

FRANCK, (Johann) Ludwig, German veterinarian; b. Mogger, nr. Mupperg, Thuringia, Germany, Mar. 7, 1834; s. Friedich Carl and Katharina Margarete (Hein) F.; ed. Munich (Germany) sch. vet. medicine (now part of Munich U.), to 1854; hon. doctorate U. Munich; m. Elise Klementine Semmel, 1869; 2 daus. Served as state or mil. veterinarian, 10 years; prof. Munich sch. vet. medicine, from 1864, dir., from 1877; tchr., agrl. dept. Tech. U. Munich, from 1872. Mem. numerous vet., med. and agrl. socs. Author: Handbuch der Anatomie der Haustiere mit besonderer Berücksichtigung des Pferdes, 1871; Handbuch der tierärztlichen Geburtshilfe, 1876. Research in anatomy of domestic animals, various other aspects of vet. medicine; pioneer in animal obstetrics. Died Munich, Apr. 4, 1884.

FRANCO, Gianfranco, Italian engr.; b. Padova, Italy, Jan. 4, 1922; s. Giovanni and Fausta (Scalco) F.; Degree in Naval-Mech. Engring., U. Naples (Italy), 1947; m. Emilia Palmegiano, Mar. 29, 1948; children—Giovanni, Lorenzo, Davide. Engring. group leader CISE (Nuclear Research Center), Milan, Italy, 1954-56; chief engr. ISPRA I reactor CISE and CNRN (now CNEN), 1956-59; dir. Ispra, Varese, Italy, 1959-60; chief engr. Dragon Project, Winfrith Heath, Eng., 1960-64; dir. CSN Casaccia, Rome, 1964—. Mem. tech. bd. Italian Nuclear Ship, 1966—. Recipient Commendatore al Merito, Republic of Italy, 1965. Mem. Italian Thermotech. Assn. (com. for nuclear plants 1959—), Italian Nat. Assn. Nuclear Engring., Ordine degli Ingegneri. Author: Les Réacteurs à Thorium, vol. 2 ency. Le Progress Scientifique; also articles. Patentee nuclear instrumentation and fuel elements charging-discharging system; research on techs. and heat exchange problems in nuclear fuel elements of nuclear reactors engring. Home: 126 Via Orti della Farmesina, Rome. Office: Km. 1-300 Strada Anguillarese, Rome, Italy.*

FRANCO, Pierre, French physician; b. Turriges, France, 1500; practice medicine Provence, France, Fribourg, Germany, Lausanne, Switzerland, Bern, Switzerland, Orange, France; tchr. anatomy U. Fribourg, Lausanne. Author: Petit traité, 1556; Traité des hernies, 1561. Invented 3-valved device to grasp the head of fetus for delivery; introduced method of suprapublic lithotomy (Franco's operation), 1556; 1st description of operation for relief of strangulated hernia, 1556. Died 1561.

FRANCOEUR, Louis-Benjamin, French mathematician, astronomer; b. Paris, Aug. 16, 1773; s. Louis-Joseph Francoeur; ed. Harcourt Coll., also Navarre Coll., Paris, l'École polytechnique. Became prof. math. Charlemagne Lycée, 1804; prof. algebra Paris Faculty Scis., from 1809. Decorated Legion of Honor. Mem. French Acad. Scis., 1842, Soc. Agr., Soc. for Encouragement Nat. Industry. Author: Traité de mécanique, 1803; Enseignement du dessin lineaire, 1820; Goniométrie, 1820; Géodésie, 1835; Cours complet de mathematiques pures, 1837; Astronomie pratique, 1840; Mémoire sur l'aréométrie, 1842; Calendrier, 1843; Traité d'arithmétique appliquée à la banque, au commerce, à l'industrie . . . , 1845; Uranographie, pub. 1853; also elementary treatise on astronomy, 1812. Specialized in pure math.; introduced teaching of linear design in pub. schs.; studied flora of Paris region; readability of his math. works led to their wide circulation. Died Paris, Dec. 15, 1849.

FRANÇOIS, Jean-Charles, French inventor, engraver; b. Nancy, France, Mar. 4, 1717; royal engraver France, from 1740. Decorated Graveur des dessins du Cabinet dur Roi; inventor engraving technique from a type of pencil sketch, 1740. Died Paris, France, Mar. 21, 1769.

FRANÇOIS-FRANCK, Charles-Albert, French physician; b. 1849; ed. Collège de France; asst. dir. lab. physiology, prof. physiology Collège de France. Research on normal and path. physiology, blood circulation including cardiac lesions, valvulary affections, vasomotor nerves, also functioning of sympathetic system. Died Paris, 1921.

FRANCOTTE, Charles-Polydore, Belgian naturalist; b. Wavre, Belgium, Nov. 21, 1851; ed. under Van Benedon; Docteur ès scis. naturelles, U. Liège (Belgium); became prof. Namur (Belgium) Athenaeum; prof. animal embryology Brussels (Belgium) Faculty Scis. Mem. French Acad. Scis., 1910, Acad. Scis. Belgium. Author: Traité de technique microscopique. Studied embryology, especially that of lizard and blindworm, also maturation and fertilization of turbellaria egg. Died Saint-Josse-tenNoode, Belgium, Apr. 21, 1916.

FRANK, Adolph, German chemist; b. Klötze/Altmark, Germany, Jan. 20, 1834; s. Salomon and Ulrike (Wolffstein) F.; student, Berlin, 1855-57; Ph.D., Göttingen, Germany, 1862; dr.engring. (hon.) Dresden, Germany; m. Meta Warburg, 1866; 2 sons, Paul, Albert, 1 dau. Became pharmacy apprentice, 1848; chemist Bennecke, Hecker & Co., Stassfurt, Germany, 1857-60; opened potassium salt factory, Stassfurt, 1861; became gen. dir. Vereinigte Chemische Fabriken, Leopoldshall, Germany, 1872; became tech. dir. bottle factory, Charlottenburg, Germany, 1876; cons. chemist, civil engr., from 1885; apptd. adviser Nobel-Dynamit-AG, Gillwärder, Germany, 1896; founder (with other cos.) Cyanid-Gesellschaft, 1898. Recipient Liebig Meml. medal. Mem. German Acetylene Assn. (co-founder, chmn.), Assn. German Engrs. (hon.), Assn. German chemists (hon.), Assn. for Promotion of Industry (hon.). Developed method for extracting potassium salts from local abraum-salts for use in soil fertilization; introduced brown beer bottles for light and heat protection; introduced Thomas basic slag for artificial fertilizer; recognized value of peat as heat source; promoted devel. of German cellulose and acetylene industries; inventor processes, also promoter industry of potassium nitrate in Germany; experimented (with H. Caro and C. von Linde) on prodn. of hydrogen gas for airships. Died Berlin-Charlottenburg, May 30, 1916.

FRANK, Albert Bernhard, German biologist, botanist; b. Dresden, Germany, Jan. 17, 1839; s. Franz Bernhard and Caroline Agnes (Wolfram) F.; Ph.D., Leipzig, Germany, 1865; 1 dau. Curator univ. herbarium, Leipzig, became mem. faculty, 1867, asso. prof., 1878; named prof. plant physiology Agrl. U. Berlin Inst. for Plant Physiology and Plant Protection, 1881; head new biol. dept. for agr. and forestry State Bd. Health, from 1899. Author: Über die Entstehung der Intercellularräume der Pflanzen, 1867; Beitrage zur Pflanzenphysiologie, 1868; Pflanzentabellen zur . . . Bestimmung der hoheren Gewächse Nord- und Mitteldeutschlands, 1869; Die Krankheiten der Pflanze, 1880; Grundzüge der Pflanzenphysiology, 1882; Über die Pilzsymbiose der Leguminosen, 1890; Lehrbuch der Pflanzenphysiologie, 1890; (with P. Sorauer) Pflanzenschutz . . . , 1892; Lehrbuch der Botanik . . . , 2 vols., 1892; Pflanzenkunder für niedere und mittlere Landwirtschaftsschulen, 1894; Kamfbuch gegen die Schädlinge unserer Feldfruchte, 1897; (with F. Krüger) Schildlausbuch, 1900; also many articles. Research on plant physiology, transversalgeotropism, heliotropism, also plant diseases of sweet cherry, beets, grains, potatoes; expert on plant protection; demonstrated that mycorrhiza fungus symbiotes are necessary for germination of some plants. Died Berlin, Sept. 27, 1900.

FRANK, Andreas, Austrian physician; b. Petzeukirchen, Austria, Dec. 5, 1904; s. Andreas and Pauline (Vollgruber) F.; M.D., U. Vienna (Austria), 1929; m. Erika Welzenbacher, Mar. 1958; children—Andreas, Heidi, Christian, Albert. Staff radiology dept. Sophien Hosp., Vienna, 1934-54; head radiology Rudolf Hosp., Vienna, 1954—; faculty radiology U. Vienna, 1959—. Mem. Congress for Radiology in Vienna (past pres.), Soc. Physicians in Vienna, German, Austrian (past pres.) socs. radiology. Research, publs. on gastroenterology; pioneered handling of tumors with radiation, removal of stones from ureter with radiation; research on reaction skin to radiation. Home: Mühlbachergasse 7, Vienna 13. Office: Maria-hilferstrasse iD, Vienna 6, Austria.*

FRANK, Evelyn, Am. research mathematician, educator; b. Chgo.; Ph.D., Northwestern U. Instr., Northwestern U., Evanston, Ill., 1942-46; prof. math. U. Ill., Chgo., 1946—. Mem. Am. Math. Soc., Math. Assn. Am., Soc. for Indsl. and Applied Math., Phi Beta Kappa, Sigma Xi. Research, publs. on spl. fields of analysis, continued fractions, spl. functions, stability theory, zeros of polynomials, number theory, numerical analysis. Address: P.O. Box 361, Evanston, Ill.*

FRANK, George Barry, pharmacologist; b. Bklyn., Feb. 1, 1929; s. Max L. and Bertha (Handelman) F.; B.S., Coll. City N.Y., 1950; M.S., Ohio State U., 1952; Ph.D., McGill U., Montreal, Que., Can., 1956; m. Rheva Adelson, Sept. 18, 1951; children—Murray, Albert, Tema. Spl. lectr. physiology, USPHS fellow McGill U., 1956-57; faculty U. Man. (Can.), 1957-65, prof. pharmacology and therapeutics, 1965; prof. pharmacology U. Alta. (Can.), Edmonton, 1965—. USPHS Spl. fellow U. Coll. London, Lund U., Sweden, 1963-64. Mem. A.A.A.S., Am., Canadian physiol. socs., Am. Soc. Pharmacology and Exptl. Therapeutics, Pharmacological Soc. Can., Soc. For Gen. Physiologists, La Sociedad Peruana de Neuro-Psiquiatria. Mem. editorial bd. Jour. Pharmacology and Exptl. Therapeutics, 1962-65; asso. editor: Canadian Jour. Physiology and Pharmacology, 1965—. Research in basic mechanism of action of drugs at level of the cell, mechanism of action of anaesthetic agts. both at cellular level and in central nervous system. Home: 5524 108th St., Edmonton, Alta. Can.*

FRANK, Gleb Mikhaylovich, Russian biophysicist; b. 1904; grad. Simferopol U., 1925; D.Biol. Sci. Asso., Leningrad Physicotech. Inst., later All-Union Inst. Exptl. Medicine, 1929-43; with various labs. USSR Acad. Sci., 1943-52, head lab. biophysics of animate structures, dep. sci. dir. Inst. Biophysics, 1952-60, now dir. Inst. Biophysics; prof. chair biophysics Biopedological Faculty, Moscow U. Recipient Stalin prize, 1951. Mem. USSR Acad. Med. Sci. (corr.), USSR Acad. Sci. (corr.). Co-editor Radiobiology sect. Large Med. Ency., 2d edit. Research and numerous publs. on effects of ultra-violet and ionizing irradiation on animal organisms and biophys. reasons for nervous stimulation and muscular contraction. Address: Inst. Biophysics, Profsoyuznaya ulitsa 7, Moscow, USSR.

FRANK, Henry Sorg, Am. phys. chemist; b. Pitts., Aug. 6, 1902; s. Austin Cleis and Alma (Sorg) F.; B.Chemistry, U. Pitts., 1922, M.S., 1922; Ph.D. (du-Pont and China Med. Bd. fellow), U. Cal. at Berkeley, 1928; m. Martha Elizabeth Griggs, June 15, 1927; children—Austin Cleis, Alice Rebecca (Mrs. Edwin P. Brown), Marian Elizabeth (Mrs. Arnold Zeitlin). Faculty, Lingnan U., Canton, China, 1922-25, 28-51; lectr. chemistry U. Cal., Berkeley, 1942-45; chief China sect. div. cultural cooperation U. S. Dept. State, 1945-46; vis. prof. U. Pitts., 1939-40, prof. chemistry, 1951—, chmn. dept., 1951-63; adj. sr. fellow Mellon Inst., Pitts., 1963—. Research visitor U. Copenhagen, 1933-34, Gottingen, 1957; cons. ICA, Nat. Tsing Hua U., Taiwan, 1959; bd. trustees Lingnan U., N.Y., 1951—. Mem. Am. Chem. Soc. (Pitts. award 1964), A.A.A.S. Theoretical and exptl. studies of liquids and solutions, particularly anomalies of water and aqueous solutions; co-introducer of iceberg and flickering cluster models for water and aqueous solutions. Home: 4 Olympia Pl., Pitts. 15217. Office: care Mellon Inst., 4400 5th Av., Pitts. 15213.*

FRANK, Ilya Mikhailovich, Russian physicist; b. Leningrad, Oct. 23, 1908; s. Mikhail Lyudvigovich and Yelizaveta Mikhailovna (Gratsianova) F.; grad. Moscow U., 1930; D.Physico-Math. Sci.; m. Ella Abramnova Beilikhis, 1937; 1 son, Alexander. With State Optical Inst., 1930-34; asso. Physics Inst., staff USSR Acad. Sci., 1934—, research atomic nucleus physics, now head atomic nucleus physics lab.; dir. lab. neutron physics Joint Inst. Nuclear Research, Dubna; faculty Moscow U., prof., head of chair, 1944—. Decorated Order of Lenin; recipient Stalin prize, 1946, Lenin prize, (with P. A. Cerenkov, I. E. Tamm) Nobel prize in physics for discovery and interpretation of Cherenkov effect, 1958. Mem. USSR Acad. Scis. (corr.). Research and publs. on photoluminescence of solutions, photochemistry, optical dissociation of molecules, phys. optics, nuclear physics; explained nature and developed (with I. E. Tamm) theory of Cherenkov-Vavilov effect. Address: Institute of Physics, USSR Acad. Sci., Moscow, USSR.

FRANK, Jerome David, Am. physician; b. N.Y.C., May 30, 1909; s. Jerome W. and Bess (Rosenbaum) F.; A.B. summa cum laude, Harvard, 1929, A.M., 1932, Ph.D., 1934, M.D. cum laude, 1939; m. Elizabeth Kleeman, Jan. 4, 1948; children—Deborah, David, Julia, Emily. Practice medicine, specializing in psychiatry, Balt., 1940—; faculty Sch. Medicine Johns Hopkins, 1942—, prof. psychiatry, 1959—. Fellow Am. Psychiat. Assn.; mem. Am. Psychol. Assn., Am. Coll. Psychiatrists, Phi Beta Kappa, Sigma Xi, Alpha Omega Alpha. Author: (with Florence Powdermaker) Group Psychotherapy: Studies in Methodology of Research and Therapy, 1953; Persuasion and Healing: A Comparative Study of Psychotherapy, 1961. Studies, numerous publs. on effects of expectations of psychiatric patients on responses to placebos and brief psychotherapy. Home: 603 W. University Pkwy., Balt. 21210. Office: Phipps Clinic, Johns Hopkins Hosp., Balt. 21205.*

FRANK, Johann Peter, physician; b. Baden, Germany, 1745; student Baden, also Metz and Pont-à-Mousson, France; med. degree, Heidelberg, Germany, 1766; Ph.D. Practiced in Bitsch, Lorraine for 2 years; became dist. med. officer, Baden; apptd. physician in ordinary to Margrave of Baden-Baden; later became urban and rural physician in service of Prince-Bishop of Spires, Bruchsal, became physician-in-ordinary, 1775; named prof. practical medicine, Göttingen, Germany, 1784; clin. tchr., chief hosp., Pavia, Italy, from 1785; protophysicus Austrian Lombardy and duchy of Mantua; named dir. Gen. Hosp., Vienna, Austria, 1795; clinician U. Vilna, 1805-06; physician-in-ordinary to Tsar, St. Petersberg, Russia, 1806-09; lived in Freiburg, Germany, 1809-11; practiced in Vienna, 1811-21. Mem. French Acad. Scis. Author: System einer vollständigen medizinischen Polizei, 1779-1817. Pioneered devel. pub. hygiene; reformed teaching and control of midwives; reformed med. faculty at U. Pavia. Died 1821.

FRANK, Karl, neurophysiologist; b. Toronto, Ont., Can., Aug. 5, 1916; s. Leslie Carl and Ethelwyn (Harris) F.; M.S. in Physics, Cornell U., 1941; Ph.D. in Physiology, U. Chgo., 1951; m. Margaret McCoy, Sept. 1, 1941; children—Kathleen (Mrs. Alfred Ochs), Eric, Caroline. Physicist, Naval Ordnance Lab., Washington, 1941-46; biophysicist USPHS Narcotic Hosp., Lexington, Ky., 1946-51; neurophysiologist NIH, Bethesda, Md., 1951-56, head, spinal cord sect. Nat. Inst. Neurol. Diseases and Blindness, 1956-63, acting asso. dir. intramural research, 1963—. Recipient Meritorious Civilian Service award USN, 1945; Superior Service award U. S. Dept. Health, Edn. and Welfare, 1965. Asso. editor Exptl. Neurology, 1960—. Research on basic mechanisms of information transfer from one nerve cell to another. Home: 4628 Chestnut St., Bethesda 20014. Office: 9000 Rockville Pike, Bethesda, Md. 20014.*

FRANK, (Friedrich Wilhelm Ferdinand) Otto, German physiologist; b. Gross-Umstadt, Germany, June 21, 1865; s. Georg and Mathilde (Linderborn) F.; ed. Munich, Kiel, Heidelberg, Glasgow, Munich, Strasbourg, Leipzig; state exam, Munich, 1889; M.D., 1892; m. Theres Schuster; 1 son. Became asst. to Carl Ludwig, Liepzig, 1892, to Carl Voit, Munich, 1894; became mem. faculty Munich, 1895, asso. prof., 1902, successor to Voit, 1908; apptd. prof., Giessen, Germany, 1905; forced by Nazis to retire, 1934. Mem. Bavarian Acad. Scis. Author: Thermodynamik des Muskels, 1904. Research on heart, elasticity of blood vessel walls, pulse theory; demonstrated that heart muscle is similar to skeleton muscles; built manometer with which he registered 1st perfect pulse curves; perfected other measuring instruments. Died Munich, Nov. 12, 1944.

FRANK, Peter Wolfgang, ecologist; b. Mainz, Germany, Sept. 24, 1923; s. Paul M. and Elizabeth (Spiegel) F.; came to U. S., 1937, naturalized, 1945; B.A., Earlham Coll., 1944; Ph.D., U. Chgo., 1951; m. Marian Hawke Burton, July 5, 1946; children—Elizabeth, Russell, Richard. Seessel fellow Yale, 1951-52; asst. prof. zoology U. Mo., Columbia, 1952-57; faculty U. Ore., Eugene, 1957—, prof. biology, 1963—; acting dir. Ore. Inst. Marine Biology, 1960-65. Mem. Ecol. Soc. Am., Am. Soc. Naturalists, Am. Soc. Zoologists, A.A.A.S., Am. Inst. Biol. Scis. Editor; Ecology, 1964—. Research, publs. on competition between species animals, role density in regulating population numbers, significance survivorship patterns. Home: 2009 Elk Dr., Eugene 97403.*

FRANKE, Wilhelm, German chemist; b. Munich, Germany, May 28, 1903; s. Wilhelm and Emilie (Scherer) F.; Ph.D., U. Munich; m. Anne Heiming, Oct. 20, 1942. Asst. at Munich, 1926-31, prof. agrégé, 1934; Rockefeller scholar, Stockholm, 1931-34; asso. prof. at Würzburg, 1941, at Cologne, 1950. Recipient Carl Duisberg prize V.D.C., 1941. Mem. Assn. German Chemists, German Soc. Lipid Scis., Soc. Physiol. Chemistry, German Soc. Hygiene and Microbiology, German Soc. Physics and Medicine. Contbr. articles to profl. publs. Research in biochemistry, particularly of higher plants and microorganisms. Home: Dieringhauserstrasse 18, Cologne-Bruck. Office: Institute for Biochemistry, Gyrhofstrasse 15, Cologne, Germany.

FRÄNKEL, Karl (name changed to Fraenken, 1912), German bacteriologist; b. Berlin-Charlottenburg, Germany, May 2, 1861; s. Maximilian and Marie Mathilde

Auguste (Nolte) Fränkel; studied medicine, Berlin, Heidelberg, Leipzig, Freiburg; Ph.D., Leipzig, 1884; m. Irmgard Frieda Helene Lahs (div.). Became asst. to Robert Koch, hygiene inst. U. Berlin, 1885, mem. faculty, 1888; asso. prof., dir. Hygiene Inst., Königsberg, Germany, 1890; named prof., Marburg, Germany, 1891, Halle, Germany, 1895; ret., 1915. Author: Grundriss der Bakterienkunde, 1886; Studien über die Mikroorganismen in verschiedenen Bodenschichten, 1887; Über das Verhalten der Kohlensäure gegenüber den Mikroorganismen, 1889; also articles. Co-founder jour. Hygienische Rundschau, 1891. Research on biology of Tb and diphtheria bacilli, 1890-1907, cholera vibrios, 1892, 94, gonococcus, 1898, meningococcus intracellularis, 1899, anthrax bacillus, 1901, staphylococcus, 1905, Bordet-Gengou's pertussis bacillus, 1908, also on purification of waste water and drinking water. Died Hamburg, Germany, Dec. 29, 1915.

FRANKEL, Max, organic chemist; b. Czechoslovakia, Oct. 31, 1900; s. Siegfried and Rose (Chajes) F.; student Tech. U., U. Vienna; Dr.phil., U. Vienna (Austria), 1923; m. Helen Hammermann, 1930. Asst., Inst. Chemistry, also Inst. Chem. Tech., Vienna U., 1920-24; sr. research chemist Verein fuer Chemische und Metallurgische Produktion, Aussig, Czechoslovakia, 1924-25; faculty Hebrew U., Jerusalem, Israel, 1925—, prof. organic chemistry, dir. dept., 1953—. Fellow Chem. Soc. London; mem. Am. Chem. Soc., Chem. Soc. Israel. Author numerous books, monographs and articles on topics in chemistry. Research on amino acids, synthesis of polypeptides, polymers and organometallic compounds, sterochemistry. Home: 13 Ramban St., Jerusalem, Israel.*

FRANKEL, Milton Bernard, Am. chemist; b. Lawrenceburg, Ind., Sept. 10, 1919; s. Isaac and Bessie (Dorn) F.; B.A., U. Cin., 1941; M.S., Pa. State U., 1947; Ph.D., Stanford, 1949; m. Shirley Ellen Besbeck, June 28, 1953; children—Terri, Pamela. Research chemist Aerojet Gen. Corp., Azusa, Cal., 1949-59; sr. chemist Stanford Research Inst., Menlo Park, Cal., 1959-65; prin. scientist Rocketdyne, Canoga Park, Cal., 1965—. Mem. Am. Chem. Soc., Sigma Xi, Phi Veloz Av., Tarzana, Cal. 91356. Office: 6633 Canoga thesis of new high energy organic compounds for use as explosives and propellant ingredients. Home: 5008 Av., Canoga Park, Cal. 91304.*

FRANKENHAEUSER, Carl Bernhard, physiologist; b. Borga, Finland, Mar. 26, 1915; s. Carl and Hanna (Astrom) F.; physician U. Helsingfors, Finland, 1946; M.D., Karolinska Inst., Stockholm, Sweden, 1949; m. Valborg Marianne von Wright, Apr. 7, 1946; 1 dau., Carola. Brit. Council's scholar Oxford, Eng., 1946-48; asst. Med. Nobel Inst., Karolinska Inst., 1948-50, asst. prof. neurophysiology, 1950-62, asso. prof., 1963-65, prof., 1965-66; prof. physiology Umea (Sweden) U., 1966-67, prof. neurophysiology Inst., 1967—; dir. Nobel Inst. for Neurophysiology, 1967—. Mem. Physiol. Soc. Research, numerous publs. on neurophysiology and biophysics of peripheral nerve; analyzed effect of calcium on ionic currents in nerve membrane; analyzed quantitatively ionic currents in vertebrate nerve fibre and given quantitative description of impulse in myelinated nerve fibre. Home: 16 Vitsippsvagen, Saltsjobaden, Sweden. Office: Nobel Inst. for Neurophysiology, Karolinska Inst., Stockholm 60, Sweden.*

FRANKENHAEUSER, Mariann von Wright, exptl. psychologist; b. Helsinki, Finland, Sept. 30, 1925; d. Tor and Ragni (Alfthan) von Wright; Diploma in Psychology, Oxford (Eng.), 1947; B.A., U. Helsinki, 1950; M.A., U. Stockholm (Sweden), 1954; Ph.D., U. Uppsala (Sweden), 1959; m. Bernhard Frankenhaeuser, Apr. 7, 1946; 1 dau., Carola. Clin. psychologist depts. neurosurgery and neurology Serafimer Hosp., Stockholm, 1951-55; research psychologist Lab. Aviation and Naval Medicine, Karolinska Inst., Stockholm, 1955-60; asst. prof. psychol. labs. U. Stockholm, 1960-63; research psychologist Swedish Council for Social Sci. Research, Stockholm, 1963-65; prof., head exptl. psychology unit Swedish Med. Research Council, Stockholm, 1965—; head physiol. psychology unit psychol. labs. U. Stockholm, 1963—. Mem. Swedish Psychol. Assn. Author: Estimation of Time: An Experimental Study, 1959; also articles in exptl. psychology. Research on human physiol. psychology especially emotional and psychol. stress, quantitative relations between physiol. behavioral and subjective reactions in various emotional states. Home: 16 Vitsippsvagen, Saltsjöbaden, Sweden. Office: 95 Drottninggatan, Stockholm, Sweden.*

FRANKENHEIM, Moritz Ludwig, German physicist, crystallographer; b. Brauschweig, Germany, June 29, 1801; m. Friderike; 2 daus.; became asso. prof. U. Breslau (now Wroclaw, Poland), 1827, prof. physics, dir. phys. cabinet, 1850-66. Author: Die Lehre von der Cohäsion . . . und die Krystallkunde, 1835; Zur Krystallkunde, I. Charakteristik der Krystalle, 1869. Forerunner of modern structural theory of crystals; 1st to show spatial lattices are geometrically possible, also showed their symmetry relationships are identical with those of crystals; opposed J. N. von Fuchs' theory of amorphy with one of aggregates of invisible crystals (important in theory of colloids); recognized relation between firmness and symmetry of crystals. Died Dresden, Germany, Jan. 14, 1869.

FRANKFORTER, George Bell, Am. chemist; b. Potter, O., Apr. 22, 1859; s. Andrew and Elizabeth (Clunk) F.; A.B., U. Neb., 1886, A.M., 1888; chem. course Bergakademie, Berlin; Ph.D., U. Berlin, 1893; m. Mary Spalding Carter, 1898; children—Alice Sylvia, Mary Elizabeth (Mrs. Charles Christian Hewitt), Eleanor Evans, George William Carter. Instr. chemistry U. Neb., 1885-87, prof. chemistry, 1893-94; high sch., Lincoln, Neb., 1887-88; dean Sch. Chemistry, dir. chem. lab. U. Minn., 1894-1917, prof. organic and indsl. chemistry, 1920-25, prof. emeritus, from 1927; prof. chemistry Stanford, 1925-26. U. S. Mint commr., 1900. Fellow A.A.A.S.; mem. Am. Chem. Soc., Am. Electrochem. Soc., Soc. Chem. Industry, Ordnance Assn., Phi Beta Kappa, Sigma Xi. Author series of bulls. on explosives, also papers on chemistry. Died Sept., 1947.

FRANKL, Viktor Emil, Austrian psychiatrist; b. Vienna, Austria, Mar. 26, 1905; s. Gabriel and Elsa (Lion) F.; M.D., U. Vienna, 1930, Ph.D., 1949; m. Eleonore Katharina Schwindt, July 18, 1947; 1 dau., Gabriele. Editor jour. Man in Everyday Life, 1927; founder, head Youth Adv. Centers, Vienna, 1928-38; staff, Neuropsychiatric Univ. Clinic, 1930-38; specialist neurology and psychiatry, 1936—; head neurological dept. Rothschild Hosp., Vienna, 1940-42; head Neurol. Poliklinik Hosp. of Vienna, 1946—; asso. prof. neurology and psychiatry U. Vienna, 1947-55, prof., 1955—; visiting professor Harvard Summer School, 1961; founder sch. logotherapy or existential analysis. Lectr. Israel, Europe, S.A., U. S., India, Australia. Mem. bd. internat. cons. Religion in Edn. Found. Imprisoned in concentration camps, 1942-45. Recipient Austrian State prize for pub. edn. Mem. Austrian Soc. Psychother. (pres. 1950—), Acad. Human Rights (adv.), Internat. Union Cultural Coop.; hon. mem. Argentine Soc. Med. Anthropology, Peruvian Soc. Neuropsychiatry and Legal Medicine, Peruvian Soc. Geriatrics, Spanish Society of Clinical and Exptl. Hypnosis. Author 10 books including: The Doctor and the Soul: From Psychotherapy to Logotherapy, 1955; From Death-Camp to Existentialism, a Psychiatrist's Path to a New Therapy (Colby Coll.'s Book of the Year, 1961-62), 1959; Man's Search for Meaning, 1962; An Introduction to Logotherapy, 1963; Psychotherapy and Existentialism, 1967. Editor: (with V. E. von Gebsattel, J. H. Schultz) Ency. of Psychotherapeutics (5 vols.). Originator of school of psychotherapy known as logotherapy or existential analysis. Home: Mariannengasse 1, Vienna IX. Office: Poliklinik Hosp. of Vienna, Vienna, Austria.*

FRANKLAND, Sir Edward, chemist; b. Churchtown, Eng., Jan. 18, 1825; apprentice to chemist, Lancaster, Eng., circa 1840; student Mus. Practical Geology, London, Eng., 1845; student under Bunsen, Marburg, Germany, 1847, Ph.D., 1849; D.C.L., Oxford, Eng., 1870; LL.D., Edinburgh, Scotland, 1884; m. Sophie Fick, Feb. 27, 1851; 3 sons, including Fredrick William, Percy Faraday, also 2 daus.; m. 2d, Ellen Grenside, 1875; 2 daus. Prof. chemistry Putney Coll. for Civil Engring., 1850, Owens Coll., Manchester, Eng., 1851; lectr. chemistry St. Bartholomew's Hosp., London, Eng., 1857; prof. chemistry Royal Instn., 1863-68, Royal Coll. Chemistry, 1865; mem. royal commn. on river pollution, from 1868. Fellow Chem. Soc. (pres. 1871-73), Royal Soc., 1853 (Royal medal 1857, Copley medal 1895); mem. Royal Acad. Scis. Bavaria; fgn. asso. French Acad. Scis., 1895; pres. Inst. Chemistry, 1877-80. Author: A Course of Lectures on Gas-Lighting, 1867; Chemical Lecture Notes, 1870-72; How to teach Chemistry, 1875; Experimental Researches in Pure, Applied, and Physical Chemistry, 1877; (with F. R. Japp) Inorganic Chemistry, 1884. Contbr. papers to sci. jours., also to English Cyclo. First to study organo-metallic compounds; used zinc alkyl reaction to synthesize hydrocarbons, 1849; made exptl. studies of influence of atmospheric pressure on combustion, 1861-68; studied acetoacetic ester, 1864; helped in transition from type theory to valence theory, also expressed modern concept of valence, 1852; recognized trivalence and pentavalence of nitrogen; discovered (with Lockyer) helium in sun; authority on river polution and sanitation; (with Tyndall) studied flame and luminoscity. Died Golaa, Norway, Aug. 9, 1899.

FRANKLAND, Percy Faraday, English chemist; b. London, Oct. 3, 1858; s. Edward Frankland; ed. U. Coll. Sch., London, Royal Sch. Mines, Würzburg (Germany) U.; Ph.D., M.Sc.; LL.D., St. Andrews (Scotland), Birmingham (Eng.) univs.; Sc.D., Dublin (Ireland), Sheffield (Eng.) univs.; m. Grace, 1882; 1 son. Demonstrator, lectr. chemistry Royal Sch. Mines, 1880-88; prof. chemistry Univ. Coll., Dundee, Scotland, 1888-94, Mason Coll., Birmingham, 1894-1900; prof. chemistry U. Birmingham, from 1900, dean Faculty Sci., from 1913, prof. emeritus, from 1919. Examiner in chemistry London U., other univs.; mem. Admiralty Inventions Bd., 1915, Anti-Gas and Chem. Warfare Com.; dep. insp. High Explosives Birmingham Area. Recipient Davy medal, 1919. Fellow Royal Soc., 1891 (council 1903-05, 16-18, v.p. 1917-18, exec. war com., chmn. chem. sect. war com., chmn. reserved occupations com.), Imperial Coll. Sci. and Tech., Royal Inst. Chemistry (pres. 1906); asso. Royal Sch. Mines; mem. Chem. Soc. (pres. 1911). Author: Agricultural Chemical Analysis, 1883; Our Secret Friends and Foes, 1894; Microorganisms in Water, 1894; Life of Pasteur, 1897; articles on fermentation and water Thorpe's Dictionary of Tech. Chemistry; numerous original memoirs to sci. publs. Inaugurated monthly systematic bacteriological exams. of London Water Supply for local govt. bd., 1885. Died Loch Awe, Scotland, Oct. 28, 1946.

FRANKLIN, Alan Douglas, Am. chemist; b. Glenside, Pa., Dec. 10, 1922; s. Benjamin and Adrienne (Kenyon) F.; A.B., Princeton, 1945, Ph.D. in Chemistry, 1949; m. Phoebe Perry Taylor, May 8, 1943 (div.); children—Adrienne K., Christopher P.; m. 2d., Katherine Ann McMurdie, Apr. 15, 1960; 1 dau., Mary L. Chief magnetics sect. Franklin Inst. Labs., Phila., 1949-55; group leader ferroelectricity group Nat. Bur. Standards, Washington, 1955-61, chief inorganic materials div., 1961-63, asst. to dir. Inst. for Materials Research, 1963—. Fellow Am. Phys. Soc.; mem. Am. Ceramic Soc., Am. Chem. Soc., Internat. Union Pure and Applied Chemistry, Phi Beta Kappa, Sigma Xi. Research in movement and displacement of atoms in crystals. Home: 6510 Ridge Dr., Washington 20016. Office: Nat. Bur. Standards, Washington 20234.

FRANKLIN, Benjamin, Am. statesman, scientist, philosopher; b. Boston, Jan. 17, 1706; s. Josiah and Abiah (Folger) F.; self-educated; M.A. (hon.), Yale, 1753, Harvard, 1753, Coll. William and Mary, 1756; LL.D., St. Andrews Coll., 1759; D.C.L. (hon.) Oxford (Eng.) U., 1762; took Deborah Read as common law wife, Sept. 1, 1730, later married, 2 children—Francis Folger, Sarah; 2 illegitimate children including William. Contbr. articles to New Eng. Courant (newspaper pub. and printed by his bro. James to whom he was apprenticed), Boston, until 1723; in London, 1724-26; owner Pa. Gazette, 1730-50; established 1st circulating library in Am. at Phila., 1731; wrote and pub. Poor Richard's Almanac (one of 1st Am. lit. prodns. to attain internat. renown, largely because of his common-sense philos. aphorisms), 1732-57; clk. Pa. Assembly, 1736-51, mem. from Phila., 1751-64; dep. postmaster Phila., 1737-53; aided Brit. in French and Indian War; founder Am. Philos. Soc., 1743, Phila. City Hosp., 1751; a founder Acad. for Edn. of Youth, 1751, inc., 1753 (later Coll. of Phila., now U. Pa.); joint dep. postmaster gen. Am. colonies, 1753-74; Pa. rep. at Albany (N.Y.) Congress, 1754; submitted "Plan of Union" adopted by Congress but vetoed by colonial legislatures (later important in drafting Articles of Confedn. and U. S. Constn.); polit. agt. Pa. Assembly, sent to Eng. to present case against Penn family for refusing to support defense expenditures, 1757-62, returned to Eng. to obtain recall of Pa. Charter, 1766; questioned before Ho. of Commons during debates on repeal of Stamp Act, 1766; colonial agt. of Ga., 1768, N.J., 1769, Mass., 1770; aided William Pitt in fruitless conciliation efforts, in Eng.; mem. 2d Continental Congress, 1775, sketched plan of union for colonies, organized post office, became 1st postmaster gen.; mem. com. to draft Declaration of Independence, also a signer, 1776; del. of mission to persuade Can. to join Am. cause; commr. to negotiate treaty with France (only Am. with whom French Prime Minister Vergennes would deal), 1776, signed final commerce and defense alliance treaties, 1778; apptd. minister to France, 1778; apptd. a commr. to negotiate peace with Gt. Britain, 1781, assumed responsibility for preliminary talks, peace treaty signed, 1783; returned to Phila., 1785; pres. Pa. Exec. Council, 1785-87; mem. U. S. Constl. Conv., 1787, largely responsible for representation compromise actually incorporated in Constn.; signed meml. to Congress for abolition of slavery (last public act), Feb. 12, 1790. Named charter mem. Hall Fame for Gt. Ams., 1900. Mem. French Acad. Scis. (1st Am. fgn. mem.; Fellow Royal Soc., (Copley medal 1753). Author: Experiments and Observations on Electricity, 1750; Opinions and Conjectures concerning the Properties and Effects of the Electrical Matter, 1750; Experiments and Observations on Electricity carried out at Philadelphia, 1751; Edict by the King of Prussia, Rules by Which a Great Empire May be Reduced to a Small One (both satires), early 1770's; Observations on the Increase of Mankind, of the Peopling of Countries, 1775; Autobiography (never finished); also essays (some of earliest signed Silence Dogood). Convinced himself of identity of lightning with electricity and suggested kite expt., 1752; from his expts. with Leyden jar (from 1747), deduced that lightning conductor could be used for safe discharge of low thunderclouds (1st mentioned, 1749, recommended to pub., 1753); developed duFay's discovery that there are 2 kinds of electricity; formulated influential one fluid theory of electricity, according to which all bodies possess normal quantity of an electric fluid and are able to produce elec. effects whenever this normal quantity is decreased or increased; suggested that an excess (or deficiency) of fluid be called positive (or negative) electricity, and held that electric fluid flows from positive to negative; implied important principle of conservation of charge; attempted to work out course of storms over N. Am.; 1st to study Gulf Stream; early investigator of population growth; early advocate of fresh air as adjunct to hygiene; inventor Pa. Fireplace (known as Franklin stove), circa 1727, Fergusson's clock, circa 1744, also bifocals. Died Phila., Apr. 17, 1790.

FRANKLIN, Beryl Cletis, Am. zoologist; b. Colly, Ky., Oct. 29, 1923; s. Andrew Kelly and Annis (Blair) F.; A.B., Ky. Wesleyan Coll., 1948; M.S., U. Ky., 1950; Ph.D., Ohio State U., 1957; m. Nancy Belle Coons, Sept. 19, 1947. Faculty, Ohio State U., 1950-57, supr. grad assts., 1954-57; instr. De Mar Coll., Corpus Christi, Tex., 1957-59; asst. prof. zoology La. State U., 1959-60; faculty N.E. La. State Coll.,

Monroe, 1960—, prof. biology, 1967—. Mem. A.A.A.S., Am. Inst. Biol. Scis., Am. Soc. Zoologists, La. Acad. Sci. (pres. 1967), Assn. Southeastern Biologists Sigma Xi. Research, publs. on reproductive physiology mammals, endocrinology reprodn. Home: 402 Beasley St., Monroe, La. 71201.*

FRANKLIN, Edward Curtis, Am. chemist; b. Geary City, Kan., Mar. 1, 1862; s. Thomas Henry and Cynthia Ann (Curtis) F.; B.S., U. Kan., 1888, M.S., 1890; student U. Berlin, 1890-91; Ph.D., Johns Hopkins, 1894; D.Sc., Northwestern, 1923, Western Res. U., 1926; LL.D., Wittenberg Coll., 1927; m. Effie June Scott, July 22, 1897; children—Mrs. Anna Comstock Barnett, Charles Scott, John Curtis. Asst. in chemistry, U. Kan., 1888-93, asso. prof., 1893-99, prof. phys. chemistry, 1899-1903; asso. prof. Stanford, 1903-06, prof. organic chemistry, 1906-29. Prof. chemistry and chief div. chemistry USPHS, 1911-13. Mem. U. S. Assay Commn., 1906; mem. adv. bd. U. S. Bur. Mines, 1917-18; phys. chemist U. S. Bur. Standards, 1918; cons. chemist Ordnance Bur., U. S. Army, 1918. Recipient Nichols medal, 1924; Willard Gibbs medal, 1931. Fellow A.A.A.S., Am. Acad. Arts and Scis.; guest Brit. Assn. Advancement Science, Capetown and Johannesburg, 1929. Research on ammonia system of compounds, liquid ammonia as an electrolytic solvent; expert on chem. warfare. Died Palo Alto, Cal., Feb. 13, 1937.

FRANKLIN, Joe Louis, Jr., Am. chemist; b. Natchez, Miss., Aug. 11, 1906; s. Joe Louis and Katherine (Balfour) F.; B.S., U. Tex., 1929, M.S., 1930, Ph.D., 1934; m. Mildred Louise Selkirk, Dec. 22, 1935; children—William Selkirk, James Balfour, Robert Ainsworth. Research chemist Humble Oil & Refining Co., Baytown, Tex., 1934-38, sect. head, 1938-45, asst. div. head, 1945-47, research asso., 1947-63; Robert A. Welch prof. chemistry Rice U., Houston, 1963—. Guest scientist Nat. Bur. Standards, 1957-58; Wiley lectr. Purdue U., 1957; mem. adv. council dept. chem. engring. Princeton, 1960. Fellow Am. Phys. Soc.; mem. Am. Chem. Soc. (S.W. award 1962), Am. Inst. Chem.E. (Publs. award S. Tex. sect. 1949, Distinguished Service award 1962), Am. Soc. Testing Materials, Faraday Soc., Combustion Inst., Sigma Xi, Tau Beta Pi, Phi Lambda Upsilon. Author: (with F. H. Field) Electron Impact Phenomena and the Properties of Gaseous Ions, 1957. Research, numerous publs. on chem. behavior ions in gases, electron impact phenoma, measurements of strengths of chem. bonds, reaction kinetics. Home: 3627 S. Braeswood St., Houston 77025.*

FRANKLIN, Joel Nicholas, Am. mathematician; b. Chgo., Apr. 4, 1930; s. Jacob Nick and Anne (Goldberg) Esau; B.S. in Math., Stanford, 1950, Ph.D., 1953; 1 dau., Sarah. Instr., N.Y. U., 1953-55; asst. prof. U. Wash., Seattle, 1955-56; sr. mathematician Burroughs Corp., Pasadena, Cal., 1956-57; asso. prof. math. Cal. Inst. Tech., Pasadena, 1957-65, prof. applied sci., 1965—, asso. dir. Computing Center, 1963—. Cons. math. and computer applications. 1953—. Mem. Am. Math. Soc., Assn. for Computing Machinery, Phi Beta Kappa, Sigma Xi. Research in math. concepts and methods for application of computers to sci. and engring., particularly, basic computational theory of random processes. Home: 1763 Alta Crest Dr., Altadena, Cal. 91001. Office: Cal. Inst. Tech., Pasadena, Cal. 91109.*

FRANKLIN, Sir John, explorer; b. Spilsby, Eng., Apr. 16, 1786; hon. doctorate U. Oxford (Eng.); explored N.Am. arctic coastline aboard HMS Trent, 1818-22; mapped littoral between Coppermine and Mackenzie rivers, 1825; made voyage to Mediterranean, 1835; gov. Tasmania, 1836-46; made arctic expdn., 1847. Created baron, 1829. Recipient Gold medal Paris Geog. Soc. Fellow Royal Soc., 1823; mem. French Acad. Scis., 1846. Author: Narrative of a Journey to the Shores of the Polar Sea in the Years 1819-22 (pub. 1823), in the Years 1825-27 (pub. 1828). Explored arctic, also S. Sea regions; discovered, named Kendall and Parry islands, also Peel River; founded sci. coll. in Tasmania (now Hobartown Royal Soc.). Died June 11, 1847.

FRANKLIN, Kenneth Linn, Am. astronomer; b. Alameda, Cal., Mar. 25, 1923; s. Myles A. and Ruth L. (Huston) F.; A.A., U. Cal., Berkeley, 1943, A.B., 1948, Ph.D., 1953; m. Beverly J. Mattson, Nov. 29, 1950 (dec. Mar. 1956); children—Kathleen Lenore, Christine Louise; m. 2d, Charlotte W. Windmuller, May 18, 1958; 1 dau., Julie Ellen. Lab. asst. Lick Obs., Mt. Hamilton, Cal., 1948-50; research asst. Leuschner Obs., Berkeley, Cal., 1953-54; research fellow dept. terrestrial magnetism Carnegie Instn., Washington, 1954-56; astronomer Am. Mus.-Hayden Planetarium, N.Y.C., 1956—, asst. chmn., 1968—. Sr. scientist Lamont Geophys. Obs., Columbia, N.Y.C., 1957-59; lectr. N.Y. U., 1958-65; instr. Coll. City N.Y., 1957, 65; adj. prof. Cooper Union, 1968—; cons. bus. firms, 1957—. Fellow A.A.A.S.; mem. Astron. Soc. Pacific, Am., Royal astron. socs., I.E.E.E., Internat. Sci. Radio Union, Sigma Xi. Author booklets: Space Age Astronomy, 1964; Birth and Death of Stars, 1964. Research on binary stars, especially Capella, galactic motion, radio astronomy, especially Jupiter, participated in discovery of radio radiation from Jupiter. Office: 81st St. and Central Park W., N.Y.C. 10024.*

FRANKLIN, Richard Harrington, Brit. surgeon; b. London, Eng., Apr. 3, 1906; s. Percival Charles and Winifred (Harrington) F.; M.B.B.S., U. London, 1930, F.R.C.S., 1934; m. Helen Margaret Kimber, Oct. 4, 1933; children—Richard Kimber, Peter James. Staff, St. Thomas Hosp., London, 1930-33; faculty Brit. Postgrad. Med. Sch., 1936-39, sr. lectr. in surgery, 1945—; cons. surgeon various hosps., also Royal Navy. Examiner in surgery U. Cambridge, 1958—; hon. surg. tutor Royal Coll. Surgeons, 1962-67, mem. council, 1965—. Mem. Royal Soc. Medicine (v.p. sect. surgery 1960-61), Assn. Surgeons Gt. Britain and Ireland, Brit. Soc. Gastroenterology, Grey Turner Surg. Club, Med. Soc. London. Author: Surgery of the Oesophagus, 1952; also articles. First successful anastomosis for congenital atresia of oesophagus with tracheo-oesophageal fistula outside U. S. Home: Wolsey House, 4 Montpelier Row, Twickenham, Middlesex, Eng. Office: 148 Harley St., London, Eng.*

FRANKLIN, Richard Morris, Am. biophysicist; b. Medford, Mass., Oct. 16, 1930; s. Morris and Esther (Anderson) F.; B.S., Tufts U., 1951; Ph.D., Yale, 1954; m. Ilse Bünning, Dec. 6, 1958; children—Sergei Tibor, Anna Victoria. Research fellow Cal. Inst. Tech., Pasadena, 1954-56; Fulbright fellow Max-Planck Institut für Virusforschung, Tübingen, Germany, 1956-57, research asso., 1957-59; asst. prof. Rockefeller Inst., N.Y.C., 1959-63; asso. prof. dept. pathology U. Colo. Sch. Medicine, Denver, 1963-65, prof., 1965-66; mem. Pub. Health Research Inst. City of N.Y., 1966—; research prof. microbiology N.Y. U., Sch. Medicine, 1966—. Asso., Lab. for Quantitative Biology, Cold Spring Harbor, N.Y., 1960-62. Recipient Pasteur medal Institut Pasteur, Paris, France, 1965. Mem. Biophys. Soc., Am. Soc. Microbiology. Editorial bd. Jour. Virology. Research, publs. on biology and molecular biology of animal virus multiplication, correlation of molecular events in animal virus multiplication with morphological and ultrastructural changes, description of enzymes and spl. molecular forms of virus ribonucleic acid concerned with replication of viral RNA. Home: 110 Bleecker St., N.Y.C. 10012. Office: 455 1st Av., N.Y.C. 10016.*

FRANKLIN, Thomas Chester, Am. chemist; b. Birmingham, Ala., Feb. 5, 1923; s. Chester S. and Irene (Tibbetts) F.; B.S. with honors, Howard Coll., 1944; Ph.D., Ohio State U., 1951; m. Nellie Louise Friel, June 26, 1946; children—Irene Elise, Margaret Elaine, Janice Carol, Thomas Edward. Instr., Howard Coll., 1946-48; asst. instr. Ohio State U., 1948-51; asst. prof. U. Richmond, Va., 1951-54; research asso. Va. Inst. for Sci. Research, Richmond, 1951-53; faculty Baylor U., Waco, Tex., 1954—, prof., 1964—. Mem. Am. Chem. Soc., Electrochem. Soc., Am. Electroplaters Soc., Internat. Com. Electrochem. Thermodynamics and Kinetics, Sigma Xi. Author: (with John Xan) A Lab Manual for Semimicroqualitative Analysis, 1948. Studies, publs. of electrochem. and catalytic properties of metal surfaces in contact with solutions using absorption of hydrogen as a measure of active area; studies of electro-deposition, catalytic and electrolytic oxidations and reductions of organic compounds. Home: 1312 Guthrie Dr., Waco, Tex. 76703.*

FRANKLIN, William Suddards, Am. physicist; b. Geary City, Kan., Oct. 27, 1863; s. Thomas Henry and Cynthia Ann (Curtis) F.; B.S., U. Kan., 1887, M.S., 1888; D.Sc., Cornell, 1901; student U. Berlin, 1890-91; holder Morgan fellowship, Harvard, 1891-92; m. Hattie F. Titus, Aug. 14, 1888; children—Curtis, Kellogg. Asst. prof. physics U. Kan., 1887-90; prof. physics and elec. engring. Ia. State Coll., 1892-97, Lehigh U., 1897-1903, prof. physics, 1903-15; prof. physics Mass. Inst. Tech., 1917-29; prof. physics Rollins Coll., Winter Park, Fla., from 1929. Hon. mem. Kan. Acad. Scis. Joint author: Elements of Physics, 3 vols.; Elements of Alternating Currents; Elements of Electrical Engineering, 2 vols., 1906; Dynamo Laboratory Manual; Elements of Electricity and Magnetism; Practical Physics, 3 vols.; Electric Waves. Died Wilmington, N.C., June 6, 1930.

FRANKS, Cyril Maurice, psychologist; b. Neath, Wales, U.K., July 26, 1923; s. Harry and Cecelia (Zeiler) F.; B.Sc., U. Wales, 1943; diploma in Edn., U. London, 1947, diploma in Ednl. Psychology, 1949, diploma in Abnormal Psychology, 1951, Ph.D., 1954; M.A., U. Minn., 1952; m. Violet Greenberg, Apr. 29, 1952; children—Steven Laurence, Sharon Leora. Came to U. S., 1957, naturalized, 1961. Jr. research engr. His Master's Voice, Hayes, Middlesex, U.K., 1943-46; instr. London Nautical Sch., 1947-50; clin. intern Maudsley Hosp., London, 1950-51; research asst. Inst. Child Welfare U. Minn., 1951-52; lectr. Inst. Psychiatry, U. London, 1952-57; dir. Psychology Service and Research Center, Neuro-Psychiat. Inst., Princeton, N.J., 1957—; adj. faculty Rutgers U., 1959—; lectr. Seton Hall Coll. Medicine and Dentistry, 1965—; vis. lectr. Princeton 1967—. Fellow Am. Psychol. Assn., Brit. Psychol. Soc.; mem. Pavlovian Soc. Am., Soc. For Psychophys. Research, InterAm. Soc. Psychology, Eastern Psychol. Assn. Author: Conditioning Techniques in Clinical Practice and Research, 1964. Research, numerous publs. on fundamental parameters in conditioning, modification of abnormal and normal behavior, and behavior therapy. Home: 315 Prospect Av. Office: Neuro-Psychiat. Inst., Princeton, N.J. 08540.*

FRANQUÉ, Otto Friedrich Wilhelm Paul von, see von Franqué, Otto Friedrich Wilhelm Paul.

FRÄNZ, Johannes, German physicist; b. Neurahnsdorf, Sept. 28, 1899; s. Georg and Alwine (Stiebert) F.; Ph.D., U. Berlin; m. Hedwig Windhorn, Sept. 1, 1926; children—Jürgen, Peter. Dir., later prof. Fed. Inst. Tech. Physics, Berlin-Braunschweig, Germany, 1925-65; hon. prof. Technische Hochschule, Brunswick, 1949—. Mem. German Soc. Physics. Collaborator: Handbuch der Physik; Praktische Physik. Research and publs. on atomic and nuclear physics; radioactivity; radiation dosimetry. Home: Sulzbacherstrasse 50. Office: Bundessallee 100, Braunschweig, Germany.

FRANZ, Shepherd Ivory, Am. psychologist; b. Jersey City, N.J., May 27, 1874; s. D. M. William and Frances Elvira (Stoddard) F.; A.B., Columbia, 1894, Ph.D., 1899; U. of Leipzig, 1896; hon. M.D., George Washington U., 1915; LL.D., Waynesburg Coll., 1915; m. Lucie Mary Niven, of London, Ont., Can., June 18, 1902; children—Theodora Niven, Elizabeth Knox, Patricia Wilderspin. Asst. in psychology, Columbia, 1897-99; asst. in physiology, Harvard, 1899-1901; instr. physiology, Dartmouth Med. Sch., 1901-04; pathol. psychologist, McLean Hosp., Waverley, Mass., 1904-06; prof. physiology, 1906-21, exptl. psychology, 1906-24, George Washington U.; psychologist, 1907-24, scientific dir., 1910-19, dir. labs., 1919-24, St. Elizabeth's Hosp. (govt. hosp. for insane), Washington, D.C.; instr. in neurology, Naval Med. Sch., 1920-24; lecturer psychology, U. of Calif. at Los Angeles, 1924-25, prof., from 1925, research lecturer, 1926; chief of psychol. and ednl. clinic, Children's Hosp., Hollywood, from 1924. Lecturer in psychology, U. of Chicago, 1922, Johns Hopkins, 1922-23. Editor Psychol. Bull., 1912-24, Psychol. Monographs, 1924-27. Awarded Butler medal, Columbia University, 1924. Author: Handbook of Mental Examination Method, 1912, 2d edit. 1919; Nervous and Mental Re-education, 1923; Persons One and Three, 1933. Devised new techniques for animal exptl. psychophysiology in testing learning and discrimination; significant work on function of frontal lobes and location of mental processes in brain. Died Los Angeles, Oct. 14, 1933.

FRANZINETTI, Carlo, Italian physicist; b. Rome, Italy, Mar. 31, 1922; s. Guido and Ada (Guastalla) F.; Laurea in Fisica, U. Roma, 1945, Libera Docenza in Fisica Nucleare, 1956; m. Joan M. J. Rees, Dec. 29, 1951; children—Victoria, Guido, Giulio. Research staff U. Bristol (Eng.), 1947-50; prof. physics U. Trieste, 1958-59, U. Pisa (Italy), 1959; sr. physicist CERN, 1962—; prof. physics U. Turin (Italy), 1966—. Mem. Italian Phys. Soc. Author: (with G. Morpurgo) Introduction to the Physics of the New Particles, 1957; also articles. Research on cosmic rays using nuclear emulsion technique, devel. spark chambers, properties of hyperons and K-mesons, neutrino interactions at high energy; contbr. to discovery of anti-protons. Home: 37/3-Q Strada del Nobile, Turin, Italy.*

FRANZINI, Tito, Italian physicist; b. Genoa, Italy, Dec. 10, 1902; s. Giacomo and Clotilde (Cipollina) F.; Physics D., U. Genoa, 1926; m. Margherita Agnoli, July 4, 1932; children—Paolo, Carlo, Marco, Clara (Mrs. Clay Armstrong). Asst. prof. U. Pavia (Italy), 1928-34, libero docente exptl. physics, 1934-38; prof. theoretical physics U. Firenze (Italy), 1938-42; prof. Naval Acad. Livorno (Italy), 1942—; prof. nuclear physics High Sch. for Nuclear Engring., U. Pisa (Italy), 1957—. Mem. sci. and tech. com. Euratom, 1958—; sci. dir. CAMEN, 1956-60. Author numerous books including Lezioni di Fis. sperimentale, 1944; Energia atomics, 1946; Tavole fisica nucleare, 1948; also numerous articles. Research on occlusion and difflusion of gases in metals, ions sources, thermionic and photoelectric emission, fission of uranium, physics of nuclear reactors. Home: Viale Italia 131, Livorno, Italy.*

FRANZIUS, Ludwig, German hydraulic engr.; b. Wittmund, Mar. 1, 1832; s. Carl Egbert and Charlotte Friederica (Bütemeister) F.; entered Hanover (Germany) Polytechnikum, 1848; D.Eng. (hon.), Tech. U. Berlin, 1901; m. Caroline Elisabeth Marie Uslar; 4 sons, 1 dau. Employed in Harburg, Stade, Neuhaus/Oste (all Germany), also on canal and locks, Papenburg/Ems, Germany; became asst., gen. hydraulics mgmt., Hanover, 1865; named to ministry commerce, industry and pub. works, Berlin, 1867; tchr. Berlin Constrn. Acad.; named chief constrn. dir. City of Breman, Germany, 1876. Mem. Commn. for Prevention of Flood Dangers; mem. Reich Bd. Health. Recipient Gold Grashot Meml. medal. Mem. Prussian Acad. Constrn. Author: Korrektion der Unterweser, 1882-95; Die Korrektion der Unterweser, 1888; Neue Hafenanhagen zu Bremen, 1889; Korrektion der Aussenweser, 1889; also articles. Made corrections in lower Weser River necessary for Bremen harbor; planned correction in outer Weser below Bremerhaven, also canal in Weser from Bremen to Hamlin. Died Bremen, June 23, 1903.

FRASCH, Herman, chem. engr., inventor; b. Würtemberg, Germany, Dec. 25, 1851; s. Johannes Frasch; m. Romalda Berks; m. 2d, Elizabeth Blee. Came to U. S., 1868. Student, employee Phila. Coll. Pharmacy; moved to Cleve., 1877, opened chem. lab. to study petroleum refining; organized Empire Oil Co., London, Ont., Can., 1885; developed Frasch process for desulfurization of crude petroleum oil; received 21 U. S. patents for refining of petroleum, 1887-94, sold patents to Standard Oil Co.; received patent for mining sulfur by means of superheated water, 1891; organizer, pres. Union Sulfur Co., 1892; pres. Internat. Sulfur Refineries of Marseilles; worked out a process for refining paraffin; recipient Perkin Gold medal in chemistry Am. sect. Soc. Chem. Industry, 1912. Died Paris, France, May 1, 1914.

FRASER, Alexander Stewart, geneticist; b. London, Eng., Dec. 24, 1923; s. Thomas Stewart and Christina (Allan) F.; B.S., U. New Zealand, 1943, M.S., 1944; Ph.D., U. Edinburgh (Scotland), 1950; m. Ella Anne Soper, Mar. 15, 1950; children—Alan, Andrew, Alexis, Annette, Alistair, Aileen. Came to U. S., 1962. Research asst. New Zealand Inst. Wheat, Christchurch, 1944-45; asst. chemist New Zealand Dept. Agr., Hamilton, 1945-46; research asst. Massey Agrl. Coll., Palmerston, New Zealand, 1946-49; research fellow Inst. Animal Genetics, Edinburgh, Scotland, 1949-52; sr. prin. research officer Commonwealth Sci. and Indsl. Research Orgn., Sydney, Australia, 1952-62; prof. genetics U. Cal. at Davis, 1962-67, chmn. dept., 1963-65; head dept. biol. scis. U. Cin., 1967—. Fellow Australian Acad. Scis. Author textbook on genetics, 1966. Research on factors of wool growth; use of computers to imitate biol. systems; use of major genes to disrupt canalized genetic systems. Home: 110 Hosea Av., Cin. 45220.*

FRASER, Frank Clarke, geneticist; b. Norwich, Conn., Mar. 29, 1920; s. Frank W. and Annie (Clarke) F.; B.Sc., Acadia University, Nova Scotia, 1940, D.Sc., 1967; M.Sc., McGill University, Montreal, Quebec, Canada, 1941, Ph.D., 1945, M.D., C.M., 1950; m. Beryl de Blois, Dec. 28, 1948; children—Norah, Noel, Alan, Scott. Faculty, McGill U., 1950—, prof. human genetics sector, 1960—; dir. dept. med. genetics Montreal Children's Hosp., 1950—; clin. fellow Royal Victoria Hosp., Montreal, 1950—. Mem. adv. com. on human genetics WHO, 1962—; mem. expert com. on occurrence congenital anomalies Canadian Dept. Nat. Health and Welfare, 1961—. Fellow Royal Soc. Can.; mem. Am. Soc. Human Genetics (past pres.), Teratology Soc. (past pres.), Genetics Soc. Can., Genetics Soc. Am., Canadian Paediatric Soc., Am. Eugenics Soc., Sigma Xi. Research, numerous publs. on genetic counselling, causes congenital malformations by collection family and prenatal data malformed children and use teratogens on mice. Home: 124 Percival Av., Montreal, Que., Can.*

FRASER, George Robert, human geneticist; b. Uzhorod, Czechoslovakia, Mar. 3, 1932; s. Stephen and Helen (Erdei) F.; B.A., Cambridge (Eng.) U., 1953, M.B., B.Chir., 1956, M.A., 1960, M.D., 1966; Ph.D., London U., 1960; m. Alonsina Ortega Velazquez, June 19, 1963; children—Elizabeth Jean Pascoe, Helen Denise Pascoe. Sci. officer Med. Research Council Population genetics research unit, Oxford, Eng., 1959-61; research asso. U. Wash., Seattle, 1961-63, asso. prof. medicine (med. genetics), 1968—; research scientist dept. ophthalmology Royal Coll. Surgeons, London, Eng., 1963-65; asso. prof. genetics, U. Adelaide, Australia, 1966-68; hon. asso. physician genetics Adelaide Children's Hosp., 1966-68; Fellow Royal Soc. Medicine, Genetical Soc. Gt. Britain. Research, numerous publs. on role genetic factors in thyroid disease, causes of pediatric deafness and blindness, genetics of hereditary blood disorders. Home: 5520 17th St. N.E., Seattle 98105.*

FRASER, Havelock Frank, Am. pharmacologist; b. Carrievale, Sask., Can., Nov. 21, 1903; s. John Alexander and Mary Jane (Morrison) F.; A.B., U. Wash., 1925; M.D., Cornell U., 1932; m. Dorothy C. Cassel, June 3, 1936; children—Robert Henry, Dorothy J. (Mrs. Robert Carpenter), Thomas Reid. Commd. asst. surgeon USPHS, 1935, advanced through grades med. dir., 1952; officer U. S. Penitentiary, Atlanta, 1934-35, med. officer in charge, 1935-36; research staff NIH, Bethesda, Md., 1937-43, asst. dir. Indsl. Hygiene Lab., 1944-46; with State Dept., Stuttgart, Germany, 1946-49; research in drug addiction, Lexington, Ky., 1949-51; asso. dir. Addiction Research Center, Nat. Inst. Mental Health, NIH, Lexington, 1952-63; ret., 1963; physician adminstrv. research Eli Lilly & Co., Indpls., 1963—. Asst. prof. dept. psychiatry Emory U. Sch. Medicine, 1935-36; lectr. pharmacology U. Louisville, 1956-63, Coll. Medicine, U. Cin., 1956-63; asst. prof. U. Ky. Coll. Medicine, Lexington, 1961-63. Mem. expert adv. panel on addiction-producing drugs WHO, 1963—. Recipient USPHS Meritorious Service medal, 1962. Mem. Am. Chem. Soc., Am. Soc. for Pharmacology and Exptl. Therapeutics, A.M.A. Contbr. articles to tech. jours. Developed new methods for evaluating addictiveness of analgesics and sedatives for man. Home: 609 E. 75th St., Indpls. 46240. Office: Eli Lilly & Co., Indpls. 46206.*

FRASER, John Stiles, physicist; b. Wonsan, Korea, June 23, 1921; s. Edward James Oxley and Margaret (McIntosh) F.; B.Sc., Dalhousie U., Halifax, N.S., Can., 1942; Ph.D., McGill U., Montreal, Que., Can., 1949; m. Dorothy Muriel Whalley, Jan. 12, 1944; children—Michael Edward, Charlotte Ann. With Atomic Energy Can. Ltd., Chalk River, Ont., Can., 1949—, sr. research officer, 1962—; vis. scientist AERE, Harwell, Eng., 1959-60. Mem. Canadian Assn. Physicists (chmn. secondary sch. sci. edn. com. 1959—), Am. Phys. Soc. Research, publs. on physics fission process, electronic devices for nuclear physics research, angular correlation studies in radioactive decay, resonant neutron scattering, computer programming. Home: 7 Cartier Circle, Deep River, Ont. Office: Atomic Energy Can. Ltd., Chalk River, Ont., Can.*

FRASER, Sir Thomas Richard, Brit. pharmacologist; b. Calcutta, Feb. 5, 1841; M.D., U. Edinburgh, LL.D.; LL.D., Aberdeen U., Glasgow U.; ScD., Cambridge U.; M.D. (hon.), Dublin; m. Susanna Margaret Duncan, 1874; 4 sons, 3 daus. Asst. physician Royal Infirmary, 1869-74; examiner in materia medica U. London, 1870-75, in pub. health, 1876-79; mem. admiralty com. Sir George Nares' Arctic Expdn., 1877; prof. materia medica U. Edinburgh, 1877-1918, prof. clin. medicine, 1878-1918, emeritus prof., from 1918, Univ. rep. on Gen. Med. Council, 1905-20, dean Faculty Medicine, 1880-1900. Med. adviser His Majesty's Prison Commn. for Scotland; cons. physician Standard Life Assurance Co.; pres. Indian Plague Commn., 1898-1901; hon. physician to king in Scotland. Named laureate Inst. France. Fellow Royal Soc., 1877, Royal Soc. Edinburgh (Macdougall-Brisbane and Keith prize), Royal Coll. Physicians Edinburgh (pres. 1900-02); mem. Assn. Physicians Gt. Britain and Ireland (pres. 1908-09), Medico Chirurg. Soc. Edinburgh (pres. 1901-03). Research and numerous publs. on practical medicine, action and therapeutic uses of medicinal substances on serpent's venom, immunization against cobra venom; 1st to study (with Alexander Crum Brown) relationship between chem. structure of drugs and physiologic action. Died Jan. 4, 1920.

FRASSETTO, Fabio, Italian anthropologist; b. Sassari, Feb. 13, 1876; s. Antonio and Maddalena (Musso) F.; Ph.D. in natural scis. and math. Former dir. Inst. Gen. and Applied Anthropology, U. Bologna; founder, dir. Mus. Anthropology, Bologna. Pres., Internat. Com. for Unification of Eugenic Anthropologic Methods and Biometry. Research on biometry, methodology, crainiology, morphology; inventor numerous anthropometric instruments. Address: via Indipendenza 31, Bologna, Italy.

FRAUENFELDER, Hans, physicist; b. Neuhausen, Switzerland, July 28, 1922; s. Otto and Emma (Ziegler) F.; Diploma, Swiss Fed. Inst. Tech., Zurich, 1947, Dr. sc.nat., 1950; m. Verena Anna Hassler, May 12, 1950; children—Ulrich Hans, Katherine Ann. Came to U. S., 1952, naturalized 1958. Asst., Swiss Federal Inst. Tech., 1946-52; faculty U. Ill., Urbana, 1952—, prof., 1958—, asso. dir. Coordinated Sci. Lab., 1962—; vis. prof. U. Hamburg, 1957; vis. scientist European Orgn. for Nuclear Research, Geneva, Switzerland, 1958, 59, 63. Cons. Los Alamos Sci. Lab., 1960—. Guggenheim fellow, 1958-59. Fellow Am. Phys. Soc.; mem. Swiss Phys. Soc. Author: The Mössbauer effect, 1962. Research in nuclear physics, surface physics, Mössbauer effect. Home: 8 Hagan Blvd., Urbana, Ill. 61801.*

FRAUNHOFER, Joseph von, German physicist; b. Straubing, Bavaria, Mar. 6, 1787; apprenticed as optician. Employed by Utzschneider Optical Institute, Benedictbeuern, from 1806; manager 1818; apptd. conservator of physical cabinet at Munich, 1823; received civil order of merit from King of Bavaria, 1824. Skilled maker of achromatic lenses and various optical instruments; discovered and carefully studied dark lines (adsorption lines) in solar spectrum (Fraunhofer lines); mapped 576 of them, denoting principal lines by letters A-G; determined their wavelengths; concluded they originate in light source; first to use diffraction grating extensively; invented machine for polishing achromatic lenses, a form of heliometer, achromatic microscope, stage-micrometer. Died Munich, Germany, June 7, 1826.

FRAUSTO DA SILVA, J. J. R., Portuguese chemist; b. Tomar (Portugal), Aug. 30, 1933; s. António Antunes Silva and Lisette W. R. Fraústo; student, Lisbon Inst. Tech., D.Sc., 1965; Ph.D., U. Oxford (Eng.), 1962; m. Georgina Madeira Marques, Mar. 14, 1960; children—Marina, Filipe. Sr. research officer C.E.E.N., 1962; faculty Lisbon (Portugal) Inst. Tech., 1962—, prof. inorganic and analytical chemistry, 1967—. Cons. bus. firms, govt. agys.; Portuguese del. OECD Com. Sci. Policy, 1965—. Recipient Artur Malheiros prize, Portuguese Acad. Scis., 1963. Mem. Portuguese Soc. Chemistry, Chem. Soc. (London), Ordem dos Engenheiros. Research, numerous publs. in coordination chemistry, synthesis of new ligands with analytical applications, invention of method for producing hydrosoluble tetracycline derivatives of high potency. Address: 3- 2°- Dt° Rua Francisco Franco, Lisbon, Portugal.*

FRAUTSCHI, Steven Clark, Am. physicist; b. Madison, Wis., Dec. 6, 1933; s. Lowell Emil and Grace (Clark) F.; B.A., Harvard, 1954; Ph.D., Stanford, 1958. Research fellow Kyoto U., Japan, 1958-59, U. Cal., Berkeley, 1959-61; faculty Cornell U., Ithaca, N.Y., 1961-62; faculty Cal. Inst. Tech., Pasadena, 1962—, prof. physics, 1966—. Mem. Am. Phys. Soc. Author: Regge Poles and S-Matrix Theory, 1963. Research, publs. on Regge poles, bootstrap theory.*

FRAZER, Alastair Campbell, Brit. biochemist; b. July 26, 1909; s. Wilson and Grace (Robbs) F.; ed. St. Mary's Hosp. Sch. Medicine; M.D., U. Birmingham; M.B., B.Sc., Ph.D., D.Sc., U. London; m. Hilary Garrod; children—Adrian, Graham, Stephen, Cynthia. Head ofcl. prof. dept., physiologist; with Med. Sch. of St. Mary Hosp., 1932-42; prof. biochem. medicine and pharmacology U. Birmingham, 1943. Cons., Atomic Energy Research Establishment, Harwell, to Brit. Army on metabolic illnesses; mem. Research Com. on Toxic Chem. Products in Agr.; mem. Com. for Safety of drugs Ministry Health; pres. Com. Toxicology; asso. mem. com. experts on manufactured additives FAO/WHO; pres. commn. manufactured additives IUPAC. Fellow Royal Coll. Physicians, Indsl. Research Assn. for Manufactured Canned Foods of Britain (pres.), Assn. Brit. Biol. Indsl. Research (pres.), Sci. Soc. Paris (corr.), Royal Flemish Acad. Scis. (hon.), Gastroent. Soc. Belgium (hon.), N.Y. Acad. Scis. (life). Research and numerous publs. on lipides, intestines, toxicology, physiology, pharmacology, exptl. biology, gastroenterology, nutrition. Address: 7 Carpenter Rd., Birmingham 15, Eng.

FRAZER, Benjamin Chalmers, Am. physicist; b. Birmingham, Ala., July 19, 1922; s. Luther Chalmers and Hallie (McCary) F.; B.S. Engring Physics, Ala. Poly. Inst. (now Auburn U.), 1947, M.S. Physics, 1948; Ph.D. in Physics Pa. State U., 1952; m. Ylia María Saldarriaga, Dec. 21, 1951; children—Hallie Mariana, Rowena Isabel, John Chalmers. Asso. physicist Brookhaven Nat. Lab., Upton, N.Y., 1952-55, physicist 1958-67, sr. physicist, dep. chmn. dept. physics, 1967—; asso. physicist Westinghouse Research Labs., Pitts., 1955-58, cons. physicist 1958-62. Guest scientist P.R. Nuclear Center, Mayaguez, 1962-63. Fellow Am. Phys. Soc.; mem. Am. Crystallographic Assn., Tau Beta Pi, Phi Kappa Phi, Eta Kappa Nu, Sigma Pi Sigma, Theta Chi. Studies in Neutron Diffraction of crystal and magnetic structures. Home: 44 Bieselin Rd., Bellport, N.Y. 11713. Office: Brookhaven Nat. Lab., Upton N.Y. 11973.*

FRAZER, Sir James (George), Scottish anthropologist; b. Glasgow, Scotland, Jan. 1, 1854; ed. Glasgow U.; Trinity Coll., Cambridge (Eng.) U., fellow; D.C.L., Oxford, LL.D., Glasgow, St. Andrews, Litt.D., Cambridge, Durham, Manchester, Ph.D., Athens, Dr. Honoris causa, univs. Paris, Strasbourg; m. Lilly Grove, 1896. Zarahoff lectr. on Condorcet Oxford, 1933. Fellow Royal Soc., 1920, Brit. Acad., Middle Temple (barrister-at-law, hon. bencher); mem. Inst. France (asso.), Prussian Acad. Sci. (corr.), Royal Netherlands Acad. Sci. (extraordinary). Author: Totemism, 1887, The Golden Bough, 1890; Pausanias' Description of Greece, translated with commentary, 1898; Pausanias and other Greek Sketches, 1900; Lectures on the Early History of Kingship, 1905; Adonis, Attis, Osiris, Studies in the History of Oriental Religion, 1906; Questions on the Customs, Beliefs, and Languages of Savages, 1907; Psyche's task, a Discourse concerning the influence of Superstition on the Growth of Institutions, 1909; Totemism and Exogamy, 1910; The Dying God, 1911; The Belief in Immortality and the Worship of the Dead, 3 vols., 1913, 22, 24; The Worship of Nature, 1926; The Gorgon's Head, and other literary pieces, 1927; The Fasti of Ovid, 5 vols., 1929; Myths of the Origin of Fire, 1930; The Fear of the Dead in Primitive Religion, 3 vols., 1933, 34, 36; Anthologia Anthropologica, a selection of Passages for the Study of Social Anthropology, 4 vols., 1938-39; also others. Known principally for extensive study on comparative folklore, magic and religion, totemism and taboo; defined concepts of imitative magic (using assumption that like produces like) and contagious magic (proposing that things which have once had contact with each other continue to affect one another after physical contact has been severed). Died Cambridge, Eng., May 7, 1941.

FRAZER, John Fries, Am. physicist, chemist; b. Phila., July 8, 1812; s. Robert and Elizabeth (Fries) F.; grad. U. Pa., 1830, M.A., 1833; LL.D., Harvard, 1857; m. Charlotte Cave, Sept. 1, 1838; 3 children, including Persifor. Experimented (with Alexander Dallas Bache) on determination of daily variations of magnetic needle in Am., circa 1828; participated in geol. survey of Pa., 1836; prof. chemistry and natural philosophy U. Pa., 1844-72, vice provost, 1855-68; lectr. Franklin Inst., editor its jour., 1850-66; a founder Nat. Acad. Scis., 1863; collaborator in study of interaction of aurora borealis and magnetic forces. Died Phila., Oct. 12, 1872.

FRAZER, Joseph Christie Whitney, Am. chemist; b. Lexington, Ky., Oct. 30, 1875; s. Joseph George and Mary Jane (Filson) F.; B.S., Ky. State U., 1897, M.S., 1898; Ph.D., Johns Hopkins, 1901; Sc.D., Kenyon Coll., 1926; m. Grace Carvill, Sept. 16, 1903; children—Joseph Hugh, Grace Carvill, Jean Cameron, Jeanne Henry. Asst. and asso. in chemistry Johns Hopkins, 1901-07, prof. chemistry since 1911, chmn. dept. of chemistry, 1916-36, B. N. Baker, prof. from 1921; chemist U. S. Bur. Mines, 1907-11. Mem. Soc. of Arts and Scis. (fgn., Utrecht), Phi Beta Kappa. Research work on osmotic pressure and vapor tension of solutions, catalysis and chem. behavior of surfaces. Died July 28, 1944.

FRAZER, Joseph Hugh, Am. physicist; b. Balt., May 13, 1905; s. Joseph Christie and Grace (Carvill) F.; A.B., Johns Hopkins, 1925, Ph.D., 1928; m. Catherine Lenore Leatherbury, Apr. 3, 1926; 1 son, Joseph Hugh. Nat. research fellow NRC, Washington, 1928-29; research asst. U. Bristol (Eng.), 1930; lectr. Victoria U., Manchester, Eng., 1930-31; instr. chemistry U.

Buffalo, 1931-33, asst. prof., 1933-41, asso. prof., 1942; chief interior ballistics lab. Ballistic Research Labs., Aberdeen (Md.) Proving Ground, 1945—. Mem. Am. Chem. Soc., Am. Phys. Soc., Army Ordnance Assn., Phi Beta Kappa, Sigma Xi. Patentee explosive compositions and processes for producing explosions on surfaces. Home: Box 107, R.D. 2, Havre de Grace, Md. 21078. Office: Interior Ballistics Lab., Ballistic Research Labs., Aberdeen, Md.*

FRAZER, Persifor, Am. geologist, chemist; b. Phila., July 24, 1844; s. John Fries (LL.D.), and Charlotte (Cave) F.; grad. U. Pa., 1862; postgrad. Atlantic squadron, 1862-63; Saxon Sch. Mines, Freiberg, Germany, 1866-69; D.-ès-Scis. Naturelles, U. France (1st foreigner so honored); m. Isabella Nevins Whelen. With U. S. Coast Guard Survey, S. Atlantic squadron, 1862-63; mineralogist and metallurgist U. S. Geol. Survey, 1869-70; instr. and prof. chemistry U. Pa., 1870-74; asst. geol. survey Pa., 1874-82. Fellow Geol. Soc. Am. A.A.A.S. Recipient John Scott legacy medal city of Phila., 1905. Author: Tables for the Determination of Minerals, 5 edits., 1874-1901; Reports of Subcom. of the Am. Com., Internat. Geol. Congress, 1888; Bibliotics, or the Study of Documents, 3 edits., 1894-1901. A founder, editor Am. Geologist, 1888-1905. Inventor colorimeter for determining relative intensity and color value of ink marks in handwriting; elucidated geology of southeastern Pa. Died 1909.

FRAZER, William Robert, Am. physicist; b. Indpls., Aug. 6, 1933; s. William Jay and Mildred (Dahlman) F.; A.B., Carleton Coll., 1954; Ph.D., U. Cal. at Berkeley, 1959; postgrad. U. Utrecht, Netherlands, 1956-57; m. Jane Zaiser, July 31, 1959; children—Bruce, Katherine. Mem. Inst. for Advanced Study, Princeton, N.J., 1959-60; faculty U. Cal., San Diego, 1960—, prof. physics, 1967—. Mem. Am. Phys. Soc. Author: Elementary Particles, 1966; also articles. Research on theory of interactions of elementary particles, principally strong interactions responsible for nuclear force. Home: 4917 Pacifica Dr., San Diego 92109. Office: U. Cal. at San Diego, La Jolla, Cal. 92038.*

FRAZIER, Charles Harrison, Am. surgeon; b. Germantown, Phila., Apr. 19, 1870; s. William W. and Harriet (Harrison) F.; A.B., U. Pa., 1889, M.D., 1892; postgrad., Berlin, 1895; m. Mary Spring Gardiner, Aug. 19, 1901. Prof. clin. surgery U. Pa., 1900-22, dean med. dept., 1902-09, John Rhea Barton prof. surgery, from 1922; on staff Univ. Hosp. Fellow A.C.S. Introduced (with William Gibson Spiller) operation for relief of trigeminal neuralgia by intracranial div. of sensory root of trigeminal nerve, circa 1902. Philadelphia, Pa. Died Aug. 26, 1936.

FRAZIER, William Carroll, Am. microbiologist; b. Madison, Wis., Sept. 26, 1895; s. William Sumner and Evalee (Palmer) F.; B.S., U. Wis., 1917, Ph.D., 1924; m. Sada Brown, May 22, 1924; 1 son, William R.; m. 2d, Hildegarde Martha Schultz, June 20, 1962. Faculty, U. Wis., Madison, 1919-24, 34—, prof. bacteriology, 1934-66, prof. emeritus, 1966—, chmn. dept. bacteriology, 1943-53; bacteriologist Bur. Dairy Industry U. S. Dept. Agr., 1924-34. Fellow A.A.A.S., Am. Acad. Microbiology; mem. Am. Soc. Microbiology, Am. Dairy Sci. Assn., Inst. Food Technologists, Internat. Assn. Milk, Food Environmental Sanitarians, Wis. Acad. Arts, Letters and Scis. Author, editor: (with others) Fundamentals of Dairy Science, 1935; author: (with others) Microbiology, General and Applied, 1951-56; Food Microbiology, 1958-67. Research, numerous publs. primarily in field of microbiology of cheeses, milk, other Foods. Home: 231 Westmorland Blvd., Madison, Wis. 53705.*

FREAS, Thomas Bruce, Am. chemist; b. nr. Newark, O., Nov. 2, 1868; s. Andrew and Mary (Bruce) F.; A.B., Stanford, 1896; Ph.D., U. Chgo., 1911; m. Mary Kuhn, Dec. 28, 1898; children—Royal Bruce, Joseph Kuhn. Prin. high sch., Hiawatha, Kan., 1896-97; chemist Western Electric Co., Chgo., 1897-98; asst. in chemistry U. Chgo., 1898-1903, instr. chemistry, 1904-11; mgr. Ernst Leitz apparatus house, Chgo., 1903-04; instr. chemistry, U. of Chicago, 1904-11; successively asst. prof., asso. prof. and prof. chemistry Columbia, from 1911; pres. Thermo Electric Instrument Co., Irvington, N.J. Fellow A.A.A.S., Am. Geog. Soc., N.Y. Acad. Scis. Died Mar. 15, 1928.

FRECH, Fritz Daniel, German geologist; b. Berlin, Mar. 16, 1861; s. Friedrich and Anna (Friedländer) F.; Ph.D., Berlin, 1885; m. Veronika Klopsch, 1894; Became mem. faculty U. Halle (Germany), 1887; named asso. prof. U. Breslau (now Wroclaw, Poland), 1891, prof., 1897; chief staff geologist, army command on Syrian front, World War I. Succeeded von Zittel as pres. Internat. Commn. for Palaeontologia Universalis. Mem. Paleontol. Soc. (v.p. 1912). Contbr. articles to sci. jours. Research on paleozoic and triassic fauna, corals from German Devonian period, also from Araxes area, China, alpine triassic; studied mussels, brachioposs, amnonites; made studies in hist. geology, especially that of Devonian period. Died Aleppo, Syria, Sept. 28, 1917.

FRÉCHET, Maurice René, French mathematician; b. Maligny, France, Sept. 2, 1878; s. Jacque Gédéon and Zoé (Cellier) F.; ed. Lycée Buffon, 1893-97, Lycée Saint Louis, 1898-99; D.Sc., 1906; m. Suzanne Carrive, Sept. 1908; children—Hélène, Henri, Denise, Alain. Prof. Besançon (France) schs., 1907; prof. Nantes, France, 1908; instr. U. Rennes (France); prof. mechanics U. Poitiers (France), 1910-19; prof. higher calculus U. Strasbourg (France), 1920-27, lectr. calculus of probabilities, 1928-33, prof. gen. math., 1933-35, prof. differential and integral calculus, 1935-40, prof. calculus of probabilities, 1941-48, ret., 1948. Decorated Legion Honor. Mem. French, Polish, Dutch acads. sci., Internat. Acad. Philosophy Sci. (past pres.). Author: Les Espaces Abstraits, 1928; Recherches Théoriques Modernes sur le Calcul des Probabilités, 1937-38; Les Probabilités Associées à un Système d'Évenements Compatibles et Dépendants, 1939-43; Pages Choisies d'Analyse Générales, 1953; Les Mathématiques et le Concret, 1955. Contbr. numerous articles to publs. First definition and theory of abstract sets; numerous studies in topology, calculus of probability, math. statistics, differential and integral calculus. Home: 2 rue Émile Faguet, Paris 14, France.*

FRED, Edwin Broun, Am. bacteriologist; b. Middleburg, Va., Mar. 22, 1887; s. Samuel Rogers and Catherine (Broun) F.; B.S., Va. Poly. Inst., 1907, M.S., 1908; Ph.D., U. Gottingen, Germany, 1911; LL.D., Lawrence Coll., 1945, Northwestern U., 1947, Mich. State U., 1955; Sc.D., Marquette U., 1945, Beloit Coll., 1946, U. N.C., 1946, Northland Coll., Ashland, Wis., 1946, U. Wis., 1958; m. Rosa Helen Parrott, June 21, 1913; children—Ann Conway, Rosalie Broun (Mrs. Thomas Moffatt). Faculty, U. Wis., Madison, 1913—, prof., 1918-58, emeritus prof., 1958—, dean Grad. Sch., 1934-43, dean Coll. Agr., dir. Agrl. Expt. Sta., 1943-45, pres., 1945-58, pres. emeritus, 1958—. Mem. U. S. NSF Board, 1950-56; member of the national advisory infectious disease council of NIH, 1956-57; mem. adv. com. for biology and medicine AEC, 1956-57, Am. Council on Edn. Commn. on Edn. and Internat. Affairs, 1958-61, Internat. Devel. Adv. Bd., 1959. Recipient Golden diploma U. Göttingen, 1961. Fellow A.A.A.S.; mem. Nat. Acad. Sci., Am. Philos. Soc., Am. Acad. Microbiology, Phi Beta Kappa, Sigma Xi, others. Author: (with F. Lohnis) Textbook of Agricultural Bacteriology, 1923; (with S. A. Waksman) Laboratory Manual of Microbiology, 1928; (with Baldwin and McCoy) Root Nodule Bacteria and Leguminous Plants, 1932. Home: 10 Babcock Dr., Madison, Wis. 53706.*

FRED, Mark, Am. chemist; b. Richmond, Ind., June 26, 1911; s. Samuel and Hannah (Simon) F.; B.S., U. Chgo., 1933, M.S., 1934, Ph.D., 1937; m. Louise Shuttles, Jan. 2, 1941; children—Susan (Mrs. Michael Schmerling), Jonathan. Chemist, Manhattan Dist. metall lab. U. Chgo., 1942-45; physicist Armour Research Found., Chgo., 1945-48; sr. chemist Argonne (Ill.) Nat. Lab., 1948—. Fellow Optical Soc. Am.; mem. Am. Chem. Soc. Research, numerous publs. in spectrochem. analysis, atomic spectra and structure of heavy elements. Home: 32 Lockport St., Plainfield, Ill. 60544. Office: 9700 S. Cass St., Argonne, Ill. 60439.*

FREDA, Vincent John, Am. physician; b. New Haven, Dec. 16, 1927; s. Matthew and Genevieve (Freda) F.; A.B., Columbia, 1948; M.D., N.Y. U., 1952; m. Carol Lynn Ury, Oct. 31, 1953; children—Pamela Ury, Andrew Giles. Clin. fellow Am. Cancer Soc., Francis Delafield Hosp., 1959; asst. obstetrics and gynecology Presbyn. Hosp., N.Y.C., 1960-62, asst. attending obstetrician-gynecologist, 1962—; faculty Columbia Coll. Phys. and Surg., 1960—, asst. prof., 1962—; career scientist Health Research Council City N.Y., 1962—. Diplomate Nat. Bd. Med. Examiners, Am. Bd. Obstetrics and Gynecology. Fellow Miami Obstet. and Gynecol. Soc. (hon.), N.Y. Obstet. Soc., Kansas City Gynecol. Soc. (hon.); mem. Am. Coll. Obstetricians and Gynecologists, N.Y. State, N.Y. County med. socs., A.M.A., Soc. for Study Blood, Internat. Soc. Blood Transfusion, N.Y. Acad. Scis., Soc. Alumni Sloane Hosp. for Women, Transplantation Soc. Research, numerous publs. on ABO and Rh hemolytic disease newborn, pathogenesis, antepartum mgmt., using amniocentesis and spectrophotmetric scanning amniotic fluid; pioneered field fetology, transfusion unborn baby in womb, performed 1st complete exchange blood transfusion during pregnancy on human unborn baby; co-originator anti-Rh gamma-immuno globulin for prevention Rh hemolytic disease newborn. Home: Warren Lane, Alpine, N.J. 07620. Office: 630 W. 168th St., N.Y.C. 10032.*

FREDENHAGEN, Karl Hermann Heinrich Philipp, German phys. chemist; b. Loitz, Pomerania, May 11, 1877; s. August and Caroline (Schultz) F.; student, Hanover, Darmstadt, Leipzig (all Germany); Ph.D. (under W. Nernst), Göttingen, Germany; m. Adelheid Bender, 1911; 2 sons, 2 daus. Asst. to H. Simon, Göttingen, to Wilhelm Ostwald and Des Coudres, Leipzig; became mem. faculty Leipzig, 1906, venia legendi for physics, 1907; prof. phys. chemistry, Griefswald, Germany, from 1923. Author: Grundlagen für den Aufbau einer Theorie der Zweistoffsysteme, 1950. Discovered causes of solubility and electrolytic dissociation in liquid solutions by taking a chem. approach; made successful exptl. studies of distbn. of solute, also of electromotive series in non-liquid solvents, liquid hydrogen fluoride as solvent. Died Griefswald, Apr. 4, 1949.

FREDERIC, Léon, Belgian physiologist; b. Grand, Belgium, Aug. 24, 1851; prof. physiology U. Liège (Belgium); mem. French Acad. Scis., 1926. Pioneer in physiology; investigated respiration and circulation; discovered heomycyanine in mollusks; wrote on autotomy of crustaceans; successfully experimented on cross circulation, 1887. Died Sept. 2, 1935.

FREDERICK II OF HOHENSTAUFEN, natural philosopher; b. Iesi, near Ancona, Italy, Dec. 26, 1194; s. Henry VI and Constance, heiress of Sicily; ed. under direction of his guardian, Innocent III; m. 1st, Constance of Aragon; m. 2nd, Yolande of Brienne, 1225 (d. 1227); m. 3rd, Isabella of England, 1235; children include: Henry, Conrad IV, Manfred, Enzio. King of Sicily, from 1198; Emperor of Holy Roman Empire, from 1220; King of Jerusalem, from 1229; undertook crusades to Holyland, 1227, 1229; put down revolt of eldest son, Henry, 1235; excommunicated twice by Gregory IX, by Innocent IV, 1245. Aided integration of Islamic thought and science into Christian west; tolerant toward Jews and Muslims; founded U. Naples, 1224; caused works of Aristotle and Ibn Rushd to be translated; prepared elaborate treatise on falconry containing a number of facts on bird anatomy not recorded previously (e.g., pneumaticity of bones, form of sternum, structure of lungs, runp glands) and remarks on mechanical conditions of flight and on bird migrations; experimented with artificial incubation of eggs; tried to determine whether vultures find food by sight or smell; established menageries including elephants, dromidaries, camels, panthers, etc.; took delight in disproving superstitions by experiment or common sense; gave set of laws to the Sicilies, 1231; promulgated 1st elaborate regulation of medical practice, 1240. Died Lucerna, Italy, Dec. 13, 1250.

FREDERICKSON, Arman Frederick, geologist; b. Winnepeg, Man., Can., May 5, 1918; s. Albert F. and Ethel May (Wilton) F.; came to U. S., 1923, naturalized, 1940; B.S. in Mining Engring., U. Wash., 1940; M.S. Metall. Engr., Mont. Sch. Mines, 1942; Sc.D., Mass. Inst. Tech., 1947; m. Mary Maxine Stubblefield, Sept. 23, 1943; children—Mary (Mrs. Jorge de la Torre), Clover (Mrs. James Niel), Penny, Kimberly, Sigrid. Mining engr., chief geologist Cornucopia Gold Mines (Ore.), 1940-41; instr. Mont. Sch. Mines, 1941-42, Mass. Inst. Tech., 1942-43; prof. geology Washington U., St. Louis, 1947-55; organizer, supr. geol. research Pan Am. Petroleum Co., Tulsa, 1955-60; prof., chmn. dept. earth and planetary scis. U. Pitts., 1960-65, dir. Center for Oceanography, 1964-66; v.p., dir. research King Resources, Denver, 1965—; pres. Denver Research Corp., N.Am. Fertilizer Corp.; exec. v.p. Rocky Mountain Mineral Corp.; Fulbright Research prof., Norway, 1951-52. Sec. clay minerals com. Nat. Acad. Sci.-NRC, 1951-52, pres., 1963; cons. to mining, metall. and petroleum cos. on oil and mineral exploration, mineralogy, geochemistry. Fellow Geol. Soc. Am., Am. Inst. Mining and Metall. Engrs.; mem. Am. Assn. Petroleum Geologists, Mineral. Soc., Geochem. Soc., Soc. Econ. Paleontologists and Mineralogists, Underwater Soc. Am. Patentee fertilizers, geochemistry, chem. engring. Research, publs. on fertilizers, application geochemistry to solution of exploration problems, earth sci. and oceanography, application computer techniques to petroleum exploration. Home: 5563 Southmoor Circle, Englewood, Colo. 80110. Office: Security Life Bldg., Denver 80202.*

FREDERICQ, Ecuyer Henri Robert Antoine, Belgian physiologist; b. Liege, June 11, 1887; s. Leon and Bertha (Spring) F.; M.D., U. Liege; m. Madeleine Nagels. Prof., U. Ghent, 1919-21; prof. U. Liege, 1921-56, dean, 1947-50, prof. emeritus, 1956—. Mem. Royal Acad. Medicine of Belgium (hon., past pres.), Royal Acad. (class of scis.), Nat. Acad. Medicine Paris (corr.), Acad. Scis. Bologne (fgn.), Acad. Medicine Mexico (fgn.), also numerous other Belgian and fgn. socs. Author: Aspects Actuels de la Physiology du Myocarde, 1927; Principles de Physiologie Générale, 4 edits., 1934-49; Traité Elémentaire de Physiologie Humaine, 3 edits., 1942-52. Research and over 200 publs. on exptl. physiology. Address: 13e, Place Xavier-Neujean, Liege, Belgium.

FREDERIK, Willem Steven, physician; b. Utrecht, The Netherlands, Mar. 28, 1928, 1913; s. Peter Willem and Johanna (Wilten) F.; M.D., Med. Sch. Utrecht, 1938; Ph.D., U. Utrecht, 1939; M.S., Harvard Sch. Pub. Health, 1950; m. Johanna C. A. Sanders, Dec. 28, 1945; children—Hanneke C. M. Willem S. P. Came to U. S., 1951, naturalized, 1956. Practice gen. medicine, Utrecht, 1940-42; physiologist Philips Radio Works, Eindhoven, Netherlands, 1942-44; research dir. KLM Royal Dutch Airlines, Netherlands, 1946-49; dir. research Liberty Mut. Ins. Co., Boston, 1951-61; research asso. dept. physics Harvard Sch. Pub. Health, Boston, 1951-54, lectr., 1954-61, biostatistician health service, Cambridge, Mass., 1962-66; cons. Cass Research Assos., Inc., Cambridge, 1951-62, dir. research, 1962-66; pres. Pharmatech Inc., Waltham, Mass., 1966—. Recipient A.M.A. Hull award, 1965. Mem. Am. Coll. Angiology, Am. Coll. Clin. Pharmacology and Chemotherapy, Am. Med. Writers Assn., Am. Pub. Health Assn., Internat. Acad. Law and Sci., Internat. Coll. Angiology, N.Y. Acad. Scis., Human Factor Soc. Research, numerous publs. on clin. pharmacology, safety engring., human fatigue, instrumentation, biostatistics; inventor differential flamoxymeter, respiron, alertomatic, sliptester, performance indicator. Home: 106 Suffolk Rd., Wellesley, Mass. 02181. Office: 223 Crescent St., Waltham, Mass. 02154.*

FREDERIKSE, H. P. R., physicist; b. The Hague, Netherlands, July 13, 1920; s. Jacobus A. R. and Sara (Malenbroek) F.; candidate U. Leiden (Netherlands), 1941, pre-doctorate, 1945, Ph.D., 1950; m. Yolanda Rossi, July 6, 1952; children—Julia, Peter, Thomas. Came to U. S., 1950, naturalized, 1956. Vis. lectr. physics Purdue U., West Lafayette, Ind., 1950-53; solid state physics sect. Nat. Bur. Standards, Washington, 1953-55, sect. chief, 1955—. Fulbright fellow, 1950-52; Guggenheim fellow, 1961-62. Recipient Exceptional Service award U. S. Dept. Commerce, 1963. Fellow Am. Phys. Soc. Important work includes research in absorption of helium at low temperatures, electronic transport in semiconductors (germanium, 3-5 compounds, oxides). Home: 9625 Dewmar Lane, Kensington, Md. Office: Solid State Physics Sect., Nat. Bur. Standards, Washington 20234.*

FREDERIKSEN, Eyvind, Danish engr.; b. Copenhagen, Denmark, Dec. 31, 1919; s. Laur and Helga (Bohm) F.; Ph.D. in Tech., Tech. U. Denmark; m. Gudrun Velmar; children—Svend, Lars. Mech. engr. Burmeister & Wain, Copenhagen, 1945-47; researcher Tech. U. Denmark, 1947-54, Stanford, 1954-56; with Titan firm, Copenhagen, 1956-58; prof. Tech. U. Denmark, 1958—. Mem. Danish Assn. Engrs. Research and numerous publs. on mech. vibrations; noise in machinery, especially reducing noise levels by suppressing devices; lubrication; rotors; bearings. Home: Almindingen 39, Soborg. Office: Technical University of Denmark, Rigensgade 13, Copenhagen K, Denmark.

FREDGA, Arne, Swedish chemist; b. Uppsala, Sweden, July 18, 1902; s. Carl and Elin (Cassel) F.; B.Sc., Uppsala U., 1924, M.Sc., 1929, D.Sc., 1935; m. Märta Brita Öhlin, Oct. 10, 1931; children—Sven, Karl, Kerstin, Märta (Mrs. Eric Granefelt). Asst. Chem. Inst., Uppsala U., 1930-35, asst. prof. chemistry, 1935-39, prof. organic chemistry, 1939—. Mem. Nobel Com. for Chemistry, 1944—, Swedish Natural Sci. Research Council, 1952-54. Recipient Celsius medal, 1963. Mem. Royal Swedish Acad. Scis., Belgian (hon.), Finnish (hon.) chem. socs., Swedish History of Sci. Soc. (pres. 1967—). Research, publs. on organic compounds of sulphur and selenium; contbns. in stereochemistry, especially steric correlations (quasiracemate method); investigations synthetic growth regulators. Home: 3A, Börjegatan, Uppsala. Office: Kemikum, Uppsala, Sweden.*

FREDHOLM, Erik Ivar, Swedish mathematician; b. Stockholm, Sweden, Apr. 7, 1866; student Stockholm Poly., 1885-86; student U. Uppsala, 1886-88, Ph.D., 1898; student U. Stockholm, 1888-93. Became prof. math. and physics U. Stockholm, 1906. Recipient Wallmark prize Swedish Acad. Sci. Corr. mem. French Acad. Scis. (Poncelet prize). Author memoir on class of functional equations, which treats integral equation named after him, and inaugurated modern theory of integral equations, 1903. Died Mörby, Sweden, Aug. 17, 1927.

FREDIANI, Harold Arthur, Am. chemist; b. N.Y.C., Dec. 23, 1911; s. Hugo J. and Mary (Ceruti) F.; B.A., State U. Ia., 1934, M.S., 1935; Ph.D., La. State U., 1937; m. Lois C. Hough, Sept. 5, 1935; children—Marita I. (Mrs. William H. Herbold), Judith L. (Mrs. Alan Yousten), Harold Arthur, Dale Steven. Faculty, La. State U., 1937-40; chief chemist Fisher Sci. Co., N.Y.C., 1940-47; asst. dir. control Merck & Co., Inc., Rahway, N.J., 1947-56; exec. dir. control Bristol Labs., Syracuse, N.Y., 1956—; Fulbright prof., Rome, Italy, 1951-52; supr. Town of Pompey, N.Y., 1964-. Recipient award La. Acad. Sci., 1937. Fellow Am. Inst. Chemists; mem. Am. Chem. Soc., Am. Pharm. Assn., N.Y. Acad. Scis., Chemists Club, Alpha Chi Sigma, Phi Lambda Upsilon. Contbr. numerous articles to profl. jours. Research in instrumentation and automation in chem. analysis. Home: R.D. 2, Pompey Center Rd., Manlius, N.Y. 13104. Office: P.O. Box 657, Syracuse, N.Y. 13201.*

FREDRICKSON, Donald Sharp, Am. physician; b. Canon City, Colo., Aug. 8, 1924; s. Charles Arthur and Blanche (Sharp) F.; student U. Colo., 1942-43; B.S., U. Mich. 1946, M.D., 1949; m. Henriette Priscilla Dorothea Eekhof, Sept. 5, 1950; children—Eric Henderikus, Rurik Charles. Clin. asso. NIH, Bethesda, Md., 1953-55, mem. sr. research staff lab. cellular physiology and metabolism, 1955-61, clin. dir., 1961—, head sect. on molecular disease lab. metabolism, 1962-66, chief lab. molecular diseases, 1966—, dir. NIH, 1966—; faculty Sch. Medicine George Washington U., 1956—, spl. lectr. internal medicine, 1959—; lectr. preventive medicine Georgetown U., 1963—. Mem. com. fats Food and Nutrition Bd. NRC, 1961—; mem. med. adv. bd. FAA, 1965—. Mem. Am. Heart Assn., Am. Oil Chemists Soc., Am. Physiol. Soc., Am. Soc. Clin. Investigation, Am. Soc. Human Genetics, Assn. Am. Physicians, Am. Coll. Cardiology, Med. Soc. Sweden, Phi Beta Kappa, Phi Kappa Phi, Alpha Omega Alpha. Research, numerous publs. in fat transport in circulation and in genetically determined diseases of fat metabolism, elucidation of several new diseases, including Tangier disease, devel. of new methods of diagnosis and study of effects of medications on fat metabolism. Home: 6615 Bradley Blvd. Office: NIH, Bethesda, Md. 20014.*

FREE, Alfred Henry, Am. biochemist; b. Bainbridge, O., Apr. 11, 1913; s. Alfred H. and Alice (Clymer) F.; A.B. magna cum laude, Miami U. (O.), 1934; M.S., Western Res. U., 1936, Ph.D., 1939; m. Helen Mae Murray, Oct. 18, 1947; children—Charles Alfred, Jane (Mrs. Richard Linderman), Barbara, Eric Scot, Penny Alene, Kurt Allen, James Jacob, Bonnie Anne, Nina Joann. Faculty, Western Res. U., 1935-46; cons. Ben Venue Labs., 1943-46; head biochem. sect. Miles Labs., Miles-Ames Research Lab., Elkhart, Ind., 1946-59, dir. Ames Research Lab., 1959-64, dir. tech. services Ames Co. div., 1964—. Mem. Am. Chem. Soc., A.A.-A.S., Am. Assn. Cereal Chemists, Am. Inst. Nutrition, Am. Inst. Chemists, Am. Inst. Vitamin Chemists, Am. Soc. Biol. Chemistry, Am. Diabetes Assn., Assn. Clin. Sci., N.Y. Acad. Scis., Soc. Exptl. Biology and Medicine, Am. Pub. Health Assn., Am. Soc. Med. Tech., Am. Bd. Clin. Chemistry, Phi Beta Kappa, Sigma Xi. Contbr. to Gould's Med. Dictionary, 1948. Editor sect. biol. abstracts Blood, 1938-67. Research, numerous publs. in simplification of urinalysis; introduced new methodology; devel. of lyophilization of human plasma for field transfusion; isolation estrogen from pregnant mare urine. Home: 3764 E. Jackson Blvd. Office: Ames Co., Elkhart, Ind. 46514.*

FREE, John Brand, English entomologist; b. Cambridge, Eng., Aug. 21, 1927; s. Frederick Charles and Gladys (Crosier) F.; B.A., Jesus Coll., Cambridge U., 1951, M.A., 1956; Ph.D., London (Eng.) U., 1954, D.Sc., 1967; m. Nancy Wilson Spiers, June 6, 1953; children—Anthony John, Mark Christopher, Nicola Caroline. Staff, Rothamsted Exptl. Sta., Harpenden, Herts., Eng., 1951. NRC Can. fellow Ont. Agrl. Coll., 1958-59. Author: (with C. G. Butler) Bumblebees, 1959; also numerous articles. Research on biology of bumble bee colonies, social orgn. of honeybees, beekeeping, pollination requirements of various hort. and agrl. crops, efficient mgmt. honey bees. Home: 2 Manland Av. Office: Rothamsted Exptl. Sta., Harpenden, Herts., Eng.*

FREED, Meier Ezra, Am. chemist; b. Phila., Oct. 20, 1925, s. Leon and Celia (Freed) F.; B.Sc., Pa. State Coll., 1948, M.Sc., 1949; Ph.D., U. Pa., 1960. With Wyeth Labs., Inc., Radnor, Pa., 1951—, research chemist, 1960—. Mem. Am. Chem. Soc., N.Y. Acad. Sci., Phila. Organic Chemists Club. Contbr. articles to sci. jours. Developed syntheses of a number heterocyclic ring systems; research related to penicillin salts, bis-acetamides, polycyclic indoles, spiroamines. Numerous patents in field. Home: 5423 Wynnefield Av., Phila. 19131. Office: Wyeth Labs., Radnor, Pa. 19131.*

FREEDBERG, A(braham) Stone, Am. physician; b. Salem, Mass., May 30, 1908; s. Hyman and Rachel Leah Freedberg; A.B., Harvard, 1929; M.D., Rush Med. Coll., U. Chgo., 1935; m. Beatrice Gordon, Aug. 29, 1935; children—Richard Gordon, Leonard Earl. Staff Beth Israel Hosp., Boston, 1938—, physician, 1964—, dir. cardiology unit, 1964—; faculty Harvard Med. Sch., 1941—, asso. prof. medicine, 1958—. Cons., mem., thyroid uptake calibration com. Med. Div. Oak Ridge Inst. Nuclear Studies, 1955-56. Diplomate Am. Bd. Internal Medicine. Mem. Am. Heart Assn. (dir. 1965—), A.A.A.S., Am. Soc. Clin. Investigation, Am. Physiol. Soc., Royal Soc. Medicine (London), Assn. Am. Physicians, Am. Thyroid Assn. (1st v.p. 1965-66). Studies, publs. on coronary heart disease and effects of altered thyroid function on heart; application of radioisotopes to study of thyroid function and measurement of thyroid-binding serum protein. Home: 30 Gardner Rd., Brookline, Mass. 02146. Office: 330 Brookline Av., Boston, Mass. 02215.*

FREEDMAN, Alfred Mordecai, Am. physician; b. Albany, N.Y., Jan. 7, 1917; s. Jacob A. and Pauline (Hoffman) F.; A.B., Cornell U., 1937; M.B., U. Minn., 1941, M.D., 1942; m. Marcia Kohl, Mar. 24, 1943; children—Paul, Daniel. Med. physiologist Army Chem. Center, Edgewood, Md., 1947-48; practice medicine, specializing in psychiatry, N.Y.C., 1948—; dir. psychiatry Flower and Fifth Av., Met., Bird S. Coler Hosps., 1960—; prof. psychiatry, chmn. dept. N.Y. Med. Coll., N.Y.C., 1960—; dir. Community Mental Health Center, Met. Hosp., N.Y.C., 1965—. Recipient Henry Wisner Miller award Manhattan Soc. Mental Health, 1964. Diplomate Am. Bd. Psychiatry and Neurology. Fellow Am. Orthopsychiat. Assn. (dir. 1962-64, editorial bd. 1964—), Am. Psychiat. Assn., Am. Acad. Child Psychiatry, N.Y. Acad. Medicine; mem. Am. Psychopath. Assn. (chmn. membership com.), Collegium Internationale Neuro-Psychopharmacologicum, Am. Coll. Neuro-Psychopharmacology, N.Y. Soc. Clin. Psychiatry (pres. 1967-68). Author: (with H. I. Kaplan) Comprehensive Textbook of Psychiatry, 1967. Research in narcotics addiction, childhood psychoses, brain damage in children, drugs and behavior; orgn. mental health services, particularly community mental health center. Home: 161 W. 86th St., N.Y.C. 10024. Office: 5 E. 102d St., N.Y.C. 10029.*

FREEDMAN, Daniel Xander, Am. psychiatrist; b. Lafayette, Ind., Aug. 17, 1921; s. Harry L. and Sophia (Feinstein) F.; B.A. cum laude in Social Relations, Harvard, 1947; M.D., Yale, 1951; m. Mary Catherine Neidigh, Mar. 20, 1945. Faculty, Yale U. Sch. Medicine, 1955-66, prof. dept. psychiatry, 1964-66, chief biol. scis. sect. dept. psychiatry, dir. grad. research tng. program in psychiatry and neurobehavioral scis., 1958-66; prof., chmn. dept. psychiatry U. Chgo., 1966—; staff Grace-New Haven Community, West Haven VA, Fairfield Hills, Connecticut Valley hosps., Yale Psychiat. Inst. Career investigator Nat. Inst. Mental Health, 1957—, guest worker, 1957-58. Recipient Keese prize for research Yale U. Sch. Medicine. Author: (with F. C. Redlich) The Theory and Practice of Psychiatry, 1966; also numerous articles. Research on action antihistamines in brain stem, role brain mechanisms in allergy, relationship psychosis brain function and allergy indicating altered central regulation peripheral autonomic function in schizophrenia, mechanism action psychotomimetic drugs showing linkage to endogenous substances in brain, drug abuse, LSD and related substances, causes and treatment. Home: 4950 S. Chicago Beach Dr., Chgo. 60615. Office: 950 E. 59th St., Chgo. 60637.*

FREEDMAN, H. H., Am. chemist; b. Malden, Mass., Mar. 5, 1924; s. Abraham and Minnie (Mintz) F.; B.S., Tufts U., 1949, M.S., 1950; Ph.D., Boston U., 1956; m. Marilyn J. Stone, May 23, 1951; children—David H., Michael A., Daniel J. Chemist, Aerovox Inc., New Bedford, Mass., 1950-51, Ionics, Inc., Cambridge, Mass., 1951-52; indsl. fellow Boston U., 1952-55; with Eastern Research Lab., Dow Chem. Co., Wayland, Mass., 1956—, asso. scientist, 1962—. Mem. Am. Chem. Soc., Sigma Xi. Research, publs. on stable carbonium ions and carbonions, synthesis and properties of small organic molecules of theoretical interest. Home: 141 Jackson St., Newton, Mass. 02159. Office: Box 400, Wayland, Mass. 01778.*

FREEDMAN, Lawrence Raphael, Am. physician; b. N.Y.C., Dec. 1, 1927; s. Hyman and Hanna (Epstein) F.; B.S., Yale, 1947, M.D., 1951; m. Rina Stahl, Apr. 3, 1955; children—Julia Ann, Leora Dina. Asst. in pathology Johns Hopkins, 1955-56; chief resident in medicine Yale-New Haven Hosp., 1956-57, asst. attending physician, 1961—; faculty Yale, 1957—, asso. prof. medicine, 1964—; cons. West Haven VA Hosp., 1964—. Chief of medicine Atomic Bomb Casualty Commn., Hiroshima-Nagasaki, 1962-64. Brown Jr. fellow, 1948-49; Markle scholar in med. scis., 1957-62. Mem. Infectious Diseases Soc. Am., Am. Fedn. for Clin. Research, Council on Epidemiology, Am. Heart Assn., Am. Soc. for Clin. Investigation, Sigma Xi, Alpha Omega Alpha. Editor-in-chief Yale Jour. Biology and Medicine, 1965—. Research on mechanisms underlying devel. of bacterial infections, particularly of kidney. Home: Rimmon Rd., Woodbridge, Conn. 06525. Office: 333 Cedar St., New Haven 06510.*

FREEDMAN, Lawrence Zelic, Am. psychiatrist; b. Gardner, Mass., Sept. 4, 1919; s. Samuel and Bessie (Lyon) F.; B.S., Tufts U., 1941, M.D., 1944; grad. N.Y. Inst. Psychoanalysis, 1955; m. Dorothy Robinson, Feb. 15, 1954; children—Bart J., Matthew E., Joshua E., Johanna J., Thomas L. Faculty med., law schs. Yale, 1946-58, fellow in psychiatry, 1946, physician in chief Psychiat. Clinic, 1951, chmn. study unit in psychiatry and law, 1953-58; fellow Center for Advanced Study in Behavioral Scis., Stanford, Cal., 1959-61, vis. scholar, 1968; 1st Founds.' Fund Research prof. in psychiatry U. Chgo., 1961—, chmn. bd. adult edn., 1966. Asst. étranger U. Paris (France) Salpetrière Hosp., 1950; faculty Cambridge (Eng.) U., 1958; chmn. research sect. Pan-Am. Congress for Psychoanalysis, 1964. Mem. Group for Advancement Psychiatry (chmn. psychiatry and law com. 1953), Am. Psychiat. Assn. (chmn. legal aspects psychiatry com. 1958), A.M.A. (chmn. panel legal aspects community mental health 1958), Am. Orthopsychiat. Assn., Sigma Xi, Phi Beta Kappa. Editor: (with H. D. Lasswell) Law, Conformity, and Psychiatry; (with others) Moral Problems in Psychoanalysis. Research, publs. on relationship between anxiety, aggression and sexuality in infrahuman primates (experimentally determined), psychosomatic relationship between human anxiety and pain (clinically ascertained); determined typology, etiology of non-conformist behaviors. Home: 1120 E. 48th St. Office: 1453 Drexel Av., Chgo. 60637.*

FREEDMAN, Ronald, sociologist; b. Winnipeg, Man., Can., Aug. 8, 1917; s. Issador and Ada (Greenstone) F.; came to U. S., 1924, naturalized, 1938; B.A., U. Mich., 1939, M.A. 1940; Ph.D., U. Chgo., 1947; m. Deborah Gail Selin, May 4, 1941; children—Joseph Selin, Jane Ilene. Faculty, U. Mich., Ann Arbor, 1946—, prof. sociology, 1957—, dir. Population Studies Center, 1960—, research asso. Survey Research Center, acting chmn. dept. sociology, 1950, 56. Cons. to population programs. Guggenheim fellow, 1957-58; Fulbright fellow, 1957-58. Recipient Class of 1919 award for excellence in teaching U. Mich., 1952. Mem. Population Assn. Am. (past. pres.), Internat. Population Union (v.p. 1965—), Am. Sociol. Assn., Sociol. Research Assn., Am. Statis. Assn. Author (with A. Hawley, H. Miner, W. Landecker, C. Lenski) Principles of Sociology, 1957; (with Whelpton, Campbell) Family Planning, Sterility, Population Growth, 1959; also articles. Editor: Population: the Vital Revolution, 1964. Research on factors affecting variations in birth rate and use family planning in both developed and underdeveloped countries, research on factors related to types and amounts internal migration. Home: 1404 Beechwood Rd., Ann Arbor, Mich. 48103.*

FREEMAN, Bruce Clark, Canadian geologist; b. Brighton, Ont., Can., Feb. 8, 1900; B.A., McMaster U., 1924; Ph.D., U. Chgo., 1932. Tchr. high schs., Ont., 1925-28; instr. geology U. Chgo., 1932, Ohio State U., 1934-40; mining geologist Ont.-Que. gold belt, 1933; geologist Geol. Survey Can., 1935-36.

Contbd. knowledge of Canadian pre-cambrian petrology, origin of metalliferous deposits. Died 1940.

FREEMAN, George Fouche, Am. plant geneticist; b. Maple Grove, Ala., Nov. 4, 1876; s. George W. and Laura C. (Nuckols) F.; B.S., Ala. Polytechnic, 1903; Sc.D., Harvard, 1917; m. L. Adelle Blachly, Kan., 1906; children—Eleanor Adelle (Mrs. G. C. Wilson), George Donald. Prin. Downer Inst., Beech Island, S.C., 1899-1901; asst. instr. in botany. Mass. Coll. Agr., 1903-04; asso. botanist. Kan. State Agr. Coll., 1904-06; plant breeder. Ariz. Agr. Expt. Sta., 1906-18; acting dean. Coll. Agr., dir. Expt. Sta., Ariz., 1915-16; chief plant breeding. Soc. Sultanienne d'Agr., Cairo, Egypt, 1918-22; chief, dept. cotton breeding Tex. Agrl. Expt. Sta., 1922-23; ofcl. agriculturist and economist mission for French Govt. to Indo-China, 1923; dir. gen. Service Technique d'Agriculture et de l'Enseignement Professionnel, Republic of Haiti, 1923-—. Died 1930.

FREEMAN, Gordon Russel, Canadian chemist; b. Hoffer, Sask., Can., Aug. 27, 1930; s. Winston Spencer Churchill and Aquila (Chapman) F.; B.A., U. Sask., 1952, M.A., 1953; Ph.D., McGill U., 1955; D.Phil., Oxford (Eng.) U., 1957; m. Phyllis Joan Elson, Sept. 8, 1951; children—Michele Leslie, Mark Russel. Postdoctoral fellow Centre D'Etudes Nucleaires De Saclay, France, 1957-58; faculty U. Alta., Edmonton, Can., 1958-—, prof. chemistry, 1965-—, head radiation research center, 1964-—, chmn. div. phys. chemistry, 1965-67. Fellow Chem. Inst. Can.; mem. Faraday Soc., Radiation Research Soc. Research, numerous publs. on mechanisms chem. reactions induced by high energy radiation, by high temperatures, and by electric discharge through nitrogen gas. Home: 14619-78 Av., Edmonton, Alta., Can.*

FREEMAN, Howard Edgar, Am. sociologist; b. N.Y.C., May 28, 1929; s. Herbert M. and Rose (Herman) F.; B.A., N.Y. U., 1948, M.A., 1950, Ph.D., 1956; m. Sharon Kleban, Aug. 7, 1953; children—Seth Richard, Lisa Jill. Social scientist RAND Corp., 1955-56; research asso. Harvard Sch. Pub. Health, 1956-62; prof. social research Florence Heller Grad. Sch. Advanced Studies Social Welfare, Brandeis U., 1960-—, dir. Research Center, 1960-68. Cons. to govt. and communities, 1956-—; cons. Russell Sage Found., N.Y.C. Mem. profl. adv. com. Mass. Mental Health Assn., 1961-—. Bd. dirs. Mass. Health Research Found., 1963-—. Fellow WHO, Eng., 1962; recipient Hofheimer prize Am. Psychiat. Assn., 1963. Author: The Mental Patient Comes Home, 1963; The Clinic Habit, 1967. Editor: The Handbook of Medical Sociology, 1963. Asso. editor Am. Sociol. Rev., 1962-67. Social Problems, 1961-67, Jour. Community Mental Health, 1963-—, Jour. Health and Social Behavior, 1967-—. Author numerous articles in field. Studies of former mental patients and sociol. problems of health care. Home: 15 Park Av., Newton, Mass. 02158. Office: Brandeis U., Waltham, Mass. 02154.*

FREEMAN, John Clinton, Jr., Am. meteorologist; b. Houston, Aug. 7, 1920; s. John Clinton and Ann (Dodson) F.; B.A., Rice Inst., 1941; M.S., Cal. Inst. Tech., 1942; postgrad. Brown U., Inst. Advanced Study; Ph.D., U. Chgo., 1952; m. Marjorie Schaefer, June 14, 1947; children—John Clinton III, Walter, Jill, Cathryn, Helen, Paul. Research asst. Brown U., 1946-48; meteorologist U. S. Weather Bur., Washington, 1948-49; mem. Inst. Advanced Study, Princeton, N.J., 1949-50; research asso. U. Chgo., 1950-52; sr. engr. Cook Research Lab., Skokie, Ill., 1951-53; owner Gulf Cons., Houston, 1955-62; sr. scientist Nat. Engring. Sci. Co., Houston, 1962-66; asso. prof. Tex. A. and M. Coll., 1952-55, U. Houston, 1958; prof., dir. research Inst. Storm Research, U. St. Thomas, Houston, 1958-—. Recipient Meisinger award Am. Mem. Am. Meteorol. Soc. (Meisinger award 1951, citation for outstanding work in establishment Tornado Warning Radar Network 1961), Am. Geophys. Union, Marine Tech. Soc., Sigma Xi. Research, publs. primarily in field of fluid dynamics, applied meteorology, oceanography; discovered finite waves on stable layers in atmosphere, methods to increase accuracy of balloon trajectories, new methods to find height storm tides, non-linear wave cyclone theory, theory of atmospheric front formation, theory of wave resulting from breaking dam, theory of motions of atmospheric jet stream. Home: 2516 Commonwealth St., Houston 77006.*

FREEMAN, John Ripley, Am. civil and mech. engr.; b. W. Bridgeton, Me., July 27, 1855; s. Nathaniel D. and Mary Elizabeth (Morse) F.; B.S., Mass. Inst. Tech., 1876; Sc.D., Brown U., 1904, Tufts, 1905, Sachs Tech. Hochschule, 1926, U. Pa., 1927, Yale, 1931; m. Elizabeth Farwell Clark, Dec. 27, 1887. Prin. asst. engr. Water Power Co., Lawrence, Mass., 1876-86; prin. asst. to Hiram F. Mills, cons. engr., 1878-86; chief engr. Associated Factory Mut. Ins. Cos., 1886-96; cons. engr. on water power and mill constrn. to varicus large mfg. corps., 1886-—; made extensive studies of water supply for Greater N.Y., for finance dept., 1899-1900; chief engr. investigations Charles River Dam, Boston Harbor, 1903; cons. engr. Boston Met. Park Commn. on san. and drainage problems, 1903-04; water commr., Winchester, Mass., 1882-86; engr. mem. Mass. Met. Water Bd., 1895-96; mem. R.I. Met. Park Commn., 1904; mem. spl. commn. additional water supply, N.Y.; planned water power devels. Feather River, Cal., 1904-05, St. Lawrence River, Long Sault, 1905, regulation of Gt. Lakes, 1924-26; cons. engr. water supplies of Nashua, Los Angeles, San Diego. Balt., City of Mexico; cons. engr. N.Y. Bd. Water Supply since 1905; in charge water power investigations N.Y. State Water Supply Commn., 1906-07; cons. engr. Isthmian Canal locks and dams, 1907, cons. engr. San Francisco water supply, from 1910 (planned Hetch Hetchy water supply); cons. Canadian Govt. on water power conservation, 1910 chmn. Nat. Adv. Com. for Aeronautics, 1918-19; cons. engr. for Chinese Govt., Grand Canal Improvement Bd., 1917-—. Twice recipient Normal medal Am. Soc. C.E. for papers contbd. to its trans. Fellow Am. Acad. Arts and Scis. Died Oct. 6, 1932.

FREEMAN, Leonard, Am. surgeon; b. Cin., Dec. 16, 1860; s. Zoeth and Ellen (Ricker) F.; B.S., U. Cin., 1882; M.D., Med. Coll. Ohio, 1886; postgrad, univs. Göttingen, Berlin and Vienna; M.A., U. Denver, 1901; m. Amanda Frank, 1894 (dec.); m. 2d, Jeanne Wright, July 27, 1905; children—Frank, Paul R., Leonard. Prof. surgery U. Colo.; surgeon St. Joseph's Hosp., Denver. Fellow A.C.S., A.M.A., Am. Surg. Assn. Author: Skin Grafting, 1912. Died Dec. 27, 1935.

FREEMAN, Ruth Benson (Mrs. Anselm Fisher), Am. nurse; b. Methuen, Mass., Dec. 5, 1906; d. Wilbur and Amy (Lawson) Freeman; R.N., Mt. Sinai Hosp. Sch. Nursing, N.Y.C., 1927; B.S., Columbia, 1934; M.A., N.Y. U., 1939, Ed.D., 1951; m. Anselm Fisher, 1927; 1 dau., Nancy R. Faculty, Johns Hopkins Sch. Hygiene and Pub. Health, Balt., 1950-—, prof. pub. health adminstrn., 1962-—. Mem. expert panel nursing WHO, 1958; pres., dir. Nat. Health Council, 1959-60; chmn. nursing research study sect. Dept. Health, Edn. and Welfare, NIH, 1959-62. Recipient Pearl McIver award, 1958; Nursing award N.Y. U., 1958. Fellow Am. Pub. Health Assn. (past dir.); mem. Am. Nurses Assn., Nat. League Nursing (past pres., dir.), Am. Assn. U. Profs., Acad. Polit. Sci., Am. Mgmt. Assn., Royal Soc. Health, Am. Nat. Council for Health Edn. of Pub., Sigma Xi, Sigma Theta Tau. Author: Techniques of Supervision in Public Health Nursing, 1945; (with Ramona Todd) Health Care of the Family, 1946; Public Health Nursing Practice, 1950, rev. edit., 1963; (with E. M. Holmes) Administration of Public Health Services, 1960. Home: 616 Massachusetts Av. N.E., Washington 20002. Office: 615 N. Wolfe St., Balt. 21205.*

FREEMAN, Smith, Am. physician; b. Joplin, Mo., Aug. 14, 1906; s. Arthur Bell and Anna (Smith) F.; M.S., Northwestern, 1930, Ph.D., 1933, M.D., 1938; m. to Willie Mae Clifton, September 1937 (deceased in 1958); children—Smith, Ann, Peter Burns. Instructor in the Northwestern University Med. Sch., 1939-40, asst. prof., 1940-42, asso. prof., 1943-45, prof. of physiology, 1945-46, prof. of exptl. medicine, 1947-51, prof. biochemistry, 1951-—, chmn. dept., 1951-65; head biochemistry section Mayo Clinic, 1946-47; cons. in physiology Hines Hosp. for Veterans, Chgo. 1944-46, 47-—, director of medical research 1947-58, chairman Hospital Research and Education Committee, 1958-—; cons. Wesley Meml. Hosp., Childrens Meml. Hosp. Mem. med. and sci. socs. Contbr. articles pertaining to medicine in med. and sci. jours. Specialist in applied physiology and biochemistry; research on bone and calcium metabolism; kidney, liver and endocrine function. Home: 247 E. Chestnut St., Chgo. 60611. Office: 303 E. Chicago Av., Chgo. 60611.

FREEMAN, Thomas, civil engr., explorer; b. Ireland. Came to Am., 1784; surveyed for new capital of U. S. 1794-96, surveyed entire northern portion of dist.; started 1st topog. survey of Washington, D.C., resigned to accept. commn. as U. S. surveyor to chart boundary between U. S. and Spanish possessions; explored Red and Arkansas rivers, 1806; surveyed boundary between Tenn. and Ala., 1807; commd. U. S. surveyor pub. lands south of Tenn., 1811; 1st to chart accurately course of lower Red River. Died Huntsville, Ala., Nov. 8, 1821.

FREEMAN, Walter, Am. physician; b. Phila., Nov. 14, 1895; s. Walter Jackson and Corinne (Keen) F.; A.B., Yale, 1916; M.D., U. Pa., 1920; M.S., Georgetown U., 1929, Ph.D., 1931; m. Marjorie Lorne Franklin, Nov. 3, 1924; children—Lorne (Mrs. Donald Canter), Walter Jackson III, Franklin, Paul, Robert FitzRandolph. Practice medicine, specializing in neurology, psychiatry, Washington, 1926-54, Los Altos, Cal., 1955-61, Sunnyvale, Cal., 1961-67; ret.; cons. neurology Walter Reed Army Hosp., 1944-54, various Cal. State Hospitals, 1956-—; Diplomate Am. Bd. Psychiatry and Neurology (past pres.). Mem. Royal Medico-Psychol. Assn., A.M.A., Am. Neurol. Assn., Am. Psychiat. Assn., Am. Assn. Neuropathologists, Soc. Biol. Psychiatry. Author: Neuropathology, 1933; (with J. W. Watts) Psychosurgery, 1942, 50; (with M. F. Robinson) Psychosurgery and the Self, 1954; The Psychiatrist, 1967. Research, numerous publs. on diseases of nervous system, weights of organs of patients dying with mental diseases, brain operations for treatment of mental disorders and pain. Home: 425 Collingwood St., San Francisco 94114.*

FREER, Otto (Tiger), Am. laryngologist; b. Chgo., Aug. 8, 1857; s. Joseph Warren and Catharine (Gatter) F.; M.D., Rush Med. Coll., 1879, followed by a semester each in univs. Munich, Vienna and Heidelberg; m. Agnes (Rand) Lee, May 18, 1911. Intern Cook County Hosp., 1879-80; specialist in diseases of nose, throat and ear; attending laryngologist Henrotin Meml. Hosp., Chgo.; prof. laryngology Chgo. Policlinic; Died April 21, 1932.

FREGE, (Friedrich Ludwig) Gottlob, German mathematician, philosopher; b. Wismar, Germany, Nov. 8, 1848; s. Karl Alexander and Auguste (Bialloblotzky) F.; student, Jena, Germany; Ph.D., Göttingen, Germany, 1873; m. Margarete Lieseberg; 1 adopted son. Became mem. faculty U. Jena, 1874, asso. prof., 1879, hon. prof., 1896, emeritus, 1918. Author: Begriffsschrift, eine der arithmetischen nachgebildete Formelsprache des reinen Denkens, 1879; Die Grundlagen der Arithmetik, 1884; Grundgesetze der Arithmetik, begriffschriftlich abgeleitet (on a theory of classes of numbers, some of which are mems. of classes they describe, others not; Frege abandoned the theory after Bertrand Russell pointed out a paradox), 2 vols., 1893-1903. also articles, lectures. Founder modern formal logic; studied relation of logic to math.; developed new set of math. definitions and symbols. Died Bad Kleinen, Germany, July 26, 1925.

FREGLY, Melvin James, Am. physiologist; b. Patton, Pa., May 26, 1925; s. Frank and Nellie (Wellings) F.; B.S., Bucknell U., 1949, M.S., 1949; Ph.D., U. Rochester, 1952; m. Marilyn Sumner Southwick, May 30, 1956. Instr., Harvard Med. Sch., 1952-56; faculty U. Fla. Coll. Medicine, Gainesville, 1956-—, prof. physiology, 1965-—, asst. dean grad. edn., 1967-—. Consultant, Strasenburgh Pharm. Company, in Rochester, N.Y., 1965-67. Recipient Internat. Physiol. Congress travel award, Brussels, Belgium, 1956, Buenos Aires, Argentina, 1959, Leiden, Netherlands, 1962, Tokyo, Japan, 1965. Mem. Am. Physiol. Soc., Endocrine Soc., Am. Thyroid Assn. (Travel award Rome Italy 1965), Am. Soc. Zoologists, A.A.A.S., N.Y. Acad. Scis., Soc. for Exptl. Biology and Medicine (sec. Southeastern sect. 1963-65), Sigma Xi. Research, numerous publs. on role endocrine glands in devel. and maintenance exptl. high blood pressure, role adrenal cortex and thyroid glands in body temperature regulation, physiol. and environmental factors affecting salt intake and water balance in animals. Home: 6581 N.W. 20th Pl., Gainesville, Fla. 32601.*

FREHSE, Helmut, German chemist; b. Cologne, Germany, June 7, 1927; s. Jakob and Maria (Hermanns) F.; Ph.D. in Enzyme Chemistry, U. Cologne, 1954. Sci. asst. Inst. for Fermentation and Enzyme Chemistry, U. Cologne, 1954-56; sci. asst. German Inst. for Lipid Research, Münster, Germany, 1956-57; residue chemist Farbenfabriken Bayer A.G., Leverkusen, West Germany, 1957-—, head residue dept., 1965-—. Contbg. author: Ency. of Plant Physiology, vol. VII 1957; Textbook of Chem. Toxicology, vol. II. Research, publs. on enzymatic oxidation and dehydrogenation of saturated and unsaturated higher fatty acids by plants, electrophoretic studies on milk proteins; analysis, behavior and metabolism of pesticides in plants and animals, devel. and application of analytical methods for pesticide residues in food. Home: 43, Hans-Sachs-Str., 509 Leverkusen 5. Office: Biol. Inst., Farbenfabriken Bayer A.G., 509 Leverkusen-Bayerwerk, West Germany.*

FREI, Emil, III, Am. physician; b. St. Louis, 1924; M.D., Yale, 1948. Intern St. Mary's Group Hosps., St. Louis, 1948-49; resident pathology Barnes Hosp., St. Louis, 1952-53; resident internal medicine St. Louis U., 1953-54, VA Hosp., St. Louis, 1954-55; chief gen. medicine br. Nat. Cancer Inst., Bethesda, Md., 1955-65; head developmental therapeutics, asso. dir. clinical research, MD Anderson Hosp., Houston, 1965-—. Diplomate Am. Bd. Internal Medicine, 1957. Mem. A.M.A.; Am. Assoc. for Cancer Research; Am. Soc. Hematology; Am. Soc. Clinical Oncology; chmn., Antineoplastic Disease Panel, Drug Efficacy Study of Nat. Acad. Scis. Identified new agents and better methods of using conventional agents in treating leukemia, cancer; clinical pharmacologic studies of such agents; studies on complications such as hemorrhage in cancer patients. Home: 4939 Valkeith St., Houston 77035. Office: 6723 Bertner, Houston 77025.*

FREI, Walter Carl, Swiss veterinarian; b. Rietheim, Nov. 12, 1882; s. Carl and Caroline (rudolf) F.; ed. U. Zurich, U. Munich; D.V.M.; D.V.M. honoris causa U. Vienna; m. Carolina Garn; 1 dau., Gerda. Asst., Inst. Vet. Physiology, Zurich, prof. vet. path. and bacteriology and dir., Vet. Pathol. Inst., U. Zurich, 1911-52; Agronomic Acad. of Bonn-Poppelsdorf, Pasteur Inst., Brussels, Inst. Vet. Research of Pretoria, Higher Vet. Inst., Berlin, Robert Koch Inst., Berlin. Mem. Med. Soc. Zurich, Soc. Natural Scis. Zurich, Soc. Swiss Agronomists, Soc. Museums Zurich; hon. mem. Soc. Swiss Veterinarians, Soc. Veterinarians of Aargau, Royal Coll. Vet. Surgeons (London). Author: Prophylaxis der Tierseuchen, 1921; Beiträge Milchdrüse und Weibl. Geschlechtsorgane (in Handbuch der Path. Anatomie), 1925; Sterrilität der Weibl. Haustiere, 1927; Tierseuchen, 1949. Research in biochemistry and path. physiology. Address: Hohestrasse 68, Zolikon, Switzerland.

FREI, Wilhelm S., German dermatologist; b. Neustadt (now Prudnik), Upper Silesia, Sept. 5, 1885; with Robert Koch Inst., Berlin, Inst. Hygiene, Göttingen, Dermatol. Clinic U. Breslau (all Germany); became pvt. docent dermatology Municipal Hosp.,

Berlin-Spandaus, 1923, prof., 1926, chief physician, dermatol. dept., 1929; resident physician in dermatology Montefiori Hosp., N.Y.C., 1937-38. Made studies of exptl. syphilis and allergic diseases of skin; devised Frei test (intradermal injection of antigen made of causative virus) for venereal lymphogranuloma (Frei's disease), 1925. Died Jan. 1943.

FREIBERG, Samuel R., Am. biologist; b. S.I., N.Y., Apr. 14, 1924; s. Abraham and Dora (Weinstein) F.; student Rugers U., 1942-43, U. N.H., 1943-44, Biarritz (France) U., 1945-46; B.Sc. cum laude, Rutgers U., 1948, Ph.D. in Plant Physiology, 1951; m. Marcia Louise Krasny, Sept. 17, 1949; children—Raymond Lawrence, Nancy Carol. Asst. dir. research United Fruit Co., Norwood, Mass., 1960-62, dir. research Central Labs., 1962-65; dir. research IRI Research Inst., N.Y.C., 1965—, v.p., 1966—. Vis. fellow Cornell U., 1957-58. Mem. Am. Inst. Biol. Scis., Agrl. Research Inst., Am. Soc. Hort. Sci., Am. Soc. Plant Physiologists, Am. Soc. Agronomy, Inst. Food Technologists. Contbg. author: Fruit Nutrition, 1966. Research, publs. on nitrogen and urea metabolism, plant nutrition and growth, polyphenols and polyphenoloxidases, postharvest fruit physiology, plant and animal food productivity, fruit flavor volatiles. Home: 138 Herrontown Rd., Princeton N.J. 08540. Office: 1 Rockefeller Plaza, N.Y.C. 10020.*

FREIBERGER, Walter Frederick, mathematician; b. Vienna, Austria, Feb. 20, 1924; s. Felix and Irene (Tagany) F.; B.A., U. Melbourne (Australia), 1947, M.A., 1949; Ph.D., U. Cambridge (1953); A.M. (Ad eundem), Brown U., 1958; m. Christine Mildred Holmberg, Oct. 6, 1956; children—Christopher Alan, Andrew James. Came to U. S. 1955, naturalized 1962. Sci. research officer Aero. Research Labs., Melbourne, 1947-49, sr. research officer, 1953-55; faculty Brown U., Providence, 1955—, dir. computing lab., 1963-—, prof. applied math., 1964——. Mem. adv. com. Mass. Inst. Tech. Computation Center. Mem. Inst. Math. Statistics, U. Stockholm, Sweden, 1962-63. Australian govt. overseas fellow U. Cambridge, 1949-53; Fulbright fellow, 1955-56; Guggenheim fellow, 1962-63. Mem. Am. Math. Soc., Inst. Math. Statistics, Soc. Indsl. and Applied Math., Assn. Computing Machinery, Am. Meteorol. Soc. (profl.). Editor-in-chief International Dictionary of Applied Mathematics, 1960; asso. editor Math. Revs., 1957-62; mng. editor Quar. Applied Math., 1966—. Study of time-series analysis; math. theories of plasticity and elasticity; numerical analysis. Home: 85 Keene St., Providence 02906.

FREIDLINA, Rakhil Khatskelevna, Russian organic chemist; b. Sept. 20, 1906; grad. Moscow U., 1930. With Sci. Research Inst. Insectofungicides, 1930-34; mem. staff Inst. Organic Chemistry, USSR Acad. Scis., 1935-39, 41-54, chief lab. inst. Organo-Elemental Compounds, 1954——; faculty Moscow Inst. Fine Chem. Tech., 1938-41. Mem. USSR Acad. Scis. (corr.). Author: Synthetic Methods in the Field of Organometallic Compounds of Arsenic, 1945; The Telomerization Reaction and Chemical Changes in Telomers, 1957; The Telomerization Reaction and New Systematic Materials, 1960; co-author: The chemistry of Quasicomplex Organometallic Compounds and the Phenomena of Dynamic Allotropy, 1947; Study of the Chemistry of the Conversion of Polychloride Hydrocarbons and Related Compounds, 1956. Research and synthesized structure and properties of organic compounds of mercury, arsenic, tin, antimony, lead, titanium, silicon, zirconium, boron, florine, chlorine; discovered homolytic isomerization of organic compounds in solutions; reactions of telomerization of olefins, chem. transformations of telomers. Home: 1-aya Cheremushkinskaya 3. Office: Institute of Organoelemental Compounds, Leninskii Prospect 31, Moscow, USSR.

FREIDSON, Eliot Lazarus, Am. sociologist; b. Boston, Feb. 20, 1923; s. Joseph and Grace (Backer) F.; student U. Me., 1940-41; Ph.B., U. Chgo., 1947, M.A., 1950, Ph.D., 1952; m. Judith Lorber, Aug. 9, 1966; children by previous marriage—Jane Beatrice, Oliver Eliot. Ford postdoctoral research fellow U. Ill., 1952-54; Social Sci. fellow Russell Sage Found., N.Y.C., 1955-56; faculty Coll. City N.Y., 1956-60; prof. sociology N.Y. U., N.Y.C., 1960—. Mem. Am. (chmn. med. sociology sect. 1963-64), Internat. (chmn. research subcom. on med. sociology 1967—) sociol. assns., Am. Anthrop. Assn., A.A.A.S., Am. Assn. U. Profs., Soc. Study Social Problems. Author: Patients' Views of Medical Practice, 1961; Medicine: Sociological Analysis of a Profession, 1969. Editor: Student Government, Student Leaders and The American College, 1955; The Hospital in Modern Society, 1963; asso. editor Social Problems, 1961-67; editor Jour. Health and Social Behavior, 1966——. Study of profl. practice as a form of orgn.*

FREILE, Alfonso José, geographer; b. Valparaiso, Chile, Feb. 15, 1917; s. Alfonso Jose and Berta (Cordovez) F.; Chemist, Universidad de Chile, 1944; Geologist, Universidad Central de Venezuela, 1947, Licenciado en Geografía, 1962; Geographer, Universidad de Chile, 1949; mem. Debora Gebaldon, Nov. 20, 1946; 1 dau., Deborah. Curator mineralogy Museo de Historia Natural, Santiago, Chile, 1940-42; prof. chemistry Instituto Pedagogico, Caracas, Venezuela, 1945-48; geologist Minsterio de Fomento, Caracas, 1947-49; prof. geomorphology Universidad de

Chile, Santiago, 1949-50, prof. geography, 1956-57; vis. prof. Universidad de Panama, 1952-54; head dept. geology Universidad de Oriente, Santiago de Cuba, Cuba, 1954-56; expert in hydrogeology UN, La Paz, Bolivia, 1958-59; prof. geomorphology and climatology Universidad Central de Venezuela, Caracas, 1959-—. Mem. Pan Am. Inst. Geography and History, Instituto Nacional de Geografia (past mem. nat. com.), Am. Soc. Photogrammetry, Am. Geol. Soc., Am. Geophys. Union, Assn. Am. Geographers, Meteorol. Soc. Am., Col. de Ingenieros de Venezuela, Inst. de Ingenieros de Minas de Chile, Soc. Venezolana de Géologos, Soc. Venezolana de Ciencias Naturales. Author: Geologia, 1956; Geologia y Geomorfología, 1957; Meterología y Climatología Tropical y de Venezuela, 1962; Provincias fisiográficas de Venezuela, 1964; Bibliografia Geográfica de Venezuela, 1964; also articles. Developed new methods climatic research in tropical meteorology and climatology. Home: Edificio Naiguata, Av. Buenos Aires, Los Caobos, Caracas, Venezuela.*

FREIMAN, David Galland, Am. pathologist; b. N.Y.C., July 1, 1911; s. Leopold and Dorothy (Galland) F.; A.B., Coll. City N.Y., 1930; M.D., L.I. Coll. Medicine, 1935; M.A. (hon.), Harvard, 1962; m. Ruth Schein, Sept. 2, 1949; children—Nancy (Mrs. Stephen Schultz), Leonard. Asst. pathologist Mass. Gen. Hosp., 1944-50; faculty U. Cin. Coll. Medicine, 1950-56; attending pathologist Cin. Gen. Hosp., Drake Meml. Hosp., 1952-56; pathologist Beth Israel Hosp., Boston 1956—; instr. pathology Tufts Coll., 1947-48, Harvard, 1949-50; clin. prof. pathology Harvard Med. Sch., Boston, 1956-62, prof., 1962——; cons. Boston VA Hosp. Mem. Am. Assn. Pathologists and Bacteriologists, Am. Soc. Clin. Pathologists, Histochem. Soc., Am. Soc. Exptl. Pathology, Internat. Acad. Pathology, A.A.A.S., Phi Beta Kappa, Sigma Xi. Author: (with W. T. Longcope) A Study of Sarcoidosis, 1952; also numerous articles. Editorial bd. Am. Jour. Pathology, 1961—, Circulation, 1962-67, Atlas of Tumor Pathology, 2d series, 1966——. Research in diseases of heart, lungs and blood vessels, histochem. studies on tissue enzymes in normal and cancerous tissues and on lipid metabolic disorders. Home: 182 Homer St., Newton Centre, Mass. 02159. Office: Beth Israel Hosp., 330 Brookline Av., Boston 02215.*

FREIND, John, Brit. physician, chemist; b. Croton, Eng., 1675; student Christ Church Coll.; Oxford (Eng.) U., B.M., 1691; M.D., 1703 apptd. reader chemistry, Oxford, 1704; army physician, Spain, 1705-07; named physician to George I, 1711; named physician to Queen Caroline, 1727. Fellow Royal Soc., 1711; mem. Royal Coll. Physicians; M.P. (from Launceston), 1722; imprisoned in Tower of London, Mar.-Nov., 1723 because of his outspoken views. Author: Emmenologia in qua fluxus muliebris . . . , 1703; Praelectiones chymicae . . . , 1709; History of Physick, 1725-26; Opera omnia medica, 1733. Expanded Keill's applications of mech. principles to chem. phenomena; divided chem. changes into 2 classes, dissociation and composition; gave specific gravities of solids and liquids; believed there is an attractive force between particles which acts over a very small distance and decreases with distance faster than the inverse square, that the attraction may be greater on one side of particle than other. Died London, July 26, 1728.

FREINKEL, Norbert, physician; b. Mannheim, Germany, Jan. 26, 1926; s. Adolf and Veronica (Kahn) F.; A.B., Princeton, 1945; M.D., N.Y. U., 1949; m. Ruth Kimmelstiel, June 19, 1955; children—Susan Elizabeth, Andrew Jonathan, Lisa Ann. Research fellow medicine Throndike Meml. Lab. Harvard, 1952-55, dir. Diabetes and Metabolism div., 1957-66, faculty Med. Sch., 1956-66; Kettering prof. medicine, chief endocrinology and metabolism Northwestern U. Med. Sch. 1966——; cons. Army Research Inst. Environmental Medicine, Natick, Mass., 1957-66; USPHS Hosp., Boston, 1958-66; cons. div. occupational health USPHS, 1966——, metabolism study sect., 1967——; mem. nat. adv. com. Nat. Inst. Metabolic Diseases, National Institutes of Health, 1967——. Recipient of the Eli Lilly award of the Am. Diabetes Assn., 1966. Mem. Endocrine Soc., Am. Physiol. Soc., Am. Goiter Soc., Am. Soc. Clin. Investigation, Assn. Am. Physicians, Am. Heart Assn., New Eng. Diabetes Assn. (dir.), Phi Beta Kappa, Sigma Xi, Alpha Omega Alpha. Editorial bd. Jour. Clin. Endocrinology and Metabolism, 1960——, Jour. Lab. and Clin. Medicine, 1966——. Contbr. numerous articles to profl. jours. Demonstrated that intermediary metabolism of glands such as thyroid and pituitary may delimit their hormonogenic function, physiol. significance of protein-binding in transport and transcellular exchange of thyroid hormone; delineation of conceptus as a new site for abstraction of maternal hormones and substrates; elucidation of syndrome of alcohol hypoglycemia and factors regulating hepatic prodn. of glucose. Home: 938 Edgemere Ct., Evanston, Ill. 60202. Office: 303 E. Chicago Av., Chgo. 60611.*

FREIREICH, Emil J, Am. physician, hematologist; b. Chgo., Mar. 16, 1927; s. David and Mary (Klein) F.; B.S., U. Ill., 1947, M.D. with honors, 1949; m. Haroldine Lee Cunningham, Mar. 13, 1953; children—Debra Ann, David Alan, Lindsay Gail, Thomas Jon. Commd. s.a. surgery USPHS, 1955, advanced through grades to med. dir., 1963; research asso. Mass. Meml.

Hosps., Boston, 1953-55; sr. investigator Nat. Cancer Inst., NIH, Bethesda, Md., 1955-64, head leukemia service, 1964-65; resigned, 1965; prof. medicine, asst. head dept. developmental therapeutics, chief research hematology U. Tex., M.D. Anderson Hosp., Houston, 1965——; Mem. A.M.A., Am. Fedn. Clin. Research, Am. Assn. for Cancer Research, Internat. Hematology Soc., Am. Hematology Soc., A.A.A.S., Am. Soc. for Clin. Oncology, Alpha Omega Alpha. Research, publs. on devel. of platelet replacement transfusion therapy to control hemorrhage; 1st successful leukocyte transfusion and bone marrow homografts from peripheral blood cells in man; defined clin. syndromes of intracranial hemorrhage from leukemia and meningeal leukemia; introduced new drugs to chemotherapy of acute leukemia. Home: 810 Monte Cello St., Houston 77024.*

FREIRE-MAIA, Newton, Brazilian geneticist; b. Boa Esperança, Brazil, June 29, 1918; s. Belini Augusto Maia and Maria Castorina Freire; D.D.S., Sch. Pharmacy and Dentistry, Alfenas, Brazil, 1945; Ph.D., Fed. U. Rio de Janeiro, 1960; m. Flávia Leite Naves, Sept. 27, 1948; children—Regina Flávia, Maria de Fátima, Newton, Marco Domiciano. Teaching asst. U. Sao Paulo, 1946-47, tchr., researcher, 1948-51; chief researches Fed. U. Paraná, 1951—; prof. Cath. U., Paraná, 1953-55; lectr. various univs. Cons. WHO. Mem. Brazilian Soc. Genetics (pres. 1960-62), Brazilian Soc. for Advancement Sci. Author: (with Ademar Freire-Maia) Medical Genetics, 2 vols., 1966. Numerous publs. on studies of inbreeding levels in Brazilians and the effect on morbidity and precocious mortality; congenital malformations; mortality and morbidity among children of physicians exposed to radiation. Office: Fed. U. Paraná, Curitiba, Brazil.*

FREKE, John, Brit. physician; b. London, 1688; s. John Freke; m. Elizabeth Blundell (dec. 1741). Apprenticed to Mr. Blundell; elected asst. surgeon St. Bartholomew's Hosp., 1726, became 1st curator soon after, surgeon, 1729-55. Fellow Royal Soc., 1729. Author: An Essay to Show the Cause of Electricity and Why Some Things are Non-electricable in which is also Considered its Influence in the Blasts on Human Bodies in the Blights on Trees, in the Damps in Mines, and as it may Affect the Sensitive Plant; Essay on the Art of Healing, 1748. Described disease in which muscles gradually become bony tissue, 1736; invented instrument for reduction of dislocations of shoulder joint, 1743; research on electricity (believed electricity the cause of fertilization in plants), acute rheumatism. Died 1756.

FREMONT, John Charles, Am. explorer; b. Savannah, Ga., Jan. 21, 1813; s. Jean Charles and Ann Whiting (Pryor) F.; attended Charleston Coll., 1829-31 (expelled); m. Jessie Benton, Oct. 19, 1841; 1 dau., Elizabeth. Commd. 1d lt. Topog. Corps, U. S. Army explored Des Moines River; helped make surveys in Carolina mountains, also in Ga.; accompanied expdn. of J. N. Nicollet to plateau between Upper Mississippi and Missouri rivers; mapped parts of Ia. Territory, 1841; made 1st important exploration of Wind River, Chain of the Rockies (with Kit Carson as guide), 1842; published Report of the Exploring Expedition of the Rocky Mountains, 1843; explored west to Ore. and South to Santa Fe, N.M., 1843-44; made survey of Central Rockies, Gt. Salt Lake Region, also part of Sierra Nevada; participated prominently in conquest of Cal.; helped capture Los Angeles during Mexican War, 1846; apptd. civil gov. Cal., 1847; involved in Stockton-Kearny mil. quarrel, court martialed, Washington, 1847-58, found guilty of mutiny, penalty remitted by Pres. Polk; resigned from U. S. Army to lead an expdn. subsidized by pvt. interests in Cal. to locate railroad passes from Upper Rio Grande into Cal., largely unsuccessful, 1848-49; mem. U. S. Senate from Cal., 1850-51; led expdn. in search of southern route for Pacific R.R., 1852-54; 1st Republican and Nat. Am. (Know Nothing) Party candidate for U. S. Pes., 1856; in quartz mining bus., Mariposa, Cal.; served as maj. gen. U. S. Army in charge of Dept. of West during Civil War, removed because of his ordering of confiscation of rebel property (including slaves); comdr. Mountain Dept. in Western Va., 1862; Radical Republican nominee for U. S. Pres. (withdrew before elections), 1864; pres., promoter Memphis & El Paso R.R.; circa 1865-73; lost fortune in railroad ventures, 1870; territorial gov. Ariz., 1878-83; restored to U. S. Army as maj. gen., 1890; called Pathfinder. Died N.Y.C., July 13, 1890.

FRÉMY, Edmond, French chemist; b. Versailles, France, Feb. 28, 1814; student of Gay-Lussac; prof. chemistry École Polytechnique, later at Mus. Natural History; mem. French Acad. Scis., 1857. Author treatise on synthesis of rubies. Editor: Encyclopedie chimique, 40 vols., 1882-94. One of the 1st to prepare flourine, but could not collect it; prepared anhydrous hydrogen flouride, artificial rubies; research on ferrates, stannates, plumbates, coloring of flowers, saponification of fats. Died Paris, Feb. 3, 1894.

FRENCH, Aaron, Am. inventor; b. Wadsworth, O., Mar. 23, 1823; s. Philo and Mary (McIntyre) F.; m. Euphrasia Terrill, 1848; m. 2d, Caroline B. Speer. Learned blacksmith trade, 1835; with Ohio Stage Co., Cleve., 2 years, Gayoso House, Memphis, Tenn., 1 year; lived in St. Louis, 1844-45; built wagons for Peter Young, Carlyle, Ill., 1845; returned to Ohio as semi-invalid, 1845-49; with Cleve. & Pitts. R.R., 1853; supt. blacksmithing Racine & Miss. R.R., Racine, Wis., until 1861; sheriff Racine County, 1861-62; partner (with Calvin Wells) in manufacture of 1st steel springs for railroad cars, Pitts., 1862; invented coiled and elliptic springs (revolutionized railroad industry); owned A. French Spring Co. (merged with Ry. Steel Spring Co.), Pitts.; left some money to Ga. Sch. Tech. Died Mar. 24, 1902.

FRENCH, C(harles) Stacy, Am. plant physiologist; b. Lowell, Mass., Dec. 13, 1907; s. Charles Ephraim and Helena (Stacy) F.; student Loomis Inst., Windsor, Conn., 1921-25; B.S., Harvard, 1930, A.M., 1932; Ph.D., 1934; m. Margaret Wendell Coolidge, Dec. 10, 1938; children—Helena Stacy, Charles Ephraim. Asst. in gen. physiology Harvard, 1930-33; research fellow Cal. Inst. Tech., 1934-35; guest worker with Otto Warburg, Kaiser Wilhelm Inst., Berlin-Dahlem, Germany, 1935-36; Austin teaching fellow in biochem. Harvard Med. Sch., 1936-38; instr. (research) in chemistry with James Franck, U. Chgo., 1938-41; asst. prof. dept. botany U. Minn., 1941-45, asso. prof., 1945-47; dir. div. plant biology Carnegie Instn. of Washington at Stanford U. since 1947. Mem. Am. Soc. Plant Physiologists (chmn. Western section, 1954), Bot. Soc. Am., National Academy of Sciences, Am. Acad. Arts and Sciences, Am. Soc. Biol. Chemists, Society General Physiologists (vice pres. 1954, pres. 1955-56), A.A.A.S., Biophys. Soc., Deutsche Akademii der Naturforscher Leopoldina. Contbr. sci. jours. on plant physiology. Study of photosynthesis of purple bacteria; characteristics, spectroscopy, and functions of plant pigments; cellular respiration; leaves and algae. Home: 11927 Rhus Ridge Rd., Los Altos Hills, Cal. Office: Carnegie Institution, Stanford, Cal.

FRENCH, Dexter, Am. biochemist; b. Des Moines, Feb. 23, 1918; s. Raymond Albert and Minnie (Ormerod) F.; A.B., U. Dubuque, 1938, D.Sc., 1960; Ph.D., Ia. State U., 1942; m. Mary Catherine Martin, June 17, 1939; children—Alfred, David, Walter, Barbara (deceased), Jean, Nancy, Carol. Fellow in physical chemistry Harvard Medical School, 1942-44; worked as chemist to the Corn Products Company, Argo, Ill., 1945; faculty Ia. State U., Ames, 1946—, prof. biochemistry, 1960—, chmn. dept. biochemistry and biophysics, 1963—. Mem. Am. Chem. Soc. (Hudson award div. carbohydrate chemistry 1964), Am. Soc. Biol. Chemists, Sigma Xi, Phi Kappa Phi. Research, numerous publs. on characterization of helical starch configuration; mechanism of amylase action; structure and biochemistry of starch and cyclodextrins. starch, starch reactions. Home: 3521 Ross Rd., Ames, Ia. 50010.*

FRENCH, Frederick Alexis, Am. chemist; b. Berkeley, Cal., Mar. 19, 1917; s. Richard Slayton and Alice (Grace) (Stone) F.; A.B. in Chemistry, U. Cal. at Berkeley, 1942; m. June Cameron Frisbie, July 16, 1937; children—Valerie Gale (Mrs. Ronald Lee Hunt), Douglas Allan. Jr. chemist Shell Devel. Co., 1940-42, chem. engr., 1942-45; research asso. Mt. Zion. Hosp., San Francisco, 1944-50, cons. in exptl. cancer chemotherapy, 1951-56, asso. dir. cancer chemotherapy research, 1956-61, dir., 1961—; radiol. chemist U. S. Naval Radiol. Def. Lab., San Francisco, 1951-56. Nat. Cancer Inst. grantee. Mem. Am. Chem. Soc., Am. Assn. for Cancer Research, Royal Soc. Medicine, N.Y. Acad. Scis., A.A.A.S. Research antibacterial and antitumor chemotherapy, medicinal chemistry, phys. chemistry, drug design theory. Home: 2 Le Roy Av., Portola Valley, Cal. 94026. Office: 2493 Pulgas Av., Palo Alto, Cal. 94303.*

FRENCH, Sir James (Weir), engr.; b. Apr. 1876; s. Andrew Gordon French; ed. Bearsden Acad., Glasgow U., Glasgow Tech. coll., U. Berlin; D.Sc.; m. Jasmine Wallace Johnstone; 2 sons. Former chmn. Barr and Stroud, Ltd., engrs., instrument makers, optical glass mfrs., Glasgow Royal Tech. Coll.; expert to Inter-Allied Commn. Control in Germany, after 1918. Fellow Inst. Physics, Phys. Soc., Soc. Glass Technicians, Soc. Antiquaries. Author books on gen. engring., machine tools, optical computation and sci., articles on physiol. and practical optics, geology, optical glass history and manufacture, sci., engring., indsl. econs. and politics. Patentee in field range and height finders, submarine periscopes, inclinometers, gunnery fire control (surface and antiaircraft), airplane and other sights, radar equipment. Died Jan. 14, 1953.

FRENCH, John, English physician; b. Broughton, nr. Banbury, Eng., 1616; s. John French; B.A., New Inn Hall, Oxford, Eng., 1637; M.A., 1640; army surgeon; one of 2 physicians to army; made doctor of physic, 1648, also physician to Savoy Hosp. Author: The Art of Distillation, 1651; The Yorkshire Spaw, or a Treatise of Four Famous Medicinal Wells . . . , 1652, 54. Translator: The New Light of Alchymy . . . (Michael Sendevogius), with Nine Books of Paracelsus on the Nature of Things, 1650; Three Books of Occult Philosophy (H. C. Agrippa), 1651; Philosophical Furnaces (J. R. Glauber), 1651. Died Oct. or Nov. 1657.

FRENCH, John Douglas, Am. neurosurgeon, neurophysiologist; b. Los Angeles, Apr. 11, 1911; s. John Rollin and Effie (Douglas) F.; A.B., U. Cal. at Los Angeles, 1933; M.D., U. So. Cal., 1937; m. Dorothy Kirsten, July 18, 1955. Asst. prof. neurol. surgery U. Rochester Sch. Medicine and Dentistry, 1943-46; staff U. Ill. Neuropsychiat. Inst., 1946-47; chief neurosurgery VA Hosp., Long Beach, Cal., 1948-58, asst. dir. profl. services for research, 1950-58, chief cons. neurosurgery, 1958—, clin. prof. surgery-neurosurgery U. Cal. Sch. Medicine, Los Angeles, 1949—, prof. anatomy, 1960—, dir. Brain Research Inst., 1960—. Mem. various coms., adv. bds. NIH, 1958-63, Office Space Sci. and Application, 1960—, Nat. Acad. Scis.-Armed Forces-NRC, 1958-62; City of Hope, 1964—. Recipient Distinguished Service award U. Cal., Los Angeles, 1965. Diplomate Am. Bd. Surgery, Am. Bd. Neurol. Surgery. Mem. Am. Neurol. Assn., Am. Acad. Neurol. Surgery, Am. Acad. Neurology, A.M.A., Harvey Cushing Soc., Pacific Coast Surg. Assn., Soc. Neurol. Surgeons, Soc. Univ. Surgeons, So. Cal. (pres. 1957-58;, Western neurosurg. socs., Sigma Xi, Alpha Omega Alpha. Author, editor: Frontiers in Brain Research, 1962. Editor: (with Editorial bd. Internat. Jour. Brain Research. R. W. Porter) Basic Research in Paraplegia, 1962. Contbr. to books. Research in physiology of brain function; contbr. studies clarifying basic mechanisms of sleep, epilepsy, and visceral disorders. Home: 10800 Chalon Rd., Los Angeles 90024. Office: Health Scis. Center, Univ. California, Los Angeles 90024.*

FRENCH, Lyle Albert, Am. neurol. surgeon; b. Worthing, S.D., Mar. 26, 1915; s. Leslie V. and Bernice (MacKenney) F.; B.S., U. Minn., 1935, M.D., 1940, M.S., 1946, Ph.D., 1947; m. Gene F. Richmond, Sept. 13, 1941; children—Frederick, Eldridge, Barbara. Faculty, U. Minn., Mpls., 1947—, prof. neurosurgery, 1956—, dir. div. neurosurgery, 1960—. Cons. VA Hosp., Mpls., 1948—, U. S. Army, 1962—, NIH, 1964—. Mem. Soc. U. Surgeons, Neurosurg. Soc. Am. (past pres.), Mpls. Acad. Medicine (past pres.), Am. Bd. Neurol. Surgery, Harvey Cushing Soc., Soc. Neurol. Surgery, Am. Acad. Neurosurgery, Minn. Soc. Neurol. Scis. (past pres.). Contbg. author: Clinical Neurology, 1955, 60; Pediatric Neurosurgery, 1959; Clinical Neurosurgery, 1963, 64. Research, numerous publs. on brain neoplasms in children, cerebrovascular permeability. Home: 85 Otis Lane, St. Paul 55104. Office: Univ. Hosps., Mpls. 55455.*

FRÉNICLE DE BESSY, Bernard, French mathematician; b. Paris, 1605; gave up math. to study theology, 1660. Author: Traité de triangles en nombre, 1676; Sur les quarrés magiques; Méthode pour trouver la solution des problèmes par exclusion. Mem. French Acad. Scis. Demonstrated that the number of magic sqs. increases enormously by writing down 880 magic sqs. of the order 4; developed method of solving indeterminate equations through whole numbers (determination method). Died Jan. 17, 1675.

FRENKEL, Jacob Karl, pathologist; b. Darmstadt, Germany, Feb. 16, 1921; s. Karl E. and Anna M. (Kaufmann) F.; came to U. S., 1940, naturalized, 1943; A.B., U. Cal. at Berkeley, 1942, M.D., 1946, Ph.D. in Comparative Pathology, 1948; m. Rebecca Lou Reese, Sept. 3, 1954; children—Lisa Marie, Linda Dawn, Carl David. Pathologist, Rocky Mountain Lab., Hamilton, Mont., 1948-50, NIH, Bethesda, Md., 1950-51; instr. U. Tenn. Sch. Medicine, 1951-52; faculty U. Kan. Sch. Medicine, Kansas City, 1952—, prof. pathology, 1960—; vis. prof. pathology U. Nacional Autonoma de Mexico, 1963-64. Cons. in pathology VA, 1955—. Fulbright fellow, Mexico, 1963. Mem. Internat. Congress Comparative Pathology (chmn. U. S. nat. com. 1965—), Am. Assn. Pathologists and Bacteriologists, Am. Soc. for Exptl. Pathology, Soc. for Exptl. Biology and Medicine, Internat. Acad. Pathology, N.Y. Acad. Scis., Endocrine Soc., Infectious Diseases Soc. Am., Soc. for Psychol. Study Social Issues, Wildlife Disease Assn., A.A.A.S., Reticuloendothelial Soc., Deutsche Gesellschaft Herpetologie. Author: (with S. Friedlander) Toxoplasmosis, 1951; also numerous articles. Research on pathogenesis toxoplasmosis in central nervous system and eye, transmission toxoplasmosis, effects corticosteroids on infections, cellular and antibody related factors in immunity; developed toxoplasmin skin test. Home: 10030 El Monte Lane, Shawnee Mission, Kan. 66207. Office: U. Kan. Med. Center, Kansas City, Kan. 66103.*

FRENKIEL, François Nafatali, physicist; b. Warsaw, Poland, Sept. 19, 1910; s. Icek and Hinda (Eilenberg) F.; Mech. Engr., U. Ghent, Belgium, 1933, Aero. Engr., 1937; Ph.D. in Physics, U. Lille, France, 1946; m. Barbara Mann, 1962. Came to U. S., 1947, naturalized, 1952. With Service Technique de l'Aeronautique, Rhode-Saint-Genese, Belgium, 1937-38, Inst. Mecanique des Fluides, U. Lille, France, 1939-40, Group Francais Devel. Recherches Aero., Toulouse, France, 1940-43, 1945-47; grad. sch. aero. engring Cornell U., 1947-48, U. S. Naval Ordnance Lab., White Oak, Md., 1948-50, Applied Physics Lab., Johns Hopkins, Silver Spring, Md., 1950-60; cons. David Taylor Model Basin, Washington, 1956-60, cons. in applied math., Naval Ship Research and Devel. Ceter, 1960—; prof. in fluid dynamics and atmospheric physics U. Minn., Mpls., 1963-67. Lectr., Inst. Fluid Dynamics and Applied Math., U. Md., 1950; cons. U. S. Weather Bur., 1952-60; research master grad. council George Washington U., 1952—; cons. Courant Inst. Math. Sci., N.Y. U., 1962-63, chmn. U. S. Nat. Com. Theoretical and Applied Mechanics, 1965-66. Recipient Prix Scientifique Interfacultaire Louis Empain, Belgium, 1939. Fellow Am. Phys. Soc. (chmn. div. fluid dynamics 1954, 57, 60, 62, 65), Am. Geophys. Union, A.A.A.S.; mem. Am. Math. Soc., Am. Inst. Aeros. and Astronautics, Philos. Soc. Washington (pres. 1963), Washington Acad. Scis. (pres. 1964), Sigma Xi. Editor: The Physics Fluids, 1957. Research in fluid dynamics, atmospheric diffusion, turbulence and applied math. Home: 4545 Connecticut Av. N.W., Washington 20008. Office: Applied Math. Lab., Naval Ship Research and Devel. Center, Washington 20007.*

FRERE, John, English geologist; b. Norfolk, Eng., Aug. 10, 1740; s. Sheppard and Sussanna (Hatley) F.; B.A., Caius Coll., 1763; M.A., 1766; m. Jane Hookham, 1768; children—John Hookham, William, Bartholomew, James Hatley, Temple. Elected mem. Parliament for Norwich, 1799. Fellow Royal Soc., 1771; v.p. Marine Soc., 1785. Contbr. papers on flint weapons of Hoxne. Discovered animal remains and implements which he dated from prehistoric times, Hoxne, Suffolk, Eng., thus providing 1st suggestion of man's coexistence with now extinct animals. Died July 12, 1807.

FRERE, Sheppard Sunderland, Brit. archeologist; b. Graffham, Sussex, Eng. Aug. 23, 1916; s. Noel Gray and Barbara (Sunderland) F.; student Lancing Coll., 1930-35; B.A., Magdalene Coll. Cambridge, 1938, M.A., 1944; M.A., Oxford U., 1966; m. Janet Cecily Hoare, July 3, 1961; children—Sarah Barbara, Bartle Henry. Schoolmaster, Lancing Coll., 1945-54; lectr. archaeology Manchester U., 1954-55; reader archaeology Roman provinces London U. Inst. Archaeology, 1955-63, prof., 1963-66; prof. archaeology Roman Empire, Oxford U., 1966—; fellow All Souls Coll., 1966—; dir. Canterbury Excavation, 1946-60; dir. Verulamium Excavations, 1955-61. Mem. Royal Commn. on Hist. Monuments Eng., Ancient Monuments Bd. Eng., 1966—. Fellow Soc. Antiquaries London. Author: Britannia, A History of Roman Britain, 1967; also numerous articles. Editor: Problems of the Iron Age in Southern Britain, 1961. Contbd. to knowledge of Roman Britain especially by excavation and study of its towns. Home: Netherfield House, Marcham, Abingdon, Berkshire, Office: All Souls Coll., Oxford, Eng.*

FREREJACQUE, Marcel, French chemist; b. Villers-les-Pots, France, Aug. 24, 1893; ed. Ecole Normale Supérieure de St Cloud (France), U. Paris; m. Lemercier, Apr. 25, 1925; children—Daniel, Gerard. Asst., French Mus. Natural History, 1925-45, under-dir., 1945-58, hon. under-dir., 1958—. Insp. gen. Classified Establishments, 1925-58. Decorated Officer of Merit for research Legion of Honor; laureat French Acad. Scis. Mem. Chem. Soc. France. Author: Traité de Chimie Organique de Grignard. Research, publs. in area of carbon hydrates and glucosides, especially digitalic glucosides. Home: 1, rue Monticelli Paris 14e. Office: 63, rue de Buffon, Paris Ve, France.*

FRERICH, Friedrich Theodor von, see von Frerich, Friedrich Theodor.

FRESENIUS, Karl Remigius, German chemist; b. Frankfort/Main, Germany, Dec. 28, 1818; ed. Bonn; studied under Liebig at Giessen, Germany; founder lab. and sch. analytical chemistry, Wiesbaden, Germany, 1848. Author: Anleitung zur qualitativen chemischen Analyse, 1814; Anleitung zur quantiv quantitativen chemische Analyse, 1846. Founder Zeitschrift für analytische Chemie. Continued Berzelius's work; prepared tables for quantitative and qualitative analysis still used in rev. edits. Died Wiesbaden, June 11, 1897.

FRESHFIELD, Douglas William, Brit. explorer; b. London, Apr. 27, 1845; s. H. R. and Jane (Crawford) F.; ed. Univ. Coll., Oxford; m. Augusta Charlotte Ritchie; 4 daus. Called to bar, 1870; mem. council Royal Geog. Soc., 1878-94, 1905, pres., 1914-17; pres. Alpine Club, 1893-95; pres. geog. sect. Brit. Assn., 1904; pres. Assn. Geog. Tchrs., 1897-1910; chmn. Soc. Authors, 1908-09. Author: Travels in the Central Caucasus and Bashan, 1869; The Italian Alps, 1875; The Exploration of the Caucasus, 1896; Round Kangchenjunga, 1903; Hannibal Once More, 1914; Unto the Hills, 1914; Life of H. B. de Saussure, 1920; Below the Snow Line, 1923; Editor Alpine Jour., 1872-80; editor Switzerland (Murray), 2 edits.; joint editor Hints to Travellers (Royal Geog. Soc.), 1883, 89, 93. Explored Alps, Apennines, mountains of central Caucasus, 1868, 87, 89, Sikkim, Nepal, 1899; 1st to climb Kazbek and Mt. Elborus, 1868; attempted Mt. Ruwenzori, 1905. Died Forest Row, Eng., Feb. 9, 1934.

FRESNEL, Augustin Jean, French physicist, engineer; b. Broglie, Normandy, May 10, 1788; s. Jacques F. and Augustine (Merimess) F.; ed. École Centrale, Caen, circa 1801-04, École Polytechnique, Paris, 1804-05, Écoles des Ponts et Chaussées, Paris. Road and bridge engr., Vendée, Drôme, l'Ille-et-Vilaine; mem. engring. corps in Paris, 1815; after 2d Restoration nominated tutor at École Polytechnique, 1815; mem. lighthouse commn., 1819, sec., 1824; mem. gen. phys. sect. French Acad. Scis., 1832; fgn. mem. Royal Soc., 1825, Rumford medal, 1824. Author: La diffraction de la lumière, 1815, others. Supplied final decisive evidence in support of transverse wave theory of light, constructed thorough theoretical and math. basis for this theory; invented biprism that produced fringes

(Fresnel fringes), explained them in terms of interference; proved by experimentation that white light is composed of waves whose length decreases from red to violet; held that oscillation of light waves is transverse; showed how theory of transverse light waves explains double refraction through Iceland spar and polarization of light; produced circularly polarized light by means of rhomb of glass; invented Fresnel's zones, 1829; worked out Fresnel's reflection formula which gives fraction of incident light reflected at surface of transparent medium; built projector with echelon lenses; 1st to replace lighthouse reflectors with compound lenses, 1819. Died Ville-d'Avray, July 14, 1827.

FRETTER, William Bache, Am. physicist, univ. dean; b. Pasadena, Cal., Sept. 28, 1916; s. William A. and Dorothy (Bach) F.; A.B., U. Cal. at Berkeley, 1937, Ph.D., 1946; m. Grace Powles, Jan. 1, 1939; children—Travis D., Gretchen, Richard Brian. Research asso. Mass. Inst. Tech., 1941; research engr. Westinghouse Electric Corp., 1941-45; faculty physics U. Cal. at Berkeley, 1946—, prof., 1955—, dean Coll. Letters and Sci., 1962—. Named chevalier de la Legion d'Honneur, France, 1964. Fellow Am. Phys. Soc. Author: Introduction to Experimental Physics, 1954. Research in high-energy nuclear reactions, particle physics, measurement of mass of muon. Home: 1120 Cragmont Av., Berkeley, Cal. 94708.*

FREUD, Geza, Hungarian mathematician; b. Budapest, Hungary, Jan. 4, 1922; s. Frigyes and Renée (Vajda) F.; degree, Tech. U. Budapest, 1950; candidate math. sci. U. Budapest, 1954, dr. math. sci., 1956; m. Anna Kastner, Jan. 8, 1964; children—Clara, Beata, Ester. Staff, Inst. Math. Research, Hungarian Acad. Scis., Budapest, 1952—; vis. prof. U. Giessen (Germany), 1961, U. Rostock (E. Germany), 1963, Syracuse U., N.Y., 1966. Recipient Grünwald prize Hungarian Math. Soc., 1954, Kossuth prize Govt. of Hungary, 1959; fellow Isituto Nazionale di Alta Matematica, Rome, Italy, 1959. Author textbooks. Research, publs. on remainder estimate in Tauberian theorems, continuity of best approximating polynomial in dependence of approximated function, test function method in linear approximation, smooth functions, approximation of derivate of a function by derivate of an approximating trigonometrical or rational polynomial, orthogonal polynomials and expansions. Office: 13-15 Reáltanoda u., Budapest V, Hungary.*

FREUD, Sigmund, psychoanalyst, psychiatrist; b. Freiberg, Moravia, May 6, 1856; s. Jacob and Amalia (Nathanson) F.; M.D., U. Vienna, 1881; LL.D., Clark U., 1909; m. Martha Bernays, 1886; 6 children including Anna. Studies psychol. aspects of hysteria under Jean Charcot, Paris, 1885-86; with physiol. lab. of Ernst Bruck, 1876-82, Inst. Cerebral Anatomy, Vienna Gen. Hosp.; lectr. U. Vienna, from 1885; studied nervous diseases of children, Kassowitz Inst., Vienna; specialist on nervous diseases, Vienna, from 1886; collaborator (with Josef Breuer) on use of hypnosis for treatment of hysteria; founder (with Eugen Bleuler and C. G. Jung) Jahrbuch für psychoanalytische und psychopathologische Forschungen, 1908; lectr. Clark U., Worcester, Mass., 1909; formed Internat. Psychoanalytical Assn., 1910; recipient Goethe prize, 1930; Fellow Royal Soc., 1936; moved to London after annexation of Austria by Germany, 1938. Author: On the Psychological Mechanism of Hysterical Phenomena, 1893; (with J. Breuer) Über Hysteria, 1895; Die Traumdeutung, 1900; The Psychopathology of Everyday Life, 1904; Totem and Taboo, 1913; The History of the Psychoanalytic Movement, 1914; Thoughts for the Times on War and Death, 1915; Beyond the Pleasure Principle, 1921; Group Psychology and the Analysis of the Ego, 1922; The Ego and the Id, 1923; Inhibitions, Symptoms and Anxiety, 1925; Autobiography, 1925; The Future of an Illusion, 1927; Civilization and its Discontents, 1929; (with A. Einstein) Why War? 1933; Moses and Monotheism, 1939. Father of psychoanalytic sch. of psychiatry; evolved concepts of dynamic subconscious, resistance and repression; discovered aetiological significance of sexual life and importance of infantile experiences in human devel.; stated that fundamental rule of psychoanalysis is to bring into consciousness repressed material held back by resistances; developed psychotherapy based on free association, interpretation of dream symbolism and patient-doctor transference; propounded theories of Odeipus complex and guilt; later divided mental apparatus into id (instinctive urges), ego, and superego (ethical standards derived from Oedipus complex); introduced concept of libido (energy of sexual instincts); conceived idea of self-preservation (eros) and death instincts; applied his theories to analyses of artistic creativity, folk culture, religion, war. Died London, Sept. 23, 1939.

FREUDENBERG, Karl Johann, German chemist; b. Weinheim, Baden, Germany, Jan. 29, 1886; s. Hermann Ernst and Helene (Siegert) F.; student U. Bonn and Berlin (Germany), 1904-10; Dr. phil. U. Berlin (Germany), 1910; m. Doris Nieden, July 30, 1910; 5 children. Asst., Emil Fischer, Berlin, 1910-14; asst. prof. U. Kiel (Germany), 1914, U. München (Germany), 1920-21; asso. prof. U. Freiburg, 1921-22; prof., dir. Chem. Inst., Tech. U., Karlsruhe, Germany, 1922-26; dir. Chem. Inst., U. Heidelberg (Germany), 1926-56; dir. Research Inst. for Chemistry Wood and Polysaccharides, Heidelberg, 1956—; Carl Schurz Meml. prof. U. Wis., Madison, 1931. Fellow Royal Soc., 1963. Author: Die Chemie der Natürlichen Gerbstoffe, 1920; (with W. Kuhn) Drehung der Polarisationsebene des Lichtes, 1932; Tannin, Cellulose, Lignin, 1933; Stereochemie, 1933; Organische Chemie, 1938; (with H. Plieninger) Organische Chemie, 1958; also numerous articles. Research in stereochemistry, sugar, and plant chemistry including cellulose, starch, lignin, proteins, natural polymers. Home: 34 Wilckensstrasse, 6900 Heidelberg, Germany.*

FREUDENTHAL, Alfred Martin, civil engr.; b. Stryj, Poland, Feb. 12, 1906; s. Simon and Gustava (Mueller) F.; C.E., Tech. U., Prague, 1929, D.Sc., 1930; M.Sc., Charles U., Prague, 1930; 1 son, Pierre Simon. Came to U. S., 1947, naturalized, 1953. Faculty, Hebrew Inst. Tech., Haifa, Israel, 1937-47; vis. prof. U. Ill., Urbana, 1947-49; prof. civil engring. Columbia, N.Y.C., 1949—. Cons. USN, USAF, industry. Recipient medal Swedish Aero. Soc., 1956. Mem. Am. Soc. C.E. (Norman medal 1948, 57), Permanent Internat. Assn. Nav. Congresses, Soc. Rheology, Internat. Assn. Bridge and Structural Engrs., Am. Inst. Aeros. and Astronautics, Sigma Xi. Author: Verbund Stuetzen, 1933; Inelastic Behavior of Materials and Structures, 1950; Solid Mechanics, 1966; Structural Reliability, 1968; also numerous articles. Developed theory of fatigue of metals, of structural reliability, mechanics of inelastic continuum and inelastic structures. Office: 624 S.W. Mudd, Columbia, N.Y.C. 10027.*

FREUDENTHAL, Hans, mathematician; b. Luckenwalde, Germany, Sept. 17, 1905; s. Josef and Elsbeth Freudenthal; ed. U. Berlin, U. Paris; Ph.D. in math.; D. honoris causa, U. Humboldt, Berlin, 1960; m. Susanna Johanna Catharina Lutter; children—Jedidja, Matthijs, Thomas, Mirjam. Asst., U. Amsterdam, 1930, rector, 1937; prof. math. U. Utrecht, 1946—, Assn. for Math. (hon.), Royal Netherlands Acad. Scis., dean, 1963-64; reader Yale, 1960-61. Mem. Belgian Internat. Commn. on Math. Instruction (pres.). Author: Logique Mathématique Appliquée, 1958; Lie Groups in Foundations of Geometry, 1964; Probability and Statistics, 1965; The Language of Logic, 1966; Mathematics Observed, 1967; (with M. de Vries) Linear Lie Groups, 1968. Research in statistics; topology; math. linguistics. Home: Fr. Schubertstraat 44. Office: Budapestlaan, Utrecht, Netherlands.

FREUDENTHAL, Poul Jhalmar, Danish surgeon; b. Copenhagen, Denmark, Feb. 27, 1895; s. Carl Friderich Wilhelm and Elisa (Sorensen) F.; M.D., U. Copenhagen; m. Eli Glud; children—Marianne, Litten. Specialist in surgery, gynecology, obstetrics, path. anatomy. Author: Experimental Rickets and the Growth Promoting Vitamin A; The Vitamin Content of Algae; About Artificial Insemination; Kvinden-Moderen; Sund Leveris. Home: H. C. Orstedsvej 10. Office: Raadhusplads 14, Copenhagen, Denmark.

FREUND, August, chemist; b. Kéty, Galacia, July 30, 1835; studied under Pebal; worked in Kéty, then Vienna. Discovered 1st polymethylene; developed synthesis for ketones. Died Lemberg, Galacia, Feb. 28, 1892.

FREUND, John Ernst, statistician; b. Berlin, Germany, Aug. 6, 1921; s. Alfred and Ida (Cahn) Lewin; came to U. S., 1939, naturalized 1952; B.A., U. Cal. at Los Angeles, 1943, M.A., 1944; Ph.D., U. Pitts., 1952; m. Maxine I. Henville, Aug. 26, 1949; children—Douglas Eric, John E. Jr. Prof. math. Alfred (N.Y.) U., 1946-54; prof. statistics Va. Poly. Inst., Blacksburg, 1954-57; prof. math. Ariz. State U., Tempe, 1957—. Fellow Am. Statis. Assn., A.A.A.S.; mem. Inst. Math. Statis., Am. Math. Soc., Math. Assn. Am. Author: Modern Elementary Statistics, 3d edit., 1967; Mathematical Statistics, 1962; Modern Business Statistics, 1958; A Modern Introduction to Mathematics, 1956; Elementary Business Statistics: The Modern Approach, 1964; Probability and Statistics for Engineers, 1965, translator French, German books. Home: 7035 N. 69th Pl., Scottsdale, Ariz. 85251.*

FREUND, Leopold, physician; b. Miskowitz, Bohemia, Apr. 5, 1868; Asst., lab. E. Finger Clinic, 1897-1913, dir., 1913-26; became pvt. docent med. radiology 1904, prof., 1914; introduced deep X-ray therapy (recorded 1897). Died 1944.

FREUNDLICH, Herbert Max Finlay, chemist; b. Berlin-Charlottenburg, Germany, Jan. 28, 1880; s. Friedrich Philipp Ernst and Ellen Elizabeth (Finlayson) F.; student, Munich; Ph.D., Leipzig, Germany, 1903; Ph.D. (hon.), Utrecht, Netherlands, 1936; m. Marie Bertha Mann, 1908; 1 son, 2 daus.; m. 2d, Maria Helene Gellert. Became mem. faculty U. Leipzig, 1906; named asso. prof. phys. chemistry and inorganic tech. Tech. U. Brunswick (Germany), 1911; joined Kaiser Wilhelm Inst. for Phys. Chemistry, Berlin-Dahlem, 1916, apptd. dep. dir., 1919; named hon. prof. U. Berlin, 1923, Tech. U. Berlin-Charlottenburg, 1925; invited to Univ. Coll., London, 1933; research prof. colloid chemistry, Minn., from 1938. Fellow Royal Soc., 1939; Chem. Soc. (hon.), Kaiser Wilhelm Soc. Author: Kapillarchemie, 1909, 4th edit. (with J. Bikerman), 2 vols., 1930-32; Grundzüge der Kolloidlehre, 1924; Kolloidchemie und Biologie, 1924; also articles. Investigated colloid chemistry, capillary chemistry and physiology; fundamental research in electrokinetics, rheology, thixotropy, adsorption; eponym of Freundlich's adsorption isotherm. Died Mpls., Mar. 31, 1941.

FREUNDLICH, Martin Morris, elec. engr.; b. Goerlitz, Germany, Nov. 23, 1905; s. Salomon and Rosa (Miodowski) F.; student Brunswick Inst. Tech., 1924-26; Dipl. Engr. Berlin Inst. Tech., 1929, Eng. Dr. summa cum laude, 1933; m. Edith Blum, Dec. 13, 1947; 1 dau., Evelyn. Came to U. S., 1936, naturalized, 1942. Research asst. Berlin Inst. Tech., 1933-34, Royal Tech. Coll., Glasgow, Scotland, 1935; staff Pye Radio Ltd., Cambridge, Eng., 1935-36, CBS, N.Y.C., 1936-44, 45-49, N. Am. Philips, Dobbs Ferry, N.Y., 1944-45; staff Airborne Instruments Lab. div. Cutler-Hammer, Deer Park, N.Y., 1949—, asst. supr., 1949-58, cons., 1958—. Mem. I.E.E.E., Am. Astronautical Soc. (dir. N.Y. sect. 1965-67, chmn. N.Y. sect. 1967), Am. Phys. Soc., A.A.A.S., Am. Vacuum Soc. Research, publs. on devel. of cathode ray oscillographs, cathode ray tubes for radar and tv, space lubrication, materials in space. Home: 16 Suydam Dr., Melville, N.Y. 11746. Office: Comac Rd., Deer Park, N.Y. 11729.*

FREWEN, Thomas, English physician; b. Eng., 1704; M.D., before 1755; practiced as surgeon, apothecary, Rye, Sussex, Eng.; later practiced medicine, Lewes, Eng. Author: Physiologia, 1780; (essays) The Practice and Theory of Inoculation, 1749, Reasons against an opinion that a person infected with the Small-pox may be cured by Antidote without incurring the Distemper, 1759. Pioneer in smallpox inoculation. Died June 14, 1791.

FREY, Albert, Swiss botanist; b. Künacht, Zurich, Switzerland, Nov. 8, 1900; s. Hans and Klara (Hoepfner) F.; ed. univs. Geneva, Paris, Lena; diploma in natural sci.; Ph.D. in Sci.; D.honoris causa univs. Utrecht, Munster, Rennes, Vancouver; m. Margrit Wyssling; children—Regula, Manuel, Barbara, Urs. Prof. agrege ETH, Zurich, 1927, dean, 1957-61; botanist Medan-Sumatra Exptl. Sta., Indonesia; asst. Instr. Lab. Gen. Botany and Vegetable Physiology, 1932-38; prof. gen. botany and vegetable physiology, 1958—. Named Marcel Benoist laureate. Mem, Kgl. Akad. van Wetensch Amsterdam, Royal Soc. (London). Author: Die Stoffausscheidung der Höheren Pflanzen; Submikroskopische Morphologie des Protoplasmas; Die Submikroskopische Struktur des Zytoplasmas; Macromolecules in Cell Structure; Die Pische Struktur des Zytoplasmas; Maoflanzliche Zellwand, 1959. Home: Schiltrain 15, Meilen ZH. Office: Universitatstrasse 2, Zurich 6, Switzerland.

FREY, Charles N., Am. biochemist; b. Hopkins, Mich., Oct. 21, 1885; s. August C. and Maria Anna (Grossklaus) F.; B.S., Mich. State U., 1911; D.Sc., 1946; M.S., U. Wis., 1915, Ph.D., 1920; m. Julia L. Leary, Dec. 28, 1920; children—Martha Anne, Charles Frederick, Florence Elizabeth (Mrs. David C. Hass). Applied sci. tchr., biologist, 1911-17; faculty U. Wis., 1916-20; fellow Mellon Inst., Pitts., 1920-22; research chemist Ward Baking Co., 1922-24; with Fleischmann Labs., N.Y.C., 1924-26, dir. research, 1926-29; dir. research Standard Brands, Inc., N.Y.C., 1929-44, dir. sci. relations, 1945-51; lectr. Mass. Inst. Tech., Cambridge, 1952-65, Columbia, 1957-63. Cons. chem. and pharm. industry, 1951—. Recipient Distinguished Alumni award Mich. State U., 1960. Mem. Am. Chem. Soc. (councilor 1949-53, 57—, named one of 10 most productive chemists Chgo. sect. 1947, Distinguished Service award 1966), Am. Assn. Cereal Chemists (pres. 1943), N.Y. Acad. Sci., Am. Soc. Bacteriology, Inst. Food Technologists (pres. 1953, Babcock Hart award 1953, Nicholas Appert award 1954), Am. Inst. Chemists (Honor Scroll 1956), Soc. Chem. Industry, N.A.M., Am. Inst. Baking (hon.). Contbr. to books in field. Studies, numerous publs. on yeast growth, anti-fungal action of acetic acid series, sorbitan monolaurate oleate, effect of high dosages of vitamin A on devel. of hypoprothrombinemia in rats and its prevention by vitamin K, fermentation factor in yeast, B1 thiamine excretion in man, influence of oxygen on fermentation of maltose and galactose, vitamin determination methods, composition, synthesis, proteolytic activity of yeast, enriched yeast as an enricher of foods; patentee in field. Home: 45 Cambridge Rd., Scarsdale, N.Y. 10583.*

FREY, David Grover, Am. limonologist; b. Hartford, Wis., Oct. 10, 1915; s. Grover Cleveland and Henrietta (Zimmerman) F.; B.A., with high honors, U. Wis., 1936, M.A., 1938, Ph.D., 1940; m. Sarah Elizabeth Jones, Jan. 24, 1948; children—Karl Frederick, Barbara Louise, Katharine Elizabeth. Jr. aquatic biologist U. S. Fish and Wildlife Service, Seattle, 1940-42, asst. aquatic biologist, College Park, Md., 1942-43, asso. aquatic biologist, 1943-45; asso. prof. zoology U. N.C., 1946-50; faculty Ind. U., Bloomington, 1950—, prof. zoology, 1955—; dir. Ind. Lake and Stream Survey, 1950-53. Guggenheim-Fulbright fellow, Austria, 1953-54; NSF Research grantee, Eng., 1962-63. Mem. Am. Soc. Limnology and Oceanography (past pres.), Internat. Assn. Limnology, Internat. Union Quaternary Research, Ecol. Soc. Am., Brit. Ecol. Soc., Soc. Systematic Zoology, Am. Soc. Zoology, Am. Micros. Soc., Marine Biol. Assn. India. Translator: (with F. E. J. Fry) Fundamentals of Limnology, 1953, 63. Editor: Limnology in North America, 1963; (with H. E. Wright) The Quaternary of the United

States, 1965, International Studies on the Quaternary, 1965; Limnology and Oceanography, 1955-59. Research on ecology carp and cisco in lakes Wis. and Ind., salmon in Columbia River, oysters in Chesapeake Bay, paleolimnology lakes in Germany, Austria, Denmark, and N.Am. based on remains microscopic animals in their sediments, taxonomy, distbn. and ecology chydorid Cladocera on world-wide basis. Home: Route 3, Box 75 Smith Rd., Bloomington, Ind. 47401.*

FREY, Emil Karl, German surgeon; b. July 27, 1888; s. Friedrich Karl and Regina (Geyrhalter) F.; ed. Wilhelms Gymnasium, Munich, U. Munich; M.D., Ph.D. in natural scis.; m. Paula von Hoesslin; children—Elisabeth Wahl-Wellhausen; children—Gitta, Otto, Rudolf. Agrégé of surgery, 1924; asso. prof. Acad. Medicine, Dusseldorf, 1927, full prof., 1930; full prof. surgery U. Munich, 1943. Recipient Bergmann medal. Mem., hon. mem. numerous German and fgn. sci. socs. Author: Der Cardiospasmus; Die Chirurgie des Herzens; Kallikrein; Die Operationen an der Lunge. Address: Arberstrasse 16, Munich 27, Germany.

FREY, Frederick Ward, Am. polit. scientist; b. Cleve., June 16, 1929; s. Frederick H. W. and Helen (Simpson) F.; A.B., Western Res. U., 1951; postgrad. (Rhodes scholar) Balliol Coll., Oxford (Eng.) U.; Ph.D., Princeton, 1962; m. Patricia Ann Evans, Dec. 16, 1967. Member of the faculty Mass. Inst. Tech., Cambridge, 1960-—, prof. polit. sci., 1966-—. Mem. Am. Polit. Sci. Assn., Am. Sociol. Assn., Am. Assn. for Pub. Opinion Research, Middle E. Inst. Author: The Turkish Political Elite, 1965; also articles. Research on comparative politics especially of developing socs., social power, pub. opinion, polit. socialization. Home: 92 Chestnut St., Boston 02108. Office: Mass. Inst. Tech., Cambridge, Mass. 02139.*

FREY, (Johann Friedrich) Heinrich Konrad, anatomist, zoologist; b. Frankfort/Main, Germany, June 15, 1822; s. Johann Peter and Maria Theresa (Griesinger) F.; studied medicine, Bonn, Berlin, Göttingen (all Germany); m. Doris Clemens, 1852; 3 sons. Became mem. faculty U. Göttingen, 1847; named mem. faculty U. Zurich, 1848, prof. med. faculty, 1851; apptd. prof. new Polytechnikum, 1855; dir. microscopic-anat. inst. Recipient service medal Vienna World Exhbn., 1873, meml. document, med. faculty, 1889. Author: Die Tineen und Pterophoren der Schweiz, 1856; Histologie und Histochemie des Menschen, 1859; Das Mikroskop und die mikroskopische Technik, 1863; Grundzüge der Histologie, 1875; Die Lepidopteren der Schweiz, 1880. Discovered and described about 80 small butterflies, many larva stages. Died Zurich, Jan. 17, 1890.

FREY, Henry Montague, English chemist; b. London, Eng., Feb. 27, 1929; s. Herman and Sadie (Steingold) F.; B.A., Balliol Coll., Oxford (Eng.) U., 1953, M.A., 1953, D.Phil., 1955; m. Leila Ann Finn, Aug. 25, 1955; children—Jeremy Graham, Colin Lionel. Commonwealth Fund fellow U. Cal. at Berkeley, 1955, Harvard, 1956; lectr. in phys. chemistry Southampton (Eng.) U., 1957-66; prof. phys. chemistry Reading (Eng.) U., 1966-—. Fellow Chem. Soc. (Corday Morgan medal 1965); mem. Faraday Soc. Research, numerous articles on gas kinetics especially unimolecular reactions, reactions of methylene and vibrationally excited molecules. Home: 10 Butts Hill Rd., Woodley, Reading. Office: Chemistry Dept., White Knights Park, Reading U., Reading, Eng.*

FREY, Max(imilian) (Ruppert Franz) von, see von Frey, Max(imilian) (Ruppert Franz).

FREY, Rudolf, German anesthesiologist; b. Heidelberg, Germany, Aug. 22, 1917; s. Alfred and Martha (Beuttel) F.; student U. Munich, 1939, U. Freiburg, 1940, U. Tübingen, 1941, U. Strassburg, 1942; M.D., U. Heidelberg, 1944; m. Heidi Immendörfer, July 24, 1943; children—Barbara, Irene, Irmgard, Martin. Staff, Field Hosp., 1944, Surg. Clinic, U. Heidelberg, 1945-49, anesthesiology dept. U. Basel (Switzerland), 1950, Mayo Clinic, Rochester, Minn., 1951; head dept. anesthesia U. Hosp., Heidelberg, 1950-59; dir. Inst. Anesthesiology, U. Mainz (Germany), 1960-—, prof., 1960-—; faculty U. Heidelberg, 1952-60, prof., 1956-60. Mem. German Soc. Anesthesia (cofounder; hon. sec. 1952-—), World Fedn. Socs. Anesthesiologists (cofounder, also coeditor News Letter 1955-—); hon. mem. French, Greek, Swiss socs. anesthesiology, Ass. des Anesthesiologists Europeans (hon. pres. 1959-—); corr. mem. Neederlandse Anaesthesisten Vereniging, Assn. Anaesthetists Gt. Britain and Ireland, Austrian Soc. Anesthesiology. Author: (with Hügin, Mayrhofer) Lehrbuch der Anaesthesiologie, 1955; also numerous articles. Co-editor: Textbook on Anaesthesiology, 1955; (with Kern, Mayrhofer) series Anaesthesiology and Resuscition, 26 vols., 1962-—; Research on curare and heart-lung resuscitation. Home: Langenbeckstrasse 1, 65 Mainz-Rhein, Germany.*

FREYCINET, see de Freycinet.

FREYDLINA, Rakhil Khatskelevna, Russian organic chemist; b. 1906; grad. Moscow U., 1930; D.Chem. Sci. With Research Inst. Insecticides and Fungicides, 1930-34; with Inst. Organic Chemistry, USSR Acad. Sci., 1935-39, 41-54, head lab. Inst. Elemental-Organic Compounds, 1954-—; instr. Moscow Inst. Fine Chem. Tech., 1938-41. Mem. USSR Acad. Sci. (corr.). Research and publs. on synthesis, structure and properties of organic compounds of mercury, arsenic, tin, antimony, lead, titanium, silicon, zirconium, boron, fluorine. Address: Inst. Elemental-Organic Compounds, USSR Acad. Sci., 28 Vavilov str., Moscow V-312, USSR.*

FREYER, Sir Peter Johnston, surgeon; b. Clifden, Ireland, July 21, 1857; s. Samuel and Celia (Burke) F.; grad. in arts Queen's U., Ireland, M.D., 1874, M.S., 1874; studied medicine under Robert McDonnell, Stevens Hosp., Dublin, Ireland; M.A., Royal U. Ireland, 1886. Joined Indian Med. Service, 1875; commd. surgeon-maj. Bengal Army, 1887, lt. col., 1895; ret. 1896, rejoined, 1914; cons. surgeon to Queen Alexandra's Mil. Hosp., London; ret. as col., 1919; named surgeon St. Peter's Hosp. for Stone, London, Eng., 1897. Recipient Arnott medal, 1904. Mem. Royal Soc. Medicine (1st pres. sect. urology). Author: The Modern Treatment of Stone in the Bladder by Litholapaxy, 1886; Stricture of Urethra and Prostatic Enlargement, 1901; Surgical Diseases of the Urinary Organs, 1908. Developed operation for total removal of prostate gland through suprapubic incision of bladder, 1900. Died London, Sept. 9, 1921.

FREYGANG, Walter Henry, Jr., Am. physiologist; b. Jersey City, Dec. 27, 1924; s. Walter Henry and Marie (Neuman) F.; M.E., Stevens Inst. Tech., 1945; M.D., U. Pa., 1949; children—Katherine Ann, Walter Nicholas Frederick. Asst. in neurology N.Y. Neurol. Inst., N.Y.C., 1951-52; fellow Nat. Found. for Infantile Paralysis, Marine Biol. Lab., Woods Hole, Mass., 1952; with NIH, Bethesda, Md., 1952-—, chief, sect. on membrane physiology, 1960-—. Fellow A.A.A.S.; mem. Am. Physiol. Soc., Sigma Xi, Soc. Gen. Physiologists, Marine Biol. Assn. Research, publs. in description and definition of elec. fields around single nerve cells in brain, mechanism of elec. coupling between elec. and mech. activity of muscle. Office: NIH, Bethesda, Md. 20014.*

FREYHAN, Fritz Adolf, psychiatrist; b. Berlin, Germany, Nov. 24, 1912; s. Max and Clara (Gotschalk) F.; M.D., U. Berlin, 1937. Came to U. S., 1937, naturalized, 1943. Mem. staff Del. State Hosp., Farnhurst, 1942-50, clin. dir., dir. research, 1950-60; dir. dept. psychiatry, neurology Del. Hosp., Wilmington, 1954-61; asso. to adj. asso. prof. dept. psychiatry U. Pa. Sch. Medicine, 1950-61; cons. psychiatry VA Hosp., Wilmington, 1954-61; dep. chief in charge clin. studies Clin. Neuropharmacology Research Center, Nat. Inst. Mental Health, Washington, 1961-66; dir. clin. studies Behavioral and Clin. Studies Center, St. Elizabeth's Hosp., Washington, 1961-66; clin. prof. psychiatry George Washington U., 1960-66; asso. clin. prof. psychiatry N.Y. U., 1966-—; dir. research dept. psychiatry St. Vincent's Hosp., N.Y.C., 1966-—. Councilor, Collegium Internationale Neuro-Psychopharmacologicum, 1964-—. Mem. Am. Psychiat. Assn., Am. Coll. Neuropsychopharmacology, A.A.A.S., A.M.A., Societe Medico-Psychologique, Am. Psychopathological Assn. (pres.-elect). Editor: Comprehensive Psychiatry, 1960-—. Contbr. numerous articles to med. jours. Clin., exptl. work on somatic treatments of mental disorders, psychopharmacology, schizophrenia, personality disorders, conceptual and clin. framework social psychiatry. Home: 4740 Connecticut Av. N.W., Washington 20032. Office: 153 W. 11th St., N.Y.C. 10011.*

FREYTAG, Gustav, cartographer; b. Neuhaldensleben, nr. Magdeburg, Germany, Jan. 23, 1852; s. Johann Ernst Friedrich and Elisabeth (Köke) F.; began studies in cartography and lithography with uncle F. Köke, Vienna, Austria, 1866; also studied in London; m. Anna Ertelt; 1 son; m. 2d, Camilla Dienstbuber, 1908; 2 daus. Became cartographer with F. A. Bockkhaus, Leipzig, Germany, 1872; with topog. dept. Prussian Gen. Staff, Berlin; returned to Köke, Vienna, 1876; established (with Wilhelm Berndt) cartographic pub. house Freytag und Berndt, 1879, continued as adviser after later merger with other firms. Author: Die Wirkung der Farben in der Geländedarstellung auf Landkarten. Pub. atlases and tourist maps with exact road delineations; used color relief to represent terrain; designed color scale still in use; helped open up Ennstaler Alps and Gesäuse Mountains. Died Bad Ischl, Austria, Dec. 19, 1938.

FREYTAG, Theodor, German engr.; b. Schweinfurt, Germany, Aug. 10, 1865; s. Heinrich Oskar and Maria Magdalene Franziska (Stepf) F.; ed. Tech. U. Munich; state exam for pub. constrn., 1892. Asst., road and river constrn. offices, Landshut and Deggendorf (both Germany); named state constrn. asst., also contracted to rebuild Kochelsee-Walchensee road, 1893; became mem. staff road and river constrn. office, Schweinfurt, Germany, 1894, chmn., 1901; named chmn. railroad constrn. office, Wurzburg, Germany, 1909; apptd. chmn. state constrn., Kochel, Germany, 1911; became chmn. dept. for utilization of water power and electricity supply Bavarian Ministry Interior, 1918, ministerial dir. highest constrn. adminstrn., 1929. Author: Das Walchenseewerk, 1922; also articles. Played important part in developing Bavaria's water power; designed and built Walchensee works, Central Isar works, other water power installations; improved navigability of Main River; built numerous roads and bridges. Died Munich, Oct. 30, 1933.

FREY-WYSSLING, Albert Friedrich, Swiss botanist; b. Küsnacht, Zürich, Switzerland, Nov. 8, 1900; s. Hans and Clara (Höpfner) F.; Dr.sc. nat. Swiss Fed. Inst. Tech., Zürich, 1924; Dr.h.c. U. Utrecht (Netherlands), 1951, U. Münster (Germany), 1954, U. Rennes (France), 1955, U. B.C., Vancouver, Can., 1961; postgrad. U. Jena (Germany), Sorbonne, Paris, France; m. Margrit Wyssling, July 26, 1928; children—Regula (Mrs. Erwin Berg), Manuel, Barbara (Mrs. Charles Bowlus), Urs. With Swiss Fed. Inst. Tech., 1926-28, 32-—, prof. gen. botany and plant physiology, 1938-—, rector, 1957-61; plant physiologist Rubber Exptl. Sta., Medan, Sumatra, 1928-32. Mem. Royal Soc. (fgn. mem.), Internat. Assn. Wood Anatomists (sec.-treas. 1957-—). Author: Submikroskopische Morphologie, 1938; Submicroscopic Morphology, 1953; Die Pflanzliche Zellwand, 1959; (with Mühlethaler) Ultrastructural Plant Cytology, 1965; also articles. Initiated ultrastructural research; established ultrastructures before introduction electron microscope; research on physiology latex flow in rubber tree; discovered Frey-Wyssling bodies in latex. Home: 15 Schiltrain, 8706 Meilen, Zurich, Switzerland.*

FRÉZIER, Amédée François, French mathematician, engr.; b. Chambréry, France, 1682; student math., pyrotechnics, Paris, France. Sent on mission to Peru, Chile, 1712; chief engr. fortifications built in Brittany, France, after 1740; became chief engr., Santo Domingo, 1719, 28. Mem. Marine Royal Acad. Author: Traité des feux d'artifice, 1706; Relation du voyage de la mer du Sud aux côtes du Perou et du Chili, 1716; Éléments de stéréotomie, 1759-60; Sur les différents ordre d'architecture, 1738. Research on stone-cutting and its application to architecture; imported Chilian strawberry to France; prepared 1st map of Santo Domingo; verified points of Patagonian coast; precursor of descriptive geometry in his treatment of conic surface and cylinders. Died Oct. 16, 1773.

FRIAUF, James Joseph, Am. zoologist; b. Toledo, Feb. 15, 1914; s. James and Nettie (Scheidler) F.; B.A., U. Toledo, 1936; M.S., U. Mich., 1937; Ph.D., U. Fla., 1942; m. Frances C. Quigley, July 8, 1944; children—Thomas Allen, William Roger. Instr., asso. prof. zoology Vanderbilt U., Nashville, 1946-—, sec. Natural Sci. Div., 1947-50, faculty council, 1955-61; pres. Highlands Biol. Sta., Inc., 1957-59. Ford Found. for Advancement Edn. fellow, 1954. Fellow A.A.A.S., Sigma Xi; mem. Soc. Systematic Zoology, Am. Soc. Zoologists, Am. Micros. Soc., Am. Inst. Biol. Scis., Assn. Southeastern Biologists. Editor: Jour. Tenn. Acad. Sci., 1963. Research in taxonomy and systematics of Dictyoptera or cockroaches of deserts, systematics, ecology and zoogeog. distbn. of interstitial fauna of marine beach sediments. Home: Ashby Dr., Route 5, Franklin, Tenn. 37064. Office: Box 1612, Sta. B, Vanderbilt U., Nashville 37203.*

FRICK, George, Am. ophthalmologist; b. U. S., 1793; student of Beer; grad. med. dept. U. Pa., 1815; postgrad. in ophthalmology, Eng.; became 1st physician to practice ophthalmology in U. S., with office in Balt., from 1819. Author: A Treatise on Diseases of the Eye; 1st complete textbook on ophthalmology, 1823; also many articles pub. in med. jours. Died 1870.

FRICK, M(atti) Heikki, Finnish physician; b. Pielisjarvi, Finland, June 10, 1931; s. Mathias and Irja (jokinen) F.; M.D., Helsinki U., 1957, Dr. Med. Sci., 1962; m. Eija Mattsson-Knuts, Feb. 20, 1951; children—Anne, Tiina, Tomi. Research worker Wihuri Research Inst., Helsinki, 1957-62; asst. in medicine Helsinki U., 1962-65, lectr., 1965-—, asso. prof., 1966-—, physician-in-chief cardiovascular lab. Central Hosp., 1965-—; lectr. cardiology Oulu U. Med. Sch., 1965-—; capt. Finnish Def. Forces, 1958-62; attending physician Inst. Occupational Health, Helsinki, 1963-64. Mem. hosp. planning com. Helsinki U. Finnish Med. Assn., 1964-—; sec. Finnish Med. Fund, 1965-—. Mem. Finnish Soc. Internal Medicine, Finnish Physiol. Soc., Am. Coll. Chest Physicians. Publs. on studies of role of biogenic amines especially serotonin in various circulatory states; facets of congenital heart disease especially coarctation of aorta; circulatory and hematological effects of acute phys. exercise; circulatory effects of long-term phys. exercise including its coronary implications and correlations between hemodynamics and fitness in normal subjects and patients with congenital heart disease. Home: 3 G 88, Suonotkontie, Helsinki 63. Office: 4 Haartman Street, Helsinki 29, Finland.*

FRICKE, Robert Elmer, Am. physician; b. Milw., May 10, 1892; s. William A. and Elma (Winegardner) F.; B.A., Johns Hopkins, 1916, M.D., 1920; m. Gertrude B. Hax, Apr. 10, 1924. Cons. in charge radium therapy sect. therapeutic radiology Mayo Clinic, Rochester, Minn., 1927-57; faculty U. Minn., Grad. Sch. Medicine, Rochester, 1929-—, asso. prof. therapeutic radiology, 1943-57, emeritus, 1957-—. Cited by U. Minn. for teaching, 1957. Diplomate Am. Bd. Radiology. Fellow Am. Coll. Radiology; mem. Zumbro Valley Med. Soc., (exec. sec. 1957-—), A.M.A., Am. Therapeutic Soc., Am. Assn. for History Medicine, A.A.A.S., Alumni Assn. Mayo Found., Sigma Xi. Contbg. author: Gynecology, 1928; Science of Radiology, 1933; Surgical Clinics of North America, 1947, 51; Medical Clinics of North America; also several other textbooks, numerous articles. Research on devel. high intensity divided dose technique in radium treatment cancer uterine cervix and uterus, radon ointment in treatment non-malignant conditions. Home: 410 6th Av. S.W. Office: Harwick Bldg., Mayo Clinic, Rochester, Minn. 55901.*

FRICKER, Alfons, chemist; b. Leupolz, Germany, Sept. 8, 1924; s. Matthaus and Pauline (Welte) F.; diploma in chemistry U. Tübingen (Germany), 1951; Dr. Agr., U. Mainz, 1965; m. Hilde Horn, May 28, 1955; children—Sabine, Benno, Christoph. Asst. Milk Research Sta., Wangen, Allgau, Germany, 1951-53; asst. Agrl. High Sch., Hohenheim, 1961—; research U. Mainz, 1962-65, lectr., 1964—; dir., prof. German Fed. Research Centre for Food Preservation, Karlsruhe, 1966—, head Inst. for Chemistry and Tech., 1966—. Research, numerous publs. on analysis of milk, butter, cheese, especially lipid constituents, nutritional properties of fish oils, parenteral nutrition with fat emulsions, deep-freezing of foods, conservation of foods, keeping quality of foods. Home: 12 a, Ringelberstrasse, 7501 Groetzingen, Germany. Office: 20 Engesserstrasse 75, Karlsruhe, Germany.*

FRICKER, Peter Emil, geologist; b. Mannedorf, Switzerland, Feb. 14, 1932; s. Emil O. and Berta (Raschle) F.; Ph.D., U. Zurich (Switzerland), 1959; m. Marie-Louise Altmann, Mar. 7, 1959; children—Daniela, Michael. Research geologist Swiss Andes Expedition to Peru, 1959; with Swiss Federal Tech. Inst., Zurich, 1959-60; Boese postdoctoral fellow dept. geology Columbia, 1960-61; sr. geologist Jacobsen-McGill Arctic Research Expdn., McGill U., Montreal, Que., Can., 1961-63; Palmer postdoctoral fellow dept. earth scis. Stanford, 1962-63; Nat. Acad. Scis.-NRC resident research asso. Ames Research Center, NASA, Moffett Field, Cal., 1963—. Mem. Am. Geophys. Union, Geol. Soc. Am., Schweizerische Geologische Gesellschaft. Contbr. articles to tech. jours. Research on stratigraphy and crustal structure of Alps, Andes, and Canadian Arctic Archipelago, thermal history terrestrial planets, origin planetary atmospheres. Home: 1211 College Av., Palo Alto, Cal. 94306. Office: Ames Research Center, NASA, Moffett Field, Cal. 94035.*

FRIDENSHTEIN, Alexander Yakovlevich, biologist; b. Kiev, USSR, June 14, 1924; s. Yakov Borisovich and Raisa (Bichovskajoi) F.; student Moscow Med. Inst., 1941-46, M.D., 1960; m. Helen Luria, Nov. 10, Mem. staff Gamalaya Inst. Epidemiology and Microbiology, Acad. Med. Scis. USSR, 1950—. Author: Experimental Ectopic Osteogenesis, 1963; (with others) Radiational Immunity, 1965; also articles. Research on mechanisms of extraskeletal bone formation; cellular basis of immunity. Home: 13 Frurise. Office: 2 Gamalaya, Moscow, USSR.*

FRIED, Bernard, Am. engr.; b. Chgo., Apr. 18, 1912; s. Joseph Leo and Ethel (Finelson) F.; B.S. in Engring. Physics, U. Ill., 1934; M.S., Ohio State U., 1937, Ph.D. in Physics, 1939; m. Bernice Armin, Feb. 23, 1936; children—Joel, Mike. Research engr. Ohio State U. Engring. Expt. Sta., 1938-41; prof. dept. mech. engring. Wash. State U., Pullman, 1941-55; cons., research engr. in structure Vultee Aircraft, San Diego, 1944-45; sr. engring. specialist AiResearch Mfg. Co., Phoenix, 1955-67; cons. engr., 1967-68; research specialist Boeing Co., Kent, Wash., 1968—. Mem. Am. Phys. Soc., Am. Soc. for Engring. Edn., Sigma Xi. Contbr. articles to profl. jours. Studies in procedures for solution partial differential equations with high speed electronic computers, extension of photoelastic method of stress analysis to inelastic range, vibration analysis of beams and plates. Home: 15235 28th St. S.W., Seattle 98166. Office: Boeing Co., Kent, Wash. 98188.*

FRIED, (Johann) Jakob, German obstetrician; b. Strasbourg (now in France), Apr. 21, 1689; s. Ulrich and Margaretha (Dietrich) F.; M.D., Strasbourg, 1710; m. Euphrosina Sophia Steinheil, 1718; 6 sons, including Georg Albrecht, 8 daus. Dir. municipal sch. for midwives (founded by Praetor Franz Joseph von Klinglin), from 1728. Mem. Leopoldina. Author: Anfangsgründe der Geburtshilfe, edited by son Georg Albrecht, 1769. Contbr. to med. compendiums, archives. Stressed importance of anatomy in obstetrics; taught many students and academicians at his sch., the prototype of other birth clinics, beginning with that founded by his student J. G. Roedever at U. Göttingen (Germany), 1751. Died Strasbourg, Sept. 3, 1769.

FRIED, Morton Herbert, Am. anthropologist; b. N.Y.C., Mar. 21, 1923; s. Norton and Sally (Solomon) F.; B.S., Coll. City N.Y., 1942; Ph.D., Columbia U., 1951; m. Martha Nemes, June 22, 1945; children—Nancy Eileen, Elman Steven. Lectr. sociology and anthropology, Coll. City N.Y., 1949-50; instr. anthropology Columbia U., 1950-53, asst. prof., 1953-57, asso. prof., 1957-61, prof., 1961—; dir. Social Sci. Research Council, 1966—; dir. summer seminar in Chinese art, culture and soc. Bd. Fgn. Scholarships U. S. Dept. State; vis. prof. U. Mich., 1960-61, Nat. Taiwan U., 1963-64, Yale 1965-66. Mem. Am. Anthropol. Assn., Am. Ethnol. Soc. (councillor 1956-59), Assn. Asian Studies. Author: Fabric of Chinese Society, 1953; Readings in Anthropology, 2 vols., 1959; The Evolution of Political Society; An Anthropological View, 1967. Contbr. articles to sci. jours. Studies in Ch'u Hsien, Anhwei Province, China; Chinese in British Guiana; clan associations in Taiwan. Office: Dept. Anthropology, Columbia U., N.Y.C. 10027.*

FRIEDBERG, Charles Kalman, Am. physician; b. Bklyn., Sept. 14, 1908; s. Meyer and Sarah (Nussbaum) F.; A.B., Columbia, 1925, M.D., 1929; m. Gertrude Tonkonogy, Mar. 21, 1933; children—Richard, Barbara. Practice medicine specializing in cardiology, N.Y.C., 1933—; attending physician, chief cardiology Mt. Sinai Hosp., N.Y.C.; faculty Columbia Coll. Phys. and Surg., 1947-67, asso. clin. prof., 1956-67; clin. prof. medicine Mt. Sinai Sch. of Medicine, 1966—; cons. physician Elmhurst City Hospital; cons. cardiologist Horton Hosp. Diplomate Am. Bd. Internal Medicine and Cardiovascular Disease. Fellow A.C.P.: Mem. Am. Heart Assn. (fellow council clin. cardiology and epidemiology), Am. Fedn. Clin. Research, Author: Diseases of the Heart, 1949, 3d edit., 1966; (with Libman) Subacute Retinal Endocarditis, 2d edit., 1948. Editor: Heart, Kidney and Electrolytes, 1962; Radioisotopes in Cardiovascular Disease, 1962; Medical Electronics in Cardiovascular Disease, 1963; Modern Trends in Coronary Disease, 1964; editor-in-chief Progress in Cardiovascular Disease. Research, numerous publs. on arrhythmias endocarditis, rheumatic fever, electrolytes in heart failure, exptl. heart failure, heart block, coronary heart disease. Home: 1185 Park Av., N.Y.C. 10028. Office: 1125 Fifth Av., N.Y.C. 10028.*

FRIEDBERG, Felix, biochemist; b. Copenhagen, Denmark, Apr. 3, 1921; s. Abram and Lina (Schwarz) F.; came to U. S., 1938, naturalized, 1943; B.S. in Chemistry, U. Denver, 1944; Ph.D. in Biochemistry, (Abraham Rosenberg Research fellow) U. Cal., 1947. Faculty, Howard U., Washington, 1948—, prof. biochemistry, 1961—. Vis. lectr. Cath. U. Am., 1950-52. Recipient Lederle Med. Faculty award, 1956. USPHS Research fellow, 1947-48. Mem. Am. Soc. Biol. Chemists, Washington Acad. Scis., Phi Beta Kappa, Sigma Xi, Phi Lambda Upsilon. Author: Thoughts About Life, 1954. Research on incorporation of radioactive amino acids in biosynthesis of proteins; effects of gamma radiation on protein structure. Home: 3636 16th St. N.W., Washington 20010.*

FRIEDBERG, Simeon Adlow, Am. physicist; b. Pitts., July 7, 1925; s. Emanuel B. and Lillian (Adlow) F.; A.B., Harvard, 1947; M.S., Carnegie Inst. Tech., 1948, D.Sc., 1951; m. Joan Brest, Sept. 4, 1950; children—Elizabeth B., Aaron L., Susan A. Fulbright grantee U. Leiden, Netherlands, 1951-52; research physicist Carnegie Inst. Tech., Pitts., 1952-53, faculty, 1953—, prof. physics, 1962—. Westinghouse fellow, 1950-51; Alfred P. Sloan Found. research fellow, 1957-61; Guggenheim fellow Imperial Coll., London, Eng., 1965-66. Fellow Am. Phys. Soc.; mem. Sigma Xi, Tau Beta Pi, Phi Kappa Phi, Pi Mu Epsilon. Contbr. chpt. to Methods of Experimental Physics, 1959. Research, numerous publs. in low temperature solid state physics, thermal and magnetic properties of coupled spin systems, thermal and transport properties in certain metals, semi-conductors, insulators. Home: 1220 S. Negley Av., Pitts. 15217.*

FRIEDE, Reinhard L., neuropathologist; b. Jaegerndorf, Czechoslovakia, May 12, 1926; s. Reinhard and Hilda (Roesner) F.; M.D., U. Vienna, 1951; m. Editha R. Franzen, Dec. 22, 1953; children—Reinhard H., Gerd R. Came to U. S., 1957, naturalized, 1962. Practice medicine, specializing in neuropathology, Dayton, O., 1957-59, Ann Arbor, Mich., 1959-65, Cleve., 1965—; with Aero. Med. Labs. Wright Air Devel. Center, 1957-59; faculty U. Mich., 1959-65, chief neuromorphology sect. Mental Health Research Inst., 1960-65, mem. exec. com., 1963-65; prof. neuropathology Western Res. U., Cleve., 1965—. Recipient Meritorious Civil Service award, 1961. Diplomate of the American Board of Pathology. Mem. Am. Assn. Neuropathologists. Author: A Histochemical Atlas of Tissue Oxidation in the Brain Stem of the Cat, 1961; Topographic Brain Chemistry, 1966. Research, numerous publs. on neuropathology, exptl. neuropathology, histochemistry of normal and diseased nerve tissue, chem. mappings of brain. Home: 2923 Drummond Rd., Shaker Heights, O. 44120. Office: Inst. Pathology, 2085 Adelbert Rd., Cleve. 44106.*

FRIEDEL, Charles, French chemist; b. Strasbourg, France, Mar. 12, 1832; successively at École des mines, École normale; prof. mineralogy, Sorbonne, Paris, France, 1876-84, prof. organic chemistry, 1884-99. Recipient Davy medal, 1880. Mem. French Acad. Scis. Determined vapor densities and molecular weights of chlorides of aluminum, iron, and gallium; prepared (with James Mason Crafts) compounds of silicon, also described synthetic reaction for producing aromatic homologues (Friedel-Crafts reaction); synthesized (with R. D. da Silva) glycerine; produced artificial minerals, including diamonds; studied pyroelectric phenomena of crystals, ketones, aldehydes. Died Montauban, France, Apr. 20, 1899.

FRIEDEL, Georges, French mineralogist; b. Mulhouse, France, July 19, 1865; gen. insp. mines; dir. Geol. Inst., prof. mineralogy and crystallography, Strasbourg, France. Mem. French Acad. Scis., 1917. Author: Leçons de cristallographie, 1926. Research on optical properties of molecular structure of certain substances under heat or pressure by transformation from crystalline to amorphous state. Died Dec. 11, 1933.

FRIEDEL, Jacques, French physicist; b. Paris, Feb. 11, 1921; s. Edmond and Jeanne (Bersier) F.; student École Polytechnique, 1944-46, École des Mines, Paris, 1946-48; Ph.D., Bristol, 1952; Thèse, Paris, 1953; m. Mary Horder, June 2, 1952; children—Jean, Paul. Mining engr. Centre de Recherches Metallurgiques, also École des Mines de Paris, H. H. Wills Phys. Lab., Bristol, 1946-56; maitre de confs. Faculté Scis. de Paris, 1956-59; prof. Faculté des Scis. d'Orsay (France), 1959—. Mem. Am. Acad. Scis. (fgn. hon.). Author: Dislocations, 1956, English edit., 1964. Research on electronic structure of metals and alloys, defects in crystals, including vacancies, dislocations, surfaces. Home: 2, rue J. F. Gerbillon, Paris, France. Office: Physique des Solides, Faculté des Scis., Orsay 95, France.*

FRIEDEL, Robert Augustine, Am. chemist; b. Kenton, O., Aug. 14, 1917; s. C. F. and G. (Trout) F.; A.B., Ohio State U., 1939, M.S., 1940; D.Sc., Carnegie Inst. Tech., 1943; m. Margaret McCredie, July 27, 1942; children—Patricia, Teresa. With Shell Oil Co., 1943-45; Coal research U. S. Bur. Mines, Pitts., 1945—, research coordinator, 1960—; guest scientist Los Angeles Air Pollution Lab., 1950, Australian CSIRO coal div., 1964-65. Mem. coms. on spectral absorption data NRC, 1951—, physico-chem. standards, 1960—; cons. on molecular structure, 1953—; participant various confs. Guggenheim fellow, 1964-65. Recipient Gold medal U. S. Dept. Interior, 1966. Fellow A.A.A.S.; mem. Am. Soc. for Testing Materials (past chmn. com. on mass spectrometry, member of the joint committee on spectral data National Bureau of Standards), Coblentz Society, (past chmn.), Am. Chem. Soc. (past chmn. Pitts. sect., Storch award 1966), Spectroscopy Soc. Pitts. (award 1964), Optical Soc. Am., Photoelectric Spectrometry Group, Geochem. Soc., Soc. Applied Spectroscopy, Japanese Mass Spectroscopy Soc. Author: (with M. Orchin) Ultraviolet Spectra of Aromatic Compounds, 1951; also numerous articles, chpt. in book. Editorial bd. Applied Spectroscopy Revs. Research on infrared, ultraviolet, electron and nuclear magnetic resonance and mass spectrometry chems., prediction isomers in catalysis also crude oils; correlated color coal and free radical content. Home: 302 Locust Lane, Pitts. 15228. Office: U. S. Bur. Mines, Pitts. 15213.*

FRIEDEMANN, Ulrich, bacteriologist, serologist; b. Berlin, Germany, May 7, 1877; s. Edmund and Auguste (Szkolny) F.; studied medicine in Stettin, Poland, Strasbourg, France, Berlin, Germany; m. Gertrud Morgenroth. Asst. to P. Ehrlich, 1903-04, to M. Rubner and C. Flugge, 1905-10; became dir. bacteriol. dept. municipal hosp., Berlin-Moabit, 1911; chief physician mil. epidemic sect. Virchow Hosp., during World War I; faculty Berlin, 1908-33, asso. prof. infectious diseases, 1920-33; with Nat. Inst. for Med. Research, London, Eng., 1933-36; head bacteriol. sect. Jewish Hosp., Bklyn., from 1936. Mem. Robert Koch Inst. Did fundamental research in infectious diseases, especially scarlet fever, diptheria, tetanus, also in immunity, anaphylaxis, theory of Wassermann reaction, virus diseases, capillary permeability, blood liquor limits of bacteria. Died N.Y., Nov. 16, 1949.

FRIEDEN, Alexander, chemist; b. Lithuania, Oct. 5, 1895; s. Abraham and Sarah (Rubin) F.; B.S., U. Va., 1919, M.S., 1919; M.A., Columbia, 1920, Ph.D., 1923; m. Evelyn Gutman, Nov. 13, 1920; children—Julian, Carl. Came to U. S., 1912, naturalized, 1919. Tchr. Charlottesville (Va.) High Sch., 1917-18, Columbia, 1920-23; with cons. labs. of Dr. Raymond F. Bacon, 1923-33; ind. cons., 1933-36; lectr. food and nutrition coll. New Rochelle (N.Y.), 1935-36; tech. dir. food products dept. Stein, Hall & Co., Inc., N.Y.C., 1936-39, tech. dir., 1939-41, v.p., 1941-46; dir. research Pabst Brewing Co., Milw., 1946-53, v.p. research, from 1953. Mem. food engring. council Ill. Inst. Tech., from 1949. bd. visitors Agrl. Research Inst., N.R.C. Mem. Cerevisiae Yeast Inst. (dir.), Am. Chem. Soc., A.A.A.S., Am. Soc. Cereal Chemists, Am. Soc. Brewing Chemists, Inst. Food Technologists, Textile Research Inst., Tech. Assn. Pulp and Paper Industry, Am. Soc. Textile Chemists and Colorists, N.Y. Acad. Sci., Chemists Club N.Y., Sigma Xi. Contbr. to sci. and tech. publs., chpts. to sci. books. Patentee in field. Died Apr. 21, 1956.

FRIEDEN, Carl, Am. biochemist; b. New Rochelle, N.Y., Dec. 31, 1928; s. Alexander and Evelyn (Gutman) F.; B.A., Carleton Coll., 1951; Ph.D., U. Wis., 1955; m. Sari Anne Schneider, Dec. 20, 1953; children—Amy, Eric, Karen. Faculty dept. biochemistry Washington U., St. Louis, 1955—, professor, 1963—; chmn. St. Louis Biochemistry Group, 1961-62. Mem. Am. Soc. Biol. Chemists, Am. Chem. Soc., A.A.A.S., Sigma Xi. Editorial bd. Jour. Biol. Chemistry, 1963-68. Research, publs. on mechanism of enzyme action including correlation of protein structure to catalytic function; devel., application of kinetic theory with respect to enzymes. Home: 7452 Wellington Way, St. Louis 63105.*

FRIEDEN, Edward Hirsch, Am. biochemist; b. Norfolk, Va., Jan. 4, 1918; s. Simon and Sarah (Bluestein) F.; A.B., U. Cal. at Los Angeles, 1939, M.A., 1941, Ph.D., 1942; m. Betty Barnett, June 29, 1941; children—Ray Allan, Jeanne E., Robert E., Rober S., Joyce S. Lalor Found. fellow U. Tex., 1942-43, instr., research asso. Med. Sch., 1943-46; research fellow Harvard, 1946-52, instr. Med. Sch., 1948-52; faculty Tufts U. Med. Sch., 1952-64, asso. prof. biochemistry, 1962-64; research coordinator, dir. Rotch Lab., Boston Dispensary, 1957-64; prof. chemistry Kent (O.) State U., 1964—. Biochem cons. Hynson, Westcott, &

Dunning, Balt., 1950—. Guggenheim fellow U. Cal. at Los Angeles, 1953-54. Mem. Am. Chem. Soc., Am. Soc. Biol. Chemists, Endocrine Soc., Soc. for Exptl. Biology and Medicine, A.A.A.S. Research, numerous publs. on equilibria between amino acids and formaldehydes, biochemistry penicillin, chem. and physiol. properties ovarian protein hormones, relationship between structure and activity biol. active proteins, mechanism action steroid hormones. Home: 359 Wilson Av., Kent, O. 44240.*

FRIEDERICH, Ernst Otto, physicist, chemist; b. Biedesheim/Rhine, Germany, June 2, 1883; s. Friedrich Ludwig and Marie (Hammann) F.; ed. Tech. U. Darmstadt (Germany), U. Heidelberg (Germany); m. Else Schmidt, 1911; 1 son, 1 dau. Chemist, Lessing firm, Nuremberg, Germany; with incandescent lamp factory AEG Gluhlampenbavrik (name changed to Osram 1919), 1909-45, later dir. Contbr. articles to profl. jours. Research on various aspects of electric incandescent lamp; improved durability of carbon filament and tungsten; made theoretical studies in physics of solid bodies; developed Friederich's law of conductivity by solids (fundamental to modern transistor research). Died Oxnard, Cal., Jan. 13, 1951.

FRIEDERICHS, Karl Paul Theodor, German entomologist; b. Wismar, Meckenbourg, Oct. 13, 1878; s. Adolf and Emilie (Paulsen) F.; Ph.D. in Entomology, Sch. of Wismar; m. Karla Bastmann; 1 dau., Angelika. Made study trips to Indies and Madagascar; agrege at Rostock; titular prof. Dutch colonial service; dir. Entomology Lab., U. Rostock; vis. prof. U. Minn.; prof. at Pozen; prof. emeritus U. Göttingen. Mem. German Acad. Natural Scis. of Halle, Russian, German (Karl Eschwig medal) socs. entomology, Soc. Zoology and Botany of Vienna (hon.), Am. Assn. Econ. Entomologists (hon.). Author: Die Grundfragen der Land und Forstwissenschaftl. Zoologie; Okologie als Wissentung des Lebendigen; Lebensdauer, Altern und Tod. Address: Am Sölenborn 14, 34 Göttingen, Germany.

FRIEDERICHSEN, Max, German geographer; b. Hamburg, Germany, June 21, 1874; s. Ludwig and Elisabeth (Kauffmann) F.; pupil of Theobald Fischer and F. von Richthofen; m. Marianne Martius, 1907; 2 sons, 1 dau. Traveled through Urals, Caucasus, Russian Armenia, after 1897, in France, 1900, in Tienshal, Ala-tua in Central Asia (with Russian botanist Saposchnikow), 1902; became mem. faculty U. Göttingen (Germany), 1903; named asso. prof., Rostock, Germany, 1906; became prof., Bern, Switzerland, 1907, Greifswald, Germany, 1909, Königsberg, Germany, 1917, Breslau, Germany, 1923; with geog. commn. German Gen. Govt., Warsaw, Poland, during World War I. Attended Internat. Congress Geologists, 1897. Author: Moderne Methoden der Erforschung, Beschreibung und Erklärung geographischer Landschaften, 1914; Die Grenzmarken des Europäischen Russlands . . . , 1915; Landschaften und Städte Polens und Litauens, 1918; Finnland, Estland und Lettland, Litauen, 1924; also articles. Investigated geography of eastern Germany and bordering countries, Baltic nations, Poland, Russia. Died Hamburg, Aug. 22, 1941.

FRIEDHEIM, Carl, chemist; b. 1858; pupil, then asst. of Rammelsberg; prof., Bern, Switzerland. Editor: Handbuch der anorganischen Chemie (Gmelin-Kraut), 7th edit. Research on complex phosphomolybdic and tungstic acids, separation of chlorine, bromine and iodine, determination of vanadium, arsenic, chromium and molyddenum, use of hydrogen peroxide and hydroxylamine in analysis. Died 1909.

FRIEDHOFF, Arnold Jerome, Am. psychiatrist, neuro-chemist; b. Johnstown, Pa., Dec. 26, 1923; s. Abraham and Stella (Beerman) F.; B.A., U. Pa., 1944, M.D., 1947; m. Frances L. Wolfe, Feb. 24, 1946; children—Lawrence, Nancy, Richard. Faculty, N.Y. U., Sch. Medicine, 1956—, head psychopharmacology research unit, 1956-62, prof. psychiatry, sci. dir Program For Study Psychotic Disorders, 1962—; mem. acad. faculty Conn. Dept. Mental Health, 1965. Recipient Career Devel. award Nat. Inst. Mental Health, 1961, Career Scientist award, 1967. Diplomate of the American Board of Psychiatry and Neurology. Fellow Am. Psychiat. Assn., Am. Coll. Clin. Pharmacology and Chemotherapy, Am. Coll. Neuropsychopharmacology; mem. Harvey Soc., Assn. For Research Nervous and Mental Diseases, A.A.A.S., Am. Chem. Soc., Inter-Am. Soc. Psychology, Royal Medico-Psychol. Assn., Am. Psychopath. Assn., Psychiat. Research Soc., Soc. Biol. Psychiatry, Sigma Xi. Research, numerous publs. on relationships between biol. abnormalities and behavioral or mental disturbances. Home: 32- 25 168th St., Flushing, N.Y. 11358. Office: N.Y. U. Sch. Medicine, 550 1st Av., N.Y.C. 10016.*

FRIEDKIN, Morris Enton, Am. biochemist; b. Kansas City, Mo., Dec. 30, 1918; s. Benjamin and Anna (Lapatuchin) F.; B.S., Ia. State Coll., 1940, M.S., 1941; Ph.D., U. Chgo., 1948; m. Roberta Vanocur, Sept. 19, 1943; children—Noah, Susanna, Deborah. Postdoctoral fellow U. Copenhagen (Denmark), 1948-49; faculty Washington U. Sch. Medicine, St. Louis, 1949-58, asso. prof. pharmacology, 1956-58; prof., chmn. pharmacology Tufts U. Sch. Medicine, Boston, 1958-67, prof., chmn. biochemistry, 1967—. Mem. Am. Acad. Arts and Scis., Am. Soc. Biol. Chemists, Am. Soc. for Pharmacology and Exptl. Therapeutics. Research and publs. on elucidation of various aspects of metabolism catalyzed by enzymes involved in nucleic acid synthesis; devel. new anticancer agts. Home: 83 Centre St., Brookline, Mass. 02146. Office: 136 Harrison Av., Boston 02111.*

FRIEDLAND, Fritz, physician; b. Berlin, Germany, Jan. 2, 1910; s. Arthur and Henriette (Baum) F.; M.D., Friedrich Wilhelm Universitaet, Berlin, 1934; m. Hilda Berta Rosenthal, June 20, 1938. Came to U. S., 1938, naturalized, 1944. Practice medicine, specializing in phys. medicine, rehab., Boston, 1939—; chief phys. medicine, rehab. service VA Hosp., Boston, 1953—; asso. prof. Sargent Coll. Boston U., 1949-56, lectr. med. rehab., 1957—; asst. prof. Tufts U., Sch. Medicine, 1954-61, asso. prof., 1961—; cons. Boston Dispensary, 1965-67; asso. staff New Eng. Med. Center Hosps., Boston, 1967—; cons. Lemuel Shattuck Hosp. Recipient John Eisele Davis Ann. award Assn. Phys. and Mental Rehab., 1957. Fellow A.C.P.; mem. A.M.A., New Eng. Soc. Phys. Medicine (award merit 1964), Am. Congress Rehab. Medicine, Assn. Phys. and Mental Rehab. (hon.), Am. Acad. Phys. Medicine and Rehab., Assn. Mil. Surgeons U. S. Contbr. chpts. to Physical Medicine in General Practice, 1952; Cyclopedia of Medicine, Surgery and Specialities, 1955, rev., 1959. Pioneer in med. rehab. of patients with spinal cord injuries; introduced into U. S. ultrasound as a phys. therapeutic modality. Home: 21 Hammond Pond Pkwy., Chestnut Hill, Mass. 02167. Office: 150 S. Huntington Av., Boston 02130.*

FRIEDLANDER, Gerhart, chemist; b. Munich, Germany, July 28, 1916; s. Max O. and Bella (Forchheimer) F.; B.S., U. Cal. at Berkeley, 1939, Ph.D., 1942; m. Gertrude Maas, Feb. 6, 1941; children—Ruth Ann, Joan Claire. Came to U. S., 1936, naturalized, 1943. Instr., U. Ida., Moscow, 1942-43; staff Los Alamos (N.M.) Sci. Lab., 1943-46; research asso. Gen. Electric Co. Research Lab., Schenectady, 1946-48; vis. lectr. Washington U., St. Louis, 1948; chemist Brookhaven Nat. Lab., Upton, N.Y., 1948-52, sr. chemist, 1952—, chmn. chemistry dept., 1968—. Chairman of the Gordon Research Conference on Nuclear Chemistry, 1954; mem. adv. com. for chemistry Oak Ridge Nat. Lab., 1966—. Fellow Am. Phys. Soc.; mem. Am. Chem. Soc. (chmn div. nuclear chemistry and tech. 1967, award for nuclear applications in chemistry 1967), A.A.A.S. Author: (with J. W. Kennedy) Introduction to Radiochemistry, 1949; Nuclear and Radiochemistry, 1955, (with J. M. Miller), 64; also articles. Asso. editor Ann. Rev. Nuclear Sci., 1958-67. Research on chem. effects of nuclear transformations, properties of radioactive isotopes, mechanisms of nuclear reactions, especially those induced by protons of very high energies. Home: 18 Arthur Av., Blue Point, N.Y. 11715. Office: Brookhaven Nat. Lab., Upton, N.Y. 11973.*

FRIEDLÄNDER, Paul, chemist; b. 1857; asst. to Baeyer in studies on indigo; prof., Karlsruhe, Germany, Vienna, Austria, Damrstadt, Germany. Patentee 235 compounds and processes of thio-indigo and indigoid dyes; began Friedlander serial collection of dye patents, 1888; studied isatin and its derivatives; synthesized thionaphthene; found Tyrian purple dye from murex mollusk to be dibromindigo; contbd. to expansion of dye industry. Died 1923.

FRIEDMAN, Abraham Solomon, Am. chemist; b. N.Y.C., Oct. 25, 1921; s. Israel Hyman and Sarah (Cohen) F.; A.B., Bklyn. Coll., 1943; Ph.D., Ohio State U., 1950; m. Diana Elena Scott, July 4, 1952; children—Danielle Suzanna, Rebecca Lea, Abigail, Michelle Miriam. Chem. engr. Manhattan Project, U. S. Army, Decatur, Ill., Oak Ridge, Metall. Lab., Chgo., 1944-46; research fellow, research asso. Ohio State U., 1946-51; Fulbright Research fellow Van der Waals Lab., Amsterdam, Netherlands, 1951-52; phys. chemist Nat. Bur. Standards, Washington, 1952-55; with U. S. AEC, 1955—, sci. rep. Paris, France, 1962-65, dep. dir. div. internat. affairs, Washington, 1966—. Mem. Am. Chem. Soc., Am., Netherlands phys. socs., Washington Acad. Sci., Philos. Soc. Washington. Author: (with L. Haar, C. W. Beckett) Ideal Gas Thermodynamic Functions and Isotope Exchange Functions, 1961; also articles, chpts. and sects. in books. Research on theory and techniques of isotope separation especially hydrogen, helium, uranium, prodn. and separation of transuranium elements, thermodynamics, statis. mechanics, cryogenics, transport phenomena; active in nuclear safeguards, internat. agreements for cooperation in peaceful applications of nuclear energy. Home: 6305 Phyllis, Lane, Bethesda, Md. 20034. Office: Div. Internat. Affairs, U. S. AEC, Washington 20545.*

FRIEDMAN, Avner, mathematician; b. Petah-Tikvah, Israel, Nov. 19, 1932; s. Moshe Simcha and Hanna (Rosenthal) F.; M.Sc., Hebrew University, Jerusalem, Israel, 1954, Ph.D., 1956; m. Lillia Lynn Kelly, July 7, 1959; children—Alissa, Joel, Naomi, Tamara. Came to the United States of America, in 1956, naturalized, 1963. Research asso. Kan. U., Lawrence, 1956-57; lectr. U. Ind., Bloomington, 1957-58; vis. asst. prof. U. Cal. at Berkeley, 1958-59; asso. prof. U. Minn., Mpls., 1959-61; vis. asso. prof. Stanford, 1961-62; prof. Northwestern U., Evanston, Ill., 1962—; vis. prof. Tel Aviv (Israel), U., 1966-67. Alfred P. Sloan fellow, 1962-65; Guggenheim fellow, 1966-67. Author: Generalized Functions and Partial Differential Equations, 1963; Partial Differential Equations of Parabolic Type, 1964; also numerous articles. Research in differentiability of solutions of partial differential equations, a priori bounds and existence theorems, asymptotic behavior of solutions, Free boundary problems for parabolic equations, ordinary differential equations and integral equations in Banach space, control theory. Home: 2669 Orrington Av., Evanston, Ill. 60201.*

FRIEDMAN, Bernard, Am. mathematician; b. Bklyn., Aug. 19, 1915; s. Nathan and Sarah (Moskowitz) F.; B.S., Coll. City N.Y., 1934; M.S., Mass. Inst. Tech., 1935, Ph.D., 1936; m. Dorothy Miller, Sept. 3, 1939; children—Neal James, Andrea Susan. Blumenthal fellow N.Y. U., 1936-37, research mathematician, 1943-53, asst. prof., 1946-53, prof., 1953-57; instr. U. Wis., 1937-40, Wilson City Coll., Chgo., 1940-43; prof. U. Cal., Berkeley, 1957—, chmn. math. dept., 1960-62. Recipient Navy award for wartime research in applied math. OSRD, 1946. Mem. Am. Math. Soc., Math. Assn. Am., A.A.A.S. (v.p. sect. A 1965—), Soc. Indsl. and Applied Math. (vis. lectr. 1960-65), N.Y. Acad. Scis., Internat. Sci. Radio Union, Inst. Math. Statistics, Phi Beta Kappa, Sigma Xi, Pi Mu Epsilon. Author: Principles and Techniques of Applied Mathematics, 1956; Intrinsic Calculus, 2d vol., 1965. Research in application of methods of abstract algebra to problems in electromagnetic wave propagation and crystal statistics. Home: 1201 Brewster Dr., El Cerrito, Cal. 94532.*

FRIEDMAN, Emanuel A., Am. physician; b. N.Y.C., June 9, 1926; s. Louis and Pauline (Feldman) F.; A.B., Bklyn. Coll., 1947; M.D., Columbia, 1951, Sc.D., 1959; m. E. Judith Salomon, June 6, 1948; children—Lynn, Meryl, Lee. Practice medicine, specializing in obstetrics, gynecology, N.Y.C., 1957-63, Chgo., 1963—; asst., asso. attending obstetrics, gynecology Columbia-Presbyn. Hosp., 1957-63; chmn. dept. obstetrics, gynecology Michael Reese Hosp., 1963—, chmn. Research Inst., 1965-67; faculty Columbia, 1957-63, coordinator Child Devel. Program, 1958-63; prof., chmn. dept. obstetrics, gynecology Chgo. Med. Sch., 1963—. Recipient Joseph Mather Smith prize Columbia, 1958, Bklyn. Coll. Alumni Honors award, 1964. Diplomate Am. Bd. Obstetrics, Gynecology. Fellow Am. Coll. Obstetrics, Gynecology, A.C.S., Internat. Coll. Surgeons, N.Y. Acad. Medicine; mem. Am. Inst. Biol. Sci., Assn. Am. Med. Colls., Soc. Exptl. Biol. Medicine, A.A.A.S., N.Y. Acad. Scis., Soc. Gynecol. Investigation, Am. Assn. U. Profs., A.M.A., Alpha Omega Alpha. Author: Labor: Clinical Evaluation and Management, 1966. Research, numerous publs. on graphicostatistical analysis of labor and factors that influence course of labor. Home: 1000 Wildwood Lane, Highland Park, Ill. 60035. Office: 2929 S. Ellis Av., Chgo. 60616.*

FRIEDMAN, Herbert, Am. astrophysicist; b. N.Y.C., June 21, 1916; s. Samuel and Rebecca (Seligson) F.; B.A., Bklyn. Coll., 1936; Ph.D. in Physics, Johns Hopkins, 1940; m. Gertrude Miller, 1940; children—Paul, Jon. With U. S. Naval Research Lab., 1940—, supt. atmosphere and astrophysics div., 1958—, chief scientist E. O. Hulburt Center Space Research, 1963—; part-time prof. physics U. Md.; spl. research V-2 rocket, satellite launchings, solar cycle variations X-ray and ultra-violet radiations from sun; produced 1st X-ray and ultra-violet photographs, also discovered hydrogen geocorona, measured ultraviolet fluxes of early-type stars. Chmn. COSPAR working group II, Internatl. Quiet Sun Year. Recipient Distinguished Service award Navy Dept., 1945, Distinguished Achievement in Sci. award 1962; medal Soc. Applied Spectroscopy, 1957; Distinguished Civilian Service award Dept. Def., 1959; Janssen medal French Photog. Soc., 1962; Eddington medal Royal Astron. Soc., 1964; Presdl. medal for distinguished fed. service, 1964; Space Sci. award Am. Inst. Aeros. and Astronautics, 1963. Fellow Am. Phys. Soc., Am. Optical Soc., Am. Geophys. Union, Am. Astron. Soc., Am. Inst. Aero. and Astronautics; mem. Nat. Acad. Scis. (exec. com. space sci. bd.), Am. Acad. Arts and Scis., Internat. Acad. Astronautics, Am. Philos. Soc. Pioneered in devel. rocket and satellite astronomy; directed expts. which detected X-ray and ultraviolet radiations from sun; produced 1st astron. photographs made in X-ray wavelengths; discovered hydrogen corona around the earth. Home: 2643 N. Upshur St., Arlington, Va. Office: Code 7100, U. S. Naval Research Lab., Washington 20390.*

FRIEDMAN, Leo, Am. food toxicologist; b. Bklyn., May 18, 1915; s. Isidore and Esther (Schub) F.; student Coll. City N.Y., 1932-35; B.S., George Washington U., 1941, postgrad.; M.S., Georgetown U., 1946, Ph.D., 1949; m. Anna Nash, Feb. 20, 1937; 1 dau., Sonia Meryl. Chemist, biochemist div. nutrition FDA, 1938-49, chief nutrition research br., 1949-59, dir. research dir. nutrition, 1959-62, acting chief nutrition research br., 1961-62; faculty Mass. Inst. Tech., Cambridge, 1962—, prof. nutrition and food safety dept. nutrition and food sci., 1965—. Mem. adv. com. on protocol for safety evaluations FDA, 1966—. Recipient Dept. Health, Edn., and Welfare Unit Superior Service award, 1960, 61. Fellow Washington Acad. Sci., A.A.A.S.; mem. Am. Chem. Soc., Am. Inst. Nutrition, N.Y. Acad. Scis., Soc. for Exptl. Biology and Medicine (chmn. nat. council 1956-57), Animal Nutrition Research Council (past chmn), Assn.

Ofcl. Agrl. Chemists, Am. Oil Chemists' Soc., Inst. Food Technologists. Research, publs. on vitamin and amino acid assays, biol. value of proteins, amino acid-sugar reaction, vitamin requirements and stability, growth-promoting action of antibiotics, nutritional effects of emulsifying agts., nutritive value of heated fats, dietary influences in carbohydrate metabolism, nutritional value of pancreatic enzymes, evaluation of food safety, physiol. effects food substances, carcinogenesis. Home: 5 Battle Green Rd., Lexington, Mass. 02173. Office: Dept. Nutrition and Food Sci., Mass. Inst. Tech., Cambridge, Mass. 02139.*

FRIEDMAN, Lester, Am. chemist; b. N.Y.C., Sept. 14, 1928; s. Joseph and Sally (Greenman) F.; B.S., Purdue U., 1951; postgrad. Ohio State U., 1951-54, Ph.D., 1959; m. Carolyn Ruth Rosen, May 27, 1956; children—Renee Frances, Robert Mark. Instr., Capital U., Columbus, O., 1952-53; staff U. S. Army Med. Research Lab., Ft. Knox, Ky., 1954-56; asst. prof. N.Y. U., N.Y.C., 1957-61; faculty Case Western Res. U. (formerly Case Inst. Tech.), Cleve., 1961—, asso. prof. chemistry, 1965—. Cons. to pvt. cos. Fellow A.A.A.S.; mem. Am. Chem. Soc. (chmn. elect Cleve. sect.), Chem. Soc. (London), N.Y. Acad. Scis. Editor series: (with G. A. Olah) Reactive Intermediates in Organic Chemistry, 1968—; also articles. Research, patents on chemistry and synthesis of reactive, transient intermediates, arynes, carbanions, carbenes, carbonium ions, and diazonium ions; effect of environment on chem. reactivity; correlation of chem. structure with biol. activity; devel. of processes for manufacture of organophosphorus compounds, new organophosphorus structures. Home: 3618 Concord Dr., Beachwood, O. 44122. Office: Case Western Res. U., University Circle, Cleve. 44106.*

FRIEDMAN, Lewis, Am. chemist; b. Spring Lake, N.J., Aug. 8, 1922; s. Joseph Henry and Tillie (Wainer) F.; B.A., Lehigh U., 1943; M.A., Princeton, 1945, Ph.D., 1947; m. Dorothy Kaplan, Aug. 24, 1948; children—Robert S., Beth, Jan. With Brookhaven Nat. Lab., Upton, L.I., N.Y., 1948—, sr. chemist, 1961—; guest scientist F.O.M. Lab. for Mass Separation, Amsterdam, The Netherlands, 1960-61. Mem. Am. Soc. Testing and Materials. Contbr. numerous articles to sci. jours. Asso. editor Jour. Chem. Physics, 1966—. Pioneer in use of mass spectrometers as a research tool for study of structures of complex organic and inorganic molecules; fundamental studies of rates of ion-molecule reactions; demonstrated use of ion molecule reactions as a tool for study of kinetic to internal energy process in reactive systems. Home: 78 Everett St., Patchogue, N.Y. Office: Chemistry Dept., Brookhaven Nat. Lab., Upton, L.I., N.Y.*

FRIEDMAN, Meyer, Am. physician; b. Kansas City, Kan., July 13, 1910; s. Joseph and Eva (Werby) F.; A.B., Yale, 1931; M.D., Johns Hopkins, 1935; m. Macia Campbell, Sept. 5, 1942; children—Joyce, Joseph, Mark. Practice medicine, specializing in cardiovascular disease, San Francisco, 1939—; dir. Harold Brunn Inst. Mt. Zion Hosp. and Med. Center, 1935—; asst. clin. prof. Med. Sch. Stanford, 1939-60. Chmn., Council For High Blood Pressure Research Am. Heart Assn., 1959-60; cardiac cons. U. S. VA, 1945-64; research cons. Riker Pharm. Labs., 1958—. Mem. Am. Soc. Clin. Investigation, Am. Physiol. Soc., Soc. Exptl. Biology and Medicine, Western Assn. Physicians, Am. Heart Assn. Author: Functional Cardiovascular Disease, 1947. Research, numerous publs. on measurement of digitalis compounds in human body fluids; discovered that overprodn. of urate induces gout, role of stress and certain behaviour patterns in elevating blood cholesterol and triglycerides, increasing incidence of coronary artery disease, exact cause of a coronary thrombus, cholesterol elevating propensities of phosphatides and fats. Home: 160 San Carlos Av., Sausalito, Cal. 94965. Office: 1600 Divisadero St., San Francisco 94115.*

FRIEDMAN, Milton, Am. physician; b. Newark, Sept. 13, 1903; s. Samuel and Sarah (Goldberg) F.; student Newark Jr. Coll., 1920-22; M.D., George Washington U., 1926; m. Elna Linborg, Dec. 21, 1947; 1 dau., Susan. Dir. radiation therapy Newark Beth Israel Hosp., 1933-38; radiation therapist Bellevue Hosp., N.Y.C., 1933—, Hosp. for Joint Diseases, N.Y.C., 1934—; attending radiotherapist N.Y. U. Hosp., N.Y.C., 1949—; prof. clin. radiology N.Y. U. Sch. Medicine, 1949—. Mem. Internat. Commn. Radiation Units, 1965—; cons. to govt. agys., hosps. Recipient numerous prizes for sci. exhibits of researches. Fellow Am. Coll. Radiology; mem. Radiol. Soc. N.Am., N.Y. Cancer Soc. (past pres.), Am. Radium Soc. (pres. 1967). Author: Roentgens, Rads and Riddles, 1959; also numerous articles, chpts. in books. Research on microscopic effects radiation on various tumors; invented many instruments and apparatus in field radium, X-ray and isotope treatment cancer; pioneered treatment cancers with supervoltage X-ray machines and cobalt; developed new radiation treatment techniques, methods for combining irradiation with cancer chemotherapy. Home: 566 1st Av., N.Y.C. 10016.*

FRIEDMAN, Milton, Am. economist; b. N.Y.C., July 31, 1912; s. Jeno Saul and Sarah (Landau) F.; A.B., Rutgers U., 1932; M.A., University Chicago, 1933; Ph.D., Columbia, 1946; LL.D., Rikkyo Univ.; 1963; LL.D., Kalamazoo College, 1968; m. Rose Director, June 25, 1938; children—Janet, David. Asso. economist Nat. Resources Com., Washington, 1935-37; mem. research staff Nat. Bur. Econ. Research, N.Y.C., 1937-45 (on leave 1940-45), 48—; prin. economist div. tax research U. S. Treasury, Washington, 1941-43; asso. dir. statis. research group Columbia, N.Y.C., 1943-45; faculty U. Chgo., 1946—, prof. econs., 1948—, Paul Snowden Russell Distinguished Service prof. econs., 1963—. Vis. Fulbright lectr. Cambridge U., Eng., 1953-54; Wesley Clair Mitchell vis. research prof. Columbia, 1964-65. Fellow Am. Statis. Assn., Econometric Soc., Inst. Math. Statistics; mem. Am. Philos. Soc., Am. Econ. Assn. (John Bates Clark medallist 1951, pres. 1967), Royal Econ. Soc., Econometrica. Author books, including A Theory of the Consumption Function, 1957; (with A. J. Schwartz) A Monetary History of the United States, 1867-1960, 1963. Devel. new theory explaining what determines fraction of income individuals spend or save; fuller explanation of role money plays in course of econ. fluctuations and long period devel.; new statis. techniques useful in econs.; contbns. to selected topics in econ. theory. Home: 5825 S. Dorchester Av., Chgo. 60637.*

FRIEDMAN, Orrie Max, chemist; b. Grenfell, Sask., Can., June 6, 1915; s. Jack and Gertrude (Shulman) F.; B.Sc., U. Man., 1935; B.Sc., McGill U., 1941, Ph.D., 1944; m. Laurel Ethel Leeder, Jan. 2, 1959; children—Mark David, Gertrude Jane, Hugh Robert. Came to U. S., 1946, naturalized, 1963. Jr. research chemist NRC, 1944-46; research fellow in chemistry Harvard, 1946-49, asst. prof. chemistry Med. Sch., 1952-53; asso. in surg. research Beth Israel Hosp., Boston, 1949-52; faculty Brandeis U., Waltham, Mass., 1953-65, prof. chemistry, 1960-65, adj. research prof. chemistry, 1965—; pres. Collaborative Research, Inc., Waltham, 1962—; v.p. United Chem. Co. Ltd., Montreal, Que., Can., 1965—. Cons. Harvard Med. Sch., 1953-54, 56-57; spl. cons. Nat. Cancer Inst., 1963—; mem. adv. coms. NIH. Fellow A.A.A.S.; mem. Am. Chem. Soc., Am. Assn. for Cancer Research, Assn. Harvard Chemists, Radiation Soc., Sigma Xi. Research, numerous publs. in field of cancer chemotherapy, devel. new cancer drugs, chemistry of high explosives, chemistry of nucleic acids. Home: 49 Warren St., Brookline, Mass. 02146. Office: 1365 Main St., Waltham, Mass. 02154.*

FRIEDMAN, Sydney Murray, Canadian anatomist; b. Montreal, Que., Can., Feb. 17, 1916; s. Jacob and Minnie (Signer) F.; B.A., McGill U., 1938, M.D., 1940, Ph.D., 1946; m. Constance Livingstone, Sept. 23, 1940. Faculty, McGill U., 1944-50; prof., head anatomy U. B.C., Vancouver, Can., 1950—. Recipient Premier award for research in aging Ciba Found., 1955. Fellow Royal Soc. Can., Am. Coll. Cardiology; mem. Am., Canadian (pres.) assns. anatomists, Canadian Physiol. Soc., Am. Heart Assn., Sigma Xi. Author: Visual Anatomy, Head and Neck, 1950; Visual Anatomy, Thorax and Abdomen, 1952. Contbr. numerous articles to profl. jours. Developed reasonable exptl. counterpart of human essential hypertension in lab. rat; demonstrated necessary role of sodium and potassium in blood pressure regulation by their direct involvement in regulation of calibre of blood vessels; participated in devel. glass electrodes specifically sensitive to sodium and potassium and initiated their application to biol. problems. Home: 4916 Chancellor Blvd., Vancouver 8, B.C., Can.*

FRIEDMANN, Herbert, Am. biologist, ornithologist; b. N.Y.C., Apr. 22, 1900; s. Uriah M. and Mary (Behrmann) F.; B.Sc., Coll. City N.Y., 1920; Ph.D., Cornell U., 1923; m. Karen Juul Veglo, May 7, 1937; 1 dau., Karen Alice (Mrs. Herbert Phillips Jones). Post-doctoral fellow NRC, Harvard, 1923-26; instr. zoology Brown U., 1926-27, Amherst Coll., 1927-29; curator birds Smithsonian Instn., Washington, 1929-57, head curator zoology, 1957-61; dir. Los Angeles County (Cal.) Mus. Natural History, 1961—. Research fellow NSF, Guggenheim Found. Recipient Leidy medal Acad. Natural Sci. Phila., 1955; Sci., 1959; Mem. Nat. Acad. Sci. (Elliott medal 1959), Am. Soc. Naturalists, Am. Soc. Zoologists, A.A.A.S. (pres. sect. zoology 1959), Am. Ornithol. Union (pres. 1939-40, Brewster medal 1964), Cooper Ornithol. Soc. (pres. So. div. 1964), So. Cal. Acad. Sci.; hon. mem. Deutsche Ornithol. Gesellschaft, S. African Soc. Ornithology, Soc. Poey Cuba, Soc. Ornithology La Plata (Argentina). Author: The Cowbirds, 1929; Birds of North and Middle America, 3 vols., 1946-50; Birds of Mexico, 2 vols., 1950, 57; Birds of Ethiopia and Kenya, 2 vols., 1930, 37; Symbolic Goldfinch, 1946; The Honey-guides, 1955; Parasitic Weaverbirds, 1961; Host Relations of Parasitic Cowbirds, 1963; also numerous articles. Ornithol. expdns. to S.Am., S.E., Central Africa, Europe; research in brood parasitism in birds, systematics of birds of Africa, N.Am., S.Am., bacterial symbiosis and break-down of waxes, symbolism of animals and plants in European medieval and renaissance art. Home: 350 S. Fuller Av., Los Angeles 90036. Office: Los Angeles County Mus. Natural History, 900 Exposition Blvd., Los Angeles 90007.*

FRIEDMANN, I(mrich), pathologist; b. S.P. Nová, Czechoslovakia, June 4, 1907; s. Arnold and Bertha (Braun) F.; M.D., U. Prague 1931; M.R.C.S., L.R.C.P., U. London, 1942, D.Sc., 1967; m. Joan Margaret Drew, June 12, 1943. Pathologist, Bata Hosp., Moravia, 1936-39; demonstrator in pathology Postgrad. Med. Sch., London, 1943; pathologist Med. Mission, Middle East, Czechoslovakia, 1944-46; asst. pathologist, reader in bacteriology U. London, 1949-53, dir. dept. pathology, Inst. Laryngology and Otology, 1952—, prof. pathology, 1963—. Medicine, 1963. Fellow Royal Soc. Medicine (Harrison prize in otology 1963), Pathol. Soc. Gt. Britain and Ireland, Coll. Pathologists (a founder), Assn. Clin. Pathologists, Brit. Soc. Cell Biology. Publs. on research in tissue culture studies of embryonic inner ear and electromicroscopy of inner ear; toxic effect of certain antibiotics on inner ear; electronmicroscopy of cancer of ear, nose, throat. Home: 46 Gordon Av., Stanmore, Eng. Office: 330 Gray's Inn Rd., London, Eng.*

FRIEDREICH, Nikolaus, German physician; b. Würzburg, Germany, July 31, 1826; s. Johann (Jean) Baptist and Catharina (Bolle) F.; ed. Würzburg, Heidelberg, Germany; M.D., 1850; m. Josephine Lauck, 1854. Dir. (with Gegenbauer) Marcus' clinic Juliusspital, Würzburg, 1850-53; became mem. faculty U. Würzburg, 1853, succeeded Virchow as prof. path. anatomy, 1856; became prof. pathology and therapy, dir. med. clinic U. Heidelberg, 1859. Contbr. articles to med. jours. Research on muscle atrophy, muscle hypertrophy, vein pulse, heart diseases; described hereditary ataxia, 1863, paramyoclonus multiplex (Friedreich's disease), 1881. Died Heidelberg, July 6, 1882.

FRIEDRICH, Ernst, German geographer; b. Klein-Lichtenau, W. Prussia, Mar. 27, 1867; s. Wilhelm and Emilie (Tornier) F.; student, Leipzig, Germany; Ph.D., Konigsberg, Germany, 1894; m. Martha Canzler, 1897; 1 son. Became mem. faculty Leipzig, 1901, asso. prof., 1906, prof. econ. geography, 1921; also tchr., comml. coll. Author: Handels- und Produktenkunde, 1898; Einführung in die Wirtschaftsgeographie, 1908; Allgemeine und spezielle Wirtshaftsgeography, 2 vols., 1909; Geography des Welthandels und Weltverkehrs, 1911; also sects. in handbooks and textbooks, maps, atlases. A founder of German econ. geography; considered climate the most important econ. factor, then the activity of man in 4 stages: reflexive, inductive, customary-traditional, economic-tech. Died Leipzig, June 17, 1937.

FRIEDRICH, Hans Joachim, German phys. chemist; b. Munich, Jan. 26, 1933; s. Franz Ignaz and Elisabeth (Roth) F.; Dipl. Chem., U. Munich, 1959, dr., 1961; m. Barbara Dorothee Wöbker, Dec. 19, 1965; 1 dau., Julia Maria. Asst. dept. organic chemistry U. Würzburg, 1962-65; research specialist molecular spectroscopy Farbwerke Hoechst C.L., Frankfurt/Main-Höchst, 1966—. Recipient pub. service awards, Bavaria, univs. Munich, Würzburg, 1961-65. Mem. Soc. German Chemists. Research and publs. on chemistry of cyanine dyes and related color bases and complex compounds; spectroscopic investigations of tautomerism and hydrogen bonds, of steric features of molecules; nuclear magnetic resonance. Home: 20 Eppsteinerstrasse, 6233 Kelkheim/Taunus. Office: Farbwerke Hoechst AG, 6 Frankfurt-Höchst, West Germany.*

FRIEDRICH, Hermann, German zoologist; b. Essen, Germany, June 5, 1906; s. Jakob and Helene (Kleinschmidt) F.; Dr. phil. U. Marburg, 1928; a.p.l. prof. U. Kiel, 1941; m. Anneliese Riemann, Aug. 25, 1931; children—Gesine (Mrs. Kurt Schmidt), Hans, Jochen, Hartmut, Gertruade (Mrs. Martin Walsdorff), Jörg, Jens. Staff Zool. Inst., U. Kiel (Germany), 1930-42, staff Inst. Marine Research, 1942-51; dir. Inst. Marine Research, Bremerhaven, Germany, 1951-62; dir. Ubersee-Museum, Bremen, Germany, 1962—. Author: Meeresbiologie, 1965; also numerous articles. Research on marine worms like Nemertreans and Polychaets. Home: 6 Nancystr. Office: 13 Bahnhofplatz, Bremen, Germany.*

FRIEDRICH, Wilhelm Karl, German hydrologist; b. Bremen, Germany, Feb. 12, 1900; s. Johann and Betty (Masemann) F.; Ph.D., U. Gottingen (Germany), 1925; m. Hildegard Pietsch, July 5, 1930; children—Christa (Mrs. Ferdinand Kuhn-Regnier), Inge (Mrs. Helmut Pein), Klaus. With German Weather Bur., 1925-26; with Fed. Inst. for Hydrology, Koblenz, Germany, 1926-65, chief research div., 1949-65; prof. hydrology U. Mainz, Germany, 1965—. Chmn., German nat. com. Internat. Hydrological Decade, 1966—. Mem. German Union Geodesy and Geophysics (chmn. German sect. hydrology 1949—), Internat. Assn. for Sci. Hydrology (past pres.), German Meteorol. Soc., German Geophys. Soc., Am. Geophys. Union. Author: Evaporation from Water Surface, 1930; Variations of Groundwater Levels, 1933; Foresthydrology, 1958; also numerous articles. Research in hydrometeorology, water-balance watersheds, foresthydrology, Lysimeter techniques. Home: 80 Brentanostr., Koblenz 54, Germany.*

FRIEDRICHS, Kurt Otto, mathematician; b. Kiel, Germany, Sept. 28, 1901; s. Karl E. and Elisabeth (Entel) F.; Ph.D. in Math., U. Gottingen, Germany, 1925; m. Nellie H. Bruell, Aug. 11, 1937; children—Walter, Liska, David, Christopher, Martin. Came to U. S., 1937, naturalized, 1944. Asst., U. Gottingen, 1925-27, privat-dozent, 1929-30; asst., privatdozent Techn. Hochschule, Aachen, Germany, 1927-29; prof. math. Tech. Hochschule, Braunschweig, 1930-37; vis. prof. applied math., N.Y. U., 1937-39, asso. prof., 1939-43, prof., 1943—, asso. dir. Courant Inst. Math. Scis., 1958-66, dir., 1966-67. Fellow Am. Acad. Arts and Scis.; mem. Nat. Acad. Sci., Mitglied Akad. der Wissenschaften (corr.), Mitglied der Braunschweigischen Wissenschaftlichen Gesellschaft (mem.). Home: 24 Lester Pl., New Rochelle, N.Y. 10804.*

FRIEMAN, Edward Allan, Am. physicist; b. N.Y.C., Jan. 19, 1926; s. Joseph and Belle (Davidson) F.; B.S., Columbia, 1946; M.S., Poly. Inst. Bklyn., 1948, Ph.D., 1951; m. Ruth Rodman, June 19, 1949; children—Jonathan Paul, Michael Rodman, Joshua Adam. Faculty, Poly. Inst. Bklyn., 1945-52; faculty Princeton, 1952—, dir. plasma physics program, 1959—, prof. astrophys. scis., 1961—, asso. dir. plasma physics lab., 1964—. Cons., Jason div. Inst. for Def. Analyses, 1960—, Aero. Research Assos. Princeton, 1960—, also dir. NSF fellow, 1964. Mem. Am. Phys. Soc., Am. Astron. Soc., A.A.A.S., Sigma Xi, Tau Beta Pi. Research in theoretical aspects of plasma physics, magnetohydrodynamics, and controlled thermonuclear fusion; maj. contbns. in these fields to stability theory and description of kinetic behavior; fundamental research in founds. nonequilibrium statis. mechanics including new methods of deriving equations of evolution. Home: 70 Heather Lane, Princeton, N.J. 08540.*

FRIEND, Albert Wiley, Am. elec. engr., physicist; b. Morgantown, W.Va., Jan. 24, 1910; s. Lemuel Elsworth and Louesa Gertrude (Michael) F.; B.S. in Elec. Engring., W.Va. U., 1932, M.S. in Physics, 1936; Sc.D. in Communications Engring., Harvard, 1948; m. Evelyn Augusta Hall, Aug. 6, 1931; children—Albert Wiley, Evelyn Joyce (Mrs. William Carter Everett), John Robert. Faculty physics W.Va. U., 1934-44; instr. Harvard, 1939-41, electronic research Cruft Lab., 1946-48; research asso., mem. staff radiation lab., Mass. Inst. Tech., 1941-42, tech. dir. heat research lab., 1942-44; research staff RCA, 1944-51; dir. engring. Daystrom, Inc., 1951, Magnetic Metals Co., 1951-53; cons. engr., physicist, 1953—; pres., chief engr. Amicon Corp., Phila., 1961—. Recipient RCA Labs. award, 1950; Nat. Electronics Conf. award, 1955. Fellow I.E.E.E. (fellowship award 1954), A.A.A.S., Am. Phys. Soc.; mem. Acoustical Soc. Am., Electrochem. Soc., Am. Meteorol. Soc., Am. Inst. Physics, Am. Geophys. Union, Nat. Soc. Profl. Engrs., Sigma Xi, Tau Beta Pi. Contbr. numerous articles to profl. jours. Patentee in field. Discovered radio wave pulse echoes from meteorol. boundaries, 1935; invented, developed heat seeking bomb (Felix) guidance principles, 1943-44, magnetic rec. components, disc rec. head, high efficiency TV scanning and high voltage systems, color TV scanning and electron beam control systems; research, consultation on electronic countermeasures equipment for aircraft, control systems, telemetering and data processing in missile systems, satellite tracking radars, large microwave antenna systems, command control communications system, Lunar Excursion Module, Lunar Exploration Systems, magnetic materials and components, electromagnetic-acoustic isolation facilities, archtl. acoustics, noise and vibration reduction. Address: 5903 City Av., Phila. 19131.*

FRIES, Elias Magnus, Swedish botanist; b. Femsjö, Sweden, Aug. 15, 1794; ed. U. Lund (Sweden); became docent U. Lund, 1814, adj. prof., 1819, prof. botany, 1824; named prof. practical economy U. Uppsala (Sweden), 1834, also botany, 1852, ret., 1859, continued as dir. garden and Bot. Mus., until 1863. Fellow Royal Soc., 1875; mem. Swedish Acad., Acad. Stockholm. Genus Freesia named in his honor. Author numerous publs. including Novitiae florae suecicae, 1814-23; Observationes mycologicae, 2 vols., 1815-18; Systema mycologicum, 3 vols., 1821-23; Elenchus fungorum, 2 vols., 1828; Lichenographica europeae reformata, 1831; Botaniska utflygter, 3 vols., 1843-64; Summa vegetabilium Scandinaviae, 1846-49; Novae symbolae mycologicae, 1851; Monographia hymonomycetum Sueciae, 2 vols., 1857-63, Icones selectae Hymenomycetum, 2 vols., 1867-84. Research on botany; introduced system of classification based on morphology and biology. Died Feb. 8, 1878.

FRIES, Jacob Friedrich, German philosopher; b. Germany, 1773; became prof. philosophy and elementary math., Heidelberg, Germany, 1905; prof. physics, Jena, Germany, 1816-43. Author: Philosophy as an Evident Science, 19 1804; Neue Kritik der Vernunft, 3 vols., 1807; System of Logic, 1811. Asserted that a scientific psychology was needed to develop adequate theory of knowledge. Died 1843.

FRIES, Lorenz (or Friesz, Frisius, Frise, Phries, Phryes, Phrisius), physician, astrologer, geographer; b. perhaps in Colmar, Germany, circa 1490; ed. probably in Vienna, Austria; m. Barbara Thun; practiced medicine, Colmar, 1518; moved to Strasbourg (now France), 1519; municipal physician, Freiburg, Switzerland; returned to Strasbourg and became citizen by marriage, 1520; settled in Metz, France, 1526. Author: Spiegel der Artzny (perhaps oldest book on internal medicine, noted for well-executed engravings), 1518; Tractat der Wildbeder Natuer, Wirkung und Eigenschafft, 1519; Synonyma (dictionary of medicines in several langs.), 1519; Ein kurtzer Bericht wie man die Gedechtnisz stercken . . . mag, 1523, Artis memorativae . . . tradition (both on hygiene and mnemonics); Sudoris anglici . . . praeservatio et cura, 1529; Defensio medicorum principis Avicennae (devense of Arabic medicine), 1529. Died perhaps in Grüningen, circa 1530-32.

FRIES, Nils Thorsten Elias, Swedish plant physiologist; b. Uppsala, Sweden, July 17, 1912; s. Thoralf and Agnes (Christensson) F.; Ph.D., U. Uppsala, 1939; m. Lisbeth Jonasson, July 19, 1941; children—Karin Elisabeth (Mrs. Torgny Hastad), Agnes Ulla Veronica, Erik Olof Elias. Faculty, U. Uppsala, 1939—, prof. plant physiology, 1956—, head dept. plant physiology, 1956—. Mem. Swedish Nat. Sci. Research Council, 1959-64. Author: Introduktion till växtfysiologin, 1966; also numerous articles. Research on physiology of fungi, especially vitamin requirements, multipolar sexuality and spore germination; physiol. genetics, especially induction and enrichment of mutations in fungi; invented filtration technique; discovered (with B. A. Kihlman) mutagenic effect of caffeine; studied importance of vitamins and other metabolites in seedling devel., induction of heat sensitivity in plant cells. Home: 15 Bergagatan, 75238 Uppsala, Sweden.*

FRIESE-GREENE, William, English inventor; b. Bristol, Eng., Sept. 7, 1855; S. James Greene; ed. Queen Elizabeth's Hosp., Clifton, Eng.; m. Victoria Marina Friese, 1874; m. 2d, Edith Jane Harrison, 1897; 5 sons. Travelling photographer; set up photog. bus., Picadilly, London, with branches in Bath, Bristol, Plymouth, Eng.; bankrupt, 1891; re-established bus., 1892. Recipient Daguerre medal, Vienna, Austria, 1888. Experimented (with John A. Roebuck Rudge) in photographing motion, 1882-84; showed his 1st motion picture, 1889; 1st to substitute sensitized celluloid ribbon-film for glass plate base for photog. emulsion, 1889; holder 1st patent for intermittent-motion device for shielding film movement between exposures by means of rotating shutter, 1889; patentee motion picture camera and projector, 1890; worked with stereoscopic and color films. Died London, May 5, 1921.

FRIESER, Hellmut, chemist; b. Franzenthal, Czechoslovakia, Aug. 22, 1901; s. Alfred and Therese (Mattausch) F.; dr.engring. Technische Hochschule Dresden (Germany), 1928; m. Ilse Anders, Oct. 25, 1930; children—Sabine (Mrs. Rhein), Veronika (Mrs. Masing), Edith (Mrs. De Bruyn). Sci. asst. Siemens & Halske, Berlin, Germany, 1930-34, Zeiss-Ikon, Berlin, 1934-36; prof. Technische Hochschule Dresden, 1936-46; sci. specialist in USSR, 1946-52; dir. Lab. Agfa, Leverkrusen, 1953-58; dir. sci. photog. Inst. Technische Hochschule, Munich, Germany, 1958—. Recipient gold medal Wiener Photographische Gesellschaft, 1955. Mem. Photog. Scientists and Engrs. (hon.), Deutsche Gesellschaft für Photographie in Cologne (chmn. research sect. 1955—), Deutsche Gesellschaft für Photographie, Deutsche Bunsengesellschaft, Münchner Chemische Gesellschaft, Deutsche Physikalische Gesellschaft, Deutsche Gesellschaft für angewandte Optik, Deutsche Kinotechnische Gesellschaft für Film und Fernsehen. Research, numerous publs. on information storage by means of photo. films and theory of photog. process. Home: 8 Hauberrisserstrasse, Munich, Germany.*

FRINGS, Hubert William, Am. sensory physiologist; b. Phila., Jan. 1, 1914; s. Martin and Myrtle (Gauweiler) F.; B.S., Pa. State U., 1936; M.S., U. Okla., 1937; Ph.D., U. Minn., 1940; m. Mable Ruth Smith, June 9, 1936; 1 son, Carl Frederick. Asst. prof. Luther Coll., Decorah, Ia., 1939-40; head dept. sci. Monett (Mo.) Jr. Coll., 1940-41; head dept. natural scis. Snead Jr. Coll., Boaz, Ala., 1941-43; head dept. physics W.Va. Wesleyan Coll., Buckhanon, 1943-46; head dept. biology Gustavus Adolphus Coll., St. Peter, Minn., 1946-47; prof. zoology Pa. State U., University Park, 1947-61; prof. zoology U. Hawaii, Honolulu, 1961-66, coordinator undergrad. honor programs, 1963-66. Mem. A.A.A.S., Am. Soc. Zoologists, Entomol. Soc. Am., Soc. Protozoologists, Am. Ornithologists Union, Acoustical Soc. Am., Internat. Com. on Biol. Acoustics (chmn. 1956—). Author: (with Mable Frings) Sound Production and Sound Reception by Insects—A Bibliography, 1960; Animal Communication, 1964; also numerous articles. Research on locations and physiology taste organs insects including new theory taste, hearing and sound prodn. by insects, mechano-reception and chemoreception invertebrates; discovered and devel. use recorded communication signals of birds for control bird depredations.

FRINGS, Mable Ruth Smith, Am. sensory physiologist; b. Shawville, Pa., Apr. 11, 1912; d. Harvey and Clara (Straw) S.; B.S., Pa. State U., 1936; postgrad. U. Okla.; m. Hubert Frings, June 9, 1936; 1 son, Carl Frederick. Instr., Snead Jr. Coll., Boaz, Ala., 1942-43, Gustavus Adolphus Coll., St. Peter, Minn., 1946-47; research asso. Pa. State U., University Park, 1949-61; research asso. zoology U. Hawaii, Honolulu, 1961-66. Mem. A.A.A.S., Am. Soc. Zoologists, Entomol. Soc. Am., Internat. Com. on Biol. Acoustics (sec. 1956—). Author: (with Hubert Frings) Sound Production and Reception by Insects—a Bibliography, 1960, Animal Communication, 1964; also numerous articles. Research on locations taste organs insects, sound prodn. and hearing by insects, chemoreception and mechanoreception invertebrates; discovered use communication signals for pest control particularly of birds. Address: Univ. of Oklahoma, Norman, Okla. 73069.*

FRINK, Orrin, Am. mathematician; b. Bklyn., May 31, 1901; s. Orrin and Elizabeth (Romeyn) F.; A.B., Columbia, 1922, M.A., 1923, Ph.D., 1926; postgrad. U. Chgo., Princeton; m. Aline Huke, June 3, 1931; children—Orrin 3d, Peter Hill, John Allen, Elizabeth. Instr., Princeton, 1925-26; NRC fellow U. Chgo., also Princeton, 1926-28; faculty Pa. State U., University Park, 1928—, prof. math., 1933—, head math. dept., 1949-60; asst. chief engr. Spl. Projects Lab., Wright Field AFB, Dayton, O., 1944-45. Fulbright lecturer at University College, Dublin, Ireland, 1960-61, 1965-66. Recipient National Science Found. award, 1964-65. Mem. Am. Math. Soc., Math. Assn. Am., Assn. for Symbolic Logic, A.A.A.S., Am. Assn. U. Profs., Pi Mu Epsilon (past nat. vice dir.), Sigma Xi, Phi Beta Kappa, and Sigma Pi Sigma. Contbr. articles to tech. jours. Proved theorems about Boolean algebras, lattice theory, orthogonal polynomials, gen. topology, calculus of variations, projective geometry, functional analysis; co-originator study Bessel polynomials. Home: 706 Sunset Rd., State College, Pa. 16801. Office: McAllister Bldg., Pa. State U., University Park, Pa. 16802.*

FRISBY, Edgar, astronomer; b. Great Easton, Eng., May 22, 1837; s. Thomas and Noah F.; B.A., U. Toronto, 1863, M.A., 1864; m. Laura Virginia Ebert, Aug. 6, 1872. Tchr. in Can., 1863-67; afterward acting prof. math. Northwestern U.; later asst. astronomer U. S. Naval Obs., Washington, and prof. math. USN, 1878; ret., 1899. Observed for Govt. several eclipses and made noteworthy computation of orbit of comet of 1882. Died Jan. 7, 1927.

FRISCH, Anton von, see von Frisch, Anton.

FRISCH, Karl von, see von Frisch, Karl.

FRISCH, Johann Leonhard, naturalist, entomologist; b. Sulzbach, nr. Nuremberg, Germany, Mar. 19, 1666; s. Johann Christoph and Sabina (Fecher) F.; student univs. Altdorf (Switzerland) 1683, Jena (Germany), 1686, Strasbourg (now France), 1688, Nuremberg; m. Sophie Elisabeth Dornmann, 1699; 8 children, including Philipp Jacob, Ferdinand Helffrich, Jodocus Leopold. Became asst. preacher, Hungary, 1691; returned to Germany to farm, 1693; traveled in Holland; became tchr. Gymnasium zum Grauen Kloster, Berlin, 1699, asst. headmaster, 1708, headmaster, 1727-43. Mem. Leopoldina, Berlin Soc. Scis. Author: Französisch-Teutsches und Teutsch-Französisches Wörterbuch, 1712; Beschreibung von allerlei Iusekten in Deutschland, 1720-38; Historia linguae sclavonicae, 1727-36; Teutschlateinisches Wörterbuch, 2 vols., 1741; Vorstellung der Vögel Deutschlands, 1763. Pioneer in Slavic studies in Germany, also in comparative linguistics; described insects and birds of Germany; founder silk industry, Brandenburg, Germany; 1st to cultivate mulberry tree in Prussia; friend of Liebniz. Died Berlin, Mar. 21, 1743.

FRISCH, Otto Robert, physicist; b. Vienna, Austria, Oct. 1, 1904; s. Justinian and Gusti (Meitner) F.; Dr.phil. Vienna U., 1926; m. Ursula Blau, Mar. 14, 1951; children—Monica Eleanor, David Antony. Research in Berlin, 1927-30, Hamburg, Germany, 1930-33, London, Eng., 1933-34, Copenhagen, Denmark, 1934-39, Birmingham, Eng., 1939-40, Liverpool, Eng., 1940-43, Los Alamos, N.M., 1943-45, Harwell, Eng., 1945-47; prof. Cavendish Lab., U. Cambridge, Eng., 1947—. Recipient Order British Empire, Medal of Freedom, 1946. Fellow Royal Soc., 1948, Phys. Soc. (London). Author: Meet the Atom, 1947; Atomic Physics Today, 1961; Working with Atoms, 1965. Editor: Progress in Nuclear Physics, 1950-64. Research on molecular beams, nuclear physics. Home: Trinity Coll., Cambridge, Eng.*

FRISCHEN, Carl Ludwig, German elec. engr.; b. Bremen, Germany, July 20, 1830; s. Johann and Anna Franziska Sophia (Leonhard) F.; student Hanover (Germany) Polytechnikum, 1849-51; m. Anna Busse, 1855; 2 daus. Entered telegraph service Hanover state ry., 1851, became telegraph engr., 1854; named head Hanover Telegraph Adminstrn., 1865; apptd. chief engr. Telegraph Adminstrn. N. German Bund, 1867, Telegraphen-Bauanstalt Siemens & Halske, 1870. Contbr. to tech. jours. Inventor (independently of W. Siemens) switch for telegraphing in both directions simultaneously on same wire; supported Siemens in 1st Bell telephone expts. in Germany; inventor ry. safety devices, including signalling sect., 1870-71, rail deflection contact, 1879; instrumental in laying cables for telegraph system in Germany. Died Berlin, May 8, 1890.

FRISE, Lorenz, see Fries, Lorenz.

FRISELL, Wilhelm Richard, biochemist; b. Two Harbors, Minn., Apr. 27, 1920; s. Olof Wilhelm and Thyra (Falk) F.; B.A., St. Olaf Coll., 1942; M.A., Johns Hopkins, 1943, Ph.D., 1946; m. Margaret Jane Fleagle, Mar. 6, 1948; children—William Richard, Robert Benjamin. Faculty, Sch. Medicine U. Colo., Denver, 1951—, prof., 1964—, asso. dean Grad. Sch., 1958—. Mem. fgn. fellowships com. Office International Research of the National Institutes of Health (USPHS), 1961-66, 67—. John G. Berquist fellow Am.-Scandinavian Found., Uppsala, Sweden, 1949-50. Fellow A.A.A.S.; mem. Am. Chem. Soc., Am. Soc. Biol. Chemists, Phi Beta Kappa, Sigma

Xi. Research, publs. on metabolism and enzymology of single-carbon compounds and sulfur amino acids, relationships between function and ultrastructure of mitochondria, mechanisms of photodynamic reactions of coenzymes. Home: 3401 E. 7th Av., Denver 80206.*

FRISH, Sergey Eduardovich, Russian physicist; b. June 19, 1899; grad. Petrograd U., 1921. With State Optical Inst., 1919-39; instr. Leningrad U., 1924-33, prof., 1933—. Del., Internat. Conf. on Spectroscopy, Lucerne, Switzerland, 1959. Mem. USSR Acad. Sci. (corr.). Author: The Analysis of Complex Spectra, 1932; Atomic Spectra, 1933; The Technique of Spectroscopy, 1936; A Course in General Physics, 1957. Research and publs. on atomic spectra, superfine structure of spectral lines, determination of nuclear moments, spectroscopy of gas discharge and spectral analysis of gases; elementary process of exciting atoms by electron impact and impacts of 2d order, 1953—. Address: Leningrad University, Universitetskaya n. 7-9, Leningrad, USSR.

FRISI, Paolo, Italian astronomer, mathematician; b. Milan, Italy, Apr. 13, 1728; ed. Monastary of Barnabites, circa 1746. Mem., Barnabite Order; prof. philosophy at Casale, Novara, and Collegio Alessandro, Milan, 1853-56; prof. philosophy, U. Pisa, 1756-64; prof. mathematics, Scuôla palatina, Milan, 1764-66. Fellow, Royal Soc., 1757; mem., French Acad. Sci., 1753; Acads. of St. Petersburg, Paris, Berlin. Works include: Disquisituo mathematicia, 1751; Del Modo di regolari fiumi e torrenti, 1762; De Gravetate universali, 1768; Cosmographia physica et mathematica, 1774-5; Opuscoli filosofici, 1781; others. Investigated physical causes of shape and size of earth; 1st introduced lightning rod into Italy; authority on hydraulics; research in math and astronomy. Died Milan, Italy, Nov. 22, 1784.

FRISIUS, Gemma, see Gemma Frisius, Regnier.

FRISIUS, Lorenz, see Fries, Lorenz.

FRISON, Theodore Henry, Am. entomologist; b. Champaign, Ill., Jan. 17, 1895; s. Joseph and Helen (O'Neal) F.; B.A., U. Ill., 1918, M.A., 1920, Ph.D., 1923; m. Ruby Gertrud Dukes, Aug. 22, 1919; children—Theodore Henry, Jr., Patricia Ann. Began as asst. Wis. state entomologist, 1920; asst. entomologist U. S. Bur. Entomology, 1921-22; asst. in dept. of entomology U. Ill., 1922-23; systematic entomologist Ill. State Natural History Survey, 1923-30, acting chief, 1930-31, chief, 1931—. Dir. Central States Forest Exptl. Sta., Central States Forestry Congress. Fellow A.A.A.S., Entomol. Soc. Am. (v.p. 1936); mem. Am. Assn. Econ. Entomologists (1st v.p. 1945). Ecol. Soc. Am. (rep. on NRC 1937), Ill. Audubon Soc. (dir. 1942), Limnological Soc. Am., Soc. Wildlife Specialists, Am. Wildlife Inst., Am. Soc. Naturalists, Wilderness Soc., Ill. Acad. Scis. (pres.), other sci. socs., Sigma Xi. Author: Fall and Winter Stoneflies, or Plecoptera, of Illinois, 1929; The Plant Lice, or Aphiidae, of Illinois (with Frederick Hottes), 1931; The Stoneflies, or Plecoptera, of Illinois, 1935; and over eighty other scientific articles. Compiler: List of Insect Types in the Collections of Illinois State Natural History Survey and University of Illinois, 1927. Editor Jour. Econ. Entomology, 1936-40; mem. editorial bd. Ecology, 1938-40. Died Dec. 9, 1945.

FRISTROM, Robert Maurice, Am. phys. chemist; b. Portland, Ore., May 26, 1922; s. Ivar and Irene (Andersen) F.; student Albany Coll. (now Lewis and Clark Coll.), 1941-42; B.A., Reed Coll., 1943; M.A. with honors, U. Ore., 1945; Ph.D., Stanford, 1949; m. Geraldine Ramona Ashour, June 11, 1957; 1 son, Robert Raymond. Research fellow Harvard, 1948-51; chemist sr. staff Applied Physics Lab., Johns Hopkins, Silver Spring, Md., 1951-61, chemist prin. staff, 1961—, lectr. chem. engring., 1961-62. Vis. lectr. chem. engring. Cal. Inst. Tech., 1966; vice chmn. Gordon Conf. on Molecular Beams, 1965. Parsons fellow, 1961. Recipient Hillebrand award chem. Soc. Washington, 1967. Fellow Random Soc. Washington; mem. A.A.A.S., Am. Chem. Soc., Am. Phys. Soc., Am. Vacuum Soc., Combustion Inst. (Silver medal 1964), Sigma Xi. Author: (with A. A. Westenberg) Flame Structure, 1965. Editor: Fire Research Abstracts and Revs. (Nat. Acad. Sci.-NRC), 1966—. Contbr. articles to profl. lit. Developed exptl. techniques for studying structure of flame fronts and used them to quantitatively characterize the phys. and chem. processes; studies in electrochemistry, microwave spectroscopy, jet propulsion, and molecular beams. Home: 10610 Mantz Rd., Silver Spring, Md. 20903.*

FRITSCH, Arnold Rudolph, Am. chemist; b. Passaic, N.J., Mar. 28, 1932; s. Arno Rudolph and Martha (Werner) F.; B.S., U. Rochester, 1953; Ph.D., U. Cal. at Berkeley, 1957; m. Betsey Rorapaugh, June 13, 1953; children—Kristin, Kerry, Arnold Read, Paul Frederick. With Westinghouse Atomic Power Dept., Pitts., 1956-59; br. chief div. internat. affairs U. S. AEC, Washington, 1959-61, spl. asst. to chmn., 1961—. Mem. Am. Phys. Soc., Am. Chem. Soc., Am. Nuclear Soc., A.A.A.S., Phi Beta Kappa, Sigma Xi. Research, publs. on nuclear spectroscopic studies neutron deficient lead isotopes, heavy ion fission uranium, surface chemistry thoria. Home: 26723 Ridge Rd., Damascus, Md. 20750. Office: U. S. AEC, Washington 20545.*

FRITSCH, Bruno, economist; b. Prague, Czechoslovakia, July 24, 1926; s. Josef and Rosa (Novak) F.; student U. Prague, 1946; Dr.rer.pol., U. Basle (Switzerland), 1952, postgrad. philosophy, 1952-57; m. Jadwiga Przybyl, Oct. 23, 1953; children—Martin, Caroline. Dir., Basel Center Econ. and Financial Research, 1958; prof. econs. Inst. Tech., Karlsruhe, Germany, 1959-63; prof. econs. U. Heidelberg, 1963-65, dir. econ. dept. S. Asia Inst., 1963-65; prof. econs. Swiss Fed. Inst. Tech., Center for Econ. Research, Zürich, 1965—; vis. prof. Coll. Europe, Bruges, Belgium, 1962—. Mem. Schweiz. Gesellschaft für Statistik und Volkswirtschaft, Am. Econ. Assn., Gesellschaft für Wirtschafts und Sozialpolitik, Verein für Socialpolitik, List Gesellschaft, Internat. Inst. Differing Civilizations, Philosophische Gesellschaft der Schweiz, Econometric Soc. Author: Geschichte und Theorie der amerikanischen Stabilisierungspolitik, 1933; Die Geld- und Kredittheorie von Karl Marx, 1954; also articles. Studies in econ. devel., especially Asia; empirical research in plan-implementation procedures in India; nat. planning and internat. devel. Home: Langackerstr. 122b, 8704 Herrliberg, Switzerland. Office: Universitätsstr. 14, 8006, Zürich, Switzerland.*

FRITSCH, Felix Eugen, botanist; b. Hampstead, Apr. 26, 1879; s. H. Fritsch; D.Sc., LL.D., U. London; Ph.D., U. Munich; m. Hedwig Lasker, 1905; 1 son. Asst. bot. dept. U. Munich, 1899-1901; asst. bot. dept. Univ. Coll., London, 1902-06, asst. prof., 1906-11; lectr. Birkbeck Coll., 1905-06, East London Coll., 1907-11; head dept. botany Queen Mary Coll., U. London, 1911-48, emeritus prof., 1948. Chmn., Council Freshwater Biol. Assn. of Brit. Empire. Fellow Royal Soc., 1932; mem. Brit. Assn. (pres. bot. sect. 1927), Linnean Soc. (pres. 1949-52), Inst. Biology (pres. 1953); fgn. mem. K. Fysiogr. Sälisk. (Lund), Norwegian Acad. Sci. and Letters, Regional Soc. Scientists (Uppsala); hon. mem. Indian, Vienna bot. socs., Royal Bot. Soc. Belgium, K. Natuur. Genootsch. Dodonaea (Gent); corr. mem. Bot. Soc. Am., Phila. Acad. Sci. Author: Plant Form and Function; An Introduction to the Study of Plants; British Fresh-water Algae; The Structure and Reproduction of the Algae; also numerous articles on algae, Antarctic algae. Died May 23, 1954.

FRITSCH, Gustav Theodor, German anatomist, anthropologist; b. Cottbus, Germany, Mar. 5, 1838; s. Luswig and Sophie (Kramsta) F.; studied natural scis. and medicine, possibly in Berlin; m. Helene Hirt, 1871; 1 son, 1 dau. Became asst., anat. inst., Berlin, 1867, prof., 1874, later head histological sect., physiol. inst.; named hon. prof., 1900. Mem. Berlin Soc. Anthropology. Author: (article with E. Hitzig) Ueber die elektrische Erregbarkeit des Grusshirns, 1870; Die Eingeborenen Sudafrikas ethnographisch und anatomisch beschrieben, 1872; Agyptische Volkstypen der Jetztzeit, 1904; Über Bau und Bedeutung d. Area Centralis des Menschen, 1908; Das Haupthaar und seine Bildungsstatten bei Menschen, 1912. Studied questions of anthropology, ethnography, astronomy and zoology on expdns. to S. Africa, 1863-66, Aden, 1868, Egypt and Levant, 1881-82, around world, 1904-04; investigated races and race mixtures; studied (with Hitizig) localization of motor functions in cerebral cortex through elec. stimulation of specific areas, 1870. Died Berlin, June 12, 1927.

FRITSCH, Heinrich, German gynecologist; b. Halle/Saale, Germany, Dec. 5, 1844; s. Gustav and Wilhelmine (Hartmann) F.; student, Tübingen, Würzburg (both Germany); M.D., Halle, 1869; m. Elisabeth Goedecke, 1874; 3 sons, Karl, Bernhard, Hans, 3 daus., including Juli Brauer, Anna Stoeckel. Asst. under von Olshausen at obstet. clinic, Halle; became mem. faculty Halle, 1873, asso. prof., 1877; also practiced medicine; named prof. obstetrics, Breslau, Germany, 1882; succeeded G. von Veit, Bonn, Germany, 1893; ret., 1910. Author: Klinik der alltäglichen geburtshilflichen Operationen, 1875; Die Krankheiten der Frauen, 1881; Grundzüge der Pathologie und Therapie des Wochenbetts, 1884; Gerichtsärztliche Geburtshilfe, 1901; Geburtshilfe, eine Einführung in die Praxis, 1904. Founder (with H. Fehling) jour. Zentrallblatt für Gynäkologie. A founder of modern gynecology and gynecol. abdominal surgery; performed operations on fistulas in female urethra; improved instruments and antisepsis. Died Hamburg, Germany, May 12, 1915.

FRITSCH, Karl Georg Wilhelm von, see von Fritsch, Karl Georg Wilhelm.

FRITSCH, Volker John William, geophysicist; b. Bruenn, Moravia, Austro-Hungary, Apr. 15, 1905; s. William and Margaret (Hutter) F.; student Tech. U., Bruenn, 1925-32; Diplomingenieur, Tech. U. Prague (Czechoslovakia), 1932, D.Sc., 1938; m. Margaret Kauder, Feb. 19, 1938; Engr. for electrotechnics, 1932; asst. Tech. U., Prague, 1937-39; cons. engr., 1939-45; honorar docent Tech. U. Vienna, 1946; authorized cons. engr., 1946-66; chief Research Sta. for Geoelectricity and Lightning Protection Vienna, 1957-66; prof. for geoelectrics Tech. U. Ilmenau, Germany, 1966—. Chmn., Austrian Com. and Normcom. for Lightning Protection, 1950—. Recipient Masaryk-Denis prize, 1937; D. Korner prize Vienna, 1957, Gold medal Honour Austrian, 1962. Mem. Am. Geophys. Union, European Assn. Exptl. Geophysics. Author numerous books including: Angewanate Geoelektric, 1949; Blitzschutztechnik, 1951; Geopathogene Erscheinungen Geophysik, 1955; Räumliche Leiter, 1960; Geoelektrische Baugrunduntersuchegen, 1960; also numerous articles. Research on subsoils and bldg. ground lightning research, lightning protection. Home: 25 Seidengasse, Vienna. Office: Arsenal Objekt 3, Vienna, Austria.*

FRITZ, Irving Bandas, Am. physiologist; b. Rocky Mount, N.C., Feb. 11, 1927; s. Henry Norman and Rose (Bandas) F.; student (Settle scholar) U. Richmond, 1943-48; D.D.S., Med. Coll. Va., 1948; Ph.D. (Zoller fellow), U. Chgo., 1951; postgrad. U. Copenhagen Inst. Physiology (Denmark); m. Helen Romaine Bridgman, Aug. 20, 1950; children—David, Jonathan, Winston, Rachel. Instr., Harvard Sch. Dental Medicine, 1951; USPHS fellow U. Copenhagen, 1953-55; asst. dir. dept. metabolic and endocrine research Michael Reese Hosp., Chgo., 1955-56; faculty dept. physiology U. Mich., Ann Arbor, 1956-68, prof., 1964-68; chmn. Banting and Bert Dept. Med. Research, University Toronto (Ont., Can.), 1968—. Member of the Am. Physiol. Soc., Endocrine Soc., A.A.A.S., Am. Assn. U. Profs., Sigma Xi. Contbg. author: Advances in Lipid Research, I, 1963; also articles, abstracts. Editorial bd. Jour. Lipid Research, 1961-64. Research on regulation and control metabolic systems, control rates fatty acid oxidation and synthesis in mammalian tissues; discovered a functional role for carnitine in fatty acid metabolism. Address: Banting and Bert Dept. Med. Research, U. Toronto, Toronto, 5, Ont., Can.*

FRITZ, James Sherwood, Am. chemist; b. Decatur, Ill., July 20, 1924; s. William Lawrence and LeOra Mae (Truster) F.; B.S., James Millikin U., 1946; M.S., U. Ill., 1946, Ph.D., 1948; m. Helen Joan Houck, Apr. 20, 1949; children—Barbara Lisa, Julie Ann, Laurel Joan, Margaret Ellen. Asst. prof. Wayne State U., 1948-51; faculty Ia. State U., Ames, 1951—, prof. chemistry, 1960—; sr. chemist Ames. Lab., U. S. AEC, 1951—. Mem. Am. Chem. Soc. Author: (with George S. Hammond) Quantitative Organic Analysis, 1957; (with George Schenk) Quantitative Analytical Chemistry, 1966; also numerous articles. Regional editor Internat. Chem. Jour., Talanta, 1967—. Research in analytical chemistry, including acid-base titrations in nonaqueous solvents, methods for determination of organic functional groups, complex-formation titrations, separation of metal ions by ion-exchange, solvent extraction and chromatography, theory of analytical processes. Home: 2018 Greenbriar St., Ames, Ia. 50010.*

FRITZ, Sigmund, Am. meteorologist; b. Bklyn., June 9, 1914; s. Morris and Celia (Berger) F.; B.A., Bklyn., 1934; M.S., Mass. Inst. Tech., 1941, Sc.D., 1953; m. Ann Robinson, Fritz, June 30, 1946; children—Maurene N., Lawrence C. Observer, research meteorologist U. S. Weather Bur., 1938-65; dir. meteorol. satellite lab. Nat. Environmental Satellite Center, Suitland, Md., 1965—. Recipient Gold medal U. S. Dept. Commerce, 1965; Distinguished Achievement award Thomas Jefferson High Sch., 1964. Mem. Am. Meteorol. Soc., Am. Geophys. Union, Internat. Union Geodesy and Geophysics, Internat. Assn. Meteorol. and Atmospheric Physics, (v.p. 1963—). Research, numerous publs. on solar radiation, scattering and absorption by atmosphere, clouds and earth surface, atmospheric ozone, interpretation pictures and radiation data from meteorol. satellites. Office: Fed. Office Bldg. 4, Suitland, Md. 20233.*

FRITZE, (Johann) Friedrich, German physician; b. Magdeburg, Germany, Oct. 3, 1735; M.D., Halle, Germany, 1756; m. Friederike Philippine Witte, 1765; 2 sons, including Friedrich Wilhelm, 2 daus. Named prof. therapy Collegium Medico-Chirurgicum, Berlin, 1764; founder clin. inst., 1789; became 2d directing physician Charité hosp., 1798. Author: Handbuch über venerische Krankheiten, 1790. Lectured on clin. practice, fever, mil. diseases, venerology; believed gonorrhea and chancre were different diseases. Died Berlin, Apr. 9, 1807.

FRITZSCHE, Carl Julius, German chemist; b. Neustadt, nr. Stolpen, Saxony, Oct. 29, 1808; Dr. Phil., Berlin, Germany, 1833; practiced pharmacy before gaining degree; asst. to Mitscherlich, Berlin; mgr. Struve's mineral-water works, St. Petersburg c. 1838, Russia; mem. Acad. Scis., St. Petersburg. Studied potassium bormate, purpuric acid, pollen, crystalline ammonium sulphides, osminidium, vanadian compounds; discovered harmine, also compounds of hydrocarbons, picnic acid, gray tin; rediscovered aniline (which he called anilin). Died Dresden, Germany, June 20, 1871.

FRITZSCHE, Carl Hellmut, German geologist; b. Gelsenkirchen, July 18, 1895; s. Emil and Berta (Lindes) F.; Ph.D. in Engring., Ph.D. in Geology, Tech. U., Berlin; m. Anni Jordan; children—Ursula, Heinz, Hellmut. Cons. in geology and exploration of mines Ministry of Industry of Chili; head enterprises, cons. engr., Chili, Bolivia; chmn. exploration Steinkohlenbergbauverein mines, Essen; prof. exploration of mines. Author: (with Potts) Horizon Mining, 1954; Lehrbuch der Bergbaukunde, 2 vols., 10 edits., 1961-62. Home: Grüne Matte 8, Essen-Bredeney. Died Mar. 15, 1968.

FROBENIUS, Ferdinand Georg, German mathematician; b. Berlin, Germany, 1849; doctorate, Berlin, 1870; named asst. prof. U. Berlin, 1874, prof., 1892;

co-ordinator, clarifier various aspects of algebra; developed method for solving linear homogeneous differential equations; working on theory of finite groups, he created theorems bearing his name, also theory of group characters. Died 1917.

FROBENIUS, Leo Viktor, German ethnologist; b. Berlin, Germany, June 29, 1873; s. Hermann and Mathilde (Bodinus) F.; m. Editha Brandt, 1901; 1 dau. Began ethographic collection (basis of Africa archives), Berlin, 1893; 12 expdns. to Africa, 1904-35; Africa archives renamed Research Inst. for Cultural Morphology, moved to Munich, 1922, moved to Frankfort/Main, Germany, 1925; hon. prof., lectr. on ethnology U. Frankfort, 1932; became dir. municipal ethnol. mus., Frankfort, 1934. Author: Und Afrika sprach, 3 vols., from 1912; Atlantis, Volksmärchen und Volksdichtungen Afrikas, 12 vols., 1921-28; Atlas Africanus, 1922-30; Erlebte Erdteile, 7 vols., 1925-30; (with H. Obermaier) Hadschra Maktuba, Urzeitliche Felsbilder Kleinafrikas, 1925; Erythräa, Länder und Zeiten des heiligen Königsmordes, 1931; Madsimu Dsangara, Südafrikanische Felsbildershronik, 2 vols., 1931; Kulturgeschichte Afrikas, 1933; Ekade Ektab, Die Felsbilder Fezzans, 1937. Helped establish sci. approach to ethnology; pioneer of hist. direction in ethnology; developed concept of culture circles (main tenet of Viennese sch.); worked in culture morphology; specialist on primitive tribes of Africa. Died Biganzolo on Lago Maggiore, Aug. 9, 1938.

FRÖBERG, Carl-Erik Torild, Swedish computer scientist; b. Slätthög, Sweden, June 23, 1918; s. Axel W. and Ruth (Andersson) F.; Ph.D., Lund (Sweden) U., 1949; m. Louise H. E. Wachtmeister, Aug. 30, 1952; children—Hans Erik, Bengt, Lars. Staff electronic computer project Inst. Advanced Study, 1947-48; with Nat. Bur. Standards, Washington, 1951; docent Lund U., 1949-56, asso. prof., 1956-65, prof. numerical analysis, 1965—. Vis. prof. U. Texas, Austin, 1965, U. Wash., Seattle, 1968. Mem. Royal Physiographic Soc. Lund. Author: Numerical Analysis, 1963-65; (with T. Ekman) ALGOL, 1964; (with others) COBOL, 1966. Editor: Scandinavian Computer Jour. BIT, 1961—. Research, publs. on software design for SMIL computer at Lund; numerical analysis; numerical computation of Coulomb Wave functions (useful in problems of nuclear physics), numerical aspects of number theory; computation of eigenvalues of arbitrary matrices. Home: 30 Fakirens Väg, Lund. Office: 14 Sölvegatan, Lund, Sweden.

FROBISHER, Sir Martin, navigator; b. Doncaster, Eng., circa 1536; became acquainted with Sir Humphrey Gilbert; attempted to discover N.W. Passage; made 3 voyages to New World (1576, 77, 78) under patronage of Queen Elizabeth and various nobles; discovered bay that is named for him on 1st voyage; mem. Sir Francis Drake's expdn. to W.I., 1585; served in sea battle against Spanish Armada, 1588; created knight for services against Spain, 1588. Died Plymouth, Eng., Nov. 7, 1594.

FROCHT, Max Mark, mech. engr., physicist; b. Warsaw, Poland, June 3, 1894; s. Meyer Loeb and Eva (Egerwald) F.; came to U. S., 1912, naturalized, 1922; B.S. in Mech. Engring., U. Mich., 1922, Ph.D. in Mechanics, 1931; M.S. in Physics, U. Pitts., 1926; hon. degree Ill. Inst. Tech., 1968; m. Dora Lipkin, Oct. 16, 1918. Faculty mechanics dept. Carnegie Inst. Tech., 1922-46; prof. mechanics Ill. Inst. Tech., Chgo., 1946-47, research prof., dir. exptl. stress analysis lab., 1947-64, emeritus prof., 1964—. Supr. research sponsored by govt. and other agys.; cons. to pvt. cos.; organized Internat. Symposium on Photoelasticity, editor Proc., 1961, 63. Fellow A.A.-A.S., Am. Soc. M.E.; mem. Soc. for Exptl. Stress Analysis (hon., William Murray lectr. 1959, M. M. Frocht award for outstanding achievement in edn. 1968), Soc. Engring. Edn., Am. Assn. U. Profs., Sigma Xi, Tau Beta Pi, Pi Tau Sigma. Author: Photoelasticity, vol. I, 1941, vol. II, 1948; Strength of Metals (translated into Russian, Spanish, Chinese), 1951; Selected Scientific Papers of M. M. Frocht on Photoelasticity, 1968; also articles. Developed new equipment and techniques; introduced new methods for stress concentrations, surface stresses and for separation of interior prin. stresses; demonstrated reliability of photoelasticity by comparisons of photoelastic, strain gage, and math. results; extended scope of photoelasticity to 3-dimensional problems, to photoplasticity and to dynamic problems; obtained solutions for complicated 3-dimensional problems; demonstrated potentialities of scattered light. Home: 4940 S. E. End St., Chgo. 60615.*

FROELICH, Ernest Julius, veterinarian; b. Vienna, Austria, Dec. 7, 1912; s. Otto and Elsa (Moller) F.; D.V.M., U. Zagreb, 1937; m. Edith Moller, Feb. 8, 1947; 1 dau., Deborah. Came to U. S., 1946, naturalized, 1949. Asst., Central Vet. Inst., Belgrade, 1938-39; head veterinarian Vet. Serum Lab., Zemun, Yugoslavia, 1939-40; with Biol. Control Lab., Zemun, 1940-41; fellow UNRRA, 1946-47; research asso. Sterling Winthrop Research Inst., Rensselaer, N.Y., 1948-50, head chemotherapy sect., 1951-62; dir. research and tech. services Winthrop Labs. Vet. Dept., Rensselaer, 1963—. Fellow N.Y. Acad. Scis.; mem. Am. Soc. Microbiology, Am. Vet. Med. Assn., U. S. Livestock San. Assn., N.Y. Acad. Scis. Developed screening techniques for study of chem. compounds against infectious diseases in exptl. animals, coordinated research leading to new antibiotics, developed stabilizer for viral vaccines, also new vaccines. Home: 174 Tampa Av., Albany 12208. Office: Sterling-Winthrop Research Inst., Rensselaer, N.Y. 12144.*

FROESCHELS, Emil, speech therapist; b. Vienna, Austria, Aug. 24, 1884; s. Siegmund and Johanna (Tintner) F.; M.D. in Speech Therapy, U. Vienna; m. Gertrude Töpfer. Instr., U. Vienna, 1914, asst. prof., 1926, head physician otol. dept. Speech-Voice Clinic, 1909-38; prof. research Washington U., St. Louis, 1939-40; head physician Speech-Voice Clinic, various hosps. in N.Y.C. Mem. Argentine Soc. Logopedics and Phonetics (hon.), Italian Soc. Phonetics, Am. Speech Correction Assn., Assn. for Advancement Psychotherapy, Gesellschaft der Arzte Wiens (corr.), Osterreichsche Oto-Laryngol. Gesellschaft (corr.), Assn. Am. Physicians, Rudolf Virchow Soc., Individual-Psychol. Soc. Am. Author 2 books, over 340 articles. Address: 133 E. 58th St., N.Y.C. 10022.

FRÖHLICH, Alfred, pharmacologist; b. Vienna, Austria, Aug. 15, 1871; s. Sigmund and Franziska (Hermann) F.; M.D., U. Vienna, 1895; m. Jenny Fröhlich, 1899. With H. Hothnagel's clinic, Vienna, 1895-98, neurol. ambulatory of L. Frankl-Hochwart's clinic, 1901; named to pharmacological inst. U. Vienna, 1906, mem. faculty, 1910, asso. prof. pharmacology and toxicology, 1912, prof., 1919-38; came to U. S., asso. May Inst. for Med. Research, Jewish Hosp., Cin. Author: Tumor of the Hypophysis without Acromegaly (Fröhlich's Syndrome), 1901. Research in vegetative innervation and hormones; 1st to describe pituitary tumor with obesity and sexual infantilism, 1901; recognized (with A. Holst) scurvy in guinea pigs fed on deficient diet, 1907. Died Cin., Mar. 22, 1953.

FRÖHLICH, Friedrich Wilhelm, physiologist; b. Vienna, Austria, May 28, 1879; s. Stefan and Bertha (Löw-Beer) F.; student of Max Verworn; m. Elisabeth Margarethe Ruhmann, 1911; 1 dau. Became mem. faculty U. Göttingen (Germany), 1907; accompanied Verworn to U. Bonn., 1910; served in war, 1914-20, then Russian prisoner of war; prof. physiology, Rostock, Germany, from 1927. Contbr. to med., physiol. jours. and handbooks. Co-editor Zeitschrift für allgemeine Physiologie. Demonstrated that frequency and strength of oscillation depends on strength of impulse also in optic nerves, also that measurable time elapses between effect of light impulse and its perception; research became important in increasing velocity of flight. Died Rostock, Nov. 8, 1932.

FROHLICH, Herbert, physicist; b. Rexingen, Germany, Dec. 9, 1905; s. Julius and Frida (Schwarz) F.; Ph.D., U. Munich (Germany), 1930; D.Sc. (hon.), U. Rennes (France), 1955; m. Fanchon Aungst, June 26, 1950. Asst. prof. U. Freiburg (Germany), 1932; asso. prof. U. Bristol (Eng.), 1935-48; prof. theoretical physics, head dept. U. Liverpool (Eng.), 1948—. Fellow Royal Soc. London; mem. Inst. Physics, Phys. Soc. Author: Elektronentheorie der Metalle, 1936; Theory of Dielectrics, 1948; also numerous articles. Devel. theory of solid state, introduction of field theory into solid state physics, prediction of isotope effect superconductors, early devel. meson theory. Office: Dept. Theoretical Physics, U. Liverpool, Eng.*

FRÖHNER, (Friedrich) Eugen, German veterinarian; b. Hirsau, Württemberg, Germany, Mar. 11, 1858; s. Ludwig Friedrich and Luise (del Marco) F.; ed. Stuttgart, Munich, Tübingen; M.D.; hon. dr. vet. medicine, Vienna, Giessen, Munich, Zurich; m. Bertha Paret, 1881; 2 children. Became prof., sch. vet. medicine, Stuttgart, Germany, 1882, vet. medicine univ., Berlin, 1886; named dir. med. clinic and br. forensic vet. medicine, 1894. Author: (with F. Friedberger, W. Zwick) Lehrbuch der speziellen Pathologie und Therapie der Haustiere, 2 vols., 1886; Lehrbuch der Arzneimittellehre für Tierärzte, 1889; Lehrbuch der Toxikologie fur Tierärzte, 1890; (with F. Friedberger) Lehrbuch der klinischen Untersuchungsmethoden für Tierärzte, 1892; Lehrbuch der allgemeinen Therapie für Tierärzte, 1893; (with J. Bayer) Handbuch der tierärztlichen Chirurgie und Geburtshilfe, 1896-1911; Kompendium der speziellen Chirurgie für Tierärzte, 1898; Chirurgische Diagnostik der Krankheiten des Pferdes, 1902; Lehrbuch der gerichtlichen Tierheilkunde, 1905; (with W. Zwink) Kompendium der speziellen Pathologie und Therapie für Tierärzte, 1912. Founder of modern vet. med. lit.; research in all fields of vet. medicine, pharmacology, forensic medicine. Died Berlin-Steglitz, Germany, June 21, 1940.

FROIDMOND, see Fromondus.

FROISSART, Marcel, French physicist; b. Paris, Dec. 20, 1934; Ingenieur, Ecole Poly., Paris, 1953-55; Ingenieur des Mines, École des Mines, Paris, 1957-58; m. Christine A. Knoll, Dec. 20, 1956; children—Olivier, Priscille, Jean-Marc, Fabrice, Marie-Thérèse. Staff, Centre D'Études Nucléaires de Saclay, Gif, France, 1959—; U. Cal. at Berkeley, 1960-61, Princeton, 1961-62, 65-66. Author: (with R. Omnes) Mandelstam Theory and Regge Poles, 1963. Research in high-energy particle scattering theory. Home: 34, Rue de Lozere, 91-Orsay, France. Office: C.E.N. Saclay, 91-Gif, France.*

FROLICH, Per Keyser, chemist, chem. engr.; b. Kristiansand S, Norway, June 29, 1889; s. Johan Keyser and Beate (Corneliussen) F.; came to U. S., 1922, naturalized, 1929; B.Sc., Norway Inst. Tech., 1921; M.Sc., Mass. Inst. Tech., 1923, D.Sc., 1925; D.Engring., Lehigh U., 1943; D.Sc., Rutgers U., 1943; m. Astrid Fronsdal, Mar. 7, 1927; children—Elizabeth Ann (Mrs. Robert B. Bachman), Astrid (Mrs. Danil R. Hancock). Instr., Kristiansand Bus. Coll., 1921-22; faculty Mass. Inst. Tech., Cambridge, 1922-29, asso. prof., 1929; with Esso Research & Engring. Co. (formerly Standard Oil Devel. Co.), 1929-46, dir. chem. research labs., 1936-46; with Merck & Co., Rahway, N.J., 1946-54, v.p., 1948-54, sci. dir. chem. div., 1953-54; dep. chief chem. officer for sci. activities U. S. Army Chem. Corps., Washington, 1954-60; tech. cons., Westfield, N.J., 1960—. Mem. adv. bd. Research Council, Rutgers U., 1944-63; mem. adv. com. dept. chem. engring. Princeton, 1948-56; mem. chem. div. NRC, 1958-60, exec. com., 1941-45. Mem. Am. Chem. Soc. (past chmn. N.J. sect., past pres. nat. soc., past dir.), Am. Inst. Chem. Engrs., Am. Inst. Chemists, Soc. Chem. Industry (Brit., Grasselli medal Am. sect. 1930), Am. Ordnance Assn., Norwegian Chem. Soc. (hon.), Norwegian Tech. Acad. Sci. Patentee in field. Research, publs. on colloids and applications to electroplating and electrorefining of metals, hydrocarbon cracking, synthesis of petrochems., rubbers, plastics. Address: 1160 Wychwood Rd., Westfield, N.J. 07090.

FROMAN, Darol Kenneth, Am. physicist; b. Harrington, Wash., Oct. 23, 1906; s. James Henderson and Eva (Wallace) F.; B.Sc., U. Alta., 1926, M.Sc., 1927, LL.D., 1964; Ph.D., U. Chgo., 1930; m. Ethel Norris, June 10, 1931; children—Kay Joyce (Mrs. Ralph P. Johnson, Jr.), Eva May (Mrs. Edwin K. Tucker). Lectr., U. Alta., Edmonton, Can., 1930-31; lectr., asst. prof. Macdonald Coll., McGill U., Quebec, Can., 1931-39; asst. prof. McGill U., Montreal, 1939-41; prof., dir. high altitude lab., U. Denver, 1941; group leader San Diego Radio and Sound Lab., 1942; research asso. metall. lab. U. Chgo., 1942-43; with Los Alamos Sci. Lab., 1943-62, asst. dir. weapons devel., 1951-53, asso. tech. dir. 1953-62; own cons. bus., Espanola, N.M., 1962—; dir. First Nat. Bank, Rio Arriba, N.M. Cons. prof. physics, U. N.M., Albuquerque, 1953-61, adv. summer sci. program, 1964—; adviser Sec. Def. Com. on Ballistic Missiles, 1955-61; sci. dir. Douglas Aircraft Co., Santa Monica, Cal., 1963—; mem. gen. adv. com. to U. S. AEC, 1964-66. Fellow Am. Phys. Soc., Am. Nuclear Soc.; mem. Sigma Xi. Research, publs. on ultrasonics, X-rays, cosmic rays, and nuclear physics; work on nuclear reactors and control; patentee in field. Home: Box 428, Rte. 1, Espanola, N.M. 87532.*

FRÖMAN, Per Olof, Swedish physicist; b. Högsby, Sweden, Apr. 27, 1926; s. Hugo and Judith (Kratz) F.; Ph.D., U. Uppsala (Sweden), 1957; m. Nanny Ingeborg Johansson, Mar. 26, 1959; 1 son, Nils Gunnar. Docent theoretical physics U. Uppsala, 1957-60, prof. theoretical physics, 1964—; prof. mechanics Royal Inst. Tech., Stockholm, Sweden, 1960-64. Author: (with N. Fröman) JWKB-Approximation, Contributions to the Theory, 1965. Research, publs. on theory of alpha decay, theory of neutron diffraction in crystals, other fields of theoretical physics. Home: 21 Skolgatan, Uppsala. Office: 18 Trädgardsgatan, Uppsala, Sweden.*

FROMENT, Paul-Gustave, French engr.; b. Paris, 1815; ed. École polytechnique, Paris; made precision instruments; built pendulum and gyroscope used by Foucault, telegraphic apparatus of Hughes; one of 1st to use electric motor for indsl. purposes, 1844. Died 1865.

FROMM, Erich, psychoanalyst; b. Frankfort/Main, Germany, Mar. 23, 1900; s. Naftali and Rosa (Krause) F.; Ph.D., U. Heidelberg, 1922; postgrad. U. Munich, 1925; m. Annis Grover, Dec. 18, 1953. Came to U. S., 1934, naturalized, 1940. Mem. staff Inst. Social Research, U. Frankfort, 1928-33; guest lectr. Columbia, 1934-38; Terry lectr. Yale, 1950; lectr. New Sch. Social Research, 1946-56; faculty Bennington Coll., 1940-50, William Alanson White Inst. Psychiatry, Psychoanalysis and Psychology, N.Y.C., 1948—; prof. Mich. State U., 1957-61; prof. psychology Mexican Nat. Autonomous U., Mexico, 1950—, N.Y. U., 1962—. Dir. Mexican Psychoanalytic Inst., 1962—. Author: Escape from Freedom, 1941; Man for Himself, 1947; Psychoanalysis and Religion, 1950; The Forgotten Language, 1951; The Sane Society, 1955; The Art of Loving, 1956; Sigmund Freud's Mission; (with D. I. Suzuki and R. de Martino) Zen Buddhism and Psychoanalysis, 1960; Marx's Concept of Man, 1961; May Man Prevail, 1961; Beyond the Chains of Illusion, 1962; The Dogma of Christ and Other Essays, 1963; The Heart of Man, 1964; You Shall Be as Gods, 1966; also articles. Research consisting of devel. of a psychoanalytic theory in its application to ethics, to philosophic anthropology, to sociology and philosophy and history (this research based on clinical data on psychoanalysis and devel. of psychoanalytic theory on basis of Freud's findings from a materialistic-mechanistic frame of reference to a humanist-existentialist one). Address: 180 Riverside Dr., N.Y.C. 10024; also 748-5 Patricio Sanz, Mexico 12, D.F. Mexico.*

FROMM, Erika (Mrs. Paul Fromm), psychologist, educator; b. Frankfort, Germany, Dec. 23, 1910; d. Siegfried and Clementine (Stern) Oppenheimer;

Ph.D. magna cum laude, U. Frankfurt, 1933; postgrad. child care program Chgo. Inst. for Psychoanalysis; m. Paul Fromm, July 20, 1928; 1 dau., Joan. Came to U. S., 1938, naturalized, 1944. Faculty, Med. Sch. Northwestern U., 1954-61; professorial lectr. U. Chgo., 1961——. Recipient award for best sci. paper in exptl. hypnosis Soc. For Clin. and Exptl. Hypnosis, 1965. Diplomate Am. Bds. Examiners Profl. Psychology, Clin. Hypnosis. Fellow Am. Psychol. Assn., Am. Orthopsychiat. Assn. (past dir.), A.A.A.S., Soc. Clin. and Exptl. Hypnosis (exec. sec.), Am. Soc. Clin. Hypnosis, Soc. For Projective Techniques. Author: (with L. D. Hartman) Intelligence—A Dynamic Approach, 1955; (with Thomas M. French) Dream Interpretation, 1964. Research in psychoanalytic ego psychology, projective techniques, dream interpretation, and hypnosis. Home: 5715 S. Kenwood Av., Chgo. 60637.*

FROMMANN, Carl, German anatomist; b. Jena, Germany, May 22, 1831; s. Friedrich Johannes and Wilhelmine (Günther) F.; ed. Jena, Göttingen (both Germany), Prague, Czechoslovakia, Vienna, Austria; M.D., 1854; asst. in clinic at Jena, 1856-58; staff German Hosp. of London, 1858-60; at Weimar, Germany, 1861-70; pvt. docent, Heidelberg, Germany, 1870-72; pvt. docent, Jena, 1872-75, became prof., 1875. Author: Untersuchungen über die normale und pathologische Histologie des centralen Nervensystems, 1876; Untersuchungen über die Gewebsveranderungen bei der multiplen Sclerose, 1878; Beobachtungen über Structur und Bewegungserscheinungen des Protoplasma der Pflanzenzellen, 1880. Discovered (with silver nitrate stain) transverse lines or striae (Frommann's lines) in axis cylinder of myelinated nerve fibers. Died Apr. 22, 1892.

FROMMANN, Johann Christian, German physicist; b. Coburg, Germany, circa 1640; ed. Königsberg, Tubingen, Leipzig, Germany; state physicist, Coburg; prof. gymnasium, 1669. Author: Tractatus de Fascinatione, 1675; Tract de haemorrhoidibus, 1677.

FROMMEL, Edouard, Swiss physician; b. Geneva, Switzerland, June 9, 1895; s. Gaston and Madeleine (Thomas) F.; ed. U. Geneva, U. Paris; M.D.; m. A. M. von Schulthes Rechberg; children—Marie-Claire, Dominique, Monique. Intern, head clinic; asst. physician Canton Hosp. and Polyclinic, Geneva; asst. U. Zurich; pvt. practice internal medicine; prof. exptl. therapy; dir. Inst. Exptl. Therapy, Geneva. Mem. Soc. Biology (corr.), Nat. Acad. Medicine of Paris (past pres.), Soc. Physics and Natural History (Geneva), Med. Soc. Geneva, Swiss Soc. Physiology, Swiss Acad. Med. Scis. (past senator). Research and over 450 publs. on internal medicine, pharmacodynamics, exptl. medicine, especially on nervous system. Address: 5 Plateau de Frontenex, Geneva, Switzerland.

FROMONDUS (or FROIDMOND), Libertus, scholar; b. Haccourt, Belgium, 1587; Catholic divine; chief prof. philosophy Coll. of Falcon, Louvain, Belgium; succeeded Jansen as prof. divinity U. Louvain, circa 1635. Author: Liberti Fromondi S.Th.L., Collegii Falconis in academia lovainensi philosophiae professoris primarii meteorologicorum libri sex, 1627. Defender of Ptolemaic system; accepted Aristotelian meteorology. Died 1653.

FRONACHER, Carl, German geneticist; b. Landshut, Mar. 8, 1871; prof., Hanover, also Berlin, Germany. Author: Grundzüge der Züchtungsbiologie, 1913; Allgemeine Tierzucht, 6 vols., 1916-27; Neuzeitliche Vererbungslehre und Tierzucht, 1924; Züchtungslehre, 1929. A founder of sci. animal breeding; authority on animal diseases. Died Munich, Apr. 9, 1938.

FRONAEUS, Sture Adolf, Swedish chemist; b. Boras, Sweden, Oct. 5, 1916; s. Gustaf A. Frann and Maria (Karlsson) F.; fil. mag. U. Lund (Sweden), 1941, fil. lic., 1947, fil. dr., 1949; m. Kerstin Molin, June 26, 1951 (dec. 1966); children—Margareta, Gunnar. Asst., U. Lund, 1943-47, faculty, 1947——, prof. chemistry, 1958——, prof. inorganic chemistry, 1964——, head chem. inst., 1962-67. Mem. Royal Physico-graphical Soc. Contbr. articles to jours. Developed methods for investigations of metal complex formation in solution; research on phys. chemistry of ion exchangers, on mechanisms of electrode reactions and of redox reactions. Home: 9 S:t Jörgens Park, Lund, Sweden.*

FRONDEL, Clifford, Am. mineralogist; b. N.Y.C., Jan. 8, 1907; s. George and Martha (Kindermann) F.; B.S. in Geol. Engring., Colo. Sch. Mines, 1929; M.A., Columbia, 1936; Ph.D., Mass. Inst. Tech., 1939; m. Eleanor Travis, Sept. 9, 1941; 1 d., Dana L.; m. 2d, Judith Weiss, Nov. 26, 1949; 1 dau., Barbara. Teaching fellow crystallography Mass. Inst. Tech., 1937-39; research asso. Harvard, 1939-42; sr. civilian physicist War Dept., 1942-43; dir. research Reeves Sound Labs., 1943-45; asso. prof. mineralogy Harvard, 1946-54, prof., 1954——, chairman department geological scis., 1965——. Recipient of Roebling medal Mineral. Soc. Am., 1964; Distinguished Achievement medal from Colorado School of Mines, 1964. Fellow of the American Academy of Arts and Scis., Mineral. Soc. Am., Geol. Soc. Am.; mem. Am. Crystallographic Assn., mineral. socs. Great Britain, Germany, France, Italy; corr. member of Natural History Museum of Vienna, American Museum of Natural History, Deutsche Akademie der Naturforscher. Author: Dana's System of Mineralogy (3 vol., with C. Palache and H. Berman), 7th edition 1951; Descriptive Mineralogy of Uranium and Thorium, 1958. Contributor profl. jours. Principal work in descriptive mineralogy; research on geochemistry of rare elements; mineral synthesis; meteoritics. Home: 20 Beatrice Circle, Belmont, Mass.*

FRONGIA, Luigi, Italian surgeon, dentist; b. Nurri, Apr. 8, 1905; s. Gaetano and Laodice (Pisano) F.; Dr. medicine and surgery; m. Lydia Occhialini; children—Maria Vittoria, Maria Luisa, Franco. Specialist in dentistry and radiology; prof. Dental Clinic, U. Cagliari, 1962. Mem. Assn. Italian Drs. of Dentistry, Italian Soc. Dental Study and Maxillo-facial Surgery, Cultivation of Med. Scis. of Cagliari. Contbr. numerous articles to profl. publs. Home: 12 via Carbonia. Office: 23 via Dettori, Cagliari, Italy.

FRONTALI, Gino, Italian pediatrician; b. Alexandria, Egypt, July 19, 1889; s. Facondo and Ernesta (Vissich) F.; M.D. Asst., asst. master Pediatric Clinic, Florence; prof., 1922; prof. of univ., 1925, full prof., 1928; titular prof. Pediatric Clinic, U. Cagliari, U. Pavia, U. Padua, U. Rome, 1943; affiliate prof. U. Rome, 1959——; dir. Center for Studies on Infantile Nutrition, Nat. Council Research. Author: Pediatria Clinica-Lezioni Clinich, 2 vols., 1952; L'Alimentazione Infantile, 4th edit., 1958. Dir., Manuala di Pediatria, 2 vols., 3d edit., 1962, Archivio Ital. di Pediatria e Puercultura, Pediatria Internaz. Research and numerous publs. on nutritional lipides, optimal absorption of trioleine, pellagra, Mediterranean anemia, life duration of red corpuscles. Address: 30 via Adige, Rome, Italy.

FRONTINUS, Sextus Julius, Roman engr.; b. circa 30; gov. Britain, 74-78; placed in charge Roman water system by Emperor Nerva, 97; author books on mil. sci., land surveying, other branches of applied sci. (none extant), also 2 vol. work on Roman aqueducts (probably most informative work on ancient engring. extant). Died 104.

FRORIEP, August Friedrich von, see von Froriep, August Friedrich.

FROSCH, Paul, pathologist; b. 1860; recognized (with F. A. J. Loeffler) cause of foot-and-mouth disease (pub. 1897), thus showing for 1st time that agt. of animal disease may be a filtrable virus. Died 1928.

FROST, Arthur Atwater, Am. chemist; b. Onarga, Ill., Aug. 3, 1909; s. Henry Hoag and Mary (Tuttle) F.; B.S., U. Cal. at Berkeley, 1931; Ph.D., Princeton, 1934; m. Faye Hibbard, Sept. 14, 1934; children—Barbara (Mrs. James W. Cernohlavek), Sylvia (Mrs. Gordon Geiger), Linda (Mrs. Don Gilbert). Research fellow Harvard, Cambridge, Mass., 1934-36; faculty Northwestern U., Evanston, Ill., 1936——, prof. chemistry, 1954——, chmn. dept., 1957-62. Fellow Am. Phys. Soc.; mem. A.A.A.S., Am. Chem. Soc. Author: (with R. G. Pearson) Kinetics and Mechanism, 1953. Research, publs. in chem. kinetics, surface elec. potentials and molecular quantum mechanics. Home: 606 Juniper Rd., Glenview, Ill. 60025. Office: Northwestern U., Evanston, Ill. 60201.*

FROST, Douglas Van Anden, Am. nutritionist; b. Pitts., Oct. 31, 1910; s. Donald Karne and Amy (Craig) F.; B.A., U. Ill., 1933; M.A., U. Wis., 1938, Ph.D., 1940; m. Muriel Louise Newkirk, Aug. 10, 1940; children—Nancy Newkirk (Mrs. James Kroening) Melodie Louise Craig, Roy Craig, Constance Manning. Analytical chemist Pacini Labs., Chgo., 1934-35, Rival Packing Co., Chgo., 1935-36; research biochemist head nutrition research dept., research specialist in nutrition Abbott Labs., North Chicago, Ill., 1940-66; research asso. Dartmouth Med. Sch., Brattleboro, Vt., 1966——. Vice-pres. Agrl. Research Inst., Nat. Acad. Sci., 1966-67; cons. Selenium-Tellurium Devel. Assn., 1962——. Fellow N.Y. Acad. Scis.; mem. Animal Nutrition Research Council (past pres.), Assn. Vitamin Chemists (past pres.), Am. Inst. Nutrition (past treas.), Am. Soc. Biol. Chemists, Am. Chem. Soc., Poultry Sci. Assn., Am. Soc. Animal Sci., Soc. for Exptl. Biology and Medicine, Metric Assn. Contbg. author: Hawk's Physiological Chemistry, 1965; Protein and Amino Acid Nutrition, 1959; Symposium on Medicated Feeds, 1956; also numerous articles. Research on vitamins, amino acids, concept nutritional relativity; one of discoverers nutrient role nicotinic acid. Home: 48 High St. Office: Brattleboro Hosp., Brattleboro, Vt. 05301.*

FROST, Edwin Brant, Am. astronomer; b. Brattleboro, Vt., July 14, 1866; s. Carlton P. and Eliza A. (DuBois) F.; A.B., Dartmouth, 1886, A.M., 1889, D.Sc., 1911; D.Sc. (hon.) Cambridge U., 1912; studied physics and astronomy, Princeton, Strassburg (Germany), Astrophys. Obs., Potsdam, Germany; m. Mary E. Hazard, Nov. 19, 1896; children—Katharine Brant, Frederick Hazard, Benjamin DuBois. Instr. physics and astronomy Dartmouth, 1887-90, asst. prof. astronomy, dir. obs., 1892-95, prof. astronomy, 1895-98, non-resident instr. 1898-1902; prof. astro-physics U. Chgo., from 1898, dir. Yerkes Obs., 1905-32, emeritus. Fellow Am. Acad. Arts and Scis., A.A.A.S. Translated, revised and enlarged A Treatise on Astronomical Spectroscopy (Dr. J. Scheiner), 1894. Author: An Astronomer's Life, 1933; Let's Look at the Stars, 1935. Asst. editor Astrophys. Jour., 1895-1901, editor, 1902. Developer stellar spectrograph; discoverer asteroid Frostia. Died Chgo., May 14, 1935.

FROST, Harold Maurice, Am. physician; b. Boston, May 21, 1921; s. Harold Maurice and Lucy (Church) F.; B.A., Dartmouth, 1943; M.D., Northwestern U., 1945; m. Elsa Claudius, Oct. 27, 1956; children—Harold Maurice III, Mary, Michael, Margaret, Patricia, Robert, Erik. Practice medicine, specializing in orthopedic surgery, Buffalo, New York, 1953-55, New Haven, Connecticut, 1955-57, Detroit, Mich., 1957——; clin. instr. orthopedic surgery U. Buffalo, 1953-55; asst. prof. Yale, 1955-57; asso. orthopedic surgeon Henry Ford Hosp., 1957——, chmn. dept. orthopedic surgery, 1966——; vis. prof. dept. anatomy Sch. Medicine U. Utah, 1963——. Mem. A.M.A. (Hektoen Gold medal 1963, certificate of merit 1966), Am. Assn. Orthopedic Surgeons, A.A.A.S., Assn. Bone and Joint Surgery, Sigma Xi. Author: Bone Remodelling Dynamics, 1963; Laws of Bone Structure, 1964; Mathematical Elements of Lamellar Bone Remodeling, 1964; Introduction to Biomechanics, 1966; Bone Dynamics in Osteoporosis and Osteomalacia, 1966; Bone Biodynamics, 1964. Research, numerous publs. in bone physiology and bone diseases. Home: 992 Dowling, Bloomfield Hills, Mich. Office: Henry Ford Hosp., Detroit 48202.*

FROST, John, English botanist; b. London, 1803; studied under Dr. Wright, apothecary of Bethlehem Hosp.; m. Harriet Yosy. Became sec. Royal Humane Soc., 1824; later practiced medicine, Berlin, Germany. Fellow Soc. Antiquaries, Royal Soc. Edinburgh, Linnean Soc.; mem. Royal Asiatic Soc., Medico-Bot. Soc. (founder). Author: Orations. Research on medicinal properties of plants. Died Mar. 17, 1840.

FROST, Paul Davis, Am. metallurgist; b. Beverly, Mass., Aug. 9, 1916; s. George Elbridge and Pearl (Doble) F.; B.S., Rensselaer Poly. Inst., 1940; postgrad. Ohio State U., 1946-50; m. Ruth Sackrider, Apr. 17, 1943; children—Francesca (Mrs. Bruce Stevenson), Jonathan James. Metallurgist, Republic Steel Corp., 1940-43, instr. ferrous metallurgy Cornell U., 1941-45, research metallurgist Cornell Aero. Lab., 1943-46 (all Buffalo); sr. tech. adviser Battelle Meml. Inst., Columbus, O., 1946——, Mem. Am. Foundrymens Soc., Am. Inst. Mining, Metall. Engrs., Am. Soc. Metals, Am. Ordnance Assn., Soc. for Die Casting Engrs., Magnesium Assn. Research and publs. on alloys in titanium, magnesium, aluminum and copper, devel. of processes for melting, casting and fabricating metals; discovered omega phase in titanium; developed heat treatments for titanium; direct research on problems of corporate and divisional mgmt. Home: 175 Webster Park Av., Columbus 43214. Office: 505 King Av., Columbus, O. 43201.*

FROST, William Dodge, Am. bacteriologist; b. Lake City, Minn., Sept. 13, 1867; s. Benjamin Cutler and Lucy Jane (Dodge) F.; B.S., U. Minn., 1893, M.S., 1894; Ph.D., U. Wis., 1903; Dr. P.H., Harvard, 1913; m. Jessie H. C. Elwell, Jan. 1, 1895; children—Herbert Cutler, Russell Elwell, Theodore Dodge. Biologist in lab. Minn. Bd. Health, 1894-95; asst. in bacteriology, U. Wis., 1895-98; instr., 1898-1902, asst. prof., 1902-07, asso. prof., 1907-15, prof. agrl. bacteriology, 1916-38. Pres. Morningside Sanatorium. Fellow A.A.A.S., Am. Pub. Health Assn., A.M.A. (asso.); mem. Am. Assn. Med. Milk Commns. (pres.), Soc. Am. Bacteriologists, Nat. Soc. Study and Prevention Tb, Wis. (dir.), Madison anti-Tb assns., (dir.), Wis. Acad. Scis., Royal Inst. Pub. Health (London), Sigma Xi. Author: Laboratory Guide in General Bacteriology, 1901; General Bacteriology (with E. G. McCampbell), 1911; The streptococci (with Mildred A. Englebrecht), 1940; also collaborator, Marshall's Microbiology. Originator cellular test for pasteurized milk and little plate method of counting bacteria in milk; research and publs. on bacteriology, especially hemolytic streptococci of milk. Died Jan. 15, 1957.

FROUDE, William, English engr.; b. Dartington, Eng., Nov. 28, 1810; ed. Oxford, Eng.; became civil engr. on ry., later naval projects; investigated water resistance of ship hulls, 1872-74; built log and pendulum equipment to measure ships' rolling; introduced exptl. tank work with model ships. Fellow Royal Soc., 1870. Inventor hydraulic break with dynamometer, 1858; discoverer law of comparison (named for him) which permits calculation of the force needed to tow an object in water against the wave raised by its own progress. Died Simonstown, S. Africa, May 4, 1879.

FRUMKIN, Alexander Naumovich, Russian physicochemist; b. Kishinev, Russia, Oct. 12, 1895; s. Naum and Margaret (Frumkin); u. diploma Physico-Math. Faculty, Odessa (Russia) U.; Dr.chem., U. Moscow (Russia), 1932; Dr.rer.nat., Technische U., Dresden, Germany, 1957; m. Amalya Davydovna Bogdanovskaja, June 1926. Chemist analytical lab., Odessa, 1915-17; faculty Odessa U., 1917-22; scientist Karpov Inst. Phys. Chemistry, Moscow, 1922-28, dep. dir., 1930-46; vis. prof. colloid chemistry U. Wis., Madison, 1928-29; dir. Inst. Phys. Chemistry, Acad. Scis., Moscow, 1946-49, head electrochemistry dept., 1949-57, dir. Inst. Electrochemistry, 1958——; prof., head chair electrochemistry Moscow State Lomonossov U., Moscow, 1930——. Recipient Red Banner Order, 1943, 45, 49; Lenin Order, 1945-65; Lenin prize, 1931; State prize 1st class, 1941; Hero of Socialist Labour, Hammer and Sickle medal, 1965; Palladium medal Am. Electrochem. Soc., 1959. Mem. Acad. Scis.

USSR, acads. of Warsaw, Sofia, Halle, Amsterdam; corr. mem. acads. of Berlin, Zagreb, Leipzig; hon. mem. Mendeleyev, Belgian chem. socs., Nat. Inst. Scis. India. Author: Electrocapillary Phenomena and Electrode Potentials, 1919; (with V. Bagotzky, S. Jofa, B. Kabanov) Kinetics of Electrode Processes, 1952; also numerous articles. Developed theory of double layer at interface between metals and electrolyte solutions; established connections between potential differences at different interfaces; introduced notion of potential of zero charge; demonstrated role of double layer structure in electrode kinetics; research on mechanism of various electrochem. processes and role of adsorption phenomena in electrochemistry. Home: Moscow, V-71, Leninsky Prospekt, 13, fl. 97., USSR. Office: Inst. Electrochemistry, Acad. Scis. USSR, Moscow, V-71, Leninsky Prospekt, 31., USSR.*

FRUTON, Joseph Stewart, biochemist; b. Czestochowa, Poland, May 14, 1912; s. Charles and Ella (Eisenstadt) F.; came to U. S., 1923, naturalized, 1929; B.A. with honors in Chemistry, Columbia, 1931, Ph.D. in Biochemistry, 1934; M.A. (hon.) Yale, 1950; m. Sofia Simmonds, Jan. 29, 1936. Faculty, Rockefeller Inst., N.Y.C., 1934-45; faculty Yale, 1945—, chmn. dept. biochemistry, 1951—, Eugene Higgins prof. biochemistry, 1957—, dir. div. sci., 1959-62. Mem. numerous coms. NRC; sci. advisor NIH, 1950-52, Anna Fuller Fund, 1951—. Rockefeller Found. spl. fellow, 1948; Commonwealth Fund fellow, 1962-63; Benjamin Franklin fellow Royal Soc. Arts, 1964; Dakin lectr., 1962. Fellow A.A.A.S.; mem. Nat. Acad. Scis., Internat. Commn. Biochem. Nomenclature, Am. Acad. Arts and Scis., Am. Soc. Biol. Chemists (council 1959-62), Am. Chem. Soc. Eli Lilly award biol. chemistry 1944), Chem. Soc. London, Biochem. Soc. Eng., Harvey Soc. (lectr. 1955), History of Sci. Soc. (council 1951-54). Author: (with S. Simmonds) General Biochemistry, 1953; 2d edit., 1958. Asso. editor Biochemistry, 1960—; mem. editorial bd. Jour. Biol. Chemistry, 1948-58. Contbr. articles to chem., biochem. jours. Research on amino acids, peptides, proteins and enzymes. Home: 123 York St., New Haven 06511.*

FRY, David Lloyd George, Am. physicist; b. Detroit, Sept. 22, 1918; s. Fred and Linda (Fry) F.; B.A., Kalamazoo Coll., 1940; M.S., Ohio State U., 1942; m. Alma Loriene Jones, Mar. 21, 1942; children—David Edward, Lawrence Bryan, Randall Keith. With Research Labs., Gen. Motors Corp., Warren, Mich., 1942—, supervisory physicist, 1952—. Mem. Am. Phys. Soc., Optical Soc. Am., A.A.A.S., Soc. Applied Spectroscopy, Engring. Soc. Detroit, Am. Soc. Testing Materials. Editor: Methods for Emission Spectrochemical Analysis, 1960. Publs. on research in emission spectroscopy, infrared spectroscopy, gases-in-solids, mass spectroscopy, semi conductor surfaces; nuclear magnetic resonance. electron spin resonance, electro-optics, magneto mech. damping, glass tech. Home: 685 Princeton St., Berkeley, Mich. 48072. Office: Research Labs., Gen. Motors Corp., Warren, Mich. 48090.*

FRY, Donald Lewis, Am. physiologist; b. Des Moines, Dec. 29, 1924; s. Clair V. and Maudie (Long) F.; student Marquette U., 1943-45; M.D., Harvard, 1949; postgrad. U. Minn., 1950-53, U. Md., George Washington U., 1953-56; m. Virginia Milne, Sept. 13, 1947; children—Donald Stewart, Ronald Sinclair, Heather Elise, Laurel Virginia. Sr. asst. surgeon Gen. Medicine and Therapeutics br. Nat. Heart Inst., 1953-56, surgeon, 1956-57, sr. surgeon cardiodynamics sect., 1957-61, med. dir., head sect. clin. biophysics Cardiology br., 1961—; asst. prof. medicine Georgetown U., 1960—. Cons. bio-engring. program Cath. U., 1965—. Mem. Am. Soc. Clin. Investigation, Biophys. Soc., Am. Physiol. Soc., Am. Fedn. Clin. Research, A.A.A.S. Contbr. chpts. to Pulsatile Blood Flow, 1964; Methods in Medical Research, 1966. Research, numerous publs. on lung, heart and vascular biophysics and physiology, measure of pressures, flows, displacements, strains, rheology, structure and dynamics of these systems in both man and animals; three dimensional analysis of pulmonary mechanics, flow vol. curve of lung, pressure gradient method for measurement of instantaneious blood flow in human subject, interrelationship of blood hydro dynamics to vessel wall behavior. Home: 5512 Lincoln St., Bethesda 20034. Office: NIH, Bethesda, Md. 20014.*

FRY, Donald William, Brit. physicist; b. Weymouth, Dorset, Eng., Nov. 30, 1910; s. William Joseph and Mary (Symonds) F.; M.Sc., King's Coll., London, 1932; m. Jessica Florence Wright, July 7, 1934; children—David, John, Peter. Research physicist Gen. Electric Research Labs., Wembley, Eng., 1932-40; with Air Ministry Research Establishment (later Telecommunications Research Establishment), 1940-42, Atomic Energy Research Establishment, 1946—, head gen. physics div., Harwell, Eng., 1950-54, chief physicist, 1954-58, dep. dir., 1958, dir. Atomic Energy Establishment, Winfrith, Dorchester, Eng., 1959—. Recipient Phys. Soc. Duddell medal. Fellow Inst. Physics and Phys. Soc., Inst. Elec. Engrs., I.E.E.E. Contbr. numerous articles to sci. jours. Demonstrated (with others) a new principle for accelerating particles-travelling wave linear accelerator. Home: Coveway Lodge, Overcombe, Weymouth, Dorset. Office: U.K.A.E.A., A.E.E., Winfrith, Dorchester, Dorset., Eng.*

FRY, Frederick Ernest Joseph, zoologist; B. Woking, Eng., Apr. 17, 1908; s. Ernest and Mabel (Holmes) F.; Ph.D., U. Toronto, 1936; m. Irene Marguerite Stewart, Oct. 19, 1935; children—John, James, Frederick. Mem. faculty U. Toronto, Ont., Can., 1936—, prof. zoology, 1956—; cons. to Fisheries of Province of Ont., 1960—; mem. Fisheries Research Bd. Can., 1965—. Recipient citation, Wildlife Soc., 1950; mem. Order Brit. Empire, 1944; Flavelle medal Royal Soc. Can., 1962. Guggenheim fellow, 1959. Mem. Am. Fisheries Soc. (pres. 1966), Am. Soc. Limnology and Oceanography (pres. 1951), Am. Soc. Zoologists, Canadian Soc. Zoologists (pres. 1966), Internat. Soc. Pure and Applied Limnology (v.p. 1962), Ecol. Soc. Am. Research and publs. on vital statistics of exploited populations of fish; influences of environment on activity of fish. Home: 10 Riverlea Rd., Weston, Ont., Can.*

FRY, Harry Shipley, Am. chemist; b. Cin., Oct. 24, 1878; s. Harry Oliver and Emma Elizabeth (Richards) F.; B.A., U. Cin., 1901, M.A., 1902, Ph.D., 1905; m. Corinne Angele Lacroix, June 16, 1904; 1 son, Harry Lacroix. With faculty U. Cin., since 1901, prof., head dept. chemistry since 1918, prof. emeritus 1945. Fellow A.A.A.S.; mem. NRC (div. chemistry and chem. tech.), Nederlandsche Chemische Vereeniging, Nat. Inst. of Social Scis., Am. Chem. Soc., Sigma Xi. Author: The Electronic Conception of Valence and the Constitution of Benzene, 1921. Contbr. many articles in Am. and foreign tech. publs. First to propose and apply terms electrometer and electronic tautomerism as fundamental concepts in electronic conception of valence and explanation of chem. reactions of compounds of carbon. Died May 18, 1949.

FRY, John, physician; b. June 16, 1922; s. Ansel and Balbina (Minton) F.; M.B., B.S., U. London, 1944, M.D., 1955; m. Joan Lilian Sabel, Feb. 4, 1944; children—James, Dimity (Mrs. Michael Dawson). Gen. practice medicine, Beckenham, Kent, Eng., 1947—. Cons., WHO, 1961—, Brit. Army, 1964—; trustee Nuffield Provinical Hosps. Trust, 1957—; councillor Coll. Gen. Practitioners, 1953—. Milbank fellow, 1965; recipient Gold medal Hunterian Soc., 1955, 56; Sir Charles Hastings prize Brit. Med. Assn., 1960-64; Rogers prize U. London, 1964; James Mackenzie prize Coll. Gen. Practitioners, 1965. Mem. Royal Soc. Medicine (pres. sect. gen practice 1962-63). Author: The Catarrhal Child, 1961; Presenting Symptoms in Childhood, 1962; Diseases of Respiratory Tract, 1962; (with N. C. Oswald) Profiles of Disease, 1966. Publs. on clin. and operational research in field of family medicine and community care. Address: 138 Croydon Rd., Beckenham, Kent, Eng.*

FRY, Joshua, surveyor; b. Crewkerne, Eng., circa 1700; s. Joseph Fry; ed. Wadham Coll., Oxford (Eng.) U., 1718; m. Mary (Micou) Hill, circa 1720. Came to Va., before 1720; prof. math. and natural philosophy Coll. William and Mary, 1731; 1st presiding justice Albemarle County; justice Ct. Chancery; county surveyor; mem. Ho. of Burgesses from Albemarle County until death; county lt., 1745; mapmaker (with Peter Jefferson) Map of the Inhabited Parts of Virginia; surveyor (with P. Jefferson), ran part of Va.-Carolina boundary, 1749; commr. to Six Nations, 1752, aided in drawing up Treaty of Logstown; col., comdr.-in-chief of a regt. Va. Militia, started for Ohio in expdn. against French, 1754, died enroute, succeeded by George Washington. Died Ft. Cumberland, Md., May 31, 1754.

FRY, Thornton Carl, Am. mathematician; b. Findlay, O., Jan. 7, 1892; s. William Watson and Elizabeth Hannah (Dingle) F.; A.B., Findlay Coll., 1912, D.Sc., 1958; A.M., U. Wis., 1913, Ph.D., 1920; m. Louise Brooks Notter, Sept. 19, 1945; children—Dinah Elizabeth (Mrs. Charles E. Jenkins). Instr., U. Wis., 1912-16; mathematician Western Electric Co., 1916-24; with Bell Telephone Labs, N.Y.C., 1924-56, math. research dir., 1940-44, dir. switching research 1944-47, dir. switching research and engring., 1947-49, asst. to exec. v.p., 1949-51, asst. to pres., 1951-56; communications cons. Internat. Tel. & Tel., N.Y.C., Paris, France, 1956-57; sr. cons. Univac div. Sperry-Rand Corp., N.Y.C., 1956-57, v.p., dir. Univac engring Remington Rand div., 1957-60, v.p. research and engring, 1960-61; cons. on orgn. and mgmt. sci. research. Lectr. elec. engring Mass. Inst. Tech., 1927; lectr. math. Princeton, 1929-30; mem. div. 7 NDRC, 1940-44, chief sect. 7.2, 1942-44, dep. chief applied math. panel 1942-45; cons. to dir. Nat. Center for Atmospheric Research, Boulder, Colo., 1961-—, Granville-Phillips Co., Boulder, 1964—, Boeing Sci. Research Labs., Seattle, 1964—. Recipient Presdl. certificate of merit for NDRC achievements, 1948. Author: Elementary Differential Equations, 1929; Probability and Its Engineering Uses, 1928. Research, publs. on math. theory of probability; especially as applied to sci. and engring. Home: P.O. Box 5966, Carmel, Cal.*

FRY, William Frederick, Am. physicist; b. Carlisle, Ia., Dec. 16, 1921; s. William C. and Flossie (Parsons) F.; B.S., Ia. State Coll., 1943, Ph.D., 1952; m. Virgie Eastburn, June 20, 1943; children—David A., Diane E. AEC postdoctoral fellow U. Chgo., 1952-53; faculty U. Wis., Madison, 1953—, prof., 1957—; Fulbright fellow, Italy, 1956-57. Fellow Am. Phys. Soc.; mem. Italian Phys. Soc. Contbr. numerous articles to profl. jours. Discovered radiative Pi-M decays, active in bubble chamber devel. and usage in studying fundamental particles. Home: 4030 Mandan Crescent, Madison, Wis. 53711.*

FRY, William James, Am. vascular surgeon; b. Ann Arbor, Mich., Mar. 21, 1928; s. Lynn Ward and Inez (Hayes) F.; M.D., U. Mich., 1952; m. Martha Ann Earl, June 18, 1949; children—Richard E., William R. Faculty, U. Mich. Med. Sch., Ann Arbor, 1957—, prof. surgery, 1967—, head sect. gen. surgery, 1967-—; chief surgery Ann Arbor VA Hosp., 1961-64, cons., 1964—. Recipient Distinguished Service award U. Mich. Med. Sch., 1963, Class of 1964 Sr. award, 1964. Diplomate Am. Bd. Surgery. Fellow A.C.S.; mem. A.M.A., Internat. Cardiovascular Soc., Soc. for Surgery Alimentary Tract, Soc. Univ. Surgeons, Assn. for Acad. Surgery, Frederick A. Coller Surg. Soc., Central, Western surg. assns., others. Research and publs. on vascular physiology, aortic resections, renal artery and femoral bypasses for treatment of arteriosclerosis, intermittent claudication, other vascular insufficiencies, operations for duodenal and gastric ulcers using vagotomy and pyloroplasty, gen. surg. procedures involving surgery of abdomen. Home: 1434 Warrington Dr., Ann Arbor, Mich.*

FRYE, John C(hapman), Am. geologist; b. Marietta, O., July 25, 1912; s. Harley Edgar and Maude Vesta (Chapman) F.; A.B., Marietta Coll., 1934, D.Sc., 1955; O. State U., 1935; M.S., U. Ia., 1937, Ph.D., 1938; m. Ruth L. Heizer, Aug. 29, 1936; children—Sally Jean, John Douglas, Terri Ruth. Grad. asst., research asst. U. Ia., 1935-38; geologist U. S. Geol. Survey, 1938-42; asst. dir. Kan. State Geol. Survey, 1942-45, exec. dir., 1945-54; asst. state geologist, 1942-45, state geologist 1952-54; asst. prof. U. Kan., 1942-45, asso. prof., 1945-52, prof. geology, 1952-54; chief Ill. State Geol. Survey, 1954—; prof. geology U. Ill., 1963—; spl. research geologist Bur. Econ. Geology, U. Tex., summer 1955—. Adv. com. to sec. interior for U. S. Geol. Survey, 1960—; del. to 19th Internat. Geol. Congress, Algiers, 1952. Fellow Geol. Soc. Am. (councillor 1958-61), A.A.A.S.; mem. Am. Inst. Mining, Metall. and Petroleum Engrs., Am. Assn. Petroleum Geologists, Soc. Econ. Geologists, Ill. Acad. Sci. (pres. 1962-63), Soc. Econ. Paleontologists and Mineralogists, Am. Com. on Stratigraphic Nomenclature (chmn. 1957-58), Assn. Am. State Geologists (editor 1956-57, sec.-treas. 1957-58, pres. 1960), NRC (com. Atomic Waste Disposal, 1955—, rep. Soc. Econ. Paleontologists and Mineralogists), Am. Geophys. Union, Ill. Mining Inst., Sigma Xi, Sigma Gamma Epsilon, Alpha Sigma Phi. Contbr. sci. articles profl. jours. Studies in classification, dating and correlation of deposits nr. surface, especially glacial deposits in interior U. S. A.; applications to search for water and mineral deposits. Home: 708 W. Vermont St., Urbana 61801. Office: State Geol. Survey, U. Ill. Campus, Urbana, Ill. 61803.

FRYE, Royal Merrill, Am. physicist; b. Milford, N.H., May 27, 1890; s. Frank Barton and Elsie Willetta (Merrill) F.; A.B., Boston U., 1911, A.M., 1912, Ph.D., 1934; student New Eng. Conservatory of Music, 1906-08, Harvard U. Grad. Sch., 1912-13, (intermittently) Mass. Inst. Tech., 1916-27, Am. Inst. Normal Methods, 1928; m. Louise Alexander, June 11, 1915. Instr. physics, Boston U., Mass. Inst. Tech., Worcester Poly. Inst. Northeastern U., 1913-36; asst. prof. physics Boston U. Grad. Sch., 1936-42, prof. physics, 1942-50, in charge research, 1931-50, chmn. dept., 1942-50; prof. physics and head dept. Simmons Coll., Boston, 1950-59; prof. physics, dean Coll. Advanced Sci., Canaan, N.H., 1959-63; prof. physics, pres. Belknap Coll., Center Harbor, N.H., 1963—. Sci. cons. USAAF, Kwajalein, Bikini, and White Sands, N.M., 1946; mem. physics panel NSF, 1965, 66. Pres. Boston Center for Adult Edn., 1950-53. Mem. Am. Phys. Soc., Am. Assn. Physics Teachers, Am. Assn. U. Profs. (nat. councillor 1950-53), Phi Beta Kappa. Author: Practical Physics (with Robert E. Hodgdon), 1935; Graphical Mathematics, 1941; Graphical Introduction to the Harmon (with Esther W. Tipple), 1942; Essentials of Applied Physics (with Hodgdon), 1947, 2d edit., 1954. Contbr. to numerous sci. jours. Research on effects of magnetic field upon spectra of certain gases; infrared studies, measurement of wave lengths in region where micro electric waves and very long infra red waves merge; other studies in relativity, sound, structure of upper atmosphere, structure of individual spectrum lines, crystal structure of minerals, determination of temperatures of atomic bomb explosions. Address: Belknap Coll., Center Harbor, N.H.

FRYE, William Wesley, Am. physician; b. North English, Ia., July 26, 1903; s. Cyrus Alexander and Martha E. (Sheetz) F.; B.S., Ia. Wesleyan Coll., 1926, M.S., Ia. State Coll., 1927, Ph.D., 1931; M.D., Vanderbilt Univ., 1939; Sc.D., Iowa Wesleyan College, 1957; m. Lillian Emily Brown, Apr. 5, 1929; children—William Wesley, Emily Ann (Mrs. Robert L. Bisso), Cynthia Brown (Mrs. Charles Wynn), Martha Lois (Mrs. Donald Drury), Jane Ellen. Instr. zoology, entomology Ia. State Coll., 1928-31; research asst. dept. preventive medicine, pub. health Vanderbilt U., 1931-37; instr., 1937, intern pediatrics Commonwealth Fund fellowship, 1939-40, asst. prof. dept. preventive medicine, pub. health, asst. clin. medicine, 1940-42, asso. prof., asst. clinical obstet., 1942-45, prof., head dept. preventive medicine, pub. health, 1945-48, dir. Sch. Pub. Health,

1946-48; asst. dean, dir. div. grad. medicine, prof. tropical medicine, pub. health, Tulane U., 1948-49; dean, prof. tropical medicine, pub. health, La. State U., 1949-59, vice pres., dean School of Medicine, 1959-65, chancellor La. State U. Medical Center, 1965—; field trip tropical diseases Central Am., 1943; mem. Cholera Commn. China, 1945; spl. cons. USPHS, 1946—, chmn. tropical disease study sect., research grants div., 1951-56, chairman advisory com. Epidemiology and Biometry, 1956—; chmn. combined deans com. VA Hosp., New Orleans, 1949—; dep. dir. commn. enteric infections Armed Forces Epidemiol. Bd., 1950-56; nat. adv. allergy and infectious diseases council NIH, Bethesda, Md., 1958—; mem. adv. sci. bd. Gorgas Meml. Inst. Tropical Medicine and Hygiene, 1961—. Recipient Ben Witt Key prize by Vanderbilt U., 1939. Diplomate Am. Bd. Preventive Medicine and Public Health. Fellow A.M.A., Am. Pub. Health Association, American College Physicians, 1951; member American Society Tropical Medicine and Hygiene (pres. 1960-61), So. Medical Assn., Pan-Am. Med. Assn. (N. Am. chmn. sect. tropical medicine 1960—), Am. Gastroenterological. Assn., Am. Cancer Soc. (mem. adv. com. Institutional Grants 1958-60), American Society of Parasitologists, American Academy of Preventive Medicine, Omicron Delta Kappa, Delta Omega, Sigma Xi, Phi Kappa Phi, Alpha Omega Alpha. Co-author tropical medicine manual. Research and publs. on parasitic diseases, particularly pathogenesis, treatment, agt. of amebiasis. Home: 14 Audubon Blvd., New Orleans 70118. Office: 1542 Tulane Av., New Orleans 70112.

FUCHS, Erich Ernst Edmund, German physician; b. Göttingen, Dec. 20, 1921; s. Wilhelm and Hella (Tiemann) F.; ed. univs. Kaliningrad, Königsberg, Vienna, Friburg, Göttingen; M.D.; m. Margret Jähne; children—Thomas, Andreas, Werner. Asst., Med. Clinic of Lübeck; chief physician Inst. Research Allergy, dir. Asthma Clinic, Bad Lippspringe, Westphalia, 1955—. Mem. German Soc. Internal Medicine, German Soc. Allergy Research, Interasma, Coll. Internat. Allergology, Internat. Soc. Allergology, European Acad. Allergology, Am. Acad. Allergy. Author: Sinusphlebitis-Myélitis-Allergie, 1945; Seide als Allergen, 1955; (with W. Gronemeyer) Der Inhalative Antigen-Pneumometrie Test als Standardmethode, 1959; Coffee an Occupational Vapour Allergen, 1959; Zur Beeinflussbarkeit des Exogenallergischen Bronchial-Asthma durch Operative Eingriffe am Vegetativen Nervensystem, 1961; Lehrbuch der Klinischen Allergie, 1967; Praktische Allergie-Diagnostik, 1968. Research on diagnosis and therapy of allergic diseases; insect, drug, and occupational allergies; asthma. Home: Kurparkstrasse 3, 4792 Bad Lippspringe. Office: Institute of Research of Allergy, Asthma Clinic, 4792 Bad Lippspringe, Germany.

FUCHS, Ernst, Austrian ophthalmologist; b. Vienna, June 14, 1851; prof., Liege, Belgium, also Vienna. Author: Lehrbuch der Augenheilkunde, 1839. Described peripheral optic nerve atrophy, 1885; gave 1st description of epidemic keratoconjunctivitis, 1889. Died Vienna, Nov. 21, 1930.

FUCHS, Fritz Friedrich, Danish gynecologist; b. Nov. 27, 1918; s. Josef and Sofie Elise (Petersen) F.; ed. U. Copenhagen, Caroline Inst., Stockholm; M.D.; m. Anna-Ritta Olsson; children—Anneli, Martin, Peter, Lars. Instr. obstet. surgery and gynecology, 1945-58; head gynecologist Commune Hosp. of Copenhagen, 1958; prof., asso. dir. obstet. and gynecol. sect. Cornell U. Med. Coll., N.Y. Hosp., 1965—. Mem. Danish Med. Soc., Danish Soc. Obstetrics and Gynecology, Danish Soc. Endocrinology, Nordic Obstet. and Gynecol. Soc., Soc. for Study Fertility, Internat. Assn. Fertility. Author numerous works, including: Studies on the Passage of Phosphate between Mother and Foetus in the Guinea, 1957; Endocrinology of Pregnancy and Labour, 1962. Home: Wilhelm Smidtsvej 12, Gentofte. Office: Commune Hospital, Copenhagen K, Denmark.

FUCHS, George, Austrian physician; b. Vienna, Austria, Oct. 25, 1908; Ph.D., U. Vienna; M.D. Chief physician, dir. Inst. Radiology, Emperor Francis-Joseph Hosp. of City of Vienna, 1946—. Mem. Soc. Doctors (Vienna), Austrian Soc. Radiotherapy. Author 8 books, including: Röntgentherapie, 1958; Strahlenschäden und Strahelnschutz, 1959; Wir und die Atombombe, 1960; Atomkrieg-Strahlenkrankheit-Strahlentod, 1964, also over 150 articles. Home: Lincke Wienzeile 158, Vienna VI. Office: Kundrastrasse 3, Vienna X, Austria.

FUCHS, Johann Nepomuk von, see von Fuchs.

FUCHS, Klaus, physicist; b. Russelheim, Germany, 1911; ed. univs. Leipzig, Kiel, Berlin (all Germany); D.Sc., U. Edinburgh (Scotland); Ph.D., U. Bristol (Eng.). With Tube Alloys, Birmingham, Eng., 1941-43; mem. Brit. Atomic Energy Diffusion Mission, N.Y., 1943-44; mem. Brit. Atomic Team, Los Alamos, 1944-46; head Theoretical Physics Div., Harwell, Eng., 1946-49; sentenced to 14 years imprisonment for disclosing nuclear information to Russians, 1950, released, 1959; became citizen E. Germany.

FUCHS, Laszlo, Hungarian mathematician; b. Budapest, Hungary, June 24, 1924; s. David Raphael and Theresia (Rosenberg) F.; M.degree, Ph.D., U. Budapest, 1947; Venia Legendi, 1951; D.Math.Scis., Hungarian Acad. Scis., 1954. High sch. tchr. Teacher's Tng. Inst., Budapest, 1947-49; asst. L. Eötvös U., Budapest, 1949-52, docent, 1952-54, prof. math., 1954—; vis. prof. Tulane U., New Orleans, La., 1961-62, U. New South Wales, Kensington, Australia, 1965, U. Miami, Coral Gables, Fla., 1966—; head algebra sect. Math. Research Inst., Budapest, 1963—. Recipient Kossuth prize, 1953. Mem. Bolyai Math. Soc. (treas. 1951-63, sec. gen. 1963-66), Am. Math. Soc., Deutsche Mathematiker Vereinigung. Author: Abelian Groups, 1958; Partially Ordered Algebraic Systems, 1963; Teilwelse Geordnete Algebraische Strukturen, 1966; also numerous articles. Research on abstract algebra, theory of commutative groups. Office: Dept. Math., U. Miami, Coral Gables, Fla. 33124.*

FUCHS, (Immanuel) Lazarus, mathematician; b. Moschin, nr. Poznan, Poland, May 5, 1833; s. Rafael and Caecilie (Katz) F.; Ph.D., U. Berlin, 1858; m. Marie Anders, 1869; 4 sons, including Richard, 2 daus., including Clara (Mrs. L. Schlesinger). Tchr. various schs., including Friedrich Werder indsl. sch., arty. and engring. sch.; became mem. faculty U. Berlin, 1865, asso. prof., 1866, succeeded Weierstrass, 1884; named prof., Griefswald, Germany, 1869, Göttingen, Germany, 1874, Heidelberg, Germany, 1875; Mem. Prussian, Bavarian, French acads. scis. Author: Collected Works I-III, edited by Richard Fuchs and Ludwig Schlesinger, 1904-09. Research on linear differential equations, theory of functions of complex variable hypergeometric series; Fuchsian differential equations have single valued coefficients and solutions with no point of indeterminateness; laid founds. for Poincaré's automorphic functions. Died Berlin, Apr. 26, 1902.

FUCHS, Leonhard, German physician, botanist; b. Wembdingen, Germany, Jan. 17, 1501; s. Johannes and Anna (Dentenorus) F.; ed. univs. Heilbron, Erfurt, circa 1513; B.A., 1514; M.A., 1521; M.D., U. Ingolstadt, 1524; m. Anna Friedberger, 1525; 4 sons, 6 daus.; m. 2d, 1565. Practiced medicine, prof. medicine, probably Munich, Germany, 1526-33; prof. U. Tübingen (Germany), from 1535. Author: De historia stirpium (based in part on Dioscorides' herbal), 1542; Neu Kreutterbuch, 1543; De humani Corporis fabrica (anat. atlas), 1551; others. Pioneer in botany; one of 1st to utilize bot. nomenclature; pub. bot. works with excellent outline illustrations; name given to fuchsia plant which he described; wrote most comprehensive work of 16th century on herbs and vegetable drugs, 1542. Died Tübingen, May 10, 1566.

FUCHS, Remaclus, Belgian physician; b. Liège, Belgium, 1510; Author: Historia omnium aquarum (manual of distilled waters in daily use at that time), 1542. Died Brussels, Belgium, 1587.

FUCHS, (Maximilian Ernst) Richard, German mathematician; b. Greifswald, Germany, Dec. 5, 1873; s. Lazarus and Marie (Anders) F.; Ph.D., U. Berlin, 1897; m. Käthe Kiepert, 1903; 1 son, 1 dau. Became tchr. math. Bismarck Gymnasium, Berlin-Wilmersdorf, 1901; apptd. mem. faculty Tech. U. Berlin, 1906, asso. prof., 1922; sci. adviser to aircraft industry, during World War I; with German Exptl. Inst. for Aviation, Berlin-Adlershof, 1924-36, German Research Inst. for Aviation, Braunschweig, Germany, 1936-45. Author: (with L. Hopf) Aerodynamics, 3 vols., 1922. Edited and pub. father's works, promoted his theory of linear differential equations with complex variables; improved lateral and longitudinal stability of aircraft. Died Rostock, 1945.

FUCHS, Sir Vivian Ernest, Brit. geologist; b. Feb. 11, 1908; s. E. and Violet Anne (Watson) F.; ed. Coll. Brighton; St. Johns Coll., Cambridge U.; M.A., Ph.D in geology. LL.D. Honoris causa, Edinburgh, D.Sc. Honoris causa, Durham; m. Joyce Connell; 2 children. Geologist with Cambridge-East Greenland expdn., 1929, Cambridge expdn., also dir. Lake Rudolf Rift Valley expdn., 1930-31, Royal Geog. Soc., Cuthbert Peek grant, 1936, dir. Lake Rukwa expdn., 1937-38; became mem. staff coll. at Camberley, 1943; mem. Falkland Islands Dependencies Survey, 1947-50; dir., after 1950. Knighted, 1958. Recipient Special Gold medal Royal Geog. Soc. (founder Gold medal), Gold medal geog. socs. Scotland, Paris, London, Chicago, Washington, Polar medal, Hubbard Medal, Nat. Geog. Soc., also numerous others. Contbr. articles to jours. Led (with Sir E. Hillary) 1st expedition to make complete overland crossing of Antarctica. Address: 78 Barton Rd., Cambridge, Eng.

FUCHS, Vladimír, Czechoslovakian gynecologist, obstetrician; b. Prague, Czechoslovakia, June 15, 1922; s. Oto and Ema (Kettner) F.; M.D., Charles U., Prague, 1949; candidate Sci., 1961; m. Jane Pros-ová, Apr. 30, 1957; 1 son, Vladimír. Staff Gynaecological-obstet. clinic, Hradec Králove, Czechoslovakia; staff III. Gyn.-obste. clinic Charles U., faculty. Author: The Influence of Toxoplasmosis in Obstetrics, 1961; Textbook of Gynecology, 1966; also articles. Research on influence of infection in obstetrics especially toxoplasmosis, infertility, operative gynecology. Home: 22. Na Porící, Prague, Czechoslovakia.*

FUCHS, Walter Maximilian, chemist; b. Vienna, June 8, 1891; s. Adolf and Anna (Fischer) F.; student, Vienna, 1909-14, Ph.D.; m. Frieda Wienstock, 1939. Worked with M. Hönig, Brno, Czechoslovakia; lectr. Tech. U. Brno, 1919-26; became chmn. dept. Kaiser Wilhelm Inst. for Coal Research, Mulheim, Germany, 1927; prof. chem. tech. Tech. U. Aachen (Germany), 1932; went to London, 1933; prof. Rutgers U., 1934; prof. fuel tech. Pa. State Coll., 1935-42; indsl. adviser, N.Y., till 1949; returned to Aachen, 1949. Author: Der gegenwärtige Stand des Gärungsproblems, 1922; Die Chemie des Lignins, 1926; Die Chemie der Kohle, 1931; When the Oil Wells Run Dry, 1946; also numerous articles. Research on utilization of sulfite lyes, theoretical problems of phenol chemistry, theory of coal and cellulose formation, cellulose chemistry, humic acid, nitrilsynthesis, number of carbon atoms in montanic acid, problems of softening boiler water, corrosion. Died Aachen, Aug. 30, 1957.

FÜCHSEL, Georg Christian, German geologist; b. Ilmenau, Germany, Feb. 14, 1722; s. Georg and Christiane (Holzey) F.; student, Jena, Leipzig; doctorate, Erfurt, Germany, 1762; in Rudolstadt, Germany, several years; began med. practice, 1756; apptd. to organize collections of prince (-to-be) Friedrich Carl von Schwarzburg-Rudolstadt, 1757, ct. physician, 1767, librarian, 1770. Mem. Erfurt Acad. Contbr. to sci. publs. Follower of J. G. Lehmann; tried to create fundamental concepts for stratigraphy, based partly on mining terminology; made geol. map of part of Thuringia (1st geol. map of German area). Died Rudolstadt, June 20, 1773.

FUCHSIG, Paul, Austrian surgeon; b. Scharding a. Inn, Austria, Mar. 3, 1908; s. Ernst and Maria (Kanzler) F.; grad. Innsbruck U. Med. Sch., 1931; m. Anny Thavonat, Dec. 20, 1938. Mem. staff U. Clinic, Vienna, Austria, 1932-56, 61—, head Surg. Clinic, 1961—; surgeon in chief Vienna Med. Hosp., 1957-61. Mem. Internat. Soc. Surgery, Internat. Coll. Surgeons, German. Austrian socs. surgery, French Acad. Surgeons, others. Research, numerous publs. on surgery and pathology of thyroid gland, goiter, trachea, duodenum, rectum, blood transfusion, etiology of fat embolism as a complication of traumatic shock and diminuation of blood volume. Home: 3 Ferstelg. Office: 4 Alserstrasse, Vienna IX, Austria.*

FUDENBERG, H. Hugh, Am. immunologists, hematologist; b. N.Y.C., Oct. 24, 1928; s. Nathan and Frances Chackowich) F.; B.A., U. Cal. at Los Angeles, 1949; M.D., U. Chgo., 1953; M.A., Boston U., 1957; m. Betty Roof, Apr. 13, 1955; children—Drew Douglas, Brooks Roberts, David Melton, Hugh Haskell. Research asso. Rockefeller Inst., 1958-60; faculty U. Cal., 1960—, prof. medicine Sch. Medicine, San Francisco, 1966—, prof. bacterology and immunology, Berkley, 1966—. Mem. expert coms. WHO; Decennial lectr. 10th Anniversary, Brit. Soc. Immunology, 1966. Recipient Pasteur medal Inst. Pasteur, Paris, France, 1962. Mem. A.A.A.S., Assn. Immunologists, Am., Netherlands (hon.), Internat. socs. hematologists, Am. Soc. Human Genetics, Am. Soc. Clin. Investigation, Am. Rheumatism Assn. Contbr. numerous articles to tech. jours. Editorial bd. Jour. Immunology and Immunochemistry, 1964—, Vox Sanglinis, 1965—, Clin. and Exptl. Immunology, 1966—, Blood, 1967—, Am. Jour. Human Anatomy, 1968—, Biochem. Genetics, 1968—. Research on antigentic, biologic, and physico-chem. properties antibody molecules and related proteins, genetic control synthesis proteins. Home: 220 Corte Madera Av., Mill Valley, Cal. Office: U. Cal. Sch. Medicine, San Francisco 94122.*

FUERST, Harold Theodore, Am. physician; b. N.Y.C., May 21, 1909; s. Sigmund and Edith (Guenzberg) F.; A.B., Cornell U., 1929; M.D., Jefferson Med. Coll., 1933; M.P.H., Columbia, 1958; m. Edith Wiener, June 14, 1940; 1 dau., Barbara Ann (Mrs. Joel J. Beaman). With Bur. Preventable Diseases, N.Y.C., 1941-42, 46-64, chief epidemiologist, 1955-60, dir., 1960-64; asst. commr. N.Y.C. Dept. Health, 1964-67; asst. clin. prof. pediatrics N.Y. Med. Coll., N.Y.C., 1958-67, clin. prof. preventive medicine, 1964-67, prof. preventive medicine, 1968—; adj. asst. prof. epidemiology Columbia U., 1960-67; lectr. N.Y. U. Grad. Med. Sch., 1957-67. Fellow Am. Pub. Health Assn.; mem. A.M.A., N.Y. Acad. Scis., A.A.A.S., Assn. Am. Med. Colls. Contbr. articles to tech. jours. Research in community adminstrn. communicable disease and chronic disease control programs, epidemiologic studies in infectious diseases, childhood lead poisoning, systemic lupus erythematosus, hosp. epidemiology. Home: 510 E. 77th St., N.Y.C. 10021.*

FUETER, Eduard, Swiss mathematician, historian; b. Zurich, Switzerland, May 16, 1908; s. Eduard and Jenny (Weber) F.; ed. univs. Zurich, Hamburg, Munich, Geneva, Paris, London; Ph.D. in letters. Swiss delegate, Com. Scientific Research, OECD, 1958-66; dir. Central Office, U. Switzerland; proponent of and lectr. on internat. research organization and research potentials, Switzerland Inst. Internat. Relations and Research. Mem. Dutch Soc. Natural Sci., Swiss Soc. Math., History Social Sci; Henry E. Sigerist Soc. for History of Medicine and Science (chmn.). Author: Grosse schweizer Forscher, 2d edit., 1941; Geschichte der Exakten Wissenschaften in der Schweizer; Aufklärung, 1941; Forschungsorganisation und Forschungsaufwendungen in der Schweiz Zurich, 1960; Probleme des Wissenschaflichen und technischen Nachwuchses, 1960; Materialprüfung und Versuchswesen in der Schweiz und im Ausland, 1965. Editor: Rev. of U. Switzerland. Internat. specialist on sci.

recruiting. Study of history of sci., historiography, universal history, marine history; university and research organization. Address: Neugut 95, Wädenswil, Zurich, Switzerland.

FUETER, (Karl) Rudolf, mathematician; b. Basel, Switzerland, June 30, 1880; s. Eduard and Jenny (Weber) F.; student, Basel; Ph.D., Göttingen, Germany, 1903; postgrad., Paris, Vienna, London; hon. doctorate, Oslo, Norway; m. Amélie von Heusinger, Mar. 1, 1908; 1 dau. Lectr., U. Marburg (Germany), also Mining Acad. Clausthal, 1907-08; named prof. math., Basel, 1908, Tech. U. Karlsruhe (Germany), 1908; apptd. to faculty U. Zurich (Switzerland), 1916, rector, 1920-22. Mem. Swiss Math. Soc. (cofounder 1910, 1st pres.), Swiss Natural Sci. Soc. (pres. Euler Commn), Leopoldina, Bavarian Acad. Scis., Sci. Inst. Coimbra. Author: Synthetische Zahlentheorie, 1917; Vorlesungen über die singulären Moduln und die komplexe Multiplikation der elliptischen Funktionen . . . , 1924; Das matematische Werkzeug des Chemikers, Biologen und Statistikers, 1926; Analytische Geometrie der Egene und des Raumes, 1945; Editor: Comentarii Mathematici Helvetici. Research on quarternion functions, also on number theory in connection with theory of functions. Died Brunnen, Switzerland, Aug. 9, 1950.

FUHLROTT, (Johann) Carl, German paleontologist; b. Leinefelde, Germany, Dec. 31, 1803; s. Philipp and Magdalene (Nussbaum) F.; student univs. Bonn, Munster (Germany); Ph.D., U. Tübingen (Germany), 1835; m. Amalie Kellner, 1835; 6 daus. Secondary sch. tchr., Heiligenstadt, 1 year; tchr. Realgymnasium, Elberfeld (now Wuppertal-Elberfeld), Germany, from 1830; became prof., 1862. Meml. plaque placed in his honor by Soc. German Scientists and Physicians, Neanderthal, 1926. Co-founder Natural History Assn. Prussian Rhineland and Westphalia, 1843, chmn. Dusseldorf br. Author: Der fossile Mensch aus dem Neanderthal, 1865; Die Hohlen und Grotten in Rheinland und Westfalen, 1869; also articles. Studied caves in Rhineland and Westphalia; 1st to assert that bones in a cave in Neanderthal were those of a prehistoric man (Homo Neanderthalensis); his theory was confirmed after his death. Died Elberfeld, Oct. 17, 1877.

FUHN, Ion Eduard, Rumanian zoologist; b. Bucharest, Rumania, Jan. 23, 1916; s. E. Edgar and Didina (Finkels) F.; LL.B., U. Bucharest, 1937, Licentiate in Philosophy, 1938, LL.D., 1946; m. Eugenia Simian, Feb. 14, 1955; 1 dau., Iris. Tchr. philosophy high schs., 1939-44; lawyer, 1944-46; diplomatic officer, 1946-47; editor sci. periodicals, 1947-54; research worker Inst. Biology of Acad. Scis., Bucharest, 1954-58, tech. asst., 1958—. Mem. Soc. Systematic Zoology, Am. Soc. Ichthyologists and Herpetologists, Senckenbergische Naturf. Gesellschaft (corr. fellow). Author: Amphibians of Rumania, 1960; (with Stefan Vancea) Reptiles of Rumania, 1961. Research, publs. on systematics, biology and ecology of amphibians and reptiles of Rumania and Balkans, revision of genus Ablepharus (Scincidae), systematics of Rumanian spiders. Home: 12 M. Glinka. Office: 296 Splaiul Independentii, Bucharest, Rumania.*

FÜHNER, Hermann Georg, German pharmacologist, toxicologist; b. Pforzheim, Germany, Apr. 10, 1871; s. Hermann and Julie (Armbruster) F.; Ph.D., Geneva, Switzerland, 1895; postgrad. in pharmacy, Berlin; M.D., Strasbourg, Germany (now France), 1902; m. Isa Keipert, 1913; 1 son, 3 daus. Became lectr., Freiburg im Bresgau, Germany, 1907, asso. prof., 1913; prof. pharmacology, Königsberg (now Kaliningrad, USSR), from 1915; in Leipzig, Germany, 1921-24, Bonn, from 1924; ret., 1937. Mem. Leopoldina. Author: Lithotherapie, historische Studien über die medizinische Verwendung der Edelsteiner, 1902; Nachweis und Bestimmung von Giften auf biologischen Wege, 1911; Pharmcologie fur Pharmazeuten, 1937; Medizinische toxikologie, 1943. Founder, editor Sammlung von Vergiftungsfallen, from 1930. Research on varied problems of pharmacology and toxicology; developed materia medica of exptl. pharmacology. Died Bonn, Jan. 11, 1944.

FUHRMAN, Frederick Alexander, Am. physiologist; b. Coquille, Ore., Aug. 13, 1915; s. Cyrus J. and Josie (Lyons) F.; B.S., Ore. State U., 1937, M.S., 1939; postgrad. U. Freiburg, Germany, U. Wash.; Ph.D., Stanford U., 1943; m. Geraldine Jackson, Nov. 12, 1942. Faculty, Stanford Med. Center, Palo Alto, Cal., 1941—, prof. physiology, 1957-59, prof. exptl. medicine, dir. Max C. Fleischmann Labs. of Med. Scis., 1959—. Cons. in lab. design and operation; cons. tech. information div. Library of Congress, 1955-56. Guggenheim Found. fellow U. Copenhagen, 1951-52; NSF sr. fellow, 1958-59; Commonwealth Fund fellow, 1965-66. Fellow A.A.A.S., N.Y. Acad. Scis.; mem. Am. Physiol. Soc., Soc. Exptl. Biology and Medicine, Sigma Xi. Contbr. numerous articles to profl. jours., books, encys. Asso. editor Stanford Med. Bull., 1950-53, Ann. Rev. of Physiology, 1953-62. Research on effects of high and low temperatures and chem. inhibitors on metabolic processes in animal tissues, pathophysiology and therapy of frostbite, exptl. hypothermia, animal toxins especially isolation and pharmacology of tetrodotoxin from amphibians. Home: 695 Coronado Av., Stanford, Cal. 94305.*

FUJII, Isao, radiochemist; b. N.Y.C., June 7, 1926; s. Shigeru and Takeko (Ito) F.; B.S., U. Tokyo (Japan), 1952, D.Sc., 1962; m. Mihoko Yoshida, Nov. 9, 1951; 1 son, Tetsuya. Chief researcher Central Research Lab., Tokyo Shibaura Electric Co., Ltd., Kawasaki, Japan, 1965—. Mem. steering com. Symposium for Radio-Chemistry in Japan, 1962—. Recipient prize Dir. Sci. and Technics Agy., Japanese Govt., 1963. Mem. Japan Soc. for Analytical Chemistry. Research, publs. on devel. neutron activation analysis method and applications of particles accelerators to analytical problems in industries, determination of rare earth elements, activation analysis by a fast neutron. Home: 408 Kishine-Cho Kohoku-Ku Yokohama-City, Kanagawa. Office: 1 Komukai Toshiba-Cho, Kawasaki-City, Kanagawa, Japan.*

FUJII, Kenjirô, Japanese cytologist; b. Kanazawa, Japan, 1866; grad., Tokyo U., 1892, D.Sc., 1913; studied in Germany and England. Asst. prof., prof., emeritus prof., Tokyo U. Edited internat. journal, Cytologia, from 1929. Mem., Japan Acad.; awarded Culture Order, 1950. Expert on cytological heredity. Died Jan. 11, 1952.

FUJII, Ryochi, Japanese pediatrician; b. Nagoya, Japan, Feb. 13, 1917; s. Seiei and Uno (Ikeda) F.; M.D., U. Tokyo (Japan), 1940, D.M.Sc., 1948; m. Shizuko Fukami, Nov. 23, 1950; 1 child, Yoshihiko. With Naval Inst. for Submarine Hygiene, 1943-45; staff Tokyo U. Hosp., 1945-54, lectr., 1954-55; faculty U. Tokyo, 1956—, head, dept. pediatrics, asso. prof. Tokyo U. Br. Hosp., 1956—; mem. vis. staff U. Cal., San Francisco Med. Center, 1959-60. Mem. com. Japanese Ministry of Welfare, 1956—. Mem. Japanese Med. Assn., Japan Soc. Pediatrics, Japan Soc. Chemotherapy, Soc. Japanese Virologists, others. Author: (with others) Text Book of Pediatrics, 1960, A Guide to Antibiotic Therapy, 1963; others. Chief editor Chemotherapy; editor Jour. Antibiotics, Japanese Jour. Pediatrics. Research, publs. on 1st case report on primary atypical pneumonia in Japan, 1948; etiological analysis for viral colds among children, 1956-65; measles and varicella studies with fluorescent antibody technique; 1st isolation of human cytomegalovirus in Japan, 1965; inventor micromethod for determining antibiotic concentration. Home: 33-2 2 chome, Higashinakano, Nakanoku, Tokyo. Office: 120 Zoshigaya, Bunkyoku, Tokyo, Japan.*

FUJIMORI, Bun-ichi, Japanese neurophysiologist; b. Matsumoto City, Japan, Nov. 23, 1910; s. Tomoji and Shige (Tanaka) F.; M.D., Tokyo U., 1935; D.Med. Scis., Tohoku U., 1943; m. Noriko Tanaka, Nov. 5, 1937. Chief, Inst. for Aviation Medicine, Naval Inst., 1943-45; chief clin. labs. 2d Tokyo Nat. Hosp., 1945-55; chmn. dept. physiology Hokkaido U. Sch. Medicine, Sapporo, 1955—. Mem. Internat. Brain Research Orgn. Author: Electroencephalography (with T. Tokizane), 1955; (with S. Itoh) Textbook of Physiology, 1964; also numerous articles. Editor: Physiology of Motor System, 1966. Research on galvanic skin response and autonomic nervous system, analysis of electroencephalogram and steady potential of brain, functions of muscle spindle and neuronal mechanisms of spinal reflexes. Home: Kita-28, Kigas hi-3, Sapporo, Japan.*

FUJIMOTO, Haruyoshi, Japanese geologist, paleontologist; b. Nara, Japan, 1897; grad. Tohoku U., 1924, Ph.D., 1938. Became instr. Tokyo Higher Normal Sch., later prof.; instr. Tokyo U., from 1928, apptd. prof., 1942; prof. Tokyo Edn. U., since 1948; chmn. Japan Phys. Geography Research Soc. Recipient prize in sci. research Ministry Edn. Author: Research on the Stratum in Japan's Palaezoic Period; Theory of Geology; The Geological Record of Japanese Localities. Research, articles on palazoic stratum of Japan, also on basic rock of Hida area in central Japan.

FUJIMOTO, James Masao, Am. pharmacologist; b. Vacaville, Cal., May 10, 1928; s. Tsuneki and Nui (Kimura) F.; A.B., U. Cal. at Berkeley, 1951, Ph.D., 1956. USPHS fellow U. Cal., San Francisco, 1954-56; faculty dept. pharmacology Tulane Med. Sch., New Orleans, 1956—, prof., 1966—. Mem. Am. Soc. for Pharmacology and Exptl. Therapeutics, Soc. Toxicology, Soc. Exptl. Biology and Medicine, A.A.A.S. Research, publs. on metabolism drugs, increases in liver function produced by drugs, inhibition drug metabolism, mechanism devel. tolerance to narcotic drugs. Home: 1483 Granada Dr., New Orleans 70122.*

FUJINAGA, Taicthiro, Japanese chemist; b. Kobe, Japan, Jan. 22, 1919; s. Taiti and Ikuko F.; grad. Kyoto (Japan) Imperial U., 1941; D.Sc., 1956; m. Kieko Fujinaga, May 20, 1946; 1 son, Kaoru. With Kyoto U. (formerly Kyoto Imperial U.), 1941, 46—, prof. dept. chemistry Faculty Sci., 1960—; research fellow U. Minn., 1956-57. Recipient prize Kinki Soc. Indsl. Chemistry, 1953. Mem. Chem. Soc. Japan (exec. com. 1965—), Am. Chem. Soc., Japan Soc. for Analytical Chemistry (prize 1963), Internat. Union Pure and Applied Chemistry (titular mem. analytical chemistry div.). Contbg. author: Progress in Polarography, 1962. Research and numerous publs. on devel. new instrumental methods in electro-analytical chemistry including sq.-wave polarography, current-scan polarography, electrolytic chromatography, new methodologies in related fields. Home: 26, Tanaka-sekidencho, Sakyo-ku, Kyoto, Japan.

FUJINAMI, Kan, Japanese physician; b. Aichi Prefecture, Japan, 1870; s. Mantoku Fujinami; M.D. Prof. Kyoto (Japan) U. Recipient Imperial Acad. prize, 1918. Noted for discovering cause of endemic disease by means of post-mortem exam. of victim of Japanese parasitic disease (parasitic organisms found in liver and their ova in uterus of corpse). Died 1934.

FUJINO, Tsunesaburo, Japanese med. bacteriologist; b. Fukui Prefecture, Japan, Jan. 28, 1907; s. Kotaku and Towa (Iwasaki) F.; grad. Faculty Medicine, Osaka (Japan) U., 1931, D.Med. Sci., 1936; m. Reiko Mori, Oct. 1, 1946; 1 dau., Michiko. Researcher dept. bacteriology Faculty Medicine, Osaka U., 1931-34, asst., 1934-36, faculty, 1936—, prof. bacteriology and serology Research Inst. for Microbial Diseases, 1948—, dir. Research Inst. for Microbial Diseases, 1955-58, univ. councilor, 1955—; lectr. Nagoya U., 1954-60. Recipient Asahi award for discovery new food poisoning bacteria Vibrio parahaemolyticus Asahi Press Co., 1965. Mem. Japan Soc. Bacteriologists (Asakawa award 1966), Japanese Soc. Virology, Japanese Soc. Med. Mycology, Japanese Soc. Infectious Diseases. Author (with Fukumi) Vibrio parahaemolyticus, 1963, Vibrio parahaemolyticus; also numerous articles. Research on pox virus, 1935, exptl. chemotherapy for monkey malaria, 1942-43, mode of action of sulfonamides, 1942-50, pathogenic fungi, 1948-63, food poisoning bacteria (Vibrio parahaemolyticus), 1950—. Home: 70 Sakura, Minoo, Osaka. Office: Yamada-kami, Suita, Osaka, Japan.*

FUJISE, Shin-ichiro, Japanese biochemist; b. Singapore, 1899; grad. Tohoku U., 1923, D.Sc., 1929; studied in Germany. Asst. researcher, Rikagaku Kenkyusho, 1926; prof., Tohoku U., from 1932. Mem., Japan Academic Council; awarded Majima prize, Japan Chem. Acad. Specialist in organic biochemistry; research on flavanon; methods of measuring melting points of small amounts of materials.

FUJITA, Francisco Eiichi, physicist; b. Hong Kong, Oct. 4, 1925; s. Seishichi and Akiko (Shimizu) F.; student Naval Acad. Japan, 1943-45; Dr. Sci., Tokyo U., 1956; m. Yoko Nagaike, Apr. 29, 1954; children—Sayuri, Masumi, Kiyoshi. Asst., Inst. for Iron and Steel, Tohoku U., Sendai, Japan, 1950-56; research asso. Japan Atomic Energy Research Inst., Tokai, 1956-63; prof. physics dept. Faculty Engring. Sci., Osaka (Japan) U., 1963—; vis. physicist U. Ill., 1956-57. Vis. physics asso. Brookhaven (N.Y.) Nat. Lab., 1959-62. Mem. Japan Inst. Metals (research paper award 1954, Merit award 1963), Phys. Soc. Japan, Japan Atomic Energy Soc., Electron Microscope Soc. Japan, Am. Phys. Soc. Author: (with others) Dislocation Theory and its Application to Metals, 1957. Research, publs. in fields of dislocation, fracture, radiation damage, alloy phase diagrams, precipitation, others; among which are known such as Fujita model for fracture, a new kinetic theory of precipitation, others. Home: 18-5 Taninoshita, Nakayama, Takarazuka, Japan. Office: Faculty Engring. Sci., Osaka U., Toyonaka, Osaka, Japan.*

FUJITA, Shigeji, physicist; b. Oita City, Japan, May 15, 1929; s. Shigeto and Makio (Eyama) F.; B.S., Kyushu U., 1953; Ph.D., U. Md., 1960; m. Sachiko Fujise, Sept. 30, 1958; children—Michio, Isao, Yoshiko. Research asso. Northwestern U., 1958-60; sr. research asso. Universite Libre de Bruxelles, 1960-64; vis. asso. prof. physics Pa. State U., 1964-65, U. Ore., 1965-66; asso. prof. physics State U. N.Y., Buffalo, 1966—. Mem. Am. Phys. Soc., Am. Assn. U. Profs., Phys. Soc. Japan. Author: Introduction to Non-Equilibrium Quantum Statistical Mechanics, 1966. Research on kinetic theory of gases and plasmas, solid-state theory and polymer physics; developed statis. mech. theory which treats transport and optical properties of matter on rigorous basis. Home: 74 Mahogany Dr., Williamsville, N.Y. 14221. Office: State U. N.Y., Buffalo.*

FUJITA, Yoshio, Japanese astronomer; b. Fukui Prefecture, Japan, 1908; grad. Tokyo U., 1931, D.Sc. Asst. engr., Tokyo Astronomical Observatory; instr., Tokyo Women's Higher Normal School; asst. prof., prof., Tokyo U. Author: The Astronomical Spectroscopy; Development of Astronomical Physics. Research on long periodic stars; spectroscopy of low temperature stars and sun's atmosphere.

FUJIWARA, Shizuo, Japanese chemist; b. Urawa, Japan, July 17, 1920; s. Seisaku and Miki (Iwasaki) F.; B.S., U. Tokyo, 1944, Ph.D., 1949; postdoctoral work U. Ill., 1953-55; m. Miyoko Kobori, Oct. 10, 1950; children—Kitao, Makio, Kikuo. Asst. prof. U. Electrocommunications, Chofu, 1949-55, prof., 1956-60; prof. U. Tokyo, 1960—; vis. prof. Ryukyu U., 1965, Temple U., Phila., 1966. Mem. Japanese (Shimpo Sho 1950), Am. chem. socs., Phys. Soc. Japan. Author: High Resolution NMR, 1962. Contbr. numerous articles to sci. jours. Investigation of separation and determination of chemically similar elements; applied results of chem. microanalysis to chem. prospecting of metal mines; magnetic resonance as one of pioneering workers in this field. Home: 364, Negishi, Urawa, Japan. Office: Dept. Chemistry, Faculty Sci., U. Tokyo, Tokyo, Japan.*

FUJIWARA, Takeo, Japanese physicist; b. 1897; grad., Kyoto U., 1924; D.Sc., 1930; studied in Europe and Am. Mem., Tóshiba Research Institute; prof., Hiroshima Bunrika U., 1933; dir., Sci. Dept., Hiroshima U., 1949—. Mem., Japan Academic Council; awarded prize of Japan Academy, 1948. Author: Researches on the Single Crystal of Aluminium and Iron, and Its X-ray Analysis; numerous papers. Specialist in experimental physics; research on single crystal of iron, tungsten, aluminium.

FUKUDA, Riken, Japanese mathematician; b. Osaka, Japan, 1815; tutor Tsuchimikado family of court nobles; tchr. sci. of Chinese calendar, circa 1858; with Bur. Astronomy, after 1868; founder, tchr. Junten Kyugosha (sch.); author numerous books on calendar sci. Died 1889.

FUKUDA, Tokuzo, Japanese economist; b. Tokyo, Japan, 1874; grad. Tokyo Higher Commerce Sch., 1894; student Leipzig, Munich (Germany) univs., 1898-1900; Ph.D., 1900. Lectr. econs. Tokyo Higher Commerce Sch. (later Tokyo U. Commerce), from 1901. Mem. 6th Internat. Acad. Conf., Brussels, Belgium, 1925. Mem. Japanese, French acads. Pioneer econ. sci. in Japan, introduced Marxism to Japan. Died 1930.

FUKUI, Saburo, Japanese chemist; b. Siga Prefecture, Japan, July 31, 1919; s. Yahei and Sada (Sichiri) F.; Dr.Engring. Kyoto U., 1953; m. Michiru Odajima, Oct. 17, 1943; 1 son, Yasukuni. Lectr. dept. indsl. chemistry Faculty Engring. Kyoto U., 1947, asst. prof., 1948-54, prof. indsl. biochemistry, 1961—; prof. Himeji Inst. Tech., 1954-61. Recipient Japan Vitamin Soc. prize, 1958; Japan Fermentation Technol. prize, 1958. Mem. Japan Soc. Fermentation tech. (exec. mgr. 1958), Japan Vitamin Soc. (sec. 1961), Japan Chem. Soc. (mem. council 1958-60), Japan Biochem. Soc. (mem. council 1966), Am. Chem. Soc., Am. Soc. Microbiology, Inst. Food Technologists, Japan Agrl. Chem. Soc. Author: Vitamins, 1957; Flavor Enhancer, 1966; (with Y. Hasitani) Yeast, 1967; also numerous articles. Research on chemistry and biochem. functions of several vitamins; nutrients and metabolism of fermentation microorganisms; improvement of Japanese fermentation technology. Home: Kyoto U. Lodge No. 24, Kumano Marutamachi, Sakyo, Kyoto, Japan. Office: Yosida, Sakyo, Kyoto, Japan.*

FUKUMOTO, Tsugio, Japanese chemist; b. Kagoshima City, Japan, Jan. 10, 1925; s. Saichi and Sumi (Fukuiri) F.; B.Sc., Osaka Sci. and Engring. U., 1948; Ph.D., Osaka U., 1960; m. Reiko Nanko, Jan. 20, 1952; children—Nobumi, Noriko. Asst., Sci., Indsl. Research Osaka U., 1948-60, asst. prof., 1960—, chief, synthetic chemistry Radiation Center Osaka Prefecture, 1960—. Studies, publs., contbns. to books on mechanisms of organic reaction; synthesis of amino acid; radiation chemistry including graft polymerization. Patentee field synthesis of Lysin. Home: No. 20, Fudai-takusha, 23, Ohonoshiba-cho, Osaka. Office: 704, Shinke-cho, Sakai, Osaka, Japan.*

FUKUOKA, Jiro, Japanese phys. oceanographer; b. Kanazawa City, Japan, Dec. 3, 1922; s. Kiyo and Mizuho (Kanauchi) F.; M.A., Tokyo (Japan) Imperial U., 1945, D.Sc., 1961; m. Chizuko Hirai, May 16, 1948; children—Yoshimasa, Kei. With Central Meteorol. Obs., Japan, 1946-56, chief hydrographical br., oceanographical sect., 1951-56; with Meteorol. Research Inst., Japan, 1956-62, chief oceanographer, 1959-62; with Fundacion La Salle, Venezuela, 1962-64, Japan, 1964-65; chief dept. phys. oceanography U. Orient, Cumana, Venezuela, 1965—; prof. oceanography Coll. Meteorology, Japan, 1964—. Mem. Oceanographical Soc. Japan (past editor), Japan Meteorol. Soc. Research, publs. on mechanism of appearance of large cold water mass in sea south of Japan; discovered that this appearance has close relation with variations of current speed of Kuroshio and bottom topography. Home: 208 Seijo Setagaya, Tokyo, Japan. Office: U. Orient, Cumana, Venezuela.*

FUKUSHIMA, Naoshi, Japanese geophysicist; b. Tokyo, Japan, Jan. 19, 1925; s. Shiro and Shizu (Yoshimoto) F.; D.Sc., U. Tokyo, 1953; m. Midori Okuno, Dec. 30, 1956; children—Hitoshi, Eri. Faculty, U. Tokyo, 1952—, prof. geophysics research lab., 1965—; research asso. Geophys. Inst., U. Göttingen (Germany), 1954-55, theoretical div. NASA, Washington, 1960-61. Sec. ionosphere research com. Sci. Council Japan, 1954—, acting sec. nat. com. for IGY, 1955-60. Mem. Soc. Terrestrial Magnetism and Electricity Japan (Tanakadate prize), Phys. Soc. Japan, Am. Geophys. Union. Research, publs. on time-variation in geomagnetic field, geomagnetic disturbance; formulated a conception on phys. mechanism of polar magnetic storms. Home: Wakabacho 1-1, Chofu-shi, Tokyo.*

FUKUTO, Tetsuo Roy, Am. chemist; b. Los Angeles, Dec. 15, 1923; s. Morito and Yaeno (Tanaka) F.; B.S., U. Minn., 1946; Ph.D., U. Cal., Los Angeles, 1950; m. Sumi Nakaki, Nov. 1, 1953; children—Jay Robert, Jon Morio, Margaret Mia, Mary Elizabeth. Postdoctoral research asst. U. Ill., 1950-51; research chemist Aerojet Engring. Corp., Azusa, Cal., 1951-52; prof. entomology and chemistry, insect toxicologist U. Cal., Riverside, 1952—; cons. toxicology study sect. USPHS, 1962-66, Stauffer Chem. Co., 1966—. Mem. Am. Chem. Soc., Entomol. Soc. Am. Contbr. numerous articles to profl. jours. Research in mode of action of organophosphorus esters, mode of action of insecticides, metabolism of insecticides, structure-activity relations in insecticides. Home: 144 E. Broadbent Dr., Riverside, Cal. 92507.*

FUKUZUMI, Kazuo, Japanese chemist; b. Nagoya, Japan, Dec. 14, 1915; s. Hideichi and Mitsuru (Yoshikawa) F.; B.Engring. Tohoku Imperial U., 1940; Dr.Engring. Nagoya U., 1958; m. Keiko Yoshimura, Nov. 30, 1943; children—Michiko, Shunichi. Lectr., Tohoku Imperial U., 1941; research officer Nagoya br. Govt. Tokyo Indsl. Chem. Research Inst., 1944-51; asst. prof. Nagoya U., 1951-66, prof., 1966—. Mem. Am. Oil Chemists Soc., Chem. Soc. Japan, Japan Oil Chemists Soc. Contbr. numerous articles to sci. jours. Contbns. to fat chemistry, new and execellent fatsplitting agts., new mechanism of fat splitting, important information about autoxidation and hydrogenation of fats, cause of atherosclerosis and cancer generation from viewpoint of fat chemistry. Home: 11 Nunoike-cho, Higashi-ku, Nagoya. Office: Nagoya U., Furo-cho, Chikusa-ku, Nagoya, Japan.*

FULFORD, Margaret Hannah, Am. botanist; b. Cin., June 14, 1904; d. Alfred Thompson and Lottie (Holloway) Fulford; B.A., U. Cin., 1926, B.E., 1927, M.A., 1928; Ph.D., Yale, 1935. Faculty, U. Cin., 1927—, prof., 1954—, fellow Grad. Sch., 1957; mem. staff U. Mich. Biol. Sta., 1947-53. Guggenheim fellow, 1951. Fellow A.A.A.S., Ind., Ohio acads. sci.; mem. Am. (curator Hepaticae 1933—), Brit. bryological socs., Internat. Bot. Congress (chmn. spl. com. for bryophytes internat. commn. for nomenclature 1950, 54, 59, 64, hon. v.p. bryological sect. 1954), Internat. Assn. Plant Taxonomists, Bot. Soc. Am., Am. Soc. Plant Taxonomists, Internat. Soc. Plant Morphologists, Soc. for Study Evolution, A.A.A.S., Am. Inst. Biol. Scis., Sigma Xi. Author: The Genus Bazzania in Central and South America, 1946; Ann. Cryptogamica et Phytopathologici, 1946; Manual of the Leafy Hepaticae of Latin America I-III; also numerous articles. Asso. editor Bryologist, 1947—. Research on structure, methods reproduction, morphogenesis, taxonomy, distbn. and significance liverworts. Home: 1372 Sutton Av., Mt. Washington, Cin. 45230.*

FULKERSON, Samuel Cole, Am. clin. psychologist; b. Cedar Rapids, Ia., Jan. 5, 1923; s. Samuel Cole and Ruth (DeSilva) F.; B.A., State U. Ia., 1948, M.A., 1951; Ph.D., Tex. U., 1955; m. Katharine Alger James, May 8, 1947; children—Gregory, Heidi. Psychologist, USAF Sch. Aviation Medicine, Randolph Field, Tex., 1954-58; asst. prof. U. Pitts. Med. Sch., 1958-61; asso. prof. U. Louisville, 1961-66, prof., 1966—, acting head, 1964-65, dir. clin. tng. program, 1961—. Diplomate Am. Bd. Examiners in Profl. Psychology. Fellow Am. Psychol. Assn.; mem. Ky. Psychol. Assn., Louisville Psychol. Soc., Am. Assn. U. Profs., Sigma Xi. Research, publs. on test constrn. and devel., cognition, psychiat. prognosis, drug effects. Home: Route 2, Jeffersontown, Ky. 40299. Office: U. Louisville, Louisville 40208.*

FULKS, Watson B., Am. mathematician; b. White County, Ark., Jan. 24, 1919; s. Bryan and Nora (Langford) F.; B.S., Ark. State Tchrs. Coll., 1940; M.S., U. Ark., 1941; postgrad. Brown U.; Ph.D., U. Minn., 1949; m. Gloria Besser, Aug. 7, 1943; children—Lyle Bryran, Wayne Mahlon. Research asso. Cal. Inst. Tech., 1949-50; asst. prof. U. Minn., 1950-56, asso. prof., 1956-60; prof. Ore. State U., Corvallis, 1960-63; prof., U. Colo., Boulder, 1963—. Author: Advanced Calculus, 1961; also articles. Home: 540 Pleasant St., Boulder, Colo. 80302.*

FULLER, Calvin Souther, Am. chemist; b. Chgo., May 25, 1902; s. Julius Quincy and Bessie (Souther) F.; B.S., U. Chgo., 1926, Ph.D., 1929; m. Willmine Works, Sept. 17, 1932; children—Robert W., Stephen S., John W. Chemist, Gen. Chem. Co., Chgo., 1920-22, Chgo. Tribune, 1924-30; phys. chemist Bell Tel. Labs., N.Y.C., 1930-42, Murray Hill, N.J., 1944-67; chief polymer research Office of Rubber Dir., Washington, 1942-44. Adviser Research and Devel. Bd., Washington, 1945-49. Recipient John Scott medal, 1956, John Price Wetherill medal, 1963. Fellow A.A.A.S. (chmn. Gordon confs. 1945), Am. Chem. Soc. (chmn. plastics div. 1944), Sigma Xi. Research, numerous publs. on x-ray structure of polymers and synthetic rubber, polymer synthesis, properties of semiconductors, diffusion in solids, solid state chemistry; inventions relating to polymeric and semiconductor materials; co-inventor Bell Solar Battery. Home: 2916 Eagle Trail, Vero Beach, Fla. 32960.*

FULLER, F. Brock, Am. mathematician; b. Eugene, Ore., July 8, 1927; s. Lon Luvois and Florence (Thompson) F.; A.B., Princeton, 1949, M.A., 1950, Ph.D., 1952; m. Alison Clark, July 4, 1957; 1 dau., Lynn Dorcas. Instr., Princeton, 1951-52; faculty Cal. Inst. Tech., 1952—, prof., 1966—. Mem. Am. Math. Soc. Contbr. articles to sci. jours. Research on theory of fixed points and periodic orbits of ordinary differential equations. Home: 1959 Meadowbrook Rd., Altadena, Cal. 91001. Office: Math. Dept., Cal. Inst. Tech., Pasadena, Cal. 91109.*

FULLER, John Langworthy, Am. biologist; b. Brandon, Vt., July 22, 1910; s. John H. and Joyce (Langworthy) F.; B.S., Bates Coll., 1931; Ph.D., Mass. Inst. Tech., 1935; m. Ruth I. Parsons, Sept. 2, 1933; children—Mary Jean (Mrs. Richard C. Farrington), Sarah Ann. Faculty, Sarah Lawrence Coll., 1935-36, Clark U., 1936-37, U. Me., 1937-47; staff Jackson Lab., Bar Harbor, Me., 1947—, sr. staff scientist, 1959—, asso. dir., 1963—. Vis. lectr. Harvard Med. Sch., 1964-65. Guggenheim fellow, 1955-56. Mem. Genetics Soc. Am., Am. Soc. Zoologists, Animal Behavior Soc., Am. Psychol. Assn. Author: Nature and Nurture, 1954; (with W. R. Thompson) Behavior Genetics, 1960; Motivation, 1962; (with J. P. Scott) Genetics and Social Behavior of the Dog., 1965; also numerous articles. Research on relationship between inherited characters of mammals and their behavior, early experience; discovered that deleterious effects psychol. deprivation are function emotional disruption rather than failure maturation. Home: Hulls Cove, Me. 04644. Office: Jackson Lab., Bar Harbor, Me. 04609.*

FULLER, R(ufus) Clinton, Am. microbiologist; b. Providence, Mar. 5, 1925; s. Rufus C. and Alice (Anthony) F.; B.A., Brown U., 1945; M.A., Amherst Coll., 1948; Ph.D., Stanford, 1952; M.A., Dartmouth, 1961; m. Carol June Seager, Sept. 14, 1946; children—David Cushman, Katherine Anne, Marilyn Gail, Jonathan Ames. Research microbiologist U. Cal. Lawrence Radiation Lab., 1952-55; asso. plant physiologist Brookhaven Nat. Lab., 1955-57, plant biochemist, 1957-59; NSF Sr. Postdoctoral fellow Oxford U., 1959-60; prof. microbiology Med. Sch. Dartmouth, 1960-66, chmn. dept., 1960-65; vis. prof. dept. life scis. U. Cal., Riverside, 1966; dir. U. Tenn.-Oak Ridge Grad. Sch. Biomed. Scis., 1966—. Cons. U. S. Surgeon Gen.; mem. com. Nat. Inst. Gen. Med. Scis., 1952-65, mem. cell biology study sect. div. research grants NIH, 1965—. Mem. Am. Soc. Biol. Chemists, Am. Soc. Microbiology, Soc. Gen. Physiologists, A.A.A.S., Am. Soc. Plant Physiologists, Am. Assn. U. Profs., Am. Soc. Cell Biology, Sigma Xi. Research, numerous publs. in comparative aspects of study of photosynthesis, understanding of simplest organisms and cellular systems that can convert light to chem. energy. Home: 105 Wiltshire Dr. Office: Oak Ridge Nat. Lab., Biology Div., P.O. Box Y, Oak Ridge 37830.*

FULLER, Richard Buckminster, Am. engr.; b. Milton, Mass., July 12, 1895; s. Richard Buckminster and Caroline Wolcott (Andrews) F.; student Milton (Mass.) Acad., 1904-13, Harvard, 1913-15, U. S. Naval Acad., 1917; Dr. Design, U. N.C., 1954; Dr. Arts, Mich. U., 1955; D.Sc., Washington U., 1957; Dr. Arts, So. Ill. U., 1959; H.H.D., Rollins Coll., 1960; Litt. D., Clemson U., 1964; A.F.D., N.M. University, 1964; Doctor of Science, Colo. Univ., 1964; m. Ann Hewlett, July 12, 1917; children—Alexandra Willets (dec.), Allegra (Mrs. Robert Snyder). Apprentice machine fitter Richards, Atkinson & Haserick, cotton mill machinery importers, Boston, 1914; in various apprentice positions Armour & Co., N.Y.C., 1915-17, asst. export mgr., 1919-21; nat. account sales mgr. Kelly-Springfield Truck Co., 1922; pres. Stockade Bldg. System, 1922-27; founder 4-D Co., Chgo., 1927, pres., 1927-32; asst. dir. research Pierce Found.-Am. Radiator-Standard San. Mfg. Co., 1930; founder Dymaxion Corp., Bridgeport, Conn., 1932, dir., chief engr., 1932-35; asst. to dir. research and devel. Phelps Dodge Corp., 1936-38; tech. cons. Fortune Mag., 1938-40; v.p., chief engr. Dymaxion Co., Inc., Del., 1941-42; chief mech. engring. sect. Bd. Econ. Warfare, 1942-44; spl. asst. to dir. Fgn. Econ. Adminstrn., 1942-44; chmn. bd. adminstrv. engr. Dymaxion Dwelling Machines, 1944-46; chmn. bd. trustees Fuller Research Found., Wichita, Kan., 1946—; pres. Geodesics, Inc., Raleigh, N.C., 1954—, Plydomes, Inc., Des Moines, Iowa, 1957—; chmn. bd. Tetrahelix Corp., Hamilton, O., Buckminster Fuller Inst., Carbondale; cons. Ford Found., Calcutta (India) Planning Orgn., 1961—; cons. space age devel. to gov. N.C., 1962—; prof. generalized design sci. exploration So. Ill. U., Carbondale, 1959—; Charles Eliot Norton prof. Harvard, 1961-62; vis. prof., lectr., critic Cornell U., Yale, Mich. U., Mass. Inst. Tech., Princeton, U. Minn., Washington U., U. Cal. at Berkeley, U. Ill., others; numerous lectrs., exhibits at major instns. throughout world. Recipient award of merit N.Y. chpt. A.I.A., 1952, USMC, 1954; Gran Premio, Trienniale de Milano, 1954, 57 (Italy); Centennial award Mich. State U., 1955; Gold medal scarab Nat. Archtl. Soc., 1958; Gold medal Phila. chpt. A.I.A., 1960; Frank P. Brown medal, Franklin Inst., 1960; Allied Professions Gold medal A.I.A., 1963; honored with Buckminster Fuller Recognition Day, U. Colo., 1963; Benjamin Franklin life fellow Royal Soc. Arts, 1960 (Eng.); Plomado de Oro award Soc. Mexican Architecture, 1963. Life fellow A.A.A.S.; fellow Inst. of Gen. Semantics (hon. trustee); mem. A.I.A. (hon. life), Nat. Inst. Arts and Letters (life mem.), Harvard Engring. Soc., Am. Soc. Profl. Geographers, Archtl. League N.Y., Am. Assn. U. Profs., Soc. Archtl. Historians. Author: Nine Chains to the Moon, 1938; No More Second Hand God, 1962; Education Automation, 1962; (poetry) Unfinished Epic of Industrialization, 1963; Charles Eliot Norton 1961-62 Lectures at Harvard University, 1963; Ideas and Integrities, 1963; (co-author) New Worlds in Engineering, 1940. Author, editor several mags. Inventor Dymaxion House, Chgo., 1927, Dymaxion 3 wheeled automobile, 1932-35; designer Dymaxion Deployment Unit steel igloo Kansas City, 1940; patentee Dymaxion World Map, 1943; designer geodesic structure for Ford Motor Company Rotunda Dome, Dearborn, Mich.; USMC advance base shelters, Dept. Commerce Internat. Trade Fair Dome Pavillions, 1956—, all D.E.W. Line Radomes, USN Geodesic Storage Domes, Antarctica, 1956, Union Tank Car repair shop domes, Baton Rouge, La., also Wood River, Ill., 1958-59; designer Kaiser aluminum domes, Hawaii, 1957, Virginia Beach, 1958, Oklahoma City, Borger, Tex., Abilene, Kan.,

Okla. Bank 1959; designer Golden Dome for Am. exhibit, Moscow, USSR, 1959, Geodesic Dome nat. hdqrs. Am. Soc. Metals, Climatron for Mo. Bot. Gardens, St. Louis, 1960; domes for U. S. Sci. Halls, Boeing Co. Spaceorama Dome, Ford Motor Dome, Encyclopedia Britannica Dome (all at Seattle World's Fair 1962), Sports Palace, Paris, France, 1960, Cinerama Theatre, Hollywood, 1963, Yomiuri Star dome, Tokyo, Japan, 1964, also dome for N.Y. World's Fair Pavilion, 1964. Inventor-discoverer energetic-synergetic geometry, geodesic structures, tensegrity structures. Address: 407 Forest St., Carbondale, Ill.

FULLMER, Harold Milton, Am. oral pathologist, histochemist; b. Gary, Ind., July 9, 1918; s. Howard and Rachael Eva (Tiedge) F.; B.S., Ind. U., 1942, D.D.S., 1944; m. Marjorie Lucile Engel, Dec. 31, 1942; children—Angela Sue, Pamela Rose. Prin. investigator Nat. Inst. Dental Research NIH, Bethesda, Md., 1953-66, chief sect. on histochemistry, 1966—; profl. lectr. dept. anatomy Georgetown U., 1964—; mem. dental study sect. USPHS, 1966—. Recipient Fulbright award U. Adelaide, 1962. Diplomate Am. Bd. Oral Pathology. Fellow Am. Coll. Dentists, A.A.-A.S.; mem. Internat. Assn. Pathologists, Am. Acad. Oral Pathology, Internat. Assn. Dental Research, Histochem. Soc., Am. Soc. Cell Biology, Am. Coll. Dentists, Biol. Stain Commn., Am. Dental Assn., Omicron Kappa Upsilon. Research, numerous publs. on pathology, histochemistry, biochemistry of connective tissues, bones, teeth; discovered oxytalan connective tissue fiber, connective tissue disorder in amyotrophic lateral sclerosis, enzyme collagenase in tissues of man, relationship between detectable collagenase and certain neuromuscular diseases. Home: 6333 Lenox Rd., Bethesda 20034. Office: NIH, Bethesda, Md. 20014.*

FULMER, Clyde Benson, Am. physicist; b. nr. Valdosta, Ga., Nov. 7, 1924; s. Michael McDuffie and Lenora (West) F.; student S. Ga. Coll., 1941-42; B.S., Berry Coll., 1948; M.S., Emory U., 1949; Ph.D., N.C. State U., 1956; m. Frieda Clarke, Mar. 20, 1954; children—Linda Faye, Clyde Benson. Asst. prof. physics Jacksonville (Ala.) State Coll., 1949-51; group leader Guided Missile Center, Redstone Arsenal, Huntsville, Ala., 1951-52; instr. physics N.C. State U., Raleigh, 1952-56; physicist Oak Ridge Nat. Lab., 1956—. Mem. Am. Phys. Soc., Am. Assn. Physics Tchrs., Sigma Xi, Sigma Pi Sigma, Tau Beta Pi. Research, publs. on equilibrium charge of fission fragments in gases, mechanism of (p,a) reactions; measurement of range-energy of fission fragments in various materials, proton elastic scattering; estimates of residual radiation levels due to induced radioactivity in charged particle accelerators, nuclear reactions. Home: 105 Mason Lane, Oak Ridge 37830. Office: P.O. Box X, Oak Ridge 37830.*

FULMER, Ellis Ingham, Am. phys. chemist; b. Gibbon, Neb., Apr. 12, 1891; s. Clark Adelbert and Evalena Anna (Ingham) F.; A.B., Neb. Wesleyan U., 1912; A.M., U. Neb., 1913; postgrad. U. Pa., 1913-14; Ph.D., U. Toronto, 1919; D.Sc., Neb. Wesleyan U., 1944; m. Ruth Emma Files, June 15, 1915; children—Norman Clark, Robert Ellery. Chemist Agr. Expt. Sta., U. Neb., summers 1913-14; in charge chemistry dept. Leander Clark Coll., Toledo, Ia., 1914-15, Friends U., Wichita, Kan., 1915-16; chemist Aetna Explosives Co., 1917; chemist, expt. farms, Ottawa, Can., 1918; asst. prof. biophys. chemistry Ia. State Coll. Agr. and Mechanic Arts, 1919-20, asso. prof., 1920-23, prof., from 1923, asst. to dir. Inst. Atomic Research, from 1947. Fellow A.A.A.S., mem. Am. Chem. Soc., Sigma Xi, Phi Kappa Phi. Died Feb. 10, 1953.

FULMER, John Leonard, Am. economist; b. Little Mountain, S.C., Sept. 24, 1911; s. Charlie Clifton and Lillian (Sease) F.; B.S., Clemson Coll., 1933; M.S., Cornell U., 1937; Ph.D., U. Va., 1939; postgrad. Am. U., Va. Poly. Inst.; m. Mary Adelaide Wise, June 30, 1936; children—June Wise (Mrs. Clyde M. Fortson, Jr.), John Leonard. Faculty, Clemson U., 1934-37, U. Va., 1939-42, 45-51; bus. economist OPS, 1951-53; faculty Emory U., 1953-56; prof. Ga. Inst. Tech., 1956-64, 68—; professor, director Bur. Bus. Research University of Kentucky, Lexington, 1964-67. Consultant to Ga. Dept. of Labor, 1957-64, Ga. Dept. Agr., 1961-63, U. S. Dept. Labor, 1961-66. Mem. Am., So. econ. assns., Am. Statis. Assn. Author: Agricultural Progress in the Cotton Belt since 1920, 1950; Regional Income, Studies in Income and Wealth, 1957; Essays in Southern Economic Development, 1964. Research in field of econ. devel., population growth of areas is function of job creation, depressed areas export people but import remittances from relatives who migrated, employment multiplier increases with urban ratio and size of urban center. Home: 1064 N. Jamestown Rd., Decatur, Ga. 30033. Office: Sch. Indsl. Mgmt., Ga. Inst. Tech., Atlanta 30332.*

FULTON, Forrest, English bacteriologist; b. Aug. 13, 1913; s. Leonard Jessupp F.; ed. Pembroke College, Oxford, B.A., 1935, M.A., 1939, D.M., 1945; London Hospital, B.M., B.Ch., 1939. Emergency staff, Public Health Service, Oxford, 1939; worked for Med. Research Council, London, 1942; taught Yale U.; reader in bacteriology and immunology, London U., 1945; prof. virology, London School of Hygiene and Tropical Medicine, 1959—. Author papers to various journals and Vol. V, Advances in Virus Research. Research on typhus vaccines; microbiology, immunology; cultivated causative agent of scrub fever in lungs of rodents; with L. Joyner, developed vaccine for scrub fever, 1945. Address: London School of Hygiene and Tropical Medicine, Keppel Street, London W.C. 1, England.

FULTON, George Pearman, Jr., Am. physiologist; b. Milton, Mass., June 3, 1914; s. George P. and Lottie (Fulton) F.; B.S., Boston U., 1936, M.A., 1938, Ph.D., 1941; m. Miriam Alice Hunt, August 1942 (div. 1964); children—Margaret, Susan, Peter Herrick, George Pearman III. Instr. biology Boston U., 1941-42; indsl. physiologist Arthur D. Little, Inc., Boston, 1946-47; asst. prof. biology Boston U., 1947-49, asso. prof., 1949-53, prof., 1953—, chmn. dept., 1956—, chmn. Biol. Grad. Sch., 1956—, Shields Warren prof., 1959—; staff mem. Children's Cancer Research Found., 1964—; vis. prof. Stanford Med. Sch., 1958. Cons. div. nursing Pub. Health Services. Director of the Greater Boston Zoo Association, 1964—. Fellow Am. Academy of Arts and Sciences, also of American Col. Angiology Gerontology Soc.; mem. Am. Physiol. Soc., Soc. Exptl. Biology and Medicine, Am. Assn. Anatomists, Am. Soc. Zoologists, Radiation Research Soc., Microcirculatory Conf., Sigma Xi, Phi Beta Kappa. Asso. editor Angiology. Author numerous papers on vascular physiology, also motion picture films. Research on physiology of small blood vessels; platelet function in control of bleeding; role of nervous system in control of vascular sphincters and smooth muscle of small blood vessels. Office: 2 Cummington St., Boston 15.*

FULTON, John Farquhar, Am. physiologist; b. Nov. 1, 1899; s. John Farquhar and Edith Stanley (Wheaton) F.; B.S., M.D., Harvard; B.A., M.A., D.Phil., D.Sc., Oxford (Eng.) U.; several hon. degrees; m. Lucia Pickering Wheatland, 1923. Demonstrator physiology Oxford U., 1923-25; asso. neurol. surgeon Peter Bent Brigham Hosp., Boston, 1928; fellow Magdalen Coll., Oxford, 1928; prof. physiology Yale, 1929-31, Sterling prof., 1931—; curator Osler Library, McGill U., 1950—. Recipient George Sarton medal History of Sci. Soc., 1958. Fellow Royal Coll. Physicians, London; mem. Am. Philos. Soc. Author: Muscular Contraction, 1926; Physiology of the Nervous System, 3d edit., 1949; Bibliog. Robert Boyle, 2d edit., 1959; Harvey Cushing, 1946; other works. Died May 29, 1960.

FULTON, Robert, Am. civil engr., inventor; b. Little Britain, Pa., Nov. 14, 1765; s. Robert and Mary (Smith) F.; m. Harriet Livingston, Jan. 8, 1808; 4 children. Devised successful manually operated mechanism to propel a boat by paddle wheels, 1779; engaged in a variety of engring. ventures connected with devel. of inland waterways; turned to painting landscapes, 1782-86; went to Eng.; worked and lived in Europe, 1786-1806; secured British patent for double inclined plane for raising and lowering canal boats; patented machine for sawing marble for which he received medal Soc. for Encouragment of Arts, Commerce and Mfg.; invented dredging machine or power shovel for cutting canal channels; published A Treatise on the Improvement of Canal Navigation, 1796; proposed constrn. of cast iron aqueducts to Bd. Agr. Gt. Britain, 1796, ultimately used; experimented unsuccessfully with self-propelling torpedo; painted L'Incendie de Moscow (thought to be 1st panorama), Paris, France; built diving boat Nautilus, 1800-01, to prove its worth he was to destroy Brit. shipping, but was unsuccessful in sinking ships, later attempted to prove worth to Brit., failed to sink French ships (torpedo failure), 1801-03; entered legal agreement with Robert Livingston (minister to France), to construct steamboat for purpose of navigating Hudson River between Albany (N.Y.) and N.Y.C., 1802; successful in expts. in France, 1803; returned to U. S., 1806; Steamboat Clermont began voyage up Hudson River, Aug. 17, 1807; engaged in establishment, mgmt. steamboat lines, constrn., 1807-15; commissioned by U. S. Congress to test torpedoes and submarine explosions, 1810; constructed steam war vessel authorized by U. S. Congress, 1814; dir. co. which produced a commercially successful steamboat; ensured introduction, continued operation of steamboats. Author: Treatise on the Improvement of Canal Navigation, 1796. Invented a machine to cut cables of ships at anchor. Died N.Y.C., Feb. 24, 1815.

FULTON, Robert Watt, Am. plant pathologist; b. Sistersville, W.Va., Jan. 29, 1914; s. Lawrence Wilson and Laura (Watt) F.; A.B., Wabash Coll., 1935; Ph.D., U. Wis., 1940. Faculty, U. Wis., Madison, 1947—, prof. plant pathology, 1960—. Mem. A.A.A.S., Am. Phytopath. Soc. Asso. editor Virology, 1958-63, editor, 1965—; editorial bd. Phytopathology, 1958-60, editor-in-chief, 1960-63. Research, numerous publs. on devel. of methods for mechanically transmitting viruses of woody plants, purification and characterization of these viruses. Home: 829 High St., Madison 53715. Office: Russell Labs., U. Wis., Madison, Wis. 53706.*

FULTON, Weston Miller, Sr., Am. inventor; b. Stewart, Ala., Aug. 3, 1871; s. William Frierson II and Mary Brown (Hudson) F.; student Howard Coll., Birmingham, 1887-89; B.S. with highest honors, U. Miss., 1892; postgrad. in engring., Tulane U., 1896-98; M.S., U. Tenn., 1902; m. Barbara S. Murrian, Aug. 17, 1910; children—Weston Miller, Jr., Barbara Alexander (Mrs. Fenton Gentry), Robert William, Jean Hudson (Mrs. James Talley III), Mary Helen (Mrs. C. J. Hartley). With U. S. Weather Bur., Vicksburg, Miss., 1892-98, Knoxville, 1898-1903; inventor Sylphon (a seamless metal bellows used in radiators, refrigerators, atomic plants, diesel, automobile and airplane engines; used in anti submarine depth bombs World War I), 1904; organizer Fulton Co. (later became Fulton Sylphon Co., then Fulton Sylphon div. Robertshaw Fulton Controls Corp.), Knoxville, 1904, prin. owner, 1904-28; pres., owner W. J. Savage Co., 1930-36, Royal Mfg. Co., 1940-45 (both Knoxville). Prof. meteorology U. Tenn., 1898-1903. Recipient Modern Pioneer award N.A.M., 1940; certificate of appreciation Pres. Roosevelt, 1945. A separate sect. U. S. Patent Office set aside to house his 125 patents which include the sylphon and other thermostatic devices. Died Knoxville, May 16, 1946.

FULTZ, Dave, Am. meteorologist; b. Chgo., Aug. 12, 1921; s. Harry T. and Ora L. (Voyles) F.; B.S., U. Chgo., 1941, certificate in meteorology, 1942, Ph.D., 1947; m. Jean L. McEldowney, Apr. 6, 1946; children—Martha M., David L., Katherine R. Asst., U. S. Weather Bur., Chgo., 1942, research asso., Chgo., P.R., 1942-44; operations analyst USAF, Guam, Marianas 1945; faculty U. Chgo., 1946—, prof. meteorology, 1960—. Cons. geophysics panel sci. adv. bd. USAF, Washington, 1959-65; mem. nat. com. on fluid mechanics films Ednl. Services, Inc., Watertown, Mass., 1963—. Guggenheim fellow, 1950-51; NSF fellow, 1957-58; recipient of the Rossby Research Medal, 1967. Fellow A.A.A.S., Am. Geophysical Union, Am. Meteorol. Soc. (Meisinger award 1950); mem. Am. Astron. Soc., Sigma Xi, Phi Beta Kappa. Author: Upper Air Trajectories and Weather Forecasting, 1945; (with R. Long, G. Owens, W. Bohan, R. Kaylor, J. Weil) Studies of Thermal Convection in a Rotating Cylinder with Some Implications for Large-Scale Atmospheric Motions, 1959. Research on fluid mech. expts. on rotating fluid systems, especially in presence of thermal convection; discoveries in conditions for applications of such expts. as scale models of atmospheric and oceanic circulations; discovery of new types of spontaneous instability for such rotating fluid systems. Home: 5516 S. Kenwood Av., Chgo. 60637.*

FUMI, Fausto Gherardo, Italian physicist; b. Milan, Italy, Aug. 22, 1924; s. Riccardo and Elfrida (Fischer) F.; M.Sc. in Phys. Chemistry, U. Genova (Italy), 1946; D.Sc. in Physics, U. Genova, 1948; m. Lina Buiatti, Jan. 26, 1951; children—Renata, Elena. Postdoctoral fellow Carnegie Inst. Tech., 1948-49, U. Ill., 1949-51; lectr. U. Milan (Italy), 1951-55; prof. U. Palermo, (Italy), 1955-57, U. Pavia (Italy), 1957-59; vis. research prof. Cornell U., Ithaca, N.Y., 1960; sr. physicist Argonne (Ill.) Nat. Lab., 1961-65; prof. Northwestern U., Evanston, Ill., 1962-66; prof., dir. Physics Inst., U. Palermo (Italy), 1966—. Cons., Italian Nat. Com. for Nuclear Research, 1956-58. Recipient Poma Prize, Accademia Nazionale dei Lincei, 1951. Fellow Am. Phys. Soc.; mem. Italian Phys. Soc. Author: (monograph) Quantochimica, 1951; also numerous articles. Editorial adv. bd. Internat. Jour. Physics and Chemistry Solids, 1957—. Research on solid state theory including study of effect of symmetry on phys. tensor properties, calculations of energy parameters for point imperfections in alkali halides and monovalent metals, phenomenological theory of pressure transitions of alkali halides, improvements on repulsive interaction between ion cores in alkali halides with proposal of new ionic crystal radii, validity of approximate equations of state in different temperature and volume ranges. Home: Via Enrico Albanese 112, Palermo, Italy.*

FUNAOKA, Seigo, Japanese anatomist; b. Nara Prefecture, Japan, 1890; ed. Kyota U., doctorate, 1921. Prof., Kyoto U., from 1923; prof. emeritus; prof., Gifu Prefectural Medical College. Author: Organism. Research on anatomy of peripheral nervous system; tuberculosis; lymph and lymphatic vessels.

FUNDERBURK, William Henry, Am. pharmacologist; b. Nevada, Mo., Mar. 12, 1917; s. William Henry and Blanche (Phillips) F.; B.S., U. Ky., 1939; M.S., U. Ill., 1950, Ph.D., 1952; m. Dorothy D. Jagoe, Dec. 16, 1938; children—Dianne (Mrs. Amiram Trauber), Jennifer, William Henry III, Kathryn. Electroencephalographer, U. Chgo., 1942-48, Traverse Cith (Mich.) State Hosp., 1951-57; pharmacologist Miles Research Labs., Elkhart, Ind., 1957-59; with Charles Pfizer Co., Groton, Conn., 1959-61; asso. dir. pharmacology A. H. Robins Co., Richmond, Va., 1961—. Lectr. pharmacology U. Ill., 1956-57, Med. Coll. of Va., 1961—. Mem. Soc. Pharmacology, Soc. Toxicology, Sigma Xi. Research, publs. in neuropharmacology, study of anticonvulsants, muscle relaxants, analeptics, tranquilizers, anti-depressants, analgesics, anti-inflammatory agts., immunosuppressant agts. Home: 13361 Kingsmill Rd., Midlothian, Va. 23113. Office: 1211 Sherwood Av., Richmond, Va. 23220.*

FUNK, Carl Friedrich, German dermatologist; b. Nuremberg, Jan. 27, 1897; s. Andreas and Babette (Herbolsheimer) F.; M.D., U. Erlangen; m. Gerda Johanna Elisabeth Lenz; children—Thomas, Joachim Alexander, Friedrich Carl. Asst., Dermatology Clinic, U. Berlin; 1st asst. Rudolf-Virchow Hosp., Berlin; with Ministry Health, Berlin, 1928-29; chief physician Therapeutic Sta., Müncheberg, 1930-45; med. dir.

Municipal Dermatology Clinic, Ratisbonne, 1949—; cons. specialist in skin Tb for dists. of Bavaria, Upper Palatinate. Mem. Assn. Chief Drs. Municipal Dermatology Clinics West Germany (sec.), Union Dermatology Practitioners of Upper Palatinate and Lower Bavaria (pres.), Med. Assn. Paris (corr.), Dermatology Assn. Argentina (corr.), German, Berlin (hon.) dermatol. assns., Assn. for Plastic Surgery (hon.). Author: Klinische Formen der Haut-Tuberkulose, 1934 (with others) Ergebnisse der Medizinischen Grundlagenforschung, vol. 1, 1956; Die Dermatologen der Deutschen Sprache, 1955. Research and publs. on skin Tb, Boeck and corrective dermatology. Home: 24 Dechbetternerstrasse, Regensburg, Bavaria. Office: 24 Maximilianstrasse, Regensburg 84; also Städtische Hautklinik, 4 Greflingerstrasse, Regensburg 84, Bavaria, Germany.

FUNK, Casimir, biochemist; b. Warsaw, Poland, Feb. 23, 1884; s. Jacques and Gustava (Zysan) F.; Ph.D., U. Berne, Switzerland, 1904; D.Sc., U. London, Eng., 1913; m. Alix Denise Schnidesch, June 19, 1914; children—Ian C., Doriane Jacqueline (Mrs. Henri Coenen). Biochemist, Pasteur Inst. Paris, 1904-06, U. Berlin, 1906-10; with Cancer Hosp., London, 1913-15, Metz & Co., N.Y.C., 1915-20; asso. Columbia Med. Sch., 1918-23; chief dep. biochemistry Rockefeller Found., State Inst. Hygiene, Warsaw, 1923-27; owner Casa Biochemica, France, 1927; biochemist Roussel Co., France, 1927-36; research cons. N.Y. Vitamin Corp., N.Y.C., 1936—; pres. Funk Found. for Med. Research, N.Y.C., 1953—. Fellow Am. Inst. Nutrition; mem. Am. Chem. Soc., Brit., French biochem. socs., Soc. Exptl. Biology and Medicine. Author: Vitamins, 1914, translated to English, 1922; also numerous articles. Research in vitamins and nutrition, male and female hormones, gonadotropic hormones, antagonistic substances in diet, importance of oncotine and oncostimuline in treatment and prevention of cancer, antagonistic substances in liver and yeast in anemia. Home: 186 Riverside Dr., N.Y.C. 10024.*

FUNK, Paul, Austrian engr.; b. Vienna, Austria, Apr. 14, 1886; s. Ignaz and Berta Funk; ed. univs. Tübingen, Vienna, Göttingen; Ph.D. in engring.; m. Grete Bender. Instr., German U. of Prague, 1915; prof. German Tech. U. of Prague, 1922, U. Vienna, 1957; prof. emeritus Tech. U. of Vienna, 1945. Mem. Austrian Acad. Scis., Austrian, Am. math. socs. Author: Die Linearen Differenzengleichungen und Ihre Andwendung in der Baumechanik; Beiträge zur Theorie der Kugelfunktionen; Beiträge zur Zweidimensionalen Finslerschen Geometrie; Die Laplace-Transformation und Ihre Anwendungen; Variationsrechnung und Ihre Anwendungen in Physik und Technik, 1962. Address: Seidengasse 25, Vienna VII, Austria.

FUNKE, Gosta Werner, Swedish physicist; b. Stockholm, Oct. 27, 1906; s. Oscar Werner and Sofia (Carlsson) F.; Fil.mag., Stockholm U., 1931, Fil.lic., 1934, Fil.dr., 1937; m. Gunborg Maria Axelina Blomquist, July 17, 1935; children—Ingela, Ann. Lectr. physics Stockholm U., 1937-41; prof. Tech. Coll., Norrkoping, 1939-43, Bromma Coll., 1943-45; sec. gen. Swedish Atomic Research Council, 1945—, Swedish Natural Sci. Research Council, 1946—. Pres., European Orgn. for Nuclear Research, also European So Obs. Decorated Kommendör av Nordstjärneordesn; Riddare av Vasaordesn. Author: Nkiskolan, 1940; Sverige inför atomaldern, 1956; Introduktion till naturvetenskaplig forskning i Sverige, 1963; also articles. Research in molecular spectroscopy; nuclear physics; applications of science to society. Home: 53 Thallavagen. Office: 166 Sveavagen, Stockholm, Sweden.

FUNKE, Otto, physician; b. Chemnitz, Germany, Oct. 27, 1828; M.D., Leipzig, Germany, 1851; became prof. physiol. chemistry, Leipzig, 1856; prof. physiology and zoology, Fribourg, Switzerland, from 1860. Author: Atlas der physiologischen Chemie, 1853, 58; Lehrbuch der Physiologie, 1855-57. Discovered hemoglobin, 1851; isolated hemoglobin in crystalline form, 1853. Died Fribourg, Aug. 16, 1879.

FUNKE, Werner Ernst, chemist; b. Stuttgart, Germany, July 31, 1928; s. Ernst Otto and Emilie (Braendle) F.; Dipl.Chem., U. Stuttgart, 1954, Dr.-degree, 1956; m. Gertrud Haegele, July 23, 1960; children—Mathias, Michaela, Marcus. Dozent for organic and polymer chemistry Technische Hochschule Stuttgart, 1963—. Cons. chemist Forschungs-institut fuer Pigmente und Lacke, 1956—. Recipient Research award paint and pigment sect. Gesellschaft Deutscher Chemiker, 1965; Richard-Zsigmondy award Kolloidgesellschaft, 1965. Research, publs. on formation and structure of crosslinked polymers, anionic polymerisation of strongly activated monomers, internal interfacial behavior of paint films in presence of water and solvents, evaluation of polymer film properties and devel. testing methods. Office: U. Stuttgart, Western Germany.*

FUNKHOUSER, William Delbert, Am. zoologist; b. Rockport, Ind., Mar. 13, 1881; s. Hugh Clark and Laura Josephine (Mobley) F.; A.B., Wabash Coll., 1905, ScD., 1929; M.A., Cornell U., 1912, Ph.D., 1916, honor fellow, 1916-17; m. Josephine H. Kinney, June 29, 1910. Instr. biology, high sch., Brazil, Ind., 1905-07; high sch., Greencastle, Ind., 1907; headmaster, high sch., Ithaca, N.Y., 1908-14; prin. Cascadilla Sch., 1915-18; head dept. zoology and entomology U. Ky., since 1918, prof. anthropology and dean Grad. Sch. from 1925. Fellow A.A.A.S., Entomol. Soc. Am. (pres. 1940); mem. N.Y., Bklyn. entomol. socs., Ky. Acad. Scis., Wilson Orinthol. Club, Am. Zoöl. Soc., Am. Anthrop. Soc., Ky. Research Club (pres. 1922-23), Ky. Archeol. Soc. (pres. 1933), Phi Beta Kappa, Sigma Xi. Author: Biology of Membracidae of Cayuga Lake, 1917; Outlines of Zoology, 1919; Wild Life in Kentucky, 1923; Birds of Kentucky, 1925; Catalogue of Membracidae, 1927; Kentucky Prehistory, 1931; Autobiobraphy of an Old Man, 1940; Ethnology Behind the War, 1943; Portraits of Kentuckians, 1943; Dead Men Tell Tales, 1944; The Days That are Gone, 1945; also articles in entomol. jours. Died June 9, 1948.

FUOSS, Raymond Matthew, Am. chemist; b. Bellwood, Pa., Sept. 28, 1905; s. Jacob Zachariah and Bertha May (Zimmerman) F.; Sc.B., Harvard, 1925; Ph.D., Brown, 1932; M.A. (hon.), Yale, 1945; m. Rose E. Harrington, July 25, 1926; 1 dau., Patricia Rose; m. 2d, Ann M. Stein, Mar. 1, 1947. Sheldon research fellow, Munich, 1925-26; Austin teaching fellow, Harvard, 1926-27; consultant with Skinner, Sherman & Esselen, Boston, 1927-30; student with C. A. Kraus, Brown U., 1930-32; research instr., Brown U., 1932-33, asst. prof., 1933-36; Internat. research fellow (on leave from Brown U.), Leipzig, Jena and Cambridge univs., 1934-35; research chemist, Gen. Elec. Co., Schenectady, 1936-45; Sterling prof. of chemistry, Yale, since 1945; Consultant in chemistry since 1945. Received Am. Chem. Soc. award in pure chemistry, 1935. Priestley lectr. Pa. State Coll., 1948. Mem. Nat. Acad. Scis., Am. Chem. Soc., Nat., N.Y., Conn. academies of science, American Academy of Arts and Sciences, Sigma Xi, Phi Beta Kappa. Contbr. articles on electrolytes, polymers and dielectrics in various scientific jours. Home: 57 Mill Rock Rd., New Haven 11, Conn.*

FURBERG, Sven Verner, Norwegian chemist; b. Sande, Norway, Apr. 16, 1920; s. Sven Verner and Anna (Larsen) F.; Cand.Real, U. Oslo, 1946; Ph.D., U. London, 1949; m. Borghild Elna Kristiansen, Dec. 29, 1952; children—Berit, Hilde, Sven. Faculty, U. Oslo (Norway), 1949-51, 57—, prof., 1966—; reader U. Bergen (Norway), 1952-56; vis. prof. U. Wash., 1966-67. Expert, UNESCO, Uruguay, 1956. Mem. Norwegian Acad. Scis. Research, publs. on X-ray crystallographic studies of molecules of biol. interest, especially nucleic acid components and carbohydrates; 1st to propose helical structure of nucleic acids, 1949. Home: 18 Roaveien, Oslo, Norway.*

FÜRBRINGER, Max Carl, German anatomist, ornithologist; b. Wittenberg, Germany, Jan. 30, 1846; s. Karl and Hermine (Gumprecht) F.; student, Jena, Germany; Ph.D., Berlin, 1869; m. Fanny Bassermann, 1878; 1 son, 1 dau., Elisabeth Braus. Asst. to C. Gegenbaur, Jena, then in Heidelberg, Germany, 1873; became mem. faculty in anatomy, Heidelberg, 1877, asso. prof., 1879, succeded Gegenbaur, 1901; named prof., Amsterdam, Netherlands, 1879; succeeded Hertwig in anatomy, Jena, 1888; ret., 1912. Author: Die Knochen und Muskeln der Extremitäten beiden schlangenühnlichen Sauriern, 1870; Zur Entwicklung der Amphibienniere, 1877; Untersuchungen zur Morphologie und Systematik der Vögel, 2 vols., 1888; Beitrag zur Systematik und Genealogie der Reptilien, 1900; Zur Frage der Abstammung der Säugetiere, 1904. Research on evolution of kidney organs; made comparative studies of breast, shoulder and wings of birds which added to knowledge of phylogeny of birds; proved that flat breasted birds (such as ostrich) had secondarily become unable to fly. Died Heidelberg, Mar. 6, 1920.

FÜRBRINGER, Paul Walther, German physician; b. Delitzsch, nr. Leipzig, Germany, Aug. 7, 1849; s. Max and Fanny (Bassermann) F.; ed. Jena, Berlin, (both Germany); m. Rose Baumgärtel, 1878; m. 2d, Ella Faulhaber, 1908; 1 son, 1 dau., Luise Reinhardt. Asst. physician, Franco-Prussian War, 1870-71; asst. Path. Inst., Jena, 1872-74; asst. to Friedreich, Internal Clinic, Heidelberg, until 1877; became mem. faculty in pharmacology and med. chemistry, Heidelberg, 1876; asso. prof., head dist. polyclinic, Jena, 1879; became chief ofcl. physician, Jena, 1880; dir. Friedrichshain municipal hosp., Berlin, 1886-1903; mem. (expert on accident medicine, indsl. diseases, phys. exercise) med. collegium Berlin and March Brandenburg, 1890-21. Author: Zur Wirkung der Salizylsäure, 1875; Die gebräuchlichsten Rezeptformeln der medizinischen Klinik zu Heidelberg, 1877; Über Spermatorrhoe und Protatorrhoe, 1881; Die Krankheiten der Harn- und Geschlechtsorgane, 1884; Untersuchung und Vorschrift über die Desinfektion der Hände des Arztes . . . , 1888; Die Storung der Geschlechtsfunktion des Mannes, 1899. Made clin. studies of infectious diseases, treatment of dropsy, climate therapy, balneological problems, phys. exercise; research on disinfection of physicians' hands; studied male impotence; showed diagnostic value of spinal puncture, 1895. Died Berlin, July 21, 1930.

FURCOLOW, Michael Leo, Am. physician; b. Alliance, O., June 8, 1907; s. Michael Lawrence and Mary (Raucci) F.; A.B., Mt. Union Coll., 1929, Sc.D., 1964; M.D., Yale, 1934; M.A., U. Cin., 1938; m. Margaret Carolyn Rowe, Nov. 28, 1942; children—Carol Ann, Michael L., John Rowe, Connie Sue. Research fellow in pediatrics U. Cin., 1937-38; officer USPHS, 1938-64, chief Kansas City (Kan.) Field Sta. Epidemiology Br., 1950-64; prof. epidemiology U. Ky. Med. Sch., Lexington, 1964—. Diplomate Am. Bd. Pediatrics, Preventive Medicine and Microbiology. Mem. A.M.A., Am. Pub. Health Assn., Am. Epidemiology Soc. Research, numerous publs. in Tb and fungus diseases of lung. Home: 3543 Tates Creek Rd., Lexington, Ky. 40502.*

FURFEY, Paul Hanly, Am. sociologist; b. Cambridge, Mass., June 30, 1896; s. James Arthur and Margaret (Connell) F.; A.B., Boston Coll., 1917; Ph.D., Cath. U. 1926; LL.D., Duquesne U., 1944; postgrad. Berlin, Frankfurt, Germany. Faculty, Cath. U., Washington, 1925—, prof. sociology, 1940-66, emeritus, 1966—, head dept., 1934-63, dir. Bur. Social Research, 1959—. Asst. dir. Juvenile Delinquency Research Project, N.Y.C., 1956-62. Recipient Papal medal Pro Ecclesia et Pontifice, 1958; named Domestic Prelate, 1958; named hon. prof. Universidad de America, Bogota, Colombia, 1956. Mem. Am., Am. Cath. sociol. socs., Am., Am. Cath. psychol. socs., Soc. for study Social Problems, A.A.A.S., Cath. Bibl. Assn. Am., Cath. Art. Assn., Cath. Anthrop. Conf., others. Author numerous books including: The Scope and Method of Sociology, 1953; (with M. E. Walsh) Social Problems and Social Action, 1958; (with G. Selwe, W. Gaughan) An Introduction to Sociology, 1958. Address: Mullen Library, Catholic U., Washington 20017.*

FURLONG, Eustace L., Am. geologist; b. San Francisco, Mar. 22, 1874; ed. U. Cal.; m. Ida Hopper, 1905; children—Edna (Mrs. Leslie M. Smith), Dorothy (Mrs. Gordon M. Wenz). Asst., dept. paleontology U. Cal., until 1910, curator paleontology, 1914-27; employed by land co. Sacramento Delta, 1910-14; curator vertebrate paleontology Cal. Inst. Tech., 1927-45. Outstanding contbr. field work in paleontology, tertiary deposits of Great Basin and Mexico, many pleistocene localities. Died Davis, Cal., Jan. 19, 1950.

FURMAN, Deane Philip, Am. entomologist; b. Richardton, N.D., June 4, 1915; s. Raymond Walter and Estelle (O'Harrow) F.; B.S., U. Cal., Berkeley, 1937, Ph.D., Davis, 1942; m. Katherine McKeehan, Dec. 17, 1938; children—Philip Deane, Bryan Dale, Lynne Anne. Entomologist, USPHS, 1946; faculty U. Cal., Berkeley, 1946—, prof. parasitology, entomologist Expt. Sta., 1958—, chmn. div. parasitology, 1963—. Trustee Alamedo County Mosquito Abatement Dist., 1961-64; mem. microbiology fellowship review panel NIH, USPHS, 1963-66; spl. fellow NIH with U. S. Naval Med. Research Unit 3, Cairo, Egypt, 1964-65. Fellow A.A.A.S.; mem. Entomol. Soc. Am., Am. Soc. Parasitologists, Am. Soc. Tropical Medicine and Hygiene, Wildlife Disease Assn., Pacific Coast Entomol. Soc. Author: Manual of Medical Entomology, 1961; Fundamentals of Applied Entomology, 1962. Research, numerous publs. primarily in field of biology, life history and classification of insects and mites parasitic on animals. Home: 235 Lake Dr., Berkeley, Cal. 94708.*

FURMAN, N(athaniel) Howell, Am. chemist; b. Lawrenceville, N.J., June 22, 1892; s. Nathaniel Higgins and Caroline H. (Howell) F.; student Lawrenceville Sch., 1904-09; B.S., Princeton, 1913, A.M., 1915, Ph.D., 1917; D.Sc. (hon.), Boston U., 1950; R.S. Brooks fellowship for travel and study in Europe, 1927-28; m. Hannah Scovel Hendrickson, Aug. 23, 1919; children—Carolyn Louise (Mrs. Samuel Kirkwood), Richard Howell. Instr. Stanford, 1917-19; asst. prof. chemistry Princeton, 1919-27, asso. prof., 1927-37, prof., 1937-45, Russell Wellman Moore professorship since 1945. Served as pvt. C.W.S., U. S. Army, at Am. U., Washington, 1918. Group leader on the analytical and process chemistry of uranium under O.S.R.D. and later Manhattan project, Princeton, 1942-46; cons. A.E.C. on analytical chemistry since 1942. Recipient Fisher award, Am. Chem. Soc. in analytical chem., 1948. Fellow A.A.A.S.; mem. Am. Chem. Soc. (pres. 1951), Am. Assn. U. Profs., Electrochem. Soc., Am. Soc. Testing Materials, N.Y. Acad. Sci., Optical Soc. Am., Phi Beta Kappa, Sigma Xi. Author: Elementary Quantitative Analysis. (with H. H. Willard), 1940; Poentiometric Titrations (with I. M. Kolthoff), 1926. Editor: Scott's Standard Methods of Chemical Analysis, 1939; mem. editorial bd. Analytical Chemistry, 1936-42, asso. editor since 1952; asso. editor Jour. Am. Chem. Soc., 1942-51, Chem. Revs., 1948-51. Responsible for work on analytical separation of uranium. Home: 201 Prospect Av. Office: Frick Chemical Lab., Princeton, N.J.*

FURMAN, Robert Howard, Am. physician; b. Schenectady, Oct. 23, 1918; s. Howard Blackall and Jane Blessing (MacChesney) F.; A.B., Union Coll., Schenectady, 1940; M.D., Yale, 1943; m. Mary Frances Kilpatrick, Feb. 10, 1945; children—Carol Kilpatrick, Jane Christine, Robert Howard, Hugh Patrick. Asst. prof. medicine Vanderbilt U. Sch. Medicine, 1950-52; asso. prof. research medicine Sch. Medicine U. Okla., 1952-65, prof., 1966—; head cardiovascular sect. Okla. Med. Research Found. and Hosp., 1952—, asso. dir., 1957—; attending cardiologist VA Hosp., Oklahoma City, 1953. Chmn. com. USPHS Area Cardiovascular grant U. Okla. Med. Center, 1957-64; com. mem. Nat. Heart Inst. NIH. Diplomate Am. Bd. Internal Medicine. Fellow A.C.P., Am. Coll. Chest Physicians; mem. A.A.A.S., Am. Clin. and Climatol. Assn., Am. Fedn. Clin Research, Am. Heart Assn. (dir.), Am.

Physiol. Soc., Central Soc. Clin. Research, N.Y. Acad. Scis., Soc. Exptl. Biology and Medicine, Southwestern Assn. Naturalists, Wilson Ornithol. Soc., Sigma Xi, Alpha Omega Alpha, Delta Upsilon. Bd. editors Jour. Lab. and Clin. Medicine, 1958-65, Circulation, 1966-—, Jour. Chronic Diseases, 1966-—. Research, numerous publs. on atherosclerosis, the basic disorder underlying heart attacks and strokes; disclosure of dietary, hormonal and enzyme mechanisms regulating amount, composition and phys. state of blood fats in health and disease. Home: 1300 Bedford Dr., Oklahoma City 73116. Office: 825 N.E. 13th St., Oklahoma City 73104.*

FURMAN, Sydney Clark, Am. chemist; b. Bakersfield, Cal., Mar. 22, 1923; s. Sydney V. and Marian F. (Ellis) F.; B.S., U. Cal. at Los Angeles, 1948, Ph.D. in Chemistry (Coffin fellow), 1951; m. Eva D. Crawshaw, May 14, 1955; children—Linda, Elyse, Robert, Michele. Research asso. Knolls Atomic Power Lab., Gen. Electric Co., 1951-57, chemist Vallecitos Atomic Lab., 1957-61, manager nuclear materials processing, Pleasanton, California, 1961-—. Research on vanadium solution chemistry, complexions, radiochemistry, isotopic exchange, fission product chemistry, metal-water reactions, gas chromatography, hot-cell exam., devel. and prodn. of isotopes, devel. track etching, membrane filtration. Home: 396 Teddy Dr., Union City, Cal. 94587. Office: P.O. Box 846, Pleasanton, Cal. 94566.*

FURNAS, Clifford Cook, Am. educator, research adminstr.; b. Sheridan, Ind., Oct. 24, 1900; s. Thomas Chalmers and Clara (Spray) F.; B.S., Purdue U., 1922, D.Eng. (hon.), 1946; Ph.D., U. Mich., 1926; LL.D., Alfred U., 1958; D.Sc. (hon.), Thiel Coll., 1960, U. National de Asuncion, Paraguay, 1963; m. Sparkle Brunes Moore, Apr. 12, 1925; 1 dau., Beatrice Louise (Mrs. Carl B. Pollock, Jr.). Faculty, Yale, 1931-42; tech. aide NDRC, Washington, 1941-43; dir. research Curtiss-Wright Research Lab., Buffalo, 1943-46; exec. v.p., dir. Cornell Aero. Lab., Inc., Buffalo, 1946-54; chancellor U. Buffalo, 1954-62; asst. sec. Def. for Research and Devel., Washington (on leave from U. Buffalo) 1955-57; pres. State U. N.Y., Buffalo, 1962-66, pres. emeritus, 1966-—; pres. Western N.Y. Nuclear Research Center, 1966; dir. Mfrs. & Traders Trust Co., Carborundum Co., Cornell Aero. Lab., Aerospace Corp. Cons. AVCO Corp., 1958-65, Bell Aerospace Corp., 1967-—. Recipient Vincent Bendix award Am. Soc. for Engring. Edn., 1956; Frank J. Tone medal Am. Inst. Mining, Metall. and Petroleum Engrs., 1957; Golden Cross of Order of Phoenix, 1963. Mem. Am. Chem. Soc., Am. Inst. Chem. Engrs. Author: The Next Hundred Years, 1936; The Storehouse of Civilization, 1939; (with S. M. Furnas) Man, Bread, and Destiny, 1937; The Engineer, 1966. Editor: Roger's Manual of Industrial Chemistry, 1942; Industrial Research - Its Origin and Management (Indsl. Research Inst.), 1948. Research on gas flow, heat transfer, and rates of reduction in iron blast furnace; studies on conservation of energy and mineral resources; inventions relating to control of gas flow through iron blast furnaces. Home: 651 LeBrun Rd., Eggertsville, N.Y. 14226.*

FURNESS, Caroline Ellen, Am. astronomer; b. Cleve., June 24, 1869; d. Henry Benjamin and Caroline Sarah (Baker) F.; A.B., Vassar, 1891; Ph.D., Columbia, 1900. Began as asst. Vassar Coll. Obs., Alumnae Maria Mitchell prof., from 1915. Vol. research asst. Yerkes Obs.; research worker, Groningen, Holland; visited oriental sci. instns. throughout world, 1926-27; del. 3d Pan-Pacific Congress, Japan, 1926. Author: Catalog of Stars Within 1° of North Pole, 1900; Catalog of Stars Within 2° of North Pole, 1906; Introduction to Study of Variable Stars, 1915. Editor Observations of Variable Stars Made at Vassar College (1901-12), 1913. Died Feb. 9, 1936.

FURNIVALL, Percy, Brit. surgeon; b. London, Apr. 5, 1868; s. F. J. Furnivall; ed. Univ. Coll., St. Bartholomew's Hosp.; m. Olive Mary Butlin. Named Hunterian prof. pathology and surgery Royal Coll. Surgeons Eng., 1901, also fellow; cons. surgeon London Hosp. Contbr. articles on abdominal surgery and malignant disease to jours. Editor: (with F. J. Furnivall) Vicary's Anatomy of the Body of Man. Died May 3, 1938.

FURON, Raymond, French geologist; b. Beaumont-de-Roger, Eure, Mar. 29, 1898; s. Charles and Thérèse (Renout) F.; Ph.D. in Sci., U. Paris; m. Madeleine Souveryn. Asst., sub-dir Nat. Mus. Natural History; head lectr., prof. geology Faculty Scis., Paris. Mem. Geol. Soc. France (past pres.), Soc. Biogeography. Author over 340 works, including La Paléogéographie; Manuel de Préhist. Générale; Géologie du Plateau Iranien; Géologie de l'Afrique; Le Problème de l'Eau dans le Monde. Home: 87, blvd. General-Leclerc, Cliche, Seine. Office: 1, rue Guy-de-la-Brosse, Paris 5, France.

FURRY, Wendell Hinkle, Am. physicist; b. Prairieton, Ind., Feb. 18, 1907; s. John Henry and Effie (Hinkle) F.; A.B., Depauw U., 1928; A.M., U. Ill., 1930, Ph.D., 1932; A.M. (hon.), Harvard, 1943; m. Elizabeth Josephine Sawdey, Dec. 27, 1931; children—Ellen Jane (Mrs. William F. Brewer, Jr.), Mary Susan. Asst. in physics, U. Ill., Urbana, 1928-31, NRC fellow U. Cal., Berkeley, Cal. Inst. Tech., Pasadena, 1932-34; instr. to asso. prof. Harvard, 1934-43, 45-62, prof. physics, 1962-—; research asso. Radiation Lab., Mass. Inst. Tech., 1943-45. Guggenheim fellow Inst. Theoretical Physics, Copenhagen, Denmark, 1950. Fellow A.A.A.S., Am. Phys. Soc., Am. Acad. Arts Scis.; mem. Phi Beta Kappa, Sigma Xi. Research in molecular energies, positron theory, microwave propagation, kinetic theory, quantum mechanics. Author: (with E. M. Purcell, J. C. Street) Physics for Science and Engineering students, 1952. Home: 17 Frost Rd., Belmont, Mass. 02178. Office: Dept. Physics, Harvard U., Cambridge, Mass. 02138.*

FURST, Arthur, Am. chemist, pharmacologist; b. Mpls., Dec. 25, 1914; s. Samuel and Dora (Cole) F.; A.A., City Coll. Los Angeles, 1935; A.B., U. Cal. at Los Angeles, 1937, M.A., 1940; Ph.D., Stanford, 1947; m. Florence Wolovitch, May 24, 1940; children—Carolyn (Mrs. Rodney M. Smith), Adrianne (Mrs. Herbert B. Chermside, III), David Michael, Timothy Daniel. Tchr., Pacific Mil. Acad., 1939-40; chemistry instr. San Francisco City Coll., 1940-47; faculty U. San Francisco, 1947-61, dir. Inst. Chem. Biology, 1961-—; research asso. Mt. Zion Hosp., San Francisco, 1949-—; faculty Stanford, 1952-61, prof. med. chemistry, pharmacology dept., 1957-61. Fellow A.A.A.S.; mem. Am. Chem. Soc., N.Y. Acad. Scis., Cal. Chemistry Tchrs., Am. Soc. Pharmacology and Exptl. Therapeutics, Western Pharmacology Soc., Am. Assn. Cancer Research, Pharm. Soc. Japan, Sigma Xi, Phi Lambda Upsilon. Author: Chemistry of Chelation in Cancer, 1963; also numerous articles. Research in chem. synthesis, cancer research, psychopharmacology and chemotherapy of Tb. Home: 3736 La Calle Ct., Palo Alto, Cal. 94306. Office; U. San Francisco, San Francisco 94117.*

FURTH, Hans Gerhard, psychologist; b. Vienna, Aust., Dec. 2, 1920; s. Hugo and Julia (Schindler) F.; B.A. in Philosophy, Charterhouse, Sussex, Eng., 1950; M.A. in Clin. Psychology, U. Ottawa (Ont., Can.), 1954; Ph.D. in Exptl. Psychology, U. Portland, 1960; m. Madeline B. Steen, May 22, 1954; children—Sonia, Peter, Julia, Daniel, David, Paul, Catherine. Came to U. S., 1955, naturalized, 1961. Research psychologist mental health project for deaf N.Y. State Psychiat. Inst., New York, 1955-57; sch. psychologist Ore. State Sch. for Deaf, Salem, 1958-60; asst. to prof. dept. psychology Cath. U. Am., Washington, 1960-—, dir. Center for Research in Thinking and Language, 1964-—. Research asso. Children's Hearing and Speech Center, Washington, 1960-—; Mem. Am. Psychol. Assn., Am. Assn. U. Profs. Author: Thinking without Language, 1966; Piaget's Theory of Intelligence, 1968. Research and publs. on intellectual processes, particularly of deaf persons, and linguistic deficiency; concluded that language is not necessary or intrinsic to intellectual devel. Home: 3224 Northampton St. N.W., Washington 20015.*

FURTH, Harold Paul, physicist; b. Vienna, Austria, Jan. 13, 1930; s. Otto and Gertrude (Harteck) F.; came to U. S., 1941, naturalized, 1947; grad. Hill Sch., 1947; A.B., Harvard, 1951, Ph.D., 1960; postgrad. Cornell U.; m. Alice May Lander, June 19, 1959; 1 son, John Frederick. Physicist, Lawrence Radiation Lab., Livermore, Cal., 1956-67, group leader, 1965-67; prof. dept. astrophys. scis., co-head exptl. div. Plasma Physics Lab. Princeton, 1967-—; cons. Advanced Kinetics, Inc., 1961-—; lectr. Livermore-Davis Sch. Applied Sci., 1964-67. Mem. Am. Phys. Soc. (past mem. exec. com. plasma physics div.). Editorial bd. Physics of Fluids, 1965-—; Nuclear Fusion, 1965-—. Research, numerous publs. on high-temperature plasmas; discovered new classes of plasma instabilities and basic stabilizing methods; devised theory and technique for transient prodn. of million-gauss magnetic fields. Home: 55 Locust Lane. Office: Princeton Plasma Physics Lab., Princeton, N.J. 08540.*

FÜRTH, Otto von, see von Fürth, Otto.

FURTWÄNGLER, (Johann) Adolf Michael, German archeologist; b. Freiburg, Germany, June 30, 1854; s. Wilhelm and Christiane (Schmidt) F.; ed. Freiburg, Leipzig; Ph.D., U. Munich (Germany), 1874; m. Adelheid Wendt, 1884; 4 children, including Wilhelm, Märis Edith Scheler. Traveled in Italy, Greece, 1876-78; participated in excavation of Olympia; became mem. faculty U. Bonn (W. Germany), 1879; with Berlin Ancient Art Collection, 1880; named asso. prof., Berlin, 1884; succeeded Brunn as prof., Munich, 1894; with adminstrn. of Glyptothek; excavated in Aegina, Orchomenos, Amyclae, from 1901. Mem. Bavarian Acad. Scis. Author: Mykenische Tongefässe, 1879; Beschreibung der Vasensammlung im Antiquarium der königlichen Museen zu Berlin, 2 vols., 1885; Bronzefunde von Olympia, 1890; Meisterwerke der geiechischen Plastik, 1893; Die antiken Gemmen . . . , 1900; (with K. Reichhold) Griechische Vasenmalerei, 1900; Aigina, der Tempel der Aphaia, 1906; other works. Authority on ancient vases and gems; made important excavations at Olympia, Aegina, Orchomenos. Died Athens, Greece, Oct. 10, 1907.

FURTWÄNGLER, (Friedrich Pius) Philipp, mathematician; b. Elze, Germany, Apr. 21, 1869; s. Wilhelm and Mathilde (Sander) F.; student, Göttingen, Germany, 1889-94; m. Ella Buchwald, 1903; 1 dau.; m. 2d, Emilie Schön, 1929. Asst. in physics and geodesy, Göttingen; prof. math., agrl. acad., Bonn-Poppelsdorf, then at Tech. U. Aachen (Germany); prof. U. Vienna (Austria), from 1912. Recipient Ernst Abbe Meml. medal. Mem. several sci. acads. Research in number theory, algebra; studied reciprocity laws, Fermat's theorem; gave proof of main ideal theorem. Died Vienna, May 19, 1940.

FUSAYAMA, Takao, Japanese dentist; b. Mino-city, Gifu, Japan, Aug. 7, 1916; s. Hideo and Shin (Yamaguchi) F.; D.D.S., Tokyo Med. and Dental U., 1938, Dr. Med. Sci., 1955; m. Setsuko Mori, Nov. 9, 1947; children—Hideko, Makoto, Akira. Faculty, Tokyo Women's Dental Coll., 1946-50; faculty Tokyo Med. and Dental U., 1950-—, prof., 1960-—; Fulbright research scholar Ind. U. Sch. Dentistry, 1956-57. Mem. Japanese Research Soc. Dental Materials and Appliances (dir. 1961-63, chmn.), Japanese Soc. Conservative Dentistry (dir., sec.-treas. 1963-65), Japanese Assn. Dental Sci. (dir. 1960-65). Author: Metal Inlay, 1952; Plastic Fillings, 1953; Acrylic Fillings, 1961; High Speach Cutting and Treatment, 1962; Cavity Preparation, 1962; New Amalgam Restoration, 1966. Research, numerous publs. on new technique of dental cast restoration; techniques for acrylic and amalgam restorations. Home: Akabanejutaku RB505, Akabanedai 2-2, Kitaku, Tokyo, Japan.*

FUSON, Reynold Clayton, Am. chemist; b. Wakefield, Ill., June 1, 1895; s. John Alvin and Jennie (Chesnut) F.; A.B., U. Mont., 1920, D.Sc., 1946; M.A., U. Cal. at Berkeley, 1921; Ph.D., U. Minn., 1924; D.Sc., U. Ill., 1966. Nat. research fellow Harvard, 1924-26, instr., 1926-27; faculty U. Ill., Urbana, 1927-—, prof. chemistry, 1932-63, prof. emeritus, 1963-—; mem. Center Advanced Study, 1959-63; vis. prof. Rice U., 1947-48; vis. prof. U. Nev., Reno, 1963-66; professor emeritus; with OSRD, 1944. Recipient Distinguished Achievement award from the University of Minnesota, 1960. Member Am. Chem. Soc. (Nichols medal N.Y. sect. 1953), A.A.A.S. (sec. chem. sect. 1930), Phi Beta Kappa, Sigma Xi, Alpha Chi Sigma (John Keubler award 1964), Sigma Sigma Kappa, Phi Lambda Upsilon (hon.), Gamma Alpha, Kappa Sigma. Author: (with L. C. Behr, H. R. Snyder) A Brief Course in Organic Chemistry; (with R. L. Shriner, D. Y. Curtin), Systematic Identification of Organic Compounds, 1956; Advanced Organic Chemistry, 1950; Reactions of Organic Compounds, 1962; also numerous articles. Research on behavior hindered ketones, enediols, vinyl alcohols, cholormethylation, alkylation, and acylation aryl ketones, coupling action Grignard reagent. Home: 1442 Hillside Dr., Reno 89503.*

FUTAKI, Kenzo, Japanese physician; b. Akita Prefecture, Japan, 1873; s. Juntai Higuchi; M.D., Tokyo U.; studied U. Munich, Germany. Vice-dir., Komagome Hospital, 1908; engr., Infectious Diseases Research Institute; dir., Komagome Hospital; prof. Tokyo U., 1919; prof., Nihon Medical College, 1930; instr., Tokyo Dental College and Nippon Women's U. Pres., Infectious Diseases Soc., 1926; mem., Japan Acad., 1951 (prize 1929). Author: Collected Theses and Lectures of Dr. Futaki. Research on dysentery bacilli; isolated 2 kinds of dysentery germs (named Komagone Bacillus A and B), 1903; studied diseases caused by rat-bite; advocated abdominal respiration, eating unhulled rice and only one meal each day.

FUTCHER, Palmer Howard, Am. physician; b. Balt., Sept. 13, 1910; s. Thomas Barnes and Marjorie (Howard) F.; A.B., Harvard, 1932; M.D., Johns Hopkins, 1936; m. Mary Viola Rightor, Nov. 21, 1942; children—Marjorie Rightor, Jane Pillow. Practice medicine, specializing in internal medicine, Balt., 1948-66, Phila., 1967-—; asso. prof. medicine Sch. Medicine Johns Hopkins, 1948-66, asst. dean Sch. Medicine, 1959-62; dir. Personnel Health Clinic, Univ. Health Service Johns Hopkins Med. Instns., 1962-66; asso. prof. medicine Sch. Medicine, U. Pa., 1967-—. Exec. dir. Am. Bd. Internal Medicine, 1967-—. Mem. Am. Soc. Clin. Investigation, Harvey Soc., Endocrine Soc., Am. Diabetes Assn., A.M.A. Author: Giants and Dwarfs, 1933. Research in impaired renal excretion of sodium chloride in congestive heart failure, abrupt transition in depth of pigmentation in upper arm of Negroes, ednl. needs of fgn.-trained physicians. Home: 273 S. 3d St., Phila. 19106. Office: 3930 Chestnut St., Phila. 19104.*

FUTTERER, Karl Joseph Xaver, German geologist; b. Stockach, Germany, Jan. 2, 1866; s. Xaver Futterer; student, Berlin, Germany; Ph.D., Heidelberg, 1889; m. Melanie Kaiser. Asst. geol. insts., Freiburg and Berlin; became mem. faculty Berlin, 1892; asso. prof. Tech. U. Karlsruhe (Germany), 1895, prof., 1897, also dep. dir. natural history collection, from 1899. Author: Durch Asien, 3 vols., 1901-11. Traveled through Turkestan, Central Asia, Tibet, China, 1897-99; one of 1st important Asian experts in Germany; added to work of Sven Hedin; made geol. and paleontol. studies of Baden, E. African Jura, Alps, earthquakes, morphology, deep sea sediment, geology of Urals, wind erosion in deserts. Died Karlsruhe, Feb. 18, 1906.

FYE, Paul McDonald, Am. oceanographer; b. Johnstown, Pa., Aug. 6, 1912; s. Orlando G. and Jennie (McDonald) F.; B.S., Albright Coll., 1935, Sc.D., 1955; Ph.D., Columbia, 1939; m. Ruth Heym, Apr. 26, 1942; children—Kenneth Paul, Elizabeth Ruth. Asst. prof. Hofstra Coll., 1939-41; research asso. Carnegie Inst. Tech., 1941-42; research supr., research dir. Underwater Explosives Research Lab., Woods Hole, Mass., 1942-47; asso. prof. chemistry U. Tenn., 1947-48; dep. chief, explosives research dept. U. S. Naval Ordnance Lab., Silver Spring, Md., 1948-52, chief, 1952-56, asso. dir. for research Woods Hole Oceanographic Inst., 1956-58, dir. instn., 1958—, pres., 1961—. Mem. ad hoc group for long range research and devel. Polaris, 1960-65; mem. USN Undersea Warfare Research and Devel. Planning Council, 1959—. Trustee Bermuda Biol. Sta. for Research. Recipient Bur. Ordnance Devel. award, 1946; Presdl. Certificate of Merit, 1948; USN Meritorious award, 1951; Distinguished Alumni award Albright Coll., 1951. Mem. Am. Chem. Soc., Am. Phys. Soc., Am. Geophys. Union, Am. Soc. Limnology and Oceanography, Marine Tech. Soc., A.A.A.S., Sigma Xi, Epsilon Chi, Phi Lambda Upsilon. Research on oceanography, explosives, photochemistry, gas kinetics, gas purification, underwater photography, liquid state. Home: 21 Challenger Dr. Office: Woods Hole Oceanographic Instn., Woods Hole, Mass. 02543.*

FYFE, William Sefton, geologist; b. Ashburton, New Zealand, June 4, 1927; s. C. A. and I. (Pullar) F.; Ph.D., U. Otago (New Zealand), 1952; div.; children—Christopher David, Catherine Mary. Reader chemistry U. Otago, 1952-59; prof. geology U. Cal. at Berkeley, 1959-66; Royal Soc. prof. U. Manchester (Eng.), 1966—. Mem. Am., Brit. chem. socs., Geol. Soc. Am., Mineral. Soc. Am. (award 1964), Mineral. Soc. Britain. Author: (with Turner, Verhoogen) Memoir, 73, 1958; Geochemistry of Solids, 1965; also articles. Exptl. and thermodynamic studies of chem. processes in crust of earth. Home: Handforth, Eng. Office: Geology Dept., Manchester U., Manchester, Eng.*

G

GAAFAR, Sayed Mohammed, vet. parasitologist; b. Tanta, Egypt, United Arab Republic, Jan. 18, 1924; s. Mohammed Hegab and Bahia (Salama) G.; B.V.-Sc., Cairo (Egypt) U., 1944; M.S., Kan. State U., 1949, Ph.D., 1950; D.V.M., Tex. A. and M. U., 1955; m. Irma Eileen Bird, Aug. 30, 1949; children—Joseph Omar, Wayne Samir, Daniel Sherief, Gail Magda. Came to U. S., 1951, naturalized, 1956. Asst. parasitologist Vet. Path. Lab., Egypt, 1944-47, parasitologist, 1950-51; veterinarian Rutherford Vet. Hosp., Dallas, 1952-54; faculty Tex. A. and M. U., College Station, 1955-58, asst. prof., 1956-58; faculty Purdue U., Lafayette, Ind., 1958—, prof. parasitology, 1964—. Mem. Am. Vet. Med. Assn., Am. Soc. Parasitologists, Am. Assn. Vet. Parasitologists (past pres.), World Assn. for Advancement Vet. Parasitology (sec.-treas. 1963—), Am. Soc. Tropical Medicine and Hygiene. Research, publs. on effect trace minerals on parasitism, pathogenesis ascarids, pathology parasitic diseases, hypersensitivity demodectic mange. Home: 2620 Newman Rd., West Lafayette, Ind. 47906. Office: Vet. Pathology Bldg., Purdue U., Lafayette, Ind. 47906.*

GAAL, Steven Alexander, mathematician; b. Budapest, Hungary, Feb. 22, 1924; s. István and Aranka (Gáspár) Gal; Ph.D., Pázmany Péter U. Budapest, 1947; m. Ilse Lisl Novak, Aug. 24, 1952; children—Barbara Sandra, Dorothy Janet. Came to U. S., 1950, naturalized, 1963. Instr., U. Budapest, 1945-47, U. Szeged, Hungary, 1947-48; asst. prof. Inst. Tech., Budapest, 1948; research asso. Centre Nat. Research Sci., Paris, France, 1948-50; mem. Inst. for Advanced Study, Princeton, N.J., 1950-52; instr. Cornell U., Ithaca, N.Y., 1953, asst. prof., 1954-59; research asso. Yale, 1958-60; faculty U. Minn., Mpls., 1960—, prof., 1963—. Principal investigator Army Research Office, Durham, N.C., 1959-60; dir. project NSF, Mpls., 1963-64; lectr. on number theory advanced grad. students, mathematicians, Mpls., 1961, point set topology, N.Y.C., 1964. Research in number theory and topology. Home: 51 Barton Av., Mpls. 55414.*

GABB, William More, Am. paleontologist; b. Phila., Jan. 20, 1839; s. Joseph H. and Christiana Gabb; B.A., Central High Sch., 1857; studied with James Hall, Albany, N.Y., 1857-60; became mem. Acad. Natural Scis., 1860; paleontologist Geol. Survey Cal., 1861-67, Classified Cretaceous, tertiary fossils; recognized as Am.'s leading expert on cretaceous marine paleontology (at age 22); made report on area of Lower Cal., made geol. map which gave true structure on Mexican peninsula, 1867; made topog. and geol. survey of Santo Domingo, 1868-71, Province of Talamanca (for Costa Rica), 1873-76; mem. Nat. Acad. Scis. Author: sects. 1, 4 of 1st vol., and entire 2d vol. of Whitney's Geological Survey of California, 1864; On the Topography and Geological Survey of San Domingo, 1873; On the Indian Tribes and Languages of Costa Rica, 1876; also monographs, papers on gen. paleontol. studies. Stricken with jungle fever in Costa Rica and unable to finish his reports. Authority on marine forms of Cretaceous period; described several new forms. Died Phila., May 30, 1878.

GABBIANI, Giulio Giuseppe, physician; b. Cremona, Italy, Mar. 19, 1937; s. Alceste and Rosita (Grisi) G.; M.D., U. Pavia Sch. Medicine (Italy), 1961; Ph.D., U. Montreal, Can., 1965; m. Francoise Chatel, Sept. 20, 1963; children—Fabrizio, Francesca. With Inst. Exptl. Medicine and Surgery, U. Montreal, 1961—, asst. prof., 1965—. Mem. Canadian Physiol. Soc. Research, numerous publs. on mechanism of physiologic and path. calcification and heavy metal poisoning. Home: 3982 Maplewood Av., Montreal. Office: Institut de Medecine et de Chirurgie expérimentales Université de Montréal, Montreal, Que., Can.*

GABOR, Dennis, physicist, engr.; b. Budapest, Hungary, June 5, 1900; s. Berthold and Ady (Jacobovits) G.; student Technol. U., Budapest., 1918-20; diploma Technische Hochschule, Charlottenburg, Germany, 1924, Dr.-Ing., 1927; D.Sc., U. London (Eng.), 1964; m. Marjorie Louise Butler, Aug. 8, 1936. Research engr. Siemens & Halske, Berlin, Germany, 1927-33, Brit. Thomson-Houston Co., Rugby, Eng., 1933-48; faculty Imperial Coll., London, 1949—, prof. applied electron physics, 1958—, staff cons. CBS Labs., Stamford, Conn., 1957—. Fellow Royal Soc., 1956, TV Soc., Inst. Physics; mem. Hungarian Acad. Scis. (hon.), Inst. Elec. Engrs. Author: The Electron Microscope, 1945; Inventing the Future, 1963; also numerous articles. Research and devel. high speed cathode ray oscillograph, shrouded magnetic lens, theory of communication, information theory, phys. optics, predicting machines, plasmas, gas discharges, tv; inventor holography. Home: 78 QueensGate, London, S.W. 7, Eng.*

GABRIEL, David Samuel, Am. mech. engr.; b. Lakewood, O., Oct. 1, 1919; s. John Ward and Lucy (Cirby) G.; student Ohio No. U., 1938-39; B.Mech. Engring., Akron U., 1943; postgrad. Case Inst. Tech., 1943-46; m. DeEtte A. Rowe, Sept. 3, 1966; children by former marriage—Sharon, Georgiana, Charles. With Lewis Research Center, NASA (formerly Lewis Flight Propulsion Lab., NACA), Cleve., 1943-66, asst. chief altitude wind tunnel br., 1951, chief engines br., 1951-53, asst. chief engine research div., 1953-55, asso. chief, 1955-58, chief propulsion systems div., 1958-61, chief, nuclear systems div., 1961-63, mgr. Centaur project, 1963-66; asst. chief engr. propulsion systems Bell Aerosystems Co.; Buffalo, 1966—. Mem. Am. Inst. Aeros. and Astronautics. Research on turbomachinery leading to discovery of Reynolds number effect on performance, research on all types of powerplants for aircraft and rockets; dir. devel. Centaur launch vehicle. Home: 352 Edgewater Dr., Tonawanda, N.Y. 14150. Office: P.O. Box 1, Buffalo.*

GABRIEL, Siegmund, German chemist; b. Berlin, Nov. 7, 1851; s. Aron and Golde (Pollnow) G.; studied at U. Berlin, Heidelberg, Germany; Ph.D. under Bunsen, 1874; m. Anna Fraenkel, 1886. Became asst. to A. W. v. Hofmann, U. Berlin, 1874, joined faculty, 1880, named asso. prof., 1886, hon. prof., 1913. Mem. German Chem. Soc. (chmn.). Research and publs. on synthesis and qualitative analysis of ring compounds, especially those containing nitrogen; synthesized phenylisochinolin (an isochinoline derivative), free unsubstituted isochinoline, phthalazine and its homologs; research on diazines, pyridazine, pyrimidine, pyrazine, amino ketones; discovered preparation of primary amines. Died Berlin, Mar. 22, 1924.

GABRIELSON, Ira Noel, Am. biologist; b. Sioux Rapids, Ia., Sept. 27, 1889; s. Frank August and Ida (Jansen) G.; B.A., Morningside Coll., 1912, LL.D., 1942; D.Sc., Ore. State U., 1936, Middlebury Coll., 1959; m. Clara Speer, Aug. 7, 1912; children—Clara June Martin, Iris Virginia Nesbitt (dec.), Jeany (Mrs. A. D. Holmes), Gail (Mrs. N. S. Ferris). Tchr. pub. schs., 1912-15; with Biol. Survey, U. S. Dept. Agr., 1915-31, regional dir., 1931-35, chief, 1935-40; dir. U. S. Fish and Wildlife Service, Dept. Interior, Washington, 1940-46; pres. Wildlife Mgmt. Inst., Washington, 1946—. Pres., World Wildlife Fund., 1961—; chmn. No. Va. Regional Park Authority, 1961—; cons. state, nat. govt. depts. Recipient Distinguished Service medal Dept. Interior, 1964, Conservation Service, 1964; Audubon medal Nat. Audubon Soc., 1949; Leopold medal Wildlife Soc., 1953; Hugh H. Bennett medal Friends of the Land, 1949; others. Fellow Am. Ornithol. Union; mem. Cooper, Wilson ornithol. socs., Soc. Mammalogists, Soc. Systematic Zoology, Washington Acad. Scis. Author: Western American Alpines, 1932; (with S. G. Jewett) Birds of Oregon, 1940; Wildlife Conservation, 1941; Wildlife Refuges, 1943; Wildlife Management, 1951; (with F. C. Lincoln) Birds of Alaska, 1959; also numerous articles. Home: 2500 Leeds Rd., Oakton, Va. 22124. Office: 709 Wire Bldg., Washington 20005.*

GABRILOVE, Jacques Lester, Am. endocrinologist; b. N.Y.C., Sept. 21, 1917; s. Benjamin and Pauline (Levine) G.; B.S. magna cum laude, Coll. City N.Y., 1936; M.D., N.Y. U., 1940; m. Hilda Roslyn Weiss, May 19, 1946; children—Sandra L. and Janice L. Staff Mt. Sinai Hosp., N.Y.C., 1940—, attending physician, 1968—; faculty State U. N.Y., N.Y.C., 1957—, clin. prof. medicine, 1966—; faculty Mt. Sinai Sch. Medicine, 1966—, clin. prof. medicine, 1968—. Mem. panel metabolic and rheumatoid diseases U. S. Pharmacopeia XVL, 1956—; cons. endocrinology VA Hosp., East Orange, N.J., 1958—, Elizabeth A. Horton Meml. Hosp., Middletown, N.Y., 1961—. Diplomate Am. Bd. Internal Medicine. Fellow A.C.P., N.Y. Acad. Medicine; mem. Harvey Soc., Endocrine Soc., Am. Diabetes Assn., Am. Fedn. Clin. Research, Alpha Omega Alpha (prize 1940), Phi Beta Kappa. Publs. on 1st delineation of male adult adrenogenital syndrome and 11-beta-hydroxylase deficiency in an adult female, Addison's disease, myxedema, Cushing's syndrome, feminizing adrenocortical tumors, effects of hormone deficiency especially adrenal cortical hormones. Home: 25 E. 86th St., N.Y.C. 10028. Office: 79 E. 79th St., N.Y.C. 10021.*

GABUZDA, George Joseph, Am. physician; b. Freeland, Pa., Jan. 26, 1920; s. George Joseph and Anna May (Silvasi) G.; A.B. summa cum laude, Lehigh U., 1941; M.D. cum laude, Harvard, 1944; m. Marion Emma Jarvis, Apr. 2, 1946; children—Anne Therese, George Joseph III, Denise Carmen. Practice medicine, specializing in internal medicine, Cleve., 1954—; faculty Western Res. U., 1954—, prof. medicine, 1964—; asso. dir. dept. medicine Cleve. Met. Gen. Hosp., 1966—. Cons., sr. attending physician Cleve. VA Hosp., 1954—; cons. Luth. Hosp., Cleve., 1964—; mem. sci. adv. bd. Cleve. Diabetes Fund, 1962—. Welch fellow in internal medicine NRC, 1951-54. Mem. Am. Soc. Clin. Investigation, Central Soc. Clin. Research, Am. Assn. Study Liver Disease (past pres.), Soc. Exptl. Biology and Medicine, Am. Inst. Nutrition, Am. Soc. Clin. Nutrition, Am. Fedn. Clin. Research, A.A.A.S. Research, numerous publs. on metabolic aspects of liver diseases and their complications, abnormalities in ammonia, nitrogen, potassium metabolism to hepatic coma, fluid and electrolyte metabolism in ascites and edema formation and moblzn., renal functional alterations asso. with severe liver disease. Home: 1854 Langerdale Blvd., South Euclid, O. 44121. Office: 3395 Scranton Rd., Cleve. 44109.*

GABY, William Laurence, Am. microbiologist; b. Hot Springs, N.C., June 15, 1917; s. Edward Lee and Bertha (Rufty) G.; B.A., U. Tenn., 1939, M.S., 1940; Ph.D., St. Louis U., 1946; m. Virginia Cox., June 10, 1940; children—Nancy Sue, Virginia Lee, Frances Ann. Sr. bacteriologist Tenn. Dept. Pub. Health, Nashville, 1940-41; bacteriologist Winthrop Chem. Co., Syracuse, N.Y., 1941-42; sr. bacteriologist Bristol Labs., Inc., Syracuse, 1946-49; asso. prof. Hahnemann Med. Coll., Phila., 1949-64; prof. microbiology, chmn. dept. health scis. E. Tenn. State U., Johnson City, 1964—. Recipient Comml. Solvents award Soc. for Microbiology, 1952. Diplomate Am. Bd. Microbiology. Fellow Am. Acad. Microbiology; mem. Am. Soc. Biol. Chemists, Am. Assn. Immunologists, Am. Soc. Microbiology, A.A.A.S., Am. Pub. Health Assn., Sigma Xi. Research, publs. on role of phospholipids in metabolism, Antigen-Antibody reactions, transport of amino acids. Home: 1408 College Height Dr., Johnson City, Tenn. 37601.*

GABY-CAZALOT, Antoine, engr.; b. Ariège, France, 1796; prof. math., later engr.; studied compressibility of liquids, 1827; built a steam engine streetcar, 1830, also engine with oscillating cylinders, shortened pressure gauge; described steel prodn. process similar to that patented by Bessemer a few months later, 1855. Died 1869.

GADAMER, Johannes Georg, German pharmaceutist; b. Waldenburg, Germany, Apr. 1, 1867; s. Oscar and Anna (Puschmann) G.; began studies in Marburg, 1891; Ph.D., 1895; hon. M.D., Breslau, 1927; m. Johanna Gewiese, 1897; 2 sons, including Hans-Georg; m. 2d, Hedwig Hellich, 1905. Joined faculty U. Marburg, 1897; named prof., dir. Pharm. Inst., U. Breslau (now Wroclaw, Poland); returned as successor to E. Schmidt, Marburg, 1919. Became mem. Reich Bd. Health, 1915. Author: Lehrbuch der Chemischen Toxikologie, 1909; also numerous articles. Collaborater: Deutsches Arzneibuch, 6th edit., 1928. Co-editor, Archiv der Pharmazie, 1921. Research on identification, qualitative analysis, and synthesis of natural substances, especially plant family with alkaloid content, also forensic chemistry. Died Marburg, Apr. 15, 1928.

GADDIS, Adam Marr, Am. chemist; b. nr. Upper Marlboro, Md., Feb. 10, 1911; s. James Pinkney and Edith (Marr) G.; B.S., Duke, 1933; m. Waneta Motsinger, Jan. 19, 1944; children—Dorothy Anna, John Marr. Chemist, U. S. Dept. Agr., Beltsville, Md., 1936—. Mem. Am. Chem. Soc., Inst. Food Tech. Research in synthesis steroids, meat research on preservation and processing, autoxidation of lipids, chromatographic methods, chemistry of rancidity. Home: R.F.D. Box 1722, Upper Marlboro, Md. 20870. Office: Meat Lab., Eastern Utilization Research and Devel. Div., U. S. Dept. Agr., Beltsville, Md. 20705.*

GADNER VON GARNECK, Georg, German cartographer; b. Landshut, Germany, 1522; s. Hyllbrant Gadner von Garneck; studied law Ingolstadt, Germany; later studied in Cologne, Germany, Tübingen, Germany; doctorate, 1550 or 51; m. between 1546 and 48; 2 children; m. 2d, Anna Ochsenbach, Apr. 9, 1561; 1 dau. Legal counsel, Ingolstadt, 1551-54; also fgn. negotiations for Duke of Bavaria; joined service Duke Christoph of Württemberg, 1555, duties included supervision of borders with neighboring states, legal disputes involving forestery and mining. Author: Chorographica Ducatus Wirtembergici, 1596. Charted forests of Württemberg, beginning 1587. Died Freudental, Germany, May 2, 1605.

GADOLIN, Johan, Finnish chemist; b. Abo, Turku, Finland, June 5, 1760; ed. Turku, Uppsala, Sweden; studied under Bergman in Sweden; prof. chemistry U. Uppsala; mem. Peterburg Acad. Sci. Boisbaudran named the rare earth element gadolinium in his honor. Isolated yttrium from the oxide, 1794; his research led to discovery of entire series of rare earths. Died Abo, Aug. 15, 1852.

GADOW, Hans Friedrich, zoologist; b. Altkrakow, Pomerania, Mar. 8, 1855; s. M. L. Gadow; began study natural scis. in Berlin, 1875; Ph.D., Jena, Germany, 1878; also studied at Frankfurt, Germany, Heidelberg, Germany; m. Clara Maud Paget. Worked with C. Gegenbauer in Heidelberg; became asst. in zool. dept. Brit. Mus., London, 1880; named curator Strickland Found. for Birds, U. Zool. Mus., Cambridge, 1882; became lectr. King's Coll., Cambridge, 1884, and reader on morphology of vertebrates in 1920; naturalized Brit. citizen, 1884. Mem. Zool. Soc. London, Royal Soc., 1892. Author: Zur vergleichenden Anatomie der Muskulatur des Beckens und des hinteren Gliedmassen der Ratiten, 1880; (with A. Newton) A Dictionary of Birds, 1893-96; A Classification of Vertebrata, Recent and Extinct, 1898; In Northern Spain, 1897; Through Southern Mexico, 1908; contbr. author: The Cambridge Natural History VIII, 1901. Research on comparative anatomy, structure and devel. of spinal column and inner ear of mammals, also geog. distbn., color change and living habits of amphibians, reptiles and birds of Spain and Mexico; continued the work of M. Fürbinger on morphology and taxonomy of birds. Died Cambridge, May 16, 1928.

GAEBLER, Oliver Henry, Am. biochemist; b. Swiss, Mo., Nov. 14, 1895; s. Franklin G. and Louisa (Grauer) G.; A.B., Central Wesleyan Coll., 1917; A.M., U. Mo., 1920; Ph.D., U. Toronto, 1922; M.D., Cornell U., 1931; m. Charlotte M. Stroetker, Sept. 17, 1928; children—Ella Louise, John Franklin. Teaching fellow U. Toronto, Ont., Can., 1920-22; asst. in physiology U. Rochester, 1923-24; asso. in biochemistry U. Ia., 1924-25, asst. prof. biochemistry, 1925-27; asso. in charge chemistry Henry Ford Hosp., Detroit, 1928-47; profl. lectr. Wayne State U., 1932-65; chmn. biochemistry dept. Edsel B. Ford Inst. Med. Research, Detroit, 1947-65, cons., 1965—. Recipient Ernst Bischoff award, 1959. Fellow A.A.A.S., Am. Inst. Chemists; mem. Am. Assn. Clin. Chemists (pres. 1958), Am. Bd. Clin. Chemistry (v.p. 1961-67), Canadian, Detroit (pres. 1942) physiol. socs., Arthritis Found. (med., sci. adv. coms.), Mich. Cancer Found., Am. Soc. Biol. Chemists, Soc. for Exptl. Biology and Medicine, Endocrine Soc., Am. Chem. Soc. Editorial bd. Clin. Chemistry, 1960-65. Research, numerous publs. on composition of blood and urine, metabolic effects of hormones, effects of x-radiation on metabolism of nitrogenous compounds. Home: 18410 Northlawn Av., Detroit 48221. Office: 2799 W. Grand Blvd., Detroit 48202.*

GAEDE, Wolfgang, German physicist; b. Lehe (now Bremerhaven-Lehe), Germany, May 25, 1878; s. Karl and Amalie (Ruef) G.; Ph.D., U. Freiburg (Germany), 1901. Asst. to F. Himstedt, U. Freiburg, until 1907; founder Tech. Physics Inst. (1st of its kind); joined faculty, 1909, named asso. prof., 1913; became prof. Karlsruhe, Germany, 1919; ret., 1934; continued research in pvt. lab.; collaborated with firm Leybold's Nachfolger, Cologne, Germany. Recipient Siemens-Ring, Elliot Cressons medal Franklin Inst., State of Pa., 1913, Dudell medal Phys. Soc., London, 1933. Mem. Leopoldina. Research and numerous publs. on devel. new apparati including rotary mercury air pump, 1905, molecular pump, 1912, diffusion pump, gas ballast pump (led to new devels. in cathode rays, X-rays, spectra of glowing gases, photoeffect). Died Munich, Germany, June 24, 1945.

GAERTNER, Gustav, see Gärtner, Gustav.

GAERTNER, Joseph, see Gärtner, Joseph.

GAERTNER, William, instrument maker; b. Merseburg, Germany, Oct. 24, 1864; s. Karl and Louise (Pippel) G.; ed. pub. sch. and Tech. Sch. for Instrument Makers, Berlin; m. Belva Eleanora Boosinger, June 14, 1917. Apprentice in instrument shop, Halle, at 16; worked for various firms in Germany, later in London, and Vienna; came to U. S., 1889, naturalized, 1896; instrument maker for Coast and Geodetic Survey, 1890-93; with Smithsonian Instn., 1893-96; opened shop in Chgo., 1896, later William Gaertner & Co. and from 1924, Gaertner Sci. Corp., of which is pres. and treas.; mfr. of interferometer for Prof. Albert A. Michelson; zenith tube, for Internat. Geodetic Assn., to determine variations of latitude, etc.; has practically solved problem of eliminating error in accurate precision screws. Awarded Howard N. Potts gold medal for notable achievements as a designer and maker of scientific instruments, Franklin Inst., 1924. Mem. Am. Astron. Soc., Army Ordnance Assn., Active in developing new instruments and improving old designs for U. S. Air Corps. Died Dec. 3, 1948.

GAERTTNER, Erwin Rudolf, Am. nuclear physicist; b. Denver, Colo., Feb. 27, 1911; s. Rudolf and Pauline Karoline (Groezinger) G.; B.S. in Elec. Engring., U. Denver, 1932; Ph.D. in Physics, U. Mich., 1937; m. Dorothy Mary Polcar, Dec. 21, 1940; children—Robert E., Martin R. NRC fellow physics Cal. Inst. Tech., 1937-38, Horace H. Rackham fellow physics, 1938-39; from instr. to asst. prof. physics Ohio State U., 1939-46; on leave to radiation lab. Mass. Inst. Tech., 1942-46; research asso. research lab. Gen. Electric Co., 1946-51, mgr. physics and exptl. equipment development Knolls Atomic Power Lab., 1951-58; professor physics Rensselaer Polytech. Institute, 1958-60, professor nuclear engineering and science, head of department, 1960—. Fellow Am. Phys. Soc., Am. Nuclear Society; member American Society for Engineering Education, U. S. Power Squadron. Phi Beta Kappa, Sigma Xi. Research in nuclear structure using betatrons, van de Graaff accelerators, reactors and linear accelerators; experimental physics. Home: 10 Kenworth Av. Office: Linear Accelerator Lab., Troy, N.Y. 12181.*

GAFFKY, Georg Theodor August, German bacteriologist; b. Hanover, Germany, Feb. 17, 1850; s. Georg Friedrich Wilhelm and Emma Wilhelmine Mathilde (Schumacher) G.; M.D., U. Berlin, 1873. Mil. physician assigned to Robert Koch, Kaiserliches Gesundheitsamt, 1880; mem. Koch's cholera expdn. to Egypt, India, 1883-84; named prof. hygiene U. Giessen (Germany), 1888; led govt. commn. to study plague in India, 1897; became dir. Inst. for Infectious Diseases, U. Berlin (renamed Robert Koch Inst. 1912), 1904-13. Author: Zur Ätiologie des Abdominaltyphus, 1884; Über die Gefahren der Serumkrankheit bei der Schutzimpfung mit Diphtherieserum, 1913. Research on etiology typhus abdominalis, rabbit septicemis, cholera, anthrax, bacteriological methods, steam disinfection, bacillus botulinus, plague epidemic of horses; 1st pure culture of typhus bacillus which was discovered by K. Eberth (Eberth-Gaffky bacillus), 1880; originated formula for prognostic classification of sputum specimens in Tb. (Gaffky's Scale). Died Hanover, Sept. 23, 1918.

GAGE, Simon Henry, Am. biologist; b. Otsego County, N.Y., May 20, 1851; s. Henry V. and Lucy (Grover) G.; B.S., Cornell U., 1877; studied in Europe, 1889; m. Susanna Phelps, Dec. 15, 1881 (dec. Oct. 1915); 1 son, Henry Phelps; m. 2d, Clara C. Starrett, Apr. 14, 1933. Instr., Cornell U., 1878-81, asst. prof., 1881-89, asso. prof. physiology, 1889-93, asso. prof. anatomy, histology and embryology, 1893-95, prof., 1895-96, prof. histology and embryology, 1896-1908, apptd. prof. histology and embryology, emeritus, 1908, after 25 yrs'. service to undertake spl. investigations, on allowance from Carnegie Found. for Advancement of Teaching. Chmn. section embryology, Internat. Congress Arts and Sciences, St. Louis, 1904. Fellow A.A.A.S. (v.p., 1885, 1892, 1899); mem. Assn. Am. Anatomists, Am. Soc. Naturalists, Am. Micros. Soc., Am. Soc. Zoölogists, Am. Soc. of Amateur Microscopists (pres. 1939), Royal Soc. of Arts, London. Author: The Microscope and Microscopic Methods, 1881, 17th edit., 1941 (leading textbook on use of microscope in biol. research in Am.); History of Microscopy in America, 1943 (in manuscript); Anatomical Technology (with Prof. Burt G. Wilder); Optic Projection with the Magic Lantern, the Reflecting Lantern, the Projection Microscope, and the Moving Picture Machine (with Henry Phelps Gage, Ph.D.), 1913-14; also numerous papers on biol. subjects; collaborator or contbr. to Foster's Encyclopaedia Medical Dictionary, Wood's Reference Handbook of the Medical Sciences, Johnson's Cyclopedia. Research in comparative histology and embryology. His collection of microscope preparations, lantern slides and other records of his research are noteworthy and valuable in tng. med. students. Died Interlaken, N.Y., Oct. 20, 1944.

GAGEL, (Friedrich August Wilhelm), German geologist; b. Heiligenbeil, Germany, Feb. 7, 1865; s. Ferdinand and Johanne (Hein) G.; state exam U. Königsberg (Germany), 1889, Ph.D., 1890; m. Lina Pieper, 1890; 3 sons, 2 daus. Employed in soils dept. Prussian Geol. Inst., Berlin, 1890, later dist. geologist, state geologist, apptd. dir. collection, 1924; mil. geologist on Eastern front, during World War I.; dir. Lüneburg Limestone Works after death of father-in-law. Research and numerous publs. on volcanic constrn. of Madeira, repeated ice formation in Eastern front during World War I; charted numerous areas of No. Germany and Scandinavia; drew geol. and morphological parallels to Island of La Palma, Canary Islands; discovered lower Permian sandstone in N. Germany nr. Lieth-Stade. Died Lüneburg, Germany, Jan. 22, 1927.

GAGER, C(harles) Stuart, Am. botanist; b. Norwich, N.Y., Dec. 23, 1872; s. Charles Carroll and Leora Josephine (Darke) G.; A.B., Syracuse U., 1895; Pd.B. and Pd.M., N.Y. State Normal Coll., Albany, 1897; Ph.D., Cornell U., 1902; D.Sc., Syracuse, 1920; Pd.D., N.Y. State Coll. for Teachers, 1921; m. Bertha Woodward Bagg, June 25, 1902; children—Benjamin Stuart (dec.), Ruth Prudence (Mrs. Kenneth G. Bucklin). Prof. biol. scis. and physiography N.Y. State Normal Coll., 1897-1905; dir. labs. N.Y. Bot. Garden, 1906-08; prof. botany U. Mo., 1908-10; dir. Bklyn. Botanic Garden, 1910-43. Mem. various commns. NRC; mem. and dir. Corp. of Bermuda Biol. Sta. for Research. Fellow A.A.A.S., N.Y. Acad. Scis.; mem. Bot. Soc. Am. (pres. 1936), Soc. Exptl. Biology and Medicine, Torrey Bot. Club (sec. 1905-08; v.p. 1911, 1917-31), Am. Soc. Biol. Chemists, Am. Soc. Naturalists, Nat. Inst. Social Scis. (pres. 1932-35), Svenska Linné Sallskapet, Société Linnéene de Lyon, Royal New Zealand Inst. Horticulture, Phi Beta Kappa, Sigma Xi, Delta Upsilon, Gamma Sigma Delta. Author: Errors in Science Teaching, 1901; Effects of the Rays of Radium on Plants, 1908; Fundamentals of Botany, 1901; Laboratory Guide for General Botany, 1916; Heredity and Evolution in Plants, 1920; The Relation between Science and Theology, 1925; General Botany with Special Reference to Its Economic Aspects, 1926; The Plant World, 1931; also numerous papers in scientific and ednl. jours. Abstractor Biol. Abstracts, 1926-31; bot. editor and contbr. Nat. Cyclo., 1932. Contbr. Ency. Brittanica, Standard Cyclo. Horticulture, Cyclo, Edn. Translator: (from the German of de Vries) Intracellular Pangenesis, 1910. Research on effect of radium rays on plants; 1st to demonstrate that radium rays under suitable conditions accelerate growth and other physiol. activities as well as retard such functions. Died Aug. 9, 1943.

GAGER, William Atkins, Am. educator; b. Cold Spring, Pa., Dec. 23, 1897; s. David Decater and Mary (Vail) G.; B.S. in San. Engring., Pa. State U., 1919, M.S., 1923; Ph.D. in Math. and Psychology, George Peabody Coll., 1940; m. Neva Marie Bemiss, Aug. 17, 1921; 1 son, William Atkins. San. engr. Johnstown Water Co. (Pa.), 1923-26; tchr. math., sci. St. Petersburg (Fla.) High Sch., 1926-27; head dept. math. St. Petersburg Jr. Coll., 1927-42; asso. prof. math. U. Fla., Gainesville, 1942-49, prof., 1949—, dir. NSF Summer Insts., 1961-68. Cons. math. U. S. Office Edn., various states, counties in Fla. Mem. Nat. Council Tchrs. Math. (dir. 1951-57), Phi Kappa Phi, Tau Beta Pi, Phi Delta Kappa, Kappa Delta Pi, N.E.A., Math. Assn. Am. Research on efficiency of liquid chlorine vs bleaching powder, efficiency open and closed sprinkling filters, chem. methods of cleaning reconditioning water meters. Directing author Functional Math. Series, 1953-61; co-author Scribner Arithmetic Series, 1955-57; author Contemporary College Algebra and Trigonometry, 1968. Contbr. articles to profl. jours. Home: 2616 S.W. 4th Pl., Gainesville, Fla. 32601.*

GAGNÉ, Robert Mills, Am. psychologist; b. North Andover, Mass., Aug. 21, 1916; s. Alphonse Francis and Alice (Mills) G.; A.B., Yale, 1937; Ph.D., Brown U., 1940; m. Harriet Nash Towle, Nov. 26, 1942; children—Samuel T., Ellen D. Faculty, Conn. Coll. For Women, 1940-41, 46-49, Pa. State Coll., 1945-46; research dir. Air Force Personnel and Tng. Research Center Air Research and Devel. Command, 1949-53, tech. dir. maintenance lab., 1953-58; prof. psychology Princeton, 1958-62; dir. research Am. Inst. Research, Pitts., 1962-65; prof. ednl. psychology U. Cal. at Berkeley, 1966—. Cons. Dept. Def. Fellow Am. Psychol. Assn., A.A.A.S.; mem. Am. Ednl. Research Assn. Author: (with E. A. Fleishman) Psychology and Human Performance, 1959; The Conditions of Learning, 1965. Editor: Psychological Principles in System Development, 1962. Research, numerous publs. primarily in field of method of measuring transfer of learning, applications psychol. prins. to tech. tng., and edn.; developed methods of task analysis for defining tng. objectives. Home: 28 Contra Costa Pl., Oakland, Cal. 94618. Office: Dept. Edn., U. Cal., Berkeley, Cal. 94720.*

GAGNEBIN, Albert Paul, Am. metallurgist; b. Torrington, Conn., Jan. 23, 1909; s. Charles A. and Marguerite E. (Huguenin) G.; B.S. in Mech. Engring., Yale, 1930, M.S., 1932; m. Genevieve Hope, October 26, 1935; children—Anne (Mrs. John D. Coffin), and Joan (Mrs. David O. Wicks). With Internat. Nickel Co., 1932, successively staff research lab., research ferrous metallurgy, development ductile iron, staff development and research div., 1932-55, sales dept., 1955-61, mgr. primary nickel dept., 1955-61, v.p., 1958-64, exec. v.p., 1964-66, pres., 1967—, mem. exec. com. dir. Internat. Nickel Co. Can., Ltd., 1965, pres., 1967—, mem. exec. com.; dir. Toronto-Dominion Bank, Abex Corp., Internat. Nickel Benelux S.A., France S.A., (Italia) S.p.A., Iberia Ltd. Bd. dirs. Am. Com. for Inst. Advanced Study-Europe, Inc., Sterling Forest Bd. of Design, 1961—, Albert Gallatin Assos. of N.Y.U., 1962—. Recipient Ann. award Ductile Iron Soc., 1965. Mem. Am. Soc. M.E., Am. Inst. Mining and Metall. Engrs., Am. Soc. Metals, American Foundrymens Soc. (hon. life, Peter L. Simpson gold medal award 1952), Am. Iron Ore Assn. (dir.), Yale Engring. Assn. (dir.), Sigma Xi. Author: The Fundamentals of Iron and Steel Castings. Co-inventor ductile iron. Home: 143 Grange Av., Fair Haven, N.J. Office: 67 Wall St., N.Y.C. 10005.

GAHN, Johan Gottlieb, Swedish chemist, mineralogist; b. Voxna, Gävleborg, Sweden, Aug. 19, 1745; studied mineralogy under Bergman; began career as miner; later owned and managed mines; mem. Swedish legislature; trained Berzelius to use the blowpipe for qualitative analysis; supplied copper to Am. ships during Am. Revolution; became prof. Coll. of Mines, Stockholm, 1784. The mineral, gahnite, was named in his honor. Discovered (with Scheele) phosphoric acid in bones, 1770; 1st to isolate metallic manganese, 1774; discovered selenium in the flue dust of one of his sulfuric acid plants. Died Stockholm, Dec. 8, 1818.

GAILLARD, Edwin Samuel, Am. surgeon; b. Charleston, S.C., Jan. 16, 1827; grad. S.C. Coll. (now U. S.C.), 1845; M.D., Med. Coll. State of S.C., Charleston M.A. (hon.) also LL.D. (hon.), U. N.C., 1873; m. Jane Marshall Thomas, 1856; m. 2nd, Mary Elizabeth Gibson, 1865; 4 children. Practiced medicine, Fla., 1855-57; after trip to Europe settled in Balt., 1861; prof. principles and practice medicine and gen. pathology Med. Coll. Va., Richmond, 1865; founded Richmond Med. Jour. (changed name to Richmond and Louisville Med. Jour., 1868), pub., 1865-79; prof. medicine Ky. Sch. Medicine, 1868; an organizer, 1st dean, prof. gen. medicine and pathology Louisville Med. Coll., 1869; established Am. Med. Weekly, Louisville, 1874, editor, 1874-83; moved to N.Y.C., 1879; published Gaillards's Med. Jour., until 1883. Author numerous papers including Ozone: Its Relation to Health and Disease (received Fiske Fund prize 1861), essay on diphtheria (received Ga. Med. Assn. prize 1866). Died Feb. 1885.

GAILLARD, Pieter Johannes, Dutch histologist; b. Rotterdam, Netherlands, Jan. 18, 1907; s. Johannes L. J. and Johanna C. (de Kok) G.; M.D., U. Leiden (Netherlands), 1931, Dr.Med., 1931, Licentiate ex.-Med., 1933; Dr.h.c., U. Bordeaux (France), U. Liège (Belgium), 1967; m. Johanna W. Schouten, Mar. 23, 1933; children—Johannes L. J., Gottlieb C. Asst., Leiden U., 1926-33, chief asst., 1933-47, prof. exptl. histology, 1947-60, prof. cell biology and histology, 1960—; dir. pub. health City of Leiden, 1945-46; dir. Cell Biology and Histology Inst., 1954—. Decorated Class of Merit Netherlands Red Cross; created Knight Order of Netherlands Lion, 1966. Mem. Royal Netherlands Acad. Scis. (pres. 1966—), Royal Soc. London (affiliate), Royal Flemmish Acad. Medicine (fgn.). Author: Hormones, Growth and Differentiation in Embryonic Explants, 1942; (with others) Alg. Celleer, 1957; also numerous articles. Editor: (with R. V. Talmage, Ann M. Budy) The Parathyroid Glands, 1965. Research on effects of hormones and vitamines on organ cultures of embryonic tissues; 1st proof of parathyroid hormone effect on living bone. Home: 24 Nassaulaan, Oegstgeest, Netherlands. Office: 10 Rynsbuygerweg, Leiden, Netherlands.*

GAIND, Kidar Nath, Indian pharm. chemist; b. Pusrur, India, Sept. 28, 1911; s. Lal C. and Purandevi (Sarpal) G.; M.Sc., Allahabad U., 1930-32; Asso. (Lady Tata Meml. Trust Research fellow), Indian Inst. Sci., Bangalore, 1935; Ph.D., Punjab U., 1938; m. Nirantra Taxali, Feb. 13, 1940; children—Arun K., Bhuwnesh. Faculty, Punjab U., 1935-39, 45—, head dept. pharmacy, 1959—, prof. pharmacy, Chandigarh, India, 1964—; chief investigator Pfizer Pvt. Ltd., Bombay, India, 1960-62, Inst. of History of Medicine and Med. Research, Delhi, India, 1963-64. Mem. Indian (council 1947—, past v.p.), Am. pharm. assns., Pharmacy Council India (chmn. edn. regulation com. 1960-65), Punjab Pharmacy Council. Author: Hospital Pharmacopoeia of Medical College, Amritsar and Attached Hospitals, 1953; also articles. Research on synthetic local anesthetics, preservative for pharms., indigenous med. plants; synthesized numerous antiseptic sulphamylazo phenolic dyes. Patentee local anesthetic, antiseptic dye. Home: 84 C/19-A, Chandigarh Union Ter., India.*

GAINES, George Loweree, Jr., Am. chemist; b. New Haven, Mar. 7, 1930; s. George Loweree and Anna (Ekman) G.; B.S., Yale, 1950, M.S., 1952, Ph.D., 1954; m. Margaret Earl Greene, Aug. 21, 1954; children—Barbara Hartley, George Loweree III, Elizabeth Wolcott. Chemist electrochems. dept., E. I. duPont de Nemours & Co., Niagara Falls, 1950-51; phys. chemist Gen. Elec. Research Lab., Research and Devel. Center, Schenectady, 1954—. Mem. Faraday Soc., Chem. Soc. London, Sigma Xi, Phi Beta Kappa. Author: Insoluble Monolayers at Liquid-Gas Interfaces, 1966. Editorial bd. Jour. Colloid and Interface Sci., 1966—. Exptl. studies in surface chemistry, ion exchange, phys. chemistry of aluminosilicate minerals. Patentee in field. Home: Box 94, R.D. 2, Scotia, N.Y. 12302. Office: P.O. Box 8, Schenectady 12301.*

GAINES, Sidney, Am. microbiologist; b. Cleve., July 15, 1917; s. Lippo and Florence (Margolin) Ginsburg; B.A., Ohio State U., 1941, M.Sc., 1943, Ph.D., 1950; m. Eva Peyton, Mar. 3, 1960. Commd. 2d lt. U. S. Army Med. Service, 1943, advanced through grades to col., 1965; bacteriologist-serologist, various hosps., Europe, U. S. A., 1944-48; research bacteriologist Walter Reed Army Inst. Research, 1952-55, chief dept. microbiology, 1956-61, asst. to dir., 1964-65; chief dept. bacteriology SEATO Med. Research Lab., Bangkok, Thailand, 1961-63; chief enteric bacteriology lab. Army Med. Research Team, Saigon, Vietnam, 1965—. Diplomate Am. Bd. Microbiology. Fellow Am. Acad. Microbiology; mem. Am. Assn. Immunologists, Am. Soc. Microbiology, Phi Beta Kappa, Sigma Xi. Research, publs. various areas in bacteriology, with emphasis on endotoxins, immunizing agts., exptl. typhoid fever in chimpanzees, bacterial virulence, enterobacteriaceae. Home: 658 Kennebec Av., Takoma Park, Md. 20012. Office: Walter Reed Army Inst. Research, Washington 20012.*

GAIRDNER, William, Brit. physician; b. Mt. Charles, Eng., Nov. 11, 1793; s. Robert Gairdner; M.D., U. Edinburgh (Scotland), 1813; postgrad. London; m. Jan. 12, 1822; 1 dau. Physician to Earl of Bristol; settled in London, 1822. Licentiate Coll. Physicians. Author: On Gout, its History, its Causes, and its Cure, 1849; Essay on the Effects of Iodine on the Human Constitution, 1824. Advocated internal use of iodine for cure of goitre and other glandular enlargements, and described ill effects of over doses. Died Avignon, France, Apr. 28, 1867.

GAIRDNER, Sir William Tennant, Scottish physician; b. Edinburgh, Scotland, Nov. 8, 1824; s. John Gairdner; ed. Edinburgh U.; M.D., LL.D.; m. Helen Bridget Wright, 1870; 4 sons, 4 daus. Became resident med. officer Royal Infirmary, Edinburgh, 1846, pathologist, 1848, physician, 1853; extra-mural lectr. in Edinburgh on practice of medicine and clin. medicine, until 1862; chief med. officer City of Glasgow, 1863-72; prof. medicine U. Glasgow, 1862-1900; hon. physician in ordinary to King Edward VII in Scotland; hon. cons. physician Western Infirmary, Glasgow. Fellow Royal Soc., 1893; pres. Brit. Med. Assn., 1888. Author: Public Health in Connection with Air and Water, 1862; Clinical Medicine, 1862; The Physician as Naturalist, 1889; also numerous articles on angina pectoris, aneurism of aorta, other med. topics. Died June 28, 1907.

GAITAN, Mario, Colombian physician; b. Neiva, Colombia, June 28, 1919; s. Anselmo and Laura (Yanguas) G.; Bachiller, Colegio del Rosario, Bogotá, Colombia, 1936; Médico-Cirujano, Nat. U., 1942; m. Clara Eugenia Gaitán, Feb. 2, 1946; children—Armando, Victoria Eugenia, Marcela, Mauricio. Practice gen. medicine, 1942-43; with Shell Oil Co., Colombia, 1944; radiotherapist Nat. Cancer Inst., 1948, head radiotherapy dept., 1949-56, dir., 1957—; faculty Nat. U., 1949-58, prof. biophysics, 1951-58; prof. cancerology Javeriana Sch. Medicine, Bogotá, 1959—. Bd. dirs. Nat. Inst. Nuclear Affairs, Bogotá. Recipient medal of physiology, 1939; Carlos Esguerra prize, 1943. Fellow Am. Coll. Radiology (hon.); mem. numerous cancer, radiology, health physics, nuclear medicine socs. in Colombia, U. S., various Latin-Am. countries. Author: Carcinoma del Seno, 1958; Alberto Verjarano, 1958; Temas de Cancerología y Radioterapia, 1946; also numerous articles. Developed new techniques for radiol. epilation of scalp; research on treatment, epidemiology, physiology, and diagnosis of cancer. Home: Carrera 16-A, Number 85-92. Office: Carrera 12, 20-69, Bogotá, Colombia.*

GAITONDE, B(hikaji) B(alwant), Indian pharmacologist; b. Banda, India, Dec. 29, 1926; s. Balwantrao S. and Annapurna (Sabnis) G.; grad. Bombay (India) Inst. Sci., 1945, M.B.B.S., 1950, M.Sc., 1955, M.D., 1957; m. Sudha B. Sabnis, June 25, 1953; children—Mangala, Uday, Girish. House officer K.I.M. Hosp., Bombay, 1950-53; asst. prof. pharmacology T.N. Med. Coll., Bombay, 1955-56; prof. pharmacology Grant Med. Coll., Bombay, 1959—; vis. scientist Dartmouth Med. Sch., 1963—. Sec. expert com. I.C.M.R.; mem. C.C.A.R., Govt. India. WHO fellow, 1954; Rockefeller fellow, 1963. Author: General Pharmacology of Drug Action, 1967; Tutorials in Pharmacology, 1967; also articles. Editorial com. Jour. Assn. Physiol and Pharmacology India. Research in indigenous drugs in India, mechanism of emetic action of digitalis.*

GAJDUSEK, Daniel Carleton, Am. pediatrician; b. Yonkers, N.Y., Sept. 9, 1923; s. Karl A. and Mahtil (Dobroczki) G.; B.S. summa cum laude in Biophysics, U. Rochester, 1943; M.D., Harvard, 1946; postgrad. in phys. chemistry Cal. Inst. Tech., 1947-49; 1 son, Ivan Mbagintao. Asst. pediatrics and virus research Harvard, 1949-52; vis. investigator Institut Pasteur, Teheran, Iran, 1954, Walter and Eliza Hall Inst. Med. Research, Melbourne, Australia, 1955-57; dir. study child growth and devel. in primitive cultures and lab. of slow, latent and temperate virus infections NIH, Bethesda, Md., 1959—. Recipient E. Meade Johnson award Am. Acad. Pediatrics, 1963. Diplomate Am. Bd. Pediatrics. Mem. Am. Pediatric Soc., Soc. for Pediatric Research, Am. Soc. Human Genetics, Am. Acad. Neurology, Phi Beta Kappa, Sigma Xi. Author: Hemorrhagic Fevers and Mycotoxicosis in the USSR, 1953. Co-editor: Slow, Latent and Temperate Virus Infections, 1965. Research, numerous publs. on learning and behavior, child growth and devel. in primitive cultures, genetic variability in brain of man, virology, pediatrics, genetics, theory of cyphers and notation for coding sensory and motor data for study neurol. patterning and learning, codiscoverer and 1st med. description of kuru disease in New Guinea. Home: 4 Laurel Pkwy., Chevy Chase, Md. Office: Nat. Inst. Neurol. Diseases and Blindness, NIH, Bethesda, Md. 20014.*

GAJEWSKI, Waclaw, Polish geneticist; b. Cracow, Poland, Feb. 28, 1911; s. Waclaw and Wanda (Landau) G.; Ph.D., U. Warsaw (Poland), 1927; m. Anna Bogdani, July 27, 1938; children—Mary, Jacek. Asst., Botanic Garden, U. Warsaw, 1926-39, prof. botany, 1945-50, prof. genetics, 1950—. Mem. Polish Acad. Scis. (head dept. gen. genetics 1956—), Polish Bot. Soc. (editor publs. 1947—), Polish Genetical Soc. (pres. 1965—). Author: A Cytogenetic Study of Genus Geum, 1957; also numerous articles. Research on evolution of plants based on genetic and cytological analysis, genetic structure of fungi. Address: 4 ul. Ujazdowskie, Warsaw, Poland.*

GALAMBOS, Robert, Am. physiologist; b. Lorain, O., Apr. 20, 1914; s. John and Julia (Petti) G.; A.B., Oberlin Coll., 1935, M.A., 1936; A.M., Harvard, 1938, Ph.D., 1941; M.D., U. Rochester, 1946; m. Jeannette Wright, Dec. 30, 1939; children—Joan B., Katherine W., Ann J. Instr. physiology Harvard Med. Sch., 1942-43, jr. investigator for OSRD, 1942-43, tutor biochemistry, 1941-42, 47-48; asst. prof. anatomy Emory U. Med. Sch., 1946-47; research fellow psycho-acoustic lab. Harvard, 1947-51; chief dept. neurophysiology Walter Reed Army Research Inst., 1951-62; prof. psychology, physiology Yale U., 1962—; spl. research hearing and learning in nervous system. Mem. Am. Physiol. Soc., Acoustical Soc. Am., Nat., Am. acads. sci. Editorial bd. Am. Jour. Physiology, Jour. Applied Physiology, 1956-59, Jour. Neurophysiology, 1958-59. Research on neurophysiology of learning; obstacle avoidance by bats; hearing. Home: 35 Tokeneke Dr., Hamden 18, Conn. Office: 333 Cedar St., New Haven.

GALANIN, Nikolay Fedorovich, Russian hygienist; b. 1893; grad. 1st Leningrad Med. Inst., 1923; D.Med. Sci. Head physics lab. Leningrad Inst. Labor Hygiene and Occupational Diseases, 1927-49; instr. Kirov Mil. Med. Acad., Leningrad, 1926-38, prof., 1938-42, head chair gen. hygiene, 1942-55; dir. Inst. Radiation Hygiene, Leningrad, 1956—; chmn. Leningrad dept. ultraviolet irradiation sect. learned council Inst. Biophysics, USSR Acad. Sci. Mem. USSR Acad. Med. Sci. (corr.), Leningrad (chmn.), All-Union (bd. mem.), All-Russian (bd. mem.) socs. hygienists and pub. health officers. Author over 100 works including Radiation and Its Importance in Hygiene, 1952; Health Climatology, 1960; co-author: Military Hygiene, 1936. Mem. editorial council Hygiene and Sanitation; co-editor Hygiene sect. Large Med. Ency., 2d edit. Address: Inst. Radiation Hygiene, ulitsa Mira 6-8, Leningrad, USSR.

GALANT, Ivan Borisovich, Russian psychiatrist; b. 1893; grad. Med. Faculty, Basel U., 1917; D.Med. Sci., 1917. Staff mem. psychiat. clinics and hosps., Moscow, Smolensk, Leningrad, 1921-35; prof., head chair psychiatry Khabarovsk Med. Inst., 1935—; chief psychiatrist Khabarovsk Kray. Mem. Khabarovsk Kray Soc. Neuropathologists and Psychiatrists (chmn.). Author: Spinal Reflex—A New Reflex in Children, 1917; Neologisms of the Mentally Sick, 1919; The Years of Research on Reflexes, 1917-1927. Research and publs. on pathology and psychopathology of Far Eastern tickborne encephalitis and summer-fall mosquito-borne encephalitis, schizophrenia, epilepsy, stomatogenic psychoses, chronic alcoholism, alcoholic psychoses, regional neuropathology and psychopathology, described various reflexes. Address: Khabarovsk Med. Inst., ulitsa Karla Marksa 32, Khabarovsk, RSFSR, USSR.

GALASSO, Francis Salvatore, Am. chemist; b. Monson, Mass., Apr. 26, 1931; s. Paul Francis and Rubino (Cirillo) G.; B.S., U. Mass., 1953; M.S., U. Conn., 1957, Ph.D., 1960; m. Lois Ernestine Wood; children—Cynthia Jean, Gary Paul. Research asst., lab. instr. U. Conn., 1955-60; with United Aircraft Research Labs., East Hartford, Conn., 1960—, supr. material synthesis group, 1963—. Mem. Sci. Research Soc. Am., Am. Chem. Soc., Sigma Xi, Phi Lambda Upsilon (v.p. 1958). Contbr. articles to sci. jours. Contbd. to understanding of ordering in oxides; prepared, studied structure and properties of superconducting, pyrolytic, laser, ferroelectric and ferro-magnetic materials; contbd. to bringing boron fibers from research to prodn. stage; compiled preparation, structure and properties of perovskite-type compounds; arranged crystallographic structures into groups in systematic manner. Home: 13 Green Manor Rd., Manchester, Conn. 06040. Office: United Aircraft Research Labs., East Hartford, Conn. 06108.*

GALBREATH, Edwin Carter, Am. paleontologist; b. Ashmore, Ill., Mar. 18, 1913; s. Walter Edwin and Ina (Pepper) G.; B.Ed., Eastern Ill. U., 1941; Ph.D., U. Kan., 1951; m. Janet Elizabeth Bute, Aug. 30, 1955; 1 son, Walter Franklin. Asst. prof. dept. anatomy U. Kan., 1952-57; prof. dept. zoology So. Ill. U., Carbondale, 1957—. Fellow Geol. Soc. Am.; mem. Soc. Vertebrate Paleontology, Soc. Study Evolution. Editor: Ill. Trans. Acad. Sci., 1963-67. Research, numerous publs. on morphology of vertebrates and stratigraphy of middle N.Am. Cenozoic deposits. Home: 608 Skyline Dr., Carbondale, Ill. 62901.*

GALE, David, Am. mathematician; b. N.Y.C., Dec. 13, 1921; s. Henry and Therese (Strauss) G.; A.B., Swarthmore Coll., 1943; M.A. in Math., U. Mich., 1947; Ph.D. in Math., Princeton, 1949; m. Julie Brigitte Skeby, Apr. 2, 1954; children—Kirsten, Karen, Katharine. Mem. staff radiation lab. Mass. Inst. Tech., 1943-45; Henry B. Fine instr. math. Princeton, 1949-50; mem. faculty Brown U., 1950-66, chmn. dept. math., 1960-66, prof. math., 1961-66; prof. math. and indsl. engring. U. Cal. at Berkeley, 1966—. Fulbright research scholar U. Copenhagen (Denmark), 1953-54; cons. RAND Corp., 1957-58; vis. prof. U. Osaka (Tokyo), 1962-63; spl. research math. econs., theory of games, geometry convex sets. Guggenheim fellow, 1962-63. Fellow Econometric Soc.; mem. Am. Math. Soc., Math. Assn. Am., Econometric Soc., Phi Beta Kappa, Sigma Xi. Author: The Theory of Linear Economic Models, 1960; also articles. Research on math. theory of econ. models including problems in theory of econ. equilibrium, econ. growth and nat. econ. planning, geometry of convex bodies, combinatorial problems in geometry. Home: 791 Hilldale Av., Office: Berkeley, Cal. 94708.*

GALE, Ernest Frederick, English chem. microbiologist; b. Luton, Eng., July 15, 1914; s. Ernest Francis and Nelly (Tomlin) G.; B.A., U. Cambridge (Eng.),

1936, Ph.D., 1939, Sc.D., 1944; B.Sc., U. London, 1935; m. Eiry Mair Jones, Aug. 28, 1937; 1 son, David Anthony. Faculty, U. Cambridge, 1936—, prof. chem. microbiology, 1960—, fellow St. John's Coll. Fellow Royal Soc., 1953; mem. Soc. Gen. Microbiology (internat. rep. 1961-67, pres. 1967—), Biochem. Soc. Author: The Chemical Activities of Bacteria, 1947, 51; Synthesis and Organisation in Bacterial Cell, 1959. Publs. on discovery of decarboxylases. roles, devel. of enzymes, biochem. sites of action of antibiotics, work on need of RNA, DNA, role of lipids in protein synthesis. Home: 25 Luard Rd., Cambridge, Eng.*

GALE, Henry Gordon, Am. physicist; b. Aurora, Ill., Sept. 12, 1874; s. Eli Holbrook and Adelaide (Parker) G.; A.B., U. Chgo., 1896, postgrad., 1896-97, fellow in physics, 1897-99, Ph.D., 1899; m. Agnes Spofford Cook, Jan. 5, 1901; 1 dau., Beatrice Gordon, Asst. in physics U. Chgo., 1899-1900, asso., 1900-02, instr., 1902-07, asst. prof., 1907-11, asso. prof. physics, 1911-16, prof., 1916-40, dean in Jr. Colls., 1908-40, dean sci. in Colls., 1912-40, dean Ogden Grad. Sch. Sci., 1922, chmn. dept. physics, 1925, dean div. phys. scis., 1931, emeritus; physicist Solar Obs., Mt. Wilson, Cal., 1906; research asso. Carnegie Inst. at Mt. Wilson, 1909, 10, 11; editor Astrophys. Jour. 1912-42. Fellow A.A.A.S. (v.p. 1934), Am. Phys. Soc. (v.p. 1927-29; pres. 1929-31), Am. Optical Society; mem. Sigma Xi, and Gamma Alpha. Author: (with R. A. Millikan) A First Course in Physics, 1906; A Laboratory Course in Physics, 1906; Practical Physics, 1920; Elements of Physics, 1926; (with R. A. Millikan and C. W. Edwards) A First Course in College Physics, 1928. Known for measuring velocity of light, testing rigidity of earth and testing ether-drift theory (in collaboration with A. A. Michelson). Died Chgo., Nov. 16, 1942.

GALE, Hoyt Stoddard, Am. geologist; b. Cleve., Dec. 9, 1876; s. George Rodney and Helen Maria (Richardson) G.; A.B., Harvard, 1900, S.B., 1902; m. Almira Miller, June 18, 1902; 1 son, Hoyt Rodney. Asst. geologist U. S. Geol. Survey, 1902-10, geologist, 1910-52; in charge sect. non-metalliferous deposits, 1912-20; chief geologist various fgn. subsidiaries Gulf Oil Corp., 1921-23; in charge Gulf Co.'s operations in Cal. and on Pacific coast, 1923-29; survey oil possibilities South Africa, 1938-39; later cons. geologist. Fellow Geol. Soc. Am. (council 1937-39), Soc. Economic Geologists, A.A.A.S. Author of U. S. Geol. Survey bulletins on coal fields, of northwestern Colo. borax, potash and nitrate deposits in U. S., and miscellaneous contbrns. on geologic subjects, including Geology of Southern California, guidebook for 16th Internat. Geological Congress, 1933, Geology of the Kramer borate deposits, Calif., 1945, etc. Research on gen. stratigraphic and structural geology, occurrence of oil and non-metallic minerals. Died July 6, 1952.

GALE, Thomas, Brit. surgeon; b. London, 1507; apprenticed to Richard Ferris, barber surgeon; a son, Thomas; served in army of Henry VII in France, 1544, under Philip II of Spain at the siege of St. Quentin, 1557; established med. practice, London, 1559; became master Barber Surgeons Co., 1561; wrote a book on surgery in 4 treatises which contained a prescription for styptic powder. Introduced term syphilis into English. Died 1587.

GALE, Walter Frederick, astronomer; b. Sydney, Australia, Nov. 27, 1865; s. Henry and Susannah Gordon (Windeyer) G.; ed. Paddington Hourse Sch.; trained in ins. and comm., offices; m. Violet Marion Birkenhead, 1899; 2 sons, 4 daus. With Savs. Bank of New S. Wales (later amalgamated with Govt. Savs. Bank), 38 years. Fellow Royal Astron. Soc.; founder (with R. T. A. Innes) New S. Wales br. Brit. Astron. Assn., 1st sec., also pres. Contbr. to astron. jours. and publs. Built a 7-inch reflecting telescope, circa 1884, later others up to 12 inches; began observations of Mars and other planets, 1886; discovered oases, also some uncharted canals, 1892; concluded that Mars can bear life; discovered 3 new comets, some double stars and nebulae; held that Argo Nebula is the center of galaxy and its 2-armed spiral character. Died June 1, 1945.

GALEAZZI LISI, Riccardo, Italian physician, surgeon; b. Rome, Italy, July 26, 1891; s. Goffredo and Emma (Lisi) G. L.; M.D. in surgery; m. Antonietta Del Giudice; children—Antonello, Lorenzo, Elisabetta, Annamaria. Prof., U. Studies, Rome; prin. physician 2 hosps. in Rome. Mem. Papal Acad. Scis., Mainz Acad. (founder, pres.). Research and numerous publs. on social sci., medicine, hygiene. Address: via Stafano Jacini 41, Rome, Italy.

GALEN, Claudius, Hellenistic physician; b. Pergamum, Asia Minor, circa 129; s. Nicon; ed. sch. philosophy in Pergamum, at Smyrna, Corinth, med. sch. in Alexandria; traveled extensively through eastern provinces of Roman Empire, circa 149-157; physician to gladiators in Pergamum, 157-161; lectr. in Rome, 162; became physician in ordinary to family of Flavian Boethius; returned famous to Pergamum, 166-168; physician to emperor Marcus Aurelius, Rome, 169. Author: De sectis ad eos qui introducuntur; De optima doctrina; De natura hominis commentaria III; De usu partium corporis humani; De anatomicis administrationibus; De elementis secundum Hippocratem, De temperamentis; De facultatibus naturalibus; De usu respirationis; De sanitate tuenda; Synopsis de pulsibus; De victu attenuante; De compositione medicamentorum; others; wrote over 500 treatises (about 100 extant). Systemized and unified Greek anat. and med. knowledge and practice; excelled in anatomy and physiology; performed careful dissections of many animals including Barbary ape, drew conclusions about human anatomy; gave account of fetal circulation; described arterial canal and foramen ovale; demonstrated that arteries carry blood instead of air; postulated there were small holes in septum of heart to allow blood to ooze from one side to other; determined mechanism of respiration and pulsation; analyzed functions of various glands, ducts, cardiac structures; one of 1st to understand value of pulse as diagnostic technique; believed that mind is located in brain (in contrast to Aristotelians); contributed to knowledge of brain, spinal cord, and nerves; explained singleness of binocular vision by referring to fact that some of optic fibers cross at chiasma and some do not; based his pathology on doctrine of four humors; many of terms he coined still used in modern anatomy; exerted great influence on devel. of medicine throughout Middle Ages; his works remained standard texts until superceded by Vesalius in anatomy and Harvey in physiology. Died probably in Sicily, circa 200.

GALENSON, Walter, Am. economist; b. N.Y.C., Dec. 5, 1914; s. Louis Peter and Libby (Mishell) G.; A.B., Columbia, 1934, Ph.D., 1940; m. Marjorie Spector, June 26, 1940; children—Emily, Alice, David. Prin. economist OSS, 1943-44; officer U. S. Fgn. Service, 1944-46; asst. prof. econs. Harvard, 1946-51; prof. U. Cal. at Berkeley, 1951-66, Cornell U., Ithaca, N.Y., 1966—. Dir. com. on economy of China, Social Sci. Research Council, 1962—; cons. ILO, 1962—. Mem. Am. Econ. Assn. Author numerous books including: CIO Challenge to the AFL, 1960; Labor in Developing Economies, 1962; Trade Union Democracy in Western Europe, 1961; Primer on Employment and Wages, 1966; The Quality of Labor Economic Development, 1964. Research, publs. on internat. labor affairs, econ. devel., analysis of planned econs. Home: 104 Homestead Circle, Ithaca, N.Y. 14850.*

GALESKI, Jozef Boguslaw Prawdzic, Polish sociologist; b. Warsaw, Poland, Feb. 22, 1921; s. Jozef and Zofia (Chienkin) G.; mgr. degree Warsaw U., 1951, Dr.Ph., 1951, dozent degree Inst. Agrl. Econs., 1958; m. Stanislawa Zabrzydowska, Nov. 13, 1943 (div. 1961); children—Maria Alina, Jan Jakub. Head div. rural social structure Inst. Agrl. Econs., 1954—; prof., head rural sociology div. Inst. Philosophy and Sociology, Polish Acad. Scis., 1961—; faculty Warsaw U., 1962—; Vice dir. rural edn. dept. Ministry Agr., 1950-51. Mem. Presidium of Com. for Indsl. Regions, Polish Acad. Scis., 1964—. Recipient Golden Merit award. Mem. European Soc. for Rural Sociology (council), Polish Sociol. Soc. (council), Polish Econ. Soc. Author: Spoleczna Struktura Wsi, 1958; Chlopi i Zawod Rolnika, 1963; Sociologia Wsi, Pojecia podstawowe, 1966. Editorial bd. Studia Sociologiczne, Roczniki Socjologii Wsi (chief editor). Research, publs. on system of rural sociology, concepts of transformation of peasantry into occupational category, determination of rural social regions in Poland, types and changes in social structure in rural areas. Home: 12 Gorczewska m.43, Warsaw. Office: 72 Nowy Swiat, Palac Staszica, Warsaw, Poland.*

GALEZOWSKI, Xavier, ophthalmologist; b. Lipowice, Poland, Jan. 5, 1833; studied medicine, St. Petersburg; M.D., Paris, 1865; m. Mlle. Tamberlieb. Author: Échelles typographiques et chromatiques pour l'examen de l'acute visuelle; Traité des maladies des yeux, 1872; Diagnostic et traitement des affections occulaires, 1874. Made detailed studies of eye anatomy; described (with Henri Parinaud) tuberculous conjunctivitis occurring in animals and man (Parinaud's oculo-glandular syndrome), 1889. Died circa, late 19th century.

GALFREDUS DE MELDIS, see Geoffroi of Meaux.

GALILEO (Galilei), Italian astronomer, mathematician, physicist; b. nr. Pisa, Italy, Feb. 15, 1564; s. Vincenzio and Giulia (Ammanati of Pescia) G.; ed. by monks at Vallombroso; studied medicine and physics U. Pisa, 1581-85; tutored in mathematics by Ostilio Ricci, 1583; children—Virginia, Livia, Vicenzio, by Marina Gamba. Prof. math. U. Pisa, 1589-92; prof. math. U. Padua, 1592-1610; returned to Florence; apptd. extraordinary prof. U. Pisa, also chief mathematician and philosopher to Grand Duke, 1610. Mem. Accademia dei Lincei Rome, 1611. Author: Sidereus nucius, 1610; Dialogo sopra i due massimi sistemi del mundo, tolemaico e copernico, 1632; Discorsi e dimostraziono matematiche intorno a due nuove scienze, 1638; many others. Discovered isochronism of pendulum, circa 1583; invented hydrostatic balance, 1586, used it to determine the specific gravity of bodies, also developed theorems on centers of gravities of solids; devised geometric and mil. compass, circa 1597; best known for his astron. contbns. made after he developed the first telescope of practical astron. value; used telescope to discover that moon has rough surface with mountains and valleys, 1610; correctly attributed the phosphorescence of dark portion of moon to sunlight reflected onto the moon from earth; discovered 4 satellites of Jupiter and charted their movements; named them Medicean planets; noted phases of Venus; discovered numerous stars previously unknown; observed spots on sun and showed that sun rotates; later studied libration of moon; endeavoured to erect a math. physics valid on a moving earth; realized distinction of gravity and levity was erroneous; demonstrated that velocity of free falling bodies is not proportional to their weights; argued that in a vacuum all objects would fall at same uniformly accelerated rate; found math. expression describing free fall; analyzed projectile motion into its components, and found path to be parabolic; invented apparatus for measurement of temperature, circa 1600; perfected compound microscope, 1624. Devoted much time to an attempt to keep his church from condemning Copernicanism, 1612-16, but failed when Copernicus' book was placed on index of forbidden books, 1616; tried by the Holy Office (Roman Catholic Church), found guilty for teaching Copernicanism and forced to make an abjuration of his beliefs, then sentenced to indefinite imprisonment, 1633; his Dialogo was placed on the index (removed only in 1822). Died a prisoner of Inquisition, under house arrest, at Arcetri, nr. Florence, Jan. 8, 1642.

GALISSARD DE MARIGNAC, Jean Charles, see de Marignac, Jean Charles Galissard.

GALL, Edward Alfred, Am. physician; b. N.Y.C., June 10, 1906; s. Julius E. and Eva (Fleischl) G.; student Coll. City N.Y., 1927; M.D., Tulane U., 1931; m. Phyllis H. Rivard, Sept. 17, 1933; children—Eric Papineau, Thomas Monroe. Instr. pathology Tufts Med. Sch., 1935-40, Harvard Med. Sch., 1940-41; dir. labs. Bethesda Hosp., Cin., 1941-48; faculty U. Cin., 1942—, Mary M. Emery prof. pathology, chmn. dept., 1948—; dir. pathology labs. Cin. Gen. Hosp., 1948—; pathologist in chief Children's Hosp., Cin., 1948—; cons. VA, Drake, Dunham, Bethesda, Booth hosps., Armed Forces Inst. Pathology, Office Surgeon Gen. Decorated grand officer Order of Daniel E. Carrion, Peru, 1965. Hon. mem. Sociedad Peruana de Anatomia Pathologica, Japanese Am. Soc. Pathologists, Kansas City Soc. Medicine, Dallas So. Clin. Soc., Western Germany Armed Forces Med. Soc.; mem. Am. Assn. Pathologists and Bacteriologists (pres. 1964), Am. Soc. Exptl. Pathology, Internat. Acad. Pathology (mem. council 1962-65, v.p. 1967-68), Ohio Soc. Pathologists (pres. 1950), Am. Soc. Clin. Pathologists (dir. 1962—), Soc. Exptl. Biology and Medicine, Central Soc. Clin. Research, Am. Assn. for Study Liver Disease. Editor in chief Am. Jour. Pathology, 1957-66. Publs. research on disorders of liver, blood forming organs, neplastic diseases. Home: 101 Lafayette Circle, Cin. 45220.*

GALL, Franz Joseph, German physiologist; b. Tiefenbrunn, Baden, Germany, Mar. 9, 1758; ed. Baden, Strasbourg, France, Vienna; began practice medicine, Vienna, 1785; lectr. phrenology and physiognomy, Vienna, 1796-1802; lectr. on tour, 1805-07; worked as lectr., physician, writer assisted by Johann Kaspar and Spurzheim, Paris, 1808-13. Author: Anatomie et physiologie du système nerveux en général, 4 vols., 1809-19; Sur les fonctions du cerveau, 6 vols., 1822-25. Founder phrenology; tried to establish the relationship between mental abilities and the shape of the brain and skull; believed different parts of the brain are asso. with different functions of the body; discovered the difference between gray matter (the active part) and white matter (connective tissue) of the brain; research on the anatomy of the nervous system. Died Montrouge, nr. Paris, Aug. 22, 1828.

GALL, Joseph Grafton, Am. biologist; b. Washington, Apr. 14, 1928; s. John Christian and Elsie (Rosenberger) G.; B.S., Yale, 1949, Ph.D., 1952; m. Dolores Marie Hogge, Sept. 17, 1955; children—Lawrence, Barbara. Faculty, U. Minn., 1952-63, prof., 1963; prof. biology Yale, 1963—. Mem. cell biology study sect. NIH, 1963-67. Mem. A.A.A.S., Am. Soc. Cell Biology (president 1967-68), Genetics Soc. Am., Am. Soc. Zoologists. Research, publs. on structure chromosomes, nucleic acid metabolism, cell fine structure by electron microscope. Home: 3 Crestview Dr., North Haven, Conn. 06473. Office: Dept. Biology, Yale, New Haven 06520.*

GALLAGHER, John Joseph, Am. zoologist; b. Phila., Mar. 30, 1914; s. Patrick Joseph and Catherine (Dowling) G.; B.A., U. Pa., 1949, Ph.D., 1955; m. Anna Helen Giordano, Aug. 2, 1947. Rotifer cons. Acad. Natural Scis. Phila., 1951, research asso. U. Pa., Phila., 1955-57, Ida. State U., Pocatello, 1957-60, N.E. La. State Coll., Monroe, 1960—. Mem. Ida. Sci. Acad. (mem., chmn. coms., sec.-treas. 1958-60), Am. Micros. Soc. (membership com. 1961—), A.A.A.S., N.W. Sci. Assn., Am. Inst. Biol. Scis., Am. Soc. Limnology and Oceanography, La., Pa. acads. sci., Internat. Limnology Assn. Contbr. articles to profl. jours., chpts. in books. Demonstrated complexity of cyclomorphosis; brought together shore-influenced methods of rotifer collecting; originated theory of rotifer population variations; discovered new rotifer species of Ia. and La. Address: 860 20th St., Knoxville, Tenn. 37916.*

GALLAGHER, Patrick Kent, Am. chemist; b. Waukegan, Ill., Mar. 17, 1931; s. George Francis and Florence (Jorgenson) G.; B.Sc., U. Wis., 1952, M.S., 1954, Ph.D., 1959; m. Marianne Ruth Maske, Aug. 29, 1953; children—Michael Kent, John Patrick. Mem. tech. staff Bell Tel. Labs., Inc., Murray Hill, N.J., 1959—. Mem. Am. Chem. Soc., Am. Ceramic Soc., A.A.A.S. Research, publs. on optical and Mossbauer spectroscopy applied to structure and bonding within inorganic materials; physiochem. studies of complexion equilibria in solution, preparation of finely divided and reactive ceramic raw materials of high purity; studies of thermal decomposition of various inorganic materials. Home: 69 Harrison Brook Dr., Basking Ridge, N.J. 07920. Office: Bell Tel. Labs., Murray Hill, N.J. 07974.*

GALLAGHER, Thomas Francis, Am. biochemist; b. Chgo., Dec. 29, 1905; s. Thomas Francis and Catherine Margaret (Regan) G.; A.B., Fordham, 1927; Ph.D., U. Chgo., 1931; m. Beatrice Marie Sheehan, July 21, 1930; children—Thomas Francis, Brian Boru, Michael Jerome. Successively research asso., asst. prof. and asso. prof. biochemistry U. Chgo., 1930-47; Gen. Edn. Bd. fellow Free City of Danzig and Berlin, Germany, 1936-37; mem. Sloan-Kettering Inst., and chief divs. steroid bio-chemistry and steroid metabolism, 1947-63; chief Inst. for Steroid Research, Montefiore Hosp. and Med. Center, 1963—. Mem. Am. Soc. Biol. Chemists, Am. Chem. Soc., Am. Assn. Cancer Research, Soc. Exptl. Biology and Medicine. Contbr. articles profl. jours. Research in biochemistry of pituitary and testicular hormones, bile acids, synthesis of adrenal cortical hormones; 1st to isolate androsterone and testosterone (with C. R. Moore, F. C. Koch), 1929. Home: 136 E. 64th St., N.Y.C. Office: 210th St. and Bainbridge Av., Bronx, N.Y.*

GALLATIN, (Abraham Alfonse) Albert, ethnologist, economist; b. Geneva, Switzerland, Jan. 29, 1761; s. Jean and Sophie Albertine (Rolaz) G.; grad. Geneva Acad., 1779; m. Sophie Allegre, 1789; m. 2d, Hannah Nicholson, 1793; at least 2 sons, Albert, Francis. Came to U. S., 1780; tutor French, Harvard, 1781; leader of settlers to Western Pa., 1784; mem. Harrisburg Conf. to revise U. S. Constrn., 1788; mem. Pa. Constl. Conv., 1789; mem. Pa. Ho. of Reps. from Fayette County, 1790-93; chiefly responsible for quelling Whiskey Rebellion of 1794 (thus preventing civil war in Pa.); mem. U. S. Ho. of Reps. from Pa., 4th-6th congresses, 1795-1801, Republican minority leader, caused creation of standing com. on finance (now ways and means com.); U. S. sec. treasury under Jefferson and Madison, 1801-14; during War of 1812, visited St. Petersburg (Russia) to attempt to secure Russian mediation to end war with Eng., 1813; peace commr. to Eng. negotiating Treaty of Ghent, 1814, concluded (with Adams and Clay) favorable comml. treaty with British, 1815; U. S. minister to France, 1815-23, to Eng., 1826-27; pres. Nat. (later Gallatin) Bank of N.Y.C.; a founder, 1st pres. council U. City N.Y. (now N.Y.U.), 1831; founder Am. Ethnol. Soc., 1842; pres. N.Y. Hist. Soc, 1843. Author: Considerations on the Currency and Banking System of the United States, 1831; Memorial of the Committee Appointed by the "Free Trade Convention" Held in Philadelphia in . . . 1831, published 1832. Instrumental in establishing state public edn.; proposed state paper money and Bank of Pa.; favored natural growth of industry, opposed excessive govt. taxation, reduced mil. appropriations and public debt, believed that Fed. money should be used to further an expanding internal economy. Died Astoria, L.I., N.Y., Aug. 12, 1849.

GALLAVARDIN, Louis, French physician; b. Lyons, France, Aug. 20, 1875; became licensed physician Lyons, 1902; prof. clin. medicine. Mem. French Acad. Scis. Research on heart diseases, inflammation of chest; introduced electrocardiograph in treatment of mitral lesions of heart. Died 1957.

GALLE, Andreas Wilhelm Gottfried, German geodesist; b. Breslau, Germany (now Wroclaw, Poland), June 22, 1858; s. Gottfried and Marie (Regenbrecht) G.; studied math. and natural scis., Göttingen, Berlin; Ph.D., Breslau, 1883; m. Clara Paetsch, 1895; 1 dau. Asst. to father Breslau Obs., 1880-83; became asst. Geodetic Inst., Potsdam, Germany, 1884, later became observer, named dept. chmn., 1911; lectr. geodesy Technische U. Berlin-Charlottenburg, 1900-10, became titular prof., 1902. Author: Die Polhöhe von Potsdam, 1898; Geodäsie, 1907; Lotabweichung im Harz und in seiner weiteren Umgebung, 1908; Mathematische Instrumente, 1912; C.F. Gauss als Zahlenrechner, 1918; also numerous articles, monographs. Studied latitude variations, longitude determination, gravity measurement; geodesic studies. Died Potsdam, Apr. 8, 1943.

GALLE, (Johann) Gottfried, German astronomer; b. Pabsthaus, Germany, June 9, 1812; s. Johann Gottfried and Henriette (Pannier) G.; student U. Berlin, 1930-33; Ph.D., 1845; m. Marie Regenbrecht, 1856; 2 sons, including Andreas. Began as secondary sch. tchr., Guben, Berlin; became asst. to Encke, Berlin Obs., 1835, later observer; dir. Prof Breslau (now Wroclaw, Poland) Obs., 1851-97. Contbr. articles to tech. jours. Discovered 3 new comets, 1839-40, innermost ring of Saturn (Krepp ring), (with Leverrier) the planet Neptune (before Adams and Challis); suggested parallax of planetoids be used to determine scale of solar system; research on meteors, earth magnetism, polar light phenomena, parhelions. Died Potsdam, Germany, July 10, 1910.

GALLEGLY, M(annon) E(lihu), Jr., Am. plant pathologist; b. Mineral Springs, Ark., Apr. 11, 1923; s. Mannon Elihu and Mary Virginia (Baber) G.; B.S. in Agr., U. Ark., 1945; M.S., U. Wis., 1946, Ph.D., 1949; m. Mary Elizabeth Smith, June 7, 1947; children—Michael Elioll, Susan Jane, Thomas William. Faculty, W.Va. U., Morgantown, 1949—, prof. plant pathology, 1960—. Recipient Campbell award A.A.A.S., 1960. Mem. Am. Phylopath. Soc., Potato Assn. Am., European Potato Assn., Mycol. Soc. Am., Am. Inst. Biol. Scis., W.Va. Acad. Sci., Sigma Xi, Gamma Sigma Delta, Alpha Zeta, Omicron Delta Kappa. Research, publs. on genetics and physiology of late blight disease of potato and tomato, description of pathogenic races of Phytophthora infestans; discovered sexual stage of this fungus and relations of sexual stage to other species of Phytophthora; devel. late blight resistant tomato varieties and potatoe lines. Home: 1292 Parkview Dr., Morgantown, W.Va. 26501.*

GALLEGO, Peter, Spanish translator; b. Galicia, Spain; flourished 1236-67; provincial of Castile, 1236; confessor to Alfonso el Sabio; head Ch. of Cartagena, 1242-50; 1st bishop of Cartagena, 1250-67; translated Aristotelian zoology from an Arabic abridgement, also treatise on economy which was ascribed to Galen.

GALLETTI, Pierre Marie, physiologist; b. Monthey, Switzerland, June 11, 1927; s. Henri and Yvonne (Chamorel) G.; M.D., U. Lausanne (Switzerland), 1951, Ph.D., 1954; m. Sonia Aidan, Dec. 31, 1959. Research fellow Inst. for Med. Research Cedars of Lebanon Hosp., Los Angeles, 1957-58; faculty Emory U., Atlanta, 1958—, prof. physiology, 1962—. E. Roosevelt fellow Internat. Union Against Cancer, U. Palermo, Italy, 1964-65. Fellow Am. Coll. Cardiology; mem. Am. Physiol. Soc., A.A.A.S., Swiss Physiol. and Pharmacological Soc., Am. Heart Assn., I.E.E.E., Am. Soc. Artificial Internal Organs. Author: (with G. A. Brecher) Heart-Lung Bypass: Principles and Techniques of Extra-corporeal Circulation, 1962; also numerous articles. Research on design, devel., technique of application artificial hearts, lungs, kidneys, placentas, enhancement drug action. Home: 308 Vickers Dr. N.E., Atlanta 30307.*

GALLI, Giuseppe, Italian surgeon; b. Rovato, Brescia, June 21, 1892. Prof., U. Modena, 1949, full prof., 1952, now full prof. gen. clin. surgery and surg. therapy; former instr. spl. surg. pathology and clin. propaedeutics U. Cagliari, U. Modena. Mem. Italian, Internat. socs. surgery, Italian Soc. Urology, Italian Soc. Radio Neurosurgery, Italian Soc. Gastroenterology, Italian Soc. Anesthesia, Italian Soc. Analgesia, Italian Soc. Exptl. Biology, Internat. Soc. Microbiology, Internat. Soc. Electroradio Biology, Med.-Surg. Soc. Modena (pres.). Contbr. numerous articles to profl. publs. Address: Univ. degli Studi, Modena, Italy.

GALLIE, William Edward, Canadian surgeon; b. Barrie, Ont., Can., Jan. 29, 1882; s. William and Annie M. (Gray) G.; M.D., U. Toronto, 1903; Sc.D. honoris causa, McGill U., 1948; m. Janet Louise Hart, Mar. 3, 1914; children—Alan Edward, Marion Louise, Hugh Richmond. Intern Hosp. for Sick Children, Toronto, 1903-04, resident surgeon, 1906-07, jr. surgeon, 1907-12; asso. surgeon, 1912-19, surgeon-in-chief, 1919-29, cons. surgeon 1929-59; intern Toronto Gen. Hosp., 1904-05, jr. surgeon, 1909-12, surgeon-in-chief, 1929-47, intern Hosp. for Ruptured and Crippled, N.Y. City, 1905-06, hon. surgeon-in-chief, 1937; with Faculty of Medicine, U. Toronto, as jr. demonstrator, 1906, demonstrator anatomy, 1906-09, demonstrator pathology, 1909-12, asst. prof. surgery, 1919-29, prof. surgery and head dept., 1929-47, emeritus prof. surgery, 1947-59, dean Faculty Medicine, 1936-46; Hunterian prof., Royal Coll. Surgeons, Eng., 1924. Fellow Royal (hon. gold medal 1947), Royal Coll. of Surgeons of Can., A.C.S. (past pres.); hon. fellow Royal Soc. Medicine, Western Surg. Assn., Acad. Orthopaedic Surgeons, Assn. Surgeons of Gt. Britain and Ireland; mem. Am. Orthopedic Assn. (pres.), Am. Surg. Assn. (pres. 1948), Central Surg. Assn., Am. Interurban Surg. Soc., Am. Interurban Orthopaedic Soc., Can. Physiol. Soc., Canadian Med. Assn., Canadian Soc. Clin. Surgs., Brit. Orthopaedic Soc. (hon.), Internat. Orthopaedic Soc., Internat. Soc. Surgeons, Author papers and lectures on med. and surg. subjects. Died Sept. 1959.

GALLIEN, Louis, French biologist; b. Cherbourg, France, Jan. 2, 1908; s. Louis Victor and Eugénie (Moret) G.; agrégé, 1930; D. Sc., 1935; D. honoris causa, U. Louvain (Belgium); m. Andrée Gueguen, Sept. 22, 1932; children—Suzanne, Claude Louis. Asst. Paris (France) Faculty Sci., 1932-37, prof., 1945—, dir. Embryology Labs., 1954—; lectr. Toulouse, Rennes, Caen (all France), 1938-44; prof. Caen Faculty Sci. Decorated Legion Honor, Palms of Acad. Mem. Inst. France, Am. Soc. Zoologists. Author: La Sexualité, 1941; Le Parasitisme, 1942; La Sélection Animale, 1943; L'Insémination Artificielle, 1946; Problemes et Concepts de L'Embryologie Expérimentale. Research, publs. on parasites, amphibian embryological expts., action steroid hormones, amphibian cytogenetics. Home: 31 Gasan. Office: 9 Quai St. Bernard, Paris, France.*

GALLIHER, Edgar Wayne, Am. geologist; b. Mcpherson, Kan., Sept. 10, 1907; A.B., Stanford, 1929, A.M., 1931, Ph.D., 1932; cons. Barnsdall Oil Co., 1932-35, geologist, 1935-37, apptd. chief geologist, 1937; research asso. Stanford, 1935; fellow Geol. Soc.; research on recent marine sedimentation, petrology of glauconite, diagenesis of marine sediments; died 1945.

GALLO, Duane Gordon, Am. biochemist; b. Aberdeen, S.D., May 15, 1926; s. Joseph William and Cora (Heidner) G.; B.S. with honors in Chemistry, U. N.D., 1951, M.S., 1953, Ph.D., 1955; m. Janice P. Lorenzen, Aug. 18, 1952; children—Roger, Rosanne, Douglas, Paula, Gregory, Celia, Christopher, Laura, Sara. Chemist, Mead Johnson Research Center, Evansville, Ind., 1955-57, asso. sr. chemist, 1957-58, sr. chemist, 1958-61, group leader, 1961—. Mem. Am. Oil Chemists Soc., N.Y. Acad. Scis., A.A.A.S., Research Soc. Am., Soc. Exptl. Biology and Medicine. Contbg. author: Encyclopedia of Chemistry, 1966. Research, publs. on factors influencing exptl. atherosclerosis, lipid synthesis, oral hypo-glycemic agts. Home: 10221 Upper Mt. Vernon Rd., Evansville, 47712. Office: 2402 Pennsylvania Av., Evansville, Ind. 47721.*

GALLOIS, Jean, French mathematician; b. Paris, June 14, 1632; abbé St. Martin de Cure, Autun diocese; protoge of Colbert, after whose death he became keeper Royal Library, also prof. Greek Coll. Royal; mem. French Acad. Scis., 1669, also French Acad.; a founder Jour. des Savants, editor, 1666-74. Died Apr. 19, 1707.

GALLONI, Ernesto Enrique, Argentine physicist; b. Buenos Aires, Mar. 5, 1906; s. Jose and Carolina (Galimberti) G.; C.E., U. Buenos Aires, 1930; m. Nelida Pedretti, Dec. 24, 1932; children—Ernesto, Marianelida (Mrs. Carlos Balmaceda), Maria Marta, Carlos, Jorge. Faculty, U. Buenos Aires, 1930—, prof. physics Faculty Engring., 1953—. Dir. Argentine Nat. Commn. on Atomic Energy, 1955-58, 63—. Mem. Academia Nacional de Ciencias Exactas Fisicas y Naturales de Buenos Aires, Academia Nacional de Ciencias Exactas Fisicas y Naturales de Cordoba, Real Academia de Ciencias de Madrid, Am. Phys. Soc., Asociacion Fisica Argentina, Mineral. Soc. Am. Author: (with J. S. Fernandez) Fisica Elemental, 2 vols., 1940; Trabajos Practicos de Fisica, 1943; (with R. Busch) Nociones Elementales de Fisico Quimica, 1952; Fisica Mecanica, 1962. Research, publs. on photoelasticity, crystal structure by X-ray and electron diffraction. Home: 1763 Yerbal, Buenos Aires, Argentina.*

GALLOWAY, Thomas, Brit. mathematician; b. Symington, Eng., Feb. 26, 1796; s. William and Janet (Watson) G.; ed. U. Edinburgh, Scotland; studied math. under Prof. Wallace; M.A.; m. dau. of Prof. Wallace, 1831. Faculty, Edinburgh, 2 years; became tchr. math. Royal Mil. Coll., Sandhurst, 1823; registrar, actuary Amicable Life Assurance Co. London, 1833-51. Fellow Royal Soc., 1834 (Royal medal 1848), Astron. Soc.; mem. Royal Astron. Soc. (became v.p. 1837, 48, fgn. sec. 1842). Contbr articles to tech. jours. on double stars, precession of equinoxes, chronology, probability theory; shape of the earth. Died London, Nov. 1, 1851.

GALLOWAY, Sir William, Brit. mining engr.; b. Paisley, Scotland; s. William Galloway; ed. U. Giessen (Germany); Bergakademie, Freiberg, Germany; Univ. Coll., London, Eng.; D.Sc. (hon.), U. Wales; m. Christiana Maud Mary Gordon; 2 sons; m. 2d, Mary Gwenap Douglas. Insp. mines W. Scotland, S. Wales dists.; prof. mining Univ. Coll. S. Wales and Monmouthshire. Mem. panel referees under Coal Mines Act, 1911; central examiner apptd. by Bd. Mining Exams.; external examiner in mining U. Birmingham, other univs. Fellow Geol. Soc., Inst. Dirs.; mem. Instn. Mining Engrs. (hon., medal for researches into action of coal-dust in colliery explosions), S. Wales Inst. Engrs. (hon., pres.). Contbr. papers on causes of colliery explosions, flying fish, mining, others. Suggested use of stone-dust as means of preventing colliery explosions, 1896 (suggested again in 1908, after which it was universally adopted); experimented with different kinds of explosives fired into mixtures of gas, air and coaldust; determined relation between height of firedamp cap and proportion of firedamp in air of mines; invented guides for sinking pits; described external capillarity, 1926. Died Nov. 2, 1927.

GALLUS (Gaius Sulpicius), Roman astronomer, orator; flourished in Rome, 166 B.C.; children include Quintus Sulpicius. Earliest of the Roman astronomers; orator, Greek scholar; served as mil. tribune under Aemilius Paullus; apptd. consul in Liguria, 166 B.C.; became envoy to Pergamum, 164 B.C.; wrote or adapted astron. work from Greek sources; predicted eclipse of the moon before the battle of Pydna, June 21, 168 B.C.

GALOIS, Evarist, French mathematician; b. Bourg-la-Reine, France, Oct. 25, 1811; s. Nicholas Gabriel and Adelaide-Marie (Demante) G.; ed. Lycee Louis-le-Grand, 1823-29, Ecole Normale, 1830. Involved radical politics, twice imprisoned for Republican sympathies. Evolved several original concepts on theory of algebra, before 1829; made significant contbns. to theories of equations, numbers and functions; research on resolubility of algebraic equations by radicals; often considered founder of theory of groups in algebraic equations; one of 1st to apply groups of substitutions to question of reducibility of algebraic equations (now known as Galois' theory); wrote paper containing most of his fundamental discoveries (not recognized until 1846, when it was pub. by Liouville in his Journal). Died in polit. duel, Paris, May 31, 1832.

GALPIN, Charles Josiah, Am. sociologist; b. Hamilton, N.Y., Mar. 16, 1864; s. Leman Quintilian and Frances Cordelia (Look) G.; A.B., Colgate U., 1885,

A.M., 1888 (Litt.D., 1919); A.M., Harvard, 1895; studied Clark U., 1898; U. Wis., 1908; m. Zoe N. Wickwire, of Hamilton, N.Y., June 22, 1887. Prof. history, Kalamazoo (Mich.) Coll., 1888-91; prin. Union Acad., Belleville, N.Y., 1891-1901; Baptist univ. pastor, U. Wis. 1905-11; asst. prof. and asso. prof. agrl. economics, U. Wis., 1911-19; economist in charge Div. of Farm Population and Rural Life, U. S. Dept. Agr., Washington, D.C., from 1919 to 1934. Editor of Century Rural Life Books. Studied rural problems in Europe, 1896, 1914, 1926; U. S. del. Gen. Assembly Internat. Inst. Agr., Rome, 1926. Mem. Am. Sociol. Assn., Delta Kappa Epsilon, Am. Country Life Assn. (v.p.); asso. mem. Internat. Inst. Sociology; corr. member Czechoslovak Acad. Agriculture, 1926. Awarded agrl. decoration of 1st class by King of Belgium, 1927. Author: The Social Anatomy of an Agricultural Community, 1915; Rural Life, 1918; Rural Social Problems, 1924; Empty Churches, 1925; My Drift into Rural Sociology, 1938. Joint editor of Systematic Source Book in Rural Sociology, 1931. Lecturer on social problems of the farmer; known for field studies of rural communities. Died 1947.

GALSTON, Arthur William, Am. biologist; b. N.Y.C., Apr. 21, 1920; s. Hyman and Frieda (Saks) G.; B.S., Cornell U., 1940; M.S., U. Ill., 1942, Ph.D., 1943; m. Dale J. Kuntz, June 27, 1941; children—William Arthur, Beth Dale. Research fellow Cal. Inst. Tech., 1947-50, sr. research fellow, 1947-50, asso. prof. biology, 1951-55; instr. Yale, 1946-47; Guggenheim fellow, Sweden, Eng., 1950-51; prof. biology Yale, 1955—, dir. div. biol. scis., 1965-66. Cons. central research dept. E.I. du Pont de Nemours & Co., 1957—. Mem. A.A.A.S., Inst. Biol. Scis., Am. Soc. Biol. Chemists, Am. Soc. Plant Physiologists, Bot Soc. Am. Author: The Life of the Green Plant, 1961, 2d edit., 1964; (with James Bonner) Principles of Plant Physiology, 1952; also numerous articles. Research on biochem. mechanisms in control plant growth and devel. by hormones and light; proposed new theories on pigments active in light absorption in phototropism, significance hormone destruction in aging. Home: 307 Manley Heights, Orange, Conn. 06477. Office: Dept. Biology Yale, New Haven. Conn. 06520.*

GALT, John Kirtland, Am. physicist; b. Portland, Ore., Sept. 1, 1920; s. Martin H. and Elsie (Lee) G.; A.B., Reed Coll., 1941; Ph.D., Mass. Inst. Tech., 1947; m. Marguerite Van Nest, Dec. 29, 1949; 1 son, James Michael. Teaching fellow physics dept. Mass. Inst. Tech., Cambridge, 1941-43, research asst. D.I.C. Project, 1943-44, electronics lab. research asso., 1945-47; research asso. radio research labs. OSRD, Harvard, Cambridge, 1944-45; NRC fellow U. Bristol, Eng., 1947-48; mem. tech. staff Bell Telephone Labs., Murray Hill, N.J., 1948-57, dept. head, 1957-61, dir. solid state electronics lab., 1961—. Mem. I.E.E.E., Am. Phys. Soc., Phi Beta Kappa, Sigma Xi. Study of band structures of metals; magnetic and mech. properties of solids. Home: 51 High St., Summit, N.J. Office: Mountain Av., Murray Hill, N.J.*

GALTON, Sir Francis, English anthropologist; b. Birmingham, Feb. 16, 1822; s. Samuel Tertius and Anne Violetta Galton; grandson of Erasmus Darwin and a cousin of Charles Darwin; studied medicine King's Coll., London, 1839-40, Trinity Coll., Cambridge, 1840-44, degree, 1844; hon. doctorates from Oxford, 1894, Cambridge, 1895, 1902; m. Louisa Jane Butler, Aug. 1, 1853. Traveled through Sudan, 1845-46; made one of 1st explorations of Damaraland and Ovampo country in S.W. Africa, 1850-52; visited Spain, 1860; founder Eugenics Lab., London; left bequest to establish eugenics chair U. London; mem. Meteorol. Council, 1868-1901. Recipient Darwin-Wallace Celebration medal Linnean Soc., 1908. Fellow Royal Soc. 1860 (Gold medal 1886, Darwin medal 1902); mem. Royal Geog. Soc. (council, Gold medal 1853), Anthropol. Inst. (pres. 1885-88, Huxley medal 1901), Brit. Assn. (gen. sec. 1863-68, pres. geog. sect. 1862, 72, pres. anthropol. sect. 1877, 85). Knighted, 1909. Author: Tropical South Africa, 1853; Meteorographica, 1863; Hereditary Genius, 1869; English Men of Science, Their Nature and Nurture, 1874; Human Faculty, 1883; Natural Inheritance, 1889; Finger Prints, 1893; Fingerprint Directory, 1895; Herbert Spencer Lecture, 1907. Cons. editor: Biometrika, from 1902. Best known for work in heredity; 1st to stress importance of studying identical twins to differentiate influence of environment from heredity; opposed theory of inheritance of acquired characteristics; investigated thought-processes of men of differing intellectual ability, thus a founder of mental testing; conducted statis. inquiries on inheritability of genius for over 40 years; credited with being founder of statis. school of genetics; devised correlational calculus; investigated capacity for recalling vivid mental images in individuals of varying mental ability and differing degrees of edn.; propounded law of filial regression (suggests that offspring of parents with unusual characteristics tend to regress to average), 1869; considered founder of eugenics; also made important contbns. to meteorology; founded modern technique of weather-mapping; invented term anticyclone for high-pressure areas; devised system of fingerprint identification, 1892; invented Galton's whistle to determine upper limit of hearing ability with regard to high frequency tones. Died Haslemere, Surrey, Jan. 17, 1911.

GALTSOFF, Paul Simon, Am. biologist; b. Moscow, Russia, Mar. 28, 1887; s. Simon Peter and Sofia (Poletaeva) G.; diploma Imperial Moscow U., 1910; Ph.D., Columbia, 1924; m. Eugenia Troussoff, Nov. 5, 1911. Came to U. S., 1921, naturalized, 1926. Sr. zoologist Imperial Acad. Scis., Sebastopol, 1915-19; privatdocent U. Crimea, Yalta, 1917-19; with U. S. Bur. Fisheries, Washington, 1922-64, chief sect. shellfish, 1928-50, dir. shellfish lab., Woods Hole, Mass., 1960-64, ret., 1964. Recipient Distinguished Service award, Gold medal Moscow U., 1911, U. S. Dept. Interior, 1963; Soc. of Friends of Natural History (Moscow) prize, 1915. Fellow N.Y. Acad. Sci.; mem. Am. Soc. Zoology (emeritus), A.A.A.S. (emeritus), Am. Soc. Limnology and Oceanography, Soc. Gen. Physiology (emeritus), Internat. Oceanographic Found. (trustee 1963—). Author: Pearl and Hermes Reef, 1932; Gulf of Mexico, Its Origin, 1955; History of the Fisheries Laboratory, Woods Hole, 1962; The American Oyster, 1964; also numerous articles. Established regeneration in sponges from separated cells, provocation of sexual activities of oysters and other bivalvers by temperature and chem. stimulation; discovered axial body in oyster sperm; established existence of toxic water soluble substances in crude oil and waste of pulp mill industry. Home: Box 167, Morgan Rd. Office: Fisheries Lab., Water St., Woods Hole, Mass. 02543.*

GALVANI, Luigi, Italian physiologist; b. Bologna, Italy, Sept. 9, 1737; studied medicine; m. Lucia Galeazzi (dec. 1790); named prof. anatomy, Bologna, 1762; also tchr. Inst. Scis.; resigned post, 1798. Author: De viribus electricitatus in motu musculari commentarius, 1791; also articles on kidneys and hearing organs of birds. Founder galvanism; observed legs of a frog touched with scalpel during dissection began to twitch, and that elec. shock on crural nerves or contact with 2 different metals produced same effect (which he called condensor effect); galvanic battery named after him; galvanism used as a term for manifestations of electric current. Died Bologna, Dec. 4, 1798.

GAMA, Vasco da, navigator, explorer; b. Sines, Portugal, circa 1460. Named to command expdn. to complete sea route to India around Southern Africa, 1497; rounded Cape of Good Hope, Nov. 22, 1497; landed at various places on both Eastern and Western African coasts during voyage; took on an Arab pilot to conduct him to India, at Malindi; reached India, May 20, 1498; left to return to Portugal, Aug. 1498; 1st European to have also landed at Mozambique and Zanzibar; returned to Portugal, Sept. 1499; commanded another expdn. to India, 1502-03; created count of Vidiguerira, 1519; viceroy of India, 1524. Opened trade route to India; discovered Amirante Islands. Died Cochin, India, Dec. 24, 1524.

GAMACHES, Étienne-Simon, French mathematician; b. Meulan, France, 1672; became clergyman; studied and wrote on astronomy, linguistics, philosophy; mem. French Acad. Scis., 1732; author: les Agréments de langage réduit a ses principes, 1718; Astonomie physique, 1740; Système du philosophe chrétien, 1746. Died Paris, Feb. 18, 1756.

GAMAIN, Bernard Jean, French physician; b. Charleville, France, Mar. 27, 1926; s. Henri and Madeleine (Deville) G.; Baccalauréat, Lycée de Charleville; student Faculty of Medicine, Paris, 1943-53; m. Jacquat Denise, Aug. 6, 1949; children—Claude, Anna, Francois, Pascal. Chief of lab. Faculty of Medicine, Paris, 1954-66; asst. biologist Hosps. of Paris, 1962—; pvt. practice medicine. Mem. French Soc. Thoracic Pathology. Author: (with Etienne Bernard) Clinical Bronchography, 1961; also numerous articles. Research on bronchology, lung function, respiratory insufficiency. Home: 105 Av. Morizet, Boulogne 92, France. Office: Hopital Cochin Rue de Faub. St. Jacques, Paris 14, France.*

GAMALIEYA, Nikolay Fyodorovich, Russian bacteriologist; b. Odessa, Russia, Feb. 5, 1859; s. Fiodor Gamalieya; studied at Mil. Med. Acad., St. Petersburg, Russia; student natural scis. U. Odessa, 1880; private docent pathology, Odessa; student and collaborator of Pasteur; in charge Jenner Inst., Leningrad, 1910-29; received chair microbiology 2d Med. Inst., Moscow, 1838; became collaborator J. J. Mechnikov, 1884. Mem. Soviet Sci. Acad. Author: Cholera and How to Combat It; Vaccination against Smallpox, 1917; Bacterial Poisons, 1893; Principles of Immunology, 1928; Filterable Viruses, 1930; Biological Processes in the Destruction of Bacteria, 1934; Infections and Immunity, 1939; Grippe and how to Combat it, 1941; Textbook of Medical Microbiology, 1943; Recent Progress in the Study of Malignant Tumors, 1946. Father of Russian med. microbiology; first to inoculate against rabies in Russia; research in etiology, epidemiology, and prophylaxis of grippe and Tb, 1940-44, methods of controlling the water supply to protect against outbreaks of cholera, plague; introduced method of rat control, small pox vaccination to Russia; inoculated for anthrax using Pasteur's methods. Died Mar. 24, 1949.

GAMBART, (Jean-Felix-) Adolphe, French astronomer; b. Cette, France, May 12, 1800; studied under Bouvard at Paris; became master astronomer; sent to Marseille (France) Obs. as adj. astronomer by Bur. Longitudes, 1819, apptd., dir., 1822. Mem. French Acad. Scis. (corr. mem. astron. sect.). Numerous observations of eclipses of satellites of Jupiter; discovered 13 comets and calculated their elliptical or parabolic orbits, 1822-34; 1st to calculate parabolic elements of Biela comet. Died Paris, July 23, 1836.

GAMBEY, Henri-Prudence, French inventor; b. Troyes, France, Oct. 8, 1787; master craftsman in Compiègne workshops; mem. French Acad. Scis., 1837, Bur. Longitudes. Recipient Gold medal Expn., 1819. Built precision instruments; perfected the cathetometer, heliostat, theodolite; invented dipping needle. Died Paris, Jan. 28, 1847.

GAMBLE, Frederick William, English zoologist; b. Manchester, Eng., July 13, 1869; s. William Gamble; ed. Manchester, Leipzig (Germany) univs.; B.Sc., 1891, M.Sc., 1893, D.Sc., 1900; m. Ellen Bamford, 1904. Asst. dir. zool. lab., lectr. zoology U. Manchester; prof. zoology U. Birmingham, from 1909. Fellow Royal Soc., 1907; mem. Brit. Assn. (pres. zoology sect. Toronto 1924). Author: Animal Life; The Animal World; also papers. Editor: Practical Zoology (Marshall and Hurst), 5th-7th edits. Died Sept. 14, 1926.

GAMBS, John Sake, economist; b. Saxoc, Guatemala, May 14, 1899; s. Gustave Adolphe and Caroline (Herrmann) G.; came to U. S., naturalized, 1904; A.B., George Washington U., 1920, M.A., 1924; Ph.D., Columbia, 1932; m. Alice L. Chase, July 7, 1938; children—Louise Martha, John Frederick. High sch. tchr. French, Washington, 1924-29; with various U. S. govt. agys., 1934-36; mem. permanent U. S. delegation Internat. Labor Office, Geneva, Switzerland, also substitute U. S. mem. governing body Internat. Labor Office, 1938-40; asso. prof. social scis. La. State U., 1940-42; mediation officer Nat. War Labor Bd., 1942-44; internat. labor adviser U. S. Dept. Labor, 1944-46; prof. econs. Hamilton Coll., Clinton, N.Y., 1946-67, Leavenworth prof., 1955-67, chmn. dept. econs., 1946-67. Mem. Am., N.Y. State (past pres.) econ. assns., Assn. for Evolutionary Econs. (a founder, pres. 1967—), Author: Decline of I.W.W., 1932; Beyond Supply and Demand, 1946; Man, Money and Goods, 1952; (with S. Wertimer, J. Komisar) Economics and Man, 1958; also articles, book revs. Assisted in devel. instl. econ. theory. Home: 81 College St., Clinton, N.Y. 13323.*

GAMGEE, Arthur, Brit. physician; b. Oct. 10, 1841; ed. Edinburgh U., also LL.D.; M.D.; D.Sc. (Hon.), Manchester, Eng., 1908; m. Mary Louisa Clark, 1875. Asst. to prof. med. jurisprudence Edinburgh U., 1863-69; named 1st Brackenbury porf. physiology Owens Coll., 1873; Fullerian prof. physiology Royal Instn. Gt. Britain, 1882-85; asst. physician, lectr. materia medica St. George's Hosp., London; examiner for several univs. Fellow Royal Coll. Physicians, Royal Soc., 1872 (Croonian lectr. 1902). Author: Text-book of the Physiological Chemistry of the Animal Body, 1880-93. Translator, editor: Human Physiology (Hermann), 1875, 2d edit., 1878. Contbr. papers on physiology and physiol. chemistry to jours. Research on continuous photog. and quasi-continuous ink registration of diurnal curve of man's temperature. Died Mar. 29, 1909.

GAMGEE, Joseph Sampson, surgeon; b. Leghorn, Italy, Apr. 17, 1828; s. Joseph Gamgee; ed. Italy; entered Royal Vet. Coll., London, 1847; m. Marion Parker, 1860; 7 children. Apptd. surgeon Brit. Italian Legion, 1855; in charge hosp. at Malta during Crimean War; surgeon to Queen's Hosp., Birmingham, Eng., 1857-81. Recipient Liston prize U. Coll., 1853. Mem. Royal Coll. Surgeons (Eng.). Author: On the Advantages of the Starched Apparatus in the treatment of Fractures and Diseases of the Joints, 1853; Reflections on Petit's Operations and on Purgatives after Herniotomy, 1855; On the Treatment of Wounds and Fractures, 1883; On Absorbent and Antiseptic Surgical Dressings, 1880. Developed several surg. devices including cotton wool absorbent pads, gauze tissue, millboards and paper splints; improved manufacture of cotton wool in an antiseptic condition. Died Sept. 18, 1886.

GAMMEL, John Ledel, Am. physicist; b. Austin, Tex., July 9, 1924; s. John Ledel and Elsie (Spletts-toesser) G.; B.S., U. Tex., 1944, M.A., 1946; Ph.D., Cornell U., 1950; m. Mary Juanita Hill, Feb. 2, 1947; children—George, Cheryl, John, Tinka, Peter. Prof., Tex. A. and M. U., 1963-67; mem. staff Los Alamos Sci. Lab., 1950-63, 67—. Fellow Am. Phys. Soc., A.A.A.S.; mem. Am. Math. Soc. Research, numerous publs. on nucleon interaction, nuclear matter, polarization phenomena in low energy nuclear physics, applications of Pade approximant to various problems. Home: 421 Estante St. Office: Los Alamos Sci. Lab., Los Alamos 87544.*

GAMO, Toshioki, Japanese biologist; b. Ibaraki, Japan, 1893; grad. Ueda Sericultural Prof. Sch.; post-grad. sci. dept. U. Tokyo, 1913. Became prof. Ueda Sericultural Profl. Sch., also at Shinshu U.; mem. Japan Acad. Council, since 1950. Recipient prize Japan Sericultural Acad. Author text-book on raising silk worms. Authority on anat. physiology of silk worms; extensive research on breathing trouble of silk worm, also relations between silk worm growth and vitamin C intake.

GAMOW, George, physicist; b. Odessa, Russia, Mar. 4, 1904; s. Anthony and Alexandra (Lebedinzeva) G.; student Normal Sch., Odessa, 1914-20, U. Leningrad, 1922-26 (Ph.D., 1928); m. Loubov Wochminzewa, Nov. 1, 1931 (div.); 1 son, Igor; m. 2d, Barbara Perkins, Oct. 11, 1958. Fellow U. Göttingen (Germany), summer, 1928; U. Copenhagen, (Denmark), 1928-29; Rockefeller fellow, Cambridge, Eng., 1929-30; asst. U. Copenhagen, 1930-31; master in research Acad. Scis., Leningrad, 1931-33; lectr. U. Paris and London, winter, 1933-34, U. Mich., summer, 1934; prof. physics, George Washington U., 1934-56, U. Colo., 1956—; lectr. Stanford, 1936; vis. lectr. Venzuelian Assn. for Advancement Sci., 1956. Participated Convegnio Fisica Nucleare, Rome, 1931; Solvay Congress, Brussels, 1933; Internat. Phys. Congress, London, 1934, Warsaw, 1938. Recipient Kalinga Price award UNESCO, 1956. Mem. Am. Phys. Soc., Washington Philos. Soc., Internat. Astron. Union, Am. Astron. Soc., Nat. and Royal Danish acads. scis. Author numerous books, latest being: Atomic Energy in Cosmic and Human Life, 1946; One, Two Three . . . Infinity, 1947; Creation of the Universe, 1952; Mr. Tompkins Learns the Facts of Life, 1953; The Moon, 1953; Puzzle-Math, 1958; Matter, Earth and Sky, 1958; (with J. Cleveland) Physics: Foundations and Frontiers, 1960; Biography of Physics, 1961; Gravity, 1962; A Planet Called Earth, 1963; A Star Called the Sun, 1964; Thirty Years That Shook Physics, 1965; Mr. Tompkins in Paperback, 1967; (with M. Icas) Mr. Tompkins Inside Himself, 1967; also articles. His early nuclear fluid hypothesis of atomic nuclei led to present theory of nuclear fission and fusion; formulated Gamow-Teller selection rule for beta emission; applied nuclear physics to problems of stellar evolution; proposed neutrino theory of supernovae, 1939, shell model of red giant stars, 1942, theory of origin of chem. elements by process of successive neutron capture, 1948, triplet-system of protein coding, 1955. Home: 785 6th St., Boulder, Colo. 80302.*

GANAPATI, P. N., Indian zoologist; b. Manjeri, India, July 15, 1910; s. P. G. and Anantlakshmi Ganapati; B.A., Maharajas Coll., Ernakulam, India, 1931; M.A., Presidency Coll., Madras, India, 1934; D.Sc., U. Madras, 1942; m. Sita Ganapati, May 21, 1939; children—Anandi, Janaki, Ramnarayan, Anantram. Postdoctoral research scholar Molteno Inst. Biology and Parasitology, U. Cambridge (Eng.), 1947-49; prof., head dept. zoology Andhra U., Waltair, India, 1949—. Mem. internat. adv. com. on marine scis., UNESCO. Fellow Indian Acad. Sci., Nat. Inst. Scis., Zool. Soc. India. Research, numerous publs. on marine biology, biol. oceanography and parasitology. Home: 6-261, Karakachettu Rd., Waltair Uplands, Waltair, India.

GANDEVIA, Bryan Harle, Australian physician; b. Melbourne, Australia, Apr. 5, 1925; s. Eric Neville and Vera Hannah (Hannah) G.; M.B., B.S., Melbourne U., 1948, M.D., 1953; m. Dorothy Virginia Murphy, Aug. 25, 1950; children—Simon Charles, Robin Harle. Various appointments in gen. and thoracic medicine, Melbourne, London, Eng., 1951-55; Royal Australasian Coll. Physicians Travelling scholar chest diseases, 1956; research fellow Postgrad. Med. Sch. and Inst. for Diseases of Chest, London, 1956-57; sr. fellow occupational medicine U. Melbourne, also other appointments as physician, thoracic specialist; 1957-62; asso. prof. thoracic medicine, dir. thoracic div. U. New S. Wales Teaching Hosps., Sydney, Australia, 1963—. Hon. curator Mus. Med. History, Australian Med. Assn., Victoria, Australia, 1957-62. Recipient Armytage prize for research in medicine U. Melbourne, 1954. Fellow Royal Australasian Coll. Physicians, Australian Soc. Allergists (hon.). Author: Annotated Bibliography of the History of Medicine in Australia, 1957; also numerous articles. Editor: Australian Jour. Physiotherapy, 1954-62, cons. editor, 1962—. Research on application of lung functions studies to clin. practice in thoracic medicine, epidemiology of chronic bronchopulmonary diseases, interaction of occupational and other inhalants in producing abnormal bronchial reactivity, history of medicine in Australia. Home: 45a Archbold Rd., Roseville, Sydney 2069. Office: Div. Thoracic Medicine, Prince Henny Hosp., Little Bay, Sydney 2036, N.S.W., Australia.*

GANDY, Harold Wells, Am. physicist; b. San Francisco, June 19, 1923; s. Luther Clark and Roma (Wells) G.; A.B., U. Cal. at Berkeley, 1949; student U. Okla., 1949-50; Ph.D., U. Mo., 1953; m. Jacqueline Wilson, Jan. 12, 1946; children—Lynn Eileen, Pamela Rae. With U. Mo., 1950-53, RCA fellow, 1952-53; research physicist Westinghouse Research Labs., East Pittsburgh, Pa., 1953-56; physicist Gen. Elec. Electronics Lab., Syracuse, N.Y., 1956-59; research physicist U. S. Naval Research Lab., Washington, 1959-67; phys. scientist U. S. Govt., 1967—. Mem. Am. Phys. Soc., Research Engring. Soc. Am., Sigma Xi, others. Research, publs. solid state physics, optical, electronic and light emitting properties of crystalline insulators, optical properties of glasses, energy transfer phenomena in glasses; Laser materials research; discoverer, patentee ytterbium glass laser; stimulated emission processes in multiply-activated glasses. Home: 2000 S. Eads St., Arlington, Va. 22202.*

GANELIUS, Tord Hjalmar, Swedish mathematician; b. Stockholm, Sweden, May 23, 1923; s. Gustaf Hjalmar and Ebba (Bejbom) G.; Fil. Dr., U. Stockholm, 1953; m. Aggie, Oct. 6, 1951; children—Per G., Truls O., M. Svante; E. Aggie S. Docent, U. Lund, Sweden, 1953-57; prof., U. Göteborg, since 1957; vis. prof., U. Washington, Seattle, 1962, Cornell U., Ithaca, 1967-68. Author: Introduction to Mathematics (in Swedish), 1966. Work in real, complex and functional analysis. Home: 19 Kallebäcksvägenm, Göteborg, Sweden. Office: U. Göteborg, Vasaparken, Göteborg, Sweden.

GANGADHARAM, P. R. J., Indian biochemist, bacteriologist; b. Vijayawada, S. India, Sept. 22, 1930; s. Jogarao Venkata and Kameswaramma (Gangadharabhatla) Pattispu; B.Sc. with honors, Andhra (India) U., 1950, M.Sc., 1951; Ph.D., Bombay U., 1954; m. Sakunthala Eranki, June 11, 1955; children—Ramakrishna, Jogarao, Kameswari. Lady Tata Research fellow pharmacology lab. Indian Inst. Sci., Bangalore, S. India, 1953-55; asst. bacteriologist Tb Chemotherapy Centre, Madras, S. India, sr. research officer; sr. research fellow microbiology Nat. Jewish Hosp., Denver, 1961-63; co-ordinator nat. drug resistance survey Indian Council Med. Research; prin. investigator, hon. project dir. biochem. studies in Tb, Service Research, USPHS. Mem. Am. Thoracic Soc., Am. Soc. for Microbiology, Internat. Union Against Tuberculosis (mem. sub com. on bacteriology and immunology), Assn. Microbiologists India, Soc. Biol. Chemists India. Research, publs. on mechanism of action of isoniazid, biochemistry and bacteriology of Tb of Indian Tubercle bacilli; developed tests for drugs in urine; established importance of peak levels of drugs in serum and scientific basis of intermittency in chemotherapy of Tb; discovered artificial man model to study drug-parasite-host relationships. Home: 6, 5th St., Dr. Thirumurthi Nagar, Madras-34. Office: Tb Chemotherapy Centre, Spur Tank Rd., Madras-31, South India.

GANGULI, Nripendra Chandra, biochemist; b. Churain, Dacca, East Pakistan, Aug. 1, 1927; s. H. C. and Manorama Ganguli; B.Sc. with honors, Dacca U., 1947; M.Sc., Calcutta (India) U., 1949, D.Sc., 1955; postgrad. U. Cal. at Berkeley; m. Dipali Ganguli, July 20, 1956; children—Sarbari, Sarmila. Research Scholar dept. applied chemistry Calcutta U., 1950-56; vis. scientist dept. agrl. biochemistry U. Cal. at Berkeley, 1956-58; asst. research officer, hon. lectr. dept. applied chemistry Calcutta U., 1958-61; asso. prof. biochemistry Nat. Dairy Research Inst., Karnal, Punjab, 1961-63, now dairy chemist, head dairy div. Fellow Indian Chem. Soc.; mem. Am., Indian (past sec.) dairy sci. assns., Indian Sci. Congress Assn., Soc. Biol. Chemists, Sigma Xi. Research, numerous publs. on devel. analytical methods applicable in biochem. research; demonstrated presence of certain enzyme in animal tissue for first time: disclosed cerbasic defects in metabolism in vitamin C. deficiency; demonstrated protein synthesis in freshly secreted milk for first time; developed methods to distinguish rennet preparations from different sources; research on proteins in milk. Home: and office: Nat. Dairy Research Inst., Karnal, Haryana, India.*

GANIS, Sam Eugene, mathematician, educator; b. Comiso, Italy, Apr. 4, 1907; s. Blase and Carmela (Bellassai) Gangarosa; came to U. S., 1907, naturalized, 1920; A.B., U. Rochester, 1931, M.A., 1932; M.S., U. Mich., 1936; J.D., John Marshall Law Sch., 1947, M.P.L., 1948; m. Marion L. Rupright, July 3, 1933; children—David R., John B. Admitted to Ill. bar, 1948; various positions in ins. and engring. fields, Ill., 1937-47, 48-51; faculty Rochester (N.Y.) Collegiate Center, 1932-36, Wilson Jr. Coll., Chgo., 1947-48; project engr. product devel. Dominion Elec. Co., Mansfield, O., 1951-54; prof. math. Ohio Wesleyan U., Delaware, 1954—. Cons. in math., ins., law fields. NSF fellow, 1955-63. Mem. Math. Assn. Am., Chgo. Bar Assn., Alpha Phi Alpha, Kappa Phi Kappa, Delta Omicron Phi, Pi Mu Alpha. Contbr. articles to ins. and math. lit. Patentee mech. field. Home: 710 W. Leonardsburg Rd., Leonardsburg, O. 43034. Office: Ohio Wesleyan U., Delaware, O.*

GANN, Thomas William Francis, English archeologist; b. Murrisk County Mayo, Ireland, May 13, 1867; s. William and Elizabeth R. (Garvey) G.; ed. King's Sch., Canterbury, Eng., Med. Sch., Middlesex Hosp., London; m. Mary Hazlemere, 1930. Practiced medicine London, Yorkshire; apptd. dist. med. officer Brit. Honduras, 1894, ret., 1923; lectr. Central Am. archaeology U. Liverpool (Eng.), until 1938; adviser to Brit Mus. expdns. to Brit. Honduras. Mem. Soc. Antiquaries. Author numerous books and articles on archeol. subjects, travel and adventure. First to excavate and describe eccentric flints and obsidians typical of lapidary art; authority on ancient Mayan civilization; discovered Coba (1926), Ichpaatum (1926), and Tzibanche (1927) in Quintana Roo, Mexico; discovered murals, burials and mortuary furniture, contbg. to history of Yucatan; presented collections to Brit. Mus. and Heye Found., N.Y. Died London, Feb. 24, 1938.

GANNETT, Henry, Am. geographer; b. Bath, Me., Aug. 24, 1846; s. Michael Farley and Hannah (Church) G.; S.B., Lawrence Sci. Sch. (Harvard), 1869; M.E., Hooper Mining Sch. (Harvard), 1870; LL.D., Bowdoin, 1899; m. Mary E. Chase, Nov. 24, 1874. Asst., Harvard Obs., 1870-71; topographer Hayden Survey, 1872-79; geographer U. S. Geol. Survey, 1882-1914. Geographer 10th, 11th and 12th censuses; asst. dir. of census of P.I., 1902, of Cuba, 1907-08; geographer Conservation Commn., 1908-09; asso. editor Bull. Am. Geog. Soc.; helped found U. S. Geog. Bd. of Geographic Names, 1890; chmn., 1890-1910; Pres. Nat. Geog. Soc. Author: Manual of Topographic Surveying; Statistical Atlases 10th, 11th and 12th Censuses; Commercial Geography; Dictionary of Altitudes: Stanford Compendium of Geography; Census of Cuba (part); Census of Porto Rico (part); Census of Philippine Islands (part); The Contour Map of U. S.; Magnetic Declination in U. S. Advanced the sciences of geography and cartography in America. Died Nov. 5, 1914.

GANONG, William Francis, Am. physiologist; b. Northampton, Mass., July 6, 1924; s. William Francis and Anna (Hobbet) G.; A.B. cum laude, Harvard, 1945, M.D. magna cum laude, 1949; m. Ruth M. Jackson, Feb. 22, 1948; children—William Francis III, Susan B., Anna H., James E. Faculty, U. Cal. Sch. Medicine, San Francisco, 1955—, prof., 1964—; cons. Cal. Dept. Mental Hygiene, Kaiser Found. Hosp. Recipient Boylston Med. Soc. prize Harvard, 1949. Mem. Am. Physiol. Soc., Endocrine Soc., Soc. Zoologists, Soc. For Exptl. Biology and Medicine (editorial bd. Proc. 1962—), N.Y. Acad. Scis., A.A.A.S., Internat. Brain Research Orgn. Author: Review of Medical Physiology, 1967. Editor: (with L. Martini) Neuroendocrinology, 2 vols., 1966-67. Mem. editorial bd. Am. Jour. Physiology, 1961-66, Endocrinology, 1961—. Research, numerous publs. in neuroendocrine regulatory mechanisms, especially those regulating secretion of adrenal cortex, control of aldosterone secretion. Home: 710 Hillside Av., Albany, Cal. 94706. Office: Dept. Physiology, San Francisco Med. Center, U. Cal., San Francisco 94122.*

GANS, Carl, biologist; b. Hamburg, Germany, Sept. 7, 1923; s. Samuel S. and Else (Leeser) G.; came to U. S., 1939, naturalized, 1945; B.M.E., N.Y. U., 1944; M.S., Columbia, 1950; Ph.D., Harvard, 1957; m. Kyoko Andow, Nov. 18, 1961. With Babcock & Wilcox Co., N.Y.C., Pitts., 1947-55; research asso. Carnegie Mus., Pitts., 1953—; research fellow U. Fla., 1957-58; faculty biology State U. N.Y. at Buffalo, 1958—, prof., 1966—; research asso. Ames Museum Nat. History, N.Y.C., 1959—. Research fellow zool. lab. U. Leiden, Holland, 1965-66. Guggenheim fellow, 1953-54. Fellow A.A.A.S.; mem. Am. Soc. Zoology (chmn. sect. vertebrate morphology 1961-64), Am. Soc. Ichthyologists and Herpetologists (gov. 1959-—). Author: (with T. S. Parsons) A Photographic Atlas of Shark Anatomy, 1964; also numerous articles. Research on functional anatomy of feeding and locomotor apparatus of vertebrates, particularly reptiles. Home: 17 Pelham Dr., Buffalo 14214.*

GANS, Leo Ludwig, German chemist, industrialist; b. Frankfort/Main, Germany, Aug. 4, 1843; s. Ludwig Aaron and Rosette (Goldschmidt) G.; studied chemistry Karlsruhe (Germany) Polytechnikum; Ph.D. under E. Erlenmeyer, Heidelberg, Germany, 1863; hon. M.D.; dr. rer. nat. (hon.), Frankfort; m. Luise Sander, 1876; 2 sons, 1 dau. Became asst. to H. Kolbe, Marburg, Germany, 1864; employed by chem. factory, Wyl/St. Gallen, Switzerland, 1865, later by Manufactur de Javel, Paris; founder chem.-tech. lab., Frankfort, 1868, which combined with aniline dye factory Teerfarbenfabrik Leopold Casella & Co., 1870; co. merged with Farbwerke Hoechst, 1904, and eventually merged into IG Farbenindustrie AG., 1925, sr. mem. bd. dirs. Participated in establishing Internat. Aviation Exhbn., Frankfort, 1909. Named hon. senator, hon. citizen of Frankfort. Mem. Frankfort Aviation Assn. (chmn.), Phys. Assn. Frankfort /Main (chmn.). Pioneered in dye chemistry and its indsl. application, especially naphthalin dyes; prodn. series azo dyes. Died Frankfort/Main, Sept. 14, 1935.

GANS, Richard Martin, physicist; b. Hamburg, Germany, Mar. 7, 1880; s. Martin and Johanna (Behrens) G.; studied electrotechnics, Technische U., Hanover, Germany; later math. and physics, Strasbourg, France; Dr.rer.nat., 1901; m. Leonie Buttmann, 1913; 2 sons. Asst. to G. Quincke, Heidelberg, Germany; went to Tübingen, 1902; joined faculty Tübingen, 1903, Strasbourg, 1911; named prof. physics, dir. Phys. Inst., U. La Plata (Argentina), 1912, became prof., 1947; apptd. dir. II Phys. Inst., U. Königsberg (Germany), 1925; various positions during war including AEG adviser, manual labor, worked on Germany's 1st betatron; became prof. theoretical physics U. Munich, 1946; named prof., Buenos Aires, 1951. Author: Einführung in die Vektroanalysis, 1905; Einführung in die Theorie des Magnetismus, 1908; also articles. Research on magnetism-molecular distbn., crystal magnetizing, localization of hysteresis loss along magnetizing curve, Einstein-Smoluchovsky theory, Brownian movement, diffusion liquids, optical activity, electro-optical phenomena; 1st quantitative treatment significance elastic tension on magnetization. Died City Bell/La Plata, Argentina, June 27, 1954.

GANSSEN, (Ernst Alwin) Robert, German chemist; b. Magdeburg, Germany, Mar. 7, 1865; s. Heinrich and Dorothea Alwinne Elisabeth (Lüder) Gans; student chemistry, Göttingen; doctorate, 1888; m. Gertrud Lichtenberg, 1894; 1 son, Robert, 1 dau. Became asst. Research Inst. for Cultivation Wine, Fruit, and Gardens, Geisenheim/Rhine, Germany, 1889; employed by chem. factory of T. Schuchardt, Görlitz, 1890; joined Prussian Geol. Inst., Berlin, 1891, named head, soils lab., 1899, head, all chem. labs., 1916, dept.

head, 1921; joined faculty, apptd. asso. prof. Tech. U. Berlin, 1909. Recipient Elliot Gresson Gold medal Franklin Inst., Phila., 1916. Research and publs. on properties of cultivated soils, natural aluminum silicates such as zeolites; patented process for synthesis zeolites called permutites (used in water filtration, softening hard water). Died Berlin, Aug. 24, 1940.

GANSWINDT, Hermann, German inventor, engr.; b. E. Prussia, June 12, 1856; ed. U. Berlin, 1881; 23 children. Founder small factory, Berlin. Author: The Dirigibility of Aerostatic Airships, 1883. Built pedal driven boat, also car; built dirigible airship, circa 1891, type of heliocopter, 1901; developed theory of establishing fulcrum in airless space to travel to other worlds. Died Oct. 25, 1934.

GANT, Frederick James, English surgeon; b. Kingsland, Eng., 1825; s. John C. Gant; ed. U. Coll., London; m. Matilda Crawshay, 1859; 1 son (dec.). Became cons. surgeon, London, 1852; lectr. anatomy and physiology Royal Free Hosp.; clin. tchr. surgery London Sch. Medicine for Women, 1878-90; Lettsomian lectr. Med. Soc. London, 1871. Mem. Royal Coll. Surgeons, Med. Soc. London, Royal Med. and Chirug. Soc. (v.p.), Med. Soc. London (pres. 1880-81). Author: Principles of Surgery, 1864; Diseases of the Bladder, Prostate Glands and Urethra, 1884; The Students Surgery, 1890; also numerous articles on surgery. Earliest description of antiseptic treatment during surgery, 1871; introduced operation for correction of abnormal bone condition (Gant's operation), circa 1865. Died 1905.

GANTT, Henry Laurence, Am. management scientist; mechanical engr.; b. Calvert Co., Md., May 20, 1861; s. Virgil and Mary Jane (Steuart) G.; McDonogh Sch., Baltimore Co., Md.; A.B., Johns Hopkins, 1880. Teacher, McDonogh Sch., 1880-3; M.E., Stevens Inst. Tech., 1884; m. Mary Eliza Snow, of Fitchburg, Mass., Nov. 29, 1899. Mech. engr., 1884-—. Mem. Am. Soc. Mech. Engrs., Soc. Naval Architects and Marine Engrs., Am. Geog. Soc. Author: Work, Wages and Profits, 2d edit., 1913; Industrial Leadership, 1915; Organizing for Work, 1919. Specialized in installing modern methods of mfg.; pioneer in human-relationship aspect of management; developed "task and bonus" wage plan, "Gantt chart" of production control. Died Pine Island Farm, N.Y., Nov. 23, 1919.

GANTZ, George Martin, Am. chemist; b. Troy, O., Jan. 31, 1915; s. Maurice A. and Marjorie (Cross) G.; A.B. cum laude, Oberlin Coll., 1937; Ph.D., U. Rochester, 1941; m. Ruth W. Leutner, June 30, 1938; children—Janet Carol (Mrs. Gilbert A. Falcone), S. Barbara, John F., Susan E., George R. Product engr., asso. dir. Gen. Aniline & Film Corp., Easton, Pa., N.Y.C., 1951-65; tech. dir. dystuff div. Geigy Chem. Corp., Ardsley, N.Y., 1965-—. Recipient Meritorious Civilian Service award USN, 1945. Mem. Am. Chem. Soc., Am. Assn. Textile Chemists and Colorists, A.A.A.S., Textile Research Inst. Research, publs. on chemistry of bleaching, dyeing, finishing of textiles, chem. warfare, product devel. and market research, application research mgmt. in dyes and chems. for textile, paper, leather, detergent, cosmetic and plastic industries. Home: 31 Rambling Brook Rd., Upper Saddle River, N.J. 07458. Office: Saw Mill River Rd., Ardsley, N.Y. 10502.*

GANZ, Albert Frederick, elec. engr.; b. Elberfeld, Germany, Apr. 25, 1872; s. Albert and Helen Theresa (Brinkmann) G.; ed. Coll. City N.Y., 1886-87; Cooper Inst. Night Sch., N.Y., 1887-91; M.E., Stevens Inst. Tech., 1895; m. Antonia Christina Stursberg, June 21, 1902. Instr. gen. physics and applied electricity Stevens Inst. Tech., 1895-97, asst. prof., 1897-1902, prof. elec. engring., and head dept. elec. engring., 1902-17. Patent expert and cons. engr., specializing in electric lighting and investigation of and remedies for electrolysis from stray electric currents. Fellow Am. Inst. E.E., A.A.A.S. Died July 27, 1917.

GAPOSCHKIN, Sergei Illarionovich, Am. astronomer; b. Eupatoria, Crimea, Russia, July 12, 1898; s. Illarion Michael and Katherine (Zolautmina) G.; Ph.D. in Arts, U. Berlin, 1928, Ph.D. in Astronomy, 1932; m. Cecilia Helena Payne, Mar. 5, 1934; children—Edward, Katherine (Mrs. John Haramundanis), Peter John Arthur. Came to U. S., 1933, naturalized, 1939. Astronomer, Harvard Coll. Obs., 1933-—. Author numerous books including: Small Magellanic Cloud, vol. 9, 1966; also numerous articles. Discovered eleven double stars providing an accurate idea of their weight, intrinsic brightness and way of life; invented method for measuring brightness of stars; research on stars in other stellar worlds. Home: 74 Shade St., Lexington, Mass. 02173. Office: Harvard Coll. Obs., Cambridge, Mass. 02138.*

GAPOSCHKIN, Cecilia Helena Payne, see Payne-Gaposchkin, Cecilia Helena.

GARABEDIAN, Henry Leslie, Am. mathematician; b. Dorchester, Mass., Nov. 22, 1901; s. Samuel A. and Hannah (Lindquist) G.; B.S., Tufts U., 1922; M.A., Harvard, 1923; Ph.D., Princeton, 1930; m. R. Geraldine Sawyer, Mar. 15, 1947; 1 dau., Nancy Ellen Senter. Faculty, Harvard, 1923-26, U. Rochester, 1926-27, Princeton, 1928-30, Northwestern U., 1927-46; prin. physicist Oak Ridge Nat. Lab., 1946-48; chief research reactors sect. div. Reactor Devel. U. S. AEC, Washington, 1948-49; cons. scientist Bettis Atomic Power div. Westinghouse Electric Corp., Pitts., 1949-56; head math. dept. research labs. Gen. Motors Corp., Warren, Mich., 1956-67; prof. math., energy engring. U. Ill. at Chgo., 1967-—. Sr. research mathematician Nat. Def. Research Council, 1944-45; research asso. Brown U., 1945-46. Fellow mem. Am. Math., Am. Nuclear Soc.; Phi Beta Kappa, Sigma Xi. Editor: Approximation of Functions, 1965. Research, publs. primarily in field of nuclear reactor physics, matrix and integral transformations. Home: 201 E. Chestnut St., Chgo. 60611. Office: U. Ill. at Chgo. Circle, Chgo. 60680.*

GARABEDIAN, Paul Roesel, Am. mathematician; b. Cin., Aug. 2, 1927; s. Carl A. and Margaret (Roesel) G.; B.A., Brown U., 1946; Ph.D., Harvard, 1948; Faculty, Stanford, 1950-1959, prof. math. 1955-59; prof. math. N.Y.U., 1959-—. Mem. Am. Math. Soc., Am. Acad. Arts and Sci. Study of partial differential equations; functions of a complex variable; hydrodynamics. Home: 100 Bleecker St., N.Y.C. 10012.

GARAND, John Cantius, engr., inventor; b. St. Remi, Que., Can., Jan. 1, 1888; s. Jean Baptiste and Elizabeth Edwidge (Oligny) G.; m. Nellie B. Shepard, Sept. 6, 1930; children—Janice Kay, Richard Norman. Began at age of 12 yrs. to repair, design and fabricate machines and machine parts, 1900-18; engaged in ordnance design engring., specializing in light arms, from 1918; perfected a machine gun, 1917-19; consulting engr., U. S. Armory, Springfield, Mass., 1919; inventor of Garand rifle, M1, standard equipment of U. S. Army. Became U. S. citizen, 1920. Recipient of William Pynchon award, 1939; Modern Pioneer in Frontier of Am. Industry award N.A.M., 1940; Holley medal Am. Soc. M.E., 1941; first Gen. John H. Rice medal Army Ordnance Assn., 1941; Lord and Taylor Am. Design award, 1942; spl. award Am. Soc. for Metals, 1942; John Skott medal, 1943; Medal for Merit from U. S. Govt., 1944. Mem. Am. Soc. Metals. Home: 25 Wilton St. Office: Springfield Armory, Springfield, Mass.

GARBER, David H(arrison), Am. physicist; b. Norfolk, Va., June 18, 1918; s. Maurice H. and Goldie H. G.; A.B., U. Pa., 1940, A.M., 1941; postgrad. U. Rochester; Ph.D., Stanford, 1953; m. Barbara Marie Buchanan, July 30, 1949; children—Richard Lincoln, Robert Carl. Devel. engr. Mpls.-Honeywell Regulator Co., Phila., 1941-42; physicist USN Electronics Lab., San Diego, 1944-45; instr. U. Wash., Seattle, 1949-51; research physicist, project engr. Cornell Aero. Lab., Buffalo, 1951-55; sr. staff scientist, asst. to v.p. program dir. Centaur, Gen. Dynamics/Convair, San Diego, 1955-65; program mgr. Aeronutronic div. Philco. Corp., Newport Beach, Cal., 1965-66; spl. asst. to dir. sci. advance space and launch systems Douglas Missile and Space Systems div. McDonnell Douglas Corp., Huntington Beach, Cal., 1966-—. Asso. fellow Am. Inst. Aeros. and Astronautics (past chmn. San Diego sect., mem. tech. com. on spacecraft 1966-—); mem. A.A.A.S., Am. Assn. Physics Tchrs., Am. Phys. Soc., Sigma Xi, Phi Beta Kappa, Pi Mu Epsilon. Contbg. author: Space Age Astronomy, 1962. Asso. editor Jour. Spacecraft and Rockets, 1967-—. Research, numerous articles on spacecraft sci. payloads, space physics and biology, lunar and planetary environments and sci. exploration, space vehicle elec. propulsion, molecular and ionic impact, infrared physics, zero and reduced gravity studies, vacuum friction, atmospheric optics and electricity, phys. meteorology, phys. oceanography, nuclear and electron magnetic resonance. Home: 2306 Arbutus St., Newport Beach, Cal. 92660. Office: 5301 Bolsa Av., Huntington Beach, Cal. 92647.*

GARBER, Harold Jerome, Am. chem. engr.; b. Cleve., Mar. 12, 1913; s. Israel and Bessie (Epstein) G.; Chem.E., U. Cin., 1935, postgrad.; m. Mary Ann Touff, Dec. 17, 1944; children—Harry Kenneth, Richard Ian, Thomas Robert, Ellen Alice, Sally Jane. Faculty, U. Cin., 1936-47, asso. prof., 1947; prof. chem. engring. U. Tenn., Knoxville, 1947-55, adminstrv. head Chem. Engring. Grad. Programs, 1947-55; mgr. chem. devel. Westinghouse Atomic Power, Pitts., 1955-59; dir. plutonium lab. Nuclear Materials and Equipment Corp., Apollo, Pa., 1959-61, tech. asst. to pres., 1959-—, dir. advanced projects, 1961-—; staff Oak Ridge Sch. Reactor Tech., 1948-55. Recipient Hochstetter prize U. Cin. 1935. Cons. to pvt. cos., govt. agys. Mem. Am. Inst. Chem. Engrs., A.A.A.S., Am. Soc. for Engring. Edn., Am. Nuclear Soc., Sigma Xi, Tau Beta Pi, Phi Lambda Upsilon, Phi Kappa Phi. Author: Chemical Engineering Problems, 1940; Slurry Engineering Handbook, 1955; also articles. Designed maj. plutonium labs.; solved multicomponent fractional crystallization problems; developed rocket fuels, radio-isotopic power generator, shrinkproofing process; research in infrared radiation heating design procedures, elucidation swirling flow hydrodynamics, gamma irradiator design, mass transfer and applied chem. reaction kinetics, catalytic recombination hydrogen and oxygen; devel. high altitude oxygen gas drying systems, design synthetic ammonia plants. Home: 5515 Darlington Rd., Pitts. 15217. Office: Nuclear Materials & Equipment Corp., Apollo, Pa. 15613.*

GARBER, Morris Joseph, Am. biometrician; b. N.Y.C., Nov. 6, 1912; s. Isadore and Ethel (Shevack) G.; B.S. in Zoology, Columbia, 1933; Ph.D., Tex. A. and M. Coll., 1951; m. Gloria Ruth Routman, Mar. 7, 1943; children—David Ira, Diana Lee. Faculty, Tex. A. and M. Coll., College Station, 1947-56, asst. prof. genetics, 1951-56; faculty U. Cal. at Riverside, 1956-—, prof. biostatistics, biometrician, 1968-—, dir. Computing Center, 1963-—. Cons. U. S. Dept. Agr. Forest Service Fire Lab., Riverside, 1965-—. Fellow A.A.A.S., Tex. Acad. Sci., Assn. for Computing Machinery (past chmn. Arrowhead chpt.), Biometric Soc., Am. Statis. Assn., Am. Genetic Assn., Sigma Xi. Research, numerous publs. on design expts., applied statistics, statis. inference in agrl. disciplines. Home: 3504 Bryce Way, Riverside, Cal. 92506.*

GARBUNY, Max, Am. physicist; b. Koenigsberg, Germany, Nov. 22, 1912; s. Efim and Elise (Sadunischker) G.; Dipl.Ing., Inst. Tech., Berlin, Germany, 1936, Dr.-Ing., 1938; m. Melitta Lowy, June 15, 1947; children—Vivian Joyce, Carole Evelyn, Ellen Virginia. Came to U. S., 1938, naturalized, 1944. Research physicist Allen-Bradley Co., Milw., 1939-43; instr. Princeton, 1943-44; research physicist Westinghouse Electric Co., Pitts., 1944-52, mgr. optical physics sect. Westinghouse Research Labs., 1952-60, cons. low temperature and radiation physics, 1960-—. Sci. adviser panel U. S. Army, 1964-—. Recipient Westinghouse Most Meritorious Patent award, 1952. Fellow Am. Phys. Soc. Author: Science of Science (with R. Fox., R. Hooke), 1963; Optical Physics, 1965; (with M. Gottlieb, W. Emmerich) Seven States of Matter, 1966; also articles. Research on optical physics, infrared tech., superconducitivity, solid state physics, nuclear instrumentation, theory and tech. ultrahigh frequencies and microwaves; developed an absolute ion speed gauge, reflex resonatron, thermal image converters, and detectors. Patentee in fields. Home: 2305 Marbury Rd., Pitts. 15221. Office: Westinghouse Research Labs., Pitts. 15235.*

GARBY, Lars Elias, Swedish physician; b. Lidköping, Sweden, July 20, 1924; s. Karl Elias and Ragnhild (Grindal) G.; M.D., U. Uppsala (Sweden), 1953, Ph.D., 1957; div.; 1 son, Jens. Asst. prof. dept. physiology U. Uppsala, 1957; research fellow U. Wash. Sch. Medicine, Seattle, 1957-58; head Swedish Med. Research Council Unit Exptl. Hematology, Uppsala, 1958-—; vis. prof. U. Fla., Gainesville, 1962, U. Liverpool (Eng.) Sch. Tropical Medicine, 1966. Tech. asst. expert IAEA, Cairo, Egypt, 1963, Bangkok, Thailand, 1965. Research, publs. on physiology, especially permeability, iron metabolism, and red blood cells. Home: 24 Döbelnsgatan, Uppsala. Office: University Hosp., Uppsala, Sweden.*

GARCIA, Amando, Spanish physicist; b. Alcoy, Spain, Jan. 3, 1934; s. Armando and Enriqueta (Rodríguez) G.; B.Sc., Faculty Scis., U. Valencia (Spain), 1957, Ph.D. in Physics, 1960; m. Manuela García, Oct. 19, 1960; children—Jorge, Ana. Staff, Instituto de Física Corpuscular, U. Valencia, 1957-—, faculty, 1957-—, adjoint prof., lectr. gen. physics, 1961-—; lectr. physics Agrl. Tech. Engring. Sch., 1961-—. Recipient Alfonso el Sabio prizes, 1961, 64. Mem. Spanish Physics and Chemistry Royal Soc. Research, publs. on low energy nuclear physics especially nuclear reactions, polarization and nuclear spectroscopy using photog. nuclear emulsion as detectors, environmental radioactivity measurements. Home: 155 Av. Primado Reig. Office: Instituto de Física Corpuscular, U. Valencia, Valencia, Spain.*

GARCIA, Manuel, music educator, inventor; b. Catalonia, Spain, Mar. 17, 1805; s. Manuel del Pholo Vicente and Joaquina (Sitches) G.; studied harmony under Fétis, Paris; m. Cecile E. Mayer, Nov. 22, 1832; children—Manuel, Gustav, Maria, Eugénie. Sang in operas, Madrid, 1820; toured Am. and Mexico; studied medicine in Paris; became prof. Paris Conservatoire, 1840; went to London, 1848; prof. singing Royal Acad. Music, London; ret. in 1895; Pvt. tchr.; tchr. of Jenny Lind. Decorated Commdr. Royal Victorian Order; Order of Alphonso XII (Spain); recipient Gold medal for Screnee (Germany). Author: Mémoire sur la voix humaine, 1840; Traité complet de l'art du chant, 1847; Observations on the Human Voice, 1855. Research on physiology of the voice; invented laryngoscope to examine the larynx of his pupils (laid founds. of modern study of the larynx.), 1855. Died London, July 1, 1906.

GARCIA DE FIGUEROLA, Luis, Spanish geologist; b. Mar. 27, 1922; s. Juán Clemente and Ramona De Figuerola; Ph.D.in Sci., U. Madrid; m. Maria del Carmen Paniagua; children—Luis, Miguel, Belén, Pablo. Asst., C.S.I.C.; asst. prof. geology; researcher Atomic Energy Commn.; prof. petrology. Recipient Alfonso de Herrera prize Higher council Sci. Research. Mem. Spanish Royal Soc. Natural History, Internat. Assn. Sedimentology. Research and publs. on geology. Home: 7 Ave. de Galicia. Office: Faculty of Sciences, Oviedo, Spain.

GARCIA HERRERA, Ernesto, Mexican physician; b. Mexico City, Nov. 7, 1925; s. Francisco H. and Eloisa (Herrera) Garcia; student Nat. Preparatory Sch. for Biol. Scis., Mexico City, 1940-41; M.D., Nat. U. Mexico, 1950; m. Blanca Rubi Neri, Feb. 12, 1942; children—David, Blanca Eloisa, Alda Patricia, Ernesto. Attending physician Children's Hosp., Mexico City, 1951-53, 59; asst. prof. Superior Sch. Rural Medicine, 1952, 58-59; residencies in pediatrics Sea View Hosp., S.I., N.Y., 1954-55, Monmouth Med.

Center, Long Branch, N.J., 1955-56, Marlboro (N.J.) Hosp., 1956-57; resident, chief resident, instr. nursery Dr. Hazard Meml. Hosp., Long Branch, 1957-58; surgeon, chief surg. service Cruz Blanca Neutral, 1959-60; med. dir. Mead Johnson of Mexico, 1960; attending pediatrician Children's Hosp. of Moctezuma Zone, 1961; head babies dept. Hosp. of Huipulco, Tlalpan, 1962—; asst. prof. Nat. U. Mexico, 1959. Mem. Soc. United Med. Professions (pres. 1963-64), Am. Med. Writers (organizer, past pres. Mexican chpt.) Author: Poisoning in Children; The Premature Baby; Pediatric Emergencies; The Mead Johnson Manual; Medical Science Among the Aztecs. Editor Actualidades Pediátricas, 1959; editor, dir. Revista de Pediatria, 1960. Devised hand pump to nebulize drugs in pulmonary patients, machine for gastric washing in cases of poisoning, new method of bronchial aspiration in children and of bronchography. Home: Alfonso 78, Mexico 13, D.F. Office: 2996 Calz. de Tlalpan, Mexical 21, D.F., Mexico.*

GARCIA-RAMOS, Juan, Mexican biophysicist; b. Queretaro, Mexico, July 26, 1915; s. Pantaleon Garcia and Simona Ramos; M.D., Army Med. Sch., Mexico City, 1936; Sc.D., Centro de Investigacion del IPN, Mexico City, 1961; m. Nelly Horsman, Nov. 11, 1957; children—Sonia, Juan. Asst. prof. Nat. Inst. Cardiology, 1944-53; head dept. physiology Nat. Inst. Pneumology, 1953-56; prof. physiology U. Mexico, Med. Sch., 1957-58, Army Med. Sch., 1958-61; prof. dept. physiology and biophysics Centro de Investigacion del IPN, 1961—. Recipient Tech. Merit medal Mexican Army, 1947; Carnot prize Nat. Acad. Medicine, 1952. Mem. Sociedad Mexicana de Ciencias Fisiológicas, Sociedad Argentina de Biologia. Research, numerous publs. on auricular flutter and ventricular fibrillation mechanisms, pulmonary function by oximetry, nervous control respiration, orgn. cerebral cortex. Home: 91 Morelos, Mexico, D.F., Mexico.*

GARCIN, Laurent, French botanist, surgeon; b. Grenoble, France, 1683; docteur en médecine; navy surgeon, 1720-29; made voyages to India, 1728-30. Mem. French Acad. Scis., 1730. Honored by Linnaeus in tree species named Garcinia. Wrote extensive descriptions of Indian trees. Died Neuchatel, Switzerland, Apr. 18, 1751.

GARCKE, (Christian) August (Friedrich), German botanist; b. Bräunrode nr. Mansfeld, Germany, Oct. 25, 1819; s. Johann August Christian and Catherine Magdalene (Hesse) G.; began study theology, Halle, Germany, 1840; state exam., 1844; Ph.D., Jena, Germany, 1844. Independent research in botany, Halle; came to Berlin to study under A. Braun, 1851; named chief asst. Royal Herbarium (later Royal Bot. Mus.), 1856, curator, 1865; joined faculty botany and pharmacognosy, 1869; became asso. prof., 1871. Mem. Leopoldina. Author: Flora von Nordund Mittel-deutchland, 1849; Flora Hallensis, 1848-56. Editor, Linnaea. Research in bot. taxonomy, Malvaceae. Died Berlin, Jan. 10, 1904.

GARDEN, Alexander, naturalist, physician; b. Aberdeenshire, Scotland, circa 1730; s. Rev. Alexander Garden; M.D., Marischal Coll., Aberdeen, Scotland, 1753; m. Elizabeth Peronneau, Dec. 24, 1755; 1 son, Maj. Alexander. Came to U. S., 1754; began med. practice, Charleston, S.C., 1755; in London, from 1783. Fellow Royal Soc., 1773 (later v.p.); mem. Royal Soc. Uppsala (Sweden), 1763. Discovered vermifugal properties of pink-root (Spigelia marilandica); discovered Congo snake, mud eel; corresponded with Am. naturalists and with noted European naturalists such as Linnaeus; instrumental in sending 1st electric eels to Europe; flower gardenia named after him. Died London, Apr. 15, 1791.

GARDINER, John Stanley, Irish zoologist; b. Belfast, Ireland, Jan. 24, 1872; s. Jephson Gardiner; ed. Marlborough Coll.; M.A., Gonville and Caius Col. Cambridge (Eng.) U.; m. Rachel Florence Dening, 1900; m. 2d, Edith Gertrude Willcock, 1909; 2 daus. Fellow Gonville and Caius Coll., Cambridge U., from 1898, demonstrator animal morphology, 1902-08, dean, 1903-09, sr. proctor, 1907-08, lectr. zoology, 1909, emeritus prof. zoology, 1909-37. Mem. Treasury com. on fishery investigations, 1907-08, govt. adv. com. on fisheries, 1913; dir. sci. investigations Ministry of Fisheries, 1920; trustee Brit. Mus., from 1931; mem. standing commn. on museums and galleries, from 1942. Recipient Agassiz medal Nat. Acad. Scis. (U. S.) Fellow Royal Soc. (Darwin medal 1944), 1908, Linnean Soc. (Linnean medal), Royal Geog. Soc. (Murchison award 1902). Author: Coral Reefs and Atolls, 1931. Editor: The Fauna and Geography of the Maldive and Laccadive Archipelagoes, 1902-06. Made numerous expdns. including Coral Reef Boring to Funafuti, 1896, Maldive and Laccadive, 1899-1901, Indian Ocean, 1905, Syechelles, 1908; research, publs. on study of animal form in relation to function; authority on madreporian corals and distbn. marine animals; made devel. reorgn. of Balfour library. Died Feb. 28, 1946.

GARDNER, Arthur Duncan, English bacteriologist; b. Staffs, Eng., Mar. 28, 1884; ed. U. Coll., (fellow) 1903-08, St. Thomas Hosp., 1911-12; m. Violet Newsam, 1918; 3 children. House surgeon, casualty officer St. Thomas' Hosp., 1911-12; lectr. in pathology, research student, until 1914; bacteriologist Standards Labs., Oxford, 1915-36, apptd. dir., 1925; reader in bacteriology U. Oxford, became prof., 1936; mem. Army Pathology Advisory Com., 1932, governing body, also chmn. Bagley Wood Dist. Nursing Assn. Fellow Royal Coll. Surgeons, Royal Soc. Medicine; mem. Path. Soc. Gt. Britain and Ireland, Med. Research Club London. Author: Microbes and Ultramicrobes, 1931; (with Chain, Florey and others) Penicillin as a Chemotheraputic Agent, also Furhter Observations on Penicillin, 1940-41. Introduced (with others) penicillin as a therapeutic agent. Address: School of Pathology, Oxford, Eng.

GARDNER, Clarence Ellsworth, Am. physician; b. Bucyrus, O., Feb. 27, 1903; s. Clarence Ellsworth and Anna (Startzman) G.; A.B., Wittenberg U., 1924, D.Sc., 1950; M.D., Johns Hopkins, 1928; m. Beatrice Ina Lockwood, June 8, 1928; 1 dau., Jane L. Asso. in surgery Johns Hopkins, 1929; asso. prof. Duke, 1932-37, prof., 1937-68, chmn. dept., 1960-64. Mem. Durham-Orange County Med. Socs., So. Surgeons Club (pres. 1940), So. (v.p. 1962), Am. (v.p. 1964) surg. assns., Am. Bd. Surgery (bd. examiners 1951-57), Am., So. med. assns., Internat. Surg. Soc., Soc. U. Surgeons, A.C.S. Research, numerous publs. on anomalies of intestinal rotation, intravenous alimentation, chest injuries, surg. mgmt. of peptic ulcer, early recognition and treatment carcinoma of breast, localization of fgn. bodies in soft tissues, use of transhepatic cholangiography in surgery. Home: Blue Creek Rd., Astor, Fla. 32002.*

GARDNER, Eldon John, Am. geneticist; b. Logan, Utah, June 5, 1909; s. John W. and Cynthia (Hill) G.; B.S., Utah State U., 1934, M.S., 1935; Ph.D., U. Cal. at Berkeley, 1939; m. Helen Richards, Aug. 21, 1939; children—Patricia (Mrs. Jerome Mahrt), Donald E., Betty Ann (Mrs. George E. Morrison), Cynthia, Alice, Mary Jane. Instr., dean Salinas (Cal.) Jr. Coll., 1939-46; with U. S. Dept. Agr. Bur. Plant Industry, 1942-44; asso. prof. U. Utah, 1946-49; prof. zoology Utah State U., Logan, 1949-62, dean Coll. Sci., prof., 1962-67, dean Sch. Grad. Studies, 1967—. Faculty Honor Research lectr., 1953. Recipient Distinguished Service award Utah Acad. Sci., Arts and Letters, 1957, Utah Sci. Tchrs. Assn., 1965. Mem. Genetics Soc. Am., Am. Soc. Human Genetics, Am. Soc. Naturalists, Genetics Soc. Utah, Sigma Xi, Phi Kappa Phi. Author: Principles of Genetics, 3d edit., 1968; History of Biology, 1965; also numerous articles. Research on human genetics especially inheritance abnormal growths and cancer, inheritance and biochem. aspects of abnormal growth in fruit flies. Home: 369 N. 5 East St., Logan, Utah 84321.*

GARDNER, Frank Herbert, Am. physician; b. San Bernardino, Cal., Sept. 21, 1919; s. Frank M. and Ernestine (Herbert) G.; B.S., Northwestern U., 1941, M.D., 1945; m. Theo Wood, June 17, 1949; children—Frank Wood, Peter Christopher, William Edward. Dir. Sprue team U. S. Army Tropical Research Lab., San Juan, P.R., 1953-55; practice medicine, specializing in hematology, Boston, 1949-66, Phila., 1966—; attending hematologist West Roxbury VA Hosp., 1952-60; mem. staff Peter Bent Brigham Hosp., 1949-66; faculty Med. Sch. Harvard, 1949-66; prof. medicine U. Pa., 1966—. Mem. Am. Fedn. Clin. Research, Am. Soc. Clin. Investigation, Am. Soc. Hematology, Am. Clin. and Climatol. Assn. Research, numerous publs. in mechanism of hemolytic anemias, including therapy with corticoids, mechanism of blood platelet lifespan and physiology, classification of thrombocytopenia, introduction of androgens to treat a variety of anemias and lymphoma. Home: 411 Gilpin Rd., Phila. 19072. Office: 51 N. 39th St., Phila. 19104.*

GARDNER, George Henry, Am. physician; b. Osborn, O., Aug. 18, 1897; s. Clarence E. and Anne Mae (Startzman) G.; A.B., Wittenberg Coll., 1917, S.C.D., 1950; M.D., Johns Hopkins, 1921; m. Marion E. Nelson, May 31, 1924; children—George Henry (dec.), Elizabeth Ann (Mrs. Richard Labahn), Mary Louise (Mrs. John Francis Ahearn, Jr.). Practice medicine, specializing in gynecology, Chgo., 1926—; mem. staffs Passavant Meml. Hosp., 1929-46, Chgo. Wesley Meml. Hosp. 1946—; faculty Northwestern U. Med. Sch., 1927-65, chmn. dept. obstetrics and gynecology, 1947-65, prof. emeritus, 1965—. Diplomate Am. Bd. Obstetrics and Gynecology. Mem. A.C.S. (life), Am., Chgo. gynecol. socs., Am. Assn. Obstetricians and Gynecologists, Am. Coll. Obstetricians and Gynecologists, Inst. Medicine, A.M.A., Ill., Chgo. med. socs., Johns Hopkins Surg. Soc., Phi Gamma Delta. Research, publs. on clinco-path. endometriosis; ovarion tumor and broad ligament. Home: 1724 Asbury Av., Evanston, Ill. 60201. Office: 720 N. Michigan Av., Chgo. 60611.*

GARDNER, John, English physician; b. Great Coggeshall, Eng.; 1804; M.D., Giessen U., 1847; m. Julia Emily Moss, 1832; several children. Settled in London, 1829; a founder, prof. chemistry and materia medica Gen. Apothecaries Co. Licentiate Royal Coll. Physicians Edinburgh, Apothecaries Soc. Mem. Royal Coll. Chemistry (a founder 1844, sec. until 1846), Chem. Soc. London, Ethnological Soc. London. Author: The Great Physician, the Connection of Diseases and Remedies with the Truths of Revelation, 1843; Household Medicine, 9th edit., 1878; Longevity: the Means of Prolonging Life after Middle Age, 5th edit., 1878; Hymns for the Sick and Convalescent, 2d edit., 1879. Introduced several drugs from Am., especially prodophyllin. Died London, Nov. 14, 1880.

GARDNER, Lytt Irvine, Am. pediatrician; b. Reidsville, N.C., Oct. 1, 1917; s. Lytt Irvine and Bettie Sue (Jones) G.; A.B., U. N.C., 1938, M.A., 1940; postgrad. U. Mich.; M.D., Harvard, 1943; m. Mary Locksley Long, June 20, 1942; children—Susan (Mrs. James E. Winfrey), Lytt Irvine, Locksley Anne, Rosalyn McAden, Miriam Clarke. Practice medicine, specializing in pediatrics, Syracuse, N.Y., 1952—; faculty State U. N.Y. Upstate Med. Center, 1952—, prof. pediatrics, 1956, 57—; prof. pediatrics Yale, 1956-57; chmn. Physicians Forum, 1966-67. Recipient E. Mead Johnson award Am. Acad. Pediatrics, 1961. Mem. Endocrine Soc., Soc. Exptl. Biology and Medicine, Am. Pediatric Soc., Soc. Pediatric Research (past pres.), Soc. Latino-Americana de Investigaçao Pediátrica (hon.). Author: Adrenal Function in Infants and Children, 1955; Molecular Genetics and Human Disease, 1961; (with R. G. Patton) Growth Failure in Maternal Deprivation, 1963; Endocrine and Genetic Diseases of Childhood, 1968. Editorial bd. Pediatrics, 1958-61; The Medical Letter, 1961-66; Jour. Clin. Endocrinology and Metabolism, 1963-67. Research, publs. on increased intracellular hydrogen ion concentration in potassium deficiency by direct chem. analysis of muscle tissue, high phosphate content of cow's milk as maj. etiologic factor in tetany of human newborn, adrenocorticolytic drug DDD, urinary excretion of dehydroepiandrosterone in children with virilizing adrenal tumors and virilizing adrenal hyperplasia, assn. between maternal deprivation and growth failure; site of biochem. error in Gaucher's disease; chromosomal abnormalities in children, internat. child health. Office: Dept. Pediatrics, Upstate Med. Center, 750 E. Adams St., Syracuse, N.Y. 13210.*

GARDNER, Riley Wetherell, Am. psychologist; b. Ree Heights, S.D., Oct. 31, 1921; s. Hugh Howard and Marguerite (Speicher) G.; A.B., Yankton Coll., 1945; Ph.D., U. Kan., 1952; m. Ruth Janssen, Aug. 27, 1950; children—Helen, Mark. Resident to sr. psychologist Menninger Found., Topeka, 1951—, dir. Cognition Research Project, 1958—. Mem. Am., Kan. (pres.), Topeka psychol. assns., N.Y. Acad. Scis. Author: (with others) Cognitive Control, 1959; (with D. N. Jackson, S. Messick) Personality Organization in Cognitive Controls and Intellectual Abilities, 1960. Research, numerous publs. on individual differences in personality orgn.; diagnosis of brain-damage by means of psychol. tests; group process in psychiat. evaluations. Home: 207 S. Broadmoor St., Topeka 66606. Office: Box 829, Topeka 66601.*

GARDNER, William James, Am. physician; b. McKeesport, Pa., June 12, 1898; s. William James and Sarah (Gongaware) G.; A.B., Washington and Jefferson Coll., 1920; M.D., U. Pa., 1924; m. Ann Ray Kieffer, Dec. 1927, 1941; children—William James III, June Patricia, Hugh Blaine. Faculty, U. Pa., 1926-29; practice medicine, specializing in neurol. surgery, Cleve., 1929—; head dept. neurol. surgery Cleve. Clinic, 1929-63, sr. cons., 1962-64; pvt. practice, 1963—; head dept. neurol. surgery Fairview Park Hosp., 1965—; asso. staff Huron Rd. Hosp.; cons. staff St. Alexis, Lutheran, St. John's, Mt. Sinai, St. Lukes, Euclid-Glenville hosps. Diplomate Am. Bd. Neurol. Surgery, Nat. Bd. Med. Examiners. Mem. A.M.A., A.C.S., Soc. Neurol. Surgeons (pres. 1956), Harvey Cushing Soc. (v.p. 1962). Research, numerous publs. on surg. removal of part of the brain; spinal cord defects; brain tumors. Home: 13700 Shaker Blvd., Cleve. 44120. Office: Hanna Bldg., Cleve. 44115.*

GARDNER, William U., Am. endocrinologist; b. Kinbrae, Minn., Nov. 11, 1907; s. James A. and Josephine (Ullmna) G.; B.S. S. D. State Coll., 1930, D.Sc., 1960; M.A., U. Mo., 1931, Ph.D., 1933; m. Katherine Homsley, July 15, 1934. NRC fellow Yale, 1933-35, research staff, 1935-41, faculty, 1941—, prof., chmn. dept. anatomy, 1943—. Mem. bd. sci. counselors USPHS; bd. sci. advisers Jane Coffin Childs Meml. Fund, Anna Fuller Fund. Mem. Internat. Union Against Cancer (v.p.), Eleanor Roosevelt Internat. Cancer Fellowship (chmn.), Sigma Xi. Contbr. numerous articles on different aspects of anatomy endocrinology and hormones in relation to cancer to tech. jours. Home: 985 Orange Center Rd., Orange, Conn. Office: 333 Cedar St., New Haven.*

GARENGEOT, see De Garengeot, Rene-Jacques Croissant.

GARG, Jagadish Behari, physicist; b. Kanpur, India, July 7, 1929; s. Mukandilal and Roopkumari (Agarwal) G.; B.S., Allahabad U., 1948; M.S., Lucknow U., 1951; D.Sc., Paris U., 1958; m. Pushpa Govila, Feb. 15, 1955; children—Suruchi, Ajay. Asst. physicist Indian AEC, New Delhi, 1951-55; research fellow College de France, Paris, 1955-58; Turner and Newall fellow U. Manchester (Eng.), 1958-61; sr. research asso. Columbia, 1961-66; prof. physics, dir. Nuclear Accelerator Lab. State U. N.Y., Albany, 1966—. Fellow Am. Phys. Soc., Brit. Inst. Physics, Phys. Soc.; mem. Société Française de Physique. Research, numerous publs. in structure of nuclei by means of slow and fast neutrons, properties of large number or resonance levels were determined by using high resolution techniques. Home: 381 Highland Dr., Schenectady 12303. Office: Western Av., Albany, N.Y. 12203.*

GARGRAVE, George, Brit. mathematician; b. Leyburn, Eng., 1710; ed. by his uncle; became asso. Joseph Randall in mgmt. acad. at Heath, Eng., 1745;

mem. faculty Wakefield (math. sch.) until 1768. Research and publs. on transit of Venus, eclipse of moon. Died Dec. 7, 1785.

GARIDEL, Pierre-Joseph, French botanist; b. Manosque, France, Aug. 1, 1658; prof. botany U. Aix en Provence; mem. French Acad. Scis., 1699; author Histoire des plantes qui naissent aux environs d'Aix, 1715; devoted most of life to study of flora in southern France, traveling, collecting specimens and making sketches; died June 6, 1937.

GARIN, J. P., French physician, biologist; b. Lyon, France, Nov. 18, 1922; s. Charles and G. (Darnat) G.; med. thesis, U. Lyon, 1953; became aggregate prof. parasitology and exotic pathology Lyon Faculty Medicine, 1958, now titular prof. Mem. Soc. Medicine Lyon, Soc. Exotic Pathology Paris. Author: Étude sur le toxoplasmose humaine acquise, 1953; also articles. Research on human toxoplasmosis, chemotherapy of exptl. toxoplasmosis, application of immunology to human paludism (malaria). Home: 47, cours Fr. Roosevelt, Lyon 69, France.*

GARLAND, Carl Wesley, Am. chemist; b. Bangor, Me., Oct. 1, 1929; s. Cecil G. and Blandena (Couillard) G.; B.S., U. Rochester, 1950; Ph.D., U. Cal., Berkeley, 1953; m. Joan A. Donaghy, July 30, 1955; children—Leslie J., Andrew E. Instr. chemistry U. Cal., Berkeley, 1953; faculty Mass. Inst. Tech., Cambridge, 1953-—, asso. prof. chemistry, 1959-—. Fellow Am. Acad. Arts and Sci., N.Y. Acad. Scis.; mem. Am. Chem. Soc., Am. Phys. Soc. Author: (with D. P. Shoemaker) Experiments in Physical Chemistry, 2d edit., 1967. Editor Optics and Spectroscopy, 1960-—. Research, publs. on infrared spectra of chemisorbed molecules to characterize structure and bonding in such species, low temperature elastic constants measured to provide information for lattice dynamical calculations of properties of solids, ultrasonic studies on order-disorder phenomena and critical points. Home: 4 Edward St., Belmont, Mass. 02178. Office: Mass. Inst. Tech., Cambridge, Mass. 02139.*

GARLAND, John, grammarian, alchemist; flourished 1230; student Oxford, Eng.; studied under Alain de Lille, Paris, France; joined faculty U. Toulouse (France), 1229; returned to Paris, 1232. Author: Compendium alchymiae cum ditionario ejusdem artis, 1560; Liber de mineralibus, 1560; Synonyma; Libellus de Praeparatione elixir; Computum; Tabula principalis, contra tabula de festis mobilitus et tabula terminorum paschalium; also poetry, works on grammar, music. Died May 1252.

GARLAND, L(eo) Henry, Am. physician; b. Dublin, Ireland, Mar. 30, 1903; s. John Peter and Mary (Martin) G.; M.B., B.Ch., B.A.O., U. Coll., Dublin, 1924, M.D. (hon.), 1960, F.F.R., 1963; m. Edith Isabel Dohrmann, July 6, 1928; children—Edith M. (Mrs. James Merrifield, Jr.), Isabel Ann (Mrs. Victor Caglieri), Judith M. (Mrs. Richard Harrington), Sheila (Mrs. Earl Reeves), Michael H. Came to U. S., 1925, naturalized, 1931. Faculty Stanford Med. Sch., San Francisco, 1929-60, clin. prof. radiology, 1948-60; clin. prof. radiology U. Cal. at San Francisco, 1960-—; practice radiology, San Francisco, 1927-—. Recipient Bronze medal Am. Cancer Soc., 1958; Gold medals Radiol. Soc. N.Am., 1960, Am. Coll. Radiology, 1961. Diplomate Am. Bd. Radiology. Mem. Cal., Am. med. assns., Am. Roentgen Ray Soc., Cal. Acad. Medicine, Cal. Radiol. Soc., Northwestern Med. Assn. Author: (with H. C. Hinshaw) Disease of the Chest, 1956, 64; also chpts. in books, numerous articles. Research on natural growth rate and duration of human cancer, methods of treating cancer using radiation, methods of improved X-ray diagnosis. Died Oct. 31, 1966.

GARLICK, George Frederick John, English physicist; b. Tipton, Staffs., Eng., Feb. 21, 1919; s. George Robert and Martha E. (Davies) G.; B.Sc., U. Birmingham, 1940, Ph.D., 1943, D.Sc., 1955; m. Dorothy Mabel Bowsher, Mar. 6, 1943; 1 dau., Elizabeth Astrid. Faculty, research physicist U. Birmingham (Eng.); prof. physics, dir. dept. Hull U., Yorkshire, Eng., 1956-—, dean faculty sci., 1957-59. Fellow Inst. Physics and Phys. Soc. Eng. Author: Luminescent Materials. Research, publs. on luminescence in solids and related phenomena. Home: 8 Parkside Close, Cottingham, Yorkshire, Eng.*

GARLICK, Theodatus, Am. surgeon, sculptor; b. Mar. 30, 1805; s. Daniel and Sabra (Kirby) G.; grad. U. Md. Med. Sch., 1834; m.3d, Mary Chittenden, 1845. Practiced surgery, Youngstown, O., 1834-52, had reputation as plastic surgeon; inventor new splints, surg. instruments; made models of surg. and pathol. anatomy; constructed camera which took daguerreotypes (photographing out of direct sunlight for 1st time), 1840; experimented in artificial trout-breeding (1st of kind in Am.). Died Dec. 9, 1884.

GARLICK, William Lynnewood, Am. physician; b. Emporia, Va., July 19, 1912; s. John Robert and Lizzie G. (Harding) G.; A.B., Emory U., 1933; M.D., George Washington U., 1937; m. Hillis Reid Morris, June 21, 1947; children—William Lynnewood, Hillis Morris, Lynn Harding, John Christopher. Faculty, U. Md. Med. Sch., Balt., 1942-—, asso. prof. thoracic surgery, 1951-—, cons. thoracic surgery, 1946-—, chief service, dir. thoracic surgery Mercy Hosp. Div., 1946-—, dir. grad. med. edn., 1953-62, dir. research lab., 1958-62; surgeon on staff Ch. Home and Hosp., Univ. Hosp., 1941-—. Fellow A.C.S.; mem. Southeastern Surg. Congress (past v.p.), A.M.A., Soc. Med. Consultants to Armed Forces, N.Y. Acad. Scis., Council Social Agys., Am. Cancer Soc., Md., Balt. med. socs. Research and numerous publs. in anticoagulant therapy and malignant tissue cultures. Home: 1866 Circle Rd., Ruxton, Md. 21204. Office: 700 N. Charles St., Balt. 21201.*

GARMAN, Samuel, Am. naturalist; b. Indiana County, Pa., June 5, 1843; s. Benjamin and Sarah Ann (Griffith) G.; grad. Ill. State Normal U., 1870; spl. student of Louis Agassiz in natural history, 1812-73; S.B. (hon.), Harvard, 1898; A.M., 1899; m. Florence Armstrong, Sept. 2, 1895. Prin. Miss. State Normal Sch., 1870-71; prof. natural scis., Ferry Hall Sem., Lake Forest, Ill., 1871-72; asst. in herpetology and ichthyology Mus. of Comparative Zoölogy, Harvard, from 1873. Mem. Maj. Powell's 1st expdn. in Colo.; with Alexander Agassiz in South Am. expdns. Mem. Linnaean Soc. Author: Deep Sea Fishes, 1899; The Chimaeroids. Chismopnea, 1904; New Plagiostomia, 1906; New Plagiostomia and Chismopnea, 1907; The Reptiles of Easter Island, 1908; Plagiostomia (sharks, skates and rays), 1913. Died Sept. 30, 1927.

GARMANN, C. F., German physician; b. Mersbourg, Jan. 19, 1640. Author: Dissertatio de nutritione infantum ad vitam longam, 1667; De miraculis mortuorum, 1667; Homo ex ovo, seu de ovo humano, 1672; Epistolaria centuria, 1714. Died July 15, 1708.

GARNER, Cifford Symes, Am. chemist; b. Newark, Oct. 4, 1912; s. Albert J. and Dorothy (Weiss) G.; B.S., Cal. Inst. Tech., 1935, Ph.D., 1938; m. Ellen Louise Sanderhoff, Aug. 7, 1937. Noyes Research fellow Cal. Inst. Tech., Pasadena, 1938-39; faculty U. Tex., Austin, 1939-46, asst. prof., 1941-46; research asso. plutonium project U. Cal. at Berkeley, 1942-43; group leader chemistry and metallurgy div. Manhattan Project, Los Alamos Sci. Lab., 1943-46; faculty U. Cal. at Los Angeles, 1946-—, prof. chemistry, 1953-—. Cons. in phys., inorganic and nuclear chemistry to various indsl. corps. John Simon Guggenheim Meml. fellow, 1959. Mem. Am. Chem. Soc., Am. Phys. Soc., Am. Inst. Physics, A.A.A.S., Sigma Xi, Phi Lambda Upsilon. Author: (with Yost, Russell) The Rare-Earth Elements and Their Compounds, 1949; Radioactivity Applied to Chemistry, 1951; also numerous articles, chps. in books. Research, patents in inorganic reaction mechanism and thermodynamics, nuclear and radiochemistry. Office: Chemistry Dept., Los Angeles 90024.*

GARNER, Harry Hyman, Am. psychiatrist; b. Chgo., Jan. 19, 1910; s. Louis and Clara (Barasch) G.; B.S., U. Ill., 1932, M.D., 1934; m. Eleanore E. Hetherington, Apr. 5, 1940; children—Edward A., Larry B. Practice medicine, specializing in psychiatry, Chgo., 1939-—; cons. neurologist, psychiatrist Oak Forest Infirmary, 1945-—; prof., chmn. dept. psychiatry and neurology Chgo. Med. Sch., 1948-—; attending psychiatrist, chmn. dept. psychiatry and neurology Mt. Sinai Hosp., 1952-—; attending physician in psychiatry, chmn. deans subcom. on psychiatry VA West Side Hosp., 1953-—. Fellow Am. Psychiat. Assn.; mem. Acad. Psychoanalysis, Central Neuropsychiat. Assn., Acad. Neurology, A.M.A., Acad. Forensic Sci., Ill. Psychiat. Soc. (past pres.), Alpha Omega Alpha. Author: Emotional Reactions to Divorce, 1955; A Confrontation Technique Used in Psychyotherapy, 1959; (with others) Progress in Psychotherapy, 1960; Treatment of Nascent Schizophrenia with a Confrontation Technique, 1962; (with others) Personality and Behavioral Aspects, 1962; The Confrontation Technique, 1963; Psychosomatic Management of the Patient with Malignancy, 1966; also numerous articles. Coeditor Unfinished Tasks in the Behavioral Sciences, 1964. Research in somatic therapies of psychoses, contbg. to technique and theory of psychotheraphy. Home: 433 W. Roscoe St., Chgo. 60657. Office: 6 N. Michigan Av., Chgo. 60602.*

GARNER, Reuben John, radiobiologist; b. Oundle, Northamptonshire, U.K., Feb. 4, 1921; s. John Henry Ebbutt and Alice (Horsley) G.; B.A. with honors, Downing Coll., U. Cambridge (Eng.), 1942, M.A. with honors, 1946; M.R.C.V.S., Royal Vet. Coll., London, 1945, F.R.C.V.S., 1952; M.V.Sc., U. Liverpool (Eng.), 1952, D.V.Sc., 1961, A.R.I.C., 1956; m. Daphne Muriel Gascoyne, June 15, 1942; children—Karen Lesley, Julian Guy. Biochemist, Col. Vet. Service, VOM, N. Nigeria, 1946-50; lectr. Liverpool U., 1950-53; sr. lectr., Bristol. (Eng.) U., 1953-56; head radiobiology dept. Inst. for Research in Animal Diseases, Agrl. Research Council, Compton, Berkshire, Eng., 1957-60; head pub. health sect. radiol. protection div. Authority Health and Safety Br., U.K. Atomic Energy Authority, Harwell, Berkshire, 1960-65; dir. pub. health service Colo. State U. Collaborative Radiol. Health Lab., Ft. Collin, 1965-—. Cons., Internat. Atomic Energy Agy., Vienna, Austria, intermittently, 1960-65. Mem. Biochem. Soc., Health Physics Soc., Vet. Research Club, Assn., Vet. Tchrs. and Research Workers, Brit. Vet. Assn., Inst. Biology, Sigma Xi. Author: Veterinary Toxicology, 1957, 2d edit., 1961; (with Clarke, Clarke, Papworth) Veterinary Toxicology, 3d edit., 1966; contbg. author: Animal Health, Production and Pasture, 1963; Radioactivity and Human Diet, 1966; also numerous articles. Research on effects radiation and radionuclides on animals, metabolism nuclides important to man, particularly in farm animals, applications to problems environmental contamination with radioactive materials. Home: 1625 Country Club Rd., Ft. Collins, Colo. 80521.*

GARNER, Wendell Richard, Am. psychologist; b. Buffalo, Jan. 21, 1921; s. Richard Charles and Lena (Cole) G.; A.B., Franklin and Marshall Coll., 1942; A.M., Harvard, 1943, Ph.D., 1946; m. Barbara Chipman Ward, Feb. 18, 1944; children—Deborah Ann, Peter Ward, Elinor Elizabeth. Research asso. Harvard, 1943-46; faculty Johns Hopkins, Balt., 1946-67, prof., 1955-67, dir. psychol. labs. Inst. Coop. Research, 1949-55, chmn. dept. psychology, 1954-64; vis. scientist Applied Psychology Research Unit, Cambridge, Eng., 1966-67; James Rowland Angell prof. psychology Yale, 1967-—; vis. asso. prof. Stanford, 1952-53. Cons. govt. agys. Mem. A.A.A.S., Am. Psychol. Assn. (Distinguished Sci. Contbn. award 1964), Psychonomic Soc., Acoustical Soc. Am., Soc. Exptl. Psychologists, Nat. Acad. Scis. Author: (with A. Chapanis, C. T. Morgan) Applied Experimental Psychology, 1949; Uncertainty and Structure as Psychological Concepts, 1962; numerous articles. Demonstrated that audibility and loudness of very short tones is related to Fourier components of tones; introduced information theory as technique for solving problems psychol. judgment; formulated methods, basic concepts for understanding nature of form and pattern perception. Home: 48 Yowago Av., Branford, Conn. 06405. Office: Dept. Psychology, Yale, New Haven 06510.*

GARNER, William Edward, English chemist; b. Dec. 5, 1889; s. William Garner; ed. univs. Birmingham (Eng.), Göttingen (Germany), Univ. Coll., London; D.Sc., Birmingham. Lectr. chemistry U. Birmingham, 1919; lectr. Univ. Coll., London, 1919-25, reader phys. chemistry, 1925-27, fellow, 1930; Leverhulme prof. phys. chemistry U. Birston (Eng.), 1927-54, dir. chem. labs., 1936; pro-vice-chancellor, 1952-54. Corr. councillor Patronato Alfonso el Sabio, 1959. Fellow Royal Soc., 1937; mem. Faraday Soc. (council), pres.), Chem. Soc. (council). Editor: Chemistry of Solid State, 1955; Chemisorption, 1957. Contbr. papers on flame, adsorption, solid decomposition, detonation of solids, long chain organic compounds to sci. publs. Died Mar. 4, 1960.

GARNERIN, André Jacques, French physicist; b. Paris, Jan. 31, 1769; studied under Alexandre Charles, physicist; received task of flying over enemy positions, 1793; invented parachute and (with his bro. Jean Baptiste Oliver Garnerin) later improved it; numerous balloon ascents beginning in 1790 in which he stayed up as long as 24 hours; jumped with parachute from a height of 1000 meters on a flight in Parc Monceau, 1797. Died Paris, Aug. 18, 1823.

GARNETT, Thomas, Brit. physician, chemist; b. Casterton, Eng., 1766; studied under Black; M.D., U. Edinburgh, Scotland, 1788; postgrad. surgery, London; m. Catherine Grace Cleveland, Mar. 1795 (dec. 1798); 2 daus., including Mrs. Catherine Grace Godwin. Lectr., Manchester, Eng.; practiced in Harrogate, Eng.; prof. Anderson's Coll., Glasgow, Scotland; became 1st prof. chemistry Royal Instn., 1799. Author: Experiments and Observations on the Horley-Green Spaw, near Halifax, with a short account of two other Mineral Waters in Yorkshire, 1790; Experiments and observations on the Crescent Water at Harrogate, 1791; Treatise on the Mineral Waters of Harrogate: History, Chemical Analysis, Medicinal Properties, and Directions for their Use, 1792. Research on mineral waters including first sci. analysis of Harrogate waters; Died London, June 28, 1802.

GARNHAM, Percy Cyril Claude, English protozoologist; b. London, Eng., Jan. 15, 1901; s. Percy Claude and Edith (Masham) G.; M.B., B.S., U. London (Eng.), 1923, M.D. with gold medal, 1928, D.Sc., 1952; Dip.Med. Malariol., U. Paris (France), 1932; Docteur Honoris causa, U. Bordeaux (France), 1965; C.M.G., 1964; hon. F.R.C.P., Edinburgh, 1966, London, 1967; m. Esther Long Price, Dec. 31, 1924; children—Diana Myfanwy, Isolde (Mrs. Christopher Meyrick), Cicely Mary (Mrs. John Alden), John Claude, Jasper Meredith, Carolyn Ismea (Mrs. Charles Stephenson). Dir. div. insect borne diseases Med. Research Lab., Kenya Med. Service, Nairobi, 1925-47; reader med. parasitology London Sch. Hygiene and Tropical Medicine, 1947-51, prof. med. protozoology, 1951-—; dir. dept. parasitology London Sch. Hygiene and Tropical Medicine, 1951-—. Chmn. tropical pesticides research com. Ministry Overseas Devel., 1959-—. Recipient Darling medal and prize, 1951, Bernhardt Nocht medal, 1957, Gaspar Vianna medal and decoration, 1962; Manson medal, 1964. Fellow Royal Soc., 1964; mem. Royal Soc. Tropical Medicine and Hygiene (pres. 1967-—), Am. Soc. Tropical Medicine, Internat. Fedn. Parasitologists (v.p. 1960-—); hon. mem. Soc. Protozoologists, Mexican, Brit., Polish socs. parasitologists, Belgian Soc. Tropical Medicine, Soc. Path. Exot. Author: (with Pierce, Roitt) Immunity of Protozoa, 1963; Malaria Parasites, 1966. Discovered 3d cycle malaria parasites in liver of monkeys and man, ultrastructures in malaria parasites and other parasitic protozoa. Home: Southernwood, Farnham Common, Buckinghamshire, Eng. Office: Imperial Coll. Field Sta., Silwood Park, Sunninghill, Berkshire, Eng.*

GARNIER, Jean-Guillaume, mathematician; b. Rheims, France, Dec. 13, 1766; prof. math. at

Rheims, Colmar; prof. U. Grand, Belgium, 1817-35; founder jour., Correspondance mathématique. Research on differential equations. Died Brussels, Belgium, Dec. 20, 1840.

GARNIER, René Édouard Louis Marie, French mathematician; b. Chalon-sur-Saône, France, Jan. 16, 1887; s. Edouard and Jeanne (Boutonnet) G.; ed. Saint-Michel at Moulins; Ph.D. in math. scis.; m. Germaine Queyrat, Oct. 21, 1913; children—Paule-Marie, Edouard, Michel, Danielle, Denise. Prof., Poitiers Faculty Scis., 1920; prof. gen. math. Sorbonne, 1936; prof. geometry, 1946; examiner École polytechnique, 1952. Mem. French Acad. Scis., 1952, French Math. Soc. (past pres.). Research on theory of differential equations, resolution of Riemann and Plateau problems, extension of Savary's formula to apply to gen. movement of a solid. Address: 21, rue Decamps, Paris, France.*

GARNIR, Henri Georges, Belgian mathematician; b. Jemeppe-sur-Meuse, Belgium, Sept. 13, 1921; s. Georges Marie and Marie (Lempereur) G.; license in phys. sci. U. Liege (Belgium); D.Sc. in Math.; m. Noelly Pierre, July 22, 1948; children—Henri, Dominique. Asst. U. Liege, 1945-50, head of works, 1950-54, agrégé, 1954-58, instr., 1958-60, prof., head Service Math. Analysis and Algebra, 1960—; dir. Liege sect. Belgian Center Research in Functional Analysis. Mem. Royal Soc. Scis. Liege, Am. Math. Soc., Math. Assn. Am., math. socs. Belgium, France. Author: Problèmes aux limites de la Physique Mathématique, 1958; Fonctions de variables réeles, 2 vols., 1963-65; Fonctions d'une variable complexe, 1965; Analyse fonctionnelle, 1968. Research, publs. on group representation theory and its applications to phys. problems; boundary value problems for partial differential equations; math. theory of wave propagation; constructive theory of functional analysis. Home: 14, rue Joiret, Angleur, Belgium. Office: Institut de Mathematique de l'Universite de Liege, 15, avenue des Tilleuls, Liege, Belgium.*

GARRELS, Robert Minard, Am. geologist; b. Detroit, Mich., Aug. 24, 1916; s. John C. and Margaret A. (Gibney) G.; B.S., U. Mich., 1937; M.S., Northwestern U., 1939, Ph.D., 1941; M.A. (hon.), Harvard, 1955; m. Jane M. Tinen, Dec. 21, 1940; children—Joan F., James C., Katherine G. From instr. to asso. prof. geology Northwestern U., 1941-52; geologist U. S. Geol. Survey, 1952-55; asso. prof. geology Harvard, 1955-57, prof., 1957-65, chmn. dept. geol. scis., 1963-65; Henri Speciael prof. sci. U. of Brussels (Belgium), 1962-63; prof. geology Northwestern U., Evanston, Ill., 1965—. Trustee Bermuda Biol. Sta. Recipient of Arthur L. Day medal Geol. Soc. Am., 1966. Fellow A.A.A.S., Geol. Soc. Am., Mineral. Soc. Am.; mem. Geochem. Soc. (pres. 1962), Nat. Acad Scis., Soc. Econ. Geologists, Geol. Soc. Washington, Am. Acad. Arts and Sci., American Chem. Soc., Sigma Xi. Author: Textbook of Geology, 1951; Mineral Equilibria, 1959; (with C. L. Christ) Minerals, Solutions, and Equilibria, 1965. Research on chemistry of seawater; crystallization of minerals; reactions among minerals; electrode techniques; ionic diffusion through rocks. Office: Locy Hall, Dept. Geology, Northwestern U., Evanston, Ill.

GARREN, Kenneth H(oward), Am. plant pathologist; b. Asheville, N.C., Nov. 26, 1912; s. Samuel and Annie (Baity) G.; A.B., Duke, 1934, M.A., 1937, Ph.D., 1938; m. Lena Oates, Aug. 12, 1945; 1 dau., Kenna Schahar. Field asst. U. S. Forest Service, Asheville, N.C., 1938; asst. prof. biology Jacksonville State Coll., 1939-41; asso. research botanist Ga. Expt. Sta., 1941-47; asso. prof. botany Auburn U., 1947-54; agrl. research adviser U. S. AID Mission to El Salvador, 1954-55; research plant pathologist Crops Research div. Agrl. Research Service, U. S. Dept. Agr., Holland, Va., 1955—, project leader peanut disease investigations, 1959. Mem. Am. Phytopath. Soc., A.A.A.S., Sigma Xi, Phi Sigma, Xi Sigma Pi. Contbr. articles to sci. jours. Investigated means by which fungi decay wood; conditions under which important diseases of a number of woody plants develop, including woody flowering plants such as camellia, forest trees of U. S., tropical plants such as coffee; discovered that an important disease of peanuts could be controlled by use of a herbicide and deep burial of trash followed by cultivation which keeps soil out of the row; studies of peculiar microbiology of peanut pod, only commercially important fruit which develops as root. Home: R.F.D. 2. Office: Tidewater Research Sta., Holland, Va. 23391.*

GARRETSON, James Edmund, (pseudonym John Darby) Am. oral surgeon; b. Wilmington, Del., Oct. 4, 1828; s. Jacob M. and Mary (Powell) G.; grad. Phila. Coll. Dental Surgery, 1856; M.D., U. Pa., 1859; m. Beulah Craft, Nov. 10, 1859; 2 daus. With Phila. Dental Coll., 1874-95, prof. anatomy and surgery, 1878, dean faculty, 1880; prof. clin. surgery Medico-Chirurg. Coll. of Phila.; pres. Med. and Chirurg. Soc. Phila., 1883. Author: A treatise on the Diseases and Surgery of the mouth, Jaws and Associated Parts (1st standard work in its field), 1869; A System of Oral Surgery, 1873; Man and His World, 1889. Originator of oral surgery as a splty. of dentistry; 1st surgeon to employ dental engine as modified for surg. operations, 1882; whenever possible, limited incisions to interior of mouth to avoid facial scars. Died Lansdowne, Pa., Oct. 26, 1895.

GARRETT, Alfred Benjamin, educator; b. Glencoe, O., June 28, 1906; s. Robert E. and Margaret (McMaster) G.; B.S., Muskingum Coll., 1928, Sc.D. (hon.), 1960; M.S., Ohio State U., 1931, Ph.D., 1932; Sc.D. (hon.), Ohio Wesleyan U., 1962, Denison U., 1966; m. Jessie Campbell, Sept. 1, 1934; children—Carol Lynn (Mrs. Alan Fisher), John Calvin, Lois Nancy. Tchr. math. Harding High Sch., Aliquippa, Pa., 1928-29; faculty Kent State U., 1932-35; faculty Ohio State U., Columbus, 1935—, prof. chemistry, 1944—, v.p. research, 1962—. Recipient Gov.'s award Ohio Newspaper Assn., 1964. Mem. Am. Chem. Soc. (chem. edn. award 1963), A.A.A.S. Studies, publs. on hydrides, low temperature batteries, trace elements, Grignard and Friedel-Craft systems. Author: Batteries of Today, 1957; The Flash of Genius, 1963; (with Evans, Sisler) Semimicro Qualitative Analysis, 1942, rev., 1946; (with Mack, Haskins, Verhoek) Textbook of Chemistry, 1949; (with Haskins, Sisler) Essentials of Chemistry, 1951; (with Richardson & Kieffer) Chemistry, 1960, rev., 1966. Home: 162 Erie Rd., Columbus, O. 43214.*

GARRETT, Edward Robert, Am. pharm. chemist; b. N.Y.C., Apr. 9, 1920; s. Morray and Stella (Abrams) G.; B.S., Mich. State U., 1941, M.S., 1948, Ph.D. (Hinman fellow), 1950; m. Irene Brewer, July 31, 1941; children—Jan Edward, Terry Lee, Kurt Lane. Sr. research scientist Upjohn Co., 1950-61; grad. research prof. U. Fla., Gainesville, 1961—; vis. prof. U. Wis., 1958, U. Buenos Aires, 1965. Cons. Smith, Kline & French, 1962—; pres. Symposium on Drug Stability, Buenos Aires, 1962; v.p. Internat. Symposium on Quality Control Drugs, Lima, Peru, 1965. Recipient Upjohn award, 1959; Research Achievement award Am. Pharm. Assn., 1963. Mem. Am. Chem. Soc., Am. Pharm. Assn. (Ebert prize for sci. publs. 1963), A.A.A.S., Am. Soc. Microbiology, Argentina, Chile socs. indsl. pharmacy and biochemistry, Sigma Xi, Rho Chi, Alpha Chi Sigma, Sigma Pi Sigma, Pi Mu Epsilon. Research, numerous publs. on kinetics and mechanisms of reaction, salicylates, steroids, nucleosides, alkaloids, prediction of stability of pharms. and antibiotics, kinetics and mechanisms of antibiotic action on microbial growth, pharm. and instrumental analysis, statistics and quality control of pharm. products, dosage forms and assays, analog computation applied to drug transformation in vitro and absorption, distbn., metabolism and excretion of drugs in vivo, pharmacokinetics. Home: 1826 N.W. 26th Way, Gainesville, Fla. 32601.*

GARREY, Walter Eugene, Am. physiologist; b. Reedsville, Wis., Apr. 7, 1874; s. John Eugene and Harriet (Anderson) G.; B.S., Lawrence U., 1894; studied U. Berlin, 1898; Ph.D., U. Chgo., 1900; M.D., Rush Med. Coll., 1909; m. Charlotte Eaton, Dec. 31, 1901; 1 son, Walter Eaton. Extension instr. in zoölogy U. Chgo., 1894-98, fellow in physiology, 1898-1900; prof. physiology Cooper Med. Coll., San Francisco, 1900-10; asso. prof. physiology Washington U., 1910-16; prof. physiology Tulane U., 1916-25; prof. physiology Vanderbilt U. Med. Sch., 1925-44, prof. emeritus, 1944. Lectr. on gen. physiology Marine Biol. Labs., Woods Hole, Mass., also mem. research staff. Mem. A.M.A., A.A.A.S., Am. Physiol. Soc. (pres. 1937, 38), Am. Soc. Biol. Chemists, NRC, San Francisco Acad. Medicine (hon.), Phi Beta Kappa, Sigma Xi. Contbr. many articles, original research, chiefly on the heart. Died June 15, 1951.

GARRISON, John Dresser, Am. physicist; b. Salt Lake City, Aug. 9, 1922; s. Lloyd and Evelyn (Dresser) G.; B.A., U. Cal. at Los Angeles, 1947, M.A., 1948; Ph.D., U. Cal. at Berkeley, 1954; m. Betty Bernhardt, Jan. 17, 1968. children—Jeffrey, Eric, Jan. Instr., Yale, 1953-56; vis. scientist Brookhaven Nat. Lab., 1955, vis. asso. physicist, 1962-63; prof. physics San Diego State Coll., 1956—. Cons. Gen. Atomic div. Gen. Dynamics, 1957—. Mem. Am. Phys. Soc., Sigma Xi. Contbr. articles to tech. jours. Analysis neutron cross sect. data; measurement proton-proton scattering cross sects. Home: 5181 College Gardens Ct., San Diego 92115.*

GARROD, Sir Alfred Baring, English physician; b. Ipswich, Eng., May 13, 1819; M.D., U. Coll. London, 1843; m. Elizabeth Ann Colchester, 1845; 4 sons including Alfred Henry, Archibald; 2 daus. Prof. therapeutics U. Coll. Hosp., 1851-63, King's Coll. Hosp., 1863-74; named physician extraordinary to Queen Victoria, 1896; Gulstonian lectr. 1857; Lumleian lectr., 1883. Created knight, 1887. Fellow Royal Soc. (v.p.), 1858. Author: Treatise on Gout and Rheumatic Gout, 1859; Essentials of Materia Medica and Therapeutics, 1855. Research on gout including discovery of uric acid in blood of patients with gout; used lithia to treat gout. Died Dec. 28, 1907.

GARROD, Alfred Henry, English zoologist; b. London, May 18, 1846; s. Sir Alfred Basing G.; ed. U. Coll. Sch., London; med. scholar, King's Coll., London, 3 times; sr. in natural sci. tripos, Cambridge, Eng., 1871; became fellow St. John's Coll., Cambridge, 1873; prof. comparative anatomy King's Coll., London, 1874-79; named Fullerian prof. physiology Royal Instn., 1875. Fellow Royal Soc., 1876; mem. Cambridge Zool. Soc. (became prosector 1871). Published collection of his papers, 1881. Contbr. author: Natural History (Gassell). Research on anatomy and myology of birds and ruminants. Died Oct. 17, 1879.

GARROD, Sir Archibald Edward, English physician; b. Nov. 25, 1857; s. Alfred B. Garrod; ed. Marlborough; Christ Coll., Oxford U., D.M., M.A.; St. Bartholomew's Hosp.; LL.D., Glasgow, Aberdeen; hon. M.D., Dublin, Malta; Dr. Honoris causa, Padua; m. Laura E. Smith, 1886; 1 dau., 3 sons. Cons. physician, dir. med. unit St. Bartholomew's Hosp., apptd. full physician, 1912; cons. physician Hosp. for Sick Children; regius prof. medicine, Oxford U.; lectr., examiner at various univs. Fellow Royal Coll. Physicians, Royal Soc., 1910, Hunterian Soc.; hon. mem. Assn. Physicians Gt. Britain and Ireland, Assn. Am. Physicians. Author: Inborn Errors of Metabolism, 1923; The Inborn Factors in Disease, 1931; also articles. Research on genetically determined human diseases, cryptonuria, alkaptonuria, fructosuria and albinism (first direct study of these problems). Died Mar. 28, 1936.

GARSTANG, John, Brit. archeologist; b. Blackburn, Eng., 1876; s. Walter and Mathilda Mary (Wardley) G.; B.A., Jesus Coll., Oxford; M.A., D.Sc., 1896-99; LL.D., Aberdeen; m. Marie Louise Berges, 1907; 1 son, 1 dau. Prof. archeology U. Liverpool (Eng.), 1907-41, emeritus, 1942-56; dir. Brit. Sch. Archaeology, Jerusalem, 1919-26; pres. Brit. Inst. Archaeology, Ankara, from 1949. Corr., Institut de France. Author: Roman Ribchester; El Arabeh; Mahasna and Bêt Khallaf; The Third Egyptian Dynasty, 1904; Burial Customs of Ancient Egypt, 1907; The Land of the Hittites, 1910; Meroë, 1911; The Hittite Empire, 1929; The Foundations of Bible History, Joshua, Judges, 1931; The Heritage of Solomon, 1934; Prehistoric Mersin, 1953; also reports on excavations. Excavated Roman sites in Britain, 1897; excavated in Egypt, Nubia, Asia Minor, N. Syria, 1900-08, Meroë, 1909-14, Askalon, Palestine, 1920-21, Jerico, 1930-36; dir. Neilson expdn. to Nr. East, 1936-52, including excavation of Mersin, 1937-47. Died Sept. 12, 1956.

GARSTIN, Sir William Edmund, engr.; b. India, Jan. 29, 1849; s. Charles and Agnes Helen (Mackenzie) G.; ed. King's Coll., London; m. Mary Isabella North, 1888; 1 son, 1 dau. Entered Indian Pub. Works Dept., 1872, sent to Egypt, 1885, ret., 1892; insp.-gen. irrigation, Egypt, 1892; under-sec. state for pub. works, 1892; adviser to ministry pub. works in Egypt, 1904-08; Brit. govt. dir. Suez Canal Co., from 1907. Author: Report on the Basin of the Upper Nile, 1904. Responsible for plans and bldg. of Aswan Dam and barrages of Asyait and Esna; compiled 2 reports on hydrography of Upper Nile; initiated geol. survey of Egypt, 1896; erected new bldgs. of Nat. Mus. Egyptian Antiquities, 1902. Died Jan. 8, 1925.

GÄRTEL, Carl Wilhèalm, German biologist; b. Staulmare, Nov. 17, 1920; s. Karl and Margarete (Hennig) G.; Ph.D. in Pedological Sci., Higher Inst. Pedology, Vienna; m. Annemarie Zurowski; children—Peter, Maria, Helene, Gabriele, Sabine. Sci. cons., dir. Inst. Diseases of Vineyards, Fed. Biol. Inst. Agrl. Sci. and Forestry. Author: Untersuchungen über die Bedeutung des Bors für Rebe under Besonderer Berücksichitigung der Befruchtung, also numerous works on oligo-elements and non-parasitic diseases of vineyards. Address: Brüningstrasse 84, Bernkastel-Kues, Germany.

GARTEN, (Ernst Heinrich) Siegfried, German physiologist; b. Kieritzsch nr. Leipzig, Germany, June 29, 1871; s. Alexander and Elisabeth (Schmiedt) G.; studied physiology, Leipzig; m.; 1 son. With U. Leipzig most of life, apptd. prof., 1916; with U. Giessen (Germany) for several years, named prof., 1908; study trip Zool. Sta., Naples, Italy. Mem. Saxon Acad. Scis. Research and numerous articles. in neuro- and sense physiol. problems, especially optical-bleaching of visual purples in retina, formation of visual yellow and orange; studied electric organs in fish; representation mech. adaptation of retina to light and dark; improved registration technique; developed 1st photocymographion, 1st electrically registering manometer. Died Leipzig, Aug. 7, 1923.

GARTENHAUS, Solomon, Am. physicist; b. Kassel, Germany, Jan. 3, 1929; s. Leopolt and Hanna (Brandler) G.; came to U. S., 1937, naturalized, 1943; B.S., U. Pa., 1951; M.S., U. Ill., 1953, Ph.D., 1955; m. Johanna Lore Weisz, Aug. 30, 1953; children—Michael M., Kevin M. Instr., Stanford, 1955-58; faculty physics Purdue U., Lafayette, Ind., 1958—, prof., 1963—. Cons. Lockheed, summers 1958-60, Advanced Research Corp., 1961-65. Mem. Am. Phys. Soc., Phi Beta Kappa. Author: Elements of Plasma Physics, 1964; also articles. Theoretical research in nuclear physics, plasma physics, many-particle systems, nuclear interactions based on meson fields and condensation phenomena at low temperatures. Home: 444 Littleton St., West Lafayette, Ind. 47906.*

GARTH, John Shrader, Am. zoologist; b. Los Angeles, Oct. 3, 1909; s. James Gray and Jessie (Imlach) G.; B.Mus., U. So. Cal., 1932, M.S., 1935, Ph.D., 1941; postgrad. Cornell U., U. Pa.; m. Isla Lora Detter, June 25, 1940; 1 dau., Linda Jean. Faculty, U. So. Cal., Los Angeles, 1935—, curator Allan Hancock Found., 1962—, prof. biology, 1967—; civilian instr. Santa Ana Army Air Base, 1942-44. Expdn. leader Hancock Found. Expdn. to Ariz. desert, 1942, 46-48; asso. marine biologist Eniwetok Marine Biol. Lab., Marshall Islands, summers 1957, 59; mem. U. S. Program in biology Internat. Indian Ocean Expdn.

UNESCO, 1964. Fellow A.A.A.S., Cal., So. Cal. acads. sci.; mem. Western Soc. Naturalists, Am. Soc. Limnology and Oceanography, Soc. Systematic Zoology (past pres. Pacific sect.), Sigma Xi. Author: Butterflies of Grand Canyon Nat. Park, 1950; Brachyura of the Pacific Coast of America, 1958; (with J. W. Tilden) Yosemite Butterflies, 1963. Research, publs. on crabs of Chile and Galapagos Islands, spider crabs of Pacific Coast of Am. Home: 1426 N. Detroit St., Hollywood, Cal. 90046. Office: Allan Hancock Found., U. So. Cal., Los Angeles 90007.*

GARTH, Thomas Russell, Am. psychologist; b. Paducah, Ky., Dec. 24, 1872; s. Robert and Jane Elizabeth (Campbell) G.; B.A., U. Denver, 1909, M.A., 1910; Ph.D., Columbia, 1917; m. Ethel Nadine Tucker, 1909; children—Thomas Russell, Francis Marion, and Ethel Nadine (adopted). Psychologist N.Y. Post-Grad. Sch., 1912-13; asst. in edn. State Normal Sch., Farmville, Va., 1913-15; asst. prin. N.Y. Parental Sch., 1915-16; prin. Barton Heights Sch., Richmond, Va., 1916-17; head dept. of edn. State Normal Sch., Canyon, Tex., 1917-19; adj. prof. psychology U. Tex., 1919-22; prof. edn. U. Denver, 1922-29, prof. ednl. psychology, 1929-30, prof. exptl. psychology, from 1930; prof. Summer Sch., U. of Colo., 1922, U. Tex., 1923. Asst. Rockefeller Hookworm Investigation, 1912; mem. Univ. Race Commn., 1920; head expdn. to study color blindness of Indians, 1930-31, to study foster Indian child in white homes, 1935-37, also with various other expdns. to study psychology of Indian, 1919—; specialist with U. S. Govt. in Indian Service Sch., 1936, 37. Author: Mental Fatigue During Continuous Exercise, 1918; Race Psychology, 1931; Educational Psychology, 1937; Life of Henry Augustus Buchtel, 1937. Died Apr. 20, 1939.

GARTLEIN, Carl Witz, Am. astrophysicist; b. Connersville, Ind., Nov. 13, 1902; B.A., DePauw U., Indiana, 1924; Ph.D., Cornell U., 1929; hon. D.Sc., Colgate U., 1965; m. Helen Hart, 1929; children: Christopher, Caroline, Delight. Prof.; dir., Visual Aurora subcenter, World Data Center A, Cornell. Mem., Am. Physical Soc., Optical Soc. Am., Am. Geophysical Union, Am. Assn. Variable Star Observers, Phi Beta Kappa, Sigma Xi. Mem., Optical Standards Com. of Nat. Bureau Standards; Upper Atmosphere subcom. of Nat. Advisory Com. for Aeronautics; other advisory coms. Research on light, especially Aurora Borealis; built one of fastest auroral spectrographs, 1939; developed All Sky Camera for continuous photography of aurora; for Canada, dir. 1st spectrographic triangulation measurements of auroral spectrum; proved that hydrogen atoms enter Earth's atmosphere during aurora. Deceased.

GÄRTNER, August Anton Hieronymus, German hygienist, bacteriologist; b. Ochtrup, Germany, Apr. 18, 1848; s. Johannes and Jenny (Dahme) G.; ed. Friedrich Wilhelm Med.-Surg. Acad., Berlin; Ph.D., Münster, Germany, 1915; m. Lilly Pross, 1878; 2 sons, including Wolfgang, 2 daus. Served in Navy, 1874-86; assigned to Kaiserliches Gesundheitsamt, 1884, asst. to Robert Koch for 2 1/2 years; named asso. prof. hygiene U. Jena (Germany), 1886, prof., 1887; traveled through Russia, U. S. studying water supplies of large cities; served as hygienic adviser for reserve hosps. XI Corps, World War I; apptd. san. insp. prisoner of war camps, 1915. Named hon. citizen, Jena. Author: (with E. Tiemann) Handbuch der Untersuchung und Beurteilung der Wässer, 1889; Die chemische und mikroskopischbackteriologische Untersuchung des Wassers, 1889; Leitfaden der Hygiene, 1892; Die Hygiene des Trinkwassers, 1896; Die Hygiene des Wassers, 1915; also numerous articles. Research on water supplies especially water problems and typhus epidemics in Ruhr; described bacillus of paratyphus group in study of mass food poisoning (enteritiditis Gärtner), 1888; proved infection of Tb in mammals and birds through female; research on joint stiffness of miners. Died Jena, Dec. 21, 1934.

GÄRTNER, Gustav, pathologist; b. Pardubitz, Bohemia, Sept. 28, 1855; s. Alois and Josephine (Liebermann) G.; M.D., U. Vienna, 1879; m. Melanie Schalek, 1897; 1 son; 1 dau., Hanna. First worked in various depts. Vienna Allgemeines Krankenhaus; then worked with Salomon Stricker, Inst. for Gen. and Exptl. Pathology; became lectr. U. Vienna, asso. prof., 1890, prof., 1918; pvt. practice, beginning in 1890. Research and publs. on splanchnic innervation of kidney, kidney secretion, innervation of brain vessels, circulation in brain; developed and introduced various elec. examination and treatment methods, including tonometer (one of 1st instrument to measure blood pressure). Died Vienna, Nov. 4, 1937, Vienna.

GÄRTNER, Joseph, German botanist; b. Calw, Württemberg, Germany, Mar. 12, 1732; s. Joseph and Eva Maria (Wagner) G.; student law, 1750; M.D., U. Tübingen, Germany, 1753; 1 son, Karl Friedrich. Prof. anatomy U. Tübingen, 1761-68, botany, natural history, St. Petersburg, Russia, from 1768; dir. bot. garden, natural history collection, St. Petersburg, til 1770. Fellow Royal Soc., 1761; mem. St. Petersburg Acad. Scis. Author: De fructibus et seminibus plantarum, 1788-91; Supplementum carpologiae, 1805-07; Contbr. articles to sci. publs. Founder carpology. Died Calw, Württemberg, 1791.

GARVEY, Gerald Thomas-John, Am. physicist; b. N.Y.C., Jan. 21, 1935; s. John Thomas and Anne (Williams) G.; B.S., Fairfield U., 1956; postgrad. Boston Coll., 1958; Ph.D. (NSF fellow), Yale, 1962; m. Doris Carol Burmester, June 6, 1959; children—Deirdre Anne, Gerald Thomas-John, Victoria Elizabeth. Research asso. Yale, 1962-63, asst. prof., 1964-66; faculty Princeton, 1963-64, 66-67, asso. prof. physics, 1967—. Alfred P. Sloan Found. fellow, 1967—. Mem. Am. Physics Soc. Research, publs. on interactions between complex nuclei revealing structure of light nuclei; studied consequences of charge-independence of nuclear forces in nuclei using nuclear reaction. Home: 54B Western Way, Faculty Rd., Princeton, N.J. 08540.*

GARVEY, William D., Am. psychologist; b. Richmond, Va., Jan. 17, 1923; s. John Wickoff and Birdie (Bique) G.; B.A., U. Richmond, 1947; M.A., U. Va., 1949, Ph.D., 1951; m. Catherine Jane Jones, Dec. 31, 1957; 1 dau., Stephanie Kate. Head engring. psychology research sect. Naval Research Lab., Washington, 1951-59; dir. project on sci. information exchange in psychology Am. Psychol. Assn., Washington, 1961-66; prof. psychology, dir. Center for Research in Sci. Communication, Johns Hopkins, Balt., 1966—. Mem. coms. Nat. Acad. Scis.-NRC, 1965—. Fellow Am. Psychol. Assn.; mem. Eastern Psychol. Assn., Psychonomic Soc., Research Engring. Soc. Am., Sigma Xi. Produced detailed and quantitative description system of sci. information in psychology. Home: 6031 Hollins Av., Balt. 21210.*

GARWIN, Richard Lawrence, Am. physicist; b. Cleve., Apr. 19, 1928; s. Robert and Leona (Schwartz) G.; B.S., Case Inst. Tech., 1947; M.S., U. Chgo., 1948, Ph.D., 1949; m. Lois E. Levy, Apr. 20, 1947; children—Jeffrey L., Thomas M., Laura J. Instr. physics U. Chgo., 1949-51, asst. prof., 1951-52; staff mem. IBM Watson Lab., Columbia, N.Y.C., 1952—, asso. dir., 1960-64, adjunct prof. physics, Columbia, 1957—. Cons. Los Alamos Sci. Lab., 1950—; mem. Pres.'s Sci. Adv. Com., 1962-65. Mem. Am. Phys. Soc., Sigma Xi. Important work includes establishment (with L. M. Lederman), of non-conservation of parity for mu-mesons, extensive research on mu-mesons, liquid and solid helium-three. Office: 612 W. 115th St., N.Y.C. 10025.*

GASCOIGNE, William, Brit. inventor; b. Leeds, Eng., circa 1612; s. Henry and Margaret (Cartwright) G.; studied astronomy under his father. Author treatise on optics. Invented methods of grinding glass; 1st to use 2 convex lenses in telescope; original inventor of wire micrometer and its application to telescope, application of telescope to quadrant. Died in battle of Marston Moor, July 2, 1644.

GASKELL, Walter Holbrook, English physiologist; b. Naples, Nov. 1, 1847; s. John Dakin and Anne Gaskell; ed. Trinity Coll., Cambridge, Univ. Coll. Med. Sch.; M.A., M.D.; LL.D., Edinburgh, 1894, McGill U., 1897; m. Catharine Sharpe Parker, 1875; 1 son, 2 daus. Univ. lectr. in physiology Cambridge U., from 1883; fellow, prelector in natural sci. Trinity Hall, named fellow, 1889. Fellow Royal Soc., 1882 (Gold medal 1889, Baly medal 1895), Medico-Chirurg. Soc. (hon.). Contbr. to sci. jours. Died Sept. 7, 1914.

GASPAR, Max Raymond, Am. physician; b. Sioux City, Ia., May 10, 1915; s. Edgar Mathias and Mabel (Teefey) G.; A.B., Morningside Coll., 1936; B.S. in Medicine, U. S.D., 1938; M.D., U. So. Cal., 1940; m. Virginia Hunter, June 2, 1938; children—Karen (Mrs. William Stivers), Thomas, James, Susan, Mary Ann. Practice medicine specializing in vascular surgery, Long Beach, Cal., 1948—; clin. prof. surgery U. So. Cal. Mem. A.C.S. (past pres. So. Cal. chpt., gov. 1961—), Internat. Cardiovascular Soc., Soc. for Vascular Surgery, Pacific Coast, Western surg. assns. Research, publs. in intestinal anastomosis, peripheral vascular surgery. Home: 4215 E. 2d St., Long Beach 90803. Office: 1777 Bellflower Blvd., Long Beach, Cal., 90815.*

GASPAR, Rezso, Hungarian physicist; b. Ersekvadkert, Hungary, Feb. 7, 1921; s. Rezso and Ilona (Racskay) G.; Diploma High Sch. Tchr. in Math. and Physics, Pazmany Peter U., Budapest, Hungary, 1943, Ph.D., 1946, Candidate of Scis., 1952, D.Sc., 1956; m. Hedvig Kosa Szabo, Aug. 8, 1943; children—Rezso, Hedvig. Asst., sr. asst., docent U. for Tech. Scis., Budapest, 1945-53; docent Kossuth Lajos U., Debrecen, Hungary, 1954-55, prof., head Inst. for Theoretical Physics, vice dean faculty natural scis., 1958—; head dept. research group for theoretical physics Hungarian Acad. Scis., Budapest, 1953—; vis. fellow Joint Inst. for Lab. Astrophysics, U. Colo., 1963-64. Recipient State prize for activity in quantum-chemistry, 1965. Mem. Eotvos Lorand Phys. Soc. (pres. sect. Debrecen). Research, numerous publs. on electronic structure of hydrides by aid of united atom model, proposed energy band spectrum for semiconductors selenium and tellurium, inventor universal potential field for atoms which approximates self-consistent potential field with exchange. Home: 5 Doczy Jozsef. Office: 1 Egyetem-ter, Debrecen, Hungary.*

GASPARIN, Adrien-Étienne-Pierre de, see De Gasparin, Adrien-Étienne-Pierre.

GASSENDI, Pierre, French philosopher, mathematician; b. Champtercier, Provence, Jan. 22, 1592; ed. Coll. of Digne, U. Aix; D.Theology from Avignon, circa 1616; lectr. theology at Digne, 1612; took holy orders, 1617; called to chair philosophy at Aix, 1617; traveled to Flanders and Holland, 1628, returned to France, 1631; provost of Cathedral of Digne, circa 1633; accepted chair math. Collège Royale, Paris, 1645-48, illness compelled him to give up lectures, 1648; lived in Toulon, 1648-50; returned to Paris, 1653; statue erected in his honor at Digne, 1852. Author: Syntagma Philosophiae Epicuri, 1649, also biographies of Tycho Brahe and Copernicus, many others; his works collected in Opera Omnia, 1658. Revived atomic theory of Democritos and Epicuros, built qualitative but mechanistic physics on it; held that atoms differed in size, weight and shape, believed that atomic motion is sole cause necessary to explain phys. phenomena; held that gaseous pressure is caused by atomic collision; suggested that light is composed of atomic particles; revived old theory of vaccum (said it was immense, immobile, incorporeal, and necessary); held that time is unlimited, incorporeal, uncreated, and continuously passing; defined motion as simple translation from one place to another; argued that gravity is not inherent property of bodies but bestowed on them by attraction of earth; supported Copernican theory; carried out expts. to refute old objections to moving-earth hypothesis (even after Church's condemnation of Galileo's work); believed that motions can be perpetual if uniform and not acted on by any outside forces, thus arrived at principle of inertia; measured velocity of sound; showed that sound of musket shot and cannon ball discharge move forward with same speed; observed transit of Mercury at time predicted by Kepler, 1631, one of 1st to observe planetary transit; gained fame (and enemies) by his unfavorable commentaries on Aristotle; opposed Descartes' philosophy and Harvey's theory of blood circulation. Died Paris, Oct. 24, 1655.

GASSER, Herbert Spencer, Am. physician; b. Platteville, Wis., July 5, 1888; s. Herman and Jane Elizabeth (Griswold) G.; A.B., U. Wis., 1910, A.M., 1911, D.Sc., 1941; M.D., Johns Hopkins, 1915; postgrad. in Europe, 1923-25; many hon. degrees. Asst. and instr. in physiology U. Wis., 1911-13; instr. pharmacology Washington U., St. Louis, 1915-16; instr. in physiology, 1916-18, asso., 1918-20, asso. prof., 1920-21; prof. pharmacology, 1921-31; prof. physiology, Cornell Univ. Med. Coll., N.Y., 1931-35; dir. Rockfeller Inst., 1935-53, mem. emeritus, 1953—. Recipient (with Erlanger) Nobel prize in physiology, 1944; Kobr medal Assn. Am. Physicians, 1954. Fellow A.A.A.S., Am. Acad. Arts and Scis., fellow Royal Soc. Edinburgh (hon.); mem. Royal Soc. London (fgn.), Nat. Acad. Scis., Am. Physiol. Soc., Am. Soc. Pharmacology and Exptl. Therapeutics, Soc. Exptl. Biology and Medicine, Am. Neurol. Assn., Assn. for Research in Nervous and Mental Disease, Harvey Soc., Am. Philos. Soc., Assn. Am. Physicians, Physiol. Soc. Eng. (hon.), Sigma Xi, other fgn. sci. socs. Editor Jour. Exptl. Medicine, 1936—. Author: (with J. Erlanger) Electrical Signs of Nervous Activity, 1937; also papers. Research on blood coagulation; investigated (with Joseph Erlanger) electrophysiology of nerves, especially action currents of phrenic nerve, also discovered conductivity difference of different groups of nerve cells, by combining electronic amplifiers and cathode-ray oscillographs, 1924. Died N.Y.C., May 11, 1963.

GASSIOT, John Peter, Brit. sci. writer; b. London, Apr. 2, 1797; ed. at Lee; m., 1818; 9 sons, 3 daus.; midshipman, Royal Navy; mem. firm Martinez, Gassiot & Co., wine mchts., London, Oporto; an endower, chmn. Kew Obs.; endowed Cowper St. Middle Class Sch.; founder Royal Soc. Sci. Relief Fund; magistrate of Surrey, Eng. Fellow Royal Soc., 1840; mem. Chem. Soc. (a founder). Contbr. articles to tech. jours. Proved, using Grove's cells, that static effect of a battery increases with its chem. action, 1844, using delicate micrometers, that Grove's arguments against the contact theory were correct, 1844; discovered stratification of electric discharge, 1852. Died Aug. 15, 1877.

GASSMANN, Fritz, Swiss geophysicist; b. Zurich, Switzerland, July 27, 1899; s. Fritz and Sophie (Kagi) G.; diploma in Math. and Physics, Swiss Fed. Inst. Tech., 1923, Ph.D. in Math., 1925; m. Rosa Deuber, Mar. 21, 1925; children—Heidi (Mrs. Rolf Steffen), Margrit (Mrs. Rudolf Brennenstuhl), Verena (Mrs. Niklaus Appenzeller). Tchr. math. Cantonal Coll., Aarau, Switzerland, 1928-37, head, 1937-42; lectr. gen. and applied geophysics Swiss Fed. Inst. Tech., Zurich, 1928-42, asso. prof., 1942-52, prof., 1952—, dir. Inst. Geophysics, 1942—. Vis. prof. Purdue U., 1952, U. Ill., 1962. Mem. Schweiz Naturforschende Gesellschaft, European Assn. Exploration Geophysicists, Soc. Exploration Geophysicists, Am. Geophys. Union. Author: (with Max Weber) Einführung in die angewandte Geophysik, 1960. Research, numerous publs. on pure and applied geophysics. Home: 30 Pestalozzistrasse. Office: 33 Leonhardstrasse, Zurich, Switzerland.*

GASSMANN, George Joseph, Am. physicist; b. Hanau, Germany, Nov. 2, 1913; s. Heinrich and Eva (Klein) G.; student U. Wuerburg (Germany). U. Vienna (Austria), 1933, U. Munich (Germany), 1933-34; Dr.nat. scis., U. Goetingen (Germany), 1939; m. Maria E. Reis, July 11, 1942; children—Ursula (Mrs. Helmut Walter), Bertwin. Came to U. S., 1952, naturalized, 1957. Physicist, Deutsche Versuchsanstalt fuer Luftfahrt Berlin, 1939-43, Reichstelle fuer Hochfrequenzforschung, 1943-45; engr. Bayerische Elektromechanische Werkstaetten, 1945-46; cons. engr., Bad-Orb, Germany, 1947-52; supervisory research physicist Air Force Cambridge Research Center, Boston, 1952—. Mem. Am. Geophys. Union, Union Radio Sci. Internat. (mem. commn. 3 U. S. nat. com. 1954—). Editor: The Effects of Disturbances of Solar Origin in Communications, 1963. Research, publs. on Arctic radio propagation, upper atmosphere tide motions, communications effects natural and manmade disturbances, propagation measurements by airborne techniques. Home: 33 Whipple Rd., Lexington, Mass. 02173. Office: Air Force Cambridge Research Labs., Cambridge Research Center, Hanscom Field, Bedford, Mass. 01730.*

GASSNER, (Johann) Gustav, German botanist; b. Berlin, Jan. 17, 1881; s. George and Luise (Voigt) G.; doctorate U. Berlin, 1906; Ph.G. (hon.), U. Göttingen (Germany); m. Lili Fassler-Farnkopf, 1910; 3 sons, 1 dau. Named prof. botany and phytopathology Agrl. U., Montevideo, Uruguay, 1907; returned to Germany, 1910; joined faculty U. Kiel (Germany), 1911, named lectr., 1912; became asso. prof., Rostock, Germany, 1915; named prof. botany, Braunschweig, Germany, 1918; dir. Turkish plant Protection Service, Ankara, 1934-39; dir. biol. dept. firm Fahlberg-List AG, Magdeburg, Germany; apptd. rector Technische U., Braunschweig, 1945, revived sci. work in univ.; gathered insts. of former Reich Biol. Inst. into a central inst., pres., 1947-51. Named hon. pres. Assn. Applied Botany, hon. councillor Technische U. Braunsweig, hon. prof. U. Montevideo. Mem. Leopoldina, Swedish Acad. Agr., also numerous other sci. socs. Author: Mikroskopische Untersuchung pflanzlicher Nahrungs- und Genussmittel, 1931; also numerous articles. Research on applied botany, especially phytopathology, blight diseases; studied devel. change caused by cold; applications in devel. and germination physiology. Died Lüneburg, Germany, Feb. 5, 1955.

GAST, (Adolf Emil) Paul, geodesist; b. Wiesbaden, Germany, Sept. 1, 1876; s. Adolf and Elise (Meyer) G.; studied surveying Agrl. U. Berlin, also Agrl. U. Bonn, 1896-1900; doctorate astronomy-geodesy, Heidelberg, 1903; m. Berta Backes, 1904; 1 dau. Joined faculty Tech. U., Darmstadt, Germany, 1904; sci. adviser Mil. Geog. Inst., also acad. dir. Mil. Acad., Buenos Aires, Argentina, 1906-10, 21-26; became prof. geodesy Tech. U. Aachen, Germany, Tech. U. Hanover, Germany, 1927; founder German-S.Am. Inst., Aachen, 1910, Ibero-Am. Inst., Berlin, 1910. Author: Vorlesungen über Photogrammetrie, 1930; Unsere neue Lebensform, 1932; also articles. Pioneered trilateration performed with phys. aids; 1st to recognize significance of photogrammetry and to develop it; invented Gast's optical pyramid for spanning a space without fixed points using aerotriangulation. Died Innsbruck, Austria, Aug. 19, 1941.

GAST, P(aul) R(upert), Am. physicist, biologist; b. Fitchburg, Mass., Jan. 27, 1897; s. Paul August and Francesca Elsenore Pelieu (Gressenich) G.; Ph.B., Brown U., 1920; M.S., N.Y. State Coll. Forestry, 1922; D.Sc., Harvard, 1927; m. Charlotte Anna Mikalson, June 30, 1924. Instr. botany Brown U., 1919-20, N.Y. State Coll. Forestry, 1920-22; instr. biophysics, forestry Harvard, 1922-24; agt. U. S. Forest Service, Northeastern Forest Expt. Sta., Amherst, Mass., instr. Harvard Forest, Petersham, Mass., 1924-29; NRC (Rockefeller Found.) fellow State Forest Research Inst., Stockholm, Sweden, 1929-30; asst. prof. forestry Harvard, 1930-47; biol. cons. U. S. Testing Co., Hoboken, N.J., 1948; physicist USAF Cambridge Research Labs., L. G. Hanscom Field, Bedford, Mass., 1948-66. Fulbright research scholar Inst. des Recherches Agronomique, Versailles, France, 1950-51. Mem. Am. Acad. Arts and Scis., Phi Beta Kappa, Sigma Xi. Research, publs. on growth and devel. of plants influenced by microclimate and mineral nutrition, measurement of solar irradiance and energy exchange by radiation between earth and atmosphere layers. Home: Red Acre Rd., Stow, Mass. 01775.*

GASTAUT, Henri Jean-Pascal, French biologist; b. Monaco, Apr. 5, 1915; s. Jean Baptiste and Marie Louise (Manceau) G.; M.D., Marseille (France) U., 1945, Ph.D., 1950; m. Claire Yvette Reynaud, Oct. 2, 1935; children—Danielle Charpy, Jean-Albert, Jean-Louis. Faculty, Faculty Medicine, Marseille, 1935—, prof. path. anatomy, 1952—, dean Faculty Medicine, 1967—; chief neurobiologist pub. hosps., Marseille, 1953—; dir. Regional Center for Epileptic children, 1960—; dir. neurobiol. research unit Nat. Inst. Health, 1961—. Recipient Prix Monthyon, French Acad. Scis., 1957—. Mem. Internat. Fedn. Clin. Neurophysiology (past pres., sec. 1949—), Internat. League against Epilepsy (past pres., sec. 1953—), Am. Acad. Neurology, Royal Soc. Medicine (U.K.), Société de Neurologie. Author numerous books, monographs, articles. Research on functional exploration of brain as applied to electroencephalography, epilepsy. Home: 87 Blvd., Perier Marseille 8, France.*

GASTON PHOEBUS (or Gaston III, Count of Foix, Viscount of Bearn), French naturalist; b. Foix, 1331; s. Gaston II, Count of Foix, and Eleonora of Comminges; m. Agnes, dau. of Philip of Evreuy, King of Navarre, 1348; 1 son. Author: Le Miroir de Phoebus des déduiz-de-la-chasse des bested sauvaiges et des oiseaulx de proye (sci. treatise on game animals and their habits, breeding and tng. of animals used for hunting; most popular book on the chase in Middle Ages), begun 1387. Died Orthez, France, 1391.

GATCH, Willis Dew, Am. surgeon; b. Aurora, Ind., 1878; s. Oliver C. and Susan Lindsay (Speidel) G.; A.B., Ind. U., 1901; M.D., Johns Hopkins, 1907; m. Jean McIntosh, Dec. 27, 1911; 1 dau., Susan. Asst. resident surgeon Johns Hopkins Hosp., 1907-11; surgeon Washington U. Hosp., St. Louis, Mo., 1911; prof. surgery Ind. U., 1911-47; dean Sch. Medicine, 1931-40; vis. surgeon Indianapolis City, Methodist, St. Vincent hosps. Fellow A.C.S.; mem. A.M.A., Am., Western, So. surg. assns. Research on bowel obstruction, burns, surg. shock, exptl. and clin. surgery, diseases of biliary passages; devised surg. bed with adjustable frame for maintaining patient in sitting position, circa 1909. Died 1954.

GATENBY, James Bronte, cytologist; b. New Zealand, 1892; s. R. Mackenzie and Catherine Jane (Bronte) G.; ed. St. Patrick's Coll., Wellington, New Zealand, Jesus Coll., Oxford; B.A., B.Sc., D.Phil., Oxford U.; M.A., Ph.D., Dublin; D.Sc., U. London; m. Enid Kathleen Mary Meade, 1922 (dec. 1950); 2 sons, 2 daus.; m. 2d, Constance Harris, 1951. Demonstrator zoology and human embryology Oxford U., 1916-19, sr. demonstrator Magdalen Coll., 1917; sr. asst. zoology and comparative anatomy Univ. coll., London, 1919; lectr. cytology U. London, 1920; prof. zoology and comparative anatomy Trinity coll., Dublin, 1921-59, prof. cytology, 1959-60. Theresa Seessel fellow Yale, 1930-31; Found. Univ. lectr. univs. Louvain, Ghent, Brussels, 1933; vis. prof. zoology King Farouk U., Alexandria, 1951-52; research asso. emeritus Argonne (Ill.) Nat. Lab., 1958; research cytologist Victoria U., Dominion Phys. Lab., Lower Hutt, New Zealand, Sydney U., 1959; OEEC sr. vis. fellow U. Paris, 1960; lectr. human histology Sch. Medicine. Recipient St. Michael medal Brussels U., 1933. Fellow Royal Micros. Soc. (London) (hon.), Acad. Zoology India (hon.); mem. Royal Soc. New Zealand (hon.), Internat. Soc. for Cell Biology (hon.). Author: Biological Technique, 1937. Biol. sub-editor Sci. Progress, 1919-27; editor, reviser Microtomist's Vade-Mecum (Bolles Lee), 1950, also numerous articles. Died July 20, 1960.

GATES, David Murray, Am. botanist, physicist; b. Manhattan, Kan., May 27, 1921; s. Frank Caleb and Margaret (Thompson) G.; B.S., U. Mich., 1942, M.S., 1944, Ph.D., 1948; m. Marian Francis Penley, June 4, 1944; children—Murray Penley, Julie Mary, Heather Margaret, Marilyn Jean. Sci. dir., liaison officer London br. Office Naval Research, 1955-57; asst. chief Upper Atmosphere and Space Physics div. Nat. Bur. Standards, Boulder, Colo., 1957-61; cons. atmospheric physics to dir., 1962-65; vis. prof. U. Mich. Biol. Sta., 1964; prof. natural history U. Colo., 1965; sr. fellow Center for Biology Natural Systems Washington U., St. Louis, 1965, prof. botany, 1965—; dir. Mo. Bot. Garden, St. Louis, 1965—. Fellow Optical Soc. Am., A.A.A.S.; mem. Ecol. Soc. Am., Am Geophys Union, Bot. Soc. Am., Royal Meteorol. Soc., Research Engring. Soc. Am., Sigma Xi, Phi Kappa Phi, Sigma Phi Sigma. Author: Energy Exchange in the Biosphere, 1962. Research, numerous publs. in response of plants to environment, biophys. ecology, temperature regulation in plants, energy budget of animals, atmospheric spectroscopy, atmospheric water vapor distbn. Home: 2361 Tower Grove Av. Office: 2315 Tower Grove Av., St. Louis 63110.*

GATES, Elmer, psychologist; b. nr. Dayton, O., 1859; s. Jacob and Phoebe Foetz; ed. in common and normal schools, but mostly by pvt. tutors, followed by spl. courses in several colleges; m. Phebe Edson, 1895. Prof. psychology, Pa. Sch. Industrial Art, Phila. Mus. Author: Psychurgy, or The Art of Using the Mind; Art of Mind-Building, etc. Evolved a practical art of brain or mind-building by systematic means, which causes an increase in the structural elements of the brain-cells, fibers and whole nervous system, increases mental capacity and skill; has made numerous other discoveries in exptl. psychology out of which he has evolved an art of using the mind more efficiently in the processes of discovery, invention, etc.; has done original work in electric meteorology, higher temperatures, and made a number of successful electric mining inventions; has laboratories for experimental research in psychology, psychurgy and in the other sciences. Died Dec. 3, 1923.

GATES, Marshall DeMotte, Jr., Am. chemist, educator; b. Boyne City, Mich., Sept. 25, 1915; s. Marshall DeMotte and Virginia (Orton) G.; B.S., Rice U., 1936, M.A., 1938; Ph.D., Harvard, 1941; D.Sc., MacMurray Coll., 1963; m. Martha L. Meyer, Sept. 9, 1941; children—Christopher D., Catharine L., Marshall DeMotte III, Virginia A. Asst. prof. chemistry Bryn Mawr (Pa.) Coll., 1941-46, asso. prof., 1946-49; lectr. chemistry U. Rochester (N.Y.), 1949-52, prof., part-time 1952-60, prof., 1960—. Max Tishler lectr. Harvard, 1953; Welch Found. lectr., 1960; mem. com. on drug addiction and narcotics div. med. scis. NRC, 1957—; tech. aide NDRC, 1943-46. Recipient Armed Services certificate of appreciation, 1946. Fellow Am. Acad. Arts and Scis.; mem. Nat. Acad. Scis. Research on gen. organic synthesis; 1st synthesis morphine. Asst. editor Jour. Am. Chem. Soc., 1949-62, editor, 1963—. Home: 41 West Brook Rd., Pittsford, N.Y. Office: Dept. Chemistry, U. Rochester, Rochester, N.Y. 14627.*

GATES, R(eginald) Ruggles, botanist, geneticist, anthropologist; b. Nova Scotia, May 1, 1882; s. A. B. and Elizabeth (Ruggles) G.; B.Sc., McGill, M.A., Mt. Allison, Ph.D., U. Chicago; hon. D.Sc., U. London, LL.D., Mt. Allison; m. Laura Greer, 1955. Demonstrator botany, McGill U., 1905; sr. fellow, U. Chicago, 1906-08, asst. 1909; research, Missouri Botanical Gardens, 1910-11; lectr. biology, St. Thomas Hospital, London, 1912-14; lectr. cytology, Bedford College, London, 1912, 1914; lectr. heredity, Oxford U., 1914; asso. prof. zoology, U. California, 1915-16; reader botany, King's College, U. London, 1919-21; prof. botany. 1921-42; emeritus, 1943; DeLamar lectr., Johns Hopkins, 1932; lecture tour U. S., 1940-42; research fellow, Harvard U., 1946-50. Recipient Mendel Medal, 1911, Huxley Medal and Prize, Imperial College of Sci., London, 1913. Fellow Royal Soc., 1931; council mem., Royal Anthropological Institute, 1927-33, 35-37, Linnean Soc., 1928-32 (vice-pres. 1931-32), Royal Microscopical Soc. (sec. 1928-30) pres., 1930-32; fellow Soc. Experimental Biology (sec. 1923-28); hon. mem., Japan Botanical Soc.; life mem., Brit. Institute of Biology. Author: The Mutation Factor in Evolution, 1915; Mutations and Evolution, 1921; Heredity and Eugenics, 1923; A Botanist in the Amazon Valley, 1927; Heredity in Man, 1929; Human Genetics (2 vol.), 1946; Human Ancestry, 1947; Pedigrees of Negro Families, 1949; numerous papers. Investigations of Ainu and race mixing in Japan, 1954; scientific expeditions in Africa, Cuba, Mexico, Australia, India, Far East; research in cytology, genetics, anthropology, botany. Died Aug. 12, 1962.

GATEWOOD, Buford Echols, Am. aero. and astronautical engr.; b. Byhalia, Miss., Aug. 23, 1913; s. Robert P. and Irene (Echols) G.; B.S. in Mech. Engring., La. Poly Inst., 1935; M.S., U. Wis., 1937, Ph.D., 1939; m. Margaret Murphy, June 28, 1939; 1 dau., Marianne. Faculty, La. Poly Inst., 1939-42, Air Force Inst. Tech., 1947-60; with McDonnell Aircraft Corp., 1942-46, Beech Aircraft Corp., 1946-47; prof. aero. and astronautical engring. Ohio State U., Columbus, 1960—. Cons. on structural design and analysis, structural fatigue, problems in dynamics, thermal problems to various cos., 1949—. Mem. Inst. Aeros. and Astronautics, Soc. Exptl. Stress Analysis, Math. Assn. Am., Am. Soc. M.E., Am. Soc. for Engring. Edn., Sigma Xi. Author: Thermal Stresses, 1957. Research, publs. on thermal stresses and inelastic structures for flight vehicle structures. Home: 2150 Waltham Rd., Columbus, O. 43221.*

GÄTKE, Heinrich, ornithologist; b. Pritzwalk, Germany, May 19, 1814; s. Johann A.F.W. and Sophia (Wenzel) G.; began career as businessman, then became artist; went to Helgoland to study bird migration and remained for 60 years; govt. sec. for English adminstrn., Helgoland. Author: Die Vogelwarte Helgoland, 1891. Research and publs. on migration of European birds; determined and confirmed flight patterns; proved Palmen's theory of inherited experience in direction was incorrect; compiled collection which was purchased by Prussia for N. Sea Mus., 1891. Died Helgoland, Jan. 1, 1897.

GATLING, Richard Jordan, inventor; b. Winton, N.C., Sept. 12, 1818; grad. Ohio Med. Coll., 1850; m. Jemima T. Sanders, 1854; 2 sons, 1 dau. As a boy assisted his father in perfecting machine for sowing cotton-seed; later invented machine for sowing rice, moved to St. Louis, where in 1844 he adapted it to sowing wheat and patented it; attended med. lectures in Cin.; lived in Indpls., 1850; inventor revolving gun known as Gatling gun (accepted by U. S. Ordnance Dept. 1866), 1862, a new gun metal, composed of steel and aluminum, 1866; Congress voted him $40,000 for proof expts. in a new method of casting cannon; inventor hemp-breaking machine, 1850; a steam plow, 1857; studied ordnance and ballistics. Died N.Y., Feb. 26, 1903.

GATTERER, Alois, Austrian astrophysicist, spectrochemist; b. Reichraming, Austira, Jan. 28, 1886; s. Anton and Emilie (Kuffrath) G.; studied theology, Innsbruck, Austria, philosophy, Pressburg, Czechoslovakia; doctorate in chemistry and physics U. Innsbruck, 1922. Joined Soc. of Jesus, 1905; ordained priest, 1915; tchr. natural scis. Institutum philosophicum, Innsbruck; lectr. natural philosophy theol. faculty U. Innsbruck, 1924-26; studied for a short period at Oxford, beginning in 1928; went to assist in reorgn. Vatican Obs., Castel Gandolfo, Rome, 1931, founder astrophys. lab., 1933, and was its head until his death. Mem. Papal Acad. Scis. Author: Il Laboratorio astrofisico della specola Vaticana, 1935; (with J. Junkes) Spark Spectrum of Iron . . . , 1935, Arc Spectrum of Iron . . . , 1935; Atlas der Restlinien, 3 vols., 1937-49; Grating Spectrum of Iron, 1951; Molecular Spectra of Metallic Oxides, 1957. A founder Spectrochimica Acta, 1939, co-editor of 1st vol. Improved numerous instruments and methods of spectrochemistry, including universal tripod for sparks and arcs, application of carbon flame to spectroscopy; spectroscopic proof of halogenes and other non-metals using ultra high frequency waves. Died Innsbruck, Feb. 17, 1953.

GATTERMANN, (Friedrich August) Ludwig, German chemist; b. Goslar, Germany, Apr. 20, 1860; s. Heinrich Friedrich Wilhelm and Marie Dorothea Louise (Creutzburg) G.; student chemistry, Leipzig, Heidelberg, Berlin, (all Germany), from 1880; Ph.D., Göttingen, Germany, 1885; m. Käthe Krausse, 1893 (div. 1918); 1 dau. Began as asst. to V. Meyer, Göttingen, joined faculty, 1886; named asso. prof., dep. dir. inst. (formerly under Bunsen) Heidelberg, 1889; named dir. chem. inst. U. Freiburg (Germany), 1900. Author: Die Praxis des organischen Chemikers, 1894-1961; also articles. 1st successful isolation and analysis of nitrogen trichloride, 1887-88; synthesized aromatic carboxylic acids; 1st synthesis and analysis of thionaphthene, thioanilide; modified Sandmeyer reaction; research on anthracene derivatives; originated Gattermann synthesis of aromatic aldehydes. Died Freiburg, June 20, 1920.

GATTESCHI, Luigi, Italian mathematician; b. Pelago, Italy, July 15, 1923; s. Gattesco and Angiolina (Masotti) G.; D.Marh., U. Florence (Italy), 1945, Libero docente, 1956; m. Marcella Alberta de Bernardi, Sept. 11, 1947; children—Gianluca, Stefano, Alessandro. Asst. prof. U. Florence, 1948-50; vis. research asso. Stanford, 1951-52; asst. prof. U. Bari, 1952-56; faculty U. Turin (Italy), 1956—, prof. incaricato higher math. analysis and numerical calculus, 1961-66, prof., 1967—. Mem. Unione Matematica Italiana, Am. Math. Soc., Associazione Italiana per il Calcolo Automatico. Author: (with Tino Zeuli) Introduzione ella Analisi Numerica, 1965; also articles. Reviewer, Zentralblatt für Mathematik, 1964—. Research on asymptotic expansions of spl. functions especially Bessel functions and orthogonal polynomials, upper explicit bound of error term. Home: Corso Re Umberto, 40, Torino, Italy.*

GATTI, Emilio, Italian physicist; b. Turin, Italy, Mar. 18, 1922; s. Aldo and Emilia (Sacchi) G.; Doctor in Electronics Engring., U. Padova (Italy), 1946; m. Laura Semenza, Sept. 10, 1948; children—Gabriella, Aldo, Carlo, Anna Paola. Research worker Istituto Galileo Ferraris, Turin, 1947-48, Cise Labs., Milan, Italy, 1948-49; head electronic lab. CISE, Milan, 1949—; faculty Politecnico, Milan, 1951—, prof. physics, 1957—. Mem. AEI It. (assn. elec. and electronics engrs.) (editor Alta Frequenze 1961—, Gold medal Bianchi 1956), SIF (Italian soc. physics) (vice dir. Il Nuovo Cimento 1963-67), I.E.E.E. (Pres. Italian sect. 1965-67). Author: (with P. Manfredi, A. Rimini) Teoria della Reti Lineari, 1966; also articles. Developed statis. theory of scintillation counter; studied limits to accuracy of timing of nuclear events; electronic instrumentation for nuclear physics research, especially vernier time sorter, subnano second digital time sorter, streamer chamber. Home: 18 Lambro Lesmo, Milan, Italy.*

GATZ, Arthur John, Am. anatomist; b. Winona, Minn., Dec. 18, 1907; s. John J. and Ida (Hagemann) G.; B.A., Carleton Coll., 1931; A.M., U. Minn., 1933, Ph.D., 1936; m. Jean L. Wells, Sept. 3, 1936; children—Margaret J., Arthur John. Instr. zoology Carleton Coll., 1935-41, asst. prof., 1941-43; asst. prof. anatomy Loyola U., Chgo., 1943-48, asso. prof. 1948-56; asso. prof. micro-anatomy Med. Coll. Ga., 1956-61, prof., acting chmn. gross anatomy, Augusta, 1961-62, chmn., 1962-64, prof., chmn. dept. anatomy, 1964—. Fellow A.A.A.S.; mem. Am. Soc. Anatomy, Am. Soc. Zoologists, Soc. Exptl. Biology and Medicine, So. Assn. Anatomists, Ga., N.Y. acads. sci., Assn. Am. Med. Colls., Sigma Xi. Author: Outline Manual of Histology, 1941; Essentials of Neuroanatomy and Neurophysiology, 1961; also articles. Research on pathology of vitamin E deficiency in animals, human cytogenetics; chromosome anomalies in Down's syndrome, Kleinfelter's syndrome, Arginosuccinic-aciduria and satellited 18 chromosome. Home: 2440 Apricot Lane, Augusta, Ga. 30904.*

GAUB, Jerome David (Gaubius), German pathologist; b. Heidelberg, Germany, 1705; studied under Brerhaaue, also Franke. Fellow Royal Soc., 1764. Author: Institutiones pathologiae medicinalis, 1758; De regimine mentis quod medicorum est, 1747. Wrote 1st work in gen. pathology. Died Leyden, Netherlands, 1780.

GAUBIL, Father Antoine, mathematician, astronomer; b. Gaillac, France, July 14, 1689. Became Jesuit missionary in China, 1723; interpreter at Imperial Ct.; mem. Acad. St. Petersburg, French Acad. Scis., 1750. Author: Traité historique et critique de l'astronomie chinoise, 1739. Compiler Chinese astron. works. Calculated and verified eclipses reported by Chinese scholars. Died Peking, China, July 24, 1759.

GAUBIUS, see Gaub, Jerome David.

GAUBY, David, dermatologist; b. Hungary, 1810; M.D., Vienna, 1839; settled in London, then in Paris; tchr. physiology and pathology (pupils included Claude Bernard); pvt. practice, Paris; physician to A. Dumas, Chopin, Liszt. Pioneer in research on parasitic diseases of man; discovered and described fungus causing avus, 1841; discovered parasitic nature of ringworm, 1844, also causative agts. of many other parasitic diseases. Died 1898.

GAUDICHAUD-BEAUPRÉ, Charles, French botanist; b. Angouleme, France, Sept. 4, 1789; student of Robiquet in Paris; studied botany under Louis-Claude Richard. Botanist Freycinet's Sci. Expdn., 1816-20; mem. French Acad. Scis. Author: Flore des iles Malouines, 1825; Botanique du voyage autor du monde exécuté pendant les années, 1836-37; Mémoires sur la physiologie des végétaux, 1851; also bot. sect. of work based on Freycinet's sci. expdn. Research on physiology of vegetables; defended phyton theory of stem tip growth. Died Paris, Jan. 16, 1854.

GAUDIN, Marc Antoine Augustine, French scientist; b. Saintes, Charente-Inférieure, France, Apr. 5, 1804; studied under Dumas and Ampère; calculator Bur. des longitudes; author: l'Architecture du monde des atoms, 1873; Vadumecum du photographie, 1861; also numerous articles on astronomy, mineralogy, photography, gun cotton, fermentation, molecular structure of minerals, Avogadro's hypothesis. 1st research on silica fused with oxyhydrogen blowpipe, using this method was 1st to make artificial rubies and sapphires; made a malleable alloy of iridium and platinum; 1st research on silver chloride and silver iodide emulsions in photography, 1861. Died Paris, France, Aug. 2, 1880.

GAUDRY, (Jean) Albert, French paleontologist; b. Saint-Germain-en-Laye, France, Sept. 15, 1827; s. Joachim-Antoine-Joseph Gaudry; docteur es sciences; began explorations in Cyprus and Greece, 1852; became asst. to A. d'Orbigny, Mus. Natural History, Paris, 1853, received chair of paleontology, 1872; lived in Greece, 1855-60; presided at 8th Internat. Congress Geology, Paris, 1900. Fellow Royal Soc., 1895; mem. French Acad. Scis., 1882 (v.p. 1902, pres. 1903). Author: les Animaux fossiles et la géologie de l'Attique; les Enchainements du monde animal dans le temps géologiques, 1878; Essai de la paléontologie philosophique, 1896. Founder paleontology in France; research and discoveries on fossil mammals which supported evolutionary theory. Died Paris, Nov. 27, 1908.

GAUDY, Anthony Francis, Jr., Am. civil engr.; b. Jamaica, N.Y., June 16, 1925; s. Anthony Francis and Catherine (Ford) G.; B.S. cum laude in Civil Engring., U. Mass., 1951; M.S. in San. Engring., Mass. Inst. Tech., 1955; Ph.D. (USPHS fellow), U. Ill., 1959; m. Elizabeth Thomas, June 11, 1955. Civil engr. E. F. Carlson, Inc., Springfield, Mass., 1951-52, Capuano Constrn., Inc., West Springfield, 1952-53; research asst. sanitary engring. Sedgewick Labs., Mass. Inst. Tech., 1953-55; research engr. Nat. Council for Stream Improvement, Va. Poly. Inst., Blacksburg, 1955, regional engr. Ore. State U., Corvallis, 1955-57; faculty U. Ill., 1959-61; faculty Okla. State U., Stillwater, 1961—, prof. civil engring., 1963—, dir. Center for Water Research in Engring., 1965—, acting head Sch. Civil Engring., 1966-67, E. R. Stapley prof. civil engring., 1968—. Dir. Water Pollution Control Fedn., 1962-65 (Service award 1965); research commn., 1959—. Recipient Eddy award for noteworthy research, 1967. Mem. Am. Soc. C.E., Am. Chem. Soc., Am. Water Works Assn., A.A.A.S., Nat. Soc. Profl. Engrs., Am. Soc. Engring. Edn., Am. Soc. Microbiology Research, publs in response of biol. waste treatment processes to changes in environment, kinetics and mechanisms of waste water purification by activated sludge process, sequential removal of organic pollutants by heterogeneous microbial populations, metabolic control mechanisms in natural populations; devel. bioengring. approach to re-use of water resource. Home: 1017 W. McElroy St., Stillwater, Okla. 74074.*

GAUGAIN, Jean-Mothée, physicist; b. Sulby, Calvados, France, 1810; ed. L'école polytechnique, dir. several mining cos. in France and Belgium, 1832-49; improved electrodynameo meter; inventor tangent magnetic compass, electrometer that permitted comparison of capacity of condensors by means of slower charge and jerky discharge. Died St.-Martin-des-Entrées, France, 1880.

GAULARD, Lucien, elec. engr.; b. Paris, France, July 16, 1850; mfr. explosives; electrician in London for several years; in charge of elec. installations at Turin (Italy) Expn., 1884; invented a thermoelectric generator, 1881, 1st alternating current transformer, 1884; installed electric lighting in subway system, London. Died Nov. 26, 1888.

GAULD, Ross Laurier, Am. epidemiologist; b. Mimico, Ont., Can., Oct. 12, 1900; s. George Robert and Janet Rome (MColl) G.; M.B., U. Toronto (Ont.), 1924, M.D., 1947; C.P.H., Johns Hopkins, 1935, Dr.PH., 1936; m. Ethel Maude Gilpin, Oct. 22, 1927; children—John Ross, Godfrey Robert. Came to U.S., 1930, naturalized, 1937. Practice medicine, Maxwell, Ont., 1927-30; with Tenn. Dept. Pub. Health, 1931-34; acting dir. Tb Study, Franklin, Tenn., 1936-37; asso. epidemiology Johns Hopkins Sch. Hygiene and Pub. Health, 1937-43; with Walter Reed Inst. Research, Washington, 1946—, chief dept. epidemiology, 1950-57, dir. div. preventive medicine, 1957—. Mem. various coms. Armed Forced Epidemiological Bd. Diplomate Am. Bd. Preventive Medicine. Fellow Am. Pub. Health Assn.; mem. Am. Epidemiological Soc. (past pres.), A.A.A.S., Constantinian Soc., Acad. Medicine, Washington. Research, numerous publs. on epidemiology infectious hepatitis, other viral and rickettsial infections, tb, rheumatic fever. Home: 8300 Thoreau Dr., Bethesda, Md. 20034. Office: Div. Preventive Medicine, Walter Reed Army Inst. Research, Washington 20012.*

GAULT, Robert Harvey, psychologist; b. Ellsworth, O.; s. Andrew Robinson and Martha (McCullough) G.; A.B., Cornell, 1902; Clark U., 1902-03; Ph.D., U. of Pa., 1905; m. Anne Lee, 1907 (died 1937); m. 2d, Mary Louise Woseczek, 1939. Prof. psychology, Washington (Md.) Coll., 1905-09; instr. and asst. prof. psychology, Northwestern U., 1909-13, asso. prof. 1913-17, prof. since 1917; temporarily on leave with Nat. Research Council, Washington, D.C., 1921-27; research asso. Carnegie Inst., Washington, 1927-29. Mem. Nat. bd. advs. Assn. for Psychiatric Treatment of Offenders. Editor Jour. Criminal Law and Criminology since 1911, Criminal Science Monographs since 1914; dir. of Vibro-Tactile Research Laboratory, 1925-40; administrative asst. Ill. State Dept. of Pub. Welfare during 1940; agent Ill. State Dept. Public Welfare relating to the deaf-blind; counsellor on edn. and training Internat. Harvester Co., 1944-45; psychol. cons., Sadler, Hafer & Assos., personnel management counsel. Specialized writer, indsl. history, sales promotion, product information and technical handbooks for industry, also consultant on employee management relations and on training and education in industry. Member board directors Chicago Crime Commission since 1919. Fellow Am. Acoustical Soc., A.A.A.S.; mem. Am. Psychologic Assn., Illinois Social Hygiene League (president, 1917-24), Sigma Xi, Phi Eta, Alpha Pi Zeta. Clubs: University (Chicago), University (Evanston). Author: Social Psychology, 1923; Criminology, 1932. Co-Author: An Outline of General Psychology. Contbr. chapters in "Recent Developments in The Social Sciences," and "Abnormal Minds and the Law", articles and reviews on psychology and education, including numerous papers relating to research on the development of a language sense and the use of speech by the aid of the organs of touch; numerous articles and editorials on criminology; report of Chicago City Council Committee on Crime, 1915. Author of A Plan for Intellectual Cooperation with Latin America, in the Area of Criminology and Contributory Sciences, 1942. Author: Phonemanship; Rewrites of Naval Torpedo Manuals, 1944. Similar Plans for U.N., interchange of information among mems. nations; chapters in Twentieth Century Applied Psychology. Home: 504 Lee St., Evanston, Ill. Office: 357 E. Chicago Av., Chicago 11, Ill.

GAULTIER DE CLAUBRY, Henry-François, French chemist; b. Paris, 1792; became prof. toxicology Paris Sch. Pharmacy, 1859; mem. Council Pub. Health. Author: (with Briand and Chaudé) Manuel de médecine légale, 1852; translator: Élements de chemie experimentale (H. William), 1832. Studied (with Colin) effect of iodine on organic substances in Gay-Lassac's Lab; discovered iodine formed blue color with starch; research on aligarin and purpurin. Died Paris, 1878.

GAUNT, Robert, Am. endocrinologist; b. Macon, Miss., Apr. 13, 1907; s. Robert Earl and Mary (Summers) G.; B.A., U. Tulsa, 1929; M.A., Princeton, 1930, Ph.D., 1932; m. Josephine Lindenkohl Howland, July 8, 1933; 1 son, Robert Howland. Prof. biology Coll. Charleston, 1932-35; from asst. to asso. prof. biology N.Y. U., 1935-45; prof. zoology, chmn. dept. Syracuse U., 1945-51; dir. endocrine research CIBA Pharm. Co., Summit, N.J., 1951-57, dir. biol. research, 1957-66, dir. basic biol. sci., 1967—. Mem. endocrinology panel, cancer chemotherapy Nat. Service Center, NIH, 1957-61. Guggenheim Meml. fellow, 1943. Fellow A.A.A.S., N.Y. Acad. Scis.; mem. The Endocrine Soc. (council 1945-49), Phila. Endocrine Soc., Am. Physiol. Soc., Am. Assn. Anatomists, Am. Soc. Zoologists, Soc. Exptl. Biology and Medicine, Royal Soc. Medicine (London), Sigma Xi. Author: Hormones and Body Water (with J. H. Birnie). Editor: Adrenal Cortex, 1949; editorial bd. Jour. Physiology, Jour. Applied Physiology, 1956-59, Endocrinology, 1953-56. Research, publs. on function mechanisms, effects of hormones and drugs; teaching of biology. Home: 35 Hilltop Terrace, Chatham, N.J. 07928. Office: CIBA Pharm. Co., Summit, N.J. 07901.*

GAUPP, Robert Eugen, German psychiatrist, neurologist; b. Neuenburg, Germany, Oct. 3, 1870; s. Robert and Julia (Faber) von G.; student medicine, Tübingen, Germany, Geneva, Strasbourg, France; state exam, Tübingen, 1893-94; M.D., 1894; m. Oktavia Hasse, 1901; 2 sons, 3 daus. Worked with psychiatrist, K. Wernicke, Breslau (now Wroclaw, Poland), 1894-99; then practiced as neurologist, Breslau, became lectr. under Kraepelin, U. Heidelberg, 1901; followed Kraepelin to Munich, 1904; named prof., Tübingen, 1906; ret., 1936. Research and publs. on progressive paralysis, paranoia, hysteria, depression, suicide, mass murder, homosexuality. Died Stuttgart, Germany, Aug. 30, 1953.

GAUS, Wilhelm Karl Friedrich, German chemist; b. Braunschweig, Germany, Oct. 26, 1876; s. Wilhelm and Anna (Ternedde) G.; student chemistry, Braunschweig, Berlin, Breslau (now Wroclaw, Poland); Ph.D., Breslau, 1900; dr. engring. (hon.) Karlsruhe, 1931; m. Helene Lina Schumann, 1906; 1 son, 1 dau. Mil. service, 1900-02; asst. to Abegg; became physico-chemist Badische Anilin- & Soda-Fabrik (BASF), Ludwigshafen am Rhein, Germany, 1902; became head BASF after it merged with other chem. firms into I. G. Farbeinindustrie AG, 1931, mem. bd. I. G. Farben, dep. chmn. tech. com., 1937. Research and publs. on devel. indigo synthesis, especially prodn. of aux. and by-products, 1902-10; developed various process, including continuous alkali chloride electrolysis, prodn. active iron, phosgene and hydrocyanic acid from methane and nitrogen; devel. (with Am. oil industry) carbonyl iron; studied ammonia synthesis. Died Gut Schmalzhof nr. Starnberg, Germany, Nov. 20, 1953.

GAUSS, Carl Joseph, German gynecologist; b. Rittergut Lohen, Germany, Oct. 29, 1875; s. Carl and Anna (Ebmeyer) G.; studied in Tübingen, Erlangen, Kiel, Würzburg, Munich (all Germany); M.D., Munich, 1898; state exam., 1899; m. Magdalene Bingel, 1919; 2 stepchildren; 2 adopted children. Asst. to Orth and Esmarch, Göttingen Germany; worked under Olshausen and Bessel-Hagen, Berlin, under Krönig and Optiz, Freiburg, Germany; joined faculty, Freiburg, 1909, named asso. prof., 1913; became dir. gynecol. dept. Diakonissen Hosp., 1921; named prof., Würzburg, 1923, dir. univ. women's clinic and midwifery sch., until 1945; head physician obstet.-gynecol. dept. St. Elisabeth Hosp., Bad Kissingen, Germany, 1947-55. Mem. German Soc. for Obstetrics-Gynecology (hon. mem.), Paracelsus Soc. (hon. pres.). Author: (with Lembcke) Röntgentiefentherapie, ihre therapeutische Grundlagen und ihr klinischen Erfolge, 1912; (with B. Wilde); Die deutschen Geburtshelferschulen, 1956; (with R. Schmiemann) Atals der geburtshilflichen Röntgen-Diagnostik, 1959; also articles. Introduced (with Krönig) narcotics to obstetrics; developed (with Wieland) narcylene narcotic; described pregnancy sign named after him (abnormally strong movement of uterus in isthmus); studied radiation therapy in women's diseases, gynecol. urology and gonorrhea. Died Bad Kissingen, Feb. 11, 1957.

GAUSS, Friedrich Gustav, German geodesist; b. Bielefeld, Germany, June 20, 1829; s. Johann Philipp and Johanna Sophie (Westermann) G.; certification exam in surveying, Minden, Germany, 1848; Ph.D. (hon.), Strasbourg, France, 1899. With Rhenish-Westphalian Land Tax Registry Office, 1848-58; called to prepare for land tax survey of Prussia and eastern provinces Prussian Finance Ministry, 1859, organized it 1861, administrated it for 3 1/2 years; named gen. insp. Prussian Land Tax Bur., 1872; set up land tax system when Schleswig-Holstein, Hanover, and Hessia-Nassau were annexed; held many high offices in Prussian Civil Service. Author: Die Gebäudesteuer in Preussen, 1866; Fünfstellige vollständige logarithmische und trigonometrische Tafeln, 1871; Fünfstellige logarithmisch-trignometrische Tafeln für Dezimalteilung der Qudranten, 1873; Die trigonometrischen und polygonometrischen Rechnungen in der Feldmesskunst, 1876; Die Teilung der Grundstücke, 1878; Die Ergänzungssteuer in Preussen, 1894; Fünfstellige vollständige trigonometrische und polygonometrische Tafeln für Maschinenrechnen, 1901. Land survey and tax assessment in Prussia; prepared logarithm tables, trigonometric and polygonometric tables for surveying. Died Berlin, June 26, 1915.

GAUSS, Karl Friedrich, German mathematician; b. Braunschweig, Germany, Apr. 30, 1777; s. Gerhard Diederich and Dorothea (Benz) G.; ed. Caroline Coll., U. Göttingen, 1795-98; doctorate (in absentia) U. Helmstedt, 1799; m. Johanna Osthoff, 1905; 3 children; m. 2d, Minna Waldeck, 1810; 3 children. Math. studies under patronage of Duke Ferdinand; offered professorship at St. Petersburg, declined; prof. astronomy, dir. obs. U. Göttingen, 1807-55; in charge of govt. project for triangulation of Hanover, 1821; built (with W. Weber) one of 1st observatories for magnetic studies at Göttingen, 1833. Recipient Lalande medal, 1810. Organizer Magnetischer Verein (soc. for research on magnetism); mem. Königliche Gesellschaft der Wissenschaften, Göttingen; corr. geometry sect. French Acad. Scis., 1804, fgn. mem., 1820, also mem. most leading sci. socs. in Europe; Fellow Royal Soc., 1804. Author: Disquisitiones arithmeticae (which covered indeterminate analysis or transcendental arithmetic), 1801; Theory of Numbers; Analysis; Geometry and Method of Least Squares; Mathematical Physics; Astronomy; Theoria motus corporum coelestium, 1809; (all pub. by Royal Soc. Göttingen, 1863-71); Intensitas Vis Magneticae Terrestris, 1833; Dioptrische Untersuchungen, 1841; Untersuchungen über Gegenstände der höheren Geodesie, 1844; Fundamente der Geometrie usw., 1900; Geodatische Nachtrage zu Band IV, 1903. Gave 1st rigorous proof of fundamental theorem of algebra (every algebraic equation must have at least one root, real or imaginary), 1799; proved fundamental theorem of arithmetic (every natural number can be represented as product of primes in one and only one way); devised rigorous proof of binomial theorem, 1792-93; demonstrated that circle can be divided into 17 equal arcs by classical methods of geometric constrn.; one of 1st to prove impossibility of performing certain constrn. by these methods; gave 1st good proof of law of quadratic reciprocity; studied complex numbers, elliptic functions, convergence of infinite series; studied Fermat numbers; provided demonstration of Fermat's theorem concerning triangular numbers; contbd. much to modern number theory; did basic work on theory of surfaces; a pioneer on non-Euclidean geometry, analytic functions and topology; originated vectorial representation of complex numbers; initiated theory of algebraic numbers in his definition of what is now called Gaussian integers; determined fundamental regions for certain modular functions; discovered Cauchy's theorem a few years before Cauchy; originated and established theories of arithmetic forms and cyclotomy; calculated orbits of planetoids Ceres and Pallas by new method; discovered method of least squares; developed theory of observational error, 1795; pioneer in electromagnetism; considered founder of math. theory of electricity; discovered famous theorem in math. of electricity (named after him); studied methods of calculating terrestrial magnetism from limited but known number of observations; originated study of goedesy; designed bifilar magnetometer, declination needle, heliotrope; unit of magnetic field named gauss in his honor. Died Göttingen, Feb. 23, 1855.

GAUSSEN, Henri Marcel, French botanist; b. Cabrières d'Aigues, July 14, 1891; s. Ulysse and Catherine (Bouchard) G.; Ph.D. in Sci., U. Toulouse; agrege; D.honoris causa, U. Madrid; m. Jeanne Chausson (dec.); 1 son, Maurice (dec.). Prof. at lycée; instr., head of work, lectr., prof.; titular prof., hon. prof. Faculty Scis., Toulouse; dir. Registry Vegetation of France; dir. Inst. Internat. Registry Vegetable Cover; pres. French Com. Cartography. Mem. Acad. Scis. of Paris (corr.), Acad. Agr. France. Author over 450 works on botany, geography, mountain economy, forests, climatology, cartography. Home: 21, rue Raymond-IV. Office: Faculty of Sciences, Allée J.-Guesde, Toulouse, France.

GAUSTER, Wilhelm Friedrich, physicist, engr.; b. Vienna, Austria, Jan. 10, 1901; s. Friedrich and Gabriele (Deltsch) G.; Dipl. Ing., U. Technology, Vienna, 1923, Dr. Techn., 1924; Dr. Habil., 1927; m. Marietta (Countess Belrupt), Feb. 24, 1940; children—Wilhelm Belrupt, Christian Belrupt. Came to U. S., 1950, naturalized, 1954. Mgr., Elin Corp. for Elec. Industrie, Austria, 1924-41; research for German Navy, 1941-45; prof. U. Techn., Vienna, 1945-50; prof. N.C. State Coll., 1950-57; dir. magnet lab. Oak Ridge Nat. Lab., 1957—. Recipient German KVK I and II medal merit; Austrian Gt. Silver Insignia of Honor. Fellow I.E.E.E.; mem. Am. Phys. Soc., Austrian Soc. Elec. Engring. Contbr. numerous sci. publs. Research in theoretical mechanics, underwater acoustics and applied electrophysics, magnetics for thermonuclear research, superconductivity; developer new types of mercury arc rectifiers. Home: 104 Seymour Lane, Oak Ridge 37831. Office: Oak Ridge Nat. Laboratory, P.O. Box Y, Oak Ridge 37830.*

GAUTIER, (Émile Justin) Armand, French physician, chemist; b. Narbonne, France, Sept. 23, 1837; Dr.-ès sci. et méd., Paris; named asso. prof. Faculty Medicine, Paris, 1869, apptd. prof. med. chemistry, 1884, also chief of lab.; practiced medicine, Nantes, France. Mem. French Acad. Scis. (became v.p. 1910, pres. 1911), Acad. Medicine. Author: La Chimie et la cellule vivante, 1894; Cours de chimie minérale et organique, 1895-96; Les toxines microbiennes et animales, 1896; Leçons de chimie biologique, 1897; L'alimentation et les régimes chez l'homme sain et chez le malade, 2d edit., 1904; also numerous articles. Determined hydrogen in atmosphere, fluorine in vegetable and animal tissues; research on photo-chem. union of hydrogen and chlorine, speed of reaction of hydrogen and oxygen; developed 1st adequate device for distilling seawater into drinkable water; studied fixation of nitrogen, 1888, organic arsenical compounds (led to modern arsenotherapy); discovered carbylamines, 1866; proved existence of compounds now known as ptomaines, 1872. Died Cannes, France, July 27, 1920.

GAUZE, Georgii Frantsevich, Russian microbiologist; b. Dec. 27, 1910; grad. Biol. Faculty, Moscow U., 1931; D.Biol. Sci. With Moscow U., 1931-40; dep. dir. studies, head lab. for procurement and cultivation of producers Research Inst. for New Antibiotics, USSR Acad. Sci., 1940—; prof., 1940—; head antibiotics procurement lab. Inst. Pharmacology and Exptl. Chemotherapy and Chemoprophylaxis, USSR Acad. Med. Sci. Del., 2d Internat. Congress on Immunological and Microbiol. Standardization, Italy, 1956; lectr. U. Chgo., Yale, Cornell U., 1959. Recipient Stalin prize, 1946. Mem. USSR Acad. Med. Sci. (corr.). Author: Lectures on Antibiotics, 1st edit., 1949, 2d edit., 1953, 3d edit., 1959; Gramicidin C and Its Uses, 1953; Means of Discovering New Antibiotics, 1958; co-author, editor: Menomycin and Its Clinical Use, 1962. Co-editor Microbiology sect. Large Med. Ency., 2d edit.; mem. editorial bd. Antibiotics. Research and publs. on action of antibiotics. Address: USSR Acad. Med. Sci., Solyanka 14, Moscow, USSR.

GAVAN, James Anderson, Am. anthropologist; b. Ludington, Mich., July 17, 1916; s. James B. and Mary (Anderson) G.; B.A., U. Ariz., 1939; M.A., U. Chgo., 1949, Ph.D., 1953; m. Margaret Sheninger, Dec. 17, 1945; children—Margaret Jean, James Charles. Research staff Yerkes Labs. Primate Biology, Orange Park, Fla., 1950-53; asst. prof. anatomy Med. Coll. S.C., 1953-60, asso. prof., 1960-62; asso. prof. anatomy, anthropology U. Fla., 1962-67; prof. anthropology U. Mo., 1967—. Mem. A.A.A.S., Am. Assn. Anatomy, Am. Anthropol. Assn., Am. Soc. Human Genetics, Am. Assn. Phys. Anthropology, Sigma Xi. Editor: The Non-Human Primates and Human Evolution, 1955. Contbr. articles to sci. jours. Studies on postnatal growth and devel. of non-human primates in order to increase our understanding of process of human and primate evolution. Home: 1503 Wilson Av., Columbia, Mo. 65201.*

GAVARD, Hyacinthe, French anatomist; b. Montmelian, France, 1753; studied medicine, Paris; follower of Desault. Author: Traité complet d'ostéologie suivant la méthode de Desault, 1791; Splanchnologie, 1800. Gave 1st description of oblique muscle fibers of stomach wall (Gavard's muscle), circa 1792. Died Paris, 1802.

GAVRILENKO, Boris Sergeevich, Russian surgeon, traumatologist; b. Oposhnya (now Poltava Oblast), 1903; grad. Dnepropetrovsk Med. Inst., 1926; Cand. Med. Sci., 1937. Asst., Ukrainian Inst. Orthopedics and Traumatology, Kharkov, 1932-34; head orthopedics and traumatology sect. Zaporozhe City Hosp., 1934-39, 46-50; chief orthopedist and traumatologist Zaporozhe Oblast, 1946—; lectr., 1950; lectr. dept. orthopedics and traumatology Ukrainian Postgrad. Med. Inst., Kharkov, 1950-55; head chair orthopedics and traumatology Zaporozhe Postgrad. Med. Inst., 1955—. Head expdn. to study effect of high mountain climate on treatment of osteo-articular Tb, Bakhmaro, Georgia, 1932. Mem. Zaporozhe Oblast Soc. Traumatologists and Orthopedists (chmn. 1956—). Research and numerous publs. on treatment of osteo-articular Tb, gunshot fractures of hip, prophylaxis of traumatism. Address: Zaporozhe Plstgrad. Med. Inst., prospect Lenina 226, Zaporozhe, Ukraine SSR, USSR.

GAY, Claude, botanist; b. Draguignan, France, Mar. 18, 1800; made voyage to Greece and Asia Minor; left to study Chilean flora, 1828, and remained 11 years; visited Russia, Morocco, Poland, U.S.A., 1856-58; mem. French Acad. Scis., 1856. Author: Historia fisica y politica de Chile, 24 vols., 1843-51. Explorations in meridional Am.; research on flora and fauna of Chile; publs. on meteorology, geology and magnetism. Died Nov. 29, 1873.

GAY, Frederick Parker, Am. pathologist, bacteriologist; b. Boston, July 22, 1874; s. George Frederick and Louisa Maria (Parker) G.; A.B., Harvard, 1897; M.D., Johns Hopkins, 1901; Sc.D. George Washington U., 1932; m. Catherine Mills Jones, Oct. 18, 1904; children—Louisa Parker, Lucia Chapman, Frederick P., William. Asst. on Johns Hopkins Med. Commn. to Philippines, 1899; asst. demonstrator pathology U. Pa., 1901-03; fellow Rockefeller Inst. for Med. Research, 1901-03; research student Pasteur Institute, Brussels, 1903-06; bacteriologist Danvers Insane Hosp., 1906-07; asst. and instr. in pathology Harvard Med. Sch., 1907-10; prof. pathology U. Cal., 1910-21, prof. bacteriology, 1921-23; prof. bacteriology Columbia, from 1923. C.R.B. exchange prof. to Belgian univs., 1926-27. Fellow A.M.A., A.A.A.S.; mem. Nat. Acad. Sci., Assn. Am. Physicians, Am. Assn. Pathologists and Bacteriologists, Soc. Exptl. Biology and Medicine Assn. Am. Bacteriologists, Am. Assn. Immunologists. Author: Studies in Immunity, 1909; Typhoid Fever, 1918; Agents of Disease and Host Resistance (with others) 1935; The Open Mind—a Life of Elmer Ernest Southard; also contbr. to sci. jours. on bacteriology, immunology and pathology. Died New Hartford, Conn., July 14, 1939.

GAY, Helen, Am. biologist; b. Pittsfield, Mass., Aug. 30, 1918; d. Ulrich E. and Alice (Gonnet) G.; B.A., Mt. Holyoke Coll., 1940; M.A., Mills Coll., 1942; Ph.D. (Lalor fellow), U. Pa., 1955. With Carnegie Inst., 1942—, cytogeneticist, 1962—; prof. zoology U. Mich., 1962—; lectr. cytology biology dept. Adelphi Coll., 1959-62; jr. profl. asst. NIH Bethesda, Md., 1943-45. Mem. Am. Soc. Zoologists, Internat. Soc. Cell Biology, A.A.A.S., Genetics Soc. Am., Am. Soc. Naturalists, Soc. Study of Growth and Devel., Am. Soc. Cell Biology, Sigma Xi. Asso. editor cytology sect. Biology Abstracts. Contbr. numerous articles on cytology and cytogenetics to sci. jours. Home: 2650 Heatherway, Ann Arbor, Mich. 48104.

GAY, John, Brit. surgeon; b. Wellington, Eng.; ed. St. Bartholomew's Hosp., London, Eng. Named surgeon Royal Free Hosp., for 18 years from 1836; 1 dau., 2 sons; became surgeon Gt. No. Hosp., 1856, sr. surgeon until 1885. Author: On Femoral Rupture, its Anatomy, Pathology and Surgery, 1848; On Indolent Ulcers and their Surgical Treatment, 1855; On Varicose Diseases of the Lower Extremities and its Allied Disorders, 1868; On Haemorrhoidal Disorder, 1882; also articles. Described new method of operating on femoral rupture, 1848. Died Sept. 15, 1885.

GAY, Leslie Newton, Am. physician; b. Shamokin, Pa., Nov. 7, 1891; s. Harry Scott and Sarah (Batdorf) G.; Ph.B., Lafayette Coll., 1909-13; M.D., Johns Hopkins, 1917; D.Sc., Lafayette Coll., 1948; m. Julia Adele Griffith, June 4, 1919; 1 son, Leslie Newton. Faculty, Johns Hopkins Sch. Medicine, 1919—, asso. prof. medicine emeritus, 1958—, founder Allergy Clinic, 1922, dir., 1922-58. Cons. USPHS, 1940—, Walter Reed Hosp., U. S. Army, 1950—; a co-founder Asthma Research Unit, Cardiff, Wales, 1948. Mem. A.M.A., So. Med. Assn., A.C.P., Am. Assn. Immunologists, Am. Acad. Allergy, A.A.A.S., Phi Beta Kappa. Author: The Diagnosis and Treatment Bronchial Asthma, 1946; also numerous articles. Pioneered use ephedrine sulphate; developed use epinephrine in oil for treatment bronchial asthma; research on antihistamine drugs; discovered and proved value of dimenhydrinate for motion sickness and labyrinthian disease; research on prolonged use steroids in treatment chronic bronchial asthma. Home: Hollins Av., Balt., 21210. Office: 1114 St. Paul St., Balt. 21202.*

GAYDON, Alfred Gordon, English physicist; b. London, Eng., Sept. 26, 1911; s. Alfred Bert and R. J. G. (Gaydon) G., B.Sc. in Physics, Imperial Coll., London U., 1932, Ph.D., 1937, D.Sc., 1942; hon. doctorate U. Dijon (France), 1957; m. Phyllis Maude Gaze, July 27, 1940; children—Julie Hazel, Bernard Gordon. Various research fellowships, 1936-47; Warren research fellow Royal Soc., 1947-—, prof. molecular spectroscopy, Imperial Coll., London U., 1961-—. Recipient Bernard Lewis Gold medal Combustion Inst., 1960. Fellow Royal Soc., 1953 (Rumford medal 1960), Inst. Physics; mem. Royal Instn. Author: (with R. W. B. Pearse) Identification of Molecular Spectra, 1941; Spectroscopy and Combustion Theory, 1942; Dissociation Energies and Spectra of Diatomic Molecules, 1947; (with H. G. Wolfhand) Flames, their Structure, Radiation and Temperature 1953; The Spectroscopy of Flames, 1957; (with I. P. Hurle) The Shock Tube in High Temperature Chemical Physics, 1963; also numerous articles. Research on molecular spectra and application of spectroscopy in various fields especially chemistry of flames, energies of dissociation, study of phys. and chem. processes in shock tubes. Home: 43 Surbiton Hill Park, Surbiton, Surrey. Office: Chem. Engring. Dept., Imperial Coll., London S.W. 7, Eng.*

GAYDUKEVICH, Viktor Frantsevich, Russian archaeologist; b. 1904; D.Hist. Sci. Prof., Leningrad U.; in charge Bosphorus archeol. expdn. to study towns and settlements of Bosphorus Empire, 1932-—. Author: Ancient Ceramic Kilns: The Excavations in Kerch and Fanagoriya in 1929-31, 1934; The Bosphorus Empire, 1949; The Burial Grounds of Certain Bosphorus Cities, 1959; Further Data on Savmak's Revolt, 1962; The Mirmeky Excavations, 1963. Research and publs. on ancient history and archaeology of No. Black Sea area; excavated towns on Tiritaka, Mirmeky and Ilurat; found ancient Uzbek culture specimens at Farkhad hydropower project, 1942-43. Address: Leningrad University, Universitetskaya n. 7-9, Leningrad, USSR.

GAY-LUSSAC, Joseph Louis, French chemist, physicist; b. St. Léonard, Haute-Vienne, France, Dec. 6, 1778; s. Antoine Gay; ed. École Polytechnique, Paris, 1797-1800; m. 1808; numerous children. Asst. to C. L. Berthollet, École des Points et Chaussées, 1800; apptd. demonstrator to A. F. Fourcroy at École Polytechnique, 1802; traveled with A. von Humboldt, 1805-06; prof. chemistry École Polytechnique, 1809; prof. physics Sorbonne, 1808-32; supt. govt. gunpowder factory, 1818; chief assayer of Mint, 1829; prof. chemistry Jardin des Plantes at Mus., 1832-50; became mem. Chamber of Deputies, Haute Vienne, 1831; elected to Chamber of Peers, 1839; an original mem. Societé d'Arcueil, 1807; elected mem. sect. gen. physics 1st class French Acad. Scis., 1806, v.p., 1821, 33, pres., 1822, 34. Author: (with Thénard) Récherches physico-chemiques, 1811, also over 150 articles, others written jointly with Humboldt, Thénard, Welter, Liebig. Discoverd law which states that all gases expand equally for equal increments of temperature, 1802; enunciated Gay-Lussac's law (law of combining volumes), states that volumes of gases involved in chem. reaction are in ratio of small whole numbers to each other and to new gas produced, 1809; made 2 balloon ascents (1st with Biot) to investigate effects of altitude on terrestrial magnetism and composition of air, 1804; experimented (with L. J. Thénard) on voltaic pile, discovered presence of hydrogen in alkalis, isolated boron (with Thénard), discovered process for preparing potassium from fused potash; studied prussic, sulfuric and oxalic acids; proved that prussic acid contains hydrogen but not oxygen; analyzed composition of water; determined density of steam; verified law of capillary action; studied fermentation and improvement of organic analysis, 1810; discovered cyanogen, 1815; introduced method of volumetric analysis, circa 1815; isolated boron from boric acid; cyanogen; improved process of manufacturing oxalic acid and sulphuric acid. investigated solubility of salts in water at different temperatures; gave solubility curves of hydrosulfides; recognized law of mass action, 1839; research on manufacture of bleaching chlorides and on assaying of silver; invented hydrometer, alcoholometer, portable barometer, maximum and minimum thermometer, steam injector pump, spirit blow-lamp. Died Paris, May 9, 1850.

GAYRE OF GAYRE AND NIGG, Robert, Scottish anthropologist; b. Ireland, Aug. 6, 1907; s. Robert and Clara (Hull) G.; ed. U. Edinburgh, Exeter Coll., Oxford; M.A., D.Phil., D.Pol. Sc., D.Sc.; m. Nina Mary Terry; 1 son, Reinold John Robert. Prof. anthropology, dir. postgrad. research dept. anthrogeography, U. Saugor India, 1954-56; talkland Pursuivant Extraordinary, 1958. Mem. Internat. Inst. Sociology, Inst. Polit. Studies Madrid (corr.). Editor: The Armorial, 1959; The Mankind Quar., 1960; Mankind Monographs, 1961. Research and publs. on ethnology and genealogy. Home: 115 The Strand, Gzira, Malta. Office: 1 Darnaway St., Edinburgh 3, Scotland.*

GAZIN, C(harles) Lewis, Am. geologist; b. Colorado Springs, Colo., June 18, 1904; s. Charles Edward and Jani Frances (Nicklaus) G.; B.S., Calif. Inst. Tech., 1927, M.S., 1928, Ph.D., 1930; married Alice Van Deusen, Jan. 29, 1927 (divorced Oct. 15, 1942; children—Margaret A. (Mrs. H. T. Schellhous), Chester L., Barbara J. (Mrs. H. G. Neubauer); m. 2d, Elisabeth Parker Hobbs, May 11, 1943. Asst. and teaching Fellow, Calif. Inst. Tech., 1927-30, also geol. and paleontol. field work in Calif., Ore., Nev. and Ariz.; jr. geologist, U. S. Geol. Survey, 1930-32 (assisted in geol. studies in Mont., Ida., Calif); asst. curator, vertebrate paleontology, Smithsonian Instn., U. S. Nat. Mus. 1932-42, asso. curator 1942-46, curator, 1946-—, and museum geologist, 1956-62, supervisory museum geologist, 1962-— (also led various expeditions to the western states and Central Am. for exploration and collection of fossil vertebrates); mem. div. earth sciences, Nat. Research Council 1948-51, 57-60; incorporator, Am. Geol. Inst. rep. Soc. Vertebrate Paleontology, 1947, dir., 1956-58. Decorated Legion of Merit, 1946. Fellow Geol. Society Am. (editorial board 1946-48), Paleontol. Society; member Society Vertebrate Paleontology (pres. 1949), Am. Soc. Mammalogists, Wash. Acad. Sci. (editor 1938-40, sec. 1947-48), Geol. Society Washington, Am. Soc. Zoologists, Soc. for Study Evolution, American Geographical Society, Asociación Paleontológica, Argentina, Sigma Xi, Tau Beta Pi, Pi Alpha Tau. Recipient Geol. Soc. Am. (Cordilleran Sect.) prize, 1930. Extensive studies of newly discovered or previously undescribed Cenozoic mammalian faunas, including new genera and species; monographic studies of extinct groups of mammals including early Tertiary tillodonts, artiodactyls, primates and condylarths; Cenozoic continental stratigraphy and correlation. Home: 6420 Broad St., Brookmont, Md. 20016. Office: U. S. National Museum, Washington 20560.

GEARIEN, James Edward, chemist; b. Peoria, Ill.; s. Grover C. and Anna (Sperry) G.; B.S., U. Ill., 1941; M.S., U. Mich., 1942, Ph.D., 1950; m. Helen G. Gray, Feb. 14, 1948; children—James Edward, Anne J. Faculty, U. Ill., Chgo., 1948-—, prof., head chemistry dept., 1958-—. Mem. Am. Chem. Soc. (chmn. div. medicinal chemistry), Am. Pharm. Assn. Research, publs. primarily on synthetic organic chemistry. Home: 1606 Courtland St., Park Ridge, Ill. 60068. Office: 833 S. Wood St., Chgo. 60612.*

GEBALLE, Ronald, Am. physicist; b. Redding, Cal., Feb. 7, 1918; s. Oscar and Alice (Glaser) G.; B.S., U. Cal. at Berkeley, 1938, M.S., 1940, Ph.D., 1943; m. Marjorie Louise Cohn, Oct. 31, 1940; children—Margaret F., Thomas R., Leslie A., Daniel T., Robert O., Jonathan L., Emily R., Anthony J. Teaching asst. physics U. Cal. at Berkeley, 1938-42, physicist radiation lab., 1943; physicist Applied Physics Lab., U. Wash., 1943-46; mem. faculty U. Wash., 1946-—, prof., chmn. dept. physics, 1959-—. Cons. NSF, Army Research Office: mem. research adv. com. electrophysics NASA; member Commission College Physics, 1966-—. Member citizens committee education Wash. State Legislature, 1960. Fellow Am. Phys. Soc.; mem. Am. Assn. U. Profs., Am. Assn. Physics Tchrs., A.A.-A.S., Fedn. Am. Scientists, Am. Civil Liberties Union, Pacific N.W. Assn. Coll. Physics (chmn. bd. dirs. 1965-—), Phi Beta Kappa, Sigma Xi. Research on atomic collision processes in gases and in single encounters. Home: 7516 28th Av. N.E., Seattle 98115. Office: Dept. of Physics, U. of Washington, Seattle, Wash. 98105.*

GEBALLE, Theodore Henry, Am. physicist; b. San Francisco, Jan. 20, 1920; s. Oscar and Alice (Glaser) G.; B.S., U. Cal., Berkeley, 1941, Ph.D., in Phys. Chemistry, 1949; m. Frances Koshland, Oct. 19, 1941; children—Gordon, Alison, Adam. Head solid state and atomic physics research dept. Bell Telephone Labs., Murray Hill, N.J., 1954-—. Mem. Am. Phys. Soc., Am. Chem. Soc. Investigation of low temperature physics; transport properties of semiconductors; properties of metals at low temperatures with emphasis on superconductivity. Home: 204 Springfield Av., Summit, N.J. 00781. Office: Bell Telephone Labs., Murray Hill, N.J. 07971.

GEBBIE, Hugh Alastair, Brit. physicist; b. Galashiels, Scotland, Jan. 5, 1922; s. Hugh Alexander and Martha (Wallace) G.; student Edinburgh (Scotland) U., 1940-42, Reading (Eng.) U., 1949-51; m. Katharine Blodgett, June 22, 1957; Faculty, Purdue U., 1952-53, Mass. Inst. Tech., 1953-55, Johns Hopkins, 1955-56; staff Nat. Phys. Lab., Teddington, Eng., 1957-—, head advanced instrumentation unit, 1966-—. Recipient Duddell medal Phys. Soc. London, 1965. Asso. Inst. Physics. Research, publs. on infrared transmission of atmosphere, interferometric spectroscopy of far infra-red; developed submillimeter maser using water and hydrogen molecules. Home: 74 Cadogan Sq., London, Eng. Office: Nat. Phys. Lab., Teddington, Eng.*

GEBER, see Jabir ibn Haiyan.

GEBHARD, Bruno Frederic William, physician; b. Rostock, Germany, Feb. 1, 1901; s. Fritz and Meta (Ross) G.; M.D., U. Rostock, 1924; m. Gertrude Adolph, Apr. 7, 1927; children—Susanne (Mrs. Alvin Goodman), Dorothy (Mrs. Bernard McCabe), Ursula (Mrs. Raymond Fink). Came to U. S., 1937, naturalized, 1944. Dir. Cleve. Health Mus., 1940-65, dir. emeritus, 1965-—; former asso. in health edn. Western Res. U.; cons. Armed Forces Med. Mus. (Washington). Recipient Prentiss award, 1965; Golden Door award, 1967. Diplomate Am. Bd. Preventive Medicine. Hon. fellow Internat. Coll. Dentistry; mem. A.M.A., Am. Pub. Health Assn., Am. Mus. Assn., Royal Soc. Health (London). Author: Wunder des Lebens, 1934; Das Leben der Frau, 1937; also numerous articles. Research in social conditions on morbichity and mortality, devel. of visual methods in health edn.; established first health mus. in U. S. Home: 3276 Braemar, Shaker Heights, O. 44120.*

GEBHARD, Kurt Alfred Thomas, German chemist; b. Elberfeld, Germany, June 27, 1881; s. Franz and Aline (Jordans) G.; studied chemistry, Munich, Leipzig, (both Germany); doctorate, Marburg, Germany, 1908; m. Natalie Mayer, 1908; 1 son, 1 dau. Employed by Photogravür-AG, Siegburg; later worked for Farbwerke Hoechst; named chem. dir. L. Marks' Sci. Inst., Frankfurt/Main, Germany, 1913. Contbr. articles to tech. jours. Pioneered exptl. research on chem. processes of dyeing, especially bleaching effect of sunlight and methods of preventing it; discovered that phosphorous, tungsten or molybdenum salts produced increased resistance to bleaching, especially with cotton dyes using cationic dyestuffs. Died in Battle of Marne, Sept. 10, 1914.

GEBHARD, Louis August, Am. radio researcher; b. Buffalo, June 11, 1896; s. August and Caroline (Deuter) G.; LL.B., Georgetown U., 1924; B.S. in Elec. Engring., George Washington U., 1930; m. Marguerite A. Strauss, Aug. 6, 1931; children—Katharine (Mrs. Harlan Q. Stevenson), Paul. With U. S. Naval Research Lab., Washington, 1923-—, successively head radio transmitter sect. radio div., supt. radio engring. div., supt. aircraft div., asst. supt. radio vis., asst. supt. charge devel., 1923-45, supt. radio div., 1945-—. Recipient Presidential Certificate of Merit for achievements in radar, 1946. Fellow Inst. Elec. and Electronic Engrs., Am. Phys. Soc., A.A.A.S.; mem. Research Soc. Am., Am. Inst. Physics, Armed Forces Communication and Electronics Assn., U. S. Naval Inst. Patentee electronics and radio. Investigation of electronics for official U. S. time and frequency; centralized electronic control; navigation; radio communication; countermeasures. Home: 2142 Branch Av. S.E., Washington 20. Office: U. S. Naval Research Lab., Washington.

GEBHARD, Paul Henry, Am. anthropologist; b. Rocky Ford, Colo., July 3, 1917; s. Paul Adam and Eva (Baker) G.; B.S. cum laude, Harvard, 1940, M.A., 1942, Ph.D., 1947; m. Agnes Elizabeth West, May 19, 1939; children—Mark, Jan., Karla. Research asso. Inst. for Sex Research, Ind. U., Bloomington, 1946-55, faculty, 1947-—, exec. dir., 1956-—, prof. anthropology, 1967-—. Fellow Am. Anthropol. Assn., Am. Sociol. Assn., Ind. Acad. Sci. Author: (with others) Sexual Behavior in the Human Female, 1953; Pregnancy, Birth and Abortion, 1958; Sex Offenders: an Analysis of Types, 1965; also articles. Research on human sexual behavior. Home: 1610 Dorchester St., Bloomington, Ind. 47401.*

GEBHARDT, Louis Philipp, Jr., Am. physician; b. Jackson, Cal., Dec. 20, 1905; s. Louis P. and Ellen (Merkel) G.; A.B., Stanford, 1929, M.S., 1934, Ph.D., 1937, M.D., 1942; m. Johann Else Burket, Sept. 11, 1938; children—Laurence P., Michael J., Karl A. Research asso. Stanford, 1937-40, acting asst. prof., 1941-42; faculty U. Utah, Salt Lake City, 1942-—, prof., head dept. bacteriology, 1943-—. Cons. to govt. agys., pvt. cos. Mem. Am. Soc. Microbiology, Am. Assn. Immunologists, Am. Pub. Health Assn., Soc. for Exptl. Biology and Medicine. Author: (with D. A. Anderson) Microbiology, 1954; also numerous articles. Research in med. bacteriology, mycology, epidemiology, virology; discovered overwintering host Western Equine encephalitis virus. Home: 2194 S. 19th E. St., Salt Lake City 84106.*

GEBHARDT, (Franz August Max) Walter, German anatomist; b. Breslau (now Wroclaw, Poland), Mar. 22, 1870; s. Johann Amand Sylvius and Eva Henriette (Troplowitz) G.; student medicine, Berlin; M.D., Breslau, 1894; m. Ida Adele Fränkel, 1898; 2 sons, including Werner. Asst., U. Breslau; with Path. Inst., 1894-95, Surg. Clinic, 1895-96, histological dept. Physiol. Inst., 1896-97; with optical firm Zeiss, Jena, Germany, 1897-99; became asst. to W. Roux, Anat. Inst., U. Halle (Germany), 1899, joined faculty, 1901, became histological prosector, 1903, titular prof., 1906, asso. prof., 1907, dir. histological and developmental history dept. under Roux, until 1918; surgeon Mil. Hosp., Halle, during World War I. Recipient Georg Hermann von Meyer prize Senckenberg Natural Sci. Soc., Frankfort, 1917. Mem. Leopoldina. Research and publs. on constrn. and devel. of bones in relation to their dependence on function; studied fine structure of vertebrate skeleton. Died Halle, Mar. 3, 1918.

GÉCZY, István, Hungarian chemist; b. Budapest, Hungary, Mar. 10, 1921; s. József and Margit (Fischer) G.; M.Sc., U. Budapest, 1943, Ph.D., 1944, Cand. of Sci., 1956; m. Klára Györk, Dec. 2, 1948; children—András, István. Asso. prof. Tech. U., Budapest, 1943-59; sr. researcher Textile Research Inst., Budapest, 1959-—. Cons. engr., mem. adv. com. plastics dept. Hungarian Acad. Scis., 1963-—. Author: (with Zoltán Csürös, Rudolf Balló, Zoltán Bruckner) Müanyagok, 1956; also numerous articles. Research on kinetics of heterogeneous catalytical reactions, kinetics of high polymer reactions, physicochemistry of oligomers and polymers. Home: 1 Kuny D., Budapest. Office: 86 Gyömröi, Budapest, Hungary.*

GED, William, Scottish inventor; b. Edinburgh, Scotland, 1690; began as goldsmith, jeweller; invented process of making stereotype plates for printing, 1725. Died Edinburgh, Oct. 19, 1749.

GEDALIA, Itzhak, human biochemist; b. Vienna, Austria, Jan. 1, 1917; s. Haim and Adele (Gottesmann) student high chem., tech. schs. Prague, Czechoslovakia, Bucharest, Rumania, 1935-41; Ph.D. in Chemistry, Hebrew U., Jerusalem, 1952; m. Miryam Stern, Jan. 19, 1950. Faculty, Hebrew U.-Hadassah Med. Sch., Jerusalem, 1942——, sr. lectr., head lab. oral chemistry and fluoride Research, 1965——; vis. scientist dept. biochemistry, U. Minn., Mpls., 1965. WHO fellow to study fluoridation in preventive dentistry in various European countries; IADR fellow Harvard Sch. Dental Medicine, 1967. Mem. European Orgn. for Fluorine Research and Dental Caries Prevention, Tooth and Bone Soc. London, Internat. Assn. for Dental Research. Author: (with others) WHO Monograph on Fluoride in Human Health, 1967; also numerous articles. Research on metabolic fluoride research and calcification of teeth and bones, placental transfer of fluoride, epidemiologic fluoride research, effect of fluoride on osteoporosis. Home: 21 Balfour, Jerusalem, Israel.*

GEDDES, Sir Patrick, Scottish biologist, sociologist; b. Ballater, Scotland, Oct. 20, 1854; s. Alexander and Janet (Stevenson) G.; ed. Perth Acad., Royal Sch. Mines, Univ. Coll., London, Sorbonne, U. Edinburgh, Montpellier; student of Huxley; m. Anna Morton, 1886 (dec. 1917); 2 sons, 1 dau.; m. 2d, Lillian Brown, 1928. Influenced by Darwin's work and its application to society; lectr. natural history Sch. Medicine, Edinburgh; with zool. stas., bot. gardens; travelled to Mexico, Cyprus, Palestine, India, U. S.; prof. botany at U. Coll., Dundee, 1883-1920; prof. sociology at U. Bombay, 1920-23; dir. Scots Coll., Montpellier; developer summer schs., Brit. residence halls at French univs. Knighted, 1932. Author: (with Arthur Thomson) The Evolution of Sex, 1889; Evolution, Sex, Biology, and Life; (with Sir J. Arthur Thomson) Outlines of General Biology; Chapters in Modern Botany; City Development; Cities in Evolution; The Life and Work of Sir Jagadis C. Bose, 1920; (with V. V. Branford) The Coming Polity, Our Social Inheritance; (with Gilbert Slater) Ideas at War. Strove to synthesize evolution, morality and biology, history and sociology; pioneer in town and regional planning, university planning; social, acad. and econ. reform; drew up original plans for Hebrew U. at Jerusalem, 1919. Died Montpellier, France, Apr. 17, 1932.

GEE, Edwin Austin, Am. chem. engr.; b. Washington, Feb. 19, 1920; s. Edwin Stanton and Marie (Junghams) G.; B.S. George Washington U., 1941, M.S., 1944; Ph.D., U. Md., 1948; m. Genevieve Riordan, Aug. 26, 1944; children—John Michael, William Stanton, David Stephen. Instr.—George Washington U., 1940-41; chemist U. S. Bur. Mines, 1942-46, asst. chief metall. div., 1946-48; with E. I. du Pont de Nemours & Co., Wilmington, Del., 1948—, asst. dir. devel. dept., 1960-63, dir. devel. dept., 1963-68, gen. mgr. photo products dept., 1968; dir. Block Engring., Inc., Cambridge, Mass., Dana Labs., Irvine, Cal. Mem. Am. Chem. Soc., Am. Soc. Metals, Am. Inst. Mining and Metall. Engrs., Del. Soc. Profl. Engrs. Contbr. articles on titanium metal to tech. jours. Research on organic kinetics; purification of salts; pigments; rare metals; semiconductor materials; extraction of ores. Home: Box 3960, Greenville, Del. 19807. Office: Nemours Bldg., Wilmington, Del. 19898.

GEER, Jack Charles, Am. pathologist; b. Galesburg, Ill., Sept. 19, 1927; s. John Charles and Ruth (McGee) G.; B.S. in Chemistry, La State U., 1950, M.D., 1956; m. Sarah Kathleen Williamson, Feb. 16, 1951; children—Charles, Richard, John, Cynthia, Michael. With La. State U. New Orleans, 1954—prof. pathology, 1965-66; prof. pathology U. Tex. S. Tex. Med. Sch., 1966-67; prof., chmn. dept. pathology Ohio State U., 1967—; vis. investigator Rockefeller Inst., 1960-61. Recipient Outstanding Sr. award Omicron D, Delta Kappa-Mortar Bd., 1956; J.A. Majors award for scholastic achievement in pathology, 1956; George W. McCoy award La. State U., 1956; USPHS Research career Devel. award, 1959-66. Mem. Am. Heart. Assn., Soc. for Exptl. Pathology and Bacteriology, A.A.A.S., Internat. Acad. Pathology, Am. Soc. Clin. Pathologists, Council Arteriosclerosis. Contbg. author: Atherosclerosis and its Origin, 1963; (with McGill) Evolution of the Atherosclerotic Plague, 1963; (with Freeman) Cellular Fine Structure, Student Atlas, 1964; also numerous articles. Research on morphology atherosclerosis, chem. changes with atherosclerosis, pulmonary vascular changes with hypertension. Home: 4391 Mumford Dr., Columbus, O. 43221.*

GEERTZ, Clifford James, Am. educator; b. San Francisco, Aug. 23, 1926; s. Clifford J. and Lois (Brieger) G.; B.A., Antioch Coll., 1950; Ph.D., Harvard, 1956; m. Hildred Storey, Oct. 30, 1948; children—Erika Storey, Benjamin Warren. With Harvard, 1956-59, instr. anthropology, 1956-57, fellow Center for Advanced study in Behavioral Scis., 1958-59; asst. prof. U. Cal., Berkeley, 1959-60; faculty U. Chgo., 1960—, prof., 1964——. Fellow Am. Acad. Arts and Scis.; mem. Am. Anthrop. Assn., Am. Soc. Study Religion, Asian Studies Assn. Author: Religion of Java, 1960; Agricultural Involution, 1963; Peddlers and Prices, 1963; The Social History of an Indonesian Town, 1965. Studies, publs. in sociology of religion, theory of culture, econ. devel. of Java, Bali, Morocco. Home: 1452 E. Park Pl., Chgo. 60637.*

GEGENBAUER, Leopold Bernhard, Austrian mathematician; b. Asperhofen, Austria, Feb. 2, 1849; s. Viktorin and Amalie (Zeitzem) G.; studied 1st history and linguistics, then math. and physics, Vienna; teaching certification, 1869; hon. Ph.D., Czernowitz, Rumania (now USSR) 1879; m. Helen Schuler von Libloy, 1877; a son, Viktor. Tchr. secondary schs., Waidhofen on Thaya, Krems, Austria, for several years; named fellow, Berlin, 1873; named asso. prof. math. U. Czernowitz, 1875; joined U. Innsbruck (Austria), 1878, became prof., 1881; succeeded J. Petzval, U. Vienna, 1893. Mem. ins. adv. bd. Ministry Interior. Mem. Acad. Scis. Vienna (corr.), Leopoldina. Research and publs. on number theory especially theory of prime numbers, on algebra, especially theory of symmetric function and theory determinants, on theory of functions, especially Bessel's and spherical functions. Died Vienna, June 3, 1903.

GEGENBAUR, Carl, German anatomist, zoologist; b. Würzburg, Germany, Aug. 21, 1826; s. Franz Joseph and Elisabeth Karoline (Roth) G.; M.D., U. Würzburg, 1851; m. Anna Margarete Emma Streng, 1863; 1 dau.; m. 2d, Ida Arnold, 1869; 1 son, 1 dau. Asst., Juliusspital, Würzburg, for 1 1/2 years; went with Kölliker to study invertebrate marine animals at Messina, 1852; joined med. faculty U. Würzburg, 1854; named asso. prof. zoology U. Jena (Germany), 1855, prof. anatomy and zoology, 1858; became prof. anatomy and comparative anatomy U. Heidelberg (Germany), 1873, also dir. Anat. Inst., emeritus 1901; tchr. Ernest Haeckel. Fellow Royal Soc., 1884. Author: Vergleichende Anatomie der Wirbeltiere mit Berücksichtgung der Wirbellosen, 2 vols., 1898-1901; Lehrbuch der Anatomie des Menschen, 2 vols., 1883-89; Grundzüge der vergleichenden Anatomie, 1859. Founder, Morphologische Jahrbuch, 1875, also editor for several years. One of 1st to study anatomy from evolutionary viewpoint; research on comparative anatomy of vertebrates; his research on fishes supported Huxley in refutation of theory of origin of skull from expanded vertebrae; studied Schwann's hypothesis and established that all vertebrates' eggs and sperm are single cells, 1861; demonstrated how embryonic parts which form gill apparatus in fish, are used for other organs, such as Eustachian tubes and thymus glands, in land vertebrates. Died Heidelberg, June 14, 1903.

GEHLEN, Adolf Ferdinand, German chemist; b. Bütow, Germany, Sept. 15, 1775; s. Jacob Ferdinand and Friederike Elisabeth (Engelmann) G.; studied pharmacy, natural scis., linguistics U. Königsberg (Germany), for 3 years; Ph.D., Halle, Germany, 1806. Worked at Rose Pharmacy, Berlin; became lectr. U. Halle, also zoo-chemist Reil's Inst., 1806; named acad. chemist Acad. Scis., Munich, Germany, 1807; died from poisoning during expt. with arsenic hydride. Monument dedicated to him; Gehlenit, a mineral, named in his honor. Mem. Pharm. Assn. Bavaria (founder), Munich Acad. Scis. Editor, (with Rose) Neues Berliner Jahrbuch für die Pharmazie, 1803-06; Neues allgemeines Journal der Chemie (now Jour. für Chemie und Physik), 1806-10; founder Repertorium der Pharmacie, 1815. Research and numerous publs. on preparation of woad and indigo, oil and extract of hops, preparation of brandy. Died Munich, July 15, 1815.

GEHLEN, Arnold, German sociologist; b. Leipzig, Germany, Jan. 29, 1904; s. Max and Margarete (Ege) G.; ed. Us. Leipzig, Cologne; m. Baroness Veronika von Wolff, 1937. Lectr., philosophy, Leipzig, 1930; prof., U. Leipzig, 1934; prof., U. Königsberg, 1938; U. Vienna, 1940; prof., sociology, Speyer Institute of Administration, 1947; prof., Aachen Technical U., 1962. Corr. mem., Austrian Acad. Scis. Author: Theorie der Willensfreiheit, 1933; Der Menschlichen Seine Natur und seine Stellung in der Welt, 1940; (with H. Schelsky) Soziologie, 1955; Urmensch und Spätkultur, 1956; Zeit-Bilder, 1960. Studies in cultural anthropology. Home: 34 Am Kupferofen, Aachen, West Germany.

GEHLEN, Heinz Johannes Georg, German chemist; b. Posen, Mar. 2, 1902; s. Bruno and Lucie (Koperski) G.; Ph.D. in Chemistry, U. Berlin. Asst. Inst. Chemistry, U. Berlin, 1927-29; with U. Königsberg, 1929-34; collaborator Gmelins Handbucher Anorg. Cehmie, 8th edit., 1934-40, Forschungsinstitut d. Vereinigten Glanzstoffwerke, Berlin, Halle, 1940-45; with Council of Govt. Dept. Edn., Darmstadt, 1945-47; head asst. Inst. Chemistry, U. Frankfort, 1947-50; dir. Inst. Inorganic Chemistry and Physics, Higher Pedagogic Sch., Potsdam, 1950-67; prof. agrégé U. Humboldt, Berlin, 1951; titular prof. Higher Pedagogic Sch. of Potsdam-Sanssouci, 1953. Recipient Dr. Theodor-Neubauer gold medal, 1961; Arndt medal, 1965. Research and publs. on inorganic, phys. and organic chemistry. Home: Seeblick 1, Gross-Glienicke of Potsdam. Office: Higher Pedagogic School, Am Neuen Palais, Potsdam-Sanssouci, Germany.

GEHLER, (Gustav) Willy, German constrn. engr.; b. Leipzig, Germany, Sept. 5, 1867; s. Gustav and Maria (Carl) G.; studied math. and natural scis., Leipzig, constrn. engring. Tech. U. Dresden, Germany; hon. dr.rer.techn., German Tech. U., Brno, Czechoslovakia, 1933; m. Elisabeth Müller, 1914; 1 son, 1 dau. Became constrn. mgr. Saxon R.R., 1905; asst. to Grübler and Mehrtens, Tech. U. Dresden; engr. Dyckerhoff & Widmann, 1905-13, also tech. dir. Dresden br.; named prof. tech. mechanics, material strength and iron bridge constrn., 1913, also prof. bldg. materials, 1918; became dir. tech. constrn. dept. State Materials Testing Bur., Dresden, 1918. Co-founder, Standards Com. adv. mem. German Concrete Assn. Recipient DIN hon. ring German Standards Commn., 1942, Emil Mörsch Meml. medal German Concrete Assn., 1950. Mem. Prussian Acad. Constrn., 1925. Author: Nebenspannungen eiserner Fachwerkbrücken, 1910; der Rahmen, 1913; (with H. Amos) Versuche mit kreuzweise bewehrten Platten, 1932; Versuche über Elastizität, Plastizität und Schwinden von Beton, 1934; Versuche an Plattenblaken mit Bewehrungen mit hoher Streckgrenze, 1937; Erläuterungen zur Bemessung von Knickstäben, 1939. Pioneered reinforced concrete, especially devel. materials such as binding substances and design; work in superstructures and bridges. Died Dresden, Apr. 13, 1953.

GEHLHOFF, Georg Richard, physicist; b. Adlig Rauden, Germany, Feb. 7, 1882; s. Eduard and Marta (Fuhlbrügge) G.; studied physics Tech. U. Danzig (Poland), then U. Berlin; doctorate, 1907; hon. dr.rer.techn., Dresden, 1927; m. Margarete Canditt; 1 son, 1 dau. Became asst. to H. Rubens, Berlin, 1911, J. Zenneck, Danzig, 1911, also lectr.; scientist optical instrument firm C. P. Goerz, Berlin-Friedenau, 1913; named dir. Leipzig-Leutzsch reflector factory taken over by Goerz, 1907, Sendlinger Optischen Glaswerke GmbH, Berlin-Zehlendorf, 1920; named dir. glass works Osram GmbH, Weisswasser, 1922, later at machine glass works, Berlin-Siemensstadt; became asso. prof. Tech U. Berlin, 1923. Golden Gehlhoff Ring donated in his honor German Tech. Glass Soc. Mem. Soc. for Tech. Physics (co-founder 1919, chmn. until 1931). Editor: Lehrbuch der Technischen Physik, 3 vols., 1924-26. Co-editor, Zeitschrift für technischen Physik, 1920-31. Research and publs. on tech. optics and lights, especially arcs and reflectors; pioneered (with Osvam) work on glass; studied phys. properties of glass in solid and liquid states, problems of glass manufacture. Died Siders, Switzerland, Mar. 12, 1931.

GEHMAN, Harry Merrill, Am. mathematician, educator; b. Norristown, Pa., Jan. 15, 1898; s. Abner Haring and Barbara (Clemens) G.; A.B., U. Pa., 1919, A.M., 1920, Ph.D., 1925; postgrad. U. Tex., m. Marian Barr, Sept. 2, 1922; children—Margery Barr (Mrs. Horace E. Dodge III), Jean Virginia (Mrs. Floyd E. Adamson), Harry Merrill. Instr. math. U. Pa., 1920-25; faculty Yale, 1926-29; prof., head math. dept. U. Buffalo, 1929-62; prof. State U. N.Y., Buffalo, 1962-—; head math. dept. U. S. Army U., Shrivenham, England, 1945. Fellow A.A.A.S.; mem. Math. Assn. Am. (sec.-treas. 1949-58, treas., exec. dir. 1959-——, Distinguished Service award for Math. 1966), Am. Math. Soc., Canadian Math. Congress, Am. Soc. Engring. Edn. (chmn. math. div. 1951-52), Soc. for Indsl. and Applied Math., Phi Beta Kappa, Sigma Xi. Research on continuous curves, irredundant sets of postulates in logic. Home: 163 Winspear Av., Buffalo 14215.*

GEHRCKE, Ernst Johann, German physicist; b. Berlin, July 1, 1878; s. Hermann and Marie (Budweg) G.; Ph.D., U. Berlin, 1901; m. Gertrud Schröder. With State Phys.-Tech. Inst., 1901-46, named dir. optical sect., 1926; also asso. prof. U. Berlin; joined U. Jena (Germany), 1946; then with German Bur. Weights and Measures, Berlin. Mem. German Soc. for Tech. Physics (a founder). Author: Anwendung der Interferenzen in der Spektroskopie, 1904; Die Relativitätstheorie eine wissenschaftliche Massensuggestion, 1920; Theorie der Atomkerne, 1920; Die Strahlen der positiven Elektrizität, 1922; Kritik der Relativitätstheorie, 1924. Invented Glimmlicht oscillograph, 1904, Lummer-Gehrcke plate, (with Lau) multiplex interference spectroscope; discovered (with Reichenheim) anode rays; observed (with Seeliger) spectral dependence of light on electron speed; research in physiol. and med. fields; discovered therapeutic value of dust particles; opposed theory of relativity. Died Birkenwerder, Germany, Jan. 25, 1960.

GEHRING, Frederick William, Am. mathematician, educator; b. Ann Arbor, Mich., Aug. 7, 1925; s. Carl Ernest and Hester (Reed) G.; B.S. in Math., U. Mich., B.S. in Elec. Engring., 1946, M.A. in Math., 1949; Ph.D. in Math., Cambridge (Eng.) U., 1952; m. Lois Caroline Bigger, Aug. 29, 1953; children—Kalle B., Peter M. Benjamin Peirce instr. Harvard, 1952-55, vis. prof., 1964-65; faculty math. U. Mich., Ann Arbor, 1955——, prof., 1962—. Fulbright and Guggenheim fellow U. Helsinki (Finland), 1958-59; NSF fellow U. Zurich (Switzerland), 1959-60. Vis. prof. Stanford, 1964; editor Van Nostrand Co., 1965——. Mem. Am., London math. socs., Schweizerische Mathematische Gesellschaft, Suomen Matemaattinen Yhdistys. Asso. editor Proc. Am. Math. Soc., 1962-65; analysis editor Duke Math. Jour., 1963——; mem. editorial bd. Jour. Math. and Mechanics, 1966——. Research, publs. on boundary behaviour of harmonic and analytic functions, applications of Tauberian theorems, and study of quasiconformal mappings. Home: 2001 Shadford Rd., Ann Arbor, Mich. 48104.*

GEIDUSCHEK, Ernest Peter, Am. biologist; b. Vienna, Austria, Apr. 11, 1928; s. Sigmund and Frida

(Tauber) G.; A.B., Columbia, 1948; A.M., Harvard, 1950, Ph.D., 1952; m. Joyce Barbara Brous, Aug. 25, 1955; children—Jeremy, Jonathan. Came to U. S., 1945, naturalized, 1946. Instr. chemistry Yale, 1952-53, 55-57; asst. prof. U. Mich., 1957-59; faculty U. Chgo., 1959—, prof. biophysics, 1964—. Cons. USPHS, 1963—. Recipient Research award Am. Postgrad. Med. Assn., 1962; Guggenheim fellow, 1964-65. Mem. Am. Chem. Soc., Biophys. Soc., Am. Soc. Biol. Chemists, A.A.A.S., Phi Beta Kappa, Sigma Xi. Bd. editors Jour. Molecular Biology, 1965—. Research, publs. on macromolecular analysis of proteins and nucleic acids, chem. genetics, synthesis and function of nucleic acids. Office: Dept. Biophysics, U. Chgo., Chgo. 60637.*

GEIGER, Johannes (Hans) Wilhelm, German physicist; b. Neustadt, Germany, Sept. 30, 1882; s. Wilhelm and Marie (Plochmann) G.; studied U. Munich; Ph.D., U. Erlangen (Germany), 1906; m. Elisabeth Heffter, 1920; 3 sons. Worked with E. Rutherford, E. Marsden, Cavendish Lab., Phys. Inst., U. Manchester (Eng.), 1906-12; named head radioactivity lab. Reich Phys.-Tech. Inst., Berlin, 1912; served in World War I, 1914-18; named prof. Phys. Inst., U. Kiel (Germany), 1925; became prof. Tübingen, Germany, 1929; named prof. Tech. U., Berlin, 1936; worked on atomic energy project, 1944-45. Recipient Hughes medal Royal Soc., 1938, Dudell medal Phys. Soc., London, 1938, Arrhenius prize U. Leipzig (Germany), Mem. Prussian Acad. Scis., Leopoldina. Author: (with W. Makower) Messmethoden auf dem Gebiete der Radioaktivität, 1920; also articles. Research (with Marsden) on dispersion of alpha particles (led to Rutherford's dispersion formula and atomic model); discovered relationship between range of alpha particles and radioactive period of radiator; confirmed character of radioactive disintegration, 1908; determined (with Rutherford) radioactive period of radium; continued devel. ionization chamber to detect single alpha particles; perfected counter for beta or cosmic ray particles (Geiger-Müller counter). Died Potsdam, Germany, Sept. 24, 1945.

GEIGER, Klaus Wilhelm, physicist; b. Berlin, Germany, Apr. 26, 1921; s. Hans and Elisabeth (Heffter) G.; Dipl. Phys., Tübingen U., 1949; Dr.rer.nat., U. Mainz, 1951; m. Charlotte Schmidt, July 26, 1947; children—Bernhard, Thomas. Research officer div. applied physics NRC, Ottawa, Ont., Can., 1954—. Mem. Can. Assn. Physicists. Research in radioactive disintegrations, nuclear fission, cosmic radiation, neutron standardization, neutron spectra. Home: 865 Chapman Blvd., Ottawa 8. Office: Div. Applied Physics, NRC, Ottawa 7, Ont., Can.

GEIGER, Philipp Lorenz, German pharmaceutist; b. Freinsheim nr. Frankenthal, Germany, July 29, 1785; s. Johannes and Louise (Hecht) G.; Ph.D., Heidelberg, Germany, 1817; M.D. (hon.), U. Marburg (Germany), 1828; m. Anna Barbara Folz Sachs, 1811; 2 stepchildren; m. 2d, Auguste Ernestine Rinck, 1826; 2 sons, including Friedrich, 4 daus. Became pharmacy apprentice, Adelsheim, 1799; journeyman, Heidelberg, Rastatt, Karlsruhe, (all Germany), until 1811; owner U. Pharmacy, Heidelberg, 1814-21; became pvt. lectr., 1816; joined faculty U. Heidelberg, 1818, named asso. prof. med. faculty, 1824. Author: Handbuch der Pharmacie, 2 vols., 1824-29; Universal pharmakopöe, 1836-45. Founder, Magazine für Pharmazie (now Annalen der Chemie). Research in phytochemistry, especially devel. alkaloid chemistry; 1st isolation chemically pure coniine, 1831; isolation (with Hesse) of atropine, acontine, colchicine, hyoscyamine, 1833. Died Heidelberg, Jan. 19, 1836.

GEIGY, (Johann) Rudolf, Swiss chem. industrialist; b. Basel, Switzerland, Mar. 4, 1830; s. Carl and Sophie (Preiswerk-Bischoff) G.; trained in father's dye-chem. factory for 3 years; Ph.D. (hon.), Basel, 1910; m. Maria Merian, 1855; children—Johann Rudolf G.-Schlumberger, Carl Alphons G.-Hagenbach; 2 daus. Became partner father's firm Johann Rudolf Geigy, 1854; supervised devel. aniline dyes; ret. from active mgmt., 1891; Swiss parliamentary leader in econ. problems especially comml. treaties with Germany, France, Italy, 1880-86. Author: Unsere Handels- und zollpolitischen Beziehungen, 1893; also articles. Devel. Basel chem. industry, especially aniline dyes; devel. social ideas. Died Basel, Feb. 17, 1917.

GEIKIE, Sir Archibald, Scottish geologist; b. Edinburgh, Scotland, Dec. 28, 1835; s. James Stuart and Isabella (Thom) G.; ed. Edinburgh U., LL.D.; D.C.L. (hon.), Oxford (Eng.) U.; D.Sc. (hon.), Cambridge (Eng.) U., Dublin (Ireland) U.; LL.D., Glasgow, Aberdeen, St. Andrews (all Scotland) univs., Durham, Birmingham, Sheffield, Liverpool (all Eng.) univs.; m. Alice Gabrielle Pignatel, 1871; 1 son, 3 daus. With Geol. Survey, from 1855; dir. Geol. Survey Scotland, from 1867; 1st Murchison prof. geology and mineralogy Edinburgh U., 1871-82; dir.-gen. Geol. Survey U.K., dir. Mus. Practical Geology, 1882-1901. Trustee Brit. Mus.; mem. Royal Commn. for Exhbn. of 1851, Council Brit. Sch. at Rome, Italy; gov. Harrow Sch., 1892-1922; chmn. Royal Commn. on Trinity Coll., Dublin, 1920. Recipient Livingston Gold medal Royal Geog. Soc. Scotland, Gold medal Inst. Mining and Metallurgy, Macdougal-Brisbane medal (twice) Royal Soc. Edinburgh; Order of Merit, 1914. Fellow Royal Soc., 1865 (fgn. sec. 1890-94, sec. 1903-08, pres. 1908-13, Royal medal), 1865; mem. Geol. Soc. (pres. 1891-92, 1906-08, Wollaston medal, Murchison medal), Brit. Assn. (pres. 1892), Lincei, French Acad. Scis. (corr. 1891, fgn. asso. 1917), acads. Berlin, Vienna, Petrograd, Belgium, Stockholm, Turin, Naples, Munich, Christiania, Göttingen, others, Phila. Acad. Natural Sci. (Hayden Gold medal), Nat. Acad. Sci. U. S., others. Author: The Story of a Boulder, 1858; Chronology of the Trap Rocks of Scotland, 1861; (with Murchison) Geological Map of Scotland, 1862; Glacial Drift of Scotland, 1863; The Scenery of Scotland viewed in connection with its Physical Geology, 1865; Basaltic Plateaux of Ireland, West of Scotland and Iceland, 1867; Old Red Sandstone of Western Europe, 1878; Carboniferous Volcanic Rocks of the Firth of Forth, 1879; Geological Sketches at Home and Abroad, 1882; Text-book of Geology, 1882; History of Volcanic Action during the Tertiary Period in Britain, 1888; Field Geology, 5th edit., 1900; New Geological Map of Scotland, with descriptive notes, 1892; The Ancient Volcanoes of Britain, 2 vols., 1897; The Founders of Geology, 1897; Geological Map of England and Wales, with descriptive notes, 1897; Types of Scenery and their Influence on Literature, 1898; The Geology of Eastern Fife, 1902; Scottish Reminiscences, 1904; Landscape in History, 1905; Charles Darwin as Geologist, 1909. Research, publs. on past volcanic history of Gt. Britain, glaciers, stratigraphical geology; encouraged micros. petrography and volcanic geology; many original contbns. to sci. Died Halsemere, Surrey, England, Nov. 10, 1924.

GEIKIE, James, Scottish geologist; b. Edinburgh, Aug. 23, 1839; ed. Edinburgh U.; LL.D., D.C.L.; m. Mary Simson Johnston; 4 sons, 1 dau. Entered H.M. Geol. Survey, 1861, became dist. surveyor, 1869; Murchison prof. geology Edinburgh U., 1882-1915; also dean faculty sci. Fellow Royal Soc., 1875, Royal Soc. Edinburgh (pres., Brisbane medal), Geol. Soc. (Murchison medal, London), Royal Scottish Geog. Soc. (a founder, pres., Gold medal). Author: The Great Ice Age (contains 1st suggestion of multiple glaciation), 1874; Prehistoric Europe, 1881; Outlines of Geology, 1884; Fragments of Earth Lore, 1892; Songs and Lyrics by Heinrich Heine . . . , 1887; Earth Sculpture, 1898; Structural and Field Geology, 1904; Mountains, their Origin, Growth, and Decay, 1913; The Antiquity of Man in Europe, 1913. Specialist in glacial geology. Died Edinburgh, Mar. 1, 1915.

GEIL, William Edgar, Am. explorer; b. nr. Doylestown, Pa.; s. Samuel and Elizabeth (Seese) G.; student Lafayette Coll.; several hon. degrees; spent 6 months making archaeol. studies in Western Asia, 1896; m. L. Constance Emerson. Began journey for comparative study of primitive races and ind. observation of missions of world, 1901; crossed China and Africa; went farther into pigmy forest than Stanley, in journey of 4 years; explored Gt. Wall of China; visited all 19 capitals of China, traveled 120,000 miles; explored Wu Yo, or 5 sacred mountains of China, 1919; lectr. in Australia, Japan, China, India, Gt. Britain, U. S. Life fellow Royal Geog. Soc., Royal Astron. Soc. Author: A Yankee on the Yangtze, 1904; The Man of Galilee, 1904, 1906; A Yankee in Pigmyland, 1905; The Men on the Mount, 1905; The Automatic Calf, 1905; The Great Wall of China, 1909-1911; Eighteen Capitals of China, 1911; Adventures in the African Jungle Hunting Pigmies, 1917. Died Apr. 1925.

GEINITZ, Hanns Bruno, German geologist, paleontologist; b. Altenburg, Germany, Oct. 16, 1814; s. Christian Traugott and Johanna Friederike (Klötzner) G.; studied chemistry, then mineralogy, geology, Berlin; doctorate, Jena, Germany, 1837; m. Luise Pusch, 1843; m. 2d, Margareta Will, 1846; 3 sons, including Eugen, 3 daus. Faculty Tech. Inst. (later Tech. U.), Dresden, Germany, 1838-94, named prof. geology and mineralogy, 1850; founded Mineral. Mus., Dresden, and became dir., 1857. Mem. Isis Natural Sci. Soc., German Geol. Soc., Leopoldina. Author: Charakteristik der Schichten und Petrefakten der sächsisch-böhmischen Kreidegebirges, 1839; Die Versteinerungen der deutschen Zechsteingebirges, 1848; Die Versteinerungen der Grauwackenformation in Sachsen, 1852; Darstellung der Flora des Hainichen-Ebersdorfer und des Flöha'er Kohlenbassins, 1854; Die Versteinerungen der Steinkohlenformation in Sachsen, 1855; Geognostische Darstellung der Steinkohlenformation in Sachsen, 1856; Die Leitpflanzen des Rothliegenden und des Zechsteingebirges in Sachsen, 1858; Das Elbthalgebirge in Sachsen, 2 vols., 1871-75. Research in geology and paleontology, especially phytopaleontology, coal formation, Saxon-Bohemian chalk mountains; an advisor in discovery of coal in Saxony. Died Dresden, Jan. 28, 1900.

GEISLER, Gerhard Paul Hermann Wilhelm, German plant physiologist; b. Berlin, Germany, Aug. 30, 1927; s. Wilhelm and Charlotte (Puzicha), G.; Diplom-Landwirt, U. Berlin, 1950; Dr.agr., Technische Hochschule Munich, Germany, 1953; m. Barbara Moeller, Jan. 12, 1950; children—Juliane, Ulrike, Michael Wolfgang. Plant breeder, sect. leader root-stock breeding Fed. Inst. Grape Vine Breeding, 1950-59; sr. research officer CSIRO, Australia, 1960-64; univ. dozent U. Hohenheim, Stuttgart, Germany, 1964—. Mem. Am. Soc. Plant Physiologists, Deutsche Botanische Gesellschaft, Gesellschaft für angewandte Botanik, Pflanzenbaugesellschaft. Research, publs. on function of plant root system in supplying plant with water and nutrients, morphogenetic effects of soil factors on roots. Home: 68 Karlfriedrich-Rumpp, Nürtingen, Germany. Office: 23 Fruwirth & 7 Stuttgart-Hohenheim, Germany.*

GEISSER, Seymour, Am. math. statistician; b. Bronx, N.Y., Oct. 5, 1929; s. Leon and Rose (Kielmanowicz) G.; B.A., Coll. City N.Y., 1950; M.A., U. N.C., 1952, Ph.D., 1955; m. Mary Lee George, Jan. 30, 1955; children—Mindy Sharon, Dan Levi, Georgia Lynn, Adam Dov. With NIH, Bethesda, Md., 1955-65, chief biometry sect. Nat. Inst. Arthritis and Metabolic Diseases, 1961-65; prof., chmn. dept. math. statistics, State U. N.Y., at Buffalo, 1965—. Fellow Am. Statis. Assn. (past dir.), Inst. Math. Statistics; mem. Biometic Soc. (mem. regional com. 1964—), Royal Statis. Soc., Math. Assn. Am., London Math. Soc. Research, publs. on theory and methodology of multivariate analysis and statis. inference. Home: 117 Meadowview Lane, Williamsvile, N.Y. 14221. Office: Dept. Math. Statistics, State U. N.Y., Buffalo 14214.*

GEISSLER, Ernst Dietrich, Am. aero. engr.; b. Chemnitz, Germany, Aug. 4, 1915; s. Johann and Elsa (Heckel) G.; B.S. in Physics and Math., Tech. U. Dresden (Germany), 1936, M.S. in Tech. Physics, 1939; D. in Applied Mat., U. Darmstadt (Germany), 1951; m. Gerda Stricker, Aug. 16, 1941; children—Barbara Ogden (Mrs. Stokes), Katharina. Came to U. S., 1945, naturalized, 1954. Physicist for V2 rocket German govt., Peenemünde, Germany, 1940-43, sect. chief Wasserfall Anti-aircraft missile, 1943-45; group leader research and devel. office U. S. Army, Ft. Bliss, Tex., 1945-50; dir. aeroballistics lab. Army Ballistic Missile Agy., Huntsville, Ala., 1950-60; dir. aero-astrodynamics lab. Marshall Space Flight Center, NASA, Huntsville, 1960—, mem. research adv. com. on missile and space vehicle aerodynamics, 1959—. Asst. prof. Inst. for Physics Tech., U. Dresden, 1939. Recipient Exception Civilian Service decoration U. S. Army, 1959; Exceptional Sci. Achievement award NASA, 1963. Asso. fellow Am. Inst. Aeros. and astronautics. Research, publs. on rocket stability and control theory, also homing methods. Patentee in field. Home: 3604 Mae Dr. S.E., Huntsville 35801. Office: R-AERO-Dir., Bldg. 4200, Marshall Space Flight Center, Huntsville, Ala. 35812.*

GEISSLER, Heinrich,, German inventor, physicist; b. Igelshieb, Saxe-Meiningen, Germany, May 26, 1814; ed. as glassblower; Ph.D. (hon.), U. Bonn (Germany), 1868; founded shop for sci. instruments, Bonn, 1854; (with Plücker) discovered maximum density of water; invented mercury pump for producing a higher vacuum, 1858; invented glass tube with a rarefied gas and 2 electrodes (Geissler tube) and demonstrated the phenomena of incadescence with electric current; proved the color depended on the gas used (neon signs a later form of Geissler tube); (with Vogelsang) demonstrated the presence of liquid carbon dioxide in cavities in quartz and topaz, 1869; invented vaporimeter, thermometer and aerometer. Died Bonn, Germany, Jan. 24, 1879.

GEIST, Lorenz Melchior, German physician; b. Nuremberg, Germany, Jan. 20, 1807; s. Johann Peter and Barbara (Hermann) G.; studied 1st philosophy, math., philology, history, then medicine, Erlangen, Germany; M.D., Munich, Germany, 1830; state exam, 1832; m. Margaretha Baumann, 1836; 1 dau. Practice medicine, Nuremberg, 1833-67; chief surgeon Nuremberg Hosp., until 1844; then became resident Holy Spirit Hosp. and Home for Aged; named head med. dept. Municipal Hosp. Recipient Monthyon prize Paris Acad. Scis. Mem. Nuremberg Local Physicians Assn. (became founding chmn. 1852). Author: De luxatione processus odontoidei, 1830; (with E. v. Bibra) Über die Krankheiten der Arbeiter in den Phosphorzündholzfabriken, 1847; Über die Regeneration des Unterkiefers nach totaler Nekrose durch Phosphordämpfe, 1852. Pioneered in indsl. diseases and geriatrics; proved (with E. v. Bibra) phosphor vapors cause necrosis of lower jaw in match factory workers and that fumes entered through tooth cavities; wrote 1st German work on geriatrics. Died Nuremberg, Oct. 20, 1867.

GEISTBECK, Alois, German geographer; b. Friedberg nr. Augsburg, Germany, Sept. 26, 1853; s. Michael and Therese (Boniberger) G.; ed. Tchrs. Coll., Freising; began studies Tech. U. Munich, 1876; teaching certificate, 1879; Ph.D., Erlangen, Germany, 1885; Pauline Reingruber Kühn, 1906; 2 step-daus. Organizer, adminstr. elementary sch. system, Ludwigshafen/Rhein, 1885-92; 1st prin., then supt. upper schs., Augsburg, 1892-99, Munich, also Neuburg/Donau, 1900-06, Kitzingen, Germany, 1906-19. Author: Die Seen der deutschen Alpen, 1885. Developed new methods of geography edn.; research and numerous publs. on Bavarian and Palatine geography, especially Bavarian lakes. Died Kitzingen, Nov. 19, 1925.

GEISTBECK, Michael, German geographer; g. Friedberg nr. Augsburg, Germany, Mar. 1, 1846; s. Alois and Therese (Boniberger) G.; studied German, history and geography; Ph.D., circa 1877; m. Anna Probst, 1875; m. 2d, Anna Lachenmaier, 1902; 4 sons, 3 daus. Engaged in tchr. edn.; supt., sem. dir., Freising, Germany. Author: Geographie für Volksschulen, 1877; Geographie für höhere Lehranstalten; Geschichte der Methodik der geographischen Unterrichts,

1877; Der Weltverkehr, 1886; (with H. Fischer) Erdkunde für höhere Lehranstalten, 1907. Advanced geography edn. in schs. Died Freising, Germany, Mar. 30, 1918.

GEITEL, Hans Friedrich, German physicist; b. Braunschweig, Germany, July 16, 1855; s. Karl Friedrich Wilhelm Theodor and Meta (Pauli) G.; studied under Bunsen, Heidelberg, Germany, Quincke, Berlin; secondary sch. teaching certificate, 1879; hon. Ph.D., Göttingen, 1899; hon. dr. engring., Braunschweig, Germany, 1915; m. Marie Scholz, 1922. Became tchr. Grosse Schule, Wolfenbüttel, Germany, 1879, sr. tchr., 1889, prof., 1896; named hon. prof. Tech. U. Braunschweig. Research and publs. (with J. Elster) on ionization in atmosphere; built 1st cathode tube; discovered selective photoelectric effect, free energy left in atom after transformation of radioactive elements; built photometer; invented photocell; formulated law of radioactive fallout; originated concept of atomic energy. Died Wolfenbüttel, Aug. 15, 1923.

GEITLER, Lothar, Austrian botanist; b. Vienna, Austria, May 18, 1899; s. Rudolf and Betty (Fürst) G.; Dr.phil., U. Vienna, 1922; m. Grete Drahowzal, July 27, 1936. Faculty, Bot. Inst., U. Vienna, 1921—, prof., 1937—, head Bot. Inst., dir. Bot. Garden, 1948—; asst. Kaiser-Wilhelm-Inst. for Biology, Berlin, Dahlem, Germany, 1929. Recipient Erzherzog-Rainer medal Zool.-Bot. Soc. Vienna, 1935. Corr. mem. Austrian Acad. Sci. Vienna, Acad. Sci. Lit. Mainz; fgn. mem. Linnean Soc. London, Bot. Soc. Am. Author: Der Formwechsel der pennaten Diatomeen, 1932; Grundriss der Cytologie, 1934; Chromosomenbau, 1938; Schnellmethoden der Kern- und Chromosomenuntersuchung, 1949; Morphologie der Pflanzen, 1953; Endomitose und endomitotische Polyploidisierung, 1953; also numerous articles. Editor: Österr. Bot. Zeitschr., 1946—. Research on life microorganisms, problems of cells and chromosomes of plants and animals, propagation of species, symbiosis, taxonomy; discovered endomitosis in animal and plant cells. Office: Botan. Institut, III. Rennweg 14, 1030 Vienna, Austria.*

GEITNER, Ernst August, German chemist; b. Gera, Germany, June 12, 1783; s. Johann Gottfried and Johanna Friederike Sophia (Dreher) G.; studied 1st theology, then medicine and natural scis., Leipzig, Germany; M.D., 1809; m. Charlotte Oppe, 1810; children—Ernst Hermann, Alfred, Gustav, Agnes (Mrs. Lange). Began practice medicine, Lössnitz, 1809; founder factory for chem. products, especially dyes, Lössnitz, 1810, moved factory to Schneeberg, Germany, after Napoleonic wars; founder argentan factory, Auerhammer, 1829; a founder AG Mechanische Weberei Auerhammer, 1837; built tropical gardens at Planitz; botanist, horticulturist. Author: Familie West, oder Unterhaltungen über die wichtigsten Gegenstände der Chemie und Technologie, 2 vols., 1806; Versuche über das Balufärben wollener Zeuge ohne Indigo, 1809. Developed yellow dye for cotton, linen, and wool from lead acetate and potassium chromate, 1819, green copper dye; developed tech. manufacture of Chinese packfong, an alloy of copper-nickel-zinc (argentan, new sliver, alpaka) which is used for flatware. Died Schneeberg, Oct. 24, 1852.

GELB, Adhémar Maximilian Maurice, psychologist; b. Moscow, Russia, Nov. 18, 1887; s. Maximilian and Wilhelmine Christine Ludovica (Stahl) G.; began study philosophy, Munich, Germany, 1906; Ph.D., Berlin, 1910; m. Nelly Achenbach, 1912; 1 son. Asst., Psychol. Inst., U. Berlin, 1909-12, Psychol. Inst., Acad. for Social Scis. (became part of U. Frankfort, 1914), Frankfort/Main, Germany, 1912-14; joined Frankfort Hosp. for Brain Damaged, 1915; joined faculty, Frankfort, 1919, named asso. prof., 1924, dir. (with M. Wertheimer), Psychol. Inst., prof. philosophy and psychology, 1929; became prof. philosophy U. Halle (Germany), 1931, also dir. Psychol. seminar, until 1933; gave occasional guest lectures until 1936, especially at U. Lund (Sweden), 1935. Research and numerous publs. on analysis of disturbances in sense perception, speech and recognition processes after brain damage, disturbances in color perception, path. spacial perception, aphasia, agnosia; discovered Gelb phenomenon (space perception depending on time experienced), 1914. Died Schömberg, Germany, Aug. 7, 1936.

GELBART, Abe, Am. mathematician; b. Paterson, N.J., Dec. 22, 1913; s. Wolf and Pauline (Landau) G.; B.S., Dalhousie U., Halifax, N.S., 1938; Ph.D., Mass. Inst. Tech., 1940; m. Sara Goodman, July 2, 1939; children—Carol Marie, Judith Sylvia, William Michael, Steven Samuel. Research asso. Brown U., Providence, 1941-42; asst. prof. math. Syracuse (N.Y.) U., 1943-47, prof., 1956-58; mem. Inst. for Advanced Study, Princeton, N.J., 1947-48; Fulbright lectr. Norway, 1951-52; dean Belfer Grad. Sch. Sci., Yeshiva U., N.Y.C., 1958—. Mem. N.Y.C. Mayor's Com on Scholastic Achievement, 1962—. Mem. Am. Math. Soc., Math. Assn. Am. Editor, Scripta Mathematica, 1957—. Cofounder (with Lipman Bers) of theory of pseudo-analytic functions. Studied existence theorems in integral equations; methods of generalizing complex function theory; functions of several complex variables; non-linear partial differential equations; fluid dynamics. Home: 140 W. End Av., N.Y.C. 23.

GELBAUM, Bernard Russell, Am. mathematician; b. N.Y.C., Feb. 26, 1922; s. Harry and Regina (Kratka) G.; A.B., Columbia, 1943; M.A., Princeton, 1947, Ph.D. (NRC Predoctoral fellow), 1948; m. Beatrice Lerner, Nov. 14, 1942; children—Daniel, David, Martin, Ethan. Instr., Princeton, 1947; faculty U. Minn., Mpls., 1948-64, prof., 1957-64; prof., chmn. dept. math. U. Cal., Irvine, 1964—; vis. mem. Inst. for Advanced Study, Princeton, N.J., 1960. Cons. Inst. for Def. Analyses, Washington, 1962—; cons. editor W. B. Saunders Co., Phila., 1964—. Pulitzer scholar, 1939; NSF research grantee, 1959—. Mem. Am. Math. Soc., Am. Assn. U. Prof., Am. Civil Liberties Union, Phi Beta Kappa, Sigma Xi. Author: (with J. M. H. Olmsted) Counterexamples in Analysis, 1964. Contbns. to functional analysis, study of topological groups, theory of games. Home: 2301 Arbutus St., Newport Beach, Cal. 92660.*

GELDARD, Frank Arthur, Am. psychologist; b. Worcester, Mass., May 20, 1904; s. Arthur and Margaret Hardy (Gordon) G.; A.B., Clark U., 1925, M.A., 1926, Ph.D., 1928; m. Jeannette Manchester, June 20, 1928; children—Deborah Rea (Mrs. Wallace Emmett Tobin, III). Faculty, U. Va., Charlottesville, 1928-62, prof. psychology, 1937-62, dean Grad. Sch. Arts and Scis., 1960-62; Stuart prof. psychology Princeton, 1962—. Research chief, human resources div. Hdqrs. USAF, 1949-50; sci. liaison officer Office Naval Research, London, 1956-57; cons. to mil., NIH; mem. div. commn. on biol. and med. scis. NSF, 1953-58, chmn. NATO adv. group on human factors, 1959-65. Decorated Legion of Merit. Recipient Pres.'s and Visitors' Research prize U. Va., 1958. Fellow A.A.A.S. (v.p. 1958-59), Am. Psychol. Assn., Assn. Exptl. and Mil. Psychology (pres. divs.), Royal Soc. Medicine; mem. Soc. Exptl. Psychology, Optical Soc. Am., Sigma Xi, Phi Beta Kappa, Phi Sigma, Chi Psi. Author: The Human Senses, 1953; Fundamentals of Psychology, 1962; also numerous articles. Editor: Defence Psychology, 1962; Communication Processes, 1965. Research on visual processes with spl. reference to phenomena of flicker and color blindness, skin sensitivity and utilization of cutaneous channels of communication, sleep motility in aviation trainees. Home: 551 Lake Dr., Princeton, N.J. 08540.*

GELFAN, Samuel, physiologist; b. Russia, Jan. 16, 1903; s. Khaim and Dora (Taradasch) G.; came to U.S., 1913, naturalized, 1927; A.B., U. Cal., Berkeley, 1925, Ph.D., 1927; m. Harriet Lucy Moore, Nov. 26, 1943; children—Caroline Moore, Peter Barrett, Deborah Elizabeth, Janet Isabel, Michael Emery, Stephanie Anna. Asst. prof. physiology Sch. Medicine Yale, 1946-51, dir. aeromed. research unit, 1947-51; research asso. neurology N.Y. Med. Coll., 1952-60, prof. neurophysiology, 1960—. Recipient award for participation in work of Div. War Research Columbia, 1945. Fellow A.A.A.S.; mem. Am. Physiol. Soc., Assn. Research Nervous and Mental Diseases, Soc. Exptl. Biology and Medicine, Harvey Soc., N.Y. Acad. Scis. Collaborating editor Textbook of Physiology, 1950. Research, numerous publs. on micro-electrodes for penetration and study of electrical properties of single living cells, localized excitation of single skeletal muscle fibers in situ; aviation physiology; functional orgn. of spinal cord; neurophysiol. alterations in exptl. rigidity. Home: 6 Murray Hill Rd., Scarsdale, N.Y. 10583. Office: Fifth Av. at 106th St., N.Y.C. 10029.*

GELFAND, Henry Morris, Am. epidemiologist; b. N.Y.C., Jan. 7, 1920; s. Michael and Wilma (Schopper) G.; B.S., Cornell U., 1940; M.D., U. Chgo., 1950; M.P.H., Tulane U., 1956; m. Jane Louise Banker, Aug. 24, 1946; children—Richard, Alaric, Christopher, Wendy. Med. entomologist Liberian Inst. Tropical Medicine, 1951-53; asso. prof. epidemiology Tulane U., New Orleans, 1953-59; chief enterovirus unit Communicable Disease Center, Atlanta, 1959-63, chief West African operations Smallpox Eradication Program, 1965—; cons. Nat. Inst. Communicable Diseases, Delhi, India, 1963-65. Mem. Am. Pub. Health Assn., Am. Soc. Tropical Medicine and Hygiene, Am. Assn. Immunologists, Am. Epidemiological Soc., Soc. Exptl. Biology and Medicine, A.A.A.S. Research, publs. on longitudinal long-term studies of epidemiology of poliomyelitis, other enterovirus infections; transmission of live attenuated poliovirus vaccine strains, epidemiology and eradication of smallpox. Home: 3006 Silvapine Trail, Atlanta 30329. Office: Communicable Disease Center, Atlanta 30333.*

GELFAND, Izrail Moiseevich, Russian mathematician; b. Krasnye Okny, Odessa Oblast, Aug. 20, 1913; D.Physico-Math. Sci., 1940. Instr., Moscow U., 1932-43, prof., 1943—; asso. Math. Inst., USSR Acad. Sci., 1939—. Recipient Stalin prize, 1951. Mem. USSR Acad. Sci. (corr.). Author over 80 works including Lectures on Linear Algebra, 1948; (with M.A. Neimark) Unitary Representations of Classic Groups, 1950; Some Functional Analysis Problems, 1956; The Integration in Functional Spaces and Its Uses in Quantum Physics, 1956. Research on theory of unitary infinite-dimensional representations of continuous series, theory of normalized rings (linear normalized spaces whose elements can be multiplied in usual manner), generalized functions and their application in differential equations. Address: Math. Inst., USSR Acad. Sci., Leninsky Prospect 19, Moscow, USSR.

GELFAND, Michael, internist; b. Wynberg Cape, Dec. 26, 1912; s. Louis and Ethel (Salkow) G.; M.B.Ch.B., U. Cape Town, 1936, M.D., 1948; D.M.R., (Eng.) 1939, D.Ph. (London) 1950; m. Esther Kollenberg, Jan. 26, 1937; children—Joy Phyllis (Mrs. Thomas Gould Phillips), Isabelle (Mrs. Simon Wapnick), Anne. Specialist physician Rhodesia Govt. and fed. govt. service, 1939-63; prof. medicine with spl. reference to Africa, U. Coll. Rhodesia, Salisbury, 1963—. Decorated officer, comdr. Brit. Empire. Fellow Royal Coll. Physicians London. Author: Sick African, 1943, 47, 57; Medicine in Tropical Africa, 1950; Schistosomiasis, 1950; Medicine and Magic of the Hashona, 1953; Tropical Victory, 1953; Lakeside Pioneers, 1964; others. Editor: Central African Jour. Medicine, 1955—. Research, publs. on clin. parasitology, med. anthropology and history. Home: 11 Lawson Av., Salisbury 5. Office: Med. Faculty, U. Rhodesia, 1/Bay 167, Salisbury, Rhodesia.*

GELFOND, Aleksandr Osipovich, Russian mathematician; b. St. Petersburg, Oct. 24, 1906; grad. Moscow U., 1927; D.Physico-Math. Sci., 1935. Prof., Moscow U., 1931—; asso. Math. Inst., USSR Acad. Sci., 1933—. Mem. USSR Acad. Sci. (corr.). Author: Transcendental Numbers, 1934; Transcendental and Algebraic Numbers, 1952; Elementary Methods in the Analytical Theory of Numbers, 1962. Research and publs. on theory of numbers, theory of functions of complex variable; established basic methods of studying transcendental numbers. Address: Moscow University, Leninskie gory, Moscow, USSR.*

GELL, Charles Fredric, Am. cons. physiologist; b. Chgo., June 16, 1907; s. Herman August and Bertha Anne (Meier) G.; B.S., Ill. Inst. Tech., 1932; B.S., Loyola U., Chgo., 1934, M.D., 1936; M.S., U. Pa., 1953, D.Sc., 1956; m. Edna Anne Leddin, June 30, 1937; children—Mada-Anne, Carl L., Michael F., Shaun-Marie. Commd. lt. M.C., USN, 1937, advanced through grades to capt., 1953; dir. USN Acceleration Lab., Johnsville, Pa., 1950-55; dir. air crew equipment lab., Phila., 1955-58, spl. asst. to chief naval research, Washington, 1958-60; ret., 1960; vis. prof. physiology U. Pa., 1955-60; dir. life scis. LTV Aerospace Corp., Astronautics Div., Dallas, 1960-66; clin. prof. physiology Southwestern Med. Sch., U. Tex., Dallas, 1960—. Cons. Dept. Def.; exec. council com. on hearing, bioacoustics and biomechanics, NRC, 1962—; chmn. biodynamics com. Aerospace Med. Panel, Adv. Group for Aerospace Research and Devel., NATO, 1959—. Recipient citations Sec. Navy, 1946, Surgeon Gen. USN, 1960; John Jeffries award Inst. Aerospace Scis., 1953; Lyster award Aerospace Med. Assn., 1957; Boynton award Am. Astronautical Soc., 1958. Fellow A.A.A.S., Aerospace Med. Assn., Am. Coll. Preventive Medicine; mem. Internat. Acad. Aerospace Medicine, A.M.A., Am. Physiol. Soc., Am. Bd. Preventive Medicine, Naval Inst., Nat. Geog. Soc. Contbg. author: Harry G. Armstrong's Textbook of Aerospace Medicine, 1962. Research, publs. in aerospace medicine involving effects on man of acceleration, extreme altitudes, explosive decompression, Aviator's Bends, long-term confinement, effect of micrometeoroidpenetration of space cabins, devel. of space suits, simulation of space environments, toxicology problems in aerospace devel., safety and survival gear for astronauts. Home: 4431 Hockaday Dr., Dallas 75229.*

GELLERT, Christlieb Ehregott, German mineralogist, chemist, metallurgist; b. Hainichen, Germany, Aug. 11, 1713; s. Christian and Johanna Salome (Schütz) G.; student U. Leipzig (Germany), 1734-36, chemistry, physics, under Euler, St. Petersburg, Russia, 1736-47, metallurgy, Freiberg, Germany. Secondary sch. tchr., St. Petersburg, 1736-37; asst. St. Petersburg Acad. Scis., 1737-57; assessor F. Foundry Works, Freiberg; founder pvt. inst. for metall. chemistry; named insp. mines and smelting, Freiberg, 1753; became chief foundry adminstr., dir. all Freiberg foundries, 1762; named prof. metall. chemistry Mining Acad., Freiburg, 1766; Mem. Acad. Scis. St. Petersburg, Soc. Mining, Schemnitz (Hungary), Econ. Soc. Leipzig (hon.). Author: Anfangsgründe der Probirkunst, 1755; Anfangsgründe der metallurgischen Chimie, 1750; also articles. Translator: Elementa artis docimasticae (J. A. Cramer), 1739. Improved mining machines and blast furnaces; believed all rocks and earths to be smeltable; discovered melting temperatures of mixtures are lower than those of single components; invented amalgamation process of extracting precious metals for ores which were poor in lead and copper sulfide (cold extraction process); discovered height to which water rose in a capillary tube was inversely proportional to square root of area of section. Died Freiberg, May 18, 1795.

GELLHORN, Ernst, neurophysiobiologist; b. Breslau, Germany, Jan. 7, 1893; s. Moritz and Hulda (Stein) G.; M.D., U. Heidelberg, (Germany), 1919; Ph.D., U. Muenster (Germany), 1919; m. Hilde Obermeier, Aug. 1, 1925; children—Irene (dec.), Helen (Mrs. Sven Hartmann), Ernest, Joyce (Mrs. Ralph Greene). Came to U. S., 1929, naturalized, 1935. Faculty, U. Ore., 1929-32, U. Ill. Med. Sch., 1932-43; prof. neurophysiology U. Minn., Mpls., 1943-60, prof. emeritus, 1960—. Recipient A. Cressy Morrison prize N.Y. Acad. Scis., 1930; Alvarenga prize Coll. Physicians Phila., 1934; Outstanding Achievement in Med. Research medal Carbon Dioxide Research Assn., 1957. Mem. Am. Physiol. Soc., A.A.A.S., Electroencephalograph Soc., Pavlovian Soc. (hon.), Epilepsy Soc. (hon.). Author: Autonomic Regulations, 1943; Physiological Foundations of Neurology and Psychiatry, 1953; Autonomic Imbalance and the Hypothalamus, 1957; Emotions and Emotional Disorders (with G. Loofbourrow), 1957; Autonomic-Somatic Integrations, 1967; also numerous articles. Research on physiology hypothalamus, functional basis of emotions and emotional disorders, integrative functions organism. Home: 2 Fellowship Circle, Santa Barbara, Cal. 93105.*

GELLHORN, George, gynecologist; b. Breslau, Germany (now Wroclaw, Poland), Nov. 7, 1870; s. Adolph and Rosalie (Pincus) G.; ed. Gymnasium, Ohlau, Germany, 1876-90; M.D., U. Würzburg, 1894; m. Edna Fischel, Oct. 21, 1903; children—George, Walter F., Martha E., Alfred A. Came to U. S., 1899. Asst. in clinics at univs. Berlin, Jena and Vienna, 1895-99; practiced medicine, St. Louis, from 1900; prof. gynecology and obstetrics, dir. dept. St. Louis U. Sch. Medicine, 1922-32; prof. clin. obstetrics and gynecology, Washington U. Sch. Medicine, from 1932; gynecologist Barnard Free Skin and Cancer Hosp.; gynecologist and obstetrician St. Luke's, City hosps., asso. gynecologist and obstetrician Barnes, St. Louis Maternity hosps.; cons. gynecologist and obstetrician Jewish, St. Louis County hosps. Designed single-stem plunger type pessary for uterine prolapse, 1908. Died Jan. 25, 1936.

GELLIBRAND, Henry, English astronomer, mathematician; b. London, Nov. 17, 1597; s. Henry Gellibrand; B.A., Oxford (Eng.) U., 1619, M.A., 1623; took holy orders and served at Chiddingstone, Kent, Eng.; became prof. astronomy at Gresham Coll., 1627; author: A Discourse Mathematical of the Variation of the Magneticall Needle, . . . , 1635; Epitome of Navigation. Discovered the earth's magnetic field changes with time. Died London, Feb. 16, 1636.

GELLIS, Sydney Saul, Am. physician; b. Claremont, N.H., Mar. 6, 1914; s. Morris A. and Minna (Bernstein) G.; A.B. magna cum laude, Harvard, 1934, M.D., 1938; m. Matilda Lichter, Mar. 7, 1939; children—Beth Louise, Stephen. Instr. pediatrics Johns Hopkins Sch. Medicine, Balt., 1941-46; asst. prof. pediatrics Harvard Med. Sch., Boston, 1946-57; prof., chmn. dept. pediatrics Boston U. Sch. Medicine, 1957-65, acting dean Sch. Medicine, 1963-65; dir. pediatrics Boston City Hosp., 1957—; prof., chmn. dept. pediatrics Tufts U. Sch. Medicine, Boston, 1965—, pediatrician-in-chief Tufts-New Eng. Med. Center, Boston, 1965—. Mem. Am., (French (corr.), pediatric socs., Soc. Pediatric Research (pres.), Am. Acad. Pediatrics. Author: Year Book of Pediatrics, 1952—; (with Benjamin Kagan) Current Pediatric Therapy, 1964; also numerous articles. Research in liver disease of children, uses of Gamma Globulin in prevention of infectious hepatitis. Home: 77 Alderwood Rd., Newton, Mass. 02159. Office: 20 Ash St., Boston 02111.*

GELL-MANN, Murray, Am. physicist; b. N.Y.C., Sept. 15, 1929; s. Arthur and Pauline (Reichstein) G.-M.; B.S., Yale, 1948, Sc.D., 1959; Ph.D., Mass. Inst. Tech., 1951; m. J. Margaret Dow, Apr. 19, 1955; children—Elizabeth, Nicholas. Mem. Inst. for Advanced Study, Princeton, 1951; faculty physics U. Chgo., 1952-55; asso. prof. theoretical physics Cal. Inst. Tech., Pasadena, 1955-56, prof., 1956—. Vis. prof. Collège de France, Paris, 1959-60; mem. Jason div. Inst. for Def. Analyses, 1961—. Mem. Am. Phys. Soc. (Dannie Heineman prize 1959), Nat. Acad. Scis., Am. Acad. Arts and Scis. Author: (with Y. Ne'eman) The Eightfold Way, 1964. Research on strangeness, weak interactions, strong interaction symmetries, dispersion relations. Home: 3637 Canyon Crest, Altadena, Cal. Office: Cal. Inst. Tech., Pasadena, Cal.*

GEMANT, Andrew, phys. chemist; b. Nagyvarad, Hungary, July 27, 1895; s. Eugene and Vilma Berkow) Gyemant; M.D.; U. Budapest, 1919; Ph.D., U. Berlin, Germany, 1922; m. Sophie Ida Staap, Dec. 28, 1933. Came to U. S. 1938, naturalized 1944. Physicist, Siemens-Schuckert, Berlin, 1925-32; privatdocent Tech. U., Berlin, 1928-33; research fellow Oxford U., Eng., 1934-37; research asso. U. Wis., 1938-39; staff physicist Detroit Edison Co., 1940-60; research asso. Grace Hosp., Detroit, 1961—. Mem. conf. elec. insulation NRC. Fellow Am. Phys. Soc., A.A.A.S.; mem. Electrochem. Soc., German Phys. Soc., Sigma Xi. Author: Colloid Physics, 1925; Electrophysics of Insulating Materials, 1930; Liquid Dielectrics, 1933; Frictional Phenomena, 1950; Nature of the Genius, 1961; Ions in Hydrocarbons, 1962; also numerous papers phys. and chem. jours. U. S., Eng., Germany, Holland, France, Japan. Research in colloid physics, dielectrics, electrets, internal friction in solids, electrochemistry of hydrocarbons, cholesteral, carcinogenesis. Home: 4501 W. Outer Dr., Detroit 48235. Office: 4160 John R, Detroit 48201.*

GEMELLI, Edoardo Agostino, Italian psychologist; b. Milan, Italy, Jan. 18, 1878; M.D., U. Pavia; degrees in histology and philosophy U. Louvain. Became physician in Italian army; then converted to Catholicism, joined Franciscan order, ordained, 1908; served as priest and physician Italian army, during World War 1; founder Sacred Heart U., Milan, 1921, pres. until 1959. Mem. Pontifical Acad. Scis. (pres). Author: L'enimiga della vita e i nzovi orizzonti della biologia, 1909; Sui Rapporti tra scienza e philosophia, 1911; L'origine subcosciente dei Fatti mistici, 1912; Metode, compiti e liniti della psicologia nello studio e nella prevenzione della delinquenza; Antropologia e psicologia, 1940; La psicotecnica applicata all'industria, 1944; Psychologie de l'enfant à l'homme, 1950; also founder Vita e Pensiero monthly, 1914. Eminent for studies in exptl. psychology, especially criminology. Died July 15, 1959.

GEMINOS OF RHODES, Greek astronomer, mathematician; lived after Posidonios but before Alexander of Aphrodisios; flourished 70 B.C.; a stoic; influenced mainly by Posidonios. Author: Introduction to Phenomena, elementary work on astronomy (main source of Proclos' commentary on 1st book of Euclid); On the Arrangement (or Theory) of Mathematics, 6 books. Explained ancient astronomy, chiefly from Hipparchos' point of view; stressed fundamental notions; classified math. scis., such as arithmetic, geometry, astronomy, optics; geodesy, mechanics; musical harmony; and practical calculation; investigated fundamental principles, axioms, definitions, and postulates; presented proofs for special properties of so-called uniform lines—straight line, circle, and cylindrical helix.

GEMINUS, Thomas, Brit. physician, engraver; b. circa 1540. Author: A Compendium of Anatomy; (published with his own copperplate engravings). Compendiosa totius anatomie delineatio (an abridgement of Vesalius's work of 1543), 1545; printed works for Leonard Digges; engraved portrait of Queen Mary, 1559. Died 1560.

GEMMA, Cornelius, physician, astronomer; b. Louvain, Netherlands, 1535; s. Regnier Gemma; M.D., 1570; prof. medicine, Louvain. Author: De arte cyclognomica; also treatises on abscesses and on plague. Studied comet of 1577. Died 1577.

GEMMA FRISIUS, Regnier (Reinerus), mathematician; b. Dockam, Holland, Dec. 8, 1508; M.D., 1542; 1 son, Cornelius; became prof. medicine and math. at U. Louvain, Belgium, 1541. Author: De principiis astronomiae, 1530; Libellus de locorum describendorum ratione, 1533; De radio astronomico et geometrico liber, 1545; De astrolabio, 1556; Arithmeticae practicae methodus facilis, 1540. Founder of Dutch sch. geography; devel. method of determining longitude using differences in local time; principle tchr. of Mercator; invented various astron. instruments including the plane table and new astrolabe; pioneer in devel. of modern methods of triangulation; Died Louvain, May 25, 1555.

GEMMILL, Chalmers Laughlin, Am. pharmacologist; b. Cresson, Pa., Nov. 24, 1901; s. Benjamin McKee and Clara (Genso) G.; grad. William Penn Charter Sch., 1918; B.S., Lafayette Coll., 1922; M.D., Johns Hopkins, 1926; m. Vivienne Angeline Warry, Jan. 12, 1938; 1 dau., Daphne DeJersey. Faculty, Johns Hopkins Sch. Medicine, 1926-45; prof., chmn. dept. pharmacology U. Va. Sch. Medicine, Charlottesville, 1945—. Commonwealth Fund fellow, Eng., 1965. Mem. Soc. for Exptl. Pharmacology and Therapeutics, Am. Physiol. Soc., Am. Biochem. Soc., Biochem. Soc. (Eng.), Am. Chem. Soc., A.A.A.S., Endocrine Soc. Author: Physiology in Aviation, 1943; also numerous articles. Research on metabolism in muscular exercise, action insulin on muscles, action of thyroxine; physiology high altitudes, effects drugs on enzymes. Home: 19 Farmington Dr., Charlottesville, Va. 22901.*

GENDLIN, Eugene Tovio, psychologist, philosopher; b. Vienna, Austria, Dec. 25, 1926; s. Leo D. and Sylvia (Tobell) G.; came to U. S., 1938, naturalized, 1944; student Drexel Inst. Tech., 1943-44, Cath. U. Am., 1946-48; M.A., U. Chgo., 1950, Ph.D., 1958; m. Frances Oshlag, Dec. 25, 1954; children—Judith Ann, Edward Jacob. Research coordinator Psychotherapy Research Group U. Wis., 1958-62; asso. prof. depts. philosophy, psychology U. Chgo., 1963—. Mem. Am. Psychol. Assn., Am. Philos. Assn., Am. Acad. Psychotherapists, Am. Assn. U. Profs., Psychologists Interested in Advancement Psychotherapy (pres.). Author: Experiencing and the Creation of Meaning, 1962. Contbr. chpts. to Personality Change, 1965. Invitation to Phenomenology, 1965, Existential Child Therapy, 1966. Editor: Psychotherapy-theory, Research and Practice (quar. jour.). Home: 5710 S. Dorchester St., Chgo. 60637.*

GENERALES, Constantine Demosthenes John, physician, educator; b. Athens, Greece, Nov. 10, 1908; s. Demosthenes John and Urania (Tselepis) G.; student Harvard, 1925-28, U. Athens, 1928-29, U. Heidelberg (Germany), 1930-31, U. Zurich (Switzerland), 1931, U. Paris (France), 1931-32; M.D., U. Berlin (Germany), 1936, D.Phil., 1937; m. Johanna Turon-Schellekens, Dec. 10, 1938. Asst. Inst. Sexual Pathology, Berlin, 1931, U. Women's Clinic, Charité, Berlin, 1936-37; research asst. Inst. Genetics and Animal Breeding, Berlin-Dahlem, 1936-37; practice medicine specializing in internal medicine, chest diseases, space medicine, N.Y.C., 1939—; staff Bellevue Hosp., 1939-42, Met. Hosp., 1939-62, Flower and Fifth Av. Hosp., 1939-62, 67—, Mt. Sinai Hosp., 1952—, N.Y. Cancer Inst., 1939-42, Bird S. Coler Hosp., 1947-62, French Hosp., 1965— (all N.Y.C.); faculty N.Y. Med. Coll., 1939, 67—, asst. prof. space medicine, 1960-62, coordinator space medicine program, 1960-62. Cons. in space medicine affairs David Sarnoff Research Center, Princeton, N.J., 1961—; lectr. univs., instns. Recipient awards Congl. Record, 1962, Am. Bill of Rights award Fellow, Mayor N.Y.C., 1963, Gold medallion City of Thessalonika, Greece, 1965. Am. Heart Assn., N.Y. Acad. Medicine, N.Y. Acad. Sci., Brit. Interplanetary Soc., Am. Geriatrics Soc., Am. Coll. Angiology (asso.); mem. Med. Soc. State N.Y. (chmn. sec. on space medicine 1960—), N.Y. Cardiological Soc. (pres.), Council on Hosp. Automation (dir.), Aerospace Med. Assn., A.A.A.S., Am. Assn. for History Medicine, Am. Astronautical Soc. (sr.), Am. Coll. Chest Physicians, Am. Geol. Soc., Am. Geophys. Union, Am. Inst. Aeros. and Astronautics, A.M.A., German Genetic Soc., I.E.E.E., Instrument Soc. Am., Soc. for Biol. Rhythms, others. Author: New Biometric Investigations of Spermatozoa and Fertility, 1937; (with H. Stiasny) Hereditary Disease and Fertility, 1937; An Abridged Concept for a Positively Curved Universe Embodying Gravitational, Thermonuclear, Electromagnetic and Biopoietic Fundamentals, 1959; Weightlessness: its Physical, Biological and Medical Aspects, 1963; China's Legacy to the Exploration of Space, 1967; also numerous articles. Research on space medicine; invented biocylothanatron; surg. aspiration syringe. Home: 2211 Broadway, N.Y.C. 10024. Office: 115 Central Park W., N.Y.C. 10023.*

GENEST, Jacques, Canadian physician; b. Montreal, Que., Can., May 29, 1919; s. Rosario and Annette (Girouard) G.; B.A., Jean de Brebeuf Coll., Montreal, 1937; M.D., U. Montreal, 1942; LL.D., Queen's U., Kingston, Ont., 1966; m. Estelle Deschamps, Oct. 3, 1953; children—Paul, Suzanne, Jacques, Marie, Hélène. Research fellow renal physiology Johns Hopkins Hosps., Balt., 1945-48; research asst. Rockefeller Inst. for Med. Research, 1948-51; dir. clin. research dept. Hotel-Dieu Hosp., Montreal, 1952-66, physician-in-chief, 1964; dir. Clin. Research Inst. Montreal, 1966—; chmn. dept. medicine U. Montreal, 1964—. Mem. com. on clin. research Med. Research Council Can., 1966—; chmn. Med. Research Council Que., 1963—. Recipient Nadeau award Montreal Med. Soc., 1957, 60; Gairdner award Toronto, 1963, Médaille Archambault A.A.A.S., 1965. Mem. Royal Coll. Physicians and Surgeons of Can., A.C.P., Royal Soc. Can. (Flavelle award 1968), Royal Soc. Medicine Eng., Assn. Am. Physicians. Research, numerous publs. on role of kidneys and adrenal glands in human physiopathology, clin. pharmacology new drugs for treatment of high blood pressure; discovery relationship of aldosterone to hypertension and renin-angiotensin to adrenal secretion of aldosterone. Home: 1171 Mont-Royal Blvd., Montreal 8. Office: 110 Pine Av. W., Montreal, Que. 18, Can.*

GENGERELLI, Joseph Anthony, Am. psychologist; b. Glouster, O., Feb. 2, 1905; s. Nugent and Filomena (Leonetti) G.; A.B., Ohio U., 1925; A.M., U. Wis., 1927; Ph.D., U. Pa., 1928; postgrad. Yale; m. Carmen Noguero Cierco, Aug. 27, 1942; 1 dau., Carmen Anna Maria. Faculty, U. Cal., Los Angeles, 1945—, asso. prof., prof., chmn. dept. psychology, 1950-55, now prof. research cons. VA Hosps., 1950—. Mem. Am., Western (past pres.) psychol. assns., N.Y. Acad. Scis., Sigma Xi. Research, numerous publs. on neurophysiol. theory of learning process, magnetic field of the nerve impulse, statis. procedures for analyzing psychol. data, remote stimulation of brain in intact animals. Home: 2001 Linda Flora St., Los Angeles 90024.*

GENKEL, Pavel Aleksandrovich, Russian plant physiologist, microbiologist; b. St. Petersburg, 1903; grad. biology dept. Perm U., 1924; Cand. Biol. Sci., 1935; D.Biol. Sci., 1946. Asso., Perm U., 1924-25, asst., 1925-30, lectr., 1930-31, prof., head chair physiology and anatomy of plants, 1931-33, former dean Biology Faculty, dir. Biol. Research Inst.; head chair plant physiology and microbiology Ural Agrl. Inst., until 1931; head salt-tolerance lab. Timiryazev Inst. Plant Physiology, USSR Acad. Sci., 1939—; prof. Moscow Oblast Pedagogical Inst., 1947—. Mem. All-Russian Bot. Soc., Moscow Naturalists Soc. Author: Lichen Symbiosis, 1938; The Story of Plant Life, 1947. Developer method of adapting plants to drought before sowing, method of increasing salt-tolerance of wheat and cotton by treating seeds before sowing. Address: Timiryazev Inst. Plant Physiocology, USSR Acad. Sci., Leninsky prospect 33, Moscow, USSR.

GENOVESE, Sebastiano, Italian hydrobiologist; b. Messina, Italy, Feb. 25, 1926; s. Angelo and Lucrezia (Garofalo) G.; dott. in Chemistry, U. Messina, 1949, dott. in Natural Sci., 1951; m. Teresa De Martino, Apr. 26, 1958; children—Lucrezia, Angelo. Asst., Inst. Hydrobiology, U. Messina, 1951-57, prof. hydrobiology, 1958—, dir. Inst. Hydrobiology, 1962—. Mem. Internat. Com. for Sci. Exploration of Mediterranean, 1957—. Mem. Società Peloritana sc.fis.-mat.nat. (pres. 1959—), Unione Zoologica Italiana, Am. Soc. for Microbiology, Societas Internationalis Limnology. Research, publs. in ecol. microbiology in marine environment and in brackish waters especially phenomenon of red water bacteria, biology of Thunnus thynnus. Home: 89, N. Bixio, Messina, Italia.*

GENSOUL, Joseph, surgeon; b. Lyons, France, 1797; chief surgeon Hôtel Dieu, Lyons. Author: Nouveau procédé pour opérer les polypes de la matrice; Lettre chirurgicale sur quelques maladies graves du sinus maxillaire et de l'os maxillaire inférieur. Perfected several new surg. techniques; 1st to make resection of superior maxillary nerve, 1827. Died 1858.

GENTAKU, see Otsuki, Bansui.

GENTH, Frederick Augustus (original name Friedrich August Ludwig Karl Wilhelm Genth), chemist; b. Wachtersbach, Hesse-Cassel, Germany, May 17, 1820; s. George Fredrich and Karoline (Freyin von Swartzenau) G.; attended U. Heidelberg, U. Giessen (both Germany); Ph.D., U. Marburg (Germany), 1845; m. Karolina Jager, 1847, 3 children; m. 2d, Minna Fischer, 1852, 9 children. Asst. to Robert Wilhelm Bunsen, Marburg, 3 years; to 1848; came to U. S., 1848; opened analytical chem. lab. in Phila., 1848; prof. chemistry U. Pa., 1872-88; re-opened pvt. lab., 1888; chemist Pa. Bd. Agr., 1877-84. Mem. Nat. Acad. Scis., A.A.A.S. Contbr. numerous papers on mineral chemistry to jours., 1842-93. In mineral chemistry, discovered 23 new mineral species, genthite (nickelgymnite) named in his honor; best example of work is paper Corundum, its Alterations and Associated Minerals, 1873; studied ammonia-cobalt bases, 1847-56. Died Phila., Feb. 2, 1893.

GENTIL, Louis-Emile, geologist; b. Algiers, Algeria, July 15, 1868; prof. geology and phys. geography Sorbonne, Paris; participated in Segonzac exploration of Morocco. Mem. French Acad. Scis., 1923. Author: Dans le Bled-es-Siba, 1906; Au centre de l'Atlas, 1910; Maroc physique (first geol. map of Morocco), 1920. Research on geology of Tafna Valley, Algeria. Died Paris, June 12, 1925.

GENTILE DA FOLIGNO, see da Foligno, Gentile.

GENTNER, Walter Andrew, Am. plant physiologist; b. Washington, Apr. 22, 1922; s. Walter Andrew and Rosa (Meyers) G.; A.B., George Washington U., 1951, M.A., 1952, Ph.D., 1962; m. Evelyn Margaret Gifford, July 30, 1948. Mng. botanist R/W Maintenance Corp., Buffalo, 1952-55; plant physiologist U. S. Dept. Agr., Beltsville, Md., 1955—. Mem. Am. Soc. Plant Physiologists, Weed Soc., Sigma Xi. Contbr. numerous articles to sci. jours. Evaluation of new chems. for their herbicidal properties; correlation of molecular structure to herbicidal activity. Home: 10522 Edgemont Dr., Adelphi, Md. 20783. Office: Plant Industry Sta., Beltsville, Md. 20705.*

GENTNER, Wolfgang, German physicist; b. Frankfort am Main, Germany, July 23, 1906; s. Carl and Luise (Klomp) G.; Ph.D. in Natural Sci., U. Frankfort, 1930; m. Alice Pfaehelr; children—Ralph, Doris. Fellow Lab. Curie, Inst. Radium, Sorbonne, 1933-35; Carnegie Found. fellow; sci. asst. Kaiser-Wilhelm Inst. Physics, Heidelberg, 1936-46; lectr. physics U. Frankfort, 1937-41; fellow Radiation Lab., U. Cal. at Berkeley, 1938-39; lectr. physics U. Heidelberg, 1941-45, prof., 1945, 58—; prof. physics, dir. Phys. Inst., U. Freiburg, 1946-58, prorektor, 1947-49; dir. Synchocyclotron, C.E.R.N., Geneva 1955-59; dir. Max-Planck Inst. for Nuclear Physics, Heidelberg, 1958—. Mem. com. of sci. and tech. Euratom, Brussels. Mem. Acad. Scis. Heidelberg, Bavarian Acad., Leopoldina Halle Acad., Max Planck Soc. Co-author Atlas of Typical Expansion Chamber Photographs, 1954. Research and publs. on biophysics, radioactivity, nuclear physics; research on age determinations. Home: Im Bäckerfeld 6. Office: Max Planck Institute, for Nuclear Physics, Saupfercheckweg, Heidelberg, Germany.*

GENTRY, John Tilmon, Am. physician; b. St. Louis, Dec. 31, 1921; s. John Tilmon and Ethel Marie (Hambley) G.; A.B., Washington U., St. Louis, 1944, B.S., M.D., 1948; M.P.H., Harvard, 1951; m. Geraldine Evelyn Heyne, Feb. 5, 1949; children—John A., Kristine A., David T., Laurie J., Glenn S. Intern U. Chgo. Clinics, 1948-49; resident pub. health N.Y. State Dept. Health, 1949-50; asst. to chief epidemiology br. Communicable Disease Center, USPHS, Atlanta, 1951-52; health officer, Anchorage, Alaska, 1952-53; dist. health officer N.Y. State Dept. Health, Syracuse, 1954-57, regional dir., Syracuse, 1957-64; clin. asst. prof. preventive medicine State U. N.Y., Upstate Med. Center, Syracuse, 1955-63, asso. prof., 1963-64; asst. dean for program devel., asso. prof. dept. pub. health administrn. U. N.C. Sch. Pub. Health, 1964—; cons. div. radiol. health USPHS, 1959-62; chief med. edn. br., dept. chief health div. AID mission to India, New Delhi, 1961-63, acting chief health div., 1961-62. Mem. vital statistics com. USPHS, 1960-61; chmn. regional planning com. health White House Conf. Children and Youth, 1960, mem. regional planning com. 1961 Conf. Aged; exec. com. N.Y. Regional Hosp. Planning Council, 1957-61, trustee, 1963-64; chmn. maternal and child health com. N.Y. State Conf. City, County and Dist. Health Officers, 1956; vice chmn. Syracuse regional N.Y. State Regional Mental Health Planning Com., 1963-64, N.Y. State Regional Interdepartmental Rehab. Com., 1963-64; mem. population panel White House Conf. Internat. Coop., 1965. Bd. dirs. Onondaga County unit Am. Cancer Soc., 1954-60, Onondaga County United Cerebral Palsy Assn., 1954-58, Onondaga Health Assn., 1954-64, Cayuga Health Assn., 1954-58; bd. dirs. council aging Council Social Agencies Onondaga County, 1954-60, legislative committee, 1955-64, chmn. planning com. health and hosp. div., 1955-60; self-study cons. Nat. Commn. on Community Health Services, Charleston, S.C., 1964-65; cons. Forsyth County Citizens Planning Council, 1964—; instl. rep., mem. exec. com. N.C. State Adv. Com. Devel. Regional Medical Complex, 1965—. Diplomate Am. Bd. Preventive Medicine. Fellow Am. Coll. Preventive Medicine (charter), Am. Pub. Health Assn., mem. A.A.A.S., N.Y. Acad. Scis., Am. Trudeau Soc., A.M.A., Population Assn. Am. N.C. Academy Preventive Medicine and Public Health, N.C. State Public Health Association, New York Acad. Preventive Medicine (charter), N.Y. State Pub. Health Assn., Phi Beta. Contbr. articles profl. jours. Epedemiological studies associating residence in sandy soil areas of U. S., esp. Southeastern coastal plain, with disease Sarcoidosis; association of increase in human congenital malformations with consumption of ground water from New York State glacial deposit areas containing above normal quantities of radioactive materials. Home: 2018 N. Lake Shore Dr., Chapel Hill, N.C.*

GENTZEN, Gerhard Karl Erich, German mathematician; b. Greifswald, Germany, Nov. 24, 1909; s. Johannes and Melanie (Bilharz) G.; Ph.D., Göttingen, Germany, 1933. Asst. to D. Hilbert for several years; served in mil. for 2 years; became lectr. math. German U. Prague, 1943. Research and publs. on formal logic and math. founds., especially theory of predicate logic and proof of freedom from contradiction of pure number theory. Died Prague, Aug. 4, 1945.

GEOFFROI OF MEAUX (Galfredus de Meldis), French astrologer, physician; flourished 1310-48; took degrees in arts and medicine; became one of masters and bachelors chosen to examine Ars brevis of Ramon Lull, 1310; one of 6 royal physicians witnessing coronation of Charles IV; possibly taught astrology in Paris and Oxford, Eng. Wrote several tracts which illustrated astrology of the period.

GEOFFROY, Claude Joseph, French chemist; b. Paris, 1685; studied under Tournefort; master apothecary. Fellow Royal Soc., 1715; mem. French Acad. Scis., 1707. Author: Observations sur les huiles essentielles, 1707; Examen des différents vitriols, 1728. Collected medicinal plants to form a bot. garden; research on properties of borax, sodium potassium tartate, Prussian blue, vegetable oil. Died Paris, Mar. 9, 1752.

GEOFFROY, Étienne-François (the Elder), French chemist, physician; b. Paris, Feb. 13, 1672; master apothecary; prof. Coll. France, 1712-31; dean Paris Faculty Medicine; prof. chemistry Royal Garden. Fellow Royal Soc., 1698; mem. French Acad. Scis., 1689 (dir. 1721); Author: Sur la preparation du bleu de Prusse; Table des différents rapports en chimie (1st table of chem. affinities), 1718; Tractatus de materia medica, 3 vols., 1741. Studied preparation of Prussian blue; studied med. botany; added to knowledge of chemistry in medicine; attacked practitioners of alchemical frauds; discussed properties of alum, boric acid; determined composition of sodium sulphate. Died Paris, Jan. 6, 1731.

GEOFFROY, Étienne Louis, French physician; b. Paris, Oct. 2, 1725; s. Étienne-Francois; docteur en médecine, 1748; mem. French Acad. Scis., 1798. Author: Histoire abrégée des insects, 1762; Dissertation sur l'organe de l'ouie de l'homme, des reptiles des poissons, 1778. Research on hearing in man, reptiles and fish, classified the insect order, Coleoptera. Died Aug. 12, 1810.

GEOFFROY SAINT-HILAIRE, Étienne, French naturalist; b. Étampes, France, Apr. 15, 1772; ed. Coll. Navarre, Paris; studied under Brisson, Haüy; 1 son, Isidore. Became prof. zoology Jardin des plantes, Paris, 1793; went to Egypt with Napoleon, 1798; a founder, Inst. Cairo; prof. zoology Faculty of Scis., Paris, 1809. Mem. French Acad. Scis., 1807 (became pres. 1833). Author: Philosophie anatomique 2 vols., 1818-22; (with F. Cuvier) l'Histoire naturelle des mammiferes, 1820-42; Principes de philosophie zoologique, 1830; Sur le principe de l'unité de composition organique, 1828; also numerous articles. Believed a single type of structure exists in the animal kingdom; did not believe in modification of existing species; founded new sci. of teratology. Died Paris, June 19, 1844.

GEOFFROY SAINT-HILAIRE, Isidore, French biologist; b. Paris, Dec. 16, 1805; s. Étienne Geoffroy Saint-Hilaire; m. Louise Blocque, 1830; a son, a dau. Named prof. Paris Mus. Natural History (Jardin des Plants); became prof. Bordeaux (France) Faculty Scis., 1837. Mem. French Acad. Medicine, French Acad. Scis., 1833 (became pres. 1856), Nat. Soc. Acclimatization (founder 1854). Author: Traité de teratologie, 1832-36; Histoire générale et particulière des anomalies de l'organisation chez l'homme et les animaux, 3 vols., 1832-36; Histoire naturelle des insectes et des mollusques, 2 vols., 1841; also memoir which led to discovery a species of Am. bat, Nyctinomus brasiliensis, 1824. Popularized eating of horse meat in France; enlarged his father's work on teratology; studied domestication of animals. Died Paris, Nov. 10, 1861.

GEORG, Lucille Katharine, Am. microbiologist; b. Ann Arbor, Mich., Oct. 9, 1912; d. Conrad and Katharine (Haller) G.; B.S., U. Mich., 1933, M.S., 1934; Ph.D., Columbia, 1948; m. William Leonard Pickard, Sept. 15, 1950. Research asst. dept. dermatology Columbia Physicians and Surgeons, 1946-49; mycologist Communicable Disease Center USPHS, Atlanta, 1949—, instr. dept. microbiology Emory U. Med. Sch., 1952—. Mem. Mycol. Soc. Am., Internat. Soc. Human and Animal Mycology, Med. Mycol. Soc. Am., Am. Soc. Microbiology, Sigma Xi, Phi Sigma. Author: (with L. Ajello, W. Kaplan, L. Kaufman) Laboratory Manual for Medical Mycology, 1963. Contbr. numerous articles to sci. jours. Developed nutritional tests for identification of dermatophyte species, selective medium for isolation of pathogenic fungi; improved methods for isolation and identification of anaerobic Actinomyces. Home: 2673 Pangborn Rd., Decatur, Ga. 30033. Office: 1600 Clifton Rd., Atlanta 30333.*

GEORGE, Donald William, Australian engr.; b. Adelaide, South Australia, Nov. 22, 1926; s. Horace William and Claudia (Cole) G.; B.Sc., U. Sydney, 1947, B.E. with First Class honors, 1949, Ph.D., 1966; m. Lorna Mildred Davey, Jan. 21, 1950; children—Christopher William, Lynden Anne. Lectr. elec. engring. U. Tech., N.S.W., Australia, 1949-53; research officer U.K. Atomic Energy Authority, Harwell, Eng., also AEC, Lucas Heights, Australia, 1954-60; asso. prof. elec. engring. U. Sydney, N.S.W., Australia, 1960—. Mem. Inst. Engrs. Australia (asso.). Research, publs. in direct energy conversion from fossil and nuclear fuels, elec. phenomena in gases at very high temperatures. Home: 31 Yellambie, Yowie Bay, N.S.W. Office: U. Sydney, Sydney, N.S.W. Australia.*

GEORGE, Eric Paul, physicist; b. London, Eng., May 1, 1914; s. Alfred William and Blanche (Skelton) G.; B.Sc., London U., 1935, Ph.D., 1941; M.Sc., U. New South Wales, Australia, 1965, D.Sc., 1966; m. Mabel Helen Swan; children—Colin Rory, Martin Anthony. Communication engr. G.E.C. Research Labs., Eng., 1935-46; research fellow London U., 1946-50; vis. prof. physics U. Rochester, N.Y., 1950-51; sr. lectr. Imperial Coll., 1951-53; reader in physics U. Sydney, Australia, 1953-55; dir. dept. physics St. Vincents Hosp., Sydney, 1955-63; prof. physics U. New South Wales, Kensington, 1963—, master Philip Baxter Coll. Fellow Australian Inst. Physics (chmn. biophysics group 1966-67); mem. Australian Physiol. Soc., Endocrine Soc. Australia. Author: (with S. T. Butler and C. Carey) High School Science; also numerous articles. Research on fundamental particles with intermediate mass, cosmic radiation, properties of cell membranes; devel. techniques in nuclear medicine. Home: Philip Baxter Coll., Box 24, P.O. Kensington, Australia.*

GEORGE, Henry, Am. economist; b. Phila., Sept. 2, 1839; s. Richard Samuel Henry and Catherine (Vallance) G.; m. Annie Fox, Dec. 3, 1861. Compositor on Home Jour., c. 1860; a founder and publisher Evening Jour.; returned to San Francisco as printer on newly established Times, 1866, became reporter editorial writer, then mng. editor; 1st of his articles appeared in Overland Monthly, Oct. 1868; revisited East as agt. for San Francisco Herald, late 1868; established independent news service; returned to Cal. became editor Oakland Transcript, 1869; became partner and editor Daily Evening Post, 1871; state insp. of gas meters, 1876; advanced theory of the single tax on unearned increment of land values in book Progress and Poverty, 1879; became leader of reform element in Am. and traveled throughout country; founded single tax party; probably most important polit. reformer of early 1880's; pub. The Irish Land Question (became very popular, had heavy influence on Fabian Socialists), 1881; correspondent for Irish World in N.Y., 1881-82; lectured in Gt. Britain under auspices of Land Reform Union, 1883-84, for Scottish Land Restoration League, 1884; ran for mayor of N.Y.C. on social reform platform, lost to Tammany leader Abrams Hewitt (but ran ahead of Theodore Roosevelt) 1886; ran for mayor again, 1897; published weekly Standard, 1887-92; lectured in Gt. Britain, 1888, 89, Australia, 1890; The Science of Political Economy, 1897. Author: Our Land and Land Policy (contained essentials of econ. philosophy which he later expanded), 1871; Progress and Poverty, 1879; Social Problems (originally series of articles in Frank Leslie's Newspaper), 1883; Protection or Free Trade, 1886; An Open Letter to the Pope, 1891; A Perplexed Philosopher, 1892. His land-tax theory influenced laws enacted in several countries, including Australia, Can. and U. S. Died N.Y.C., Oct. 29, 1897.

GEORGE, James Henry Bryn, chemist; b. Swansea, Wales, Feb. 5, 1929; s. Brinley Clifford and Kate (Rees) G.; B.A., Oxford U., Eng., 1949, M.A., 1952, D. Phil., 1952; S.M. in Chem. Engring., Mass. Inst. Tech., 1953; m. Brigitte Böck, Aug. 10, 1963. Came to U. S., 1952, naturalized, 1960. Faculty, Mass. Inst. Tech., 1953; chem. engr. Ionics, Inc., Cambridge, Mass., 1953-54; sr. phys. chemist Arthur D. Little, Inc., Cambridge, 1954-60, group leader phys. chemistry, 1960-66, sect. head, 1966—. Mem. Am. Chem. Soc., Soc. Chemistry and Industry. Investigations of ion-solvent interactions in solution, transport properties in ion-exchange membranes; tech. and econ. studies in fuel cells, saline water conversion and radio-isotopes. Patentee in solvent extraction, sealed rechargeable battery tech. Home: 31 Linnaean St., Cambridge 02138. Office: Acorn Park, Cambridge, Mass. 02140.*

GEORGE, John Caleekal, Indian zoologist, physiologist, ornithologist; b. Kerala, India, June 16, 1921; s. Caleekal John and Annamma (Varghese) G.; B.Sc. with honors, Wilson Coll., U. Bombay, India, 1942, Ph.D., 1947; postgrad. (Fulbright Smith-Mundt fellow), U. Pa., 1953-54; m. Achamma Mathew, June 12, 1950; children—Vinod Caleekal, Manoj Caleekal, Anuppa Caleekal. Demonstrator, Ismail Yusuf Coll., Bombay, 1945-48, lectr. Inst. Sci., Bombay, 1948, vertebrate zoologist, dept. anthropology Govt. of India, 1948-50; faculty Maharaja Sayajirao U., Baroda, India, 1950—, prof. zoology, head dept., 1957—. Dutch Govt. Research scholar, 1958; Fulbright scholar and lectr., 1961-62; research grantee Muscular Dystrophy Assns. Am., 1963-65. Fellow Zool. Soc. India; mem. All India Congress Zoology (pres. sect. 1962), Soc. Animal Morphologists and Physiologists India (hon. sec.). Author: (with A. J. Berger) Avian Myology, 1966. Founder editor: Pavo; editor: Jour. Animal Morphology and Physiology; editorial bd. Indian Jour. Exptl. Biology. Research, publs. on biology of avian muscle. Home: Adhyapak Nivas. Office: Dept. Zoology, Maharaja Sayajirao U., Faculty of Sci., Baroda 2, India.*

GEORGE, John Lothar, Am. wildlife ecologist; b. Milw., Apr. 17, 1916; s. John T. and Laura (Sauer) G.; student U. Wis., 1934-36; B.S. in Forestry, U. Mich., 1939, M.S. in Zoology, 1941, Ph.D. (Rackham Spl. fellow), 1952; m. Jean C. Craighead, Jan. 28, 1944; children—Carolyn L., John C., Thomas L.; m. 2d Janice Chennault, June 17, 1967; 1 son, David N. Coordinator grad. conservation program Vassar Coll., Poughkeepsie, N.Y., 1950-57; curator mammals Bronx Zoo, N.Y.C., 1957-58; research specialist U. S. Fish and Wildlife, Washington, 1958-63; prof. wildlife mgmt. Pa. State U., University Park, 1963—. Chmn. ecol. effects chem. controls Internat. Union for Conservation Nature and Natural Resources, Lucerne, Switzerland, 1963—. Recipient Auriane award for best children's book on humane treatment of animals, 1956. Mem. Wilderness Soc., Wildlife Soc., Nature Conservancy, Soc. Mammalogists, Am. Ornithologists Union, Wilson Ornithol. Soc., N.E. Bird Banding Assn., Sigma Xi, Phi Sigma, Sigma Pi, Phi Epsilon Phi. Research, publs. on role of pesticides as oral intoxicants to birds, global contaminants, food poisoners. Author: (with J. C. George) Vulpes, the Red Fox, 1948, Vison the Mink, 1949, Masked Prowler, the Story of a Raccoon, 1950, Meph, the Pet Skunk, 1952, Bubo, the Great Horned Owl, 1954, Dipper of Copper Creek, 1956. Home: 685 Westerly Pkwy., State College, Pa. 16801.*

GEORGE, Joseph Johnson, Am. meteorologist; b. West Plains, Mo., June 20, 1909; s. Will A. and Bess (Johnson) G.; student U. Cal. at Los Angeles, 1926-29, Cal. Inst. Tech., 1933-34; m. Mary Beale Sasscer, Sept. 16, 1934; children—Mary Beale (Mrs. Zolly Derryberry), Margaret Lynn (Mrs. Dennis Barré), Penelope (Mrs. James Robinson), Joseph S. Chief meteorologist Western Air Express, Los Angeles, 1929-34; supt. meteorology Eastern Air Lines, Atlanta, 1934-42, dir. meteorology, 1946—. Recipient Losey award Inst. Aero. Scis., 1943. Mem. Am. Meteorol. Soc. (past councilor, past v.p. Meisinger award 1941, award for applied meteorology 1956). Author: Weather Forecasting for Aeronautics, 1960; also articles. Research on weather forecasting as applied to aeros., fog, physics and forecasting. Home: 2698 Piney Wood Dr., East Point, Ga. 30044. Office: Eastern Air Lines, Atlanta Airport, Atlanta, 30320.*

GEORGE, (Johann Friedrich) Leopold, German philosopher, psychologist; b. Berlin, Germany, Aug. 14, 1811; s. Johann Gottlieb and Friederike (Bartz) G.; studied 1st theology, then Oriental langs.; studied under Schleiermacher; m. Auguste Bührig, 1844; 2 sons, 3 daus. Joined philosophy faculty U. Berlin, 1834, lectr., secondary sch. tchr., until 1856; named prof. philosophy U. Greifswald (Germany). Author: System der Metaphysik, 1844; Die fünf Sinne, 1846; Lehrbuch der Psychologie, 1854; Die Logik als Wissenschaftslehre, 1868; also others on philosophy and theology. Tried to reconcile religious faith with sci. knowledge; studied devel. and theory of sense organs; believed psychology is a sci. base for metaphysics. Died Berlin, May 24, 1873.

GEORGE, Philip, biophys. chemist; b. Maidstone, Kent, Eng., Jan. 30, 1920; s. Walter and Frances Alice (Brook) G.; B.A., Christ's Coll., Cambridge, Eng., 1941, M.A., 1944, Ph.D., 1945; m. Kathleen Margaret Hoff, Sept. 27, 1946; children—Francis, Sarah, Emma, Simon, Hannah, Edwin. Came to U. S., 1955. Research Molteno Inst. for Parasitology, Cambridge, 1945-47; lectr. phys. chemistry Leeds (Eng.) U., 1947-49; asst. dir. research dept. colloid sci. Cambridge U., 1949-55; prof. biophys. chemistry U. Pa., 1955—, chmn. group com. in molecular biology, dir. tng. program, 1960-61, dir. gen. honors program, 1961-63, chmn. dept. history and philosophy of sci. Grad. Sch. Arts and Sci., 1964—. Cons. Hartford Found. project Presbyn. Hosp., Phila., 1960—; mem. biophys. scis. study sect. Nat. Inst. Gen. Med. Scis., 1963-66. Mem. Am. Chem. Soc., Soc. Biol. Chemists, Biophys. Soc., Am. History Sci. Soc., A.A.A.S., Franklin Inst., Renaissance Soc. Chem. Soc. (Eng.), Biochem. Soc. (Eng.), Faraday Soc. (Eng.), Brit. Soc. for Histr. Sci., Phi Beta Kappa, Sigma Xi. Contbr. profl. jours. Studies on the mechanism of oxidation-reduction reactions and the relationship between the structure and function of hemoglobin and the hemoprotein enzymes that catalyse oxidation-reduction processes; the investigation of fundamental thermodynamic aspects of the conservation and utilization of energy in cellular metabolism. Home: 4 Herford Pl., Lansdowne, Pa. 19050. Office: Dept. of Chemistry, University of Pa., Phila. 19104.*

GEORGE, Thomas Neville, Brit. geologist; b. Swansea, Wales, U.K., May 13, 1904; s. Thomas Rupert and Elizabeth (Evans) G.; B.Sc., U. Wales, 1924, M.Sc., 1925, D.Sc., 1932; Ph.D., U. Cambridge (Eng.), 1928; postgrad. U. London (Eng.), 1930-31; D.Sc. (hon.), U. Rennes (France), 1956; m. Sarah Hannah Davies, Aug. 25, 1932. Demonstrator, U. Wales, Swansea, 1928-30, prof. geology, 1933-46, dean sci., 1936-39; geologist Geol. Survey Gt. Britain, 1930-33; prof. geology U. Glasgow (Scotland), 1947—, dean sci., 1956-59; Woodward lectr. Yale, 1956; Sr. Fgn. Sci. fellow NSF, Northwestern U., 1964; vis. prof. U. Witwatersrand, Cape Town, Natal, 1967. Fellow Royal Soc., 1963, Royal Soc. Edinburgh (past v.p.), Geol. Soc. London (pres.; Lyell medal 1963); mem. Paleontol. Assn. (past pres.), Scottish Field Studies Assn. (pres.), Brit. Assn. (geology-past pres.), Geol. Soc. Glasgow (past pres.), Edinburgh Geol. Soc. (hon.), Assn. U. Tchrs. (past pres.). Author: Evolution in Outline, 1951; Regional Geology of North Wales, 1961, South Wales, 1968; British Caledonides (in part), 1963; University Instruction in Geology, 1965; Geology of Scotland (in part), 1965; also numerous articles. Research on regional geology of Britain, stratigraphy and paleogeography of Carboniferous rocks mainly in Wales and Ireland, structural growth of palaeozoic Britain, geomorphology of Wales, Ireland and Scotland, processes of evolution especially invertebrate fossils, fossil brachiopods. Home: 1 Princes Terrace, Glasgow, W.2, Scotland, Office: U. Glasgow, Glasgow, Scotland.*

GEORGESCU-ROEGEN, Nicholas, econometrician; b. Constantza, Rumania, Feb. 4, 1906; s. Stavru and Maria (Niculescu) G-R.; Lic. Math., Bucharest U., 1926; D.Stat., U. Paris, Sorbonne, 1930; m. Otilia Busuioc, Sept. 2, 1934. Came to U. S., 1948, naturalized, 1954. With Bucharest U. Sch. Statistics, 1932-47; asst. dir. Central Statis. Inst. Rumania, 1932-38; econ. adviser Finance Ministry Rumania, 1938-39; dir. Bd. Trade, Rumania, 1939-44; sec. gen. Rumanian Armistice Commn., 1944-45; research asso., lectr. Harvard, 1948-49; prof. econs. Vanderbilt U., Nashville, 1949—. Rockefeller vis. prof., Japan, 1962-63; cons., vis. prof. Ford Found., Brazil, 1964, 66. Rockefeller fellow, 1934-36; Guggenheim fellow, 1958-59; Fulbright scholar, 1958-59. Fellow Econometric Soc., Internat. Inst. Sociology, Rumanian Acad. Moral Scis.; mem. Am., So. econ. assns., Société de Statistique de Paris. Author: Metoda Statistica, 1933; Analytical Economics: Issues and Problems, 1966; (with T. C. Koopmans et al) Activity Analysis of Production and Allocation, 1951. Asso. editor Econometrica, 1951—. Research in theory of consumer's behavior covering measurability of utility, consistent ordering of preferences, and distinction between risk and uncertainty; analytical properties and generalizations of input-output systems; econ. devel.; overpopulated agrarian econs., breakdown of capitalism, inflaction-lock of Latin Am. Home: 2614 Hemingway Dr., Nashville 37215.*

GEORGI, Carl Eduard, Am. microbiologist; b. Milw., Feb. 18, 1906; s. Herman Emil and Ottilie (Memmler) G.; B.S., U. Wis., 1930, M.S., 1932, Ph.D., 1934; m. Marjorie Clare Womelsdorff, Aug. 20, 1936; children—Liesl Andrea (Mrs. Benjamin Vrana), Todd Anthony. Asst. instr. chemistry U. Wis., 1934-35; faculty U. Neb., Lincoln, 1935—, prof. microbiology, 1947—, chmn. dept., 1953—, mem. staff dept. biochemistry and nutrition, 1955—. Fulbright research scholar U. Paris, 1951-52; Murray Longworth Prof. microbiology, U. Neb., 1964—. Diplomate Am. Bd. Microbiology. Fellow Am. Acad. Microbiology, A.A.A.S.; mem. Am. Soc. Microbiology (pres. Missouri Valley br. 1942-43, chmn. div. gen. microbiology 1966), Am. Chem. Soc. (chmn. Neb. 1942, 47), Neb. Acad. Sci. (pres. 1944-45), Am. Inst. Biol. Scs. (vis. coll. lectr. 1960—), Am. Soc. Biol. Chemists, Am. Soc. Cell Biology, Soc. Exptl. Biology and Medicine, Soc. Gen. Microbiology, Electron Microscope Soc. Am., Soc. Indsl. Microbiology, Sigma Xi (pres. Neb. 1948-49). Author: Laboratory Manual for General Microbiology, 5th edit., 1959; also numerous articles. Research on fixation of atmospheric nitrogen by leguminous plants, thermophilic bacteria, microbial conversion of agrl. products to other uses, structure and composition of bacterial cells; enzyme systems. Home: 3033 Georgian Ct., Lincoln, Neb. 68502.*

GEORGI, Felix, Swiss neurologist; b. Zurich, Switzerland, Sept. 17, 1893; s. Georg Cohn and Margarethe (Levin) G.; ed. Us. Zurich, Frieburg/Br., Berlin, Munich; M.D., Freiburg, 1919; hon. dr., U. Münster, 1963; m. Mildred Arbenz, Dec. 1920; children: Walter, Mareile, Helmut. Asst., Psychiatric U. Hospital, Frankfort am Main, 1919-20; asst., Institute of Experimental Cancer Research, Heidelburg, 1920-23; chief-asst., U. Clinic, Breslau, 1923-33; dir., Bellevue Clinic, Yverdon, 1933-46; asst. Psychiatric U. Clinic, Bern, 1933-46; chief-asst., Psychiatric U. Clinic, Basel, 1946-65, dir., Neurological U. Clinic, Basel, 1951; Prof., 1955. Pres. of Medical Advisory Committee of Swiss MS Soc.; pres., C. Barell-Fund for Cerebral-Palsyed. Hon. mem., French Neurological Soc.; N.Y. Acad. Scis.; mem., Swiss Neurological Soc., German Neurological Soc., World Fed. Neurology. Author: Körperbau und seelische Anlage, 1928; Humoralpathische der Nervenkrankh, 1935; Zur Problematik der MS, 1961. Pioneer of psychosomatic medicine; research in humoral-pathology of mental and nervous diseases; specialist in geomedicine especially multiple sclerosis. Died Basel, Feb. 21, 1965.

GEORGIADE, Nicholas George, Am. physician; b. Lowell, Mass., Dec. 25, 1918; s. George N. and Stefanie (Englisch) G.; student Fordham U., 1937-40; D.D.S., Columbia, 1944; B.S. in Medicine, M.D., Duke, 1949; m. Ruth Catherine Saver, Sept. 21, 1942; children—Greg, Robert, Nancy. Practice medicine specializing in plastic and maxillofacial surgery Duke Med. Center, Durham; faculty Duke Sch. Medicine, 1954—, prof. plastic, maxillofacial and oral surgery, 1961—; chief plastic and maxillofacial surgery Watts Hosp., Durham, 1956—; staff Lincoln Hosp., Durham. Cons. to hosps., govt. agys. Diplomate Am. Bd. Plastic Surgery, Am. Bd. Oral Surgery. Fellow A.C.S.; mem. Am. Soc. Head and Neck Surgeons, Plastic Surgery Research Council, Am. Soc. Plastic and Reconstructive Surgery, Am. Soc. Maxillofacial Surgeons (past pres.), Am. Assn. Plastic Surgeons, Southeastern Soc. Plastic and Reconstructive Surgeons, Societe Internationale Chirurgie, Am. Assn. U. Profs., Am. Soc. Cryobiology, Tissue Cultures Assn., Cleft Palate Assn., Transplantation Soc., Sigma Xi. Contbg. author: Textbook of plastic and Maxillofacial Surgery, 1964; Textbook Burns, 1965. Research, numerous publs. on tissue preservation, clin. and investigative work in mgmt. burns, including microbiology burns, facial growth and devel. Home: 2523 Wrightwood Av., Durham 27705. Office: Duke U. Sch. Medicine, Durham, N.C. 27706.*

GEORGIEV, Vladimir Ivanov, Bulgarian linguist; b. Gabare, Bulgaria, Feb. 3, 1908; s. Ivan Georgiev and Janka (Zeljazkova) Grazdanina; student U. Sofia (Bulgaria), 1926-30; Dr., U. Vienna (Austria), 1934; Dr.hon. causa, U. Berlin (Germany), 1960; m. Magdalina Alexandrova Obroimova, July 3, 1953. Faculty, U. Sofia, 1931—, prof. linguistics. Recipient Nat. Prize sci. Mem. Bulgarian Acad. Scis. (corr., academician), Académie des Inscriptions et Belles Lettres (corr.), French Acad. Scis. (corr.), Assn. Internationale des Études de Sud-Est Européen (pres.), Internat. Com. Slavists (v.p.). Author: Vorgriechische Sprachwissenschaft, 1941; La Langue thrace, 1957; Issledovanija pr sravniteljno-istoriceskomu jasykoznanija, 1958; Introduzione alla storia della lingue indeuropee, 1966; also numerous articles. Home: 11 Oboriste, Sofia, Bulgaria.*

GERALDINUS, Johannes, astronomer, natural philosopher; flourished early 17th century. Author: De meteoris tractus . . . , 1613. Believed comets were of elementary, not celestial nature and that they had power to signal some events; speculated on origin of springs and elevation of water.

GERARD, John, English surgeon, herbalist; b. Nantwich, Cheshire, Eng., 1545; practicing barber-surgeon for 20 years; supt. of gardens of Lord Burghley, London, Eng.; Linnaeus named the genus Gerardia in his honor; author: Catalogus arborum, fruticum, ac plantarum . . . (1st catalog of any garden), 1596; Herball, 1597. Had medicinal garden with more than 1000 species of plants. Died London, Feb. 1612.

GERARD, Ralph Waldo, Am. physiologist; b. Harvey, Ill., Oct. 7, 1900; s. Maurice and Eva (Teitelbaum) G.; B.S., U. Chgo., 1919, Ph.D., 1921; M.D., Rush Med. Coll., 1924; D.Sc. (hon.), U. Md., M.D. (hon.), U. Leiden (Holland), LL.D., U. St. Andrews (Scotland), Litt.D., Brown U.; D.Sc., McGill U.; m. Margaret Wilson, June 15, 1922; m. 2d, Leona Bachrach, Jan. 1, 1955; 1 son, James. Faculty, S.D. U., 1921-22; nat. research fellow, Europe, 1925-27; faculty U. Chgo., 1927-52, 54-55, U. Ill., 1952-55; prof. neurophysiology Mental Health Research Inst., U. Mich., Ann Arbor, 1955-64, cons. sr. scientist, 1964—; dean grad. div., dir. spl. studies, prof. biol. scis., U. Cal., Irvine, 1964—; also lectrs. Mem. coms. NRC. Recipient medal Charles U. (Prague, Czechoslovakia), 1946, Order of White Lion, Govt. of Czechoslovakia, 1946; Alumni award U. Chgo., 1967. Ford Found. fellow, 1954-55. Mem. Nat. Acad. Scis., Am. Psychiat. Assn. (hon.), Am. Acad. Arts and Scis., Physiol. Soc. (pres. 1951-52), Assn. Research Nervous and Mental Disease, Physiol. Soc. (Gt. Britain), Biochem. Soc. (Gt. Britain), Am. Neurol. Assn., Am. Naturalists, A.A.A.S., Pan Hellenic Med. Assn. (hon.), Am. Assn. U. Profs., Soc. Exptl. Biology and Medicine, Soc. Gen. Physiology, Nat. Soc. Med. Research, Am. Neurol. Assn., Soc. Electroencephalogy, Internat. Brain Research Orgn., Acad. Psychoanalysis, Soc. Biol. Psychiatry (pres. 1964), Phi Beta Kappa, Sigma Xi, Alpha Omega Alpha, others. Author: Unresting Cells, 1940; The Body Functions, 1941; Food for Life, 1952; Mirror to Physiology; A Self-Survey of Physiological Science, 1958. Editor: Methods in Medical Research, 1950; Concepts of Biology, 1958; (with Cole) Psychopharmacology, Problems in Evaluation, 1959; (with Duyff) Information Processing in the Nervous System, 1964; Computers and Education, 1967. Research, publs. in chemistry, mitochondria metabolism, electrophysiology of nervous system, mechanisms of memory fixation, measurement of muscle, brain activity, organistic theory, respiratory irritant gases, relation of biology to social and ethical problems. Home: 1007 Goldenrod Av., Corona del Mar, Cal. 92625. Office: U. Cal., Irvine, Cal. 92664.*

GERARD OF BRUSSELS, physicist; probably Flemish; flourished 13th century; attempted in a treatise written in Euclidean style, De moto, to solve difficulties which were removed later by notion of angular rotation.

GERARD OF CREMONA, translator; b. Cremona, Italy, circa 1114; worked under ch. auspices in Toledo, Spain; translated math., astron. and med. works from Arabic to Latin, including: Almgest (Ptolemy); Elements (Euclid); also parts of Aristotle, Archimedes, Avicenna, Theodosius, Galen, al-Kindi, al-Razi, al-Khwarizmi. Died Toledo, 1187.

GERARD OF SABBIONETA, Italian astrologer; flourished circa 1255-59. Author: Geomantiae astronomiae libellus (treatise on geomancy); Theorica planetarum (summary of Ptolemaic astronomy as explained by al-Battani and al-Farghani).

GERASIMOV, Innokentii Petrovich, Russian geographer, pedologist; b. Dec. 9, 1905; grad. Leningrad U., 1929; D.Geog. Sci. Asso., USSR Acad. Sci., 1929-47, with Dokuchaev Inst. Soil Research, 1947-51, dir. Inst. Geography, 1951—, bur. mem. geology and geography sci. dept., 1960—, organizer, dir. Inst. Geography, Siberian Dept., 1958-60. Head Soviet delegation Internat. Geography Congress, Stockholm, 1960. Decorated Order of Lenin. Mem. USSR, Bulgarian (hon.) acads. sci., USSR Geog. Soc. (pres.). Co-author: The Soils of Bulgaria, 1948. Research and numerous publs. on soils, geomorphology, paleogeography, geology of quaternary deposits and pedology of Central Asia, Kazakhstan, Urals, Western Siberia, Caucasus, Mongolia and Bulgaria; devel. of natural physico-geographical zones, geomorphological zoning, cartography and classification of soils. Address: Inst. Geography, USSR Acad. Sci., Staromonetny p. 29, Moscow, USSR.

GERASIMOV, Mikhail Mikhaylovich, Russian anthropologist, archeologist, sculptor; b. 1907. With Archeol. Commn., USSR Acad. Sci. Recipient Stalin prize, 1950. Author: The Principles of Facial Reconstruction on the Basis of Skull Formation, 1949; Facial Reconstruction on the Basis of Skull Formation: Modern Fossilized Man, 1955; The Malta Paleolithic Settlement and Its Place among the Paleolithic Relics of Siberia, 1961. Research and publs. on facial reconstruction on basis of skull formation; discovered Malta settlement (west of Irkutsk), relic of Upper Paleolithic period, 1927; does sculptures of anthrop. types populating USSR from Paleolithic times to present, beginning with Pithecanthropos and Sinanthropos Address: Archeol. Commn., USSR Acad. Sci., Leninsky prospect 14, Moscow, USSR.

GERASIMOV, Yakov Ivanovich, Russian phys. chemist; b. Sept. 23, 1903; grad. Moscow U., 1925. Asst., Moscow U., 1927-31, docent, 1931-41; prof., 1941—. Del., 13th Internat. Congress on Theoretical and Applied Chemistry, Stockholm, 1953. Mem. USSR Acad. Sci. (corr.). Research and publs. on thermodynamic properties of metal compounds and alloys. Address: Moscow University, Leninskie gory, Moscow, USSR.*

GERBER, Niklaus, Swiss chemist; b. Thun, Switzerland, June 8, 1850; s. Niklaus and Anna Barbara (Steinmann) G.; studied chemistry, Bern, Switzerland; Ph.D., Zürich, 1874; postgrad. under Gautier, Paris, v. Pettenkofer, and v. Voit, Munich; m. Mathilde Theresia Weber, 1880; children—Walo Niklaus, Max, Viktor. Worked on problems of milk condensation and baby food prodn. with uncles' firm, beginning in 1874, Glocykenthal nr. Thun, until 1878, Little Falls, N.Y., 1880-83; returned to Switzerland, 1883; founder 1st dairy in Zurich, 1887; founder (with Hagershoff) Dr. N. Gerber's Co.m.b.H. for manufacture of apparatus for milk fat determination, Leipzig-Zurich, 1904. Author: Zur Ernährung der Kinder und das Kindenahrungsmittel, 1875; Chemisch Physikalische Analyse der verschiedenen Milch-Arten und Kindermehle, 1880; Anleitung zur praktischen Milchprüfung, 1881, pub. under title Die praktische Milchprüfung, 1954; Die natürliche Preservation der Kulnmilch und die Milchverproviantierung der Zukunft, 1883; Verbessergungen im Schweizer Molkereiwesen, 1886; Die Acid-Butyrometrie als Universal-Fettbestimmungsmethode, 1892. Research in milk chemistry, differences in human and cow's milk, milk condensation methods, manufacture of baby foods; invented acid-butyro-metry process for determining fat content of milk; pioneered establishment of hygienic dairies. Died Zurich, Feb. 9, 1914.

GERBERT OF AURILLAC (Pope Sylvester II), French mathematician, scholar; b. Aurillac, Auvergne, France, circa 940; ed. in Moslem schs. in Spain; in Barcelona, Spain, for several years; began teaching in Reims, France, 972; became 1st French pope, 999. Initiated rebirth of European learning; reintroduced abacus; built several clocks and astron. instruments; possibly 1st Christian to give sci. description of Spanish-Arabic numerals (although without zero). Died Rome, Italy, May 12, 1003.

GERBILLON, Father Jean, French mathematician; b. Verdun-sur-Meuse, France, June 4, 1654; Jesuit missionary sent to China by Louis XIV; mem. French Acad. Scis., 1699. Author: Traité de géométrie; wrote and translated treatises on geometry into Chinese and Tartar. Died Pekin, China, Mar. 27, 1707.

GERBODE, Frank Leven Albert, Am. surgeon; b. Placerville, Cal., Feb. 3, 1907; s. Frank A. and Anna (Leven) G.; A.B. cum laude, Stanford, 1932, M.D., 1936; M.Surgery (hon.), Nat. U. Ireland; M.D. (hon.), U. Thessaloniki, Greece, U. Uppsala, Sweden; m. Martha Barker Alexander, Dec. 24, 1931; children—Wallace Alexander, Maryanna (Mrs. Joseph Lionel Shaw), Frank Albert, Penelope, John Philip. Faculty, Stanford Med. Sch., Palo Alto, Cal., 1945—, clin. prof. surgery, 1959—; clin. prof. surgery U. Cal. Med. Sch., San Francisco, 1963—, chief cardiovascular surgery dept. Presbyn. Med. Center, San Francisco, 1959—. Hon. student St. Bartholomew's Hosp., London, Eng., 1949; lectr., guest prof. various instns., univs. U. S. and fgn. countries; cons. to USPHS-NIH. Named comdr. Brit. Order St. John of Jerusalem, 1965. Diplomate Am. Bd. Surgery, Am. Bd. Thoracic Surgery. Mem. A.A.A.S., Am. Assn. Thoracic Surgery, Am. Assn. U. Profs., A.C.S., Am. Heart Assn., A.M.A., Western, Am. surg. assns., Deutsche Gesselschaft Fur Chirurgie (corr.), Excelsior Surg. Soc., Internat. Cardiovascular Soc., Internat. Soc. Surgery, Soc. Clin. Surgery, Soc. Thoracic Surgeons, Soc. U. Surgeons, Soc. Vascular Surgery, So. Surg. Soc., Sociedad Medico Quirungice del Guayas, Guayaquil, Ecuador (hon.), Argentina Assn. Surgery (hon.), Alpha Omega Alpha, Sigma Xi, others. Editorial bd. Annals of Surgery, Surgery, Rev. Surgery, Surg. Sci.; sci. counsel Annales de Chirurgie Thoracique et Cardio-Vasculaire. Research, publs. on cardiovascular diseases. Home: 2560 Divisadero St., San Francisco 94115. Office: Presbyn. Med. Center, San Francisco 94115.*

GERDIEN, Hans, German physicist; b. Königsberg, Germany, May 13, 1877; s. Hugo and Emmy (Prange) G.; studied physics, math., chemistry, Munich, Göttingen, Germany; Ph.D., Göttingen, 1903; m. Alice von Wedel, 1909; 2 sons, 1 dau. Asst. to Wiechert, Geophys. Inst., Göttingen; asst. to E. Ricke, Inst. for Exptl. Physics, also univ. lectr., 1906-08; with firm Siemens, 1908-44, named dir. research lab Siemens & Halske AG and Siemens-Schuckertwerke, 1912, later dir. Mem. Soc. for Tech. Physics (hon.), German Phys. Soc. Patented numerous inventions, especially in electronics; pioneered ultrasonic research, especially in liquids. Died Brenke nr. Göttingen, Feb. 1, 1951.

GERDING, Harm, Dutch chemist; b. Sept. 10, 1899; s. Hendrik and Roelofje (Palthe) G.; Ph.D. in Sci., U. Amsterdam; m. Reincutje Kroon; children—Hendrike Hermance, Jacobus Joan Theophiel, Else Denise, Roelfine, Jahannes Constantijn. Asst., prof. chemistry H.B.S., Amsterdam; prof. natural sci. Gem. Kweekschool voor Onderwijzers, Amsterdam; head asst. U. Amsterdam, prof. agrégé, curator, reader, prof., lectr. electro-chemistry, photochemistry, spectroscopes and organic chemistry. Mem. Koninklijke Nederl. Chemische Vereiniging (past pres.), Am. Chem. Soc., A.A.-A.S., French Soc. Physics, Faraday Soc. Home: Michel Angelostraat 26. Office: Nwe Achtergracht 123, Amsterdam, Netherlands.

GERGEN, James Bernard, Am. chemist; b. Hastings, Minn., Nov. 19, 1922; s. Bernard G. and Leola (Thill) G.; B.S., Coll. St. Thomas, 1943; M.S. in Chemistry, U. Minn., 1949; m. Elizabeth Jane Duncan, Jan. 26, 1944; children—Kathryn Mary (Mrs. Patrick A. Coury), James Richard, Robert Joseph, Susan Marie. Chemist, Socony-Mobil Oil Co., St. Paul, 1942-43; research chemist AEC, Los Alamos, 1944-46; instr. Coll. St. Thomas, St. Paul, 1947-49; new products supr., tech. mgr. film products group Minn. Mining & Mfg. Co., St. Paul, 1949-60, program coordinator, projects mgr. microfilm products div., 1962—; pres., gen. mgr. Space Products, Inc., Chanhassen, Minn., 1960-62. Mem. Am. Chem. Soc. Invented 3M makeready system. Investigation of inorganic chemistry of plutonium and uranium; saponification of highly sterically hindered esters; promotion and devel. of new products. Home: 2570 Olson Lake Rd., St. Paul 55109. Office: 3M Center, St. Paul 55101.

GERGEN, John Jay, Am. mathematician; b. St. Paul, Apr. 17, 1903; s. John Andrew and Cora (Johnson) G.; B.A., U. Minn., 1925, M.A., 1926; Ph.D., Rice U., 1928; m. Aubigne Munger Lermond, June 11, 1931; children—John Andrew, Kenneth Jay, Stephen Lermond, David Richmond. NRC fellow U. Cal., Berkeley, Princeton, Oxford U., Clermont U., 1928-30; Benjamin Peirce instr. Harvard, 1930-33; asst. prof. U. Rochester, 1933-36; faculty Duke, 1936—, prof., 1939—, chmn. dept. math., 1937—. Acting dir. math. scis. div. Office Ordnance Research, U. S. Army, 1951-61. Recipient Dept. Army Outstanding Civilian Service medal, 1959. Contbr. articles to math. jours. Study of harmonic functions; partial differential equations; complex variables; Fourier series. Died Jan. 16, 1967.

GERGONNE, Joseph Diaz, French mathematician; b. Nancy, France, June 19, 1771; became officer arty. French Army, 1816; named prof. math. Lyceum in Nimes, 1795; prof. physics and astronomy, Montpellier, France, became rector, 1831. Mem. French Acad. Scis. Founder, editor, contbr. numerous articles Annales de mathematiques pures et appliquées, 1810-31. Originated terms polar, 1810, class of curve, 1827; formulated principle of duality as ind. principle of geometry (debated with Poncelot over priority of discovery); perfected analytic geometry of Descartes; solved Apollonian problem; believed analytic methods superior to synthetic methods. Died Montpellier, May 4, 1859.

GERHARD, (Friedrich Wilhelm) Eduard, archeologist; b. Poznan, Nov. 29, 1795; s. Johann David and Sophie (Nösselt) G.; student U. Breslau; Ph.D., U. Berlin, 1915; m. Emilie von Scheurnschloss, 1942; Worked in Italy, mainly Rome, 1818-33; founder Instituto di Corrispondenza Archeologica (became German Archeol. Inst. 1929), Rome; became archeologist in charge enlarging collection Berlin Mus., 1933; apptd. prof. U. Berlin, 1844. Mem. Berlin Acad. Scis. Author: Antike Bildwerke, zum erstenmal bekanntgemacht, 1828-44; Neapels antike Bildwerke, 1828; Beschreibung Roms, 1830-40; Auserlesene griechische Vasnbilder, 4 vols., 1939-58; Etruskische Spiegel, 4 vols., 1939-69; Trinkschalen und Gefässe, 2 vols., 1843-48; Griechische Mythologie, 2 vols., 1954-55; Gesamte Abhandungen, 2 vols., 1867. Founder Archäologische Zeitung, 1943. Pioneer in application of sci. methods to archeology; created basis for research by making sources available through collection and publs. of material. Died Berlin, May 12, 1867.

GERHARD, William Wood, Am. physician; b. Phila., July 23, 1809; s. William and Sarah (Wood) G.; A.B., Dickinson Coll., 1826; M.D., U. Pa., 1830; m. Miss Dobbyn, 1850; 3 children; went to Paris to study Asiatic cholera epidemic of 1831-32; resident physician Pa. Gen. Hosp., Phila., 1834-68; prof. physiology U. Pa., 1838-72. Gave 1st accurate description of tuberculosis meningitis, 1833. Wrote thesis on endermic application of medicaments, 1830; papers include those on pathology of smallpox, 1833, pneumonia in children 1834; wrote on Cerebral Affections of Children (study of tuberculosis meningitis), 1834; published his most important paper, On the Typhus Fever, which Occurred at Philadelphia in 1836 . . . showing the Distinction between this Form of Disease and . . . Typhoid Fever with Alteration of the Follicles of the Small Intestine (clearly distinguished typhus from typhoid fever for 1st time; 1837; published paper epidemic meningitis, 1863. Author: On the Diagnosis of Diseases of the Chest, 1836; Lectures on the Diagnosis, Pathology, and Treatment of the Diseases of the Chest, 1842; Diagnosis of Thoracic Diseases, 1835. Editor: Grave's System of Clinical Medicines. Specialized in pulmonary diseases. Died Phila., Apr. 28, 1872.

GERHARDT, Carl Jakob Christian Adolph, German physician; b. Speyer, Germany, June 5, 1833; s. Abraham and Clementine (Kolb) G.; began study medicine, Würzburg, Germany, 1850; m. Wanda von Barby; 7 children, including Dietrich. Became asst. to v. Rinecker, Med. Polyclinic, Würzburg, 1856; asst. to Griesinger, Med. Clinic, Tübingen, 1858-60; joined faculty U. Würzburg, 1860; named prof., dir. med. clinic, Jena, 1862, Würzburg, 1872; succeeded Friedrichs in Berlin, 1885. Author: Der Kehlkopfcroup, 1859; Der Stand des Diaphragmas, 1860; Lehrbuch der Kinderkrankheiten, 1861; Lehrbuch der Auscultation und Perkussion, 1866; Handbuch der Kinderkrankheiten, 6 vols., 1877-96. Introduced iron chloride test for acetic acid in urine; demonstrated presence of peptone in urine; originated concept of paralysis of larynx; research on laryngology, Tb., children's diseases, malaria, parasitology; described erythromelalgia (Gerhardt's disease), 1892. Died Gamburg nr. Mosback, Germany, July 21, 1902.

GERHARDT, Charles (Carl) Frédéric, chemist; b. Strasbourg (now France), Aug. 21, 1816; s. Samuel and Charlotte Henreiette (Lobstein) G.; student Polytechnikum, Karlsruhe, Germany, with O. Erdmann at U. Leipzig (Germany), with Liebig at U. Giessen (Germany), 1836-37; doctorate, U. Montpellier, 1841; m. Jane Megget Sanders, 1844; 2 sons, 1 dau. Worked in Persoz' lab., Strasbourg, 1837; asst. to J. B. Dumas, Paris, 1838, then worked with Chevreul; prof. chemistry, Montpellier, France, 1844-48; returned to Paris, 1848; founder (with Laurent) pvt. teaching lab., Paris, 1851; became prof. U. Strasbourg, 1855. Author: (pub. under pseudonyms Boyreau and Pellet) Manuel de manipulation à l'usage des personnes qui suivent les cours de la Sorbonne et de l'École de médecine, 1841; Précis de chimie organique, 1844-45; Introduction à l'étude de la chimie par le système unitaire, 1848; Traité de chimie organique, 4 vols., 1853-56; (with G. Chancel) Précis d'analyse chimique qualitative, 1855; Précis d'analyse chimique quantitative, 1859; Developed theory that all organic compounds derive from 4 types—hydrogen, hydrogen chloride, water, ammonia; developed theory of multi-basic acids built around remainder theory; confirmed theory of homologous series, also concept of heterology; developed unitary system from homologies; discovered anhydrides of monobasic acids, phenol, quinoline, acetanilide, cymol proposed a classification of organic compounds; contributed to devel. of atomic-weight theory. Died Strasbourg, Aug. 19, 1856.

GERHARDT, Dietrich, German physician; b. Jena, Germany, Feb. 16, 1866; s. Carl and Wanda (v. Barby) G.; studied medicine, Heidelberg, Würzburg, Berlin (all Germany); M.D., state exam, 1889; m.

Franziska Reye, 1898; 4 children. Asst. to Rindfleisch Würzburg, 1889-92; became asst. to Naunym, Strasbourg, France, 1892, joined faculty, 1894, became asso. prof., 1900; named asso. prof., dir. med. polyclinic, U. Erlangen (Germany), 1903, Jena, Germany, 1905; became prof., dir. med. clinic, Basel, Switzerland, 1907, Würzburg, 1911; ret. 1919. Author: Über Entstehung und diagnostischen Bedeutung der Herztöne, 1898; Über Herzmuskelerkrankungen, 1902; Über einige neuere Gesichtspunkte für die Diagnose und Therapie der Hierenkrankheiten, 1906; Über Anpassungs- und Ausgleichsvorgänge bei Krankheiten, 1908; Herzklappenfehler, 1913; Die Endokarditis, 1914; also articles. Research on internal medicine, especially heart, lung and pleura diseases; considered neurology a part of internal medicine; developed hypotheses important for modern views on heart disease. Died Meiningen, Germany, July 31, 1921.

GERHARDT, Ulrich Karl Friedrich Kurt Eduard, German zoologist; b. Würzburg, Germany, Oct. 11, 1875; s. Carl and Wanda (v. Barby) G.; studied medicine, natural scis., then zoology; m. Renate Zittelmann, 1904; 1 son, Dietrich; m. 2d, Renate Rauch; 1 dau., Eva-Maria (Mrs. Brandt). Joined faculty U. Breslau (now Wroclaw, Poland), 1905, named asso. prof., 1922; became prof. anatomy and physiology domestic animals, Halle, Germany, 1924, also dean natural sci. faculty, prorector during rebldg. of univ. after 1945. Mem. Leopoldina, German Zool. Soc. Research and numerous publs. on anatomy of kidneys and reproductive organs of mammals, especially sexual biology of spiders and snails; developed theory of orthogenesis. Died Halle, June 8, 1950.

GERISCHER, Heinz, German phys. chemist; b. Wittenberg, Germany, Mar. 31, 1919; s. Oskar and Amalie (Scheuer) G.; diploma U. Leipzig, 1943, Dr. rer. nat., 1946; m. Renate Gersdorf, Oct. 22, 1948; children—Cornelia, Ulrike, Christiane, Bettina. Asst. prof. U. Berlin, 1946-48; research asso. Max Planck Inst. for Phys. Chemistry, Göttingen, Germany, 1949-53; sr. research fellow Max Planck Inst. Metals, Stuttgart, Germany, 1954-61, now external mem.; asso. prof. electrochemistry Tech. U., Munich, Germany, 1962-64, prof. phys. chemistry, dir. Inst. for Phys. Chemistry and Electrochemistry, 1964—. Recipient Bodenstein award German Bunson Soc., 1953. Contbr. numerous articles to profl. jours. Research on methods for investigating fast electrode reactions, mechanism of electrolytic deposition of metals, electrochemistry of semiconductors, fast reactions in homogeneous solution, heterogeneous catalysis. Home: 20 Alpenstrasse, Ebersberg, Bavaria 8019. Office: 21 Arcisstrasse, Munich, Bavaria 8000, Germany.*

GÉRKE, Peter Iakovlevich, Russian histologist, embryologist; b. Russia, July 12, 1904; grad. Byelorussian U., Minsk, 1927. Mem. staff Minsk Med. Inst., 1930-52; prof., since 1937; dir. Inst. Exptl. Medicine, Latvian SSR Acad. Scis. (mem), since 1952. Made cytological studies of nematodes, amphioxus lanceolutus, and several mammals; also writings on comparative embryology of mammals.

GERKING, Shelby Delos, Jr., Am. zoologist, educator; b. Elkhart, Ind., Nov. 16, 1918; s. Shelby Delos and Fezon (Churchill) G.; A.B. with distinction DePauw U., 1940; student U. Mich., 1939, 41; Ph.D., Ind. U., 1944; m. Louisa B. Pfretzschner, Dec. 28, 1943; children—Shelby Delos III, Timothy Churchill, Andrew Alfred. Research asso. physiology Ind. U., 1944-46, faculty zoology, 1946-67, prof., dir. U. Biol. Sta., 1959-67; prof., chmn. dept. zoology Ariz. State U., Tempe, 1967—. Dir. Ind. Aquatic Research Unit; 1959-67; research asso. lake and stream survey Ind. Dept. Conservation, 1946-53; asso. dir. Water Resources Research Center; mem. freshwater productivity com. Internat. Biol. Program; mem. U. S. nat. com. Internat. Biol. Program. Fellow A.A.A.S., Ind. Acad. Sci., Am. Inst. Fishery Research Biologists; mem. Ecol. Soc. Am., Am. Inst. Biol. Scis., Internat. Assn. Limnology, Am. Fisheries Soc., Am. Soc. Limnology and Oceanography, Am. Soc. Ichthyologists and Herpetologists, Am. Soc. Zoologists, Wildlife Soc., N.Y. Acad. Scis., Sigma Xi. Author: Laboratory Manual for Man and the Biological World, 1958, 2d edit., 1963. Editor: Biological Basis of Fresh Water Fish Production, 1967. Research, publs. on lake and stream populations and relation to food supply, factors behavioral and nutritional important in fish prodn. Home: 418 E. Alameda St., Tempe, Ariz. 85281.*

GERLACH, Arch Clive, Am. geographer; b. Tacoma, May 12, 1911; s. William Henry and Kathryn Alice (Cooper) G.; B.A., San Diego State Coll., 1933; M.A., U. Cal. at Los Angeles, 1935; Ph.D., U. Wash., 1943; m. Arlene Marie Schmiedeman, Dec. 31, 1935. Faculty, Los Angeles City Coll., 1939-42, U. Wis., 1946-50; chief map div. U. S. Dept. State, 1945-46; chief geography and map div. Library of Congress, Washington, 1950-67; chief Nat. Atlas Project, coordinator for geog. applications of satellite data U. S. Geol. Survey, 1962-67, chief geographer, 1967—; chmn. adv. com. on geography U. S. Dept. State, Nat. Acad. Scis., 1956-61. Hon. fellow Am. Geog. Soc.; mem. Internat. Geog. Union (v.p. 1964—), Pan Am. Inst. Geography and History (v.p. 1965—), Assn. Am. Geographers (pres 1962-63), Spl. Libraries Assn. (Outstanding Achievement award 1961), Nat. Council for Geog. Edn., Assn. Pacific Geographers, Am. Congress Surveying and Mapping. Editor: National Atlas of the United States. Contbr. articles to profl. publs. Developer thematic maps for intelligence studies in OSS; research on land use, resource inventories, urban devel., coastal morphology. Home: 5615 Newington Rd., Washington 20016. Office: U. S. Geol. Survey, Washington 20542.*

GERLACH, Eckehart, German physiologist; b. Göttingen, Germany, Apr. 2, 1927; s. Walter and Elisabeth (Küch) G.; student U. Göttingen, 1946-52, U. Heidelberg, 1952-53, M.D., 1954; m. Ingrid Bues, Dec. 18, 1954; children—Imke, Evelin, Jörn Tilman. Mem. staff dept. pharmacology U. Heidelberg (Germany), 1954-56; staff dept. physiology U. Freiburg/Breisgau (Germany), 1956—, prof. physiology, 1966—; chmn. physiology faculty medicine Tech. U. Aachen (Germany), 1966—. Mem. Deutsche Physiologische Gesellschaft, Deutsche Gesellschaft f. Biologische Chemie, Nephrologische Gesellschaft, N.Y. Acad. Scis. Research, publs. on cell physiology, application of modern methods of biochemistry in combination with radioisotopes; devel. of new analytical procedures for micro-determination of organo-phosphates in living cells and tissues; basic studies on energy and phosphate metabolism in erythrocytes kidney, brain and heart; studies on significance of naturally occurring compounds for regulation of coronary blood flow. Home: 49 Okenstrasse, Freiburg. Office: 7 Herm.-Herder-Str., Freiburg/Br., Germany.*

GERLACH, Joachim, German surgeon; b. Breslau, Germany (now Wroclaw, Poland), Mar. 30, 1908; s. Ernst and Gertrude (Jahn) G.; ed. univs. Breslau, Munich, Frankfort am Main; M.D.; m. Hanna Cruse; children—Anneliese, Brigitte, Bernhard, Ernst. With Anat. Inst., U. Lena, 1933-34, surg. dept. Bethany Diaconic Clinic, Liegnitz, 1934-37, Kaiser Wilhelm Inst. for Study Brain, 1938-45, neurosurg. dept. Clinic of Schleswig-Stadfeld, 1945-48, U. Würzburg, 1948-63. Mem. German Soc. Neurosurgery, German Soc. Surgery, Phys.-Med. Soc. Würzburg. Home: 11 Essigkrug, 8702 Versbach. Office: Luitpoldkrankenhaus, 87 Würzburg, Germany.

GERLACH, Max, agrl. chemist; b. Prenzlau, Germany, May 28, 1861; s. Friedrich Wilhelm Ferdinand and Wilhelmine Natalie Ida (Keitz) G.; student chemistry, agr.; doctorate, Halle, Germany, 1888; hon. dr. Agr. U. Berlin, 1932. Became asst. to M. Maerker at Agrl. Exptl. Sta., Halle, 1887; named dir. exptl. sta., Poznan (now Poland), 1893; apptd. prof., 1904; named dir. Kaiser Wilhelm Inst. for Agr., Bromberg, Germany, 1906; made expdn. to Turkey, 1917; dir. Exptl. Inst. for Cultivation of Plants and Feed, 1923-30. Became mem. sci. council Reich Bur. Interior, 1912. Mem. Assn. Agrl. Exptl. Stas. in German Reich. Research and publs. on soils, plant prodn., plant protection, fertilizer, feed; co-discoverer nitrogen-fixing bacterium, azotobakter. Died Berlin-Steglitz, Mar. 30. 1940.

GERLACH, Walther, German physicist; b. 1889; ed. U. Tübingen. Lectr., U. Tübingen, 1916, prof. physics, 1925; lectr. U. Göttingen, 1917; lectr. U. Frankfurt, 1920, extraordinary prof., 1921-25; prof. physics Munich U., 1929-57. Mem. Bavarian Acad. sci., Göttingen Acad. Sci., Acad. Leopoldina Halle. Author: Grundlagen der Quantentheorie, 1921; Atombau und Atomabbau, 1923; Materie, Elektrizität, Energie, 1923; Die chemische Spektralanalyse, 1930, Part II, 1933, Part III, 1936; Magnetismus, 1931; Foundations and Methods of Chemical Analysis by the Emission Spectrum, 1934; Methoden der naturwissenschaftlichen Erkenntis, 1936; Max Planck—Werk und Wirkung, 1948; Akademische Provinz, 1949; Humaniora und Natur, 1950; Physik des täglichen Lebens, 1956; Physik (Fischer Lexikon), 1960; Humanität und Naturwissenschaftliche Forschung, 1962; Die Sprache der Physik, 1962; Physik in Geistesschichte und Paedagogik, 1964. Research and publs. on magnetism; made important expts. on spatial aspects of quantum theory. Address: 15 Franz Joseph Strasse, Munich 13, Germany.

GERLAND, (Anton Werner) Ernst, physicist; b. Kassel, Germany, Mar. 16, 1838; s. Althasar and Wilhelmine (Grandidier) G.; ed. Polytechnikum, Karlsruhe, Germany; began study math. and physics U. Marburg, 1859; teaching certificate 1863; Ph.D., 1864; m. Henriette Doussin, 1876; a dau., Margarete (Mrs. Hellmuth van Steinwehr). Asst. tchr. secondary sch. in Kassel, 1863-67; became asst. to Wüllner, U. Bonn (Germany), 1867; asst. to Rijke, then mem. faculty in physics U. Leyden (Netherlands); participated in Franco-Prussian War, 1870-71; became tchr. physics and math. Upper Indsl. Sch., Kassel, 1872; named lectr. Mining Acad., Clausthal, Germany, 1888, prof. physics and electrotechnics, 1892. Author: Leibnizens und Hugens' Briefwechsel mit Papin, nebst der Biographie Papins, 1881; Licht und Wärme, 1883; Die Anwendung der elektrizität bei registrierenden Apparaten, 1887; Leibnizens nachgelassene Schriften physikalischen, mechanischen und technischen Inhalts, 1906; Geschichte der Physik, part I, 1913; (with F. Traumüller) Geschichte der physikalischen Experimentierkunst, 1899. Research on exptl. physics and plant physiology; noted for hist. studies on phys. apparatus, Leibniz, Papin. Died Clausthal, Mar. 22, 1910.

GERLAND, Georg Cornelius Karl, German geographer, geophysicist; b. Kassel, Germany, Jan. 29, 1833; s. Balthasar and Wilhelmine (Grandidier) G.; studied classical philology, German lit., anthropology in Berlin and Marburg, Germany; Ph.D., Marburg, 1859; m. Wilhelmine Henke, 1864; 1 son, Heinrich, 4 daus. Secondary sch. tchr., Kassel, also Hanau, Magdeburg, Halle, 1856-75; apptd. prof. geography U. Strasbourg (France), 1875, remained for 35 years lecturing on theology, ethnology, math., geography, cartography, geography of organisms, descriptive geography, geophysics.; helped found Meteorol. Inst. of Alsace-Lorraine, also Seismographic Research Sta., Strasbourg, 1899. Mem. Internat. Seismol. Soc. (a founder). Author: Über den Aussterben der Naturvölker, 1868; Atlas der Völkerkunde, 1892; Geographischen Abhandlungen aus dem Reichslande Elsass-Lothringen, 1892; Das Reichland Elsass-Lothringen, part I, 1898-1901; completed 5th vol. of Weitz's work on anthropology of primitive peoples and wrote 6th vol., Die Völker der Südsee, 1872. Founder Beiträge zur Geophysik (now Gerlands Beiträge zur Geophysik), 1887. Research on anthropology and ethnology; a founder of geophysics; believed study of man derived from geography. Died Strasbourg, Feb. 16, 1919.

GERLING, Christian Ludwig, German mathematician, physicist, astronomer; b. Hamburg, Germany, July 10, 1788; s. Christian Ludwig and Charlotte (Helmer) G.; studied in Helmstedt, Göttingen (both Germany); Ph.D., 1812; studied under C. E. Gauss; m. Christian Wilhelmine Suabedissen, 1814; children— 1 son, 3 daus., including Emma (Mrs. Karl Winkelblech), Marie (Mrs. Herman Platner). Prof., Lyceum, Kassel; named prof. math., physics and astronomy U. Marburg (Germany), 1817-64, built math.-phys. inst. with obs., 1841. Mem. Göttingen Acad. Scis. Author: Die Ausgleichrechnung der praktischen Geometrie oder die Methode der kleinsten Quadrate mit ihrer Anwendung auf geodätische Aufgaben, 1843; also numerous articles. Determined solar parallax, star magnitude; measured geomagnetic variation; performed triangulation of Kur-Hessig. Died Marburg, Jan. 15, 1864.

GERMAIN, Jean Eugene, French chemist; b. Boujan, France, Jan. 28, 1922; s. Martial and Fernande (Feral) G.; B.Sc., Sorbonne, Paris, France, 1939, M.Sc., 1945, Ph.D., 1952; postgrad. Northwestern U.; m. Dorothy Lamar Moore, Oct. 1951; children—Richard Jean, Elisabeth Dorothy. Faculty, U. Lille, Faculty Scis., 1952-66, prof., 1954-66, dir. Sch. Chem. Engring., 1963-66; prof. Faculty Scis., U. Lyon (France), 1966—, dir. Sch. Indsl. Chemistry, 1966—. Mem. French (council 1965-66), Am. chem. socs., Soc. Indsl. Chemistry (past pres. No. sect.), Soc. Phys. Chemistry (council 1963-66), Sigma Xi, Alpha Chi Sigma. Author: Heterogeneous Catalysis, 1959. Research, numerous publs. on heterogeneous catalysis and catalytic conversion of hydrocarbons. Address: 43 Blvd. du 11 Novembre, 69 Lyon-Villeurbanne, France.*

GERMAIN, Paul Marie, French physicist; b. St. Malo, France, Aug. 28, 1920; s. Paul Albert and Elisabeth (Frangeul) G.; ed. Faculty Scis. Paris, École Normale Supérieure; agrégé de l'Université (Math.), 1942; Dr. ès Scis., Paris, 1948; hon. doctorate U. Louvain (Belgium); m. Marie Antoinette Gardent, Oct. 10, 1942; children—Marie Hélène, François. Researcher, C.N.R.S., 1942-46; research engr. Office Nat. d'Études et de Recherches Aeronautiques (ONERA), 1946-49, dir. gen., 1962-67; prof. Faculty Scis. Poitiers (France), 1949-53, Lille (France), 1954-58, Paris, 1958—. Vis. prof. Brown U., 1953-54, Cal. Inst. Tech., 1957-58. Mem. French Acad. Scis. (corr.), Internat. Acad. Astronautics, Am. Acad. Arts and Scis. (fgn. hon.). Author: Mécanique des milieux continus, 1962; also articles. Research on supersonic and transonic flows, problems in theory of partial differential equations of mixed type, shock waves, magneto fluid dynamics, shock wave and shock structure of plasma physics. Home: 3 Av. de Champaubert, Paris 15, France.*

GERMAIN, Sophie, French mathematician; b. Paris, Apr. 1, 1776; d. Ambroise François and Marie (Bruguelu) G.; corr. study at École polytechnique, Paris. Recipient prize Institut de France, 1816. Author: Mémoire sur vibrations des lames élastiques, 1816; Recherches sur la théorie des surfaces élastiques, 1821. Proved impossibility of solving Fermat's last equation if x, y, and z are not divisible by an odd prime.; gave theoretical explanation of Chaloni's vibrating plates. Died Paris, France, June 27, 1831.

GERMANI, Gino, sociologist; b. Rome, Italy, Feb. 2, 1911; s. Luigi and Lina (Catalini) G.; student Faculty Econ. Scis., U. Rome, 1930-34; Licenciado in Philosophy, U. Buenos Aires (Argentina), 1943; M.A. (hon.), Harvard, 1966; m. Celia Carpi, Nov. 12, 1954; children—Louis Sergio, Anna Alejandra. With U. Buenos Aires, 1942-45, prof. sociology, dir. Inst. Sociology, 1955-66, chmn. dept. sociology, 1957-62; prof., dir. seminar sociology Colegio Libre de Estudios Superiores, Buenos Aires, Rosario, Argentina, 1946-55; Monroe Gutman prof. Latin Am. affairs dept. social relations Harvard, 1966—; vis. prof. U. Chgo., 1959, U. Cal. at Berkeley, 1961-62, Columbia, 1964-65. Fellow Am. Social. Soc.; mem. Asociación Sociológica Argentina, Associazione Italiana di Scienze Sociali, Am. Acad. Arts and Scis. (fgn. hon.). Author: Sociología Cientifica, 1956; Estudios de Psicología Social,

1956; Estructura Social de la Argentina, 1955; Politica e Mass, 1962; Politica y Sociedad en una Epoca de transición, 1964; Estudios de Sociología y Psicología Social, 1966; also articles. Editor: (with T. Di Tella, J. Graciarena) Argentina, Sociedad de Masas, 1965. Analysis of social structure of Argentina, theory of polit. and social change in Argentina and Latin Am., characteristics of process of mobilization theory and application to Latin Am. Home: 201 Highland St., West Newton, Mass. 02165. Office: William James Hall, Harvard Cambridge, Mass. 02138.*

GERMANN, Frank Erhart Emmanuel, Am. physicist, chemist; b. Peru, Ind., Dec. 6, 1887; s. Gustave Adolph and Mary (Miller) G.; A.B. cum laude, U. Ind., 1911; postgrad. U. Berlin, U. Lausanne, Switzerland, U. Neuchatel, Switzerland; Dr. ès Sc. Phys., U. Geneva, Switzerland, 1913; m. Martha Marie Knechtel, July 25, 1916; children—Richard Paul, Lois Marie (Mrs. Horace Jones). Docent, U. Geneva, 1912-13; faculty U. Ind., 1913-14; Carnegie teaching fellow Cornell U., 1914-18; prof. Colo. Sch. Mines, 1918-19; prof. phys. chemistry U. Colo., Boulder, 1919-56, prof. emeritus, 1956—; chemist Nat. Bur. Standards, Boulder, 1956—. Fellow Am. Phys. Soc., A.A.A.S. (pres. S.W. div. 1936-38); mem. Am. Chem. Soc. (pres. Colo. sect. 1934-36), Colo.-Wyo. Acad. sci. (pres. 1930-31). Contbr. articles to Am., internat. sci. mags., textbook. Research in field of phys. constants of cryogenic materials. Home: 1800 Sunset Blvd. Office: care Nat. Bur. Standards, Boulder, Colo. 80302.

GERMAR, Ernst Friedrich, German entomologist, paleontologist, geologist; b. Glauchau, Saxony, Feb. 3, 1786; s. Johann Ernst Eberhardt and Susanne Magdalene (Zergiebel) G.; began study natural scis. in Freiburg, Germany, 1804; studied law in Leipzig: Ph.D., Halle, Germany, 1810; m. Wilhelmine Keferstein, 1815. Visited Dalmatia, 1811; joined faculty U. Halle, 1812, named asso. prof., 1817, prof. mineralogy, 1824. Author: Systematis glassatorum prodromus, 1810; Fauna insectorum Europae, 24 pamphlets, 1812-48; Insectorum species novae aut minus cognitae I, Coleoptera, 1824; Die Versteinerungen des Mansfelder Kupferschiefers, 1840; Die Versteinerungen des Steinkohlengebirges von Wettin und Löbejün, 8 pamphlets, 1844-52. Editor, Magazin der Entomologie, 4 vols., 1813-21, Zeitschrift für die Entomologie, 5 vols., 1839-44. Research on Coleopterae and Hemipterae, fossil insects, fossil flora of Wettin coal; made entomol. collection 2d in size in Germany only to that of Berlin Mus. Died Halle/Salle, July 8, 1853.

GERMER, Lester Halbert, Am. physicist; b. Chgo., Oct. 10, 1896; s. Hermann G. and Marcia (Halbert) G.; A.B., Cornell U., 1917; M.A., Columbia, 1922. Ph.D., 1927; m. Ruth Woodard, Oct. 2, 1919; children—Emily (Mrs. V. W. Samms), John Halbert G. With engrng. dept. Western Electric Co., 1917-25, research physicist, 1925-53; tech. staff, research physicist Bell Telephone Labs., 1925-61, Cornell U., Ithaca, N.Y., 1961—. Recipient Elliot Cresson medal, 1931. Fellow A.A.A.S., N.Y. Acad. Sci., Am. Phys. Soc. (chmn. N.Y. sect. 1944); mem. Soc. X-Ray and Electron Diffraction (v.p. 1943, pres. 1944), Sigma Xi. Author sci. articles. Discoverer (with Dr. C. J. Davisson) of diffraction of electrons by crystals, 1927; other studies in thermionics, erosion of metals, contact physics, plating of molyodenum and tungsten by thermal decomposition of carbonyls. Home: Long Hill Rd., Millington, N.J. Office: Clark Hall, Cornell U., Ithaca, N.Y.

GERNEZ, Désiré-Jean-Baptiste, French chemist; b. Valenciennes, France, Apr. 24, 1834; prof. gen. chemistry École normale, Paris. Mem. French Acad. Scis., 1906. Collaborator on Annales scientifiques de l'école Normale. Research on saturated salt solutions, crystallization of super saturated salts; studied (with Pasteur) wine and silk worm diseases; used blow-hole to investigate speed of steams; studied rotary power of various liquids, boiling, supercooling. Died Paris, Oct. 31, 1910.

GERNGROSS, Otto, chemist; b. Vienna, Austria, Feb. 26, 1882; s. Alfred and Emma (Sichel) G.; ed. univs. Friburg, Bern, Berlin; Ph.D. in chemistry; m. Erna Hecht, 1910; children—Edith, Susi, Veronika. Asst., agrégé, asso. prof., full prof. Technische Hochschule, Berlin; full prof. Sch. for Higher Agrl. Studies, Ankara, Turkey; prof., dir. Inst. Tech. Chemistry, U. Ankara. Recipient Karl Auer von Welsbach medal. Mem. Austrian Tech. Assn. Leather (hon.), Am. Chem. Soc. of Leather. Editor: Chemie und Technologie der Leim und Gelatine Fabrikotien. Research and publs. on organic chemistry, proteins, gelatin, ferments, and leather. Died Jan. 23, 1966.

GERO, Jan, Czechoslovakian physiologist; b. Trnava, Czechoslovakia, Mar. 21, 1921; s. Ignac and Vilma (Hoffmanova) G.; M.D., Comenius U. Med. Sch., 1950; C.Sc., Charles U., Prague, 1955; m. Maria Gerova, Feb. 1, 1926; 1 dau., Zora. Asso. prof. dept. social scis. Comenius U. Med. Sch., 1950-51; postgrad. fellow dept. pharmacology Charles U. Med. Sch., Prague, 1951-54, asst. prof., 1954-55; sr. investigator dept. cardiovascular physiology Inst. Normal and Path. Physiology, Slovak Acad. Scis., Bratislava, Czechoslovakia, 1955—, chief dept., mem. bd., 1960—. Mem. Czechoslovakian Med. Soc. Sects. Physiology, Pharmacology, Cardiology. Author: (with Maria Gerova) Elasticity of Sinocarotid Region, 1961. Research, numerous publs. on reflex regulation of blood circulation, role of central nervous system as influenced by drugs, hemodynamic parameters inducing stimulation of blood vessels smooth muscle activity, phys. properties of blood vessels, elasticity of blood vessels and their determinant factors, tension of vessel wall as determinant of reactivity, sympathetic regulation of vascular smooth muscle. Home: 6 Belehradska. Office: 1 Sienkiewiczova, Bratislava, Czechoslovakia.*

GERRARD, John Watson, physician; b. N. Rhodesia, Apr. 14, 1916; s. Herbert Shaw and Doris (Watson) G.; B.M., B.Ch., D.M., Oxford U., Eng., Birmingham U., Eng., m. Lilian Elisabeth Whitehead, Aug. 28, 1941; children—Jonathan, Peter, Christopher. Lectr. pediatrics U. Birmingham, 1948-55; prof. pediatrics U. Sask., Saskatoon, Can., 1955—. Recipient John Scott award, 1962. Fellow Royal Coll. Physicians. Contbr. numerous articles to sci. jours. Assisted in discovery of low phenylalanine diet for phenylketonuria; research in celiac disease and food allergies. Home: 809 Colony St. Office: Univ. Hosp., Saskatoon, Sask., Can.*

GERRARD, William, English chemist; b. Tyldesley, Eng., Feb. 24, 1900; s. Thomas and Mary Ann (Hampson) G.; Ph.D., U. London (Eng.), 1932, D.Sc., 1947; postgrad. Wigan Mining and Tech. Coll., Battersea Poly.; m. Janet Hilton Smith, Dec. 28, 1927; children—Sheilah Mary (Mrs. Raynham Herbert Sawyer), John Peter Duncan, Paul Michael. Lectr. in charge chemistry and physics London Coll. Pharmacy, 1925-28; head dept. sci. Norwood Tech. Coll., London, 1928-34; lectr. No. Poly., London, 1934-47, head dept. chemistry, maths., biology and geology, 1947-65, asso. lectr. for research, 1965—. Cons. and control work in rubber tech., milk analysis, organophosphorus and organoboron chemistry, 1934—. Fellow Royal Inst. Chemistry; mem. Chem. Soc., Am. Chem. Soc., Soc. Chem. Ind., Royal Instn. Author: The Organic Chemistry of Boron, 1961; also numerous articles. Research on interaction of hydroxy compounds with halides, preparation of applicable inorganic polymers, prevalence of rearrangement in simple alkyl groups, significance of solubility of hydrogen halides in compounds. Home: 15 Oakroyd Close, Potters Bar, Hertfordshire, Eng. Office: No. Poly., Holloway Rd., London N.7., Eng.*

GERRETSEN, Johan Cornelis Hendrik, Dutch mathematician; b. Winschoten, Netherlands, May 20, 1907; s. Frederik Willem and Carolina (de Vries) G.; ed. U. Groningen (Netherlands), 1925-30; m. Clasina Willemse, Nov. 30, 1933; children—Frederik, Livinus, Johannes, Carolina (Mrs. Catharinus Heierman). Tchr. math., 1930-43; prof. math. U. Groningen, 1946—; dir. Math. Inst. Mem. Wiskundig Genootschap. Author: The Topological Foundations of the Enumerative Geometry, 1939; Descriptive Geometry, 1940; Non-Euclidean Geometry, 1949; (with G. Sansone) Lectures on the Theory of Functions of a Complex Variable, vol. I, 1960, vol. II, 1968; Lectures on Tensor Calculus and Differential Geometry, 1962; Tangente und Flächeninhalt, Dutch edit., 1964, German edit., 1966. Research, publs. on founds. of enumerative geometry, founds. of hyperbolic trigonometry, more-dimensional projective geometry (Segre-manifolds), inequalities in triangles, functional equations. Home: 31 Gratamastraat, Groningen. Office: 4 Reitdiepskade, Groningen, Netherlands.*

GERRITSEN, Alexander Nicolaas, physicist; b. Den Haag, Netherlands, Nov. 29, 1913; s. Sander and Gerrie (Vander Willik) G.; Candidate, Leiden U. (Netherlands), 1933, Doctorandus, 1937, Doctor, 1948; m. Jacqueline Voolhaas, Sept. 27, 1943; children—Rob, Jeroen. Head instr. lab. courses Kamerlingh Onnes Lab., Leiden U., 1943-47; sr. scientist Netherlands Orgn. for Basic Research on Matter, 1947-55; vis. prof. Purdue U., Lafayette, Ind., 1954, asso. prof. 1956-59, prof. 1960—. Sec., Netherlands Orgn. Sci. Investigators, 1948-54. Fellow Am. Phys. Soc.; mem. Netherlands Phys. Soc., Ind. Acad. Scis., Am. Assn. U. Profs. Work in separation electronic and lattice thermal conductivity, 1936; ion transport in liquefied gases, 1949; anomalous electron transport behavior paramagnetic alloys, 1951-56; susceptibility of semiconductors, 1962; transport properties magnesium, 1964. Home: 100 Wheeler Lane, West Lafayette, Ind. Office: Dept. Physics, Purdue U., Lafayette, Ind. 47907.*

GERSCHENKRON, Alexander, economist; b. Odessa, Russia, Oct. 1, 1904; s. Paul and Sophie (Kardow) G.; Doctor Rerum Politicarum, U. Vienna, 1928; m. Erica Matschnigg, 1928; children—Helga-Susanna, Maria-Renate. Came to U. S., 1938, naturalized, 1945. Research asso. Austrian Inst. for Bus. Cycle Research, 1937-38, dept. econs. U. Cal., Berkeley, 1938-42, lectr., 1942-44; staff mem. Bd. Govs. Fed. Res. System, 1944-48, chief fgn. areas sect., 1946-48; asso. prof. econs. Harvard, 1948-51, prof. econs., 1951-55, Walter S. Barker prof. econs., 1955—, dir. econ. history, 1959—, Frank W. Taussig research prof. econs., 1961-62; Guggenheim fellow, 1954-55; director economic projects Russian Research Center 1949-56; Ford Research prof. U. Cal., Berkeley, 1958; vis. fellow St. Catherine's Coll., Oxford, Eng. Mem. Am. Statis. Assn., Am. Hist. Assn., Am. Econ. Assn., Econ. History Assn. (pres. 1966-68), Econ. History Soc. (Eng.), Phi Beta Kappa (hon.). Author: Bread and Demoracy in Germany, 1943; Economic Relations with the U.S.S.R., 1945; A Dollar Index of Soviet Machinery Output, 1951; Economic Backwardness in Historical Perspective, 1962; Continuity in History, 1968; also articles econ. jours. Developed approach to indsl. devel. Europe in 19th century in terms of relative backwardness of areas concerned; hist. analysis of index number problem with regard to measurement indsl. output in periods rapid econ. growth. Home: Shady Hill Sq., Cambridge, Mass. 02138.*

GERSH, Isidore, Am. anatomist, b. Bklyn., Oct. 6, 1907; s. Charles and Rebecca (Kaplan) G.; B.S., Cornell U., 1928; Ph.D., U. Chgo., 1932; m. Eileen Mary Sutton, Dec. 4, 1944; children—Frank S., Ilona M. Instr., asso. Johns Hopkins, 1933-46; asso. prof. U. Ill. Coll. Medicine, 1946-49; faculty dept. anatomy, U. Chgo., 1949-64, prof., 1950-64; research prof. dept. animal biology U. Pa., Phila., 1963—; vis. prof. U. Birmingham (Eng.), 1951, U. Oslo (Norway), 1957, U. Santiago (Chile), 1959, U. Va., 1958, 61, 62. Mem. Am. Assn. Anatomists, Biophys. Soc., Am. Soc. for Exptl. Pathology, A.A.A.S. Research, numerous articles. on freeze-dry method for morphological and cytochem. studies with light and electron microscopes, structure of protoplasm, nature of connective tissues. Home: 4037 Baltimore Av., Phila. 19104.*

GERSHENFELD, Louis, Am. microbiologist; b. Phila., Dec. 25, 1895; s. George and Jennie (Stupe) G.; B.S., Phila. Coll. Pharmacy and Sci., 1917, Ph.M., 1920, D.Sc., 1940; postgrad. U. Pa. Jefferson Med. Coll.; m. Bertha Miller, Nov. 17, 1918; children—George, Marvin Aaron. Dir. dept. bacteriology Phila. Coll. Pharmacy and Sci., 1917—, prof. bacteriology, 1920—, also charge clin. chemistry labs.; dir. Gershenfeld Lab., 1919—. Bacteriologist, sci. cons. Upper Darby Twp. Dept. Health, 1930-66; cons. to instns., indsl., chem., pharm. firms; mem. sterile adv. bd. U. S. Pharmacopeia XII, XIII, XIV; chmn. clin. lab. prep. Nat. Formulary VII; pub. com. chmn. Nat. Biol. Stain Commn., Nat. Council Pharm. Research. Fellow Am. Pub. Health Assn., A.A.A.S.; mem. Am. Pharm. Assn., Am. Soc. Microbiologists, N.Y. Acad. Scis., Am. Acad. Polit. and Social Sci., many others. Author: Bacteriology and Sanitary Science, 1929, 33; The Jew in Science, 1934; Biological Products, 1939; Bacteriology and Allied Subjects, 1945, 47; Urine and Urinalysis, 1933, 43, 48. Research, publs. in numerous fields including disinfectants and insecticides, sterilization of medicants and injectible preparations, surface active agts., snake venom, air and atmospheric conditions, urinalysis and biol. products, others. Home: 1101 N. 63d St., Phila. 19151. Office: 43d and Kingsessing Av., Phila. 19104.*

GERSHENOVICH, Zoundel Semenovich, Russian biochemist; b. Tremgen, Belorussia, USSR, May 25, 1905; s. Semen and Elizabeth (Kaganovich) G.; D.Biochemistry, Krasnodar Med. Inst., 1928, Prof., 1940; m. Klavdia Bocharova, Nov., 1930; children—Oleg, Alexander, Natalia. Staff, Inst. Biochemistry, Kiev, USSR, 1929-41; head biochemistry dept. Rostov-on-Don (USSR) State U., 1945—. Decorated Krasnaya Zvedzda and medals. Mem. Biochemistry Soc. N. Caucasus (pres. 1964—). Research, numerous publs. on chem. activity of brain in different conditions. Home: 247 Gorky St., Rostov-on-Don, USSR.*

GERSHON-COHEN, Jacob, Am. physician, radiologist; b. Phila., Jan. 9, 1899; s. Abraham and Dora (Starkman) Cohen; M.D., U. Pa., 1924, D.Sc., 1936; m. Sara Eskin, Mar. 26, 1921. Practiced medicine, specializing in radiology, Phila., 1928—; dir. div. radiology Albert Einstein Med. Center, 1949-66, now dir. emeritus; prof. radiol. research Sch. Medicine Temple U.; asst. prof. radiology Grad. Sch. Medicine U. Pa., 1941—. Recipient Alverenga prize, 1934, Gold medal Internat. Coll. Radiology, 1937, Man of Year award Phila. 4-8-2 Square Club, 1959, Medallion of Achievement Golden Slipper Square Club, 1960, Mankind and Medicine award Albert Einstein Med. Center, 1966. Diplomate Am. Bd. Radiology and Nuclear Medicine. Author: (with H. Ingleby) Comparative Anatomy, Pathology, and Roentgenology of the Breast, 1960; (monograph) Thermography, 1965. Research, numerous publs. on mammography, thermography, role of viruses in cancer, osteoporosis, physiology of the G-I tract, x-ray telometry, contrast and color enhancement, telognosis, ultrasonics and virology. Home: 2401 Pennsylvania Av., Phila. 19130. Office: 255 S. 17th St., Phila. 19103.

GERSON, Nathaniel Charles, Am. physicist; b. Boston, Oct. 15, 1915; s. Benjamin Kolman and Julia (Bluemnthal) G.; B.S. magna cum laude, U. P.R., 1944; M.S., N.Y. U., 1948; m. Sareen Ruth Epstein, Aug. 26, 1945; children—Donald F., Stanton L., Richard K., Martha B., Stephanie L. Asst. chief tech. investigations sec. U. S. Weather Bur., 1944-46; asst. on propagation LF Loran system Watson Labs., USAF, 1946-48; chief Ionospheric Physics Lab., USAF Cambridge Research Labs., 1948-56; cons. and research physicist, Lincoln, Mass., 1956—. Sec., U. S. Nat. Com., IGY, sec. exec. com., 1953-57. Cons. Mass. Inst. Tech. Lincoln Lab., Lexington, 1957, Advanced Research Projects Agy., Washington, 1958-65. Mem. Research Soc. Am., Canadian Assn. Physicists, Arctic

Inst. N.Am., Am. Meteorol. Soc., Am. Geophys. Union, A.A.A.S. Editor: Radio Wave Absorption in the Ionosphere, 1962. Research, numerous publs. in physics and movements in upper atmosphere particularly ionosphere and aurora. Address: Trapelo Rd., Lincoln, Mass. 01773.*

GERSTEL, Dan Ulrich, geneticist; b. Berlin, Germany, Oct. 23, 1914; s. Alfred and Else (Flato) G.; came to U. S., 1938, naturalized, 1944; B.S., U. Cal. at Davis, 1940; M.S., U. Cal. at Berkeley, 1942, Ph.D., 1945; m. Eva Krojanker, Jan. 10, 1937; children—David, Naomi, Asso. genetics U. Cal. at Berkeley, 1944-46; asso. geneticist rubber plant research sta. U. S. Dept. Agr., Saliner, Cal., 1946-49; research fellow Cal. Inst. Tech., 1949-50; faculty N.C. State U., Raleigh, 1950—, prof. dept. crop sci., 1956-63, William Neal Reynolds prof., 1963—; vis. prof. Weizmann Inst. Sci., Rehvoth, Israel, 1961-62. Mem. Genetics Soc., Evolution Soc., Am. Soc. Botany, Am. Naturalists, A.A.A.S., N.C. Acad. Sci. Contbg. author: Chromosomes Today, 1966; Chromosome Manipulation and Plant Breeding, 1966; also numerous articles. Research on hereditary mechanism of self-sterility of sunflower family, asexual reprodn. in plants, hybridization between different species, transfer of genes from one species to another, abnormalities which arise from this transfer, origins cultivated plants. Dept. Crop Science, N.C. State U., Raleigh, N.C. 27607.*

GERSTEN, Christian Ludwig, German mathematician; b. Giessen, Germany, 1701; ed. Geissen; became prof. math., Giessen, 1733. Fellow Royal Soc., 1733. Described a calculating machine he had designed (never built), 1735; imprisoned for a letter written to Landgrave at Darmstadt, 1748-60. Died 1762.

GERSTEN, Jerome William, Am. physician; b. N.Y.C., Apr. 20, 1917; s. Louis and Bessie (Abrams) G.; B.S. magna cum laude, Coll. City N.Y., 1935; M.D., N.Y. U., 1939; M.S., U. Minn., 1949; m. Rhoda Rich, Nov. 8, 1941; children—Steven Paul, Wendy Lee, Christopher Carl, Dennis John, Madeleine Lou. Practice medicine, specializing in phys. medicine, Denver, 1949—; faculty Sch. Medicine U. Colo., 1949—, prof., chmn. dept. phys. medicine and rehab., 1957—. Vice pres. Am. Congress Rehab. Medicine. Mem. Am. Physiol. Soc., Soc. Exptl. Biology and Medicine, A.M.A., A.A.A.S., Am. Acad. Phys. Medicine and Rehab., Am. Assn. EMG and Electrodiagnosis (past pres.), Am. Inst. Ultrasonics in Medicine (past pres.), Am. Rehab. Found., Phi Beta Kappa, Sigma Xi, Alpha Omega Alpha. Contbr. chpt. to Ultrasonic Energy, Biological Investigations and Medical Applications, 1965. Editorial bd. Am. Jour. Phys. Medicine, 1961—. Research, numerous publs. on physiol. effects of phys. agts. with spl. emphasis on ultrasound, exercise physiology in normal individuals, and in patients with spinal cord trauma, home care and clinic care programs for chronically ill and homebound individuals. Home: 1370 Forest St. Office: 4200 E. 9th Av., Denver 80220.*

GERSTENHABER, Murray, Am. mathematician; b. N.Y.C., May 6, 1927; s. Joseph and Pauline (Rosenzweig) G.; B.S., Yale, 1948; M.S., U. Chgo., 1949, Ph.D., 1951; m. Ruth P. Zager, June 3, 1956; children—Jeremy J., David E., Rachel R. Jewett postdoctoral fellow Harvard, 1951-52, Inst. for Advanced Study, Princeton, N.J., 1952-53; faculty math. U. Pa., Phila., 1953—, prof., 1961—. NSF sr. postdoctoral fellow Inst. for Advanced Study, 1958-59. Mem. Am. Math. Soc. (organizer and 1st chmn. com. on math. in life scis., editor bull.). Research, publs. on algebra; introduced deformation theory of algebras. Home: 64 Clover Lane, Princeton, N.J. 08540. Office: Dept. of Math., U. Pa., Phila. 19104.*

GERSTER, Arpad Geyza (Charles), surgeon; b. Kassa, Hungary, Dec. 22, 1848; s. Nicholas and Caroline (Schmidt-Adamkovich) G.; grad. Acad. Kassa; grad. in medicine U. Vienna, 1872; m. Anna Barnard Wynne, Dec. 14, 1875. Came to U. S., 1873. Asst. surgeon, Austrian Army, 1872-73; surgeon Lenox Hill Hosp., N.Y.C., 1878—, Mt. Sinai Hosp., N.Y.C., from 1879; prof. surgery, N.Y. Polyclinic, 1882-94; prof. clin. surgery Columbia, from 1916; practiced medicine, N.Y.C., Bklyn., 1873-82. Author: Rules of Aseptic and Antiseptic Surgery, 1888; Recollections of a New York Surgeon, 1917; also papers. Tchr. of the Mayo brothers. Pioneer in antiseptic surgery. Died Mar. 11, 1923.

GERSUNY, Robert, surgeon; b. Teplitz-Schönau, Bohemia, Jan. 15, 1844; asst. to Theodore Billroth, Vienna, Austria; dir. Rodolfinerhaus, Vienna; dir. surgery Karolinen-Hosp. for Children. Author: Ärtzt un Patient, 1884; Bodensatz des Lebens, 1906; Theodor Billroth, 1922. Noted for operation for fecal incontinence (Gersuny's operation), also for work in gynecol. and plastic surgery. Died Oct. 31, 1924.

GERTHSEN, Christian, German physicist; b. Hörup auf Alsen, Germany, Nov. 21, 1894; s. Nis and Helen (Jörgensen) G.; studied physics, Heidelberg, Germany, Munich, until 1914; studied in Göttingen, beginning 1919; Ph.D., Kiel, Germany, 1922; m. Elisabeth Tönsfeldt, 1922; 2 sons. Named lectr., Tübingen, Germany, 1928; apptd. prof. exptl. physics, Giessen, Germany, 1932; named prof. physics (inst. destroyed in war), Berlin, 1939; became prof. Tech. U. Karlsruhe (Germany), 1948. Mem. Acad. Scis. Heidelberg. Author: (with K. Bechert) Atomphysik, 1938; (with M. Pollermann) Einführung in das Physikalische Praktikum für Studierende der Medezin und anderer Fächer, 1941; Physik, Ein Lehrbuch zum Gebrauch neben Vorlesungen, 1948; various articles. Research on stimulation of X-rays by protons, alpha particles, dispersion of protons on atom nuclei; investigated canal rays, transference of protons and helium rays, method of energy multiplication using transference. Died Karlsruhe, Dec. 8, 1956.

GERTLER, Menard M., physician; b. Saskatoon, Sask., Can., May 21, 1919; s. Franklin and Clare (Delman) G.; B.A., U. Sask., 1940; M.D., McGill U., 1943, M.Sc., 1946; D.Sc., N.Y. U. Postgrad. Med. Sch., 1958; m. Anna Paull, Sept. 4, 1943; children—Barbara, Stephanie, Jonathan. Came to U. S., 1946, naturalized, 1953. Exec. dir. coronary research project Mass. Gen. Hosp., Boston, 1947-50; staff Columbia Coll. Phys. and Surg., 1951-54; Nat. Heart Inst. Research fellow N.Y. U., 1954-56, Vocational Rehab. Adminstrn. Sr. fellow, 1957-58, faculty Sch. Medicine, 1958—, prof. exptl. medicine Inst. Rehab. Medicine, 1966—, dir. cardiovascular research, 1958—, staff U. Hosp., 1960—; med. dir. Sinclair Oil Corp., N.Y.C., 1958—. Cons.-lectr. U. S. Naval Hosp., St. Albans, N.Y., 1958—. Recipient Founder's Day award N.Y. U., 1958. Mem. A.M.A., N.Y. Acad. Scis., Am. Heart Assn., A.C.P., Am. Chem. Soc., Am. Fedn. for Clin. Research, Harvey Soc. N.Y., Soc. for Exptl. Biology and Medicine. Author: (with others) Coronary Heart Disease in Young Adults, 1954; also numerous articles. Developed concept early detection and prevention coronary heart disease; exptl. prodn. heart failure in guinea pigs. Home: 1000 Park Av., N.Y.C. 10028.*

GERVAIS, (François-Louis-) Paul, French zoologist; b. Paris, Sept. 26, 1816; M.D., Dr. Sci.; named dean Montpellier (France) Faculty Scis., 1856; prof. natural history, Paris Faculty Scis., 1868-79; mem. French Acad. Scis., 1874 Author: Histoire naturelle des insectes aptères, 1847; Historie des mammifères, 1855; Zoologie et paléontologie générales, 1867. Studied paleontol. zoology, relation of zoology to medicine, evolution of mammals, life of wingless insects. Died Feb. 10, 1879.

GERVASE OF CANTERBURY, scholar; b. Tilbury, Essex, Eng., ed. Bologna; tchr., Bologna; in Venice, 1177; in service of William II the Good (king of Siciles 1169-89); in Salerno, 1190-91; created marshal of kingdom of Arles by Otto III (emperor 1209-18); author (for Otto IV) Otia imperialia, a geog. and hist. olla-podrida in which he discusses topography of Rome and mentions asbestos or salamander skin, circa 1211.

GESCHWIND, Norman, Am. physician; b. N.Y.C., Jan. 8, 1926; s. Morris and Anna (Blau) G.; B.A., Harvard, 1946; M.D., Harvard, 1951; m. Patricia Dougan, Sept. 8, 1956; children—Naomi, David, Claudia. Moseley Travelling fellow Nat. Hosp., London, Eng., 1952-53, USPHS Research fellow, 1953-55; research fellow Mass. Inst. Tech., 1956-58; staff neurologist Boston VA Hosp., 1958-62, chief neurology service, 1962-66; asso. prof. neurology Boston U., 1962-66, prof., chmn. dept., 1966—. Fellow Am. Acad. Neurology; mem. Acad. Aphasia, Am. Neurol. Assn. Research, publs. on anat. basis higher functions nervous system, explanation disturbances higher function on basis damage to pattern cortico-cortical connections in animals and man. Home: 137 Clinton Rd., Brookline, Mass. 02146. Office: 80 E. Concord St., Boston.*

GESNER, Abraham, Canadian geologist; b. Cornwallis, N.S., Can., May 2, 1797. Studied medicine in London; M.D., Guy's Hospital Author: On the Mineralogy and Geology of Nova Scotia, 1847. Discovered and named kerosene (introduced into U. S. 1853); patented process for making it from crude petroleum, 1852. Died Apr. 29, 1864.

GESNER, Conrad, botanist, zoologist; b. Zurich, Mar. 26, 1516; ed. Strasbourg, Bourges, Paris, Basel; 1541. Prof. of Greek, Lausanne, 1537-40; also of physics and natural history, studied medicine further at Montpellier, M.D. degree, U. Basel, 1541, Zurich, town physician, Zurich, from 1554. Author: Historia Plantarum, 1541; Bibliotheca universalis (gives list and summary of all Hebrew, Greek, Latin books known to him, 20 vols., 1545-49; Historia animalium (a cornerstone in devel. of zoology), 5 vols., 1551-58, 87; De secretis Remediis, 2 parts, 1552, 59; Opera Botanica (which presents 1st attempt at scientific classification in botany), collected, pub. posthumously, 1753-59. Among first to present illustrations of fossils (though he did not recognize their significance); collected about 500 previously unrecognized species of plants; attempted to describe all known animals; work on distillation includes significant section on analysis of aqueous solutions. Died Zürich, Dec. 13, 1565.

GESSEL, Stanley Paul, Am. forester; b. Providence, Utah, Oct. 14, 1916; s. Gottleib and Esther (Heyrend) G.; B.S. in Forestry, Utah State Agrl. Coll., 1939; Ph.D., U. Cal. at Berkeley, 1950; m. Persis Annette Craig, Aug. 9, 1947; children—Paula, Susan. Engr., Army Engrs., Honolulu, 1941-42; faculty Coll. Forestry, U. Wash., Seattle, 1948—, prof., 1958—, asso. dean research, 1964—; vis. prof. U. Wis., 1955. Mem. Soc. Am. Foresters, Am. Geophys. Union, A.A.A.S., N.W. Sci. Assn., N.W. Forest Soils Council (sec. 1950—), Soil Sci. Soc. Am. (asso. editor proc.), Nat. Geog. Assn., Sigma Xi, Xi Sigma Pi, Gamma Alpha. Research on forest soils classification, forest growth, silviculture, tree nutrition, forest soil fertilization. Home: 2026 N.E. 120th St., Seattle 98125.*

GESSLER, A(lbert) E(dward), chemist; b. Metzingen, Wurtt, Germany, May 8, 1885; s. Edward Albert and Marie Louise (Leuze) G.; B.S., U. of Stuttgart, 1905; Ph.D., U. of Berlin, 1907; m. Mildred B. Murray, Feb. 2, 1915; children—Isolde (Mrs. Craig P. Smith), Albert; married 2d, Helen Yarnall, Mar. 31, 1932; 1 daughter Sally (Mrs. F. G. Appleton). Came to United States, 1908, became naturalized, 1922. Chemist, G. Siegle Co., Rosebank, N.Y., 1908, vice pres. and mem. bd., 1914-18; partner and vice pres. Ultro Chem. Corp., 1918, firm consol. with Zinsser & Co., 1926; chief chemist, mem. bd. and exec. com. Zinsser & Co., 1926-34; dir. research Interchem. Corp., N.Y. City, 1934-44, vice pres. and dir. research, 1944-52, director emeritus of research since 1952; now engaged in private practice as chemical cons. Recipient certificate awarded for effective research in connection with atomic bomb, 1945; certificate awarded for effective service in work on camouflage organized through Nat. Defense Research Council, Mar. 1, 1945; recipient grant for cancer research from Lillia Babbitt Hyde Found., 1944-53; Adult Award, National Assn. Printing Ink Makers. Fellow N.Y. Acad. Sci.; mem. Am. Chem. Soc. (councilor), Am. Assn. Cancer Research, A.A.A.S., Electron Microscope Society. Assn. Research Dirs. Rep. Contributor of papers to chemical and to medical publications. Holder 50 U. S. patents on pigments; dyes; high speed publication printing; key inventions to mass printing in black and in natural colors for majority of present day magazines since 1934; U. S. patent on fast isolation of pure viruses (for use in vaccines). Address: 100 Sands Point Rd., Sarasota, Fla. 33577.*

GESSNER, Johannes, Swiss physician, natural scientist; b. Wangen, Switzerland, Mar. 18, 1709; s. Christoph and Esther (Maag) G.; ed. privately under Johann Jakob Scheuchzer, Zurich; studied in Leyden, Netherlands, Paris, France, Basel, Switzerland; m. Katharina Escher, 1738. Became Scheuchzer's successor at Carolinum, Zurich, 1738; prof. natural scis.; founder Bot. Garden, Zurich; responsible for founding obs. in Zurich. Mem. Leopoldina, Phys. Soc. Zurich (founder, later named Nat. Sci. Soc.), acads. of Berlin, Göttingen, St. Petersburg, Stockholm, Uppsala. Made Zurich sci. edn. center of eastern Switzerland even before univ. was founded; few of his publs. are extant. Died Zurich, May 6, 1790.

GETMAN, Frederick Hutton, Am. chemist; b. Oswego, N.Y., Feb. 9, 1877; s. Charles Henry and Alice (Peake) G.; ed. Rensselaer Poly. Inst., Lehigh U.; grad. chem. dept. U. Va., 1896; Ph.D., Johns Hopkins, 1903; m. Ellen M. Holbrook, Nov. 26, 1906. Instr. chemistry and physics Stamford High Sch., 1897-1901, instr., 1905-06; fellow by courtesy, Carnegie research asst. in phys. chemistry Johns Hopkins, 1903-04; lectr. phys. chemistry Coll. City of N.Y., 1904-05; lectr. physics, Columbia, 1907-08; asso. prof. chemistry Bryn Mawr (Pa.) Coll., 1909-15; dir. Hillside Lab., Stamford, Conn.; pres. The Getman & Judd Co. Fellow A.A.A.S., Am. Inst. Chemists, mem. Am. Electrochem. Soc., History of Scis. Soc., London, Am. chem. socs., Nederlandsche Chemische Vereeniging, Phi Beta Kappa. Author: Blowpipe Analysis, 1899; Laboratory Exercises in Physical Chemistry, 1904; Introduction to Physical Science, 1909; Outlines of Theoretical Chemistry, 1913; Electrochemical Equivalents (with Dr. Carl Hering); Life of Ira Remsen; also scientific papers on various problems in phys. chemistry and chem. biography. Died Dec. 2, 1941.

GETOFF, Nikola, radiation chemist; b. Lom, Bulgaria, Dec. 18, 1922; s. Avram P. and Ganka (Ivanova) G.; diploma engring. and chemistry Tech. U. Vienna, 1950, Dr.Tech., 1952; m. Edith Thiel, Dec. 6, 1952; children—Doris, Roman. Asst., Tech. U. Vienna, 1952-56; research chemist, then head radioisotope lab. Austrian Nitrogen Works Ltd., Linz, 1956-59; research Inst. Radium Research and Nuclear Physics, Vienna, 1959—; lectr. radiation chemistry and hot atom chemistry U. Vienna, 1965—. Research cons. Austrian Nuclear Reactor Center, 1961—. Mem. Austrian Chem. Soc., Faraday Soc., others. Author: Kurzes Radiochemisches Praktikum, 1961; (with others) Isotope in der Landwirtschaft, 1960; (with others) Strahlenchemie, 1967; also numerous articles. Synthesis of amino acids and simple organic substances from carbon dioxide, ammonia and water under the influence of ionizing radiation and ultraviolet light; pulse radiolysis of organic substances; chem. transformations following neutron capture; primary processes in matter caused by radiation. Home: 5/II/12 Flurschützstrasse, A-1120 Vienna, Austria.*

GETOOR, Ronald Kay, Am. mathematician; b. Royal Oak, Mich., Feb. 9, 1929; s. K.G. and Ruth (Eberle) G.; A.B., U. Mich., 1950, M.S., 1951, Ph.D., 1954; m. Ann Broman, Mar. 19, 1959; 1 dau., Lise. Fine instr. Princeton, 1954-56; faculty U. Wash., Seattle, 1956-—, prof., 1964-—, postdoctoral fellow, NSF, 1959-60. Vis. prof. German Sci. Found., U. Hamburg, 1964. Mem. Am. Math. Soc., Inst. Math. Statistics. Contbr. articles to math. jours. Study of probability theory, particularly general theory of Markov processes and their associated potential theory. Home: 4810 38th N.E., Seattle 98105.*

GETTENS, Rutherford John, Am. chemist; b. Mooers, N.Y., Jan. 17, 1900; s. Daniel Patterson and Clara (Rutherford) G.; B.S., Middlebury (Vt.) Coll., 1923; M.A., Harvard, 1929; m. Katharine Hilliard Covelle, Nov. 26, 1930; 1 dau., Rebecca Hilliard. Instr., Colby Coll., Waterville, Me., 1923-27; staff Fogg Mus. Art, Harvard, 1928-51, chief Mus. Tech. Research, 1949-51, lectr. fine arts Harvard, 1948-51; staff Freer Gallery Art, Smithsonian Inst., 1951-—, head curator, Freer Gallery Lab., 1961-—. Cons. fellow N.Y. U. Conservation Center, 1960-—. Mem. Archeol. Inst. Am., Internat. Inst. for Conservation Historic and Artistic Works (pres. 1968-—, editor Abstracts 1958-65), Am. Chem. Soc., Am. Assn. Museums. Author: (with G. L. Stout) Painting Materials, 1942, 66; also numerous articles. Sect. editor Chem. Abstracts, 1962-—. Research on materials and processes art and archeology, conservation cultural arts, history techn. based on studies material cultures ancient civilizations. Home: 6011 Broad Branch Rd., N.W., Washington 20015. Office: Freer Gallery of Art Smithsonian Instn., Washington 20560.*

GETTING, Ivan Alexander, Am. physicist; b. N.Y.C., Jan. 18, 1912; s. Milan A. and Hariet (Almasy) G.; B.S., Mass. Inst. Tech., 1933; Ph.D., Oxon, 1935; D.Sc. (hon.), Northeastern U., 1955; m. Dorothea Louise Gracy, Oct. 2, 1937; children—Nancy Louise G. (Mrs. Richard Resch), Ivan Craig, Peter Alexander. Jr. fellow Soc. Fellows, Harvard, 1935-40; staff mem. Radiation Lab., Mass. Inst. Tech., 1940-45, prof. elec. engring., 1945-50; asst. for devel. and planning Dep. Chief Staff/Devel., USAF, Washington, 1950-51; v.p. engring. and research Raytheon Co., Waltham, Mass., 1951-60; pres. Aerospace Corp., El Segundo, Cal., 1960-—. Recipient medal for Merit, 1946, Air Force Distinguished Service medal, 1958; Bur. of Ordnance Devel. award, 1945, Fellow A.A.A.S., Am. Phys. Soc., A.I.E.E.; mem. Am. Assn. Rhodes Scholars, World Affairs Council (dir. Los Angeles, 1960-—). Research in nuclear physics, astrophysics; particle accelerators; gaseous discharges; radar; automatic tracking of targets by radar, rapid scanning radar antennas; fire control. Home: 605 Tigertail Rd., Los Angeles 90049. Office: 2350 E. El Segundo Blvd., El Segundo, Cal. 90045.*

GEULINCX, Arnold, Flemish philosopher; baptized Antwerp, Belgium, Jan. 31, 1624; studied philosophy and theology at U. Louvain; professorships there, 1646, 1652; dismissed because Jansenist; became Calvinist; M.D., 1658; lectured privately in Leyden; prof. extraordinary of philosophy and ethics, U. Leyden, 1665-69; became prof., Leyden, Belgium, 1665. Author: Quaestiones quod libeticae, 1653; Logica Restituta, 1662; Methodus inveniendi argumenta, 1663; De virtute et primis ejus proprietatibus, 1665; Gnothi seauton, sive . . . ethica, 1675; Physica vera, 1688; Annotata praecurrentia ad R. Cartesii principia, 1690; Annotata majora, 1691; Metaphysica vera et ad mentem peripateticum, 1691; Collegium oratorium, 1696. Early follower of Descartes; founder metaphys. theory called occasionalism. Died Nov. 1669.

GEUTHER, (Johann Georg) Anton, German chemist; b. Neustadt nr. Coburg, Germany, Apr. 23, 1833; s. Christian Friedrich and Anna Cordula (Eichhorn) G.; studied in Jena, beginning 1852, Berlin, Göttingen (all Germany), 1853-54; Ph.D., Göttingen, 1855; m. Amalie Agnes Sindram, 1863; 1 son, 1 dau. Asst. to F. Wöhler, Göttingen, named lectr., 1857, asso. prof., 1862; prof., Jena, from 1863. Mem. Chem. Soc. London (hon.). Author: Kurzer Lehrgang der chemischen Analyse, 1867; Lehrbuch der Chemie, gegründet auf die Wertigkeit der Elemente, 1870; also articles. Research on make-up of various double compounds, reduction of nitrobenzene to aniline, acetic acid and its compounds; isolated nitrosamine (nitro-diethylene) and dimethylene carbon-ethylene-ether; discovered an isomeric acid of crotonic acid; developed new synthesis methods. Died Jena, Aug. 23, 1889.

GEVANTMAN, Lewis Herman, Am. chemist; b. N.Y.C., Sept. 12, 1921; s. Benjamin and Ida (Goldberg) G.; B.Engring., Johns Hopkins, 1942; Ph.D. in Phys. Chemistry, U. Notre Dame, 1951; m. Leatrice Black, Aug. 22, 1948; children—Sandra Cay, Janis Mara. Chem. operator Johns Hopkins, also Bethlehem Steel Co., 1942-43; research chemist Clinton Labs., Manhattan Project, 1943-46; supervisory research chemist U. S. Naval Radiol. Def. Lab., San Francisco, 1951-56, acting head applied research br., 1956-59, head radiation chemistry br., 1959-61, sci. research adminstr., head chem. tech. div., 1961-64; sr. sci. adviser U. S. mission IAEA, 1964-—. Cons. Nuclear Sci. and Engring. Corp., 1956-59; lab. rep. to 3 TV programs dealing with radioactivity. Mem. Am. Chem. Soc. (chmn. civil def. com. No. Cal. sect. 1961-64), Am. Soc. Testing Materials (ad hoc com. dosimetry), A.A.A.S., Radiation Research Soc., Sigma Xi. Author articles, patentee in field. Research on chem. and phys. methods to study mechanism of reaction caused by irradiation of selected chem. systems (demonstrated role of free radical entities in such reactions); applied chem. systems with well-known behavior to radiation to measure dose delivered to mock biol. systems to demonstrate relationship between energy absorbed and biol. effect. Home: Gymnasiumstrasse 45. Office: Schmidgasse 14, Vienna, Austria.*

GEWÜRZNÄGELEIN, see Carion, Johannes.

GEY, Karl Friedrich, biochemist; b. Leipzig, Germany, Sept. 7, 1925; s. Karl Paul and Franziska (Möller) G.; student Jena (Germany) U., 1945-48; M.D., Basel (Switzerland), 1952; postgrad. chemistry Göttingen (Germany) U. m. Heidi Ruth Wydler, Apr. 1, 1954; children—Christoph, Matthias. Research fellow biochem. sect. Max-Planck Inst., Göttingen, 1953-54, dept. biochemistry Oxford (Eng.) 1953-54; research worker dept. exptl. medicine F. Hoffmann-La Roche & Co., Ltd. Basel, 1955-—, head biochem. sect., 1961-—, procuration, 1963-—; lectr. biochemistry Bern (Switzerland) U., 1967-—. Mem. Biochem. Soc. Switzerland, Gt. Britain, France, Germany. Research, numerous publs. on animal atherosclerosis, cholesterol biosynthesis, metabolism of fatty acids, carbohydrates, monoamines and pharmacological alterations. Home: 54 Ob. Rebbergweg, 4153 Reinach/Bl, Switzerland. Office: 1234 Grenzacherstrasse 4002, Basel, Switzerland.*

GEYLER, Hermann Theodor, German botanist, paleobotanist; b. Schwarzbach nr. Gera, Germany, Jan. 15, 1835; s. Hermann Gustav and Adelgunde Schiller (von Schillershausen) G.; studied botany, Leipzig, Jena, (both Germany), 1857-61; Ph.D., 1860; m. Anna Theresia Krahmer, 1871; 1 son. Worked under C. Cramer, Zürich, Switzerland, 1864-67; prof. Senckenberg Med. Inst., Frankfort/Main, Germany, 1867-89; lectr. on anatomy and physiology of plants, specialized botany; adminstr. bot. and phytopaleont. sect. Senckenberg Natural Sci. Soc.; 2d dir. Senckenberg, 1873-75, 77-79. Mem. Leopoldina, Acad. Natural Scis. Phila. (corr.). Author: Über fossile Pflanzen aus Borneo (early work on plant fossils from tropical regions), 1875; various articles. Research on phytopaleontology of tertiary and other periods, vascular tissue in conifer leaves. Died Frankfort/Main, Mar. 22, 1889.

GHENT, Arthur Warren, biologist; b. Toronto, Ont., Can., Sept. 8, 1927; s. Percy Parker and Mildred (Warren) G.; B.S., U. Toronto, 1950, M.A., 1954; Ph.D., U. Chgo., 1960; m. Jocelyn Kathryn Maynard, Oct. 14, 1961; children—Jennifer Barbara, Warren David, Tove Jacqueline. Agrl. research officer Forest Biology div. Canadian Dept. Agr., 1950-59; asst. prof. zoology U. Okla., 1960-64; asst. prof. Sch. Life Scis. U. Ill., Urbana, 1964-65, asso. prof., 1965-—. Mem. Canadian Entomol. Soc., Soc. For Study Evolution, Am. Inst. Biol. Scis., Sigma Xi. Research, publs. on hymenopteran behavior, tribolium behavior relative to conditioning of flour medium; biostatis. studies, time series correlations, non-parametric methods. Home: 203 E. Dodson Dr., Urbana, Ill. 61801.*

GHEORGHESCU, Benedict, Rumanian physician; b. Dobresti, Arges, Rumania, Jan. 14, 1924; s. Vasile and Florica (Marinescu) G.; Physician, Faculty Medicine, Bucharest, Rumania, 1948, Primar Physician, 1959, D.Med. Scis., 1962; m. Florica Radu, June 17, 1952; 1 dau., Ruxandra Mihaela. Asst., Gastroenterology Clinic, 1951-62; chief asst. Gastroenterology Center, 1962-—; chief radioisotope lab. Gastroent. Center, Bucharest, 1958-—. Fellow Internat. Atomic Energy Agy., Vienna, Austria, 1964-65, Rome, Italy, 1964-65. Mem. Internat. Soc. Gastroenterology. Author: (with I. Brasla) Diagnostic with Radioisotopes in Medical Practice, 1964; also numerous articles. Co-inventor method of color scintigraphy of liver; research on metabolism of Vitamin B12 in chronic hepatitis, pathology of resected stomach. Home: 6 Aprodu Purice, Bucharest. Office: 37-39 Bd.1 Mai, Bucharest, Rumania.*

GHEORGHIU, Ion S., Rumanian elec. engr.; b. 1885; prof. Bucharest (Rumania) Poly. Inst.; mem. Acad. Socialist Republic Rumania; planned 1st electrification of some ry. lines in Rumania, 1914-15. Author: Electrical Machines, 1957-61; numerous works on electro-technique.

GHERARDI, Bancroft, Am. telephone engr.; b. San Francisco, Apr. 6, 1873; s. Bancroft and Anna Talbot (Rockwell) G.; B.Sc., Poly. Inst. Bklyn., 1891, Dr. Eng., 1933; M.E., Cornell U., 1893, M.M.E., 1894; m. Mary Hornblower Butler, June 15, 1898. Became engr. asst. N.Y. Telephone Co., 1895, traffic engr., 1900; chief engr. N.Y. and N.J. Telephone Co., 1901-06, asst. chief engr. N.Y. and N.J. Telephone Co., 1906-07; also asst. chief engr. N.Y. Telephone Co., 1906-07; equipment engr. Am. Tel. & Tel., 1907-09, engr. plant, 1909-18, acting chief engr., 1918-19, chief engr., 1919-20, v.p., chief engr., 1920-38. Recipient Awarded Edison medal, 1932, for contbns. to telephone engring. and devel. of elec. communication. Died Aug. 14, 1941.

GHETALDI, Marino, mathematician, physicist; b. Dubrovnik (now Yugoslavia), 1566; studied at Rome, Anvers (Belgium), Paris. Author: Archimedes promotus (contains oldest specific weight of any metal), 1603; De resolutione et compositione mathematica, pub. 1630. A precursor of students of analytical geometry; used algebra in the solution of geometric problems; determined the specific weight of seven of the most important metals. Died 1626

GHILAROV, Mercury Sergeyevich, Russian zoologist, entomologist; b. Kiev, USSR, Mar. 6, 1912; s. Sergei A. and Elisabeth A. (Ivanova) G.; Dipl. zoologist, U. Kiev, 1929-33; Cand. Agrl. Sci., Timiriazev Acad. Agr., Moscow, 1937; D. Biol. Sci., Acad. Scis. USSR, 1947; m. Irene I. Blokhintseva, Mar. 18, 1936; 1 son, Alexei M. Jr. entomologist Sugar Beet Research Inst., Kiev, 1933-34; sr. entomologist, lab. leader All-Union Rubber Plants Research Inst., Moscow, 1934-44; sr. scientist Inst. Animal Morphology, Acad. Scis. USSR, 1944-56, chief lab soil zoology at lab., 1956-—; docent soil zoology U. Moscow, 1946-51; prof. invertebrate zoology State V. I. Lenin Pedagogical Inst., Moscow, 1949-—. Pres. Nat. Com. Soviet Biologists. Recipient A. N. Severtzov prize Acad. Scis. USSR, 1947; State prize USSR, 1951; Badge of Honor, USSR, 1953; Silvestri Golden medal F. Silvestri Found. (Italy), 1965; Gustav Kraatz Plakette Acad. Agrl. Scis., Berlin, Germany, 1966. Corr. mem. Acad. Scis. USSR; mem. Entomol. Soc. USSR, Acad. Zoology Agra, India (v.p.), F.A.Z (v.p.); hon. mem. entomol. socs. France, Finland, Czechoslovakia. Author: Insect Pests of koksaghyz, 1943; Peculiarities of the Soil as Environment and its Significance in Insect Evolutions, 1949; (with others) Keys to Soil-Dwelling Insect Larvae, 1964; Zoological Methods in Soil Diagnostics, 1965; also numerous articles. Studied soil arthropods—their adaptation to soil life and evolution from aquatic life to the terrestrial through soil during phylogenesis; proved significance of soil forming animal activity in different soil types; elaborated systems of soil pest control by means of agrl. measures; described many larval forms of insects, compiled keys to insect larvae; studied evolution of internal insemination in land arthropods. Address: Inst. Animal Morphology, 12-2 Vavilov St., Moscow W-133, USSR.*

GHINI, Luca (di Croara d'Imola), physician, botanist; b. Croara d'Imola, circa 1490; s. Leonora (Ravaglia di Molinella) G.; m. Gentile Sarti, Dec. 3, 1528; 1 son, Galeazzo. Prof., Bologna, Italy, 1527-44; called to Pisa by Granduke Cosimo I and remained there 1544-54; taught at Bologna, 1554-56. Author: Lectionum de herbis ex luca ghini epitome, pub. 1657. Founder bot. gardens, Pisa, later Florence, circa 1550. One of 1st to collect and classify plants; studied cure for yellow fever. Died May 4, 1556.

GHIRARDELLI, Elvezio, Italian zoologist; b. Orasso, Novare, Jan. 30, 1918; s. Vincenzo and Angela (Mazza) G.; Ph.D. in natural sci.; m. Zoé Ambrosini, 1946; children—Lia, Paola. Prof. zoology Inst. Zoology, Trieste. Mem. Adriatic Soc. Scis. Trieste (pres.). Research and publs. on chetognaths and gonads, determination of differentiation of gonads. Address: Zoological Institute, 32 via A. Valeria, Trieste, Italy.

GHON, Anton, Austrian bacteriologist, path. anatomist; b. 1866; demonstrator Inst. Path. Histology and Bacteriology, Vienna; asst. Path. Inst., 1894-1910; became privat docent path. anatomy U. Vienna, 1899, prof., 1902; named prof. path. anatomy German U. Prague, 1902. Described primary lesion of pulmonary Tb (Ghon Tubercle), 1912, devel. primary Tb in children; recognized difference between lesions of 1st infection and those from reinfection. Died 1936.

GHOSH, Jagal Jiban, Indian biochemist; b. nr. Calcutta, India, June 19, 1925; s. Jugal Kishore and Nalini (Bose) G.; B.S., Calcutta U., 1945, M.Sc., 1947, D.Phil., 1950, D.Sc., 1965; m. Dipti Deb, Apr. 17, 1962; 1 dau., Kum Kum. Postdoctoral fellow McGill U., Montreal, Que., Can., 1953-55, N.Y. U. Coll. Medicine, 1955-56, Columbia Coll. Phys. and Surg., 1959-60, Harvard, 1960-61; faculty Calcutta U., 1957-—, Centenary prof. biochemistry, 1965-—. Panelist neurochemistry sect. Internat. Brain Research Orgn., UNESCO, Paris, 1960-—. Mem. Indian Brain Research Assn. (editor-in-chief Brain News, 1965-—), Internat. Soc. Neurochemistry (mem. organizing com. 1966-—), A.A.A.S., Am. Inst. Biol. Scis. Research, numerous publs. on brain biochemistry, mode of action of narcotic, convulsant, and tranquilizer drugs at level of nucleoprotein metabolism in brain tissue, nutritional biochemistry. Home: P-10 Grey St. Extension, Calcutta-4, West Bengal, India.*

GHOSH, S. N., Indian physicist; b. Calcutta, India, Feb. 1, 1918; s. A. C. and Urmila (Biswas) G.; B.Sc., St. Xaviers Coll., 1937; M.Sc., Calcutta U., 1942; D.Sc., U. Calcutta, 1948; m. Shiela Majumdar, June 18, 1949; children—Santanu, Atanu. Adair-Dutt Research scholar, Calcutta, 1943-45; research asst., physicist Council Sci. and Indsl. Research, New Delhi, 1945-49; research fellow Nat. Inst. Sci., New Delhi, 1949-50; post-doctoral research fellow Duke, 1950-52, Harvard, 1952-54; physicist Wentworth Inst., Boston, 1954-56, Geophys. Research Center, Boston, 1957-58; vis. prof. U. Mich., Ann Arbor, 1966-—; faculty Allahabad U., India, 1956-57, prof., head dept. applied physics, 1959-—. Mem. adv. bd. Def. Sci. Orgn., Govt. India, 1962-—; vis. com. India U. Grants Commn., 1965-—; fgn. sec. Nat. Acad. Sci. of India, 1960-64, treas., editor Procs., 1965-—. Research, publs. on upper atmosphere, micro-wave and collisional problems, atmospheric phenomena, processes in collisions between ion-ion and ion-neutral particles. Home: 12 Lake Temple Rd., Calcutta, India.*

GHOSH, Sailaja Prasad, Indian inorganic chemist; b. Bihar, India, Jan. 6, 1914; s. Bagala Kinkor and Nihar Bindu (Sen) G.; M.Sc., Sci. Coll., Patna, India, 1938; D.Phil., Calcutta (India) U.; Ph.D., Durham (Eng.) U.; m. Mamata Das, July 7, 1949; children—Debi, Jyoti. Faculty, Sci. Coll., Patna, 1943—, prof. Bihar Ednl. Service Class I, 1959-66, univ. prof., head chemistry, 1966—. Fellow Indian Chem. Soc., Royal Inst. Chemistry London, Inst. Fuel London (asso.). Author: (with B. N. Sahai) Handbook of Practical Chemistry for B.Sc. Students; also articles. Research on complex compounds, preparation, properties, reactions and structure of transitional elements, new analytical methods, tridentate ligands, graphitization of pitch, potash salts from molasses. Address: Sci. Coll. Campus, Patna-5, India.*

GHURYE, Sudhish G., mathematician; b. Bombay, India, Nov. 10, 1924; s. G. S. and Sajubai (Aroskar) G.; M.Sc., U. Bombay, 1947; Ph.D., U. N.C., 1952; m. Charlotte Wolf, Jan. 28, 1953. Came to U. S., 1956. asst. prof., Univ. of Chicago, 1956-59; assoc. prof., Northwestern Univ., 1959-61; assoc. prof., Univ. of Minnesota, 1961-62; prof. math. Ind. U., 1962—, chmn. dept., 1964—. Study of mathematical statistics; probability theory. Home: Route 1, Audubon Dr., Bloomington, Ind. 47405.

GIACCONI, Riccardo, physicist; b. Genoa, Italy, Oct. 6, 1931; s. Antonio and Elsa (Canni) G.; doctorate in physics U. Milan, 1954; m. Mirella Manaira, Feb. 16, 1957; children—Guia, Anna Lee, Marc. Came to U. S., 1956, naturalized, 1960. Asso. prof. physics Milan U., 1954-56; research asso. Ind. U., Bloomington, 1956-58, Princeton, 1958-59; sr. scientist Am. Sci. & Engring., Inc., Cambridge, Mass., 1959-60, chief space physics div., 1963, v.p. space research and systems div., 1960—. Fulbright fellow, 1956-58. Mem. A.A.A.S., Am. Geophys. Union, Italian Phys. Soc., Am. Astronom. Soc. (Helen B. Warner prize 1966). Publs. on research in cosmic ray physics; cloud chamber studies of high energy nuclear interactions; scintillation and Cherenkov counter study of elementary particles; scintillation chambers technique; heavy particles in Van Allen belts; X-ray astronomy; dir. group that discovered 1st extra solar X-ray sources 1962; galactic and solar X-ray astronomy; invented telescopes for X-ray astronomy. Home: 14 Pollywog Lane, Weston, Mass. 02193. Office: 11 Carleton St., Cambridge, Mass. 02142.*

GIACOMELLO, Giordano, Italian chemist, pharmacist; b. Montereale Cellina, July 26, 1910; s. Pietro and Luigia (Torresin) G.; ed. in chemistry and pharmacology; m. Maria Romeo, 1944; children—Pierluigi, Alessandro, Francesco. Titular head, dir. Inst. Chemistry, Pharmaceutics and Toxicology, U. Rome; dir. Higher Inst. Pub. Health, Rome, Center Studies for Nuclear Chemistry, Nat. Center Applied Radiobiology and Agrl. Problems. Mem. council direction Nat. Com. for Chemistry; v.p. Sci. Com. of Euratom; pres. Center Studies for Protection against Radion, E. Paterno Found. Contbr. numerous articles to sci. publs. Home: 93 viale Ippocrate. Office: 1st Chemical Farm, University of Rome; also 299 viale Regina Elena, Rome, Italy.

GIACOMINI, Amedeo, Italian physicist, mathematician; b. Cuneo, Mar. 11, 1905; s. Cesare Amedeo and Anna (Riccardi); Ph.D. in Phys. Scis., U. Pisa; m. Rina Mantelli, 1932; children—Anna Maria, Amedeo Cesare. Dir., Faculty Phys. Scis. and Math., U. Perouse; dir. Nat. Inst. Ultra Acoustics, Nat. Council Research, Rome. Mem. Italian Soc. Physics, Acoustical Soc. Am. Research and numerous publs. on physics. Home: 19 viale Pinturicchio. Office: 4 Piazzale delle Scienze, Rome, Italy.

GIACOMINI, Carlo, Italian anatomist; b. Tortona, Italy, Nov. 15, 1840; s. Vincenzo and Felicita (Alvigini) G.; degree medicine, U. Turin (Italy), 1864; became asst. Inst. Anatomy, Turin, Italy, 1866, head, 1871; became prof. anatomy, 1880. Mem. Acad. Medicine Turin, Acad. Sci. Turin. Author: Guida allo studio delle circonvoluzioni cerebrali, 1878; Varietà delle circonvoluzioni cerebrali dell'uomo, 1881; Cervelli dei Microcefali; Topografia cranio-cerebrale; Sull'anatomia del Negro e delle scimmie anthropomorfe; Annotazioni soptra l'a atomia del Negro, 1884. Developed method of hardening brain to preserve form and proportion; collected 1000 brains including those from criminals, normal individuals, idiots, and colored races; perfected method of making microscopic sects. of entire cerebral hemisphere, also colored sects., sectioning congealed cadavers and fixing organs; described external morphology of brain; proved that character, morals, intelligence or other traits could not be told from superficial observation of brain; his research enabled surgeons to determine exact location of cerebral cortex and center single cortical regions. Died July 5, 1898.

GIANNINI, Gabriel Maria, industrialist, physicist; b. Rome, Italy, Oct. 21, 1905; s. Torquato and Maria (Laccetti) G.; D.Physics, U. Rome, 1929; m. Luisa Casazza, July 18, 1931; children—Maria Laura (Mrs. Gerald F. Madigan), Valerio; m. 2d, Olga Harrington, September 27, 1964; one dau., Gabriella-Carla. Came to the United States, 1930, naturalized, 1938. Research in acoustics RCA, also Curtis Inst. Music, Phila., 1931-35; research in acoustics and telephony Transducer Corp., N.Y.C., 1936-40; engring. mgmt. Lockheed Aircraft Co., 1941-44; founder, 1945, pres. until 1957, Giannini Controls Corp., Pasadena, Cal.; founder, 1959, since chairman bd. Giannini Voltex; president Giannini Inst. Associate Cal. Inst. Tech., 1957—. Asso. fellow Am. Inst. Aero. and Astronautics; senior member I.E.E.E.; mem. Società Italiana Di Fisica, Instrument Society Am., Am. Phys. Soc., Am. Optical Soc., Acoustical Soc., Am. Soc. M.E., Soc. Automotive Engrs., Assn. Computing Machinery, Solar Energy Soc., Marine Tech. Soc., Royal Aero. Soc., British Interplanetary Soc., U. S. Naval Inst. Approximately 50 patents, including acoustic networks; telephone devices; aircraft instruments; high temperature gas sources; plasma and electrostatic generators. Home: Valmaria Ranch, Indio, Cal. 92201; also Lungo Tevere Mellini 24, Rome, Italy.*

GIANOLA, Umberto Ferdinando, physicist; b. Birmingham, U.K., Oct. 29, 1927; s. Ferdinando and Elizabeth (Worrall) G.; B.Sc. in Physics, U. Birmingham, 1948, Ph.D., 1951; m. Mary Randle Daffern, Aug. 16, 1952; children—Francis, Bruce, Tracy. With Royal Aircraft Establishment, Farnborough, U.K., 1951; post-doctoral fellow U. B.C., Can., 1951-53; mem. tech. staff Bell Telephone Labs., Inc., Murray Hill, N.J., 1953—. head, fundamental memory components dept., 1963—. Co-chmn. Conf. on Magnetism and Magnetic Materials, 1965. Sr. mem. I.E.E.E.; mem. Am. Phys. Soc., Research Soc. Am. (pres. Summit br. 1960). Research and devel. electronic and solid state devices, especially conception and application of magnetic devices to digital memory and logic. Patentee in field. Home: 3 Hancock Dr., Florham Park, N.J. 07932. Office: Bell Telephone Labs., Inc., Murray Hill, N.J. 07971.*

GIANOTTI, Ferdinando, Italian physician, dermatologist; b. Corsico, Italy, Aug. 22, 1920; s. Giuseppe and Rosa (Sala) G.; grad. in Medicine, U. Milan (Italy), 1947, postgrad. faculty, U. Milan, 1950—, prof. dermatology, 1956—, prof. in charge allergic skin diseases, 1966—. Recipient several Italian sci. awards. Mem. Società Italiana Dermatologia e Sifilografia, Association des Dermatologistes et syphiligraphes de langue francaise. Author: La sifilide, 1966; also numerous articles. Discovered infantile papular acrodermatitis characterized by papular dermatitis on face and limbs, anicteric hepatitis and histiocitic polylymphadenitis; research on virology of pemphiguspemphigoid diseases group; first recognized in Italy schistosome dermatitis and pustolosis of rice pickers, enteropathic acrodermatitis, lichen striatus, juvenile xanthogranuloma and granulomatosis of milkers' hands. Home: 26 Cimarosa, Milan, Italy.*

GIANTURCO, Cesare, physician; b. Naples, Italy, Feb. 12, 1905; s. Emanuel and Remigia (Guargiglia) G.; M.D., U. Naples, 1924; postgrad. U. Rome, U. Berlin; M.S., U. Minn., 1934; m. Verna Daily, Nov. 9, 1934; children—Paola, Michael. Came to U. S., 1930, naturalized, 1936. Practice medicine, specializing in radiology, Urbana, Ill., 1934—; radiologist Carle Hosp. Clinic, Urbana, 1934-67; prof. clin. physiology U. Ill., 1967—. Mem. A.M.A., Am. Roentgen Ray Soc., Radiol. Soc. N.Am., Am. Coll. Radiology, Italian (hon.), Inter Am., Panamanian (hon.) radiol. socs., Sigma Xi. Research on mech. factors of gastric digestion, shifting of blood under various conditions, permanent visualization of organs toward roentgen rays, combination of various radiol. examinations, speed of progress of intestinal tubes. Home: 101 W. Meadows St. Office: 602 W. University Av., Urbana, Ill. 61801.*

GIARD, Alfred, French biologist; b. Valenciennes, France, Aug. 8, 1846; prof. at Lille (France) Faculty Scis., 1873-87; named prof. Paris Faculty Scis., 1888; worked at Wimereux Lab., Paris, founder Lille Zool. Sch. Mem. French Acad. Scis., 1900. Founder jour. Bulletin biologique de France et de Belgique. Research on parasitism, parasitic castration, anhydrobiosis, proecilogony; a supporter of the theory of evolution. Died Orsay, France, Aug. 8, 1908.

GIARDINI, Armando Alfonzo, Am. mineralogist; b. Salamanca, N.Y., June 5, 1925; s. Giardino B. and Rose (Ferrara) G.; B.S., U. Mich., 1951, M.S., 1953, Ph.D., 1956; m. Anne Morton Johnston, Aug. 21, 1950; children—Michele, Richard, Melissa, Peter. Phys. scientist electrotech. lab. U. S. Bur. Mines, Norris, Tenn., 1952; sr. research engr., research and devel. div. Carborundum Co., Niagara Falls, N.Y., 1953-55; research asso., teaching fellow U. Mich., Ann Arbor, 1956-57; physicist, project leader high pressure research U. S. Army Electronics Command, Ft. Monmouth, N.J., 1957-65; prof. geology U. Ga., Athens, 1965—. Cons. on high pressure instrumentation design, engring., geophysics, geochemistry, materials synthesis, gems, and gem materials, 1956—. Recipient several awards and citations for outstanding sci. contbns. U. S. Dept. Army. Mem. Am. Mineral. Soc., Am. Geophys. Union, Am. Soc. M.E. (chmn. research com. on pressure tech. 1965—), Geochem. Soc., A.A.A.S., Sigma Xi. Editor: (with Lloyd) High Pressure Measurement, 1963; High Pressure Technology, 1965. Contbr. numerous articles to tech. jours. Research on optical properties single crystals, hardness vectors in single crystals, mineralogy and geochemistry, instrumentation exptl. and pressure measuring techniques for ultra high pressure research and prodn. purposes, diamond synthesis, syntheses new materials and phys. measurements under high pressures and temperatures. Home: 180 Lanier Ct., Athens, Ga. 30601.*

GIAUQUE, William Francis, chemist; b. Niagara Falls, Ont., May 12, 1895; s. William Tecumseh Sherman and Isabella Jane (Duncan) G. (parents U. S. citizens); B.S., U. Cal., 1920, Ph.D., 1922, LL.D., 1963; D.Sc., Columbia, 1936; m. Muriel Frances Ashley, July 19, 1932; children—William Francis Ashley, Robert David Ashley. Instr. of chemistry, U. Cal., Berkeley, 1922-27, asst. prof., 1927-30, asso. prof., 1930-34, prof. chemistry, 1934—. Recipient prize of A.A.A.S. (Pacific Div.), 1929, for discovery (with H. L. Johnston) of the second and third oxygen isotopes; Chandler medal Columbia, 1936, Elliott Cresson medal Franklin Inst., 1937, for discovery of adiabatic demagnetization method of producing temperatures approaching absolute zero; Nobel Prize for chemistry, 1949; G. N. Lewis medal, 1956. Fellow Am. Phys. Soc., Am. Acad. Arts and Scis.; mem. Am. Chem. Soc. (Gibbs medal 1951), Am. Assn. U. Profs., Nat. Acad. Scis., Am. Philos. Soc., Sigma Xi, Phi Lambda Upsilon (hon.). Proved 3d law of thermodynamics; determined accurately the entropy of numerous substances at absolute zero; other research on low temperature calorimetry, especially on condensed gases, cryogenic apparatus, conversion of ortho to para hydrogen, isotopes and band spectra, free energy and entropy from spectroscopy. Home: 2643 Benvenue Av., Berkeley 4, Cal.

GIBB, Cecil A., Australian social psychologist; b. Sydney, Australia, Aug. 18, 1913; s. Harry A. and Sophia (Renner) G.; B.A., U. Sydney, 1935, B.Ed., 1939, M.A., 1940; Ph.D., U. Ill., 1949; m. Margaret V. Young, Aug. 12, 1939. Dep. dir. Australian Army Psychol. Service, 1942-46; sr. lectr. U. Sydney, 1937-42, 46-47, 49-50; vis. prof. Dartmouth, 1951-55; prof., head social psychology dept. Australian Nat. U., Canberra, 1955—, chmn. bd. sch. gen. studies, 1966—; vis. research asso. Princeton, 1962. Fellow Am., Australian psychol. socs.; mem. Social Research Council Australia (sec. 1964-66). Research, publs. on interaction theory of leadership which proved influential in moving emphasis from search for leadership traits toward consideration of conditions under which leader behavior occurs. Home: 256 Laperouse St., Canberra, A.C.T., Australia.*

GIBB, Jack Rex, psychologist; b. Magrath, Alta., Can., Dec. 20, 1914 (parents Am. citizens); s. John Lye and Ada (Dyer) G.; B.A., Brigham Young U., 1936, M.A., 1937; postgrad. U. Chgo.; Ph.D., Stanford, 1943; m. Lorraine Miller, Dec. 29, 1951; children—Lawrence Henry, Blair Bradford, John Randolph. Faculty, Brigham Young U., 1937-46, asso. prof., 1945-46; asso. prof. psychology Mich. State U., East Lansing, 1946-49; asso. prof. psychology U. Colo., Boulder, 1949-54, prof., 1954-56; research prof. Fels Group Dynamics Center, Newark, Del., 1956-59; dir. research Nat. Tng. Lab., Washington, 1959-61, mem. nat. bd., 1953-59, 65—; cons. psychologist, pvt. practice psychology, Newark, 1961-63; resident fellow Western Behavioral Scis. Inst., La Jolla, Cal., 1963-67; pvt. practice cons. psychologist; 1965—. Cons. Dow Chem. Co., Am. Tel. & Tel., State Dept. Faculty fellow Fund for Advancement of Edn., Ford Found., 1951-52. Fellow Am. Psychol. Assn., Am. Sociol. Assn.; mem. Soc. for Advanced Mgmt. (v.p. research 1959-61), Soc. for Psychol. Study Social Issues (nat. bd. 1956-59), Am. Personnel and Guidance Assn., Am. Assn. for Humanistic Psychology (pres. 1967-68), Nat. Soc. for Study Communication, Internat. Soc. for Study Gen. Semantics. Author: (with Leland P. Bradford, Kenneth Benne) T.-Group Theory and Laboratory Method, 1964; (with Grace N. Platts, Lorraine F. Miller) Dynamics of Participative Groups, 1951; also numerous articles. Derived theory of social orgn. and group mgmt. using high-trust, low-control model; research on small group orgn. using different variables. Address: 8475 La Jolla Scenic Dr., La Jolla, Cal. 92037.*

GIBBON, John Heysham, Jr., Am. physician; b. Phila., Sept. 29, 1903; s. John Heysham and Marjorie (Young) G.; A.B., Princeton, 1923, Sc.D., 1961; M.D., Jefferson Med. Coll., 1927; Sc.D., U. Buffalo, 1959, U. Pa., 1965; m. Mary Hopkinson, Mar. 1931; children—John, Mary (Mrs. John Clarke), Alice (Mrs. Christopher Boehm), Marjorie (Mrs. Robert Shepard). Research fellow in surgery Harvard Med. Sch., 1930-31, 1934-35; fellow in medicine U. Pa. Sch. Med., 1936-42; asst. surgeon Bryn Mawr Hosp., 1936-46; surgeon Pa. Hosp., 1937-50; chief surg. service Mayo Gen. Hosp., Galesburg, Ill., 1945; asst. prof. surgery U. Pa., 1945-46; prof. surgery, dir. surg. research Jefferson Med. Coll., 1946-56, Samuel D. Gross prof. surgery, head dept. surgery, 1956—. Recipient John Scott award, 1953; Am. Heart Assn. Research Achievement award, 1965; Internat. Soc. Surgery Distinguished Service award, 1959; Phila. award, 1964; Rudolph Matas award Tulane U., 1958. Mem. A.C.S. (bd. govs., 1950-64), Am. Surg. Assn. (pres. 1953), Am. Assn. Thoracic Surgery (pres. 1960), Coll. Physicians of Phila. (pres. 1964-66), Heart Assn. Southeastern Pa. (pres. 1958), Soc. Vascular Surgery (pres. 1964), Soc. Clin. Surgery (pres. 1953), Phila. Acad. Surgery (pres. 1956-58). Editor: Surgery of the Chest, 1962. Contbr. numerous articles to med. jours. Inventor heart-lung apparatus; pioneer to employ artificial ventilation to prevent respiratory failure in chest surgery patients; inventor Jefferson Ventilator. Home: 2103 N. Providence Rd., Lynfield Farm, Media, Pa. 19063.*

GIBBONS, Norman Edwin, Canadian bacteriologist; b. Niagara Falls, Ont., Can., Apr. 8, 1906; s. William Edwin and Maud (Perry) G.; B.A., Queen's U., 1927, M.A., 1928; Ph.D., Yale, 1932; m. Marian Alice Bennie, Sept. 22, 1931; children—Janith C. (Mrs. Donald Gordon Ross), Diana M. (Mrs. Terrance Joseph Grinnell). Asst. bacteriologist Fisheries Exptl. Sta., Halifax, N.S., Can. 1931-37; with NRC, Ottawa, Ont., 1937—, asst. dir. div. bioscis., 1959—. Sec.-gen. VIII Internat. Congress for Microbiology, 1962; chmn. Canadian Com. on Culture Collections, 1947; trustee Biol. Abstracts, Bergey's Manual Trust. Mem. Canadian Soc. Microbiologists (pres. 1959-60), Internat. Assn. Microbiol. Socs. (sec.-gen.), Royal Soc. Can., Am. Soc. Microbiology, Soc. Applied Bacteriology, Soc. Gen. Microbiology, Canadian Inst. Food Tech., Sigma Xi. Editor: Recent Progress in Microbiology, VIII. Studies, numerous publs. on microbiology of fresh and salt fish, cured meats, shell and dried eggs, particularly in relation to storage; research on physiology of halophilic bacteria, particularly why they require high concentrations of sodium chloride for growth and survival. Office: Div. Bioscis., NRC, Sussex Dr., Ottawa 7, Ont., Can.*

GIBBS, Erna Leonhardt, med. scientist; b. Bad Homburg, West Germany, Mar. 5, 1904; d. Jean Emil and Ida (Schneider) Leonhardt; student pvt. schs.; m. Frederic Andrew Gibbs, Dec. 16, 1930; children—Erich L., Frederic Andrews. Came to U. S., 1927, naturalized, 1939. Research asst. Boston City Hosp., also Harvard Med. Sch., 1927-30, Johnson Found. for Med. Physics, U. Pa., 1930-31, Harvard Med. Sch., 1931-44; founder (with husband) 1st Am. lab. clin. electroencephalography Boston City Hosp., 1934; research asst. U. Ill. Sch. Medicine, Chgo., 1944—. Recipient Eminent Achievement award Ill. Interprofl. Council, 1956; award woman's council Brain Research Found., 1967; named Woman of Year, Am. Woman's Assn., 1958. Author: (with F. A. Gibbs) Atlas of Electroencephalography, vol. 1, 1950, vol. 2, 1955, vol. 3, 1965; also articles. Determined clin. significance of numerous electroencephalographic patterns. Home: 1427 N. Astor St., Chgo. 60610.*

GIBBS, Frederic Andrews, Am. physician; b. Balt., Feb. 9, 1903; s. Rufus Macqueen and Cornelia (Andrews) G.; A.B., Yale, 1925; M.D., Johns Hopkins, 1929; hon. doctorate U. Montpellier (France), 1965; m. Erna Leonhardt, Dec. 16, 1930; children—Erich Leonhardt, Frederic Andrews. Asst. neuropathology Harvard Med. Sch., 1929-30, research fellow, 1933-37, instr., 1937-44; Johnson Found. research fellow U. Pa., 1930-32; asso. prof. psychiatry Ill. Neuropsychiat Inst., U. Ill. Med. Sch., Chgo., 1944-51, prof. neurology, dir. div. electroencephalography and epilepsy Neuropsychiat. Inst., 1951—; staff St. Luke's Hosp., Chgo., 1944-60, dir. electroencephalographic lab., 1947-60; cons. neurology and psychiatry Presbyn.-St. Luke's Hosp., Chgo., 1960—. Cons. to various state and fed. agys. Recipient Mead Johnson award Am. Acad. Pediatrics, 1938; Lasker award Am. Pub. Health Assn., 1952. Mem. Physiol. Soc., Am. Neurol. Assn., Am. Acad. Neurology, Acad. Cerebral Palsy, Brain Research Found. (trustee), Am. Epilepsy Soc., Epilepsy Assn. Am., Am. Med. Electroencephalographic Assn. Author: (with E. L. Gibbs) Atlas of Electroencephalography, 1941, 2d edit., vol. 1, 1950, vol. 2, 1952, vol. 3, 1964; (with F. W. Stamps) Epilepsy Handbook, 1958; (with M. S. Sadove and D. Becka) Electroencephalography for Anesthesiologists and Surgeons, 1967; (with F. L. Gibbs) Medical Electroencephalography, 1967; also numerous articles. Research on electroencephalogram; thermoelectric blood flow recorder in needle form, dye injection measurement of cerebral blood flow. Home: 1427 N. Astor St., Chgo. 60610. Office: 720 N. Michigan Av., Chgo. 60611.*

GIBBS, Gordon Everett, Am. physician; b. Cordova, Ill., Sept. 25, 1911; s. George E. and Mabel (Ewell) G.; A.B., U. Redlands, 1932; Ph.D., U. Cal. at Berkeley, 1939, M.D., 1942; m. Gertrude Fleischmann, Dec. 19, 1941; children—Gale (Mrs. Michael Roeder), Gwen (Mrs. Randal Stone), Gerald, Greta. NRC fellow U. Cal. at San Francisco, 1947-49, U. Ill., 1949-50; asso. prof. pediatric research U. Md., Balt., 1950-54; asso. prof. pediatrics U. Neb., Omaha, 1954-56, prof., chmn. dept., 1956-66, research prof. pediatrics, 1966—. Mem. gen. med. and sci. adv. council Nat. Cystic Fibrosis Research Found. Mem. Am. Pediatric Soc., Soc. Pediatric Research, Soc. Exptl. Biology and Medicine, Am. Acad. Pediatrics, Am. Diabetes Assn., Am. Coll. Chest Physicians, Sigma Xi. Contbr. chpts. to books, articles to profl. jours. Developed ultra-micro volumetric diffusion method for urea and ammonia; demonstrated effect of pancreas feeding to restore severity of diabetes in depancreatized dogs; initiated clin. use of aerosol neomycin; refined technique of duodenal aspiration for diagnosis of pancreatic deficiency in children; demonstrated incomplete pancreatic deficiency in cystic fibrosis; initiated use of glucagon to treat insulin hypoglycemia in diabetic children, produced glomerulosclerosis and retinal microaneurisms in long term alloxan-diabetic monkeys; assayed metabolic enzymes in isolated sweat gland tissue in cystic fibrosis. Home: 1334 S. 94th St., Omaha 68124.*

GIBBS, Jack Porter, Am. sociologist, educator; b. Brownwood, Tex., Aug. 26, 1927; s. Mayfield and Catherine (Porter) G.; B.A., Tex. Christian U., 1950, M.A., 1952; Ph.D., U. Ore., 1957; m. Sylvia Kroenlein, Dec. 23, 1961; children—Laura Kathleen, Douglas. Lectr., research sociologist internat. population and urban research U. Cal., Berkeley, 1957-59; faculty sociology, U. Tex., Austin, 1959-65, prof., 1963-65, 67—, asso. dir. Population Research Center, 1961-63; prof., chmn. dept. sociology Wash. State U., Pullman, 1965-67. Fulbright fellow, 1952-53; Russell Sage fellow, 1964-65. Mem. Pacific (pres. 1967-68), Am. sociol. assns., Population Assn. Am. Author (with Kingsley Davis, et al) The World's Metropolitan Areas, 1959; (with Walter T. Martin) Status Integration and Suicide, 1964. Editor Urban Research Methods, 1961; Suicide, 1968. Contbr. articles to profl. jours. Formulation of theory on suicide rates, theory on urbanization, conceptual schemes for classifying normative phenomena. Office: U. Tex., Austin, Tex.

GIBBS, James Ethan Allen, Am. inventor; b. Rockbridge County, Va., Aug. 1, 1829; s. Richard and Isbella (Poague) G.; m. Catherine Givens, 1883; m. 2d, Margaret Craig, 1893; at least 3 children. Asso. with father's machine carding bus., Rockbridge County, until 1846; made unsuccessful attempt to establish wool-carding bus. utilizing machine of own design, Mill Point, Pocahontas County, W. Va., 1846; engaged in agr., 1846-50; built a sewing maching (from pictures of sewing machines appearing in advertisements), patented 2 improvements (a forerunner of automatic tensions system and a material feeding device), 1856; patented several chain and lock stitch machines, 1857; invented twisted loop rotary hook machine, 1857; formed partnership with James Willcox, introduced Willcox and Gibbs sewing machine, 1858; patented a lock and clutch driven bicycle; mfr. gunpowder for Confederate Army during Civil War; ret. from bus., 1890; traveled around U. S. and to Europe; gave name to Town of Raphine (from Greek: to sew), Rockbridge County. Died Raphine, Nov. 25, 1902.

GIBBS, Josiah Willard, Am. mathematical physicist; b. New Haven, Conn., Feb. 11, 1839; s. Prof. Josiah Willard and Mary Anna (Van Cleve) G.; grad. Yale U., 1858; Ph.D., Yale U., 1863; studied in Paris, 1866-67; Berlin, 1867, Heidelberg, 1868; hon. doctorates from Erlangen, 1893; Williams Coll., 1893; Princeton, 1896; tutor in Latin, 1863-65, tutor in natural philosophy, 1865-66, Yale College; prof. of mathematical physics, Yale, 1871-1903. Mem., Nat. Acad. Scis.; Fellow Royal Soc., 1897 (Copley medal, 1901); elected to Hall of Fame for Great Americans, 1950. Author: Graphical methods in the thermodynamics of fluids, 1873; A method of geometrical representation of the thermodynamic properties of substances by means of surfaces, 1873; On the equilibrium of heterogeneous substances, 1876-78; Elementary principle in statistical mechanics, 1902. Applied thermodynamics to chemistry (by application of 1st and 2nd laws of thermodynamics to heterogeneous substances, he established theoretical basis for physical chemistry); discovered Gibbs phase rule; began development of vector analysis and applied it to problems of crystallography and to computation of planetary and cometary orbits; also conducted research in optics and statistical mechanics; patented railroad brake. Died New Haven, Conn., Apr. 20, 1903.

GIBBS, Julian Howard, Am. phys. chemist; b. Greenfield, Mass., June 24, 1924; s. Howard Brown and Judith Martha Bassett (Hemenway) G.; B.A., Amherst Coll., 1947; M.A., Princeton, 1949, Ph.D., 1950; m. Cora Lee Gethman, July 27, 1946; children—James Hemenway, Judith Maxwell, Jeffrey Stephen, Jonathan Myles. Instr. phys. chemistry U. Minn., 1951-52; with research lab. Gen. Electric Co., 1952-55, Am. Viscose Corp., 1955-60; mem. faculty Brown U., 1960—, prof. chemistry, 1963—, chmn. dept., 1964—. Fulbright fellow Cambridge (Eng.) U., 1950-51; Fulbright research scholar Max Planck Inst. for Phys. Chemistry, Göttingen, Germany, 1967-68; Guggenheim fellow, 1967-68. Fellow Am. Phys. Soc. (chmn. div. high polymer physics 1963, exec. com. div. 1962-65, high polymer physics prize 1967), Am. Chem. Soc., Phi Beta Kappa, Sigma Xi, Theta Delta Chi. Mem. adv. bd. Biopolymers. Author articles in field. Research on statistical mechanics; dipole moments; valency theory; biol. macromolecules; glasses; polymers. Home: White Birch Lane, Barrington, R.I. 02806. Office: Brown Univ., Providence 02912.

GIBBS, Martin, Am. biologist; b. Phila., Nov. 11, 1922; s. Sam and Rose (Sugarman) G.; B.S., Phila. Coll. Pharmacy and Sci., 1943; Ph.D., U. Ill., 1947; m. S. Karen Kvale, Oct. 11, 1950; children—Janet H., Laura J., Steven J. Michael S., Robert K. Scientist, Brookhaven Nat. Lab., 1947-56; prof. biology Cornell U., Ithaca, N.Y., 1956-64; prof., chmn. biology dept. Brandeis U., Waltham, Mass, 1964—; vis. prof. U. Pa., 1954. Cons. NSF, 1961-64, NIH, 1964—. Mem. Am. Soc. Plant Physiologists, Am. Soc. Biol. Chemists, Biochem. Soc. Editor-in-chief Plant Physiology, 1963—. Research, numerous publs. on photosynthesis, respiration in plants. Home: 32 Slocum Rd., Lexington, Mass. 02173. Office: Dept. Biology, Brandeis U., Waltham, Mass. 02154.*

GIBBS, Peter (Godbe), Am. physicist; b. Salt Lake City, Dec. 7, 1924; s. Lauren Worthen and Mary (Godbe) G.; student Berea Coll., summer 1943, 43-44, U. Mich., summer 1944, 44-45; B.S., U. Utah, 1947, M.S., 1949, Ph.D., 1951; m. Miriam Starling Kvetensky, July 12, 1953; children—Laurence Kay (Doon), Victoria Emmeline, Nicholas, Peter. Research asso. U. Ill., Urbana, 1951-52, instr., 1952-54; Fulbright lectr. theoretical physics U. Ceylon, Colombo, 1954-55; faculty U. Utah, Salt Lake City, 1956—, prof. physics, 1962—. Fulbright lectr. solid state physics Sao Carlos Engring. Sch., U. Sao Paulo (Brazil), 1963. Recipient Ross Coffin Purdy prize Am. Ceramic Soc., 1962. Mem. Am. Phys. Soc., Am. Assn. Physics Tchrs., Sigma Xi. Contbg. author: Physical Chem—An Advanced Treatise, 1967; also articles. Research on properties of dislocations especially in aluminum oxide, motions of positive ions in quartz, effects of high pressure on mech. properties of metals especially creep and internal friction, origin of brittle fracture. Home: 79 Laurel St., Salt Lake City 84103.*

GIBBS, William Francis, Am. naval architect, marine engr.; b. Phila., Aug. 24, 1886; s. William Warren and Frances Ayers (Johnson) G.; student Harvard, 1906-10, Sc.D., 1947; LL.B. and M.A., Columbia, 1913; E.D. (hon.), Stevens Inst. Tech., 1938, N.Y.U., 1955; D.Sc., Bowdoin Coll., 1955; m. Mrs. Vera Cravath Larkin, 1927; children—Francis C., Christopher L.; (step-son) Adrain C. Larkin. Organizer Gibbs Bros., Inc., 1922, Gibbs & Cox, Inc., 1929 (pres.). Controller of shipbuilding, WPB, 1942-43; chmn. Combined Shipbuilding Com. of Combined Chiefs of Staff, 1943; rep. Office War Mblzn. on Procurement Rev. Bd. of Navy, 1943. Recipient Am. Design award, 1943, David W. Taylor gold medal Soc. Naval Architects and Marine Engrs., 1946. Presdl. Certificate of Merit, 1947, Holland Soc. of N.Y. Distinguished service gold medal, 1951, Franklin Gold medal Franklin Inst., 1953; Elmer A. Sperry award, 1955; Michael Pupin medal, Columbia Engring. Alumni Assn., 1959; Allied Professions medal A.I.A., 1960; William S. Newell Meml. award United Seaman's Service, 1962. Fellow Royal Soc. Arts, Am. Soc. M.E. (hon.), (asso.) Inst. Aero. Scis.; mem. Soc., N.Y. Acad. Scis., Naval Architects and Marine Engrs. (v.p.), Instn. Naval Architects, Am. Soc. Naval Engrs., Am. Bur. Shipping (tech. com.), N.E. Coast Inst. Engrs. and Shipbuilders, U. S. Naval Inst., Nat. Acad. Scis., N.Y. Bar, Phi Beta Kappa. Designed faster, more efficient ships for mil. and comml. use, notably the S.S. United States. Home: 945 Fifth Av. Office: 1 Broadway, N.Y.C. 4.

GIBERT, Rene Justin Joachim, French phys. chemist; b. Tarascon, France, Mar. 20, 1908; s. Auguste and Louise (Moutardier) G.; Docteur ès Sciences physiques, U. Clermont, Ferrand (France), 1941; student Faculty Scis. de Marseille (France), Facultes des Sciences, Paris; m. Valentine Pressoiras, June 24, 1933; children—Huguette, Raymonde. With U. Clermont-Ferrand, Nancy, France, 1933—, prof. chem. engring., 1949-58, prof. phys. chemistry, 1958—. Decorated officer l'ordre des Palmes académiques, Mem. Société chimique de France, Société de Chimie physique de France. Author: Heart Transfer, 1963; Fluid Mechanics, 1964; also articles. Research on physico-chem. interpretation of properties of irritable cell. Home: 42 rue de la Foucotte, Nancy 54, France.*

GIBIAN, Heinz Rudolf, German chemist; b. Munich, Germany, Mar. 30, 1916; s. Rudolf and Emmi (Wieneke) G.; ed. Technische Hochschule, Munich; Ph.D. in engring.; diploma in chem. sci.; m. Margarete Jänicke, 1947; 1 son, Christoph. With Schering AG, Berlin, 1942—; dir. Govt. Dept. Biology, 1961—, Govt. Chem. and Pharm. Dept., 1963—. Mem. Soc. German Chemists, German Soc. Biol. Chemistry, Soc. German Doctors and Sci. Researchers. Author: Mucopolysaccharide und Mucopolysaccharidasen. Editor: Deuticke Vienne, 1959, also articles. Home: Bergstrasse 20, 1 Berlin 39. Office: Mullerstrasse 170-172, 1 Berlin 65, Germany.

GIBLETT, Eloise Rosalie, Am. physician; b. Tacoma, Jan. 17, 1921; d. William Richard and Rose (Godfrey) Giblett; B.S., U. Wash., 1942, M.S., 1947, M.D., 1951. Practice medicine, specializing in hematology, Seattle, 1953—; research fellow U. Wash., Postgrad. Med. Sch. London, 1953-55; faculty U. Wash., Seattle, 1953—, research prof. medicine, 1967—; asso. dir. King County Central Blood Bank, 1955—. Mem. Am., Internat. socs. hematologists, Am. Soc. Human Genetics, Am., Brit. socs. immunology, Am. Fedn. Clin. Research, Western Soc. Clin. Research, Western Assn. Physicians, N.Y. Acad. Scis., Sigma Xi, Alpha Omega Alpha. Asso. editor: Transfusion, 1963—; Am. Jour. Human Genetics, 1964—; Am. editor Brief Reports, Vox Sanguinis, 1963—. Studies, numerous publs. on splenic mechanisms, iron kinetics, red cell destruction due to isoantibodies, detection of variants in blood group antigen, serum protein, red cell enzyme genetic systems, changes in red cell antigen asso. with marrow stress. Home: 6533 53d St. N.E., Seattle 98115. Office: Terry at Madison, Seattle 98104.*

GIBSON, Alexander George, English physician; b. Hull, Eng., Sept. 21, 1875; s. William and Elizabeth Gibson; ed. Univ. Coll., Aberystwyth, Christ Ch., Oxford, St. Thomas's Hosp., London; D.M.; m. Constance Muriel Jones; 2 sons, 1 dau. Cons. physician Radcliffe

Infirmary and County Hosp., Oxford, Eng. Fellow Royal Coll. Physicians. Author: The Radcliffe Infirmary, 1926; The Heart, 1926; Clinical Methods, 1927; The Mycoses of the Spleen, 1930; The Physician's Art, 1933. Gave 1st description of mid-diastolic wave in jugular pulse during slow pulse (b-wave), 1907. Died Jan. 11, 1950.

GIBSON, Arnold Hartley, English engr.; b. Yorkshire, Eng., July 26, 1878; s. W. H. Gibson; B.Sc., Manchester (Eng.) U., 1903, D.Sc., 1909; LL.D. U. St. Andrews; m. Amy Quarmby, 1905; 3 sons. Head math. dept. Salford Tech. Inst., 1903-04; asst. lectr. engring. and hydraulics Manchester U., 1904-09, later prof. engring.; prof. engring. U. St. Andrews, 1909-20; mem. Bd. Trad. Com. on Water Power of Brit. Isles; mem. Air Ministry Engine Research Com.; mem. Severn Barrage Com. Mem. Inst. Civil Engr. (Ewing medal 1939), Inst. Mech. Engrs. Author: Hydraulics; Water Hammer in Hydraulic Pipe Lines; Natural Sources of Energy; A Study of the Circular-Arc Bow Girder; Hydro-Electric Engineering; also papers. Died Feb. 16, 1959.

GIBSON, Arthur, Canadian entomologist; b. Toronto, Ont., Can., Dec. 23, 1875; LL.D., Queen's U., 1935. Asst. entomologist exptl. farms br. Canadian Dept. Agr., 1899-1905, asst. entomologists, 1905-08, chief asst. entomologist, 1908-14, chief div. field crop and garden insects, 1914-20, dominion entomologist, head entomol. br., 1920-42. Fellow Entomol. Soc. Am. (past pres.), Royal Soc. (Can.), Entomol. Soc. London; mem. Assn. Econ. Entomologists (past pres.). Research on control of cutworms, army worms, cabbage maggot and flea beetles, flower and common garden, household, greenhouse insects, lepidoptera. Died 1959.

GIBSON, Charles Stanley, English chemist; b. Feb. 8, 1884; s. Joshua Gibson; B.Sc. with honors in chemistry, Oxford (Eng.), U., 1905; M.A., Oxford, Cambridge univs.; Sc.D., Cambridge U.; M.Sc., Tech. U. Manchester (Eng.). Sr. research student Goldsmiths' Co., Oxford U., 1909-12; prof. chemistry Maharaja's Coll., Trivandrum, S. India, 1912-19; prof. chemistry Egyptian Govt. Sch. Medicine, Cairo, 1919-20; prof. chemistry London U., Guy's Hosp. Med. Sch., 1921-39, emeritus, from 1939. Asst. lectr., demonstrator Chem. Lab., Cambridge U., mem. Sidney Sussex Coll.; hon. adviser Chem. Warfare Com., Ministry of Munitions, 1916-19; sr. gas adviser S.E. Eng., 1939-45. Fellow Royal Soc., 1931; Royal Inst. Chemistry; mem. Brit. Assn. (pres. chemistry sect. 1938), Société de Chimie Industrielle (hon.), Chem. Soc. (hon. sec. 1924-33). Author: Essential Principles of Organic Chemistry. Contbr. papers on stereo-chemistry, chemistry of natural products, organic chemistry of arsenic, sulphur, selenium and gold, prodn. films of gold, chemotherapy, others. Worked with W. J. Pope on resolution of benzoylalanine and on alkyl derivatives of gold and dichlorodiethyl sulfide; also work on organic arsenic and gold compounds, setting of Plaster of Paris, others. Died London, Mar. 24, 1950.

GIBSON, David, Canadian psychologist; b. Galt, Ont., Can., Apr. 6, 1926; s. John William and Martha (Sweeney) G.; B.A., U. Toronto, 1950, M.A., 1952, Ph.D., 1960; m. Kathleen May Sudbury, Sept. 9, 1961; children—Martha, Jennifer, Judith. Chief psychologist Ont. Hosp. Schs., 1956-61; research psychologist Childrens Psychiat. Research Inst., London, Ont., 1961-63; prof. U. Calgary (Alta., Can.), 1963—; research dir. Vocational Research Inst., Calgary, 1967—. Fellow Am. Assn. on Mental Deficiency; mem. Am., Canadian (dir.) psychol. assns., Council for Exceptional Children, Psychologists Assn. Alta. (pres. 1966-67). Editor: Canadian Psychologist, 1966—. Research, publs. on genetic and physiol. factors for understanding and prediction of human behavior especially with mentally retarded and disturbed children, physiology of human thinking. Home: 2411 Usher Rd. N.W., Calgary, Alta., Can.*

GIBSON, David Mark, Am. biochemist; b. Kokomo, Ind., Aug. 7, 1923; s. Carl Banta and Marie (Loop) G.; A.B., Wabash Coll., 1944; M.D., Harvard, 1948; m. Margaret Lockhart, June 2, 1951; children—Carl, John, Shauna, Heather, Mark. Research asso. U. Ill., 1950-53; research asso. Enzyme Inst. U. Wis., 1953-55, asst. prof, 1955-58; faculty Ind. U. Sch. Medicine, 1958—, prof., 1961—, chmn. dept., 1967—; investigator Am. Heart Assn., 1957-62. Mem. Am. Soc. Biol. Chemistry, Am. Chem. Soc., A.A.A.S., Am. Assn. U. Profs., Am. Diabetes Assn., Am. Soc. Cell Biology, Sigma Xi. Contbr. numerous articles in field to sci. jours. Research in enzymes concerned with formation and utilization of fatty acids, control of these enzyme systems. Home: 3436 Brisbane Rd., Indpls. 46208.*

GIBSON, Frank Alexander, Irish marine biologist; b. Lurgan, Ireland, May 18, 1924; s. Gibson and Christina (Paton) G.; grad. with 1st class honors U. Coll., Cork, 1945; M.Sc., U. Coll. Cork, 1947, Ph.D., 1953; m. Alice Mary Howe, Sept. 21, 1951; children—Patricia, Lesley, Peter, Jillian. With Dept. Agr. and Fisheries, 1948—, insp. in charge marine fishery research, Dublin, 1959—. Del. Internat. Council for Expln. of Sea; mem. council Irish Specimen Fish Com. Mem. Inst. Biology (founder mem.), Inst. Profl. Civil Servants. Research and publs. on bionomics of shellfish prodn., life history of lobsters and prawns; formulation of control measures designed to promote maximum yield from fishing; design of storage units for comml. purposes; promotion of farming techniques for harvesting mussels and oysters. Home: 13 Barton Dr., Dublin 14. Office: 3 Cathal Brugha St., Dublin 1, Ireland.*

GIBSON, Frank Curry, Am. physicist; b. West Homestead, Pa., Nov. 23, 1916; s. Francis S. and Jennie (Curry) G.; B.S., U. Pitts., 1938; m. Ruth Edna Frey, Aug. 17, 1945; children—Robert Francis, Carol Ruth. Physicist fuel and explosives br. U. S. Bur. Mines, Pitts., 1945-53, supervisory physicist explosives and phys. scis. div., 1954-59, project coordinator Explosives Research Center, 1960—. Mem. Am. Inst. Physics, I.E.E.E., Am. Inst. Aeros. and Astronautics. Research in detonation and combustion in explosives and propellants and hazardous chems. leading toward greater safety in mfg., storage and usage of such materials, low-velocity detonation regime in liquid explosives systems. Home: 915 Gibson Lane, Pitts. 15236. Office: 4800 Forbes Av., Pitts. 15213.*

GIBSON, George, Am. chemist; b. Yonkers, N.Y., Mar. 20, 1909; s. John and Louisa (Sutherland) G.; B.S. in Chemistry, Bklyn. Poly. Inst., 1932, M.S., 1935, Ph.D., 1942; m. Victorine Robinson, Aug. 17, 1940; children—Janet (Mrs. William Teagardin), David George. Chemist, Chemco Photoproducts Co., Inc., 1932-33, Charles Pfizer Co., 1936-37; research chemist E. I. duPont de Nemours & Co., Inc., 1937-42; from asst. prof. to prof. chemistry Ill. Inst. Tech., 1942-60; prof. chemistry Bklyn. Coll. City U. N.Y., 1960—, chmn. dept., 1962—. Asso. chemist Argonne Nat. Labs., 1949-50. Mem. Am. Chem. Soc., Sigma Xi, Phi Lambda Upsilon. Contbr. profl. jours. Research on chemistry of hydrazine; chemistry of uranium; coordination compounds; nonaqueous solvents. Home: 636 Baldwin Av., Baldwin, L.I., N.Y. 11510. Office: Bklyn. Coll., Bedford Av. and Av. H, Bklyn. 11210.*

GIBSON, George Alexander, Scottish mathematician; b. Greenlaw, Scotland, Apr. 19, 1858; s. Robert Gibson; ed. Glasgow (Scotland), Berlin (Germany) univs.; M.A., LL.D.; m. Nellie Stenhouse Hunter; 2 sons, 1 dau. Asst., lectr. in math. Glasgow U., 1883-95, prof. math. 1909-27; prof. math. Glasgow and W. of Scotland Tech. Coll., 1895-1909. Author: Treatise on the Calculus; Introduction to the Calculus; Treatise on Graphs. Died Apr. 1, 1930.

GIBSON, John Ebert, physicist; b. Gibson, N.C., Dec. 25, 1914; s. John Shaw and Edna (Ebert) G.; B.S. in Physics, U. N.C., 1939; Physicist, U. S. Naval Research Lab., Washington, 1940-65. Fellow Washington Acad. Sci.; mem. Am. Astron. Soc., Research Soc. Am., Union Radio Sci. Internat. (mem. U. S. com. V 1955—). Contbr. articles to tech. jours. Patentee push-pull oscillator, tunable slot coupling. Originated microwave triode coaxial circuits; determined microwave solar brightness; confirmed theory stratified lunar surface by microwave eclipse observations; investigated planetary radiation; measured terrestrial atmospheric absorption microwaves. Home: P.O. Box 96, Gibson, N.C. 28343.*

GIBSON, R(alph) E(dward), physical chemist; b. Kings Lynn, Norfolk, Eng., Mar. 30, 1901; s. John and Jane (Ferry) G.; student George Watson's Boys' Coll., Edinburgh, 1914-19; B.S., U. Edinburgh, 1922; Ph.D. (Carnegie Research scholar 1922-24), 1924; m. Elizabeth Burnham Derby, Apr. 4, 1927; children—John D. Southmayd, Anne K. (Mrs. W. Kumm), Ronald Malcolm Eustace. Came to U. S., 1924, naturalized, 1940. Mem. staff Geophys. Lab., Carnegie Instn. of Washington, 1924-46 (on leave of absence 1941-46); lectr. in chem., George Washington U., 1929-39, adjunct prof., 1932-45; vice chmn. sect. H, div. 3, Nat. Defense Research Com., 1941-44; dir. research Allegany Ballistics Lab., 1944-46; mem. of the Applied Physics Lab., Johns Hopkins, 1946—, acting dir., 1947-48, dir. since 1948. Past mem. numerous nat. coms. on weapons and rocketry. Bd. dirs. Found. Advanced Edn. in Scis., NIH. Awarded Crum Brown medal in chemistry, 1920, Hope prize in chemistry, 1921, Hillebrand prize, Chem. Soc. Washington, 1939, Edward Orton Jr. fellow lectr. Am. Ceramic Soc., 1947; President's Certificate of Merit, 1948; Navy Distinguished Pub. Service award, 1958; Captain David Dexter Conrad award, 1960. Fellow Am. Inst. Aeros. and Astronautics; mem. Am. Ordnance Assn., Am. Geophys. Union, Am. Chem. Soc. (chmn. Chem. Soc. of Washington, 1931, councilor, 1932, 33, 1935-36, 1938-41; chmn. div. phys. and inorganic chemistry, 1942-43, councilor, 1942-43), Am. Phys. Soc., Washington Acad. Scis. (pres. 1956), Armed Forces Chem. Assn. (v.p. 1953-56), Philos. Soc. Washington (pres. 1940), Sigma Xi, Sigma Tau. Research, publs. in phys. chemistry, solid propellants, guided missile devel., missile systems, space tech., research adminstrn. Home: 3607 Dunlop St., Chevy Chase 15, Md. Office: The Applied Physics Lab., Johns Hopkins U., Silver Spring, Md.

GIBSON, William, Am. surgeon; b. Balt., Mar. 14, 1788; s. John Gibson; ed. St. John's Coll., Annapolis; M.D., U. Edinburgh (Scotland), 1809; studied with Sir Charles Bell in Eng.; m. Sarah Hollingsworth; 8 children, including Charles Bell. Prof. surgery U. Md., 1811-19; 1812; prof. surgery U. Pa., 1819-55. Author: The Institutes and Practice of Surgery (prin. publ.), 1824. Did much to advance knowledge and practice surgery; tied common iliac artery for aneurism (1st time done in Am.), 1812; performed caesarean section twice on same patient who lived 50 years after 1st operation (most striking surg. success). Died Savannah, Ga., Mar. 2, 1868.

GICLAS, Henry Lee, Am. astronomer; b. Flagstaff, Ariz., Dec. 9, 1910; s. Eli and Hedwig (Leissling) G.; B.S., in Astronomy, U. Ariz., 1937; postgrad. U. Cal. at Berkeley, 1939-40; m. Bernice Francis Kent, May 23, 1936; 1 son, Henry Lee. Research asst. Lowell Obs., Flagstaff, 1931-44, astronomer, 1944—, exec. officer, 1952—. Dir., Raymond Endl. Found., 1963, sec.-treas., 1964—. Fellow A.A.A.S.; mem. Internat. Astron. Union, Am. Astron. Soc., Astron. Soc. Pacific (past dir.), Ariz. Acad. Sci. Contbr. articles to tech. jours. Research on accurate position measurement moving objects such as comets and minor planets including search and recovery periodic ones, stellar photometry, proper motion survey 13-inch Pluto discovery telescope to discover hi-velocity stars, moving groups including many new white dwarf, double and spl. degenerate stars. Home: 120 E. Elm St. Office: Lowell Obs., Flagstaff, Ariz. 86001.*

GIDDINGS, Franklin Henry, Am. sociologist; b. Sherman, Conn., Mar. 23, 1855; s. Rev. Edward J. and Rebecca Jane (Fuller) G.; A.B., Union Coll., 1877, A.M., 1899, hon. Ph.D., 1897; LL.D., Columbia, 1929; m. Elizabeth P. Hawes, Nov. 8, 1876; children—Henry Starr, Elizabeth Rebecca (Mrs. Ralph Abercrombie), Lorinda Margaret. Engaged in journalism, 1877-83; prof. Bryn Mawr Coll., 1888-94; lecturer sociology, 1891-94, prof. sociology, 1894-1906, sociology and history of civilization, 1906—, Columbia. Fellow Am. Statis. Assn.; mem. Nat. Inst. of Arts and Letters; history Am. Sociol. Soc. (pres. 1910-11). Author: The Modern Distributive Process (with J. B. Clark), 1888; The Theory of Sociology, 1894; The Principles of Sociology, 1896 (also French, German, Russian, Spanish, Hebrew, Czech and Japanese transls.); The Theory of Socialization, 1897 (Italian translation, 1898); The Elements of Sociology, 1898; Democracy and Empire, 1900; Inductive Sociology, 1901; Descriptive and Historical Sociology, 1906; Pagan Poems, 1914; The Western Hemisphere in the World of Tomorrow, 1915; The Responsible State, 1918; Studies in the Theory of Human Society, 1922; The Scientific Study of Human Society, 1924; The Mighty Medicine—Superstition and Its Antidote, 1929; Civilization and Society, 1932. Laid groundwork for sci. investigation of social phenomena, basing his studies on idea of consciousness of kind, and using inductive and statis. methods in sociology. Died Scarsdale, N.Y., June 11, 1931.

GIDDINGS, J(ohn) Calvin, Am. chemist; b. American Fork, Utah, Sept. 26, 1930; s. Luther W. and Berniece (Crandall) G.; B.S., Brigham Young U., 1952; Ph.D., U. Utah, 1954; m. Jennifer Jean Hill, Dec. 26, 1957; children—Steven B., Michael C. Faculty U. Utah, 1956—, prof., 1966—. Mem. chemistry research evaluation panel Air Force Office Sci. Research, 1964—. Recipient Am. Chem. Soc. award in Chromatography and Electrophoresis, 1967. Mem. Am. Chem. Soc., A.A.A.S., Sigma Xi. Author Dynamics of Chromatography, 1965; Editor: Separation Science, (with R. A. Keller) Chromatographic Science, Advances in Chromatography, Vols. 1, 2, 3, 4, 5, 1965, 66, 67, 68. Contbr. numerous articles in field to sci. jours. Devel. of new theoretical concepts in chromatography, electrophoris, snow physics, diffusion processes, chem. kinetics. Home: 904 Military Dr., Salt Lake City 84108.*

GIDEON, Peter Miller, Am. pomologist; b. Woodstock, O., Feb. 9, 1820; s. George and Elizabeth (Miller) G.; m. Wealthy Hull, Jan. 2, 1849. Moved to land claim Gideon's Bay, Lake Minnetonka Minn., 1858; head Minn. exptl. fruit farm, 1878; engaged in developing varieties of fruit to withstand Northern climates for 41 years; developed Wealthy apple from seeds of Siberian crabtree, also Peter and Gideon variety of apples; originated new varieties of crab apples. Author: Growing Hardy Fruits, 1885; Our Seedling and Russian Apples, 1887. Died Oct. 27, 1899.

GIEBE, Erich, German physicist; b. Brandenburg, Germany, Dec. 20, 1877; studied electrotechnics, then physics and math. Tech. U. Berlin, 1897-1902; Ph.D., U. Berlin, 1903. Joined strong current lab. Reich Phys.-Tech. Inst., 1904, became dir. dept. II of electricity and magnetism, 1928. Research and publs. on devel. high frequency technics; developed capacity and induction norms, bifilar bridge (GiebeZickner bridge); built precision high frequency gauges; discovered piezoelec. illumination phenomenon; set up series law on regular behavior of elastic individual frequencies of quartz crystals and isotopes. Died Berlin-Charlottenburg, June 22, 1940.

GIEBEL, Christoph Gottfried Andreas, German paleontologist, zoologist; b. Quedlinburgh, Germany, Sept. 13, 1820; s. Christoph and Johanna (Keilholz) G.; Ph.D., U. Halle (Germany), 1845. Became mem. faculty U. Halle, 1848, asso. prof., 1858, prof. zoology, 1861. Author: Allgemeine palaeontologie, 1852; Odontographie, 1854; Die Versteinerungen im Muschelkralk von Lieskau bei Halle, 1856; Die Fauna der Braunkohlenformation von Latdorf bei Bernburg, 1864; Vogelschutzbuch, 1868; (monograph) Insecta epizoa, 1874; also numerous articles. Research on paleontology, osteology of mammals, protection of birds and plants. Died Halle/Saale, Nov. 14, 1881.

GIEMSA, (Berthold) Gustav (Carl), chemotherapeutist; b. Blechhammer, Germany, Nov. 20, 1867; s. Gustav and Franziska G.; student pharmacy, Leipzig, Germany, 1892-94, later chemistry, mineralogy, and bacteriology; student organic and physiol. chemistry at U. Berlin, beginning 1898; hon. M.D.; m. Eugenie Mayer, 1916. Govt. pharmacist and chemist German E. Africa, from 1895; founder (with B. Nocht) Inst. for Ship and Tropical Diseases, Hamburg, Germany, named dir. chem. dept., 1900. Contbr. articles to med. jours. and publs. of his inst. Research in tropical medicine; discovered the Giemsa stain (mixture of methylene-azure which stained protozoan blood parasites), 1903; studies on quinine derivatives, bismuth preparations. Died Biberwier, Austria, June 10, 1948.

GIER, Herschel Thomas, Am. embryologist; b. Hepler, Kan., May 11, 1907; s. Delta L. and Jennie Smart) G.; B.S., Kan. State Tchr.'s Coll., 1931; Ph.D., Ind. U., 1907; m. Wilma Hobson, Sept. 4, 1932; children—Donald, Ronald, Harold. Postdoctoral fellow Harvard, 1936-37; faculty Ohio U., Athens, 1937-47, asso. prof., 1946-47; faculty Kan. State U., Manhattan, 1947—, prof. embryology, 1961—. Fellow A.A.A.S.; mem. Am. Assn. Anatomists, Am. Soc. Zoologists, Soc. for Study Reprodn. Author: Coyote in Kansas, 1957; also articles. Research on taxonomy and distbn. of amphibians; numbers, food, diseases (especially rabies) in foxes; biology and econs. of coyotes; reprodn. in dogs; reprodn. and embryology of dairy cattle. Home: 1123 Vattier St., Manhattan, Kan. 66502.*

GIERKE, Edgar Otto Konrad von, see von Gierke, Edgar Otto Konrad.

GIES, William John, Am. biol. chemist; b. Reisterstown, Md., Feb. 21, 1872; s. John Jr., and Ophelia Letitia (Ensminger) G., B.S., Gettysburg, 1893, M.S., 1896, Sc.D., 1914; Ph.B., Yale, 1894, Ph.D., 1897; 1940; postgrad. U. Berne, 1899, Marine Biol. Lab., Woods Hole, Mass., 1901, 02; several hon. degrees; m. Mabel Loyetta Lark, May 24, 1899; children—John, James Tressler, Robert Henry, Mary. Asst. in physiol. chemistry Yale, 1894-98, in zoölogy, 1895, tutor in physiology, 1896-98; instr. physiol. chemistry Columbia, 1898-1902, adj. prof., 1902-05, prof., 1905-07, prof. biol. chemistry, 1907-55, sec. faculty Coll. Phys. and Surg., 1905-21; prof. physiol. chemistry, N.Y. Coll. Pharmacy, 1904-22, Tchrs. Coll. (Columbia), 1909-28; cons. chemist N.Y. Bot. Garden, 1902-21; cons. pathol. chemist Bellevue Hosp., 1910-22. Dental awards and fellowships founded in his honor; Recipient awards N.Y.C. and State dental socs.; plaque erected to him as founder Columbia U. Dental Sch., 1952. Fellow A.A.A.S. (organizer), Am. Coll. Dentists, Am. Acad. Periodontol., N.Y. Acad. Dentistry (v.p., pres. pres., 1st hon. fellow), N.Y. Acad. Medicine (asso.); mem. Am. Philos. Soc., Soc. Exptl. Biology and Medicine (pres. 1917-19), Am. Soc. Biol. Chemists (sec. 1906-10), Internat. Assn. Dental Research (pres. 1939-40), Pan-Am. Odontol Assn. (pres. 1943-44), Dental Council (hon.), Am. Physiol. Soc., Soc. Pharmacology and Exptl. Therapeutics, Phi Beta Kappa, Sigma Xi, other sci. socs. Author: Biochemical Researches, 8 vols., 1903-27; Text-Book of General Chemistry, 1904; Text-Book of Organic Chemistry, 1905, 09; Laboratory Work in Biological Chemistry, 1906; Bull. on Dental Edn., for Carnegie Found., 1926. Editor Am. Dental Assn. vol. on Dental Caries, 1939, 2d edit., 1941, Wells Meml. vol., 1948; editor Am. Coll. of Dentists vol. on Dental Care Under Clinical Conditions, 1943; editor vol. for N.Y. Inst. Clin. Oral Path., on Fluorine in Dental Public Health, 1945. Founder, editor Proc. Soc. Exptl. Biology and Medicine, 1904-10, Proc. Am. Soc. Biol. Chemists, 1907-10, Bio-chem. Bull., 1911-16; Jour. Dental Research, 1919-36, editor emeritus since 1936; Jour. Am. Coll. of Dentists, 1934-38; asst. editor, 1938-40; editor dept. of biol. chemistry Chem. Abstracts, 1911-21; asso. editor N.Y. Jour. Dentistry, 1935-36; contbg. editor Annals of Dentistry, 1936-42, mem. editorial bd., 1947-52; editor History of First Three Decades of N.Y. Acad. of Dentistry 1953. Died May 20, 1956.

GIESE, Arthur Charles, Am. biologist; b. Chgo., Dec. 19, 1904; s. Theodore and Bronice (Bombinska) G.; B.S., U. Chgo., 1927; Ph.D., Stanford U., 1933; m. Raina Ivanoff, Aug. 3, 1928; 1 son, Arthur Theodore. Mem. faculty Stanford (Cal.) U., 1933—, prof. biology, 1947—; staff Biol. Lab., Cold Spring Harbor, N.Y., summer 1935, Woods Hole, Mass., summer 1942, 44, 46; vis. prof. Cal. Inst. Tech., 1950. Rockefeller Found. fellow, 1939-40; Guggenheim fellow, 1947, 59. Mem. Soc. Gen. Physiologists, Am. Soc. Zoologists, Am. Soc. Protozoology, Western Naturalists, A.A.A.S., Am. Inst. Biol. Scis. Author: Cell Physiology, 1957, 62, 68; also articles. Editor: Photophysiology, 2 vols., 1964, 68. Research in radiation biology, physiology of regeneration of single cells, reproductive physiology of marine invertebrates. Home: 792 Santa Maria St., Stanford, Cal. 94305.*

GIESEL, Friedrich Otto (Fritz), German chemist; b. Winzig, Germany, May 20, 1852; s. Johannes and Auguste (Freitag) G.; student Berlin Indsl. Acad., 1872-74; Ph.D., Göttingen, Germany, 1867; hon. dr.-engring., Braunsweig, Germany, 1916; m. Martha Schwormstädt, 1884; 1 son, 2 daus. Named asst. Exptl. Inst. German Alcohol. Mfrs., Berlin, 1875; Joined C. Libermann at Berlin Indsl. Acad., 1876; employee Chininfabrik Braunschweig (quinine factory), from 1878; hon. prof. Tech. U. Braunschweig; Mem. Leopoldina. Contbr. numerous articles to tech. jours. Pioneer in radiology and radiochemistry in Germany; 1st in Germany to produce radium salts commercially; discovered that a magnetic field deflects beta rays, also discovered radioactive element emanium (closely related to actinium, discovered by Debierne, 1900), 1903, actinium-X (an isotope of radium). Died of lung cancer caused by radiation, Braunschweig, Nov. 14, 1927.

GIFFARD, Henri, French engr., inventor; b. Paris, 1825; invented, built first motor-powered balloon aircraft, 1852, invented automatic injection pump, 1858; built dirigible displayed at World's Fair 1878. Died 1882.

GIFFORD, Ray Wallace, Jr., Am. physician; b. Westerville, O., Aug. 13, 1923; s. Ray W. and A. Marie (Wagoner) G.; B.S., Otterbein Coll., 1947; M.D., Ohio State U., 1947; M.S., U. Minn., 1952; m. Mary E. Morris, May 23, 1947; children—Peggy Elizabeth, Cynthia Ann, Susan Jane. Practice medicine, specializing in internal medicine, Rochester, Minn., 1953-61, Cleve., 1961—; cons. internal medicine, mem. staff Mayo Clinic, 1953-61; staff mem. dept. hypertension, renal disease Cleve. Clinic Found., 1961—; faculty Mayo Found., 1953-61. Recipient Alcorn Ophthalmology prize, 1947, Alumni Achievement award, 1962 (both Ohio State U.). Fellow A.C.P., Am. Coll. Chest Physicians, Am. Coll. Clin. Pharmacology and Chemotherapy, Am. Geriatrics Soc.; mem. A.M.A., Central Soc. Clin. Research, Am. Fedn. Clin. Research, Am. Heart Assn., Heart Assn. Northeastern Ohio (pres. elect), Am. Therapeutic Soc., Am. Soc. Nephrology, Alpha Omega Alpha. Contbr. numerous articles to profl. jours. Developed phentolamine test for pheochromocytoma; pioneer new antihypertensive drugs; evaluated effect of long term antihypertensive therapy on course of severe and malignant hypertension, results of operations in mgmt. of renal vascular hypertension; studies natural history of Raynaud's disease, mechanism of hypotensive action of diuretics. Home: 2504 Newbury Dr., Cleve. 44118. Office: 2020 E. 93d St., Cleve. 44106.*

GIFFORD, Sanford Robinson, Am. ophthalmologist; b. Omaha, Jan. 8, 1892; s. Harold and Mary (Millard) G.; A.B., Cornell U., 1913; M.A., M.D., U. Neb., 1918; m. Alice Carter, July 11, 1917; children—Sanford R., Carter. Bacteriologist, rank of 1st lt. U. S. Army Base Hosp., in France, 1918-19; in practice with father, Omaha, 1919-29; instr. in ophthalmology U. Neb. Med. Sch., 1919-24, asst. prof. ophthalmology, 1924-29; prof. ophthalmology, Northwestern U. Med. Sch., also ophthalmologist to allied hosps. from 1929; attending ophthalmologist Cook County Hosp. since 1932. Mem. Am. Ophthal. Soc., Am. Acad. Ophthalmology and Oto-Rhino.-Laryngology, Chgo. Inst. Medicine, Sigma Xi. Author: Handbook of Ophthalmic Therapeutics, 3d edit., 1942; Textbook of Ophthalmology, 2d edit., 1941. Asso. editor Archives of Ophthalmology since 1928. Corr. editor Klinische Monatsblätter für Augenheilkunde, 1928-40. Contbr. articles on bacteriology of the eyes, especially diseases due to fungi and higher bacteria. Reported (with J. M. Patton) probable etiological agt. of hitherto unknown disease, agrl. conjunctivitis. Died Feb. 25, 1944.

GIGAS, Gunter, physicist; b. Dürrenberg, Germany, Aug. 2, 1920; s. William Felix and Irmgard (Behrend) G.; came to U. S., 1930, naturalized, 1936; B.Sc., U. Nev., 1950; M.Sc., U. So. Cal., 1959, Ph.D., 1962; m. Joan E. Brankman, Feb. 20, 1954; children—Mark, Marina. Group leader High Altitude Lab., Airesearch Co., Los Angeles, 1951-60; with Litton Industries, Canoga Park, Cal., 1962; faculty U. So. Cal., 1960-62; research scientist Atomics Internat., Canoga Park, 1962—, supr. radiation effects, 1964—. Mem. Am. Phys. Soc., Sci. Research Soc. Am., Am. Geophys. Union, Sigma Xi, Phi Kappa Phi. Research, publs. in radiation effects in dielectric materials, temperature control coatings, elastic and inelastic scattering of protons, measurement of excitation functions, optical model analytical study of inelastic scattering, neutron spectral determination. Home: 1313 Ramona Dr., Camarillo, Cal. 93010. Office: 8900 Desoto Av., Canoga Park, Cal. 91304.*

GIGAS, Johann, German cartographer; b. Lügde, Germany, circa 1582; student medicine, math., Helmstedt, Germany, 1597-99, from 1600, Wittenberg, Germany, 1599; m. Maria von Dorsten, circa 1607; 8 children. Named prof. math. and medicine U. Burgsteinfurt, Germany, 1607, also in charge pharmacy, lectr. on physics; went to Münster, Germany, 1614; became personal physician to prince bishop; practiced medicine. Author: Prodromus geographicus (atlas archdiocese of Cologne, Germany, and its bishoprics), 1620; also numerous maps of Westphalian territories, some of which were based on his own surveying. Died Münster, 1637.

GIGNOUX, Maurice Irénée Marie, French geologist; b. Lyons, France, Oct. 19, 1881; Ph.D., 1913; m. Miss Garel, 1909; 6 children. Became asst. to prof. Kilian in lab. Grenoble, France, 1909, succeeded Kilian, 1926; joined meteorol. research dept. French Army, 1913; joined U. Strasbourg (France), 1918; Mem. French Acad. Scis., 1932. Recipient Penrose medal Geol. Soc. Author: Stratigraphical Geology, 1926; Geology of Dams, 1955; also publ. which is basis of all Mediterranean Pliocene and Quarternary stratigraphy, 1913. Died Grenoble, Oct. 20, 1955.

GILBERD, William, see Gilbert, William.

GILBERT, Benjamin, chemist; b. Felixstowe, Eng., Sept. 27, 1929; s. William Richard and Dorothy O. (Naylor) G.; B.Sc., U. Bristol (Eng.), 1950, Ph.D., 1954; m. Maria Elisa Alentejano, Sept. 27, 1959; children—William Richard, Peter Alentejano. Ministry Aviation, Waltham Abbey, Essex, Eng., 1953-57; staff Wayne State U., Detroit, 1957-58, Instituto de Quimica Agricola, Rio de Janeiro, Brazil, 1958-62; faculty Universidade Fed. do Rio de Janeiro, 1962—. Fellow Chem. Soc., Am. Chem. Soc. Contbg. author The Alkaloids, 1965; also articles. Research on conformational studies of large rings, synthesis of isoflavone, isolation and structure of many indole alkaloids and some triterpenes. Office: Faculdade de Farmacia, Av. Wenceslau Braz, 49 fundos, Rio de Janeiro, ZC 82, Brazil.*

GILBERT, Charles Henry, Am. zoologist; b. Rockford, Ill., Dec. 5, 1859; s. Edward and Sarah (Bean) G.; B.S., Butler U., 1879; M.S., Ind. U., 1882, Ph.D., 1883; m. Julia R. Hughes, Aug. 7, 1883. Asst. in natural sciences and modern langs. Ind. U., 1880-84, prof. zoology, 1889-91; prof. natural hist. U. Cin., 1884-89; prof. zoology Stanford, 1891—. Asst. to U. S. Fish Commn., 1880-98; naturalist in charge U. S. Fish Commn. steamer Albatross, 1889-90; naturalist in charge Hawaiian Explorations of U. S. Fish Commn. steamer Albatross, 1902, and explorations in northwest Pacific and Japan, 1906; asst. Internat. Fisheries Commn., 1909; in charge expert salmon investigations for U. S. Bur. of Fisheries and British Columbia Fisheries Dept., 1909-27. Author: Synopsis of the Fishes of North America (with David Starr Jordan), 1882; The Deep-Sea Fishes (of the Hawaiian Islands), 1905. Died Apr. 20, 1928.

GILBERT, Daniel Lee, Am. physiologist; b. Bklyn., July 2, 1925; s. Louis and Blanche (Lutz) G.; A.B., Drew U., 1948; M.S., State U. Ia., 1950; Ph.D., U. Rochester, 1955; m. Claire Plunguian, July 26, 1964. Instr., U. Rochester, 1955-56; faculty Albany Med. Coll., 1956-60, asst. prof., 1959-60; faculty Jefferson Med. Coll., Phila., 1960-63, asso. prof., 1962-63; physiologist NIH, Bethesda, Md., 1962-63, head sect. on cellular biophysics Lab. Biophysics, Nat. Inst. Neurol. Diseases and Blindness, 1963—. Research cons. grad. council biophysics George Washington U., 1965—. Mem. corp. Marine Biol. Lab., Woods Hole, Mass. Fellow A.A.A.S.; mem. Am. Chem. Soc., Am. Physiol. Soc. (Bowditch lectr. 1964), Biophys. Soc., Soc. Exptl. Biology and Medicine, Soc. Gen. Physiologists, Sigma Xi. Research and publs. on toxic effects oxygen in biol. systems, physiology of biol membranes by measuring ion permeabilities and elec. properties of muscle and nerve membranes, significance of atmospheric oxygen in biol. evolution. Home: 4879 Battery Lane, Bethesda 20014. Office: Lab. Biophysics, Nat. Inst. Neurol. Diseases and Blindness, NIH, Bethesda, Md. 20014.*

GILBERT, Everett Eddy, Am. chemist; b. Ithaca, N.Y., May 5, 1914; s. Allan H. and Katharine (Everett) G.; A.B., Yale, 1935, Ph.D., 1938; m. Norma Hyde, June 17, 1944; children—Steven M., Susan K. Chemist, Tidewater Asso. Oil Co., 1938-41; with Allied Chem. Corp., Morristown, N.J., 1941—, research asso., 1962—; spl. research insecticides, agrl. chems., fluorine chemistry, sulfur chemistry, sulfonates, and detergents. Mem. Phi Beta Kappa, Sigma Xi. Author: Sulfonation and Related Reactions, 1965; also numerous articles on detergent mfr. and fluorine chems. Patentee in field. Home: 7 Frederick Pl., Route 12, Morristown 07960. Office: P.O. Box 405, Morristown, N.J. 07960.*

GILBERT, François-Hilaire, veterinarian; b. Chatellerault, France, Mar. 18, 1757; prof. Alfort Vet. Sch.; mem. French Acad. Scis., 1795, Soc. Agr. Author: Recherches sur les causes des maladies char bonneuses dans les animaux, 1795; la Propagation des betes à laine de race d'Espagne, 1797. Introduced merino sheep to France; studied symptomatic anthrax in animals. Died Signoriolano, Spain, Sept. 6, 1800.

GILBERT, Grove Karl, Am. geologist; b. Rochester, N.Y., May 6, 1843; s. Grove Sheldon and Eliza (Stanley) G.; A.B., U. of Rochester, 1862, A.M., 1872, LL.D., 1898, U. of Wis., 1904, U. of Pa., 1907; m. Fannie L. Porter, Nov. 10, 1874. Asst. in Ward Mus., Rochester, 1863-68; geologist in Ohio survey, 1868-70, Wheeler survey, 1871-74, Powell survey, 1875-79, U. S. Geol. Survey, from 1879 (chief geologist, 1889-92). Spl. lecturer, Cornell, 1886, Columbia, 1892, Johns Hopkins, 1895 and 1896. Walker grand prize, Boston Soc. Natural History, 1908. Fellow Royal Soc., 1918; mem. Nat. Acad. Scis. Editor geol. and phys. geography depts., Johnson's Encyclopaedia. Author: Report on the Geology of the Henry Mountains, 1877; Lake Bonneville, 1890; Introduction to Physical Geography, 1902; Glaciers and Glaciation, vol. 3, Harriman Alaska Expdn., 1904. First to recognize a laccolithic mountain group and to introduce the ideas of erosion, glaciation and river devel. to geomorphology; notable concepts on relationship between formation of Great Lakes and glacial action; studied Niagara River and Niagara Falls, hydraulic mining debris in Sierra Nevada; morphology and glaciation of Sierra Nevada. Died Jackson, Mich., May 1, 1918.

GILBERT, Gustave M(ark), Am. psychologist; b. N.Y.C., Sept. 30, 1911; s. Mark and Ethel (Nierenberg) G.; B.A., Coll. City N.Y., 1932; M.A., Columbia, 1936, Ph.D., 1939; m. Matilda Safran, June 15, 1941; children—Robert, John, Charles. Instr., Conn. Coll. For Women, 1939-40, Bard Coll., Anandale, N.Y., 1940-42; prison psychologist Internat. Mil. Tribunal, Nuremberg, Germany, 1945-46; asso. prof. Princeton, 1947-50; chief psychologist VA Hosp., Northport, N.Y., 1950-51; asso. prof. Mich. State U., 1951-58; cons. Mich. Dept. Corrections, 1953-58; prof., chmn. psychol. dept. L.I. U., Bklyn., 1958—. Cons. Peace Corps., 1963—. Recipient Soc. for Psychol. Study Social Issues award for best study internat. tensions, 1950. Diplomate Am. Bd. Expt. Profl. Psychology. Mem. Am. Psychol. Assn., Interam. Soc. Psychology (past pres.). Author: Nuremberg Diary, 1947; The Psychology of Dictatorship, 1950; also articles. Research on relationship between psychopathology and social pathology, advancement biosocial theory personality. Home: 6 Crystal Dr., Great Neck, N.Y. 11021. Office: L.I. U., Bklyn. 11201.*

GILBERT, Harry, Am. chemist; b. Cleve., Aug. 16, 1918; s. Max and Myrtle (Lemel) G.; B.S. in Chem. Engring., Case Inst. Tech., 1941; m. Esmine Brillis, Dec. 31, 1944; children—Karen, Natalie, Paul, Sally, Diana. Research asso. Case Inst. Tech., Cleve., 1941-42, U. Ill., 1944-45; prodn. supr. Plumbrook TNT Plant, Sandusky, O., 1942-44; research chemist B.F. Goodrich Co., Akron, O., 1946-60; supr. polymer research Hercules Powder Co., Allegany Ballistics Lab., Cumberland, Md., 1960—. Recipient NDRC award, 1945. Mem. Am. Chem. Soc. Patentee in field. Pioneered synthesis vinylidene cyanide which was made into synthetic fiber, purification procedure for poisoned water; developed filament-wound ablative nozzle, exit cone for ultra-high acceleration rockets; developed improved solid propellants for rocket motors. Home: Av. B, Potomac Park. Office: Box 210 Cumberland, Md. 21502.*

GILBERT, James Freeman, Am. geophysicist; b. Vincennes, Ind., Aug. 9, 1931; s. James Freeman and Gladys (Paugh) G.; B.S., Mass. Inst. Tech., 1953, Ph.D., 1956; NSF postdoctoral fellow, U. Cambridge (Eng.), 1956-57; m. Sally Bonney, June 19, 1959; children—Cynthia, Sarah, James Sherwood. Research asso. Mass. Inst. Tech., 1956-57; successively asst. research geophysicist, asst. prof., asso. prof. U. Cal. at Los Angeles, 1957-59; sr. research geophysicist Tex. Inst., Dallas, 1959-61; prof. geophysics U. Cal. at San Diego, 1961—, chmn. dept. earth scis., 1963-64. Guggenheim fellow, 1964-65. Fellow Am. Acad. Arts and Scis., Am. Geophys. Union; mem. Am. Phys. Soc., Am. Math. Soc., Seismological Soc. Research on seismology; diffraction theory; communication theory; elastodynamics. Home: 650 Rimini Rd., Del Mar, Cal. 92014. Office: P.O. Box 109, La Jolla, Cal. 92038.

GILBERT, Sir Joseph Henry, English chemist; b. Hull, Eng., Aug. 1, 1817; ed. Glasgow U., Univ. Coll., London; student of Liebig, Ph.D., Giessen, Germany; M.A., Oxford U.; Sc.D., Cambridge U.; LL.D., Glasgow, Edinburgh; m. Eliza Laurie, 1850 (dec. 1853); m. 2d, Maria Smith, 1855. (with John Bennet Lawes) dir. Rothamsted Lab., from 1843; Sibthorpean prof. rural economy Oxford U., 1884-90. Fellow Royal Soc., 1860, Chem. Soc. (pres. 1882), Linnaean Soc.; corr. French Acad. Scis. Research on nitrogen fertilizers. Died Harpenden, Eng., Dec. 23, 1901.

GILBERT, Lawrence Irwin, Am. biologist; b. N.Y.C., Jan. 24, 1929; s. Charles and Matilda (Bronznick) G.; B.S., L.I. U., 1950; M.S., N.Y. U., 1955; Ph.D., Cornell U., 1958; m. Doris Paule Millstein, Oct. 26, 1952; children—Scott David, Daniel Todd, Joanne Robin. Faculty Northwestern U., Evanston, Ill., 1958—, prof., biol. scis., chmn. dept., 1965—, sec.-treas. div. comparative endocrinology. NSF Sr. fellow Univerisität Bern (Switzerland), 1964-65. Mem. Soc. Developmental Biology, Soc. Exptl. Biology, Am. Soc. Zoologists, Am. Soc. Cell Biology, Entomol. Soc. Am., Soc. Gen. Physiologists. Research and numerous publs. on hormonal control growth and devel. in insects, effects hormones on lipid metabolism. Home: 937 Sutton Dr., Northbrook, Ill. 60062. Office: Northwestern U., Evanston, Ill. 60201.*

GILBERT, Ludwig Wilhelm, German physicist; b. Berlin, Aug. 12, 1769; s. Johann Ludwig and Dorothea Sophia (Jäner) G.; began study of geography and physics in Halle, Germany, 1786; Ph.D., 1794; M.D., Greifswald, Germany, 1808. Became observer, asso. prof., Halle Obs., 1795; lectr. math; lectr. physics, 1798-99; became librarian, 1798; named prof. physics and chemistry at Halle; apptd. prof. physics, Leipzig, Germany, 1811. Author: Die Geometrie nach Legendre, Simpson van Swinden, Gregor a St. Vincentio und den Alten dargestellt, 1798; also wrote introductions, critical supplements to phys. essays. Editor, Annalen der Physik, 76 vols., 1799-1824. Research on acids and electricity; discovered dry lime does not absorb dry chlorine. Died Leipzig, Mar. 7, 1824.

GILBERT, Philippe, mathematician; b. Beauvaing, France, 1832; prof. Catholic U., Louvain, Belgium, 1855-92; mem. French Acad. Scis. Developed analytical formula for components of acceleration, 1888; studied geometrical analysis, mechanics, history of scis. Died Louvain, 1892.

GILBERT, Robert Pertsch, Am. mathematician; b. N.Y.C., Jan. 8, 1932; s. Ralph H. and Ruth (Pertsch) G.; B.S., Bklyn. Coll., 1952; M.S. in Physics, Carnegie Inst. Tech., 1955, M.S. in Math., 1955, Ph.D. in Math., 1958; m. E. Eileen Manton, Oct. 28, 1955. Faculty, U. Pitts., 1957-60, Mich. State U., 1960-63; research asst. prof. Inst. for Fluid Dynamics and Applied Math., U. Md., 1961-64, research asso. prof., 1964-65; prof. dept. math. Georgetown U., Washington, 1965-66, Ind. U., 1966. Cons. spl. coal research div. U. S. Bur. Mines, 1958-60, Naval Ordnance Lab., 1961-64. Mem. Am. Inst. Physics, Am. Math. Soc., Soc. for Indsl. and Applied Math., Washington Acad. Scis., Sigma Xi, Pi Mu Epsilon. Research and publs. on analysis, especially harmonic functions, boundary problems, math. physics, partial differential equations. Home: 2201 Queens Way, Bloomington, Ind. 47401.*

GILBERT, Rufus Henry, Am. physician, inventor; b. Guilford, N.Y., Jan. 26, 1832; s. William Dwight Gilbert; attended Coll. Phys. and Surg., N.Y.C.; m. Miss Maynard; m. 2d, Miss Price; 2 children. Began practice medicine, Corning, N.Y., circa 1853; went to Europe, circa 1857, became convinced that pub. health problems could best be solved by rapid transp. facilities to permit urban residents to live outside cities in cleaner atmosphere; surgeon to Duryée Zouaves, 1861, later served as med. dir. XIV Corps, U. S. Army during Civil War; to implement his ideas on rapid transp., became asst. supt. Central R.R. of N.J.; obtained patents for pneumatic tube system, 1870; instrumental in incorporation of Gilbert Elevated R.R. Co., 1872 (opened for travel 1878), forced out of mgmt. of co., circa 1878. Died N.Y.C., July 10, 1885.

GILBERT (or GILBERD), William, English physician, natural philosopher; b. Colchester, Eng., May 24, 1544; s. Hierom Gylberd; A.B., St. John's Coll., Cambridge, 1560, M.A., 1564, M.D., 1569; unmarried. Fellow, St. John's Coll., 1561, sr. fellow, 1569; settled in London, practiced medicine, 1573; apptd. physician to Queen Elizabeth, 1601, to James I, 1603; became fellow Royal Coll. Physicians, circa 1576, apptd. censor, 1581-88; treas. of Coll., 9 years, elected pres., 1600. Author: De magnete, magneticisque corporibus, et de magno magnete tellure, 1600; also one posthumous work (edited by his brother) De mundo nostro sublunari philosophia nova, 1651. Pioneer in study of magnetism; showed that compass needle when freely suspended points downward toward earth (magnetic dip); showed also that there is dip effect in neighborhood of spherical magnet; held that earth was great spherical magnet and that compass points not to heavens but to magnetic poles of earth; unit of magneto-motor force named gilbert; studied other natural attractive forces; showed that amber is not only substance which will attract light objects when rubbed, grouped together substances which exhibit such attractive power calling them electrics; 1st to use terms electric force, electric attraction and magnetic pole, thus is considered father of electricity; held modern notions on structure of universe; agreed with Copernicus that earth rotated on its axis; concluded that fixed stars are not all at same distance from earth; believed that form of magnetic attraction held planets in their orbits. Died London or Colchester, Nov. 30, 1603.

GILBERT THE ENGLISHMAN (Gilbertus Anglicus, or Dr. Desideralissimus), physician, med. writer; b. Eng.; flourished 13th century; studied in Eng. and abroad; became chancellor of Montpellier, France, 1250. Author: Compendium medicinae; Commentairii in versus aegidii de urinis; Practica medicinae; Experimenta magistri Gilberti cancellarii Montepessulani; Compendium super librum aphorismorum Hippocratis; Eorundem expositio; Antidotarium; De viribus aquarum et specierum; De proportione fistularum; De judicio patientis; De re herbaria; De tuenda valetudine; De particularibus morbis; Thesaurus pauperum. First practical English med. writer; 1st to recognize contagious nature of smallpox; advocated use of distilled water for travelers, fruit for sea travelers; emphasized surg. treatment of cancer.

GILBERTUS ANGLICUS, see Gilbert the Englishman.

GILBOE, David Dougherty, Am. biochemist, physiologist; b. Richland Center, Wis., July 13, 1929; s. Harvey Bernard and Margaret (Dougherty) G.; B.A., Miami U., Oxford, O., 1951; M.S., U. Wis., 1955, Ph.D., 1958; m. Myrtle Marie Kroll, Aug. 18, 1951; children—Andrew John, Sarah Ann. Research asst. in biochemistry U. Wis., Madison, 1955-58, Wis. Alumni Research Found. fellow, 1958-59, instr. surgery, 1959-61, asst. prof. surgery and physiology, 1961—. Mem. Am. Chem. Soc., Am. Soc. for. Exptl. Biology and Medicine, A.A.A.S., Sigma Xi. Research, publs. in devel. hypertonic solutions for use in reducing intracranial pressure in human patients; synthesized cellulose ion exchange material for protein chromatography which did not involve substitution of prosthetic group in cellulose chain; 1st to successfully perform surg. isolation of dog brain and keep it alive for studies of brain metabolism and physiology of cerebral blood flow. Home: 5001 Tokay Blvd., Madison, Wis. 53711.*

GILBRETH, Frank Bunker, Am. engr., management scientist; b. Fairfield, Me., July 7, 1868; s. John Hiram and Martha (Bunker) G.; grad. English High Sch., Boston, 1885; LL.D., U. of Me., 1920; m. Lillian Evelyn Moller, Oct. 19, 1904. Contracting engr., Boston, 1895-1904, N.Y., 1904-11; cons. engr., from 1911; pres. Frank B. Gilbreth (Inc.). Lectr. at Am. and European univs.; dir. Summer Sch. of Mgmt. for Profs. of Engring., Psychology and Economics; organized Soc. Promotion Sci. of Mgmt. (afterwards Taylor Soc., 1st of its kind); founder internat. museums for elimination of unnecessary fatigue of workers in industries. Author: Field System, 1908; Concrete System, 1908; Bricklaying System, 1909; Motion Study, 1911; Primer of Scientific Management, 1911; also with wife, Time Study; Fatigue Study, 1916; Applied Motion Study, 1917; Motion Study for the Handicapped, 1919. Inventor of micro-motion and chronocyclegraph processes for determining fundamental units and methods of indsl. edn.; 1st to apply motion picture camera to recording and analysis of operations; developed methods for fitting crippled soldiers for indsl. life. Died June 14, 1924.

GILBRETH, Lillian Moller, Am. engr.; b. Oakland, Cal., May 24, 1878; d. William and Annie (Delger) Moller; B.Litt., U. Cal., 1900, M.Litt., 1902, LL.D., 1933; Ph.D., Brown, 1915, Sc.D., 1931; M. Engring., U. Mich., 1928; Dr. Engring., Rutgers Coll., 1929, Stevens Inst. Tech., 1950, Syracuse U., 1952; Sc.D., Russell Sage Coll., 1931, Colby Coll., 1951, Lafayette Coll., 1952; LL.D., Smith Coll., 1945, Mills Coll., 1952; L.H.D., Temple U., 1949; m. Frank Bunker Gilbreth, Oct. 19, 1904; children—Anne Moller (Mrs. Robert E. Barney), Mary Elizabeth (dec.), Ernestine Moller (Mrs. Charles E. Carey), Martha Bunker (Mrs. Richard E. Tallman), Frank Bunker, William Moller, Lillian Moller (Mrs. Donald D. Johnson), Frederick Moller, Daniel Bunker, John Moller, Robert Moller, Jane Moller (Mrs. G. Paul Heppes, Jr.). Pres. of Gilbreth, Inc., constrn. engrs. in mgmt.; dir. courses in motion study and utilization of technol. progress; prof. mgmt. Purdue U., 1935-48; courses for the disabled, from 1948; chmn. dept. personnel relations, Newark Coll. Engring., 1941-43; prof. mgmt. U. Wis., 1955; lectr. on tech. and human relations problems in mgmt., Asia, Australia, Can., Europe, Mexico, U. S., 1955—. Mem. com. on coll. women students and the war, Am. Council on Edn.; mem. com. ednl. advisers, Office War Information; mem. sub-com. on edn., War Manpower Commn. Recipient Henry Lawrence Gareth Medal (with Frank Gilbreth), Nat. Inst. Social Scis., Wallace Clark Internat. award, Washington award, 1954; hon. fellow Brit. Inst. Mgmt., 1951; Allan R. Cullimore medal, 1959. Mem. Am. Assn. U. Women, Am. Mgmt. Assn. (hon.), Inst. of Mgmt., Soc. for Advancement Mgmt. (hon.), Acad. Masaryk, Am. Psychol. Assn., A.S.M.E. (hon. mem. 1950), Engring. Inst. Can. (hon. mem. 1949), Am. Home Econs. Assn. (hon. mem. 1952), Soc. Indsl. Engrs. (hon.), Inst. for Sci. Mgmt. of Poland, Women's Engring. Soc. of London, Phi Beta Kappa. Author: Psychology of Management, 1912; also with husband, Time Study, Fatigue Study, 1916. Applied Motion Study, 1917, Motion Study for the Handicapped, 1919; The Home Maker and Her Job, 1927; Living With Our Children, 1928; (with Edna Yost); Normal Lives for the Disabled (with Alice Rice Cook), 1944; The Foreman and Manpower Management, 1947; Living With Our Children, 1951; also papers on edn., mgmt., psychology and research; contbd. chpt. "Work and Leisure," in Toward Civilization (edited by Charles A. Beard); article on Scientific Management in New Internat Ency. Home: 30 The Crescent, Montclair, N.J.; also The Shoe, Nantucket, Mass.

GILCHRIST, Ebenezer, Scottish physician; b. Dumfries, Scotland, 1707; student medicine, Edinburgh, Scotland, London, Eng., Paris, France; grad. Reims, France; practiced in Dumfreis, 1732-74. Author: Essays Physical and Literary, vol. III, 1770. Revived certain ancient methods of treatment; defended inoculation for small-pox. Died June 12, 1774.

GILCHRIST, John Dow Fisher, zoologist; b. 1866; s. Andrew Gilchrist; ed. Madras Coll., St. Andrews, univs. in Edinburgh, Munich, Zurich; M.A., D.Sc., Ph.D.; m. Elfreda Ruth Raubenheimer; 1 son, 1 dau. Marine biologist Govt. of Cape of Good Hope, 1895; asst. zool. dept. U. Edinburgh, 1898; prof. zoology South African Coll., Cape Town, 1905-17, U. Cape Town, 1918-26. Made marine biol. and fisheries survey of Cape Coasts, 1896-1904, for Natal Govt., 1902; discovered trawling ground on Agulhas Bank, 1898; chmn. Fishery Bd. Cape Province, 1908; marine biol. and fisheries adviser for Cape Province, 1912; hon. dir. South Africa Fishing and Marine Biol. Survey, 1920. Fellow Linnean Soc.; mem. Royal Soc. South Africa (pres. 1918), Soc. Centrale d'Acquiculture, South African Assn. for Advancement Sci., South African Philos. Soc. (pres. 1903-04). Author: Annual Reports of Marine Biologist to Cape Government, 1895-1904; South African Zoology, 1911; Marine Biological Reports to Cape Province, from 1913; Reports on the Fisheries and Marine Biological Survey of South Africa, 1920-26. Editor, co-author: Marine Investigations in South Africa, 5 vols., 1902-08; Joint editor Science in South Africa, 1905, also articles on fishes, mollusca, crustacea, hemichordata, temperatures and currents of South African seas. Died Oct. 1926.

GILCHRIST, Percy Carlyle, Brit. metallurgist; b. Lyme Regis, Eng., Dec. 27, 1851; s. Alexander and Anne (Burrows) G.; ed. Royal Sch. Mines; m. Nora Fitzmaurice, 1877; 1 son, 1 dau. Asso., Royal Sch. Mines; fellow Royal Soc., 1891; mem. Soc. Chem. Industry, Royal Soc. Arts, Instn. Mech. Engrs., Instrn. Civil Engrs.; v.p. Iron and Steel Inst. Founder (with Sidney G. Thomas) "Basic" or Thomas-Gilchrist method for making steel from phosphoric pig iron. Died Dec. 15, 1935.

GILCHRIST, Richard Kennedy, Am. surgeon; b. Canadian, Tex., Jan. 20, 1904; s. Charles E. and Mabel E. (Boyle) G.; B.S., U. Chgo., 1926; M.D., Rush Med. Coll., 1930; m. Madeline Wenger, Mar. 25, 1936; children—Kennedy W., Lynn E. Faculty U. Ill., Chgo., 1934—, clin. prof. surgery, 1950—; attending surgeon Presbyn.-St. Luke's Hosp., Chgo., 1935—. Diplomate Am. Bd. Surgery. Mem. A.M.A., A.C.S. (treas. 1955-62), Am. (sec. 1953-57) Central (pres. 1951-52), Western, So. surg. assns., Soc. Clin. Surgery (pres. 1954-65), Internat. Soc. Surgeons, Chgo. Surg. Soc. (pres. 1961-62), Chgo. Med. Soc., Billings Med. Club, Contbr. articles to med. jours. Research in lymphatic spread cancer, constrn. substitute bladder and urethra, electromagnetic heating lymph nodes. Home: 2430 Lakeview Av., Chgo. 60614. Office: 122 S. Michigan Av., Chgo. 60603.*

GILCHRIST, T(homas) Caspar, physician; b. Crewe Cheshire, Eng., June 15, 1862; s. Robert and Emma (Weiss) G.; ed. Owen's Coll. (Victoria U.), Eng.; intermediate M.B., U. London, 1886; licentiate in medicine, surgery and midwifery, London, 1887; hon. M.D., U. Md., 1907; m. Annie McKerrow Hall, 1894. Came to U. S., 1890. Clin. prof. dermatology U. Md., from 1897; clin. prof. dermatology Johns Hopkins U., also dermatologist Johns Hopkins Hosp., from 1898. Mem. Royal Coll. Surgeons, London. Described generalized blastomycosis caused by Blastomyces dermatitidis (Gilchrist's disease), 1896. Died Nov. 14, 1927.

GILDEMEISTER, (Georg Eduard) Martin, German physiologist; b. Wangerin, W. Prussia, Feb. 21, 1876; s. Georg Eduard and Marie (Borchmann) G.; student natural scis., medicine, Würzburg, Munich, Germany; doctorate, Berlin, Germany, 1898; m. Marie Bauck, 1908; 1 son, Hermann; 2 daus. including Marie Anna (Mrs. Horst Frunder). Asst. in physiology U. Königsberg, Germany, 1899-1904, faculty, 1904-07; lectr. Strasbourg, France, 1907-10, titular prof., 1910-17; asso. prof., dir. phys. and sense physiol. dept. Physiol. Inst., U. Berlin, 1917-24; named prof. in Leipzig, Germany, 1924. Research on relationship between stimulus and response, effect of induction stimulus, electro-physiology of skin; tried to develop a gen. theory of response especially on hearing threshhold and limits and its dependence on age; invented numerous instruments for measuring physiol. reactions. Died Leipzig, Oct. 13, 1943.

GILES, Norman Henry, Am. geneticist; b. Atlanta, Aug. 6, 1915; s. Norman Henry and Alice (Guerard) G.; A.B., Emory U., 1937; M.A., Harvard, 1938, Ph.D., 1940; M.A. (hon.), Yale, 1951; m. Dorothy Lunsford, Aug. 26, 1939; children—Annette Guerard, David Lunsford. Instr. botany Yale, 1941-45, asst. prof., 1945-46, asso. prof., 1946-51, prof., 1951-61, Eugene Higgins professor of genetics, 1961—; prin. biologist Oak Ridge Nat. Lab., 1947-50. Cons. AEC, 1954-64. Parker fellow Harvard, 1940-41; Fulbright and Guggenheim fellow U. Genetics Inst., Copenhagen, 1959-60; mem. genetics study sect. Nat. Insts. Health, 1960-64, mem. genetics tng. com., 1966—; Guggenheim fellow Australian Nat. U., Canberra, 1966. Fellow Am. Acad. Arts and Sciences, Am. Assn. Advancement Sci.; mem. Nat. Acad. Scis., Genetics Soc. Am. (treas. 1954-56), Bot. Soc. Am., Am. Soc. Naturalists, Radiation Research Soc., Am. Ornithologists Union, Phi Beta Kappa, Sigma Xi. Mem. editorial bd. Radiation Research, 1953-58, Am. Naturalist, 1961-64. Research on cytology and genetics. Home: 85 Jackson Rd., Hamden, Conn. 06517. Office: Dept. Biology, Kline Biology Tower, Yale U., New Haven 06520.

GILES OF CORBEIL (Aegidius Corboliensis), physician, humanist; studied in Salerno; student of Peter of Musanda; became canon of Notre Dame, archiater to Philip Augustus (king 1180-1223). Author med. poems in Leonine verse (main source of salernitan information for Parisian doctors); works include: De urinis (most popular textbook on uroscopy in Christian West until 16th century); De pulsibus; De laudibus et virtutibus compositorum medicaminum; Viaticus (treatise on pathology); De physiognomiis (continued physiognomic tradition).

GILES OF LESSINES (Aegidius a Lessinia), philosopher, astronomer; b. Lessines in Hainaut, Belgium, circa 1230; studied under Albert the Gt., St. Thomas; joined Dominican Order; assigned to abbey of St. James, Paris. Author: De concordia temporum (on chronology); De unitate formae (on unity of forms), 1278. Died circa 1304.

GILETTI, Bruno John, Am. geochemist; b. N.Y.C., Dec. 6, 1929; s. John M. and Rita (Baltera) G.; A.B., Columbia, 1951, B.S., 1952, M.A., 1954, Ph.D., 1957; m. Dody Marshak, June 12, 1953; children—Ann, Laura. Research asso. Lamont Geol. Obs., Columbia, 1957-58; research asso. dept. geology, mineralogy Oxford U., 1958-60; faculty dept. geol. scis. Brown U., Providence, 1960—, prof., 1967—. Fellow Geol. Soc. Am; mem. Am. Geophys. Union, Am. Chem. Soc., Geochem. Soc. Research, publs. on chem. and phys. processes which metamorphose rocks, use of radioactive isotopes to determine ages of rocks with a view to understanding process of mountain and continent formation. Home: 188 Bowen St., Providence 02906.*

GILFORD, Hastings, English physician; b. Melton Mowbray, Eng., 1861; m. Lilian Adele Hope, 1889; 2 sons, 2 daus. Hunterian prof., fellow Royal Coll. Surgeons Eng.; surgeon in charge Sutherland War Hosp., Hosp. for Pensioners. Fellow Royal Soc. Medicine. Author: Disorders of Post-natal Growth and Development, 1911; Tumors and Cancers: A Biological Study, 1925; Cancer, Civilization, Degeneration, 1932; The Cancer Problem and its Solution, 1934; The Hysterozoa, 1936; also various articles. Believed to have introduced term progeria (also known as Hutchinson-Gilford disease), a form of infantilism which he described, 1904. Died Sept. 6, 1941.

GILGENKRANTZ, J. M., French physician; b. Nancy, France, Aug. 24, 1926; s. R. G. and S. (Rouche) G.; M.D., U. Nancy, 1958; m. Simone Pointet, July 20, 1954; children—Claire, Helene. Faculty, U. Nancy Sch. Medicine, prof., 1963—; with cardiological dept. Central Hosp., Nancy. Mem. French Soc. Cardiology, Author: (with Faivre) Épreuve d'effort et hypoxie, 1960; (with Faivre, Renaud) l'entrainement électrique du coeur, 1964; Stimulation et chocs électriques, 1965; also numerous articles. Research on elec. stimulation of heart, hemodynamics of heart, coronary diseases. Home: 79 rue de Villers, Vandoeuvre 54. Office: Service de Cardiologie, C.H.U. de Nancy 54, France.*

GILIBERT, Jean Immanuel, French botanist, physician; b. Lyons, France, 1741; planted a botanic garden, Grodno, circa 1775; returned to Lyons, 1783. Author: L'anarchie médicinale, 3 vols., 1772; Flora Lithuanica, 1781; History of the Plants of Europe or Elements of Practical Botany, 2 vols., 1798; Abridgment of the Natural System of Linnaeus, 1802. Died 1814.

GILL, Atticus James, physician; b. Okmulgee, Okla., June 8, 1914; s. X. R. and Martha (Trotter) G.; student Duke, 1931-33, M.D., 1938; m. Lucille Hodge, Nov. 8, 1941; children—Frank H., Mary L., James H. Faculty, U. Tenn. Sch. Medicine, 1941-43, asst. prof., 1942-43; asst. prof. Southwestern Med. Coll., Dallas, 1943-47, asso. prof., 1947-49; faculty U. Tex. Southwestern Med. Sch., Dallas, 1949—, prof. pathology 1950—, asso. dean, 1950-51, asst. dean, 1952-54, dean, 1955-67. Diplomate Am. Bd. Pathology. Fellow Am. Coll. Pathologists, Am. Soc. Clin. Pathology, A.C.P.; mem. A.M.A., So., Tex. med. assns., Dallas County Med. Soc., Contbr. articles to tech. jours. Research on inflammatory reaction in malignant diseases.*

GILL, Sir David, Brit. astronomer; b. Aberdeen, Scotland, June 12, 1843; s. David and Margaret (Mitchell) G.; ed. Marischal Coll., also under Clerk Maxwell at U. Aberdeen; several hon. degrees; m. Isobel Black, 1870. Began career as watchmaker, Aberdeen, also established obs. with David Thompson; dir. pvt. obs. of Lord Lindsay (later Earl of Crawford), Dunecht, Scotland (also organized Lindsay's transit of Venus expdn. to Mauritius, measured base line for geodetic survey of Egypt, nr. Cairo, connected longitudes of Berlin, Malta, Alexandria, Suez, Aden, Seychelles, Mauritius, Rodriquez), 1873-76; head expdn. to determine solar parallax by observations of Mars, from Ascension Island, 1877; royal astronomer, Cape of Good Hope, 1879-1907; organized Transit of Venus expdns., S. Africa, 1882; dir. Geodetic Survey of Natal and Cape Colony (which he had earlier proposed and organized), 1896; emissary Brit. Cononial Office to work on boundary survey between Brit. Bechuanaland and German S.W. Africa (later completed under his dir.), Berlin, 1896; proposed, organized Geodetic Survey of Rhodesia, 1897; Brit. rep. to com. Internat. Bur. Weights and Measures. Fellow royal socs. London, 1883 (Royal medal 1903), Edinburgh (hon.); mem. Royal Astron. Soc. (Gold medal 1882, 1907, pres. 1909-11), Inst. Marine Engrs. (pres. 1910-11), Brit. Assn. (pres. 1907-08), Bur. Longitudes, Paris, France, Research Def. Soc. (pres.), Nat. Acad. Sci. (Watson Gold medal 1900), Washington, also many other fgn. acads. scis.; corr. Inst. France (Valz prize Acad. Scis. 1882). Author: Six Months in Ascension, 1878; Catalogues of Stars for the Equinoxes 1850, 1860, 1885, 1890, and 1900, from observations made at the Royal Observatory, Cape of Good Hope; The Cape Photographic Durchmusterung; Geodetic Survey of South Africa, Vols. I-V; A History and Description of the Royal Observatory, Cape of Good Hope, 1913; also many papers. Photographed Great Comet of 1882; also pointed out advantages of using photography for complete cataloging of stars; devised most accurate method of the time for measuring sun's distance from earth; accurately determined distances of 22 stars; determined mass of Jupiter; made survey showing magnitude and projection of 400,000 stars, 1885-98; extended (with Kapteyn) Argelander's star chart to s. celestial pole. Died London, Eng., Jan. 24, 1914.

GILL, James Edward, Canadian geologist; b. Nelson, B.C., Can., Jan. 16, 1901; s. James and Catherine Jane (Walton) G.; student U. B.C., 1917-20; B.Sc. in Mining Engring., McGill U., Que., 1921; Ph.D., Princeton, 1925; m. Florence Edna Drysdale, Sept. 9, 1925; children—Robert James, Claire Ann (Mrs. Rhodes Bethune Evans). Engr., Granby Consol. Mining, Smelting & Power Co., 1921-22; asst., fellow, Procter fellow, Princeton, 1923-25; faculty U. Rochester (N.Y.), 1925-29; faculty McGill U., 1929—, prof., 1949—, Dawson prof., 1957—, chmn. dept. geol. scis., 1959—. Sr. lectr. Can., USSR, 1961. Recipient Leonard medal Engring. Inst. Can., 1943. Fellow Royal Soc. Can. (Miller medal 1957), Geol. Assn. Can., Mineral Assn. Can., Geol. Soc. Am., A.A.A.S.; mem. Soc. Ecol. Geologists, Canadian Inst. Mining and Metallurgy (Barlow Meml. prize 1939, Distinguished Service medal, 1964), Am. Geophys. Union, Geochem. Soc., Corp. Profl. Engrs. Editor, contbg. author: The Proterozoic in Canada, 1957; also numerous articles. Research on Gunflint iron-formation, mineral deposits, properties metallic sulphides, early history earth; discovered high grade iron ore in Labrador, three gold ore bodies in Que. Home: 5 Lilac Av., Dorval, Que. Office: Dept. Geol. Scis., McGill U., 3450 University St., Montreal, Que., Can.*

GILL, Jocelyn Ruth, Am. astronomer; b. Flagstaff, Ariz., Oct. 29, 1916; d. Thomas B. and Sarah (Bailey) Gill; A.B. in Math., Wellesley Coll., 1938; S.M. in Astronomy and Astrophysics, U. Chgo., 1941; Ph.D. in Astronomy, Yale, 1959. Lab. asst., instr. astronomy Mt. Holyoke Coll., 1940-42; staff mem. radiation lab. Mass. Inst. Tech., 1942-45; from instr. to asst. prof. astronomy Smith Coll., 1945-52; grad. work, teaching asst. U. Cal. at Berkeley, 1946-48; asst. prof., then acting chmn. dept. astronomy Mt. Holyoke Coll., 1952-57; asso. prof. astronomy and math. Ariz. State Coll., 1959-60; vis. lectr. Wellesley Coll., 1960-61; staff scientist Office Astronomy and Solar Physics, Office Space Sci., NASA, Washington, 1961-63, with manned space sci. div., 1963—, chief in-flight scis. br., 1963—; made solar eclipse flight, 1963. Fellow A.A.A.S.; mem. Am. Astron. Soc., Nantucket Maria Mitchell Assn., Am. Assn. Variable Star Observers, Sigma Xi. Work on Gemini Sci. program. Investigation of motion of Neptune's satellite, Triton; study of celestial mechanics; numerical analysis of satellite orbits. Home: 560 N. St. S.W., Washington 20024. Office: 400 Maryland Av. S.W., Washington 20546.

GILL, Piara Singh, Indian physicist; b. Chela, Punjab, India, Oct. 28, 1911; s. Basant Singh and Partap Kaur (Bains) G.; A.B., U. So. Cal., 1935, M.S., 1936; Ph.D., U. Chgo., 1940, postgrad. (research fellow); m. Chambeli Hukam Chand, Feb. 17, 1942; children—Nishtha, Surishtha. Lectr. physics Forman Christian Coll., Lahore, 1940-47; prof. exptl. physics TATA Inst. Fundamental Research, Bombay, India, 1947-48; vis. scientist dept. Terrestrial magnetism and electricity Carnegie Instn., 1948; prof. physics, dean faculty sci. Aligarh (India) Muslim U., 1949-63; dir. Gulmarg (Kashmir, India) Research Obs., 1951—; dir. Central Sci. Instruments Orgn., Chandigarh, Punjab 1963—; vis. prof. Wash. State U., Pullman, 1961-62; vis. scientist Nat. Center for Atmospheric Research, Boulder, Colo., part-time 1963-65. Cons. Nat. Bur. Standards, Washington, 1949; Mem. nat. sci. adv. council Inst. for Comprehensive Medicine, editorial bd. Jour. Fellow Nat. Inst. Scis. (India), Nat. Acad. Scis. (India), Am., Indian phys. socs., Sigma Xi. Research in cosmic rays, nuclear physics. Home: 69 Deputy Minister's, Sector 7, Chandigarh. Office: Bay Bldg., Sector 17, Chandigarh, Punjab, India.*

GILL, Richard C(ochran), Am. research-explorer; b. Washington, Nov. 22, 1901; s. William Tignor Gill and Flora May (Allen) G.; B.A., Cornell U., 1924; grad. student Columbia, N.Y. U., 1926-27; m. Ruth Lenfest, Apr. 19, 1926. Instr. English, Lafayette Coll., Easton, Pa., 1925-28; field mgr. in Ecuador, Peru, Bolivia, B. F. Goodrich Rubber Corp., 1928-30; pres. Gill, Miller Co., ranching, Ecuador, from 1930; pres. Gill, Dundas and Co., Palo Alto. Cal., leader Gill-Merrill expdn., other S.Am. (upper-Amazonian) expdns. in ethno-botany, tropical Am. pharmacognosy; pres. Gill-Merrill Participants; pres., founder S.Am. mfg. base Transandino Co., curare and other tropical drugs; founded tech. library, over 5000 items on history, botany, pharmacognosy, pharmacology and clin. reprints and curare, curare synthetics, anti-curare agts. Fellow Am. Geog. Soc., Am. Polar Soc.; mem. Hakluyt Soc., Explorers Club, A.A.A.S., Am. Acad. Polit. and Social Sci.; Cal. Acad. Scis., Am. Pharm. Assn., Torrey Bot. Club. Author: Manga, 1937; Volcano of Gold, 1938; Kalu, 1939; White Water and Black Magic (a history of curare), 1940; Paco Goes to the Fair (juvenile), 1940; The Other America, 1941; The Flying Death (Curare): A Manga Book, 1942; Francisco de Oreliana, a Biography; Scientific Bibliography of Curare (1595-1945); Clinical Employments and History of Modern Curarization; Physico-Chemical Determination of d-Tubocurarine in Therapeutic Formulations; South American juvenile books, Contbr. articles to Ency. Brit., also to jours. Research in ethnobotany of drug curare, its application in treatment of spastic paralysis, mental diseases, also as anti-convulsant and anesthesia; investigated evolution and clin. application of therapeutic curare variant of higher alkaloidal potency, also evolution of econ. prodn. technique for chemically pure d-tubocurarine, its clin. use in acute anterior poliomyelitis, obstet. anesthesia, co-adminstrn. with intravenous anesthetics. Died July 7, 1958.

GILL, Theodore Nicholas, Am. zoologist; b. N.Y.C., Mar. 21, 1837; s. James Darrell and Elizabeth (Vosburgh) G.; ed. in pvt. schs. and under spl. tutors; A.M. (hon.), Columbian (now George Washington U., 1865, M.D., 1866, Ph.D., 1870, LL.D., 1895; grantee Wagner Free Inst. of Sci., Phila. Mem. expdn. to collect fish specimens in W.I., 1858; adj. prof. physics and natural history George Washington U., 1860-61, lectr. natural history, 1864-66 73-84, prof. zoology, 1884-1910. Librarian, Smithsonian Instn., 1865-67; asst. librarian Library of Congress, 1866-75; asso. in zoology, U. S. Nat. Mus. Fellow A.A.A.S. (pres. 1897); mem. Nat. Acad. Scis., Am. Philos. Soc., Acad. Natural Sci. (corr., Phila.), Philos. Soc. Washington, Biol. Soc., Zool. Soc. London (fgn.). Author: Catalogue of the Fishes of the East Coast of North America, 1861-73; Arrangement of the Families of Fishes, 1872; Arrangement of the Families of Mammals, 1872; Parental Care Among Fresh-Water Fishes, 1906; also contbns. to Life-histories of Fishes, Vol. I, 1909. Asso. editor Johnson's New Univ. Ency., Century Dictionary and Standard Dictionary; editor The Osprey. Died 1914.

GILLES, André Édouard, Belgian geneticist; b. Fosses, Belgium, Apr. 21, 1922; s. Nicolas and Anna (Bodart) F.; licence U. Louvain (Belgium), 1946, D.Sc., 1948; m. Francine Van de Putte, Mar. 20, 1948; children—Pierre, Jean, Anne, Claire. Asst. U. Louvain, from 1946, head of works, from 1949, lectr., from 1950, prof., 1954-—, sec. Faculty Sci., 1961-—. Mem. biology com. EURATOM, 1961-—. Mem. Belgian Soc. Genetics (gen. sec.). Author: (with A. Boterberg) Botanique, 1964. Research, publs. on hybrid sterility; effects of radiation and microbeams on pollen mother cells; variability of contents of DNA following irradiations and mutagen actions; bacterial genetics; correlation between methylation of s-RNA and differentiation in Arabidopsis thaliana, others. Home: 158, Tiensevest, Louvain. Office: 24, Vaartstraat, Louvain, Belgium.*

GILLES, Herbert Michael, physician; b. Port Saïd, Egypt, Oct. 9, 1921; s. Joseph and Clementine (Farrugan) G.; B.Sc., Royal U. Malta, 1943, M.D., 1946; B.Sc., U. Oxford (Eng.), 1951; m. Wilhelmina Antoinette Caruana, Feb. 5, 1955; children—Michael, Robert, Marisa, Anthony. Jr. clin. appointments United Oxford Hosps., 1949-54; sci. staff Med. Research Council, 1954-58; lectr. trop. medicine Liverpool (Eng.) Sch. Trop. Medicine, 1958-62; prof. preventive and social medicine U. Ibadan (Nigeria), 1963-65; sr. lectr. trop. medicine U. Liverpool, 1965-—; vis. prof. U. Lagos Med. Sch., Nigeria, 1966-—. Cons. physician U. Coll. Hosp., Ibadan, Nigeria, 1963-65, Liverpool Regional Hosp. Bd., 1965-—. Mem. Royal Coll. Physicians. Author: Akufo—An Environmental Study of a Nigerian Village Community, 1964; also articles. Research on immunity to malaria, epidemiology of hookworm infection and schistosomiasis; environmental study of village community in Nigeria. Office: Liverpool Sch. Tropical Medicine, Liverpool 3, Eng.*

GILLES D'ALBI, Pierre, see d'Albi, Pierre Gilles.

GILLETTE, King Camp, Am. inventor; b. Fond du Lac, Wis., Jan. 5, 1855; s. George Wolcott and Fanny Lamira (Camp) G.; ed. pub. schs., Chgo.; m. Alanta Ella Gaines, July 2, 1890; 1 son, King G. Inventor Gillette razor, 1895 (marketed 1903); organizer Gillette Safety Razor Co., 1901, pres. until 1931. Author: Human Drift, 1894; Gillette's Social Redemption, 1900; Gillette's Industrial Solution, 1900; World Corporation, 1906; The People's Corporation, 1924. Died July 9, 1932.

GILLETTE, Philip Roger, Am. physicist; b. Mount Vernon, Ia., May 12, 1917; s. Clinton Edgar and Celia (Rogers) G.; B.A., Cornell Coll., Ia., 1937; B.S., U. Ill., 1938, M.S., 1939, Ph.D., 1942; m. Bettelaine Dunbar, Apr. 26, 1947; children—Kenneth Lee, Sandra Jo. Mem. staff radiation lab. Mass. Inst. Tech., Cambridge, 1942-45; project engr. Sperry Gyroscope Co., Great Neck, N.Y., 1945-48; physicist Hanford Works, Gen. Electric Co., Richland, Wash., 1948-50; research engr., sr. physicist Stanford Research Inst., Menlo Park, Cal., 1950-64, Washington, 1964-—. Mem. Am. Phys. Soc., Am. Inst. Aeros. and Astronautics, I.E.E.E. (sr. mem.). Contbr. to Pulse Generators, 1947; also articles, classified reports. Research in theory and methodology of nuclear radiation shielding, nuclear reactor operation, magnetic behavior of thin ferromagnetic films, operational behavior, design, test and application of microsecond-duration pulse transformers and pulse-forming networks, propellant selection and performance prediction for ballistic missiles and space vehicles, analysis of mil. reconnaissance, command control and mgmt. problems. Home: Crofton, Md. 21113. Office: Stanford Research Inst., 1000 Connecticut Av., Washington 20036.*

GILLIAM, David Tod, Am. surgeon; b. Hebron, O., Apr. 3, 1844; s. William and Mary Elizabeth (Bryan) G.; ed. pub. schs., Bartlett's Commercial Coll., Cin.; M.D., Med. Coll. Ohio, 1871; m. Lucinda E. Mintun, Oct. 7, 1866; 1 son, Earl M. Enlisted 2d Va. (Union) Cav., Aug., 1861; elected corporal Co. I; with Garfield in march against Humphrey Marshall on Big Sandy River, Ky.; sent to Wheeling, W. Va., as recruiting officer; later ascended Kanawha River and took part in many skirmishes; with Crook in battle of Lewisburg, Va.; wounded and taken prisoner nr. Gauley, Va., by Gen. Loring; escaped 5 weeks later; sent to parole camp; discharged spring of 1863. In practice medicine, Columbus, O., from 1868; emeritus prof. gynecology, Med. Dept., Ohio State U. (trustee); gynecologist to St. Anthony's and St. Francis hosps.; pathologist Columbus Med. Coll., from 1877; choir physiology Starling Med. Coll., from 1879, prof. gynecology, obstetrics, from 1885. Hon. fellow Am. Assn. Obstetricians and Gynecologists (v.-p., 1905-6); mem. A.M.A., Ohio State Med. Assn., hon. mem. Northwestern Ohio Med. Assn.; mem. Pan.-Am. Med. Congress, World's Med. Congress. Author: Pocket Book of Medicine, 1882; Essentials of Pathology, 1883; Practical Gynecology, 1903; The Rose Croix, 1906; The Righting of Richard Devereux. Collaborator, Randall & Ryan's History of Ohio, 5 vols., 1912. Originated operations for suspension of uterus (called Gilliam operation), for cystocele, for incontinence of urine in the female; devised many surg. instruments and the Gilliam operating table. Died 1923.

GILLILAND, Edwin R., Am. chem. engr.; b. El Reno, Okla., July 10, 1909; s. Owen Edwin and Elsie (Kelly) G.; B.S., U. Ill., 1930; M.S., Pa. State Coll., 1931; Sc.D., Mass. Inst. Tech., 1933; D.Eng. (hon.), Northeastern U., 1948; m. Ann F. Miller, June 15, 1938; 1 dau., Gail Ann. Faculty, Mass. Inst. Tech., Cambridge, 1934-—, prof. chem. engring., 1944-—, head, dept. chem. engring., 1961-—. Mem. chmn. various coms. NACA, NDRC, WPB, Pres.' Sci. Adv. Com., others; cons. Office Sci. and Tech., 1965-—; mem. ad hoc adv. com. Office of Saline Water, Dept. Interior, 1965-—. Mem. Am. Inst. Chem. Engrs. (Profl. Progress award in chem. engring. 1950, William H. Walker award 1954, Warren K. Lewis award in chem. engring. edn. 1965, dir. 1958-60), Am. Chem Soc. (Bakeland medal and award for achievement in chemistry 1944, Indsl. and Engring. Chemistry award 1959, chmn. canvassing com. for awards 1962), Nat. Acad. Scis., Nat. Acad. Engring., Am. Acad. Arts and Scis., Soc. Chem. Industry, Sigma Xi (pres. chpt. 1961-62), Tau Beta Pi. Author: Elements of Fractional Distillation, 1950; (with Walker, Lewis, McAdams) Principles of Chemical Engineering, 1937. Research, publs. on synthetic rubber manufacture, fluidized solid system for catalytic reactors, distillation systems; developed new methods for demineralization of saline water. Home: 95 Longmeadow Rd., Belmont, Mass. 02178. Office: 77 Massachusetts Av., Cambridge, Mass. 02139.*

GILLIN, John P(hilip), Am. anthropologist; b. Waterloo, Ia., Aug. 1, 1907; s. John Lewis and Etta (Shaffner) G.; A.B., Univ. of Wis., 1927, A.M., 1930; A.M., Harvard, 1931, Ph.D., 1934; student Univ. of Berlin, 1928, U. London, 1928; m. Helen Norgord, Mar. 29, 1934; 1 son, John Christian. Engaged in anthropol. field work in Algeria, 1930, Europe, 1930, New Mex., 1931, British Guiana, 1932-33, Ecuador and eastern Peru, 1934-35, Utah, 1936-37, Wis., 1938-39, Guatemala, 1942, 46, 48, Peru, 1944-45, Colombia 1946, Cuba, 1948, various parts of Latin Am. since, Europe, 1958; staff Peabody Mus., 1934-35; faculty Sarah Lawrence, 1933-34 Univ. of Utah, 1935-37; Ohio State U., 1937-41, Duke, 1941-46; prof. anthropology and research U. N.C., 1946-59; dean div. social sciences, prof. anthropology, University of Pittsburgh, 1959-62, research professor, 1962-—; visiting professor Columbia, 1957-58, U. Hawaii, summer 1956; hon. fellow in psychology, Yale, 1940-41; mem. bd. of econ. warfare, U. S. Embassy, Lima, Peru, 1942-44; Smithsonian rep., Peru, 1944-45; research asso. Carnegie Instn. of Washington, 1942, 46. Cons., AID, 1962-—. Fellow A.A.A.S., Am. Anthrop. Assn. (exec. bd. 1949-52), Center for Advanced Study in Behavioral Scis., 1954-55; mem. American Sociol. Soc., Nat. Research Council (chmn. com. Latin Am. 1945-51), Soc. for Applied Anthropology (pres. 1959-60), Sigma Xi, Phi Kappa Phi, Alpha Kappa Delta, Alpha Kappa Lambda. Author, 1933-, —among latest publs. The Ways of Men, 1948; Moche: A Peruvian Coastal Community, 1947; Cultural Sociology (with J. L. Gillin), 1948, The Culture of Security in San Carlos, 1951; For A Science of Social Man (editor, co-author), 1954; Integration Social de Guatemala (with others), 1957; San Luis Jilotepeque, 1958; (with others) Social Change in Latin America Today, 1961. Research and publs. on basic values of modern Latin Am. culture; ethnology of Caribs in Brit. Guiana, of Pokomam (Maya) Indians in Guatemala, of peasants in Peru; anthropometric relations of Caribs with other forest tribes, of Ecuador Quichua Indians with other tribes; archaeology of Utah and no. periphery of Pueblo area of U. S.; theory of psychology of cultures, learning and practice; theory of modern complex sociocultural systems. Office: Cathedral of Learning, U. Pitts., Pitts. 13.

GILLINGS, Barrie Roderick D'Arcy, Australian dentist; b. Newcastle, New South Wales, Australia, June 17, 1934; s. Richard John and Esme (Salisbury) G.; B.D.S. with honors, U. Sydney, 1954; M.Sc. in Dentistry, U. Rochester, 1963; m. Margaret Tingle, Nov. 10, 1956; children—Michael, John, Anthony, David, Christopher, Jane. Teaching fellow U. Sydney Faculty Dentistry, 1954-56, lectr., sr. lectr., 1963-—; pvt. practice dentistry, London, Eng., 1957, 61; research asso. Eastman Dental Dispensary, U. Rochester (N.Y.), 1957-63. Council mem., chmn. field research com. Dental Health Found., 1962-—. Mem. Internat. Assn. for Dental Research (sec. Australian sect. 1966). Contbr. articles to profl. jours. Research on tooth morphology, dental materials, tooth enamel dissolution, effects of foods on tooth enamel; inventor, developer thin sect. machine for hard materials, photoelectric mandibulograph (device to record and display 3D movements) of jaws during function, originator 1st intraoral tooth-sized radio transmitter for studying jaw contacts. Home: 121 Bannockburn Rd., Turramurra, New South Wales. Office: Faculty of Dentistry, 2 Chalmers St., Sydney, New South Wales, Australia.*

GILLIS, James Melville, Am. astronomer; b. Georgetown, D.C., Sept. 6, 1811; s. George and Mary (Melville) G.; attended U. Va., 1833; m. Rebecca Roberts, Dec. 1837. Entered USN, 1826; commd. midshipman, 1833; studied in Paris for 6 months, 1835, ordered back to Washington (D.C.), assigned to Depot of Charts and Instruments; in charge of depot, 1837; commd. to make astron. observations in Washington necessary for evaluation of longitude observation of Lt. Charles Wilkes' expdn., 1838; pointed out inadequacy of existing building and equipment for astron. research to Bd. Naval Commrs., 1841 (led to act of Congress providing for establishment U. S. Naval Obs., Washington); visited Europe in interests of obs., circa 1842-1843, authorized by Congress to go to Santiago (Chile) to observe Venus and Mars, 1849-52; became supt. U. S. Naval Obs., 1861; mem. Nat. Acad. Sci. Died Feb. 9, 1865.

GILLIS, Jan, Belgian chemist; b. Arlon, Aug. 8, 1893; s. Louis and Pauline (De Keghel) G.; ed. at Athens, Ghent, Antwerp; Ph.D. in Chem. Sci. and Botany, U. Ghent; D.honoris causa, U. Geneva; m. Paula Vande Velde, 1920; 1 son, Paul. Prof. analytic chemistry, 1923-—, prof. emeritus, 1961, dean, 1953-57, hon. dean, 1957-—. Mem. Koninkl. Vlaamse Academie voor Wetenschappen, van Belgie. Author: Reagents for Qualitative Inorganic Analysis; Electrochimie; Spectrochemie; Analyse par Activation. Address: Graaf de Smet de Nayer Plein 11, Ghent, Belgium.

GILLIS, Marvin Bob, biochemist; b. Soperton, Ga., Apr. 5, 1920; s. Bob Lee and Pearl (Gillis) G.; B.S.A., U. Ga., 1940; Ph.D., Cornell U., 1947; m. Helen Reed, Dec. 23, 1946; children—Margaret Susan, Marvin Reed, Kenneth Robert. Research chemist Internat. Minerals & Chem. Corp., Skokie, Ill., 1947-50, supr. nutrition research, 1950-51, supr. biol. research, 1951-54, mgr. organic and biol. research, 1954-56, asst. dir. research, 1956-57, dir. research and devel., 1957-64, dir. animal health and nutrition, 1964-66; Div. vice pres. 1966-—; research asso. Cornell U., 1947-51. Chmn. bd. Cal. Cattle Supply Co., Bellflower, Calif., 1964-—. Pres. Agrl. Research Inst., Nat. Acad. Scis-NRC, 1962-63, mem. agrl. bd., 1962-—. Mem. Bd. Dir. Animal Health Inst. Recipient Poultry Sci. Research prize Poultry Sci. Assn., 1948. Mem. Am. Chem. Soc.; Am. Inst. Nutrition; N.Y Acad. Scis.; Poultry Science Assn.; Am. Soc. Animal Sci., Sigma Xi, Phi Kappa Phi, Alpha Zeta. Contbr. articles to tech. jours. Research on role vitamins and minerals in nutrition, research on prodn. amino acids by fermentation and chem. synthesis, processing and refining nonmetallic minerals, devel. improved agrl. chems. and fertilizers. Patentee in field. Home: 2116 Larkdale Dr., Glenview, Ill. 60025. Office: 5401 Old Orchard Rd., Skokie, Ill. 60076.

GILLISSEN, Günther Josef, German biochemist, physician; b. Stuttgart, Germany, Oct. 6, 1917; s. Josef and Pauline (Gröber) G.; ed. univs. Heidelberg, Wurtzburg, Lena, Danzig; M.D., Ph.D. in natural sci.; m. Ilse Wotschke, 1954; children—Annette, Adrian, Olivier, Evelyne. Prof., Upper Sch., specialist in microbiology. Mem. Soc. Biol. Chemists (Paris), Soc. Research on Nature and Medicine, German Soc. Internal Medicine, Soc. Hygiene and Microbiology, German Soc. Pathology. Contbr. over 100 articles to profl. publs. Home: Lennebergstrasse 15, Mainz-Gonsenheim. Office: Institute for Microbiological Medicine, University of Mainz, 1 Langenbeckstrasse, Mainz, Germany.

GILLIUS, Petrus, naturalist; b. Alby, Languedoc, 1490; patronized by Bishop Armagnac (a liberal); commd. by King Francis I to visit the Levant; enlisted in Turkish army; escaped to France, 1550; pub. a Latin edit. of Aelian, identifying species when possible; wrote on topography of Constantinople; described an elephant sent from Persia to the sultan. Died 1554.

GILLMAN, Leonard, Am. mathematician; b. Cleve., Jan. 8, 1917; s. Joseph Moses and Etta (Cohen) G.; diploma (fellow in piano 1933-38) Julliard Grad. Sch., 1938; B.S., Columbia U., 1941, A.M., 1945, Ph.D., 1953; m. Reba Parks Marcus, Dec. 24, 1938; children—Jonathan Webb, Michal Judith. Asst., lectr. Columbia, 1941-43; research asso. Tufts, 1943-45, Mass. Inst. Tech., 1945-51; operations analyst Navy Dept., 1943-51; instr. math. Purdue U., 1952-53, asst. prof., 1953-56, asso. prof., 1956-60; prof., chmn. dept. math. U. Rochester (N.Y.), 1960-—; mem. Inst. for Advanced Study, Princeton, N.J., 1958-60. Carnegie Corp. fellow in math. statistics, 1942-43; Guggenheim fellow in math., 1958-59; NSF sr. post-doctoral fellow in math., 1959-60. Author: (with

Meyer Jerison) Rings of Continuous Functions, 1960. Home: 368 Council Rock Av., Rochester, N.Y. 14610.*

GILLULY, James, Am. geologist; b. Seattle, June 24, 1896; s. Charles E. and Louisa (Briegel) G.; B.S., U. Wash., 1920; Ph.D., Yale, 1926; Sc.D., Princeton, 1959; m. Enid Adelaide Frazier, June 30, 1925; children—Molly (Mrs. David L. Shaw), Sally (Mrs. Duane L. Jones). Prof. geology U. Cal. at Los Angeles, 1938-50; with U. S. Geol. Survey, 1921-38, 50-66, research geologist, Denver, 1950-66. Geol. cons. to chief engr. S.W. Pacific War Theater, 1943-44. Recipient Distinguished Service medal U. S. Dept. Interior, 1960. Mem. Geol. Soc. Am. (Penrose medal 1958, pres. 1948), Nat. Acad. Scis., Am. Acad. Arts and Scis., Soc. Econ. Geologists (mem. council 1938-40), Seismol. Soc. Am., Am. Assn. Petroleum Geologists, Am. Mineral. Soc., Colo. Sci. Soc., Geol. Soc. London (fgn. mem.). Author: (with A. C. Waters and A. O. Woodford) Principles of Geology, 1951, 59; also numerous articles. Research on econ. geology of copper, especially in S.W. U. S., structural geology, distbn. of mountain bldg. and granitic masses in geologic time. Home: 975 Estes St., Lakewood, Colo. 80215. Office: U. S. Geol. Survey, Denver 80225.*

GILMAN, Alfred, Am. pharmacologist; b. Bridgeport, Conn., Feb. 5, 1908; s. Joseph and Frances (Zack) G.; B.S., Yale, 1928, Ph.D., 1931; m. Mabel J. Schmidt, Jan. 11, 1934; children—Joanna, Alfred Goodman. Research fellow biochemistry Yale, 1931-32, research asso., asst. prof. pharmacology, 1933-43; asso. prof. pharmacology Columbia, 1946-48, prof., 1948-55; prof. pharmacology, chmn. dept. Albert Einstein Coll. Medicine, Yeshiva U., 1956—. Spl. cons. USPHS; mem. sci. and edn. council Am. Found. Allergic Disease; mem. exec. com. Medical division NRC. Hon. fellow Am. Acad. Allergy; fellow N.Y. Acad. Sci.; mem. Am. Physiol. Soc., Am. Soc. Pharmacology and Exptl. Therapy (pres. 1960), Harvey Soc., Soc. Exptl. Biology and Medicine, N.Y. Acad. Medicine (asso.), Nat. Acad. Scis., Sigma Xi, Alpha Omega Alpha. Author: (with L. S. Goodman), Pharmacological Basis of Therapeutics, 1941. Mem. editorial bd. Am. Jour. Physiology, Jour. Applied Physiology, Pharmacol. Rev. Introduced (with F. S. Philips) use of nitrogen mustards in treatment Hodgkin's disease, 1946.*

GILMAN, Henry, Am. chemist; b. Boston, May 9, 1893; s. David and Jane (Gordon) G.; B.S., Harvard, 1915, M.A., 1917, Ph.D., 1918; postgrad. Polytechnikum, Zurich, Switzerland, 1915, Sorbonne, Paris, France, 1916, Oxford (Eng.) U., 1916; m. Ruth Shaw, June 20, 1929; children—Jane Gordon, Henry Shaw. Instr., Harvard, 1918; asso. U. Ill., 1919; prof. chemistry Ia. State U., Ames, 1919—, Distinguished prof. scis. and humanities, 1962; research project dir. Manhattan project, OSRD, NSRC, 1940-45. Trustee Carver Research Found. Recipient Frederic Stanley Kipping award in organosilicon chemistry Am. Chem. Soc., 1962; Firestone Internat. Lecture award in organometallic chemistry, 1968. Hon. fellow Chem. Soc. (Gt. Britain), Phi Lambda Upsilon; mem. Am. Chem. Soc. (councilor-at-large 1939-45), A.A.A.S. (v.p. 1930—), Nat. Acad. Scis. (ofcl. rep. in Russia 1963), Phi Beta Kappa, Sigma Xi, Phi Lambda Upsilon. Author: Organic Chemistry—An Advanced Treatise, 4 vols., 1938, 43, 53; Organic Syntheses, Vol. 6, 1926; co-author Organic Substituted Cyclo-Silanes; Catenated Organic Compounds of Group IV-B Elements; also numerous articles, chpts. and sects. in books, encys. Editorial bd. Advances in Organometallic Chemistry, Organic Syntheses, Jour. Organometallic Chemistry, Organmetallic Reactions, Sci. Citation Index, Organometallic Syntheses. Research in organometallic chemistry, organosilicon chemistry, furans, correlations between physiol. action and chem. constitution, long-chained aliphatic compounds, heterocyclic chemistry, thermally stable fluids and lubricants. Home: 3221 Oakland St., Ames, Ia. 50012.*

GILMAN, John Ellis, Am. physician; b. Harmar, nr. Marietta, O., July 24, 1841; s. John Salvin and Elizabeth C. (Fay) G.; student medicine and surgery under his father and brother, and Dr. George Hartwell, of Toledo, O.; M.D., Hahnemann Med. Coll., 1871; m. Mary D. Johnson, 1860. First physician to offer his services for relief of sufferers at time of Chicago Fire, 1871, and apptd. by the Relief and Aid Soc. as sec. of its com. on sick and hosps.; prof. physiology, san. sci. and hygiene, later prof. materia medica Hahnemann Med. Coll., Chgo., 1884-1904. Introduced X-ray in therapeutic use, 1906; afterwards used it in treating cancers. Died June 21, 1916.

GILMARTIN, Malvern, Am. oceanographer; b. Los Angeles, Nov. 14, 1926; s. Malvern and Gladys (Hall) G.; A.A., Chaffey Coll., 1950; B.A., Pomona Coll., 1954; M.S., U. Hawaii, 1956; Ph.D., U. B.C., 1960; m. Amy Jean Finch, June 8, 1954; children—Malvern, Dale Moana, Sheila Ann, Ian Harvey. Faculty, U. Hawaii, 1954-56, 64-66; faculty U. B.C., 1956-60; sr. scientist oceanography Inter-Am. Tropical Tuna Com., 1960-64, lab. dir., Ecuador, 1961-64; prof. biol. oceanography Stanford, 1966—. Cons. Instituto Nacional de Pesca, Ecuador, 1961-64, Empresa Puertos de Colombia, 1963; coordinator Inter-Am. El Nino Project, 1962-64. Mem. Am. Soc. Limnology and Oceanography, Phycol. Soc. Am., A.A.A.S., Am. Inst. Biol. Scis., U. S. Naval Inst. Research on oceanography of estuaries, mediating influence of marine phys. conditions on primary prodn. of organic material by plants in sea, measurement of Peru-Chile Under-current. Office: Hopkins Marine Sta., Pacific Grove, Cal. 93950.*

GILMORE, Charles Whitney, Am. vertebrate paléontologist; b. Pavilion, N.Y., Mar. 11, 1874; s. John Edward and Caroline M. (Whitney) G., B.S., U. Wyo., 1901; m. Laure Contant, Oct. 11, 1902; children—Eloise Elizabeth, Dorothy Caroline, Helen Rosalie. Student asst. to prof. W. C. Knight; dept. vertebrate paleontology Carnegie Mus., 1901-04; with U. S. Nat. Mus., since 1904, as preparator dept. vertebrate paleontology until 1909, custodian, 1908-11, asst. curator 1911-18, asso. curator, 1918-23, curator from 1923. Mem. Paleontol. Soc. (pres. 1938), Biol. Soc. Washington, Geol. Soc. Washington, Geol. Soc. Am., Paleontol. Soc. Washington (pres. 1935), Soc. Vertebrate Paleontol. (pres. 1943), Phi Beta Kappa. Specialist in extinct reptiles, especially fossil lizards, mem. 16 major expdns., 1907-45, in charge of 15. Died Sept. 27, 1945.

GILMORE, Melvin Randolph, Am. ethnobotanist; b. Valley, Neb., Mar. 11, 1868; s. John Randolph and Mary Louisa (Concannon) G.; A.B., Cotner Coll., 1904; M.A., U. Neb., 1909, Ph.D., 1914. Prof. biol. scis. Cotner Coll., Neb., 1905-11; curator Mus. Neb. State Hist. Soc., 1911-16; curator State Hist. Soc. N. D., 1916-23; mem. sci. staff Mus. of Am. Indian, Heye Found., N.Y., 1923-28; curator of ethnology Mus. of Anthropology, U. Mich., from 1929. Mem. teaching staff Am. Sch. Wild Life Protection, McGregor, Ia., from 1922, Nature Training Sch., Gardner Lake, Conn., 1929; rep. of gov. of N.D. on Nat. Cons. on State Parks, Des Moines, Ia., 1921; conducted expdn. of Mus. Am. Indian to record ancient ritualistic ceremonies of Arikara tribe, 1924; originated proposed project of Am. Ethnobotanical Garden of Mus. of Am. Indian; established,, ethnobot. lab. Mus. of Anthropology, U. Mich., 1931. Fellow Am. Geog. Soc. Author Uses of Plants by Indians of the Missouri River Region (in Ann. Report Bur. Am. Ethnology) 1919. Died July 25, 1940.

GILMORE, Paul Carl, mathematician; b. Lethbridge, Can., Dec. 5, 1925; s. John A. and Emma (Mueller) G.; B.A., U. B.C., 1949; B.A., Cambridge U., 1951, M.A., 1954; Ph.D., Amsterdam U., 1953; m. Marijke Gerda Worp, June 12, 1954; children—Ian Alexander, Karen Marijke. Came to U. S., 1955, naturalized, 1964. Research asso. U. Toronto, Ont., Can., 1953-55; asst. prof. Pa. State U., 1955-58; staff mathematician IBM Research, Yorktown Heights, N.Y., 1958—; adj. prof. Columbia U., 1966—. Recipient Operations Research Soc. Am. Lancaster prize, 1964. Mem. Am. Math. Soc., Canadian Math. Congress, Assn. Symbolic Logic, Soc. for Indsl. and Applied Math., Operations Research Soc. Am. Developed methods for handling very large linear programming problems. Home: 39 Oak Hill Rd., Chappaqua, N.Y. 10514. Office: IBM Research, Box 218, Yorktown Heights, N.Y. 10598.*

GILREATH, Esmarch Senn, Am. chemist; b. North Wilkesboro, N.C., Sept. 21, 1904; s. Frank Hackett and Mamie (Williams) G.; A.B., U. N.C., 1926, M.A., 1927, Ph.D., 1945; m. Sara Taylor, Oct. 17, 1936. Faculty, Ga. Inst. Tech., 1927-29, high sch., Burlington, N.C., 1933-42, U. N.C., 1942-44; research chemist Am. Enka Corp., Enka, N.C., 1945-46; faculty Washington and Lee U., Lexington, Va., 1946—, prof., head, dept., chemistry, 1954—. Mem. Am., Brit. chem. socs., Brit. Soc. Chem. Industry, Va. Acad. Sci., Phi Beta Kappa, Sigma Xi, Alpha Chi Sigma. Author: Qualitative Analysis, 1954, Fundamental Concepts of Inorganic Chemistry, 1958. Contbr. articles to chem. jours. Devised a semi-micro scheme of qualitative analysis for cations without using hydrogen sulfide. Home: 420 Honeysuckle Hill, Lexington, Va. 24450.*

GILVARG, Charles, Am. biochemist, educator; b. N.Y.C., June 13, 1925; s. Hyman and Rose (Kreitzer) G.; B.Chem. Engring., Cooper Union, 1948; Ph.D., U. Chgo., 1951; m. Frieda Marie Mueller, June 21, 1949; children—Karyn Marie, David, Martin Howard, Gail Elizabeth. Postdoctoral fellow U. Chgo., 1951-52, N.Y.U., 1952-53, Cornell U., 1953-54; faculty dept. biochemistry N.Y.U., 1954-64; prof. dept. chemistry Princeton (N.J.), 1964—. Cons. biochemistry study sect. USPHS, 1960—. Guggenheim fellow, 1964. Mem. Am. Soc. Biol. Chemists, Am. Chem. Soc. (Paul Lewis award in enzyme chemistry 1963), Harvey Soc., A.A.A.S., Sigma Xi. Editor: Jour. Biol. Chemistry, 1965—. Contbr. articles to sci. jours. Research on intermediary metabolism of microorganisms. Home: 240 Hartley Av., Princeton, N.J. 08510.*

GILVARRY, John James, physicist; b. Manchester, Eng., July 15, 1917; s. John and Honoria (Kelly) G.; B.S. in Physics, Coll. City N.Y., 1939; M.A., Princeton, 1943, Ph.D., 1943. Came to U. S., 1924, naturalized, 1932. Physicist, OSRD at Princeton, 1942-43, Naval Ordnance Lab., Washington, 1943-46, N.Am. Aviation Co., Los Angeles, 1946-48; staff Rand Corp., Santa Monica, Cal., 1948-56, sr. physicist, 1950-56; sr. scientist Allis-Chalmers Mfg. Co., Milw., 1956-61; sr. staff scientist, leader plasma, atomic and molecular physics group Space Scis. Lab., Gen. Dynamics Corp., San Diego, 1961-64; Nat. Acad. Scis.-NRC sr. postdoctoral resident research asso. theoretical studies br. space scis. div. Ames Research Center, NASA, Moffett Field, Cal., 1964—; staff Livermore Radiation Lab., U. Cal., 1952. Fellow Am. Phys. Soc., Phys. Soc. (Gt. Britain), A.A.A.S.; mem. Italian Soc. Physics, Am. Geophys. Union, Am. Astron. Soc., Am. Math. Soc., Math. Assn. Am., Am. Assn. Physics Tchrs., Astron. Soc. Pacific, Phi Beta Kappa, Sigma Xi. Editorial adviser Fracture Processes in Polymeric Solids (B. Rosen), 1964. Research and numerous publs. on relativity effects on planets; thermodynamic functions from statis. atom model, solid state physics; determination of interior temperature of the earth; meteorites and asteroids; escape of planetary atmospheres; lunar maria; terrestrial oceans. Home: 1735 Woodland Av., Palo Alto, Cal. 94303. Office: Ames Research Center, Moffett Field, Cal. 94035.*

GIMBERNAT, Antonio de, see de Gimbernat, Antonio.

GIMBERT ROURA, José, Spanish scientist; b. Barcelona, Spain, Dec. 19, 1921; s. José and Carlota G.; ed. univs. Grenada, Barcelona, Cambridge; Ph.D. in pharmacy; m. Teresa Ràfois, 1953; children—José Maria, Javier, Rosa Mariá. Collaborator, chief microbiology, 1951-57; prof. biochemistry, pharm. insp. dir. lab., Frumtost. Recipient Nat. prize of pharmacy, Victor de Plata for profl. merit, prize City of Barcelona. Mem. Royal Acad. Pharmacy, Acad. Med. Sci., Am. Chem. Soc., others. Author: Antibiotiques, Revision de la Conception Actuelle; Antibiotiques Vieux; Les Antibiotiques en Ophthalmology; Stabilité et Diffusibilite de la Chlortétracycline. Home: 214 calle General Mitre. Office: 9-11 calle Suiza, Barcelona 6, Spain.

GIN, Winston, Am. engr.; b. San Francisco, Aug. 24, 1928; s. Kwock H. and Shee (Yee) G.; B.A., U. Ariz., 1950, B.S., 1951; M.A., U. Cal. at Los Angeles, 1955, M.S., 1959; m. Ellen Ong, Aug. 20, 1960; 1 son, Robert. Mem. tech. staff Hughes Aircraft Co., Culver City, Cal., 1953-57; research engr. Jet Propulsion Lab., Cal. Inst. Tech., Pasadena, 1957-59, research group supr., 1959-63, sect. mgr. solid propellant engring., 1963—. Mem. NASA Adv. Group on Solid Rockets, 1964—. Mem. Am. Inst. Aeros. and Astronautics, Phi Beta Kappa, Sigma Xi, Phi Kappa Phi, Pi Mu Epsilon. Research and publs. on chem. reactions in supersonic flow, devel. of solid propellant rocket engines; inventor thrust determination device for solid propellant rocket engines. Home: 1418 Star Ridge Dr., Monterey Park, Cal. 91754.*

GINGERICH, Owen Jay, Am. astronomer; b. Washington, Iowa, Mar. 24, 1930; B.A., Goshen College, 1951; M.A., Harvard U., 1953; Ph.D. astronomy, Harvard, 1962; m. 1954; 3 children. Dir. observatory, 1955-58, instr., 1955-57, asst. prof., 1957-58, Am. U., Beirut; lectr. astronomy, Wellesley College, 1958-59; astrophysicist, Smithsonian Astrophysical Observatory, since 1961; lectr. general education, from 1960, in hist. sci. 1962-63, 1965, in astronomy from 1964, Harvard U.; asso. prof. astronomy and hist. sci., 1968—; Mem., Harvard Obs. eclipse expedition to Ceylon, India, 1955; mem., Harvard expedition to observe occultation of Regulus by Venus, 1959; astronomy consultant, Harvard Project Physics, since 1964; dir., central telegram bureau, Internat. Astronomical Union, 1965. Research and publications on model stellar atmospheres and in history of astronomy. Office: Smithsonian Asteophysical Observatory, Cambridge, Mass.

GINGRICH, Curvin Henry, Am. astronomer; b. York, Pa., Nov. 20, 1880; s. William Henry and Ellen (Kindig) G.; B.A., Dickinson Coll., 1903, M.A., 1905; Ph.D., U. Chgo., 1912, postgrad. Yerkes Obs., 1911-12; Sc.D., Dickinson College, 1941; m. Mary Ann Gross, Aug. 10, 1915; 1 dau., Gertrude. Maryville Sem., Mo., 1903-05, Northwest Mo. Coll., Albany, 1905-07, Baker U., Baldwin, Kan., 1907-09; instr. math. Carleton Coll., 1909-12, prof. math. and astronomy from 1912, acting dean, 1914-15, dean, 1915-17; asst. to pres. registrar, 1917-19; at Mt. Wilson Obs., 1921-22; in charge courses in astronomy; Columbia, summers, 1929, 30; lectr. Adler Planetarium, summers, 1931, 32, 33; research asst. McCormick Obs., summer 1935. Mem. Am. Astron. Soc., Math. Assn. Am., Phi Beta Kappa, Sigma Xi. Asso. editor Popular Astronomy, 1912-26, editor from 1926. Astron. work at Goodsell Obs., Carleton Coll., principally micrometric measures of comet positions and double stars, also celestial photography and photog. determinations of positions of asteroids; stellar photometry. Died June 17, 1951.

GINGRICH, Newell Shiffer, Am. physicist; b. Orwigsburg, Pa., Jan. 29, 1906; s. Felix Moyer and Minnie (Shiffer) G.; A.B., N. Central Coll., Naperville, Ill., 1926; M.A., Lafayette Coll., 1927; Ph.D., U. Chgo., 1930; m. Fern Priscilla Riedel, June 5, 1928; children—Phillip Riedel, Katherine Ann (Mrs. Thomas Allan Brady). Faculty physics Lafayette Coll., Easton, Pa., 1927-28, Mt. Allison U., Sackville, N.B., Can., 1930-31, Mass. Inst. Tech., Cambridge, 1931-36; faculty U. Mo., Columbia, 1936—, prof. physics, 1943—. Physicist, Oak Ridge Nat. Lab. 1952, summer 1953, Argonne Nat. Lab. (Ill.), summers 1956, 57, 58. Recipient Citation of Merit OSRD, 1945; named Distinguished Prof. U. Mo., 1962. NSF fellow Mass. Inst. Tech., 1959-60. Fellow Am. Phys. Soc., A.A.A.S.; mem. Am. Assn. Physics Tchrs. (nat. and local offices), Am. Crystallographic Assn., Am. Assn. U. Profs., Sigma Xi, Phi Beta Kappa (hon.). Reviser: Physics—A Textbook for Colleges, 1950, 57. Cons. editor McGraw-Hill Ency. Sci. & Tech., 1960—. Contbr. articles to sci. jours., 1930—. Research on neutron diffraction; Compton effect in x-rays; structure in crystals; atomic distribution in liquids. Home: 313 E. Brandon Rd., Columbia, Mo. 65201.

GINI, Corrado, Italian economist, social scientist; b. Motta di Livenza, May 23, 1884; s. Lavinia Locatelli; D. Law and Econs., U. Bologna; D. honoris causa, U. Milan, U. Geneva, Harvard; m. Valentina Poggioli, 1920; children—Paola, Renata. Prof. statistics emeritus U. Cagliari, U. Bologna; prof. statistics and sociology U. Rome. Recipient Royal prize Acad. dei Lincei, Marzotto Nat. prize. Mem., hon. mem., pres. 20 internat., Italian and fgn. sci. socs. Author 20 books, over 1000 articles. Home: 39 via Adige. Office: 10 via Terme di Diocleziano, Rome, Italy.

GINSBERG, Donald Maurice, Am. physicist; b. Chgo., Nov. 19, 1933; s. Maurice J. and Zelda (Robbins) G.; B.A., U. Chgo., 1952, B.S., 1955, M.S., 1956; Ph.D., U. Cal., Berkeley, 1960; m. Joli D. Lasker, June 10, 1957; children—Mark D., Dana L. Faculty U. Ill., Urbana, 1959—, prof. physics, 1966—. Mem. Am. Phys. Soc., Am. Assn. Physics Tchrs., Sigma Xi, Phi Beta Kappa. Research in elec. and magnetic properties of metals at very low temperatures. Home: 1804 Bellamy St., Champaign, Ill. 61820.*

GINSBERG, Stewart Theodore, Am. psychiatrist; b. St. Paul, Apr. 18, 1906; s. Jacob and Mollie (Baldkind) G.; B.S., B.M., U. Minn., 1932, M.D., 1933; m. Ada Leach, Aug. 31, 1930; children—Barbara (Mrs. Sam Tisherman), Janet Mary (Mrs. Alvin Klein), Mark Bruce. Chief, psychiatry div. VA central office, Washington, 1955-57; commr. mental health state Ind., Indpls., 1957-66; dir. VA Hosp., Lyons, N.J., 1966—. Faculty Georgetown U., 1956-57, Ind. U., Indpls., 1948-66, prof. psychiatry, 1957-66; asst. examiner Am. Bd. Psychiatry and Neurology, Inc., 1950—. Recipient Distinguished Service award Ind. Pub. Health Assn., 1964, Award of Merit Tri-State Hosp. Assembly, 1964. Fellow Am. Psychiat. Assn., Coll. Am. Psychiatrists, Internat. Coll. Dentists (hon.); mem. A.M.A., Ind. Med. Assn., Ind. Neuro-psychiat. Assn. Contbr. articles to profl. jours. Research in clinical psychiatry; improving utilization of personnel in mental health specialties; utilization of community resources in psychiatric treatment. Address: VA Hosp., Lyons, N.J. 07939.

GINSBURG, Nathan, Am. physicist; b. Casey, Ill., Aug. 25, 1910; s. Louis and Dora (Brachman) G.; B.A., Ohio State U., 1931, M.A., 1932; Ph.D., U. Mich, 1935; m. Ruth Ostrow, Aug. 25, 1942; 1 dau., Susan E. Engring. research fellow U. Mich., 1935-36; Johnston scholar Johns Hopkins U., 1936-38, research asso., 1938-41; research asso. Carnegie Inst. Dept. Embryology, 1941-42; asst. prof. U. Tex., 1942-46; asso. prof. Syracuse U., N.Y., 1946-52, prof., 1952—, chmn. dept. physics, 1965—. Fellow Am. Phys. Soc., Optical Soc. Am.; mem. A.A.A.S., N.Y. Acad. Scis. Research in structure of molecules by spectroscopic methods, optical constants of single crystals, far infrared spectroscopy. Home: 989 James St., Syracuse, N.Y. 13203.*

GINSBURG, Seymour, Am. mathematician, computer scientist; b. Bklyn., Dec. 12, 1927; s. William and Bessie (Setomer) G.; B.S., Coll. City N.Y., 1948; M.S., U. Mich., 1949, Ph.D., 1952; m. Eleanor Shore, June 13, 1954; children—Diane, David. Asst. prof. math. U. Miami (Fla.), 1951-55; engr. Northrop Corp., Hawthorne, Cal., 1955-56; sr. research engr. Nat. Cash Register Co., Hawthorne, 1956-59; sect. head Hughes Aircraft Co., Culver City, Cal., 1959-60; sr. mathematician System Devel. Corp., Santa Monica, Cal., 1960—; prof. dept. elec. engring. U. So. Cal., 1966—. Mem. I.E.E.E., Am. Math. Soc., Assn. Computing Machinery, Soc. Indsl. and Applied Math., Math. Assn. Am. Author: An Introduction to Mathematical Machine Theory, 1962; The Mathematical Theory of Context Free Languages, 1966; also articles. Asso. editor Jour. Computer and System Scis., 1967—; area editor Jour. Assn. for Computing Machinery, 1968—. Pioneered automata theory including input-output behavior of machines, minimization of internal memory of machines; introduced models of data processing devices; pioneered formal programming lang. theory including devel. theory context free langs; introduced number different formal langs. Home: 14031 Margate St., Van Nuys, Cal. 91401. Office: 2500 Colorado Av., Santa Monica, Cal. 90406.

GINTL, Wilhelm Friedrich, chemist; b. Vienna, Austria, Aug. 5, 1843; s. Wilhelm and Anna Maria (Gullich) G.; student chemistry Vienna; doctorate Prague, Czechoslovakia 1867; m. Amalie Rosenbach, 1869; 4 sons. Indsl. chemist; became asst. to F. Rochleder at U. Prague, 1865, joined faculty in gen. and applied chemistry, 1868; prof. gen., analytical chemistry German Poly. Inst. of the Kingdom of Bohemia (later German Tech. U.), Prague, 1870-1908, rector 4 times; became mem. State (Bohemia) Bd. Health, 1870; advisor to comml. ct.; mem. Bohemian parliament, 1878-87; named ct. chemist, 1879; joined adminstrn. Austrian Assn. for Chem. and Metall. Prodn., 1887, became mem. exec. com., 1889, v.p., 1896, pres., 1898; mem. various govt. adv. bds.; became mem. Herrenhaus, 1902. Mem. Assn. for Promotion of Chem. Industry in Austria (founder, pres., v.p.). Author: Die Mineralwasserquellen von Bilin in Böhmen, 1898; Der österreichische Verein für chemische und metallurgische Produktion, 1856-1906, 1906. Research in analytical and inorganic chemistry, 1865-70, also in phytochemistry. Died Prague, Feb. 26, 1908.

GINTRAC, Élie, French physician; b. Bordeaux, France, Nov. 9, 1791; M.D., Paris, 1814; dir. Bordeaux Sch. Medicine; prof. internal medicine, Bordeaux; mem. French Acad. Medicine, French Acad. Scis., 1864, French Acad. Surgery. Author: Observations et recherches sur la cyanose, 1824; De l'influence de l'hérédite sur la production de la surexcitation nerveuse, 1845. First description of scleroderma; research on cyanosis, heredity in mental disturbances. Died Bordeaux, Dec. 10, 1877.

GINZBERG, Albert Semenovich, Russian geologist; b. Apr. 8, 1883; prof. Gerstsen (Hertzen) Pedagogical Inst., Leningrad. Author Lectures in Experimental Petrography (standard text in USSR), 1938; principal work in the field of mineralogy and petrography.

GINZBURG, Vitaly Lazarevich, Russian physicist, astrophysicist; b. Moscow, USSR, Oct. 4, 1916; s. Lazar and Augusta (Wildauer) G.; grad. Moscow U., 1938, postgrad., 1938-40; postgrad. Phys. Inst., USSR Acad. Sci., 1940-42; m. Nina Ezmakova, Sept. 30, 1946; 1 dau., Irina (Mrs. L. I. Dorman). Head subdept. theoretical physics Phys. Inst., USSR Acad. Sci., Moscow, 1942—. Vis. prof. Gorky U., 1945-59, cons., 1959—. Recipient Mandelstam prize, 1947, State prize, 1953, Lomonsov prize, 1962, Lenin prize, 1966. Mem. USSR Acad. Scis. Author: Propagation of Electromagnetic Waves in Plasma, 1960, 64, 67; (with Syrovatskii) Origin of Cosmic Rays, 1964; Superconductivity, 1946; (with Agronovich) Crystalloptics, 1967. Research on elementary particles, quantum theory of Cherenkov radiation, transition radiation, superconductivity theory, superfluidity theory, theory of ferroelectric phenomena, theory of cosmic rays origin, radioastronomy, theory of electromagnetic waves propagation. Address: P.V. Lebedev Phys. Inst., 53 Leninsky Prospect, Moscow, USSR.*

GINZBURG, Vulf Venyaminovich, Russian anthropologist; anatomist; b. Kursk, 1904; grad. Leningrad Inst. Med. Sci., 1926; postgrad. in anthropology USSR Acad. Sci., 1931-35; Cand. Biol. Sci., 1935; D.Med. Sci., 1944. Asst. dept. normal anatomy 1st Leningrad Med. Inst., 1929-39; jr., later sr. instr. dept. normal anatomy Kirov Mil. Med.Acad., Leningrad, 1937-59; head Leningrad group anthropologists and archeologists, dept. head Miklukho-Maklaya Mus. Anthropology and Ethnography, USSR Acad. Sci., 1937—; hean anthropology course Leningrad U., 1938—; lectr. anthropology, 1939-49; prof., 1949—. Mem. archeol. and anthrop. expdns. to Pamira, Tadzhikistan, Kazakhstan, Turkmenia, Kareliya-on-Don, South Russian Steppes, 1932, 33, 39; del. 25th Internat. Congress Orientalists, Moscow, 1960. Mem. All-Union (bd. mem.), Leningrad (bd. mem.) socs. anatomists, histologists and embryologists. Author: The Mountain Tadzhiks: Date on the Anthropology of the Tadzhiks of Karategin and Darvaza, 1937; Human Races and the Reactionary Nature of Racist Theories, 1958; The Lymphatic System of Human Lower Limbs, 1959; Anthropological Characteristics of the Peoples of Central Asia, 1962; The Elements of Anthropology for Physicians, 1963; co-author: History of Anatomy, Histology and Embryology in St. Petersburg over 250 Years, 1957. Mem. editorial bd. Archives Anatomy, Histology and Embryology. Address: Leningrad University, Universitetskaya n. 7-9, Leningrad, USSR.

GINZEL, Friedrich Karl, Austrian astronomer; b. Reichenberg, Bohemia, Feb. 26, 1850. Author: Handbuch der mathematischen und technischen Chronologie, 3 vols., 1906-14; Spezieller Kanon der Finsternisse, 1899. Studied astron. problems of chronology. Died Berlin, June 29, 1926.

GINZEL, Karl Heinz, pharmacologist; b. Reichenberg, Czechoslovakia, June 1, 1921; s. Otto and Irmgard (Weyde) G.; M.D., U. Vienna, 1948; m. Christine Böck-Greissau, Apr. 22, 1947, (div. July 1954); 1 son, Hellmut; m. 2d, Gerda Schutte, Mar. 21, 1958. Came to U. S., 1960. Research asst. Pharmacol. Inst. U. Vienna, 1948-52; World Health Orgn. fellow U. Oxford (Eng.), 1952-53; sr. scientific officer Glaxo Labs., Greenford (Eng.), 1954; sr. research fellow U. Birmingham (Eng.), 1955-57; sr. lectr. neuropharmacology Inst. Neurology Nat. Hosp. Nervous Diseases, U. London, 1957-60; sr. pharmacologist Riker Labs., Los Angeles, 1960—. Mem. Brit. Physiol. Soc., Am., German pharmacol. socs. Contbr. numerous articles in field to profl. jours. Contbr. field neuromuscular blocking agts. culminating in introdn. Succinyl-Bis-Choline into clin. use as a muscle relaxant in anesthesia and electroconvulsive therapy; research neuropharmacology. Home: 9956 Rudnick Av., Chatsworth, Cal. 91311. Office: Riker Labs., Northridge, Cal. 91326.*

GINZTON, Edward Leonard, engineer; b. nr. Ekaterinoslavsk, Russia, Dec. 27, 1915; s. Leonard Luis and Natalia P. (Philipova) G.; came to U. S., 1929, derivative citizen; B.S., U. Calif., 1936, M.S., 1937; E.E., Stanford, 1938, Ph.D., 1940; m. Artemas A. McCann, July 12, 1939; children—Anne, Leonard, Nancy, David. Research engr. Sperry Gyroscope Co., N.Y.C., 1940-46; asst. prof. applied physics and elec. engineering Stanford University 1946-47, associate prof., 1947-50, prof., 1951; dir. Microwave Lab., 1949-59; bd. dirs. Varian Assos., 1948—, chmn. bd., chief exec. officer, 1959-64, chairman of board and president, 1964—; expert consultant of research, devel. bd. Nat. Mil. Establishment, 1947-50; mem. commn. 1, U. S. Nat. Com. of Internat. Sci. Radio Union; dir. Stanford U. Project M., 1957-60; dir. Stanford Bank. Chmn. adv. bd. Sch. Engring., Stanford. Recipient Morris Liebmann Meml. prize, I.R.E., 1958. Fellow I.E.E.E., mem. Nat. Acad. Scis., Nat. Acad. Engring., mem. Sigma Xi. Author: Microwave Measurements, 1957. Contbr. articles to tech. jours. Patentee in field. Research on radar; circuit development; microwave tube development; linear electron accelerators. Home: 28014 Natoma Rd., Los Altos Hills, Cal. Office: 611 Hansen Way, Palo Alto, Cal. 94303.

GIOLITTI, Giovanni, Italian surgeon, veterinarian; b. Rome, Italy, Mar. 24, 1918; s. Giuseppe and Marie (Tami) G.; M.D. in surgery and vet. medicine; m. Carla Appiani, 1946; children—Giuseppe G., Andrea V. Prof., U. Milan. Mem. Am. Chem. Soc., Internat. Soc. for Advancement Sci. Research and publs. on antibiotics, prevention of illness. Home: Conca del Naviglion 4. Office: via Celoria 10, Milan, Italy.

GIORDANI, Francesco, Italian chemist; b. Naples, Italy, July 5, 1896; s. Giulio and Maria (Rossi) G.; Ph.D. in Chemistry. Became prof. gen. and inorganic chemistry Faculty Sci., U. Naples, 1957; dir. lab. electrochemistry Engring. Soc., also Inst. Gen. Chemistry; pres. tech. cons. com. Inst. Indsl. Reconstrn.; pres. Italian Nat. Research Com.; pres. Italian Nat. Com. for Nuclear Research. Mem. Nat. Acad. Lincei, Nat. Soc. Sci., Letters and Arts Italy. Developed theory of electrolytic diaphragm and circulation of alkaline chloride; studied (with Armand, Etzel) methods and speed with which atomic energy could be produced in 6 nations of European Coal and Steel Community. Died 1961.

GIORDANO, Alfonso, Italian physician; biologist; b. Lercara, Palermo, Dec. 10, 1910; s. Luigi and Vincenzina (Bongiovanni) G.; M.D.; Ph.D. in biol. sci.; M.D. honoris causa, U. Louvain; Dr. honoris causa dei Lincei Acad.; Ph.D. honoris causa in sci. and letters; m. Maria Teresa Scarlata, 1938; children—Pierluigi, Ferdinando, Paolo. Prof., U. Bari, 1938-40, U. Pavia, 1943-56, U. Milan, 1956—. Mem. Italian Path. Soc., Belgian Soc. Pathology. Author: Sifilide Renale, Disencefalie, Patologia Ereditoria. Home: via Madre Cabrini 10. Office: via Francesco Sforza 38, Milan, Italy.

GIORGI, Maurizio, Italian geophysicist; b. Albano Laziale, Italy, Jan. 19, 1914; s. Isidoro and Annunziata (Borgognoni) G.; Dr. Math. and Physics, Rome U., 1937; m. Amelia d'Andrea, Jan. 8, 1942; children—Maria Angela, Rita, Marta, Isidoro. Geophysicist, Nat. Inst. Geophysics, Rome, 1937-51, sr. geophysicist, 1951-54; researcher NRC, Rome, 1954-56, sr. researcher in geophysics, 1956-57, dir. study and devel. office, 1960—; prof. geophysics and meteoroogy Navy Hydrographical Inst., Genoa, Italy, 1957—; asst. prof. geomagnetism Rome U., 1957—; dir. Nat. Center Atmospheric Physics and Meteorology, Rome, 1961—. Decorated knight Order for Italian Republic, 1961. Mem. Italian Geophys. Assn. (sec. 1951—), European Soc. Seismology, Italian Inst. Navigation, Am. Geophys. Union, others. Author: Geofisica, 1958; Il campo magnetico terrestre, 1960; Lezioni di magnetismo terrestre e magnetismo navale, 1960; also numerous articles. Research on geomagnetism, seismology, meteorology. Home: 112, Viale Africa. Office: 31 Piazzale L. Sturzo, 00144 Rome, Italy.*

GIOVANNI DA SANTA SOFIA, see De Sancta Sophia, Giovanni.

GIOVANNOZZI, Renato, Italian engr.; b. Florence, Italy, July 21, 1921; s. Aldo and Maria (Bettoni) G.; ed. Classical Sch. and Univ.; Ph.D. in engring.; m. Maria Cecilia Ferrari, 1940; children—Francesco, Paolo, Maria, Elisabetta, Elena. Full prof. applied physics Poly. Inst., Turin. Mem. Acad. Scis. of Turin. Author: Trattoto di Costruzione de Machine, 2 vols. Research and publs. on applied mechanics, aerodynamics, constrn. of machines. Home: via Susa 32. Office: Faculty Engring., Politecnico di Torino, Turin, Italy.

GIPPRICH, John L., Am. physicist; b. Balt., Jan. 5, 1880; s. Anton and Mary (Hopf) G.; A.B., Loyola Coll., 1900; studied Johns Hopkins, 1908-09. Joined Soc. of Jesus, 1900; ordained priest R.C. Ch., 1913; prof. physics Holy Cross Coll., 1903-05, Boston Coll., 1906-07; prof. math. Fordham, 1907-08; prof. physics Georgetown U., 1914-29, became regent med. and dental schs., 1929; prof. physics St. Joseph's Coll., Phila., 1935-36; pastor So. Md. Chs., 1936-44; mem. faculty Georgetown U., from 1944. Mem. Washington Acad. Sci., Royal Astron. Soc., London. Author: Laboratory Manual of Mechanics, Heat and Sound, 1927. Died Mar. 7, 1950.

GIRALDES, Joaquim Pedro Casado, surgeon, geographer; b. Portugal; 1808; consul to France; worked in Paris. Author: Tableau des colonies et possessions anglaises dans les quatre parties du monde, par un patriote portugais, 1814; Tratado completo de cosmographia e geographia historica, phisica e commercial, antiqa e moderna, offrecada a S.M.F. e Senhor D. Joao VI, 1825. Considered 1st geographer of Portugal; described paradidymis (organ of Giraldes), 1859. Died 1875.

GIRARD, Aimé, French chemist; b. Paris, Dec. 2, 1830; joined lab. of Pelouze, 1851; conservator sci. collections l'École polytechnique, 1858-71; prof. chemistry l'École supérieure de commerce, 1858-69; named prof. indsl. chemistry Conservatoire des arts et métiers, 1871, Inst. agronomique, 1876; mem. French Acad. Scis., 1894, Société d'agriculture. Research on refining sugar, manufacture of sulfuric acid and hydrocelluloses; improved processes in wine, bread making, distillation and preparation of vegetable fibers. Died Paris, Apr. 12, 1898.

GIRARD, Albert, mathematician; b. Lorraine, France, 1595. Author: Invention nouvelle en algebre (contained 1st use of brackets, recognized imaginary roots, and stated that the number of roots of an equation is equal to its degree), 1629; Oeuvres mathematiques de Stevin, 1634; also book on trigonometry in which the contractions sin, tan, and sec were used for the first time, 1629. Applications of algebra to geometry; showed that in cubic equations that lead to irreducible case there are always 3 roots, 2 positive and 1 negative or vice versa; found areas of spherical triangles and figures traced on the surface of a sphere by arcs of a circle; studied negative roots (which Descartes later developed); 1st publs. on solid angles and their measurement. Died 1632.

GIRARD, Charles Frédéric, zoologist; b. Mulhausen, Upper Alsace, France, Mar. 9, 1822; ed. Neuchatel Switzerland, circa 1843-47; M.D., Georgetown (D.C.) Coll., 1856. Came to U. S. with Louis Agassiz, 1847, lived in Cambridge (Mass.) until 1850, published papers on flatworms and fish; moved to Washington, 1850, served as asst. to Spencer Baird at Smithsonian Instn. until 1860, in this capacity helped plan U. S. Nat. Mus., 1857; pub. many reports on fish and reptiles, especially dealing with collections made by exploring and survey parties in Far West during 1850's; went to France, 1860; became Confederate sympathizer, 1861, travelled through Va. and Carolinas as Confederate agt. for drug and surg. supplies, 1863; practiced medicine, Paris, France, 1865-85; pub. Herpetology of Wilkes expdn., 1858; pub. essay on fish in reports of railroad explorations to the Pacific (Vol. X), 1859; pub. paper on typhoid fever during Franco-Prussian War, 1872, paper on fishes, also bibliography of his works, 1888, paper on N.Am. flatworms, 1891. Died Neuilly-sur-Seine, France, Jan. 29, 1895.

GIRARD, Louis Joseph, Am. opthalmologist; b. Spokane, Wash., Mar. 29, 1919; s. Harry and Bonita Inez (Crosshay) G.; B.A., Rice U., 1941; M.D., U. Texas, 1944; N.Y. U. Post Graduate School of Medicine. Coordinator, 1948-49, lectr., 1951-53, N.Y. U. Post Graduate Med. School; dir., Chronic Infection Project, N.Y. Eye and Ear Infirmary, 1949-52; assoc. mng. dir., Ophthalmological Found. Inc., 1951-55; assoc. dir., Dept. Research, N.Y. Eye and Ear Infirmary, 1952-53; clinical asst. prof., U. Texas, 1953-57, lectr. 1957; clinical assoc. prof., 1953-56; assoc. chmn., Ophthalmology, 1956-58, clinical prof. and head Ophthalmology, 1958-61; prof. and chmn., 1961—, Baylor U. College of Medicine; exec. dir., The Eyes of Texas Inc., 1960—; numerous hospital appointments. Recipient many honors. Mem., Am. Acad. Ophthalmology and Otolaryngology; Pam-Am. Assn. Ophthalmology; Am. College Surgeons; Assn. for Research in Ophthalmology; N.Y. Acads. of Sci. and Medicine; French Soc. of Ophthalmology; Assn. Am. Physicians and Surgeons, others. Author and co-author numerous publications in field. Research on glaucoma, ocular motility, corneal lenses, transplantation, ocular allergy, cornea, amblyopia, and other problems in opthalmology. Home: 3205 Del Monte, Houston, Texas 77019. Office: 1200 Moursund, Houston, Texas 77025.*

GIRARD, Philippe de, see de Girard.

GIRARD, Pierre-(Simon), French engr.; b. Caen, France, Nov. 4, 1765; head engr. Dept. Civil Engring.; accompanied Napoleon on his Egyptian expdn.; mem. French Acad. Scis., 1815 (became pres. 1829), Egyptian Inst., Soc. Agr. Author: Oeuvres complètes, 1830-32. Directed constrn. of Ourog Canal; determined level of Nile, and studied formation of its alluvial deposits. Died Paris, Nov. 30, 1836.

GIRARDIN, Jean-(Pierre-Louis), French agronomist; b. Paris, Nov. 16, 1803; named prof. applied chemistry, Rouen, France, 1828; prof., dean Lille (France) Faculty Scis.; rector Clermont Acad. Recipient Gold medal Paris Sch. Pharmacy, 1824, St. Petersburg Acad., 1860. Mem. French Acad. Scis., 1842, Central Agrl. Soc., Paris Acad. Medicine. Author: Leçons de chémie élémentaire, 1835, 60; Technologie de la garance, 1844. Popularized study of chemistry; studied agrl. chemistry especially perfection of fertilizers and madder dye. Died Jan. 24, 1842.

GIRAUD, Raoul Gaston, French physician; b. Privas, Oct. 10, 1888; s. Gaston and Anita (Rourin) G.; ed. Coll. of Privas, Faculty Sci. and Medicine, Montpellier; M.D., lic. in sc.; m. Germaine Villard, 1930; children—Marie-Claude, Jean-François (dec.), Isabelle. Agrégé of med., 1923; prof. therapeutic hydrology and climatology, 1928-60; prof. path. and clin. propaedeutic medicine, 1932-37; prof. clin. medicine, 1937-60; dean Faculty of Montpellier, 1941-60, now hon. dean. Mem. Nat. Acad. Medicine France, Inst. France (corr.), French Soc. Cardiology (past pres.), Mediterranean Med. Union (pres. med. div.). Author: Annales de Cardiologie, 10 vols.; Hypotension Artérielle; Pratique Médico-Chirurgicale; Cardiologic, Hydrologie; Traité de Médecine; Traité de Climatologie, also articles. Address: 5 Enclos Tissié, Montpellier, France.

GIRDLER, Ronald, Brit. geophysicist; b. Reading, U.K., Aug. 2, 1930; s. George William and Annie (Sinkins) G.; B.Sc., Reading U., 1955; Ph.D., Cambridge (Eng.) U., 1958; Research asso. Lamont Geol. Obs., Columbia, U. N.Y.C., 1959-60; Imperial Chems. Industry fellow U. Durham (Eng.), 1960-63; lectr. physics U. Newcastle (Eng.), 1963—. Mem. Royal Astron. Soc., Geol. Soc. London, Am. Geophys. Union, Math. Assn. Research and publs. on Rift valleys especially Red Sea and Gulf of Aden., palaeomagnetism and its application to study horizontal motions of Earth's crust, seismicity of Earth, evolution of ocean basins, statis. studies terrestrial heat flow, convection in Earth's mantle. Home: 304 Kidmore Rd., Caversham, Reading, U.K. Office: Sch. Physics, Univ., Newcastle upon Tyne, U.K.*

GIRDNER, John Harvey, Am. physician, surgeon; b. Cedar Creek, Tenn., Mar. 8, 1856; s. William and Mary Ann (Link) G.; A.B., Tusculum Coll., Tenn., 1876; M.D., Univ. Med. Coll., N.Y. U., 1879; m. Adela Pratt, Sept. 23, 1886. Intern Bellevue Hosp., 1879-80; lectr. surgery N.Y. Post-Grad. Med. Sch. and Hosp. (mem. corp.). Fellow N.Y. Acad. Medicine. Author: Newyorkitis, 1901; also many essays on med. and other subjects. First to graft skin successfully from dead body onto the living; inventor telephonic bullet probe, 1887, also phymosis forceps. Died Nov. 25, 1933.

GIRIFALCO, Louis A., Am. physicist; b. N.Y.C., July 3, 1928; s. Anthony and Santa (Compagnino) G.; B.S., Rutgers U., 1950; M.S., U. Cin., 1952, Ph.D., 1954; m. Catherine A. Lyons, Sept. 3, 1950; children—Sandra, Anthony, Mary, John, Robert, Theresa, Stephen, Dorothea. Research chemist E. I. du Pont de Nemours & Co., Inc., 1954-55; solid state physicist NASA, 1955-61; with U. Pa., 1961—, prof., 1965—; cons. Frankford Arsenal; cons. editor John Wiley & Sons. Mem. Am. Phys. Soc., A.A.A.S., Am. Inst. Metall. Engrs. Author: Atomic Migration in Crystals, 1964. Research and publs. in crystal imperfections; statist. mechanics of crystals; properties of surfaces; thermal and electronic phenomena in crystals. Home: 155 Union Av., Bala-Cynwyd, Pa. 19004.*

GIROD, Christian Alphonse, physician; b. Bône, Algeria, Aug. 31, 1930; s. Lucien Alphonse and Ernestine (Guy) G.; Dr. degree, Med. Faculty Algiers, 1957; m. Raymonde Vaille, Dec. 19, 1955; children—Michèle, Monique, Anne-Marie, Geneviève, Isabelle. Asso. prof. histology embryology, 1961-62; asso. prof. histologyembryology U. Lyon (France), 1962—; biologist Hôpitaux, Lyon, 1962—. Mem. Societé de Biologie Algiers, Société de Biologie Lyons, Société d'Endocinologie, Association des Anatomistes de Langue francaise, Société Francaise de Cytologie Clinique, Société Française de Microscopic électronique, Société Francaise d'Histochimie, Société Francaise de Biologie Medicale, Research, numerous publs. on exptl. endocrinology, including influence of endocrine glands on leukocytes, cytology of hypophysis using light or electron microscopy, exptl. cytology of Langerhans' islets in monkey, hypophyseal-genital correlations in monkey. Home: 51 Avenue Rockefeller Lyon-3e, Rhône, France.*

GIROUS DE BUZAREINGUES, (Louis-François) Charles, French inventor, physiologist; b. St. Geniez, France, May 1, 1773; mem. French Acad. Scis., 1826, Soc. Agriculture. Author: Essai sur les mérinos, 1812; Physiologie agricole, 1849; Précis de morale, 1852. Invented micrometer for measuring delicacy of wool; research and publs. on Merino sheep, agr., plant physiology and anatomy. Died Aveyron, France, July 25, 1856.

GIRTANNER, Christoph, physician, chemist; b. St. Gallen, Switzerland, Dec. 7, 1760; s. Hieronymus and Barbara Felicitas (Wegelin) G.; student Lausanne, Switzerland, Strasbourg, France; M.D., Göttingen, Germany, 1782; postgrad. chemistry Paris, Edinburgh, Scotland; postgrad. medicine London, Eng., m. Catherine Marie Erdmann, 1790; 2 daus. Practiced medicine specializing in pediatrics, St. Gallen, for a short period; returned to Göttingen, 1787; traveled to Eng., Holland, France, 1788-89; then settled in Göttingen. Mem. Med. Soc. Edinburgh (hon.), Lit. and Philos. Soc. Manchester (hon.). Author: Abhandlung über die venerische Krankheit, 3 vols., 1783-89; Neue chemische Nomenklature für die deutsche Sprache, 1791; Anfangsgründe der antiphlogistischen Chemie, 1792; Historische Nachrichten und politische Betrachtungen über die französische Revolution, 13 vols., 1793-1803; Abhandlung über die Krankheiten der Kinder und über die physische Erziehung derselben, 2 vols., 1794; Über das Kautische Prinzip für die Naturgeschichte, 1796; Ausführliche Darstellung des Brownischen Systemes der praktischen Heilkunde . . . , 1798; Ausführliche Darstellung des Darwinschen Systemes der praktischen Heilkunde . . . , 1799. Research on irritability as a life principle and oxygen as a principle of irritability; proved venous blood turns bright red when oxygen is added; opposed phlogiston theory in chemistry; publs. on medicine of Brown and Darwin; studied French Revolution. Died Göttingen, May 17, 1800.

GITLIN, David, Am. physician, educator; b. N.Y.C., Aug. 7, 1921; s. Simon and Yetta (Rubin) G.; B.S., Coll. City N.Y., 1942; M.D., N.Y. U., 1947; m. Geraldine Mary Wadsworth, Nov. 22, 1944; children—Susan Jane, Jonathan David. Faculty dept. pediatrics Harvard Med. Sch., Boston, 1952-63; prof. pediatrics U. Pitts. Sch. Medicine, 1963—. Research collaborator Brookhaven Nat. Lab., 1954—; cons. to U. S. surgeon-gen., 1960—; mem. human embryology and devel. study sect. NIH, 1960-64. Recipient E. Mead Johnson award for research in pediatrics, 1956. Guggenheim fellow, 1958-59. Mem. Am. Soc. Clin. Investigation, Soc. Pediatric Research, Am. Acad. Pediatrics (Borden award 1963), Am. Assn. Immunologists. Contbr. numerous articles to sci. jours. Research on pathophysiology of immunological deficiency diseases and devel. of immunity in children, kinetics of synthesis and degradation of plasma proteins in man. Home: Wagner Rd., R.D. 3, Allison Park, Pa. 15101. Office: 125 DeSoto St., Pitts. 15213.*

GITSCH, Eduard, Austrian physician; b. Wels, Austria, Aug. 3, 1920; s. Wilelm Max and Maria (Wallmann) G.; M.D., U. Vienna, 1944; m. Irena Zambrzycka, Apr. 26, 1949; children—Jolanta, Gerald. Staff, U. Women's Clinic, Vienna, 1947—, prof., 1966—. Research grant Duke, 1957, Harvard, 1957. Author: Prophylaxe und Therapie postoperativer Komplikationen nach gynükologischen Laparotomien durch Langzeit-Cholinesterase hemmer, 1965; also numerous articles. Research on endocrinology, obstetrics, gynecology; new method to prevent postoperative urol. complications after abdominal radical operation of cervical cancer. Home: 17 Feldgasse, 1080, Vienna VIII, Austria.*

GIUFFI, Giovanni Antonio, Italian astrologer; flourished 17th century. Lived in Palermo. Author: Tractatus de eclipsibus, 1623. Wrote astrological treatise on eclipses; discussed effects, length and when influence of eclipses would be at height and how to find "dispositor" of eclipse; Ptolemy and Haly are cited.

GIUFFRIDA-RUGGERI, Vincenzo, Italian anthropologist; b. Catania, Sicily, Feb. 1, 1872; m. Lillian C. Strachan, 1914; 2 sons. Physician, Rome, 1896; libero docente of anthropology; asst. prof. anthropology U. Rome, 1900-05; prof. anthropology U. Pavia, 1906-07; prof. ordinary anthropology, dir. Anthrop. Inst., Royal U., Naples, from 1907; prof. ethnology Royal Oriental Inst. Naples, from 1914. Fellow Royal Anthrop. Inst. Gt. Britain and Ireland (hon.); corr. mem. anthrop. socs. Paris, Vienna, Moscow, Lyon, Brussels, Porto, Liège; mem. R. Accad. delle scienze fis. e matem. di Napoli, Italian Soc. Anthropology and Ethnology (v.p.), Roman Soc. Anthropology (v.p.), Swiss Inst. Gen. Anthropology, Anthropology Sch. Paris. Author: Sulla dignità morfologica dei segni degenerativi, 1897; Homo sapiens, Einleitung zu einem Kurse der Anthropologie, 1913; L'Uomo attuale, Una specie collettiva, 1913; Antropologia dell 'Africa orientale, 1915; A Sketch of the Anthropology of Italy, 1918; Antropologia sistematica dell 'Asia, 1919. Observed abnormal shallowness of glenoid fossa, 1897 (called Giuffrida-Ruggeri's stigma). Died 1922.

GIUL, Kasum K., Russian geographer; b. Shemakha, Baku, Russia; s. Kasum Meshady Kasum and Soltan-Khanum Hadjy (Husein) G.; candidate sci., 1947, dr. of geographical sci., 1951, Leningrad Institute of Water Transport; m. Anna Avetovna Bostanjan. Tchr., Azerbaidjan State U., from 1936; vice-rector, 1954-57; dir., Geographical Institute of Acad. of Scis., 1957-62; head dept. of Caspian Sea. Chmn., Geographical Soc. of Azerbaidjan SSR; vice-chmn., Caspian Commn., Acad. of Scis. Azerb. SSR; chrm., Aral-Caspain section of Oceanographical Commn. Author: Marine Terminological Dictionary, 1940; Hydrology of Marine Oil Fields of Western Coast of Caspian Sea, 1955; The Caspian Sea, 1956; The Influence of the Caspian Sea Level Fluctuations on the People's Economy, 1960; The Atlas of Agriculture of the USSR, 1960; The Composite Atlas of the Azerbaidjan SSR; textbooks and articles pub. in Polish, Bulgarian, Chinese. Specialist in study of Caspian Sea.

GIULANDINO, Melchiorre, botanist, explorer; b. Konigsberg, Germany, circa 1520; studied Latin, Greek, medicine; traveled in Europe and Nr. East, Palestine, Egypt: taken prisoner by Algerian pirates, ransomed and returned to Venice; traveled in N. Africa; became head, bot. (herb) garden, Padua, Italy, 1561; became mem. faculty botany, Padua, 1564; prof. for life U. Padua. Author: De stirpium, 1557; Hortus Patavinus, 1600. Contbd. to knowledge of Egyptian flora; described rare herbs; offered an hypothesis of how plants take in water. Died Padua, Jan. 8, 1589.

GIULIANO, Landolino, Italian mathematician; b. Catane, July 17, 1914; s. Giuseppe and Carmela (Caruso) G.; Ph.D. in math. scis.; m. Maria Luisa Cagnoni, 1950; children—Rita, Carmela, Gabriella, Giuseppe, Luigi, Francesco. Prof. math. analyses Naval Acad., Livorne; prof. math. U. Pisa and Upper Normal

Sch., Pisa. Research and publs. on math. analyses. Home: 59 via Ulvi Liegi. Office: University of Pisa, Upper Normal School, Pisa; also Naval Academy, Livorne, Italy.

GIULOTTO, Luigi, Italian physicist; b. Mantua, May 23, 1911; s. Virgilio and Antonietta (Perini) G.; Ph.D. in scis. and physics; m. Gilda Olivelli, 1948; children—Laura Elena, Enrico. Asst., Poly. Inst., Milan; asst., instr. spectroscopics U. Pavia, titular prof. higher and gen. physics, 1949. Named Dalla Ricca laureate, Somaini laureate for physics. Mem. Lombardo Inst. Scis. Letters, Foratom, Ampere, Radiofrequency Spectroscopy Group, Lombard (v.p.), Italian socs. physics, Am. Phys. Soc. Contbr. numerous articles to sci. jours. Home: 12 viale della Liberta. Office: Institute of Physics, University of Pavia, Pavia, Italy.

GIURGEA, Margareta Alexandru Valeriu, Rumanian physicist; b. Bucharest, Rumania, Aug. 19, 1915; d. Alexandru L. and Natalia (Simionescu) Valeriu; student U. Bucharest, 1933-36, D. phys. scis., 1943; m. Gheorghe D. Giurgea, July 13, 1952. Staff, U. Bucharest, 1937—, prof. physics, 1962—; chief dept. spectroscopy Inst. Physics, Rumanian Acad., Bucharest, 1949—. Recipient Ordinul Muncii, 1965. Mem. Fédération internationale des hommes de science. Author: (with D. Barca-Galateanu, I. Iova, V. Sahini, R. Titeica, A. Trutia) Introducere in Spectroscopie, 1966; also articles. Research on elec. discharges in gases, spectrophotometry, light scattering in liquids and crystals. Home: 27A Lazureanu, Bucharest, Rumania.*

GIVENS, James Wallace, Jr., Am. computer scientist; b. Alberene, Va., Dec. 14, 1910; s. James Wallace and Mamie (Hughes) G.; B.S. cum laude, Lynchburg Coll., 1928, D.Sc., 1965; postgrad. U. Ky., 1929; M.S. in Math., U. Va.; Ph.D. in Math., 1936, Princeton; postgrad. Inst. for Advanced Study, Princeton, 1935-37; m. Virginia Catherine Shelton, Sept. 16, 1937; children—James Wallace III, Brian Hughes, Barry Shelton. Faculty, Cornell U., 1937-41, Northwestern U., Ill., asst. prof., 1941-46; asso. prof. Ill. Inst. Tech., 1946-47; prof. U. Tenn., 1947-56; prof., chmn. math. dept. Wayne State U., Detroit, 1956-60; prof. engring. scis. Northwestern U., Evanston, 1960-64, prof. math., 1960—; asso. dir. applied math. div. Argonne Nat. Lab., Ill., 1962-64, dir., 1964—. Cons. math. panel, Oak Ridge Nat. Lab., 1951-62; sr. sci. Inst. Math. Scis., N.Y. U., 1953-55. Author: (with O. Venlen, A. H. Taub) Geometry of Complex Domains, 1936; Numerical Computation of the Characteristic Values of a Real Symmetric Matrix, 1954. Home: 1926 Orrington Av., Evanston, Ill. 60201. Office: Argonne Nat. Lab., Argonne, Ill. 60439.*

GIVENS, Miles Parker, Am. physicist; b. Richmond, Va., June 9, 1916; s. Charles Watson and B. Mellie (Richardson) G.; B.S., U. Richmond, 1937; Ph.D., Cornell U., 1942; m. Mary Imogene Morgan, Sept. 17, 1941; children—Roger Wayne, Robert Parker, Jean Frances. Instr. Pa. State U., 1942-46; physicist Johns Hopkins Applied Physics Lab., Silver Spring, Md., 1946-47; with U. Rochester, 1947—, prof. optics, 1957—, asso. dir. Inst. Optics, 1963-64. Fellow Optical Soc. Am.; mem. Am. Phys. Soc., Am. Assn. Physics Tchrs., Am. Soc. Engring. Edn., Phi Beta Kappa, Sigma Xi, Phi Kappa Phi. Research and publs. on band structure in solids by soft x-ray spectroscopy and measurement of optical properties; exptl. work in holography. Home: 657 Ridge Chapel Rd., Williamson, N.Y. 14589.*

GJESSING, Dag Tryveson, Norwegian geophysicist; b. Talvik, Norway, Feb. 24, 1930; s. Trygve R. and Ruth (Lofting-Hansen) G.; student Bergen (Norway) U., 1949-50; B.Sc. in Engring., London (Eng.) U., 1954; Dr. Philos., Oslo (Norway) U., 1964; m. Toril Johansen, Sept. 15, 1958; 1 son, Trygve D. Staff, Norwegian Def. Research Establishment, Kjeller, Norway, 1954—, research asso., 1955-65, sr. research asso., 1965—; vis. research asso. Stanford, 1960-61. Recipient premium Instn. Elec. Engrs., London, 1963. Mem. Norwegian Instn. for Engrs., Norwegian Geophys. Union, Internat. Sci. Radio Union. Contbg. author: Radiometeorologie en vue de L'Application aux Telecommunications, 1966. Research, publs. on fine structure of troposphere with regard to wind velocity, refractive index, temperature and humidity with radio, radar and direct methods. Home: 54 Skogfaret, Skedsmokorset, Norway. Office: P.O. Box 25, Kjeller, Norway.*

GJESSING, Leiv Rolvssön, Norwegian physician; b. Alten, Norway, Oct. 24, 1918; s. Rolv R. and Susanne (Milberg) G.; M.D., U. Oslo (Norway), 1947, M.Sc., 1948, Dr.med., 1965; m. Avis Margareth Swimm, 1948; children—Jan, Rolv, Kai, Geir. Research fellow U. Toronto (Ont., Can.), 1948-50; practice medicine, specializing in psychiatry and biochemistry, 1950-59; chief lab., asst. med. supt., physician in charge psychiat. patients Dikemark Hosp., Asker, Norway, 1959—. Mem. Internat. Soc. for Neurochemistry, Biochem. Soc. (London, Eng.), A.A.A.S. Editor: Symposium on Tyrosinosis in Honour of Dr. Grace Medes, 1966. Research, publs. on metabolism of periodic psychosis, metabolic diseases as phenylketonuria and tyrosinosis, and tumors of sympathetic nervous system. Home: 39 Ingolfs vei, Engelsrud, Asker, Norway. Office: Dikemark Hosp., Asker, Solberg p.a., Norway.*

GLACET, Charles, French chemist; b. St. Vaast, France, Feb. 17, 1911; s. Arthur and Berthe (Lequet) G.; Licencé ès Scis., Faculté Scis. Lille (France), 1934; Dr. ès Scis., 1947; m. Marguerite Helene Butez, Apr. 8, 1939; children—Hélène, Lucette. With Faculté Scis. Lille, 1938—, attaché au Centre Nat. de la Recherche Scientifique, 1938-47, prof. sans chaire, 1958-59, prof. titulaire à titre personnel, 1960-65, titular prof. organic chemistry, 1965—. Mem. Chem. Soc. France (prix Adrian 1945, prix R. Berr 1954). Author: (with J. Wiemann) Cours de chimie physique, 1947; also articles. Research on alpha-hydroxytetrahydrofurans, amines, alpha-aminotetrahydrofurans, pyrans, delta-aminoalcohols, unsaturated aminoalcohols, polyaminopolyalcohols, hydroxides of piperidiniums. Home: 16, rue Victor Lelievre, Mons en Baroeul, France. Office: Faculté des Scis., Lille, B.P. 36, France.*

GLADSTONE, John Hall, English chemist; b. Hackney, London, Eng., Mar. 7, 1827; ed. Univ. Coll., London, Giessen (Germany) U.; Ph.D., D.Sc., 1847; m. May Tilt, 1852; m. 2d, Margaret King, 1869. Fullerton prof. chemistry, Royal Inst., 1847-50; lectr. St. Thomas's Hosp., 1850-52; mem. Royal Commn. on Lights, Buoys, and Beacons, 1858-61; mem. gun cotton com. War Office, 1864-68; Fullerian prof. chemistry Royal Instn., 1874-77. Fellow Royal Soc., 1853 (Davy medal 1897); pres. Chem. Soc., 1877-79, Phys. Soc., 1874-76. Author: Life of Michael Faraday, 1872; Spelling Reform from an Educational Point of View, 1878; Chemistry of Secondary Batteries, 1883; also papers on laws of chem. combination, relations of chem. and optical sci. Pioneer in field of phys. chemistry; research (with Dale) on refraction; proposed formula for molecular refractivity of organic compounds; found that state of equilibrium is produced in action of thiocyanates on ferric salts in solution; investigated colorless and colored salt solution reactions; worked on drawing up tables of relative affinities; correctly explained H. Rose's observation that solution of barium sulfate in hydrochloric acid can be precipitated by sulphuric acid and barium chloride as well; studied dissociation of certain gases; spectroscopy; chemistry in relation to optics; electric batteries. Died London, Oct. 6, 1902.

GLADYSHEVSKII, Evhen Ivanovich, see Hladyshevsky, Evhen Ivanovich.

GLAISHER, James, English meteorologist; b. London, Eng., Apr. 7, 1809; s. James Glaisher; m. Cecilia Louise Belville, 1843; 2 sons, 1 dau. With Ordnance Survey Ireland, 1829-30; apptd. asst. Cambridge U. Obs., 1833; joined Royal Obs., Greenwich, Eng., 1835, chief magnetic and meteorol. dept., 1838-74; helped Daily News organize its 1st daily weather report. Fellow Royal Soc., 1849; mem. Royal Astron. Soc., Royal Micros. Soc. (pres. 1865-68), Photog. Soc. (pres. 1869-72), Royal Meteorol. Soc. (co-founder, sec. 1850-67, 68-72, pres. 1867-68), Aero. Soc. (co-founder, became 1st treas. 1866), Brit. Assn. (com. on math. tables). Chmn. exec. com. Palestine Exploration Fund, 1880. Author: Hygrometrical Tables adapted to the Use of the Wet and Dry Bulb Thermometer, 1847; Travels in Air, 1867. Made observations on position of Haley's comet at its return, 1835; organized system of precise meteorol. observations; tried to discover relation between weather and cholera epidemics in London, 1832, 49, 53-54; made 8 ascents in a balloon, 1862; computed smallest factor of every number not divisible by 2, 3, or 5, of the 4th, 5th and 6th millions. Died Croydon, Surrey, Eng., Feb. 7, 1903.

GLAISHER, James Whitbread Lee, English mathematician; b. Lewisham, Eng., Nov. 5, 1848; s. James and Cecilia (Belleville) G.; grad. Trinity Coll., Cambridge, 1871; M.A.; Sc.D. Cambridge, 1887, Dublin, 1892, Victoria U., 1902. Tutor Trinity Coll., 1883-93, lectr., 1871-1901, also fellow. Fellow Royal Soc., 1875 (Sylvester medal 1913); pres. Cambridge Philos. Soc., 1882-84, London Math. Soc., 1883-84 (De Morgan medal 1908), Royal Astron. Soc., 1885-87, 1900-02, sect. A. Brit. Assn., 1890, Cambridge Antiquarian Soc., 1899-1901. Editor Messenger of Math., from 1871, Quar. Jour. Pure and Applied Math., from 1878. Contbr. papers on pure math to jours. Died Dec. 7, 1928.

GLAISTER, John, Scottish surgeon; b. Lanark, Scotland, Mar. 9, 1856; s. Joseph and Marion Hamilton (Weir) G.; M.B., M.D., LL.D., U. Glasgow; D.P.H., Cambridge U.; m. Mary Scott Clark; 2 sons, 4 daus. Named lectr. med. jurisprudence and hygiene Royal Infirmary Sch. Medicine, 1881; became prof. med. jurisprudence and pub. health St. Mungo's Coll., Glasgow, 1888; prof. forensic medicine U. Glasgow, 1898-1931; police divisional surgeon, Glasgow; examiner for diplomas in pub. health Royal Colls. Physicians and Surgeons, Edinburgh, Faculty Physicians and Surgeons, Glasgow, 1883, 88; mem. referee in indsl. diseases for 6 Scottish counties; medico-legal examiner in Crown cases. Fellow Royal Soc. Edinburgh, Coll. Surgeons, Royal Faculty Physicians and Surgeons, Glasgow. Author: William Smellie and his Contemporaries: a Contribution to the Literature of Midwifery of the Eighteenth Century, 1894; A Manual of Hygiene for Students, 3d edit., 1920; Text-Book of Medical Jurisprudence, Toxicology, and Public Health, 2 vols., 5th edit., 1921; Gas Poisoning in Mining and other Industries, 1914. Died Dec. 18, 1932.

GLANVILL, Joseph, English natural philosopher; b. Plymouth, Eng., 1636; s. Nicholas Glanvill; B.A., Exeter Coll., Oxford, (Eng.) U., 1655; M.A., Lincoln Coll., Oxford U., 1658; m. Mary Stocker; 2 children, including Maurice; m. 2d, Margaret Selwyn; children—Sophia, Henry, Mary. Chaplain to Francis Rons, lord of Cromwell and provost of Eton; later with Oxford U.; named rector Abbey Ch., Bath, Eng., 1666. Fellow Royal Soc., 1664; founder (with Henry More) assn. for psychical research. Author: The Vanity of Dogmatizing, 1661; Lux orientalis, 1662; Saepsis scientifica, 1665; Philosophical Considerations touching Witches and Witchcraft, 1666; Plus Ultra, or the Progress and Advancement of Knowledge since the Days of Aristotle, 1668; An Earnest Invitation to the Lord's Supper, 1673; Essays on Several Important Subjects, 1676; The Zealous and Impartial Protestant, 1681. Tried to find empirical ground for belief in supernatural; defended witchcraft. Died Bath, Nov. 4, 1680.

GLASER, Christopher, Swiss chemist; b. Basel, Switzerland, circa 1628; demonstrator Jardin du roi; tchr. of Nicholas Lemery, who succeeded him; apothecary to Louis XIV of France. Mineral glaserite (aphthitalite) named in his honor. Author: Traité de la chimie, 1663 (which went through 12 editions in French, German, and English to 1710. Discovered a new method for making potassium sulphate, which is also called Glaserite. Died circa 1672.

GLASER, Donald A(rthur), Am. physicist; b. Cleve., Sept. 21, 1926; s. William Joseph Glaser; B.S., Case Inst. Tech., 1946, Sc.D. (hon.), 1959; Ph.D., Cal. Inst. Tech., 1949; m. Ruth Louise Thompson, Nov. 28, 1961; children—Louise, William. Prof. physics U. Mich., 1949-59, U. Cal. at Berkeley, 1959-64, prof. physics and molecular biology, 1964—. NSF fellow, 1961; Guggenheim fellow, 1961-62; recipient Henry Russel award U. Mich., 1955; Charles V. Boys prize Phys. Soc. London, 1958, Nobel prize in physics, 1960. Fellow Am. Physics Soc. (prize 1959); mem. Nat. Acad. Scis., Sigma Xi, Tau Kappa Alpha, Theta Tau. Constructed 1st bubble chamber for visual demonstrations of movements of high-energy atomic particles; study of cosmic rays; application of methods of physics to problems of molecular biology. Office: Molecular Biology Dept., U. Cal., Berkeley, Cal.*

GLASER, Gilbert H., Am. neurologist; b. N.Y.C., Nov. 10, 1920; s. Burnard R. and Sidelle (Rogers) B.; A.B., Columbia, 1940, M.D., 1943, Sci. D., 1951; M.A., Yale, 1963; m. Morfydd M. Pugh, Mar. 17, 1946; children—Gareth Evan, Sara Elizabeth. Faculty, Columbia U., Coll. Phys. and Surg., 1948-52, asso., 1951-52; head sect. neurology, dir. Electroencephalograph Lab., faculty Yale Sch. Medicine, also Yale-New Haven Med. Center, 1952—, prof. neurology, 1963—; vis. prof. neurology Inst. Child Health and U. Coll., London, Eng., 1965-66. Cons. neurologist U. S. Vets. Hosp., West Haven, Conn., 1955—; mem. coms. USPHS, 1956-60, 68—. Fellow Am. Acad. Neurology (past trustee), Royal Soc. Medicine; mem. Am. Neurol. Assn., Am. Epilepsy Soc. (past pres.), Am. Electroencephalographic Soc. (past mem. council), EEG Soc. (Eng.), Eastern Electroencephalograph Assn. (past pres.), Phi Beta Kappa, Alpha Omega Alpha, Sigma Xi. Author: EEG and Behavior, 1963; also numerous articles. Research in electrophysiol. and metabolic mechanism in epilepsy, effects hormones on nervous system, relationships between ionic environment and neural excitability. Home: 205 Millbrook Rd., Hamden, Conn. 06518. Office: 333 Cedar St., New Haven 06510.*

GLASER, Johann Heinrich, Swiss physician; b. Basel, Switzerland, Oct. 6, 1629; prof. anatomy and botany, Basel; reformed clin. instrn.; introduced bedside tng. and post-mortem examinations. Died Basel, Feb. 5, 1675.

GLASER, Karl Andreas, German chemist; b. Kirchheimbolanden, Germany, June 27, 1841; s. Friedrich Wilhelm and Regina Luise (Giessen) G.; student, poly. schs., Nuremberg, Germany, Munich, Germany, 1856-59; engring. degree; student chemistry U. Erlangen (Germany); Ph.D., U. Tübingen (Germany), 1864; m. Anna Jakobine Doflein, 1874; 2 sons, 1 dau.; m. 2d, Elisabeth Kern, 1885; 1 son, 1 dau. Asst. to Kekulé in Ghent, Belgium; went to Bonn, Germany, 1867; joined faculty U. Bonn, 1869; joined firm Badische Anilin- & Soda-Fabrik, Ludwigshafen (BASF), 1869, founded br. dye factory, Russia, 1877-78, apptd. dep. dir. and dir. entire dye prodn. operation, 1877-78, became mem. mng. bd. dirs., 1883, ret. from mgmt., 1895, dir. until 1920, became chmn., 1912. Research and publs. on aromatic compounds; developed process for synthetic prodn. of Alizarin dyes; discovered carbazol and phenanthren. Died Heidelberg, Germany, July 25, 1935.

GLASER, Kurt, physician; b. Vienna, Austria, Feb. 16, 1915; s. Richard and Hedwig (Schiller) G.; student U. Vienna Med. Sch., 1933-38; M.D., U. Lausanne Med. Sch., 1939; M.Sc., U. Ill. Med. Sch., 1948; m. Susanne Stein, Dec. 6, 1946; children—Richard, Benjamin, David, Dan. Came to U. S., 1939, naturalized, 1945. Instr. pediatrics U. Ill. Med. Sch., 1945-50; tchg. fellow Cook County Hosp., Chgo.,

1945-50; asst. chief physician, instr. pediatrics Hadassah U. Hosp. Hebrew U., Jerusalem, Israel, 1950-54; with U. Md., 1954—, asso. prof. pediatrics, 1962—, asst. clin. prof. psychiatry, 1965—; clin. dir. Rosewood State Hosp., 1961—; asso. cons. pediatrics and psychiatry Sinai Hosp. Recipient Raymond B. Allen Instructorship award U. Ill., 1949. Mem. Am. Acad. Pediatrics, Am. Psychiatric Assn., Am. Assn. Mental Deficiency, Am. Acad. Mental Retardation, A.M.A., Md. Psychiat. Soc., Balt. City Med. Soc., Med. and Chirurgical Soc., Sigma Xi. Contbr. chpt. Suicide in Children and Adolescents in: Acting Out Theoretical and Clinical Aspects, 1965; book translation (with Susanne Glaser) Physiology & Pathology of Infant Nutrition, 1955. Research and publs. on drug adminstrn. to children; bone marrow devel., growth patterns in premature infants, psychiat. problems of children and adolescents. Home: 6114 Biltmore Av., Balt. 21215. Office: Rosewood State Hosp., Owings Mills, Md. 21117.*

GLASER, (Johann) Ludwig (Valentin), German biologist; b. Grünberg (Hessia), Germany, Feb. 9, 1818; s. Johann and Friederike Magdalene Marie (Vigelius) G.; student natural scis., philology, Giessen, Germany, 1837-39, Ph.D., 1842; student Polytechnikum Darmstadt, until 1842; m. Anna Margarete Luise App, 1848; 2 sons, 3 daus. Became dir. pvt. sch. Biedenkopf, Germany, 1842; became tchr., dir. secondary schs. in Biedenkopf, 1846, in Friedberg, 1856, in Worms, 1858, in Bingen, 1874; named titular prof., 1872; ret., 1879. Author: Der neue Borkhausen oder hessisch-rheinische Falterfauna, 1863; Landwirtschaftliche Ungeziefer, dessen Feinde und Vertilgungsmittel, 1867; Taschenwörterbuch der Botanik, 1885; Die Kleintiere in ihrem Nutzen und Schaden, 1886; Catalogus etymologicus coleopterorum et lepidopterorum, 1887. Described flora and fauna of Hessia; research on defense adaption of phys. characteristics of insects to the plants on which they live (mimicry); popularized biology; adaptation in animal protection and pest control. Died Mannheim, Germany, Jan. 20, 1898.

GLASER, Otto (Charles), biologist; b. Wiesbaden, Germany, Oct. 13, 1880; s. Charles and Eleanore (Blum) G.; brought to U. S. in infancy; A.B., Johns Hopkins, 1900, Ph.D., 1904; postgrad. U. Budapest, 1911-12; m. Dorothy Gibbs Merrylees, Sept. 1, 1909; children—Comstock, Victoria; m. 2d, Anita Gibson Glaenzer, 1934 (dec. 1940); m. 3d, Dorothy Wrinch, Aug. 20, 1941. Demonstrator in comparative anatomy, embryology and biology Coll. Physicians and Surgeons (now U. Md.), Balt., 1901-03; fellow Johns Hopkins, 1903-05; investigator oyster culture N.C. Geol. Survey and U. S. Bureau Fisheries, 1901-02; in charge oyster culture Gulf Biol. Station, La., 1903-04; instr. zoölogy and embryology Marine Biol. Lab., Woods Hole, Mass., 1905-07; with U. Mich., 1905-08, advancing to asso. prof. biology; prof. biology Amherst (Mass.) Coll., from 1918, Harkness prof., from 1939; lectr. on biology New Sch. for Social Research, N.Y.C. Fellow A.A.A.S.; mem. Am. Soc. Naturalists, Am. Soc. Zoölogists, Am. Physiol. Soc., Soc. Exptl. Biology and Medicine, N.Y. Acad. Sci., Soc. Growth and Devel., Phi Beta Kappa, Sigma Xi. Contbr. to sci. jours. on developmental physiology and growth. Died Feb. 7, 1951.

GLASER, Walter, mathematician, physicist; b. Oberbaumgarten, Bohemia, July 31, 1906; s. Franz and Maria (Riebl) G.; Ph.D., German U., Prague, 1930; studied in Vienna several semesters; degree in library sci.; m. Maria Scholz; 1 son, 2 daus. Became asst. to Philipp Frank at inst. for theoretical physics, German U., Prague, 1929, joined faculty, 1933, named dir. inst., 1936; became prof. theoretical physics, 1938; joined inst. for theoretical physics U. Vienna, 1947; became asso. prof. applied physics Tech. U. Vienna, 1949, dir. inst. for applied physics, 1952, prof., 1953, chmn. inst. for theoretical and applied physics, 1958; worked with B. von Borries & E. Ruska of Siemens & Halske AG, Berlin, 1938-54; chief physicist Farrand Optical Co., N.Y., 1954-58. Mem. Austrian Acad. Scis. (corr.). Author: Grundlagen der Elektronenoptik, 1952; Elektronen and Jonenoptik, 1954; articles on electron optics. Worked on devel. of magnetic electron microscope, differential and integral equations, vector analysis, vector algebra and non-Euclidian geometry, statis. mechanics and wave mechanics, theory of ideal gas and electron theory of metals. Died Vienna, Feb. 3, 1960.

GLASGOW, Arthur Graham, Am. engr.; b. Buchanan, Va., May 30, 1865; s. Francis Thomas and Anne Jane (Gholson) G.; M.E., Stevens Inst. Tech., 1885; E.D., 1928; D.Sc., Washington and Lee U., 1930; LL.D., Wabash Coll., 1950; m. Margaret Elizabeth Branch, 1901. Joined United Gas Improvement Co., 1885; sec., mgr. Lewiston (Me.) Gas Light Co., 1886; engr. Kansas City (Mo.) Gas Light & Coke Co., 1888; gen. insp. United Gas Improvement Co., 1890; gen. mgr. and chief engr. Standard Gas Light Co. of City of N.Y., 1891; founder firm Humphreys & Glasgow (installations of Humphreys-Glasgow plant and processes aggregate daily capacity over 2,500 millions cubic ft.), London, Eng., 1892; chmn. and pres. Bldg. Supplies Corp. Norfolk, Va. U. S. World's Gas Congress, Paris, 1900; chmn. Com. on Electrolysis of Am. Gas Light Assn., 1903-06; rep. Am. Gas at Paris Conf. for Standardizing Threads, 1908; fixed-nitrogen adminstr. U. S. War Dept., founder U. S. Fixed-Nitrogen Lab., later administered by Dept. of Agr., 1919. Recipient Gold medal Am. Gas Assn., 1910, Franklin Inst., 1928. Life mem. Inst. Civil Engrs., Inst. Mech. Engrs., Am. Soc. M.E., Am. Soc. C.E.; hon. mem. Inst. Gas Engrs. (Gt. Britain). Contbr. to jours. Patentee in gas tech. Died Oct. 28, 1955.

GLASHOW, Sheldon Lee, physicist; b. N.Y.C., Dec. 5, 1932; s. Lewis and Bella (Rubin) G.; A.B., Cornell U., 1954; A.M., Harvard, 1955, Ph.D., 1958. NSF fellow U. Copenhagen, 1958-60; research fellow Cal. Inst. Tech., 1960-61; faculty Stanford, 1961-62, U. Cal., Berkeley, 1962-66; faculty Harvard, 1966—, prof., 1967—; cons. Brookhaven Nat. Lab., 1966—. Alfred P. Sloan fellow, 1962-66. Mem. Am. Phys. Soc. Research, numerous publs. in theory of elementary particles. Home: 84 Prescott St., Cambridge, Mass. 02138.*

GLASNER, Abraham, chemist; b. Cluj, Rumania, Mar. 3, 1910; s. Simon and Sarah (Hoffmann) G.; B.Sc. in Chemistry, U. London, 1935; Ph.D., Hebrew U., Jerusalem, Israel, 1940; m. Leah Shapira, Mar. 1937; children—Yael, Samuel, Simon. Faculty, Hebrew U., 1940—, prof. chemistry, 1967—. Vis. lectr. Princeton, 1961-62; lectr. analytical chemistry U. Tel-Aviv, 1954-56; sci. dir. Labs. of Security Office, Jerusalem, 1950-52. Mem. Israel, Am. chem. socs., Chem. Soc. (London), Sigma Xi. Author textbook on qualitative analytical chemistry, 1944, gen. and inorganic chemistry, 1948 (in Hebrew). Research and publs. on kinetics of concentrated solutions, peroxides of transition metals, thermal decomposition of perchlorates, rare earth oxalates, diffusion of metal ions in alkali halides, color centers and spectra of metal ions substituted in alkali halides, trace analysis by spectrophotometric methods, crystallization and preparation of highly pure potassium chloride. Home: 9, Aza Rd., Jerusalem, Israel.*

GLASOE, Gynther Norris, Am. physicist; b. Northfield, Minn., Oct. 29, 1902; s. Paul Maurice and Gena (Kirkwold) G.; B.A., St. Olaf Coll., 1924; M.A., U. Wis., 1926, Ph.D., 1930; m. Nora Boraas, Aug. 22, 1928; 1 son, Paul John. Instr. physics St. Olaf Coll., Northfield, Minn., 1928-29, U. Wis., 1930-31, Columbia, 1931-41; asso. group leader, staff mem. radiation lab. Mass. Inst. Tech., Cambridge, 1941-46; asso. prof. Rensselaer Poly. Inst., Troy, N.Y., 1946-47, prof., 1947-52; asso. chmn. physics dept. Brookhaven Nat. Lab., Upton, N.Y., 1952-62, asst. dir. tech. services, 1962—. Fellow Am. Phys. Soc.; mem. Am. Assn. Physics Tchrs. Research in photoelectricity, nuclear fission, neutron physics, radar physics, time of flight techniques. Co-editor, contbg. author Pulse Generators. Home: 29 Livingston Rd., Bellport, N.Y. 11713. Office: Brookhaven Nat. Lab., Upton, L.I., N.Y. 11973.*

GLASS, H(iram) Bentley, biologist; b. Laichowfu (now Yehsien), Shantung, China, Jan. 17, 1906 (parents Am. citizens); s. Wiley B. and Eunice (Taylor) G.; student Decatur (Tex.) Bapt. Coll., 1923-25; A.B, Baylor U., 1926, M.A., 1929, LL.D., 1958; Ph.D, U. Tex., 1932; NRC fellow U. Oslo, Norway, 1932-33, Kaiser-Whlhelm Inst. Biologie, Kaiser-Wilhelm Hirnforschung, Berlin, Germany, 1933, U. Mo., 1933-34; Sc.D., Washington Coll., 1957, Western Res. U., 1962, Cornell Coll., 1965, Western Maryland Coll., 1966; m. Suzanne G. Smith, Aug. 10, 1934; children—Lois Anne (Mrs. R. S. Edgar), Alan Bentley. Tchr. high sch., Timpson, Tex., 1926-28; teaching fellow Baylor U., 1928-29; instr. Stephens Coll., 1934-38; research asso. bur. ednl. research in sci. Columbia Tchrs. Coll., 1936-37; from asst. prof. biology to prof. Goucher Coll., Balt., 1938-47; asso. prof. biology Johns Hopkins, 1947-52, prof., 1952-65; acad. v.p., Distinguished prof. biology State University of New York at Stony Brook, 1965—. Mem. Internat. Genetics Congress, 1932, 48, 53, 58, 63; chmn. adv. com. for biology and medicine AEC, 1962-63, mem., 1956-63; mem. continuing com. Pugwash Confs. on Sci. and World Affairs, 1958-66; mem. Md. Gov.'s Adv. Com. on Nuclear Energy, 1959-65; ofcl. U. S. del. Internat. Union Biol. Socs., 1953, 55. Pres., Md. br. Am. Civil Liberties Union, 1955-65. Trustee Biol. Abstracts, 1954-60, pres., 1958-60; trustee Cold Spring Harbor Lab. for Quantitative Biology, 1965—, Eastern regional Inst. Edn., 1966—; president Fund for Overseas Research Grants and Education, 1966—. Fellow Am. Acad. Arts and Scis.; mem. Am. Inst. Biol. Scis. (pres. 1954-56; chmn. biol. sci. curriculum study 1959-65), Nat. Acad. Sci. (com. on genetic effect atomic radiation; chmn. com. sci. edn. Pacific Sci. Bd. 1967—), Am. Philos. Society (mem. council 1966—), A.A.A.S. (dir., 1959-66), Am. Soc. Zoologists, Am. Soc. Naturalists (pres. 1965), Am. Genetic Assn. (council 1952—), Genetics Soc. Am. (v.p. 1960), Soc. Study Evolution, also American Society of Human Genetics (president 1967), also the American Soc. Phys. Anthropology, Conf. Biol. Editors (chmn. 1957-59), History Sci. Soc., Eugenics Soc. (dir. 1958—), Am. Assn. U. Profs. (chmn. spl. com. on acad. freedom and tenure, pres. 1958-60), Phi Beta Kappa (senator 1963-66, pres. 1967—), Sigma Xi. Author: Genes and the Man, 1943; Science and Liberal Education, 1959; Science and Ethical Values, published in 1966. Editor: McCollum-Pratt Symposia vols. 1-9 (with W. D. McElroy), Forerunners of Darwin, 1959; Survey of Biological Progress, vol. 3-4. Editor Quar. Rev. Biology. Mem. editorial bd. Human Biology; Isis. Research on human genetics; genetics of Drosphila; suppressor genes; history of genetics; Rh blood types. Home: Box 65, East Setauket, N.Y. 11733. Office: State U. N.Y. at Stony Brook, Stony Brook, N.Y. 11790.

GLASSER, Arthur Charles, Am. pharm. chemist; b. Pitts., June 19, 1921; s. Herbert G. and Louise (Lindner) G.; B.S. in Pharmacy, Duquesne U., 1949; Ph.D. in Pharm. Chemistry, Ohio State U., 1953; m. Marjorie Ellen John, May 6, 1943; children—Stephan, Ellen, David. Faculty U. Ky., Lexington, 1953—, prof., chmn. dept. pharm. chemistry, 1959—, acting dean Coll. Pharmacy, 1964—. Mem. Am. Pharm. Assn., Am. Chem. Soc. Research and publs. on relationship between chem. structure organic sulfur containing molecules and their physiol. activity. Home: 951 Stonewall Rd., Lexington, Ky. 40504.*

GLASSER, Mervyn Lawrence, Am. physicist; b. Crookston, Minn., Oct. 5, 1933; s. Abraham N. and Estelle (Berger) G.; B.A., U. Chgo., 1953, M.S., 1955; Ph.D., Carnegie Inst. Tech., 1961; m. Judith Jay Sensibar, Aug. 30, 1956; children—Trina, Daniel, Joshua, Tanya. Faculty U. Miami, Coral Gables, Fla., 1956-58, U. Wis., 1963-64; physicist Battelle Meml. Inst., Columbus, O., 1962-63, sr. staff scientist, 1964—. Mem. Math. Assn. Am., Am. Phys. Soc., Sigma Xi. Research in physics of solids, effect of crystal structure on electronic properties of metals, magnetic properties of ferromagnetic insulators, math. of symmetry. Home: 475 Clinton Heights, Columbus 43202. Office: 505 King Av., Columbus, O. 43201.*

GLASSER, Otto, biophysicist; b. Saarbruecken, Germany, Sept. 2, 1895; s. Alexander and Lina (Gentsch) G.; Ph.D., U. Freiburg (Germany), 1919; m. Emmy von Ehrenberg, July 19, 1922; 1 dau., Hannelore. Came to U. S., 1922, naturalized, 1929. Instr. Radiol. Inst., U. Freiburg, 1919-21, U. Frankfurt (Germany), 1921-22; biophysicist Howard Kelly Hosp., Balt., 1922-23, Cleve. Clinic, 1923-25; asst. prof. biophysics N.Y. Postgrad. Med. Sch., Columbia, 1925-27; head dept. biophysics Cleve. Clinic Found., 1927-60, prof. biophysics Cleve. Clinic Ednl. Found., 1937-60, prof. emeritus, 1961-64. Cons. biophysicist Western Res. U. Hosp., Cleve., 1934-46; cons. radiology VA, Washington, 1946-55; mem. com. phys. medicine A.M.A., 1951-56, vice chmn. com. med. physics, 1956-62. Recipient Honor-Roentgen plaque Roentgen-Mus., Lennep, 1951, John Stanley Coulter plaque Am. Congress Phys. Medicine, 1953, Comdrs. Cross, Order of Merit Fed. Republic Germany, 1960, Otto Glasser award Ohio Radiol. Soc., 1962, also 9 awards for sci. exhibits, 1927-58. Diplomate Am. Bd. Radiology (examiner). Fellow Ohio Acad. Sci., A.A.A.S., Am. Phys. Soc., Am. Coll. Radiology; mem. Radiol. Soc. N.Am. (gold medal achievement award 1936), Am. Radium Soc. (Janeway medal 1950), Deutsche Roentgen Gesellschaft (hon.), Am. Roentgen Ray Soc., A.M.A., Sigma Xi. Author: Wilhelm Conrad Roentgen and the Early History of the Roentgen Rays, 1931; Dr. W. C. Roentgen, 1945. Editor: The Science of Radiology, 1933. Editor, collaborator: Physical Foundations of Radiology, 1944. Editor, compiler: Medical Physics, Vol. I, 1944, Vol. II, 1950, Vol. III, 1960. Contbr. numerous articles to sci. jours. Research on measurements radiant and atomic energies; inventor condenser dosimeter for measurement radiant energies. Died Dec. 11, 1964.

GLASSER, Richard Lee, Am. physiologist; b. Balt., Mar. 26, 1927; s. Abraham I. and Rose (Plaine) G.; A.B., Johns Hopkins, 1949; Ph.D. (J. F. B. Weaver fellow), U. Md., 1957; m. Florence Coplan, Aug. 8, 1950; children—Mardi Coplan, Scott Mitchell, Rebecca Ann. Health physicist Army Chem. Center, Md., 1949-52; faculty U. N.C., Chapel Hill, 1957—, asso. prof. physiology, 1965—; asst. prof. Duke, 1959-61. Mem. Am. Physiol. Soc., A.A.A.S., Am. Inst. Biol. Scis., N.Y. Acad. Scis., N.C. Acad. Sci. Contbr. articles to tech. jours. Discovered cardiovascular augmenter and depressor areas in brain stem; localized brain stem inhibitory system; research on brain stem facilitatory systems which regulate respiratory, cardiovascular and somatic postural activity, role of cerebellum in control respiration. Home: 503 Morgan Creek Rd., Chapel Hill, N.C. 27514.*

GLASZIOU, K. T., Australian biologist; b. Sydney, Australia, Jan. 16, 1923; s. Victor and Elsie (Gelding) G.; B.Sc., U. Sydney, 1954, M.Sc., 1955, Ph.D., 1957; m. Betty Burnett Mitchell, Nov. 20, 1945; children—John D., Douglas K., Paul P., Jan. M. Plant physiologist Colonial Sugar Refining Co., Sydney, 1956-60, head David North Plant Research Centre, Brisbane, Australia, 1960—. Mem. Australian, Am., Scandinavian socs. plant physiologists, Australian Biochem. Soc. Research, publs. on metabolic blockages in bacterial mutants; biochemistry of regulatory mechanisms of plant growth and particularly sugar cane plant. Home: 52 Meirs Rd., Indooroopilly, Qld., Australia. Office: 50 Meirs Rd., Indooroopilly, Qld., Australia.*

GLATT, Max Meier, psychiatrist; b. Berlin, Germany, Jan. 26, 1912; s. Chaim and Gilly (Joachimsmann) G.; M.D., U. Leipzig, 1937; D.P.M., Royal Coll. Physicians and Surgeons, London, 1950; m. Gizela Irom, Sept. 4, 1960; 1 son, Julian Charles. Clin. asst. U. Berlin, 1937; staff Jewish Hosp., Berlin, 1937-38; med. officer Cane Hill Hosp., Coulsdon, Eng., 1942-44; registrar St. Lawrence's Hosp., Caterham, Eng., 1944-51; sr. hosp. med. officer Warlingham Park Hosp., Warlingham, Eng., 1952-58; cons. psychiatrist

St. Bernard's Hosp., Southall, Eng., 1958—; cons. in charge Regional Alcoholism and Addiction Unit, N.W. Met. Regional Hosp. Bd., Southall, 1964—; hon. cons. physician dept. psychol. medicine U. Coll. Hosp., London, hon. lectr. Med. Sch., 1967—. Cons. psychiatrist to hosps.; mem. coms. WHO; exec. bd. Internat. Council on Alcohol and Alcoholism; vice chmn. Med. Council on Alcoholism; mem. bds., coms. for treatment of alcoholism. Fellow Am. Group Psychotherapy Assn., Royal Soc. Medicine; mem. Council of Soc. for Study of Addiction, Royal Medico Psychol. Assn., numerous others. Co-author: The Drug Scene in Gt. Britain, 1967; contbr. chpts. to books, numerous articles in profl. jours. Established 1st alcoholic unit in Great Britain, 1952, 1st drug dependence unit; research and promotion of group therapy and therapeutic community as treatment for alcoholism and drug dependence; studies of alcoholism as an illness. Home: 16 Southbourne Crescent, Finchley, London, N.W. 4, Eng. Office: St. Bernard's Hosp., Southall, Middlesex, Eng.*

GLATZEL, (James Moritz) Bruno, German physicist; b. Berlin, Sept. 24, 1878; s. Paul and Adelheid (Meyer) B.; studied math. and physics at U. Berlin, also Tech. U. Berlin for 3 semesters; Ph.D., U. Erlangen (Germany), 1901; later studied electronic engring. Electronic engr. firm Carl Flohr; made study trip to Eng. and Scotland; became lectr. Tech. U. Berlin, Mil. Tech. Acad., 1908; engr. firm Carl-Lorenz AG, also dir. Berlin Phototelegraphy Sta. Author: Erzeugung von Hochfrequenzenergie, 1913; Elektrische Methoden der Momentphotographie, 1915; also numerous articles. Editor: (with A. Korn) Handbuch der Phototelegraphie und Teleautographie, 1911. Research on phototelegraphy, ballistic high frequency cinematography, telegraphy with weakly attenuated waves, prodn. of high frequency currents, behavior of selenium cell and photo cells. Died Verdun, France, Oct. 8, 1914.

GLATZEL, Hans, physician; b. Göppingen, Germany, Aug. 22, 1902; s. Friedrich and Helen (Landerer) G.; student U. Tubingen (Germany), 1921-23, U. Konigsberg (Germany), 1924, U. Wien (Austria), 1924-25; Med.exam., U. Berlin (Germany), 1926; promotion U. Hamburg (Germany), 1926; inaugur., U. Gottingen (Germany), 1936; m. Marianne Handtmann, May 11, 1951; children—Regina Beate (Mrs. Rainer Schmidt-Ott), Enno Richard, Johann Christian, Sabine Désirée, Martina Juliane. Asst. med. clinic Heidelberg, Gottingen, 1927-38; registrar Kaiser-Wilhelm Inst. für Anthropologie, Berlin, 1927-38; registrar U. Kiel (Germany), 1938-46; chief internal dept. Diakonissenanstalt Felnsburg, 1946-47; practice medicine, Fensburg, Germany, 1946-57; dir. clin.-physiol. dept. Max-Planck-Inst. für Ernahrungsphysiologie, Dortmund, Germany, 1957—. Recipient Willmar-Schwabe prize, 1966. Author: Nahrung und Ernahrung, 1955; Krankenernahrung, 1953; also numerous articles. Research on mineral metabolism, psychosomatic investigation on ulcus pepticum and ischaemic heart diseases, radiol. and biochem. investigation on digestibility of different food, effect of low fat and high fat diets on man, alimentary influence on stroke volume, frequency and blood pressure. Home: 12 Duwelssiepen, Dortmund. Office: 201 Rheinlanddamm, Dortmund, Germany.*

GLAUBER, Johann Rudolph, German chemist; b. Karlstadt/Main, Germany, 1603 or 1604; s. Rudolf and Miss (Gertraut) G.; self-taught, apparently had some form of tng. in Salzburg, Paris, Basle, 1626-44; m. Rebecca Jacobs, 1636; no children; m. 2d, Helene Cornelis, 1641; 8 children. Dir., Furstliche Apothek, Giessen; built true chem. lab. in his house, Amsterdam, designed spl. furnaces and other equipment for it, 1648-50; dir. lab. in Kitzingen, 1651; returned to his lab., Amsterdam, 1655/56-68. Author numerous works, including De auri tinctura, 1646; Furni novi philosophici, 1646-50; Miraculum mundi, 1653; Miraculum mundi continuatio, 1657; Pharmacopoea spagyrica, 1654-68; Des Teutschlands-Wohlfahrt, 1656-61; Tractatus de natura salium, 1658; Novum lumen chimicum, 1664; collected works in German, 1658-59, French, 1695, English, 1689. One of 1st to have clear ideas concerning formation of salts from bases by action of acids, recognized that salts contain acid and base constituent; discovered "sal mirabile" (sodium sulphate or Glauber's salt) in water of a mineral spring, Vienna, 1625, introduced it 1st as cathartic, later as cure-all; prepared other chem. compounds and sold them for medicinal purposes; used anhydrous sodium sulfate to remove water from oils and mineral acids; gave clear descriptions of preparation of sulphuric, hydrochloric, nitric acids; related order in which mercury dissolves metals from their ores; concerned with indsl. uses of chemistry; prepared hydrochloric acid, fuming nitric acid, arsenic chloride, ammonium nitrate, chlorides, nitrates, sulfates, acetates of various metals, studied their properties; paid careful attention to crystal forms; described number of organic compounds, mostly in crude and impure form; recognized all mercury compounds are poisonous; described several qualitative analytic reactions; contbd. to knowledge of agr., medicine, paints, explosives, glass prodn., porcelain making; theoretical views strongly influenced by Paracelsus; improved and invented many kinds of stills and furnaces, used a chimney; understood that nation's natural resources could be exploited for nat. power and improved living conditions; opposed Germany's excessive exportation of raw materials to Vienna and France. Died Amsterdam, Netherlands, 1668/70.

GLAUBER, Roy J(ay), Am. theoretical physicist; b. N.Y.C., Sept. 1, 1925; s. Emanuel B. and Felicia (Fox) G.; B.S. summa cum laude, Harvard, 1946, M.A., 1947, Ph.D., 1949; mem. Inst. Advanced Study, 1949-51; research fellow, Swiss Fed. Poly. Inst., Zürich, 1950; m. Cynthia Marshall Rich, July 26, 1960; 1 son, Jeffrey Marshall. Mem. staff theoretical physics div. Los Alamos Sci. Lab., 1944-46; lectr. Cal. Inst. Tech., 1951-52; mem. faculty Harvard, 1952—, prof. physics, 1962—; vis. lectr. Ecole d' Été de Physique Théorique, Les Houches, France, 1954, 64, U. Cal. at Berkeley, 1955, 57, 63, U. Colo., 1958, 61, U. Wash. 1960, Brandeis U., 1961, U. Leningrad (USSR), 1964; Fulbright lectr., France, 1954; spl. research nuclear physics, quantum theory radiation, statis. mechanics. NRC predoctoral fellow, 1946-49; AEC postdoctoral fellow, 1949-50; Frank B. Jewett fellow, 1950-51; Guggenheim fellow, 1959; also NSF Sr. Postdoctoral fellow, 1966, 67. Mem. Am. Phys. Soc., Am. Acad. Arts and Scis., Phi Beta Kappa, Sigma Xi. Contbr. articles profl. jours. Bd. editors Jour. Math. Physics, 1961-63. Investigation of the theory of diffraction by molecules; theory of neutron scattering by crystals and by molecules; theory of radiative capture of orbital electrons by nuclei; theory of high energy collisions of elementary particles with nuclei; quantum theory of optical coherence and photon counting; quantum theory of electromagnetic amplification. Home: 221 Pleasant St., Arlington, Mass. 02174. Office: Lyman Lab. Physics, Harvard Univ., Cambridge, Mass. 02138.*

GLAUCOS OF CHIOS (Glaucos of Samos); made a stand of iron supporting a silver bowl for Alyttes of Lydia (reigned 617-560 B.C.); invented welding iron.

GLAUERT, Hermann, Brit. physicist; b. Sheffield, Eng., Oct. 4, 1892; s. Louis Glauert; ed. Trinity Coll., Cambridge, Eng.; M.A.; m. Muriel Baker, 1922; 2 sons, 1 dau. Fellow Trinity Coll., 1920-26; researcher in aerodynamics Royal Aircraft Establishment, Farnborough, Eng., from 1916, became prin. sci. officer. Fellow Royal Soc., 1931, Royal Aero. Soc. Author: Elements of Aerofoil and Airscrew Theory; also numerous papers. Died Aug. 4, 1934.

GLAVIANO, Vincent Valentino, Am. physiologist; b. Frankfort, N.Y., July 19, 1920; s. Salvatore and Josephine (Manzo) G.; B.S., City Coll. N.Y., 1950; Ph.D., Columbia, 1954; m. Eleanor Spargimino, July 18, 1943; children—Joan J., Vincent S. Faculty, Columbia, 1951-53, fellow, 1954-56; instr. Hunter Coll., N.Y.C., 1952-54; asst. prof. physiology U. Ill. Coll. Medicine, Chgo., 1956-60; asso. prof. physiology Loyola U. Sch. Medicine, Chgo., 1960-64, prof., 1964—. Cons., Cook County Hosp. Cardio Pulmonary Lab., Abbott Labs. Postdoctoral research fellow N.Y. Heart Assn., 1954-56; travel awards Nat. Acad. Scis., 1962, 65. Mem. Am. Physiol. Soc., Soc. Exptl. Biology and Medicine, Am., Chgo. heart assns., Harvey Soc., N.Y. Acad. Sci., Sigma Xi. Research, numerous publs. on heart function in hypotension, role of cardiac neurohormones in states of stress, mechanism of tranquilizer drugs on central nervous system, metabolism of heart. Home: 121 N. Charles Av., Villa Park, Ill. 60817.*

GLAZEBROOK, Sir Richard Tetley, English aero. physicist; b. West Derby, Liverpool, Eng., Sept. 18, 1854; s. Nicholas Smith Glazebrook; ed. Liverpool Coll., Trinity Coll., Cambridge; M.A., D.Sc., Oxford U.; LL.D., Edinburgh U.; Sc.D., Victoria U.; D.Sc., U. Heidelberg; m. Frances Gertrude Atkinson, 1883; 1 son, 3 daus. Prin., Univ. Coll., Liverpool, 1898-99; dir., Nat. Phys. Lab., 1899-1919, chmn. exec. com., 1925-32; chmn. Aero. Research Com., 1908-33; Zaharoff prof. aviation, dir. dept. aeros. Imperial Coll. Tech., 1920-23; Univ. lectr. math.; asst. dir. Cavendish Lab., Cambridge, Rede lectr., 1917. Recipient Hopkins prize Cambridge Philos. Soc., 1886, Royal medal, 1931; Gold medal Royal Aero. Soc., 1933, Albert medal Royal Soc. Arts. Fellow Royal Soc., 1882 (fgn. sec. 1926-29, Hughes medal 1909); mem. Instn. Civil Engrs. (hon.), Instn. Mech. Engrs. (life), Inst. Elec. Engrs. (hon., pres. 1906), Brit. Assn. (sec. com. on elec. standards). Author: Text-Book of Physical Optics; (with W. N. Shaw) Text-Book of Practical Physics; Laws and Properties of Matter; Clerk-Maxwell and Modern Physics; Science and Industry, 1917; textbooks on heat, light, mechanics, electricity; Cambridge Natural Science Manuals. Editor: Dictionary of Applied Physics, also numerous articles. Research on theory of light, expts. in elec. measurements and thermometry, wind tunnel expts. for aircraft devel., airplane engines; adopted scientific apparatus for military purposes and in aeronautical engring. Died Limpsfield Common, Eng., Dec. 15, 1935.

GLAZER, Harold, Am. mathematician, systems analyst; b. Phila., Apr. 29, 1929; s. Barney and Alice (Chazen) G.; A.B. in Math., Boston U., 1949, A.M. in Math., 1950, Ph.D., 1963; m. Estelle Rosen, Aug. 7, 1955; children—Jo-Anne B., Stephen W., Amy L. Applied mathematician Harvard Obs., 1953-55, cons., 1955-56; sr. engr. Sylvania Research Labs., Waltham, Mass., 1955-57; sr. engr. systems analysis Raytheon, Sudbury, Mass., 1958-61, mgr. computer systems and analysis, 1961; mem. tech. staff systems analysis Mitre Corp., Bedford, Mass., 1961-68; mgr. Datafacts Group, Viatron Computer Systems Corp., Burlington, Mass., 1968—. Instr. math. Boston U. evening div., 1958-59. Fellow A.A.A.S.; mem. Inst. Math. Statistics, Am. Statis. Assn., Inst. Mgmt. Sci., Operations Research Soc. Am., Assn. for Computing Machinery. Devel. methods for cost/effectiveness analysis, including computerized cost models; applications to mil. planning; system and subsystem evaluations, computerized techniques and modeling, data processing methods. Home: 9 Spring Lane, Framingham, Mass. 01701. Office: 105 Terrace Hall Av., Burlington, Mass. 01803.*

GLAZER, Nathan, Am. sociologist; b. N.Y.C., Feb. 25, 1923; s. Louis and Tillie (Zacharevich) G.; B.S.S., Coll. City N.Y., 1944; M.A., U. Pa., 1944; Ph.D., Columbia, 1962; m. Ruth Slotkin, Sept. 26, 1943 (div. 1958); children—Sarah, Sophie, Elizabeth; m. 2d, Sulochana Raghavan, Oct. 4, 1962. Mem. editorial staff Commentary mag., 1945-53, Doubleday-Anchor Books, 1954-55; Walgreen lectr. U. Chgo., 1955; mem. staff Communism in Am. Life project Fund for Republic, 1956-57; vis. lectr. U. Cal. at Berkeley, 1957-58, prof. sociology, 1963—; instr. Bennington Coll., 1958-59; vis. asso. prof. Smith Coll., 1959-60; fellow Joint Center Urban Studies, Harvard-Mass. Inst. Tech., 1960-61; study and travel in Japan, 1961-62; urban sociologist HHFA, Washington, 1962-63. Guggenheim fellow, 1954. Author: (with D. Riesman and R. Denney) The Lonely Crowd, 1950; (with D. Riesman) Faces in the Crowd, 1952; American Judaism, 1957; The Social Basis of American Communism, 1961; (with D. P. Moynihan) Beyond the Melting Pot (Anisfield-Wolf award Sat. Rev. 1964), 1963. Research on Am. ethnic groups, analysis of their devel. in and relationship to Am. soc., devel. of social policy and planning. Home: 1508 LaLoma St., Berkeley, Cal. 94708.*

GLAZKO, Anthony Joachim, Am. biochemist; b. San Francisco, Aug. 15, 1914; s. Joachim Nicholas and Josephine (Ruthkovska) G.; A.B., U. Cal. at Berkeley, 1935, Ph.D., 1939; m. Margaret Jean Hemans, Dec. 17, 1959; children—John Nicholas, Susan Jean, Mary Janet. Research asso. in pharmacology U. Mich., 1939-41; served with med. research unit USNR, 1941-45; asst. prof. biochemistry Emory U. Med. Sch., Atlanta, 1945-47; with research dept. Parke, Davis & Co., Detroit, 1947—, lab. dir. in chem. pharmacology, 1961—. Mem. Am. Chem. Soc., Am. Coll. Clin. Pharmacology and Chemotherapy, Am. Fedn. Clin. Research, Am. Soc. Pharmacology and Exptl. Therapeutics, Am. Physiol. Soc., N.Y. Acad. Sci., Soc. Exptl. Biology and Medicine, Sigma Xi. Contbr. to Remington's Pharmaceutical Sciences, 1965, Shirkey's Pediatric Therapy, 1966; also numerous articles. Research in fields of absorption, tissue distbn., excretion and metabolic disposition of new drugs, species differences in metabolism, pediatric dosage, devel. of new analytical procedures based on colorimetry, fluorescence, gas chromatography, radio-isotope techniques. Home: 1245 Fair Oaks Pkwy., Ann Arbor, Mich. 48104. Office: Research Labs., Parke, Davis & Co., Ann Arbor, Mich. 48106.*

GLAZUNOV, Mikhail Fedorovich, Russian path. anatomist, oncologist; b. 1896; grad. Petrograd. Mil. Med. Acad., postgrad., until 1923; D.Med. Sci., 1935. Instr. path. anatomy dept. Leningrad Mil. Med. Acad., 1924-29, dir. chair path. anatomy, 1944—; dir. path. dept. Inst. Oncology, USSR Acad. Sci., 1929-41, asso., head path. and morphological lab., 1944—. Decorated Order of Lenin. Mem. USSR Acad. Med. Sci., All-Union Soc. Path. Anatomists (dep. chmn. presidium). Author: Instructions on Pathological Anatomical Service; Malignant Tumors, 1947; Tumors of the Ovaries, 1954, 61. Mem. editorial bd. Problems of Oncology, also many-vol. handbook on path. anatomy; mem. editorial council Pathology Archives; co-editor pathology and morphology sects. Large Med. Ency., 2d edit. Research and numerous publs. on nature of malaria pigment, morphology and histogenesis of various tumors. Address: Leningrad Mi-. Med. Acad., ulitsa Lebedeva 6, Moscow, USSR.

GLEASON, Andrew Mattei, Am. mathematician; b. Fresno, Cal., Nov. 4, 1921; s. Henry Allan and Theodalinda (Mattei) G.; B.S. Yale, 1942; postgrad. (Soc. fellows), Harvard, 1946-50, M.A. (hon.), 1953; m. Jean Berko, Jan. 26, 1959; children—Katherine Anne, Pamela, Cynthia. Asst. prof. Harvard, 1950-53, asso. prof., 1953-57, prof., 1957—. Served to lt. comdr. USNR, 1942-46, 1950-52. Recipient, Newcomb Cleve. prize, A.A.A.S., 1951. Mem. Nat. Acad. Sci., Am. Math. Soc., Math. Assn. Am., Soc. Math. de France, Am. Acad. Arts and Sci. Club: Cosmos (Washington). Contbd. to solution of Hilbeart's fifth problem; research and publs. on finite projective planes, founds. of quantum mechanics, function algebras, reform of elementary and secondary math. curriculum. Home: 110 Larchwood Dr., Cambridge, Mass. 02138.*

GLEASON, Elliott Perry, Am. inventor; b. Westmoreland, N.H., June 27, 1821; common school edn.; a pioneer in manufacture of gas burners in early stages of industry; inventor regulating Argand burner and many other standard devices; also identified with devel. of electric lighting; pres., and prin. owner E. P. Gleason Mfg. Co., N.Y., mfrs. gas and electric lighting appliances, also Gleason & Bailey Mfg. Co., Seneca Falls, N.Y., mfrs. fire dept. rolling stock. Died 1901.

GLEDHILL, John Alan, physicist; b. Littleborough, Eng., Jan. 1, 1920; s. Albert and Winifred (Harrison) G.; B.Sc., Rhodes U. Coll., 1938, M.Sc., 1942, Ph.D., 1947; Ph.D., Yale, 1949; m. Eily E. A. Archibald, Apr. 25, 1953; children—Irvy. Faculty, Rhodes U. Coll., Grahamtown, South Africa, 1943——, prof., head dept. physics, 1954——. Nuffield fellow, 1955; sr. postdoctoral research asso. Nat. Acad. Scis.-NASA, 1966-67. Fellow Inst. Physics (London); mem. Faraday Soc., South African Inst. Physics, Sigma Xi. Research, publs. on determination of solubilities of sparingly soluble salts in water by conductimetry, theory of ionospheric layer formation, effect of solar eclipses on ionosphere, interaction between radiation belts and ionosphere, Jupiter's magnetosphere and radio emission. Home: 18 Hare St., Grahamstown, South Africa.*

GLEDITSCH, John Gottleib, German botanist; b. Leipzig, Germany, Feb. 5, 1714; M.D., Frankfort/Oder, 1740; lectr. botany, physiology, materia medica, Frankfurt/Oder; prof. anatomy, dir. bot. garden Berlin Acad. Sci. Author: Systematische Einleitung in der neuere Forstwissenschaft, 1774-75; Vermischte physikalisch-botanisch-ökonomische Abhandlungen, 1765-66; Methodus fungorum, 1753; Systema plantarum a staminum situ, 1764. Mem. Berlin Acad. Sci. Died Berlin, Oct. 5, 1786.

GLEICHEN, Alexander Wilhelm, German physicist; b. Niederschöneweide, Germany, Sept. 23, 1862; studied math. and natural scis., Berlin; Ph.D., Kiel, Germany, 1889. Tchr., Friedrich Wilhelm Gymnasium, 1887-97; tchr. math. Helene Lange's 2d sch. courses for women; tchr. Kaiser Wilhelm Realgymnasium, 1897-1904; joined faculty Technische U. Berlin, 1902; tester patent applications in class 42h (instrumental optics) Reich Patent Office, 1904-18; joined optical co. C. P. Goerz, Berlin-Friedenau, 1919; sci. tech. editor Centralblatt für Optik und Mechanik, 1915-19; founder German Sch. for Optics and Phototechnics, Berlin, 1922. Mem. London Optical Soc. Author: Lehrbuch der geometrischen Optik, 1902; Einführung in die medizinische Optik, 1904; Vorlesungen über photographischen Optik, 1905; Leitfaden der praktischen Optik, 1906; Die Grundgesetze der naturgetreuen photographischen Abbildung, 1910; Die Theorie der modernen optischen Instrumente, 1911; Die Optik in der Photographie, 1911; Grundriss der photographischen optik auf physiologischer Grundlage mit elementarmathematischer Begründung, 1913; (with E. Klein) Schule der Optik; Übungsaufgaben aus der geometrischen Optik, 1917; also numerous articles. Research on geometric optics, correction of optical image imperfections, requirements of true photog. image, ophthal. optics. Died Berlin, Oct. 21, 1923.

GLEICHEN, Friedrich Wilhelm von, see von Gleichen, Friedrich Wilhelm.

GLEISBERG, Walther, German botanist; b. Wroclaw, Breslau, Mar. 29, 1891; s. Maximilian and Anna (Lange) G.; ed. univs. Breslau, Lena, Zurich; Ph.D. in botany; m. Charlotte Baude, 1918; children—Johann George, Christian Friedrich. Asst. scientist Bot. and Pomol. Inst., Proskau, 1919; prof., dir. Hort. Inst., Pillnitz, 1928-33, Inst. Fruit Culture, Vineyard and Horticulture, 1933-39, Inst. Cult. Horticulture, U. Posen, 1939-45; prof. emeritus Faculty Math. and Physics, Hamburg. Mem. Soc. Botany, Soc. Applied Botany, Soc. Horticulture, Soc. Agrl. Sci. Editor, founder: Die Gartenbauwissenschaft. Research and publs. on botany and applied botany. Address: la Flurweg, Braunlage-Harz, Germany.

GLEISS, Jörn, German pediatrician; b. Bremen, Germany, May 25, 1919; s. Hans-Werner and Ria (Lucker) G.; student medicine, U. Leipzig, U. Greifswald (Germany), State Bd., U. Hamburg (Germany), 1944, Promotion, 1944. Staff, Children's Clinic, Med. Sch., 1946——; with U. Duesseldorf (Germany), asst. prof., 1963——. Author: Thalidomide Embryopathie; Preventive Pediatrics; also numerous articles. Research on newborn, mature and immature infant (neonatology), hematological studies in infants, problems of preventive pediatrics, including infant mortality, feeding of prematures, malformations, pharmacotoxic studies. Home: 404 Heine-13, Neuss, West Germany. Office: 4000 Mooren-5, Düsseldorf, West Germany.*

GLEISSBERG, Wolfgang, German astronomer; b. Breslau, Dec. 26, 1903; s. Carl and Erna (Wollstein) G.; ed. U. Berlin, U. Breslau; Ph.D. in astronomy; m. Charlotte Michael, 1934; 1 dau., Ingrid. Asst., Breslau Obs., 1927-33; sci. asso. Istanbul Obs., 1934-47, dir., 1948-58; hon. prof. U. Frankfurt am Main, 1958——; dir. Astron. Inst., 1960——. Mem. Turkish Soc. Astronomy (hon. pres.), Internat. Astron. Union, Astronomische Gesellschaft, Ver. der Sternfreunde, Physikalischer Verein zu Frankfurt. Author: Die Häufigkeit der Sonnenflecken, 1952, also numerous articles. Home: 12 Buchenweg, 6375 Oberstedten. Office: 23 Senckenberg-Anlage, 6 Frankfurt am Main, Germany.

GLEMSER, Oskar Max, German chemist; b. Stuttgart, Germany, Nov. 12, 1911; s. Karl and Amalie (Gogel) G.; Dipl. Ing., Tech. Hochschule, Stuttgart, 1934, Dr. Ing., 1935; m. Ida-Marie Glemser, May 21, 1938; children: Christel, Rainer. Dozent, Techn. Hochschule, Aachen, 1941; full prof., U. Göttingen, since 1952; dir., Institute of Inorganic Chemistry, U. Göttingen. Pres., Acad. of Scis., Göttingen; mem., Leopoldina, Halle; Ges. Deutscher chem.; Bunsengesellschaft; Am. Chem. Soc.; Am. Mineralogical Soc. Published 235 articles. Research in inorganic and organic fluorine chemistry; hydroxides, hydrates, oxides, isopolianions. Home: 10 Ricard-Zsigmondy-Weg, Göttingen, Germany. Office: 8-9, Hospitalstrasse, Göttingen, Germany.

GLEN, Robert, entomologist; b. Paisley, Scotland, June 20, 1905; s. James Allison and Jeanie Blackwood (Barr) G.; B.Sc., U. Sask., 1929, M.Sc., 1931, LL.D., 1959; Ph.D., U. Minn., 1940; D.Sc., U. Ottawa, 1960; m. Margaret Helen Cameron, June 30, 1931; children—Robert Cameron, Ian Robert. Econ. entomologist Dominion Entomol. Lab., Saskatoon, Sask., 1928-45, charge wireworm investigations, Prairie Provinces, 1936-45; research coordinator, entomology div. Science Service, Can. Dept. Agr., Ottawa, Ont., 1945-50, chief entomology div., 1950-57, asso. dir. Science Service, 1957-59; dir.-gen. research br. Can. Dept. Agr., 1959-62, asst. dep. minister, 1962——. Mem. entomol. research panel Def. Research Bd., 1947-55, chmn., 1955-59; adv. com. faculty sci. Ottawa U., 1961——. Recipient Outstanding Achievement award U. Minn., 1960; Outstanding Achievement award Entomol. Soc. Can., 1964; certificate of merit Entomol. Soc. Am., 1964. Fellow Agrl. Inst. Can. (pres. Eastern Ont. br. 1950), Royal Soc. Can. (mem. council 1961-62); mem. Entomol. Soc. Can. (pres. 1957), Entomol. Soc. Am. (pres. 1962), Entomol Soc. Ont., Agrl. Econ. Research Council (vice president of governing bd. 1962-66), Science Council of Canada, Ontario Society Profl. Agrologists, Profl. Inst. Can., Sigma Xi. Editorial bd. Ann. Review Entomology, 1955-60, Canadian Entomologist, 1959-62. Provided characters for identifying some 30 pest species of wireworms occurring in Canada and methods for assessing their numbers, injury to crops, and economic importance; contributed to the organization and development of agricultural research in Canada and procedures for its management. Home: 832 Riddell Av. Office: Confederation Bldg., Ottawa, Can.*

GLENN, Frank, Am. surgeon; b. Marissa, Ill., Aug. 7, 1901; s. Charles and Minnie (McMurdo) G.; M.D., Washington U., St. Louis, 1927; m. Esther Wheelwright Child, Jan. 1938; children—Gardner, Prudence (Mrs. Donald Harris), Frank. Practice medicine, specializing in surgery, N.Y.C., 1932——; mem. staff N.Y. Hosp., 1932——, surgeon-in-chief, 1947——; faculty Med. Coll. Cornell U., 1932——, Lewis Atterbury Stimson prof. surgery, 1947——; surg. cons. Sch. Aerospace Medicine USAF. Bd. dirs. Medico. Diplomate Am. Bd. Surgery. Mem. A.C.S. (past pres.), Soc. Med. Cons. to Armed Forces (past pres.), N.Y. Acad. Medicine (past pres.), Am. Assn. Surgery Trauma, Am. Assn. Thoracic Surgery, Am. Geriatrics Soc. (past pres.), Am. Heart Assn. (past dir.), A.M.A., Am., So. surg. assns., Gerontol. Research Found., Harvey Soc., Internat. Soc. Surgery, Pan Am. Med. Assn., Soc. Clin. Research, Soc. Exptl. Biology and Medicine, Soc. U. Surgeons (past pres.), Alpha Omega Alpha. Author: (with Harold J. Stewart) Mitral Valvulotomy, 1959; Surgery in the Aged, 1960; (with George E. Wantz, Jr.) Problems in Surgery, 1961; Atlas of Biliary Tract Surgery, 1963. Research and publs. on hypertension, cardiovascular surgery, chemotherapy, surgery of gastrointestinal tract, hernia, trauma, med. edn. Home: 809 Wolf's Lane, Pelham Manor, N.Y. 10803. Office: 525 E. 68th St., N.Y.C. 10021.*

GLENN, James Francis, Am. physician; b. Lexington, Ky., May 10, 1928; s. Cambridge Francis and Martha (Morrow) G.; student U. Ky., 1947; B.A., U. Rochester, 1949; M.D., Duke, 1952; m. Gale Brooke Morrison, Dec. 29, 1948; children—Cambridge Francis, Sara Brooke, Nancy Carrick, James Morrison Woodworth. House officer gen. surgery Peter Bent Brigham Hosp., 1952-54; asst. prof. urology Yale U. Sch. Medicine, 1959-61; asso. prof. urology Bowman Gray Sch. Medicine, 1961-63; prof. urology, chmn. dept. Duke U. Med. Center, 1963——. Cons. urology Durham VA, Watts, Lincoln hosps., all Durham, N.C.; sci. adv. bd. Nat. Kidney Found. Mem. A.C.S., Am. Urol. Assn., Internat. Urol. Soc., A.M.A., Soc. Pediatric Urology, Am. Assn. Genitourinary Surgeons, Soc. Pelvic Surgeons, Sigma Xi, Alpha Omega Alpha. Author: Diagnostic Urology, 1964. Research and publs. in kidney disorders causing hypertensive vascular disease, diseases of the adrenal glands, congenital malformations of the urinary tract, combined treatments of cancer. Home: 27 Oak Dr., Forest Hills, Durham 27707. Office: Box 3707 Duke Hosp., Durham, N.C. 27706.*

GLENN, Leonidas Chalmers, Am. geologist; b. Crowder's Creek, N.C., Sept. 9, 1871; s. William Davis and Sarah P. (Torrence) G.; A.B., U. S.C., 1891; postgrad. Harvard, 1895; Ph.D., Johns Hopkins, 1899; m. Nellie Louise McCullough, Sept. 12, 1900; children—William David, Hugh Wilson. Taught in secondary schs., 1891-94; supt. town schs., Darlington, 1894-96; adj. prof. biology and geology, S.C. Coll., Columbia, 1899-1900; adj. prof. geology Vanderbilt U., 1900-03, prof., 1903-42, head div. of sci., 1928-42, emeritus from 1942. With N.C., Ky., Tenn., U. S. geol. surveys and U. S. Forest Service; mem. faculty George Peabody Coll. for Tchrs., 1914-15; oil geologist Sinclair Oil Co., 1916-17, 18; acting state geologist of Tenn., 1918; spl. agt. U. S. Internal Revenue Dept. oil and gas valuation works, 1918-19; investigated changes in Red River, Tex. Okla. boundary, for U. S. Dept. Justice, 1919-21; studied mining for oil and oil shale industry in Europe, 1923; mapping W. Ky. coal field for Ky. Geol. Survey since 1924; cons. geologist TVA, also U. S. Army Engrs. Fellow A.A.A.S.; mem. Geol. Soc. Washington, Am. Inst. Mining and Metall. Engrs., Geol. Soc. Am., Scis. Soc. Am., Tenn., Ky. acads. sci., Phi Beta Kappa, Sigma Xi. Contbr. to sci. periodicals. Died Jan. 11, 1951.

GLENN, William Wallace Lumpkin, Am. surgeon; b. Asheville, N.C., Aug. 12, 1914; s. Eugene Byron and Elizabeth Elliot (Lumpkin) G.; B.S., U. S.C., 1934; M.D., Jefferson Med. Coll., 1938; M.A. (hon.), Yale, 1962; m. Amory Potter, May 15, 1943; children—William Lumpkin, Elizabeth Amory. Asst. physiology Harvard Sch. Pub. Health, 1941-43; asso. surgery Jefferson Med. Coll., 1946-48; mem. faculty Yale Med. Sch., 1948——, prof. surgery, 1962——, chief cardiovascular surgery, dir. surg. labs., 1948——; attending surgeon Grace-New Haven Hosp.; cons. VA Hosp., W. Haven, Derby Hosp., Ansonia and Meridan (Conn.) Hosp. Cons. Surgeon Gen. Com. Environmental Medicine, 1962-64; mem. com. cardiovascular systems NRC, 1955-56. Mem. Am. Heart Assn. (chmn. council cardiovascular surgery 1960-62, v.p. 1961-64, Merit award 1966), Conn. Soc. Med. Research (pres. 1962), Internat. Surg. Group (pres. 1964), Am. Surg. Assn., A.C.S. (chmn. Conn. adv. com. 1963——), Am. Assn. Thoracic Surgery, Vascular Surg. Soc., Halstead Soc., Société Chirurgie Internat. (treas. 1967). Co-author: Thoracic and Cardiovascular Surgery, 1962. Contbr. profl. jours. Cons. editor: Cardiac Pacemakers, 1964. Developer operation to bypass right side of heart (caval-pulmonary artery shunt), developer radio frequency pacemaker. Home: 685 Forest Rd., New Haven 06515.*

GLENNER, George Geiger, Am. physician; b. Bklyn., Sept. 17, 1927; s. Francis Richard and Jennie (Geiger) G.; A.B., Johns Hopkins U., 1949, M.D., 1953; m. Joyce M. Saunders, June 26, 1954; children—Jonathan, Amanda. Pathologist, Harvard Sch. Legal Medicine, Boston, 1955; research pathologist NIH, Bethesda, Md., 1955-58, chief sect. histochemistry, 1959——; asst. pathologist Johns Hopkins Hosp., Balt., 1957-58; asso. prof. pathology Georgetown U. Sch. Medicine, Washington, 1965——. Mem. Am. Assn. Pathologists and Bacteriologists, Am. Soc. Exptl. Pathologists, Am. Soc. Cell Biology. Research and numerous articles on enzyme activity in breakdown of tissues in pathologic conditions, study of enzyme kinetics in histochemical systems; discovered norepinephrine in tumors of carotid body; definition of ultrastructure of a pathologic protein material, amyloid. Home: 6209 Lone Oak Dr., Bethesda 20034. Office: NIH, Bethesda, Md. 20014.*

GLEY, Émile, French biologist; b. Epinal, France, 1857; aggregate prof.; named prof. gen. biology Coll. de France, 1908; mem. French Acad. Medicine. Author: Essai de philosophie et d'histoire de la biologie, 1900; (with M. Duval) Traité élémentaire de physiologie, 1909. Described parathyroid glands and showed their essential function, 1891; research on nervous system, physiology of glands, psychol. physiology, hemoglobin, physiology of liver. Died 1930.

GLICK, David, Am. histochemist; b. Homestead, Pa., May 3, 1908; s. Max and Anne (Lasday) G.; B.S., U. Pitts., 1929, Ph.D., 1932; m. Ruth Mueller, Sept. 16, 1929; children—David, Peter; m. 2d, Irena Ross, Sept. 2, 1945; children—Jonathon Michael, Jeffrey Alan. Hernsheim fellow Mt. Sinai Hosp., N.Y.C., 1932-34; head chemistry dept. Mt. Zion Hosp., San Francisco, 1934-36; Rockefeller fellow Carlsberg Lab., Copenhagen, Denmark, 1937; head chemistry Beth Israel Hosp., Newark, 1937-42; chemist Office Sci. Research and Devel., Mt. Sinai Hosp., N.Y.C., 1942-43; head research Russell Miller Co., Mpls., 1943-46; cons. U. S. Army, 1945-46; asso. prof. biochemistry U. Minn., Mpls., 1946-50, prof., 1950-61; Commonwealth fellow European Labs., 1949, 58-59; prof. pathology Stanford, Palo Alto, Cal., 1961——. Recipient Phillips Medal, U. Pitts., 1929; Career award USPHS, 1962. Diplomate Am. Bd. Clin. Chemistry. Mem. Histochem. Soc. (v.p. 1950, pres. 1951), Am. Chem. Soc., A.A.A.S., Am. Soc. Biol. Chemists, Am., Internat. socs. cell biology, Internat. Histochem. Soc., Am. Assn. U. Profs., Roy Microscopal Soc., Royal Danish Acad. Sci. and Letters. Author: Techniques of Histo and Cytochemistry, 1949; Quantitative Chemical Techniques of Histo- and Cytochemistry, 2 vols., 1961, 63; Black and White and Other Poems, 1946; also articles. Editor: Methods of Biochemical Analysis; editorial bd. Jour. Histochem. and Cytochem., 1953-56, 65——. Research in quantitative histochemistry by elaboration of new techniques and methods of analysis by microchem., microphys., microbiol. means. Home: 680 Foothill Rd., Stanford, Cal. 94305.*

GLICKMAN, Irving, Am. oral pathologist; b. N.Y.C., Jan. 17, 1914; s. Nathan and Rose (Gurland) G.; B.S., Bklyn. Coll., 1933; D.M.D., Tufts Coll., 1938, postgrad.; m. Violeta Arboleda, Mar. 13, 1954; children—Alan, Denise. Faculty, Tufts U., Boston, 1938——, research prof. oral pathology, prof., chmn. dept. periodontology Sch. Dental Medicine, 1960——. Cons., mem. adv. groups to govt. agys., hosps.; mem. cons. staff hosps.; dir. Berkshire Conf. in Periodontology and Oral Pathology, 1950—. Recipient Distinguished Service award U. S. Jr. C. of C., 1949; Distinguished Service award Tufts U. Alumni Assn., 1955; Samuel Charles Miller Meml. award in oral medicine, 1965; Basic Research in Periodontology award Internat. Assn. Dental Research, 1966. Diplomate Am. Bd. Periodontology, Am. Bd. Oral Medicine, Fellow Am. Acad. Dental Sci., A.A.A.S., Am. Coll. Dentists, Internat. Coll. Dentists; mem. Am. Acad. Periodontology, Am. Acad. Dental Medicine, Am. Assn. Anatomists, Internat. Assn. Dental Research, Internat. Acad. Oral Pathology, numerous other Am., fgn. socs., Sigma Xi, Omicron Kappa Upsilon. Author: Clinical Periodontology, 1964; contbg. author others; also numerous articles. Research on causes and treatment of pyorrhea, including effects of diabetes, vitamin C deficiency, vitamin A deficiency, estrogen deficiency. Home: 24 Manor House Rd., Newton Center, Mass. Office: 483 Beacon St., Boston.*

GLIDDEN, Joseph Farwell, Am. inventor; b. Charleston, N.H., Jan. 18, 1813; s. David and Polly (Hurd) G.; attended sem., Lima, N.Y.; m. Clarissa Foster, 1837; m. 2d, Lucinda Warne, 1851; at least 1 child. Worked way West from N.H. as thresher, 1842-44; purchased land in De Kalb County, Ill., also a cattle ranch in Tex.; sheriff of De Kalb County, 1852-53; patented improvement on barb wire fences, 1874; organized (with Isaac L. Ellwood) Barb Wire Fence Co., De Kalb County, 1875; sold his half interest to Washburn & Moen Mfg. Co., 1876. Died Oct. 9, 1906.

GLINOS, Andre Dimitri, biomed. scientist; b. Athens, Greece, Oct. 18, 1919; s. Dimitri Alexander and Anna (Chronis) G.; M.D., U. Athens, 1941; m. Loni Anna Wallschuetzky, Dec. 28, 1959. Came to U. S., 1946, naturalized, 1952. Research fellow, mem. faculties U. Paris, Harvard, Johns Hopkins, 1945-53; chief growth physiology sect. Walter Reed Army Inst. Research, 1953-61, dept. cellular physiology, 1961—; research prof. U. Md., 1964—; research collaborator Brookhaven Nat. Lab., 1964-66. Mem. Am. Physiol. Soc., Am. Gerontol. Soc., Am. Assn. Cancer Research, Tissue Culture Assn., N.Y. Acad. Scis., Internat., Am. socs. cell biology. Research publs. on cellular growth control mechanisms, liver regeneration, carcinogenesis, tissue culture, cellular effects of radiation, physiopathology of aging, methodology of med. and biol. research. Home: 4977 Battery Lane, Bethesda, Md. 20014. Office: Walter Reed Army Inst. Research, Washington 20012.*

GLISSON, Francis, English physician; b. Dorsetshire, Eng., 1597; s. William Glisson; B.A., Cambridge, (Eng.) U., 1621, M.A., 1624, M.D., 1634; incorporated M.A. Oxford, 1627. Reguis prof. physics Cambridge U., 1636-77, named anatomy reader, 1639. Fellow Royal Soc. (charter), 1663, Coll. Physicians (pres. 1667-69). Author: Tractatus de rachitide (one of 1st English med. monographs), 1650; Anatomie hepatitis, 1654; Tractatus de ventriculo et intestinis, 1677; De natura substantiae energetica, 1672. Described rickets (Glisson's disease); original description of infantile rickets which he assumed was caused by poor nutrition; 1st description liver stroma (sheath of connective tissue accompanying the vessels and bile ducts known as Glissons Capsule), 1654; introduced concept of irritability (as opposed to sensibility) as a property of living tissue, circa 1675; 1st to prove muscles contract when brought into action; distinguished between direct tissue excitability and tissue excitability transmitted to brain by nerves. Died Oct. 16, 1677.

GLOCK, Waldo Sumner, Am. dendroclimatologist; b. Vinton, Ia., Aug. 20, 1897; s. John and Lena (Biebescheimer) G.; B.A., U. Ia., 1922; Ph.D., Yale, 1924; m. Betty Wellman, Mar. 14, 1924; 1 son, Waldo Sumner. Instr. Ohio State U., 1924-31; mem. research staff div. plant biology Carnegie Inst. Washington, 1931-38; mem. faculty Tex. Technol. coll., 1938-42, 45-48; prof., dir. tree-ring research lab. Macalester Coll., St. Paul, 1948-66, chmn. dept. geology, 1948-64; vis. prof., dir. tree-ring research lab. U. N.C., 1966—. also lectr. and cons. Fellow A.A.A.S., Geol. Soc. Am.; mem. Ecol. Soc. Am., Am. Geophs. Union (life), Internat. Soc. Biometeorology, Phi Beta Kappa, Sigma Xi. Author: Principles and Methods of Tree-ring Analysis, 1937; Tree Growth and Rainfall, 1950; Classification and Multiplicity of Growth Layers in Branches of Trees, 1960; (with Paul J. Germann, Sharlene Agerter) Uniformity Among Growth Layers of Three Ponderosa Pine, 1963; (with Sharlene Agerter) Annotated Bibliography of Tree Growth and Growth Rings-1950-62, 1966. Research and publs. on growth patterns and environmental influences on growth of forest trees; use of algae and tree growth as climatic indicators.*

GLOCKER, Richard, German physicist; b. Calw, Germany, Sept. 21, 1890; s. Richard and Helen (Maler) G.; student U. Berlin (Germany), 1909, Technische U. Stuttgart (Germany), 1910; Dr.phil., U. Munich (Germany), 1914; Dr.med. h.c. U. Tübingen (Germany), 1955; m. Elisabeth Stribeck, Sept. 24, 1925; 1 dau., Renate (Mrs. Leins). Faculty, Technische U. Stuttgart, 1919—, prof. X-ray tech., 1923-60, dir. Röntgeninstitut, 1920-60; dir. inst. für metallphysik Max-Planck-Institut für Metallforschung, Stuttgart, 1934-60. Hon. mem. Deutsche Röntgengesellschaft, Deutsche Gesellschaft für Biophysik, Deutsche Gesellschaft für Metallkunde. Author: Materialprüfung mit Röntgenstrahlen, 1957; (with Macherauch) Röntgen- und Kernphysik für Mediziner und Biophysiker, 1965; also numerous articles. Home: Stuttgart-N, Robert Bochstr. 10, Germany.*

GLOERSEN, Per, Am. physicist; b. Washington, Pa., Dec. 19, 1927; s. Frede and Froydis (Lund) G.; M.A., Johns Hopkins, 1952, Ph.D., 1956; m. Barbara Ann Bentley, July 4, 1953; children—William Bentley, Peter Frede, Kathryn Ann. Research asso. Johns Hopkins, Silver Spring, Md., 1952; physicist U. S. Army Ballistic Research Lab., Aberdeen Proving Ground, Md., 1952-54; with Gen. Electric Co., 1956—, project leader exptl. plasma physics Space Scis. Lab., Phila., 1959-63, group leader physics missile and space div. Space Scis. Lab., King of Prussia, Pa., 1963—. Mem. Am. Phys. Soc., Am. Inst. Aeros. and Astronautics (mem. tech. com. on electric propulsion 1965-67). Contbr. articles to tech. jours. Research in atomic, molecular and plasma physics with emphasis on spectrometric investigations. Home: 93 Flamehill Rd., Levittown, Pa. 19056. Office: P.O. Box 8555, Phila. 19101.*

GLOGER, Constantine Wilhelm Lambert, German zoologist, ornithologist; b. Kasischka/Brottkau, Germany, Sept. 17, 1803; s. Franz and Johanna (Klar) G.; studied natural scis. in Berlin, 1824-25; Ph.D., Breslau (now Wroclaw, Poland), 1830. Schooltchr., Breslau; moved to Berlin, 1843; asst. editorial bd. Jour. für Ornithologie; received 3 year grant from Prussian Ministry of Agr. to work out bird protection law. Mem. Leopoldina. Author: Wirbeltierfauna von Schlesien, 1833; Das Abädern der Vögel durch Einfluss des Klimas, 1833; (with H. Lichtenstein) Vollständiges Handbuch der Naturgeschichte der Vögel Europas, part I, 1834; many articles. Research on purposefulness in animal life, geog. variation and climatic modifications, bird conservation and protection. Died Berlin, Dec. 30, 1863.

GLOOR, Pierre, physician, physiologist; b. Basel, Switzerland, Apr. 5, 1923; s. Fritz and Marie (Meier) G.; B.A., Humanistisches Gymnasium Basel, 1942; postgrad. U. Lausanne (Switzerland); M.D., U. Basel, 1949; Ph.D., McGill U., Montreal, Que., Can., 1957; m. Luba Genush, Sept. 17, 1954; children—Irene Mary, Daniel Victor. Faculty, McGill U., 1954—, asso. prof. clin. neurophysiology, 1962—, chief Lab. Electroencephalography and Clin. Neurophysiology, Montreal Neurol. Inst., 1961—, Chief Lab. Neurophysiology, 1966—. Recipient Robert Bing prize Swiss Acad. Med. Scis., 1962. Mem. Canadian Physiol. Soc., Canadian Neurol. Soc., Canadian Soc. Electroencephalographers, Am. Electroencephalograph Soc., Am. Epilepsy Soc., Eastern Assn. Electroencephalographers, N.Y. Acad. Scis. Contbg. author: Hypothalamo-hypophyseal Interrelationships, 1956; Handbook of Physiology, Vol. III, 1960; Les grandes activités du Rhinencéphale Masson, 1960; Physiologie und Pathophysiologie des vegetativen Nervensystems, 1963; also numerous articles. Research on problems pressure cerebro-spinal fluid, organization limbic system, its electrophysiology, its anatomy in human brain, role of amygdala and hippocampus in epilepsy, analysis complex seizure problems in man using intracarotid drug injections. Home: 17 Strathcona Dr., Montreal 16, Que., Can.*

GLOVER, Townend, entomologist; b. Rio de Janeiro, Brazil, Feb. 20, 1813 (parents Brit. citizens); s. Henry and Mary (Townend) G.; ed. in Eng.; m. Sarah T. Byrnes, Sept. 1840; 1 adopted dau.; studied art in Germany under Mattenheimer, 1834-35; came to New Rochelle, N.Y., 1836; took his collection of modelled fruits to Washington, 1853-54; govt. entomologist, 1854-59; prof. natural sci. Md. Agrl. Coll., 1859-63; 1st ofcl. entomologist U. S. Agrl. Dept., 1863-78, also with U. S. Patent Office until 1878; had plans for work illustrating insects of U. S., never completed; studied insect pests of citrus fruit, cotton, other crops of Am. South.; ret. to Balt., 1878. Died Balt., Sept. 7, 1883.

GLOWINSKI, Mieczyslaw, physician; b. Bydgoszcz, Poland, May 26, 1899; s. John and Otylia (Calubecka) G.; Medicinae univ. d., U. Lwow (Poland), 1924; m. Wera Jäckel, Jan. 27, 1934 (dec. 1955); 1 dau., Nina (Mrs. Kazimierz Juzwa); m. 2d, Helena Batachowska, Jan. 24, 1958. Asst. Med. Faculty Poznan, 1924-28; staff Primarius Regional Hosp., Rybnik, Poland, 1929-39, 45-51; I asst. dept. histology and embryology Silesian Med. Acad., 1951-54; dir. Districtal Gynecol. Hosp., Siemianowice, Poland, 1954-55; faculty Gynecol. Clinic Silesian Med. Acad., Zabrze, (Poland), 1962-65. Cons. on gynecology Polish Rys., 1945-65; co-redactor Ginekologia Polska, 1950-65. Recipient Medal Independence, 1938, Cross of Merit II, 1938, Cross Gt. Polen Independence, 1957, medal for extraordinary merit in Health Service, 1963, medal for merit in devel. Katowice Dist. in gold, 1958. Mem. Polish Gynacologists (mem. nat. exec. com. 1956-65), Polish Soc. Histochemistry (mem. nat. exec. com. 1962-65), Polish Soc. Endocrinology, Polish Soc. Oncology. Research and numerous publs. on surg. methods for correction congenital gynecol. defects, pathology pregnancy and causes perinatal deaths of the newborn. Died Katowice, Poland, Nov. 12, 1965.

GLOYNA, Earnest Frederick, Am. engr., educator; b. Vernon, Tex., June 30, 1921; s. Herman E. and Johanna M. (Riethmayer) G.; B.S. in Civil Engring. Tex. Tech. Coll., 1946; M.S. in Civil Engring., U. Tex., 1949; D.Eng., Johns Hopkins, 1952; m. Mary Agnes Lehman, Feb. 17, 1946; children—David F., Lisa M. Engr., 1942-46; faculty U. Tex., 1947—, prof. civil engring., 1959—, dir. environmental health engring., 1960—, Center for Research in Water Resources, 1963—; sr. nuclear engr. Convair Corp., 1956, cons., 1959-61. Cons. numerous govt. agys., industries and municipalities, 1952—. Recipient Water Resources Div. award Am. Water Works Jour., 1959, Harrison Prescott Eddy medal Water Pollution Control Jour., 1959. Mem. Southwestern Soc. Nuclear Medicine (hon.), Am. Assn. Profs. San. Engring., Am. Acad. Environmental Engrs., Am. Soc. Engring. Edn., Am. Soc. C.E., U. S.-Mexico Border Pub. Health Assn., Am. Inst. Chem. Engring. Research, publs. on water pollution control and treatment; handling of indsl. waste. Home: 3317 River Rd., Austin, Tex. 78703.*

GLOZMAN, Osip Sergeevich, Russian pathophysiologist; b. Vilno, 1900; grad. Med. Faculty, Saratov U., 1922; D.Med. Sci. Asst. dept. path. physiology Saratov Med. Inst., 1926-32, head chair path. physiology, 1932-46, dep. dir. for sci. work and studies, 1938-41; dir. Saratov Regional Malaria Sta., 1932-34; charge endocrinology sta. Saratov Oblast Dept. Health, 1934-36; head chair path. physiology Alma-Ata Med. Inst., 1946—. Chmn. learned med. council Kazakhstan Ministry Health, 1948-53. Mem. Kazakhstan Soc. Gerontologists (chmn.), Kazakhstan Soc. Pathologists (chmn.), Kazakhstan Znanie Soc. (Presidium mem. med. sect.). Mem. editorial council Path. Physiology and Exptl. Therapy. Research and numerous publs. on hematology, hemorrhage, shock, tissue cultures, biochemistry, silicosis, lead poisoning, pathology; discovered enhanced absorption of amino acids by red cells in anemic states, 1930; 1st described (with Planeles and Kasatkins morphine shock, 1942; described (with Kasatkina) post-transfusion anemia, 1948-50; proposed method of transfusion in cases of acute endotoxicosis and exotoxicosis, 1950. Address: Alma-Ata Med. Inst., Komsomolskaya ulitsa 96, Alma-Ata, Kazkhstan, USSR.

GLUECK, Bernard Charles, Am. physician; b. Balt., Aug. 26, 1914; s. Bernard D. and Josephine (Stransky) G.; A.B., Columbia, 1934, certificate in Psychoanalytic Medicine, 1951; M.D., Harvard, 1938; m. Marie Louise Howard, Sept. 19, 1936; children—Susan Howard, Charles David. Practice medicine, specializing in psychiatry, Mpls., 1955-60, Hartford, Conn., 1960—; prof., dir. research Med. Sch. U. Minn., 1955-60; dir. research Inst. Living, 1960—; lectr. psychiatry Sch. Medicine Yale, 1965—. Harry Armstrong lectr. Aerospace Med. Assn., 1967. Fellow Am. Psychiat. Assn. (chmn. research com.); mem. Am. Psychopath. Assn. (pres.), Assn. For Psychoanalytic Medicine, Am. Coll. Neuropsychopharmacology, Sigma Xi, Alpha Omega Alpha. Research, numerous publs. in psychodynamics of sexual delinquency, devel. of qualitative personality assessments, application of computer techniques to psychiat. hosps. and psychiat. practice.*

GLUECK, Eleanor Touroff, Am. research criminologist; b. N.Y.C., Apr. 12, 1898; d. Bernard Leo and Anna (Wodzislawski) Touroff; A.B., Barnard Coll., 1920; diploma N.Y. Sch. of Social Work, 1921; Ed.M., Harvard, 1923, Ed.D., 1925, Sc.D., 1958; m. Sheldon Glueck, Apr. 16, 1922; 1 dau., Joyce Glueck Rosberg (dec.). Head worker Dorchester (Mass.) Welfare Center, 1921-22; research in criminology, dept. social ethics, Harvard U., 1925-28; research asst. Harvard Law Sch. Crime Survey, 1928-30; research asst. in criminology, Harvard Law Sch., 1930-53, research asso. in criminology, 1953—. Trustee, exec. com. Judge Baker Guidance Center. Recipient August Vollmer award, Am. Soc. Criminology, 1961; (with husband) Beccaria gold medal German Criminological Soc., 1964; gold medal Inst. of Criminal Anthropology, University Rome, Italy, 1964. Fellow Am. Acad. Arts and Scis.; mem. Am., German, Internat. socs. criminology, Nat. Assn. Social Workers, Nat. Conf. Social Welfare, Internat. Conf. Social Work, Internat. Soc. Social Def., World Fedn. Mental Health, Med. Correctional Assn., United Prison Assn. (mem. corp.), Harvard Found. for Advanced Study and Research, Acad. Certified Social Workers. Am. Assn. U. Women, League of Women Voters, Assn. N.Y. Sch. Social Work, Pi Gamma Mu. Author (with Sheldon Glueck) books including: After-Conduct of Discharged Offenders, 1945; Unraveling Juvenile Delinquency, 1950; Delinquents in the Making, 1952; Physique and Delinquency, 1956; Ventures in Criminology, 1964; also articles in profl. jours. Coeditor Preventing Crime, 1936; Predicting Delinquency and Crime, 1959; Family Environment and Delinquency, 1962; mem. adv. editorial bd. Internat. Jour. Social Psychiatry. Pioneer (with husband) in various aspects of criminological research, particularly studies of effectiveness of penocorrectional treatments, etiology of delinquency, crime and recidivism; established 1st prediction system in history of criminology, Glueck Social Prediction Table. Office: 3 Garden St., Cambridge 38, Mass.

GLUECK, Nelson, Am. archaeologist; b. Cincinnati, June 4, 1900; s. Morris and Anna (Rubin) G.; A.B., U. Cincinnati, 1920; B.H.L., Hebrew Union Coll., 1918, Rabbi, 1923; student U. of Berlin and Heidelberg U., 1923-24; Ph.D., U. of Jena, Germany, 1926;

Morgenthau fellow Am. Sch. of Oriental Research, Jerusalem, 1928-29; LL.D., U. of Cincinnati, 1936, U. Pa., 1960, Miami U., 1962; D.H.L., Jewish Theol. Sem., 1947, Jewish Inst. of Religion, 1947; L.H.D., Brandeis U., 1961, Wayne State University, 1962, New York University, 1963; Litt.D., Dropsie Coll., 1947; D.S.L., Kenyon Coll., O., 1955; D.D., Drake U., Ia., 1956; m. Helen Ransohof Iglauer, Mar. 26, 1931; 1 son, Charles Jonathan, Instr. Hebrew Union Coll., 1929-31, asst. prof., 1932-33, asso. prof., 1934-35, professor Bible and Biblical archaeology, 1936-—, pres., 1947-50; pres. Jewish Inst. of Religion, 1949-50; pres. combined Hebrew Union Coll.-Jewish Inst. of Religion, Cin., N.Y., Los Angeles, Jerusalem, 1950-—; lecturer Biblical literature U. Cin., 1932-36; annual prof. Am. Sch. of Oriental Research, Baghdad, 1933-34; dir. Am. School of Oriental Research, Jerusalem, Palestine, 1932-33, 1936-40, 43-47; field dir. Baghdad, 1942-47. Recipient Cin. Fine Arts Award, 1940, Ohiana Career medal, 1956; Ohiana non-fiction book award, 1960. Mem. Am. Philos. Soc., Central Conf. Am. Rabbis, Archaeol. Inst. Am., Am. Oriental Soc., Israel Exploration Soc., Am. Sch. Oriental Research, Phi Beta Kappa. Author: Das Wort Hesed im alttestamentlichen Sprachgebrauche, 1927; Explorations in Eastern Palestine, Vol. 1, 1934, 2, 1935, 3, 1939, 4, 1951; The Other Side of the Jordan, 1940; The River Jordan, 1946; Rivers in the Desert: A History of the Negev, 1959. Contbr. articles on archaeology and Bible to mags., books, encys. Made excavations and important archaeological discoveries in Palestine and Transjordan, 1932-47; archaeol. explorations in Negev, 1952-—. Home: 162 Glenmary Av., Cin. 45220.

GLUECK, Sheldon, criminologist; b. Warsaw, Poland, Aug. 15, 1896; s. Charles and Anna (Steinhardt) G.; brought to U. S., 1903, naturalized, 1920; student Georgetown U. Law Sch., 1914, 15; A.B., George Washington U., 1920, S.S.D., 1963; LL.B., LL.M., Nat. U. Law School, 1920; postgrad. Harvard Law Sch., 1926; A.M., Harvard, 1922, Ph.D., 1924, Sc.D., 1958; LL.D., U. Thessalonika (Greece), 1948; m. Eleanor Touroff, Apr. 16, 1922; 1 dau., Anitra Joyce Rosberg (dec.). Instr. criminology, penology Dept. Social Ethics, Harvard, 1925-29; asst. prof. criminology Harvard Law Sch., 1929-31, prof., 1931-—, Lowell lectr., 1935, Roscoe Pound prof. law, 1950-63, emeritus, 1963-—. Ofcl. U. S. del. Internat. Prison Congress, Prague, 1930, Paris, 1950; mem. Adv. Com. on Rules of Criminal Procedure, Supreme Ct. of U. S., Am. Law Inst. for Youth Correction Authority Model Bill and for Model Penal Code; adviser to Justice Robert H. Jackson on law governing War Crime Trials, Nuremburg. Recipient Isaac Ray award Am. Psychiat. Assn., 1961; August Vollmer award Am. Soc. Criminology, 1961; (with wife) Beccaria gold medal German Criminological Soc., 1964; gold medal Inst. Criminal Anthropology, U. Rome (Italy), 1964. Fellow Am. Acad. Arts & Sci., Am. Psychiat. Assn. (hon.; Isaac Ray award 1961); mem. Am. Soc. Criminology (past v.p.), Am. Bar Assn. Author numerous books 1925-—, including: The Nuremberg Trial and Aggressive War, 1946; Crime and Corrections: Selected Papers, 1952; Co-author (with Eleanor T. Glueck): After-conduct of Discharged Offenders, 1945; Unraveling Juvenile Delinquency, 1950; Delinquents in the Making, 1952; Physique and Delinquency, 1956; Predicting Delinquency and Crime, 1959; Family Environment and Delinquency, published 1962; Ventures in Criminology, published 1964; (sole author) Law and Psychiatry: Cold War or Entente Cordiale?, 1962; others. Editor publs., including: The Welfare State and The National Welfare, 1952; The Problem of Delinquency, 1958; Roscoe Pound and Criminal Justice, published 1965. Mem. editorial bd. Fed. Probation, Internat. Jour. Social Psychiatry. Contbr. profl. jours. Pioneered in devel. of relationship of law and psychiatry, war criminal law, and in teaching in field of juvenile delinquency; pioneered (with wife) in various aspects of criminological research, particularly studies of effectiveness of penocorrectional treatments, etiology of delinquency, crime and recidivism; established 1st prediction system in history of criminology, Glueck Social Prediction Table. Office: Langdell Hall, Harvard Law Sch., Cambridge 38, Mass.

GLUECKSOHN-WAELSCH, Salome, geneticist; b. Germany, Oct. 6, 1907; d. Ilja and Nadia Gluecksohn; Ph.D., U. Freiburg, 1932; m. Heinrich Waelsch, Jan. 8, 1943; children—Naomi, Peter. Came to U. S., 1933, naturalized, 1939. Univ. asst. U. Berlin, 1932-33; research asso., lectr. zoology Columbia, 1936-52, research asso. in obstetrics Coll. Phys. and Surg., 1952-55; asso. prof. anatomy Albert Einstein Coll. Medicine, Bronx, N.Y., 1955-58, prof. genetics, 1958-—. Mem. Am. Soc. Zoologists, Am. Assn. Anatomists, Genetics Soc., Soc. Developmental Biology. Research and numerous articles on developmental genetics, role of genes in normal and abnormal growth and differentiation, genetic and developmental aspects of congenital malformations, mammalian genetics. Home: 90 Morningside Dr., N.Y.C. 10027. Office: Dept. of Genetics, Albert Einstein College of Medicine, Bronx, N.Y. 10461.*

GLUSHCHENKO, Ivan Evdokimovich, Russian biologist; b. Jan. 28, 1907; grad. Kharkov Agronomical Inst.; D.Agrl. Sci. Asso. of Lysenko; dir. lab. plant genetics Inst. Genetics, USSR Acad. Sci., 1939-—; mem. All-Union Lenin Acad. Agrl. Sci., 1956-—. Recipient Stalin prize, 1943, 50. Research and publs. on genetic properties of tissues of plantal organism, vegetational hybridization of plants, biology of fertilization in remote and close cross breeding of cross-pollinating plants. Address: Inst. Genetics, USSR Acad. Sci., Leninsky prospect 33, Moscow, USSR.

GLUSHOV, Viktor Mikhaylovich, Russian mathematician; b. Rostov-on-Don, 1923; grad. Rostov-on-Don U., 1948; D.Physico-Math. Sci., 1956. With Urals Timber Inst., 1948-56; with Inst. Math., Ukrainian Acad. Sci., 1956-57, dir. Computer Center, 1957, now dir. Inst. Cybernetics; prof., 1957-—. Mem. Ukrainian Acad. Sci. (v.p. 1962-—). Research and publs. on theory of infinite discrete values, topological bicompact groups, specialist in computer engring. and cybernetics. Address: Inst. Cybernetics, Ukrainian Acad. Sci., Kiev, Ukraine SSR, USSR.

GLUSKER, Donald Leonard, Am. chemist; b. Chgo., Oct. 6, 1930; s. Albert Joseph and Ann (Goldsmith) G.; B.S., U. Cal. at Berkeley, 1951; D.Phil. (Rhodes scholar) Oxford (Eng.) U., 1954; m. Jenny Pickworth, Dec. 18, 1955; children—Ann, Mark, Katharine. NSF Postdoctoral fellow Cal. Inst. Tech., 1954-56; sr. chemist Rohm and Haas Co., Phila., 1956-61, lab. head, 1961-—. Mem. Am. Chem. Soc., Sigma Xi. Contbr. articles to tech. jours. Research on infra-red spectra charge transfer complexes (intermolecular interactions), mechanisms organic reactions, polymers and polymerization mechanisms. Home: 1011 Annac Rd., Huntingdon Valley, Pa. 19006. Office: Research Labs., Rohm and Haas Co., Norristown and McKean Rds., Spring House, Pa. 19477.*

GLUUD, (Friedrich) Wilhelm, German chemist; b. Bremen, Germany, Apr. 12, 1887; s. Johann Friedrich and Clara (Hoffmann) G.; began study chemistry in Munich, Freiburg im Breisgau, Berlin (all Germany), 1905; Ph.D. (under Emil Fischer), Berlin, 1909; m. Hildegard Lechler, 1920; 1 adopted dau. Asst. to Fischer in Berlin, until 1912, then at chem. inst. U. Bonn (Germany); mem. staff Davy Faraday Lab., London, until 1914; joined Kaiser Wilhelm Inst. for Coal Research, Mühlheim, Germany, 1915; became mem. faculty U. Münster (Germany), 1928, named asso. prof., 1928; became dir. Gesellschaft für Kohlentechnik m.b.H., Dortmund, Germany, 1919; named head Bergwerksverband zur Verweitung von Schutzrechten, der Kohlentechnik G.m.b.H., Dortmund, 1928. Author: Tieftemperaturverkokung der Steinkohle; Berichten der Gesellschaft für Kohlentechnik, vols., 1-4, 1921-35; also articles in chem. jours. Editor: Handbuch der Kokerei, 2 vols., 1927/28. Research on coal and coke chemistry; improved thermic purification of bituminous coal, also improved use of coal as fuel and raw material for chem. processes. Died Münster, Aug. 9, 1936.

GLYNN, Ernest E., English pathologist; s. T. R. Glynn; ed. Clare Coll., Cambridge, Univ. Coll., Liverpool; M.A., M.D., Cambridge; Asst. physician Chest Hosp., 1902-03; asst. physician Liverpool Royal Infirmary, 1904-06, later hon. pathologist and bacteriologist; prof. pathology U. Liverpool. Fellow Royal Coll. Physicians, London. Author: Study of Disease in Domesticated Animals, 1913; Microbes, the War, and Science, 1916; also articles on pathology of sex characters, other path. studies. Died Sept. 22, 1929.

GMEINER, Josef Anton, Austrian mathematician; b. Bizau (Vorarlberg), Austria, July 12, 1862; s. Josef and Margaretha (Feuerstein) G.; teaching diploma, 1890; Ph.D., U. Innsbruck (Austria), 1895; m. Maria Rosa Feuerstein, 1896. Tchr. intermediate schs. in Graz, Austria, Flume, Yugoslavia, Klagenfurt, Vienna (both Austria), 1890-95; tchr. secondary sch., Pola, Yugoslavia, and in Vienna, 1899; became lectr. math. U. and Tech. U. Vienna, 1900; named asso. prof. math. German U. Prague, 1901; named prof., 1904; apptd. prof. in Innsbruck, 1906. Author: (with O. Stolz) Theoretische Arithmetik I, 1900, II, 1902; Einleitung in die Funktionen Theorie I, 1904, II, 1905; treatise on number theoretical reduction of binary quadratic forms; numerous articles. Research in number theory, algebra, theory of series, integral calculus and theory of functions; a founder Austrian number theory-algebraic sch. Died Innsbruck, Jan. 11, 1927.

GMELIN, Christian Gottlob, German chemist; b. Tübingen, Germany, Oct. 12, 1792; travelled in France, Eng., Norway, Sweden; worked with Berzelius in Sweden; became prof. chemistry and pharmacy Tübingen, 1817. Author: Einleitung in die Chemie, 2 vols., 1833-37. Translator: Jahres-Bericht (Berzelius), 1st 3 vols. Research on lithium compounds, mineral analysis; independently discovered artificial ultramarine; mentioned tumeric paper test for boric acid. Died Tübingen, May 13, 1860.

GMELIN, Johann Friedrich, German chemist, physician; b. Tübingen, Germany, Aug. 8, 1748; s. Philipp Friedrich Gmelin; a son, Leopold. Became asso. prof. medicine Tübingen, 1772; became prof. medicine and chemistry, Göttingen, Germany, 1775, full prof., 1780-1804. Author: Onomatologia botanica completa, 1771-79; Allgemeine Geschichte der Gifte, 1776-78; Grundsätze der technischen Chemie, 1786; Grundriss der allgemeine Chemie, 1789, Geschichte der Chemie, 3 vols., 1797-99. Editor: Systema natura (Linneaus), 13th edit., 1788-93. Research on vegetative poisons, chemistry and history of chemistry; numerous publs. on botany, chemistry and other natural scis. Died Göttingen, Nov. 1, 1804.

GMELIN, Johann Georg the Elder, German apothecary; b. 1674; student of Hiärne, m. Barbara Haas; children include Johann Conrad, Johann Georg, Philipp Friedrich. Owner, Gmelinschen pharmacy, Tübingen, Germany; tchr. pvt. classes in chemistry U. Tübingen. Analyzed mineral water; developed tests for metals and minerals; prepared mercuric acetate (sperma mercurii, pub. by son Johann Georg). Died Tübingen, 1728.

GMELIN, Johann Georg the Younger, German botanist, naturalist, chemist; b. Tübingen, Germany, Aug. 10, 1709; s. Johann Georg and Barbara (Haas) G.; began study medicine and natural scis. U. Tübingen, 1722; M.D., 1728; m. Barbara Frommann, 1749; 3 sons, including Christian von G., Eberhard. Traveled to St. Petersburg after his tchr. G. B. Bilfinger, 1727; granted yearly fellowship Imperial Acad., 1728, teaching contract, 1730, named prof. chemistry and natural history, 1731; explored Siberia, 1733-43; returned to Germany, 1747; apptd. prof. botany and chemistry, Tübingen, 1749. Author: Flora Sibirica, sive historia plantarum Sibiriae, 4 vols., 1747-69; Reise durch Sibirien, 4 vols., 1751/52. Pioneer in modern sci. study of Siberia; described many new plant species; research on composition of plant ashes, dye extraction, drug effects, volcanos, ammonites, heat conditions of water. Died Tübingen, May 20, 1755.

GMELIN, Leopold, German chemist; b. Göttingen, Germany, Aug. 2, 1788; s. Johann Friedrich and Rosine Luise (Schott) G.; studied in Tübingen, Germany; doctorate, Göttingen, 1812; studied medicine in Italy, 1813; m. Luise Maurer, 1816; 1 son, Adolf; 3 daus., including Auguste (Mrs. Theodor von Dusch). Began work under von Jacquin in Vienna, Austria, 1811; joined faculty U. Heidelberg (Germany), 1813, became asso. prof., 1814, prof., 1817-51. Author: Handbuch der Chemie, 2 vols., 1817-19, enlarged to 13 vols., 1843-70. Translated: Handbook of Chemistry (H. Watts), 19 vols., 1848-72. A founder of physiol. chemistry; research on chemistry of digestion; discovered potassium ferrocyanide (Gmelin's salts), 1822; discovered (with F. Tiedemann) various organic substances; developed Gmelin's test for presence of bile pigments; introduced terms ester and ketone, 1848. Died Heidelberg, Apr. 13, 1853.

GMELIN, Samuel Gottlieb, German botanist, naturalist, explorer; b. Tübingen, Germany, July 4, 1744; s. Johann Conrad and Veronika (Erhardt) G.; studied medicine and natural scis. U. Tübingen; M.D., 1763; m. Anna de Chappuzeau, 1772. Practiced medicine, Leyden, The Hague (both Netherlands), Paris, then returned to Tübingen; called to St. Petersburg, Russia, as mem. Acad., 1767; named prof. botany U. Tübingen, 1768, but continued travels instead; entered service of Empress Catherine, 1768, and explored Russia, Woronesch, Tscherkask, Zarizyn, Astrakhan; explored Persia, 1770-72, 73. Author: Historia fucorum (on seaweed, marine algae), 1768; Reise durch Russland zur Untersuchung der drey Hatur-Reiche, 4 vols., 1770-84. Described flora, fauna, geography, ethnology, history of regions hitherto unexplored. Died Achmetkent, Caucasus, Russia, July 27, 1774.

GNEDENKO, Boris Vladimirovich, Russian mathematician; b. Ulyanovsk, Jan. 1, 1912; grad. Saratov, U., 1930; postgrad. Moscow U., 1937; D.Math. Sci. With Ivanovo Textile Inst., 1930-34; instr. Moscow U., 1934-42, prof., 1942-45; prof. Lvov U., 1945-50, Kiev U., 1950-—; dir. Inst. Math., Ukrainian Acad. Sci., 1955-58. Mem. Ukrainian Acad. Sci. Author: A Course on the Theory of Probabilities, 1954; Electronic Computers, 1957; Some Problems of the Theory of Probabilities, 1957; The Role of Mathematical Methods in Biological Research, 1959; The Achievement of Ukrainian Mathematicians, 1957; co-author: Works of A. N. Kolmogorov on the Theory of Probabilities, 1963. Research on theory of probabilities. Address: Kiev University, ulitsa Vladmirskaya 58, Kiev, USSR.

GOADBY, Sir Kenneth Weldon, Brit. bacteriologist; b. Gravesend, Eng., Mar. 7, 1873; s. J. J. Goadby; ed. Univ. Extension Coll., Reading, Guy's Hosp., 1899-1902; m. Constance Eva Olding, 1898; 1 son. Lectr. bacteriology Nat. Dental Hosp., from 1904; lectr. oral hygiene London Sch. Tropical Medicine; Erasmus Wilson lectr. Royal Coll. Surgeons, 1907; Hunterian prof., 1911; hon. v.p., later pres. Sci. Film Assn.; med. referee for indsl. poisoning County of London, 1913; mem. med. adv. com. Ministry Mines; hon. bacteriological specialist for vaccine therapy Royal Herbert Hosp., Woolwich, during War; mem. War Office Com. for Study Tetanus; rep. med. sci. adv. com. to sec. of state for mines. Author: Mycology of the Mouth, A Textbook of Oral Bacteria; The Vaccine Treatment of Pyorrhoea Alvelorasis; The Relation of Diseases of the Mouth to Rheumatism; Report to the Home Office on the Working of Special Rules in Force in Match Factories; Lead Absorption and Lead Poisoning; (with T. M. Legge) Manual of Industrial Lead Poisoning; Text-Book on Diseases of Mouth and Oral Mucous Membrane, 3d edit., 1927, also articles. Research on causes and pathology of lead poisoning, on pathology and bacteriology of diseases of mouth and upper air passages, on rheumatoid arthritis. Died Aug. 10, 1958.

GOARLE, David, see van Goorle, David.

GOBINEAU, Count Joseph Arthur de, see de Gobineau, Count Joseph Arthur.

GOBLE, Frans Cleon, Am. biologist; b. Chgo., July 11, 1913; s. Leroy Truman and Winifred (McKee) G.; student U. Colo., 1930; B.S., Battle Creek Coll., 1933; M.S., U. Mich., 1934, Sc.D., 1939; postgrad. Rice Inst., 1934-35; m. Bernice L. Dertinger, Dec. 30, 1953. Pathologist, N.Y. State Bur. Game, 1938-45, Sterling Winthrop Research Inst., Rensselaer, N.Y., 1945-51, Abbott Labs., North Chgo., Ill., 1951-52; biol. cons. Charlotte Amalie, V.I., 1952-53; dir. parasitology Ciba Pharm. Products, Inc., Summit, N.J., 1954-60, dir. chemotherapy, 1961—. Cons. Chagas disease Pan Am. Health Orgn., WHO, 1962—, mem. expert adv. panel on parasitic diseases, 1964—. Mem. Am. Soc. Trop. Medicine and Hygiene, Royal Soc. Trop. Medicine and Hygiene, Am. Soc. Parasitologists, Am. Assn. Pathologists and Bacteriologists, N.Y. Soc. Trop. Medicine (past pres.), Soc. Protozoologists, Soc. Exptl. Biology and Medicine, Reticuloendothelial Soc. Contbr. numerous articles to tech. jours. Classification lung parasites mammals and birds; research on pathology and chemotherapy of infectious diseases, immunology Chagas Disease; toxicity studies on pharm. preparations. Home: Carnach, Rocky Run Rd., Glen Gardner, N.J. Office: 556 Morris Av., Summit, N.J. 07901.*

GOBRECHT, Heinrich Friedrich Hermann, German physicist; b. Bremen, Germany, July 20, 1909; s. Heinrich and Caroline (Oesterhelweg) G.; student U. Hanover (Germany), 1929-31, U. Göttingen (Germany), 1931-33, U. Marburg (Germany), 1933-34; Dipl.Ing., U. Dresden (Germany), 1936, Dr.Ing., 1937, Dr.Ing.habil., 1939; m. Christa Schubbe, Aug. 25, 1938; children—Klaus, Juergen, Jens. Chief engr. tv dept. Loewe Opta Berlin, 1938-45; chief engr. Siemens Radio Arnstadt, 1946-48; prof. physics Tech. U., Berlin, Germany, 1948—, dir. II Phys. Inst., 1952—. Mem. Deutsche Physikalische Gesellschaft, Deutsche Bunsengesellschaft für Physikalische Chemie, Gesellschaft Deutscher Chemiker. First analysis of spectra of rare earths, 1937; research, numerous publs. on luminescence and semiconductor physics since 1948. Home: 42 Marinesteig, Berlin 38, Germany.*

GOBRONIDZE, Yevtikhiy Georgievich, Russian psychiatrist; b. 1903; grad. Med. Faculty, Tbilisi U., 1928; D.Med. Sci. Asst., lectr. dept. psychiatry Tbilisi Med. Inst., 1931-37, prof., head chair psychiatry, 1937—; asso. Inst. Psychiatry, Georgian Ministry Health, 1931-44, dep. dir. for sci. work, 1944—. Decorated Order of Lenin. Mem. All-Union (bd. mem.), Tbilisi (chmn.) socs. neuropathologists and psychiatrists. Author: Insulin Therapy of Schizophrenia. Research and numerous publs. on regional pathology, history of psychiatry, clin. psychiatry, psychopathology and psychotherapy, postnatal psychoses, psychic disorders of infectious and intoxication origin. Address: Tbilisi Med. Inst., ulitsa Milikashvili 16, Tbilisi, Gruz. SSR, USSR.

GOCHOLASHVILI, Mariya Mikievna, Russian plant physiologist; b. Tiflis, 1904; grad. Agrl. Faculty, Tiflis Poly. Inst., 1927. With dept. plant physiology All-Union Inst. Tea Cultivation, 1927-29, All-Union Inst. Phytoculture, 1929; head lab. plant physiology All-Union Research Inst. Subtropical Crops, 1929-31; with So. Inst. for Combatting Plant Pests, 1932-34; head dept. plant physiology Batumi Subtropical Bot. Gardens, 1935-46, now asso. Co-author: Biological Principles of Tea Plant Cultivation in Georgia, 1963. Research and publs. on winter-hardiness and vegetative reproduction of subtropical crops. Address: Batumi Subtropical Botanical Gardens, Batumi, Gruz. SSR, USSR.

GOCHT, (Moritz) Hermann, German radiologist, orthopedist; b. Köthen, Germany, Feb. 3, 1869; s. Hermann and Minna (Niemann) G.; M.D., U. Erlangen (Germany), 1893; m. Margarete Kassler, 1897; 1 foster dau. Mem. staff Allgemeine Krankenhaus, Hamburg-Eppendorf, Germany, 1895; worked under Albert Hoffa, Orthopedic Clinic, Würzburg, Germany, beginning 1897; founder pvt. clinic for orthopedic surgery, Halle, Germany, 1900; named dir. home for cripples in Merseburg, 1909; named prof., 1910; asso. prof. at Orthopedic Inst., U. Berlin, during World War I; named prof., 1927. Author: Lehrbuch der Röntgen-Untersuchung zum Gebrauch für Mediziner, 1898, under title of Handbuch der Röntgenlehre, 1921; Orthopädische Technik, Anleitung zur Herstellung orthopädische Verbandapparate, 1901; (with R. Radike, F. Schede) Künstliche Glieder, 1907; Die Röntgen-Literatur, 1-15, 7 vols., 1911-34. Editor (with F. König) Archiv für orthopädie und Unfall-Chirurgie, 16-29, 1909-30. Pioneered in X-ray technique; research on orthopedic technique, amputation stumps and artificial limbs; research on path. anatomy of congenital hip luxation. Died Berlin, May 18, 1938.

GOCKEL, Albert Wilhelm Friedrich Eduard, meteorologist; b. Stockach, Baden, Germany, Nov. 27, 1860; s. Albert and Wilhelmine (Trolle) G.; studied physics, Freiburg/Breisgau, Würzburg, Karlsruhe; Ph.D., Heidelberg, 1885; m. Paula Baumhauer, 1902. Secondary sch. tchr.; asst. to Joseph von Kowalski, Fribourg, Switzerland, 1895, became mem. faculty, Fribourg, 1901, prof. cosmic physics, 1909. Author: Das Gewitter, 1895; Die Luftelektrizität, 1908; Die Radioaktivität von Boden und Quellen, 1914. Researched relationship between atmospheric elec. elements and meteorol. phenomena through numerous measurements all over Switzerland; studied natural radioactivity of earth, storm electricity, atmospheric interference of radio waves; his observations on balloon flights led to discovery of cosmic rays, 1910. Died Fribourg, Mar. 4, 1927.

GOCLENIUS, Rodolphus (Göckel, Rudolph the Younger), German physician, mathematician; b. Wittenberg, Germany, 1572; studied at Marburg, Copenhagen, Padua; M.D., Marburg, Germany, 1601; became prof. physics, Marburg, 1608; prof. medicine, 1611; prof. math., 1613. Author: Tractatus de Magnetica Curatione Vulneris, 1609 (defense of the weapon-salve); other works on augury, chiromancy, astrology and medicine. Defended Paracelsian medicine; engaged in controversy with Roberti on magnetic cure of wounds. Died Marburg, Mar. 2, 1621.

GODAL, Hans Christian, Norwegian physician; b. Dröbak, Norway, Dec. 8, 1922; s. Arne and Ellen Godal; M.D., U. Copenhagen (Denmark), 1952; m. Ragnhild Nordbö, May 20, 1949; children—Aslak, Ellen, Inge Christian. Hosp. service, 1953-56; staff Rikshospitalet, Oslo, Norway, 1957, Inst. for Protein Chemistry, Bern, Switzerland, 1958; research fellow Inst. for Thrombosis Research, Rikshospitalet, Oslo, 1958-60; hematologist dept. IX, Oslo City Hosp., 1961—, sci. leader Hematological Research Lab. Mem. European Soc. Hematology. Author: Precipitation and Coagulation of Fibrinogen, 1961; also articles. Research on fibrinogen, fibrinogen-fibrin conversion and fibrionolysis; lipids and lipoproteins; paraproteinemias. Home: 116 Ullevalsveien, Oslo-3. Office: Ulleval sykehus, Dept. IX, Oslo-3, Norway.*

GODDARD, David Rockwell, Am. biologist; b. Carmel, Cal., Jan. 3, 1908; s. Pliny Earle and Alice (Rockwell) G.; A.B., U. Cal. at Berkeley, 1929, A.M., 1930, Ph.D., 1933; m. Doris Martin, Aug. 21, 1933; (dec.); children—Alison (Mrs. John Elliott), Robert Martin; m. 2d, Katharine Evans, Feb. 2, 1952. NRC fellow Rockefeller Inst., 1933-35; faculty U. Rochester, 1935-46, chmn. dept. biology, 1941-46; prof. botany U. Pa., Phila., 1946—; chmn. dept. botany, 1953-57, dir. div. biology, 1957-61, provost, 1961—; Walker-Ames prof. U. Wash., 1955. Guggenheim fellow U. Chgo., 1942-43, Columbia, 1950. Recipient Stephen Hales award Am. Soc. Plant Physiology, 1948. Mem. Nat. Acad. Scis., Am. Philos. Soc., Am. Acad. Arts and Scis. Author: (with Hober) The Physical Chemistry of Cells and Tissues, 1945; contbg. author: Plant Physiology; also numerous articles. Research on plant physiology especially cellular metabolism and respiratory enzymes, early demonstrator existence cytochrom c in plants. Home: 490 E. Abington Av., Phila. 19104.

GODDARD, Jonathan, Brit. physician; b. Greenwich, Eng., circa 1617; s. Henry Goddard; M.B., Christ's Coll., Cambridge, (Eng.) U., 1638; M.D., Catharine Hall, 1643; Gulstonian lectr., 1648; as physician-in-chief Brit. army accompanied Cromwell to Ireland, 1649, Scotland, 1650; warden Merton Coll., Oxford (Eng.) U., 1651-60; became mem. Little Parliament and Council of State, 1653; apptd. Gresham prof. physic, 1655. Fellow Royal Coll. Physicians; mem. Royal Soc. (became mem. council 1663). Author: Discourse Concerning Physick, 1668; Discourse on the Unhappy Condition of the Practice of Physick, 1670; also articles. 1st Englishman to make telescopes; improved optical instruments; research on the refining of gold with antimony. Died Mar. 24, 1675.

GODDARD, Robert Hutchings, Am. physicist; b. Worcester, Mass., Oct. 5, 1882; s. Nahum Danford and Fannie Louise (Hoyt) G.; B.Sc., Worcester Poly. Inst., 1908; A.M., Clark U., 1910, Ph.D., 1911, Sc.D., 1945; m. Esther Christine Kisk, June 21, 1924. Instr., Worcester Poly. Inst., 1909-11, Princeton, 1912-13; instr. and fellow, physics Clark U., 1914-15, asst. prof., 1915-19, prof., 1919-43, also dir. phys. labs.; leave of absence, 1930-32, 34-42, engaged in rocket research, under Daniel and Florence Guggenheim Found. grants; dir. research, bur. aeros. Navy Dept., 1942-45; cons. engr. Curtiss-Wright Corp., 1943-45. Fellow A.A.A.S.; mem. Am. Phys. Soc., Am. Meteorol. Soc., Inst. Aero. Scis., Nat. Aero. Assn., Geophys. Union, Sigma Xi, Sigma Alpha Epsilon. Author: A Method of Reaching Extreme Altitudes, 1914; Rocket Development: Liquid-Fuel Rocket Research, 1929-41 (pub. posthumously, 1948). Elaborated the fundamental theory of rocket flight, 1914-16; built 2-stage solid fuel rocket, 1914; developed rocket engine fueled with gasoline and liquid oxygen, 1923; launched 1st liquid fuel rocket (4 feet high, 6 inches diameter), 1926; launched 1st rocket to carry instruments (barometer, thermometer, camera), 1929; engaged in rocket research for reaching high altitudes, developed self-cooling combustion chambers, 1st automatic steering systems, multistage rockets, 1924-42; designed small rockets to help Navy airplanes take off from carriers, during World War II; developed 1st smokeless powder rocket; proved experimentally efficiency of rocket propulsion in vacuum; an early invention perfected as the bazooka; holder 214 patents in rocketry. Died Balt., Aug. 10, 1945.

GODEAUX, Lucien, Belgian mathematician; b. Morlanwelz, Belgium, Oct. 11, 1887. s. Auguste and Leontine Godeaux; ed. Ecole des Mines de Mons, U. Liège (Belgium); m. M. H. Luthers, Jan. 29, 1919; children—Jean, Paul. Prof., Ecole militaire, 1920-25; prof. U. Liège, 1925-58, emeritus, 1958—. Mem. Royal Acad. Belgium, Royal Soc. Scis. Liège, Soc. Scis. Arts and Letters Hainaut; mem. of corr. acads. Padua, Bologna, Milan, Bordeaux, Lima, math. socs. Amsterdam, France. Studies in differential projective geometry and algebraic geometry. Office: Univ. of Liège, Liège, Belgium.*

GÖDEL, Kurt, logician; b. Bruenn, Czechoslovakia, Apr. 28, 1906; s. Rudolf and Marianne (Handschuh) G.; Ph.D., U. Vienna, 1930; Litt.D., Yale, 1951; Sc.D., Harvard, 1952; Sc.D., Amherst, 1967; m. Adele Porkert, Sept. 20, 1938. Came to U. S., 1940, naturalized, 1948. Faculty, U. Vienna, 1933-38; mem. Inst. for Advanced Study, Princeton (N.J.), 1933, 35, 38-52, prof., 1953—. Co-recipient Einstein award, Lewis and Rosa Strauss Meml. Fund, 1951. Mem. Am. Philos. Soc., Am. Math. Soc., Assn. Symbolic Logic, Nat. Acad. Scis., Am. Acad. Arts and Scis. Author: The Consistency of the Continuum Hypothesis, 1940. Contbr. articles to profl. jours. Gave proof of completeness of predicate logic; method of finding, for any given formalized axiom system of math., a question of Diophantine analysis undecidable in that system; proof that consistency of system cannot be proved in same system, proof of consistency of axiom of choice and of Cantor's continuum hypothesis with currently assumed axioms of set theory, constrn. of rotating universes on basis of Einstein's theory of gravitation. Address: Inst. for Advanced Study, Princeton, N.J. 08540.*

GODFRAIND, Théophile, Belgian pharmacologist; b. Bande, Belgium, Feb. 18, 1931; s. René and Edith (Tasiaux) G.; M.D., U. Louvain (Belgium), 1955, Agrégé Enseignement Supérieur, 1958; m. Anne De Becker, Dec. 12, 1957; children—Pierre, Catherine. Research fellow Laboratoire Therapeutique Expérimentale, U. Louvain, 1955-58, prof. exptl. pharmacology, 1964—; prof. pharmacology U. Lovanium, Leopoldville, 1960-64. Author: L'auto-intoxication après brulure, 1958. Research and publs. on activation of proteolytic enzymes after extensive burns; pharmacology of cardiac glycosides, angiotensin antagonists. Home: 4 Hofter Bekelaan, Winksele, Belgium. Office: 4 rue Van Even, Louvain, Belgium.*

GODFREY, Ambrose (Godfrey-Hanckwitz); m. Mary; children—Boyle, Ambrose, John; operator lab. of Robert Boyle, London. Fellow Royal Soc., Contbr. articles to tech. jours. Analysed water of medicinal spring, Nottington, Eng., 1719; invented and patented machine for extinguishing fires by explosion and suffocation, 1761. Died Jan. 15, 1741.

GODFREY, Thomas, Am. inventor, mathematician; b. Bristol Twp., Pa., 1704; s. Joseph Godfrey; m.; at least 1 child, Thomas. Apprenticed to glazier; invented improved quadrant for ascertaining latitude, 1730; gave quadrant to Joshua Fisher to test in Delaware Bay; James Hadley of Eng. claimed the invention, Godfrey's cause pleaded unsuccessfully by Gov. Logan of Pa. Died Dec. 1749.

GODIN, Louis, French astronomer, mathematician; b. Paris, Feb. 28, 1704; conducted measurement of degree of meridian for French Acad. Scis., 1735; prof. math., Lima, Peru; dir. Naval Acad., Cadiz, Spain. Fellow Royal Soc., 1735; mem. French Acad. Scis., 1725. Author: History of the Academy of Sciences from 1680-1699, 11 vols. Died Cadiz, Sept. 11, 1760.

GODING, Frederic Webster, Am. entomologist; b. Hyde Park, Mass., May 9, 1858; s. Alphonso Landon and Lydia Mehitable (Chandler) G.; M.D., Northwestern U., 1882; Ph.D., Bethel Coll., Tenn., 1890; m. Ella Blanche Phelps, June 8, 1880; children—Hazle Vera (Mrs. H. B. Ames), Frederic Landon; m. 2d, Jessie E. Ayre, May 12, 1913. Taught in pub. schs., Ill., 5 yrs.; prof. science Loudon (Tenn.) Coll., 1 yr.; practiced medicine, 1882-98; spl. asst. to state entomologist of Ill., 1885; mayor of Rutland, Ill., 10 yrs.; consul at Newcastle, N.S.W., 1898-1908; coal insp. U. S. Army, Newcastle, 1905-08; consul at Montevideo, Uruguay, 1908-13; in charge Am. Legation, Montevideo, 1911; consul gen. at Guayaquil, Ecuador, 1913-24. Mem. First Ecuadorian Med. Congress, Med. Surg. Soc. of Guayaquil. Contbr. articles and monographs on entomology and biology, ethnology, also comml. reports; began work on Cercopidae and Membracidae, 1885; published extensive descriptions of S.Am. species of Membracidae, also classification and monograph of Membracidae of the World. Died May 5, 1933.

GODLEE, Sir Rickman John, Brit. surgeon; b. Apr. 15, 1849; nephew of Lister; ed. Univ. Coll., London; B.A., M.S.; LL.D., M.D., Dublin, Ireland; m. Juliet Mary Seebohm, 1891. Prof. clin. surgery, fellow Univ. Coll., London; surgeon Univ. Coll. Hosp.; hon. surgeon in ordinary to King. Fellow Royal Coll. Surgeons (pres. 1911-13); pres. Royal Coll. Medicine, 1916-18. Author: Lord Lister, 1917; also numerous works on med. topics. Performed (with A. H. Bennett) 1st surg. removal of brain tumor, 1884. Died Whitchurch, Eng., Apr. 20, 1925.

GODLEWSKI, Henry Gabriel, Polish histochemist; b. Lwow, Poland, July 23, 1914; s. Tadeus F. and Janina Z. (Dzieslewska) G.; student Agrl. Faculty, Jagiellonian U., Cracov, Poland, 1932-33, physician, 1939, M.D., 1946; Docent in Histology and Embryology, Gdansk Med. Sch., 1951; Prof., Polish Acad. Scis., Warsaw, Poland, 1961; m. Alice F. Ziomek, Sept. 4, 1939; children—Eva, Marek. Vol., histology and embryology dept. Warsaw U., 1935-37; physician Cracov dist., 1945-46; dir. town hosp., head Malbork br. Polish Red Cross. 1945-46; asst. dept. histology and embryology Gdansk Med. Sch., 1946-50; organized chair histology and embryology Bialystok Med. Sch., 1951; head histochem. lab. dept. tumour biology Inst. Oncology, Gliwice, Poland, 1951-59; fellow Brit. Council, dept. histology Liverpool (Eng.) U., 1958; head histochem. lab. Inst. Exptl. Pathology, Polish Acad. Scis., Warsaw, 1960-67; vis. scientist Yerkes Primate Center, Emory U., Atlanta, 1965-66; sci. dir. Pharm. Inst., Warsaw, 1967—. Mem. Polish Histochem. Soc. (pres. 1961—), Copernicus's Soc. Naturalists and Physicians, Internat. Cell Biology Soc., French Histochem. Soc. Research and numerous publs. on cancer, glycogen in processes of tissue growth and differentiation, precancerous lesions, histochemistry of transplantable tumors. Home: 23/8 Warskiego, Warsaw 87. Office: 8 Rydygiera, Warsaw 86, Poland.*

GODLOVE, Isaac Hahn, Am. color physicist; b. St. Louis, June 13, 1892; s. Lewis and Lillie G.; B.S., M.A., Washington U., 1915; Ph.D., U. Ill., 1926; m. Esther Alice Hurlbut, Dec. 22, 1923; 1 son, Terry Francis; m. 2d, Margaret Noss, Aug. 6, 1949. Prof. chemistry Mo. State Normal Sch., 1915-16; asso. prof. U. Okla., 1921-26; research dir. Munsell Color Co., 1926-30; dir., exhbn. color N.Y. Mus. Sci. and Industry, 1930-31; color editor Webster's New Internat. Dictionary, 1931-32; propr. Color Service Labs., 1932-35; chemist and physicist DuPont Co., 1935-43, Gen. Aniline & Film Corp., from 1943. Mem. Optical Soc. Am. (com. colorimetry), Am. Assn. Textile Chemists and Colorists (chmn. color com.), Inter-Soc. Color Council (chmn. 1948-49, editor), Sigma Xi. Author articles on color physics and psychology. Joint author: The Science of Color, 1953. Co-author: The Smithsonian Tables of Physical Constants, 1954. Contbr. articles on color physics and psychology. Died Aug. 14, 1954.

GODMAN, Frederick Du Cane, Brit. entomologist; b. 1834; s. J. Godman; D.C.L.; m. Edith Mary Elwes (dec. 1875); m. 2d, Alice Mary Chaplin; (with Osbert Salvin) increased knowledge of insect fauna in Americas; trustee Brit. Mus. (which houses his collections of 50,000 bird specimens, (120,000 Coleoptera, 30,000 Lepidoptera, 18,000 Diptera, 11,000 Hemiptera, 5000 Hymenoptera). Fellow Royal Soc.; mem. Royal Instn. Author: Natural History of the Azores, 1870; Monograph of the Petrels, 1910; Biologia Centrali-Americana, 1916. Died Feb. 19, 1919.

GODMAN, John Davidson, Am. naturalist, anatomist; b. Annapolis, Md., Dec. 20, 1794; s. Capt. Samuel and Anna (Henderson) G.; M.D., U. Md., 1818; m. Angelica Kauffman Peale, Oct. 6, 1821. Practiced medicine, New Holland, Pa.; moved to village nr. Balt.; gave series of lectures on anatomy and physiology, Phila.; became prof. surgery Med. Coll. of Ohio, Cin.; head Phil. Sch. Anatomy; prof. anatomy Rutgers Med. Coll., N.Y.C., 1826-27; ret. to Germantown, Pa. Editor 1st issue Quar. Reports of Med., Surg. and Natural Science (1st med. jour. published West of Allegheny Mountains), 1822; editorial bd. Phila. Jour. Med. and Phys. Scis. (became Am. Jour. Med. Scis., 1827), 1825-27. Author: Anatomical Investigations, 1824; American Natural History, 3 vols., 1826-28; Rambles of a Naturalist, 1833. Died Apr. 17, 1830.

GODNEV, Tikhon Nikolaevich, Russian plant physiologist; b. Apr. 5, 1893; grad. Physico-Math. Sch., Moscow U., 1916; Ph.D. in Biol. Scis. Prof., Belorussian U., Minsk, 1927—. Belorussian del. Internat. Conf. Peaceful Uses Atomic Energy. Decorated Order of Lenin, 1961. Mem. Belorussian Acad. Scis. Author: The Formation of Organic Matter in Green Foliage, 1934; The Structure of Chlorophyll Molecules, 1936; The Structure of Chlorophyll and Possible Means by which It Is Formed in Plants, 1947. Research on photosynthesis, possibilities for reproducing chlorophyll under artificially created conditions. Address: Belorussian University, Minsk, Belorussian SSR, USSR.

GODRON, Alexandre, French naturalist; b. Hayange, France, Mar. 25, 1807; ed. Stanislas Coll., Paris; M.D., Strasbourg, France, 1833; practiced medicine at Nancy, France; prof. natural history Nancy Sch. Medicine; prof. natural scis. Nancy Faculty Scis., 1854-76; dir. Nancy Bot. Gardens. Mem. French Acad. Scis., 1877. Author: La Flore de la Lorraine; De l'espèce et des races. Research on botany especially hybrids, geology and zoology especially flora and fauna of Lorraine, France. Died nr. Nancy, Aug. 16, 1880.

GODSKE, Carl Ludvig Schreiner, Norwegian meteorologist; b. May 20, 1906; s. Wilhelm and Hildur (Schreiner) G.; Cand. Real., Oslo (Norway) U., 1931, Ph.D., 1933. Sci. asst. to Vilhelm Bjerknes, 1929-39; prof. U. Bergen (Norway), 1946—. Mem. Norsk Geofysisk Forening, others. Author: (with T. Bergeron, J. Bjerknes, R. Bundgaard) Dynamic Meteorology and Weather Forecasting, 1957; books on statistics and popular meteorology in Norwegian. Research, publs. on diverse investigations in dynamical meteorology, local meteorology, climatology and statis. meteorology. Home: 20 Professorvegen, Minde, Norway. Office: Geofysisk Institutt, Allegaten 70, Bergen, Norway.*

GODSON, Warren Lehman, Canadian meteorologist; b. Victoria, B.C., Can., May 4, 1920; s. Walter Ernest Henry and Mary Edna (Lehman) G.; B.A., U. B.C., 1939, M.A., 1941; M.A., U. Toronto, 1944, Ph.D., 1948; m. Merl Ellen Hotson, Dec. 26, 1942, (dec. Sept. 1965); children—Elliott Fyfe, Marilyn Ivy, Murray Alexander, Ralph Cecil, Ellen Florence; m. 2d, Ruth Margaret Clarke, Sept. 16, 1967. Research meteorologist Meteorol. Service Can., Toronto, Ont., 1943—, supt. atmospheric research sect., 1954—; spl. lectr. U. Toronto, 1951-61. Recipient Buchan prize, Presidents' prize, Darton prize, all from Royal Meteorol. Soc. Fellow Royal Soc. Can., Royal Meteorol. Soc., Am. Meteorol. Soc.; mem. Canadian Meteorol. Soc., Canadian Assn. Physicists, Internat. Assn., Meteorology and Atmospheric Physics (sec. 1963—), World Meteorol. Orgn. (Internat. Union Geodesy and Geophysics liaison officer 1963—). Contbr. numerous articles to sci. jours. Discovery of Arctic stratospheric jet stream and of final warming process in polar winter stratosphere and relation of these to large-scale and seasonal changes in atmospheric ozone; Curtis-Godson approximation technique used in atmospheric infrared radiation studies to handle vertical variability of atmosphere; inventor Ozonogram diagram used for ozonagram representation. Home: 47 Winsdale Rd., Etobicoke, Ont. Office: 315 Bloor St. W., Toronto 5, Ont., Can.*

GODWIN, John Thomas, Am. physician; b. Social Circle, Ga., Dec. 2, 1917; s. Hubert Olie and Georgie Ann (Adams) G.; B.S., Emory U., 1938, M.D., 1941; m. Sara Elizabeth Moak, Mar. 5, 1948; children—Elizabeth, Thomas Adams, Patricia Ann. Practice medicine, specializing in pathology, Atlanta, 1955—; pathologist, dir. labs. St. Joseph's Infirmary, 1955—; prof. pathology Dental Sch. Emory U., 1958—; spl. research scientist Ga. Inst. Tech., 1959—. Mem. Gov.'s Radiation Control Council, Ga. Sci. and Tech. Commn. Recipient Silver medal Am. Soc. Clin. Pathologists, 1952, Gold medal, 1953, 1st prize Radiol. Meeting N.Am., 1953, A.M.A., 1955, 1st prize So. Med. Assn., 1958, Med. Assn. Ga., 1959. Mem. A.M.A., Am. Assn. Pathologists and Bacteriologists, Am. Goiter Assn., Internat. Assn. Pathologists, Am. Soc. Clin. Pathologists, Coll. Am. Pathologists, Southeastern Surg. Congress, So. Med. Assn., Am. Pub. Health Assn., James Ewing Soc., Am. Soc. Exptl. Pathology, N.Y. Acad. Scis., Fulton County Med. Soc. (pres. 1966). Contbr. numerous articles to profl. jours. Research and publs. on use of radioactive isotopes in diagnosis, treatment of various tumors; effects of ACTH on mouse trichinosis. Home: 1164 Springdale Rd. N.E., Atlanta 30306. Office: 265 Ivy St. N.E., Atlanta 30303.*

GODWIN-AUSTEN, Henry Haversham, English explorer; b. Teignmouth, Devonshire, Eng., July 6, 1834; s. Robert A. C. and Maria (Godwin) G.-A.; ed. Royal Mil. Coll. Sandhurst; m. Pauline Plowden, 1861; 1 son; m. 2d, Jessie Robinson, 1881. Gazeteer, Her Majesty's 24th Regt. of Foot, 1851; went to India, 1852; named topog. asst. Trigonometrical Survey of India, mem. Kashmir Survey party (surveyed areas of Kashmir and Baltistan, glaciers at head of Shigar river and Hunza Nagar frontier, discovered Baltoro glacier, which comes partly from 2d highest mountain in Himalayas, later named after him), 1857; surveyed country of Rupshu and Zaskar in Ladakh, 1862; investigated topography of Changchingmo to eastern end of Pang Kong Lake, nr. Rudok, 1863; with last mission to Bhutan (mapped area between Darjelling and Punakha), 1863-64; with Bhutan Field Force, 1864-65; in charge survey in Garo, Khasi, Jaintia, N. Cachar, Naga Hills, Manipur (during expdn. against Dafla tribe at base eastern Himalayas he mapped large area and fixed various distant peaks 1874), 1866-76. Fellow Royal Soc., 1880, Zool. Soc., Royal Geog. Soc. (Founder's medal 1910); pres. sect. E. Brit. Assn., 1883, Malacological Soc., 1897-09, Conchological Soc. Gt. Britain and Ireland, 1908-09. Author: On the Land and Freshwater Mollusca of India, 1882-1920; (with W. T. Blanford) The Fauna of British India, vol. Mollusca, 1908. Contbr. papers on geology, phys. features, natural history, ethnology to sci. jours. Died Nore, Godahmung, Dec. 2, 1923.

GOEBEL, Charles James, Am. physicist; b. Chgo., Dec. 16, 1930; s. Harry Wilhelm and Gladys (Ibach) G.; Ph.B., U. Chgo., 1949; Ph.D., 1956; m. Belle Catherine Gorman, Oct. 27, 1951; children—George Harry, John Philip. Research asso. Lawrence Radiation Lab., Berkeley, Cal., 1954-56; faculty U. Rochester (N.Y.), 1956-61, asso. prof. physics, 1959-61; faculty U. Wis., Madison, 1961—, prof. physics, 1964—; mem. staff Inst. for Advanced Study, Princeton, 1960-61. Fellow Am. Phys. Soc. Research, publs. in quantum field theory of elementary particles especially using analytic properties (dispersion relations), solution of strong coupling limit of static models with analytic and group theoretic methods. Home: 10 N. Roby Rd., Madison, Wis. 53705.*

GOEBEL, Fritz, German pediatrician; b. Bielefeld, Germany, June 3, 1888; s. Karl and Emilie (Freudenberg) G.; studied medicine, Bonn, Freiburg (both Germany); M.D., Munich, 1913; m. Gisela von Held, 1915; 1 son; 2 daus. Asst., Gisela Children's Hosp., Munich, also Path. Inst., U. Berlin; became mem. faculty U. Jena (Germany), 1922, apptd. asso. prof., 1924, mem. staff at children's clinic, until 1925; apptd. prof., head children's clinic, Halle, Germany, 1925; named prof. Düsseldorf (Germany) Med. Acad., 1937; apptd. dir. municipal hosps., 1938. Mem. Leopoldina. Research on protein metabolism, intestinal parasites, anemia, digestive disturbances, especially measles, polio, and Tb, other infectious diseases; introduced streptomycin therapy for Tb in Germany after World War II. Died Munich, Sept. 1, 1950.

GOEBEL, Karl Immanuel Eberhard von, see von Goebel, Karl Immanuel Eberhard.

GOEBEL, Walther Frederick, Am. biochemist; b. Palo Alto, Cal., Dec. 24, 1899; s. Julius Ludwig and Kathryn (Vreeland) G.; A.B., U. Ill., 1920, A.M., 1921, Ph.D., 1923; postgrad. U. Munich (Germany); D.Sc., Middlebury Coll., 1959; m. Cornelia Van Rensselaer Robb, Oct. 23, 1930; children—Cornelia van Rensselaer (Mrs. Nathaniel Bronson, II), Anne Kathryn, Bruce Barkman. With Rockefeller Inst. for Med. Research, N.Y.C., 1924-64, mem., 1944-64, prof. biochemistry, 1957-64; prof., head dept. biochemistry Rockefeller U., 1964—. Mem. Am. Chem. Soc., Am. Soc. Microbiology, Am. Assn. Immunologists, Am. Soc. Biol. Chemists, Harvey Soc., Nat. Acad. Scis., Phi Beta Kappa, Sigma Xi, Phi Lambda Upsilon, Phi Eta. Research, numerous publs. on relationship between structure carbohydrates and their ability to induce immunity to disease; pioneered demonstration prodn. immunity to penumoccocal infection with antigen synthetic origin; discovered nature receptor for certain bacteriophages on cell surface susceptible microorganisms; discovered chem. nature colicines, potent bactericidal agts. produced by certain enteric microorganisms. Home: Vineyard Lane, Greenwich, Conn. Office: The Rockefeller U., N.Y.C. 10021.*

GOEDDE, Heinz Werner, German biochemist, geneticist; b. Lippstadt, Germany, July 9, 1927; s. Heinrich and Maria (Storck) G.; Vordiplom in Chemistry, U. Munich (Germany), 1951, diploma of chem., 1954; Ph.D., U. Hamburg (Germany), 1957; m. Gisela Schweins, Aug. 16, 1958. Research asst. Inst. Biochemistry U. Freiburg (Germany), 1957-61, chief asst., dept. to dir. Inst. Human Genetics, head dept. biochemistry, 1961-66, lectr. biochemistry and human genetics, 1963—. Mem. Gesellschaft Deutscher Chemiker, Deutsche Gesellschaft für Physiol. Chemie, Deutsche Gesellschaft für Klin. Chemie, Deutsche Gesellschaft für Anthropologie. Author: (with K. Altland, A. Doenicke), Serumcholinesterasen: Pharmacogenetik- Biochemi- Klinik, 1966; also numerous articles. Research on mechanisms of enzyme oxidation of oxo-acids and intermediates, biochemistry of lipoic acid; pharmacogenetics and enzyme protein polymorphisms, inborn errors of metabolism. Home: 40 Schlossbergstreet, 78 Freiburg, Baden-Württemb., Germany.*

GOEDKOOP, Jacob Adriaan, Dutch phys. chemist; b. Velsen, Netherlands, July 4, 1921; s. Jacob and Cornelia (Baljet) G.; student U. Leyden (Netherland), 1939-42; student U. Amsterdam (Netherlands), 1942-48, D.degree, 1952; postgrad. Pa. State Coll.; m. Elly de Goede, July 15, 1952; children—Mark Jacob, Caroline Astrid, Jeroen Björn, Karin Elisabeth. Asst., U. Amsterdam, 1947-52; dir. physics Netherlands-Norwegian Joint Establishment for Nuclear Energy Research, Kjeller, Norway, 1952-59; dir. physics Reactor Centrum Nederland, Petten, Netherlands, 1959-61, mng. dir. for research, 1961—; asso. prof. U. Leyden, 1958—. Sci. sec. 1st Conf. on Peaceful Applications of Atomic Energy, UN, 1955. Created Knight 1st class Order of St. Olaf (Norway). Mem. Royal Netherlands Chem. Soc., Netherlands Phys. Soc., Norwegian Phys. Soc., Am. Crystallographic Assn. Author: Theoretical Aspects of X-ray Crystal Structure Analysis, 1952; also articles. Research on crystal structure determinations by means of X-ray and neutron diffraction. Home: 14, Irenelaan, Bergen, N.H., Netherlands. Office: Reactor Centrum Nederland, Petten, N.H., Netherlands.*

GOEPPERT-MAYER, Maria, physicist; b. Kattowitz, Poland, June 28, 1906; d. Friedrich and Maria (Wolff) Goeppert; Ph.D., U. Göttingen, Germany, 1930; D.Sc., Russel Sage Coll., 1960, Mt. Holyoke Coll., 1961, Smith Coll., 1961, U. Portland, 1968; m. Joseph E. Mayer, Jan. 18, 1930; children—Marianne, Peter Conrad. Came to U. S., 1930, naturalized, 1933. Vol. asso. Johns Hopkins, 1931-39; lectr. Columbia, 1939-46, Sarah Lawrence Coll., 1942-45; physicist SAM Labs., 1942-45; sr. physicist Argonne Nat. Lab., 1946-60; vol. prof. Enrico Fermi Inst. Nuclear Studies, U. Chgo., 1946-59, prof., 1959-60; prof. sch. sci. and engring. Revelle Coll., U. Cal. San Diego at La Jolla, 1960—. Recipient (with Jensen and Wigner) Nobel prize in physics, 1963. Mem. Akademie der Wissenschaften of Heidelberg, Am. Acad. Arts and Scis., Am. Phys. Soc., Nat. Acad. Scis., Sigma Xi. Author: (with Joseph E. Mayer) Statistical Mechanics, 1940; (with J. H. D. Jensen) Elementary Theory of Nuclear Shell Structure, 1951. Developed shell theory of atomic nucleus, with protons and neutrons arranged in shells as electrons are in the outer atom. Home: 2345 Via Siena, La Jolla, Cal.

GOERCKE, Johann, German physician; b. Sorquitten (Sensburg), Germany, May 3, 1750; s. Johann Friedrich and Anna Elisabeth (Apfelbaum) G.; studied surgery under his uncle (Apfelbaum), at Tilsit; student Königsburg, Germany, Berlin; state exam., circa 1786-87; hon. M.D., Erlangen, 1795; m. Wilhelmine Lehmann, 1799. Became under regt. surgeon, Gerlach, Königsberg, 1766; became co. surgeon, Königsberg, 1767, in Crown Prince's Regt., Potsdam, Germany, 1774; named surgeon King's Own Co., Potsdam, 1778; became surgeon Hosp. for Disabled, Berlin, 1784; traveled to Vienna, Italy, France, Eng., Scotland, Netherlands, 1787-89; named dept. gen. staff surgeon under Theden, 1789, co-dir. all Prussian field hosps., gen. staff surgeon, 1797; at hdqrs. in East Prussia, 1806-09. Mem. Leopoldina, Acad. Surgery Copenhagen. Author: Kurze Beschreibung der bei der Königlich-Preussichen Armee Stattfindenden Kranken transportmittel . . . , 1814. Important in the devel. Prussian mil. medicine; organized flying field hosp.; introduced ambulance wagons with spl. springs; improved edn. in mil. medicine. Died Potsdam, June 30, 1822.

GOERTZEL, Gerald, Am. physicist; b. N.Y.C., Aug. 18, 1919; s. Martin B. and Edith (Wilson) G.; M.E., Stevens Inst. Tech., 1940, M.S., 1940; student N.Y. U., 1940-41, 1942-44, Ph.D., 1947; m. Martha Bendheim, June 15, 1941; children—Frances, Marion, Susan. Engr., Republic Aviation Corp., 1941; instr., aero. engr. N.Y. U., 1942-44; physicist Kellex Corp., 1944-45, Oak Ridge Nat. Lab., 1946-48; asso. prof. physics N.Y. U., 1948-52; tech. dir. Nuclear Devel. Corp. Am., 1953-59; v.p. Sage Instruments, Inc., 1960-63, sr. engr. IBM Corp., Yorktown Heights, N.Y., 1964—. Fellow, Am. Phys. Soc., A.A.A.S., N.Y. Acad. Scis.; mem. I.E.E.E., Instrument Soc. Am., Sigma Xi, Tau Beta Pi. Author: Some Mathematical Methods of Physics, 1960. Research and publs. in the fields of theoretical physics, nuclear reactor theory, numerical analysis, instrumentation. Home: 7 Sparrow Circle, White Plains, N.Y. 10605.*

GOESSMANN, Charles Anthony, chemist; b. Naumburg, Germany, June 13, 1827; prepared in gymnasium, Fritzlar, Germany, Ph.D., U. Göttingen, 1853; LL.D., Amherst, 1899; m. Miss M. A. Kinny, 1862. Came to U. S., 1857. Became asst. to Friedrich Wöhler, 1853; invited by Eastwich Bros., Phila., to improve sugar refining process; chemist, mgr. sugar refinery, Phila., 1857-61; chemist Onondaga (N.Y.) Salt Co., 1862-69; prof. chemistry Rensselaer Poly. Inst., 1866-68, chemistry, Mass. Agrl. Coll., 1869—. Chemist Mass. State Bd. Agr.; state insp. fertilizers; analyst Mass. State Bd. Health; dir. Mass. Agrl. Expt. Sta.; made hon. dir. Hatch Expt. Sta., 1895; hon. rep. Dept. of Agr. to study certain sci. matters in Germany, France, 1899. Pres. Assn. Ofcl. Agrl. Chemists. Author many reports, monographs and papers on chem. subjects. Discovered arachic acid in peanuts; described leucine (1st known amino acid); studied chemistry of sorghum and of sugar beet. Died 1910.

GOETHALS, George Washington, Am. engr.; b. Bklyn., June 29, 1858; student Coll. City of N.Y., 1873-76; grad. U. S. Mil. Acad., 1880; grad. Army War. Coll., 1905; LL.D., U. Pa., 1913, Princeton, 1915. Apptd. 2d lt. engrs., 1880, discharged as lt. col., 1898; maj. engr. corps. 1900, maj. gen., 1915, ret., 1916; instr. in civil and mil. engring., U. S. Mil. Acad., several yrs. until 1888; in charge Muscle Shoals Canal constrn., on Tenn. River; chief of engrs. during Spanish-Am. War; mem. Bd. of Fortifications (coast and harbor defense); chief engr. Panama Canal, 1907-14; 1st civil gov. Panama Canal Zone, 1914-16; chmn. bd. apptd. to report on Adamson 8-hour law, 1916; apptd. state engr., N.J., 1917; gen. mgr. Emergency Fleet Corp., 1917; active duty U. S. Army, 1917-19; head, consulting engring. firm of George W. Goethals and Co., 1923-28. Received thanks of Congress, 1915, for service in constructing Panama Canal; recipient medals, Nat. Geog. Soc., Civic Forum, Nat. Inst. Social Scis. Goethals Bridge between Staaten Island and N.J. named for him. Died N.Y.C., Jan. 21, 1928.

GOETHE, Friedrich Walter, German biologist; b. Kiel, June 30, 1911; s. Walter and Maria Goethe; ed. univs. Friburg, Basle, Munster; Ph.D. in biology; m. Elisabeth Peters, 1936; children—Bernhart, Swanhild, Burkhart. Asst., Inst. Animal Biology; mus. asst.; mem. Commn. for Preservation Nature; asst., dir. Vogelwarte Helgoland Inst. Ornithology, Wilhelmshaven. Mem. Soc. German Ornithology, German Soc. Mammalogy, Finnish Assn. Ornithology, Am. Ornithol. Union. Author: Biologie der Silbermöwe, 1937; Die Vogelinsel Mellum, 1939; Die Silbermöwe, 1956; Ethologie der Musteliden. Home: 19 Kirchreihe. Office: Vogelwarte Helgoland, 294 Wilhelmshaven-Rüstersiel, Germany.

GOETHE, Johann Wolfgang von, German poet, naturalist, anatomist, physicist; b. Frankfort/Main, Germany, Aug. 28, 1749; s. Johann Kaspar and Katharine Elisabeth (Textor) G.; student U. Leipzig (Germany), 1765-68, U. Strasbourg (France), 1770-71; m. Christiane Vulpius, 1806; several children. Author various sci. works, including: Metamorphose der Pflanzen, 1790; Farbenlehre, 1805-10. Held that all plant structures are modifications of a type-leaf; maintained an evolutionary view of devel. of plants and animals from original archetypes; studies of bone structure resulted in discovery of recognizable os intermaxillare in human beings, 1784; founded and named science of morphology; a neptunist in geology; attacked Newton's theory of formation of colors from white light; held that light is one and indivisible. Died Weimar, Germany, Mar. 22, 1832.

GOETHERT, Bernhard Hermann, aero. engr.; b. Hannover, Germany, Oct. 20, 1907; s. Bernhard August and Elsie (Rickmeyer) G.; B.S., Tech. U., Hannover, 1930; M.S., Tech. U., Danzig (Poland), 1934; Ph.D. cum laude Tech. U., Berlin, Germany, 1938; m. Hertha Tod., Mar. 29, 1935; children—Hella (Mrs. D. A. Lacy), Winfried, Wolfhart, Reinhard. Came to U. S., 1945, naturalized, 1954. With DVL, Berlin, Germany, 1934-45, dept. chief high speed aerodynamics, 1939-45, cons. as rep. USAF, 1958-59; cons. USAF, Wright Aeronautical Devel. Center, Dayton, O., 1945-59; chief wind tunnel test activities Wright Air Devel. Center, Dayton, 1949-52; with ARO, Inc., 1952-64, chief engine test facility, Tullahoma, Tenn., 1956-59, dir. engring., 1959-63, research v.p., chief scientist, 1963-64, dir., 1959-64; chief scientist Air Force Systems Command USAF, Andrews AFB, Washington, 1964-66; prof., dir. U. Tenn. Space Inst., Tullahoma, 1964—. Mem. various adv. groups NATO; mem. research adv. com. on fluid mechanics, NASA, 1963—; cons. Nat. Acad. Sci. Recipient USAF Scroll of Appreciation, Chief of Staff, 1959; named hon. prof. Tech. U., Aachen, Germany, 1961. Fellow Am. Rocket Soc., Inst. Aerospace Scis. (asso.), Am. Inst. Aeros and Astronautics. Author: Transonic Testing, 1961; also numerous articles. Research and devel. rocketry and constrn. wind tunnels; directed constrn. first high speed wind tunnel. Home: 1703 Sycamore Circle, Manchester, Tenn. 37388. Office: U. Tenn. Space Inst., Tullahoma, Tenn. 37388.*

GOETSCH, Emil, Am. physiologist; b. Davenport, Ia., Jan. 23, 1883; B.S., U. Chgo., 1903, Ph.D., 1906; M.D., Johns Hopkins, 1909; postgrad. Harvard, 1912-15. Asst. in anatomy U. Chgo., 1906-07, asso., 1907-08, asst. in exptl. therapeutics, 1908-09; surgeon in charge Hunterian Labs., Johns Hopkins, 1909-10, asso. surgeon, 1915-18, asso. prof., 1918-19; asst. surgeon Harvard Med. Sch., resident surgeon Peter Bent Brigham Hosp., 1912-15; prof. L.I. Coll. Medicine, 1919-48, emeritus prof. Coll. Medicine, State U. N.Y., 1948—; surgeon in chief Columbia Hosp., 1919-48, attending surgeon, 1948-55. Mem. Am. Surg. Assn., Assn. for Study Goiter. Research on diseases of thyroid gland, physiology and pathology of goiter, chronic thyroditis, cancer of thyroid, hygroma, Riedel's struma, structure of mammalian esophagus, surgery of ductless glands, interrelation of thyroid and adrenal glands; developed test to differentiate hyperthyroidism from functional nervous disorders by use of subcutaneous infection of epinephrine.

GOETTE, Alexander Wilhelm, zoologist; b. St. Petersburg, Russia, Dec. 31, 1840; s. Ernst Bernhard and Natalie (Bagh) G.; studied medicine in Dorpat, Estonia, beginning 1860; M.D., Tübingen, Germany, 1866; m. Maria Hoerschelmann, 1867; 3 sons including Ernst; m. 2d, Ida Peters; 2 daus. Joined faculty, also became asst. to O. Schmidt at zool. inst. U. Strasbourg (France), 1872, became asso. prof., 1877; named dir. zool. collection Municipal Mus., 1880; prof. zoology, dir. zool. inst. U. Rostock (Germany), 1882-86; prof. in Strasbourg, 1886-1918. Author: Die Entwickelungsgeschichte der Unke (Bombinator igneus) als Grundlage einer vergleichenden Morphologie der Wirbeltiere, 1875; Abhandlungen zur Entwicklungsgeschichte der Tiere, 1882-90; Über Vererbung und Anpassung, 1898; Lehrbuch der Zoologie, 1902; Die Entwicklung der Kopfnerven bei Fischen und Amphibien, 1914; Die Entwicklungsgeschichte der Tiere, 1921. Tried to understand orgn. of animals through ontogeny; research on morphology of vertebrates, regeneration, evolution of worms, scyphopolypa, molluscs; prepared way for founding of evolution mechanics by his student Wilhelm Roux. Died Handschuhsheim, Germany, Feb. 5, 1922.

GOETTELMAN, Robert Clement, Am. physicist; b. Grand Junction, Colo., May 26, 1929; s. Clement George and Kathleen (Plunkett) G.; Asso. Sci., Mesa Jr. Coll., 1950; B.A., U. Colo., 1953; postgrad. U. So. Cal., 1954, Stanford, 1956-57; m. Beverly Ann Klein, June 9, 1951; children—Sharon Lee, Timothy Brian, Leslie Brook, Kelly Marie. Physicist, Cryogenic Lab., Nat. Bur. Standards, Boulder, Colo., Washington, 1951-53, Douglas Aircraft Co., El Segundo, Cal., 1953-54; physicist Stanford Research Inst., Menlo Park, Cal., 1954—. Mem. Research Soc. Am., Am. Phys. Soc. Contbr. articles to tech. jours. Research on physics instrumentation, explosive phenomena. Patentee remote icing detector. Home: 413 Dracena Lane, Los Altos, Cal. 94022. Office: 333 Ravenswood St., Menlo Park, Cal. 94025.*

GOETZ, Alexander, physicist; b. Wiesbaden, Germany, Nov. 17, 1897; s. Carl and Ellinor (Elbers) G.; Ph.D., U. Goettingen (Germany), 1921, Ph.D. in Physics and Phys. Chemistry, 1923; m. Sylvia Scott, June 29, 1936; children—Alexander Franklin Hermann, Ellinor Margaretta. Came to U. S., 1927, naturalized, 1940; Privatdozent, U. Goettingen, 1923-29, prof., 1929-41; Rockefeller fellow Cal. Inst. Tech., Pasadena, Cal., 1927-29, asso. prof., 1930—, research dir. Rare Metals Inst., 1937-47, project dir. research contracts and grants, 1948—; guest prof. Imperial Univs., Japan also Tsin Huan U., China, 1929-30; sr. staff cons. Nat. Center for Atmospheric Research, Boulder, Colo., 1964—. Cons. to govt. state and city agys., industry. Mem. Am. Phys. Soc., German Phys. Soc., Am. Pub. Health Assn., A.A.A.S. Sigma Xi. Author: Physik und Technik des Hochvakuums, 1922, 25. Research and publs. on emergency desalting of seawater; devel. membrane filter prodn. and application to microbiol. detection; invented aerosol spectrometer and micro-analytic instrumentation for determination of aerocolloidal matter. Home: 1317 Boston St., Altadena, Cal. 91001.*

GOETZ, Charles Albert, Am. chemist; b. Lockwood, Mo., Jan. 7, 1908; s. Jacob George and Mary (Ripper) G.; student U. Wis., 1926-31; B.S., U. Ill., 1932, M.S., 1934, Ph.D., 1938; m. Sidonia Helen Heck, Feb. 24, 1934; children—Charles Albert II, Roger Melvin. Grad. asst. U. Ill., 1933-38; with Cardox Corp., Chgo., 1938-46, v.p., 1943-46; dir. engring. Brunswick Corp., Chgo., 1946-48; faculty Ia. State U., Ames, 1948—, prof., 1950—, head dept. chemistry, 1950-65; sr. chemist Ames Lab., AEC, 1950—, chief metallurgy div., 1950-61, chief chem. div., 1950-65; v.p., cons. Aeration Processes, Inc., Columbus, O. Mem. Am. Chem. Soc., Electrochem. Soc., Nat. Fire Protection Assn., Sigma Xi. Contbr. numerous articles to field to sci. jours. Inventor process whipping cream by saturation with nitrous oxide under pressure. Home: 822 Ash Av., Ames, Ia. 50010.*

GOETZ, Robert Hans, surgeon; b. Frankfort/Main, Germany, Apr. 17, 1910; s. Johan K. and Emmy (Lesch) G.; Staatexamen, M.D., U. Frankfort, 1935, M.D., 1935; M.D., U. Berne (Switzerland), 1936; M.B., Ch.B., U. Capetown (S. Africa), 1944; m. Verena M. Bluntschli, June 26, 1937; children—Sylvia (Mrs. Eugene Perle), Stephen, Lionel, Angela. Came to U. S., 1957, naturalized, 1965; Marais Meml. Research scholar U. Cape Town, 1937-40, asso. prof. surgery, chief dept. surg. research, 1945-57, chief vascular investigation service Groote Schuur Hosp., 1952-57; faculty Albert Einstein Coll. Medicine, 1957—, prof. surgery, 1964—; attending surgeon Bronx Municipal Hosp. Center; asso. attending surgeon Bronx-Lebanon Hosp. Center, Beth-El Hosp; Health Research Council Career scientist, 1962—. Carnegie Traveling fellow, 1955. Fellow Am. Coll. Angiology (hon.), S. African Coll. Physicians and Surgeons (founding), A.C.S. (sec. Bronx chpt.); mem. N.Y. Soc. for Med. Research (v.p.), Rudolf Virchow Med. Soc. (pres.), Am. Coll. Chest Physicians, Am. Coll. Cardiology, N.Y. Soc. for Cardiovascular Surgery, Internat. Cardiovascular Soc., Royal Soc. Medicine, Physiol. Soc. (London), Royal Soc. S. Africa, Brit. Med. Assn., A.M.A., Deutsche Gesellschaft fur Kreislaufforschung, N.Y. Surg. Soc., Brazilian Coll. Angiology (hon. corr.). Research, numerous publs. on vascular and cardiac physiology, vascular surgery, comparative physiology. Home: 80 Vernon Dr., Scarsdale, N.Y. 10585. Office: Dept. Surgery, Albert Einstein Coll. Medicine, Bronx, N.Y. 10461.*

GOETZ, (Johann Konrad) Wilhelm (Friedrich Eduard), German geographer; b. Schnabelwaid, Germany, July 27, 1844; s. August and Elisabeth (Tretzel) G.; studied theology and philology, Erlangen, also Leipzig, Germany, 1861-65; state exam in theology, 1870, in history, geography, 1874; Ph.D., Tübingen, Germany, 1882; m. Johanna Kiderlin, 1872; 2 daus., 2 sons, including August. Became prison pastor in Lichtenau, 1867, in Sulzbach, 1871; began teaching in Munich, 1876; prof. Bavarian mil. edn. instns., 1890-1909; joined faculty of Tech. U., Munich, 1886; named hon. prof., 1900. Recipient Silver Prince Ludwig medal Munich Geog. Soc., 1904. Author: Die Verkehrswege im Dienste des Welthandels, 1888; Lehrbuch der wirtschaftlichen Geographie für Handels-, Real- und Gewerbeschulen, 1891; Geographische-Historisches Handbuch von Bayern, 2 vols., 1895-98; Landeskunde des Königreiches Bayern, 1904; Historische Geographie, Beispiele und Grundlinien, 1904. First to promote econ. and comml. geography in Germany; father of nat. (Bavarian) hist. geography; research on hist. geography of southeastern, eastern and no. Europe. Died Munich, Mar. 26, 1911.

GOEZE, Johann August Ephraim, German zoologist; b. Aschersleben, Germany, May 28, 1731; s. Johann Heinrich and Catherine Margarete (Kirchhoff) G.; student theology, Halle, Germany; m. Leopoldine Maria Keller, 1770; 4 children; Became pastor in Aschersleben and Quedlinburg, 1751, at Blasius Ch., 1762; named 1st ct. deacon of Quedlinburg Sem., 1787. Author: Entomologische Beiträge zu des Rr Linné zwölften Ausgabe des Natursystems, 1777-83; Versuch einer Naturgeschichte der Eingeweidewürmer thierischer Körper, 1782. 1st research on insects and small animal life in water; 1st descriptions of numerous entomol. types; described 1st tardigrades, 1773; research in helminthology, arthropods. Died Quedlinburg, Germany, June 27, 1793.

GOFF, Emmet Stull, Am. horticulturist; b. Elmira, N.Y., Sept. 3, 1852; s. Gustavus A. and Mary (Stull) G.; m. S. Antoinette Carr, Oct. 2, 1880. Fruit grower and farmer, nr. Elmira, 13 years; horticulturist

N.Y. Exptl. Sta., Geneva, 1882-89, experimented on culture of many economically important plants; 1st prof. horticulture U. Wis., 1899-1902. Author: Principles of Plant Culture, 1897; Lessons in Commercial Fruit Growing, 1902. Did his most notable research in field of differentiation of flower buds of fruit plants; pioneered in using sprays to combat fungi and harmful insects; studied root systems and budding of plants; developed new, hardy varieties of plums. Died June 6, 1902.

GOFF, John Alonzo, Am. mechanical engineer; b. Colorado City, Colo., Oct. 19, 1899; s. Joseph Randolph and Anna Bell (Cadwell) G.; student Colo. Agrl. Coll., 1917-19; B.S. in M.E., U. of Ill., 1921, M.S., 1924, Ph.D., 1927; m. Virginia Brands Hanawalt, Jan. 27, 1933; children—John Randolph, Susan Virginia. Asst. in mech. engring., U. of Ill., 1921-23, instr., 1923-25, asso., 1925-27; geophysicist Sun Oil Co., Dallas, Tex., 1927-28; engr. Linde Air Products Co., New York, 1928-29; research engr. Westinghouse Research Labs., 1929-30; asso. prof. thermodynamics, U. of Ill., 1930-35, prof. of thermodynamics, 1935-38; dean Towne Sci. Sch. and dir. mech. engring., 1938-50; prof. mech. engring. U. of Pa. since 1950; consultant on thermodynamics Westinghouse Electric & Mfg. Co., 1930-40, Div. 6. O.S.-R.D., 1942-44; dir. U. Pa. Thermodynamics Research Lab., 1945-51; cons. thermodynamics U. S. Nat. Bur. Standards, Mechanical Systems Section, 1961—. Fellow Am. Soc. M.E. (chmn. bd. honors 1964-65); mem. Am. Math. Soc. Author: Notes on Thermodynamics, 4th edit. 1946; also tech. articles and reviews. Research on thermodynamic properties of moist air (the Goff-Gratch formulation), 1950; thermodynamic analysis of the limits imposed by chemical equilibrium on the performance of the torpedo powerplant, 1945; investigated the Caratheodory axiomatic approach in the construction of thermodynamic theory. Home: 623 Righters Mill Rd., Narberth, Pa. 19072. Office: U. of Pa., Phila. 19104.*

GOGLIA, Mario Joseph, Am. mech. engr.; b. Hoboken, N.J., Mar. 30, 1916; s. Frederick Louis and Roza (Coppola) G.; M.E., Stevens Inst. Tech., 1937, M.S., 1941; Ph.D., Purdue U., 1948; m. Juanita Dixon, June 8, 1940; children—David, Rozanne. Faculty, U. Ill., 1938-45, asst. prof. mech. engr., 1944-45; application engr. Republic Flowmeter Co., Chgo., 1945-46; faculty Purdue U., 1946-48; prof. Ga. Inst. Tech., Atlanta, 1948-55, Regents prof., 1955-58, dean grad. div., 1960-66; dean engring. U. Notre Dame, 1958-60; vice chancellor for research Univ. System of Ga., 1966—. Cons., Office Edn., Bur. Higher Edn. Recipient Naval Ordnance award, 1945. Mem. Am. Soc. M.E., Am. Soc. Engring. Edn., Sigma Xi, Tau Beta Pi, Pi Tau Sigma, Pi Mu Epsilon, Phi Kappa Phi. Author: Thermodynamics, 1955; also articles. Research on producing smoke screens by airplanes, parachute cloth phys. changes as air passes thru fabric; explained mechanics fluid flow in channel with uniformly porous walls. Patentee design modification of gaseous diffusion plant. Home: 3066 Arden Rd. N.W., Atlanta 30305.*

GOGUEL, Jean, French geologist; b. Paris, France, Jan. 2, 1908; s. Maurice and Jeanne (Nyegaard) G.; ed. Poly. Sch., Paris Sch. Mines; D.Sc., 1937; m. Micheline Vernes, Mar. 20, 1931; children—Arian, Alain, Claude, Sylvie, Bernard, Béatrice. Prof. Paris Sch. Mines; engr. Mines Corps Service; dir. French Geol. Map Service, from 1953. Head, Geophysics Service, Overseas Office Sci. and Tech. Research. Decorated Legion Honor. Mem. London Geol. Soc. (hon.), Am. Geology Soc. (hon.), Am. Acad. Arts and Scis. Author: Traité de Tectonique; Géologie de la France, 1950; Application de la Géologie aux Travaux de L'Ingenieur, 1959. La Graviométrie, 1963. Research, publs. on geol. field work in French Alps; studies in gen. tectonics, mechanics tectonic deformation; work in geophysics. Home: 100 rue du Bac, Paris 7. Office: 62 Blvd. St. Michel, Paris 6, France.*

GOHL, Johann Daniel, (pseudonym: Ursinus Wahrmund), German physician; b. Berlin, July 26, 1674; s. Johann and Sophia (Weiland) Gool; ed. in langs., law and medicine U. Halle, M.D., 1698; m. Eva Victoria Eisener, 4 children, including Daniel August, Eleaonore Catherine (m. Manitius). Practiced medicine, Berlin, 1698, Halle, from 1699, also lectr. in medicine; later practiced again in Berlin; physician in Freienwalde, 1711; ofcl. physician Oberbarnim dist., Wriezen, from 1721. Mem. Soc. Scis. Berlin, Soc. Scis. Leopoldina. Author: Gedancken vom gesunden und langen Leben der Menschen, 1709; Gantz generale Instruction von der Tugend und Gebrauch der Freyenwalder Gesundbrunnens, 1716; Medicina practica, clinica et forensis, 1735. Pioneer in med. statistics; pub. Acta medicorum Berolinensuim (probably 1st German statistics on death causes), from 1717; considered mental illness within scope of medicine, tried to treat it scientifically. Died Wriezen/Oder, Apr. 2, 1731.

GOHORY, Jacques (Gohorry, also Sauvius, Leo; Solitarius), French naturalist; b. Paris, early 16th century; Author: Discours responsif . . . , 1575; Theophrasti Paracelsi philosophiae et medicinae . . . , 1568; published def. of ancient medicine and philosophy. Translated the book of Lemnius into French; also other works. Recommended antimony and potable gold as medicines; publs. on tobacco. Died Paris, Mar. 5, 1576.

GOIDANICH, Gabriele, Italian agronomist; b. Aosta, Aug. 30, 1912; s. Pier Gabriele and Itala (Gasperini) G.; prof. vegetable pathology; m. Piera Pozzoli, 1937. Dir., Exptl. Lab. and Obs. for Study Diseases of Plants, U. Bologna; former prof. vegetable pathology U. Cal. Mem. Nat. Acad. Agr. (v.p.). Author: Avversita delle Plante Agrarie; Alterazioni Crittogamiche del Legno; Manuale di Patologia Vegetale. Address: 29/III, via Gandino, Bologna, Italy.

GOIN, Coleman Jett, Am. biologist; b. Gainesville, Fla., Feb. 25, 1911; s. Newbold Loecher and Mariam (Jett) G.; B.S., U. Fla., 1939, M.S., 1941, Ph.D., 1946; m. Olive Lynda Bown, June 7, 1940; children—Olive Lynda, Coleman Jett. Mem. faculty U. Fla., Gainesville, 1945—, prof. biol. sci., 1956—; research asso. Carnegie Mus., Pitts., 1950—. Panelist, NSF, 1964—. Recipient certificate of appreciation Fla. Blue Key, 1959. Fellow Herpetologists League; mem. Am. Soc. Ichthyologists and Herpetologists (pres. 1966), Am. Soc. Naturalists, Evolution Soc. Author: (with A. F. Carr) Guide to Reptiles, Amphibians and Fresh Water Fishes of Florida, 1955; (with O. B. Goin) Introduction to Herpetology, 1962, Comparative Vertebrate Anatomy, 1965. Research and publs. on speciation, evolution, ecology of amphibians. Home: 626 N.W. 36th Av., Gainesville, Fla. 32601.*

GOKIELI, Levan Petrovich, Russian logician, mathematician; b. Kutaisi, 1901; grad. Math. Faculty, Tbilisi U., 1924; D.Physico-Math. Sci., 1935. Head chair gen. math. Tbilisi U., 1934—; asso. Georgian Acad. Sci., 1935—; prof., 1936—. Mem. Georgian Acad. Scis. (corr.). Author: The Mathematics of Possibility and the Mathematics of Reality, 1939; The Concept of Existence in Mathematics, 1944; The Axiomatization of Logic, 1947; The Concept of Numbers, 1951; The Paradoxes of the Theory of Sets, 1957; The Nature of the Logical, 1958; Induction in Mathematics, 1960; The Philosophical Importance of Non-Euclidean Geometry, 1963. Address: Tbilisi University, prospect Chavchavadze 45, Tbilisi, Gruz, SSR, USSR.

GOKSOYR, Jostein, Norwegian microbiologist; b. Koparrik, Norway, June 28, 1922; s. Harald and Aslaug (Nilsen) G.; mag.scient., U. Oslo, 1950, Dr.-philos., 1955; m. Eva-Liisa Laitinen, June 19, 1952; children—Matti Erik, Harald, Anders. Asst., U. Oslo, 1948-51; with Central Inst. for Indsl. Research, Oslo, Norway, 1952-55; faculty U. Bergen (Norway), 1956—, prof. microbiology, 1966—, vice dean Faculty Sci., 1966-67. Mem. Norwegian Soc. for Microbiology (past pres.), Norwegian Chem. Soc., Norwegian Biochem. Soc., Scandinavian Soc. for Plant Physiology, Soc. for Gen. Microbiology. Author: Kroppen var, 1965; also articles. Norwegian editor Physiologia Plantarum, 1967—. Research on modes of action of fungicides, especially sulfur-containing compounds, wood-decomposing fungi, especially dry rot fungus Merulius lacrymans. Home: 18 Solbakken, Bergen, Norway.*

GOLAND, Martin, Am. engineer; b. N.Y.C., July 12, 1919; s. Herman and Josephine (Bloch) G.; M.E., Cornell U., 1940; m. Charlotte Nelson, Oct. 16, 1948; children—Claudia, Lawrence, Stewart. Instr. mech. engring. Cornell U., 1940-42; sect. head structures dept. research lab., airplane div. Curtiss-Wright Corp., Buffalo, 1942-46; chmn. div. engring. Midwest Research Inst., Kansas City, Mo., 1946-50, dir. for engring. scis., 1950-55; v.p. Southwest Research Inst., San Antonio, 1955-57, dir., 1957-59, pres., 1959—; prof. research (honoris causa) St. Mary's U., San Antonio. Chmn. subcom. vibration and flutter NACA, 1952-60; chmn. research adv. com. on aircraft structures NASA, 1960—; sci. adv. com. Harry Diamond Labs., U. S. Army Materiel Command, 1955—; adv. panel com. sci. and astronautics Ho. of Reps., 1960—; mem. high speed ground transp. panel Dept. Commerce 1966—, nat. inventors council, 1966—; sci. adv. panel Dept. Army, 1966—; chmn. U. S. Army Weapons Command Advisory Group, 1966—. Chmn. space age industry devel. com.; mem. Greater San Antonio Devel. Com. Bd. govs. St. Mary's U., San Antonio; grad. council Trinity U., San Antonio; vis. com. dept. aerospace scis. University of Texas; research advisory committee coordinating board Texas College and University System, 1966—; director of the San Antonio Symphony. Recipient Spirit of St. Louis jr. award Am. Soc. M.E., 1945, jr. award, 1946, Alfred E. Nobel prize Am. Soc. C.E., 1947 Fellow Am. Soc. M.E. (dir., mem. bd. tech., mem. tech. devel. com.; v.p. communications), Am. Ordnance Assn., A.A.A.S.; mem. C. of C. (dir.), Soc. Exptl. Stress Analysis, Am. Inst. Aero. and Astronautics (dir.; treas.), Research Soc. Am., Sigma Xi. Editor Applied Mechanics Review, 1952-59, editorial adviser, 1959—. Research on aircraft design; applied mathematics; structures; aerodynamics; dynamics; engineering analysis. Home: 211 Five Oaks St., San Antonio 78209. Office: 8500 Culebra Rd., San Antonio 78206.*

GOLD, Bela, economist; b. Kolozsvar, Hungary, Jan. 30, 1915; s. Leo and Esther (Ludwig) G.; came to U. S., 1920, naturalized, 1927; B.S. in Mech. Engring., N.Y. U., 1934; Ph.D. (Univ. fellow), Columbia, 1948; m. Sonia Steinman, July 5, 1938; 1 son, Robert. Research cons. Life Ins. Sales Research Bur., Hartford, Conn., 1938-39; asst. head, div. program surveys U. S. Bur. Agrl. Econs., Washington, 1939-43; econ. cons. U. S. Senate subcom. on war moblzn., 1943-44; econ. adviser U. S. Fgn. Econ. Adminstrn., U. S. Dept. Commerce, 1944-46; prof. indsl. econs. U. Pitts. Grad. Sch. Bus., 1947-66; Timken prof. indsl. econs., dir. research program indsl. econs., prof.-in-charge econs. Case Inst. Tech., Cleve., 1966—. Indsl. cons., 1950—; cons. research univs. Pa., Pitts., Temple, 1957-59; vis. prof. Imperial Coll. Sci. and Tech., London, 1967. Social Sci. Research Council fellow, 1937-38; Ford Found. fellow, 1961-62; vis. fellow Nuffield Coll., Oxford, 1964; Ford Found. grantee, 1966-67. Mem. Am. Econ. Assn., Inst. Mgmt. Scis., Internat. U. Contact for Mgmt. Edn., Nat. Assn. Bus. Economists, Acad. Mgmt., Am. Assn. U. Profs., A.A.A.S. Author: Wartime Economic Planning in Agriculture, 1949, How Is Higher Education Financed?, 1959, Foundations of Productivity Analysis, 1955, Long Term Iron and Steel Manufacturing Costs, 1956; (with J. O. Stalson) Organizing for Sales, 1939. Research in indsl. costs, growth patterns, and tech. effects; productivity analysis concepts and tools covering multiple inputs, changes in product-mix and different aggregation levels; studies in potential, planned and actual performance in interrelated sectors of econ. moblzn. Home: 2300 Overlook Rd., Cleve. 44106.*

GOLD, Harry, Am. physician; b. Bialestok, Russia, Dec. 25, 1899; s. Samuel and Naomi (Katz) G.; came to U. S., 1903, naturalized, 1910; B.A., Cornell U., 1919; M.D., 1922; m. Bertha Goldman, Aug. 26, 1926; children—Naomi (Mrs. Norbert Steinberger), Stanley, Muriel (Mrs. Alan Morris). Faculty, Cornell U. Med. Coll., 1922-65, prof. clin. pharmacology, 1947-65, prof. emeritus, 1965—; practice medicine specializing in cardiology, N.Y.C., 1922—; chief cardiovascular research unit Beth Israel Hosp., N.Y.C., 1931-64; chief cardiac clinic Hosp. for Joint Diseases, N.Y.C., 1930-63, cons., 1963—. Mem. revision com. U. S. Pharmacopaeia, 1940-57; lectr. to med. colls., U. S., Europe, 1930—. Fellow N.Y. Heart Assn., A.M.A., Am. Coll. Clin. Pharmacology, N.Y. Acad. Sci., N.Y. Acad. Medicine; mem. Am. Pharmacological Soc., Biometric Soc., Phi Beta Kappa, Alpha Omega Alpha; hon. mem. Cardiology Soc. Brazil, Argentine Med. Assn. Author: Quinidine in Disorders of the Heart, 1950; also numerous articles on pharmacology and cardiology. Research on animal lab. pharmacology, human hosp., pharmacology. Address: 7 E. 82d St., N.Y.C. 10028.*

GOLD, Marvin Harold, Am. chemist; b. Buffalo, June 23, 1915; s. Max and Jennie (Franklin) G.; B.A., U. Cal., Los Angeles, 1937; Ph.D., U. Ill., 1940; postgrad. Cal. Inst. Tech., U. Cal., Los Angeles; m. Sophye Mendelson, Aug. 31, 1940; children—Judith May, Norman Charles. Head organic chemistry The Visking Corp., Chgo., 1942-48; with Aerojet-Gen. Corp., Sacramento, 1948—, sr. scientist, 1963—. Recipient Navy Meritorious Pub. Service citation, 1962; Anna Fuller Fund Cancer fellow Northwestern U., 1940-42. Mem. Research Soc. Am. (br. pres.), Am. Inst. Aeros. and Astronautics, Nat. Security Indsl. Assn., Am. Chem. Soc., Am. Ordnance Assn., Sigma Xi, Phi Lambda Upsilon. Research, publs., patents in field of polynitro aliphatic chemistry, catalysis, pyrolysis, reaction kinetics. Home: 4849 Alexon Way, Sacramento 95841. Office: P.O. Box 15847, Sacramento 95813.*

GOLD, Thomas, astronomer, physicist; b. Vienna, Austria, May 22, 1920; s. Max and Josephine (Martin) G.; B.A., Cambridge U., Eng., 1942, M.A., 1945, fellow Trinity Coll., Cambridge, 1947; M.A. (hon.), Harvard, 1957; m. Merle Eleanor Tuberg, June 21, 1947; children—Linda, Lucy, Tanya. Came to U. S., 1956. Lectr. physics Cambridge U., 1948-52; chief asst. to Astronomer Royal, Gt. Britain, 1952-56; prof. astronomy Harvard, 1958, Robert Wheeler Willson prof., 1958-59; prof. astronomy, chmn. dept. astronomy, dir. Center Radio-physics and Space Research, Cornell U., 1959—. Mem. lunar and planetary missions bd., cons. NASA; mem. space panel Pres.'s Sci. Adv. Com. Trustee Asso. Univs., Inc. Fellow Am. Acad. Arts and Scis., Royal Soc. London; member Internat. Acad. Astronautics, Royal Astron. Soc. (past councillor), Nat. Acad. Scis., Am. Astron. Soc., Am. Geophys. Union. Contbr. articles profl. jours. Study of astronomy; geophysics; biophysics; the steady state theory of the expanding universe; dynamical problems in the solar system; such as the motion of the earth's axis of rotation, the spin of the planet Mercury; the constitution of the surface of the moon and its derivation; theories of Quasars based on star collisions; discussions of the nature of time and of problems arising out of relativity theory; theories of the physics of sense organs, particularly the inner ear. Address: 414 Cayuga Heights Rd., Ithaca, N.Y.*

GOLD, Victor, chemist; b. Vienna, Austria, June 29, 1922; s. Oscar and Emmy G.; student King's Coll., London, Eng., 1939-40; B.Sc., U. Coll., London, 1942, Ph.D., 1945; D.Sc., London U., 1958; m. Jean Sandiford, Mar. 27, 1954; children—Elizabeth Helen, Martin Peter. Faculty, King's Coll., U. London, 1944—, prof. chemistry, 1964—; vis. prof. Cornell U., 1962, 65, research fellow, 1951-52. Mem. Chem. Soc. (mem. primary jours. com. 1966-67), Faraday Soc.,

Am. Chem. Soc. Author: pH Measurements, 1956; (with D. Bethell) Carbonium Ions, 1967; also numerous articles. Editor: Advances in Physical Organic Chemistry. Research in reaction mechanisms in solution, theory and application of kinetic hydrogen isotope effects; basicity of unsaturated hydrocarbons and acidity of aromatic nitro-compounds and structures of carbonium ions and carbanions formed. Office: King's Coll., Strand, London W.C.2., Eng.*

GOLDA, Veroslav, Czechoslovak, physician; b. Police, Czechoslovakia, Feb. 11, 1927; s. Frantisek and Stepánka (Zajíckova) G.; Ph.D., Palacky U., 1953, M.D., 1959; C.Sc., Komensky U., 1964; m. Anna Urbáskova, Oct. 13, 1952; 1 dau., Hana. Staff, Inst. Exptl. Medicine, Czechoslovakian Acad. Scis., Bratislava, 1960, dept. physiology, Prague, 1961; staff dept. physiology Hungarian Acad. Scis., Pécs, 1961-62, Nencki Inst. Exptl. Biology, Polish Acad. Scis., Warsaw, 1965, Inst. Exptl. Medicine, USSR Acad. Scis., Leningrad, 1966; head working group, tchr. Inst. Psychology, Olomouc, Czechoslovakia, 1966—. Mem. Czechoslovakian Physicians J. E. Purkyne. Research, publs. on orgn. of motor system, laterality of motor functions in cats, morphological and electrophysiol. asymmetry of hemispheres in cats, restitution of motor functions after partial destruction of CNS. Home: 1 Cyrilometod Sq. Office: 3 Hnevotinská Str., Olomouc, Czechoslovakia.*

GOLDANSKII, Vitalii Iosifovich, Russian chemist, physicist; b. Vitebsk, U.S.S.R., June 18, 1923; s. Joseph E. and Judith (Melamed) G.; grad. Moscow (USSR) U., 1944, Candidate chemistry, 1947, D.Sc., 1954; m. Ljudmila N. Semenova, Oct. 12, 1947; children—Dmitrij, Andrej. Staff, Inst. Chem. Physics, Acad. Scis. U.S.S.R., Moscow, 1944-52, 61—, sr. scientist, 1949-52, head lab. nuclear and radiation chemistry, 1961—, sr. scientist, head div. lab. photomesonic processes Lebedev Phys. Inst., Acad. Scis. USSR, 1952-61; prof. Moscow Inst. Phys. Engring. 1951—. Mem. U.S.S.R. Acad. Sci. (corr.; Mendelejiev prize 1966). Author: (with A. M. Bladin, I. L. Rozenthal) Kinematics of Nuclear Reactions, 1959; (with A. V. Kutzenko, M. I. Podgoretzkii) Counting Statistics of Nuclear Particles Recording, 1959; Mössbauer Effect and its Applications in Chemistry, 1963; Physical Chemistry of Positron and Positronium, 1968; also numerous articles. Research on ionic catalysis in polymolecular adsorption layers, high energy nuclear and photonuclear reactions; theory of Cerenkov radiation of cosmic rays in atmosphere; exptl. determination of electric polarizability of proton; prediction and theoretical description of 2-proton radioactivity; predictions of properties of neutron-excessive and neutron-deficient nuclei; devel. applications of Mössbauer effect in different brs. chemistry; positron annihilation and devel. its chem. applications; solid phase polymerization induced by radiation and shock waves. Address: Vorobjevskoje Shausse, 2b, Moscow V-334, U.S.S.R.*

GOLDBACH, Christian, mathematician; b. Konigsberg, Prussia, Mar. 8, 1690; ed. Oxford, Eng., 1712; made frequent study trips through Europe, during early 1720's; studied under A. Duhre, Sweden; apptd. Prussian ambassador to St. Petersburg, 1725; mem. staff Russian Ministry of Fgn. Affairs, from 1742; prof. math.; mem. St. Petersburg Acad. Scis (sec.). Author: Recueil de l'Academie de Saint Petersburg, 1728-38. Among the few Russian mathematicians of time; formulated Goldbach's theorem dealing with number theory, which states that every even number is sum of 2 prime numbers, 1742; also dealt with curves, series theory, and their application to integration of differential equations; corresponded with Daniel Bernoulli and Euler. Died St. Petersburg, Nov. 20, 1764.

GOLDBERG, Abraham, Scottish physician; b. Edinburgh, Scotland, Dec. 7, 1923; s. Julius and Rachel (Varinovsky) G.; M.B. Ch.B., U. Edinburgh; M.D. with gold medal, 1956; D.Sc., U. Glasgow, 1966; m. Clarice Cussin, Sept. 3, 1957; children—David, Jennifer, Richard. Nuffield research fellow Med. Sch. London, Eng., 1952-54; Med. Research Council travelling fellow U. Utah, Salt Lake, City, 1954-56; lectr. medicine U. Glasgow (Scotland), 1956-60, sr. lectr., 1960-65, reader in medicine, 1965-67, prof., 1967—. Dir. group on iron and Porphyrin metabolism Med. Research Council. Sydney Watson Smith lectureship, 1964. Recipient Fleck award, 1967. Fellow Royal Coll. Physicians (Glasgow, Edinburgh and London); mem. Assn. Physicians, Scottish Soc. Physicians, Med. Research Soc. Author: (with Rimington) Diseases of Porphyrin Metabolism, 1962. Editor: Scottish Med. Jour., 1962-63. Research, publs. on clin. and exptl. studies in haemoglobin, porphyrin and iron metabolism and asso. aspects of clin. medicine. Home: 15 Baronald Dr., Glasgow W.2. Office: Western Infirmary, Glasgow, Scotland.*

GOLDBERG, Edward David, Am. geochemist; b. Sacramento, Aug. 2, 1921; s. Edward Davidow and Lillian (Rothholz) G.; B.S., U. Cal., Berkeley, 1942; Ph.D., U. Chgo., 1949; m. Betty Jean Anderson, Feb. 23, 1945; children—David Wilkes, Wendy Jean. Faculty, Scripps Instn. Oceanography, La Jolla, Cal., 1949—, prof. chemistry, 1960—; provost Revelle Coll. U. Cal., San Diego, 1965-66. Guggenheim fellow, 1961. Mem. Am. Geophys. Union, A.A.A.S., Geochemical Soc., Sigma Xi. Author: (with J. Geiss) Earth Sciences and Meteoritics, 1964. Contbr. numerous articles to profl. jours. Research, publs. primarily in chem. composition of sea water, sediments, marine organisms; dating techniques in marine environment; radioactive dating techniques applied to glaciers; co-discoverer silicon-32 in nature. Home: 2614 Ellentown Rd., La Jolla 92037. Office: P.O. Box 109, La Jolla, Cal. 92037.*

GOLDBERG, John Edward, Am. civil engr.; b. Seattle, Sept. 29, 1909; s. Max A. and Ida (Hanock) G.; B.S. in Civil Engring. with highest distinction, Northwestern U., 1930, C.E., 1931; Ph.D., Ill. Inst. Tech., 1955; m. Dorothy M. Long, June 16, 1944; 1 dau., Jane. Engr., City of Chgo., 1931-42; structures engr. Waco Aircraft Co., Troy, O., 1942-43; structures engr. Convair, San Diego, Cal., 1943-47; instr. engring. U. Cal. at San Diego, 1944-47; asst. prof. mechanics, dir. fundamental mechanics research Ill. Inst. Tech., Chgo., 1947-50; faculty Purdue U., Lafayette, Ind., 1950—, prof. structural engring. 1955—, head dept., 1965—. Dir. Midwest Applied Sci. Corp., Lafayette. Cons. nuclear engring., structural engring.; mem. ship hull research com. Nat. Acad. Scis. Mem. Am. Soc. C. E., Am. Soc. Engring. Edn., Internat. Assn. Bridge and Structural Engring., Sigma Xi, Tau Beta Pi, Chi Epsilon. Asso. editor Applied Mechanics Revs., 1947-50. Research and publs. in fields of multistory bldg. structures, shell theory, stability theory, elasticity, vibrations, engring. seismology. Home: 1805 Western Dr., West Lafayette, Ind. 47906.*

GOLDBERG, Leo, Am. astronomer; b. Bklyn., Jan. 26, 1913; s. Harry and Rose (Ambush) G.; S.B., Harvard, 1934, A.M., 1927, Ph.D., 1938; m. Charlotte B. Wyman, July 9, 1943; children—Suzanne, David Henry, Edward Wyman. With U. Mich., 1938-60, asso. prof., chmn. dept. astronomy, dir. obs., 1946-60; Higgins prof. astronomy Harvard, 1960—, chmn. dept., dir. Coll. Obs., 1966—; staff Smithsonian Astrophys. Obs., Cambridge, Mass., 1960-66. Mem. sci. adv. bd. USAF, 1959-62; mem. def. sci. bd. U. S. Dept. Def., 1962-64. Mem. solar physics sub-com. NASA, 1962-65, sci. and tech. adv. com., 1964—, ad hoc sci. adv. com., 1966—. Recipient Bowdoin Essay prize Harvard, 1938, Navy award for exceptional service to naval ordnance devel., 1946. Fellow Am. Acad. Arts and Scis.; mem. Nat. Acad. Scis. (past mem. space sci. bd., dir. Benjamin Apthorp Gould Fund 1959—), Internat. Astron. Union (mem. U. S. nat. com. 1956—, past chmn. com., pres. com. 44 1961-67), Am. Philoso. Soc., Am. Phys. Soc., Am. Astron. Soc. (pres. 1964-66), Royal Astron. Soc. (fgn. asso.), Societe Royale des Sciences de Liege (fgn. asso.), Internat. Astron. Union (v.p. 1958-64), Amales d'Astrophysique (fgn. corr.), Sigma Xi. Author: (with L. H. Aller) Atoms, Stars and Nebulae, 1943; also numerous articles. Collaborating editor Astrophys. Jour., 1949-51, chmn. editorial bd., 1954; editor Ann. Rev. Astronomy and Astrophysics, 1961—; editorial bd. Space Sci. Revs., 1961—, Solar Physics, 1966—. Research on temperature, density and chem. composition sun and other stars with spl. emphasis on atomic processes of importance to astrophysics; observations from space vehicles. Home: 33 Ledgewood Dr., Weston, Mass. 02193.*

GOLDBERG, Richard Robinson, Am. mathematician; b. Chgo., Sept. 6, 1931; s. Berthold L. and Sara (Robinson) G.; B.S., Northwestern U., 1951, A.M., Harvard, 1952, Ph.D., 1956; m. Evelyn Kelman, Aug. 9, 1953; children—Deborah, Daniel. Mathematician, Atomic Power div. Westinghouse Corp., Pitts., 1955-57; instr. Northwestern U., Evanston, Ill., 1957-58, asst. prof., 1958-61, asso. prof., 1961-67, prof., 1967—. Author: Fourier Transforms, 1961; Methods of Real Analysis, 1964. Contbr. articles to sci. jours. Research in harmonic analysis; integral transforms. Home: 320 Wesley St., Evanston, Ill. 60202.*

GOLDBERGER, Isidore Harry, Am. physician; b. N.Y.C., Aug. 24, 1888; s. Herman and Rose (White) G.; M.D., N.Y. U., 1910; m. Minnie Snow, Feb. 12, 1913; children—Eleanor (Mrs. Robert S. Frank), Marjorie (Mrs. Edmund Grasheim). Practice medicine, specializing in pediatrics, N.Y.C., 1911—; co-founder, cons. pediatrician Morrisania City Hosp.; dir. health edn. Bd. Edn., N.Y.C., 1914-58, dir. emeritus, 1958—; faculty Sch. for Oral Hygiene, Columbia, 1916-47; clin. prof. pediatrics, 1911-41, emeritus, 1941—; cons. coll. physician Bd. Higher Edn., N.Y.C., 1964—. Recipient award for services to children State N.Y. and Nation, N.Y. State Assn. Health, Phys. Edn. and Recreation, 1958; trophy for promotion mut. good will and understanding Interfaith Movement, 1958; Silver scroll N.Y.C. Bd. Edn., 1958, scroll for distinguished and exceptional pub. service City, N.Y., 1941, 56, 59. Fellow A.M.A.; mem. Am. Pub. Health Assn., A.A.H.P.E.R. (Anderson award 1963), Am. Cancer Soc. (dir. N.Y.C.), Am. Sch. Health Assn., Royal Soc. For Promotion Health (Eng.), N.Y. Acad. Medicine, Bronx Pediatric Soc. (founder, past pres.), others. Author: (with J. Mace Andress, A. K. Aldinger) Health Essentials, 1928; (with J. Mace Andress) Health School on Wheels, 1933, Broadcasting Health, 1933; Spic and Span, 1939; The Health Parade, 1939; Growing Big and Strong, 1941; Safety Every Day, 1941; Health and Physical Fitness, 1943; Doing Your Best for Health, 1945; Building Good Health, 1945; Helping the Body in Its Work, 1949; The Healthy Home and Community, 1949; Understanding Health, 1950; (with Grace T. Hallock) Health For Life, 1964. Introduced diphtheria toxoid for mass prophylaxis, use of Salk vaccine, chest X-ray, Sch. Health Day, audiometer testing, other health programs in N.Y.C.; one of discoverers of adding Vitamin D in milk to eliminate rickets, co-worker use of measles convalescent serum. Home: 2420 Sedgwick Av., Bronx 10468. Office: 2625 Grand Concourse, Bronx, N.Y. 10468.*

GOLDBERGER, Joseph, physician; b. Austria-Hungary, July 16, 1874; s. Samuel and Sarah (Gutman) G.; student Coll. City of N.Y., 1890-92; M.D., Bellevue Hosp. Med. Coll. (N.Y. U.), 1895; m. Mary Humphreys Farrar, Apr. 19, 1906; children—E. Farrar, Joseph H., B. Humphreys, Mary Humphreys. Resident physician Bellevue Hosp., 1895-97; pvt. practice, Wilkes-Barre, Pa., 1897-99; commd. asst. surgeon USPHS, 1899, passed asst. surgeon, 1904, surgeon, 1912, sent to study yellow fever and typhus, Cuba, 1899, both of which he contracted; Chief work, research in preventive medicine, infectious diseases; dir. field nutrition investigations; published results of original investigations of trematodes, the straw itch, yellow fever, dengue fever, measles, typhus fever, cholera media, diphtheria carriers; showed pellagra is a dietary deficiency disease, 1915. Died Washington, D.C., Jan. 17, 1929.

GOLDBERGER, Marvin Leonard, Am. physicist; b. Chgo., Oct. 22, 1922; s. Joseph and Mildred (Sedwitz) G.; B.S., Carnegie Inst. Tech., 1943; Ph.D., U. Chgo., 1948; m. Mildred Ginsburg, Nov. 25, 1945; children—Samuel M., Joel S. Research asso. Radiation Lab., U. Cal., 1948-49, now cons.; research asso. Mass. Inst. Tech., 1949-50; asst. prof. U. Chgo., 1950-55, prof., 1955-57; Higgins prof. physics Princeton, 1953-54, 57—. Cons. Los Alamos Sci. Lab., Brookhaven Nat. Lab. Fellow Am. Phys. Soc., Am. Acad. Arts and Scis.; mem. Nat. Acad. Scis. Author: (with W. Watson) Collision Theory, 1954. Worked with Manhattan Project on reactor design, 1944; developed (with Fermi) theory of high-energy nuclear reactions, involving predictions of energy and angular distbn. of particles emerging from a nucleus, 1948; developed (with Gell-Mann) theory of collision processes, 1952, dispersion theory (now basic to theoretical physics), 1962. Home: 125 Fitz-Randolph Rd., Princeton, N.J.*

GOLDBLATT, Harry, Am. pathologist; b. Ia., Mar. 14, 1891; s. Philip and Jennie (Spitz) G.; A.B., McGill U., 1912, M.D., 1916; m. Jean Elizabeth Rea, June 25, 1929; children—David, Peter Jerome. Beit Meml. research fellow Lister Inst. Preventive Medicine, London, 1921-23; with dept. physiology Univ. Coll., London, 1923-24; asst. prof. pathology Western Res. U., 1924-27, asso. prof., 1927-35, prof. exptl. pathology, 1935—, asso. dir. Inst. Pathology, 1929—. Mem. A.M.A., Am. Assn. Pathologists and Bacteriologists, Soc. for Exptl. Biology and Medicine, Soc. Clin. Research, Am. Heart Assn., Path. Soc. Gt. Britain and Ireland. Research on vitamins, ultraviolet light, peritonitis, hypertension; invented (with L. Gross) autotechnicon (device for automatic fixation, dehydration and paraffin impregnation of tissue specimens), 1929.

GOLDBLITH, Samuel Abraham, Am. food scientist; b. Laurence, Mass., May 5, 1919; s. Abraham and Fannie (Rubin) G.; S.B., Mass. Inst. Tech., 1940, S.M., 1947, Ph.D., 1949; m. Diana Greenberg, Apr. 27, 1941; children—Errol D. (dec.), Judith Ann, Jonathan Mark. Faculty, Mass. Inst. Tech., Cambridge, 1952—, prof. food sci., exec. officer dept. nutrition and food sci., 1959—. Chmn. com. radiation NRC, 1961—; mem. com. on radiation preservation foods Am. Inst. Biol. Scis.-AEC, 1964—. Mem. Am. Chem. Soc., Inst. Food Tech. (Monsanto award 1953), N.Y. Acad. Scis., A.A.A.S., Soc. for Cryobiology, Soc. Am. Microbiology. Author: (with M. A. Joslyn, J. Nickerson) An Introduction to Thermal Processing, 1961; (with M. A. Joslyn) Milestones in Nutrition, 1964; also numerous articles. Research on food preservation techniques, including radiation, freeze-dehydration methods. Home: 228 Upham St., Melrose, Mass. 02176. Office: 27 Massachusetts Av., Cambridge, Mass. 02139.*

GOLDEMBERG, J., Brazilian physicist; b. Rio Grande Sul, Brazil, May 27, 1928; s. Jacob and Bertha (Tessler) G.; grad. U. Sao Paulo (Brazil), 1950, Ph.D., 1954; widower; children—Clovis, Ricardo, Sergio. Faculty, U. Sao Paulo, 1950—, prof. exptl. physics, 1957—; research asso. prof. Stanford, 1962-64; asso. prof. U. Paris (France), 1965. Mem. Academia Brasileira de Ciencias, Am. Phys. Soc. Research and publs. on systematics of photonuclear reactions; established upper limit for electric dipole moment of electron; helped establish new method for electron scattering measurements. Home: Rua Itambé, Sao Paulo, Brazil.*

GOLDEN, Alfred, Am. pathologist; b. N.Y.C., Aug. 4, 1908; s. Bernard and Rheba (Dryer) G.; B.S., U. Wis., 1934, M.S., 1935; M.D., Washington U., St. Louis, 1938; m. Libby Siegel, Sept. 8, 1965. Pathologist, exec. officer, cons. Armed Forces Inst. Pathology, 1940-44; lectr. Sch. Trop. Medicine, Army Med. Center, Washington, 1942-45; med. officer-in-charge tropical disease investigation unit Office Intern-Am. Affairs, Washington, 1945-46; asso. prof. pathology U. Tenn. Sch. Medicine, 1946-50; asso. prof. U. Buffalo Sch. Medicine, 1950-55; asso. prof. Wayne State U. Coll. Medicine, Detroit, 1956-57; dir. labs. Jennings Meml. Hosp., also Alexander-Blain Hosp. and Clinic, Detroit, 1955——. Cons. VA, Detroit, 1955-57, McGregor Found., Detroit, 1959-63. Recipient Alexander Berg prize in bacteriology Washington U. Sch. Medicine, 1938, Army Commendation award, 1945. Fellow A.C.P., Am. Coll. Pathologists, Am. Soc. Clin. Pathologists, A.M.A.; mem. Am. Soc. Exptl. Pathology, Internat. Acad. Pathology, Sigma Xi. Research and publs. on tropical ulcer, viral pneumonia, injuries to adrenal glands, diffuse fibrosis of lungs, fungus disease of central nervous system, primary tumor of genital system. Home: 26764 York Rd., Huntington Woods, Mich. 48070. Office: 7815 E. Jefferson St., Detroit 48214.*

GOLDENBERG, Ira Stovin, Am. physician; b. Bridgeport, Conn., Feb. 23, 1925; s. Edward and Gertrude (Stovin) G.; B.A., U. Mich., 1947; M.D., Boston U., 1951; m. Rena Sasson, Oct. 5, 1952; children—Deborah, Beth, Nancy, Ellen. Resident, Yale-New Haven Hosp., 1951-56, now attending surgeon; practice medicine, specializing in surgery, New Haven, 1957——; faculty Sch. Medicine Yale, 1957——, asso. prof. surgery, 1964——; attending surgeon West Haven VA Hosp. Diplomate Am. Bd. Surgery. Fellow A.C.S.; mem. A.A.A.S., Soc. U. Surgeons, Endocrine Soc., Am. Assn. Cancer Research, Sigma Xi, Alpha Omega Alpha. Research and publs. on delineation of metabolic interrelations in the patient undergoing surgery, hormonal and chemotherapy of breast cancer. Home: 76 Shepherd Lane, Orange, Conn. 06477. Office: 333 Cedar St., New Haven 06510.*

GOLDENSOHN, Eli Samuel, Am. neurologist; b. N.Y.C., June 26, 1916; s. Matthias and Bessie (Mencher) G.; A.B., George Washington U., 1937, M.D., 1940; m. Betty Brown, June 16, 1941; children—Ellen, Richard, Martin. Asst. prof. physiology U. Colo., neurology U. Pa., Phila., 1963-67; prof. neurology Columbia, 1967——; cons. neurology Bronx VA Hosp.; attending physician Grad. Hosp. U. Pa. Career investigator N.Y.C. Health Council. Mem. Am. Neurology Assn., Am. Acad. Neurology, Am., Eastern (pres.) EEG socs., Am. Epilepsy Soc. (pres.), Sigma Xi. Research, numerous publs. field epilepsy, electroencephalography, brain function and characteristics of single neurones. Home: Route 9, West Nyack, N.Y. 10960. Office: 710 W. 168th St., N.Y.C. 10032.*

GOLDENWEISER, Alexander, anthropologist, sociologist; b. Kiev, Russia, Jan. 29, 1880; s. Alexander S. and Sofia G. (Munstein) G.; Kiev Gymnasium, 1896-1900; Harvard, 1900-01; A.B., Columbia, 1902, A.M., 1904, Ph.D., 1910; m. Anna Hallow, July 31, 1906; 1 dau., Alice Rosalind; m. 2d, Ethel Cantor, Jan. 31, 1930. Lectured on anthropology, Columbia, 1910-19; lecturer on anthropology and sociology, New Sch. for Social Research, New York, 1919-26; lecturer on anthropology and psychology, Rand School of Social Science, 1915-29; prof. thought and culture, Ore. State System of Higher Edn., Portland Extension, 1930——; visiting prof. sociology, Reed Coll., Ore., 1933-39; visiting prof. anthropology, U. of Wis., 1937-38; prof. anthropology, U. of Wash., summer 1923; prof. sociology, U. of Ore., summer 1925; prof. sociology, Stanford U., summer 1935, U. of Buffalo, summer 1936; public lecturer. On editorial staff, Encyclopedia of the Social Sciences, 1927-28. Editor: (with W. F. Ogburn) The Social Sciences and Their Inter-relations, 1927. Author: Totemism, an Analytical Study, 1910; Early Civilization, 1922; Robots or Gods, 1932; History, Psychology and Culture, 1933; Anthropology, an Introduction to Primitive Culture, 1937; (co-author) American Indian Life, 1924; Our Changing Morality, 1924; Political Theories—Recent Times, 1925; History and Prospects of the Social Sciences, 1925; Population Problems, 1925; Sex in Civilization, 1929. Known for comparative studies of primitive cultures from which general characteristics of folk-culture were drawn. Died July 6, 1940.

GOLDFUSS, (Georg) August, German zoologist, paleontologist; b. Thurnau, Germany, Apr. 18, 1782; s. Johann August and Margarete (Wächter) G.; student natural scis. Erlangen, Germany; m. Eleonore Oelhafen von Schöllenbach, 1815; 5 sons, 6 daus. Faculty zoology U. Erlangen, from 1804, named lectr., 1810; became prof. zoology, mineralogy, Bonn, Germany, 1818, also dir. Zool. Mus. and the paleontol. collection. Mem. Leopoldina (librarian). Author: Naturbeschreibung der Säugetiere, 2 vols., 1809-12; Die Umgebungen von Muggendorf, 1810; (with K. G. C. Bischoff) Physikalisch-statistische Beschreibung des Fichtelgebirges, 1816; Naturhistorischer Atlas, 1824-43; Petrefacta Germaniae, 3 vols., 1826-35, 42, 44; Beiträge zur vorweltlichen Fauna des Steinkohlengebirges, 1847; also a handbook, 1820. Research on African beetles, paleontology especially fossilized cave animals of Germany; introduced concept of protozoa although he included animals other than single celled organisms, 1818. Died Bonn-Poppelsdorf, Oct. 2, 1848.

GOLDHABER, Gerson, physicist; b. Chemnitz, Germany, Feb. 20, 1924; s. Charles and Ethel (Frisch) G.; M.Sc., Hebrew U., 1949; Ph.D., U. Wis., 1950; m. Sulamith Low, May 8, 1947 (dec. Dec. 1965); 1 son, Amos Nathaniel. Came to U. S., 1948, naturalized, 1953. Faculty, Columbia, N.Y.C., 1950-53; faculty U. Cal., Berkeley, 1953——, prof. physics, 1964——, co-group leader Lawrence Radiation Lab., 1963——. Ford Found. fellow, 1960-61. Mem. Am. Physics Soc., Sigma Xi. Exptl. study of elementary particle interactions and properties; exptl. techniques in photographic emulsions, hydrogen bubble chamber; co-discoverer antiproton anihilation process, A mesons.*

GOLDHABER, Gertrude Scharff (Mrs. Maurice Goldhaber), physicist; b. Mannheim, Germany, July 14, 1911; d. Otto and Nelly (Steinharter) Scharff; Ph.D., U. Munich, 1935; m. Maurice Goldhaber, May 24, 1939; children—Alfred S., Michael H. Came to U. S., 1939, naturalized, 1944. Research asso. in physics Imperial Coll., London, Eng., 1935-39; physicist U. Ill., Urbana, 1939-48, asst. prof., 1948-50; asso. physicist Brookhaven Nat. Lab., Upton, L.I., N.Y., 1950-58, physicist, 1958-62, sr. physicist, 1962——. Cons., Argonne Nat. Lab., 1946-56, Los Alamos Sci. Lab., 1953——; mem. adv. panel on nuclear data project NRC, 1959-64. Fellow Am. Phys. Soc.; mem. Sigma Xi. Research on identity of beta particles with atomic electrons, ferromagnetism, neutron physics, photoneutrons, neutrons from spontaneous fission of uranium, nuclear isomers, systematics of nuclear levels, tests of parity violation in strong interactions. Home: 92 S. Gillette Av., Bayport, L.I., N.Y. 11705. Office: Physics Dept., Brookhaven Nat. Lab., Upton, L.I., N.Y. 11073.*

GOLDHABER, Maurice, physicist; b. Lemberg, Austria, Apr. 18, 1911; s. Charles and Ethel (Frisch) G.; student Berlin (Germany) U., 1930-33; Ph.D., Cambridge (Eng.) U., 1936; m. Gertrude Scharff, May 24, 1939; children—Alfred S., Michael H. Came to U. S., 1938, naturalized, 1944. Charles Kingsley Bye fellow Magdalene Coll., Cambridge, 1936-38; faculty physics U. Ill., Urbana, 1938-50, prof., 1945-50; sr. scientist Brookhaven Nat. Lab., Upton, L.I., N.Y., 1950-60, chmn. dept. physics, 1960-61, dir., 1961——. Morris Loeb lectr. Harvard, 1955; vis. lectr. physics Weizmann Inst., Italian Summer Sch. Physics, Scandinavian Inst. Theoretical and Atomic Physics; cons. Los Alamos Sci. Lab.; mem. adv. com. for physics Oak Ridge Nat. Lab., Nat. Acad. Scis.; mem. N.Y. State Adv. Council for Advancement Indsl. Research and Devel. Fellow Am. Phys. Soc.; mem. Nat. Acad. Scis. Asso. editor Phys. Rev., 1951-53; editorial com. Il Nuovo Cimento. Contbr. numerous articles on neutron physics, radioactivity, nuclear isomers, nuclear photoelectric effect, nuclear models, fundamental particles to profl. jours. Home: 91 S. Gillette Av., Bayport, N.Y. 11705. Office: Brookhaven Nat. Lab., Upton, L.I., N.Y. 11973.*

GOLDHABER, Sulamith, physicist; b. Vienna, Austria, Nov. 11, 1923; d. Abraham and Toni (Reinisch) Low; M.A. Hebrew U., Jerusalem; Ph.D., U. Wisconsin, 1951; m. Gerson Goldhaber, May 8, 1947; son, Amos Nathaniel. Asst. inorganic chem., Hebrew U., 1946-47; radio chem., U. Wisconsin, 1948-51; res. assoc., nuclear chem. Columbia, 1941-53; research physicist, Radiation Lab., California, 1954-58; physicist, Lawrence Radiation Lab., California, 1959-65; Ford Foundation Fellow, 1960-61; Gugenheim Fellow, 1965. Mem., Am. Physical Soc., Sigma Xi. Author of numerous articles. Experimental study of elementary particle interactions and properties using photographic emulsions, 1950-58, and Hydrogen Bubble Chamber; studied anti-proton annihilation process; K plus nuclear interaction; Sigma plus — Sigma minus electromagnetic mass difference; dynamics of elementary particle resonance and double resonance formation; co-discoverer of A mesons. Died Madras, India, Dec. 8, 1965.

GOLDIAMOND, Israel, psychologist; b. Ukraine, Nov. 1, 1919; s. Samuel and Clara (Rothenburg) G.; came to U. S., 1923, naturalized, 1929; B.A., Bklyn. Coll., 1942; Ph.D., U. Chgo., 1955; m. Betty Johnson, Feb. 28, 1946; children—Lisa Catherine, Joe David, Susannah. Faculty, dir. perception, conditioning lab. So. Ill. U., 1955-60; prof. psychology Ariz. State U., Tempe, 1960-63; exec. dir. Inst. For Behavioral Research, Silver Spring, Md., 1963-68; asso. prof. Johns Hopkins, 1966-67, prof. psychiatry, 1967-68; prof. psychiatry and psychology U. Chgo., 1968——. Recipient Research Career Devel. award Nat. Inst. Mental Health, 1963——. Fellow Am. Psychol. Assn., A.A.A.S.; mem. Optical Soc. Am., Am. Ecol. Soc., Psychonomic Soc. Research and publs. on perception and perceptual methodology, exptl. analysis, programming and systematic alteration of behavioral repertoires of ednl., social and clin. relevance. Home: 5555 S. Everett Av., Chgo.*

GOLDICH, Samuel Stephen, Am. geologist; b. Grand Forks, N.D., Jan. 17, 1909; s. Louis and Fannie (Lazar) G.; A.B., U. Minn., 1929, Ph.D., 1936; M.A., Syracuse U., 1930. Faculty, U. Minn., 1948-59, prof., 1949-59; geologist U. S. Geol. Survey, 1959-64, br. chief isotope geology, 1960-64; prof. Pa. State U., 1964; prof. N.Y. State U., Stony Brook, 1965——. Recipient Distinguished Service award Dept. Interior, 1965. Fellow Am. Geophys. Union, Geol. Soc. Am., Mineral. Soc. Am.; mem. A.A.A.S., Am. Chem. Soc., Assn. Petroleum Geologists, Geochem. Soc., Mineral. Soc. Can., Soc. Econ. Paleontology and Mineralogy. Research, numerous publs. in area geology Trans-Pecos, Tex., origin laterite and bauxite, precambrian geology and geochronology, Minn.-Ont. border region, isotope geology. Home: 449 Lincoln Blvd., Hauppauge, N.Y. 11788. Office: Dept. Earth and Space Scis., State U. N.Y., Stony Brook, N.Y. 11790.*

GOLDIN, Abraham, Am. pharmacologist, cancer research specialist; b. N.Y.C., Nov. 10, 1911; s. Hyman E. and Anna (Rosansky) G.; B.S., Bklyn. Coll., 1933; M.A., Columbia, 1935, Ph.D. in Zoology, 1942; m. Jessica Wolfe, Mar. 29, 1942; children—Deborah, Laura. Lab. asst. in bacteriology, physiology Bklyn. Coll., 1933-38, instr. zoology, 1945-46; lab. instr. biology Queens Coll., Coll. City N.Y., 1938-42; biologist, chief biol. sect. Army Chem. Center, 1946-49; asst. Johns Hopkins, Balt., 1947-48, 50-52, research asso., 1954; biologist, clin. research unit Nat. Cancer Inst., Marine Hosp., Balt., 1949, Chemistry and Pharmacy Lab., 1952, head, biochem. pharmacy sect., 1956-63, chief, drug evaluation br., 1963-65, asso. chief lab. research Cancer Chemotherapy Nat. Service Center, Bethesda, Md., 1966——; staff USPHS Hosp., Balt., 1949. Vis. prof. biochemistry Brandeis U., Waltham, Mass., 1957——; research adviser to grad. council George Washington U., 1960——. Mem. Societa Italiana di Cancerologia (fgn. corr.), A.A.-A.S., N.Y. Acad. Sci. (asso.), Am. Physiol. Soc., Soc. Pharmacology and Exptl. Therapeutics; Royal Soc. Medicine (affiliate), Internat. Soc. Chemotherapy, Am. Soc. Microbiology. Research, publs. in exptl. cancer chemotherapy, including drug evaluation, synergism, metabolite-antimetabolite relationships. Home: 6910 Hillmead Rd., Bethesda. Office: Nat. Cancer Inst., NIH, Bethesda, Md.*

GOLDIN, Abraham Samuel, Am. chemist; b. Bklyn., Apr. 22, 1917; s. Samuel and Jennie (Greenberg) G.; A.B., Columbia U., 1937, A.M. in Chemistry, 1941; Ph.D., U. Tenn., 1951; m. Shirley May Schlick, Aug. 16, 1945; children—Stephen Lloyd, David John, Lorraine Susan. Chemist, Eastern Wine Corp., 1941-42, Mut. Chem. Co., 1942; research asso. SAM Labs., Columbia U., 1943-45; tech. engr. Union Carbide Corp., Oak Ridge, 1945-46, chemist-physicist, 1946-50; radiochemist USPHS, Cin., 1951-60, dep. officer in charge, dir. research, Winchester, Mass., 1962-68; chem. dir. Nat. Lead Co., Inc., Winchester, 1960-61; asso. prof. indsl. medicine N.Y. U., 1961-62; lectr. Yale, 1967——; asso. prof. radiochemistry Harvard Sch. Pub. Health, 1968——. Internat. Panel on Standardization Low-Level Radiochem. Measurement. Mem. Am. Chem. Soc., A.A.A.S., Am. Pub. Health Assn., Health Physics Soc., Sigma Xi. Contbr. numerous articles to tech. jours. Research on measurement radionuclides in humans, radioactive waste disposal, environmental surveillance; developed analytical methods in radiochemistry. Home: 15 Carriage Lane, Winchester, Mass. 01890. Office: 665 Huntington Av., Boston 02115.*

GOLDMAN, Alan J., Am. mathematician; b. N.Y.C., Mar. 2, 1932; s. Leonard and Sylvia (Schachter) G.; B.A., Bklyn. Coll., 1952; M.A., Princeton, 1954, Ph.D., 1956; m. Cynthia G. Timberg, June 25, 1955; 1 son, Peter H. Instr., Princeton, 1955-56; mathematician Nat. Bur. Standards, Washington, 1956-60, chief operations research sect., 1961——, dep. chief applied math. div., 1966——; lectr. Am. U., 1956-57, Cath. U. Am., 1957-60. Recipient Dept. Commerce Silver medal for meritorious service, 1966. Mem. Am. Math. Soc., Math. Assn. Am., Soc. for Indsl. and Applied Math., Operations Research Soc. Am., Phi Beta Kappa, Sigma Xi, Pi Mu Epsilon. Research, publs. into and applications of math. methods and models useful in studying complex systems of modern tech. especially weapons, transport, communications. Home: 4822 Chevy Chase Dr., Chevy Chase, Md. 20015. Office: Nat. Bur. Standards, Washington 20234.*

GOLDMAN, Charles Remington, Am. limnologist; b. Urbana, Ill., Nov. 9, 1930; s. Marcus Selden and Olive (Remington) G.; B.A., U. Ill., 1952, M.S. in Zoology, 1955; Ph.D., U. Mich., 1958; m. Shirley Ann Aldous, Apr. 4, 1953; children—Christopher Selden, Margaret Blanch, Olivia Remington, Ann Aldous. Field asst. Ill. State Natural History Survey, 1954-55; biologist Engring. Research Inst., U. Mich., part-time, 1956-57; faculty U. Cal. at Davis, 1958——, prof. zoology, dir. Inst. Ecology, 1966——. Cons. to pvt. cos., govt. agys. NSF fellow, Italy, 1964; Guggenheim fellow, 1965. Fellow A.A.A.S.; mem. Am. Soc. Limnology and Oceanography (pres. elect 1966-67, pres. 1967-68, editorial bd. 1964-67), Ecol. Soc. Am., Internat. Assn. Theoretical and Applied Limnology. Editor: Primary Productivity in Aquatic Environments, 1965. Editorial bd. Ecology, 1966-68. Contbr. articles to tech. jours. Developed carbon-14 bioassay method for detecting nutrient factors limiting biol. productivity red salmon nursery lakes in Alaska; discovered molybdenum limiting photosynthesis in a lake; evaluated contbn. nitrogen fixing alder trees to productivity mountain lake. Home: 626 Elmwood Dr., Davis, Cal. 95616.*

GOLDMAN, David Eliot, Am. biophysicist; b. Boston, Aug. 11, 1910; s. Hiram R. and Sophie (Loman) G.; A.B., Harvard, 1931; Ph.D., Columbia, 1943; m. Jeanne Loewenstam, Jan. 2, 1938; 1 son, James. Faculty, Coll. Phys. and Surgs. Columbia, 1941, mem. war research staff OSRD, 1943; commd. lt. (j.g.) USN, 1944, advanced through grades to capt., 1963; mem. aviation physiology div. Naval Med. Research Inst., Bethesda, Md., 1944-47, head biophysics div., 1947-55, 57——; sci. liaison officer Office Naval Research, London, Eng., 1955-57; com. mem. NRC, Office Sci. and Tech. NIH, Am. Standards Assn. Mem. Biophys. Soc., Am. Physiol. Soc., Am. Phys. Soc., Acoustical Soc. Am., N.Y. Acad. Scis., A.A.A.S.

Contbr. numerous articles to profl. jours. Research, publs. on theoretical, exptl. analysis elec., mech., chem. behavior nerve membranes. Home: 7020 Wilson Lane, Bethesda 20034. Office: U. S. Naval Med. Research Inst., Bethesda, Md. 20014.*

GOLDMAN, Edward Alphonse, Am. naturalist; b. Mt. Carroll, Ill., July 7, 1873; s. Jacob Henry and Laura Carrie (Nicodemus) G.; ed. pub. schs. and under pvt. tutors; m. Emma May Chase, June 23, 1902; children—Nelson Edward, Orville Merriam, Luther Chase. With U. S. Biol. Survey since 1892; much of time, 1892-1906, in biol. investigations in Mexico; in biol. survey of Panama, 1911-12, of Ariz., 1913-17; in charge div. biol. investigations, 1919-25; in charge div. of game and bird reservations, 1925-28; sr. biologist since 1928. Fellow A.A.A.S.; mem. Am. Ornithologists' Union, Am. Soc. Mammalogists, Washington Acad. Sci, Biol. Soc. Washington (pres. 1927-29), Am. Forestry Assn., Cooper Ornith. Club Cal., Washington Biologists' Field Club. Author: Revision of Wood Rats of Genus Neotoma, 1910; Revision of Spiny Pocket Mice (Heteromys and Liomys), 1911; Plant Records of an Expedition to Lower California, 1916; Rice Rats of North America (Oryzomys), 1918; Mammals of Panama, 1920; The Wolves of North America (with Stanley P. Young); also numerous shorter papers, mainly on mammals and birds and conservation of wild life. Deceased.

GOLDMAN, Leon, Am. surgeon; b. San Francisco, Feb. 14, 1904; s. Samuel and Lilly (Kalfin) G.; A.B., U. Cal., Berkeley, 1926; M.D., U. Cal., San Francisco, 1930; M.S. in Physiology, Northwestern U., 1939; m. Betty Rosenburg, June 19, 1939; children—Dianne (Mrs. Bertram Fienstein), Yvonne (Mrs. Walter Banks), Lynn (Mrs. Carey Klippsten). Faculty U. Cal. Sch. Medicine, San Francisco, 1935—, prof. surgery, 1950—. Mem. Am. Surg. Assn., Am. Coll. Surgery, Am. Thyroid Assn., Am. Gastroent. Assn., Assn. for Surgery Gastrointestinal Tract, A.M.A., Western, Pan Pacific, Pacific Coast (past pres.), San Francisco (past pres.) surg. socs., Soc. U. Surgeons, Internat. Soc. Surgery. Research in surgery of thyroid and parathyroid glands; surgery of gastrointestinal tract. Home: 1050 North Point St., San Francisco 94109.*

GOLDMAN, Leon, Am. physician; b. Cin., Dec. 7, 1905; s. Abraham and Fannie (Friedman) G.; B.S., U. Cin., 1927, M.D., 1929; m. Belle Hurwitz, Aug. 23, 1936; children—John, Steven, Carol. Practice medicine, specializing in dermatology, Cin., 1932—; prof., chmn. dept. dermatology Coll. Medicine U. Cin., 1946—, dir. Laser Labs. Med. Center, 1962—. Cons. USPHS. Diplomate Am. Bd. Dermatology. Mem. Am. Dermatol. Assn., Sigma Xi, Alpha Omega Alpha. Author: Laser in Cancer Research, 1966; Biomedical Applications of the Laser, 1967. Contbr. numerous articles to profl. jours. Research and publs. on laser, dermatology, insect control, trop. medicine, cancer, history medicine.*

GOLDMAN, Oscar, Am. mathematician; b. Bklyn., Feb. 2, 1925; s. Isaac and Esther (Schwartz) G.; B.S., Coll. City N.Y., 1944; A.M., Princeton, 1946, Ph.D., 1948; m. Madge Rosenbaum, Aug. 8, 1949. Benjamin Peirce instr. Harvard, 1948-51; mem. faculty Brandeis U., 1951-61, prof. math., 1961-62, chmn. dept., 1956-61; prof. math. U. Pa., 1962—, chmn. dept., 1963—; mem. Inst. Advanced Study, 1960-62. NSF Sci. faculty fellow, 1960-61. Mem. Am. Math. Soc., Math. Assn. Am., Math. Soc. Japan, Sigma Xi. Author research papers algebra, theory numbers. Asso. editor Jour. Franklin Inst., 1963—. Office: Math. Dept., University of Pa., Phila. 19104.*

GOLDMANN, Edwin E., surgeon; b. Burgersdorp, Cape Colony, Nov. 12, 1862; s. B. N. Goldmann; ed. Breslau, Freiburg (both Germany) univs.; m. Lorna Larrence Bosworth Smith, 1906; 1 son, 1 dau. Asst., Univ. Hosp., Freiburg, 11 years; asso. (under Ehrlich) at inst. for exptl. therapeutics, Frankfort/Main, Germany; prof. surgery Freiburg U., from 1892; sr. surgeon Deaconess House, Freiburg, from 1898. Fellow Royal Soc. Medicine, London, Eng. Contbr. papers to med. jours., from 1887. Investigated malignant growth, use of X-rays for cancer diagnosis, external and internal secretions revealed by vital staining, new operative methods for surgery of esophagus, breast and stricture of urethra. Died Aug. 12, 1913.

GOLDMARK, Peter Carl, physicist; b. Budapest, Hungary, Dec. 2, 1906; s. Alexander and Emmy G.; student U. Vienna, 1925-31, B.S., Ph.D.; children—Peter Carl. Frances C., Christopher W., Andrew G., Jonathan, Susan. Came U. S., 1933, naturalized, 1937. Television engr. in charge dept. Pye Radio, Ltd., Cambridge, Eng., 1931-33; cons. engr., N.Y. City, 1934-35; chief engr. television dept., CBS, N.Y.C., 1936-44, dir. of engring. research and devel., 1944-50, v.p. Engring. Research and Devel. Dept., 1950—; also pres. CBS Labs. div. CBS; doing war research part time, Radio Research Lab. Harvard U.; vis. prof. med. electronics Med. Sch., U. Pa. Recipient Morris Liebman Meml. prize for electronic research, 1946; Vladimir K. Zworykin award for devel. and utilization electronic television, 1961. Fellow Am. Inst. E.E., I.R.E., Soc. Motion Picture Engrs., TV Soc. (London). Contbr. to profl. jours. Invented field sequential system color TV, 1940; developed long-playing microgroove phonograph record; numerous patents in field of television and radio. Office: CBS Labs., High Ridge Rd., Stamford, Conn.

GOLDNER, Joseph Leonard, Am. physician; b. Omaha, Nov. 18, 1918; s. Oscar Charles and Rose (Wolf) G.; student U. Minn., 1936-39; B.S., U. Neb., 1941, M.D., 1943; m. Eunice Ruth Kensinger, June 3, 1944; children—Richard Douglas, Steven Craig. Practice medicine, specializing in surgery, Durham, N.C., 1950—; faculty Duke Med. Center, 1950—, prof., 1957—, chief Orthopaedic Amputee Clinic, 1955—, chief orthopaedic hand surgery, 1955—, chmn. div. orthopaedic surgery, 1967—; instr. Am. Acad. Orthopaedic Surgeons, 1960—; instr. Am. Acad. Cerebral Palsy, 1960—; vis. prof. Royal Childrens Hosp., Melbourne, Australia, 1965; cons. various hosps. Traveling fellow Orthopaedic Assn., Britain, 1955. Diplomate Am. Bd. Orthopaedic Surgery (examiner). Mem. A.M.A., So. Med. Assn. (chmn. exec. com.), Am. Orthopaedic Assn., Am. Acad. Orthopaedic Surgeons, Am. Soc. Surgery of Hand, Am. Acad. Cerebral Palsy, Pan-Pacific Surg. Soc., Alpha Omega Alpha. Contbr. chpts. to Hand Surgery, 1966. Contbr. numerous articles to profl. jours. Research on reconstructive surgery of cerebral palsy in the upper extremity, surgery of poliomyelitis, reconstructive surgery of the hand in all aspects; contbr. to mgmt. of Volkmann's ischaemia in orthopaedic surgery. Home: 602 E. Forest Hills Blvd., Durham 27707. Office: Duke Med. Center, Durham, N.C. 27706.*

GOLDRING, William, Am. physician; b. Anniston, Ala., May 8, 1898; s. Paul and Rebecca (Rosenfeld) G.; student Fordham U., 1916-19; B.S., N.Y. U., M.D., 1922; m. Helen Heffner, Sept. 24, 1928; 1 dau., Roberta (Mrs. Robert S. Coles). Faculty, N.Y. U. Sch. Medicine, N.Y.C., 1926—, prof., 1960—; staff Bellevue Hosp., N.Y.C., 1926—, N.Y. U. Hosp., 1949—. Cons. WHO, 1951-53, also hosps. and brs. of govt.; mem. Macy Found. Confs. on Hypertension; mem. med. adv. bd. Ruth Papier Nephrosis Found. Recipient Alumni Sci. award N.Y. U. Sch. Medicine, 1961, Med. Sch. Alumni medallion, 1965. Diplomate Am. Bd. Internal Medicine. Fellow N.Y. Acad. Medicine, A.C.P.; mem. Harvey Soc., Am. Physiol. Soc., Soc. Exptl. Biology and Medicine, Am. Heart Assn., Alpha Omega Alpha, Sigma Xi, others. Research, publs. on secondary hypertension and renal pathology; insulin effects on man. Home: 325 E. 79th St., N.Y.C. 10021.*

GOLDSCHEIDER, (Johann Karl August Eugen) Alfred, German physician, physiologist; b. Sommerfeld, Germany, Aug. 4, 1858; s. Max G.; M.D., Friedrich-Wilhelm Inst., Berlin, Germany, 1881; m. Margarete Alexandra Fischer, 1902; 1 son, 2 daus. Mem. staff, faculty, med. clinic U. Berlin; became directing physician Moabit Hosp., 1894, Virchow Hosp., Berlin, 1906; dir. polyclinic, 1910-33; became prof., prof. emeritus, 1926. Author: (with E. v. Leyden) Spezifische Energie der Sinnesnerven, 1881; Diagnostik der Krankheiten des Nervensystem, 1893; Über den Schmerz, 1894; Gesamte Abhandlungen, 1898; Die Bedeutung der Reize für Pathologie und Therapie im Lichte der Neuronlehre, 1898; (with E. Flatau) Normale und pathologische Anatomie der Nervenzellen, 1898; Anleitung zur Übungstherapie der Ataxie, 1899; Über die spinalen Sensibilitätsbezirke der Haut, 1917; Tafeln der spinalen Sensibilitätsberzirke der Haut, 1918; Das Schmerz problem, 1920; Therapie innerer Krankheiten, 1929. Became editor Zeitschrift für physikalisch-diätetische Therapie, 1898, (with K. Eberth) Fortschritte der Medizin, 1894; became coeditor Zeitschrift für klinische Medizine, 1911. Research on sense physiology, percussion and auscultation, internal medicine especially therapy; described epidermolysis bullosa (Goldscheider's disease), 1882; discovered (independently of Blix) separate sensory spots in the skin for heat, cold, pressure, 1883-84. Died Berlin, Apr. 10, 1935.

GOLDSCHMIDT, Bertrand Léopold, French chemist; b. Paris, France, Nov. 2, 1912; s. Paul and Naomi (de Rotschild) G.; Ph.D. in Engring., U. Paris; children—Paul, Emma. Asst., Curie Lab., Paris, 1935-40; div. head Anglo-Canadian Atomic Energy Project, Montreal, Chalk River, 1942-45, div. head chemistry, 1946; dir. chemistry Atomic Energy Commn., 1946-59, dir. external relations and programs, 1959—. Contbr. articles to profl. publs. Home: 24 rue Galilée, Paris 16. Office: 29-33 rue de la Fédération, Paris 15, France.

GOLDSCHMIDT, Hans, German chemist; b. Berlin, Jan. 18, 1861; studied under Bunsen. Invented alumino-thermic process for reducing metals from their oxides and for welding (Goldschmidt's process), 1905; produced carbon-free chromium, manganese and cobalt using his process. Died Baden-Baden, Germany, May 20, 1923.

GOLDSCHMIDT, Hermann (Hayum) Mayer Salomon, astronomer; b. Frankfort/Main, Germany, June 17, 1802; s. Mayer Salomon and Helen (Cassel) G.; ed. as artist, Munich, Germany; m. Adelaide Pierrette Moreau, 1861; 2 children. Painter of hist. scenes, Paris, France, 1834; amateur astronomer, from 1847. Recipient numerous honors for discoveries. Contbr. articles to tech. jours. Discovered 14 asteroids, 1852-61; observed solar eclipse, July 18, 1860. Died Fontainebleau, France, Sept. 19, 1866.

GOLDSCHMIDT, (Emil August) Johannes, German physicist, meteorologist, climatologist; b. Dresden, Germany, Aug. 24, 1894; s. August and Minna (Schiller) G.; state exam in physics, pure and applied math. Technische U. Dresden, 1924; Dr.rer.techn., 1925; m. Gertrud Lehmann, 1925; 2 daus. Asst. under E. Alt at Saxon Weather Bur., dir. research dept.; dir. obs., Wahnsdorf, Germany, 1936-45, 51-52; lectr. meteorology Technische U. Dresden; became lectr. forestry faculty at Tharandt, Germany, 1945; lectr. Hort. Inst., Pillnitz, also U. Leipzig (Germany). Research and numerous publs. on atmospheric radiation and electricity, climatic situations of Saxony; made long range weather predictions; developed techniques for measuring the upper ozone. Died Dresden, Nov. 1, 1952.

GOLDSCHMIDT, Leontine, biochemist; b. Arad, Austria, Mar. 9, 1913; d. Max and Anna (Michel) G.; Ph.M., Vienna (Austria) U., 1935, Ph.D., 1937. Research asst. Boston U. Sch. Medicine, 1942-45; research biologist U. S. Naval Radiol. Def. Lab., San Francisco, 1947-50; dir. biochem. research Creedmoor Inst. Psychobiologic studies, Queens Village, N.Y., 1954—; asso. research scientist N.Y. State Dept. Mental Hygiene, 1965—. Fellow A.A.A.S.; mem. Am. Chem. Soc., N.Y. Acad. Sci. Contbr. articles to tech. jours. Research on physico-chem. changes brought about in red blood cells by aging process, atomic radiations, changes in blood constituents during different states mental disease. Home: P.O. Box 40, Sta. 60, 80-45 Winchester Blvd. Office: Creedmoor Inst., Queens Village, N.Y. 11427.*

GOLDSCHMIDT, Richard Benedikt, zoologist, geneticist; b. Frankfort/Main, Germany, Apr. 12, 1878; s. Salomon and Emma Rosette (Flürscheim) G.; student Munich, Germany, from 1898; Ph.D., Heidelberg, Germany, 1902; M.D. (hon.), U. Kiel, Germany, 1928; Sc.D. (hon.), Madrid, 1935; m. Else Kühnlein, 1906; 1 son, 1 dau. Asst. to Hertwig at U. Munich, 1903, lectr., 1904-09, asso. prof., from 1909; with Kaiser Wilhelm Inst., Berlin, from 1914, 2d dir., 1919-24; prof., Tokyo, Japan, 1924-26; prof. genetics, cytology U. Cal. at Berkeley, 1936-46, prof. emeritus, 1946-58. Mem. Leopoldina, also numerous acads. and profl. socs. Author: Einführung in die Vererbungslehre, 1911; Physiologische Theorie der Vererbung, 1927; The Material Basis of Evolution, 1940; Theoretical Genetics, 1955; Die quantiativen Grundlagen von Vererbung und Artbildung, 1920; Mechanismus und Geschlechtsbildun, 1920; Gen und Ausseneigenschaft, 1928-35; Die sexuellen Zwischenstufen, 1931; Physiological Genetics, 1938. Research and numerous publs. on heredity and genetics, X-chromosomes using butterflies; developed theory that serial pattern of chromosome and chem. configuration of chromosome molecules rather than qualities of individual genes determine heredity; studied effect of environmental changes on phenotype. Died Berkeley, Apr. 24, 1958.

GOLDSCHMIDT, Victor Mordechai, German mineralogist; b. Mainz, Germany, Feb. 10, 1853; s. Salomon and Josephine (Porges) G.; student metallurgy Indsl. Acad., Berlin, also Mining Acad., Freiberg, Germany; Ph.D., U. Heidelberg, Germany 1880; m. Leontine Porges, 1888. Research, Vienna, Austria, 1880; faculty U. Heidelberg, 1888; founder Mineral.-Crytallographic Inst., Heidelberg. Author: Atlas der Kristallformen, 1913-23; Index der Kristallformen, 1886-91; Krystallographische Winkeltabellen, 1897; Farben in der Kunst, 1919; Über Harmonie und Complication, 1901; Research on molecular structure of solids, intensity and means of effect of molecular power; measurement of crystals. Died Salzburg, Germany, May 8, 1933.

GOLDSCHMIDT, Victor Moritz, mineralogist, geochemist; b. Zurich, Switzerland, Jan. 27, 1888; s. Heinrich and Amalie (Köhne) G.; student mineralogy, geology, chemistry Oslo, Norway, from 1905, Ph.D., 1911; student Vienna, Austria, 1908-09; Dr. rer. nat. h.c., Freiburg, Germany, 1931. Faculty, U. Oslo, from 1912, asso. prof., dir. Mineral. Inst., 1914-29, dir. Geol. Mus., 1935-42, 1946-47; prof. Göttingen, Germany, 1929-35; with Macaulay Inst., Aberdeen, Scotland, Agrl. Exptl. Inst., Rothamsted, 1942-46. Author: Geochemische Vereilungsgesetze IX, Die Mengenverhältnisse der Elemente und der Atomarten, 1938. Contbr. articles to sci. publs. Founder modern geochemistry, crystalchemistry; new basis for theory of transformation of minerals from pressure, temperature changes; exptl. studies on occurrence of elements in rock, minerals and others. Died Vestre Aker, Norway, Mar. 20, 1947.

GOLDSCHMIDT, Walter (Rochs), Am. anthropologist; b. San Antonio, Feb. 24, 1913; s. Hermann and Gretchen (Rochs) G.; B.A., U. Tex., 1933, M.A., 1935; Ph.D., U. Cal. at Berkeley, 1942; m. Beatrice Lucia Gale, May 27, 1937; children—Karl Gale, Mark Stefan. Social scientist Bur. Agrl. Econs., 1940-46; faculty U. Cal. at Los Angeles, 1946—, prof. anthropology, 1956—, chmn. dept. anthropology, 1964—; dir. Ways of Mankind Radio Program, 1950-53; dir. Culture and Ecology in E. Africa, 1960-67. Fulbright Research scholar, U.K., 1953; Social Sci. Research Council grantee, 1953; Wenner-Gren Found. Postdoctoral Research grantee, 1953; NSF Postdoctoral fellow, 1964; fellow Center for Advanced Study in Behavioral Scis., 1964-65. Cons., Ginn and Co., 1963-67. Fellow Am. Anthrop. Assn., African Studies Assn. (founding; past dir.); mem. A.A.A.S., Soc. for Applied Anthropology, Southwestern Anthrop. Assn. (past pres.), Am. Ethnol. Soc., Internat. African Inst., Am. Assn. U. Profs., Phi Beta Kappa, Sigma Xi. Author: Small Business and the Community, 1946; As You

Sow, 1947; Nomlaki Ethnography, 1951; Ways to Justice, 1953; Man's Way, 1959; Exploring the Ways of Mankind, 1960; Comparative Functionalism, 1966; Sebei Law, 1967; Kambuya's Cattle; the Legacy of an African Herdsman, 1968; also articles. Editor: The United States and Africa, 1958, rev., 1963; The Anthropology of Franz Boas, 1959. Editor: Am. Anthropologist, 1956-59; spl. editor Aldine Pub. Co., 1966—. Dir., producer recs., Ways of Mankind, Albums I, II, 1953. Research on adaptation of social instns. and behavioral attributes of native peoples in adjustment to econ. conditions, especially shift between pastoralism and hoe farming. Home: 978 Norman Pl., Los Angeles 90049.*

GOLDSCHMIEDT, Guido, chemist; b. Trieste, May 29, 1850; s. Siegmund and Henriette (Herzfeld) G.; student chemistry Vienna, Austria, 1869-71; Ph.D., Heidelberg, Germany, 1872; m. Angelika von Herzfeld, 1886; 1 dau., Guida (Mrs. Otto Loewi). Worked under A. v. Baeyer and P. H. v. Groth, Strasbourg, France, 1872-74, under F. C. Schneider and L. v. Barth in Vienna, 1874; joined faculty of Vienna, 1875, became asso. prof., 1890; named prof. Agrl. U., 1891; apptd. prof. German U. Prague, 1891, rector, 1907-08; became dir. II chem. lab. U. Vienna, 1911. Recipient Lieben prize, 1892. Mem. Austrian Acad. Scis. Research and publs. on qualitative analysis of natural substances, potassium melt, papaverine and structure of other opium alkaloids, transition of an unsaturated fatty acid of high carbon number into the corresponding saturated fatty acid; 1st to demonstrate the presence of an isoquinoline ring in an alkaloid; isolated pyrene. Died Vienna, Aug. 6, 1915.

GOLDSHTEYN, Dmitriy Yefimovich, Russian roentgenologist; b. 1899; grad. med. faculty, 1924; D.Med. Sci. Asst. lectr. Kazan Postgrad. Med. Inst., 1929, now prof., head chair roentgenology and radiology. Mem. All-Russian (bd. mem.), Tatar (dep. chmn.) socs. roentgenologists and radiologists Znanie Soc. (Presidium mem. Tatar br.). Mem. editorial council Vestnik Rentgenologii. Research and numerous publs. on bone pathology, diseases of digestive tract, biol. effect of x-rays, lymphatic system. Address: Kazan Postgrad. Med. Inst., ulitsa Komleva 11, Kazan, RSFSR, USSR.

GOLDSMID, Hiroshi Julian, English physicist; b. Southsea, Eng., Aug. 11, 1928; s. David Koichi and Nina (Goldsmid) Nishikawa; B.Sc. with spl. honors in physics Queen Mary Coll., U. London (Eng.), 1949, Ph.D., 1958, D.Sc., 1966; m. Joan May Mortlock, Aug. 9, 1952; children—Jonathan Mark, Jane Felicity, Sarah Elizabeth. Sci. staff Hirst Research Centre, Central Research Labs., Gen. Electric Co. Ltd., Wembley, Eng., 1951-61, prin. sci. staff, 1961-64; reader solid state physics Bath U. Tech., Bath, Eng., 1964—. Recipient Lightfoot medal Inst. Refrigeration, 1958-59. Fellow Inst. Physics. Author: Applications of Thermoelectricity, 1960; (with J. R. Drabble) Thermal Conduction in Semiconductors, 1961; Thermoelectric Refrigeration, 1964; Thermal Properties of Solids, 1965. Publs. on discovery of bismuth telluride as first practicable material for thermoelectric refrigeration; 1st observation of heat conduction by electron-hole pairs; invention of thermoelectric direct-current transformer; research on thermoelectricity and thermal conductivity. Home: 20, Bramble Dr., Bristol 9, Eng.*

GOLDSMITH, Eli David, Am. biologist; b. N.Y.C., Apr. 10, 1907; s. Morris and Rose G.; B.S., Coll. City N.Y., 1926; A.M., Harvard, 1928, Ph.D., 1934; M.S., N.Y. U., 1930; m. Gertrude Alper, Dec. 21, 1940; 1 dau., Cathy Ellen. Faculty Harvard, 1928-29, 31-34, Washington Sq. Coll., 1929-31, Coll. City N.Y., 1934-45; faculty N.Y. U., N.Y.C., 1934—, prof. histology Grad. Sch. Arts and Sci. and Coll. Dentistry, 1951—, research coordinator Coll. Dentistry, 1948—; dir., prin. investigator research projects Am. Cancer Soc., USPHS, and other agys., 1946—. Dir. USPHS Grad. Research Tng. Program, 1958—, cons. to surgeon gen., 1959—. Commonwealth Fund grantee, 1944. Recipient Gold medal Alumni Research award Columbia Dental Alumni, 1961. Fellow Am. Coll. Dentists, N.Y. Acad. Scis., Royal Micros. Soc., A.A.A.S., Gerontol.; asso. fellow N.Y. Acad. Medicine; mem. Sci. Research Soc. Am. (pres. N.Y. U. Coll. Dentistry br. 1952-53), Am. Acad. Dental Medicine (hon.), Aerospace Med. Assn., Am. Assn. Anatomists, Am. Assn. Cancer Research, Am. Soc. for Zoologists, Am. Micros. Soc., Am. Soc. for Naturalists, Am. Physiol. Soc., Am. Chem. Soc., Endocrine Soc., Internat. Assn. for Dental Research (sec.-treas. N.Y. sect. 1955—), Soc. Exptl. Biology and Medicine, Soc. Study Devel. and Growth, Entomol. Soc. Am., Harvey Soc., Phi Beta Kappa, Sigma Xi, Beta Lambda Sigma. Author (with Gerrit Bevelander): Laboratory Directions in Histology, 1948. Editor: (with R. I. Dorfman) The Influence of Hormones on Enzymes, 1951. Research and publs. on growth and devel.; mechanisms involved in chemotherapy of antimetabolites and their use in animal sterilization: hazards of drugs to fetuses, prodn. blood diseases; discovered insect chemosterilants.*

GOLDSMITH, George Jason, Am. physicist; b. Newburyport, Mass., Mar. 29, 1923; s. Albert Abe and Bessie (Segal) G.; B.S. in Chemistry, U. Vt., 1944; M.S. in Chemistry, Purdue U., 1948, Ph.D. in Physics, 1955; m. Sonia Perkins, June 10, 1945; children—Lynn, Peter, Robert, Laurie. Instr., Purdue U., Lafayette, Ind., 1948-55; mem. tech. staff RCA Labs., Princeton, N.J., 1955-68; mem. staff plasma physics lab. Princeton (on loan from RCA), 1965-68; asso. prof. physics Boston Coll., 1968—. Mem. Am Phys. Soc., A.A.A.S., Sigma Xi, Sigma Pi Epsilon. Author (with E. Bleuler): Experimental Nucleonics, 1952. Research, publs. on basic research in nuclear physics, solid state physics, ferroelectrics, solid state plasmas. Home: 27 Longview Dr., Princeton, N.J. 08540.*

GOLDSMITH, Grace A., Am. internist-nutritionist; b. St. Paul, Apr. 8, 1904; d. Arthur William and Arabell Louise (Coleman) Goldsmith; B.S., U. Wis., 1925; M.D. (Isidore Dyer medal), Tulane, 1932; M.S., U. Minn., 1936; D.M.S. (hon.), Woman's Med. Coll. Pa. Mem. faculty Tulane, 1936—, prof. medicine, 1949—, dir. nutrition-metabolism sect. dept. medicine, 1946-67, dean Sch. Pub. Health and Trop. Medicine, 1967—; cons. to hosps., brs. of govt. Mem. bd. NRC, 1948—; mem. U. S. nat. com. Internat. Union Nutritional Scis., 1960—; trustee Am. Freedom from Hunger Found. Recipient Outstanding Achievement award U. Minn., 1964. Diplomate Am. Bd. Internal Medicine Am. Bd. Nutrition, Pan Am. Med. Assn. Fellow A.A.A.S., A.C.P., A.M.A. (Goldberger award 1965), Am. Pub. Health Assn., N.Y. Acad. Scis.; mem. Assn. Am. Physicians, Am. Diabetes Assn., Am. Inst. Nutrition (Osborne and Mendel award 1959, pres. 1963-64), Am. Soc. Clin. Investigation, Am. Soc. Clin. Nutrition, Fedn. Am. Socs. Exptl. Biology, Internat. Soc. Hematology, Soc. Exptl. Biology and Medicine, So. Med. Assn., So. Soc. Clin. Research, Alpha Omega Alpha, Sigma Xi, Delta Omega. Author: Nutritional Diagnosis, 1959. Mem. editorial bd. Am. Jour. Atherosclerosis Research, Jour. Am. Med. Women's Assn., Archives Internal Medicine. Research and publs. on role of corn diets in pellagra; nutritional value of ascorbic acid and B vitamins; lipid metabolism. Diplomate Pan Am. Med. Assn. Home: 1621 Peniston St., New Orleans 70115.*

GOLDSMITH, Julian Royce, Am. geochemist; b. Chgo., Feb. 26, 1918; s. Mitchel and Ceclia (Kallis) G.; S.B., U. Chgo., 1940, Ph.D., 1947; m. Ethel J. Frank, Sept. 4, 1940; children—Richard Norman, Susan Jean, John Elmore. Research chemist Corning Glass Works, 1942-46; research asso. geochemistry U. Chgo., 1947-51, faculty, 1951—, asso. prof. geochemistry, 1958—, asso. dean div. phys. scis., 1960—, acting dean, 1962, asso. chmn. dept. geophys. scis., 1961-62, chmn., 1963—. Mem. earth sci. panel NSF, 1958-61, chmn., 1960-61; mem. Nat. Sci. Bd., 1964—. Fellow Geol. Soc. Am., Mineral. Soc. Am. (past councillor), A.A.A.S., Am. Acad. Arts and Scis.; mem. Am. Chem. Soc., Am. Crystallographic Assn., Geochem. Soc. (pres. 1966), Am. Ceramic Soc., Mineral. Soc. Gt. Britain, Am. Geophys. Union, Phi Beta Kappa, Sigma Xi. Cons. editor Ency. Brit., 1955—, McGraw-Hill Ency. Sci. and Tech., 1957—. Research, publs. on chem. inter-relations at high temperatures and pressures, substances in earth's crust, crystallization phenomena minerals as related to atomic structure. Home: 5631 Blackstone Av., Chgo. 60637.*

GOLDSMITH, Middleton, Am. surgeon; b. Port Tobacco, Md., Aug. 5, 1818; s. Alban and Talia Ferro (Middleton) Smith; attended Hanover (Ind.) Coll.; grad. Coll. Physicians and Surgeons, N.Y.C., 1840; m. Frances Swift, June 1843; 2 daus. a founder N.Y. Path. Soc., 1844; prof. surgery Castleton (Vt.) Med. Coll., 1844-45; pres. Vt. Med. Soc., 1851; prof. surgery Ky. Sch. Medicine, Louisville, 1856; served in Civil War; wrote pamphlet A Report on Hospital Gangrene, Erysipelas, and Pyaemin as Observed in the Departments of the Ohio and the Cumberland: With Cases Appended, 1863; pioneer in antiseptic surgery; established Rutland (Vt.) Free Dispensary; spl. commr. to investigate state insane asylum, Vt., which resulted in its improvement and reform; drew up game laws of Vt. Introduced (with father) practice of lithority (method of crushing bladder stones) in Am. Died Nov. 26, 1887.

GOLDSMITH, William Noel, Brit. dermatologist; b. London, Eng., Dec. 26, 1893; s. Ernest and Bertha Fanny (Feist) G.; ed. univs. Cambridge, Wroclaw, Vienna; M.A., M.D. With Cambridge U., U. Breslau, St. John's Hosp.; dermatologist Univ. Coll. Hosp., London, 1932-59. Fellow Royal Coll. Physicians; mem. Brit. Assn. Dermatology (hon., pres.), St. John's Hosp. Dermatol. Soc. (pres.), Royal Soc. Medicine (pres. dept. dermatology); hon. mem. socs. dermatology Austria, Germany, Berlin, Am., Hungary, Denmark, Sweden, France. Author: Recent Advances in Dermatology, 1936, ed edit., 1954, also articles. Editor: Brit. Jour. Dermatology, 1939-49. Home: 129 Whitehall Ct., London S.W.1. Office: 6 Upper Wimple St., London W.1, Eng.

GOLDSPOHN, Albert, Am. physician; b. Dane County, Wis., Sept. 23, 1851; B.S., Northwestern Coll., Naperville, Ill., 1875; M.D., Rush Med. Coll., 1878; studied in Germany 2 yrs.; m. Cornelia E. Walz, 1887. Resident physician and surgeon. Cook County Hosp., Chgo., 1878-79; practiced at Des Plaines 6 yrs.; settled in Chgo.; attending surgeon former German Hosp., 1888-1904; prof. diseases of women and abdominal surgery. Post-Grad. Med. Sch. and Hosp., 1890-1922; surgeon in chief Evang. Deaconess Hosp. Devised a fundamental operation for injuries sustained during childbirth, also 2 major operations for displacements and prolapse of womb. Died Sept. 1, 1929.

GOLDSTEIN, Allen Abbey, Am. mathematician; b. Balt., Jan. 7, 1925; s. Joseph and Sophia (Sklar) G.; B.A., St. John's Coll., 1947; M.A., Georgetown U., 1952, Ph.D., 1954; m. Martha D. Svendsen, May 21, 1945; children—Stephen, Laurence, Julie, Walter, Phyllis, Pauline. With Nat. Bur. Standards, 1951-53, Georgetown U., 1954-55, Convair Astronautics, San Diego, 1955-60, Mass. Inst. Tech., 1960-63; asso. prof. math. U. Tex., Austin, 1963-64; asso. prof. U. Wash., Seattle, 1964-65, prof. math., 1965—; guest prof. U. Hamburg (Germany), 1966-67. Cons., Raytheon Corp., 1960-63, Boeing Sci. Research Labs., 1963—, Livermore Lab., U. Cal., 1964—. Mem. Am. Math. Soc. Author: Constructive Real Analysis, 1967. Contbr. articles to profl. jours. Research on theory of numerical solution of problems of applied math., constructive methods for finding roots of equations and systems of equations and extreme values of functions, theory and constructive methods for approximating functions by other functions, problems of optimal control theory. Office: Dept. of Math., University of Wash., Seattle 98105.*

GOLDSTEIN, Avram Shalom, Am. pharmacologist; b. N.Y.C., July 3, 1919; s. Israel and Bertha (Markowitz) G.; A.B., Harvard, 1940, M.D., 1943; m. Dora Benedict, Aug. 29, 1947; children—Margaret, Daniel, Joshua, Michael. Intern Mt. Sinai Hosp., N.Y.C., 1944; successively instr., asso., asst. prof. pharmacology Harvard, 1947-55; prof., exec. head dept. pharmacology, Stanford, 1955—. Mem. Am. Soc. Pharmacology and Exptl. Therapeutics, A.A.A.S., Am. Soc. Biol. Chemists, Am. Assn. Cancer Research, Genetics Soc. Am. Numerous investigations on the biochemical and molecular basis of drug action and the principles underlying the distribution and fate of drugs in the body; mechanisms of drug tolerance and addiction. Home: 735 Dolores St., Stanford, Cal.*

GOLDSTEIN, Eugen, German physicist; b. Gleiwitz, Germany, Sept. 5, 1850; s. Julius and Bertha (Neumann) G.; Ph.D., U. Berlin, Germany, 1881; m. Laura Baer. Research, U. Berlin from 1872; with Berlin Obs., from 1878, asst., from 1888; guest worker Reich Phys.-Tech. Inst., 1890-96; independent research, Berlin-Schöneberg, 1896-1927. Research and publs. on elec. discharge in rarified gases; discovered canal rays, 1886, elec. deflection of cathode rays, spark spectra of simple ionized atoms and band spectrum of helium molecule. Died Berlin, Dec. 25, 1930.

GOLDSTEIN, Henri, Swiss chemist; b. Paris, France, Sept. 14, 1897; s. Raphäel and Marie (Sachs) G.; Ph.D. in Sci., U. Lausanne; m. Giorgina Brociner, 1942. Asst., U. Lausanne, prof. organic chemistry, hon. prof.; asst. Technische Hochschule, Zurich; chemist Sandoz Mfg. Co., Basle. Mem. Swiss Soc. Chemistry (pres. 1946-48). Contbr. articles to profl. publs. Address: Bonne-Esperance 30, Lausanne, Switzerland.

GOLDSTEIN, Jacob Herman, Am. chemist; b. Atlanta, Dec. 18, 1915; s. David and Jennie (Levine) G.; A.B., Emory U., 1942, M.S., 1944; A.M., Harvard, 1947, Ph.D., 1949; m. Audrey Jones, Dec. 26, 1952. Faculty, Emory U., 1949—, prof. chemistry, 1959—, Charles H. Candler prof., 1960—. NRC postdoctoral fellow Harvard, 1946-49. Fellow Am. Phys. Soc.; mem. Am. Chem. Soc., Soc. Nuclear Medicine, Am. Assn. U. Profs., Phi Beta Kappa, Sigma Xi, Omicron Delta Kappa. Special research on molecular spectroscopy and structure, valence theory.

GOLDSTEIN, Kurt, psychologist; b. Kattowitz, Germany, Nov. 6, 1878; M.D., U Breslau, 1903. Prof., Konigsberg, 1912; head dept. neurology Neurol. Inst., U. Frankfurt, 1915, prof., dir. Neurol. Inst., 1922; head dept. neurology Moabit Hosp., Berlin, 1929; prof. U. Berlin, 1930; Rockefeller fellow, Amsterdam, 1933-34; with Psychiat. Inst., Columbia, 1934-35, clin. prof. neurology, lectr. psychopath. dept., 1936-40; head neurophysiol. lab. Montefiore Hosp., 1936-40; clin. prof. neurology Tufts Coll. Med. Sch., 1940-45; vis. prof. psychology City Coll. N.Y., 1950-55, New Sch. Social Research, 1955—; James lectr. Harvard, 1938-39; head neurol. lab. Boston Dispensary, 1940-45. Fellow N.Y. Acad. Scis.; mem. Am. Neurol. Assn., Am. Psychiat. Assn., Am. Psychol. Assn. Asst. editor: Jour. Nervous and Mental Diseases. Research on psychopathology, comparative normal and path. anatomy, speech and optic sphere disorders, injuries and tumors of brain, schizophrenia.

GOLDSTEIN, Lester, Am. biologist; b. Bklyn., June 28, 1924; s. Charles and Gussie (Silverman) G.; B.A., Bklyn. Coll., 1948; Ph.D., U. Pa., 1953; m. Margaretta Gilboy, Aug. 29, 1964. Research fellow U. Cal. at Berkeley, 1953-55; research asso. U. Cal. Med. Center, San Francisco, 1955-59; faculty U. Pa., 1959-67, prof. biology, 1964-67; prof. U. Colo., Boulder, 1967—. Harrison fellow U. Pa., 1950-51; USPHS fellow, 1952-53, 53-55; Damon Runyon fellow, 1955-56, Lalor Found. fellow, 1959; USPHS Spl. fellow, 1965-66. Mem. A.A.A.S., Am. Inst. Biol. Scis., Am. Soc. Cell Biology, Internat. Soc. Cell Biology. Editor: Readings in Cell Biology, 1966; The Control of Nuclear Activity, 1967; editor Biosci. jour., 1966—. Research and publs. on relationship between nuclear RNA and cytoplasmic RNA; discovered behavior of proteins of cell nucleus. Home: Sunshine Canyon, Boulder, Colo. 80302.*

GOLDSTEIN, Louis, physicist; b. Dombrad, Hungary, Mar. 25, 1904; s. Morris M. and Josephine (Stern) G.; Licence es Sciences, Sorbonne, Paris, France, 1926, Doctorat es Sciences, 1932; m. Ella Trammer, Nov. 14, 1933; 1 son, John C. Came to U. S., 1939, naturalized, 1943. Research worker Sorbonne, 1928-32; research asso. Institut Henri Poincare, U. Paris, 1932-39; research worker physics dept. N.Y. U., 1939-41; instr. physics Coll. City N.Y., 1942-44; mem. wave propagation group Div. War Research, Columbia, 1944-45; physicist Fed. Telecommunications Lab., Internat. Tel. & Tel., 1946; mem. staff Los Alamos Sci. Lab., 1946-—. Mem. Am. Phys. Soc. Mem. B'nai B'rith. Author: Conservation Theorems in the Theory of Electronic Collisions, 1933. Research and publs. in statis. mechanics, statis. thermodynamics of quantum systems, including atoms, atomic nuclei, dense phases of helium isotopes. Home: 1300 Canyon Rd., Los Alamos 87544. Office: Los Alamos Sci. Lab., P.O. Box 1663, Los Alamos 87544.*

GOLDSTEIN, Menek, biochemist; b. Kolomya, Poland, Apr. 8, 1924; s. Jakob and Cylia, (Hirsch) G.; Ph.D., U. Berne (Switzerland), 1955. Came to U. S., 1956, naturalized, 1961. Asst. in biochemistry, U. Berne, 1953-56; research staff Worcester Found. for Exptl. Biology, Shrewsbury, Mass., 1956-57; faculty N.Y. U. Med. Center, N.Y.C., 1958-—, asso. prof. biochemistry, 1963-—. Mem. Am. Chem. Soc., Am. Soc. Biol. Chemists, Am. Soc. Pharmacology and Exptl. Therapeutics, Inc. Research and publs. on biosynthesis and metabolism of adrenergic neurotransmitters. Home: 333 30th St., N.Y.C. 10016.

GOLDSTEIN, Richard J., Am. mech. engr.; b. N.Y.C., Mar. 27, 1928; s. Henry and Rose (Steierman) G.; B.Mech. Engring., Cornell U., 1948, M.S. in Mech. Engring., 1950, M.S. in Physics, 1951; Ph.D., U. Minn., 1959; m. A. Nancy Klein, Sept. 5, 1963; children—Arthur Sander, Jonathan Jacob. Instr., research fellow U. Minn., 1948-51; devel. engr. Oak Ridge Nat. Lab., 1951-54; fellow, instr. U. Minn., 1956-58, asso. prof., 1961-65, prof., mech. engring., 1965-—; asst. prof., Brown U., Providence, 1959-61. Cons., Gen. Electric Co., 1965-—. Mem. Am. Soc. M.E., Minn. Acad. Sci., Am. Assn. U. Profs., Tau Beta Pi, Pi Tau Sigma, Sigma Xi. Research and publs. on fluid mechanics and heat transfer by free and forced convection, evel. exptl. techniques for measurements of velocity and temperature; measurements of transport properties. Home: 520 Janalyn Circle, Golden Valley, Minn. 55416. Office: Mech. Engring. Dept., U. Minn., Mpls. 55455.*

GOLDSTEIN, Sidney, Am. sociologist; b. New London, Conn., Aug. 4, 1927; s. Max and Bella (Hoffman) G.; B.A., U. Conn., 1949, M.A., 1951; Ph.D., U. Pa., 1953; m. Alice Dreifuss, June 21, 1953; children—Beth Leah, David Louis, Brenda Ruth. Instr. sociology U. Pa., 1953-55; faculty Brown U., Providence, 1955-—, prof., 1960-—, chmn. dept. sociology and anthropology, 1963-—. Cons., Inst. Neurol. Diseases, NIH, 1960-—, population div. UN, 1965-66; mem. adv. com. U. S. Bur. Census, 1965-—. Harrison fellow U. Pa., 1951-53; Fulbright scholar, Denmark, 1961-62; Guggenheim fellow, 1961-62; Social Sci. Research Council fellow, 1961-62; A.M. Dushkin Hadassah fellow, 1965-66. Mem. Am. Statis. Assn., Am. Sociol. Assn., Population Assn. Am. (dir. 1966-68), Internat. Union for Sci. Study Population, Gerontol. Soc., Assn. for Jewish Demography (dir. 1965-—), Eastern Sociol. Soc., Phi Beta Kappa. Author: Patterns of Mobility, 1910-50, 1958; Migration and Economic Development in Rhode Island, 1958; Consumption Patterns of the Aged, 1960; (with Kurt B. Mayer) The First Two Years, 1961; The Norristown Study, 1961; People of Rhode Island, 1963; (with Calvin Goldscheider) Jewish Americans: Three Generations in a Jewish Community, 1968; also articles. Devel. techniques for measuring population mobility; evaluation of characteristics of migrants and relation between mobility and econ. opportunity; research on changing age structure and changing levels of income, expenditures and savs. of aged. Home: 95 Kiwanee Rd., Warwick, R.I. 02888. Office: Brown U., Providence, R.I. 02912.*

GOLDSTEIN, Sydney, applied mathematician; b. Hull, Eng., Dec. 3, 1903; s. Joseph and Hilda (Jacobs) G.; student U. Leeds (Eng.), 1921-22; M.A., Ph.D., Cambridge (Eng.) U., 1928; D.Eng. (hon.), Purdue U., 1967; D.Sc. (hon.), Case Inst., 1967; m. Rosa R. Sass, Mar. 23, 1926; children—David John, Ruth Hilda (Mrs. Joakim J. Donner). Rockefeller Found. fellow U. Gottingen (Germany), 1928-29; lectr. Manchester (Eng.), 1929-31, Beyer prof. applied math., 1945-50; fellow St. John's Coll., Cambridge, 1929-32, 33-45; staff aerodynamics div. Nat. Phys. Lab., 1939-45; chmn. Brit. Aero. Research Council, 1946-49; prof. applied math., chmn. dept. aero. engring. Technion, Haifa, Israel, 1950-55, v.p., 1951-54; Gordon McKay prof. applied math. Harvard, 1955-—. Recipient Adams prize, 1935, Timoshenko medal Am. Soc. M.E., 1965; hon. fellow St. John's Coll., U. Cambridge, 1965; Leverhulme fellow Cal. Inst. Tech., 1938-39. Fellow Royal Aero. Soc., Royal Soc. Arts, Royal Soc., 1937; mem. Internat. Acad. Astronautics, Am. Acad. Arts and Scis., Royal Netherlands Acad. Arts and Sci. (fgn. mem.). Author: Lectures on Fluid Mechanics, 1960; also numerous articles. Editor: Modern Developments in Fluid Dynamics, 1938. Home: 28 Elizabeth Rd., Belmont, Mass. 02178. Office: Div. Engring. and Applied Physics, Pierce Hall, Harvard U., Cambridge, Mass. 02138.

GOLDSTINE, Herman Heine, Am. mathematician; b. Chgo., Sept. 13, 1913; s. Isaac Oscar and Bessie (Lipsey) G.; B.S., U. Chgo., 1933, M.S., 1934, Ph.D., 1936; m. Adele Katz, Sept. 15, 1941 (dec.); children—Madlen, Jonathan; m. 2d, Ellen Watson. With Inst. Advanced Study Princeton, 1946-57; with IBM Corp., 1958-—, cons. to dir. research, 1967-—; adj. prof. City U. N.Y.; mem. bd. sci. advisers Cornell U./Sloan-Kettering Inst.; mem. adv. council dept. math. Princeton; mem. numerous coms. Army Math. Steering Com., Air Force Office Sci. Research, NRC. Mem. Am. Math. Soc., Circolo Mathematico di Palermo, Math. Assn. Am., Soc. Indsl. and Applied Math., Sigma Xi. Contbr. numerous articles to profl. jours. Research on computer devel.; collaborated with von Neumann on design and devel. of first computer. Home: 18 Hayrake Lane, Chappaqua, N.Y. 10514. Office: IBM Corp., P.O. Box 218, Yorktown Heights, N.Y. 10598.*

GOLDTHWAIT, James Walter, Am. geologist; b. Lynn, Mass., Mar. 22, 1880; s. James Wesley and Olive Jane (Parker) G.; A.B., Harvard, 1902, A.M., 1903, Ph.D., 1906; D.Sc., U. N.H., 1945; m. Edith Dunnels Richards, June 25, 1906; children—Richard Parker, Lawrence, Teaching fellow Harvard, 1902-04; tchr. Radcliff Coll., 1904; asst. prof. geology Northwestern U., 1904-08; asst. prof. geology Dartmouth, 1908-11, prof. geology since 1911. Engaged during summers in geologic field work for state surveys of Wis. and Ill., for U. S. Geol. Survey and for Geol. Survey of Can.; geologist N.H. Hwy. Dept. since 1917; cons. Gt. Lakes diversion suits. Fellow Geol. Soc. Am., Am. Acad. Arts and Sciences; mem. Phi Beta Kappa, Sigma Xi. Author: Abandoned Shorelines of Eastern Wisconsin (Wis. Geol. Survey), 1906; Physiography of Nova Scotia (Geol. Survey Can.), 1925; Geology of New Hampshire (N.H. Acad. Science Handbook No. 1), 1925; also numerous reports and papers, dealing with extinct shorelines, earth movements, river floods, glacial and physiographic studies in N.E. and Can. Died Dec. 31, 1947.

GOLDTHWAIT, Richard Parker, Am. geologist; b. Hanover, N.H., June 6, 1911; s. James Walter and Edith (Richards) G.; B.A., Dartmouth, 1933; M.A., Harvard, 1937, Ph.D., 1939; m. Katherine Davenport Burnham, June 12, 1937; children—Jane (Mrs. Robert Oldham), Susan Betsy, Thomas Burnham. Asst. prof. Brown U., 1939-43; faculty Ohio State U., Columbus, 1946-—, prof. geology, 1948-—, dir. Inst. Polar Studies, 1960-65, chmn. dept. geology, 1965-—; mem. expdns. to Chinese Tibet, Antarctica, Greenland, Baffin Island. Recipient Award of Merit, USAAF, 1946; named one of Ten Outstanding Men of Year, Columbus Citizen Jour., 1962; Antarctica Service medal 1966. Fellow Geol. Soc. Am.; Am. Geog. Soc., A.A.A.S., Ohio Acad. Sci.; mem. Am. Geophys. Union, Glaciological Soc. Author: Geology of New Hampshire, Part I Surficial Geology, 1951; also numerous articles. Research on glaciers in action to explain glacial records with application to Pleistocene history several regions; design and supervision team research projects in geology, soils, ecology, glaciology in arctic-subarctic regions. Home: 452 Colonial Av., Worthington, O. 43085. Office: Dept. Geology, Ohio State U., 125 S. Oval Dr., Columbus, O. 43210.*

GOLDWASSER, Edwin Leo, Am. physicist; b. N.Y.C., Mar. 9, 1919; s. Israel Edwin and Edith (Goldstein) G.; B.A., Harvard, 1940; student Columbia, 1941; Ph.D., U. Cal. at Berkeley, 1950; m. Elizabeth Weiss, Oct. 27, 1940; children—Michael, John, Katherine, David, Richard. Physicist, U. S. Navy Bur. Ordnance, 1941-45; sr. physicist 12th Naval Dist., also U. S. Navy Yard, Mare Is., Cal., 1943-45; teaching asst., research asst. U. Cal. at Berkeley, 1948-50, research asso., 1950-51; faculty U. Ill., 1951-67, prof. physics, 1959-67, mem. phys. scis. study com., 1956-61; dep. dir. Nat. Accelerator Lab., Oak Brook, Ill., 1967-—; spl. research primary cosmic radiation, energy loss charged particles, photoprodn. of pi mesons, interactions of strange particles. Mem. physics survey com. NRC, 1964-—, chmn. div. phys. scis., 1966-—; mem. panel high energy accelerator physics gen. adv. com. of AEC and President's Sci. Adv. Com., 1962-63. Westinghouse Fellow, 1949-50; Fulbright fellow to Italy, 1957-58; Guggenheim fellow, 1957-58. Fellow Am. Phys. Soc.; mem. Fedn. Am. Scientists, Sigma Xi. Author: Optics, Waves, Atoms and Nuclei, 1965. Author numerous articles, contbr. to books. Research on proton and neutron components of cosmic radiation at sea level, energy loss of electrons in rare and dense materials, distbn. of energy losses and density effect, elementary particle interactions at high energies, photoprodn. of Pi mesons with spl. emphasis on threshold region. Home: 1755 E. 55th St., Chgo. 60615. Office: Nat. Accelerator Lab., 1301 W. 22d St., Oak Brook, Ill. 60521.*

GOLDWATER, Leonard John, Am. physician; b. N.Y.C., Jan. 15, 1903; s. Abraham Lincoln and Belle (Delmar) G.; A.B., U. Mich., 1924; M.D., N.Y. U., 1928, Sc.D. in Medicine, 1936; M.S. in Pub. Health, Columbia, 1941; m. Charlotte von der Heyde, Dec. 15, 1953. Faculty N.Y. U. Coll. Medicine, 1932-36, 1938-41, 46; faculty Columbia, N.Y.C., 1946-—, prof. occupational medicine, 1952-—; sr. indsl. hygiene physician N.Y. Dept. Labor, N.Y.C., 1936-38; praelector St. Andrews U., Scotland, 1966; vis. scholar Duke, 1967-68. Practice internal medicine, N.Y.C., 1933-41; cons. in occupational medicine AEC, WHO, and others, 1946-—. Hon. fellow Royal Inst. Pub. Health and Hygiene; mem. A.M.A., N.Y. Acad. Medicine, Am. Pub. Health Assn., Am. Acad. Occupational Medicine (pres. 1959), Indsl. Med. Assn., Am. Indsl. Hygiene Assn., Am. Conf. Govtl. Indsl. Hygienists, Med. Soc. N.C., Finnish (hon.), Egyptian (hon.) socs. indsl. medicine. Contbr. to Dangerous Properties of Industrial Materials (N. I. Sax), 1963. Organized cardiac work classification unit for selective placement of cardiacs in employment, 1941; basic studies in toxicology of benzene, glycols, mercury, and other indsl. chems., especially mercury. Home: Route 3, Chapel Hill, N.C. 27514.*

GOLDZIEHER, Joseph William, Am. physician; b. Budapest, Hungary, Sept. 21, 1919; s. Max A. and Margaret (Strasser) G.; A.B., Harvard, 1940; M.D., N.Y.U., 1943; m. Maria Trinidad Garcez, June 10, 1955; 1 dau., Michele Alexandra (Mrs. Allan Shedlin, Jr.). Practice medicine, specializing in endocrinology, N.Y.C. and San Antonio, 1947-—; head endocrine sect. S.W. Found. For Research and Edn., 1953-57, chmn. dept. endocrinology, 1957-—, dir. div. clin. scis., 1966-—; research prof. Trinity U., 1959-—; nat. cons. U. S. Army dept. medicine Brooke Gen. Hosp., Ft. Sam Houston, 1956-—, USAF dept. medicine Wilford Hall Hosp., Lackland AFB, Tex., 1960-—. Diplomate Am. Bd. Clin. Chemistry. Fellow Am. Coll. Clin. Pharmacology and Chemotherapy, Am. Coll. Obstetricians and Gynecologists, N.Y. Diabetes Assn.; mem. Endocrine Soc., Mexican Soc. Nutrition and Endocrinology, Laurentian Hormone Conf., Am. Soc. Study Sterility, Pacific Coast Sterility Soc., Am. Assn. Clin. Chemists, Assn. Harvard Chemists, A.M.A. Contbr. numerous articles to profl. jours. Research, publs. on steroid methodology, organ disorders, metabolism; clin. and exptl. reproductive physiology. Home: 9717 Blanco Rd. Office: S.W. Found. For Research and Edn., San Antonio 78206.*

GOLENHOFEN, Klaus Winfried Franz, German physiologist; b. Breslau, Germany, Oct. 19, 1929; s. Paul and Elisabeth (Nickel) G.; student U. Erlangen (Germany), 1949-52, U. Göttingen (Germany), 1952, Medizinische Akademie Düsseldorf (Germany), 1952-53; Dr.med., U. Heidelberg (Germany), 1955; m. Renate Neff, Nov. 24, 1956; children—Rainer-Anke, Martina Nikola. Faculty, Inst. Physiology U. Marburg/Lahn (W. Germany), 1955-—, prof. physiology, 1966-—; vis. researcher dept. pharmacology U. Oxford (Eng.), 1963-64. Mem. Deutsche Physiologische Gesellschaft, Deutsche Gesellschaft fur Kreislaufforschung, Gesellschaft Deutscher Naturforscher und Ärzte, Physiol. Soc. (asso.). Author: (with H. Hensel, G. Hildebrandt) Durchblutungsmessung mit Wärmeleitelmenten, 1963; also numerous articles. Research on muscle blood flow in man, local heat clearance technique for measuring blood flow, shivering during cold stress, coordination of circulatory rhythms, oxygen consumption of smooth muscle, electrophysiology of smooth muscle. Home: 15 Calvinstrasse. Office: Physiologisches Institut, Deutschlansstrasse 2, Marburg/Lahn, West Germany.*

GOLGI, Camillo, Italian cytologist; b. Corteno, Brescia, Lombardy, July 7, 1843 (or 1844); med. tng. U. Padua, grad. 1865; numerous hon. degrees. Did some work at psychiat. clinic of Cesare Lombroso, then at lab. of G. Bizzozero, histologist, Pavia; took post at small hosp. for incurables at Abbiategrasso, 1872; prof. anatomy U. Siena, 1879-80; prof. histology and gen. pathology U. Pavia, 1880. Recipient (with Ramon y Cajal) Nobel prize in medicine and physiology, 1906, numerous other awards. Developer silver-impregnation method to stain nervous tissues, 1873; described 2 main types of nerve cells (since known as Golgi cells type I and II), 1873; described large nerve cells of granular layer of cerebellum, 1874; described structure of olfactory bulb, internal reticular apparatus of cells which still bears his name, 1885; discovered muscle spindles or nervous musculotendinous end organs; demonstrated that chorea is associated with definite lesions in nervous system; showed microscopic differentiation between sarcomas and gliomas; showed that different forms of malaria caused by different protozoans, also that severity of malaria attack depends on number of parasites in blood and that malarial paroxysms are coincident with sporulation of parasites, 1886-89; did work on structure of kidney and other organs; made valuable observations on pellagra and causation of mental disease. Died Pavia, Jan. 21, 1926.

GOLINEVICH, Elena Mikhailovna, Russian microbiologist; b. 1901; grad. Kiev Med. Inst., 1925; M.D., 1953. Lab. dir. in hosp., 1926-31; research worker Inst. Exptl. Medicine, Leningrad, 1931-34, then at All-Union Inst. Exptl. Medicine, Moscow, 1934-44; with Gamalci Inst. Exptl. Medicine Acad. Med. Sci. USSR, since 1946, apptd. lab. dir., 1954. Recipient (with P. F. Zdrodovskii) Lenin prize, 1959, for The Study of Rickets and Ricketsiosa (pub. 1956).

GOLL, Friedrich, Swiss physician, neurohistologist; b. Zofingen, Switzerland, Mar. 1, 1829; s. Johann Ulrich and Sophie (Herosé) G., student natural scis., Zurich, Switzerland, M.D., 1853; student Würzburg, Germany. Worked under C. Bernard, Paris,

1853-55; began practice medicine, Zurich, 1855; named lectr. materia medica, 1862, asso. prof., 1885; dir. Med. Polyclinic, 1863-69. Mem. Soc. Physicians Canton Zurich (pres. 1885-95). Author: Über die feinere anatomie des Rückenmarkes, 1860. Histological research on the spinal cord with more than 2,000 tissue samples (used chromic acid and Gerlach's carmine dye); prepared tables; described parts of architectonics of nerve fibers; median portion of dorsal funiculus of spinal cord was named after him. Died Zurich, Nov. 12, 1903.

GOLOMB, Michael, Am. mathematician; b. Munich, Germany, May 3, 1909; s. Moritz and Miriam (Margulies) G.; abiturium U. Würzburg, 1928-29; Ph.D., U. Berlin, 1933; m. Dagmar Racic, Feb. 19, 1939; children—Miriam Wanda, Deborah. Came to U. S., 1939, naturalized, 1946. Lectr., Zagreb, Yugoslavia, 1934-39; research asso., instr. Cornell U., Ithaca, N.Y., 1939-42; faculty Purdue U., Lafayette, Ind., 1942—, prof., 1950—; chief analysis sect. Research Lab., Franklin Inst., Phila., 1944-46; vis. prof. Math. Research Center, U. Wis., Madison, 1957-58, summer 58; research asso., lectr. div. applied math. Argonne Nat. Labs., Chgo. summers 1946-48, 60, 62; lectr. Ecole d' Été, Ablis, France, 1963. Cons. Franklin Inst., Phila., Naval Ordnance Plant, Indpls., 1950-52, Argonne Nat. Lab., Chgo., 1961—. Fellow A.A.A.S.; mem. Am. Math. Soc., Math. Assn. Am., Soc. Indsl. and Applied Math., Oesterreichishe Math. Gesellschaft, Ind. Acad. Sci., Soc. for Natural Philosophy, Sigma Xi. Author: (with M. E. Shanks) Elements of Ord. Differential Equations, 1950; Lectures on Theoretical Mechanics, 1960; Lectures on Theory of Approximations, 1963. Research in integral equations, differential equations, control systems, approximation theory. Home: 1407 Woodland Av., West Lafayette, Ind. 47906.*

GOLOVIN, Igor Nikolaevich, Russian exptl. physicist; b. 1913; grad. Moscow U., 1936. Mem. staff Inst. Atomic Energy Acad. Scis. USSR, since 1946, apptd. dir. research, dep. dir., during 1950-57. Recipient Lenin prize, 1958 for research on powerful impulse discharges in gas to receive high-temperature plasma. Specialist in electronics. Address: USSR Moskva, Leninsky prosp., Institut atomney energu, Aw USSR.

GOLTZ, Friedrich Leopold, physiologist; b. Poznan, Poland, Aug. 14, 1834; s. Heinrich and Leopoldine Friederike (von Blumberg) G.; attended lectures of Helmholtz, Königsberg, Germany, however mainly self-taught; m. Agnes Simon, 1868. Anat. prosector U. Königsberg; named prof. physiology, Halle, Germany, 1870, Strasbourg, France, 1872. Author: Beiträge zur Lehre von den Functionen der Nervencentren des Frosches, 1869; Gesamte Abhandlungen über die Verrichtungen des Grosshirns, 1881; also numerous articles. Research on central nervous system, especially reflex processes, localization of cerebrum function in dogs (kept dogs alive without cerebrums for up to 3 years); observed that vestibule of inner ear functioned in balance, not for hearing. Died Strasbourg, May 4, 1902.

GOLTZ, Robert William, Am. physician; b. St. Paul, Minn., Sept. 21, 1923; s. Edward Victor and Clare (O'Neill) G.; B.S., U. Minn., 1943, M.B., 1944, M.D., 1945; m. Patricia Ann Sweeney, Sept. 27, 1945; children—Leni, Paul. Practice medicine specializing in dermatology, Mpls., 1951-65; faculty div. dermatology U. Minn., Mpls., 1951-65, asst. prof., 1958-61, asso. prof., 1961-65; prof., dir. div. dermatology U. Colo. Med. Center, Denver, 1965—. Am. Cancer Soc. grantee, 1956-58; NIH grantee, 1958-66. Mem. A.M.A., Am. Acad. Dermatology, Soc. for Investigative Dermatology, Histochem. Soc., Am., Pacific dermatol. assns., Am. Soc. Dermopathology. Research and publs. in histopathology and histochemistry of skin, elastic tissue physiology and pathology. Home: 5735 E. Oxford Av., Englewood, Colo. 80110. Office: U. Colo. Med. Center, Denver 80220.*

GOLUB, Morton Allan, Am. chemist; b. Montreal, Que., Can., June 11, 1925; s. Mike and Fanny (Gold) G.; B.Sc., McGill U., Montreal, 1944; M.Sc., U. N.B., 1947; Ph.D., U. Mo., 1951; m. Joanna Belle Hoffmaster; children—Kevin King, Eric Norman. Came to U. S., 1947, naturalized, 1954. Sci. instr. Ross Sch., Montreal, 1945-46; staff B. F. Goodrich Co., Brecksville, O., 1951-60, research asso., 1956-60; sr. polymer chemist Stanford Research Inst., Menlo Park, Cal., 1960-67; instr. math. Foothill Jr. Coll., Los Altos Hills, Cal., 1963-67; NRC research asso. Ames Research Center, NASA, Moffett Field, Cal., 1968—. DuPont postgrad. fellow U. Mo., 1948-49. Mem. Am. Chem. Soc., A.A.A.S., Research Soc. Am. Research and publs. on chem. reactions and structure analysis of macromolecules; isomerization and cyclization of natural rubber; discovered cis-trans isomerization of polybutadiene. Home: 1224 Thorpe Ct., Los Altos, Cal. 94022. Office: Ames Research Center, NASA, Moffett Field, Cal. 94035.*

GOLUBTSOV, Vyachoslav Alekseevich, Russian power engr.; b. Apr. 10, 1894; grad. Leningrad Electro-Tech. Inst., 1925. Chief engr. Kashmira State Electric Power Plant, then Chelyabinsk, 1934-36; with Moscow Power Inst., apptd. prof., 1945; chief of lab. USSR Acad. Scis. Energy Inst., since 1955. Recipient Stalin prize. Mem. USSR Acad. Scis. (corr.). Author: The Operation of Boiler Installations at Electric Power Plants, 1950; Refractory Materials and Slags in Power Engineering, 1953; Some Problems of the Efficient Use of Fuels, 1955; Handbook on Chemistry in Power Engineering, 1960; also others. Participated in devel. of methods for softening water for indsl. boilers; research on air-preheating, deaeration, preparation of pulverized coal and ash utilization (using ash from Moscow coal for aluminum oxide prodn.). Address: USSR Acad. Scis. Energy Institute, Moscow, Russia.

GOMBAS, Paul, Hungarian physicist; b. Selegszántó, Hungary, June 5, 1909; s. János and Maria (Gräf) G.; Ph.D., U. Budapest (Hungary), 1933; m. Ida Krepil, Sept. 12, 1931. Dozent, U. Budapest, 1938-39; prof. U. Szeged (Hungary), 1939-40, U. Kolozsvár, 1940-44; U. Tech. Scis., Budapest, 1944—. Recipient Kossuth prize, 1948, 50. Mem. Hungarian Acad. Scis. (v.p. 1948-58). Author: Die statistische Theorie des Atoms und ihre Anwendungen, 1949; Theorie und Lösungsmethoden des Mehrteilchenproblems der Wellenmechanik, 1950; Die statistische Behandlung des Atoms, 1956; Pseudopotentiale, 1967; also numerous articles. Editor, Acta Physica Hungarica, 1949—. Research on statis. theory of atoms and atomic nuclei, theory of metals, especially theory of cohesion; devel. method of pseudopotentials, 1935. Home: 83-85 Villányi ut, Budapest, Hungary.*

GOMBERG, Moses, chemist; b. Elizabetgrad, Russia, Feb. 8, 1866; s. George and Marie Ethel (Resnikoff) G.; ed. Elizabetgrad Gymnasium, 1878-84; B.S., U. Mich., 1890, M.S., 1892, Sc.D., 1894, LL.D., 1937; postgrad., U. Munich, 1896-97, U. Heidelberg, 1897; hon. Sc.D., U. Chgo., 1929, Poly. Inst., Bklyn., 1932. Instr. chemistry U. Mich., 1893-99; asst. prof. organic chemistry, 1899-1902, jr. prof., 1902-04, prof., 1904-36; also chmn. dept. of chemistry, 1927-36. Cons. chemist Bur. Mines, 1917-18. Recipient Nichols medal Am. Chem. Soc., 1914, Willard Gibbs medal, 1925, Chandler medal, 1927. Fellow A.A.A.S., (v.p., chmn. sect. C, 1935); mem. Nat. Acad. Sciences, Am. Philos. Soc., Am. (pres. 1931), Netherlands (hon.) chem. socs., Am. Inst. Chemists. Contbr. to jours. Discovered trivalent carbon, 1910; developed 1st satisfactory antifreeze for automobiles. Died Ann Arbor, Mich., Feb. 12, 1947.

GOMES, Bernardino Antonio, Portuguese physician botanist; b. Arcos, Portugal, 1769; M.D., Coimbra, 1793. Practiced medicine, Lisbon, 1793; became surgeon to Portuguese navy, 1797; went to Brazil as ship's surgeon; fought typhoid in Portuguese fleet at Gibralter; physician to Hospital Real, 1805; dir., Hospital of San Lázaro; took another trip to Brazil. Author: Memoria sobre la ipecacuana gris del Brazil, 1801; Método de Tratamiento del Tifus u otras fiebres malignas contagiosas, por la afusión del agua fria, 1806; Ensayo demografico, 1820; Sobre los medios de disminuir la elefantiasis en Portugal y de perfeccionar el conocimiento y la curacíon de las enfermedades cutaneas, 1821; Carta á los médicos Portugueses sobre la elefantiasis, 1821. Developed cold water treatment of fever credited with saving many lives; 1st to isolate medicine from cinchona bark (calling it cinchonino), 1810. Died Lisbon, Portugal, Jan. 13, 1824.

GOMOT, Lucien, French zoologist; b. La Geneytouse, France, Dec. 16, 1929; s. Etienne and Maria (Mousset) G.; D.Sc., Clermont-Ferrand (France) Faculty Sci.; m. Paulette Chastrette, July 8, 1954; children—Philippe, Annette. Scholar, Limoges (France) Ednl. Sch., 1947-51; tchr. Clermont-Ferrand Ednl. Sch., 1951-55, research asst. Faculty Sci., 1955-57, asst. in chief, Faculty Sci., 1960-61; research asst. Nat. Center Sci. Research, 1955-57, research asso., 1957-59, research chief, 1959-60; lectr. U. Besançon (France), 1961-63, prof. zoology Faculty Sci., 1963—, chair zoology, 1964—. Chevalier, Ordre des Palmes Académiques. Mem. Zool. Soc. France, Biology Soc. France. Research, publs. on organogenesis of uropygiene gland of birds, mammary gland of rabbit, mouse; (with A. Propper) study interspecific duck hybrids; (with P. Guyard) study of differentiation of hermaphrodite gland of snail. Home: Villa le Grillon, Franois. Office: Laboratoire de Zoologie, Place Maréchal Leclerc, Besançon (Doubs) 25, France.*

GOMPERS, Samuel, management scientist; b. England, Jan. 27, 1850; s. Solomon and Sarah (Root) G.; m. Sophia Julian, Jan. 27, 1867 (died 1920); m. 2d, Gertrude Gleaves Neuscheler, Apr. 16, 1921. Cigarmaker by trade; helped develop the Cigarmakers Internat. Union, becoming an officer, 1887; one of the founders of the Federation of Trades and Labor Unions, organized in 1881, of which was pres. 3 yrs.; one of founders Am. Federation of Labor, 1886, and continuously served as pres. excepting 1895, also editor of the American Federationist; 1st v.p. Nat. Civic Federation; mem. Advisory Com. Council Nat. Defense, 1917-19; rep. of A. F. of L. at Peace Conf., Paris, France, 1918-19; pres. Internat. Commn. on Labor Legislation, at the Peace Congress; chmn. delegates from A. F. of L. to Conv. of Internat. Federation of Trades Unions, Amsterdam, 1919; mem. President's First Industrial Conf., 1919, President's Unemployment Conf., 1921, President's Advisory Disarmament Com., 1921—, President's Agricultural Conference, 1921. Pres. Pan-Am. Federation of Labor; member of Sulgrave Inst. Author: Labor in Europe and America; American Labor and the War; Labor and the Common Welfare; Labor and the Employer; Out of Their Own Mouths; pamphlets on labor question and labor movement. Advocate of rights of labor; exerted great influence on development of Am. labor movement. Died Dec. 13, 1924.

GOMPERTZ, Benjamin, English mathematician, astronomer, actuary; b. London, Mar. 5, 1779; self-educated; m. Abigail Montefiore, 1816; actuary Alliance Assurance Co., 1824-28. Fellow Royal Soc., 1819 (became mem. council 1832); mem. Astron. Soc. (mem. council 1821-31), Math. Soc. Spitalfields (pres.). Prepared (with Francis Baily) a catalogue of stars, 1822; devised new series of mortality tables for Royal Soc.; publs. on imaginary quantities of prisms, 1817-18; developed law of human mortality (Gompertz's law). Died London July 14, 1865.

GONÇALVES DE LIMA, Oswaldo, Brazilian chemist; b. Recife, Pernambuco, Brazil, Nov. 7, 1908; s. Vicente Gonçalves Ourem de Lima and Júlia de Lima; ed. Colégio Arquidiocesano de Olinda; Celégio Nóbrega do Recife; Escola de Engenharia de Pernambuco; Escola Nacional de Química do Rio de Janeiro; m. Honorina de Souza Lima, Mar. 7, 1936; children: Clausius, Sônia, Clarissa. Chief chemist, Usina Professor Portela, Minas Gerais, 1930; chief chemist, Usina Agua Branca, 1932; chief chemist, Industrias Carlos de Brito & Cia., 1937-44; 1st dir., Escola Superior de Química, 1947-58; researcher, 1950-52, visiting prof., 1951 and 1958, Us. in Mexico; holder of professorships in Brazil; now dir., Institute of Antibiotics, U. Pernambuco. Recipient, Comenda de Pavia, 1956; hon. prof. U. Pernambuco, 1960; Order of Merit, State of Pernambuco, 1963. Titular of Brazilian Acad. of Scis., correspondent, Soc. of Pharmacy and Chemistry, Sao Paulo; mem., Soc. of Biology of Pernambuco; Soc. Mexicana de Quimica. Author: El Maguey y el pulque en los codices mexicanos, 1956; about 150 articles. Research in industrial and phytochemical microbiology; studies on various S. Am. native drinks; isolation of actinomycin, biflorin, ussamycin, dalbergion I and II; investigations of antibiotic and antimicrobic action of various drugs. Home: Rua Silveira Martins, 30, Rio de Janeiro, Brazil. Office: Instituto de Antibioticos da Universidade Federal de Pernambuco, Cidade Universitaria, Recife, Brazil.*

GONIN, Jules, Swiss ophthalmologist; b. 1870; ed. U. Lausanne (Switzerland); chief ophthalmic clinic, prof. ophthalmology U. Lausanne. Research and publs. on a method of treating detachment of retina using ignipuncture of retinal fissure through an incision in the sclera, 1927. Died 1935.

GONNARD, Pierre, French biochemist; b. Paris, France, Dec. 30, 1911; s. Jean and Elise (Garas) G.; Pharm. D., U. Paris, 1937, M.D., 1947, D.Sc., 1948; m. Alice Laurette, Aug. 12, 1944; children—Claude, Isabelle. With Med. Faculty, U. Paris, 1939—, asso. prof., 1955-64, prof. biochemistry, 1964—; dir. lab. Maison Départementale, Nanterre, France, 1941—. Recipient French Acad. Medicine prize. Mem. Soc. Biol. Chemistry, Internat. Soc. Neurochemistry, N.Y. Acad. Scis., Brazilian Inst. Investigation Tuberculosis. Research, numerous publs. on enzymology and applications to physiology, pathology, pharmacology, on melanins, pyridoxal phosphate, effects of pharmacological agts. on enzyme reactions, especially on cerebral metabolism. Home: 64 de Bellechasse, Paris 7e, France. Office: Maison Départementale de Nanterre 92, France.*

GONSER, Bruce Winfred, Am. metallurgist; b. Hudson, Ind., Sept. 9, 1899; s. Robert M. and Lillian A. (Bonbrake) G.; B.S. in Chem. Engring., Purdue, 1923, Chem.E., 1929; M.S. in Metallurgy, 1924; Sc.D., Harvard, 1933; m. Helen M. Vincent, Mar. 18, 1925; children—Gretchen (Mrs. John E. Cumming), Diana (Mrs. Richard P. DeVere), Galen Craig. Research chemist, metallurgist Am. Smelting and Refining Co., Colo., N.J., Utah, Cal., Tex., Neb., 1924-31; metallurgist Nat. Radiator Corp., Johnstown, Pa., 1933-34; chief nonferrous div. Battelle Meml. Inst., Columbus, O., 1934-52, asst. dir., tech. dir., 1952-64, cons. 1964—; expert tech. asst. missions, Buenos Aires, Argentina, 1965. Trustee Tin Research Inst., Columbus. Mem. Am. Soc. for Testing and Materials (past dir. award of merit 1956, Gillett lectr. 1966), Wire Assn. (past dir., Best Paper award 1951, Mordical Meml. lectr. 1963), A.A.A.S. (past mem. council), Am. Inst. Mining, Metall. and Petroleum Engrs. (past chmn. extractive metallurgy div.), Am. Soc. for Metals, Electrochem. Soc. Author: (with C. F. Powell and I. E. Campbell) Vapor Plating, 1955; also numerous articles. Editor: (with E. M. Sherwood) Columbium, 1958; Rhenium, 1962. Research in nonferrous metallurgy particularly tin, zinc, lead, antimony, titanium, zirconium. Home: 1301 Arlington Av., Columbus 43212. Office: 505 King Av., Columbus, O. 43201.*

GONZAGA, Arcadio C. Philippine vet. physiologist; b. Tanauan, Batangas, Philippines, Jan. 12, 1902; s. Ambrosio M. and Geronima (Castillo) G.; D.V.M., U. Philippines, 1926; Ph.D., Cornell U., 1933; m. Lourdes Piansay, Dec. 7, 1938; children—Nemesio, Rodeo. Instr., Coll. Vet. Medicine U. Philippines, 1928-39, asst. prof., 1939-52, asso. prof., 1952-56, prof., 1956-67, emeritus, 1967—, chmn. dept., 1949-67, dean Coll., 1961-63; professorial lectr. Araneta U. Found., Philippines, 1967—. Mem. Philippine Soc. Animal Sci., Philippine Assn. for Advancement Sci. Information and Communication, Philippine Vet. Med. Assn., NRC Philippines, Soc. Advancement Research, Phi Kappa Phi, Phi Sigma, Phi Zeta. Editor: Philippine Jour. Vet. Medicine. 1962—. Contbr. numerous articles to sci. jours. Pioneer in blood physiol. studies on domestic animals; semen preservation and artificial insemination. Home: 12 Maginhawa, U. P. Village, Quezon City, P.I. Office: Inst. Vet. Medicine, Araneta U. Found., Malabon, Rizal, P.I.*

GONZALEZ, Jose Oliver, parasitologist; b. Lares, P.R., Sept. 21, 1912; s. Jose Oliver-Lugo and Maria (Gonzalez) G.; B.A., U. P.R., 1938; M.S., U. Chgo., 1939, Ph.D., 1941; m. Maria Isabel Silva, June 29, 1935; children—Jose Enrique Oliver-Silva, Felipe Oliver-Silva, Jorge Oliver-Silva. With U. P.R. Sch. Medicine, 1928—, prof., head dept. med. zoology, 1961—, prof. parasitology, 1955—. Recipient Howard T. Rickett prize, 1941, Bailey K. Ashford award, 1947, Purdue Frederick prize, 1957. Mem. Am. Soc. Trop. Medicine and Hygiene, A.A.A.S., Am. Acad. Microbiology, Am. Soc. Parasitologists, Soc. Biology and Exptl. Medicine, Helminthol. Soc. Washington. Contbr. numerous articles in field to sci. jours. Developed tests for serological diagnosis of bilharziasis; contbd. to knowledge of antigenicity of parasitic forms. Home: 13 Bucare St., Santurce, P.R. 00913.*

GOOCH, Sir Daniel, Brit. engr.; b. Bedlington, Northumberland, Eng., Aug. 24, 1816; s. John and Anna (Longridge) G.; apprenticed as practical engr., Newcastle-upon-Tyne, Eng.; m. Margaret Tanner, Mar. 22, 1838; 4 sons, including Henry Daniel, 2 daus.; m. 2d, Emily Burder, Sept. 14, 1870. Became locomotive supt. Gt. Western Ry., 1837, later chmn.; chmn. Telegraph Constrn. and Maintenance Co.; dir. Anglo Am. Co.; mem. Parliament for Crickdale; justice of peace for Berkshire, 1865-85. Pioneered in telegraphic media; built 1st fast locomotive, 1846; set up 1st telegraphic connection to U. S., 1866; research on atmospheric resistance of trains and internal and rolling friction. Died nr. Windsor, Oct. 15, 1889.

GOOCH, Frank Austin, Am. chemist; b. Watertown, Mass., May 21, 1852; s. Joshua Goodale and Sarah Gates (Coolidge) G.; A.B., Harvard, 1872, A.M., Ph.D., 1877; student of Wolcott Gibbs, also in Europe until 1878; hon. M.A., Yale 1887; m. Sarah Elisabeth Wyman, Aug. 12, 1880; 1 dau. Meredyth (Mrs. John Downes Whiting). Asst. in chem. lab. under Prof. Josiah P. Cooke, 1872-1875; engaged in analytical work at Newport for U. S. 10th Census, 1879-81; chemist No. Transcontinental Survey, 1881-84, U. S. Geol. Survey, 1884-86; prof. chemistry Yale, 1885-1918, also dir. Kent Chem. Lab. Fellow or mem. Nat. Acad. Scis., A.A.A.S., Am. Philos. Soc. Author: Analyses of Waters of the Yellowstone Park (with J. E. Whitfield), 1888; Research Papers from the Kent Chemical Laboratory of Yale Univ. (2 vols.), 1901; (with C. F. Walker) Outlines of Inorganic Chemistry, 19051, Laboratory Experiments, 1905; Outlines of Qualitative Chemical Analysis (with P. E. Browning), 1906; Methods in Chemical Analysis, 1912; Representative Procedures in Quantitative Chemical Analysis, 1915. Developed Gooch filtering crucible, also method of rapid electrolytic estimation of metals, distillation method of estimating boric acid, method of quantitative separation of lithium from other alkalai metals; contbd. to iodometric methods. Died Aug. 12, 1929.

GOOD, Adolphus Clemens, naturalist; b. West Mahoning, Pa., Dec. 19, 1856; s. Abram and Hannah (Irwin) G.; grad. Washington and Jefferson Coll., 1879, Western Theol. Sem., 1882; m. Lydia Walker, June 21, 1883. Ordained to ministry Presbyn. Ch., 1882; active in missionary work, Baraka, French Congo, Africa, 1882-85, Kangwe, French Congo, 1885-92; worked in Bulu Country in German Territory, 1892-94 (penetrated further inland than any other white man up to that time); prepared a Bulu primer; translated Gospels into Bulu; reputedly added more to knowledge of insect forms of Africa than any other single collector. Died Efulen, German Territory, North of French Congo, Dec. 13, 1894.

GOOD, Richard Albert, Am. mathematician; b. Ashland, O., Sept. 24, 1917; s. Charles Wesley and Eva (Davis) G.; A.B., Ashland Coll., 1939; M.A., U. Wis., 1940, Ph.D., 1945; m. Josephine Gardner, Sept. 5, 1946; children—Mary J., Catherine A., Margery L. Asst. prof., U. Md., College Park, 1945-54, asso. prof., 1954-60, prof., 1960—. Mem. Soc. for Indsl. and Applied Math. (editor publs. 1962—). Study of groups; matrices; theory of clusters. Home: 6908 Wells Pkwy., Hyattsville, Md. 20782.

GOOD, Roland Hamilton, Jr., theoretical physicist; b. Toronto, Ont., Can., Oct. 22, 1923; s. Roland Hamilton and Marie (Smith) G.; came to U. S. 1948, naturalized, 1950; B.M.E., Lawrence Inst. Tech., 1944; MAE, Chrysler Inst. Engring., 1946; M.S., U. Mich., 1948, Ph.D., 1951; m. Ferol Hendrickson, May 7, 1944; children—Roland H. III, Patricia Gail, Sue Marie. Engr., Chrysler Corp., Windsor, Ont. and Highland Park, Mich., 1942-47; instr. U. Cal., Berkeley, 1951-53; faculty Pa. State U., 1953-56; from asso. prof. to prof. physics Ia. State U., Ames, 1956—, physicist, sr. physicist, Ames Lab. of U. S. AEC, 1956—. Vis. asso. prof. U. Colo. summer 1958; NSF sr. postdoctoral fellow, Inst. Advanced Study, Princeton, 1960-61; vis. prof. Inst. Math. Sci., Madras, India, 1968. Fellow Am. Phys. Soc. Research and publs. theoretical physics, especially, relativistic wave equations, polarization of elementary particles, metallic binding, electron emission from medals, and spectroscopy of rare earths. Home: 1724 Meadow Lane, Ames, Ia. 50010.*

GOOD, Walter Amos, Am. physicist; b. Hillsdale, Mich., Apr. 25, 1916; s. Lester O. and Fern (Hallett) G.; A.B. magna cum laude, Kalamazoo Coll., 1937; M.S., U. Ia., 1939, Ph.D., 1941; m. Joyce M. Force, Oct. 30, 1942; children—Ginnie F., Terry G. Grad. asst. physics dept. U. Ia., 1937-41; physicist dept. terrestrial magnetism Carnegie Instn. Washington, 1941-42; physicist Applied Physics Lab., Johns Hopkins U., Silver Spring, Md., 1942—, chmn. Bumblebee Guidance Panel, 1956-58, mem. SAE flight controls com., 1960—, mem. APL Adv. Bd., 1960—. Fellow Acad. Model Aeronautics; mem. Fedn. Internat. Aeronautique, Am. Phys. Soc., Am. Inst. Aeros. and Astronautics, Sigma Xi. Contbr. articles in field to sci. jours. Participant in development of Proximity Fuze for Navy Shells and Gun Fire Control Systems; specialist in control systems for guided missiles; developed early radio-controlled model airplanes. Patentee missile motion simulators. Home: 9802 Parkwood Dr., Bethesda, Md. 20014. Office: 8621 Georgia Av., Silver Spring, Md. 20910.*

GOODALE, Fairfield, Am. pathologist; b. Framingham, Mass., May 4, 1923; s. Fairfield and Anna P. (Perkins) G.; student Harvard, 1941-42; postgrad. Adelbert Coll., Western Res. U., 1945-46, M.D., 1950; m. Mary Margaret Lyman, Aug. 19, 1945; children—Tad, Nan, John, Susan, Tim. Research fellow Am. Cancer Soc., 1954-55, London, Eng., 1955-56, USPHS, Oxford, Eng., 1956-57, Boston, 1957-58; teaching fellow Harvard, 1954-55, asst. pathology, 1957-58; asst. prof. pathology Dartmouth Med. Sch., 1958-60; asso. prof. Albany Med. Sch., 1960-63; prof., chmn. dept. pathology Med. Coll. Va., Richmond, 1963—. Fellow Am. Soc. Exptl. Pathology; mem. New Eng. Soc. Pathologists, A.M.A., Internat. Acad. Pathologists, Am. Soc. Clin. Pathologists, Soc. Exptl. Biology and Medicine. Contbg. author Etiology of Myocardial Infarction, 1963; also articles. Research on causes of fever in humans, causes of atherosclerosis; isolated protein released from white blood cells which produces fever. Home: 1217 Rothesay Circle, Richmond, Va. 23221.*

GOODALE, George Lincoln, Am. botanist; b. Saco, Me., Aug. 3, 1839; s. Stephen Lincoln and Prudence Aiken (Nourse) G.; A.B., Amherst, 1860, A.M., 1866; M.D., Harvard, 1863; Bowdoin, 1863; several hon. degrees; m. Henrietta Juel Hobson, 1866; 5 children, including Joseph Lincoln, Francis Greenleaf. Practiced medicine, Portland, Me., 3 yrs.; Josiah Little prof. natural science and prof. mineralogy, botany and applied chemistry. Bowdoin Coll., 1868-72; instr. botany, lectr. vegetable physiology, Harvard, 1872-73, asst. prof., 1873;78, prof. botany, 1878-88, Fisher prof. botany, 1888-1909, Fisher prof. natural history emeritus, 1909, curator Bot. Mus., 1879-1909, hon. curator, 1909—. Author: Wild Flowers of North America, 1882; Vegetable Physiology, 1885; Vegetable Histology, 1885; also articles. Research in econ. botany, tropical agr., improvement of sugar cane; obtained fgn. plants for culture in U. S. Died Cambridge, Mass., Apr. 12, 1923.

GOODALE, Hubert Dana, Am. zoologist, geneticist; b. Troy, N.H., June 5, 1879; s. David Wilder and Mary (Reed) G.; A.B., Trinity Coll., 1903, A.M. (scholar), 1904; Ph.D. (fellow 1906-07), Columbia U., 1913; m. Lottie Ann Merrell, June 20, 1906; children—Hazel Margaret (Mrs. James Edwin Bullock), Marion Putnam, Wendell Merrell (dec.). Researcher pvt. genetic sta., 1907-11; staff Carnegie Instn., Cold Spring Harbor, N.Y., 1911-13, Mass. Agr. Expt. Sta., Amherst, 1913-22; geneticist Mt. Hope Farm, Williamstown, Mass., 1922-62. Recipient Mass. Soc. Promoting Agr. medal, 1942; New Eng. fellow Agrl. Adventurers, 1954; named to Poultry Hall of Fame, 1955. Fellow A.A.A.S., Poultry Sci. Assn.; mem. Am. Zool. Soc., Am. Soc. Naturalists. Contbr. numerous articles to tech. jours. Research on intravitam staining amphibian eggs; demonstrated control female plumage by ovary; research on factors controlling egg prodn., applications of research resulting in doubling egg prodn., genetics of milk prodn., demonstrated effective way progeny test selection on two strains mice. Home: 257 W. Main St., Williamstown, Mass. 01267.*

GOODALL, David William, botanist; b. Edmonton, Middlesex, Eng., Apr. 4, 1914; s. Henry William and Isabel Blanche (Harlow) G.; B.Sc., Imperial Coll. Sci. and Tech., London, Eng., 1935, Diploma Imperial Coll., Ph.D., 1941; D.Sc., U. Melbourne (Australia), 1953; m. Audrey Veronica Kirwin, Aug. 31, 1940 (div. Aug. 1949); 1 son, Patrick Thompson; m. 2d, Muriel Grace King, Sept. 15, 1949; children—Jan Peter, Glyn, Karen. Research officer Research Inst. Plant Physiology, 1939-46; plant physiologist W. African Cacao Research Inst., Tafo, 1946-48; sr. lectr. botany Melbourne U., 1948-52; reader botany U. Coll. Gold Coast, 1952-54; prof. agrl. botany Reading (Eng.) U., 1954-56; dir. C.S.I.R.O. Tobacco Research Inst., Mareeba, Queenland, Australia, 1956-61; sr. prin. research scientist div. math. statistics, Nedlands, Western Australia, 1961—. Hon. reader in botany U. Western Australia, Nedlands, 1964—. Recipient Forbes Meml. medal Imperial Coll., 1935; Beit Sci. Research fellow, 1937-39. Fellow Inst. Biology, Linnean Soc. London; mem. Brit. Ecol. Soc., Ecol. Soc. Am., Assn. Applied Biologists, Soc. Exptl. Biology, Internat. Soc. Biometerology, Internat. Soc. Trop. Ecology, Australian Statis. Soc. Author: (with F. G. Gregory) Chemical Composition of Plants as an Index of Their Nutritional Status, 1947; also numerous articles. Research on analysis of plant growth, diagnosis of mineral deficiencies in plants, plant numerical taxonomy. Home: 117 Dalkeith Rd. Office: C.S.I.R.O. Lab., Wembley, Western Australia.*

GOODALL, McChesney, Am. physician; b. Staunton, Va., Nov. 10, 1916; s. McChesney and Julia (Ellison) G.; B.S., U. Va., 1939; M.D., Med. Coll. Va., 1948; Ph.D., Karolinska Inst., Stockholm, Sweden, 1952; m. Wayne Stokes, Feb. 28, 1952; children—Bettine Marshall (Mrs. Scott Weiss), Eugenia Ellison, McChesney, Pendleton, William Stokes. Asst. prof. physiology Yale Sch. Medicine, 1952-54; asso. prof. physiology, surgery Sch. Medicine Duke, 1954-58; asst. dir. Research Center and Hosp. U. Tenn., 1958-61, med. dir., 1961-64; research dir. Shriners Burns Inst. Med. br. U. Tex., 1965—. Dir. Va. Trout Co., Hobb's House Am., Calbiochem. Mem. So. Med. Assn., Am. Physiology Soc., Sigma Xi. Author: Adrenaline-Noradrenaline in the Mammalian Tissues, 1951. Research, publs. on sympatho-adrenal medullary system, including physiology, biosynthesis and metabolism of the hormones, adrenaline and noradrenaline, cardiovascular. Home: 5128 Av. T., Galveston 77550. Office: Shriners Burns Inst., Galveston, Tex. 77551.*

GOODE, George Brown, Am. naturalist, govt. ofcl.; b. New Albany, Ind., Feb. 13, 1851; s. Francis Collier and Sarah (Crane) G.; grad. Wesleyan U., Middletown, Conn., 1870; m. Sarah Lamson Ford Judd; 4 children. Moved to N.Y., 1857; in charge Orange Judd Mus. Natural History, 1871-77; mem. staff Smithsonian Instn., 1873, asst. sec., 1887; employed in Atlantic Coast explorations of Fish Commn.; U. S. commr. fish, 1887-88; supervised Smithsonian exhibits at Phila. Centennial Expn., 1876; U. S. commr. at fisheries exhbns., Berlin, Germany, 1880, London, Eng., 1883; conducted survey Am. fisheries for 10th census, 1880. Mem. or fellow Am. Acad. Arts, Nat. Acad. Scis., A.A.A.S., Anthrop. Soc. Washington, Geog. Soc. Washington. Author: Catalogue of the Fishes of the Bermudas, 1876; Oceanic Ichthyology (added 156 new species of fish from Atlantic), 1895; An Account of the Smithsonian Institution, 1895, and The Smithsonian Institution 1846-96, 1897 (best known hist. treatises); Virginia Cousins (his own family record), 1887; The Game Fishes of North America; American Fishes, 1888; The Beginnings of American Science; The Origin of the Scientific and Educational Institutions of the United States, 1890; The Museums of the Future, 1891. Died Washington, Sept. 6, 1896.

GOODE, William Josiah, Am. sociologist; b. Houston, Aug. 30, 1917; s. William Josiah and Lillian R. (Bare) G.; B.A., U. Tex., 1928, M.A., 1939; Ph.D., Pa. State U., 1946; m. Josephine M. Canuizzo, Dec. 24, 1937 (dec. Feb. 1947); children—Brian Erich, Rachel Tara (dec.), Barbara Nan (Mrs. Nathaniel Platen Baldwin); m. 2d, Ruth Siegel, Oct. 21, 1950; 1 son, Andrew Josiah. Instr., Pa. State U., 1940-43; social sci. analyst Latin Am. Statis. Inst.; asst. prof. Wayne State U., 1946-50; staff Columbia, 1950—, prof. sociology, 1956—. Vis. prof. Free U. Berlin, 1954; research asso., bd. govs. Bur. Applied Social Research, 1958—; U. S. del. UN Conf. on Underdeveloped Nations, 1962. Recipient MacIver prize in sociology, 1966. Mem. Am. Anthrop. Assn., Am (v.p. 1967), Eastern (pres. 1966-67) sociol. assns.,

Sociol. Research Assn. (pres. 1967—), Social Sci. Research Council (dir.). Author: Religion Among the Primitives, 1951; (with Paul K. Hatt) Methods in Social Research, 1952; After Divorce, 1956; Die Struktur der Familie, 1960; World Revolution and Family Patterns, 1963; The Family, 1964; Readings on the Family and Society, 1964; Dynamics of Modern Society, 1966; also articles. Developed theory of illegitimacy in New World; research on relations between role theory and instnl. theory through role strain, class and divorce rates over time, cross-culturally, inputs-outputs between religion and other institutions, social structure of professions, morale and productivity. Home: 44 Franklin St., Piermont, N.Y. Office: 403 Fayerweather St., Columbia U., N.Y.C. 10027.*

GOODENOUGH, John Bannister, Am. physicist; b. Jena, Germany, July 25, 1922 (parents Am. citizens); s. Erwin Ramsdell and Helen (Lewis) G.; grad. Groton Sch., 1940; A.B., Yale Coll., 1943; Ph.D., U. Chgo., 1952; hon. degree U. Bordeaux, 1967; m. Irene Johnston Wiseman, June 16, 1951. Research engr. Westinghouse Research Lab., 1951-52; research physicist Lincoln Lab., Mass. Inst. Tech., 1952—, group leader, 1958—; Fellow Am. Phys. Soc. (mem. exec. com. div. solid state physics); Japanese Phys. Soc., Neurosciences Research Program (trustee 1962—). Author: Magnetism and the Chemical Bond, 1963. Asso. editor Materials Research Bull., 1966—. Contbr. numerous articles in field to sci. jours. Research in tech. magnetism, particularly domain-wall theory and its application to digital-computer components; fundamental magnetism, particularly magnetic-exchange interactions and conditions for localized-electron vs collective-electron magnetism; theory of electrons in solids, particularly relationship between state of outer electrons and electron-induced phase transformations. Home: South Hill Rd., New Boston, N.H. 03070. Office: Lincoln Lab., Mass. Inst. Tech., Lexington, Mass. 02173.*

GOODENOUGH, Ward Hunt, Am. anthropologist; b. Cambridge, Mass., May 30, 1919; s. Erwin Ramsdell and Miriam (Lewis) G.; grad. Groton Sch.; A.B., Cornell U., 1940; Ph.D., Yale, 1949; m. Ruth A. Gallagher, Feb. 8, 1941; children—Hester (Mrs. Steven Gelber), Deborah L., Oliver R., Garrick G. Mem. field staff, research br., information and edn. div. War Dept., 1942-45, case analyst, correction br., 1946; instr. U. Wis., 1948-49; mem. faculty U. Pa., Phila., 1949—, prof. anthropology, 1962—; vis. lectr. Cornell U., 1950, vis. prof., 1961-62; vis. prof. Bryn Mawr Coll., 1955, Swarthmore Coll., 1955, U. Hawaii, 1959. Cons. editor Bobbs-Merrill Co. Fellow Center for Advanced Study in Behavioral Scis., 1957-58. Fellow Am. Anthrop. Assn. (editor 1966—), A.A.A.S., Royal Anthrop. Inst., N.Y. Acad. Scis.; mem. Am. Ethnol. Assn. (pres. 1962), Soc. Applied Anthropology (pres. 1963), Linguistics Soc. Am., Am. Oriental Soc., Polynesian Soc., Phi Beta Kappa, Sigma Xi, Phi Kappa Phi. Author: Property, Kin and Community on Truk, 1951; Native Astronomy in the Central Carolines, 1953; Cooperation in Change, 1963. Editor: Explorations in Cultural Anthropology, 1964; also articles. Expdns. to Truk, Gilbert Islands, New Guinea; research in anthropology, ethnology of Micronesia and Melanesia, comparative study of Malayo-Polynesian langs., methodology of ethnography and descriptive semantics, theory of culture and its application to problems of community devel. Home: 204 Fox Lane, Wallingford, Pa. 19086. Office: Univ. Mus., 33d and Spruce Sts., Phila. 19104.*

GOODEVE, Charles Frederick, phys. chemist; b. Neepawa, Man., Can., Feb. 21, 1904; s. Frederick William and Emma (Hand) G.; B.Sc., U. Man., M.Sc., U. Coll., London, Eng., 1927, D.Sc., 1934; m. Janet I. Wallace, Sept. 17, 1932; children—Peter Julian, John Anthony. Lectr., later reader U. Coll., London, 1928-39; dep. dir. dept. miscellaneous weapon devel. Admiralty, 1940-42; dep. controller for research and devel., 1942-45; dir. Brit. Iron and Steel Research Assn., London, 1945—; dir. Indsl. and Comml. Finance Corp., Ltd., London. Gov., Imperial Coll., London. Decorated Order Brit. Empire; knighted, 1946; recipient U. S. medal of Freedom with silver palm, 1945; Bessemer Gold medal, 1962; Carl Leug Gold medal, 1962; Silver medal Operational Research Soc., 1964. Fellow Royal Soc., 1940, Royal Inst. Chemistry, Inst. Metals; mem. Inst. Transport. Contbg. author Iron and Steel Productivity Report. Research in molecular spectroscopy, reaction kinetics, visual process, operational research. Home: 38 Middleway, London N.W. 11. Office: 24 Buckingham Gate, London S.W.1, Eng.*

GOODEY, Tom, English zoologist; b. July 28, 1885; s. Thomas and Hannah Goodey; B.Sc., U. Birmingham (Eng.), 1908, M.Sc., 1909, D.Sc., 1915; Mackinnon student Royal Soc. London, 1910-11; m. Constance Lewis; 1 son, 3 daus. Protozoologist Rothamsted Expt. Sta., Harpenden, Eng.; 1912-13, plant heminthologist, 1920, head nematology dept.; protozoologist, lab. agrl. zoology U. Birmingham, 1913-19; mem. research staff, dept. helminthology London (Eng.) Sch. Tropical Medicine, later Inst. Agrl. Parasitology, Winches Farm, St. Albans, Eng.; with govt. service in India, 1919-39. Fellow Royal Soc., 1947. Died Sept. 29, 1951.

GOODIER, James Norman, Am. mech. engr.; b. Preston, Eng., Oct. 17, 1905; s. James and Martha (Grimshaw) G.; B.A., Cambridge (Eng.) U., 1927, Ph.D., 1931; Sc.D., U. Mich., 1931; m. Marina Timoshenko, July 17, 1931; 1 son, Peter N. Came to U. S., 1938, naturalized, 1946. Fellow, Ont. Research Found., Toronto, Can., 1931-38; prof. mechanics Cornell U., 1938-47; prof. engring. mechanics Stanford (Cal.) U., 1947—, chmn. div. engring. mechanics, 1954—. Cons. Stanford Research Inst., 1948—. Recipient George Westinghouse award Am. Soc. Engring. Edn., 1946; Fellow Am. Soc. M.E., (Timoshenko medal 1961). Author: (with S. Timoshenko) Theory of Elasticity, 1951; (with P. G. Hodge) Elasticity and Plasticity, 1958. Contbr. numerous research articles in applied mechanics to tech. jours. Study of stress waves, theory of elasticity; elastic-plastic instabilities. Home: 506 Mayfield Av., Stanford, Cal. 94305.

GOODLETT, Vernon Wilson, Am. chemist; b. Greenville, S.C., May 16, 1934; s. Claud Bernard and Mildred (Wilson) G.; B.S., Wofford Coll., 1956; Ph.D., Vanderbilt U., 1959; m. Sara Jacqueline Murphy, June 30, 1956; children—Carol Elizabeth, Steven Wilson. Research chemist Tenn. Eastman Co., Kingsport, 1959-61, sr. research chemist, 1962—. Mem. Am. Chem. Soc. (past sec.-treas. local sect.), Sigma Xi. Contbr. articles to tech. jours. Research on high-resolution visible and ultraviolet spectrum aluminum monoxide molecule, nuclear magnetic resonance. Home: 2305 Woodridge Av., Kingsport, Tenn. 37662.*

GOODMAN, Bruce Bailey, English physicist; b. Taunton, Eng., Apr. 22, 1928; s. Stuart and Nora (Bailey) G.; B.A., Cambridge U., 1948, Ph.D., 1952; m. Claude Suzanne Esménard, Feb. 22, 1963; 1 son, Patrick Michael. Research physicist Westinghouse Electric Corp., Pitts., 1953-57; prof. asso. Université de Grenoble (France), 1957; dir. research Centre National de la Recherche Scientifique, 1963-66; chief physicist Brit. Oxygen Co., London, 1966—; vis. prof. physics U. Sussex, 1967—. Recipient Silver medal Centre National de la Recherche Scientifique, 1966, Prix Ancel French Phys. Soc., 1966. Fellow Phys. Soc. London; mem. French, Am. phys. socs. Research and publs. on low temperatures, especially superconductors and hyperfine specific heats of rare earth metals; discovered superconductivity in osmium and rubidium; pioneered work on energy gap in superconductors and discovery and properties of type II superconductors. Home: 16 Courtfield Rd., London S.W. 7, Eng. Office: Brit. Oxygen Co., Deer Park Rd., London S.W. 19, Eng.*

GOODMAN, Charles David, Am. physicist; b. N.Y.C., May 9, 1928; s. Jacob and Libby (Freed) G.; A.B., Clark U., 1949; postgrad. Amherst Coll., 1949-50; Ph.D., U. Rochester, 1955; m. Joan Louise Wright, June 11, 1952; children—Henry Nicholas, Diana Ruth. Physicist, Oak Ridge Nat. Lab., 1955—, group leader, 1958—. Louis Lipsky fellow Weizmann Inst. Sci., Rehovoth, Israel, 1966. Fellow Am. Phys. Soc.; mem. Sigma Xi. Research and publs. in field of structure atomic nuclei; developed apparatus for distinguishing particle types in nuclear reactions, techniques for automatic computer processing exptl. nuclear data. Home: 100 Scenic Dr. Office: P.O. Box X, Oak Ridge 37830.*

GOODMAN, Clark Drouillard, Am. nuclear physicist; b. Memphis, Sept. 9, 1909; s. J. Alma and Naomi (Clark) G.; B.S., Cal. Inst. Tech., 1932; Ph.D., Mass. Inst. Tech., 1940; m. Mary Ellen Hohiesel, Aug. 8, 1933; children—Gaye Ellen, Alan Clark. Tech. aide div. 16 and 17, OSRD, 1942-45; sr. physicist Oak Ridge Nat. Lab., 1945-46; asso. prof. physics Mass. Inst. Tech., 1947-58; asst. dir. div. reactor devel. AEC, Washington, 1955-58; v.p. technique Schlumberger Ltd., Houston, 1958-62; v.p. and tech. dir. Prengle Dukler and Crump, Inc., 1962—, Houston Research Institute, 1962—; professor of physics, University of Houston, 1963—, chmn. physics dept., 1966—. Sec. com. radio-activity NRC, 1939-47. Cons. Joint Congl. Com. Atomic Energy, 1961-62, Dept. of Def., 1961—, NASA Hdqrs., 1966—. F. Am. Phys. Soc., Geol. Soc. Am.; mem. Am. Rocket Soc., Am. Nuclear Soc., Sigma Xi. Author: Science and Engineering of Nuclear Power, vols. I and II, 1947, 48; Introduction to Nuclear Power, 1955; Atomic Energy (in Japanese), 1957. Basic research in low energy nuclear physics; applied research in nuclear engineering and geophysics; study of space science. Home: 12511 Old Oaks Dr., Houston 77024. Office: 5417 Crawford St., Houston 77004; also 6001 Gulf Freeway, Houston.*

GOODMAN, DeWitt Stetten, Am. physician; b. N.Y.C., July 18, 1930; s. Max and Jennie (Katz) G.; A.B., Harvard, 1951, M.D., 1955; m. Ann Bregstein, July 7, 1957; children—Daniel W., Elizabeth. Investigator Nat. Heart Inst., Bethesda, Md., 1956-58, 1960-62; vis. fellow Med. Research Council Exptl. Radiopathology Research Unit, Hammersmith Hosp., London, England, 1959-60; asst. prof. medicine Columbia U., 1962-67, asso. prof. medicine, 1967—; asso. editor Jour. Clin. Investigation, 1967—; mem. editorial bd. Jour. Lipid Research, 1965—; mem. metabolism study sect. Nat. Insts. Health, 1966—; asst. attending physician The Presbyn. Hosp., N.Y.C., 1962-67, asso. attending physician, 1967—. Recipient Health Research Council of the City N.Y. Career Scientist award, 1964; Soc. for Exptl. Biology and Medicine Meltzer award, 1963. Mem. Am. Soc. for Clin. Investigation, Am. Soc. Biol. Chemists, Council on Arteriosclerosis Am. Heart Assn., Harvey Soc., Am. Oil Chemists Soc., A.A.A.S., Phi Beta Kappa, Alpha Omega Alpha. Contbr. numerous articles in field to sci., med. jours. Research, publs. contbg. to the understanding of biosynthesis and metabolism of cholesterol and cholesterol esters; the metabolism of B-Carotene and Vitamin A; the biosynthesis of Vitamin A from B-Carotene; the transport of lipids in blood plasma and the interaction of lipids with proteins. Office: 630 W. 168th St., N.Y.C. 10032.*

GOODMAN, Donald C., Am. neuroanatomist; b. Chgo., Nov. 24, 1927; s. Alexander and Freda (Mermelstein) G.; student U. Chgo., 1944-45; B.S., U. Ill., 1949, M.S., 1950, Ph.D., 1954; m. Adelle Bortz, Apr. 21, 1949; children—Brian and Eric (twins), Michael and Susan (twins), Elaine. Instr. anatomy U. Pa., 1954-56; postdoctoral fellow Inst. Neurol. Scis., 1954-55; with U. Fla., 1956—, prof. anatomy, 1962—, co-dir. Center for Neurobiol. Scis., prof. medicine, 1964—; chmn. dept. anatomical Scis., 1966—. Recipient Fla. chpt. Sigma Xi Annual Research award, 1962. Mem. Am. Assn. Anatomists, A.A.A.S., Am. Inst. Biol. Scis., So. Soc. Anatomists. Author (with Fisher) Myology of the Whooping Crane, 1956, Functional Anatomy of the Feeding Apparatus in Water Fowl, 1962. Research and publs. on the exptl. determination of the evolution of cerebellar function; research on the plasticity and recovery activities of the nervous system. Home: 2034 N.E. 9th Terrace, Gainesville, Fla. 32603. Office: J. Hillis Miller Health Center, U. Fla., Gainesville, Fla. 32603.*

GOODMAN, George J(ones), Am. botanist; b. Evanston, Wyo., Nov. 5, 1904; s. Arthur Duane and Mary Elizabeth (Jones) G.; A.B., U. Wyo, 1926-29; M.S., Washington U., St. Louis, 1930, Ph.D., 1933; m. Marcia McCay, Dec. 19, 1948; Faculty, U. Okla., Norman, 1933-36, prof. botany, curator Herbarium, 1945—, curator botany Mus., 1945—; faculty Ia. State Coll., 1936-45, asso. prof., 1944-45; plant taxonomist Okla. Biol. Survey, 1950—. Mem. A.A.A.S., Am. Assn. U. Profs., Am. Fern Soc., Am. Soc. Plant Taxonomists, Bot. Soc. Am., Cal. Bot. Soc., New Eng. Bot. Club, Internat. Assn. Plant Taxonomists, Soc. Study Evolution, Torrey Bot. Club, Southwestern Assn. Naturalists (past pres.), Colo.-Wyo. Acad. Sci., Ia. Acad. Sci., Okla. Acad. Sci. (past pres.), Phi Beta Kappa, Phi Kappa Phi, Phi Sigma, Sigma Xi. Author: Spring Flora of Central Oklahoma, 1958; also numerous articles. Research on flora Okla.; taxonomic revisional work in flowering plants N.Am. Home: 1229 Avondale Dr., Norman, Okla. 73069.*

GOODMAN, John D., Am. physician; b. Annapolis, Md., Dec. 20, 1794; ed. U. Md., 1815-18, M.D.; unmarried. Participated in defense of Fort Henry; practiced medicine, Md., 1818-21; became prof. anatomy and surgery Ohio Med. Coll., 1821; taught privately in Phila.; apptd. prof. anatomy Rutgers, 1826; contbr. articles to Physiol. and Path. Anatomy, 1825, Am. Natural History, 3 vols., 1826-28, Am. Quarterly Review; editor Phila. Jour. Med. Scis.; research on physiology, pathology, anatomy, also natural history, compiled catalogue of plants; died 1830.

GOODMAN, Leo A., Am. statistician, sociologist; b. N.Y.C., Aug. 7, 1928; s. Abraham J. and Mollie (Sacks) G.; A.B. summa cum laude Syracuse U., 1948; M.A. in Math., Math. Statistics, Princeton, 1950, Ph.D. in Math., Math. Statistics, 1950; m. Ann Haven Davidow, Aug. 28, 1960; children—Andrew Martin, Thomas Henry. Research asst. in math. statistics Princeton, 1948-49; asst. prof. statistics and sociology, U. Chgo., 1950-53, asso. prof. 1953-55, prof., 1955—; vis. prof. math. statistics and sociology, Columbia, 1960-61. Mem. com. on statistics, div. math. Nat. Acad. Scis., NRC, 1961-64, chmn., 1963-64. Research tng. fellow, Social Sci. Research Council, Princeton, 1950, Advanced Research scholarship, Fulbright award, honorary research tng. fellow, Soc. Sci. Research Council, Cambridge U., 1953-54, sr. postdoctoral fellow, NSF, Guggenheim fellow, Cambridge U., London Sch. Economics and Polit. Sci., 1959-60. Fellow Inst. Math. Statistics (mem. council 1955-57), Am. Statis. Assn., A.A.A.S., Am. Sociol. Assn., Royal Statis. Soc.; mem., Am. Math. Soc., Econometric Soc., Biometric soc., Am. Assn. U. Profs., Phi Beta Kappa, Sigma Xi. Club: Quadrangle, Princeton (Chgo.). Contbr. articles to tech. jours. Research and publs. on topics in math. statistics, sociology, math., applied statistics, social research, econometrics, psychometrics, human genetics, math. biophysics, biometrics, marketing, behavioral sci., sociometrics, demography, applied probability.*

GOODMAN, Leon, Am. chemist; b. Livingston, Mont., Dec. 16, 1920; s. Samuel and Sadie (Kopald) G.; student Fresno State Coll., 1937-40; B.S., U. Cal. at Berkeley, 1941; Ph.D., U. Cal. at Los Angeles, 1950; m. Marilyn Shear, Feb. 1, 1956; children—Laura Elizabeth, Andrew Bentley. Chemist, Office of Sci. Research and Devel., Pitts., 1942-45; chemist Los Alamos Sci. Lab., 1950-53; research asso. U. So. Cal., Los Angeles, 1953-54; chmn. dept. bio-organic chemistry Stanford Research Inst., Menlo Park, Cal., 1955—. Mem. Am. Chem. Soc., Chem. Soc. London, Sigma Xi. Research and numerous publs. in field of carbohydrate chemistry and nucleic acid chemistry as related to cancer research; research in synthesis of analogs of folic acid; participated in 1st syntheses of a natural purine 2-deoxynucleoside. Home: 1795 Guinda St., Palo Alto, Cal. 94030. Office: Stanford Research Inst., Menlo Park, Cal. 94025.*

GOODMAN, Louis S., Am. pharmacologist; b. Portland, Ore., Aug. 27, 1906; s. Charles William and Dora (Hurwitz) G.; A.B., Reed Coll., 1928; M.D., M.A., U. of Ore., 1932; grad. work Johns Hopkins, 1932-33; Dr. of Scis. (hon.), Univ. Manitoba, 1965; m. Helen Ricen, Dec. 14, 1933; children—Carolyn, Debora. Teaching asst. psychology, Reed Coll., 1927-29; house officer in medicine Johns Hopkins Hosp., 1932-33; Nat. Research Council fellow in pharmacology, Yale, 1934, instr. pharmacology and toxicology, 1935-37, asst. prof., 1937-43; prof. and chmn. dept. pharmacology and physiology, U. of Vt., 1943-44; prof. and head dept. pharmacology, U. of Utah, 1944—. Mem. nat. bd. med. examiners Pharmacology Test Com., 1955-59; mem. nat. adv. Neurological Diseases and Blindness Council, 1954-58; mem. adv. council Life Ins. Med. Research Fund, 1956-59; mem. adv. com. Pharmacology Service Center, Nat. Inst. Mental Health, 1957-63; chmn. pharmacol. tng. com. Nat. Inst. Health, 1958-62; nat. adv. com. Mental Health Council, 1963-66, nat. adv. council health research facilities, 1966—; mem. council on drugs A.M.A., 1958-61; mem. grad. council, research grants com. U. Utah, 1957—; mem. med. bd. Myasthenia Gravis Found., 1953-59; mem. sci. bd. Nat. Neurological Research Found., 1957-64; mem. sr. postdoctoral fellowship evaluation com. Nat. Sci. Found., 1955-58. Fellow N.Y. Acad. Sci.; mem. Nat. Acad. Scis., Am. Academy of Neurology, Acad. Anesthesiology (hon.), Am. Soc. Pharmacol. and Exptl. Therapeutics (mem. bd. publication trustees 1949-61), Soc. Exptl. Biology and Medicine, American Assn. Advancement Science, Am. Physiol. Soc., Phi Beta Kappa, Sigma Xi. Author: Pharmacological Basis of Therapeutics (with Gilman) 1965. Editor: Pharmacological Revs., 1949-53; asso. editor: Proc. Soc. Exptl. Biology and Medicine, 1947-51; mem. editorial bd. Annual Review of Pharmacology, 1959-65, Journal of Exptl. Psychiatry, 1959—; contbr. to sci. publs. Research on pharmacodynamics; anticonvulsant and sympathetic drugs. Home: 2926 Crestview Dr., Salt Lake City 8, Utah.

GOODMAN, Mary Ellen Hoheisel, Am. anthropologist; b. Los Angeles, Aug. 8, 1911; d. August and Grace (Goold) Hoheisel; B.E., U. Cal. at Los Angeles, 1932; M.A., Radcliffe Coll., 1943, Ph.D., 1946; m. Clark D. Goodman, Aug. 8, 1933; children—Gaye Ellen, Alan Clark. Faculty, Wellesley (Mass.) Coll., 1946-54, asst. prof. sociology and anthropology, 1947-54; social sci. analyst U. S. Pub. Housing Adminstrn., Washington, 1957-58, U. S. Dept. Health, Edn. and Welfare, Washington, 1958-59; lectr., asso. prof. U. Houston, 1961-65; lectr. prof. Rice U., Houston, 1963—; dir. early childhood studies Tufts U., 1956-59. Coordinator studies White House Conf. on Children and Youth, 1958-59; intergroup relations cons. Antidefamation League U. S., 1959-60. Fulbright Research scholar, Japan, 1954-55; named Nat. Woman of Year, Delta Zeta, 1963; recipient Matrix award Theta Sigma Phi, 1962. Fellow Am. Anthrop. Assn., Am. Sociol. Assn., Soc. Research in Child Devel; mem. N.Y. Acad. Scis., Phi Beta Kappa, Sigma Xi. Author: Race Awareness in Young Children, 1952, rev., 1964; Individual and Culture, 1967; also articles, weekly column in newspaper. Described and documented genesis of race attitudes in young children, nature of social concepts, attitudes and values in presch. and elementary sch. age children in U. S. and Japan; documented nature of process of intra-family transmission of values to young child; research on life ways of Negroes in U. S. and Latin Ams. in Houston. Home: 2701 Bellefontaine, Houston 77025.*

GOODMAN, Morris, Am. immunologist; b. Milw., Jan. 12, 1925; s. Benjamin and Sarah (Bratt) G.; B.S., U. Wis., 1948, M.S., 1949, Ph.D., 1951; m. Selma Kessler, Apr. 5, 1946; children—Louise, Julia, David B. Research fellow Cal. Inst. Tech., 1951-52; research asso. U. Ill., 1952-54, Detroit Inst. Cancer Research, 1954-58; research asso. Sch. Medicine Wayne State U., 1958-60, faculty, 1960—, prof. anatomy, 1966—; dir. research Plymouth State Home and Tng. Sch., Northville, Mich., 1966—. Fellow A.A.A.S.; mem. Am. Soc. Immunologists, Am. Soc. Zoologists, Am. Acad. Neurology, Am. Assn. Phys. Anthropologists, Am. Soc. Animal Sci., Am. Assn. Anatomists, Soc. Study Devel. and Growth, Internat. Soc. Primatology. Contbr. numerous articles to profl. jours. Research and publs. on molecular evolution of primates, in particular establishing genetic relationship of man to other primates. Home: 24211 Oneida St., Oak Park, Mich. 48237. Office: 15480 Sheldon Rd., Northville, Mich. 48167.*

GOODMAN, Murray, Am. chemist; b. N.Y.C., July 6, 1928; s. Louis and Frieda (Bercun) G.; B.S., Bklyn. Coll., 1949; Ph.D., U. Cal. at Berkeley, 1952; m. Zelda Silverman, Aug. 26, 1951; children—Andrew, Joshua, David. Postdoctoral fellow Mass. Inst. Tech., 1952-55; research fellow U. Cambridge (Eng.), 1955-56; faculty Poly. Inst. Bklyn., 1956—, prof. chemistry, 1964—. Recipient Distinguished Alumnus medal Bklyn. Coll., 1965. Mem. Am. Chem. Soc. Editor, Biopolymers Jour., 1963—. Research, publs. on structure of macromolecules; synthesized models of biopolymers and analyzed their structure and properties using spectroscopic methods. Home: 116 Willow St., Bklyn. 11201.*

GOODMAN, Robert Norman, Am. plant pathologist; b. Yonkers, N.Y., Dec. 15, 1921; s. Sidney William and Margaret (Fried) G.; B.S., U. N.H., 1948, M.S., 1950; Ph.D., U. Mo., 1952; m. Phoebe Newman, Sept. 4, 1949; children—Joyce Beth, Rachael Lea, Janet Faith. Grad. asst. U. N.H., 1948-50; grad. asst. U. Mo., 1950-52, asst. prof., 1952-55, asso. prof., 1955-61, prof., 1961—; post-doctorate fellow Swiss Fed. Inst. Tech., 1958-59, U. Leeds, Eng.; 1965-66. Guggenheim Found. fellow, 1958-59; Lalor Found. fellow, 1958-59; NIH spl. fellow, 1965-66; recipient U. Mo. Jr. Faculty Research award of merit, 1955. Mem. Am. Phytopathol. Soc., Am. Soc. Microbiology, Sigma Xi, Gamma Sigma Delta. Author: Advances in Pest Control Research, 1961; Antibiotics in Agriculture, 1963; author (with H. S. Goldberg) Biochemistry of Physiology of Infectious Plant Disease, 1967. Contbr. numerous articles to sci. jours. Research on chemotherapy of bacterial plant diseases; penetration and absorption of chemotheraputants by plants; nature of virulence of and resistance to bacterial plant pathogens; electronmicroscopy in bacterial plant diseases. Home: 605 Crestland St., Columbia, Mo. 65201.*

GOODNIGHT, Clarence James, Am. biologist; b. Gillespie, Ill., May 30, 1914; s. Charles A. and Phoebe (Perseneus) G.; A.B., U. Ill., 1936, M.A., 1937, Ph.D., 1939; m. Marie L. Ostendorf, Aug. 25, 1940; children—Ann Marie, Charles James. Faculty, U. Ill., 1939-40, 42-44; instr. Bklyn., 1940-42; asst. prof. Jersey City State Tchrs. Coll., 1944-46; faculty Purdue U., West Lafayette, Ind., 1946-65, prof., 1955-65; prof., head dept. biology Western Mich. U., Kalamazoo, 1965—. Mem. Am. Assn. U. Profs., A.A.A.S., Am. Inst. Biol. Scis., Am. Micros. Soc., Am. Soc. Zoologists, Ecol. Soc. Am., Nat. Assn. Biology Tchrs., Soc. Systematic Zoology. Author: (with M. L. Goodnight) Zoology, 1954; (with M. L. Goodnight, R. Armacost) Biology, An Introduction to the Science of Life, 1962; (with P. Gray, M. L. Goodnight) General Zoology, 1964; also numerous articles. Research on physiol. adaption animals to various environmental factors, taxonomy, distbn. and evolution arachnids. Home: 1633 Chevy Chase Blvd., Kalamazoo 49001.*

GOODPASTURE, Ernest William, Am. pathologist; b. Montgomery County, Tenn., Oct. 17, 1886; s. Albert Virgil and Jennie Wilson (Dawson) G.; B.A., Vanderbilt, 1907; hon. M.S., Yale; M.D., Johns Hopkins U., 1912; D.Sc. (hon.), U. of Chgo., Washington U., LL.D., Tulane U., 1957; m. Sarah Marsh Catlett, Aug. 11, 1915; 1 dau., Sarah; m. 2d, Frances Katharine Anderson, May 25, 1945. Rockefeller fellow in pathology Johns Hopkins U., 1912-14; faculty Harvard, 1913-22; dir. Singer Meml. Research Lab., 1922-24; prof. pathology Vanderbilt U., 1924-55, dean sch. medicine, 1945-50; sci. dir. dept. pathology, Armed Forces Ints. Pathology, 1955—; mem. Army Epidemiol. Bd., 1941-46; cons. health div. TVA, 1942-55; Sci. dir. Internat. Health div. Rockefeller Found., 1938-40, 42-44; Com. on Growth, Panel on Virus, NRC, 1946-48; mem. Nat. Inst. Health, virus and rickettsia section, 1946-49. Adv. com., div. biology, medicine AEC, 1947-52. Recipient Sedgwick Meml. medal. Am. Pub. Health Assn., 1944; John Scott award City of Philadelphia, 1945; Passano Found. award, 1946; Howard Taylor Ricketts award U. Chgo., 1955; Kovalenko medal. Nat. Acad. Sci., 1958; gold headed cane award Am. Assn. Pathol. and Bacteriol., 1958. Fellow A.A.A.S. (v.p. 1940); mem. Nat. Acad. Sciences (council 1944-52), Am. Philos. Soc., Assn. Am. Physicians (Kober medalist 1943), So. Med. Assn. (research medalist 1937), Gorgas Soc. (hon. Ala.), Am. Assn. Pathol. and Bacteriol. (council since 1942; pres. 1948-49), Am. Soc. Exptl. Pathol. (pres. 1939-40), A.C.P. (John Phillips lectr. 1948), Harvey Soc., Phi Beta Kappa, Sigma Xi. Acting Editor Am. Jour. Pathology, 1957—. Contbr. to jours. Research in infectious diseases, etiology, pathogenesis, chick embryo techniques, viruses; isolated (with Claud D. Johnson) virus causing mumps, circa 1934; studied formation of fibrinogen; fibrinolysis in hepatic insufficiency. Died Sept. 20, 1960.

GOODRICH, Edwin Stephen, English zoologist; b. 1868; Merton Coll., M.A., D.Sc., LL.D., Oxford (Eng.) U.; m. Helen L. M. Pixell, 1913. Aldrichian demonstrator in comparative anatomy; Linacre prof. zoology and comparative anatomy U. Mus., Oxford. Fellow Royal Soc., 1905 (Royal medal 1936); mem. Linnean Soc. (Gold medal 1932). Author: Vertebrate Craniata, 1909; Living Organisms, 1924; Structure and Development of Vertebrates, 1930; also other publs. Research on morphology and comparative anatomy; distinguished between the nephridium and coelomoduct; established the differences between various types of fish scales. Died Jan. 6, 1946.

GOODRICH, Ernest Payson, Am. engr.; b. Decatur, Mich., May 7, 1874; s. Edward Payson and Mary Isabelle (Hall) G.; B.Pd., Mich. Normal Coll., 1898, M.Ed., 1936; B.S., U. Mich., 1898; C.E., 1901, D. Eng., 1935; D.Eng., Polytechnic Institute, 1955; m. Mildred Louise Weed, May 18, 1899; 1 son, Ernest Weed. Commd. civil engr. (lt. jr. grade), USN, 1899; resigned, 1903; chief engr. Bush Terminal Co. and affiliated cos., N.Y., 1903-07; pvt. practice since 1907, designing harbors in many parts of world, surveying zoning and planning cities and regions, including new capital of Nanking Whampoa, port of Canton, China, Bogota, Colombia, S.A., Los Angeles, Portland, Ore., Newark; cons. eng. N.Y.C. Govt., 1910-16; mng. dir. N.Y. Bur. Municipal Research; cons. Regional Plan of N.Y. and Environs, also Cin. Norfolk, Newark, Springfield, New Haven; dep. commr., chief engr. Dept. Sanitation, N.Y.C., later commr., 1933-34; port engr., later cons. Albany Port Commn., 1924-1933; prof. engring. economics. N.Y. U. 1934-35; Fellow A.A.-A.S.; mem. Am. Soc. C.E. (dir., Collingswood prize 1905), Am. Inst. Cons. Engrs. (pres. 1951), Soc. Terminal Engrs. (dir.), Am. Inst. Planning (dir.), Inst. Traffic Engrs. (pres.), Lectr. and writer on tech. subjects; inventor prog. system of electrical light signal street traffic control; discoverer of laws of population distbn.; made extended studies of application of sunlight to building orientation and city planning. Translator; Der Eisenbetonbau (Concrete Steel Construction), by Prof. Emil Mörsch, 1909. Died Oct. 7, 1955.

GOODRICH, Max, Am. physicist; b. Calhoun, Mo., Dec. 11, 1905; s. Henry Charles and Elma (Shafer) G., B.A., Westminster Coll., 1927; Ph.D., U. Minn., 1936; m. Marian Jeanette Guyer, Aug. 24, 1929; children—Mary Lee Ann (Mrs. Roy T. Matthews), Marna Jean. Instr. Salt Lake Collegiate Inst., Westminster Coll., 1927-29; asst. instr. U. Minn., 1929-36; faculty La. State U., Baton Rouge, 1936—, asso. prof., 1944-50, prof. physics, 1950—, dean grad. sch. 1961—. Instr. Pa. State Extension summer 1941; physicist War Research Lab., U. Tex., summer 1945; sr. physicist Oak Ridge Nat. Lab., 1949-50, summers 1952, 54. Fellow Am. Phys. Soc.; mem. Am. Assn. Physics Tchrs., A.A.A.S., Am. Assn. U. Profs., La. Acad. Scis., Sigma Xi, Sigma Pi Sigma, Omicron Delta Kappa. Determination of properties and energy structure of atomic nuclei made radioactive by artificial transmutation through exposure to neutrons of nuclear reactors. Home: 1708 Hood Av., Baton Rouge, La. 70808.*

GOODRICKE, John, astronomer; b. Groningen, Netherlands, Sept. 17, 1764; s. Henry and Levina (Sessler) G. Fellow Royal Soc., 1786 (Copley medal 1783). Research and publs. on variability of stars; discovered period and law of star Algol's changes, 1732. Died York, Eng., Apr. 20, 1786.

GOODSIR, John, anatomist, pathologist; b. Anstruther, Scotland, Mar. 20, 1814; s. John and Elizabeth (Taylor) G.; ed. at Coll. St. Andrews (Scotland), Edinburgh; practiced with his father, Anstruther, 1835-40; moved to Edinburgh, 1840; named curator Coll. Surgeons, 1841; curator, demonstrator anatomy, U. Mus., 1843-46; prof. anatomy, 1846-67. Virchow dedicated his publ. Cellular-Pathologie to him, 1859. Fellow Royal Soc., 1846. Author: Anatomical and Pathological Observations, 1845; also articles. Research in anatomy and physiology, especially growth of teeth, endocrine glands, embryology, nutrition and bone growth; discovered vegetable spores (sarcina ventriculi) in the stomach of animals and man, 1842. Died Mar. 6, 1867.

GOODSTEIN, Leonard David, Am. psychologist; b. N.Y.C., Jan. 11, 1927; s. Moses and Stella (Warshar) G.; B.S., City Coll. N.Y., 1948; M.A., Columbia U., 1948, Ph.D., 1952; m. Ruth Einhorn, Dec. 18, 1948; children—Richard E., Steven M. Instr. Hofstra Coll., 1948-51; with U. Ia., 1951-64, prof. psychology, 1961-64; prof. psychology U. Cin., 1964—; cons. VA, 1953—, Peace Corps, 1963—. Diplomate Am. Bd. Clin. Psychology. Fellow Am. Psychol. Assn.; mem. Am. Personnel and Guidance Assn., Midwestern, Ohio psychol. assns., Ia. Acad. Sci. Publs. on exptl. studies of psychopathology, especially schizophrenia and stuttering; exptl. and theoretical analysis of psychotherapy. Home: 373 Compton Rd., Cin. 45215.*

GOODSTEIN, Reuben Louis, English mathematician; b. London, Eng., Dec. 16, 1912; s. Alexander and Sophia (Fisher) G.; ed. Magdalene Coll., Cambridge U.; B.A., Ph.D., D.Litt.; m. Louba Etkin, 1938; children—Peter David, Margaret Ann Sophia. Reader, U. Reading, 1935-48; prof. math. U. Leicester, 1948—. Mem. Math. Soc. London, Math. Assn., Am. Math. Soc., Assn. for Symbolic Logic. Author: Mathematical Logic; Axiomatic Projective Geometry; Recursive Number Theory; Recursive Analysis; Boolean Algebra. Editor: Math. Gazette. Home: 16 Clarendon Park Rd. Office: University of Leicester, Leicester, Eng.

GOODWIN, Harry Manley, Am. physicist; b. Boston, Apr. 18, 1870; s. Richard D. and Sarah (Clisby) G.; S.B., Mass. Inst. Tech., 1890; postgrad. Harvard, 1890, 91; Ph.D., U. Leipzig, 1893; U. Berlin, 1894; m. Mary B. Linder, Apr. 16, 1906; 1 son, Richard Hale. Asst. in physics Mass. Inst. Tech., 1890-92, instr., 1892-97, asst. prof., 1897-1903, asso. prof. physics, 1903-06, prof. physics and electrochemistry in charge dept. of electrochemistry, 1906-34, dean of grad. sch., 1932-40, prof. emeritus, 1940—. Fellow Am. Acad. Arts and Scis., Washington Acad. Scis., A.A.A.S., Am. Phys. Soc.; mem. Am. Astron. Soc., Sigma Xi. Author: Physical Laboratory Manuals; The Precision of Measurements and Graphical Methods. Contbr. of papers on physics and electrochemistry to sci. jours. Died June 26, 1949.

GOODWIN, John Forrest, English physician; b. Ealing, London, Eng., Dec. 1, 1918; s. William Richard and Myrtle (Forrest) G.; ed. Cheltenham Coll., 1932-

37, St. Mary's Hosp. Med. Sch., U. London, 1937-42; m. Barbara Robertson, Oct. 27, 1943; children—Jennifer, Martin. Med. registrar St. Mary's Hosp., London, 1943; physician Anglo-Iranian Oil Co., Iran, 1945; med. 1st asst. Sheffield Royal Infirmary, 1946-49; lectr. physician Postgrad. Med. Sch., U. London, 1949-59, sr. lectr., 1959-63, prof. clin. cardiology, 1963—. Mem. council Brit. Heart Found., 1964—. Fellow Royal Coll. Physicians London, Royal Soc. Medicine, Am. Coll. Cardiology; mem. Brit. (treas. 1958—), Italian cardiac socs., Assn. Physicians Gt. Britain and Ireland, Med. Research Soc., N.Y. Acad. Scis. Author: (with R. Daley and R. Steiner) Clinical Disorders of the Pulmonary Circulation, 1960; also numerous articles. Research on disorders of pulmonary blood vessels in heart disease, diseases of heart muscle, congenital heart disease, cardiac function and surg. treatment of disease of heart valves, electrocardiography in congenital and acquired heart disease, cardiac tumours. Home: 18 Augustus Rd., Wimbledon Park S.W. 19, Eng. Office: Royal Postgrad. Med. Sch., London W.12, Eng.*

GOODWIN, Melvin Harris, Am. pub. health biologist; b. Thomasville, Ga., Jan. 9, 1917; s. Melvin Harris and Munson (Beverly) G.; B.S., U. Ga., 1941, M.S., 1951; Ph.D., Emory U., 1955; m. Virginia Peacock, Oct. 12, 1942; children—Melvin Harris, III, Phyllis V. Dir. Field St., Emory U., Atlanta, 1939-57, asso. preventive medicine Med. Sch., 1944-57; commd. lt. (j.g.) USPHS, 1944, advanced through grades to capt., 1957, asst. chief tech. br. Communicable Disease Center, Atlanta, 1953-57, chief Phoenix Field Sta., 1957-66; ret., 1966; dir. div. preventive med. services Ariz. Dept. Health, Phoenix, 1966—. Mem. expert com. on insecticides WHO, 1957—. Mem. Ga. Acad. Sci. (past pres.), Nat. Malaria Soc. Research and publs. on ecology of malaria-carrying mosquitoes contbg. to control by use of residual insecticide; developed rationale for investigation and control of enteric infections, epidemiol. procedures. Home: 327 W. Orchid Lane, Phoenix 85021. Office: 1624 W. Adams St., Phoenix 85007.*

GOODWIN, Richard Hale, Am. botanist; b. Brookline, Mass., Dec. 14, 1910; s. Harry Manley and Mary (Linder) G.; A.B., Harvard, 1933, M.A., 1934, Ph.D. (U. Scholar, fellow), 1937; fellow Atkins Inst. Arnold Arboretum, Cuba, 1935, U. Copenhagen, 1937-38, Am. Scandinavian Found., E. Africa, 1937; m. Esther Bemis, Oct. 12, 1936; children—Mary (Mrs. Bruce K. Wetzel), Richard Hale. Instr. botany U. Rochester (N.Y.), 1938-41, asst. prof., 1941-44; prof. botany Conn. Coll., New London, 1944—; dir. Conn. Arboretum, New London, 1944-65. Pres., Conservation and Research Found., 1953—, Nature Conservancy, 1956-58, 64-66; commr. Conn. Geol. and Natural History Survey, 1945—; mem. biology council NRC, 1954-57. Recipient award Bklyn. Bot. Garden, 1963. Fellow A.A.A.S., Am. Acad. Arts and Scis.; mem. Am. Inst. Biol. Scis. (governing bd. 1963—), Bot. Soc. Am., Am. Soc. Plant Physiologists, Soc. Study Devel. and Growth (sec. 1949-53), Ecol. Soc. Am., Am. Soc. Plant Taxonomy, Soc. Study Evolution, Torrey Bot. Club, New Eng. Bot. Club, Soc. Gen. Physiology. Research and numerous publs. in cytogenetics and morphogenesis of interspecific Solidage hybrids, morphogenetic effects of light on plant growth, fluorescent substances in plants, morphogenesis of roots, long-range vegetation studies in natural areas. Address: Conn. Coll., New London, Conn. 06320.*

GOODWIN, Trevor Walworth, English biochemist; b. Neston, Eng., June 22, 1916; s. Arthur Walworth and Agnes (Jones) G.; B.Sc., Birkenhead Inst., U. Liverpool (Eng.), 1938, M.S., 1939, diploma in edn. with distinction, 1940, D.Sc., 1950; m. Kathleen Sarah Hill, Dec. 28, 1944; children—Jane, Clare, Ann. Research worker Brit. Ministry of Food, 1940-44; faculty U. Liverpool, 1944-59, Johnston prof. biochemistry, 1966—; prof. biochemistry and agrl. biochemistry U. Wales, Aberystwyth, 1959-66. Fellow Royal Soc., Royal Inst. Chemistry; mem. Biochem. Soc. (symposium organizer), Internat. Union Photochemistry (vice chmn.), Internat. Union Biol. Scis. (sec. commn. on biochemistry), Chem. Soc., Phytochem. Soc., Author: Comparative Biochemistry, 1952; Recent Advances in Biochemistry, 1959; Biosynthesis of Vitamins, 1962; also numerous articles. Editor: Biochemistry of Plant Pigments, 1965. Research on biochemistry of carotenoid pigments in relation to photosynthesis in plants and their conversion to vitamin A in animals, biochemistry of chlorplasts (photosynthesizing units in plants), manufacture of vitamins by microorganisms. Home: The Beeches, Stoveron Rd., Birkenhead, Eng. Office: Biochemistry Dept., The Univ., Liverpool 3, Eng.*

GOODWIN, Willard E., Am. urologist; b. Los Angeles, July 24, 1915; s. Willard and Olive (Belt) G.; A.B., U. Cal. at Berkeley, 1937; M.D., Johns Hopkins, 1941; m. Mary Pearson Josephs, Feb. 21, 1942; children—Mary Devereux (Mrs. Lloyd Jones), Peter Colt, Willard, II. Faculty, Johns Hopkins Med. Sch., 1948-51, asst. prof. urology; faculty U. Cal. at Los Angeles, 1951—, prof., 1953—; chief urology Wadsworth VA Hosp., Los Angeles. Diplomate Am. Bd. Urology. Fellow A.C.S.; mem. A.M.A., Am. Urol. Assn., Pacific Coast Surg. Assn., Am. Surg. Assn. Research and numerous publs. on urol. surgery, kidney transplantation. Home: 254 Bronwood Av., Los Angeles 90049.*

GOODWIN, William, Brit. chemist; b. Macclesfield, Scotland, 1873; s. George James Goodwin; ed. Owens Coll., Manchester, Glasgow U., Glasgow Agrl. Coll., U. Göttingen (Germany), Lab. for Vegetable Physiology, Paris, Queen's Coll., Galway, Eire; M.Sc., Ph.D.; m. Helene Käthe Elizabeth Zyska, 1908; 2 daus. Demonstrator and asst. lectr. Queen's Coll., Galway; lectr. agrl. chemistry, Harper-Adams Agrl. Coll.; lectr. agrl. chemistry, head chemistry dept., adv. and research chemist South-Eastern Agrl. Coll., Wye; prin. Midland Agrl. and Dairy Coll., Sutton Bonington and Kingston-on-Soar; ofcl. agrl. analyst Notts and Lindsey county councils; sec. Notts War Agrl. Com.; examiner various univs. Research and publs. on sci. feeding of animals, methods of examining milk and dairy products, other agrl. topics. Died Dec. 30, 1953.

GOODWIN, William Lawton, chemist, geologist; b. Baie Verte, N.B., Can., Apr. 30, 1856; s. Edward Chappell and Margaret (Carey) G.; student, LL.D., Mt. Allison U., Sackville, N.B.; B.Sc., London, Eng.; D.Sc., Edinburgh (Scotland) U.; postgrad., Heidelberg, Germany; m. Christina Murray, 1885; 1 son, 3 daus. Demonstrator chemistry U. Edinburgh, 1879-80, with William Ramsay at Univ. Coll., Bristol, Eng., 1881-82; prof. chemistry, Mt. Allison, 1882-83; prof. chemistry Queen's U., Kingston, Ont., Can., 1883-1921, dean Sch. Mining, 1893-1913, dean Faculty Applied Sci., 1913-12. Author: A Text-book of Chemistry, 1886; Handbook for Prospectors, 1924, 29, 32; Manuel de Prospecteur, 1930; Geology and Minerals of New Brunswick, 1928; Geology and Minerals of Ontario, 1929; Geology and Minerals of Quebec, 1929; Geology and Minerals of Manitoba, 1930; Everyday Science, 1931; also papers in sci. jours. Died Jan. 1941.

GOODY, Richard Mead, atmospheric physicist; b. Welwyn, Eng., June 19, 1921; s. Harold Earnest and Lilian (Rankine) G.; B.A., Cambridge U., 1942, M.A., 1946, Ph.D., 1949; A.M. (hon.), Harvard, 1958; m. Elfriede Koch, July 11, 1947; 1 dau., Brigid R. Came to U. S., 1958, naturalized, 1964. With Brit. Ministry Aircraft Prodn., 1942-46; fellow St. John's Coll., Cambridge, Eng., 1950-53; reader Imperial Coll., London, Eng., 1953-58; Abbot Lawrence Rotch prof. dynamical meteorology, dir. Blue Hill Obs., Harvard, Cambridge, Mass., 1958—. Cons. to Brit., U. S. govt. agys. Mem. Royal, Am. meteorol. socs., Am. Astron. Soc., Am. Acad. Arts and Scis., Phys. Soc. of London. Author: Physics of the Stratosphere, 1954; Atmospheric Radiation, 1964; also articles. Research in theory and measurement of role of radiation in earth's atmosphere, lab. and solar spectroscopy on atmospheric gases, atmospheres of Mars and Venus, infra-red emission of solar photosphere, interaction of radiation and fluid motions. Home: 805 Brush Hill Rd., Milton, Mass. 02186.*

GOODYEAR, Charles, Am. inventor; b. New Haven, Conn., Dec. 29, 1800; s. Amasa and Cynthia (Bateman) G.; ed. pub. schs.; apprentice in merchandising; m. Clarissa Beecher, Aug. 24, 1824; m. 2d, Fanny Wardell, 1854; 7 children. Author: Gum Elastic and Its Varieties, 2 vols., 1853. Began expts., 1834; obtained 1st patent for acid and metal coating to destroy the adhesive properties of rubber, 1837; discovered (with N. M. Hayward, whom he employed to expt. with effects of sulphur on rubber) what became vulcanization process, patented 1844; went to Europe to extend his patent, 1851; obtained fgn. patents in all countries but Eng.; spent years 1851-58 in Europe; sold mfg. licenses and thus was largely responsible for establishment of rubber industry in Europe; unable to profit financially from his discovery since his patents were constantly infringed upon. Died in poverty, N.Y.C., July 1, 1860.

GÖPEL, Adolph, German mathematician; b. 1812; prof., gymnasium, nr. Potsdam, Germany. Author: Theoriae transcendentium primi ordinis adumbratio levis (extended Jacobi's work on Abelian and theta functions), 1847. Died 1847.

GOPPELSROEDER, (Christoph) Friedrich, Swiss chemist; b. Basel, Switzerland, Apr. 1, 1837; s. George Friedrich and Emma (von Speyr) G.; student chemistry Basel, Switzerland, 1855, Berlin, 1856, Heidelberg, beginning in 1857; Ph.D. under R. Bunsen, 1858; m. Rosina La Roche, 1863; 2 sons, 2 daus. Joined firm Koechlin, Baumgartner und Cie., Lörrach, 1858; became dept. pub. chemist, Basel, 1860; named lectr. on chemistry U. Basel, 1861, asso. prof., 1869; dir. Chem. Sch., Mulhouse France, 1872-80; independent research, Basel, 1880-98. Research and numerous publs. on capillary analysis (which led to the discovery of paper chromatography, 1944), food chemistry and physiology; began bacteriological testing of water; restored oil paintings by chemistry; discovered reaction between aluminum and alcoholic morin solution; extraction of dyestuffs by electrolytic reactions. Died Basel, Oct. 14, 1919.

GORA, Edwin Karl, physicist; b. Bielsko, Poland, Oct. 22, 1911; s. Joseph F. F. and Klara (Foerster) G.; student Jagellonian U., Cracow, Poland, 1929-34; D.Sc., U. Leipzig, Germany, 1942; m. Erika E. Buschmann, Jan. 18, 1945; children—Michael, Evelyn, Monica, Angela, Sonia. Faculty, Coll. Steubenville (O.), 1948-52; faculty Providence Coll., 1949—, prof., 1957—. Mem. Am. Phys. Soc. Research in quantum theory of radiation damping, partitioning technique, asymptotic methods in asymmetric rotor and angular momentum theory. Home: 86 Sandringham Av., Providence 02908.*

GORBACH, Georg, Austrian engr.; b. Hohenems, Vorarlberg, Mar. 13, 1901; s. Karl and Berta (Vogel) G.; ed. Technische Hochschule, Graz; Ph.D. in tech. sci.; m. Ilse Hemmer, 1931; children—Brigitte, Gerhild, Jörg Wolfgang, Karl Heinz. Asst., Technische Hochschule, Graz, 1927, titular prof., 1935, full prof. chemistry, 1948. Named laureat of Pregl prize, 1953. Mem. Assn. Austrian Chemists, Austrian Soc. Microchemistry, Austrian Soc. Microbiology, German Soc. Research on Fatty Matters, Internat. Soc. Sci. on Fatty Matters, Internat. Soc. Pure and Applied Chemistry. Home: 67 Brucknerstrasse. Office: 9 Schloegelgasse, Graz, Austria.*

GORBACHEV, Tinofei Fedorovich, Russian mining engr.; b. Troitskaya (now Tambov Oblast), USSR, June 23, 1900; grad. mining sch. Tomsk Poly. Inst. Chief engr. Kuzbas Coal, 1946-50; dir. Kemerovo Mining Inst., 1950-54; head Lab. Mining Pressure, Inst. Mining, USSR Acad., Siberia. Mem. USSR Acad. Scis. (corr., v.p. Siberian br. 1961—). Author: Working Thick Seams in the Kuzbas, 1944; Ways of Improving Systems of Working Thick, Steeply-Dipping Seams in the Southern Kuzbas, 1949; Preliminary Results of Observations of Coal Undercutting, Mine Workings, Installations and Water Sources in the Kuzbas, 1951; A Combined System of Shield Mining, 1954; Experience with Thick-Seam Working in the USSR and Abroad, 1957; (with others) Working the Kuzbas Coal Deposits, 1959. Editor: Scientists of Siberia to the Kuzbas, 1961; chief editor: Problems of Mining Pressure, 1962. Developed self-propelling mining machine using water. Office: Siberian Branch of USSR Acad. Sciences, Novosibirsk, Siberia, USSR.

GORBMAN, Aubrey, Am. endocrinologist; b. Detroit, Dec. 13, 1914; s. David and Esther (Korenblit) G.; A.B., Wayne State U., 1935, M.S., 1936; Ph.D., U. Cal. at Berkeley, 1940; m. Genevieve D. Tapperman, Dec. 26, 1938; children—Beryl Ann, Leila Harriet, Claudia Louise, Eric Jay. Faculty, U. Cal. at Berkeley, 1940-41, Wayne State U., Detroit, 1941-44, Yale, 1944-46; faculty Barnard Coll. and Columbia, 1946-62, prof., 1953-62, chmn. zoology dept., 1952-55; prof., chmn. zoology dept., U. Wash., Seattle, 1963—; vis. prof. chemistry Nagoya (Japan) U., 1955; vis. prof. zoology Tokyo (Japan) U., 1960. Cons. biologist Brookhaven Nat. Lab., Upton, N.Y., 1949-58; mem. panel on preparation of biology tchrs. Commn. Undergrad. Edn. in Biol. Sci., 1965—. Fulbright fellow Collège de France, Paris, 1950; Guggenheim fellow U. Hawaii, 1955. Mem. Am. Soc. Zoologists (chmn. div. comparative endocrinology), Am. Assn. Anatomists, Endocrine Soc., Soc. for Exptl. Biol. Medicine. Author: Comparative Endocrinology, 1959; (with H. A. Bern) A Textbook of Comparative Endocrinology, 1962; also numerous articles. Editor-in-chief Gen. and Comparative Endocrinology, 1960—. Research in comparative endocrinology, ecology radioactive iodine in environment, action hormones on nervous system. Home: 4218 55th Av. N.E., Seattle 98105.*

GORBOVITSKY, Samuil Yefimovich, Russian dermatologist, venerologist; b. 1900; grad. 1st Leningrad. Med. Inst., 1925, postgrad. dept. skin and venereal diseases, 1925-29; D.Med. Sci. Asst. dept. skin and venereal diseases Leningrad Postgrad. Med. Inst., 1929-30, lectr., 1931-35; chief physician Tarnovsky Skin and Venereal Diseases Hosp., Leningrad, 1929-30; dir. Leningrad Skin and Venereal Diseases Inst., 1930-51; head chair dermatology and venerology 3d Leningrad Med. Inst., 1935-40; prof., 1935—; head chair skin and venereal diseases Naval Med. Acad., 1940-56; dep. head chair skin and venereal diseases Kirov Mil. Med. Acad., Leningrad., 1956—. Chmn. venerology council Leningrad City Dept. Health, 1930-51. Mem. All-Union Soc. Dermatologists and Venerologists (bd. mem. 1937—), Leningrad Dermatol. Soc. (bd. mem. 1930-35, dep. chmn. 1935—). Author: Manual for Naval Surgeons on Treatment and Prophylaxis of Venereal and Infectious Skin Diseases, 1943; Chemotherapy and Criterions for Pronouncing Complete Cure of Syphilis, 1944. Mem. editorial council Vestnik Dermatology and Venerology. Address: Kirow Mil. Med. Acad., ulitsa Lebedeva 6, Leningrad, USSR.

GORCZYNSKI, Wladyslaw Jozef, Polish meteorologist; b. Bramki, nr. Warsaw, Poland, Mar. 19, 1879; s. Franciszek and Marja (Higersberger) G.; ed. U. Warsaw; D.Sc., U. Montpellier, 1906; m. Stanislawa Bejt, 1914. Dir., Meteorol. Service of Poland, ret., 1928; chief Meteorol. Bur., Warsaw, 1903-19; 1st dir. Central Meteorol. Inst. Poland, 1919-27. Made several voyages abroad to measure solar radiation (actionometry), 1904-14, 19-28. Mem. Polish Acad. Scis., Societas Scientiarum Varsovien. Author 3 books on climatology of Poland and world. Research and over 250 publs. on actionometry; invented solarimeters and pyrheliometers (actionometers) for solar radiation measurements.

GORDAN, Gilbert Saul, Am. endocrinologist; b. San Francisco, July 8, 1916; s. Gilbert S. and Sadie (Joseph) G.; B.A., U. Cal., Berkeley, 1937, Ph.D., 1947; M.D., U. Cal., San Francisco, 1941; m. Jane Rafter, Dec. 5, 1965. Faculty Sch. Medicine, U. Cal. San Francisco, 1948—, prof. medicine, 1962—; Commonwealth Fund research fellow Harvard, 1947-48. Mem. Am. Soc. Clin. Investigation, Royal Soc. Medicine (London), Sigma Xi, Alpha Omega Alpha. Author: Year Book of Endocrinology, 1951-63. Contbr. numerous articles to profl. jours. Research and publs. primarily in field of anabolic steroids, bone metabolism and parathyroid physiology. Home: 100 Pemberton Pl., San Francisco 94114.*

GORDAN, Paul, German mathematician; b. Breslau, (now Wroclaw, Poland), Apr. 29, 1837; s. Daniel and Friedericke (Friedenthal) G.; studied math. in Breslau, Königsberg, Germany, Berlin; Ph.D., Berlin, 1862; hon. dr., Dublin; m. Sophie Deurer, 1869; a son, Paul. Joined faculty U. Giessen (Germany), 1863, named asso. prof., 1865; became prof. Erlangen, Germany, 1874, emeritus, 1910. Named hon. mem. U. Dorpat (Estonia). Mem. Bavarian Acad. Scis., French Acad. Scis., 1904. Author: (with A. Clebsch) Theorie der Abelschen Funktionen 1866; Über des Formensystem binärer Formen, 1875; Dr. Paul Gorden's Vorlesungen über Invariantentheorie, 2 vols. (edited by G. Kerschensteiner), 1887; also articles. Research on theory of elliptical function and their application to equation theory, form theory of linear transformations; Gordan's theorem proves finiteness of any invariant system for single binary form; proved fundamental theorems for algebraic forms using algorithms. Died Erlangen, Dec. 21, 1912.

GORDON, Albert Saul, Am. physiologist; b. Bklyn., Aug. 8, 1910; s. David and Lydia (Riover) G.; B.S., Coll. City N.Y., 1930; M.S., N.Y. U., 1931, Ph.D., 1934; m. Ruth Beitler, Dec. 6, 1935; 1 dau., Enid Doris (Mrs. Barry Ira Bruck). Faculty, N.Y. U., 1931—, prof., 1954—; faculty Newark Inst. Arts and Scis., 1934-35; mem. study sect. hematology USPHS, 1958-64; mem. erythropoietin com. Nat. Heart Inst. NIH, 1964—. Recipient Cressy Morrison award in natural sci., 1948, 66, grants USPHS, 1957, Am. Cancer Soc., 1951, USAF, 1952, Damon Runyon Meml. Fund For Cancer Research, 1961, Commonwealth Fund 1942, NSF, 1955, numerous other grants. Dazian Found. Med. Research fellow, 1949-51. Fellow N.Y. Acad. Scis.; mem. Am. Assn. U. Profs., Am. Soc. Zoologists, Am. Assn. Anatomists, Am. Physiol. Soc., Soc. Exptl. Biology and Medicine, Endocrine Soc., Harvey Soc., Soc. for Study Blood, Internat., Am. socs. hematology, Reticulo Endothelial Soc., (founding editor-in-chief jour.), Royal Soc. Medicine, A.A.A.S. Contbr. numerous articles to profl. jours. Research, publs. on exptl. hematology and endocrinology, blood cell formation and destruction in higher animals, mechanisms underlying blood diseases like anemias, leukemias. Home: 424 Forest Pl., West Hempstead, N.Y. 11552. Office: N.Y. U., 100 Washington Sq. E., N.Y.C. 10003.*

GORDON, Alexander, Scottish obstetrician; b. 1752; practiced medecine, Aberdeen, Scotland. Author: History, Pathology and Treatment of Puerperal Fever, 1786; Treatise on the Epidemic Puerperal Fever of Aberdeen (advocated that physicians and nurses who had attended puerperal fever patients fumigate themselves and their clothes), 1795. Formulated theory on the contagiousness of puerperal fever, 1795. Died 1799.

GORDON, Andrew, physicist; b. Gofforach, Scotland, June 15, 1712; ed. at Ratisbon; student law, Salzburg, Austria; traveled in Austria; joined Order St. Benedict; ordained priest; prof. philosophy, Erfurt, Germany 1737; mem. Acad. Scis. Paris (corr.). Author: Programma de studii philosophici dignitate et utilitate, 1737; De concordandis mensuris, 1742; Phaenomena electricitatis exposita, 1744; Dissertatio de spectris, 1746; Physicae experimentalis elementa, 1751-52. Invented elec. fountain, elec. turnstile, elec. whirl, earliest known electrostatic reaction motor, electric chimes (earliest application of electric convection). Died Erfurt, Aug. 22, 1751.

GORDON, Charles Henry, Am. geologist; b. Caledonia, N.Y., May 10, 1857; s. John and Ann (McKinnon) G.; B.S., Albion Coll., 1886, M.S., 1890; Ph.D. U. Chgo., 1895; postgrad. U. Heidelberg, 1897-98; m. Mary E. Hydorn, June 22, 1887; children—Irene Hydorn (Mrs. Burton Ashton Gaskill), Helen Garnett (Mrs. Don Carlos Ellis), Isabel (Mrs. Hugh Sevier Carter). Instr. high sch., Keokuk, Ia., 1886-87; prin. Wells Sch., Keokuk, 1887-90; instr. natural history, Northwestern U., Ill., 1890-93; supt. schs., Beloit, Wis., 1895-97, Lincoln, Neb., 1899-1903; lectr. U. Neb., 1901-03; acting prof. geology U. Wash., 1903-04; prof. geology and mineralogy N.M. Sch. Mines, 1904-05; field asst. U.S. Geol. Survey, 1905-06, asst. geologist, 1906-13; prof. geology U. Tenn., 1907-31. Asso. state geologist, Tenn., 1910-14; dir. and supt. Dept. of Minerals, Nat. Conservation Expn., Knoxville, 1913; connected with Mo., Ia., Mich. surveys. A founder Knoxville Geol. Soc., Tenn. Acad. Scis.; fellow A.A.A.S., Geol. Soc. Am. Research on rock deposits, marble formations, iron ore, copper. Died June 12, 1934.

GORDON, Cyrus Herzl, Am. archaeologist; b. Phila., June 29, 1908; s. Benjamin Lee and Dorothy (Cohen) G.; A.B., U. Pa., 1927, M.A., 1928, Ph.D., 1930; m. Joan Elizabeth Kendall, Sept. 22, 1946; children—Deborah, Sarah, Rachel, Noah, Dan. Instr., U. Pa., Phila., 1930-31; archaeologist Am. Sch. Oriental Research, Baghdad, 1931-35; fellow Johns Hopkins, Balt., 1935-38; lectr. Smith Coll., Northampton, Mass., 1938-39, 40-41; mem. Inst. for Advanced Study, Princeton, N.J., 1939-40, 41-42; prof. Dropsie Coll., Phila., 1946-56; prof. Brandeis U., Waltham, Mass., 1956—. Mem. Am. Inst. Archaeology, Am. Oriental Soc., Am. Philol. Assn., Soc. Bibl. Lit., Am. Hist. Assn., Am. Assn. U. Profs. Author: Ugaritic Literature, 1949; Hammurapi's Code, 1957; Adventures in the Nearest East, 1957; Smith College Tablets, 1952; Ugaritic Textbook, 1965; Ancient Near East, 1965; Common Background of Greek and Hebrew Civilizations, 1965; Ugaritic and Minoan Crete, 1966; Evidence for the Minoan Language, 1966; Forgotten Scripts, 1968; also numerous articles. Discovered common background of Greek and Hebrew civilizations; deciphered Minoan and Eteocretan inscriptions, formulated 1st detailed grammar and lexicon of Ugaritic. Home: 130 Dean Rd., Brookline, Mass. 02146. Office: Brandeis U., Waltham, Mass. 02154.*

GORDON, Edgar Stillwell, Am. physician; b. Chgo., Nov. 6, 1907; s. Edgar Bernard and Edna (Stillwell) G.; B.S., U. Wis., 1927, M.A., 1929; M.D., Harvard, 1932; m. Lola A. Gray, June 27, 1936; children—Joan Elizabeth (Mrs. Arthur Wayne Owens), Stuart Gray, Robert Bruce. Practice medicine, specializing in endocrinology and metabolism, Madison, Wis., 1938—; faculty U. Wis., 1938—, prof. medicine, 1952—; cons. AEC, Oak Ridge Inst. Nuclear Studies, NASA; mem. program project com. Inst. Arthritis and Metabolic Diseases, NIH. Mem. Assn. Am. Physicians, A.C.P., Am. Soc. Clin. Investigation, Central Soc. Clin. Research, Endocrine Soc., Am. Inst. Nutrition, Am. Diabetes Assn., Am. Clin. and Climatol. Assn. Author: Nutritional and Vitamin Therapy, 1942. Editor: Steroid Hormones, 1950. Research and publs. in metabolic diseases, endocrinology and nutrition, including diabetes, obesity, atherosclerosis, coronary heart disease, nutritional deficiency, physiology of energy metabolism, physiology of growth, physiology of exercise. Home: 1520 Wood Lane, Madison 53705. Office: University Hosp., Madison, Wis. 53706.*

GORDON, Eugene I., Am. physicist; b. N.Y.C., Sept. 14, 1930; s. Sol and Gertrude (Lassen) G.; B.S., City Coll. N.Y., 1952; Ph.D., Mass. Inst. Tech., 1957; m. Barbara Young, Aug. 19, 1956; children—Laurence Mark, Peter Elliot. Research asso. Mass. Inst. Tech., 1957; physicist Bell Telephone Labs., Inc., Murray Hill, N.J., 1957—. Fellow I.E.E.E.; mem. Am. Phys. Soc., Phi Beta Kappa. Asso. editor Jour. Quantum Electronics. Research and publs. on phys. mechanisms and fabrication of various types of gas lasers, particularly first continuously operating blue-green gas laser; techniques for modulating and deflecting light beams by the use of high frequency acoustic waves; new types of microwave amplification. Home: 14 Braidburn Way, Convent, N.J. 07961. Office: Bell Telephone Labs., Inc., Murray Hill, N.J. 07971.*

GORDON, Francis Byron, Am. microbiologist; b. Fairbury, Ill., Mar. 15, 1905; s. Marshall and Bertha (Patton) G.; B.S., Ill. Wesleyan U., 1927; Ph.D., U. Chgo., 1936, M.D., 1937; m. Mary Elizabeth Vrooman, Aug. 30, 1930; children—John Marshall, Ellen Jane (Mrs. Paul Eric Soherr), James Byron, William Ethan. From instr. to prof. dept. microbiology U. Chgo., 1936-48; div. chief Chem. Corps Biol. Labs., Fort Detrick, Md., 1948-54; spl. lectr. virology George Washington U., 1951—; lectr. bacteriology Coll. Spl. and Continuation Studies U. Md., 1951-53, vis. prof., 1958-60; head div. virology Naval Med. Research Inst., Nat. Naval Med. Center, 1954-62, dir. dept. microbiology, 1962—; cons. Sec. War, 1942-45. Recipient Distinguished Service award Med. Alumni Assn. U. Chgo., 1952. Mem. Am. Acad. Microbiology (bd. govs.), Am. Soc. Microbiology, Soc. for Exptl. Biology and Medicine, A.A.A.S., N.Y. Acad. Scis., Am. Assn. Immunologists, Am. Inst. Biol. Scis. Mng. editor: Jour. Infectious Diseases, 1941-48, adv. editor, 1948—. Contbr. chpts. to Jordan-Burrows' Textbook of Bacteriology, 1941, 45, 49, 54, 58. Research, numerous publs. on poliomyelitis, its epidemiologic pattern, nature of exptl. disease in monkeys, arthropod-borne diseases of central nervous system, agts. of diseases of trachoma-psittacosis group, effect of artificial atmospheres and alterations in barometric pressure. Home: 17410 New Hampshire Av., Ashton, Md. 20702. Office: Naval Med. Research Inst., NNMC, Bethesda, Md. 20014.*

GORDON, George Phineas, Am. inventor; b. Salem, N.H., Apr. 21, 1810; s. Phinias and Mary (White) G.; m. Sarah Cornish, 1846; m. 2d, Lenore May, 1856; 1 child. Apprenticed to printer, N.Y.C.; opened small job-printing office, N.Y.C.; began experimenting on improved press for card printing, 1835; obtained 1st patent for Yankee job press, 1851; received over 50 patents during his career; introduced Firefly job press, 1854, turned out 10,000 cards per hour; built more than 100 kinds of presses; established factory in Rahway, N.J., with offices in N.Y.C., 1872. Died Norfolk, Va., Jan. 21, 1878.

GORDON, Glen Everett, Am. chemist; b. Keokuk, Ia., Oct. 13, 1935; s. Scott Robert and Sara (Perry) G.; student Western Ill. State Coll., 1952-54; B.S., U. Ill., 1956; Ph.D., U. Cal. at Berkeley, 1960; m. Constance Hereshoff, May 30, 1958; children—Karl, Christine. With Mass. Inst. Tech., 1960—, asso. prof. chemistry, 1964—. Research in nuclear chemistry and geochemistry, including nuclear decay properties, nuclear fission, trace-element distbns. Home: 141 Oxford St., Cambridge, Mass. 02140.*

GORDON, Harry Haskin, Am. physician; b. Bklyn., Aug. 4, 1906; s. Samuel and Ida (Haskin) G.; B.A., Cornell U., 1926, M.D., 1929; m. Fayga Halpern, June 8, 1948; children—Charles, Deborah. Faculty Yale, 1931-32, Cornell U., 1932-46, U. Colo. Sch. Medicine, 1946-52, Johns Hopkins, 1952-62; prof. pediatrics Albert Einstein Coll. Medicine, Bronx, N.Y., 1962—, dean, 1967—; sr. specialist maternal and child health Children's Bur. U. S. Dept. Labor, N.Y.C., 1937-42; dir. Rose F. Kennedy Center for Research in Mental Retardation and Human Devel., Albert Einstein Coll. Medicine-Bronx Municipal Hosp. Center, 1965—. Cons. to Surgeon Gen. U. S. Army, 1947—, U. S. Children's Bur., 1948-51, USPHS, 1949—; vice chmn. research adv. bd. Nat. Assn. for Retarded Children, 1963—. Recipient Borden award Am. Acad. Pediatrics, 1944; Career Scientist award Health Research Council N.Y.C., 1962; Grover F. Powers Distinguished Prof. Nat. Assn. Retarded Children, 1963. Diplomate Am. Bd. Pediatrics. Fellow A.A.A.S.; mem. A.M.A., Am. Acad. Pediatrics, Am. Inst. Nutrition, Am. Pediatric Soc. (past pres.), Am. Soc. Clin. Investigation, Harvey Soc., Soc. for Pediatric Research (past pres.), Soc. Exptl. Biology and Medicine, N.Y. Acad. Scis., Bronx County Med. Soc., Sigma Xi, Alpha Omega Alpha. Research on devel. of prematurely born infants and in trop. diseases. Home: 369 Orienta Av., Mamaroneck, N.Y. 10543. Office: care Albert Einstein Coll. Medicine, Bronx, N.Y. 10461.*

GORDON, Louis, Am. analytical chemist; b. N.Y.C., Sept. 3, 1914; s. Harry and Gussie (Feinstein) G.; B.S., U. Ky., 1937; M.S., U. Mich., 1939, Ph.D., 1947; m. Ruth Levy, June 13, 1940; 1 son, Michael. Asst. prof. Ohio State U., Columbus, 1946-48; faculty Syracuse U. (N.Y.), 1948-50, prof., 1956-57; prof. chemistry Case Inst. Tech., Cleve., 1957—, dean Grad. Studies, 1961—; Centennial prof. U. Ky., Lexington, 1965. Mem. Am. Chem. Soc., Am. Assn. U. Profs. Author: (with M. L. Salutsky, H. H. Willard) Precipitation from Homogeneous Solution, 1959; also numerous articles. Research in analytical chemistry aspects of precipitation phenomena particularly nucleation, coprecipitation, devel. new precipitation processes. Died Cleve., Oct. 21, 1966.

GORDON, Maxwell, organic chemist; b. USSR, Feb. 13, 1921; s. Abraham and Sara (Gordon) G.; came to U. S., 1923, naturalized, 1935; B.Sc., Phila. Coll. Pharmacy and Sci., 1941; M.S., U. Pa., 1946, Ph.D. in Organic Chemistry, 1948; D.I.C., Imperial Coll. London (Eng.), 1951; m. Ethel Mayer, June 5, 1949; children—Alan Michael, Sandra Lynn. With Squibb Inst. Med. research, New Brunswick, N.J., 1951-55; resarch and devel. chemist Smith Kline & French Lab., Phila., 1955-57, head phys. scis., 1957—. Mem. com. on modern methods handling chem. information Nat. Acad. Sci./NRC, 1965—. Mem. A.A.A.S., Am., Swiss, German, Austrian chem. socs., Chem. Soc. (London, Eng.), Pharm. Soc. Japan. Editor, author: Psychopharmacological Agents, vol. I, 1964, vol. II, 1966; also numerous articles. Patentee in field. Discovered various pheonthiazine tranquilizers, diuretics, analgetics; invented chem. typewriter, phototypesetting system for automatic photocomposition chem. structures; research on mechanism biosynthesis antibiotics and metabolism drugs. Home: 1919 Chestnut St., Phila. 19103. Office: 1500 Spring Garden St., Phila. 19101.*

GORDON, Milton Paul, Am. biochemist; b. St. Paul, Feb. 8, 1930; s. Abraham and Rebecca (Ryan) G.; B.A. summa cum laude, U. Minn., 1950; Ph.D., U. Ill., 1953; m. Elaine Travis, Jan. 1, 1955; children—David, Karen, Nancy. Upjohn Co. fellow U. Ill., 1950-51; Am. Cancer Inst. fellow Sloan-Kettering Inst. for Cancer Research, N.Y.C., 1953-55, research asst., 1955-57; lectr. Bklyn. Coll., 1955-57; asst. research biochemist Virus Lab., U. Cal. at Berkeley, 1957-59; faculty U. Wash., Seattle, 1959—, prof., 1966—. Sec.-treas. Pacific Slope Biochem. Conf., 1964-68, pres., 1968. Mem. Am. Chem. Soc., Harvey Soc. A.A.A.S., Am. Soc. Biol. Chemists, Am. Inst. Aeros. and Astronautics, Am. Soc. Plant Pathologists, Research and publs. on multiplication of plant viruses. Home: 8255 45th Av. N.E., Seattle 98115.*

GORDON, Neil Elbridge, Am. chemist; b. Spafford, N.Y., Oct. 7, 1886; s. William James and Ella C. (Mason) G.; Ph.B., Syracuse U., 1911; M.A., 1912, Pd.B., 1921; Ph.D., Johns Hopkins, 1917; m. Hazel A. Mothersell, June 29, 1915; children—Neil Elbridge, Fortuna Lucille. Asst. prof. chemistry. Goucher Coll., Balt., 1917-19; prof. phys. chemistry U. Md., 1919-21, dir. chem. dept., 1921-28; state chemist of Md., 1921-28; Francis P. Garvan prof. chem. edn. Johns Hopkins U., 1928-36; head chemistry dept. Central Coll., Fayette, Mo., 1936-42; chmn. chemistry dept., Wayne U., from 1942. Fellow A.A.A.S. (organizer Gibson Island Research Confs.); mem. Am. Chem. Soc., Mo. Acad. Sci., Faraday Soc., Royal Soc. Arts and Scis., Sigma Xi. Organizer and editor Jour. Chem. Edn., 1924-33; dir. Hooker Scientific Library 1936-46; dir. Friends of the Kresge-Hooker Scientific Library, since 1946; organizer and editor. Record of Chem. Progress, since 1939. Author: Project Study of Chemistry, 1925; Introductory College Chemistry,

1926, 2d edit., 1941; Introductory Chemistry, 1927, revised 1940; College Chemistry, 1928; Record Book for Introductory Chemistry, 1928. Contbr. numerous articles in sci. jours. Research on soil colloids, absorption theory, potash availability in mixed fertilizers, solubility of liquids in liquids; patentee thermoregulator for constant bath temperature. Died May 20, 1949.

GORDON, Paul, Am. physical metallurgist; b. Hartford, Conn., Jan. 1, 1918; s. Charles Dana and Anne Mabel (Hirshberg) G.; student Wesleyan U., Middletown, Conn., 1935-37; B.S. in Metallurgy, Mass. Inst. Tech., 1939, M.S., 1940, Sc.D., 1949; m. Evelyn Rubin, Oct. 16, 1941; children—Dana Charles, Jane Ellen. Research asso. metallurgy Mass. Inst. Tech., 1941-42, group leader Manhattan Project, 1942-47; mem. faculty Ill. Inst. Tech., 1949-50, 54-—, prof. metall. engring., 1957-—, chmn. dept., 1966-—; asst. prof. Inst. Study Metals, U. Chgo., 1951-54. Mem. Am. Soc. Metals, Am. Inst. Mining and Metall. Engrs. (Mathewson gold medal 1957), Inst. Metals, A.A.A.S., Am. Soc. Engring. Edn., Engrs. Council Profl. Devel., Sigma Xi. Contbr. profl. jours., chpts. to books. Research on metal failures; surface reactions; transformations in metals; grain growth; recrystallization; precipitation; order-disorder phenomena. Home: 5648 S. Harper Av., Chgo. 60637.

GORDON, Sir Robert, inventor; b. Elginshire, Scotland, Mar. 7, 1647; s. Sir Ludovick and Elizabeth (Farquhar) G.; m. Margaret, widow of Alexander, 1st Lord of Duffus, Feb. 23, 1676 (dec. Apr. 1677); 1 dau.; m. 2d, Elizabeth Dunbar; 3 sons, 4 daus. Represented Sutherlandshire, Scotch parliament, 1672-74; sat in Conv., 1678, 1681-82, 85-86. Fellow Royal Soc., 1686. Author: Recipe to Cure Mad Dogs, or Men and Beasts bitten by Mad Dogs, 1687. Invented machine or pump for raising water. Died 1704.

GORDON, Robert Boyd, Am. physicist; b. East Orange, N.J., Dec. 25, 1929; s. Myron Boyd and Catherine (Rote) G.; B.S., Yale, 1952, D.Eng., 1955; m. Joan Parke Ruttiger, Sept. 13, 1952; children—Penelope, Margaret. Asst. prof. Sch. Mines, Columbia U., 1955-57; faculty Yale, 1957-—, asso. prof. applied sci., 1960-—. Mem. Am. Phys. Soc., Am. Geophys. Union, Am. Inst. Mining and Metall. Engrs., Sigma Xi. Author: (with R. M. Brick, A. Phillips), Structure and Properties of Alloys, 1965; also articles. Research on mech. properties of solids particularly at high pressures and related to geophysics and properties of earth. Home: 121 Boston St., Guilford, Conn. 06437. Office: Kline Geology Lab., Yale, New Haven 06520.*

GORDON, Robert Edward, Am. biologist; b. N.Y.C., June 20, 1925; s. Lewis Francis and Claire (McEvoy) G.; A.B., Emory U., 1949; M.S., U. Ga., 1950; Ph.D., Tulane U., 1956; m. Catherine Tigner, Sept. 16, 1948; children—Claire Catherine, Martha Lee. Faculty N.E. La. State Coll., Monroe, 1954-58; faculty U. Notre Dame (Ind.), 1958-—, prof. biology, 1966-—, head, dept. biology, 1964-67, asso. dean, 1967-—. Mem. UNESCO Working Parties on Sci. Publs., Phila., 1963, Paris, 1964; mem. U. S. panel U. S./Japan Co-op, Sci. Program, Primary Jour. Editors, Tokyo, 1965, 67; chmn. council biol. scis. information Nat. Acad. Scis., 1967-—. Fellow A.A.A.S., Herpetologists League; mem. Am. Inst. Biol. Sci., Am. Soc. Ichthyologists & Herpetologists, Am. Soc. Zoology, Animal Behavior Soc., Conf. Biol. Editors (sec. 1963-—), Herpetological Soc. Japan, Soc. Study Amphibians and Reptiles, Sigma Xi. Editor Am. Midland Naturalist, 1958-64, sect. on amphibia and reptilia Biol. Abstracts, 1963-—. Research and publs. on behavior, ecology of lower vertebrates; problems in sci. communication, information. Home: 19551 Oakdale Av., South Bend, Ind. 46637.*

GORDON, Sheffield, Am. chemist; b. Chgo., Feb. 10, 1916; s. William and Edna (Astrahan) G.; B.S., U. Chgo., 1937, postgrad., 1941-42; Ph.D., U. Notre Dame, 1949; m. Marceline L. Dzurus, Nov. 28, 1964. Chemist, Alton R.R. Co., 1938-40; research asso. metall. lab. U. Chgo., 1946-49; staff Argonne (Ill.) Nat. Lab., 1950-—; collaborateur etranger Centre Etude Nucleaire, Saclay, France, 1965-66. Mem. Am. Chem. Soc., Am. Phys. Soc., Faraday Soc., Radiation Research Soc., A.A.A.S. Contbg. author Nuclear Reactor Experiments, 1958. Research and publs. in radiation chemistry and photochemistry of liquid and gaseous systems, elucidation of interaction of ionizing radiation with matter, reactions unstable intermediate species produced by these interaction. Home: 5000 East End Av., Chgo. 60615. Office: Argonne Nat. Lab., 9700 Cass Av., Argonne, Ill. 60439.*

GORDON, Walter, physicist; b. Apolda, Germany, Aug. 3, 1893; s. Arnold and Bianca (Braun) G.; student math., physics, Berlin, 1915-21; Ph.D. under M. Planck, 1921; m. Gertrud Lobbenberg, 1932. Successively asst. to M. v. Laue, Berlin. W. Bragg, Manchester; worked at Kaiser Wilhelm Inst. for Fiber Chemistry, Berlin-Dahlem, Germany; joined U. Hamburg (Germany), 1926, became mem. faculty, 1929, named asso. prof., 1930; moved to Stockholm, Sweden, 1933; worked at Inst. for Mechanics and Math. Physics, U. Stockholm. Research on atom mechanics, quantum mechanics of wave mech. form, relativistic Dirac equation of the spin electron, (independently of Klein) Compton effect; originated Klein-Gordon equation. Died Stockholm, Dec. 24, 1939.

GORDON, William Edwin, Am. radiophysicist; b. Paterson, N.J., Jan. 8, 1918; s. William and Mary (Scott) G.; B.A., Montclair (N.J.) State Coll., 1939, M.S., 1942; M.S., N.Y. U., 1946; Ph.D., Cornell U., 1953; m. Elva Freile, June 22, 1941; children—Larry Scott, Nancy Lynn. Faculty, Cornell U., Ithaca, N.Y., 1948-66, prof., 1959-65, Walter R. Read prof. engring., 1965-66, dir. Arecibo Ionon. Obs., 1959-65; dean engring. and sci. Rice U., Houston, 1966-—. Com. chmn. Internat. Sci. Radio Union, 1957-60, Internat. Council Sci. Unions, 1954-60; trustee U. Corp. For Atmospheric Research, 1965. Recipient Air Force Cambridge Research Lab. commendation, 1963; Balth van der Pol Gold medal for distinguished research in radio sci., 1966. Fellow I.E.E.E.; mem. Am. Meteorology Soc., Am. Geophys. Union, A.A.A.S., Sigma Xi, Tau Beta Pi, Kappa Delta Pi, Sigma Kappa Nu, Phi Kappa Phi. Research on troposphere, especially role of tropospheric scattering of electromagnetic waves in radio communication; establishment of Arecibo ionospheric obs. Home: 12422 Mossycup, Houston 77024.*

GORDON, William St. Clair, Am. physician; b. Raleigh, N.C., Mar. 28, 1858; s. James and Mary St. Clair (Cooke) G.; M.D., Med. Coll. Va., 1879; spl. courses, Jefferson Med Coll., Phila., U. Pa.; m. Kate Blanks Gordon, Oct. 16, 1890. Practiced at Richmond, Va.; a founder University Coll. Medicine, Richmond, 1893, also prof. physiology, later asso. prof. clin. medicine; prof. medicine Med. Coll. of Va., 1913-14; physician and mem. bd. dirs. Laurel (Va.) Reformatory. Author: Recollections of the Old Quarters (dialect, prose and verse), 1902. Investigated outbreak of typhoid in Richmond, 1884, submitted extensive report, 1893. Died Apr. 24, 1924.

GORDONOFF, Toni, physician, b. Novosybkov, Russia, Feb. 3, 1893; s. Moise and Eida (Kasarnovsky) G.; M.D., U. Bern (Switzerland), 1922; postgrad. U. Nancy, U. Bern; m. Eugenie de Sboeff, Aug. 6, 1926. Faculty U. Berne Faculty Medicine, 1922-63, docent, 1926, prof. pharmacology and toxicology, 1946-63. Hon. mem. bd. govs. U. Jerusalem. Corr. mem. Soc. de Thérapie (Paris, France), Med. Soc. Vienna, Avard Intern. Price Buergi; mem. Pharmakol. Soc. Germany, Pharmakol. Soc. Switzerland, Eurotox, Med. Soc. Berne. Author: Textbook of Pharmacology, 1956; Rezeptierkunde, 1936, Handbuch d. Therapie, 1946; Grundriss der gebräuchl. Arzneimittel, 1965; also numerous articles. Editor in chief Internat. Rev. Vitamin Research, 1932-—, Therapeutische Umschau, 1944-—. Research on expectorants, pain, and vegetative nervous system. Home: 23 Ankerstr., Bern, Switzerland.*

GORDON-TAYLOR, Sir Gordon, Brit. surgeon, anatomist; student Gordon's Coll., Aberdeen, Scotland; M.A., Aberdeen U.; M.S., B.Sc., London (Eng.) U.; med. student Middlesex Hosp.; several hon. Brit. and fgn. degrees; m. Florence Mary Pegrume, 1920. Demonstrator anatomy King's Coll.; Hunterian prof. surgery Royal Coll. Surgeons Eng., 1929, 42, 44, mem. council, 1932-48, v.p., 1941-43, also examiner in anatomy; lectr. on surgery, hon. demonstrator anatomy Middlesex Hosp.; examiner in surgery at univs. Cambridge (Eng.), London, Belfast (Ireland), Leeds (Eng.), Durham (Eng.), Edinburgh (Scotland); Moseley prof. surgery Peter Bent Brigham Hosp., Harvard U., 1941, 46; postgrad. prof. surgery U. Cairo (Egypt), 1947; surgeon to out-patients Royal No. Hosp.; surgeon Royal Scottish Hosp. and Corp.; cons. surgeon to Royal Navy, also to Middlesex Hosp., Alfred and St. Vincent's Hosp., Melbourne, Australia. Lectr. at various univs. and sci. socs. Recipient Triennial Gold medal Royal Soc. Medicine, 1956. Fellow Royal Coll. Surgeons; pres. Med. Soc. London, 1941-42, Royal Soc. Medicine, 1944-46, Assn. Surgeons Gt. Britain and Ireland, 1944-45; chmn. Horatian Soc.; hon. fellow, mem. many Brit. and fgn. med. socs. Author books and papers contbd. to med. jours. on anatomy, abdominal cancer, mil. surgery. Died Sept. 3, 1960.

GORDY, Walter, Am. physicist; b. Lawrence, Miss., Apr. 20, 1909; s. Walter Kalin and Gertrude (Jones) G.; B.A., Miss. Coll., 1932, LL.D., 1959; M.A., U. N.C., 1933, Ph.D., 1935; Dr. honoris causa U. de Lille (France), 1955; m. Vida Brown Miller, June 19, 1935; children—Eileen (Mrs. George Kohut), Walter Terrell. Asso. prof. math., physics Mary Hardin-Baylor Coll., Belton, Tex., 1935-41; Nat. Research fellow Cal. Inst. Tech., Pasadena, 1941-42; staff. Radiation Lab., Mass. Inst. Tech., 1942-46; asso. prof. physics Duke, Durham, N.C., 1946-48, prof. 1948-—, James B. Duke prof. 1958-—. Vis. prof. physics U. Tex., spring 1958; mem. phys. sci. div. NRC 1954-57. Recipient Science Research award Oak Ridge Inst. Nuclear Studies, 1948. Fellow Am. Phys. Soc. (chmn. Southeastern Sect. 1953-54), A.A.A.S. (council 1955); mem. Radiation Research Soc. (council 1961-64), Nat. Acad. Scis., Sigma Xi. Author: (with W. V. Smith, R. F. Trambarulo) Microwave Spectroscopy, 1953; (with others) Chemical Applications of Spectroscopy, 1956. Asso. editor Jour. Chem. Physics, 1955-58, Spectrochimica Acta, 1957-60. Research interests in infrared spectroscopy, microwave spectroscopy, magnetic resonance, millimeter and submillimeter wave radiation, nuclear moments, molecular structures, chem. physics. Home: 2521 Perkins Rd., Durham, N.C. 27706.*

GORE, George, Brit. electrochemist; b. Bristol, Eng., Jan. 22, 1826; s. George Gore; LL.D., U. Edinburgh (Scotland), 1877; m. Hannah Owen, 1849; 1 son, 1 dau. Apprentice to cooper, circa 1847; migrated to Birmingham, Eng., 1851; became time keeper Soho works, subsequently practitioner in med. galvanism, chemist in phosphorus factory; lectr. chem. and phys. sci. King Edward's Sch., Birmingham, 1870-80; became head Inst. Sci. Research, Birmingham, 1870. Fellow Royal Soc., 1865. Author: The Art of Electrometallurgy, 1877; The Art of Scientific Discovery, 1878; The Scientific Basis of National Progress, 1882; The Electrolytic Separation and Refining of Metals, 1890; The Scientific Basis of Morality, 1899; also papers. Investigated properties of electro deposited antimony, also of liquid carbonic and hydrofluoric acids; improved art of electroplating. Died Dec. 23, 1908.

GORE, Greenville D., Am. mathematician, astronomer; b. Freeman, Mo., June 5, 1899; s. Greenville Francis and Eliza Ellen (Foley) G.; B.A., William Jewell Coll., 1923; S.M., U. Chgo., 1925, Ph.D., 1932; m. Mary Fenley Bryan, Dec. 31, 1932; 1 son, Bryan Frank. Instr., S.D. State Coll., 1925-27; chmn. dept. math. and engring. sci. Central YMCA Coll. (later became Roosevelt U.), Chgo., 1935-45, 1945-—; research asso. Dearborn Obs., Northwestern U., 1945. Staff Northwestern U. Navy V 12 program, 1942-44, dir., 1944. Mem. Math. Assn. Am. (past pres. Ill. sect.), Chgo. Men's Math. Club (past pres.), Sigma Xi, Sigma Nu. Author: (with H. Simmons) Plane and Spherical Trigonometry, 1945. Research, publs. in geometry, especially trigonometry and transformation of surfaces; participated in astronomical research, particularly, cataloguing of faint red stars. Home: 1707 Gilbert Av., Downers Grove, Ill. 60515. Office: 430 So. Michigan Av., Chgo. 60605.*

GORE, John Ellard, Irish astronomer; b. Athlone, Ireland, June 1, 1845; s. John Ribton Gore; ed. privately, engring. degree Trinity Coll., Dublin, Ireland. Joined Indian govt. as asst. engr., constn. Sirkhind Canal, 1869, ret., 1879; astron. observer, Dublin and Sligo, Ireland, from 1879. Fellow Royal Astron. Soc.; mem. Royal Irish Acad. (council), Royal Astron. Soc. Can. (corr.), Brit. Astron. Assn. (v.p., dir. variable star sect.), Liverpool Astron. Soc. Author: Star Groups; Astronomical Lessons; The Visible Universe; The Worlds of Space; The Stellar Heavens; Studies in Astronomy; Catalogue of Known Variable Stars, 1884; Planetary and Stellar Studies, 1888; The Scenery of the Heavens, 1890; An Astronomical Glossary, 1893; Southern Stellar Objects (gives results of own observations with achromatic telescopes of three and four inch apertures). Translator: Popular Astronomy (Flammarion); sidereal portion of Concise Astronomy. Contbr. articles to profl. jours. Discovered several variable stars, computed orbits of numerous binary stars. Died July 18, 1910.

GOREV, Nikolai Nikolaevich, Russian pathophysiologist; b. Apr. 21, 1900; grad. Med. Faculty, Irkutsk U., 1926; D.Med. Sci. Prof. chair gen. pathology Irkutsk U., 1926-31; head chair pathophysiology Khabarovsk Med. Inst., 1931-34; head dept. exptl. pathology Kiev Inst. Exptl. Biology and Pathology, 1934-53; head lab. blood circulation and respiration Bogomolets Inst. Physiology, Ukrainian Acad. Scis., 1953-—; head chair path. physiology Stomatological Faculty, Kiev Bogomolets Med. Inst.; head pathophysiol. lab. Yanovsky Ukrainian Tb Research Inst.; dir. Research Inst. Gerontology and Exptl. Pathology, USSR Acad. Med. Scis., Kiev, 1958-—. Del. 4th European Congress on Allergy, London, 1959. Decorated Order of Lenin. Mem. USSR Acad. Med. Scis. Author: An Outline Study of Hypertonia, 1957. Mem. editorial bd. Physiol. Jour. of Ukrainian Acad. Scis.; mem. editorial council Path. Physiology and Exptl. Therapy. Research and numerous publs. on pathology of cardiovascular system. Address: Gosudarstvenny Med. Inst., b. Tarasa Shevchenko 13, Kiev, Ukraine SSR, USSR.

GORGAS, William Crawford, Am. surgeon; b. Mobile, Ala., Oct. 3, 1854; s. Josiah and Amelia (Gayle) G.; A.B., U. of South, 1875; M.D., Bellevue Hosp. Med. Coll. (N.Y. U.), 1879; several hon. degrees; m. Marie Cook Doughty, Sept. 15, 1885. Intern Bellevue Hosp., 1878-80; apptd. surgeon U. S. Army, 1880; chief san. officer of Havana, 1898-1902; applied methods of combating yellow fever which eliminated that disease in Havana; col. asst. surgeon gen., by spl. act of Congress, for work at Havana, 1903; major gen., surgeon gen. U. S. Army, Mar. 4, 1915; ret., 1918; dir. yellow fever research Rockefeller Found. Apptd. chief san. officer Panama Canal, 1904, eliminated malaria and yellow fever, thus making possible work on the canal; mem. Isthmian Canal Commn., 1907-—. Recipient Mary Kingsley medal Liverpool Sch. Tropical Medicine, 1907; Gold medal Am. Mus. Safety, 1914; elected to Hall of Fame for Gt. Americans, 1950. Worked to control yellow fever in Ecuador, Panama, and Cuba. Died July 4, 1920.

GÖRGES, Johannes (Hans) Friedrich Heinrich, German physicist; b. Lüneburg, Germany, Sept. 21, 1859; s. Christian Wilhelm Ferdinand and Marie (Meyer) G.; studied math. and physics, Göttingen, Germany; hon. dr. engring. Technische U. Berlin; m. Marie Krichel-dorff; 2 daus. Joined firm Siemens & Halske, Berlin, 1884, rose to position of dir. Charlottenburg works; became lectr. electrotechnics Technische U., Dresden, Germany, 1901, remained until 1930, his Inst. for Strong Current Tech. was a model of its type; relocated Aue, Germany, 1945. Named hon. citizen Technische U. Stuttgart, Germany. Author: Grundzüge der

Elektrotechnik, 1913; also articles. Used math. and geometry for single phase and multiphase alternating current; introduced Görges vector diagram of transformer; research on field distbn. in A.C. motors, parallel action of synchronous engines, motors especially single phase induction motor. Died Aue, Oct. 7, 1946.

GORHAM, John, Am. physician, chemist; b. Boston, Feb. 24, 1783; s. Stephen and Molly (White) G.; B.A., Harvard, 1801, M.B., 1804, M.D., 1811; studied exptl. chemistry under Friedrich Accum, London, Eng.; m. Mary Warren, June 2, 1808. Adj. prof. chemistry and materia medica Harvard, 1809-15, Ewing prof. chemistry and mineralogy, 1815-27; librarian Mass. Med. Soc., 1814-18; ret., 1827. Fellow Am. Acad. Arts and Scis. Author: The Elements of Chemical Science (1st systematic chemistry textbook written by an Am. and published in U. S.), 2 vols., 1819-20; also articles on medicine and chemistry. Died Mar. 27, 1829.

GORIA, Carlo Angelo Secondo, Italian chemist; b. Cheri, Turin, Italy, Nov. 3, 1910; s. Francesco and Emma Goria; Ph.D. in chemistry; m. Carla Repetto, 1938; children—Francesco, Carlo-Alberto. Prof. applied chemistry, 1939, U. Palermo, 1954-56; prof. gen. and applied chemistry Poly. of Turin, 1956. Research and numerous publs. on applied chemistry, constrn. materials. Home: 13 via Torricelli. Office: Castello del Valentino, Turin, Italy.

GORIAEV, Mikhail Ivanovich, Russian biochemist; b. Sept. 30, 1904. Prof. Zoo-Vet. Inst. Alma-Ata, 1938—, also head Biochem. Lab.; dir. Inst. Metallurgy, Chemistry and Bldg. Materials, Kazakh Affiliate, USSR Acad. Scis., 1942—, v.p. Kazakh SSR Acad., 1946-56, also chief lab. for plant chemistry, Inst. Chem. Scis., 1946-56. Work in bio-chemistry of milk and milk products and biochemistry of plants.

GORINI, Costantino, Italian microbiologist; b. Rimini, Italy, Jan. 9, 1865; M.D., 1890; aggregate physician U. Pavia (Italy); prof. hygiene and bacteriology Agrarian Faculty of Milan (Italy); microbiologist Rome Dept. Health. Mem. French Acad. Scis., 1939. Research on morphology, culture, physiology of microbes in milk. Died Milan, Sept. 3, 1950.

GORINOV, Aleksandr Vasilevich, Russian engr.; b. Aug. 4, 1902; ed. Moscow Inst. Communication Engrs. Chief-constrn. engr. Moscow-Donbass Railroad; with Leningrad Inst. Railroad Engrs., 1931-46, Military-Transport Acad., 1932-38, Moscow Inst. of Railroad Engrs.; became prof., 1941. Corr. mem. USSR Acad. Scis. Author: Classification of USSR Railroads, 1946; The Planning of Railroads, 1948; A Powerful Reserve for Increasing the Weight of Trains Along Entire Lines, 1954; Problems of Planning Railroads with Electric and Steam Traction, 1959; The Development of All Forms of Transport as component Parts of the Integral Transport Network of the USSR, 1962; also others. Research on complex designs of railroads, theory of inertia calculations. reserve use of train's kinetic energy, improvement of transportation with gradually increasing railroad power. Address: Moscow Inst. of Railroad Engineers, Moscow, Russia.

GORIZONTOV, Petr Dmitrievich, Russian pathophysiologist; b. 1902; grad. Omsk Med. Inst., 1927; D.Med. Sci. Asst. path. physiology 2d Moscow Med. Inst., 1928-33; asst., later lectr. chair path. physiology Sechenov 1st Moscow Med. Inst., 1934-38, prof., 1939-52; prof., head chair path. physiology Moscow Central Postgrad. Med. Inst., USSR Ministry Health, 1953—; head pathophysiol. lab., dep. dir. for sci. work Inst. Biophysics, USSR Acad. Med. Sci. Decorated Order of Lenin, 1961. Mem. USSR Acad. Med. Sci., All-Union Soc. Pathophysiologists (chmn.). Author: The Pathological Physiology of Radiation Lesions, 1955. Dep. editor Archives Pathology; mem. editorial bd. Med. Radiology, Path. Physiology and Exptl. Therapy; co-editor Pathology and Morphology sects. Large Med. Ency., 2d edit. Research and publs. on significance of brain in cholesterol metabolism. Address: Moscow Central Postgrad. Med. Inst., pl. Vosstaniya 1-2, Moscow, USSR.

GORKIN, Vladimir Zinovievich, Russian biochemist; b. Kharkov, USSR, Nov. 2, 1927; s. Zinovii Davidovich and Frida (Guerbich) G.; ed. Kharkov Med. Sch., 1944-49; postgrad. student Inst. Biol. and Med. Chemistry, lab. of Prof. A. E. Braunstein, 1949-52; candidate med. sci., 1952; D.Biol. Scis., 1964; m. Lyna Levitina, Sept. 3, 1950; 1 son, Alexander. Mem. staff biochem. lab. Inst. Orthopaedics and Traumatology, 1953-56; research fellow Inst. Biol. and Med. Chemistry, Acad. Med. Scis. USSR, 1956-59, sr. research fellow, 1959-61, head dept. biochemistry, 1961—, vice dir., inst., 1964—; lectr. selected topics of enzymology Moscow State U., 1963—. Mem. Moscow br. All-Union Biochem. Soc. Author articles, chpts. in books. Research on biochem. aspects of inflammation; methodological works on zonal electrophoresis; research biochemistry and biochem. pharmacology of biogenic monoamines (nature of mitochondrial monoamine oxidases, mechanism of action of monoamine oxidase inhibitors). Home: 57/65 Novoslobodskaya, Moscow A-55. Office: 10 Pogodinskaya, Moscow, USSR.*

GORLAEUS, see van Goorle, David.

GORLENKO, Mikhail Vladimirovich, Russian phytopathologist, mycologist; b. Vladimir, USSR, June 12, 1908; s. Vladimir Mikhailovich and Josephina Gorlenko; diploma Voronezh (USSR) State U., 1930, candidate of sci., 1936; D.Sc., Moscow State U., 1947; m. Tretyakova Elena Gorlenko, Dec. 15, 1932; children—Josephina, Vladimir. Sr. research phytopathologist Voronzh Research Sta. for Plant Protection, Voronezh, 1929-41; dir. Moscow Research Sta. Plant Protection, 1941-55; head dept. lower plants Moscow State U., 1955—; cons. Ministry of Agr. Decorated Order Lenin, Order Red Banner Labor, Badge of Honor. Mem. Moscow Soc. Naturalists, All-Union Bot. Soc., All-Union Microbiology Soc. Author: Rust of Cereals, 1948; Diseases of Wheat, 1951; Diseases of Plants and Environments, 1950. Plant Immunity, 1962; Bacterial Diseases of Plants, 1966. Research on modes of wintering of wheat powdery mildew, biology of pathogens of wheat black bacteriosis; control measures against rust of cereals; control measures and biology of pathogens of potato blackleg; bacteriosis cabbage and cucumbers; discovered that majority of plant pathogenic bacteria could not survive in soil and died out under the action of antagonists; regularities of plant pathogen distbn. Home: Bldg. U, 58, Lenin Hills, Moscow B-234. Office: Dept. Lower Plants, Moscow State U., Moscow B-234, USSR.

GORLIN, Richard, Am. cardiologist; b. Jersey City, June 30, 1926; s. S. G. and Henrietta (Bernfield) G.; M.D., Harvard, 1948; m. Marjorie Shore, Apr. 16, 1960; children—Wendy Elizabeth, William Barry, Douglas James. Faculty Harvard Med. Sch., 1956—, asst. prof. medicine, 1961—; physician Peter Bent Brigham Hosp., Boston, 1967—, dir. cardiovascular research lab., 1957—; dir. cardiovascular unit, 1967—. Diplomate Am. Bd. Internal Medicine, Nat. Bd. Med. Examiners. Fellow Am. Coll. Physicians, Am. Coll. Cardiology (trustee); mem. Am. Fedn. Clin. Research (past nat. councillor), Am. Heart Assn., Am. Physiol. Soc., Am. Soc. Clin. Investigation, Assn. Am. Physicians, Assn. U. Cardiologists, Internat. Soc. Internal Medicine, New Eng. Cardiovascular Soc. (pres.), Royal Soc. Medicine. Research, publs. on pathophysiology diagnosis, treatment of coronary arterial disease, congestive heart failure; influence of drugs and exercise on cardiac hemodynamics and body metabolism; measurement of heart function in normal and disease states. Home: 87 Gray Cliff Rd., Newton, Mass. 02159.*

GORMSEN, Svend Theodore, mathematician, educator; b. Odense, Denmark, Feb. 24, 1909; s. Hans Peder and Anna (Jensen) G.; came to U. S. 1927, naturalized 1932; B.S., Ohio State U., 1935; M.S., U. Fla., 1949, Ph.D., 1953; m. Virginia Margaret Warden, June 25, 1936; children—James Warden, Gayle Patricia. Tchr. pub. schs. N.Y., Ohio, 1935-43; asst. prof. math. U. Fla., 1947-54; prof. Rollins Coll., 1954 56, Va. Poly. Inst., 1956—. Served to Lt. Cdr. USNR, 1943-46. Mem. Math. Assn. Am., Am. Math. Soc., A.A.U.P., Pi Mu Epsilon, Sigma Xi. Made maps of certain algebraic curves invariant under cyclic involutions of periods three-five-seven. Home: 702 Preston Av., Blacksburg, Va. 24060.*

GORNALL, Allan Godfrey, Canadian biochemist; b. River Hebert, N.S., Can., Aug. 28, 1914; s. Herbert Thomas and Lucy A. (Markham) G.; B.A., Mt. Allison U., 1936; Ph.D., U. Toronto, 1941; m. Mary Elizabeth Sheila Stewart, Dec. 27, 1941; children—William Stewart, Douglas Allan, Thomas Herbert, Catherine Ann. Faculty, U. Toronto (Ont., Can.), 1946—, prof., 1963—, chmn. dept. path. chemistry, 1966—, cons. clin. biochemistry to hosps. in Faculty Medicine. Served to lt. comdr. Royal Canadian Navy, 1942-46. Recipient Reeve prize, 1941; Nuffield scholar, 1949. Fellow Royal Soc. Can.; mem. Am. Chem. Soc., Endocrine Soc., Canadian Biochem. Soc., Canadian Physiol. Soc. (past treas.), Canadian Soc. Clin. Chemists, Canadian Soc. Clin. Investigation, Canadian Fedn. Biol. Socs. (past hon. treas.), Biochem. Soc. Gt. Britain. Research and publs. on intermediates in urea synthesis, liver function tests, nutrition and liver injury, methods for serum proteins, urinary corticosteroids, aldosterone; aldosterone prodn. in pregnancy, hypertension, liver disease; aldosterone in exptl. hypertension, responses to renin and angiotensin, metabolism of heart muscle. Home: 135 Hanna Rd., Toronto 17, Ont., Can.*

GORODETSKY, Aleksandr Afanasevich, Russian radiologist; b. Novo-Kostichi (now Kuybyshev Oblast), 1897; grad. Med. Faculty, Saratov U., 1924; D.Med. Sci., 1940. Physician, Urals Oblast, 1924-34; instr. radiology Bashkir Med. Inst., Ufa, 1935-41; dep. sci. dir. Kiev Inst. Radiology and Oncology, later head radiology dep. Bogomolets Inst. Exptl. Biology and Pathology, 1944-49; chief radiologist Ukrainian Ministry Health, 1945-50, now dir. radiology dept. Inst. Exptl. Biology and Pathology, also chmn. radiology commn.; dir. chair radiology Kiev Postgrad. Med. Inst., 1944—; head biophysics lab. Bogomolets Inst. Physiology, Ukrainian Acad. Sci., 1953—. Decorated Order of Lenin. Mem. Ukrainian Acad. Sci. (corr.), Ukrainian (chmn. 1956—), Kiev (chmn. 1945—) socs. radiologists. Mem. editorial council Vestnik Roentgenology and Radiology. Research and numerous publs. on therapy and radiology, developer methods of treatment including radiotherapy of bullet wounds with diagnostic device, single large dose irradiation of patients with breast cancer, also combined treatment of breast cancer, radiation lesions, refined indications for radioactive phosphorus treatment of tumors. Address: Kiev Postgrad. Med. Inst., ulitsa 9-go yanvarya, Kiev, Ukraine SSR, USSR.

GORODETZKY, Serge, French physicist; b. Montpellier, France, Apr. 16, 1907; s. Grégoire and Perel (Krementchoutsky) G.; Dr. ès sc., U. Paris, Sorbonne; m. Reine Guesnon, Mar. 26, 1936; children—Philippe, Françoise, Elisabeth. Staff, Laboratoire de l'Institut de Radium, Laboratoire de Broglie; dir. Institut de Recherches Nucléaires, Strasbourg, France; examiner student l'École Polytechnique. Decorated officer la Légion d'Honneur; commandeur l'Ordre National du Mérite; officier de l'Instruction Publique. Mem. Am. Phys. Soc., Société Française de Physique. Research, numerous publs. on radioactivity, cosmic radiation, especially on measurement of mass of cosmic ray muons, nuclear physics. Home: 4 rue Jacques Kablé, Office: Institut de Recherches Nucléaires, B.P. 16, 67-Strasbourg-Gronenbourg, France.*

GORSKY, Ivan Ivanovich, Russian paleontologist; b. Sept. 12, 1893. Became prof. Leningrad Mining Inst., 1935; dir. All-Union Sci. Research Inst. Geology, 1943-47; chmn. Karelo-Finnish br. USSR Acad. Scis., 1947-52; dir. USSR Acad. Scis. Lab. on Geology, 1950; chmn. All-Union Paleontol. Soc., 1954. Corr. mem. USSR Acad. Scis. Author: Corals from the Lower Coal Deposits of the Kirghiz Steppe, 1932; Geological Survey of the Kizel Region, 1932; Geotectonic Conditions for the Formation of the Ural Coal Deposits and Allied Features of the Geological Structure of the Deposits, 1943; The History of Coal Formation on USSR Territory, 1956; also numerous others. Research on geology of Urals, particularly coal deposits, also stratigraphy and tectonics of Urals, Kazakhstan, other regions of USSR; compiled maps of various parts of USSR. Address: Dept. Geology and Geography, Presidium, USSR Acad. Scis., Lenin Prospekt, Moscow, USSR.*

GORTER, Cornelis Jacobus, Dutch physicist; b. Utrecht, Netherlands, Aug. 14, 1907; s. Harmanus Johan and Anne (van Eck) G.; Ph.D., U. Leyden (Netherlands), 1932; D.Sc., U. Grenoble (France), 1955, U. Paris (France), 1963, U. Nancy (France), 1966; LL.D., U. Halifax (N.S., Can.), 1960; m. Lilla Catharina Elisabeth Charlotte von Krogh, June 27, 1938; children—Fridtjof, Herman, Annekari, Lilla. Conservator, Teyler's Stichting, Haarlem, Netherlands, 1931-36; reader U. Groningen (Netherlands), 1936-40; prof. U. Amsterdam (Netherlands), 1940-46; prof. physics U. Leiden, 1946—, dir. Kamerlingh Onnes Lab., 1946—. Pres., Found. for Research on Matter, 1954-60. Decorated chevalier Lion of Netherlands, commdr. Order of Merit in Research and Invention (France). Mem. Nat. Acad. Scis. (U. S.), Royal Netherlands Acad. Scis. (pres. 1906-66), Royal Swedish Acad. Sci., Royal Flemish Acad. Scis., Finnish Acad. Scis., Am. Acad. Arts and Scis. Author: Paramagnetische Eigenschaften von Salzen, 1932; Paramagnetic Relaxation, 1947; also numerous articles. Editor: Progress in Low Temperature Physics, I, 1955, II, 1957, III, 1961, IV, 1964, V, 1967. Research on thermodynamical treatment of superconductivity; discovered paramagnetic relaxation, antiferromagnetic resonance. Home: 3 Burggravenlaan, Leyden, Netherlands. Office: 18 Nieuwsteeg, Leyden, Netherlands.*

GORTNER, Ross Aiken, Am. biol. chemist; b. nr. O'Neill, Neb., Mar. 20, 1885; s. Joseph Ross and Louisa E. (Waters) G.; B.S., Neb. Wesleyan U., 1907; M.A., U. Toronto, 1908; Ph.D., Columbia, 1909; Sc.D., Lawrence Coll., 1932; m. Catherine V. Willis, Aug. 4, 1909 (dec. 1930); children—Elora Catherine, Ross Aiken, Willis Alway, Alice Louise; m. 2d, Rachel Rude, Jan. 12, 1931. Research asst. in agrl. chemistry U. Neb., 1906-07; asst. in chemistry Faculty Arts, U. Toronto, 1907-08; resident investigator in biol. chemistry Sta. for Exptl. Evolution, Carnegie Instn. of Washington, Cold Spring Harbor, N.Y., 1909-1914; asso. prof. soil chemistry U. Minn., 1914-16, asso. prof. agr. biochemistry, 1916-17, prof. agrl. biochemistry and chief of div. of agrl. biochemistry U. Minn. and Minn. Agrl. Expt. Sta., since 1917. Cons., Chem. Warfare Service, U. S. Army, since 1926; mem. NRC; chmn. U. S. com. Internat. Com. on Biochem. Nomenclature of Union of Pure and Applied Chemistry, 1930-37. Fellow A.A.A.S.; mem. Am. Chem. Soc. (councillor 1918-25, 29-39; vice chmn. and sec. 1919; chmn. biol. div. 1920; sec. colloid div., 1929; chmn. 1931), Nat. Acad. Scis., Am. Soc. Biol. Chemists, Soc. Exptl. Biology and Medicine, Am. Soc. Naturalists (pres. 1932), Sigma Xi (exec. com. 1936-40). Author: Outlines of Biochemistry, 1929, 2d edit., 1938; J. Arthur Harris, Botanist and Biometrician (with others), 1936; Selected Topics in Colloid Chemistry, 1937; also extensive contbr. on topics pertaining to biol. chemistry. Asso. editor Jour. Am. Chem. Soc., Jour. Phys. Chemistry, 1929-30, 34-35; asst. editor Chem. Abstracts (zoology). Died Sept. 30, 1942.

GORTNER, Ross Aiken, Jr., biochemist; b. Cold Spring Harbor, L.I., N.Y., June 2, 1912; s. Ross Aiken and Catherine (Willis) G.; student Oberlin Coll., 1929-30; A.B., U. Minn., 1933, M.A., 1934; Ph.D., U. Mich., 1937; M.A. (hon.), Wesleyan U., 1948; m. Mary Priscilla Cahill, Dec. 20, 1938; children—Catherine Clarke (Mrs. David R. Singleton), Douglas Ross.

Grad. teaching asst. U. Mich., 1935-37; faculty Wesleyan U., Middletown, Conn., 1937—, prof. biochemistry, 1948—. Fulbright lectr., Copenhagen, Denmark, 1954-55; vis. research prof. U. Giessen, Max. Planck Inst. Biochemistry, Munich, Germany, 1961-62; mem. bd. control Conn. Agr. Expt. Sta., New Haven, 1964—. Program dir. sci. curriculum improvement NSF, 1966-67. Fellow A.A.A.S.; mem. Am. Chem. Soc., Am. Inst. Nutrition, Conn. Nutrition Council, Phi Beta Kappa, Sigma Xi, Phi Lambda Upsilon, Gamma Alpha, Alpha Chi Sigma. Author: (with W. A. Gortner) Outlines of Biochemistry, 1949; (with P. E. Marsh) Federal Aid to Science Education-Two Programs, 1963. Contbr. articles to sci. jours. Research on nutritional aspects of selenium poisoning, action of acidic foods and beverages on dental hard tissues, protein nutrition, biochem. aspects of Vitamin E deficiency. Home: 261 Washington Terrace, Middletown, Conn. 06457.*

GORTNER, Willis Alway, Am. biochemist; b. Cold Spring Harbor, N.Y., Dec. 20, 1913; s. Ross Aiken and Catherine (Willis) G.; B.A. magna cum laude, U. Minn., 1934; Ph.D., in Biochemistry, U. Rochester, 1940; m. Susan Leet Reichert, Aug. 25, 1960; children—Willis Alway, II, David Allen, Catherine Willis, Frederick Aiken. Research chemist, Gen. Mills, Inc., Mpls., 1934-37, 40-42; faculty Cornell U., Ithaca, 1943-48; head chem. dept. Pineapple Research Inst. Hawaii, Honolulu, 1948-64; dir. human nutrition research div. Agrl. Research Service, U. S. Dept. Agr., 1964—; with Bikini Sci. Resurvey Team, 1947, Bjorksten Research Found., Madison, Wis., 1953, NRC-Nat. Acad. Scis., 1957, Nat. Canners Assn. Research Lab., 1960-61, dept. nutritional scis. U. Cal. at Berkeley, 1963; affiliate grad. faculty U. Hawaii, 1956-64. Recipient Thomas Andrews award, U. Minn., 1934. Fellow A.A.A.S.; mem. Am. Soc. Biol. Chemists, Am. Inst. Nutrition, Am. Chem. Soc., Inst. Food Technologists, Sigma Xi, Phi Lambda Upsilon, Alpha Chi Sigma. Author: (with F. S. Erdman, N. K. Masterman) Principles of Food Freezing, 1948; also numerous articles. Coeditor, author Outlines of Biochemistry, 1949. Patentee plant enzyme preparations, fruit senescence inhibitors. Research on effect freezing and storage conditions on nutritional value and quality foods; established needs enzyme system involved in plant hormone metabolism; demonstrated usefulness biochem. basis for more precise and meaningful horticultural terminology on fruit devel. Home: 12701 Lacy Dr., Meadowood, Silver Spring, Md. 20904. Office: Agrl. Research Center, Beltsville, Md. 20705.*

GORZ, Herman Jacob, Am. geneticist; b. Eagle River, Wis., Nov. 22, 1920; s. Louis and Katherine (Bartold) G.; student LaCrosse (Wis.) State Tchrs. Coll., 1938-39; B.S., U. Wis., 1942, M.S., 1948, Ph.D., 1951; m. Jeanette Muriel Gundlach, Sept. 8, 1951; children—Marily Jean, Jean Marie. Faculty, N.D. State U., 1951-54; research geneticist Crops Research div. Agrl. Research Service, U. S. Dept. Agr., prof. agronomy U. Neb., Lincoln, 1954—. Mem. Genetics Soc. Am., Am. Soc. Agronomy, Sigma Xi, Gamma Sigma Delta, Alpha Zeta. Contbr. chpts. to Germ Plasm Resources, 1961; Forages, 1962; Advances in Agronomy, 1965; Proc. of 10th Internat. Grassland Congress, 1966. Research in genetic control of coumarin biosynthesis in Melilotus, barriers to interspecific hybridization within the genus Melilotus, resistance to insects in Melilotus, inheritance of various characters in sweetclover, selection for improved characteristics in lines of sweetclover. Home: 6126 Leighton Av., Lincoln, Neb. 68507.*

GOSAR, Peter, Yugoslavian physicist; b. Ljubljana, Yugoslavia, Oct. 15, 1923; s. Andrej and Antonija (Mihevc) G.; B.S. in Physics, U. Ljubljana, 1951, Ph.D., 1956; m. Danila Nemec, June 30, 1956; children—Alenka, Tomaz, Andrej. Research asso. Solid State Lab., Inst. for Telecommunications, Ljubljana, 1953-57; postdoctoral fellow Laboratoire du Magnétisme, CNRS, Bellevue, France, 1957-59; sr. scientist J. Stefan Inst., Ljubljana, 1959-64; research asso. physics dept. U. N.C., Chapel Hill, 1964-66; prof. physics U. Ljubljana, 1966—. Mem. Sigma Xi. Research, publs. on surface photo-effect in semiconductors, theory of propagation of light through inhomogeneous media, proton motion in ice crystals, impurity conduction and other relaxation phenomena in solids. Home: 21 Mirje, Ljubljana, Yugoslavia.*

GOSLAR, Hans Günter, German anatomist; b. Cologne, Germany, Dec. 28, 1918; s. Julio and Christine (Waimann) G.; ed. U. Cologne, U. Bonn; M.D.; m. Hilde Fortmann, 1947; 1 dau., Ingeborg. Prof. anatomy U. Bonn. Mem. Royal Soc. Medicine (Eng.), Soc. Anatomy, Soc. Histochemistry, Soc. Chemists of Germany, German Soc. Endocrinology, European Soc. Comparative Endocrinology. Research and publs. on cytology of neuro-vegetative secretion, active tissues of thymus, thymus and endocrinology, histochem. research on moulting cycle of reptiles, histochem. research methods, histochemistry of reproductive system. Home: 23 Drachenfelsstrasse 23, Duisdorf bei Bonn. Office: Anatomisches Institut, University of Bonn, Nussallee 10, 53 Bonn, Germany.

GOSS, Charles Mayo, Am. anatomist; b. Peoria, Ill., Feb. 16, 1899; s. Charles E. and Frances Wade (Mayo) G.; A.B., Yale, 1921, M.D. cum laude, 1926; m. Josephine Cowell, Aug. 14, 1928; children—Elizabeth Cowell (Mrs. Henry Chodkowski), Frances Mayo (Mrs. Luis J. Vergne), Marianna Cowell. Faculty Yale, 1926-29, Columbia, 1929-38; prof., head anatomy U. Ala., 1938-47; prof., head dept. anatomy La. State U., New Orleans, 1947-65, prof. anatomy, 1965—; vis. prof. anatomy George Washington U., 1966—. Guggenheim fellow, 1956. Distinguished Alumnus Bradley U., 1956. Mem. Am. Assn. Anatomists (past pres.), Am. Assn. Phys. Anthropologists, Am. Assn. History Medicine, Sigma Xi, Alpha Omega Alpha, Omicron Delta Kappa. Editor: Gray's Anatomy, 1948, 54, 59, 66; mng. editor Anat. Record, 1948—. Contbr. numerous articles to profl. jours. Research and publs. primarily in field of devel. of heart in very early mammalian embryos, particularly rats and squirrel monkeys; translations of Ancient Greek texts. Home: 7809 Moorland Lane, Bethesda, Md. 20014. Office: 1335 H St., Washington 20005.*

GOSSE, Henri-Albert, Swiss chemist, pharmacist; b. Geneva, Switzerland, May 25, 1753; master apothecary, Geneva; recipient Gold medal Paris Coll. Pharmacy, 1781; mem. French Acad. Scis., Geneva Soc. Physics (founder). Studied diseases of metal workers; 1st substitution of hydrogen gas for caloric gas in aerostat; invented artificial mineral water. Died Geneva, Feb. 1, 1816.

GOSSE, Philip Henry, Brit. ornithologist; b. Worcester, Eng., Apr. 6, 1810; s. Thomas Gosse; ed. Poole and Blandford, Eng.; became clk. in whaler's office; farmed in Can.; traveled in U. S.; taught in Ala.; returned to Eng., 1839; sent to Jamaica on ornithol. expdn. by Brit. Mus., 1844-46. Fellow Royal Soc., 1865. Author numerous books on zoology including: The Canadian Naturalist, 1840; Introduction to Zoology, 1843; Birds of Jamaica, 1847-49; A Naturalist's Sojourn in Jamaica, 1851. Died Aug. 23, 1888.

GOSSELIN, Athanase Léon, French surgeon; b. Paris, June 16, 1815; M.D., 1843. Surgeon, Lourcine Hosp., from 1851; prof. pathology Paris Faculty Medicine, from 1858. Mem. French Acad. Scis., 1874 (pres. 1887), Acad. Medicine, Soc. Surgery. Author: Hernies étranglées; hémorroides de l'orchite parenchymateuse, 1852. Devised new treatment for hernia and hemorrhoids, studied orchitis in smallpox and diseases of eye. Died Paris, Apr. 30, 1887.

GOSSELIN, Richard P., Am. mathematician; b. Springfield, Mass., June 29, 1921; s. A. Edmond and Grace (Pettengill) G.; B.S., U. Chgo., 1944, Ph.D., 1951; M.A., U. Rochester, 1948; m. Jean R. Wenneis, May 13, 1944; children—Philip W., Janet R. Prof., Youngstown Coll., 1952-55; asst. prof. U. Conn., 1955-58, asso. prof., 1958-61, prof., 1961—. Mem. Am. Math. Soc., Math. Assn. Am. Contbr. articles to sci. jours. Research in Fourier analysis; trigonometric series; localization theory. Office: Dept. Math., U. Conn., Storrs, Conn. 06268.*

GOSSELIN, Robert Edmond, Am. pharmacologist; b. Springfield, Mass., Sept. 2, 1919; s. A. Edmond and Grace (Pettengill) G.; A.B., Brown U., 1941; Ph.D., in Physiology, U. Rochester, 1945, M.D., 1947; m. Ruth Lackmann Smith, June 26, 1948; children—Peter G., Andrea L. Instr., U. Rochester Sch. Medicine, 1948-52, asst. prof. pharmacology, 1954-56; jr. scientist Rochester Atomic Energy Project, 1948-52, 54-56; prof. pharmacology, chmn. dept. Dartmouth Med. Sch., 1956—; clin. toxicologist, dir. poison information center Mary Hitchcock Meml. Hosp., Hanover, N.H. Mem. toxicology study sect. USPHS, 1964—; mem. adv. com. on safety evaluation FDA, 1966—. Mem. Soc. Pharmacology and Exptl. Therapeutics, Am. Physiol. Soc., A.A.A.S., N.Y. Acad. Scis., Toxicology Soc., Am. Assn. Poison Control Centers. Author: (with Gleason, Hodge) Clinical Toxicology of Commercial Products, 1957; also numerous articles. Research in clin. toxicology, devel. exptl. antidotes, control of ciliary motion. Home: Elm St., Norwich, Vt. 05055. Office: dept. Pharmacology, Dartmouth Med. Sch., Hanover, N.H. 03755.*

GOSSET, Montague, Brit. surgeon; b. Tanner's End, Edmonton, Eng., July 1, 1792; s. Daniel Gosset; diploma Guy's Hosp., 1814; 8 children. Apprentice to Stocker, Guy's Hosp., from 1809; lived in Scotland, 1815-1817; cons. surgeon, Great George Street, Westminster, Eng., 1819, London, for 34 years. Hon. fellow Royal Coll. Surgeons. 1st to detect renal aneurysm, 1829; described improved tonsil iron, 1835; demonstrated use of nitric acid for destruction of nevi; an introducer of 2 instruments for dividing strictures of urethra, 1818. Died, Oct. 21, 1854.

GOSSET, William Sealy (known as Student), Brit. indsl. statistician; b. Canterbury, Eng., June 13, 1876; s. Frederick and Agnes (Vidal) G.; ed. Winchester Coll., New Coll., Oxford U., Karl Pearson's lab. U. Coll., London; m. Marjory Surtees Phillpotts, 1906; 1 son, 2 daus. Sci. staff Arthur Guinness, Son and Co., Dublin, Ireland; charge Guinness Brewery, London. Author: Student's Collected Papers (pub. in Biometrika, edited by E. S. Pearson, J. Wishart), 1942. Applied statis. methods to exptl. and routine work of large-scale industry; introduced new math. approach to decision-making by analysis of relatively small numbers; performed (with R. A. Fisher and E. S. Beaven) agrl. expts. involving barley. Died Beaconsfield, Eng., Oct. 16, 1937.

GÖSSWALD, Karl, German zoologist; b. Würzburg, Jan. 26, 1907; s. Max and Margarete (Geyer) G.; ed. Univ. Gymnasium, Würzburg; Ph.D. in zoology; m. Johanna Traumann, 1937; children—Karl, Barbara. Asst., U. Munich, Munich Exptl. Sta. Arboriculture, Viticulture and Horticulture, Newstadt; sci. asst., head service biology in Berlin, 1935; mem. governing council, dept. head Exptl. Sta. Forestry Econs., Eberswalde, 1942; prof. agrégé, 1947; asso. prof., council mem., dir. Inst. Visible Zoology, U. Würzburg, 1948. Head Office Control of 1st Matters; mem. dir.'s council German dept. Internat. Union for Study Social Insects. Mem. Internat. Soc. Bioclimatology and Biometeorology (sci. council), Internat. Soc. Nutritional Research. Author: Die Rote Waldemeise, Bedeut., Nutz. und Zucht in Forstw., 1949; Aculeata. Editor: Waldhygiene Jour.; co-editor: Angew. Zool., Insectes Sociaux, Jour. Bioclimatology and Biometeorology, also numerous articles. Home: 14a Scheffelstrasse. Office: 10 Rontgenring, Würzburg, Germany.

GOSTING, Louis Joseph, Am. phys. chemist; b. Kildare, Okla., Nov. 10, 1921; s. Ralph Louis and Jessie (McKinney) G.; A.B., Southwestern Coll., Winfield, Kan., 1943, D.Sc., 1967; Ph.D., U. Wis., 1948; m. Dorothy Mae Clark, Sept. 21, 1947. Project asso. U. Wis., 1947-48, faculty 1950—, prof. Inst. for Enzyme Research, 1963—; NRC fellow Rockefeller Inst. for Med. Research, N.Y.C., 1948-49; du Pont fellow Yale, 1949-50. Recipient Research Career award NIH, 1963. Fellow A.A.A.S.; mem. Am. Chem. Soc., N.Y. Acad. Scis., Am. Assn. U. Profs., Wis. Acad. Scis., Arts, and Letters, Sigma Xi. Publs. on measurement and interpretation of transport of substances in solution by diffusion, electrophoresis and sedimentation. Home: 4830 S. Hill Dr., Madison, Wis. 53705.*

GOSZCZYNSKI, Stefan, Polish chemist; b. Radomsko, Poland, Apr. 14, 1924; s. Tadeusz and Zofia (Nowak) G.; grad. Inst. Tech., Gliwice, Poland, 1950, Ph.D., 1960; m. Hanna Jaroslawska, June 28, 1953; children—Peter, Thomas. Sci. worker Silesian Inst. Tech. dept. organic chemistry, 1948—; postdoctoral Brit. Council scholar Birmingham (Eng.) U., 1960-61; reader organic chemistry Inst. Tech., Gliwice, 1964—. Mem. Polish Chem. Soc. Research, publs. on new method of synthesis of quinoline nucleus by cyclization (ring closure) of beta-aryl-alpha, beta-unsaturated oximes resulting in formation of carbon-nitrogen bond, analytical problems in polyamide fibres manufacture. Home: 42 Dworcowa, Gliwice, Poland.*

GOTAAS, Harold Benedict, Am. civil engr.; b. Mellette, S.D., Sept. 3, 1906; s. Halfdan and Emma (Cady) G.; B.S., U. S.D., 1928, D.Sc., 1955; M.S., Ia. State Coll., 1930; M.S., Harvard, 1937, Sc.D., 1942; m. Alice E. McLaughlin, Apr. 11, 1931; 1 son, Richard M. Faculty, U. Cal., Berkeley, 1946-57; prof. civil engring., dean Technol. Inst. Northwestern U., Evanston, Ill., 1957—, Walter P. Murphy prof., 1967—; cons. WHO, 1954—; chmn. panel on closed ecol. systems NRC, 1959-61; mem. engring. panel NSF, 1960-63. Recipient Kenneth Allen award, 1946; Harrison P. Eddy medal, 1954; James R. Croes medal, 1958; Rudolph Hering medal, 1958; named Engr. of Year, Ill., 1961. Mem. Nat. Acad. Engring., Am. Soc. C.E., Am. Pub. Health Assn., Am. Water Works Assn., Am. Soc. Engring. Edn., Inter-Am. Assn. San. Engring. (past pres. U. S. sect.), Nat. Soc. Profl. Engrs., Water Pollution Control Fedn. (past pres. Cal. sect.), Am. Acad. Environmental Engrs., Western Soc. Engrs., Sigma Xi, Phi Kappa Phi, Tau Beta Pi, Delta Omega, Chi Epsilon. Author: Composting of Organic Wastes for Sanitary Disposal and Reclamation, 1956. Contbr. numerous articles to profl. jours. Research in environment control, especially san. and municipal engring. and planning, photosynthetic use of organic matter in waste waters; studies of econ. progress in under-developed countries. Home: 618 Colfax St., Evanston, Ill. 60201.*

GOTCH, Francis, English physiologist; b. Bristol, Eng., 1853; s. Rev. Dr. Gotch; B.A., B.Sc., London (Eng.) U.; D.Sc., M.A., U. Oxford (Eng.); LL.D., St. Andrews; D.Sc. (hon.), Liverpool, Eng.; m. Rosamund B. Horsley, 1887. Became demonstrator in physiol. lab. Oxford, 1883; named Holt prof. physiology Univ. Coll., Liverpool, 1891; Waynflete prof. physiology, Oxford, from 1895; fellow Magdalen Coll.; Oxford. Croonian lectr., 1891; mem. departmental com. on sight tests for mercantile marine Bd. Trade, 1910-12. Fellow Royal Soc., 1892; mem. Royal Coll. Surgeons, Soc. Biology (corr. Paris, France). Contbr. papers on physiology of nerve, retina muscles, elec. organs to sci. jours. Demonstrated (with Victor Alexander Haden Horsley) that elec. currents are produced by mammalian brain in action, also recorded them with string galvanometers, 1891. Died July 17, 1913.

GOTH, Robert William, Am. plant pathologist; b. Phillips, Wis., May 10, 1927; s. William Edward and Rose (Dolezalek) G.; B.S., Wis. State Coll., 1954; M.S., U. Minn., 1957, Ph.D. (Tozer Found. fellow), 1961; m. Joyce Marie Nelson, Dec. 29, 1954; children—Valerie Dianne, Robert William. Faculty, U. Minn., 1959-60; plant pathologist Plant Industry Sta. U. S. Dept. Agr., Beltsville, Md., 1961—; vis. scientist dept. plant pathology U. Cal., Davis, 1966. Mem. Am. Phytopath. Soc., Bot. Soc. Washington,

Sigma Xi, Sigma Gamma Delta, Gamma Alpha. Research on compatability relationship between nuclear components of inter-specific crosses of sorghum smuts, effects of virus infection upon growth, seed yield, flower prodn., predisposition, to adverse environmental conditions of red and white clover; discovered techniques for studying bacterial pathogens of beans. Home: 4213 Wicomico Av., Beltsville 20705. Office: Crops Research Div., Agrl. Research Sta., U. S. Dept. Agr. Plant Industry Sta., Beltsville, Md. 20705.*

GOTHAN, Walter Ulrich Eduard Friedrich, German paleobotanist; b. Woldegk (Mecklenburg), Germany, Aug. 26, 1879; s. Heinrich and Minna (Heinrichs) G.; began study geology and mining Mining acads. of Clausthal, Germany, Berlin, 1899, botany, chemistry and philosophy, U. Berlin, 1903-04; Ph.D., Jena, Germany, 1905; m. Marie Schmidt, 1923; m. 2d, Hertha Müller, 1948. Asst. Prussian Geol. Inst., 1903-10, full asst., 1910-13, collection curator 1913-27, dist. geologist, 1927-29, state geologist, 1929-38, sect. head, From 1938, titular prof., from 1919; named lectr. Berlin Mining Acad. (merged with Technische U., 1914), 1908, unofcl. asso. prof., 1926; named hon. prof. U. Berlin, 1927, prof. paleobotany, 1947. Recipient Leopold von Buch plaque, 1948, Orville Derby medal, 1951, medal Société botanique de France, 1954. Mem. German Acad. Scis. Re-edited: Lehrbuch der Paleobotanik (Potonié), 1921, 3d edit., 1954. Research and numerous publs. on morphological, taxonomic description of soft coal flora, especially of Tertiary; founder coal petrography; developed a comparative stratigraphy, especially of Carboniferous and Permian periods; developed unified nomenclature of coal layers and their paleontol. limits. Died Berlin, Dec. 30, 1954.

GOTO, Hidehiro, Japanese chemist; b. nr. Osaka City, Japan, Sept. 11, 1907; s. Shugaku and Chido (Okada) G.; grad. Sch. Chemistry, Faculty Sci., Tohoku U., Sendai, Japan, 1933, D.Sc., 1940; m. Tae Sagawa, Apr. 26, 1935; children—Kazuhiro, Yukihiro, Akihiro. Faculty, Tohoku U., 1933—, prof., 1945—. Chmn., Japanese Indsl. Standard Com. for Chem. Analysis of Iron, Steel and Other Metals, 1951—. Recipient prizes Japan Inst. Metals, 1947, Iron and Steel Inst. Japan, 1945, Chem. Soc. Japan, 1965. Mem. Chem. Soc. Japan, Japan Inst. Metals, Iron and Steel Inst. Japan, Japan Soc. for Analytical Chemistry, Atomic Energy Soc. Japan. Research and numerous publs. on analytical chemistry, including fluorescence analysis, catalytic analysis, spectrophotometry; studies in metal analysis in Japan. Home: 14-25 3 chome, Mukaiyama, Sendai. Office: Research Inst. for Iron, Steel and Other Metals, Tohoku U., 75 Katahira-cho, Sendai, Japan.*

GOTO, Toshio, Japanese chemist; b. Gifu, Japan, Apr. 23, 1929; s. Morio and Fujie (Yanagihara) G.; B.Sc., Nagoya U., 1952, Ph.D., 1960; m. Izumi Gorokawa, Apr. 22, 1960; children—Mikio, Kaoru. Instr. faculty sci. Nagoya U., 1954-61, asst. prof., 1961-66, prof. faculty agr., 1966—. Mem. Pharm. Soc. Japan, Chem. Soc. Japan, Chem. Soc. (London). Research and publs. on structure determination and synthesis of biologically active natural products, steroid studies, studies on bioluminescent substances, organic instrumental analysis. Home: 10, Yagumadori 1, Nakagawaku, Nagoya-shi. Office: Furocho, Chikusaku, Nagoya-shi, Japan.*

GOTRIAN, Walter Robert Wilhelm, German astrophysicist; b. Aachen, Germany, Apr. 21, 1890; s. Otto and Luise (Dieckmann) G.; student physics, Aachen; Ph.D., Göttingen, Germany, 1914; m. Eva Merkel, 1921. Research staff U. Göttingen; asst. physics and spectroscopy Astrophys. Obs., Potsdam, Germany, 1922-54; joined faculty U. Berlin (Germany), 1923, became asso. prof., 1927; became prof. astrophysics Humboldt U., Berlin, also dir. Potsdam Obs., 1951; mem. solar eclipse expdn. to Sumatra, 1929. Mem. German Acad. Scis. Author: Elektronenstoss und geschichtete Entladung, 1921; Graphische Darstellung der Spektren von Atomen und Ionen mit ein, zwei und drei Valenzdektronen, 1928; also articles, chpts. in books. Research on nature of solar corona and solar magnetic fields; suggested explanation of corona spectrum, 1939. Died Potsdam, Mar. 3, 1954.

GOTS, Joseph Simon, Am. microbiologist; b. Phila., Oct. 12, 1917; s. Solomon D. and Lillian (Shlomoff) G.; A.B., Temple U., 1939; M.S., U. Pa., 1941, Ph.D., 1948; m. Selma Sheinbeck, Jan. 24, 1941; children—Ronald Eric, Lynne Suzanne. Field agt. U. S. Dept. Agr., 1941-42; served to lt. col. San. Corps, U. S. Army, 1942-46; Abbott fellow Sch. Medicine, U. Pa., Phila., 1946-48, mem. faculty, 1948—, prof. microbiology, 1963—, mem. council of Grad. Sch. Arts and Scis., 1964—. Cons. Parke Davis & Co., Inst. Energy Conversion. Mem. Am. Soc. Microbiology, Soc. Gen. Microbiology, Am. Acad. Microbiology, Am. Soc. Biol. Chemists, A.A.A.S., N.Y. Acad. Sci., Am. Soc. Gen. Physiology, Am. Assn. Cancer Research, Soc. Exptl. Medicine, Sigma Xi. Research and numerous publs. in bacterial genetics and enzymology, action of genes, metabolism of microorganisms, antibiotics, mechanisms of drug action and antibiotic resistance, biochem. genetics, control mechanisms, regulatory action of biosynthesis, biosynthesis of proteins and nucleic acids. Home: 1209 Greenhill Rd., Flourtown, Pa. 19031.*

GOTSCHLICH, Emil Carl Anton Constantin, German hygienist; b. Beuthen, Germany, Mar. 28, 1870; s. Emil and Maria (Kinne) G.; began study medicine, Breslau (now Wroclaw, Poland), 1889; M.D., 1894; m. Rosa Kaufmann, 1898; 1 son, 1 dau. Asst. to C. Flügge; named dir. Municipal Bd. Health, Alexandria, Egypt, 1896; became dep. dir. Hygiene Inst., U. Halle (Germany), 1915, Saarbrücken, 1915; named prof. hygiene, Giessen, Germany, 1917; named prof. hygiene, Heidelberg, Germany, 1926, emeritus, 1935-41, tchng. prof., 1941-49; dir. Central Hygiene Inst., Ankara, Turkey, until 1941. Recipient Goethe medal, 1940. Author; (with W. Schürmann) Leitfaden der Mikroparasitologie, 1920; Erfahrungen über Pest in Egypten, 1923; Die Bedeutung der Hygiene für die Erneuerung des Deutschen Volkes, 1924; Die Variabilität der Mikroorganismen in allgemeiner biologischer Hinsicht, 1924; Kommen und Gehen der Epidemien, 1928; Hygiene Zivilisation und Kultur, 1929; Fabrikspeisung, Die wissenschaftlichen Grundlagen der Volksernährung, 1930; also numerous articles. Editor: Handbuch der hygienischen Untersuchungsmethoden, 1926-29. Founder, Turkish Bull. Hygiene and Exptl. Biology. Research on morphology and biology of pathogenic microorganisms, prophylaxis for infectious diseases, disinfection, gen. theory of epidemics; organized pub. health and combated plague, 1899, cholera, 1902, Egypt. Died Heidelberg, Dec. 19, 1949.

GOTSHALL, William Charles, Am. engr.; b. St. Louis, May 9, 1875; s. Daniel H. and Minnie Wortmann Gotshall. Began as an elec. expert with Mo. Electric Light & Power Co.; U. S. Govt. engr. in charge of work of riprapping and protecting banks of 150 miles of Mississippi River; in charge location St. Louis & Eastern R.R.; rebuilt and operated Cairo (Ill.) Electric Ry.; built Belleville (Ill.) Electric Ry.; built Marshalltown (Ia.) ry. and lighting plant, Muncie (Ind.) Electric Ry., Grand Av. Ry., St. Louis; apptd. chief engr. Union Depot Ry. Co., St. Louis, rehabilitated entire system and made pioneer intro. of and operated 3-wire system on electric rys.; converted 2d Av. (horse) Ry., N.Y., into a conduit electric ry., 1897-98; as pres. and chief engr. N.Y. & Portchester R.R. Co., accomplished pioneer work in design and devel. of high-speed electric traction produced 1st high speed electric ry. in U. S. located entirely on pvt. right of way, with no grade crossings; engaged in design and devel. of new and in purchase and rehabilitation of existing railroads in U. S., Europe, Near and Far East, Africa. Author: Electric Railway Economics (textbook), 1914. Active in organizing and directing Near East archaeol. excavations Died Aug. 20, 1935.

GOTSIRIDZE, Othary Alexandrovitch, Russian surgeon, physiologist; b. Tbilissi, Georgian SSR, Jan. 3, 1932; s. Alexander Mikhailovich and Maria (Varlamishvili) G.; physician Tbilissi State Med. Inst., 1956, surgeon, 1958, obstetrician-gynecologist, 1962; m. Leila Grigorievna Gogitchaishvili, Feb. 8, 1953; children—Alexander, Nino. Surgeon, Regional Child Hosp., also head clin.-diagnostical lab., Tkibuli, 1956-57; surgeon Regional Child Hosp., also head lab. blood transfusions and blood conservation, Tkibuli, 1957-58; sci. worker lab. exptl. surgery Inst. Female Physiology and Pathology, Tbilissi, 1958-59, sci. worker, 1959-60, head lab., 1960-63; head lab. exptl. surgery Inst. Obstetrics and Gynecology, Tbilissi, 1963—. Mem. bd. Georgian br. Assn. USSR-Italy; mem. bd. med. sect. Georgian Soc. Cultural Relations with Fgn. Countries; mem. Georgian br. Soc. USSR-Iran; mem. Georgian Com. Solidarity African and Asian Countries. Recipient hon. diploma IVth World Congress Fertility and Sterility. Fellow Georgian Assn. Surgeons, Georgian Assn. Obstetrics and Gynecology; mem. Internat. Assn. Fertility, Georgian Assn. Fertility and Sterility (gen. sec. surg. sect.). Author: Problems of Physiology and Pathology, 1958; also numerous articles. Research on study of reflectory afterdisage; original method of operation for recanalization of fallopian tubes; operative interventions for intensification of blood supply to female and male genitalia; new methods of transplantations of ovaries, uterus, fallopian tubes, female internal sexual organs (uterus and its appendages); revascularization of sclerotized auditory nerve. Address: 25 Atonely str., Tbilissi-26, Georgian SSR.*

GOTTARDI, Vittorio, Itlaian chemist; b. Venice, Italy, Oct. 9, 1923; s. Francesco and Ida (Folin) G.; Degree in chem. engring., U. Padua (Italy), 1947; m. Maria Forti, Jan. 30, 1956; children—Pietro, Guido. Asst. prof. applied chemistry U. Padua, 1950-57, fellow applied chemistry, metallurgy, metallography, 1957—, prof. materials sci. Faculty Engring., 1955-67; dir. Stazione Sperimentale del Vetro, Venice, 1957—. Exec. com. Internat. Glass Commn., 1964—. Mem. Istituto Veneto per le Scienze, Lettere ed Arti, Am. Ceramic Soc., Società Italiana di Metallurgia, Soc. Glass Tech., Deutsche Glastechnische Gesellschaft, Union Scientifique Continentale du Verre. Research, publs. on applied chemistry, especially glass, silicates and metallurgy. Home: 251 S. Croce, Venice. Office: Stazione Sperimentale del Vetro- 10, Via Briati, Murano (Venice), Italy.*

GOTTFRIED, Kurt, physicist, educator; b. Vienna, Austria, May 17, 1929; s. Salomon and Augusta (Werner) G.; B.Eng., McGill U., 1951, M.Sc., 1952; Ph.D., Mass. Inst. Tech., 1955; m. Sorel B. Dickstein, June 26, 1955; children—David M., Laura S. Jr. fellow Soc. Fellows Harvard U., 1955-58; research fellow Inst. Theoret. Physics, Copenhagen, 1958-59; research fellow Harvard, 1959-60, asst. prof. physics, 1960-64; asso. prof. physics Cornell U., 1964-68, prof. physics, 1968—. Mem. Am. Phys. Soc. Author: Quantum Mechanics, 1966. Research on the theory of complex nuclei, behaviour of fluids at low temperatures, high energy reactions between elementary particles. Address: Lab. Nuclear Studies, Cornell U., Ithaca, N.Y. 14850.*

GOTTHARDT, Ernst, German engr.; b. Kassel, Apr. 26, 1908; s. Otto and Helene (Faillard) G.; ed. U. Göttingen, U. Berlin; Ph.D. in engring.; m. Elfriede Frevert, 1940; children—Rolf, Volker. Instr., U. Berlin; prof. U. Stuttgart. Author: Ermittlung von Korrelationen; Erfahrungen mit dem Geodimeter; Geodätisches Rechnen mit der Rechenanlage Zusa Z 22; Genauigkeitsfragen zur Analytischen Orientierung von Luftbildern. Home: 42 Schottstrasse. Office: 11 Keplerstrasse 11, Stuttgart, Germany.

GOTTLIEB, Jacques Simon, Am. psychiatrist; b. Trinidad, Colo., Feb. 2, 1907; s. David Hart and Sara (Sanders) G.; student State U. Colo., 1923-25; B.S., Harvard, 1928, M.D., 1932; m. Helen Mae White, Dec. 29, 1934; children—Marilyn Lee, Jacquelyn (Mrs. John McCall), David Hart. Mem. faculty Ia. State Psychopathic Hosp., also State U. Ia., Iowa City, 1936-53, asst. dir. hosp., prof. psychiatry; dir. Inst., Jackson Meml. Hosp., Miami, Fla., 1953-55, also prof., chmn. dept. psychiatry U. Miami, 1953-55; dir. Lafayette Clinic, Detroit, 1955—; prof. psychiatry Wayne State U., Detroit, chmn. dept., 1961—. Dir., Am. Bd. Psychiatry and Neurology, 1959-67, pres., 1966; mem. adv. com. to Psychiatric, Neurol. and Psychol. Service, VA, 1965-67, com. on pharmacology and chemistry Nat. Inst. Mental Health, 1965-67. Mem. A.M.A. (council on med. edn. and hosp. residency rev. com.), Am. Coll. Psychiatrists, Am. Psychiat. Assn., Mich. State Med. Soc., Wayne County Med. Soc. Research and numerous publs. in field of interdisciplinary studies of schizophrenia. Home: 1712 Lafayette Towers West, Detroit 48207. Office: 951 E. Lafayette St., Detroit 48207.*

GOTTLIEB, Otto Richard, organic chemist; b. Brno, Czechoslovakia, Aug. 31, 1920; s. Adolf and Dora (Ornstein) G.; indsl. chemist degree Fed. U. Rio de Janeiro, 1945; Sc.D., Rural U. Brazil, 1966; m. Franca Cohen, Oct. 8, 1947; children—Hugo Emilio, Raul, Marcel Bernardo. Prodn. chemist Ornstein & Cia, Rio De Janeiro, 1946-54; research asso. Inst. Agrl. Chemistry, Ministry Agr. and Nat. Research, Rio de Janeiro, 1955-58, research head, 1959-63; prof. organic chemistry U. Brasilia, 1964-65; prof. organic chemistry Rural U. Brazil, 1966—. Research fellow Weizmann Inst. Sci., Israel, 1960; prof., adviser Fed. U. Minas Gerais, 1962—, Fed. U. Pernambuco, 1966—; vis. prof. U. Sheffield, Eng., 1964. Mem. Brazilian Acad. Scis., Chem. Soc. Gt. Britain. Contbr. numerous articles to profl. jours., chpts. to books. Research in analytical chemistry, gasometric titrations, natural products chemistry, pyrones and nitro-derivative from Lauraceae species, xanthones and biphenyl-derivative from Guttiferae species, neoflavanoids and cinnamylphenols from Leguminosae species. Home: 323 Cinco de Julho, Rio de Janeiro, Brazil.*

GÖTTLING, Johann Friedrich August, German chemist, pharmaceutist; b. Derenburg, Germany, June 5, 1753; s. Johann Friedrich and Christian (Würtzler) G.; studied chemistry, Göttingen, 1785-87; Ph.D., Jena, Germany, circa 1789; studied pharmacy under Wiegelb, Langensalza; m. Sophie Schultze, 1789; 1 son, Karl Wilhelm, 1 dau. Employed by W.H.S. Buchholz, owner of Hofapotheke, Weimar, Germany, 1775; became asso. prof. gen. and pharm. chemistry U. Jena, 1789, hon. prof., 1799, prof., 1809. Mem. Leopoldina. Author: Einleitung in die pharmazeutische Chemie, 1778; Tabelle über die Lehre von den Salzen und ihren mittelsalzartigen Verbindungen, 1784; (with J. D. Brandis) Technologisches Taschenbuch für Künstler, Fabrikanten und Metallurgen, 1784; Vollständiges chemisches Probier-Cabinet . . . , 1790; Versuch einer physikalischen Chemie, 1797; Beiträge zur Berichtigung der antiphlogiston Chemie, 1794-98; Handbuch der theoretischen und praktischen Chemie, 3 vols., 1798-1800; Physikalisch-chemische Enzyklopädie . . . , 3 vols., 1804-07; Die Sirup- und Zuckerbereitung aus Runkelruben, 1808; Elementarbuch der chemischen Experimentierkunst, 2 vols., 1808. Editor: Almanach oder Taschen-Buch für Scheidekünstler und Apotheker (first pharm. jour.), 1779-1802. One of 1st in Germany to oppose phlogiston theory; wrote one of oldest books on phys. chemistry; research camphor, resins and various ethers, on compounds of sulphur, arsenic, phosphorous, mercury, improved chem. apparatus; 1st in Thuringia to extract beet sugar. Died Jena, Sept. 1, 1809.

GOTTLOB, Rainer Maximilian Ewald, Austrian physician; b. Vienna, Austria, Dec. 17, 1918; s. Kurt and Hedwig (Bierhoff) G.; M.D., German U. Prague; m. Hedwig Knapp, 1952; children—Georg Bernhard, Irene Veronika. Head physician in office of Prof. Mandl, until 1957; with surg. dept. Kaiser Franz Josef Hosp.; dir. exptl. surg. dept. Univ. Clinic Surgery, Vienna. Named laureat of Theodor Körner prize (2). Mem. Internat. Coll. Surgeons, European Soc. Cardiovascular Surgery, German Soc. Surgery, Order Physicians Vienna. Author: Angiographie und Klinik, 1956. Research and numerous publs. on surgery, surg. treatment of diseases of vascular system, angiology, pathology of vascular system. Home: 28 Kirchengasse, Vienna VII. Office: Surgical Clinic, 4 Alserstrasse, Vienna IX, Austria.

GOTTMANN, Jean (Iona), French geographer; b. Kharkov, Russia, Oct. 10, 1915; s. Elie and Sonia (Ettinger) G.; Bacc.es Litt., Sorbonne, Paris, France, 1932, Dipl.Et.Sup., 1934, Lic. es Litt., 1937; m. Bernice Adelson, Aug. 11, 1957. Research asst. human geography, Sorbonne, 1937-41, prof. Ecole Pratique des Hautes Etudes, 1960-—; asso. prof. Johns Hopkins, Balt., 1943-48; prof. Inst. d'Etudes Politiques, U. Paris, 1948-56; research dir. Twentieth Century Fund, N.Y.C., 1956-61; mem. Inst. for Advanced Study, Princeton, intermittently, 1942-65. Cons. Fgn. Econ. Adminstrn., Washington, 1942-44; adviser French Ministry Nat. Economy, Paris, 1945-46; chmn. internat. Com. on Regional Planning, 1949-52; dir. research and studies UN Secretariat, N.Y.C., 1946-47. Recipient Bonaparte-Wyse award, Paris, 1962, Sully-Olivier de Serres award, Paris, 1965. Mem. Assn. Am. Geographers, (Charles P. Daly medal 1964), Societe de Geographie, Am., Royal Netherlands geog. socs., World Soc. for Ekistics, Prospective Paris. Author: L'Amerique, 1949; Ageography of Europe, 1950; La Politique des Etats et leur geographie, 1952; Megalopolis, 1961; also other books, numerous articles. Description and analysis of N. Atlantic civilization particularly in Western Europe and Eastern U. S.; research on concept of megalopolis with gen. theory of urbanization, role geog. factors in modern soc. and internat. relations. Home: 9 rue Sedillot, Paris 7, France.*

GOTTSCHALK, Alexander, Am. radiologist; b. Chgo., Mar. 23, 1932; s. Louis R. and Fruma (Kasdan) G.; B.A., Harvard, 1954; M.D., Washington U., St. Louis, 1958; m. Jane P. Rosenbloom, Aug. 13, 1960; children—Rand, Karen, Amy. Research asso. Donner Lab., Lawrence Radiation Lab., U. Cal. at Berkeley, 1962-64; faculty U. Chgo., 1964-—, asso. prof. radiology, 1966-—, dir. sect. nuclear medicine dept. radiology, 1964-—; dir. Argonne Cancer Research Hosp., 1967-—; faculty research asso. Am. Cancer Soc. Diplomate Am. Bd. Radiology; mem. Soc. Nuclear Medicine, Radiation Soc. N.Am., Chgo. Roentgen Soc., A.M.A., Sigma Xi. Contbr. articles to tech. jours. Research on use radioisotopes in diagnosis, instrumentation in nuclear medicine, use accelerated heavy particles for therapy, exptl. radiobiology. Home: 5543 S. Harper Av., Chgo. 60637.*

GOTTSCHALK, Alfred, Australian biochemist; b. Aachen, Germany, Apr. 22, 1894; s. Ben Carl and Rosa (Kahn) G.; student U. Munich (Germany), 1912-13, U. Freiburg (Germany), 1913; M.D., U. Bonn (Germany), 1920; postgrad. Kaiser-Wilhelm Inst. Biochemistry, Berlin, Germany, 1921-26; m. Elizabeth Oygler, Aug. 2, 1923; 1 son, Rudolf. Dir. biochem. dept. Gen. Hosp., Stettin, Germany, 1927-35; sr. biochemist Walter and Eliza Hall Inst. Med. Research, Melbourne, Australia, 1939-59; hon. fellow Australian Nat. U., Canberra, Australia, 1959-63; guest prof. Max-Planck Inst. Virus Research, Tübingen, Germany, 1963-—; hon. prof. U. Tübingen, 1966-—; lectr. U. Melbourne, 1949-59. Recipient David Syme prize, also medal U. Melbourne, 1950, H. G. Smith Meml. medal Royal Australian Chem. Inst., 1953. Fellow Royal Inst. Chemistry, Royal Australian Chem. Inst., A.A.A.S., Australian Acad. Sci., Max Planck Inst. (fgn. sci. mem.). Author: Chemistry and Biology of Sialic Acids, 1960; Glycoproteins, Their Composition, Structure and Function, 1966, numerous articles. Research and publs. on molecular structure of neuraminic acid, reaction mechanism of enzyme neuraminidase, molecular structure of glyoproteins, especially of carbohydrate group, elucidation of chemistry underlying initial phases of influenza virus infections. Office: Max Planck Inst. of Virus Research, Spemannstr. 35, 74 Tübingen, West Germany.*

GOTTSCHALK, Louis August, Am. med. scientist; b. St. Louis, Aug. 26, 1916; s. Max Wilhelm and Kelmie (Mutrux) G.; A.B., Washington U., 1940, M.D., 1943; certificate in adult and child analysis Chgo. Inst. for Psychoanalysis, 1951; postgrad. Washington Psychoanalytic Inst., 1951-53; m. Helen C. Reller, July 24, 1944; children—Guy H., Claire A., Louise H., Susan E. Staff psychiatrist, Dir. electroencephalography lab. USPHS Hosp., Ft. Worth, 1946-48; instr. dept. psychiatry Southwestern Med. Sch., Dallas, 1947-48; research asso., asst. chief child psychiatry div. Inst. for Psychosomatic and Psychiat. Research and Tng., Michael Reese Hosp., Chgo., 1948-51; research psychiatrist Nat. Inst. Mental Health, NIH, Bethesda, Md., 1951-53, cons. psychiatrist, 1955-—; faculty U. Cin. Coll. Medicine, 1953-67, research prof. psychiatry, coordinator research dept. psychiatry, 1960-68; tng. and supervising analyst Chgo. Inst. for Psychoanalysis, 1957-68; staff Cin. Gen. Hosp., 1953-67; prof., chmn. dept. psychiatry and human behavior Coll. Medicine, U. Cal. at Irvine, 1967-—; dir. residency tng. Orange County Med. Center, Orange, Cal., 1967-—. Vice pres. bd. trustees Cin. Speech and Hearing Center. Recipient Research Career award USPHS, 1961-68. Fellow Am. Psychiat. Assn., Am. Coll. Psychiatry, A.A.A.S., Am. Coll. Neuropsychopharmacology; mem. Am. Psychosomatic Soc., Assn. for Research in Nervous and Mental Disease, Acad. Medicine Cin., Cin. Soc. Neurology and Psychiatry (past pres.), Group for Advancement of Psychiatry (mem publs. com. and research com. 1960-—), Phi Beta Kappa, Sigma Xi, Phi Eta Sigma, Alpha Omega Alpha, Omicron Delta Kappa. Author: Psycholinguistic Analysis of Two Psychotherapeutic Interviews, 1961; (with A. H. Auerbach) Methods of Research in Psychotherapy, 1966; (with G. C. Gleser) Measurement of Psychological States Through the Content Analysis of Verbal Behavior; also numerous articles. Editorial bd. Psychosomatic Medicine, 1958-—; cons. editor Science, 1963-—. Research on relation emotions to epilepsy, effects on emotions of tranquilizing and anti-depressant medication, phases of menstrual cycle and hypnotic suggestion, devel. brief verbal behavior analysis method measuring magnitude transient emotional states, effects anxiety and hostility on blood levels of adrenalin, free fatty acids, corticosteroids, methods research in psychotherapy. Home: 4607 Perham Rd., Corona del Mar, Cal. 92625. Office: Coll. Medicine, U. Cal., Irvine, Cal. 92664.*

GOTTSCHALK, Walter Helbig, Am. mathematician; b. Lynchburg, Va., Nov. 3, 1918; s. Carl and Lula (Helbig) G.; B.S., U. Va., 1939, M.A., 1942, Ph.D. in Math., 1944; M.A. (hon.), Wesleyan U., Middletown, Conn., 1964; m. Margaret Hemsworth, Aug. 27, 1952; children—Healther, Steven. First instr. to prof. math. U. Pa., 1944-63, chmn. dept., 1955-58; prof. math. Wesleyan U., 1963-—, chmn. dept., 1964-—; mem. Inst. Advanced Study, Princeton, 1947-48; research asso. Yale, 1960-61. Mem. Am. Math. Soc. (asso. editor proc. 1954-56), Math. Assn. Am., Soc. Indsl. and Applied Math., Assn. Symbolic Logic, Am. Assn. U. Profs., Phi Beta Kappa, Sigma Xi. Author: (with G. A. Hedlund) Topological Dynamics, 1955; also articles. Mem. editorial bd. Math. Systems Theory, 1967-—. Home: 271 Washington Terrace, Middletown, Conn. 06457.

GOTTSCHALK, Werner Max, German geneticist; b. Marienberg, Germany, May 15, 1920; s. Max Paul and Elsa (Neubauer) G.; Dr.rer.nat., U. Freiburg, 1950; Scholar, Bot. Inst., U. Freiburg, 1950-51; asst. Inst. Forage Crops, U. Giessen (Germany), 1952-55; privatdozent Faculty Natural Scis., U. Giessen, 1953-56; asst. Inst. Plant Breeding, U. Göttingen (Germany), 1956; faculty U. Bonn (Germany), 1957-—, prof., head Inst. Genetics, 1965-—. Mem. mutation adv. group div. atomic energy in agr. Joint FAO/IAEA, 1965-—. Mem. Deutsche Botanische Gesellschaft, Vereinigung für angewandte Botanik. Author: The Cytology of the Cultivated Tomato and its Wild-growing Relatives, 1956; The Action of Mutant Genes on Morphology and Function of Plant Organs, 1964; also articles. Research in chromosomal fine structure, prodn. and cytological evaluation of highly polyploid plants, genetic control of germ cell formation, alteration of genes and analysis of gene action, prodn. of mutants useful for plant breeding, radiation genetics, cytogenetics and evolution research. Home: 10 Lengsdorfer, Bonn, Germany.*

GOTTSCHALK, Winston Malcolm, Am. physicist; b. Rolla, Mo., Feb. 2, 1908; s. Victor Hugo and Kathrine (Cox) G.; B.S., Princeton U., 1929; M.S., Cal. Inst. Tech., 1932; Ph.D., Harvard, 1946; m. Ruth Strong Bradley, June 22, 1929; children—Ruth Patricia (Mrs. Robert L. Kepner), W. Bradley. Tchr., St. Mark's Sch., Southboro, Mass., 1937-41, Exeter Acad., 1941-42; instr. Harvard, 1942-45; head microwave research div. Raytheon Co., Waltham, Mass., 1945-58; prof. elec. engring. U. Del., Newark, 1958-61; leader in exptl. physics RCA, Moorestown, N.J., 1961-63; staff mem. Mitre Corp., Bedford, Mass., 1963-—. Mem. Am. Phys. Soc., Sigma Xi. Research and publs. on devel. of microwave tubes, plasma physics; participant underground nuclear tests; specialist in nuclear weapons effects; theoretical studies of atomic spectra. Home: 68 Claflin St., Belmont, Mass. 02178. Office: Mitre Corp., Bedford, Mass. 01730.*

GÖTZ, (Friedrich Wilhelm) Paul, geophysicist, meteorologist; b. Heilbronn, Germany, May 20, 1891; s. Paul and Maria (Mühlschlegel) G.; studied astronomy, math., physics; m. Margarethe Beversdorf, 1932. Began as tchr. in Davos, Switzerland; accepted contract to build inst. for climate and radiation research, Arosa, 1921, and worked there until his death; joined faculty U. Zurich, 1931, named titular prof. meteorology, 1940. Mem. Internat. Assn. Meteorology and Atmospheric Physics, Internat. Assn. Geomagnetism and Aeronomy. Author: Das Strahlungsklima von Arosa, 1926; Klima und Wetter in Arosa, 1954; also numerous articles and monographs on ozone. Research on light, ultra-violet radiation, upper atmosphere, aurora borealis and other light effects, dimming of atmosphere, climate of Arosa; a leader in ozone research; discovered reversal effect known as Götz effect (led to 1st practical method of determining vertical distbn. of ozone in atmosphere). Died Chur, Switzerland, Aug. 29, 1954.

GOUAN, Antoine, French botanist; b. Montpellier, Nov. 15, 1733; pupil of Boissier de Sauvages; M.D., 1752. Prof. botany Montpellier Faculty Medicine, until 1803; served at mil. hosps. Mem. French Acad. Scis., 1796. Author: Hortus regius Montpelliensis, 1762; Flora Montpelliaca, 1765; Historia piscium 1770; Nomenclature botanique, 1803; Traité de botanique, 1812. Introduced Linnean classification into France. Died Montpellier, Sept. 1, 1821.

GOUBEAU, Joseph Anton, German chemist; b. Augsburg, Mar. 31, 1901; s. Anton and Franziska (Nestl) G.; Ph.D. in Chemistry, U. Munich; m. Helene Müller, 1930; children—Elisabeth, Irmingard, Agnes, Helene, Andreas. Asst., U. Munich, 1926-28, U. Friburg, 1928-29, Bergakademie of Clausthal, 1929-35; instr., prof. U. Göttingen, 1935-51; with U. Stuttgart, Germany, 1951-—. Recipient Adolf-Stock prize German Chem. Assn., 1953. Mem. Acad. Heidelberg, Acad. Göttingen, Acad. Leopoldina Halle, Acad. Bologna, Spanish Assn. Physics and Chemistry (hon.), Higher Council Sci. Investigation Madrid (hon. councilor). Research and publs. on analysis of middle of spectrum of Raman, vibration spectra of inorganic compounds. Home: 5 Hallimaschweg 7000 Stuttgart-Schonberg. Office: 26 Schellingstrasse, 7000 Stuttgart, Germany.*

GOUDEMAND, Maurice, French physician; b. Meudon, France, Aug. 30, 1919; s. Charles and Clémence (Defrance) G.; Docteur en Médecine, Faculté de Lille (France), 1945; Professeur Agrégé d'Hematologie, 1958; m. Gillette Fontaine, May 27, 1944; children—Michel, Jenny. Adj. dir. Centre Régional de Transfusion Sanguine, Institut Pasteur, Lille, 1947-58; prof. medicine Faculté de Médecine, Lille, 1958-—. Mem. French, Internat. socs. hematology, Internat. Soc. Blood Transfusion. Research, numerous publs. on hemorrhagic diseases, especially hemophilia, immunohematology (auto-immune anemias). Home: 28, rue Alexandre Leleux 59, Lille, France.*

GOUDET, Georges, French physicist; b. Dijon, Cote-d'Or, June 2, 1912; ed. Higher Normal Sch.; agrégé in phys. sci.; Ph.D. in sci. Head of service Central Lab. Telecommunications, until 1955, then dir. gen.; with Nat. Center Studies of Tele-communications; prof. U. Nancy; dir. Nat. Higher Sch. Electronics and Mechanics, Nancy; prof. Higher Sch. Physics and Indsl. Chemistry; dir. gen. Gen. Co. Telephone Constrns. Adm. of C.G.C.T., L.T.T., M.T.I., Metrix, Sicopel, S.A.C.T. Research and numerous publs. on electronic optics, electromagnetism, centimetric waves. Address: 251 rue de Vaugirard, Paris, France.

GOUDSMIT, Samuel Abraham, physicist; b. The Hague, Netherlands, July 11, 1902; s. Isaac and Marianne (Gompers) G.; student U. Leiden and Amsterdam; Ph.D., U. Leiden, 1927; D.Sc., Case Inst. Tech., 1958; m. Jaantje Logher, Jan. 19, 1927 (div. 1960); 1 dau., Esther Marianne; m. 2d, Irene Bejach Rothschild, 1960; Came to U. S., 1927. Joined Faculty U. Mich., 1927, prof. physics, 1932-46; Rockefeller fellow, 1926, Guggenheim fellow, 1938; vis. prof. Harvard, 1941; prof. physics Northwestern U., 1946-48. On leave to Radiation (Radar) Lab. Mass. Inst. Tech. and Eng., 1942-46, sr. scientist, Brookhaven Nat. Lab., 1948-—. Detailed to War Dept. as chief Sci. Intelligence Mission in Europe, 1944-45. Awarded Medal of Freedom; Officer Order Brit. Empire; recipient Research Corp. Scientific Award, 1954; Max Planck medal German Phys. Soc., 1965. Fellow Am. phys. Soc. (mng. editor), Netherlands Phys. Soc., Nat. Acad. Scis., Am. Nuclear Soc.; mem. Am. Philos. Soc. Editor Phys. Review Letters. Author: (with Pauling) The Structure of Line Spectra, 1930; (with Bacher) Atomic Energy States, 1931; Alsos, 1947; also articles. Discovered (in studies with George E. Uhlenbeck) that all electrons spin around an axis (electron spin), 1925. Office: Brookhaven National Lab., Upton, L.I., N.Y.*

GOUGH, Denis Ian, geophysicist; b. Port Elizabeth, S. Africa, June 20, 1922; s. Frederick W. and Ivy C. (Hingle) G.; B.S., Rhodes U., 1943, M.S., 1947; Ph.D., U. Witwatersrand, 1953; m. Winifred I. Nelson, June 1, 1945; children—Catherine V., Stephen W. Research officer, sr. research officer S. African Nat. Phys. Lab., 1947-58; lectr., sr. lectr. U. Coll. Rhodesia and Nyasaland, 1959-63; asso. prof. S.W. Center for Advanced Studies, Dallas, 1964-66; prof. U. Alta, Edmonton, Can., 1966-—. Fellow Royal Astron. Soc. (London); mem. Am. Geophys. Union, European Assn. Exploration Geophysicists. Author: (with A. L. Hales) Measurements of Gravity in Southern Africa, 1950. Research of crustal structure of earth through gravity and magnetic surveys; devel. of instrument for shallow seismic surveys using hammer and electronic delay timing; studies related to polar wander and continental drift, terrestrial heat flow. Home: 11747 83d Av., Edmonton, Alta., Can.*

GOUGH, Harrison Gould, Am. psychologist; b. Buffalo, Minn., Feb. 25, 1921; s. Harry Betzer and Aelfreda (Gould) G.; B.A. summa cum laude, U. Minn., 1942, M.A. (Social Sci. Research Council fellow), 1947, Ph.D., 1949; m. Kathryn H. Whittier, Jan. 23, 1943; 1 dau., Jane Kathryn. Faculty U. Minn., 1948-49; faculty U. Cal., Berkeley, 1949-—, prof. psychology, 1960-—, asso. dir. Inst. Personality Assessment and Research, 1964-67, chmn. dept. psychology, 1967-—. Cons. clin. psychology U. S. VA, 1951-—, Letterman Army Hosp., 1957-—; mem. research adv. com. Cal. Dept. Corrections, 1957-64; mem. research adv. com. Cal. Dept. Mental Hygiene, 1964-—; dir. Cons. Psychologists Press, 1956-—. Fulbright fellow, Italy, 1958-59, 65-66; Guggenheim Found. fellow, 1965-66. Mem. Internat. Assn. Applied Psychology, Am., Western, Cal. (pres. 1960-61) psychol. assns Originator of adjective check list and Cal. psychol. tests; contbr. sci. articles on test devel. and interpretation, crime and delinquency, social maturity, scholastic achievement and progress, performance in med. tng. and practice, polit. attitudes, and perception of geometric forms and illusions. Home: 10 Florida Av., Berkeley, Cal. 94707.*

GOUGH, Jethro, Brit. pathologist; b. Mountain Ash, U. K., Dec. 29, 1903; s. Jabez and Ellen (Mortimer) G.; B.Sc., Welsh Nat. Sch. Medicine, 1924, M.D., 1930; m. Annie Mary Thomas, Aug. 11, 1934; children—John, William. Faculty, Welsh Nat. Sch. Medicine, 1927, 1930—, prof. pathology, 1948—; demonstrator pathology U. Manchester (Eng.), 1929-30; dir. Inst. Pathology, Cardiff, 1954—. Cons. adviser Nat. Health Service, 1948—. Fellow Coll. Pathologists. (mem. council 1962—); hon. mem. Harvey Soc. N.Y., Am. Thoracic Soc. Research and publs. on diseases of lungs; developed (with J. E. Wentworth) new technique for comparing changes in lungs with abnormalities in X-ray films. Home: 22 Park Rd., Whitchurch, Cardiff, U.K.*

GOUJAUD, Aimé Jacques Alexandre, see Bonpland, Aimé Jacques Alexandre.

GOULARD, Thomas, French physician; b. St. Nicolas de la Grave, France, 1720. Maj. surgeon Montpellier Mil. Hosp. Author: Oeuvres de chirurgie avec un traité sur les effets des préparation de plomb, 1770. Discovered extract of Saturn or white-water (eau de Goulard); described urethral and venereal diseases. Died 1790.

GOULD, Augustus Addison, Am. physician, conchologist; b. New Ipswich, N.H., Apr. 23, 1805; s. Nathaniel Duren and Sally (Prichard) G.; grad. Harvard, 1825, M.D., 1830; m. Harriet Sheafe, Nov. 25, 1833. Became a leading physician in Mass.; influen-in devel. of conchology in Am.; wrote Report on the Invertebrata of Mass., 1841, Mollusca and Shells, 1852 (his most important contbns. to Am. science); editor The Terrestrial-Air-Breathing Mollusks of the United States and the Adjacent Territories of North America; introduced such subjects as principles of classification, geog. distbn. of genera and species, geol. relationships, and anatomical structures; mem. Boston Soc. Natural History, pres., several years; an original mem. Nat. Acad. Arts and Scis. Author: Invertebrate Animals of Mass., 1841; Mollusca and Shells of the United States Exploring Expedition under Capt. Wilkes, 1852; A System of Natural History Containing Scientific and Popular Descriptions of Various Animals, 1833; (with Louis Agassiz) Principles of Zoology, 1848. Research in zoology; study of insects and mollusks. Died Boston, Sept. 15, 1866.

GOULD, Benjamin Apthorp, Am. astronomer; b. Boston, Sept. 27, 1824; s. Benjamin Apthorp and Lucretia (Goddard) G.; grad. Harvard, 1884, LL.D. (hon.) 1885; Ph.D., under Gauss, U. Gottingen (Germany), 1848; LL.D., Columbia, 1887; m. Mary Quincy, 1861; 5 children. Established and conducted Astron. Jour., 1849-51, re-established, 1886; in charge of longitude dept. U. S. Coast Survey, 1852-67; organized and directed Dudley Obs., Albany, N.Y., 1855-59; guaged (by aid of submarine cable) difference in longitude between Am. and Europe, 1860; prepared standard catalogue, applying for 1st time systematic corrections to various star catalogues, 1862; built pvt. obs., nr. Cambridge, Mass. 1862; dir. Nat. Obs., Cordoba, Argentina, 1870-circa 1884, did greatest work in observation of stars of southern hemisphere; established meteorol. stas. as far south as Tierra del Fuego, 1872; 1st astronomer to use telegraph in geodetic work, made 15 determinations before method introduced in Europe. Fellow U. Chile, Royal Soc., 1891, Acad. of Sci., Paris, Imperial Acad., St. Petersburg, Bur. Longitudes, Paris, Astron. Gesellschaft, Berlin, Am. Acad. Arts and Scis. (v.p., Watson medal), Nat. Acad. Scis. (charter). Founder, editor, Astronomical Jour., 1846-61; 1886-96. Author: Investigation of the Orbit of the Comet U., 1847; Reports on the Discovery of the Planet Neptune, 1850; Uranometria Argentina, 1879; Resultados Del Observatorio Nacional Argentino en Cordoba (most important work, contains zone catalogues giving positions of 73,160 stars and gen. catalogue of 32,448 stars in So. hemisphere), 1879-96. Perfected methods for determining longitude telegraphically. Died Cambridge, Nov. 26, 1896.

GOULD, Charles Newton, Am. geologist; b. Lower Salem, O., July 22, 1868; s. Simon G. and Arvilla A. G.; grad. Southwestern Coll., Winfield, Kan., 1899; spl. studies geology and paleobotany; A.M., U. Neb., 1900, Ph.D., 1906, D.Sc., 1928; U., LL.D., Oklahoma City U., 1933; m. Nina Leola Swan, Sept. 24, 1903; children—Lois Hazel, Donald Boyd. Prof. geology U. Okla., 1900-11; resident hydrographer U. S. Geol. Survey, 1902-06; dir. Okla. Geol. Survey, 1908-11; cons. geologist, engaged chiefly in petroleum investigations, 1911-24; dir. Okla. Geol. Survey, 1924-32, with rank of dean U. Okla. Dir. structural materials survey Okla., Fed. Emergency Relief Adminstrn., 1934; state dir. mineral survey Okla. Geol. Survey, WPA, 1935; regional geologist Nat. Park Service, 1936-40. Chmn. Am. Com. on Revision of Rules of Stratigraphic Nomenclature, 1929-33. Member Am. Inst. Mining and Metall. Engrs., Geol. Soc. Am., Am. Assn. State Geologists, Paleontol. Soc. Am., A.A.A.S., Internat. Geol. Congress, Inst. Petroleum Technologist, Okla. Acad. Sci. (twice pres.), Am. Assn. Petroleum Geologists, Royal Soc. Arts, Okla. Bd. Geographic Names, Am. Mining Congress (v.p.), Am. Geophys. Union, Am. Geog. Union, Okla. Hall Fame. Author: Geology and Water Resources of Oklahoma, 1905; Geography of Oklahoma, 1909; Petroleum and Natural Gas in Oklahoma, 1912; Index to the Stratigraphy of Oklahoma,, 1926; Travels Through Oklahoma, 1928; Humanizing Geology, 1928; Oklahoma Place Names, 1933; Covered Wagon, Geologist, 1947; also numerous bulls. and articles. Died Aug. 13, 1949.

GOULD, George Milbry, Am. physician, ophthalmologist; b. Auburn, Me., Nov. 8, 1848; s. George Thomas and Eliza A. (Lapham) G.; A.B., Ohio Wesleyan U., 1873, A.M., 1892; S.T.B., Harvard 1874; postgrad. univs. Paris, Leipzig. Berlin. M.D., Jefferson Med. Coll., 1888; m. Harriet Fletcher Cartwright, Oct. 15, 1876; m. 2d, Laura Stedman, Oct. 3, 1917. Pastor, prop. book and art store, Chillicothe, O., 1874-85; began practice, Phila., 1888, splty. ophthalmology; ophthalmologist, Phila., Almshouse, 1892-94. Recipient 1st Doyne medal Ophthal. Congress, Oxford, Eng. Editor Med. News, 1891-95, Phila. Med. Jour., 1898-1900, Am. Medicine, 1901-06. Author: The Student's Medical Dictionary, 1890; An Illustrated Dictionary of Medicine, Biology and Allied Subjects, 1894; An American Yearbook of Medicine and Surgery, 1896-1905; A Cyclopedia of Practical Medicine and Surgery, 1900; other books on ophthalmology. Devised cemented bifocal lenses which came into wide use, 1888-89; described constitutional effects of eyestrain; demonstrated that even small errors of refraction may cause psychol. reaction of nervous irritation, in report of 1888. Died Atlantic City, N.J., Aug. 8, 1922.

GOULD, Howard Ross, Am. geologist; b. Adrian, W.Va., Nov. 10, 1921; s. S. R. and Grace (Harris) G.; B.A., U. Minn., 1943; postgrad. U. Cal. Scripps Instn., La Jolla, 1946-47; Ph.D. in Geology, U. So. Cal., 1953; m. Marilyn Bradley, Feb. 14, 1948; children—Bradley, Suzanne. Spl. cons. USN, also civilian with OSRD, 1943-46; marine geologist, oceanographer U. Cal. at San Diego, 1943-46; geologist U. S. Geol. Survey, 1947-54; asst. prof. oceanography and marine geology U. Wash., 1953-56; coordinator facies geology research Humble Oil & Refining Co., Houston, 1956-63, chief geologic research sect., 1964; mgr. statigraphic and structural geology div. Esso Prodn. Research Co., Houston, 1964—. Fellow Geol. Soc. Am.; mem. Am. Assn. Petroleum Geologists, Soc. Econ. Paleontologists and Mineralogists, Gulf Coast Assn. Geol. Socs., Am. Geophys. Union, A.A.A.S., Geochem. Soc., Houston Geol. Soc., Am. Petroleum Inst. Research and publs. on sedimentation, especially in marine and lake environments, gen. geology of the sea floor, petroleum geology. Home: 13043 Kimberly Lane, Houston 77024. Office: P.O. Box 2189, Houston 77001.*

GOULD, Ira A., Jr., Am. dairy technologist; b. Atchison, Kan., Sept. 28, 1905; s. Ira Alfred and Waunettia H. (Adams) G.; B.S.A., W.Va. U., 1931; M.S., Mich. State Coll., 1933; Ph.D., U. Wis., 1938; m. Genevieve M. Wilson, May 4, 1929; 1 son, Kaye Harter. Staff dairy manufacturers Mich. State Coll., 1933, instr., 1934-38, asst. prof., 1938-42, asso. prof. 1942-44; prof. dairy manufacturers U. Md., 1944-49; chmn. dept., prof. dairy tech. Ohio State U., 1949—; cons. various dairy and sci. orgns. U. S. del., XIV Internat. Dairy Congress, Rome, 1956, adviser U. S. delegation XV Congress, London, Eng., 1959; lecturer XVI Congress, Copenhagen, 1962; UNICEF cons. dairy edn., Near and Middle East, 1960; cons. dairy programs UNICEF-FAO, 1966. Bd. dirs. Columbus Milk Council. Recipient Borden award for research milk chemistry Am. Chem. Soc., 1946; award of honor Am. Dairy Sci. Assn., 1966. Fellow American Assn. for Advancement Science; mem. Am. Dairy Sci. Assn. (dir. 1951-53, v.p., pres. 1955-56), Inst. Food Tech., Wash. Acad. of Sci. Internat. Assn. Milk, Food Sanitarians. Author sci. articles. Asso. editor of Jour. Dairy Sci., 1941-51. Research on chemical changes in milk caused by heat, enzymes, and microbial action; study of dairy technology and administration. Home: 2216 Dorset Rd., Columbus 21, O.*

GOULD, John, English ornithologist; b. Lyme Regis, Dorsetshire, Eng., Sept. 14, 1804; s. gardener at Windsor (Eng.) Castle; m. Miss Coxen, 1829; 1 son, Charles. Became taxidermist to Zool. Soc., 1827; travelled in Australasia, 1838-40. Fellow Royal Soc., Author numerous folios on birds including: A Century of Birds from the Himalayan Mountains, 1932; Birds of Europe, 1832-37; Birds of Australia, 1840-48, supplement, 1851-69; Birds of Asia, 1850-80; Birds of Great Britain, 1862-73; also numerous articles. Observations and collections from Australasia, including hummingbirds. Died London, Eng., Feb. 3, 1881.

GOULD, Sylvester Emanuel, Am. physician; b. Detroit, July 31, 1900; s. Jude and Sarah (Stolarsky) G.; A.B., U. Mich., 1920, M.D., 1924, M.S., 1939, D.Sc., 1942; m. Minna Blumenthal, July 22, 1926; children—Joyce (Mrs. Ronald Marvin Rothstein), Carol (Mrs. Berton Alan Leon) (dec.), Mark. Practice medicine, specializing in pathology, Detroit, 1932-63, Miami, Fla., 1963-66, 67—; dir. pathology Wayne County Gen. Hosp., 1932-63; asso. in pathology U. Mich., 1952-56; faculty Wayne State U., 1957-63, prof. emeritus, 1964—; vis. prof. pathology U. Miami, 1963-66, adj. prof., 1966—; chief research in pathology Atomic Bomb Casualty Commn., 1966-67; trustee Detroit Inst. Cancer Research, 1946-64, treas., 1956-60; vice chmn. Internat. Commn. on Trichinosis, 1960—. Mem. Am. Assn. Pathologists and Bacteriologists, Coll. Am. Pathologists, Am. Soc. Clin. Pathology, Am. Soc. Parasitologists, Japan Pathology Soc. (councillor). Author: Trichinosis, 1945; Pathology of Heart, 1953, 60, 67, 68; Microscopic Pathology, 1964; The Acute Abdomen, 1966. Editor: Bull. Coll. Am. Pathologists, 1957-60, Bull. Pathology, 1965-66, 67—; editor-in-chief Am. Jour. Clin. Pathology, 1946-55. Contbr. numerous articles to profl. jours. Research on prevention of trichinosis by exposure of pork to ionizing radiation, trichinosis, pathology of heart, microscopic pathology. Home: 801 Venetian Way, Miami 33139. Office: Jackson Meml. Hosp., Miami, Fla. 33136.*

GOULDNER, Alvin Ward, Am. sociologist; b. N.Y.C., July 29, 1920; s. Louis and Estelle (Fetbrandt) G.; B.B.A., Coll. City N.Y., 1941; M.A., Columbia, 1945, Ph.D., 1952; children—Richard Lee, Alan Jeremy, Andrew Ward; m. 2d, Janet Lee Walker, Feb. 5, 1966. Asso. prof., prof. dept. sociology U. Ill., 1954-59; prof. dept. sociology Washington U., St. Louis, 1959-67, Max Weber research prof. social theory, 1967—, chmn. dept., 1959-64. Recipient Aux. Research award Social Sci. Research Council, 1957. Fellow Center For Advanced Study Behavioral Scis., 1961-62. Mem. Am. Sociol. Assn., Soc. Study Social Problems (past pres.), Soc. Psychol. Study Social Issues, Soc. Sociol. Research. Author: Patterns of Industrial Bureaucracy, 1954; Wildcat Strike, 1954; (with R. Peterson) Notes on Technology and Moral Order, 1963; (with H. P. Gouldner) Modern Sociology, 1963; Enter Plato, 1965. Editor: Studies in Leadership, 1950; (with S. M. Miller) Applied Sociology, 1965; founder Trans-Action mag., 1963. Research, publs. on sociol. studies of trade union leadership, strikes, indsl. bureaucracy; problem of causal inference; theory of applying sociology for policy purposes; hist. study Greek antiquity, social theory. Home: 9 Washington Terrace, St. Louis 63112. Office: Dept. Sociology, Washington U., St. Louis 63130.*

GOURMELEN, Étienne, French surgeon; b. Finistère, France. Succeeded Akakia as prof. of surgery; known for work during plague, 1581; contbd. to devel. of French surgery. Author: Synopseos chirurgiae libri sex, 1566; Avertissements et conseils à messieurs de Paris pour se prèserver de la peste, 1581; Guide des chirurgiens, 1634. Died Melun, France, 1593.

GOURSAT, Édouard Jean Baptiste, French mathematician; b. Lanzac, France, May 21, 1858; began teaching at Toulouse, France, 1881; named prof. math. analysis U. Paris, 1897. Mem. French Acad. Scis. Author: Leçons sur l'integration des équations aux dérivées partielles, 1890-98; Théorie des fonctions algébriques et leurs intégrales, 1894; Cours d'analyse mathématique, 1902-05; Leçons sur le problème de Pfaff, 1922. Research on infinitesimal analysis; studied theory of functions, pseudo- and hyperelliptic integrals, differential equations, invariants and surfaces. Died Paris, Nov. 25, 1936.

GOUTAREL, Robert, French chemist; b. Dôle, France, Mar. 15, 1909; s. Mathieu and Andrée (Rouget) G.; Baccalaureat Sc., U. Rennes (France), 1927; pharm. Faculty Pharmacy, Paris, 1932; Dr. med., Faculty Medecine, Paris, 1939; Dr. es Sc., Faculty Sci., Paris, 1954; m. Berthe Pontich, Nov. 10, 1942. Chef de travaux Faculty Pharmacy, Paris, 1946-60; head research Centre National de la Recherche Scientifique, Gif-sur-Yvette, France, 1956-61, sci. dir., 1961—, sub-director, Institute de Chimie des Substances naturelles, 1961—. Decorated commandeur Ordre des Palmes Académiques. Mem. Société Chimique de France, Am. Chem. Soc. Author: Les alcaloides stéroidiques des Apocynacées, 1964; also numerous articles. Research on chemistry of natural products, especially indol and steroid alkaloids. Home: 47 Madeleine, Crenon Sceaux 92, France. Office: CNRS, ICSN, Gif-sur-Yvette 91, France.*

GOUTIER, Roland, Belgium radiobiologist; b. La Louvière, Belgium, Feb. 10, 1927; s. Raoul and Mariette (Bacq) G.; M.D., U. Liège (Belgium), 1951; m. Marguerite Pirotte, July 11, 1953; children—Jean-Michel, Philippe. Fonds Nat. de la Recherche Scientifique belge scholar U. Liège, 1953-57, asst., 1958; head sect. physiol. chemistry, radiobiology dept. Centre d'Etude de l'Energie Nucléaire, Mol, Belgium, 1959—. Decorated chevalier Ordre de la Couronne. Mem. Soc. belge de Biochimie, Soc. Physiologistes de langue française, Soc. de Biologie, Soc. European de Radiobiologie, Soc. belge de Physiologie et Pharmacologie, Biochem. Soc. U.K. Research and publs. demonstrating action of caffeine and related compounds on neuromuscular system, separation of 2 aliesterases in blood serum, activation of deoxyribonuclease in irradiated tissues resulting from lesions of lysosomes and changes in cell populations. Home: 24 Frankrijklaan, Geel, Belgium. Office: Centre d'Etude de l'Energie Nucléaire, Boeretang, Mol, Belgium.*

GOUY, Georges, French physicist; b. Vals-les-Bains, Feb. 19, 1854; prof. U. Lyons (France); mem. French Acad. Scis., 1901; studied photometrically the total intensity of spectral lines in flames, 1879; studied (with G. Chapuis) osmotic equilibrium, 1888, also refraction and diffraction of X rays, 1896; set up an expression for thermodynamic efficiency which considered change in inner and kinetic energy of a system, heat quantity, accomplished outer work, change in volume, 1889. Died Vals-les-Bains, Jan. 27, 1926.

GOVAERTS, Jean Marie Lambert Theo, Belgian nuclear chemist; b. Belgium, July 15, 1913; s. Theo and Paula Govaerts; Licencie, 1936, Dr. es Sc., 1939, Agrégé Enseignement Supérieur, 1955; m. Bindelle Govaerts, Nov. 1941; children—Michele, Dominique. Faculty, U. Liège (Belgium), now prof. Decorated chevalier l'Ordre de Leopold II; Fulbright fellow, 1949. Author: Introduction a la Chimie Nucleaire, 1961; La Gamma Scintigraphie, 1965; also numerous articles. Research in nuclear chemistry and radioactive tracers. Home: 7 Quai de la Boverie, Liège, Belgium.*

GOVIER, William Miller, Am. pharmacologist; b. Defiance, O., Mar. 26, 1915; s. Charles William and Clarissa (Miller) G.; B.A., Kalamazoo Coll., 1935; M.D., Vanderbilt U., 1939; Intern, USPHS, 1939-40; Post Doctorate, Vanderbilt 1941-43, Pharmacology; m. Beverly Newell, Dec. 27, 1932; children—William Charles, Mercedes Joyce, Suzanne Ellen, Deborah Ann. Asst. prof. pharmacology and physiology Bowman Gray Sch. Medicine, 1943-44; pharmacologist Sharp & Dohme, 1944-47, Upjohn Co., 1947-54; dir. dept. pharmacology Schering Corp., Bloomfield, N.J. 1954-58; head dept. pharmacology McNeil Labs., Phila., Pa., 1958-60; dir. biol. scis. Warner-Lambert Research Inst., Morris Plains, N.J. 1960-65; v.p. Ethical Drug Research, Morris Plains, N.J., 1965——. Fellow Am. College Clin. Pharmacology and Chemotherapy; mem. A.A.A.S., Am. Soc. Pharmacology and Exptl. Therapeutics, Can. Pharm. Soc., N.Y. Acad. Sci., A.C.S. Research and numerous publs. in fields of enzymic aspects of shock and mechanism of action of sympathomimetic amines, vitamin E and digitoxin, isolation of coenzyme A, effect of barbituates on acetylation, choline cycle in cardiac decompensation, plasma substitutes, serum heart disease, barbiturate anaesthetics, tranquilizers, serotonin metabolism, antihistamines, muscle relaxants. Home: 47 Cobb Rd., Mountain Lakes, N.J. 07046. Office: 170 Tabor Rd., Morris Plains, N.J.

GOWANS, James Learmonth, immunologist; b. May 7, 1924; s. John Gowans and Selma Josephina (Ljung) G.; ed. Trinity Sch., Croydon; King's Coll. Hosp.; Lincoln Coll., Oxford; M.B., B.S., London, 1947; M.A., Oxford, 1953; m. Moyra Leatham, 1956; 1 son, 2 daus. Med. Research Exchange Council scholar Pasteur Inst., Paris, 1953; Staines med. research fellow Exeter Coll., Oxford, 1955-60; Henry Dale research prof. Royal Soc., since 1962. Fellow Royal Soc., 1963, St. Catherine's Coll., Oxford. Research, publs. on physiology of lymphoid tissue, lymphocytes, and cellular immunology.*

GOWDEY, Charles Willis, Canadian pharmacologist; b. St. Thomas, Ont., Can., Sept. 3, 1920; s. William Charles and Myrtle (Craford) G.; B.A., U. Western Ont., 1944, M.Sc., 1946; D.Phil., U. Oxford, 1948; m. Madelon Craig Gilmour, Sept. 7, 1946; children—David, Katherine, Kevin, Sheila. Faculty, U. Western Ont., London, Can., 1948——, prof. pharmacology, head, dept., 1960——. Mem. Canadian Physiol. Soc., Am. Soc. for Pharmacology and Exptl. Therapeutics, Pharmacological Soc. Can. Research on blocking agts., insecticides, shock, anaemia, oxygen toxicity, psychopharmacology, decompression sickness and hyperbaric procedures. Home: 428 Wortley Rd., London, Ont., Can.*

GOWEN, John Whittemore, Am. biologist; b. Evinston, Fla., Sept. 5, 1893; s. Charles Hayes and Gertrude (Whittemore) G.; B.S., U. Me., 1914, M.S., 1915; Ph.D., 1917; m. Marie Helena Stadler, Sept. 10, 1917; children—Elaine Stadler (Mrs. Jay Townsent Wakeley), Helen Marie (Mrs. Reid Anderson Cameron). Biologist-in-charge Biol. Dept., Me. Agrl. Exptl. Sta., Orono, Me., 1917-26; asso. mem. Rockefeller Inst. for Med. Research, Princeton, N.J., 1926-37; prof. genetics Ia. State U., Ames, 1937-64, head dept., 1948-59; prof. radiology and radiation biology Colo. State U., Ft. Collins, 1964——. Cons. AEC, 1948-64, NIH, 1954-64. Mem. A.A.A.S., Am. Soc. Naturalists, Genetics Soc. Am. (pres. 1952), Am. Soc. Human Genetics, Biometric Soc., Inst. Math. Statistics, Gerontol. Soc., Am. Genetics Assn., Radiation Research Soc., Am. Inst. Biol. Scis., Am. Soc. Zoologists (v.p. 1945——), Ia. Cancer Soc., Phi Beta Kappa, Sigma Xi, Phi Kappa Phi. Research, numerous publs. on factors important to inheritance, physiology of quantity and constituents of milk; growth of cattle; mouse and fowl diseases; biology of irradiation in man, mouse, bacteria, viruses; sex determination; mutation, gerontology, genetic theory. Died Sept. 14, 1967.

GOWERS, Sir William Richard, English neurologist; b. London, Mar. 20, 1845; s. William and Ann (Venables) G.; ed. Christ Church Coll. Sch., Oxford, Univ. Coll., London; M.D.; LL.D., Edinburgh U.; M.D. (hon.), Dublin; m. Mary Baines, 1875; 2 sons, 2 daus. House physician, pvt. sec. to William Jenner, 1867-70; med. registrar Univ. Coll., London, 1870, asst. physician, 1872, physician, later prof. clin. medicine, 1883, ret., 1888; asst. physician hosp. for paralyzed and epileptic Queen Square, London, 1873; cons. physician, from 1888. Fellow Royal Soc., 1887, Royal Coll. Physicians, Royal Coll. Physicians (Ireland) (hon.); mem. Royal Coll. Surgeons, Royal Soc. Sci. (Upsala), Soc. for Internal Medicine (Vienna), Société des Medicines Russes de St. Petersburg (hon.). Author: Manual and Atlas of Medical Ophthalmoscopy, 1879; Diagnosis of Diseases of the Spinal Cord, 1880; Epilepsy and Other Chronic Convulsive Diseases, Their Causes, Symptoms and Treatment, 1881; Diagnosis of Diseases of the Brain, 1885; Manual of Diseases of the Nervous System, 1886; The Dynamics of Life, 1894, also works on diseases of nervous system, heart, others. Research on blood vascular system; described retinal changes in patients with chronic nephritis, 1876; introduced term haemocytometer for instrument developed by Malassey in 1874, 1877; invented 1st practical hemoglobinometer, 1878; described Gowers' tract (bundle of nerve fibers in spinal cord), 1880; 1st to describe tetanic nature of epileptic convulsions, 1881, local panatrophy, 1886; 1st (with V. A. H. Horsley) to remove tumor of spinal cord successfully, 1888; described Gowers' sign (irregular contraction of pupil in tabes dorsalis in response to light), 1895; described progressive form of muscular atrophy (distal myopathy), 1902. Died London, May 4, 1915.

GOWLAND, William, English metallurgist; b. Sunderland, Eng., 1842; s. George Thompson Gowland; ed. Royal Coll. Chemistry, Royal Sch. Mines; m. Joanna Macaulay, 1890 (dec. 1909); 1 dau.; m. 2d, Maude Margaret Connacher, 1910. Chemist, metallurgist Broughton Copper Co., 1870-72, chief metallurgist, 1889-91; chemist, assayer, fgn. head Imperial Japanese Mint, 1872-88; metall. adviser to War Dept. Japan; asso. in mining and metallurgy Royal Sch. Mines, London, emeritus prof. metallurgy. Gov. Imperial Coll. Sci. and Tech., Sch. Metalliferous Mining, Camborne, Eng.; examiner in metallurgy Bd. Edn., London U. Recipient Murchison and De la Beche medals. Fellow Royal Soc., 1908; asso. Royal Sch. Mines; mem. Inst. Mining and Metallurgy (pres., Gold medal), Royal Anthrop. Inst. (pres.), Inst. Metals (pres.), Soc. Antiquaries, Inst. Chemistry, Chem. Soc., Am. Inst. Mining and Metall. Engrs., Soc. Chem. Industry, Royal Soc. Arts, Royal Instn. Author: The Metallurgy of the Non-Ferrous Metals; Imperial Mint Technical Reports; contbr. articles to tech. publs. Died June 10, 1922.

GOY, Robert William, Am. psychologist; b. Detroit, Jan. 25, 1924; s. George Frederick and Charlotte (McDowell) G.; B.S., U. Mich., 1947; Ph.D., U. Chgo., 1953; m. Barbara Elaine Perry, Nov. 13, 1948; children—Michael Frederick, Peter William, Elizabeth Ruth. Faculty dept. anatomy U. Kan., Lawrence, 1956-63, asso. prof. psychiatry, 1962-63; asso. scientist dept. reproductive physiology and behavior Ore. Regional Primate Research Center, Beaverton, 1963-65, scientist, chmn. dept., 1965——; asso. prof. med. psychology U. Ore. Med. Sch., Portland, 1964-66, prof., 1966——. Mem. Am. Assn. Anatomists, Animal Behavior Soc., Psychonomic Soc., Soc. for Study Fertility, Endocrine Soc. Contbr. articles to profl. jours. Research on gonadal hormones on influence of behavior of individual, importance of heredity and early experience as factors limiting actions of hormones, influence of hormones present prior to birth on devel. of masculine or feminine psychosexual orientation, demonstrator that hormones present before birth influence mode of conduct in adulthood leading to concept of hormonal orgn. of central nervous tissues into masculine and feminine forms. Home: 14465 S. W. Hargis Rd., Beaverton 97005. Office: 505 N.W. 185th Av., Beaverton, Ore. 97006.*

GOYAN, Frank Mayer, Am. phys. chemist; b. Placerville, Cal., June 22, 1908; s. Frank James and Mary (Mayer) G.; B.S., U. Cal. at Berkeley, 1930, M.S., 1931, Ph.D., 1937. Faculty, U. Cal. Sch. Pharmacy, San Francisco, 1932——, prof. chemistry, pharm. chemistry, 1956——, acting dean students, 1963-65. Mem. Am. Chem. Soc., Am. Pharm. Assn., Faraday Soc., Electrochem. Soc., A.A.A.S. Contbr. chpts. to books, numerous articles to sci. jours. Pioneer with glass electrode pH meter, ophthalmic solutions, thermoelectric methods for determination of colligative properties. Home: 625 Ashbury St., San Francisco 94117.*

GOZZINI, Adriano, Italian physicist; b. Florence, Italy, Apr. 13, 1917; s. Pilade and Amalia (Chiosso) G.; Ph.D. in physics; D. honoris causa, U. Clermont-Gerrand; m. Giulia Favilla, 1952; children—Andrea, Silvia, Alfredo, Carlo. Prof. exptl. physics U. Pisa. Research and numerous publs. on physics, spectroscopics, radio frequency. Home: 5 via Zerboglio. Office: 2 Piazza Torricelli, Pisa, Italy.

GRAAF, Regnier de, see de Graaf, Regnier.

GRABAU, Amadeus William, Am. paleontologist; b. Cedarburgh, Wis., Jan. 9, 1870; S.B., Mass. Inst. Tech., 1896; S.M., Harvard, 1898; Sc.D., 1900; m. Mary Antim, Oct. 5, 1901; one dau., Josephine Esther (Mrs. Kenneth Ross). Asst. instr. palaeontology Mass. Inst. Tech., 1892-97; prof. geology Rensselaer Poly. Inst., Troy, N.Y., 1899-1901; successively lectr., adj. prof., prof. paleontology Columbia U., 1901-19; became prof. paleontology Nat. U., Peking, China, 1919; chief paleontologist Chinese Geol. Survey; research asso. 3d Asiatic Expdn., Am. Mus. Natural History. Fellow Geol. Soc. Am.; mem. Paleontol. Soc. Am., N.Y. Acad. Scis. Author: (with H. W. Skinner) North American Index of Fossils, 1909-10; Principles of Stratigraphy, 1913; Text Book of Geology, 2 vols., 1920, 21; Silurian Fossils of Yunnan, 1920; Ordovician Fossils of North China, 1921; Palaeozoic Corals of China, 1921; Stratigraphy of China, vol. 1, 1924-25; Migration of Geosynclines, 1924; Silurian Fossils of Yunnan, 1926; Early Permian Fossils of China, 1934; The Rhythm of the Ages, 1940; also numerous articles. Applied principles of evolution to certain groups of invertebrates; developed pulsation and polar control theories that crustal features of earth were caused by rhythmic rise and fall of sea level as shown by its transgressions and regressions; collected information on world stratigraphic deposits and applied it to earth's history. Died Peiping, China, Mar. 20, 1946.

GRABOWSKI, Casimer Thaddeus, Am. biologist; b. Cleve., Aug. 16, 1927; s. Michael and Catherine (Alinski) G.; B.S., Western Res. U., 1950; Ph.D., Johns Hopkins, 1954; m. Marian Elizabeth Zinger, Aug. 7, 1954; 1 son, Robert. Faculty, U. Pitts. Sch. Medicine, 1954-60, asst. prof. anatomy, 1958-60; faculty U. Miami, Fla., 1960——, prof. biology, 1967——. Mem. Am. Soc. Anatomists, Am. Soc. Zoologists, N.Y. Acad. Sci., A.A.A.S., Teratology Soc., Soc. for study Growth and Devel., Phi Beta Kappa, Sigma Xi. Research, publs. on organizer center of chick embryo, mechanism of action of agts. producing birth defects, effects of oxygen lack on embryo, blood chemistry of embryo. Home: 1925 S.W. 70 Av., Miami 33155.*

GRACE, James Thomas, Jr., Am. physician, surgeon; b. Troy, Ala., July 16, 1923; s. James Thomas and Anne (Salter) G.; B.S., Yale, 1945; M.D., Harvard, 1948; m. Betty Bryant Thornton, Nov. 21, 1951; children—Elizabeth Anne, Mary Day, John Bryant, Patricia Merrill. Practiced medicine, specializing in surgery, Huntsville, Ala., 1950-51; Asst. resident surgery Vanderbilt-VA Hosp., Nashville, Tenn., 1953-56, resident surgeon, 1956-57; mem. staff Roswell Park Meml. Inst., Buffalo, 1957——, asst. dir., 1959-67, dir. viral oncology sect., 1963——, dir. inst., 1967——; faculty Vanderbilt U., 1956-57; faculty U. Buffalo Sch. Medicine, 1958——, asso. research prof. surgery, 1962——; research prof. microbiology State U. N.Y., Buffalo, 1964——. Mem. cancer virology panel NIH, 1961-62; mem. human cancer virus task force USPHS, 1962-64. Recipient Billings medal A.M.A., 1961. Diplomate Am. Bd. Surgery. Mem. A.M.A., Soc. U. Surgeons, A.C.S., Halsted Soc., N.Y. Acad. Scis., Am. Fedn. Clin. Research (councillor/Eastern sect. 1960-62), Soc. Exptl. Biology and Medicine (councillor Western N.Y. sect. 1959-64), A.A.A.S., Am. Assn. Cancer Research (dir. 1966——), Am. Soc. Mammalogist, Surg. Investigators Exchange, Am. Cancer Soc. (pres. Erie Co. div. 1965——), Nat. Inst. Gen. Med. Scis. (clin. research tng. com. 1965——), Sigma Xi. Editor surg. sect. Yearbook of Cancer, 1960——; editorial bd. Rev. Surgery, 1964——. Contbr. articles to surg. jours. Studies of host defense factors in cancer; immunology of cancer; relationship of viruses to malignancy; surg. physiology; devel. of surgical techniques for management of cancer. Established immunological relationship between cancer and coexisting dermatomyositis. Home: 4848 Smiley Terrace, Clarence, N.Y. 14031. Office: 666 Elm St., Buffalo, N.Y. 14203.*

GRACE, Oliver Davies, Am. educator; b. Washington, Dec. 21, 1914; s. Oliver Joseph and Gladys (Davies) G.; D.V.M., Colo. State U., 1940; M.S., U. Ill., 1951; m. Vera M. Hanawalt, July 17, 1948; children—Kerstin Elaine, Edward Oliver. Field worker U. S. Dept. Agr., N.D., 1940-41, lab. worker, Va., 1942, lab. worker, N.H., 1942-46; asst. prof. vet. hygiene Vet. Medicine br. Food and Drug Adminstrn., 1946-53; asst. prof. vet. sci. U. Ill., 1948-53; head vet. med. dept. Baxter Labs., Inc., Morton Grove, Ill., 1953-55; ext. vet. hygienist U. Neb., 1955-57, asso. prof. vet. sci., 1957-62, prof., 1962——; mem. Basic Sci. Exam. Bd., 1958——. Mem. N. Central Vet. Diagnosticians (chmn.), N. Central Poultry Disease Conf. (chmn.), Am., Neb. vet. medicine assns., U. S. Livestock San. Assn., Conf. Research Workers in Animal Diseases. Contbr. articles in field to sci. jours. Research cause, methods of prevention, treatment of diseases affecting domestic animals. Home: 1720 Donald Circle, Lincoln, Neb. 68505.*

GRACIAN, Zerahiah (Zerahiah ben Isaac ben Shealtiel Gracian), philosopher, physician; b. Barcelona or Toledo, Spain; flourished in Rome, circa 1277-1288. Author commentaries on Job, also on Proverbs, 1288, and Moreh Webakim (Maimonides). Translator (from Arabic to Hebrew): Aristotle's Physics, Metaphysics, De coelo et mundo, and Ibn Rushd's middle commentaries on them; De anima (Aristotle); Commentary on De coelo (Themistios); De causis (a neo-Platonic treatise); Treatise on the Nature of the Soul (al-Farabi); De causis et symptomatibus, 3 chpts. of De compositione medicamentorum secundum genera (both Galen); books 1 and 2 of Quanum (Ibn Sina); Maimonides' aphorisms, 1277, and his treatise on coitus.

GRAD, Arthur, mathematician; b. Austria, Jan. 31, 1918; s. Herman and Helen (Selinger) G.; B.S., Coll. City N.Y., 1938; A.M., Columbia, 1939; Ph.D., Stanford, 1948; m. Irene Smiley, June 21, 1946; children—Susan, Laura. Asst. sci. aid Nat. Bur. Standards, 1941; mathematician U. S. Coast and Geodetic Survey, 1941-46; research asso., acting instr. Stanford, 1946-48; mathematician Office Naval Research, 1948-53; lectr. U. Md., 1949-53; mathematician AEC computing facility N.Y. U., 1953-54; head math. br. Office Naval Research, 1954-59; program dir. math. scis., head math. scis. sect. NSF, 1959-63; asso. dean grad. div. Stanford, 1963-64; dean Grad. Sch., Prof. math. Ill. Inst. Tech., 1964——. Rep. undersea warfare group NRC, 1950, mem. div. math., 1959-63; adv. council on automatic data processing Bur. of Budget, 1962-63. Mem. Am. Math. Soc., Math. Assn. Am., A.A.A.S. Contbr. numerous articles to profl. jours. Study of confomal mapping; schlicht functions; functions of a complex variable. Home: 1359 E. 55th Pl., Chgo. 60637.

GRAD, Harold, Am. mathematician, educator; b. N.Y.C., Jan. 14, 1923; s. Herman and Helen (Selinger) G.; B.E.E., Cooper Union, 1943; M.S., N.Y. U., 1945, Ph.D., 1948; m. Betty Miller, Jan. 23, 1949; children—Hilary Lynn, Michael. Engr., Westinghouse Electric Corp., 1943-44; research asso. N.Y. U., N.Y.C., 1944-48, asst. prof. math., 1948-53, asso. prof., 1953-57, prof., 1957—, dir. magneto-fluid dynamics div. Courant Inst. Math. Sci., 1960—; dir. Space Scis. Inc., Waltham, Mass. Mem. adv. com. on thermonuclear research Oak Ridge Nat. Lab., 1964-67. Fellow Am. Phys. Soc. (chmn. div. fluid dynamics 1963); mem. Soc. Engring. Sci. (dir.), Am. Math. Soc. Research in kinetic theory, magnetohydrodynamics, plasma physics, statis. mechanics. Home: 248 Overlook Rd., New Rochelle, N.Y. Office: 251 Mercer St., N.Y.C. 10012.*

GRADE, Hans Gustav Bernhard, German aviation engr.; b. Köslin, Germany, May 17, 1879; s. Wilhelm and Anna (Walter) G.; ed. Tech. U. Charlottenburg (Germany); m. Käthe Grotum, 1910. Founder Grade-Motorenwerke GmbH. (mfr. of 2-cycle engines for bicycles and boats), Magdeburg, Germany, 1905, engine and airplane factory, Bork. Germany, 1909, flying sch. (most important in Germany), 1910; engaged in repair work, World War I; factory taken over by Aviatikflugzeug-Werke, 1916; built automobiles, 1919-28; various engring. positions, Bork, from 1928. Recipient prize donated by Karl Lanz for flight in German-built plane, 1909. Contbr. articles to aviation jours. Pioneer in German aviation; 2d licensed pilot in Germany; 1st to fly self-built airplane made completely in Germany; produced light, econ. sport planes with 2-cycle motors. Died Borkheide, Germany, Oct. 22, 1946.

GRADMAN, Robert Julius Wilhelm, German geographer, botanist; b. Lauffen/Neckar, Germany, July 13, 1865; s. Adolf and Pauline (Hörlin) G.; student theology, Tübingen, Germany, 1883-87, Ph.D. in geography and botany, 1898, hon. Ph.D., 1941; m. Julie Tritschler 1891; 1 son, Hans, 1 dau. Vicar Öhringen and Kuchen, Germany, 1889-90; became pastor, Forchtenberg, Germany, 1891; named librarian, univ. library, Tübingen, 1901, joined faculty, 1909; made expdns. to Alps, also Algeria, 1913; became prof. Erlangen, Germany, 1919; rector, Erlangen, 1925-26; expdn. to Syria, Palestine, Egypt, 1933; became emeritus, 1934; worked independently in Tübingen and Sindelfingen until 1950; founder Franconian Geog. Research, Erlangen. Chmn., Central Com. for Sci. Geography Germany, for several years. Recipient Carl Ritter Gold medal, 1931; named hon. senator U. Erlangen, 1943. Author: Pflanzenleben der schwäbischen Alb, 1898; Dinkel und die Alemannen, 1901; Geschichte des Getreidebaus im römischen und deutschen Altertum, 1909; Siedlungsgeographie des Königreichs Württemberg, 1913; Länderkunde von Süddeutschland, 2 vols., 1931; Wörterbuch deutscher Ortsnamen in den Grenz- und Auslandsgebieten, 1925. Research on geography of settlement, plant geography of Swabian Alb, relation between desert and steppe, steppes of Middle East; originated theory of influence of climatic and vegetation changes on prehistoric settlements; combined viewpoints of botanist, plant geographer, geomorphologist, historian. Died Sindelfingen, Germany, Sept. 16, 1950.

GRAEBE, Carl, German chemist; b. Frankfort/Main, Germany, Feb. 24, 1841; s. Carl and Emmeline (Boeddinghaus) G.; grad., Heidelberg, Germany, 1862; studied under Kolbe; m. Albertine Burgdorfer, 1896. Lecture asst. to Bunsen for 3 sessions; joined Baeyer's Lab., 1865; went to Leipzig, Germany, 1869; named prof. Königsberg, Germany, 1870, Geneva, 1878-1906; returned to Frankfort after retirement. Recipient Perkin medal, Lavoisier medal, Berthollet medal. Mem. French Acad. Scis., 1913, Bavarian Acad. Scis. Author: Geschichte der organischen Chemie I, 1920. Research on organic dyes, salicylic acid, quinones; discovered formula for phthalic acid, (with Caro) acridine in coal tar; synthesized (with Glaser) carbazole and phenanthrane, (with Lieberman) alizarin dye, 1869; introduced terms, ortho, meta, and para for disubstituted benzene derivatives; studied relationship between color and structure. Died Frankfort, Jan. 19, 1927.

GRAEFE, Albrecht Friedrich Wilhelm Ernst von, see von Graefe, Albrecht Friedrich Wilhelm Ernst.

GRAEFE, (Karl) Alfred, German ophthalmologist; b. Martinskirchen, Germany, Nov. 23, 1830; s. Friedrich Heinrich and Florentine (Stephan) G.; studied medicine in Halle, Heidelberg, Würzburg, Leipzig (all Germany), Prague, 1850-54; M.D., Halle, 1854; m. Marie Charlotte Colberg, 1857; 5 sons, including Felix, 5 daus. Asst. physician in his cousin's (Albrecht von Graefe) pvt. eye clinic, Berlin, 1855-58; joined faculty U. Halle, 1858, became lectr., 1859, asso. prof., 1864, prof., 1873, dir. eye clinic, 1884, became emeritus and moved to Weimar, Germany, 1892; founder pvt. clinic for eye diseases, Halle, 1859. Author: Klinische Analyse der Motilitätstörungen des Auges, 1858. Editor: (with T. Sämisch) Handbuch der gesamten Augenheilkunde, 7 vols., 1874-80. Research on motility defects of eye; 1st description of Ischaemia retinae; 1st to perform tear duct extirpation; introduced Lister's antisepsis to ophthalmology. Died Weimar, Apr. 12, 1899.

GRAEFE, Karl Ferdinand von, see von Graefe, Karl Ferdinand.

GRAEFE, Richard Edmund, German chemist; b. Dresden, Germany, Dec. 18, 1876; s. Emanuel and Emma (Funk) G.; studied chemistry Technische U., Dresden, 1897-1901; Ph.D., Basel, Switzerland, 1901; hon. dr. engring. Freiberg, 1925; m. Johanna Sommer, 1905; 1 son, Lothar, 1 dau. Joined firm Riebecksche Montanwerke beginning as lab dir., Webau nr. Halle, Germany, 1901, then tech. dir.; made study trips to U. S., Scotland, 1905-07; joined Tex. Co., 1910; a founder Deutsche Trinidad-Asphalt Gesellschaft, 1911, also bus. mgr.; worked for Colas GmbH, Dresden-Reick; ind. chemist, 1933-45; a founder lignite found. Freiberg Mining Acad., 1917; became titular prof., 1918. Author: Laboratoriumsbuch für die Braunkohlenindustrie, 1908; also numerous articles. Editor: Einführung in die chemische Technologie der Brennstoffe, 1927. Developed processes in fuel research, especially tar (chief product of distillation); invented (with R. v. Walther) process of heat-pressure splitting of heavy lignite oil to extract benzine; studied fuel chemistry, bituminous hwy. materials, regeneration used lubricating oil; suggested extraction of creosote, resins and asphalt with alcohol; studied refining of mineral oil. Died Dresden, Feb. 20, 1945.

GRAETZ, Leo, German physicist; b. Breslau (now Wroclaw, Poland), Sept. 26, 1856; s. Heinrich and Marie (Monasch) G.; studied math. and physics, Berlin, Strasbourg, France; Ph.D., Breslau, 1880; m. Emilie Heller, 1883; 1 son, 1 dau., Leonie (Mrs. Ernst von Seuffert). Became asst. to A. Kundt, Strasbourg, 1881; named lectr., Munich, 1883, asso. prof., 1893, prof. physics with Röntgen, 1908. Author: Die Elektrizität und ihre Anwendungen, 1883; Kompendium der Physik, 1887, 5th edit. under title, Lehrbuch der Physik, 1922; Kurzer Abriss der Elektrizität, 1897; Das Licht und die Farben, 1900; Die Atomtheorie in ihrer neuesten Entwicklung, 1918; Die Physik, 1919. Editor: Handbuch der Elektrizität und des Magnetismus, 5 vols., 1912-28. Research on heat conduction, heat radiation, friction, electricity, elec. waves, including normal and abnormal dispersion, X-rays and cathode rays. Died Munich, Nov. 12, 1941.

GRAF, Engelbert, German chemist; b. Steinheim, Germany, June 15, 1922; s. Peter Engelbert and Ida (Kämmerer) G.; Apotheker, Frankfurt (Germany) U., 1942, Jena (Germany) U., 1945; Dr.rer.nat., Würzburg (Germany) U., 1950; m. Ingeborg Steiff, Oct. 28, 1960. Sci. asst. Würzburg (Germany) U., 1948-55, privatdozent Inst. für Pharmacy and Food Chemistry, 1956-60; asso. prof. Tübingen (Germany) U., 1960—, leader dept. pharm. tech., 1967—. Mem. German Pharm. Soc. (v.p. Bavaria and Württemberg 1958—), Am., German chem. socs., German Soc. Medicinal Plant Research, Physic.-Med. Soc. Würzburg. Author: (with F. R. Preuss) Gadamers Lehrbuch d. chemisch. Toxikologie, 1966; also articles. Editor: Mitteil. Deutsche Pharmaz. Gesellschaft, 1965—. Research on analytical methods, including determinations of alkaloids and other drugs, elucidations of chem. structures; transformations of eburicoic acid. Home: 18 Philosophenweg, Tübingen 74, Germany.*

GRAFF, Frederic, Am. civil engr.; b. Phila., May 23, 1817; s. Frederick and Judith (Sawyer) G.; m. Elizabeth Mathieu. Asst. engr. Phila. Water Dept., 1842-47, chief engr., 1847-56, 66-62, reorganized dept. by combining dist. works with prin. city works, planned and directed constrn. of 3 reservoirs and 1 dam, modernized Phila. water system; park commr. City of Phila., 1851, established Fairmount Park System; experimented with pumping machinery, suggested water-supply systems for many of larger cities in East; pres. Am. Soc. C.E. 1885, dir. several years; pres. Engrs. Club of Phila., 1880; pres. Franklin Inst. Died Phila., Mar. 30, 1890.

GRAFF, Frederick, Am. engr.; b. Phila., Aug. 27, 1774; s. Jacob Graff; m. Judith Sawyer; 1 son Frederic. Became draftsman Phila. Water Works (1st steam-powered water works in U. S.) 1797, supt., 1805; selected (with John Davis) Mt. Morris (now Fairmount), Pa. as site for new reservoir, 1810; designed mains, over 113 miles of which were laid by 1842, also connections, stopcocks, and fire plugs for water system of Phila. (1st efficient hydraulic water system in nation), chief engr. Phila. Water Dept. until 1847; mem. Franklin Inst. Died Apr. 13, 1847.

GRAFF, Kasimir Romuald, astronomer; b. Próchnowo, Poland, Feb. 7, 1878; s. Stanislaus and Valentine (Rother) G.; Ph.D., U. Berlin, 1901; m. Frida Hoffmann, 1905; m. 2d, Maria Frank, 1943. Asst., Urania Obs., Berlin, 1897-1902; joined Hamburg (Germany) Obs., 1902; became observer new obs., Bergedorf, Germany, 1909; named prof. U. Hamburg, 1917; geodesist, cartographer during World War I; became prof., dir. obs. U. Vienna, 1928; ret., 1948; made study trip to Pulkowo, Moscow, Tashkent, 1906-07; mem. solar eclipse expdn., Crimea, 1914. Mem. Austrian Acad. Scis., Academia Pontificia Vaticana, Commn. for Internat. Geodesy, German Astron. Soc. Author textbooks including: Elemente der Astronomie, 1910; also numerous articles. Research on unstable stars and photometric comparison stars nr. them, planets, including surfaces of Jupiter, Staturn and especially Mars; measured star colors, nebulae, various constellations; invented instruments for measuring star brightness. Died Breitenfurt, nr. Vienna, Feb. 15, 1950.

GRAFF, Theobald Louis, German ophthalmologist; b. Wiesbaden, Germany, Oct. 16, 1899; s. Jacob and Bertha (Lehmann) G.; Ph.D. in Natural Sci., U. Frankfort on Main; m. Ruth Glokke, 1927. With obs. U. Frankfort, 1922, instr. ophthal. optics, 1949—; with optical industry, Rathenow, 1923-45; prof. U. Berlin, 1946-47; with adminstrv. council optical research, Rathenow, 1947-49. Recipient Duncker medal German Community Optical Visual Research. Mem. German Soc. Applied Optics, Ophthal. Soc. Germany, Union Med. Ophthalmologists Rhein and Main. Research and publs. on bifocals, refraction of human eye, ophthalmoscope. Home: 21 Kirchhainerstrasse. Office: University Eye Clinic, 14 Ludwig-Rehnstrasse, Frankfort on Main, Germany.

GRAFF DE PANCSOVA, Ludwig Bartholomäus, zoologist; b. Pancsova, Yugoslavia, Jan. 2, 1851; s. Hermann and Elisabeth (Zoldy de Zold) G. de P.; began study medicine U. Vienna, 1868, zoology U. Graz (Austria), 1871; Ph.D., Strasbourg, France, 1873; m. Eugenie Schorisch, 1874; 2 sons, 2 daus., including Rosa (Mrs. Alfred Filz von Reiterdauk). Became asst. to C.T. v. Siebold, U. Munich (Germany), 1873, joined faculty, 1874; named prof. forestry acad., Aschaffenburg, Germany, 1876; became prof. zoology U. Graz, 1884, dean, 1888-89, rector, 1896-97; mem. expdns. to Ceylon, Java, 1893-94, Arctic sea, 1902, N.Am., 1907. Hon. pres. VIII Internat. Congress Zoologists, Graz, 1910. Mem. Soc. Morphology and Physiology (co-founder), German Zool. Soc. (a founder), Austrian Acad. Scis. Author: (with F. L. Leuckart) Das Genus Myzostoma, 1877; Monographie der Turbellarien, 1882-99; Das Schmarotzertum in Tierreich und sein Bedeutung für die Artbildung, 1907; also articles. Research on animal parasites, turbellarians (flatworms), including taxonomy, morphology, Died Graz, Feb. 6, 1924.

GRAFFAR, Marcel, Belgian physician; b. Huy, Belgium, July 14, 1910; s. Joseph and Marie (Ramlot) G.; ed. Med. Sch., Free U. Brussels (Belgium), 1928-35, Harvard Sch. Pub. Health, 1936-37; m. Antoinette Fuss, Apr. 29, 1939; 1 child, Françoise. Asst. in pediatric service St. Pierre U. Hosp., 1946-53, agrégé of advanced instruction pediatrics, from 1950, instr. social medicine Med. Sch., from 1952, also head pediatric clinic; asso. prof. Free U. Brussels, 1952-55, prof., from 1955, dir. research Sociology Inst.; head pediatrics service Brugmann Hosp., from 1956, also head pediatric clinic; dir. Nat. Center Study Child Growth and Devel., 1960—; asst. dir. Inst. Hygiene and Social Medicine; v.p. Sch. Pub. Health. Mem. Belgian Soc. Pediatrics (past pres.), Belgian Assn. Hygiene and Social Medicine. Author: (with others) Cinq cents familles d'une commune de l'agglomération bruxelloise, 1957. Research, publs. dealing with growth and devel. of normal children, effects of socio-econ. conditions on health and devel., epidemiology of infectious and non-infectious diseases in children. Home: 52, rue Hector Denis, Brussels 5. Office: Faculté de Médecine et de Pharmacie, Laboratoire de Médecine Sociale, 7 rue Heger-Bordet, Brussels 1, Belgium.*

GRÄFFE, Karl Heinrich, mathematician; b. Braunschweig, Germany, Nov. 7, 1799; s. Dietrich Heinrich and Johanna Friederike (Moritz) G.; ed. Carolineum in Braunschweig, 1821; studied math. U. Göttingen (Germany); Ph.D., 1825; m. Lucie Sulzer, 1830; 2 sons, including Eduard, 3 daus. Became tchr. Tech. Inst., Zurich, 1828; named prof. Upper Indsl. Sch., 1833; named lectr. U. Zurich, 1833, asso. prof., 1860. Recipient prize for dissertation, 1825, prize for Gräffe method Berlin Acad. Scis. Author. Die Geschichte der Variationsrechnung vom Ursprung der Differential-und Integralrechnung bis auf die heutige Zeit, 1825; Die Auflösung der höheren numerischen Gleichungen, 1837. Invented method of computing imaginary roots (Gräffe method). Died Zurich, Feb. 17, 1873.

GRAFFI, Dario, Italian physicist; b. Rovigo, Italy, Jan. 10, 1905; s. Michele and Amalia (Tedeschi) G.; Dr.Physics, U. Bologna (Italy), 1925, Dr.Math., 1927; m. Lina Cavazza, Dec. 11, 1937; children—Mariella, Paola, (Mrs. Colombo), Sandro, Giorgio. Asst., Faculty Engring., U. Bologna, 1925-35; prof. math., and physics Lyceum of Cagliari, 1935-36; asso. prof. rational mechanics U. Torino (Italy), 1936-38; prof. U. Bologna, 1939—; asso. prof. U. Paris, 1937-38. Recipient Medagli 'oro dei benemeriti Scuola della cultura dell'arte. Mem. Acad. Lincei, Acad. Bologna, Acad. Torino, Istituto Lombardo Scienze e Lettere. Author: Onde elettromagnetiche, 1965; also numerous articles. Research in electromagnetism, mechanics of fluids, nonlinear mechanics, reciprocal theorems, operational calculus, adiabatic invariants, hereditary phenomena. Home: 9 A. Murri, Bologna, Italy.*

GRAFFUNDER, Walter, physicist; b. Frankfort/Main, Germany, Jan. 7, 1898; s. Wilhelm and Anna (Brauer) G.; studied natural scis. Technische U., Darmstadt, Germany, U. Frankfort; Ph.D., Frankfort,

1922; m. Clara Borsari, 1940; 1 dau. Asst. to G. Schmaltz, K. W. Meissner, U. Frankfort, 1922-34; with tube devel. lab. Telefunken G.m.b.H., 1934-45; settled in Switzerland, 1946, and became asst. to F. Dessauer; named lectr. U. Freiburg, (Germany), 1948, asso. prof., 1950. Research and publs. on Debeye's theory of dielectric properties of liquids, temperature dependence of dielectricity constants of organic liquids and binary liquid mixtures, optical problems, including photoelectric effect, ion variation in atmosphere, linear accelerators; developed numerous measurement methods in field of electronic tubes and high frequency technics. Died Engadin, Switzerland, Aug. 12, 1953.

GRAHAM, A. Stephens, Am. physician; b. Augusta, Ga., Nov. 3, 1900; s. Frank T. and Estelle (Lewis) G.; student U. Fla., 1919-21; M.D., U. Va., 1925; M.S. in surgery U. Minn., 1929; m. Nancy Hart Gordon, June 28, 1926; children—Nancy Gordon (Mrs. Oliver Hitch), Elizabeth Lindsay (Mrs. Brenton Halsey). Practice medicine, specializing in gen. surgery, Richmond, Va., 1933—; faculty surgery Med. Coll. Va., 1933—, asso. prof., 1938-66, emeritus prof., 1966—; sr. surg. cons. VA Hosp., Richmond, 1947-67; v.p., 1960-66, chief surgeon Stuart Circle Hosp., 1946—, also bd. dirs. Mem. Am., So. surg. assns., Internat. Soc. Surgeons, A.C.S., A.M.A., Soc. Cons. to Armed Forces, Sigma Xi. Author: (with Fred Rankin) Cancer of Colon and Rectum, 1939; Christopher's Text Book of Surgery, 1940; Livingston and Pack's Treatment of Cancer, 1940; Bancroft's Abdominal Surgery, 1941; Fields Fundamentals in Cancer Practice, 1963; Ochsnes's Cyclopedia of Medicine, 1964; also numerous articles. Research on lymphatics, physiology of liver including developing of shunt maneuvers (with F. Mann) later used in surgery treatment of cirrhosis of liver in man; developed (with Rankin) operations for cancer of colon, particularly obstructive resection and spl. instrument for its implementation and also for aseptic union of colon after removal of cancer. Home: 811 Arlington Circle, Richmond 23229. Office: 1805 Monument Av., Richmond, Va. 23220.*

GRAHAM, Angus Frederick, Am. microbiologist; b. Toronto, Ont., Can., Mar. 28, 1916; s. Frederick J. and Mary (Ball) G.; came to U. S., 1958, naturalized, 1963; B.A.Sc., U. Toronto (Ont.), 1938, M.A.Sc., 1939; Ph.D., U. Edinburgh (Scotland), 1942, D.Sc., 1952; m. Jacqueline Francoise Poirier, July 3, 1955; children—Robert J., Andrew D., Paul F. Lectr. biochemistry Carnegie Teaching fellow U. Edinburgh, 1942-47; research asso. Connaught Med. Research Labs., U. Toronto, 1947-58, asso. prof. microbiology, 1953-58; mem. Wistar Inst. Anatomy and Biology, Wistar prof. microbiology U. Pa., Phila., 1958—. Eleanor Roosevelt Internat. Cancer fellow Institut du Radium, Paris, France, 1964-65. Mem. Canadian Soc. Microbiology, Am. Soc. Microbiologists, N.Y. Acad. Scis., A.A.A.S. Editor-in-chief, Jour. Cellular Physiology, 1965—; editorial bd. Jour. Virology, 1964—. Research and publs. on chemistry of enzymes and viruses, especially mechanism of multiplication of viruses in living cells. Home: 148 David Dr., Havertown, Pa. 19083.* Office: Wistar Inst., 36th and Spruce Sts., Phila.

GRAHAM, Clarence Henry, Am. psychophysiologist; b. Worcester, Mass., Jan. 6, 1906; s. Robert Samuel and Ann Jane (Gillespie) G.; A.B., Clark U., 1927, A.M., 1928, Ph.D., 1930; M.A., Brown U., 1943, D.Sc. (hon.), 1958; m. Elaine R. Hammer, Sept. 6, 1949. Instr. psychology Temple U., Phila., 1930-31; Nat. Research fellow Johnson Found. for Med. Physics, U. Pa., 1931-32; asst. prof. Clark U., Worcester, 1932-36; mem. faculty Brown U., 1936-45, prof., 1941-45; prof. psychology Columbia U., N.Y.C., 1945—. Sci. liaison officer Office Naval Research, London, 1952-53, mem. psychology panel, 1947-55; participant Kyoto (Japan) Seminars in Am. Studies, 1952; mem. Armed Forces-NRC Vision Com., 1946-58, exec. council, 1956-58. Recipient Presidential Certificate of Merit, 1948; certificate of appreciation Office Naval Research, 1961. Mem. Soc. Exptl. Psychologists (Howard Crosby Warren medal 1941), Optical Soc. Am. (Tillyer medal 1963), Am. Psychol. Assn., Am. Physiol. Assn., Nat. Acad. Scis., Am. Philos. Soc., Am. Acad. Arts and Scis., Sigma Xi. Editor: (with others) Vision and Visual Perception, 1965. Contbr. numerous articles to profl. jours. Research in data and theory of brightness vision, color vision, visual perception, elec. responses from eyes of lower organisms, visual functions of person colorblind in one eye and normal in other, space and movement perception, analysis of "illusion" of movement. Home: 70 Haven Av., N.Y.C. 10032.*

GRAHAM, Eugene Alexander, Jr., Am. engr.; b. Phila., Mar. 25, 1925; s. Eugene Alexander and Irma (Meyer) G.; E.E., Mass. Inst. Tech., 1954; Ph.D., U. Turin (Italy), 1957; m. Luciana Angela Fornara, Aug. 22, 1957; children—Alexander Eugene, Carl Albert. Prof.-researcher scholar Galileo Ferraris Nat. Advanced Studies Inst. of Polytechnic, and U. Turin, 1955-57, fellow, vis. prof., 1962; sr. systems research engr. Nat. Co., Inc., 1958-59; dir. research Microwave Assos., Inc., 1959-60; sr. staff engr. Raytheon Co., 1960-63; staff cons. Space-Gen. Corp. subsidiary Aero-jet Gen. Corp., 1963—; prof.-researcher Université de Paris, École Normale Supérieure, Lab. de Physique, 1966—. Decorated Silver Star with cluster, Bronze Star. Mem. I.E.E.E., Am. Phys. Soc., Am. Math. Soc., A.A.A.S., Soc. for Indsl. and Applied Math., Nat. Geog. Soc. Research and publs. in communications, including statis. communication and detection theory, synthesis of systems, network synthesis, microwave through millimeter wave to optical and solid state circuitry and systems, upper atmosphere physics, plasmas. Home: 68, rue du Moulin de la Pointe, Paris 13e. Office: 24, rue Lhomond, Paris Ve, France.*

GRAHAM, Evarts Ambrose, Am. surgeon; b. Chgo., Ill., Mar. 19, 1883; s. David Wilson and Ida Anspach (Barned) G.; A.B., Princeton, 1904; M.D., Rush Med. Coll., 1907; spl. chem. studies U. Chgo., 1913, 14; many hon. degrees; m. Helen Tredway, Jan. 29, 1916; children—David Tredway, Evarts Ambrose. Intern Presbyn. Hosp. Chgo., 1907-08, asst. attending surgeon, 1912-15; asst., also instr. surgery Rush Med. Coll., 1909-15; mem. staff Otho. S. A. Sprague Meml. Inst., Rush, Chgo., 1912-15; chief surgeon Park Hosp., Mason City, Ia., 1915-17; prof. surgery Washington U. Sch. of Medicine, 1919-51, emeritus since 1951; surgeon in chief Barnes Hosp., St. Louis Children's Hosp.; surgeon in chief Peter Bent Brigham Hosp., 1925; temp. prof. surgery St. Bartholomew's Hosp., London, 1939. Mem. NRC, 1925-46; mem. Nat. Bd. Med. Examiners, 1924-33 chmn. Am. Bd. Surgery, 1937-41. Recipient others; Gross prize in surgery, 1920; Leonard prize Am. Roentgen-Ray Soc., 1925; Gold medal Am. Radiol. Soc., 1925, for devel. of cholecystography; Gold medal and certificate of merit St. Louis Med. Soc., 1927; Gold medal So. Med. Assn., 1934; John Scott medal City of Phila., 1937; St. Louis Award, 1942; Lister medal Royal Coll. Surgeons, Eng., 1942; Roswell Park medal, 1949; Am. Coll. Chest Physician medal, 1949; Miss. Valley Med. Soc. medal, 1949; distinguished service medal A.M.A., 1950; Bigelow medal Boston Surg. Soc., 1951, Charles Mickle Fellowship, Univ., Toronto, 1943. Fellow A.C.S. (pres.); mem. A.M.A., Am. Assn. Thoracic Surgery (pres.), Soc. Clin. Research, Am. Philos. Soc.; mem., hon. mem. other Am. and fgn. med. socs. Author: Empyema Thoracis, 1925; Diseases of the Gall-Bladder and Bile Ducts, 1928; also sect. in Medical and Surgical History of World War, 1924. Editor: Surgical Diagnosis, 1930; Yearbook of Surgery, 1926-27. Co-editor Archives of Surgery, 1920-45, Annals of Surgery, 1935-45; editor Jour. Thoracic Surgery, 1931—. Described disturbed mechanics of respiration and circulation when normal intrathoracic pressures are altered; developed (with W. H. Cole) Graham-Cole test, or method for cholecystography, an X-ray visualization of gall-bladder, 1924; devised new treatment for chronic abscess of the lung; contbd. to pathology and treatment of carcinoma of bronchus, explanation of particular toxicity of choloroform and similar anaesthetic agts.; performed (with J. J. Singer) 1st successful total pneumonectomy, 1933. Died Mar. 4, 1957.

GRAHAM, Frances Keesler (Mrs. David Tredway Graham), Am. psychologist; b. Canastota, N.Y., Aug. 1, 1918; d. Clyde C. and Norma (Van Surdam) Keesler; B.A., Pa. State U., 1938; Ph.D., Yale, 1942; m. David Tredway Graham, June 14, 1941; children—Norma, Andrew, Polly. Acting dir. St. Louis Psychiat. Clinic, 1942-44; instr. Barnard Coll., 1948-51; instr. Sch. Medicine Washington U., St. Louis, 1942-48, 53-55, research asso., 1953-57; research asso. U. Wis., Madison, 1957-64, asso. prof., 1964—; cons. NINDB Perinatal Research Br. Mem. Am. Psychol. Assn., Soc. Research Child Devel., Soc. Psychophysiol. Research, A.A.A.S., Phi Beta Kappa, Sigma Xi. Mem. editorial bd. Jour. Exptl. Child Psychology, 1964—, Child Devel., 1966—. Research and publs. primarily in field of developmental changes in autonomic responsiveness to stimulation; developed psychol. tests for study brain injury. Home: 2927 Harvard Dr., Madison 53705. Office: U. Hosps., Madison, Wis. 53706.*

GRAHAM, Frank Dunstan, Am. aero. engr.; b. Princeton, N.J., Aug. 17, 1922; s. Frank Dunstone and Mary Louise (Power) G.; B.S. in Aero. Engring., Princeton, 1943, M.S., 1947; m. Arline Marie Roesch, Dec. 23, 1944; children—Bruce, Geoffry. Draftsman, Fleetwings, Inc., Bristol, Pa., 1943; aerodynamicist Boeing Airplane Co., Seattle, 1947-48; flight research engr. Cornell Aero. Lab., Buffalo, 1948-50; gen. supervisory engr. All-Weather Flying div. USAF, Wright-Patterson AFB, Ohio, 1950-54; chief engr. flight controls Lear, Inc., Grand Rapids, Mich., 1954-59; asso. prof. aero. engring. Princeton, 1959-67, prof., 1967—. Mem. USAF B-52 Structural Modifications Rev. Com., 1965-66; guest lectr. U. Cal., Los Angeles, 1965-67; tech. dir. Systems Tech., Inc., Hawthorne, Cal., 1959—. Mem. I.E.E.E., Am. Inst. Aeros. and Astronautics, Am. Assn. U. Profs., Sigma Xi. Author (with D. T. McRuer): Analysis of Nonlinear Control Systems, 1961. Contbr. articles to sci. jours. Devel. integral criteria for servomechanism performance; expts. on human dynamic response and formulation of a predictive model of human operator behavior in control tasks. Home: 54 Maclean Circle, Princeton, N.J. 08540.*

GRAHAM, George, Brit. inventor; b. Horsgill, Eng., 1675; apprenticed to watchmaker, London, 1688; succeeded to business of Tompion; built astron. instruments for Halley, Bradley, French Acad. Scis.; Fellow Royal Soc., 1720. Contbr. articles to tech. jours. Invented mercurial pendulum, 1726, dead beat escapement; also invented and improved many astron. instruments; discovered daily variation of magnetic declination. Died Nov. 20, 1751.

GRAHAM, Horace Delbert, Am. food technologist; b. Camaguey, Cuba, Aug. 7, 1925; s. Ernest Elisha and Rose Jane (Duncan) G.; B.S., McGill U., 1950; M.S., Mich. State U., 1952; Ph.D., U. Ill., 1958; m. Dorothy Jean Goins, May 23, 1962; children—Dexter, De-Albert Yolande, Ernesto Jonathan Fitzgerald. Research asst. Mich. State U., 1950-52; research asst. U. Ill., 1955-58; research asso. Carver Found. Tuskegee Inst. (Ala.), 1959-62; head applied biochemistry unit Sci. Research Council Govt. Jamaica, 1962-63; asso. prof. biology U. P.R., Mayaquez, 1963-66, prof. biology, chief scientist P.R. Nuclear Center, 1966—. Mem. Inst. Food Technologists, A.A.A.S., Am. Dairy Sci. Assn., Am. Pharm. Assn., Sigma Xi. Contbr. numerous articles to profl. jours. Research on devel. of methods for isolation and quantitative determination of food gums, polyoxyethylene and polyoxypropylene compounds, devel. of methods for determination of piperine and gibberellic acid, interactions of food gums and irradiation of foods, anionic detergents. Address: P.O. Box 1326, Mayaquez, P.R. 00709.*

GRAHAM, John, surgeon; b. Glasgow, May 12, 1879; s. Daniel Graham; B.Sc., 1902; M.B., Ch.B., Glasgow U., 1904; postgrad. in anatomy and surgery in Berlin and London. House surgeon Victoria infirmary, 1905; asst. surgeon Royal Samaritan Hosp. for Women, Glasgow; prof. anatomy Anderson Coll. Medicine, Glasgow, 1919-46; examiner in anatomy Royal Faculty Physicians and Surgeons Glasgow; ret., 1946. Fellow Royal Faculty Physicians and Surgeons Glasgow; mem. Glasgow So. Med. Soc. (pres. 1922-23). First to report case of trench fever under name relapsing febrile illness, 1915. Died Dec. 15, 1958.

GRAHAM, John Borden, Am. pathologist; b. Goldsboro, N.C., Jan. 26, 1918; s. Ernest Heap and Mary (Borden) G.; B.S., Davidson Coll., 1938; M.D., Cornell U., 1942; m. Ruby Barrett, Mar. 23, 1943; children—Charles Barrett, Virginia Borden, Thomas Wentworth. Asst., Cornell U., 1943-44; faculty U. N.C., Chapel Hill, 1946—, Alumni Distinguished prof. pathology, 1966—. Mem. genetics tng. com. USPHS, 1962-66; mem. pathology test com. Nat. Bd. Med. Examiners, 1963-67, chmn., 1967—; mem. Internat. Com. Hamostasis and Thrombosis, 1963—; cons. Environmental Health Center, USPHS. Markle scholar in med. sci., 1949-54. Mem. A.M.A., So. Med. Assn., A.A.A.S., Elisha Mitchell Sci. Soc., Am. Assn. U. Profs., Soc. Exptl. Biology and Medicine, Am. Soc. Exptl. Pathology, Assn. U. Pathologists, Am. Assn. Pathologists and Bacteriologists, Am. Soc. Human Genetics, Internat. Soc. Hematology, Sigma Xi. Publs. on co-discovery of Stuart Factor in blood clotting; among first to map genes on human X-chromosome; descriptions of inherited diseases in humans. Home: 108 Glendale Dr., Chapel Hill, N.C. 27514.*

GRAHAM, John Warren, Am. geoscientist, educator; b. Boston, Aug. 4, 1918; s. Clarence Flack and Mary Jarvis (Fish) G.; A.B., Johns Hopkins, 1940, Ph.D., 1949; m. Helen-Stuart Link, June 7, 1941; children—Kristianne (Mrs. Peter Bumpus), Peter C. Fellow, Carnegie Instn. of Washington, 1947-51, staff mem. dept. terrestrial magnetism, 1951-58; staff Woods Hole Oceanographic Instn., 1958-63; prof. geoscis. S.W. Center for Advanced Studies, Dallas, 1963—. Adj. prof. So. Methodist U., 1963—. Fellow Geol. Soc. Am.; mem. Am. Geophys. Union, Geochem. Soc., Sigma Xi. Asso. editor Jour. Geophys. Research, 1965—. Contbr. articles to profl. jours. Research on origin and significance of magnetic properties of rocks, history of earth's magnetic field in geologic time, biol. origin of manganese modules on sea floor, deformation mechanics of sedimentary rocks, high sensitivity instruments for magnetic measurements. Home: 6626 Northaven, Dallas 75230. Office: P.O. Box 30365, Dallas 75230.*

GRAHAM, Owen Hugh, Am. med. entomologist; b. Thorndale, Tex., Apr. 18, 1917; s. Owen Daniel and Clara (Clymore) G.; B.S., Tex. A. and M. U., 1938, M.S., 1940, Ph.D., 1962; m. Joy Sanderson, Oct. 4, 1940; children—Georgia, Owen Hugh, Elena, Andrew B. Entomologist, Bur. Entomology and Plant Quarantine, U. S. Dept. Agr., Brownwood, Tex. and Yuma, Ariz., 1939-42, Menard, Tex., 1942, 46, entomologist, Kerrville, Tex., 1947-50; entomologist C.E. U. S. Army, Canal Zone, 1950-56; entomologist, asst. sta. leader entomology research div. Livestock Insects Lab., U. S. Dept. Agr. Kerrville, 1956-63, investigations leader, 1963—. Adviser to U. S. delegation to UN FAO Conf., Jamaica, 1958; cons. OIRSA, 1964—. Recipient Agrl. Research Service Incentive award, 1958. Mem. A.A.A.S., Entomol. Soc. Am., Am. Inst. Biol. Scis., Am. Mosquito Control Assn., Sociedad Mexicana de Entomologia, Sigma Xi. Contributed to improved procedures for control malaria and oriental schistosomiasis; devel. systemic insecticides; studies of biology and control of cattle fever ticks and other ticks, sterile male eradication of Dermatobia hominis. Home: 212 Woodcrest Dr. Office: P.O. Box 232, Kerrville, Tex. 78028.*

GRAHAM, Richard Hugh, Am. physicist; b. Pitts., May 25, 1921; s. Herbert W. and Izetta (Ewens) G.; student Rensselaer Poly. Inst., 1938-40; B.S., U. Wash., 1948; m. Margaret Murnane, May 17, 1946; children—Roger W., Robin A., Rosalyn L. Physicist,

Allis Chalmers Mfg. Co., Milw., 1948-51, Cal. Research and Devel. Co., Livermore, 1951-54; sect. head U. S. AEC, Washington, 1954-56; staff scientist Lockheed Missile and Space Div., Palo Alto, Cal., 1957-59; mgr. uranium devel. operation nuclear energy div. Gen. Electric Co., Santa Clara, Cal., 1959-67; staff Gulf Gen. Atomic Inc., 1967——. Cons. to AEC, Holmes & Narver, Inc. Served with weather service USAAF, 1942-45. Mem. Am. Nuclear Soc., Am. Phys. Soc., A.A.A.S. Research and publs. in nuclear reactor fuel cycles; nuclear reactor physics; nuclear reactor safety experimentation; med. physics including artificial kidney devel.*

GRAHAM, Robert Lockhart, Canadian physicist; b. Peterboro, Ont., Can., Feb. 17, 1921; s. Hugh Henry and Evelyn (Davis) G.; B.A., McMaster U., 1943, M.A., 1945; Ph.D., U. London (Eng.), 1949, D.I.C., 1949; m. Charlotte Thomas, Sept. 2, 1944; children—Diana, Evan, Hugh. Asso. research officer Atomic Energy Can., Ltd., Chalk River, Ont., 1953-62, sr. research officer, 1962——; vis. physicist Lawrence Radiation Lab. U. Cal., Berkeley, 1962-63. Cons. Med. Research Council, London, 1946-49, Imperial Chem. Industries, Widnes, Eng., 1947-49. Mem. Am. Phys. Soc., Canadian Assn. Physicists. Contbr. numerous articles to profl. jours. Developed mass spectrometer techniques for studying mass distbn. of fission fragments, first techniques for measuring nanosecond nuclear lifetimes; studies of nuclear structure, beta and gamma rays from radioactive isotopes, internal conversion electrons using 100cm radius beta-spectrometer. Home: 4 Tweedsmuir Pl., Deep River, Ont. Office: Physics Div., Atomic Energy Can., Ltd., Chalk River, Ont., Can.*

GRAHAM, Ruth Moore (Mrs. John B. Graham), Am. cytologist; b. Paris, Ida., Mar. 11, 1917; d. Charles O. and Josephine (Durkee) Moore; B.S., U. Mich., 1938; diploma Lab. Technique, Simmons Coll., 1939; Sc.D. (hon.), Women's Med. Coll. Pa., 1954; m. John B. Graham, June 15, 1940; children—Michael, Susan. Dir., Vincent Cytology Lab., clin. cytologist Mass. Gen. Hosp., research asso. dept. gynecology Harvard Med. Sch., Boston, 1951-56; asso. cancer research scientist Roswell Park Meml. Inst., Buffalo, 1957; clin. asso. in gynecology and pathology State U. N.Y., Buffalo, 1963——. Cons. Walter Reed Gen. Hosp., Washington, 1952-55. Recipient Annual Leonard Wien award Cancer Inst. Miami, 1955. Mem. Am. Soc. Cytology (award for outstanding achievement in cytology 1963), Am. Assn. Cancer Research. Author: Cytologic Diagnosis of Cancer, 1950, 12th edit., 1963; also articles. Research in early diagnosis of uterine cancer; mode of action of radiation. Office: 666 Elm St., Buffalo 14203.*

GRAHAM, Thomas, Brit. chemist; b. Glasgow, Scotland, Dec. 20, 1805; M.A., Glasgow, 1824; postgrad. in Edinburgh, Scotland; D.C.L., U. Oxford (Eng.), 1853; tchr. math. and chemistry in pvt. lab., Glasgow, 1825-29; asst. Mechanics' Instn., 1829; prof. chemistry Andersonian U., Glasgow, 1830-37; successor to Turner at Univ. Coll., London, Eng., 1837-55; master of mint, 1855-69. Bakerian lectr., 1850, 54; v.p. chem. jury for exhbn. of 1851. Recipient Keith prize Royal Soc. Edinburgh, 1834. Fellow Royal Soc., 1836 (twice v.p., Gold medal 1840, 50); corr. mem. French Acad. Scis., 1847; 1st pres. Chem. Soc., 1840, Cavendish Soc., 1846. Author: Elements of Chemistry, 1842; also 63 sci. papers. Pioneer in phys. chemistry; founder of colloid chemistry; differentiated between 2 classes of substances (which he called colloids and crystalloids) that pass through a membrane at different rates; discovered process of dialysis; formulated Graham's Law (speeds of diffusion of different gases are inversely proportional to sq. roots of their densities), 1834; introduced term osmosis; investigated alcohol and salt compounds; his investigations of phosphoric acid led to modern concept of polybasic acids. Died London, Sept. 11, 1869.

GRAHAM, Wallace Harry, Am. surgeon; b. Highland, Kan., Oct. 9, 1910; s. James Walter and Elizabeth Marie (Veneman) G.; student Central Mo. State, Warrensburg, Mo., 1928-32; B.S., Creighton Univ., 1934, M.D., 1936; student Univ. of Vienna, 1937, Univ. of Budapest, 1938, Royal Coll. of Surgeons, Edinburgh, Scotland, 1939; m. Velma Ruth Hill, Sept. 15, 1935; children—Wallace Scott, Heather Ellen, Bruce. Served externship, Mass. Gen. Hosp., Boston, 1935; internship and pres. house staff, Kansas City Gen. Hosp., Kansas City, Mo., 1936; personal physician to the President of the United States, September 12, 1945; professorial lecturer in surgery George Washington Univ. Sch. of Medicine and associate surgeon George Washington U., Washington; surg. staff Bapt. Meml., Research, Menorah, St. Margarets hosps. (Kansas City, Mo.), Walter Reed Gen. Hosp. (Washington); founder Westport Med. Center (Kansas City, Mo.); hon. mem. faculty U. of Santo Domingo, Dominican Republic; research worker in cancer; special asst. to the surgeon general United States Air Force, senior flight surgeon. Member of Presidential Spl. Mission to Guatemala City and the Middle East; United States Air Force delegate to International Coll. Medicine and Pharmacy, Mexico City, Internat. Symposium of High Altitude Biology, Lima, Peru. Entered active service with U. S. Army as 1st lt., advanced through the grades to maj. gen., 1951; chief of 24th Evacuation Hospital Surgical Service in combat, 1942-45. Section chief of surgery Walter Reed Army Hosp. Decorated Bronze Star, Purple Heart (U. S.); Chevalier of the Legion of Honor, Croix de Guerre with palm (France); Distinguished Service Order, Hon. Companion (Great Britain); Officer Order of Orange-Nassau with crossed sabres and rosette (Netherlands); Officer Order of Leopold with palm and attrition; Order of Finlay, Grade of Comdr. (Cuba); Order Italian Star of Solidarity, 2d Class; Croix de Guerre with palm (Belgium); Comdr., Order of Mil. Merit (Brazil); Cross of Comdr. of the Royal Order of the Phoenix (Greece). Elected to Hall of Fame, Creighton U., 1945. Certified fellow Internat. Coll. Surgeons. Fellow A.C.S.; mem. Aero Med. Assn., Royal Soc. Medicine, A.M.A., Mo., Jackson County (Mo.), D.C. med. socs., Mo., D.C., George Washington U. surg. socs., Pan Am. Surg. Assn. Contbr. numerous articles to med. and surgical publs. Home: 5157 Ward Parkway. Office: 1815 E. 63d St., Kansas City, Mo.

GRAHN, Douglas, Am. geneticist; b. Newark, Apr. 25, 1923; s. Victor F. and Greta (Franzen) G.; B.S., Rutgers U., 1948; M.S., Ia. State Coll., 1950, Ph.D., 1952; m. Sally Linn Smythe, Dec. 21, 1946; children—Frederick S., Catherine L., Alice A. Faculty, Rutgers U., 1948, Ia. State Coll., 1948-52; asso. biologist div. Biol., Med. Research Argonne (Ill.) Nat. Lab., 1953-58, 61-62, asso. div. dir., 1962-66, sr. biologist, 1966——; geneticist U. S. AEC, Washington, 1958-61; mem. radiation control adv. bd. Md., 1960-61; co-chmn. space radiation study panel Space Sci. Bd. Nat. Acad. Sci.-NRC, 1962——; cons. McDonnell Aircraft Co., St. Louis, 1963-64, Office Manned Space Flight NASA, 1963——. Mem. Genetics Soc. Am., A.A.A.S., Am. Soc. Naturalists, Radiation Research Soc., Am. Soc. Human Genetics, Am. Genetics Assn., Am. Soc. Zoology, Sigma Xi. Research and publs. primarily in field of determination of genetic basis of radiation sensitivity in mice, environmental factors affecting neonatal death rate and birth weight in U. S. population, life shortening and other pathol. effects of radiation and the influence of genetic factors on life expectancy. Home: 5720 Carpenter St., Downers Grove, Ill. 60515. Office: 9700 S. Cass Av., Argonne, Ill. 60439.*

GRAINGER, John, Brit. plant pathologist; b. nr. Huddersfield, Yorkshire, Eng., Oct. 2, 1904; s. Edward and Lillian (Pugh) G.; student Huddersfield Tech. Coll., 1922-23; B.Sc., Leeds (Eng.) U., 1928, Ph.D., 1930; m. Mary Hewlett, Apr. 8, 1931. Hon. fellow plant pathology U. Wis., 1927-28; fellow for research virus diseases of plants Leeds U., 1928-30, asst. lectr. hort. botany, 1930-33; dir. Tolson Meml. Mus., Huddersfield, 1933-43; head dept. plant pathology W. of Scotland Agrl. Coll., Auchincruive, 1943——; hort. adviser Huddersfield Corp., 1940-43; Brit. Council visitor to New Zealand, 1960. Recipient Silver medal award for new machines Royal Highland Show, Aberdeen, Scotland, 1951, Royal Agrl. Show, Oxford, 1959. Mem. Am. Phytopath. Soc., Brit. Mycol. Soc., Assn. Applied Biologists, Soc. European Nematologists, European Assn. for Potato Research, Yorkshire Naturalists' Union. Author: Virus Diseases of Plants, 1934; Garden Science, 1935; (with F. A. Mason) A Catalogue of Yorkshire Fungi, 1937. Research, publs. on discovery of Cp/Rs, a measure of a host plant's physiol. capacity to become diseased, costless control of crop disease, new machines for control of pathogens infesting whole soil mass, self-calculating instrument for forecasting potato late blight, econ. assessments of crop disease on local, nat. and internat. scales. Home: White Gables, St. Quivox, Ayr. Office: Dept. Plant Pathology, W. of Scotland, Agrl. Coll., Auchincruive, Ayr, Scotland.*

GRAINGER, Richard Dugard, Brit. anatomist, physiologist; b. Birmingham, Eng., 1801; s. Edward G.; cadet Mil. Acad., Woolwich, Eng., later surgeon; lectr. anatomy; apptd. lectr. St. Thomas Hosp. Sch., 1842; ret. 1860; named insp. to inquire into cholera Bd. Health, 1849. Fellow Royal Soc., 1846 mem. Christian Med. Assn. (a founder 1854), Royal Coll. Surgeons (became mem. council 1845). Author: Observations on Structure and Functions of the Spinal Chord, 1837. Developed theory of functions of sympathetic nervous system. Died Feb. 1, 1865.

GRAM, Hans Christian Joachim, Danish bacteriologist; b. Copenhagen, Denmark, Sept. 13, 1853; ed. Copenhagen, Berlin, Strasbourg, France, Marburg, Germany; M.D., 1883. Named lectr. U. Copenhagen, 1883, prof. pharmacology, 1891, prof. pathology and therapy, 1900-23; dir. clin. medicine Frederiks and Rigs-Hospitalet, Copenhagen, 1892-1923. Author: Undersogeler over de Rode Blodlegemers Storelse hos Menesket, 1883; Laegemidlernes Egenkaber og doseri Tabelform, 1897; Klinisk-Therapeutiske Foreslaninger, 1902-09; The Results of a New Method for Determining the Fibrin Percentage in Blood and Plasma, 1922. Research on physiology of anemia in pregnancy, hemoblasts in pernicious anemia; determined cell vol. and elec. conductivity of blood; introduced differential staining method for dividing bacteria into gram negative and gram positive groups, 1884; described method for determining fibrin percentage in blood and plasma, 1922. Died Copenhagen, Nov. 14, 1938.

GRAMATZKI, Hugh Ivan (John), physicist; b. Shillong, India, Aug. 12, 1882; s. Emil Ludwig and Martha Catherine (Hensel) G.; grad. Tech. U., Karlsruhe, Germany, 1907; Ph.D., Berlin, 1937; m. Elsa Suchland, 1932. Owned pvt. obs., Kleinmachnow, nr. Berlin. Mem. Astro-Gesellschaft (co-founder), Richard Wagner Soc. (2d chmn.). Author: Leitfaden der astronomischen Beobachtung, 1928; Probleme der konstriktiven Optik und ihre mathematische Hilfsmittel, 1954; also astron. articles, novels, novellas, radio plays. Research on non-Archimedian math., geometric optics, modern photometry; developed sensitive photo and projection objective lenses, variable focus-length lens, photometer equipment for star measurement; studied planets, especially observations of Mars, 1924; color photographs of lunar eclipse, 1942. Died Kleinmachnow, Mar. 14, 1957.

GRAMMATEUS, Heinriel (Schreyber, Heinrich), mathematician; b. Erfurt, Germany, before 1496; ed. U. Vienna, 1507; studied in Cracow, Poland, 1514-17; With U. Vienna until 1521 (when univ. was closed due to plague); then went to Nuremberg, later to Erfurt; returned to Vienna as examiner of baccalaureat candidates and procurator of nation of Saxony. Author: Algorithmus proportionum . . . , 1514; Libellus de compositione regularum pro vasorum mensuratione . . . , 1518; Ayn new kunstlich Buech . . . , 1521; Behend und Khunstlich Rechnung . . . , 1521; Ein kunstreich und behend Instrument . . . , 1522; Eyn kurtz newe Rechnen und Visyrbuech leynn . . . , 1523; Algorismus de integris . . . , 1523. First in Germany to have expositions on algebraic problems (binomials, quadratic equations) appear in print; contributed to spread of new algebra; improved math. symbolic language. Died Vienna, winter 1525/26.

GRAMME, Zénobe-Théophile, engr., inventor; b. Jehay-Bodegnée, Belgium, Apr. 4, 1826; studied at École industrielle in Luttich; model worker in Société l'alliance; specialized in constrn. of elec. apparatus; worked for French engr., Bazin; worked as a turner; entered into bus. with Hippolyte Fontaine, 1871; founder (with Fontaine) Societe des machines magneto-electriques Gramme. Recipient Volta prize from Louis Napoleon, 1852; decorated Commdr. Order of Leopold, 1898. Developed (with Hippolyte Fontaine) magneto-electric machines; invented alternating current machines, 1867; improved Pacinotti's machine; in 1869 invented continuous current dynamo which was used in electrometallurgy and to produce electric light, in 1871; Gramme ring, armature, and machine named after him. Died Jan. 20, 1901.

GRAMONT. Arnaud de, see de Gramont, (Antoine-Alfred) Arnaud-Xavier-Louis.

GRANATA, Angelo, Italian physician; b. Varese, Italy, Oct. 29, 1922; s. Attilio and Eugenia (Cantu) G.; Med.Dr., Pavia U., 1947; Haematology specialization, Haemat. Sch., Pavia, 1949; Cardiology specialization, (French Govt. scholar), Paris Med. Sch., 1952. Asst. med. dept. Pavia U., 1947-49; asst. to chair chest diseases Milan (Italy) U., 1953-55; asst. to occupational medicine dept. Messina (Italy) U., 1956-59, asst. prof. medicine, 1959——. Fellow Italian Histochemistry Soc. (founding), Italian Occupational Medicine Soc., Italian Hepatology Soc. Contbg. author: Trattato Italiano di Medicina Interna, 1966. Research and numerous publs. on cytomorphological and cytochem. modifications of blood from exogenic noxae, analysis of biol. and toxicological activity of metals, phagocytosis of blood cells from metals, prophylaxis of heart diseases, organic and cytological reactions from ascorbic acids, cytochemistry of vaginal smears; developed cytochem. method for detecting vitamin C in cells and tissues. Home: Via Bruschetti 18 Milan, Italy. Office: U. Messina, Messina, Italy.*

GRANATO, Andrew Vincent, physicist; b. Cleve., May 9, 1926; s. Salvatore and Francesca (Polizzi) G.; student U. Rochester, 1944-46; B.S., Rensselaer Poly. Inst., 1948, M.S., 1950; Ph.D., Brown U., 1955; m. Pauline Brassard, July 21, 1956; children—Samuel Charles, Andrea Marguerite, Sarah Francesca, Ann Vivian. Research asso. Brown U., 1955-57; research asst. prof. U. Ill., Urbana, 1957-59, asso. prof., 1961-64, prof., 1964——; vis. prof. Technische Hochschule, Aachen, Germany, 1960-61. Guggenheim fellow, 1960. Mem. Am. Phys. Soc., Acoustical Soc. Am., Am. Assn. Physics Tchrs. Research, publs. primarily in field of solid state physics. Home: 1702 W. Green St., Champaign, Ill. 61820.*

GRANCHER, Jacques Joseph, French physician; b. Felletin, France, 1843; agrégé, 1873; Physician in hosps., 1879; became clinician of diseases of infants, 1885. Confirmed Villemain's work showing that tubercular virus is specific and inoculable, 1873; described splenization of lung in lobar pneumonia (Grancher's pneumonia, Desnos's pneumonia, splenophneumonia), 1883; studied children of tubercular mothers; introduced Grancher's system of boarding children from tuberculous households, 1903. Died Paris, 1907.

GRAND'EURY, (François) Cyrille, French paleo-botanist; b. Houdreville, France, Mar. 9, 1839; prof. St. Etienne Sch. Mines; mem. French Acad. Scis., 1885. Author: Sur la flore carbonifère des environs de St. Etienne, 1876; Mémoire sur la formation de la houille, 1882; Géologie et paléontologie, 1890. Studied coal formation, also carboniferous flora of Loire region; identified Westphalian formations as Carboniferous. Died Malzeville, France, July 22, 1917.

GRANDI, Guido, Italian entomologist; b. Vigevano, Pavia, Italy, Mar. 3, 1886; s. Giuseppe and Elisabetta (Mainardi) G.; Ph.D. in natural sci. Full prof. entomology U. Bologna, also dean Faculty Agrarian Scis., dir. Bollettino dell' Instituto de Entomologia; mem.

Higher Council Pub. Instrn. and Higher Council Diseases of Plants. Mem. Italian Nat. Acad. Entomology (perpetual pres.), Acad. dei Lincei, Acad. Scis. Bologna, also numerous Italian and fgn. sci. socs., entomol. socs. Berlin, Brussels, Zurich, Paris. Author: Intorduzione allo Studio dell Entomologia, 2 vols., 1951; Studi di un Entomologo sugli Invenotteri Superiori, vol. 1, 1961. Research and numerous publs. on problems of specialized biology, geneology of morphology. Address: 6 via Filippo Re, Bologna, Italy.

GRANDI, Luigi Guido, Italian mathematician; b. Cremona, Oct. 7, 1671. Prof. philosophy U. Pisa, 1700, prof. math., 1714; mem. Order of Camaldolites. Fellow Royal Soc., 1709. Author: On Series and Infinitesmals; On Sound; Flores Geometrici, 1728; Quadratura Circuli et Hyperbolae, 1703, 10. Made important contbns. to devel. of geometric theory; wrote especially on analogies of circle and equilateral hyperbola, logarithmic curve, curves of double curvature on sphere and quadrature of certain parts of spherical surface; also studied problems of sound and conics. Died Pisa, July 4, 1742.

GRANDJEAN, François Alfred, French engr.; b. Lyons, France, Oct. 17, 1882; s. Jules and (Bonidal) G.; ed. École nationale supérieure des mines; École polytechnique; m. Hélèn Ouvre, June 4, 1914; children—Suzanne, Marthe, Bernard. Became engr.; apptd. prof. École nationale supérieure des mines, Saint-Etienne, then at Paris, later hon. prof.; inspector-gen. of mines; dir. Service for Geol. Map of France; retired. Mem. Nat. Mus. Natural History (asso.), French Acad. Scis. Research, publs. on liquid crystals. Home: 15, quai de l'Ile, Geneva, Switzerland. Office: 61, rue Buffon, Paris 5, France.

GRANDJEAN, P. Étienne, Swiss physician; b. Bern, Feb. 24, 1914; s. Fritz and Marie (Kindler) G.; ed. univs. Bern, Vienna, Oxford, Harvard; M.D.; m. F. Siebenmann, 1941; children—Christine, Dominique. Prof. agrégé U. Lausanne, 1948; titular prof. hygiene and physiology E.T.H. of Zurich, 1950. Mem. Internat. Assn. Human Engring. (hon. sec.). Author over 200 works, including: Physiologische Arbeitsgestaltung. Home: 115 Susenbergstrasse. Office: Eidg. Techn. High School, Zurich, Switzerland.

GRANGER, Walter Willis, Am. paleontologist; b. Middletown Springs, Vt., Nov. 7, 1872; s. Charles H. and Ada (Haynes) G.; studied in high sch. for 2 years; hon. D.Sc. Middlebury Coll., 1931; m. Anna Dean, Apr. 7, 1904. Mem. staff Am. Mus. Natural History, from 1890, successively asst. in taxidermy, field collection in zoology, asst. curator, apptd. asso. curator, 1927; 2d in command Asiatic expdn., 1924-31. Mem. Soc. Vertebrate Paleontology, Explorer's Club, Paleontol. Soc., Geol. Soc. Contbr. articles to jours. Contbd. to knowledge of Eocene stratigraphy and fauna, also faunas of eastern Asia; assisted in collecting various Sauropod dinasaurs (great Brontosaurus skeleton being most famous); credited with classifying and describing Paleocene epoch. Died Sept. 6, 1941.

GRANICK, Sam, Am. biochemist; b. N.Y.C., Feb. 16, 1909; s. Aaron and Dora (Ustin) G.; B.S., U. Mich., 1929, M.S., 1933, Ph.D., 1938; m. Elsa Bachman, 1938; children—Donna, Joel Leonor. Mem. staff Rockefeller Univ., 1938-—, mem., prof. biochemistry, 1964-—, Harvey lectr., 1949. Mem. Nat. Acad. Scis., Am. Soc. Naturalists (v.p. 1963), Am. Soc. Biol. Chemists, Am. Chem. Soc., Society for Developmental Biology (pres. 1967), Am. Soc. Plant Physiologists, Bot. Soc. Am. Studies on chloroplasts: their isolation, structure, and inheritance; studies on free radicals of redox dyes; on iron metabolism; studies on heme and chlorophyll biosynthesis and the control mechanisms involved. Home: 43-17 48th St., Long Island City, L.I., N.Y. 11104. Office: Rockefeller Univ., York Av. and 67th St., N.Y.C. 10021.*

GRANIT, Ragnar Arthur, physiologist; b. Helsinge, Finland, Oct. 30, 1900; s. Arthur Wilhelm and Bertie (Malmberg) G.; Mag.phil., Swedish Normallyceum, 1923, M.D., 1927; M.D. (hon.), U. Oslo (Norway), 1951; D.Sc., Oxford (Eng.) U., 1956, U. Hong Kong, 1961; m. Marquerite M. Bruun, Oct. 2, 1929; 1 son, Michael W. Th. Docent physiology Helsingfors (Finland) U., 1929, prof. 1937-40; fellow med. physics U. Pa., Phila., 1929-31; Rockefeller fellow Oxford U., 1932-33; faculty Med. Sch., Karolinska Inst., Stockholm, Sweden, 1940-—, chmn. neurophysiology, 1946-—; dir. Nobel Inst. for Neurophysiology, Stockholm, 1945-—; vis. prof. Rockefeller U., 1956-66. Silliman lectr. Yale, 1954. Recipient Donders medal Utrecht U., 1957; Retzius Gold medal Oslo U., 1957; Jahre prize, 1961; III Internat. St. Vincent prize Tourin Acad. Medicine, 1961; Sherrington gold medal, 1967; Nobel prize in physiology and medicine, (with H. K. Hartline and G. Wald) 1967. Fellow Royal Soc., 1960; mem. Nat. Acad. Scis. (U. S.), Royal Swedish Acad. Sci. (pres. 1963-65), Physiol. Soc. (Eng.) (hon.), Am. Physiol. Soc. (hon.), Am. Assn. Neurologists (hon.), Indian Acad. Scis. (hon.), Am. Philos. Soc., others. Author: Sensory Mechanisms of the Retina, 1947; Receptors and Sensory Perception, 1955; Charles Scott Sherrington: An Appraisal, 1966. Discovered inhibition in retina and color specific modulators and dominators in optic nerve, existence of tonic motoneurones of various types, control of gamma motor nerves for muscle spindles and their relevance for rigidites and spasticities; intracellular research on naturally and artificially stimulated motoneurones. Home: 14 Eriksbergsgatan. Office: Med. Nobel Inst., Stockholm 60, Sweden.*

GRANSTROM, Marvin Leroy, Am. civil engr.; b. Anaconda, Mont., Sept. 25, 1920; s. Carl August and Alida (Eckstrom) G.; B.S., Morningside Coll., 1942; B.S. in Civil Engring., Ia. State U., 1943; M.S., Harvard, 1947, Ph.D., 1955; m. Ruth Maybelle Olsen, Jan. 1, 1944; children—David Marvin, Kay Ruth, Chris Carl. Instr., Case Inst. Tech., 1947-49; asso. prof. U. N.C., 1949-58; prof., chmn. dept. civil engring. Rutgers U., New Brunswick, N.J., 1958-—. Cons., Nat. Univ. for Engrs., Lima, Peru, 1955-57; mem. sub-com. on san. engring. Nat. Acad. Sci.-NRC. Mem. Sigma Xi, Delta Omega, Tau Beta Pi. Research on coagulation, chlorine and chlorine dioxide chemistry, indsl. wastes, radioactive wastes. Contbr. articles to tech. jours. Home: 1796 Watchung Av., Plainfield, N.J., 07060. Office: Rutgers U., New Brunswick, N.J.*

GRANT, Harold Johnson, Jr., Am. zoologist; b. Bronx, N.Y., Nov. 4, 1921; s. Harold Johnson and Esther (Blake) G.; A.B., U. Colo., 1950, M.A., 1952, Ph.D., 1962; m. Margaret Louise Smith, Dec. 20, 1947; children—Lucinda Borrows, Laurel Smith. Grad. fellow U. Colo., 1953-54, instr. biology, 1953; asst. curator Acad. Natural Scis., Phila., 1954-60, asso. curator, 1960-63, curator, chmn. dept. insects, 1963-66. Fellow Royal Entomology Soc. London; mem. Am. Entomology Soc. (pres., editor, cons.), A.A.A.S., Soc. Systematic Zoology, Am. Soc. Zoologists. Author: Orthoptera of North America, 1960. Contbr. numerous articles to profl. jours. Research, publs. on classification, evolution, distbn. and bionomics of the Order Orthoptera (grasshoppers and allies). Home: 628 Conestoga Rd., Berwyn, Pa. 19312. Office: 19th and Parkway, Phila. 19103. Died Feb. 27, 1966.

GRANT, Sir James Alexander, physician; b. Inverness, Scotland; s. James and Jane (Ord) G.; went to Can., 1831; ed. Queen's Coll., Kingston, Ont., also London, Edinburgh; M.D., McGill U., 1854; m. Maria Malloch, 1856. Began practice medicine, Ottawa, Can.; physician to gov.-gen. of Ottawa, 1867-1905; mem. Canadian Parliament, 1865-78, 92-96. Recipient Gold medal for med. sci., Palermo, Sicily. Fellow Royal Coll. Physicians, London, Royal Coll. Surgeons, Edinburgh; mem. Royal Coll. Surgeons, London; pres. Canadian Med. Assn., 1872, Internat. Congress of Hygiene for Can., 1909, Canadian sect. 4th Internat. Hygienic Congress, Buffalo, 1913, Medico-Chirurg. Soc. (hon. Ottawa). Contbr. to med. and surg. jours. Discoverer of serum therapy, 1861; advocated elec. treatments for prolonging life, 1909. Died Feb. 1920.

GRANT, James William, astronomer; b. Wester Elchies, Scotland, Aug. 12, 1788; s. Robert G.; m. Margaret Wilson; 3 sons, 4 daus. With E. India Co., 1805-49, beginning as writer and held successively more important posts in Bengal until retirement; built granite obs. at Elchies. Fellow Royal Astron. Soc. Detected companion of Antares 2 years before observed by Mitchel, 1844. Died Sept. 17, 1865.

GRANT, Louis Strathmore, West Indian microbiologist; b. Jamaica, W.I., Mar. 6, 1913; s. Henry Augustus and Catherine (Clucas) G.; M.B., Ch.B., Edinburgh U., 1934; M.P.H., U. Mich., 1941; diploma London U., 1949; m. Pauline Phillips, June 7, 1947; 4 children. Med. officer Govt. Bacteriol. and Path. Lab., Jamaica, faculty UCWI, sr. lectr.; reader U. W.I., Mona, Jamaica, 1960-64, head dept. microbiology, 1960-—, prof., 1964-—, mem. med. faculty bd., 1953-54, 56-—. Mem. adv. com. to Minister Health, 1955-—; mem. Expert Panel on Virus Diseases, WHO, 1951-—; mem. Internat. Com. for Phage Typing, 1950-—. Fellow Royal Soc. Medicine London, Am. Pub. Health Assn., Coll. Pathologists; mem. Am. Pub. Health Assn., Brit. Med. Assn. (pres. Jamaica), Soc. Clin. Pathologists. Research, numerous publs. on Leptospirosis, Arbovirus disease, poliomyelitis, syphilis, Salmonella and Klebsiella infections, viral encephalitis. Home: 8 College Common, Mona, Kingston 7, Jamaica, W.I.*

GRANT, Nicholas John, Am. metallurgist; b. South River, N.J., Oct. 21, 1915; s. John and Mary (Sudnik) G.; S.B., Carnegie Inst. Tech., 1938; Sc.D., Mass. Institute of Technology, 1944; married to Anne T. Phillips, Sept. 12, 1942 (deceased April 1957); children—Anne P., William D., Nicholas P.; m. 2d, Susan Mary Cooper, Aug. 1963; son, Jonathan. Metallurgist Bethlehem Steel Co., 1938-40; mem. faculty Mass. Inst. Tech., 1942-—, prof. metallurgy, 1955-—; pres. N.E. Materials Lab., Inc., 1954-—; tech. dir. Investment Castings Inst., 1954-—; cons. industry, 1947-—. Vice pres. Nitralloy Corp. U. S. A.; dir. General Diode Corporation, Loomis-Sayles Mutual Fund. Member materials committee NASA, 1958-—. Recipient distinguished service award Investment Castings Inst., 1956. Mem. Am. Soc. Metals, Am. Inst. Mining, Metall. & Petroleum Engrs., Inst. Metals (London), Sigma Xi. Contbr. articles profl. jours., chpts. in books. Contributions to high temperature understanding of deformation and fracture; to high temperature understanding of alloy theory; development of new class of alloys: oxide dispersion strengthened; holder of 22 U. S. patents and numerous issues in Europe. Home: 10 Leslie Rd., Winchester, Mass. 01890. Office: Massachusetts Inst. Technology, Cambridge 39, Mass.*

GRANT, Robert, astronomer; b. Grantown-on-Spey, Scotland, June 17, 1814; ed. King's Coll., Aberdeen, Scotland; student Paris, France, 1845-47; M.A., U. Aberdeen, 1855, LL.D., 1865; m. Elizabeth Emma Davison, Sept. 3, 1874; 1 son, 3 daus. Editor, Monthly Notices, 1852-60; apptd. prof. astronomy, dir. obs. Glasgow U., 1859. Fellow Royal Astron. Soc. (Gold medal 1856), Royal Soc., 1865. Author: History of Physical Astronomy from the Earliest Ages to the Middle of the Nineteenth Century, 1852; Catalogue of 6415 Stars for the Epoch 1870, 1883; also articles. Observed chromosphere and prominences from his sta. nr. Vittoria, 1860; observed Leonid meteors of 1866, 68, Andromed of 1872, 85, ingress of Venus at transit of 1882; identified positions of thousands of stars. Died Oct. 24, 1892.

GRANT, Robert Edmond, Scotch comparative anatomist; b. Edinburgh, Nov. 11, 1793; s. Alexander Grant; M.D., U. Edinburgh, 1814; studied medicine and natural history, Paris, other continental univs., 1815-20. Explored the coasts of Scotland, Ireland and adjacent islands; lectr. on comparative anatomy of invertebrates for Dr. John Barclay, 1824; prof. comparative anatomy and zoology London U., 1827-74; Fullerian prof. physiology, 1837-40; Swiney lectr. on geology Brit. Mus. Fellow Royal Soc., 1836, Royal Soc. Edinburgh. Author: Lectures, 1833-34; Outlines of Comparative Anatomy, 1835-41; also articles. Publs. on sponges which showed their animal nature; believed in the transformation of species. Died Aug. 23, 1874.

GRANT, Ronald Thomson, Brit. physician; b. Glasgow, Scotland, Nov. 5, 1892; M.B., Ch.B., Glasgow (Scotland) U., 1915, M.D., 1921. Staff cardiac dept. U. Coll. Hosp., London, Eng., 1921-34; dir. clin. research unit Med. Research Council, Guy's Hosp., London, 1934-57, researcher, Med. Sch., 1957-—. Decorated Order Brit. Empire. Fellow Royal Soc., 1934, Royal Coll. Physicians. Research and publs. on cardiovascular system in health and disease. Home: Farley Green Cottage, Albury, Guildford, Surrey, Eng. Office: Guy's Hosp. Med. Sch., London S.E.1., Eng.*

GRANT, Sir Thomas Tassell, inventor; b. 1795; joined service, 1812; became comptroller victualling and transport service, 1850; ret. 1858. Recipient medals for invention biscuit machine French Crown, Soc. Arts. Fellow Royal Soc., 1840. Invented steam machinery for mfg. biscuits, 1829, life-buoy, feathering paddle wheel, Grant's patent fuel (used in navy); proposed distillion of fresh water from sea water for drinking and cooking, 1834. Died Oct. 15, 1859.

GRANT, Ulysses S., IV, Am. geologist, petroleum engr.; b. Salem Center, N.Y., May 23, 1893; s. Ulysses S. and Fannie (Chaffee) G.; A.B. cum laude, Harvard, 1915; Ph.D., Stanford, 1929; m. Frances Dean, Sept. 27, 1950. Faculty, U. Cal., Los Angeles, 1931-59, prof. geology, 1940-59, chmn. geology dept., 1937-45; dir. U. S. Grant Investment Co., San Diego, 1949-—; cons. indsl. firms, govt. agys. Fellow Geol. Soc. Am.; mem. Am. Soc. C.E., Am. Assn. Petroleum Geologists (past pres. Pacific sect.). Research, publs. on geology, paleontology, engring; study of Tertiary and Quaternary mollusks of Cal. Home and office: 121 Groverton Pl., Los Angeles 90024.*

GRANT, Ulysses Sherman, Am. geologist; b. Moline, Ill., Feb. 14, 1867; s. Lewis Addison G. and Mary Helen (Pierce) G.; B.S., U. Minn., 1888; Ph.D., Johns Hopkins, 1893; m. Avis Winchell, Oct. 1, 1891; children—Addison Winchell, Lois, Avis Harriet (Mrs. E. E. Swick), Willard Winchell. Asst. state geologist, Minn., 1893-99; instr. geology U. Minn., 1897-98; prof. geology and curator of Mus., 1899-—; acting dean Coll. Liberal Arts, Northwestern U., 1907-08, 16-19. Geologist on Geol. and Natural History Survey of Wis., 1899-1907, U. S. Geol. Survey, 1904-25, Ill. Geol. Survey, 1906-20, Ore. Bur. Mines and Geology, 1913-15. Author: Preliminary Report on the Copper-bearing Rocks of Douglas County, Wis., 1900, 2d edit., 1901; Vols. IV and V, Final Report of the Geol. and Natural History Survey of Minn. (with N. H. Winchell), 1899-1900; Report on the Lead and Zinc Deposits of Wis., 1906; Copper and Other Mineral Resources of Prince William Sound, Alaska, 1906, 1910; Description of the Lancaster and Mineral Point Quadrangles, Wis., 1907; Glaciers of Prince William Sound and Kenai Peninsula, Alaska, 1910-12; Mineral Resources of Kenai Peninsula, Alaska, 1910-12. Asso. editor Am. Geologist, 1897-1906. Died Sept. 22, 1932.

GRANT, Verne Edwin, Am. biologist; b. San Francisco, Oct. 17, 1917; s. Edwin E. and Bessie (Swallow) G.; A.B., U. Cal. at Berkeley, 1940, Ph.D., 1949; m. Karen Alt, Nov. 2, 1960. Teaching asst. in botany U. Cal. at Berkeley, 1946-49; NRC fellow Carnegie Inst., Stamford, Cal., 1949-50; geneticist Rancho Santa Ana Bot. Garden, Claremont, Cal., 1950-—; asst. prof. botany Claremont Grad. Sch., 1951-53, asso. prof. 1953-57, prof., 1957-67; prof. biology Inst. Life Sci., Tex. A. and M. U., College Station, 1967-—. Mem. A.A.A.S., Genetics Soc., Soc. for Study Evolution, Am. Soc. Naturalists. Author: Natural History of the Phlox Family, 1959; The Origin of Adaptations, 1963; The Architecture of the Germplasm, 1964; (with Karen Grant) Flower Pollination in the Phlox Family, 1965; also numerous articles. Adv. editor for genetics and evolution Ency. Am., 1955-64. Research on genetics and evolution of plants. Home: Superior, Ariz.*

GRANT, William West, Am. surgeon; b. Russell County, Ala., Nov. 15, 1846; s. Dr. Thomas McDonough and Mary J. (Benton) G.; student Jefferson Med. Coll.; Bellevue Hosp. Med. Coll.; M.D., L.I. Med. Coll., 1868; postgrad., hosps. Berlin, Vienna, London; m. Mary A. Mosely, 1878 (dec. 1888); m. 2d, Nanny Craig Green, 1895. Practiced at Davenport, Ia., 1870-88; post surgeon Rock Island (Ill.) Arsenal, 1885-88; in Europe, 1888-89, Denver, 1890—; pres. surg. staff St. Joseph's Hosp.; surgeon St. Luke's Hosp.; local surgeon C.,R.I.&P. Ry., Denver, 1890—; surgeon gen., Colo., 1899-1903. Performed 1st complete operation for facial paralysis by anastomosis of facial and spinal accessory nerves; originator of standard operation for diseases and deformity of the mouth; performed early operation for appendicitis, Jan. 4, 1885. Died Jan. 8, 1934.

GRANVILLE, Joseph Mortimer, Brit. physician; b. Eng., 1833. Author: (with Andrew Wynter) The Borderlands of Insanity, and Other Papers; The Cure and Cause of the Insane; being the Report of the Lancet Comm. on the Lunatic Asylums, 1875-77, for Middlesex, London, and Surry, 1877; Nerve-vibration and Excitation as Agents in the Treatment of Functional Disorders and Organic Disease, 1883; Gout and Its Clinical Aspects, 1885; Devised hammer for giving vibratory massage to treat functional and organic nervous disorders, circa 1882. Died 1900.

GRAPPIN, Guy, French stomatologist; b. Berck, France, Sept. 9, 1917; s. François and Lovely (Christen) G.; ed. med. schs. U. Bordeaux, U. Montpellier; m. Josette Ollivier-Pallud, Mar. 7, 1942; 1 dau., Joycelyn. Physician, Chad, 1945-56, Congo, 1947-48, Sénégal, 1950-53; prin. stomatologist Far Eastern Ground Forces, Vietnam, 1954-56; insp. stomatology services, Madagascar, 1957-60; chief physician stomatology service Hôpital Principal; dir. dental instrn. Faculté de médecine, Dakar, Sénégal; dir. Institut d'Odontologie et de stomatologie tropicale, 1966—. Mem. Med. Soc. French Lang. in Negro Africa, French Soc. Stomatology. Research, publs. on paradontosis in Africa, dysmorphosis of African children, stomatological pathology among rural African regions, problem of fluorine in Sénégal. Home: 22, ave. de la République, Dakar, Sénégal, W. Africa.*

GRÄSBECK, Ralph Gustaf, Finnish biochemist; b. Helsinki, Finland, July 6, 1930; s. Armas Rafael and Mary (Lehtio) G.; M.D., U. Helsinki, 1953, M.Sc.D., 1956; m. Christina Ebba Strömberg, Sept. 11, 1954; children—Svante Gustav, Jerker Sigurd. Research fellow biochemistry Johns Hopkins, Balt., 1954-55, U. Helsinki, 1956-59, Med. Nobel Inst., Karolinska Inst., Stockholm, 1957; docent clin. chemistry U. Helsinki, 1959—, sr. scientist Minerva Inst. Med. Research, 1959—; chief physician lab. dept. Maria City Hosp., Helsinki, 1960—. Recipient Jahre prize U. Oslo, 1966. Mem. Am. Chem. Soc., Finska Läkaresällskapet, Danish Soc. Internal Medicine, Finska kemistsamfundet. Research, publs. on metabolism of vitamin B12; mechanism of mitosis in lymphocytes, developed radiochem. ultramicromethod for protein assay, fish tapeworm studies. Home: Gäddvik, Mattby (nr. Helsinki). Office: Minerva Inst., 12 Toeloegatan, Helsinki 10, Finland.*

GRASHCHENKOV, Nikolai Ivanovich, Russian neurologist, neurophysiologist; b. Zabor (now Smolensk Oblast), Mar. 26, 1901; grad. Med. Faculty, Moscow U., 1926, Red Prof. Inst. Philosophy and Natural Sci., 1932; D.Med. Sci., 1935. With Med. Faculty Moscow U. (later Sechenov 1st Moscow Med. Inst.), later All-Union Inst. Exptl. Medicine, 1926-33; dir. All-Union Inst. Exptl. Medicine, 1933-34; dir. Inst. Neurology, USSR Acad. Med. Sci., 1944-48; researcher on sensatory organs and electrophysiology of nervous system univ. labs., Cambridge, Eng., N.Y.C., New Haven, Boston, Montreal, Que., Can., 1935-57; head chair nervous diseases Moscow Central Postgrad. Med. Inst., 1951—, Sechenov 1st Moscow Med. Inst., 1958—; head clin. physiol. lab., USSR Acad. Med. Scis., 1951—. Chmn. learned med. council USSR Ministry Health, 1951-58; v.p. European regional com. WHO; del. Internat. Congress Neuropathologists, London, 1955. Decorated Order of Lenin. Mem. USSR (corr.), Belorussian (pres. 1947-51) acads. scis., USSR Acad. Med. Scis. Author over 160 works including textbooks, monographs; co-author: Manual of Nervous Diseases, 1939; author: Diagnosis and Treatment of Injuries of the Peripheral Nerves, 1942; Anaerobic Infections of the Brain, 1944; An Outline of Viral Lesions of the Central Nervous System, 1951; co-author, editor multivolume Manual on Neurology, 1957. Dep. chief editor Large Med. Ency., 2d edit.; mem. editor council Problems of Neurosurgery, Korsakov Jour. Neuropathology and Psychiatry. Research and numerous publs. on electrophysiology of central nervous system, physiology and pathology of sensatory organs, traumatic lesions, infectious diseases of nervous system. Died Oct. 8, 1965.

GRASHOF, Franz, German engr.; b. Düsseldorf, Germany, July 11, 1826; s. Karl and Lisette (Bruggemann) G.; studied math., physics, mech. engring., Royal Indsl. Inst., Berlin; hon. doctorate Rostock, Germany, 1860; m. Henriette Nottebohm, 1854; 1 son, 2 daus., including Elisabeth (Mrs. Karl Hoffacker). Joined navy; became tchr. math. and mechanics Royal Indsl. Inst., Berlin, 1854; also named dir. Standards Bur., 1856; became prof. theoretical mechanics Tech. U., Karlsruhe, Germany, 1863-93. Mem. Assn. German Engrs. (a founder, 1st dir., editor jour. until 1867, annual award named in his honor). Author: Die Festigkeitslehre mit besonderem Rücksicht auf die Bedürfnisse des Maschinenbaues, 1866; Theoretische Maschinenlehre, 3 vols., 1875-90; also articles. Introduced math.-sci. methods into tech. field; a founder sci. tech. especially in mech. engring. Died Karlsruhe, Oct. 26, 1893.

GRASMANN, Eustachius, forestry scientist; b. Inning, Germany, Feb. 19, 1856; s. Karl and Josefa (Zintner) G.; studied forestry, Forestry Acad., Aschaffenburg, Germany, 1875-78; Ph.D., U. Munich, 1878; m. Anna Henle, 1887; children—Karl, Max. Asst., U. Munich, 1878-82, 83-87; at estates of future Bavarian Queen Maria Theresia in Moravia and Hungary, 1882-83; lectr. politics and econs. forestry Agrl. and Forestry Sch., Tokyo, Japan, from 1887, also agr. dept. Imperial U.; 1896, joined Bavarian Forestry Adminstrn., Germany, 1896, govt. dir. Chamber of Forests, until 1922; became forestry advisor to Royal House, 1923. 1st German forestry scientist in Japan; pioneered modern Japanese forestry policy, developed reform plans for Japanese forests. Died Munich, June 8, 1935.

GRASSBERGER, Roland, Austrian physician; b. Salzburg, Austria, Nov. 26, 1867; s. Karl and Maria (Daniel) G.; M.D., U. Vienna, 1892; m. Mathilde Rabl; 3 sons, Hans, Alfred, Roland, 1 dau. Worked in Franz-Josef Hosp., Vienna; became asst Hygiene Inst., U. Vienna, 1897, lectr., 1902, asso. prof., 1906, titular prof., 1917, prof., chmn. inst., 1924; ret., 1936. Mem. Austrian Acad. Scis. Author: (with A. Schattenfroh) Über Beziehungen von Toxin und Antitoxin, 1904; (with M. Grassberger) Die Desinfektion in Theorie und Praxis für Arzte, Chemiker und Ingenieure, 1913; Über Krankheitsübertragung durch Nahrungsmittel, 1929; also handbooks, articles. Research on anaerobic sporogenesis, hygienic problems in milk and drinking water, hygiene of home, sch., and industry, disinfection, vermin control. Died Vienna, Dec. 4, 1956.

GRASSE, Pierre Paul, French zoologist; b. Perigueux, Dordogne, France, Nov. 27, 1895; s. Jules and Suzanne (Focke) G.; ed. Perigneux Sch.; Ph.D. in sci.; D. honoris causa, U. Brussels, U. Basle; m. Madeleine Pierre, 1947; children—Monique, Claude, Isabelle. Asst., Faculty Scis., Montpellier, 1920-29; prof. Sci. Faculty Clermont-Ferrand, 1929-37; prof. Sci. Faculty, Paris, titular chair evolution organic beings. Mem. Acad. Scis., Royal Acad. Scis. and Letters Belgium, also numerous others. Author over 300 works, including: Traité de Zoologie, 24 vols.; Parasites and Parasitisme; Précis de Biologie Animale; Précis de Zoologie. Home: 61 blvd. Saint-Michel, Paris 5. Office: 105 blvd. Raspail, Paris, France.

GRASSELLI, Eugene Ramiro, chemist; b. Strasburg, Jan. 31, 1810; s. Giovanni Angelo G.; ed. U. Strasburg, U. Heidelberg; m. Fredericka Eisenbarth, circa 1838; 8 children, including Lucretia (Mrs. Daniel Bailey), Caesar Augustin. Came to U. S., 1837; worked in Phila., until 1839; established his 1st factory for manufacture sulfuric acid, white alum, soda ash and Glauber's salt, Cin., 1839; built new plant for expanded manufacture of chems. and pharm. preparations, Cleve., 1865. A pioneer in Am. heavy chem. industry; designed and built apparatus for manufacture of chloroform; innovated sulfuric acid chambers; built railroad sulphuric acid tank cars. Died June 31, 1882.

GRASSET, Joseph, French physician; b. Montpellier, Mar. 18, 1849. Prof. pathology and gen. theraputics at Montpellier. Author: Des localisations dans les maladies cérébrales, 1878; Anatomie clinique des centres nerveux, 1900; Traité de pysiopathologie clinique; Les limites de la Biologie, 1902. Authority on neural diseases; investigated diseases of nerves, brain and spinal cord, also hypnotism, spiritism and occultism (contbg. to broader understanding of neural disease). Died Montpellier, July, 1918.

GRASSI, Giovanni Battista, Italian zoologist; b. Rovellasca, May 27, 1854; M.D., U. Pavia, 1878; studied zoology at Heidelberg, Würzburg. Prof. zoology, Catania, 1883; prof. comparative anatomy, Rome, 1895. Author: I Chetognati, 1883; I progenitori degli insetti e dei miriapodi, l'Japyx e la Campodia, 1886; Studi di uno zoologo sulla malaria, 1900; Flagellati viventi nei termiti, 1917. Research on life histories of intestinal worms, eels, termites and Protozoa; discovered transitional types of eel in Straits of Messina, 1896, led to solution of problem of eel migration; studied sporozoan malarial parasite of mosquito, demonstrated that Anopheles mosquito carries plasmodium of malaria in digestive tract, 1898; demonstrated that parasite causing malaria goes through sexual phase of its life cycle only in Anopheles mosquito (with Amico Bignami), 1899. Died Rome, May 4, 1925.

GRASSMAN, Wolfgang, German biochemist; b. Munich, Germany, Feb. 20, 1898; s. Karl and Auguste (Rothmund) G.; D.Phil., U. Munich, 1923; m. Elfriede von Hörmann, 1930; children—Irmtraud (Mrs. Arnold Nordwig), Peter. Prof., Dresden Inst. Tech., 1934-45; dir. Kaiser Wilhelm Institut für Lederforschung; expert for Bavarian Ministry Econs., 1945; lectr. in chemistry and physiology Regensburg Coll. of Philosophy and Theology, 1947-56; dir. Max Planck Institut fur Eiweiss und Lederforschung, Regensburg, 1948-56; dir. Max Planck Institut fur Eiweiss und Lederforschung, Munich, 1956—; hon. prof. U. Munich, 1957—. Recipient Stiasny medal. Mem. Bayerische Akademie der Wissenschaften, Am. Chem. Soc., Gesellschaft Deutscher Chemiker, others. Author: Methoden und Ergebnisse der Enzymforschung, 1928; (with T. W. Bertho) Biochemisches Praktikum, 1936; also numerous articles. Research on analytical methods in biochemistry; electrophorese; proteins; leather; enzyme chemistry; peptides. Home: 8036 Herrsching Lochschwab, Gachenaustrasse 21. Office: 46 Schillerstrasse, Munich 15, Germany.*

GRASSMANN, Hermann Günther, German mathematician; b. Stettin, Prussia, Apr. 15, 1809; s. Justus Günther and Johanne (Medenwaldt) G.; studied theology U. Berlin, 3 years, diplomas, 1834, 39; no formal math. study, passed state exam., Berlin, 1831; Ph.D. (hon.), U. Tübingen, 1876; m. Therese Knappe, 1849; 7 sons, 4 daus., including Justus, Hermann, Richard. Tchr. math. secondary sch., Stettin, from 1831, 35, Berlin Indsl. Sch., 1834; tchr. Friedrich-Wilhelm Sch., Stettin, 1842; prof. Marienstift Sch., 1852; pub. newspaper with brother Robert, 1848-49; devoted himself to intensive Sanskrit studies, from 1849. Author: Die Wissenschaft der extensiven Grössen oder die Ausdehnungslehre, 1844; Die Ausdehnungslehre, 1862; Wörterbuch zur Rig-Veda, 1875; Gesamte mathematische und physikalische Werke, 3 vols., 1894-96, 1902-04, 11 (collected works edited by son Hermann and others), also works on Sanskrit philology. Pioneer modern vector analysis; created new algebra of n-dimensional space, called the theory of extension, 1840-44; formulated linguistics law which is named after him; research on theory of expansion derived from studies of tides; his work at 1st not recognized, turned to study of Sanskrit. Died Stettin, Sept. 26, 1877.

GRASSUS, Benevenutus, oculist, physician; flourished 12th century; most famous non-Muslim oculist of medieval times; author: Practica oculorum (most popular Latin text on eye diseases, (became 1st printed book on the subject). 1st printed edit., 1474.

GRASTY, John Sharshall, Am. geologist, mining engr.; b. Versailles, Ky., Mar. 15, 1880; s. Thomas Percy and Mattie Virginia (White) G.; A.B., Johns Hopkins, 1902, Ph.D., 1908; postgrad. Washington and Lee U., Mass. Inst. Tech.; Sc.D., Washington Coll., Md., 1912; m. Elizabeth Montgomery Cochran, Nov. 9, 1909; children—Thomas P., John Sharshall. Asst., U. S. Geol. Survey, 1905; geologist Md. Geol. Survey, 1906-08; asst. state geologist, Va., 1909-16; adj. prof. econ. geology U. Va., 1908-13, asso. prof., 1913-16; prof. mining geology, Washington and Lee U., 1916-18; investigations for B.&O. Ry., Southern Ry. and C.&O. Ry.; spl. geologist Ala. Geol. Survey; chem. engr. Ordnance Dept., U. S. Army, 1918; oil geologist Mid-Continent Fields, 1919-20. Inventor fire sentinel apparatus; specialist on rock slides and foundations. Author: Limestones of Maryland, 1909; The Slate Deposits of Virginia, 1917; Origin of Caverns in Relation to Structure, 1925. Died June 5, 1930.

GRASTYAN, Endre, Hungarian neurophysiologist; b. Öriszentpéter, Hungary, Feb. 25, 1924; s. Joseph and Gizella (Siska) G.; Med. doctor, U. Pécs, (Hungary), 1951, Candidate Med. Scis., 1958; m. Anna Horváth, Apr. 16, 1953; 1 son, Andras. Neurophysiologist, Inst. Physiology, U. Pécs Med. Sch., 1951-57, faculty, 1957—, docent, 1962—. Research, publs. on behavioral, psychic processes with neurophysiol. techniques, electrophysiology. Home: Dr. Berze Nagy u.l, Pécs, Hungary.*

GRATAROLO, Guglielmo, physician, alchemist; b. Bergamo, Lombardy, 1516; studied philosophy and medicine, Padua. Moved to Basel to avoid persecution for his Protestant beliefs, 1555; taught at Marburg, 1562; returned to Basel to practice medicine, 1563. Author: De praedictione morum naturarumque hominum cum ex inspectione partium corporis tum aliis modus, 1554; Pestia descriptio, 1555; De peste these, 1568. Used diagnostic procedure based in part on continuing acceptance of occultism, involving observation of patient's posture and facial expression. Died Basel, Switzerland, circa 1563-68.

GRATIOLET, Louis Pierre, French physiologist; b. Sainte Foy, France, July 6, 1815; ed. Paris. Preparer Mus. Natural History, Paris, asst. naturalist, 1854-63, prof. anatomy, from 1863. Author: Mémoire sur les plis cérébraux de l'homme et des primates, 1854; Anatomie comparée du système nerveux, 1858; Recherches sur les systèmes vasculaires de la sanguine médicinale et de l'anlastomeraroce, 1864. Brain specialist; made comparative studies of brain lobes of man and primates, studies on bloodstone, vascular system; a fiber tract of the brain, radiatio optica, is also called Gratiolet's optic radiation after him. Died Paris, Feb. 16, 1865.

GRATZL, Erwin, Austrian veterinarian; b. Vienna, Austria, Mar. 25, 1902; s. Ignaz and Helene (Bernhard) G.; M.D. in Vet. Medicine; m. Hildegard Hönlinger, 1938 (dec.); m. 2d, Cornelia Zelinger, 1960; children—Angela, Dorothea. Asst., Med. Clinic of Vet. Faculty, until 1939; titular prof. internal medicine and study clin. epidemics Faculty Vet. Medicine, U. Giessen, 1939, full prof. Med. Clinic, prof. Faculty Vet. Medicine, 1940; full prof. Med. Clinic of Vet.

Faculty, Vienna, 1948, dean Vet. Faculty, 1953-56; hon. prof. Faculty Vet. Medicine, U. Munich, Free U. Berlin. Mem. Austrian Soc. Veterinarians, Austrian Soc. Microbiology and Hygiene, Austrian Soc. Biochemistry, Austrian Soc. Pure and Applied Biophysics, World Vet. Poultry Assn., Audio-Visual Vet. Assn. Collaborator: Zentralblattes für Veterinärmedizin. Editor-in-chief: Wiener Tierärztliche Monatsschrift; co-editor Monatshefte für Tierheilkunde; editorial cons. com. Jour. Small Animal Practice, Veterinarian, Deutsche Tierärztliche Wochenschrift. Address: Medical Clinic of Tierärztlichen High School, 11 Linke Bahngasse, Vienna III, Austria.

GRAUE, Louis Charles, Am. mathematician, educator; b. Louisiana, Mo., Dec. 23, 1923; s. Louis J. and Ruth (Forman) G.; B.S., U. Chgo., 1947, M.S., 1948; Ph.D., Ind. U., 1950; m. Patricia J. Hock, June 21, 1949; children—Geoffrey, Nancy. Instr., asst. prof. math. Sacramento State Coll., 1950-56; asso. prof. math. Coe Coll., Cedar Rapids, Ia., 1956-59; asso. prof. math. Bowling Green (O.) State U., 1959-64, prof., 1964—, chmn. dept. math., 1965—. Served with USNR, 1944-46. Mem. Math. Assn. Am., Am. Math. Soc., Phi Gamma Delta. Contbr. articles to profl. jours. Research in differential geometry, algebra, bird orientation. Home: Route 3, Bowling Green, O. 43402.*

GRAUL, E. H., German biophysicist; b. Zeitz, Germany, Dec. 29, 1920; s. Emil and Kathe Graul; M.D., 1946; Ph.D., 1948; m. Ruth Menzel, Aug. 9, 1950; children—Eva, Sybille, Udo. Head radiobiol. lab. and x-ray sect. U. Munster, Germany, 1948-53; head dept. radiobiology and isotope research U. Marburg (Germany), 1954—, prof. nuclear medicine, biophysics, space medicine, 1954—; vis. scientist Brit. Atomic Energy Center, 1954, Atomic Research Center, Mol, Belgium, 1957; asso. prof. Argonne Nat. Lab., Chgo., 1956-58; Lectr., Japan, 1960; mem. space research com. German Fed. Ministry Transport; sci. adviser German Def. Ministry, German Med. Assn. Mem. German Soc. Rockets and Space Travel (DGRR) (bd. dirs., sci. adviser), German Soc. Aviation and Space Medicine (v.p.). Author books, numerous articles. Editor-in-chief Atompraxis; co-editor various internat. jours. Research on radiobiology, uses of radio-active isotopes in research and practice, space medicine. Home: 355 Marburg/Lahn, Körnerstr. 19. Office: 355 Marburg/Lahn, Lahnstr. 4 a, Germany.*

GRAUNT, John, Brit. statistician; b. Hampshire, Eng., Apr. 24, 1620; s. Henry and Mary Graunt; m. Mary; several children; apprenticed to haberdasher. Fellow Royal Soc. (charter), 1663. Author: Natural and Political Observations . . . Made upon the Bills of Mortality, 1662. Father vital statistics; 1st application math. to integration vital statistics; demonstrated that population can be estimated from death rates, that life in the country was healthier than in city; recognized principle of uniformity of vital facts in large groups; advocated the establishment of govt. dept. for collecting demographic data. Died London, Apr. 18, 1674.

GRAUSTEIN, William Caspar, Am. mathematician; b. Cambridge, Mass., Nov. 15, 1888; s. Adolf Henry and Julia (Caspar) G.; A.B., Harvard, 1910, A.M., 1911; traveling fellow Harvard; Ph.D., Bonn, 1913; m. Mary Florence Curtis, June 10, 1921. Instr. math. Harvard, 1913-14, lectr., 1919, asst. prof., 1919-26, asso. prof., 1926-33, prof., 1933—; instr. math. Rice Inst., Houston, 1914-15, asst. prof., 1915-18; lecturer in mathematics, Harvard, 1919, asst. prof., 1919-26, asso. prof., 1926-33, prof. 1933—. Author: (with William F. Osgood) Plane and Solid Analytic Geometry, 1921; Introduction to Higher Geometry, 1930; Differential Geometry, 1935. Editor Trans. Am. Math. Soc., 1936—. Died Jan. 22, 1941.

GRAVENHORST, (Heinrich Ludwig Diedrich) Friedrich, German civil engr.; b. Weide, Germany, Jan. 3, 1835; s. Carl Friedrich and Margaretha Christina (Cirzovius) G.; grad. Polytechnikum, Hanover, Germany, 1860; m. Luise Lehmann, 1867; 4 sons, including Otto, 5 daus. Joined rd. bldg. adminstrn. firm Schleswig-Holstein, 1860; became dist. engr., Ottendorf, 1872; named hwy. insp., dir. State Office, Stade, Germany, 1873. Invented a plaster formed of cube-shaped granite stones, 4 to 6 centimeters on a side, which was used as a rd. surface, all over Europe, 1885. Died Stade, June 11, 1915.

GRAVENHORST, Johann Heinrich, chemist; b. 1719; manufactured sal ammoniac, Glauber's salt, red alum, Brunswick green. Author: (with brother Christoph Julius Gravenhorst) Einige Nachrichten am das Publikum . . . , 1769. Inventor (with brother) pigment Brunswick green (copper oxychloride). Died 1781 or 82.

GRAVENHORST, Johann Ludwig Christian Carl, German zoologist; b. Braunschweig, Germany, Nov. 14, 1777; s. Johann Conrad Wilhelm and Johanna Maria Sophia (Oldendorp) G.; began study law, Helmstedt, Germany, 1797, Ph.D. in Entomology, 1801; student natural scis., Göttingen, Germany, 1799; m. Marianne Charlotte Elsner, 1811. Made study trip to Paris, 1802; later became ind. scholar, Braunschweig; joined faculty natural history U. Göttingen, 1804, became asso. prof., asst. Acad. Mus., 1808; named prof. natural history, dir. bot. garden U. Frankfort/Oder (Germany), 1810; moved with univ. to Breslau (now Wroclaw, Poland); founder with pvt. collection Breslau Zool. Mus., 1814. Author: System der Natur, 1804; Vergleichende Übersicht der Linné'schen und einiger neueren zoologischen Systeme, 1807; Grundzüge der systematischen Naturgeschichte, 1817; Ichneumonologia europaea, 1829; Das zoologische Museum der Universität Breslau, 1832; Vergleichende Zoologie, 1843; Das Tierreich nach den Verwandtschaften und Übergängen in den Klassen und Ordnungen deselben dargestellt, 1845. Studied ichneumon fly, infusoria, amphibians, reptiles, marine animals, descriptive taxonomy. Died Breslau, Jan. 14, 1857.

GRAVENSTEIN, Joachim Stefan, Am. physician; b. Berlin, Germany, Jan. 25, 1925; Dr.med., U. Bonn (Germany), 1951; M.D., Harvard, 1958; m. Alix Trutschler, Aug. 27, 1949; children—Nikolaus, Alix, Frederike, Stefan, Ruprecht, Dietrich, Constanze. Came to U. S., 1952, naturalized, 1959. Clin. fellow anesthesia Mass. Gen. Hosp., 1952-56; research fellow Harvard, 1952-56, clin. asso., 1956-58; asso. anesthetist Mass. Gen. Hosp., 1958; faculty U. Fla., Gainesville, 1958—, prof. anesthesiology, 1963—, chief anesthesia, 1958—. Mem. Assn. U. Anesthetists, Am. Soc. Anesthesiologists, Fla. Soc. Anesthesiologists, Am. Soc. Pharmacologists, A.M.A., Internat. Anesthesia Research Soc., Fla. Med. Assn., Alachua County Med. Soc., Alpha Omega Alpha. Research and numerous publs. in drug actions in anesthesia. Address: 101 N.W. 44th St., Gainesville, Fla. 32601.

GRAVES, Alvin C(usman), Am. physicist; b. Washington, Nov. 4, 1909; s. Herbert C. and Clara Edith (Walter) G.; B.S., U. Va., 1931; postgrad. Mass. Inst. Tech., 1932; Ph.D., U. Chgo., 1939; m. Elizabeth Riddle, Sept. 27, 1937; children—Marilyn Edith, Alvin Palmer, Elizabeth Anne. Instr. physics U. Tex., 1939-41, asst. prof., 1941, asso. prof. (on leave), 1942—; with U. Chgo. Metall. Lab., 1942-43; staff mem. Los Alamos (N.M.) Sci. Lab., U. Cal., 1943-45, group leader, 1945-47, asso. div. leader, 1947-48, div. leader, 1948—; dep. sci. dir. Pacific Proving Grounds operations, 1947-48, sci. dir., 1948—; test dir. Nev. Proving Grounds operations, 1951-54, sci. advisor Nev. Test Site operations, 1955—. Recipient Exceptional Civilian Service award, Air Force, 1951, Certificate of Achievement, Army, 1954; Distinguished Service award FCDA, 1955. Fellow Am. Phys. Soc.; mem. Am. Inst. Physics, Com. of Sr. Reviewers AEC, Sigma Xi, Gamma Alpha, Tau Beta Pi. Compared packing fractions of 14 metals, calculated two atomic weights. Author: (with R. L. Walker) A Method for Measuring Half-Lives, 1947; (with others) Spin and Magnetic Moment of Tritium, 1947, Relative Moments of H-1 and H-3, 1947; The Packing Fraction Difference Among Heavy Elements. Home: 1459 46th St. Office: P.O. Box 1663, Los Alamos Sci. Lab., Los Alamos, N.M.

GRAVES, John Thomas, mathematician, jurist; b. Dublin, Dec. 4, 1806; s. John Crosbie Graves; B.A., Dublin 1827, M.A., 1832; M.A., Oxford, 1831; m. Miss Tooke, 1846. Dean of Ardagh; became barrister Inner Temple, 1831; named prof. jurisprudence U. Coll., London, 1889; poor-law insp., 1847-70. Fellow Royal Soc., 1839. Contbg. author: Dictionary of Greek and Roman Biography (Smith). Research and publs. on imaginary logarithms (led to Sir William Rowan Hamilton's discovery of quaternions); bequeathed math. library to U. Coll. Died Mar. 29, 1870.

GRAVES, Lawrence Murray, Am. mathematician; b. Topeka, Aug. 7, 1896; s. William James and Sarah (Cowgill) G.; A.B., Washburn Coll., 1918, Sc.D., 1941; M.S., U. Chgo., 1920, Ph.D., 1924; m. Josephine Mary Wells, Aug. 27, 1924; children—Robert Lawrence, John Lowell, Anne Lowell. Faculty, Washington U., St. Louis, 1920-22; NRC fellow Harvard, 1924-26; faculty U. Chgo., 1926—, prof., 1939-61, prof. emeritus 1961—; vis. prof. Ind. U., 1947-48, Ill. Inst. Tech., 1961-66; chmn. editorial com. Internat. Congress Mathematicians, Cambridge, Mass., 1950. Recipient Presidential Certificate of Merit, 1948. Mem. Am. Math. Soc., Math. Assn. Am., A.A.A.S., Sigma Xi. Author: Theory of Functions of Real Variables, 1946, rev., 1956. Research and publs. primarily in field of calculus of variations, differential equations, abstract functional analysis. Home: 5842 Stony Island Av., Chgo. 60637.*

GRAVES, Robert James, Irish physician; b. 1797; s. Richard Graves; M.B., Dublin, 1818; studied on continent; settled in Dublin, 1821 and founded Park St. Sch.; became prof. Royal Coll. Dublin, 1827; named physician Meath Hosp., 1821. Fellow Irish Coll. Physicians (pres. 1843-45), Royal Soc., 1850. Author: Lectures on the Functions of the Lymphatic System, 1828; System of Clinical Medicine, 1843; Clinical Lectures on the practice of Medicine, 2 vols., 1848; Studies in Physiological and Medicine, 1863. Founder (with Stokes) Dublin Jour. Medicine and Chem. Sci. 1st to use stimulants and food instead of dieting at the beginning of typhoid and other fevers; 1st to explain exophalmic goiter (Grave's disease). Died Dublin, Mar. 20, 1853.

GRAVES, William Washington, Am. neuropsychiatrist; b. La Grange, Ky., Nov. 13, 1865; s. David William and Julia Ann (Crockett) G.; M.D., St. Louis Coll. Physicians and Surgeons, 1888; postgrad., London, Heidelberg, Berlin and Vienna, 1901-04; m. Helena J. Sessinghaus, June 9, 1891; 1 dau., Helen. With St. Louis U. Med. Sch., 1905—, prof. nervous and mental diseases, 1914—, dir. dept., 1923—. Mem. A.M.A., A.C.P., St. Louis Med. Soc. (certificate of merit, Gold medal for classification of Scapulae and discovery of age-incidence principle of investigation 1939), Am. Neurol. Assn., Am. Psychiatric Assn. Died Apr. 18, 1949.

GRAVESANDE, Willem Jacob Storm van s', Dutch mathematician; b. Herzogenbusch, Netherlands, Sept. 27, 1688; student law U. Leiden (Netherlands). Practiced law, The Hague, circa 1707; sec. to embassy sent to Eng., 1715-17; prof. math., astronomy U. Leiden, from 1717. Fellow Royal Soc., 1715. Author: Physices Elementa Mathematica Experimentis Confirmata, Sive Introduction ad Philosophiam Newtonianam, 2 vols., 1720-21; Oevres Philosophiques et Mathématiques, 1774; Nouvelle Théorie sur le Choc des Corps; Remarques sur la Construction des Machines Pneumatiques. Invented 1st heliostat, 1719; popularized sci. ideas of Newton at U. Leiden, new theories of calculus; made various discoveries in physics; demonstrated that when ring expands, internal diameter increases as external diameter increases. Died Leiden, Feb. 28, 1742.

GRAVESEN, Johannes, Danish physician; b. Ringkobing, Jan. 3, 1889; s. Carl and Johanne (Friis) G.; M.D., U. Copenhagen; m. Ellen Seidelin Larsen, 1915; children—Lars, Lizzie, Olaf, Niels, Kirsten, Per. Physician sanatorium, India, 1921-23; med. dir. Sanatorium of Wejlefjord, 1933-44; chief physician dept. Tb, Hosp. of Frederiksberg and Pulmonary Clinic of Frederiksberg, Copenhagen, 1944-59. Mem. Brit. Tb Assn. (hon.). Author: On Collapse Therapy of Pulmonary Tuberculosis, 1920; Surgical Treatment of Pulmonary Tuberculosis, 1925, also numerous articles on Tb of lungs and its treatment. Address: 41 Tesdorpfsvej, Copenhagen f, Denmark.

GRAVIER, Charles-Joseph, French zoologist; b. Orleans, France, Mar. 4, 1865; Docteur ès scis., École normale supérieure de St. Cloud, 1896; mem. Charcot's Antarctic expdn., 1903-05; named prof. natural sci. École normale, Grenoble, France, 1915; prof. zoology Paris Mus. Natural History, from 1917; mem. French Acad. Scis. Author: Annélides polychètes, 1903; Les récifs de coraux, 1911. Research on Bristle worms of order Archiannelida from Antarctic, coral formation, crustaceans, parasitic crustaceans. Died Paris, Nov. 15, 1937.

GRAWITZ, Paul Albert, German pathologist; b. Zerrin, Germany, Oct. 1, 1850; s. Wilhelm G. and Agnes (Fischer) G.; ed. U. Halle (Germany), U. Berlin; M.D., 1873; m. Anna Cunitz, 1879; 1 dau., Lotte (Mrs. Otto Busse). Asst. to Virchow, Berlin, 1875; faculty pathology and path. anatomy U. Berlin, 1884; asso. prof. U. Greifswald (Germany), 1886; prof. path. anatomy, from 1887; dir. Path. Anatomy Inst., U. Greifswald. Research on mold fungi, congenital hips laceration, jaundice in newborn infants, lung infarct; and on infection; proved chem. substances can also cause infection; proposed sleeping cell theory, worked on cell cultures; described hypernephroma. (known as Grawitz's tumor), 1884. Died Greifswald, Germany, June 27, 1932.

GRAY, Andrew, natural philosopher; b. Scotland, 1847; s. John Gray; M.A., Glasgow U.; D.Sc., LL.D.; m. Annie Gordon; 3 sons, 4 daus. Pvt. sec., asst. to Sir W. Thomson, 1875-80; Eglinton fellow in math. Glasgow U., 1876, ofcl. asst. to prof. natural philosophy, 1880-84, prof. natural philosophy, 1899-1923, prof. emeritus, 1923-25; prof. physics Univ. Coll., North Wales, 1884-89. Fellow Royal Soc., 1896. Author: Absolute Measurements in Electricity and Magnetism, 1883; Theory and Practice of Absolute Measurements in Electricity and Magnetism, Vol. I, 1888, Vol. II (parts 1 and 2), 1893, new edit., 1921; (with G. B. Mathews) A Treatise on Bessel Functions, 1895, rev. (with MacRobert), 1922; Magnetism and Electricity, Vol. I, 1898; Dynamics and Properties of Matter, 1901; The Scientific Work of Lord Kelvin, 1908; (with J. G. Gray) A Treatise on Dynamics, 1911, rev., 1920; A Treatise on Gyrostatics and Rotational Motion, 1919, also articles. Died Oct. 10, 1925.

GRAY, Sir Archibald Montague Henry, English dermatologist; b. Ottery St. Mary, Devonshire, Feb. 1, 1880; s. Frederick Archibald and Louisa Frances (Waterworth) G.; ed. Cheltenham Coll., Univ. Coll., London; M.D., B.S.; LL.D., U. London; m. Elsie Cooper, 1917; children—John A. B., Sybil M. W. Physician, dermatol. service Univ. Coll. Hosp. and Hosp. for Sick Children; hon. cons. Ministry Health; dean Faculty Medicine. Fellow Royal Coll. Physicians, Royal Coll. Surgeons; mem. Royal Soc. Medicine (past pres.), Brit. Assn. Dermatology, also numerous fgn. dermatol. socs. Author: Sclerema Neonatorum, 1926; Haematoporphyria Congenita, 1926; Dermatology from the Time of Harvey, 1951; The Founders of Modern Dermatology, 1953. Address: 7 Alvanley Gardens, London N.W.6, Eng.

GRAY, Asa, Am. botanist; b. Sauquoit, N.Y., Nov. 8, 1810; s. Moses and Roxana (Howard) G.; M.D., Fairfield Med. Sch., Jan. 25, 1831; hon. degrees Oxford and Cambridge (Eng.), Aberdeen (Scotland) univs.; A.M. (hon.), Harvard, 1844; LL.D., Hamilton Coll., 1860; m. Jane Lathrop Loring, May 4, 1848.

Mem. U. S. exploring expdn. under command of Capt. Charles Wilkes, 1834-37; prof. botany U. Mich., 1838; Fisher prof. natural history Harvard, 1842-73, 79, organized dept. botany; recognized authority in area of plant geography; wrote famous study of Japanese botany and its relationship to that of N.Am.; described N. American plants in hundreds of publs.; a founder Nat. Acad. Arts and Scis.; pres. A.A.A.S., 1863-73; regent Smithsonian Instn., 1874-88; mem. Hall of Fame. Fellow Royal Soc., 1873. Author: North American Gramineae and Cyperaceae, 1835; Elements of Botany, 1836; First Lessons in Botany and Vegetable Physiology, 1857, revised edit., 1868; How Plants Grow, 1858; Field, Forest and Garden Botany, 1868, 70; How Plants Behave, 1872; Elements of Botany, 1887; Botanical Textbook, 1842, Natural Science and Religion, 1880. A teleogist; believed species were differentiated according to a preordained plan in the mind of a creator; his work influenced Darwin, helped propagate Darwinian ideas in U. S.; helped revise the taxonomic procedure of Linnaeus on basis of a more natural classification, based primarily on fruit anatomy rather than on gross morphology. Died Cambridge, Mass., Jan. 30, 1888.

GRAY, Edward George, English biologist; b. Pontypool, Eng., Jan. 11, 1924; s. William Godfrey and Charlotte (Atkinson) G.; B.Sc., Univ. Coll. of Wales, Aberystwyth, 1952, Ph.D., 1955; m. May Eine Kyllikki, Feb. 14, 1953; children—Michael Timothy, Peter Marlow. Research asst. dept. anatomy Univ. Coll., London, Eng., 1955-57, lectr., 1957-61, reader in cytology, 1961-67, prof. cytology, 1967—. Mem. Anat. Soc. Gt. Britain and No. Ireland, Soc. Exptl. Biology (Gt. Britain). Contbr. articles to profl. jours. Research on observations with electron microscope on structures in brain and spinal cord, on octopus leech and insect nervous systems, analysis of such structures by isolation with ultracentrifuge, devel. of stereoscopy for electron microscope. Home: 23 Bradgate, Cuffley, Eng. Office: Anatomy Dept., Gower St., London S.C.1, Eng.*

GRAY, Elisha, Am. electrician, inventor; b. nr. Barnesville, O., Aug. 2, 1835; ed. Oberlin Coll.; m. M. Delia Shepard, 1862; learned blacksmithing, carpentry and boatbuilding; pursued spl. studies in phys. sci. Oberlin Coll.; D.Sc., 1874; studied acoustics abroad; began career as electrician, 1865; inventor a self-adjusting telegraph relay, 1867; established as mfr. of electric apparatus, Cleve., 1869; perfected type-writing telegraph, telegraph repeater, telegraphic switch, annunciator; organized Western Electric Mfg. Co., 1872, ret., 1874; inventor speaking telephone, 1876 (his claim to invention of telephone beaten by A. G. Bell in patent litigation), telautograph, 1888; established Gray Electric Co., Highland Park, Ill.; organized Congress of Electricians, in connection with World's Columbian Expn., 1893, served as chmn. Author: Experimental Researches in Electro-Harmonic Telegraphy and Telephony, 1878; Elementary Talks on Science. Died Newtonville, Mass., Jan. 21, 1901.

GRAY, Ernest Paul, Am. physicist; b. Vienna, Austria, Mar. 12, 1926; s. Cornel and Alice (May) G.; B.A., Cornell U., 1947, Ph.D., 1952; m. Miriam Neuberger, June 27, 1954; children—Peter L., Robert L. Jr. engr. Los Alamos Sci. Lab., 1946; mathematician Nat. Bur. Standards, 1947; sr. staff Johns Hopkins U. Applied Physics Lab., Silver Spring, Md., 1951-58, prin. profl. staff mem., 1958—, chief theoretical plasma physics staff, 1966—, faculty Johns Hopkins, 1957—, prof. 1958—. AEC fellow, 1948-49; William S. Parson fellow, 1957-58. Mem. Am. Phys. Soc., Philos. Soc. Washington, Washington Acad. Scis. Contbr. articles to tech. jours. Research on scattering electromagnetic radiation from ocean surface enabling radar to track a target instead of image; single-particle trajectories in complex magnetic fields, dispersion relation when electron beam penetrates and interacts with plasma; devised method analyzing electron density data in decaying discharge so as to make sure whether diffusion or recombination control electron loss; formulated theory inelastic collisions in such a way as to make clear which part cross sect. reflects interaction and which part is statis. in sense reflecting only fact that strong interaction has taken place. Home: 12516 Davan Dr., Silver Spring 20904. Office: 8621 Georgia Av., Silver Spring, Md. 20910.*

GRAY, Fenton, Am. soil scientist; b. Santa Clara, Utah, Aug. 12, 1916; s. Alden and Matilda (Stucki) G.; student Dixie Jr. Coll., 1934-36; B.S., U. Utah, 1938; Ph.D., Ohio State U., 1951; m. Evelyn Snow, Dec. 21, 1938; children—Fenton Michael, William Alden, Stephen S., Philip A., Marilyn. Faculty, Okla. State U., Stillwater, 1951—, prof., 1959—; sr. soil scientist, acting project mgr. FAO Spl. Fund N.E. Brazil, 1961-62. Mem. Am. Soc. Agronomy, Soil Sci. Soc. Am., A.A.A.S., Soil Conservation Soc. Am., Clay Minerals Soc., Okla. Acad. Sci., Sigma Xi, Gamma Sigma Delta. Author: Soils of Oklahoma, 1959. Contbr. numerous articles to profl. jours. Research on morphology and genesis of reddish prairie soils Okla. soils, their classification, use and productivity, large scale irrigation possibilities in FAO Francisco River Project. Home: 1017 Orchard Lane, Stillwater, Okla. 74075.*

GRAY, Frank Davis, Jr., Am. physician; b. Marshall, Minn., Aug. 24, 1916; s. Frank Davis and Nettie (Urbach) G.; B.S., Northwestern U., 1938; M.D., Columbia, 1943; m. Frieda Gersh, June 27, 1941. Mem. faculty Yale U. Med. Sch., New Haven, 1949-68, asso. prof. medicine, 1957-68; dir. sect. chest diseases Yale-New Haven Med. Center, 1949-68; prof. medicine Jefferson Med. Coll., 1968—; dir. div. medicine Lankenau Hosp., Phila., 1968—; cons. VA Hosp., West Haven, Conn., Laurel Heights Hosp., Shelton, Conn. Fellow A.C.P., Am. Coll. Chest Physicians (pres. New Eng. 1966—); mem. Conn. Heart Assn. (pres. 1958-60), Conn. Thoracic Soc. (pres. 1964-65), A.M.A., Am. Soc. Clin. Investigation, Am. Fedn. Clin. Research, History of Sci. Soc., Sigma Xi. Author: Pulmonary Embolism, 1966; also articles. Research in circulation to lung including studies of normal control of lung blood flow, factors influencing loss of fluid into lung air spaces, changes resulting from obstruction to lung blood flow. Home: 775 Mill Creek Rd., Gladwyne, Pa. 19035. Office: Lankenau Hosp., Phila. 19151.*

GRAY, George Robert, English entomologist; b. Chelsea, Eng., July 8, 1808; s. Samuel Frederick Gray; studied at Merchant Taylor's Sch.; became asst. zoology dept. Brit. Mus., 1831. Fellow Royal Soc., 1865; mem. Academia economico-agraria dei georgofili Florence. Author: Entomology of Australia, 1833; Genera of Birds, 3 vols., 1837-49; Handlist of the Genera and Species of Birds, 3 vols., 1869-71; Catalogue of the British Birds in the Collection of the British Museum, 5 vols., 1848-63; His catalogs listed more than 11,000 species. Died London, May 6, 1872.

GRAY, James Gordon, Scotch physicist; b. Glasgow, Scotland, 1874; s. Andrew Gray; ed. U. Coll. N. Wales, U. Glasgow; became mem. teaching staff natural philosophy dept. U. Glasgow, 1904; Cargill prof. applied physics, 1920, (with A. Gray) planned bldg. and equipping of Natural Philosophy Inst. Author: A Treatise on Dynamics, 1911, rev. edits., 1920, 31; also numerous articles, patent specifications. Fellow Royal Soc. Edinburgh; mem. Instn. Elec. Engrs. Invented motor gyrostats and accessories, several new spinning-tops, including series of animated gyrostats and of devices for use in aerial and marine navigation, nat. def. including stabilisers, artificial horizons, cloud-leveling apparatus, steering devices; pioneered invention of inductor compass used by Lindbergh in Atlantic flight. Died Nov. 6, 1934.

GRAY, John Edward, Brit. naturalist; b. Walsall, Staffordshire, Eng., Feb. 12, 1800; s. Samuel Frederick Gray; m. Maria Emma Smith, 1826; entered lab. of a chemist, Cripplegate, 1844; entered med. schs. of St. Bartholomew's, Middlesex hosps.; studied under Mr. Taunton at Hatton Garden and Maze Pond; Ph.D., U. Munich (Germany), 1852. Lectr. botany Borough Sch. Medicine; became asst. zool. keeper at Brit. Mus., 1824, keeper, 1840-74. Fellow Royal Soc., 1832; mem. Zool. Soc. (v.p.), Bot. Soc. (pres.), Entomol. Soc. (pres.). Author: Handbook of British Waterweeds, 1864; Synopsis of British Mollusks, 1852; also numerous articles. Created largest zool. collection in Europe, 1852. Died Mar. 7, 1875.

GRAY, John Purdue, Am. physician, alienist; b. Center County, Pa., Aug. 6, 1825; s. Peter B. and Elizabeth (Purdue) G.; A.M., Dickinson Coll., 1846; M.D., U. Pa., 1849; LL.D., Hamilton Coll., 1874; m. Mary B. Wetmore, Sept. 6, 1854. Resident physician Blockley Hosp., Phila., 1849; 1st asst., acting med. supt. N.Y. State Lunatic Asylum, Utica, 1853; med. supt. Mich. State Lunatic Asylum, 1853; full editor Am. Jour. Insanity, 1854; prof. psychol. medicine and med. jurisprudence Bellevue Hosp. Med. Coll., 1874-82, Albany Med. Coll., 1876-82; pres. Assn. Med. Supts. of Am. Instn. for the Insane; revolutionized asylum constrn., introduced steam heat and ventilation; abolished as far as possible mech. restraint and solitary feeding for patients. Died Utica, Nov. 29, 1886.

GRAY, John Stephens, Am. physiologist; b. Chgo., Aug. 11, 1910; s. Joseph W. and Carrie (Weston) G.; B.S., Knox Coll., 1932; M.S., Northwestern U., 1934, Ph.D., 1936, M.D., 1946; m. Elma Nash, June 15, 1935; children—Ann (Mrs. Stanley R. Riggs), Virginia (Mrs. Daniel D. Spence). Instr. Northwestern U. Med. Sch., 1936-39, asst. prof., 1940-44, asso. prof., 1945-46, prof., chmn. dept., 1946—. Recipient Knox Coll. Alumni Achievement award, 1957; Guggenheim fellow, 1962. Mem. Am. Physiol. Soc., Soc. Exptl. Biology and Medicine, A.A.A.S. Author: Pulmonary Ventilation and Its Physiological Regulation, 1950. Contbr. numerous articles field physiology to sci. jours. Research, publs. on the purification and biological properties of gastro-intestinal hormones and quantification of gastric acid secretion; early application of quantitative mathematical analysis to biological control systems, especially the respiratory system. Home: 422 15th St., Wilmette, Ill. 60091.*

GRAY, Louis Harold, English radiobiologist; b. London, Eng., Nov. 10, 1905; s. Harry and Amy (Bowen) G.; ed. Christ's Hosp., Trinity Coll., Cambridge U.; honors in natural sci.; D.Sc., U. Leeds; m. Frieda Picot; children—John, Giles. Researcher, Rouse Ball, 1927-30; fellowship instr. Trinity Coll., Cambridge U., 1930-34; Prophet fellow Royal Coll. Surgeons, 1934-39; chief physician Mt. Vernon Hosp., 1934-47, dir. Brit. Empire cancer campaign research unit in radiobiology, 1953—; asso. dir. M.R.C. Radiotherapeutic Research Unit, Hammersmith Hosp., 1947-53. Fellow Royal Soc., 1961; mem. Royal Soc. Medicine, Brit. Inst. Radiology, Faculty Radiologists, Radiation Research Soc. U. S., Assn. for Radiation Research, Hosp. Physicists Assn. Research and numerous publs. on nuclear physics, radiology, radiobiology with reference to radiotherapy of cancer and risks asso. with use of ionium radiation. Home: 5 St. Mary's Av., Northwood. Office: British Empire Cancer Campaign Research Unit of Radiobiology, Mt. Vernon Hospital, Northwood, Middlesex, Eng.

GRAY, Peter, biologist; b. London, Eng., June 4, 1908; s. Oscar and Dorothy (Selby) G.; B.S., U. London, 1929; Ph.D., Imperial Coll. Sci., 1931; m. Freda Ethel Dolman, July 29, 1933. Faculty, U. Pitts., 1939—, prof., 1943-45, chmn., 1945-63, Andrey Avinoff prof. biology, 1964—; cons. editor Reinhold Pub. Corp. Fellow A.A.A.S., Royal Micros. Soc.; mem. Am. Assn. Anatomists, Soc. Am. Zoologists, Am. Micros. Soc. (past pres.), Soc. For Indsl. Microbiology (past dir.), Phi Beta Kappa, Sigma Xi, Alpha Epsilon Delta. Author: French Grammar for Science Students, 1931; Basic Microtechnique, 1952; Microtomists's Formulary and Guide, 1954; The Mistress Cook, 1956; (with Freda Gray) Bibliography of Works in Microtechnique in Latin Alphabet Languages, 1956; Handbook of Basic Microtechnique, 1952, 2d edit., 1958, 3d edit., 1964; French Grammar for Science Students, 1959; (with Clarence and Marie Goodnight) General Zoology, 1964; How to Use the Microscope, 1967; Dictionary of the Biological Sciences, 1967. Editor: Ency. of Biol. Scis., 1961. Research, numerous publs. on microscopy preparation, interrelations of invertebrate parasites and symbiotes; tropical fungus protection methods. Home: 5131 Ellsworth Av., Pitts. 15232.*

GRAY, Peter Rygaard, Am. physicist; b. Ishpeming, Mich. Nov. 7, 1928; s. William Arthur and Mary (Bilkey) G.; B.S., Mich. Coll. Mining and Tech., 1950, M.S., 1951; Ph.D., U. Cal. at Berkeley, 1955; m. Norma Joyce Doolittle, July 15, 1951; children—Peter, David, Eric, Nancy. Chemist, Argonne Nat. Lab., Lemont, Ill., 1950-52, Dow Chem. Co., Midland, Mich., 1955-61; mgr. chem. physics Phillips Petroleum Co., Bartlesville, Okla., 1961—. Mem. Am. Phys. Soc. Research, publs. in areas of interactions of radiation, primarily electrons and positrons with matter; spectroscopic studies, primarily positranium and X-Ray. Home: 937 Sandstone Dr. Office: Phillips Petroleum Co., Radiation Lab., Bartlesville, Okla. 74003.*

GRAY, Seymour Jerome, Am. physician; b. Rochester, N.Y., Nov. 30, 1911; s. H. Lewis and Eva (Bobry) G.; B.A., U. Rochester, 1933; M.D., U. Pa., 1936; Ph.D., U. Chgo., 1943; m. Ruth Hart, July 2, 1935; children—Alfred Hart, Roger. Faculty U. Chgo., 1937-43; faculty Harvard Med. Sch., Boston, 1946—, asso. clin. prof. medicine, 1957—; mem. staff Peter Bent Brigham Hosp., Boston, 1947—. Vis. prof. nutrition Mass. Inst. Tech., 1963—; cons. various govt. agys. and hosps. Recipient Order of Hipolito Unanue Peru; named Comendador de la Orden al Merito Bernardo O'Higgins Chile; Hon. Prof. Medicine U. Chile, 1956. Diplomate Am. Bd. Internal Medicine. Mem. A.C.P., Am. Soc. Clin. Investigation, Am. Gastroent. Assn., A.A.A.S, Central Soc. Clin. Research, Soc. Exptl. Biology and Medicine, Am. Fedn. Clin. Research, A.M.A., Internat. Soc. Internal Medicine, Am. Physiol. Soc., Am. Soc. Clin. Nutrition, also hon. mem. many fgn. socs., Sigma Xi. Contbr. to med. textbooks, 1954—. Publs. on first preparations of radioactive chromium compounds and their use in blood volume measurements; stress, hormone effects on gastrointestinal tract. Home: 282 Warren St., Brookline, Mass. 02146. Office: 1180 Beacon St., Brookline, Mass. 02146.*

GRAY, Stephen, English electrician; b. circa 1696; pensioner of Charterhouse; Fellow Royal Soc., 1732. Research and publs. on elec. conduction, 1720-36; discovered basic principles of electric conductivity, 1729; 1st to divide materials into electrics and non-electrics; discovered non-electrics could be transformed into electric state by contact with active electrics; discovered electricity can be carried any distance perpendicularly by a thread; his research led the way for the invention of the Leyden phial, electric batteries. Died Feb. 25, 1736.

GRAY, Stephen Wood, Am. anatomist; b. Oakland, Cal., Apr. 27, 1915; s. Roy Barnett and Ethel (Graham) G.; B.A., Lake Forest Coll., 1936; M.A., U. Ill., 1937, Ph.D., 1939; m. Netta E. Steinhaus, Sept. 9, 1938. Instr., U. Ill., 1939-42, asst. prof. zoology, 1946; faculty Emory U., Atlanta, 1946—, prof. anatomy, 1961—. Mem. Ga. Acad. Sci. (exec. council 1964—), Am. Assn. Anatomists, Am. Physiol. Soc., Am. Soc. Zoologists, Sigma Xi. Author: (with Skandalakis, Shepard, Bourne) Smooth Muscle Tumors of the Alimentary Tract, 1962; also numerous articles. Research on natural history of smooth muscle tumors, effects of gravity on plant growth, effects of weightlessness on growth; origin, frequency and diagnosis of congenital malformations in humans. Home: 3008 Wadsworth Mill Ct., Decatur, Ga. 30032. Office: Dept. Anatomy, Emory U., Atlanta 30022.*

GRAY, Thomas, elec. engr.; b. Lochgelly, Scotland, Feb. 4, 1850; grad. Glasgow (Scotland) U., 1878; tchr. engring. Imperial Coll. of Engring., Tokyo, Japan, 1879-81; asst. in elec. engring. work to Lord Kelvin, 1881-88; represented him and Prof. Fleming Jenkin during manufacture and laying of Mackay-

Bennett system of Transatlantic cable; dir. dept. mech. and elec. engring. Rose Poly. Inst., Terre Haute, Ind. Contbr. Directions for Seismological Observations in Brit. Admiralty Manual of Sci. Enquiry, also articles on telegraphs and telephones in Ency. Brit.; prepared Smithsonian Tables; mem. expert staff Century Dictionary in dept. electricity. Died 1908.

GRAY, Thomas Cecil, English anesthesiologist; b. Liverpool, Mar. 11, 1913; s. Thomas and Ethel (Unwin) G.; ed. Ampleforth Coll., York; M.D.; m. Marjorie Hely, 1937; children—David, Beverley. Cons. anesthesiologist to council adminstrn. Meml. Regional Hosp., Liverpool; reader U. Liverpool, 1947-59, now prof. anesthesiology, dean postgrad. med. studies; dean Faculty Anaesthesiologists, Royal Coll. Surgeons Eng., 1964, 67. Mem. Assn. Anaesthesiologists Gt. Britain and Ireland (pres. 1957-60), Argentine, Brazilian, Sheffield, Australian socs. anesthesiology, Lisbon Soc. Medicine (corr.), Biol. Soc. Trinity Coll. of Dublin (hon.), Royal Soc. Medicine. Cons. editor Brit. Jour. Anaesthesia; Modern Trends in Anaesthesia and General Anaesthesia. Home: 1 Acrefield Rd., Liverpool 25. Office: 48 Bedford St., Liverpool, Eng.

GRAYBIEL, Ashton, Am. physician; b. Port Huron, Mich., July 24, 1902; s. William G. and Lucy Ann (Young) G.; A.B., U. So. Cal., 1924, A.M., 1925; M.D. Harvard, 1930; m. Moira Barkley Martin, Mar. 23, 1934; children—Ashton Lynd, Ann Martin. Research asso. Fatigue Lab., Harvard, 1936-42, instr., 1940-42; staff Naval Aerospace Med. Inst., Pensacola, Fla., 1942—, dir. research, 1945—; lectr. medicine U. Ala. Med. Sch., 1958—. Decorated Legion of Merit. Recipient Theodore C. Lyster award Aerospace Med. Assn., 1950, Liljencrantz award, 1961, Tuttle award, 1965; Navy League Adm. William S. Parsons award, 1960; Jeffries award Inst. Aerospace Scis., 1962; Boynton award Am. Astronautical Assn., 1962; Groedel medal Am. Coll. Cardiology, 1962; Conrad award Dept. Navy, 1965. Fellow A.A.A.S.; mem. Am. Coll. Cardiology (past pres.), Aerospace Med. Assn. (past. pres.), Aeromed. Assn. (past chmn. space medicine br.), Am. Heart Assn., Am. Physiol. Soc., Assn. Mil. Surgeons, Am. Otological Soc. (asso.), Am. Coll. Sports Medicine, Internat. Acad. Aviation Medicine, Internat. Acad. Astronautics, Am. Inst. Aeros. and Astronautics. Author: Clinical Electrocardiography, 1950; (with P. D. White, L. Wheeler, C. Williams) Electrocardiography in Practice, 1951; also numerous articles. Research on cardiovascular problems including long-term follow-up; role of vestibular organs under ordinary and stressful circumstances. Home: P.O. Box 4063, Bayshore, Warrington, Fla. 32507. Office: Naval Aerospace Med. Inst., Pensacola, Fla. 32512.*

GREATHOUSE, Glenn Arthur, Am. nuclear chemist, physicist; b. West Salem, Ill., Aug. 16, 1903; s. Chester A. and Alta (Brown) G.; B. Ed., Ill. State Normal U., 1927; M.S., U. of Ill., 1929; Ph.D., Duke U., 1931; m. Edith Mary Bennett, Aug. 22, 1925; children—Rosemary, Glenna Lu. Teacher, Arenzville, Ill. High Sch., 1925-26; asst. in research, University of Ill., 1927-30; fellow in chemistry, physics and biochemistry, Duke Univ. 1930-31; asst. prof. biophysics, Univ. of Md., 1931-36; physiologist, U. S. Dept. of Agr. Washington D.C., 1936-44, lecturer, U. S. Dept. of Agriculture Grad. Sch., 1935-44; dir. Army-Navy Nat. Defense Research Corp. Information Center, 1944-45; dir. Prevention of Deterioration Center, Nat. Research Council, Nat. Acad. of Sciences 1945-55, science adv. 1955—; interim prof. chemical engineering, U. of Florida, 1951-56; prof. Nuclear Engineering, 1956-60, head dept., 1957-60, pres., treas., Nuclear Research Chem., Inc., 1960-67; spl. cons. Mallincrodt Chem. Works, St. Louis, 1967—. Mem. sci. com., Office of Science Research and Development, 1944-45; special consultant, Office of the Chief of Ordnance, Army Dept., Washington, D.C., 1944-45; head nuclear cons. Commonwealth of Mass., 1959—; mem. U. S. British Science Mission to England, Europe, Africa and Panama, 1945; United States delegate to the second World Conference on peaceful use of the atom, Geneva, 1958. Awarded National Defense Research Committee-Award; Bureau of Ordnance Development Award U. S. Navy; His Majesty's Medal for Service in the Cause of Freedom British Government. Member of American Nuclear Society, Am. Chemical Soc., A.A.A.S., Wash. Acad. of Science, Sigma Xi. Author: (with Wessel) Deterioration of Materials—Causes and Preventive Techniques, 1954. Contbr. many articles to various profl. publs. Research on nuclear chemistry; mechanisms of chemical action by C-14 tracers; isolation and classification of natural chemical products. Home: The Towers Apt., Dayton Beach, Fla. 32020. Office: Spl. Cons., Mallincrodt Chem. Works, St. Louis.*

GREATOREX, Ralph, math. instrument maker; b. 1625; apprenticed to Elias Allen, instrument-maker; designed and made precise instruments; Samuel Pepys and Robert Boyle were among his customers; helped in the surveying of London after the Great Fire. Died 1712.

GREAVES, Harold Richard Goring, educator, polit. scientist; b. Naples, Italy, Nov. 17, 1907; s. Harold Frederick and Beatrice Violet Heather Greaves; student U. London, 1925-29, Grad. Inst., Geneva, Switzerland, 1929-30. Faculty, U. London (Eng.), 1931—, prof. polit. sci., 1961—, dean Faculty Econs. Author numerous books including: Foundations of Political Theory; The Civil Service in the Changing State; The British Constitution; Reactionary England; The League Committees; Democratic Participation and Public Enterprise. Lit. editor Polit. Quar.; English editor Internat. Polit. Sci. Abstracts. Address: London Sch. Econs., Houghton St., London, W.C.2, Eng.*

GREAVES, John, Brit. mathematician, astronomer; b. Hampshire, Eng., 1602; s. John Greaves; ed. Balliol Coll., Oxford, (Eng.) U., 1st became Gresham prof. geometry, London, then Savilian prof., Oxford; traveled in Near E. Author: Pyramidographia, 1646; Elementa linguoe Persicae, 1649; Philosophical Transactions; A Discourse on the Roman Foot and Denarius; miscellaneous works were published by Dr. Thomas Birch, 1737. Made instrumental survey of the Gt. Pyramid, 1638; research on longitude and navigation. Died 1652.

GREAVES, Joseph Eames, Am. biochemist; b. Logan City, Utah, Nov. 2, 1880; s. Joseph C. and Catherine (Eames) G.; B.S., Utah Agrl. Coll., 1904; M.S., U. Ill., 1907; Ph.D., U. Cal., 1911; m. Pernecy Dudley, June 10, 1907 (dec. May 1918); children—Joseph D., Florence D., Pernecy D., Vera D., Mary Oretta; m. 2d, Ethelyn Oliver, May 5, 1920; children—Marguerite Oliver, Thelma Mae, Oliver. Began as instr. chemistry Utah Agrl. Coll., 1907, asst. prof., 1908-10, asso. prof. physiol. chemistry, 1911-13, prof. bacteriology and physiol. chemistry, 1913-27, prof. bacteriology and pub. health, 1927-29, prof. bacteriology and biochemistry since 1929; bacteriologist Utah Exptl. Sta. Mem. Logan City Bd. of Health. Mem. Am. Chem. Soc., Am. Bacteriol. Soc., A.A.A.S., Am. Pub. Health Assn., Am. Soc. Biochemists, Utah Acad. Sci., Arts and Letters (pres. 1947). Author: Agricultural Bacteriology, 1922; Bacteria in Relation to Soil Fertility (with E. O. Greaves), 1925, Elementary Bacteriology, 5th edit., 1945; also articles in profl. jours. Cons. editor Soil Science; contbg. editor Americana Ann. Died June 6, 1954.

GREAVES, William Michael Herbert, astronomer; b. Barbados, W.I., Sept. 10, 1897; s. E. C. Greaves; ed. Codrington Coll., Barbados, St. John's Coll., Cambridge, M.A.; m. Caroline Grace Kitto, 1926; 1 son. Isaac Newton student Cambridge U., 1921-23, fellow St. John's Coll., 1922-25; chief asst. Royal Obs., Greenwich, 1924-38; royal astronomer for Scotland, prof. astronomy U. Edinburgh, 1938-55. Recipient Tyson medal in astronomy. Fellow Royal Soc., 1943; mem. Royal Astron. Soc. (past sec.), Royal Soc. Edinburgh (v.p.), Brit. Assn. for Advancement Sci. (sec., recorder sect. A). Contbr. articles to profl. publs. Died Dec. 24, 1955.

GREBE, Ernst Wilhelm, German mathematician, physicist; b. Michelbach nr. Marburg, Germany, Aug. 30, 1804; s. Friedrich and Jacobine Henriette (Bock) G.; studied philology and math., Bonn, Leipzig, Germany, 1821-24, theology, Marburg, 1824-26; Ph.D., 1829; m. Caroline Creuzer, 1831; 3 children, including Leonhard. Became lectr. U. Marburg, 1829; secondary sch. tchr. successively in Rinteln, Marburg, Kassel, and again in Marburg; named rector Realschule, Kassel, Germany, 1855. Research and publs. on devel. topology founded by L. Euler; discovered independently Grebe's point (also Lemoine's point) which is point on the surface of a triangle for which the sum of the sqs. of distances from sides of triangle is the smallest possible. Died Kassel, Jan. 14, 1874.

GREBE, John Josef, chemist, inventor; b. Uerzig, Rhineland, Germany, Feb. 1, 1900; s. Carl and Gertrude (Erbes) Grebe; B.S., Case Sch. Applied Sci., 1924, M.S., 1927, D.Sc., 1935; m. Hazel Amanda Holmes, March 2, 1929; children—Ruth Elaine (Mrs. Jos. T. Davis), Joanne Hazel (Mrs. Dwight T. Hendricks), John Holmes, Carolyn (Mrs. Wm. Moloney), James Carl. Came to U. S. 1914, naturalized, 1921. With Dow Chem. Co., Midland, Mich., 1924—, as dir. Phys. Research Lab.; on loan from Dow Chem. Co. to Oak Ridge Nat. Lab., 1946-47, to Army Chem. Corps as chief tech. adviser, 1948-49; research counselor Dow Chem. Co., 1949-53, dir. nuclear research and devel., 1953-58, dir. nuclear and basic research, 1958—; developed electrometric control. Recipient certificate of merit for devel. of sun screen, Franklin Inst., 1942; Chem. Ind. medal, 1943; John Wesley Hyatt award, 1947. V.p. and dir. Dowell; dir. Ewing Development Co. Civilian observer Bikini bomb tests, 1946. Fellow A.A.A.S., Am. Inst. Chemists; mem. Am. Chem. Soc., Am. Inst. Chem. Engrs., Am. Physical Soc., Am. Soc. M.E., Engring. Soc. of Detroit, Mich. Acad. Scis., Soc. Chem. Industry, Found. Study of Cycles, Atomic Industrial Forum, Am. Nuclear Soc., Am. Ordnance Assn., Armed Forces Chem. Assn., Franklin Inst., N.Y. Acad. Sci., Sigma Xi. Contbr. articles to sci. jours. Known for investigation of quantum theory; formulated a period table for fundamental particles; credited with making possible the wartime prodn. of styrene for synthetic GR-S rubber; patentee in areas of power generation, electrochemistry, air conditioning and plastics; contbd. to chem. process of recovering magnesium from sea water. Home: 1505 W. St. Andrews Dr. Office: Dow Chem. Co., Midland, Mich.

GREBE, Leonhard, German radiologist; b. Elberfeld, Dec. 15, 1883; s. Friedrich and Josefine (Krienen) G.; Ph.D. in Radiology, U. Bonn; m. Gertrud Klose, 1913; children—Gertrud, Hilde, Friedrich. Inst. Research, Bonn, 1923-45. Recipient plaque Roentgen Mus., Remscheid-Lennep. Hon. mem. Roentgen Assn. Rhine and Westphalia, German Roentgen Mus. Co-author: Dosimetrie der Röntgenstrahl. Research and publs. on spectroscope, theory of relativity. Address: 8 Maltererstrasse, Bonn, Germany.

GREBENSCHCHIKOV, Ilya Vasilievich, Russian physicist; b. June 21, 1887. Organized State Optical Inst., Leningrad. Decorated Order of Lenin; recipient Stalin prize, 1942. Mem. USSR Acad. Scis. Author several sci. publs. Established founds. for silicate industry, 1933; produced optical glasses. Died Feb. 8, 1953.

GREBINSKY, Sergey Orestovich, Russian plant physiologist, biochemist; b. Fergana (now Uzbek SSR), 1905; grad. Biol. Faculty, Leningrad U., 1935, postgrad., until 1938; D.Biol. Sci., 1942. Lectr., head chair plant physiology and microbiology Kazakhstan U., 1938-45; head sect. plant physiology and biochemistry Bot. Inst., Kazakhstan br. USSR Acad. Sci., 1944-45; prof., head chair plant anatomy and physiology Lvov U., 1945—. Mem. Ukrainian Bot. Soc. (Lvov br.). Author: The Biochemistry of Tobacco and Nicotiana rustica, 1938; The Biochemical Features of Alpine Plants, 1939; The Biochemistry of Citrus Plants, 1940; The Physiological and Biochemical Features of Mountain Plants, 1944; The Physiology and Biochemistry of Plants in Kazakhstan, 1945; Plant Respiration in the Light of Modern Findings, 1945; Organic Acids, 1948. Research in plant metabolism. Address: Lvov University, Marshalovskaya 1, Lvov, Ukraine SSR, USSR.

GRECO, Donato, Italian physicist; b. Cervinara, Avelino, Oct. 5, 1923. Prof., U. Bari, later full prof. math., algebraic and infinitesimal analyses Faculty Math., Phys. and Natural Scis.; instr. higher math., dir. Inst. Analysis and Inst. Geometry; instr. higher analysis Faculty Scis., U. Naples. Address: Institute of Mathematics, University of Naples, Naples, Italy.

GREDLER, Vincenz Maria (baptized Ignaz), natural scientist; b. Telfs, Austria-Hungary, Sept. 30, 1823; s. Johann Gredler; teaching certificate Innsbruck, Austria, 1852. Joined Franciscan Order, Salzburg, Austria, 1841; ordained priest, 1846; tchr. secondary sch., Hall, Austria, 1848-49; became tchr. natural history Sch. Franciscans, Bozen, Italy, 1849; founder, dir. Ordensgymnasium, 1872-98; continued to teach until 1901. Author: Die Käfer von Tirol nach ihrer horizontalen und vertikalen Verbreitung, 2 vols., 1863-66; also numerous articles. Studied distbn. of beetles, ants, diptera, heteroptera, molluscs in Tirol, especially land and fresh water molluscs, plant pests; pioneered geol. study ice age in Austria; studied material from Carinthia, Styria, Herzegorina, Central Africa, Borneo, Summatra, Japan and especially China. Died Bozen, May 4, 1912.

GREEBLER, Paul, Am. physicist; b. Buffalo, Dec. 13, 1922; s. Ben and Rebecca (Yager) G.; B.S., U. Colo., 1944; M.S., Rutgers U., 1952, Ph.D., 1954; m. Doris Harriet Schier, Oct. 20, 1946; children—Veronica Phyllis, Carol Susan, Arlene Joy. Physicist, Johns Manville Corp. Research Center, Manville, N.J., 1946-53, sr. research physicist, 1953-55; research physicist Knolls Atomic Power Lab., Gen. Electric Co., Schenectady, N.Y., 1955-56, with Atomic Power Equipment dept., San Jose, Cal., 1956—, mgr. advance reactor physics sub-sect. advanced products operation, 1964—. Mem. Am. Phys. Soc., Am. Inst. Physics, Am. Nuclear Soc., Sigma Xi, Tau Beta Pi, Sigma Pi Sigma. Editor: (with Ernest Henley) Advances in Nuclear Science and Technology series, 1963-68; contbr. numerous articles to profl. jours. Theoretical and exptl. research on heat transfer mechanisms in insulating materials; developed physics methods for theoretical calculations on nuclear reactors; contbr. to nuclear design and evaluation of atomic reactors; patentee. Home: 1461 Cherry Garden Lane. Office: 310 De Guigne Dr., Sunnyvale, Cal. 94086.*

GREEFF, Richard, German zoologist; b. Elberfeld, Germany, Mar. 14, 1829; s. Peter and Sophie (Bredt) G.; student medicine, Würzburg, Heidelberg, (both Germany); M.D., Berlin, 1857; m. Maria Esch; children include Richard, Maria (Mrs. Hugo Vogel). Began practice medicine, Elberfeld, 1859; left med. profession and joined faculty zoology U. Bonn, 1863; became prof. zoology and comparative anatomy U. Marburg (Germany), 1871; traveled in Canary Island, Cape Verde Islands, Guinea Islands, Adriatic Coast, Naples, Lisbon. Research and publs. on annelids, structure and reproduction conditions of single-celled animals, especially rhizopods. Died Marburg, Aug. 30, 1892.

GREEFF, (Karl) Richard, German ophthalmologist, med. historian; b. Elberfeld, Germany, June 18, 1862; s. Richard G. and Maria (Esch) G.; student Leipzig, Berlin; M.D., Marburg, 1888. Asst. to H. Schmidt-Rimpler, Marburg, C. Schweigger, Berlin, C. Weigert, Senckenberg Inst., Frankfort/Main, Germany; joined faculty U. Berlin, 1894, asso. prof. in charge charity patients as new univ. eye clinic was established, 1897-1928. Author: Über das ophthalmoskopische Aussehen und den Bau der Chorioidea, 1897; Anleitung zur mikroskopischen Untersuchung des Auges, 1898; Rembrandts Darstellung der Tobiasheilung, 1907; Erfindung der Augengläser, 1921; Das menschliche Auge, 1935; Kurze Geschichte der Brille und des Optiker-

Handwerks, 1938. 1st research on path. anatomy of eye; studied trachoma, history of eye therapy and eye glasses in art, practical optics, sci. edn. of opticians; collected hist. glasses, ophthal. instruments, pictures and prints on history of glasses. Died Berlin, Nov. 4, 1938.

GREELEY, Andrew M(oran), Am. sociologist; b. Oak Park, Ill., Feb. 5, 1928; s. Andrew T. and Grace (McNichols) G.; A.B., St. Mary of Lake Sem., 1950, S.T.L., 1954; M.A., U. Chgo., 1961, Ph.D., 1961. Ordained priest Roman Catholic Ch., 1954; asst. pastor Ch. of Christ the King, Chgo., 1954-64; sr. study dir. Nat. Opinion Research Center, Chgo., 1961—; lectr. sociology U. Chgo., 1961—; cons. Archdiocesan Office Urban Affairs, Hazen Found. Commn. Recipient Cath. Press Assn. award for best book for young people, 1965; Thomas Alva Edison award for radio broadcast, 1963. Mem. Am. Sociol. Assn., Am. Cath. Sociol. Soc. (pres.), Soc. for Sci. Study Religion, Religious Research Assn. Author: The Church and the Suburbs, 1959; Strangers in the House, 1961; Religion and Career, 1963; (with Peter H. Rossi) Education of Catholic Americans, 1966. Research and publs. on influence religious, ethnic groups on Am. social structure with emphasis on effects of Cath. edn. Home: 450 E. 78th St., Chgo. 60615. Office: 5720 S. Woodlawn Av., Chgo. 60637.*

GREELY, Adolphus Washington, Am. arctic explorer; b. Newburyport, Mass., Mar. 27, 1844; s. John Balch and Frances (Cobb) G.; grad. Newburyport High Sch., 1860; m. Henrietta H. C. Nesmith, June 20, 1878; children—Antoinette, Adola, John Nesmith, Rose Ishbel, Adolphus W., Gertrude Gale. Served in Civil War, 1861-65, pvt. to capt., and bvt. maj. vols. (thrice wounded); apptd. 2d lt. 36th U. S. Inf., Mar. 7, 1867; 1st lt. 5th Cav., May 27, 1873; capt., June 11, 1886; brig. gen. chief signal officer U. S. A., Mar. 3, 1887; maj. gen., Feb. 10, 1906. First vol. pvt. soldier of Civil War to reach grade of brig. gen. U. S. A. Constructed 2,000 miles mil. telegraph in Tex., Dak. and Mont., 1876-79; in pursuance of recommendation of Hamburg Internat. Geog. Congress (1879) was placed, 1881, in command of U. S. expdn. to establish one of a chain of 13 circumpolar stations; his party of 25 reached further north (83°-24′) than any previous record; discovered new land N. of Greenland and crossed Grinnell Land to the Polar Sea; two relief expdns. failed to reach the party, which retreated S. to Cape Sabine, where, relief still failing, the party largely perished of starvation, only 7 survivors being found by 3d expdn. under Capt. Winfield S. Schley. During mil. operations abroad (1898-1902) there were built and operated under his direction 1,000 miles of telegraph in P.R., 3,800 miles in Cuba, 250 miles in China, and 13,500 miles of lines and cables in P.I.; installed system of 3,900 miles of telegraph lines, submarine cables and wireless in Alaska, 1900-04, the wireless section of 107 miles, from Nome to St. Michael, being the first successful long-distance wireless operated regularly as part of a commercial system. Mem. bd. to regulate wireless telegraphy in U. S., 1904; mem. bd. to report on coast defenses of U. S., 1905; U. S. del. Internat. Telegraph Conf., London, 1903, Internat. Wireless Telegraph Conf., Berlin, 1903. Comdg. Pacific Div. and in charge relief operations, San Francisco earthquake sufferers, Apr.-Aug. 1906; comdg. Northern Div. 1906, Dept. Columbia, 1907; retired by operation of law, 1908. Gold medalist, Royal. Am. and French geog. socs. Author: Isothermal Lines of the United States, 1881; Chronological List of Auroras, 1881; Diurnal Fluctuations of Barometric Pressure, 1891; Three Years of Arctic Service, 2 vols., 1885; Proceedings of Lady Franklin Bay Expedition, 1888; American Weather, 1890; American Explorers, 1894; Handbook of Arctic Discoveries, 1896; Rainfall of Western States and Territories, 1888; Climate of Oregon and Washington, 1889; Climate of Nebraska, 1890; Climatology of Arid Region, 1891; Climate of Texas, 1891; Public Documents First Fourteen Congresses of United States, 1900; Handbook of Polar Discoveries, 1909; Handbook of Alaska, 1925; True Tales of Arctic Heroism, 1912; Reminiscences of Travel and Adventure, 1927; Polar Regions in Twentieth Century, 1928. Work on arid lands, meteorology and climatology; his explorations and books added important knowledge about polar regions. Died Washington, Oct. 20, 1935.

GREEN, Alex E. S., physicist; b. N.Y.C., June 2, 1919; s. Joseph M. and Celia (Kahn) G.; B.S., Coll. City N.Y., 1940; M.S., Cal. Inst. Tech., 1941; Ph.D., U. Cin., 1948; m. Freda Kaplowitz, June 2, 1946; children—Bruce, Deborah, Marcia, Linda, Tamara. From research asst. to physicist Cal. Inst. Tech., Pasadena, 1940-43; from instr. to asso. prof., U. Cin., 1946-53; asso. prof., acting chmn. physics, sci. dir. Tandem Van de Graff Program, Fla. State U., Tallahassee, 1953-59; chief physics Convair Div., Gen. Dynamics, San Diego, 1959-62, mgr. Space Sci. Lab., Astronautics Div., 1962-63; grad. research prof. physics U. Fla., Gainesville, Fla., 1963—. Research participant Oak Ridge Nat. Lab., summer 1956; vis. staff scientist, group leader T-11, Los Alamos Sci. Lab., 1957-58; cons. Jet Propulsion Lab. Pasadena, Cal., 1963—, Marshall Space Flight Center, Huntsville, Ala., 1964—, Inst. for Def. Analysis, Washington, 1963—. Decorated medal of Freedom. Fellow Am. Phys. Soc.; mem. Am. Assn. Physics Tchrs., Am. Geophys. Union, Am. Rocket Soc. (pres. San Diego chpt. 1962-63), Am. Inst. Aeros. and Astronautics (vice chmn. space and atmospheric physics com. 1965—, chmn. tech. com. 1965-66). Author: Nuclear Physics, 1955; Atomic and Space Physics, 1965. Editor: The Middle Ultraviolet in Space, 1966. Research, publs. in nuclear physics, atomic physics, atmospheric and space physics. Home: 3535 N.W. 7th Place, Gainesville, Fla. 32601.*

GREEN, Arthur George, Brit. chemist; b. London, Feb. 27, 1864; student of Williamson; prof. color chemistry and dyeing in Leeds, Eng., 1903-16; organizer dyestuffs research lab. Manchester (Eng.) Coll. Tech.; cons. firm Levinstein Ltd. Fellow Royal Soc., 1915. In 1887 discovered primuline, later found to be a thiazole derivative; originated ingrain process of developing a dye on a fabric; research on sulfur and stibene dyes, aniline black, oxonism compounds, sulfanilamide; (with A. G. Perkin) opposed Ostwald's theory of indicatiors and believed a quinonoid form of the undissociated molecule caused the red color of alkaline phenolpthalein. Died London, Sept. 12, 1941.

GREEN, Charles, Brit. inventor; b. London, Eng., 1785; s. Thomas Green; m. Martha Morrell; a son, George. Went into father's fruit bus.; made 1st balloon ascent, 1821; subsequently made 526 ascents; proved coal-gas can be used to inflate balloons, 1821; invented guide-rope to regulate ascent and descent of balloon. Died 1870.

GREEN, David Ezra, Am. biochemist; b. N.Y.C., Aug. 5, 1910; s. Herman and Jennie (Marrow) G.; B.A., N.Y. U., 1930, M.A., 1932; Ph.D., Cambridge (Eng.) U., 1934; m. Doris Cribb, Apr. 15, 1935; children—Rowena (Mrs. Larry Matthews), Pamela (Mrs. Joseph Baldwin, Jr.). Beit Meml. Research fellow Cambridge, Eng., 1934-40; fellow Harvard, 1940-41; with enzyme lab. Coll. Phys. and Surg., Columbia, 1941-48; asso. prof. biochemistry Columbia, 1947; co-dir. Inst. Enzyme Research, U. Wis., 1948—, prof. enzyme chemistry, 1948—. Recipient Paul-Lewis Labs. award enzyme chemistry, 1946. Fellow Am. Acad. Arts and Scis.; fgn. fellow Royal Flemish Acad. Arts and Scis.; mem. Am. Soc. Biol. Chemists, Nat. Acad. Scis., Am. Chem. Soc., Harvey Soc., Biochem. Soc., Am. Soc. Cell Biology, Phi Beta Kappa, Sigma Xi. Author: Mechanisms of Biological Oxidations, 1939; Molecular Insights into the Living Process, 1967. Organized enzyme systems; reconstructed fatty acid oxidizing system; study of structure and function of membrane systems; mechanism of oxidative phosphorylation; mitochondrial structure and function. Home: 1525 Sumac Dr., Madison, Wis. 53705.*

GREEN, Gabriel Marcus, Am. mathematician; b. N.Y.C., Oct. 19, 1891; grad. Coll. City N.Y., 1911; A.M., Columbia, 1912, Ph.D., 1913. Tchr. math. Coll. City N.Y. for a year; became instr. Harvard, 1914. Recipient prizes for math. ability, prize for highest rank in all coll. subjects. Wrote work on projective differential geometry; also articles in math. and geometry. Research on projective differential geometry of triple systems of surfaces, one-parameter family of space curves and conjugate nets on a curved surface. Died Jan. 24, 1919.

GREEN, George, Brit. mathematician; b. Sneinton, Eng., July 14, 1793; B.A., Caius Coll., Cambridge U. (4th wrangler), 1837; named Perse fellow Caius Coll., 1839; became pensioner of Caius Coll., Cambridge, 1833. Author: Essay on the Application of Mathematical Analysis to the Theory of Electricity and Magnetism; also articles on laws of equilibrium of fluids, attractions in n-dimensional space, motion of a fluid agitated by vibrations of a solid ellipsoid. Generalized and extended Poisson's research; introduced term, potential; originated theorem to study potential in the case of fields. Died Sneinton, Mar. 31, 1841.

GREEN, Harold Alfred John, economist; b. Birmingham, Eng., Feb. 13, 1923; s. William Harold and Winifred (Joynes) G.; B.A., Oxford U., 1947, M.A., 1948; Ph.D., Mass. Inst. Tech., 1954; m. Barbara Ingalls Robertson, June 3, 1950; children—Catherine Jennifer, David William Robertson. Faculty, Clark U., 1948-50, Brown U., 1952-54, U. Manchester, Eng., 1954-55, U. Keele, Eng., 1955-58, U. Cal. at Santa Barbara, 1958-59; faculty U. Toronto, Ont., Can., 1959—, prof. econs., 1963—; Simon vis. prof. U. Manchester, 1965-66. Mem. Am. Econ. Assn., Econometric Soc., Royal Econ. Soc., Canadian Polit. Sci. Assn. Author: Aggregation in Economic Analysis, 1964; also articles. Research on econ. theories of consumer behavior, allocation of resources, growth and tech. progress, choice over time and under uncertainty. Home: 30 Queen Mary's Dr., Toronto 18, Ont., Can.*

GREEN, Harold David, Am. physiologist; b. Zanesville, O., Aug. 11, 1905; s. Louis Harold and Laura (Gobel) G.; B.S., Wooster Coll., 1927, D.Sc., 1957; M.D., Western Res. U., 1931; m. Bonnie Louise McClung, June 30, 1934; children—Barbara Holden (Mrs. Jerry A. Trivette), David Louis Darragh. Faculty, Western Res. U. Med. Sch., 1933-35, 36-39, 40-45, Yale Med. Sch., 1935-36, Mass. Inst. Tech., 1939-40; prof. physiology and pharmacology, dir. dept. Bowman Gray Sch. Medicine, Winston-Salem, N.C., 1945-60, asso. dept. internal medicine, 1945—, Gordon Gray prof., 1960—, chmn. dept. physiology and pharmacology, 1960-63, chmn. dept. physiology, 1963—, asso. in pharmacology, 1963—; staff mem. N.C. Baptist Hosp. Mem. physiology study sect. NIH, USPHS, 1959-62; mem. sci. council Inst. for Advancement Med. Communications, 1961-64. Fellow A.C.P.; mem. A.M.A., Am. Fedn. for Clin. Research, Am. Heart Assn. (council for high blood pressure research, sect. on circulation), Am. Coll. Clin. Pharmacology and Chemotherapy, Am. Physiol. Soc., Am. Soc. for Pharmacology and Exptl. Therapeutics, A.A.A.S., Biophys. Soc., Soc. for Exptl. Biology and Medicine, Central Soc. for Clin. Investigation, So. Soc. for Clin. Research, Phi Beta Kappa, Sigma Xi, Alpha Omega Alpha, Theta Chi Delta. Research and numerous publs. on measurement of coronary blood flow, peripheral circulatory studies in normal and abnormal conditions, hormonal and autonomic nervous control of circulation in various vascular beds, hypertension, shock. Home: 3619 Dewsbury Rd., Winston-Salem, N.C. 27104.*

GREEN, Harry, Am. biochemist; b. Phila., Sept. 7, 1917; s. Samuel and Mary (Bogatin) G.; A.B., U. Pa., 1938, M.S., 1939, Ph.D., (Harrison fellow), 1942; m. Harriett Borten, Oct. 6, 1945; children—Ann Ellen, Jane Merle. Sr. organic chemist Lion Oil Refining Co., El Dorado, Ark., 1941-44, Pa. Salt Mfg. Co., Phila., 1944-47; sr. research asso. U. Pa. Med. Sch. Phila., 1947-52; head biochemistry research Wills Eye Hosp., Phila., 1952-58; dir. biochemistry-macrobiology Smith, Kline & French Lab., Phila., 1958—; asst. prof. biochemistry U. Pa. Med. Sch., 1954—. Mem. Am. Soc. Biol. Chemist, Am. Soc. Pharmacy and Exptl. Therapeutics, Assn. Research Ophthalmology, N.Y. Acad. Scis., Assn. for Research in Nervous and Mental Diseases, Sigma Xi. Research and publs. on synthesis organic compounds related to indsl. uses of petroleum products, synthesis organic insecticides and agrl. chems., biochemistry cancer, eye disease, nervous and mental diseases, hypertension, obesity, biochemistry behavior. Home: 6232 N. 12th St., Phila. 19141. Office: 1500 Spring Garden St., Phila. 19101.*

GREEN, Herbert Sydney, math. physicist; b. Ipswich, Gt. Britain, Dec. 17, 1920; s. Sydney and Violet (Hindry) G.; B.Sc., Royal Coll. Sci., 1941; Ph.D., U. Edinburgh (Scotland), 1947, D.Sc., 1949; D.Sc., U. Adelaide (Australia), 1952. Research fellow U. Edinburgh, 1945-49; mem. Inst. for Advanced Study, Princeton, N.J., 1949-50; vis. prof. Dublin (Ireland) Inst. for Advanced Studies, 1950-51, 58; prof. math. physics U. Adelaide (Australia), 1951—. Fellow Australian Acad. Sci.; mem. Australian Math. Soc., Australian Inst. Physics. Author: (with M. Born) General Kinetic Theory of Liquids, 1950; Molecular Theory of Fluids; Encyclopaedia of Physics, Vol. 10, 1959; (with C. A. Hurst) Order-Disorder Phenomena, 1964; Matrix Mechanics, 1965; also articles. Statis. mechanics of fluids especially irreversible processes, theoretical nuclear and particle physics, plasma physics, lattice statistics, spinors in gen. relativity. Office: U. Adelaide, Adelaide, S. Australia, Australia.*

GREEN, Horace, Am. laryngologist; b. Chittenden, Vt., Dec. 24, 1802; s. Zeeb and Sarah (Cowee) G.; M.D., Castleton Med. Coll., 1825; m. Mary Butler, Oct. 20, 1829; m. 2d, Harriet Douglas, Oct. 27, 1841; 11 children. First Am. physician to specialize in diseases of throat; wrote Treatise on Diseases of the Air Passages: Comprising an Inquiry into the History, Pathology Causes and Treatment of those Affectations of the Throat Called Bronchitis, Chronic Laryngetis, Clergyman's Sore Throat (created controversy by advocating application of local medication in larynx, written before invention of laryngscope); prof. medicine and pres. Castleton Med. Coll., 1840-43; a founder N.Y. Med. Coll., 1850, prof. medicine, 1850-60; founded Am. Med. Monthly, 1854; contbr. numerous articles to med. jours. Died Ossining, N.Y., Nov. 29, 1866.

GREEN, Irving Joseph, Am. virologist; b. Bayonne, N.J., Sept. 10, 1923; s. Louis and Sadie (Ran) G.; student San Francisco City Coll., 1941-42; A.B., U. Cal. at Berkeley, 1948; student Stanford U., 1949-50; M.S., U. Wis., 1952-53; postgrad. U. Cal. at Los Angeles, 1963-65; m. Matilda Cassorla, Dec. 30, 1951; children—Louis, Alan, Sharon. With U. S. Army Med. Dept., 1943-46; virology dept. 6th Army Area Med. Lab., 1948-49; epidemic disease control officer 1st Marine Div., Korea, 1950-51; research virologist Navy Med. Research Unit 4, Great Lakes, Ill., 1951-56, Unit 1, U. Cal. at Berkeley, 1956-60, head virology and tissue culture dept. Unit 2, Taipei, Taiwan, 1960-63; officer in charge U. S. Navy Biol. Lab., Naval Supply Center, Oakland, Cal., 1965—. Mem. A.A.A.S., Tissue Culture Assn., Am. Soc. Microbiology. Research and publs. in viruses in tissue culture, preservation of tissue culture cells by freezing, isolation of a new influenza B virus, epidemiology of Japanese B encephalitis virus. Home: 1016 Ashmount Av., Oakland 94610. Office: U. S. Naval Biol. Lab., Oakland, Cal. 94625.*

GREEN, Jack Peter, Am. research physician; b. N.Y.C., Oct. 4, 1925; s. Maurice and Ethel (Herman) G.; B.S., Pa. State U., 1947, M.S., 1949; Ph.D., Yale, 1951, M.D., 1957; m. Arlyne Frank, Oct. 25, 1958. Asso. prof. pharmacology Yale U. Sch. Medicine New Haven, 1957-65, Cornell U. Med. Coll., N.Y.C., 1966-68; prof., chmn. dept. pharmacology

Mt. Sinai Sch. Medicine, N.Y.C., 1968—. Mem. A.A.A.S., N.Y. Acad. Sci., Harvey Soc., Am. Soc. Pharm. and Exptl. Therapeutics, Am. Chem. Soc. Editorial bd. Jour. Pharmacology and Exptl. Therapeutics, Clin. Pharmacology and Therapeutics. Research and numerous publs. in chemistry of brain, metabolism of neurotransmitters and hormones, studies of mastocytes, application of quantum mechanics to biology. Home: 435 E. 70th St., N.Y.C. 10021. Office: Dept. Pharmacology, Mt. Sinai Sch. Medicine, Fifth Av. and 100th St., N.Y.C.

GREEN, James W(eston), Am. physiologist; b. Elkins, W.Va., May 16, 1913; s. James Weston and Adah (Harshberger) G.; B.S., Davis-Elkins Coll., 1935; student U. Pa., 1940-41; Ph.D., Princeton, 1948; m. Erika Roth, July 8, 1961; children—James Philip, Stephen Henry. Mem. faculty Rutgers U., 1948—, prof. physiology, 1961—, chmn. dept. physiology and biochemistry, 1962-67, chairman of department of physiology, 1967—. Ford Found. fellow, 1953-54. Mem. Am. Physiol. Soc., Soc. Gen. Physiologists. Editor: New Developments in Tissue Culture, 1961. Research on the physical functioning of the surface membranes of cells—showing that hydrogen ions, various chemical agents, and radiations have marked effects on the rate of penetration of ions into cells; study of the effects of these agents on enzyme systems in cells. Home: 409 Grant Av., Highland Park, N.J. 08904. Office: Rutgers Univ., New Brunswick, N.J. 08903.*

GREEN, John W(illie), Am. mathematician; b. Hearne, Tex., Mar. 8, 1914; s. John Willie and Maude (Owen) G.; B.A., Rice Inst., 1935, M.A., 1936; Ph.D., U. Cal. at Berkeley, 1938; m. Catherine Alden Reid, Aug. 30, 1938; children—Harriet Alden, John Owen. Instr. Harvard, 1938-39; instr. U. Rochester, 1939-41, asst. prof., 1941-43; mathematician Aberdeen Proving Ground, Md., 1943-45; asst. prof. U. Cal. at Los Angeles, 1945-48, asso. prof., 1948-54, prof., 1954—; statis. investigator Nat. Sci. Found., 1955-56. Mem. Am. Math. Soc. (sec. 1957—), Math. Assn. Am., Soc. Indsl. and Applied Mathematics. Contributions to potential theory, differential equations, convex sets, applied mathematics. Home: 612 17th St., Santa Monica, Cal. Office: University of California, Los Angeles.*

GREEN, Jonathan, physician, surgeon; b. circa 1788; M.D., Heidelberg, Germany, 1834; surgeon in navy; fellow Royal Med. and Chirurg. Soc.; mem. Royal Coll. Surgeons. Author: The Utility and Importance of Fumigating Baths illustrated; or a Series of Facts and Remarks, shewing the Origin, Progress and final Establishment of the practice of Fumigations for the Cure of various Diseases, 8 vols., 1823; A short Illustration of the Advantages derived by the use of Sulphurous Fumigating Hot air, and Vapour Baths, 8 vols., 1825; Some Observations on the utility of Fumigating and other Baths . . . With a Summary of . . . Cases, 1831; A Practical Compendium of the Diseases of the Skin with Cases, 8 vols., 1835; On the Utility and Safety of the Fumigating Bath as a remedial agent in Complaints of the Skin, Joints, Rheumatism, 1847; An Improved Method of Employing Mercury by Fumigation to the Whole Body, 8 vols., 1852. Patented portable vapour bath. Died Feb. 23, 1864.

GREEN, Joseph, English chemist; b. London, Eng., Aug. 13, 1920; s. Louis and Eva (Herman) G.; B.Sc., Kings Coll., London U., 1941, Ph.D., 1944, D.Sc., 1965; m. Olive Pauline Jones, Mar. 17, 1950. Head chem. research Vitamins Ltd., London, 1944—. Fellow Royal Inst. Chemistry; mem. Chem. Soc., Biochem. Soc., Nutrition Soc. Research, numerous publs. on theory of chromatography, fat-soluble vitamins, especially chemistry and biochemistry of vitamin E. Home: 19, Vineyard Hill Rd., London, S.W.-19, Eng. Office: Walton Oaks Exptl. Sta., Tadworth, Surrey, Eng.*

GREEN, Joseph Reynolds, physiologist; b. Stowmarket, Suffolk; B.Sc., U. London, 1880; M.A., Cambridge U., 1888; Sc.D., 1894. Sr. demonstrator physiology Cambridge U., 1885-87, fellow, lectr. Downing Coll., 1902; prof. botany Pharm. Soc. Gt. Britain, 1887-1907; Hartley lectr. vegetable physiology U. Liverpool. Recipient Rolleston prize Oxford U., 1890. Fellow Royal Soc., 1895, Linnean Soc.; mem. Brit. Assn. (pres. botany sect. 1902). Author: A Manual of Botany, 1895; The Soluble Ferments and Fermentation, 1899; Introduction to Vegetable Physiology, 1900; Primer of Botany, 1910; History of Botany from 1860-1900, 1910, also articles. Died June 3, 1914.

GREEN, L. L., English physicist; b. Leicester, Eng., Mar. 30, 1925; s. Leonard and Victoria (Hughes) G.; M.A., King's Coll., Cambridge U., Ph.D., 1948; m. Helen Therese Morgan, Apr. 29, 1952; children—Sarah Elizabeth, Paul Nicholas. Faculty, U. Liverpool (Eng.), 1948—, prof. exptl. physics, 1964—. Fellow Phys. Soc.; mem. Inst. Physics (sec. nuclear physics sub-com.). Research, publs. on secondary particles in fission, nuclear spectroscopy, nuclear reactions. Home: 1 De Grouchy St., West Kirby, Cheshire, Eng. Office: Chadwick Lab., U. Liverpool. Liverpool, Eng.*

GREEN, Louis Craig, Am. astrophysicist; b. Macon, Ga., Feb. 2, 1911; s. Edward Melvin and Ann Field (Craig) G., A.B., Princeton, 1932, M.A., 1933, Ph.D., 1937; m. Elizabeth Hazard Ufford, July 27, 1940. Tchr. math. and astronomy Allegheny Coll., 1937-41; faculty Haverford (Pa.) Coll., 1941—, prof. astronomy, 1953—; tchr. math. Swarthmore Coll., 1944; tchr. physics Bryn Mawr Coll., 1944-46; dir. Strawbridge Meml. Obs., Haverford Coll., 1942—, chmn. physics, 1963-65, provost, 1965; vis. prof. Max Planck Inst. Physics and Astrophysics, Munich, Germany, summer 1959; mem. Inst. for Advanced Study, Princeton, N.J., 1962-63. Guggenheim fellow 1955-56. Fellow Am. Phys. Soc.; mem. Am. Astronom. Soc., Phi Beta Kappa, Sigma Xi. Contbr. articles to profl. jours. Research on computations of atomic constants including transition probabilities of astro physical interest. Home: 791 College Av., Havefford, Pa. 19041.*

GREEN, Melville Saul, Am. physicist; b. Jamaica, N.Y., June 9, 1922; s. Maurice S. and Ella (Prichep) G.; B.A., Columbia, 1944; M.A., Princeton, 1947, Ph.D., 1952; m. Vivian Grossman, Feb. 12, 1950; children—Aliza, Joel. Faculty U. Chgo., 1947-51; research asso. Inst. Fluid Dynamics and Applied Math. U. Md., 1951-54; chief, statis. physics sect. Nat. Bur. Standards, Washington, 1954-68; prof. physics Temple U., Phila., 1968—. Vis. lectr. Weitzmann Inst. Sci., Rehovoth, Israel, 1958; with Office of Asst. Sec. Commerce for Sci. and Tech., Washington, 1963; OAS lectr. in statis. mechanics Inst. Polytechnico Nacional de Mexico, 1964. Fulbright grantee, 1957, Guggenheim fellow, 1957; recipient Gold Medal for Distinguished Achievement in Fed. Service U. S. Dept. Commerce, 1965. Fellow Am. Phys. Soc.; mem. Washington Philos. Soc., Washington Acad. Scis. Editor (with J. V. Sengers): Critical Phenomena: The Report of a Conference, 1965; bd. editors Phys. Rev., 1944—, Jour. Math. Physics, 1953-65, Jour. Physics of Fluids, 1966—. Research in statis. mechanics, especially theory of irreversible processes, theory of dense systems, kinetic theory of gases, theory of light scattering. Address: Temple U., Phila.*

GREEN, Morton, Am. paleontologist; b. Bklyn., Oct. 25, 1917; s. Leon and Nessie (Lackowitzkaia) G.; A.B., U. Kan., 1940, M.A., 1942; Ph.D., U. Cal. at Berkeley, 1954; m. Elizabeth Ann Griffith, June 6, 1946; children—Leon Morton, Joel Griffith, Julia Louise. Faculty, S.D. Sch. Mines and Tech., Rapid City, 1950—, prof. biology, head dept., curator vertebrate paleontology Mus. Geology, 1962—; prof. associe Faculte des Scis. Universite de Montpellier (France), 1965-66. Mem. Geol. Soc. Am., Soc. Vertebrate Paleontology, Paleontology Soc., Am. Inst. Biol. Sci., Soc. Systematic Zoology, Soc. Study Evolution, Soc. Geology France, Sigma Xi. Research on fossil mammals especially of S.D., ages of rocks in which they have been found. Home: 3810 Brookside Dr., Rapid City, S.D. 57701.*

GREEN, Peter Shaw, English botanist; b. Rochester, Eng., Sept. 11, 1920; s. John William and Elizabeth E. (Hainsworth) G.; B.Sc. with honors, King's Coll., U. London, 1948; m. Winifred E. S. Brown, Aug. 26, 1946; children—Fiona Mary, Andrew Shaw, Marian Ruth, Jane Elizabeth. Faculty, U. Birmingham (Eng.), 1948-52; sr. sci. officer Royal Bot. Garden, Edinburgh, Scotland, 1952-61; hort. taxonomist Arnold Arboretum, Harvard, 1961-66; sr. sci. officer Royal Botanic Gardens, Kew, Eng., 1966—. Fellow Linnean Soc.; mem. Internat. Assn. Bot. Gardens (past sec., past treas.), Am. Soc. Plant Taxonomists, Bot. Soc. Brit. Isles. Research, publs. on taxonomy of Gleaceae and of cultivated plants. Home: 47 Holmesdale Rd., Teddington, Middlesex. Office: Royal Botanic Gardens, Kew, Richmond, Surrey, Eng.*

GREEN, Robert Holt, Am. physician; b. Charleston, S.C., Oct. 31, 1911; s. Walter Guerry and Daisie (Holt) G.; B.A., U. South, 1933; postgrad. U. N.C., 1933-34; M.D., Johns Hopkins, 1938; M.A. (hon.), Yale, 1967; m. Audrey Greet Johnston, Apr. 29, 1943; children—Robert Holt, Barbara Johnston, William Guerry, NRC fellow Rockefeller Inst. Med. Research, 1941-42, 46-47; faculty Yale, 1947-60, 67—, prof. medicine, 1967—; asso. sci. dir. Health Research Council City N.Y.; faculty N.Y. U., 1960-67, prof. medicine, 1965-67; affiliated Bellevue Hosp. Center, Univ. Hosp., N.Y.C., 1961-67; chief med. service VA Hosp., N.Y.C., 1964-67; asso. chief staff for research VA Hosp., West Haven, Conn., 1967—. Cons. internal medicine Willowbrook State Sch., Staten Island, N.Y., 1961-67. Diplomate Am. Bd. Internal Medicine. Mem. Am. Assn. Immunologists, Am. Fedn. Clin. Research, A.M.A., Am. Soc. Clin. Investigation, Harvey Soc., Soc. Exptl. Biology and Medicine, Phi Beta Kappa, Sigma Xi. Research and publs. on infectious diseases of man emphasizing viral agts. of smallpox, influenza, polio, rubella. Home: 363 Post Rd., R.F.D. 4, Madison, Conn. 06443.*

GREEN, William Thomas, Am. surgeon; b. Waucoma, Ia., Aug. 29, 1901; s. William L. and Jessie A. (Scott) G.; A.B., Ind. U., 1921, A.M., 1922, M.D., 1925, Sc.D., 1960; M.A., Harvard, 1962; m. Gladys M. Griffith, Dec. 18, 1930; children—William Thomas, Janet (Mrs. Henry Vaillant), Elisabeth Ann (Mrs. John Fogarty). Practice medicine specializing in surgery, Boston; Harriet M. Peabody prof. orthopedic surgery, co-head dept. Harvard Med. Sch., 1962—; orthopedic surgeon-in-chief Children's Hosp. Med. Center, 1946—; chief orthopedic surgeon Peter Bent Brigham Hosp., 1946—; cons. VA, West Roxbury hosps., 1953—, Robert Breck Brigham Hosp., 1964—. Recipient Ravdin medal. Diplomate Am. Bd. Orthopedic Surgery (past pres.). Mem. Am. Acad. Orthopedic Surgeons (past pres.), Am. Acad. Cerebral Palsy (past pres.), Am. Acad. Pediatrics, Am. Orthopedic Assn., Internat. Soc. Orthopedic Surgery and Traumatology, Assn. Bone and Joint Surgeons, Soc. Pediatric Research, Orthopedic Research Soc., New Eng. Surg. Soc., Soc. Research Child Devel., A.M.A., Am. Rheumatism Assn., Societa Italiana Di Ortopedia E Traumatologia, Phi Beta Kappa, Sigma Xi, Alpha Omega Alpha. Research, numerous publs. on skeletal growth and devel., methods of predicting growth and correcting discrepancies in length of extremities, paralytic disease, poliomyelitis and cerebral palsy, surg. procedures in rehab., factors producing skeletal deformity and their correction, localized lesions of bone. Home: 126 Prospect St., Belmont, Mass. Office: 300 Longwood Av., Boston 02115.*

GREENBERG, J. Mayo, Am. astrophysicist, educator; b. Balt., Jan. 14, 1922; s. Henry and Ree (Goldenberg) G.; Ph.D., Johns Hopkins, 1948; m. Naomi Slovin, June 21, 1947; children—Toby, Joshua, Shelly, Jonathan. Physicist NACA, Langley Field, Va., 1944-46; asst. prof. U. Del., Newark, 1948-51; research asso. Inst. for Fluid Dynamics and Applied Math., U. Md., 1951-52; asst. prof. physics Rensselaer Poly. Inst., Troy, N.Y., 1952-56, asso. prof., 1956-57, prof., 1957—, mem. Inst. Advanced Study, Princeton, N.J., 1965-66. OEEC sr. vis. fellow U. Leiden (Holland), 1961. Fellow Am. Phys. Soc.; mem. Internat. Astron. Union, Am. Astron. Soc. Developed and applied theories of scattering of light by small particles, theories of potential scattering, studies on interstellar matter. Home: 2126 Union St., Schenectady 12309. Office: Dept. Physics and Astronomy, Rensselaer Poly. Inst., Troy, N.Y.*

GREENBERG, Joseph, Am. microbiologist; b. Revere, Mass., Dec. 20, 1918; s. Louis and Fannie (Walker) G.; A.B. magna cum laude, Harvard, 1940, M.A., 1941, Ph.D., 1947; m. Lucille Charlotte Stein, Dec. 14, 1946; children—David, Martha, Douglas, Russell. Scientist dir. USPHS, NIH chief of sect. on chemotherapy Lab. Tropical Diseases, AID, 1947-58; program dir. microbiology Stanford Research Inst., Menlo Park, Cal., 1958-60; chief, microbiology div. Palo Alto (Cal.) Med. Research Found., 1960—. Recipient Bailey K. Ashford award in tropical medicine, 1954; Career Devel. award Nat. Cancer Inst., 1963. Fellow Am. Soc. Tropical Medicine and Hygiene, A.A.A.S.; mem. Am. Soc. Parasitologists, N.Y. Acad. Scis., Soc. for Exptl. Biology and Medicine. Research on chemotherapy of malaria; microbial genetics; amebiasis and cancer; radiation biology. Home: 270 Kellogg Av. Office: 860 Bryant St., Palo Alto, Cal. 94301.*

GREENBERG, Joseph H., Am. anthropologist; b. Bklyn., May 28, 1915; s. Jacob and Florenze (Pilzer) G.; A.B., Columbia, 1936; Ph.D. in Anthropology (Social Sci. Research Council fellow), Northwestern U., 1940; m. Selma Berkowitz, Nov. 23, 1940. Faculty, U. Minn., 1946-48; asst. prof. Columbia, 1948-53, asso. prof., 1953-57, prof. anthropology, 1957-62; prof. Stanford, 1962—; vis. prof. Summer Linguistic Inst., Mich. U., 1957, U. Minn., 1960; mem. panel anthropology and philosophy and history of sci. NSF, 1959-61; vis. prof. summer inst., U. Colo., 1961; dir. West African Langs. Survey, 1959—. Served with Signal Intelligence Corps, AUS, 1940-45. Stanford fellow, 1958-59; Ford Found. grantee, 1952, 57-62; recipient Demobilization award Social Sci. Research Council, 1945-46, Guggenheim award, 1954-55. Mem. Nat. Acad. Scis., Am. Anthro. Assn. (rep. to gov. bd. Internat. Inst. 1955—), Linguistic Soc. Am. (exec. com. 1953-55), West African Linguistics Soc. (chmn. 1965-66), African Studies Assn. (exec. com. also com. on langs. and linguistics; 1959—, pres. 1964-65), Phi Beta Kappa. Author: Languages of Africa, Essays in Linguistics, 1957; Universals of Language, 1963. Co-editor, Word, 1950-54; adv. editor, Am. Anthropologist, 1952-54; asso. editor Jour. of African Langs., 1962—; editorial bd. Ency. Social Scis., 1961—. Study of the ethnology of Africa; classification of African languages; general linguistics. Home: 860 Mayfield St., Stanford, Cal. 94305.

GREENBERG, Louis Donald, Am. biochemist; b. Pueblo, Colo., July 14, 1905; s. Samuel and Ida (Cooperman) G.; A.B., U. Cal. at Berkeley, 1930, Ph.D. in Biochemistry, 1936; m. Helen H. Lepon, May 29, 1939; 1 dau., Carolyn. Biochemist, So. Pacific Gen. Hosp., San Francisco, 1935-39; research asst. U. Cal. Sch. Medicine, San Francisco, 1936-42, faculty, 1942—, prof. pathology, 1960—, chmn. grad. group in nutrition, 1963—. Cons., Oakland (Cal.) Vets. Hosp., 1959-62; research div. Napa State Psychiat. Hosp., Imola, Cal., 1961-62. Grantee, Nat. Vitamin Found., 1952-55, USPHS, 1957—, Nutrition Found., Inc., 1957-63, Spl. Dairy Bd. Nat. Dairy Council, 1963-66. Fellow A.A.A.S.; mem. Am. Soc. Biol. Chemists, Am. Chem. Soc., Soc. Exptl. Biology and Medicine, Western Soc. Clin. Research, Pacific Slope Biochem. Conf. Contbr. numerous articles to tech. jours. Research in biochemistry and metabolism vitamins, pathology of exptl. nutritional deficiencies, exptl. arteriosclerosis. Home: 1644 Monterey Blvd., San Francisco 94127.*

GREENBERG, Milton, Am. geophysicist; b. Carteret, N.J., Apr. 21, 1918; s. David and Eva (Salzer) G.; student Coll. City N.Y., 1934-40; B.A., N.Y.U., 1943; M.P.A., Harvard, 1954; Sc.D., College of Advanced Science, 1962; m. Maxine Carol Baer, June 30, 1948; children—Eve Diane, David Max, Alan Baer. Research and development planner Air Force Cambridge Research Center, 1947-49, dep. dir. operations and planning Geophysics Research div., 1947-54, dir. Geophysics Research Directorate, 1954-58; president Geophysics Corporation of America (name changed to GCA Corp.), 1958—. First chairman tech. mgmt. council, Air Research and Development Command U. S. delegate to XIth Gen. Assembly, Internat. Union Geodesy & Geophysics, 1957; mem. Upper-Air Rocket & Satellite Research Panel; mem. central radio propagation laboratory advisory panel of National Academy of Sciences, 1963-1966. Recipient exceptional civilian service citation USAF, 1957. Trustee Coll. Advanced Science, Canaan, N.H. Mem. A.A.A.S., Am. Geophys. Union, Am. Meteorol. Soc., Internat. Assn. Geomagnetism & Aeronomy, Internat. Assn. Meteorology and Atmospheric Physics, Internat. Union Geodesy and Geophysics, U. S. Rocket and Satellite Research Panel, American Institute of Aeronautics and Astronautics, Scientific Research Soc. of Am. Editor-in-chief Planetary and Space Science, 1957-62, editorial adv. bd., 1962—. Research on geophysics; physics of the atmosphere and space. Home: 46 Sagamore Dr., Andover, Mass. 01810. Office: GCA Corp., Burlington Rd., Bedford, Mass. 01730.

GREENBERG, Samuel M., Am. biochemist; b. Louisville, Nov. 10, 1915; s. Louis and Tillie (Shapiro) G.; Ph.D., U. So. Cal., 1951; m. Mildred E. Elfman, July 9, 1943; children—Susan, Jane. Staff dir. research and devel. dept. metabolism Smith Kline & French Labs., Phila., 1954-67; dir. sci. information div. McNeil Labs., Ft. Washington, Pa., 1967—. Mem. Am. Chem. Soc., A.A.A.S., Am. Inst. Nutrition, Biochem. Soc. Great Britain, Soc. Exptl. Biology and Medicine, Sigma Xi. Contbr. numerous articles in field to sci. jours. Studied factors affecting vitamin and mineral absorption; assay of essential fatty acids. Home: 203 Conshohocken State Rd., Bala-Cynwyd, Pa. 19004. Office: McNeil Labs., Ft. Washington, Pa. 19034.*

GREENBLATT, Robert Benjamin, Am. physician; b. Montreal, Que., Can., Oct. 12, 1906; s. Louis and Hannah (Robbins) G.; B.A., McGill U., 1928, M.D., C.M., 1932; m. Gwen Lande, June 23, 1932; children—Nathaniel, Edward, Deborah. With Ga. Med. Coll., 1935—, prof., chmn. dept. endocrinology, 1946—. Adv. bd. scope panel on endocrinology U. S. Pharmacopoeia; cons. council on drugs A.M.A., also govt. agys. Recipient Crawford W. Long Gold medal, 1941; Phi Lambda Kappa Gold medal award, 1955; Rubin award, 1960; Billings Silver medal, 1964. Diplomate Am. Bd. Obstetrics and Gynecology. Hon. mem. obstet. and gynecol. socs. numerous states and fgn. countries, Queen's Gynecol. Soc., Canadian Soc. Study Sterility; mem. Endocrine Soc., Am. Coll. Obstetricians and Gynecologists, Endocrine Soc., Royal Soc. Medicine (asso.). Author: Office Endocrinology, 4th ed., 1952, Search The Scriptures, 1963; Ovulation, 1966; Progress in Conception Control, 1966. Asso. editor Geriatrics; bd. editors Obstetrics and Gynecol. Rev. Numerous publs. on gynecol. and endocrinol. problems; discovered that Clomiphene will stimulate ovulation in non-ovulatory women; an originator of sequential contraceptive pill. Home: 3011 Bransford Rd., Augusta, 30904. Office: Medical Arts Bldg., 1467 Harper St., Augusta, Ga. 30902.*

GREENE, Edward Forbes, Am. chemist; b. N.Y.C., Dec. 29, 1922; s. Roger Sherman and Kate (Brown) G.; A.B., Harvard U., 1943, M.A., 1947, Ph.D., 1949; m. Hildegarde Forbes, June 11, 1949; children—Susan Curtis, Judith Elizabeth, David Forbes, Roger Cobb. Jr. chemist Shell Oil Co., Wood River, Ill., 1943-44; mem. staff Los Alamos (N.M.) Sci. Lab., 1949; research asso. Brown U., 1949-51, instr., 1951-53, asst. prof., 1953-57, asso. prof., 1957-63, prof., 1963—; vis. prof. Tougaloo Coll., 1965. NSF Sr. Postdoctoral fellow U. Bonn, Germany, 1959-60, Cal. Inst. Tech., 1966-67. Fellow Am. Phys. Soc.; mem. Am. Chem. Soc., Combustion Inst. Author: (with J. P. Toennies) Chemical Reactions in Shock Waves, 1964. Contbr. articles in field to sci. jours. Research in phys. chemistry, particularly on energy transfer between molecules and the kinetics of chemical reactions; studies in shock waves, molecular beams. Home: 10 Patterson St., Providence, R.I. 02906.*

GREENE, Edward Lee, botanist; b. Hopkinton, R.I., Aug. 20, 1843; s. William M. and Abby (Crandall) G.; student of Knure Kumlein; Ph.B., Albion (Wis.) Coll., 1866; degree Jarvis; LL.D., U. Notre Dame, 1895. Episcopal clergyman, 1871-85; then R. C. layman; prof. botany U. Cal., 1885-95, Cath. U. Am., 1895-1904; asso. in botany Smithsonian Instn., 1904—. Pres. Internat. Congress of Botanists, Chgo. Expn., 1893. Author: Illustration of West American Oaks, 1887; Pittonia, 5 vols., 1887-1903; Flora Franciscana, 1891; Manual of Botany for the region of San Francisco Bay, 1894; Plantae Bakerianae, 1901; Leaflets of Botanical Observation, 2 vols., 1903-09; Landmarks of Botanical History, 1909. Discovered and named 5000 new specimens; established that Cesalpino (not Linnaeus) was founder of sci. botany. Died Washington, Nov. 10, 1915.

GREENE, Frederick Davis, Am. chemist; b. Glen Ridge, N.J., July 9, 1927; s. Phillips Foster and Ruth (Altman) G., B.A., Amherst Coll., 1949; Ph.D., Harvard, 1952; m. Theodora Elizabeth Whatmough, June 5, 1953; children—Alan, Carol, Elizabeth, Phillips. Postdoctoral fellow with D. J. Cram, U. Cal., Los Angeles, 1952-53; instr. Mass. Inst. Tech., 1953-55, asst. prof., 1955-58, asso. prof., 1958-62, prof., 1962—. Alfred P. Sloan fellow, 1958-62; NSF Sr. Postdoctoral fellow, 1965-66. Mem. Am. Chem. Soc., Chem. Soc. London, Am. Acad. Arts and Scis., Phi Beta Kappa. Editor, Jour. Organic Chemistry, 1962—. Contbr. articles to sci. jours. Studies of mechanisms of organic reactions. Office: Dept. Chemistry, Dreyfus Lab., Mass. Inst. Tech., Cambridge, Mass. 02139.*

GREENE, Harry Sylvestre Nutting, Am. pathologist; b. Woonsocket, R.I., Sept. 22, 1904; s. George Wellington and Gertrude (Earl) G.; student Wilbraham (Mass.) Acad., 1917-21; student Brown U., 1921-25; M.D., C.M., McGill U., 1930; A.M. (hon.), Yale, 1943; m. Helen May Davis, Sept. 27, 1930; 1 dau., Judith Ann; m. 2d, Jean Barnes, Dec. 18, 1954; 2 daus., Susan, Melissa. Inst. pathology, Path. Inst., McGill U., 1930-31; asst. in pathology, Rockefeller Inst. for Med. Research, N.Y. City, 1931-35; asso. in pathology, Rockefeller Inst., Princeton, N.J., 1935-41; asso. prof. of pathology and surgery, Yale U. Sch. of Medicine, 1941-43, prof. of pathology, 1943-50. Anthony N. Brady prof. of pathology since 1950. Recipient Borden award, 1956. Mem. Am. Assn. Advancement Sci., Am. Assn. Pathologists and Bacteriologists, Am. Genetic Assn., Soc. for Exptl. Biology and Medicine, Soc. for Study of Growth and Development, Am. Assn. for Cancer Research, Harvey Soc. of N.Y., Am. Acad. of Arts and Scis., N.E., Cancer Soc., Sigma Xi. Research on cancer-transplantation. Home: Podunk Rd., Guilford, Conn. Office: 310 Cedar St., New Haven 11.*

GREENE, James Sonnett, Am. physician; b. N.Y.C., Dec. 25, 1880; s. Jacob J. and Doris (Harrow) G.; M.D., Cornell U., 1902; postgrad, U. Berlin, Allerheilegen Hosp. (Breslau), U. Jena, 1906-12; m. Emilie Josephine Wells, Aug. 27, 1919. Began practice at N.Y.C., 1902; founded Nat. Hosp. for Speech Disorders (devoted to diagnosis and treatment of voice and speech disorders), N.Y.C., 1916, and since med. dir.; prof. speech Coll. Dental and Oral Surgery, N.Y.C., 1916-18; lectr. on voice and speech disorders Grad. Sch. Med., U. Pa. Cons. on speech disorders N.Y. Eye and Ear Infirm., Meml. Hosp., N.Y.C. Recipient spl. Gold medal Am. Laryngol., Rhinol. and Otol. Soc., 1940. Fellow N.Y. Acad. Medicine; mem. Med. Soc. Co. of N.Y., A.M.A., Am. Acad. Ophthalmology and Otolaryngology, Am. Laryngol., Rhinol. and Otol. Soc., Am. Group Therapy Assn., A.A.A.S., Nat. Assn. Tchrs. Speech (sustaining), Am. Soc. for Research in Psychosomatic problems. Author: The Cause and Cure of Speech Disorders (with E. J. Wells), 1927; I Was a Stutterer, 1932; Straight Talk (essays) 1948, also many monographs on speech and voice disorders. Editor, Talk (pub. by Nat. Hosp. for Speech Disorders); asso. editor Better English in Speech and Writing. Contbg. editor Year Book of the Eye, Ear, Nose and Throat, 1938. Contbr. to med. jours. Died Sept. 17, 1950.

GREENE, Joseph Lee, Jr., chemist; b. Montgomery, Ala., May 5, 1924; s. Joseph Lee and Annie C. (Gibson) G.; student Cornell U., 1944-45, U. S. Mil. Acad., 1945; B.S., Auburn U., 1948, M.S., 1949; Ph.D., Emory U., 1957; m. Jane Fackler, June 7, 1948; children—Janet, William Lee. Research chemist Tenn. Eastman Co., Kingsport, 1950-55; research asso. Emory U., 1955-57; research chemist Shell Devel. Co., Emeryville, Cal., 1957-58; sr. chemist, sect. head Thiokol Chem. Co., Huntsville, Ala., 1958-60; sr. chemist So. Research Inst., Birmingham, Ala., 1960—; lectr. Birmingham So. Coll., 1961-62; instr. U. Ala., 1966—. Recipient Pi Alpha Research award Emory U., 1957. Mem. Am. Chem. Soc., Sigma Xi, Phi Lambda Upsilon. Research on chemistry of heterocyclic compounds, terpenes, organo-aluminum compounds and organic sulfur compounds, autoxidations of hydrocarbons and thermal degradation of high polymers. Home: 2732 Cherokee Ct., Birmingham 35216. Office: 2000 9th Av. S., Birmingham, Ala. 35205.*

GREENE, Leon Charles, Am. physiologist, pharmacologist; b. Rochester, N.Y., June 1, 1925; s. Floyd and Florence (Dittman) G.; B.Sc. in Biology and Chemistry, Denison U., 1950; M.S. in Gross and Microscopic Anatomy (Baxter fellow), U. Pa., 1951, Ph.D. in Physiology, 1957; m. Grace Smith, Dec. 30, 1950; children—Jennifer, Jeffrey, James. Physiologist, U. S. Govt., Johnsville, Pa., 1954-58; faculty U. Pa., Phila., 1957—, asso. in physiology, 1958—; sr. physiologist/pharmacologist Smith Kline & French Labs., Phila., 1958-60, group leader pharmacology sect., 1960, asst. sect. head, dir. gen. pharmacology, 1960-62, sect. head, gen. therapeutics dept., dir. biomed. investigations, 1962-64, dir. biomed. activities, 1965, asso. dir. research and devel., mgr. gen. therapeutics dept., 1966—. Mem. Am., Phila. physiol. socs., N.Y. Acad. Sci., A.A.A.S., Sigma Xi. Basic research in fields of pain and temperature sensation; applied research in cardiovascular and respiratory areas. Home: 32 Beechfern Lane, Willingboro, N.J. 08046. Office: 1500 Spring Garden St., Phila. 19101.*

GREENE, Nicholas Misplee, Am. physician; b. Milford, Conn., July 11, 1922; s. Joseph N. and Nanine W. (Pond) G.; B.S., Yale, 1944, M.A., 1955; M.D., Columbia, 1946; m. Elizabeth R. Miller, May 21, 1946; m. Elizabeth R. Miller, May 21, 1946; children—Nicholas P., Cynthia R., Joseph Nathaniel II. Intern Presbyn. Hosp., N.Y.C., 1946-47; resident anesthesiology Mass. Gen. Hosp., Boston, 1949-51; vis. fellow U. Edinburgh and Royal Infirmary, Edinburgh, 1951; asst. anesthetist Mass. Gen. Hosp., also instr. anesthesia Harvard Med. Sch., 1951;53; asso. prof. anesthesiology, asst. prof. pharmacology U. Rochester Sch. Medicine, also dir. anethesia Strong Meml. Hosp., 1953-55; prof. anesthesiology, lectr. pharmacology Yale Sch. Medicine, also director anesthesia at Yale-New Haven Hospital, 1955—. Diplomate Am. Bd. Anesthesiology. Mem. Assn. Univ. Anesthetists, Am. Soc. Anesthesiologists, Internat. Anesthesia Research Soc., N.E. Soc. New York Acad. Scis., Sigma Xi. Author articles in field. Research on effect of anesthetics on carbohydrate metabolism; rationale and safe use of spinal anesthesia. Home: 1220 Ridge Rd., Hamden, Conn. 06517. Office: 789 Howard Av., New Haven 06504.*

GREENE, Norbert Dennis, Jr., Am. materials engr., b. Rochester, N.Y., Sept. 7, 1931; s. Norbert Dennis and Kathryn (Downes) G.; B.Ch.E., U. Rochester, 1953; M.S., Ohio State U., 1954, Ph.D., 1957; m. Simone Blumenthal, Sept. 21, 1955; children—Sharon, Clifford. Research asso. Engring. Expt. Sta. Ohio State U., 1955-57; research metallurgist Metals Research Labs. Union Carbide Corp., Niagara Falls, N.Y., 1957-59; with Rensselaer Poly. Inst., 1959—, prof., 1965—. Recipient Young Author's award, Nat. Assn. Corrosion Engrs., 1960, Geisler Meml. award, Am. Soc. Metals, 1962. Mem. Nat. Assn. Corrosion Engrs., Electrochem. Soc., Am. Soc. Metals, Electrochem. Soc. Japan, Welding Research Council, A.A.A.S., Internat. Com. for Electrochem. Thermodynamics and Kinetics, Sigma Xi. Author: Experimental Electrode Kinetics, 1965. Research and publs. on metal oxidation, fuel cells, med. materials, surg. implants and dental materials; basic research on corrosion, metallic passivity, alloy devel. Home: Box 197, R.D. #4, Troy, N.Y. 12181.*

GREENE, William Friese, see Friese-Greene, William.

GREENE, William Houston, Am. chemist; b. Columbia, Pa., Dec. 30, 1853; s. Stephen and Martha (Mifflin) G.; A.M., Central High Sch., Phila.; M.D., Jefferson Med. Coll., 1873. Asst. prof. chemistry, Jefferson Med. Coll., 1870-77, demonstrator chemistry, 1875-77; pursued original research in lab. of Adolph Wurtz, Paris, also in pvt. lab., Phila., 1877-79; demonstrator chemistry U. Pa., 1879-80; prof. chemistry Central High Sch., 1880-92. Author: A Hand-book of Medical Chemistry, 1880; Lessons in Chemistry, 1884. Translator, editor Wurtz's Elements of Modern Chemistry, 1880, 84, 87. Am. editor Paul Berts' First Steps in Scientific Knowledge. Died Aug. 8, 1918.

GREENEWALT, Crawford Hallock, Am. chemist; b. Cummington, Mass., Aug. 16, 1902; s. Frank Lindsay and Mary Elizabeth (Hallock) G.; ed. William Penn Charter Sch., 1914-18, B.S. in Chem. Engring., Mass. Inst. Tech., 1922; D.Sc., Univ. of Delaware, 1940; D.Sc., Northeastern University, 1950; E.D. (hon.), Rensselaer Polytechnic Inst., 1952; LL.D., Columbia U., 1953; LL.D., Williams Coll., 1953; Sc.D., Boston U., 1953; D.C.S., N.Y.U., 1954; D.Eng., Poly. Institute of Brooklyn, 1954; D.Sc., Phila. Coll. Pharmacy and Sci., 1955, Drexel Inst. Tech., 1961; LL.D., Kenyon Coll., 1958, Kan. State U., Temple U., 1960, U. Pa., 1961, Swarthmore Coll. 1961, Univ. of Notre Dame, 1965, Bowdoin Coll., 1965; L.H.D., Jefferson Med. College, 1960; m. Margaretta Lammot du Pont, June 4, 1926; children—Nancy Crawford Frederick, David, Crawford Hallock. With E. I. du Pont de Nemours & Co. since 1922, asst. dir. exptl. station, chem. dept., 1939, chem. dir. Grasselli chem. dept., 1942, tech. dir. explosives dept., 1943, asst. dir. development dept., 1945; asst. gen. mgr. pigments dept. Sept. 1945; vice pres. June 1946-48, pres., 1948-62, chmn. bd., 1962—; dir. Equitable Trust Company, 1935-43, E.I. du Pont de Nemours & Company, Christiana Securities Co., Morgan Guaranty Trust Company, The Boeing Co. Member of the Business Council. Member board of directors Wilmington Music Sch., Winterthur Corp., Massachusetts Institute of Technology, Philadelphia Orchestra Assn. New School of Music, Trustee Longwood Foundation, Carnegie Instn., Regent Smithsonian Instn. Recipient John Fritz Medal, Am. Institute of Chemical Engrs., 1962. Mem. Am. Inst. Chem. Engrs., Am. Inst. Chemists, Soc. Chem. Industry, Am. Acad. Arts and Scis., Nat. Acad. Scis., Am. Chem. Soc., A.A.A.S., Am. Philos. Soc., National Geographic Society (trustee). Author: The Uncommon Man, 1959; Hummingbirds, 1960. Research on catalysis; high pressure reactions; the partial pressure of water out of aqueous solutions of sulfuric acid; the separation of gaseous hydrocarbons; absorption of water vapor by sulfuric acid solutions. Home: Greenville, Del. Office: Du Pont Bldg., Wilmington, Del. 19898.

GREENFIELD, Moses A., Am. physicist, educator; b. Bklyn., Mar. 8, 1915; s. Benjamin and Goldie (Seewald) G., B.S., Coll. City N.Y., 1935; M.S. in Physics, N.Y. U., 1937, Ph.D. in Physics, 1941; m. Sylvia Sorkin, June 13, 1937; children—Richard,

Carolyn. Grad. asst. physics N.Y. U., 1936-37; tutor physics dept. Coll. City N.Y., 1937-39; mathematician U. S. Coast and Geodetic Survey, Washington, 1939-40; examiner in physics and chemistry U. S. Civil Service Commn., Washington, 1940-41; sr. physicist David Taylor Model Basin, Bur. Ships, U. S. Navy, Washington, 1941-46; research specialist N.Am. Aviation, Los Angeles, 1946-48; chief spectroscopy sect. Atomic Energy Project, U. Cal. at Los Angeles, 1948-49, chief tolerance sect. and x-ray physics unit, 1948-51, asso. clin. prof. radiology Sch. Medicine, 1949-51, asso. prof. radiology, radiation physicist, 1951-56, prof. radiology, radiation physicist, 1956—. Cons. physicist Atomics Internat., Canoga Park, Cal., 1948—, VA Wadsworth Hosp., 1953—, Harbor Gen. Hosp., Los Angeles County, 1954—, U. S. Naval Hosp., San Diego, 1959—; Internat. Ednl. Exchange fellow U. S. Dept. State, Rio de Janeiro, Brazil, 1957—, Am. specialist to S.Am., 1960—, S.E. Asia, 1964-65. Diplomate Am. Bd. Radiology. Fellow Am. Phys. Soc.; mem. Am. Coll. Radiology, N.Y. Acad. Scis., Radiation Research Soc., Am. Nuclear Soc. (past pres. Los Angeles chpt.), Soc. Nuclear Medicine (past pres. Los Angeles chpt.), Phi Beta Kappa, Sigma Xi. Contbg. author Isotope Dosimetry, 1965. Research in nuclear medicine including isotope dosimetry, low level counting, applications of isotopes to medicine and industry. Home: 1955 Mandeville Canyon Rd., Los Angeles 90049.*

GREENFIELD, Stanley Marshall, Am. geophysicist; b. N.Y.C., Apr. 16, 1927; s. Harry William and Millie (Cahan) G.; B.S., N.Y. U., 1950; Ph.D., U. Cal. at Los Angeles, 1967; m. Rhoda Claire Barish, Sept. 1, 1951; children—Diane Robin, David H. Research asst. N.Y. U., 1948-50; staff scientist RAND Corp., Santa Monica, Cal., 1950-59, asso. dept. head planetary scis. dept., 1962-64, head dept. geophysics and astronomy, 1964—; sci. adviser to Dep. Chief of Staff for Devel. USAF, Washington, 1959-61. Mem. geophysics panel sci. adv. bd. USAF, 1964—; mem. space scis. com. NASA-DOD, 1960-61; mem. com. on upper atmosphere rocket research Space Sci. Bd., NASA, 1959-61. Mem. N.Y. Acad. Sci., Am. Geophys. Union, Am. Meteorol. Soc. (Spl. award 1961), A.A.A.S., Sigma Xi, Tau Beta Pi. Author chpt. on satellite meteorology in Space Handbook, 1958; also articles. Asso. editor ICARUS, 1962—. Pioneer in research establishing idea, concept and feasibility of a meteorol. satellite, which idea was implemented as TIROS weather satellite; demonstrated through research feasibility of utilizing balloons as exploration tools on planets; theoretical investigation which showed how molecule size particles of radioactive material could be scavanged from atmosphere; demonstrated how solar flare effects on geomagnetic field could be used to gain knowledge on dynamics of region between 80-140 km. Office: 1700 Main St., Santa Monica, Cal. 90406.*

GREENHILL, Sir (Alfred) George, English mathematician; b. London, Nov. 29, 1847; math. lectr. Cambridge prof. math. Arty. Coll., Woolwich, Eng., 1876-1906. Fellow Royal Soc., 1888; mem. French Acad. Scis. (corr.). Author: Differential and Integral Calculus with Applications, 1885; Applications of the elliptic Function, 1892; Hydrostatics, 1894; Notes on Dynamics, 1908; Report 19, Theory of a Stream Line with Application to an Aeroplane, 1910, 16; Dynamics of Mechanical Flight, 1912; Gyroscopic Theory, 1914. Research effect of the rifling of the bore on the flight of projectiles, dynamics mech. flight, applications of elliptic function. Died London, Feb. 10, 1927.

GREENHILL, J(acob) P(earl), Am. gynecologist and obstetrician; b. N.Y.C., Feb. 28, 1895; s. Charles and Fanny (Pearl) G.; B.S., Coll. City N.Y., 1915; M.D., Johns Hopkins, 1919; m. Olga B. Hess, Mar. 16, 1929. Resident house officer Johns Hopkins Hosp., 1919-20; asst. resident Sinai Hosp., Balt., 1920-21; first resident Chgo. Lying-in Hosp., 1921-23, later sr. attending obstet.; sr. attending obstet., gynecol. Michael Reese Hosp., Chgo., 1931—; attending gynecol. Cook County Hosp., Chgo., 1925—, chmn. dept., 1941-47; prof. gynecol. Cook County Grad. Sch. Medicine, 1936—; asso. obstet. Northwestern U., 1922-32; prof. obstet., gynecol., vice chmn. dept. Loyola U., Chgo., 1933-47; hon. prof. Nat. U., Peru. Mem. White House Conf. Com., 1930. Decorated chevalier Legion of Honor, 1957; conseiller d'Honneur, Inst. Endocrinologie, Haiti, 1960; recipient of the Fulbright Travel Award, 1962. Diplomate Am. Bd. Obstet. and Gynecol. (charter). Fellow A.C.S. (life), Am. Coll. Obstet. and Gynecol. (charter), Am. Acad. Psychomatic Medicine, Chgo. Inst. Medicine; hon. fellow Internat. Coll. Surgeons (treas., trustee), Internat. Fertility Soc. (charter, v.p.); A.M.A. (chmn. com. female genital system for standard nomenclature diseases and operations), A.A.A.S., Am. Assn. Anatomists, Venereal Disease Assn., Johns Hopkins Med. and Surg. Assn., Ill., N.Y. acads. scis., Am. Assn. Study Internal Secretions, Am. Soc. Cancer Control, Soc. Sci. Study Sex (charter), Pan-Pacific Surg. Assn.; hon. mem. obstet. and gynecol. socs. So. Africa, Dominican Republic, Chile, Argentine, Panama, Brazil, Uruguay, Venezuela, Portugal, Algeria, Turkey, Guatemala, gynecol. socs. France, Germany, Brazil, Peruvian Acad. Surg., Argentine Soc. Fertility, fertility socs. Brazil, Portugal. Author: Obstetrics for the General Practitioner, 1935; Office Gynecology, 8th edit., 1965; Obstetrics in General Practice, 4th edit., 1948; Obstetrics, 13th edition published, 1965; Surgical Gynecology, 3d edition published 1965; Analgesia and Anesthesia in Obstetrics, 1952, 2d edit., 1962. Editor: (with J. B. DeLee) Year Books of Obstetrics, 1923-31; Year Book of Gynecology, 1931-42; Year Book of Obstetrics and Gynecology, 1942—; book rev. editor: Fertility and Sterility, 1949-67; editor in chief Ob/Gyn Digest. Book reviewer, mem. editorial bd. numerous med. jours. U. S., Europe, S.A. Editor Hosp. Publs., Inc., 1964. Research on principles and practice of obstetrics; surgical gynecology; fertility and sterility; anesthesia and analgesia in obstetrics. Home: 190 E. Pearson St., Chgo. 60611. Office: 55 E. Washington St., Chgo. 60602.

GREENHILL, William Laurence, Australian agrl. engr.; b. Tasmania, June 18, 1907; s. William Alexander and Florence (Atkinson) G.; B. Engring., U. Tasmania, 1927, M.Engring., 1934; m. Agnes May Fieldwick, Dec. 19, 1931; children—William Michael, David Laurence. Civil engr. State Electricity Commn., Victoria, Melbourne, 1927-29; with Commonwealth Sci. and Indsl. Research Orgn., 1929—, Colombo Plan expert for Applied Sci. Research Corp., Thailand, Bangkok, 1964—. Mem. Grassland Soc., Agrl. Engring. Soc. Research, numerous publs. on nature and means of removal of collapse in timber, effect of temperature on strength of timber, aspects of haymaking and ensilage; devel. methods of processing bast fibers. Home: 108 Haydens, Beaumaris, Victoria. Office: 314 Albert, East Melbourne, Victoria, Australia.*

GREENHOUSE, Samuel W., Am. math. statistician; b. N.Y.C., Jan. 13, 1918; s. Joseph and Lena (Rubin) G.; B.S., Coll. City N.Y., 1938; M.A., George Washington U., 1954, Ph.D., 1959; m. Selma Simon, Jan. 16, 1944; children—Joan E., Richard N., Joel B., Robin S. Statistician, U. S. Bur. Census, Washington, 1940-42; statis. analyst UNRRA, Washington, 1945-48; math. statistician Nat. Cancer Inst., NIH, 1948-54, chief statistics and math. sect. biometry br. Nat. Inst. Mental Health, 1954-66, chief epidemiology and biometry br. Nat. Inst. Child Health and Human Devel., Bethesda, Md., 1966—; vis. prof. statistics Stanford, 1960-61; professorial lectr. statistics George Washington U., Washington, 1947—. Mem. research adv. council Office of Aviation medicine Feb. Aviation Agy., 1959-67. Fellow Am. Statis. Assn.; mem. Am. Statis. Assn., Inst. Math. Statistics, Math. Assn. Am., Biometric Soc. (pres. elect 1968), A.A.A.S., Royal Statis. Soc., Internat. Assn. for Statistics in Phys. Scis., Am. Psychopathology Assn., Washington Statis. Soc. Author, editor (with J. E. Birren, R. N. Butler, L. Sokoloff, M. Yarrow) Human Aging, 1963. Asso. editor Jour. Am. Statis. Assn., 1967—. Research and publs. on statis. methods and inference particularly multivariate problems, multivariate data and times series, Bayesian analysis of sequential trials. Home: 1724 Ladd St., Silver Spring, Md. 20902. Office: Nat. Insts. Health, Bethesda, Md. 20014.*

GREENLY, Edward, geologist; b. Bristol, Dec. 3, 1861; s. Charles Hickes and Harriett (Dowling) G.; ed. Clifton Coll., Univ. Coll., London; D.Sc., Wales. With His Majesty's Geol. Survey, Scotland, 1889, made re-survey of Anglesey, 1895-1919, surveyed country between Menai Strait and mountains. Fellow Geol. Soc. (Lyell medal); hon. mem. Edinburgh, Liverpool (medal 1933) geol. socs., Anglesey Antiquarian Soc. Author: The Geology of Anglesey, 2 vols., 1919; The One Inch Geological Map of Anglesey, 1919; The Earth, 1927; (with Williams) Methods in Geological Surveying, 1930, also numerous articles and chpts. Died Mar. 4, 1951.

GREENMAN, Jesse More, Am. botanist; b. North East, Pa., Dec. 27, 1867; s. James William and Clarissa (More) G.; B.S., U. Pa., 1893; M.S., Harvard, 1899; Ph.D., U. Berlin, 1901; m. Anne Louise Turner, Sept. 20, 1902; children—Jesse More, Milton Turner. Asst. in botany, U. Pa., 1890-92, instr., 1893-94; asst. Gray Herbarium, Harvard, 1894-99; Harvard U. Kirkland fellow U. Berlin, 1899-1901; instr. Harvard, 1902-05; asst. curator botany Field Mus., Chgo., 1905-13; asst. prof. bot. U. Chgo., 1908-13; asso. prof. botany, 1913-16, prof., 1917-45, prof. emeritus 1945-51; in charge grad. Henry Shaw Sch. Botany, Washington U., St. Louis, 1927; emeritus curator herbarium Mo. Bot. Garden, 1948-51. Editor taxonomy vascular plants, Bot. Abstracts, 1917-27. Fellow A.A.A.S.; mem. Bot. Soc. Am., Ecol. Soc. Am., Am. Soc. Naturalists, N.E. Bot. Club, Am. Assn. U. Profs., N.Y. Acad. Scis., Phi Beta Kappa, Sigma Xi, Phi Sigma. Contbr. numerous articles on flora of N. Am. and Mexico, also monograph on Genus Senecio. Died Jan. 20, 1951.

GREENOUGH, Geoffrey Blakeley, metallurgist; b. Warrington, Eng., Nov. 17, 1921; s. Arthur and May (Blakeley) G.; B.A., Queens' Coll., Cambridge (Eng.) U., 1943, M.A., 1947, Ph.D., 1948; m. Eleanor Violet Joyce Scott, Aug. 21, 1948; children—Susan Eleanor, Charles Geoffrey. Head X-ray and physics sects metallurgy dept. Royal Aircraft Establishment, Farnborough, Eng., 1943-45, 47-51; sr. lectr. phys. metallurgy U. Sheffield (Eng.), 1951-56; staff Reactor Devel. Lab., U.K. Atomic Energy Authority, Windscale, 1956-61, dep. head lab. Reactor Fuel Lab., 1961—. Mem. adv. council on metallurgy U. Sheffield, 1966—. Fellow Phys. Soc.; mem. Inst. Metals. Research and publs. on changes in crystal lattice dimensions of metals during deformation, cleavage fracture of metals, behaviour of graphite and fuel elements in gas-cooled reactors. Home: 22 Elmhurst Rd., St. Annes, Lancashire, Eng. Office: Springfields Works, Salwick, Preston, Eng.*

GREENOUGH, George Bellas, English geographer, geologist; b. London, Eng., 1778; ed. Peterhouse, Cambridge, also Göttingen, Freiburg (where Werner was his teacher). Mem. Parliament, 1807-12; sec. to Royal Instn. Fellow Royal Soc., 1807; organizer Geol. Soc. London, 1807, 1st pres. 1811; 1st pres. Geog. Soc., 1839-40. Author: A Critical Examination of the First Principles of Geology, 1819; Geological Maps of the United Kingdom, 1820; Geological Map of England and Wales, 2d ed., 1839; Geological Map of India, 1854. Died Apr. 2, 1855.

GREENSPAN, Donald, Am. mathematician; b. N.Y.C., Jan. 24, 1928; s. Louis and Jessie (Scholnick) G.; B.S., N.Y.U., 1948; M.S., U. Wis., 1949; Ph.D., U. Md., 1956; m. Ruth Lucas, July 3, 1957; children—James, Marc, Rona. Research engr. Hughes Aircraft, Culver City, Cal., 1956-57; asso. prof. mathematics Purdue U., 1957-62; permanent member Math Research Center, U. Wis., Madison, 1962—. Served to 2nd Lt. USAF, 1953-54. Mem. Assn. Computing Machinery, Math. Assn. Am. (referee), Am. Math. Soc. Author: Theory and Solution of Ordinary Differential Equations, 1960; Introduction to Partial Differential Equations, 1961; Introductory Numerical Analysis of Elliptic Boundary Value Problems, 1965; also numerous research papers. Asso. editor Soc. Indsl. and Applied Math. Review, 1960—; reviewer Am. Math. Soc., 1960—. Research on applications of computers to nonlinear boundary value problems from physics and engring. Home: 817 Hiawatha Dr. Office: Mathematics Research Center, U. Wis., Madison, Wis.*

GREENSTEIN, Jesse Leonard, Am. astronomer; b. N.Y.C., Oct. 15, 1909; s. Maurice and Leah (Feingold) G.; A.B., Harvard U., 1929, A.M., 1930, Ph.D., 1937; m. Naomi Kitay, Jan. 7, 1934; children—George Samuel, Peter Daniel. Nat. Research fellow Yerkes Observatory U. Chgo., 1937-39. faculty, 1939-48; research asso. U. Tex., 1939-48; asso. prof. Cal. Inst. Tech., 1948-49, prof., chmn. dept., 1949—; staff mem., mem. obs. com. Mt. Wilson and Palomar Observatories, 1948—, exec. officer astronomy, 1961—; cons. USAF, NSF, Office Naval Research, RAND Corp., Hycon Corp.; vis. prof. Princeton U.; staff Owens Valley Radio Obs., 1966—. Fellow Inst. Advanced Study; mem. Nat. Acad. Scis., Am. Acad. Arts and Scis., Am. Astron. Soc., Royal Astron. Soc., Internat. Astron. Union, Royal Acad. Scis. Liège. Research and publs. on interstellar absorption, nature of interstellar matter and its interaction with stars, presence of magnetic fields in space, composition of stars and resulting devel. of concept that composition of universe is related to intrastellar nuclear processes, radio astronomy. Home: 2057 San Pasqual St., Pasadena, Cal. 91107. Office: 1201 E. California St., Pasadena, Cal.*

GREENSTEIN, Jesse P(hilip), Am. biochemist; b. N.Y.C., June 20, 1902; s. Louis and Lena (Birnbaum) G.; B.S., Poly. Inst. Bklyn., 1926; Ph.D., Brown U., 1930; m. Lucy Louise Mitchell, May 19, 1933; children—Louise (Mrs. Warren Brill), Michael Efrem. NRC fellow Harvard, 1930-31, instr., 1933-39; NRC fellow Kaiser Wilhelm Inst., Dresden, Germany, 1931-32; instr. U. Cal., 1932-33, vis. prof., 1948; chief biochemist Nat. Cancer Inst., NIH, Bethesda, Md., 1939—, chief biochemistry lab., 1945—. Mem. com. sci. advisors Inst. Microbiology, Rutgers U., 1958-59; mem. Am. delegation Cancer Colloquium, Rome, 1949. Recipient Neuberg medal in biochemistry, 1950; Distinguished Service award U. S. Dept. Health, Edn. and Welfare, 1954; Hillebrand prize Wash. chpt. Am. Chem. Soc., 1957. Hon. mem. Japanese Found. for Cancer Research, Japanese Biochem. Soc.; mem. Am. Chem. Soc. (chmn. div. biol. chemistry), Am. Soc. Biol. Chemistry, Am. Assn. Cancer Research, NRC (com. biochemistry, 1957-59, chmn. sub-com. on amino acids). Author: Biochemistry of Cancer, 2d edit. 1954. Editor: Cancer Research Advances in Cancer Research, Archives of Biochemistry; co-author: Chemistry of the Amino Acids, 1960. Died Feb. 12, 1959.

GREENSTEIN, Julius Sidney, Am. biologist; b. Boston, July 13, 1927; s. Samuel and Helen (Shriber) G.; B.A., Clark U., 1948; M.S., U. Ill., 1951, Ph.D., 1955; m. Joette Mason, Aug. 23, 1954; children—Gail Susan, Jodi Beth, Jay Mason, Blake Jeffrey, Joette Elise. Faculty, U. Mass., 1954-59; faculty Duquesne U., Pitts., 1959—, chmn. dept. biol. scis., 1961—, prof., 1964—; vis. lectr. Am. Inst. Biol. Scis., 1966—. Mem. Am. Assn. Anatomists, Internat. Fertility Assn., A.A.A.S., Am. Soc. Zoologists, Am. Fertility Soc., Soc. Study Fertility, Council Biol. Editors, N.Y. Acad. Sci., Soc. Study Reprodn., Am. Assn. U. Profs. (pres. Duquesne chpt.), Sigma Xi. Editor: Internat. Jour. Fertility, 1958—; editorial bd. Pa. Acad. Sci., 1963—. Contbr. to understanding of causes and prevention of reproductive failure in mammals by studying early developmental stages of embryo, nature of male and female reproductive organs and asso. hormone-secreting glands at various stages before, during and after conception; developed new techniques for studying tissues under the microscope; investigated relationship of specific diseases to normal reproductive performance. Home: 1628 Worcester Dr., Pitts. 15243.

GREENWALD, Gilbert Saul, Am. anatomist; b. N.Y.C., June 24, 1927; s. Morris and Celia (Levy) G.; A.B. in Zoology, U. Cal. at Berkeley, 1949, Ph.D., 1954; m. Pola Gorsky, Sept. 8, 1950; children—Susan, Elizabeth, Douglas. Asso., U. Cal. at Berkeley, 1954; USPHS fellow Carnegie Instn., Washington, Balt., 1954-56; faculty U. Wash., Seattle, 1956-61, asst. prof., 1959-61; faculty U. Kan. Med. Center, Kansas City, 1961—, research prof. human reprdn., since 1964—, prof. anatomy, co-dir. Ford Found grant for research in reprodn., 1965—. Mem. Brit. Soc. Fertility, Am. Assn. Anatomists, Endocrine Soc., Sigma Xi. Contbr. articles to tech. jours. Research on hormonal regulation egg transport thru oviduct, mode action various fertility control techniques, factors regulating ovarian follicular devel. and mammalian corpus luteum. Home: 3108 W. 79th Terrace, Prairie Village, Kan. Office: Dept. Obstetrics-Gynecology, U. Kan. Med. Center, Kansas City, Kan. 66103.*

GREENWALT, Tibor Jack, Am. physician; b. Budapest, Hungary, Jan. 23, 1914; s. Bela William and Irene (Foldes) G.; came to U. S., 1920, naturalized, 1923; B.A., N.Y. U., 1933, M.D., 1937; m. Shirley Alma Johnson, Aug. 6, 1960; 1 son, Peter H. Intern, Mt. Sinae Hosp., N.Y.C., 1937-38; rotating intern Kings County Hosp., Bklyn., 1938-40; resident in medicine Montefiore Hosp., N.Y.C., 1940-41; research asst. hematology New Eng. Med. Center, Boston, 1941-42; med. dir. Milw. Blood Center, 1947-66; prof. medicine Sch. Medicine, Marquette U., Milw., 1948-66; med. dir. blood program Am. Nat. Red Cross, 1966—; clin. prof. medicine George Washington U., 1966—. Cons. in hematology VA Hosp., Wood, Wis., 1946-66, Milw. County Gen. Hosp., 1946-66; chmn. com. on blood and transfusion problems Nat. Acad. Sci.-NRC, 1963-66. Served to maj., M.C., AUS, 1942-46. Recipient Gold medal, Caduceus Soc., Washington Sq. Coll., N.Y. U., 1933. Diplomate Am. Bd. Internal Medicine. Mem. A.M.A., A.C.P., A.A.A.S., Am. Assn. Blood Banks (Elliott award 1966), Am. Soc. Clin. Pathologists, Am. Fedn. Clin. Research, Central Soc. Clin. Research, Am. Fedn. Clin. Research, Central Soc. Clin. Research, Am. Assn. Immunologists, Internat. Soc. Blood Transfusion (pres.), Am. Soc. Human Genetics, Internat. Soc. Hematologists, Am. Soc. Hematologists, Alpha Omega Alpha. Author: (with W. Dameshek, R. J. Tat, C. Dreyfus) Hemolytic Syndromes, 1942; (with S. A. Johnson) Blood Coagulation and Transfusion in Clinical Medicine, 1965; Advances in Immunogenetics, 1967. Research, publs. in immunochematology; hemolytic disease of the newborn; discovery of new red blood cell groups; inventor of filter to remove white cells. Home: 4617 Kenmore Dr., N.W., Washington 20007. Office: 1730 E. St. S.W., Washington 20006.*

GREENWOOD, Delbert A., Am. biochemist, educator; b. Lehi, Utah, Oct. 15, 1904; s. Thomas A. and Mary (Losee) G.; B.S., Brigham Young U., 1926, M.S., 1930; postgrad. Ia. State U.; Ph.D., U. Chgo., 1946; Maxine S. Clayton, Sept. 12, 1934; children—Joan (Mrs. Leonard Dahle), Rae (Mrs. John Payne). Instr. chemistry Brigham Young U., 1926-30; instr. Ia. State U., 1930-36; research biochemist Am. Meat Inst., Chgo., 1936-46, research asso. pharmacology, 1939-46; prof. biochemistry, pharmacology Utah State U., Logan, 1946—. Mem. subcom. fluorosis in livestock prodn. NRC. Mem. Am. Chem. Soc., Am. Forestry Assn., Food Tech. Soc., Utah Acad. Scis., Arts and Letters (pres. 1949-50), numerous others. Author (with others): Methods of Vitamin Assay, 1947. Studies, numerous publs. on physiol. effects of fluorine compounds; effects of storage, thermal processing and variety on nutritive values of meats and vegetables; toxicity of DDT and other pesticides to farm animals; effect of fertilizers on nutritive value of plants; toxicity of organic phosphorous compounds, carbamates and polynuclear aromatic hydrocarbons. Home: 601 River Heights Blvd., Logan, Utah 84321.*

GREENWOOD, Fred Laurel, Am. chemist; b. nr. Mt. Holly Springs, Pa., July 31, 1911; s. Irvin E. and Orii (Bishop) G.; B.S., Dickinson Coll., 1933; M.S., Pa. State U., 1936; Ph.D., U. Minn., 1940; m. V. Louise Shafer, Sept. 4, 1944; Instr. U. Minn., 1941-46; faculty Tufts U., Medford, Mass., 1946—, prof., 1958—; fellow U. Ill., 1949-50. NSF Faculty fellow U. Cal. at Berkeley, 1959-60. Mem. Am. Chem. Soc., Phi Beta Kappa, Sigma Xi. Contbr. articles to tech. jours. Research in reaction of alkenes with ozone; synthesis of conjugated dienes; Grignard reactions; pyrolysis of allylic esters. Home: 112 Arlington St., Winchester, Mass. 01890. Office: Dept. Chemistry, Tufts U., Medford, Mass. 02155.*

GREENWOOD, John, Am. dentist; b. Boston, May 17, 1760; s. Isaac and Mary (I'ans) G.; m. Elizabeth Weaver, Mar. 22, 1788. Apprenticed to cabinetmaker; served as rifleman and scout during Revolutionary War; became dentist, N.Y.C., 1785; credited with originating foot-power drill, also springs which held plates of false teeth in position, and use of porcelain in manufacture of false teeth; George Washington was one of his patients (found Greenwood to be most satisfactory of his numerous dentists). Died N.Y.C., Nov. 16, 1819.

GREENWOOD, Peter Humphry, English ichthyologist; b. Redruth, Eng., Apr. 21, 1927; s. Percy Ashworth and Joyce (Wilton) G.; B.Sc. with honors, U. Witwatersrand, Johannesburg, South Africa, 1949, D.Sc., 1962; m. Marjorie George, Jan. 10, 1950; children—Pamela, Jennifer, Nan, Philippa. Research officer East African Fisheries Research Orgn., Jinja, Uganda, 1951-57; sr. research fellow Brit. Mus. Natural History, London, Eng., 1958-59, asst. keeper dept. zoology, curator fishes, 1959—. Research asso. Am. Mus. Natural History, 1965—; mem. Brit. subcom. on productivity of freshwaters Internat. Biol. Programme, 1964—. Fellow Linnean Soc. (council, editorial com.), Zool. Soc. London (Sci. medal 1963); mem. Inst. Biology. Author: Fishes of Uganda, 1958, rev., 1966; reviser A History of Fishes (J. R. Norman), 1963. Contbr. numerous articles to profl. jours. Research on biology, taxonomy and evolution of east African freshwater fishes, problems relating to rapid and prolific speciation of cichlid fishes in African lakes; relationship and evolution of living bony fishes especially their comparative anatomy, studies of African Pleistocene fossil fishes and breeding biology of certain species. Home: 20 Cromer Villas Rd., London, S.W.18. Office: British Museum of Natural History, Cromwell Rd., London, S.W.7, Eng.*

GREENWOOD, Robert E., Am. mathematician; b. Navasota, Tex., June 21, 1911; s. Robert E. and Lula (Lewis) G.; B.A., U. Texas, 1933; postgrad. Brown U.; M.A., Princeton, 1938, Ph.D., 1939; m. Mary Maud Brown, Dec. 15, 1951; 1 dau., Barbara Frances. Faculty, U. Texas, Austin, 1938—, now prof. math. Contbr. reports for various research agys., articles to math. jours., Collier's Ency. Study of probability; numerical methods; combinatory analysis. Home: 3203 Breeze Terrace, Austin, Tex. 78722.*

GREER, Monte Arnold, physician; b. Portland, Ore., Oct. 26, 1922; s. William Wallace and Rose (Rasmussen) G.; student Ore. State U., 1940-43; A.B., Stanford, 1944, M.D., 1947; m. Margaret Johnson, Dec. 31, 1943; children—Susan Elizabeth, Richard Arnold. Research asso. in endocrinology New Eng. Med. Center Hosp., Boston, 1950-51; sr. investigator, sr. asst. surgeon USPHS, Nat. Cancer Inst., NIH, Bethesda, Md., 1951-55; chief radioisotope unit VA Hosp., Long Beach, Cal., clin. asst. prof. U. Cal. at Los Angeles, 1955-56; faculty, head div. endocrinology, U. Ore. Med. Sch., Portland, 1956—, prof. medicine, 1962—. Recipient Ciba award Endocrine Soc., 1958; Research Career award NIH, 1962—. Mem. Am. Fedn. for Clin. Research (chmn. Western sect. 1958-59), Western Soc. for Clin. Research (past v.p.), Endocrine Soc. (mem. council 1965—), Am. Thyroid Assn., Am. Soc. for Clin. Investigation, Soc. for Exptl. Biology and Medicine, Portland Acad. Medicine, Internat. Brain Research Orgn. Research, numerous publs. on isolation and mechanism action naturally-occurring goitrogens, thyroid physiology, neuroendocrinology. Home: 2706 Glen Eagles Rd., Lake Oswego, Ore. 97034. Office: U. Ore. Med. Sch., Portland, Ore. 97201.*

GREER, Scott, Am. sociologist, polit. scientist; b. Sweetwater, Tex., Oct. 25, 1922; s. A. A. and Mary (Scott) G.; B.A., Baylor U., 1946; M.A., U. Cal., Los Angeles, 1951, Ph.D., 1952; m. Dorothy Dewey, 1946; 1 dau., Eve. Faculty, U. Cal., Santa Barbara, 1951-52, Occidental Coll., 1952-55; chief sociologist, asst. research dir. Met. St. Louis Survey, 1956-57; faculty Northwestern U., Evanston, Ill., 1957—, prof. sociology, polit. sci., dir. Center Met. Studies, 1962—. Recipient Ford Found. Pub. Affairs div. grant, 1961-62. Mem. Am., Midwest sociol. assns., Sociol. Research Assos., Internat. City Mgrs. Assn. Author: Social Organization, 1959; Last Man In: Racial Access to Union Power, 1959; (with others) Exploring the Metropolitan Community, 1961; The Emerging City: Myth and Reality, 1962; Governing the Metropolis, 1962; (with Norton E. Long) Metropolitics: The Study of Political Culture, 1963; (with Robert F. Winch) The Family and Voluntary Organizations in the Post-Thermonuclear Attack Situation, 1963; Via Urbana and Other Poems, 1963; Urban Renewal and American Cities: The Dilemma of Democratic Intervention, 1965. Editor: (with others) The New Urbanization in New Nations and Old, 1966. Contbr. numerous articles to profl. jours. Studies, publs. on urbanism and social and polit. structures; housing and redevelopment in the metropolis; problems of social change. Home: 904 Michigan St., Evanston, Ill. 60201.*

GREGG, Donald Crowther, Am. chemist; b. Marlboro, N.H., June 25, 1913; s. Arthur E. and Ida May (Crowther) G.; B.S., U. Vt., 1935; M.S., U. N.H., 1937; Ph.D., Columbia, 1941; m. Florence Bentley Green, May 19, 1943; children—Bentley Crowther, Fulton Mills. Asst. chemistry U. N.H., 1935-37, Columbia, 1937-39; research chemist Wallace & Tiernan Products, Belleville, N.J., 1940-41; faculty Harvard, 1940; vis. lectr., summer 1946; instr., asst. prof. Amherst (Mass.) Coll., 1941-46; prof. U. Vt., 1952—, Pomeroy prof. chemistry, 1963—; vis. prof. U. Fla., 1962-63; sci. faculty fellow Nat. Sci. Found., 1962-63. Mem. Am. Chem. Soc. (chmn. Western Vt. sect. 1953), Am. Assn. U. Profs. (pres. Vt. chpt. 1953-55), New Eng. Assn. Chemistry Tchrs., N. New Eng. Acad. Sci. (pres. 1965-66), Sigma Xi (pres. Vt. chpt. 1952-53), Alpha Chi Sigma, Phi Lambda Upsilon, Sigma Delta Xi. Author: Principles of Chemistry, 2d edit., 1963; College Chemistry, 2d edit., 1965; Chemistry in the Laboratory, 1966. Research on activity of enzymes found in common mushrooms, especially their ability to catalyze oxidations of certain organic compounds; structure and oxidation of some organic sulfides of unique structure. Home: 60 University Terrace, Burlington, Vt. 05401.*

GREGG, Donald Eaton, Am. physiologist; b. Bridgeport, Conn., Mar. 24, 1902; s. Hugh Gilmore and Julia (Ober) G.; B.S., Colgate U., 1924; M.S. (Porter fellow), U. Rochester, 1929, Ph.D., 1930, M.D., 1946; m. Maria Grana, Jan. 8, 1927; children—James Alan, William Gilmore, John Bruce. Faculty, Western Res. U., 1930-44; staff med. research lab. M.C., U. S. Army, Ft. Knox, Ky., 1946-50; chief dept. cardiorespiratory diseases Walter Reed Army Inst. Research, Washington, 1950—; lectr., U. S., fgn. countries. Recipient Exceptional Service award Sec. Army, 1959; Distinguished Civilian Service award Sec. Def., 1961; Dept. Army Certificate of Achievement, 1961; Presdl. award Distinguished Fed. Civilian Service, 1962. Mem. Am. Physiol. Soc., Am. Heart Assn. (Research Achievement award 1963), A.A.A.S., Am. Coll. Cardiology, Am. Coll. Chest Physicians, Am. Soc. Study Arteriosclerosis, Soc. Exptl. Biology and Medicine, Inter-Am., Peruvian Nat. cardiological socs., Sigma Xi, Phi Beta Kappa, Alpha Omega Alpha, others. Author: The Coronary Circulation in Health and Disease, 1950. Research, publs. on blood flow and methods of measurement, energetics of heart and cardiovascular system, regulation of coronary flow and coronary collateral circulation. Home: 3535 Chevy Chase Lake Dr., Chevy Chase, Md. 20015. Office: Walter Reed Army Inst. Research, Washington 20012.*

GREGG, Robert Arden, Am. chemist; b. Dundee, Mich., Mar. 8, 1918; s. Marvin C. and Ethel (Ball) G.; A.B. magna cum laude, Adrian Coll., 1937; M.S., U. Mich., 1940, Ph.D., 1943; m. Alberta Jean Westerman, Mar. 31, 1951; children—Timothy Arden, Deborah Ann. Instr., Adrian Coll., 1937-38; teaching fellow U. Mich., 1941-42; with U. S. Rubber Co., 1942—, research planning, research and devel. dept. Research Center, Wayne, N.J., 1964, personnel recruiting, 1965—. Mem. Am. Chem. Soc., Phi Beta Kappa, Sigma Xi, Tau Kappa Alpha, Beta Pi Theta. Patentee in rubber chemistry. Research and publs. on mechanisms by which monomers are converted to polymers; discovered elastic polyurethane fiber; improvements in elastomer and fiber products and processes. Home: 475 Laurel Lane, Kinnelon, N.J. 07405. Office: Research Center, U. S. Rubber Co., Wayne, N.J.*

GREGG, Willis Ray, Am. meteorologist; b. Phoenix, N.Y., Jan. 4, 1880; s. Willis Perry and Jennie E. (Ray) G.; A.B., Cornell U., 1903; Sc.D., Norwich U., 1937; m. Mary Chamberlayne Wall, Oct. 15, 1914; 1 dau., Ruth Marguerite. With U. S. Weather Bur., 1904-38, at Mt. Weather Obs., Va., 1907-14, Washington, 1915-38, in charge Aerological div., 1917-34; chief, 1934-38; spl. meteorol. adviser trans-Atlantic flight for NC seaplanes (U. S. Navy) at Trepassey, Newfoundland, May 1919, and for Brit. dirigible R34, at Mineola, N.Y., July 1919, engaged in organizing weather service for comml. airways activities, 1926—. Mem. NACA (chmn. exec. com. and of subcom. on meteorol. problems), Internat. Meteorol. Orgn., Internat. Meteorol. Com. (pres. commn. on projections of meteorol. maps), Daniel Guggenheim Com. on Aero. Meteorology, Interdepartmental Com. on Coordination of Meteorol. Service for Aeronautics, etc. Fellow Am. Meteorol. Soc. (treas. 1923-35). Co-Author: Introductory Meteorology, 1918; Meteorology, 1931. Author: Aeronautical Meteorology, 1925, 2d edit., 1930; Aerological Survey of the United States (monograph), 1922 and 1926. Died Sept. 14, 1938.

GREGOR, James Wyllie, Scottish botanist; b. Innerwick, Scotland, Jan. 14, 1900; s. Charles Edward and Janet (Wyllie) G.; Ph.D., U. Edinburgh (Scotland), 1926, D.Sc., 1939; m. Mary Joanne Farquharson Wilson, Feb. 14, 1929. Chief asst. Scottish Soc. Research in Plant Breeding, 1926-49; dir. Scottish Plant Breeding Sta., Pentlandfield, Roslin Midlothian, 1950-65. Decorated comdr. Order Brit. Empire. Fellow Royal Soc. Edinburgh, Linnean Soc. London, Inst. Biology. Research and publs. on genecology, particularly nature and patterns of ecotypic variation relating to interactions, formulation (with J. S. L. Gilmour) of Deme terminology of microevolutionary description. Home: Old Mill House, Balerno, Scotland.*

GREGOR, William, English chemist, mineralogist; b. Trewarthenick, Cornwall, Eng., Dec. 25, 1761; s. Francis and Mary (Copley) G.; B.A., St. John's Coll., Cambridge (Eng.) U., 1784, M.A., 1787; m. Charlotte Anne Gwatkin, 1790; 1 dau. Rector of Diptford, Devonshire, Eng., 1787-93, of Creed, Cornwall, Eng., 1794-1817; pub. sermons and pamphlets. Discovered titanium (sometimes called gregorite); research on zeolite, wavelite and other substances. Died Creed, July 11, 1817.

GREGORAS, Nicephoros, Byzantine astronomer; b. Heraclea Pontica, (now Turkey), 1295; studied in Heraclea Pontica, also Constantinople, Turkey; studied astronomy under Theodoros Metaochites; flourished under the Palaeologoi, 1261-1453. Wrote 2 treatises on astrolabe, one before 1335, the other circa 1350; treatise on reform of calendar, 1324. Predicted 2 lunar and 1 solar eclipses in 1330. Died 1359.

GREGORY, Christopher, Am. physicist; b. Cleve., June 6, 1916; s. Thomas and Catherine (Schwan) G.; B.S., Cal. Inst. Tech., 1938, M.S., 1939, Ph.D. in Physics, 1941; m. Rose Ching, Aug. 8, 1942; children—Cheryl, Christopher Thomas, Charles Michael, Christina, Deeling. Faculty, U. Hawaii, Honolulu, 1941-43, 1945—, prof., chmn. dept., 1952—, chmn. grad. faculty, 1960—; physicist Bur. Standards, 1944-45. Mem. Am. Phys. Soc., Am. Mathematica Soc. Bd. editors, Pacific Sci., 1951-54. Research on absorption spectra, relativistic cosmology, non-linear invariants and problem of motion, field equations, non-local field theory, problem of motion in extra-dimensional spaces, extra dimensional effect research. Home: 3633 Woodlawn Terrace Pl., Honolulu 96822.*

GREGORY, David, astronomer, mathematician; b. Kinnairdie, Banffshire, Scotland, June 24, 1661; s. David Gregory; ed. Marischal Coll., Aberdeen, Scotland; M.A., U. Edinburgh, 1683; M.A., M.D., Oxford (Eng.) U., 1692; m. Elizabeth Oliphant, 1695; 4 children including David. Prof. math. Edinburgh, 1683-91; apptd. Savilian prof. astronomy Oxford, 1691, master commoner of Balliol Coll. Fellow Royal Soc., 1692, Royal Coll. Physicians Edinburgh (hon.). Author: Astronomiae Physicae et Geometricae Elementa (1st textbook on gravitational principles, tried to remodel astronomy in conformity with phys. theory), 1702; also articles. 1st to suggest possibility of an achromatic combination of lenses. Died Maidenhead, Berkshire, Eng., Oct. 10, 1708.

GREGORY, Duncan Farquharson, mathematician; b. Edinburgh, Scotland, Apr. 13, 1813; s. James Gregory; ed. U. Edinburgh, U. Geneva (Switzerland) also Trinity Coll., Cambridge (Eng.) U. became fellow, 1840, fifth wrangler, 1837; M.A., 1841; asst. to chemistry prof.; 1st editor Cambridge Math. Jour. Author: Examples of the Processes of the Differential and Integral Calculus, 1841; A Treatise on the Application of Analysis to Solid Geometry; Mathematical Writings (editor W. Walton), 1865. Wrote 1st treatise in which the system of solid geometry is developed using symmetrical equations. Died Edinburgh, Feb. 23, 1844.

GREGORY, Herbert E(rnest), Am. geologist; b. Middleville, Mich., Oct. 15, 1869; s. George Albert and Anne (Bross) G.; A.B., Yale, 1896, Ph.D., 1899; D.Sc. hon., Doane Coll., 1934; m. Edna Earle Hope, June 30, 1908; 1 dau., Anne Cutts (Mrs. John L. Scarlett). Asst. in botany Yale, 1896-98, instr. phys. geography, 1899-01, asst. prof. physiography, 1901-04, Silliman prof. geology, 1904-36, emeritus since 1936; asst. geologist U. S. Geol. Survey, 1900-09, geologist, 1909-48; supt. Conn. Geol. and Natural History Survey, 1916-20; acting dir. Bernice P. Bishop Museum, Honolulu, T.H., 1919-20, dir., 1920-36, emeritus dir. since 1936; chmn. bd. regents U. Hawaii, 1934-42; organized 1st Pacific Sci. Congress, 1920; del. European sci. congresses, 1948, 1950. Mem. bd. water supply, Honolulu, 1928-36. Trustee Palama Settlement, Honolulu, 1934-42. Served as maj., supervisor sci., com. edn. and special training War Dept., 1918. Fellow Geol. Soc. Am., Assn. Am. Geographers, Am. Acad. Arts and Scis., Washington Acad. Sci., Am. Philos. Soc.; mem. NRC. Contbr. geol. articles profl. jours. Asso. editor Am. Jour. Sci., 1904-28. Spl. study of glacial geology of Alps, 1902; extensive exploration of New Zealand and Australia, 1915-16. Died Honolulu, Jan. 23, 1952.

GREGORY, Jack Norman, Australian chemist; b. Melbourne, Aug. 29, 1920; s. Martin Norman and Gertrude (Treyvaud) G.; B.S., Melbourne U., 1942, M.S., 1943, D.Sc., 1955; m. Sheila Muriel Ewan, Dec. 6, 1945; children—Ian Raymond, Susan Margaret. Research officer, div. tribophysics Commonwealth Sci. and Indsl. Orgn., 1942-48, sr., prin. research officer attached U.K. Atomic Energy Research Establishment, Harwell, Eng., 1948-53, sr., prin. research officer Australian AEC attached U.K. Atomic Energy Research Establishment, 1953-56; chief isotope div. Australian AEC Research Establishment, Lucas Heights, Sydney, Australia, 1956—. Mem. New S. Wales Radiol. Adv. Council, 1957-64, Adv. Panel Hunter Valley Research Found., Newcastle, 1958—. Recipient Grimwade research prize in chemistry, 1947. Fellow Royal Australian Chem. Inst. Author: The World of Radioisotopes, 1966; also numerous articles. Research on friction and lubrication - phys. chemistry of sliding interface, solid state by diffusion of radioactive tracers, chem. problems of materials and design of advanced nuclear reactors, application of radioisotopes and radiation. Home: 13 Nellella St., Blakehurst. Office: New Illawarra Rd., Lucas Heights, New South Wales, Australia.*

GREGORY, James, Scottish mathematician; b. Drumoak, Aberdeen, Scotland, Nov. 1638; s. John and Janet (Anderson) G.; ed. Marishal Coll., Aberdeen; m. Mary Jameson, 1669; 2 daus.; 1 son, James. Math. studies at Padua, Italy, 1664-67; became math. prof. at St. Andrews, Scotland, 1668; named 1st prof. math. at Edinburgh, 1674; became blind from amaurosis. Fellow Royal Soc., 1688. Author: Optica promota, 1663; Vera circuli et hyperbolae quadratura, 1667; Exercitationes geometricae, 1668. Invented reflecting telescope (Gregorian telescope), 1661; discovered a solution of the Keplerian problem by infinite series, method of drawing tangents to curves geometrically, rule founded on the principle of exhaustion for the direct and inverse method of tangents, series from which pi can be computed; 1st to distinguish between convergent and divergent series; studied use of the transits of Mercury and Venus to calculate the distance of the sun; developed the photometric method of estimating distances of the stars. Died Edinburgh, Oct. 1675.

GREGORY, John Walter, Brit. geologist; b. London, Jan. 27, 1864; s. John James Gregory; LL.D., D.Sc.; m. Audrey Chaplin, 1895; 1 son, 1 dau. Travelled in Rocky Mountains and Great Basin of western states for geol. purposes, 1891; explored Brit. East Africa, 1882-93; naturalist Sir Martin Conway's Expdn. across Spitzbergen, 1896; asst. geol. dept. Brit. Mus., 1887-1900; dir. civilian sci. staff Antarctic Expdn., 1900-01; dir. geol. survey Mines Dept., Victoria, 1902-04; head Lake Eyre Expdn., 1901-02, Expdn. to Cyremica, 1908, to So. Angola, 1912, Alps of Chinese Tibet, 1922; prof. geology and mineralogy U. Melbourne, 1900-04; prof. geology U. Glasgow, 1904-29. Recipient Victoria medal Royal Geog. Soc., 1919, Gold medal Scottish Geog. Soc., 1922, Royal Soc. Edinburgh, 1924, Gallois medal Société Geographique de Paris, 1922. Fellow Royal Soc., 1901, Geol. Soc. (pres. 1928-30, Bigsby medal); mem. Instn. Mining and Metallurgy, Brit. Assn. (pres. geol. sect. 1907, geog. sect. 1924), New Zealand Inst. (hon.), Geog. Soc. Australia (hon.), Geol. Soc. South Africa, Acad. Sci. (Halle), Geol. Soc. Belgium. Author: The Great Rift Valley: A Narrative of a Journey to Mount Kenya and Lake Baringo, 1896; Catalogue of Fossil Bryozoa in British Museum, Vol. 1, 1896, Vol. 3, 1908; The Mount Lyell Mining Field, Tasmania, 1904; The Foundation of British East Africa, 1901; The Dead Heart of Australia, 1906; The Rift Valleys and Geology of East Africa, 1921; Elements of Economic Geology, 1928; (with C. J. Gregory) To the Alps of Chinese Tibet, 1923; also over 200 articles on geology of Alps, Mediterranean Basin, W.I., Africa, Australia, fossil corals, echinoderms, Bryozoa, Eozoon, mining geology. Drowned in the Urubamba River, Peru, June 2, 1932.

GREGORY, Joseph Tracy, Am. paleontologist; b. Eureka, Cal., July 28, 1914; s. Frank C. and Edith (Tracy) G.; A.B., U. Cal. at Berkeley, 1935, Ph.D., 1938; postgrad. Inst. Meteorology, U. Chgo., 1943-44; m. Jane Everest, Feb. 21, 1949; 2 children. Lectr., Columbia U., 1939; technician, supr. paleontology lab. bur. econ. geology U. Tex., Austin, 1939-41; instr., curator fossil vertebrates U. Mich., Ann Arbor, 1941-46; asst. prof. geology Yale, 1946-52, asso. prof., 1952-60, curator fossil vertebrates Peabody Mus., 1946-60; prof. paleontology, curator lower vertebrates Mus. Paleontology U. Cal. at Berkeley, 1960—, chmn. dept. paleontology, 1960-65. Mem. Am. Soc. Mammalogists, Am. Soc. Zoologists, A.A.A.S., Geol. Soc. Am., The Paleontol. Soc., Soc. for Study Evolution, Soc. Vertebrate Paleontology. Contbr. articles to tech. jours. Described anatomy fossil skeletons various amphibians and reptiles Carboniferous, Permian, Triassic age. Office: Dept. Paleontology, U. Cal., Berkeley, Cal. 94720.*

GREGORY, Paul W., Am. geneticist; b. Frankfort, Ky., June 3, 1898; s. Robert Lee and Annie (Wallace) G.; B.S., U. Ky., 1922; M.S., Kan. State U., 1924; M.S., Harvard, 1927, Sc.D., 1928; m. Lucille Minerva Dean, May 29, 1924; children—Dean Wallace, Milton Lee. George H. Emerson fellow Harvard, 1926-28; asso. prof. biology La. State Normal Coll., 1924-26; prof. zoology Baker U., 1928-30; faculty U. Cal. at Davis, 1930—, prof., 1947—; staff Marine Biol. Lab., Woods Hole, Mass., 1927; spl. agt., cons. beef breeding research, U. S. Dept. Agr., Denver, 1952-62. Spl. adviser to breeders and breed assns., 1948-62. Fellow Genetics Soc. Am., Am. Genetics Assn., Am. Soc. Zoologists, Soc. for Exptl. Biology and Medicine; mem. A.A.A.S., Sigma Xi. Research and numerous publs. on early embryology of rabbit, embryological basis of size inheritance, hereditary prolonged gestation in cattle and dam-fetus relationships, inbreeding in cattle and gene specific types of deterioration, achondroplastic deterioration in cattle resulting from selection enforcing a directional evolution, bovine cytology, erythrocyte chimera, Klinefelter's syndrome. Home: 26 College Park, Davis, Cal. 95616.*

GREGORY, Philip Herries, Brit. biologist; b. Exmouth, Devon, Eng., July 24, 1907; s. Herries Smith and Muriel (Eldridge) G.; B.Sc., Brighton Tech. Coll., 1928; Ph.D., Imperial Coll. Sci. and Tech., 1931; m. Margaret Fearn Culverhouse, Feb. 5, 1932; children—Andrew Herries, Rachel Helen. Mycologist, Man. Med. Coll., Winnipeg, Can., 1931-34, Seale-Hayne Agrl. Coll., Devon, 1935-40; plant pathologist Rothamsted Exptl. Sta., Harpenden, Eng., 1940-54; prof. botany Imperial Coll. Sci. and Tech., London, 1954-58; head plant pathology dept. Rothamsted Exptl. Sta., 1958—. Fellow Royal Soc., 1962, Linnean Soc.; mem. Inst. Biology, Brit. Mycol. Soc. (pres. 1951), Assn. Applied Biologists, Brit. Allergy Soc., Indian Phytopathol. Soc., Hertfordshire Natural History Soc. (pres. 1952, 53, 64-65). Author: Microbiology of the Atmosphere, 1961. Publs. on behavioral studies of pathogenic microbes, especially airborne fungi; aerobiology. Home: 11 Topstreet Way, Harpenden. Office: Rothamsted Exptl. Sta., Harpenden, Eng.*

GREGORY, Raymond Leslie, Am. physician; b. Beeville, Tex., Feb. 20, 1901; s. Charles Hardy and Molly (Keller) G.; B.A., U. Tex., 1922, M.A., 1923; Ph.D., U. Minn., 1927, M.D., 1929; m. Lois Clarissa Null, Sept. 7, 1927; children—Charles H., Raymond Frederick, Mary Edith (Mrs. Peyton Hawes). Faculty, U. Ia., 1934-36, La. State U., 1936-37; prof., head dept. medicine Howard U. Med. Sch., 1937-39, U. Ark. Sch. Medicine, 1939-40; faculty U. Tex. Med. Br., Galveston, 1940—, prof. internal medicine, 1943—, dir. Endocrine Clinic, 1947-59, dir. cardiac catheterization unit, 1954-59, chmn. dept. internal medicine, 1961—. Diplomate Am. Bd. Internal Medicine. Mem. A.M.A., Am. Physiol. Soc., Am. Council for High Blood Pressure, A.A.A.S., Soc. for Exptl. Biology and Medicine, Central, So. socs. clin. research, A.C.P., Endocrine Soc., N.Y. Acad. Scis., Am. Goiter Assn., Am. Diabetes Assn., Sigma Xi, Phi Beta Pi, Alpha Omega Alpha, Gamma Alpha, Mu Delta. Contbr. numerous articles to profl. jours. Research on metabolic and endocrine diseases; metabolism of alcohol; azotemia associated with massive gastrointestinal bleeding; exptl. hypertension. Home: 3111 Av. O, Galveston, Tex. 77550.*

GREGORY, Sir Richard (Arman), English astronomer; b. Bristol, Eng., Jan. 29, 1864; s. John Gregory; D.Sc., U. Leeds, U. Bristol; LL.D., St. Andrews U.; m. Kate Florence Dugan Pearn, 1888 (dec. 1926); 1 son, 1 dau.; m. 2d, Dorothy Mary Page, 1931. Asst., Phys. Lab., Clifton Coll., 1882-85; tchr. in training at Royal Coll. of Sci., 1886-87; sci. demonstrator His Majesty's Dockyard Sch., Portsmouth, 1887-88; computer to Solar Physics Com., asst. to Sir Norman Lockyer, 1889-93; asst. editor Nature, 1893-1919, editor, 1919-39; emeritus prof. astronomy Queen's Coll., London; Ext. lectr. Oxford U., 1890-95; chmn. council Norman Lockver Obs. Corp. (now inc. in Univ. Coll. of S.W., Exeter), 1920-48. Fellow Royal Soc., 1933, Royal Astron. Soc., Royal Meteorol. Soc. (pres. 1928-29), Inst. Physics; mem. Brit. Assn. for Advancement Sci. (pres. 1940-46), South Eastern Union Sci. Socs. (pres. 1924), Geog. Assn. (pres. 1923-24), Brit. Assn. (pres. ednl. sci. sect. 1922). Author several textbooks on phys. geography, physiography, physics, chemistry, gen. exptl. sci.; The Vault of Heaven, 1893; Discovery, or the Spirit and Service of Science, 1916; Cultural Contacts of Science, 1938; Religion in Science and Civilization, 1940; British Scientists; Science in Chains; Education in World Ethics and Science. Died Middleton, Eng., Sept. 15, 1952.

GREGORY, Robert Todd, Am. mathematician, educator; b. Owensboro, Ky., Mar. 19, 1920; s. Richeson Todd and Jennie (Howard) G.; student Georgetown (Ky.) Coll., 1937-39; B.S., U. S. Naval Acad., 1939-42; M.S., Ia. State U., 1948; Ph.D., U. Ill., 1955; m. Margaret Bentzinger, Dec. 29, 1944; children—Rosalie, Carl. Mathematician, U. S. Naval Proving Ground, Dahlgren, Va., 1949; instr. Fla. State U., 1949-50; research asst. Digital Computer Lab., U. Ill., Urbana, 1950-55, vis. asso. prof., summer 1960; asst. prof. U. Cal., Santa Barbara, 1955-59, vis. prof., Berkeley, summer 1964; asso. prof. math. U. Tex., Austin, 1959-63, sr. research mathematician Computation Center, 1959—, prof. math., 1963—. Cons. U. S. Naval Air Missile Test Center, Point Mugu, Cal., 1956, Space Tech. Labs., Los Angeles, 1956-58, Argonne Nat. Lab., 1962-63; chmn. numerical analysis com. Coop. Orgn. Users Control Data 1604 Computers, 1964. Mem. Am. Math. Soc., Math. Assn. Am., Soc. for Indsl. and Applied Math. (vis. scientists panel 1961-62), Tex. Acad. Sci. (vis. lectr. 1961-65), Assn. for Computing Machinery. Author: Numeral Systems, 1963. Reviewer, Computing Revs., 1961—. Contbr. articles to profl. publs. Study of computational problems in matrix algebra; numerical analysis; numerical solutions of partial and ordinary differential equations. Home: 2703 Mountain Laurel Dr., Austin, Tex. 78703.*

GREGORY, William, Scottish chemist; b. Edinburgh, Dec. 25, 1803; studied under Liebig, 1835; became prof. chemistry Anderson's Coll., Glasgow, Scotland, 1837; named prof. medicine and chemistry Aberdeen, Scotland, 1839; prof. chemistry Edinburgh, 1844. Author: A Hand-Book of Inorganic Chemistry, 3d edit., 1853; Elementary Treatise on Chemistry, 1855; Outlines of Chemistry for the Use of Students, 1845, an enlarged version called A Handbook of Organic Chemistry, 3d edit., 1852. Revised (with Liebig Elements of Chemistry (Turner), 1842. Translated several of Liebig's books. Discovered nitrogen sulfide; research on preparation of pure morphine from opium, analysis of opium, meconic acid, methyl mercaptan, derivatives of uric acid, preparation of potassium permanganate, distillation of rubber. Died Edinburgh, Apr. 24, 1858.

GREGORY, William K(ing), Am. paleontologist, morphologist; b. N.Y.C., May 19, 1876; s. George and Jane (King) G.; student Sch. Mines, Columbia, 1894-96; A.B., Columbia, 1900, A.M., 1905, Ph.D., 1910; D.Sc., Witwatersrand, 1938; m. Laura Grace Foote, Dec. 4, 1899; m. 2d, Angela DuBois, 1938. Research asst. to Henry Fairfield Osborn, 1899-1913; asst. curator dept. vertebrate paleontology Am. Museum Natural History, 1911-14, asso. in paleontology, 1914-26, curator dept. comparative anatomy, 1921-44, emeritus, 1944—; curator dept. ichthyology, 1925-44, emeritus, 1944—; lectr., asst. and asso. prof., prof. vertebrate paleontology, Da Costa prof., Columbia, 1943-45, emeritus, 1945—. Fellow N.Y. Acad.

Scis. (pres. 1932-33), N.Y. Zool. Soc., A.A.A.S. (v.p. sect. II, 1931); mem. Am. Soc. Naturalists (v.p. 1936). Am. Assn. Anatomists, Geol. Soc. Am., Paleontol. Soc. Am., Am. Soc. Mammalogists, Am. Philos. Soc., Nat. Acad. Scis., Am. Acad. Arts and Scis., Am. Assn. Physical Anthropology (pres. 1941-42), Am. Soc. Ichthyology and Herpetology (pres. 1936-38); fgn. fellow London Zool. Soc., Geol. Soc. London, Linnean Society London, Royal Soc. Queensland, Royal Soc. Sci. Upsala, State Russian Paleontol. Soc. Leningrad. Acad. Hon., Museo de la Plata. Author: The Orders of Mammals, 1910; On the Structure and Relations of Notharctus, an American Eocene Primate, 1920; The Origin and Evolution of the Human Dentition, 1922; Our Face from Fish to Man, 1929; Fish Skulls—A Study of the Evolution of Natural Mechanisms, 1933; A Half Century of Trituberculy—The Cope-Osborn Theory of Dental Evolution, 1934; In Quest of Gorillas, 1937; Studies on the Origin and Early Evolution of Paired Fins and Limbs, Parts I-IV (with H. C. Raven) 1941; The Monotromes and the Palimpest Theory, 1947; also numerous revs. and tech. papers. Research on Eocene primates; evolution and morphology of vertebrate skull and locomotor systems; origin of man; emergent evolution; evolution of mammalian molar teeth; conchology; phylogeny of fish skulls; authority on gorillas. Home: 235 W. 76th St., N.Y.C. 10023.

GREGORY OF NYSSA, St., Greek natural philosopher; b. circa 331; younger bro. of St. Basil; m. Theosebia. Bishop of Nyssa, 371; attended great council at Constantinople; composed sermons, lives, letters, commentaries on scripture (an edit. of his works prepared by Froton du Duc, 1615); believed that creation was potential, with God imparting to matter its fundamental properties and laws, and the objects and completed forms of universe developing gradually from chaotic material. Died circa 396.

GREGORY OF RIMINI (Gregorio Novelli da Rimini; Gregorius de Arimino, Ariminesis), natural philosopher; b. Rimini, Italy; studied in Rimini, Paris, 1323-29. Augustinian hermit, theologian, Occamist; magister regens, Paris, 1345; principalis lector in Augustinian House, Rimini, 1351, gen. of Augustinian order, 1357. Author: (treatises) De intentione et remissione formarum; Quaestiones metaphysicales; De usuris. Wrote commentary on 4 books of Sentences, 1344 (contains discussions of math. continuity and infinity. Died Vienna, Nov. 1358.

GREHANT, Nestor, French physician; b. Lyons, France, 1838. Author; Manuel de physique medicale 1869; Les poisons de l'air, 1890; Recherches physiologiques, 1903; Le gaz du sang, 1894; others. Devised a method (with C. E. Quinquaud) to determine volume of blood by use of carbon monoxide, 1882. Died Paris, 1910.

GREIDER, Kenneth Randolph, Am. physicist; b. Cleve., Feb. 10, 1929; s. Clarence Edwin and Cornelia (Widney) G.; B.S. in Engring., U. Mich., 1950; M.S., U. N.M., 1954; Ph.D., U. Cal., Berkeley, 1958; m. Jean Marie Foley, Dec. 22, 1956 (dec. Dec. 1967); children—Mark, Carolyn. Faculty U. Cal., Berkeley, 1958-59, La Jolla, Cal., 1959-62, Yale, 1962-65; faculty U. Cal., Davis, 1965—, prof. physics, 1967—. Cons. Lawrence Radiation Lab., U. Cal., Berkeley, 1959-61, Inst. for Def. Analyses, Washington, 1962-63, Los Alamos Sci. Lab., N.M., 1964-67. Mem. Am. Phys. Soc. Contbr. articles to sci. jours. Research in theoretical investigations reaction mechanisms in nuclear physics, gen. scattering theory elementary particles. Home: 1202 Bucknell Dr., Davis, Cal. 95616.*

GREIF, Roger Louis, Am. physiologist; b. Balt., Aug. 23, 1916; s. Leonard L. and Amy (Frederleicht) G.; B.S., Haverford Coll., 1937; M.D., Johns Hopkins U., 1941; m. Carol Clement Prince, July 24, 1950; children—Peter Clement, Nicholas Peabody, Matthew Payson. Fellow in medicine Johns Hopkins Hosp., 1946-47; asst. physician Rockefeller Inst. Hosp., N.Y.C., 1947-53; faculty Cornell U. Med. Coll., N.Y.C., 1953—, prof. physiology, 1965—; cons. Health Research Council N.Y.C., chmn. metabolic disease panel, 1962-64. Mem. Harvey Soc. (mem. council 1962-65), Soc. Exptl. Biology and Medicine (sec. N.Y.C. chpt. 1962), Am. Physiol. Soc., Am. Fedn. Clin. Research, Soc. Gen. Physiologists, Endocrine Soc., Phi Beta Kappa, Alpha Omega Alpha. Research, numerous publs. on albuminuria in nephrosis, in treatment of malaria, in hormone-enzyme relationships, comparative physiology of kidney; mechanism of action of thyroid hormones and of hormone synthesis by thyroid gland. Home: 534 E. 87th St., N.Y.C. 10028.*

GREIG, Edward David Wilson, physician; b. Edinburgh, Scotland, Mar. 20, 1874; s. David Greig; M.B., C.M., U. Edinburgh, 1895, B.Sc., 1898; M.D., D.Sc. Asst. to prof. pathology Univ. Coll., London, 1898; entered Indian Med. Service, 1899; plague researcher, Bombay, 1902-03, mem. sleeping sickness commn. Royal Soc., Uganda, 1902-03, also dir.; mem. Enteric fever research com. Govt. India, 1907-08, also dir.; research for Govt. India on etiology of dysentery, 1908-09, epidemic dropsy, 1909-11, cholera, 1912-16; dir. med. research, India, 1921-23; dir. Pasteur Inst., Shillong; physician-cons. on tropical diseases Royal Infirmary, Edinburgh, 1929-39; lectr. tropical diseases Edinburgh U., 1924-39. Fellow Royal Coll. Physicians Edinburgh, Royal Soc. Edinburgh; mem. Soc. Exotic Pathology (corr.). Author reports, papers; joint editor Indian Jour. Med. Research, 1921-23. Died Apr. 13, 1950.

GREIG, Margaret Elizabeth, pharmacologist; b. Cumberland, Ont., Can., Mar. 12, 1907; s. John Graham and Jane Eva (Brodie) G.; B.A., McGill U., 1928; M.A., U. Sask., 1928-30; Ph.D., McGill U., 1932. Demonstrator chemistry U. Sask., 1928-30; NRC Can., fellow, 1930-32; research asst. dept. cellulose chemistry McGill U., 1932-35; chemist Biochem. Research Found., Franklin Inst., Phila., 1935-42; faculty Vanderbilt U., 1942-53, asso. prof., 1948-53; with OSRD, 1942-44; pharmacologist, sr. scientist Upjohn Co., Kalamazoo, Mich., 1953—. Fellow N.Y. Acad. Scis., A.A.A.S.; mem. Am. Soc. for Pharmacology and Exptl. Therapeutics, Sigma Xi. Contbr. numerous articles to tech. jours. Research on difference in metabolism between normal and tumor tissues, shock from hemorrhage, action of stimulants and depressants on brain metabolism, etryptamine, allergy and anaphylaxis. Home: 2923 Memory Lane, Kalamazoo 49007. Office: Upjohn Co., Henrietta St., Kalamazoo 49001.*

GREIM, Georg Heinrich, German geographer, meteorologist; b. Offenbach, Germany, July 15, 1866; s. Friedrich Wilhelm and Mathilde (Heddaeus) G.; Ph.D. in Natural Scis., Giessen, Germany, 1887; further studies in Göttingen, Germany, Munich, Germany. Joined faculty Tech. U., Darmstadt, Germany, 1891, became lectr. phys. geography and mineralogy, 1897; placed in charge erecting weather stas. in Hessia, 1898; later dir. Hessian Weather Service; became chmn. State Bur. for Meteorology and Hydrology, 1917; became prof. geography Tech. U. Munich, 1920; ret., 1931. Author: Landeskunde des Grossherzogtums Hessen . . . , 1908; Beiträge zur Anthropogeographie des Grossherzogtums Hessen, 1912; Physikalische Geographie, 1927; Frankenland, 1933. Research in phys. geography, including continuation Finsterwalder's work of exact photogrammetric measurement of alpine glaciers, meteorol. conditions in Hessia; set up water gauge sta. in Jambach, 1893, and collected data on flow, speed and temperature for 25 years. Died Jugenheim, Germany, Apr. 5, 1946.

GREINACHER, Heinrich, Swiss physicist; b. Switzerland, May 31, 1880; ed. univs. Zurich, Geneva (Switzerland), Berlin, Heidelberg (Germany); Ph.D. Became prof. physics U. Berne (Switzerland), dir. Phys. Inst., also Meteorol. Obs., since 1924. Mem. Swiss Phys. Soc. Recipient Theodore Kocher prize U. Berne, 1948. Author: Die neueren Strahlen, 1909; Einführung in die Ionen- und Elecktronenlehre der Gase, 1923; Physik in Streifzügen, 2d edit., 1943; Ergänzungen zur Experimentalphysik, 3d edit., 1953. Research on radioactivity, ionology, electronics; invented Greinacher-connection, cascade generator, diffusion hygrometer, spark counter, vibration electromer, static voltmeter. Address: Alpeneggstr. 17, Berne, Switzerland.

GREINEL, Hermann Paul, physicist; b. Nuremberg, Germany, Sept. 12, 1909; s. Wilhelm Anton and Frida (Hofmann) G.; Ph.D., Friderico-Alexandrina U., Erlangen, Germany, 1937; m. Antonia M. H. Stark, Sept. 12, 1942; children—Irmela M., Herbert R. C. Came to U. S., 1956, naturalized, 1962. Asst. chief of dept. Hagenuk, Kiel, Germany, 1937-42; group leader DFS Ernst Udet, Darmstadt-Freilassing, Germany, 1942-45; research engr. Arsenal de l'Aéronautique, Chatillon-sous-Bagneux (Seine), France, 1946-49, Centre National d'Etudes des Télécommunications, Palaiseau (Seine-Oise) France, 1949; subdir. Institute Laboratorios Metalúrgicos de Alta Temperatura, San Carlos de Bariloche (T. N. Rio Negro), Argentina, 1950; sect. head Indústrias Aeronáuticas y Mecánicas del Estado, Instituto Aerotécnico, Córdoba, Argentina, 1950-56; asso. scientist Republic Aviation Corp., Farmingdale, N.Y., 1956-58; aero. research engr. AFMDC, Holloman AFB, N.M., 1958-59; research scientist Lockheed Missiles & Space Co., Palo Alto, Sunnyvale, Cal., 1959—. Mem. Wissenschaftliche Gesellschaft fuer Luft-und Raumfahrt (Germany), Am. Phys. Soc., Am. Inst. Aeros. and Astronautics, Optical Soc. Am. Research in geometrical and phys. optics, infrared instrumentation and sensors, guidance and control, nuclear and plasma physics, flutter and vibration, flight mechanics. Home: 2246 Deodara Dr., Los Altos, Cal. 94022. Office: P.O. Box 504, Sunnyvale, Cal. 94088.*

GREINER, Franz Ferdinand, glass industrialist; b. Stützerbach, Germany, Apr. 3, 1808; s. Johann Günther and Sophie Johanna Dorothea (Handlin) G.; m. Caroline Güntherine Wilhelmine Lattermann, 1828; 3 daus. Learned glass blowing and set up glass factory in father's mill, 1830. 1st to blow glass thermometers and utilize process commercially in Thuringian Forest; manufactured 64 different aerometers and thermometers, also phys. apparatus, chem. glass articles. Died Stützerbach, June 9, 1855.

GREINER, Friedrich Eberhard (Fritz), foundry engr.; b. Weingarten, Germany, Sept. 7, 1867; s. Christian Friedrich and Ursula (Hiplinger) G.; ed. Indsl. Sch., Stüttgart, Germany; hon.dr.engring. Tech. U., Stuttgart, 1925; m. Viktoria Rauscher, 1891. Worked in Magdeburg, Braunschweig, Ettlingen, Germany, then operations engr. iron casting firm Gebrüder Bolze, Mannheim, became foundry dir. engine firm G. Kuhn, Stüttgart, 1901, placed in charge of all foundry work when firm merged with Maschinenfabrik Esslingen, 1905, planned and built firm's new model foundry, Esslingen-Mettingen, 1912-18; ret., 1929. Recipient Siegfried Werner Meml. medal, 1924. Mem. Assn. German Iron Foundries (chmn. Württemberg). Made transition from ribbed cylinder to hollow constrn. in steam engines; devel. cast cylinders for automobiles and airships; studied acid resistant cast iron, cast aluminum, bearing metals; developed Greiner-Klingenstein cast iron diagram, alloy additives for cast iron. Died Stuttgart, Aug. 26, 1936.

GREINER, Walter, German physicist; b. Neuenbau, Germany, Oct. 29, 1935; s. Albin and Elsa (Fischer) G.; Diplom physiker, Technische Hochschule Darmstodt, 1960; Dr.rer.nat., U. Freiburg, 1961; m. Baerbel Chun, Aug. 27, 1960; children—Martin, Carsten. Asst. prof. U. Md., 1962-64; dir. Inst. Theoretical Physics, prof. physics U. Frankfurt (Germany), 1965—; guest prof. U. Melbourne (Australia), 1966; guest scientist NBS, Washington, 1966. Mem. German Phys. Soc. Research, publs. in nuclear physics, collective motion in nuclei, dynamic collective theory of giant resonances, eigenchannel theory of nuclear reactions. Home: 3468 Gundelhardt 44, Kelkheim, Germany. Office: 77064/2331 Robert Mayer Str., Frankfort/Main, Germany.*

GREISHEIMER, Esther Maud, Am. physiologist; b. Chillicothe, O., Oct. 31, 1891; s. William and Elizabeth (Andre) G.; B.S. in Edn., Ohio U., 1914; M.A., Clark U., 1916; Ph.D., U. Chgo., 1919; M.D., U. Minn., 1923. Faculty, U. Minn., Mpls., 1918-35, asso. prof., 1931-35; prof. physiology Woman's Med. Coll. Pa., 1935-43; prof. Temple U. Sch. Medicine, Phila., 1944-56, prof. emeritus, 1956—. Mem. A.M.A., Am. Med. Woman's Assn. (named Med. Woman of Year br. 25 1954), Am. Physiol. Soc., Am. Assn. U. Profs., A.A.A.S., Phi Beta Kappa, Sigma Xi, Alpha Omega Alpha, Sigma Delta Epsilon. Author: Physiology and Anatomy, 1932, 8th edit., 1963; (with E. E. Chaffee) Basic Physiology and Anatomy, 1964; also numerous articles. Research in physiology anesthesia. Home: 5450 Wissahickon Av., Phila. 19144.*

GREMELS, (Karl Felix) Hans, German pharmacologist; b. Grossbreitenbach, Germany, Dec. 9, 1896; s. Paul and Helene (Lincke) G.; student natural scis. and medicine; later student Hamburg, Berlin, Munich, (all Germany), London; m. Gisela Fuchs, 1943; 1 son, 2 daus. Joined faculty path. physiology and pharmacology U. Munich, 1932; named prof. pharmacology U. Marburg (Germany), 1938. Research and publs. on physiology and pharmacology of urine, effect of acetylcholin, adrenalin, digitalis glycoside and strophanthin on heart metabolism, pharmacology of parasympathetic nervous system, control of energy metabolism through nutritive substances, especially monosaccharide and amino acids. Died Freiburg, Germany, Mar. 25, 1949.

GREN, Friedrich Albert Carl, German chemist, physicist; b. Bernburg, Germany, Apr. 29, 1760; s. Johann Magnus and Dorothea Elisabeth (Sieber) G.; studied under v. Grell, Helmstedt, from 1782; M.D., Halle, Germany, 1786, Ph.D., 1787; m. Johanna Sophie Karsten, 1788; 1 dau. Began as pharmacist; became asso. prof. pharmacology U. Halle, 1787, prof. philosophy, 1788, later prof. medicine; lectured on natural history, chemistry, pharmacology, physics. Author: Systematisches Handbuch der gesamten Chemie, 3 vols., 1787-90; Dissertation inauguralis physico-medica sistens observationes et experimenta circa genesin aëris fixi et phlogisticati, 1786; Grundriss der Naturlehre; Grundriss der Pharmacologie . . . , 1790; Handbuch der Pharmacologie oder der Lehre von den Arzneimitteln, 1791; Grundriss der Chemie zum Gebrauche akademischer Vorlesungen, 2 vols., 1796. Editor, Jour. der Physik, 1790-94, Neues Jour. der Physik, 4 vols., 1795-97, Annalen der Physik. Wrote 1st practical sci. textbook in chemistry; 1st defended phlogiston theory, then changed to combustion theory. Died Halle, Nov. 26, 1798.

GRENACHER, (Georg) Hermann, German zoologist; b. Lipburg, Germany, Mar. 18, 1843; s. Johann Georg and Katherina Barbara (Claiss) G.; studied math. and chemistry Tech. U., Karlsruhe, later zoology U. Würzburg (Germany), U. Göttingen (Germany); Ph.D., 1867; m. Louisa Ruprecht, 1873; 1 son, 3 daus., including Else (Mrs. Paul Holdefleiss), Hedwig (Mrs. Karl Heldmann). Prosector under A. Kölliker, Würzburg; joined faculty, Göttingen, 1869; mem. research expdn. to Canary and Cape Verde Islands, 1871-72; became prof. Forestry Acad., Hanover-Münden, 1872, U. Rostock (Germany), 1873, U. Halle, 1882; emeritus, 1909. Author: Untersuchungen über das Sehorgan der Arthropoden, insbesondere der Spinnen, Insecten und Crustaceen, 1879; also articles. Introduced borax carmine and alum carmine dyes; discovered central body heliozoa; studied morphology and evolution of eyes of arthropods and molluscs especially cephalopods. Died Halle, Apr. 25, 1923.

GRENANDER, Ulf, mathematician; b. Västervik, Sweden, July 23, 1923; s. Sven and Maza (Dersson) G.; Fil. Dr., U. Stockholm (Sweden), 1950; m. Emma-Stina Hallquist, Dec. 22, 1946; children—Sven, Angela, Charlotte. Docent, U. Stockholm, 1950-51, 54, prof., 1959-65; vis. asst. prof. U. Chgo., 1951-52; vis. asso. prof., U. Cal., 1953; prof. Brown U. Providence, 1957-58, 66—. Fellow Inst. Math. Statistics; mem. Royal Acad. Scis., Stockholm, Internat. Inst. Author: Probabilities on Algebraic Structures, 1963; other books, also articles. Research on statis. inference in stochastic processes, time series analysis, ins. math., pattern theory, operations research. Home: 26 Barberry Hill, Providence 02906.*

GRENELL, Robert Gordon, Am. neurobiologist; b. N.Y.C., Apr. 3, 1916; s. Max and Lee (Gordon) G.; B.A., Coll. City N.Y., 1935; M.S., N.Y. U., 1936; Ph.D., U. Minn., 1943; m. Freda Zierler, June 18, 1943. With Yale U. Sch. Medicine, 1943-47, instr. neuroanatomy, 1945-47; USPHS sr. fellow E. R. Johnson Found. for Research in Med. Physics U. Pa., dept. biophysics Johns Hopkins U., 1947-49; with U. Md. Sch. Medicine, 1950—, prof. neurobiology in psychiatry, 1959—; cons. NASA, 1963-64; cons. neurophysiologist Balt. City Hosps.; cons. Mass. Gen. Hosp.; prof. pharmacology U. S. AID, Trivandrum, India, 1961-62. Fellow A.A.A.S.; mem. Am. Physiol. Soc. (cons. survey physiology), Assn. Research Nervous and Mental Disease, Am. Physiol. Soc., Biophys. Soc., Soc. Biol. Psychiatry, Internat. Soc. Cell Biology, Am. EEG Assn., Soc. Exptl. Biology and Medicine, Md. Psychiat. Soc. Author (with L. J. Mullins) Molecular Structure and Functional Activity of Nerve Cells, 1956, Neural Physiopathology, 1962. Research on chem. and physiol. properties of brain cells, mechanisms by which they function; reactions occuring at and across the cell surface; relationships between phys. and chem. mechanisms in cerebral nerve cells and certain aspects of behavior. Home: 210 E. Highfield Rd., Balt. 21218.*

GRENFELL, Sir Wilfred Thomason, English surgeon; b. Mostyn House, Parkgate, nr. Chester, Eng., Feb. 28, 1865; s. Algernon Sydney and Jane Georgiana (Hutchinson) G.; ed. Marlborough Coll., Oxford U. and London Hosp.; hon. LL.D., Williams Coll., 1909; hon. M.A., Harvard, 1909; hon. M.D., Toronto U., 1911; L.H.D., U. of N.Y., 1928; LL.D., McGill U., Montreal, Can., 1928, Middlebury Coll., Vt., 1928, Princeton U., Bowdoin, St. Andrews, 1929, Berea Coll., Ky., 1930; D.Sc., U. of Louisville, 1933; m. Anne MacClanahan, 1909; (died 1938); children—Wilfred Thomason, Kinloch Pascoe, Rosamond Loveday. House surgeon London Hosp., 1890-91; entered med. service of Royal Nat. Mission to Fishermen, 1889; fitted out first hosp. ship for North Sea fisheries; cruised with fishermen, established houses and mission vessels for them; went to Labrador, 1892, and built 5 hosps., 7 nursing stas., 2 orphanages, 2 large schools, coop. stores, and inaugurated industrial work and child welfare work; surgeon in charge hospital steamer, Strathcona II, of which is master, and cruises each year on coasts of northern Newfoundland and Labrador; owned and operated S.S. Maraval, S.S. Zavorah, M.V. Jessie Goldthwait, M.-V. George B. Cluett, and yawls in connection with hosps. F.R.C.S., L.R.C.P., London; fellow Am. Coll. Surgeons, 1915, Royal Coll. Surgeons, 1920; mem. Royal Instn. of Gt. Britain; awarded Murchison Bequest, Royal Geog. Soc., 1911; gold medal Nat. Acad. Social Sciences (U. S.), 1920; Livingston gold medal Royal Scottish Geog. Soc., 1930; lord rector St. Andrews U., 1929-31. Hon. fellow of Queen's Coll., Oxford, England, 1936. Author: Harvest of the Sea; A Man's Faith, 1908; Labrador, 1909; Adrift on an Ice-Pan, 1909; Down to the Sea, 1910; What Life Means to Me, 1910; What the Church Means to Me, 1911; Down North on the Labrador, 1911; The Adventure of Life, 1912; Tales of the Labrador, 1916; Labrador Days, 1919; A Labrador Doctor (autobiography); Northern Neighbors, 1923; Yourself and Your Body, 1925; Labrador Looks at the Orient, 1928; Forty Years for Labrador, 1932; The Romance of Labrador, 1934. Died Oct. 9, 1940.

GRENGG, Roman, Austrian mineralogist; b. Stein, Danube, Austria, Dec. 1, 1884; s. Roman and Maria Grengg; ed. Technol. U., U. Vienna; Ph.D. in engring.; Ph.D. in tech., commandant agrege; m. Hedwig Mirau Vve Haarmann, 1942; children—Roman, Ida Christine. Asst., Technol. U., Vienna, prof., dir. Inst. Applied Mineralogy and Petrography, 1925—; instr. mineralogy and applied geology; prof. Indsl. Mus. Tech.; head of research in tech. chemistry, legal expert; researcher, councillor in chemistry, 1st materials, geology and hygiene in constrn., 1947—. Mem. German Assn. Mineralogy, Viennese Assn. Geology. Research and numerous publs. on modernization of sci. methods and practices of research of minerals, rocks, mortar, concrete, disposition of strata, sources of mineral water, water problems, terrain for constrn. Home: 8 Liesing Jos Kutschagasse, Vienna XXIII, Austria.

GREPPI, Enrico, Italian physician; b. Bologna, Italy, July 22, 1896; s. Luigi and Eugenia (Rossetti) G.; M.D., titular prof.; m. Nella Giampiccolli, 1930; children—Eugenio, Daisy, Annalisa, Donatella, Claudio. Dir., Medico-Univ. Clinic, Florence, Italy. Mem. Italian and fgn. med. assns., Italian Gerontology Soc. (pres.). Author monographs on illnesses of rat, hypertension, plethora, diabetes of the elderly, senility, senile illnesses, cephalalgy. Home: 13 via Buonvicini. Office: Medical Clinic, viale Morgagni, Florence, Italy.

GRESHAM, Geoffrey Austin, pathologist; b. Wrexham, North Wales, Nov. 1, 1924; s. Thomas Michael and Harriet Anne (Richards) G.; M.C. in Pathology, Cambridge (Eng.) U., 1946; M.B., B.Chir., Kings Coll. Hosp., London, Eng., 1947, M.A., 1949, M.D., 1957; m. Gweneth Margery, July 1, 1950; children—Christopher, Diana, Andrew, Robert, Susan. Faculty, Cambridge U., 1952—, lectr., 1957-63, univ. morbid anatomist and histologist, 1963—. cons. pathologist Brit. Army, 1949—; fellow Jesus Coll., Cambridge, 1965—. Mem. com. on comparative cardiology WHO, 1962—. Fellow Zool. Soc.; mem. Pathol. Soc., European Atherosclerosis Group. Author: An Introduction to Comparative Pathology (with A. R. Jennings), 1962; (with D. G. Chalmers) Biological Aspects of Occlusive Vascular Disease, 1964. Research and publs. on factors leading to coronary arteriosclerosis, particularly role of dietary fats. Home: 18 Rutherford Rd., Cambridge, Eng.*

GRESHAM, Sir Thomas, Brit. economist; b. London, 1519; s. Sir Richard Gresham; ed. Caius Coll., Cambridge (Eng.) U.; m. Mrs. William Read, 1544; 1 son; also illegitimate dau., Anne (Mrs. Nathaniel Bacon). Admitted to the Mercers' Co., 1543; went to the Low Countries to became mcht., 1543; acted as agt. for Henry VIII; became financial agt. of the crown; temporary ambassador at the court of the Duchess of Parma under Queen Elizabeth; various diplomatic missions; founder Gresham Coll., London, 1596; founder Royal Exchange. Created knight, 1559. Described the tendency of the inferior of 2 forms of currency to circulate more freely (Gresham's law). Died London, 1579.

GRESKY, Alan Tolstoy, Am. chemist; b. Piper, Ala., Sept. 29, 1917; s. John Leon and Luda (Harper) G.; student Ala. Coll., summer 1936; A.B. in Chemistry, U. Ala., 1939; postgrad. Ala. Med. Sch., U. Tenn.; m. Charlotte Ruth Almgren, Nov. 26, 1938; children—Frederic Paul, Ruth Alane (Mrs. Donald Ray Privett), Mary Lou (Mrs. Edward James Spitzer II). With Tenn. Coal & Iron Co. ore div. U. S. Steel Co., Birmingham, Ala., 1939-42, DuPont Chem. Co. Ordnance Works, Childersburg, Ala., 1942-45, Carbide and Carbon Chem. Co., Oak Ridge, Tenn., 1945-46, Monsanto Chem. Co. Clinton Labs., Oak Ridge, 1946-48; chemist, chem. tech. div. Union Carbide Chem. Co. Oak Ridge Nat. Lab., 1948—. With information and exhibits div. Oak Ridge Inst. Nuclear Studies, part time 1955-60; lectr. nuclear fuel cycle course Oak Ridge Sch. Reactor Tech., 1958-65. Mem. Am. Chem. Soc., Am. Soc. Testing and Materials. Author: New Laws of Nature, 1964. Research and publs. on chem. processing of n-irradiated fuels; long-range planning studies of nuclear fuel cycle; phys., chem. theories relating the micro and macroworlds; proposed unified equation of particle mechanics, particle wave mechanics, particle interaction. Home: 113 Kingsley Rd., Oak Ridge 37832. Office: P.O. Box X, Oak Ridge 37831.*

GRESSITT, Judson Linsley, entomologist; b. Tokyo, Japan, June 16, 1914; s. James Fullerton and Edna (Linsley) G.; student Stanford, 1932-34; B.S., U. Cal., Berkeley, 1938, M.S., 1939, Ph.D., 1945; m. Margaret Kriete, Mar. 20, 1941; children—Sylvia Anne, Rebecca Louise, Edna Carolyn, Ellyn Elizabeth. Asst. zoologist U. Cal., Berkeley, 1938-39, research fellow, 1944-45, asst. entomologist, Riverside, 1947-50; faculty Lingnan U., Canton, China, 1939-43, 46-51; entomologist Pacific Sci. Bd. NRC, 1951-52; with Bishop Mus., Honolulu, 1953—, chmn. entomology dept., 1955—, L.A. Bishop Distinguished chair zoology, 1964—; faculty U. Hawaii, 1955—; cons. NIH, Nat. Acad. Sci. Fellow Entomol. Soc. Am., Cal. Acad. Sci., A.A.A.S.; mem. Pacific Coast, Hawaiian, Japan, Royal entomol. socs. Author: Zoogeography of Pacific and Antarctic Insects, 1961; (with S. Kimoto) Chrysomelidae of China and Korea, 1961; (with others) Insects of Campbell Island, 1964. Editor: Pacific Basin Biogeography, 1964; Pacific Insects, 1959—, Jour. Med. Entomology, 1964—. Research and publs. primarily in zoogeography of Pacific, Antartic insects; described, named 1000 species beetles; revised Oriental zoogeographical region to include New Guinea and S. Pacific islands. Home: 1053-A Ilima St., Honolulu 96817. Office: Bishop Mus., Honolulu 96819.*

GRESSLY, Amanz, Swiss geologist; b. Bärschwil, Switzerland, July 17, 1814; s. Xaver and Margaritha (v. Glutz-Ruchti) G.; began med. studies in Strasbourg, France; later studied geology and paleontology, Porrentruy. Worked with Thurmann, Porrentruy; invited to Musée de Neuchâtel (France) where he worked with Desor and Vogt, 1839; Author: Observations gélogiques sur le Jura soleurois, 1838-41. Research on geology and paleontology of Jura; introduced concept of various compositions of sediments of equal age. Died Bern, Switzerland, Apr. 13, 1865.

GRESSON, Richard Arbuthnot Reynell, Irish cytologist; b. Dunlavin, Ireland, Apr. 17, 1899; s. William and Charlotte (Brush) G.; B.Sc., U. Edinburgh, 1926, D.Sc., 1941; Ph.D., U. St. Andrews, 1930; m. Helen Docherty, Sept. 24, 1926; 1 dau., Deirdre (Mrs. Ivan F. Nelson). Asst. in zoology U. Coll., Dundee, 1926-31; lectr. in cytology, then sr. lectr. zoology U. Edinburgh, 1931-49; prof. zoology Queen's U., Belfast, N. Ireland, 1949-64, prof. emeritus, 1964, dean faculty sci., 1951-54. Fellow Royal Soc. Edinburgh, Inst. Biology; mem. Royal Irish Acad., Soc. Exptl. Biology. Author: Essentials of General Cytology. Research and publs. in light and electron microscopy of animal cells. Home: Strand Rd., Castlegregory, County Kerry, Republic of Ireland.*

GRETE, (Ernst) August (Heinrich), agrl. chemist; b. Celle, Germany, Sept. 29, 1848; s. Heinrich and Wilhelmina (Bockemeyer) G.; studied philology, chemistry, Göttingen, Germany; Ph.D., 1875; m. Emma Bäbler, 1895; 1 son. Asst. to P. Zöller, Agrl. Inst., Vienna; named dir. Swiss Agrl. Chem. Research Inst., Zurich, 1878; also lectr. Fed. Polytechnikum, Zurich. Author: Die Conservierung und Vorbesserung der Gülle und des Stallmistes durch Phosphorsäure, 1898; also numerous articles. Improved agrl. chem. control research; determined phosphoric acid using titration with molybdic acid; pioneered application of fertilizers to Swiss agr. Died Zurich, Mar. 26, 1919.

GREULACH, Victor August, Am. botanist; b. Convoy, O., Dec. 6, 1906; s. John Adam and Margaret (Giessler) G.; A.B., DePauw U., 1929; M.Sc., Ohio State U., 1933, Ph.D., 1940; m. Elizabeth Dunnells, Oct. 6, 1935; children—Dorothy (Mrs. Roger G. Herbert), Susan (Mrs. J. Laurent Scharff), Vicki (Mrs. Bruce Marsh). Instr., Muskingum Coll., New Concord, O., 1933-35; faculty U. Houston, 1935-46, asso. prof., 1940-46, chmn. dept. biology, 1944-46; asso. prof. plant physiology Tex. A. and M. U., 1946-49; faculty U. N.C. Chapel Hill., 1949—, prof. botany, 1951—, chmn. dept. botany, 1960—. Cons. NSF, 1958-62; exec. dir. Commn. on Undergrad. Edn. in Biol. Scis., 1964-65. Rector scholar DePauw U., 1925-29; U. scholar Ohio State U., 1938-39. Fellow A.A.A.S.; mem. Bot. Soc. Am., Am. Soc. Plant Physiologist, Assn. Southeastern Biologists (past pres.), N.C. Acad. Sci. (past pres.), Sigma Xi (past pres. N.C. chpt.). Author: Plants: An Introduction to Modern Botany (with J. E. Adams), 1962, 67; also numerous articles. Research on photoperiodic responses plants, influences maleic hydrazide, gibberellic acid, and other substances on plant growth and devel. Home: 304 Laurel Hill Rd., Chapel Hill, N.C. 27514.*

GRÉVIN, Jacques, French poet, physician; b. Clermont (Beauvoisis), France, 1539; ed. U. Paris; mem. Paris Med. Sch.; became physician and councillor to Margaret of Savoy, 1561. Author several med. texts including: Traité d'anatomie, 1562; Traité des venins. Translated med. texts from Latin. Expert on anatomy. Died 1570.

GREW, Nehemiah, English physician, botanist; b. Mancetter Parish, Warwickshire, Eng., 1641; s. Obadiah Grew; studied at Pembroke Hall, Cambridge U.; grad. B.A., 1661; M.D., Leyden, 1671; became noted physician in London. Fellow Royal Soc., 1671 (sec. 1677); mem. Royal Coll. Physicians. Author: Disputatio medicophysics . . . de liquore nervoso, 1671; Anatomy of Vegetables Begun, 1673; Idea of Phytological History, 1673; Comparative Anatomy of Stomachs and Guts, 1681; Anatomy of Plants, 1682; Cosmologia sacra, 1701. Pioneer microscopist; observed cellular structure of plants; reputedly 1st to recognize that flowers are sexual organs of plants; studied stamens and pistils; 1st to distinguish pollen grains; coined term comparative anatomy, 1675; Linnaeus named genus of trees Grewis in his honor. Died London, Mar. 25, 1712.

GRIENBERGER, Christoph, astronomer, mathematician; b. Hall, Austria, July 2, 1561; studied theology for 3 years; studied math. under Clavius, Collegium Romanum. Joined Soc. of Jesus, 1590; went to Graz, Austria, 1597; returned to Rome, and worked with Clavius, became his successor, 1612; in charge math. and astronomy edn., Jesuit China Mission. Author: Prospectiva nova coelestis, 1612; De speculo ustorio elliptico, 1613; also contributed to catalog of fixed stars. Studied practical optics; followed beliefs of Copernicus and Galileo. Died Rome, Mar. 11, 1636.

GRIESBACH, Carl Ludolph, Austrian geologist; b. Vienna, Dec. 11, 1847; s. George Ludolph and Caroline (Skriwanek) G.; ed. U. Vienna; m. Emma Griesbach, 1869; 1 son, 1 dau. Joined German Geol. Inst., 1867; mem. German expdn. to Natal and Portuguese E. Africa, 1869-70; made paleontol. drawings for Brit. Mus., London, beginning in 1871; became officer Royal Fusiliers, 1874; named asst. supt. Geol. Survey of India, Calcutta, 1878, dir., 1894-1903; mem. Commn. for Arbitration Russian-Afghanistan Border Dispute, 1884-86; at ct. of Emir Abdurrahman of Afghanistan. Research and publs. on geology of Central Himalayas, Jurassic cliffs of Vienna, carpathian cliff zone in Arr, seismology; 1st studies on Safid-Kuh, Afghan-Turkestan, Upper Burma; believed in tectonic origin of earthquakes. Died Graz, Austria, Apr. 13, 1907.

GRIESBACH, Hermann Adolf, German hygienist; b. Schwartau nr. Lübeck, Germany, Apr. 9, 1854; s. Georg Christoph Alexander and Emilie Augusta Wilhelmine (Böhme) G.; studied biology and chemistry, then medicine, Marburg, Göttingen, Berlin, Würzburg (all Germany); Ph.D., Leipzig, 1876; M.D., Heidelberg, 1883; m. Amalie von Heimburg, 1895; 1 son, Rolf, 1 dau. Joined faculty U. Basel (Switzerland), 1883, became lectr. histology, 1885; prof. biology and hygiene, Strasbourg, France, 1890-1918; also tchr. natural scis., Mulhouse; named hon. prof. hygiene, Giessen, Germany. Mem. German Assn. for Sch. Hygiene (founder, also founder jour. Gesunde Jugend, 1900). Author: Physikalisch-chemische Propädeutik, 2 vols.,

1895-98; Gesundheit und Schule, 1902; Physiologie und Hygiene der Ernährung . . . , 1915; Arteriosklerose und Hypertonie . . . , 1923; Medizinisches Wörter- und Nachschlagebuch, 1927; Persönliche Hygiene und schulhygienische Richlinien an Pädegogischen Hochschulen, 1930; also numerous articles. Research in zoology, embryology, microscopic techniques, histochemistry, morphology and coagulation of blood, physiology of senses; pioneered improvement sch. hygiene, including phys. exercises, sex edn., regular phys. examinations. Died Schwartau, June 23, 1941.

GRIESBACH, Rolf, German physician; b. Mullhouse, Alsace, July 22, 1899; s. Hermann and Heimburg Amalie Griesbach; ed. univs. Strasbourg, Marbourg, Giessen; M.D.; m. Herta Rüdinger, 1939; children—Erika, Ulla Renate, Ruth Marion, Gert. Chief physician, pres. Office Treatment of Tb, Augsburg; chief physician Sanatorium of Wasach; sec. gen. Central Commn. of Germany of Fight against Tb; med. specialist in illnesses of lungs. Mem. Fed. Union German Thorocologues (pres.), Austrian and South German Soc. Tb (hon.), Chilean Soc. Tb (hon.). Author 4 books on Tb, also over 200 articles. Editor, dir.: Der Tuberkuloscarzt. Address: 5 Frohsinn, Augsburg, Bavaria, Germany.

GRIESINGER, Wilhelm, German physician, psychiatrist; b. Stuttgart, Germany, July 29, 1817; s. Gottfried Ferdinand and Karoline Luise (Dürr) G.; studied medicine, Tübingen, Germany, Zurich, Switzerland; M.D., 1838; studied under F. Magendie, Paris, for 1 year; m. Josephine von Rom, 1850. Pvt. practice, Lake Constance; became asst. physician to A. Zeller, Winnenthal; joined Wunderlich's med. clinic, U. Tübingen, 1843, named asso. prof. pathology, materia medica, history medicine, 1847; joined U. Kiel (Germany), 1849; became dir. Med. Sch., Cairo, Egypt, 1850; returned to Germany, 1852; apptd. dir. Tübingen Med. Clinic, 1854; became prof., Berlin, 1864; chmn. psychiat. dept. Charité; dir. Univ. Poly-clinical Inst. for Internal Medicine, until 1867. Author: Lehrbuch der Psychiatrie, die Pathologie und Therapie der psychiatrischen Krankheiten, 1845; Gesammelte Abhandlungen, 2 vols., 1872. Research on psychic reflex actions, psychic tonus equaling character or soul, reflex energy controlled by inhibition, typhus, rheumatism; described pseudohypertrophic muscular dystrophy (Duchenn-Griesinger disease, Erb's paralysis or dystrophy, 1865; 1st to describe infantile splenic anemia, 1866. Died Berlin, Oct. 26, 1868.

GRIESS, Johann Peter, chemist; b. Kirchhosbach, Hesse-Cassel, Germany, Sept. 6, 1829; s. Johann Heinrich and Catherine Elisabeth (Gliem) G.; ed. U. Jena, Germany, 1850-51, U. Marburg, Germany, 1851-58; Ph.D. (hon.) U. Munich, 1877; studied under Kolbe; m. Louisa Anna Mason, 1869; 2 sons, 2 daus. With Oehler chem. factory, Offenbach/Main, Germany, 1856-57; became asst. to Hofmann, Royal Coll. Chemistry, London, 1858; became dir. brewery firm Allsopp & Sons, Eng., 1862. Fellow Royal Soc., 1868. Research on azo and diazo compounds, decomposition of diazonium salts with the liberation of nitrogen and the formation of phenols; discovered alpha-naphthol, 1862; discovered action of nitrous acid on amines (a step in making certain dyes). Died Bournemouth, Eng., Aug. 30, 1888.

GRIFFEN, Allen Beattie, Am. cytogeneticist; b. Corsicana, Tex., Jan. 6, 1914; s. Allen William and Theodora (Schiermann) G.; A.B., U. Tex., 1935, Ph.D., 1939; m. Elizabeth Ann Lindsey Taylor, June 10, 1947—children—Julia Ann, Margaret Allen, Robert Dana. Student asst. zoology U. Tex., 1932-35, tutor, 1935-36, research asst., 1936-39, instr., research asso., 1939-41, asso. prof. zoology, 1941-45; asso. prof. U. Mo., 1945-47, prof., 1947-50; research asso. Jackson Lab., Bar Harbor, Me., 1950-57, staff scientist, 1957-67, sr. staff scientist, 1967——. Mem. Am. Soc. Zoologists, Genetics Soc. Am., Phi Beta Kappa, Sigma Xi. Research and numerous publs. on gene position effects, chromosome mapping, speciation in drosophila; germinal and tumor cell cytology, chromosome mapping, radiation cytogenetics in the mouse. Home: Barberry Mews, Barberry Lane, Bar Harbor, Me. 04609. Office: The Jackson Lab., Bar Harbor, Me. 04609.*

GRIFFIN, Amos Clark, Am. biochemist; b. Newton, Utah, May 16, 1917; s. Amos R. and Della (Petersen) G.; B.S., Utah State U., 1939; M.S., Mich. State U., 1941; student U. Cal., Berkeley, 1941-42; Ph.D., U. Wis., 1947; m. Cleo Lundstrom, Aug. 16, 1939; children—Douglas, Paul, Brent, Robert, Janis. With Stanford, 1948-55, prof. dept. chemistry, 1950-55; head dept. biochemistry U. Tex. M.D. Anderson Hosp. and Tumor Inst., 1955-62; chmn., prof. dept. biochemistry Baylor U. Coll. Medicine, 1956-61; Am. Cancer Soc. prof. biochemistry U. Tex. M. D. Anderson Hosp. and Tumor Inst., 1962——. Mem. Am. Assn. Cancer Research, Am. Soc. Biol. Chemists, Biochem. Soc. (London), Am. Chem. Soc. Research, publs. on etiology, behavior, biochem. changes, protein synthesis, induction of tumors; role of B-vitamins in liver tumor origin. Home: 3843 Tartan Lane, Houston 77025. Office: M. D. Anderson Hosp. and Tumor Inst., Dept. Biochemistry, Houston 77025.*

GRIFFIN, David Michael, botanist; b. Plymouth, Eng., Oct. 18, 1929; s. Stanley James and Ada (Allen) G.; B.A., Emmanuel Coll., Cambridge, Eng., 1953, M.A., 1956, Ph.D., 1957; m. Margaret Isobel Burford, Aug. 17, 1957; children—Stephen David, Michael Julian, Philip Mark, Anne Margaret. Research student Forestry Commn., Botany Sch., Cambridge U., 1953-55; reader plant pathology dept. agrl. botany U. Sydney (New South Wales, Australia), 1955——. Mem. Brit. Mycol. Soc., Assn. Applied Biologists, Am. Phytopathol. Soc., Mycol. Soc. Am., Linnean Soc. of New South Wales. Contbr. articles to profl. jours. Research on ecology of fungi in soil with reference to soil phys. factors as temperature, moisture, aeration. Home: 17 Denman St., Turramurra, New South Wales. 2074. Office: Dept. of Agrl. Botany, University of Sydney, Sydney, New South Wales 2006, Australia.*

GRIFFIN, Donald Redfield, Am. biologist; b. Southampton, N.Y., Aug. 3, 1915; s. Henry Ferrand and Mary (Redfield) G.; B.S., Harvard, 1938; M.A., 1940, Ph.D., 1942; D.Sc., Ripon Coll., 1966; m. Ruth Castle, Sept. 6, 1941 (div. Aug. 1965); children—Nancy Jean (Mrs. Rex Jackson) Janet Redfield (Mrs. Freeland Abbott), Margaret Louise, John Hadley; m. 2d, Jocelyn Crane, Dec. 16, 1965. Research asso. Psychoacoustic Lab., Fatigue Lab. and Biol. Labs., Harvard, 1942-45; faculty Cornell U., 1946-53, prof. zoology, 1952-53; prof. zoology Harvard, 1953-65, chmn. dept. biology, 1962-65; prof. Rockefeller U., N.Y.C., 1965——; sr. research zoologist N.Y. Zool. Soc., 1965——; dir. Inst. for Research in Animal Behavior, Rockefeller U.-N.Y. Zool. Soc., 1965——. Recipient Prize in Sci., Phi Beta Kappa, 1964. Mem. Am. Soc. Mammalogists, Am. Ornithologists Union, Am. Soc. Zoologists, Am. Physiol. Soc., Ecol. Soc. Am., Am. Acad. Arts and Scis., Nat. Acad. Scis. (Elliott medal 1962). Author: Listening in the Dark, 1958; Animal Structure and Function, 1962; Echoes of Bats and Men, 1959; Bird Migration, 1964; also articles. Research on animal behavior and comparative physiology, animal orientation particularly bird navigation and echolocation in bats. Home: 1 W. 67th St., N.Y.C. 10023.*

GRIFFIN, Harriet Madeline, Am. educator; b. Bklyn, Apr. 6, 1903; d. Harry and Madeline (Gully) G.; A.B., Hunter Coll., 1925; M.A., Columbia, 1929; Ph.D., N.Y.U., 1939. Tutor, instr., Hunter Coll., N.Y.C., 1926-30; instr. Bklyn. Coll., 1930-41; asst. prof., 1941-50, asso. prof., 1950-56, prof., 1956-66, emeritus, 1966——; adj. prof. Molloy Coll., 1966——. Fellow A.A.A.S.; mem. Am. Math. Soc., Math. Assn. Am., N.Y. Acad. Sci., Albertus Magnus Guild, Phi Beta Kappa, Pi Mu Epsilon, Sigma Xi, Key Pin Soc. of N.Y. U. Author: The Concepts of Theory of Numbers, 1949; Elementary Theory of Numbers, 1954. Research in abstract algebra and theory of numbers. Home: 3609 Farragut Rd., Bklyn. 11210.*

GRIFFIN, James Bennett, Am. anthropologist; b. Atchison, Kan., Jan. 12, 1905; s. Charles Bennett and Maude (Bostwick) G.; Ph.B., U. Chgo., 1927, A.M., 1930; Ph.D., U. Mich., m. Ruby Fletcher, Feb. 14, 1936; children—John B., David M., James C. Research asso. U. Mich. Mus. of Anthropology, 1936-41, asst. curator archaeology, 1937-42, asso. curator, 1942-45, asso. prof. anthropology, 1945-49, prof., 1949——, curator, 1945——, dir. Museum Anthropology, 1946——. Recipient Viking Fund medal, award, 1957. Fellow A.A.A.S. (v.p. 1953-54), Soc. Am. Archaeology; mem. Pan-Am. Inst. Geog. and History (pres. com. on anthropology 1954-59), Soc. Am. Archaeology (pres. 1951-52), Internat. Union Prehistoric and Protohistoric Scis. (permanent council 1950——, exec. com. 1962——), Am. Anthropol. Assn., Sociedad Mexicana de Antropologia, Nat. Acad. Sci. Author The Fort Ancient Aspect, 1943. Editor, co-author Archeology of Eastern United States, 1952. Research and publs. on prehistoric ceramic cultures east of the Rocky Mountains, helping to order relationships between units and aligning them in correct sequence; studies, publs. on prehistoric connections of Eastern U. S. with Middle Am. and N.E. Asia and on the correlation of cultural and ecol. changes in the northeast and on possible effects of minor climatic changes on prehistoric cultures. Office: 4017 Museums, Ann Arbor, Mich. 48104.*

GRIFFIN, Martin Luther, Am. cons. chem. engr.; b. Northampton, Mass., May 21, 1859; s. John and Naomi (Estabrook) G.; A.B., Amherst Coll., 1883, A.M., 1886; m. Ada Riggs, Mar. 28, 1894; children—Arthcher Estabrook, Carroll Riggs. Expert in pulp and paper-making processes and raw materials therefor, also in textile finishing of cotton, silk and woolen fabrics, combustion of fuels, destructive distillation, deportment of gases, prcesses of evaporation and drying, the electrolytic cell for production of caustic soda and chlorine and application of their products to industry, treatment of water and trades wastes in industry. Mem. various chem. socs. and author of tech. papers. Died Aug. 28, 1942.

GRIFFITH, Frederick Reece, Jr., Am. physiologist; b. St. Louis, June 30, 1891; s. Fred R. and Elizabeth Hannah (Welch) G.; A.B., Washington U., St. Louis, 1914, M.A., 1915; Ph.D., Harvard, 1923; m. Varina Davis O'Hara, June 15, 1921; children—Anne Elizabeth (Mrs. Lincoln C. Aldridge), James O'Hara, Davis Walker. Asst. prof. U. Miss., 1915-16, So. Meth. U., 1916-19; instr. Harvard, 1919-23; faculty U. Buffalo, 1923-56, prof., head dept. physiology, 1935-56; prof. biology D'Youville Coll., Buffalo, 1960——. Mem. Am. Physiol. Soc., Am. Inst. Nutrition, Soc. Exptl. Biology and Medicine, A.A.A.S., Phi Beta Kappa, Sigma Xi. Author: Adrenaline, Adrenergic-Sympathomimetic, and Adrenolytic-Sympatholoytic Drugs, 1956; also numerous articles. Research on mechanisms calorigenic action adrenaline, variability in human physiol. functions. Home: 48 Eagle St., Williamsville, N.Y. 14221. Office: Dept. biology D'Youville Coll., Buffalo 14201.

GRIFFITH, Wendell Horace, Am. biochemist; b. Churdan, Ia., Nov. 7, 1895; s. George William and May (Fowler) G.; B.S., Greenville (Ill.) Coll., 1917; M.S., U. Ill., 1919, Ph.D., 1923; m. Harriet Isabel Leas, Aug. 31, 1922; 1 son, Wendell Horace. Chmn., prof. dept. biol. chemistry U. Cal., Los Angeles, 1951-63, prof. emeritus, 1963——; dir. Life Scis. Research Office Fedn. Am. Socs. Exptl. Biology, Bethesda, Md., 1962——; nutrition adviser in India FAO, 1959-60; mem. food and nutrition bd. NRC, 1950-62; com. chmn., cons. Office Surgeon Gen. USPHS; 1st dir., Life Scis. Research Office, Am. Soc. Exptl. Biol.; mem. sci. adv. com. Nutrition Found. Fellow Am. Pub. Health Assn., A.A.A.S., N.Y. Acad. Scis.; mem. Am. Soc. Biol. Chemists, Am. Chem. Soc., Inst. Food Technologists, Soc. Exptl. Biology and Medicine, Sigma Xi. Contbr. numerous articles to profl. jours. Research, publs. on mechanisms of biochemical detoxification of choline in maintenance of renal and hepatic tissues, amino acids in animals and man, dietary lipids; med. and pub. health edn., mil. feeding, nutritional recovery of malnourished prisoners of war and relation of nutrition and other environmental factors to human performance. Died Feb. 55, 1968.

GRIFFITH, William, Brit. botanist; b. Ham Common, Eng., Mar. 4, 1810; s. Thomas Griffith; ed. medicine; grad. U. Coll., London; m. Miss Henderson, Sept. 1844; Mem. various expdns., many in service of E. India Co.; apptd. to Malacca on med. duty; dir. Bot. Garden., Calcutta, India; named prof. botany Med. Coll., Calcutta, 1842. Contbr. articles to tech. jours. Drew up a gen. flora of India; extensive collection of bot. specimens. Died Feb. 9, 1845.

GRIFFITHS, Ernest Howard, Brit. physicist; b. Brecon, Eng., June 15, 1851; s. Henry and Mary (Blake) G.; ed. Owens Coll., Manchester, Eng., Sydney Sussex Coll., Cambridge (Eng.) U.; hon. degrees from Aberdeen, Scotland, Manchester, Liverpool, Eng.; m. Elizabeth Martha Clark, 1877. Became pvt. tutor, univ. coach, 1873; named fellow Sydney Sussex Coll., 1897, named hon. fellow, 1904; became prin. U. Coll. of S. Wales and Monmouthshire, Cardiff, 1901; became fellow at Jesus Coll., Oxford (Eng.) U., 1905, 09, 13, 17; ret., 1918; named gen. treas. Brit. Assn., 1920; ret., 1928. Mem. exec. com. Nat. Phys. Lab. Recipient Hughes medal. Fellow Royal Soc., 1895; mem. Brit. Assn. (became pres. of phys. sect. 1906, of ednl. sci. sec., 1913). Developed elec. method to determine mech. equivalent heat; made platinum resistance thermometers; (with H. L. Callendar) redetermined the boiling point of sulphur, improved resistance boxes for precision measurements; research on latent heat for evaporation of benzene and water; (with Ezer Griffiths) studied thermal capacities of metals from liquid air temperatures up to 100° C. and compared results with quantum theories. Died Cambridge, Mar. 3, 1932.

GRIFFITHS, Ezer, English physicist, b. Wales, Eng., Nov. 28, 1888; s. Abraham Lincoln and Anne G.; ed. U. Coll., Cardiff, 1908-14. Fellow, U. Wales; with Nat. Phys. Lab., from 1914, apptd. principal officer, 1948. Pres. com. on insulating materials Internat. Inst. Refrigeration, also 2d com. on phys.-tech. problems of idsl. refrigeration. Recipient Moulton medal Instn. Chem. Engrs. Fellow Royal Soc., 1926, Phys. Soc. (v.p.); mem. Brit. Assn. (recorder section A), Inst. of Refrigeration (pres., v.p.). Author: Methods of Measuring Temperature; Pyrometry; also numerous papers, reports on thermal measurement, refrigeration and insulation. Died Feb. 14, 1962.

GRIFFITHS, George Motley, Canadian physicist; b. Thorold, Ont., Can., Dec. 12, 1923; s. George Ewart and Valeska (Motley) G.; B.A., U. Toronto, 1949; M.A., U. B.C., 1950, Ph.D., 1953; m. Joyce Margaret Craig, Sept. 18, 1948; children—Robert A., Lloyd O., Donald I., David O., Brenda Lee. Rutherford Meml. fellow Cavendish Lab., Cambridge, Eng., 1953-55; faculty U. B.C., Vancouver, 1955——, prof. physics, 1963——; sr. research fellow Cal. Inst. Tech., 1962-63. Mem. Canadian Assn. Physicists, Am. Phys. Soc., Inst. Physics, Phys. Soc. (London). Research in low energy nuclear physics with particular emphasis on elecctromagnetic interactions in nuclei and their application to processes involving energy generation and element formation in the stars. Home: 4645 Langara Av., Vancouver 8, B.C., Can.*

GRIFFITHS, James Edward, Am. chemist; b. Ft. Frances, Ont., Can., June 1, 1931; s. Edward and Marie (Grouette) G.; B.Sc. with honors U. Man., 1955, M.Sc., 1956; Ph.D., McGill U., 1959; m. Mary C. Lytwynka, Sept. 2, 1954; 1 son, James A. Research asso. U. So. Cal., 1958-60; mem. tech. staff Bell Telephone Labs., Inc., Murray Hill, N.J., 1960——. Mem. Am. Chem. Soc., The Chem. Soc. (London). Contbr. articles in field to sci. jours. Research on molecular structures of inorganic materials using infrared, Raman, nuclear magnetic resonance and microwave spectroscopy; photoelectric detection techniques, intermolecular interactions in the liquid and gas states and the chemistry of volatile compounds of the Group IV, V and VI elements. Home: 77 Hansell Rd. Office; Bell Telephone Labs., Murray Hill, N.J. 07971.*

GRIFFITHS, John Willis, Am. naval architect; b. N.Y.C., Oct. 6, 1809; s. John Griffiths. Wrote series of articles on naval architecture in Portsmouth Adv., 1836; delivered 1st formal lecture on naval architecture, N.Y.C.; editor Am. Ship, 1878-82; one of 1st to specialize in designing; designed Rainbow (1st extreme clipper ship), Sea Witch; developed improved form of rivet; invented machine for bending timber into crooked forms used in shipbldg.; designed New Era (1st ship with mechanically bent timber), 1870. Author: Treatise on Marine and Naval Architecture, 1850; Ship Builder's Manual, 1853; Progressive Ship-Builder, 1875. Died Bklyn., Mar. 30, 1882.

GRIFFITHS, Robert, inventor; b. Vale of Clwydd, Dec. 13, 1805; carpentry apprentice; later pattern maker engine works, Birmingham, Eng. Patentee improved method screw propulsion (adopted by navy), 1853, with separate blades and less vibration, 1858, automatic electric hairbrush, (with M. C. W. Copeland) damper for steam boilers and method preventing scale in boilers, (with Mr. Bouill) improvement in atmospheric ry. (used vacuum on one side and plenum on other to act on piston, also closing of atmospheric pipe), (with Samuel Evers) machine improving manufacture hexagon nuts, 1837; improved machinery for making bolts, ry. spikes, rivets; invented rivet machine, 1835, (with John Gold) glass-grinding and polishing machine, 1836. Died June 1883.

GRIFFITTS, James John, Am. physician; b. Springfield, Ill., Dec. 13, 1912; s. Thomas H. D. and Elizabeth (Glynn) G.; B.S., U. Va., 1933, M.D., 1937; m. Leola Horton, June 13, 1940; children—Susan, Dickson, Sally Lee, Sharon Lynn, Shelly. Commd. asst. surgeon USPHS, 1939, advanced through ranks to sr. surgeon, 1947; with NIH, Bethesda, Md., 1940-49; asso. dir. Blood bank Dade County, Fla., 1949-65; pres. Dade Reagents, Inc., Miami, Fla., 1954—. Dir. Am. Hosp. Supply Corp., Evanston, Ill. Recipient John Elliott award in blood banking Am. Assn. Blood Banks, 1963. Mem. A.M.A., Am. Soc. Clin. Pathologists, Alpha Omega Alpha. Contbr. numerous articles to tech. jours. Research in immunology, infectious diseases, human nutrition, blood transfusion. Home: 1213 N.E. 94th St., Miami Shores, Fla. 33138. Office: 1851 Delaware Pkwy., Miami, Fla. 33152.*

GRIGG, Charles Meade, Am. sociologist, coll. dean; b. Richmond, Va., Nov. 1, 1918; s. Joseph W. and Nellie A. (Chockley) G.; B.S., Richmond Profl. Inst., Coll. William and Mary, 1947; M.A., U. N.C., 1950, Ph.D., 1952; m. Virginia E. Caffee, Aug. 23, 1947; children—Charles M., John W., Joseph G., Ruth E. Faculty, Brown U., 1952-55; faculty Fla. State U., Tallahassee, 1955—, prof. sociology, dir. Inst. Social Research, 1960—, asso. dean Coll. Arts and Scis., 1966—. Mem. com. on research and edn. Fla. Tb and Health Assn., 1960—; mem. adv. panel co-op. research and demonstration grants program Welfare and Social Security Adminstrn., 1966—; mem. Fla. Com. on Law Enforcement and Adminstrn. Justice, 1966—. Fellow Am. Sociol. Assn.; mem. So. Sociol. Soc. (exec. com. 1962—), Population Assn. Am. Author: (with L. M. Killian) Racial Crisis in America, 1964; Recruitment to Graduate Study, 1965; Graduate Education, 1965. Editor: (with C. N. Millican) The Setting for Higher Education in Florida, 1965; (with A. M. Hartsfield and M. A. Griffin) The Impact of NASA on Brevard County, 1966; (with K. S. Miller) Mental Health and the Lower Social Classes, 1966. Research contbns. to understanding of process of conflict, particularly as it relates to community orgn.; adaptation of different socioeconomic groups to urban environment. Home: 1559 Cristobal Dr., Tallahassee, Fla. 32303. Fla. State U., Tallahassee 32305.*

GRIGGS, David Tressel, Am. geophysicist; b. Columbus, O., Oct. 6, 1911; s. Robert Fiske and Laura (Tressel) G.; B.A., Ohio State, 1932, M.A., 1933; postgrad Harvard, 1934-41; m. Helen Irene Avery, May 4, 1946; children—Nicola, Stephen Fiske. Research asso. Radiation Lab., Mass. Inst. Tech, Cambridge, 1941-42; expert cons. office sec. war, Washington, 1942-46; chief nuclear energy sect. project RAND, Santa Monica, Cal., 1946-48; prof. geophysics Inst. Geophysics and Planetary Physics, U. Cal. at Los Angeles, 1948—; chief sci. USAF, 1951-52; dir. FMA, Inc., Inglewood, Cal., 1959-67. Decorated Purple Heart. Recipient Presidential medal of Merit, 1946, award for exceptional civilian service USAF, 1952. Mem. Nat. Acad. Scis., Am. Geophys. Union (pres. sect. of tectonophysics, 1964-67). Research, publs. on earthquake analysis. Home: 190 Granville Av., Los Angeles, 90049.

GRIGGS, William, Brit. inventor; b. Bedfordshire, Eng., Oct. 4, 1832; m. Elizabeth Jane Gill, 1851 (dec. 1903); 2 sons. Tech. asst. to dir. Indian Mus., until 1855; photo-lithographer India Office, 1855-1885; set up photo-lithographic works, Peckham, Eng., 1868. Pioneer in diffusion of color and half-tone block making; produced plates for Forbes Watson Textile Manufactures, India, 1866; invented photo-chrome lithography, circa 1868; reproduced facsimile editions of old manuscripts, including 43 vols. of Shakespeare, 1881-91. Died Worthing, Eng., Dec. 7, 1911.

GRIGNARD, (François Auguste) Victor, French chemist; b. Cherbourg, France, May 6, 1871; s. Theophile Henri and Marie (Herbert) G.; Ph.D., U. Lyons (France), 1901; student of Philip Barbier; became asst. chemistry dept., Lyons, 1900, asso. prof., 1908, prof. from 1919; named lectr., Besancon, France, 1905, Nancy, France, 1906; prof., Nancy, from 1910. Recipient (with Paul Sabatier) Nobel prize in chemistry, 1912. Author: Thèses sur les combinaisons organomagnésiennes mixtes et leurs applications à des synthèses, 1901; Traité de chimie organique, 15 vols., 1935. Discovered organomagnesium compounds used for organic synthesis (Grignard's reagent), 1900. Died Lyons, Dec. 13, 1935.

GRIGNASCHI, Victor José, Argentine physician; b. Buenos Aires, Argentine, Feb. 3, 1917; s. Victor and Teresa M.N. (Sardá) G.; Physician, U. Buenos Aires, 1950, M.D., 1953, honor certificate, 1955; m. María Zulema Palvecino, Jan. 9, 1943; 1 dau., Maria Zulema. Titular prof. hematology Salvador U. Med. Sch., 1964—, chief hematologic div., 1957—; dir. lab. Nat. Direction Scholar Sanity, 1962—; physician Hosp. Rivadavia, Buenos Aires, 1953—; ofcl. investigator mongolism Nat. Exptl. Inst. Mongolism, 1963—; docent collaborator Ofcl. U. Buenos Aires, 1960—. Ministry Health fellow to conduct studies on tropical anemias at Bolivian Amazonic basic, 1961, 62; Fundaleu fellow, 1962; recipient First award (internat.) for best work on cancerology, 1963, 1st Nat. award in medicine for book, 1963. Fellow Argentine Med. Assn.; mem. Argentine (directive commn.), Internat. socs. hematology. Author: (with M. A. Etcheverry) Leukaemia and Cancer as Molecular Disease, 1963; also articles. Discovered spontaneous mutation in leukemic cells by alterations of nucleic acids, 1961; originated (with M. A. Etcheverry) theory of molecular origin of leukemia and cancer; confirmed theory of enzymatic deletion in neoplastic cells (Potter, Haddow); studied alterations of cytochemistry in neoplastic cells. Home: 315. Avda. Parral. Office: Hospital Rivadavia, Bustamente y Las Heras, Buenos Aires, Argentina.*

GRIGNON, Pierre-Clément, French metallurgist; b. Saint-Dizier, France, Aug. 24, 1723; dir. Bayard Forge; mem. French Acad. Scis., 1768. Author: Sur l'art de fabriquer le fer, 1775; publs. on analysis of iron ore. Research on iron prodn.; built cannons and arty. Died Bourbonne, France, Aug. 1784.

GRIGOLYUK, Eduard Ivanovich, Russian mechanics specialist; b. 1923; grad. Moscow Aviation Inst., 1944. Instr., Moscow Aviation Inst., 1944-46, Moscow Higher Tech. Sch., 1946-50; with Exptl. Design Bur., 1948—; asso. Inst. Mechanics, USSR Acad. Scis., 1953—. Mem. USSR Acad. Scis. (corr.). Editor: Mechanics, 1952—. Research and publs. on theory of shells, theory of elasticity and plasticity. Address: Inst. Mechanics, USSR Acad. Sci., Leningradsky prospect 7, Moscow, USSR.*

GRIGORIEFF, Wladimir W., chemist; b. Hankow, China, July 11, 1908; s. Wladimir J. and Olga (Anchougova) G.; Chem.E., Tech. Sch., Zurich, Switzerland, 1932; Ph.D., U. Chgo., 1939; m. Lucile W. Pfaender, July 30, 1935 (div. 1952); 1 son, Paul; m. 2d, Lilian Armstrong, Feb. 14, 1953. Came to U. S. 1933, naturalized, 1939. Chief chemist Danziger Refineries, Inc., Pampa, Tex., 1934-35; rsrch. chemist Gen. Electric Co., Pittsfield, Mass., 1939-45; dir. Ordark, U. Ark., 1946-53, inst. sci. and tech., 1947-53, council mem. Oak Ridge Inst. Nuclear Studies, 1949-53, chmn. univ. relations div. of Inst., 1952-64, asst. to exec. dir. spl. projects, 1964—; sr. officer Internat. Atomic Energy Agy., Vienna, 1958-59. Fellow A.A.A.S.; mem. Tenn. Acad. Sci., Am. Chem. Soc., Am. Soc. Engring. Edn. (chmn. com. internat. engring. edn.), Am. Assn. U. Profs., World Congress on Engring. Edn. (program chmn. 1965), Am. Nuclear Soc., Inc. (exec. sec., 1955-58), Sigma Xi. Contbr. articles profl. jours. Research on pyrotechnics; glass compositions of low-coefficient of thermal expansion; synthetic electrical insulation; glass to metal seals. Home: 104 Ogden Circle, Oak Ridge.

GRIGORIEV, Andrei Aleksandrovich, Russian geographer; b. Nov. 1, 1883; student Heidelberg, Berlin univs. (both Germany); grad. in natural scis. St. Petersburg (now Leningrad) U. Lectr. geography High Pedagogical Coll., St. Petersburg, 1914-19; prof. Geog. Inst., Leningrad, 1918-36; dir. Inst. Geomorphology, Leningrad, 1931-34, Inst. Geography, Moscow, 1934-50; chief Dept. History of Geography, Inst. Geography, 1950—. Mem. USSR Acad. Sci., Geog. Soc. USSR. Author: Analytical Study of the Composition and Structure of the Earth's Physico-Geographical Crust, 1937; Soviet Geography up to the 18th Congress of the All-Union CP, 1939; Soviet Geography During the Second Five-Year Plan, 1939; The Characteristics of the Basic Types of Physico-Geographical Environment, 1946; Some Problems of Physical Geography, 1951; Subarctic, 1956; Development of the Theoretical Problems of Soviet Physical Geography, 1965; Regularities of Composition and Development of Geographical Environment, 1966. Mem. editorial bd. Gt. Soviet Ency. Made 2 expdns. to Bolshezemelskaya tundra, 1904, 21, also expdns. to So. Urals, 1923, Kola Penninsula, 1928-29, 31; introduced dynamic geography (would convert study from being merely descriptive to establishing gen. laws of physico-geog. processes. Office: Inst. Geography, Acad. Sciences USSR, Staromonetny pereulok 29, Moscow, USSR.*

GRILLY, Edward Rogers, Am. physicist; b. Cleve., Dec. 30, 1917; s. Charles B. and Julia (Varady) G.; B.A., Ohio State U., 1940, Ph.D., 1944; m. Mary Hedwig Witholter, Dec. 14, 1942; children—David, Janice. Chemist Carbide and Carbon Chems. Corp., Oak Ridge, 1944-45; asst. prof. chemistry U. N.H., 1946-47; mem. staff Los Alamos Sci. Lab., U. Cal., 1947—. Mem. Am. Phys. Soc. Contbr. articles to profl. jours., also chpts. to books. Research on transport properties of gases, thermodynamic properties of liquid hydrogen, liquid and solid helium, melting properties of permanent gases. Home: 1467 42d St. Office: Los Alamos Sci. Lab., P.O. Box 1663, Los Alamos 87544.*

GRIM, Ralph Early, Am. geologist, mineralogist; b. Reading, Pa., Feb. 25, 1902; s. Harry W. and Etta C. (Early) G.; Ph.B., Yale, 1924; Ph.D., State U. Ia., 1931; m. Frances E. Reed, Jan. 9, 1964. Asst. state geologist State of Miss., 1926-30; asst. prof. U. Miss., 1926-30; geologist Ill. State Geol. Survey, 1931-46, prin. geologist, 1946-50; research prof. geology U. Ill., Urbana, 1950—. Chmn., exec. com. Internat. Com. for Study Clays, 1950-60; chmn. clay minerals com. NRC, 1950-60. Fellow Acad. Sci. India, Geol. Soc. Am., Mineral Soc. Am., Am. Ceramic Soc.; mem. Ceramic Soc. France, Ceramic Soc. Brazil, Brit. Ceramic Soc., Soc. Econ. Geology. Author: Clay Mineralogy, 1953; Applied Clay Mineralogy, 1962; also numerous articles on composition, properties, occurrence clay materials. Home: 704 W. Florida Av., Urbana, Ill. 61801.*

GRIMALDI, Albert-Honoré-Charles, see Albert I.

GRIMALDI, Francesco Maria, Italian physicist, astronomer; b. Bologna, Italy, Apr. 2, 1618; joined Soc. of Jesus, 1632; ordained, 1638; prof. U. Bologna. Author: Physico-mathesis de lumine, 1665. Discovered the diffraction of light; research on interference and the dispersion of the sun's rays through a prism, moon's surface; one of 1st to formulate a wave theory of light (led to Newton's research on optics); studied and named dark areas of moon; made lunar map, 1650. Died Bologna, Italy, Dec. 28, 1663.

GRIMAUX, Louis Édouard, French chemist; b. Rochefort-sur-Mer, France, July 3, 1835; M.D., Faculty Medicine, Paris, 1865; grad. in chemistry, 1866; Tchr. chemistry Faculty of Medicine, 1869, 71, 73; became tchr. École polytechnique, Paris, 1876; named prof. gen. chemistry Institut agronomique, 1876; prof. chemistry École polytechnique, 1881-98. Recipient Jecker prize Acad. Scis. Paris, 1870, 75. Mem. French Acad. Scis., 1894. Author: Chemie organique élémentaire, 1872; Chemie inorganique élémentaire, 1874; Theories et notations chimiques, 1884; Lavoisier, 1888; wrote biography of Gerhardt, 1900. Research on nitriles, aromatic glycols, 1870, acid ureides, 1871, allantoin, 1876, synthesis of citric acid, 1881; prepared quinine and its salts, also other alkaloids, 1892; prepared codeine from morphine and showed it was a methyl ether of morphine. Died Paris, May 2, 1900.

GRIMES, James Stanley, Am. natural philosopher; b. Boston, May 10, 1807; probably son of Andrew (or Joseph) and Polly (or Sally) (Robbins) Grimes. Practiced law, Boston and N.Y.C.; prof. med. jurisprudence Castleton Med. Coll., also Willard Inst.; one of 1st Am. evolutionists; became interested in phrenology, 1832; in mesmerism; published Phreno-Geology (which he claimed to be 1st essay on theistic evolution), 1851; took part in series of 8 debates with Leo Miller, Boston, 1860; moved to Evanston, Ill. Author: A New System of Phrenology, 1839; Etherology, 1845; The Mysteries of Human Explained, 1857; Phreno-Physiology, 1893. Died Evanston, Sept. 27, 1903.

GRIMINGER, Paul, nutritionist; b. Vienna, Austria, Aug. 29, 1920; s. Sigmund and Frieda (Klein) G.; student Agrl. Coll., Vienna, 1950-51; B.S., U. Ill., 1952, M.S., 1953, Ph.D., 1955; m. Olga Maria Egger, Aug. 20, 1954; children—Jeannette Nurith, Elaine Kathrine, Karen Anne, Andrew Bert. Research asst. U. Ill., 1952-53, research fellow, 1953-55; asst. prof. U. Neb., 1955-57; faculty Rutgers U., 1957—, prof., 1966—. Guggenheim Found. fellow, 1964. Mem. Am. Inst. Nutrition, Poultry Sci. Assn., Soc. Exptl. Biology and Medicine, A.A.A.S., N.Y. Acad. Scis., Am. Nutrition Research Council. Contbr. numerous articles to sci. jours. Investigation of nutritional factors in atherosclerosis, in particular of blood cholesterol, lowering agts. such as pectin; determination of calcium needs of laying hens and investigation of various other nutritional requirements of poultry; investigation of interaction of Vitamin K and anticoagulants, and certain phases of Vitamin K metabolism; research on passage of vitamins from mother to offspring. Home: 325 N. 4th Av., Highland Park, N.J. 08904. Office: Thompson Hall, Rutgers U., New Brunswick, N.J. 08903.*

GRIMM, Carl Albert, Am. mathematician; b. Cin., Apr. 1, 1926; s. Carl Hugo and Alberta (Kumler) G.; B.A., U. Cin., 1950, M.A., 1952; m. Jeanne Irma Blase, Sept. 10, 1948; children—Eric Christopher, David Conrad, Jeffrey Hamilton. Instr., South Dakota Sch. Mines and Tech., Rapid City, 1952-55, asst.

prof., 1955-58, asso. prof., 1958-65, prof., 1965—. Tchr. for high sch. math. tchrs., NSF, 1958—. Mem. Math. Assn. Am., Am. Math. Soc., Phi Beta Kappa, Delta Phi Alpha, Sigma Xi. Presbyn. (deacon). Contbr. articles to tech. jours. Home: 2709 Evergreen Dr., Rapid City, S.D., 57701.*

GRIMM, Hans August Georg, German phys. chemist; b. Hamburg, Germany, Oct. 20, 1887; s. Johannes and Rosette (van Dieman) G.; studied food chemistry, U. Munich (Germany); Ph.D., 1911; studied phys. chemistry, beginning in 1918; m. Hedwig Weidinger; 1 son. On western and so. fronts ending as capt. in Alpine Corps Staff, 1914-18; joined faculty U. Munich (Germany), 1923; became asso. prof. phys. chemistry U. Würzburg, 1924, prof., 1927; apptd. dir. research lab. firm Badische Anilin-und Soda-Fabrik, Oppau; also hon. prof. Würzburg, Germany; ret. to Diessen, Germany, and concentrated on philos. problems, 1938; named hon. prof. U. Munich, 1949. Mem. Göttingen Acad. Scis. Research and publs. on geometric similarities in crystals; increased number of hard, diamond-like compounds; developed Grimm's hydride displacement theorem on systematic properties of inorganic hydrogen compounds; used (with others) X-rays to measure electron distbn. in compounds which demonstrated various types of bonds. Died Gauting nr. Munich, Oct. 25, 1958.

GRIMMER, Heinz Helmut, physician; b. Duala, Cameroon, Dec. 12, 1913; s. Alfred and Johanna (Bittmann) G.; doctorate, U. Berlin (Germany), 1940; m. Gisela Smuda, Dec. 24, 1947; children—Klaus, Joachim. Faculty skin and venereal diseases U. Berlin, 1951-62, prof., 1957-62; staff skin clinic Free U. Berlin, 1946-62, head physician 1949-62; head physician skin clinic Municipal Hosps., Wiesbaden, Germany, 1962—; prof. U. Heidelberg (Germany), 1965—. Mem. Polnische Dermatologische Gesellschaft (hon.). Author: (with H. Rieth) Krankheiten durch Schimmelpilze bei Mensch und Tier, 1965; also numerous articles. Research on mycology, eczema, histology, andrology. Home: 82 Bierstadterstrasse. Office: 81 Schwalbacherstrasse, Wiesbaden, Germany.*

GRIMMER, Walter Eugen, German dairy chemist; b. Böhlen nr. Leipzig, Germany, June 28, 1878; studied chemistry Tech. U., Dresden, Germany, U. Göttingen (Germany); Ph.D., 1904; m. Margarete Weise; 2 sons (killed in war), 1 dau. Mem. staff physiol.-chem. inst. Dresden Sch. Vet. Medicine, returned and joined faculty, 1912; joined Dairy Inst. Greifswald, Germany, 1907; named dir. research and teaching inst. dairying Agrl. Chamber E. Prussia, 1916; also joined faculty U. Königsberg, named prof., 1924. Author: Chemie und Physiologie der Milch, 1910; Rationelle Milchwirtschaft, 1913; Leitfaden der Milchhygiene, 1922; Milchwirtschaftliches Praktikum, 1926. Editor: (with H. Weigmann, W. Winkler) Handbuch der Milchwirtschaft, 1930-36. Editor jour. Milchwirtschaftliche Forschungen. Research on dairy chemistry including milk enzymes, ripening of cheeses, chem. reactions, biochemistry of microscopic life, mycology of cheese, decomposition of protein, isolation of decomposition products. Died Königsberg, Sept. 24, 1944.

GRIMPE, (Johann) Georg, German zoologist; b. Leipzig, Germany, Feb. 16, 1889; s. Georg and Hedwig (Werner) G.; studied natural scis., especially zoology, comparative anatomy U. Leipzig; Ph.D., 1912; m. Elisabeth Bückmann, 1916. Asst., Leipzig Zool. Garden, until 1920; became asst. Zool. Inst. of U., 1915, lectr., 1922, asso. prof., 1928; worked at marine biol. insts. in Naples, Italy, Ville Franche-sur-Mer, Helgoland, Monaco. Editor: Tierwelf der Nord- und Ostsee. Editor internat. jour. Der Zoologische Garten, 1927. Research and publs. on marine animals, especially cephalopods; prepared collections of zool. gardens. Died Leipzig, Jan. 22, 1936.

GRIMSON, Keith Sanford, Am. physician; b. Munich, N.D., Apr. 21, 1910; s. Gudmundur and Ina (Sanford) G.; B.A., U. N.D., 1930, B.S., 1931; M.D., U. Chgo., 1934; m. Mozelle A. Johnson, Oct. 16, 1934; children—Roger Connell, Baird Sanford, Keith Sanford. Instr. dept. surgery U. Chgo., 1940-42; Belgian Am. Ednl. Found. Research fellow with C. Heymans, Ghent, Belgium, 1939-40; asst. prof. surgery Duke U. Sch. Medicine, 1943-48, asso. prof., 1948, prof., 1949—; cons. A.M.A., U. S. Dept. Health, Edn. and Welfare, Am. Heart Assn., surg. and med. socs.; past chmn. cardiovascular study sect. USPHS. Recipient distinguished service medals U. N.D., 1958, U. Chgo., 1956; Modern Medicine award, 1958. Diplomate Am. Bd. Surgery. Fellow A.C.S.; mem. Am., Soc. surg. assns., Am. Physiol. Soc., Am., So. socs. clin. research, Am. Heart Assn. (chmn. council for high blood pressure research 1959), Am. Soc. Pharmacology and Exptl. Therapeutics, Soc. U. Surgeons, Internat. Soc. for Angiology (past v.p.), Internat. Cardiovascular Soc. (v.p. 1962), A.A.A.S., Soc. for Exptl. Biology and Medicine, A.M.A. Sigma Xi, Alpha Kappa Kappa. Mem. bd. editors Modern Medicine, 1957—, Am. Surgeon, 1959—. Research in fields of pharmacology, physiology, medicine and surgery, including initial devel. of 8 new drugs for peptic ulcer or hypertension; establishment of principles concerning reflex control of blood pressure, respiration and heart rate; mgmt. of constipation, peripheral vascular disorders, high blood pressure and duodenal ulcer; devel. of operations for pancreatitis, obstipation, hypertension, arterial diseases and gastric or duodenal ulcer. Home: 3313 Devon Rd., Hope Valley, Durham, N.C.*

GRINBERG, Aleksandr Abramovich, Russian inorganic chemist; b. Apr. 20, 1898; grad. Leningrad U., 1924. Prof., Leningrad Tech. Inst., 1936—. Recipient Stalin prize, 1946. Mem. USSR Acad. Sci. Author: Introduction to the Chemistry of Complex Compounds, 1951. Research and publs. on chemistry of complex compounds, structure of platinum salts, isomerism of platinum and palladium divalent derivatives, acid-base and redox properties of complex compounds, balance in their aqueous solutions, use of tracer atoms in chemistry of complex compounds. Address: Leningrad Tech. Inst., Zagorodny pr. 49, Leningrad, USSR.

GRINBERG, Georgi Abramovich, Russian physicist; b. Leningrad, June 16, 1900; grad. Leningrad Poly. Inst., 1923; D. Physico-Math. Sci., 1935. With Leningrad. Inst. Roentgenology and Radiology, later Leningrad Physicotech. Inst., 1919-30; instr. Leningrad Polytech. Inst., 1924-30, prof., 1930-55; with Leningrad Svetlana Plant, 1929-41; asso. Leningrad Electro-Phys. Inst., 1930-36, Physicotech. Inst., USSR Acad. Sci., Leningrad, 1941—. Recipient Stalin prize, 1949. Mem. USSR Acad. Sci. (corr.). Author: Selected Problems on the Mathematical Theory of Electric and Magnetic Phenomena, 1948. Research and publs. on theoretical electronics, theory for focusing effect of electric and magnetic fields, theory of electromagnetic wave propagation and math. physics. Address: Physicotech. Inst., USSR Acad. Sci., Politekhnicheskaya ulitsa, Leningrad, USSR.

GRINBERG, Raul, physician, endocrinologist, encologist; b. Buenos Aires, Argentina, Aug. 15, 1922; s. David and Anna (Tabachnicoff) G.; M.D., Faculty Medicine, U. Buenos Aires, 1946, Ph.D., 1952; m. Claire C. Rabow, Aug. 16, 1965; children—George, Ricardo, Stephen, Diego, Andrew. Came to U. S., 1958, naturalized, 1967. Asso., Roffo Inst. Exptl. Medicine, Buenos Aires, 1948-50; teaching staff Rivadavia Hosp., Buenos Aires, 1952-58; asso. research medicine Columbia, 1958-61; sr. internist Roswell Park Meml. Inst., Buffalo, 1961-65; practice medicine specializing in endocrinology, oncology, Buenos Aires, 1952-58, Binghamton, N.Y., 1965—; vis. prof. endocrinology Cornell U., Ithaca, N.Y., 1966. Inst. Exptl. Medicine fellow, Santiago, Chile, 1948; Traveler grantee Francis Delafield Hosp., N.Y.C., 1958. Fellow Am. Geriatrics Assn., A.M.A.; mem. Endocrine Soc., Am. Assn. Cancer Research, A.A.A.S., N.Y. Acad. Scis., Buffalo Acad. Medicine. Research, publs. on combined chemotherapy of solid tumors; devel. new concepts in relationship between thyrotropic hormone and lymphomas. Home: 537 Clark St., Waverly, N.Y. 14892. Office: Morrison Hall, Cornell U., Ithaca, N.Y.; also 86 Walnut St., Binghamton, N.Y.*

GRINDELL-MATTHEWS, Harry, Brit. inventor; b. Mar. 17, 1880; s. Daniel and Jane Rymer (Grindell) Matthews; ed. Mcht. Venturers' Coll., Bristol, Eng. Volunteered for service in Boer War, 1899; research in broadcasting, wireless telephone; established wireless telephonic conversation with airplane in flight, also sent 1st press message by radio telephone, 1911; developed automatic pilot for flying; demonstrated wireless telephone between automobiles, 1912; controlled boats with searchlight, 1914; worked on submarine detection devices, 1915-17, recording pictures and sound simultaneously on same film, 1918-23, sky projector, 1925, luminaphone (organ played by light), 1926; cons. on sound film prodn. Warner Bros., U. S., 1926-27; research on def. from air and underwater attack, from 1930. Died Sept. 11, 1941.

GRINDON, Joseph, Sr., Am. dermatologist; b. St. Louis, Aug. 20, 1858; s. Arthur St. Leger and Kelis (Chérot-Dupavillon) G.; M.D., St. Louis Med. Coll., 1879; Ph.B., St. Louis U., 1884, Sc.D., 1943; m. Lina Boislinière, Sept. 30, 1903 (dec.); children—Pauline C., Joseph B. (dec.), Dorothy M. (Mrs. A. B. Murphy, Jr.), Joseph B. Began practice, 1879; lecturer on diseases of skin, 1886-95, prof. physiology, 1894-95, prof. dermatology, 1895-1900, St. Louis Med. Coll.; prof. clin. dermatology and syphilology, Washington U., 1900-12; prof. dermatology, St. Louis U., 1912-44, prof. emeritus 1944-50; physician St. Louis Smallpox Hosp., 1881-83; formerly dermatologist O'Fallon Dispensary; dermatologist St. John's, St. Mary's, Desloge and St. Louis City hospitals. Diplomate Am. Bd. Dermatology and Syphilus. Mem. A.M.A., Am. Dermatol. Assn. (pres. 1928), Dermatol. Conf. Mississippi Valley, Am. Acad. Dermatology, Soc. for Investigative Dermatology; corr. mem. Société Francaise de Dermatologie et de Syphiligraphie. Author: Diseases of the Skin, 1902; also several chapters in American Text-Book of Genito-Urinary Diseases, Syphilis and Diseases of the Skin, 1898; Handbook of Cutaneous Therapeutics (with Dr. W. A. Hardaway), 1907. Contbr. to med. jours. Died Apr. 1, 1950.

GRINGS, William Washburn, Am. psychologist; b. Superior, Wis., Mar. 19, 1918; s. William W. and Jessie (Washburn) G.; student U. Ia., 1936-38; B.A. magna cum laude, U. Dubuque, 1940; M.A., U. Ia., 1941, Ph.D., 1946; m. Hilda C. Balster, Aug. 27, 1942; children—Carol Ann, Janet Marie, Steven Frederick, Elaine Ethel. Asst. prof. U. Denver, 1946-47; faculty U. So. Cal., Los Angeles, 1947—, prof. psychology, chmn. dept. psychology, 1960—. Cons. VA, 1948—, NIH, 1966—. Mem. Soc. for Psychophysiol. Research (pres. 1967-68), Am., Cal. (dir.), Los Angeles County (pres. 1960-61;, Western psychol. assns., Human Factors Soc., A.A.A.S., Psychometric Soc. Author: Laboratory Instrumentation in Psychology, 1954; also articles. Research on learning (conditioning) of bodily responses asso. with emotional experience, especially effect of knowledge, perception and verbalizations; changes in skin impedance (galvanic skin response) as index of autonomic behavior. Home: 8020 Agnew Av., Los Angeles 90045.*

GRINKER, Roy Richard, Am. neuropsychiatrist, psychoanalyst; b. Chgo., Aug. 1900; s. Julius and Minnie (Friend) G.; S.B., U. Chgo., 1919; M.D., Rush Med. Coll., 1921; m. Mildred Barman, July 24, 1924, 1 son, Roy Richard. Instr. neurology Northwestern U., 1925; instr. neurology, U. Chgo., 1927-29, asst. prof., 1929-31, asso. prof. 1931-35, asso. prof. psychiatry, 1935-36, and chief of psychiatric div., 1935-36, lectr. in psychiatry Social Service Adminstrn., 1936-50; chmn. dept. of neuropsychiatry, and dir. Inst. for Psychosomatic and Psychiatric Research and Training, Michael Reese Hosp., 1946—; clin. prof. psychiatry U. Ill. Med. Sch., 1951-56, 61—. Chmn. psychopharm. study sect. USPHS; chmn. State Ill. Research and Tng. Authority. Fellow A.A.A.S., N.Y. Acad. Sciences, American College of Neuropharmacology; mem. Am. Psychopathol. Assn., Acad. Psychoanalysis (president 1961), American Assn. Research in Nervous and Mental Diseases, Am Assn. Neuropathologists, Am. Neurol. Assn., Am. Psychiat. Assn., Am. Psychoanalytic Soc., A.M.A. (editor-in-chief archives neurology, psychiatry 1956-59; archives of gen. psychiatry 1959—), Sigma Xi. Author: Neurology, 1934; Psychosomatic Research; and numerous sci. publs.; co-author: (with Spiegel) Men Under Stress, 1945; War Neuroses, 1945; Anxiety and Stress; (with others) The Phenomena of Depression, 1961; Psychiatric Social Work. Editor: Mid-Century Psychiatry, Toward a United Theory of Human Behavior, 1956. Research and publs. on degenerative, infectious, toxic, vascular diseases of brain; psychotherapy; war neuroses. Home: 910 N. Lake Shore Dr. Office: Michael Reese Hosp., 29th and Ellis Av., Chgo. 60616.

GRINNELL, Frederick, Am. mech. engr., inventor; b. New Bedford, Mass., Aug. 14, 1836; s. Lawrence and Rebecca S. G.; prep. edn. Friends' Acad., New Bedford; grad. Rensselaer Poly. Inst., Troy, 1855, as civ. and mech. engr.; m. Mary B. Page, Feb. 17, 1874. Was successively supt. Corliss Steam-engine Works, Providence; mgr. Jersey City Locomotive Works; supt. motive power Atlantic & Great Western R.R.; from 1869, pres., mgr. and mech. engr. Providence Steam and Gas Pipe Co. Introduced and did much to perfect automatic fire extinguisher and alarm, taking out about 40 patents in connection with it. Died 1905.

GRINNELL, George Bird, Am. ethnologist, explorer; b. Bklyn., Sept. 20, 1849; s. George Blake and Helen (Lansing) G.; A.B., Yale, 1870, Ph.D., 1880, Litt.D., 1921; m. Elizabeth Curtis Williams, Aug. 21, 1902. Six months in unmapped West, 1870; in bus., N.Y., 1871-74; asst. in osteology, Peabody Mus., Yale, 1874-80; naturalist with Gen. Custer's expdn. to Black Hills, 1874, and with Col. William Ludlow's reconnaissance to Yellowstone Park, 1875; an editor Forest and Stream, 1876-1911; pres. Forest and Stream Pub. Co., 1880-1911, Bosworth Machine Co., 1887—, dir., 1886; mem. Harriman Alaska Expedition, 1899; commr. to treat with Blackfoot and Ft. Belknap Indians, 1895. Mem. adv. bd. Federal Migratory Bird Law; student of N. Am. ethnology. Fellow Am. Ornithologists Union; mem. Nat. Parks Assn. (pres.). Author: Pawnee Hero Stories and Folk Tales, 1889; Blackfoot Lodge Tales, 1892; The Story of the Indian, 1895; Jack the Young Ranchman, 1899; Jack Among the Indians, 1900; Jack in the Rockies, 1904; Jack the Young Canoeman, 1906; Jack, the Young Trapper, 1907; Jack the Young Explorer, 1908; Trails of the Pathfinders, 1911; The Indians of Today (to 1910), 1911; Jack, the Young Cowboy, 1913; Beyond the Old Frontier, 1913; Blackfeet Indian Stories, 1913; The Fighting Cheyennes, 1915; When Buffalo Ran, 1920; The Cheyenne Indians (2 vols.), 1923; Bent's Old Fort and Its Builders, 1923; By Cheyenne Campfires, 1926; Two Great Scouts, 1929. Co-editor: American Big Game Hunting, 1893; Hunting in Many Lands, 1895; Trail and Campfire, 1897; Harper's Camping and Scouting, 1911. Editor: American Big Game in Its Haunts, 1904; Hunting at High Altitudes, 1913; Hunting and Conservation, 1925; Hunting Trails on Three Continents, 1933. Discovered glacier in Montana which now bears his name, 1885; influential in legislation which led to establishment of Glacier Nat. Park, 1910. Died N.Y.C., Apr. 11, 1938.

GRINSHTEIN, Aleksandr Mikailovich, Russian neuropathologist; b. Tula, Russia, Aug. 10, 1881; grad. med. faculty Moscow U., 1904. Intern nervous diseases clinic Moscow U., 1904-06, asst. from 1906; prof., head nervous diseases dept., Veronezh, then Kharkov Med. Inst., from 1921; head nervous disease dept. Second Moscow Med. Inst., also cons. Kremlin Med. Adminstrn., 1940-55; became cons. Fourth Main Adminstrn. USSR Ministry of Health; mem. Presidium Coordinating Com., also bur. mem. clin. dept. Recipient medal Order of Red Banner of Labor, named Honored Sci. Worker; mem. USSR Acad. Med. Scis. Author: Routes and Centers of the Nervous System;

also numerous articles. Research on influence of cerebral cortex on autonomic innervation; established that one symptom of lesions in hypothalamus area is path. intense feeling of hunger, also that gen. asiposity is sign of medullary lesions, 1927; then showed that path. intense feeling of hunger also can indicate epileptic aura in cases of lesions of hypothalamus area, 1931; proposed preganglionis sympathectomy as treatment for causalgia, convulsive reflexes, other diseases.

GRINSTEAD, Robert Russell, Am. chemist; b. Sacramento, Apr. 15, 1923; s. Allen Ray and Emily (Poppe) G.; B.S., U. Cal. at Berkeley, 1946; Ph.D., Cal. Inst. Tech., 1950; m. Helen Janney Stabler, Oct. 29, 1949; children—James Russell, Charles Miller, Catherine Roberta. Research chemist Dow Chem. Co., Pittsburg, Cal., 1949-52, project leader, 1952-65, sr. research chemist, 1965—. Mem. Am. Chem. Soc. (exec. com. Cal. sect. 1962—), A.A.A.S., Phi Beta Kappa, Sigma Xi, Alpha Chi Sigma. Contbr. articles to profl. jours. Patentee in field. Research on ion exchange and solvent extraction methods for recovery of uranium, other elements; model enzyme systems. Home: 1016 Minert Rd. Office: 2800 Mitchell Dr., Walnut Creek, Cal. 94598.*

GRINTER, Linton Elias, Am. engr.; b. Kansas City, Mo., Aug. 28, 1902; s. Linton Earl and Mary Mandeville (Masterson) G.; B.S., U. of Kan., 1923; M.S., U. of Ill., 1924, Ph.D., 1926; C.E. U. of Kan., 1930; LL.D. (honorary), Arizona State Univ., 1962; m. Constance Louise Hall, Oct. 19, 1926; children—Mary Constance, Lawrence Edward. Part-time draftsman and engr., 1923-26; engr., designer, Standard Oil Co. of Ind., Whiting, 1926-28. Asso. prof. civil engring., Tex. A. and M. Coll., 1928-29, prof. structural engring., 1929-37; dean grad. div., dir. civil engring., Armour Inst. Tech., Chicago, 1937-39, v.p. and dean grad. div., 1939-40; v.p. and dean Grad. School, Illinois Institute Tech., 1940-46; research prof., Ill. Inst. Tech., 1946-52; dean grad. sch. and dir. research U. Fla. since 1952. Cons. War Manpower Commn., 1943; coordinator 6th Service Command of Army Specialized Training Program, 1943-44. Cons. to Editor of Encyclopedia Britannica on technical articles since 1944; consultant to bd. control So. Regional Edn. since 1950 and mem. commn. on grad. studies. Del. International Tech. Congress, and to Internat. Congress for Applied Mechanics, Paris, France, 1946; International Congress Engring. Edn. Zurich, 1954. Cons., chmn. Panel on Heavy Equipment, Research and Development Bd., Office Sec. of Defense, since 1949; mem. bd. univ. representatives Argonne Nat. Lab. Awarded Univ. scholarship, 1923-24, fellowship, 1924-26. U. Ill. Recipient Lamme medal for engring. edn., 1958. Mem. Am. Soc. Civil Engrs. (chmn. exec. com. mechanics, 1950-51). Am. Soc. Mech. Engrs. Western Soc. Engrs. (chmn. civic com. 1948-50, mem. bd. dirs. 1950-52). American Concrete Inst. Internat. Assn. of Bridge and Structural Engrs., Am. Soc. for Engring. Edn. (chmn. Ill.-Ind. section 1947-48, pres. 1953-54, chmn., com. on evaluation of engring. edn., 1952-56), Engrs. Council for Profl. Devel. (pres. 1965-67), Sigma Xi. Author numerous books since 1936; co-author of Engineering Preview, 1945; Numerical Methods of Analysis in Engineering, 1949; Engineering Mechanics, 1952; numerous papers and monographs. Developed 1st practical method of computing wind stresses in skyscrapers or other multi-story buildings; research on vibrations of tower structures due to wind pulses in hurricane. Home: 2256 N.W. 4th Pl., Gainesville, Fla. 32601.*

GRIOLI, Giuseppe, Italian mathematician, physicist; b. Messina, Apr. 10, 1912; s. Antonio and Caterina (Villari) G.; Ph.D. in math. and physics. Researcher, Inst. for Applications Calculus, Nat. Council Research, 1938-49; asso. prof. U. Padua, 1949, full prof. rational mechanics, 1952—. Research and numerous publs. on cinematics, math. analyses, theory of elasticity, dynamics of rigid bodies. Address: 16 via Luzzatti, Padua, Italy.

GRIPONISSIOTIS, Basil, Greek neurosurgeon; b. Levadia, Oct. 10, 1910; s. John and Joe (Kalis) G.; M.D., U. Athens; m. Maria Dames, 1950. Dir., Neurosurg. Center, Evangelism Hosp.; prof. agrege U. Athens. Mem. Assn. Greek Surgeons, Assn. Neurology and Psychiatry Athens, Internat. Congress Neurol. Surgeons. Author: Ossifying Chronic Subdural Hematoma, 1955; Hydatid Cyst of the Brain, 1957; The Brachiale Vertebralis Angiography, 1957. Home: 25 Od Fok. Negri. Office: 26 Od. Patriarchou Loakim, Athens, Greece.

GRISAUNT, William, physician; b. Eng.; flourished 1350; studied medicine at Montpellier, France; became student or fellow at Merton Coll., 1299; possibly father of Pope Urban V; tchr. philosophy Oxford (Eng.) U.; practiced medicine, Marseilles, France. Author: Speculum astrologiae; De qualitatibus astorum; De magnitudine solis; De quadratura circuli; De motu capetis; De significatione astrorum; De causa ignorantiae; De judicio patientis; De urina non visa; Ne ignorantiae nel poltius invidiae.

GRISCOM, John, Am. chemist; b. Hancock's Bridge, N.J., Sept. 17, 1774; s. William and Rachel (Denn) G.; m. Abigail Hoskins, 1800; m. 2d, Rachel Denn, Dec. 13, 1843; children—Abigail, John. First Am. educator to teach chemistry and give lectures on subject to classes, 1803; prof. chemistry and natural history Queens Coll. (now Rutgers U.), 1812-28; organized N.Y. High. Sch. for Boys 1825; instituted Lancasterian system of monitorial instrn.; prin. Friends' Sch., Providence, R.I.; prof. chemistry Columbia; made known med. properties of cod-liver oil and value iodine in treatment of goiter. Discourse on Character and Education, 1823; Monitorial Instruction, 1825. Died Burlington, N.J., Feb. 26, 1852.

GRISCOM, Ludlow, Am. ornithologist; b. N.Y.C., June 17, 1890; s. Clement Acton and Genevieve Sprigg (Ludlow) G.; A.B., Columbia, 1912; A.M., Cornell U., 1915; m. Edith Sumner Sloan, Sept. 14, 1926; children—Edith Rapallo (Mrs. P. O. Daley), Andrew, Joan Ludlow. Instr. elementary biology Cornell U., 1915-16; asst. Am. Mus. Natural History, 1917-20, asst. curator ornithology, 1921-27; research curator zoölogy Mus. Comparative Zoölogy, Harvard, 1927-48, and research ornithologist, 1948-59. Mem. zool. exploration parties in Panama, 1924, 27, Yucatan, 1926, Nicaragua, 1917, Guatemala, 1930; vol. asst. with Gray Herbarium Expdn. to Arctic Newfoundland, 1925, and Gaspé Peninsula, 1923; del. 8th Internat. Ornithol. Congress, Oxford, Eng., 1934. Recipient Conservation medal, 1956. Fellow A.A.A.S., Am. Ornithologists Union (pres. 1956), N.Y. Acad. Scis., Linnaean Soc. N.Y. (pres. 1927); mem. Ecol. Soc. Am. Brit. Ornithologists Union, Nat. Audubon Soc. (chmn. bd. dirs.), Boston Soc. Natural History (trustee, pres. 1948, hon. curator birds), Am. Mus. Natural History, Mass. Audubon Soc. (dir.), Sigma Xi. Author: Birds of the N.Y. City Region, 1923; Distribution of Bird Life in Guatemala, 1932; Ornithology of the Republic of Panama, 1935; A Monographic Study of the Red Crossbill, 1937; Modern Bird Study, 1945; Birds of Nantucket, 1948; Birds of the Concord Region, a Study in Population Trends, 1949. Origin and Distribution Birds of Mexico, 1940; Distributional Check-List Birds of Mexico, Part I, 1940. Birds of Massachusetts, 1955, Annotated and Revised Checklist, with Dorothy E. Snyder, 1955. Contbr. articles to ornithol. and bot. jours.; spl. field of research birds of C.A., field identification of N. Am. birds, conservation. Contbg. editor Nat. Audubon Mag.; asso. editor, Audubon Field Notes. Died May 28, 1959.

GRISEBACH, August Heinrich Rudolf, German botanist; b. Hanover, Germany, Apr. 17, 1814; became prof. at Göttingen, Germany, 1847; made sci. trips to Turkey, 1839, the Pyrenees, 1850, Norway, 1842; Author: Die Vegetation der Erde, 1872; also other publs. on Oriental and S.Am. plants. A founder of plant geography. Research on vegetation of Am., W.I., and the Orient. Died Göttingen, Germany, May 9, 1879.

GRISHKO, Nikolai Nikolavich, Russian geneticist; b. Jan. 4, 1901; grad. Poltava Agrl. Inst., 1925, Kiev Agrl. Inst., 1926. Dir., Inst. Botany, Ukranian Acad. Scis., 1933-34, dir. bur. dept. bipl. and agrl. sci., 1940-48, prof., since, 1944, dir. bot. gardens, 1944-59. Mem. Ukranian SSR Acad. Scis. Recipient Order of Lenin. Author: Course of General Genetics, 1933; New Developments in Hemp Selection, 1935; Michurin's Methods of Plant Acclimatization, 1955; also others. Specialist in acclimatization of plants of southern origin in northern Ukrania, also in selection of floral plants; bred new types of uniformly ripening hemp suitable for mech. harvesting. Address: Botanichesky sad AN, Kiev, Ukranian SSR.

GRISOLIA, Santiago, chemist; b. Valencia, Spain, Jan. 6, 1923; s. Santiago and Concepcion (Garcia) G.; B.A., Inst. Nat. Cuenca, Spain, 1939; postgrad. Med. Sch., Madrid, Spain; M.D., Med. Sch., Valencia, 1944; m. Frances L. Thompson, Aug. 16, 1949; children—William, James. Faculty, Med. Sch., Valencia, 1942-45, asst. prof. physiology, 1944-45; fellow chem. pharmacology N.Y. U., 1946; vis. asso. prof. biochemistry Chgo. U., 1946-47; research asso. physiol. chemistry U. Wis., 1947-51, asst. prof., 1951-54; faculty U. Kan., Kansas City, 1954—, prof. medicine and biochemistry, 1959—, chmn. dept., 1962—. Established investigator Am. Heart Assn., 1943-58; cons. VA Hosp., Kansas City, Mo., 1959—, U. Guadalajara, Mexico, 1963—. Mem. Am., Spanish socs. biol. chemists, Sigma Xi, Alpha Omega Alpha. Research, numerous publs. on phosphoglycerate metabolism, carbamyl and acetyl amino acids metabolism, substrate induced enzyme inactivation. Home: 2900 W. 48th Terrace, Shawnee Mission, Kan. 66205. Office: 39th and Rainbow Blvd., Kansas City, Kan. 66103.*

GRISSOM, Robert Leslie, Am. physician; b. Decatur, Ill., Mar. 5, 1917; s. Leo L. and Ruth (English) G.; student Millikin U., 1934-37; B.S., M.S., M.D., U. Ill., 1937-41; m. Virginia B. Beal, Oct. 28, 1944; children—Nancy, Carol, Leslie, Timothy. Faculty, U. Ill., 1946-53, asst. prof., 1950-53; faculty U. Neb., Omaha, 1953—, prof. internal medicine, chmn. dept., 1955—. Markle scholar, 1950-56. Mem. Sigma Xi. Contbr. articles to tech. jours. Cardiovascular research. Home: 5521 Harney St., Omaha 68132.*

GRISWOLD, Herbert Edward, Am. physician; b. Kansas City, Kan., Apr. 15, 1917; s. Herbert Edward and Zula (Green) G.; B.A. in Chemistry, Reed Coll., 1939; M.S. in Physiology, U. Ore., M.D., 1943; m. Norma Georgeia Walker, Dec. 24, 1943; children—Cheryll Ann (Mrs. Patrick Scanlon), David, Bartley, Thomas. Fellow pediatrics Johns Hopkins Hosps., Balt., 1947-48; faculty U. Ore. Med. Sch., Portland, 1949—, prof., medicine, 1958—, head div. cardiovascular renal disease, 1955—; with Nat. Heart Hosp., Inst. Cardiology, U. London (Eng.), 1957-58. Mem. Ore., Multnomah County med. socs., Ore. Heart Assn., Western Soc. for Clin. Research, Western Assn. Physicians, A.C.P. Research, numerous publs. on hemodynamic adjustments in congenital and acquired heart disease. Home: 3526 S.E. Henry St., Portland 97202. Office: 3181 S.W. Sam Jackson Park Rd., Portland, Ore. 97201.

GRIVSKY, Eugene Michael, chemist; b. Pskov, Russia, Dec. 20, 1911; s. Michael Theodore and Alexandra (Gemchuzhina) G.; B.S., U. Brussels, 1936, M.S., 1938, D.Sc. summa cum lauda, 1940; m. Helen Vlasova, Oct. 27, 1935; children—Michael Eugene, Tatiana. Sr. research chemist, group leader research Labs. Organic Chemistry, pharm. div. Union Chimique Belge, Brussels, 1941-56; sr. research organic chemist Burroughs Wellcome & Co., Inc., Wellcome Research Labs., Tuckahoe, N.Y., 1957—; abstractor chem. abstracts Russian, German, French, Estonian, Latvian sci. lit., 1958—. Mem. chem. socs. Belgium, France, Eng., Am. Chem. Soc., Pharm. Soc. Japan, A.A.A.S. Research and publs. in organic and medicinal chemistry, mechanism of reactions, synthesis of various novel medicinal compounds; synthesis of several new chemotherapeutic agts. Home: 32 River Rd., Scarsdale, N.Y. 10583. Office: Wellcome Research Labs., 1 Scarsdale Rd., Tuckahoe, N.Y. 10707.*

GRIZE, Jean-Blaise, Swiss mathematician; b. Les Verrières, Mar. 16, 1922; s. Jean and Louise (Dällenbach) G.; ed. U. Neuchatel, U. Louvain; Ph.D. in Sci.; m. Gertrude Keiner, 1946; children—Florence, François, Michele. Prof. math. Sch. Commerce, Neuchatel, 1947-59; instr. Faculty Scis., U. Geneva, 1958; asso. prof. Faculty Letters, U. Neuchatel, 1960, asso. prof., 1961. Mem. Internat. Center Genetic Epistemology, Swiss Soc. Logic and Philosophy Scis., Dutch Soc. Natural Scis., Romand Soc. Philosophy, Assn. for Symbolic Logic. Author: Essai sur le rôle du temps en analyse math., 1954, also articles. Home: 1 rue Chantemerle. Office: University of Neuchatel, Neuchatel, Switzerland.

GRJOTHEM, Kai Gudbrand, Norwegian chemist; b. Aasnes, Norway, July 13, 1919; s. Erland and Kaya (Haarbye) G.; siv.ing., Tech. U. Norway, 1950, dr.techn., 1955; m. Jorunn Synnöve Andersen, June 28, 1943; 1 son, Steinar. Research asst. Silicate Inst., 1951; faculty Tech. U. Norway, Trondheim, 1952—, prof., head inorganic chemistry, 1959—; head phys. chemistry Tech. U. Denmark, 1961-62; vis. research fellow U. Toronto (Ont., Can.), 1956-58; distinguished vis. prof. metallurgy Pa. State U., 1964-65; acting dir. Inst. Chemistry, Norwegian Tchrs. Coll., 1959-64. Cons. in light metals chemistry and metallurgy. Fellow Royal Norwegian Soc. Sci., Norwegian Tech. Acad.; mem. C.I.T.C.E. (nat. sec. for Norway 1962—, mem. council (1963—), Am. Chem. Soc., Electrochem. Soc., Am. Inst. Mining, Metall. and Petroleum Engrs., Sigma Xi. Author: Contribution to the Theory of Aluminum Electroylsis, 1956; (with T. Förland, K. Motzfeldt, S. Urnes) Selected Topics in High-Temperature Chemistry, 1965; also numerous sci. articles. Research on light metals chemistry, thermodynamics of ionic melts, equilibria in fluoride and fluoride-oxide systems resulting in new concepts on constitution of cryolite melts used in electrolysis of aluminum and new theories on molten salts. Home: Biskop Skaars gt. 2, Trondheim, Norway.*

GROAT, Benjamin Feland, Am. cons. engr.; b. Hannibal, Mo., Oct. 18, 1867; s. Peter Benjamin and Ann Garnett (Ritter) G.; B.Sc. in Engring., U. Minn., 1901, LL.B., 1908, LL.M., 1910; m. Harriet Grace Mitchell, June 25, 1906; 1 dau., Lucy Mitchell (Mrs. George Ashmun Morton). Identified with r.r. service, various branches, for a number of yrs.; instr. physics, 1895, prof. in charge mechanics and mathematics, U. of Minn. Sch. of Mines, 1898-1910, admitted to bar, 1908; hydro-electric engr. Aluminum Co. Am., 1910-20; cons. practice 1910-49; made alloy lead pipe coupling with adjusted expansion properties; developed precise turbine tests, measuring water chemically, 1914; planned and directed dredging and power improvements of Grasse River, N.Y., 1914; inventor-patentee method of automatic ice diversion; originated plan and wrote application for permit to install ice diversion, St. Lawrence River, which was granted by Internat. Joint Commn.; advocate of Nat. Hydraulic Laboratory (citing its value in planning means to stop erosion of soils); Awarded silver medal, Engrs. Soc. Western Pa., 1915; Norman medal, Am. Soc. Civil Engrs., 1917. Former fellow A.A.A.S.; mem. Am. Soc. C.E., Soc. for Promotion Engring. Edn., Am. Math. Soc., Am. Inst. E.E. (asso.), Sigma Xi. Author sci. papers relating to force of New Richmond tornado; summation of differences; inversions and determinants; back-water slopes; chemihydrometry; ice diversion; rod float theory and tables; dimensional theory; similarity and models (founding proposed new branch of accurate engring. design); gas flow with frictional and received (rejected) heat; rules of quadrature, Generalized Maxwell's Viscosity Theory, disclosing error not previously recognized. Died June 16, 1949.

GROAT, William Avery, Am. physician; b. Canastota, N.Y., Nov. 9, 1876; s. William Robert and Elizabeth Morgan (Avery) G.; B.S., Syracuse U., 1897, M.D., 1900; m. Nellie Nichols Bacon, Oct. 2, 1901; children—William Avery, Robert Andrews, Elsie

(Wade). In practice Syracuse, N.Y., 1901-45; faculty, Coll. Medicine, Syracuse U., 1902-45, prof. clin. pathology, 1911-45; sr. attending physician and dir. Hazard Lab., Meml. Hosp.; sr. attending physician diseases of metabolism and dir. Jacobson Meml. Lab., St. Joseph Hosp.; cons. Univ., City and Psychopathic hosps. and Syracuse Free Dispensary. Chmn. advisory com. on pub. health, City of Syracuse. Diplomate Am. Bd. Internal Medicine. Fellow A.A.A.S.; mem. A.M.A., N.Y. State Med. Soc. (chmn. bd. trustees, past pres., mem. house of delegates), Am. Assn. Immunologists, Am. Assn. Clin. Pathologists, Am. Assn. for Study of Goitre, Am. Assn. for Diseases of Internal Secretions, Sigma Xi, Nu Sigma Nu, Alpha Omega Alpha, Phi Kappa Phi, Phi Kappa Alpha. Contbr. articles and reports of researches, particularly diseases of blood and metabolism, to med. publs. Died Sept. 9, 1945.

GROB, Cyril A., chemist; b. London, Eng., Mar. 12, 1917; s. Albert John and Lydia (Hongler) G.; Ph.D., Fed. Inst. Tech., Zurich, Switzerland, 1942; m. Maria Rainer, Dec. 28, 1943; children—Christoph, Michael. Research asso. U. Basel, Switzerland, 1942-47, faculty, 1948—, prof., 1954—, dir. Inst. Organic Chemistry, 1960—; distinguished vis. prof. Fordham U., N.Y. Rockefeller Found. fellow U. Cal., Los Angeles, 1950-51. Mem. Swiss (pres. 1964-66), Am. chem. socs. Chem. Soc. London. Research, numerous publs. on syntheses of natural products, structure-reactivity relationships in organic chemistry, electrostatic effects and principles of fragmentation in organic chemistry. Home: 66 Lerchenstrasse, Basel 4000. Office: Institut fur Organische Chemie, U. Basel, Switzerland.*

GROB, David, Am. physician; b. N.Y.C., Feb. 23, 1919; s. Hyman and Fannie (Baumwall) G.; B.S., Coll. City N.Y., 1937; M.D., Johns Hopkins, 1942; m. Elizabeth Nussbaum, Dec. 24, 1949; children—Charles, Susan, Emily, Philip. Fellow, Johns Hopkins Sch. Medicine, Balt., faculty, 1948-58, asso. prof. medicine, 1955-58; prof. State U. N.Y. Coll. Medicine, Bklyn., 1958—; physician Johns Hopkins Hosp., Balt., 1951-58; dir. med. services Maimonides Hosp., Bklyn., 1958—, dir. research and edn., 1960—; asst. dean State U. N.Y. Coll. Medicine, 1962—. Cons. various govtl. agys.; mem. med. adv. bd. Myasthenia Gravis Found., 1953—, chmn., 1961-63. Recipient Townsend Harris medal Alumni Assn. Coll. City N.Y., 1964. Diplomate Am. Bd. Internal Medicine. Fellow A.C.P.; mem. A.M.A., Am. Soc. for Clin. Investigation, Am. Fedn. for Clin. Research, Am. Physiol. Soc., Am. Soc. for Pharmacology and Exptl. Therapeutics, Am. Acad. Neurology, Kings County Med. Soc., Bklyn. Soc. Internal Medicine (past treas.). Research, numerous publs. on clin. pharmacology, diseases of muscle, mechanism of neuromuscular transmission in normal subjects and in various diseases, mechanism of action of neuromuscular blocking drugs and anticholinesterase compounds, mechanism of action of poisons, mgmt. hypertension, diseases of muscle. Home: 20 Fern Dr., Roslyn, N.Y. 11576. Office: 4802 10th Av., Bklyn. 11219.*

GROBBEN, Karl, zoologist; b. Brno, Czechoslovakia, Aug. 27, 1854; s. Ludwig and Bertha (Fischer) G.; studied natural scis, especially zoology, U. Vienna; Ph.D., 1877; m. Ida Tschermak von Seysenegg, 1885; 1 dau. Joined faculty U. Vienna, 1879, asso. prof., 1884, prof., dir. Zootomic Inst., 1893; became chmn. 1st Zool. Inst., U. Vienna, 1896; mem. Pola expdn. to E. Mediterranean, sponsored by Vienna Acad. Scis., 1889. Mem. Vienna Acad. Scis. Author: Zur Kenntnis des Stammbaumes und des Systems der Crustaceen, 1892; Die systematische Einteilung des Tierreichs, 1908; also publs. in zool. jours. Re-edited: Lehrbuch der Zoologie (Claus). Research on anatomy, histology and evolution of lower crustaceans, morphology of molluscs, including medusae and worms, evolution of sexual organs of moina and cetochilus; reclassified and improved system of crustaceans. Died Salzburg, Austria, Apr. 13, 1945.

GROBER, Julius Johann August Armin, German physician; b. Laucha, Unstrut, Nov. 27, 1895; s. Gustav and Anna (Ahomeyer) G.; ed. univs. Bonn, Strasbourg, Lena; M.D.; m. Else Greve, 1901 (div.); m. 2d, Marie Luise Schmidt, 1959; children—Inge, Renate, Eva. Instr., prof., full prof., dir. Physio-Therapy Inst. of Lena and Tartu. Recipient Bernhardt-Nocht medal. Research and publs. on dietetics, acclimation, phys. therapy, tropical medicine, colds. Address: Hans Gudrun, Bodendorf, Ahr über Remagen, Germany.

GROBMAN, Arnold B(rams), Am. biologist; b. Newark, Apr. 18, 1918; s. Samuel H. and Sophia (Brams) G.; B.S., U. Mich., 1939; M.S., U. Rochester, 1941, Ph.D., 1943; m. Hulda Gross, Feb. 20, 1944; children—Marc Ross, Beth Allison. Faculty, U. Rochester, 1943-44, research asso. Manhattan Dist., 1944-46; faculty U. Fla., 1946-59, dir. Fla. State Mus., 1952-59; dir., biol. scis. curriculum study U. Colo., 1959-65; dean Coll. Arts and Scis., Rutgers U., New Brunswick, N.J., 1965—. Cons. numerous colls. Recipient A. Cressy Morrison prize N.Y. Acad. Scis., 1944. Mem. A.A.A.S. (council 1961-64), Am. Assn. Museums, Am. Inst. Biol. Scis. (exec. com. 1953-59), Nat. Assn. Biology Tchrs. (pres. 1965, exec. com. 1964—), NRC, Nat. Acad. Scis., Soc. Study Evolution, Am. Assn. U. Profs., Am. Ednl. Research Assn., Am. Soc. Naturalists, Am. Soc. Zoologists, Assn. Am. Med. Colls., Assn. Higher Edn., Assn. Southeastern Biologists, Assn. Supervision and Curriculum Devel., Genetics Soc. Am., Nat. Assn. Research in Sci. Teaching, N.E.A., Nat. Sci. Tchrs. Assn., Nature Conservancy, Soc. Systematic Zoology, Soc. Vertebrate Paleontology, Am. Soc. Ichthyologists and Herpetologists (Fred H. Stoye prize 1940, bd. govs. 1945—, pres. 1963), several fgn. sci. assns., Sigma Xi, Phi Sigma, Alpha Epsilon Delta. Studies, publs. on geog. distbn. and systematics of N.Am. reptiles and amphibians; genetic effects of radiation on mice; biol. edn. Office: Rutgers U., New Brunswick, N.J. 08901.*

GROBSTEIN, Clifford, Am. biologist; b. N.Y.C., July 20, 1916; s. Aaron Joshua and Birdie (Vurdin) G.; B.S., Coll. City N.Y., 1936; M.A., U. Cal. at Los Angeles, 1938, Ph.D., 1940; m. Rose Gruver, Aug. 6, 1938; children—Paul, Joan; m. Ruth Hirsch Beloff, June 12, 1966. Instr. of zoology Ore. State Coll., 1940-43; sr. research fellow USPHS, 1946-47; biologist Nat. Cancer Inst., 1947-57; prof. biology Stanford, 1957-65, exec. head dept. biol. scis., 1963-65; chmn. dept. biology U. Cal. San Diego, 1965—. Cons. NSF, NIH, Am. Cancer Soc. Recipient Brachet award Royal Acad. Scis. Belgium, 1959. Fellow Am. Acad. Arts and Scis.; mem. Am. Soc. Zoologists, Am. Soc. Cell Biologists, Internat. Inst. Embryology, Soc. Study Devel. and Growth (past pres.), Nat. Acad. Scis. Research on embryonic induction in vitro: studies of mechanisms; demonstration of occurrence across an interspace; control of cytodifferentiation. Home: 8295 Prestwick Dr., La Jolla, Cal. 92307.*

GRODINS, Fred Sherman, Am. physiologist; b. Chgo., Nov. 18, 1915; s. Abe E. and Minnie (Levine) G.; B.S., Northwestern U., 1937, M.D., 1942, Ph.D., 1944; m. Sylvia Johnson, Mar. 28, 1942. Faculty, U. Ill. Coll. Medicine, 1946-47, asso. prof., 1947; faculty Northwestern U. Med. Sch., Chgo., 1948—, prof. physiology, 1950-67; prof. elec. engring. and physiology U. So. Cal., 1967—. Mem. physiology tng. com. NIH, 1964—; cons. math. dept. Rand Corp., Santa Monica, Cal., 1964—. Recipient Research career award USPHS, 1962. Mem. Am. Physiol. Soc., A.A.A.S., Soc. for Exptl. Biology and Medicine (mem. simulation council 1961—), Phi Beta Kappa, Sigma Xi, Phi Lambda Upsilon, Alpha Omega Alpha. Author: Control Theory and Biological Systems, 1963; also numerous articles. Editorial bd. Circulation research, 1960-65. Research on biol. control systems, math. models for control pulmonary ventilation and cardiac output. Home: 26 Chuckwagon Rd., Rolling Hills, Cal. 90274. Office: Dept. Elec. Engring., U. So. Cal., Los Angeles 90007.*

GRODZINS, L., Am. physicist; b. Lowell, Mass., July 10, 1921; s. David Melvin and Taube (Bialoblotsky) G.; B.S., U. N.H., 1946; M.S., Union Coll., 1948; Ph.D., Purdue U., 1954; m. Lulu F. Anderson, Dec. 16, 1956; children—Dean David, Henry Jacob. Research asst. Gen. Electric Research Lab., 1946-48; instr. Purdue U., 1954-55; with Brookhaven Nat. Lab., 1955-59; faculty Mass. Inst. Tech., Cambridge, 1959—, prof. physics, 1966—. Cons. USN, industry. Guggenheim fellow, 1964-65. Mem. Am. Phys. Soc. Research, publs. on symmetry properties, particularly parity non-conservation and nuclear spectroscopy, particularly electromagnetic moments of excited nuclear states. Home: 92 Churchill Av., Arlington, Mass. 02174.*

GROEBBELS, Franz Maria, German ornithologist, physiologist; b. Sigmaringen, Germany, Sept. 1, 1888; s. Johann and Laura (Salm) G.; studied medicine, Munich, Germany, Heidelberg, Germany, 1907-12; further studies at Frankfurt, Germany, Munich; m. Magda Emma Struss, 1921; 1 dau. Joined faculty physiology U. Hamburg (Germany), 1921; prof., chief physician Physiol. Inst., Hamburg-Eppendorf; ret., 1953. Author: Der Vogel, 2 vols., 1932, 37; Der Vogel in der deutschen Landschaft, 1938; (with F. Moebert, H. Kirchner) Ornithologische Hilfstabellen, 1938; also numerous articles on physiology and ornithology, 1907-60. Studied functional morphology and histology, physiology and biology; combined field obs. with lab. research in ornithology. Died Mölln, Germany, Nov. 7, 1960.

GROEDEL, Franz Maximilian, radiologist, cardiologist; b. Bad Nauheim, Germany, May 23, 1881; s. Isidor Maximilian and Rosa (Klopter) G.; ed. U. Munich; Dr. Rer. Nat. (hon.), U. Frankfurt, 1951. Head x-ray dept. Hosp. zum Heiligen Geist, Frankfurt am Main, 1909; with father's pvt. sanatorium, Bad Nauheim, from 1921; faculty dept. radiology and phys. therapy U. Frankfurt, 1920, asso. prof., 1926; founder William G. Kerckhoff heart Research Inst., Bad Nauheim, 1931; remained in U. S. after lecture tour; cons. cardiologist several N.Y. hosps.; faculty Fordham U. Author: (with H. Liniger and H. Lossen) Materialiensammlung der Unfälle und Schäden in Röntgenbetrieben, 1925; (with Klopfer) Gesetzbuch und ärztlicher Röntgenbetrieb, 1925; Die biologische Wirkung der Röntgenstrahlen speziell im Lichte der modernen Kapillarforschung und der Modernen Entzündungslehre, 1925; Das Extremitäten-, Thorax- und Partial-Elektrokardiogramm des Menschen, 2 vols., 1934; The Chest Leads in Pericarditis, 1941; the Venous pulse and Its Graphic Recording, 1946; (with P. R. Borchardt) Direct Electrocardiography of the Human Heart and Intrathoracic Electrocardiography, 1948, also over 300 articles. Pioneer in radiology and cardiology, x-ray gastrointestinal diagnostics; modified Levy-Dorn orthodiagraph, combined x-ray and electrocardiagram; devised method of direct x-ray cinematography; worked on elec. manifestations of heart; discovered difference between electrocardiograms of left and right hearts, 1932; made graphic reprodn. of vein pulse, 1946, direct electrocardiographic tapping from surface of human heart, 1948. Died N.Y.C., Oct. 12, 1951.

GROEN, Pier, Dutch physicist, oceanographer; b. Sneek, Netherlands, Dec. 6, 1912; s. Johannes and Tjitske (De Jong) G.; student Free U., Amsterdam, Netherlands, 1931-37, D.Sc., 1942; m. Arga Wilhelmina N. de Haan, Aug. 6, 1942; children—Johannes B., Annemarie T., Bastiaan A., Tjitske M., Maaike W. Asst. for theoretical Physics Free U., Amsterdam, 1938-41; meteorologist, phys. oceanographer Koninklijk Nederlands Meteorologisch Inst., De Bilt, Netherlands, 1942-64; prof. oceanography and meteorology Free U., Amsterdam, Netherlands, 1952—; prof. oceanography State U., Utrecht, Netherlands, 1964—. Sci. adviser Netherlands Meteorol. Inst., De Bilt, 1964—; chmn. sci. com. Netherlands Inst. Sea Research, 1965—. Mem. Netherlands Assn. Meteorology and Astronomy, Royal Netherlands Geog. Assn. Christian Assn. Sci. and Medicine. Author: (with Dorrestein) Zeegolven, 1949, 58; De Wateren der Wereldzee, 1951, 61; The Waters of the Sea, 1967; also articles. Research on theories of order and disorder, nocturnal cooling of earth's surface, internal waves, surface waves of sea, eddy motions, storm surges, exchange of suspended particles by currents. Home: 42 Wilhelminalaan, De Bilt, Netherlands. Office: Inst. v. Aarwetenschappen, De Boelelaan 1085, Amsterdam, Netherlands.*

GROENEWOLD, Hilbrand Johannes, Dutch theoretical physicist; b. Muntendam, Netherlands, June 29, 1910; s. Johannes and Bernardina (Gatsonides) G.; cand. Groningen U., 1930, drs., 1934; postgrad. Cambridge (Eng.) U., 1934-35; dr., Utrecht (Netherlands), 1946; m. Geertje Balder, Apr. 7, 1942; children—Anita, Aart, Ank. With U. Groningen (Netherlands), 1935-37, 43-47, 51—, prof. theoretical physics and founds. sci., 1956—; theoretical physics asst. Leyden (Netherlands) U., 1937-42; tchr. Gymnasium, the Hague, 1940-43; theoretical physicist Royal Netherlands Meteorol. Inst., Bilt, 1947-51. Research, publs. on founds. and interpretation of quantum theory, measuring process in quantum theory, relativistic classical and quantum theories of particles and fields, arrow of time, founds. and methodology of phys. sci., role of observation, logical operations and hypothesis formation, ethics and sci., social implications of sci. and social responsibility of scientists. Home: 222 van Houtenlaan, Groningen, Netherlands.*

GROENMAN, Sjoerd, Dutch sociologist; b. Roosendaal, The Netherlands, Nov. 28, 1913; s. Berend and Anna (Joustra) G.; D.Litt. Arts, U. Amsterdam, 1947; m. Lucie Limborgh Meijer, Sept. 16, 1939; children—Louisa (Mrs. Frederick C. Samuel), Anne Marten, Berend Sjoerd, Tertius. Journalist, 1937-38, 39-40; sociologist Community of Emmen, 1938-39, Provincial Inst. for Promotion of Industrialization, 1940-43; sociologist Authority of Reclaimed Polders, 1943-48; prof. sociology Utrecht State U., 1948—, dean Faculty Social Scis., 1963—. Chmn. adv. com. Dutch Ministry of Social Work, Culture and Recreation; chmn. Internat. Social Sci. Council, Paris, 1961—. Mem. Union Sci. Study Population, Internat. Com. for Documentation Social Scis., European Center for Social Sci. (dir.). Author: Staphorst, A Closed Society, 1947; Methods of Social Research, 1950, 66; Colonisation on New Land, 1953; Our Part in Space, 1960. Numerous publs. on social research for phys. planning. Home: 188 Anna Paulownalaan, Zeist, Netherlands. Office: 2 Varkenmarkt, Utrecht, Netherlands.*

GROGAN, Clarence Orval, Am. educator; b. Grogan, Mo., June 1, 1921; s. John Thomas and Maria (Schull) G.; B.S., U. Mo., 1946, M.A., 1949, Ph.D., 1951; m. Cora Alice West, June 21, 1945; children—Virginia, Richard, Wendell. Corn breeder Mo. Farmers Assn., Marshall, 1946-48; asst. instr. U. Mo., 1949-51; in charge of maize research Orange Free State, Union S. Africa, 1951-54; agronomist U. S. Dept. Agr., also research asso. U. Mo., 1954-59; agronomist U. S. Dept. Agr., also prof. agronomy Miss. State U., 1959-66; prof. Cornell U., 1966—. Chmn., N.C. Corn Improvement Conf., 1958, So. Corn Improvement Conf., 1964. Mem. Am. Soc. Agronomy, Am. Genetic Assn., Sigma Xi, Gamma Sigma Delta, Alpha Zeta. Research, numerous publs. on genetic, chemical, pathological and entomological investigations of maize. Home: 1493 Ellis Hollow Rd., Ithaca, N.Y. 14850.*

GROISSMAYR, Fritz Béla, meteorologist; b. Attnang, Austria, Apr. 4, 1894; s. Franz and Leontine (Mayr) G.; began study chemistry Polytechnikum, Arnstadt, Germany, 1914; m. Marianne Hulak, 1937. Lived several years in Hungary; became dir. Reich Weather Service Sta., Passau, Germany, 1932; engr. Töging Aluminum Works, 1942-44. Author: Die säkulare Klimgwende um 1940 und das Katastrophenjahr 1947 in Zentraleuropa, 1949; also articles. Research on meteorology; attempted to discover correlations between weather phenomena of large, widely separated areas to prove world significance of meteorol. manifestations. Died Passau, Sept. 16, 1948.

GROLLMAN, Arthur, Am. physician; b. Balt., Oct. 20, 1901; s. Simon and Bessie (Karu) G.; A.B., Johns Hopkins, 1920, Ph.D., 1923, M.D., 1930; m. Anna Louise Costello, Mar. 26, 1926; children—Arthur Patrick, Catherine Ann, Evelyn Frances. Faculty, Johns Hopkins, 1923-41, asso. prof. pharmacology and therapeutics, 1932-41; prof. research medicine Bowman Gray Sch. Medicine, Winston-Salem, N.C., 1941-44; prof. medicine Southwestern Med. Sch., U. Tex., Dallas, 1944-50, chmn. dept. exptl. medicine, 1950—. Cons. to Surgeon Gen., USAF. Mem. Am. Physiol. Soc., Am. Pharm. Soc., A.M.A., A.C.P. Author: Cardiac Output, 1932; The Adrenals, 1936; Clinical Physiology, 1960, 2d edit., 1963; Clinical Endocrinology, 1964; Pharmacology and Therapeutics, 1965; also numerous articles. Research in fields of cardiac output, adrenal glands, acute renal failure, hypertension, endocrinology. Home: 3501 Princeton Av., Dallas 75205. Office: 5323 Harry Hines Blvd., Dallas 75235.*

GROMASHEVSKII, Lev Vasilevich, Russian epidemiologist; b. Nikolaev, Ukraine, Oct. 13, 1887; grad. Med. Faculty, Novorossiyak U., 1912; D.Social Med., 1926. Asst., lectr. Odessa Inst. Health Bacteriology, 1920-23, prof., dir., 1923-27; head Odessa Guberniya Dept. Health, sr. asst., later prof. epidemiology and rector Odessa Med. Inst., 1920-28; prof., 1923—; prof., head chair epidemiology Med. Inst. and Postgrad. Med. Inst., Dnepropetrovsk, 1928-31; dir. dnepropetrovsk Inst. Health Bacteriology, 1928-30; founder, dir. Moscow Central Inst. Epidemiology and Microbiology, 1931-48; founder Moscow Central Research Lab. Hygiene and Epidemiology, 1933; head Moscow Lab. Hygiene and Epidemiology, 1933-39; prof. Moscow Central Postgrad. Med. Inst., 1939-48; dir. dept. epidemiology Moscow Mechnikov Inst. Epidemiology and Microbiology, 1943-47; founder, dir. Kiev Inst. Infectious Diseases, USSR Acad. Med. Sci., 1948-51, bur. mem. dept. hygiene and epidemiology, 1944-48; head chair epidemiology Kiev Med. Inst., 1951-62; dep. dir. for sci. work Kiev Inst. Epidemiology, Microbiology and Hygiene, 1953—. Chmn. epidemiological commn. learned med. council Ukrainian Ministry Health. Mem. USSR Acad. Med. Sci., Kiev Soc. Epidemiologists, Microbiologists and Infectionists (chmn.). Co-author: Cholera in Odessa in 1918-22, 1929; Specialized Epidemiology, 1947; Dysentery, 1956; author: Principles in the Classification of Infectious Diseases, 1947; General Epidemiology, 3d edit., 1949. Co-editor Epidemiology and Infectious Diseases sects. Large Med. Ency., 2d edit.; mem. editorial council Jour. Microbiology, Epidemiology and Immunology, Med. Affairs. Research and numerous publs. on epidemiology of infectious diseases, cholera, typhus, measures against infections, formulated system of classifying infectious diseases based on mechanism of transmission, analyzed epidemiology of typhus, determined nature of septic angina caused by consumption of food containing grain which had remained in field during winter and developed measures against it, research on helminthiasis in southwestern Ukraine. Address: Inst. Epidemiology, Microbiology and Hygiene, ulitsa S. Razina 4, Kiev, Ukraine SSR, USSR.

GROMOV, Vitaliy Viktorinovich, Russian otorhinolaryngologist; b. 1901; grad. Med. Faculty, Kazan U., 1924; postgrad. Saratov Research Inst. Physiology of Upper Respiratory Tracts, 1933-36; Cand. Med. Sci., 1936; D.Med. Sci., 1944. Mil. physician, supernumerary asso. ear, nose and throat diseases clinic Kazan Med. Inst., 1925-31; otorhinolaryngologist, Tadzhikistan, 1932 33; asst. dept. otorhinolaryngology Kazan Postgrad. Med. Inst., 1937-41, head chair, 1949—, prof., 1950—. Hon. physician of Tatar ASSR, 1948—. Research and publs. on devel. of human larynx, relationship between gastric secretion and respiration. Address: Kazan Postgrad. Med. Inst., ulitza Komleva 12, Kazan RSFSR, USSR.

GRONAU, Karl Ludwig, German meteorologist; b. Berlin, June 7, 1742; s. Johannn Hermann and Luise (von Bergen) G.; m. Johanne Hermann, 1777; children include Johann Carl Ludwig. Pastor, Parochial Ch., Berlin. Author: Versuch einger Beobachtungen über die Witterung in der Mark Brandenburg, besonders in der Begend um Berlin seit ältester Zeit, 1794. Made meteorol. observations, 1756-1826, using barometer and thermometer, beginning in 1774; collected weather observations. Died Berlin, Dec. 8, 1826.

GRONDAHL, Lars Olai, Am. physicist; b. Hendrum, Minn., Nov. 27, 1880; s. Peter Elias and Herborg (Huglen) G.; student Concordia Coll., 1997-98; B.S., St. Olaf Coll., 1904, M.S., 1905, D.Sc., 1940; Ph.D., Johns Hopkins, 1908; postgrad. U. Chgo. summer 1903, 1909, U. Berlin, summer 1914; m. Grace Elizabeth Fuller, Sept. 11, 1907; 1 son, Martin (dec.) Instr., St. Olaf Coll., 1904-05; lectr., asst. Johns Hopkins, 1906-8; prof. math. and physics Spokane Coll., 1908-09; instr. physics U. Wash., 1909-12; from instr. to asso. prof. physics Carnegie Inst. Tech., 1912-20; dir. research Union Switch and Signal Co., 1920-37, dir. research and engring., 1937-47, cons., 1947-49; mem. Navy Electronics Lab. Coop Research Group, San Diego, 1951; mem. staff research project for army engrs. corps, dept. engring. Pa. State Coll., 1953. Staff, Naval Cons. Bd., 1917-19; mem. NRC, 1933-36; chmn. exec. com. Council on Applied Physics, Am. Inst. Physics, 1936; sect. mem. NDRC, 1940-42; chief sect. 52 OSRD, 1942-45; cons. Westinghouse Air Brake Co., 1947-53, Navy Electronics Lab., San Diego, 1949-53. Recipient Potts medal Franklin Inst., 1938, Modern Pioneers award N.A.M., 1940, Feorge R. Henderson medal Franklin Inst., 1947, Pres.'s medal for Merit, 1948, Pitts. Physics award, 1952. Fellow A.A.A.S., Am. Phys. Soc., Am. Inst. E.E.; mem. Humanist Assn., Phi Beta Kappa, Sigma Xi, Sigma Pi Sigma. Discovered phenomenon of asymmetric conduction in solids; invented and developed copper oxide rectifier, also railway inductive train communication system, also numerous other elec., optical and mech. devices and systems. Home: 2025 Wightman St., Pitts. 15217.*

GRONEMEYER, Wilhelm, German physician; b. Paderborn, May 9, 1912; s. August and Helene (Baumann) G.; ed. U. Tübingen, U. Kiel; M.D.; m. Hanne Pöhler, 1939; children—Uwe, Steffen, Alke, Martin. Med. asst. Univ. and Med. Clinic, specialist in internal medicine, chief physician, Lubeck; chief physician Asthma and Allergy Inst. Research and Asthma Clinic, Bad Lippspringe, 1952; instr. allergy U. Göttingen, 1959. Mem. European Acad. Allergology (v.p.), Interasma, German Soc. for Internal Medicine (editor), German Soc. for Allergy (editor), Deutsche Gesellschaft für Abeitsmedizin. Collaborator: Handbuch der Allergie; Lehrbuch der Innere Medizin; Lehrbuch der Allergie; collaborating editor: Internat. Archives Allergy, also numerous articles on asthma and allergy. Home: 10 An der Jordanquelle. Office: 5 Arminiuspark, Bad Lippspringe, Germany.

GRONOVIUS, Johann Friedrich, Dutch botanist; b. 1690; s. Jakob Gronovius; studied law; children include Lorenz Theodor. Magistrate of Leyden; friend of Linnaeus. Author: Flora Virginica, 2 parts, 1739, 43; Flora orientalis, 1775 (mentioned coffee). Died 1760.

GROOM, Dale, Am. physician; b. Tulsa, Nov. 6, 1912; s. F. H. and Mary (Dale) G.; A.B., Hiram Coll., 1936; M.D., Med. Coll. Va., 1943; M.S. in Medicine, U. Minn., 1948; m. Marjorie Tweed, Jan. 26, 1944; children—Shelley Anne, Lincoln Dale, Randall Tweed. Fellow in medicine Mayo Found., Rochester, Minn., 1945-49; practice medicine specializing in cardiology, Miami, Fla., 1949-52; faculty Med. Coll. S.C., Charleston, 1952—, asso. prof. medicine, dir. postgrad. edn., 1960—, chmn. student health, 1958—, asst. dean, 1966—. Nat. cons. cardiology USAF, 1966—. Diplomate Am. Bd. Internal Medicine. Fellow A.C.P. (chmn. bd. govs. 1968—); mem. A.M.A. (mem. council on postgrad. programs 1965—), Am. Heart Assn., Alumni Assn. Mayo Found., Sigma Xi, Alpha Omega Alpha (hon.). Author: Clinics in Electrocardiography, 1960; also numerous articles. Research in fields coronary heart disease, cardiovascular sounds and electrocardiography; pioneered use TV broadcasting for postgrad. med. edn. Home: Stono Battery, John's Island, S.C. 29455. Office: Med. Coll. S.C. Hosp., Charleston, S.C. 29401.*

GROOM, Percy, Brit. botanist; b. Sept. 12, 1865; s. T. R. Groom; B.Sc., Mason Coll., Birmingham; postgrad. U. Bonn; M.A., Trinity Coll., Cambridge; M.A., D.Sc., Oxford U.; m. Mary Harrop; 1 dau. Frank Samrk student of botany Gonville and Caius Coll., 1888-89; prof. botany and arboriculture Imperial Coll., Whampoa, China, 1889-92; resident Oxford, 1892-98, mem. Exeter Coll.; univ. lectr. plant physiology, Edinburgh, 1898; head biol. depts. Royal Indian Engeing. Coll., Coopers Hill, 1899-1905, also Univ. Coll., Reading; lectr. botany No. Poly. Inst., Holloway, 1907-08; asst. prof. botany Imperial Coll. Sci. and Tech., London, 1908-11, prof. of tech. of woods and fibers, 1911-31. Fellow Royal Soc., 1924, Linnean Soc. Author: Trees and their Life Histories; Elementary Botany; also papers. Editor various sci. works and transls. Died Sept. 16, 1931.

GROOMBRIDGE, Stephen, Brit. astronomer; b. Goudhurst, Kent, Eng., Jan. 7, 1755; a dau., Mrs. Newton Smart; from apprentice to linendraper, London; West Indian mcht. until 1815; built obs. at Goudhurst. Fellow Royal Soc., 1812; mem. Astron. Soc. (a founder), Acad. Naples. Author: A Catalogue of Circumpolar Stars, deduced from the Observations of Stephen Groombridge, F.R.S.; reduced to Jan. 1, 1810 (contained 4,243 star places). Observed fastest moving known star, eclipses of sun, 1816, 20. Died Mar. 30, 1832.

GROOS, Friedrich, German physician, philosopher; b. Karlsruhe, Germany, Apr. 23, 1768; s. Immanuel and Sophie (Gerzog) G.; ed. Tübingen, Karlsruhe, Stuttgart, Pavia; M.D., Freiburg; m. Christiane Theilacker, 1806; 12 children; m. 2d, Jakobine Schippel, 1831. Asst. to city physician, Karlsruhe; became town physician, Stein/Pforzheim, Gochscheim-Odenheim, 1809; ct. physician, Schwetzingen, 1813; became directing physician, asylum for insane and infirm, Pforzheim, 1814 (became insane asylum in Heidelberg 1826); ret., 1836. Author: Untersuchunge über die moralischen und organischen Bedingungen des Irrseins und die Lasterhaftigkeit, 1826; Entwurf einer philosophischen Grundlage für die Lehre von dem Geisteskrankheiten, 1828; Ideen zur Begründung eines obersten Princips für die psychische Legalmedicin, 1829; Der Geist der psychischen Arzneiwissenschaft in nosologischer und gerichtlicher Beziehung, 1831; Die geistige Natur des Menschen, Bruchstücke zur einer psychischen Anthropologie, 1834; Der unverwesliche Leib, als Organ des Geistes und Sitz der Seelenstörungen, 1837; other works on philosophy, also numerous articles. Dealt with problems of philos. nature, also some problems of psychology; sought to find basis of insanity in devel. and life of a person as a whole; contbd. to knowledge of criminal psychology; research on theoretical psychiatry. Died Eberbach/Neckar, June 15, 1852.

GROOS, Karl Theodor, German philosopher, psychologist; b. Heidelberg, Germany, Dec. 10, 1861; s. Julius and Sophie (Koopman) G.; Ph.D., U. Heidelberg; m. Emma Kraut, 1884; 1 adopted dau. Lectr. U. Giessen (Germany), 1889-92, asso. prof. philosophy, 1892-98, 1901-11; prof. U. Basel (Switzerland), 1898-1901; prof. philosophy and psychology U. Tübingen (Germany), 1911-1929, emeritus, from 1929. Author: Einleitung in die Asthetik, 1892; Die Spiele der Tiere, 1896, English edit., 1898; Die Spiele der Meuschen, 1899, English edit., 1901; Das Seelenleben des kindes, 1903. Work was mainly of philos. nature, attempted synthesis of biol., psychol. and metaphys. aspects; studies on psychol. esthetics; intensive studies on games (play) of man and animals, promoted theory that games and play are tng. for future serious pursuits; also research characterological and developmental psychology themes. Died Tübingen, Mar. 27, 1946.

GROOTEN, Christian s' (Schrotenius, Sgrothenius, Schrootz, Scroot), German cartographer; b. Sonsbeck nr. Xanten, Germany, circa 1530; 1 son, 1 dau. Worked as cartographer, Kalkar, most of life; most of his maps were on contract to Spain; contracted by Duke of Alba to make atlas of N.W. Europe, 1568, presented to Philipp II of Spain, 1573, continued contracts to improve atlas until 1576; moved to Jesuit Coll., Cologne, 1590. Author: Atlas I (38 maps, now in Brussels), 1573; Atlas II (38 maps, now in Madrid), 1592. Oldest cartographic survey of states, including Duchy of Westphalia, Bishopric Paderborn; 1st atlas with roads and sea routes, was oldest comml. atlas of No. Europe. Died Kalkar, 1603/04.

GROS, (Franz Jakob) Oskar, German pharmacologist; b. Werneck nr. Würzburg, Germany, Mar. 13, 1877; s. Peter and Mathilde (Bischoff) G.; studied chemistry, Würzburg, 1896-98; Ph.D., Leipzig, Germany, 1901, M.D., 1908; m. Helene von Werthern, 1915; 1 dau. Joined faculty U. Leipzig, 1909; became prof. pharmacology U. Halle (Germany), 1915, Cologne, Germany, 1919, Kiel, Germany, 1922, Leipzig, 1925; became emeritus, 1943. Mem. Saxon Acad. Scis. Research and publs. on application of phys. chemistry to pharmacological reactions, especially hemolysis, silver salt reactions, local anesthesia. Died Uffing am Staffelsee, Germany, Aug. 3, 1947.

GROSCH, Herbert Reuben John, computer scientist; b. Saskatoon, Sask., Can., Sept. 13, 1918; s. Reuben John and Bessie (Adams) G.; came to U. S., 1928, naturalized, 1934; B.Sc., U. Mich., 1938, Ph.D., 1942; m. L. Joyce Labots-Misbeek, Dec. 23, 1966. Jr. astronomer U. S. Naval Obs., 1941-43; optical designer USN Dept., Sperry Gyro., Farrand Optical, 1943-45; dir. computing Watson Sci. Computing Lab., IBM-Columbia, 1945-50; mgr. tech. computing bur. IBM, Washington, 1951, asst. to dir. sales services, White Plains, 1958, mgr. space program, N.Y.C., 1959; head logical design research group Whirlwind II, Mass. Inst. Tech., 1952; mgr. computer lab. G.E. Evendale, 1952-56, mgr. applications computer dept. G.E. Phoenix, 1956-57, also mgr. Deacon project G.E., Santa Barbara, Cal., 1965-67; cons. computer applications U. S. and W. Europe, 1959-65; dir. Center for Computer Scis. and Tech., Nat. Bur. Standards, Washington, 1967—, Tchr., Columbia, 1946-50, Ariz. State U., 1957. Fellow Am., Inst. Aeros. and Astronautics (past nat. pres.); mem. Am. Astron. Soc., Am. Optical Soc. Assn. for Computing Machinery (charter), I.E.E.E., Brit. Computer Soc., Assn. for Machine Transl. and Computational Linguistics, Sigma Xi. Research, numerous publs. on devel. and application computers especially analysis orbits, mechanized optical design, math. tables, polynomial approximation. Home: 9 Carderock Ct., Bethesda, Md. 20034. Office: Nat. Bur. Standards, Washington 20234.*

GROSCHWITZ, Eberhard Gustav Paul, German physicist; b. Freiburg, Germany, Feb. 3, 1915; s. Gustav and Klara (Barbenheim) G.; Dr.rer.nat., U. Freiburg, 1951, Dipl.Phys., 1944; m. Luise Lotte Thiele, Oct. 26, 1940. Sci. collaborator Siemens AG, WWB, München, Germany, 1952—. Mem. Deutsche Physikalische Gesellschaft. Research, publs., inventions on semiconductors, including semiconductor plasma and amplification effects, surface currents, theory of point contact rectifier, physics of maser and laser, nonlinear effects, thermodynamics of electronic semiconductor parameters, problems of charge carrier-motions. Home: Eduard-Schmid-Str. 4, 8 München 90. Office: Balanstr. 73, 8 München 8, Fed. Republic Germany.*

GROSJEAN, Carl Clement, Belgian theoretical physicist; b. Courtrai, Belgium, Sept. 5, 1926; s. Carl Adolphe and Germaine (Lecompte) G.; B.S. in Math., State U. Ghent (Belgium), 1947, Lic.Sc. in Physics, 1949; D.Sc., 1951, Geaggr. H.O., 1955; Ph.D., Columbia U., 1951; m. Andrea Emma Bernolet, Aug. 7, 1965. Fellow, Watson Sci. Computing Lab., 1949-50; research asst. Institut Interuniversitaire des Sciences Nucleaires, Brussels, Belgium, 1951; research asso.,

1952-58; prof. applied math. and nuclear reactor theory State U. Ghent, 1958—, also prof. advanced math. analysis, 1965—, dir. computing lab., 1960. Vis. fellow Princeton, 1956-58; mem. adv. com., neutron phys. div. Centre d'Etudes de l'Energie nucleaire, Mol, Belgium, 1960-64; vis. prof. Univ. Coll. London, 1963, hon. research asso., 1963. Laureate, Concours interuniversitaire, 1950; recipient Empain prize, 1953. Mem. Belgian Math. Soc., Belgian Phys. Soc. Author: Formal Theory of Scattering Phenomena, 1960; (with W. Bossaert) Table of absolute gamma-ray detection efficiencies, 1965; (with J. Meeus and W. Vanderleen) Canon of Solar Eclipses, 1966; also articles. Research and publs. on multiple scattering of elementary particles; extended theory of scattering phenomena to case of completely arbitrary anisotropic scattering laws; contbd. to theory of slowing down of neutrons, particularly in media subjected to thermal motion; collaborated in 1st exptl. observation of dislocations in pure metallic crystals. Home: 19 Recollettenlei. Office: 6 Rozier, Ghent, Belgium.*

GROSS, Alfred Otto, Am. ornithologist; b. Atwood, Ill., Apr. 8, 1883; s. Henry and Sophia (Gross) G.; A.B., U. Ill., 1908; Ph.D., Harvard, 1912; D.Sc., Bowdoin Coll., 1952; m. Edna Grace Gross, July 2, 1913; children—William Albert, Thomas Alfred, Louise Edna (Mrs. Otis Northrop Minot). Ornithologist, field dir. Ill. Natural History Survey, 1906-09; investigator Bermuda Biol. Sta., 1910-12; ornithologist Roosevelt Wild Life Expt. Sta., Ithaca, N.Y., 1926-27; dir. heath hen investigation State of Mass., Boston, 1923-35; dir. Ruffed Grouse Investigation of New Eng., 1925-35; dir. Wis. Prairie Chicken Investigation, Madison, 1928-29; ornithologist Bowdoin MacMillan Arctic Expdn., 1934; dir. Bowdoin Sci. Sta., Kent Island, New Brunswick, Can., 1935-53; prof. ornithology U. Mich. Biol. Sta., 1928; faculty dept. biology Bowdoin Coll., 1912-53, prof. ornithology, 1922-53. Fellow Am. Ornithologist's Union, Cooper Ornithol. Club, Wilson Ornithol. Soc. Author: The Heath Hen, 1928; Wisconsin Prairie Chicken, 1930; also numerous articles. Research on life histories of N.Am. birds, albinism and melanism of plumages, albinism and erythrism of eggs of N.Am. birds. Home: 11 Boody St., Brunswick, Me. 04011.*

GROSS, Bernhard, physicist; b. Stuttgart, Germany, Nov. 12, 1905; s. Wilhelm and Sophie (Hirsch) G.; Dr.rer.nat., Tech. Highsch., Stuttgart, 1956; Diplomingenieur, Tech. High Sch., Stuttgart, 1932; m. Gertrud Gunz, Dec. 31, 1935; children—Antonio, Roberto. Asst., Inst. Physics, Tech. Highsch., Stuttgart, 1932-33; staff Nat. Inst. Tech., Rio de Janeiro, Brazil, 1934—, dir. div. electricity, 1946—; prof. Cath. U., Rio de Janeiro, 1955—; dir. physics Brazilian Nat. Research Council, 1950-54; dir. div. sci. and tech. information IAEA, Vienna, Austria, 1961-67, mem. sci. adv. com., 1958-60. Brazilian rep. sci. adv. com. UN, 1958-60. Elec. Research Assn. fellow, London, Eng., 1949-50; fellow Yale, 1954-55. Mem. Am. Phys. Soc., Brazilian Acad. Scis. Author: Mathematical Structure of Theories of Viscoelasticity, 1953; Singularities of Linear System Functions (with E. P. Braga), 1961; Charge Storage in Solid Dielectrics, 1964. Research on theory absorption, math. theory of behavior of linear viscoelastic systems, dispersion equations for elec. networks, electrets, irradiation effects in solid dielectrics; established Gross equation in theory of cosmic rays for transformation from isotropical into unidirectional incidence, two-charge theory of electrets. Patentee Compton dosimeter. Home: Rua Nascimento Silva 178. Office: Comissao Nacional de Energia Nuclear, Rua General Severino 90, Rio de Janeiro, Brazil.*

GROSS, Desiderio, physician; b. Budapest, Hungary, Nov. 4, 1902; s. Emanuel and Ana (Weiss) G.; student German U. Prague, 1920-22, Med. Faculty Pozsony, Hungary, 1923, Med. Faculty, Pécs, Hungary, 1925-57; M.D., U. Pécs, 1927, U. Santiago de Chile, 1930. Prof. agregé clin. electrocardiography, Santiago de Chile, 1935-36; practice medicine, specializing in cardiology, Santiago de Chile, since 1930 —. Fellow Am. Coll. Cardiology; mem. Sociedad Chilena de Cardiologia, Research and publs. dealing with problems of clinical electrocardiography. Address: 418 San Antonio, Santiago de Chile.*

GROSS, Edward, sociologist; b. Czibo, Rumania, Oct. 18, 1921; s. Samuel and Dora (Levi) G.; B.A., U. B.C., Vancouver, Can., 1942; M.A., U. Toronto (Ont., Can.), 1943; Ph.D., U. Chgo., 1947; m. Florence Rebecca Goldman, Feb. 18, 1943; children—David Philip, Deborah Laura. Came to U. S., 1945, naturalized, 1951. Faculty, Wash. State U., Pullman, 1947-51, 53-60, asso. prof., 1955-60; faculty U. Wash., Seattle, 1951-53, prof. sociology, 1965-66, 67—; prof. U. Minn., 1960-65, 66-67. cons. to local and nat. govt. agys.; mem. panel on counseling and selection Nat. Manpower Adv. Com., 1965—. Recipient Gov.-Gen's Gold medal, 1942. Mem. Am. Sociol. Assn., Pacific Sociol. Soc., Soc. for Study Social Problems, A.A.A.S. Author: Work and Society, 1958; Industry and Social Life, 1965; The Professionals, 1968; Academic Adminstrators and University Goals, 1968; also numerous articles. Research on importance of small groups and informal attachments to organizational performance, structure of univs. Office: Dept. Sociology, U. Wash., Seattle 98105.*

GROSS, Evgenii Fedorovich, Russian physicist; b. Oct. 20, 1897; grad. Leningrad U., 1924. Staff, Leningrad U., 1924; prof. from 1938; staff, Physico-Technical Institute, USSR Acad. Scis., from 1944. Recipient Stalin Prize, 1946. Corr. mem., USSR Acad. Scis., 1946—. Author: Light Scattering, 1934; Debye Transverse Thermal Waves and Light Scattering in Crystals, 1940; Light Scattering and Relaxation, 1940; Entropy Fluctuation in a Liquid and the Rayleigh Line, 1946; The Excitone Structure of the Spectrum Curves of the Internal Photoelectrical Effect in Crystals, 1956; The Linear and Quadratic Zeeman Effects and Diamagnetism of the Copper Oxide Excitone, 1956. Research in spectroscopy of solids; discovered fine Rayleigh line structure in light scattering in crystals and liquids, 1930; proposed method of determining orientation relaxation time of molecules from anisotropic scattering spectrum, 1940; discovered optical spectrum of excitones, 1951; studied Zeeman and Stark effects in excitones; excitone dissociation; radiation spectrum. Address: Fiziko-tekhnichesky institut AN SSR, Sosnovka 2, Leningrad, USSR.

GROSS, Johann Friedrich, physicist; b. 1732; prof. exptl. physics Karl's Sch., Stuttgart, Germany; discovered elec. pause (zones in an electrically charged body in which no sparks can be produced), 1776. Died 1795.

GROSS, Ludwik, Am. physician; b. Krakow, Poland, Sept. 11, 1904; s. Adolf and Augusta Silbiger (Alexander) G.; M.D., Iagellon U., Krakow, 1929; m. Dorothy Nelson, Oct. 7, 1943; 1 dau., Augusta Helene. Came to U. S., 1940, naturalized, 1943. Cancer research Pasteur Inst., Paris, 1932-39, Christ Hosp., Cin., 1941-43; cons., 1953-56; asso. scientist Sloan-Kettering Inst., Meml. Center, N.Y.C., 1957-60; chief cancer research VA Hosp., Bronx, N.Y., 1945—. Recipient Prix Chevillon, Acad. Medicine, Paris, 1937; R. R. DeVilliers Found. award for leukemia research, N.Y.C., 1953; Walker prize Royal Coll. Surgeons, London, 1962; Pasteur Silver medal Pasteur Inst., Paris, 1962; Lucy Wortham James award James Ewing Soc., 1962; WHO UN prize, 1962; Bertner Found. award U. Tex., 1963; Polish Med. Alliance Distinguished Achievement award, 1964; Albert Einstein Med. Center's Centennial medal, Phila., 1965. Diplomate Am. Bd. Internal Medicine. Fellow A.C.P., A.A.A.S., Internat. Soc. Hematology; mem. A.M.A., Am. Soc. Hematologists, Am. Assn. Cancer Research, Assn. Mil. Surgeons U. S., Soc. for Exptl. Biology and Medicine, N.Y. State, Bronx County med. socs. Author: Oncogenic Viruses, 1961; also numerous articles on exptl. cancer and leukemia. Research on transmission of mouse leukemia by filtrates, isolation mouse leukemia virus, isolation oncogenic virus causing parotid tumors and other neoplasms in newborn mice. Home: 29 Ramona Ct., New Rochelle, N.Y. 10804. Office: VA Hosp., 130 W. Kingsbridge Rd., Bronx, N.Y. 10468.*

GROSS, Paul, physician; b. Berlin, Germany, June 8, 1902; s. Martin and Julie (Baumgarten) G.; A.B., Western Res. U., 1924, M.D., 1927; M.A., 1929; m. Dorothy J. V. Mulac, Aug. 4, 1930; children—Julianne (Mrs. A. Burt Sauvageot), Paul James, Peter Martin, John Edwin. Came to U. S., 1913, naturalized, 1921. Dir., Research Lab., Indsl. Hygiene Found., Mellon Inst., Pitts., 1948—, sr. fellow Mellon Inst., 1954—; instr. Western Res. U., 1931-35, curator mus., 1933-35; adj. prof. Grad. Sch. Pub. Health, U. Pitts., 1960—. Mem. Am. Soc. Clin. Pathology, Am. Indsl. Hygiene Assn., Indsl. Med. Assn., A.C.P., Am. Coll. Chest Physicians, Internat. Acad. Pathologists, Coll. Am. Pathologists, Am. Soc. Exptl. Pathology, Am. Assn. Pathologists, and Bacteriologists, Am. Soc. for Exptl. Biology and Medicine. Author: (with Mellon, Cooper) Sulfanilamide Therapy of Bacterial Infections, 1938; (with T. F. Hatch) Pulmonary Deposition and Retention of Inhaled Aerosols, 1964; also numerous articles. Research on mechanisms chronic pulmonary inflammations, lung clearance mechanism. Home: 207 Iola St., Glenshaw, Pa. 15116. Office: 4400 5th Av., Pitts. 15213.*

GROSS, Paul Magnus, Am. physical chemist; b. N.Y. City, Sept. 15, 1895; s. Magnus and Ellen (Sullivan) G; B.S., Coll. City New York, 1916; M.S., Columbia, 1917, Ph.D., 1919; grad. study U. of Leipzig; m. Gladys Cobb Petersen, Aug. 4, 1918; children—Paul Magnus, Beatrix Cobb. Instr. in chemistry, Coll. of the City of New York, 1916-18; asst. prof. chemistry, Trinity Coll. (now Duke University), 1919-20, prof., 1920-25, Duke, 1925-66, professor emeritus, 1966—, chairman of the department of chemistry, 1921-48, dean Grad. Sch., 1947-52, v.p., 1949-60, dean of university, 1952-58; president Oak Ridge Inst. Nuclear Studies since 1949. Served as 2d lt., C.W.S., 1918; vice chmn. bd. National Sci. Foundation, 1955-62; sci. advisor U. S. delegation to 5th conf. U.N.E.S.C.O., Paris, 1949; cons. NASA, 1963—; mem. N.C. bd. Sci. and Space Tech.; mem. N.C. Gov.'s Sci. Adv. Com., nat. adv. environmental health com. USPHS, army adv. panel to Sec. Army, panel toxicol. information Pres.'s Sci. Adv. Com.; chmn. munitions adv. group U. S. Army Munitions Command; former member of the board American Cancer Adv. Council NIH. Treas. Council So. Univs., 1954—; trustee Woodrow Wilson National Fellowship Found., 1960—. Awarded President's Medal for Merit, 1948, medal from Southern Assn. Sci. and Industry, 1951; Townsend Harris Award, 1953; Carnegie Manship Award, 1954; Distinguished Civilian Service award AUS, 1963; Comdr. Civil Order British Empire. Fellow Am. Phys. Soc., N.Y. Acad. Sci.; mem. Am. Chem. Soc. (Herty medal Ga. sect. 1945; medal Fla. sect. 1952), A.A.A.S. (pres. 1962, chmn. of board of directors 1963), American Chemical Society, American Assn. U. Professor, National Research Council, Sigma Xi. Phi Beta Kappa. Author: Elements of Physical Chemistry (with J. M. Bell), 1929. Contbr. to chem. and phys. jours. Research in the thermodynamics of solutions; biochemistry of tobacco; electrical properties of non-aqueous solutions and of organic molecules through dipole moment measurements. Home: Hope Valley, Durham, N.C. 27707.*

GROSS, Robert Edward, Am. surgeon; b. Balt., July 2, 1905; B.A., Carleton Coll., 1927, D.Sc. (hon.), 1951; M.D., Harvard, 1931; M.D. (hon.), Louvain (Belgium) U., 1959, Turin (Italy) U., 1961; D.Sc. (hon.), Suffolk U., 1962, U. Sheffield (Eng.), 1963. George Groham Peters travelling fellow Peter Brent Brigham Hosp., Boston, 1937, resident in surgery, 1938-39, mem. staff, 1939—, sr. asso. in surgery, 1946—; asso. vis. surgeon Children's Hosp., Boston, 1939-46, surgeon in chief, 1947—; faculty Harvard Med. Sch., Boston, 1934—, Ladd prof. children's surgery, 1947—. Cons. numerous hosps. Diplomate Am. Bd. Surgery, Am. Bd. Thoracic Surgery. Mem. Am. Assn. Pathologists and Bacteriologists, Mass. Med. Soc., A.M.A., Boston Med. History Club, New Eng. Pediatric Soc., Am. Assn. Thoracic Surgery, A.C.S., Am. Soc. Exptl. Pathology, Soc. U. Surgeons, Soc. Pediatric Research, Soc. Clin. Surgery, New Eng., Boston surg. socs., Am., New Eng., Mass. (past pres., dir. 1965) heart assns., Am. Surg. Assn., Soc. Vascular Surgery, Am. Acad. Pediatrics, Am. Acad. Arts and Scis., Am. Coll. Cardiology, Pan Am. Med. Assn., Brit. Assn. Pediatric Surgeons. Author: (with W. E. Ladd) Abdominal Surgery in Infancy and Childhood, 1941; Surgical Treatment for Abnormalities of the Heart and Great Vessels, 1947; The Surgery of Infancy and Childhood, 1953; Trattato di Chirurgia Infantile, 1956; Cirugia Infantil, 1956; also numerous articles. Research on surgical correction of coarctation of aorta; patent ductus arteriosus; ventricular septal defects; atrial septal defects; abnormalities of heart and great vessels; surgical repair of tetralogy of Fallot; pump-oxygenator for maintaining circulation during open-heart surgery. Home: 25 Wayside Inn Rd., Framingham, Mass. Office: 300 Longwood Av., Boston 02115.*

GROSS, Rudolf Josef, German crystallographer, mineralogist; b. Gaustadt nr. Bamberg, Germany, Oct. 22, 1888; s. Benedict and Anna (Stürzer) G.; student U. Jena, U. Leipzig (all Germany); Ph.D., U. Rostock (Germany); m. Nora Blassmann, 1919; 2 daus. Asst. to Rinnes, U. Leipzig; joined faculty U. Greifswald (Germany), 1918, became prof., 1922, rector, 1949-50; named asso. prof., Hamburg, Germany, 1919. Research and publs. on growth of crystals, especially nucleus formation in crystallization, collective crystallization, oriented deformation of isotypic crystals, also hardening of metals, grinding techniques; one of 1st to apply X-ray diffraction to determination of mineral structure. Died Greifswald, July 12, 1954.

GROSS, Samuel David, Am. surgeon; b. Easton, Pa., July 8, 1805; s. Philip and Johanna (Brown) G.; grad. Jefferson Med. Coll., 1828; D.C.L., Oxford (Eng.) U.; LL.D., Cambridge (Eng.), U., U. Edinburgh (Scotland), U. Pa.; m. Louisa Weissell, 1828. Apptd. demonstrator anatomy Med. Coll., O., 1833; prof. path. anatomy Cin. Med. Coll., 1835; prof. surgery U. Louisville, 1840; Jefferson Med. Coll., 1856; a founder A.M.A.; founder Phila. Path. Soc., Phila. Acad. Surgery, Am. Surg. Soc.; established Acad. Surgery prize for original articles; presided over Internat. Congress Surgeons, 1876; v.p. German Surg. Soc.; inventor numerous instruments; translator from French and German med. texts. Author: Elements of Pathological Anatomy (1st work on subject in English lang.), 1839; A Practical Treatise on Foreign Bodies in the Air Passages, 1854; A System of Surgery, Pathological, Diagnostic, Theraputive and Operative, 2 vols., 1859. Published 1st account of the use of adhesive plaster as a means of extension in treatment of fracture, 1830; systematically studied morbid anatomy. Died Phila., May 6, 1884.

GROSS, Samuel Weissell, Am. surgeon; b. Cin., Feb. 4, 1837; s. Samuel David and Louisa (Weissell) G.; grad. Jefferson Med. Coll., 1857; m. Grace Linzee Revere, Dec. 28, 1876. Served as surgeon U. S. Army during Civil War; commd. lt. col. M.C., 1865; surgeon Phila. Hosp; surgeon hosp. of Jefferson Med. Coll.; founded, developed present-day radical operation for cancer; prof. principles of surgery and clin. surgery Jefferson Med. Coll., 1882; one of 1st physicians in Phila. to use antiseptic surgery. Author: Practical Treatise on Tumors of the Mammary Gland, 1880; Practical Treatise on Impotence, Sterility and Allied Disorders of the Male Sexual Organs, 1881, Died Phila., Apr. 16, 1889.

GROSS, Sidney William, Am. physician; b. Cleve., Aug. 28, 1904; s. Joseph and Frieda (Weiss) G.; A.B., Western Res. U., 1924, M.D., 1928; m. Evelyn E. Swanson, Apr. 27, 1937; 1 son, Samuel Warren; m. 2d., Molly Harr, 1966. Dir. neurol. surgery Mt.

Sinai Hosp., N.Y.C., 1955—, City Hosp., N.Y.C., 1936—; practice neurol. surgery, N.Y.C., 1934—; prof. neurosurgery Mt. Sinai Sch. Medicine, 1965—. Fellow A.C.S.; mem. Am., N.Y. neurol. socs., Harvey Cushing Soc., N.Y. Soc. Neurosurgery, Phi Beta Kappa, Alpha Omega Alpha. Diplomate Am. Bd. Neurol. Surgery. Author: (with William Ehrilich) Diagnosis and Treatment of Head Injuries, 1940; also numerous articles. Research on use organiciodid for cerebral augiography, opaque ventriculography, surgery brain tumors and head injuries. Address: 44 E. 81st St., N.Y.C. 10028.*

GROSS, Walter, German pathologist; b. Baden, Germany, Jan. 12, 1878; s. Ferdinand and Bertha (Doll) G.; student medicine Lausanne, Switzerland, Heidelberg, Germany, Berlin; state exam., 1903; M.D., Heidelberg, 1905; m. Margarethe Schmitzdorf, 1909; 2 sons. Asst. to Vierordt, Pawlow, F. Müller, P. Ernst; joined faculty, 1911; became lectr. forensic medicine, 1912; army pathologist in field during World War I; became asso. prof. U. Dorpat (Estonia), 1917; named prof. U. Greifswald (Germany), 1921, U. Münster (Germany), 1924. Research on relationship between histological changes and functional disturbances of kidneys, influence of vital coloring on cells and on dye secretion, histotech. coloring, glomerulonephritis, encephalitis, infections. Died Münster, Nov. 14, 1933.

GROSS, Wilhelm, Austrian mathematician; b. Molln, Austria, Mar. 24, 1886; s. Wilhelm and Anna (Maurer) G.; Ph.D., U. Vienna, 1910; postgrad. U. Göttingen (Germany), 1910-12. Became asst. U. Vienna, 1912, joined faculty math., 1913; named asso. prof. U. Czernowitz, Rumania (now USSR), 1918. Recipient Richard Lieben prize Acad. Scis., Vienna. Research in differential equations, theory of functions, variation calculus, theory of invariants, theory of quantities, geometry, especially singularities of analytic functions, isoperimetric problem in double integrals, expansion of theory of invariants; demonstrated smallest surface of quantities of points. Died Vienna, Oct. 22, 1918.

GROSSBERG, Sidney Edward, Am. physician; b. Miami, Fla., Nov. 13, 1929; s. Lazar and Anita (Mandell) G.; B.S., Emory U., 1951, M.D., 1954; m. Josette Brugerolle, May 20, 1959; children—Daniel Eliot, Leslie David. Fellow, Sch. Medicine Johns Hopkins, 1958-59; practice medicine, specializing in virology and infectious diseases, Mpls., 1959-62, N.Y.C., 1962-66; faculty Med. Sch. U. Minn., 1959-62; asst. prof. microbiology Med. Coll. Cornell U., 1962-66; asst. attending N.Y. Hosp., 1964-66; prof., chmn. dept. microbiology Marquette U. Sch. Medicine, Milw., 1966—; cons., attending staff Milw. County Gen. Hosp., 1966—. Vis. investigator Pasteur Inst., Paris, 1964-65; USPHS fellow Nat. Inst. Allergy and Infectious Diseases, 1959-61. Recipient Research Career Devel. award USPHS, 1961; John and Mary Markle scholar, 1965—. Mem. Am. Assn. Immunologists, Am. Fedn. Clin. Research, Am. Soc. Microbiologists, Infectious Disease Soc. Am., Soc. Exptl. Biology and Medicine, Sigma Xi. Research, publs. on inhibition of viral RNA replication by interferon, mechanisms viral interference at a distant embryonic organ site, sequelae of encephalitis, cell biology cells; discovery viral-induced hyperlipemia. Home: 3005 E. Kenwood Blvd., Milw. 53211. Office: 561 N. 15th St., Milw. 53233.*

GROSSE, Aristid V., chemist; b. Riga, Russia, Jan. 4, 1905; s. Victor G. and Ella (Lieven) G.; Dr. Engring., Technische Hochschule, Berlin-Charlottenburg, Germany, 1927; came to U. S., 1930, naturalized, 1937; m. Irene Lieven, Mar. 3, 1932; 1 son, Aristid. Research chemist, Kaiser Wilhelm Inst. for Chemistry, Berlin-Dahlem, Germany, 1927-28, Med. Analysis Labs., Shanghai, China, 1928-29; research asso. Technische Hochschule, Berlin-Charlottenburg, Germany, 1929-32; vis. asst. prof., Dept. of Chemistry, U. Chgo., 1931-40; research asso., Universal Oil Products Co., Chgo., 1930-35, asso. dir. research, 1935-40; John Simon Guggenheim Research fellow, Dept. of Physics, Columbia U., 1940-41; asso. with H. C. Urey in war research labs., Columbia U. (Manhattan Project), 1942-43; cons. on synthetic rubber, Houdry Process Corp., Phila., since 1942; chief cons. on synthetic rubber, WPB, Washington, 1942-43; dir. research, Houdry Labs., Houdry Process Corp. of Pa., 1943-48. Pres. Research Inst. of Temple U., Phila., since 1948. Mem. Am. Rubber Mission to USSR, 1942-43. Mem. Am. Chem. Soc. (sec. com. on foreign compendia), A.A.-A.S., Am. Phys. Soc., Soc. Am. Mil. Engrs., Am. Inst. Aeros. and Astronautics, Sigma Xi. Author: (with E. Krause) Chemie der Metallorganischen Verbindungen, 1937. Specialized in catalytic chemistry of hydrocarbons; discovered (with V. N. Ipatieff) reactions of paraffins with olefins, aromatics, dehydrogenation of paraffins to olefins; with W. Mattox and J. C. Morrell, cyclization of straight chain hydrocarbons to aromatics; (with J. C. Morrell and J. Mavity) dehydrogenation of n-butane and n-butenes to butadiene; (with C. Linn) hydrogen fluoride catalytic alkylation process for production of aviation gasoline; radioactivity; isolated element 91, protactinium, 1927; proved (with A. Nier, E. Booth and J. R. Dunning) slow neutron fission of uranium 235, 1940; developed (with J. R. Dunning and E. Booth), 1940-43, fundamentals of diffusion process for separation of U235; discovered (with W. Libby) cosmic ray carbon, 1947; cosmic ray tritium, 1950; high temperature research; flame studies; rocket propulsion fuels; ozone studies; containment of liquid metallic substances up to 5000° K and properties of liquid metals up to critical temperature; noble gases; solidification and efflux time of soap bubbles. Home: 456 Glynwynne Av., Haverford, Pa. 19041. Office: Research Inst. of Temple Univ., Phila. 19144.*

GROSSE-BROCKHOFF, Franz, German physician; b. Osterfeld, Germany, Nov. 26, 1907; s. Heinrich and Elisabeth (Niehaus) G.-B.; student univs. Würzburg, Leipzig, Berlin, Kiel, Cologne, Graz, Bonn; degree, Bonn, 1932; m. Maria Lenz, Oct. 5, 1939; children—Ursula, Rita, Rudolf, Hans-Heinrich. Asst., univs. Bonn, Göttingen, 1932-36; asst. prof. Med. Clinic, U. Bonn, 1936-54; dir. I. Med. Clinic, U. Düsseldorf, 1954— , prof., 1966—. Mem. Fed. Health Bd., Acad. Leopoldina. Author: Pathologische Physiologie, 1950; also numerous articles. Research on exptl. physiology of circulation, clin. cardiology, path. physiology of heart and lungs. Address: 4 Moorenstrasse, Düsseldorf, West Germany.*

GROSSELET, Jules Auguste Alexandre, French geologist; b. Cambrai, France, Apr. 19, 1832; Doctuer ès Sciences, 1860. Prof., Quesnoy Coll., 1852; asst. Sorbonne, 1853; prof. geology Lille (France) Faculty Scis., from 1864, dean, 1902. Founder Geol. Soc. World; mem. French Acad. Scis., 1913. Author: L'ardenne geologie des terrains primaires de Belgique (complete geol. description of Ardennes region), 1860. Authority on geology applied to indsl. problems. Died Lille, Mar. 20, 1916.

GROSSER, Otto, anatomist; b. Vienna, Austria, Nov. 21, 1873; s. Anton and Marie (Persina) G.; M.D., U. Vienna, 1899; M.D. (hon.), Breslau (now Wroclaw, Poland), 1936; m. Maria von Winiwarter; 5 daus. Joined faculty U. Vienna, 1902, asso. prof., 1907; named prof. anatomy German U. Prague (Czechoslovakia), 1909-45; 2d chmn. Internat. Embryol. Inst., Utrecht, Holland. Mem. German Soc. Arts and Scis. Prague (became chmn. 1918), Berlin Acad. Scis. (corr.), Vienna Acad. Scis. (corr.), Leopoldina. Author: Vergleichende Anatomie und Entwicklungsgeschichte der Eihäute und der Placenta . . . , 1909; Die Wege der feralen Ernährung innerhalb der Säugetierreihe, 1909; Frühentwicklung, Eihautbildung und Placenta . . . , 1909; Die Wege der fetalen Ernährung innerhalb der Säugetierreihe, 1909; (with G. Politzer) Frühentwicklung, Eihautbildung und Placentation der Entwicklungsgeschichte des Menschen, 1944; Vorlesungen über topographische Anatomie des Menschen, 1950; also articles. Research on embryology, evolutionary basis for body defects, devel. human intestine and respiratory tract, devel. trophoblasts, nourishment in pregnant animals; developed new classification of placenta. Died Thumersbach, Mar. 23, 1951.

GROSSET, Antonin Louis Charles Sébastien, French surgeon; b. Fécamp, France, Jan. 20, 1872. Prof. clin. surgery Paris Faculty Medicine, from 1916; mem. Acad. Medicine, Acad. Surgery, French Acad. Scis., 1934. Author: Nouvelle pratique medio-chirurgie; (with Duplay and Reclus) Traité de chirurgie. Studied anatomy of uterus and kidney; credited with new surg. techniques of gastrotomy, gastroenterostomy and duodenotomy. Died Paris, Oct. 24, 1944.

GROSSETESTE, Robert, Brit. philosopher; b. Stradbroke, Suffolk, Eng., circa 1175; ed. Oxford and Paris; 1st chancellor U. Oxford; 1st rector Oxford Franciscan Sch., 1224; bishop of Lincoln, 1235-53; chief organizer philos. studies at Oxford; taught Roger Bacon. Author: Treatise on the compotus, circa 1232; Compendium sphaerae; also commentaries on the Posterior Analytics, Aristotle's Physics; wrote on natural sci. including sound, motion, light, heat. Made first complete translation of Nicomachean Ethics into Latin, circa 1240-43. Research on mirrors and lenses; formulated explanation for the rainbow; held by some to be pioneer of inductive method. Died Buckden, Huntingdon, Eng., Oct. 9, 1253.

GROSSFELD, Johann Gerhard, German food chemist; b. Bentheim nr. Osnabrück, Germany, Feb. 25, 1889; s. Johann Heinrich and Maria (Lüfolding) G.; student chemistry U. Freiburg (Germany), U. Munich (Germany); Ph.D., U. Münster (Germany), 1913; m. Elisabeth Masbaum, 1919. Food chemist Chem. Research Bur., Recklinghausen, Germany, 1914-27; with State Food Research Inst., Berlin, (later Reich Inst. for Food and Pharm. Chemistry), 1927-44. Author: Anleitung zur Untersuchung der Lebensmittel, 1927; Handbuch der Eierkunde, 1938; also numerous articles. Improved methods of research, especially analytic processes for investigating foods; studied fats, oils. Editor, Zeitschrift für Untersuchung der Lebensmittle; co-editor Handbuch der Lebensmittelchemie, IV-IX, 1939-42. Died Munich, June 3, 1944.

GROSSKREUTZ, Joseph Charles, Am. physicist; b. Springfield, Mo., Jan. 5, 1922; s. Joseph Charles and Helen (Mobley) G.; B.S., Drury Coll., 1943; M.A., Wash. U., 1948, Ph.D., 1950; m. Mary Catherine Schubel, Sept. 7, 1949; children—Cynthia Lee, Barbara Helen. Research asst. Wash. U., 1947-50; research physicist Cal. Research Corp., La Habra, Cal., 1950-52; with U. Tex., 1952-56, research scientist Nuclear Physics Lab., 1952-56; sr. physicist Midwest Research Inst., Kansas City, Mo., 1956-59, prin. physicist, 1959-63, sr. advisor, 1963—. Fellow Am. Phys. Soc.; mem. Am. Soc. Testing and Materials, Am. Inst. Mech. Engrs., Electron Microscopy Soc. Am., Sigma Xi. Research on micro-mechanisms of fatigue fracture; electron microscopy of metal surfaces; small angle x-ray scattering from metals; phys. structure of wheat protein, nuclear energy level spectra; seismic surface waves. Home: 8221 Linden Dr., Prairie Village, Kan. 66208. Office: 425 Volker Blvd., Kansas City, Mo. 64110.*

GROSSMAN, Andrzej, Polish chemist; b. Warsaw, Poland, Apr. 6, 1908; s. Wladyslaw and Maria (Sawrymowicz) G.; m. Henryka Grabowska, Aug. 25, 1942; children—Andrzej, Anna. Engr. coke-oven plant, Orzegow, Upper Silesia, Poland, 1934-36, gas work, Warsaw, 1937-39; tech. dir. carbon electrodes plant, Raciborz, Poland, 1945-46; dir. carbon electrodes plant, Starogard, Poland, 1947-48; tech. dir. trust of coking plants, Zabrze, Poland, 1949-53; prof., acting vice rector Tech. U. Silesia, Gliwice, Poland, 1954—. Recipient Golden Cross of Merit, 1959; Knight Cross Order Polonia Restituta, 1967. Mem. Polish Soc. Chemistry. Author: (with Balczewski) Technological Control in Coking Plants, 1956; also articles. Research on coal carbonization, graphite for nuclear purposes, radioactivity of water, purification of wastewaters. Home: 366 Wolnosci, Zabrze, Poland. Office: 23 Konarskiego, Gliwice, Poland.*

GROSSMAN, Burton Jay, Am. physician; b. Chgo., Nov. 27, 1924; s. Paul and Neva (Sonnenschein) G.; B.S., U. Chgo., 1946, M.D., 1949. With U. Chgo. Sch. Medicine, 1948—, prof. dept. pediatrics, 1961—; med. dir. La Rabida Sanitarium, 1962—. Mem. Am., Ill. acads. pediatrics, Soc. Pediatric Research, Midwest Soc. Pediatric Research (mem. council, 1959-62), Nat. Soc. Med. Research, Chgo. Pediatric Soc., Chgo., Ill., Am. heart assns., Am. Bd. Pediatrics, Chgo. Rheumatism Soc., Soc. Med. History of Chgo., Ill. Assn. Maternal and Child Health, Am. Pediatric Soc., Sigma Xi. Contbr. numerous articles to med. jours. Research on optimum therapy for children with rheumatic fever and rheumatic heart disease, physiol. effects of naturally occurring mucopolysaccharides especially in field of lipid coagulation. Home: 5050 East End Av., Chgo. 60615. Office: La Rabida Sanitarium, Chgo. 60649.*

GROSSMAN, Moses, Am. physician; b. Kiev, Russia, Oct. 14, 1921; s. Gregory and Klara (Kaufman) G.; came to U. S., 1941, naturalized, 1944; A.B., U. Cal. at Berkeley, 1943; M.D., U. Cal. at San Francisco, 1946; m. Verle Anne Campbell, July 14, 1951; children—Deborah, Pamela, David, Daniel. Faculty, U. Cal. at San Francisco, 1951—, prof. pediatrics, 1964—; chief pediatrics and infectious diseases San Francisco Gen. Hosp., 1959—; asso. dean. U. Cal. Sch. Medicine, 1964—. Cons. to A.R.C. for vaccine immune globulin program, 1959. Mem. Am. Acad. Pediatrics, Soc. for Pediatric Research, Infectious Disease Soc. Am. Research, publs. in infections in hosps., infections in the neonate. Home: 1001 Ulloa St., San Francisco. 94127.*

GROSSMAN, Sebastian Peter, psychologist, educator; b. Coburg, Bavaria, Jan. 21, 1934; s. Otto and Arnet (Peipers) G.; B.A., U. Md., 1958; M.S., Yale, 1959, Ph.D., 1961; m. Lore Bensel, June 30, 1955. Asst. prof. psychology U. Ia., 1961-64; asso. prof. psychology U. Chgo., 1964-67, prof., 1967—. Mem. Am. Psychol. Assn., Am. Physiol. Soc., Royal Soc. Medicine, Sigma Xi, Phi Kappa Phi. Author: A Textbook in Physiological Psychology, 1967. Regional editor Jour. Physiology and Behavior, 1965—. Publs. on devel. technique for direct injection of drugs into brain for study of brain-governed activities. Home: 1156 E. 56th St., Chgo. 60637.*

GROSSMANN, Gustav, X-ray physicist; b. Budapest, Hungary, Aug. 10, 1878; s. Leopold and Karoline (Pick) G.; student U. Budapest, 1895-97; mech. engring. degree Fed. Tech. U., Zürich, Switzerland, 1900; Ph.D., U. Zürich, 1903; m. Thea Pohl, 1918. Lectr. physics and electrotechnics Fed. Tech. U., Zürich, 1900-5; lab. engr. various electro-tech. firms, including AEG Berlin, 1905-10; joined firm Siemens & Halske AG Berlin, 1911, with Vienna (Austria) works, 1915-18, named dir. electro-med. dept., Berlin-Siemensstadt, 1919; bd. dir. mut. undertaking of firms Siemens-Reininger-Feifa GmbH Berlin; Reiniger, Gebbert & Schall AG Erlangen; Phönix-Röntgenröhrenfabriken AG Rudolstadt (became Siemens-Reiniger-Werke AG Berlin 1932), 1925-32; ind. scientist, 1933-42; became adviser Hungarian Ministry Health, 1942; became chmn. X-ray group Assn. for Communications Tech., 1951; named chmn. sci. dept. Budapest Inst. for Oncology, 1954. Author: Einführung in die Röntgentechnik, 1912; Physikalische und technische Grundlagen der Röntgentherapie, 1925. Under his direction Siemens & Halske developed new practical examination apparatus; developed direct radiation measurement devices, X-ray generators with valve rectifiers, device for making X-ray films; research on protection from radiation and high tension. Died Budapest, Jan. 16, 1957.

GROSSMANN, Louis Adolf, meteorologist; b. Bklyn., Sept. 24, 1855; s. Louis and Hulda (Foerster) G.; student U. Berlin (Germany), Breslau (now Wroclaw, Poland), Ph.D., 1880; m. Cornelie Doerth, 1892; 3 sons, including Hans, 1 dau. Asst. to A. Müttrich, Eberswalde (Germany) Forestry Acad., 1880-86; joined weather service dept. German Marine Obs., 1886, named prof., 1904; became head dept., 1907. Research and publs. on theoretical math. formulations, especially statis. studies of climate, weather of German coast, including storms and tidal waves; corrected and expanded weather prognosis rules of Guilbert for alternating displacement of high and low pressure zones (Guilbert-Grossmann rules); devel. weather service. Died Hamburg, Germany, Feb. 9, 1917.

GROSSMANN, Marcel Hans, mathematician; b. Budapest, Apr. 9, 1878; s. Jules and Henriette (Lichtenhahn) G.; studied math. Zürich Polytechnikum; Ph.D., U. Zürich, 1902; m. Anna Keller, 1903; 1 son, Marcel, 1 dau. Became asst. to W. Fiedler, Polytechnikum, Zürich, 1900; named prof. Frauenfeld Canton Sch., 1901, Oberrealschule, Basel, Switzerland, 1905; became prof. geometry Fed. Tech. U., Zurich, 1907. Mem. Swiss Math. Soc. (hon., co-founder). Author: (with A. Einstein) Entwurf einer verallgemeinerten Relativitätstheorie und einer Theorie der Gravitation, 1913; also articles. Worked with A. Einstein and introduced him to theory of invariants; studied theories of relativity and gravitation; advanced reform of Swiss secondary sch. edn. Died Zürich, Sept. 7, 1936.

GROSSMANN, Vojtech, Czechoslovakian pharmacologist; b. Vlcovice, Czechoslovakia, Feb. 2, 1922; s. Vilém and Josefa (Skrabalova) G.; M.D., Charles U., Prague, Czechoslovakia, 1950; m. Vera Cespivova, Dec. 28, 1947; children—Irena, Vojtech. Mem. staff Charles U., 1950-52, faculty, 1955—, prof. pharmacy, head dept. pharmacy of Med. Faculty, Hradec Králové, Czechoslovakia, 1963—; leader Pharm. Inst., Mil. Acad. J. E. Purkyne, Hradec Králové, 1952-55. Mem. com. Ministry of Health, mem. State Problem Commn. CSSR. Recipient Meritorious diploma for bldg. health service in East Bohemian Country. Mem. Soc. Medicine J. E. Purkyne, Assn. Nuclear Medicine and Radiohygiene, Assn. Pharmacy, European Soc. Radio Biology, European Soc. Research of Drug Toxicity. Author: Textbook of Pharmacy for Nurses, 1956, rev., 1959, 60; (with Votava, Rasková) Pharmacology, 1959. Research and publs. on change of reactivity of irradiated organisms to applied drugs; estimated changes or resorption from gastrointestinal tract; stronger binding on plasma proteins and slower penetration through the hematoencephalic barrier for drug-cation in contrast to drug-anions. Home: 1041 Svendova Hradec Králové, Czechoslovakia.*

GROSSWALD, Emil, mathematician; b. Bucharest, Rumania, Dec. 15, 1912; s. Paul and Elsa (Iscovitsch) G.; B.Sc., Seminarul Pedagogic U., Bucharest, 1929; M.Sc., U. Bucharest, 1932; M.E.E., Poly. Sch. Bucharest, 1936; Engr., Ecole Supériore d'Electricité, Paris, France, 1940; Ph.D., U. Pa., 1950; m. Elizabeth Ronald, Nov. 20, 1950; children—Blanche, Vivian. Came to U. S., 1946, naturalized, 1952. Engr., Gen. Electric Co., Milano, Italy, 1937; with Co. Electro-Technica, Bucharest, 1937-39; contractor, Havana, Cuba, 1942-43; faculty Academia de Estudios U., Havana, 1943-46, U. P.R., Rio Piedras, 1947-48; faculty U. Pa., Phila., 1948-50, 1952—, prof. math., 1960—; lectr. U. Sask., Saskatoon, Can., 1950-51; mem. Inst. for Advance Study, Princeton, 1951-52, 1959-60; vis. prof. U. Paris (Sorbonne), 1964-65. Mem. Am., French math. socs., Math. Assn. Am. (gov. Phila. sect. 1966—), A.A.A.S., Sigma Xi. Author: Topics from the Theory of Numbers, 1966. Research, publs. on Jacobi polynomials and Mean Value Theorem, Bessel polynomials and their Galois group; study of zeta-function in critical strip, average order of arithmetical functions, oscillation theorems for arithmetical functions, neutron transport, others. Home: 318 S. St. Bernard St., Phila. 19143.*

GROSVENOR, Gilbert Hovey, geographer; b. Constantinople, Turkey, Oct. 28, 1875; s. Edwin Augustus and Lilian H. (Waters) Grosvenor; A.B., magna cum laude, Amherst Coll., 1897, A.M., 1901, Litt.D., 1926; LL.D., Georgetown U., 1921, Coll. William and Mary, 1930; Sc.D., S.D. State Sch. Mines, 1935, George Washington U., 1952; Litt. D., U. Md., 1938, U. Miami, 1944; LL.D., Lafayette Coll., 1938; m. Elsie May Bell, Oct. 23, 1900 (dec. 1964); children—Melville Bell, Gertrude Hubbard (Mrs. Samuel A. Gayley), Mabel, Lilian Waters (Mrs. Joseph Marion Jones), Alexander Graham Bell (dec.), Elsie Alexandra Carolyn (Mrs. Walter K. Myers), Gloria (Mrs. Torfinn Oftedal). Tchr., Englewood (N.J.) Acad., 1897-98; asst. editor Nat. Geog. Mag. 1899-1900, mng. editor, 1900-02, editor-in-chief, 1903-54, also dir. Nat. Geog. Soc., 1899-1920, pres., 1920-54 (during his directorship members increased from 900 to 1,950,000), chmn. Bd. Trustees, 1954-66. Del., 17th Internat. Geog. Congress, 1952. Mem. bd. mgrs. Am. Assn. to Promote Teaching of Speech to Deaf, Walter Reed Meml. Assn., Save the Redwoods League; trustee George Washington U., U. Miami. Recipient numerous honors and awards, including Culver medal (Chgo. Geog. Soc.), 1927, Bryant medal (Geog. Soc. Phila.), 1941, Grosvenor medal (Nat. Geog. Soc.), 1949, Samuel Morse award (Am. Geog. Soc.), 1952, Explorers medal (Explorers Club), 1964. Fellow Cal. Acad. Sci., A.A.A.S.; mem. Am. Antiquarian Society, Assn. of Am. Geographers, Audubon Soc., Sierra Club, Washington Acad. Scis. (v.p.), Columbia Hist. Soc., Washington Nat. Monument Soc. (hon. v.p.), Telephone Pioneers Am.; hon. corr. mem., geog. socs. of Australasia, Bolivia, Edinburgh, Guatemala, Uruguay, Rio de Janeiro, Michoacan and Lima; mem. Phi Beta Kappa, Sigma Xi. Author: Young Russia, 1914; The Land of the Best, 1916; Flags of the World (with Byron McCandless), 1917; (with W. J. Showalter), 1934; The Hawaiian Islands, 1924; Discovery and Exploration, 1924; A Maryland Pilgrimage, 1927; History of National Geographic Society, 1936; Maps for Victory, 1942; The Society's Maps of Europe, 1944; also numerous articles for mags. Asso. editor Proc. 8th Internat. Geog. Congress, 1905; Scientific Report of the Ziegler Polar Expedition of 1905-06. Editor: Scenes from Every Land, 1907, 2d series, 1909, 3d series, 1912, 4th series, 1917; Book of Birds (with Alexander Wetmore), 1937; Cumulative Index Nat. Geog. Mag., 1899-1948, 52; Review and Place Name Index of Nat. Geog. Soc. Maps of Continents and Oceans, 1944. Lake 28 miles long, discovered in Alaska, 1919, named Grosvenor Lake in recognition of his encouragement of Alaskan explorations; Gilbert Grosvenor Range in Antarctica discovered and so named by Adm. Byrd, 1929; Grosvenorflellet (mountain in Spitzbergen) named by British-Spitzbergen Expedition, 1952. Died Baddeck, N.S., Feb. 4, 1966.

GROSVENOR, William Mason, Am. chem. engr.; b. St. Louis, Oct. 5, 1873; s. William and Ellen (Sage) G.; B.S., Poly. Inst. Bklyn., 1893; postgrad. 1893; Johns Hopkins, Ph.D., U. Pa., 1898; m. Marie Dexter, Apr. 9, 1901; children—Mary Dexter (Mrs. Ralph O. Ellsworth), William Mason. Served as indsl. investigator N.Y. Tribune, 1895; was chemist Mathieson Alkali Works, 1896; Millview Mining Co., 1898; engr. and asst. treas. Ampere Electrochem. Co., 1899; asst. supt. Gen. Chem. Co., 1900-02, asst. mgr. investigating dept., 1903-04; supt. Contact Process Co., Buffalo, 1905; engr., treas. George F. Westcott Co., 1906; engr., sec. Dryer Engring. Co., 1907; cons. practice and chem. expert, 1907-44. Mem. Am. Chem. Soc., Soc. Chem. Industry (chmn. 1915), Am. Inst. Chem. Engrs. (charter), Electrochem. Soc., Am. Inst. Chemists, Profl. Engrs. Assn., A.A.A.S., Soc. de Chimie et Industrie. Contbr. to profl. jours. First to introduce high speed moving picture and projection to analysis of rapid motion in industrial work. Holder of numerous U. S. patents. Died May 30, 1944.

GROTE, Augustus Radcliffe, entomologist; b. Aigburt, Eng., Feb. 7, 1841; came to Am. as a child and lived in N.Y.; asso. with E. T. Gresson, G. H. Horn, E. Norton, C. R. Ostin Sacken, James Riddings, P. R. Uhlen; with Buffalo Soc. Natural History, 1873-82, curator; lived in New Brighton, N.Y., 1882-84; then left for Bremen, Germany. Became editor N.Am. Entomologist, 1879. Research on Lepidoptera; 1st in Am. to study Noctuidae; (with Coleman T. Robinson) described numerous species of moths; described over 1,000 new species of Lepidoptera and named the most important including Euxoa excellens, Western armyworm, army cutworm, olive green cutworm, yellow-striped armyworm, brassy cutworm, Zimmerman pine moth; numerous publs. on Lepidoptera. Died Sept. 12, 1903.

GROTE, Claus Willi Walter, German physicist; b. Bückeburg, Germany, Aug. 8, 1927; s. Heinrich F. and Hedwig (Sawartowski) G.; Diplomphysiker, Humboldt U., Berlin, Germany, 1958, Dr.rer.nat., 1962; m. Hilde Rausch, Jan. 16, 1954; children—Heinz, Jürgen. Asst., Deutsche Akademie der Wissenschaften zu Berlin, Zeuthen Kernphysikalisches Institut, 1958-60, head asst., 1960-62; sci. staff Forschungsstelle für Physik hoher Energien, Zeuthen, Germany, 1962—; lectr. exptl. methods in elementary particle physics Humboldt U., 1966—. Chmn. track chamber com. Joint Inst. for Nuclear Research, Dubna, USSR, 1966. Mem. Physikalische Gesellschaft in der DDR. Research, publs. on high energy physics, properties of very short-lived elementary particles (resonant states) using bubble chamber experiments; co-discoverer A1, A2, and L-Meson. Home: Volkswohlstrasse 120, DDR 1199, Berlin, Germany. Office: Platanenallee 6, DDR 1615 Zeuthen, Germany.*

GROTE, Louis Radcliffe, German physician; b. Bremen, Germany, Apr. 19, 1886; s. Augustus Radcliffe and Gesa Maria (Ruyfer) G.; studied first music, then medicine, Freiburg, Rostock, Munich, Göttingen, Berlin (all Germany); m. Ida Behrmann, 1911; 1 son, 3 daus. Joined U. Med. Clinic, Halle, Germany, 1914, became mem. faculty, 1922, later prof.; became dir. sanitoriums and hosps., Dresden, Germany, 1924, Frankfurt/Main, Germany, 1928, Zwickau, 1933, Dresden, 1934, Wetzlar, 1945, Glotterbad, 1952-57; named dir. Karlsruhe (Germany) Therapy Week, 1956. Mem. Europeaum Medicum Collegium (became pres. 1958). Author: Grundlagen ärztlicher Betrachtung, 1921; (with A. Brauchle) Gespräche über Schulmedizin und Naturheilkunde, 1935; Wege zur Verständnis der Naturheilkunde, 1936; Grundzüge der Individualtherapie, 1939; Allgemeine Therapeutik, 1948; also numerous articles. Research on theory of constn. and regulation pathology, phys. therapy, dietetics, metabolic problems, especially diabetes; one of 1st to study music therapy; studied ethics of med. treatment; developed circulation capacity tests; studied significance of blood pressure diagnosis; exponent of natural healing. Died Siensbach, Germany, Mar. 15, 1960.

GROTEFEND, Georg Friedrich, German philologist, archeologist; b. Münden, Prussia, June 9, 1775; s. Johann Christian and Sophie (Wolff) G.; ed. U. Göttingen, 1795-97; m. Christine Bornemann, 1805; 5 sons, 2 daus. Received an asst. mastership Göttingen gymnasium, 1797; apptd. prorector Frankfort/Main gymnasium, 1803; then corrector, 1803-21; dir. Hanover gymnasium, 1821-49. Pub. numerous papers on language including: Rudimenta Linguae Umbricae, 1835-38; Neue Beiträge zur Erläuterung der persepolitanischen Keilschrift, 1837; Rudimenta Linguae Osae, 1839; Zur Geographie und Geschichte von Altitalien, 1840-42; Neue Beiträge zur Erläuterung der babylonischen Keilschrift, 1840. Known in his lifetime as Latin and Italian philologist; most significant contbn. was first successful, though partial diciphering of Persepolitan cuneiform inscriptions (laid found. for interpretation of old Persian form of trilingual achaemendic cuneiform alphabet), 1802; founder soc. for investigation of German lang., 1815; attempted to explain extant remains of Umbrian dialect, 1835-38; important work on coins of Bactria. Died Hanover, Dec. 15, 1853.

GROTH, Paul Heinrich von, see von Groth, Paul Heinrich.

GROTH, Wilhelm, German phys. chemist; b. Hamburg, Jan. 9, 1904; s. Wilhelm and Anna (Hamdorf) G.; student Munich Inst. Tech., univs. Munich, Tübingen; Dr.rer.nat., 1927; m. Margot Effenberger, Dec. 26, 1945; children—Thomas, Astrid. Faculty, U. Hamburg, 1932-50, asso. prof., 1948-50; prof. U. Bonn, 1950—, dir. Inst. Phys. Chemistry; dir. Inst. Phys. Chemistry, Nuclear Research Installation, Jülich. Mem. German AEC. Mem. German Bunsen Soc., Soc. German Chemists, German Phys. Soc., Leopoldina, German Atomic Forum. Author: (with K. Beyerle, P. Harteck, H. Jensen) Über Gaszentrifugen, 1950; also numerous articles. Research on photochemistry in far ultraviolet, reaction kinetics, mass spectrometry, isotope separation, reactions in upper atmosphere, radiation chemistry; developed new method for dehydration. Address: 38 Melbweg, Bonn, West Germany.*

GROTIUS, Hugo (Huigh de Groot), scholar; b. Delft, Netherlands, Apr. 10, 1583; ed. U. Leyden; LL.D., U. Orleans, 1598; m. Maria van Reigersberch, 1608. Accompanied Dutch statesman, Johan van Oldenbarnevelt on mission to France, 1598; practiced law, from 1599; fiscal advocate for Province of Holland, 1607; Dutch spokesman at conferences with English on colonial affairs, 1613-15; pensionary of Rotterdam, 1613, represented it in States of Holland; represented Holland in States General of United Provinces; captured by Stadtholder Prince Maurice during revolution of 1618, sentenced to life imprisonment in Loevenstein Castle; escaped 1621 to Paris where he lived under protection of Louis XIII (until 1631); returned to Holland, 1631, but not allowed to remain; went to Hamburg, 1632; became Swedish ambassador to Paris, 1635. Author: De veritate religionis christianae, 1622; De iure belli ac pacis, 1625 (pub. as extension of De iure praedae; De iure belli ac pacis considered 1st definitive text on internat. law). Attempted to make warfare more humane; stated it was wrong to wage war but for certain reasons; believed natural law prescribed rules of conduct for nations; derived specific content of internat. law from Bible and classical history; borrowed from earlier scholars, especially Gentilli; founder of internat. law. Died Rostock, Mecklenburg-Schwerin, Aug. 28, 1645.

GROTJAHN, Alfred, German social hygienist; b. Schladen, Germany, Nov. 25, 1869; s. Robert and Emma (Frey) G.; M.D., U. Berlin, 1894; state exam., 1896; 2 sons, including Martin, 1 dau. Pvt. practice medicine, Berlin; became prof. U. Berlin, 1912; apptd. chmn. social hygiene dept. med. bur. City of Berlin, 1915; became prof. social hygiene U. Berlin, 1920; mem. Reichstag, 1921-24. Author: Der Alkohollismus nach Wesen, Wirkung und Verbreitung, 1898; Über Wendlungen in der Volksernährung, 1902; (with I. Kaup) Krankenhauswesen und Heilstättenbewegung im Lichte der Sozialen Hygiene, 1912; Geburtenrückgang und Geburtenregelund, 1914; Die Hygiene der menschlichen Fortpflanzung, 1926; (with F. Goldmann) Die Leistungen der deutschen Krankenversicherung im Lichte der Sozialen Hygiene, 1928; Arzte als Patienten, 1929. Founder of sci. of social hygiene, including definition of the concept, and establishment of research methods; worked for laws on youth welfare and venereal disease. Died Berlin, Sept. 4, 1931.

GROTTHUSS, Christian Johann Dietrich (Theodor) von, physicist; b. Leipzig, Germany, Jan. 20, 1785; s. Ewald Dietrich and Elisabeth Eleonore V. G.; student chemistry U. Leipzig, from 1803, École Polytechnique, Paris, France, also Italy. Mgr. Geddutz estate, Kurland, Germany, also ind. research, from 1808. Author: Physisch-chemische Forschungen, 1820; Abhandlungen über Elektrizitat und Licht, 1906. Formulated theory decomposition liquids by galvanic current, basic thought was that polarity is formed between hydrogen, oxygen when current is sent through water (water particles assume polar arrangement which results in regrouping through entire liquid when hydrogen, oxygen given off at electrodes, also that reactions at electrodes are reactions of oxidation, reduction), 1805; formulated photochem. absorption law, 1819. Died Geddutz, Russia, Mar. 14, 1822.

GROUPÉ, Vincent, Am. virologist; b. Phila., Sept. 13, 1918; s. Andrew Vincent and Georgia (Patterson) G.; B.A., Wesleyan U. (Conn.), 1939; Ph.D., U. Pa., 1942; m. Gerry Finley Nash, Mar. 30, 1942; children—David Vincent, Lawrence Nash. Faculty, U. Conn., 1947-49; prof. virology Rutgers U. Inst. Microbiology, New Brunswick, N.J., 1949-—. Cons. Wallace Labs., Cranbury, N.J., 1960-—, Nat. Cancer Inst., NIH, Bethesda, Md., 1961-—. Recipient citation Wesleyan U., 1959. Fellow Am. Acad. Microbiology, N.Y. Acad. Medicine; mem. Am. Assn. Cancer Research, Soc. Exptl. Biology and Medicine, Am. Assn. Immunologists. Author: (with others) Avian Tumor Viruses, 1965. Research, numerous publs. in exptl. antibiotic and chemotherapy of viral diseases, epidemic typhus vaccine, cancer producing viruses. Home: 151 Cedar Lane, Princeton, N.J. 08540. Office: Inst. Microbiology, Rutgers U., New Brunswick, N.J. 08903.*

GROVE, Andrew S., Am. engring. physicist; b. Budapest, Hungary, Sept. 2, 1936; B.Chem. Engring. magna cum laude, City U. N.Y., 1960; Ph.D., U. Cal. at Berkeley, 1963; m. Eva Kastan, June 8, 1958; children—Karen, Roberta. Came to U. S., 1957, naturalized, 1962. Asst. dir. research and devel. Fairchild Semiconductor, Palo Alto, Cal., 1963-—. Lectr., U. Cal. at Berkeley, 1966-—. Mem. Am. Phys. Soc., I.E.E.E., Am. Inst. Chem. Engrs., Tau Beta Pi. Author: Physics and Technology of Semiconductor Devices, 1967; also articles. Research on nature of high Reynolds number separated flows, properties of semiconductor surfaces and their influence on semiconductor device characteristics, devels. leading to first stable surface field-effect transistor and to high-voltage planar transistors. Home: 3011 Ross Rd., Palo Alto, Cal. 94303.*

GROVE, William Johnson, Am. physician; b. Ottawa, Ill., Mar. 23, 1920; s. J. Roy and Florence (Johnson) G.; B.S., U. Ill., 1942, M.D., 1943, M.S., 1949; m. Betty Pedigo, Mar. 23, 1944; children—William Johnson, Pamela J., Holly L. Practice medicine, specializing in surgery, Chgo.; faculty U. Ill. Coll. Medicine, 1951-—, prof., 1964, dean, 1968-—. Recipient Raymond B. Allen Instructorship award, 1960. Mem. Soc. U. Surgeons, A.C.S., A.M.A., Assn. Am. Med. Colls., Am., Western surg. assns., Central Surg. Soc., Soc. Clin. Surgery, Warren H. Cole Soc., Alpha Omega Alpha. Research, numerous publs. in transplantation of blood vessels and reconstructive cardiovascular surgery, cancer, med. edn. Home: 664 58th St., Hinsdale, Ill. 60521. Office: 1853 W. Polk St., Chgo. 60612.*

GROVE, Sir William Robert, Brit. physicist; b. Swansea, Wales, July 11, 1811; s. John and Anne (Bevan) G.; M.A., Brasenose Coll., Oxford, 1835; D.C.L., 1875; LL.D., Cambridge, 1879; m. Emma Maria Powles, May 27, 1837; 1 dau., Mrs. William Edmund Hall. Became barrister Lincoln's Inn, 1835; named prof. exptl. philosophy, London Instn., 1847; named queen's counselor, 1853; apptd. mem. Royal Commn. on Law of Patents, 1864; became judge Queen's Bench, 1880; named privy councillor, 1887. Fellow Royal Soc., 1840 (Royal medal 1847); mem. Royal Instn. (became v.p. 1844), Chem. Soc. (charter), Accademia dei Lincei (Rome). Author: Correlation of Physical Forces, 1846; also articles. Research on electrolytic decompositions, including 1st demonstration of dissociation of water; invented Grove gas voltaic battery, 1839; originated theory of mut. convertibility of forces, 1846. Died Aug. 1, 1896.

GROVER, Horace John, Am. physicist; b. Cayuga, N.Y., May 14, 1909; s. Frank B. and Mary (Wiley) G.; A.B., U. Rochester, 1929, M.A., 1931; Ph.D., Cornell U., 1937; m. Elizabeth Hutchinson, July 2, 1938; 1 dau., Anne Elizabeth. Physicist, Renssalaer Poly. Inst., 1936-42, Battelle Meml. Inst., Columbus, O., 1942-—. Mem. com. on fatigue research and tech. NASA, 1962-—. Mem. Am. Soc. Testing Material (award of merit 1964), Am. Phys. Soc., Soc. Exptl. Stress Analysis. Author: (with S. A. Gordon and L. R. Jackson) Fatigue of Metals and Structures, 1954. Research, numerous publs. in mech. properties of structures, especially those of metals and alloys, resistance to fatigue under repeated stressing. Home: 584 Fox Lane, Worthington, O. 43085. Office: 505 King Av., Columbus, O. 43201.*

GROVES, Leslie Richard, Am. engr.; b. Albany, N.Y., Aug. 17, 1896; s. Leslie Richard and Gwen (Griffith) G.; student, U. Wash., 1913-14, Mass. Inst. Tech., 1914-16; B.S., U. S. Mil. Acad., 1918; grad. Army Engr. Sch., 1921, Command and Gen. Staff Sch., 1936, Army War Coll., 1939; LL.D., St. Ambrose Coll., U. Cal., Hamilton Coll., St. Ambrose; D.S.C., Lafayette, Williams, Hobart, Ripon and Pa. Mil. Colls.; m. Grace Hulbert Wilson, Feb. 10, 1922; children—Richard Hulbert, Gwen (Mrs. John A. Robinson). Commd. 2d lt. U. S. Army, 1918, advanced through grades to lt. gen., 1948, retired 1948; various assignments U. S., Hawaii, Europe, Nicaragua, dep. chief construction, Corps Engrs. 1941; headed Manhattan Atomic Devel. Project, 1942-47, in responsible charge all phrases of project; v.p. Remington div. Sperry Rand Corp., 1948-61; dir. Cooper-Bessemer Corp. Decorated D.S.M., Legion of Merit (U. S.); also fgn. decorations. Mem. Am. Soc. C.E., Am. Soc. M.E. Author: Now it can be Told, The Story of the Manhattan Project. Home: 2101 Connecticut Av. N.W., Washington 20008.

GROVES, Marion Herbert, Am. psychologist, univ. dean; b. Medicine Lodge, Kan., Feb. 18, 1910; s. Herbert St. Clair and Martha Mae (Pepoon) G.; B.A., Phillips U., 1933; Ph.D., U. Chgo., 1950; m. Arleen Elizabeth Daffer, Nov. 21, 1943 (div. 1959); m. 2d, Mildred Ann (Blakesley) Mitchell, June 1, 1963; children—Marshall Ray, Mariann Jama. Ordained to ministry Disciples of Christ, 1933; chief psychologist Behavior Clinic Cook County Criminal Ct., 1943-45; psychologist VA counseling service U. Chgo., 1945, instr., 1945-47; lectr., dir. exptl. psychology lab. U. Cal. at Santa Barbara, 1947-49; faculty Ill. Inst. Tech., Chgo., 1950-—, asso. prof. psychology, 1954-—, asst. dean Grad. Sch., 1951-58, asso. dean, 1958-—. Cons. Ventura (Cal.) Juvenile Ct., 1947-49; gov. Internat. House; lectr. neurology and psychiatry Northwestern U. Med. Sch., 1954-—. Mem. A.A.A.S., Am., Midwestern psychol. assns., Am. Soc. for Engring. Edn., Ill. Soc. for Personality Study (past pres.), Internat. Platform Assn., Nat. Assn. for Fgn. Student Affairs, Sigma Xi. Author: Methods and Principles of Psychological Research, 1948; also articles. Research on effects various diseases and other sustained phys. and psychol. stresses on mental and phys. health humans and lower animals, neurol. theory learning. Home: 3100 S. Michigan Av., Chgo. 60616.*

GRUBB, Sir Howard, Brit. engr.; b. 1844; ed. Trinity Coll., Dublin, Ireland; M. Engring. (hon.), U. Dublin; m. Mary Hester Walker, 1871; established factory for astron. and other precise instruments, Dublin; astron, instrument maker, contractor to Brit., colonial, fgn. govts.; sci. adviser to Commrs. Irish Lights, from 1913. Recipient Cunningham Gold medal, 1881, Boyle medal, 1912. Fellow Royal Soc., 1883, Royal Astron. Soc.; hon. mem. Royal Inst. Engrs., Ireland; gov. Royal Dublin Soc. Contbr. papers to sci. publs. Died Sept. 16, 1931.

GRUBB, Thomas, Irish optician; b. Kilkenny, Ireland, 1800; a son, Sir Howard Grubb. Abandoned merc. business to open workshops in Dublin; fellow Royal Soc., 1864, Royal Astron. Soc.; mem. Royal Irish Acad. Contbr. articles to tech. jours. Built reflecting telescopes, including 15 inch, Armagh, 1835, 20 inch Glasgow (Scotland) Obs. reflector, 4 foot Melbourne (Australia) reflector, 1867. Died Sept. 19, 1878.

GRUBBE, Emil Herman, Am. radiologist; b. Chgo., Jan. 1, 1875; s. Albert and Bertha (Reets) G.; tchrs. and pharmacy degrees Valparaiso U.; 1890-92, B.S., 1893, A.M., 1894, Ph.D., 1895, chemist and physicist, 1895-96; M.D., Gen. Med. Coll., Chgo., 1898, Chicago Coll. Medicine and Surgery, 1910; postgrad. Bellevue Hosp., N.Y.C., 1901. Specialist x-ray therapy, 1896-—; pioneer application x-ray for treatment of disease, for cure recurrent carcinoma; treatment cancer with x-ray, 1896, Tb of skin, 1896; pioneered use of lead as protection untoward effects x-ray; pioneered design, establishment of hosp. x-ray dept.; dir. Ill. X-Ray and Electro-Therapeutic Lab.; lectr. chemistry and physics Gen. Med. Coll., 1895, adj. prof. chemistry, physics and x-rays, 1896-97, prof. roentgenology, electrotherapeutics dept. Hahnemann Hosp., Chgo., 1896-1919; prof. radiology, x-ray therapeutics and electrophysics Ill. Postgrad. Sch. Electro-Therapeutics, 1899-1921; roentgenologist Chgo. Bapt. Hosp., 1900-16, Pine Sanitarium, 1911-16; cons. staff Peekskill Sanitarium, 1900-06; prof., head dept. roentgenology and electrotherapeutics Chgo. Coll. Medicine and Surgery, 1910-20; cons. physician Frances E. Willard Hosp., 1910-24; prof. roentgenology, phys. therapeutics Jenner Med. Coll., 1914-17; hon. cons. roentgenologist Streeter Meml. Hosp., Chgo., 1938. Recipient award for pioneer work x-ray therapy and electrotherapy Am. Inst. Medicine; citation Chgo. Med. Soc., 1946; award Walter Reed Soc., 1952; citation scroll Chgo. Roentgen Soc., 1956. Diplomate Am. Bd. Radiology, 1937. Fellow A.C.P.; mem. Am. Roentgen Ray Soc., Radiol. Soc. N.A. (founder), A.M.A., Assn. Am Physicians, Nat. Acad. Scis., Am. Assn. Cancer Research, Nat. Soc. Phys. Therapeutics (pres. 1912), A.A.A.S., Assn. Approved Radiology, Am. Philos. Soc. Author: A System of Inorganic Chemical Analysis, 1898; X-Ray Treatment—Its Origin, Birth and Early History, 1949. Editor Archives of Electrology and Radiology, 1904-08; asso. editor Am. Electro-Therapeutic and X-Ray Era, 1901-13. Author numerous med. articles, monographs. Died Mar. 26, 1960.

GRUBBS, Samuel Bates, Am. physician; b. Indpls., Feb. 11, 1871; s. Daniel Webster and Matilda (Miller) G.; A.B., U. Mich., 1893; M.D., Columbia, 1896; postgrad. Paris, Vienna; m. Mary Evelyn Noble, June 17, 1903; 1 son, Daniel Dean. With USPHS, 1897, specializing in yellow fever, bubonic plague, typhus fever and meningitis prevention, advanced through ranks to med. dir. ret., 1933; service in Europe, Mexico, South America, P.I. and Orient; was chief quarantine officer, P.R., 1908-12; vis. physician Presbyn. Hosp., San Juan, P.R., 1910-12; chief extra cantonment san. area, Newport News, Va., 1917-18, san. insp. U. S. Army, post of embarkation, N.Y.; served in France; chief quarantine officer, Panama Canal, 1919-20; health officer, Port of N.Y. 1921-25; chief fgn. quarantine div., 1925-27; dir. Great Lakes dist., 1928; adviser to Los Angeles Health Dept., 1929; chief quarantine officer, Hawaiian Islands, 1929-33. San. insp. on board transport Sedgwich, Spanish-Am. War. Am. del. orgn. office Internat. d'Hygiene Publique. Mem. A.M.A., A.A.A.S., Nu Sigma Nu, Delta Upsilon. Contbr. to Pub. Health Reports. Originator of vacuum cyanide method of disinfecting clothing, ratproofing of ships and cheopis index for bubonic plague. Died Sept. 19, 1942.

GRUBE, (Adolf) Eduard, German zoologist; b. Königsberg (now Kaliningrad, USSR), May 18, 1812; s. Karl Eduard and Ernestine Luise (Mertens) G.; student natural scis., Königsberg; Ph.D., 1834; m. Josephine Schäfer, 1839; 2 sons, including Max, 2 daus. Joined faculty U. Königsberg, 1837, became asso. prof., 1843; named prof. zoology, Dorpat, Estonia, 1844; became prof., Breslau (now Wroclaw, Poland), 1957; worked on devel. Zool. Mus., Breslau. Mem. numerous socs., including Leopoldina. Author: Zur Anatomie und Physiologie der Kiemenwürmer, 1838; Actinien, Echinodermen und Würmer des adriatischen und Mittelmeers, 1840; Untersuchungen über die Entwicklung der Clepsinen, 1844; Die Familie der Anneliden, 1851; Die Insel Lussin und ihre Meeresfauna, 1864; Die Anneliden der Novara-Expedition, 1868. Research on lower marine animals, including structure, devel. and taxonomy of annelids; named the onychophora (group of animals intermediate between annelids and arthropods); discovered radiation figures of dividing cells, 1844; described numerous new species of annelids. Died Breslau, June 28, 1880.

GRUBE, Hermann, physician; flourished late 17th century. Author: Commentarius de modo simplicium medi camentorum facultates cognoscendi, 1669; De arcanis medicorum non arcanis commentatio, 1673; De transplantatione morborum, 1674; De ictu tarantulae, 1679. Discussed beneficial results of Harvey's discovery; wrote on medicinal simples; discussed symptoms of tarantula poisoning; criticized polypharmacy of his time; attacked belief in universal medicine, transplantation of diseases, weapon ointment, doctrine of signature in plants, gathering of herbs under certain constellations, use of letters, words and characters.

GRUBENMANN, (Johann) Ulrich, Swiss mineralogist, petrographer; b. Trogen, Switzerland, Apr. 15, 1850; s. Hans Caspar and Katharina (Eugster); student Fed. Poly. Sch., Zürich, Switzerland, until 1874, U. Munich, 1875-76, Heidelberg, Germany, 1886; Ph.D., U. Zürich, 1886; m. Ida Carolina Baumer, 1876; 1 son; m. 2d, Lisette Augusta Fisch, 1881; 1 son, 1 dau. Became tchr. canton sch., Frauenfeld, Switzerland, 1874, rector, 1886-88; joined faculty U. Zürich, 1888, became prof., 1893; named prof. Fed. Tech. U., Zürich, 1893, rector 1909-11; emeritus, 1921. Mem. Swiss Natural Sci. Soc. (pres. geotech. commn. 1894-1924). Author: Tabellen zur Bestimmung der Mineralien . . . , 1900; Die kristallinen Schiefer, 2 parts, 1904/07, 3d edit. (with P. Niggli) published as Die Gesteinsmetamorphose, 1924; Die schweizerische Tonlagerstätten, 1907; also articles. Founder, editor Schweizer Mineralogische-petrographische Mitteilungen, 1921-24, Studies in crystalline state, rock metamorphosis; one of 1st to recognize significance of polarization microscopic methods and chem.-analytical methods in mineralogy. Died Zürich, Mar. 16, 1924.

GRUBER, Charles Michael, Am. physician; b. Hope, Kan., Mar. 11, 1887; s. John Nicholas and Barbara (Ehrsam) G.; A.B., U. Kan., 1911, A.M., 1912; Ph.D., Harvard, 1914; M.D., Washington U., St. Louis, 1921; m. Hermione Archer Sterling, June 6, 1912 (dec. 1959); children—Barbara (Mrs. Joseph Wrigley), Charles Michael; m. 2d, Mildred Farquhar Alexander, 1960. Instr., U. Pa., 1914-15; prof. physiology and pharmacology Albany Med. Coll., 1915-17; faculty U. Colo., 1917-20, prof., head dept. pharmacology, 1918-20; asso. physiology Washington U., St. Louis, 1920-21, asso. prof., 1921-32; prof., head dept. pharmacology Jefferson Med. Coll., Phila., 1932-53, prof. emeritus, 1953-—; prof., head dept. pharmacology Loma Linda U., 1953-57; vis. prof. biology U. Redlands (Cal.), 1957-63. Fellow A.M.A.; mem. Pa. State, Philadelphia County med. socs., Phila. Coll. Physicians, Am. Coll. Cardiology, Am. Physiol. Soc., Am. Soc. for Pharmacology and Exptl. Therapeutics (past pres.), Soc. for Exptl. Biology and Medicine, Phila. Phys. Soc. (past pres.), A.A.A.S., Sigma Xi, Phi Beta Pi, Alpha Omega Alpha, Beta Beta Beta, Phi Sigma. Author: Handbook of Treatment, 1948; also numerous articles. Research on actions barbiturates and similar compounds, effects change in rates interruption elec. impulses on blood pressure. Home: 1313 College Av., Redlands, Cal. 92373.*

GRUBER, Elbert Egidius, Am. chemist; b. Cin., Sept. 11, 1910; s. Leo J. and Bertha (Kuhlmann) G.; B.S. in Chemistry, Xavier U., 1932; M.S., U. Ill., 1934, Ph.D., 1937; m. Mary I. Stumpf, Aug. 4, 1934; children—Thomas Leo, David Paul, Mary Christine, Carolyn Marie. With research div. B.F. Goodrich Co., 1937-50, research supr., 1946-50; with Central Research Labs., Gen. Tire & Rubber Co., Akron, O., 1950-—, asst. dir. exploratory research, 1955-62, dir. research and devel., 1962-—. Mem. solid com. large rocket com. Aerojet-Gen. Corp., Azusa, Cal., 1949-—; chmn. Gordon Research Conf. on Elastomers, 1957. Mem. Am. Chem. Soc. (past chmn. Akron sect.), Akron Polymer Group (past chmn.), Soc. Plastic Engrs., Indsl. Research Inst. Research, publs. on organic synthesis related to sex hormones, local anesthetics, rubber antioxidants, catalyst for polymerization and polymerization modifiers; co-inventor polyurethan rubbers, stabilizer for plastics; directed research and devel. new synthetic rubber based on propylene oxide, binder for propellants. Home: 1604 Highbridge Rd., Cuyahoga Falls, O. 44223. Office: P.O. Box 951, Akron, O. 44309.*

GRUBER, Georg Benno Otto, German pathologist; b. Munich, Germany, Feb. 22, 1884; s. Max-Emmanuel and Anna (von Waechter) G.; M.D., U. Munich; D.V.M. honoris causa; m. Maria Schafer, 1918; children—Irmingard, Haans (dec.), Elisabeth. Physician, prof. agrege U. Strasbourg, 1913; with Municipal Hosp. of Mayence, 1917; full prof. path. anatomy U. Innsbruck, 1923; full prof. gen. pathology and path. anatomy U. Göttingen, 1928; instr. history of medicine, 1941. Recipient Paracelsus medal German Assn. Doctors. Mem. Sci. Acad. Göttingen, Acad. Physicians South Halle. Collaborator: Morphologie der Missbildungen, Einführung in Geist und Studium der Medizin, Von Arzlicher Ethik, 50 Jahre Pathologie in Deutschland. Editor: Zentralblatt für Allgemeine Pathologie und Pathologische Anatomie. Research and publs. on path. anatomy, teratology. Address: 8 Planckstrasse, Gottingen 34, Germany.

GRUBER, Max, biochemist; b. Dresden, Germany, Nov. 9, 1921; s. Hermann and Gisella (Schnek) G.; doctorandus U. Amsterdam (Netherlands), 1949; Ph.D., U. Utrecht (Netherlands), 1952; m. Susanne Heynemann, May 6, 1952. From instr. to asst. prof. U. Utrecht, to 1952; prof. biochemistry U. Indonesia, Bandung, 1952-55; Rockefeller Found. fellow Yale, 1955-56; prof. biochemistry, head dept. biochemistry Groningen (Netherlands) U., 1956—. Adv. bd. Biochimica et Biophysica Acta, later editor, 1962-64, asso. mng. editor, 1964—; Netherlands del. Council Internat. Union Biochemistry, 1967—; gov. Netherlands Found. Chem. Research, 1965—. Recipient Shell prize for chemistry, 1962. Mem. Netherlands Soc. Biochemistry (chmn. 1965-67), A.A.A.S., Royal Netherlands Chem. Soc., European Molecular Biology Orgn. Research, publs. on elucidation of vitamin B1-sparing action of fat; structural studies on immunoglobulins (antibodies); structure-function relationship of enzyme from different animal species; isolation and localization of intracellular animal proteolytic enzymes; regulation of RNA and protein synthesis; studies on structure and shape of biol. macromolecules, especially haemocyanin; elucidation of circular structure of mitochondrial DNA. Home: 19 Emmalaan, Haren (Gr), Netherlands. Office: 10 Bloemsingel, Groningen, Netherlands.*

GRUBER, Max(imilian) Franz Maria von, see von Gruber, Max(imilian) Franz Maria.

GRUEBER, Johann, explorer; b. Linz, Austria, Oct. 28, 1623; student, Vienna, Austria, Leoben, 1644-47; student theology, Graz, Austria, 1651-55; Joined Soc. Jesus, 1641; prof. secondary schs., Graz, Ödenburg, Leoben, 1647-51; assigned to discover land route to China (successively with P. Bernhard Diestel, Albert d'Orville, P. Heinrich Roth), 1656; sailed to Smyrna from India to Macao; arrived in Peking, China, 1659; with Imperial Astronomy Bur., Peking, until 1661; began trip westward through China, Tibet, Himalayas, Nepal, India, 1661; then through Persia, Asia Minor; arrived in Rome, 1664; joined Jesuit Mission, Hungary, 1665. First European to explore many areas of E. Asia; 1st European to cross all of Tibet. Died Oct. 28, 1623.

GRUEN, Johann Ludwig, German bacteriologist, hygienist; b. Crefeld, May 19, 1919; s. Johann and Margarethe (Lünger) G.; ed. univs. Bonn, Vienna, Dusseldorf, Danzig; M.D.; m. Elisabeth Gruen, 1945. Chief physician Inst. for Hygiene and Microbiology, Acad. Medicine Dusseldorf. Research and numerous publs. on bacteriology and hygiene, clin. and practical study of staphylococtus. Home: 42 Boverter Kirchweg, 4151 Osterath. Office: 111 Witzelstrasse, Dusseldorf, Germany.

GRUENEBERG, Hans, geneticist; b. Elberfeld, Germany, May 26, 1907; s. Levi and Else (Steinberg) G.; Ph.D., U. Berlin, 1929; M.D., U. Bonn (Germany), 1932; D.Sc., U. London (Eng.), 1948; m. Elsbeth Capell, July, 1933 (dec.); children—Reuben N., Daniel S.; m. 2d, Hannah Blumenfeld, Nov., 1946 (dec.). Hon. research asst. dept. genetics U. Coll. London, 1933-38, Moseley Research student Royal Soc., 1938-42, reader dept. animal genetics, 1946-55, prof., 1956—. Hon. dir. exptl. genetics research unit Med. Research Council, 1955—. Fellow Royal Soc., Inst. Biology, Internat. Inst. Embryology; mem. Genetical Soc., Soc. for Exptl. Biology, Soc. for Developmental Biology. Author: The Genetics of the Mouse, 1943, 52; Animal Genetics and Medicine, 1947; The Pathology of Development, 1963; also numerous articles. Research into mechanisms of gene action in mouse. Home: 66A Belsize Park Gardens, London N.W. 3. Office: Dept. Animal Genetics, Wolfson House, 4 Euston Bldgs., London N.W. 1, Eng.*

GRUENWALD, Peter, Am. physician; b. Schoenwald, Czechoslovakia, Mar. 12, 1912; s. Julius and Paula (Loewy) G.; M.D., U. Vienna (Austria), 1936; m. Eva Glas, Sept. 15, 1951; 1 son, Thomas. Came to U. S., 1938, naturalized, 1944. Asst., U. Vienna, 1938; faculty State U. N.Y., Bklyn., 1945-54, asst. prof., 1949-54; pathologist Margaret Hague Maternity Hosp., Jersey City, 1955-58; asso. prof. Johns Hopkins, 1958—; research pathologist Sinai Hosp. Balt., 1961—. Mem. Am. Assn. Anatomists, Am. Assn. Pathologists and Bacteriologists, Am. Acad. Pediatrics, Am. Coll. Obstetricians and Gynecologists, Internat. Acad. Pathology, Am. Soc. Human Genetics, N.Y. Acad. Scis. Contbg. author: Resuscitation of the Newborn Infant, 1966. Research, numerous publs. on embryology of urogenital tract, congenital malformations, normal and abnormal growth and devel. of human fetus, normal and pathologic structure of human placenta, role of surface tension in expansion of lungs in newborn and its abnormalities. Home: 4001 Fords Lane, Balt. 21215. Office: Sinai Hosp., Balt. 21215.*

GRUHLE, Hans Walther, German psychiatrist; b. Lübben, Germany, Nov. 7, 1880; s. Franz Karl Heinrich and Thekla (Schumann) G.; M.D. under Kraepelin, U. Munich, 1905; m. Atha Nodnagel, 1922; 1 son, 1 dau. Began as asst. to Nissel, U. Heidelberg Clinic, joined faculty, 1912; named provisional dir. Bonn (Germany) Neurol. Clinic, 1934, but not accepted because of politics; dir. sanatorium, first in Zwiefalten, then in Weissenan; became dir. psychiat. and neurol. clinic, Bonn, 1946, emeritus, 1952, provisional dir., 1955. Author: Psychiatrie für Arzte, 1918; Geisteskrankheiten und Strafrecht, 1927; Psychologie der Schizophrenie, 1929; Psychopathologie der Schizophrenie, 1932; Theorien der Schizophrenie, 1932; Grundriss der Psychiatrie, 1937, 15th edit., 1948; Selbstmord, 1940; Das Portrait, 1948; (collected works) Verstehen und Einfühlen, 1953; also numerous articles. Research on psychopathology, psychiatry, criminal psychology, criminality of youth, mental disease and penal law, psychopathology of schizophrenia, suicide, history of psychology. Died Bonn, Oct. 3, 1958.

GRUITHUISEN, Franz Paula von, German physician, astronomer; b. Schloss Hallenberg am Lech, Bavaria, Mar. 19, 1774; s. Peter and Marie Rosina G.; student in a surgeon's sch.; studied natural scis. and medicine, Landshut; doctorate, 1808; m. Antoine Neuner, 1820; 1 son, 1 dau. Became hosp. orderly Austrian army in Turkish war, 1788; joined service of Ct. of Elector Karl Theodor of Bavaria, 1793; tchr. at sch. for rural physicians, Munich, from 1808; became asso. prof. U. Munich, 1826, prof. astronomy, 1830. Mem. Acad. Leopoldina. Author: Organozoonomie, 1811. Developed theory of infection in genetic sense as a productive step backward to polp stage; discovered differentiation of form elements of blood; attempted lithotrity and chem. dissolution of kidney and gall stones; studies in astronomy, geology and geography; proposed theory that moon craters were caused by meteors; studies on weather prognosis, influence of sunspots, measurement of gravitation and tides, horizontal pendulum. Died Munich, June 21, 1852.

GRUMBACH, Arthur Samuel, Swiss physician; b. Zurich, June 25, 1895; s. Jacques and Berthe (Rothschild) G.; ed. univs. Zurich, Bern, Lausanne; M.D.; m. Gurti Dagmar, 1939. Asst., Inst. Path. Anatomy, Geneva; fgn. asst. to Prof. F. Widal, Paris, Pasteur Inst. Paris; with Rockefeller Inst., N.Y.C.; 1st asst. Inst. Hygiene of Zurich; prof. agrege; prof., dir. Inst. Microbiology. Mem., hon. mem. nat. and internat. socs. Author: Das Handskelett im Lichte des Rontgenstrahlen, 1921; Die Besredkov'sche Lehre von Antivirus und der Lokalen Immuniteit, 1928; Die Lehre von der Fokalen Infektion, 1934, 37; Die Infektions Krankhetten des Menschen und ihre Erreger, 2 vols., 1958. Home: 20 Rutistrasse. Office: 32 Gloriastrasse, Zurich, Switzerland.

GRUMBACH, Eugene Albert, French physicist; b. Paris, France, May 2, 1880; s. Gustave and Sophie (Lippmann) G.; ed. Faculty Scis., Paris; Ph.D., agrégé in sci.; m. Henr. Rueff, 1920; children—Marcel, Jacqueline. Prof., Lycée of Poitiers, 1906; asst. Faculty Scis., Paris, 1911, lectr., prof. without chair, 1913; prof. physics Faculty Scis., Poitiers, 1928-36, dean, 1938-48. Mem. French Soc. Physics, Soc. Chem. Physics. Research and publs. on photovoltaic batteries, discontinuity of potential photocapillary phenomena, free circulation of liquids. Home: 7 route de Nouaillé, Poitiers. Office: Physics Lab., Faculty of Sciences, Poitiers, Vienna, France.

GRUMBACH, Melvin Malcolm, Am. physician; b. N.Y.C., Dec. 21, 1925; s. Emanuel and Adele (Weil) G.; A.B., Columbia, 1945, M.D., 1948; m. Madeleine F. Butt, Dec. 1, 1951; children—Ethan Malcolm, Kevi Lawrence, Anthony Havemeyer. Vis. fellow Oak Ridge Inst. Nuclear Studies, 1952; fellow, asst. in pediatrics Johns Hopkins Sch. Medicine, 1953-55; faculty Columbia Coll. Physicians and Surgeons, 1955-66, asso. prof. pediatrics, 1961-66; with Babies Hosp. Presbyn. Hosp., N.Y.C., 1955-66, asso. attending pediatrician, 1961-66; head pediatric endocrinology div. and postdoctoral tng. program Columbia Presbyn. Med. Center, 1955-66; prof. pediatrics, chmn. dept. U. Cal. Sch. Medicine, San Francisco, 1966—; pediatrician-in-chief U. Cal. San Francisco Med. Center, 1966—. Cons., mem. human embryology and devel. study sect. NIH, USPHS, 1962—; mem. Nat. Bd. Med. Examiners, 1964—. Recipient Joseph Mather Smith prize Columbia, Career Scientist award Health Research Council City N.Y., 1961-66. Mem. Am. Soc. Clin. Investigation, Pediatric Soc., Endocrine Soc., Am. Acad. Pediatrics, Soc. Pediatric Research, Am. Soc. Human Genetics, Teratology Soc., Harvey Soc., N.Y. Acad. Scis., Sigma Xi, Alpha Omega Alpha. Mem. editorial bd. Jour. Clin. Endocrinology and Metabolism, 1957—, asso. editor, 1963-66; editorial cons. Jour. Pediatrics, 1966—. Research, publs. on hormonal effects on growth and maturation, function of human sex chromosomes, sex chromosome abnormalities and disorders of sex differentiation, genetic factors in endocrine disease, characterization, function and determination of human placental hormone with growth hormone-like and lactogenic activity; pathogenesis and treatment of growth disorders, determinants of fetal growth. Home: 230 Santa Clara, San Francisco 94122. Office: Dept. Pediatrics U. Cal. Med. Center, San Francisco 94122.

GRUMM, Hans Josef, Austrian physicist; b. Melk, Austria, Oct. 25, 1919; s. Josef Josef and Theresia (Bauer) G.; Dr.Phil., U. Vienna (Austria), 1949; postgrad. Techn. U. Stuttgart (Germany), 1957; m. Elfriede Altmann, Sept. 27, 1947; 1 son, Hans Richard. Research asst. Techn. U. Vienna, 1950-53, dozent nuclear physics, 1961—; head electron optical research div. Goerz Optical Co., Vienna, 1953-55; staff Central Inst. for Nuclear Physics, Dresden, Germany, 1956; sci. asst. Inst. for Reactor Physics, Techn. U. Stuttgart, 1957; head reactor dept. SGP Co., Vienna, 1958-64; mgr. Reactor-I.G., Vienna, 1961-66; reactor engring. dept. NUKEM, Hanau, Germany, 1964-66; head Inst. for Reactor Engring., SGAE, Vienna-Seibersdorf, 1961—. Mem. Brit. Inst. Nuclear Engrs., Am. Nuclear Soc. Author: (with K. A. Höcker) Lineare Reaktorkinetik und Störungstheorie, 1958; (with W. Glaser) Kernreaktortheorie, 1962; (with H. Thirring) Kernenergie, 1967; also articles. Research on theory of corpuscular optics especially electron microscope, reactor physics, theory of reactor fuel cycles. Home: 7-9 Joanelligasse 1060, Vienna. Office: 10 Lenaugasse 1082, Vienna, Austria.*

GRÜMMER, Gerhard, German plant pathologist; b. Görlitz, Aug. 20, 1926; s. Richard and Erna (Heitzsch) G.; Ph.D., U. Jena, 1951; Sc.D., U. Greifswald, 1954; m. Gertraud Ranis, June 5, 1954; children—Gerald, Harald. Faculty U. Greifswald, 1952—, asso. prof., 1962—, head dept. plant pathology, 1963—; vis. prof. Indian Statis. Inst., Calcutta, 1964-65. Author: Die gegenseitige Beeinflussung höherer Pflanzen-Allelopathie, 1955; Tiere und Pflanzen tropischer Länder, 1966. Research on diseases of poppies, toxic secretions of plants, Phytophthora infestans on potatoes and tomatoes, ecology of weeds, diseases of rice, toxic effects of insecticides on plants, control of grassland weeds. Address: 5 Robert-Blum-Strasse, Greifswald, East Germany.*

GRUMMITT, Oliver Joseph, Am. chemist; b. Cleve., Jan. 16, 1910; s. Joseph James and Hannah (Gleine) G.; A.B., Oberlin Coll., 1932; M.A., Western Res. U., 1934, Ph.D., 1936; DuPont postdoctoral fellow chemistry, Cornell U., 1936-38. From instr. to asso. prof. chemistry Western Res. U., 1938-54, prof., 1954—, chmn. dept. chemistry, 1958-62; research cons. Sherwin-Williams Co., 1939—. Chem. investigator in Germany, U. S. Army, Dept. Commerce, 1946. Fellow A.A.A.S.; mem. Am. Chem. Soc., Am. Assn. U. Profs., Am. Assn. Oil Chemists, Sigma Xi, Phi Beta Kappa. Alpha Chi Sigma. Synthesis and mechanism studies in various areas of organic chemistry. Home: 15949 Cleviden Rd., East Cleveland, O. 44112.*

GRÜNBAUM, Albert S. F. (Leyton, A. S. F.), Brit. bacteriologist; b. 1869; ed. Cambridge, Eng., St. Thomas' Hosp., London, Eng.; worked with Gruber, Vienna, Austria; became physiologist, Liverpool, Eng., 1904; prof. pathology, Leeds, Eng. First to study agglutinating reaction of serum in enteric fever (F. Widal published before him). Died 1921.

GRÜNBAUM, Branko, mathematician; b. Osijek, Yugoslavia, Oct. 2, 1929; s. Vlado and Margueritte (Banderier) G.; student U. Zagreb, Yugoslavia, 1948-49; M.Sc., Hebrew U., Israel, 1954, Ph.D., 1958; m. Zdenka Bienenstock, June 30, 1954; children—Ram, Daniel. Mem. Inst. for Advanced Study, Princeton, 1958-60; vis. asst. prof. U. Wash., 1960-61; faculty Hebrew U., Jerusalem, 1961-65; vis. prof. Mich. State U., 1965-66; prof. math. U. Wash., Seattle, 1966—. Mem. Am. Math. Soc. Author: Convex Polytopes, 1967, also articles. Editor, Israel Jour. Math., 1963—. Research in combinatorial geometry, convexity, graph theory, and convex polytopes. Home: 5512 N.E. 63d St., Seattle 98115.*

GRÜNBERG, Hans, geneticist; b. Wuppertal, Germany, May 26, 1907; s. Levy and Else (Steinberg) G.; ed. Wuppertal-Elberfeld, univs. Bonn, Friburg, Berlin, London; m. Elsbeth Capell, 1933; children—Reuben N., Daniel S. Research asst. Univ. Coll., London, 1933-38, instr. genetics, 1946-55, prof., 1956; Moseley research student Royal Soc., 1938-42; hon. adminstr. exptl. genetics research unit Med. Research Council, 1955. Fellow Royal Soc.; mem. Internat. Inst. Embryology, Inst. Biology. Author: The Genetics of the Mouse, 1943, 52; Animal Genetics and Medicine, 1947; The Pathology of Development, 1963, also numerous articles. Home: 66A Belsize Park Gardens, London N.W.3. Office: University College, Gower St., London W.C.1, Eng.

GRUND, Alfred Johannes, geologist, geographer; b. Prague, Czechoslovakia, Aug. 3, 1875; s. Otto and Udalrike (Wischniowsky) G.; Ph.D., U. Vienna, 1899; postgrad., Berlin; m. Camilla Bischof, 1907; 1 son, 2 daus. Began as asst. Geog. Inst., U. Vienna, joined faculty, 1904; became asso. prof. Inst. Oceanography,

U. Berlin, 1907; named prof. geography German U., Prague, 1910; dir. hydrographic work on joint Austrian-Italian undertaking for research in Adriatic Sea, 1911-14. Author: Landeskunde von Österreich-Ungarn, 1903; Beiträge zur Morphologie des Dinarischen Gebirge, 1910; also articles. Research on morphology and hydrography of Dinaric chalk formations, subterranean hydrography; applied W. M. Davis' morphological cycle theory to chalk formations. Died Temes-Kubin nr. Belgrade, Yugoslavia, Nov. 11, 1914.

GRUND, (Franz) Friedrich (Alexander), German hydraulic engr.; b. Heinrichau, Germany, May 5, 1814; surveyor's exam. Constrn. Trades Sch., Breslau (now Wroclaw, Poland), 1836; degree Gen. Constrn. Sch., Berlin, 1841; m. Worked in Siegtal, Coblenz; successively became state insp. land and waterways, 1844, dir. harbor constrn., Cologne, Germany, 1847, dir. waterways, Cochem, waterways insp. Province of Rhine, Germany, 1854, state soil enrichment insp., Düsseldorf, Germany, 1856, govt. counsel, Stettin, 1860; joined Ministry Trade Industry and Pub. Works, Berlin, 1862; became commr. for border regulation between Prussia and Oldenburg, 1869; ret., 1887. Mem. numerous commns. Mem. Berlin Acad. for Constrn. (dir. 1866-73). Laid found. for agrl. waterway constrn. and water supply; designed Rhein-Maas Canal, harbors of Ruhrort, Oberlahnstein, Emmerich; responsible for canalization of upper Saar. Died Berlin, May 16, 1892.

GRUNDMANN, Franz Henrich, German metallurgist; b. Quelle, Germany, Jan. 1, 1808; s. Franz Hermann and Anna Margarete Elisabeth (Kröger) G.; student Tchr.'s Sem., Soest, 1928-30; state exam. for tchrs. natural scis. in indsl. schs., 1854. Became tchr. Rectorate Sch., Lüdenscheid, Germany, 1940, provincial indsl. sch., Hagen, 1847, indsl. sch., Schweidnitz, Germany, 1854, mining sch., Tarnowitz, 1857; ret., 1875. Author: Sind die englischen Steinkohlen besser als die schlesisichen?, 1864. Studies on ore preparation, especially puddling process; 1st chem. studies of coal in Upper Silesia (helped eliminate favoritism toward English coal). Died Tarnowitz, Feb. 23, 1887.

GRÜNEBERG, Hermann Julius, German chem. industrialist; b. Stettin, Germany, Apr. 11, 1827; s. August Wilhelm and Henriette Caroline (Breslich) G.; trained as pharmacist; student chemistry, Berlin, Paris; Ph.D., Leipzig, Germany, 1860; m. Emilie Schmidtborn, 1860; 3 sons, 1 dau. Founder (with Klee) saltpeter factory, 1854; made study trips to S. France, Eng., Scotland; founder (with Julius Vorster) potassium nitrate factory Vorster und Grüneberg, Kalk, nr. Cologne, Germany, circa 1858. Mem. Orgn. Chem. Industry Germany (co-founder 1877). Founder German potassium industry; developed processes for preparation of potash using abraum salts; produced artificial fertilizers, custom-mixed for particular purposes. Died Cologne, June 7, 1894.

GRÜNEISEN, Eduard, German physicist; b. Giebichenstein, Germany, May 26, 1877; s. Eduard and Elisabeth (Dryander) G.; ed. U. Halle (Germany), U. Berlin (Germany); Ph.D., 1900; m. Charlotte Bruns, 1910; 3 sons, 2 daus. Began as asst. to F. Kohlrausche, State Phys.-Tech. Inst., Berlin-Charlottenburg, became permanent colleague, 1904, head weak current lab., 1911, dir. dept. electricity and magnetism, 1919; joined faculty U. Berlin, 1905; named prof. exptl. physics, dir. Phys. Inst., U. Marburg (Germany), 1927. Research and numerous publs. on laws of solid state, crystalanisotrophy of elastic properties and thermic expansion of metal monocrystals, precision measurement of speed of sound in gases; developed Gustav Mie's solid state theory, including application of quantum theory to a comprehensive thermodynamic theory culminating in Grüneisen's Relation between heat expansion, specific heat, atomic vol. and compressibility. Died Marbury, Apr. 5, 1949.

GRUNER, Christian Gottfried, German physician; b. Sagan, Nov. 8, 1744; s. Balthasar and Maria Dorothea (Golisch) Grunerth; ed. Leipzig, Germany, Halle, Germany; M.D., 1747; m. Christina Margarete Haase, 1777; 4 sons, 4 daus. Practice medicine Breslau (now Wroclaw, Poland); became prof. med. theory U. Jena (Germany), 1773. Mem. Leopoldina. Studied writings of Hippocrates, 1772; publs. on theory of disease, pharmacology; wrote work on morbus gallicus, 1793. Died Jena, Germany, Dec. 5, 1815.

GRUNER, Gottlieb Sigmund, Swiss natural scientist; b. Trachselwald (Bern), Switzerland, July 20, 1717; s. Johann Rudolf and Anna Magdalene (Kastenhofer) G.; Ph.D.; m. Maria Rosina Schnell, 1755; m. 2d, Katherina Esther Delosea, 1765; 2 sons, including Gottlieb Siegmund, 3 daus. Became archivist Langrave of Hessia-Homburg, 1741; became pvt. tutor to Prince Christian von Anhalt-Schaumburg, 1743; named vice ofcl. scribe in Thorberg, Switzerland, 1749, advocate, 1755, state scribe of Landshut and Fraubrunnen, 1764. Mem. Econ. Soc. Bern (6 prizes). Author: Die Eisgebirge des Schweizerlandes, 3 vols., 1760-62; Naturgeschichte Helvetiens in der alten Welt, 1773; Versuch eines Verzeichnisses der Mineralien des Schweizerlandes, 1775; Reisen durch die merkwürdigsten Gegenden Helvetiens, 2 vols., 1778. Research on cultivation of bees, mining; published 1st map of mineral locations in Switzerland; described rocks, minerals and fossils of Switzerland; 1st to classify Swiss minerals; recognized origin of erratic boulders. Died Utzerstorf, Switzerland, Apr. 10, 1778.

GRUNER, (Franz Rudolf) Paul, Swiss theoretical physicist; b. Bern, Switzerland, Jan. 13, 1869; s. Friedrich August and Hermine (von Lerber) G.; ed. Bern, Strasbourg, France; Ph.D., Zürich, Switzerland, 1893; m. Bertha Bettina Fanny Bovet, 1896; 1 son, 3 daus. Tchr. physics and math. Freies Gymnasium, Bern, 1893-1903; became lectr. U. Bern, 1894, titular prof., 1903, asso. prof., 1906, asso. prof. theoretical and math. physics, 1913; co-founder, colleague Jungfraujoch Research Sta. Mem. Christian Student Assn. Switzerland and Germany, Swiss Nat. Sci. Soc. (v.p. 1917-22), Fed. Meteorol. Commn. (pres.). Author: (with Kleinert) Dämmerungserscheinungen, 1927; also sci. and popular articles. Research on optics of opaque media especially twilight phenomena, light effect, photometry, electron theory of metals, specific theory of relativity, thermodynamics, radioactivity. Died Bern, Dec. 11, 1957.

GRUNERT, Johann August, German mathematician, physicist; b. Halle/Saale, Germany, Feb. 2, 1797; s. Johann Friedrich August and Dorothea Charlotte (Krünitz) G.; student, Göttingen, Germany; Ph.D., Halle, Germany, 1820; m. Amalie Therese Bergener, 1825; m. 2d; at least 1 son, 2 daus. Secondary sch. tchr., Torgau, 1820-28; in Brandenburg, Germany, 1828-33; prof. U. Griefswald (Germany), from 1833; also tchr. Agrl. Inst., Eldena, from 1838. Mem. Leopoldina, acads. scis. of Munich, Vienna, Stockholm, Uppsala, Pest. Author. Mathematische Abhandlungen, 1822; Die Kegelschnitte, 1823; Statikfester Körper, 1825; Spheroidische Trigonometrie, 1833; Beiträge z. slenen und angen Mathematik, 1838-40; Lehrbuch der Mathematik und Physik, 1841-57; Neue Methode zu Bestimmen die Polhöhe, 1844; Über die mittlere Entfernung einer Figur von einke einem Puncte, 1848; Oxodromische Trigonomertir optische Untersuchungen, 1846-51; Beitrage zu meteorlogishe Optik, 1850; Analytische Geometrie d. Ebene und d. Raumes, 1857; also completed Klügel-Mollwide math. dictionary, 1831-36. Founder, editor Archiv der Mathematik und Physik, 1841-57. Gave a proof of Descartes' rule of signs; revived interest in math. work of E. W. Tschirnhaus. Died Griefsward, July 7, 1872.

GRÜNFELD, Josef, physician; b. Györke, Czechoslovakia, Nov. 19, 1840; s. Samuel and Rosa (Roth) G.; student medicine Pest, also Vienna, Austria; M.D., 1867; m. Sofie Schneider, 1874; 1 son, 3 daus. Asst. to K. L. Sigmund von Ileanor, U. Vienna, joined faculty dermatology and syphilidology, 1881; chmn. dept. Vienna Gen. Polyclinic, 1885-1907. Publs. on devel. endoscope which made it possible to detect normal and path. changes in urinary tract and bladder. Died Vienna, May 14, 1910.

GRÜNHUT, Jacques-Leo, food chemist; b. Vienna, Austria, May 22, 1863; s. Adolf and Amalie (Fränkel) G.; teaching degree U. Leipzig (Germany), 1886, Ph.D., 1886, certifying exam for food chemistry, 1892; m. Betty Epstein. Lectr., chmn. dept. Fresenius Chem. Lab., Wiesbaden, Germany, 1895-1918; prof., chmn. dept. Research Inst. for Food Chemistry, Munich, 1918-21. Mem. Assn. German Food Chemists (mem. adminstrv. com.). Author: Trinkwasser und Tafelwasser, 1920; also articles. Research in food chemistry, especially analysis of foods; studies of wine, drinking water, baking powder, meat extracts, mineral water, biochem. processes in treatment of foods with sulphurous acid, food control. Died Munich, Jan. 5, 1921.

GRUNSKY, Helmut, German mathematician; b. Aalen-Württ, July 11, 1904; s. Heinrich and Lydia (Stahl) G.; ed. Technische Hochschule, Stuttgart, Berlin, U. Berlin; Ph.D., Dipl. Ing.; m. Irma Schenck, 1935; children—Wolfgang, Hiltrud, Eberhard. Asst. scientist Prussian Acad. Scis., Berlin, 1930-39; with Higher Sch., 1945-49; instr. U. Tübingen, 1949-51; Pullman vis. prof. Wash. State Coll., 1950-51; asso. prof. U. Mayence, 1951-58; full prof. U. Würzburg, 1958—. Mem. German Soc. Mathematicians. Editor: Jahrbuch über die Fortschritte der Mathematik, 1935-39; Jahrbericht der Deutschen Mathematiker Ver., 1957-63. Research and publs. on conformable diagrams, theory of potential and differential equations. Home: 4 Scheffelstrasse, Würzburg 87. Office: 6 Klinikstrasse, Würzburg 87, Germany.

GRUNSTEIN, Nathan David, Am. public mgmt. scientist; b. Ashland, O., Sept. 19, 1913; s. Samuel Lewis and Rose (Kolinsky) G.; B.A., Ohio State U., 1935, M.Sc., 1936; Ph.D., Syracuse U., 1943; LL.B., George Washington U., 1951; m. Dorothy Deborah Davis, Nov. 12, 1938; children—Miriam R. (Mrs. Bruce Richard Levin), Margaret J., Leon D., Robert H. Legal research asst., office head atty. U. S. Dept. Agr., Washington, 1939-40; adminstrv. asst. to asst. commr. FDA, Washington, 1940-41; adminstrv. officer, exec. asst. to vice chmn. for labor prodn. WPB, Washington, 1941-47; prof. pub. law and adminstrn. Wayne State U., Detroit, 1947-58; prof. adminstrn. U. Pitts. Grad. Sch. Pub. and Internat. Affairs 1958-64; dir. grad. program in pub. mgmt. sci. Western Res. U., Cleve., 1964—. Dir. pub. exec. devel. programs for fed., urban, state govts., 1951-64; cons. CONSAD Research Corp., 1961—; lectr. Center for Advanced Study Orgn. Sci., U. Wis., Milw., 1963—. Mem. Inst. Mgmt. Sci., Soc. for Gen. Systems Research, Am. Soc. for Polit. and Legal Philosophy, Internat. Assn. for Philosophy of Law and Social Philosophy, Mich., Allegheny County bar assns. Author: (with Ashley Sellers) Regulatory Procedure Plant Quarantine Act, 1941, Federal Food Drug and Cosmetic Act, 1941; General Management of Michigan State Government, 1951; (with J. F. Davison) Administrative Law, 1952; Presidential Delegation of Authority in Wartime, 1961; Executive Development in Business and Government, 1961; (with J. F. Davison) Administrative Law and the Regulatory System, 1966; also articles. First explication of feasible method for simulation and gaming of an urban community; first applications of systems theory to law and procedures of regulatory agys.; initial devel. hist. and conceptual founds. for an urban mgmt. sci.; related human learning and devel. theory to exec. devel. and innovated in design of exec. devel. programs. Home: 2872 Washington Blvd., Cleveland Heights, O. 44118. Office: Western Res. U., Hayden Hall, Cleve. 44106.*

GRÜNWALD, Josef, Czechoslovakian mathematician; b. Prague, Czechoslovakia, Apr. 11, 1876; s. Anton Karl Grünwald; Ph.D., 1898; joined faculty U. Vienna, Austria, 1903; became asso. prof. German U., Prague, 1906. Research and publs. on basis of dual numbers —opposite to Riemann's number sphere; studied basis of complex numbers, geometric interpretation of dual numbers, diagram principle coupling plane geometry and cinematics with solid geometry; Grünwald's cone is a cone of 2d order in projective 2 dimensional space. Died Prague, July 1, 1911.

GRUNZKE, Marvin Elwood, Am. psychologist; b. Matawan, Minn., Nov. 1, 1923; s. Edward A. and Alvina (Gabriel) G.; B.S., Trinity U., San Antonio, 1954; M.A., U. Tex., 1958; Ph.D., Baylor U., 1965; m. Eunice M. Bebler, Aug. 25, 1949; 1 son, Paul M. Enlisted USAF, 1947, advanced through grades to lt. col., 1967; dir. human resources research detachment Sampson AFB, N.Y., 1952-53; devel. job performance examinations Personnel Research Lab., Lackland AFB, Tex., 1953-57; research psychologist comparative psychology div. Aeromed. Research Lab., Holloman AFB, N.M., 1958-66, chief, 1966; staff scientist directorate personnel plans Hdqrs USAF, Pentagon, 1966—. Mem. Am. Psychol. Assn. Research on validation of Air Force Airman classification test battery; developed apparatus for behavioral analysis of chimpanzees in space flight program; invented zero-gravity water and food dispensers used in animal space flight programs. Home: 8818 Skokie Lane, Vienna, Va. 22180. Office: Hdqrs USAF, PDPL, Washington 20330.*

GRÜNZWEIG, Carl Otto, German chemist; b. Schorndorf, Germany, Dec. 31, 1845; s. Carl and Luise Christine (Guapp) G.; student chemistry Munich, Germany, Stuttgart, Germany; Ph.D., Tübingen, Germany, 1872; m. Wilhelmine Krämer, 1879; 4 sons, 1 dau. Indsl. chemist Ultramarin-Fabrik Marienberg, nr. Bensheim, Germany, 1872-78; founder (with Paul Hartmann) Grünzweig & Hartmann, Ludwigshafen, Germany, 1878; became mem. city council, 1889, hon. mayor, Ludwigshafen, 1891-96. Founder insulation materials industry; originated and produced cork brick from cork waste with binding substance for insulation (patented 1880); developed and modified cork brick for various purposes; invented cork expansion process, 1906; discovered insulating properties of Kieselguhr. Died Ludwigshafen, July 9, 1913.

GRUPE, Oskar, German geologist; b. Einbeck, Germany, Apr. 14, 1878; s. Hermann and Louise (Böhlecke) G.; student Tech. U. Hanover; Ph.D., U. Göttingen (Germany), 1901; m. Elly Kunz, 1906. Geologist on expdn. to Sumatra, Konikl. Nederlandsche Petroleum Maatschappij; became asst. geologist Prussian Geol. Inst., Berlin, 1902, permanent colleague, 1904; named dist. geologist, 1914, state geologist, 1927; apptd. dir. State Exam. Com. for Qualifying Geologists, 1934. Research and numerous publs. on salt, brown coal, asphalt beds; created unified representation of Triassic for large areas of Prussia; worked on 35 spl. geol. maps of Prussia. Died Berlin-Schmargendorf, Germany, Feb. 26, 1940.

GRUPP, Gunter, Am. physiologist; b. Esslingen, Germany, Feb. 6, 1920; s. Alois and Martha (Baechle) G.; M.D., U. Freiburg, 1948, Dozent for Pharmacology, 1953; m. Ingrid L. Bettinger, Oct. 5, 1958; children—Michael, Stephan, Lily, Deborah, Jacqueline, Karen. Came to U. S., 1958, naturalized, 1963. Faculty pharmacology U. Freiburg, Germany, 1948-58; faculty U. Cin., 1958—, prof. exptl. medicine, 1965—, dir. exptl. pharmacology lab., cardiac lab., 1962—. Mem. Am. Soc. for Pharmacology and Exptl. Therapeutics, German Pharmacology Soc., Soc. for Exptl. Biology and Medicine. Research, publs. on correlation of heat prodn. of kidney to aerobic and anaerobic metabolism; analysis of mechanism of autoregulation of organ blood flow; electrolyte transport across cell membranes and relation to contractile force of heart; cardiac and circulatory problems. Home: 5052 Collinwood Pl., Cin. 45227.*

GRUSON, Hermann Jacques August, German metallurgist; b. Magdeburg, Germany, Mar. 13, 1821; s. Louis Abraham and Caroline (Bodenstein) G.; studied mech. engring. under August Borsig, Berlin, 1840-45; m. Emma Lemelson, 1847; 1 son, Hermann, 2 daus.; m. 2d, Helene Hildebrandt, 1889. Mech. engr. Berlin-Hamburg R.R., 6 years; chief engr. Wöhlert Engine Factory, Berlin, 3 years; tech. dir. Hamburg-Magdeburg S.S. Co., 1 year; founder shipyard with mech. shop and iron foundry, 1855; built new plant on

Magdeburg-Halberstadt R.R., 1869-72; asso. with Krupp firm, beginning in 1870; bought out by Krupp, 1893. Created cast iron of superior hardness by mixing various types of pig iron; developed chill casting process for r.r. parts, armored plates, gun carriages; produced cast steel. Died Magdeburg, Jan. 31, 1895.

GRUSS, Gerhard Christian, German mathematician; b. Berlin, Mar. 16, 1902; s. Emil and Martha (Kühne) G.; certification diploma Tech. U. Berlin, 1925, Ph.D., 1927; postgrad. U. Göttingen, Germany; m. Charlotte Glasenapp, 1928; 2 sons, 2 daus. Asst. to R. Rothe, Tech. U. Berlin, joined faculty, 1929; became prof. Freiberg (Germany) Mining Acad., 1935. Research and publs. on differential geometry, including Levi-Civitas parallelism in 3-dimensional gen. metric space, also variation calculus, theory of complete and multiple monotonic functions, applications in mechanics. Died Freiberg, May 20, 1950.

GRÜTZNER, Paul von, see von Grützner, Paul.

GRYNAEUS, Simon, theologian, geographer; b. Veringen, Swabia, Germany, 1493. Prof. Greek, Heidelberg, 1523, Basel, 1536; attended Diet of Worms, 1540; discovered last 5 works of Livy, pub. Almagest of Ptolemy in Greek, 1538; translated Plato into Latin; councillor to Duke Ulrich of Württemberg. Author: The New World of Regions and Islands Unknown to the Ancients, 1532 (contains narratives of Marco Polo and numerous other travellers). Died Basel, Switzerland, 1541.

GRYTING, Harold Julian, Am. chemist; b. Belview, Minn., Dec. 31, 1919; s. Reier Elling and Julia (Olsen) G.; B.A., St. Olaf Coll., 1941; postgrad. N.D. State U., 1941-42; Ph.D., Purdue, 1947, postdoctoral fellow, 1947; m. Barbara Jean Ruggles, June 25, 1954; children—Corrine Suzanne, Paul Julian. Chemist, E.I. DuPont de Nemours & Co., 1942-43; with Naval Ordnance Test Sta., China Lake Cal., 1947—, head explosives research br., 1952-65, tech. asst. to head explosives and pyrotechnics div., 1965—, rep. to Tech. Coop. Program, 1959, 63, 65. Fellow A.A.A.S.; mem. Am. Chem. Soc., Am. Ordnance Assn. (govt. cons. 1959—), Nat. Tng. Lab., Research Soc. Am., Sigma Xi, Phi Lambda Upsilon. Collaborator: Outline of Organic Nitrogen Compounds, 1945; Outline of Organic Chemistry, 1961; (with E. F. Degering, others) Fundamental Organic Chemistry, 1948; also tech. reports. Research in synthesis of numerous compounds in alkylphenol and quaternary ammonium series for germicides, research in new explosives compositions; inventor manometer based on vapor pressure of constantly boiling liquid, plastic bonded explosives formulations, high temperature resistant explosives. Home: P.O. 875, 1900 Linda Vista, Ridgecrest, Cal. 93555. Office: Code 45402, U. S. Naval Weapons Center, China Lake, Cal. 93557.*

GRZIMEK, Bernhard, German zoologist; b. Neisse/Schlesien, Germany, Apr. 24, 1909; s. Paul Franz and Margot (Wanke) G.; studied veterinary medicine and zoology; Ph.D.in zoology; m. Hildegard Prüfer; children—Rochus, Michael, Thomas. Became dir. zool. gardens of Frankfurt/Main; hon. trustee Tanganyika Nat. Parks. Recipient prize German Republic, Oscar prize for documentary film Serengeti Shall Not Die. Author: Das Eierbuch; Handbuch der Geflügel-Krankheiten; Kein Platz für wilde Tiere; Serengeti darf nicht sterben; Aus Nashömer gehoren allen Menschen. Research on animal physiology, also poultry diseases. Address: Zoological Garden, Frankfurt am Main, Germany.

GSCHEIDLEN, Richard, German physiologist; b. Augsburg, Germany, 1842; became asst. to Heidenhain, Breslau (now Wroclaw, Poland), 1869; later dir. Bd. Health, Breslau; author work opposing doctrine of spontaneous generation, 1874, also work (with M. Traube) on putrefaction. Died 1889.

GSCHNEIDNER, Karl Albert, Jr., Am. metallurgist; b. Detroit, Nov. 16, 1930; s. Karl and Eugenie (Zehetmair) G.; B.S. in Chemistry, U. Detroit, 1952; Ph.D. in Phys. Chemistry, Ia. State U., 1957; m. Melba E. Pickenpaugh, Nov. 4, 1957; children—Thomas, David, Edward, Kathryn. Staff, Los Alamos Sci. Lab., 1957-63, sect. leader, 1961-63; vis. asst. prof. U. Ill., Urbana, 1962-63; asso. prof. metallurgy Ia. State U., Ames, 1963-67, prof., 1967—, dir. rare earth information center, 1966—. Mem. Metall. Soc. Am. (chmn. com. on alloy phases, 1965-68), Am. Chem. Soc., Am. Soc. Metals, Am. Crystallographic Assn., Am. Inst. Mining, Metall. and Petroleum Engring., A.A.A.S., Sigma Xi, Phi Lambda Upsilon. Author: Rare Earth Alloys, 1961; also articles, chpts. in books, reports. Editor: (with M. T. Hepworth, N. A. Parlee) Metallurgy at High Pressures and High Temperatures, 1964. Research on phys. metallurgy of rare-earth metals and alloys, electronic transformation of cerium, theory of alloy formation especially rare earth metals. Home: 2216 Duff Av., Ames, Ia. 50010.*

GUADAGNI, Dante George, Am. chemist; b. Healdsburg, Cal., Jan. 7, 1920; s. Raffaello and Ida (Gilardi) G.; B.S., U. Cal. at Berkeley, 1942; m. Lorraine Ridolfi, Nov. 28, 1942; children—Richard, Thomas, Marie. Research chemist Stauffer Chem. Co., San Francisco, 1942; research chemist Western Regional Research Lab., Albany, Cal., 1946—, head food appraisal investigations, 1955—. Mem. Inst. Food Technologists, Am. Chem. Soc. Contbr. numerous articles in field to sci. jours. Devel. of improved processes for freezing preservation of fruits; established role of phenolic compounds in browning of frozen fruits; devel. of objective techniques for measurement of quality loss in frozen fruits; sensory procedures and conditions for stability of frozen fruits; relation between instrumental and sensory techniques in aroma evaluation.*

GUAZZI, Gian Carlo, Italian physician; b. Reggio Emilia, Italy, Aug. 27, 1931; s. Giovanni and Ines (Iemmi) G.; M.D. cum laude, Parma (Italy) U., 1956. Asst. in neurology, psychiatry Parma U., from 1956, asso. prof. neurology, psychiatry, from 1962; research chief, neuropathology dept. Bunge Inst., Berchem-Antwerp, Belgium, from 1962, later neuropathology cons.; head neuropathology dept. Born-Bunge Found., Berchem-Antwerp, 1966—. Mem. Belgian, French neurology socs., Belgian Neuro-Pathologie Société, Italian Neuropath. Soc., French Electronic Microscopy Soc. Research, publs. on nonexptl. human and comparative neuropathology, demyelinating diseases, leuco- and polio-encephalitis, cerebral, systemic thesaurismosis. Home: 17 Venusstrasse, Antwerp. Office: Dept. Neuropathology, Born-Bunge Found., 59 F. Williotstr., Berchem-Antwerp, Belgium.*

GUBERNIEV, Mikhail Alekseevich, Russian biochemist; b. 1900; grad. 2d Moscow U., 1926; postgrad. All-Union Inst. Exptl. Medicine, Inst. Biol. and Med. Chemistry; D.Biol. Sci. Tchr. organic, inorganic and analytical chemistry at secondary and higher schs., 1926-54; former asso. All-Union Inst. Exptl. Medicine, Inst. Biol. and Med. Chemistry; head lab. biochemistry, dir. All-Union Antibiotics Research Inst., 1954—. Dep. chmn. antibiotics com. USSR Acad. Med. Sci. Mem. editorial bd. Antibiotics. Research and numerous publs. on animo and nucleic acids, phosphoric compounds and antibiotics; developer antibiotic levomitsetin, 1949; worked on prodn. of major amino acids. Patentee in field. Address: All-Union Antibiotics Research Inst., Nogatinskoe sh., Moscow, USSR.

GUBLER, Adolphe, French physician; b. Metz, France, 1821; studied under Trousseau; aggregate physician; became intern, 1844; named head Paris Clinic, 1848; apptd. prof. therapeutics, 1868; recipient Gold medal Paris Hosps., 1848; mem. Soc. Biology (pres.), Acad. Medicine. Described a form of paralysis affecting parts on opposite sides of body (alternate or crossed hemiplegia, Gubler's paralysis, Weber-Gubler syndrome, or Millard-Gubler paralysis), 1856; opposed homeopathy; research on curare. Died Toulon, 1879.

GUBLER, Clark Johnson, Am. chemist; b. LaVerkin, Utah, July 14, 1913; s. Henry W. and Susanna (Pickett) G.; A.B., Brigham Young U., 1939; M.S., Utah State U., 1941; Ph.D., U. Cal. at Berkeley, 1945; m. Maurine Kjar, Sept. 21, 1938; children—David Clark, Kathleen, Anne, Ronald Kjar. Faculty, U. Utah, 1946-56; spl. research fellow USPHS U. Wis., 1956-58; faculty Brigham Young U., Provo, Utah, 1958—, prof. chemistry, 1960—, established investigator Am. Heart Assn., 1960-65. Mem. Am. Soc. Biol. Chemists, Am. Inst. Nutrition, N.Y. Acad. Scis., Am. Inst. Clin. Nutrition, A.A.A.S., Sigma Xi. Research, numerous publs on copper and iron functions and metabolism, Wilson's Disease, anemias and blood disorders; functions, mode of action and metabolism of thiamine. Home: 466 N. 550 E., Orem, Utah 84057. Office: Dept. Chemistry, Brigham Young U., Provo, Utah 84601.*

GUCKER, Frank Thomson, Am. phys. chemist; b. Phila., Apr. 8, 1900; s. Frank Thomson and Louise (Fulton) G.; A.B., Haverford Coll., 1920, M.A., 1921, LL.D., 1966; Ph.D., Harvard, 1925; m. Eleonore Dubois Harris, June 17, 1925; children—Frank Fulton, Katharine Harris (Mrs. Herbert H. Hand). Research fellow Harvard, 1924-25, 27-28; Nat. Research fellow Cal. Inst. Tech., 1925-27; research chemist E. I. DuPont de Nemours & Co., Wilmington, Del. 1928-29; faculty Northwestern U., 1929-47; faculty Ind. U., Bloomington, 1947—, prof., chmn. dept. chemistry, 1947-51, dean Coll. Arts and Scis., 1951-65, research prof. chemistry, 1965—. Chief tech. aide NDRC, 1941-42. Fellow Carnegie Inst. Washington, 1940-50. Fellow A.A.A.S., Ind. Acad. Scis.; Mem. Am. Chem. Soc., A.A.U.P., Phi Beta Kappa, Sigma Xi, Alpha Chi Sigma, Phi Lambda Upsilon. Author: (with W. Buell Meldrum) Introduction to Theoretical Chemistry, 1936; (with W. Buell Meldrum) Physical Chemistry, 1942; (with Ralph L. Seifert) Physical Chemistry, 1966. Research on aerosols, aqueous solutions of electrolytes and amino acids, precise measurement of heat capacities, heats of dilution, sonic velocities in liquids. Home: 1125 E. Hunter Av., Bloomington, Ind. 47401.*

GUDDEN, (Johann) Bernhard Aloys von, see von Gudden, (Johann) Bernhard Aloys.

GUDDEN, Bernhard Friedrich Adolf, German physicist; b. Pützchen, Germany, Mar. 14, 1892; s. Clemens and Elisabeth (Fick) G.; student natural scis. and math. Bonn, Würzburg, Göttingen (all Germany); Ph.D., Göttingen, 1919; m. Clara Bohnert, 1921; 3 sons, 3 daus. Became asst. to R. W. Phol, U. Göttingen, 1919, joined faculty exptl. physics, 1921, became unofcl. prof., 1924; named prof. exptl. physics U. Erlangen (Germany), 1926; became dir. Phys. Inst., German U., Prague, Czechoslovakia, 1939; arrested at end of World War II. Author: Die lichtelektrischen Erscheinungen, 1928; also articles. Research on photo-electric phenomena, conduction of electricity in semi-conductors, photoelectric processes in diamond, selenium, phosphoric sulfides, alkali halogenides (confirmed quantum equivalents); a founder solid state physics; introduced hypothesis of dissociating interference points; separation of primary and secondary processes in photo-electric conduction. Died Prague, Aug. 3, 1945.

GUDERMANN, Christoph, German mathematician; b. Vienenburg, Germany, Mar. 25, 1798; s. Joseph and Anna Maria (Eilers) G.; student U. Göttingen (Germany), 1820-21; math. tchr.'s certification, Berlin, Germany, hon. doctorate, 1832; m. Magdalena Flohr, before 1832; 5 children. Became math. tchr. secondary sch., Cleves, 1823; named asso. prof. Theol. and Philos. Acad., Münster, Germany, 1832, prof., 1839; Taught K. Weierstrass. Author: Grundriss der analytischen Sphärik, 1830; Theorie der Potenzial-oder der cyklisch-hyperbolischen Functionen, 1833; Lehrbuch der niederen Sphärik, 1835; Theorie der Modular-Functionen und der Modular-Integrale, 1844; Über die wissenschaftliche Anwendung der Beagerungs-Geschütz, Nebst einer Anhange: Von der Prall- (Ricochet-) Schüssen, 1850; also articles. Research on geometry of spheres and elliptical functions (modular functions); introduced terminology for elliptical functions; studied astronomy, mechanics, determination of latitude, proof of sectional elevation point in triangle; function of hyperbola amplitude named after him; calculated 1st tables for hyperbolic functions. Died Münster, Sept. 25, 1851.

GUDGER, Eugene Willis, Am. ichthyologist; b. Waynesville, N.C., Aug. 10, 1866; s. James Cassius Lowry and Mary Goodwin (Willis) G.; student Emory & Henry Coll., Va., 1883-87; B.S., U. Nashville, 1892, M.S., 1893; Ph.D., Johns Hopkins, 1905. Instr. sci. Asheville Coll., 1894-59; lab. asst. Gen. Biol. Lab., Johns Hopkins, 1902-04; prof. biology N.C. Coll. for Women, Greensboro, 1905-19; investigator U. S. Bur. Fisheries, Beaufort, N.C., 1902-11; research asso., Tortugas Lab., Carnegie Inst., 1912-15; editor Vol. III, Bibliography of Fishes (Am. Museum), 1919-23; asso. in ichthyology, Am. Museum, 1921, bibliographer in ichthyology, 1923-38, asso. curator, 1935-38; hon. asso. in ichthyology Am. Mus. and librarian Dean Meml. Library. Fellow A.A.A.S., N.Y. Zoöl. Soc.; mem. Am. Mus. Natural History (life), Am. Soc. Zoölogists, Am. Soc. Naturalists, History of Sci. Soc., hon. corr. mem. Salmon-Trout Assn. of Gt. Britain; corr. mem. Zool. Soc. London, 1939. Author numerous works, including: Natural History of the Whale Shark, 1915; Structure and Habits of Barracuda, 1918; The Candiru, the Only Vertebrate Parasite of Man, 1930; Beginnings of Fish Teratology (1555-1642), 1936; Oral Breathing Valves in Fishes (1685-1935), 1946. Editor of Bashford Dean Memorial Volume; editor for ichthyological terms in 2d edit. Webster's New Internat. Dictionary, 1935. Died Feb. 19, 1956.

GUDNELFINGER, Sigmund, mathematician; b. Kirchberg, Feb. 14, 1846; s. Salomo and Julie (Simon) G.; student Tübingen, Germany, Heidelberg, Germany, Königsburg (now Kalingrad, USSR); Ph.D., Giessen, Germany, 1867; m. Amalie Gunz, 1878; 2 sons, including Friedrich. Joined faculty, Tübingen, 1869, became lectr. on analytic geometry and algebra, 1872, asso. prof. math., 1873; named prof. math. Polytechnikum (later Tech. U.), Darmstadt, Germany, 1879. Co-recipient Steiner prize Prussian Acad. Scis., 1895; recipient gold medal Bene merenti Bavarian Acad. Scis., 1897. Research and publs. on invariant theory of algebraic forms and application to geometry of algebraic curves, especially of 2d and 3d degree, analytic geometry, transformation of quadratic form of n variables into a sum of squares of linear functions of variables; tables, including logarithms and real roots. Died Darmstadt, Dec. 13, 1910.

GUEFT, Boris, Am. pathologist; b. Cannes, France, Nov. 10, 1916; s. Amshel and Nina (Oussoltseff) G.: came to U. S., 1917, naturalized, 1927; A.B., Columbia, 1938; M.D., N.Y. U., 1941; m. Eula Mae Respess, June 25, 1943; children—Nina, Esther, Michael. Research asst. USPHS, New Britain Gen. Hosp., also instr. Yale, 1946-47; resident, fellow in pathology Mt. Sinai Hosp., N.Y.C., 1947-50; pathologist Fairfield State Hosp., also instr. Yale, 1950-55; asst. prof. pathology U. Cin., pathologist VA Hosp., Cin., 1955-58; asso. prof. pathology Albert Einstein Coll. Medicine, N.Y.C., 1958-66, prof., 1966—. Mem. Am. Assn. Pathologists and Bacteriologists, Electron Microscope Soc. Am., Internat. Acad. Pathology. Publs. on new descriptions and understandings of tissue changes in systemic lupus, relation of electron microscopic findings in hepatitis to viral agts., description of ultra structural components of amyloid deposits. Home: 128 Vernon Dr., Scarsdale, N.Y. 10583.*

GUENTHER, Konrad Eduard Franz, zoologist; b. Riga, Latvia, May 23, 1874; s. Hermann and Barbara Julie (Behrens) G.; ed. Bonn, Leipzig, Freiburg (all Germany); m. Eva Fehsenfeld, 1902; 2 sons, including Ekke Wolfgang. Faculty, U. Freiburg, 1902-47, became asso. prof., 1913; invited to Brazil, Argentina, 1923-24; traveled to Mesopotamian Steppes, tropical forests of India and Brazil. Author: Der Darwinismus, 1904; Erhaltet unserer Heimat die Vogelwelt, 1906; Natur und Mensch, 1907; Atlas zur Abstammungs- und Entwicklungsgeschichte des Menschen, 1909; Tierleben unserer Heimat, 1923; Das Antlitz Braziliens,

1927; Sprache der Natur seit der Vorzeit unserers Volkes, 1930; Natur als Offenbarung, 1933; Naturbuch von Schwarzwald, 1942. Pioneered nature protection, especially bird sanctuaries; founder biol. pest control in tropical S. Am.; one of 1st to use nature tours as teaching device; research on bird forms and voices. Died Freiburg, Jan. 26, 1955.

GUENTHER, William Charles, Am. statistician, educator; b. Stewartville, Minn., Dec. 17, 1921; s. Clayton A. and Helen (Forney) G.; B.A., state U. Ia., 1943, M.S., 1946; Ph.D., U. Wash., 1952; m. Norma Berry, Aug. 4, 1955; children—Eda, Clark, Paul, Lee Ann. Math. statistician Naval Ordnance Lab., Corona, Cal., 1952-55; asst. prof. Ariz. State U., Tempe, 1955-57, Fresno (Cal.) State Coll., 1957-59; faculty dept. statistics U. Wyo., Laramie, 1959-—, prof., 1963-—. With Martin Co., Denver, summers 1960-61, Westat Research Analysts, Denver, summer 1962. Decorated D.F.C., Air medal. Mem. Inst. Math. Statistics, Am. Statis. Assn., Am. Math. Assn., Phi Beta Kappa, Sigma Xi. Author: Analysis of Variance, 1964; Concepts of Statistical Inference, 1965; Concepts of Probability, 1968; also articles. Home: 1515 Kearney St., Laramie, Wyo. 82070.*

GUERICKE, Otto von, German physicist, engineer, natural philosopher; b. Magdeburg, Prussian Saxony (Germany), Nov. 20, 1602; s. Hans and Anna (von Zweydorff) Gericke; ed. U. Leipzig, from 1617, U. Helmstadt, 1620; studied law at Jena, 1621, and math and mechanics at Leyden, 1623; traveled in France, England, returning to Madgeburg, 1626; m. Margarethe Alemann, 1626 (d. 1645); 2 sons, 1 dau.; m. 2nd. Dorothea Lentke, 1652. Elected alderman, 1627; when Madgeburg captured by Flemish troops, went to Braunschweig, became engineer under Gustavus II Adolphus of Sweden; as such, returned to Madgeburg, 1632; served as engineer-in-chief for Saxony and as member of new city council; Bürgermeister of Magdeburg and magistrate for Brandenburg, 1646-81; resigned and retired to Hamburg, 1681. Author: Experimenta nova (ut vocantur) magdeburgica de vacuo spatio, 1672; Geschichte der Belagerung und Eroberung von Magdeburg. Invented a water-pump, and then 1st air pump, circa 1650, with which he studied role of air in combustion and respiration; showed that light is and sound is not propagated in a vacuum. In a famous series of public experiments, G. placed together 2 hollow copper hemispheres (Magdeburg hemispheres), the edges of which fitted smoothly together, evacuated air from inside resulting sphere (about 14 inches in diameter), and showed it held together by air pressure on outside of hemispheres and that a team of horses could not pull them apart; this demonstrated before Emperor Ferdinand III at Imperial Diet assembled at Regensburg, 1654; invented an electrical machine that generated static electricity by friction; invented a gygrometer; studied magnetism; predicted periodic return of comets. Died Hamburg, Germany, May 11, 1686.

GUERIN, Alphonse F. M., French surgeon; b. Ploermel, France, circa 1816; M.D., 1847. Interne, Paris hosps., 1840, aide in anatomy, 1843-46, prosector, 1847-53, surgeon, 1850. Author: Maladies des organes génito-externes de la femme; Sur la fièvre purulente, 1847; Traité de chirurgie opératoire, 1855; Mémoire sur les rétrécissements de l'urèthre, 1857; Du pansement ouaté et de son application à la thérapeutique chirurgicale, 1885. Described fold of mucous membrane sometimes found in roof of fossa navicularis of urethra, (known as Guerin's valve or fold) 1864. Died 1895.

GUERIN, Henri, French chemist; b. Paris, Aug. 3, 1906; s. Francis and Marthe (Dury) G.; Engr., Sch. Physics and Indsl. Chemistry; D.Sc.; diploma in advanced law studies; m. Roberte Bonny, Aug. 7, 1933; children—Marthe, Chantol, Françoise, Jean-Claude, Philippe, Jacques, Béatrice. Attaché, then research master Nat. Center for Sci. Research, 1930-43; prof. Faculty Scis. Nancy (France), 1953-58, Faculty Scis. Paris, 1958-65, Faculty Scis. Paris-Orsay, 1965-—. Mem. Chem. Soc. France, Soc. Chem. Physics, Am. Chem. Soc. Author: Reactivité des combustibles, 1945; Manipulations et Analyse des Gaz, 1951; Chimie industrielle, 1962; Chimie, 1962, 65, 67; also articles. Research on arsenites and phosphates corresponding notably to establishment of equilibrium diagrams; studies of gasification of solid combustibles and their reactivity. Home: 95, Boulevard Jourdan, Paris 15, France. Office: Faculté des Scis., Orsay 91, France.*

GUERIN, Jules René, French surgeon; b. Boussu, France, 1801; M.D., Paris, 1826. Founded Gazette Médicale de Paris, 1830; dir. orthopedic services l'Hôpital des Enfants, 1839-49. Recipient med. prize French Acad. Scis., 1837. Author: Essais sur la méthode sous-cutanée, 1841; Essai de physiologie générale 1843; Pansement des plaies par l'occlusion pneumatique, 1878; Etude sur l'intoxication purulente, 1879; Recherches sur les difformités congénitales, 1880-82. Establishes treatment of osseous deformities as a specialty; introduced subcutaneous incision in orthopedic operations. Died 1886.

GUERRANT, Nollie Burnham, Am. biochemist; b. Fulton, Mo., Aug. 15, 1899; s. John Wesley and Ocie (Craighead) G.; A.B., Westminster Coll., 1921; M.S., N.D. State Coll., 1923; postgrad. Ia. State Coll.; Ph.D., Mo. U., 1925; m. Prudence Lorraine Fennel, June 7, 1924; children—Martha (Mrs. Early Lee Files), Nollie Burnham, Ronald Pierre. Asst. research chemist N.D. Agr. Expt. Sta., Fargo, 1921-23; asst. research chemist Mo. U., Columbia, 1924-25; asst. prof. biochemistry Okla. State U., Stillwater, 1925-27; asso. biochemist Auburn (Ala.) U., 1927-30; asso. prof. Pa. State U., University Park, 1930-37, prof. biochemistry, 1937-64, head nutrition lab., 1930-64, prof. emeritus, 1964-—. Mem. vitamin adv. bd. U. S. Pharmacopeaia, 1948-—; mem. food and nutrition bd. NRC, 1949-52, vitamin standards com. WHO, 1949-—. Mem. A.A.A.S., Biochem. Soc. London, Am. Chem. Soc., Am. Inst. Nutrition, Am. Soc. Biol. Chemists, Soc. for Exptl. Biology and Medicine, Inst. Food Technologists, Sigma Xi, Phi Kappa Phi, Alpha Chi Sigma, Phi Lambda Upsilon, Gamma Sigma Delta. Research, numerous publs. on biochemistry in health, food and deficiency diseases, vitamin requirements in men and animals, factors influencing vitamin stability and vitamin content of foods, influence of certain antibiotics on vitamin requirements. Address: 400 S. Ravine St., Fulton, Mo. 65251.*

GUERTLER, William Minot, German metallurgist; b. Hanover, Germany, Mar. 10, 1880; s. Alexander and Grace (Sedgwick) G.; student Hanover, Munich, Göttingen (all Germany); Ph.D., 1904; m. Felicitas de la Porte, 1908; 2 daus. Became asst. to Tamann, Göttingen, 1904; with Tech. U., Berlin, 1907-45, joined faculty, 1908, became asso. prof., 1917, prof., dir., Inst. for Applied Metallurgy, 1933, emeritus, 1945; became research asso. Mass. Inst. Tech., 1908; became prof., dir. Inst., Tech. U., Dresden, Germany. Numerous lecture tours to India, Japan, U. S. Mem. German Soc. for Metallurgy (a founder). Author: Einführung in die Metallkunde, 2 vols., 1943, under title Metallkunde, 1954; also numerous articles. Editor, contbg. author: Metallographie, Ein aufürhliches Lehr- und Handbuch der Konstitution und der physikalischen, chemischen und technischen Eigenschaften der Metelle und metallischen Legierungen, 3 vols., 1912-35. Research and numerous patents on qualitative analysis of alloys, including their classification and nomenclature, theoretical found. and investigation methods, especially aluminum, non-ferrous metals; contbr. to devel. of metallurgy as ind. sci. Died Mar. 21, 1959.

GUEST, Howard Russell, Am. chemist; b. Scranton, Pa., Oct. 11, 1914; s. Halton W. and Mabel (Boorem) G.; B.S., Pa. State U., 1936; m. Hallie M. Hughart, Dec. 20, 1947; children—Susan, Stephen. With Union Carbide Corp., 1936-—, asso. dir. research and devel. Olefins Div., South Charleston, W.Va., 1964-—. Mem. Am. Chem. Soc., Entomol. Soc. Am., Am. Phytopath. Soc. Publs. patentee, research in catalytic process studies for new techn. in petrochems.; devel. new pesticides; discovery new commercially important chems. by oxidation, hydrogenation, condensation, Home: 845 Lower Chester Rd., Charleston, W.Va. 25302. Office: P.O. Box 8361, South Charleston, W.Va. 25303.*

GUETTARD, Jean-Étienne, French geologist, mineralogist; b. Étampes, France, Sept. 22, 1715; studied medicine, Paris; keeper natural history collections of Duke of Orleans; mem. French Acad. Scis. Author: Mémoire et carte mineralogique; Sur la nature et situation des terrains qui traversent la France et l'Angleterre, 1746; Atlas et description mineralogiques de la France, 1780; Mémoire sur la transpiration insensibledes plantes. Made 1st geol. survey and map of France; originated idea that minerals and rocks around Paris are arranged in concentric bands of sand, marl, and schist which continue to other side of channel; 1st to identify trilobite fossils, mammalian remains and nummucites in Paris area; studied weathering by rain and wind, 1770; deduced that land and sea floor rise to compensate for weathering loss; suspected volcanic origin of many mountains in central France. Died Paris, Jan. 6, 1786.

GUEVARA POZO, Diego, Spanish pharmacist; b. Tinola, Almeria, Apr. 15, 1909; s. Francisco and Carolina P. Guevara Pozo; M.D. in pharmacy; m. Augustias Benitez Alahija, 1930. Prof. parasitology Faculty Pharmacy, Grenada, later vice dean; dir. Lopez Neyra Inst. Parasitology, C.S.I.C., also mem. council; v.p. delegation of council of higher sci. research, Granada. Mem. Spanish Assn. Natural History, Spanish Soc. Physics. Mem. Royal Acad. Medicine Grenada, Assn. Parasitology (founder, v.p.). Dir., Spanish Rev. Parasitology. Contbr. articles to profl. publs.

GUFFY, Joseph Claude, Am. chemist; b. Rosston, Okla., Jan. 12, 1920; s. Adolphus Claude and Roxie (Burke) G.; B.S. summa cum laude, Southwestern Inst. Tech., Okla., 1940; M.S., U. Wis., 1943, Ph.D., 1947; m. Mary Adrienne, Tack, Dec. 20, 1943; children—Lynn P., David J. Research chemist OSRD, 1943-45, asst., 1945-47; chemist Shell Oil Co., 1947-48; instr. U. Cal. at Berkeley, 1948-51; with Chevron Research Co. (formerly Cal. Research Corp.) div. Standard Oil Co., Cal., Richmond, Cal., 1951-—, group supr., 1959-62, supervising research chemist, 1962-66, sr. research asso., 1966-—. Mem. Am. Chem. Soc., A.A.-A.S., Soc. Applied Spectroscopy, Sigma Xi, Alpha Xi Sigma. Research, publs. on spectroscopy applied to data composition matter. Home: 624 Santa Barbara St., Berkeley, Cal. 94707. Office: 576 Standard Av., Richmond, Cal. 94802.*

GUGGENBÜHL, (Johann) Jakob, Swiss physician; b. Meilen, Switzerland, Aug. 13, 1816; s. Hans Jakob and Maria (Hottinger) G.; student medicine, Zürich, Geneva, Bern (all Switzerland); M.D. Physician, Sernftal, 1837-39; joined Hofwil Tng. Instn., 1840; opened instn. for cretins and other retarded children, on Abendberg nr. Interlaken, 1841. Author: L'Abendberg, Premier Rapport, 1844; Briefe über den Abendberg und die Heilanstalt für Cretinismus, 1846; Senschreiben an Lord Ashley, 1851; Die Heilung und Verhütung des Cretinismus und ihre neuesten Fortschritte, 1853; Die Erfroschung des Cretinismus und Blödsinns, 1860. Worked with cretins and retarded children by first increasing body strength, then by work therapy (gardening, crafts) and edn. Died Montreux, Feb. 2, 1863.

GUGGENHEIM, Edward Armand, Brit. phys. chemist; b. Manchester, Eng., Aug. 11, 1901; s. Armand and Marguerite (Simon) G.; M.A., Cambridge U., 1927, Sc.D., 1937; m. Simone Ganzin, Sept. 20, 1934; m. 2d., Ruth Helen (Clarke) Aitken, Nov. 3, 1955. Tchr., Danish Royal Agr. Coll., 1928-29, Stanford, 1932-33, Reading U., 1933-35, London U., 1936-39; mem. sci. staff Admiralty, 1939-44; staff Montreal Lab. Atomic Energy, 1944-46; prof. chemistry Reading U., 1946-66, ret., 1966. Fellow Royal Soc., 1946, Royal Inst. Chemistry; fgn. mem. Royal Danish Acad. Author: Modern Thermodynamics, 1933; (with R. H. Fowler) Statistical Thermodynamics, 1939; Thermodynamics for Chemists and Physicists, 1949; Mixtures, 1952; (with J. E. Prue) Physico-Chemical Calculations, 1954; Boltzmann's Distribution Law, 1955; Elements of the Kinetic Theory of Gases, 1960; Applications of Statistical Mechanics, 1966; Elements of Chemical Thermodynamics, 1966; Elements and Formulas of Special Relativity, 1967. Research, publs. in applications of thermodynamics and statistical mechanics to properties of gases, mixtures, electrolyte solutions. Home: Selborne, 71 Peppard Rd, Caversham, Reading, Berkshire, Eng.*

GUGGENHEIMER, Heinrich W., Am. mathematician; b. Nurnberg, Germany, July 21, 1924; s. Siegfried and Marguerite (Bloch) G.; dipl. Swiss Fed. Inst. Tech., 1947, Sc.D., 1950; m. Eva Auguste Horovitz, June 6, 1947; children—S. Michael, Esther H., Tobias I. S., Hanna Y. Came to U. S., 1959, naturalized, 1965. Lectr., Hebrew U. of Jerusalem, 1954-56; prof. Bar Ilan U., Israel, 1956-59; asso. prof. Wash. State U., 1959-60; asso. prof. U. Minn., 1960-62, prof., 1962-67; prof. Bklyn. Poly. Inst., 1967-—. Speaker, Internat. Colloquium Differential Geometry, Italy, 1953, Founds. of Geometry, Holland, 1959, Algebraic Geometry, Rome, 1965, Differential Geometry, Italy, 1967. Mem. Math. Assn. Am. (vis. lectr. 1963-65), Am., Swiss math. socs. Author: Differential Geometry, 1963; Plane Geometry And Its Groups, 1967; also numerous articles. Research in connections between differential geometry and topology - algebraic geometry; differential geometry and convexity. Home: 426 Wilson St., West Hempstead, N.Y. 11552. Office: 333 Jay St., Bklyn. 11201.*

GUGLIELMINETTI, Ernest, Swiss physician; b. Glis Wallis, Switzerland, Nov. 4, 1862; s. Anton and Luise (Furrer) G.; student medicine, Bern, Switzerland; state exam, 1885; m. Bertha Cecilie Oppenheimer, 1899. Asst. to Hugo Kronecker; colonial physicians for Netherlands Govt. in Java, Sumatra, Borneo, 1886-90; physiologist Jansen Expdn. to Mont Blanc; physician in Monaco, 40 years. Named hon. citizen Brig, 1938. Mem. Internat. Rd. Congress (founder), Internat. League for Combating Dust (founder). Originated method of putting tar on surface of roads, 1902; studied height physiology in balloon flights. Died Geneva, Switzerland, Feb. 1943.

GUGLIELMINI, Domenico, Italian mathematician, physician; b. Bologna, Italy, Sept. 27, 1655; M.D., 1678; became gen. intendant water, Bologna, 1688; named prof. math. U. Bologna, 1690, prof. hydrometry, 1694; prof. medicine U. Padua (Italy), 1702-10. Fellow Royal Soc., 1697; mem. French Acad. Scis. (fgn. asso.). Author: Volantis flammae epitropeia, 1677; De cometarum natura et ortu, 1681; Riflessioni philosophiche, 1688; Aquarum fluentium mensura, 1690; Della natura de'fiumi, 1697; Exercitatio physico-medica, 1701. Research on hydraulics; originated law of constant interfacial angles for salt crystals. Died Padua, July 11, 1710.

GUGLIEMINI, Giovanni Battista, Italian astronomer; b. Bologna, Italy, Aug. 16, 1763; faculty U. Bologna, 1794-1817, prof. astronomy, 1801-17, rector, 1814-15; dir. Bologna's waterworks, 1802-10. Research and publs. on earth's rotation; studied objects falling from tower in Bologna to show deviation to east, 1791-92. Died Bologna, Dec. 15, 1817.

GUHL, Alphaeus Matthew, Am. zoologist; b. Cleve., Aug. 14, 1898; s. Matthew and Katherine (Haller) G.; B.A., N. Central Coll., Ill., 1922; M.S., U. Chgo., 1939, Ph.D., 1943; m. Della A. Schuelke, Aug. 23, 1924; children—Kathryn J. (Mrs. James M. Campbell). High sch. tchr., prin., 1922-37; faculty Kan. State U., Manhattan, 1943-—, prof. 1954-—. Fellow Poultry Sci. Assn., Animal Behavior Soc., A.A.A.S.; mem. Kan. Acad. Sci. (past sec., pres.), Ecol. Soc. Am. (past chmn. sect. animal behavior and sociobiology), Am. Inst. Biol. Scis. (gov. bd. 1959-65), Am. Soc. Zoologists, Cooper Ornithol. Club, Poultry Sci. Assn., Wilson Ornithol. Soc., World's Poultry Sci. Assn., Sigma Xi, Gamma Sigma Delta, Phi Kappa Phi (nat. pres.). Contbg. author: Sex and Internal Secretions, 1961; Behavior of Domestic Animals, 1951. Research on significance dominance orders in chickens, sexual behavior and effects social rank, genetic selections for levels of aggressiveness, gonadal hormones and agonistic behavior, social inertia and integration flocks, devel. social behavior and orgn. in chicks. Home: 1744 Leavenworth St., Manhattan, Kan. 66502.*

GUIART, Jean, French anthropologist; b. Lyon, France, July 22, 1925; s. Jules and Helen (Pierret) G.; LL.D., U. Paris; diploma Sch. of Living Oriental Langs., Paris; m. Josephine Soot-Calimbre, Oct. 10, 1951; children—Michel, Armand, Rene. With Mus. of Man, Paris, 1945-47; research work Office of Fgn. Sci. Research and Tech., 1947-57; prof. Practical Sch. of Higher Studies, Paris, 1957—; dir. Centre Documentaire pour l'Oceanie, Paris. Mem. Nat. Com. Sci. Research, 1958—, Planning Commn. for Overseas Terrs., 1964—. Recipient Silver medal Nat. Center Sci. Research, Paris. Mem. Polymesoan Soc., Soc. Oceanic Studies, Soc. Oceanistes (v.p.). Author: Un siecle et demi de contacts culturels a Tanna, Nouvelles Hebrides, 1957; (with H. Deschamps) Tahiti, Nouvelle Caledonie, Nouvelles Hebrides, 1958; Espiritu Santo, 1959; The Arts of the South Pacific, 1963; Structure de la Chefferie en Melanesie du Sud, 1964; Nouvelles Hebrides, 1966; Mythologie du masque en Nouvelle Caledonie, 1966; also articles. Devel. inventory method for study of formalized social structures such as land-tenure and chieftainship. Home: 147 Av. J. B. Clement, Clamart 92, France. Office: 293 Av. Daumesmil, Paris 12, France.*

GUIBAL, Jules, French mining engr., inventor; b. Toulouse, France, 1813. Prof., L'École des Mines, Hainaut. Conducted 1st expts. in ventilating mines; determined composition of mine air, 1840; built widely accepted centrifugal mine ventilator, 1860; unit of conductivity of mine ventilation is named for him. Died Paris, 1888.

GUIDI, Guido (Vidius, Vidus), physician, anatomist; b. Florence, Italy, 1500; came to France, circa 1542; physician to François I, later to Duke Cosme I; prof. Collège de France; prof. philosophy, then of medicine U. Pisa (Italy). Author: Chirurgia e Graeco in Latinum a se conserva, 1544; De anatomia corporus humani, 7 vols., 1611; Opera omnia medica, chirurgico-anatomica, 1667. Described pterygoid canal of vidian artery and nerve (Vidian canal), also nerve of pterygoid canal (Vidian nerve), circa 1555, internal maxillary artery (Vidian artery), circa 1559; restored Greek medicine and surgery in France. Died 1569.

GUIGNARD, (Jean Louis) Leon, French biologist; b. Mont-sous-Vaudrey, France, Apr. 13, 1852; prof. botany, dir. Ecole supérieure, Paris; prof., dean Faculty Pharmacy. Mem. French Acad. Scis. (pres. 1911), Acad. Medicine, Acad. Agr. Author: Anatomie végétale, 1886; L'appareil sexuel et la double fécondation dans les tulipes, 1893. Research on double fertilization in tulips, corn, and angiosperms in gen., cellular multiplication, plant immunology, enzymes and glucosides in plants, male gametes of Crytogams. Died Paris, Mar. 7, 1928.

GUILCHER, André Julien, French oceanographer; b. Brest, France, May 19, 1913; s. Julien Pierre and Mathilde (Thoreux) G.; student Louis Le Grand Coll., Paris, France, 1931-34; Licence and Agrégation in History and Geography, U. Paris, 1935, 36, Doctorate, 1948; m. Yvonne Mailharrou, July 17, 1937; children—Goulven and Rozenn (Mrs. René Battistini) (twins). Faculty, Brest Coll., 1937-40, Nantes Coll., 1940-47; prof. geography U. Nancy (France), 1947-57; prof. marine and fluvial hydrology U. Paris, 1957—. Mem. French Nat. Council for Sci. Research, 1952—, Comite Central d'Océanographie et d'Etude des Cotes, 1953—, French Com. Geodesy and Geophysics, 1951—. Decorated Croix de Guerre, 1939-45. Mem. Assn. French Geographers (gen. sec. 1962—), Internat. Geog. Union (mem. commn. coastal geomorphology 1952—), Société Géologique de France, Brit. Assn. Geographers, Société des Océanistes. Author: Morphologie Littorale and sous-marine, 1954; (with J. Beaujeu) Les Iles Britanniques, 1963; Précis d'Hydrologie Marine et continentale, 1965; also numerous articles. Research in coastal geomorphology, coral reefs, marine and continental hydrology, submarine geomorphology and geology of continental shelf off Brittany. Home: 2 Boulevard Joffre Bourg la Reine, Hauts de Seine, France. Office: 191 rue Saint Jacques, Paris 5e, France.*

GUILFORD, Joy Paul, Am. psychologist; b. Marquette, Neb., Mar. 7, 1897; s. Edwin Augustus and Arvilla (Monroe) G.; A.B., U. of Neb., 1922, A.M., 1924; Ph.D., Cornell, 1927; LL.D. (honorary), University of Nebraska, 1952; Sc.D., University Southern California, 1962; married Ruth S. Burke, Sept. 8, 1927; 1 dau., Joan S. Instr. psychology U. of Ill., 1926-27; asst. prof. psychology U. of Kan., 1927-28; asso. prof., prof. psychology U. of Neb., 1928-40, dir. Bur. Instructional Research, 1938-40; prof. psychology U. of So. Calif. since 1940 (on leave of absence, 1942-45). Served with U.S.A.A.F. as aviation psychologist, 1942-45; dir. psychol. research units No. 2 and No. 3; chief field research unit, Air Force Training Command Hdqrs.; chief Dept. Records and Analysis, Sch. Aviation Medicine; disch. rank col. Awarded Legion of Merit. Fellow A.A.A.S., Am. Psychol. Assn. (pres., 1949-50); mem. Psychometric Society (president 1937-38), National Academy of Sciences, Society of Experimental Psychologists, Midwestern Psychol. Assn. (pres., 1939-40), Western Psychol. Assn. (pres., 1946-47), So. Calif., Cal. State psychol. assns., Soc. Multivariate Exptl. Psychologist, Cal. Color Soc. (pres. 1948-49), Blue Key, Phi Beta Kappa, Sigma Xi. Minor contributions to applied statistics, psychophysical methods, scaling methods, and test methods; two psychophysical laws; laws of color preference; discovery of fundamental dimensions (traits) of personality, including intellectual abilities; development of the structure-of-intellect theory and model, with derivations of implications for general psychological theory. Author: Psychometric Methods, 1936, 1954. Other books. Home: P.O. Box 1288, Beverly Hills, Cal. 90213. Office: U. So. Cal., Los Angeles 90007.*

GUILLAIN, Georges Charles, French physician; b. Rouen, France, Mar. 3, 1876; M.D., 1902. Prof. neurology Paris Faculty Medicine, 1923-44. Mem. French Acad. Scis., 1951, Acad. Medicine. Authority on spinal column; studied cephalorhachidian liquid, spinal cord marrow, pathology of head; described polyradicular syndrome. Died May 21, 1951.

GUILLAUME, Charles Edouard, Swiss meteorologist, physicist; b. Fleurier, Switzerland, Feb. 15, 1861; ed. Fed. Poly. Sch., Zurich, Switzerland, 1878-1882; Ph.D., 1883; D.Sc. (hon.), Geneva (Switzerland) U., Neuchatel (Switzerland) U., U. Paris; m. A. M. Taufflieb, 1888; 3 children. Asst., Internat. Bur. Weights and Measures, from 1883, asso. dir., from 1902, dir., from 1905. Recipient Nobel prize in physics, 1920, Duddell medal, 1929. Mem. French Acad. Scis. (corr.), Swedish (fgn.), Russian acads., Royal Soc. Uppsala, Brit. Assn. (corr.), Institut Genevois, Soc. Arts Geneva, Phys. Soc. London, Royal Inst. London, Soc. for Physics (pres.), others. Author: Traité pratique de la thermométrie de précision, 1889; Units and Standards, 1893; Les radiations nouvelles, 1896; Les applications des aciers au nickel, 1898; Metrical Convention, 1902; Applications of Nickel-Steels, 1904; Les récents progrès du système metrique, 1907-21; Initiation to Mechanics, 1909. Translator: Boys, Soap Bubbles: La Création du Bureau International des Poids et Mesures et son Oeuvre, 1927. Redetermined volume of liter; investigated sources of error in mercury-in-glass thermometer and corrections to be used, 1889; investigated properties of nickel-steel alloys; discovered INVAR (low-expansion coefficient) and ELINVAR (low elasticity coefficient), 2 inexpensive alloys used for high precision instruments and measurement standards, 1899. Died Paris, June 13, 1938.

GUILLEMEAU, Jacques, French physician; b. Orléans, France, 1550; ed. l'Hôtel-Dieu, Paris; a son, Charles. Served in army and hosps. in Flanders; named dir. Coll. Surgery, 1595; chief surgeon Hôtel Dieu; surgeon to kings Charles IX, Henry III, Henry IV. Author: Traité des maladies de l'oeil, 1588; De la grossesse et accouchement des femmes, 1621; Oeuvres de chirurgie, 1602; La chirurgie francaise, 1594; L'heureaux accouchement des femmes, 1609. First to suggest inorganic material for tooth fillings, 1594; described pyorrhea alveolaris, 1594; made 1st reference to making of artificial teeth, 1594. Died 1613.

GUILLEMIN, Ernest Adolph, Am. engineer; b. Milwaukee, Wis., May 8, 1898; s. Victor and Erna (Jacobsen) G.; B.S., U. of Wis., 1922; S.M., Mass. Inst. Tech., 1924; Ph.D., U. of Munich, 1926; m. Mary Lanier Moran, May 29, 1929 1 dau., Mary Grace. Asst. in elec. engring., Mass. Inst. Tech., 1922-24, instr., 1926-28, asst. prof., 1928-36, asso. prof. 1936-44, professor, 1944-60; Webster prof. of electrical engring., 1960—; Saltonstall traveling fellow 1924-26; consultant to Radiation Laboratory, 1940-44, Raytheon Manufacturing Co., 1946—, Edgerton, Germeshausen & Greer, Inc., 1965—; member advisory bd. Hycon Eastern, Cambridge. Awarded President's Certificate of Merit 1948; medal of honor Inst. Radio Engineers, 1961. Fellow Institute Radio Engineers, American Inst. Elec. Engrs., Am. Acad. Arts, Scis., Royal Society Arts (London); Sigma Xi. Author: Communication Networks, Vols. I and II, 1931 and 1935; Mathematics of Circuit Analysis, 1949; Introductory Circuit Theory, 1953; Synthesis of Passive Networks, 1957; Linear Physical Systems, 1963; collaborator: Electric Circuits, 1940. Applied Electronics, 1943. Significant contributions to the design procedure of electrical networks for prescribed response, especially for specified behavior in the time domain; pulse-forming networks for radar (during war); design of Loran transmitter for navigational aid. Home: 41 Woodlawn Av., Wellesley Hills 82, Mass. Office: Mass. Inst. Tech., Cambridge, Mass.*

GUILLEMIN, Roger, Am. physiologist; b. Dijon, France, Jan. 11, 1924; s. Raymond and Blanche (Rigollot) G.; B.A., U. Dijon, 1941, B.Sc., 1942; M.D., Faculty of Medicine, Lyons, France, 1949; Ph.D., U. Montreal, 1953; m. Lucienne Jeanne Billard, Mar. 22, 1951; children—Chantal Claude Marie, Francois Jean Marie, Claire Marguerite Marie, Helene Marie, Elizabeth Marie, Cecile Marie. Came to U. S., 1953, naturalized 1965. Asso. dir., asst. prof. U. Montreal Inst. Exptl. Medicine and Surgery, 1951-53; asso. dir. dept. exptl. endocrinology Coll. France, Paris, 1960-63; faculty Baylor U., Houston, 1953-60, 63—, prof. physiology Coll. Medicine, 1963—. Mem. Am. Physiol. Soc., Endocrine Soc., Soc. Exptl. Biology and Medicine, A.A.A.S., Soc. Biology of Paris, N.Y. Acad. Scis. Research, numerous publs. on physiol. studies and isolation of hypothalamic hormones. Home: 17 W. Shady Lane, Houston 77042.*

GUILLEMIN, Victor, Jr., Am. physicist; b. Milw., Feb. 10, 1896; s. Victor and Erna (Jacobsen) G.; B.A., U. Wis., 1923; M.A., Harvard, 1925; Dr.Phil., U. Munich, Germany, 1926; m. Eileen W. Whall, June 14, 1936; children—Victor W., Robert C., Richard E., Marie-Jean. Instr. physics Harvard, 1930-31, lectr. 1931-35, research asso. Fatigue Lab., 1935-41; sr. physicist USAAF, Dayton, O., 1941-48, cons. USAF, 1948-52; prof. biophysics U. Ill., 1948-59; lectr. physics Harvard, 1959-66, hon. research asso., 1966—. Mem. com. on med. scis. Nat. Mil. Establishment, 1948-53; mem. com. on bioacoustics NRC 1953-56. Harvard Sheldon fellow, 1926; NRC fellow, 1927-28; recipient USAF Meritorious Service award, 1948. Mem. A.A.A.S., Am. Phys. Soc., Am. Physiol. Soc., Am. Assn. Physics Tchrs., Aeromed. Assn., Biophys. Soc., Phi Beta Kappa, Sigma Xi. Author: The Story of Quantum Mechanics, 1968. Research and publs. on atomic and molecular structure; biol. and aeromed. research; mil. aviation apparatus and equipment. Home: 61 Foster Rd., Belmont, Mass. 02178.*

GUILLET, James Edwin, Canadian chemist; b. Toronto, Ont., Can., Jan. 14, 1927; s. Edwin C. and Mary E. (Scott) G.; B.A., U. Toronto, 1948; Ph.D., Cambridge U., 1955; m. Helen Ann Bircher, July 4, 1953; children—Edwin Louis, Barbara Lynn, Patricia Ann, Carolyn Jean. Research chemist Eastman Kodak Co., Rochester, N.Y., 1948-50; research chemist Tenn. Eastman Co., Kingsport, 1950-52, sr. research chemist, 1955-62, research asso., 1963; asso. prof., U. Toronto, 1963—; cons. Glidden Co. of Can., 1964—, Imperial Oil Enterprises Ltd., 1965—; vis. prof. polymer chemistry Vanderbilt U., 1964, 66. Chem. Inst. of Can. fellow, 1966. Mem. Chem. Inst. Can., Faraday Soc., Am. Chem. Soc. Research, publs. on kinetics and mechanism of polymerization and degradation of polymers, establishment of mechanisms of Photodegradation of various polymeric materials, relations between molecular structure and phys. properties of polymers. Numerous patents in field. Home: 31 Sagebrush Lane, Don Mills, Ont. Office: Dept. Chemistry, U. Toronto, Toronto 5, Ont., Can.*

GUILLET, Leon Alexandre, French metall. engr.; b. St. Nazaire, France, July 11, 1873; Docteur ès Sciences, 1902. Prof., Conservatory Arts and Crafts, from 1906; with Central Sch. Mfg., from 1911, dir., 1923. Mem. French Acad. Scis., 1925. Research primarily on alloys; studied mech. properties of spl. steels and thermal processing of metals. Died Paris, May 9, 1946.

GUILLIERMOND, (Marie-Antoine) Alexandre, French botanist; b. Lyons, France, Aug. 19, 1876; Docteur ès scis., 1913; prof. botany Paris Faculty Scis., from 1927; mem. French Acad. Scis., 1935. Author: Recherches cytologiques, 1902; Les Levures, 1912. Added to knowledge of cytology; made classic studies on chondrin; research on reprodn. of mushrooms. Died Apr. 1, 1945.

GUILLOTIN, Joseph Ignace, French physician; b. Santes, France, 1738; ed. Jesuit Coll., Bordeaux. Held various polit. offices, including dep. to States-General, 1789. Author: Petition des citoyens domociliés à Paris, 1788. Defended capital punishment and proposed on humanitarian grounds use of beheading machine (named for him), 1789; re-established Acad. Medicine, after French Revolution. Died Paris, 1814.

GUINAND, Pierre Louis, Swiss optician; b. Corbatière, Switzerland, 1848; mfr. wood clockcases; later cast bells for clocks; began expts. in optical glass, 1768; built full-scale furnace, 1775; glass mfr. (with Josph von Benediktbenern of Utzschneider works), Benekiktbenern, Germany; returned to Switzerland, 1814. Discovered importance of stirring in mfg. optical glass, 1798; invented new stirrer which was universally used by 1805; studied manufacture of flint glass, crown glass. Died 1824.

GUINTERIUS, Johannes (Johann Winther of Andernach, Gunther of Andernach), physician; b. Andernach, Germany, 1505; M.D., U. Paris, 1528. Mem. faculty U. Paris; practiced medicine in Metz, France; became prof. Greek, Strasbourg (now in France), 1544, also physician to Elector of Palatinate. Author: Medical Knowledge and Practice in Ancient and Modern Times, 1571. Translator (into Latin): De re medica (Paul of Agina), 1532; De compositione medicamentorum, 1533, On Anatomical Procedure, 1536, Anatomical Institutions, 1536 (Galen). Died 1574.

GUIRAUD, Paul Louis Emile, French psychiatrist; b. Cessenon, Hérault, Aug. 4, 1882; s. Elie and Louise (Grasset) G.; M.D., Sch. and Faculty Medicine, Montpellier; m. Suzanne Berquez, 1913. Intern hosps. of Montpellier, Charenton; med. dir. of Saint-Dizier, 1908; chief physician psychiat. hosps. of Seine, 1922-52. Mem. Medico-Psychol. Soc. (past pres.), Psychiat. Evolution (past pres.), Congress Psychiatrists and Neurologists (past pres.). Author: (with M. Dide) Psychiatry Practicing Medicine, 1922-29; Psychiatrie Generale, 1950; Rapporteur au Premier Congrès Mondial de Psychiatrie, 1950; Psychiatrie Clinique, 1956. Address: 4 rue Marie-Deraismes, Paris 17, France.

GUISNÉE, French mathematician; b. France; flourished 1705. Engr. to King Louis XIV; prof. math. Master Gervais Coll. Mem. French Acad. Scis., 1702. Author: Méthode de démontrev par l'algèbre les théorèms do geometrie, 1705. Applied algebra to solving geometric problems; constructed equations according to methods of Descartes.

GUITERAS, Gregarlo Maria, physician; b. Matanzas, Cuba, Mar. 12, 1863; s. Eusebio and Josefa (Gener) G.; A.B., LaSalle Coll., Phila., 1880, A.M., 1881, M.S. (hon.), 1917; M.D., U. Pa., 1885; m. Maria Hortensia Aranguren, Sept. 11, 1897; children—Nestor Ramon, Blanche Maria, John Raoul, Maria Hortense, Matilde Maria, George Gustavus, Mary Louise. Pvt. practice medicine, Charleston, S.C., 1886-87; asst. surgeon, advancing through grades to sr. surgeon, U. S. Marine Hosp. Service (now USPHS), 1888-1927, ret. as lt. col., 1927. Del. to Pan-Am. Med. Congress, Havana, 1902, Guatemala, 1907, Santiago de Chili, 1911, Montevideo, Uruguay, 1920. Served in all yellow fever epidemics in U. S. and considered as authority on this disease. Died July 5, 1934.

GUITERAS, Juan, Cuban physician; b. Matanzas, Cuba, Jan. 4, 1852; s. Eusebio and Josefa (Gener) G.; ed. at La Empresa, Matanzas; M.D., U. Pa., 1873 (Ph.D.); m. Dolores Gener, May 5, 1883. Resident, vis. phys. Phila. Hosp., 1873-79; marine hosp. service, 1879-89; served as expert on yellow fever in all epidemics, 1881-—; prof. medicine Charleston Med. Sch., 1884-88; prof. pathology U. Pa., 1889-99; on staff Gen. Shafter as yellow fever expert in Santiago campaign, 1898; prof. gen. pathology and tropical diseases U. Havana, 1900-21; dir. pub. health, Cuba, 1909-21, pres. Nat. Bd. Health; sec. Pub. Health and Charities, Cuba, 1921-22, resigned. Pres. 2d Nat. Med. Congress of Cuba; mem. Yellow Fever Commn. of Internat. Health Bd., Rockefeller Found., 1916-25. Mem. Assn. Health Officers N.Am. (v.p.), Am. Pub. Health Assn. Editor: La Revista de Medicina Tropical. Discovered filaria Bancrofti in the U. S. and uncinaria in Cuba; verified the cause of yellow fever independently of Maj. Walter Reed's army bd., 1901. Died Oct. 28, 1925.

GULBRANSEN, Earl Alfred, Am. chemist; b. Seattle, Jan. 20, 1909; s. Christian A. and Ellen (Johnson) G.; B.Ch.E., Wash. State Coll., 1931; Ph.D., U. Pitts., 1934; m. Margery Evernden, July 2, 1938; children—Karen, Kristin, David. Nat. Research Council fellow U. Cal., Berkeley, 1934-35, research asso., 1935-36; instr. chemistry, chem. engring. Tufts U., 1936-40; research chemist Westinghouse Research Labs., Pitts., 1940-47, adv. chemist, 1947-66, cons. scientist, 1966-—. Recipient Inst. of Metals award, 1949; Nat. Assn. Corrosion Engrs. Whitney award, 1952; Am. Chem. Soc. Pitts. award, 1961; Electrochem. Soc. Acheson award and prize, 1964. Mem. Am. Chem. Soc., Internat. Conf. Surface Reactions (chmn. 1948), Internat. Union of Pure and Applied Chemistry, Chemists Club Pitts. (chmn. 1956-57), Corrosion Research Council (chmn. 1963-64), Am. Inst. Mining and Metall. Engrs., Electron Microscope Soc. Contbr. numerous articles in field to sci. jours. Developed sensitive microbalance for use in high vacuum and gas reaction systems for study of surface reactions, electron diffraction and electron microscope techniques for study of surface reactions; study of reactions of metals and alloys with oxygen, hydrogen and water at high temperature; spl. basic studies of materials for use in atomic power plants and in space applications; thermochem. studies of high temperature materials; studies on geochem. processes. Home: 63 Hathaway Court, Pitts. 15235. Office: Westinghouse Research Labs., Pitts. 15235.*

GULDBERG, Cato Maximilian, Norwegian chemist, mathematician; b. Christiana, (Oslo), Norway, Aug. 1, 1836; ed. U. Christiana; tchr. Royal Mil. Sch., 1860; apptd. prof. applied math. Royal Mil. Acad., Oslo, 1862; named lectr. U. Christiana, 1867, apptd. prof. applied math., 1869. Author: Études sur les affinités chimiques, 1867. Research on thermodynamics and chem. equilibrium; (with Peter Waage) developed chem. law of mass action which deals with the speed of reaction and the relative concentrations of reactants, 1864. Died Christiana, June 14, 1902.

GÜLDENSTÄDT, Johann Anton, explorer; b. Riga, Latvia, Apr. 26, 1745; s. Anton and Dorothea (von Virgin) G.; student medicine, Berlin, Germany; M.D., Franfort/Oder, Germany, 1767. Invited to join 1768 expdn. program of St. Petersburg (Russia) Acad., 1767; traveled through Caucasus lands, sources of Duna, Dnieper, Don and Volga, via Moscow, Stalingrad to Astrachan to No. Foreland of Caucasus, in mountain's southward, then to Don, Asow Sea, over Kiev, Moscow to St. Petersburg, 7 years; practiced medicine, St. Petersburg. Author: Reisen durch Russland und im kaukasischen Gebirge (editor P. S. Pallas), 2 vols., posthumously 1787-91; also articles. Observations on plant and animal geography; descriptions of rocky subsoil and covering soil; 1st publs. on black soil and its origin. Died St. Petersburg, Mar. 23, 1781.

GULDIN, Paul (Habakuk), mathematician; b. Mels, nr. St. Gallen, Switzerland, June 12, 1577; ed. Collegium Romanum, under Clavius. Trained as goldsmith; joined Soc. Jesus, Freising, 1597; prof. math. Jesuit Colls., successively in Rome, Graz, Austria, Vienna, Sagan, and Vienna. Author: Refutatio elenchi a S. Calvisio conscripti, 1616; Paralipomena, 1616; Problema arithmeticum de rerum combinationibus, 1622; Dissertatio de motu Terrae, 1622; Problema geographicum de discrepantia in numero ac denominatione dierum, 1633; Centrobaryca, 4 vols., 1635-41. Introduced centrobaric method (Guldin's theorem) which gives relation between volume and area of a revolving figure, 1635. Died Graz, Nov. 3, 1643.

GÜLDNER, Hugo, German mech. engr.; b. Herdecke, Germany, July 18, 1866; s. Gustav and Ida (Erdmann) G.; student Upper Profl. Sch., Hagen, Germany; hon. dr. engring., Karslruhe, Germany; m. Adele Benecken, 1891; 2 sons, 2 daus. Became engr., Magdeburg, Germany, 1890; joined firm H. Laas & Co., Magdeburg, 1895; founder motor factory, Magdeburg, 1897, went bankrupt, 1897; became chief engr. factory Gebrüder Pfeiffer, Kaiserslautern, Germany, 1898; chief engr. R. Diesel's firm Allgemeine Gesellschaft für Dieselmotoren AG, Augsburg, Germany, 1899, left firm, 1901; became chief engr., tech. dir. firm Maschinenbaugesellschaft München, 1903; also founder Güldner-Motoren-Gesellschaft, Munich, Germany, 1903; moved plant to Aschaffenburg, Germany, 1906-07. Author: Monteurschalen ein Bedürfnis des practischen Maschinenbaues, 1895; Das Entwerfen und Berechnen von Verbrennungskraftmaschinen, 1903; Untersuchungen über den Einfluss der Betriebswärme auf die Steuerungseingriffe der Verbrennungsmaschinen, 1924; also articles. Patentee improvements in combustion engines, after 1890, 2-cycle motor, 1897; invented generator system also designed 4-cycle motor, 1902; manufactured oil motors which reached capacity of 1200 horsepower, beginning 1906; worked (with Diesel) on impulse oil motors and diesel ship's engines. Died Frankfort/Main, Germany, Mar. 12, 1926.

GULEKE, Nicolai Gustav Hermann, surgeon; b. Pernau, Estonia, Apr. 25, 1878; s. Reinhold and Ludmilla (Weiser) G.; student medicine, Berlin, Münster, Bonn, Germany, Rostock, Germany, Strasbourg, France; M.D., 1902; m. Agathe von Kapff, 1913; 3 daus. Asst. to Bergmann, Berlin, to Madelung, Strasbourg; joined faculty Strasbourg; ship's physician; became prof. surgery, 1913; named prof. Marburg, Germany, 1918, Jena, Germany, 1919; advising surgeon both world wars. Hon. mem. German, Austrian, Swedish socs. surgery; mem. Internat. Coll. Surgeons (master surgeon), Leopoldina. Author: Die Chirrugie der Nebenschildrüsen, 1913; Kriegschirurgiescher Röntgenatlas (with H. Dietlen), 1917; (with O. W. Gross) Die Erkrankungen des Pankreas, 1924; Die Chirrugie der Hirngeschwülste, 1936; Kreigschirurgie und Kriegschirurgen im Wandel der Zeiten, 1945; Der A ärztliche Kunstfehler, 1955; 50 Jahre Chirurgie, 1955; Die bösartigen Geschwülste des Dickdarms und Mastdarms, 1957; also monographs, articles. Studied neurosurgery, urology, orthopedics, radiology, tumor operations. Died Wiesbaden, Germany, Apr. 4, 1958.

GULISASHVILI, Vasilii Zakharevich, Russian sylviculturist, ecologist; b. Apr. 23, 1903; grad. Forestry Faculty, Leningrad Forestry Inst., 1927, Chem. Tech. of Timber Faculty, 1931; D.Agrl. Sci., 1936. Asso., Exptl. Forestry Sta., Leningrad Agrl. Inst., 1926-31; former dep. dir. Transcaucasian Forestry Research Inst.; acad. dir. Tbilisi Timber Tech. Inst., until 1937, head dept. gen. forestry, 1961; dir. Forestry Inst., Georgian Acad. Scis., 1944-—. Mem. Georgian Acad. Sci. Author: Textbook on Sylviculture, 1944; Sylviculture and the Principles of Plant Geography and Ecology, 1944; General Sylviculture, 1957. Research and publs. on biology, ecology and ecol. geography of tree species. Address: Forestry Inst., Georgian Acad. Sci., Tbilisi, Gruz. SSR, USSR.

GULKANIAN, Vartan Oganesovich, Russian geneticist; b. Feb. 28, 1902; grad. Yerevan U., 1927. Mem. staff Armenian branch USSR Acad. Scis., from 1927, asso. Armenian Plant Protection Sta., also Biol. Inst., until 1943, asso. Inst. Genetics and Plant Selection, 1950-56; asso. Inst. Farming, Armenian Ministry of Agrl., since 1956. Mem. Armenian Acad. Scis. (v.p. 1943-50). Author: The Nature of the Splitting of Wheat Hybrids Obtained by Zonal Pollination, 1948; Deep Chopping of Cotton, 1949; Progress in Agrobiological Science in the Armenian SSR, 1949; Agrobiological Science and Wheat Seed Culture, 1962; also others. Co-developed new wheat type (Artashat 42) for Armenia, 1949; research on nutrition related to plant growth, yield and heredity; derived ramose wheat types using local strains; gave method of deep chopping cotton. Address: Armenian SSR Acad. Scis., Yerevan, Armenian SSR.

GULL, Sir William Withey, Brit. physician; b. Colchester, Eng., Dec. 31, 1816; s. John Gull; ed. Guy's Hosp.; M.B., London U., 1841; M.D., 1846; D.C.L., Oxford, 1868; LL.D., Cambridge and Edinburgh, 1880; m. Miss Lacey, 1848; 1 son, William Cameron; 1 dau. Lectr. natural philosophy, 1843-47; lectr. physiology and comparative anatomy Guy's Hosp., 1846-56, asst. physician, 1851-56, full physician from 1856, joint lectr. on medicine, 1856-65, cons. physician, 1865-90; Fullerian prof. physiology Royal Instn., 1847-49; censor Coll. Physicians, 1859-61, 72-73, councilor, 1863-64; mem. Gen. Med. Council, 1871-83, 86-97; attended the Prince of Wales in his illness of 1871; named physician extraordinary to Queen, 1872, in ordinary, 1887. Fellow Royal Coll. Physicians, Royal Soc., 1869. Contbr. articles to Guy's Hosp. Reports. Described disease, usually of middle age, with renal hemorrhage but no lesion which is known as renal hemophilia, essential renal hematuria, or angioneurotic hematuria, 1866; described path. lesions in tabes dorsalis, 1856; research on use of static electricity for treatment of nervous disorders, 1849; (with Henry Gawen Sutton) first description of arteriosclerotic atrophy of the kidney (Gull-Sutton disease), 1872; described myxedema with atrophy of the thyroid gland (Gull's disease), 1874. Died Jan. 29, 1890.

GULLIKSEN, Harold, Am. psychologist; b. Washington, July 18, 1903; s. Charles and Signe (Engebretsen) G.; B.A., U. Wash., 1926, M.A., 1927; postgrad. Ohio State U.; Ph.D., U. Chgo., 1931; m. Dorothy Eleanor Palmer, Sept. 6, 1930; children—Eleanor Louise (Mrs. John William McLauchlin), Katherine Jean (Mrs. Leon Goodrich). Instr., Ohio State U., 1927-29; faculty U. Chgo., 1929-31, 33-45; research asso. Mooseheart Lab. for Child Research, 1931-33; prof. psychology Princeton, 1945-—. Research sec. Coll. Entrance Examination Bd., 1945-48; research adviser Ednl. Testing Service, 1948-—. Fellow Am. Statis. Assn., Am. Psychol. Assn.; mem. Inst. Math. Statistics, Psychometric Soc. (past pres.), Am. Assn. U. Profs., Eastern Psychol. Assn., Phi Beta Kappa, Sigma Xi. Author: Theory of Mental Tests, 1950; also articles. Editor: (with S. Messick) Psychological Scaling—Theory and Applications, 1960; co-editor, contbg. author Mathematical Psychology, 1964. Research on formulation math. equation to express psychol. theories; developed theories on math. formulations applicable to mental tests, to curves produced during process learning, to measurement intangibles such as an individuals values, attitudes and preferences. Home: 12 Aiken Av., Princeton, N.J. 08540.*

GULLINO, Pietro Michele, pathologist; b. Saluzzo, Italy, Mar. 24, 1919; s. Antonio and Olimpia (Camisassi) G.; M.D. summa cum laude, U. Torino, 1943; m. Marisa Irene Bigo, Feb. 1, 1956. Came to U. S., 1959, naturalized, 1963. Vis. scientist Nat. Cancer Inst., Bethesda, Md., 1955-59, staff mem., 1959-—. Recipient Ganassini award, 1953; award Italian Council for Research, 1953; award Italian League Against Cancer, 1954. Mem. Am. Assn. Cancer Research, Am. Soc. Exptl. Pathology, Italian Soc. Cancer, Italian Soc. Pathology. Research, numerous publs. on lymphatic tissue, exptl. reprodn. of morphologic patterns found in diseases, amino-acid toxicity, physiopathology of tumors. Home: 832 Melody Ct., Bethesda 20034. Office: Nat. Cancer Inst., Bethesda, Md. 20014.*

GULLIVER, George, Brit. anatomist, physiologist; b. Banbury, Eng., June 4, 1804; ed. St. Bartholomew's Hosp., London; 1 son, George. Prosector to Abernethy; dresser to Lawrence at St. Bartholomew's Hosp; became Hunterian prof. comparative anatomy and physiology, 1861; surgeon to royal Horse Guards. Fellow Royal Soc., 1839, Royal Coll. Surgeons. Editor: (English translation) General and Minute Anatomy of Man and the Mammalia (Gerber), 1842; The Works of William Hewson, F.R.S., 1846. Research and publs. on formation and repair of bone; 1st to give tables of measurement and full observations on shape and structure of red blood corpuscles in man and many vertebrates; demonstrated prevalence of cholesterine and fatty degeneration in several organs; in botany showed important varieties of character in raphides, pollen and some tissues. Died Nov. 17, 1882.

GULLOTTA, Filippo, neuropathologist; b. Catania, Italy, Feb. 4, 1931; s. Giovanni and Adelaide (Boratto) G.; M.D., U. Catania, 1955; m. Helmtrud Goeken, May 11, 1963; 1 son, Giovanni. Asst. neurologist U. Catania, 1955-56; asst. instr. pathology and anatomy U. Modena, Italy, 1957-60; asst. instr. U. Bonn, Germany, 1960, asst. instr., 1963-66, asst. prof. neuropathology, 1966-—; asst. Deutsche Forschungsanstalt für Psychiatrie, Munich, Germany, 1961-63. Mem. Soc. Ital. Neurologia, Deutsche Gesellschaft für Neuropathologie, Soc. Ital. di Neuropatologia. Author: Patologia dei Gliomi Encefalici (with G. Spigolon), 1961; Anatomia Patologica del Sistema Nervoso, 1965; Das sogenannte Medulloblastom, 1967; also articles. Research on morphology and histochemistry of brain tumors, pathology of nervous system. Home: 57 Gudenauerweg, Bonn-Ippendorf, Germany. Office: Inst. für Neuropathology, 7 Wilhelmsplatz, Bonn, Germany.*

GULLSTRAND, Allvar, Swedish ophthalmologist; b. Landskrona, Sweden, June 5, 1862; ed. univs. Uppsala (Sweden), Vienna, Stockholm; M.D., 1888. Became lectr. ophthalmiatrics Carolina Inst., Stockholm, 1891; prof. ophthalmiatrics U. Uppsala, 1894-1927, emeritus, 1927-30. Recipient Nobel prize for physiology and medicine, 1911. Improved slit lamp, 1911, spectacle lenses, methods for estimating astigmatism and abnormal shapes of cornea and for locating paralysed muscles, corrective glasses after removal cataractous lens; studied formation of optical images. Died Stockholm, July 28, 1930.

GULVI, Maksim Fedotovich, Russian biochemist; b. Mar. 3, 1905; grad. Kiev Vet. and Zootech. Inst., 1929; D.Biol. Sci. Asso., Inst. Biochemistry, Ukrainian Acad. Sci., 1932-41, 44-—, head lab. tissue proteins, 1950-—, also chmn. editorial and publs. council; prof. Kiev Vet. Inst., Ukrainian Acad. Agri. Sci., 1944-—. Mem. Commn. for Coordination Cancer Research, 1959. Recipient Stalin prize, 1952. Mem. Ukrainian Acad. Sci. (v.p. 1958-62). Author: The Biological Activity of Some Purified and Crystalline Proteins, 1954; The Tricarbon Oxydation Cycle and its Physiological Significance, 1957; Ribonucleic Acid, the Agent and its Role as the Matrix in the Biosynthesis of Protein, 1962. Research and publs. on biochemistry of muscular activity, methods of preparing highly purified and crystalline tissue proteins and their functional interrelations during metabolism. Address: Ukrainian Acad. Sci., Vladimirskaya ulitsa 54, Kiev, USSR.

GUMBEL, Emil J., mathematician; b. Munich, Germany, July 18, 1891; s. Hermann and Flora Gumbel; Ph.D. in Math., U. Munich; m. Marie-Louise Czettritz, 1930 (dec.); 1 son, Harald. With U. Heidelberg; head research U. Lyons; asso. prof. New Sch. for Social Research, N.Y.C., 1940-42; prof. French U., N.Y.C., 1942-46; asso. prof. Bklyn. Coll.; adj. prof. indsl. engring., Columbia U., 1953—. Fellow Am. Statis. Assn., Inst. Math. Statistics; mem. Internat. Inst. Statistics. Author: Verräter Verfallen der Ferne, 1930; La Durée Extreme de la Vie Humaine, 1937; Statistical Theory of Extreme Values and Applications, 1954; Statistics of Extremes, 1958. Research on math. theory of population, statistics; meteorological phenomena, droughts, floods, fatigue and breaking strength failures. Home: 441 Ocean Av., Bklyn. 11226. Office: S.W. Mudd Bldg., Columbia, N.Y.C. 10027.

GÜMBEL, Ludwig Karl Friedrich, German naval engr.; b. St. Julian, Germany, Mar. 12, 1874; s. Karl Ludwig and Julie (Koch) G.; student Tech. U. Berlin, Germany, 1894-98; doctorate, 1910; m. Olga Catherine Dretz, 1902; 3 sons, 1 dau. Joined shipyard Schichau-Werft, Elbing, 1898; adminstrv. head mech. dept. Hamburg-Amerika Line; named dep. dir. Atlas-Werke, Bremen, Germany, 1906; became prof. Tech. U. Berlin, 1910; co. comdr. in heavy combat during World War I; then called to build up submarine fleet. Mem. Soc. Naval Engrs. Research and publs. on engine lubrication leading to rules for constrn. of bearings, transversal vibration of ship, torsion vibration of waves, hydrodynamics (friction resistance of ship), theory of screw propellor. Died Berlin-Charlottenburg, Germany, Feb. 8, 1923.

GÜMBEL, (Wilhelm) Theodor, German botanist; b. Dannenfels, Germany, May 19, 1812; s. Johann Friedrich and Charlotte (Roos) G.; student forestry, Würzburg, Germany; degree, Munich, Germany, 1835; postgrad. philosophy, 1 year; m. Salomea Amalie Mohr, 1844; 1 son, Theodor, 3 daus. Became tchr. Indsl. Sch., Zweibrücken, Germany; became tchr. Landau, Germany, 1843, rector, 1853. Mem. Leopoldina. Worked on Bryologia Europaea (Philipp Schimper). Contbr. articles to sci. jours. Discovered (at the same time as H. Koch) regenerative ability of leaves and rhizoids of mosses (formed basis for genetic work of F. Wettstein). Died Landau, Feb. 10, 1858.

GÜMBEL, (Carl) Wilhelm von, see von Gümbel, (Carl) Wilhelm.

GUMLICH, Ernst Carl Adolf, German physicist; b. Ahorn nr. Coburg, Germany, Apr. 23, 1859; s. Ernst and Sophie (Tritschler) G.; student physics, math., natural scis., Jena, Germany, Tübingen, Germany, Berlin, Germany; teaching degree, Berlin, 1883; Ph.D., Jena, 1885; m. Else Hein, 1890; 2 sons, 1 dau. Tchr. for several years; joined Reich Phys. Tech. Inst., Berlin, 1887, dir. magnetic lab., 1898-1924, became prof., 1898. Research and publs. on magnetic materials under alternating magnetization, iron-silicon alloys, magnetic and elec. properties of pure iron, coal steels and ferro-alloys of silicon, aluminum, manganese, measuring methods for determination magnetic properties; improved Epstein process, Joch-Isthmus method; studied location and time changes in ferromagnetic properties; showed parallelism with elec. conduction capacity. Died Berlin, Feb. 12, 1930.

GUMMERE, John, Am. mathematician; b. Willow Grove, Pa., 1784; s. Samuel and Rachel (James) G.; m. Elizabeth Bugby, 1808, at least 1 son Samuel James. For most part self-educated; taught math., Horsham, Pa., 1803-05; taught in Rancocas, N.J., 1806-11; opened, operated boarding sch., Burlington, N.J., 1814-33; elected to Am. Philos. Soc., 1814, began contbg. articles to their Transactions; taught mathematics Haverford (Mass.) Sch., 1833-44, supt., several years; reestablished (with son Samuel James) sch., Burlington, 1843-45; recognized as one of ablest mathematicians in U. S. at that time. Author: A Treatise on Surveying, 1814; Elementary Treatise on Astronomy, 1822. Died May 31, 1845.

GUMPLOWICZ, Ludwig, sociologist, polit. scientist; b. Cracow, Austrian Poland, Mar. 9, 1838; studied at Cracow and Vienna. Practiced law, then turned to sci. work; became docent U. Graz, 1875, prof. extraordinarius, 1882, prof. ordinarius, 1893-1908. Author: Rasse und Staat, 1875; Der Rassenkampf, 1893; Grundriss der Soziologie, 1885; Die Soziologische Staatsidee, 1892. Adopted some aspects of the sociology of Herbert Spencer; replaced struggle for survival among individuals with struggle for survival among social groups; studied antagonisms between racial groups; defined society as a group centering about some 1 or 2 common interests; proposed that social phenomena of any order are due to interaction of hetrogeneous elements; defined sociology as discipline concerned with formulation and illustration of universal laws governing social phenomena. Died Graz, Austria, Aug. 9, 1909.

GUND, Konrad, physicist, inventor; b. Vienna, Austria, Apr. 25, 1907; s. Robert and Henrike (Krafft von Dellmensingen) G.; student Tech. U. Vienna; Dr. rer. nat., Göttingen, Germany, 1946; m. Déjanire Caurairy, 1932. Became X-ray engr. Siemens & Halske, Vienna, 1931; with Siemens-Reiniger-Werke, Erlangen, Germany, from 1936, dir. design offices, from 1949. Mem. commn. European Bd. for Nuclear Physics Research. Contbr. articles to sci. publs. Pioneer in devel. betatron for med. radiation with electron rays and X-rays of high quantum energy, from 1941; developed a 6 MeV betatron for physics research, 1946, 1st used for med. purposes, 1948; developed a 15 MeV betatron, also a portable clin. application of it. Died Göttingen, May 31, 1953.

GUNDISALVO, Domingo, (also: Dominicus Gundissalinus) Spanish philosopher, translator; fl. early 12th century; translated in collaboration with John of Seville. Archdeacon of Segovia. Author: De divisione philosophiae; De immortalitate animae; De processione mundi; De unitate; De anima. His philosophy was Muslim Aristotelian and neo-Platonism modified by Christian theology and individualism; Muslim-Jewish philosophy was introduced into Latin Christendom by his writings and translations.

GUNDLACH, Johannes Christoph, German zoologist; b. Marburg, Germany, July 17, 1810; s. Johannes Christoph and Christine (Rethberg) G.; Ph.D., Marburg, 1837. Invited to Dutch Guiana (Surinam) financed by Assn. Natural Sci. in Kassel, 1838; remained in Cuba and spent life there; made journeys to P.R. Author: Contribución a la Ornithologia Cubana, 1873-76; also articles. Pioneered research on fauna of Gt. Antilles, including many new forms; studied systematic classification, distbn. ecology, biology of birds, reptiles, amphibians, molluscs. Died Havana, Cuba, Mar. 15, 1896.

GUNN, John Battiscombe, physicist; b. Cairo, Egypt, May 13, 1928; s. Battiscombe George and Lilian (Meacham) G.; B.A., Trinity Coll., Cambridge (Eng.) U., 1948; m. Freda Elizabeth Pilcher, Aug. 5, 1950; children—Janet P., Gillian R. Research engr. Elliott Bros. (London), Ltd., Borehamwood, Eng., 1948-53; jr. research fellow Royal Radar Establishment, Malvern, Eng., 1953-56; asst. prof. physics dept. U. B.C., Vancouver, Can., 1956-59; staff mem. IBM Thomas J. Watson Research Center, Yorktown Heights, N.Y., 1959—. Fellow Am. Phys. Soc., I.E.E.E. Research, publs., patents in hot electron phenomena in semiconductors, semiconductor devices; discovered microwave oscillations of current and asso. travelling domains of elec. field in gallium arsenide. Home: Box 23, Route 1, Mt. Kisco, N.Y. 10549. Office: Box 218, Yorktown Heights, N.Y. 10598.*

GUNN, Ross, Am. physicist; b. Cleve., May 12, 1897; s. Ross Delano Aldrich (M.D.) and Lora Arletta (Conner) G.; B.S. in E.E., U. Mich., 1920, M.S., 1921; Ph.D., Yale, 1926; m. Gladys Jeannette Rowley, Sept. 8, 1923; children—Ross, Andrew Leigh, Charles Rowley, Robert Burns. Instr. in engring. physics U. Mich., 1920-22; radio research engr. U. S. Air Service, 1922-23; instr. physics Yale, 1923-27; research physicist U. S. Naval Research Lab., 1927-33, tech. adviser, 1933-47, supt. mechanics and electricity div., 1938-46; supt. aircraft electrical div., 1943-46; supt. physics div., 1946-47; tech. dir., Army-Navy Precipitation Static Project, 1943-46; Army-Navy Atmospheric Electricity Project, 1946-47; cons. Nat. Adv. Com. for Aeronautics, 1943-59; Nat. Defense Research Com., 1942-43, Research and Development Bd., 1946-48; dir. phys. research, U. S. Weather Bur., 1947-57; also asst. chief bur., 1955-56; research prof. physics, Am. U., Washington, 1958—; cons. AEC, 1958-60; dir. Air-Force-Weather Bur. Cloud Physics Project 1947-50; mem. Sci. Adv. Bd., chief of staff, U. S. A. F., 1948-53; cons. C. F. Kettering Found. 1951-54. Cited by sec. of navy for exceptionally distinguished service in connection with devel. atomic bomb, 1945; Distinguished Service award, Flight Safety Found., 1951, Robert M. Losey Award, Inst. Aero. Scis., 1956, Gold medal for Exceptional Service, U. S. Dept. Commerce, 1957, Distinguished Alumnus award U. Mich., 1953. Fellow Am. Phys. Soc., Inst. Elec. and Electronic Engrs., Geophys. Union; mem. Nat. Acad. Scis., Am. Meteorol. Soc., Sigma Xi. Contbr. many sci. and tech. articles. Inventor, organizer 1st work on atomic powered submarine; invented elec. devices for mil. and naval use; investigated solar and terrestrial electricity and magnetism. Died Oct. 15, 1966.

GUNNING, Harry Emmet, Canadian research chemist; b. Toronto, Ont., Can., Dec. 16, 1916; s. Lorenzo Edward and Ledo Beryl (Shangraw) G.; B.A. with honors in Chemistry 1st class, U. Toronto, 1939, M.A., 1940, Ph.D., 1942; m. Donna Marie Beahan, Jan. 30, 1943; 1 dau., Judith Beryl. Fellow, Harvard, 1942-43; staff div. pure chemistry NRC, Ottawa, Ont., 1943-46; asst. prof. chemistry U. Rochester, 1946-48; faculty Ill. Inst. Tech., 1948-57, prof., 1955-57; prof., head dept. chemistry U. Alta., Edmonton, Can., 1957—. Chmn. adv. com. on chem. research Def. Research Bd. Can.; mem. adv. council NRC Can.; cons. in phys. chemistry. Fellow Royal Soc. Can., A.A.A.S., Chem. Inst. Can. Research, numerous publs. on atomic and free radical chemistry, intermediates in photochem. reactions, flashphotolysis and kinetic mass spectrometry, separation isotopes using hyperfine excitation, identification and characterization unpairedspin species by ESR spectroscopy. Home: 13119 Grandview Dr., St., Edmonton, Alta., Can.*

GUNNING, Robert Clifford, Am. mathematician; b. Longmont, Colo., Nov. 27, 1931; s. Clifford Henry and Inez (Wilhelm) G.; A.B., U. Colo., 1952; M.A., Princeton, 1953, Ph.D., 1955; m. Wanda S. Holtzinger, 1966. Fellow, U. Chgo., 1955-56; faculty Princeton, 1956—, prof. math., 1966—; asst. dir. studies in math. St. Catharine's Coll., Cambridge, Eng., 1959-60. Mem. Am. Math. Soc., Société Mathématique de France. Author: Lectures on Modular Forms, 1962; (with Hugo Rossi) Analytic Function of Several Complex Variables, 1965; Lectures on Riemann Surfaces, 1966; also articles. Research on function theory of complex variables. Office: Fine Hall, Princeton, Princeton, N.J. 08540.*

GUNSALUS, Irwin Clyde, Am. biochemist; b. nr. Blunt, S.D., June 29, 1912; s. I. Clyde and Anna (Shea) G.; B.S., Cornell U., 1935, M.S., 1937, Ph.D., 1940; m. Carolyn Foust, June 16, 1951; children—Gene, Glen, Ann, Robert, Richard, Carolyn Kristina, Kristin Carla. Faculty, Cornell U., Ithaca, N.Y., 1937-46, prof. bacteriology, 1945-47; prof. Ind. U., 1947-50; prof. bacteriology U. Ill., Urbana, 1950-55, prof. biochemistry, 1955—, head div., 1955-66. Mem. biochemistry study sect. NIH, 1955-60, 64—, NSF, 1950-60. Guggenheim fellow, 1950, 59. Co-recipient Mead Johnson award in biochemistry Am. Inst. Nutrition, 1946. Mem. A.A.A.S., Am. Chem. Sc., Am. Soc. Biol. Chemists, Biochem. Soc. (Britain), Canadian Soc. Microbiologists, Harvey Soc. N.Y. (hon.), N.Y. Acad. Scis., Am. Soc. Microbiology, Soc. Gen. Microbiologists, Soc. Gen. Physiologists, Nat. Acad. Scis., Genetics Soc. Editor: (with R. Y. Stanier) The Bacteria, vols. I-V, 1960-64. Research, publs. on cell and enzyme regulation, oxidative energy coupling, keto acid and thioester systems, enzymatic catalysis oxygenation, dehydrogenation, transfer and polymerization, genetic recombination and variation in growth. Home: 2002 S. Race St., Urbana, Ill. 61801.*

GUNTER, Edmund, English mathematician; b. Hertfordshire, Eng., 1581; ed. at Westminster, also Christ Church, Oxford (Eng.) U., M.A., 1606; B.D., 1615. Became incumbent St. George's, Southwark, 1615; named 3d prof. astronom Gresham Coll., 1619-26. Author: Book of the Sector; Canon triangulorum or Table of Artificial Sines and Tangents, 1620; complete works edited by Samuel Foster, 1636, by William Leybourn, 1673. Invented Gunter's quadrant, 1618, surveyor's chain (Gunter's chain); 1st to use terms, cosine and cotangent, 1620; discovered variation in magnetic needle, 1622; pioneered devel. logarithms; invented Gunter's line or rule of projection on which the logarithms for numbers were printed (forerunner of the slide rule); introduced use of arithmetical complements to logarithmetical arithmetics; calculated logarithmic sines and tangents for every minute to 7 places. Died London, Dec. 10, 1626.

GUNTER, Gordon, Am. zoologist; b. Goldonna, La., Aug. 18, 1909; s. John O. and Joanna (Pennington) G.; B.A., La. State Normal Coll., 1929; M.A., U. Tex., 1931, Ph.D., 1945; postgrad., La. State U., 1934; m. Lottie Gertrude LaCour, June 6, 1932 (div. June 1957); children—Miles G., Forrest P., Charlotte (Mrs. B. G. Wood); m. Frances M. Hudgins, Sept. 7, 1957; children—Edmund O., Harry A. Biologist, U. S. Bur. Fisheries, intermittently 1931-38; marine biologist Tex. Game Fish and Oyster Commn., 1939-45; research scientist Inst. Marine Sci., U. Tex., 1945-49, dir., 1949-55; prof. zoology Marine Lab., U. Miami (Fla.), 1946-47; sr. marine biologist Scripps Instn. Oceanography, 1948-49; dir. Gulf Coast Research Lab., Ocean Springs, Miss., 1955—; prof. biology U. Miss., University, 1956—; prof. zoology Miss. State U., State College, 1956—. Vice chmn. for biology NRC, 1942-57; sci. adv. panel Gulf States Marine Fisheries Commn., 1956—; mem. bd. advisers Fla. Bd. Conservation, 1963—. Fellow Am. Inst. Fishery Research Biologists; mem. Tex. (past sect. v.p.), Miss. (pres. 1965—) acads. scis., Am. Fisheries Soc. (life), Am. Soc. Naturalists, Am. Soc. Zoologists. Research, numerous publs. on seasonal movements and distbn. marine animals of Gulf of Mexico, fisheries, seasonal cycles related to temperature. Home: Halstead Rd. Office: Gulf Coast Research Lab., Ocean Springs, Miss. 39564.*

GUNTHER, Albert Charles Lewis Gotthilf, zoologist; b. Esslingen/Neckar, Germany, Oct. 3, 1830; s. Friedrich Gotthilf and Eleonore (Nagel) G.; Ph.D., Tübingen, Germany, 1853; M.D., 1857; hon. Dr., 1903; studied under Johannes Müller, Berlin, also Bonn, Germany; m. Roberta MacIntosh, 1 son; m. 2d, Theodora D. Drake, 1879; 1 son. Naturalized Brit. citizen, 1862. Received holy orders; apptd. to staff zool. dept. Brit. Mus., 1862; keeper zool. dept., Brit. Mus., 1862; keeper zool. dept., 1875-95. Mem. Royal Soc., 1867 (v.p. 1875-76, gold medal 1878), Brit. Assn. (became pres. biol. sect. 1880), Linnean Soc. (pres. 1848-1901, gold medal 1904). Author: Handbuch der medizinischen Zoologie, 1858; Geographical Distribution of Reptiles, 1858; Catalogue of

Fishes in the British Museum, 8 vols., 1859-70; Reptiles of British India, 1864; Fische der Südsee, 3 vols., 1873-1909; Gigantic Land Tortoises, 1877; Introduction to the Study of Fishes, 1880; Deep-sea Fishes of the Challenger Expedition, 1887. Founder, 1st editor Record of Zoological Lit., 1864. Research on anatomical features, habits, life-history of lower orders of vertebrates; important in devel. Mus. Natural History collections. Died Kew, Eng., Feb. 1, 1914.

GUNTHER, Carl Oscar, German bacteriologist; b. Naumburg/Saale, Germany, 1854; dir. Prussian Inst. for Water Hygiene, Berlin; a founder Hygienische Rundschau; research on bacteriology with regard to pub. health. Died 1929.

GUNTHER, Erna, Am. anthropologist; b. Bklyn., Nov. 9, 1896; d. C.N. and Olga Joanna Elizabeth (Ehren) Gunther; B.A., Barnard Coll., 1919; M.A., Columbia, 1920, Ph.D., 1928; m. Leslie Spier, July 23, 1921 (div. Feb. 1932); children—Robert F. G., Christopher Lawrence. Prof. anthropology U. Wash., Seattle, 1929-66, dir. Wash. State Mus., 1929-62; prof. anthropology U. Alaska, College, 1966——. With U. Ariz.-Am. Assn. Mus. Inst., 1963; vis. prof. So. Ill. U., Carbondale, 1960. Research on reconstrn. of Indian life in N.W. coast in 18th century. Home: 4065 Iris Lane, Fairbanks, Alaska 99701. Office: Eielson Bldg., U. Alaska, College, Alaska. 99735.*

GÜNTHER, (Johann Heinrich) Friedrich, German veterinarian; b. Kelbra, Saxony, Germany, Dec. 6, 1794; s. Friedrich and Sybille Dorothea (Werther) G.; student medicine and botany U. Jena (Germany), student, vet. medicine, Berlin, Germany; student vet. medicine, Hanover, Germany, 1816-18; m. Dorothea Friederike Luise Dittmar; a son, Karl. Became veterinarian, Kelbra, 1818; named tchr. Sch. Vet. Medicine, Hanover, 1819, dir., 1847-58. Author: Untersuchungen und Erfahrungen im Gebiete der Anatomie, Physiologie und Tierheilkunde, 1837; Lehrbuch der praktischen Veterinärgeburtshilfe, 1838; (with son Karl Günther) Das Gangwerk der Pferde, ein Beitrag zur Beurteillungslehre und Züchtungskunde des Pferdes . . . , 1845; Lupinenbau und darauf basierte Sommer-und Winterfütterung der Schafe und übrigen Haustiere . . . , 1857. Improved vet. med. edn. in Hanover, including addition of courses in pharmacognosy, obstetrics; invented instruments for diagnosis and therapy; research on bisection of tendons and muscles under skin, tooth diseases, salivary duct, nutritional value of lupine, fever, sheep pox, pneumonia. Died Hanover, Nov. 19, 1858.

GÜNTHER, Friedrich Christian, German ornithologist; b. Kahla, Germany, Apr. 22, 1726; s. Johann Kaspar and Katherine Margarethe (Mecke) G.; M.D., Jena, Germany, 1747; m. Eleonora Sophia Johanna Trautmann, 1761; 4 children. Practiced medicine, Kahla; became mem. city council with duties of city judge, 1752; later became vice mayor; mayor, 1767-68. Author: Bemerkungen aus Naturgeschichte, 1770. Translator: Annus historica-naturalis (J. A. Scopoli) (with his own notes which were first such notes on Thuringia). Performed some of 1st ornithol. work in Germany, including ecol. conditions, and identification of species. Died Kahla, Apr. 25, 1774.

GÜNTHER, Fritz Karl, German chemist; b. Winkel, Germany, Sept. 27, 1877; s. Richard and Anna (Brustmann) G.; Ph.D., U. Munich (Germany), 1901; m. Edith Klingelhöffer, 1907; 2 sons. Became chemist Badische Anilin- & Soda-Fabrik, Ludwigshafen, Germany, later became dep. dir. main lab.; ret., 1938. Research on organic dyes, especially azo dyes; synthesis of alkylated naphthalin sulfonic acid and discovered its soap character (important in textile industry, synthetic soaps), 1917; produced new varieties of stable diazonium compounds. Died Heidelberg, Jan. 4, 1957.

GÜNTHER, Gustav, pharmacologist; b. Leipa, Bohemia, Jan. 27, 1868; s. Albert and Maria Anna (Konrad) G.; master pharmacy, 1889; M.D., U. Vienna (Austria), 1898; degree in vet. medicine, 1901; m. Maria Josepha Lorenz, 1914; 1 dau. Demonstrator, Histological Inst., U. Vienna, 1895-98; later asst. surg. clinic Rudolf Hosp., Vienna; became asst. Vienna Sch. Vet. Medicine, then joined faculty anatomy, histology and embryology, transferred to dept. pharmacology and botany, 1901, named asso. prof., 1902, prof. pharmacology Sch. Vet. Medicine, 1908, rector, 1915-17, founder ind. pharmacological Inst., Mem. Patent Office, 1915. Research and numerous publs. on histology, pharmacology, toxicology, skin resorption; 1st to perform resistance test on spermatozoa using organic substances. Died Vienna, Mar. 25, 1935.

GUNTHER, O., German immunologist; b. Hamburg, Germany, Feb. 29, 1908; s. Willy Franz and Antoinette (Schommer) G.; grad. Pub. Sch. Gymnasium Christianeum Hamburg-Altona, 1927; m. Gerda Schneider, Aug. 8, 1936; children—Gotz, Michaela. Asst. city hosps., Hamburg-Altona, Germany, 1935-39, Magdeburg, Germany, 1939-40, Essen, 1948-49; asst. Inst. Microbiology, Jena, Germany, 1946-48; staff inst. exptl. therapy Paul-Ehrlich-Inst., Frankfort/Maine, Germany, 1949——, prof., sci. mem., 1957——; teaching order immunology U. Frankfort, Sch. Medicine, 1964——. Mem. immunization com. Germany Soc. Against Poliomyelitis (immunization com. 1955——), German Soc. Hygiene and Microbiology, German Soc. Allergy, Med. Soc. Frankfort. Author: Neu Antigentabelle der Salmonellagruppe Jena, 1954. Research, publs. on allergic encephalitis after vaccination, antibody prodn., stepwise devel. of immunity. Home: 7 Binding, Str. Office: 44 Paul-Ehrlich Str., Frankfort/Main, West Germany.*

GÜNTHER, (Adam Wilhelm) Siegmund, German mathematician, geographer; b. Nuremberg, Germany, Feb. 6, 1848; s. Ludwig Leonhard and Johanna (Weiser) G.; ed. Heidelberg, Leipzig, Berlin, Göttingen (all Germany); Ph.D., Erlangen, Germany, 1870; m. Marie Weiser, 1872; 3 sons, including Adolf; 1 dau. Became mem. faculty U. Erlangen, 1872; at Poly. Sch., Munich, 1874; secondary sch. techr., 1876-86; prof. geography Tech. U. Munich, 1886-1920; mem. parliament; comdr. Bavarian field weather service during World War I. Mem. Bavarian Acad. Scis., Leopoldina. Author: Lehrbuch der Determinatentheorie, 1875; Vermischte Untersuchungen zur Geschichte der mathematischen Wissenschaft, 1876; Ziele und Resultate der neueren mathematischhistorischen Forschung, 1876; Studien zur Geschichte der mathematischen und physikalischen Geographie, 1877-79; Grundlehren der mathematischen Geographie, und elementaren Astronomie, 1878; Die Lehre von der gewöhnlichen und verallgemeinerten Hyperbelfunction, 1881; Lehrbuch der Geophysik und physikalischen Geographie, 2 vols., 1884-86, rewritten as Handbuch der Geophysik, 1897-99; Meteorologie, 1889; Handbuch der mathematischen Geographie, 1891; Geschichte der anorganischen Naturwissenschaft in 19. Jahrhundert, 1901; Entdeckungsgeschichte und Fortschritt der Geographie im 19. Jahrhundert, 1902; Geschichte der Erdkunde, 1904; Geschichte der Mathematik, 1907. Worked on theory of compound fractions, determinants and hyperbolic functions, also in areas of math. and phys. geography. Died Munich, Feb. 3, 1923.

GUNTHER OF ANDERNACH, see Guinterius, Johannes.

GUNTHEROTH, Warren Gaden, Am. physician; b. Hominy, Okla., July 27, 1927; s. Harry W. and Callie (Cornett) G.; M.D., Harvard, 1952; m. Ethel Haglund, July 3, 1954; children—Kurt, Karl, Sten. Fellow in cardiology Children's Hosp., Boston, 1954; research fellow pediatrics Harvard Med. Sch., 1954; Exchange fellow Karolinska Sjukhuset, Stockholm, 1955; spl. research fellow dept. physiology and biophysics U. Wash., 1958-59, instr. pediatrics, 1958-59, asst. prof., 1959-62, asso. prof. pediatrics, head div. pediatric cardiology, 1962——; cons. pediatric cardiology Univ., King County, Madigan Army, Pub. Health Service, Children's Orthopedic hosps., Seattle. Fellow Am. Bd. Pediatrics, Am. Bd. Pediatric Cardiology; mem. Wash. State Heart Assn. (trustee), Am. Physiol. Soc., Soc. Pediatric Research, Am. Acad. Pediatrics. Author: Pediatric Electrocardiography, 1965. Publs. on studies of hypertension, shock, congenital heart disease, arrhythmias, electrocardiogram, cardiac drugs; devel. of instruments for clin. and physiol. expts. Home: 8903 27th Av. N.E., Seattle 98115.*

GUNTZ, Antoine Nicolas, chemist; b. Wiesbaden, Germany, July 9, 1859; pupil of Berthelot. Prof. chemistry Nancy (France) Faculty Scis. Mem. French Acad. Scis., 1912. Research on fluorine, lithium, manganese, barium and their compounds; discovered effects of hydrogen and azote on lithium. Died Paris, Aug. 7, 1935.

GUNZ, Frederick Walter, physician; b. Munich, Germany, Nov. 17, 1914; s. Hugo and Johanna (Loewenfeld) G.; M.B., B.S., St. Bartholomew's Hosp., U. London, 1939, M.D., 1942; Ph.D., U. Cambridge (Eng.), 1949; m. Joan Phelps Tuckey, Sept. 23, 1944; children—Hugh Phelps, Philippa Joan, Sarah Phelps. Research scholar in leukemia Cambridge U., 1946-49; research fellow New Eng. Med. Center, Boston, 1956; hematologist N. Canterbury Hosp. Bd., Christchurch, New Zealand, 1950——; dir. cytogenetics unit Cancer Soc. New Zealand, 1961——. Fellow Royal Soc. Medicine London, Internat. Soc. Hematology; mem. Royal Coll. Physicians London, Asian-Pacific Soc. Hematology (councillor for N.Z.), N.Y. Acad. Scis., Assn. Pathologists, Path. Soc. Gt. Britain. Author: (with William Dameshek) Leukemia, 1958; also numerous articles. Research on causation of leukemia, its geog. distbn. and cellular abnormalities. Home: 32 Hawford Rd., Christchurch 2. Office: Pathology Dept., Christchurch Hosp., Christchurch, New Zealand.*

GÜNZ, Justus Gottfried, physician; b. Königstein, Germany, Mar. 1, 1714; M.D., Leipzig. Prof. anatomy and surgery U. Leipzig. Mem. French Acad. Scis., 1744. Author: De mammarum fabrica et lactis secretione, 1734. Specialist in female anatomy and childbirth; research on breast anatomy. Died Dresden, Germany, June 23, 1754.

GUPPY, Henry Brougham, Brit. botanist; b. Dec. 1854; s. T. S. Guppy; ed. Queen's Coll., Birminham, St. Bartholomew's Hosp., London; M.B., C.M., Edinburgh U., 1876; m. Annie Jordon, 1887; m. 2d, Letitia Warde, 1900. With med. service Royal Navy, 1876-85, as surgeon H.M.S. Hornet on China and Japan sta., 1877-80, H.M.S. Lark in Western Pacific, 1881-84; studied coral reef formation and plant dispersal in Keeling Islands and W. Java, 1887-88; made bot. and geol. studies in Hawaiian and Fijian islands, 1896-1900; studied littoral flora of Pacific side of S. Am., from Strait of Magellan to Panama, 1903-04; visited Teneriffe; engaged in bot. work in W.I., 1907-11; studied flora of Mt. Pico, Azores, 1913-14. Fellow Royal Soc., 1918, Royal Soc. Edinburgh, Linnean Soc. (Gold medal 1917). Author books on Solomon Islands, 1887, homes of family names, 1890, observations in Hawaii and Fiji, 1903-06, studies in seeds and fruits, 1912, studies in W.I. and Azores, 1917. Died Apr. 23, 1926.

GUPTA, Suraj Narayan, physicist; b. Punjab, India, Dec. 1, 1924; s. Lakshmi N. and Devi (Goyal) G.; M.S., U. Delhi (India), 1946; Ph.D., U. Cambridge (Eng.), 1951; m. Letty Gupta, July 14, 1948; children—Paul, Ranee. Came to U. S., 1953, naturalized, 1963. Imperial Chem. Industries fellow U. Manchester (Eng.), 1951-53; vis. prof. physics Purdue U., Lafayette, Ind., 1953-56; prof. physics Wayne State U., Detroit, 1956-61, Distinguished prof. physics, 1961——. Vis. physicist Argonne Nat. Lab., Brookhaven Nat. Lab., NRC Can. Fellow Am. Phys. Soc., Nat. Acad. Scis. (India). Contbr. articles to profl. jours. Quantized electromagnetic field; formulated method of aux. fields to isolate divergencies, renormalized by counter terms; introduced flat-space interpretation Einstein's theory gravitation; quantized gravitational field. Home: Belcrest Hotel, Detroit 48202.*

GURCHOT, Charles, chemist; b. Paris, France, July 19, 1898; s. Marcus and Rebecca (Norflus) G.; came to U. S., 1909, naturalized, 1921; B.S., Coll. City N.Y., 1921; Ph.D., Cornell U., 1927; postgrad. Ecole de Medecine, Paris; m. Georgette Helen Rossillon, May 28, 1930. Ind. cancer researcher U. Pa. Med. Sch., also Cornell U., 1933-35; faculty U. Cal. Med. Coll., San Francisco, 1937-45, asst. prof., 1941-45; research dir. John Beard Meml. Found., San Francisco, 1945-63. Fellow A.A.A.S.; mem. Am. Genetics Assn., Soc. for Exptl. Biology and Medicine, Pasteur Soc., Sigma Xi. Author: Biology—Key to the Riddle of Cancer, 1949; also articles. Research on compounds containing electro-negative bismuth for use in neurosyphilis, electro-negative arsenic in peridentitis, cyanogenetic glucosides in cancer treatment, devel. trophoblast theory of etiology of cancer. Address: 150 Palo Alto Av., San Francisco. 94114.*

GURD, Frank Ross Newman, biochemist; b. Montreal, Que., Can., Jan. 20, 1924; s. Fraser Baillie and Jessie (Newman) G.; grad. cum laude Phillips Exeter Acad., 1941; B.S., McGill U., 1945, M.S., 1946; Ph.D., Harvard, 1949; m. Ruth Sights, June 12, 1956; children—Fraser, Kathleen, Martha, Charles. Came to U. S., 1946, naturalized, 1954. Asst. dir. Bur. Med. Research, Equitable Life Assurance Soc., N.Y.C., 1955-59; asst. prof. clin. biochemistry Med. Coll. Cornell U., 1955-60; prof. biochemistry Sch. Medicine Ind. U., Bloomington, 1960-66, prof. chemistry, 1965——. John Simon Guggenheim and Helen Hay Whitney fellow dept. biochemistry Sch. Medicine Washington U., 1954-55. Mem. Am. Soc. Biol. Chemists, Am. Chem. Soc., Biophys. Soc., N.Y. Acad. Scis., A.A.A.S., Sigma Xi. Author: Chemical Specificity in Biological Interactions, 1954; (with D. J. Hanahan) Chemistry of the Lipides, 1960. Editorial bd. Jour. Biol. Chemistry, 1966——. Research, publs. on lipoprotein isolated from blood; combination of proteins with certain metal salts; identification of sites of binding and effects on conformation; modes of combination of metal ions with peptides; chem. modification of proteins to correlate structure in solution with that in crystalline state. Home: 2600 Fairoaks Lane, Bloomington, Ind. 47401.*

GURDJIAN, Elisha Stephens, neurosurgeon; b. Smyrna, Asia Minor, Apr. 18, 1900; s. Stepan and Beruke (Hagopian) G.; came to U. S., 1920, naturalized, 1930; M.S., U. Mich., 1924, M.D., 1926, Ph.D., 1927; m. Dorothy Eileen Kratz, May 29, 1933; children—Edwin S., Joan (Mrs. L. K. Cox), Ronald, Richard. Practice medicine, specializing in neurol. surgery, Detroit, 1930——; head neurosurg. service Grace Hosp., 1938——, chief of staff, 1961-64; prof. neurol. surgery Wayne State U., 1949——, chmn. dept., 1956——; chief Detroit Gen. Hosp. Neurosurg. Service, 1961——. Recipient Bronze medal Roentgen Ray Soc., 1948, Gold medal, 1961. Mem. A.M.A. (Hektoen silver medal 1967), A.C.S., Harvey Cushing Soc., Congress Neurol. Surgeons, Am. Neurol. Assn., Am. Assn. Anatomists, Am. Assn. Surgery of Trauma, Central Surg. Assn., Central Neuro-Psychiat. Assn., Soc. Neurol. Surgeons, Soc. Nuclear Medicine, Am. Assn. U. Profs., Soc. Cryobiology. Author: Operative Neurosurgery, 1952; Head Injury: Mechanism, Diagnosis and Management, 1958; also numerous articles. Studies on causes and mgmt. stroke, mechanisms, types, treatment of head injury, pial circulation. Home: 19385 Renfrew Rd., Detroit 48221. Office: 1553 Woodward St., Detroit 48226.

GUREWITSCH, Anatole Matvey, physicist; b. Chisinau, Russia, June 12, 1911; s. Matvey N. and Sonya (Kornblad) G.; Diploma in Engring., Swiss Fed. Inst. Tech., Zurich, 1935, Ph.D. in Nuclear Physics, 1953; m. Eleanor Joyce Chestnut, July 11, 1943; children—Sonya, Matthew, Kathryn. Staff, Gen. Electric Engring. and Research Labs., Schenectady, 1937-56, sci. rep. Gen. Electric Research Lab., Zürich, Switzerland, 1956-59, cons. applied sci. and tech. Gen. Electric Co., N.Y.C., 1959——. Mem. Am. Phys. Soc.,

I.E.E.E., Brit., German phys. socs., German Soc. for Electron Microscopy, Brit. Soc. Environmental Engrs., Fedn. Am. Scientists. Contbr. articles to tech. jours. Patentee in field electronics, microwaves, vacuum tech.; inventor ionic vacuum pump; a developer 80 million volt electron accelerator of synchnotron type; developed numerous microwave circuits for radar. Home: 142 Zurichbergstr., Zurich; G.E. Löwenstr. 29, Zurich, Switzerland; also 570 Lexington Av., N.Y.C. 10022.*

GURGEL, Octavio Amaral, Jr., Brazilian silviculturist; b. Sao Paulo, Brazil, May 25, 1918; s. Octavio Amaral and Francisca A. Gurgel; diploma in Agronomy, Escola Superior de Agr., Luiz de Queiroz, Brazil, 1942; D.Agronomy, U. Sao Paulo, 1953; m. Dyla Paes de Barros Amaral, Dec. 8, 1943; children—Leda Maria, Sonia Maria, Alda Maria Barros. Country agronomist, 1943-44; chief charge Sta. Exptl. Sao Simao, 1945-50; chief in charge Sta. Exptl. Santa Rita Passa Quatro (Brazil), 1951—; chief substitute Sta. Exptl. Mogi Mirim, Forest Service, 1959-60, chief substitute State Forest Mogi Guacu, 1959-60; faculty silviculture Center Research Forest Service, State of Sao Paulo, trustee Forest Research, Forest Service, 1962—, mem. commn. for reforestation; mem. movement for forest edn. Ministry Agr., Brazil. Mem. Soc. for Advancement Sci. Brazil, Brazilian Bot. Soc., Brazilian Genetical Soc. Research, numerous publs. on indigenous species of trees with econ. value; determined minimal number of trees per plot; correlations in juvenile characters which can be used for precocious selection of best trees; mgmt. indigenous forest trees, mgmt. and selection of various exotic species. Home: Hôrto Experimental. Caixa Postal 15, Santa Rita do Passa Quatro, Estado de Sao Paulo, Brazil.*

GÜRICH, Georg Julius Ernst, German geologist, paleontologist; b. Guttentag, Germany, Sept. 25, 1859; s. Eduard and Emilie (Bruckisch) G.; Ph.D., Breslau (now Wroclaw, Poland), 1883; m. Margarethe Quenstedt, 1896; 2 sons, 2 daus. Became mem. E. Flegel's Sudan expdn., 1885; joined faculty, 1887; commd. expdns. to German S.W. Africa and Venezuela, 1888-90; tchr., Breslau (now Wroclaw, Poland), 1894-1910; lectr. U. Breslau; began charting work Prussian Geol. Inst., 1901; became dir. State Mineral.-Geol. Inst., also prof. Colonial Inst., 1810; mem. expdn. to S. and E. Africa, 1914-15; became prof. geology and paleontology U. Hamburg (Germany), 1919; visited S. and S.W. Africa, 1928-29. Author: Geologische Übersichtskarte von Schlesien nebst 1 Band Erläuterungen dazu, 1890; Deutsch-Südwestafrika, Reisebilder und Skizzen, 1891; Geologische Führer in das Riesengebirge, 1900; Erdgestaltung und Erdgeschichte, 1928; also numerous articles. Research on geology and paleontology of Silesia, Central Mountains, Poland; studied quartenary period, oil of N. Germany; wrote 1st geol. report on German S.W. Africa; studied S. and S.W. African fossils, especially Kuibis quarzite fossil and fishes from Dwyka shale. Died Berlin, Germany, Aug. 16, 1938.

GURIN, Samuel, Am. biochemist; b. N.Y.C., July 1, 1905; s. Morris and Rose (Zwinig) G.; B.A., Columbia, 1928, M.S., 1930, Ph.D., 1934; D.Sc., Phila. Coll. Pharmacy, 1964, LaSalle Coll., 1965; m. Celia Zall, June 14, 1930; children—Robert Neil, Richard Stephen. Faculty U. Pa. Med. Sch., Phila., 1937—, prof., 1948-55, Benjamin Rush prof. biochemistry, chmn. dept., 1955—, dean Sch. Medicine, 1962—. Mem. NSF Panel on Metabolism, 1950-55; study sect. in physiol. chemistry NIH Commn. on Career Research and Devel. Awards, 1960-66; NIH Inst. Arthritis and Metabolic Diseases Commn. on Growth, 1950-55, bd. sci. counselors, 1955-60. Mem. Am. Chem. Soc., Am. Soc. Biol. Chemists, A.A.A.S., Soc. Exptl. Biology and Medicine, Physiol. Soc., John Morgan Soc., Sigma Xi. Research, publs. on chem. nature of vitamins, Thiazol component of vitamin B, isolation and chem. properties of chorionic gonadotrophin of human pregnancy urine, protein structure, biosynthesis and metabolic fate of fatty acids and cholesterol, formation of bile salts from cholesterol. Home: 401 S. 47th St., Phila. 19143.*

GURLAND, John, statistician; b. Ottawa, Can., Jan. 6, 1917; s. Max G.; B.A., U. Toronto, 1939, M.A., 1942; Ph.D., U. Cal. at Berkeley, 1948; m. Vera Frances Green, June 26, 1948; children—Marsha, Iva. Came to U. S., 1945, naturalized, 1955. Faculty, Ia. State U., Ames, 1952-60, prof. statistics, 1958-60; prof. U. Wis. Math. Research Center, Madison, 1960-63, prof. statistics and med. statistics, 1963—. Editor: Stochastic Models in Medicine and Biology, 1964. Home: 3521 Sunset Dr., Madison, Wis. 53705.*

GURLT, Ernst Friedrich, German veterinarian; b. Drentkau, Germany, Oct. 13, 1794; s. Johann Carl Gurlt; M.D., Breslau (now Wroclaw, Poland), 1819; m. Henriette Emilie Doniges, 1824; 3 sons, including Ernst Julius, Hermann; 1 dau., Antonie (Mrs. Robert Lehmann-Nitsche). Became tutor in zootomy and botany faculty Berlin (Germany) Sch. Vet. Medicine, 1819, named prof., 1826, dir. sch., 1849-70; ret., 1870. Author: Handbuch der vergleichenden Anatomie der Haussäugetiere, 1822, revised Ellenberger-Bauss, 18th edit., 1943; Anatomische Abbildungen der Haussäugetiere, 1824-33; Anatomie des Pferdes, 1831; Lehrbuch der pathologischen Anatomie der Haussäugetiere, 1831; Lehrbuch der vergleichenden Physiologie der Haussäugetiere, 1837; (with H. Hertwig) Chirurgische Anatomie und Operationslehr für Tierärzte, 1847; Handatlas zu dem Handbuch der vergleichenden Anatomie der Haussäugetiere, 1860. Coeditor (with C. H. Hertwig) Magazin für die gesamte Tierheilkunde, 1835-74. Studies in comparative anatomy domestic mammals, path. anatomy, physiology, surgery on domestic animals, botany. Died Berlin, Aug. 13, 1882.

GURLT, Ernst Julius, German surgeon; b. Berlin, Germany, Sept. 13, 1825; s. Ernst Friedrich and Henriette Emilie (Doniges) G.; M.D., Berlin, 1848; m. Marie Auguste Caroline Wilhelmine Büttner, 1858; m. 2d, Helen Giesche. Made study trip to Austria, France, Gt. Britain; became asst. in von Langenbeck's clinic, Berlin, 1852; joined faculty in surgery, U. Berlin, 1853, asso. prof., 1862; prof. surgery Kaiser Wilhelm Acad.; participated in mil. campaigns of 1848, 64, 66, 70-71. Hon. mem. German Soc. Surgery. Author: Beiträge zur vergleichenden pathologischen Anatomie der Gelenkkrankheiten, 1853; Über den Transport Schwerverwundeter und Kranker im Kriege, 1859; Über einige durch Erkrankung der Gelenkverbindungen verursachte Missstaltungen des menschlichen Beckens, 1854; Geschichte der Chirurgie und ihrer Ausübung, 3 vols., 1898; Handbuch der Lehre von den Knochenbrüchen, 1862-65; Leitfaden für Operationsübungen am Cadaver, 1862; Zur Geschichte der internationalen und freiwilligen Krankenpflege im Kriege, 1873; Die Krieschirurgie der letzten 150 Jahren in Preussen, 1875; Die Gelenkresektionen nach Schussverletzungen, ihre Geschichte, Statistik, und Endresultate, 1879. Research on bone and joint diseases, history of surgery, mil. surgery; anesthesia studies included proof of greater complications of chloroform over ether. Died Berlin, Jan. 9, 1899.

GURNEY, Ashley Buell, Am. entomologist; b. Cummington, Mass., May 16, 1911; s. Homer Wilder and Mary (Shaw) G.; B.S., U. Mass., 1933, M.S., 1935, Ph.D., 1940; m. Ruth Florence Peters, Nov. 4, 1941; children—John W., Richard D., Ann Elizabeth. With entomology research div. U. S. Dept. Agr., 1936—, research entomologist, Washington, 1955—. Exec. sec. Entomol. Soc. Am., 1954, rep. on NRC, 1956-59. Mem. Washington Acad. Scis., A.A.A.S., Entomol. Soc. Washington, Biol. Soc. Washington. Research, numerous publs. in identification of certain groups of insects, particularly grasshoppers, cockroaches, related insects, their distribution and identification characteristics. Home: 4606 N. 41st St., Arlington, Va. 22207. Office: Dept. Entomology, U. S. Nat. Museum, Washington 20560.*

GURNEY, Sir Goldsworthy, English inventor; b. Treator, Cornwall, Eng., Feb. 14, 1793; s. John Gurney; studied medicine under Dr. Avery, Wadebridge; m. Elizabeth Symons, 1814; a dau., Anna J. Practiced surgery, Wadebridge; moved to London, 1820; Lectr. elements chem. sci. Surrey Inst.; supervised lighting and ventilation in new houses of parliament, 1854-63. Recipient gold medal Soc. Arts. Author: Course of Lectures on Chemical Science as delivered at the Surrey Institution, 1823; Observations on Steam Carriages on Turnpike Roads, . . . with the Report of the House of Commons, 1832; Account of the Invention of the Steamjet or Blast . . . , 1859; Observations pointing out a Means by which a Seaman may identify Lighthouses, 1864. Went from London to Bath and back at 15 miles per hour in his steam carriage, 1829; his steamjet extinguished mines fire, 1st applied to steamboats, 1824; prin. of Gurney stove used in warming and ventilating old House of Commons; anticipated principles of electric telegraph; invented oxyhydrogen blowpipe; discovered Drummond Light. Died Reeds, Cornwall, Eng., Feb. 28, 1875.

GURR, Edward, English biologist, chemist; b. Leicester, Eng., Oct. 3, 1905;.s. William Stanton Collins and Violette (Clarke) G.; student Leicester Coll. Sci. and Tech., Chelsea (Eng.) Coll. Sci.; Ph.D., U. Coll., Cork, Ireland, 1962; m. Florence P. Gruner, Dec. 27, 1930. Works mgr. George T. Gurr, London, 1925-45; head lab. chems. div. May & Baker, Ltd., Dagenham, Eng., 1945-46; dir. Michrome Labs., Edward Gurr, Ltd., East Sheen, London, 1946—; hon. research asso. dept. anatomy U. Coll., Cork, 1958—. Fellow Linnean Soc., Instn. Biology, Soc. Dyers and Colourists, Royal Micros. Soc., Royal Inst. Chemistry; mem. Anat. Soc. Gt. Britain and Ireland. Author: Microscopic Staining Techniques, Numbers 1 and 2, 1950, Number 3, 1951, Number 4, 1958; Practical Manual of Medical and Biological Staining Techniques, 1953; Methods of Analytical Histology and Histochemistry, 1958; Encyclopaedia of Microscopic Stains, 1960; Staining, Practical and Theoretical, 1962; Rational Use of Dyes in Biology, 1965; also articles. Research on biol. stains in animal and plant histology and cytology; reclassification of biol. stains; collaborator in devel. new histological techniques including Falg method; studies on effect of heat on pH of water and aqueous dye solutions, nature of certain granules present on malaria parasite; synthesized new types of polychrome dyes, including Trifalgic acid, Rhodanile blue. Home: 19 Fife Rd., Office: 42 Upper Richmond Rd., London, S.W.14, Eng.*

GÜRS, Karl August, German physicist; b. Frankfort, Germany, Nov. 5, 1927; s. Karl Hermann and Kathe (Bernhardt) G.; grad. Goethe-Realgymnasium, Frankfort, 1949; diploma Goethe-U., Frankfort, 1954, Doctor's degree, 1959; m. Ursula Preuschen, Apr. 9, 1960; children—Inge Barbara, Karl Reinhard. Research asst. Physikalisches Inst., Frankfort U., 1955-59; with Lemens & Halske Research Lab., Munich, Germany, 1960—, head solid-state laser lab., 1964—; instr. physics Frankfort U., 1964—, habilitation, 1964. Mem. Deutsche Physikalische Gesellschaft, Deutscher Hochschulverband. Research, publs in semiconductor field, investigation of temporary behavior of solid-state lasers, devel. of internal laser modulation, resonant intracavity modulation and coupling laser modulation, improvement of cw solid-state laser to produce an emission like that of gas laser. Home: 40 Eucken, 8 Munich 25. Office: 8 Balan, 8 Munich 8, Germany.*

GURTIN, Morton E., Am. mathematician; b. Jersey City, Mar. 7, 1934; s. Saul G. and Irene (Hoffman) Burns; B.M.E., Rensselaer Poly. Inst., 1955; Ph.D. (Nat. Def. fellow), Brown U., 1961; m. Leatrice T. Kagan, June 12, 1955; children—Amy Lynn, William Robert. Engr., Douglas Aircraft Co., Los Angeles, 1955-56, Gen. Electric Co., Schenectady, 1956-59; faculty Brown U., 1961-66; prof. math. Carnegie-Mellon U., Pitts., 1966—. Mem. Am. Math. Soc., Soc. for Natural Philosophy, Sigma Xi. Author: (with others) Wave Propagation in Dissipative Materials, 1965. Asso. editor: Math. Revs., 1963-65, Archive for Rational Mechanics and Analysis, 1968—. Research, numerous publs. in math. theory of linear elasticity, linear viscoelasticity, math. structure of continuum thermodynamics and continuum mechanics, variational principles for linear initial-value problems, theory of wave propagation in non-linear materials with memory. Home: 5848 Elmer St., Pitts. 15232.*

GURVICH, Alexsandr Gavrilovich, Russia biologist; b. Poltawa, Russia, Sept. 13, 1874; mem. staff Inst. Chemistry of USSR Acad. Scis. (mem.) Recipient Stalin prize. Credited with discovery of mitogenic rays; theorectical studies on sphere of mitogenic irridation of blood. Address: Inst. Chemistry, USSR Academy Sciences, Moscow, USSR.

GUSBERG, Saul Bernard, Am. physician; b. Newark, Aug. 3, 1913; s. Morris and Lina (Gelfand) G.; student U. Mich., 1930-33; M.D., Harvard, 1937; D.Sc. in Medicine, Columbia, 1948; m. Dorothy Cushner, June 17, 1938; 1 son, Richard. Faculty. Colombia, 1946—, clin. prof. obstetrics and gynecology, 1962—, Jules Bache fellow in gynecol. research, 1947—; prof., chmn. dept. obstetrics and gynecology Mt. Sinai Sch. Medicine, 1965—; obstetrician, gynecologist in chief Mt. Sinai Hosp., N.Y.C., 1962—. Mem. Am. Cancer Soc. (pres. elect N.Y.C. div. 1966—), Am. Gynecol. Soc., Soc. Pelvic Surgeons, Soc. Gynecol. Investigation, Soc. Profs. Obstetrics and Gynecology, Am. Coll. Obstetrics and Gynecology, A.C.S., Am. Radium Soc. Editorial bd. Obstetrics and Gynecology, 1963—. Research, publs. on diagnosis, prevention, treatment of cancer of female reproductive organs. Home: 257 Palisade Av., Dobbs Ferry, N.Y. Office: 1176 Fifth Av., N.Y.C. 10029.*

GUSEK, Wilfried Walter, German pathologist; b. Gelsenkirchen, Nov. 14, 1928; s. Wilhelm and Hedwig Gusek; M.D., U. Lena; m. Charlotte-Luise Sanden, 1957; 1 dau., Gabriele-Charlotte. Practice gen. medicine, 1955, anat. pathology, 1956; agrégé in gen. pathology and path. anatomy U. Hamburg, 1960. Mem. German Assn. Pathology, Internat. Acad. Pathology, Ev. Akademikerschaft in Deutschland, Gesellschaft Deutsche Naturforscher und Aerzte, Deutsche Gesellschaft für Elektronenmikroskopie. Author: Elektronenoptische Untersuchungen am Zellbild des Granulationsgewebes, 1959; Zur Ultrastruktur der Epiphysics Cerebri der Ratte, 1961; Submikroskopische Untersuchungen zur Feinstruktur aktiver Bindegewebszellen, 1962; Submikroskopische Untersuchungen als Beitrag zur Struktur und Onkologie der Meningione, 1962; Neuere Ergebnisse und Aspekte zur Pathologischen Anatomie der Entzündung, 1962. Home: 135 Langer Kamp, Hamburg-Garstedt. Office: Pathological Institute, University of Hamburg, 52 Martinistrasse, Hamburg 20, Germany.

GUSEYNOV, Dzhebrail Mukhtarovich, Russian agrochemist, agronomist; b. 1913; grad. Azerbaijan Agrl. Inst.; D. Agrl. Sci. Dir., Inst. Agrochemistry and Pedology, Azerbaijan Acad. Sci., 1945—, acad. sec. agrl. dept., 1963—; prof. Recipient Stalin prize. Mem. Azerbaijan Acad. Sci. Author: Organic Mineral Fertilizers from Petroleum Industry Waste, 1944; Fertilizers from Petroleum Industry Waste, 1949; A Study of the Agrochemical Properties of Soils, 1960. Editor (symposium) Agrochemical Research in the Azerbaijan SSR, 1960. Research and publs. on producing mineral fertilizers from petroleum indsl. waste. Address: Inst. Agrochemistry and Pedology, Azerbaijan Acad. Sci., Baku, Azerbaijan, SSR, USSR.

GUSIC, Branimir, Yugoslavian physician, surgeon, anthropogeographer; b. Zagreb, Yugoslavia, Apr. 6, 1901; s. Ferdinand and Jelka (Vrbanic) G.; M.D., Med. Faculty in Zagreb, 1926; Ph.D., Faculty Sci. in Zagreb, 1929; m. Marijana Heneberg, June 30, 1927; 1 son, Ivan. Faculty oto-rino-laryngol. dept. Faculty Medicine, Zagreb, 1927—, prof., head dept., 1945—, dean, 1945-48, 60-62; mem. Yugoslav Acad. Sci. and Arts, 1947—, gen. sec., 1947-50. Recipient several high Yugoslav decorations. Hon. mem. Croatian, Czechoslovakian med. socs., Austrian, Bulgarian, Greek oto-rhino-laryngol. socs.; mem. Serbian Acad. Sci. and Arts, Slovenian Acad. Sci. and Arts, Collegium Oto-rhynolaryngologicum Amicitiae Sacrum. Author: Otologija prakticnog lijecnika, 1957; also numerous articles. Research on pathophysiology of esophagus and respiratory mucous membrane; pathophysiology and surgery of malignancies of head and neck; influence of environment on devel. of ORL diseases; man in Dinaric Alps, role of man in karstification of perimediterranean region; ethnogenesis of Dalmatian littoral towns. Home: 24 Malinova, Zagreb I, Yugoslavia.*

GUSINDE, Martin, ethnologist; b. Breslau, Germany, Oct. 29, 1886; s. Aegidius and Bertha (Hentschel) G.; Ph.D., U. Vienna. Priest, Societas Verbi Divini; prof. biology, Liceo Alemán Santiago de Chile, 1912-22; head, dept., Museo Nacional de Etnologia, prof. anthropology, Catholic U., Santiago, 1916-24; on editorial staff, Anthropos, since 1925; prof., anthropology, Catholic U., Washington, D. C., from 1949; prof. U. Vienna. Mem. several academies and assocs. Author: Bibliografia de la Isla de Pascua; Die Feuerlandindianer (3 vol.), 1931-39; Urmenschen in Feuerland; Kongo-Pygmäen, 1942; Urwaldmenschen am Ituri: Pygmäen und Waldneger; Die Twa-Pygmäen und Ptgmoiden im tropischen Afrika; Die Kleinwuchsvölker in heutiger Beurteilung. Undertook ethnological expeditions: to Araucanian Indians, 1916-17; 4 to indians in Fuegia, 1918-24; N. Am. indians, 1928-9; to pygmies in Belgian Congo and Ruanda, 1934-35; to Bushmen and Hottentots in S. Africa, 1950-51 and 1953; to Yupa Indians, Venezuela, 1954; to Ainu of Japan and Aeta of Philippines, 1955; to Ayom-pygmies, New Guinea, 1956; to Indoesia, 1960. Address: Mödling bei Wien, Austria.

GÜSSEFELD, Franz Ludwig, cartographer; b. Osterburg, Altmark, Dec. 5, 1744; s. Franz Joachim and Catherine Sophie (Albrecht) G.; pupil of Hahn in Königsberg (now Kaliningrad, USSR); m. Johanna Christiana Brunguell, 1780; 4 stepchildren. Surveyor under Hahn and Petri; forestry adviser, other positions in service Saxony-Weimar; with Landes-Industrie Comptoir, also Bertuch's Geog. Inst., Weimar, Germany. Produced 1st noteworthy Prussian map in one of Brandenburg, also others, characterized by precision, use of geog., astron. and math. data, careful orthography; produced some maps from geog. data of A. Büsching, after 1813. Died Weimar, June 17, 1808.

GUSSENBAUER, Carl Ignatz, Austrian surgeon; b. Obervellach, Austria, Oct. 30, 1842; s. Richard Joseph and Antoinie Elisabeth (Krauss) G.; M.D., U. Vienna, 1867; m. Clotilde Clara Josepha Halla, 1884; 1 son, 3 daus. Specialized in obstetrics and ophthalmology; intern Allgemeines Krankenhaus, Vienna, Rudolfspital; became asst. to Billroth, 2d surg. clinic of U., 1872; joined faculty, Vienna, 1874; became asso. prof. U. Liege (Belgium), 1875, prof., 1876; apptd. chmn. 2d. surg. clinic, Prague, Czechoslovakia, 1878; named chmn. 2d surg. clinic, Vienna, 1894. Contbr. articles to tech. jours. Built 1st practical artificial larynx, 1873; studied cancer of large intestine, bacteriology, neurosurgery; 1st surg. treatment pancreatic cysts; improved breast cancer operation; exptl. preparation for Billroth's stomach resection. Died Vienna, June 1903.

GÜSSFELDT, (Richard) Paul (Wilhelm), German explorer; b. Berlin, Germany, Oct. 14, 1840; s. Maria Sophie Friederike Heideblut; adopted by Johann Friedrich Wilhelm and Auguste Antonie (Müller) Güssfeldt; student natural scis. and math. Heidelberg, Berlin, Giessen, Bonn, (all Germany); Ph.D., 1865; m. Helene Sobernheim, 1895. Joined faculty math. U. Bonn, 1868; made expdn. to Loango Coast, W. Africa, 1873; traveled through Egypt, Arabic Desert, 1876; made expdn. to Cordilleras, S. Am., 1882-83; became prof., dir. natural sci. instruction Seminar for Oriental Langs., also tchr. geographic-astron. determination, 1892; travel companion Kaiser Wilhelm II. Author: Die Loango-Expedition, 1873-76, eine Reisewerk in 3 Abteilunge (with J. Falkenstein, E. Pechuel-Loesche), 1879; In den Hochalpen, 1886; Reise in d. Andes von Chile und Argentinien, 1888; Der Montblanc Studie im Hochgebirge vornehmlich in der Montblanc-Gruppe, 1894; Grundzüge der astronomische-geographischen Ortsbestimmung . . . , 1902. Described flora, fauna and settlement of S.Am.; bearing and height findings, geomagnetic measurements. Died Berlin, Jan. 17, 1920.

GUSTAFSON, J(ohn) K(yle), Am. mining geologist; b. Chgo., Mar. 13, 1906; s. Lewis and Irene Stoddard (Baker) G.; A.B., Washington U. St. Louis, 1927; A.M., Harvard, 1928, Ph.D., 1930; D.Sc., Mich. Coll. Mining and Tech., 1963; m. Elizabeth Brigham, June 11, 1930; children—Lewis Brigham, Judith (Mrs. Walter Lyall Milde), Andrew Baker. Geologist with various companies, Can. and Australia, 1930-39; in charge Toronto, office, Hollinger Exploration Ltd., 1939-42, 1st Hollinger expdn. to Labrador and Ungava, 1942; adviser Metals Reserve Co., Washington, 1942-44; chief geol. Newmont Mining Corp., Magma Copper Co. and affiliated cos., 1944-49; cons. Zinc Corp. Ltd., New Broken Hill Consol. Ltd., Australia, 1947; cons. geologist M. A. Hanna Co., 1950-56, dir. explorations, 1956-60; v.p. Hanna Mining Co. and predecessor corp., 1953-60; pres., chief exec. officer, dir. Homestake Mining Co., 1961—; chmn., chief exec. officer Port Costa Clay Products Co., 1964—; pres., dir. Homestake Iron Ore Co. of Australia Ltd., 1964—, Homestake Lead Co. of Mo., 1965—, Homestake Potash Co., 1965—, Little Beaver Mining Co., 1961—, Mineral Devel. Co., 1965. Dir. raw materials, U. S. AEC, 1947-48, mgr. raw materials operations, 1948-49, mem. adv. com. on raw materials, 1950-59; panel mineral exploration research NSF, 1952-59; mem. adv. council Inst. Geophysics and Planetary Physics, U. Cal., 1961—; Hugh Exton McKinstry lectr. Harvard, 1965. Fellow Geol. Soc. Am. (mem. council, 1957-60), Soc. Econ. Geologists (v.p. 1961-63); mem. Am. Inst. Mining and Metall. Engrs., Canadian Inst. Mining and Metall., Mining. Metall. Soc. Am., Am. Mining Congress (bd. govs. 1962), Societe du Cuivre de Mauritanie, Akjoujt, Mauritania (dir.), Phi Beta Kappa, Sigma Xi. Research and publs. on metamorphism and ore deposits, especially uranium. Home: 41 Alvarado Rd., Berkeley, Cal. 94705. Office: 100 Bush St., San Francisco.

GUSTAFSON, Torsten Valdemar, Sweden physicist; b. May 8, 1904; s. Albin and Hulda (Bramstang) G.; Fil.Dr. U. Lund (Sweden), 1933; m. Karin Lindskog, Jan. 5, 1935; children—Ingrid (Mrs. Jörgen Lindström), Carita (Mrs. Per Brinck), Torbjörn, Christina. Asst., Swedish Hydr-Biol. Com., 1930-34; faculty Lund U., 1933—, prof. physics, 1939—. Chmn. bd. Nordita, 1963—; mem. Gov.'s Bd. Research, 1963—, Gov.'s Adv. Bd. Atomic Energy, 1956—. Mem. acads. of Sweden, Denmark, Finland, Soc. Lund and Uppsala. Research, publs. on aerodynamics, flow round aerfoils, ocean currents, inertia currents, atomic physics, theory of light, nuclear physics. Home: Gyllenkroks allè 13, Lund, Sweden.*

GUT, Rudolf Max, Swiss mathematician; b. Zurich, July 4, 1898; s. Friedrich and Elisabeth (Weiss) G.; ed. univs. Zurich, Berlin, Paris, Yale; Ph.D. in math.; m. Gret Glaser, 1942; children—Martin, Charlotte. Prof., U. Zurich. Mem. Swiss Soc. Hem Math. (pres.). Research and publs. on theory of algebraic numbers. Home: 598 Glaernischstrasse, Herrliberg bei Zurich. Office: Mathematical Institute, University of Zurich, Zurich, Switzerland.

GUTBERLET, Louis Charles, Am. chemist; b. Chgo., Mar. 29, 1928; s. Ferdinand and Frances (Habisohn) G.; student Wright Jr. Coll., 1946-48; B.S., Ill. Inst. Tech., 1950; M.S., Purdue U., 1954; m. Patricia Ann Panka, Sept. 20, 1952; children—Eric John, Paul James. With research dept. Am. Oil Co., Whiting, Ind., 1954—, sr. project chemist, 1965—. Mem. Am. Chem. Soc., Sigma Xi, Alpha Chi Sigma. Research on heterogeneous catalysis, catalytic hydrogenation, reaction kinetics and mechanisms. Patentee petroleum processing devices. Home: 735 S. East St., Crown Point, Ind., 46307. Office: P.O. Box 431, Whiting, Ind. 46394.*

GUTBIER, (Felix) Alexander, German chemist; b. Leipzig, Germany, Mar. 21, 1876; s. Karl and Fanny (Thilo) G.; student Dresden, Germany, Munich, Germany, Zurich, Switzerland; Ph.D., Erlangen, Germany, 1899; m. Olga Fischer, 1902 (div. 1916); 3 sons, including Rolf; m. 2d, Gertrud Gaugler, 1919. Joined faculty U. Erlangen, 1902, became asso. prof., 1907; named prof. electrochem. and chem. tech. Tech. U., Stuttgart, Germany, 1912; became dir. chem. lab. U. Jena (Germany), 1922. Author: (with L. Birkenbach) Praktische Anleitung zur Massanalyse, 1905; Lietfaden der Qualitativen Analyse, 1920; Goethe, Grossherzog Carl August und die Chemie, in Jena, 1926; also articles. Research on anorganic chemistry, colloid chemistry, numerous new compounds; produced in colloid form, tellurium, silver, platinum, gold, silicon, selenium, mercury, boron, arsenic, antimony, copper, bismuth, organic substances as protective colloids; developed new methods in colloid chemistry, new method of analysis and separation in complex chemistry; determined atomic weights of lead, telerium, bismuth. Died Jena, Oct. 4, 1926.

GUTENBERG, Beno, b. Darmstadt, Germany, June 4, 1889; s. Hermann and Pauline (Hachenburger) G.; student Technische Hochschule, Darmstadt, 1907-08; Ph.D., U. Göttingen, 1911; Ph.D. (hon.), U. Uppsala, 1955; m. Hertha Dernburg, Aug. 17, 1919; children —Arthur, Stephanie. Came to U. S., 1930, naturalized, 1936. Asst. in central office Internat. Seismol. Assn., Strasbourg, 1913; prof. geophysics U. Frankfurt am Main, 1926-30, Cal. Inst. Tech., 1930-60. Recipient Prix de Physique du Globe, Acad. Royal de Belgique, William Bowie Medal, 1953; Wiechert Medal 1956. Fellow Royal Astron. Soc., Geol. Soc. (London) (fgn.), Internat. Assn. for Seismology and Physics of Interior of Earth (pres. 1951-54), Seismol. Soc. Am. (pres. 1945), Geol. Soc. Am., Am. Geophys. Union, A.A.A.S., Soc. Exploration Geophysicists, Nat. Acad. Sci.; hon. mem. Royal Soc. New Zealand; corr. mem. Finnish Geog. Soc.; fgn. mem. Academy Lincei, Rome, Swedish Acad. Sci., Finnish Acad. Letters and Science. Author: Seismicity of the Earth (with C. F. Richter), 1949. Author, co-author or editor several books pub. in Germany; also more than 150 sci. papers. Editor: Internal Constitution of the Earth. Made first exact determination of radius of earth's core; 1st to explain existence of shadow zone (where earthquake waves not felt); postulated existence of core at center of earth about 2100 miles in radius, 1913; sharp boundary between core and rocky mantle lying above it called Gutenberg discontinuity. regarded as authority on earthquakes. Died Los Angeles, Jan. 25, 1960.

GUTERMUTH, C(linton) R(aymond), Am. conservationist; b. Ft. Wayne, Ind., Aug. 16, 1900; s. Henry Christian and Alice Virtue (Zion) G.; student U. Notre Dame, 1918-19; grad. Am. Inst. Banking, 1927, postgrad.; m. Ila Bessie Horm, Mar. 4, 1922; Asst. cashier St. Joseph Valley Bank, Elkhart, Ind., 1922-34; dir. edn. Ind. Dept. Conservation, 1934-40, dir. fish and game, 1940-42; Ind. rent dir. OPA, 1942-45; exec. sec. Am. Wildlife Inst., Washington, 1945-46; v.p. Wildlife Mgmt. Inst., Washington, 1946—. Trustee, exec. sec. N.Am. Wildlife Found., 1945—; dir., treas. World Wildlife Fund, 1962—. Trustee Stronghold, Inc., Sugar Loaf Mountain, Md. Recipient Distinguished Service award Nat. Assn. Soil Conservation Dists., 1958, Nat. award Nat. Watershed Congress, 1961. Fellow A.A.A.S. (mem. council 1959—), Wildlife Soc. (trustee 1951-66, Leopold medal 1957), Am. Fisheries Soc., Am. Soc. for Range Mgmt., Soil Conservation Soc. Am. (hon.). Author: (with Wynne Thorne) Land and Water Use, 1963; Indiana Official Lake Guide, 1938; (with Ira N. Gabrielson) Fisherman's Encyclopedia, 1950; (with Bruce R. Tuttle) Standard Book of Fishing, 1950; also numerous articles. Research on better mgmt. renewable natural resources. Home: 4801 Connecticut Av. N.W., Washington 20008. Office: Wire Bldg., Washington 20005.*

GUTH, Lloyd, Am. anatomist; b. N.Y.C., Oct. 8, 1929; s. Benjamin G. and Syd (Fisher) G.; B.A. cum laude, N.Y. U., 1949, M.D., 1953; m. Josephine Rose Zalewski, Aug. 5, 1955; children—Michael Walter, Robert William. Commd. asst. surgeon USPHS, 1954, advanced through grades to med. dir., 1967; jr. scientist NIH, Bethesda, Md., 1954-61. head sect. on exptl. neurology, lab. neuroanat. scis. Nat. Inst. Neurol. Diseases and Blindness, 1961—. Mem. Am. Assn. Anatomists, Am. Physiol. Soc., A.A.A.S. Translator: Studies on Vertebrate Neurogenesis (Santiago Ramon y Cajal), 1960. Asso. editor Exptl. Neurology, 1965—. Research, publs. on peripheral nerve regeneration mechanism by which peripheral nerve fiber regulates protein synthesis muscle it innervates. Home: 6004 Kingsford Rd. Office: NIH, Bethesda, Md. 20014.

GUTH, Sylvester Karl, Am. physicist; b. Milw., Dec. 31, 1908; s. Alexander C. and Laura (Kiesslich); B.S., U. Wis., 1930, E.E., 1950; D. Ocular Sci., No. Ill. Coll. Optometry, 1953; m. Beryl K. Van Deraa, May 2, 1931. With Gen. Electric Co., 1930—, with lighting research lab., 1930-54, mgr. radiant energy effects lab., Cleve., 1955—. Del., Commn. Internationale de l'Eclairage, 1951, 55, 59, 63; lectr. Case Inst. Tech., 1950-64; also cons. various instns. Fellow A.A.A.S., Illuminating Engring. Soc., Am. Acad. Optometry; mem. Optical Soc. Am., Ohio Acad. Sci., Inter-Soc. Color Council, Assn. for Research in Ophthalmology, Illuminating Engring. Soc. (London, Eng.), Armed Forces Nat. Research Council Com. on Vision. Research, publs. on psychol. and physiol. aspects of light, vision and seeing; also effects of radiant energy on man, animals and plants. Home: 637 Quilliams Rd., South Euclid, O. 44121. Office: Gen. Electric Co., Nela Park, Cleve. 44112.*

GUTHE, Carl Eugen, Am. anthropologist; b. Kearney, Neb., June 1, 1893; s. Karl Eugen and Clara Belle (Ware) G.; B.S., U. Mich., 1914; A.M., Harvard, 1915, Ph.D., 1917; m. Grace Ethel McDonald, Sept. 12, 1916; children—Karl Frederick, Alfred Kidder, Marjorie Belle (dec.), James McDonald. Austin teaching fellow Harvard, 1915-17; asso. dir. Andover-Pecos Expdn., Phillips Acad., Andover, Mass., 1917-21; research asso. in Middle Am. archaeology, Carnegie Instn., Washington, 1921-22; asso. dir. anthropology Univ. Museums, U. Mich., 1922-29; dir. Mus. of Anthropology, U. Michigan, 1929-43; chmn. div. social scis. U. Mich., 1935-38, dir. Univ. Museums, 1936-43; dir. N.Y. State Mus., 1944-53, N.Y. State Sci. Service, 1945-53; research asso. Am. Assn. Museums, 1953-59. Chmn. Com. State Archaeol. Surveys, NRC, 1927-37; chmn. Div. Anthropology and psychology. NRC, 1938-41; chmn. bd. Lab. of Anthropology, Santa Fe, 1936-39; pres. Midwest Museum Conf., 1940-43. Has made excavations in N.M., Guatemala, and Philippine Islands. Mem. Soc. Am. Archaeology (pres. 1945-46), American Assn. Museums, A.A.A.S., Sigma Xi, Phi Kappa Phi, Delta Tau Delta. Author: Pueblo Pottery Making, 1925; So You Want a Good Museum, 1957; The Canadian Museum Movement, 1958; The Management of Small History Museums, 1959; also pamphlets and articles in jours. Home: 1407 Ferdon Rd., Ann Arbor, Mich. 48104.*

GUTHE, Hermann Adolph Wilhelm, German geographer; b. St. Andreasberg, Germany, Aug. 12, 1825; s. Friedrich Wilhelm and Wilhelmine Sophie Friederike (Wage) G.; student U. Göttingen (Germany), 1845-47, 48-49, U. Berlin, 1847-48; secondary sch. teaching certificate, 1850; Ph.D., 1856; m. Hauke-Margarete Schomerus, 1854; 7 children. Became tchr. 1849; became sr. tchr. Lyceum, Hanover, Germany, 1851; also began teaching geography to cadet corps, Hanover, 1863; named instr. math. and mineralogy Poly. Sch., Hanover, circa 1863, prof. title, 1868; became prof. geography Poly. Sch., Munich, Germany, 1873. Hermann Guthe medal (bronze) established in his honor Geog. Soc. Hanover, 1928. Author: Die Lande Braunschweig und Hannover, mit Rücksicht auf die Nachbargebiete geographische dargestellt, 1867; Lehrbuch der Geographie, 1868, continued by H. Wagner, 10th edit., 1920. Died Munich, Jan. 29, 1874.

GUTHE, Karl Eugen, physicist; b. Hanover, Germany, Mar. 5, 1866; s. Otto and Anna (Hanstein) G.; ed. Gymnasium, Hanover, Hanover Tech. Sch., univs. of Marburg, Strassburg and Berlin; passed state exam., Marburg, 1889, Ph.D., 1892; m. Clara Belle Ware, 1892. Came to U. S., 1892. Instr. physics U. Mich., 1893-1900, asst. prof., 1900-03, prof., 1909-15, dean grad. dept., 1912-15; asso. physicist, Bur. of Standards, 1903-05; prof. physics State U. Iowa, 1905-09; Fellow A.A.A.S. (v.p. 1908). Author: (with J. O. Reed) Manual of Physical Measurements, 1902, 07, 12; Laboratory Exercises with Primary and Storage Cells, 1903; (with others) Textbook of Physics, 1908, 1909; College Physics, 1911; Definitions in Physics, 1913. Died Sept. 10, 1915.

GUTHNICK, Paul, German astronomer; b. Hitdorf nr. Leverkusen, Germany, Jan. 12, 1879; s. Hermann and Anna Katherine (Panzer) G.; student math. and natural scis., Bonn, Germany; Ph.D., 1901; m. Melitta Lang, 1923; 2 sons. Asst., Berlin Obs., 1 year; then with F. G. von Bülow's pvt. obs., Bothkamp nr. Kiel, Germany; became observer Berlin Obs. (moved to Babelsberg, Germany 1913), 1906; named asso. prof. U. Berlin, 1916, prof. practical astronomy, dir. obs., 1921; obs. lost most of its instruments, 1945; continued to work independently until his death. Mem. Astron. Soc. (chmn. commn. on variable stars), Prussian, Bavarian acads. scis., Leopoldina, Vatican Pontificia Accademia delle Scienze Nuovi Lincei. Research and publs. on use of photoelectric effect, application photocell in astrophotometry, 1911; developed apparatus to measure brightness with increased precision. Died Potsdam-Babelsberg, Sept. 6, 1947.

GUTHRIE, Andrew, physicist; b. Ladysmith, B.C., Can., s. Samuel and Lempi (Jarvinen) G.; came to U. S., 1937, naturalized, 1949; B.A., U. B.C., 1934; M.S., Purdue U., 1939, Ph.D., 1941; m. Denneta McClung, Nov. 22, 1948; children—Andrew, Steven. Instr., Purdue U., Lafayette, Ind., 1941-42; asso. physicist Radiation Lab., U. Cal., Berkeley, 1942-46, cons., Oakland, 1946-48; asso. prof. San Jose (Cal.) State Coll., 1948-50; head nucleonics div. U. S. Naval Radiol. Def. Lab., San Francisco, 1950-60; prof. physics, chmn. dept. Cal. State Coll., Hayward, 1960—. Fellow Am. Phys. Soc.; mem. Am. Vacuum Soc., Internat. Orgn. Vacuum Tech. and Engrs., Sigma Xi, Sigma Pi Sigma. Author: Vacuum Technology, 1963; contbr. author: Handbook of Physics, 1956, rev., 1965; National Nuclear Energy Series, 7 vols., 1949. Contbr. articles to tech. jours. Study of nuclear and surface physics; electrons; ultra-high vacuum science. Home: 1025 Alvarado Rd., Berkeley 5, Cal. 94705.

GUTHRIE, Frederick, Brit. physicist; b. Bayswater, Eng., Oct. 15, 1833; s. Alexander Guthrie; B.A., U. Coll., London, 1855; studied chemistry under Bunsen, Heidelberg, Germany, under Kolbe, Marburg, Germany; Ph.D., 1854; m. 4 times. Asst. to Frankland, Owens Coll., to Playfair, Edinburgh; prof. chemistry and physics Royal Coll., Mauritius, 1861-67; later became prof. Normal Sch. Sci., South Kensington, Eng. Fellow Royal Soc. Edinburgh, Royal Soc., 1871; mem. Phys. Soc. London (founder 1873). Author: Elements of Heat, 1868; Magnetism and Electricity, 1873; Introduction to Physics; also articles. Discovered approach caused by vibration, 1870, cryohydrates. Died Oct. 21, 1886.

GUTHRIE, George James, English surgeon; b. London, Eng., May 1, 1785; m. Margaret Paterson; 2 sons, Lowry, Charles, 1 dau.; m. 2d. Apprenticed surgeon; asst. York (Eng.) Hosp., later lectr.; surgeon 29th Regiment, Can., 1801, Peninsula, 1808-14; performed several novel operations Waterloo, 1815, recognized as foremost English mil. surgeon of Napoleonic era; lectr. surgery before mil. officers, East India Co., Eng., 1816-46; founder infirmary eye diseases. Mem. Royal Coll. Surgeons. Author: On Gunshot Wounds, on Inflammation, Erysipelas and Mortification, on Injuries of Nerves, and on Wounds of Extremities, 1836; Injuries of Head Affecting Brain, 1842; On Wounds and Injuries of the Arteries of Human Body, 1846; Commentaries on Surgery of the War, 1808-15(pub. 1853. Performed 1st recorded successful ligation of peroneal artery, battle of Waterloo; introduced method of long incisions through skin to relieve diffused erysipelas; 1st in Eng. to use a lithotrite to crush bladder stone; credited with 1st description non-prostatic obstruction at neck of bladder, 1834. Died London, May 1, 1856.

GUTHRIE, Samuel, Am. chemist; b. Brimfield, Mass., 1782; s. Samuel and Sarah Guthrie; attended Coll. Physicians and Surgeons, N.Y.C., 1810-11, U. Pa., 1815; m. Sybil Sexton, 1804, 4 children including Alfred, Edwin. Served in med. corps, U. S. Army, War of 1812; moved from Sherburne, N.Y., to Sacketts Harbor, N.Y., 1817, practiced medicine, set up exptl. chem. lab.; said to have invented an effective priming powder called percussion pill and punch lock for exploding it which together made the flintlock musket obsolete; devised process for rapid conversion of potato starch into molasses, 1830; discovered chloric ether by distilling chloride of lime with alcohol in copper (proved to be chloroform), 1831. Author: The Complete Writings of Samuel Guthrie (collection of letters and comments), 1832. Died Sacketts Harbor, Oct. 19, 1848.

GUTMAN, Robert, Am. sociologist; b. N.Y.C., Aug. 3, 1926; s. Bert and Beatrice (Thalheim) B.; A.B., Columbia, 1946, Ph.D. (Social Sci. Research Council fellow), 1955; postgrad. Princeton, London Sch. Econs.; m. Sonya Rudikoff, Sept. 17, 1950; children—John A. D., Elizabeth C. R. Faculty, Dartmouth, 1949-57; faculty Rutgers U., New Brunswick, N.J., 1957—, prof. sociology, 1960—, Faculty Research fellow, 1965-66; vis. prof. Stanford, 1964, Bartlett Sch. Architecture, London U., 1966. Population Council fellow, 1956-58. Mem. Population Assn. Am. (dir. 1963—), Am. Sociol. Assn., Am. Statis. Assn. Author: (with others) The Mark of Oppression, 1951; The Accuracy of Vital Statistics in Massachusetts, 1842-1901, 1956; Birth and Death Registration in Massachusetts, 1960; Site Planning and Social Organization, 1964. Asso. editor Am. Sociol. Rev., 1961-63; adv. editor Am. Jour. Sociology, 1962-65. Devel. methods for evaluating hist. statistics; proposed uses of sociology in archtl. practice and design edn. Home: 180 Jefferson Rd., Princeton, N.J. 08540. Office: Rutgers U., New Brunswick, N.J. 08903.*

GUTMANN, Ernest, Czechoslovakian physiologist; b. Usti nad Labem, Czechoslovakia, July 16, 1910; s. Josef and Anna (Popperová) G.; M.D., Prague, 1936, D.Sc., 1956; Ph.D., Oxford (Eng.) U., 1943; m. Amalie Szusterová, July 1, 1936; children—Sylvie (Mrs. Gilbert), Anna (Mrs. Koffer), Tomás. Research asst. Inst. Comparative Physiology, Oxford U., 1939-45; lectr. Inst. for Brain Research, U. Prague, until 1949; head dept. physiology of neuromuscular system Physiol. Inst.—Czechoslovak Acad. Scis., Prague, 1950—; prof. pathol. physiology U. Prague. Recipient Czechoslovak State prize for medicine, 1964. Mem. Czechoslovak Physiol. Soc. (chmn.), Czechoslovak Acad. Scis. (corr.), German Acad. Natural Sci. Leopoldina, Med. Research Council Czechoslovakia. Author: Functional Regeneration of Nerves, 1958; The Denervated Muscle, 1962; also many articles. Editor (jours.) Physiologia Bohemoslovenica, 1950—, Ceskoslovenská fysiologie, 1950—. Analysis of factors affecting rate of regeneration of peripheal nerves and of recovery motor and sensory functions after nerve injury; study of influence of increased and decreased activity on metabolism and function of nerve and muscle cell; analysis of trophic functions (long term mechanisms by which nervous system influences muscle metabolism). Home: 11 Schnirchova, Prague 6. Office: 1083 Budejovická, Prague 4, Czechoslovakia.*

GUTMANN, Viktor, Austrian chemist; b. Vienna, Austria, Nov. 10, 1921; s. Viktor and Margarete (Lehmann) G.; Dr.techn., Tech. U., 1946; Ph.D., U. Cambridge, 1950, Sc.D., 1964; 1 son, Christian Meinrad. Dozent, Tech. U., Vienna, 1952, faculty, 1957—, prof., dir. Inst. Inorganic Chemistry, 1960—; dozent U. Vienna, 1952; prof. chemistry U. Baghdad (Iraq), 1952-53. Recipient sci. prize City of Vienna, 1952, Wegscheider-Prize, Austrian Acad. Sci., 1963. Mem. Chem. Soc. (London), Am. Chem. Soc., Gesellschaft Deutsche Chem. Editor: Proceedings VIII ICCC, 1964; Coordination Chemistry in Non-Aqueous Solutions, 1967; Halogen Chemistry, 1967. Research and numerous publs. on new solvents in inorganic chemistry, electrochem. and physiochem. investigations; characterization of coordinating properties of solvents by donor number; devel. new analytical methods by polarography in non-aqueous solutions; preparative studies in phosphorus-nitrogen-chemistry, boron chemistry, boron-nitrogen chemistry. Home: 16 Trinksgeldgasse, Perchtoldsdorf, A 2380 Austria. Office: 9 Getreidemarkt, A 1060 Vienna, Austria.*

GUTOWSKY, Herbert Sander, Am. chemist; b. Bridgman, Mich., Nov. 8, 1919; s. Otto and Hattie (Meyers) G.; A.B., Ind. U., 1940; M.S., U. Cal. at Berkeley, 1946; Ph.D., Harvard, 1949; m. Barbara Joan Stuart, June 22, 1949; children—Daniel Kurt, Robb Edward, Christopher Carl. Faculty, U. Ill., Urbana, 1948—, prof., 1956—, head div. phys. chemistry, 1956-62, head dept. chemistry and chem. engring., 1967—. Mem. Petroleum Research Fund Adv. Bd., 1959-61; mem. selection, scheduling com. Gordon Research Confs., 1959-63; chmn. chemistry panel NSF, 1965-66. Recipient Langmuir award in chem. physics. Guggenheim fellow, 1954. Fellow Am. Phys. Soc., A.A.A.S.; mem. Am. Chem. Soc. (chmn. div. phys. chemistry 1966-67), Nat. Acad. Scis., Faraday Soc., Phi Beta Kappa, Sigma Xi. Research, numerous publs. nuclear magnetic and electron spin resonance. Home: 508 S. Ridgeway St., Champaign, Ill. 61820. Office: 177 Noyes Lab., Urbana, Ill. 61801.*

GUTSCHICK, Raymond Charles, Am. geologist; b. Chgo., Oct. 3, 1913; s. William Anthony and Bessie (Kosatka) G.; student Morton Jr. Coll., 1932-34; B.S., U. Ill., 1938, M.S., 1939, Ph.D. in Geology, 1942; m. Alice Edna Lude, July 2, 1939; children—Alice Antoinette (Mrs. Stuart Turner), Raal Emily (Mrs. Gary Hoff). Instr., U. Ill., Urbana, 1942-43; geologist Magnolia Petroleum Co., Oklahoma City, 1943-45, Aluminum Ore Co., Rosiclaire, Ill., 1942, 45-47, Gulf Oil Corp., Oklahoma City, 1947; faculty dept. geology U. Notre Dame (Ind.), 1947—, prof., 1954—, head dept. geology, 1956—; cons. geologist, South Bend, Ind., 1962—; vis. prof. Ind. U. Geol. Field Sta. Mont., summers 1951, 52, 56-62. Recipient Lay Faculty award U. Notre Dame, 1964. Fellow A.A.A.S., Geol. Soc. Am.; mem. Am. Assn. Petroleum Geologists, Paleontol. Soc., Soc. Econ. Paleontology and Mineralogy, Nat. Assn. Geology Tchrs., Sigma Xi. Research, publs. on geol. history Mississippian rocks Western U. S., Ariz., Mont., paleontology and micropaleontology Mississippian invertebrates, arenaceous foraminifera, petroleum and nonmetallic econ. geology. Home: 53176 West Dr., South Bend, Ind. 46637. Office: Geology Bldg., U. Notre Dame, Notre Dame, Ind. 46556.*

GUTTMAN, Helene Augusta Nathan, Am. microbiologist; b. N.Y.C., July 21, 1930; d. Arthur and Mollie (Bergovoy) Nathan; B.A., Bklyn. Coll., 1951; A.M. (Andelot fellow), Harvard, 1956; M.A., Columbia, 1958; Ph.D., Rutgers U., 1960; m. Newman Guttman, Mar. 4, 1962. Research technician Pub. Health Research Inst., 1951-52; control bacteriologist Burroughs-Wellcome, Inc., Tuckahoe, N.Y. 1952-53; with Haskins Labs., N.Y.C., 1952-64, staff microbiologist, 1960-64; lectr. dept. biology Queens Coll., N.Y.C., 1956-57; research collaborator Brookhaven Nat. Lab., 1958; research asso. Goucher Coll., 1960-62; vis. asst. research prof. Med. Coll. Va., 1960-62; faculty N.Y. U., 1962-67, asso. prof., 1965-67; asso. prof. cell biology U. Ill. at Chgo. Circle, 1967—. Rep. for Soc. Protozoologists to Nat. Register Sci. and Tech. Personnel, 1965—; Dazian Found. fellow, 1956; Pres.'s fellow Soc. Am. Bacteriologists, 1957. Fellow A.A.A.S., Sigma Xi, Sigma Delta Epsilon.; mem. Am. Soc. Microbiologists, Am. Soc. for Cell Biology (chmn. ednl. policy com. 1966—), N.Y. Acad. Scis., Soc. Protozoology, Soc. Indsl. Microbiologists, Soc. Gen. Microbiology, Soc. for Exptl. Biology and Medicine and Hygiene, Am. Inst. Nutrition, N.Y. Soc. Tropical Medicine, Theobald Smith Soc. (T. Jefferson Murray prize 1959). Contbr. articles to tech. jours., chpts. in books. Co-discoverer Crithidia factor; defined nutritional requirements and metabolic relationships for parasitic protozoa of family Trypanosomatidae, cellular mode of action of drugs, vitamin control metabolism, chem. factor influencing chem. contro. membrane permeability, control kinetoplast devel., autotrophy. Home: 916 60th Pl., Downers Grove, Ill. 60515. Office: Dept. Biol. Scis., U. Ill. at Chgo. Circle, Chgo. 60680.*

GUTTMANN, Ludwig, neurosurgeon; b. Tost, Germany, July 3, 1899; s. Bernhard and Dorothea (Weissenberg) G.; student U. Breslau, U. Wurzberg; M.D., U. Freiburg, 1924; Dr. Surgery (hon.) Durham U., 1960; F.R.C.P., F.R.C.S., Eng.; m. Else Samuel, Apr. 7, 1927; children—Dennis, Eva (Mrs. Frank Loeffler). Asst. in neurology Wenzel Hancke Kronkenhaus, Breslau, 1923-28; mem. dept. psychiatry U. Hamburg, 1928-29; lectr. neurology and neurosurgery U. Breslau, 1929-33; dir. neurology and neurosurgery dept. Jewish Hosp., Breslau, 1933-39; research asst., also mem. Peripheral Nerve Injuries Centre, Wingfield Morris Hosp., Oxford, Eng., 1939-44; dir. Nat. Spinal Injuries Centre, Stoke Mandeville Hosp., Aylesbury, Eng., 1944-66, cons. to affiliate units. hon. cons. Nat. Spinal Injuries Center, 1966—. Decorated knight bachelor, comdr. Brit. Empire; decorations from govts. of Belgium, France, Germany, Holland, Japan, Spain. Mem. Internat. Sports Assn. for Disabled (pres.), Internat. Med. Soc. of Paraplegia (pres.); hon. mem. med. socs. various countries. Editor-in-chief Paraplegia jour. Publs. on studies of new concepts in treatment of spinal cord injuries; research on nerve regeneration, epilepsy, air-encephalography, myelography; discovery of Quinizarin Sweat Test. Home: Menorah, Daws Hill Lane, High Wycombe, Eng. Office: Stoke Mandeville Hosp., Aylesbury, Eng.*

GUTTMANN, Stephan, chemist; b. Carei, Rumania, Apr. 25, 1928; s. Maurice and Eve (Burger) G.; Licence ès Sciences Chim., U. Geneva (Switzerland), 1951, Doctorat ès Scis., 1953; m. Elsa Israel, Mar. 25, 1950; children—Eve, Michel, Helene. Research asso. dept. organic chemistry U. Geneva, 1953-56; research asso. dept. pharm. chemistry Sandoz Ltd., Basel, Switzerland, 1956-62, chief research team, 1963—. Mem. Chem. Soc. Geneva, Chem. Soc. Basel. Research, publs. on 1st indsl. synthesis of oxytocin and alpha melanocyte stimulating hormone, bradykinin; synthesis of angiotensin-I, analogues of oxytocin, vasopressin, bradykinin, corticotropin, aminoacyl-nucleosides; research on peptide synthesis. Home: 18 Hegenheimermattweg, Allschwil BL, 4123, Switzerland. Office: 13 Fabrikstrasse, Basle B. 4002 Switzerland.*

GUTTON, Camille Antoine Marie, French physicist; b. Nancy, France, Aug. 30, 1872. Prof. physics Nancy Faculty Scis.; dir. Nat. Lab. Radioelectricity. Mem. French Acad. Scis., 1928. Research in wireless telegraphy and telephony, ultra high frequency waves and their application; studied tech. of radio; made first radio liaisons between a plane and earth; invented high sensitivity electro-meter. Died Jan. 31, 1938.

GUTTSTADT, Albert, German physician; b. Rastenburg, Germany, Jan. 25, 1840; student medicine, Berlin; M.D., circa 1866; m. Klara Guhrauer, 1876; 2 sons. Welfare physician, Berlin; mil. physician Franco-Prussian War, 1870-71, also at smallpox hosp., Tempelhof, Berlin; joined Prussian Statis. Bur., 1872, became med. adviser, 1874, later mem.; joined faculty hygiene U. Berlin, 1875. Author: Krankenhauslexikon für das Königreich Preussen, 1885; Die naturwissenschaftlichen und medizinischen Staatsaustalten Berlins, 1886; Anstalten und Einrichtungen des offetlichen Gesundheitswesen in Preussen, 1890; Deutschlands Gesundheitswesen, 2 vols., 1890; Über die praktische Ausbildung von Ärzten an Kliniken, 1892; Krankenhauslexikon für das Deutsche Reich, 1910. Mem. Chamber Physicians. Editor numerous jours, including Deutsch Medizinische Wochenschrift, Klinisches Jahrbuch, Korrespondenzblatt Berliner Ärzte. Statis. study of pub. health; devel. new methods; research on cripples, blind, deaf and dumb, mentally ill, welfare care. Died Berlin, May 3, 1909.

GUTZEIT, (Robert Julius) Kurt, German internist; b. Berlin, Germany, June 2, 1893; s. Georg and Agnes (Fuchs) G.; studied natural scis., then medicine; tng. in splty., U. Jena (Germany); m. Erna Stintzing; 1 son, 4 daus. Head physician under Stepp, U. Breslau (now Wroclaw, Poland); joined faculty U. Jena (Germany), 1923; became prof., Breslau, 1929; named dir. Rudolf Vierchow Hosp., Berlin, 1933; prof. internal medicine, Breslau, 1934-45; adv. internist for army, 1939-45; interned, 1945-48; became dir. Herzoghöhe sanatorium, Bayreuth, 1949, Fürstenhof clinic, Bad Wildungen, 1957. Mem. Leopoldina. Author: Über gutartige Magentumoren, 1926; Die Gastroskopie, Lehrbuch und Atlas, 1929, 30, 37; Die Gastroenteritis, 1933; Die Duodenalsonde, 1935-45; (with G. W. Parade) Die Fokalinfektion, 1939; Hepatitis epidemica, 1942-44; Magenkrankheiten, 1943; Die Wirbelsäule als Krankheitsfaktor, 1951. Editor, collaborator on several handbooks. Research on digestive and metabolic diseases, neural pathology and chiropractics, chemotherapy of malignant tumors, genetic pathology, indsl. diseases; in X-ray diagnostics developed 6 valve, multiphase current X-ray apparatus; developed technique of gastroscopy. Died Bad Wildungen, Oct. 28, 1957.

GUTZMANN, Hermann Carl Albert, German physician; b. Bütow, Germany, Jan. 29, 1865; s. Albert and Charlotte (Trabant) G.; M.D., U. Berlin, 1887; m. Martha Pahl, 1891; 1 son, Hermann, 3 daus. Founder clinic for speech and voice defects (later became part Med. Polyclinic 1907, joined to Killian throat and nose clinic Charité Hosp. 1912); founder phonetic lab.; joined faculty U. Berlin (Germany), 1905, became asso. prof., 1909; named dir. pvt. clinic and instn. for speech defects, 1896. Author: Vorlesungen über die Störungen der Sprache und ihre Heilung, 1893; Stimmbildung und Stimmpflege, 1906; Physiologie der Stimme und Sprache, 1909; also numerous articles. Studies on speech and voice defects, stuttering and other speech defects of children, history of speech therapy, exptl. phonetics. Died Berlin-Lichterfelde, Nov. 4, 1922.

GUTZMER, (Carl Friedrich) August, German mathematician; b. Neuroddahn, Germany, Feb. 2, 1860; s. August and Wilhelmine (Schultze) G.; student math. U. Berlin (Germany); Ph.D., U. Halle (Germany), 1893; m. Helene Günther, 1893; 1 dau., 2 stepchildren. Asst. to E. Lampe, Tech. U. Berlin, 1894-96; joined faculty U. Halle, 1896, named asso. prof., 1899; named prof. U. Jena (Germany), 1900, Halle, 1905. Mem. German Assn. Math. (became editor ann. 1901), Leopoldina (pres. 1922-24, became editor 1921), German Soc. German Scientists and Physicians (mem. edn. com.), German com. for Edn. in Math. and Natural Scis. Author: Theorie der eindeutigen analytischen Funktionen, 1906; Die Tätigkeit der Unterrichtskommission der Gesellschaft Deutscher Naturwissenschaftler und Arzte, 1908; Die Tätigkeit des Deutchen Ausschusses für die mathematische und naturwissenschaftliche Unterricht in den Jahren, 1908-13, 1914. Co-editor, Leonhardi Euleri opera omnia, Series I., vol. 17, 1915, 18, 20. Editor, Nova Acta, 1921-24. Research on application of math. to modern sci., tech. and econ. problems. Died Halle, May 10, 1924.

GUY, Jean Claude, French physicist; b. Compiegne, France, May 14, 1919; s. Paul and Suzanne (Mosnier) G.; Licencié, Sorbonne, U. Paris, 1939, Pharmacien, 1943, Dr. ès Sci., 1948; m. Louise Mogford, Apr. 10, 1945; children—Jean-Louis, Pierre. Prof., Faculte de Pharmacie de Paris, 1950-64; prof. Sorbonne, 1964—; lectr. L'École Polytéchnique, 1957—. Laureat, Chem. Soc. France, French Acad. Medicine. Mem. Physics Soc., Chem. Soc, Phys. Chemistry Soc. France. Research, publs. on molecular physics, particularly calculations of magnetic susceptibilities and constants of magnetic shields in relation to theory of nuclear magnetic resonance. Office: 9, Quai St. Bernard, Paris, France.*

GUYE, Charles Eugène, Swiss physicist; b. St. Christophe, Switzerland, Oct. 15, 1866. Prof. exptl. physics, Geneva (Switzerland) Faculty Scis.; dir. physics lab. U. Geneva. Mem. French Acad. Scis., 1929. Studied area between living and inanimate matter; discovered (with his brother, Philippe Auguste Guye) fixation of atmospheric azote by means of electric arc, 1895; studied (with Lavanchy) electrons of cosmic rays and proved that the mass of these corpuscles grew with their speed according to theory of relativity, 1913. Died Geneva, July 15, 1942.

GUYE, Phillippe-Auguste, Swiss chemist; b. St. Christophe, Switzerland, June 12, 1862; Ph.D., Geneva, 1884; studied under Graebe and Friedel. Asst. to Carl Graebe; named privatdocent, 1885; became asso. prof. theoretical and tech. chemistry, Geneva, Switzerland, 1892, prof., 1895-1922. Recipient Davy medal, 1921. Mem. French Acad. Scis. Founder Jour. de Chimie physique, 1903, editor until 1922; a founder Helvetica Chimica Acta. Research and numerous publs. on rotary power of optically active substances, molecular constitution, polymerization in liquids, indsl. electrolysis of alkali chlorides, electrochem. synthesis of nitric acid, gas densities, compressibilities and critical data; determined atomic weights of nitrogen, chlorine, silver; determined value for R in gas equation, PV=RT, 1900. Died 1922.

GUYENOT, Emile Louis Charles, zoologist; b. Lons-le-Saunier, France, June 9, 1885; s. Desire and Marie Guyenot; M.D., Faculties Scis. and Medicine, Paris, 1909, Sc.D., 1917; hon. doctorate U. Lausanne; m. Marie Billot, July 20, 1909; 4 children. Asst. in bacteriology U. Besancon, 1905-08; with Sorbonne, 1908-17; prof. physiology and histology Dental Sch., Paris, 1912-18; prof. zoology and comparative anatomy U. Geneva, 1918—, founder, dir. Zool. Expt. Sta., 1933, also dir. Inst. Zoology and Comparative Anatomy; asso. dir. Zool. Sta. (Faculty Scis., Paris, Wimereux, 1924. Mem. French Acad. Scis. (corr.), Geneva Soc. Physics and Natural History, Helvetic Soc. Natural Scis., Swiss Soc. Genetics; hon. mem. Biol. Soc. (Paris), Neuchatel Soc. Natural Scis., Geneva Med. Soc. Author: Les Mollusques, 1919; L'Hérédité, 1924, 2d edit., 1931, 3d edit., 1942; La Variation, 1931; L'Evolution, 1931; La détermination du sexe et l'héredité, 1935; Les sciences de la vie aux XVIIe and XVIIIe siècles, 1941; L'origine des espèces, 1944; Les Problèmes de la vie, 1946. Research and over 300 publs. on genetics, regeneration, endocrinology, aseptic life, air bladder of fishes, sporazoa. Address: University of Geneva, Geneva, Switzerland.

GUYON, Jean Casimir Félix, French physician, surgeon; b. St. Denis, France, July 21, 1831; ed. U. Nantes (France), U. Paris; M.D., 1858. Urol. surgeon Necker Hosp.; became prof. surg. pathology Paris Faculty Medicine, 1877, prof. genitourol. surgery U. Paris, 1890. Mem. Acad. Medicine (became pres. 1901), Acad. Surgery (pres.), French Acad. Scis. (became pres. 1913). Author: Leçons sur les maladies des voies urinaires, 1881; (with Bazy), Atlas des maladies des voies urinaires, 1886; Leçons cliniques, 1888; Diagnostic des affections chiruricales des reins, 1891. Founder of urology; developed technique of amputation; improved Bigelow's method of litholapaxy. Died Paris, July 21, 1920.

GUYON, Jean Louis Genevière, French surgeon; b. Albert, France, Apr. 5, 1794. Made voyage to Martinique to study yellow fever, 1815, to Poland to study cholera, 1831; chief army surgeon, from 1838. Mem. French Acad. Scis., 1856. Author: Sur le traitement de la fièvre jaune, 1826; Sur la cholera en Pologne, 1832; Sur les maladies de l'afrique. Died Paris, Aug. 23, 1870.

GUYOT, Arnold Henry, geographer, geologist; b. Boudevilliers, Switzerland, Sept. 28, 1807; s. David Pierre and Constance (Favarger) G.; grad. Coll. of Neuchatel, 1825; Ph.D., U. Berlin, 1835; LL.D. (hon.), Union Coll., 1873; m. Sarah Doremus Haines, July 2, 1867. Prof. history and phys. geography Acad. of Neuchatel, 1839-48; 1st to formulate laws of structure and movement of glaciers; came to U. S., 1848; lectured under Mass. Bd. Edn. 6 years; published geography textbooks, 1866-75, made 1st definite attempt at scientific presentation of geography in Am. schs.; prof. phys. geography and geology Princeton, 1854-84. Author: Meteorological and Physical Tables, 1852; Earth and Man, 1853; Treatise on Physical Geography, 1873; Memoir of Louis Agassiz, 1883. His researches led to establishment of U. S. Weather Bur.; founded museum at Princeton; selected and equipped weather observation stations in N.Y. and Mass.; flat-topped submarine mountains (guyots) named in his honor. Died Princeton, N.J., Feb. 8, 1884.

GUYTON, Arthur Clifton, Am. physician; b. Oxford, Miss., Sept. 8, 1919; s. Billy Sylvester and Katherine (Smallwood) G.; B.A., (Taylor medal), U. Miss., 1939; M.D., Harvard, 1943; m. Ruth Alice Weigle, June 12, 1943; children—David, Robert, John, Steven, Catherine, Jean, Douglas, James, Thomas. Faculty, U. Miss., Oxford, 1947—, prof., chmn. dept. physiology and biophysics, 1948. Recipient U. S. Presdl. citation, 1956. Fellow A.A.A.S. (Gould award 1959), Am. Coll. Cardiology; mem. Am. Physiol. Soc., Biophys. Soc., Am. Fedn. Clin. Research, So. Soc. Clin. Research, Am. Heart Assn. (bd. dirs.), Miss. Acad. Sci. (bd. dirs.), Nat. Bd. Med. Examiners, Sigma Xi, Alpha Omega Alpha, Omicron Delta Kappa. others. Author: Textbook of Medical Physiology, 1956, 61, 66; Function of the Human Body, 1959, 64; Circulatory Physiology: Cardiac Output and Its Regulation, 1963. Editorial bds. Am. Jour. Physiology, Jour. Applied Physiology, Am. Jour. Cardiology, Excerpta Medica, Circulation Research. Research, publs. on intracardiac pressure, measurements in health and disease, pulmonary and systemic circulation including renal hypertension and the Goldblatt phenomenon, venous flow and return, cardiac physiology, cardiac output. Home: 234 Meadow Rd., Jackson, Miss. 39206. Office: Dept. Physiology and Biophysics, U. Miss., Oxford, Miss. 39216.*

GUYTON DE MORVEAU, Baron Louis Bernard, French chemist; b. Dijon, France, Jan. 4, 1737; s. Antoine Guyton; studied law in Dijon; m. Claudine Poulet. Advocate-gen. Dijon parliament, 1755-82; tchr. acad. Dijon, 1774; elected mem. legislative assembly, 1791, conv., 1792, Com. Pub. Safety, 1793; commissary Army of N. (dir. aerial reconnaissance balloons at Battle of Fleuros), 1794; founder (with others), dir., tchr. Ecole Polytechnique, Paris, France, 1795-1805; master mint, 1800-1814. Original mem. French Acad. Scis., 1796; Fellow Royal Soc., 1788. Author: Digressions académiques (views on phlogiston and crystallization), 1772; Éléments de chimie théorique et pratique, 3 vols., 1776-77; (with Lavoisier, Berthollet, Fourcroy) Méthode d'une nomenclature chimique (presents nomenclature based on sci. principles), 4 vols., 1787. Contbr. articles to Encyclopédie méthodique, 1786. Used chlorine and hydrochloric acid gas as disinfectant, 1773; showed that sodium sulphate could be used to make soda, 1782; concluded that combination with carbon causes conversion of iron to steel; introduced name Prussic acid; investigated gunpowder manufacture. Died Paris, Jan. 2, 1816.

GUZE, Henry, Am. psychologist, human biologist; b. Newark, June 7, 1919; s. Julius and Celia (Huberman) G.; B.A., U. Newark, 1942; Ph.D., N.Y. U., 1955; m. Vivian Segerman, June 18, 1945. Adj. prof. psychology L.I. U. Grad. Sch. Art and Scis., N.Y.C., 1948—; pvt. practice psychology, 1952—; guest lectr. N.Y. U., 1961, 62, 64, U. Kan., 1961, Phila. Mental Health Clinic, 1965; vis. prof. anthropology Drew U. Research cons. Inst. Research in Hypnosis, Westchester Bd. Health. Diplomate Am. Bd. Examiners Psychol. Hypnosis. Fellow Acad. Psychosomatic Medicine, A.A.A.S., Soc. Clin. and Exptl. Hypnosis (certificate of merit 1955); mem. Am. Acad. Psychotherapists (certificate of merit 1964), Internat. Soc. Comprehensive Medicine, Soc. Sci. Study Sex (pres.), Am. Anthrop. Assn., Am. Assn. Phys. Anthropologists, Am. Genetic Assn., Am. Psychol. Assn., Am. Psychosomatic Soc., Am. Soc. Psychosomatic Dentistry and Medicine, Am. Sociol. Assn., Med. Correctional Assn. Asso. editor: Jour. Internat. Soc. Clin. and Exptl. Hypnosis, 1954—; Voices—The Art and Sci. of Psychotherapy, 1964—; Jour. Sex Research, 1965—. Research, numerous publs. on effects of pre-weaning nursing deprivation on hoarding, maternal and sexual behavior in adult rats, anesthesia trauma as a cause of anxiety and depression, compensatory neuro-muscular behavior as a result of blocking sensory motor reactions, psychol. aspects of contraceptive choice and failure, cyclic activity and emotional response in human female, hypnosis as set for emotional expression, pain induction in study of image and hallucination, transsexual phenomena and gender research. Home: 66 Sunset Av., Montclair, N.J. 07042. Office: 10 W. 15th St., N.Y.C. 10011.*

GUZE, Samuel Barry, Am. psychiatrist; b. N.Y.C., Oct. 18, 1923; s. Jacob and Jenny (Berry) G.; student Coll. City N.Y., 1939-41; M.D., Washington U., 1945; m. Joy Lawrence Campbell, June 7, 1946; children—Jonathan David, Ann. Faculty, Washington U. Sch. Medicine, St. Louis, 1951—, prof. psychiatry, asso. prof. medicine, 1964—, asst. to dean, 1965—. Staff Barnes Hosp., St. Louis, 1951—, Renard Hosp., 1963—; asst. dir. Psychiatry Clinic, Washington U. Sch. Medicine, 1951-55, dir., 1955—. Diplomate Am. Bd. Internal Medicine, Am. Bd. Psychiatry and Neurology. Fellow A.C.P., Am. Psychiat. Assn.; mem. Am. Fedn. for Clin. Research, Central Soc. for Clin. Research, Psychiat. Research Soc., A.M.A., Am. Psychosomatic Soc., Assn. for Research in Neurol. and Mental Diseases, Am. Psychopathol. Soc., Soc. Biol. Psychiatry, Sigma Xi, Alpha Omega Alpha. Research, numerous publs. in natural history psychiat. disorders including hysteria, sociopathic personality disorder, alcoholism, organic brain syndromes. Home: 17 Ridgemoor Dr., St. Louis 63105. Office: 4940 Audubon Av., St. Louis 63110.*

GUZMAN, Miguel Angel, statistician; b. San Vincente, El Salvador, Dec. 20, 1925; s. Patrocinio and Elsa (Foresti) G.; B.A., U. Tenn., 1949; M.S., N.C. State Coll., 1956, Ph.D., 1961; m. Maria Victoria Sobalvarro, Dec. 1, 1951; children—Miguel Antonio, Ricardo Enrique, Ana Cecilia, Juan Carlos. With Instituto de Nutrición de Centro América y Panama, Guatemala, 1949—, chief labs, asst. dir., 1953-54, chief div. statistics, 1957—, chmn. publs. com., 1957-61; vis. asso. prof. biostatistics, vis. lectr. dept. nutrition and food sci. Mass. Inst. Tech., 1964—. Cons. U. S. Dept. Agr., 1961—, Dupont de Nemours, 1960—, WHO, 1959—. Mem. Am. Inst. Nutrition, Am. Statis. Assn., Biometric Soc., Am. Assn. Phys. Anthropologists, Asociacian Latinoamericana de Nutricion, Sigma Xi, Phi Kappa Phi, Sigma Pi Alpha. Research, numerous publs. on human growth and devel., math. methods in biology, computer simulation in studying effects of diseases and nutrition. Home: 21-24 Avenida Las Americas, Guatemala, Guatemal Z.13. Office: Apartado 11-88, Guatema, Guatemala, Central Am.*

GWATKIN, Ralph Buchanan Lloyd, Am. biologist; b. Newport, Eng., May 23, 1929; s. Ralph L. and Ada A. (Lennie) G.; B.A. with honors, U. Toronto, 1950, M.A. in Bacteriology, 1951; Ph.D. in Microbiology,

Rutgers U., 1954; m. Selma L. Schatz, June 20, 1954; children—Sharon, Adina. Came to U. S., 1959, naturalized, 1964. Research asso. Connaught Labs., U. Toronto, Can., 1956-59; asso. Wistar Inst. Biology, Phila., 1959-62; research asst. prof. reproductive physiology U. Pa., Phila., 1963-66; dir. Tissue Culture Research Labs., asst. dir. microbiol. and natural products research dept. Merck Sharp & Dohme Research Labs., Rahway, N.J., 1967—. Recipient Research Career Devel. award NIH, 1964. Mem. N.Y. Acad. Sci., Tissue Culture Assn., Soc. Cell Biologists, Sigma Xi. Pioneered in colony counts of animal cells growing in suspension; organ culture of fallopian tube; study of growth of viruses in mammalian eggs; nutritional requirements of blastocyst in mammals. Address: Merck Sharp & Dohme Research Labs., Rahway, N.J.*

GWINN, William Dulaney, Am. phys. chemist; b. Bloomington, Ill., Sept. 28, 1916; s. Walter E. and Allyne (Dulaney) G.; A.B., U. Mo., 1937, M.A., 1939; Ph.D., U. Cal. at Berkeley, 1942; m. Margaret Boothby, July 11, 1953; children—Robert B., Ellen, Kathleen. Faculty, U. Cal. at Berkeley, 1942—, prof. phys. chemistry, 1955—, research prof. Miller Research Inst., 1961-62. Guggenheim fellow, 1954; Sloan fellow, 1955-59; cited for merit U. Mo., 1964. Fellow Am. Phys. Soc.; mem. Phi Beta Kappa, Sigma Xi, Pi Mu Epsilon. Asso. editor Jour. Chem. Physics, 1962-64. Research, publs. on molecular structure, microwave spectroscopy, quantum mechanics, direct digital control. Home: 8506 Terrace Dr., El Cerrito, Cal. 94530. Office: Dept. Chemistry, U. Cal., Berkeley, Cal. 94720.*

GWINNE, Matthew, English physician; b. London, circa 1558; s. Edward Gwinne; B.A., St. John's Coll., Oxford, 1578, M.A., 1582; M.B., M.D., 1593; regent master in music St. John's, 1582-83, jr. proctor, 1588; attended Ambassador Sir Henry Unton in France, 1595; became 1st prof. physics Gresham Coll., 1598-1607; named physician to Tower, 1605; maintained pvt. practice. Apptd. commr. for inspecting tobacco, 1620. Licentiate, fellow Coll. physicians. Author: In assertorem chymicae sed verae medicinae desertorem (proved gold lacks medicinal value equivalent to metallic value, also that Francis Anthony's aurum potabile was without gold), 1611. Died 1627.

GYFTOPOULOS, Elias Panayiotis, nuclear engr., educator; b. Athens, Greece, July 4, 1927; s. Panayiotis Elias and Despina (Louvaris) G.; diploma mech. and elec. engring. Tech. U. Athens; Sc.D. in Elec. Engring., Mass. Inst. Tech., 1958; m. Artemis S. Scalleri, Sept. 3, 1962; children—Vasso E., Maro E., Rena E. Came to U. S., 1953, naturalized, 1963. Faculty, Mass. Inst. Tech., Cambridge, 1954—, prof. nuclear engring., 1965—. Cons. Brookhaven Nat. Lab., Phillips Petroleum Co., E. I. DuPont De Nemours & Co., others. Fellow Am. Acad. Arts and Scis., Am. Nuclear Soc.; mem. Am. Inst. Aeros. and Astronautics, Am. Phys. Soc., Am. Nuclear Soc., A.A.A.S., Am. Soc. M.E., Am. Inst. Aeros. and Astronautics. Research, numerous publs. on nuclear reactor dynamics, stability of non-linear systems, physics intermetallic surfaces, physics of low energy plasmas, thermionic conversion.*

GYGER, Hans Conrad (Geiger), Swiss cartographer, mathematician; b. Zurich, Switzerland, July 22, 1599; s. Hans Georg and Verena (Leemann) G.; m. Elisabeth Meyer, Jan. 8, 1627; 3 sons, including Johann Georg, Hans Friedrich, 3 daus. Became ofcl. Kappeler Ct., Zurich, 1647; painter, especially on glass; cartographic work Zurich Surveying Sch., later with instrument mfrs. P. Eberhart & Leonhard Zubler; commd. by state to prepare map of Zurich canton. Author: Register über Gygers Risse und Schriften; Marchenbeschreibung der Zürchergebietes von 1664; Karte der Kanton Zürich, 1667, facsimile edit., 1944. Prepared numerous maps of canton of Zurich, Switzerland; used triangulation which showed mountains in horizontal projection and shaded earth surface forms (1st relief map). Died Zurich, Sept. 25, 1674.

GYLDEN, Johan August Hugo, astronomer; b. Helsingfors, Sweden, May 29, 1841. Dir., astronomer Acad. Stockholm Obs., 1871-96; with Obs. Pulkova, Russia, 1862-65. Mem. French Acad. Scis., 1879. Studied motion of celestial bodies and libration; made series of important observations on variation of latitude, 1863-70. Died Nov. 9, 1896.

GYLLENHAAL, Leonhard, entomologist; b. Alzustorp, Germany, Dec. 3, 1752; joined army as under officer, 1769; ret. as maj., 1799; became agriculturist, Hoegberg; studied natural history under Linneaus and Thimberg, Upsalla, Sweden; presented insect collection to Upsalla Acad. Scis. Mem. Upsalla Acad. Scis., Stockholm Acad. Sci., Entomol. Soc. Paris. Author: Insecta suecica, 1808; Nova acta regiae societatis scientiarum Upsaliensis, 1799; Synonymia insectorum. Research on entomology. Died May 13, 1840.

GYLLENSTEN, Lars, Swedish histologist, physician; b. Stockholm, Sweden, Nov. 12, 1921; s. Carl and Ingrid (Rangström) G.; M.D., Karolinska Institutet, Stockholm, 1948, Ph.D. in Medicine, 1953; m. Inga-Lisa Hultén, Oct. 5, 1946; 1 dau., Katarina. Faculty, Karolinska Institutet, 1953—, asso. prof. histology, 1955—; lectr. histology Sch. Dentistry Stockholm, 1951-53. Mem. Swedish Acad. Author numerous novels, short novels, essays, poetry, sci. articles. Research on hormonal regulation of thymus and lymphnodes, seeding of lymphocytes from thymus, oxygen induced injuries of eyes, influence of function and lack of function on devel. of brain, regulative mechanisms in growth of visual centers. Home: 121 Karlavägen, Stockholm, Sweden.*

GYORGY, Paul, physician; b. Nagy Varad, Hungary, Apr. 7, 1893; M.D., U. Budapest, 1915; M.D. (hon.), Heidelberg U., 1958; m. Margaret A. John, Oct. 23, 1920. Came U. S., 1935, naturalized, 1941. Prof. pediatrics Heidelberg U., 1920-33; research fellow U. Cambridge (Eng.), 1933-35; vis. research prof. Western Res. U., 1935-44; prof. pediatrics U. Pa., 1944-58, emeritus, 1958—. Chmn. organizing com. 5th Internat. Cong. on Nutrition, Wash., 1960; chmn. protein adv. group WHO, FAO, UNICEF. Recipient Borden award Am. Inst. Nutrition and Acad. Pediatrics; Mendel award Am. Inst. Nutrition; Goldberger award A.M.A.; Howland award Am. Pediatric Soc. Markle fellowship, U. Toronto. Pioneer in discovery riboflavin, vitamin B6, biotin, growth factors in human milk, exptl. dietary liver injury; isolated (with others) vitamins biotin, riboflavin and pyridoxine. Home: 201 Curwen Rd., Rosemont, Pa. 19010. Office: Phila. Gen. Hosp., Phila. 19104.*

GYRISCO, George Gordon, Am. entomologist; b. South Hadley, Mass., Mar. 25, 1920; s. Michael and Maria (Zastoelke) G.; B.S. summa cum laude, Mass. State Coll., 1943; Ph.D., Cornell U., 1943-47; m. Valerie Maitland, Horn, June 20, 1947; children—Geoffrey Maitland, Jill Lea, Glenn Gordon. Asst. biologist Mass. Conservation Dept., 1942; faculty Cornell U., 1947—, prof. entomology, 1955—, head dept., 1962-63. Fellow A.A.A.S.; mem. Entomol. Soc. Am., Entomol. Soc. Can., Ecol. Soc. Am., Soc. for Study Evolution, Am. Daffodil Soc., Sigma Xi, Phi Kappa Phi, Gamma Alpha, Sigma Alpha Epsilon. Research, numerous publs. on behavior-diapause in alfalfa weevil, insect flight-alfalfa weevil, European chafer, circadian rhythm in insects, insecticide residues on field crops, storage and loss products in dairy cattle, virus vectors sugar beets, their behavior and control, control field crop insects, population dynamics soil insects. Home: 36 Twin Glens, R.F.D. 1, Ithaca, N.Y. 14850.*

GZHITSKY, Stepan Zenonovich, Russian biochemist; b. Ostrovets (now Ternpolo Oblast, Ukraine), 1900; grad. Lvov Acad. Vet. Med., 1929. Instr. biochemistry Lvov Acad. Vet. Medicine (name now Lvov Zootech. and Vet. Inst.), 1929—, now head chair biochemistry; prof., 1939—; head dept. biochemistry of farm animals Research Inst. Agr. and Stock-Raising in Western Regions of Ukraine SSR, 1951—. Decorated Order of Lenin. Mem. Ukrainian Acad. Agrl. Scis., Ukrainian Acad. Sci. (corr.). Research and numerous publs. on etiology and pathogenesis of chronic hematuria in cattle. Address: Lvov Zootech. and Vet. Inst., ulitsa Dragomanova 14-16, Lvov, Ukraine SSR, USSR.

H

HAAB, Otto, Swiss ophthalmologist; b. Wülfingen, Switzerland, Apr. 19, 1850; prof., Zurich, Switzerland. Author atlases of ophthalmoscopy and external diseases of eye. Discovered cortical reflex of pupil (Haab's reflex, or cerebral cortex reflex), 1891; inventor magnet for extracting metal particles from eye, 1902; built an ophthalmoscope. Died Zurich, Oct. 17, 1931.

HAACK, Wolfgang, German mathematician; b. Gotha, Apr. 24, 1902; s. Hermann and Johanna (König) H.; student Hanover Inst. Tech., univs. Jena, Hamburg; dr., 1926; m. Marianne Blumentritt, May 30, 1936. Docent, Danzig Inst. Tech., 1929-35, Berlin Inst. Tech., 1935-37; prof. Karlsruhe Inst. Tech., 1938-44; prof. Berlin Inst. Tech., 1944—; hon. prof. Free U. Berlin; dir. math. sect. Hahn-Meitner Inst. Nuclear Research, Berlin, 1959—. Cons. Telefunken AG, Ulm, Inst. Radio and Mathematics, Werthhoven, Bonn. Mem. German Assn. Mathematicians, Soc. Applied Math. and Mechanics, numerous others. Author: Differentialgeometrie I, II, 1948; Darstellende Geometrie I, II, III, 1954; Elementare Differentialgeometrie, 1955; also numerous articles. Research on shape of projectiles with minimum wave resistance, detection and tracking of air traffic by computing primary radar signals, existence theorems of systems of partial differential equations. Home: 55b Königsalle, Berlin 33. Office: Hahn-Meitner Institute for Nuclear Research, 100 Glienicker Strasse, Berlin 39, West Germany.*

HAACKE, (Johann) Wilhelm, German zoologist; b. Clenze, Germany, Aug. 23, 1855; s. Johann Wilhelm and Wilhelmine Friederike Louise (Wacker) H.; student zoology, U. Jena; Ph.D., 1878; m. Emily von Bertouch, 1883; 1 son, 1 dau. Became asst. U. Jena, 1878, U. Kiel (Germany), 1879; emigrated to New Zealand, 1881; dir. Mus., Adelaide, Australia, 1882-84; returned to Germany, 1886; zoo dir., Frankfort/Main, 1888-93; lectr. zoology Tech. U., Darmstadt, Germany, 1890-97; ind. scholar; later secondary sch. tchr. Author: Die Schöpfung der Tierwelt, 1893; Die Schöpfung des Menschen und seiner Ideale, 1895; Aus der Schöpfungswerkstatt, 1897; Grundriss der Entwicklungsmechanik, 1897; also articles. Studied morphology of corals and medusae, evolution; discovered (independently of Caldwell) that ant sea urchin (Ameisenigel) produced eggs, 1884; originated a gemmae theory in attempt to explain inheritance of acquired characteristics; used concept of orthogenesis. Died Lüneburg, Germany, Dec. 6, 1912.

HAAG, Henrique Paulo, Brazilian chemist; b. Sao Paulo, Brazil, Apr. 9, 1928; s. Paul Henry and Rosa (Raes) H.; Agronomy engr. Escola Superior de Agrl. Luiz de Queiroz, Sao Paulo, 1954, Dr., 1958, Docente-Livre, Ph.D., 1966; m. Maria Jose Martins, Sept. 4, 1955; children—Paulo Sergio, Carlos Alberto. Asst. biochemistry Escola Superior de Agrl. Luiz de Queiroz, Piracicaba, Sao Paulo, 1955-58, asst. prof., 1958-66, prof. applied biochemistry, 1966—, prof. mineral nutrition in plants, 1964-66. Mem. Am. Soc. Plant Physiologists, Brazilian Soc. Soil Sci., Brazilian Soc. Botany, Latin Am. Soc. Soil Sci. Author: (with E. Malavolta, F. A. F. de Mello, M. O. C. Brasil Sobro) On the Mineral Nutrition of Some Tropical Crops, 1962; also numerous articles. Fundamental studies in mineral nutrition in tropical crops, especially sugar cane, cotton, eucalyptus, coffee; interactions among virus and mineral nutrition. Home: 55 Av. Independencia, Piracicaba. Office: Escola Superior de Agricultura Luiz de Queiroz, Piracicaba, Sao Paulo, Brazil.*

HAAG, Jules, French mathematician; b. Flirey, France, Aug. 19, 1882; prof. spl. math., then rational mechanics Besancon (France) Faculty Scis., named dir. Chronometric Inst., Besancon, 1927. Mem. French Acad. Scis. Publs. on mechanics, chronometry, analysis, geometry, calculus of probabilities. Died Besancon, Feb. 16, 1953.

HAAG, Rudolf, German physicist; b. Tübingen, Germany, Aug. 17, 1922; s. Albert and Anna (Schaich) H.; Diplomphysiker, T. H. Stuttgart (Germany), 1948; Ph.D. in Physics, U. Munich (Germany), 1951; m. Kaethe Fues, June 17, 1948; children—Albert, Friedrich, Elisabeth, Ulrich. Asst., Inst. P. Theoret. Physik, U. Munich, 1951-53; research asso. theoretical study group CERN, Copenhagen, Denmark, 1953-54; lectr. theoretical physics Munich U., 1954-56; research asso. Max-Planck Inst., Göttingen, Germany, 1956-57; vis. prof. theoretical physics Princeton, 1957-59; prof. physics U. Ill., Urbana, 1960-66; prof. Hamburg (Germany) U., 1966—. Research, publs. on math. founds. of quantum theory of fields; devel. collision theory of particles. Home: 20 Oeltingsallee, Pinneberg 208, W. Germany. Office: 149 Luruper Chaussee, Hamburg 2, W. Germany.*

HAAGENSEN, Cushman Davis, Am. physician; b. Hillsboro, N.D., July 6, 1900; s. Edward Cornelius and Henrietta (Paulson) H.; B.A., U. N.D., 1921, D.Sc. (hon.), 1958; M.D., Harvard, 1923; D.Sc. (hon.), U. Athens (Greece), 1964; m. Alice Munro, Sept. 11, 1928; children—Alice (Mrs. R. D. Gerard), Karen (Mrs. W. L. Savage). Fellow, Meml. Hosp., N.Y.C., 1929-31; asso. Crocker Inst. Cancer Research, N.Y.C., 1931-37; asso. in cancer research Columbia, N.Y.C., 1931-34, faculty Coll. Physicians and Surgeons, 1934—, prof. clin. surgery, 1956-66, prof. emeritus, 1966—, coordinator cancer teaching, 1948-66; dir. Columbia Inst. Cancer Research, 1949-52; dir. surgery Francis Delafield Hosp., 1955-66. Recipient Harlow medal N.Y. Acad. Medicine, 1961. Mem. Halsted, Whipple, Arthur Purdy Stout, Ewing socs., Am. Assn. for Cancer Research, Am. Cancer Soc. Author: (with W. B. Lloyd) A Hundred Years of Medicine, 1943; Diseases of the Breast, 1956; also numerous monographs, articles, chpts. in med. books. Editor: Lymphatics in Cancer. Research in diseases of breast; originator Criteria of Operability, Columbia Clin. Classification; coordinator study of milk factor in mice and humans, 1934-66. Home: Woods Rd., Palisades, N.Y. 10964.*

HAAN, Dieter, German physician; b. Merzig, Germany, Sept. 13, 1928; s. George and Grete (Mennong) H.; student U. Homburg (Germany), 1947-50, U. Freiburg, (Germany), 1950-53; exam. and grad., U. Freiburg, 1953; m. Armgard Wenck, Aug. 4, 1961; 1 son, Christoph. Asst. physician clinic Grand Hôtel Bad Nauheim, 1958-60; head physician med. clinic, Darmstadt, Germany, 1960-65; head cardiological service I, med. clinic U. Hamburg (Germany), 1965—, privat dozent, 1966—. Fellow Sci. Council Internat. Coll. Angiology (past pres.); mem. Internat. Acad. Law and Sci. (past v.p.), Brasilian Coll. Angiology (corr.), Deutsche Gesellschaft für innere Medizin and Deutsche Gesellschaft für Kreislaufforschung. Author: (with A. H. Lemmerz) Praktisch-synoptische EKG-Interpretation, 1959; (with A. H. Lemmerz, W. Heigl) Herzauskultation, 1961; (with M. Ratschow, A. Halpern) Fortschritte der Angiologie, 1963; (with H. M. Hasse) Klinik der Gefässkrankheiten, 1963; Beta-Receptoren-Blockade, 1966; also articles. Editor: Documenta cardioangiologica, Angiologica; asso. editor Angiology, Vascular Diseases, Angéiologie, Angiopathias, Lex et Scientia. Research in diagnostic and therapeutic methods in peripheral, cerebral and mesenterical angiopathies; pathophysiology, diagnosis and treatment of coronary heart diseases, myocardial infarction, cardiac insufficiency and valvular heart diseases, (especially central venous pressure); coagulopathies, anticoagulants, thrombolysis; intensive care units; intoxications; arterial blood pressure. Home: 13 Woldsenweg, Hamburg 20, Germany.*

HAARDT VON HARTENTHURN, Vinzenz Karl, cartographer, ethnographer, geographer; b. Iglau, Moravia, Aug. 11, 1843; s. Karl and Maria (Sagrave) H. von H.; ed. mil. acad. and war sch., Vienna, Austria; m. Helene Juras, 1872; 2 sons, 2 daus. Commd. lt., 1862; capt. in gen. staff; participated in campaigns against Italy, in Dalmatia, 1866-71; named prof. Tech. Mil. Acad., Vienna, 1872; resigned commn., 1873; sci. dir. Geog. Inst. E. Hölzel, Vienna, 1877-96; head 1st dept. Mil. Geog. Inst., Vienna, 1897-1914. Mem. Leopoldina. Author: Geographischer Atlas der Österreichisch-Ungarischen Monarchie, 1883; Die Occupation Bosniens und der Herzegowina, 1878; also articles. Studies in cartography and sch. geography; prepared phys.-geog., polit., r.r., hist., ethnographic, lang. maps; his map of S. Pole 1895 used in preparation for expdn.; studied mil. geography of Dalmatia and Austrian coastlands. Died Vienna, Aug. 1, 1914.

HAARMANN, (Hermann) August, German metallurgist; b. Blankenstein, Germany, Aug. 4, 1840; s. Johann Heinrich and Wilhelmine Anna (Thomas) H.; ed. Berlin Indsl. Acad.; hon. dr. engring., Tech. U. Berlin; m. Elfried Tappe, 1871; 4 sons, including Allan, Hermann, Erich, 2 daus. Worked as miner for 5 years; began as puddler Steinhauser Hütte, foundry, Witten, Germany; named mgr. Heinrichshütte, Hattingen, 1870; became dir. Eisen- und Stahlwerke Osnabrück subsidiary Georg-Marien-Bergwerk- und Hütten-Verein (GMV), 1872, dir., 1911. Mem. Prussian R.R. Bd. Recipient Carl Lueg Meml. medal, 1907. Mem. Assn. German Metallurgists (became mem. bd. 1886), Assn. German Iron Foundries (dep. chmn.). Author: Das Eisenbahngeleise, Geschichtlicher Teil, 1891; Kritischer Teil, 1902. Research on devel. r.r. rails and ties, streetcars; numerous patents, including Haarmann twin rail roadbed for streetcars, ground plates, Haarmann clamp. Died Osnabrück, Aug. 7, 1913.

HAARMANN, Erich, German geologist; b. Osnabrück, Germany, June 14, 1882. Prof. Berlin, Germany. Author: Die Oszillationstheorie: eine Erklärung der Krustenbewegungen von Erde und Mond, 1930. Originator of oscillation theory.

HAARMANN, (Gustav Ludwig Friedrich) Wilhelm, German chemist; b. Holzminden, Germany, May 24, 1847; s. Heinrich Wilhelm and Augusta Luisa (Seulcke) H.; student Mining Acad., Clausthal, U. Berlin; Ph.D., U. Göttingen (Germany), 1872; dr. engring. (hon.), Tech. U. Braunschweig (Germany); m. Louise Stieren, 1876; 2 sons, 1 dau. Worked (with Ferdinand Tiermann, under A. W. von Hofmann) in chemistry, Berlin; founder Haarmann & Reimer (factory for prodn. vanillin, ionon), Holzminden. Recipient Cothenius medal. Pioneer in modern manufacture of scent; discovered conversion process of coniferin to vanillin by oxydation with chromic acid; discovered (with Reimer) method of producing vanillin from eygenol. Died Höxter, Westphalia, Mar. 6, 1931.

HAAS, Arend Maarten, Dutch engr.; b. Hoorn, The Netherlands, Dec. 25, 1898; s. Maarten and G. (Stam) H.; ed. Tech. U. Delft, N.Y. U.; Ph.D. in applied sci. and civil engring.; m. Maria Sieuwerts, 1923; 3 children. Engr. in U. S., 1923-27; civil engr. on bridges and hwys., 1927-29; engr. Hollandse Beton Mij, Indonesia, 1929-39; engr. in Netherlands, 1939-49; dir. Nederlandse Aann. Mij, 1949-53; prof. Tech. U. Delft, 1953—. Mem. Internat. Assn. Shell Structures (pres.), Netherlands Assn. for Concrete Research. Author books, including: (with Bouma) Proceedings of the Symposium on Shell Research; Design of Thin Concrete Shells I and II; (with others) Code et Manuel d'application pour le calcul et l'exécution du Béton Armé; also numerous articles. Home: 1 Danskersstraat, The Hague. Office: 25 Oostplantsoen, Delft, Netherlands.*

HAAS, Arthur, physicist; b. Brunn, Czechoslovakia, Apr. 30, 1884; s. Gustav and Gabriele (Strakosch) H.; student Göttingen (Germany) U.; Phil.D., U. Vienna, 1906; m. Emma Beatrice Huber, 1924. Privat lectr. U. Vienna, 1912, prof., from 1923; asst. prof. Leipzig (Germany) U., 1913; prof. physics U. Notre Dame, from 1936. Lectr. Univ. Coll., London, 1924, 27, 31, also guest lectr. numerous univs. in U. S. Mem. Académie Internationale d'Histoire des Sciences (corr.). Author: Introduction to Theoretical Physics; Atomic Theory; Quantum Mechanics; Quantum Chemistry; The New Physics; World of Atoms; Physics for Everybody; Cosmological Problems of Physics (works translated into English, French, German, Swedish). Explained shift in light rays toward red end of spectrum with theory that light loses energy in speeding to Earth. Died Chgo., Apr. 20, 1941.

HAAS, Cornelis, Dutch chemist; b. Amsterdam, Netherlands, July 13, 1930; s. Johannes Christiaan and Johanna (van Nigtevecht) H.; Doctorate, U. Amsterdam, 1956; m. Johanna Sluyters, July 14, 1956; children—Elisabeth, Maryke, Johannes, Christiaan. Research asso. Brown U., Providence, 1956-57; with Philips Research Labs., Eindhoven, Netherlands, 1957—. Research, publs. on infrared spectra of solids, phys. chemistry of semiconductors, magnetism. Home: 39 Neerlandstraat, Geldrop, Netherlands. Office: Philips Research Lab., Eindhoven, Netherlands.*

HAAS, Erwin, Am. med. scientist; b. Budapest, Hungary, Sept. 11, 1906; s. Jacob and Hedwig (Bass) H.; M.E., Gauss Sch., Berlin, Germany, 1928; postgrad. Kaiser Wilhelm Inst., Berlin; Ph.D., U. Chgo., 1942; m. Elisabeth Tysper, Feb. 6, 1932; children—Wolfgang, Robert. Came to U. S., 1938, naturalized, 1945. Asst. prof. chemistry U. Chgo., 1938-44; sr. mem. Worcester Found. for Exptl. Biology and Medicine, Worcester (Mass.) State Hosp., 1944-45; asst. prof. exptl. pathology Western Res. U., Cleve., 1945-46; asst. dir. Inst. for Med. Research, Cedars of Lebanon Hosp., Los Angeles, 1946-53; asst. dir. Beaumont Meml. Research Labs., Mt. Sinai Hosp., Cleve., 1953—, cons. to staff, 1953—. Recipient Julius E. Goodmand award Mt. Sinai Hosp., 1965. Mem. Am. Soc. Biol. Chemists, Am. Chem. Soc., Am. Heart Assn. (med. adv. bd. Council for High Blood Pressure Research), Sigma Xi. Contbr. numerous articles to profl. jours. Discovered new respiratory enzymes New Yellow Enzyme and Cytochrome Reductase, antibacterial enzymes Antinvasion I and II in blood plasma; research on biochemistry, physiology, immunology of hypertension associated with kidney disease. Home: 3838 Bainbridge Rd., Cleveland Heights, O. 44118. Office: 1800 E. 105th St., Cleve. 44106.*

HAAS, George Arthur, physicist; b. Vienna, Austria, Sept. 7, 1926; s. Arthur Erich and Emma (Huber) H.; came to U. S., 1935, naturalized, 1941; B.S., U. Notre Dame, 1950, Ph.D., 1953; m. Mary Jane Farrell, Dec. 27, 1951; children—Therese, Patricia, Susan, Christopher. Physicist, surface physics sect. U. S. Naval Research Lab., Washington, 1951—, sect. head, 1954—. Cons. Office Naval Research, Bur. Ships, NSF; mem. thermionic panel Interagy. Advanced Power Group, 1964—. Recipient Centennial of Sci. award U. Notre Dame, 1965. Mem. Am. Phys. Soc., Research Soc. Am., Sigma Xi. Research, publs. on electron emission, phys. and chem. properties of cathodes, electronic properties of surfaces, energy conversion. Home: 3512 Fort Hill Dr., Alexandria, Va. 22310. Office: U. S. Naval Research Lab., Washington 20390.*

HAAS, Hippolyte, German geologist; b. Stuttgart, Germany, 1855; prof. U. Kiel (Germany); investigated Swiss Alps of Vaud region, geol. formations of Slesvig-Holstein; drew geol. map of Kiel Canal; made detailed studies of Jurassic brachiopods. Died Munich, 1913.

HAAS, Howard Clyde, Am. scientist; b. N.Y.C., Oct. 31, 1920; s. Constantine and Minnie (Honig) H.; B.S., Coll. City N.Y., 1941; postgrad. Bklyn. Coll., 1941-42; Ph.D., Poly. Inst. Bklyn., 1949; m. Flora Elinor Nathan, Dec. 19, 1948; children—Judith Elinor, Donald Clyde. Research chemist Gen. Foods Corp., Hoboken, N.J., 1941-42, 45-46; research asso. USN sponsored program in copolymerization Poly. Inst. Bklyn., 1946-48; mgr. polymer research dept. Polaroid Corp., Cambridge, Mass., 1949—. Cons. govt. agys. Mem. Am. Chem. Soc. Research, numerous publs. in copolymerization, copolymer structure, free radical reactions, graft copolymer, ionic polymerization, syndiotactic polymers, peroxide decomposition, kinetics, novel monomers, rheology, synthetic gelatin substitutes. Patentee in fields, diffusion transfer photography, duplication processes, protection photog. images, vectography, new polymers and processes, light polarization. Home: 25 Marion Rd. Arlington, Mass. 02174. Office: 730 Main St., Cambridge, Mass. 02174.*

HAAS, Howard James, Am. soil scientist; b. Sharon Springs, Kan., June 10, 1913; s. Adam Darius and Carrie (Gilbert) H.; B.S., Kan. State U., 1936, M.S., 1946; postgrad. Colo. State U.; m. Mildred Evelyn Wagner, Sept. 4, 1937; children—Gordon Allen (dec.), Russell James. With Agr. Research Service, U. S. Dept. Agr., 1936—, soil scientist soil and water conservation research div. No. Great Plains Research Center, Mandan, N.D., 1946—. Recipient Sustained Outstanding Work Performance award U. S. Dept. Agr., 1962. Fellow Soil Conservation Soc. Am. (Outstanding Profl. Conservationist, N.D. state chpt. 1962); mem. Am. Soc. Agronomy, Soil Sci. Soc. Am., Western Soc. Soil Sci., Soil Sci. Soc. Can., N.D. Acad. Sci. Research, publs. on effect various crop rotations on crop yields and soil minerals, more efficient use of ltd. water in Gt. Plains. Home: 1212 N. Parkview Dr., Bismarck, N.D. 58501. Office: No. Gt. Plains Research Center, Mandan, N.D. 58554.*

HAAS, Peter Herbert, physicist; b. Frankfort, Germany, Apr. 20, 1921; s. Hugo H. and Erna (Blumenthal) H.; B.S., Columbia, 1949; postgrad. U. Md.; m. Albina D. Amoroso, June 24, 1945; children—Peter II, Robert E., Jeffery J. Came to U. S., 1937, naturalized, 1942. Supervisory physicist high frequency standards sect. Nat. Bur. Standards, 1949-54; cons. to mine fuse div., later chief nuclear vulnerability br. Harry Diamond Labs., Washington, 1954-65; asst. dept. dir. exptl. research Def. Atomic Support Agy., Arlington, Va., 1965—. Recipient Meritorious Service awards Dept. Commerce, 1954, Dept. Army, 1966. Mem. Am. Phys. Soc., Washington Acad. Scis. Research, publs. on magnetic measurements at high frequencies, radio frequency noise standards, high frequency current standard devel., radiation effects in materials, nuclear weapon effects. Home: 9232 E. Parkhill Dr., Bethesda, Md. 20014. Office: Thomas Bldg., Court House St., Arlington, Va. 20305.*

HAAS, Philipp, Austrian inventor; b. Vienna, June 7, 1791; s. Philipp and Anna (Zechbauer) H.; ed. St. Anna indsl. and drawing sch.; m. Sabine Aumann, 1829; 3 sons, including Robert, Eduard Haas von Teppichen, 1 dau. Founder textile workshop, 1810, also 1st mech. carpet loom factory in Austria, other factories by 1860. Patentee loom regulator which permitted econ. manufacture of smooth organtine, 1822, lace machine mechanisms for manufacture of bobbinet, 1923, 26; inventor chem. substance which increased clarity and body of lawn materials. Died Vöslau, Austria, June 3, 1870.

HAAS, Wilhelm, Swiss cartographer, typecaster; b. Basel, Switzerland, Aug. 23, 1741; s. Johann Wilhelm and Margreth (Christ) H.; m. Anna Münch, 1765; 1 son, Wilhelm. In charge family typecasting firm, Basel, Switzerland; gen. insp. Helvetian arty.; instr. arty. sch., St. Urban. Author: Beschreibung einer neuen Buchdruckerpresse, 1772; Erklärung einer neuerfundenen und gemeinützlicheren Einrichtung der Stücklinien und Zwischenspäne mit der dazugehörigen Tabellen, 1772; Erste typometrische Landkarte Respublica Basileensis, 1776; Carta della Sicilia, 1777. Improved printing press, including use of movable letter and symbols called typometry (after A. G. Preuschen's idea); printed maps in various langs. and large edits. for low prices. Died St. Urban, June 8, 1800.

HAASE, (Reinhold) Ernst, German mineralogist; b. Gerbstedt, Germany, Oct. 21, 1871; s. Reinhold and Christiane Emilie (Baetz) H.; student tchr.'s coll., Eisleben, 1889-92; teaching degree U. Halle (Germany), 1898, hon. dr.rer.nat., 1946; m. Hedwig Kürsten, 1901; 2 sons, 3 daus. Tchr., Belleben, then in elementary and intermediate sch., Halle, 1896, Halle gymnasium, 1918-34, prof. methods geog. edn. Edn. Acad., Halle, 1930-31, pensioned, 1934, became prof. practical edn., especially methods biology and chem. edn., 1947, emeritus, 1952. Mem. Leopoldina. Author: Lötrohrpraktikum, 1908; Die Erdrinde, 1909; Allerlei Küchenweisheit, 1912; Tiere der Vorzeit, 1916; Die Geologie in der Schule, 1918; Grundriss der Geologie, 1919; Physik des Spielzeugs, 1921; Die Erziehung zur Freude an der Natur, 1922; Die Grunlagen der sozialen Gesinnung in der kindlichen Spielgesellschaft, 1924; Die Wetterkunde in der Volksschule, 1925; Die Himmelskunde in der Volksschule, 1930; Volkesgesundung durch Volserziehung. Studied methods of teaching natural sci., detrital, gravel; heavy mineral analysis; petrographic research on Hallic porphyry. Died Halle, Dec. 13, 1959.

HAASEN, Peter, German physicist; b. Gotha, Germany, July 21, 1927; s. Herbert and Ingeborg (Sammer) H.; Dipl.Phys., U. Göttingen (Germany), 1951, Dr.rer.nat., 1953; m. Barbara Kulp, Sept. 12, 1958; children—Christine, Elisabeth. Research asso. U. Chgo., 1954-56; sci. staff Max Planck Inst., Stuttgart, Germany, 1956-58; prof. for metal physics U. Göttingen, 1959—. Mem. Deutsche Gesellschaft für Metallkunde (mem. bd. 1965—), Deutsche Forschungs Gesmeinschaft (mem. senate 1966—), Acad. Sci. Göttingen. Research, publs. on physics of defect crystals, plasticity of metals and semiconductor crystals. Home: 18 Tannenweg, Göttingen 34, Germany.*

HAAST, Sir Julius Johann Frank von, see von Haast, Sir Julius Johann Frank.

HABAS (Ahmad ben Adb Allah al-Marwazi) al-Hasib, see al-Hasib, Habas.

HABEL, Karl, Am. virologist; b. Phila., Sept. 28, 1908; s. Charles and Claire Leslie (Ward) H.; A.B., U. Pa., 1929; M.D., Jefferson Med. Coll., Phila., 1933; m. Ruth Jennette Carter, May 26, 1934; children—Kurt Carter, Gretchen Bryant (Mrs. Kenneth Hill). Intern Phila. Gen. Hosp., 1933-35, resident pediatrics, 1935-36; resident contagious medicine Phila. Hosp. Contagious Diseases, 1936-38; instr. pediatrics U. Pa., 1936-38; officer USPHS, 1938—, med. dir., 1948; chief lab. infectious diseases NIH, 1948-54, chief lab. biology viruses, 1959—. Mem. expert com. rabies WHO. Trustee Am. Type Culture Collection, U. Pa.; adv. bd. Fed. Societies Exptl. Biology. Diplomate Am. Bd. Preventive Medicine. Mem. Am. Acad. Microbiology (bd. govs.), Am. Assn. Immunologists, N.Y. Acad. Scis., Washington Acad. Medicine, Am. Coll. Preventive Medicine. Mem. editorial bd. Virology, Proc. Soc. Exptl. Biology and Medicine. Developed improved rabies vaccine, mumps vaccine; research poliomyelitis. Home: 9426 Locust Hill Rd., Bethesda 14. Office: Lab. Biology Viruses, Nat. Insts. Health, Bethesda 14, Md.

HABER, Fritz, chemist; b. Breslau, Silesia (now Wroclaw, Poland), Dec. 9, 1868; ed. Technische Hochschule, Karlsruhe; studied under Hofmann at U. Berlin, Ph.D. Prof., Technische Hochschule, Karlsruhe, 1906; 1st dir. Kaiser Wilhelm Inst., Berlin, Germany, 1911-33; fled to Switzerland, 1933. Recipient Nobel Prize in chemistry, 1918, Rumford medal, 1932. Worked in electrochemistry; devised glass electrode to measure acidity of a solution, 1909; research on thermodynamic gas reactions; developed Haber process for prodn. of ammonia from atmospheric nitrogen, this process supplied Germany with nitrates for explosives during World War I (now important source of fixed nitrogen for prodn. of fertilizer); worked on gas warfare, directed 1st use of poisonous chlorine gas, 1915. Died Basle, Switzerland, Jan. 29, 1934.

HABERER VON KREMSHOHENSTEIN, Hans, surgeon; b. Vienna, Austria, Mar. 12, 1875; s. Theodor and Elise (Seidl) H. von K.; ed. Vienna, Graz, Austria; M.D., 1900; hon.dr.jur., Cologne, Germany; M.D. (hon.), Athens, Greece; m. Hermine Rziha, 1903; 1 dau. Asst., U. Graz; became intern under Eiselsberg, 1st Surg. Clinic, Vienna, 1902; joined faculty U. Vienna, 1907; became prof., Innsbruck, Austria, 1911, Graz, Austria, 1924, Düsseldorf, 1926; rector Med. Acad. Düsseldorf, 1929-30; became dir. surg. clinic, Cologne, Germany, 1930; rector U. Cologne, 1935-36; emeritus, 1948. Hon. mem. A.C.S., Viennese Soc. Physicians, Soc. Surgeons Vienna; mem. German Soc. Surgery (became chmn. 1940). Author: Indikationsstellung bei operation Eingriffen wegen Erkrankungen des Magens und Duodenums, 1923; Rolle des Pylorus bei den Geswürskrankheiten, 1925; Die Erkrankungen der Leber und Gallenwege, 1947; also articles. Devel. of gastro-intestinal surgery; improved Billroth's termino-lateral modification; studied kidney reduction, neurosurgery; surg. treatment for stomach and duodenal ulcers, cancer, aneurysm, arteriomesenterial intestine closure; used stomach resection for treatment of old hardened ulcers; performed adrenal gland transplantations. Died Düren, Germany, Apr. 29, 1958.

HABERLAND, Ulrich Klaus Walther Werner, German indsl. chemist; b. Sollstedt, Germany, Dec. 6, 1900; s. Hermann and Martha (Könnecke) H.; Ph.D., U. Halle (Germany), 1924; hon.dr.rer.pol., Bonn, Germany, 1960; hon. dr.rer.nat., Technische U., Aachen, Germany, 1960, Cologne, 1960; m. Ilse Könnecke, 1926; 4 sons, 1 dau. Became asst. Chem. Inst., U. Halle, 1923; later with dye factory, Hanover, Germany; joined Uerdingen works IG-Farbenindustrie AG, 1928, became dept. head, 1931, dir. Uerdingen works, 1937, head Leverkusen works, 1943, head 4 works of Lower Rhein group, including Leverkusen and Uerdingen, 1943-45; re-established Farbenfabriken Bayer AG, 1951. Recipient Carl Duisberg Plaque. Chmn. fgn. commerce adv. bd. Fed. Econ. Ministry; mem. German Atomic Commn. Mem. Max-Planck Soc. Research and publs. on devel. iron oxide pigment from iron oxide (patentee 1930-38). Died Anteweiler, Germany, Sept. 10, 1961.

HABERLANDT, Gottlieb, botanist; b. Ungarisch-Altenberg, Hungary, Nov. 28, 1854; prof., Graz, Austria, also Berlin, Germany. Author: Physiologische Pflanzenanatomie, 1884; Das reizleitende Gewebesystem der Sinnpflanze, 1890; Sinnesorgane im Pflanzenreich, 1901. Research on plant sensitivity to external stimuli. Died Berlin, Jan. 30, 1945.

HABERLANDT, Ludwig, Austrian physiologist; b. Graz, Austria, Feb. 1, 1885; s. Gottlieb and Charlotte (Haecker) H.; M.D., Graz, 1909; m. Therese Brem, 1914; 2 sons, 1 dau. Asst., Physiol. Inst., U. Graz; followed father to U. Berlin; later with W. Trendelenburg, Innsbruck, Austria, became lectr., 1913, asso. prof., asst. to E. Brücke, 1919. Author: Das Herzflimmern, seine Entstehung und Beziehung zu den Herznerven, 1914; Über Stoffwechsel und Ermüdbarkeit der peripheren Nerven, 1916; Die Physiologie der Atrioventrikularverbindung des Kaltblutherzens, 1917; Über hormonale Sterilisierung des weiblichen Tierkörpers, 1924; Das Hormon der Herzbewegungen, 1927; Das Herzhormon, 1930; Die hormonale Sterilisierung des weiblichen Organismus, 1931; also numerous articles. Research on impulse formation and conduction in heart, sterilization of female animals using ovarial extracts to inhibit ovulation; pioneered myogenic theory of impulse formation; observed substance with hormone effect on frequency and intropy in heart chamber (noradrenalin); one of 1st to study hormonal contraception. Died Innsbruck, July 22, 1932.

HABERLANDT, Walter Friedrich, Austrian geneticist; b. Innsbruck, Austria, Feb. 21, 1921; s. Ludwig and Therese (Brem) H.; student U. Vienna, 1939; M.D., U. Innsbruck, 1945; m. Jutta Solarek, March 25, 1964; children—Kirsten G., Solveig A. With dept. neurology and psychiatry U. Hosp., Innsbruck, 1945-53; research asso. Inst. for Direct Analysis, N.Y.C., 1954, dept. med. genetics, Psychiat. Inst., N.Y.C., 1954-55; asst. prof. Inst. Human Genetics, U. Münster (Germany), 1956-60; asso. prof. dept. psychiatry U. Hosp., Düsseldorf, Germany, 1961-62; prof. med. genetics Inst. Anthropology and Human Genetics, U. Tübingen (Germany), 1963—. Corr. mem. Problem Commn. Neurogenetics, Am. Soc. Human Genetics, Am. Eugenics Soc., numerous others. Author: Amyotrophische Lateralsklerose: Klinischpathologische und genetisch-demographische Studie, 1964. Research in psychopathology, psychosomatics, psychotherapy, mental hygiene; twin, family, population studies in neuropsychiatry; analyses in cytogenetics and other fields of human genetics. Address: 137 Gartenstrasse, 74 Tübingen, West Germany.*

HABERLING, Wilhelm Gustav Moritz, German med. historian; b. Liegnitz, Germany, Feb. 14, 1871; s. Conrad and Emma (Hüttner) H.; student Breslau (now Wroclaw, Poland), Königsberg, Marburg, Germany; M.D., Königsberg, 1895; m. Elseluise Meyer-Becherer, 1915. Mil. physician; bacteriological and surg. tng., Berlin, Rostock, Germany; under Garrè, U. Surg. Clinic, Rostock; moved to Düsseldorf, Germany, 1900; joined faculty Med. Acad., Düsseldorf, 1914, became asso. prof., 1923, dir. of own inst., 1931; sr. govt. med. adviser, Koblenz, after World War I. Recipient Sudhoff medal, 1935. Author: Heldenliedern des Mittelalters, 1917; Die Entwicklung der Kriegsbeschädigtenfürsorge von den ältesten Zeiten bis zur Gegenwart, 1918; German Medicine, 1934; Die Geschichte der Düsseldorfer Ärtze und Krankenhäuser bis zum Jahre 1907, 1936. Studied history of mil. medicine, med. biography and bibliography, history of surgery and hygiene, local med. history. Died Düsseldorf, Aug. 22, 1940.

HABERMAN, Sol, Am. immunologist, bacteriologist; b. Chgo., Jan. 15, 1914; s. Nathan and Eva (Yankovitch) H.; Asso. Sci., N. Tex. Agrl. Coll., 1934; B.A., U. Tex., 1936, M.A., 1937; Ph.D., Ohio State U., 1941; m. Carleta Jeanne Rambo, May 14, 1948; 1 son, Hardy Kemp. Faculty, Baylor U., Dallas, 1941—, prof., chmn. dept. microbiology, dir. grad. studies Coll. Dentistry, 1963—, dir. microbiology Baylor Hosp., 1941—. Mem. adv. bd. Nat. Blood Research Found., 1952-58, med. adv. bd. Leukemia Soc., Inc., 1954-58. Recipient Merit award Tex. Soc. Pathologists, 1946. Recipient Outstanding Alumnus award Arlington State Coll. Ex-Student Assn., 1966. Diplomate Am. Bd. Microbiologists (editor newsletter Diplomate Forum 1966—). Fellow A.A.A.S., Am. Coll. Dentists (hon.); mem. Soc. Am. Bacteriologists, Am. Assn. Immunologists, Am. Soc. U. Profs., Internat. Soc. Hematology (co-editor proc. internat. congresses 1950, 52), Am. Soc. Human Genetics, Am. Acad. Microbiology, Orthodontic Alumni Assn. (hon.), Sigma Xi, Omicron Kappa Upsilon, Beta Beta Beta, Sigma Pi Sigma, others. Author: Laboratory Manual for Dental Bacteriology, 1956. Research publs. on Rh antigens and antibodies, dental bacteria, isoimmunity, antibodies to cancer, effects of plastics and their components on antibodies and other serum proteins. Home: 7223 Edgerton Dr., Dallas 75231. Office: 3500 Gaston Av., Dallas 75246.*

HABERMANN, Ernst Richard, German pharmacologist; b. Gössenheim, Bavaria, Germany, July 31, 1926; s. Alfred and Auguste (Ammersbach) H.; M.D., U. Würzburg, 1951; m. Christa Wagner, May 19, 1956; children—Peter, Ulrich, Mareile. Asst., Inst. for Pharmacology and Toxicology, U. Wurzburg, 1952-61; prof. pharmacology U. Wurzburg, 1962—; dir. Pharm. Inst., U. Giessen (Germany), 1966—. Research, numerous publs. on pharmacology and biochemistry of proteins and peptides, substances with novel structure and mode of action were identified in bee, snake and bacterial venoms, prodn., structures, actions and fate of pharmacologically active agts. in blood plasma (mems. of kinin system) have been determined. Home: 4 Tannenweg, Leihgestern, Hessen. Office: 4 R.-Büchheim Strasse, Giessen, Hessen, Germany.*

HABERMEHL, Erasmus (or Habermel), instrument maker; b. 1538; m. Susanna Solis, 1593; became ct. instrument maker for Kaiser Rudolph II, Prague, Czechoslovakia, 1594; made instruments for Tycho Brahe (then ct. astronomer), also for Franciscus de Padovanis von Forli, dukes of Rosenberg in Wittingau, others; produced instruments characterized by beautiful form, sparse decoration, clean workmanship) represented in instrument collections at Vienna, Prague, Hamburg, Dresden, Munich, Nuremberg. Died Prague, Nov. 15, 1606.

HABETLER, George Joseph, Am. mathematician; b. McKees Rocks, Pa., Oct. 31, 1928; s. Stephen and Mary (Tisenay) H.; B.A. summa cum laude, Duquesne U., 1949; D.Sc., Carnegie Inst. Tech., 1952; m. Clementine S. Williams, July 13, 1953; children—Christy S., Linda R., Mark G., Research scientist Knolls Atomic Power Lab., Gen. Electric Co., Schenectady, 1952-64; prof. math. Rensselear Poly. Inst., Troy, N.Y. 1964—. Mem. Am. Math. Soc., Am. Phys. Soc., Soc. Indsl. and Applied Math. Research, publs. on foundations and numerical analysis of neutron diffusion problems in reactor physics. Home: Lee Av., Rexford, N.Y. 12148. Office: Math. Dept., Rensselaer Poly. Inst., Troy, N.Y. 12180.*

HACATAEOS OF MILETOS, Greek geographer; b. Miletos, circa 550 B.C.; s. Hagesander. Author: Circuit of the Earth (extant in fragments); a history which is lost but may have influenced Herodotos' work. Father of geography; continued Anaximander's map of world; divided land area into northern half (Europe), southern half (Asia), drawing both as semicircles; traveled widely through Persian empire. Died circa 476 B.C.

HACHETTE, Jean Nicolas Pierre, French mathematician; b. Mézières, France, May 6, 1769; ed. U. Rheims (France); appt. asst. prof. Poly. Sch., 1794, prof. descriptive geometry, 1797-1816; named prof. Paris Faculty Scis., 1810. Mem. Soc. Agr., French Acad. Scis. Author: Supplements aux leçons de géométrie descriptive données par Monge 1795, 1811; Éléments de géométrie solide, 1817; Traité élementaire des machines, 1828; Traité de géométrie descriptive, 1828. Applied algebra to 3 dimensional geometry; gave 1st complete discussion of 2d degree surfaces; applied geometry to constrn. of farm machinery; helped disseminate descriptive geometry; studied aspects of mech. physics. Died, Paris, Jan. 16, 1834.

HACHEY, Henry Bennedict, Canadian oceanographer; b. West Bathurst, N.B., Can., June 7, 1901; s. Joseph Bennet and Sarah (Kelly) H.; B.Sc., St. Francis Xavier Coll., 1922; M.Sc., McGill U., 1925; LL.D., St. Thomas U., 1950; m. Katherine Avis Cox, Oct. 18, 1930; children—Philip Osmund, Mary Isabel, John Henry, Jane Elizabeth. Lectr., St. Francis Xavier Coll., 1922-23; tutor McGill U., 1925-26; prof. physics U. N.B., 1926-28; oceanographer Fisheries Research Bd. Can., St. Andrews, N.B., 1928-46, chief oceanographer, 1946-63. Sec., Canadian Com. on Oceanography, 1946-66; Canadian del. to numerous internat. meetings. Research, numerous publs. on phys. and chem. nature of ocean waters, especially N. Atlantic, E. Canadian Arctic. Home: 132 Edward St., St. Andrews, N.B., Can.*

HACHIHAMA, Yoshikazu, Japanese chemist; b. Tokyo, 1900; grad. Tohoku U., Japan, 1924; E.D., 1937. Asst. prof. Osaka (Japan) Tech. Coll. from 1930; asst. prof. Osaka U., prof., 1937—. Recipient Japan Chem. Soc. prize, 1950. Author: Chemistry of Lignin. Specialist applied chemistry; researched pulp, fiber, plastics.

HACHISUKA, Yoetsu, Japanese bacteriologist; b. Hiroshima, Japan, July 7, 1920; s. Einosuke and Kazu (Matushima) H.; M.D., Nagoya U., 1944, Ph.D., 1950; m. Hisako Haruki, Nov. 9, 1944; children—Yasuto, Misako, Shinzi. Lectr., Mie Med. Coll., 1951; research fellow U. Tex. M. D. Anderson Hosp. and Tumor Inst., 1955-56; asso. prof. Nagoya City U., 1958—. Author: Spores, 1962; Food Science, 1967; also articles. Research on spore germination; introduced use of changes in optical density to quantitatively measure kinetics of germination; discovery of chemicals which stimulate spore germination, including L-asparagine, DL-isoleucine-DL-serine, caramel. Home: 3-24 Nishizaka-cho, Nagoya, Japan.*

HACK, Marvin H., Am. histochemist; b. Rochester, Feb. 25, 1917; s. James W. and Dorothy (Bennett) H.; B.S., Roosevelt Coll., 1949; Ph.D., U. Chgo., 1951. Mem. faculty Tulane U., New Orleans, 1952—, prof. histochemistry, 1964—; biochemist, NIH, 1955. Mem. Am. Assn. Anatomists, Soc. Exptl. Biology and Medicine, Am. Soc. Biochemists, Sigma Xi. Research and publs. on biochem., histochem. methodology and physiology of lipids using comparative techniques. Home: 1430 Tulane Av., New Orleans 70112.*

HACKEL, Donald Benjamin, Am. pathologist; b. Boston, July 7, 1921; s. Paul Louis and Sonya (Yesner) H.; A.B., Harvard, 1943; M.D., 1946; m. Irene Goos, June 22, 1947; children—Constance Ann, Andrea Joyce, Richard Elliott. Faculty, Western Res. U., Cleve., 1948-60, asso. prof. pathology, 1952-60; prof. pathology Duke, Durham, N.C., 1960—. Recipient USPHS Career Research award, 1962—. Mem. Am. Assn. Pathologists and Bacteriologists, Am. Physiol. Soc., Soc. for Exptl. Pathology. (Park-Davis award 1961), Internat. Acad. Pathology. Research, numerous publs. on cardiovascular effects hemorrhagic shock, exptl. renal disease, exptl. diabetes mellitus. Home: 4018 Bristol Rd., Durham, N.C. 27707.*

HACKEL, Eduard, botanist; b. Haida, Bohemia, Mar. 17, 1850; s. Josef and Margaretha (Werner) H.; student Poly. Sch., Vienna (later Tech. U.), 1865-69; m. Barbara Kürner, 1877; 1 son, 1 dau. Became asst. sci. secondary sch., St. Pölten, 1869, tchr. natural history, 1871-1900; moved to Graz, Austria, 1904, Attersee, 1907; made trip to Portugal, Spain, 1876. Author: Catalogue raisonné des graminées de Portugal, 1880; Monographia festucarum europaearum, 1882; also numerous articles. Research on grasses, especially family Festuca, gramineae taxonomy; differentiated between intravaginal and extravaginal shoots; used histology of leaf layers to determine blossomless grasses; studies on lodiculae; 1st to discover function of lodiculae in flowering grasses. Died Attersee, Feb. 17, 1926.

HACKER, Viktor von, see von Hacker, Viktor.

HACKERMAN, Norman, Am. chemist; b. Balt., Mar. 2, 1912; s. Jacob and Anna (Raffel) H.; A.B., Johns Hopkins, 1932, Ph.D., 1935; m. Gene Allison Coulbourn, Aug. 25, 1940; children—Patricia Gale (Mrs. Raymond Rosenthal), Stephen Miles, Sally Griffith, Katherine Elizabeth. Asst. prof. phys. chemistry Loyola Coll., Balt., Md., 1935-39, Va. Poly. Inst., 1941-43; research chemist 1936-40, Kellex Corp., N.Y.C., 1944, Colloid Corp., asst. chemist USCG, 1940-41; faculty U. Tex., 1945—, prof. chemistry, 1950—, chmn. dept., dir. Corrosion Research Lab., 1952-61, v.p., provost, 1960-61, vice chancellor for acad. affairs, 1963-67, president U. Tex. at Austin, since 1967—. Joseph J. Mattiello lectr. Fedn. for Socs. Paint Tech., 1964; chem. Gordon Corrosion Research Conf., 1950, Intersoc. Corrosion Com., 1956-58, Gordon Research Conf. on Chemistry, 1959. Fellow N.Y. Acad. Sci.; mem. A.A.A.S., Am. Chem. Soc. (S.W. Regional award 1965), Electrochem. Soc. (tech. editor Jour., v.p. 1954-57, pres. 1957-58, Palladium medal 1965), Nat. Assn. Corrosion Engrs. (bd. dirs. 1952-55, Whitney award 1956), Faraday Soc., Sigma Xi, Alpha Chi Sigma, Phi Kappa Phi. Research on metal corrosion, surface chemistry of metals and oxides, passivity, elec. double layer at solid metal-solution interfaces. Home: 2101 Meadowbrook, Austin, Tex. 78703.*

HACKETT, Felix Edward, Irish physicist; b. Omagh, Ireland, Aug. 16, 1882; s. Daniel and Jeanne (Walsh) H.; ed. Univ. Coll., Dublin, Johns Hopkins; Ph.D. in phys. scis.; m. Mary Murnaghan, 1910. Reader of physics Royal Coll. Scis., Dublin, 1909-21; prof. Univ. Coll., Dublin, 1921-26, dean, 1922-24, prof. physics and electricity, 1926-52. Pres. administrv. council Sch. Theoretic Physics, Inst. Progress of Sci., Dublin. Mem. Royal Irish Acad., Royal Soc. Dublin (past pres.), Inst. Physics (London), Inst. Indsl. Research. Address: 20 Zion Rd., Rathgar, Dublin, Ireland.

HACKSPILL, Louis Jean Henri, French chemist; b. Paris, May 3, 1880; s. Louis-Francois and Marguerite (Franck) H.; D.Sc., Chemistry Inst., Faculty Scis., Paris; m. Marie-Thérèse Haizet, Dec. 17, 1927; children—Christian, Denys. Asst. to Henri Moissan, 1903-07; asst. Faculty Scis., Paris, 1907-12; instr. Faculty Scis., Nancy, France, 1913-14; prof. Faculty Scis., Strasbourg, France, 1919-32; prof. Faculty Scis., Paris, 1932-51; dir. Nat. Superior Sch. Chemistry, 1939-50, hon. prof., 1951——. Decorated Légion d'honneur, Palmes Academique; recipient Houzeau Prize, 1962, La Caze Prize, 1932. Mem. French Acad. Scis. (became pres. 1961, Cahours Prize 1911), French Chem. Soc. (hon. pres.), Soc. Indsl. Chemistry (hon.). Research in alkaline-earth chemistry, alkaline alloys, phosphides and hydrides; 1st preparation of pure cesium, 1905, originated new methods for preparation of liquid boron. Author: les Métaux alcalins, 1911; l'Azote, 1922; (with Rémy) la Petite industrie chimique; (with Besson, Harold) Chimie minérale, 2 vols. Died Oct. 7, 1963.

HACQUET, Belsazar (Balthasar), natural scientist; b. Le Conquet, France, 1739; student medicine, Paris, medicine, law, Vienna, Austria; m. 1799. Pressed into French fleet, 1755; surgeon in French, English, Prussian, and Austrian service; traveled to Constantinople, 1763; joined U. Vienna, 1764; assigned as surgeon mercury mining works Idria, Austria-Hungary, 1766; became prof. anatomy, physiology, Ljubljana, Austria-Hungary, 1733; named prof. natural history, Lemberg, Hungary, 1787; named prof., Cracow, Poland, 1805, dean med. faculty, 1907; ret., 1810 and moved to Vienna. Author: Oryctographia Carniolica . . . Physikalische Erdbeschreibung des Herzogtums Krain, Istrien und zum Teil der benachbarten Länder, 4 vols., 1778-80; Nachrichten von Versteinerungen von Schalthieren, die sich in ausgebrannten feuerspeienden Bergen finden, 1780; Plantae alpinae Carniolicae, 1782; Reise durch die Nordischen Alpen, physikalisches under anderen Inhalts, unternommen in den Jahren, 1784-86, 2 parts, 1791; Physikalische und technische Beschreibung der Feuersteine, 1792; Abbildung und Beschreibung der südwestlichen und östlichen Wenden, Illyrier und Slaven . . . , 1802-05; Blicke über das menschliche Wissen in der Naturkunde, 1813. Made maps and drawings on travel observations, on geology, paleontology, petrography, mining, balneology, botany, ethnology in Austria, Balkans, Turkey, Bohemia, Galicia, Poland, Germany, Scandinavia; important in opening up E. Alps, and Carparhians for geology and tourisms; studied ethnology and geography of Carniola and Austrian empire. Died Vienna, Jan. 10, 1815.

HADAMARD, Jacques Salomon, mathematician; b. Versailles, France, Dec. 8, 1865; s. Amédée and Claire (Picard) H. doctorate, Paris, France, 1892; hon. degrees from Göttingen, Yale, Oslo, Brussels. m. Louise Anna Trénel, June 30, 1892; 5 children. Prof., U. Bordeaux, from 1894, Sorbonne, until 1897, Coll. France, 1897-1935, prof. math. analysis Polytech. Sch. Paris, 1912-35, prof. Ecole Centrale des Arts et Manufactures, 1920-35; went to Princeton (N.J.), U. S., during World War II. Recipient Grand Cross, Legion of Honor. Fellow Royal Soc., 1932; mem. French Acad. Scis., London Math. Soc. (hon.). Author: La série de Taylor et son prolongement analytique, 1901; Leçons sur la propagation des ondes, 1903; Notice sur les travaux scientifiques de M. Jacques Hadamard, 1912; Lectures on Cauchy's Problem, 1923; An Essay on the Psychology of Invention in the Mathematical Field, 1945; Leçons de géométrie élémentaire; Cours d'analyse de l'école polytechnique; also editor of Annales scientifiques de l'ecole normale supérieure. Known for his work on infinitesimal calculus; rediscovered Cauchy's root test, 1892; proved the prime number theorem, 1896; developed (with Kelvin) theorem on minimum of absolute value of determinant; introduced the term functional; his work led to basis of theory of functional analysis. Died October 17, 1963.

HADDON, Alfred Cort, English ethnologist, anthropologist; b. London, May 24, 1855; s. John and Caroline (Waterman) H.; ed. Christ's Coll., Cambridge, 1875-79, M.A., Sc.D.; hon. Sc.D., Manchester, Perth univs.; m. Fanny Elizabeth Rose, 1881; 1 son, 2 daus. Prof. zoology Royal Coll. Sci., Dublin, 1880-1901; lectr. ethnology Cambridge U., 1900-09, reader, 1909-26; lectr. U. London, 1904-09; led Cambridge anthrop. expdn., studied zoology and ethnology of Torres Straits, New Guinea, Sarawak. Pres. (sect. H) Brit. Assn., 1902-05, Royal Anthrop. Inst. Gt. Britain, 1902-02. Fellow Royal Soc., 1899. Author: Introduction to Embryology, 1887; Evolution in Art: Study of Man, 1898; Head Hunters, Black, White and Brown, 1901; Magic and Fetishism, 1906; The Races of Man and Their Distribution, 1909; History of Anthropology; The Wanderings of Peoples, 1912; Migrations of Cultures in British New Guinea, 1920; (with A. H. Keane) Man, Past and Present, 1920; Practical Value of Ethnology, 1921; (Laura Start) Iban or Sea Dayak Fabrics and their Patterns, 1936; (with J. Hornell) Canoes of Oceania, 3 vols., 1936-38; Smoking and Tobacco Pipes in New Guinea, 1946. Special interest in art and tech.; important work in South Seas, where he developed field study methods in ground-breaking study among primitive cultures. Died Cambridge, Apr. 20, 1940.

HADDOW, Alexander, Brit. physician; b. Loch Linnhe, Scotland, Jan. 18, 1907; s. William and Margaret Haddow; M.D., D.Sc., Ph.D., U. Edinburgh; D. honoris causa, U. Perouse; m. Lucia Lindsay Crosbie Black, 1932; 1 son, William. Physician, Royal Infirmary, Edinburgh; reader bacteriology U. Edinburgh; dir. Chester Beatty Inst. Research, Inst. Research on Cancer, London; prof. exptl. pathology U. London. Fellow Royal Soc., 1958, Royal Soc. Edinburgh; mem. Internat. Union against Cancer (pres.), Soc. Vis. Scholars (treas.), Med. Assn. for Protection against War (pres.); hon. mem. Royal Acad. Medicine Belgium, Am. Acad. Arts and Scis., USSR Acad. Med. Scis. Contbr. numerous articles to sci. and med. jours. Home: 6 Neville Terrace, London S.W.7. Office: Chester Beatty Research Institute, Fulham Rd., London S.W.3, Eng.

HADDY, Francis John, Am. physician; b. Walter, Minn., Sept. 6, 1922; s. Thomas J. and Frances (Shaheen) H.; student Luther Coll., 1940-42; B.S., U. Minn., 1943, B.M., 1946, M.D., 1947, M.S. in Physiology, 1949, Ph.D. in Physiology (Mayo Found. fellow, Am. Heart Assn. fellow), 1953; m. Theresa Eileen Brey, Sept. 21, 1946; children—Richard, Carol, Alice. Asst. prof. Northwestern U., 1953-55, 57-61; prof., chmn. dept. physiology, asso. prof. medicine U. Okla. Med. Center, Oklahoma City, 1961-66; staff VA Research Hosp., Chgo., 1953-55, 57-61, clin. investigator, dir. div. research, 1959-61; chmn., prof. physiology Mich. State U., East Lansing, 1966——. Diplomate Am. Bd. Internal Medicine; mem. Am. Physiol. Soc., Am. Soc. Clin. Investigation. Editorial bd. Am. Jour. Physiology, 1964——, cardiovascular study sect NIH, 1964——. Research, numerous publs. on mechanism lung edema, role veins in formation edema, effect of ions on blood vessels, mechanism local regulation blood flow, characterization kidney lymph. Home: 900 Audubon Rd., East Lansing, Mich. 48823.*

HADFIELD, Sir Robert Abbott, English metallurgist; b. Sheffield, Eng., Nov. 28, 1858; s. Robert and Marianne (Abbott) H.; ed. Collegiate Sch., Sheffield; m. Frances Belt Wickersham, 1894. Entered father's firm, on father's death became chmn. and mng. dir. Fellow Royal Soc., 1909; mem. Faraday Soc. (pres. 1914-20), Iron and Steel Inst. (pres. 1905-07), French Acad. Scis. Author: Metallurgy and Its Influence on Modern Progress, 1925; Faraday and His Metallurgical Researches, 1931. Invented manganese steel (harder than ordinary steel), 1882; examined elec. properties of silicon steels, studied magnetic permeability, 1900, influence of low temperatures, 1921, metallography with X-rays, 1920, ferrous alloys; expts. on corrosion and deformation of steel at high velocities, also worked on armor piercing shells. Died Surrey, Eng., Sept. 30, 1940.

HADJIOLOFF, Assen Ivanov, Bulgarian histologist, embryologist; b. Shirokovo, Bulgaria, Jan. 19, 1903; s. Ivan Radkov and Maria (Djanamova) H.; M.D., Faculty Medicine, Sofia, Bulgaria, 1920-26; Sc.D., Faculty Scis., Lyons, France, 1929; m. Helena Atilova Silaghieva, Dec. 11, 1927; children—Assen, Kremena. Staff, Med. Faculty, Sofia, 1928——, prof., dir. inst. and chair histology and embryology, 1933——; dir. Research Inst. Morphology, Bulgarian Acad. Scis., 1953——, editor bull., 1953——, gen. sci. sec., 1953-56. Mem. nat. commn. UNESCO, 1956——, pres. sec. natural scis., 1958-66. Recipient State (Dimitrov's) Nat. Prize for Sci., 1951, Cyrill and Methodius Order, 1958, People's Republic Bulgaria Order I degree, 1963, Honorable Sci. Worker, 1963. Mem. Hungarian Acad. Scis. (hon.), Yugoslavian Acads. Scis. and Arts (corr.), Sci. Workers (past pres.), Académie internationale histoire des sciences Paris (corr.). Author: Histophysiologie du tissu adipeux et le métabolisme morphologique des graisses, Actinoluminescences des tissues, 1931; Haematology, Biology of blood tissue, 1950; Histology, 1945, 46, 54, 58, 68; Embryology, 1950, 56; also numerous articles. Research on histochemistry and morphological metabolism of lipids, fluorescent microscopy and fluorescent histochemistry, tissue cultures, electron microscopy and histochemistry of blood, sexual and nervous tissues and organs, philos. histology, history and orgn. of scis., especially natural and med. scis. Home: Komplex Istok, Block 2A, Sofia-13. Office: Inst. Morphology, Acad. Scis., Sofia-13, Bulgaria.*

HADLEY, George, English meteorologist; b. London, Feb. 12, 1685; s. George and Katherine (FitzJames) H.; ed. Pembroke Coll.; became mem. Lincoln's Inn, 1701; became barrister, 1709; in charge meteorol. observations presented to Royal Soc., 7 years. Fellow Royal Soc., 1754. Author: Account and Abstract of the Meteorological Diaries communicated for 1729 and 1730. Formulated modern theory of trade winds. Died June 28, 1768.

HADLEY, John, English mathematician, sci. mechanist; b. Eng., Apr. 16, 1682; s. George and Katherine (FitzJames) H.; m. Elizabeth Hodges, 1734; 1 son, John. Obtained his 1st success by improvement he effected in reflecting telescope, 1719-20; invented reflecting quadrant, 1730. Fellow Royal Soc., 1716 (v.p. 1728). Author (paper to Royal Soc.) Description of a new Instrument for Taking Angles by John Hadley, Vice Pres. R.S., 1731. Invented 1st practical reflecting telescope; his reflecting quadrant (Hadley's quadrant) was predecessor of sextant. Died Feb. 14, 1744.

HADORN, Ernst, Swiss zoologist; b. Forst, Switzerland, May 31, 1902; s. Christian and Elisabeth (Lehner) H.; student Tchrs. Coll., Muristalden, Bern, Switzerland, 1918-22; Ph.D., U. Bern, 1931; student U. Munich (Germany), 1927-28; Dr.med. (hon.) U. Basle (Switzerland), 1961; Dr.sci. (hon.) U. Utrecht (Netherlands), 1963; m. Marie Daepp, Oct. 21, 1930; children—Hans Beat, Irene Birnstiel-Hadorn, Marianne. Docent, U. Bern, 1935-39; prof. zoology and comparative anatomy U. Zürich (Switzerland), 1939——, rector, 1962-64. Hon. mem. Soc. Biol. exper. Italia, Báyerische Alcademie, Am. Acad. Recipient Pris Benoist, 1955, Pris Albert Brachet, 1965. Author: Developmental genetics and Lethal Factors, 1961; also numerous articles. Research on function of cell nucleus and cytoplasmic in amphibian hybrids, developmental genetics of lethal factors, pteridines in insects, culture of Drosophila cells in vivo, transdeterminations in cell cultures; discovered ring gland (center of hormone prodn. in Diptera). Home: 15 Häldeliweg 15, Zürich, Switzerland.*

HADQUAERT, Armand, Belgian mineralogist; b. Mont-Saint-Amand, Mar. 30, 1906; s. Henri and Jeanne (Monnet) H.; Ph.D. in Sci., U. Ghent; m. Renee Schepers; children—Jeanne M., Nicole M. R. Head of works, instr., full prof., former dean Faculty Scis.; former pres. Sch. Econ. Scis., U. Ghent. Vice pres. Nat. Comm. for UNESCO, also Belgian del. to gen. conf. Mem. Internat. Assn. U. Profs. (sec. gen.). Research and publs. on geology and mineralogy. Home: 43 Vaderlandstrasse. Office: Rozier, Ghent, Belgium.

HAECKEL, Ernst Heinrich, German biologist, natural philosopher; b. Potsdam, Prussia, Feb. 16, 1834; ed. under Johannes Müller, R. Virchow and R. A. Kölliker at Würtzburg, Vienna, Berlin; M.D., M.Ch., U. Berlin, 1857; m. 1862; 3 children. Practiced medicine, Berlin, 1857-61; privatdozent U. Jena, 1861, prof. extraordinary comparative anatomy, dir. Zool. Inst., 1862, chair zoology established for him, 1865-1909; made sci. expdns. to Canary Islands, 1866-67, Red Sea, 1873, Ceylon, 1881-82, Java, 1900-01, also others. Recipient Turin Bressa prize, 1901. Author: Anthropogenie oder Entwicklungsgeschichte des Menschen, 1874; Monismus als Band zwischen Religion und Wissenschaft, 1892; (in translation) Natural History of Creation, 1868; Radiolaria, 1862, 87; Siphonophora, 1869, 88; Monera, 1870; Calcareous Sponges, 1872; Deep-Sea Medusae, 1881; General Morphology, 1866; Studies on the Gastraea Theory, 1873-84; The Riddle of the Universe, 1899; The Last Link, 1898; Last Words on Evolution, 1906. Made important contbns. to biol. knowledge on radiolaria, sponges, medusae, jelly fishes and siphonophora; 1st to draw up geneal. tree relating various orders of animals with regard both to one another and to their common origins; separated animal kingdom into unicellular and multicellular organisms; formulated principle ontogeny recapitulates phylogeny (devel. of individual reflects that of species); 1st German advocate of Darwinism, he expounded on possibility of sexual selection before Darwin did; held that life evolved from non-life and that psychology was only br. of physiology; held that human intelligence developed by evolution from simplest forms of psychical activity which can be found in protozoa; proposed theory of 2-layered gastrula from which all multicellular organisms originate, 1847; exponent of materialistic monistic philosophy; asserted essential unity of organic and inorganic nature; denied existence of personal God but believed God is identical with eternal all-inspiring energy, that He is one with matter and that His will is reflected in all phys. and psychical phenomena. Died Jena, Germany, Aug. 8, 1919.

HAECKER, (Ferdinand Carl) Valentin, zoologist; b. Altenburg, Hungary, Sept. 15, 1864; s. Ludwig and Julie (Schübler) H.; student natural scis., Strasbourg, France; Ph.D., Tübingen, Germany, 1889; M.D. (hon.), Halle, Germany; m. Lucie Kühn, 1903; 1 son, 1 dau. became asst. to A. Weismann, U. Freiburg (Germany), 1890, joined faculty, 1892, became asso. prof. zoology, 1895; became prof. zoology Tech. U., Stuttgart, Germany, 1900, U. Halle-Wittenberg, Germany, 1909; also tchr. Agrl. and Vet. Insts. Mem. Leopoldina (sec.). Author: Praxis und Theorie der Zellen- und Befruchtungslehre, 1899; Der Gesang der Vögel seine anatomischen und biologischen Grundlagen, 1900; Bastardierung und Geschlechtszellenbildung, 1904; Allgemeine Vererbungslehre, 1911; Über Gedächtnis, Vererbung und Pluripotenz, 1914; Über Aufgaben der Phänogenetik, 1923; Umwelt und Erbgut, 1926; (with T. Ziehen) Über die Erblichkeit der musikalischen Begabung, 1922; Goethes morphologischen Arbeiten und die neuerer Forshung, 1927. Research on maturation and cell division processes in lower forms of crabs, chromosome reprodn. in germ cells, phaenogenetics (origin of outward characteristics of an organism in earliest states of devel.), radiolariae, plankton, bird migration and song, animal psychology; introduced term, pluripotency, for capacity of embryo cells or organs to develop differently from their types. Died Halle, Dec. 19, 1927.

HAEFELY, Emil, Swiss inventor; b. Mümliswil, Switzerland, May 22, 1866; s. Leonz and Elisabeth (Mischler) H.; dr. engring. (hon.), Tech. U. Darmstadt (Germany), 1922; m. Philomena Saner, 1908; 3 sons, 3 daus.; m. 2d, Mathilde Meier, 1887; 2 sons. Apprentice, comb factory, Mümliswil; with constrn. bur. von Rollsche Eisenwerke Klub; joined firm Alioth, Münchenstein, 1896, Brown, Boveri & Cie., 1896; founder firm for prodn. of his insulation inventions, Neuewelt, 1904, moved to Basel, Switzerland, 1914, opened br., St. Louis, France, 1921. Inventor micafolium (mica paper insulation for windings with complete impregnation), layered resin paper, both basic in elec. insulation. Died Basel, Feb. 28, 1939.

HAEFF, Andrew Vasily, electronics engr.; b. Moscow, Russia, Dec. 30, 1904; s. Vasily A. and Klavdia A. Haeff; came to U. S., 1928, naturalized 1936; E.E. and M.E., Poly. Inst., 1928; M.S. in Elec. Engring., Cal. Inst. Tech., 1929, Ph.D., 1932; m. Sonya I. Bibikoff, Oct. 22, 1936; 1 son, Andre. Spl. research fellow Cal. Inst. Tech., Pasadena, 1932-33; research engr. RCA, Harrison, N.J., 1934-41; cons. physicist U. S. Naval Research Lab., Washington, 1941-42; prin. radio engr., cons. inelectronics, head vacuum tube research br., 1941-45; mem. adv. council, head electron tube lab. Hughes Aircraft Co., Culver City, Cal., 1950-54, dir. research labs., v.p., 1954-61, cons., 1961-64; sr. scientist TRW Space Tech. Labs., Redondo Beach, Cal., 1965, director of the phys. electronics laboratory, 1966—. With vacuum tube devel. com. OSRD, 1942—; research and devel. bd. U. S. Dept. Def. 1942—. Fellow I.E.E.E. (Diamond award 1950), Am. Phys. Soc. Patentee in Field. Research in electron tubes—UHF, microwave and storage tubes; radar systems, microwave signal generators, space charge effects in electron beams, interaction of electrons with fields; invented traveling wave tube. Home 11134 Bellagio Rd., Los Angeles 90049. TRW Systems, Redondo Beach, Cal.*

HAEHN, Hugo, German chemist; b. Herzberg/Elster, Saxony, Sept. 26, 1880; s. Robert and Berta (Langhammer) H.; student, Leipzig, Germany, Basel, Switzerland, Heidelberg, Germany; Ph.D., Munich, 1904; m. Else Flader, 1920; 1 son, 1 dau. With chem.-pharm. lab. U. Königsberg (now Kaliningrad, USSR), 1904-08; with Agrl. Inst., Berlin; asst. to E. Buchner, U. Breslau (now Wroclaw, Poland); with chem. inst. Berlin Inst. Vet. Medicine, during World War I; with Inst. for Indsl. Fermentation, from 1918. Author: Biochemie der Gärungen unter besonderer Berücksichtigung der Hefe, 1952. Biochem. research on indsl. fermentation, coloration of chopped potatoes, citric acid fermentation, lactic acid fermentation, yeast, silage agrl. products, diatase formation in mold fungi; transformation of sugar into fat by microorganisms; developed theory on microbic fat synthesis from sugar. Died Berlin, Dec. 6, 1957.

HAEMPEL, Oskar, Austrian hydrobiologist; b. Malec, W. Galicia, May 12, 1882; s. Karl and Wilhelmine (Kögler) H.; student Agrl. Inst., Vienna, 1902-03, also Tech. U., U. Munich; Ph.D., 1906; m. Martha Umlauf, 1917; 2 sons. Became asst., Bavarian Exptl. Inst., 1906; joined Agrl.-Chem. Exptl. Inst., Vienna, 1908; joined faculty Agrl. Inst., Vienna, 1910; bacteriologist on E. front, also at Rainer Hosp., Vienna, World War I; became adviser fish provisioning Ministry Food, 1918; became asso. prof. hydrobiology and fishery econs. Agrl. Inst., 1920, prof., 1924, ret., 1934. Author: Fischzucht und deren Bedeutung für das wirtschaftliche Leben, 1914; Das Tier- und Pflanzenleben unserer Alpenseen, 1915; (with E. Doljan) Handbuch der modernen Fischereiwirtschaftslehre, 1921; Fischereibiologie der Alpenseen, 1930; also articles. Research on fishery biology and econs., including water pollution; studies on anatomy and physiology of fish, fish diseases, biology of alpine lakes, nourishment of carp and trout, influence of hormones on sex determination; devel. (with Glaser) Glaser-Haempel fish test for human pregnancy. Died Vienna, Jan. 2, 1953.

HAENDLER, Helmut Max, Am. inorganic chemist; b. Boston, June 19, 1913; s. Max Anton and Frieda Anna (Lange) H.; B.S. in Chem. Engring., Northeastern U., 1935; Ph.D., U. Wash., 1940; m. Mildred Chandler Bragdon, June 20, 1936; children—Blanca Louise, Steven Andrew. Instr., U. Wash., 1940-42; research chemist, supr. Substitute Alloy Materials Labs., Manhattan Project, 1942-45; faculty U. N.H., Durham, 1945—, prof. inorganic chemistry, 1952—. Cons. Brookhaven Nat. Lab., Upton, N.Y., 1958-63; chmn. Gordon Research Conf. in Inorganic Chemistry, 1961. Fellow A.A.A.S.; mem. Am. Chem. Soc., Am. Crystallographic Assn., N.Y. Acad. Scis., Sigma Xi, Tau Beta Pi. Research, numerous publs. application structural prins. to inorganic chemistry. Home: Lee Hook Rd., R.F.D., Newmarket, N.H. 03857. Office: Dept. Chemistry, U. N.H., Durham, N.H. 03824.*

HAENKE, Thaddaeus, botanist, explorer; b. Kreibitz, Bohemia, Dec. 5, 1761; s. Elias Georg Thomas and Anna Rosalia (Eschler) H.; student Karls U., Prague, Czechoslovakia; beginning in 1780, U. Vienna, 1786-89. Made numerous bot. trips, Bohemia, E. Alps; botanist on expdn. led by A. Malaspina from Santiago, Chila, along W. coast of S.Am., C.Am., and N.Am. to Alaska, across Pacific to Marianas, Philippines, E. coast of Australia, and Tonga Archipelago to Callao, 1789-94; commd. to cross S.Am. from Lima, Peru, to Buenos Aires, Argentina; settled in Cochabamba, Bolivia, 1796. Author: Descripción geográfica física e histórica de las Montañas habitadas por la nación de los Indios Yuracarées, 1796; Memorandum sobre los Rios navegables que fluyen al Marañon, 1799; Introducción a la Historia Natural de la Provincia de Cochabamba, 1798. Discovered source of Marañon; research on flora of Cal., Alaska, Botany Bay, Australia; collected flora in Guam; explored new regions in internal S.Am.; collected and described plants and their significance in agr., medicine, tech. in Cochabamba; sci. initiator Chilean saltpeter industry. Died Cochabamba, 1817.

HAENLEIN, Paul, German aero. engr.; b. Cologne, Germany, Oct. 17, 1835; s. Johann Baptist and Wilhelmine (Poirez) H.; engring. degree Poly. Sch., Karlsruhe, Germany; m. Mathilde Thanel, 1877; employed in engine and mech. factories, Germany, also Sweden, Eng., Austria, Switzerland, 1860-1903. Author: Über das jetzige Stadium des lenkbaren Luftschiffes, 1904. Patentee airship driven by Lenoir gas motor, 1865; built steerable airship filled with coal gas (Aeolus), 1872. Died Mainz, Germany, Jan. 27, 1905.

HAENNY, Charles Bertrand, Swiss physicist; b. Bassins, Switzerland, 1906; s. Charles and Angèle H.; D.Sc., U. Lausanne, 1928; D.Sc., U. Paris, 1936; m. Josette Barthelemy, March 27, 1943; children: Catherine, Suzanne, Luc. Editor, Internat. Table of Constants, 1928-30; fellow & asst., U. Paris, 1934; Swiss. Ramsay Fellow, 1934-35; Privat Docent, U. Lausanne; head, research, Nat. Center for Scientific Research, Paris, 1935-39; lectr., U. Lausanne, 1940; extraordinary prof., 1943, ordinary prof., 1951, dir. Nuclear Physics Inst., U. Lausanne, 1951. Mem. Swiss Chem. Soc., French Physical Chem. Soc.; Swiss, French, Italian, Am. Physical Socs.; mem. of honour, Soc. for Development of Internat. Tables of Constant. Author numerous articles. Research on double refraction and magnetic properties of rare earths in solution; cosmic rays physics; neutron emission by fission and chain reaction; nuclear physics low and high energy; tracers. Home: 21, Av. du Général Guisan, Pully, Switzerland. Office: Institut de Physique Nucléaire de l'Université, 19, rue César-Roux, Lausanne, Switzerland.*

HAENSEL, (Heinrich) Gustav, chemist; b. Pirna, Saxony, Nov. 22, 1841; s. Heinrich and Amalie (Hippe) H.; student comml. sch., Dresden, Germany, later Poly. Sch., Dresden, 1875-77; m. Jerta Pienitz, 1868; 3 sons, including Otto. Enlarged operation begun by father for mfg. liquor essences, Pirna, 1867; founder br. operation, Aussig, 1878. Recipient prizes for products at world exhbns., Phila., 1876, Paris, 1900. Research on manufacture of volatile oils; determined that oxygen-containing component of oil is aromatic rather than terpene; patentee apparatus for distilling cumin oil to separate carvene and isolate pure carvone, 1876; manufactured peppermint oil, carnation oil, lavender oil. Died Pirna, July 14, 1923.

HAENSEL, Vladimir, chemist; b. Freiburg, Germany, Sept. 1, 1914; s. Paul and Nina (Von Tugenhold) H.; came to U. S., 1930, naturalized, 1936; B.S., Northwestern U., 1935, Ph.D., 1941, D.Sc. (hon.), 1957; M.S., Mass. Inst. Tech., 1937; m. Mary Magraw, Aug. 28, 1939; children—Mary Ann (Mrs. Michael J. Ahlen), Katherine. With Universal Oil Products Co., Des Plaines, Ill., 1937—, v.p., dir. research, 1964—. Recipient Precision Sci. Co. award for achievement in petroleum chemistry, 1952; Profession Progress award Am. Inst. Chem. Engrs., 1957; Perkin medal, 1967. Mem. Am. Chem. Soc. (award in indsl. and engring. chemistry Esso Research & Engring. Co. 1964). Research, numerous publs. on catalysis especially conversion of hydrocarbons; discoveries leading to comml. wide scale utilization of platinum containing catalysts in refining industry. Home: 705 W. North St., Hinsdale, Ill. 60521. Office: 30 Algonquin Rd., Des Plaines, Ill. 60016.*

HAENZEL, Gerhard Karl Theodor, mathematician; b. Wollin, Pomerania, Mar. 5, 1898; s. Alwin and Margarethe (Hellwig) H.; Dr. Engring., Tech. U. Berlin, 1926; Dr. rer. nat., U. Freiburg, 1940; m. Gertrud Wichmann, 1924; 1 son, 2 daus. Became mem. faculty Tech. U. Berlin, 1929; named prof. geometry Tech. U. Karlsruch (Germany), 1933, prof. math. and math. tech., 1937; prof. math. U. Munster (Germany), 1943-44. Contbr. to math. and phys. jours. Work on line geometry, relationship between geometry and physics, theory of functions and differential equations, relationships between geometry and wave mechanics; stressed use of higher math. in engring. Killed at Lesneven, France, Mar. 6, 1944.

HAERTZEN, Charles Anthony, Am. psychologist; b. Mpls., Aug. 17, 1925; s. Ray and Laura (McFall) H.; B.A., U. Minn., 1949, M.A., 1950; Ph.D., U. Ky., 1961; m. Margaret Lucille Davis, May 14, 1955; children—John, Linda, Mark, Charlene. Psychologist, instr. Rochester (Minn.) State Hosp., 1950-53; psychologist Dept. Pub. Welfare, St. Paul, 1953-56; research psychologist Nat. Inst. Mental Health, Addiction Research Center, Lexington, Ky., 1956—. Cons. Gen. Mills, Mpls., 1963—. Mem. Am., Midwestern, Central Ky. psychol. assns., Sigma Xi, Phi Delta Kappa. Author: Follett Vest Pocket Anagram Dictionary, 1964; also articles. Co-constructor of inventory and scales used for measurement of subjective experience asso. with drugs, altered states of consciousness, alcohol or opiate withdrawal and psychiat. diagnosis; defined personality characteristics of opiate addicts, alcoholics, and criminals; statis. methodology for factor analysis, tetrachoric correlations, and constrn. of psychol. scales; constrn. anagram lists in psychol expts. Office: USPHS Hosp., Lexington, Ky. 40501.

HAESER, Heinrich, physician; b. Rome, Italy, Oct. 15, 1811; s. August Ferdinand and Dorothea (Schwabedissen) H.; M.D., U. Jena (Germany), 1834. Practiced medicine Thuringia, Germany, 1835; med. faculty U. Jena, 1836-49, prof., 1846-49; prof. Leipzig, Germany, from 1849, U. Greifswald (Germany), U. Breslau (now Wroclaw, Poland), from 1862. Author: Historisch-pathologische Untersuchungen, Als Beiträge zur Geschichte der Volkskrankheiten, 2 vols., 1839-41; Bibliotheca epidemiographica, sive catalogus librorum . . . conscriptorum, 1843; Lehrbuch der Geschichte der Medizin und der epidemischen Krankheiten, 2 vols., 1845 (standard work in med. history); Die Vaccination und ihre neuesten Gegner . . . , 1854; Dissertatione de cura aegrotorum publica a christianis oriunda, 1856; Geschichte christlicher Krankenpflege und Pflegerschaften, 1857; Über das Sittliche in dem Berufe des Arztes, 1860; Zur Geschichte der medizinischen Fakultät Greifswald, 1879; Grundriss der Geschichte der Medizin, 1884. Editor: Archiv für die gesammte Medizin. Died Breslau, Sept. 13, 1885.

HAEUSSERMANN, Carl Friedrich, German chemist; b. Stuttgart, Germany, July 24, 1853; s. Friedrich and Gottliebin (Ergenzinger) H.; student Stuttgart Polytechnikum, Poly. Inst. Munich; Ph.D., Heidelberg, 1876; m. Anna Hubertin Beiell; 2 foster daus. Became mem. faculty Stuttgart Polytechnikum, 1877; named sub-dir. Chemische Fabrik Griesheim-Elektron, 1883, also mem. bd.; commd. adv. chemist by Prussian War Ministry to Königliche Pulverfabriken Hanau (royal powder factory); named prof. chem. tech. Tech. U. Stuttgart, 1891. Author: Sprengstoffe und Zündwaren, 1894; Alfred Nobel und die Erfindung der nitroglyzerinpulver, 1904; Die Nitrozellulosen, ihre Bildungsweisen, Eigenschaften und Zusammensetzung, 1914; also articles. Research on coal tar dyes, inorganic and organic by-products, aniline and aniline salts, nitrobenzene, toluidine and its derivatives; discovered 2.4.6 trinitrotoluene (TNT); developed hexanitrodiphenylamine (underwater explosive); studied nitration of cellulose. Died nr. Ulm, Germany, July 9, 1918.

HAFEZ, Mostafa Mahmoud, Egyptian chemist; b. Cairo, Egypt, Oct. 7, 1909; s. Mahmoud and Zeinab (Ashmawi) H.; B.Sc. with honors, Sheffield U., 1938; Ph.D., London U., 1941; m. Enayat Hafez, Aug. 6, 1942; children—Mahmoud, Sherif, Khadiga. Tchr. secondary sch., 1930-35; faculty sci. Alexandria U., 1942-50; vis. fellow Princeton, 1946-48; sec. gen. Nat. Research Center, Cairo, 1950-62; state sec. Ministry Sci. Research, Cairo, 1962-65; v.p. Supreme Council Sci. Research, Cairo, 1965—. Decorated Order of the Republic, Egypt. Mem. Egyptian Chem. Soc. Author: Plastics, 1957 (in Arabic). Research, publs. on reaction kinetics, ring closure, indole formation, isocholesterol, anti-oxidants, solar energy. Home: 3 Hasba, Heliopolis, Cairo. Office: 101 Kasr el Aini, Cairo, UAR.*

HAFFKINE, Waldemar Mordecai Wolff, bacteriologist; b. Odessa, Russia, Mar. 15, 1860; asst. prof. physiology, med. sch., Geneva, Switzerland, 1888-89; asst. to Pasteur, Paris, 1889-93; physiologist, bacteriologist Indian Govt., from 1893; founder research lab., Bombay. Discovered and used method of inoculation against cholera, 1893-94; introduced into India method of inoculation against plague, 1897. Died Oct. 26, 1930.

HAFFNER, Felix, German pharmacologist, toxicologist; b. Marbach/Neckar, Germany, Oct. 18, 1886; s. Traugott and Sophie (Mozer) H.; ed. Tübingen, Munich (both Germany); m. Margatete Gudden, 1912; 1 dau. Became asst. to H. von Tappeiner, Munich Pharmacological Inst., 1912; became mem. faculty, 1922; later at U. Freiburg (Germany); named prof. U. Königsberg (Germany); 1925, Tübingen, 1927. Author: Über das Wesen der unspezifischen Therapie, 1927; Der chemische Reiz, 1936; Über die Bedeutung des Stickstoffs fur die zentrale Erregung, 1936; Belebung der Rezeptur unter Berücksichtigung des Vorschlages der Einführung von Normdosen, 1937; also articles. Research on phys. effect of light on cells and cell metabolism, limitations of chem. attraction, effect of hydrogen ions, standardization of medicines; inventor test for anodyne medicines, still used in modified form; developed standardized medicine containers. Died Tübingen, Mar. 18, 1953.

HAFNER, Everett Mark, Am. physicist; b. Bklyn., May 20, 1920; s. Mark and Frances (Cisin) H.; B.S. in Physics (Pulitzer scholar), Union Coll., Schenectady, 1940; Ph.D., U. Rochester, 1948. Physicist, Naval Ordnance Lab., 1941-45; asso. physicist Brookhaven Nat. Lab., 1948-52; faculty U. Rochester (N.Y.), 1953-

—, prof. physics, 1966—; staff physicist Commn. on Coll. Physics, 1964-65. Mem. com. on proficiency examinations in physics N.Y. State Dept. Edn., 1964—. NSF fellow, 1952-53. Fellow Am. Phys. Soc. (exec. com. N.Y. State sect 1965—); mem. Am. Assn. Physics Tchrs., A.A.A.S., Sigma Xi. Research, publs. on inelastic scattering protons by nuclei, proton-induced fission, neutron-proton scattering, scattering polarized protons and neutrons, cosmic gamma radiation, magnetometer detection submarines. Home: 460 Oakridge Dr., Rochester, N.Y. 14617.*

HÄFNER, Heinz, German neurologist; b. Munich, Germany, May 20, 1926; s. Heinrich and Elisabeth (Gerner) H.; M.D., Ph.D., U. Munich; m. Ursula von Rintelen, 1951; children—Gilbert, Gerald. Med. asst. Nymphenburger Krankenhaus and Clinic of Neurology, U. Munich, 1950-51, asst., dir. psychodiagnostic dept., 1951-54; chief physician Clinic Psychiatry and Neurology, U. Heidelberg, 1958—. Author: Schulderleben und Gewissen, 1956; Psychopathen: Daseinsanalytische Untersuchungen zur Struktur und Verlaufsgestalt von Psychopathien, 1961; Psychiatrie der Verfolgten; (with Baeyer and Kisker) Psychopathologische und Gutachliche Erfahrungen an Opfern der Nationalsoziolistischen Verfolgung und Vergleichbarer Extrembelastungen, 1964, also numerous articles on neurology, psychiatry and psychotherapy. Home: 3 Gundolfstrasse. Office: 4 Vossstrasse, Heidelberg, Germany.

HAFSTAD, Lawrence Randolph, Am. physicist; b. Minneapolis, Minn., June 18, 1904; s. Bernt Andrew and Ellen (Bruem) H.; B.S., U. of Minn., 1926; Ph.D., Johns Hopkins, 1933; m. Mary Cowen; 1 son, William A. Engr., Northwestern Bell Tel. Co., Mpls. 1920-28; physicist, Carnegie Instn. of Wash. (D.C.), 1928-42; staff Applied Physics Lab., Johns Hopkins U., Silver Spring, Md., 1942-45, dir. of research, 1945-47; dir. Inst. Cooperative Research, 1947-49; exec. sec. of Research and Development Bd., Office of Sec. of Defense, 1947-49; named first director Reactor Development Div., AEC, 1949-55; dir. Atomic Energy Div. Chase Manhatten Bank, 1955; v.p. Gen. Motors Corp., director research labs., 1955—; director Atomic Power Devel. Assos.; trustee Power Reactor Devel. Corp. Cons. exec. office of Pres.; Dept. Def. dir. research and engring.; mem. gen. adv. com. AEC. Trustee Cranbrook Inst. Sci., Johns Hopkins Carnegie Endowment for Internat. Peace; mem. adv. council Fund for Peaceful Atomic Devel. Recipient (with M. A. Tuve) of A.A.A.S. award for research and development 1,000,000-volt vacuum tube, 1931. Recipient medal for merit from Sec. Navy for maj. contribution in development of significant improvements in ordnanc for Army and Navy, 1946; Proctor prize Sci. Res. Soc. Am., 1956. Fellow of American Phys. Soc., Am. Inst. Radio Engrs.; mem. Am. Geophys. Union, Washington Philos. Soc., Washington Acad. Scis., Am., Am. Inst. E.E., Indsl. Research Inst., Soc. Automotive Engrs., Sigma Xi, Tau Beta Pi. Contbr. articles tech. jours. Research on wave propagation, atomic disintegration, artificial radioactivity. Home: 191 Marblehead Dr., Bloomfield Hills, Mich. Office: Gen. Motors Corp. Research Labs., Warren, Mich.

HAFTORN, Svein, Norwegian zoologist; b. Drammen, Norway, Jan. 30, 1925; s. Birger and Martha (Winsvold) H.; cand.real., U. Oslo, 1952, Dr.Sci., 1957; m. Eva Hamborgstrom, Sept. 22, 1951; children—Jens, Sylvi, Tone, Beate. Curator zoology Mus. Royal Norwegian Soc. Sci., Trondheim, 1952-66; prof. zoology U. Trondheim (Norway), 1966—, head Zool. Inst., 1966—. Mem. Royal Norwegian Soc. Scis., Norwegian, Am., Brit., German ornithol. unions, Norwegian Zool. Union. Author: Vare Fugler (Birds of Norway), 1962; Fjellfauna (The Alpine Fauna of Norway), 1966; also articles. Research on bird ecology, especially breeding habits, feeding behavior and food requirements of chickadees (in Norway and Alaska); distbn. of Norwegian birds. Home: Malsjoen, Klaebu, Norway.* Office: 12 Bjornsonsgt., Trondheim, Norway.*

HAGAN, Wallace Woodrow, Am. geologist; b. Griggsville, Ill., Feb. 3, 1913; s. Warren L., and Mabel Rea (Bruner) H.; B.S., U. Ill., 1935, M.S. (Grad. scholar 1935-36), 1936, Ph.D. (Grad. fellow 1936-37, 40-41), 1942; summer student U. Mo., 1937; m. Mary Elizabeth LeVan, Nov. 30, 1940; children—Karen Rae, Elizabeth Annette. Asst. petroleum geologist, J. V. Wicklund Devel. Co., Detroit, 1937-39; cons. geologist Greenville Ky. and Urbana, Ill., 1939-40; geologist charge ground water sect., div. geology Ind. Dept. Conservation, 1942-44; geologist Sohio Petroleum Co., 1945-48, Felmont Oil Corp., 1948-52; cons. geologist, Owensboro, Ky., 1952-58; dir. state geologist Ky. Geol. Survey, 1958—. Mem topographic mapping com. Ky. C. of C., 1947-51; adv. bd. Ky. Geol. Survey, 1952-58, ex officio mem., 1958—; rep. gov. Ky. research com. Interstate Oil Compact Commn., 1958-65, chmn. com., 1965-66; bd. dirs. Ky. Conservation Congress, also mem. natural resources devel. com. and mineral resources subcom., 1961-64; chmn. quality water com. Ky. Water Resources Study Co. mmn., 1959; rep. gov. Ky. Nat. Water Research Symposium, 1961. Fellow Geol. Soc. Am. (vice chmn. S.E. sect. 1957); mem. Am. Assn. Petroleum Geologists (dist. rep. Great Lakes 1954), Ky. (pres. 1966-67), Ind.-Ky. (exec. officer 1955-56), Lexington geol. socs., Paleontological Soc., Assn. Am. State Geologists (statistician 1963-66, pres.-elect 1967-68, chmn. liaison com. 1966—), Ky. Acad. Sci., Phi Beta Kappa, Sigma Xi, Phi Kappa Phi, Sigma Gamma Epsilon. Contbr. articles profl. jours. Research on exploration of water supplies by elec. resistivity methods. Home: 317 Jesselin Dr., Lexington 40503. Office: Mineral Industries Bldg., University of Ky., Lexington, Ky. 40506.*

HAGBERG, Bengt. A., Swedish physician, pediatrician; b. Gothenburgh, Sweden, Aug. 9, 1923; s. Eric and Gulli (Zachau) H.; grad. Med. Sch., Uppsala (Sweden) U., 1950, doctors degree in pediatrics, 1953; m. Gudrun Wranne, Jan. 18, 1947; children—Gunilla, Barbro, Hans, Hars, Kerstin. Faculty, Uppsala U., 1956—, asso. prof. pediatric neurology, 1964—. Author: The Iron-binding Capacity of Serum in Infants and Children, 1953; also numerous articles. Research in pediatric hematology, neurology, endocrinology. Home: Eriksbergs V.18, Uppsala, Sweden.*

HAGEDORN, Donald James, Am. plant pathologist; b. Moscow, Ida., May 18, 1919; s. Fred W. and Elizabeth (Scheyer) H.; B.S., U. Ida., 1941; M.S., U. Wis., 1943, Ph.D., 1948; m. Eloise Tierney, July 18, 1943; 1 son, James William. Faculty, U. Wis., Madison, 1948—, prof. plant pathology, 1964—. Exec. sec., presiding officer Internat. Working Group on Legume Viruses, 1961-66. NSF fellow, 1957. Fellow A.A.A.S. (Campbell award 1961); mem. Am. Phytopath. Soc., Am. Inst. Biol. Sci., Nat. Pea Improvement Assn. (v.p. 1966—), Sigma Xi, Alpha Zeta. Fundamental and practical studies on diseases of vegetable crops especially peas and beans, basic studies on legume viruses, breeding for disease resistance in processing peas resulting in devel. and release of 3 new highly resistant canning peas and several new lines for sources of useful germplasm. Home: 927 University Bay Dr., Madison, Wis. 53705.*

HAGEDORN, Fred Bassett, Am. physicist; b. Boone, Ia., June 8, 1928; s. Hans James and Christine (Johnson) H.; B.S., Ia. State U., 1952; Ph.D., Cal. Inst. Tech., 1957; m. Grace Alexander Helms, June 27, 1954; children—Martha Gray, Katherine Johanna. Mem. tech. staff Bell Telephone Labs., Murray Hill, N.J., 1957—, group supr., 1961—. Mem. Am. Phys. Soc., Sigma Xi. Research, publs. in low energy nuclear physics, magnetic materials, superconducting thin films. Home: 30 Fawn Circle, Berkeley Heights, N.J. 07922. Office: Bell Telephone Labs., Murray Hill, N.J. 07971.*

HAGEDORN, Rolf, physicist; b. Wuppertal, Germany, July 20, 1919; s. Max Paul and Selinde (Reinecke) H.; Abiturium, Realgymnasium, Wuppertal, 1937; dipl. phys., Dr.rer.nat., U. Göttingen, 1952; postdoctoral fellow Max Planck Inst. Physics, Göttingen, 1952-54; m. Anneliese Westerhaus, 1952; 1 dau., Regine. Theoretical physicist, tchr. CERN, Geneva Switzerland, 1952—; lectr. Max Planck Inst. Physics, Munich, Germany, Matscience, Madras, India, various physics schs. Author: Relativistic Kinematics, 1963. Research, publs on nuclear physics, high-energy and elementary particle physics, relativistic quantum mechanics, symmetries, statis. models of particle prodn., thermodynamic description of strong interactions; discovery of presumable existence of an absolute highest temperature. Home: 01 Sergy-St. Genis, France. Office: CERN-TH, 1211, Geneva, Switzerland.*

HAGEDORN, Werner, German surgeon; b. Eichsfeld, Germany, 1831; ed. Berlin; pupil of Johannes Müller, B. von Langenbeck; became asst., hosp. in Magdeburg, Germany, 1855, dir. surgery, 1863. Inventor flat-sided surg. needle with cutting edge nr. point, circa 1885. Died 1894.

HAGEK, Thaddäus (Hagecius ab Hayek, also Nemicus), Czechoslovakian astronomer, physician; b. Prague, Czechoslovakia, 1525; personal physician to Kaiser Maximilian II, then to Rudolph II; partially responsible for Brahe's coming to Prague, 1598; prof. math., Carolinum, Prague. Author: Oratio de laudibus geometriae, 1557; Dialexis de novae et prius incognitae stellae inusitatae magnitudinis . . . appratitione deque eiusdem vero loco constituendo, 1574; Responsio ad virulentem et maledicum Hannibulis Raymundi . . . scriptum, 1576; Descriptio cometae, que apparuit A.D. 1577 . . . , 1578; Apodixis physica et mathematica de cometis tum in genere, tum imprimis de eo, qui 1580 . . . efulsit, 1581; De cerevisia eiusque conficiendi ratione, 1585; contbr. Commentarius in Hermetis Trismegisti aphorismos sive 100 sententias astrologicas. Observed new star, 1572 and reported that it had no observable parallax, studied comets. Died Prague, Sept. 1, 1600.

HAGEN, Carl Gottfried, pharm. chemist; b. Königsberg (now Kaliningrad, USSR), Dec. 24, 1749; s. Heinrich and Maria Elizabeth (Georgesohn) H.; M.D., U. Königsberg, 1755, Ph.D., 1807; m. Johanna Maria Rabe, 1784; 9 children, including Carl Heinrich, Johann Friedrich, August, Johanna (Mrs. F. W. Bessel), Louise Florentine (Mrs. F. E. Neumann). Apprenticed at family pharmacy, Königsberg, 1763, asst., 1766-69, mgr., 1772-1816; apptd. lectr. U. Königsberg, 1775, asso. prof., 1779, prof. med. faculty, 1788, prof. physics, chemistry and natural history, philos. faculty, 1807; named med. adviser to provincial health collegium, 1800. Galeopsis Hagenii (botany), Mytilus Hagenii (zoology) named in his honor; Hagen-Bucholz Found. established to promote sci. pharm. work. Mem. numerous socs. Author: Lehrbuch der Apothekerkunst, 1778; Grundriss der Experimentalchemie zum Gebrauch bey dem Vortrage derselben, 1786; Grundriss der Experimentalpharmacie, 1790; Preussens Pflanzen, 2 vols., 1818; Chloris Borussica, 1819. Developed new methods of teaching chemistry in which students performed expts.; 1st to treat pharmacy scientifically in a textbook. Died Königsberg, Mar. 2, 1829.

HAGEN (-BISCHOFF), (Jakob) Eduard, Swiss physicist; b. Basel, Switzerland, Feb. 20, 1833; s. Karl Rudolf and Rosina (Geigy) Hagen; student, Berlin, Geneva, Paris; Ph.D., Basel, Switzerland, 1855; m. Marguerite Bischoff, 1862; children—4 sons, including August, 3 daus., including Margaretha (Mrs. J. H. W. Rupe). Became tchr. physics and chemistry Basel Indsl. Sch., 1856; named prof. math. U. Basel, 1862, prof. physics, 1863, emeritus, 1906. Organizer pub. contbns. for Bernoullianum (inst. for exact natural scis.), 1872. Mem. Swiss Natural Sci. Soc. (pres. glacier commn.). Contbr. articles to profl. jours. Research on flourescence in 30 substances; helped confirm Stokes' law; studied relation between flourescence and adsorption, also spectral analysis of flourescence light; helped measure Rhone glacier; worked on theory of structure of ice crystals; experimented on elec. spark between point and plate. Died Basel, Dec. 23, 1910.

HAGEN, (Carl) Ernst Bessel, German physicist; b. Königsberg, Germany (now Kaliningrad, USSR), Jan. 31, 1851; s. Adolf and Johanna (Bessel) H.; ed. Berlin, Heidelberg; Ph.D., 1875; m. Wilhelmine von Bezold, 1896; 2 sons. Asst. to Bunsen, 1873-75; asst. to A. Toepler, Dresden, Germany, 2 years, to Helmoholz, Berlin, 6 years; became mem. faculty U. Berlin, 1883; visited U. S., 1884; named asso. prof. applied physics, chmn. new electrotech. lab. Dresden Polytechnikum, 1884; became physicist Imperial Navy, Kiel, Germany, 1887; apptd. dir. 2d (tech.) dept. Reich Phys.-Tech. Inst., 1893; ret., Berlin-Charlottenburg, 1918. Mem. Imperial Standards Commn., 1894; asso. mem. Patent Bur., 1895-1908; sec. bd. German Mus. Author: Die elektrische Beleuchtung mit besonderer Berucksichtigung der in den Vereinigten Staaten verwandten Systeme, 1885; also articles. Studied elec. lighting in U. S.; conducted (with H. Rubens) classic expts. to confirm Maxwell's electromagnetic light theory, thus making possible determination of elec. conductivity of metals by radiation measurement alone. Died Solln, nr. Munich, Jan. 15, 1923.

HAGEN, Everett Einar, Am. economist; b. Holloway, Minn., July 5, 1906; s. John J. and Marthea (Moe) H.; B.A., St. Olaf Coll., 1927; M.A., U. Wis., 1932, Ph.D., 1941; m. Ruth Alexander, June 4, 1937. Instr., Mich. State Coll., 1937-42; with Fed. Govt., 1942-48; prof. econs. U. Ill., 1948-51, chmn. dept., 1950-51; econ. adviser Govt. Union of Burma, 1951-53; vis. prof. Mass. Inst. Tech., 1953-59, prof. 1959—, prof. polit. sci., 1964—. Cons. U. S. econ. aid. agys., intermittently 1948—, Govt. Japan, 1956, Govt. El Salvador, 1962-63. John Simon Guggenheim Meml. Found. fellow, 1962. Mem. Am. Econ. Assn., Royal Econ. Assn., Am. Assn. U. Profs., Phi Beta Kappa. Author: Handbook for Industry Studies, 1958; On the Theory of Social Change, 1963. Editor: Planning Economic Development, 1964. Research, publs. on econ. devel. low-income countries. Home: 100 Memorial Dr., Cambridge, Mass. 02142.*

HAGEN, Friedrich Wilhelm, German psychiatrist; b. Dottenheim, Germany, June 16, 1814; s. Friedrich Wilhelm and Christiane Elisabeth (Schmauss) H.; ed. Munich, Erlangen (both Germany); M.D., 1836; m. Margarethe Engerer, 1847; 3 sons, including Eduard, 1 dau. Practiced medicine, Velden, Germany; visited insane asylums, Eng., France, Germany 1844; became asst. physician new asylum, Erlangen, 1846, dir., 1859; named dir. Cloister Irsee asylum, nr. Kaufbeuren, Germany, 1849; apptd. prof. U. Erlangen, 1860. Author: Physiologische Untersuchungen, Studien in Gebiete der physiologische Psychologie, 1847; Studien auf dem Gebiete der Ärztlichen Seelenheilkunde, 1870-72; Chorinsky, eine gerichtlich-psychologische Untersuchung, 1872. Research on psychic importance of brain and nerve organs, also on constn. and temperament, mutual effect of mood shifts on physiology, body and skull measurements of insane, weights of brain and spinal chord; contbd. to modern humane treatment of insane. Died Erlangen, June 13, 1889.

HAGEN, Gotthilf Heinrich Ludwig, German hydraulic engr.; b. Königsberg, Germany (now Kaliningrad, USSR), Mar. 3, 1797; s. Friedrich Ludwig and Helene Charlotte (Reccard) H.; ed. Königsberg; surveyor's exam, 1819; masterbuilder's degree, 1822; Ph.D. (hon.); m. Wilhelmine Auguste Hagen, 1827; 4 sons, including Friedrich Ludwig, Carl Heinrich, Otto Albert, 1 dau. In state service, from 1819; made study trip, 1822-23; in Königsberg, 1924; became insp. harbor installation, Pillau, 1826; named delegate to Upper Constrn. Delegation, Berlin, 1831, chmn., 1859-75; tchr. Constrn. Acad., also arty. and engring sch., 1934-49; expert delegate to work on laws for German rivers, Frankford/Main, 1849; adviser Ministry Commerce, 1850; with Prussian admiralty, 1854-56. Recipient Gold medal for service in constrn.; Hagen Found. established in his honor. Mem. Acad. Scis., Berlin. Author: Beschreibung neurer Wasserbauwerke in Deutschland, Frankreich, den Niederlanden und der Schwiiz, 1826; Untersuchung über den Druck und die Reibung des Sandes, 1833; Grundzüge der Wahrscheinlichkeitsrechrung, 1837; Über die Bewegung des Wassers in engen zylindrischen Röhren, 1839; Handbuch der Wasserbaukunst (classic work in hydraulic constrn.), III, 1941-63; Über die Oberflüche der Flüssigkeiten, 1845; Über den Einfluss der Temperatur auf die Bewegung des Wassers in Röhren, 1854; Zur theorie der Meereswellen, 1859; Die neueren Theorien der Bwegeung dse srömtne, Wassers, 1868; Über die Bewegung des Wassers in Strömen, 1868; Über den Seitendruck der Erde, 1871; Geschwindigkeit des strömenden Wassers in verschiedenen Tieten, 1883. Dir. constrn. on Rhin, Elbe, Weser, other rivers; projects include dikes, harbor installations, dune fortifications; invented and experimented with apparatus; eponym of Hagen-Poiseuill law of viscosity measurement. Died Berlin, Feb. 3, 1884.

HAGEN, Hermann August, entomologist; b. Königsberg, Germany, May 30, 1817; s. Carl Heinrich and Anna (Linck) H.; degree in medicine U. Königsberg, 1840, Ph.D. (hon.), 1863; D.S. (hon.), Harvard, 1887; m. Johanna Maria Gerhards, 1851. Came to U. S., 1867; developed entomol. dept. Museum Comparative Zoology, Harvard; apptd. prof. entomology Harvard, (1st prof. entomology at any U. S. coll.) 1870; mem. Am. Acad. Arts and Scis., Am. Philos. Soc., Am. Entomol. Soc. Author: Monographie der Termiten, 1855-60; (at request Smithsonian Instn.) Synopsis of North American Neuroptera, 1861; Bibliotheca Entomologica, 1862-63; contbr. numerous articles to profl. pubis. Founder 1st entomol. museum in U. S.; described numerous insects, including large termite, Nevada termite, common raphidia, stigmatic snake fly; influenced entomol. study in U. S. and Europe. Died Cambridge, Mass., Nov. 9, 1893.

HAGEN, Johann Georg, astronomer; b. Bregenz, Austria, Mar. 6, 1847; s. Martin and Theresia (Schick) H.; studied math., astronomy and physics in Münster and Bonn. Entered Jesuit order, 1863, received edn. from them; taught math. Coll. Prairie du Chien (Wis.), 8 years; dir. Georgetown obs., Washington, 1888-1906; Vatican obs., nr. Rome, Italy, 1906. Author: Atlas Stellarum Variabilium, 1899-1908, 1927-34 (exhaustive study on variable stars); Synopsis der höheren Mathematik, 4 vols., 1891-1905, 1930; Die veränderlichen Sterne, 2 vols., 1913-24; La rotation de la terre, 1911. Contbr. numerous articles to profl. jours. Revitalized work on Vatican's zone of internat. photog. sky chart; revised critically Dreyer gen. catalog of constellations and nebulae; made expts. on earth's rotation using isotomeograph which he developed; discovered and observed extensive cosmic cloud fields; representative of data-collecting period of astronomy. Died Vatican City, Sept. 6, 1930.

HAGEN, John P., astronomer; b. Amherst, N.S., Can., July 31, 1908; s. John T. and Ella Bertha (Fisher) H.; B.S., Boston U., 1929, Sc.D., 1959; M.A., Wesleyan U., Middletown, Conn., 1931; postgrad. Yale, 1931-33; Ph.D., Georgetown U., 1949; Sc.D., Fairfield U., 1958, Loyola University, 1959, Adelphi College, 1959; Sc.D., Mount Allison (Canada), 1960; married to Edith W. Soderling, October 12, 1935; children—J. Peter, E. Christopher. Research asso. Wesleyan U., 1931-35; supt. atmosphere and astrophysics div., development microwave radar, radio astronomer, mem. various eclipse expdns. Naval Research Lab., Washington, 1935-58; asst. dir. space flight development Nat. Aero. and Space Adminstrn., 1958-60, dir. OUNC, 1960-62; prof. radio astronomy Pa. State U., 1962-—. Project dir. Earth Satellite Project; lectr. Georgetown U. Recipient Presdl. Certificate of Merit. Fellow Institute Radio Engrs., Am. Acad. Arts and Scis.; mem. Am. Astron. Soc., Internat. Astron. Union, Washington Acad. Scis., U.R.S.I. (chmn. nat. com.), Phi Beta Kappa, Sigma Xi. Research in properties of quartz crystals, microwave radio circuits, mil. use of microwave radio, radio astronomy. Home: 513 W. Park Av., State College, Pa. Office: 103 Whitmore, Pa. State U., University Park, Pa. 16802.*

HAGEN, Paul Beo, biochemist; b. Sydney, Australia, Feb. 15, 1920; s. Conrad James and Mary (McFadzean) von Hagen; M.B., U. Sydney, 1945, B.S., 1945; m. Jean M. Himms, Sept. 29, 1956; children—Anna Jean, Nina Jean. Med. officer New South Wales Health Dept., 1945-48; lectr. physiology U. Sydney, 1949-51; sr. lectr. U. Queensland (Australia), 1951-52; C. J. Martin fellow med. research Oxford (Eng.) U., 1952-54; asst. prof. pharmacology dept. Yale, 1954-56, Harvard, 1956-59; prof. biochemistry, head dept. U. Man., Med. Coll., Winnipeg, Can., 1959-64; prof. biochemistry Queen's U., Kingston, Ont., Can., 1964-—, head dept., 1964-67. Mem. Med. Research Council Can., 1965-—, vice chmn., 1967. Fellow Chem. Inst. Can.; mem. A.A.A.S., Am. Chem. Soc., Am. Soc. Pharmacology and Exptl. Therapeutics, Biochem. Soc., Physiol. Soc., Pharmacological Soc., Can. Biochem. Soc., Chem. Inst. Can. Editorial bd. Jour. Pharmacology and Exptl. Therapeutics, 1960-64, Biochem. Pharmacology, 1960-64, Canadian Jour. Biochemistry, 1963-67. Research, numerous publs. on mechanisms of biosynthesis and storage of adrenaline and noradrenaline, histamine and serotonin, digestive process of herbivores, regulation of metabolism. Home: 235 Tudor Pl., Ottawa 7, Ont., Can.*

HAGEN, Ulrich, German biologist; b. Frankfört/Main, Germany, Feb. 21, 1925; s. Wilhelm and Grete (Pokowski) H.; gen. certification as physician U. Munich (Germany), 1950, Ph.D., 1952, M.D., 1954; m. Malvine Seyfferth, Dec. 29, 1954; children—Martin, Stefan Wilhelm. Sci. asst. biophys. dept. Heiligenberg-Inst., Heiligenberg, Baden, Germany, 1953-61; dozent Med. Faculty, U. Freiburg (Germany), 1961-65; biochemist Inst. für Strahlenbiologie, Nuclear Research Center, Karlsruhe, Germany, 1965-—. Recipient Röntgenpreis, U. Giessen (Germany), 1965. Author: Erg. med. Strahlenforsch., 1964; also numerous articles, chpt. in book. Research on mechanism of effect of ionizing radiation on living organisms, alteration of metabolism and macromolecules in irradiated cells, biochem. studies on radiation effect on deoxyribonucleic acid. Home: 31 Berlinerstr., Office: Inst. f. Strahlenbiologie, Kernforschungszentrum, Post box 947, Karlsruhe 75, Germany.*

HAGENBACH, August, Swiss physicist; b. Basel, Switzerland, Dec. 22, 1871; s. Eduard and Marguerite (Bischoff) H.; student U. Basel; Ph.D., U. Leipzig (Germany), 1894; m. Elena Jenny Luisa Aman, 1906; 1 son, 1 dau. Asst., U. Bonn (Germany), apptd. lectr., 1898, then prof.; went to Aachen, Germany, 1904; named prof. physics U. Basel, 1906, univ. rector, 1926. Author: (with H. Konen) Atlas der Emissionsspektren der meisten Elemente, 1905; (with A. Wüllner) Lehrbuch der Experimental physik I (Allgemeine Physik und Akustik), 1907; also articles. Research in spectroscopy, elec. arc, rotation dispersion; his intro. of new optical lattice in his physics inst. led to fruitful work on spectral lines of molecules. Died Basel, Aug. 11, 1955.

HAGER, Charles Keemle, Am. physicist; b. Oak Park, Ill., Oct. 11, 1923; s. Charles Henry and Olive (Benedict) H.; student Kansas City (Mo.) Jr. Coll., 1940-43; B.S., U. Tex., 1948, M.A., 1949, Ph.D., 1952; m. Gladys Marie Reeds, Aug. 25, 1947; children—Janice Susan, Michael Kevin. Research scientist Def. Research Lab., Austin, Tex., 1949-52; project aerophysics engr. Gen. Dynamics Corp., Ft. Worth, 1952-55; research dir. Mandrel Industries, Houston, 1956-58; dir. applications research Varo, Inc., Garland, Tex., 1958-59; with LTV Electrosystems, Garland, Tex., 1959-—, now sci. cons.; teaching fellow U., Tex., 1948-49; instr. So. Meth. U., 1955. Recipient Sigma Pi Sigma award for outstanding research, 1950. Mem. I.R.E. (sec. 1961-62), I.E.E.E., Am. Astronautical Soc. Contbr. articles to profl. jours. Design, devel. fields electronics, nuclear reactors, servomechanisms, communications systems, instrumentation; math. analysis in statistics, circuits, electronic systems, analog computers, noise. Home: 7048 Cliffbrook, Dallas 75240. Office: P.O. Box 6118, Dallas 75222.*

HAGER, George Philip, Jr., Am. chemist; b. Balt., Mar. 16, 1916; s. George Philip and Marie Theresa (Zilch) H.; B.S., U. Md., 1938, M.S., 1940, Ph.D., 1942; postgrad. U. Colo., (fellow) Northwestern U.; m. Margaret Kathryn League, Dec. 24, 1938; children—George Philip III, Priscilla Jane, Deborah Ethel, Andrew DuMez. Research organic chemist Eli Lilly & Co., Indpls., 1942-45; faculty U. Md. Sch. Pharmacy, 1945-55, prof., head dept. pharm. chemistry, 1948-55; sr. scientist Smith, Kline & French Labs., Phila., 1955-57; dean, prof. U. Minn. Coll. Pharmacy, 1957-65; dean, prof. U. N.C. Sch. Pharmacy, Chapel Hill., 1966-—. Chmn. gen. research support adv. com. NIH; mem. nat. adv. com. on selection physicians, dentists, allied specialists SSS. Mem. Acad. Pharm. Scis. (pres. elect), Am. Pharm. Assn., Am. Chem. Soc., A.A.A.S., Am. Assn. Colls. of Pharmacy (past pres.), Nat. Acad. Scis. (past com. chmn.), Sigma Xi, Rho Chi. Author: (with others) Quantitative Pharmaceutical Chemistry, 1957. Research, numerous publs. primarily in field of relationships between molecular structure and biol. activity. Home: 339 Burlage Circle, Chapel Hill, N.C. 27515.*

HAGER, (Hans) Herman Julius, German pharmacist; b. Düben, Germany, Jan. 3, 1816; s. Johannes Hager; pharmacy apprentice, Salzwedel; passed state pharmacist exam, 1841; hon. dr., Jena, Germany; m. Auguste Bauer; 1 son, 1 dau. Mgr. pharmacy, Fraustadt, 1843-59; moved to Berlin; bought powder mill nr. Fürstenberg/Oder, 1871; moved to Frankfort/Oder, 1881, to Neuruppin, 1896 (all Germany). Author: Handbuch der pharmaceutischen Receptirkunst, 1850; Die neuesten Pharmakopöen Norddeutschlands, from 1854; Anleitung zur Fabrikation künstlicher Mineralwässer, 1860; Medicamenta hoeoepathica et isopathica omnia, ad id tempus a medicis aut examinata aut usa recepta, 1861; Lateinisch-deutsches Wörtenbuch zu den neuesten und auch älteren Pharmakopöen . . . , 1862; Erster Unterricht der Pharmaceuten, 1868; Untersuchungen, Ein Handbuch der Untersuchung, Prüfung und Werthbestimmung aller Handelswaaren . . . , 2 vols., 1870-74; Übersetzung des Deutschen Arzneibuches ins Lateinische, 1872; Manuale pharmaceuticum, 2 vols., 1859-73; Handbuch der pharmaceutischen Praxis, 1875. Devised new testing methods in food chemistry; invented, improved apparatus used in pharm. analysis. Died Neuruppin, Jan. 24, 1897.

HAGER, Karl Heinrich, German engr.; b. Mainz, Germany, Jan. 17, 1868; s. Theodor and Anna (Mayer) H.; ed. tech. univs. Dresden, Munich; engring. degree, 1893; m. Elsa Halm, 1897; 1 dau. With Bavarian state rys., Nuremberg, Ingolstadt, Munich; became 1st prof. reinforced concrete Tech. U. Munich, 1906, prof. engring. sci., 1908, recoor, 1917-18, 18-19; advisor Bavarian Ministry Commerce, 1907; dir. Bavarian state indsl. inst., Nuremberg, from 1919. Author: Die Berechnung rechteckiger Platten mittels trigonometrischen Reichen, 1911; Theorie des Eisenbetons, 1916; also articles. Pioneer in methods of reinforced concrete constrn.; 1st to collect measurement bases for reinforced concrete parts subject to pressure, torsion, thrust, flexibility; calculated methods for cross spans of reinforced concrete plates. Died Simbach/Inn, Dec. 25, 1946.

HAGFORS, Tor, physicist; b. Oslo, Norway, Dec. 18, 1930; s. Vidar Johan and Hanna (Edmundsson) H.; Ph.D., U. Oslo, 1959; M.A., U. Tech. Norway, 1955; m. Gillian Patricia Hart, Jan. 3, 1953; children—John, Toril, Martin, Vivien. Staff, Norwegian Def. Research Establishment, 1955-59, sr. sci. officer, 1961-63; research asso. Stanford, 1959-60; staff Mass. Inst. Tech., Lincoln Lab., Lexington, 1963-67; dir. Radio Observatorio de Jicamarca, Lima, Peru, 1967-—. Mem. Am. Geophys. Union, I.E.E.E., Am. Astron. Soc., Norwegian Phys. Soc., Norwegian Geophys. Union. Research, publs. on propagation of electromagnetic waves, lunar and planetary exploration by radio wave scattering, reflection and scattering of electromagnetic waves from plasmas, stoachastic phenomena in wave propagation. Home: Acton St., Carlisle, Mass. 01741. Office: Radio Observatorio de Jicmarca, Lima, Peru.*

HÄGG, Gunnar, Swedish chemist; b. Stockholm, Sweden, Dec. 14, 1903; s. Erik and Hertha (Trägardh) H.; Filosofie kandidat, U. Stockholm, 1925, Filosofie licentiat, 1927, Filosofie dr., 1929; m. Gunnel Silfwerbrand, Apr. 20, 1934; children—Ingemund, Erik, Birgitta (Mrs. Lars Landström), Anders, Bengt. Lectr. gen. and inorganic chemistry U. Stockholm, 1929-36; prof. gen. and inorganic chemistry U. Uppsala (Sweden), 1936-—. Pres., Swedish Nat. Com. Crystallography, 1960-—; mem. Nobel Com. for Chemistry, 1965-—. Decorated knight comdr. Order of North Star. Mem. Internat. Union Crystallography (past v.p.), royal acads. scis. in Stockholm, Uppsala, Lund, Copenhagen, Oslo and Amsterdam, Royal Acad. Engring. Scis., Stockholm, German Acad. for Scientists, Leopoldina. Author: Kemisk reaktionslära, 1940; Allmän och oorganisk kemi, 1963; also numerous articles. Research in inorganic chemistry and metallography, especially with X-ray diffraction methods; crystallography, crystal structure and crystal chemistry; metallic compounds between transition elements and non-metals, metal oxides, non-stoichiometry; constrn. and devel. of X-ray diffraction instruments and computers. Home: 24 Thunbergsvägen, Uppsala, Sweden.*

HAGGENMACHER, (Gustav) Adolf, explorer; b. Island Limmatau, Switzerland, May 3, 1845; s. Johann Jakob and Marie (Eichenberger) H.; m. Maria Cantariny, 1868; 4 children. Apprentice, comml. house, Basel, Switzerland, 1863-65; bookkeeper, piano tchr., Cairo, Egypt; traveled with caravan to Khartoum, 1868; in service Sudanese missionary soc. to explore Ethiopia, 5 years; entered (with Werner Munzinger) service of Khedive of Egypt, explored Somaliland, Gallaland; named dep. of Gov.-gen. Munzinger, Kassala, also Massaua. Helped open up Africa; gathered agrl. and handwork products, geog. and ethnographic data in unexplored regions of Somaliland and Gallaland; searched for lost Ethiopian writings in Ankober, Greek traces in Akik; collected tales and folklore of nomadic tribes; explored western Ethiopia; settled boundary dispute between Ethiopia and Egypt. Killed in mil. expdn. by Gallas, Tadjurra, nr. Aussa, Nov. 20, 1875.

HAGIHARA, Yusuke, Japanese astrophysicist; b. Osaka, Japan, Mar. 28, 1897; D.Sc. in Astronomy, U. Tokyo, 1927; m. Yuko. Joined Tokyo Astron. Obs., 1921; became asst. prof. U. Tokyo, 1923, prof., 1935-—; dir. Tokyo Astron. Obs., 1946-57; prof. Tohoku U., Sendai, Japan, 1957-60; pres. Utsonomiya U., 1960-64. Recipient Watson medal U. S. Nat. Acad. Scis., 1960. Mem. Japan Acad., Japan Sci. Council, Internat. Astron. Union (past v.p.). Author: General Astronomy, 1955; Stability in Celestial Mechanics, 1957. Research on stability of natural and artificial satellites, theory of planetary nebulae, libration phenomena in planetary and satellite motions; developed theory of astron. refraction. Home: 51 Hikawa-cho, Minatoku, Tokyo, Japan.

HAGIWARA, Takahiro, Japanese seismologist; b. Tokyo, Japan, May 11, 1908; s. Yoshio and Hisa (Kubo) H.; student Tokyo U., 1929-32, D.Sc., 1942; m. Toyoko Takémura, Mar. 28, 1946; children—Kazuyoshi, Kenji, Toru. Faculty, Earthquake Research Inst., Tokyo U., 1933-—, prof., 1944-—, dir., 1965-67. Chmn. sectional meeting earthquake prediction Geodetic Council Ministry Edn. Japan, 1963-—; leader Earthquake Prediction Research Group in Japan, 1961-—. Author: Vibration Measurements, 1945; Prediction of Earthquake, 1966; also numerous articles. Research on improvement of seismographs, instrumental study of changes in inclination and extension of earth's surface related to earthquake occurrence; invented water-tube tiltmeter, silica-tube geoextensometer, electromagnetic seismograph. Home: 3-24 Nishigahara, Kita-ku, Tokyo, Japan.*

HAGNER, Francis Randall, Am. genito-urinary surgeon; b. Washington, Feb. 19, 1873; s. Charles E. and Isabella (Davis) H.; M.D., George Washington U., 1894, D.Sc., 1939; m. Elizabeth Allemong, June 1899. Practiced medicine, Washington, 1896—; prof. genito-urinary surgery George Washington U., 1905-39, emeritus. Fellow A.C.S. Introduced open operation for relief of gonorrheal epididymitis, 1906. Died July 7, 1940.

HAGSTRÖM, Stig Bernt Matteus, Swedish physicist; b. nr. Nässjö, Sweden, Sept. 21, 1932; s. Johan and Nanny (Svanberg) H.; Fil.Dr., U. Uppsala, Sweden, 1964; m. Brita-Stina Felldin, June 23, 1957; children—Stig Anders, Mats Fredrik, Karin Maria. Staff research lab. Mass. Inst. Tech., Cambridge, 1964, Lawrence radiation lab. U. Cal. at Berkeley; univ. lectr. Chalmers U. Tech., Gothenburg, Sweden, 1966—. Mem. Am. Phys. Soc. Research, publs. on atomic and x-ray physics; contbr. to devel. of a highly accurate method for measuring atomic energy levels of core electrons by means of photoelectron spectroscopy. Home: Kungsbackag. 3. Office: Dept. Physics, Chalmers Technical U., Gothenburg S, Sweden.*

HAGSTRUM, Homer Dupre, Am. physicist; b. St. Paul, Minn., Mar. 11, 1915; s. Andrew and Sadie (Fryckberg) H.; B.E.E., U. Minn., 1935, B.A., 1936, M.S., 1939, Ph.D. Physics, 1940; m. Bonnie Doon Cairns, Aug. 29, 1948; children—Melissa Billings, Jonathan Tryon. With Bell Telephone Labs., Inc., Murray Hill, N.J., 1940—, research physicist, head, surface physics research dept. 1954—. Recipient Eta Kappa Nu prize, 1932. Fellow Am. Phys. Soc.; mem. Research Soc. Am., Sigma Xi, Eta Kappa Nu, Tau Beta Pi. Research in ionization and dissociation of diatomic gases by electron impact, microwave magnetrons (patents), phys. electronics, mass spectroscopy, electron ejection from solids by ions. Home: 27 Colony Dr., Summit, N.J. 07901. Office: Bell Tel. Labs., Murray Hill, N.J. 07971.

HAGUE, Arnold, Am. geologist; b. Boston, Dec. 3, 1840; s. William and Mary Bowditch (Moriarty) H.; Ph.B., Sheffield Scientific Sch., Yale, 1863; studied at univs. of Göttingen and Heidelberg and at Freiberg Sch. of Mines, 1863-66; Sc.D., Columbia, 1901; LL.D., Aberdeen, 1906; m. Mary Bruce Howe, Nov. 14, 1893. Asst. geologist U. S. Geol. Exploration, 40th parallel, and made investigations of mines and mining processes in Nevada, 1867-77; govt. geologist, Guatemala, 1877-78; examined mines in northern China for the govt., 1878-79; geologist of U. S. Geol. Survey, from 1879. Mem. commn. apptd. by Nat. Acad. Scis. at request of U. S. Govt., 1896, to prepare plan of Nat. forest reserves; v.p. Internat. Geol. Congresses, Paris, 1900, Stockholm, 1910, Toronto, 1913. Author: Geology of the Yellowstone National Park (with others), 1899; Atlas of Yellowstone National Park (2 plates), 1904; The Origin of the Thermal Waters in the Yellowstone National Park, 1911. Died May 14, 1917.

HAHN, Amandus Wilhelm Luitpold Emil Louis Heinrich, German physiologist; b. Düsseldorf, Germany, Jan. 16, 1889; s. Otto and Anna (Scheepers) H.; Ph.D., U. Munich, (Germany), 1911; M.D., 1914; m. Maria Gerstner, 1937; 1 son. Mil. physical, World War I; asst. to O. Frank, physiol. inst. U. Munich, became mem. faculty, 1919, asso. prof., 1924, prof., 1946. Author: Die Bedeutung des Massenwirkungsgesetzes fur die Physiologie, 1921; Grundriss der Biochemie fur Studienrende, 1922, reissued as Grundriss der physiologischen Chemie, 1946; also articles. Research on pyrole synthesis, blood and bile dyes, properties and reactions of enzymes; made studies related to Wieland's dehydrogenation theory on breakdown of succinic acid to pyroracemic acid, later recognized as part of citric acid cycle; studied nature of recovery processes, also concept of coupled reaction. Died Munich, Jan. 1, 1952.

HAHN, Dietrich, German physicist; b. Berlin, Germany, Feb. 28, 1922; s. Otto and Hanna (Gerlach) H.; ed. in physics Tech. U. Berlin; Ph.D. in engring.; m. Friederike Kubetschka, 1964; children—Hartmut, Peter. Dir., Fed. Inst. Tech. Physics; prof. Tech. U. Berlin. Mem. Assn. German Assn. Physicists, Assn. for Tech. Lighting, German Assn. Tech. Optics. Contbr. articles to profl. jours. Home: 98 Am Schlachtensee, Berlin 38. Office: 2-12 Abbestrasse, Berlin 10, Germany.

HAHN, Erwin Louis, Am. physicist; b. Sharon, Pa., June 9, 1921; s. Israel and Mary W. (Weiss) H.; B.S., Juniata Coll., 1943, Sc.D., 1966; M.S., U. Ill., 1947, Ph.D., 1949; m. Marian Ethel Failing, Apr. 8, 1944; children—David L., Deboreh A., Katherine L. NRC fellow Stanford, 1950-52, instr. 1951-52; cons. Office Naval Research, Stanford, 1950-52; research physicist Watson Sci. Computing Lab., 1952-55; asso. Columbia, 1952-55; faculty U. Cal., Berkeley, 1955—, prof., 1961—, A. C. Miller Sci. prof., 1966-67; asso. prof. Miller Inst. for Basic Research, Berkeley, 1958-59. Cons. U. S. AEC, 1955—; spl. cons. UN, 1959. Guggenheim fellow, 1961-62; NSF fellow, 1961-62. Fellow Am. Phys. Soc. (West Coast dep. sec. 1960-61). Research in nuclear and electron spin magnetic resonance, electronic instruments, electron physics, nuclear spin coupling in molecules, low temperature physics, discovered spin echoes. Home: 69 Stevenson Av., Berkeley, Cal. 94708.*

HAHN, Frank John, Am. mathematician; b. N.Y.C., Apr. 14, 1929; s. Karl Heinrich and Margaretta (Roth) H.; B.A., U. Rochester, 1951; M.A., U. Ill., 1956, Ph.D., 1959; m. Elizabeth Jacobson, Aug. 4, 1951; children—Karl, Nancy, Katherine, Amy. NSF postdoctoral fellow, vis. mem. Inst. Advanced Study, 1959-60; faculty Yale, 1960—, asso. prof., 1965—, prof., 1968—; NSF sr. postdoctoral fellow, vis. scholar U. Cal., Berkeley, 1965-66. Mem. Am. Math. Soc., Math. Assn. Am. Author: (with L. Auslander and L. Green) Flows on Homogeneous Spaces, 1963. Research on understanding of pointwise almost periodic transformations and flows, their relationship to diophantine approximations; information on transformations with a unique invariant volume has been obtained. Died Jan. 14, 1968.

HAHN, Fred Ernst, molecular biologist; b. Allenstein, Germany, Sept. 22, 1916; s. George H. J. and Ruth (Prusse) H.; Dipl.Chem., U. Kiel (Germany), 1943, Dr. rer.nat., 1948; m. Rosemarie Weyland, July 12, 1945; children—Patricia Monica H., Michael Gregory, Christina Andrea. Came to U. S., 1949, naturalized, 1958. Chief biochem. dept. Inst. Virus Research, U. Heidelberg (Germany), 1946-49; research biochemist Walter Reed Army Inst. Research, Washington, 1949-56, head chemotherapy research unit, 1956-59, chief dept. molecular biology, 1959—. Professorial lectr. molecular pharmacology George Washington U., Sch. Medicine, Washington, 1965—. Recipient Meritorious Civilian Service award U. S. Dept. Army, 1961. Mem. Am. Soc. Biol. Chemists, Am. Soc. for Microbiology, A.A.A.S. Research, numerous publs. on mode action antimicrobial substances, molecular biology protein and nucleic acid synthesis, biochem. studies on rickettsiae demonstrating cellular nature, structure-activity relationships. Home: 8309 Westmont Terrace, Bethesda, Md. 20034. Office: Dept. Molecular Biology, Walter Reed Army Inst. Research, Washington 20012.*

HAHN, Fritz Philipp Karl, German pharmacologist; b. Koenigstein, Germany, Feb. 13, 1907; s. Karl and Karoline (Stiehl) H.; Medizinischens Staatsexamen, U. Frankfurt, 1926, Dr.med., 1932; m. Margot Wedekind, Aug. 22, 1935; children—Helmut, Renate. Faculty, U. Frankfurt, 1931-33; staff serology inst. U. Heidelberg (Germany), 1933-36, Pharmakolog. Inst., U. Cologne (Germany), 1936-45; staff Medizin. Akad. Düsseldorf (Germany) Pharmakol. Inst., 1945-60 dir., 1951-60; dir. Pharmakoal. Inst., U. Freiburg (Germany), 1960—. Mem. Deutsche Pharmakolog. Gesellschaft, Deutsche Physiolog. Gesellschaft, Deutsche Biochemische Gesellschaft, Collegium internationale allergologicum. Contbg. author: Handbook of Pharmacology, 1954; Pharmacological Revs., 1960. Research and numerous articles on blood groups, immunological organ specificity, lability reactions of serum, autoantibodies, anaphylatoxin, histamine and histaminase in anaphylaxis, digitalis, central nervous system, toxicological problems of food additions. Home: 12 Stollenweg, 7801 Wittnau, Germany. Office: 29 Katharinenstrasse, 78 Freiburg i.Br., Germany.*

HAHN, George Thomas, metallurgist; b. Vienna, Austria, July 28, 1930; s. Rudolf and Stella (Honig) H.; came to U. S., 1938, naturalized, 1943; B.Mech. Engring., N.Y. U., 1952; M.S. in Metall. Engring., Columbia, 1953; Sc.D., Mass. Inst. Tech., 1959; m. Charlotte Minovitz, June 10, 1956; children—Claudia, Elizabeth. Research asst. N.Y. U., 1952; research engr. Westinghouse Research Lab., Pitts., 1952; project engr. USAF Materials Lab., Dayton, O., 1953-55; cons. Mfg. Labs., Cambridge, Mass., 1958-60; research asso. Mass. Inst. Tech., 1959-60; div. cons. Battelle Meml. Inst., Columbus, O., 1960-62, research asso., 1962-63, fellow, 1963-64, asso. div. chief, 1965—. Mem. Am. Inst. Metall. Engrs., Am. Soc. Metallurgists, Am. Soc. for Testing and Materials. Editor: (with others) Fracture, 1959. Research, publs. on yielding, cleavage of metals fracture, and fracture mechanics. Home: 786 Eastmoor Blvd., Columbus, O. 43209.*

HAHN, Hans, Austrian mathematician; b. Vienna, Austria, Sept. 27, 1879; s. Ludwig Benedikt and Emma (Blümel) H.; student, Strasbourg (now in France), Munich, Germany; Ph.D., Vienna, 1902; m. Eleonore Lilly Minor, 1909; 1 dau., Nora. Became mem. faculty U. Vienna, 1905, prof., from 1921; substitute for O. Stolz, Innsbruck, Austria, 1905-06, then lectr.; named asso. prof., Czernowitz (now USSR), 1909; became asso. prof., Bonn, Germany, 1916; prof., 1917. Recipient Lieben prize, 1921. Mem. Acad. Scis. Vienna (corr.), Calcutta Math. Soc. (hon.), German Assn. Mathematicians (com. mem. 1931-32). Author: Theorie der reellen Funktionen I, 1921; Reele Funktionen I, 1932; also articles. Investigated calculus of variations, theory of quantities, real functions; proved many theorems by gen. postulates; introduced concept of small-scale correlation and used it to elucidate theory of quantity of points; studied representation of given functions by singular integrals and their application. Died Vienna, July 24, 1934.

HAHN, Hans, psychologist; b. Nuremberg, W. Germany, Mar. 12, 1900; s. Sigmund and Nini (Schwarzbauer) H.; Ph.D., U. Heidelberg (Germany), 1923; Chief asst. Jugendkundisches, Psychology Inst., Nuremberg, 1922-24; practice psychology, Germany, 1922-33; spl. research asst. Inst. Hautes Études, Belgium, 1933; prof., dir. Inst. Exptl. Psychology, State U. Trujillo, Peru, 1938-39; prof. applied psychology and anthropology, head dept. U. Nacional Superior, San Marcos, Lima, Peru, 1939-50; head dept. psychology Transylvania Coll., Lexington, Ky., 1950-65, prof. emeritus, 1965—; dir. clin. psychology and research Woodbridge State Sch. (N.J.), 1965—; acting dir. dept. psychology N.J. State Hosp., Trenton; head dept. vocational guidance and applied exptl. psychology Brit.-Peruvian High Sch. System, 1944-50; head psychotechnician Inst. Peruvian Police Army, Navy, and Air Force, 1943-50. Fellow Acad. Psychosomatic Medicine (contbg. editor Psychosomatics), Internat. Council Psychologists, Am. Assn. for Humanistic Psychology; mem. Am., N.J., Ky., Central Ky., Southeastern psychol. assns., A.A.A.S., Am. Assn. on Mental Deficiency, German Assn. Psychologists, Research, publs. on applied exptl. psychology in industry, atitude research, empathy and perception, apperception in normal, neurotic and psychotic individuals, personality tests, accident proneness, influence of advt. on children, applied anthropology. Office: Woodbridge State Sch., Woodbridge, N.J.*

HAHN, Johann Siegmund, physician; b. Schweidnitz, Silesia, Nov. 23, 1664; s. Gottfried and Anna Maria (Hube) H.; ed. Leipzig, Germany, Leiden, Netherlands; M.D.; m. Katherine Sophie Grass; 1 son, Johann Gottfried von H.; m. 2d; 1 son, Johann Siegmund. Practiced medicine, also municipal physician, Schweidnitz; personal physician to Polish prince Jakob Sobieski. Author: Peterswalder Gesundbrunnen, 1732; Psychrolusia veterum renovata, jam decocta, 1738. Pioneer of natural water cure in Germany; cured a case of typhus by cold water cure; followed Schwerdtner's methods of water treatment. Died Schweidnitz, Oct. 6, 1742.

HAHN, Karl von, see von Hahn, Karl.

HAHN, Milton Edwin, Am. psychologist; b. Galion, O., Oct. 15, 1903; s. Dale Owen and Madge (Linnell) H.; B.A. Hamline U., 1927; M.S., U. Minn., 1938, Ph.D., 1942; m. Margaret I. Benson, Feb. 28, 1928; 1 son, Owen M. High sch. tchr., dir. student personnel, St. Paul Pub. Schs., 1931-38; instr. U. Minn., 1938-41, dir. men's activities, 1941-42; prof., dir. Psychol. Services Center, Syracuse U., 1944-48; prof., dean student U. Cal. at Los Angeles, 1948-60, prof. psychology, 1960—. Cons. to bus. and industry, govt. agys., 1940—. Mem. Am. Psychol. Assn. Author numerous books including: (with M. S. Maclean) Counseling Psychology, 1955; Psychoevaluation: Adaptation, Distribution, Adjustment, 1963; also numerous articles. Research on personality dimensions of normal adult individual including edn., mil. service, growth and devel., nonverbal communication and aging. Home: 1483 La Linda Lake, San Marcos, Cal. 92069. Office: 405 Hilgard St., Los Angeles 90024.*

HAHN, Otto, German physical chemist; b. Frankfort on Main, Germany, Mar. 8, 1879; s. Heinrich and Charlotte (Giese) II; student U. Marburg, 1897, U. Munich, 1898, Dr. deg., 1901; student Univ. Coll., London, 1904-05, Phys. Lab. McGill U., Montreal, Can., 1905-06; m. Edith Junghans, Mar. 22, 1913; 1 son, Hanno. Faculty, U. Berlin, 1907, prof. chemistry, 1910-33; mem. Kaiser Wilhelm Inst. Chemistry, 1912-28, dir., 1928-45; pres. Max Planck Soc. for Advancement Sci. (formerly Kaiser Wilhelm Soc.), 1946-60. Recipient Nobel prize for chemistry (for splitting uranium atom and discoveries in chain reaction prin.), 1944. Mem. Berlin, Göttingen, Munich, Stockholm, Vienna, Madrid, Helsinki, Lissabon, Mainz, Rome (Vatican), Allahabad, Copenhagen, Boston, Indian acads. sci. Author: Applied Radiochemistry, 1936; New Atoms, 1950; Vom Radiothor zur Uranspaltung, 1962. Did important work on nuclear fission; discovered protactinium, 1917, later discovered several transuranium elements from atomic number 94 to 96; obtained exptl. evidence of nuclear fission in uranium, noting a product to be rare gas krypton. Died Göttingen, Germany, July 28, 1968.

HAHN, Petr, physiologist; b. Berlin, Germany, Nov. 11, 1923; s. Arnold and Maria (Katz) H.; 1st B.Sc., U. Wales, Swansea, 1941; M.D., Faculty Medicine, Prague (Czechoslovakia), 1950; C.Sc., Acad. Scis., Prague, 1953, D.Sc., 1964; m. Nadezda Novozamska, July 16, 1948; children—Gena, Martin. Asst., Inst. Physiology, Prague U., 1950-51; staff Inst. Physiology, Czechoslovak Acad. Scis., Prague, 1951—, sci. worker, 1953, head lab. for developmental nutrition, 1962—. Lectr., Pediatric Faculty, Prague U., 1963—; cons. Parliamentary Com. on Nutrition, 1963—. Recipient awards for sci. achievement Czechoslovak Acad. Scis., 1956, 60, 64. Mem. Am. Oil Chemists Soc., Physiol. Soc. Prague, Pediatric Soc. Prague. Author: (with Krecek, Capek, Martinek) Development of Homeostasis, 1956; (with Mourek, Jilek, Kopecky) Resistance to Hypoxia in the Newborn, 1961; Development of Nutritional Functions, 1966; (with Koldovsky) Development of Metabolic Functions and Nutrition, 1967; also numerous articles. Research on effect of food composition on postnatal devel. of mammals and man, milk fat especially related to enzyme activity of infant, adrenal glands. Home: Nad Budankami I/13 Prague. Office: Inst. Physiology, Acad. Scis., Prague, CSSR.*

HAHN, Richard Balser, Am. chemist; b. Detroit, July 6, 1913; s. Balser P. and Hattie (Liebau) H.; B.S., Wayne State U., 1935, M.S., 1936; Ph.D., U. Mich., 1948; m. Constance L. Lake, June 25, 1938; children—Paul B., Thomas E. Tchr. high sch., Detroit, 1936-42; faculty Wayne State U., Detroit, 1942—, prof. chemistry, 1956—; instr. U. Mich., 1946-48; research chemist Oak Ridge Nat. Lab., 1950-51. Mem. Am. Chem. Soc., Phi Lambda Upsilon, Phi Kappa Phi. Author: (with F. J. Welcher) Semi Micro Qualitative Analysis, 1955, Inorganic Qualitative Analysis, 1960; also articles. Research on chemistry of zirconium, hafnium, radiochem. analysis for fission products. Home: 11917 Payton St., Detroit 48224.*

HAHN, Richard Leonard, Am. nuclear chemist; b. N.Y.C., May 25, 1934; s. Robert and Sylvia (Weingarten) H.; B.S. summa cum laude, Bklyn. Coll., 1955; M.A., Columbia, 1956, Ph.D., 1960; m. Sheila Thomas, Sept. 3, 1956; children—Sharyn Gale, Jill Karen, Pamela Diane. Research asso. Brookhaven Nat. Lab., Upton, N.Y., 1960-62; chemist Oak Ridge Nat. Lab., 1962—. Mem. orgn. com. 9th Conf. on Analytical Chemistry in Nuclear Tech., 1965. Mem. A.A.A.S., Am. Phys. Soc., Am. Chem. Soc., Phi Beta Kappa, Phi Lambda Upsilon. Author: Guide to Activation Analysis, 1964; also articles. Research in nuclear chemistry, mechanisms reactions nuclei with charged particles, nuclear properties of transuranium nuclei, new isotopes high-energy nuclear fission, decay schemes, effects nuclear decay on chem. properties, activation analysis. Office: Oak Ridge Nat. Lab., Oak Ridge 37830.*

HAHN, Stuart Hamilton, Am. mech. engr.; b. Wilkinsburg, Pa., Feb. 28, 1903; s. John Cooper and Katherine (McKee) H.; B.S., Carnegie Inst. Tech., 1927; m. Aileen Gregory, Sept. 6, 1930; 1 son, Peter Hamilton. With Timken Roller Bearing Co., 1927-28; with B. F. Goodrich Co., 1928-41, sect. head, phys. research lab., 1933-41; with Curtiss-Wright Corp., 1941-44; resident project engr. AiResearch Mfg. Co. Ariz., 1944-46; with Dalmo Victor Co. div. Textron Corp., 1947-55, engring. cons., 1950-55; engring. specialist Sylvania Electronic Systems-West div. Sylvania Elec. Products, Inc., Mountain View, Cal., 1955-58, head mech. devel. sect., 1958-63, staff mech. engr., 1963—. Mem. Soc. Automotive Engrs. Contbr. articles to profl. publs. Patentee composite elasto-meric vibration isolator, direct-reading portable air-flow resistance meter, automatic ice warning and de-icer control apparatus, light-weight laminated surface honeycomb-core constrn. for microwave reflectors, high speed inertialess electro-mech. recorder; inventor numerous other devices. Home: 600 Pennsylvania Av., Los Gatos, Cal. 95030. Office: P.O. Box 205, Mountain View, Cal. 94042.*

HAHNEMANN, (Christian Friedrich) Samuel, German physician; b. Meissen, Germany, Apr. 10, 1755; s. Christian Gottfried and Johanna Christiane (Spiess) H.; attended schs. in Vienna, Leopoldstadt; studied medicine in Leipzig, from 1775; doctorate from Erlangen, 1779; m. Henriette Küchler, 1782; 2 sons, 8 daus.; m. 2d, Mélanie d'Herville, 1835. House physician to gov. of Siebenbürgen, 1777-79; physician in Hettstedt and Gommern; gave up practice, worked in chemistry in Dresden, 1784-88; returned to medicine, moved 17 times, settled in Torgau, 1805; accepted to faculty U. Leipzig, 1811, expelled, 1821; practiced in Köthen, became internationally known; moved to Paris, 1835, had huge practice. Mem. Leopoldina. Author: Versungen über ein neues Prinzip zur Auffindung der Heilkräfte, Hufelands Journal, 1796; Fragments sur le propriétés positives des médicaments, 1805; Organon der rationellen Heilkunde, 1810; Reine Arzneimittellehre, 6 vols., 1811-20; La matiére médicale pure, 1820; Traité des maladies chroniques, 1821/28; Die chronische Krankheiten, ihre eigentümliche Natur und Heilung, 4 parts, 1828-38. Founder of homeopathy; discovered that all medicines are made soluble in water by rubbing for long time (anticipated colloid chemistry); in cholera epidemic of 1831 was only person to point to tiny form of life as cause, introduced antiseptic treatment against it. Died Paris, July 2, 1843.

HAHON, Nicholas, Am. virologist; b. N.Y.C., Mar. 24, 1924; s. Samuel A. and Catherine (Walesk) H.; B.S. cum laude, Davis and Elkins Coll., 1948; Sc.M. in Hygiene, Johns Hopkins, postgrad. (Kellogg scholar); m. Katheryn E. Harris, Jan. 31, 1948; 1 dau., Nicolette K. Virologist, So. div. U. S. Army Biol. Labs., Ft. Detrick, Frederick, Md., 1951-58, supervising virologist aerobiology div., 1958—. Recipient Sci. Research Soc. Am. award for outstanding research, 1965. Mem. A.A.A.S., Am. Soc. for Microbiology, Research Soc. Am., Nat. Registry Microbiologists, Am. Bd. Microbiology. Author: Selected Papers on Virology, 1964; also articles. Characterized smallpox and related pox virses in cell cultures and chick embryos, pathogenesis poxvirus infections and immune responses in monkeys; developed quantitative and rapid assay viruses and rickettsiae and serum-neutralization test employing fluorescent antibody techniques. Home: 803 Shawnee Dr. Office: U. S. Army Biology Labs., Ft. Detrick, Frederick, Md. 21701.*

HAID, August, German chemist; b. Horb am Neckar, Germany, July 14, 1886; s. August and Anna (Maucher) H.; D.Eng., Tech U. Stuttgart (Germany). With central inst. for sci. research, Heubabelsberg, 3 years; specialist on explosives Reich navy office, 6 years; became mem. staff Reich chem.-tech. inst. (name changed to Fed. Inst. for Materials Testing after World War II), 1920, dir. dept. explosive materials, 1928, v.p. after World War II, ret., 1946. Contbr. articles to jours. Research on detonation of explosives, hydrodynamic theory of detonation, relationship between chem. make up of material and its capacity for explosive reaction; clarified concept of explosive power; suggested intro. of pentaerythrite-tetranitrate (nitropenta), cyclotrimethylene-trinitramine (hexogen), nitroguanidine; developed pourable ammonium nitrate with hexogen additive, testing processes for explosive materials. Died Munich, Germany, Jan. 21, 1963.

HAID, Matthäus Franz, German geodesist; b. Speyer, Germany, Feb. 28, 1853; s. Johann Friedrich and Catharina (Lederle) H.; student Berlin Indsl. Acad., 1870-71, Aachen Polytechnikum, 1871-72, Munich Polytechnikum, until 1874; doctorate, Jena, Germany; m. Henriette Lichtenberger, 1880; 3 children. With Bavarian state service, 1874-77; asst. to C. M. von Vauernfeind, Munich, became mem. faculty, 1880; named asso. prof. practical geometry and higher geodesy Karlsruhe (Germany) Poly. Sch., 1882, prof., 1884, emeritus, 1917, rector, 1894-95, 1901-02. Chmn. topog. bur., Baden; asso. mem. adminstrn. roads and waterways, Baden; mem. Baden standards commn.: adviser on land survey Greek govt. Mem. Nat. Sci. Assn. (dir. seismol. commn. Karlsruhe 1903). Author: Gezeiten und Starrheitskoeffizienten der festen Erde abgeleitet an den Registrierungen des Horizontalpendel in Freiburg i.B. und Durlach, 1912; also articles. Made depth measurements in Lake Constance; founder seismol. stas. equipped with Hecker horizontal pendulums, Durlach, Freiburg. Died Speyer, Nov. 5, 1919.

HAIDER, Syed Z., chemist of Pakistan; b. Dacca, Pakistan, Sept. 1, 1927; s. Zamir and Hajera Haider; B.Sc. with honours in Chemistry, Dacca U., 1947, M.Sc., 1948; Ph.D., London U., 1958, diploma Imperial Coll., 1958; m. Saleha Haider. Mar. 1, 1950; 6 children. With Dacca U., 1950—, reader chemistry, 1961—, head. dept. chemistry; faculty Notre Dame Coll., Dacca. Recipient H.P. Royal Gold medal Dacca U., 1962. Mem. Pakistan Assn. for Advancement Sci. (asso. sec.), Pakistan Assn. Scientists and Sci. Professions (gen. sec.), Royal Inst. Chemistry (asso.). Author: A Text Book of Inorganic Chemistry; also articles. Developed new methods for synthesis of metal borate complexes, organo-boron compounds, phosphates of zirconium, niobium, tantalum, antimony; new methods for determinations of borate and phosphate in complex mixtures; established tetragonal unit cell of niobium oxy-phosphate by X-ray diffraction; synthesized new inorganic ion-exchangers; studied mechanism of reaction kinetics of inorganic complex formations, adsorption phenomena in number of co-precipitated products, magnetic properties of mixed metal/oxides; observed explosive and germicidal properties of organo-boron compounds. Home: 37/B Fuller Rd., Dacca-2, Pakistan.*

HAIDINGER, Wilhem Karl von, see von Haidinger, Wilhelm Karl.

HAIG, Charles, physiologist; b. N.Y.C., Nov. 18, 1900; s. Val Douglas and Pauline (Brown) H.; B.S., Columbia, 1927, M.A., 1928, Ph.D., 1935; m. Eleanor Mancell Hughes, Mar. 29, 1924 (dec.); m. 2d, Marjorie Douglas Faulkner Mancell, Apr. 24, 1962. Fellow in biology Coll. City N.Y., 1927-28, tutor, 1928-41; faculty Columbia, N.Y.C., 1926-27, 35-44, research fellow, 1937-44; faculty N.Y. Med. Coll., 1942-60, prof. physiology, 1954-60, dir. dept., 1954-56; sr. research fellow Nat. Council to Combat Blindness, U. Lab. Physiology, Oxford, Eng., 1961-62; sr. research fellow in physiol. optics City U., London, 1963-65; ret. Fellow A.A.A.S., N.Y. Acad. Scis., N.Y. Acad. Medicine (asso.); mem. Optical Soc. Am., Am. Phys. Soc., Am. Physiol. Soc., Am. Soc. Zoologists, Harvey Soc., Photobiology Group, Colour Group, Brit. Assn. for Advancement Sci. Publs. on research in phototropism, human vision, visual pigments, vitamin A in chronic diseases and normal requirements, diagnosis and therapy of retinitis pigmentosa, diagnosis of congenital stationary night blindness. Home: 3 Portland Terrace, The Green, Richmond, Surrey, Eng.*

HAILER, Ekkehard Eugen Reinhold, German bacteriologist; b. Horb/Neckar, Germany, Apr. 21, 1877; s. Albert and Anna (Hillerbrand) H.; ed. univs. Munich, Tübingen (both Germany); Dr. rer. nat., circa 1903. Became sci. asst. Imperial Bd. Health, Berlin-Dahlem br., 1903; named govt. advisor, mem. Reich Bd. Health, 1925, sr. govt. advisor, 1932; apptd. dir. chem. dept. Robert Koch Inst. for Infectious Diseases, 1937. Contbg. author: Die Desinfektion, in Handbuch der Hygiene VIII (Weyls), 1922; also articles in jours. and Bd. Health publs. A founder of modern disinfection techniques; worked on disinfection in Tb to combat indsl. anthrax infection; made exptl. studies of value of chem. disinfection products. Died Berlin, May 19, 1939.

HAINES, Thomas Harvey, Am. psychologist, psychiatrist; b. Moorestown, N.J., Nov. 4, 1871; s. Zebedee and Anna Philips (Harvey) Haines; A.B., Haverford (Pa.) Coll., 1896; Ph.D., Harvard, 1901; M.D., Ohio State U., 1912; spl. studies in neurology and psychiatry. Munich, Zurich and London, 1912-13; m. Helen Manley Hague, Aug. 15, 1912. Asst. prof. philosophy, prof. psychology Ohio State U., 1901-15. First asst. phys., Boston Psychopathic Hosp., 1913-14, also prof. psychology Smith Coll. (part time); clin. dir. Ohio Bur. Juvenile Research, 1914-17; prof. medicine (nervous and mental diseases) Ohio State U., 1915-20. Fied cons. and dir. mental health surveys for Nat. Com. for Mental Hygiene in Ky., Ala., Miss., La., Mo., Md., Ariz., N.D., 1917-22; dir. div. on mental deficiency Nat. Com. for Mental Hygiene, 1922-25. Psycho. examiner Camp Dix and Camp Stuart, 1917-18; mem. com. on psychol. exam. of recruits, Nat. Research Council, 1917; psychiatrist at N.Y. Hosp. Out-Patient Dept., 1932-42. Fellow A.A.A.S., Am. Psychiatric Assn. (life mem.); mem. Med. Soc., State N.Y., Med. Soc. County N.Y., Sigma Xi, Phi Beta Kappa. Author: Mental Measurement of the Blind, 1915; reports of mental hygiene studies in many states and bills offered to legislatures to improve administration in mental health fields; some fifty other titles in psychol. and med. jours. Research on mental measurements; developer of specialized tests. Died Mar. 2, 1951.

HAINES, Walter Wells, Am. economist; b. Stamford, Conn., Dec. 1, 1918; s. Thomas Kelly Peterson and Carrie Hooker (Williams); H.; B.A., U. Pa., 1940, M.A., 1941; M.A., Harvard, 1942, Ph.D., 1943; m. Hazel Ellen Maxwell, Jan. 1, 1945; children—Jennifer Jean, Deborah Lee, Pamela Ann, Christopher Alan, Liseli Ellen, Timothy Maxwell. Instr., Kenyon Coll., 1946-47; faculty N.Y. U., 1947—, prof. econs., 1960—, chmn. dept., 1956—. Fulbright lectr. U. Peshawar (Pakistan), 1962-63. Mem. Am. Econ. Assn., Am. Assn. U. Profs., Phi Beta Kappa, Pi Gamma Mu. Author: Money, Prices, and Policy, 1961; also articles. Clarification theory underlying circuit flow of money and integration into nat. income theory; critical analysis of role of imports in benzenoid chem. industry; exploration theoretical and practical problems associated with unified internat. monetary system. Home: Skyview Acres, Pomona, N.Y. 10970. Office: N.Y. U., University Heights, N.Y.C. 10453.*

HAIRSTON, Nelson George, Am. ecologist; b. nr. Mocksville, N.C., Oct. 16, 1917; s. Peter Wilson and Margaret E. (George) H.; B.A., U. N.C., 1937, M.A., 1939; Ph.D., Northwestern U., 1948; m. Martha Turner Patton, Aug. 19, 1942; children—Martha Patton, Nelson George, Margaret Elmer. Faculty, U. Mich., Ann Arbor, 1948—, prof. ecology, 1961—. Cons., WHO, 1954-56, 56, 60, 65, mem. expert adv. panel on parasitic disease, 1964—; mem. tropical medicine and parasitology study sect. NIH, 1965—. Mem. Ecol. Soc. Am., Brit. Ecol. Soc., Soc. for Study Evolution, Am. Soc. Ichthyologists, and Herpetologists, Am. Inst. Biol. Scis., Sigma Xi. Editorial bd. Ecol. Monographs, 1957-63, Ecology, 1963—. Research, publs. on ecol. distbn. salamanders, orgn. natural communities animals, interspecific relationships in Paramecium in nature, sampling techniques, population dynamics snails, schistosomes, filiarial worms, epidemiology worm parasites. Home: 1711 Arlington Blvd., Ann Arbor, Mich. 48104.*

HAISSINSKY, Moise, chemist; b. Taracha, Russia, Nov. 4, 1898; Dr.Chemistry, U. Rome, 1927; Dr.Sci., U. Paris, 1959; m. Jeanne Gay, July 12, 1930; children—Michel, Jacques. Faculty, U. Rome, 1923-28; with Curie Lab., Paris, 1930—, dir. sci. prof. chemistry U. Paris, 1957-62; Donegani prof. Accademia dei Lincei, Rome, 1953. Cons. French AEC. Recipient Hebert prize Paris Acad. Sci., 1939. Mem. Société Chimie Physique, Société Physique, Faraday Soc., Am. Chem. Soc., Am. Phys. Soc. Author: L'Atomistique Moderne, 1930; Chimie Nucleaire, 1957; (with Adloff) Radiochemical Survey of the Elements, 1965; also other books, numerous articles. Research on new methods of electrochem. deposition of radioelements, mechanism of formation of monoatomic layers, nature of radiocolloids; discovery of tetravalent Pa; mechanism of isotope exchange between ions, radiation chemistry of aqueous solutions. Office: 11 Rue Pierre Curie, Paris, France.*

HAIST, Reginald Evan, Canadian physiologist; b. Hamilton, Ont., Can., Jan. 14, 1910; s. Oscar Wesley and Anna (Leppert) H.; B.A., U. Toronto, 1933, M.D., 1936, M.A., 1937, Ph.D., 1940; m. Margaret Scott Milne, Aug. 7, 1936; children—Lynda M. (Mrs. B. D. Pullen), David W., Stephen R., Margaret C., Frances E. B. Faculty, U. Toronto (Ont.), 1940—, prof. physiology, 1949—, chmn. dept., 1965—. Fellow Royal Soc. Can.; mem. Brit., Can., Am. physiol. socs., Am. Diabetes Assn., Acad. Medicine of Toronto, N.Y. Acad. Scis. Research, numerous publs. on exptl. diabetes, function of islets of Langerhams, factors influencing synthesis and secretion of insulin and its carriage by blood, shock secondary to injury, particularly metabolic effects, hypothermia and endocrine function. Home: 56 Cheltenham Av., Toronto, Ont., Can.*

HAITINGER, Ludwig Camillo, chemist; b. Vienna, Austria, Oct. 23, 1860; s. Carl Ludwig and Anna (Wenzl) H.; ed. U. Vienna. Became pvt. asst. to A. Lieben, U. Vienna, 1880; joined firm Welsbach-Williams-Gesellschaft, Vienna-Atzgersdord, 1886; dir. Osterreichische Gasglühlicht- und Elektrizitätsgesellschaft, Vienna; set up incandescent lamp factory,

Phila., 1888; ret., 1907. Contbr. articles to handbooks and jours. Research on acid (important in heterocyclic compound chemistry), chemistry of rare earths in regard to incandescent mantle; recognized cerium oxyde as effective additive in incandescent mantle, 1886; studied emission spectra of praeseodymium and neodynium; assted in Auer von Welsbach's invention of gas incandescent mantle. Died Weidling, nr. Klosterneuburg, Dec. 28, 1945.

HAITINGER, Max, Austrian chemist; b. Vienna, Apr. 20, 1868; s. Carl Ludwig and Anna (Wenzl) H.; ed. U. Vienna, Ph.D. (hon), 1944; m. Emilie Gräf, 1896; m. 2d, Hermine Fabianitsch, 1925; 1 son, 3 daus. Joined army, 1888; tchr. math., physics, chemistry Pioneer Cadet Sch., Hainburg, from 1896, later comdt.; resigned from army as col., 1918; exptl. researcher, inst., Klosterneuburg; became asso. with E. Haschek, 2d phys. inst. U. Vienna, 1931, also with H. Eppinger, 2d med. clinic. Recipient Fritz-Pregl prize for microchemistry, 1937. Author: (with E. Haschek) Farbmessungen, 1936; Die Floureszenzanalyse in der Mikrochemie, 1937; Floureszens-Mikroskopie, ihre Anwendung in der Historlogie und Chemie, 1938; also articles. Founder fluorescense microscopy; discovered secondary fluorescense in form of fluoro-chroming process; contbd. to protein and permeability pathology; introduced fluoro-chroming process into microchemistry and microscopy; developed fluorescense microscope for firm C. Reichert. Died Vienna, Feb. 19, 1946.

HAJDU, Stephen, physiologist; b. Szombathely, Hungary, Jan. 9, 1916; s. Istavan and Elisabeth (Krisch); M.D., U. Budapest (Hungary), 1941; m. Margaret Blakey, Jan. 31, 1949; children—Nicholas, Michael. Came to U. S., 1949, naturalized, 1952. Faculty, U. Budapest, 1939-45, asst. prof., 1941-45; research prof. Hungarian Biol. Inst., Lake Balaton, 1945-47; Brit. Council fellow King's Coll., London, Eng., 1947-48, lectr., 1948-49; research asso. Inst. for Muscle Research, Woods Hole, Mass., 1949-54; surgeon Nat. Heart Inst., Bethesda, Md., 1954-59, sr. surgeon, 1959-64, med. dir., 1964—. Eszterhazy Found. fellow, 1942-43. Mem. Marine Biol. Lab. Assn. Woods Hole, Am. Physiol. Soc., Soc. Gen. Physiology. Research, publs. on nutrition, endocrinology, muscle and heart physiology. Home: 5517 Oak Pl., Bethesda 20034. Office: NIH, Rockville Pike, Bethesda, Md. 20014.*

HAJEK, Marcus, physician; b. Wershetz, Hungary, 1861; M.D., U. Vienna, 1879. Prof. laryngology U. Vienna, 1912. Author: Pathology and Theory of the Inflammatory Diseases of the Nasal Sinuses, 1899. Developed systematic and sci. approach to diagnosis of sinus ailments; improved treatment of empyema of sinuses; studied Tb of upper respiratory tract, syphilis of nose, sinuses and oral cavities; devised instruments including nasal hook for opening ethmoidal cells, forceps for opening sphenoidal sinus, several devices for resection of septum; suggested new method of radical operation of frontal sinusitis; described method of treating scars and abrasions in posterior nares and epipharynx, improved technique of extralaryngeal operations for cancer of larynx. Died 1941.

HAJNAL, András, Hungarian mathematician; b. Budapest, Hungary, May 13, 1931; s. György and Ilona (Fein) H.; D.Math. Scis., Eötvös Loránd U., Budapest, 1953; m. Emilia Márkus, Nov. 20, 1962; 1 son, Péter. Researcher, U. Eötvös Loránd; vis. research asst. U. Cal. at Berkeley, 1964. Research, publs. on relative consistency theorems concerning generalized continuum hypothesis, set-theoretical partition calculus (with P. Erdös, R. Rado); proved a conjecture of S. Ruziewicz. Home: 22/a Sziget, Budapest XIII, Hungary.*

HAJOS, Zoltan George, chemist; b. Budapest, Hungary, Mar. 3, 1926; s. Imre Henrik and Elizabeth (Teichner) H.; Dipl. Chem. Engring., Tech. U., Budapest, 1947, D.Sc. in Chem. Engring., 1949; m. Irene Edith Pal, Feb. 19, 1955. Came to U. S., 1957. Assistant professor of organic chem. tech. inst. Tech. U., 1948-52; asso. prof. organic chemistry inst. U. Veszprem (Hungary), 1952-53; asso. prof., research asso. organic chemistry inst. Tech. U., Budapest, 1953-57; research asso. dept. chemistry Princeton, 1957-60; sr. chemist dept. chem. research Hoffman-La Roche, Inc., Nutley, N.J., 1960—. Mem. Am. Chem. Soc., Chem. Soc. (London), Sigma Xi. Research, publs. on heterogeneous catalysis, synthesis heterocyclic compounds with physiol. interest, synthesis and stereochem. studies steroidal compounds. Home: 243 Highland Av., Upper Montclair, N.J. 07043. Office: 340 Kingsland Av., Nutley, N.J. 07110.*

HAKEN, John Kingsford, Australian chemist; b. Sydney, Australia, Jan. 25, 1932; s. George Robert and Dorothy (Kingsford) H.; grad. Sydney Tech. Coll., 1956, U. New S. Wales, 1957, 62; m. Violet May Lance, Dec. 24, 1955; children—Steven John, Roslyn Elizabeth. Tech. officer U. New S. Wales, Kensington, Australia, 1957-59, lectr. polymer chemistry, 1959—; vis. prof. chemistry U. So. Miss., 1965-66. Cons. Unisearch, Ltd., 1960—. Mem. Royal Australian Chem. Inst. (asso.), Oil and Colour Chemists Assn. (London). Author: The Synthesis of Acrylic Esters by Transesterification, 1967; also articles, chpts. in books. Synthesis of acrylic esters; devel. correlations of gas chromatographic retention behavior and chem. structure. Home: 7 Condor Crescent, Blakehurst, New S. Wales. Office: Dept. Polymer Sci., U. New S. Wales, Box 1, P.O. Kensington, Australia.*

HAKLUYT, Richard, English geographer; b. Herefordshire, Eng., circa 1552; B.A., Christ's Coll., Oxford, 1574, M.A., 1577; m. twice, one son. Lectr. on geography, Oxford; ordained to ministry Anglican Ch., 1578; chaplain to Sir Edward Stafford, English ambassador to France, 1583; in Paris, France, 1583-88; prebendary Bristol Cathedral, circa 1584; rector, Wetheringsett, Suffolk, Eng., circa 1589; chaplain of Savoy, archdeacon Westminster Abby, circa 1600. Author: Divers Voyages Touching the Discovery of America, 1582, enlarged to 3 vols., 1598-1600; A Particular Discourse Concerning Western Discoveries, 1584; The Principall Navigations, Voiages, and Discoveries of the English Nation . . . within the Compass of these 1500 Yeares, 1589; Virginia Richly Valued, 1609. Collected narratives of voyages and stimulated interest in exploration and colonization New World. Died London, Eng., Nov. 23, 1616.

HAKURA, Yukio, physicist; b. Osaka, Japan, Mar. 17, 1926; s. Gihei and Haruko (Kuwahara) H.; B. Eng., Kyoto U., 1951, D.Eng., 1961; m. Kimiko Tokunaga, Dec. 10, 1955; children—Akihiko, Yumiko. Chief forecaster ionospheric storms Radio Research Labs., Tokyo, Japan, 1951-61, chief, radio propagation research sect. and radio and space warning sect. Hiraiso Radio Wave Obs., 1961-63, project physicist space research, Tokyo, 1963-65, sr. research physicist, 1965-66; with theoretical div. Goddard Space Flight Center, NASA, Greenbelt, Md., 1966—. Guest worker CRPL, Nat. Bur. Standards, Boulder, Colo., 1961-62. Recipient Ministry of Telecommunications award, 1959. Mem. Soc. Terrestrial Magnetism and Electricity Japan (Tanakadate award 1965), Soc. Astrophysics Japan, Am. Geophys. Union, Internat. Union Geodesy and Geophysics, Assn. Geomagnetism and Aeronomy. Pioneer in field solar-terrestrial relationship and morphology of ionospheric storms including discovery of Polar Cap Absorption events by synoptic study of ionospheric vertical sounding data; establishment of affinity of solar cosmic radiations and type IV solar radio outbursts, resulting in improvement of radio propagation disturbance forecasting; invention of corrected geomagnetic coordinates useful for polar geophys. research; detailed analysis of ionospheric storms using corrected coordinates. Home: 6-7 Honda-4 chome, Kokubunji, Tokyo, Japan. Office: Theoretical div. Goddard Space Flight Center, NASA, Greenbelt, Md.*

HALBAN, Hans von, see von Halban, Hans.

HALBAN, Josef von, see von Halban, Josef.

HALBERSTAEDTER, Ludwig, dermatologist, radiologist; b. Beuthen, Upper Silesia, Dec. 9, 1876; s. Hermann Halberstaedter; studied surgery under Garre in Königsberg (now Kaliningrad, USSR), dermatology under Neisser in Breslau (now Wroclaw, Poland); 2 sons, 1 dau. Mem. syphilis research expdn., Java, 1907; became mem. faculty U. Berlin, 1922, prof., 1926; also dir. radiation dept. Inst. for Cancer Research, Berlin-Dahlem; fled to Palestine, 1933; dir. radiation therapy Hadassah Hosp., Jerusalem, also tchr. Hebrew U. Contbr. to jours. Demonstrated sensitivity of ovary to radiation, 1904; used thorium to treat cancer, also a spl. method for larynx cancer, Berlin; conducted expts. on effect of radiation on lower forms of life; studied biol. effect of million volt radiation, also application of cathode rays, radiation biology of tissues and cells. Died N.Y., Apr. 20, 1949.

HALBERT, Melvyn Leonard, Am. physicist; b. Phila., Aug. 12, 1929; s. Herman and Florence (Nachman) H.; A.B., Cornell U., 1950; Ph.D., U. Rochester, 1955; m. Edith H. Conrad, Mar. 25, 1951; children—Daniel C., Joel M., Alan L. Physicist, Oak Ridge Nat. Lab., 1955—. Fellow Am. Phys. Soc. Research, publs. on nuclear reactions induced by nitrogen ions, fluctuations in nuclear reactions, scattering alpha particles, X-rays from mesonic atoms, Nucleon-nucleon interactions. Home: 104 Morgan Rd., Oak Ridge 37830. Office: Oak Ridge Nat. Lab., Oak Ridge 37830.*

HALBERT, Seymour Putterman, Am. physician; b. Phila., Mar. 20, 1917; s. Morris and Fannie (Ellman) H.; student Bklyn. Coll., 1933-35; B.A., U. N.C., 1937; M.D., Johns Hopkins, 1941; m. Martha Swanson, Nov. 7, 1965; children—Lynn, Alan, Bruce. Med. research specializing in infectious diseases and ophthalmology, N.Y.C., 1949-65, Miami, Fla., 1965—; faculty Columbia Coll. Phys. and Surg., 1952-65, prof. ophthalmology, chief sect. immunology, 1961-65; prof. pediatrics U. Miami Med. Sch., 1965—. Cons. NIH, 1963—; dir. immunology Cordis Lab., 1963—. Guggenheim fellow, Helen Hay Whitney fellow, 1956-57. Mem. Am. Soc. Microbiologists, Am. Assn. Immunologists, N.Y. Acad. Scis., Harvey Soc., A.A.A.S., Assn. for Research Ophthalmology, Brit. Soc. Immunology, N.Y. Acad. Medicine, Biochem. Soc., Am. Chem. Soc., Phi Beta Kappa, Sigma Xi. Research, numerous publs. in biochem. evolution of proteins, analysis of human infections with particular reference to streptococcal diseases and rheumatic fever, significance of bacterial antibiosis in resistance to superficial infections, auto-immune diseases of the eye, and heart. Home: 12450 Rock Garden Lane, Miami 33156. Office: Nat. Children's Cardiac Hosp., 1475 N.W. 12th Av., Miami, Fla. 33134.*

HALBFASS, Wilhelm, German hydrogeographer; b. Hamburg, Germany, June 26, 1856; s. Heinrich Christina and Josiana Marie (Thomsen) H.; student, Freiburg, Würzburg; Ph.D., Strasbourg (now in France), 1880; postgrad., Halle, Germany; m. Charlotte Kühne, circa 1890. tchr. secondary sch., Neuhaldensleben, prof., 1901-10; became mem. faculty U. Jena, 1923, hon. prof., 1931, lectr. until 1936. Author: (classic articles in lake geography) Morphometrie der europäischen Seen, 1903-04, Dei Seen der Erde, 1922; Beiträge zur Kenntnis der pommerschen Seen, 1901; Das Wasser im Wirtschaftsleben des Menschen, 1911; Der gegenwärtige Stand der Seenforschung, 1912; Abseits der Heerstrassen, 1913; Das Süsswasser der Erde, 1914; Moderne Aufgaben der geographischen Seenforschung, 1914; Methoden der Seenforschung; Grundlagen der Wasserwirtschaft, 1921; Grundzüge einer vergliechenden Seekunde (standard work), 1923; Der Jahreswasserhalt der Erde, 1934. Research in morphometry and physics of lakes; sounded many German and fgn. lakes; worked toward complete inventory of lakes of world with goal of geog. comparisons; also studied gen. hydrology, water economics. Died Jena, Oct. 29, 1938.

HALBOUTY, Michel Thomas, Am. geologist; b. Beaumont, Tex., June 21, 1909; s. Tom Christian and Sodia (Monnelly) H.; B.S., Tex. A. and M. U., 1930, M.S., 1931, Geol. Engr., 1956; D.Engring. (hon.), Mont. Coll. Mining, Sci. and Tech., 1966; m. Fay Renfro, June 22, 1945; children—Thomas E. Kelly (foster child), Linda Fay. Cons. geologist, petroleum engr., Houston, 1937—. Rep. div. earth scis. Nat. Acad. Scis., NRC. Recipient Distinguished Service award Tex. Mid-Continent Oil and Gas Assn., 1965. Fellow A.A.A.S., Geol. Soc. Am., Inst. Petroleum (London), Tex. Acad. Sci.; mem. Am. Assn. Petroleum Geologists (pres.), Am. Inst. Mining, Metall. and Petroleum Engrs. (dir., v.p.), Am. Inst. Profl. Geologists (past pres. Tex. sect.), Am. Mus. Natural History, Am. Petroleum Inst., Am. Soc. Oceanography, Ind. Petroleum Assn. Am., Internat. Assn. Sedimentology, Internat. Oceanographic Found., Internat. Oil Scouts Assn., Mid-Continent Oil and Gas Assn., Mineral. Soc. Am., Nat. Water Well Assn., N.Y. Acad. Scis., Seismol. Soc. Am., Soc. Econ. Paleontologists and Mineralogists Am., Soc. Exploration Geophysicists, Soc. Ind. Profl. Earth Scientists (past dir.), Tau Beta Pi. Author: Petrographic Characteristics of Gulf Coast Oil Sands, 1937; (with James A. Clark) Spindletop, 1952; Salt Domes, Gulf Region of the United States, 1966. Research, publs. in petroleum geology, especially geology of salt domes; petrography of Gulf Coast Province; regional geology; stratigraphy. Home: 3630 Willowick Dr., Houston 77019. Office: Michel T. Halbouty Bldg., 5111 Westheimer Rd., Houston 77027.*

HALDANE, John Burdon Sanderson, geneticist; b. Oxford, Eng., Nov. 5, 1892; s. John Scott Haldane; M.A. (fellow 1919-22), New Coll., Oxford; m. 1st, Charlotte Franken, 1926 (div. 1945), 2nd, Helen Spurway, 1945. Reader in biochemistry Cambridge U., 1922-32; Fullerian prof. physiology Royal Instn., 1930-32; prof. genetics London U., 1933-37, prof. biochemistry, 1937-57; research prof. Indian Statis. Inst., 1957-61. Chmn. editorial bd. Daily Worker, 1940-49. Fellow Royal Soc., 1932; mem. Genetical Soc. (pres. 1932-36), Moscow Acad. Scis. (hon.), Société de Biologie (corr.), Deutsche Akademie der Wissenschaften zu Berlin; Nat. Inst. Scis. India; Royal Danish Acad. Scis. Recipient: chevalier Legion of Honor, 1937; Darwin medal Royal Soc., 1953; Darwin-Wallace medal Linnean Soc., 1958; Kimber medal Nat. Acad. Scis. (U. S.), 1961; Feltrinelli prize Accademia dei Lincei, 1961. Author: Daedelus, 1924; Callinicus, 1925; Possible Worlds, 1927; (with J. S. Huxley) Animal Biology, 1927; Science and Ethics, 1928; Enzymes, 1930; The Inequality of Man, 1932; The Causes of Evolution, 1933; Fact and Faith, 1934; My Friend Mr. Leakey, 1937; Heredity and Politics, 1938; Marxist Philosophy and the Sciences, 1938; Science and Every Day Life, 1939; Keeping Cool, 1939; Science in Peace and War, 1940; New Paths in Genetics, 1941; Banned Broadcast, 1947; Science Advances, 1947; Everything Has a History, 1951; Biochemistry of Genetics, 1953. Applied math. analysis to genetic phenomena; research on evolution, human physiology; developed methods to measure gene linkage; determined safe gas mixture for breathing under water (to reduce possibility of bends). Died Bhudaneswar, India, Dec. 1, 1964.

HALDANE, John Scott, physiologist; b. Edinburgh, Scotland, May 3, 1860; s. Robert and Mary (Burton-Sanderson) H.; student U. Jena; M.A., M.D., LL.D., U. Edinburgh; numerous hon. degrees; m. Louisa Kathleen Trotter, 1891; children—John B. S., Naomi Mary Margaret (Mrs. Mitchison). Demonstrator physiology Univ. Coll., Dundee, 1887, Oxford U., 1887; fellow New Coll., 1901; reader physiology, 1907, resigned 1913; dir. mining research labr., nr. Doncaster, 1912. Mem. admiralty com. on caisson disease; mem. Safety in Mines Research Bd., 1921; hon. prof. mining Birmingham U., 1921; Silliman lectr. Yale, 1916. Recipient Copley medal, 1934. Fellow Royal Soc., 1897 (Royal medal 1916), Inst. Mining Engrs. (pres. 1924-28). Author: Respiration, 1922, 35; Gases and Liquids, 1928; The Sciences and Philosophy, 1929; The Philosophical Basis of Biology, 1931; The Philosophy of a Biologist, 1935. Made important contbns. to mine safety; studied carbon monoxide and its physiol. action; invented new methods for gas analysis of blood, notably improved hemoglobinometer, 1901; discovered that respiration is regulated by amount of carbon dioxide in blood, 1905; investigated coal mine gases, determined their composition before and after explosion and their physiol. effect; in expdn. to Pikes Peak studied effects of low barometric pressure, 1911; results of his trip revolutionized ideas about respiration; devised apparatus for stage decompression for safe ascent of deep sea divers and for people afflicted with caisson disease, 1907; studied cause of heat stroke in tropics and significance of high temperatures in mines. Died Oxford, Eng., midnight Mar. 14-15, 1936.

HALDAT DU LYS, Charles de, see de Haldat du Lys.

HALDEN, Wilhelm, Austrian chemist; b. Prague, Dec. 24, 1892; s. Carl and Eugénie (Jarka) H.; Ph.D. in Chemistry, U. Graz; m. Berta Payer, 1940; children—Ingeborg, Franz-Werner, Heinz-Dieter. Asst., agrégé of med. chemistry, asso. prof., ofcl. expert for research on nutrition Inst. Chemistry, U. Graz. Austrian rep. Internat. Commn. Nutritional Sci. and Tech. Mem. Am. Assn. Pub. Health. Author: (with A. Grün) Analyse der Fette und Wachse; Grundlagen der Ernährung; Cholesterin und Ernahrung; (with L. Prokop) Sport und Ernährung. Home: 6 Josef Marxstrasse, A 8043 Graz-Kroisbach. Office: Medical Chemical Institute, University of Graz, Graz, Austria.

HALE, George Ellery, Am. astronomer; b. Chgo., June 29, 1868; s. William Ellery and Mary S. H.; S.B., Mass. Inst. Tech., 1890; studied Harvard College Obs., 1889-90, Univ. Berlin, winter 1893-94; hon. Sc.D., U. Pittsburgh, 1897, Yale, 1905, Victoria U., Manchester, 1907, Oxford, 1909, Cambridge, 1911, Chicago, 1916, Columbia, 1917, Harvard, 1921; LL.D., Beloit, 1904, U. of Calif., 1912, Princeton, 1917; Ph.D., Berlin, 1910; m. Evelina S. Conklin, June 5, 1890; children—Margaret, William Ellery. Dir. Kenwood Astrophys. Obs., 1890-96; asso. prof. astrophysics, 1892-97, prof., 1897-1905, organizer, and dir. Yerkes Obs., 1895-1905, U. Chicago; organizer, and dir. Mount Wilson Obs. of Carnegie Instn. of Wash., 1904-23, hon. dir., from 1923. Joint editor Astronomy and Astrophysics, 1892-95, Astrophysical Journal, 1895-1935. Janssen medal, Paris Acad. Sciences, 1894; Rumford medal, 1902; Draper medal, 1903; gold medal, Royal Astron. Soc., 1904; Bruce medal, Astron. Soc. of the Pacific, 1916; Janssen medal, Astron. Soc. of France, 1917; Galileo medal, Florence, 1920; Actonian prize, Royal Instn., 1921; Arthur Noble medal for civic service (Pasadena), 1927; Cresson and Franklin medals, Franklin Inst., 1926, 27; Holland Society medal, 1931; Copley medal, Royal Society, London, 1932; awarded Ives medal by Optical Soc. Am. Mem. com. on intellectual cooperation League of Nations, 1922. Author: The Study of Stellar Evolution; Ten Years' Work of a Mountain Observatory; The New Heavens; The Depths of the Universe, 1924; Beyond the Milky Way, 1926; Signals from the Stars, 1931. First to measure diameter of a star (Alpha Orionis) with interferometer (with Pease), 1920; invented spectroheliograph (at about same time as Deslandres), 1889; proved that sun's magnetic and geog. poles do not coincide; established theory of movement of sun's spots. Died Pasadena, Feb. 21, 1938.

HALE, Henry Bixby, Am. physiologist; b. Estherville, Ia., July 22, 1913; s. John Murray and Mabel (Bixby) H.; B.S., Ia. State Coll., 1936, M.S., 1939; Ph.D., U. Cin., 1944; m. Mary Nell Wood, Sept. 2, 1939; 1 son, John Murray II. Tchr. high sch., Keokuk, Ia., 1936-39; faculty Stephens Coll., Columbia, Mo., 1944, Still Coll., Des Moines, 1945-47, Okla. A. and M. Coll., 1947-49; aviation physiologist USAF Sch. Aviation Medicine (now USAF Sch. Aerospace Medicine), Brooks AFB, Tex., 1949—. Mem. Aerospace Med. Assn., Am. Physiol. Soc., Endocrine Soc., Soc. for Exptl. Biology and Medicine, Research Soc. Am., A.A.A.S., Sigma Xi. Research, publs. on environmental physiology, aviation physiology, aerospace physiology, stress, acclimatization, endocrine-metabolic responses to changes in climate. Home: 3819 Oaktrail St., San Antonio 78228. Office: Box 4273, USAF Sch. Aerospace Medicine, Brooks AFB, Tex. 78235.*

HALE, William Harris, Am. animal nutritionist; b. Richmond, Ky., Feb. 11, 1920; s. Alva T. and Sara (Winkler) H.; B.S., U. Ky., 1946, M.S., 1947; Ph.D., U. Wis., 1950; m. Margaret Smiley, Oct. 27, 1945; children—Karen, Ann, Margaret, Mary, Barbara. Faculty, U. Ill., 1950-52, Ia. State U., 1952-57; head, large animal nutrition research Charles Pfizer & Co., Terre Haute, Ind., 1957-60; prof. animal nutrition U. Ariz., Tucson, 1960—. Research in ruminant nutrition, including estrogens, trace minerals, non-protein nitrogen, forage utilization, grain processing and utilization, vitamin A, antibiotics, fat and non-protein nitrogen utilization. Home: 6549 E. Koralee St., Tucson 85710.*

HALE, William Jay, Am. chemist; b. Ada, O., Jan. 5, 1876; s. James Thomas and Emma Elizabeth (Ogle) H.; A.B., A.M., Miami U., 1897, LL.D., 1937; A.B., Harvard Univ., 1898, A.M., 1899, Ph.D., 1902; traveling fellow in chemistry, Technische Hochschule, Berlin, and U. Göttingen, 1902-03; m. Helen Dow, Feb. 7, 1917 (died Oct. 1918); 1 dau., Ruth Elizabeth (Mrs. Wiley T. Buchanan, Jr.). Research assistant, University of Chicago, 1 term, 1903; instructor, chemistry, 1904-08, asst. prof., 1908-15; asso. prof., 1915-19, U. of Mich.; dir. organic chem. research, Dow Chem. Co., Midland, Mich., 1919-1934, research consultant since 1934; president Verdurin Co., Detroit and Midland, Mich., since 1951. Chairman division of chemistry and chemical technology. Nat. Research Council, Washington, D.C., 1925-27; vis. prof. chemurgy, Conn. Coll., 1936-39. Fellow A.A.A.S., London Chem. Soc.; mem. American Chem. Soc., Société Suisse de Chimie Société Chimique de France, Deutsche Chemische Gesellschaft, Phi Beta Kappa, Sigma Xi, Phi Lambda Upsilon, Alpha Chi Sigma. Author: A Laboratory Outline of General Chemistry (with Alexander Smith), 1907; The Calculations of General Chemistry, 1909; A Laboratory Manual of General Chemistry, 1917, 2d edit. with Wm. G. Smeaton, 1930; Chemistry Triumphant, 1932; The Farm Chemurgic, 1934; Prosperity Beckons, 1936; Farmward March, 1939; Farmer Victorious, 1949; Chemivision, 1952; papers organic chem., addresses chemurgic development for industrialization agr. Patentee of new process for mfr. of phenol, aniline, acetic acid, butadiene and their derivatives. Often described as father of chemurgy. Died Aug. 8, 1955.

HALE, William Mason, Am. microbiologist; b. Nashville, Ark., Dec. 8, 1898; s. Jurdon Christopher and Nancy (Rammage) H.; B.S., U. N.M., 1924; M.D., Yale, 1929; m. Frances Laverne Hardaway, Mar. 25, 1955; step children—Beverly (Mrs. William C. Smith), Cheryl, Lynn Hale, Herbert Ellis, Gerald Wayne Hale. Faculty, Yale Sch., Medicine, 1929-38, asst. prof., 1932-38; prof., head dept. microbiology State U. Ia., 1938-49; sr. scientist, head div. bacteriology and virology Brookhaven Nat. Lab., Upton, N.Y., 1949-53; prof. microbiology U. Tenn. Coll. Medicine, 1953-64, prof. emeritus, 1964—. Med. officer U. S. FDA, 1937-38; cons. to Sec. of War, 1941-49; cons. pathology U. Tenn. Coll. Medicine, 1964—. Fellow Am. Assn. Immunologists, Ia. Acad. Sci.; mem. Soc. Bacteriologists, Soc. for Exptl. Biology and Medicine, Am. Soc. for Exptl. Pathology, Am. Pub. Health Assn., Ia. Path. Assn. Research, publs. on resistance to infections, immunization against influenza, effect of ionization on resistance and infections; discovered toxic factor in influenza virus. Home: Box 31, Rushing Route, Mountain View, Ark. 72560.*

HALEM, Friedrich Wilhelm von, see von Halem, Friedrich Wilhelm.

HALES, Stephen, English botanist, physiologist, chemist; b. Bekesbourne, Kent, Eng., Sept. 7, or 17, 1677; s. Thomas Hales; M.A., Benet Coll., Cambridge (Eng.) U., 1703; B.D., 1711; D.D., Oxford (Eng.) U., 1733. Made perpetual curate Teddington Middlesex, Eng., 1708; apptd. rector Porlock, Somerset, Eng., later Faringdon, Hampshire, Eng. Mem. com. to establish colony in Ga., 1732; apptd. almoner to princess-dowager of Wales, 1750. Fellow Royal Soc. (Copley medal 1739), 1717; fgn. asso. mem. French Acad. Scis. Inventor devices for food preservation, sea water purification, a ventilator to convey fresh air into jails and hosps.; performed earliest known expts. to determine blood pressure of animals, circa 1705 (described in Haemostaticks, 1733); credited with establishing sci. of plant physiology (because of book Vegetable Staticks, 1727); measured growth rate of plants, loss of water in evaporation, variations of rootforce at different times of day; conducted many other expts. with plants; 1st to collect gases over water with pneumatic trough (however, thought all gases so collected were air; not interested in qualitative tests). Died Teddington, Jan. 4, 1761.

HALE-WHITE, Sir William, English physician; b. London, Nov. 7, 1857; s. William Hale White (Mark Rutherford); M.D., London, Dublin; LL.D., Edinburgh, 1927; m. E. J. S. Fripp; 1 son. Entered Guy's Hosp., 1875, became asst. physician, 1885, lectr. medicine, 1899, physician, from 1890, also cons.; chmn. Queen Mary's Hosp.; vice chmn. Queen's Inst. of Dist. Nursing; councillor Brit. Red Cross Soc.; chmn., fellow Bedford Coll.; treas. Epsom Coll. Fellow Royal Coll. Physicians (Harveian orator 1927, Croonian lectr. 1897), Royal Coll. Physicians Edinburgh (hon.); pres. Royal Soc. Medicine. Author: Test-Book of General Therapeutics, 1889; Materia Medica, 26th edit., 1945; Text-Book of Pharmacology and Therapeutics, 1901; Common Affections of the Liver, 1908; Bacon, Gilbert and Harvey, 1927; Laennec, 1923; Great Doctors of the Nineteenth Century, 1935; Keats as Doctor and Patient, 1938. Died Feb. 26, 1949.

HALEY, Harold Bernard, Am. surgeon; b. Madison, Wis., Sept. 11, 1923; s. Harry Bernard and Annabelle (Buchanan) H.; student U. Notre Dame, 1940-42, U. Wis., 1942-43; M.D., St. Louis U., 1946; m. Margaret Carroll, Nov. 26, 1923; children—Claudia, Harold and Patrick (twins), Mark, Gregory. Practice medicine, specializing in surgery, Chgo., 1955—; mem. staffs Cook County, Mercy, Oak Park hosps.; professor surgery Stritch Sch. Medicine, Loyola U., Chgo. Damon Runyon fellow, 1950-51. Mem. A.C.S., Am. Assn. for Cancer Research, Am. Thyroid Assn., A.A.A.S., Am. Cancer Soc., Sigma Xi. Research in biochem. aspects of wound healing, steroid metabolism in cancer, ednl. research. Home: 843 Columbian St., Oak Park, Ill. 60302. Office: P.O. Box 1336, Hines, Ill. 60141.*

HALEY, Jay, Am. communications analyst; b. Midwest, Wyo, July 19, 1923; s. Andrew J. and Mary (Sneddon) H.; B.A., U. Cal. at Los Angeles, 1948; B.L.S., U. Cal. at Berkeley, 1951; M.A., Stanford, 1953; m. Elizabeth Kuehn, Dec. 4, 1949; children—Kathleen, Andrew, Gregory. Research asso. dept. anthropology Stanford, 1954-60, Palo Alto (Cal.) Med. Research Found., 1958-62; dir. family expt. project Mental Research Inst., Palo Alto, 1962-67; dir. family research Phila. Child Guidance Clinic, 1967—. Cons. Group for Advancement Psychiatry, 1964—. Author: Strategies of Psychotherapy, 1963; Techniques of Family Therapy, 1967; also articles. Research in animal and human communication, metacommunication and its relation to psychiatry, relationship family to psychiat. problems and social deviancy, change in psychotherapy and social systems. Office: 1700 Bainbridge St., Phila. 19143.*

HALEY, Thomas John, Am. biologist; b. Crosby, Minn., Nov. 4, 1913; s. Thomas E. and Ida May (Young) H.; student Loyola U., 1931; B.S., U. So. Cal., 1938, M.S., 1942; Ph.D., U. Fla., 1945; m. Jeanne Wall, Sept. 24, 1964; children—Kathleen, Barbara Anne. Med. dir. E. S. Miller Labs., Los Angeles, 1945-47; fellow in pharmacology and toxicology U. So. Cal., 1946-48; chief div. pharmacology and toxicology, dept. biophysics and nuclear medicine Lab. Nuclear Medicine and Radiation Biology, U. Cal. at Los Angeles Sch. Medicine, 1958-66, asso. clin. prof. medicine, 1951-60; prof. pharmacology U. Hawaii Sch. Medicine, Honolulu, 1966—. Sci. com. Air Pollution Control Dist., Los Angeles, 1955; vis. scientist Sandoz Labs., Basle, 1956. Del Amo Found. fellow, 1955; AEC fellow, 1956. Fellow Walter Reed Soc., A.A.A.S., Internat. Acad. Law and Sci.; mem. Am. Acad. Forensic Scis., Am. Heart Assn., Am. Soc. Pharmacology and Exptl. Therapeutics, N.Y. Acad. Sci., Am. Pharm. Assn., Am. Chem. Soc., Soc. Exptl. Biology and Medicine, Biometric Soc., Radiation Research Soc., Am. Indsl. Hygiene Assn., Reticulo-endothelial Soc., Soc. Toxicology Western Pharmacology Soc., Acad. Pharm. Scis. Author: Clinical Toxicology, 4th edit., 1964; Response of Nervous System to Ionizing Radiation, 1962, 64; Nuclear Hematology, 1965; also numerous articles. Research on evaluation of new drugs, effects of drugs on nervous system, effects of ionizing radiation on animals. Home: 3518 Alohea Av., Honolulu.*

HALFF, Albert H., Am. civil engr.; b. Midland, Texas, Aug, 20, 1915; s. Henry M. and Rosa (Wechsler) H.; B.S.C.E., Southern Methodist U., 1937; M.S.C.E., Ill. Inst. Tech., 1942; D.Eng., Hopkins, 1950; m. Lee Benson, Aug. 24, 1940; children: Henry M., Albert L. Asst., Tex. Hwy. Dept., 1936-37; asst. engr., Koch & Fowler, 1937-39; enstr. engrg., Texas College Arts & Indus., 1939-41; instr., Ill. Inst .Tech., 1940-42; design engr., Charles Deleuw & Co., 1942; engr., Brown & Bellows, 1942-43; sanitary engr., 1950-53; partner, Hundley & Halff, 1953-60; owner, Albert H. Halff Assoc. Engrs., 1960—. Mem., Nat. Soc. Profl. Eng.; Water Works Assn.; Soc. Civil Engr.; Am. Acad. Sanitary Engrs.; Sigma Xi; pres., Dallas Branch, ASCE. Pub. several articles. Patentee of process of production water by condensation; threaded devices with antirotational means; electric relay; direct determination of density of solids. Home: 3514 Rock Creek Dr., Dallas, Texas 75219. Office: 3636 Lemmon Ave., Dallas, Texas 75219.*

HALFORD, Joseph Olney, Am. chemist; b. Oakland, Cal., Aug. 12, 1903; s. John J. and Kathleen (Loftus) H.; B.S. in Chemistry, U. Cal. at Berkeley, 1924, Ph.D., 1927; m. Elizabeth Knudson, July 19, 1940; children—Mary E., Janine L. Faculty, U. Mich., Ann Arbor, 1926—, prof. chemistry, 1945—. Mem. A.A.-A.S., Am. Chem. Soc. Research in thermodynamics of internal rotation, stability of aromatic polyenes. Home: 1719 Morton Av., Ann Arbor, Mich. 48104.*

HALFORD, Ralph Stanley, Am. chemist, univ. dean; b. Vallejo, Cal., Apr. 21, 1914; s. John Jefferson and Kathleen (Loftus) H.; B.S., U. Cal. at Berkeley, 1935, Ph.D., 1938; m. Marion Kee Glover, Mar. 3, 1939. Instr. chemistry U. Cal. at Berkeley, 1938-40; NRC fellow Harvard, 1940-41, instr. chemistry, 1941-43, lectr., 1943-46; asso. prof. chemistry Columbia, N.Y.C., 1946-51, prof., 1952—, vice provost for projects and grants, 1959-67, dean grad. faculties, 1961-67, special assistant to the president, since 1967—. With OSRD, 1944; mem. com. adv. to Office Ordnance Research, NRC, 1951-52; mem. nat. selection com. on Fulbright awards to U.K., Inst. Internat. Edn., 1954-56; vis. com. to chemistry dept. Brookhaven Nat. Lab., 1954-56. Fellow Am. Phys. Soc.; mem. Am. Chem. Soc., N.Y. Acad. Scis., Sigma Xi, Alpha Chi Sigma. Research, publs. on structure of condensed systems, infrared spectra of liquids, absorption anisotropy in crystals. Home: 445 Riverside Dr., N.Y.C. 10027.*

HALL, Albert Mangold, Am. metallurgist; b. Bklyn., Oct. 8, 1914; s. Edgar Albert and Salena (Mangold) H.; A.B., Columbia, 1935, B.S., 1936, M.S., 1937; m. Jean Collier Lamb, Dec. 27, 1938; children—Charles Hamilton, David Arthur, Peter Albert. Research metallurgist Huntington Alloy Products div. Internat. Nickel Co., Inc., Huntington, W.Va., 1937-45; research engr. Battelle Meml. Inst., Columbus, O., 1945-47, asst. supr., 1947-53, div. chief, 1953—. Mem. Columbus Tech. Council (pres. 1956), Am. Soc. Testing and Materials (com. on stainless steels), Wire Assn., Am. Soc. for Metals, Am. Soc. for Mining and Metall. Engrs., Am. Ordnance Assn., Sigma Xi, Tau Beta Pi, Epsilon Chi. Contbr. numerous articles to profl. jours. Author: Nickel in Iron and Steel, 1954. Developed process to cadmium coat metals; identified sulfide inclusions in nickel alloys; devel. etchant to reveal oxidation in iron alloys, semi-austenitic stainless steel and manganese steel with exceptional ductility; research in metallurgy of fine wire made from iron- and nickel-base alloys, kinetics of graphite formation in steel, process of recrystallization in copper alloys. Home: 2342 Dorset Rd., Columbus 43221. Office: 505 King Av., Columbus, O. 43201.*

HALL, Sir Arnold Alexander, English aero. engr.; b. Liverpool, Eng., Apr. 23, 1915; s. Robert Alexander and Ellen Elizabeth (Parkinson) H.; ed. Clare Coll., Cambridge U.; M.A. with 1st honours in mech. sci.; m. Dianne Hall, 1946; children—Caroline, Elizabeth, Veronica. Sci. officer Royal Aircraft Establishment, Farnborough, Hampshire, 1938-45, adminstr., 1951-55; Zahuroff prof. aviation U. London, dir. aero. dept. Imperial Coll. Sci. and Tech., London, 1945-51; tech. adminstr. Hawker Siddeley Group Ltd., London, 1955-58, adminstr., 1958-63; adminstrv. del. Bristol Siddeley Engines Ltd., London, 1958-63; adminstr. del., v.p. Hawker Siddeley Engines Ltd., 1963. Mem. Govt. Com. on Problem of Noise, Engring. Adv. Council. Recipient Rex Moir prize for mechanics, John Bernard Seely prize for aeros., Ricardo prize for thermodynamics, Von Baumhauer medal for contbn. to aviation rescue Royal Netherlands Aero Club. Fellow Royal Soc., 1953; mem. Royal Aero. Soc. (pres., Gold medal), Inst. Sci. Aeros., Am. Inst. Aeros. and Astronautics, Internat. Council Aero. Sci. Home: Wakehams, Borney, North Windsor. Office: 18 St. James's Sq., London S.W.1, Eng.

HALL, Arthur Lewis, Brit. geologist; b. Birmingham, Jan. 10, 1872; s. William and Mary Ann (Smith) H.; ed. Univ. Coll., Bristol; Sc.D., Caius Coll., Cambridge, 1925; m. Rosalie Riddick Powell, 1900; 3 sons, 1 dau. Sr. sci. master Dulwich Coll., 1900-02; field geologist Geol. Survey Transvaal, 1903-15; asst. dir. Geol. Survey Union S. Africa, 1915-32. Gen. Sec. Internat. Geol. Congress, 1929. Fellow Royal Soc. London, 1935, Geol. Soc. London (Murchison medal 1930), S. Africa (pres., Draper medal 1932), Royal Soc. S. Africa; fgn. corr. Geol. Soc. Belgium. Author: Memoir on the Boshveld Igneous Comples, 1932; Memoir on the Barberton Moutain-Land; also articles. on geology of S. Africa. Died Aug. 13, 1955.

HALL, Asaph, Am. astronomer; b. Goshen, Conn., Oct. 15, 1829; worked at farming and carpentry; spl. studies at Norfolk Acad., U. Mich. LL.D., Yale, Harvard; Ph.D., Hamilton; m. Angeline Stickney. Taught school; became asst. Harvard Coll. Obs., 1857-62; aide U. S. Naval Obs., 1862, prof. math., 1863; prof. astronomy Harvard, from 1895; chief astronomer, expdn. to San Antonio, Tex., 1882. Mem. Nat. Acad. Scis.; corr. French Acad. Scis., 1879 (Lalande prize 1878, Arago medal 1879); fgn. mem. Royal Astron. Soc. (gold medal 1879). Discovered 2 satellites of Mars, Deimos and Phobos, 1877; also did work on double stars, Jupiter's satellite orbits, mass of Mars; determined period of rotation of Saturn, also period of movements of its satellite Hyperion. Died Annapolis, Md., Nov. 22, 1907.

HALL, Asaph, Jr., Am. astronomer; b. Cambridge, Mass., Oct. 6, 1859; s. Asaph and Angeline (Stickney) H.; A.B., Harvard, 1882; Ph.D., Yale, 1889; m. Mary Estella Cockrell, July 14, 1897. Asst., U. S. Naval Obs., Washington, 1882-85; asst. astronomer, Yale Obs., 1885-89, Naval Obs., 1889-92; prof. astronomy and dir. obs., U. Mich., 1892-1905; asst. Naval Obs., 1905-08; prof. math. U.S.N., from 1908. Died Jan. 12, 1930.

HALL, C(harles) William, Am. surgeon; b. Gage, Okla., Feb. 8, 1922; s. Cecil A. and Helen (Greene) H.; A.B., Kan. U., 1950, M.A., 1952, M.D., 1956; m. Betty Arlene Woodring, June 6, 1943 (div. Apr. 1962), children—Daniel C., Kendall W., Gregory A., Patrick C., Connan L.; m. 2d, Shirley Anne Thompson, Oct. 21, 1962. Fellow cardiovascular surgery Baylor U. Coll. Medicine, 1962-63, asst. prof. surgery and physiology, 1964—, dir. Artificial Heart program, 1964—; hon. prof. Cath. U., Cordoba, Argentina. Diplomate Am. Bd. Surgery. Mem. Am. Soc. Artificial Internal Organs, Am. Heart Assn., Am. Assn. for Advancement Med. Instrumentation, A.A.A.S., Internat. Cardiovascular Soc., Soc. Vascular Surgery, Sigma Xi, Phi Sigma. Research, publs. on devel. type of artificial skin; co-developer of artificial heart. Home: 9120 Kenilworth Dr., Houston 77024.*

HALL, Calvin S(pringer), Am. psychologist; b. Seattle, Jan. 18, 1909; s. Calvin S. and Dovre (Johnson) H.; student U. Wash., 1926-29; A.B., U. Cal. at Berkeley, 1930, Ph.D., 1933; m. Irene Sanborn, Nov. 10, 1932; 1 dau., Dovre (Mrs. David Busch). Prof., chmn. dept. psychology Western Res. U., 1937-57; prof., exec. officer Syracuse U., 1957-59; vis. prof. U. Miami, 1959-60; Fulbright prof. U. Nijmegen, Holland, 1960-61; dir. Inst. Dream Research, Santa Cruz, Cal., 1961—; lectr. U. Cal., Santa Cruz, 1966—. Mem. Am. Psychol. Assn. Author: The Meaning of Dreams, 1953; A Primer of Freudian Psychology, 1954; (with G. Lindzey) Theories of Personality, 1957; (with R. L. Van De Castle) Content Analysis of Dreams, 1966; also numerous articles. Research in field of psychology, genetics of behavior and psychology of dreams; devised objective quantitative methods for analyzing contents of dreams. Address: 1310 W. Cliff Dr., Santa Cruz, Cal. 95060.*

HALL, Charles Francis, Am. explorer; b. Rochester, N.H., 1821. Became interested in fate of Arctic explorer Sir John Franklin and his party, left New London, Conn. in search of relics and survivors of expdn., May 1860; landed alone at Frobisher Bay in Arctic region, July 1860; lived with and gained help of Eskimos until his return to New London, Sept. 1862, having found relics of Frobisher's expdn. of 1577-78 but none of Franklin expdn.; made similar journey, 1864-69, during which he found relics of Franklin expdn. as well as traces of possible survivors; on June 29, 1871, with backing of U. S. Congress, he undertook last voyage to North in command of ship Polaris, penetrated to 82° 11′ N and 61° W, farthest point attained up to that time by vessel; Hall died on trip, but crew members survived extremely severe winter to return to U. S.; trip proved to have important geog. results. Author: Arctic Researches and Life Among the Esquimaux, 2 vols., 1865; Narrative of the Second Arctic Expedition, 1879. Considered one of greatest Arctic explorers. Died Thank God Harbor, Greenland, Nov. 8, 1871.

HALL, Charles Martin, Am. inventor; b. Thompson, O., Dec. 6, 1863; s. Rev. Heman B. and Sophronia (Brooks) H.; A.B., Oberlin, 1885, A.M., 1893 (LL.D., 1910); unmarried. Invented the electrolytic process for manufacture of aluminum (at same time as Héroult in France), 1886; began comml. manufacture of aluminum with Pittsburgh Reduction Co., 1888 (now Aluminum Co. of America), v.p., from 1890; 1890—; U. S. court sustained Hall patent and conceded priority of invention, 1893. Recipient Perkin medal, for work in chemistry, 1911. Died Daytona Beach, Fla., Dec. 27, 1914.

HALL, Chester Moor, English inventor; b. Leigh, Eng., Dec. 9, 1703; s. Jehu and Martha (Brittridge) H.; ed. Inner Temple, became bencher, 1763; inventor achromatic telescope, 1733. Died Sutton, Surrey, Mar. 17, 1771.

HALL, Sir (Alfred) Daniel, English agriculturist; b. June 22, 1864; s. Edwin Hall; ed. Balliol Coll., Oxford; M.A.; D.Sc., Oxford U.; LL.D., Aberdeen U., Cambridge U.; m. Louisa Brooks, (dec. 1921)); 1 son; m. 2d, Ida S. A. Beaver, 1922. First prin. South-Eastern Agrl. Coll., Wye, 1894-1902; dir. Rothamsted Exptl. Sta. (Lawes Agrl. Trust), 1902-12; chmn. Commn. on Agr., Kenya Colony, 1929; former dir. John Innes Hort. Instn.; former chief sci. adviser Ministry Agr.; mem. Econ. Adv. Council; Rede lectr. Cambridge U., 1935; Leath Clark lectr. U. London, 1935; Earl Grey lectr. King's Coll., Newcastle-upon-Tyne, 1939. Fellow Royal Soc., 1909; mem. Royal Hort. Soc. (v.p., Victoria medal honor), Royal Acad. Agr. (Sweden) (fgn.). Author: The Soil, 4th edit., 1931; The Book of the Rothamsted Experiments, 1919; Fertilizers and Manures, 3d edit., 1928; The Feeding of Crops and Stock, new edit., 1936; (with E. J. Russell) The Agriculture and Soils of Kent, Surrey and Sussex, 1911; A Pilgrimage of British Farming, 1913; Agriculture after the War, 1916; (with W. B. Crane) The Apple, 1933; The Improvement of Native Agriculture, 1936; The Genus Tulipa, 1940; Reconstruction and the Land, 1941; also articles. Developed sci-agrl. edn. in Gt. Britain; conducted many studies in agr., made results available to farmers and agriculturists. Died London, July 5, 1942.

HALL, David Alan, English biochemist, gerontologist; b. Leeds, Eng., July 22, 1921; s. George and Eleanor (Hirst) H.; B.Sc. in Chemistry, Leeds U., 1942, Ph.D., 1945; m. Shirley Robinson, June 12, 1947; children—Alison, Jonathan, Helen. Research asst. U. Leeds, 1944-48, faculty, 1948—, sr. lectr. dept. medicine, 1963—. First Nuffield Gerontological Research fellow, 1954-60. Mem. Brit. Soc. for Research on Ageing (hon. sec. 1962—), Internat. Assn. Gerontology (chmn. European biol. research com. 1966—) Biochem. Soc., Author: Chemistry of Connective Tissue, 1961, Elastolysis and Aging, 1964; also numerous articles. Editor: Internat. Revs. Connective Tissue Research, 1961—, Gerontolgia 1964—. Research on purification and characterization of pancreatic enzymes, elastase, elastomucase, elastolipoproteinase, age changes in collagen and elastin; discovered cellulose in mammalian tissues. Home: 11 Tinshill Rd., Leeds, 16, Eng.*

HALL, Edward Twitchell, Am. anthropologist; b. Webster Groves, Mo., May 16, 1914; s. Edward T. and Jessie (Gilroy) H.; student Pomona Coll., 1929-30; A.B., U. Denver, 1936; M.A., U. Ariz., 1938; Ph.D., Columbia U., 1942; m. Mildred Ellis Reed, Dec. 16, 1946; children—Ellen McCoy, Eric Reed. Asst. staff archeologist Lab. Anthropology, Sante Fe, 1937; staff dendrocronologist Peabody Mus. Awatovi expdn., 1937-39; asso. prof. anthropology, chmn. dept. U. Denver, 1946-48; faculty Bennington (Vt.) Coll., 1948-51; dir. Point IV tng. program Fgn. Service Inst., U. S. State Dept., 1950-55; dir. research, dep. dir. Washington Office, Human Relations Area Files, 1955-57; pres. Overseas Tng. and Research, Inc., Washington, 1955-60; dir. communications research project Washington Sch. Psychiatry, 1959-63; prof. anthropology Ill. Inst. Tech., Chgo., 1963-67, Northwestern U., Evanston, Ill., 1967—. Dir. Columbia Governador expdn., 1941; field work Micronesia, 1946, Southwestern U. S., 1933-42, 66-67, Europe, 1952—; econ. and cultural survey, Micronesia, 1946; cons. inter-cultural relations to internat. bus. and govt., 1955—; mem. bldg. research adv. bd. NRC, 1964—; mem. small grants com. Nat. Inst. Mental Health, 1962-65. Fellow Soc. Applied Anthropology (sec. 1960-63), Am. Anthropol. Assn.; mem. Am. Enthnol. Assn., Soc. Am. Archeology, Tree Ring Soc. Author: Earl Stockaded Settlements in Governador, N.M., 1942; The Silent Language, 1959; The Hidden Dimension, 1966; also articles. Defined culture technically as primarily a system of communication; established and named field of proxemics; study man's perception of micro-space (distance maintained between individuals) as function of status, emotion and culture. Home: 1218 Astor St., Chgo. 60610.*

HALL, Edwin Herbert, Am. physicist; b. Gorham, Me., Nov. 7, 1855; s. Joshua E. and Lucy Ann (Hilborn) H.; A.B., Bowdoin, 1875, A.M., 1878; Ph.D., Johns Hopkins, 1880; LL.D., Bowdoin, 1905; m. Caroline Eliza Bottum, August 31, 1882 (dec. 1921); 1 dau., Constance Huntington. Instr. physics Harvard, 1881-88, asst. prof., 1888-95, prof. 1895-1921, emeritus prof., 1921. Recipient medal Am. Assn. of Physics Teachers, 1937. Fellow Am. Acad. Arts and Scis.; corr. mem. Brit. Assn. Adv. Science; fgn. mem. Société Hollandaise des Sciences. Author: A Textbook of Physics (with J. Y. Bergen), 1891, 3d edit., 1903; Lessons in Physics, 1900; The Teaching of Chemistry and Physics (with Alexander Smith), 1902; College Laboratory Manual of Physics, 1904; Elements of Physics, 1912; Dual Theory of Conduction in Metals, 1938. Discovered the shifting of equipotential lines of electric current under influence of magnetic field (Hall effect), 1879; contbd. to theory of thermo-electric and thermal conduction in metals; studied thermomagnetic and electromagnetic effects in soft iron. Died Nov. 20, 1938.

HALL, E(ugene) Raymond, Am. zoologist; b. Imes, Kan., May 11, 1902; s. Wilber Downs and Effie Susan (Donovan) H.; A.B., U. Kan., 1924; M.A., U. Cal. at Berkeley, 1925, Ph.D., 1928; m. Mary Frances Harkey, Aug. 9, 1924; children—William Joel, Hubert Handel, Benjamin Downs. Curator mammals Mus. Vertebrate Zoology, U. Cal. at Berkeley, 1927-44, faculty, 1930-44, asso. prof., 1937-44; dir. Mus. Natural History, prof. zoology U. Kan., Lawrence, 1944—, chmn. dept. zoology, 1944-61, Summerfield Distinguished prof. in zoology, 1958—, editor mus. publs., 1946—; dir. Kan. Biol. Survey, 1946-67. Zoologist State of Kan., 1959-67; prin. investigator research projects Office Naval Research, NSF, 1950—; mem. adv. bd. on nat. parks Sec. of Interior, 1953-59. Guggenheim fellow, 1943-44. Recipient Certificate of Merit, Nash Conservation award 1956. Mem. Am. Soc. Mammologists, (hon.; past pres.), Defenders Wildlife (exec. com., v.p.), Am. Wildlife Soc., Prairie Nat. Park Natural History Assn. (pres.), Am. Assn., Am. Ornithologists Union, Cooper Club, Deutsche Gesellschaft für Saugetierkunde, Am. Soc. Systematic Zoologists, Cal. Acad. Scis., A.A.A.S. (past v.p. sect. F). Author: Mammals of Nevada, 1946; American Weasels, 1951; Handbook of Mammals of Kansas, 1955; (with J. W. Bee), Mammals of Northern Alaska; 1956; (with K. R. Kelson) The Mammals of North America, 1959; also numerous articles. Named numerous species and subspecies N. Am. mammals; improved classification Mammalia. Home: 1637 W. 9th St., Lawrence, Kan. 66044.*

HALL, George Arthur, Jr., Am. chemist; b. Parkerburg, W.Va., June 16, 1920; s. George A. and Elizabeth (Dressel) H.; B.S., W.Va. U., 1941; Ph.D., Ohio State U., 1945; m. Tanya Gussie, Jan. 26, 1963. Instr. chemistry Ohio State U., 1944-46, U. Wis., Madison, 1946-50; faculty W.Va. U., Morgantown, 1950—, asso. prof., 1956—. Mem. Am. Chem. Soc. (past exec. sec. Midwest cooperating sect. 1962-65), A.A.A.S., Am., Brit. ornithologists unions, Wilson Ornithol. Soc. (editor Bull., 1964—). Research, publs. on solvent effect on rates of chem. reactions in solution. Home: Box 52A, Route 6, Morgantown, W.Va. 26505.*

HALL, G(ranville) Stanley, Am. psychologist; b. Ashfield, Mass., Feb. 1, 1846; s. Granville Bascom and Abigail (Beals) H.; A.B., Williams, 1867, A.M., 1870; student Union Theol. Sem., 1867-68, Berlin and Bonn, 1868-71, Berlin and Heidelberg, 1871-72; Ph.D., Harvard, 1878; Leipzig, Berlin, London, 1878-81; (LL.D., U. of Mich., 1888, Williams, 1889, Johns Hopkins, 1902); married. Prof. psychology Antioch Coll., 1872-76; instr. English, Harvard, 1876-77; lecturer, psychology, Harvard and Williams, 1880-81; prof. psychology, Johns Hopkins, 1881-88; pres.

and prof. psychology Clark U., 1889-1919. Founder and editor Am. Jour. Psychology 1887-1921; editor Pedagogical Sem., 1892—, Am. Jour. Religious Psychology and Edn., 1904-15, Jour. Applied Psychology, from 1917. Founder Am. Psychol. Assn. (pres. 1891, 1924). Author: The Contents of Children's Minds, 1883; Adolescence (2 vols.), 1904; Youth—It's Education, 1907; Educational Problems (2 vols.), 1911; Founders of Modern Psychology, 1912; Jesus the Christ, in the Light of Psychology, 1917; Morals: The Supreme Standard of Life and Conduct, 1920; Recreations of a Psychologist, 1920; Senescence, 1922; Life and Confessions of a Psychologist, 1923. Noted as pioneer in child psychology and for establishment of psychol. lab. at Johns Hopkins (1st exptl. psychology lab. in Am.); subsequent work had great influence on Am. psychology. Died Worcester, Mass., Apr. 24, 1924.

HALL, Harlow Homer, Am. microbiologist; b. East Leroy, Mich., Oct. 2, 1904; s. Fred S. and Madge (Bushnell) H.; B.S., Mich. State U., 1927, M.S., 1936, Ph.D., 1944; m. Mabel Brandt, Apr. 5, 1928; 1 dau., Janet A. (Mrs. John A. Ulland). Bacteriologist, Mich. Dept. Agr., Lansing, 1927-30; microbiologist U. S. Dept. Agr., Washington, 1930-43, New Orleans, 1943-49, Peoria, Ill., 1949—. Recipient Superior Service award U. S. Dept. Agr., 1953; Centennial award Mich. State U., 1955; Distinguished Nutritionist award Distillers Feed Council, 1964. Mem. Inst. Food Technologists, Am. Soc. for Microbiology, Entomol. Soc. Am., Am. Chem. Soc., Nat. Geog. Soc., Sigma Xi, Pi Kappa Phi. Research, numerous articles on microbiology prodn. sugar, sirups, starch, citrus products, silage fermentation, fermentative prodn. riboflavin, beta-carotene, and vitamin B12, stblzn. microbial cells, devel. microbial insecticides, investigation mycotoxins. Died Apr. 19, 1967.

HALL, Harry Reginald Holland, English archeologist; b. Fulham, Sept. 30, 1873; s. Sydney Prior and Hannah (Holland) H.; ed. St. John's Coll., Oxford, 1891, D.Litt., 1920. Asst. dept. Egyptian and Assyrian antiquities Brit. Mus., 1896, dep. keeper, 1919, keeper, 1924; excavated at Der el-Bahri, Egypt, 1903-04, winters 1905-06, Abydos, 1910, 25; dir. Museum's excavations at Ur of Chaldees and Abu Shahrain; discovered new site at Al-Ubaid; sent to Iraq by Brit. Mus. with mandate to protect antiquities from damage, World War I. Chmn., Palestine Exploration Fund, 1922. Fellow Brit. Acad.; mem. Egypt Exploration Soc., Royal Asiatic Soc., Soc. for Promotion Hellenic Studies, Soc. Antiquaries (v.p.). Author: Oldest Civilization of Greece, 1901; Coptic and Greek Texts of the Christian Period, 1905; Egyptian Scarabs, 1913; Hieroglyphic Texts from Egyptian Stelae, 1912-25; The XIth Dynasty Temple at Deir el-Bahari; The Ancient History of the Near East, 1913; The Civilization of Greece in the Bronze Age, 1923; A Seasons Work at Ur, 1930. One of foremost authorities on date, style and authenticity of Near East artifacts; 1st to try to harmonize new discoveries in Crete with previous knowledge of Aegean area; noted for work in Egypt, Mesopotamia and Greece. Died London, Oct. 13, 1930.

HALL, Herbert James, Am. physician; b. Manchester, N.H., Mar. 12, 1870; s. Marshall Parker and Susan (James) H.; M.D., Harvard, 1895; m. Eliza Pitman Goldthwait, Dec. 29, 1897. Practiced in Marblehead, Mass., from 1897; especially interested in developing manual work as a remedy in nervous disabilities; medical dir. Devereux Mansion (sanitarium), from 1912. Mem. Am. Occupational Therapy Assn. (pres. 1920-21). Editor occupational therapy and rehab. sect. The Modern Hospital. Author: The Untroubled Mind (essays), 1915; (with Mertice MacCrea Buck) The Work of Our Hands, 1915; (with same) Handicrafts for the Handicapped. 1916; War-Time Nerves, 1918. Died Feb. 19, 1923.

HALL, I. Walker, Brit. physician; b. Sept. 16, 1868; ed. Owens Coll., Manchester; M.D.; m.; 1 dau. Sr. demonstrator physiology Manchester U., 1900, became lectr. pathology, 1901; prof. pathology Bristol U., 1906-33, also dir. preventive medicine lab., curator Canynge Hall. Author: Purin Bodies of Foodstuffs, 1901; Methods of Morbid Histology, 1903; also papers. Editor: Metabolism and Practical Medicine, 1908. Died Oct. 9, 1953.

HALL, Irvin Monroe, Jr., Am. insect pathologist; b. Daly City, Cal., Sept. 30, 1921; s. Irvin Monroe and Clara (George) H.; B.S., U. Cal. at Berkeley, 1948, Ph.D., 1952; m. Maisie Maree McCann, Feb. 9, 1947; children—James Irvin, Catherine Louise, Roberta Ann. Staff, U. Cal. at Berkeley, 1952, U. Cal. at Riverside, 1953—, prof. insect pathology, also insect pathologist, 1967—, vice chmn. dept. biol. control, 1960-62, chmn., 1967—. Sr. Fulbright research scholar, New Zealand, 1963-64. Mem. Entomol. Soc. Am., Soc. Invertebrate Pathology, Phi Beta Kappa, Sigma Xi. Research, numerous publs. on insect pathology, microbial control, bacterial, fungus, virus, rickettsial, protozoan diseases of insects, diseases of phytophagous mites, devel. and use of microbial insecticides. Home: 2355 Oak Crest Dr., Riverside, Cal. 92506.*

HALL, Ivan Victor, Am. botanist; b. Parrsboro, N.S., Can., Aug. 24, 1927; s. Bradford Rhodes and Evelyn Ida (Smith) H.; B.S., Acadia U., 1948, M.S., 1949; Ph.D., Cornell U., 1953; m. Carol Wright Haff, June 17, 1953; children—Dorothy Ella, Helen Frances, Peter Wright, John Victor. Tech. officer Can. Dept. Agr. Research Sta., Kentville, N.S., 1949-52, research officer, 1952-65, research scientist, 1965—. Mem. Canadian Bot. Soc., Bot. Soc. Am., Ecol. Soc. Am., Brit. Ecol. Soc., Am. Soc. Hort. Sci., Agrl. Inst. Can., Can. Field-Naturalists Club. Author: (with E. L. Eaton) Blueberry Culture and Propagation, 1961. Research, numerous publs. on botany of lowbush blueberry from an ecol., taxonomic, genetic and physiol. standpoint; devel. of methods to cultivate and set out selected clones of the lowbush blueberry; culture of the cranberry in Can.; studies on culture, pollination and nutrition of the highbush blueberry. Home: 59 Elm Av. Office: Research Sta., Kentville, N.S., Can.*

HALL, Sir James, Brit. geologist, chemist; b. Haddingtonshire, Eng., Jan. 17, 1761; s. Sir John and Magdalen (Pringle) H.; student Christ Coll., Cambridge; m. Helen Douglas, Nov. 9, 1786; 3 sons, John, Capt. Basil, James; 3 daus. Intimate of Hutton and Playfair; tested Huttonian system by study of continental and Scottish formations; refuted Wernerian views by lab. expts. Mem. Royal Soc. Edinburgh (pres.). Fellow Royal Soc., 1806. Author: Essay on the Origin, History and Principles of Gothic Architecture, 1813; series of memoirs communicated to Royal Soc. Edinburgh. First geologist to apply directly test of lab. expts. to geol. hypotheses, beginning sci. of exptl. geology; gave account of formation of volcanic cones; invented high temperature regulating machine. Died Edinburgh, June 23, 1832.

HALL, James, Am. geologist; b. Hingham, Mass., Sept. 12, 1811; s. James and Susanna (Doordain) H. grad. Rensselaer Poly. Inst., 1832; A.M. (hon.). Union Coll., 1842; M.D. (hon.), U. Md., 1846; LL.D., Hamilton Coll., 1863, McGill U., 1884, Harvard, 1886; m. Sarah Aiken, 1838, 4 children. Asst. prof. chemistry Rensselaer Poly. Inst., 1832-36, prof. geology, 1836; began explorations, Western N.Y., 1838; state geologist Ia., 1855-58, Wis., 1857-60; dir. N.Y. State Museum, 1866-98. Vice pres. Internat. Congress Geologists, Paris, 1878, Bologna, 1881, Berlin, 1885. Recipient Wollaston medal Geol. Soc. of London, 1858, Walker prize Boston Soc. Natural History, 1884, Hayden medal Acad. Natural Scis. of Phila., 1890. Mem. Geol. Soc. Am. (1st pres.), Nat. Acad. Scis. (charter), French Acad. Scis. Author: The Paleontology of New York, Part IV Comprising the Survey of the Fourth Geological District, 1843, (classic of early geol. lit.); New York State Natural History Survey: Paleontology, 8 vols., 1847-94. Known as authority on invertebrate paleontology and stratigraphic geology. Died Albany, N.Y., Aug. 7, 1898.

HALL, John Fry, Am. psychologist; b. Phila., Apr. 24, 1919; s. Harry R. and Alta (Herner) H.; B.S., Ohio U., 1946; M.A., Ohio State U., 1947, Ph.D., 1949; m. Jean Midlam, May 14, 1943; 1 son, John. Faculty, Pa. State U., State College, 1949—, prof. psychology, 1958—. Program dir. in psychobiology NSF, Washington, 1966-67; vis. prof. U. Va., 1952, U. Wis., 1954, U. Cal. at Berkeley, 1962. Mem. Am. Psychol. Assn., Psychonomics Assn., A.A.A.S., Sigma Xi. Author: Psychology of Motivation, 1961; Psychology of Learning, 1966; Readings in the Psychology of Learning, 1967; also articles. Home: 1288 Penfield Rd., State College, Pa. 16801.*

HALL, John Scoville, Am. astronomer; b. Old Lyme, Conn., June 20, 1908; s. Nathaniel Conkling and Harriet Rose (Lance) H.; A.B., Amherst Coll., 1930; Ph.D., Yale, 1933; m. Ruth Carolyn Chandler, June 18, 1935; children—Richard Chandler, Carolyn Davison (Mrs. Richard R. Smith). Engaged as assistant in astronomy Columbia, 1933-34; instr. Sproul Obs., Swarthmore Coll., 1934-38; asst. prof., then asso. prof. astronomy Amherst Coll., 1938-48; staff mem. Radiation Lab., Mass. Inst. Tech., 1942-46; dir. astrometric and astrophys. div. U. S. Naval Obs., 1948-58; dir. Lowell Obs., Flagstaff, Ariz., 1958—; pioneer in photoelectric photometry of stars in infrared region of spectrum; co-discoverer (with W. A. Hiltner) polarization of starlight, 1947; initiated, supervised transfer 40-inch Ritchey Chrétien reflector, also 69-inch Perkins reflector to Flagstaff. Mem. adv. panel astronomy Nat. Sci. Found., 1958-61, chmn., 1960-61; mem. Carnegie Image Tube Com., 1953—. Trustee Museum No. Ariz. Recipient Boyden premium Franklin Inst., 1939. Mem. Internat. Astron. Union (pres. sub-commn. 1955-61), Am. Astron. Soc. (councilor 1944-47, 60-62, v.p. 1963-65), Astron. Society Pacific (first vice president 1963—), Phi Beta Kappa, Sigma Xi, Alpha Delta Phi. Editor: Radar Aids to Navigation, 1947. Contbr. numerous articles profl. jours. Research in astrophysics, photoelectric spectrophotometry, polarization of starlight, elec. measurement of scintillation and seeing, application of image tubes to astron. problems. Address: Lowell Observatory, Flagstaff, Ariz.

HALL, Josiah Newhall, Am. physician; b. North Chelsea, Mass., Oct. 11, 1859; s. Stephen A. and Evalina A. (Newhall) H.; B.S., Mass. Agrl. Coll., 1878; M.D., Harvard, 1882; m. Carrie G. Ayres, Apr. 12, 1885; children—Sigourney D., Oliver W. (dec.). Practiced medicine, Sterling, Colo., 1883-92, Denver, 1892-1937; prof. medicine, med. dept., U. Colo.; physician to Denver City and County, St. Joseph's, St. Anthony's hosps., and Mercy Sanitarium. Pres. State Bd. Med. Examiners, Colo., 1891; pres. Colo. Bd. Health, 1903-04; cons. in internal medicine to the 16 southwestern mil. hosps. during World War I. Author: Borderline Diseases, 2 vols., 1915. Died Dec. 17, 1939.

HALL, Lawrence Babcock, Am. med. engr.; b. Fargo, N.D., June 9, 1909; s. Lawrence J. and Lucy (Babcock) H.; B.S., N.D. State Coll., 1935, M.E., 1953; M.S., Ore State U., 1936; postgrad. Johns Hopkins; m. Sennie Marguerite Hargreaves, Sept. 20, 1941; children—Marguerite Lucy, Henrietta Elizabeth, Lawrence Joshua. Regional engr. Ga. Dept. Pub. Health, 1937-41; mem. med. mission to china, USPHS, Burma Rd., 1941-43, dep. chief engring. br. Malaria control in war areas, Atlanta, 1943-49, chief biophysics sect. Communicable Disease Center, Savannah, Ga., 1950-63; planetary quarantine officer NASA, Washington, 1963—. Cons. to Shah of Iran, 1949, WHO, 1955, 62, 64, Pan Am. Health Orgn., 1956, 58, 59, 61; chmn. expert com. WHO, 1955. Recipient Alumni Achievement award N.D. State U., 1965. Mem. Am. Pub. Health Assn., Conf. Fed. San. Engrs., Am. Acad. San. Engrs., Am. Soc. Tropical Medicine and Hygiene. Contbr. articles to profl. jours. Developed world-wide standards for malaria eradication equipment, methodology for environmental microbiology related to hosp.-acquired infections, control of microbial contamination on spacecraft. Home: 4412 Franklin St., Kensington, Md. 20795. Office: NASA, 400 Maryland Av. S.W., Washington 20546.*

HALL, Marshall, English physiologist; b. Basford, nr. Nottinghamshire, Eng., Feb. 18, 1790; s. Robert Hall; M.D., Edinburgh, 1812; visited med. schs. at Paris, Göttingen, Berlin, 1814-15; m. Nov. 1829; 1 son, Marshall. Practiced in Nottingham, 1817-25, London, 1826-53, specializing in nervous diseases; Gulstonian lectr., 1842; active in found. of Brit. Assn. Fellow Royal Soc., 1832, Royal Soc. Edinburgh, Royal Coll. Physicians; mem. French Acad. Scis. Author: Diagnosis, 1817; Diseases of Females, 1828; Lectures on the Nervous System and Its Diseases, 1836; Memoirs on the Nervous System, 1837; Practical Observations in Medicine, 1845, 46, also numerous articles to Royal Soc. Advanced theory of reflex actions in paper before Royal Soc. (on reflex function of medulla oblongata and medulla spinalis), 1833; located nervous centers for reflex actions in spinal cord; introduced new method of treatment in asphyxia; devised method of artificial respiration (named after him); devised treatment for epilepsy; 1st to note relationship between anemia and epilepsy, 1851. Died Brighton, Eng., May 11, 1857.

HALL, Maurice Crowther, Am. zoologist; b. Golden, Colo., July 15, 1881; s. George Hemingway Birtby and Marion Wallace (Crowther) H.; S.B., Colo. Coll., Colorado Springs, 1905, hon. D.Sc., 1925; M.A., U. Neb., 1906; Ph.D., George Washington U., 1915, D.V.M., 1916; m. Lola May Davis, June 18, 1906; children—Marion Millicent, Winifred Lois, Margaret Lola. Instr. botany and physiology Cutler Acad., Colorado Springs, 1904-05; instr. biology and chemistry, Cañon City (Colo.) High Sch., 1906-07; jr. zoologist U. S. Bur. Animal Industry, 1907-11, asst. zoologist, 1911-16; asst. zoologist; U. S. Insecticide and Fungicide Bd., 1915-16; prof. zoology and parasitology Coll. of Vet. Medicine, George Washington U., 1914-16; parasitologist, research lab. of Parke, Davis & Co., Detroit, 1916-18; sr. zoologist U. S. Bur. Animal Industry, 1919-36, chief of zool. div. 1925-36; asst. custodian Helminthol. Collection, U. S. Nat. Museum, 1925-31, custodian, from 1931. Pres. Permanent Internat. Commn. on Parasitology of Internat. Zool. Congress, from 1930, Internat. Commn. on Control of Parasites of Internat. Veterinary Congress, from 1934. Mem. NRC; prof. zoology and chief div. of zoology NIH-USPHS, from 1936. Proposed the carbon tetrachloride treatment, and, with Dr. J. E. Shillinger, the tetrachlorethylene treatment for hookworm disease. Died 1938.

HALL, Nathan Isaac, Am. elec. engr.; b. Elkins, W.Va., Oct. 24, 1910; s. Nathan I. and Grace (Darlington) H.; B.S. in Elec. Engring., W.Va. U., 1936, M.S., D.Sc.; E.E., Stanford, 1937; D.Sc., David and Elkins Coll.; m. Margaret Sabelstrom, July 15, 1939; children—Natalie Grace, Leonard Eric. Elec. engr. Bell Telephone Labs., N.Y.C., 1937-47; v.p. Hughes Aircraft Co., Culver City, Cal., 1947—. Named Nation's Most Outstanding Young Elec. Engr., Eta Kappa Nu, 1943. Mem. I.R.E., Inst. Aero. Scis., Am. Inst. Elec. Engrs. Contbr. articles to tech. jours. Patentee in field. Developed radar modulator, electronic telephone switching systems, radar, guided missiles and other mil. electronic devices. Home: 400 Bonhill Rd., Los Angeles, Cal. 90049. Office: Hughes Aircraft Co., Culver City, Cal. 90232.*

HALL, Peter Francis, Australian biochemist; b. Sydney, Australia, Dec. 13, 1924; s. William and Alice (Price) H.; M.B., B.S., U. Sydney, 1947, M.D., 1956; Ph.D., U. Utah, 1961. Registrar, Maudsley Hosp., London, Eng., 1952, Guy's Hosp., London, 1952-55; asst. physician Sydney Hosp., 1955-59; lectr. Sydney U., 1956-59; fellow U. Salt Lake City, 1959-62; asst. prof. physiology U. Pitts., 1962-65; prof. biochemistry U. Melbourne (Australia), 1966—. Fellow Royal Australian Coll. Physicians. Author: The Functions of the Endocrine Glands, 1959; Gynaecomastia, 1959; also numerous articles. Research on causes of gynecomastia, demonstration of an hereditary factor in etiology of myxedema, action of gonadotrophic hormones on testis and ovary, role of mitochondria in steroid biosynthesis. Home: 162 Hotham St., East Melbourne, Victoria, Australia.*

HALL, Philip, English mathematician; b. London, Apr. 11, 1904; ed. King's Coll., Cambridge, Eng.; hon. D.Sc., Tübingen, Germany, 1963. Became fellow King's Coll., Cambridge, 1927, faculty, 1933-—, Sadleirian prof. pure math., 1953-67. Recipient De Morgan medal London Math. Soc., 1965. Fellow Royal Soc., 1942 (Sylvester medal 1961). In algebra developed theory of p-groups, 1933, gen. structure theory for finite soluble groups, 1937, (with Graham Higman) theories of p-soluble groups, 1956. Office: King's Coll., Cambridge, Eng.*

HALL, Reginald, English physician; b. Durham, Eng., Oct. 1, 1931; s. Reginald P. and Maggie W. (Wilson) H.; B.Sc., U. Durham, 1953, M.B.B.S., 1956, M.D., 1963; m. Molly Hill, June 11, 1960; children—Susan M., Amanda M., John Reginald, Andrew James, Stephanie Claire. Harkness fellow Commonwealth Fund, clin. and research fellow in medicine Harvard, Mass. Gen. Hosp., Boston, 1960-62; Wellcome sr. research fellow in clin. sci. U. Newcastle upon Tyne, 1964-—, hon. lectr. medicine, 1965, now sr. lectr.; hon. cons. physician Royal Victoria Infirmary. Newcastle upon Tyne, 1965-—. Mem. Royal Soc. Med. (council endocrine sect.), Med. Research Soc., European Thyroid Assn., Scottish Soc. for Exptl. Medicine, London Thyroid Club. Research, numerous publs. on autoantibodies in iodide goiter and asthma, stimulation of purine synthesis in vitro in calf thyroid by thyroid-stimulating hormone, evidence of genetic predisposition to formation of thyroid autoantibodies. Home: 105, Moorside North, Newcastle upon Tyne 4. Office: Ward 10 Office, Royal Victoria Infirmary, Newcastle on Tyne 1, Eng.*

HALL, Richard Pinkham, Am. biologist; b. Eagle Mills, Ark., Feb. 21, 1900; s. Eugene Roy and Merta (Smith) H.; B.A., Henderson-Brown Coll., 1919; student Emory U., 1920-21; A.M., U. Cal. at Berkeley, 1922, Ph.D., 1924; m. Hazel Griggs, June 22, 1935; 1 dau., Barbara (Mrs. William Curtis Blaylock). Instr., coach Camden (Ark.) High Sch., 1919-20; instr. biology Emory U., 1920-21; teaching fellow zoology U. Cal., Berkeley, 1921-23; instr. biology Rice U., 1924-26; faculty N.Y. U., 1926-—, prof., 1938-—, acting chmn. biology, 1931-32, 38-39, 47-48. Mem. Am. Microscopial Soc. (pres., 1957), Soc. Protozoologists (pres., 1950-51), Am. Type Culture Collection (trustee 1953-61), A.A.A.S., Am. Soc. Naturalists, Am. Soc. Zoologists, Am. Soc. Tropical Medicine and Hygiene, Soc. Protozoologists, N.Y. Acad. Sci., Soc. Exptl. Biology and Medicine, Am. Inst. Biol. Sci., Am. Soc. Parasitologists, Sigma Xi, Phi Beta Kappa. Author: Protozoology, 1953; Protozoa, 1964; Protozoan Nutrition, 1965. Research, publs. on cytology of various Protozoa, nutrient requirements, resistance to drugs. Home: 1219 Post Rd., Scarsdale, N.Y. 10583. Office: N.Y. U., University Heights, N.Y.C. 10453.*

HALL, Robert Noel, Am. physicist; b. New Haven, Dec. 25, 1919; s. Harry V. M. and Clara (Kommers) H.; B.S., Cal. Inst. Tech., 1942, Ph.D., 1948; m. Dora Siechert, Aug. 2, 1941; children—Richard H., Elaine L. Lab. asst., research lab. Gen. Electric Co., 1942-46, physicist, 1948-—; spl. research semiconductor laser using gallium arsenide. Fellow Am. Phys. Soc., I.E.E.E. (David Sarnoff award in electronics 1963); mem. Electrochem. Soc. Research on semicondrs., transistors, tunnel diodes and recombination; inventor power rectifier; co-inventor injection laser. Home: 2315 Gurenson Lane, Schenectady 12309. Office: Gen. Electric Research. and Devel. Center, Box 8, Schenactady 12301.*

HALL, Robert Turner, Am. chemist; b. Front Royal, Va., Dec. 25, 1911; s. Robert Carter and Martha Adams (Sowers) H.; B.S., Va. Mil. Inst., 1931; M.S., U. Va., 1933, Ph.D., 1935; m. Alice Churchill Edwards, July 31, 1937; children—Nancy Grayson (Mrs. W. S. Colburn), Alice Pendleton (Mrs. S. W. Nixon). Chief chemist Swann Chem. Co., 1936-39; with Hercules, Inc., Wilmington, 1939-—, research adminstr., 1965-—. Mem. Am. Chem. Soc. (div. councilor 1961-64), Sigma Xi. Research and publs. on devel. methods and techniques for analysis of explosives, synthetic resins, agrl. chems, especially in trace analysis. Home: 103 Briar Lane, Newark, Del. 19711. Office: Research Center, Hercules, Inc., Wilmington 19899.*

HALL, Ross Hume, biochemist; b. Winnipeg, Man., Can., Nov. 22, 1926; s. Reginald M. and Elizabeth (Hume) H.; B.A., U. B.C., 1948; Ph.D., U. Toronto, 1950; Ph.D., U. Cambridge, 1953; m. Rachel May, Sept. 8, 1950; children—Stewart, Donald, Mary-Elizabeth. Came to U. S., 1954, naturalized, 1960. Research chemist Lederle Labs. div. Am. Cyanamid Co., 1954-58; prin. cancer research scientist Roswell Park Meml. Inst., Buffalo, 1958-—; asso. research prof. biochemistry State U. N.Y. at Buffalo, 1964-—. Mem. Am. Chem. Soc., Am. Soc. Biol. Chemists, Am. Assn. Cancer Research, A.A.A.S. Contbr. numerous articles to profl. jours. Discovered new components in structure of ribonucleic acid. Office: 666 Elm St., Buffalo 14203.*

HALL, Thomas, Am. inventor; b. Phila., Feb. 4, 1834; ed. U. Lewisburg (Pa.). Devised mechanism for printing by touching keys, and in 1867 exhibited a keyed typewriter at Paris Expn.; later studied mechanics in Europe; invented and, 1881, placed on the market the Hall typewriter; also invented several successful sewing machines, drill-grinding and other machinists' tools; later patent atty. Died 1911.

HALL, Thomas Christopher, Am. physician; b. N.Y.C., Nov. 26, 1921; s. John Clarence and Theresa (McDonald) H.; student Coll. City N.Y., 1938-40; B.S., Harvard, 1944, M.D., 1949; m. Martha Hartwell Bentley, Mar. 14, 1964; children—Christopher, Thomas, Seth, Amity, Bronwen, Nathan. Faculty, Harvard, Boston, 1949-—, asst. prof. medicine, 1965-—; asso. prof. biochemistry Brandeis U., 1961-64; chief pharmacology Children's Cancer Research Found., Boston, 1961-—. Cons. to NIH, Nat. Cancer Inst., also hosps.; mem. human cancer com. Internat. Union Against Cancer. Mem. Am. Assn. Cancer Research, A.C.P., Am. Soc. Exptl. Pharmacology and Therapeutics, Soc. for Study Growth and Devel., James Ewing Soc., Internat. Soc. Chemotherapy, Am. Therapeutic Soc., Sigma Xi, Alpha Omega Alpha. Research, numerous publs. on immunosuppressive action of drugs, radiosensitization by drugs, chemotherapy of cancer, prediction of response to cancer drugs, induction of enzymes by drugs. Home: 7 Oakland Rd., Brookline, Mass. 02146. Office: 35 Binney St., Boston 02115.*

HALL, Victor Ernest, physiologist; b. Victoria, B.C., Can., Feb. 11, 1901; s. Ernest Amos and Mary Louisa (Fox) H.; A.B., Stanford, 1922, A.M., 1925, M.D., 1928; m. Frances Marie Gould, Sept. 2, 1940; children —Elizabeth Jean (Mrs. Patrick E. O'Neil), Robert Ernest, Barbara Catherine. Faculty Stanford, 1925-51, prof. physiology, 1941-51; prof. physiology U. Cal. at Los Angeles, 1951-—, dir. Brain Information Service, 1964-—. Mem. A.A.A.S., Am. Physiol. Soc. Editor: Ann. Rev. Physiology, 1943-—; U.C.L.A. Forum in Med. Scis., 1963-—; exec. editor Handbook of Physiology, 1952-55. Research, publs. on metabolic theory of temperature regulation, aspects of information storage, retrieval and analysis. Home: 506 Hanley Pl., Los Angeles 90049.*

HALL, Warren Acker, Am. civil engr.; b. Hot Springs, S.D., Aug. 12, 1919; s. Frank L. and Zula (Acker) H.; A.A., Santa Ana Jr. Coll., 1939; B.S., Cal. Inst. Tech., 1942; Ph.D., U. Cal. at Los Angeles, 1952; m. Betty R. Nickerson, May 29, 1942; children—Beverly A., Frank L., Marilyn S. Faculty, U. Cal., 1947-—, prof., asst. dean undergrad. studies and engring., Los Angeles, 1956-60, dir. Water Resources Center, 1961-—. Mem. U. S. nat. com. Internat. Hydrological Decade, 1964-—; mem. steering com. U. S. delegation coordinating council UNESCO, 1965-—; chmn. Western Interstate Water Conf., 1965-66. Recipient Distinguished Alumni award Santa Ana Coll., 1960. Mem. Univs. Council on Water Resources (exec. bd. 1963-—, exec. sec. 1963-66, treas. 1963-—, chmn. 1966-—), Am. Geophys. Union, Am. Soc. C.E., Am. Soc. Agrl. Engrs., Am. Soc. for Engring. Edn. Contbr. articles to tech. jours. Developed theory unsteady shallow flow over ground surface, analysis criteria for artificial replenishment groundwater; derived basic equation for groundwater flow; developed methods for anlysis water resources system for best results, analytical cost allocation for multiple purpose aqueducts, optimal policies for irrigation under drought conditions; developed design criteria for border irrigation systems. Home: 4108 Beverly Glen Blvd., Sherman Oaks, Cal. 91403. Office: Water Resources Center, U. Cal. at Los Angeles 90024.*

HALL, Wendell Howard, Am. physician; b. Redfield, S.D., May 18, 1916; s. Eugene W. and Lora (Pendell) H.; B.S., U. Minn., 1938, M.B., 1940, M.D., 1941, Ph.D., 1950; m. Muriel H. Griffith, Feb. 11, 1948; children—David, Donald, Joan, John. Prof. medicine U. Minn., Mpls., 1961-—, prof. microbiology, 1964-—, prof. lab. medicine, 1966-—; chief med. services Mpls. VA Hosp., 1961-—. Fulbright lectr. Free U. Brussels, Belgium, 1961. Diplomate Am. Bd. Internal Medicine. Mem. Am. Soc. Clin. Investigation, Central Soc. Clin. Research. Research, numerous publs. in fields of infectious diseases, antibiotics, Tb, brucellosis, immunity to infections, fungus diseases. Home: 6609 Galway Dr., Mpls. 55435. Office: VA Hosp., Mpls. 55417.*

HALLBAUER, Joseph August, German engr., metallurgist; b. Zittau, Saxony, Nov. 23, 1842; s. Anton and Agnes (Breithaupt) H.; student Dresden (Germany) Poly. Sch., 1859-62; dr. engring. (hon.), Dresden; m. Elisabeth Nowotny; m. 2d, Anna Schickert; m. 3d, Helene Lochner; 1 son, 2 daus. Employed in shops Saxon R.R., Leipzig, Germany, also in mech. factory of R. Hartmann; with Matthiesen & Hegeler, Lasalle, Mich., 1866-68; became Hartmann's rep. in Russia; rep. of Alfred Krupp, Saxon-Thuringia, then St. Petersburg, Russia, 1874; named dir. G. Hartmann's Lauch factories, Lauchhammer, Germany. Introduced Siemens-Martin process for smelting scrap, Lauchhammer, also transformed power source to lower Lausitz brown coal, built power plant fed by brown coal, 1st large German power sta. supplying grid; built electric charging machine for carrying scrap to Martin furnace; introduced load lifting magnet. Died Kötzschenbroda, nr. Dresden, Apr. 18, 1922.

HALLDAL, Per Haakon Haller, plant physiologist; b. Frogn, Norway, Feb. 22, 1922; s. Haakon and Asta (Amundsen) H.; Cand.real. Marine Botany, U. Oslo (Norway), 1950, Dr.Phil. Plant Physiology, 1959; m. Kari Balke, Feb. 5, 1952; children—Knut, Inger, Dag. Docent plant physiology U. Lund, Sweden, 1958-62; lectr. plant physiology U. Göteborg, Sweden, 1963-67; prof. plant physiology U. Umea, Sweden, 1967-—; biophys. research group Phys. Inst. State U., Utrecht, Netherlands, 1953. Staff Hopkins Marine Sta., Stanford, Pacific Grove, Cal., also dept. plant biology Carnegie Instn., Washington, Stanford, Cal., 1955-57, U. Cal., Scripps Instn. Oceanography, 1965; participant Scripps Instr. expan. to Gt. Barrier Reef, 1966. Mem. Deutsche Botanische Gesellschaft. Author: Fotosyntesen, 1966; also articles. Editorial bd. Zeitschrift für Pflanzenphysiologie. Research on phytoplankton ecology and taxonomy, photobiology including phototaxis in unicellular marine algae, photoreaction, photosynthesis. Home: Trastvägen 2c, Umea, Sweden.*

HALLÉ, Jean Noël, French physician; b. Paris, Jan. 6, 1754; docteur en médecine, 1777; prof. physics Sorbonne; named prof. medicine Coll. France, 1794; 1st physician to Napoleon. Mem. French Acad. Scis. (pres. 1813); Royal Soc. Medicine. Author: Traité d'hygiène, 1806; Histoire de plusieurs vaccinations, 1806. Editor Codex medicmentarius, Vol. I. Research on effects of camphor, breast cancer, sci. hygiene; defended Lavoisier at his trial by Convention; promoted vaccination. Died Paris, Feb. 11, 1822.

HALLE, Louis, polit. scientist; b. N.Y.C., Nov. 17, 1910; s. Louis Joseph and Rita (Sulzbacher) H.; B.S., Harvard, 1932, postgrad.; postgrad. Nat. War Coll.; m. Barbara Mark, Mar. 16, 1946; children—John Joseph, Julia, Mark, Robin, Anne. With Dept. State, Washington, 1941-54, mem. policy planning staff, 1952-54; research prof. Woodrow Wilson dept. fgn. affairs U. Va., 1954-56; prof. Grad. Inst. Internat. Studies, Geneva, Switzerland, 1956-—. Recipient John Burroughs award 1941. Author: Transcaribbean, 1936; Birds against Man, 1938; River of Ruins, 1941; Spring in Washington, 1947; On Facing the World, 1950; Civilization and Foreign Policy, 1955; Choice for Survival, 1958; Dream and Reality, 1959; Man and Nations, 1962; Sedge, 1963; The Society of Man, 1965. Home: Vésenaz, Geneva, Switzerland.*

HALLE, Thore Gustaf, Swedish botanist, geologist; b. 1884; D.Sc., Uppsala (Sweden) U.; D.Sc. honoris causa, Cambridge (Eng.) U. Dir. dept. paleobotany Naturhistoriska Riksmuseum, Stockholm, Sweden. Mem. Swedish Acad. Sci., Linnean Acad., Geol. Soc. London (fgn. corr.), Royal Acad. Belgium (asso.), Geol. Soc. China (corr.). Research and publs. on geol. structure of Falkland Islands, Norwegian Lower Devonian plants, Palaeozoic plants of Central Shansi, fossilized Pteridosperms, Mesozoic flora from Graham Land. Office: Naturhistoriska Riksmuseum, Stockholm, 50, Sweden.

HALL-EDWARDS, John Francis, English roentgenologist; b. Dec. 19, 1858; s. John Edwards; ed. King Edward's Sch., Queen's Coll., Birmingham; L.M., Edinburgh; D.M.R. and E., Cambridge. Former cons. radiographer 1st So. Gen. Hosp., Guest Hosp., Dudley; radiographer 1st Birmingham War Hosp., Rubery, Mil. Orthopaedic Hosp., Hollymoor, Monyhull Mil. Hosp.; surgeon radiographer Royal Orthopaedic and Dental Hosps., Birmingham. Fellow Royal Coll. Physicians, Royal Soc. Medicine, Royal Soc. Edinburgh, Royal Photog. Soc. (hon.), Am. Electro-Therapeutic Soc. (hon.); hon. life mem. Roentgen Soc., Brit. Inst. Radiology, Am. Roentgen-Ray Soc., Birmingham Natural History and Philos. Soc., Am. Roentgen Ray Soc.; mem. Birmingham Photog. Soc., Birmingham and Midland Inst. Sci. Soc., Am. X-Ray Soc. (Brit. corr.). Author: Bullets and Their Billets, or Experiences with the X-Rays in South Africa; X-Ray Dermatitis—Its Cause, Cure and Prevention; Carbondioxide Snow: Its Therapeutic Uses; The Radiography of Metals; The Metal Industry, also numerous articles on application of electricity, photography, x-rays and radium to medicine and surgery. Pioneer in x-ray investigation (lost both hands from burns). Died Aug. 15, 1926.

HALLÉN, Erik Gustaf, Swedish physicist; b. Göteborg, Sweden, Sept. 16, 1899; s. Hans and Matilda (Richardson) H.; grad. Chalmers' Tech. U., Göteborg, 1921; Ph.D., Uppsala (Sweden) U., 1930; m. Lila Richardson, Nov. 25, 1954; Asst. prof. math. physics Uppsala U., 1931-43; faculty Royal Inst. Tech., Stockholm, 1943-—, prof. theoretical elec. engring., 1946-65, emeritus, 1965-—; lectr. high sch., 1942-46. Decorated Kommendör av Kungliga Nordstjärne Orden, 1964. Mem. Kungliga Vetenskaps och Vitterhetssamhället i Göteborg, Ingenjörsvetenskapskademien, 1951. Author: Electromagnetic Theory, 1962; Tvungna svängningar Operatorkalkyl, 1965; also articles. Research on antenna theory, theory of oscillations and transients in mech. systems and elec. systems such as coils, networks and antennas; invented broad band antenna. Home: 25 Alvägen Sollentuna, Sweden. Office: 70 Valhallavägen, Stockholm, Sweden.*

HALLENBECK, George Aaron, Am. physician; b. Rochester, Minn., June 29, 1915; s. Dorr Foster and Bessie (Graham) H.; B.S., Northwestern U., 1936, M.D., 1940; Ph.D. in Physiology, Mayo Found., U. Minn., 1943; m. Marian Mansfield, Dec. 16, 1938; children—John M., George A., Christopher G., Linda.

Cons. surgeon Mayo Clinic, Rochester, 1949-60, head surgeon Mayo Clinic, Rochester, 1949-60, head sect. surg. research, 1961—, chmn. sects. surgery, 1966-—; prof. surgery and physiology Mayo Found., U. Minn., Rochester, 1960—. Research, numerous publs. in physiology gastric and pancreatic secretion, clin. investigation in surgery gastrointestinal tract, transplantation tissue. Home: Mounted Route, 72, Rochester, Minn. 55901.*

HALLER, Albin, French chemist; b. Felleringen, France, Mar. 7, 1849; agrégé in pharmacy; docteur ès scis.; becamse prof. gen. chemistry Faculty Nancy (France), 1885; founder Chem. Inst., Nancy; named prof. Sorbonne, 1899; became dir. Paris Sch. Physics and Chemistry, 1905. Mem. French Acad. Scis., Soc. Agr. Research on influence of unsaturation and ring formation of optical activity, synthesis of cyano-fatty acids, camphor; synthesized borneol, 1891, camphor, 1895, menthol, 1905; discovered constituents of camphoric acid, 1893. Died Paris, Apr. 29, 1925.

HALLER, Albrecht von, see von Haller, Albrecht.

HALLER, Heinz, economist; b. Schwenningen, N. Germany, Mar. 19, 1914; s. Martin and Ursula (Schlenker) H.; dipl.rer.pol., U. Tübingen (Germany), 1935, Dr.rer.pol., 1936; m. Hildegard Maurer, Dec. 12, 1939; children—Gert-Rüdiger, Bettina. Accountant, Schwäbische Treuhand AG, Stuttgart, Germany, 1938-46; asst. dept. econs. U. Tübingen, 1946-48, dozent, 1948-53, asst. prof., 1953-54; prof. U. Kiel (Germany), 1954-57; prof. U. Heidelberg (Germany), 1957-67; prof. econs. U. Zürich (Switzerland), 1967-—. Mem. sci. council Fed. Ministry Finance, Bonn., 1955—, Fed. Ministry of Economy, Bonn., 1965—. Mem. Econometric Soc., Institut International de Finances Publiques. Author: Gibt es eine Lohn-theorie, 1936; Typus and Gesetz in der Nationalökonomie, 1950; Finanzpolitik, Grundlagen und Hauptprobleme, 1957; Die Steuern, 1964; Das Problem der Geldwertstabilität, 1966; also articles. Integration of pub. finance and fiscal policy in gen. frame of econ. policy; analysis of structure of rational tax and duty system, some spl. problems in econ. theory expecially theory of distbn. Home: 29 Etzelstr., Stäfa (ZH) Switzerland. Office: U. Zürich, Zürich, Switzerland.*

HALLERSTEIN, August, see Allerstein, August.

HALLER VON HALLERSTEIN, (Johann) Carl Christoph Wilhelm Joachim, archeologist; b. Hiltpoltstein, Germany, June 10, 1774; s. Carl Joachim and Sophie Amalie Luise (Imhof von Mörlach) H. von H.; ed. Karls-Acad., Stuttgart, Germany, also Berlin; built houses, painted theater decorations, Nuremberg, Germany; in Italy, 1808; went to Greece, 1810, participated in excavations; in process of bldg. bridge for Turkish govt., Greece, 1817. Excavated (with C. R. Cockerell) pediment sculptures of Aphaia temple, Aegina; excavated (with O. M. von Stackelberg) reliefs of Apollo temple, Bassae. Died Ampelakia, Greece, Nov. 5, 1817.

HALLEY, Edmund, Brit. astronomer; b. Haggerston, nr. London, Eng., Nov. 8, 1656; s. Edmund Halley; ed. St. Paul's Coll., London, Queen's Coll., Oxford; received mandamus from Charles II to Oxford U. for master's degree, 1678; D.C.L., Oxford U., 1710; m. Mary Tooke, 1682; children who survived infancy—Edmund, Margaret, Mrs. Price. Conducted stellar studies in St. Helena, 1676-78; arbitrated between Hooke and Hevelius at Danzig, 1679; traveled in Italy, 1681; dep. controller of mint, Chester, 1696-98; explored the Atlantic as comdr. Paramour Pink naval vessel, 1699-1700; at request of German emperor King Leopold I inspected Adriatic harbors, aided imperial engrs. in fortification of Trieste; apptd. Savilian prof. geometry Oxford U., 1703; became astronomer royal Greenwich obs., 1721. Fellow Royal Soc., 1678, asst. sec., also editor Philos. Trans., 1685-93, sec., 1713; fgn. asso. French Acad. Scis., 1729. Author: A Direct and Geometrical Method of Finding the Aphelia and Eccentricity of Planets, 1676; Catalogus stellarum australium, 1679; Theory of the Variation of the Magnetical Compass, 1683; Synopsis astronomiae cometicae, 1705; Astronomical Tables, 1752; also papers in Philos. Trans. Translator: Apollonius' De sectione rationis (from Arabic), also restored lost books De sectione spatii, 1706; Apollonios' Conics, 1710; Serenus' De sectione cylindri et coni, 1710; Menelaus' Spherics, 1758. Made 1st catalogue of stars of so. hemisphere, 1676-78; began study of comets and their orbits, 1680; discovered a comet (later named after him), 1682, identified it as comet which had appeared in 1531 and 1607, correctly predicted its return in 1759 (1st accurate prediction of this sort); held correctly that comets travel in very elongated orbits about the sun; pointed out that at least 3 stars had changed position since ancient times, even since Tycho Brahe's day, 1718, thus concluded that stars are not fixed but have proper motions which are perceptible only over extended periods because of their great distance from the earth; affirmed Kepler's suggestion that transits of Venus could be used to determine scale of solar system; observed complete transit of Mercury; detected acceleration of moon's mean motion; accurately predicted total solar eclipse of 1715; observed eclipse and the great aurora, 1715; made 1st detailed description of trade winds, 1786; suggested that sea is salty because of salt deposits from rivers; ascended Mt. Snowdon to test his method of determining heights by barometer; developed barometric formulas, 1685; demonstrated law connecting atmospheric elevation with density, 1686; observed constancy of temperature of boiling water, utilized this knowledge to compare expansion of water and quicksilver; devised formula for combustion and combination expansion, 1693; prepared 1st declination map, 1701; surveyed coasts and tides of English Channel, pub. map of them, 1702; did important work in neglected field of Greek geometry; discovered methods of computing logarithms by infinite series; worked on geometric solutions to numerical equations; credited with originating science of life statistics in his Breslau Table of Mortality; introduced Newton's Principia to Royal Soc., pub. the work at his own expense and corrected all the proofs, 1687; defended Newton's chronology against attacks by French and German academics. Died Greenwich, Eng., Jan. 14, 1742.

HALLIBURTON, William Dobinson, English physiologist, chemist; b. London, Eng., June 21, 1860; s. Thomas and Mary (Homan) H.; B.Sc., Univ. Coll., London, 1879; M.D. with Gold medal, 1884; postgrad. Vienna, 1885; LL.D., Aberdeen, Toronto; m. Annie Dawes, 1886. Asst. prof. physiology Univ. Coll., 1883-89; prof. physiology King's Coll., London, 1889-1928, emeritus, 1928-31, former dean med. faculty. Fellow Royal Soc., 1891, Royal Coll. Physicians (Baly medal 1911); mem. Biochemistry Soc. (hon.), Brit. Med. Assn. (pres. 1900, 06). Author: text-Book of Chemical Physiology and Pathology, 1891; Essentials of Chemical Psychology, 11th edit., 1922; Handbook of Physiology, 18th edit., 1928; Chemical Side of Nervous Activity (Croonian lectrs. Royal Coll. Physicians), 1901; Biochemistry of Muscle and Nerve, 1904; Physiology and National Needs, 1919, also numerous articles. Editor: Physiol. Abstracts, 1916. Pioneer in biochemistry; studied differentiation of proteins of blood and muscle and constitution of nerve tissue. Died Exeter, Eng., May 21, 1931.

HALLIDAY, Ian, Canadian astronomer; b. Lloydminster, Sask., Can., Nov. 10, 1928; s. Clarence Peter and Edith (Phillips) H.; B.A., U. Toronto, 1949, M.A., 1950, Ph.D., 1954; m. Norma Lillian Mobley, July 7, 1951; children—John Douglas, Janet Elizabeth. Research scientist, Dominion Obs., Ottawa, Ont., Can., 1952—. Chmn. asso. Com. on Meteorites Can., 1966-—. Recipient Gold medal Royal Astron. Soc. Can., 1949. Mem. Internat. Astron. Union, Am. Astron. Soc., Meteoritical Soc. Asst. editor Jour. Royal Astron. Soc. Can., 1964—. Research, publs. on meteor physics, meteor spectra, hot gas behind meteor, geophys. analysis of old meteorite crater, determination of diameter of planet Pluto. Home: 825 Killeen Av. Office: Dominion Obs., Dept. Energy, Mines and Resources, Ottawa, Ont., Can.*

HALLIER, Ernst, German botanist; b. Hamburg, Germany, Nov. 15, 1831; s. Johann Gottfried and Marie (Schleiden) H.; ed. Berlin, Göttingen; Ph.D., Jena, Germany, 1858; m. Agnes Schröter, 1866; 3 sons, including Hans, 1 dau., Dorothea (Mrs. R. Erdmann); m. 2d, Helene Schade, 1893; 1 son. Became asso. with H. Ludwig's pvt. chem.-pharm. inst., Jena, 1858; became mem. faculty U. Jena, 1860, asst. to Schlieden; named asso. prof., 1864; temporary dir. bot. garden, 1862; ret., 1884. Author: Der Grosherzogliche Sächsische Botanische Garten zu Jena, 1864; Die pflanzlichen Parasiten des menschlichen Körpers, 1866; Phytopathologie, 1868; Excursionbuch, enthaltend praktischer Anllitung zum Bestimmen der im Deutschen Reich heimischen Phanerogamen, 1874; Schule der systematischen Botanik, 1878; Die Pflanze und der Mensch in ihrer Wechselbeziehung, 1878; Kulturgeschichte des 19. Jahrhunderts in ihren Beziehungen zu der Entwicklung der Naturwissenschaft geschildert, 1889; Die Pestkrankhei ten (Infektionskrankheiten) der Kulturgewächse, 1895. Editor Zeitschrift für Parasitenkunde, from 1869. Pioneer in microscopic plant pathology and mycology; discovered a group of pathogenic fungi and diseases they cause on plants; isolated and cultivated microorganisms in artificial culture medium; work in plant taxonomy; studied causes of cholera, typhus, measles, smallpox; introduced hypothesis of polymorphism of disease causes. Died Dachau, nr. Munich, Germany, Dec. 20, 1904.

HALLIER, Johannes Hans Gottfried (called Hans), botanist; b. Jena, Germany, July 6, 1868; s. Ernst and Agnes (Schröter) H.; student, Munich, Germany, 1890-92; Ph.D., Jena, 1892; m. Dorothea Ludwig, 1899; 1 son, 1 dau.; m. 2d, Karla de Heus, 1911; 3 daus. Asst., Göttingen (Germany) Bot. Garden, 1892; with bot. garden, Buitenzorg, Java, 1893-97; asst., bot. lab. U. Munich, 1897; became asso. with Bot. Mus., Hamburg, Germany, 1898; traveled in Ceylon, India, Malaya, China, Japan, 1903-04; curator Rijks Herbarium, Leiden, Netherlands, 1908-22. Contbr. articles to bot. jours. Pioneer in phylogenetic taxonomy; introduced biochem. methods in classification; developed microscopic-anat. methods of grouping flowering plants; developed theory of varying characters to deal with problem of mono- or polyphylie; studied taxonomy of convolulaceae and acanthaceae, also problems of internat. nomenclature. Died Oegstgeest, nr. Leiden, Mar. 10, 1932.

HALLOPEAU, François Henri, French dermatologist; b. Paris, 1842; certified physician City of Paris, 1877; became aggregate prof. Paris Faculty Medicine, 1878; mem. Acad. Medicine. Author: (with Fouquet) Traité de syphilis; (with Leredde) Traité de dermatologie, 1900. Research on leprosy, dermatology of extremities; described mycosis fungoides; gave 1st description of dermatitis vegetans (also known as pyodermatitis vegetans, pyodermite végétante), 1889. Died Paris, 1919.

HALLOWELL, A(lfred) Irving, Am. anthropologist; b. Phila., Dec. 28, 1892; s. Edgar Lloyd and Dorothy (Edsall) H.; B.S., U. Pa., 1914, M.A., 1920, Ph.D., 1924, Sc.D., 1963; m. Maude Frame, Oct. 17, 1942. Faculty, U. Pa., 1923-44, 47—, emeritus prof. anthropology, 1963—; prof. Northwestern U., Evanston, Ill., 1944-47; vis. prof. Temple U., 1963-64, Bryn Mawr (Pa.) Coll., 1964-65. Chmn. div. anthropology and psychology NRC, 1946-49; chmn. human relations Area Files, Inc., 1957-63. Guggenheim fellow, 1940-41. Recipient Viking Fund medal Wenner-Gren Found. N.Y., 1955. Mem. Am. Anthropol. Assn. (past pres.), Am. Folklore Soc. (past pres.), Soc. for Projective Techniques (past pres.), Am. Psychopath. Assn., Nat. Acad. Scis., Am. Philos. Soc. Author: Bear Ceremonialism in the Northern Hemisphere, 1926; The Role of Conjuring in Saulteaux Society, 1942; Culture and Experience, 1955. Research on No. Ojiba of Can., other Algonkian peoples. Home: 401 Woodland Av., Wayne, Pa. 19087. Office: Univ. Mus., 33d and Spruce Sts., Phila. 19087.*

HALLWACHS, Wilhelm Ludwig Franz, German physicist; b. Darmstadt, Germany, July 9, 1859; s. Ludwig and Emilie (Hoffmann) H.; ed. Strasbourg (now France), Berlin, Germany; Ph.D., 1883; m. Marie Kohlrausch, 1890; 4 daus., including Helene (Mrs. Karl Huch-Hallwuchs). Asst. to A. Kundt, Strasbourg; asst. to F. Kohlrausch, Würzburg, Germany, 1884-86; prof. physics Tech. U. Dresden (Germany), from 1900; rector, 1921-22. Author: Die Lichtelektrizität, 1914; also articles. Discovered Hallwachs photoelectric effect (partial basis for quantum theory, also for photocell), 1888; discovered photoelectric exhaustion; built double trough refractometer, also periodic quadrant electrometer free of magnetic and after effect; Proved that under ultraviolet light negative electric metal plates discharge and uncharged plates are charged positively (effect is linked with absorption of ultra-violet light). Died Dresden, June 20, 1922.

HALM, Jakob Karl Ernst, astronomer, astrophysicist; b. Bingen, Germany, Nov. 30, 1866; s. Carl Joseph and Sabina (Dietrich) H.; Ph.D., U. Kiel (Germany), 1890; mem. editorial staff Astronomische Nachrichten; asst. Strasbourg (now in France) obs., 1889-95; 1st class asst. Edinburgh, Scotland, 1895-1907; chief asst. Cape obs., 1907-26; tchr., Stellenbosch, S. Africa, from 1926. Recipient Macdougall-Brisbane prize Royal Soc. Edinburgh, 1904-06. Contbr. articles to aston. jours. Research on theory of star atmospheres, spectroscopy of Fraunhofer lines; discovered systematic variation in solar rotation according to solar activity and heliographic width; studied star currents by radial velocity measurements, made 3d probable; determined galactic absorption zone; introduced concept of mass-luminosity relationship; indicated approached equal distbn. of energy in star systems. Died Stellenbosch, July 17, 1944.

HALMA, Nicolas, French mathematician; b. Sedan, France, 1755; prof. math. École polytechnique, Paris; prof. geography Fontainebleau Mil. Sch.; librarian Sainte-Geneviève Library, Paris. Author: Commentaires de Ptolémée, 1813-22; Hypothèses et époques des planétes de Ptolémée, 1820; Commentaires de Théon d'Alexandrie, 1822; Astrologie judiciaire de divination égyptienne, 1824. Translated complete works of Ptolemy into French. Died Paris, 1828.

HALMAGYI, Denis Francis Julius, physiologist; b. Budapest, Hungary, Jan. 18, 1921; s. Francis and Rose (Glaser) H.; M.D., U. Med. Sch., Szeged, Hungary, 1945; M.D., Sydney U., 1962, D.Sc., 1967; m. Alice J. Timar, July 18, 1945; 1 son, Gabor Michael. Asso. prof. dept. medicine U. Szeged, 1947-56; research registrar pneumoconiosis research unit Med. Research Council, Llandough Hosp., nr. Cardiff, Wales, 1956-58; Adolph Basser research fellow dept. medicine U. Sydney (Australia), 1958-63, dir. Gordon Craig Research Lab., reader exptl. surgery, 1963—, lectr. respiratory physiology, 1961-67. Mem. Royal Australasian Coll. Physicians, Australian Physiol. Soc., Australian Thoracic Soc., Cardiac Soc. Australia and New Zealand. Author: Die Klinische Physiologie des kleinen Kreislaufs, 1957. Research, numerous publs. on nervous regulation of pulmonary circulation in heart disease, physiol. mechanisms in fluid inhalation and drowning, circulatory and response mechanisms in acute lung embolism, endotoxin shock, shock after incompatible blood transfusion. Home: 174 New Canterbury Rd., Petersham, New South Wales. Office: Gordon Craig Research Lab., Dept. Surgery, U. Sydney, New South Wales, Australia.*

HALMI, Nicholas Stephen, endocrinologist; b. Budapest, Hungary, June 6, 1922; s. Nicholas and Irene (Hay) H.; M.D., U. Budapest, 1947; m. Katherine Beck, June 10, 1963; 1 son, Nicholas Alexander. Came to U. S., 1949, naturalized, 1955. Asst. in pathology St. John's Hosp., Budapest, 1945-47; asst. prof. anatomy U. Pécs (Hungary), 1947-49; instr. anatomy U. Chgo., 1949-50; faculty U. Ia., Iowa City, 1950—, prof., 1958—. Mem. endocrinology study sect. NIH, 1962-66; NSF sr. postdoctoral fellow, 1961-62. Recipient NIH Research Career award, 1962. Mem. Endocrine Soc. (Ciba award 1957), Am. Physiol. Soc., Am. Assn. Anatomists, Soc. Exptl. Biology and Medicine, Am. Thyroid Assn. Contbr. chpts. to The Thyroid, 1964; Histology, 1966. Research, numerous publs. on cells producing different hormones in pituitary gland, pituitary-adrenal relationships, regulation of thyroid function, effects of undernutrition on carbohydrate metabolism, metabolism of iodide in the thyroid, other tissues. Home: Box 256A, R.R. 1, Iowa City. Office: 356 Med. Labs., Iowa City, Ia. 52240.*

HALMOS, Paul Richard, mathematician; b. Budapest, Hungary, Mar. 3, 1916 (parents Am. citizens); s. Alexander Charles and Paula (Rosenberg) H.; B.S., U. Ill., 1934, M.S., 1935, Ph.D., 1938; m. Virginia Templeton Pritchett, Apr. 7, 1945. Faculty, U. Ill., 1938-39, 42-43; asst. Inst. for Advanced Study, Princeton, N.J., 1939-42; faculty Syracuse U., 1943-46, U. Chgo., 1946-61; prof. math. U. Mich., Ann Arbor, 1961-68; prof., chmn. dept. math. U. Hawaii, 1968—. With radiation lab. Cambridge U., 1945; vis. prof. U. Montevideo, 1951-52, U. Miami, 1965-66. Guggenheim fellow, 1947-48. Mem. Am. Math. Soc. (editor Proc. 1956-62, exec. com. council 1964-66), Math. Assn. Am. (Chauvenet prize for Math. Exposition 1946, gov. 1957-60), Assn. for Symbolic Logic. Author: Finitedimensional Vector Spaces, 1942; Measure Theory, 1950; Introduction to Hilbert Space, 1951; Naive Set Theory, 1960; Lectures on Ergodic Theory, 1961. Editor Math. Revs., 1963—. Research in ergodic theory and operator theory in Hilbert space; contbns. to math. exposition and teaching. Dept. Math., U. Hawaii, Honolulu 96822.*

HALONEN, Pentti Ilmari, Finnish physician; b. Kuopio, Finland, Apr. 27, 1914; s. Erkki and Henni (Rytkönen) H.; M.D., U. Helsinki; m. Lilli Kolmijoik, Aug. 15, 1944; 1 child, Tiina Marjaana. Became prof. internal medicine, 1941; named head 1st dept. medicine U. Helsinki, 1961. Mem. Duodecim (pres.), Finnish Heart Diseases Assn. (pres.), Finnish Soc. Internal Medicine, European Cardiology Soc., European Cardiovascular Surgery Soc., Internat. Coll. Angiology, Finnish Hematology Assn., Finnish Acad. Sci., Belgian, Brit. cardiology socs., N.Y. Acad. Scis. Contbr. numerous articles to tech. jours. Home: It. Kaivop. 15, Helsinki, Finland. Office: Unioninkatu 33, Helsinki, Finland.

HALPERN, Julius, Am. physicist; b. Norfolk, Va., Feb. 4, 1912; s. Jacob and Lena (Kanter) H.; B.S., Carnegie Inst. Tech., 1933, M.S., 1935, Sc.D., 1938; m. Phyllis E. Melnick, Feb. 4, 1940; children—Paul J., Sydney Ann. Research asso. U. Mich., 1937-40, U. Cal., Berkeley, 1940-41; with Mass. Inst. Tech. Radiation Lab., 1941-46, asso. dir. Brit. br., Malvern, Eng., 1944-45, sr. staff research lab. of electronics, 1946-47; faculty U. Pa., Phila., 1947—, prof. 1952—; guest sr. physicist Brookhaven Nat. Lab., Upton, N.Y., 1964—. Chmn. Fedn. Am. Scis., 1952. Recipient U. S. Dept. Def. Merit award, 1947. Fellow Am. Phys. Soc.; mem. Am. Assn. U. Profs., Sigma Xi, Tau Beta Pi, Beta Sigma Rho. Patentee in field. Research, publs. in nuclear and particle physics. Home: 243 S. 4th St., Phila. 19106.*

HALPERN, Otto, physicist; b. Vienna, Austria, Apr. 25, 1899; s. Heinrich and Ernestine Marie H.; Ph.D., U. Vienna, 1922; m. Hilde Berl, 1942; 1 dau., Maria Elizabeth. Came to U. S., 1930, naturalized, 1936. Prof. physics N.Y. U., 1931-41, U. So. Cal., Los Angeles, 1947-52, Columbia, 1947—; staff mem. Mass. Inst. Tech., 1941-55. Cons. Los Alamos Lab., 1946-52, Naval Research Lab., 1947—, Brookhaven Nat. Lab., 1949-53, Livermore Radiation Lab., 1959-61. Recipient Distinguished Service medal Dept. Def., 1960. Research on theory of fine spectral lines, neutron optics. Inventor dielectric (Harp). Contbr. numerous articles on theoretical physics. Address: Park Towers, 2 Brick St., London W. 1, Eng.*

HALPERT, Béla, pathologist; b. Czecze, Hungary, Aug. 31, 1896; s. Joseph and Cecilia (Schiffer) H.; M.D., German U., Prague, Czechoslovakia, 1921; m. Priscilla W. Humphrey, June 30, 1933. Came to U. S., 1924, naturalized, 1930. Chief lab. service VA Hosp., Houston, 1949-64; prof. pathology Baylor U. Med. Sch., 1949-64, prof. emeritus, 1965—; spl. asst. VA Pathology, Armed Forces Inst. Pathology, Washington, 1966—. Mem. Am. Assn. Anatomists, A.A.A.S., Am. Assn. Cancer Research, Am. Assn. Path. Bacteriologists, Am. Assn. U. Profs., Am. Coll. Gastroenterology, Am. Soc. Clin. Pathologists, Am. Soc. Exptl. Pathology, A.M.A., Internat. Acad. Pathology, Coll. Am. Pathologists, Soc. Exptl. Biology and Medicine, So. Med. Assn., Sigma Xi. Author: Necropsy, 1941. Studies, publs. on structure and function of gallbladder, morphologic studies cardiac and vascular anomalies, structural alterations in vascular grafts and prostheses in exptl. animals and man, frequency, structure and clin. behavior of benign and malignant neoplasms. Home: 9039 Sligo Creek Pkwy., Silver Spring, Md. 20901. Office: Armed Forces Inst. Pathology, Washington 20305.*

HALPHEN, Georges Henri, French mathematician; b. Rouen, France, Oct. 30, 1844; ed. Ecole Polytechnique, Paris, 1862, Ecole d'Application d'Artillerie de Metz, 1864; doctorate, 1878. Fought in Franco-Prussian war; became répétiteur Ecole Polytechnique, 1873, examinateur, 1884. Recipient Steiner prize Berlin Acad., 1881; Mem. French Acad. Scis. (Grande Prix des Sciences Mathematiques 1880). Author: Mémoire sur la classification des courbes gauches, 1882; Traité des fonctions elliptiques et de leur application, 1886-91. First to discover differential invariants in his work on differential equations; his investigations touched mainly geometry of algebraic curves and surfaces, differential invariants, Laguerre's invariants, elliptic functions and their applications; pub. 1st systematic presentation of Weierstrass' theory of elliptic functions, 1886. Died Versailles, France, May 21, 1899.

HALSBAND, Egon Gerhard, German ichthyologist; b. Bochum, Germany, Oct. 25, 1921; s. Hugo and Martha (Müller) H.; student U. Bonn; Dr., U. Marburg, 1952; m. Ingeborg Bläsing, Mar. 7, 1954; 1 dau., Ulrike. Researcher, Westphalian State Inst. Fisheries, 1952-54; sci. asst. Fed. Research Inst. Fisheries, Hamburg, Germany, 1954—. Mem. Limnological Soc. Author: (with P. F. Meyer-Waarden) Einführung in die Elektrofischerei, also numerous articles. Research on founds. and practice of electrofishing, devel. new apparatus, problems of coastal pollution by indsl. wastes, nerve physiology of fishes. Home: 49 Schmarje. Office: 9 Palmaille, Hamburg, West Germany.*

HALSEY, Frederick Arthur, Am. management scientist; b. Unadilla, N.Y., July 12, 1856; s. Gaius Leonard (M.D.) and Juliet E. (Carrington) H.; B.M.E., Cornell, 1878; m. Stella D. Spencer, May 12, 1885; children—Olga S., Marion S. Engr. Rand Drill Co., 1880-90; engr. and gen. mgr. Canadian Rand Drill Co., 1890-94; asso. editor, 1894-1907, editor, Feb. 1, 1907-May 1, 1911, editor emeritus, the American Machinist. Active in opposition to the bill for the adoption of the metric system which was before the Ho. of Rep. 1902-06; presented with a testimonial by mfrs. in recognition of these efforts, and since made commr. of Am. Inst. of Weights and Measures, organized by leading engrs. and mfrs. to oppose the metric system. Recipient gold medal Am. Soc. M.E., 1923. Author: Slide Valve Gears, 1890; Slide Rule, 1899; Worm and Spiral Gearing, 1902; Metric Fallacy, 1904, rewritten edit., 1919; (with C. F. Smith) Design and Construction of Cams, 1906; Halsey's Handbook for Machine Designers and Draftsmen, 1913; Methods of Machine Shop Work, 1914; Metric System in Export Trade, 1917; Weights and Measures of Latin America, 1918. Inventor of "premium plan" of paying labor, now recognized factor in factory management in U. S. and Europe. Died Oct. 20, 1935.

HALSKE, Johann Georg, German elec. engr.; b. Hamburg, Germany, July 30, 1814; s. Johann Hinrich and Johanna Catharina (Hahn) H.; m. Christiane Schmidt, 1845; 2 sons, including Albert, 2 daus. With firms Hirschmann, also Pistor & Martins, in Berlin, Repsold, in Hamburg; founder (with F. Bötticher) mech. firm, Berlin, 1844; founder (with W. Siemens) firm Telegraphen-Bauanstatt von Siemens & Halske, Berlin, 1847, resigned when mass prodn. became necessary, 1867; asso. with Berlin applied arts mus. Responsible for rapid growth of Siemens & Halske by careful mech. work on instruments and apparatus, including telegraph equipment, relays, crank inductors, elec. gauge instruments, signal installations, liquid gauges; laid found. of precision mechanics in electrotechnology. Died Berlin, Mar. 18, 1890.

HALSTEAD, Bruce Walter, Am. biotoxicologist; b. San Francisco, Mar. 28, 1920; s. Walter and Ethel Muriel (Shanks) H.; A.A., San Francisco City Coll., 1941; B.A., U. Cal. at Berkeley, 1943; M.D., Loma Linda U., 1948; m. Joy Arloa Mallory, Aug. 3, 1941; children—Linda (Mrs. Robert Baldwin), Sandra, David, Larry, Claudia. Research asst. in ichthyology Cal. Acad. Scis., 1935-43; instr. Pacific Union Coll., 1943-44; mem. faculty Loma Linda U., 1948-58, research asso. Lab. Neurol. Research, Sch. Medicine, 1964—; dir. World Life Research Inst., Colton, Cal., 1959—, Internat. Biotoxicological Center; research asso. in ichthyology Los Angeles County Mus., 1964—; instr. Walla Walla Coll., summers 1964—. Cons. to govt. agys., pvt. corps.; editorial staff Exerpta Medica, 1959—, Toxicon, 1962—; dir. Nat. Assn. Underwater Instrs., Internat. Underwater Enterprises, Internat. Bots., Inc. Fellow A.A.A.S., Internat. Soc. Toxinology (a founder), N.Y. Acad. Scis., Royal Soc. Tropical Medicine and Hygiene; mem. Am. Inst. Biol. Scis., Am. Micros. Soc., Am. Soc. Ichthyologists and Herpetologists, Am. Soc. Limnology and Oceanography, numerous others. Author: Poisonous and Venomous Marine Animals of the World, 3 vols., 1966; also numerous articles. Pioneer marine biotoxicology in U. S.; dir. expdns. around world in search poisonous plants, animals. Address: 23000 Grand Terrace Rd., Colton, Cal. 92324.*

HALSTEAD, Ward Campbell, Am. psychologist; b. Sciotoville, O., Dec. 31, 1908; s. Ward Beecher and Fannie (Campbell) H.; student Miami U., 1925-27; A.B., Ohio U., 1930; A.M., Ohio State U., 1931; Ph.D., Northwestern U., 1935; Nat. Research fellow, U. Chgo., 1935; m. Elizabeth Lee, Dec. 6, 1932; 1 son, Mark Beecher. Instr. in exptl. psychology, dept. medicine, U. Chgo., 1936-39; asst. prof. and asso. mem. Otho S. A. Sprague Meml. Inst., U. Chgo., 1939-43, asso. prof., 1943-46, prof. exptl. psychology, dept. of medicine, 1946—, dir. psychology sect., chmn. sect. biopsychology, 1953; vis. prof. psychology, U. Cal., Los Angeles, summer 1947, Berkeley, summer 1949; lectr. Hixon Symposium on Brain Mechanisms, Cal. Inst. Tech., 1948; James Arthur lectr. Am. Museum Natural History, 1950. Mem. com. on psychiatry, NRC, 1947—; cons. Nat. Inst. Neurol. Diseases and Blindness, 1954—. Mem. Am. Psychol. Assn., A.A.-A.S., Am. Physiol. Soc., Am. Neurol. Assn., Soc. of Biol. Psychiatry, Midwestern Psychol. Assn. Author: Medicine and the War (with W. H. Taliaferro, editor), 1945; Psychology for the Returning Serviceman (with E. G. Boring, editor), 1945; Trauma of the Central Nervous System (with J. Browder, editor), 1945; Brain and Intelligence, 1947; The Frontal Lobes (with J. F. Fulton, editor), 1948; Brain Mechanisms and Behavior (with L. Jeffress, editor), 1951; monographs: Cerebellar Functions, 1935; Brain and Behavior (editor and author with J. Katz), 1950. Contbr. various tech. papers on brain and personality functions in man and on brain functions in lower animals in profl. publs. Used factor analysis to determine nature of functions of cortex, specifically in frontal lobes. Home: 5537 University Av., Chgo. 60637.

HALSTED, William Stewart, Am. surgeon; b. N.Y.C., Sept. 23, 1852; s. William Mills, Jr. and Mary Louisa (Haines) H.; A.B., Yale, 1874; M.D., Coll. Phys. and Surg. (Columbia), 1877; postgrad. univs. of Vienna, Leipzig, Würzburg, 1878-80; (hon. F.R.-C.S., Eng., 1900, Edinburgh, 1905; LL.D., Yale, 1904, Edinburgh, 1905; Sc.D., Columbia, 1904), m. Caroline Hampton, June 4, 1890. Attending physician Charity Hosp., New York, 1881-83; attending surgeon Bellevue and Presbyn. hosps., 1885-87; asso. surgeon Roosevelt Hosp. and surgeon-in-chief, Out-Patient Dept., 1881-87; surgeon-in-chief Emigrant Hosp., N.Y., 1881-84, Johns Hopkins Hosp., Balt., from 1889; prof. surgery Johns Hopkins U., from 1889. Fellow Am. Surg. Assn., Am. Soc. Exptl. Pathology. Introduced cocaine injections for local anesthesia, 1885; 1st surgeon of importance to introduce rubber gloves into operating room, 1890; devised operative techniques based on minimal injury to tissues; devised operations for cancer of breast, hernia; did exptl. work on thyroid. Died Balt., Sept. 7, 1922.

HALTER, Samuel, physician; b. Geneva, Switzerland, Nov. 3, 1916; s. Joseph and Regina (Horowitz) H.; M.D., Free U., Brussels, 1941, D.P.H., 1946; Laureat, Concours U., 1949; m. Paule F. I., Nisen, Dec. 18, 1946; 1 dau., Regine. Research fellow embryology, 1935-49; practice medicine, 1941-42; med. officer in charge Belgian Seamens Hosp., Liverpool, Eng., 1943-45; med. officer health, Brussels, Belgium, 1945-47; chief hosp. div. Ministry Health, Brussels, 1947-56, dir. gen. pub. health, 1957—; with Free U. Brussels, 1945-60, lectr. pub. health, 1952-60; prof. Sch. Pub. Health, Brussels, 1960—; Prof. Pub. Health, Faculty Medicine, Brussels, 1965—, Prof. Pub. Health Legislation, hosp. orgn. and adminstrn., radiol. protection. Expert, WHO, Euratom, OECD, IHF; cons. WHO. Decorated Croix des Evadés; Croix prisonnier politique; chevalier couronne; officer de Leopold; commandeur de l'ordre le Couronne. Mem. Assn. Belge Hygiene et Med. Sociale, Assn. Belge de Radioprotection (hon. pres.), Health Physics Soc. Research, numerous publs. on embryology, med. care and health planning, radiation protection, air pollution. Home: 48 rue E. Bouillot. Office: Ministry Pub. Health, Brussels, Belgium.*

HALVORSON, Halvor Orin, Am. bacteriologist; b. River Falls, Wis., Mar. 26, 1897; s. Hallie and Elizabeth (Heverdahl) H.; B.S. in Chem. Engring., U. Minn., 1921, M. in Chem. Engring., 1922, Ph.D., 1928; D.Sci., St. Olaf Coll., 1948; m. Selma C. Halvorson, Aug. 22, 1922; children—Betty Jean (Mrs. Theodore Caspar), Harlyn Odell, Loren Ellis, Gayle Adair. Faculty, U. Minn., St. Paul, 1928-49, prof., 1940-49, dir. Hormel Inst., 1943-49, research prof. dept. biochemistry, 1965—; faculty U. Ill., 1949-65. Mem. adv. panels Dept. Def., NSF, Army Chem. Corps.; also cons. various govtl. brs. Named Outstanding Am. Investigator, Am. Mfg. Assn., 1940. Mem. Soc. Am. Bacteriologists (Pasteur award 1960), Am. Acad. Microbiology (chmn. 1958—), Nat. Acad. Scis., Sigma Xi, Alpha Chi Sigma, Gamma Alpha, Phi Kappa Phi, Phi Lambda Upsilon. Research, publs. in chem. san. engring.; surface tension factors in immunology; soaps; antiseptics; bacterial physiology; polio vaccine purification. Home: 1901 E. River Rd., Mpls. 55414. Office: Dept. Biochemistry, Coll. Biol. Scis., St. Paul 55101.*

HALVORSON, Harlyn Odell, Am. biologist; b. Mpls., May 17, 1925; s. Halvor Orin and Selma Elizabeth Halvorson; B.S. cum laude, U. Minn., 1948, M.S., 1949; Ph.D., U. Ill., 1952; m. Jean Ericksen, Aug. 27, 1954; children—Lisa Marie, Philip Raymond. Faculty, U. Mich. Med. Sch., 1952-56; faculty U. Wis., Madison, 1956—, prof. bacteriology, 1962—, chmn. Lab. Molecular Biology, 1966—; instr. Woods Hole Marine Biol. Sta., summers 1962-65, 67, Hebrew U., 1965. Mem. adv. bd. on q.m. research NRC-Nat. Acad. Scis.; chmn. physiology div. NIH, 1961; NIH Career prof., 1963. Mem. Am. Soc. Microbiology, Am. Acad. Microbiology, Am. Chem. Soc., N.Y. Acad. Scis., Sigma

XI, Alpha Chi Sigma. Asso. editor Bacteriol. Revs., 1961-66, Archives of Biochem. Biophysics, 1965—; editor series on molecular biology Harper & Row, 1964—. Research on regulation of enzyme synthesis in microorganisms, nature of dormancy in bacterial spores. Home: 1114 Gilbert Rd., Madison, Wis. 53711.*

HAM, Thomas Hale, Am. physician; b. Oklahoma City, Okla., July 19, 1905; s. Thomas Caverno and Lola (Trickey) H.; B.S., Dartmouth, 1927; M.D., Cornell U., 1931; m. Fanny Curtis, May 16, 1936; children—Thomas Caverno (dec.), Margaret Curtis, Lola Josephine (Mrs. James M. Minifie, Jr.). With Thorndike Meml. Lab., Boston City Hosp., 1934-50, asso. dir., 1948-50; asst. prof. medicine Harvard Med. Sch., 1943-50; prof. medicine Western Res. U., Cleve., 1950—, dir. div. research in med. edn., 1958—. Mem. Am. Soc. for Clin. Investigation (past pres.), Am. Soc. Hematology (pres. 1965), A.M.A., Assn. Am. Physicians, A.C.P., Assn. Am. Med. Colls., Am. Acad. Arts and Scis. Author: Syllabus of Laboratory Examination in Clinical Diagnosis, 1950; also articles. Editorial bd. Jour. Med. Education, 1955-60. Research on sedimentation rate of red cells as related to plasma proteins, hemolytic anemias, including paraoxysmal nocturnal hemoglobinuria, sickle cell anemia, hereditary spherocytosis, phys. characteristics of red cells. Home: 2961 Broxton Rd., Shaker Heights, O. 44120. Office: 2064 Abington Rd., Cleve. 44106.*

HAMADA, Kosaku (pseudonym: Seiryo), Japanese archeologist; b. Osaka, Japan, 1881; grad. Tokyo U.; Litt.D.; m. Tomaki Ogawa; children include Seiko. Became instr. Kyoto U., 1909, asst. prof., 1913, prof. 1907, pres., 1937; founder Asiatic Archeol. Assn. Systematized archeol. research in Japan, and contbd. to knowledge of history of fine arts and religion in Asian countries; conducted excavations in South Manchuria (publ. report on findings, which was first report of its kind in Japan), 1910, also excavations in Japan, China, Korea. Died Kyoto, 1938.

HAMADA, Shigenori, Japanese elec. engr.; b. Tokyo, Japan, Sept. 21, 1900; grad. Tokyo U. Dir. Toshiba Elec. Research Inst., Japan; later prof. Tokai U., Japan; now pres. emeritus; currently prof. engrng. dept. Tohoku U., Japan. Author: Theory of Thermo-Electron Radiation; Vacuum Tube Engineering, other publs. Research on electron and elect. communication engrng.; specialist vacuum tube engrng.

HAMAGUCHI, Hiroshi, Japanese chemist; b. Mie Pref., Japan, Jan. 18. 1915; s. Eikichi and Tomiko (Okuno) H.; student Tokyo U., 1934-37, D.Sc., 1943; m. Saito Ikuko, Nov. 23, 1942; children—Noriko, Hitoshi. Prof., Akita Mining Coll., 1938-43, The 1st Higher Sch., 1943-47; faculty sci. Tokyo (Japan) U. Edn., 1947-64, prof. radiochemistry, faculty sci., U. Tokyo, 1964—. Analytical chemist, 1952—. Mem. Am. Chem. Soc., Chem. Soc. Japan, Japanese Soc. for Analytical Chemistry, Atomic Energy Soc. Japan. Research, publs. in radiochemistry, geochemistry, analytical chemistry. Home: 3-50-7 Hayamiya Nerimaku, Tokyo, Japan.*

HAMAKER, John Charles, Jr., Am. metallurgist; b. Canton, O., Apr. 21, 1924; s. John Charles and Lucile (Hammersmith) H.; B.S. in Metall. Engring., U. Mich., 1945, M.S. in Metall. Engring., 1947, Ph.D. (Internat. Nickel Co. fellow), 1952; m. Phyllis L. Bourbonnais, June 14, 1947; children—Joanne Cynthia, John Charles III. Research engr. Internat. Nickel Co., Bayonne, N.J., 1948; cons. Foundry Services Inc., Cleve., 1950; metallurgist Gen. Iron Works, Denver, 1951-53; research metallurgist Vanadium-Alloys Steel Co., Latrobe, Pa., 1953-54, mgr. research, 1955-58, dir. research and metall. engring., 1959-61, v.p. tech., 1961-64, mem. exec. com., 1963—, also dir.; dir. Vanadium-Alloys Steel Can. Ltd.; v.p. tech., dir. Vasco Metals Corp., 1965—; pres. New Eng. Materials Lab., 1966—. Mem. Am. Soc. Metals (chmn. tech. council 1965—), Nat. Acad. Sci., Am. Standards Assn., Am. Iron and Steel Inst., Am. Inst. Mining, Metall. and Petroleum Engrs., Am. Soc. Tool and Mfg. Engrs., Soc. Automotive Engrs., Am. Soc. Testing Materials, Am. Vacuum Soc., Iron and Steel Inst. London, Sigma Xi, numerous others. Author: (with George A. Roberts) Tool Steels, 1958; (with George A. Roberts, Alan R. Johnson) Tool Steels, 1962. Contbr. numerous articles to profl. jours. Developer new steels for ultra high strength, high speed steel cutting, bearings, also forging methods and heat treatments for improved properties. Home: 919 Walnut St. Office: Vasco Metals Corp., Latrobe, Pa. 15650.*

HAMANN, Christel Bernhard Julius, German engr.; b. Hammelwarden, Germany, Feb. 27, 1870; s. Georg Wilhelm Christian and Catherine Margarethe Louise (Schumacher) H.; ed. Nautical Inst., Bremerhaven; D.Eng. (hon.), Berlin, 1933; m. Hedwig Schindler. With Math.-Mech. Inst. of A. Ott, then asso. with Carl Zeiss, Jena, Germany, Carl Bamberg, Berlin; began ind. research, Berlin, 1896; later with Mercedes Büromaschineuwerke; with Deutsche Telephonwerke-und Kabelindustrie AG, Berlin, after World War I. Recipient Gold medal Paris World Exhbn. for geodetic and math. instruments, 1900. Built and improved adding machines and calculators, from 1898, also bookkeeping machines; built Mercedes Euclid with proportional lever system; developed new operation method for hand calculators, thus making automatic div. possible (Hamann Manus). Died Berlin, June 9, 1948.

HAMAR, Martin, Rumanian zoologist; b. Iernut, Rumania, Feb. 5, 1927; s. Martin and Suzanna (Hajdu) H.; M.Sc., Lomonosov U., Moscow, USSR, 1959; student Faculty Biology, Sverdporsk, USSR, 1949-54; m. Maya Sutova, Dec. 31, 1957; 1 dau., Olga. Research worker Kinst. Agr. Research Rumania, Rumanian Acad. Sci., 1954-58; head lab. mammals Central Research Inst. for Agr., Bucharest, Rumania, 1959—. Mem. commn. for use radioisotopes in biology Rumanian Acad. Sci., since 1961—. Mem. bd. Natura rev. Mem. Rumanian Soc. Naturalists, Deutsche Gesellschaft für Säugetierkunde. Author: On Eastern Routes, 1963; Rodents of Rumania, 1967; also articles. Elaboration of methods for rodent tagging with radioisotopes; elucidation of specific composition of rodent and insectivora fauna of Rumania, Pleistocen mammal fauna; research on mechanism of Crictidae and Spalax evolution. Home: Str. Chopin 7, Office: Bd. Marasti 61, Bucharest, Rumania.*

HAMBERGER, C. A., Swedish oto-laryngologist; b. Lidköping, Sweden, Aug. 29, 1908; s. Carl Johan and Hilma (Hammarberg) H.; Med.Kand., Karolinska Institutet, Stockholm, Sweden, 1931, Med.Lic., 1936, Med.Dr., 1942; m. Astrid Gunborg Maria, May 23, 1935; children—Anders, Lars, Bertil, Kerstin. Asso. prof. oto-laryngology U. Stockholm, 1942-49; prof. U. Göteborg (Sweden), 1949-60; prof. Karolinska Institutet, 1960—; med. dir. U. Hosp., Göteborg, 1956-60, Karolinska Hosp., Stockholm, 1960—. Research, numerous publs. on otolaryngology. Home: 166 Sveavägen, Stockholm Va., Sweden.*

HAMBERGER, Georg Edhard, German physician; b. Jena, Germany, Dec. 13, 1697; s. Georg Albrecht and Sophie Catherine (Spitz) H.; ed. Jena; magister, 1717, M.D., 1721; m. Sophie Margarete Wedel, 1724; 6 sons, including Adolf Friedrich, Adolf Albrecht, 3 daus., Johanna Sophie Margarete (Mrs. C. H. Eckhard), Dorothea Elisabeth (Mrs. B. C. B. Wiedeburg), Clara Catherine Sophie (Mrs. J. C. Blasche). Named state physician, Weimar, 1724, Jena, 1729; became asso. prof. medicine and physics, 1726; named prof. math. and physics U. Jena, 1737, prof. medicine 1744, pathology, 1748, physiology, 1749, also taught botany, anatomy, surgery. Mem. Leopoldina. Author: Dissertatione de respirationis mechanismo et usu genuino, 1727; Elementa physices methodo mathematica, 1727; Dissertation sur la mécanique des sécrétions dans le corps humain, 1747; Physiologia medica, 1751; Elementa physiologicae medicae, 1757. Research on relationship between liquid and solid bodies, secretion process; tried to develop math.-phys. laws to explain phenomena of life; showed that inspiration is due to external intercostal muscles, expiration to internal intercostal muscles, 1740; gave 1st description of perforated duodenal ulcer, 1746. Died Jena, July 22, 1755.

HAMBLETON, William Weldon, Am. geologist; b. Lancaster, Pa., Sept. 10, 1921; s. Harry C. and Ella (Moore) H.; B.S. in Chemistry, Franklin and Marshall Coll., 1943; M.A. in Geology, Northwestern U., 1947; Ph.D., U. Kan., 1951; m. Nancy Jane Schnelli, Sept. 7, 1946; children—Ann Louise, Jeffrey Craig. Petrographer, U. S. Bur. Mines, 1949, Pa. Geol. Survey, 1951, 56; faculty U. Kan., Lawrence, 1951—, prof. geophysics, asso. dir. Kan. Geol. Survey, 1962—, asso. dean Grad Sch., 1967—. Geophysicist, Chevron Oil Co., 1955; vis. scientist Lamont Geol. Obs., 1959-60; research com. Interstate Oil Compact Commn., 1967—. Fellow Geol. Soc. Am., Am. Assn. Petroleum Geologists, Soc. Exptl. Geophysicists, Am. Geophys. Union, Nat. Assn. Geography Tchrs., Sigma Xi. Research, publs. on petrography of igneous rocks, clays, coals, magnetic and gravity studies, Precambrian geology. Home: 2009 Oxford Rd., Lawrence, Kan. 66044.*

HAMBOURGER, Walter E., Am. pharmacologist, toxicologist; b. Cleve., Aug. 13, 1906; s. Jacob I. and Esther (Labowitch) H.; B.S. in Chem. Engring., Case Inst. Tech., 1928; Ph.D., Western Res. U., 1931; M.D., U. Chgo., 1939; m. Reva Casler, May 24, 1936; children—Paul David, Robert Michael. Research fellow pharmacology Western Res. U., 1929-32, asst. prof., 1936-42; faculty Yale, 1932-36; chief pharmacologist G. D. Searle & Co., Skokie, Ill., 1942-63, head toxicology dept., 1963—. Mem. Soc. Toxicology, Am. Soc. Pharm. and Exptl. Therapeutics, Soc. Exptl. Biology and Medicine, Nat. Soc. Med. Research, Sigma Xi, Alpha Omega Alpha. Publs. on pharmacologic and toxicologic research in animals, including misuse of barbiturates, morphine excitement, anticholinergic drugs, probable safety of new synthetic drugs. Home: 573 Jackson Av., Glencoe, Ill. 60022. Office: 4900 Searle Pkwy., Skokie, Ill. 60076.*

HAMBRUCH, Paul, German ethnologist; b. Hamburg, Germany, Nov. 22, 1882; s. Heinrich Friedrich and Anna (Obermann) H.; student, Göttingen, Germany; Ph.D., U. Berlin, 1907; m. Paula Schmidt, 1912; 1 son. Asst. Berlin Ethnol. Mus., before 1907, dir., curator Indo-oceanic dept., after 1910; mem. Hamburg S. Seas Expdn., 1909-10; lectr. Hamburg Colonial Inst.; became mem. faculty U. Hamburg at its founding, asso. prof., 1922. Author: Ergebuisse der Hamburger Südsee-Expedition 1908-10, Vol. II, Nauru, 1914, Vol. III, Ponape, 1932-36; Südseemärchen, 1922; Faraulip, Liefeslegenden aus der Südsee, 1924; Die Irrtümer und Phantasien des Harrn Prof. Dr. Hermann Wirth, 1924; Malaiische Märchen, 1926; Ozeanische Rindenstoffe, 1926. Ethnol. research on Micronesian Islands, Nauru, Ponape; collected folklore of S. Seas, also of Lübeburger Heide region, Germany. Died Hamburg, June 23, 1933.

HAMBURGER, H. J., physiologist; b. Mar. 9, 1859; ed. Alkmaar, Netherlands, U. Utrecht (Netherlands); Sc.D., M.D., LL.D., Aberdeen, St. Andrews; m. Frédérique Coben Gosschalk; 1 son, 1 dau. Asst. to Prof. Donders, Physiol. Lab., Utrecht, 1882-88; prof. Royal Vet. Sch., Utrecht, 1988-1901; prof. physiology U. Groningen (Netherlands), from 1901. Herter lectr., N.Y., 1922; 1st Charles E. Dolme lectr. Johns Hopkins, 1922; pres. 9th Internat. Physiol. Congress, Groningen, 1913. Honored by publ. of book on biochem. contbns. of European and Am. scientists, 1908. Author: Osmotischer Druck und Ionenlehre in den medicinischen Wissenschaften, 1902-04. Introduced application of phys. chemistry to med. scis., 1883. Died Jan. 5, 1924.

HAMBURGER, Hans Ludwig, mathematician; b. Berlin, Aug. 5, 1889; s. Karl and Margarete (Levy) H.; student, Berlin, Göttingen; Ph.D., Munich, 1914; lectr., then asst. asso. prof. U. Berlin, 1919-24; became prof. U. Cologne (Germany), 1924; sought polit. asylum, Eng., 1939; prof. U. Ankara (Turkey), 1947-53. Contbr. articles to math. jours. Research on expansion of Stieltjess momentary problems, also on Riemann's functional equation and theory of zeta function; studied problems of differential geometry, including theory of spherical diagrams, linear and partial differential equations of hyperbolic type, also problems of algebraic and operation theory; worked unsuccessfully on Caratheodory's assumption on navel points of regular, closed plane. Died Cologne, Aug. 14, 1956.

HAMBURGER, Meyer, mathematician; b. Poznan, Poland, Apr. 5, 1838; student U. Berlin; Ph.D., Halle, Germany, 1865; 1 son, 4 daus. Tchr., boys' sch. in Jewish community, Berlin, 1864-1903; named lectr. Tech. U. Berlin, 1879, titled prof. algebraic analysis and algebra, 1885, lectr. on theory of functions, calculus of variations, exponential theory, from 1896. Contbr. articles to math. jours. Studied integration of partial differential equations, multiple roots of the fundamental equation of a linear differential equation, Pfaff's reduction process, singular solutions of algebraic differential equations of 1st order. Died Berlin, June 9, 1903.

HAMBURGER, Morton, Am. physician; b. Balt., Aug. 18, 1907; s. Morton and Helen (Bechhofer) H.; B.A., Johns Hopkins, 1928, M.D., 1934; m. Virginia Gray, July 23, 1943; children—Susan, Robert, John. Research asso. U. Chgo., 1937-38; instr. medicine U. Cin. Coll. Medicine, 1938-41; field dir. Commn. on Air Borne Infections, Armed Forces Epidemiology Bd., Washington, 1942-46, mem. cons. to sec. war, 1944-46; faculty U. Cin. Coll. Medicine, 1946—, prof., 1958—, dir. infectious disease div. medicine, dept. medicine, 1946—, asst. dir. dept. medicine, 1958—; cons. VA, Children's, Bethesda hosps. (all Cin.) Fellow A.C.P., mem. Central Soc. Clin. Research, Am. Soc. Clin. Investigation, Am. Clin. and Climatol. Assn., Assn. Am. Physicians, Am. Heart Assn., Am., Ohio State med. assns., Infectious Diseases Soc. Am., Cin. Soc. Internal Medicine, Soc. Gen. Microbiology, Sigma Xi. Editorial bds. Archives of Internal Medicine, Antimicrobial Agts. and Chemotherapy, 1960—. Research, publs. on transmission, carriers of streptococcal infections in army camps; studies on treatment of various bacterial infections. Home: 140 Elm Av., Wyoming, O. 45215. Office: Cin. Gen. Hosp., Cin. 45229.*

HAMBURGER, Viktor, biologist; b. Landeshut, Germany, July 9, 1900; s. Max and Else (Gradenwitz) H.; student U. of Heidelberg, 1919-20, U. of Munich, 1920-21; Ph.D., U. of Freiburg (Baden), 1924; m. Martha Fricke, July 28, 1928; children—Doris, Carola. Came to U. S., 1932, naturalized, 1940. Privatdozent, dept. zoology, U. of Freiburg, 1927-32; research fellow, dept. zoology, U. of Chicago, 1932-33, instr., 1933-35; asst. prof. dept. zoology Washington U., St. Louis, 1935-39, asso. prof., 1939-41, prof. zoology and head dept., 1941—. Fellow American Academy of Arts and Sciences; mem. Am. Assn. Anatomists, Am. Soc. Zoologists (pres. 1955), Nat. Acad. Scis., Soc. Development and Growth (past pres.), Genetics Soc. Am. Naturalists, Institut International d'Embryologie. Author: Manual of Experimental Embryology 1960. Research on embryology of behavior; experimental neuro-embryology; mode of gene action in development. Home: 740 Trinity Av., St. Louis 63130.

HAMBURGER, Walter Julian, Am. textile physicist; b. Boston, Nov. 19, 1901; s. Simon B. and Clara (Schindler) H.; B.S., Mass. Inst. Tech., 1921, M.S., 1941; Ph.D., Poly. Inst. Bklyn., 1948; M.Sc. (hon.), Lowell Technol. Inst., 1952; m. Janet V. Lambert, Sept. 12, 1927; children—Walter J., Suzanne L. (Mrs. Robert N. Thurston) Gen. engr., 1922-25; indsl. engr. paper industry, 1925-30; treas., tech. dir. H. Schindler & Co., Inc., Canton, Mass., 1930-44; chmn., chief exec. officer Fabric Research Labs., Inc., Dedham, Mass., 1942—. Lectr., prof. various technol. insts., 1941—; cons. U.S.A.A.F., World War II. Recipient Olney medal Am. Assn. Textile Chemists and Colorists, 1956; certificate Distinction, Poly. Inst. Bklyn., 1956.

Fellow Textile Inst. (Manchester, Eng., past v.p.); mem. Fiber Soc. (past pres.), A.A.A.S. (chmn. Gordon Research Confs. Textiles 1952), Am. Soc. for Testing Materials (Edgar Marburg lectr. 1955, Harold deWitt Smith medal 1959), N.Y. Acad. Scis., Inst. Textile Sci. (Can.), Am. Chem. Soc. Studies, publs. on fibrous, organic and related materials. Home: 15 Crest Dr., Dover, Mass. 02030. Office: 1000 Providence Hwy., Dedham, Mass. 02026.*

HAMBY, Wallace Bernard, Am. physician; b. Ennis, Tex., Aug. 2, 1903; s. Adrain D. and Willie Elizabeth (Moseley) H.; student Mercer U., 1920-22; A.B., U. Okla., 1926, M.D., 1928; m. Hellyn Greenwald, Aug. 15, 1934 (dec. Apr. 1964); children—Marcia Ruth (Mrs. Donald R. Calavan), Wallace Bernard II; m. 2d, Mary Jane Gutierrez Husk, Oct. 1, 1965. Asst. histology and embryology U. Okla., 1924-26; prof. neurol. surgery U. Buffalo Sch. Medicine, 1941-60; attending neurol. surgeon. head dept. Cleve. Clinic Found., 1960—. Diplomate Am. Bd. Neurol. Surgery. Mem. A.M.A., A.C.S., Harvey Cushing Soc. (v.p. 1957-58), Soc. Neurol. Surgery, Am. Acad. Neurol. Surgery (pres. 1951), Am. Neurol. Assn., Am. Acad. Neurology, Scandinavian Neurosurg. Soc., Sigma Xi, Alpha Omega Alpha. Author: Hospital Care Neurosurgical Patients, 1940, 2d edit., 1948; Intracranial Aneurysms, 1952; Case Reports and Autopsy Records of Ambroise Pare, 1960; Surgery and Ambroise Pare, 1965; Carotid-Cavernous Fistula, 1966; Ambroise Paré, Surgeon of the Renaissance, 1967; also numerous articles. Investigation, treatment of surg. diseases of nervous system; particularly intracranial aneurysms. Home: 13700 Fairhill Rd., Shaker Heights O. 44120. Office: 2020 E. 93d St., Cleve. 44106.*

HAMDALLAH, Mustawfi, Persian natural philosopher; b. Qazwin, circa 1281. Of Arabian and Shi'a ancestry; belonged to the lit. circle of il-khan Uljaii'tu, 1304-16; had unusual opportunities of meeting the learned men of his time and obtaining valuable information. Author: Zafar-nama (Book of Victory), 1335; Ta'rikh Inguzida (Select History), 1330; Nuzhat Al-Qulub (Delight of the Hearts), 1340; An Encyclopedia of Cosmography and Science. His ency. is especially concerned with med. facts, geography and customs but covers astronomy, physics, chemistry, mineralogy, botany, zoology, anatomy, human characteristics (mental, moral and physical). Died after 1340.

HAMEAU, Jean, French physician; b. La Teste, France, 1779; M.D., Montpellier, France, 1807; became health officer, Bordeaux, France, 1804. Author: Étude sur le virus, 1847. Precursor of Pasteur; introduced Jenner's vaccine into Gironde area, 1801; proved man can contract glanders; discovered pellagra, 1829. Died 1851.

HAMEKA, Hendrik Frederik, chemist; b. Rotterdam, Holland, May 25, 1931; s. Dirk C. and Johanna (Mannebeck) H.; H.Drs., U. Leiden (Netherland), 1953, D.Sc., 1956; m. Carol E. Carson, Aug. 9, 1958; children—Richard Charles, Christina Laura. Came to U. S., 1960, naturalized, 1963. Research asso. U. Rome, Italy, 1956-57; fellow Carnegie Inst. Tech., 1957-58; research physicist N. V. Phillips Lamps, Eindhoven, Netherlands, 1958-60; asst. prof. chemistry Johns Hopkins, 1960-62; asso. prof. chemistry U. Pa., Phila., 1962—; mem. editorial bd. Chem. Physics Letters. Alfred P. Sloan Research fellow, 1963—. Mem. Am. Phys. Soc. Author: Advanced Quantum Chemistry, 1965; Introductory Quantum Theory, 1967. Contbr. numerous articles to sci. jours. Research on theory of molecular structure and optical and magnetic properties of molecules; calculations of spin-orbit and spin-spin coupling; theory of resonance optical rotation, magneto-optical rotation and multiple-photon processes. Home: 617 Magill Rd., Swarthmore, Pa. 19081.*

HAMEL, Carl, German bacteriologist; b. Düren, Germany, June 19, 1870; s. Robert and Gertrud (Schmitz) H.; ed. Strasbourg (now France), Heidelberg, Berlin, Munich (all Germany); M.D., 1894; hon. dr.med.vet., Leipzig; m. Hedwig Kluxen, 1910; 2 children. With path.-anat. inst. U. Munich, 1894; with surg. dept. Heurahnsdorf hosp., 1895-98; in Friedrichshain, Berlin, 1898; in internal dept. Charlottenburg (Berlin) hosp., 1898-1901; with Charité hosp., Berlin, 1901-02; became asst. Reich Bd. Health, 1902, mem., 1906, pres., 1926-22; apptd. adviser Reich office of interior, 1918, ministerial dir. sub-dept. pub. health care, 1922. Mem. Internat. Bd. Health, Paris, Soc. Med. Officers of Health, other health orgns. Discoverer basophilic granular red corpuscles in blood in chronic lead poisoning; research on anti-Tb measures; a founder lupus sanitarium, Münchberg. Died Rhöndorf, nr. Bonn, Germany, Sept. 12, 1949.

HAMEL, Georg Karl Wilhelm, German mathematician; b. Düren, Germany, Sept. 12, 1877; s. Leonhard and Paula (Jansen) H.; ed. Tech. U. Aachen, U. Berlin; Ph.D., U. Göttingen (Germany), 1901; dr.rer.nat. (hon.) Aachen, 1954; m. Agnes Frangenheim, 1909; 3 daus. Asst. to Felix Klein, 1902; became lectr., asst. to Heun, Karlsruhe, Germany, 1903; named prof. mechanics and math. Tech. U. Brno (Czechoslovakia), 1905, Aachen, 1912, Berlin, 1919; emeritus, from 1949; vis. prof. U. Tübingen, 1946-47, U. Berlin, 1947-48; tchr. Tech. U. Munich, from 1948. Mem. Prussian, Bavarian acads. scis., German Acad. Scis. and Lit., Mainz, Leopoldina. Author: Elementare Mechanik, 1912; Integralgleichungen, 1937; Theoretische Mechanik, 1949; also articles. Research in math. mechanics and axiomatic devel. of mechanics; studied d'Alembert's principle, problem of plasticity, ground water currents, theory of cables and rods, films and shells; worked on problems of real numbers, also linear differential equations with periodic coefficients and non-linear differential equations in relation to a question of mech. stability; introduced integral equations into teaching and use. Died Landshut, Germany, Oct. 4, 1954.

HAMER, Walter Jay, Am. chemist; b. Altoona, Pa., Nov. 5, 1907; s. Jesse James and Naomi (Roland) H.; B.S., Juniata Coll., 1929, Sc.D. (hon.), 1966; Ph.D., Yale, 1932; m. Alma Robinson, Mar. 19, 1941; 1 dau., Margaret. Postdoctoral research fellow Yale, 1932-34; research asso. Mass. Inst. Tech., 1934-35; chemist Nat. Bur. Standards, Washington, 1935-50, chief electrochemistry sect., 1950—. Research chemist nat. def. research com. OSRD and Mahattan Project, 1943-44; cons. U. S. Dept. Def., 1951-53; lectr. to univs., govt. agys. Recipient certificate Merit, OSRD, 1945, Manhattan Project, 1945; U. S. Dept. Commerce Superior Acomplishment award, 1954, Gold medal exceptional service award, 1965. Fellow I.E.E.E., A.A.A.S., Washington Acad. Sci.; mem. Am. Chem. Soc., Am. Phys. Soc., Electrochem. Soc. (past pres.), Faraday Soc., Sigma Xi, Alpha Chi Sigma. Editor: The Structure of Electrolytic Solutions, 1959. Coordinator: Electrochemical Constants, 1953. Research, numerous publs. on electrolytic solutions, primary and secondary batteries, voltage or electromotive force standards; determined Faraday constant. Home: 3028 Dogwood St. N.W., Washington 20015. Office: Nat. Bur. Standards, Washington 20234.*

HAMILL, William Henry, Am. chemist; b. Oswego, N.Y., June 13, 1908; s. William Henry and Lorena (Brecht) H.; B.S., U. Notre Dame, 1930, M.S., 1931; Ph.D., Columbia, 1936; m. Angela Irma Tobia, Aug. 17, 1934; children—Mary Angela (Mrs. William Walker), Carol Rita, Irma Clare, Ann Therese, Catherine Bridget. Faculty, Fordham U., N.Y.C., 1931-38; prof. chemistry U. Notre Dame (Ind.), 1938—, Chmn., Gordon Conf. on Radiation Chemistry, 1958. Mem. Am. Chem. Soc., Faraday Soc., Sigma Xi. Author: (with R. R. Williams, Colin Mackay) Principles of Physical Chemistry, 1959, 2d edit., 1966. Asso. editor Radiation Research, 1961-63; editorial bd. Jour. Am. Chem. Soc., 1957—. Research, publs. on mechanisms chem. reactions induced by light, radiation; mass spectrometer studies, electrons, ions in irradiated organic solids. Home: 17899 Edgewood Walk, South Bend, Ind. 46635.*

HAMILTON, Albert Hine, Am. micro-chem. investigator; b. Weedsport, N.Y., Dec. 10, 1859; s. James Theodore and Clarissa (Hine) H.; Ph.G., Coll. of Pharmacy of City of N.Y. (now dept. Columbia U.), 1885; m. Jessie Eccles, Feb. 1, 1888. Pharmacist, 1887-1911; chemist, Auburn (N.Y.) Bd. of Health; frequently called as expert chemist in legal proceedings, appearing in numerous homicide, forgery, arson, burglary, bomb, assault and other cases. Discovered by test shots into human bodies, how to identify the "contact shot" in homicide cases; originated a system of examination of exhibits in circumstantial evidence cases whereby the exhibits reveal the truth, regardless of claims to the contrary; discovered by test shots, that the fine scratches on murder bullets, which have been relied upon by forensic ballistics to identify a suspected firearm, are not made by the barrel interior but by the crimping of the cartridge shell and the hot escaping exploding powder. Died July 1, 1938.

HAMILTON, Alexander, Am. statesman, polit. theorist, economist; b. Charlestown, Nevis, B.W.I., probably Jan. 11, 1755 (or 56, 57); s. James and Rachel Faucitt (Lavien) H.; came to colonies, 1772; ed. King's Coll. (now Columbia), 1773-76; m. Elizabeth Schuyler, 1780; 8 children. Wrote 3 influential and anonymous pamphlets (which were attributed to John Jay and John Adams) defending colonist's position against Brit., 1774-75; commd. capt., artillery co. formed by N.Y. Provencial Congress, 1766; apptd. lt. col., George Washington's staff, then private sec. to Washington, 1777-81; led Am. column in final assault on Brit., Yorktown, 1781; admitted to N.Y. bar, 1782, practiced law, N.Y.C., from 1783; N.Y. del., Continental Congress, 1782-83, 88-89, also to Annapolis Convention, 1786; founder Bank of N.Y., 1784; mem. N.Y. legislature, 1787; drafted call for and represented N.Y. at Constitutional Convention, Phila., 1787, signed the Constitution as an individual; wrote (with John Jay and James Madison) the Federalist Papers, a series of 85 articles favoring adoption of Constitution, 1787-88 (classic of polit. lit. which helped shape Am. polit. instns.); persuaded N.Y. Convention to ratify the Constitution, 1788; became sec. of Treasury, Washington's cabinet, 1789-95; among maj. spokesmen of Federalist Party; had many conflicts with Jefferson, which in essence were over Hamilton nationalism versus Jeffersonian democracy, and led to formation of 2 party system (Federalists versus Democrats); Am. principle of neutrality in fgn. policy largely due to his influence, 1793-94; drafted much of Washington's farewell address, 1796; founder N.Y. Evening Post, 1801; his polit., financial, diplomatic programs were directed toward strengthening federal union at expense of states; correctly defined problems inherent in Articles of Confederation, in his letters, 1779-80; advocated aristocratic, strongly centralized, rep. union with devices to give weight to influence of class and property; opposed powerful state legislatures; presented financial program to Congress in series of reports, and argued that Constitution was source of implied as well as enumerated powers; (doctrine became basis for interpreting and expanding Constitution in later years); adviser to cabinet of John Adams. Died N.Y.C., after duel with Aaron Burr, July 12, 1804.

HAMILTON, Alice, Am. physician; b. N.Y.C., Feb. 27, 1869; d. Montgomery and Gertrude (Pond) H.; M.D., U. Mich., 1893; postgrad., Germany, 1896, Johns Hopkins, 1897, Pasteur Inst., Paris, 1902; Sc.D., Tulane U., 1936; hon. degrees U. Mich., Holyoke Coll., Smith Coll., Tulane U., U. Rochester. Prof. pathology Woman's Med. Coll., Chgo., 1897-1901; asst. pathologist McCormick Inst., 1902-09; dir. Ill. Commn. to Investigate Occupational Diseases, 1908-10; with Dept. Labor, 1911-21; asst. prof. indsl. medicine Harvard Med. Sch., later Sch. Pub. Health, 1919-35, emeritus prof., 1935—. Mem. League Nations Health Com., 1924-30; cons. Internat. Labor Office, Geneva, Switzerland, 1924—; mem. U. S. Research Com. Recent Social Trends, 1923-33; pres. Hoover's Com. on Social Trends, 1930-33. Mem. A.M.A., A.A.A.S., Am. Pub. Health Assn. Author: Industrial Poisons in the United States, 1925; Industrial Toxicology, 1934; Exploring Dangerous Trades, 1943. Pioneered research in indsl. diseases. Home: Hadlyme Ferry, Hadlyme, Conn.

HAMILTON, Charles Horace, Am. sociologist; b. McLennan County, Tex., June 10, 1901; s. William Clark and Amy Jane (Whittenberg) H.; A.B., Southern Meth. U., 1924; M.S., Tex. Agrl. and Mech. Coll., 1925; Ph.D., U. of N.C., 1932; student U. of Chicago, 1925, Harvard, 1930-31; m. Maurine Phifer, Sept. 9, 1928; children—Charles Edward (dec.), Elizabeth Ayers, Mary Bellamy. Prof. edn., Lon Morris Coll., Jacksonville, Tex., 1925-26; sr. social scientist, bur. agrl. econs., U. S. Dept. Agr., Washington, 1939-40; head dept. of rural sociology, N.C. State Coll., 1940-60, William Neal Reynolds professor rural sociology, 1960—; vis. prof. Univ. of Wisconsin, 1959-60; vis. prof. sociology and biostatistics, asso. dir. Carolina Population Center, U. N.C., Chapel Hill, 1967—; on leave as dir. sociol. research, Commission on Hosp. Care, Chgo., 1945-46; mng. editor Rural Sociology, Journal of Rural Sociol. Soc. since 1940; consultant research grants div. Nat. Inst. Health, USPHS, 1946-49. Pres. N.C. Conf. for Social Service, 1961-62. Social Science Research Council Fellow, 1930-31. Sec. rural sub-com. Gov's. Commission on Hosp. and Medical Care, also chmn. statistics and publications sub-com.; mem. Adv. Com. on Population for 1950 U. S. Census. Oliver Max Gardner award, 1958. Fellow Am. Statis. Assn.; mem. Am. Rural Sociol. Soc. (pres.), Am. Assn. U. Profs., Am., So. (pres., 1957-58) sociol. socs., Population Assn. Am. (pres. 1960-61), Internat. Population Union. Author: (with W. E. Garnett) The Role of the Church in Rural Community Life in Virginia, 1929; Rural-Urban Migration in North Carolina, 1934; Recent Changes in the Social and Economic Status of Farm Families in North Carolina, 1937; (with Selz C. Mayo) Rural Population Problems in North Carolina, 1943; (with A. C. Bachmeyer and Staff) Hospital Resources and Needs in Michigan, 1946; (with A. C. Bachmeyer and Staff) Hospital Care in the United States; (with Clarence Poe) Hospital Care for All Our People, 1947; (with Herbert A. Aurbach) What's Happening to North Carolina Farms and Farmers, 1958; (with Thomas R. Ford, ed.) The Southern Appalachian Region, A Survey, 1962; (with Josef Perry) 1980 Population Projections for North Carolina Counties, 1963, (with Mildred Kantov, ed.) Mobility and Mental Health, 1965; (with John McKinney and Edgar Thompson, eds.) Continuity and Change in the South, 1965. Research in rural life, in population trends and problems, in land tenure, hospitals and medical care, population and college enrollment projections, delineation of Community College areas, derivation of statistical formula and models for the study of sociological phenomena. Home: 1515 Duplin Rd., Raleigh, N.C.*

HAMILTON, Cliff Struthers, Am. chemist; b. Blair, Neb., Nov. 23, 1889; s. S. Leigh and Lillie (Brownlee) H.; B.S., Monmouth Coll., 1912, Sc.D., 1954; postgrad. U. Minn.; Ph.D., Northwestern U., 1922; m. Frances Howe, Aug. 4, 1921; children—Robert W., Mary Frances (dec.), Martha H., Clif Struthers. Instr., Ohio Wesleyan U., 1917-19; research instr. pharmacology U. Wis., 1922-23; faculty U. Neb., Lincoln, 1923-27, prof. chemistry, 1929-58, chmn. dept., 1939-55, prof. emeritus, 1958—, asso. prof. Northwestern U., 1927-29; cons. Parke, Davis & Co., Detroit, 1927-63. Recipient St. Louis Am. Chem. Soc. Midwest medal award, 1955. Mem. Am. Chem. Soc. A.A.A.S., Sigma Xi, Phi Lambda Upsilon. Chief editor vol. 29 Organic Synthesis, 1949. Publs. on synthesis of organic compounds containing arsenic, antimony, or bismuth; study of heterocyclic compounds utilizable as drugs. Home: 2829 Van Dorn St., Lincoln, Neb. 68502.*

HAMILTON, David James, Scottish pathologist; ed. Edinburgh Vienna, Strasbourg; M.B., LL.D.; m. Catherine Wilson; demonstrator pathology Edinburgh U.;

prof. pathology Aberdeen U., to 1908; pathologist, royal infirmaries Edinburgh, Aberdeen. Licentiate Royal Coll. Physicians. Fellow Royal Soc., 1908, Royal Coll. Surgeons, Royal Soc. Edinburgh; mem. Medico-Chirurg. Soc., Edinburgh, Path. Soc. Gt. Britain and Ireland, other socs. Author: Pathology of Bronchitis . . . , 1883; Text-Book of Pathology, 1889; Board of Agriculture Report of Louping-ill and Braxy, 1906; also articles. Died Feb. 19, 1909.

HAMILTON, Donald Ross, Am. physicist; b. Hartford, Vt., Sept. 5, 1914; s. Rollo Albert and May Davina (Ross) H.; A.B., Princeton, 1935, Ph.D., Columbia, 1939; m. Eileen Mary Clare-Patton, Aug. 20, 1938; children—Erica Lynn (Mrs. Richard S. Weeder), Eleanor Patton, David Ross. Jr. fellow Soc. Fellows, Harvard, 1939-42; staff mem. Mass. Inst. Tech. Radiation Lab., 1940-46; project engr. Sperry Gyroscope Co. Research Labs., Garden City N.Y., 1941-42, research engr., 1942-45; asst. prof. physics Princeton, 1946-48, asso. prof., 1948-55, prof., 1955—, dean grad. sch., 1958-65; mem. Inst. for Advanced Study, 1952. Vis. sr. physicist Brookhaven Nat. Lab. Upton, N.Y., 1953-54; Vis. prof. physics Stanford. 1965. Fellow Am. Phys. Soc., Fedn. Am. Scientist. Author: (with J. B. H. Kuper, J. L. Knipp) Klystrons and Microwave Triods, 1947. Editorial bd.; Princeton U. Press, 1956-60. Office: Palmer Phys. Lab., Princeton U., Princeton, N.J. 08540.*

HAMILTON, Edwin Lee, Am. marine geologist; b. Sherman, Tex., Dec. 20, 1914; s. Terrel Lee and Clara (Baten) H.; B.S., Tex. A. and M. U., 1936; M.S., Stanford, 1950, Ph.D., 1952; m. Marie Hall, July 7, 1938; children—Marcia (Mrs. Istvan Mocsy), Carolyn (Mrs. George Senge), Laura. Joined USMC, 1936, ret. as col., 1947; lectr. U. Wash., 1951; research USN Electronics Lab., San Diego, Cal., 1951—, supervisory oceanographer, 1955—; research asso., lectr. U. Cal. at San Diego, 1954—; dir. Gen. Oceanographics, Inc., 1953—. Mem. Geol. Soc. Am., Am. Assn. Petroleum Geologists, Am. Geophys. Union, Paleontol. Soc. Am., Sigma Xi. Research, publs. on topography, structure, sediments of sea floor, evolution of sunken islands, formation of abyssal plains, sedimentary processes, sound velocity and soil mechanics of marine sediments. Home: 3594 Dupont St., San Diego 92106. Office: USN Electronics Lab., San Diego 92152.*

HAMILTON, Eric Ishmael, English geochemist; b. London, Eng., Oct. 30, 1931; s. Ghulam and Kathleen (Bond) H.; B.Sc., London U., 1954; D.Phil., Oxford (Eng.) U., 1957; m. Shirley Pamela Jenkins, July 30, 1955; children—Mark Jonathan Ross, Malcolm Richard Ian, Piers Lawrence Alexander, Quentin Ulrich Ivan. Geochemist, Greenland Geol. Survey, Copenhagen, Denmark, 1957-60; research fellow Oxford U., 1960-65; sect. head chemistry, chemist radiol. protection service Med. Research Council, Belmont, Eng., 1965—. Adviser, Royal Coll. Gen. Practitioners. Fellow Geol. Soc. London, Marine Biol. Assn. U.K. Author: Applied Geochronology, 1965; (with R. Farquhar) Radiometric Dating for Geologists, 1968; also articles. Editor: Earth Planetary Science Letters, 1966. Application of modern methods of chem. and phys. analysis to earth scis., chem. composition of man and his environment and pollution of the environment, neutron activation analysis, mass spectrometry, x-ray fluorescence. Home: 21 Woodcrest Walk Reigate, Surrey. Office: Radiol. Protection Service, Clifton Av., Belmont, Sutton Surrey, Eng.*

HAMILTON, Frank Hastings, Am. surgeon; b. Wilmington, Vt., Sept. 10, 1813; s. Calvin and Lucinda H.; grad. Union Coll., 1830; M.D., U. Pa., 1835; m. Mary (van Arsdale) McMurran, Oct. 15, 1834; m. 2d, Mary Gertrude Hart; Sept. 1, 1840. Licensed to practice medicine, N.Y., 1833; prof. surgery Fairfield Sch., 1839, Geneva Med. Sch., 1840-44; helped found med. dept. U. Buffalo, 1846, 1st prof. surgery until 1858; prof. surgery L.I. Coll. Hosp.; prof. clin. and mil. surgery Bellevue Hosp. Med. Coll., 1861, became full prof. surgery and surg. pathology, 1868; med. insp. U. S. Army; cons. physician to Pres. Garfield, in attendance until Garfield died after assassination. Author: A Practical Treatise on Fractures and Dislocations, 1860; Treatise on Military Surgery and Hygiene, 1862; Surgical Memoirs of the War of the Rebellion, 2 vols., 1870-71; The Principles and Practice of Surgery, 1872. Pioneered grafting in treatment of skin ulcers. Died N.Y.C., Aug. 11, 1886.

HAMILTON, Howard Laverne, Am. biologist; b. Lone Tree, Ia., July 20, 1916; s. Harry Stephen and Gertrude (Shilbley) H.; B.A. with highest distinction, U. Ia., 1937, M.S., 1938; postgrad. U. Rochester; Ph.D., Johns Hopkins, 1941; m. Alison Phillips, Dec. 22, 1945; children—Christina Helen, Phillips Howard, Martha Jayne. Faculty zoology Ia. State U., 1946-62, prof., 1953-62, acting head dept. zoology, entomology, 1960-61, chmn. dept., 1961-62; prof. biology U. Va., Charlottesville, 1962—. Mem. Am. Soc. Zoologists (mem. subcom. on edn., policy com. 1957-59), Society for the Study of Devel. and Growth, Am. Soc. Naturalists, Ia. Acad. Sci., International Inst. Embryology. Author: Lillie's Developement of the Chick, 1952, Brief edit., 1953; also numerous articles. Cons. editor McGraw-Hill Ency. Sci. and Tech., 1963-65; mng. editor Am. Zoologist, 1966—. Research, publs. on culture of viruses and rickettsiae; chemotherapy in rickettsial diseases; chem. control of organogenesis, normal stages of chick embryo, biol. effects of rare earths. Home: Jumping Branch Farm, Route 5, Garth Rd., Charlottesville, Va. 22901.*

HAMILTON, James, physicist; b. Sligo, Ireland, Jan. 29, 1918; s. Joseph and Jessie (Mackay) H.; ed. Royal Acad. Instn., Belfast, Ireland, Queen's U., Belfast, Inst. for Advanced Studies, Dublin, Manchester (Eng.) U.; m. Glen Dobbs, Aug. 8, 1945; children—Elizabeth, Andrew, Patrick. Imperial Chemistries Industry fellow Manchester U., 1945-50; lectr. math. Cambridge (Eng.) U., 1950-60; prof. physics U. Coll., London, Eng., 1960-64; prof. physics Nordita, Copenhagen, Denmark, 1964—; fellow Christ's Coll., Cambridge, 1953-60; research asso. Cornell U., Ithaca, N.Y., 1957-58. Mem. Royal Danish Acad. Author: Theory of Elementary Particles, 1959; also articles. Research elementary particle physics, including cosmic rays, radiation damping effects, unitarity, detailed balance, pertubation theory of S-Matrix, dispersion relations and analysis of structure of pion-nucleon interaction. Home: Soldalen 9, 2100 Copenhagen O, Office: Nordita, Blegdamsvej 17, 2100 Copenhagen O, Denmark.*

HAMILTON, John Frederick, Am. physicist; b. Knoxville, Tenn., Mar. 19, 1928; s. Frederick W. and Ruth (Lyon) H.; B.S., U. Tenn., 1950; children—Joseph W., Martha L., Thomas L. Physicist, Eastman Kodak Co. Research Labs., Rochester, N.Y., 1950—. Mem. Am. Phys. Soc., Electron Microscopy Soc. Am., Soc. Sigma Xi. Research, publs. in theory of silver halide photog. process, application of electron microscopy to physics of ionic crystals, including crystal growth, elec. properties, interaction with radiation. Home: 25 E. Hanford Landing Rd., Rochester 14615. Office: Kodak Co. Research Labs., Rochester, N.Y. 14650.*

HAMILTON, Leicester Forsyth, Am. chemist; b. Medford, Mass., Feb. 23, 1893; s. Frank Herbert and Janet (Hintze) H.; S.B., Mass. Inst. Tech., 1914; m. Mary Alma Nichols, Oct. 6, 1917; children—Jean (Mrs. John B. Stevens), Helen (Mrs. David S. Paulsen). Faculty, Mass. Inst. Tech., Cambridge, 1914-63, prof. chemistry, 1935-58, prof. emeritus, 1958-63, emeritus lectr., 1963—, acting head dept., 1942-46, exec. officer, 1946-63, sec. faculty, 1953-58. Mem. Am. Chem. Soc. Author (with S. G. Simpson) books on analytical chemistry. Studies in quantitative chem. analysis; application of precise measurement to quantitative analysis; calculations in analytical chemistry. Home: 100 Memorial Dr., Cambridge, Mass. 02142.*

HAMILTON, Leonard Derwent, physician, molecular biologist; b. Manchester, Eng., May 7, 1921; s. Jacob and Sara (Sandelson) H.; B.A. (County U. scholar), Balliol Coll., Oxford, 1943, B.M., 1945, M.A., 1946, D.M., 1951; M.A., Trinity Coll., Cambridge, 1948, Ph.D., 1952; m. Ann Twynam Blake, July 20, 1945; children—Jane Derwent, Stephen David, Robin Michael. Came to U. S., 1949, naturalized, 1964. USPHS research fellow dept. medicine U. Utah, 1949-50; mem. staff Sloan-Kettering Inst., N.Y.C., 1950—, head, isotope studies sect., 1957-64, asso. scientist, 1965—; mem. staff Meml. Hosp., N.Y.C., 1950-65, asst. attending physician dept. medicine, 1958-65; faculty Sloan-Kettering div. Grad. Sch. Med. Scis., Cornell U. Med. Coll., 1956-64; sr. scientist, head, div. microbiology Med. Research Center, Brookhaven Nat. Lab., Upton, N.Y., 1964—, also attending physician Hosp. Med. Research Center, 1964—. Cons. Office of Under-Secs. for Spl. Polit. Affairs, Sci. Com. on Effects Atomic Radiation, UN, 1960-62; mem. Com. Biol. Effects Atomic Radiation, Nat. Acad. Sci. -NRC, Washington, 1960-64; mem. Mayor's Tech. Adv. Com. on Radiation, N.Y.C., 1963—. Am. Cancer Soc. scholar, 1953-58; Commonwealth Fund grantee, 1955-62. Diplomate Am. Bd. Pathology. Mem. Am. Assn. Cancer Research, Am. Soc. Clin. Investigation, Am. Soc. Exptl. Pathology, Biochem. Soc., Brit. Med. Assn., Harvey Soc., Osler Soc., Soc. Protozoologists. Editor: Gerrard Winstanley, Selections From His Works, 1944; Physical Factors and Modification of Radiation Injury, 1964. Research on life-span of lymphocytes and their function in immunity; collaborator (with M. H. F. Wilkins) on proof of 3-dimensional structure of DNA; effects of various chems. and ionizing radiation on cells and man. Home: Childs Lane, Old Field, Setauket, N.Y. 11785. Office: Med. Research Center, Brookhaven Nat. Lab., Upton, N.Y. 11973.*

HAMILTON, Lyle Howard, Am. physiologist; b. Superior, Neb., June 11, 1924; s. Will B. and Iva (Reid) H.; student Boise Jr. Coll., 1943-44; A.B., Willamette U., Ore., 1950; M.S., State U. Ia., 1952, Ph.D., 1954; m. E. Eloise Nelson, Aug. 21, 1949; 1 son, Steven D. Asst. prof. physiology U. Sask., Saskatoon, 1954-57, adminstrv. asst. to dean medicine Sch. Medicine, 1956-57; chief physiology sect. research service Wood (Wis.) VA Center, 1957-61, prin. scientist, 1961—; faculty Marquette U. Sch. Medicine, Milw., 1957—, prof. physiology, dir. clin. physiology sect., 1967—. Mem. Am., Canadian phys. socs., Soc. Exptl. Biology and Medicine, Wis. Soc. for Med. Research, Sigma Xi. Research, publs. on areas of circulation, hematology and temperature regulation, area pulmonary physiology, devel. and evaluation of instruments used in pulmonary physiology or disease. Home: 1465 N. Hamilton Dr., Brookfield, Wis. 53005. Office: VA Center, Wood, Wis. 53193.*

HAMILTON, Paul Barnard, biochemist; b. Toronto, Ont., Can., Nov. 9, 1909; s. Edger Percival and Emma (Barnard) H.; B.A., U. Toronto, 1932, M.D., 1935, Ph.D., 1940; m. Suzanne Currelly, June 5, 1937; children—Paul C., John B., Rafe A., Came to U. S., 1940, naturalized, 1952. Fellow dept. biochemistry U. Toronto, 1936-40; with Rockefeller Inst. Hosp., N.Y.C., 1940-46, asso. dept. chemistry, 1945-46; chief biochemistry A.I. du Pont Inst., Nemours Found., Wilmington, Del., 1946—; adv. bd. mem. Analytical 1962-65. Mem. Am. Chem. Soc., A.A.A.S., Am. Soc. Biol. Chemistry, Harvey Soc., N.Y. Acad. Sci., Soc. Exptl. Biology and Medicine, Am. Bd. Clin. Chemistry, Ont. Coll. Physicians and Surgeons. Studies, publs. on gastric, kidney, shock physiology, amino acid chemistry, ion exchange chromatography, amino acid content of blood, urine, spinal fluid. Office: P.O. Box 269, Wilmington, Del. 19899.*

HAMILTON, Robert, Brit. physician; b. Edinburgh, Scotland, Dec. 6, 1721; attended lectures of William Hunter and Smellie in London; M.D., U. St. Andrews, 1766. Apprenticed to William Edmonston, surgeon and apothecary in Leith; entered Navy as surgeon's mate, 1741; in service until 1748; practiced medicine at King's Lynn, Eng. Fellow Royal Coll. Physicians, Edinburgh. Author: Observations on Scrophulous affections with remarks on papers in Philos. Trans. Royal Soc.; Schirrus Cancer and Rachitis, 1791; Observations on the Marsh Remittent Fever, on Water Canker and Leprosy, with Memoir of the Author's Life (pub. posthumously 1801); Letters on the cause and Treatment of the Gout (pub. posthumously 1806). Gave first modern description of orchitis as a complication of mumps, 1773. Died Nov. 9, 1793.

HAMILTON, Wallis Sylvester, Am. civil engr., b. Palmyra, N.J., Feb. 14, 1911; s. Wallis Henry and Hazel (Corle) H.; B.S., Carnegie Inst. Tech., 1935, M.S., 1939; postgrad. Colo. A. and M. U.; Ph.D., State U. Ia., 1943; postgrad. Mass. Inst. Tech.; m. Eva Blichfeldt, June 13, 1937; 1 dau., Mary Susan. Instr., Carnegie Inst. Tech., 1939-42; faculty Northwestern U., Evanston, Ill., 1943—, prof. civil engring., 1953—; vis. prof. State U. Utah, 1962. Research engr. Armour Research Found., 1946; dir. projects for seaplane testing and research Bur. Aeros., 1948-54; prin. investigator research contract Carter Oil Co., 1954-59; research engr. Aerial Measurements Lab., Northwestern Tech. Inst., 1959; civil engr. Harza Engring. Co., 1961; also cons. Mem. Am. Soc. C.E., Am. Geophys. Union, Wave Research Council, Sigma Xi, Tau Beta Pi. Research, publs. on distbn. of spray from hulls, calculation of flood movements in rivers, subsidence of Earth's surface due to pumping liquid from wells, structural model of portion of Earth, forces on bodies accelerating in liquid. Home: 4058 Fairway Dr., Wilmette, Ill. 60091. Office: Dept. Civil Engring., Technol. Inst., Northwestern U., Evanston, Ill. 60201.*

HAMILTON, Walter Clark, Am. chemist; b. Austin, Tex., Feb. 16, 1931; s. Olan Harvey; and Vida (Pearce) H.; B.S., Okla. State U., 1950; Ph.D., Cal. Inst. Tech., 1954; m. Marjorie Kimmel, May 30, 1953; children—Kim Luise, Douglas Olan, Karl Walter. NSF fellow Math. Inst., Oxford, Eng., 1954-55; staff Brookhaven Nat. Lab., Upton, N.Y., 1955—, chemist, 1959-65, sr. chemist, 1965—. Lectr., Columbia Sch. Gen. Studies, 1960-63; vis. lectr. Princeton, 1965; mem. U. S. Nat. Com. for Crystallography, 1964—; vis. prof. State U. N.Y., 1966-67; mem. NRC, 1966—. Mem. Am. Crystallographic Assn. (v.p. 1968), Am. Chem. Soc., Math. Assn. Am., Sigma Xi. Author: Statistics in Physical Science, 1964; Hydrogen Bonding in Solids, 1968; also articles. Research on theoretical and exptl. determination molecular and crystal structures, applications computers and statistics to these fields. Office: 33 Lewis St., Upton, N.Y. 11973.*

HAMILTON, William, Scottish philosopher; b. Glasgow, Scotland, Mar. 8, 1788; s. William Hamilton; B.A., Oxford U., 1911. Became mem. Scottish bar, 1813; made unsuccessful bid for chair moral philosophy Edinburgh U., 1820, apptd. prof. civil history, 1821, elected to chair logic and metaphysics, 1836-56. Author: Philosophy of the Unconditioned, 1829; Discussions in Philosophy, Literature, and Education, 1852-53; Metaphysics and Logic, 1858-60. Emphasized psychology vs. older metaphysical method; recognized importance of Aristotle, German philosophers, especially Kant; his logic anticipated later devels. in math. logic. Died May 6, 1856.

HAMILTON, Sir William, Brit. seismologist; b. Scotland, 1730; s. Archibald Hamilton; m. Emma Lyon, 1791; served with 3d regiment of foot guards, 1747-58; Brit. envoy to ct. of Naples, 1764-1800. Fellow Royal Soc., 1766. Author: (letters in Annual Register) Observations on Mount Vesuvius, Mount Etna and other Volcanoes of the Two Sicilies, 1772, Campi Phlegraei, 1776; other treatises on earthquakes and volcanoes, 1772-83. Studied action of volcanoes of Vesuvius and Etna; helped excavate Herculaneum and Pompeii. Died 1803.

HAMILTON, William John, Jr., Am. zoologist; b. Corona, N.Y., Dec. 11, 1902; s. William J. and Charlotte (Richardson) H.; B.S., Cornell U., 1926, M.S., 1928, Ph.D., 1930; m. Nellie R. Rightmyer, Oct. 12, 1928; children—Ruth E. (Mrs. Charles William Fisher), William John III, June C. (Mrs. David F. Beneway). Faculty, Cornell U., Ithaca, N.Y., 1926—, prof. zoology, 1947-63, prof. emeritus, 1964—; research asso. mammalogy Am. Mus. Nat. History, 1958—. Chmn. sci. adv. com. Edmund Niles Huyck Biol. Sta., Rensellearville, N.Y., 1939-55; mem. environmental biology panel NSF, 1956-58. Fellow A.A.A.S., N.Y. Acad. Sci.; mem. Am. Soc. Mammalogists (hon. past pres.), Ecol. Soc. Am. (past pres.), Soc. Naturalists, Am. Soc. Zoologists, Soc. Systematic Zoology, Wildlife Soc. Soc. Ichtyologists and Herpetologists, Am. Ornithol. Union, Wilson Ornithol. Club, Sigma Xi, Phi Kappa Phi. Author: American Mammals, 1939; Mammals of Eastern United States, 1943; (with A. F. Gustafson, H. Ries, C. E. Guise) Conservation in the United States, 1939; also numerous articles. Research on food and feeding habits N.Am. vertebrates, life histories N.Am. mammals, econs. fur-bearing mammals, delayed implantation reproduction. Home: 615 Highland Rd., Ithaca, N.Y. 14850.*

HAMILTON, Sir William Rowan, Irish mathematician; b. Dublin, Ireland, Aug. 4, 1805; s. Archibald and Sarah (Hutton) H.; student Trinity Coll., Dublin, 1823-27; had mastered 13 langs. by age 14; m. Helen Maria Bayley, 1833; at least 1 son. Apptd. to Andrews chair astronomy, Dublin, 1827; spent rest of life at obs. at Dunsink, nr. Dunkirk, engaged in study of math; Irish royal astronomer, 1827. Mem. Royal Irish Acad. (pres. 1837), Royal Soc. (Gold medal 1834), French Acad. Scis. Author: Theory of Systems of Rays, 1828; General Methods of Dynamics, 1834-35; Lectures on Quaternions, 1853; Elements of Quaternions, 1866, also numerous papers to Royal Irish Acad. Inventor theory of quaternions (3-dimensional algebra); invented and named hodography; predicted conical refraction with aid of math.; his work on dynamics modeled classical mechanics on lines later followed by quantum theory. Died Sept. 2, 1865.

HAMM, Franklin Albert, physicist; b. New Tripoli, Pa., Feb. 23, 1918; s. Mahlon Albert and Helen (Reimert) H.; B.S., Muhlenberg Coll., 1939; M.S., Cornell U., 1941, Ph.D., 1943; m. Frances Melba Wertz, Sept. 9, 1941; children—Michael Franklin, Terry Francis. Group leader Gen. Aniline & Film Corp., Easton, Pa., 1943-52; asst. dir. research Burroughs Corp., Phila., 1952-53; research leader Eastman Kodak Co., Rochester, N.Y., 1953-58; sect. leader Minn. Mining & Mfg. Co., St. Paul, 1958-64, lab. mgr., 1964-65, tech. mgr. new product devel. photog. film div., 1966—. Recipient Am. Chem. Soc. award, 1939. Mem. Am. Phys. Soc., Sigma Xi. Contbg. author: Physical Methods of Organic Chemistry, vol. I, part III, 1954, vol. I, part II, 1960; also tech. articles. Research on use electron microscope on dyes, fibers, polymers, photg. emulsions, photog. latent image theory. Patentee magnetic and semiconductor materials. Home: 1505 N. 2d St., Stillwater, Minn. 55082. Office: 444 McKnight Rd., St. Paul 55119.*

HAMM, Wilhelm Philipp von, see von Hamm, Wilhelm Philipp.

HAMMAR, Harald Edwin, Am. chemist; b. Worcester, Mass., Sept. 3, 1902; s. Julius Waldemar and Ida Nelson) H.; B.S.A., B.S. in Agrl. Edn., U. Fla., 1925, M.S., 1927; Ph.D., Rutgers U., 1931; m. Edith Lillian Crone, Aug. 6, 1934. Field asst., meteorologist Everglades Exptl. Sta., Belle Glade, Fla., 1927-28; research fellow Rutgers U., New Brunswick, N.J., 1928-31; research asso. chemist Nat. Bur. Standards, Washington, 1931-36; research chemist U. S. Dept. Agr., Shreveport, La., 1936-61, Albany, Ga., 1943-48, Clemson, S.C., 1961-63; fertilizer chemist Clemson U. Agr. Chemistry Service, Clemson, S.C., 1963—. Fellow Am. Inst. Chemists; mem. Am. Chem. Soc. (past chmn. Ark.-La.-Tex. sect.), Am. Soc. Plant Physiologists, Soil and Plant Soc. Fla., Sigma Xi, Phi Kappa Phi, Alpha Zeta, Kappa Delta Pi, Pi Delta Epsilon. Author: (with P. D. Trask, C. C. Wu) Source Sediments of Petroleum, 1932; also articles; contbr. chpt. to Ency. Chem. Tech., vol. 9, 1952. Analyzed oil well cores and marine sediments for carbon-nitrogen ratio; chem. research on nutrition and fertilizer requirements of edible tree nuts and peanuts; first atomic tracer studies on tree nutrition; chem. methods of inorganic analysis, minor elements, mineral analysis by atomic absorption spectroscopy. Home: 254 Grove Dr., Sunny Acres, Clemson, S.C. 29631.*

HAMMEL, Harold Theodore, Am. physiologist; b. Huntington, Ind., May 8, 1921; s. Audry H. and Ferne (Wiles) H.; B.S., Purdue U., 1943; M.S., Cornell U., 1950, Ph.D., 1953; m. Dorothy King, Dec. 29, 1948; children—Nannette, Heidi. Faculty, U. Pa., 1953-61; asso. prof. dept. physiology, Yale Sch. Medicine, 1961-68, fellow John B. Pierce Lab., 1961-68; prof. marine biology Scripps Instn. Oceanography, U. Cal. at San Diego, La Jolla, 1968—. Mem. Am. Phys. Soc., Am. Physiol. Soc., Am. Soc. Mammalogy, Am. Soc. Plant Physiology, Am. Soc. Zoologists. Research, numerous publs. in thermal physiology, regulation of internal body temperature in dogs, rats, monkeys, hibernators and lizards, responses to thermal stress in reindeer, camel and jack rabbit, cold adaptation in several ethnic groups of man, supercooling and anti-freeze in Arctic marine fish, freezing of xylem sap without cavitation in conifers. Office: Physiol. Research Lab., Scripps Instn. Oceanography, U. Cal. San Diego, La Jolla, Cal. 92037.

HAMMER, Adam, physician; b. Baden, Germany, 1818; studied medicine, Heidelberg, Germany, Paris, France; practiced medicine, Mannheim, Germany; came to St. Louis, Mo., 1848. Author: Die Anwendung des Schwefeläthers . . . , 1847. Credited with 1st diagnosis of coronary thrombosis in living patient, 1878. Died 1878.

HAMMER, Ernst Hermann Heinrich von, see von Hammer, Ernst Herman Heinrich.

HAMMER, Marie Signe, Danish naturalist; b. Copenhagen, Denmark, Mar. 20, 1907; d. Niels and Alma Jorgensen; Ph.D. in Natural Science, U. Copenhagen; m. Ole Hammer, 1936; children—Karen, Inga, Birgitte, Peder. Studied natural sci., mem. expdns. to Iceland, Greenland, Can., Mexico, Andes, Fiji Islands, New Zealand, New Guinea, Hawaii. Mem. Danish Assn. Natural History, also others. Author: Studies on the Oribatids and Collemboles of Greenland, 1944; Investigations on the Microfauna of Northern Canada, 1952-53; Investigations on the Oribatid Fauna of the Andes Mountains, 1958-62. Address: Roland, Fredensborg, Denmark.

HAMMER, Preston Clarence, Am. computer scientist; b. Rockford, Mich., Oct. 12, 1913; s. Adam and Ada (Zimmerman) H., A.B. magna cum laude (fellow), Kalamazoo Coll., 1934; A.M. (fellow), U. Mich., 1935; Ph.D., Ohio State U., 1938; m. Hilda Knight, Aug. 6, 1939; children—Phoebe Ann, Frances Jean, Nick Adam, Kathryn Molly, Arthur Stephen. Instr., U. Mich., 1938-40; instr. Ore. State Coll., 1941-42, asst. prof., 1942-44, asso. prof., 1946-47; supr. Lockheed Aircraft Corp., 1944-45; supr. quality control Los Alamos Sci. Lab., 1947-52; prof. U. Wis., 1952-65, dir. Numerical Analysis Lab., 1952-62, traveling faculty fellow Europe, 1959-60, chmn. dept. computer scis, 1961-65; prof., head computer sci. dept. Pa. State U., 1965—; vis. prof. research U. Cal., 1962-63. Cons. Argonne Nat. Lab., Sandia Corp., Los Alamos Sci. Lab., NSF. Fellow A.A.A.S.; mem. Wis. Acad. Sci. Arts and Letters, Am. Math. Soc., Math. Assn. Am. (lectr. 1960—), Inst. Math. Statisticians, Assn. Computing Machinery (lectr. 1966—) Soc. Indsl. and Applied Math. (lectr. 1962—, council 1962-65), Soc. Symbolic Logic, Am. Assn. U. Profs., Wiskundig Genootshai, Circolo Mathematico di Palermo, Sigma Xi. Editor: The Computing Laboratory in the University, 1957. Editorial bd. Mathematics of Computation, 1960-63, Communication Jour., 1962—; asso. editor Math. Systems Jour., 1966—. Research on gen. theory of formal systems called extended topology, new founds. for convexity theory, for modeling and approximation theory; convexity, computing, numerical methods, functional analysis, linguistics. Home: 331 W. Fairmount St., State College, Pa. 16801. Office: McAllister Bldg., Pa. State U., University Park, Pa. 16802.*

HAMMER, Sigmund Immanuel, Am. geophysicist; b. Webster, S.D., Aug. 13, 1901; s. Ludvig E. and Laura (Anderson) H.; B.A., St. Olaf Coll., 1924; Ph.D., U. Minn., 1929; m. Norma Lucille Johnson, Nov. 28, 1925; children—Sigmund L., Laura Blanche (Mrs. William D. Inglis II), John P., Paul L., Kirsten Norma (Mrs. T. G. Carter), Douglas J., Ludvig E. Geophysicist, Gulf Research & Devel. Co., Pitts., 1929-46, sect. supr., 1946-66, research asso., 1966-67; prof. geophysics, dept. geology and geophysics U. Wis., Madison, 1967—. Adj. mem. grad. faculty U. Pitts., 1946-67; cons. NSF; chmn. adv. com. Mem. Nat. Acad. Sci.-Nat. Acad. Engring. Joint Adv. Com. to Environmental Sci. Services Adminstrn., 1967—. Mem. Am. Phys. Soc., Phys. Soc. Pitts., A.A.A.S., Pa. Acad. Sci., Am. Assn. Petroleum Geologists, Am. Geophys. Union, Soc. Exploration Geophysicists (hon. life), European Assn. Exploration Geophysicists, Internat. Assn. Geodesy and Geophysics, Am. Geol. Inst., Assn. Mexicana de Geofisicos de Exploration, Geosci. Information Soc., Sigma Xi. Contbr. articles in field to profl. publs. Developer theories of gravitational field of earth and applications to search for mineral accumulations. Office: Dept. Geology and Geophysics, U. Wis., Madison, Wis.*

HAMMER, William Joseph, Am. elec. engr.; b. Cressona, Pa., Feb. 26, 1858; s. William Alexander and Martha A. (Beck) H.; ed. pub. schs., Newark, N.J.; m. Alice Maude White, Jan. 3, 1894; 1 dau., Mabel White (Mrs. Thomas Cleveland Assheton). Asst. to Edward Weston in Weston Malleable Nickel Co., Newark, 1878; became asst. in lab. of Thomas A. Edison, Menlo Park, N.J., 1879; chief engr. Edison Lamp Works, 1880-81; sent to Eng. by Edison as chief engr. English Edison Co., 1881, and established in London 1st central sta. in world for incandescent elec. lighting; chief engr. German Edison Co. (now Allgemeine Elektrische Gesellschaft), 1883-84; represented Edison at Franklin Inst. Elec. Expn., 1884, Crystal Palace Elec. Expn., 1882, and Paris Expn., 1889; at close of latter expn. made balloon flight over France; chief insp. central stas., Edison Elec. Lab. Co., 1884-85; became confidential asst. of pres. of parent Edison Co., 1884, and incorporator and trustee Sprague Elec. R.R. & Motor Co.; installed the 8,000-light plant of Ponce de Leon Hotel at St. Augustine, Fla., 1887; cons. and contracting engr. in connection with elec. effects at Cincinnati Expn., 1888; chief engr., gen. mgr. Boston Edison Co., 1886-87; cons. practice as elec. engr., from 1890. Served as maj. Gen. Staff, U. S. Army, World War I, in charge elec. and aero. war inventions. Recipient John Scott Legacy medal and premium from Franklin Inst.; grand prize, St. Louis Expn., 1904, for hist. collection of incandescent elec. lamps; Gold medal St. Louis Expn., and Elliott Cresson Gold medal Franklin Inst., 1906; World War medal (U. S.), 1920; Chevalier Legion of Honor (France), 1925. Fellow Am. Inst. E.E. (life mem., v.p.), Am. Phys. Soc., A.A.A.S., Am. Acoustical Soc.; mem. Edison Pioneers, (pres. 1920-21), Franklin Exptl. Club (pres.), Nat. Conf. on Standard Elec. Rules (pres. 10 years). Invented the radium luminous preparations used for watches, clocks, airplane and automobile instruments, etc.; brought 9 tubes of radium from Curie labs., 1902, and delivered 88 lectures on radium; first suggested and used radium for cancer and tumor treatment; inventor of motor-driven flashing electric sign. Died Mar. 24, 1934.

HÄMMERLING, Joachim August Wilhelm, German biologist; b. Berlin, Germany, Mar. 9, 1901; s. Auguste and Gertrud (Title) H.; ed. U. Berlin, U. Marbourg; Ph.D. in sci.; m. Charlotte Klose, 1931; children—Brititte, Wolfgang (dec.), Ulrich, Helmut, Günter. Asst., Kaiser-Wilhelm Inst. Biology, 1922-40, dept. chief, 1945-48; German dir. German-Italian Inst. for Marion Biology, Rovigno d'Istria, Italy, 1940-45; dir. Max Planck Inst. Marine Biology, Wilhelmshaven, 1948—. Mem. numerous German and internat. socs. Author: Fortpflzg. im Tier- und Pflanzenreich. Research and publs. on physiology of devel., genetics, cellular biochemistry. Address: Max Planck Institute for Marine Biology, Anton Dohrn Wag, 294 Wilhelmshaven, Germany.

HAMMETT, Louis Plack, Am. chemist; b. Wilmington, Del., Apr. 7, 1894; s. Philip Melancthon and Louise (Plack) H.; A.B., Harvard, 1916; postgrad. Eidg. Tech. Inst. Zurich (Switzerland); Ph.D., Columbia, 1923, D.Sc., 1962; m. Janet Thorpe Marriner, June 4, 1919; children—Philip Marriner, Jane (Mrs. Howard Adrian Zwemer). Chemist, Bur. Aircraft Prodn., U. S. Army, 1917-19, E. C. Worden, 1919-20; faculty Columbia, N.Y.C., 1920—, prof. chemistry, 1935-59, Mitchill prof., 1959-61, Mitchill prof. emeritus, 1961—. Research supr., research dir. Explosives Research Lab., Bruceton, Pa., 1941-45. Recipient Hon. fellow Chem. Soc. London; mem. Am. Chem. Soc. (Nichols award N.Y. sect. 1957, Norris award Northeastern sect. 1960, Priestley medal 1961, Gibbs medal, Chgo. sect. 1961, Norris award 1966), Nat. Acad. Scis. Author: Solutions of Electrolytes, 1929; Physical Organic Chemistry, 1940; Introduction to the Study of Physical Chemistry, 1952; also numerous articles. Research on correlation reaction of organic chem. substances with quantitative relationships. Address: R.D. 4, Box 310, Newton, N.J. 07860.*

HAMMITT, Frederick Gnichtel, Am. nuclear and mech. engr.; b. Trenton, N.J., Sept. 25, 1923; s. Andrew Baker and Julia (Gnichtel) H.; B.S. in Mech. Engring., Princeton, 1944; M.S., U. Pa., 1949; M.S. in Applied Mechanics, Stevens Inst. Tech., 1956; Ph.D. in Nuclear Engring., U. Mich., 1958; m. Barbara Ann Hill, June 11, 1949; children—Frederick Gnichtel, Harry Andrew, Jane Stevenson. Engr., John A. Roebling Sons Co., Trenton, 1946-48, Power Generators Ltd., Trenton, 1948-50, Reaction Motors, Inc., Rockaway, N.J., 1950-53; project engr. Worthington Corp., Harrison, N.J., 1953-55; faculty U. Mich., Ann Arbor, 1955—, prof. nuclear engring., 1961-68, prof. mech. engring., 1968—, dir. founder Lab. for Fluid Flow and Heat Transport Phenomena, 1961—. Mem. Am. Nuclear Soc. (past chmn. Mich. sect.), Am. Soc. M.E. (chmn. cavitation com. fluids engring. div. 1966—), Am. Soc. for Testing and Materials (vice chmn. com. G-2 on cavitation and impingement erosion 1965—), also numerous articles. Design and devel. highspeed spl. pumps and turbines in rocket and nuclear industry; research in cavitation. Patentee cavitation test device. Home: 1306 Olivia St., Ann Arbor, Mich. 48104.*

HAMMON, William McDowell, Am. epidemiologist; b. Columbus, O., July 20, 1904; s. William Henry and Adaline (McDowell) H.; A.B., Allegheny Coll., 1932, Doctor of Science (honorary), 1959; M.D., Harvard University, 1936, M.P.H., 1938, Dr. P.H., 1939; Certificate Army Med. Sch., 1943; m. Helen Black, Aug. 3, 1926; children—William M., Barbara Helen. Dir. Med. Dispensary, nr. Shabunda, Belgian Congo, 1926-30; instr. epidemiology school pub. health, Harvard, 1939-40; mem. faculty The Hooper Found. for Med. Research, U. of Calif., 1940-50, prof. epidemiology, 1947-50, mem. faculty, sch. pub. health, 1944-50, prof. epidemiology, 1947-50, lecturer in medicine and neurology, sch. medicine, 1942-50; prof. and head dept. epidemiology and microbiology, Grad. Sch. Pub. Health, U. Pittsburgh since 1950; cons. surgeon gen., U. S. Army 1941—, dir. virus infections commn. Armed Forces Epidemiology Bd., 1956-65, mem. bd., 1965—; cons. Communicable Disease Center, USPHS, 1948—, mem. nat. adv. cancer council, 1960-63, adv. com. live polio-virus vaccine Surgeon Gen., 1959—; mem. panel viruses and cancer Nat. Cancer Inst., 1959-62 virus panel U. S. Japan Med. Scis. Coop. Program, 1966—. Mem. Pacific Sci. Bd., NRC, 1950-56. Diplomate Am. Bd. Preventive Medicine and Pub. Health. Fellow Am. Pub. Health Assn. (mem. research and standards com. 1949, chmn. 1950-55), A.A.A.S., N.Y. Acad. Scis., Am. Acad. Microbiology (gov.); mem. Soc. of Am. Bacteriologists (pres. 1949), Am. Assn. Immunology, Am. Soc. Tropical Medicine and Hygiene (pres. 1967, editorial bd. 1951-67), Soc. Exptl. Biology and Medicine Soc. Exptl. Pathology, Am. Assn. Physicians, American Epidemiological Society (pres. 1962-63), Phi Beta Kappa, Sigma Xi, Alpha Omega Alpha, Delta Omega. Research in field and lab. aspects of arthropod-borne virus encephalitides and poliomyelitis, also arthropod transmission and immunology of virus infections. Home: 317 Sleep Hollow Rd., Pitts. 15228.*

HAMMOND, Datus Miller, Am. zoologist; b. Providence, Utah, May 20, 1911; s. Horace E. and Selina (Tibbitts) H.; B.S., Utah State Agr. Coll., 1932; M.A., U. Cal. at Berkeley, 1934, Ph.D., 1936; m. Emily Merrill, Dec. 23, 1937; children—Marie, Louise, Betty, Marilyn, Carol. Research asso. U. Cal., 1936; instr. Utah State Agr. Coll., 1936-39, asst. prof., 1939-41; asst. protozoologist zool. div. Bur. Animal Industry, U. S. Dept. Agr., Beltsville, Md., 1941-42; asso. protozoologist U. S. Regional Lab. for Animal Disease Research, Auburn, Ala., 1942-44; prof., head zoology dept. Utah State U., Logan, 1945—. Fulbright Research scholar U. Munich (Germany), 1955-56; named Utah State U. Faculty Honor lectr., 1964; guest prof. U. Bonn, Germany, 1966. Mem. A.A.A.S., Am. Soc. Zoologists, Soc. Protozoologists, Am. Soc. Parasitologists, Am. Micros. Soc. Research, publs. on internal parasites livestock particularly protozoa in cattle; Home: 479 W. Center St., Logan, Utah 84321.*

HAMMOND, E(dward) Cuyler, Am. biostatistician; b. Balt., June 14, 1912; s. Edward and Agnes (Cuyler) H.; student Gilman Country Day Sch., 1929-31; B.S., Yale, 1935, M.A., 1953; Sc.D., Johns Hopkins, 1938; m. Marian E. Thomas, Jan. 3, 1948; children—Thomas Cuyler, Richard Render, Jonathan Cuyler. Asso. statistician, div. indsl. hygiene Nat. Inst. Health, USPHS, 1938-42; cons. med. research sect. Bur. Aero., U. S. Navy, 1941-42; civilian requirements br. OQMG, 1942; dir. statis. research sect. Am. Cancer Soc., 1946—; lectr. statistics dept. pub. health Yale, 1952-53, prof. biometry, dir. grad. studies, 1953-58, also chmn. univ. exec. com. statistics; lectr. preventive and environmental medicine Albert Einstein Coll. Medicine; cons. dept. biology Brookhaven Nat. Laboratory. Member scientific advisory panel Research to Prevent Blindness, Inc. Nat. lectr. Sigma Xi, 1957-58. Recipient William R. Belknap award for excellence in biol. studies, 1935. Fellow Am. Pub. Health Assn.; mem. American Statistical Association, Inst. Math. Studies, N.Y. Acad. Sci., Biometric Soc., A.A.A.S. Suggested relationship between cigarette smoking and lung cancer; research on epidemiology of cancer, statis. cons., fatigue, psychomotor and visual tests, aviation medicine, aircraft accidents, evaluation of therapy, test validation, med. econ. studies, design of expts., indsl. hygiene, effects of radiation, human and exptl. population studies. Home: 164 E. 72d St., N.Y.C. 10021. Office: 219 E. 42d St., N.Y.C. 10017.

HAMMOND, Edwin Hughes, Am. geographer; b. Ann Arbor, Mich., Jan. 8, 1919; s. Harry Emmons and Elizabeth (Huddle) H.; B.A., U. Mo., 1939; M.A., U. Wis., 1940; Ph.D., U. Cal. at Berkeley, 1951; m. Elizabeth Mills, Dec. 28, 1940; children—Janet Elizabeth (Mrs. William P. Stilwell, Jr.), Richard Edwin, Lawrence Alan. Lectr. geography U. Cal. at Berkeley, 1946-48; instr. geography U. Neb., 1948-49; faculty U. Wis., Madison, 1949-64; prof. geography Syracuse (N.Y.) U., 1964—. Mem. Assn. Am. Geographers, Am. Assn. U. Profs., A.A.A.S., Phi Beta Kappa, Sigma Xi. Author: (with G. Trewartha, A. Robinson) Elements of Geography; Fundamentals of Physical Geography, 1961. Research, publs. in geomorphology and land-surface morphometry and mapping. Home: 1709 Euclid Av., Syracuse, N.Y. 13224.*

HAMMOND, George Simms, Am. chemist; b. Auburn, Me., May 22, 1921; s. Oswald Kenric and Marjorie (Thomas) H.; B.S., Bates Coll., 1963; M.S., Ph.D., Harvard, 1947; m. Marian Reese, June 8, 1945; children—Kenric, Janet, Steven, Barbara, Jeremy. Postdoctoral fellow U. Cal. at Los Angeles, 1947-48; faculty Ia. State Coll., 1948-58, prof. chemistry, 1956-58; vis. asso. prof. U. Ill., summer 1953; prof. organic chemistry Cal. Inst. Tech, Pasadena, 1958-63, Arthur Amos Noyes prof. chemistry, 1963—. Mem. chemistry adv. panel NSF, 1962—. Guggenheim fellow, also NSF sr. fellow Oxford (Eng.) U. and U. Basel (Switzerland), also Cal. Inst. Tech., 1956-57. Fellow Am. Acad. Arts and Scis.; mem. Nat. Acad. Scis., Am. Chem. Soc. (award in petroleum chemistry 1960), Chemistry Soc. (London), Phi Beta Kappa, Sigma Xi. Author: (with J. S. Fritz) Quantitative Organic Analysis, 1956; (with D. J. Cram) Organic Chemistry, 1958. Editor: Advances in Photochemistry, 1961. Research on chem. reactivity, photochemistry, principles of chem. dynamics. Home: 1521 E. Mountain St., Pasadena, Cal. 91104.*

HAMMOND, Sir John, English veterinarian; b. Bristol, England, Feb. 23, 1889; s. Burrell and Janett L. (Aldis) H.; ed. Coll. of Downing, Cambridge U.; M.A.; D.Sc., U. Ia., U. Durham, U. Leeds.; D.Ag.Sc., U. Vienna, U. Louvain, U. Copenhagen, U. Cracovie; m. Mercy Goulder, 1916; children—John, Christopher H. Physiologist, Inst. Animal Nutrition, Cambridge U., 1920-43, lectr. agrl. physiology, 1943-54. Guest prof. State U. Ia., 1932, U. Abadan, 1964. Mem. Royal Agrl. Acad. Sweden, Agrl. Tech. Acad., Vet. and Zool. Soc. Madrid, French Agrl. Acad., German Soc. Animal Prodn. Fellow Royal Soc., 1933. Author: Reproduction in the Rabbit, 1925; Reproduction in the Cow, 1927; Growth and Development of Mutton Qualities in the Sheep, 1932; Farm Animals, 1940; Animal Breeding, 1963. Home: 1 Luard Rd. Office: School of Agriculture, Cambridge University, Cambridge, Eng.

HAMMOND, R. Philip, Am. chem. engr.; b. Creston, Ia., May 28, 1916; s. Robert Hugh and Helen (Williams) H.; B.S., U. So. Cal., 1938; Ph.D., U. Chgo., 1947; m. Amy Louise Farmer, Feb. 28, 1941; children—Allen Lee, David Michael, Jean Phyllis, Stanley Wayne. Chief chemist, dir. lab. Lindsay Chem. Co., West Chicago, Ill., 1938-46; research fellow U. Chgo., 1946-47; asso. div. leader, reactor devel. div. Los Alamos Sci. Lab., 1947-64; dir. Nuclear Desalination program Oak Ridge Nat. Lab., 1963—. Mem. U. S. delegation advisers Geneva Conf. on Peaceful Uses of Atom, 1955, 64; mem. Internat. Atomic Energy Agy. spl. panel on desalination, 1964—; alternate mem. U. S.-Mexico-Internat. Atomic Energy Agy. study team for desalting plant. Mem. Am. Nuclear Soc., Am. Chem. Soc., Am. Assn. Cost Engrs. Contbr. chpt. Ency. Brit., 1963, Progress in Nuclear Energy, 1956; also articles. Developer nuclear powered desalination plants, chem. processes for intensely radioactive materials, methods of cost analysis and prediction for nuclear reactors, cost allocation for dual purpose plants; patentee reactors, reactor fuels, fuel-handling devices, devices for high speed r.r. trains; contbr. to devel. fast breeder reactors. Home: 879 W. Outer Dr. Office: P.O. Box Y, Oak Ridge 37830.

HAMMOND, Thomas Edwin, Brit. surgeon; b. Aug. 5, 1888; s. Edwin and Jane (Jenkins) H.; ed. Cheltenham Coll., St. Bartholomew's Hosp. Med. Sch. Surgeon, Cardiff Royal Infirmary; cons. surgeon Welsh Nat. Meml. Assn. Fellow Royal Coll. Surgeons; mem. Cardiff Med. Soc. (pres.), Royal Soc. Medicine (pres. sect. urology). Author: The Constitution and its Reaction in Health; Principles in the Treatment of Inflammation, 1934; Infections of the Urinary Tract; Vitality and Energy in Relation to the Constitution; also papers in sci. jours. Died Mar. 25, 1943.

HAMMONS, Ray Otto, Am. plant geneticist; b. Wesson, Miss., Oct. 2, 1919; s. William D. and Lou (Douglas) H.; B.S. with high honors, Miss. State U., 1947, M.S., 1948; Ph.D. in Agronomy, N.C. State U., 1953; m. Annie Ray Howell, Sept. 13, 1942; children—Mary A. (Mrs. Fred E. Harmon, Jr.), Lynda F. (Mrs. G. C. Eidson, Jr.), Beverly Sue, Ray Otto. Research instr. peanut radiation genetics N.C. State U., 1949-53, asso. genetics, 1952-53; asst. prof. forage breeding Purdue U., 1953-55; geneticist crops research div. Agr. Research Service, U. S. Dept. Agr., Tifton, Ga., 1955-63, research geneticist, 1963—; asso. prof. U. Ga., Athens, 1955—. Mem. Crop Sci. Soc. Am., Am. Soc. Agronomy, Am. Genetics Assn., Sigma Xi, Alpha Zeta. Contbr. articles to tech. jours. Developed key for identification Miss. grasses by vegetative morphology; discovered and applied new genetic technique to increase 1st generation hybrid populations in peanuts; research on radiation-induced mutants in peanuts. Home: 1203 Lake Dr. Office: P.O. Box 748, Tifton, Ga. 31794.*

HAMNER, Karl Clemens, Am. botanist; b. Salina, Kans., Oct. 15, 1908; s. William Marion and Millie (Swenson) H.; B.S., U. Cal. at Berkeley, 1931; M.S., U. Chgo., 1934, Ph.D., 1935; m. Gladys Gerner, July 24, 1941; children—Lois Karen (Mrs. Arthur Hawley Parmelle III), Melinda Lee. Asso. plant physiologist U. S. Dept. Agr., Beltsville, Md., 1936-37; faculty U. Chgo., 1937-40; plant physiologist U. S. Plant, Soil and Nutrition Lab., Ithaca, N.Y., 1940-48, dir., 1946-48; prof. botany U. Cal., Los Angeles, 1948—, chmn. dept., 1948-57. Speaker before prof. groups; cons. AEC, 1949-65. Carnegie Inst. fellow, 1949. Mem. Am. Soc. Naturalists, Bot. Soc. Am., Am. Soc. Plant Physiologists (pres. Western sect. 1950), Soc. Hort. Sci., Sigma Xi (pres. chpt. 1962). Studies, publs. of plant photoperiodism, nutrition; endogenous circadian rhythms. Home: 11917 Ayres Av., Los Angeles 90064.*

HAMON, Jacques Pierre Jean, French med. entomologist; b. Paris, France, Jan. 1, 1926; s. André and Berthe (Laberthonnière) H.; Ingénieur agronome Inst. Nat. Agronomique, Paris, 1947; Diplomé d'Entomologie médicale and vétérinaire, ORSTOM, Paris, 1949; m. Marcelle Gourdet, July 23, 1955; children—Hervé, Axelle and Murielle (twins), Patrice. Entomologist in charge Malaria Service, Corsica, 1949-50, Réunion Island, 1950-52; chief dept. med. entomology Centre Muraz, OCCGE (formerly SGHMP), 1952—. Cons. in med. entomology WHO, 1955, 57, 61, 65, mem. expert panel on insecticides, 1959—. Decorated Médaille de bronze des épidémies (France); Officier de l'Ordre National Voltaique (Haute-Volta). Mem. Soc. Pathologie exotique, Soc. entomol. France, Soc. franc. Parasitologie, Royal Entomol. Soc. London, Royal Soc. Tropical Med. Hygiene, Am. Mosquito Control Assn. Contbg. author: Practical Malariology, 1963; Evolution of Infectious Diseases, 1966. Research, numerous publs. on identity, distbn. and ecology of African Mosquitos and transmitted diseases, especially human malaria and Wuchereria filariasis; orgn. and evaluation of malaria control operation in E. and W. Africa; field and lab. studies on phenomenon of insecticide resistance in vectors. Home: 82 rue Aristide Briand, Orsay 91, France. Office: B.P. 153 Bob-Dioulasso, Haute-Volta.*

HAMOR, Glenn Herbert, Am. medicinal chemist; b. Kootenai, Ida., May 14, 1920; s. Samuel Clark and Della (Shoop) H.; B.S., U. Mont., 1941, M.S., 1947; Ph.D., U. Minn., 1952; m. Eileen Deegan, Sept. 7, 1947; children—Patricia, Ellen, Kathleen, Timothy. Instr., U. Mont., 1947-48; research asst. U. Minn., 1959-60; faculty U. So. Cal., Los Angeles, 1952—, prof., 1966—. Cons., Smith, Klin & French Labs., Phila., 1958-65, IBM, 1966; lectr. chemistry dept. Pepperdine Coll., 1955-64. Am. Found. for Pharm. Edn. fellow, 1949-51; Research Corp. research fellow, 1954; USPHS grantee, 1961-65. Fellow A.A.-A.S.; mem. Am. Chem. Soc., Am. Pharm. Assn., Sigma Xi, Phi Lambda Upsilon, Rho Chi. Mem. editorial adv. bd. Current Contents, 1962—. Contbr. articles to tech. jours. Preparation new drugs with correlation chem. structure and biol. activity, chemistry taste. Patentee drugs. Home: 6519 W. 87th St., Los Angeles 90045.*

HAMPSON, William, Brit. physician; b. 1895; s. William Hampson; M.A., Oxford U.; m. Amy Bolton. Former med. officer charge elec. and x-ray depts. Queen's Hosp. for Children, St. John's Hosp., Leicester Sq. Mem. Röntgen Soc. (council). Author: Radium Explained; Paradoxes of Nature and Science; Modern Thraldom—a New Social Gospel; also numerous articles in jours. Inventor self-intensive method of refrigeration of gases by which air 1st liquefied cheaply, and subsequently method by which hydrogen and helium were liquefied; inventor 1st apparatus for making surg. pencils of carbonic acid snow, Hampson radiometer for measuring doses of x-rays; discovered method to control beats of heart by electrically stimulated muscle contractions. Died Jan. 1, 1926.

HAMPTON, James C., Am. anatomist; b. Juliaetta, Ida., Sept. 20, 1921; s. Joseph D. and Mary (Horn) H.; student San Francisco State Coll., 1947-49; B.S., U. Ida., 1951, M.S., 1952; Ph.D., U. Wash., 1957; m. Erma Dean Law, Mar. 28, 1942; children—Susan Carol, Connie Lou, Joseph Douglas, Walter William. Faculty, U. Wash. Sch. Medicine, 1957-59, Baylor U. Coll. Medicine, 1959-61, med. br. U. Tex., Galveston, 1961-62; prof., chmn. dept. anatomy Med. Sch. Northwestern U., Chgo., 1962—. Mem. Am. Assn. Anatomists, A.A.A.S. Biophys. Soc., Electron Microscopy Soc. Am., Am. Soc. Cell Biology, Colegio Anatomico Brasileiro (hon.), Sigma Xi, Phi Beta Pi. Contbr. chpt. to Electron Microscopic Anatomy, 1964. Research, publs. on microscopic and submicroscopic anatomy, radiation biology and effects of x-rays on lining of small intestine using electron microscope, histochemistry and autoradiography. Home: 1434 Lake Av., Wilmette, Ill. 60091. Office: 303 E. Chicago Av., Chgo. 60611.*

HAMWI, George John, Am. physician; b. N.Y.C., July 23, 1914; s. Habib and Aneesa (Merhige) H.; student Columbia, 1932-35; B.S., Am. U., Beirut, 1936, M.D., 1940; M.S., Ohio State U., 1947; m. Isabel Roberts, Oct. 31, 1942; children—George John, Paul Roberts, Alan Hoffman. Practiced internal medicine, specializing in endocrinology and metabolism, Columbus, O., 1949—; faculty Ohio State U., 1949—, prof. medicine, dir. div. endocrinology and metabolism, 1958—; internal medicine cons. VA Hosp. (Dayton, O.), United Mine Workers Am.; com. mem. NIH. Diplomate Am. Bd. Internal Medicine. Fellow A.C.P., Am. Geriatrics Soc.; mem. A.M.A., Central Soc. Clin. Research, Am. Fedn. Clin. Research, N.Y. Acad. Scis., A.A.A.S., Am. Diabetes Assn. (dir.), Endocrine Soc., Am. Thyroid Assn., Am. Cancer Soc., Am. Soc. Clin. Nutrition, Am. Therapeutic Soc., Royal Soc. Medicine, Gerontol. Soc., Alpha Omega Alpha. Contbr. numerous articles on diabetes, thyroid, pituitary and adrenal diseases to profl. jours. Died Feb. 4, 1967.

HAMY, Maurice Theodore Adolphe, French astronomer; b. Boulogne-sur-mer, France, Oct. 31, 1861; Dr.-ès-sc., Paris, 1887; astronomer Paris Obs., until 1929, hon. astronomer, from 1929; pres., Bur. des Longitudes. Mem. French Acad. Scis. (pres. 1928). Research in celestial mechanics; perfected meridian circle, other astron. instruments. Died Paris, Apr. 9, 1936.

HANACK, Michael, German chemist; b. Luckenwalde/Berlin, Germany, Oct. 22, 1931; s. Georg and Elisabeth (Gimpel) H.; student Freiburg (Germany) U., 1949-50, Bonn (Germany) U., 1950-52; Dr.rer.-nat. Tübingen U., 1957; m. Ingrid Grabert, Mar. 9, 1964; 1 son, Matthias. Asst., U. Tübingen, 1957-61, dozent organic chemistry, 1962-67, prof., 1967—. Fellow N.Y. Acad. Scis.; mem. Gesellschaft Deutscher Chemiker, Chem. Soc. London, N.Y. Acad. Sci. Author: Conformation Theory, 1965; also articles. Research on chemistry of small ring compounds, rearrangement reactions of unsaturated compounds under cyclization to small ring compounds, conformational analysis, organic fluorine compounds. Home: 9 Bei der Ochsenweide, Office: Chemisches Institut, 33 Wilhelmstrasse, 74, Tübingen, Germany.*

HANAFEE, William, radiologist; b. Louisville, Mar. 21, 1926; s. John F. and Mary (Christ) H.; student Ind. Coll. Pharmacy, 1943-44; B.A., U. Rochester, 1946; M.D., U. Louisville, 1949; m. Constance Gandolph, Nov. 25, 1948; children—William N., Linda, Patrick and Michael (twins). Faculty, U. Cal., Los Angeles, 1953—, prof. radiology Sch. Medicine since

1966—, acting chmn. dept., 1966—; attending physician in neuroradiology Wadsworth Gen. Hosp., VA, Los Angeles, 1962—. Cons., Long Beach VA, 1964—, Jules Stein Eye Inst., Los Angeles, 1966—; lectr. San Diego Naval Hosp., 1966. Fellow Am. Coll. Radiology; mem. A.M.A., Am. Soc. Neuroradiology, Los Angeles Soc. Neurology and Psychiatry, Cal., Los Angeles County med. assns., N.Am., Pacific N.W., Cal. Los Angeles radiol. socs., Am. Roentgen Ray Soc., Pan Am. Med. Assn. Devel. new techniques for coronary arteriography, using percutaneous approach through the axilla, as well as new techniques for obtaining contrast of previously unexplored regions of the body, such as cavernous sinus, and the retro-orbital space; contbr. to investigation of trans-jugular cholangiography and controlled selective carotid angiography. Home: 4634 Via Apuesta, Tarzana, Cal. 91356.*

HANAHAN, Donald James, Am. biochemist; b. Springfield, Ill., May 13, 1919; s. James Francis and Clara (Schiller) H.; B.S., U. Ill., 1941, Ph.D., 1944; m. Lillian Marie Larsen, June 27, 1947; children—Douglas A., Laura J., Timothy J., Colleen J., Carolyn M. Research asso. Manhattan Project, 1944-45; postdoctoral fellow U. Cal. at Berkeley 1945-47; faculty U. Wash., Seattle, 1948-67, prof. biochemistry, 1958-67; prof., head dept. biochemistry U. Ariz., 1967—. Guggenheim Found. fellow, 1955; NIH spl. fellow, 1965-66. Mem. Am. Chem. Soc., Am. Soc. Biol. Chemists. Author: Lipid Chemistry, 1960; also numerous articles. Research in chemistry and biochemistry of simple and complex lipids, action of snake venom on lipids, nature of erythrocyte membrane, behavior of lipids in blood coagulation. Home: 90 Calle Primorosa, Tucson 85716.*

HANANI, Haim, mathematician; b. Slupca, Poland, Sept. 11, 1912; s. Michael and Sara (Lubraniecki) Chojnacki; student U. Vienna, 1929-31; M.A., U. Warsaw, 1934; Ph.D., Hebrew U., Jerusalem, Israel, 1938; m. Esther Bogatin, Feb. 16, 1938; children—Michael, Abraham, Nizza. Financial sec. Hebrew U., Jerusalem, 1951-55; faculty Technion-Israel Inst. of Tech., Haifa, 1955—, prof. math., 1962—, dean dept. math., 1961-62, v.p. for acad. affairs, 1962-65; research asso. prof. Math. Research Center, U. S. Army, Madison, Wis., 1959-60; vis. prof. U. Western Australia, Perth, 1961. Adviser on higher edn. to minister edn. and culture, 1964—. Mem. Israel Math. Union (past chmn.), Am. Math. Soc., Math. Assn. Am., Inst. Math., Statistics, Am. Statis. Assn., Israeli Assn. for Data Processing, Israeli Soc. for Logic. Research, publs. on four color problem, summability of series, theory of games, theory of balanced incomplete block designs, combinatorial analysis. Home: 63 Hanassi Blvd., Haifa, Israel.*

HANAU, Arthur, physician; b. Frankfort/Main, Germany, May 11, 1858; ed. Marburg, Bonn; M.D., 1881; asst. to E. Klebs, Zurich, Switzerland, later mem. faculty; became prosector, hosp., St. Gallen, Switzerland, 1897. Contbr. articles to med. jours. Research on prostatitis, infection of salivary glands, physiology of intestinal secretions, pathology of lung diseases, physiology of thrombi; refuted theory of viscous metamorphosis; made 1st successful cancer inoculation in rats, 1889. Died Konstanz, Germany, Aug. 2, 1900.

HANAUSEK, Thomas Franz, Austrian specialist in raw materials and commodities; b. Schloss Weitwörth, Austria, Sept. 26, 1852; s. Eduard and Josephine (Luksche) H.; teaching degree U. Vienna, 1879, Ph.D., 1881; m. Maria Theresia Kurz, 1885; 1 son, 1 dau.; m. 2d, Marie Linde, 1912. Became asst. tchr. 1875; named prof. commodities research and natural history State Sci. and Comml. Sch., Krems, 1880, Schottenfelder Sch., Vienna, 1885; also lectr. tech. microscopy Vienna Comml. Acad.; joined Ford Research Inst., Vienna, 1897; lectr. tech. raw materials and commodities research Tech. Indsl. Mus., Vienna, 1898-1902; prof. Secondary Sch., Vienna III; became sch. supt., Krems, 1902; ret., 1910. Author: Die Nahrungs-und Genussmittel aus dem Pflanzenreiche, 1884; Lehrbuch der Materialienkunde auf naturgeschitlicher Grundlage . . . I, 1887, II, 1891; Lehrbuch der Somatologie und Hygiene für Lehrer-und Lehrerinnenbildungsanstalten, 1894; Lehrbuch der technischen Mikroskopie, 1901; also articles. Research on structure and research possibilities of foods, especially flour, coffee, cocao, spices. Died Vienna, Feb. 4, 1918.

HANCKE, Johann Wenceslaus (Wenzel), German surgeon; b. Mertschütz, Germany, Mar. 16, 1770; s. Johann Georg and Maria Anna Rosina (Koschke) H.; apprenticed to barber-surgeon, Jauer; student Breslau (now Wroclaw, Poland); student surgery Pepiniere, Berlin, 1795-99; M.D., Frankfort/Oder, Germany, 1807; m. Plümicke, 1807; 2 sons; m. Schick; 1 son; m. 3d, von Krafft. Became journeyman barber, Breslau, 1786; joined army as surgeon, 1790; after studies returned to army as sr. surgeon; settled in Breslau, 1807; named physician Hosp. of Compassionate Bros., 1810; chief surgeon Bürgerwerder Mil. Hosp., 1813-14; med. adviser Provinzial-Kollegium-Medicum of Silesia, also prof. Sch. Surgery, Breslau. Hosp. in Breslau named in his honor. Author: Dissertatione inauguralis de inaccessa pericardii inflammati diagnosi memorabile quodam morbi huiusque exemplo illustrata, 1807; Über Eröffnung der Eitergeschwülste nach verschiedenen Methoden, 1829; Prophylaktische Heilverfahren bei Verletzungen von tollen Hunde, und Behandlung der eingetretenen Wuthkrankheit, 1832; Die asiatische Cholera, 1832. Recommended zinc chloride in treatment of 2d stage syphilis; fought cholera epidemic in Silesia, 1831. Died Breslau, June 22, 1849.

HAND, Cadet Hammond, Jr., marine biologist; b. Patchogue, N.Y., Apr. 23, 1920; s. Cadet Hammond and Myra (Wells) H.; B.S., U. Conn., 1946; M.A., U. Cal., Berkeley, 1948, Ph.D., 1951; m. Winifred H. Werdelin, June 6, 1942; children—Cadet Hammond III, Gary Alan. Faculty, U. Cal., Berkeley, 1953—, acting chmn. dept. zoology, 1960-61, prof. zoology, 1963—, dir. Bodega Marine Lab., 1961—; cons. NIH, NSF. NSF Sr. Postdoctoral fellow to Australia, New Zealand, 1959-60, Guggenheim fellow, 1967—. Mem. Ecol. Soc. Am., Am. Inst. Biol. Scis., Am. Soc. Limnology and Oceanography, Soc. Systematic Zoology, No. Cal. Malacological Soc., Western Soc. Naturalists, Ray Soc. (Gt. Britain). Contbr. numerous articles to profl. jours. Research in natural history, life history, distbn., zoogeography and taxonomy of coelenterates, symbiosis between coelenterates and other animals, natural history of mollusks, physiology of partnership between sea anemones and symbiotic algae. Home: 1737 Zinn St., Richmond, Cal. 94805. Office: Dept. Zoology, U. Cal., Berkeley, Cal. 94720.*

HANDEL-MAZZETTI, Heinrich Raphael Eduard von, see von Handel-Mazzetti.

HANDFORD, Stanley Wing, Am. physiologist; b. Batesville, Ark., June 28, 1906; s. Robert Stanley and Helen Margaret (Wing) H.; B.A., U. Ill., 1928; Ph.D., Yale, 1939; m. Elizabeth Gordon Shortall, Aug. 20, 1945; children—Eric S. S. W., Peter S. W. Commd. lt. (j.g.) USN, 1949, advanced through grades to capt., 1961, ret., 1966; physiologist Research Unit 3, Cairo, Egypt, 1954-56; head applied physiology Naval Med. Research Inst., Bethesda, Md., 1956-61, physiologist, head Environmental Stress div., adviser to dir. physiol. scis. dept., 1964-66; biomed. liaison scientist Office Naval Research br. office, London, Eng., 1961-64; program dir. arthritis area Nat. Inst. Allergy and Metabolic Diseases, Extra-mural program NIH, 1966—. Recipient Office Naval Research award, 1964. Mem. Am. Physiol. Soc., Am. Zool. Soc., Soc. Study Growth and Devel., N.Y. Acad. Scis., Brit., Assn. Radiation Research, Royal Soc. Medicine. Research, publs. on burn shock; use of artificial colloidal agts. for transfusion in place of blood or plasma; effects of total body radiation on gastrointestinal tract; growth of embryonic and larval transplanted eyes in amphibians; underwater physiology with respect to oxygen toxicity and inert gas narcosis. Home: 1410 Dale Dr., Silver Spring, Md. 20910. Office: Nat. Institutes of Health, Extra-mural Program, Westwood Bldg., Bethesda, Md. 20014.*

HANDIN, John Walter, Am. geophysicist; b. Salt Lake City, June 27, 1919; s. Walter H. and Dolores (Peirce) H.; A.B., U. Cal. at Los Angeles, 1941, M.A., 1948, Ph.D., 1949; m. Frances Robertson, Sept. 2, 1947; children—Diane, Katherine. Geologist, C.E., U. S. Army, 1948-49; research asso. Inst. Geophysics, U. Cal. at Los Angeles, 1949-50; with Shell Devel. Co., Houston, 1950-67, sr. research geologist, 1957-62, research asso., 1962-67; Distinguished prof. geology and geophysics, dir. Center for Tectonophysics, Tex. A. and M. U., College Station, 1967—, vis. prof. Columbia U., 1964. Cons., U. S. Army C.E., 1963—, Pres.'s Sci. Adv. Com., 1965; mem. U. S. Nat. Com. on Rock Mechanics, 1964- ; cons. to govt. and industry. Fellow Geol. Soc. Am.; Am. Geophys. Union; mem. Sigma Xi. Editor: (with D. T. Griggs) Rock Deformation, 1960; also numerous articles. Research on rock mechanics with applications to tecconophysics, structural geology, engrng. geology. Home: 2614 Melba Circle, Bryan, Tex. 77801. Office: Tex. A. and M. U., College Station, Tex. 77843.*

HANDLER, Alfred Harris, Am. pathologist; b. Boston, Apr. 4, 1923; s. Louis and Anna (Hite) H.; B.S., Providence Coll., 1943; A.M., Boston U., 1948, Ph.D., 1951; m. Barbara Shapiro, Aug. 3, 1951; children—Jane, Amy. Research fellow USPHS, Boston U., 1950-52; sr. tech. investigator Pondville Hosp., Walpole, Mass., 1952-53; research asso. in pathology Children's Hosp. Med. Center and Children's Cancer Research Found., Boston, 1952-65, also chief labs. of tumor transplantation, 1952-65; research asso. Harvard Med. Sch., 1956-65; asso. research prof. carcinogenisis Sch. Pub. Health, U. Pitts., 1965-66; lectr. in exptl. embryology Carnegie Inst. Tech., Pitts., 1965-66; asso. prof. pathology Rutgers U. Med. Sch., New Brunswick, N.J., 1966—. Fellow Royal Soc. Medicine; mem. Royal Soc. Health (hon.), A.A.A.S., N.Y. Acad. Scis., Am. Assn. Cancer Research, Am. Soc. Exptl. Pathology, Am. Assn. Pathologists and Bacteriologists, Transplantation Soc., A.M.A., Sigma Xi. Research and numerous publs. in exptl. pathology, oncology, tissue and cell transplantation; established use of cheek pouch of hamster as tool for growing human and animal tumors.*

HANDLER, Philip Am. biochemist; b. N.Y.C., Aug. 13, 1917; s. Jacob and Lena (Heisen) H.; B.S., Coll. City of N.Y., 1936; Ph.D., U. Ill., 1939; m. Lucille P. Marcus, Dec. 6, 1939; children—Mark, Eric Paul. Jr. chemist U. S. Regional Soybean Byproducts Lab., 1937-39; faculty Duke, Durham, N.C., 1939—, James B. Duke prof., chmn. dept. biochemistry, 1961—. Cons. U. S. VA, Oak Ridge Inst. Nuclear Studies, com. NASA; mem. adv. coms. mem. Nat. Adv. Health Council President's Commn. on Heart Disease, Pres.'s Sci. Adv. Com., Cancer and Stroke; Johns Hopkins, U. Notre Dame, NIH, Kettering Research Inst., Scripps Metabolic Clinic and Research Found., Pa. State U., Brandeis U.; chmn. Nat. Sci. bd. NSF. Trustee Found. Advancement Edn. in Scis., Cold Spring Harbor Lab. for Quantitative Biology. Fellow N.Y. Acad. Sci. (C. B. Mayer award 1943), A.A.A.S.; mem. Nat. Acad. Scis. (chmn. survey of life scis.), Am. Soc. Biol. Chemists (pres. 1963-64), Fedn. Am. Soc. Exptl. Biology (chmn. bd. 1965-66), Am. Chem. Soc., Am. Inst. Nutrition, Soc. Exptl. Biology and Medicine, Am. Acad. Arts and Scis. Biochem. Soc. (Great Britain), Sigma Xi. Author: (with Abraham White, Emil Smith) Principles of Biochemistry, 1954, 59, 64, 68. Mem. editorial com. Geriatrics; adv. bd. Comparative Biochemistry and Physiology, Jour. Theoretical Biology, Jour. Biol. Chemistry. Research and publs. on nicotinic acid and choline deficiency states; function of parathormone in renal secretion; oxidative-enzymes, mechanism of enzyme action, evolution; amino acid metabolism. Home: 2529 Perkins Rd., Durham, N.C. 27706.

HANDLEY, William Sampson, English surgeon; b. Eng., 1872; M.D., M.S., U. London; m. Muriel Rigby; 4 sons, 1 dau. Cons. surgeon Middlesex Hosp.; Hunterian prof. surgery and pathology Royal Coll. Surgeons; med. dir. Bournemouth Municipal Cancer Clinic. Mem. Acad. Medicine Rome, Royal Flemish Acad. Medicine, Royal Coll. Surgeons (v.p.); hon. fellow Royal Soc. Medicine (pres. surg. sec.), A.C.S., Med. Soc. London, Royal Soc. Medicine. Author: Cancer of the Breast and its Operative Treatment, 2d edit., 1922; The Genesis of Cancer, 1932; also papers and lectures. Probably first to stress the importance of the lymphatics in the spread of mammary carcinoma, 1906; research on abdominal surgery and surg. treatment of tumors. Died Mar. 18, 1962.

HANDLIRSCH, Anton, Austrian entomologist; b. Vienna, Austria, Jan. 20, 1865; s. Peter and Kosina (Kaufmann) H.; student pharmacy, then entomology U. Vienna; Ph.D. (hon.), Graz, Austria, 1923; m. Martha Alaunek, 1892; 1 son, 1 dau. Became asst. Vienna Natural History Mus., 1892, asst. curator, 1899, curator, 1906, dir., 1922; joined faculty U. Vienna, 1924, became asso. prof., 1931. Became sec. Zool.-Bot. Soc. Vienna, 1893, editor Verhandlungen, 1900, pres., 1922-28. Mem. Austrian Acad. Scis. Author: Die fossilen Insekten und die Phylogenie der rezenten Formen, 1908; Rekonstruktion paläozoischer und mesozoischer Insekten, 1912; Gegen die übermässige Zersplitterung systematischer Gruppen, 1929; also numerous articles. Pioneered change from descriptive to comparative and phylogenetic method of observation in insect taxonomy; research on fossils from Am., Belgium, Germany, Russia; developed new system using phylogeny; recognized connection between size relationships of insects and their geog. distbn. Died Vienna, Aug. 28, 1935.

HANDOVSKY, Hans, pharmacologist, toxicologist; b. Vienna, Austria, May 18, 1888; s. Leopold and Karoline (Sinaiberger) H.; doctorate U. Vienna, 1912; m. Hermine Käthe Sagebiel, 1932; 1 son, 1 dau.; m. 2d, Grit Preussler, 1951. Asst. biocolloid chem. inst. U. Vienna, 1908-10; asst., Heidelberg, Germany, 1912-13; with Wiedhowski, Prague, Czechoslovakia, 1913-14; med. officer World War I; joined Pharmacological Inst., U. Göttingen, (Germany), 1920, joined faculty, 1924, became asso. prof. pharmacology and toxicology, 1926; fled from Germany to Sorbonne, Paris, 1933; later asst. Inst. for Pharmacology and Toxicology, Geneva, Switzerland; then dir. cancer research lab. Women's Clinic, U. Geneva; ret., 1957; returned to Göttingen. Author: Fortschritte in der Kolloidchemie der Eiweisskörper, 1911; Leitfaden der Kolloidchemie für Biologen und Mediziner, 1922; Pharmacotherapie, 1952; also articles. Research on phys.-chem. problems of pharmacology, blood chemistry, cancer, especially relation between doses of vitamin E and speed of cancer devel. Died Remscheid-Lüttringhausen, Germany, Nov. 11, 1959.

HANDSCHIN-(HOFSTETTER), Eduard, Swiss zoologist, entomologist; b. Liestal nr, Basel, Switzerland, Aug. 31, 1894; s. Albert and Anna (Freyvogel) H.-H.; doctorate U. Basel (Switzerland); m. Mathilde Hofstetter, 1922; 3 sons. With exptl. sta., Harpenden, Eng., 1 year; mem. research commn. to study biology of buffalo fly Govt. of Australia, 1930-32; joined faculty U. Basel, 1921, became asso. prof., 1927, prof., 1941; colleague Natural History Mus., named pres. bd., 1946, dir., 1956-59; became mem. Sci. Nat. Park Commn., 1926, also pres. Mem. Entomol. Soc. Belgium (hon.), Entomol. Soc. Lund (hon.), Entomol. Soc. Basel (hon.), Natural Sci. Soc. Basel, Swiss League for Nature Protection; corr. mem. Bogor Bot. Garden, Zool.-Bot. Soc. Vienna. Author: Praktische Einführung in die Morphologie der Insekten, 1928; also articles. Research on earth animals, primitive terrestrial arthropods (especially Collembolae), including morphology-taxonomy combined with ecology and evolution; problems of soil biology, hydrobiology, parasitology; attempeed to rearrange taxonomy of Neuroptera; expanded entomol. collection of Basel Natural History Mus. Died Basel, Jan. 19, 1962.

HANDSCHUMACHER, Robert Edmund, Am. pharmacologist; b. Abington, Pa., Oct. 16, 1927; s. Gustave

Heinrich and Emma (Streck) H.; B.S., Drexel Inst., 1949; M.S., U. Wis., 1951, Ph.D., 1953; m. Elizabeth Grafly, Sept. 6, 1949; children—Robert Kurt, Mark Davis. Faculty Yale, 1956—, prof. pharmacology, 1964—, Career prof. in cancer research, 1964—. Sci. cons. Am. Cancer Soc., 1963; Anna Fuller Fund, 1965; Internat. Union Against Cancer, 1966. Mem. Am. Soc. Biol. Chemists, Am. Assn. for Cancer Research, Am. Soc. Pharmacology and Exptl. Therapeutics, Am. Chem. Soc., A.A.A.S., Sigma Xi. Contbg. author: Nucleic Acids, vol. 3, 1959; Methods in Medical Research, 1965; also numerous articles. Editorial bd. Cancer Research, 1962-65, Molecular Pharmacology, 1965—; asso. editor Cancer Research, 1965—. Developed antimetabolites for cancer therapy; biochem. research on metabolic control mechanisms. Home: Northford Rd., Branford, Conn. 06405. Office: 333 Cedar St., New Haven 06510.*

HANDSON, Ralph (or Raphe), mathematician; flourished Eng. 1612-30; pupil Henry Briggs. Tchr. math.; auditor, accounts of Chancery. Translator (English): Trigonometrica (Pitiscus, 1600; important nav. test), 1614.

HANES, Charles Samuel, Canadian biologist; b. Toronto, Ont., Can., May 21, 1903; s. Carmi Addison and Anastasia (Kavanagh) H.; B.A., U. Toronto, 1925; Ph.D., Cambridge U., 1931, Sc.D., 1951; m. Theodora Burleigh Auret, Mar. 7, 1931; 1 dau., Ursula Ann (Mrs. David John Fry). Lectr. biology Queens U., Kingston, Ont., 1927-28; research officer Ont. Research Found., 1932-34; research officer Low Temperature Sta. for Research in Biochemistry and Biophysics, Cambridge U., Eng., 1934-42, reader in plant biochemistry, dir. agrl. research council unit, 1947-51, professorial fellow, Downing Coll., 1947-51, hon. fellow, 1951—; prof. biochemistry U. Toronto, 1951-68, emeritus, 1968—. Sci. mem. Brit. Food Mission in N.Am., 1942-44; dir. food investigation Dept. Sci. and Indsl. Research U.K., 1943-47. Fellow Royal Soc., 1942, Royal Soc. Can. (Flavelle medal 1955); mem. Biochem. Soc. London Fedn. Am., Fedn. Canadian biol. socs. Numerous publs. on interpretation of structure of starch from mode of attack by different amylases; postulation of 1st helical structure for a macromolecule; discovered amylose phosphorylase of higher plants; studies of various enzymes; devel. (with J. T. Wong) of unrestricted kinetic theory of enzyme action. Office: Dept. Biochemistry, U. Toronto, Toronto 5, Ont., Can.*

HANF, Karl Ignaz (Blasius), Austrian ornithologist; b. St. Lambrecht, Austria, Oct. 30, 1808; s. Karl and Elisabeth (Zach); ed. U. Graz (Austria); joined Benedictine Order, St. Lambrecht Monastery, 1828; ordained priest, Mariahof, Austria, 1832; became priest in Zeutschach, 1843; priest, Mariahof, 1853-89. Recipient prizes for bird collection Vienna World Exhbn. 1873; bird sanctuary at Furtner Lake named in his honor. Research and publs. on problems of bird migration, Furtner Lake, feather change of Lagopus mutus, reproductive biology of Loxia curvirostra and Cuculus canorus; collected over 1000 birds; discovered brooding spot of Eudromias morinellus. Died Mariahof, Jan. 2, 1892.

HANGER, Franklin M., Am. physician; b. Staunton, Va., Sept. 6, 1894; s. Frank M. and Martha (McDowell) H.; B.S., U. Va., 1916; M.D., Johns Hopkins, 1920; m. Harriet Echols, Apr. 15, 1942; 1 dau., Harriet Echols. Staff Presbyn. Hosp., N.Y.C., 1940-60, cons., 1960—; prof. medicine, N.Y.C., 1947-60, prof. emeritus, 1960—; vis. prof. medicine U. Beirut (Lebanon), 1960-61. Sr. med. cons. Kingsbridge VA Hosp., N.Y.C., 1948-60. Diplomate Am. Bd. Internal Medicine. Mem. A.C.P. (past pres. regent), Am. Assn. for Study Liver Diseases (past pres.), Assn. Am. Physicians, Am. Soc. Clin. Investigation, Am. Assn. Immunologists. Author: Oxford Looseleaf Medicine; Cecil Textbook of Medicine; contbg. author: Nelson System of Medicine; Diseases of the Liver, 1956; The Thyroid, 1955; also numerous articles. Discovered specific toxicity gram negative material on capillary walls; described cat scratch disease and developed skin test antigen; originated cephalin flocculation test. Home: 605 E. Beverly St., Staunton, Va. 24401.*

HANGGI, George John, Am. mech. engr.; b. Little Rock, Dec. 6, 1922; s. Emil John and Alma (Franks) H.; student Ark. Poly. Coll., 1940-42; B.S. in Mech. Engring., Okla. A. and M. Coll., 1947, M.S., 1950; m. Nina E. Sandlin, Aug. 3, 1944; children—George Gerard, John Alan, James Rodney. Instr., Okla. A. and M. Coll., Okla. State U., Stillwater, 1947-51; engr.-designer Continental Oil Co., Ponca City, Okla., 1951-52, research engr., 1952-61, sr. research engr., 1961—. Registered profl. engr., Okla. Designed and developed lab. apparatus, high pressure equipment, automatic apparatus. Patentee, vibrators, low temperature analysis, fatigue testing device, soil injection apparatus. Home: 1605 Trio Lane, Ponca City 74601. Office: Continental Oil Co., Ponca City, Okla. 74602.*

HANKEL, Hermann, German mathematician; b. Halle/Saale, Germany, Feb. 14, 1839; s. Wilhelm Gottlieb Hankel; studied under Mobius in Leipzig, Germany, 1857 under Riemann, in Göttingen, Germany, 1860 under Weir Weierstrass and Kronecker in Berlin, 1861; Ph.D., Leipzig; m. Maria Dippe, 1867. Became mem. faculty U. Leipzig, 1863, asso. prof., 1867; named prof., Erlangen, Germany, 1876; prof., Tübingen, Germany, from 1869. Author: Zur allgemeinen Theorie der Bewegung der Flüssigkeiten, 1861; Die Erlerschen Integrale bei unbeschränkter Variabilität des Argumentes, 1863; Vorlesungen über komplexe Zahlen und ihre Funktionen I (Theorie der complexen Zahlensysteme), 1867; Zur Geschichte der Mathematik in Altertum und Mittelalter, 1874; Elemente der projektiven Geometrie in synthetischer Behandlung, 1875. Worked on complex numbers and functions of complex variable; research on history of math.; studies in projective geometry; known for complex integral representation of gamma function, solutions to Bessel's differential equation, proof of theorem that expansion of complex number system is not possible if all laws of arithmetic remain valid; developed Riemann's ideas. Died Schramberg, Germany, Aug. 29, 1873.

HANKEL, Wilhelm Gottlieb, German physicist; b. Ermsleben, Germany, May 17, 1814; grad. physics and chemistry U. Halle (Germany), 1840; m. Stegmann; children include Hermann. Became asso. prof. U. Halle, 1847; prof. physics U. Leipzig (Germany), 1849-87. Author: Grundriss der Physik, 1848; Elektrische Untersuchungen, 1856-83; also articles. Translator, Arago's works into German, 1854-60. Research on pyroelectricity of crystals, piezoelectricity and photoelectricity, electromotive series and galvanic chains, electromagnetism; built numerous measuring instruments. Died Leipzig, Feb. 17, 1899.

HANKES, Lawrence Valentine, Am. biochemist; b. Chgo., Nov. 24, 1919; s. Michael John and Matilda (Bachman) H.; A.B., DePauw U., 1942; M.S., Mich. State U., 1943; Ph.D., U. Wis., 1949, postgrad., 1949-50; m. Mary Catherine Hamm, Sept. 16, 1951; children—Lawrence, M., Catherine A., Matthew W. Instr., Mich. State U., 1942-43, Northwestern U. Dental Sch., 1944; head allergy group VA Hosp., Aspinwall, Pa., 1950; head clin. chemistry Med. Research Center, Brookhaven Nat. Lab., Upton, L.I., N.Y., 1951—. Fellow Royal Chem. Soc. (London, Eng.); mem. Am. Chem. Soc., Soc. for Exptl. Biology and Medicine, Am. Soc. Biol. Chemists, Am. Assn. Clin. Chemists, Sigma Xi, Alpha Chi Sigma, Phi Sigma. Contbr. numerous articles to tech. jours. Research on metabolism tryptophan with synthesis metabolites, clin. chemistry methods. Home: 11 Maple Rd., Strongs Neck, Setauket N.Y. 11785. Office: Brookhaven Nat. Lab., Upton, N.Y. 11973.*

HANKS, John Harold, Am. microbiologist; b. Fowlerton, Ind., Sept. 16, 1906; s. John Rufus and Augusta Mae (Smith) H.; B.S., Allegheny Coll., 1928; Ph.D., Yale, 1931; m. Julia King, June 9, 1930; children—John King, James Philip, Juliette Mae, Janet Susan. Fellow NRC, Harvard, 1931-32; faculty George Washington U., 1932-39; bacteriologist Leonard Wood Meml. Lab., Balt., 1939—, microbiologist, 1959—; lectr. Harvard, 1946-59; asso. prof. Johns Hopkins U. Sch. Hygiene, 1959-62, prof. pathobiology, 1962—. Cons. Pan Am. Health Orgn.; mem. expert com. on leprosy WHO. Mem. Internat. Leprosy Assn. (chmn. div. bacteriology and immunology), Am. Soc. Microbiology, Am. Assn. Immunologists, Soc. Exptl. Biology and Medicine, Am. Acad. Microbiology. Research, numerous publs. on cultivation of causative agt. of human leprosy; cultivation and physiology of mycobacteria; mgmt. of cell and tissue cultures; investigation of lab. and pub. health problems in leprosy. Home: 204 Ridgemede Rd., Balt. 21210. Office: 615 N. Wolfe St., Balt. 21205.*

HÄNLE, Georg Friedrich, German pharmacist; b. Lahr, Germany, Jan. 6, 1763; s. Johann Daniel and Catherine Salome (Bohnert) H.; student medicine Upper Karls Sch., Stuttgart, Germany, 1 1/2 years; m. Sybille Katherine Scholderer, 1788; 1 son, Christian Friedrich; 3 daus. Began as pharmacist apprentice, Strassbourg, France; asst., Darmstadt, also Zweibrücken; became pharmacist Untere Apotheke (family ownership, since 1666); Lahr, 1788; ret., 1815. Med. adviser State of Baden, Germany, Mem. Pharmacists Assn. No. Germany (hon.). Author: Chemische-technischen Abhandlungen, 4 vols., 1808-21; Entwurf zu einer all-gemeinen und beständigen Apothekertaxe, 1818. Founder jour. Magazin der Pharmacie (now Annalen der Pharmacie), 1823. Contributed to recognition of pharmacy as independent sci. Died Karlsruhe, Germany, June 23, 1824.

HANLE, Wilhelm, German physicist; b. Mannheim, Jan. 13, 1901; s. Adolf and Elsbeth Hanle; student univs. Heidelberg, Göttingen; Dr.phil., 1924; m. Dora Gmelin, June 8, 1927; children—Irmgard (Mrs. Scharmann), Eberhard, Hellmut. Asst. univs. Göttingen, 1924, Tübingen, 1925, Halle, 1926; unscheduled prof. univs. Jena, 1929-35, Leipzig, 1935-37, Göttingen, 1937-41; prof., dir. phys. inst. U. Giessen, 1941—. Chmn. radiation measurement group, Fed. Ministry Sci. Research. Mem. German Phys. Soc. (chmn. edn. commn.), Am. Phys. Soc. Author: Künstliche Radioaktivität; also numerous articles. Research on influence of magnetic fields on polarization of resonance radiation (Hanle effect), measuring lifetimes of luminescence of atoms, excitation function for collisions of electrons and ions with atoms and molecules, exoelectron emission, dosimetry. Address: 74 Goethestrasse, 63 Giessen, West Germany.*

HANN, Julius Ferdinand von, see von Hann, Julius Ferdinand.

HANNA, Clinton Richards, Am. elec. engr.; b. Indpls. Dec. 17, 1899; s. Walter Parks and Elinor (Vestal) H.; B.S. in Elec. Engring., Purdue U., 1922, E.E., 1926, D.Engring. (hon.), 1945; m. Dorothy R. Wilharm, June 8, 1926; children—Marilyn Louise (Mrs. John L. Williams), David Walter. With Westinghouse Electric Corp., Pitts., 1922—, mgr. electromech. dept., 1930-44, asso. dir. research, 1944-60, cons., 1960—. Recipient Westinghouse Order of Merit, 1940, WPB citation Pres. Roosevelt, 1942, Potts medal Franklin Inst, 1949. Fellow I.E.E.E., Acoustical Soc. Am.; mem. Am. Inst. Aeros. and Astronautics, Am. Inst. E.E. (Lamme medal 1955). Contbr. articles to tech. jours. Patentee in field. Inventions in fields recording on film, loudspeakers, microphones, sound motion pictures, voltage, position and speed regulation, gyro stablizn. aircraft autopilots, vehicle stablzn. and tilting. Home: 2756 N.E. 37th Dr., Ft. Lauderdale, Fla. 33308.*

HANNA, Michael George, Am. biologist; b. Cleve., July 7, 1936; s. Michael George and Camella (Karem) H.; B.Sc., Baldwin Wallace Coll., 1958; M.Sc., U. Notre Dame, 1960; Ph.D., U. Tenn., 1964; m. Barbara Ann Pearson, Sept. 6, 1958; children—Michael George, III, Christina Louise, Susanne Kathleen. AEC Research fellow, U. Notre Dame, 1959-60; Pub. Health fellow U. Tenn., 1961-62; predoctoral fellow Oak Ridge Inst. Nuclear Studies, 1962-64; research biologist Oak Ridge Nat. Lab., 1964—. Mem. Royal Soc. Medicine, Am. Soc. Zoologists, Sigma Xi. Contbr. articles to tech. jours. Research on kinetics and transformation affected cell populations during induction phase immune response in lymphatic tissue, role lymphatic tissue germinal centers in immune response, role immune response in carcinogenesis. Home: 117 Monticello Rd., Oak Ridge 37831. Office: P.O. Box Y, Biology Div., Oak Ridge Nat. Lab., Oak Ridge 37831.*

HANNAN, Edward J., Australian statistician; b. Australia, Jan. 29, 1921; s. James Thomas and Margaret (McEwan) H.; student Xavier Coll., Melbourne, Australia, 1933-36; B.Commerce, Melbourne U., 1948; Ph.D., Australian Nat. U., 1955; m. Irene D. Trott, Mar. 1, 1949; children—Christine Josephine, Jennifer Marion, Patrick Edward, David James. Clk., Commonwealth Bank of Australia, Melbourne, 1936-40, economist, 1949-53; research fellow Australian Nat. U., Canberra, 1953-58, prof. statistics, 1959—. Mem. Am. Math. Soc., Royal Statis. Soc., Econometric Soc., others. Author: Time Series Analysis, 1960; Group Representations and Applied Probability, 1965; also articles. Research on use of Fourier methods in analysis of data primarily in problems of relating time series. Home: 8 Penrhyn, Canberra, Australia.*

HANNAN, Patrick Jeremiah, Am. chemist; b. Washington, Mar. 17, 1922; s. Patrick F. and Lilian (Keefe) H.; B.S., Cath. U., 1942, M.S., 1948; m. Margaret Ellen Trewhella, Feb. 9, 1952; children—Mary Catherine, Margaret Boucher, Marilyn Louise, Thomas Patrick. Chemist, Carnegie Instn., 1942-45, U. S. Dept. Agr., 1947-48, ERDL, 1948-50, 51-56; owner Mildex Corp., 1950-51; chemist U. S. Naval Research Lab., Washington, 1956—. Mem. Am. Chem. Soc., Soc. Indsl. Microbiology. Research, publs., patents on factors controlling oxygen prodn. by mass cultures of algae, use of algae as oxygen source. Home: 3632 Veazey St. N.W., Washington 20008. Office: U. S. Naval Research Lab., Washington 20390.*

HANNAY, N(orman) Bruce, Am. chemist; b. Mt. Vernon, Wash., Feb. 9, 1921; s. Norman Bond and Winnie (Evans) H.; B.A., Swarthmore Coll., 1942; M.S., Princeton, 1943, Ph.D., 1944; m. Joan Anderson, May 27, 1943; children—Robin, Brooke. With Bell Telephone Labs., Murray Hill, N.J., 1944—, exec. dir. materials research div., 1967—. Mem. adv. coms. Nat. Bur. Standards, Dept. Def., Cornell, Princeton, Am. Cancer Soc., Materials Adv. Bd.; mem. solid state adv. panel NRC; cons. Orgn. Econ. Coop. and Devel. Fellow Am. Phys. Soc.; mem. Am. Chem. Soc., Electrochem. Soc. (chmn. electronics div.). Editor: Semiconductors, 1959. Author: Solid State Chemistry, 1967; also articles. Research on dipole moments and molecular structure; thermionic emission and oxide cathodes; mass spectroscopy; analysis of solids; semiconductor materials. Home: 29 Old Fort Rd., Bernardsville, N.J. 07924. Office: Bell Telephone Labs., Murray Hill, N.J. 07971.*

HANNI, Lucius, Austrian mathematician; b. Göfis, Austria, Mar. 31, 1875; s. Franz Xaver and Maria Kreszentia (Specht) H.; Ph.D. in Math. and Physics, U. Innsbruck (Austria), 1900. Became asst. math., librarian Tech. U., Vienna, 1904; joined faculty U. Vienna, 1906; became asso. prof. U. Graz (Austria), 1922. Research and publs. on pure math., Borel's generalization of concept of limit, Leffler's infinite series, applied math., Maxwell's differential equation in wave theory, theory of relativity, theory of vector fields variable in time. Died Waltendorf, Austria, Mar. 16, 1931.

HANNO, Carthaginian navigator; b. Carthage; flourished 500 B.C. Author: Periplus. Claimed to have circumnavigated Africa (if true he was 1st man of Mediterranean world to cross the equator).

HANNON, John Patrick, Am. physiologist; b. Richmond, Cal., May 12, 1927; s. Francis P. and A. Isabel (Krumlinde) H.; B.A., U. Cal., Berkeley, 1950, Ph.D., 1954; m. Trudy R. Eberhardt, Jan. 26, 1952; children—Christie A., Kathleen E., John Paterick. Research physiologist USAF Arctic Aeromedical Lab., Fairbanks, Alaska, 1955-58, chief physiology dept., 1958-63; asst. chief physiology div. U. S. Army Med. Research and Nutrition Lab. Fitzsimons Gen. Hosp., Denver, 1963-65, chief physiology div., 1965—; lectr. U. Alaska, 1958-60, asso. prof. zoology, mem. grad. faculty, 1960-61; gen. chmn. 12th Alaska Sci. Conf., 1961. Recipient Civilian Service Outstanding Performance award, 1960. Fellow A.A.A.S. (past pres. Alaska div.); mem. Am. Physiol. Soc., Soc. Exptl. Biology and Medicine. Editor: (with E. Viereck) Symposium: Neural Factors in Temperature Regulation, 1960, Symposium: Comparative Physiology of Temperature Regulation, 1962. Contbr. numerous articles to profl. jours. Research in responses of lower animals and man to environmental stresses of cold and altitude, biochemical and tissue enzyme changes associated with cold acclimatization, high altitude acclimatization in women. Home: 315 Pine Cone Rd., Parker, Colo. 80134. Office: Physiology Div., U. S. Army Med. Research and Nutrition Lab., Fitzsimons Gen. Hosp., Denver 80240.*

HANNOVER, Adolph, Danish pathologist, physician; b. Copenhagen, Denmark, Nov. 24, 1814; mem. French Acad. Scis. (prize 1856). Author: Sur l'importance de menstruation, 1851; Sur l'épithélioma, 1852. Investigated anatomy, physiology and pathology of eye, also constrn. and use of microscope; micrometric research on vertebrates; credited with introducing term epithelioma, circa 1843. Died Taarback, Denmark, July 17, 1894.

HANOT, Victor Charles, French physician; b. Paris, July 6, 1844; M.D., 1875. Author: Cirrhose du foie, 1875-87; . . . la Cirrhose hypertrophique avec ictère chronique (Charcot-Debove collection); Endocardite aiquë (Léauté collection); Etudes sur les maladies dis fore (with M. Gilbert), 1888; contbg. author: Traité de la phtisie pulmonaire (Hérard, Cornil), 1888. Described hypertrophic cirrhosis of liver caused by chronic cholangitis or obstruction of bile ducts (Honot's disease), 1875. Died 1896.

HANRIOT, Maurice, chemist; studied under Wurtz; prof. Acad. Medicine Paris. Research on derivatives of glycerine and substitution of hydrogen by alkali metals in hydrocarbons, (with Richet) respiratory exchange in man and animals; discovered chloralose and its isomer parachloralose by union of chloral anhydride and glucose, 1894, ferment lipase in blood, 1896; synthesized collargol. Died 1933.

HANSARD, Samuel L(eroy), nutritionist; b. Knoxville, Tenn., June 5, 1914; s. Clarence Clayton and Martha E. (Carden) H.; B.S. in Agr., U. Tenn., 1937; M.S. in Nutrition, Ohio State U., 1941; Ph.D., U. Fla., 1953; m. Erma Brobeck, Nov. 8, 1941; children—Samuel Leroy, II, Martha Louise, Peggy Lynn. Faculty U. Tenn., 1946-57, prof., sr. scientist, 1951-57; dir. radioisotope Lab., chmn. nutrition sect. La. State U., 1957-68; prof. animal sci. U. Tenn., Knoxville, 1968—. Recipient Gamma Sigma Delta Research award, 1963; Am. Feed Ingredients Assn. Calcium Carbonate award, 1965. Fellow A.A.A.S.; mem. Am. Inst. Nutrition, Soc. Exptl. Biology and Medicine, La., N.Y. acads. scis., Am. Soc. Animal Sci. (Feed mfrs. award 1966), Sigma Xi, Phi Kappa Phi, Gamma Sigma Delta, Omicron Delta Kappa, Alpha Zeta., Phi Sigma. Research and numerous publs. on mineral absorption, placental transfer, prenatal nutrition, environmental physiology, radiochem. procedures for thryoid function, mineral metabolism, body composition of farm and lab. animals. Home: 100 Eastwood Dr., Knoxville, Tenn. 37920.*

HANSCOME, Thomas Dixon, Am. physicist; b. Austin, Minn., July 2, 1914; s. William Thomas and Ethel (Dixon) H.; B.A. cum laude, U. Minn., 1938; M.A., 1939, postgrad.; m. JoAnn Elizabeth Chandler, Nov. 8, 1940; children—Thomas Chandler, Cheryl (dec.). Physicist, Naval Research Lab., Washington, 1942-59, br. head, chief scientist Chesapeake Bay Expt. Sta., 1947-48; mgr. radiation effects research dept. Hughes Aircraft Co., Fullerton, Cal., 1959—. Recipient Meritorious Civilian Service award USN, 1946. Mem. Am. Inst. Physics, Am. Phys. Soc., Gamma Alpha. Contbr. articles to profl. jours. Developed radiation effects tech. to counteract radiation in materials and electronic devices. Home: 1782 Terry Lynn Dr., Santa Ana, Cal. 92705. Office: P.O. Box 3310, Fullerton, Cal. 92634.*

HÄNSEL, Hermann, Austrian agronomist; b. Vienna, Austria, Jan. 13, 1918; s. Ludwig and Anna (Sandner) H.; ed. Central Sch. Agr., Vienna; agron. engring. dipl.; Ph.D. in engring.; m. Ingrid Hacker, 1951; children—Bernhard, Georg, Arno, Agnes. Asst., Inst. Phytology, Central Sch. Agr., Vienna, 1946-49; researcher Plant Breeding Inst., Cambridge U., 1949-50; dir. Sta. Selection of Cereals, Probstdorfer Saatzucht S.P.R.L., 1951—; agrégé, 1945; asso. prof., 1962. Recipient Theodor Körner prize, 1953, 59. Mem. Soc. Zoology and Botany Vienna, German Soc. Botany, European Soc. Phytology. Research and numerous publs. on physiology and genetics of triticum, hordeum and pisum. Home: 61/7 Lange Gasse. Office: 33 Gregor Mendelstrasse, Vienna, Austria.

HANSEL, Rudolf, German botanist; b. Feb. 5, 1920; s. Wenzel and Maria (Rudolf) H.; ed. Athenaeum of Teplitz-Schönau, U. Vienna, U. Munich; Ph.D. in natural sci.; m. Margarete Hölzlmeyer, 1947; children—Klaus-Dieter, Birgit, Ruth-Maggit. Agrege, U. Munich, 1954; asso. prof. U. Berlin, 1954, full prof., 1963. Research and numerous publs. on medicinal plants. Home: 4 Albiger Weg, 1 Berlin 38. Office: Institute for Pharmacognosy, 63 Schweinfurstrasse, 1 Berlin 33, Germany.

HANSEL, William, Am. physiologist; b. Vale Summit, Md., Sept. 16, 1918; s. John William and Helene M. (Sperlein) H.; B.S., U. Md., 1940; M.S., Cornell U., 1947, Ph.D., 1949; m. Milbrey A. Downey, Aug. 12, 1942; children—Barbara Ann, Kay Elizabeth. Faculty, Cornell U., Ithaca, N.Y., 1949—, prof., 1960—. Recipient Ann. Soc. Animal Sci. award in animal physiology, 1960, N.Y. Farmers award, 1964. Guggenheim Found. fellow U. Chgo., 1958, NSF Sr. Postdoctoral fellow, 1966. Mem. A.A.A.S., N.Y. Acad. Scis., Soc. Study Fertility, Internat. Fertility Assn., Am. Dairy Sci. Assn., Am. Soc. Animal Sci., Am. Physiol. Soc., Sigma Xi, Phi Kappa Phi. Contbr. numerous articles to profl. jours. Research, publs. on physiology of ovarian and pituitary hormones in domestic and lab. animals, causes of infertility in animals; developed techniques for artificial estrous cycle synchronization and insemination in domestic animals. Home: 1039 Hanshaw Rd., Ithaca, N.Y. 14850.*

HANSEMANN, David Paul von, see von Hansemann, David Paul.

HANSEN, Armauer Gerhard Henrik, Norwegian bacteriologist; b. Bergen, Norway, July 29, 1841; practice medicine, Bergen; dir. Leprosy Hosp., Bergen. Discovered bacillus causing leprosy, 1879. Died Feb. 12, 1912.

HANSEN, Emil Christian, Danish botanist, bacteriologist; b. Ribe, Denmark, May 8, 1842; student art, Copenhagen, Denmark; later studied sci; became head physiol. dept. Carlsburg Inst., 1879. Studied fungi; perfected methods of culture and isolation; devised way of cultivating pure strains of yeast (revolutionized brewing industry); defined morphological and physiol. characters of different species and devised system of classification. Died 1909.

HANSEN, H. G., German physician; b. Flensburg, May 6, 1922; s. G. and M. (Philipsen) H.; student univs. Berlin, Würzburg; m. Ursula Hartmann, 1944; children—Owe, Birgit. Head physician, children's clinic, U. Kiel, 1954-65; dir. children's clinic, prof. pediatrics Lübeck Med. Acad., 1966—. Recipient Med. Soc. award U. Kiel, 1958. Corr. mem. Pediatric Soc. Bela Horizonte, Brazil. Author: Phasenkontrastverfahren in der Medizin, 1953; Phasenkontrastverfahren in der Hämatologie, 1955; Die Physiologie des Lymphocytenwechsels, 1958; Differentialdiagnose von Krankheitssymptomen bei Kindern und Jugendlichen, vol. I, 1961; (with Koecher) Klinische Diagnostik für den Kinderarzt, 1964; also numerous articles. Research in pediatric hematology, clin. pediatrics, bone diseases in childhood, pediatric radiology, exptl. endocrinology. Address: 71 Kronsforder Allee, Lübeck, West Germany.*

HANSEN, Johannes, animal husbandry scientist; b. Nadelhöft, Germany, Mar. 9, 1863; s. Nicolai and Marie Dorothea (Thomsen) H.; student agrl. sci., Kiel, Germany, then Jena, Germany; doctorate, 1886; hon. dr.agr.; hon.dr.vet.med.; m. Helene Schenk, 1890; 1 son, 1 dau. Tchr. agr., Silesia and Mecklenburg, Germany, 1886; became dir. Agr. Sch., Zwätzen, Germany, 1889; named adminstr. agrl. operation, Oberglogau, Upper Silesia, 1897; joined faculty U. Jena, 1896; became prof. Agrl. Acad., Bonn-Popelsdorf, 1901; named prof., Königsberg (now Kaliningrad, USSR), 1910; prof. animal breeding, Berlin, 1922-29, emeritus, 1929—. Mem. German Agrl. Soc., German Soc. for Animal Breeding. Author: Lehrbuch der Rinderzucht, 1920, 27. Orgn. of German domestic animal breeding, especially cattle; introduced milk control assns. in Germany; recognized role of fodder in modern animal nutrition physiology; tested performance of cattle. Died Berlin, Jan. 3, 1938.

HANSEN, Max, German metallurgist; b. Bremerhaven, Jan. 27, 1901; s. Max and Helene (Petersen) H.; ed. U. Marbourg, U. Göttingen; Ph.D. in chemistry; m. Mary Tittel, 1925; children—Helga, Dagmar, Burkhard. Asst., I.G. Farbenindustrie A.G., 1924-25; dept. dir. Kaiser-Wilhelm Inst. Metall. Research, 1925-34; sci. collaborator Duerener Metallwerke A.G., 1934-36, chief metallurgist, 1936-39; dir. Research Establishment, 1939-44, 45; prof. Poly. Sch., Berlin, 1940-45; hon. prof. Ill. Inst. Tech., Chgo., 1945-47, asso. prof., 1947-49; dir. dept. metall. research Armour Research Found., Chgo., 1949-54; dir. Metallgesellschaft A.G., Frankfort on Main. Mem. German Assn. Metallurgy (Heyn medal 1963), Inst. Metals (London), Am. Soc. for Metals, Am. Inst. Mining, Metall. and Petroleum Engrs., Sigma Xi, Tau Beta Pi. Contbr. numerous articles to German and Am. sci. jours. Home: 8 Königsteinerstrasse, Kronberg Ts. Office: 14 Reuterweg, Frankfort on Main, Germany.*

HANSEN, Niels Ebbesen, horticulturist; b. nr. Ribe, Denmark, Jan. 4, 1866; s. Andrew and Bodil (Midtgaard) H.; came to U. S. with parents, 1873; B.S., Ia. Agrl. Coll., 1887; M.S., 1895; Sc.D., U. S.D., 1917; m. Emma Elise Pammel, Nov. 16, 1898 (dec. Dec. 1904); children—Eva Pammel (Mrs. David L. Gilkerson), Carl Andreas; m. 2d, Dora Sophie Pammel, Aug. 27, 1907. In practical horticulture, comml. Ia. nurseries, 1888-91; asst. prof. horticulture Ia. Agrl. Coll., 1891-95; prof. horticulture S.D. Agrl. Coll. and Expt. Sta., 1895-1937, emeritus from 1937. Conducted hort. study in 8 countries of Europe, including Russia, 1894; made exploration trip for U. S. Dept. Agr., collecting new econ. seeds and plants in Russia, Turkestan, Western China, Siberia, Transcaucasia, 1897-98; made exploration for U. S. Dept. Agr. around the world through Lapland, Finland, Russia, Siberia, Manchuria and Japan, 1906, to Siberia, Mongolia, Manchuria, Turkestan, Transcaucasia and N. Africa, 1908-09; went to North China (Manchuria) for State of S.D., 1924; del. Internat. Congress Horticulture, London, 1930; originator new fruits, especially the Hansen hybrid plums, later extensively grown in the West; introduced the Turkestan, Siberian and many other alfalfas; also introduced and named the Cossack alfalfa, later widely grown in the prairie Northwest; originated a method of field hybridization of hardy alfalfas by transplanting; made exploration for alfalfa for S.D. in Siberia, 1913; also imported the Siberian fat-rumped sheep from which James W. Wilson developed a tailless breed of sheep; made exploration tour invitation of the Soviet govt., to East Siberia, 1934; a program of expts. in horticulture and agr., covering 100 points, was completed Mar. 1935, and published by USSR, 1937. Recipient George Robert White gold medal of honor Mass. Hort. Soc., 1917; Marshall P. Wilder silver medal Am. Pomol. Soc. (for new fruits), 1929; Cosmopolitan gold medal "for public service," Sioux Falls, 1933; awarded A. P. Stevenson gold medal Man. Hort. Soc. (for new fruits), 1935; Alumni Merit award Chgo. Alumni Assn., 1942; Medal of Honor, Ia. Hort. Soc., 1944, S.D. Hort. Soc., 1946. Author: Handbook of Fruit-culture and Tree-Planting (in Danish-Norwegian), 1890; Systematic Pomology (with J. L. Budd), 1903. Died Oct. 5, 1950.

HANSEN, Peder Gregers, Danish physicist; b. Frederiksberg, Denmark, Jan. 11, 1933; s. P. G. and Musse (Lindeloff) H.; M.Sc. in Chemistry, Tech. U. Denmark, 1955; Dr.phil. U. Copenhagen (Denmark), 1965; m. Bitten Bisbjerg, May 25, 1957; children—Peder, Ole. Research scientist Danish Atomic Energy Commn., Research Establishment, Risoe, Denmark, 1956-66; prof. physics U. Aarhus (Denmark), 1966—; staff Niels Bohr Inst., Copenhagen, 1956-58. Mem. Am. Phys. Soc., Fysiskforening, Kemisk Forening. Research, publs. on radiochemistry and nuclear structure physics, especially levels of deformed nuclei and beta decay. Home: Soefoften 36, 8250 Egaa, Denmark. Office: Inst. Physics, U. Aarhus, Denmark.*

HANSEN, Peter Andreas, Danish astronomer; b. Schleswig, Denmark, Dec. 8, 1795; asst. in measuring arc of meridian Altona Obs., 1821; became Seeberg Obs., 1815, obs. moved to Gotha, Sweden, 1859. Recipient Gold medal Royal Astron. Soc., twice, Fellow Royal Soc., 1835 (Copley medal 1850). Author: Fundamenta, 1838; (with Olafsen) Tables du soleil, 1853; Tables de la lune, 1857; Darlegung, 1862-64. Mem. French Acad. Scis. Research on theory of perturbations, lunar theory. Died Mar. 28, 1874.

HANSEN, Richard Mads, Am. biologist; b. Goshen, Utah, Jan. 11, 1924; s. Hyrum and Maggie (Lauredsen) H.; B.S., U. Utah, 1950, M.S., 1951, Ph.D., 1954; m. Carol Kay, Sept. 11, 1945; children—Karl R., Stephen K., Victor L. Faculty, Colo. State U. (formerly Colo. A. and M. Coll.), Ft. Collins, 1954—, asso. biologist, 1959—. Chmn., Colo. Coop. Range Rodent Project, 1959—. Mem. Am. Soc. Mammalogists, Wildlife Soc., Herpetologists League, Ecol. Soc. Am., Am. Soc. Range Mgmt., Am. Inst. Biol. Scis., Sigma Xi, Phi Kappa Phi. Research, numerous publs. on effects of rodents and rabbits on range lands. Home: 1709 Morningside Dr., Ft. Collins, Colo. 80521.*

HANSEN, Richard Thomas, Am. astronomer; b. Balt., Mar. 14, 1928; s. John Henry and Anna Mary (Ruth) H.; student Johns Hopkins, 1945-46; B.S. in Mech. Engring., U. Colo., 1950; m. Shirley Rae Friel, June 6, 1951; children—Susan, Thomas, Katrinka. Observer, Sacramento Peak Obs., N.M., 1950-51; observer-in-charge High Altitude Obs., Climax, Colo., 1952-56, sci. staff, Boulder, Colo., 1956-62; exec. officer Joint Inst. for Lab. Astrophysics, Boulder, 1962-63; sci. staff High Altitude Obs., Haleakala, also Mauna Loa, Hawaii, 1963—. Supr., IGY World Data Center A-Solar Activity, Boulder, 1957-62; vis. colleague, Haleakala Obs., Hawaii Inst. Geophysics, U. Hawaii, 1963-65. Fellow A.A.A.S.; mem. Am. Astron. Soc., Am. Geophys. Union, Sigma Xi. Research, publs. on observation of sun especially its corona, relations between solar activity phenomena and terrestrial events. Home: P.O. Box 727, Kamuela, Hawaii 96743. Office: High Altitude Obs., Boulder, Colo. 80302.*

HANSEN, Robert Suttle, Am. chemist; b. Salt Lake City, June 17, 1918; s. Charles Andrew and Bessie (Suttle) H.; B.S. in Chemistry, U. Mich., 1940, M.S., 1941, Ph.D., 1948; m. Gilda Cappannari, Apr. 8, 1939; 1 son, Edward Charles. Faculty, Ia. State U., Ames, 1948—, prof. chemistry, 1955—, distinguished prof., 1967—, chmn. chemistry dept., 1965-68; asso.

chemist Ames Lab. Atomic Energy Commn., Ames, 1948-55, sr. chemist, 1955—, chief chemistry div., 1965-68, dir., 1968—. Cons. to various cos. NSF Sr. fellow U. So. Cal., also U. Utrecht (Netherlands), 1959-60. Fellow A.A.A.S.; mem. Am. Chem. Soc. (Kendall award in colloid chemistry 1966), Am. Phys. Soc., Faraday Soc., Phi Beta Kappa, Sigma Xi, Phi Kappa Phi, Phi Lambda Upsilon. Research and numerous publs. on instrumentation and theory for measuring wave length and damping of capillary waves; deduced surface compositions and molecular orientation from elec. properties of interfaces; application ultra-high vacuum field emission and flash filament techniques to catalytic reactions. Home: 2030 McCarthy Rd., Ames, Ia. 50010.*

HANSEN, Wilhelm Emil Leopold, German mech. engr.; b. Gotha, Germany, Aug. 28, 1832; s. Peter Andreas and Lina (Braun) H.; student mech. engring. U. Göttingen (Germany), U. Berlin; m. Ida Hornbostel, 1863; 1 son, 2 daus. Built factory for manufacture machines for gear wheels, Gotha, 1861, added iron foundry, produced steam engines and turbines, 1863. Invented movable furnace for drying casting molds in foundry, control for variable expansion in steam engines, process for calculating fall coefficients in hydraulic measurements. Died Gotha, Oct. 14, 1906.

HANSEN, William W(ebster), Am. physicist; b. Fresno, Cal., May 27, 1909; s. William George and Laura Louise (Gillogly) H.; m. Betsy Ross, Oct. 18, 1938; A.B., Stanford, 1929, Ph.D., 1932. Instr. in physics, Stanford, 1930-34; Nat. Research fellow, 1932-34; asst. prof. physics, Stanford U., 1935-37; asso. prof., 1937-42; prof., from 1942; research engr. Sperry Gyroscope Co., Garden City. N.Y. cons. to Nat. Def. Research Com., 1941-45. Recipient Liebmann prize for work in electromagnetic theory I.R.E., 1945. Fellow Am. Phys. Soc., I.R.E., A.A.A.S.; mem. Nat. Acad. Scis., Am. Assn. Physics Tchrs., Phi Beta Kappa, Sigma Xi. Contbr. articles to sci. jours. Worked on accelerating electrons by electromagnetic resonators, 1936-37 (led to devel. klystron); developed methods of microwave measurements; antenna design; radar design, especially Doppler radar, gun-laying radar; blind-landing microwave systems, during World War II; design and devel. linear electron accelators by means of klystrons, after World War II; worked with F. Bloch in devel. nuclear magnetic resonance. Died May 23, 1949.

HANSON, Adolph Melanchton, Am. surgeon; b. St. Paul, Minn., Sept. 11, 1888; s. Martin Gustav and Caroline (Runice) H.; student Red Wing (Minn.) Sem., 1903-07, Hamline U., 1907-08, U. of Minn., 1908-09; M.D., Northwestern U., 1911; grad. study, Sch. of Neurosurgery, U. of Pa., 1917; A.B., St. Olaf Coll., Northfield, Minn., 1922, M.A., 1923; m. Marie Lucile Boxrud, Nov. 26, 1914; children—Jane Lucile, Anne Marie, Adolph Martin, Patricia. Began practice at Red Wing, 1911; 1st house surgeon City Hosp., Seattle, Wash., 1912; in practice at Faribault, Minnesota, since August 1912; condr. of the Hanson Research Laboratory. Mem. A.M.A., Am. Chem. Soc., Assn. Mil. Surgeons U. S., Assn. for Study Internal Secretions, Am. Med. Editors and Authors Assn. Awarded 1st prize for work on parathyroid gland, Minn. Soc. Internal Medicine, 1927. Author: Practical Helps in the Study and Treatment of Head Injuries, 1925; also chapter, Management of Gunshot Wounds of the Head and Spine, in Vol. XI, The Med. Dept. of The U. S. Army in the World War. Contbr. numerous articles on parathyroid and thymus glands, head, spine and superficial nerve surgery, etc. Discoverer parathyroid, thymus, and pineal gland active extracts; inventor of dural separator and bone elevator used in brain surgery; research in glands of internal secretion, since 1922. Home: 16 2d Av. N.W. Office: Hanson Research Laboratory, Faribault, Minn.

HANSON, Alden Wade, Am. chemist; b. Jennings, Mich., June 19, 1910; s. Yorgan and Goldie (Fairchilds) H.; student Mich. State Normal Coll., 1928-29; B.S., Alma Coll., 1934; m. Helen Bennett, Feb. 22, 1930; children—Peter W., Helen L. (Mrs. Robert A. Brady), Chris A., Alden B. With Dow Chem. Co., 1934—, groups leader, Midland, Mich., 1942-54, dir. Nuclear and Basic Research Lab., 1954—. Cons., to regents U. Cal. at Livermore, 1956-62. Mem. A.A.A.S., Am. Chem. Soc., Am. Nuclear Soc., Am. Soc. for Testing Materials, Am. Inst. Chem. Engrs., Midland Coatings Soc., Sci. Research Soc. Am. Research and numerous patents on method of continuous cycle formation, compositions and processes in plastics and chemicals, mech. apparatus, oil shale recovery, well processes. Home: 1605 W. St. Andrews Dr. Office: 1704 Bldg., Midland, Mich. 48640.*

HANSON, Alfred Olaf, Am. physicist; b. Braddock, N.D., Sept. 26, 1914; s. George Olaf and Sina P. (Farness) H.; B.S. in Edn., U. N.D., 1936, M.S. in Math., 1938; Ph.D. in Physics, U. Wis., 1942; m. Elisabeth Marie Miller, May 16, 1942; children—Andrew J., Donald F., Ardith Marie, Craig D. Tchr., Pine River (Minn.) High Sch., 1936-37; grad. asst. U. N.D., Grand Forks, 1937-38, U. Wis., Madison, 1938-43; scientist Los Alamos (N.M.) Sci. Lab., 1943-45; mem. faculty U. Ill., Urbana, 1946—, prof. physics, 1951—. Cons. Naval Research Lab. Fulbright scholar to Inst. Physics, Turin, Italy, 1955-56; Fulbright lectr., Sao Paulo, Brazil, 1960. Fellow Am. Phys. Soc.; mem. Sigma Xi. Research in devel. and use of fast neutron source and applications to fast neutron research, patentee; also photonuclear studies with betatrons, 1946—. Home: 707 W. Iowa St., Urbana, Ill.*

HANSON, Clarence Herman, Am. research agronomist; b. DeKalb, Ill., June 20, 1913; s. Charles A. and Anna E. (Johnson) H.; B.S., U. Minn., 1940; M.A., U. Mo., 1942; Ph.D., N.C. State Coll., 1951; m. Gertrude E. Schmidt, July 5, 1941; children—Elizabeth Ann (Mrs. Richard H. Phillips), James C. Research agronomist U. S. Dept. Agr., N.C. State Coll., Raleigh, 1946-58; research leader alfalfa investigations crops research div. Agrl. Research Service, Beltsville, Md., 1958—. Chmn. Nat. Certified Alfalfa Variety Review Bd., 1962—. Fellow Am. Soc. Agronomy; mem. Am. Dehydrators Assn. (research council 1965—), Am. Genetic Assn., Sigma Xi, Alpha Zeta, Gamma Sigma Delta. Research, publs. on devel. disease and insect resistance, breeding procedures, plant estrogens, inheritance studies, alfalfa and lespedeza, cleistogamy. Home: 10726 Kinloch Rd., Silver Spring, Md. 20903. Office: Crops Research Div., Agrl. Research Service, U. S. Dept. Agr., Beltsville, Md. 20705.*

HANSON, David Lee, Am. mathematician; b. Mpls., July 14, 1935; s. Gordon Carl and Martha (Van Metre) H.; B.S., Mass. Inst. Tech., 1956; M.A., Ind. U., 1959, Ph.D., 1960; m. Alison Warren Robbins, June 1, 1956; children—Linda, Sharon, Scott, Carol. Staff mem. IBM Research Center, Yorktown Heights, N.Y., 1960-61; staff mem. Sandia Corp., Albuquerque, 1961-63; asso. prof. statistics U. Mo., Columbia, 1963—, prof. statistics and math. 1967—. Fellow Inst. Math. Statistics; mem. Am. Math. Soc., Math. Assn. Am., Am. Statis. Assn., Sigma Xi. Asso. editor Annals Math. Statistics. Research in ergodic theory, Markov processes, and probabilistic behavior of weighted sums of random observations. Home: 1112 W. Rollins Rd., Columbia, Mo. 65201.*

HANSON, Earle William, Am. plant pathologist; b. Wheaton, Minn., Oct. 18, 1910; s. Edwin W. and Esther (Lundquist) H.; B.S., U. Minn., 1933, M.S., (Firestone fellow) 1939, Ph.D., 1942; m. Maryan McIntosh, Aug. 3, 1941; children—Paula Jean, Ruth Ann. Seed analyst Minn. Seed Testing Lab., St. Paul, 1930-33; asst. plant pathologist U. Minn., 1934-35; pathologist Bur. Plant Industry, Soils & Agrl. Engring., U. S. Dept. Agr., St. Paul, 1937-46, Agrl. Research Service, U. S. Dept. Agr., Madison, Wis., 1946-51; faculty U. Wis., Madison, 1951—, prof., 1956—. Mem. Am. Phytopath. Soc. (editor 1955-58, pres. N.Central div. 1964-65), Am. Grassland Council (dir. 1962-65, OFCL. rep. 10th Internat. Congress, Helsinki, Finland, 1966), Am. Inst. Biol. Scis., Am. Soc. Agronomy, Crop Sci. Soc. Am., Nat. Geog. Soc., Sigma Xi, Alpha Zeta, Gamma Sigma Delta, Phi Sigma, Gamma Alpha. Cooperated in devel. and release of disease resistant, high yielding varieties of crop plants; developed methods for evaluating disease reactions of cereal and forage crops; research on nature of host-parasite relationships and effects of environment on disease devel. Tested fungicides as seed treatments and foliage protectants; worked with project leaders of U. S. Dept. Agr. in correlating forage improvement investigations on regional and nat. basis. Home: 4016 Hiawatha Dr., Madison, Wis. 53711.*

HANSON, George Henry, Am. chem., nuclear engr.; b. Alpena, Mich., Jan. 26, 1918; s. Anton Roine and Anna (Halvorsen) H.; B.S. in Chem. Engring., U. Mich., 1939, M.S. (Simon Mandlebaum scholar, U., Tau Beta Pi fellow), 1940, Ph.D. (Ethyl Gasoline Corp. fellow), 1942; m. Jean Marie Wahl, May 19, 1945. With research and devel. dept. Phillips Petroleum Co., 1942—, sr. chem. engr., group leader, asst. sect. chief, research div., Bartlesville, Okla., Borger, Tex., 1942-51, sr. engr., sect. chief, sr. reactor engring. scientist atomic energy div. AEC's Nat. Reactor Testing Sta., Idaho Falls, Ida., 1951—. Phillips Co. engr. on loan Oak Ridge Nat. Lab., 1947-48; guest lectr. Oak Ridge Sch. Reactor Tech., 1957; rep. Phillips Co. on various nuclear study groups. Mem. A.A.A.S., Am. Inst. Chem. Engrs., Am. Nuclear Soc., Am. Chem. Soc., Sigma Xi, Tau Beta Pi, Phi Lambda Upsilon, Iota Alpha, Phi Kappa Phi, Phi Eta Sigma. Patentee, publs., extensive research in nuclear test reactors; devel. butadiene process for synthetic rubber manufacture; inventor (with Brown, Katz) Convergence Pressure system; pioneer in engring. studies of application of atomic power to central sta. elec. power plants. Home: 444 7th St., Idaho Falls 83401. Office: P.O. Box 2067, Idaho Falls, Ida. 83401.*

HANSON, Harold Palmer, Am. physicist; b. Virginia, Minn., Dec. 27, 1921; s. Martin Bernhard and Elvida Elaine (Paulsen) H.; B.S., Superior (Wis.) State Coll., 1942; M.S., U. Wis., 1944, Ph.D., 1948; m. Mary Jean Stevenson, June 22, 1944; children—Steven Bernard, Barbara Jean. Faculty, U. Fla., 1948-54; faculty U. Tex., 1954—, prof. physics, 1961—, chmn. dept., 1962—; summer research physicist Lincoln Labs., Mass. Inst. Tech., 1953, Gen. Atomic Co., San Diego, 1964; Fulbright research scholar, Norway, 1960-61. Mem. Am. Phys. Soc., Sigma Xi, Sigma Pi Sigma. Contbr. articles to profl. jours. Research on X-ray spectroscopy, electron diffraction, atomic structure. Home: 2903 Clarice Ct., Austin, Tex. 78731.*

HANSON, Horst, German biochemist; b. Lauchhammer, Germany, May 6, 1911; s. Alfred and Elsa (Stockmar) H.; Dr.med., U. Halle (Germany), 1934, Dr.med.habil., 1941; m. Else Schroder, Feb. 26, 1938; children—Jurg, Heike (Mrs. Messerschmidt). With U. Halle (Germany), 1935-39, 46—, prof. physiol. chemistry, 1948—, dir. Inst. Physiol. Chemistry; sci. coworker Reichsgesundheitsamt Berlin (Germany), 1943-45; scientist, capt. Luftfahrtmedizininisches Forschungs Institut Berlin, 1946-48. Mem. German Acad. Scientists Nature, Saxon Acad. Scis. Leipzig. Research, numerous publs. on occurrence, specificity, activity, and distbn. of proteolytic enzymes, work on peptidases and peptide analysis. Home: 93 Strasse der DSF, Halle, Saale, Germany.

HANSON, Lars A., Swedish physician, immunologist; b. Naverstad, Sweden, Aug. 10, 1934; s. Hanson C. O. and Frideborg (Johannesson) H.; M.D., U. Göteberg, 1961; m. Monika Tunbäck, Aug. 15, 1958; children—Björn, Petra. Amanuens dept. med. biochemistry U. Göteborg (Sweden), 1954-55, dept. bacteriology, 1955-61, asst. prof. immunology, 1961-62, asso. prof., cons. pediatric clinic, 1965—; asst. research fellow Pasteur Inst., Paris, 1958; research asso. Rockefeller Inst., N.Y.C., 1962-63; asst. prof. immunology Karolinska Inst., Stockholm, 1963-65. Research, publs. on immunology of milk proteins, structure of antibodies in serum, milk and urine, immunity in urinary tract infections in childhood; devel. immunodiffusion methods, neonatal immunity. Home: 2 Västergatan, Göteborg C, Sweden.*

HANSON, Lyle Eugene, Am. microbiologist; b. Sarona, Wis., Oct. 2, 1920; s. Fred E. and Marion (Bergquist) H.; Ph.B., Northland Coll., 1942; D.V.M., Mich. State U., 1950; M.S., U. Ill., 1953, Ph.D., 1957; m. Ruth Allene Magruder, June 18, 1945; children—Bruce Lloyd, Karen Ruth, Craig Lyle, Jane Eileen. Faculty, Coll. Vet. Medicine U. Ill., Urbana, 1951—, prof. and head of the department of veterinary pathology and hygiene, 1961—. Mem. Am. Vet. Med. Assn., A.A.A.S., Soc. Am. Microbiologists, Avian Pathologists, Phi Zeta, Sigma Xi. Contbr. numerous articles to profl. jours. Research on investigation of properties and reactions produced in hosts by viruses of Newcastle disease, laryngotracheitis, fowlpox, duck hepatitis and infectious bronchitis; study of leptospirosis in cattle, swine and wild animals. Home: 1306 E. Pennsylvania St., Urbana, Ill. 61801.*

HANSON, William Bert, Am. physicist; b. Warroad, Minn., Dec. 30, 1923; s. Bert and Viola (Carlquist) H.; B.Chem.Engring., U. Minn., 1944, M.S. in Physics, 1949; Ph.D. in Physics, George Washington U., 1954; m. Wenonah Ann Dahlquist, Mar. 14, 1946; children—Bryan, Craig, David Arthur, Karen Lee. Physicist, Nat. Bur. Standards, Washington, 1949-54; sr. mem. research lab. Lockheed Missiles and Space Co., Palo Alto, Cal., 1954-62; prof. atmospheric and space scis. S.W. Center for Advanced Studies, Dallas, 1962—. Mem. subcom. on ionospheres and radio physics NASA, 1963—; U. S. Nat. Com. commn. IV Union Radio Sci. Internal.; Arecibo Evaluation Panel. Mem. Am. Geophys. Union. Author: (with others) Satellite Environment Handbook, 1961; also articles. Research on measurement properties second-sound in liquid helium, first calculations electron and ion temperatures in earth's ionosphere, measured ion concentrations and temperatures and neutral particle densities using rockets and satellites; theoretical description distbns. various constituents high atmosphere earth.

HANSON, William Roderick, Am. zoologist; b. Hanly, N.D., Nov. 25, 1918; s. John Hiram and Katherine (Munro) H.; student N.D. State Tchrs. Coll., 1937-40; B.A., U. Mont., 1943; Ph.D., Okla. State U., 1953; postgrad. U. Cal. at Los Angeles, N.C. State U.; m. Helen Louise Bordley, Aug. 26, 1946; children—Roderick, Lynne, Keith. Zoologist, N.D. Game and Fish Dept., 1948-50, Ariz. Game and Fish Dept., 1953-55, Ill. Natural History Survey, 1955-59; faculty Cal. State Coll., Los Angeles, 1959—, asso. prof., 1964—. Cons. Nat. Systems Corp. Fulbright lectr., Finland, 1967-68. Mem. Ecol. Soc. Am., Brit. Ecol. Soc., Am. Ornithologists Union, Am. Soc. Mammalogists, Wildlife Soc., A.A.A.S., Am. Inst. Biol. Scis. Research, publs. on relations of pheasants to vegetative types, food and cover plants of bobwhite quail in relation to land use by man, sex and age ratios of populations in relation to productivity, survival, abundance; estimating population density of animals. Home: 1309 Fleetwell Av., West Covina, Cal. 91790. Office: 5151 State College Dr., Los Angeles 90032.*

HANSTEEN, Christopher, Norwegian astronomer, physicist; b. Oslo, Norway, Sept. 26, 1784; student math. U. Copenhagen; dir., founder astron. obs., Oslo, 1833, magnetic obs., Oslo, 1838; joined faculty U. Christiana (now Oslo), 1814, became prof. astronomy and applied math., 1816; sent to Siberia by govt. to make observations, 1828-30; directed measurement of meridian arc between Atjik and Fuglenas, 1845-60. Fellow Royal Soc., 1839. Author: a treatise on geometry, 1835, on mechanics, 3 vols., 1834-38; Souvenirs d'un voyage en Sibérie, 1857, German edit., 1854; A Scientific Account of the Voyage, 1863; Observations de l'inclination magnétique faites pendent les anées, 1855 à 1864, 1865; Sur les variations séculaires du magnétism, 1865. Mem. French Acad. Scis. Research in terrestrial magnetism. Discovered daily variation in horizontal magnetic intensity, 1821. Died Oslo, Apr. 11, 1873.

HANSTEIN, Johannes Ludwig Emil Robert von, see von Hanstein, Johannes Ludwig Emil Robert.

HANTUSH, Mahdi Salih, Iraqi hydrologist; b. Hit, Iraq, Dec. 12, 1921; s. Salih and Rima (Ismail) H.; A.B. in Engring., Am. U., Beirut, Lebanon, 1942; M.S. Irrigation Engring., U. Cal. at Berkeley, 1945-47; Ph.D. in Civil Engring., U. Utah, 1949; m. Iqbal A. Al-Husaini, Aug. 7, 1953; children—Hudda, Layla, Ahmad Ameen, Suha, Muhammad Dhia. Supt., Iraq Irrigation Dept., 1942-44, asst. engr., 1950; faculty Coll. Engring., Baghdad, Iraq, 1951-53, dean, 1956-58, prof. hydrology, 1965—; spl. lectr. U. Utah, 1953-54; prin. hydrologist N.M. Inst. Tech., 1954-55, sr. hydrologists, prof. hydrology, head ground water hydrology dept., 1958-65. Dir. sci. missions Iraq Ministry Edn., 1951-53; cons. devel. bd. Govt. of Iraq., 1955-58. Fellow A.A.A.S.; mem. Am. Soc. C.E., Am. Geophys. Union, Soc. Math. and Physics Iraq. Contbg. author Advances in Hydroscience, vol. 1, 1964. Research, publs. on theory of leaky aquifers, hydraulic properties of aquifers, advances in unsteady motion of ground water. Home: 4 Jeem/10, Waziriah, Baghdad, Iraq.*

HANTZSCH, Arthur Rudolf, German chemist; b. Dresden, Germany, Mar. 7, 1857; s. Rudolf Georg and Clara (Bähr) H.; student chemistry Tech. U., Dresden; doctorate Würzburg, Germany, 1880; M.D. (hon.), Leipzig, Germany; Hon.dr.engring., Dresden, Germany; m. Katharina Schilling, 1883; 2 sons, 1 dau.; m. 2d, Hedwig Steiner, 1911. Asst., Berlin; asst. Phys. Chemistry Inst., U. Leipzig, joined faculty, 1883; became prof. organic chemistry Fed. Tech. U. Zurich (Switzerland), 1885; named prof., Würzburg, 1893, U. Leipzig, 1903; emeritus, 1928. Mem. Saxon Acad. Scis., Acad. Scis. Göttingen, Leopoldina. Author: Grundriss der Steroechemie, 2d edit., 1904; (with S. Reddelien) Diazoverbindungen, 2d edit., 1921; Theorie der Ionogenen Bindungen, 1923. Research and numerous publs. on synthesis of pyridine and its derivatives, stereochemistry of nitrogen compounds, phys. methods of structure analysis, nitrogen compounds, tautomeric rearrangements, acids and pseudo-acids; synthesis of heterocyclic compounds, especially thiazole (later important for vitamin B1 and penicillin); conductivity of organic compounds. Died Dresden, Mar. 14, 1935.

HANTZSCH, (Gustav Robert) Viktor, German geographer; b. Dresden, Germany, May 10, 1868; s. Adolf and Emma (Jencke) H.; student geography and history, Leipzig, Germany, 1892-95, Ph.D., 1898. Commd. to catalog map collection Dresden Royal Library, 1899, later dir. map collection; ret., 1902. Author monograph on Sebastian Münster. Bibliog. studies in hist. geography and cartography; one of 1st to support Ratzel's ideas of anthropogeography and to create relationship between modern geography and history. Died Dresden, Nov. 12, 1910.

HANTZSCHE (Willy) Walter, German mathematician; b. Koselitz, Germany, Dec. 5, 1912; s. Huge Hermann and Helena (Mehnert) H.; student math. and physics Tech. U. Dresden, Germany, 1932-36; student U. Hamburg (Germany) Dr.sc.nat., U. Halle (Germany), 1936; m. Christa Rothardt, 1942. Became aerodynamist AGO-Flugzeugwerken, Oscherslében, Germany, 1937; mem. staff Inst. for Gas Dynamics, Aeros. Research Inst., Braunschweig, Germany, 1939-44; also lectr. Braunschweig. Research and numerous publs. on topology, applied math. especially gas dynamics, diffusion of compression waves, stationary under-sound and over-sound currents; topological studies of Euclidian spatial forms. Died Braunschweig, Mar. 29, 1944.

HÄNTZSCHEL, Walter Helmut, German geologist, paleontologist; b. Dresden, Germany, Nov. 16, 1904; s. Theodor Johannes and Minna (Müller) H.; student Tech. U., Dresden, 1924-29; D., 1932; m. Marianne Krausse, May 23, 1936; children—Maria, Kristine (Mrs. Volker Nolte). Chief Senckenberg Inst. Marine Geology, Wilhelmshaven, Germany, 1934-38; custodian for paleontology State Mus. Mineralogy and Geology, Dresden, 1938-45; head custodian, hon. prof. geology and paleontology Hamburg State Inst. Geology, U. Hamburg (Germany), 1949—. Mem. Deutsche Geologische Gesellschaft, Geologische Vereinigung, Paläontologische Gesellschaft, Internat. Assn. Sedimentologists. Author: Vestigia Invertebratorum et Problematica, 1965; also numerous articles. Research on geology and paleontology of Cretaceous Formation in Germany, actualistic geol. and paleontol. investigations of tidal flat deposits off coast North Sea, trace fossils from all over world. Home: 31 Brockdorffstr., 2 Hamburg 73, West Germany.*

HANZAWA, Masao, Japanese oceanographer; b. Sendai, Japan, Oct. 17, 1922; s. Shojiro and Shigeko (Shimizu) H.; M.S., Tohoku U., 1947, D.Sc., 1962; m. Motoko Yasukawa, Oct. 17, 1949; children—Masaki, Toshimichi Hanzawa. With Japan Meteorol. Agy., Tokyo, 1947—, chief maritime meteorology br., 1956-60, research ofcl. phys. oceanography and marine meteorology, 1960—; vis. prof. Scripps Instn. Oceanography, 1965. Profl. mem. com. promotion of scis. Dept. Edn., 1957, com. on marine sci. and technology Prime Minister's Office, Japanese Govt., 1962. Recipient Sankei Grand Prix, for book, 1964. Mem. Oceanographical Soc. Japan, Meteorol. Soc. Japan, Am. Geophys. Union, A.A.A.S. Author: (with H. Htakeyama, J. Nemoto, A. Suwa) Science of the Earth, 1961; also articles. Discovered inverse relationship between temperature anomalies of seasurface along U. S. and Japanese coasts in some years. Home: 6-6-7 OOmachi, Kamakura, Japan. Office: Japan Meteorol. Agy., 1-7, Ootemachi, Chiyoda-ku, Tokyo, Japan.

HANZE, Arthur Raymond, Am. chemist; b. Canton, O., Aug. 27, 1915; s. Edward E. and Eva (Gloss) H.; B.S., Washington and Jefferson Coll., 1938; Ph.D., Ohio State U., 1942, postgrad., 1942-43; m. Janice L. Willis, June 14, 1942; children—Stephen, Douglas. Research asso. Upjohn Co., Kalamazoo, Mich., 1944—. Mem. Am. Chem. Soc., Chem. Soc. London. Contbr. articles to sci. jours. Patentee, publs. synthesis vitamin A, folic acid, adrenocortical hormones, mental health drugs, nucleic acids, sugars. Home: 2632 Taliesin Dr., Kalamazoo 49001. Office: care Upjohn Co., Kalamazoo, Mich. 49001.*

HANZLIK, Paul John, Am. pharmacologist; b. Shueyville, Ia., July 24, 1885; s. Martin and Mary (Kreysa) H.; Ph.G., State U. Ia., 1902, Ph.C., 1908; A.B., U. Ill., 1908, A.M., 1911; M.D., Western Reserve U., 1912; m. Bertha Shimek, Aug. 1909; children—Harold, Dorothy. Demonstrator in pharmacology, Western Res. U., 1912-13, instr., 1913-15, asso., 1915-17, asst. prof., 1917-20, asso. prof., 1920-21; with Pharmacol. Inst. and Physico-Chem.-Biol. Inst., U. Vienna, 1913-14; prof. pharmacology Stanford U., 1921-50, prof. emeritus from 1950; cons. pharmacologist San Francisco Dept. Health from 1934, U. S. Dept. Agr., 1936-44, Fellow A.C.P., Am. Coll. Dentists (hon.), Am. Social Hygiene Assn.; mem. Am. Med. Assn., Am. Physiol. Soc., Soc. Pharmacology and Exptl. Therapeutics, Soc. Exptl. Biology and Medicine, A.A.-A.S., Com. on Research on Syphilis, Inc., Internat. Assn. Dental Research, Sigma Xi, Phi Rho Sigma, Alpha Omega Alpha. Author: Actions and Uses of the Salicylates and Cinchophen in Medicine, 1927; (with Prof. T. Sollman), Fundamentals of Experimental Pharmacology, 1928, 2d ed., 1939; Handbook of Accepted Remedies, Symptoms and Treatment of Poisoning, Diagnostic Procedures and Miscellaneous Information, 3d edit., 1940. Contbr. to med. and scientific periodicals. Died Feb. 1, 1951.

HAPEL-LACHENAIE, Thomas-Luc-Augustin, naturalist; b. Argentan, France, Apr. 2, 1760; royal veterinarian; became head pharmacist, Guadeloupe, 1790; mem. French Acad. Scis. Research on culture and products of Guadeloupe; made analyses of spring waters, also meteorol. and climatological observations; conducted bot. studies of sugar cane of Antilles; investigated epizootic diseases. Died Sainte-Rose, Guadeloupe, May 14, 1808.

HAPPOLD, Frank Charles, English biochemist; b. Barrow, Eng., Sept. 23, 1902; s. Henry and Emma (Ley) H.; D.Sc., Ph.D., Manchester, Eng.; m. Margaret Smith, 1926; 1 son, 1 dau. Prof. biochemistry Sch. Medicine, Leeds, Eng., 1946—. Leverhulme fellow, Harvard. Research on applications of biochemistry to microbiology; enzymology, especially oxidation of amino acids; distinguished (with Anderson, McLeod, Thomson) 3 types of specific etiologic agt. of diphtheria (gravis, mitis, intermedius), 1931. Home: 6A Bainbridge Rd., Leeds 6, Eng.

HARADA, Tokuya, Japanese microbiologist; b. Osaka, Japan, Oct. 14, 1918; s. Komanosuke and Umeko (Nishimatsu) H.; B.D., Tokyo U., 1941, D.Sc., 1954; m. Michiyo Honda, Nov. 16, 1945; children—Miyoko, Akira. Faculty, Inst. Sci. and Indsl. Research, Osaka U., 1943—, prof., 1966—; lectr. Kobe Womens Coll., 1951-66. Recipient medal Agrl. and Chem. Soc. of Japan, 1958. Research and numerous publs. on gen. and applied microbiology; discovered succinoglucan (a bacterial polysaccharide). Home: 4-460 Funao-Nishi, Sasai, Osaka, Japan. Office: 3-145 Higashi-Asakayama, Sakai, Osaka, Japan.*

HARANGHY, László, Hungarian pathologist, gerontologist, biologist; b. Debrecen, Hungary, Aug. 10, 1897; s. György and Maria (Gulyas) H.; diploma in medicine, Med. U. Debrecen and Budapest, 1923; privat-docent, Pecs, Hungary, 1932; m. Leona Csór, Jan. 10, 1925; 1 dau., Márta (Mrs. André Rimon). Prosector head physician, Baja, Hungary, 1923-40; prof. pathology and forensic medicine, Kolozsvar, 1940-45, Marosvasarhely, Hungary, 1945-52; prof., Budapest, Hungary, 1952—; sci. dir. Hungarian Research Inst. Gerontology, Budapest, 1963—. Vice pres. forensic med. council Hungarian Sci. Council on Health, 1955—; pres. morphological and gerontological council Ministry of Health. Corr. mem. Hungarian Acad. Scis. (pres. med. history council, named Eminent Worker of Upper Edn., Eminent Physician); mem. Hungarian, German path. socs., Internat., Hungarian, Am. gerontological socs. Author: Gerontological Studies in Hungarian Centenarians, 1965. Research, publs. on diseases of reticuloendothelial system, origin of tumors, alterations and pathology of old age, biology of aging on invertebrates. Home: 72/c Németvölgyi, Budapest. Office: 93. Üllöi, Budapest, Hungary.*

HARASHIMA, Akira, Japanese theoretical physicist; b. Tokyo, Japan, Feb. 18, 1908; grad. Tokyo U., 1930; Sc.D., 1942. Asst. prof. Tokyo U.; later prof. 1st Higher Sch.; asst. prof. Kyushu (Japan) U., prof., 1942—; prof. Tokyo Inst. Tech., 1949—. Author: Dynamics; Force; Thermo Dynamics, other works. Research, publs. on friction and function of lubricant, theory on liquid structure; specialist statis. dynamics.

HARBERGER, Arnold C(arl), Am. economist; b. Newark, July 27, 1924; s. Ferdinand C. and Martha (Bucher) H.; student Johns Hopkins, 1941-43; M.A., U. Chgo., 1947, Ph.D., 1950; m. Ana Beatriz Valjalo, Mar. 15, 1958; children—Paul Vincent, Carl David. Asst. prof. Johns Hopkins, 1949-53; faculty U. Chgo., 1953—, prof. econs., 1959—, chmn. dept., 1964—, dir. Center Latin Am. Econ. Studies, 1965—. Vis. prof. Mass. Inst. Tech. Center Internat. Studies, New Delhi, 1961-62, Econ. Devel. Inst., World Bank, 1965; cons. to numerous govt. commns., depts. Guggenheim fellow, 1958; Fulbright scholar, 1958; Faculty Research fellow Social Sci. Research Council, 1952-55. Mem. Am. Econ. Assn., Econometric Soc., Royal Econ. Soc., Phi Beta Kappa. Editor: The Demand for Durable Goods, 1960. Publs. on integration of classical and Keynesian approaches to analysis of currency depreciation, measurement of cost to economy of various classes of distortions, theory of incidence of corp. income tax, gen.-equilibrium theory of welfare cost, dynamics of inflation, techniques of project appraisal. Home: 4840 S. Greenwood Av., Chgo. 60615.*

HARBERT, Fred, Am. physician; b. Detroit, Jan. 27, 1905; s. Frederick and Albertina (Burkshtaler) H.; A.B., Wayne State U., 1928, M.B., 1928, M.D., 1929; M.Sc., U. Pa., 1940, D.Sc., 1942; m. Frances Marie Clark, Jan. 8, 1927; children—E. Lorraine (Mrs. Fred Morey), Winifred L. (Mrs. George Keane), Barbara E. (Mrs. Lawrence Schauber), Mary J. (Mrs. John Weightman), Carol A. (Mrs. John Singleton), E. Jean. Asso. prof. U. Pa. Grad. Sch. Medicine, 1948-51, vis. lectr., 1953-64; prof. otolaryngology, head dept. Jefferson Med. Coll., Phila., 1954—. Mem. spl. med. adv. group VA, D.C., 1965—. Diplomate Am. Bd. Ophthalmology, Am. Bd. Otolaryngology. Fellow Am. Acad. Facial Plastic and Reconstructive Surgery, A.C.S., Internat. Coll. Surgeons, Am. Otorhinologic Soc. for Plastic Surgery (pres. 1964), Am. Laryngol., Rhinol. and Otol. Soc. (v.p. 1968), Am. Otol. Soc.; mem. Keorotology Group (pres. 1968), Am. Acoustical Soc., Otosclerosis Study Group, Am. Acad. Ophthalmology and Otolaryngology. Research and publs. on field otolaryngology, audiology; basic contbns. field auditory masking and adaption, especially related to Bekesy Audiometry. Home: 931 Waverly Rd., Bryn Mawr, Pa. 19010. Office: Jefferson Med. Coll. Hosp., Phila. 19107.*

HARBISON, John Stewart, Am. entomologist; b. Chenango, Pa., Sept. 29, 1826; came to Cal., 1854; purchased half-interest William Buck Apiary, nr. Sacramento, 1855; became partner R. G. Clark, bee keeper, San Diego County, Cal., 1864; moved to Harbison Canyon, San Diego County, 1874; owned 6 apiaries and 2000 hives, 1876. Author: The Beekeeper's Directory or the Theory and Practice of Bee Culture, 1861. Invented Harbison hive, 1848, improved it (later known as California hive), 1857; invented 2 pound section box, 1857-58; introduced honeybee to Sacramento Valley, 1856; built 1st honey house in Cal., 1880. Died San Diego, Oct. 12, 1912.

HARBITZ, Hans Fredrik, Norwegian surgeon; b. Oslo, Norway, Feb. 1, 1900; s. Francis and Alma (Borchgrevink) H.; M.D., U. Oslo; m. Elise Bache, 1927; children—Francis, Hans Fredrik, Thorstein. Instr. surgery, and pathology, asst. in surgery Ulleval Hosp., Oslo; dir., chief surgeon Regional Hosp. of More and Romsdal, Molde; head surg. service Hosp. of Aker, Oslo, 1938-47; prof. surgery U. Oslo, 1947- 5. Mem. Norwegian Med. Assn., Svenska Läkaresellskapet, Svensk Kirurgisk Förening, Internat. Assn. Surgery. Research and publs. on thyreotoxin, abortions, anaesthesia in operative treatment of fractures, postoperative thromoembolism, plastic thorax in pulmonary Tb and Tb empyemae, hemorrhoids and varices, parathyreodism, cancer of rectum, fatal appendicitis, hormone treatment of chest cancer, diseases of neck and chest. Address: Aker Sykehus, Oslo, Norway.

HARBOTTLE, Garman, Am. nuclear chemist; b. Dayton, O., Sept. 25, 1923; s. William Edwin and Susan (Garman) H.; B.S., Cal. Inst. Tech., 1947; Ph.D., Columbia, 1949; m. Naomi Perkiss, June 10, 1949; 1 dau., Laura. Staff, Brookhaven Nat. Lab., Upton, L.I., N.Y., 1949—, scientist, 1957—; dir. div. research and labs. IAEA, Vienna, Austria, 1965—. AEC fellow, 1951-52; Guggenheim fellow, 1957-58. Mem. Sigma Xi, Phi Lambda Upsilon, Tau Beta Pi. Research and numerous publs. on recoil atom reactions in crystals, also annealing reactions of such atoms. Home: Crane Neck Rd., Old Field, N.Y. 11785. Office: Brookhaven Nat. Lab., Upton, N.Y. 11973.*

HARCOURT, Augustus George Vernon, English chemist; b. London, Dec. 24, 1834; s. Frederick E. and Maria (Tollemache) H.; student Balliol Coll., Oxford, Eng., 1854-58, Christ Church, 1859; m. Rachel Mary Bruce, 1872; 2 sons, 8 daus. Became Lee's reader chemistry Christ Church, 1859, tutor, 1864-1902. Became met. coal-gas referee, 1872; named Bakerian lectr., 1895. Fellow Royal Soc., 1868 (mem. council 1878-80); mem. Chem. Soc. (a sec. 1865-73, became pres. 1895), Brit. Soc. (gen. sec. 14 years). Determined amount of oxygen absorbed by potassium and sodium; research on rate of chem. changes which helped (with Guildberg and Berthelots) to establish quantitatively Berthelot's law of mass action; proved (with William Esson) velocity of change varied directly with quantities of reacting substances; attempted to purify coal gas from sulphur compounds and devised sulphur test; introduced Pentane lamp for ofcl. standard of light; invented inhaler for adminstrn. chloroform; studied effect of low temperature on chem. reaction. Died 1919.

HARCOURT, William Vernon, English natural philosopher; b. Sudbury, Eng., 1789; s. Edward Harcourt; B.A., Christ Church, Oxford, Eng., 1811, M.A., 1814; m. Matilda Mary Gooch, 1824; children—Edward William, William Vernon, 5 daus. Served in navy 5 years; successively became clergyman Bishopthorpe, Eng., 1811, canon of York, Eng., 1824, rector of Wheldrake, Eng., 1824, Bolton Perry, Eng., 1837. Mem. Brit. Assn. (became gen. sec. 1831, pres. 1839), Yorkshire Philos. Soc. (1st pres.). Fellow Royal Soc., 1881. Research on effect of heat on inorganic compounds; collected glasses of definite and mutually compensative dispersions to make achromatic combinations. Died Apr. 1871.

HARCUM, Eugene Rae, Am. psychologist; b. Cambridge, Md., Mar. 1, 1927; s. Eugene Payten and Myrtle (Larmore) H.; B.S., Coll. William and Mary, 1950; M.A., Johns Hopkins, 1952; Ph.D., U. Mich., 1955; m. Phoebe Carroll Martin, Aug. 30, 1952; children—Sarah Lois, James Payten. Grad. research asst., research asso. U. Mich., 1955-58; faculty Coll. William and Mary, Williamsburg, Va., 1958—, prof. psychology, 1965—. Mem. com. on vision Nat. Acad. Sci.-NRC, 1958—. Fellow A.A.A.S.; mem. Am., Eastern, Va. psychol. assns., Am. Assn. U. Profs., Va. Acad. Sci. (J. Shelton Horsley award 1964), Phi Beta Kappa, Sigma Xi. Author: Reproduction of Linear Visual Patterns Tachistoscopically Exposed in Various Orientations, 1964. Asso. editor: Perceptual and Motor Skills, 1966. Contbr. articles to profl. jours. Demonstrator strategy of observer being important in perception and memorization of visual patterns appearing in different temporal and spatial configurations; interpreted differential accuracy in reprodn. of individual elements within the patterns, in case of spatial patterns, as alternative to notion of simple structural dominance of one cerebral hemisphere. Home: 103 Plantation Dr., Williamsburg, Va. 23185.*

HARDAWAY, Robert Morris, III, Am. surgeon; b. Camp John Hay, P.I., Jan. 9, 1916; s. Robert M. and Olive (Gray) H.; A.B., U. Denver, 1936; postgrad. U. Colo. Med. Sch., 1935-37; M.D., Washington U., St. Louis, 1939; m. Lee Harkey, June 12, 1939; children—Robert Morris IV, Joan Elizabeth, Thomas Gray, Christopher Lee. Commd. 1st lt., U. S. Army, 1940, advanced through grades to col., 1960; chief surg. service U. S. Army Hosp., Ft. Belvoir, Va., 1950-54, 97th Gen. Hosp., Frankfort, Germany, 1954-58, Martin Army Hosp., Ft. Benning, Ga., 1958-60; dir. div. surgery Walter Reed Army Inst. Research, Washington, 1960-67; comdg. officer Frankfort (Germany) Med. Service Area, 1967—. Diplomate Am. Bd. Surgery. Fellow A.C.S.; mem. A.M.A., Microcirculatory Conf., Am. Assn. for Surgery of Trauma, Am. Coll. Angiology, Assn. Mil Surgeons U. S., Alpha Omega Alpha. Author: Syndromes of Disseminated Intravascular Coagulation with Special Reverence to Shock and Hemorrhage, 1966; also numerous articles. Initiated and developed concept that wide variety of clin. syndromes are due to transient coagulation in circulating blood, gen. surgery, especially surg. physiology, etiology of shock and related syndromes. Address: Hdqrs. 97th Gen. Hosp., APO N.Y. 09757.*

HARDAWAY, William Augustus, Am. physician; b. Mobile, Ala., Jan. 8, 1850; s. William Augustus and Mary (del Barco) H.; student Westminster Coll., U. Va.; M.D., Mo. Med. Coll., 1870; A.M., St. Louis U.; LL.D. Westminster Coll.; m. Lucy Nelson Page, Jan. 8, 1877. Practiced and taught medicine, St. Louis; established (with others) St. Louis Courier of Medicine, 1879. Mem. Am. Dermatol. Assn. (pres. 1885). Author: Manual of Skin Diseases, 1890. Editor (with Dr. L. B. Bangs) of American Text-Book of Genito-Urinary Diseases, Syphilis and Diseases of the Skin, 1898; Handbook of Cutaneous Therapeutics (with Joseph Grindon), 1907. First to describe prurigo nodularis (later called Hyde's disease), 1879; introduced use of elec. needle for removal of unwanted hair. Died Mar. 3, 1923.

HARDEN, Sir Arthur, English chemist; b. Manchester, Eng., Oct. 12, 1865; doctorate in chemistry, U. Erlangen (Germany), 1888. Lectr., demonstrator chemistry Owens Coll., Manchester, 1888-97; joined Jenner Inst. Preventive Medicine (later Lister Inst.), 1897, ret. as head of biochem. sect., 1930. Recipient (with von Euler-Chelpin) Nobel prize in chemistry, 1929; Davy medal, 1935. Fellow Royal Soc., 1909. Author several books including: Alcoholic Fermentation, 1911; (with F. C. Garett) Practical Organic Chemistry. Editor, Biochem. Jour. Demonstrated structure of zymose and discovered other fermentation enzymes; showed presence of inorganic phosphates increases speed of fermentation. Died June 17, 1940.

HARDER, Hermann, German mineralogist; b. Essen, Germany, Nov. 16, 1923; s. Hermann and Carola (Tiggemann) H.; Dissertation, U. Göttingen (Germany), 1956, Habilitation, 1956. Dozent, U. Braunschweig (Germany), 1956-59; prof. mineralogy and petrography U. Münster (Germany), 1959-66; prof. sedimentpetrography U. Göttingen, 1966—. Mem. Internat. Assn. Seimentology (German re.). Research, publs. on sedimentary rocks and deposits, geochemistry, precious stones. Office: Sedimentpetrographisches Institut der Universität, 13 Lotzestrasse, Göttingen, Germany.*

HARDER, Johann Jakob, Swiss physician; b. Basel, Switzerland, Sept. 7, 1656; prof., Basel; physician to several princes. Discovered lacrimal gland which is present in many animals (Harder's gland); one of 1st to write on physiology and toxicology. Died Basel, Apr. 28, 1711.

HARDIE, Robert Howie, astronomer; b. Lachine, Que., Can., Dec. 5, 1923; s. Robert Howie and Catherine (Campbell) H.; B.Sc., McGill U., 1945, M.Sc., 1946; Ph.D., U. Chgo., 1950; m. Frances Harriett Isley, Aug. 23, 1950; children—James Alexander, Robert Stephen. Came to U. S., 1946, naturalized, 1960. Fellow, Paris (France) Obs., 1950-51; vis. asst. prof. Ohio State U., 1951-53; astronomer Lowell Obs., Flagstaff, Ariz., 1953-55; faculty Vanderbilt U., Nashville, 1955—, prof. astronomy, 1963—, dir. Dyer Obs., 1961—. Mem. Assn. Univs. for Research in Astronomy (dir.-at-large 1965-68). Mem. Assn. Univs. for Research in Astronomy (dir.-at-large 1965-68). Research and publs. on stellar brightness, variable stars, star clusters, astron. instruments and photometry; co-discoverer Pluto's rotation. Home: Oman Dr., Brentwood, Tenn. 37027. Office: Dyer Obs., Vanderbilt University, Nashville 37203.*

HARDIN, Creighton A., Am. physician; b. Clinton, N.C., July 20, 1918; s. Louis Thomas and Norma (Alves) H.; B.A., U. Wis., 1940, M.D., 1943; M.S. in Surgery, U. Kan., 1950; m. Helen E. Tank, June 12, 1943; children—Mary, Creighton A. II, Daniel. Prof. surgery U. Kan. Med. Center, Kansas City, 1965—. Diplomate Am. Bd. Surgery, Am. Bd. Plastic Surgery; mem. Am. Surg. Soc., Soc. Univ. Surgeons, Western, Central surg. socs., N.Y. Acad. Scis., A.C.S. Research and numerous publs. on field vascular surgery; transplantation of tissues and organs. Home: 8229 Nall St., Prairie Village, Kan. 66208. Office: University of Kan. Med. Center, Kansas City, Kan. 66103.*

HARDIN, Garrett, Am. biologist; b. Dallas, Apr. 21, 1915; s. Hugh and Agnes (Garrett) H.; B.Sc., U. Chgo., 1936; Ph.D., Stanford, 1941; m. Jane Swanson, Sept. 7, 1941; children—Hyla (Mrs. Timothy Fetler), Peter, Sharon, David. Staff, Carnegie Inst. Washington, 1942-46; faculty U. Cal. at Santa Barbara, 1946—, prof. biology, 1956—; vis. fellow Cal. Inst. Tech., 1952-53. Vis. prof. U. Cal. at Los Angeles, 1961, at Berkeley, 1964. Mem. A.A.A.S., (president of the Western division 1968-69), Ecol. Soc., Eugenics Soc., Soc. for Gen. Systems, Phi Beta Kappa (sci. award com. 1964-67), Sigma Xi. Author: Biology: It's Principles and Implications, 1949; Nature and Man's Fate, 1959; Population, Evolution and Birth Control, 1964. Research and numerous publs. on competitive exclusion principle, logical basis on which abortion-on-demand can be justified. Home: 399 Arboleda Rd., Santa Barbara, Cal. 93105.*

HARDIN, James Walker, Am. botanist; b. Charlotte, N.C., Mar. 31, 1929; s. Henry Grady and Olive (Walker) H.; B.S., Fla. So. Coll., 1950; M.S., U. Tenn., 1951; Ph.D., U. Mich., 1957; m. Dorthy A. Struck, Dec. 21, 1957; children—Elizabeth Ann, Patricia Lynn, Catherine Louise. Instr., U. Mich., Ann Arbor, 1956-57; faculty N.C. State U., Raleigh, 1957—, asso. prof. botany and forestry, 1961—, curator Herbarium, 1957—. Pres., chmn. bd. trustees Highlands (N.C.) Biol. Station, Inc., 1963—; vis. prof. U. Va., 1962, 64. Mem. Am. Soc. Plant Taxonomists (Cooley award for best paper 1957), Bot. Soc. Am., Internat. Assn. for Plant Taxonomists, Soc. for Study Evolution, So. Appalachian Bot. Club (past pres.), Sigma Xi, Phi Kappa Phi. Author: Workbook for Woody Plants, 1960; General Plant Morphology, A Laboratory Guide, 1965; Poisonous Plants of North Carolina, 1961. Mem. editorial bd. Brittonia, 1964—. Research and publs. on poisonous and drug plants of N.C., flora of southeastern U. S., woody plants, herbaceous angiosperms, plant taxonomy. Home: 200 Furches St., Raleigh, N.C. 27607.*

HARDING, Denys Clement Wyatt, English psychologist; b. Lowestoft, Eng., July 13, 1906; s. Clement George and Harriet (Wyatt) H.; B.A., Cambridge (Eng.) U., 1928, M.A., 1931; m. Jessie M. Ward, Dec. 18, 1930; Investigator, research staff Nat. Inst. Indsl. Psychology, 1928-33; lectr. psychology London (Eng.) Sch. Econs., 1933-38; sr. lectr. psychology U. Liverpool (Eng.), 1938-45; prof. psychology U. London, 1945-68. Fellow Brit. Psychol. Soc. (past gen. sec.). Author: The Impulse To Dominate, 1941; Social Psychology and Individual Values, 1953; Experience into Words, 1963. Editor: Brit. Jour. Psychology, 1948-54. Research and publs. on rhythm in motor skills and literature; psychology of literary criticism; social conflict; motivation and social background of complex civilized pursuits. Home: Ashbocking Old Vicarage, Ipswich, Suffolk, Eng.*

HARDING, Karl Ludwig, German astronomer; b. Lauenburg, Germany, Sept. 30, 1765; s. Carl Ludwig and Louisa Dorothea Sabine (Cordes) H.; studied theology; 1 dau. House tutor to J. H. Schroeder who had own obs., Lileinthal, Germany, beginning 1779, insp. obs., 1800-05; became prof. astronomy (with Gauss), U. Göttingen (Germany), 1805. Mem. French Acad. Scis. Fellow Royal Soc., 1806. Author: Atlas novus coelestis, 1808-26; also articles. Research on rotation of Mercury, illumination of dark side of Venus, form and visibility of Saturn ring, asteroid Ceres I; discovered asteroid, Juno, 1804, 3 comets, 1813, 24, 32, also several variable stars; tested and corrected all known star positions. Died Göttingen, Aug. 31, 1834.

HARDISON, John Robert, Am. plant pathologist; b. Yakima, Wash., Jan. 12, 1918; s. Earl James and Genevieve (Somppi) H.; B.S., Wash. State U., 1939; M.S., U. Mich., 1940, Ph.D., 1942; m. Eleanor Marie Lynch, Oct. 9, 1937; children—John Robert, Catherine Ann, Mary Eleanor, Patrick James. Asst. in forage crops investigations Ky. Agrl. Expt. Sta., Lexington, 1942-44; research plant pathologist seed prodn. investigations forage and range research br. crops research div. Agrl. Research Service, U. S. Dept. Agr., Corvallis, Ore., 1944—. Mem. Am. Phytopath. Soc., Mycol. Soc. Am., Am. Soc. Agronomy, Sigma Xi. Contbr. numerous articles to profl. jours. Developed control of blind seed, ergot, seed nematode, and dwarf bunt diseases of grasses by cultural methods; practical control of grass rust diseases by nickel fungicides; chemotherapeutic control of flag smut and stripe smut by systemic chems.; clarified causal agts. of silver top. Home: 1125 S.W. Stopp Pl., Corvallis 97330. Office: Botany Dept., Ore. State University, Corvallis, Ore. 97331.*

HARDY, Sir Alister Clavering, English zoologist; b. Nottingham, Eng., Feb. 10, 1896; s. Richard and Elizabeth (Clavering) H.; M.A., Oxford U., 1921, D.Sc., 1938; LL.D. (hon.), U. Aberdeen, 1962; D.Sc. (hon.), Southampton U., 1962, Hull U., 1963; m. Sylvia Lucy Garstang, Dec. 3, 1927; children—Michael Garstang, Alisa Belinda (Mrs. John D. Farley). Asst. naturalist Ministry Agr. and Fisheries, 1921-24; chief zoologist Discovery Expdn. to Antarctic, 1924-28; prof. zoology and oceanography U. Coll. Hull, 1928-42; Regius prof. natural history U. Aberdeen, 1942-45; Linaere prof. zoology and comparative anatomy Oxford U., 1946-61, emeritus prof., hon. fellow Merton and Exeter colls. Knighted, 1957. Fellow Royal Soc., 1940; mem. Linnean Soc., Zool. Soc., Ecol. Soc., Soc. Exptl. Biology, Marine Biol. Assn. U.K., others. Author: The Open Sea, 2 vols., 1956, 59; The Living Stream, 1965; The Divine Flame, 1966; Great Waters, 1967. Research on oceanography, including diet of herring; invented automatic plankton recorder (used to make quantitative study of ocean plankton); studies on vertical migration and bioluminescence of plankton, insect drift in the atmosphere. Home: 7 Capel Close, Oxford, Eng.*

HARDY, Claude, French mathematician; b. Le Mans, France, possibly 1600; career in law. Author: Data d'Euclide, 1625. Prepared Greek edit. of Euclid, with Latin transl.; arbitrator for Descartes in dispute with Fermat. Died Paris, 1678.

HARDY, Godfrey Harold, English mathematician; b. Cranleigh, Eng., Feb. 7, 1877; s. Isaac and Sophia (Hall) H.; ed. Trinity Coll., Cambridge. Fellow, Trinity Coll., Cambridge, 1900-19, lectr., 1906-19, Cayley lectr., 1914-19; prof. geometry U. Oxford (Eng.), 1919-31; prof. pure math. U. Cambridge, 1931-42; vis. prof. Princeton, also Cal. Inst. Tech., 1928-29. Recipient Chauvenet prize Am. Math. Assn., 1933. Fellow Cambridge Philos. Soc., Royal Soc., 1910 (Royal medal 1920, Sylvester medal 1940, Copley medal 1947); mem. French Acad. Scis., Brit. Assn. (became pres. sect. A 1922), Nat. Union Sci. Workers (pres. 1924-26), Math. Assn. (pres. 1924-26), London Math. Soc. (pres. 1926-28, 39-41, De Morgan medal 1929). Author: A Course of Pure Mathematics, 1908; (with J. E. Littlewood, G. Polya) Inequalities, 1934; (with E. Maitland Wright) Introduction to the Theory of Numbers, 2d edit., 1945; Ramanujan, 1940; A Mathematician's Apology, 1940; Fourier Series, 1944; Divergent Series, 1949; Cambridge Mathematical Tracts, numbers 2, 12, 18, 38; also numerous articles. Originated law on proportions of dominant and recessive Mendelian characters transmitted in large mixed populations (Hardy's law), 1908; (with J. E. Littlewood) studies on theory of Diophantine approximation; theory of Fourier series, Riemann zeta-function and distbn. prime numbers; solution of Waring's problem on expression of a number as sum of cubes, fourth powers. Died Cambridge, Dec. 1, 1947.

HARDY, James Daniel, Am. physiologist; b. Georgetown, Tex., Aug, 11, 1904; s. James Chappell and Lulu (Daniel) H.; A.B., U. Miss., 1924, M.A., 1925; Ph.D., Johns Hopkins, 1930; D.Sc., Kansas City Coll. Osteopathy and Surgery, 1966, Southwestern U., 1967; m. Augusta Ewing Haugh, June 8, 1928; children—James Daniel, George F. Prof-physiology U. Pa., Phila., 1953-61; with U. S. Naval Med. Acceleration Lab., Johnsville, Pa., 1953-61; dir. John B. Pierce Found. Lab., New Haven, Conn., 1961—; prof. physiology Yale, New Haven, 1961—. Chmn. com. on naval med. research NRC. Recipient Meritorious Civilian Service Award, U. S. N., 1961. Fellow Am. Phys. Soc., Aerospace Med. Assn.; mem. Am. Physiol. Soc., Biophysics Soc., Am. Soc. Heating, Refrigeration and Air Conditioning Engrs. Research on pain and temperature, developer instruments to monitor such stimuli. Home: Racebrook Rd., Woodbridge, Conn. 06525. Office: 290 Congress Av. New Haven, Conn. 06519.*

HARDY, James Daniel, Am. surgeon; b. Birmingham, Ala., May 14, 1918; s. Fred Henry and Julia (Poynor) H.; B.A., U. Ala., 1938; M.D., U. Pa., 1942, M.Med. Sci., 1951; m. Louise Scott Sams, July 1, 1949; children—Louise Scott, Julia Ann, Bettie Winn, Katherine Poynor. Practice medicine specializing in surgery, Memphis, 1951-55, Jackson, Miss., 1955—; faculty U. Tenn. Coll. Medicine, 1951-55; prof., chmn. dept. surgery, dir. surg. research U. Miss., Med. Center, 1955—; surgeon-in-chief Hosp. U. Miss.; chief surg. cons. VA Hosp., Miss. Tb. Sanatorium, Magee; cons. Oak Ridge Inst. Nuclear Scis., 1951-54, surgery study sect. NIH, 1958-62; vis. prof. numerous univs., also numerous lectures. Recipient Ann. award for distinguished service Pitts. Surg. Soc., 1962, Hektoen Silver medal for sci. research A.M.A., 1965. Diplomate Am. Bd. Surgery, Am. Bd. Thoracic Surgery. Mem. A.A.A.S., Am. Assn. Surgery of Trauma, Am. Assn. Thoracic Surgery, A.C.S., A.M.A., Am. Physiol. Soc., Am., So. surg. assns., Internat. Soc. Surgeons, Internat. Surg. Group, Soc. Surgery Alimentary Tract, Soc. Clin. Vascular Surgery, Internat. Cardiovascular Soc., Am. Cancer Soc., Alpha Omega Alpha. Author: Surgery and the Endocrine System, 1952; Fluid Therapy, 1954; Surgical Physiology of the Adrenal Cortex, 1955; Pathophysiology in Surgery, 1958; Total Surgical Management, 1959; (with others) Biopsy Manual, 1959; Surgery of the Aorta and Its Branches, 1960. Editor: (with C. P. Artz) Complications in Surgery and Their Management, 1960. Research and numerous publs. on body water and electrolytes, trauma, endocrine organs; performed 1st human lung, heart, and successful kidney autotransplants, 1st successful reproductive organ replantation in dogs. Home: 2531 Eastover St., Jackson 39211. Office: 2500 N. State St., Jackson, Miss. 39216.*

HARDY, Le Grand Haven, Am. ophthalmologist; b. Provo City, Utah, June 13, 1894; s. Milton Henry and Elizabeth (Smoot) H.; A.B., Brigham Young U., 1916; postgrad. U. Chgo., 1917; B.S., Columbia, 1919, M.D., 1921; m. Susanna Edwards Schuyler Haigh, July 9, 1923. Began practice as ophthalmologist, N.Y.C., 1922; asst. surgeon N.Y. Eye and Ear Infirmary, 1924-30; ophthalmologist Northern Dispensary, 1924-29, consultant from 1929; ophthalmic surgeon Midtown Hosp., 1925-29; dir. ophthalmology Fifth Avenue Hosp., 1929-36; prof. clin. Ophthalmology Coll. Phys. and Surg., Columbia U., attending ophthalmologist Presbyn. Hosp. and Vanderbilt Clinic; staff mem. Inst. of Ophthalmology; dir. Functional Testing and Physiologic Optics Labs., Inst. of Ophthalmology. Fellow A.C.S., A.M.A., Am. Ophthalmol. Soc., Am. Acad. Ophthalmology and Otolaryngology, A.A.A.S., Assn. Research Ophthalmologists, Am. Orthoptic Council (pres. from 1938), Inter-Society Color Council. Author: History and Technic of Scotometry, published by American Ophthal. Society, 1931; also articles on illumination as it affects the eye, orthoptics, the bases of color vision, measurements of sight. Co-author: The Geometry of Binocular Space Perception, 1953. Co-inventor: Hardy-Rand-Ritter Pseudoisochromatic Plates. Died Apr. 14, 1954.

HARDY, Miriam Pauls, Am. audiologist, speech pathologist; b. St. Louis, Feb. 4, 1912; d. Otto and Amelia (Ottesky) Pauls; A.B., Harris Tchrs. Coll., 1932; postgrad. Washington U., St. Louis; M.A., Wayne State U., 1939; Ph.D., Northwestern U., 1949; m. William G. Hardy, Aug. 3, 1960. Tchr., N.J. State Sch. for Deaf, 1934-41; instr., supr. deaf edn. Rackham Sch. Spl. Edn., Mich. State Coll., Ypsilanti, Mich., 1941-42; supr. hearing therapy Spl. Edn. Clinics, Ind. State U., Terre Haute, 1942-44; supr. aural rehab. USN Hosp., Phila., 1944-48; faculty Johns Hopkins U. and Hosp., 1949—, asso. prof. otolaryngology and environmental medicine, 1955—. Mem. Sec. of Health, Edn. and Welfare Adv. Com. on Deaf Edn., mem. communicative scis. study sect. Nat. Inst. Neural Diseases and Blindness, NIH, 1962-66. Recipient award of merit Am. Acad. Ophthalmology and Otolaryngology. Fellow Am. Speech and Hearing Assn. (past v.p.); mem. Acoustical Soc., Alexander Graham Bell Assn. Contbg. author: Hearing and Deafness. Research and publs. on differential diagnosis and mgmt. of communicative disorders in children, normal and abnormal lang. and speech devel., hearing in infants and young children. Home: 2533 Pickwick Rd., Balt. 21207. Office: Hearing and Speech Center, Johns Hopkins Hosp., Balt. 21205.*

HARDY, (Cecil) Ross, Am. biologist; b. Cleveland, Utah, Apr. 4, 1908; s. John James and Ruth Adeline (Oviatt) H.; B.S., U. Utah, 1931, M.S., 1938; Ph.D., U. Mich., 1943; m. Lucile Neff, May 29, 1934; children—Alan Ross, Kaye Marilyn (Mrs. Arnold Leonard Kaufman). Tchr. high sch., 1933-38; chmn. div. biol. scis. Dixie Jr. Coll., 1938-46; head biology dept. Weber Coll., 1946-49; prof. zoology Cal. State Coll., Long Beach, 1949—. Fellow Herpetologists League, Utah Acad. Scis., Arts and Letters; mem. Am. Soc. Mammalogists, Am. Ornithologists Union, Wilson, Cooper ornithol. socs., Soc. Systematic Zoology, Am. Soc. Icthologists and Herpetologists, Ecol. Soc. Am., Wildlife Soc., Sigma Xi, Alpha Chi Sigma, Sigma Upsilon, Phi Sigma, also others. Research and publs. on relationships between rodents and desert plants to soil; study of rodents, desert tortoise, other retiles, ecology of birds, migration. Home: 5351 Las Lomas St., Long Beach, Cal. 90815.*

HARDY, Sir William Bate, English biologist; b. Erdington, Warwickshire, Eng., Apr. 6, 1864; s. William and Sarah (Bate) H.; B.Sc., Gonville and Caius Coll., Cambridge (Eng.) U., 1888; m. Alice Mary Finch, 1898; 3 children. Fellow Gonville and Caius Coll., 1892, tutor, 1900-18, univ. lectr. in physiology, 1913; 1st chmn. food investigation bd. Dept. Sci. and Indsl. Research, 1917-28, dir. food investigation, 1917-34, supt. low temperature research sta., Cambridge, 1922-34; Croonian lectr., 1905, Bakerian lectr., 1925, Abraham Flexner lectr. Vanderbilt U., 1931. Responsible for creation Torry Research Sta. and Ditton Lab.; chmn. adv. com. on fisheries Devel. Commn., 1919-31. Fellow Royal Soc., 1902, biol. sec., 1915-25, organizer food com., 1915; pres. Brit. Assn., 1934. Author: Collected Scientific Papers (edited by E. K. Rideal), 1936. Pioneer work in colloid chemistry, molecular physics of films, surfaces and boundary conditions; research on static friction, action of lubricants; basic work in modern theory protein solutions. Died Cambridge, Jan. 23, 1934.

HARE, Alexander Paul, Am. sociologist; b. Washington, June 29, 1923; s. Alexander Paul and Lulu (Waters) H.; B.A., Swarthmore Coll., 1947; B.S., Ia. State U., 1948; M.A., U. Pa., 1949; Ph.D., U. Chgo., 1951; m. Rachel Diana Thies, June 21, 1947; children—Sharon E., Diana S., Mally M., Christopher P. Instr., Wellesley Coll., 1952-53; Nat. Inst. Mental Health fellow, instr. Harvard, 1953-59, research asso., lectr. dept. social relations, 1957-60; asso. sociologist dept. psychiatry Mass. Gen. Hosp., 1955-60; sociologist Family Guidance Center, Harvard Sch. Pub. Health, 1956-58; asso. prof. Haverford (Pa.) Coll., 1960-66, prof. sociology, 1966—, chmn. dept. sociology, 1965—. Vis. prof. Makerere U., Uganda, U. Ibadan, Nigeria, 1964-65. Mem. Am. Sociol. Assn., Pa. (pres. 1966-67), Eastern sociol. socs., Am. Assn. U. Profs., Soc. Exptl. Social Psychologists. Author: Handbook of Small Group Research, 1962; also articles. Editor: (with E. F. Borgatta, Bales) Small Groups, 1955, rev., 1965; Sociol. Inguiry, 1967—. Sumarized and integrated research on small groups from 1900 to present; research on computer simulation of behavior in small discussion groups, cross cultural (African and European) studies of group problem solving, influence of group size on interaction and consensus in small groups. Home: 715 Brooke Rd., Wayne, Pa. 19087. Office: Haverford College, Haverford, Pa. 19041.*

HARE, Hobart Amory, Am. physician; b. Phila., Sept. 20, 1862; s. William Hobart and Mary Amory (Howe) H.; B.S., U. Pa., 1885, M.D., 1884, LL.D., 1921; M.D., Jefferson Med. Coll., 1893; m. Rebecca Clifford Pemberton, May 8, 1884. Prof. children's diseases U. Pa., 1890; prof. therapeutics and diagnosis, Jefferson Med. Coll., from 1891. Editor Univ. Med. Mag., 1888-89, Med. News, 1890-91, Therapeutic Gazette, 1891-1927. Fellow Coll. Physicians (Phila.), Assn. Am. Physicians. Author: Practical Therapeutics, 1890, 20th edit., 1927; Practical Diagnos, 1896, 9th edit., 1927, also many med. essays. Editor: A System of Therapeutics, 4 vols., 1890, 1901, 11; Medical Compilations and Sequels of Typhoid and Other Fevers, 1901. Died June 15, 1931.

HARE, Robert, Am. chemist; b. Phila., Jan. 17, 1781; s. Robert and Margaret (Willing) H.; studied chemistry under James Woodhouse; A.M., M.D. (both hon.), Yale, 1806; M.D. (hon.), Harvard, 1816; m. Harriett Clark, Sept. 11, 1811; 1 son, John Clark. Prof. chemistry Coll. William and Mary, 1818, U. Pa., 1818-47; hon. mem. Smithsonian Instn. Author: Brief View of the Policy and Resource of the United States, 1810; Chemical Apparatus and Manipulations, 1836; Compendium of the Course of Chemical Instruction in the Medical Department of the University of Pennsylvania, 1840. Discovered oxy-hydrogen blow-pipe (source of highest degree of heat then known, which enabled him to fuse the most refractory substances and led to founding of platinum industry), 1801; inventor calorimeter, 1816; received 1st Rumford medal from Am. Acad. Arts and Scis., for discovery of mercury cathode in electrolysis of aqueous solutions of metallic salts, 1839; developed means of using tar for lighting; inventor deflagrator, 1820) developed chem. process for denarcotizing laudanum, also processes for isolating various elements, including boron and silicon; first to isolate calcium metal in U. S. Died Phila., May 15, 1858.

HARET, Spiru, Rumanian mathematician; b. 1851; prof. Bucharest (Rumania) U.; mem. Rumanian Acad.; minister of edn. Showed existence of secular perturbations of big axes of planetary orbits, 1878; initiated reforms which led to modern system of edn. on all levels in Rumania. Died 1912.

HARFORD, Carl Gayler, Am. physician; b. St. Louis, June 27, 1906; s. Edwin Marvin and Agnes (Gayler) H.; A.B., Amherst Coll., 1928; M.D., Washington U., St. Louis, 1933; m. Mary Broadhead Cowan, Aug. 5, 1933; children—John Gayler, and Carolyn Harford. Asst. in medicine, bacteriology, and immunology Washington U. Sch. Medicine, 1935-36, faculty, 1938—, prof. medicine, 1964—, acting head dept. microbiology, 1952-53, 59-60; fellow pathology and bacteriology Rockefeller Inst. for Med. Research, N.Y.C., 1936-38. Diplomate Am. Bd. Internal Medicine. Mem. Am. Soc. Microbiologists, Central Soc. for Clin. Research, Am. Assn. Immunologists, Am. Soc. for Clin. Investigation, Assn. Am. Physicians, A.M.A., St. Louis Soc. Internal Medicine. Research and articles in mechanisms of immunity of viruses of equine and St. Louis encephalitis, mechanisms of resistance and susceptibility of lung; electron microscopy of cells infected with viruses, electron microscopic autoradiography in viral infection. Home: 6940 Waterman St., St. Louis 63130. Office: 660 S. Euclid St., St. Louis 63110.*

HARGER, Rolla Neil, Am. chemist, toxicologist; b. Decatur Co., Kan., Jan. 14, 1890; s. William Delashmutt and Margaret Elizabeth (Neil) H.; A.B., Washburn Coll., 1915; A.M., Kan. U., 1917; Ph.D., Yale, 1922; m. Helen Harriet Dick, June 6, 1917; children—Elizabeth Ann (Mrs. John W. Stalcup), Robert William, Susan Margaret. Lab. instr. chemistry Washburn Coll., 1913-15; instr. Kan. U., 1915-17; asst. biochemist U. S. Dept. of Agr., Washington, D.C., 1917-20; Nat. Research Council fellow, Yale, 1920-22; asst. prof. biochemistry and toxicology sch. medicine Ind. U., 1922-29, asso. prof., 1929-33, prof., 1933-1960, chmn. dept., 1933-1956; prof. emeritus biochemistry and toxicology, 1960—. Research worker and consultant on poisons, alcohol, ether, vitamins; inventor "drunkometer" for testing intoxication, 1931; consultant on poisons Ind. State Bd. of Health, Ind. State Police; expert witness and consultant in many notable cases as Indianapolis "lead epidemic," 1940-45, mineral oil popcorn, 1945, Cline wives' ashes, 1946, Dupont, Ind., mercury cases, 1946, Kaadt diabetic cure, 1948; guest lecturer on chem. tests for intoxication before various regional cons. of traffic court judges and prosecutors, sponsored by American Bar Association and Northwestern University Traffic Institute, since 1947. Member committee on driver intoxication Nat. Safety Council since 1938, v. chmn. Indianapolis Mayor's Traffic Safety Com., 1941, chmn. enforcement com. Indianapolis Safety Council since 1941-1958, cons. traffic div., 1960—. Mem. Am. Chem. Soc., Am. Soc. Biol. Chemists, Ind. Chem. Soc., A.A.A.S., Sigma Xi. Home: 5015 Graceland Av. Office: Ind. Univ. School of Medicine, Indianapolis, Ind.

HARGREAVES, James, English inventor; b. nr. Blackburn, Eng., 1720; a dau., Jenny; worked as carpenter, then hand-loom weaver, Standhill, 1740-50; employed by Robert Peel, Blackburn, circa 1760. Devised spinning jenny, circa 1764; local spinners destroyed machines; moved to Nottingham 1768; built (with Thomas James) mill; received patent for spinning jenny, 1770. Died Apr. 22, 1778.

HARGREAVES, James, English chemist; b. Hoarstones, Pendle Forest, Eng., 1834. Employed by W. Gossage, soap mfr., 1856-65; built works for mfg. hydrochloric acid, 1873. Numerous patents on improvements in soap manufacture, recovery of sulphur and phosphates from waste, hydrochloric acid manufacture, thermomotor which ran on gas-tar (forerunner of Diesel engine), (with Bird) electrolytic process for making soda. Died Widnes, Eng., Apr. 4, 1915.

HARING, Douglas Gilbert, Am. anthropologist; b. Watkins Glen, N.Y., Aug. 6, 1894; s. Leon and Emillie (Gilbert) H.; B.S., Colgate U., 1914, D.Sc., 1962; postgrad. Rochester Theol. Sem., 1915-17, B.D., 1923; A.M., Columbia U., 1923, postgrad. 1926-27; diploma Sch. Japanese Lang. and Culture, Tokyo, Japan, 1925; LL.D., Syracuse U., 1962; m. Ann Teasdale Howell, Aug. 8, 1918; 1 dau., Ruth Ann (Mrs. Leonard Lief). Lectr., Columbia, 1926-27; faculty Syracuse (N.Y.) U., 1927-66, prof. anthropology, 1946-61, prof. emeritus., 1961—. Vis. lectr. Harvard, 1944-46; field research asso. Pacific Sci. Bd., NRC, Ryukyu Islands, 1951-52. Postdoctoral fellow Viking fund, 1948. Fellow A.A.A.S. (past council mem.), Am. Anthrop. Assn., Am. Sociol. Assn.; mem. Assn. for Asian Studies (past dir.), Asiatic Soc. Japan, Japanese Soc. Ethnology, Am. Oriental Soc., Am. Folklore Soc., Far Eastern Prehistory Assn., Japan Soc., Phi Beta Kappa, Sigma Xi. Author: The Land of Gods and Earthquakes, 1929; (with M. E. Johnson) Order and Possibility in Social Life, 1940; Blood on the Rising Sun, 1943. Editor, contbg. author: Japan's Prospect, 1946; Personal Character and Cultural Milieu, 1956. Translator: (with W. Tsuneishi) Social History of Japan (Takikawa), 1948. Research and numerous publs. on Japanese and Ryukyuan culture, statis. studies of conditions during depression of 1930's, culture and personality, Okinawa. Home: 117 Euclid Terrace, Syracuse, N.Y. 13210.*

HARINGTON, Sir Charles Robert, Brit. biochemist; b. Llanerfyl, North Wales, Aug. 1, 1897; s. Charles and Audrey Emma Burges (Bayly) H.; student Malvern Coll., 1914-16; B.A., Cambridge U., 1916-19, M.A., 1927, Sc.D. (hon.), 1949; Ph.D., U. Edinburgh, 1922; Dr. (hon.), U. Paris, 1945; m. Jessie McCririe Craig, Aug. 1, 1923; children—Michael, Alison Mary, Margaret Jane. Research asst. dept. therapeutics U. Edinburgh, 1920-22; lectr. path. chemistry U. London,

1928-31, prof. chem. pathology, 1931-42; dir. Nat. Inst. Med. Research, London, 1942—. Mem. Med. Research Council, 1938-42, Agrl. Research Council, 1941-45. Created Knight, 1948; recipient Gold Medal for Therapeutics, Soc. Apothecaries, 1953. Fellow Royal Soc., 1931 (Royal medal 1944). Hon. fellow Royal Soc. of Edinburgh; mem. Chem. Soc., Physiol. Soc., Biochem. Soc., Soc. Endocrinology, Med. Research Soc., Am. Acad. Arts and Scis. (hon.). Author: The Thyroid Gland, 1933. Editor of Biochem. Jour., 1930-42. Contbr. articles profl. jours. Extensive research biochemistry; (with Barger) synthesized thyroxine, demonstrating it as derivative of tyrosine, 1927; determined structure of thyrozine. Home: Mt. Vernon House, Hampstead, London N.W. 3. Office: The Ridgeway, Mill Hill, London N.W. 7, Eng.

HARINGTON, Sir John, Brit. inventor; b. 1561; s. John Harington; ed. Cambridge, Lincoln's Inn, London; m. Mary Rogers, 1583; many children; credited with inventing 1st functional toilet in which a stream of water disposed of excreta; banished from ct. for describing the invention, 1596; accompanied earl of Essex to Ireland, 1599. Author transl. Orlando furioso (Ariosto), 1591; The Metamorphosis of Ajax, 1596. Died Kelston, Eng., Nov. 20, 1612.

HARKEN, Dwight Emary, Am. surgeon; b. Osceola, Ia., June 5, 1910; s. Conreid Rex and Edna (Emary) H.; A.B., Harvard, 1931, M.D., 1936; L.M.S.S.S., Soc. Apothecaries London, 1939; D.S., Suffolk U., 1964; m. Anne Louise Hood, Aug. 29, 1934; children—Alden Hood, Anne Louise (Mrs. Ridgway Macy Hall, Jr.). Faculty, Tufts Med. Sch., 1946-48; faculty Harvard Med. Sch., Boston, 1948—, clin. prof. surgery, 1963—; surgeon, chief dept. thoracic surgery Peter Bent Brigham Hosp., Boston, Mt. Auburn Hosp., Cambridge, Mass; cons. hosps. Malden, Waltham, Melrose, Boston Hosp. for Women, Carney, Quincy City, others. Cons. VA, U. S. Naval Hosps.; vis. prof., guest lectr. numerous instns. in fgn. countries. Recipient Susan and Theodore Cummings Humanitarian award, 1963-66, Centennial award Boston City Hosp., 1964. Diplomate Am. Bd. Surgery, Am. Bd. Thoracic Surgery (a founder 1948). Fellow Am. Coll. Cardiology (pres. 1964), A.C.S., Am. Coll. Chest Physicians; mem. Am. Assn. Thoracic Surgery, Am. Heart Assn., A.M.A., Soc. Vascular Surgery, Internat. Cardiology, Assn. Am. Med. Colls., Am. Thoracic Soc., New Eng. Surg. Soc., Soc. Thoracic Surgeons, Pan-Pacific Surg. Assn., Pan Am. Med. Assn., Sociedad Colombiana de Cirujanos, Sociedad Chilena da Cardiologia, Sociedad de Cirujanos de Chile, Sociedad Medica de Valparaiso, Sociedad Argentina de Cardiologia, Japanese Med. Soc., Japanese Soc. Artificial Organs, Alpha Omega Alpha, Sigma Xi, others. Research and publs. on cardiac and thoracic surgery; also pioneered cardiac surgery during World War II and subsequently developed cardiac and thoracic surg. operations, designed various types of surg. instruments, heart-lung machines, heart pacemakers and artificial heart valves for thoracic and cardiac surgery. Home: 4 Lowell St., Cambridge, Mass. 02138. Office: 67 Bay State Rd., Boston 02115.*

HARKER, Alfred, English petrologist; b. Kingston-upon-Hull, Eng., Feb. 19, 1859; s. Portas Hewart and Ellen (Tarbotton) H.; grad. St. John's Coll., Cambridge, 1878; hon. degrees McGill U., Montreal, Que., Can., 1913, Edinburgh, Scotland, 1919. Elected fellow St. John's Coll., Cambridge, 1885, became demonstrator geology, 1884, univ. lectr., 1904, reader petrology, 1918-31; on Geol. Survey Scotland, 1895-1905. Recipient Royal medal, 1935, Sedgwick prize, 1888, Murchison medal, 1907, Wollaston medal, 1922. Fellow Royal Soc., 1902; mem. Geol. Soc. London (pres. 1916-18). Author: Petrology for Students, 1895; Tertiary Igneous Rocks of Skye, 1904; Geology of Small Isles of Invernesshire, 1908; Natural History of Igneous Rocks, 1909; Metamorphism, 1932. Pioneered research on thermal metamorphism of rocks, 1891, 93; studies on igneous rocks. Died Cambridge, July 18, 1939.

HARKER, David, Am. crystallographer; b. San Francisco, Oct. 19, 1906; s. George Asa and Harriette (Buttler) H.; B.S., U. Cal. at Berkeley, 1928; Ph.D., Cal. Inst. Tech., 1936; m. Katherine DeSavich, July 1, 1930; children—Tatiana (Mrs. Harrison McG. Yates), Liudmila (Mrs. Ignatius A. Triolo). Faculty, Johns Hopkins, 1936-41; staff research lab. Gen. Electric Co., Schenectady, 1941-50, cons. x-ray diffraction, Milw., 1953—; faculty Poly. Inst. Bklyn., 1950-59; head biophysics dept., Roswell Park Meml. Inst. div. State U. N.Y., Buffalo, 1959—, research prof., 1960—. Cons., Carborundum Co., Niagara Falls, N.Y., 1962—. Fellow Am. Inst. Mining and Metall. Engrs., Am. Phys. Soc., N.Y. Acad. Scis.; mem. Am. Inst. Physics (adv. editor Soviet Physics-Crystallography, also mem. adv. bd. Russian transls.), NRC (U. S. A. nat. com. crystallography), Internat. Union Crystallography (del. to 7th gen. assembly Internat. Congress and Symposium, Moscow, 1966), A.A.A.S., Am. Assn. U. Profs., Am. Crystallographic Assn., Biophys. Soc., Societe Francaise de Mineralogie et de Crystallographie, Sigma Xi, others. Research and publs. on methods of finding atomic arrangement in crystals; elucidation of structures of several crystals including ribonuclease; relationship between external form of crystals and their internal symmetry; designer apparatus for measuring x-ray diffraction patterns of crystals; calculation of strength of acids from their molecular structures. Home: 23 High St., Buffalo 14203.*

HARKIN, Duncan Claire, Am. mathematician; b. Cameron, W.Va., Sept. 20, 1899; s. Daniel C. and Flora (Duncan) H.; A.B., W.Va. U., 1921, A.M., 1923; Ph.D., U. Chgo., 1927; m. Alice Mildren, Sept. 1, 1925; children—Phyllis (Mrs. Marlin Paul Willis), Duncan Alfred, Anita Rose (Mrs. Richard Oscar Peterson). Instr., W.Va. U., 1923-25, U. Fla., 1925-26; lectr. U. Man. (Can.), 1926-27; asso. prof. Ala. Poly. Inst., Auburn, 1927-37, prof., 1937-38; mem. staff Bklyn. Coll., 1938-46; head math. analysis group, research div. Bur. Ordnance, U. S. Navy Dept., 1941-46; head math. cons. Naval Research Lab., Washington, 1946-49; vis. Fulbright prof. Math. Inst., U. Oslo (Norway), 1950; cons. U. S. Govt., 1951-58, 59-63; prof. U. S. Naval Acad., Annapolis, 1958-59; head dept. math. and statistics Am. U., Washington, 1960-62. Fellow A.A.A.S.; mem. Am. Math. Soc., Math. Assn. Am., Société Mathématique de France, Gauss Gesellschaft. Phi Beta Kappa. Author: (with B. H. Crenshaw) College Algebra, 1929; Brief College Algebra, 1939; Fundamental Mathematics, 1940. Research and publs. on simplification of algebraic processes, simplification and arithmetization of elliptic functions; initiated mech. translation showing its econ. feasibility. Home: 1034 Havenhurst Dr., La-Jolla, Cal. 92037.*

HARKINS, Henry Nelson, Am. surgeon; b. Missoula, Mont., July 13, 1905; s. William Draper and Anna Louise (Hatheway) H.; B.S., U. Chgo., 1925, M.S., 1926, Ph.D. (Douglas Smith fellow), 1928; M.D., Rush Med. Coll., 1931; m. Jean Hamilton Trester, June 19, 1937; children—Pamela Jean (Mrs. John Parr), Ellen Christine, Anne Wayne, Harriet Nelson. With U. Chgo., 1931-36, chief resident surgeon, instr., 1936-38; asso. surgeon Henry Ford Hosp., Detroit, 1939-43; asso. prof. Johns Hopkins, 1943-47; prof. surgery, exec. officer, chmn. dept. surgery U. Washington, Seattle, 1947-64, prof. surgery, 1964-67; vis. prof. numerous instns., including U. Ind., 1965, U. Queensland, Brisbane, Australia, 1965, U. Lund, Sweden, 1965, U. Gothenburg, Sweden, 1966; lectr. numerous univs. Spl. cons. hematology NIH, 1946-48, surgery, 1950-55, gen. medicine, 1960-65. Guggenheim Meml. fellow, 1938-39, 1965-66. Diplomate Nat. Bd. Med. Examiners, Am. Bd. Surgery. Fellow A.C.S.; mem. A.M.A., Am. Fedn. Clin. Research, Am. Physiol. Soc., Soc. U. Surgeons, Internat. Soc. Surgery, Am. Surg. Assn., Soc. Surgery Alimentary Tract, Am. Cancer Soc., Internat. Surg. Club, Alpha Omega Alpha, Sigma Xi, numerous others. Author: The Treatment of Burns, 1942; (with H. G. Moore, Jr.) The Billroth I Gastric Resection, 1954; Editor: (with others) Surgery: Principles and Practice, 1957, 3d. edit., 1965; Surgery of the Stomach and Duodenum, 1962; Geriatric Surgical Emergencies, 1963; Hernia, 1964; Surgical Anatomy, 1964. Research and publs. on treatment of burns; originator of combined operation of vagotomy, antrectomy, and gastroduodenostomy for treatment of duodenal ulcer. Died Aug. 12, 1967.

HARKINS, William Draper, Am. chemist; b. Titusville, Pa., Dec. 28, 1873; s. Nelson Goodrich and Sarah Eliza (Draper) H.; A.B., Stanford, 1900, Ph.D., 1907; post-grad. U. Chgo., 1901, 04, Stanford, 1905-06, Institut für Physikalische Chemie, Karlsruhe, Germany, 1909; research asso. Research Lab. Phys. Chemistry, Mass. Inst. Tech., 1909-10, 11; m. Anna Louise Hatheway, June 9, 1905; children—Henry Nelson, Alice Marion. Asst. and instr. chemistry Stanford, 1898-1900; prof., head dept. U. Mont., 1900-12; chemist in charge of smelter smoke investigation Anaconda Farmers's Assn., 1902-10, Mountain Copper Co. of Cal., 1904, U. S. Dept. of Justice, 1910-12; research work for Carnegie Instn., Washington, 1911; asst. prof. gen. chemistry U. Chgo., 1912-14, asso. prof., 1914-17, prof. phys. chemistry from 1907, Andrew MacLeish Distinguished Service prof., from 1935, dir. rubber research. from 1942; lectr. Mellon Inst. Indsl. Research, 1916-17, U. Ill., 1918-19; cons. C.W.S., from 1927; cons. chemist Libbey-Owens-Ford Glass Co., Universal Oil Products Co., U. S. Rubber Co., Nat. Def. Research Com.; George Fisher Baker lectr. Cornell U., 1936-37. Mem. Internat. Com. on Atoms, from 1932. Fellow A.A.A.S. (v.p.); mem. Am. Chem. Soc. (chmn. div. phys. and inorganic chemistry 1919-20, Willard Gibbs medal 1928). Nat. Acad. Scis., Am. Philos. Soc., hon. mem. Alpha Omega Alpha. Research and publs. on heat of sun and stars, stability of atomic nuclei and physics and chemistry of surfaces; known for work on atomic structure; predicted existence of neutron and heavy hydrogen; pioneer in studying problem of relative proportions of elements in universe as a whole. Died Mar. 7, 1951.

HARKNESS, James, mathematician; b. Derby, England Jan. 24, 1864; s. John Harkness; ed. Trinity Coll., Cambridge; B.A., London U.; LL.D., McGill U., 1921; m. Katherine E. Cam, 1908; 2 sons, 1 dau. Asso., asso. prof., prof. math. Bryn Mawr (Pa.) Coll., 1888-1901; Peter Redpath prof. pure math. McGill U., Montreal, Que., Can., 1903-13. Fellow Royal Soc. Can.; mem. Am. Math. Soc. (past v.p., sub-editor Trans.). Author (with Frank Morley) 2 treatises on theory of functions, also articles. Died Dec. 1923.

HARKNESS, Robert Angus, Brit. med. biochemist; b. Stockton-on-Tees, Eng., Nov. 13, 1931; s. Robert and Evelyn (Underwood) H.; M.B., Ch.B., U. Edinburgh, (Scotland), 1955, Ph.D., 1962, M.C. Pathology, 1966; postgrad. Med. Sch. London, 1958; m. Eveline J. Roy, Oct. 10, 1962; children—Anne, Derek. Research fellow dept. biochemistry U. Edinburgh, also Med. Research Council clin. endocrinology research unit, 1958-62; sci. staff Med. Research Council, U.K., 1962-66; sr. lectr. pediatric biochemistry U. Edinburgh, 1966—. Mem. Royal Coll. Physicians Edinburgh. Research and publs. on metabolism of progesterone, devel. of steroid assays to study endocrine gland function, especially testosterone and pregnanetrial. Home: 33 Fox Spring Rise, Edinburgh 10. Office: Dept. Pediatric Biochemistry, Royal Hosp. for Sick Children, Edinburgh, Scotland.*

HARKNESS, William, astronomer; b. Ecclefechan, Scotland, Dec. 17, 1837; s. James and Jane Weild H.; studied Lafayette Coll., Pa., 1854-56; A.B., Rochester U., 1858, A.M., 1861, LL.D., 1874; A.M. Lafayette, 1865; studied medicine, New York, M.D., 1862. Apptd. aide U. S. Naval Obs., 1862; commd. prof. math. with rank of lt. comdr., 1863; served at Naval Obs. until 1865. served on U. S. monitor Monadnock, 1865-66; attached to U. S. Hydrographic Office, 1867; discovered the coronal line K 1474, during total solar eclipse of Aug. 1869; mem., 1871, and from 1882 exec. officer U. S. Transit of Venus Commn.; had charge of transit of Venus parties at Hobart, Tasmania, 1874, Washington, 1882; discovered theory of focal curve of achromatic telescopes, 1879; attached to U. S. Naval Obs., from 1862, designed most of its large instruments; astron. dir. Naval Obs., 1894-99, and dir. American Ephemeris and Nautical Almanac, 1897-99; ret. as rear adm. USN, 1899. Invented spherometer calipher, other astron. instruments. Died 1903.

HARKORT, Johann Caspar, German engr.; b. Harkorten, Germany, Jan. 22, 1817; s. Johann Caspar and Johanna (Ihne) H.; ed. indsl. sch., Hagen, comml. sch., Leipzig, Germany; m. Maria Cäcilia Pottgiesser, 1844; 5 daus. including Maria (Mrs. W. Liebe), Anna (Mrs. R. Böker), Cäcilie (Mrs. A. Böker), Johanna (Mrs. W. Funcke). With father's firm from late 1840's, expanded it to include machine factory, bridge bldg., gen. iron constrn.; founder factory, Duisburg, sold to Gesellschaft Harkort, 1872. Pioneer in use of ingot iron and rolled steel in bridges; built one of 1st privately constructed bridges, over Wupper at Ritterhausen; other bridges include those over Ruhr at Werden, over Rhine at Coblenz, or Elbe at Stendal, over Weser at Bemmen, over Danube at Vienna; also built bridges in Netherlands, Russia, Portugal, Java; responsible for constrn. for Vienna World Exhbn., 1871-72. Died Harkorten, Oct. 13, 1896.

HARLAN, Jack Rodney, Am. geneticist; b. Washington, June 7, 1917; s. Harry Vaughn and Augusta (Griffing) H.; B.S. with distinction, George Washington U., 1938; Ph.D., U. Cal. at Berkeley, 1942; m. Jean Yocum, Aug. 4, 1939; children—Sue Caryl (Mrs. Robert L. Hughes), Harry Vaughn, Sherry Ruth, Richard Edwin. Research asst. Tela R.R. Co., Honduras, 1942; geneticist Agrl. Research Service, U. S. Dept. Agr., Woodward, Okla., 1942-51, Stillwater, Okla., 1951-61; prof. agronomy Okla. State U., 1951-66; prof. plant genetics U. Ill., Urbana, 1966—. Botanist plant exploration Agrl. Research Service, U. S. Dept. Agr., Middle Eastern countries, 1948, 60-61; sr. staff mem. Iranian Prehistoric Project, U. Chgo., Iran, 1960, Turkish Prehistoric Project, Turkey, 1964. Guggenheim fellow, 1959-60. Fellow A.A.A.S., Am. Soc. Agronomy; mem. Crop Sci. Soc. Am. (pres. 1966), Bot. Soc. Am., Am. Soc. Range Mgmt., Am. Inst. Biol. Sci., Sigma Xi. Author: Theory and Dynamics of Grassland Agriculture, 1956; Plant Scientists and What They Do, 1964. Research and numerous publs. in biosystematics of grasses, particularly Dichanthium and Cynodon; improved varieties of native grasses, grass breeding, origin and evolution of cultivated plants; contbn. to cytogenetic theory, including concepts of microcenters, diffuse origins and compilospecies. Office: Agronomy Dept., University of Ill., Urbana, Ill.*

HARLAND, Walter Brian, English geologist; b. Scarborough, Eng., Mar. 23, 1917; s. Walter E. and Alice M. (Whitfield) H.; B.A., U. Cambridge (Eng.), 1938; m. Elisabeth M. E. Lewis, Jan. 3, 1942; children—Marian J., Edward B., Hilary F., Beatrice H. Asso. prof. W. China Union, U., Chengtu, 1943-46; U. Cambridge demonstrator, 1946-48, lectr., 1948—, reader, 1966—, fellow Gonville and Caius Coll., 1950—. Fellow Geol. Soc. London (sec. 1963—), Royal Geog. Soc., Royal Astron. Sci. Author: The Earth, Rocks, Minerals and Fossils, 1960; The Phanerozoic Time-scale, 1964; The Fossil Record, 1967; also numerous papers. Co-editor, Geol. Mag., 1955—. Structural, stratigraphical and geophys. investigation of Arctic, especially Spitsbergen; study ice ages and Precambrian stratigraphy. Home: 117 Grantchester Meadows, Cambridge, U.K.*

HARLESS, (Johann) Christian Friedrich, German physician; b. Erlangen, Germany, June 11, 1773; s. Gottlieb Christoph and Sophie H. C. (Weiss) H.; Ph.D., Erlangen, 1793, M.D., 1794; m. Elisabeth (Bettina) Pfauz; 1 son, Hermann. Joined faculty U. Erlangen, circa 1794, became asso. prof., 1794; became prof., co-dir. med. clinic, Erlangen, 1814; named prof. pathology and therapy U. Bonn, 1818, also founded clin. instns. Recipient numerous honors; Praemium Harlessianum, scholarship found., founded in his honor. Author: Versuch einer vollständigen Geschichte der Hirn- und Nervenlehre, 1801; Analecta de dysenteria et imprimis ejus therapia in antiquitatibus, 1801; Analecta historico-critica, De Archigene medico et de Apolloniis medicis eorumque scriptis . . . , 1816; Handbuch der ärztlichen Klinik, 3 vols., 1817-26; Die Verdienste der Frauen um Naturwissenschaften, Gesundheits- und Heilkunde . . . , 1830; Servilii Democratis quae supersunt carmina medicinalia, Graece et Latine, I, 1833; Die sämtlichen bisher in Gebrauch genommenen Heilquellen und Kurbäder des südlichen und mittleren Europas, West Asiens und Nordafrikas, I, 1, Die Heilquellen und Kurbäder Griechenlands, 1846. Research in medicine, including history of medicine and epidemics. Died Bonn, Mar. 13, 1853.

HARLEY, Robison Dooling, Am. physician; b. Pleasantville, N.J., Feb. 27, 1911; s. Halvor L. and Alice (Robison) H.; B.Sc., Rutgers U., 1932; M.D., U. Pa., 1936; Ph.D., U. Minn., 1941; m. Loyde Gochnauer, Dec. 18, 1944; children—Robison Dooling, Ardee R., Heather L., Halvor, William W. Practice medicine specializing in opthalmic surgery, Atlantic City, 1947—; prof. ophthalmology Temple U. Med. Sch.; lectr. U. Pa. Grad. Sch. Medicine, 1960—; attending surgeon Wills Eye Hosp., St. Christopher's Hosp. for Children. Fellow A.C.S. (gov. 1958—), Am. Acad. Ophthalmology and Otolaryngology. Contbg. author: Pediatrics (Nelson), 1965; Sensorimotor Anomolies of Extrinic Ocular Muscles. Research and numerous publs. on surgery, congenital defects, clin. diseases of eye.*

HARLOW, Francis Harvey, Am. physicist; b. Seattle, Jan. 22, 1928; s. Francis H. and Florence (Melvin) H.; B.S., U. Wash., 1949, Ph.D., 1953; m. Patricia Jean Nystuen, June 21, 1952; children—Catherine Jean, Carol Elizabeth, Celia Ann, Keith Francis. With Los Alamos Sci. Lab., 1953—, group leader, 1959—. Research and publs. on fluid dynamics and devel. of techniques for numerical solution of problems by high speed computer; pioneer in devel. of multi-dimensional techniques, including PIC and MAC methods. Home: 1407 11th St. Office: P.O. Box 1663, Los Alamos 87544.*

HARLOW, Harry F., Am. psychologist; b. Farifield, Ia., Oct. 31, 1905; s. Lon H. and Mable (Rock) Israel; student Reed Coll., 1923-24; A.B., Stanford, 1927, Ph.D., 1930; m. Clara Mears, 1932 (div. 1946); children—Robert M., Richard F.; m. 2d, Margaret Kuenne, 1948; children—Pamela Ann, Jonathan. Asst. prof. psychology U. Wis., 1930-38, asso. prof., 1938-44, prof., 1944-50, 1952-56, George Cary Comstock, prof. psychology, dir. primate lab., 1956—, dir. regional primate center, 1961—. Chief human resources research U. S. Army, 1950-52. Fellow American Academy of Arts and Scis.; mem. NRC (chmn. div. anthropology, psychology, 1954-56), Am. (pres. div. exptl. psychology 1950-51, pres. 1957-58), Midwest (pres. 1947-48) psychol. assns., American Philosophical Society, National Academy of Sciences, Gamma Alpha, Sigma Xi. Editor of Jour. Comparative and Physiological Psychology, 1951-63. In 35 years of expts. with rhesus monkeys, offered new hypothesis in areas of neurophysiology, motivation, love; showed that animal and human socs. are based on multiplicity of affectional bonds, rather than on sex as was previously assumed. Home: 2005 Jefferson St., Madison, Wis. 53711.*

HARMAN, Theodore Carter, Am. physicist; b. Warsaw, Ind., July 22, 1929; s. Seward W. and Bernice (Irvine) H.; A.B., Manchester Coll., 1951; M.S., Purdue U., 1953; m. Marilyn Axline, Aug. 3, 1957; children—Elizabeth Ann, Janet Elaine, Kathryn Sue, Thomas Carter. Project leader Battelle Meml. Inst., Columbus, O., 1953-58; asst. group leader Lincoln Lab., Mass. Inst. Tech., Lexington, 1959-65, 66—; asst. dir. for materials sci. Advanced Research Projects Agy., Washington, 1965-66. Mem. Am. Phys. Soc., Am. Inst. Mining, Metall. and Petroleum Engrs. (electronic materials com., 1965—), Sigma Xi. Author: (with J. M. Honig) Thermoelectricity, 1967. Invented thermal conductivity measurement technique known as Harman method; contbns. to preparation of high purity indium antimonide, mercury telluride and other intermetallic compounds, to measurement and understanding of properties of compound semicondrs.; understanding of thermoelectric devices. Home: 4 Ross Rd. Office: 244 Wood St., Lexington, Mass. 02173.*

HARMEL, Merel Hilber, Am. anesthesiologist; b. Cleve., May 19, 1917; s. Louis and Hermine (Greenbaum) H.; B.A., Johns Hopkins, 1938, M.D., 1943; m. Armide Chilcoat, July 2, 1944; children—Nancy Armide, Ruth Courtney, Priscilla Gover, Mary Louise. Practice medicine specializing in anesthesiology, Bklyn., 1952—; anesthesiologist-in-chief State U. Kings County Med. Center, 1952—, pres. med. bd., 1958-62, chmn. exec. com., 1964-65; cons. L.I. Jewish, St. Albans Naval, Maimonides, St. John's Episcopal, VA hosps.; prof., chmn. dept. anesthesiology State U. N.Y. Downstate Med. Center, 1952—. Commonwealth fellow Oxford U., 1961-62. Diplomate Am. Bd. Anesthesiology. Fellow Am. Coll. Anesthesiology; mem. A.M.A., Am. Soc. Anesthesiologists. Research and publs. on cerebral circulation and its regulation, particularly during anesthesia; effect of drugs on nervous system. Home: 6 Terrace Pl., Port Washington, N.Y. 11050. Office: Dept. Anesthesiology, State U. N.Y. Downstate Med. Center, 450 Clarkson Av., Bklyn. 11203.*

HARMEN, Raymond Andrew, Am. engr.; b. Oakland, Cal., Feb. 2, 1917; s. Charles and Rose (Alves) H.; A.B., U. Cal. at Berkeley, 1939, M.S., 1947; M.A., U. S. Naval Acad., 1943; m. Letitia Rooney, Oct. 4, 1951; 1 dau., Rosalee Pauline. Aerological officer U. S. Navy, 1944-46; meteorologist Landing Aids Expt. Sta., Arcata, Cal., 1948-50; physicist U. S. Naval Missile Center, Point Mugu, Cal., 1950-56; head environment determination br. U. S. Naval Missile Center, 1957-58, head flight reliability br., 1958-60, head reliability office, 1960—. Mem. Am. Inst. Aeros. and Astronautics (nat. tech. com. on reliability and maintainability). Research and publs. in application of math. theory of probability to engrng. systems, devel. math. models for predicting mission success of missile and astronautic systems, relating measurement to reliability, availability, performance. Home: 208 Ramona Pl., Camarillo, Cal. 93010. Office: P.O. Box 15, U. S. Naval Missile Center, Point Mugu, Cal. 93041.*

HARMER, Frederic William, Brit. geologist; b. Norwich, England, Apr. 24, 1835; s. Thomas and Emily Harmer; ed. Norwich; M.A. (hon.), Cambridge U.; m. Mary Lyon; 4 sons, 1 dau. Former chmn., mng. dir. Norwich Electricity Co. Ltd.; made 1st survey of glacial beds of east of Eng. (with Searles W. Wood, Jr.), 1864-72; investigator Pliocene and Pleistocene deposits of eastern and midland countries, Belgium, Holland. Recipient Murchison medal, 1902. Fellow Geol. Soc. (council 1896-1900), Royal Meteorol. Soc.; mem. Palaeontographical Soc. (council 1878-82), Norfolk and Norwich Naturalists Soc. (pres. 1877-79), Norfolk and Norwich Naturalists Soc.(pres. 1877-79), Geol. Soc. France, Soc. Belgian Geologists (hon.), Norfolk and Norwich Hort. Soc. Author: Pliocene Mollusca of Great Britain, 1914-21, also numerous articles on geology, paleontology, paleometeorology. Died Apr. 11, 1923.

HARMER, Sir Sidney Frederic, English zoologist; b. Norfolk, Eng., Mar. 9, 1862; s. Frederic William and Mary (Lyon) H.; B.Sc., U. Coll., London, 1880; B.A., King's Coll., Cambridge, 1883, M.A., 1886, Sc.D., 1893; m. Laura R. Howell, Sept. 2, 1891; 4 children. Became lectr. natural scis., Cambridge, 1886; asst. tutor King's Coll., 1890-1908; dir. Natural History Mus., London, 1919-27. Mem. Brit. Assn. for Advancement Sci., Royal Geographical Soc., Cambridge Philos. Soc., Linnean Soc. (pres. 1927-31, Gold medal 1934). Knighted, 1920. Became joint editor Cambridge Natural History, 1890. Author: Report on Polyzoa collected by (Dutch) Sibolga Expedition, 4 vols. Research and publs. on Polyzoa and cetacea, including reports of stranded whales. Died Oct. 22, 1950.

HARMON, Kent Midgley, Am. chemist; b. St. George, Utah, May 17, 1920; s. Irvin W. and Winnie (Midgley) H.; student Glendale Coll., 1937-40; A.B., summa cum laude, U. Redlands, 1942; M.S. in Chemistry, U. Cal. at Berkeley, 1944; Ph.D., Cal. Inst. Tech., 1949; m. LaFaye Steed, Sept. 23, 1942; children—David Kent, Claudia Faye, Kristi Ileene, Peter Irvin, Gary Louis. With Los Alamos Sci. Lab., U. Cal., 1944-46, supr. uranium recovery and purification group, 1945-46, supr. plutonium chemistry group, 1946; with Hanford Atomic Products Operation, Gen. Electric Co., Richland, Wash., 1946-64, supr. chem. research and separations chemistry groups, 1951-64; mgr. chem. separations unit chemistry dept. Battelle N.W., Richland, 1965—. Mem. Am. Chem. Soc., A.A.A.S., Sigma Xi. Contbg. author: Reactor Handbook, Vol. II, 1961; Process Chemistry,Vol. II, 1958. Research and numerous publs. in chemistry fission products, plutonium, uranium, Home: 1821 Thayer St. Office: 300 Area, Battelle N.W., Richland, Wash. 99352.*

HARMS, Jürgen Wilhelm, German zoologist; b. Bargdorf, Germany, Feb. 2, 1885; s. Wilhelm and Dorothea (Grimm) H.; student Marburg, Germany, Cambridge, Eng.; Ph.D., 1907; m. Frances Schele, 1909. Asst., U. Bonn; joined faculty U. Marburg, 1910; became prof. zoology U. Königsberg (now Kaliningrad, USSR), 1910, Tübingen, Germany, 1925, Jena, Germany, 1935; became guest worker Anat. Inst., U. Marburg, 1949; guest prof. U. Cairo (Egypt), 1951-52; made numerous research expdns., including, Lanzarole, 1913, Balearics, 1923, Java, 1926, Sunda Islands, 1928-30, Christmas Island, 1932, Sumatra, Japan, 1939, Chile, 1956. Author: Experimentelle Untersuchung über die innere Sekretion der Keimdrüsen, 1914; Individualzyklen als Grundlage für die Erforschung des biologischen Geschehens, 1924; Körper und Keimzellen, 2 vols., 1926; Wandlungen des Artgefüges unter natürlichen und Künstlichen Umweltbedingungen, 1934; also articles. Research on species transformation in evolution, endocrinology, especially effect of sex hormones in amphibians, earthworms, gephyreae, toads; thymus, problems of aging and rejuvenation; believed in inheritance of acquired characteristics. Died Marburg, Oct. 2, 1956.

HARMS, Robert Henry, Am. nutritionist; b. Dover, Ark., Sept. 27, 1923; s. Charles W. and Stella (Moore) H.; B.S., U. Ark., 1953, M.S., 1954; Ph.D., Tex. A. and M. U., 1956; m. Kathryn Sue McAlister, Apr. 29, 1944 (dec. Nov. 1961); m. 2d, Mary Eugenia Bryan, June 22, 1966; children—Carolyn Sue (Mrs. Elwin Thrasher), Robert Henry. Tchr., Springdale, Ark., 1946-51; asst. prof. U. Tenn. 1955-57; faculty U. Fla., Gainesville, 1957—, prof. nutrition, 1962—, chmn. dept., 1963—. Ralston Purina fellow, 1954. Recipient award Am. Feed Mfrs., 1965. Mem. Soc. Exptl. Biology and Medicine, Poultry Sci. Assn., Am. Inst. Nutrition, World Poultry Sci. Assn., Sigma Xi, Gamma Sigma Delta (award of merit U. Fla. 1962). Research, numerous publs. on protein, calcium and phosphorous requirements of chicks and hens; interrelationship of nutrition and disease. Home: 1421 N.W. 28th St., Gainesville, Fla. 32601.*

HARMS, Wilhelm Max Bruno, German physician; b. Berlin, Germany, Mar. 23, 1890; s. August and Emilie (Rutkowsky) H.; Ph.D., M.D., U. Berlin; m. Elisabeth Petroll, 1922; 1 son, Jurgen. Med. officer Verw. Bez. Tiergarten, Berlin, 1922-33; practice medicine, 1934-43; prof. Robert-Koch Inst. for Hygiene and Infectious Diseases, Berlin, 1949-52, and currently. Mem. German, Swiss socs. for history medicine and natural sci., Berlin Soc. for History Medicine, German Entomol. Soc., Internat. Paracelsus Soc., Gesellschaft der Naturforscher und Aerzte. Research and numerous works on hygiene and history of medicine. Address: 30 Katharinenstrasse, Berlin 37, Germany.

HARMSEN, Hans Ludwig Friedrich, German physician; b. Berlin-Charlottenburg, May 5, 1899; s. Friedrich and Therese (Hegemann) H.; ed. univs. Berlin, Marbourg, Munich, Vienna; M.D., Ph.D.; m. Annelise von Borsig; children—Hinrich, Ursula, Jörg, Dirk, Ruth, Hans-Christoph, Claas, Ernst, Arnold. Practice medicine, asst. Inst. Hygiene, Berlin, 1926-27; dir. Fedn. of Evang. Settlement of Hosps. and Treatment; dir. tech. medicine Interprofl. Control Health and Welfare, 1937-39; mem. Council Hygiene, 1939-45; head Acad. Medicine, Hamburg, 1946; full prof. hygiene and pub. health; dir. Inst. Hygiene, Hamburg. Mem. Ernst Barlach Soc. (pres.), Nor. Wissenschaftliche (pres.), Soc. for Social Hygiene, German Acad. Demographie (pres.). Editor: Städte-hygiene, 1950—. Research and numerous publs. on population, social hygiene and health. Home: 32 Ranzaustrasse, 2 Hamburg 70, Wandbeck. Office: 15-17 Gorch Fock Wall, 2 Hamburg 36, Germany.

HARNDT, Ewald Albert Heinrich, German dentist; b. Berlin, Germany, Jan. 22, 1901; s. Adolf and Emma Harndt; M.D., D.D.S., Friedrich Wilhelm U., Berlin; m. Frida Koepnik, 1928; children—Raimund, Thomas. Sci. asst. dept. dental treatment, Dental Inst., Berlin, 1927; chief physician, 1935; agrégé, 1936; instt., 1938; affiliate prof., 1944; asso. prof., dir., 1948, 50; full prof. Dental Inst., Free U. Berlin, 1956, dean Faculty Medicine, 1961. Recipient medal Gen. Assn. Dentists Belgium, Miller prize of Germany. Mem. Deutsche Gesellschaft für Zahn-Mund- und Kieferheilkunde (pres.), Deutsche Gesellschaft für Ernährung, Orca, Arbeitsgemeinschaft für Parodontopathien, Internat. Fedn. Dentists, Griechische Stomatologische Gesellschaft, Italian Stomatological Soc., Arpa Internat., Gesellschaft für Natur- und Heilkunde. Research and numerous publs. on histopathology, bacteriology and physiology. Home: 68-69 Knesebeckstrasse, Berlin 12. Office: 4-6 Assmannshauserstrasse, Berlin 41, Germany.

HARNED, Herbert S., Am. phys. chemist; b. Camden, N.J., Dec. 2, 1888; s. Thomas B. and Augusta (Traubel) H.; A.B., U. Pa., 1909, B.S., 1910, Ph.D., 1913; m. Dorothy Elizabeth Foltz, Sept. 8, 1917; children—Julia (Mrs. Chester F. Pardee), Herbert S., Louise, Eleanor. Faculty, U. Pa., Phila., 1913-28, prof., 1926-28; prof. phys. chemistry Yale, 1928-57, prof. emeritus, 1957—; group leader Manhattan Project, 1945-46. Cons., AEC, 1950—. Mem. Am. Chem. Soc., Faraday Soc., Nat. Acad. Sci. Author: (with Benton B. Owen) Physical Chemistry of Electrolytic Solutions, 1943. Research and numerous publs. on lower chloride niobium, salts in water, precise measurements of diffusion coefficients of salt in dilute solutions; determined ionization constant of water as a function of temperature; discovered rule mixture electrolytes known as Harneds rule. Home: 207 Armory St., New Haven 06511.*

HARNWELL, Gaylord P., Am. physicist; b. Evanston, Ill., Sept. 29, 1903; s. Frederick William and Anna Jane (Wilcox) H.; B.S., Haverford Coll., 1924; student Cambridge U. (Eng.), 1924-25; M.A., Princeton, Ph.D., 1927, LL.D., 1955; LL.D., U. Pa., 1953, Ursinus College, 1954, Dropsie College, 1955, U. Pitts. and Columbia U., 1957; D.Sc., Temple U., Haverford College, Hahnemann Med. College, 1954, Franklin and Marshall College, 1956, U. So. Cal., 1959, Drexel Inst., 1961; LL.D., Washington Coll., 1957, Northwestern U., 1958; Brown U., 1959, Swarthmore, 1959, William and Mary, 1960, Duke, St. Andrews (Scotland), 1963, U. Cal. at Los Angeles, Occidental Coll., 1964, Harvard U., 1965; Ph.D., LaSalle Coll., 1957; C.L.D., The Divinity School in Philadelphia, 1957; Sc.Ped. D., Elizabethtown Coll., 1959; L.H.D., Wilkes Coll., 1965; married Mary Louise Rowland, June 18, 1927; children—Mary Jane (Mrs. Harnwell Krumbhaar), Ann Wheeler (Mrs. John Ashmead), Robert Gaylord. Nat Research Council fellow, Calif. Inst. Tech., 1927-28, Princeton, 1928-29; asst. prof.

of physics, Princeton, 1929-36, asso. prof., 1936-38; prof. of physics, chmn. of the dept. and dir. of Randal Morgan Lab., U. Pa., 1938-53, pres., 1953—; leave absence, 1942-46 to act as dir. U. Cal. div. war research, U. S. Navy Radio and Sound Lab., San Diego, Cal.; dir. Philco Broadcasting Co., Pa. R.R. Co., PEN JERDEL; pres. West Phila. Corp. Member com. undersea warfare NRC. Mem. bd. mgrs. Franklin Inst.; dir. University City Sci. Center, University City Science Institute; member board of trustees Carnegie Found. for Advancement Teaching, Awarded medal for Merit. Mem. bd. dirs. United Fund of Phila., 1954—; chmn. exec. com. Greater Phila. Movement. Fellow Am. Phys. Soc., Accoustical Soc.; mem. Am. Philos. Soc. Newcomen Soc. N.A., Am. Council on Education (chmn. 1959-60), Sigma Xi, Phi Beta Kappa, Sigma Pi Sigma Alpha Epsilon Delta, Alpha Phi Omega. Cons. editor, Internat. Series in Physics, McGraw-Hill Book Co., 1946-53. Author: Principles of Electricity and Electromagnetism, 1929; Experimental Atomic Physics (with John J. Livingood), 1936; (with W. E. Stephens); Atomic Physics, 1955; Russian Diary, 1960; Educational Voyaging in Iran, 1962; (with George Legge) Physics—Matter, Energy and the Universe, 1967. Helped develop sonar; research in mass spectroscopy and collision of 2d kind in rare gases. Home: 8212 St. Martins Lane, Chestnut Hill, Phila.*

HARPER, John Lander, Brit. botanist; b. Rugby, Eng., May 27, 1925; s. John Hindley and Harriett (Archer) H.; M.A., Magdalen Coll., Oxford (Eng.) U., 1951, Ph.D., 1951; m. Borgny Lerö, Jan. 8, 1954; children—Belinda Solveig Jane; Claire Catherine Elise, Jonathan Tor. Demonstrator dept. agr. U. Oxford, 1948-59; prof. agrl. botany U. Coll. North Wales, Bangor, 1959—, head Sch. Plant Biology, 1967—. Fellow Inst. Biology; mem. Brit. Ecol. Soc. (pres. 1965-67), Soc. Exptl. Biology. Editor: Biology of Weeds, 1960. Research and publs. on weed biology, seed and seedling mortality, causes and consequences, plant competition and its ecol. and evolutionary consequences, population biology of plants, application of exptl. methods to ecol. problems. Home: Cae Groes, Glan-Y-Coed Park, Dwygyfylchi, Penmaenmawr, North Wales. Office: School Plant Biology, U. Coll. North Wales, Bangor, North Wales.*

HARPER, Paul Vincent, Am. surgeon; b. Chgo., July 27, 1915; s. Paul V. and Isabel (Vincent) H.; A.B. cum laude, Harvard, 1937, M.D., 1941; m. Phyllis Sweetser, Aug. 31, 1939; children—Stephanie Alice, Cynthia, William Rainey, David Paul. Faculty, U. Chgo., 1949—, asso. dir. Argonne Cancer Research Hosp., 1963—. Mem. Am. Physiol. Soc., Am. Surg. Assn., A.A.A.S., Soc. Clin. Surgery, Soc. Exptl. Biology and Medicine, A.C.S., Radiation Research Soc., Soc. Nuclear Medicine, Central, Chgo. (treas. 1960-63) surg. socs., Sigma Xi. Research, numerous publs. on gastric, pancreatic, pulmonary physiology, diagnosis and therapy with radioisotopes, radiative destruction of hypophysis, prodn. and use of Iodine 125 and Technetium 99m. Home: 543 Dundee Rd., Glencoe, Ill. Office: 950 E. 59th St., Chgo. 60637.

HARPER, Robert Almer, Am. botanist; b. Le Claire, Ia., Jan. 21, 1862; s. Almer and Eunice (Thomson) H.; A.B., Oberlin Coll., 1886, A.M., 1891; Ph.D., U. Bonn, 1896; m. Alice Jean McQueen, June 25, 1899; m. 2d, Helen Sherman, Jan. 2, 1918; 1 son, Robert Sherman. Prof. Greek and Latin, Gates Coll., 1886-88; instr. in acad. Lake Forest Coll., 1889-91, prof. botany and geology, 1891-98; prof. botany U. Wis., 1898-1911, Columbia, 1911-30, now emeritus. Mem. Nat. Acad. Scis., Am. Acad. Arts and Scis., Am. Philos. Soc., Bot. Soc. Am., A.A.A.S. Research on physiology of plants, cytology of fungi, fertilization in mildews, cell div. in slime molds, morphogenesis in algae. Died May 12, 1946.

HARPER, Roland M., Am. botanist, geographer; b. Farmington, Me., Aug. 11, 1878; s. William and Bertha (Tauber) H.; B.E., U. Ga., 1897, Sc.D., 1929; Ph.D., Columbia, 1905; m. Mary Susan Wigley, June 23, 1943. Aide, U. S. Nat. Herbarium, 1901-02; forestry collector Geol. Survey Ga., 1903-04; botanist, geographer Geol. Survey Ala., 1905-66, Fla. State Geol. Survey, 1908-31; asst. in agrl. geography U. S. Dept. Agr., 1917-18; head Fla. State Census, 1925; with Ala. Commn. Forestry, 1927; research prof. econs. U. Ga., 1928-29; former prof. U. Ala. Fellow A.A.A.S.; mem. Am. Eugenics Soc., Inter-Am. Soc. Anthropology and Geography, Ala. Acad. Sci., Fla. Acad. Scis., Am. Soc. Plant Taxonomists, Assn. Southeastern Biologists, N.E. Bot. Club, Torrey Bot. Club, So. Appalachian Bot. Club., Ecol. Soc. Am., Assn. Am. Geographers, So., Ga. hist. assns., Population Assn. Am. Contbr. numerous articles to tech. jours. Discovered one genus and 30 species of flowering plants. Home: 309 9th St., Tuscaloosa, Ala., 35401. Died Apr. 30, 1966.*

HARPESTRAENG, Henrik, physician; b. circa 1164; physician to Eric IV Waldemarssön (King of Denmark 1241-50); earliest Scandinavian writer on natural history and medicine; works include tract on laxative remedies, astrological papers, an herbal, articles on hygiene, diagnosis, surgery, med. tract of Salernitan type, lapidaries and cook books. Died 1244.

HARPMAN, Justo Arthur, surgeon, anatomist; b. Buenos Aires, Argentina, June 22, 1916; s. Jacob and Johanna (Vredenburg) H.; B.Sc. with 1st Class Honors in Anatomy, London U., 1937, M.B., B.S., 1940, Diploma in Laryngology and Otology, 1942, M.S., 1952; m. Annie Scott, June 7, 1941. Research asst. U. Coll., London U., 1937-39; asst. ear nose and throat surgeon Warwick Hosp., 1945-48, cons., 1948—; cons. ear, nose and throat surgeon Stratford-on-Avon, Shipston-on-Stour hosps., Warwickshire, Eng., 1948—. Postgrad. tchr. to hosp. resident; lectr. socs., orgns.; examiner Midland Inst. Otology, Birmingham, Eng., 1965-67. Fellow Royal Coll. Surgeons Edinburgh, Royal Soc. Medicine; mem. Anat. Soc. Gt. Britain, Inst. Otology. Research, publs. on pain, nerves of hearing, brain part connected with hearing, anatomy of head and neck, surgery of brain infections, diseases of nose, sinuses, especially treatment of cancer of throat and nose. Home: 32, Wellesbourne Rd., Barford, Warwick. Office: Warwick Hosp., Lakin Rd., Warwick, Eng.*

HARRAR, J. George, Am. biologist, found. exec.; b. Painesville, Ohio, Dec. 2, 1906; s. Ellwood Scott and Lucetta (Sterner) H.; A.B., Oberlin Coll., 1928; M.S., Ia. State U., 1929; Ph.D., U. Minn., 1935; m. Georgetta Steese, Jan. 11, 1930; children—Cynthia Ann (Mrs. Alvin Wilson), Georgetta Louise (Mrs. David T. Denhardt). Prof. biology U. P.R., 1929-33; instr. phytopathology U. Minn., 1934-35; prof. biology Va. Poly. Inst., 1935-41; head dept. phytopathology Wash. State U., 1941-42; with Rockefeller Found., N.Y.C., 1943—, field dir., 1943-52, deputy dir. agr., 1952-53, dir. agr., 1955-59, v.p., 1959-61, pres., trustee, 1961—. Mem. Nat. Research Council, Nat. Acad. Scis., Agy. for Internat. Devel. (Pres.'s sci. adv. com.), Inst. Nutrition Scis. (adv.bd.), Phytopathol. Soc., A.A.A.S., Am. Acad. Arts and Scis., Am. Pub. Health Assn. Author (with E. S. Harrar): Guide to Southern Trees, 1946, (with E. C. Stakman): Principles of Plant Pathology, 1957, Strategy for the Conquest of Hunger, 1963, rev. edit., 1967. Contbr. numerous articles in field to sci. jours. Pub. results of sci. contbns. in plant pathology, mycology; adminstr. program tech. assistance developing nations in Latin Am., Asia, Africa (emphasis on application of sci. and tech. to increasing food supplies). Home: 125 Puritan Dr., Scarsdale, N.Y. 10585. Office: 111 W. 50th St., N.Y.C. 10020.*

HARRASSOWITZ, Herman Ludwig Friedrich (Meyer until 1917), German geologist, paleontologist; b. Cottbus, Germany, Oct. 19, 1885; s. Hermann and Elizabeth (Harrassowitz) Meyer; student, Berlin; Ph.D., Freiburg (Germany), 1908; m. Ilse Zoch, 1918; 2 sons, 1 dau. Asst., Geol. Inst., Freiburg, 1906-08; became asst. Mineral. Inst., U. Giessen (Germany), 1908, joined faculty in geology and paleontology, 1910, became asso. prof., 1915; army geologist, Flanders and Russia, 1917-18; became prof., Giessen, 1920; forced to retire by Nazis, 1934; prof. Justus Liebig Inst., 1947-52, became emeritus, 1952. Recipient Von Reinach prize Senckenberg Natural Sci. Soc., 1921. Mem. Leopoldina. Author monograph on Eocene turtles of Messel. Research and publs. on chem. processes in geol. devel. of earth's crust, especially pioneered work in weathering, bauxite beds. Died Bad Ems, Germany, Apr. 18, 1956.

HARRELL, Everett Richard, Jr., Am. physician; b. Checotah, Okla., Apr. 5, 1922; s. Everett Richard and Golden (Duncan) H.; M.D., Duke, 1946; m. Rachel Ann McSwain, Oct. 22, 1947; children—Kathryn Reed, Patricia Ann, Carolyn Lee, Nancy West. Faculty, Med. Center, U. Mich., Ann Arbor, 1952—, prof. dermatology, 1960—, chmn. dept., 1967—. Mem. Am., Detroit (pres.) dermatol. assns., A.M.A. (sec. dermatology sect.), Am. Acad. Dermatology (bd. dirs.), Soc. for Investigative Dermatology, Mycol. Soc. Am., Internat. Soc. Tropical Dermatology, Sigma Xi, Alpha Omega Alpha. Research, publs. on fungus infections in humans; developed intravenous use of amphotericin B in systematic blastomycosis. Home: 3076 Geddes St., Ann Arbor, Mich. 48104.*

HARRELL, George Thomas, Am. physician; b. Washington, June 16, 1908; s. George Thomas and Anna (Muhlenberg) H.; A.B., Duke U., 1932, M.D., 1936; m. Janet Elliott Griffin, June 17, 1937; children—George Thomas III, Robert Griffin. Instr. medicine Duke U., 1939-41; mem. faculty Bowman Gray Sch. Medicine, Wake Forest Coll., Winston Salem, N.C., 1941-54, research prof. medicine, 1952-54; dean Coll. of Medicine, prof., U. Fla., Gainesville, 1954-64; dean Coll. Medicine, dir. Milton S. Hershey Med. Center, Pa. State U., Hershey, 1964—. Fellow Am. Coll. Physicians, Royal Soc. Tropical Medicine; mem. A.M.A., So. Med. Assn. (Silver medal 1944, Gold medal 1953), Assn. Am. Med. Colls., Pa., Dauphin County med. socs., Am. Soc. Clin. Investigation, Soc. for Experimental Biology and Medicine. Author: Medical Education Facilities, published 1964; also numerous articles. Research in alterations in capillary and cellular permeability to water, salts and proteins by isotopic techniques in infectious diseases and myxedema, clin. correlation of basic lab. studies in Rickettsial Spotted Fever, Boreck's Sarcoid and Myxedema. Home: 1141 Cocoa Av. Office: Milton S. Hershey Med. Center, 500 Univ. Dr., Hershey, Pa. 17033.*

HARRER, Gerhart, Austrian neurologist; b. Innsbruck, Austria, Jan. 28, 1917; s. Alois and Else (Göbel) H.; M.D., U. Vienna; m. Hildegund Baur, 1951; children—Stefan, Michael. Asst., Psychiat. Clinic, U. Vienna; chief physician, med. dir. Neurol. Clinic, Salzburg, Austria. Mem. Austrian Union Neurology and Psychiatry, Austrian van Swiet Soc. (pres.). Author: Vegetativendokrine Diagnostik. Research and numerous works on vegetative disturbances, traumatology of brain, psycho-pharmacology. Home: 8 Fichtenweg. Office: 79 Ignaz-Harrerstrasse, Salzburg, Austria.

HARRER, Joseph Marie, Am. elec. engr.; b. North Buffalo, N.Y., May 1, 1913; s. Joseph and Theresa (Harrer) Pschierer; B.S. in Elec. Engring., Rensselaer Poly Inst., 1934; M.S., Ill. Inst. Tech., 1952; m. Suzanne C. Litter, June 15, 1935; children—Lora (Mrs. G. Lavin), Nancy, Elizabeth Ann. Power plant engr. E.I. DuPont de Nemours, Buffalo, 1935-38; valuation engr. Republic Light Heat & Power, Buffalo, 1938-39; N.Y. State Pub. Service Commn., Rochester, 1939-40; elec. engr. Bausch & Lomb Optical Co., Rochester, 1940-43; process supr. Tenn. Eastman Corp., Oak Ridge, 1943-47; control engr. Oak Ridge Nat. Lab., 1947-48; sr. elec. engr. Argonne (Ill.) Nat. Lab., 1948—. Mem. Am. Nuclear Soc., Nat. Soc. Profl. Engrs., I.E.E.E., Research Soc. Am. Author: Nuclear Reactor Control Engineering, 1963; also articles. Research on nuclear reactor control and instrument engring., safety analysis, engring. and operation nuclear power plants. Patentee nuclear reactor control devices. Home: 265 Cottage Hill St., Elmhurst, Ill. 60126. Office: 9700 S. Cass Av., Argonne, Ill. 60439.*

HARRINGTON, Charles Dana, Am. chemist; b. White Plains, N.Y., July 22, 1910; s. Eugene Thomas and Mary (Quirk) H.; B.S. in Chemistry, Harvard, 1932, M.A., 1939, Ph.D., 1941; m. Emily Elizabeth Whitney, 1941. Chemist, Mallinckrodt Chem. Works, St. Louis, 1941-43, plant 4 mgr., 1943-46, tech. dir. uranium div., 1946-52, mgr. uranium div., 1952-60, v.p., 1960-61; v.p. United Nuclear Corp., Centreville, Md., 1961-63, sr. v.p., 1963-65; pres. Douglas United Nuclear, Inc., Richland, Wash., 1965—, dir., 1965—. Mem. Am. Chem. Soc. (Mid-West award for work in devel. uranium processing 1960), Am. Nuclear Soc. Author: (with A. E. Ruehle) Uranium Production Technology, 1959; also articles. Research in chemistry and metallurgy uranium and processing methods, spl. classified fuels for mil. application, plutonium prodn. methods. Patentee in field. Home: 722 W. Court St., Pasco, Wash. Office: Fed. Bldg., Richland, Wash. 99352.*

HARRINGTON, James Foster, Am. plant physiologist; b. Newark, N.J., Nov. 24, 1916; s. Ralph Emerson and Elizabeth (Williams) H.; B.S., Ohio State U., 1939; M.S., 1940; Ph.D., Cornell U., 1944; m. Helen Myrtle Butcher, June 18, 1939; children—Sally (Mrs. Leigh Holmes), Gordon James, Mark John. Asst. prof. horticulture, Ia. State U., 1944-46; faculty U. Cal. at Davis, 1946—, prof. vegetable crops, 1961—. Cons., U. S. AID program, Miss. Contract, Brazil, 1965-66. Fulbright Research grantee Hort. Lab. Netherlands Agrl. U., 1954-55; NSF-U. S. Dept. Agr. fellow Seed Protein Pioneering Research Lab., New Orleans, 1962-63; named Horticulturist of Year, All Am. Selections, Am. Seed Trade Assn., 1964. Mem. A.A.A.S., Am. Soc. for Hort. Sci., Am. Soc. Plant Physiologists, Assn. Ofcl. Seed Analysts, Am. Assn. U. Profs., Cal. Seed Council, Sigma Xi. Research and publs. on environmental factors which improve yield and germination of seed crops, storage and moisture resistance packaging of seed leading to improved germination of seed for planting. Home: 714 Anderson Rd., Davis, Cal. 95616.*

HARRINGTON, Joseph, Jr., Am. mech. engr.; b. Riverside, Ill., Sept. 21, 1908; s. Joseph and Cora (Dunlap) H.; S.B., Mass. Inst. Tech., 1930, Sc.D., 1932; m. Alene Louisa Smith, Sept. 17, 1932; children—Joan White (Mrs. Roger G. Smith), Joseph III, Anne Dunlap (Mrs. John F. Heider). With United Shoe Machinery Corp., Boston, 1932-55, asst. dir. research, 1945-55; sr. staff mem. in charge mech. engring. Arthur D. Little, Inc., Cambridge, Mass., 1955—. Mem. Electronic Industries Assn., Soc. for Exptl. Stress Analysis, Numerical Control Assn., A.A.A.S., Phi Eta Sigma, Tau Beta Pi. Contbr. numerous articles to profl. pubs. Inventor, designer, developer mech. products and automatic prodnl. machinery. Patentee in field. Home: 1 Cherry St., Wenham, Mass. 01984. Office: 20 Acorn Park, Cambridge, Mass. 02140.*

HARRINGTON, Mark Walrod, Am. astronomer; b. Sycamore, Ill., Aug. 18, 1848; s. James and Charlotte (Walrod) H.; student Northwestern U., 1864-66; grad. U. Mich., 1868, A.M. (hon.), 1870, LL.D., 1894; postgrad. U. Leipzig (Germany), 1876-77. Taught math. and geology U. Mich., 1868-71, 71-76, prof. astronomy, 1879-91; worked for U. S. Coastal Survey in Alaska, 1870-71; taught math. and astronomy at Fgn. Office Cadet Sch., Peking, China, 1877-78; prof. astronomy U. La., 1878-79; founder, editor Am. Meteorol. Jour., 1884-92; 1st civilian head U. S. Weather Bur., 1891-95; pres. U. Wash., 1895-97; mem. Imperial Anthrop. Soc. Moscow, Linnean Soc. of London, Austrian Meteorol. Soc.; corr. mem. Scottish Geog. Soc. Author: About the Weather, 1899; also many scientific papers. Discovered irregular periodic change of light in star Vesta, trifid character of Gt. Hercules nebula. Died Oct. 9, 1926.

HARRINGTON, Marshall Cathcart, Am. physicist, govt. ofcl.; b. Rockford, O., Aug.28, 1904; s. Marshall and Elizabeth Jane Cathcart) H.; A.B. in Physics with highest honors, Princeton, 1926, A.M. (Queen fellow in physics), 1927, Ph.D., 1932; m. Lillian Gottlieb, Nov. 7, 1958; Instr. physics Princeton, 1927-29, research asst., 1930-31; faculty physics Drew U., 1931-55, prof., 1941-55, chmn. div. sci., 1943-51; physicist David Taylor Model Basin, Dept. Navy, 1955-57, head fluid dynamics br., 1957-60, contract research administr. Bur. Ships Hydromechanics Research Program, 1960-62; project scientist gen. physics div. Directorate Phys. Scis., Air Force Office Sci. Research, Washington, 1962-63, acting chief gen. physics div., 1963-64, project scientist, 1962—. Tech. assistance expert physics UNESCO, Iraq, 1952-54; professorial lectr. physics George Washington U., 1955-57, U. Md., 1956-57. Fellow A.A.A.S., Am. Phys. Soc., Am. Inst. Aeros. and Astronautics (asso.); mem. Am. Astron. Soc., Am. Physics Tchrs. Assn., Philos. Soc. Washington, Optical Soc. Am., I.E.E.E. (sr.), Am. Geophys. Union, Sci. Research Soc. Am., N.Y. Acad. Scis., Nat. Audubon Soc., Nat. Wildlife Fedn., Wilderness Soc., Nat. Parks Assn., Phi Beta Kappa, Sigma Xi, Sigma Pi Sigma. Research in molecular and atomic physics, radio astronomy, optics. Home: 4545 Connecticut Av. N.W., Washington 20008. Office: Air Force Office Sci. Research, Arlington, Va. 22209.*

HARRINGTON, Richards Harry, Am. phys. metallurgist; b. Grand Rapids, Mich., Apr. 23, 1904; s. Harry Eli and Gretchen (Orth) H.; B.S., U. Mich., 1925, M.S., 1926, Sc.D., 1929; m. Iverna Hill, Nov. 22, 1941. Research asso. Gen. Electric Co., Schenectady, 1929-51; metall. cons., Schenectady, 1951-55; asst. prof. Engring. Sch. U. Mass., 1955-59; chief materials sci. lab. Watervliet (N.Y.) Arsenal, 1959—. Mem. Am. Soc. Metals, Am. Inst. Mining, Metall. Engrs., Sigma Xi, Phi Lambda Upsilon, Alpha Chi Sigma. Author: Modern Metallurgy of Alloys, 1949; also contbr. to three symposia. Patentee in field. Research and publs. primarily in field of new alloys, treatments of metals. Home: 1512 Barclay Pl., Schenectady 12309. Office: Watervliet Arsenal, Watervliet, N.Y.*

HARRIOT, Thomas, English mathematician, astronomer; b. Oxford, Eng., 1560; B.A., Oxford U., 1580. Math. advisor to Sir Walter Raleigh, Durham House, 1579-1603; sent as surveyor to Virginia, (U. S.), 1585; in service of Earl of Northumberland, 1607-1621. Author: A Brief and True Report of the New Found Land of Virginia (1 of earliest statis. surveys on large scale), 1588; Arts Analyticae Praxis, 1631. 1st Englishman to explore and describe natural history of N. America (brought tobacco plants and potatoes to Europe); considered founder of English sch. of algebra; gave algebra its modern form; made exhaustive study on theory of equations; 1st to factor equations; adopted small letter notation used today, also introduced symbols of inequality and used dot to signify multiplication; in navigation anticipated work of Edward Wright; observed sun's spots and Jupiter's satellites at same time as Galileo (but did not publish report); Mentioned ellipticity of planetary orbits; discussed atomism. Died July 2, 1621.

HARRIS, Albert Sidney, Am. physiologist; b. Banks, Ala.; s. Thomas Lee and Amanda (McLendon) H.; B.S., U. Ala., 1930; Ph.D., Wash. U., 1934; m. Margaret Nancy Black, Dec. 27, 1940. Instr., Ala. Inst. Deaf and Blind, 1930-31; lectr. Central Inst. Deaf, St. Louis, 1931-33; instr. physiology, pharmacology U. Ala., 1934-35, asst. prof., 1935-37; research asso. Wash. U., 1937-39; sr. instr. physiology Western Res. U., 1939-41, asst. prof., 1941-45; asso. prof. Baylor Coll. Med., 1945-47, prof., 1947-51; prof., head dept. physiology La. State U., New Orleans, 1951—; mem. Heart Program Project Com., 1963-67; vis. prof. U. Indonesia Sch. Medicine, 1957-58. Mem. Am. Physiol. Soc., Am. Heart Assn., InterAm. Cardiology Congress, World Congress Cardiology, Internat. Physiology Congress, Soc. Exptl. Biology and Medicine. Mem. editorial bd. Am. Jour. Electrocardiology, 1966—. Research, publs. on specific afferent nerve fibers in limb extension and flexion, analysis of mechanisms of initiation of ventricular fibrillation, antiarrhythmic drug actions, mechanisms of ventricular arrhythmias following occlusion of coronary arteries, effectiveness of coordinated ventricular assistance pump in myocardial infarction and other cardiovascular impairments. Home: 1525 Seville Dr., New Orleans 70122.*

HARRIS, Chapin Aaron, Am. dentist; b. Pompey N.Y., May 6, 1806; s. John and Elizabeth (Brundage) H. A.M., M.D.; m. Lucinda Heath Hawley, Jan. 11, 1826, 9 children. Licensed as dentist by Med. and Chirurg. Faculty of Md., Balt., 1833; editor Am. Jour. Dental Science (1st dental periodical), 1839; organizer Balt. Coll. of Dental Surgery (world's 1st dental coll.), 1839, 1st dean, 1st prof. operative dentistry and dental prosthesis, 1840-44, pres., 1844-60; an organizer Am. Soc. of Dental Surgeons (1st nat. dental assn.), 1840, 1st corr. sec., pres. 1844; an organizer Am. Dental Conv., 1855, pres., 1856-57. Author: The Dental Art, a Practical Treatise on Dental Surgery, 1839; translator: A Treatise on Second Dentition (C. F. Delabarre), 1845; Complete Elements of the Science and Art of the Dentist (A. M. Desiraback), 1847; editor: Natural History and Diseases of the Human Teeth (Joseph Fox). Died Balt., Sept. 29, 1860.

HARRIS, Chauncy Dennison, Am. geographer; b. Logan, Utah, Jan. 31, 1914; s. Franklin S. and Estelle (Spilsbury) H.; A.B., Brigham Young U., 1933; B.A., Oxford U., 1936, M.A., 1943; Ph.D., U. Chgo., 1940; D. Econ. (hon.), Cath. U. Chile, 1956; m. Edith Young, Sept. 5, 1940; 1 dau., Margaret. Faculty, Ind. U., 1939-41, U. Neb., 1941-43; faculty U. Chgo., 1943—, prof. geography, 1947—, chmn. Non-Western Area Programs and Internat. Studies, 1960-66, dir. Center for Internat. Studies, 1966—. Dir. Social Sci. Research Council, 1959—. Recipient Distinguished Service award Geog. Soc. Chgo., 1965. Mem. Assn. Am. Geographers (pres. 1957), Inst. Brit. Geographers, Association de Géographes Français, Internat. Geog. Union (v.p. 1956-64), Am. Assn. for Advancement Slavic Studies (pres. 1962), Am. Geog. Soc. (councilor 1962—); hon. mem. Royal Geog. Soc. London, geog. socs. Berlin, Frankfurt, Rome, Beograd. Author: Salt Lake City: A Regional Capital, 1940; (with Jerome D. Fellmann) International List of Geographical Serials, 1960. Devised functional classification of cities based on employment and occupational data; innovated method of measuring market potential with spl. reference to indsl. location; studied pattern, growth, characteristics, and interrelationships of urban centers in Soviet Union. Home: 5649 Blackstone Av., Chgo. 60637.*

HARRIS, Cyril Manton, Am. physicist; b. Detroit, June 20, 1917; s. Bernard O. and Ida (Moss) H.; B.A., U. Cal. at Los Angeles, 1938, M.A., 1940; Ph.D., Mass. Inst. Tech., 1945; m. Ann Schakne, July 12, 1949; children—Nicholas Bennett, Katherine Anne. Researcher, Carnegie Instn. Washington, 1941; research engr. Bell Telephone Labs., 1949-51; sci. cons. Office Naval Research, U. S. Embassy, London, Eng., 1951; Fulbright vis. lectr. Tech. U. Delft (Holland), 1951-52; vis. Fulbright prof. U. Tokyo (Japan), 1960; faculty Columbia U., N.Y.C., 1952—, prof. elec. engring. and architecture, 1964—. Panelist, Nat. Acad. Scis. 1965—; acoustical cons. Met. Opera House, Lincoln Center for Performing Arts, N.Y.C., John F. Kennedy Center for the Performing Arts, Washington, Powell Symphony Hall, St. Louis. Fellow I.E.E.E., Audio Engring. Soc., Acoustical Soc. Am. (past prs., asso. editor Jour. 1959—); mem. Bd. Standards and Planning, ANTA, U. S. Inst. for Theatre Tech. (dir. 1964—), Physics Inst., Phys. Soc. London, Acoustics Group, Am. Inst. Physics (mem. governing bd. 1965-66). Author: (with V. O. Knudsen) Acoustical Designing in Architecture, 1950; (with C. E. Crede) Shock and Vibration Handbook, 1961; also articles. Editor: Handbook of Noise Control, 1957. Mem. editorial adv. bd. Physics Today, 1955-66. Research in room acoustics and application to design auditoriums and concert halls; developed acoustical analysis equipment, techniques and methods noise control; research on analysis speech and techniques for synthesizing human speech. Home: 425 Riverside Dr., N.Y.C. 10025. Office: S.W. Mudd Bldg., Columbia U., N.Y.C. 10027.*

HARRIS, Dale B., Am. psychologist; b. Elkhart, Ind., June 28, 1914; s. Ward Manning and Lillian (Benner) H.; A.B., DePauw U., 1935; M.A., U. Minn., 1937, Ph.D., 1941; m. Elizabeth Saltmarsh, July 17, 1935; children—Ruthann E., James S., David B., Geoffrey M. Ednl. dir. Minn. State Tng. Sch. for Boys, Red Wing, 1936-38; research asst. Stanford Language Arts Investigation, Stanford U., 1938-39; with U. Minn., 1940-59, prof. Inst. Child Welfare, 1948-59, dir., 1954-59; prof. Pa. State U., 1959—, acting head dept. psychology, 1962-63, head dept. psychology, 1963-67; cons. U. S. VA, 1963—, U. S. Children's Bur., 1962—, U. S. Dept. Labor 1955-59; mem. edl. bd. Annual Review of Psychology, 1956-62. Fellow Am. Psychol. Assn., Soc. for Research in Child Development, A.A.A.S.; mem. Am. Ednl. Research Assn. (v.p., 1964-66). Author: Children's Drawings as Measures of Intellectual Maturity, 1963. Editor Child Development Abstracts & Bibliography. Studies, publs. in child development and adjustment including interest patterns of juvenile delinquents; parent-child relationships; character structure of children; children's cognitive and intellectual development; early childhood edn. Home: 317 Ridge Av., State College, Pa. 16801. Office: Burrowes Bldg., University Park, Pa. 16802.*

HARRIS, Frank Ephraim, Am. chemist; b. Boston, Aug. 26, 1929; s. Frank Ephraim and Wilhelmina (Sellers) H.; A.B., Harvard, 1951; Ph.D., U. Cal. at Berkeley, 1954. Instr. chemistry Harvard, 1953-56; asst. prof. chemistry U. Cal. at Berkeley, 1956-59; faculty Stanford U., 1959-68, asso. prof. chemistry, 1961-68; professor of physics U. Utah, since 1968—; cons. Lawrence Radiation Lab. U. Cal., United Aircraft Research Labs.; lectr. Internat. Summer Inst. in Quantum Chemistry; acting dir. Quantum Chemistry group Uppsala U. (Sweden), 1963-64; mem. adv. editorial bd. Internat. Jour. of Quantum Chemistry, also Quantum Theory of Matter. A.P. Sloan fellow, 1957-59. Author: Principles of Chemistry, 1963-65. Research, publs. in theoretical chemistry; theory of orientation of molecules in an electric field; dimensions and charge distbn. in charged polymer molecules; devel. of methods for the calculation of the distbn. of electrons in atoms and molecules, especially for use on large-scale digital computers. Office: Dept. Physics, U. Utah, Salt Lake City 84112.*

HARRIS, Franklin Stewart, Jr., Am. physicist; b. Logan, Utah, May 24, 1912; s. Franklin Stewart and Estella (Spilsbury) H.; A.B., Brigham Young U., 1931, M.A., 1936; Ph.D., Cal. Inst. Tech., 1941; m. Maurine Steed, June 7, 1950; children—Franklin Stewart, III, JoAnne (Mrs. Robert Franklin Moyle), Bernie McCune, Michael Glenn, Morgan James, Wendell Leon, Della Elizabeth, Chauncy Steed, Linda Maurine, Mark Dennison, Diana Marie. Research asst. Cal. Inst. Tech., 1937-41; lectr. U. B.C., Vancouver, Can., 1941-43; faculty U. Utah, 1943-63, prof. physics, 1952-63; mem. tech. staff Aerospace Corp., El Segundo, Cal., 1962—; prin. engr., cons. Sperry Utah Co., Salt Lake City, 1959-62; cons. Intermountain Weather, Inc., 1961-65. Fellow Optical Soc. Am., Phys. Soc. London, A.A.A.S.; mem. Am. Assn. Physics Tchrs., I.E.E.E., Am. Meterol. Soc., Utah Acad. Scis., Arts and Letters (past mem. council), Am. Geophys. Union, Sigma Xi. Author: Experimental Physics, Part I 1961, Part II, 1961, Part III, 1963; also articles. Research on diffraction of light, photography of patterns, measurement of far-field and wide angle patterns, polarization, cloud physics, scattering of light, heterogenous nucleation of ice, lasers, infrared satellites. Home: 15514 Tuba St., Mission Hills, Cal. 91340. Office: Aerospace Corp., P.O. Box 95085, Los Angeles 90045.*

HARRIS, Gerard Glendinning, Am. physicist; b. Rochester, N.Y., Sept. 28, 1926; s. Ellwood Mathew and Irene (Glendinning) H.; B.S., U. Rochester, 1947; Ph.D. in Physics, Princeton, 1951; m. Anita Pratt, Sept. 7, 1949 (div. Sept. 1958); children—Neal Kathleen, Paul Nathan; m. 2d, Josephine Whitmeyer, June 17, 1961; children—Helen, Jessica, Josephine. Fulbright scholar, Bristol, Eng., 1951-53; faculty Columbia, 1953-58, asst. prof., 1955-58; mem. tech. staff Bell Telephone Lab., Murray Hill, N.J., 1958—. Mem. A.A.A.S., Acoustical Soc. Am. Author: (with others) Modern Communications, 1962; also articles. Research in elementary particle physics, psychoacoustics, bioacoustics, neurophysiology; helped establish properties tau meson leading to prediction non-conservation parity; clarified function of lateral line sensory organ in fish. Home: 87 Salem Rd. Office: Bell Telephone Labs., Murray Hill, N.J. 07974.*

HARRIS, Gilbert Dennison, Am. geologist; b. Jamestown, N.Y., Oct. 2, 1864; s. Francis E. and Lydia Helen (Crandall) H.; Ph.B., Cornell U., 1886; m. Clara Stoneman, Dec. 30, 1890; 1 dau., Rebecca Stoneman. On U. S., Tex. and Ark. geol. surveys, 1887-93; investigated tertiary deposits of So. Eng. and No. France, 1894; asst. prof. paleontology and strategraphic geology Cornell U., Ithaca, N.Y., 1894-1909, prof. 1900-35. emeritus; founder Paleontol. Research Instn., 1932. State geologist of La., 1899-1909. Cons. geologist Trinidad Petroleum Devel. Co., Trinidad, B.W.I., 1919-23; paleontologist Standard Oil Co. of Venezuela, 1923-25; vis. grad. lectr. geology Tex. State U., 1927; paleontologist Oldenbergische Erdölgesellschaft, Germany, 1929. Fellow Geol. Soc. Am. (v.p. 1937); Paleontol. (pres. 1936); mem. Société Geologique de France, Paleontologische Gesellschaft v. Deutschland, Société géologique de Suisse, Acad. Nat. Sci., Phi Beta Kappa, Sigma Xi. Editor, propr. Bulls. Am. Paleontology, 1895-1952, Paleontographica Americana. Research on Eocene and Miocene periods. Died Dec. 4, 1952.

HARRIS, Gordon McLeod, chemist; b. Chungking, China, July 23, 1913; s. George Gordon and Agnes (McLeod) H.; B.Sc. with honors in Chemistry, U. Sask. (Can.), 1939, M.Sc., 1940; A.M., Harvard, 1942, Ph.D., 1943; m. Justyn Lee Montgomery, June 10, 1943; children—David, Deborah, Matthew. Came to U.S., 1952, naturalized, 1958. Research chemist NRC Can., Ottawa, Ont., 1944-45; asst. prof. chemistry U. Sask., Saskatoon, 1945-48; sr. lectr. in phys. chemistry U. Melbourne, Australia, 1948-52; research asso. U. Wis., Madison, 1952-53; faculty U. Buffalo, 1953—, prof., 1955-61, Larkin Prof. chemistry, 1961—, chmn. chemistry dept., 1956—; hon. research asso. U. Coll., London, Eng., 1961. Mem. Am. Chem. Soc. (chmn. Western N.Y. sect. 1964-65), Chem. Soc. (London), A.A.A.S., Am. Assn. U. Profs., Sigma Xi. Author: Chemical Kinetics, 1966; also articles. Research on reactions hydrogen atoms in gas phase, isotopic tracer studies chem. reactions, isotopic exchange reactions, isotopic effects on rate reaction and chem. equilibrium, mechanisms reactions inorganic complexes in solution and in solid state. Home: 128 Crosby Blvd., Buffalo 14226.*

HARRIS, Halbert Marion, Am. entomologist; b. Charleston, Miss., July 18, 1900; s. Thomas Marion and Louise Elizabeth (Ross) H.; B.Sc., Miss. A. and M. Coll., 1923; M.Sc., Ia. State U., 1925, Ph.D., 1928; m. Katherine Rentrop Day, Aug. 6, 1927; 1 son, Halbert Marion. Mem. faculty Ia. State U., Ames, 1923-61, prof. entomology, 1942-61, chmn. dept. zoology and entomology, 1947-61; state entomologist State of Ia., 1947-61, also state apiarist; vis. prof. U. Agrl. Scis., Mysore State, India, 1965-68, Mem. Nat. Plant Bd., 1950-52, Central Plant Bd., 1947-61; chmn. com. NRC, 1956-60; cons. Ford Found., 1960—. Mem. Entomol. Soc. Am., Entomol. Soc. India, Sigma Xi. Research and numerous publs. on taxonomy and biology of Nabidae, Gerridae, related families of heteropterous insects. Home: 16 Commissariat Rd., Bangalore, Mysore, India. Office: 32 Ferozshah Rd., New Delhi 1, India.*

HARRIS, Henry Clayton, Am. agronomist, plant physiologist; b. Pike Road, N.C., Dec. 12, 1898; s. Jesse Bryant and Jessie (Harrison) H.; A.B., U. N.C., 1922; Ph.D., Cornell U., 1927; postgrad. Oak Ridge Inst. Nuclear Studies; m. Martha Louise Robertson, Dec. 24, 1930; children—Henry Clayton, Ann Robertson (Mrs. Fred Copeland Hunter). Faculty, Cornell U., 1924-26, Sam Houston State Tchrs. Coll., 1927-28, U. Del., 1928-43; faculty U. Fla., Gainesville, 1943-—, prof. agronomy, 1946-50, agronomist, 1946-—, in charge agronomy isotope and biochem. lab., 1954-60. Fellow Oak Ridge Inst. Nuclear Studies, 1950, recipient Nuclear Studies Sci. award, 1950. Fellow A.A.A.S., Am. Soc. Agronomy; mem. Am. Chem. Soc., Soil Sci. Soc. Am., Internat. Soil Sci. Soc., Am. Soc. Plant Physiologists, Scandinavian Soc. for plant Physiology, Soil and Crop Sci. Soc. of Fla., Sigma Xi, Gamma Alpha, Gamma Sigma Delta Distinguished Service award Fla. 1958). Research, numerous publs. on effect of nutrient deficiencies, particularly micronutrients, on crops; demonstrated that imbalance of micronutrients greatly affected nitrogen and mineral content of crops; 1st to find boron deficiency caused internal imperfection of peanut seed; research with radioactive calcium showing that root applications of calcium failed to move adequately into peanut fruit for normal seed devel. Home: 3020 S.W. 1st Av., Gainesville, Fla. 32601.*

HARRIS, Jesse Graham, Jr., Am. psychologist; b. Jacksonville, Fla., Jan. 5, 1926; s. Jesse Graham and Mona (Woods) H.; B.A., Harvard, 1946; Ph.D., Duke, 1955; m. Julia Patricia McNamee, Sept. 5, 1953; children—Julia Kathleen, Cecilia Anne. Asst. prof. dept. psychology U. Conn., Storrs, 1958-60; faculty U. Ky., Lexington, 1960-—, prof., chmn. dept. psychology, 1963-—, dir. psychology dept. psychiatry, 1960-63. Mem. Ky. State Bd. Psychology, 1966-—, Ky. Mental Health Manpower Commn., 1963-—; cons. to hosps.; cons. Peace Corps, 1967-—. Served to lt. Med. Services, USNR, 1955-58; now lt. comdr. Res. Diplomate Am. Bd. Examiners in Profl. Psychology. Fellow Am. Psychol. Assn.; mem. Southeastern, Midwestern, Ky., Central Ky. (past pres.), Western Fla. (past pres.) psychol. assns. Contbg. author: Rorschach Psychology, 1960. Research, publs. in schizophrenia, test analysis and evaluation, clin. psychology. Home: 3356 Bellefonte Dr., Lexington, Ky. 40502.*

HARRIS, John, Brit. mathematician, clergyman; b. circa 1667; ed. Oxford; took holy orders; pvt. tchr. math.; pub. lectr. applied math.; prof. math., London. Fellow Royal Soc., 1696 (sec. 1709-10). Author: Lexicon Technicum; Collection of Voyages; Voyages and Travels; Dictionary of Arts and Sciences, 1704. Died Sept. 9, 1719.

HARRIS, John Donald, Am. psychologist; b. Brownsburg, Pa., May 12, 1914; s. Walter W. and Pearl (Graves) H.; B.A., Maryville Coll., 1931; M.A., Vanderbilt U., 1936; postgrad. U. Colo.; Ph.D., U. Rochester, 1942; m. Barbara Lillian Nelson, July 12, 1946; children—John Curtis, Ralph Dexter, Lisa Helene, Todd Nelson, Heather Anne. Instr. English, Del. Sr. High Sch., 1936-39; Univ. fellow U. Colo., 1939-40; fellow U. Rochester, 1940-42, instr. psychology, 1942-43; head auditory research br. U. S. Naval Med. Research Lab., Groton, Conn., 1943-—, chief scientist Behavioral Scis. Div., Submarine Med. Center, 1966-—; dir. research C. W. Shilling Auditory Research Center, 1957-—; adj. prof. City U. N.Y., 1965-—. Mem. A.A.A.S., Am. Inst. Physics, Acoustical Soc. Am., Am. Speech and Hearing Soc., Am. Psychol. Assn. Editor: Jour. Auditory Research, 1959-—. Author: Some Relations Between Vision and Audition, 1953. Contbr. numerous definitive articles to sci. jours. on pitch discrimination, loudness discrimination, binaural hearing, speech intelligibility testing, audiol. testing in otol. clinic; definition of primary auditory abilities; definition of auditory capabilities necessary for sonar operator. Home: 230 Prospect Hill Rd., Groton, Conn. 06340. Office: U. S. Naval Submarine Med. Research Lab., U. S. Naval Submarine Base, Groton, Conn. 06340.*

HARRIS, John Ferguson, Jr., Am. chemist; b. Stroudsburg, Pa., Apr. 15, 1925; s. John Ferguson and Georgia (Lightbourne) H.; A.B., U. Pa., 1948, M.S., 1950, Ph.D., 1953; m. Jacqueline Scott, Sept. 3, 1949; children—Mark Scott, John David, Katherine. Research chemist, central research dept. E. I. DuPont de Nemours & Co., Wilmington, Del., 1952-—. Guggenheim fellow U. Cologne (Germany), 1963-64. Mem. Am. Chem. Soc., Chem. Soc. London. Research, publs. on chemistry of organic fluorine and sulfur compounds. Home: 2541 Deepwood Dr., Foulk Woods, Wilmington 19803. Office: DuPont Exptl. Sta., Wilmington, Del. 19898.*

HARRIS, John William, Am. physician; b. Boston, Mar. 30, 1920; s. Ulysses S. and Lillian (Dennett) H.; B.S., Trinity Coll., 1941; M.D., Harvard Med. Sch., 1944; m. Stephanie Jean Bunting, Apr. 7, 1951; children—Wendy Alexandra, Stephen Bunting, Alison Dennett. Research fellow in medicine Harvard, 1948-51, research asso., 1951-52; sr. instr. medicine Western Reserve U. Sch. Med., 1952-54, Markle scholar in med. sci., 1955-60, prof. medicine, 1962-—; hematologist, vis. physician Cleve. Met. Gen. Hosp., 1952-—; attending physician Crile VA Hosp., Cleve., 1953-58; mem. com. on blood and transfusion problems Nat. Acad. Scis., 1963-66; mem. Hematology study sect. USPHS (research career award 1962), 1962-64. Mem. Am. (v.p. 1965), central socs. clin. investigation, A.C.P., Am. Fedn. Clin. Investigation, Assn. Am. Physicians, Soc. Exptl. Biology and Medicine Am. Soc. Hematology, Phi Beta Kappa, Alpha Omega Alpha. Author: The Red Cell, 1963. (with D. L. Horrigan) Case Problems in Hematology, 1963. Investigations into anemias due to defective red cell prodn., increased red cell destruction, red cell metabolism, correlations between structure, function and behavior of hemoglobin. Home: 3080 Coleridge Rd., Cleveland Heights, Ohio 44118. Office: Cleve. Metropolitan Gen. Hosp., 3395 Scranton Rd., Cleve. 44106.*

HARRIS, Lawrence Arnold, physicist; b. Toronto, Ont., Can., Jan. 25, 1923; s. William and Tillie (Shayne) H.; B.A.Sc., U. Toronto, 1946; S.M., Mass. Inst. Tech., 1948, Sc.D., 1950; m. Wilma Esser, May 19, 1946; children—Daniel Jay, Dorothy Ellen. Research asst. Mass. Inst. Tech., 1946-50; asso. prof. elec. engring. U. Fla., 1950-51, U. Minn., 1951-55; physicist Gen. Electric Research and Devel. Center, Schenectady, 1955-—. Mem. Am. Phys. Soc., A.A.A.S., N.Y. Acad. Sci., Sigma Xi. Research, publs. on electron beam dynamics, electron optics, electron gun design, vacuum techniques, electron emission phenomena, material analysis by electron emission. Home: 2259 Berkley Av., Schenectady 12309. Office: P.O. Box 8, The Knolls, Schenectady 12301.*

HARRIS, Malcolm LaSalle, Am. surgeon; b. Port Byron, Ill., June 27, 1862; s. Samuel Gedney and Frances (Greene) H.; M.D., Rush Med. Coll., Chgo., 1882; m. Rose Breckenridge, Oct. 12, 1887; children—Samuel, Florence H. (Mrs. Max von Schegell). Practiced medicine, Chgo., from 1884; prof. clin. surgery Chgo. Polyclinic, 1889-1935. Mem. Am. Med. Assn. (pres. 1929-30). Developed method for obtaining urine from each kidney separately by using a double catheter (known as Harris' segregator), 1898. Died Wauwatosa, Wis., Mar. 2, 1936.

HARRIS, Marvin, Am. anthropologist; b. Bklyn., Aug. 18, 1927; s. Irving and Sadie (Newman) H.; A.B., Columbia, 1949, Ph.D., 1953; m. Madeline Grove, Jan. 25, 1953; children—Robert Eric, Susan Lynn. Research asso. anthropology Columbia, N.Y.C., 1952-53, asst. prof., 1953-59, asso. prof., 1959-63, prof., chmn. dept. anthropology, 1963-—; exec. sec. Columbia-Cornell U.-Harvard-U. Ill. Summer Field Studies Program, 1960-—, field leader Ecuador Field Team, 1960, Brazil Field Team, 1962. Field study race relations UNESCO, Bahia, 1950-51; tech. adviser Ministry Edn., Rio de Janeiro, Brazil, 1953; research adviser Nat. Inst. Pedogogical Studies, Brazil, 1953. Mem. Am. Anthrop. Assn. Author: Town and Country in Brazil, 1956; (with Charles Wagley) Minorities in the New World, 1958; Patterns of Race in the Americas, 1964; The Nature of Cultural Things, 1964; Editor: (with Morton Fried) Studies in Anthropology, 20 vols., 1960-—. Contbr. articles to profl. publs. Home: 325 Highwood Av., Leonia, N.J. 07605. Office: Dept. Anthropology, Columbia U., N.Y.C., 10027.*

HARRIS, Milton, Am. chemist; b. Los Angeles, Mar. 21, 1906; s. Louis and Naomi (Granish) H.; B.S., Ore. State U., 1926; Ph.D., Yale, 1929; D.Sc., Phila. Coll. Textiles and Sci., 1955; m. Carolyn Wolf, Mar. 30, 1934; children—Barney Dreyfuss II (foster child), John A. Founder, Harris Research Labs., 1945, pres., 1945-61; v.p. research Gillette Co., Boston, 1957-66, dir. research, 1956-66; chm. exec. com. Sealectro Corp., 1966-—; dir. Hazelton Labs. Mem. utilization research and devel. adv. com. U. S. Dept. Agr. Pres. Yale Chemists Assn., 1961-67; mem. Yale Council, 1964-—; mem. Alumni Bd., Devel. Bd., 1964-67. Recipient Washington Acad. Sci. award, 1943, Olney medal for research in field textile chemistry, 1945, Naval Ordnance Devel. award, Honor award Am. Inst. Chemists, 1957, Harold DeWitt Smith Meml. medal, 1966. Fellow Textile Inst. Eng.; mem. Nat. Acad. Scis., NRC, Am. Chem. Soc. (chmn. bd. 1966-—), Am. Assn. Textile Chemists and Colorists, Am. Inst. Chemists, Dirs. Indsl. Research, Indsl. Research Inst., A.A.A.S., Am. Mgmt. Assn., Fiber Soc. Editor: Harris Handbook of Textile Fibers, 1951; (with H. Mark) Natural and Synthetic Fibers, 1944-61. Contbr. numerous articles to profl. jours. Patentee in field. Research, publs. on chemistry, structure, properties of textiles, chemistry of polymers. Home: 4101 Linnean Av., Washington 20008. Office: 3300 Whitehaven St. N.W., Washington 20007.*

HARRIS, Morgan, Am. zoologist; b. St. Anthony, Ida, May 25, 1916; s. Archibald Overton and Augusta Pearl (Lewelling) H.; A.B. with highest honors in Zoology, U. Cal. at Berkeley, 1938, Ph.D., 1941; m. Marjorie Ruth Mason, Aug. 10, 1940; children—Roger Mason, Ronald Morgan. Harrison postdoctoral fellow U. Pa., 1941-42; research asst. Stanford, 1942-44; instr. U. Wash., 1944-45; faculty U. Cal. at Berkeley, 1945-—, now prof. zoology, Miller Research prof., 1963-65, chmn. dept. zoology, 1957-65. Mem. cell biology study sect. NIH, 1958-60, 61-63, nat. adv. gen. med. sci. council, 1963-65. Merck sr. postdoctoral fellow U. Paris, 1953-54; Guggenheim fellow U. Cambridge (Eng.), 1960-61. Mem. Internat. Soc. for Cell Biology (exec. com. 1964-—), Am. Soc. for Cell Biology (exec. com. 1965-—), Am. Soc. Zoologists, Tissue Culture Assn., Soc. Gen. Physiologists, Soc. for Growth and Devel. Author: Cell Culture and Somatic Variation, 1964; also articles. Research on cell growth and cell genetics with cell cultures, role proteins and other nutritional factors on growth isolated cell populations, mechanisms drug resistance, chromosomal variation, thermotolerance in vitro. Home: 605 Plateau Dr., Berkeley, Cal. 94708.*

HARRIS, Philip L., Am. biochemist; b. Detroit, May 7, 1910; s. David Charles and Agnes (McCullough) H.; B.S., Alma Coll., 1931; M.S., Pa. State U., 1932, Ph.D., 1934; m. Flora Ellen Lau, Aug. 24, 1933; children—David A., Gordon G., Philip S. Research biochemist United Fruit Co., 1935-37; instr. physiol. chemistry U. S.C., 1937-40; research dir. biol. research Distilation Products Industries, Eastman Kodak Co., Rochester, N.Y., 1940-62; dir. div. nutrition Dept. Health, Edn., Welfare, Food and Drug Adminstrn., Washington, 1962-—. Asso. in nutrition U. Rochester, 1942-62; research advisor George Washington U., 1966-—; mem. Mayor's Com. on Fluoridation, Rochester, 1950-55. Mem. Am. Chem. Soc., Am. Inst. Nutrition, Am. Inst. Food Technologists, Am. Bd. Nutrition. Contbr. numerous articles to profl. jours. Determined vitamin content, nutritive values of banana; discovered effects of vitamin and nutrients on functioning of thyroid; developed analytical methods for fat-soluble vitamins; determined metabolic role for vitamins, lipids, fatty acids. Home: 3122 Birch St. N.W., Washington 20015. Office: 200 C ST. N.W., Washington 20204.*

HARRIS, Stanley Cyril, Am. pharmacologist; b. Chgo., July 2, 1916; s. Louis and Rose (Waldman) H.; B.S., Northwestern U., 1938, M.S., 1941, Ph.D., 1946; m. Mary Z. Gilmore, Mar. 28, 1942; children—Patricia, Michael. Faculty, Northwestern U. Dental Sch., Chgo., 1942-66; prof., chmn. physiology and pharmacology, 1954-66; prof., chmn. dept. pharmacology (dental), dir. div. advanced dental edn. U. Pa., Phila., 1966-—. Chmn. program project com. Nat. Inst. Dental Research, 1967-68; chmn. ad hoc com. on advanced edn. Am. Assn. Dental Schs., 1965-68. Mem. Am. Physiol. Soc., Am. Soc. Pharmacology and Exptl. Therapeutics, Chgo. Inst. Medicine, Sigma Xi, others. Research, numerous publs. on analgesics, local anesthetics, appetite depression. Home: 488 Timber Lane, Devon, Pa. 19333. Office: 4001 Spruce St., Phila. 19104.*

HARRIS, Stanton Avery, Am. chemist; b. Middlebury, Vt., Oct. 7, 1902; s. Charles Emerson and Nellie (Blakely) H.; B.S., Middlebury Coll., 1924, M.S., 1926; Ph.D., Ia. State U., 1931; m. Mary Sulser Pilgrim, June 4, 1927; children—Anna Ruth (Mrs. Kendall P. Carlson), Robert Stanton, Mary Elizabeth (Mrs. Ralph L. Chapek). Sr. chemist Merck Research Labs., 1937-50, dir. tech. information dept., 1950-54; vis. prof. depts. biochemistry, urology Coll. Medicine U. Ia., 1954-55; research asso. Merck Sharp & Dohme Research Labs., Rahway, N.J., 1955-64, research fellow, 1964-67. Recipient Mead Johnson award, 1940. Mem. Am. Chem. Soc., Am. Assn. Cancer Research, Sigma Xi. Contbr. numerous articles to profl. jours. Research, publs. on structure, synthesis of vitamin B6, pantothenic acid, biotin; penicillin chemistry. Home: 47 Arrowhead Lane, Penfield, N.Y. 14526.*

HARRIS, Thomas Maxwell, English botanist; b. Leicester, Jan. 8, 1903; s. Alexander and Lucy Frances N. Harris; Sc.D., Cambridge U.; m. Katharine Massey, 1928; children—Peter, Margaret, Frances, Elizabeth. Instr., Cambridge U.; prof. botany U. Reading, U. Ghana. Fellow Royal Soc., 1948. Recipient gold medal (botany) Linnean Soc., 1968. Author: Catalogue of the British Museum on Jurassic Flora, also numerous works on paleobotany. Address: University of Reading, Reading, Eng.*

HARRIS, Walter, English physician; b. Gloucester, Eng., 1647; B.A., New Coll., Oxford (Eng.) U., 1670; M.D., Bourges, France, 1675. Physician to Charles III, 1683, also to William III and Queen Mary. Author: Pharmacologia antiempirica or A Rational Discourse of remedies Both Chemical and Galenical (gave accounts of mercury, antimony, vitriol, iron, quinine and opium as medicines); 1683; De morbis acutis infantum, 1698. Probably 1st to use calcium salts in treatment of infantile tetany, recorded 1689. Died London, Aug. 1, 1732.

HARRIS, Walter Edgar, Canadian chemist; b. Alta., Can., June 9, 1915; s. William Ernest and Emma (Humbke) H.; B.Sc., U. Alta., 1938, M.Sc., 1939; Ph.D., U. Minn., 1944; m. Phyllis Pangburn, June 14, 1942; children—Margaret Anne, William Edgar. Postdoctoral fellow U. Minn., 1943-46; prof. analytical chemistry U. Alta., Edmonton, 1946-—. Research asso. Atomic Energy Can., 1957-58. Fellow Chem. Inst. Can.; mem. Am. Chem. Soc., Sigma Xi. Author: (with Henry W. Habgood) Programmed Temperature Gas Chromatography, 1966; also numerous articles. Research on electroanalytical chemistry of uranium, amperometric titrations, synthetic rubber, hot atom chemistry, programmed temperature gas chromatography. Address: Dept. Chemistry, U. Alta., Edmonton, Alta., Can.*

HARRIS, Sir William Snow, English inventor; b. Plymouth, Eng., Apr. 1, 1791; s. Thomas and Mary (Snow) H.; ed. U. Edinburgh; m. Elizabeth Snow Thorne, 1824. Surgeon in militia; practiced medicine, Plymouth; devoted to elec. research, from 1824; Bakerian lectr. on elementary laws of electricity, 1839; govt. grantee; apptd. sci. referee, 1860. Fellow Royal Soc., 1831 (Copley medal 1835). Author: Electricity, 1848; Magnetism, 1850-52; Galvanism . . . , 1856; Treatise on Frictional Electricity, 1867; also papers. Inventor method of arranging lightning conductors of ships, 1820, also an improved lightning conductor and mariner's compass; 1st to conceive of disc electrometer. Died Jan. 22, 1867.

HARRISON, Charles Victor, English pathologist; b. Newport, Eng., Jan. 21, 1907; s. Charles Henry and Violet (Witchell) H.; B.Sc., U. Wales, 1926, M.B., B.Ch., 1929; M.B., B.S., U. London, 1929, M.D., 1937; m. Olga Beatrice Cochrane, Sept. 8, 1937; children—Douglas Victor, Judy Olga. Sr. lectr. pathology Liverpool (Eng.) U., 1939-46; reader in morbid anatomy Postgrad. Med. Sch., London, Eng., 1946-56, prof., 1956—. Cons. in pathology Royal Air Force, 1960—. Fellow Royal Coll. Physicians, Coll. Pathology; mem. Pathol. Soc. Gt. Britain, Internat. Acad. Pathology. Author: Recent Advances in Pathology, 1960. Contbr. numerous articles to med. jours. Research on mechanism of prodn. of arterial disease, effects of dusts on lung, behavior of tumors of lymph nodes. Home: 23 Woodville Gardens, London, W.5, Eng.*

HARRISON, David Lakin, English mammalogist, physician; b. Sevenoaks, Eng., Oct. 1, 1926; s. James Maurice and Rita G. (Sorley) H.; M.A., M.B., B.Ch., 1951, Clare Coll., Cambridge (Eng.) U., Ph.D., 1967; student St. Thomas' Hosp., London, 1948-51. Practice gen. medicine, Sevenoaks, 1955—. Recipient H.H. Fellow Linnean Soc. (Bloomer medal 1966), Zool. Soc. London; mem. Am. Soc. Mammalogists; Brit. Mammal Soc., E. African Natural History Soc. Author: Footsteps in the Sand, 1959; The Mammals of Arabia, vol. I, 1964; also articles. Research on mammals of Eurasia and Africa. Address: Bowerwood House, St. Botolph's Rd, Sevenoaks, Kent, Eng.*

HARRISON, Donald Fredrick Norris, English surgeon; b. Portsmouth, Eng., Mar. 9, 1925; s. Frederick William and Florence (Norris) Rees; student Guys Hosp. Med. Sch., 1943-48; F.R.C.S., U. London, 1955, M.S., 1959, M.D., 1960; m. Audrey Clubb, Jan. 29, 1948; children—Susan Patricia, Zoë Claire. Sr. registrar Guys Hosp., 1956-60; faculty U. London, 1960—, Hunterian prof., 1962—; surgeon Royal Nat. Ear Nose and Throat Hosp., London, 1960—. Fellow Royal Coll. Surgeons (ct. examiners); corr. mem. med. socs. U. S., Australia, France, Denmark. Research, publs. on cancer of head and neck, chemotherapy, diseases affecting ear, nose and throat; developed techniques utilizing hypothermia. Home: 6 Fishers Farm, Horley, Surrey, Eng. Office: Inst. Laryngology and Otology, 330 Grays Inn Rd., London, Eng.*

HARRISON, Edgar George, Jr., Am. physician; b. St. Louis, Feb. 24, 1925; s. Edgar George and Mary F. (Shelton) H.; B.S., U. Ill., 1948; M.D., St. Louis U., 1952; M.S., U. Minn., 1959; m. Rita Catherine Delaney, June 11, 1949; children—David G., Dianne G., Brian D., Mark A., Matthew S., Rebecca A. Practice medicine, specializing in pathology, Rochester, Minn.; research pathologist St. Louis U., 1953-54; fellow pathology Mayo Found., 1954-58; cons. surg. pathology Mayo Clinic, 1958—; asst. prof. pathology U. Minn., 1964. Mem. A.M.A., Am. Soc. Clin. Pathologists, Internat. Acad. Pathologists, Sigma Xi, Alpha Omega Alpha. Contbr. numerous articles to profl. jours. Research, publs. on diseases of blood vessels, lymphoid tissues including thymus and lung as applied to surg. pathology. Home: 522 14th Av. S.W. Office: 200 First St. S.W., Rochester, Minn. 55901.*

HARRISON, Edward, English physician; b. Lancashire, Eng., 1766; ed. Edinburgh, London; practiced medicine, Horncastle, Eng. Fellow Royal Soc. Author: Pathological and Practical Observations on Spinal Diseases . . . , 1827; An Essay on the Powerful Influence of the Spinal Nerves . . . , 1831. Described Harrison's grove nr. lower border of chest, found especially in children with rickets, 1798. Died May 6, 1838.

HARRISON, Geoffrey Ainsworth, Brit. anthropologist; b. Teddington, Eng., Aug. 6, 1927; s. Harold Ainsworth and Eva (Chipchase) H.; B.A., Trinity Coll., Cambridge U., 1952; M.A., Christ Church, Oxford U., 1951, D.Phil., 1954; B.Sc., U. London; m. Mary Elizabeth Atkinson, May 8, 1946; 1 dau., Elizabeth Anne. Faculty, U. Oxford, 1953-54, reader phys. anthropology, 1963—; lectr. human biology U. Liverpool (Eng.), 1954-63. Fellow Linacre Coll., Oxford U. Fellow Zool. Soc., Royal Anthrop. Inst. Author: (with J. S. Weiner, J. M. Tanner, N. A. Barnicot) Human Biology, 1964; also articles. Research on nature of effects of heredity and environment on growth and devel., anthropology of human adaptation to desert and high altitude regions, reconstrn. of demography and genetic structure of past populations, analysis of quantitative variation in man. Home: 42, Cumnor Hill, Oxford, Eng.*

HARRISON, George Carol, Am. chemist; b. Osage, Ia., Dec. 19, 1917; s. Harry Hilton and Florence Olive (Gross) H.; B.S., Ia. State U., 1940, Ph.D., 1944; m. Ardis Marie Vaughn, June 21, 1942; children—Carol Marie, G. Carter Vaughn. With OSRD, 1941-42, G.F. Smith Chem., 1942-43; chemist Minn. Mining & Mfg. Co., St. Paul, 1944—. Mem. Am. Chem. Soc., A.A.A.S., Sigma Xi, Alpha Chi Sigma, Phi Lambda Upsilon. Fellow Nat. Geog. Soc. Contbr. articles to tech. jours. Patentee in field. Research on pressure sensitive adhesives, castable elastomers, resin tooling compound, structural adhesives, molded products, factory installed vitrified clay pipe gaskets and seals, resinous flooring compounds. Home: 32 Mid Oaks Lane, St. Paul 55113. Office: 367 Grove St., St. Paul 55101.*

HARRISON, George Ernest, English physicist; b. Birmingham, Eng., July 28, 1904; s. Joseph Ernest and Leah Marion (Turner) H.; B.Sc., U. Birmingham, 1925, Ph.D., 1927, D.Sc., 1954; m. Frances Marjorie Holdcroft, July 6, 1935; children—Alan Stuart, Christine Elizabeth, Brenda Marion. Lectr. physics U. Birmingham, 1930-46; head math. and physics dept. South-East Essex (Eng.) Tech. Coll., 1946-47; head radiochemistry sect. Med. Research Council, Radiobiol. Research Unit, Harwell, Eng., 1947—, dep. dir., 1965. Fellow Phys. Soc. (London), Inst. Physics (London). Contbr. articles to profl. jours. Research on diffusion of rare gases with reference to molecular forces at collision, metabolism of various trace elements in man and exptl. animals with reference to strontium, therapeutic means of decreasing uptake of strontium by dietary additives. Home: Newlands, Frilford Heath, Abingdon, Berkshire. Office: Radiobiological Research Unit, Harwell, Berkshire, Eng.*

HARRISON, George Russell, Am. physicist; b. San Diego, July 14, 1898; s. Ernest and Magda (Lincke) H.; B.A., Stanford, 1919, M.A., 1920, Ph.D., 1922; Sc.D., Northeastern U., 1943, St. Lawrence U., 1952; LL.D., Middlebury Coll., 1955; D.Engr., Drexel Inst., 1955; m. Florence Kent, Dec. 25, 1922 (dec. May 1955); children—Mary Lou, Nancy Kent (Mrs. Kurt Herman Grab), David Kent; m. 2d, Elizabeth Cavanna Headley, Mar. 9, 1957. Instr. physics Stanford, 1919-23, asst. prof., 1925-30; prof. physics Mass. Inst. Tech., 1930-64, dean Sch. Sci., 1942-64, dean, prof. emeritus, 1964—; trustee Babson Inst., 1956—. Dir. Colt Industries, Inc., Bausch & Lomb, 1965—. Chief optics div. NDRC, 1940-46. Recipient Ives medal, Rumford medal, Cresson medal, Mees medal, medal of Soc. Applied Spectroscopy, Pitts. Spectroscopy award, Presdl. medal of merit. Nat. Research fellow Harvard, 1925. Mem. Am. Acad. Arts and Scis. (v.p. 1944-46), A.A.A.S. (past dir. 1956), Am. Phys. Soc., Optical Soc. Am. (pres. 1946-48), Am. Inst. Physics (chmn. dirs. 1947-54). Author: Atoms in Action, 1939; How Things Work, 1941; What Man May Be, 1956; First Book of Light, 1963; First Book of Energy, 1965; (with R. C. Lord, J. Loufbourow) Practical Spectroscopy, 1948; also numerous articles. Editor: Mass. Inst. Tech. Wavelength Tables, 1939. Invented automatic wavelength measuring engines, echelle spectroscopy, interferometrically controlled ruling engine for diffraction gratings. Home: Barnes Hill Rd., Concord, Mass. 01742.*

HARRISON, Harold Edward, Am. physician; b. New Haven, July 23, 1908; s. Abraham and Rose (Chaikind) H.; B.S., Yale, 1928, M.D., 1931; m. Helen Miriam Coplan, Aug. 2, 1936; children—Stephen Coplan, Richard Gerald. Pediatrician-in-chief Balt. City Hosp., 1945—; faculty Sch. Medicine Johns Hopkins, 1945—, prof. pediatrics, 1965—; mem. nutrition study sect. NIH, 1958-63; mem. food, nutrition bd. Nat. Acad. Sci., NRC, 1964—. Recipient (with Helen C. Harrison) Mead Johnson award for research in pediatrics, 1941, Borden award Am. Acad. Pediatrics, 1960. Mem. Am. Pediatric Soc., Am. Soc. Clin. Investigation, Am. Soc. Pediatric Research, Endocrine Soc., Am. Inst. Nutrition, A.M.A., Sigma Xi, Alpha Omega Alpha. Research, publs. on increased understanding of factors which control transport of calcium and phosphate across intestinal wall, mechanism of action of Vitamin D on calcium and phosphate transport and interrelation of Vitamin D and parathyroid hormone action; developed concept of renal hypophosphatemia as cause of Vitamin D resistant rickets. Home: 5500 N. Charles St., Balt. 21210. Office: 4940 Eastern Av., Balt. 21224.*

HARRISON, James Merritt, Canadian geologist; b. Regina, Sask., Can., Sept. 20, 1915; s. Roland O. and Vera (Merritt) H.; B.Sc., U. Man., 1935, Sc.D., 1965; M.A., Queen's U., 1941, Ph.D., 1943, LL.D., 1967; D.U.C., U. Calgary, 1967; m. Herta Boehmer Sliter, May 5, 1944; 1 stepson, Norman. With Geol. Survey Can., 1943—, dir., 1956-64, asst. dep. minister Dept. Energy, Mines and Resources, 1964—; spl. lectr. Precambrian geology, econ. geolooy Queen's U., 1949-50. Pres., Internat. Union Geol. Scis., 1961-64; v.p. Internat. Council Sci. Unions, 1963-66, pres., 1966—; mem. Sci. Council Can., 1966—; Distinguished lectr. Am. Assn. Petroleum Geologists, 1948. Recipient Kemp Meml. Gold medal Columbia, 1963; Gold medal Profl. Inst. of Pub. Service Can., 1966; Blaylock medal Canadian Inst. Mining and Metallurgy, 1966. Fellow Royal Soc. Can. (v.p. 1966-67, pres. 1967—), Geol. Soc. Am. (council), Royal Canadian Geog. Soc. (dir.); mem. Geol. Assn. Can. (pres.), Soc. Econ. Geology (v.p. N.Am.), Mineral. Assn. Can., Arctic Inst. N.Am. Research, numerous publs. on Precambrian geology, banded iron-formations, ore deposition, tectonics. Home: 4 Kippewa Dr. Office: 588 Booth St., Ottawa, Ont., Can.*

HARRISON, John, English horologist, inventor; b. Yorkshire, Eng., Mar., 1693; s. Henry and Elizabeth (Barber) H.; m. Elizabeth; 1 son, William. Recipient: Copley medal Royal Soc., 1749; award Bd. of Longitudes competition from parliament (for construction of chronometers), 1763, 1773. Author: An Account of the Proceedings in Order to the Discovery of the Longitude, 1763; Narrative of the Proceedings Relative to the Discovery of the Longitude at Sea . . . , 1765; The Principles of Mr. Harrison's Timekeeper with Plates of the Same . . . , 1767. Constructed an eight-day clock, 1715, gridiron pendulum, 1726, also recoil escapement, secondary spring (going ratchet); devised new musical scale; built at least 5 chronometers (for determining longitude at sea), from 1736-73. Died Mar. 24, 1776.

HARRISON, John Christopher, geophysicist; b. Stockton-on-Tees, Eng., May 20, 1929; s. Charles Frederick R. and Anne L. (Middleton) H.; B.A., Cambridge U. (Eng.), 1950, Ph.D., 1953; m. Elaine C. P. Millar, Aug. 12, 1960; children—Kirsteen Elizabeth, Fiona Anne, Keith Thomas. Came to U. S., 1953, naturalized, 1967. Jr. research geophysicist U. Cal. at Los Angeles, 1953-55; asst. research geophysicist, 1957-61; tech. staff Hughes Research Lab., Malibu, Cal., 1961-65; asso. prof. geol. scis. U. Colo., Boulder, 1965—. Mem. Am. Geophys. Union (sec. Western nat. meeting 1964, chmn. 1967—), Royal Astron. Soc. Contbr. articles to tech. jours. Research on techniques for accurate measurement gravity, geol. interpretation gravity measurements. Home: 2305 Dartmouth Av., Boulder, Colo. 80302.*

HARRISON, John Hartwell, Am. surgeon; b. Clarksville, Va., Feb. 16, 1909; s. Isaac Carrington and Rosalie (Smith) H.; B.S., U. Va., 1929, M.D., 1932; M.A., Harvard, 1965; m. Gertrude Chisholm, June 16, 1934 (dec. Feb. 1965); children—John Hartwell, Robert C. II, Cornelia (Mrs. Curtis Scribner), Jeffrey C.; m. 2d, Mary Louise Rice Harding, July 16, 1965. Practice medicine, specializing in surgery, Boston; faculty Med. Sch. Harvard, 1935—, Elliott Carr Cutler prof. surgery, 1965—; cons. urology Mass. Hosp. Sch. For Crippled Children (Canton), Childrens Med. Center, Boston Lying-In-Hosp. (both Boston), West Roxbury, Mass. VA Hosp. Diplomate Am. Bd. Surgery, Am. Bd. Urology. Fellow A.C.S. (gov.), A.M.A.; mem. Am. Urol. Assn. (Raymon Guiteras award 1965, past sect. pres.), Am., New Eng. urol. socs., Am. Surg. Assn. (past pres.), Clin. Soc. Genito-Urinary Surgeons, Am. Assn. Genito-Urinary Surgeons, Am. Acad. Arts and Scis. (Frances Amory award 1962). Specialist in human Kidney transplants, adrenal disorders. Home: 68 Dudley St., Brookline, Mass. 02115. Office: 721 Huntington Av., Boston 02115.*

HARRISON, Ronald George, English anatomist; b. Ulverston, England, Apr. 5, 1921; s. James and Alice Hannah Harrison; B.A., M.A., B.M., BCh., D.M., Oxford U., married; four children. Physician, Radcliffe Infirmary, Oxford, 1944-45; demonstrator Oxford U., 1945-49, Ruskin Sch. Drawing, 1946-50, Pembroke Coll., 1950; prof. anatomy U. Liverpool, 1950—. Mem. Zool. Soc. London, Anat. Soc., N.Y. Acad. Scis., Soc. for Human Biology and Study Fertility, Soc. for Endocrinology. Author: Textbook of Human Embryology, 1959, 63; The Adrenal Circulation, 1960, also chpts. and articles. Home: The Stables, Fenn Hill, Upper Brighton, Wallasey, Cheshire. Office: University of Liverpool, Eng.*

HARRISON, Ross Granville, Am. biologist; b. Germantown (Phila.), Pa., Jan. 13, 1870; s. Samuel and Katherine (Diggs) H.; A.B., Johns Hopkins, 1889, Ph.D., 1894, LL.D., 1942; M.D., U. Bonn, 1899; A.M. (hon.), Yale, 1907; D.Sc., U. Cin., 1920, U. Mich., 1929, U. Dublin, 1932, Harvard, 1936, Yale, 1939, Columbia, 1940, U. Chgo., 1949; Ph.D., U. of Freiburg, 1929; M.D., U. Budapest, 1935; Dr. Rev. Nat., U. Tübingen, 1953; m. Ida Lange, Jan. 9, 1896; children—Dorothea Katharine, Elizabeth Ross, Richard Edes, Eleanor Barrington (Mrs. Rufus Putney, Jr.), Ross Granville. Lectr. on morphology Bryn Mawr Coll., 1894-95; instr. and asso. in anatomy Johns Hopkins, 1896-99, asso. prof. anatomy, 1899-1907; prof. comparative anatomy Yale, 1907-27, Sterling prof. of biology, 1927-38, prof. emeritus, from 1938. Chmn. NRC, Washington, 1938-46; pres. 6th Pacific Sci. Congress, 1939; Dunham lectr. Harvard, 1926; Croonian lectr. Royal Soc., London, 1933; Linacre lectr. St. John's Coll., Cambridge, 1939; Silliman lectr. Yale, 1949. Mem. science com. Nat. Resources Planning. Bd., 1938-43; mem. U. S. Nat. Com. for UNESCO, 1946-51. Conf. bd., Assoc. Research Councils (chmn., 1944—). Mng. editor Jour. Exptl. Zoology, 1903-46; trustee Marine Biol. Lab., Woods Hole, Mass., 1908-40; bd. dirs. L.I. Biol. Sta.; trustee Bermuda Biol. Sta. (treas. 1930-46), Woods Hole Oceanographic Inst., Science Service. Mem. adv. bd. Wistar Inst.; bd. sci. dirs. Rockefeller Inst. Med. Research (v.p., 1939—). Fellow Am. Acad. Arts and Scis., A.A.A.S. (v.p. 1936); mem. Nat. Acad. Scis. (council 1932-46), Am. Philos. Soc. (council 1941-44, v.p. 1947-50), Am. Soc. Zoologists (pres. 1924), Soc. Study Devel. and Growth (pres. 1946-47), Am. Assn. Anatomists (pres. 1912-13), Anatomische Gesellschaft (pres. 1934-35), Royal Physiog. Soc. (Lund),

Royal Soc. Science (Uppsala), Am. Soc. Naturalists (pres. 1913), Am. Neurol. Assn., Soc. Exptl. Biology and Medicine; corr., hon. or fgn. mem. many others. Fgn. corr., Acad. Science of Institute of Bologna, Italy. Awarded (1914) The Archduke Rainer medal of Imperial Royal Zool. Bot. Soc. of Vienna, 1925; John Scott medal and premium, of the City of Phila., "for the invention of devices for tissue grafting and for tissue culture." John J. Carty medal Nat. Acad. Scis., 1947; Antonio Feltrinelli Internat. prize Accademia Nazionale de'l Lincei, Italy, 1956. Author numerous scientific papers on devel. of fishes, nervous system, embryonic transplantation and cultivation of animal tissues outside the organism. In 1907 first adapted the hanging drop culture method to the study of embryonic tissues and demonstrated directly the outgrowth of the developing nerve-fiber; did important work on regeneration of peripheral nerves, and on part played by nervous system in embryonic muscle differentiation; did pioneer work in presently accepted nerve growth theory. Died Sept. 30, 1959.

HARRISON-CHURCH, Ronald James; English geographer; b. Wimbledon, Eng., July 26, 1915; s. James Walter and Jessie May (Fennymore) C.; B.Sc. in Econs., London Sch. Econs., 1936; postgrad. U. Paris; Tchr.'s Diploma, Inst. Edn., U. London, 1939; m. Dorothy Violet Harrison, Aug. 2, 1944; children—Julia Rosalind, Christopher Julian. Lectr., London (Eng.) Sch. Econs., 1944-58, reader, 1958-64, prof. geography, 1964—. Lectr. in U. S. Brazil, France, Belgium, Germany, 1946—; cons. UN Econ. Commn. for Africa, 1962. Mem. Royal Geog. Soc. (Back award 1957), Inst. Brit. Geographers, Geog. Assn., African Studies Assn. Author: Modern Colonization, 1951; West Africa, 1957, 8th edit., 1968; Environment and Policies in West Africa, 1963; Africa and the Islands, 1964, 2d edit., 1967. Contbr. articles to profl. jours. Research on phys. environment and econ. devel. of West Africa. Home: 8 Mannicotts, Welwyn Garden City, Eng. Office: London School of Economics, Houghton St., Aldwych, London W.C.2, Eng.*

HARRISS, C(lement) Lowell, Am. economist; b. Fairbury, Neb., Aug. 2, 1912; s. Riley Clement and Alice (Hunt) H.; B.S. summa cum laude, Harvard, 1934; postgrad. U. Chgo., 1936-37; Ph.D., Columbia U., 1940; m. Agnes Bennett Murphy, June 1, 1936; children—Patricia Louise, Lowell Gordon, Martha Anne, Brian Clement. Mem. faculty Columbia U., N.Y.C., 1938—, prof. econs., 1957—. Vis. prof. univs. in U. S., Europe; cons. to city, state fed. govt., spl. commns. Dir., Internat. Inst. Pub. Finance, Tax Inst. Am., Nat. Tax Assn., Lincoln Found. Mem. Am. Econ. Assn., Internat. Fiscal Assn., Internat. Inst. Pub. Finance, Royal Econ. Soc., numerous others. Author: The American Economy, 5th edit., 1965; Selected Readings in Economics, 3rd edit., 1967; History and Policies of Home Owners Loan Corporation, 1951; (with W. J. Schultz) Am. Public Finance, 8th edit., 1965; Money and Banking, 2d edit., 1965; other books, numerous articles. Research on role of govt. in econ. life, especially taxation and govt. spending. Home: 14 Plateau Circle, Bronxville, N.Y. 10708.*

HARROD, Sir Roy (Forbes), English economist; b. London, Feb. 13, 1900; s. Henry Dawes and Frances Marie Desirée (Forbes-Robertson) H.; 1st class honors in Classics, Oxford U., 1921, 1st class honors in modern history, 1922; LL.D., U. Poitiers (France), 1950, U. Aberdeen (Scotland), 1956, U. Pa., 1964, U. Glasgow, 1968, U. Warwick, 1968; m. Wilhelmine Cresswell, Jan. 8, 1938; children—Henry Mark, Dominick Roy. Tutor, Christ Church, Oxford U., 1922-67; mem. Hebdomadal Council, Oxford U., 1929-35; U. lectr. in economics, 1929-37, 1946-52; mem. governing body, Christ Church, Oxford U., 1924-67; mem. Winston Churchill's Pvt. Statis. Staff, 1940-45. Mem. sub-commn. on employment and econ. stability UN, 1947-50; research staff IMF, 1952-53. Recipient Bernard Harms prize U. Kiel (Germany), 1966; created knight, 1959; mem. Royal Economic Soc. (council, since 1933; v.p.; pres. 1963-65). Author: International Economics, 1933, rev., 1957; Trade Cycle, 1936; Britain's Future Population, 1943; A Page of British Folly, 1946; Are These Hardships Necessary, 1947; Towards a Dynamic Economics, 1948; Life of John Maynard Keynes, 1951; And So It Goes On, 1951; Economic Essays, 1952; The Dollar, 1953; Foundations of Inductive Logic, 1956; Policy Against Inflation, 1958; The Prof. (memoir of Lord Cherwell F.R.S.), 1959; Topical Comment, 1961; The British Economy, 1963; Reforming the World's Money, 1965; also numerous articles. Contbr. to theory of imperfect competition, especially concept of Marginal Revenue Curve, theory of fgn. trade, especially fgn. trade multiplier, theory of econ. growth, especially Harrod-Donar growth equations; studies in logic. Home: The Old Rectory, Holt, Norfolk, Eng. Office: 51 Campden Hill Sq., London, Eng.

HARROLD, Orville, Goodwin, Jr., Am. topologist; b. Chgo., Sept. 2, 1909; s. Orville Goodwin and Estelle (Pancake) H.; A.B., with great distinction, Stanford, 1931, M.A., 1932, Ph.D., 1936; m. Gladys Estelle Buell, June 30, 1934; children—Phillip, Jeffrey. Instr., Stanford, 1933-37, Ore. State Coll., 1937-39, Northwestern U., 1940-42; asst. prof. La. State U., 1942-43, Pomona Coll., 1943-45; vis. lectr. Princeton, 1946-47; prof. U. Tenn., 1947-64, head dept. mathematics, 1961-64; prof. topology, chmn. dept. mathematics Fla. State U., Tallahassee, 1964—; mem. Inst. for Advanced Study, 1958, 64. Cons., Union Carbide Corp., 1949-61. Nat. Research fellow, 1939-40; John Simon Guggenheim Meml. fellow Oxford (Eng.) U., 1958-59; recipient $500 prize Oak Ridge Inst. Nuclear Competition, 1949. Mem Internat. Congress Mathematicians, Am. Math. Soc. (asso. sec.), Math. Assn. Am., Phi Beta Kappa, Sigma Xi. Contbr. articles to tech. jours. Reviewer, Math. Revs. Research on theory of curves and structural properties; numerical quadrature processes; isotopy problems in space; continuous transformations. Home: 1101 Lothian St., Tallahassee 32303.*

HARSANYI, Kálmán, Hungarian chemist; b. Budapest, Hungary, Aug. 25, 1927; s. Miksa and Józsa (Frank) H.; chem. engr., Poly. U. Budapest, 1949, D.Chem. Scis., 1959; m. Edit Szabó, Apr. 2, 1958; 1 son, Gábor. Asst., Inst. Organic Chemistry, Poly. U. Budapest, 1949-54, 1st asst., 1954-60, lectr., 1962—; head research lab. Chinoin Pharm. Works, Budapest, 1960-63, head chem. research, 1963—. Mem. Hungarian Chem. Soc., Hungarian Pharmacological Soc. Research and publs. on pseudobasic amincarbinols, new isoquinoline synthesis; preparation of series of new diphenylaklkylamines, new oxadiazole derivatives, new methods for synthesis of oxadiazoles. Patentee in field. Home: 19-21 Jozsef Egri, Budapest. Office: 1-5 Tó, Budapest, Hungary.*

HARSHBARGER, Boyd, Am. math. statistician; b. Weyers Cave, Va., Feb. 15, 1906; s. John A. and Eugenia (Tutwiler) H.; A.B., Bridgewater Coll., 1928, D.Sc., 1955; M.S., Va. Poly. Inst., 1931; A.M., U. Ill., 1935; Ph.D. (Rockefeller fellow 1939-41), George Washington U., 1943; postgrad. U. N.C., 1940, Ia. State Coll., 1941-42; m. Isabelle Hoge, Sept. 5, 1935; children—John Hoge, Barbara Hume Church. Instr. Miller (Va.) Sch., 1929-30; instr. Va. Poly. Inst., 1931-37, asst. prof. mathematics, 1937-39, prof. statistics, 1941-48, head dep. statistics, dir. statis lab., 1948—; statistician Va. Agrl. Expt. Sta., Blacksburg, 1941-48; tchr. quality control courses Princeton, U. N.C., Va. Poly. Inst., 1941-43; cons. ordnance dept. U. S. Army, 1949—. Recipient J. Shelton Horsley research award for meritorious research, 1946. Fellow Am. Statis. Assn. (mem. council 1956-58); mem. Inst. Math. Statistics, Biometric Soc. (pres. 1957), Va. Acad. Sci. (pres. 1949-50), A.A.A.S., Va. Edn. Assn., Am. Assn. U. Profs., Sigma Xi. Author (booklet); Rectangular Lattices, Memoir No. L, 1947. Editor-in-chief Va. Jour. Sci., 1950-55. Contbr. articles profl. jours Research on theory and application of lattice designs. Home: 113 Country Club Dr., Blacksburg, Va. 24060.*

HARSHBARGER, Kenneth E., Am. nutritionist; b. Arcola, Ill., Nov. 12, 1914; s. Harry H. and Maude (Whipple) H.; B.S., U. Ill., 1937, M.S., 1939, Ph.D., 1960; m. Elsie J. Brown, Aug. 31, 1957; children—Lee, Keven, Kent, Karen. Faculty, U. Ill., Urbana, 1939—, prof. dairy sci., 1962—, asso. head dept., 1963—. Spl. cons. nutrition sect. Office Internat. Research, NIH. Mem. Am. Inst. Nutrition, Am. Dairy Sci. Assn., Am. Soc. Animal Sci. Research, numerous publs. on nutritive value of high-moisture corn, feeding regimens for young calves, relationship of vitamin D to milk fever, automation of dairy cattle feeding operations, concept of group feeding for milk prodn. Home: 502 E. Pennsylvania Av., Urbana, Ill. 61801.*

HARSHBERGER, John William, Am. botanist; b. Phila., Jan. 1, 1869; s. A. and Jennie (Walk) H.; A.B., Central High Sch., Phila., 1888; B.S., U. Pa., 1892, Ph.D., 1893; spl. student Harvard, 1890; traveled and botanized in Europe, Brazil, Argentine, Chile, Mexico, Alaska, Ariz., Utah, Calif., Can., W.I., Southern Fla., Northwestern and Eastern states; m. Helen B. Cole, June 28, 1907 (dec. 1923); children—Jane Yard, Elyonta Cole. Instr. botany and zoölogy U. Pa., 1892-1907, asst. prof. botany, 1907-11, prof., from 1911. Lectr. Soc. for Extension of Univ. Teaching; lectured in farmers' insts. in Pa., 1904-06; in charge nature study, Pocono Pines Assembly, summers 1903-08; in charge ecology, Marine Biol. Laboratory, Cold Spring Harbor, L.I., 1913-21; in charge botany Nantucket Maria Mitchell Assn., 1914-15. Author: Maize, a Botanical and Economic Study, 1893; The Botanists of Philadelphia, and Their Work, 1899. Bot. editor Worcester's New English Dictionary; Student's Herbarium for Descriptive and Geographic Purposes, 1901; A Phytogeographic Survey of North America, 1911, for the series of monographs, "Die Vegetation der Erde," The Vegetation of South Florida, 1914. The Vegetation of the New Jersey Pine Barrens, 1916; A Text-Book of Mycology and Plant Pathology, 1917; Colored Wall Map Vegetation of North America, 1919; Pastoral and Agricultural Botany, 1920. Bot. editor new Funk & Wagnall's College Dictionary. Died Apr. 27, 1929.

HART, Sir Andrew Searle, Irish mathematician; b. Limerick, Ireland, Mar. 14, 1811; s. George Vaughan and Maria (Hume) H.; B.A., (fellow) Trinity Coll., Dublin, Ireland, 1833, M.A., 1839; LL.D., LL.B., 1840; m. Frances MacDougall, 1840; children—George Vaughan, Henry Chichester. Mem. staff Trinity Coll., Dublin, from 1835, v.p., 1876. Knighted, 1886. Author: Elementary Treatise on Mechanics, 1844. Research on hydrostatics and mechanics; worked on proofs to various math. problems; extended Peurbach's theorem by showing circles touching 3 given circles can be distributed into sets of 4, all touched by same circle. Died Kilderry, Ireland, Apr. 13, 1890.

HART, Edward, Am. chemist; b. Doylestown, Pa., Nov. 18, 1854; s. George and Martha Longstreth (Watson) H.; B.S., Lafayette Coll., 1874, LL.D., 1924; Ph.D., Johns Hopkins, 1878; m. Jennie Darlington, Aug., 1878. Asst. and tutor chemistry Lafayette Coll., 1874-76, adj. prof., 1876-82, prof., 1882-1924, also dean Pardee Scientific Dept., 1909-24. Pres. Baker & Adamson Chem. Co., 1881-1913; propr. Chem. Pub. Co., from 1892. Recipient John Scott medal and premium Franklin Inst., Phila. Editor Jour. Analytical and Applied Chemistry, 1887-93, Jour. Am. Chem. Soc., 1893-1901. Author: Volumetric Analysis, 1876; Text Book of Chemical Engineering, 1920; The Silica Gel Pseudomorph, 1924. Inventor mineral wax bottle for holding hydrofluoric acid. Died June 6, 1931.

HART, Edward Walter, Am. physicist; b. Easton, Pa., Jan. 14, 1918; s. Abraham S. and Sara (Rosenstrauch) H.; B.S., City Coll. N.Y., 1938; Ph.D., U. Cal. at Berkeley, 1950; m. Flori L. Feder, Dec. 15, 1940; children—Enid L., Lucinda M. Physicist, elec. engr. U. S. Navy Dept., Washington, San Diego, 1940-45; physicist U. Cal. Radiation Lab., Berkeley, 1946-51; physicist Gen. Electric Research Lab., Schenectady, N.Y., 1951—; adjunct prof. physics Rensselaer Poly Inst., Troy, N.Y., 1963. Recipient U. S. Navy Meritorious Civilian Service award, 1945. Fellow Am. Phys. Soc. Research and publs. on theory of nuclear forces, study of metals, thermodynamic theory of inhomgeneous systems. Home: 1168 Hedgewood Lane, Schenectady, N.Y. 12309*

HART, Edwin Bret, Am. biochemist; b. Sandusky, O., Dec. 25, 1874; s. William and Mary (Hess) H.; student U. Mich., 1892-97, U. Marburg (Germany), U. Heidelberg (Germany), 1900-01; m. Ann Virginia De Mille, Nov. 18, 1903; 1 dau., Margaret Virginia. Asst. chemist, N.Y. Agrl. Exptl. Sta., 1897-1902, asso. chem., 1902-06; prof. agrl. chemistry U. Wis.; chemist, Wis. Exptl. Sta., from 1906. Fellow A.A.A.S.; mem. Nat. Acad. Sci., Am. Chem. Soc., Soc. Biol. Chemists, Soc. Animal Prodn., Soc. Dairy Science. Devised simple method for estimating casein in cows milk; differentiated between organic and inorganic phosphorus in foods; discovered importance of iodine for goiter prevention; helped discover vitamins. Died Mar. 12, 1953.

HART, George H(art), Am. veterinarian; b. Phila., Pa., Oct. 10, 1883; s. John Robbins and Jane (Sheard) H.; V.M.D., U. Pa., 1903; M.D., George Washington U., 1908; LL.D., U. Cal., 1958; m. Eva M. Cadman, Nov. 13, 1912 (dec. Nov. 1949); m. 2d, Theresa A. Dennis, Jan. 6, 1954. Asst. bacteriology, pathology, pathol. div., bur. animal husbandry U. S. Dept. Agr., Washington, 1903-08, field worker, 1908-10; city veterinarian Health Dept., Los Angeles, 1910-17; asso. prof. vet. sci. Coll. of Agr., U. Cal., 1917-24, prof., 1924-26, prof. and head div. of animal husbandry, 1926-48, dean Sch. Vet. Med., prof. vet. sci., 1948-54, dean emeritus, 1954—; Fulbright lectr. U. Sydney (Australia), 1955. U. S. Govt. del. to 13th Internat. Vet. Congress, Zurich, Switzerland, 1938, World Vet. Congress, Stockholm, 1953; del. NRC to 7th Pacific Science Congress, Auckland and Christ church, New Zealand, 1949; chmn. com. on animal health NRC 1943-44, 45, mem. coms. on animal nutrition, vet. services for farm animals and tng. research workers in agr., 1944-47. Recipient Internat. Vet. Congress prize, 1938; Borden award Am. Vet. Med. Assn., 1953. Mem. A.A.A.S., Am. Assn. Univ. Profs., Am. Vet. Med. Assn. (mem. research council 1941-50), Am. Soc. Animal Prodn. (pres. 1940, 1947 honor guest). U. S. Livestock San. Assn., Wilderness Soc., Soc. Exptl. Biol. and Med., Sigma Xi. Died Aug. 2, 1959.

HART, Harold, Am. chemist; b. N.Y.C., May 14, 1922; s. David and Ruth (Feldman) H.; student Bklyn. Coll., 1937-39; B.S., U. Ill., 1941; M.S., Pa. State U., 1943, Ph.D., 1947; m. Geraldine Marjorie Cohen, Oct. 4, 1942; children—Leslie Elisabeth, David Joel, Diana Louise, Ariel Linda. Faculty, Mich. State U., 1946—, prof., 1957—. Cons. Standard Oil Co. (Ind.), Amoco Chems. Corp., 1955—; mem. adv. bd. Am. Chem. Soc. Petroleum Research Fund, 1966—; mem. chem. panel NSF, 1961-65. Guggenheim fellow Harvard, 1955-56; NSF fellow, Cambridge, Eng., 1962-63. Recipient Mich. State U. Distinguished Faculty award, 1965. Mem. Am. Chem. Soc. (award in petroleum chemistry 1962), Chem. Soc. (London), A.A.A.S., Am. Assn. U. Profs., Sigma Xi. Author: Organic Chemistry, 1953, 2d edit., 1959, Korean edit., 1961, 3d edit., 1966; also numerous articles. Editorial advisor Houghton Mifflin Co., 1958—; editorial bd. Jour. of Organic Chemistry, 1965—; editor-in-chief Chem. Revs. 1967—; Studies, publs. of chemistry of hydrocarbons related to petroleum; synthesis of molecules with strained rings; discovery of new reactions from interaction of light with organic molecules. Home: 2323 E. Mt. Hope Rd., Okemos, Mich. 48864.*

HART, Hiram, Am. biophysicist; b. Bklyn., May 29, 1924; s. Max and Ruth (Meyers) H.; B.S., Coll. City N.Y., 1943; Ph.D. in Theoretical Physics, N.Y. U., 1952. Physicist, U. Camera Corp., 1943-46; with N.Y.C. Bd. Edn., 1946-53; faculty dept. physics Coll. City N.Y., 1953—, prof., since 1968—; physicist, head med. physics lab. Montefiore Hosp. and Med. Center, N.Y.C., 1952-—; research asso. med. physics dept. Brookhaven Nat. Lab., Upton, L.I., N.Y., 1964-—. Cons., Walter Reed Army Inst. Research, 1960-62. NSF Sci. Faculty fellow Yale, 1959-60. Mem. Am. Phys. Soc., Biophys. Soc., Soc. Nuclear Medicine, Royal Soc. Medicine, Am. Assn. Physicists in Medicine. Editor: Multicompartment Analysis Tracer Experiments, 1963. Contbr. articles to tech. jours. Research on analysis kinetics from results radioisotope tracer expts. in biol. and biochem. systems; developed 1st gen. single isotope solution contiguous multi-compartment systems, 1st perturbation-tracer treatment steady state systems, 1st gen. solution for non-contiguous multi-compartment systems. Home: 1710 Montgomery Av., N.Y.C. 10453.*

HART, James, English physician; b. Northamptonshire, Eng., flourished circa 1633; ed. Paris, France, circa 1608; later at Basle; M.D. Practiced medicine, Northampton, from 1612. Author: The Arraignment of Urines, by Peter Forrest, Epitomized and Translated by James Hart, 1623; The Anatomie of Urines, 1625; Diet of the Diseased (an attempt in Hippocratic tradition to prescribe, but with exclusion of drugs, proper regimen and phys. conditions such as air and exercise in both health and disease), 1633.

HART, John Lawson, Canadian biologist; b. Toronto, Ont., Can., May 8, 1904; s. John S. and Jean (Lawson) H.; B.A., U. Toronto, 1925, M.A., 1926, Ph.D., 1930; m. Olive Helen Doan, May 22, 1930; 1 dau., Kathryn Helen (Mrs. Ronald Alexander Manzer). Lectr. U. Alta., Can., 1929-30; with Fisheries Research Bd. Can., 1929-—, biologist Biol. Sta. Nanaimo, B.C., 1929-50, dir., 1950-54; dir. Biol. Sta. St. Andrews, N.B., Can., 1954-67. Mem. Am. Inst. Fisheries Research Biologists, A.A.A.S., Am. Fisheries Soc., Am. Soc. Ichthyologists Herpetologists. Contbr. numerous articles in field to sci. jours. Research life history white fish in Ont. Lakes, also population dynamics pilchards herring, lingcon in B.C. waters, food of fur seals, faunology of fishes. Address: 60 Queen St., St. Andrews, N.B., Can.*

HART, Philip D'Arcy, English med. scientist; b. London, Eng., June 25, 1900; s. Henry D'Arcy and Ethel (Montagu) H.; student Caius Coll., Cambridge (Eng.) U., 1919-22, U. Coll. Hosp. Med. Sch., London, 1922-25; M.A., 1926; M.D., Cambridge U., 1930; m. Ruth Meyer, Oct. 25, 1941; 1 son, Oliver Simon D'Arcy. Temple Cross fellow, U. S., 1934-35; asst. physician U. Coll. Hosp., London, 1934-37; mem. sci. staff Med. Research Council, 1937-65, dir. Tb. research unit, London, 1948-65; attached Nat. Inst. for Med. Research, London, 1965-—; mem. investigatory missions for Brit. Govt. on Tb., Newfoundland, 1945, West Germany, 1947. Mem. expert com. of Tb., WHO, 1947-64. Fellow Royal Coll. Physicians; mem. Brit. Soc. Immunology, Soc. Gen. Microbiology, Path. Soc. Gt. Britain. Research and publs. on dust disease of coalminers, chemotherapeutic agts. in Tb, BCG vaccination, action of certain surfactants in exptl. Tb. Home: 37 Belsize Ct., London N.W.3. Office: Nat. Inst. for Med. Research, London, N.W.7., Eng.*

HART, Theodore Stuart, Am. physician; b. Groving, Ill., Feb. 25, 1869; s. Charles Langdon and Sarah (Franks) H.; A.B., Yale, 1891, A.M., 1893; M.D., Columbia U. Coll. Phys. and Surg., 1895; m. Mary Robbins, June 12, 1901. Vis. physician Seton Hosp., 1901-11; instr. Coll. Phys. and Surg., 1903-13, asst. prof. clin. medicine, 1913-22; vis. physician Presbyn. Hosp., 1914-22, cons. physician, from 1922; cons. physician Manhattan Eye, Ear and Throat Hosp., from 1922, Neurol. Inst., from 1936. Diplomate Am. Bd. Internal Medicine. Mem. Assn. Am. Physicians, Am. Soc. Clin. Investigation, A.M.A., N.Y. Acad. Medicine, N.Y. Heart Assn. (pres. 1920-24), Am. Heart Assn. (chmn. exec. com., 1924-31), Harvey Soc., Zeta Psi, Chi Delta Theta. Author: The Diagnosis and Treatment of Abnormalities of Myocardial Function, 1917; Taking Care of Your Heart, 2d edit., 1937. Contbr. many papers on metabolic disorders and diseases of circulation to med. jours. Died Jan. 1, 1951.

HART, Thomas John, marine biologist; b. Little Shelford, Eng., Sept. 17, 1907; s. John Henry Arthur and Katharine Mary (Gwatkin) H.; L.M.S., Bakewell, 1925; B.Sc., Leeds (Eng.) U., 1928, B.Sc. with honors in Zoology I, 1929, M.Sc., 1934, D.Sc., 1935; m. Edith Angood, June 17, 1931; children—William John, Richard Henry Angood, Thomas Andrew. Zoologist sci. staff Discovery Investigations, 1929-49; staff Nat. Inst. Oceanography, Wormley, Eng., 1949-—. Recipient Polar medal, Antarctic, 1929-39. Mem. Marine Biol. Assn. U.K., Challenger Soc., Inst. Biology, Brit. Phycological Soc. Author or co-author of Discovery Reports, 1934, 35, 37, 42, 46, 60; also articles. Description and interpretation of seasonal and spatial distbn. of marine phytoplankton in relation to environment, phys. and biol., fishery problems, discolored water phenomena, amphipid crustacea, skin film diatoms of whales. Office: Nat. Inst. Oceanography, Wormley, nr. Godalming, Surrey, Eng.*

HARTE, Richard Hickman, Am. surgeon; b. Rock Island, Ill., Oct. 23, 1855; s. William H. and Mary A. (Betty) H.; M.D., U. Pa., 1878; m. Maria H. Ames, 1888; children—Katherine (Mrs. George Putnam), Helen (Mrs. R. Ellison Thompson), Richard. Emeritus surgeon, Pa. Hosp.; cons. surgeon St. Mary's, St. Timothy's, Bryn Mawr and Abington hosps.; adj. prof. surgery, U. Pa. Fellow Am. Surg. Assn. (pres. 1910-11, editor Transactions 12 years). Author: Hand Book of Local Therapeutics, 1893. Pioneer in plastic surgery. Died Nov. 14, 1925.

HARTECK, Paul, chemist; b. Vienna, Austria, July 20, 1902; s. Josef and Gabriele (Schattenfroh) H.; student U. Vienna, 1921-23; Ph.D., U. Berlin, 1926; D.Sc., U. Bonn, 1966; m. Marcella Piccino-Hay, Sept. 27, 1948; children—Claudia, Christian Wolfgang, Paul Laurence. Distinguished research prof. phys. chemistry Rensselaer Poly. Inst., Troy, N.Y., 1951-—. Recipient Wilhelm Exner medal Oester Gewerbeverein, 1961. Mem. Joachim-Jungius Gesellschaft der Wissenschaftern (past v.p.), Max Planck Soc., Societe Chimique de Belgique (Jean Servais Stas medal 1957). Author: (with K. F. Bonhoeffer) Grundlagen der Photochemie, 1933. Contbr. numerous articles to profl. jours. Research on ortho-para hydrogen-photochemistry-nuclear physics, Deuterium-Deuterium reaction producing Tritium and Helium III; chemistry of upper atmosphere, radiation chemistry, chemonuclear research. Home: Brunswick Hills, Troy, N.Y. 12180.*

HARTER, Harman Leon, Am. statistician; b. Keokuk, Ia., Aug. 15, 1919; s. Harman Theodore and Mary Josie (Hough) H.; A.B., Carthage Coll., 1940; A.M., U. Ill., 1941; Ph.D., Purdue U., 1949; m. Alice Lauretta Madden, Oct. 23, 1943. Grad. asst. math. U. Ill., 1941-43; prof. physics Mo. Valley Coll., 1943-44; instr. math. Purdue U., 1946-48, research fellow in math., 1948-49; asst. prof. math. Mich. State U., 1949-52; math. statistician Aerospace Research Labs., Wright-Patterson AFB, Ohio, 1952-—. Fellow Am. Statis. Assn. (chmn. sect. on phys. and engrng. scis. 1964); mem. Inst. Math. Statistics, Math. Assn. Am., Operations Research Soc. Am., Soc. Indsl. and Applied Math., Internat. Assn. for Statistics in Phys. Scis. Author: New Tables of the Incomplete Gamma-Function Ratio and of Percentage Points of the Chi-Square and Beta Distributions, 1964. Research, publs. on theory of order statistics and their use in testing of statis. hypotheses and in estimation of parameters of statis. distbns., especially those useful in life testing; computation of extensive and highly accurate tables, including tables of critical values for various statis. tests. Home: 32 S. Wright Av., Dayton, Ohio 45403. Office: Aerospace Research Labs., Bldg. 450, Area B, Wright-Patterson AFB, Ohio 45433.*

HARTERT, Ernst Johann Otto, German ornithologist; b. Hamburg, Germany, Oct. 29, 1859; s. Carl and Elisabeth (Raysen) H.; Ph.D. (hon.), Marburg, Germany, 1904. m. Claudia Endris, 1891; 1 son, (dec. 1916). Zoologist, Eduard Flegel's expdn. to Hausaland, 1885-86; traveled through Sumatra and India, 1887-89; began work at a Senckenberg mus., Frankfort, Germany, 1889, later with zool. mus., London; became dir. Walter Rothschild's pvt. zool. mus., Tring, Eng., 1892; ret. 1930; after retirement worked at Zool. Mus., Berlin. Author: Vorläufige Versuch einer Ornis Preussens, 1887; Aus den Wanderjahren eines Naturfurschers, 1901; Die Vögel der paläarktischen Fauna, 3 vols., 1903-23. Research and publs. on bird taxonomy, theory of origin of species (led to reform in sci. nomenclature of animals in Europe). Died Berlin, Nov. 11, 1933.

HARTFORD, Winslow Hopper, Am. chemist; b. Newton, Mass., June 1, 1910; s. James Bradley and Alice (Winslow) H.; A.B., Boston U., 1928; D.B., Mass. Inst. Tech., 1930, Ph.D., 1933; m. Mary Emily Haviland, May 27, 1939; children—Douglas Bennett, Janet Winslow. Research chemist Mut. Chem. Co. Am., Balt., 1933-45; research supr. Mut. Chem. div. Allied Chem. Corp., Balt., 1945-58, research supr. Solvay Process div., Syracuse, N.Y., 1958-63, sr. scientist, 1963-67, with the industrial chemicals division, 1967-—. Mem. Am. Chem. Soc., A.A.A.S., Soc. Testing and Materials, Leather Chemists Assn., Wood Preservers Assn., Am. Electroplaters' Soc., Sigma Xi. Contbg. author: Chromium, 1956; Treatise Analytical Chemistry, 1963; Encyclopedia Chemical Technology, 1964; also numerous articles. Research on chemistry and tech. chromium compounds, wood preservatives, analytical chemistry leather, tanning compounds, indsl. waste water, corrosion prevention, metal plating and finishing, hydrogen-chlorine explosions, organic chlorinations. Patentee in field. Home: 309 Southfield Dr., Fayetteville, N.Y. 13066. Office: P.O. Box 6, Syracuse, N.Y. 13209.*

HARTIALA, Kaarlo Jacob, physiologist; b. Hancock, Mich., Dec. 10, 1919; s. Rafael Johan and Hilja (Wargelin) Hartman; licentiate of Medicine, Helsinki (Finland), 1947, M.D., 1950; m. Airi Järvinen, Dec. 27, 1943; children—Anja, Jaakko, Marja, Kari, Eeva. Docent physiology Helsinki U., 1954; prof. physiology Turku (Finland) U., 1955-—, dean Med. Faculty, 1960-63. Med. dir. Lääke Oy, Medipolar; chmn. Finnish Sci. Council. Decorated Comdr. Cross Finland, White Lion Order, Finland Liberty Cross, Order 4 with laureate. Mem. Finnish State Sport Fedn. (sci. council 1962-—), Scandinavian Physiol. Soc. (pres.), Finland Sci. Acad., A.A.A.S. Land. editor Acta physiol. Scand. Research, numerous publs. on role of gastrointestinal mucous membrane in handling of fgn. substances (detoxification reactions), mechanism of exptl. drug or psychic stress induced peptic ulceration; pointed out insufficiency of fetal organ to cope with fgn. substances, 1954. Home: 1 A 9 Satalaistenkatu, Turku, Finland.*

HARTIG, Georg Ludwig, German forester; b. Gladenbach, nr. Biedenkopf, Germany, Sept. 2, 1764; s. Friedrich Christian and Sophie Catherine (Venator) H.; student U. Giessen (Germany), 1781-83; Ph.D. (hon.); m. Theodore Klipstein, 1787; 9 sons, including Theodor, 4 daus. Became chief forestry ofcl., Hungen, 1886; forestry ofcl. for house of Nassau-Oranien, 1797-1866; named chief forestry adviser, Stuttgart, Germany, 1806; chief state forestry ofcl. of Prussia, dir. forestry in Prussia, 25 years; lectr. forestry sci. U. Berln, named hon. prof., 1838. Author: Anweisung zur Holzzucht für Förster, 1791, 8th edit., 1818; Anweisung zur Taxation der Forste oder zur Bestimmung des Holzertrags, 1795; Grundsätze der Forstdirection, 1803; Lehrbuch für Förster und die es werden wollen, 1808, 11th edit., 1877; Die Forstwissenschaft nach irhem canzen Umfange in gedrängter Kürze, 1831; Forstliches und forst-naturwissenschaftliches Conversations-Lexikon, 1834. Pioneer (with Heinrich Cotta) in practice of classical forestry, research and teaching. Died Berlin, Feb. 2, 1837.

HARTIG, (Heinrich Julius Adolph) Robert, German botanist; b. Braunschweig, Germany, May 30, 1839; s. Theodor and Agnes (von Heidenreich) H.; studied forestry Collegium Carolinum, Braunschweig, 1861-63; other studies in Berlin; m. Adolfine Geller, 1869; 1 dau., Edith (Mrs. Karl von Tubeuf). With state forestry adminstrn., Braunschweig, also Hanover, Germany; named lectr. botany and zoology, Eberswalde, Germany, 1867; became prof. botany U. Munich, also chmn. bot. dept. Forestry Research Inst., 1878. Author: Vergleichende Untersuchungen über Wachstumsgang und Ertrag der Rotbuche und Eiche im Spessart, der Rotbuche im östlichen Wesergebirge, der Kiefer in Pommern und der Weisstanne im Schwarzwald, 1865; Über das Dickenwachstum der Waldbäume, 1874; Lehrbuch der Pflanzenkrankheiten, 1900. A founder of wood pathology; concentrated on study of fungus diseases of forest trees. Died Munich, Oct. 9, 1901.

HARTIG, Theodor, German silviculturist, entomologist; b. Dillenburg, Germany, Feb. 21, 1805; s. Georg Ludwig and Theore (Klipstein) H.; student forestry inst. U. Berlin, 1924-27; m. Agnes von Heidenreich, 1830, 1835; 4 sons, including Robert, 3 daus.; employed in forestry adminstrn.; became lectr. forestry sci., Berlin, 1830; prof., chmn. forestry dept. Collegium Carolinum, Brunswick, Germany, 1838-77; mem. Leopoldina. Author: Die Adlerflügler Deutschlands mit besonderer Berücksichtigung ihres Larvenzustandes und ihres Wirkens in Wäldern und Gärten, 1837; Lehrbuch der Pflanzenkunde in ihrer Anwendung auf Forstwirthschaft, 1940-46; Anatomie und Physiologie der Holzpflanzen, 1878; also articles. Leading forest botanist of his time; research on anatomy and physiology of wood growth, also forest entomology. Died Brunswick, Mar. 26, 1880.

HARTING, (Carl August) Johannes (Hans), German physicist; b. Rummelsburg, Germany, Feb. 15, 1868; s. Carl Friedrich Fides Ferdinand and Anna (Dittmann) H.; student Berlin, Munich; Ph.D., 1889; m. Margarete Hergesell, 1895; 1 son; m. 2d, Agathe Sohr, 1916. Joined State Phys.-Tech. Inst., 1893; became asst. to Ernst Abbe, 1897; named sci. and tech. dir. optical firm Voigtländer, Braunschweig, 1899; also lectr. Tech. U., Braunschweig; joined optical dept. Reich Patent Office, 1907, became dir., 1929, ret. but remained provisional pres., 1933; successor to Ernst Abbe, Carl Zeiss Found., 1934-41; joined Babelsberg Obs., 1941; rebuilt Zeiss firm, Jena, Germany, 1945-51. Recipient Goethe medal for art and sci., 1943, Nat. prize 1st class German Democratic Republic, 1949. Mem. German Soc. for Applied Optics (hon.; became chmn. 1923), German Acad. Scis. (hon.). Author: Optisches Hilfsbuch für Photographierende, 1909, 3d edit. under title Photographische Optik, 1948. Editor Centralzeitung für Optik und Mechanik, 1919-34. Research and publs. on telescopic and photog. objective lenses, geometric optics, microphoto optics, crystal cultivation; photoelectronic-photometric research (with P. Guthnick); devel. electron microscope; interpolation of optical refractive indices. Died Jena, Sept. 21, 1951.

HARTING, Pieter, physiologist; b. Rotterdam, Netherlands, Feb. 27, 1812; M.D., Utrecht, Netherlands, 1835; practiced medicine, Oudewater; tchr. microscopic anatomy and vegetable physiology, 1843-56, tchr. zoology, 1856-82, Utrecht. Author: Het Mikroskoop (treatise and exhaustive history of microscope), 4 vols., 1848-54. Died 1885.

HARTINGER, Hans, German physicist; b. Munich, Germany, Feb. 21, 1891; s. Michael and Karoline (Schwaiger) H.; Dr.rer.techn., Tech. U. Munich, 1916; student U. Munich; b. Irma Hammerschmidt, 1921; 2 sons, 1 dau. Joined optical firm C. Zeiss, Jena, Germany, 1917; Tchr. sch., 1914-16; joined faculty State Engrng. Sch. for Optics, 1925; became hon. prof. med. optics, U. Jena (Germany), 1937; became tchr. secondary sch., Munich, 1945; also hon. prof. Tech. U. Munich. Research and numerous publs. on med.-optical devices, including elec. hand eye mirror, photo-ceratoscope, poly-ophthalmoscope, projection perimeter, projection coordimeter; also improved numerous apparatus; participated in invention of shadow-free operation lamps. Died Munich, July 18, 1960.

HARTL, Heinrich Josef Franz, geographer, geodesist; b. Brno, Czechoslovakia, Jan. 23, 1840; student Poly. Inst., Vienna, Austria, 1856-59; Ph.D. (hon.), Vienna, 1899. Joined army, 1859; transferred to navy, 1861; assigned to Mil. Geog. Inst., 1865; named dir. trigonometric dept. Mil. Geog. Inst., 1887; commd. col., 1895; ret. from mil. service, 1898; prof. geodetics U. Vienna. Apptd. authorized del. Internat. Geodetic conf., 1882. Mem. Leopoldina. Author: Die Höhenmessung des Mappeurs, 2 vols., 1876; also articles. Astronomic location finding in Balkans; staked out Arlberg tunnel; research on meteorology, history of surveying, atmospheric refraction, 1st order triangulation of Greece; also triangulation work in Dalmatia, Bohemia, Moravia. Died Vienna, Apr. 3, 1903.

HARTL, Rudolf, veterinarian; b. Donawitz, Bohemia, May 28, 1868; studied medicine, Prague, Vienna; M.D., 1894; became lectr. pathology and bacteriology Vienna Sch. Vet. Medicine, 1902, asso. prof., 1906, prof. pathology, bacteriology, path. anatomy, forensic vet. medicine, 1909, expanded its path.-anat. mus. and created unique documentation of pathology of domestic animals; dir. sta. for diagnostic animal innoculation, mem. vet. adv. bd. Ministry Agr. Contbr. to med. jours. Research on infectious diseases and epidemics among domestic animals, also livestock trading and liability. Died Vienna, May 8, 1943.

HARTLAUB, (Carl Johann) Gustav, German ornithologist; b. Bremen, Germany, Nov. 8, 1814; s. Karl Friedrich Ludwig and Johanna (Buch) H.; studied medicine, Bonn, Berlin; M.D., Göttingen, 1838; m. Caroline H. Stachow, 1844; children—Carl, Clemens. Traveled in Austria, France, Netherlands, Gt. Britain; practiced medicine, Bremen; engaged in ornithol. research. Author: Jahresberichte über die Leistungen in der Naturgeschichte der Vögel, 1846-71; System der Ornithologie Westafricas, 1857; Ornithologie der Viti-, Samoa- und Tonga-Inseln, 1867; (with O. Finsch) Die Vögel Ostafrikas, 1870; Die Vögel Madagascars und der benachbarten Inselgruppen, 1877. Expanded bird collection of mus. Bremen Natural History Assn. to internat. significance; important work on birds of Africa and Polynesia; considered most eminent ornithologist in Germany in 2d half of 19th century. Died Bremen, Nov. 20, 1900.

HARTLES, Ronald Lewis, English biochemist; b. Widnes, Eng., Nov. 15, 1915; s. William Edgar and Bertha (Lewis) H.; B.Sc., U. Liverpool (Eng.), 1938, Ph.D., 1948, D.Sc., 1959; m. Lilian Welbon, Apr. 5, 1941; children—Richard Peter Welbon, Elizabeth Joan. Staff, H. M. Forces, 1939-46, 92d E. African Malaria Lab., 1943-46; faculty U. Liverpool (Eng.), 1948—, prof. dental sci., 1963—. Cons. Unilever, Ltd., 1956—. Mem. Biochem. Soc., Nutrition Soc., Bone and Tooth Soc., Soc. for Gen. Microbiology, Internat. Assn. for Dental Research, Brit. Dental Assn. Author: (with D. H. Goose) Principles of Preventive Dentistry, 1964; also articles. Research on nutritional factors in metabolism of bones and teeth, metabolism of oral flora, exptl. dental caries. Home: Warden's Lodge, Greenbank Lane, Liverpool 17. Office: Dept. Dental Scis., U. Liverpool, Liverpool, Eng.*

HARTLEY, Sir Charles Augustus, Brit. civil engr.; b. Hedworth, Durham, Eng., Feb. 3, 1825; s. W. A. and Lillias (Todd) H. Engr.-in-chief, cons. engr. European Commn. of Danube, 1856-1907; made engrg. plans for River Scheldt, 1867; planned enlargement of Port of Odessa, 1867; mem. bd. engrs. for improvement of Mississippi River, 1875; mem. Paris conf. for route of Panama Canal, 1879; mem. Internat. Tech. Commn. Suez Canal, 1884-1907; mem. Ribble Nav. Commn., 1889; insp. Durban Harbor, 1896. Recipient awards Royal Soc. Arts, Inst. Civil Engrs. Fellow Royal Soc. Edinburgh; mem. Inst. Civil Engrs.; asso. Royal Inst. Brit. Architects. Author: Delta of the Danube, 1873; Public Works in the United States and Canada, 1874; Inland Navigations in Europe, 1885; History of the Engineering of the Suez Canal. Planned engrng. improvements on Danube, Hugli, Don, Dnieper, Mississippi, Scheldt rivers, also harbor improvements for Odessa, Madras, Trieste, Constanza, Bourgas, Varna; made the Danube navigable for cargo ships; noted for work on Suez, Panama canals. Died London, Feb. 20, 1915.

HARTLEY, David, English psychologist, physician, philosopher; b. Armley, Yorkshire, Eng., Aug. 30, 1705; s. David and Evelyn (Wadsworth) H.; ed. Jesus Coll., Cambridge, B.A., 1726, M.A., 1729; m. 1st 1730, 2d 1735; several children, including David. Practiced medicine, Newark and Bury St. Edmunds, 1730-35, London, 1735-42, Bath, 1742-57. Fellow Royal Soc., 1736. Author: Some Reasons Why the Practice of Innoculation ought to be introduced into the Town of Bury, 1733; Ten Cases of Persons who have Taken Mrs. Stephen's Medicines . . . , 1738; A View of the present Evidence for and against Mrs. Stephen's Medicines, 1741; Observations on Man, His Frame, His Duty and His Expectations, 2 vols., 1749. Originator of physiol. psychology, also doctrine of associationism; believed the mind developed from association of sensations, which are the product of vibrations in nervous system. Died Bath, Aug. 28, 1757.

HARTLEY, Frank, Am. surgeon; b. Washington, June 10, 1856; s. John Fairfield and Mary (King) H.; grad. Princeton, 1877; M.D., Columbia U., 1880; student Heidelberg, Germany, Vienna, Austria, Berlin, Germany; LL.D., Princeton, 1909; m. Emma Allyce Parker, Aug. 1, 1897. Became cons. surgeon to N.Y. Skin and Cancer Hosp., 1893. Mem. N. Y. Surg. Soc. First to use intracranial neurectomy (of 2d and 3d divs. of trigeminal nerve) for treatment of trigeminal neuralgia, 1892. Died N.Y.C., June 19, 1913.

HARTLEY, Sir Percival, English biochemist; b. Calverley, Eng., May 28, 1881; s. W. T. and (Grimshaw) H.; ed. Tech. Coll., Bradford; U. Leeds; D.Sc., London; hon. D.Sc., Leeds; m. Olga Parnell, 1920; 2 daus. Researcher, Lister Inst., London, Eng., 1906-09, asst. in biochemistry, 1913-15, also mem. staff, from 1949; physiol. chemist to govt. of India, Muktesar, 1909-13; with Wellcome Physiol. Research Labs., 1919-21; dir. dept. biol. standards Nat. Inst. for Med. Research, 1922-46; mem. staff U. London, 1927; with London Sch. Hygiene and Tropical Medicine, 1946-48; with dept. pathology Oxford (Eng.) U., 1948-49. Knighted, 1944. Fellow Royal Soc., 1937; mem. Pathol. Soc. Gt. Britain and Ireland, Physiol. Soc., Biochem. Soc., Soc. Gen. Microbiology. Research on biochemistry, bacteriology, immunology; discovered (with J. B. Leathers) arachidonic acid; gave conditions for Hartley's broth (effective broth for production of potent diphtheria toxin). Died Feb. 16, 1957.

HARTLEY, Ralph Vinton Lyon, Am. physicist; b. Spruce, Nev., Nov. 30, 1888; s. Robert and Matilda (Hutchison) H.; A.B., U. Utah, 1909; B.A. (Rhodes scholar), Oxford (Eng.) U., 1912, B.Sc., 1913; m. Florence Vail, Mar. 21, 1916. Instr., U. Nev., Reno, 1909-10; mem. research dept. Western Electric Co., N.Y.C., 1913-25; mem. research dept. Bell Telephone Labs., N.Y.C., 1925-42, Murray Hill, N.J., 1942-50. Recipient Medal of Honor, I.R.E., 1946. Fellow Am. Phys. Soc., Acoustical Soc. Am., I.E.E.E. Author: Oscillator Circuit, 1914; Binaural Sound Location, 1919; Information Theory, 1927; Parametric Oscillations (Ramon Effect), 1928; Mechanical Interpretation of Physics, 1959. Home: Hotel Suburban, Summit, N.J. 07901.*

HARTLEY, Sir Walter Noel, chemist; D.Sc. honoris causa, Royal U. Ireland. Prof. chemistry, dean faculty Royal Coll. Sci., Dublin. Recipient Gold medal for sci. applications of photography, Silver medal in chem. arts St. Louis Expn., 1904, Longstaff medal for research in spectro-chemistry Chem. Soc., 1906, Grand Prix for spectrographic research Franco-Brit. Exhbn., 1908. Fellow Royal Soc., 1884; mem. Inst. Chemistry Gt. Britain and Ireland (past v.p.), Brit. Assn. (pres. sect. B chemistry 1903-04). Author: Air and Its Relations to Life, 1876; Water, Air and Disinfectants, 1877; Quantitative Analysis, 1887, also articles. Died Sept. 11, 1913.

HARTLINE, Haldan Keffer, Am. physiologist; b. Bloomsburg, Pa., Dec. 22, 1903; s. Daniel Schollenberger and Harriet Franklin (Keffer) H.; B.S., Lafayette Coll., Easton, Pa., 1923; M.D., Johns Hopkins, 1927; Eldridge Johnson traveling research scholar, Univs. of Leipzig and Munich, 1929-31; Doctor Science, Lafayette College, 1959; married Mary Elizabeth Kraus, Apr. 11, 1936; children—Daniel Keffer, Peter Haldan, Frederick Flanders. Nat. Research fellow in med. sciences, Johns Hopkins, 1927-29; fellow in med. physics, Eldridge Johnson Research Found., U. of Pa., 1931-36, asst. prof. biophysics, 1936-40, 1941-42; asso. prof. physiology, Cornell Univ. Med. Coll., 1940-41; asso. prof. biophysics, U. of Pa., 1943-48, prof., 1948-49; prof. biophysics and chairman department Johns Hopkins University, Baltimore, 1949-53; member and professor Rockefeller Univ., New York City, 1953—. Recipient Nobel prize (with G. Wald, R. Granit), 1967; William H. Howell award (physiology), 1927. Howard Crosby Warren medal (exptl. psychology), 1948; A.A. Michelson award case Inst. 1964. Mem. Nat. Acad. Scis., American Physiol. Soc., Am. Philos. Soc., Am. Acad. Arts and Scis., Royal Soc. (London), Biophys. Soc., Optical Soc. Am., Phi Beta Kappa, Sigma Xi. Research on neurophysiology of vision; physiology of sense receptors; electric responses of arthropod and vertebrate eyes; physiology of photoreception; research on electrical responses of nerve fibers and cells when light hits retina showed how eye differentiates shapes and sharpens contrasts. Address: Rockefeller U., 66th St. and York Av., N.Y.C. 10021.

HARTMAN, Carl Gottfried, Am. physiologist; b. Reinbeck, Ia., June 3, 1879; s. Ossiar William and Sophie (Lemvigh) H.; student U. Ia., 1896-97; B.A., U. Tex., 1902, M.A., 1904, Ph.D. in Embryology, 1915; m. Eva M. Rettenmeyer, June 23, 1919; children—Carl F., Philip E., Paul A., Grace (Mrs. Richard Leighton). Mem. faculty Sam Houston State Tchrs. Coll., 1909-12, U. Tex., 1912-25; research asso. Carnegie Inst., 1925-41; prof. zoology, head depts. zoology and physiology, U. Ill., 1941-47; staff Ortho Research Found., 1947-51, asso. dir., 1951-58, dir. emeritus, 1958—; research cons. Margaret Sanger Research Bur., N.Y.C., 1960; Recipient Squibb award Endocrine Soc., 1946; Lasker award, 1949, 1st Margaret Sanger award in medicine, 1966 Planned Parenthood Fedn. Am.; Barren medal Barren Found. Chgo., 1963; First Marshal medal Brit. Soc. Study Fertility, 1965. Wistar Inst. fellow, 1917-18. Mem. Nat. Acad., A.A.A.S., Physiol. Soc., Soc. Zoology (pres. 1948), Gynecol. Soc., Soc. Study Sterility (hon. v.p. 1959), Soc. Gynecol. Investigation (hon.), Am. Assn. Anatomists, Anat. Soc. Gt. Britain and Ireland (hon.), Brazil Soc. Study Sterility (hon.), Internat. Inst. Embryology. Author: (with L. B. Bibb) First Book of Health and Human Body and Its Enemies, 1932; Time of Ovulation in Women, 1936; Possums, 1952; Science and the Safe Period, 1963. Research and publs. on physiology of reprodn. in man, animals; opossum embryology and devel.; drugs and fertility. Home: 606 Crescent Av., Plainfield, N.J. 07060. Office: 17 W. 16th St., N.Y.C. 10011.*

HARTMAN, Frank Alexander, Am. physiologist; b. Gibbon, Neb., Dec. 4, 1883; s. George Washington and Flora (Sprague) H.; A.B., U. Kan., 1905, M.A., 1909; Ph.D., U. Wash., 1914; m. Anna Botsford, Feb. 10, 1906; children—William B., Warren E., Donald G., Flora L., Mary L. (Mrs. Merrill Barnebey). Tchr. high schs., 1906-14; Austin teaching fellow Harvard Med. Sch., 1914-15; faculty U. Toronto (Can.), 1915-19, U. Buffalo, 1919-34; prof., chmn. dept. physiology Ohio State U., Columbus, 1934-47, research prof., 1947-54, research prof. emeritus, 1954—. Recipient Chancellors medal U. Buffalo, 1932; Schoelkopf medal Western N.Y. Sect. Am. Chem. Soc., 1932. Fellow A.M.A. (Gold medal 1932); mem. A.A.A.S., Physiol. Soc., Soc. Zoology, Endocrine Soc. (pres. 1935), Ornithol. Union, Wilson Ornithol. Soc., Cooper Ornithol. Club, Phi Beta Kappa, Sigma Xi, Phi Lambda Upsilon, Alpha Omega Alpha. Author: (with Katharine A. Brownell) The Adrenal Gland, 1949; also numerous articles. Research on adrenal glands of vertebrate species; separated 1st active substance from adrenal gland effective in treatment of Addison's disease. Home: 183 W. Como Av., Columbus, O. 43202.*

HARTMAN, Leon Wilson, Am. physicist; b. Downsville, N.Y., June 18, 1876; s. Henry and Sarah Eleanor (Wilson) H.; B.S., Cornell U., 1898, A.M., 1899; Ph.D., U. Pa., 1903; postgrad. U. Göttingen, 1903-04; m. Edith Dabele Kast, July 31, 1907; children—Margaret Eleanor, Sara Louise (dec.), Paul Leon, Charles Frederick, David Kast. Asst. in physics Cornell U., 1900-01; prof. physics Kan. Agrl. Coll., 1901-02; Fraser fellow in physics U. Pa., 1902-03, Tyndale fellow, 1903-04; instr. physics Cornell U., 1904-05; asst. prof. physics U. Utah, 1905-06, asso. prof., 1906-09; prof. physics U. Nev., from 1909, acting pres., 1938, pres., from 1939. Mem. A.A.A.S., Am. Phys. Soc., Utah Acad. Sci., Illuminating Engring. Soc., Am. Assn. Univ. Profs., Sigma Xi, Phi Kappa Phi. Author: Laboratory Manual of Experiments in Physics, 1906; An Introduction to Electrical Measurements, 1930; Measurement of Coefficient of Self-Inductance in Terms of Resistance and Time (joint author), 1940. Contbr. to various scientific journals, articles on radiation, pyrometry, acetylene, Nernst lamp, spectro-photometry, visibility. Died Aug. 27, 1943.

HARTMANN, Arthur, physician; b. Germany, 1849; credited with inventing audiometer, circa 1878. Died 1931.

HARTMANN, (Wilhelm) Eugen, German elec. engr.; b. Nürtingen, Germany, May 26, 1853; s. Karl Friedrich and Christiane Karoline (Wiedenmann) H.; trained as precision mechanic; hon. D. Eng., Tech. U. Stuttgart (Germany); worked in Vienna, Austria, also Göttingen, Germany (as tech. asst. to Wilhelm Weber); began ind. practice, Würzburg, Germany, 1879; transferred firm (with Wunibald Braun) under name Hartmann & Braun to Frankfort/Main, Germany; manufactured elec. measuring instruments, mirror galvanometer with telescopic scale, 1st spring galvanometer for practical electrotech.; a founder univ. in Frankfort; a founder Electrotech. Assn.; named prof. by Prussian govt. for promotion of Physics Assn. in Frankfort. Died Munich, Oct. 18, 1915.

HARTMANN, Georg, physicist, inventor; b. Eckoltsheim, Bavaria, 1489; studied theology and mathematics; became vicar St. Sebaldus Ch., Nuremberg, Germany; served Duke Albert of Prussia, 1544. Author: Directorium, 1554. Constructed a star altimeter, compass and sundials; wrote reports on magnetic declination and inclination; observed inclination of magnetic needle, 1544. Died Nuremberg, 1564.

HARTMANN, Henri-Albert-Charles-Antoine, French surgeon; b. Paris, June 16, 1860; aggregation in Surgery, French Acad. Scis. Research on intestinal surgeon, Paris, 1892; became prof. pathology Paris Faculty Medicine, 1909. Mem. Acad. Medicine, Acad. Surgery, French Acad. Scis. Research on intestinal and gen. surgery. Died Paris, Jan. 11, 1952.

HARTMANN, Henrik Anton L., physician; b. Sandefjord, Norway, Mar. 20, 1920; s. Henrik A. and Elvira (Erlandsen) H.; M.D., U. Oslo Med. Sch., 1949; m. Anastatia Smith, Sept. 17, 1952; children—Lisa, Tony, Jeni, Arne, Signe. Came to U. S., 1950, naturalized, 1956. Pathologist, St. Clare Hosp., Monroe, Wis., 1953; faculty U. Wis., Madison, 1954—, prof. pathology, since 1967—; staff Inst. for Neurobiology, U. Goteborg (Sweden), 1958. Cons., Mendota State Hosp., Madison, Wis., 1954—. Diplomate Am. Bd. Pathology. Mem. Am. Assn. Neuropathologists, Am. Soc. for Exptl. Pathology, Am. Assn. Pathologists and Bacteriologists. Research and publs. on nervous system, correlation between abnormal behavior in animals and quantitative changes in nerve cells. Home: 10 S. Kenosha Dr., Madison, Wis. 53705.*

HARTMANN, Hermann Leon H., German chemist; b. Bischofsheim, May 4, 1914; s. Michael and Anna (Bauer) H.; ed. Higher Tech. Sch., U. Munich; Ph.D. in natural sci.; m. Else Eylitz, 1941; children—Wolfgang Friedel, Olaf-René. Asst., Inst. Phys. Chemistry, Frankfort am Main, 1942; qualified prof., 1932; with Max Planck Inst., Göttingen, 1946; prof. agrégé and dir. Inst. Phys. Chemistry, U. Frankfort on Main, 1952. Mem. Wissenschaftliche Gesellschaft der U. Frankfurt, Bunsen Soc., Faraday Soc. Author: Theorie der Chemischen Bindung auf Quatentheoretischer Grundlage, 1954. Editor: Theoretica Chimica Acta, also numerous articles on quantum and phys. chemistry. Home: 8 Flughefenstrasse 8. Office: 11 Robert Mayerstrasse, Frankfort on Main, Germany.

HARTMANN, Hudson Thomas, Am. plant physiologist; b. Kansas City, Kan., Dec. 6, 1914; s. Dale and Violet (Thomas) H.; B.S., U. Mo., 1939, M.S., 1940; Ph.D., U. Cal. at Berkeley, 1947; m. Dorothy Henson, Sept. 23, 1940; children—Carol Lee, Donald, Marilyn, Lawrence. Asso. Expt. Sta., U. Cal. at Davis, 1945-47, faculty, 1947—, prof. pomology, 1960—. Recipient Fulbright award, Australia, 1960-61, Italy, 1964-65, Greece, 1968-69. Mem. A.A.A.S., Am. Soc. Plant Physiology, Am. Soc. Hort. Sci., Scandinavian Soc. Plant Physiology, Sigma Xi. Author: (with Dale Kester) Plant Propagation: Principles and Practices, 1959, 2d edit., 1968; also articles. Research on physiology of flowering in olive tree, physiol. studies on root initiation in cuttings of various fruit species. Home: 35 Parkside Dr., Davis, Cal. 95616.*

HARTMANN, Johann, mathematician, iatrochemist; b. Amberg, Bavaria, Jan. 15, 1568; s. Adam and Caecilia (Flick) H.; studied math. at univs. Altdorf, Jena, Leipzig, Helmstedt, Wittenberg (all Germany); graduated at Marburg, 1591; Doctor of Medicine, 1606. m. Susanne Mylius, June 26, 1592; 4 sons, including Georg Eberhard, 2 daus., including Margarete (Mrs. H. Petraeus). Prof. mathematics Marburg, 1592, prof. iatrochemistry, 1609; resigned chair to become councillor and physician to Duke of Hesse-Kassel, 1616. Author: Praxis Chymiatrica, 1633 (one of best known 17th century collections of chem. recipes; went through at least 13 editions by 1690); collected Opera Omnia Chymica, 1684 (went through 3 editions). First chemistry prof. in Europe; research on opium; collected chem. formulas including antimonial and mercurial preparations. Died Kassel, Dec. 7, 1631.

HARTMANN, Johannes Franz, German astrophysicist, astronomer; b. Erfurt, Germany, Jan. 11, 1865; s. Daniel and Sophia (Evers) H.; ed. univs. Tübingen, Berlin; doctorate, U. Leipzig (Germany), 1891; m. Angelika Scherr, 1907; 2 sons, 1 dau. Asst. to L. de Ball, Vienna, Austria, to H. Bruns, Leipzig; became asst., astro-phys. obs., Potsdam, Germany, 1896, observer, 1898, prof., 1902; named prof. U. Göttingen, dir. obs., 1909; dir. La Plata obs., 1921-34. Contbr. to profl. jours. A leading astrophysicist of his time; authority on spectroscopy; discovered interpolatorial dispersion formula (named for him); helped set up internat. system of wave length norms; built a microphotometer (named after him), 1899, spectro comparator, 1904; developed method (named after him) of precision testing large objectives, used it to correct the great double refractor; discovered static calcium cloud; set up a nova theory; determined deformation of Eros. Died Göttingen, Sept. 13, 1936.

HARTMANN, John Rudolf, Am. physician; b. Everett, Wash., Jan. 2, 1924; s. Rudolf and Eugenie (Kaiser) H.; M.D., Johns Hopkins, 1947; m. Elma C. Lile, July 7, 1956; children—John L., Anne C., Allison E., Andrew P. Practice medicine specializing in pediatric hematology, Seattle, 1958—; staff Kinderspital, Zurich, Switzerland, 1956-57; dir. hematology Children's Orthopedic Hosp. and Med. Center, Seattle, 1960—; prof. pediatrics U. Wash., since 1966—. Mem. exec. com. Leukemia Task Force, Nat. Cancer Inst., NIH, 1965—. Mem. Am. Acad. Pediatrics, Am. Soc. Hematology, Phi Beta Kappa, Alpha Omicron Alpha. Research and publs. in bleeding disorders, hemophilia, leukemia, malignant diseases of children. Home: 4316 N.E. 33d St., Seattle 98105. Office: 4800 Sand Point Way, Seattle, 98105.*

HARTMANN, Max, German biologist; b. Lauterecken, Germany, July 7, 1876; m. Louise Ruecklos; 2 children. Asst., Zoology Inst., Munich, Germany, 1902-05; with U. Giessen (Germany), 1905-14; dir. Robert Koch Inst., Berlin, 1909-14; became dir. Max Planck Inst., 1944; hon. prof. zoology, Tübingen, Germany, Author: Allgemeine Biologie, 1925; Die Sexualität, 1943; Die philosophische Grundlagen der Naturwissenschaften, 1948.

HARTMANN, Miner Louis, Am. chemist; b. Hutchinson, Kan., July 14, 1889; s. Henry Philip and Anne (Bussinger) H.; B.S., U. Ariz., 1911; Ph.D., Harvard, 1915; m. Elizabeth Trowbridge, Apr. 2, 1915; 1 son, Robert Trowbridge. Instr., U. Mo., 1914-15; prof. S.D. Sch. Mines, 1915-18; dir. research Carborundum Co., 1918-26; tech. dir. Celite Co., 1926-29; cons. chem. engr., patent atty., Los Angeles and Beverly Hills, Cal., 1929—. Mem. Los Angeles Air Pollution Hearing Bd., 1946-52. Mem. Am. Chem. Soc., Los Angeles Patent Law Assn. Research, publs., inventions in fields of abrasives, high temperature refractories, radio, non-metallic mineral processing, foundry sand bonds electrochemistry, air pollution abatement; determined atomic weight of cadmium; developed high temperature heat insulating materials. Home: 300 S. Canon Dr. Office: 204 S. Beverly Dr., Beverly Hills, Cal. 90212.*

HARTMANN, Richard, mech. engr.; b. Barr, Alsace, Nov. 8, 1809; s. Johannes and Marie Madeleine (Schwartz) H.; m. Berta Augusta Schwartz, 1837; 4 sons, including Gustav, Richard, 4 daus.; m. 2d, Elise Schäffer, 1872. Learned toolsmith's trade, Barr; with mech. firm Carl Gottlieb Haubold, Chemnitz, Germany, 5 years; founder (with Illing and Götze) factory for repair and rebldg. of spinning machines, 1837, expanded it to manufacture steam engines, 1840, further expansion, 1843, built weaving machines, from 1844, locomotives, from 1846; formed (with sons) a co., 1868, sold into corp. Sächsische Maschinenfabrik zu Chemnitz, 1869, chmn. bd. until death. Inventor "Continue" spinning machine (replaced 2 others), 1839; introduced manufacture of steam engines in Saxony. Died Chemnitz, Dec. 16, 1878.

HARTNETT, James Patrick, Am. mech. engr.; b. Lynn, Mass., Mar. 19, 1924; s. James Patrick and Anna (Ryan) H.; B.S., Ill. Inst. Tech., 1947; M.S., Mass. Inst. Tech., 1948; Ph.D., U. Cal. at Berkeley, 1954; m. Shirley Carlson, July 14, 1945; children—James, David, Paul, Carla, Dennis. Engr., Gen. Electric Co., 1947-48; research engr. U. Cal. at Berkeley, 1949-54; faculty U. Minn., 1954-60; Guggenheim fellow, vis. prof. U. Tokyo (Japan), 1960; Fulbright lectr., vis. prof. U. Alexandria (Egypt), 1961; prof., chmn. mech. engring. dept. U. Del., 1961-65; prof., head energy engring. dept. U. Ill., Chgo., 1965—; cons. to industry. Mem. Am. Soc. M.E., Am. Inst. Aeros. and Astronautics, A.A.A.S., Am. Assn. U. Profs., Sigma Xi. Editor: Recent Advances in Heat and Mass Transfer, 1961; (with T. F. Irvine) Advances in Heat Transfer, 1963; (with Carl Gazley, Jr.) Internat. Jour. of Heat and Mass Transfer, 1960—. Research and publs. primarily in field of heat and mass transfer. Home: 6738 N. Talman Av., Chgo. 60645.*

HARTOG, Marcus, Brit. zoologist; s. Alphonse and Marion (Moss) H.; ed. U. Coll., London; M.A., Cambridge; D.Sc., London, Royal U. Ireland; ed. Trinity Coll. Asst. to dir. Royal Botanic Gardens, Ceylon, 1874-77; became demonstrator Owens Coll., Manchester, Eng., 1888, later lectr. natural history; prof. natural history Queen's Coll., Cork, Ireland, 1882-1909; prof. zoology U. Coll., Cork, 1909-21; fellow Royal U., Ireland. Fellow Royal Hort. Soc. (hon.), Linnean Soc. Author: Problems of Life and Reproduction, 1913; The Function of Examinations in Education, Merit Proficiencey, 1919. Research and publs. on zoology, botany, gen. biology, physics nuclear div. Died Jan. 21, 1923.

HARTREE, Douglas Rayner, physicist; b. Cambridge, Eng., Mar. 27, 1897; s. William and Eva (Rayner) H.; B.A., M.A., Ph.D., St. John's Coll., Cambridge; M.Sc., U. Manchester; m. Elaine Charlton, Aug. 21, 1923; children—Nesta Margaret (Mrs. Edward L. Booth), Oliver Penn, John Richard. Research fellow St. John's Coll., Cambridge, 1924-27. Christ's Coll., Cambridge, 1928-29; prof. applied math. U. Manchester, 1929-37, prof. theoretical physics, 1937-45; prof. math. physics Cambridge U., from 1946; acting chief, Inst. Numerical Analysis, U. S. Bur. Standards, U. Cal., 1948; vis. prof. Princeton, 1955. Served with anti-aircraft exptl. sect. Munitions Inventions Dept., Ministry Munitions, 1916-19; with sci. research sect. Ministry of Supply, 1940-45. Rockefeller fellow Inst. Theoretical Physics, Copenhagen, 1928. Fellow Royal Soc., 1932; mem. Inst. Physics, Inst. Elec. Engrs., Cambridge Philos. Soc. Author: Calculating Instruments and Machines, 1949; Numerical Analysis, 1952; The Calculation of Atomic Structure, 1957; also papers on propagation of radio waves, calculating machines, atomic structure. Died Feb. 12, 1958.

HARTSOEKER, Niklaas (Nicolaus), Dutch physicist, histologist; b. Gouda, Netherlands, Mar. 26, 1656; s. evang. minister; student U. Leyden (Netherlands), 1674-76; prof. math. and philosophy U. Dusseldorf, 1704-16; tchr., Peter the Gt., circa 1700. Mem. French Acad. Scis. (fgn. asso.), 1699. Author: Essai de Dioptrique, 1694; Principes de physique, 1696; Conjectures physiques, 2 vols., 1706-08; Eclaircissements sur les conjectures physiques, 2 vols., 1710-12; Recueil de pleusieurs pièces de physique, 1722; Cours de physique, 1730. Believed fetus was preformed in spermatozoon; published illustrations of a homunculus crouching inside head of a spermatozoon; improved microscope and telescope. Died Utrecht, Netherlands, Dec. 10, 1725.

HARTT, Charles Frederick, geologist; b. Fredericton, N.B., Can., Aug. 23, 1840; s. Jarvis William Hartt; grad. Arcadia Coll., N.S., Can., 1860. Studied at Mus. of Comparative Anatomy on invitation of Louis Agassiz, Cambridge, Mass., 1861-64; geologist on Thayer expdn. to Brazil, 1865-66; prof. geology Vassar Coll., 1866-67; prof. geology and phys. geography Columbia, 1868-75; chief Geol. Commn. of Brazil, 1875-78. Author: Geography of Brazil, 1870; Notes on the Modern Tupi of the Ana, 1872; Crustacea Collected on the Coast of Brazil, 1866-73. Discovered remains of 5 species of oldest known insects; proved that ice sheets had not extended over Amazon Valley. Died Rio de Janeiro, Brazil, Mar. 18, 1878.

HARTT, Constance Endicott, Am. plant physiologist; b. Passaic, N.J., Nov. 2, 1900; d. George leBaron and Claudine (Millington) Hartt; A.B., Mt. Holyoke Coll., 1922; S.M., U. Chgo., 1924, Ph.D. 1928. Instr., N.C. Coll. for Women, 1922-24, St. Laurence U., 1925-30; asst. prof. botany Conn. Coll. for Women, 1930-31; research staff sugarcane physiology Exptl. Sta., Hawaiian Sugar Planters Assn., Honolulu, 1931-63, sr. physiologist, 1959-63; asst. prof. botany U. Hawaii, 1932. Sec., Hawaiian Bot. Gardens Found., Honolulu, 1959—. Recipient Distinguished Service award Hawaiian Sugar Technologists, 1963. Fellow A.A.A.S.; mem. Bot. Soc. Am., Hawaiian Bot. Soc. (hon. life), Am. Soc. Plant Physiologists, Am. Assn. U. Women (research fellow 1931, past Honolulu pres.). Research and publs. on sugarcane physiology, potassium deficiency, phosphorus nutrition, mechanism of photosynthesis and factors affecting it, translocation of photosynthate and factors affecting it including temperature, intensity and quality of light. Home: 42 Coelho Way, Honolulu 96817. Office: 1527 Keeaumoku, St., Honolulu 96822.*

HARTWEG, Karl Theodore, German botanist; b. Karlsruhe, Germany, July 18, 1812; with Jardin des Plantes, Paris; sent by London Hort. Soc. to collect seeds and plants in Mexico, C.Am., northern S.Am., 1836-43, Cal., 1845-48; became dir. Grand Ducal gardens, Baden, Germany. Journal of his trip pub. in jour. and trans. of London Hort. Soc. Brought 81 new species from Cal., including Caenothus prostratus, several new species of pines from tropical Am.; amassed most extensive collection made by single individual from Mexico and tropical Am. in 1st half of 19th century (pub. in George Bentham's Plantae Hartwegianae 1839-57). Died Baden, Feb. 3, 1871.

HARTWELL, Jonathan Lutton, Am. chemist; b. Boston, Feb. 21, 1906; s. John Frederic and Mary (Lutton) H.; A.B., Harvard, 1928, M.A., 1929, Ph.D., 1935; m. Ann Kathryn Goebel, Oct. 28, 1943. Research chemist E. I. DuPont de Nemours & Co., Wilmington, Del., 1929-32, Interchem. Corp., N.Y.C., 1935-37; chemist Nat. Cancer Inst., NIH, Bethesda, Md., 1938—; head natural products sect., Cancer Chemotherapy, Nat. Service Center, Nat. Cancer Inst., 1958—. Mem. Am. Chem. Soc., Soc. for Econ. Botany, Am. Soc. Pharmacognosy, Alpha Chi Sigma. Author: Compounds Which Have Been Tested for Carcinogenic Activity, 1941; (with P. Shubik) Supplement I, 1957; also numerous articles. Research in organic chemistry, chem. carcinogenesis, chemotherapy cancer, devel. drugs for human cancer. Home: 5038 Massachusetts Av., Washington 20016. Office: Nat. Cancer Inst., Bethesda, Md. 20014.*

HARTWIG, Walenty, physician; b. Moscow, USSR, Oct. 7, 1910; s. Ludwik and Maria (Biriukow) H.; M.D., Warsaw (Poland) U., 1938; m. Danuta Rachwal, Apr. 18, 1964; children—Elzbieta, Jolanta. Lectr. gen. medicine Stomatologic Faculty, Warsaw U., 1946-55, Psychol. Hygiene Sch., Warsaw, Poland, 1949-52; in charge 1st Med. Clinic, Postgrad. Med. Sch., Warsaw, 1953—, prof., 1954—. Chmn. endocrinological Council Ministry Health, 1963—. Mem. Internat. Soc. Internal Medicine. Author: Textbook of Internal Medicine for Dentists, 1950. Contbr. numerous articles to profl. jours. Research on physiology and pathology of pituitary, adrenals and thyroid. Home: 9 Spasowskiego, Warsaw 30. Office: 80 Ceglowska, Warsaw 45, Poland.*

HARTZ, Theodore Robert, Canadian physicist; b. Saskatoon, Sask, Can., Mar. 23, 1923; s. Philip Ernest and Barbara (Gabel) H.; B.A., U. Sask., 1945, B.Ed., 1946, M.A., 1948; Ph.D., U. B.C., Vancouver, Can., 1957; m. Elizabeth Leota Irwin, Dec. 14, 1955; children—Leslie Elizabeth, Judith Margaret. Research scientist Def. Research Telecommunications Establishment, Ottawa, Ont., Can., 1952—; lectr. Carleton U., part-time 1959-60. Mem. Internat. Astron. Union, Canadian Assn. Physicists, Am. Geophys. Union. Editor: (with C. O. Hines, I. Paghis, J. A. Fejer) Physics of the Earth's Upper Atmosphere, 1965; also articles. Research on phys. processes upper atmosphere, radio astronomy; pioneered exploitation satellite radio astronomy. Home: 915 Mountainview Av., Ottawa 14. Office: Def. Research Telecommunications Establishment, Ottawa 4, Ont., Can.*

HARVALD, Bent, Danish physician; b. Copenhagen, Denmark, Feb. 11, 1924; s. Aage and Elly (Greiff) H.; qualified U. Copenhagen, 1947, Ph.D., 1954; m. Ulla Nielsen, Apr. 5, 1952; children—Ulla Birgitte, Thomas, Claes. Sci. asst. U. Inst. for Human Genetics, Copenhagen, 1949-51; specialist internal medicine at various Copenhagen hosps., 1951-61; reader clin. genetics U. Copenhagen, 1961—; asst. chief physician dept. internal medicine C. Bispebjerg Hosp., Copenhagen, 1961—; chief physician Dr. Ingrids Hosp., Godthab, Greenland, 1965-66. Mem. Danish Med. Assn., Danish Assn. Internal Medicine, Am. Eugenic Soc., World Assn. Neurol. Scis. (corr.). Author: Heredity of Epilepsy, 1954; (with T. Kemp, M. Hauge) Arvepartologi, 1962; also numerous articles. Research on heredity of epilepsy by exam. of brain wave patterns, genetic factors in cancer, diabetes and other chronic diseases using twins, kidney damage caused by analgesics. Home: 85 Hvidegardspark, Lyngby, Denmark. Office: U. Inst. Human Genetics, 14 Tagensvej, Copenhagen, Denmark.*

HARVEY, Edmund Newton, Am. physiologist; b. Phila., Nov. 25, 1887; s. William and Althea Ann (Newton) H.; B.Sc., U. Pa., 1909; Ph.D., Columbia, 1911; m. Ethel Nicholson Browne, Mar. 12, 1916; children—Edmund Newton, Richard Bennet. Instr. physiology Princeton, 1911-15, asst. prof., 1915-19, prof. 1919-33, H. F. Osborn prof., 1933—; vis. lectr. in biology Mass. Inst. Tech., 1940-41. Trustee Bermuda Biol. Sta.; v.p., trustee Marine Biol. Lab., Woods Hole, Mass. Vis. lectr. Inst. de Biofisica, Rio de Janeiro, 1946. Recipient John Price Wetherill medal Franklin Inst. Pa., 1934, Rumford medal Am. Acad. Arts and Scis., 1947; Certificate of Merit, U. S. Armed Forces. Mem. A.A.A.S., Am. Soc. Naturalists, Am. Soc. Biol. Chemists, Am. Physiol. Soc., Soc. Exptl. Biology and Medicine, Am. Soc. Zoölogists, Growth Soc., Nat. Geog. Soc., Am. Assn. U. Profs., N.Y. Acad. Scis., Soc. Am. Bacteriologists, Bot. Soc. Am., Harvey Soc., Am. Philos. Soc., Nat. Acad. Scis., Am. Acad. Arts and Sciences Boston, Internat. Soc. Cell Biology (v.p. 1947-50), NRC, Sigma Xi. Author: The Nature of Animal Light, 1920; Laboratory Directions in General Physiology, 1933; Living Light, 1940. Editor Survey of Biol. Progress; asso. editor Biol. Bull., Biol. Abstracts, Jour. of Cellular and Comparative Physiology. Research in bioluminescence, cell permeability, nerve conduction, regulation in plants, ultrasonic radiation, cell surface tension, brain potentials, decompression sickness, mechanism of wounding. Died July 21, 1959.

HARVEY, G(eorge) G(raham), Am. physicist; b. St. Louis, Jan. 25, 1908; s. William Augustus and Pauline V(irginia (Graham) H.; A.B., Washington U., St. Louis, 1928, M.Sc., 1930, Ph.D., 1932; m. Dorothy Evelyn Howe, Mar. 27, 1954. Asst. in physics Washington U., 1928-30, fellow in physics, 1930-32; NRC fellow U. Chgo., 1932-34; faculty Mass. Inst. Tech., 1934—, prof., 1961, staff Radiation lab., 1942-46, asst. dir. Research Lab. Electronics, 1950-52, asso. dir., 1952—, exec. officer dept. physics, 1952—; instr. Coll. City N.Y., 1937-38. Cons. office field service OSRD, 1944-45. Recipient certificate appreciation War Dept., 1946; fellow Inst. Physics, London, 1963. Fellow A.A.A.S., Am. Acad. Arts and Scis., Am. Phys. Soc., Phys. Soc. (London); mem., Am. Assn. Physics Tchrs., Am. Assn. U. Profs., Am. Math. Soc., Edinburgh Math Soc., London Math. Soc., Math. Assn. Am., Optical Soc. Am., Phi Beta Kappa, Phi Beta Kappa Assos., Sigma Xi, Pi Mu Epsilon. Study of theoretical physics, phys. electronics. Home: 12 Washington St., Belmont, Mass. 02178. Office: Mass. Inst. Tech., Cambridge, Mass. 02139.

HARVEY, John Arthur, physicist; b. Saskatoon, Sask., Can., Dec. 14, 1921; s. Albert F. J. and Elizabeth (Hummerstone) H.; came to U. S., 1945, naturalized, 1950; B.Sc., Queens U., Kingston, Ont., Can., 1945; Ph.D., Mass. Inst. Tech., 1950; m. Mary Therese Kelly, Jan. 29, 1949; children—Johanne Mary, William Joseph. Jr. physicist Atomic Energy Can. Ltd., Chalk River, Ont., 1945-46; physicist Brookhaven Nat. Lab., Upton, N.Y., 1951-55; physicist Oak Ridge Nat. Lab., 1955—. Fellow Am. Phys. Soc.; mem. Sci. Research Soc. Am., Tenn. Acad. Sci. Research and numerous publs. in nuclear structure physics, low energy nuclear physics, nuclear spectroscopy, neutron physics, gravitational acceleration of free neutrons, neutron electron interaction. Home: 108 Ogontz Lane, Oak Ridge 37830. Office: P.O. Box X, Oak Ridge Nat. Lab., Oak Ridge, 37830.*

HARVEY, John Collins, Am. physician; b. Youngstown, Ohio, Sept. 11, 1923; s. Joseph Paul and Mary (Collins) H.; grad. Phillips Exeter Acad., 1941; B.S., Yale U., 1943; M.D., Johns Hopkins U., 1947; m. Adele Dillon, Nov. 26, 1949; children—Elizabeth V., John Collins II, William Charles, Amy L., Margaret J. With Johns Hopkins U., 1948—, prof. medicine, 1966—; Johns Hopkins Hosp., 1948—, physician, 1952—, dir. outpatient services, 1960—; cons. medicine Ft. Howard VA Hosp., Sparrows Point, Md.; mem. council on med. care State Md. Dept. Health, 1960—, chmn., 1964—; bd. trustees Provident Hosp., Balt.; sr. cons. in medicine Balt. City Hosps.; mem. advisory panel med. edn. Vocational Rehabilitation Adminstrn. Dept. Health, Edn. and Welfare, Washington, 1964—, program review panel, neurol. and sensory disease service program div. chronic disease Pub. Health Service, 1964—; editor Tice's Practice Medicine; editor Internat. Digest. Diplomate Nat. Bd. Med. Examiners, Am. Bd. Internal Medicine; fellow Am. Coll. Physicians, Am. Pub. Health Assn.; mem. Balt. City Med. Soc., A.M.A., Am. Fedn. Clin. Research, A.A.A.S., Biophys. Soc., N.Y. Acad. Sci., So. Med. Assn., Md. Soc. Internal Medicine, Nat. Rehabilitation Assn., Am. Clin. and Climatol. Assn., Phi Beta Kappa, Sigma Xi, Alpha Omega Alpha. Author numerous book reviews to sci. jours. Contbr. numerous articles in field to sci. jours. Research on muscle diseases, especially myolonic muscular distrophy. Home: 119 Upnor Rd., Balt. 21212 and (summer) 607 Maid Marian Hill, Sherwood Forest, Anne Arundel County, Md. Office: 153 Carnegie Dispensary Bldg., The Johns Hopkins Hosp., Balt. 21205.*

HARVEY, O. J., psychologist; b. Corinne, Okla., Aug. 27, 1927; s. Joseph Marion and Nina (Little) H.; B.A., U. Okla., 1950, M.A., 1951, Ph.D., 1954; m. Mary Christine Minton, Nov. 17, 1950. Post doctoral fellow Yale U., 1954-55; asst. prof. psychology Vanderbilt U., 1955-58; with U. Colo., 1958—, prof. psychology, 1962—; fellow Center for Advanced Study in Behavior Scis., Stanford, 1965-66. Recipient award Career Devel. NIMH, 1966. Mem. Am., Rocky Mountain psychol. assns. Author (with others) Conceptual Systems and Personality Organization, 1961, (with others) Intergroup Conflict and Cooperation, 1961, (with others) Motivation and Social Interaction, 1963, Experience, Structure and Adaptability, 1966. Studies, publs. in group psychology; social perception; personality and attitude development and change. Home: 435 S. 68th St., Boulder, Colo. 80302.*

HARVEY, Paul Henry, Am. plant breeder; b. St. Paul, Neb., Aug. 2, 1911; s. William and Ella Mae (Roe) H.; B.S. in Agr., U. Neb., 1934, Ph.D. in Genetics, Ia. State U., 1938; m. Ethel Marie Larson, June 8, 1938; children—Ann (Mrs. Herbert P. Scott), Lois Kay, David Paul. Grad. asst. genetics Ia. State U., 1934-38; faculty N.C. State U., Raleigh, 1938—, prof. agronomy, 1945-56, William Neal Reynolds prof., 1955—, head dept. crop sci., 1956—, Mem. Gov. N.C. Sci. Adv. Com. Mem. Am. Soc. Agronomy, Crop Sci. Soc. Am. (pres. 1965), Sigma Xi, Phi Kappa Phi, Gamma Sigma Delta (nat. award distinguished service agr. 1956). Research on plant breeding methodology and quantitative genetics. Home: 1311 Mayfair Rd., Raleigh, N.C. 27608.*

HARVEY, Rodney Beecher, Am. plant physiologist; b. Monroeville, Ind., May 26, 1890; s. Aaron Lawrence and Mary Vandervort (Hester) H.; Ph.C., Purdue U., 1912, D.Sc., 1939; B.S., U. Mich., 1915; Ph.D., U. Chgo., 1918; postgrad. Cambridge (Eng.) U., 1927-28, U. Bonn (Germany), summer 1928; m. Helen M. Whittier, June 17, 1916; children—Hale M. Whittier (dec.), Rodney Bryce, Rhoda Beatrice, Helen Elizabeth, Eleanor Whittier. Asst. botanist Eli Lilly Co., 1912-13; asst. pharmacognosist U. S. Bur. Chemistry, 1915; asso. pharmacognosist U. S. Bur. Plant Industry, 1918, plant physiologist, 1918-20; asst. prof. botany U. Minn., 1920-21, asso. prof. plant pathology and botany, 1921-31; prof. plant physiology, agrl. botany and horticulture from 1931; dir. Fla. Citrus Expt. Sta., 1936-37; dir. div. indsl. microbiology, Gen. Mills Research Lab., 1942-43. John Simon Guggenheim fellow, 1927-28. Fellow A.A.A.S.; mem. Bot. Soc. Am., Am. Chem. Soc., Am. Soc. Plant Physiologists (pres. 1936-37), Minn. Acad. Sci. (v.p. 1938-39, pres. 1942-43), Am. Phytopathol. Soc., Ecol. Soc. Am., Bot. Soc. of Czechoslovakia (corr. mem.), Phi Lambda Upsilon, Sigma Xi, Gamma Sigma Delta, Alpha Zeta, Gamma Alpha. Author: Plant Physiological Chemistry, 1930; A Textbook of Plant Physiology (with A. E. Murneck), 1930; An Annotated Bibliography of Low Temperature Relations of Plants, 1935; Plant Physiology (with A. E. Murneck), 1938. Discoverer of ethylene process of ripening fruits; patentee fruit ripening and coloration processes. Died Nov. 4, 1945.

HARVEY, Roger Allen, Am. physician; b. Binghamton, N.Y., Mar. 7, 1910; s. Zina Austin and Alice (Finch) H.; B.S., Hamilton Coll., 1933; M.S. in Radiology, U. Rochester, 1938, M.D. with honors, 1939; m. Marjorie Harding, June 20, 1940; children—Carol (Mrs. Donald S. Thompson, Jr.), Jean. Fellow Internat. Cancer Research Found., U. Rochester, 1936-37, faculty, 1942—, asst. prof., 1945-46; prof. radiology, head dept. U. Ill., Chgo., 1946—; radiologist-in-chief U. Ill. Hosps., 1946—. Cons., USPHS, 1964—; chmn. med. adv. com. Hinsdale Health Mus., 1957—; Carman lectr. Radiol. Soc. N.Am., 1953. Recipient Distinguished Service in Cancer Control award Am. Cancer Soc., 1964. Fellow Am. Coll. Radiology; mem. Assn. U. Radiologists (past pres.), Am. Cancer Soc. (past pres. Ill.; nat. pres., dir.), A.M.A., A.A.A.S., Radiation Research Soc., Radiol. Soc. N.Am., Am. Roentgen Ray Soc., Sigma Xi. Contbr. numerous articles to tech. jours., chpts. to books. Research on visualization circulating cancer cells in movie studies, chronic radiation exposure effects on skin, effects Betaron irradiations (X-ray and electron beams) on normal tissues, and variety human cancers, methods of enhancing radiation effects on cancer, population effects of radiation. Home: 14 S. Elm St., Hinsdale, Ill. 60521. Office: 840 S. Wood St., Chgo. 60612.*

HARVEY, Walter Robert, Am. biologist; b. Tucumcari, N.M., June 19, 1919; s. Clarence R. and Nora (Burges) H.; Asso. Sci., Cameron State Agrl. Coll., 1940; B.S., Okla. State U., 1942; M.S., Ia. State U., 1947, Ph.D., 1949; m. Marie H. Zigler, Sept. 1, 1940; children—Robert Dale, Jerry Ray, Donna Marie. Instr., Ia. State U., Ames, 1947-49; asso. prof. U. Ida., Moscow, 1950-54; biometrician, Biometrical Services, Agrl. Research Service, U. S. Dept. Agr., Beltsville, Md., 1954-64; vis. prof. Ohio State U., Columbus, 1964, Distinguished prof., 1964—. Mem. Am. Soc. Animal Sci., Am. Dairy Sci. Assn., Am. Inst. Biol. Scientists, Am. Statis. Assn., Biometrics Soc., Sigma Xi, Gamma Sigma Delta. Research, publs. on statis. genetics as applied to animal breeding problems, application of computers to analysis of exptl. data. Home: 4255 Mumford Dr., Columbus, O. 43221.*

HARVEY, William, English physician, anatomist; b. Folkestone, Kent, Eng., Apr. 1, 1578; s. Thomas and Joane (Halke) H.; B.A., Gonville and Caius Coll., Cambridge (Eng.) U., 1597; M.D., U. Padua (Italy), 1602, Cambridge U., 1602; m. Elizabeth Browne, Nov. 1604. Practiced medicine, London; physician St. Bartholomew's Hosp., from 1609; Lumleian lectr. in surgery and anatomy Royal Coll. Physicians, 1615-56; apptd. physician extraordinary to James I, 1618, to Charles I, 1625, accompanied Charles I to Scotland, 1633; physician to Lord Arundel, 1636; apptd. warden Merton Coll., Oxford (Eng.) U., 1645, ret., circa 1646. Mem. Royal Coll. Physicians. Author: Exercitatio anatomica de motu cordis et sanguinis in animalibus, 1628; Exercitationes de generatione animalium, 1651. May be considered founder of modern physiology; showed experimentally that blood does not oscillate back and forth in blood vessels (as Galen had believed) but travels in 1 direction only; calculated that in 1 hour heart pumps quantity of blood equal to 3 times weight of a man; concluded that blood flows in continuous stream through closed circulatory system, moving from heart through arteries into veins and back to heart; (though unable to find connection between arteries and veins which his theory demanded) postulated correctly that these connections (capillaries) are too small to be seen by naked eye; became interested in generation and embryology; 1 of 1st since Aristotle to study carefully stages of developing chick embryo. Died London, June 3, 1657.

HARVEY, William Henry, botanist; b. Summerville, Limerick, Ireland, Feb. 5, 1811; s. Joseph Massey Harvey; ed. sch. in Ballitore County, Kildare, 1824-27; M.D. (hon.), Dublin, 1844. Asso. with father's bus., 1827; made frequent bot. and zool. excursions to County Clare; colonial treas. at Cape Town, S. Africa, 1836-42; lectr. in U. S., 1849; visited India, Australia and South Seas, 1853-56; prof. botany, 1856. Fellow Royal Soc., 1858. Author: Genera of South African Plants, 1838; Manual of British Algae, 1841; Phycologia Britannica, A History of British Seaweeds, 4 vols., 1846-51; The Seaside Book, 1849; Phycologia Australica, 5 vols., 1858-63; (with O. W. Sonder) Flora Capensis, 7 vols., 1859-1925. Discovered Hookeria laete virens (before unknown as Irish moss), 1831; chief authority on algae and South African flora. Died May 15, 1866.

HARVEY-GIBSON, Robert John, Brit. botanist; b. Nov. 2, 1860; s. R. Gibson; ed. Univs. Aberdeen, Edinburgh, Strasburg; M.A.; D.Sc. m. Eda Lawrie, 1887; 1 dau.; Demonstrator zoology, later physiology U. Edinburgh; demonstrator biology Univ. Coll., Liverpool; prof. botany U. Liverpool, 1894-1921, now emeritus prof.; former examiner in botany Univs. New Zealand, Bristol, Aberdeen, Glasgow, Durham, Nat. U. Ireland, Pharm. Soc. Gt. Britain. Fellow Royal Soc. Edinburgh. Author: Outlines of the History of Botany; Primer of Biology; British Plant Names and Their Derivations; The World of Plants; The Master Thinkers; Two Thousand Years of Science, also articles. Died June 3, 1929.

HARWOOD, Sir Busick, Brit. surgeon; b. Newmarket, Eng., possibly 1745; s. John Harwood; apprenticed to apothecary; qualified as surgeon; M.B., Christ's Coll., Cambridge, 1785; m. dau. of Sir John Peshall, 1798; apptd. to Indian service; named prof. anatomy Cambridge U., 1785, Downing prof. medicine, 1800. Fellow Royal Soc., 1784. Author: System of Comparative Anatomy and Physiology, 1796; also synopses of course lectures. Research and expts. on blood transfusion. Died Nov. 10, 1814.

HASCHE, Rudolph Leonard, Am. chemist; b. Doon, Ia., June 20, 1896; s. Carl Harmann and Clara Belle (Lemon) H.; B.S., Tarkio (Mo.) Coll., LL.D., 1953; M.S., Washington and Jefferson Coll., 1919, Ph.D., Johns Hopkins, 1924; fellow U. Cal., 1924-25, univs. Berlin and Vienna, 1925-26; m. Blanche Knox, Aug. 25, 1920; 1 dau., Blanche Geraldine (Mrs. Richard P. Clarke). Research chemist Am. Smelting & Refining Co., 1926-30, supt. research, 1930-31; dir. chem. research A. O. Smith Corp., 1931-34; supt. research and devel. div. Tenn. Eastman Corp., 1934-43, sci. counsel, 1943-52; pres., chmn. bd. Carbonic Devel. Corp., from 1931; pres. Hasche Engring. Co., from 1950. Hasche Process Co., from 1958; also pvt. cons. Cons. WPB, 1943, F.E.A., 1945, investigator chem. plants in Germany, 1945. Served in poison gas research dept., C.W.S., U. S. Army, World War, Nat. Research fellow of Rockefeller Found., 1924-25; Internat. Edn. Bd. fellow, 1925-26. Recipient certificate of appreciation Dept. of Army, 1951. Fellow A.A.A.S.; mem. Am. Gas Assn., Am. Inst. C.E., Am. Chem. Soc., Soc. Chem. Industry (London), Am. Petroleum Inst., Am. Soc. Refrigerating Engrs., Phi Beta Kappa, Sigma Xi, Gamma Alpha. Author: Plastics, Theory and Practice, 1947. Contbr. to chem. jours. Patentee Hasche process for mfg. dry ice, gas reforming and other chem. and liquefaction processes. Died Jan. 8, 1959.

HASEDA, Taiichiro, Japanese physicist; b. Tokyo, Japan, Oct. 25, 1921; s. Taizo and Fukiko (Uzawa) H.; Bachelor, Tokyo U., 1940, D.Sc., 1950; m. Kazuko Watanabe, Oct. 30, 1950; children—Yasukiko, Kayako. Staff, Tohoku U., 1941-64, asst. prof. Research Inst. for Iron, Steel and other Metals, 1951-64; prof. Kyoto (Japan) U., 1964—. Mem. Japan Phys. Soc., Japan Chem. Soc. Research, publs. on low temperature physics, including properties of glassy or amorphous state, paramagnetic compounds, magnetic ordering in low dimensional lattices and finite clusters; magnetic and thermal studies. Home: 32-46 Otowa-Isejuku-cho, Yamashima, Kyoto, Japan.*

HASEGAWA, Shuji, Japanese bacteriologist; b. Fukui Prefecture, Japan, July 5, 1898; grad. Tokyo U., 1933; M.D.; a dau., Yōko (Mrs. Nisshin Honma). Became asst. technician Infectious Diseases Research Inst., Tokyo U., 1928, technician, 1938, pres., 1949, then dir.; became prof. Tokyo U., 1941; pres. Gumma U., 1961—. Recipient Asahi Culture prize for research in chemotherapy of Tb, 1942. Studies in chemotherapy for leprosy, whooping cough.

HASEK, Milan, Czechoslovakian physician; b. Prague, Czechoslovakia, Oct. 4, 1925; s. Rudolf and Bozena (Zahalkova) H.; M.D., Faculty Medicine, Charles U., Prague, 1949, candidate sci., 1955, D.Sci., 1961; m. Vera Hoscalkova, Feb. 11, 1950; children—Zora, Jiri, Darine. Research asst. Inst. Biology Charles U., 1947-52; research worker Inst. Biology Czechoslovakian Acad. Sci., Prague, 1952-61, Inst. Exptl. Biology and Genetics, 1962—, dir., since 1962—; prof. gen. biology Charles U., 1966—. Recipient Czechoslovak State prize for sci., 1961. Mem. N.Y. Acad. Scis., Royal Soc. Medicine, Internat. Inst. Embryology. Author: Vegetative Hybridization in Animals, 1953; The Development of the Chicken, 1955; General Biology, 1966; also numerous articles. Research in exptl. and embryology and immunology; discoveries in immunological reactions; devel. some new exptl. embryological methods. Home: 935 Stallichova, Prague 4. Office: 2 Flemingovo, Prague 6, Czechoslovakia.

HASENÖHRL, Friedrich, Austrian physicist; b. Vienna, Austria, Oct. 30, 1874; Ph.D., Vienna, 1897; pupil of Stefan and Boltzmann; prof. Tech. High Sch., Vienna, 1905-07; prof. theoretical physics U. Vienna, from 1907. Research on theory of relativity, electro-magnetic theory; showed interrelationship between mass and energy, finding same results that follow from Einstein's theory of relativity. Died Vielgereuth, Austria, Oct. 7, 1917.

HASERICK, John Roger, Am. dermatologist; b. Mnpls., Sept. 23, 1915; s. Ernest and Addie (Swanson) H.; B.A., Macalester Coll., 1937; M.S., in Dermatology, U. Minn., 1947, M.D., 1940; m. Jane Fleckenstein, May 10, 1941; children—John, Jane. With Cleve. Clinic, 1948—, head dermatology dept., 1959—; asso. clin. prof. dermatology Western Reserve U., Cleve., 1959—. Fellow A.C.P.; mem. A.M.A., Am. Dermatol. Assn., Soc. for Investigative Dermatology, Cleve. Dermatol. Soc., Am. Acad. Dermatology. Author: (with R. Kellum) Primer on Lupus Erythematous, 1962; Practical Dermatopathology, 1958, rev., 1967; also numerous publs. Discovered L.E. factor in blood of patients with lupus erythematous. Home: 31200 Fox Hollow, Cleve. 44124. Office: 2020 E. 93d St., Cleve. 44106.*

HASHEM, Nemat, Egyptian pediatrician; b. Cairo, Egypt, Sept. 1923; s. Hashem Omar and Ratiba (Roushdi) H.; M.B., B. Ch., Cairo U., 1948, DCH, 1950, M.D. in Pediatrics, 1952; postgrad. Boston Children's Med. Centre, Harvard. Chief pediatrician Sch. Health, 1950-52; lectr. pediatrics Cairo U., 1953-61, supr. pediatric neurology clinic, 1956-61; prof. asst. pediatrics Ain Shams U., Cairo, 1962—, supr. genetics unit, 1964—; research visitor N.Y. U., 1962-66, Harvard U. Boston Children's Med. Centre, 1963-64, London, Ont., Can., 1963. Mem. Egyptian Endocrinology Soc., Egyptian Pediatric Assn., Egyptian Soc. Neurology and Neuropsychiatry and Neurosurgery (editorial bd. Jour. 1960—). Research, publs. on various pediatric problems and identification certain tissue extracts promoting peripheral human lymphocytes to undergo mitosis; extraction certain nucleic acid from human lymphocytes cultured with certain viruses or bacteria. Home: 5 El-Goumhouria, Cairo, Egypt.*

HASHIMOTO, Haruo, Japanese sericulturist; b. Gumma Prefecture, Japan, 1904; grad. Kyushu U., 1930; D.Agr. Dir., Maebashi br. Sericultural Expt. Sta., Agrl. and Forestry Ministry, later dir. Matsumoto br. Recipient Japan Acad. prize, 1954. Research on heredity and breeding of silk-worms, role of Z and W chromosome in sex determination.

HASHISH, Salah el Din, biologist; b. Atbara, Sudan, Oct. 16, 1926; s. El-Sayed Aly and Hanem (Khattab) H.; B.Sc. with honors in Zoology, Cairo (Egypt) U., 1948; Ph.D., London Hosp. Med. Coll., London (Eng.) U., 1954; m. Füg Ertka, Oct. 1953; children—Mona, Dina, Dia. Demonstrator physiology Abbasia Med. Faculty, 1948-51, lectr., 1954-55; asst. prof. Egyptian Atomic Energy Establishment, Cairo, 1960-64, prof., head radiobiology and radioisotopes dept., 1964—; faculty Cairo U., Ein Shams U.; supr. radioisotopic research Med. Inst., Alexandria. Mem. governing bd. Middle Eastern Radioisotope Centre for Arab Countries. Decorated Republican Order, 1956; Sci. and Arts Decoration, 1959; Sci. Medal, 1959. Mem. Radioisotopes Soc. Egypt. Author: Radioisotopes Applications in Biology, 1956. Research, publs. on introduction of radioisotopes and radiation sources in Egypt, uses radioisotopes in biology, physiology, biochemistry, radiation biology; devel. spl. apparatus and techniques for dynamic studies in biol. systems in vitro. Home: 11 El Mahalawi St., Dokki, Cairo. Office: Atomic Energy Establishment, Cairo, Egypt.*

HASHIZUME, Binroji, (real name, Megumu Hashimoto), Japanese pharmacologist; b. Nagano Prefecture, Japan, 1897; grad. Tokyo U., 1922; research and numerous publs. on synthetic chemistry of medicines.

HASIGUTI, Ryukiti Robert (Hashiguchi, Ryukichi), Japanese metallurgist; b. Yokosuka, Japan, Aug. 23, 1914; s. Yasutaka and Hide (Simana) H.; B.Engring., U. Tokyo (Japan), 1939, Dr.Engring., 1953; m. Kiku Misumi, Apr. 25, 1939. Asso. prof. U. Tokyo, 1945-54, prof., 1954—; sr. scientist Inst. Phys. and Chem. Research, Saitama-ken, 1959—; research ofcl. Nat. Research Inst. for Metals, 1957—. Mem. Phys. Soc. Japan, Japan Inst. Metals (past v.p., Achievement medal 1948), Iron and Steel Inst. Japan (v.p.), Atomic Energy Soc. Japan. Author: Theory of Dislocations in Crystals, 1955; (with others) Solid State Physics, 1961; also numerous articles. Research on crystal lattic defects in metals and semiconductors, behavior of various point defects in metals and semi-conductors; discovered internal friction peaks in metals (Hasiguti peaks). Home: 2-17-6 Taira-Machi, Meguro-ku, Tokyo, Japan.*

HASKINS, Caryl Parker, Am. biophysicist, geneticist, research scientist; b. Schenectady, N.Y., Aug. 12, 1908; s. Caryl Davis and Frances Julia (Parker) H.; Ph.B., Yale, 1930; Ph.D., Harvard University, 1935; D.Sc., Tufts Coll., 1951, Union Coll., 1955, Northeastern U., 1955, Yale, 1958, Hamilton Coll., 1959; LL.D., Carnegie Inst. Tech., 1960, U. Cin., 1960, Boston College, 1960, Washington and Jefferson College, 1961, U. Del., 1965; Sc.D., George Washington University, 1963; m. Edna Ferrell, July 12, 1940. Staff member, research laboratory, General Electric Company, Schenectady, New York 1931-35; research associate Massachusetts Institute Technology, 1935-45; pres., research dir. Haskins Labs., Inc., 1935-55, dir., 1955—; pres. Nat. Photocolor Corp. of N.Y.C., 1939-55, dir., 1939-56; research prof. Union coll., Schenectady, 1937-55; pres. Carnegie Instn. of Washington, 1956—; mem. bd. dirs. Schenectady Trust Company, 1934. Assistant liaison officer OSRD, 1941-42, sr. liaison officer, 1942-43; exec. asst. to chmn. NDRC, 1943-44, dep. exec. officer, 1944-45; sci. adv. bd. Policy Council, Research and Development Bd. of Army and Navy, 1947-48; consultant Research and Development Bd., 1947-51, Sec. of Def., 1950-60, Sec. of State, 1950—; mem. President's Science adv. com., 1955-58, cons., 1959—. Trustee Carnegie Institution, Washington, Pacific Sci. Center, Center Advanced Study Behavioral Scis., Asia Found., Carnegie Corp. of N.Y., Marlboro Coll.; Woods Hole Oceanographic Instn.; Rand Corp.; trustee Ednl. Testing Service, chmn. bd., 1960-61, 67—; regent Smithsonian Instn.; trustee, mem. corp. Yale; Awarded Certificate of Merit, (U. S.), 1948; King's Medal for Service in Cause Freedom (Gt. Brit.), 1948; Benjamin Franklin Fellow Royal Soc. Arts; Fellow Am. Phys. Soc., A.A.A.S., Am. Acad. Arts and Sciences, Royal Entomol. Soc. (Gt. Britain), Entomol. Soc. Am., mem. Nat. Acad. Scis., N.Y. Acad. Scis., Sigma Xi, Phi Beta Kappa. Author: Of Ants and Men, 1939; The Amazon, 1943; Of Societies and Men, 1950; The Scientific Revolution and World Politics, 1964; The Search for Understanding, 1967; contbr. to anthologies and tech. papers. Research on biol. effects of radiation, genetics of Drosophila, Lebistes and certain flagellates, also cellular physiology, nutritional requirements of microorganisms. Home: 1545 18th St. N.W., Washington 20036. Office: 1530 P St., N.W., Washington 20005.

HASKINS, Reginald Hinton, Canadian biologist; b. North Bay, Ont., Can., July 16, 1916; s. Alfred James and Eva (Horrell) H.; B.A., U. Western Ont., 1938, M.A., 1940; student Botany Sch., Cambridge, Eng., 1945; Ph.D., Harvard, 1948; m. Shirley Elieta Koester, Aug. 12, 1949; children—Reginald James, Susan Gay, Phillip John. Demonstrator botany U. Western Ont., 1938-40, 45-46; with Harvard, 1946-48, teaching fellow, 1946-48, tutor, 1947-48; with NRC Can., Prairie Regional Lab., Saskatoon, Sask., Can., 1948—, sr. research officer, 1958-67, head physiology and biochemistry of fungi sect., 1956—, prin. research officer, 1967—. Sec. asso. com. on animal nutrition, 1951—; asso. editor Canadian Jour. Botany, 1964—; abstracter Biol. Abstracts. Mem. Bot. Soc. Am. (chmn. microbiol. sect. 1964-65), Mycol. Soc. Am. (councilor), Canadian Soc. Microbiologists, Brit. Mycol. Soc., Am. Soc. Plant Physiologists, Internat. Assn. Plant Taxonomy, Am. Inst. Biol. Scis. Contbr. to McGraw-Hill Ency. Sci. and Tech., numerous articles to sci. jours. Clarified criteria for identification of lower fungi; discovered and developed methods for prodn. of fungal products, antibiotics of potential value with patents on ustilagic acids, lysine, arginine and glutamic acid; improved methods of long-term preservation of microorganisms; described and named new fungi; 1st showed requirement of certain fungi for sterols for sexual reprodn., growth and temperature resistance, opening new area of study of plant disease; developed new techniques for study of fungi. Home: 222 Lake Crescent. Can. Office: NRC, Prairie Regional Lab., Saskatoon, Sask., Can.*

HASKINS, Willard Theodore, chemist; b. Binghamton, N.Y., Mar. 18, 1907; s. Clothier Theodore and Valerie (Lawyer) H.; B.Chemistry, Cornell U., 1930; M.S., U. Md., 1933, Ph.D., 1936; m. Bertha May Gover, July 27, 1933; children—Karen A. (Mrs. Richard Ragland), James E. Research chemist Armstrong Cork Co., Lancaster, Pa., 1930-31; chemist Carbohydrate Lab., NIH 1936-47, Lab. Tropical Diseases, Bethesda, Md., 1947-59, Rocky Mountain Lab., Hamilton, Mont., 1959-66 (retired); commd. scientist USPHS, 1948, advanced through grades to scientist dir., 1953. Mem. Am. Chem. Soc., Sigma Xi, Tau Beta Pi. Contbr. numerous articles to tech. jours. Research on carbohydrate chemistry, chemotherapy and control tropical diseases, chem. composition immunologically active substances isolated from microorganisms, relationship chem. composition to biol. activity. Home: 612 Desta St.

HASLER, Maurice Fred, Am. physicist; b. N.Y.C., May 19, 1907; s. Fred and Johanna (Raese) H.; B.S., Cal. Inst. Tech., 1929, M.S., 1930; Ph.D., 1933; m. Helen Ott, Jan. 14, 1939; children—Andrea, Stephen. Partner, Applied Research Labs., Glendale, Cal., 1934-50, pres., Applied Research Labs., Inc., 1950-61, chmn. bd. dirs., 1961—, dir. research and devel. Hasler Research Center of Applied Research Labs., Goleta, Cal., 1963—; v.p. Bausch & Lomb, 1958-62, dir., 1962—. Recipient Beckman award in chem. instrumentation Am. Chem. Soc., 1958. Fellow Optical Soc. Am.; mem. Am. Chem. Soc., Am. Phys. Soc. Research, publs., invention in fields of spectrochem. field utilizing optical emission and X-ray techniques for analysis of materials. Home: 63 Eucalyptus Lane, Santa Barbara, Cal. 93103. Office: 95 La Patera Lane, Goleta, Cal. 93017.*

HASS, Georg Hartwig, physicist; b. Hanau, Germany, Aug. 8, 1913; s. Max Hartwig and Bertha (Roloff) H.; student U. Innsbruck (Austria), 1932-33, U. Freiburg, (Germany), 1933-34, Inst. Tech. (Danzig) 1934-43, Dr. rer. techn. Physics, 1937, Dr. habil. Physics, 1943; m. Gisela Voigt, Dec. 28, 1939. Came to U. S. 1946, naturalized 1954. Research asst., Inst. Tech., Danzig, 1936-38; asst. prof., 1938-45; cons. Siemens, Telefunken and Heraeus Platinschmelze, Germany, 1938-45; cons. U. S. Army Engr. Research & Development Labs., Ft. Belvoir, Va., 1946-52, supr. physicist, head physics research lab., 1952-66, dir. physics research tech. area night vision Lab., Electronics Command, since 1966—. Recipient Commanding General's Medal for Tech. Achievement, 1957; Research and Development Achievement award, Dept. of the Army, 1961; Fellow Optical Soc. Am. (mem. bd. dirs., bd. editors, internat. commn. optics), Am. Phys. Soc., Washington Acad. Scis. Studies of coatings for controlling the temperature of satellites developed and applied to numerous U. S. satellites. Editor and contbr. book series Physics of Thin Films, 1963-64. Research and publs. on vacuum deposited films, oxidation and optical properties of metals, and electron diffraction and electron microscopy Home: 7728 Lee Av., Alexandria, Va. 22308. Office: U. S. Army Electronics Command, Night Vision Lab., Fort Belvoir, Va. 22060.*

HASS, Hans, marine biologist; b. Vienna, Austria, Jan. 23, 1919; s. Hans and Margarete (Brausewetter) H.; ed. U. Vienna, U. Berlin; m. Lotte Baierl, 1950; 1 son by previous marriage. Recipient Internat. prize Venice Film Festival, 1951. Author: Unter Koralen und Haien, 1941; Photojagd am Meeresgrund, 1942; Beitrag zur Kenntnis der Teteporiden, 1945; Drei Jäger auf dem Meeresgrund, 1946; Menschen und Haie, 1947; Manta, 1952; Wir kommen aus dem Meer, 1957; also films. Mem. several underwater expdns.; research on various deep-sea animals. Office: Internat. Institut für submarine Forschung, Vaduz, Liechtenstein.

HASS, Henry Bohn, Am. chemist; b. Huntington, Ohio, Jan. 25, 1902; s. Frederick William and Alma Marie (Bohn) H.; student Heidelberg Coll., 1917-18; B.A., Ohio Wesleyan U., 1921, D.Sc., 1942; M.A., Ohio State U., 1923, Ph.D., 1925; LL.D., U. Chattanooga, 1945, D.Sc., L.I. U., 1965; m. Georgia May Herancourt, Sept. 20, 1921; children—Robert Henry, Charlotte Frances (Mrs. Donald E. Hudgin), Thomas William, Richard Frederick. Dir. research Balt. Gas Engring. Co., 1925-28; with Purdue U., 1928-49, head dept., 1928-49, prof. chemistry, 1937-49; mgr. research and devel. Gen. Aniline & Film Corp., N.Y.C., 1949-52; pres. Sugar Research Found., N.Y.C., 1952-60; dir. chem. research M. W. Kellogg Co., Piscataway, N.J., 1961—. Fellow Royal Soc. Arts, Internat. Congress Anesthetists; mem. Research Engring. Soc. Am., Soc. Chem. Industry, Am. Chem. Soc. (chmn. N.Y. sect.), Am. Inst. Chemists (pres.), Am. br. Societe de Chimie Industrielle (past pres.), Phi Beta Kappa, Sigma Xi, Alpha Sigma Phi, Phi Lambda Upsilon, Pi Tau Sigma. Research on synthesis of cyclopropane (led to establishment of nitro paraffin industry); discovery of rules for aliphatic chlorinations; contbns. to prodn. of atomic bomb; gas chromatographic studies. Home: 95 Fernwood Rd., Summit, N.J. 07901. Office: Kellogg Lab., Piscataway, N.J. 08854.

HASS, Karl Willi, German chemist; b. Hanau on Main, Jan. 5, 1908; s. Max and Berta (Roloff) H.; ed. U. Friburg, Technische Hochschule, Karlsruhe and Ydansk, Danzig; eng. dipl.; Ph.D. in engring.; m. Annerose Sievers, 1944; children—Renat, Ellen, Brigitte, Klaus. Asst., Technische Hochschule, Ydansk;

asst. dir. Fabrique Friedr. Krupp Essen; chemist Farbwerke Hoechst; chief prodn., Frankfort, Hoechst, Fa. Feldmuhle Papier- und Zellstoffwerke A.G., Werk Koholyt Lülsdorf; now dir. Dynamit Nobel Aktienges und Werk Feldmuhle Lülsdorf; prof. electrochemistry apparentée; pres. Fachverband Elektrokorund- und Siliziumkarbid-Hersteller E.V. Mem. Assn. Indsl. Chemists, Soc. German Chemists. Author: Stand Sowie Neueste Technische Entwicklung auf dem Gebiet der Wässrigen Chlorakali-Elektrolyse, 1962. Home: 1 Deutz-Monodorferstrasse. Office: Dynamit Nobel Aktienges, Werk Feldmühle, 5211 Ranzel, Germany.

HASSALL, Cedric Herbert, chemist; b. New Zealand, Dec. 6, 1919; s. Herbert and Lydia (Porter) H.; M.Sc., U. New Zealand, 1941; Ph.D. (Royal Commn. 1851 sr. student) Cambridge (Eng.) U., 1946-47; postgrad. (Carnegie fellow) Harvard, 1950; Sc.D., Cambridge U., 1966; m. Elisabeth Cotti, Sept. 28, 1946. children—Peter Alexander, Maureen Elisabeth. Lectr. chemistry U. Otago (New Zealand), 1942-45; 1st prof. chemistry U. W.I., Jamaica, 1948-57; prof., head dept. chemistry U. Coll. Swansea (U.K.), 1957—; dean Faculty Sci., 1964—. Mem. Caribbean Research Council, 1954-57; sci. adviser U. Jordan, since 1964—; mem. adv. com. King Abdulaziz U., Saudi Arabia, since 1966—. Rockefeller Travel fellow, 1956. Fellow Royal Inst. Chemistry. Research and numerous publs. on chemistry and biosynthesis of antibiotics and fungal metabolites, chemistry of natural products including polypeptides, cardiac glycosides, alkaloids, flavonoids. Home: Remuera, Brynfield Rd., Langland, Bay, Swansea. Office: Dept. Chemistry, Univ. Coll. of Swansea, Swansea, Gt. Britain.

HASSE, Helmut, German mathematician; b. Kassel, Aug. 25, 1898; s. Paul and Margarete (Quentin) H.; ed. univs. Kiel, Göttingen, Marburg; Ph.D. in math.; m. Clara Ohle, 1923; children—Jutta, Rüdiger. Prof. U. Halle, 1925, U. Marburg, 1930, U. Göttingen, 1934, U. Berlin, 1949, U. Hamburg, 1950. Mem. German, Austrian math. socs. Author: Höhere Algebra, Vols. I and II, 1925-26; Zahlentheorie, 1949; Vorlesungen über Zahlentheorie, 1950; Klassenzahl Abelscher Zahlkörper, 1952, also numerous articles. Home: 35 Hagener Allee, 207 Ahrensburg, Germany.

HASSELL, Odd, Norwegian chemist; b. Oslo, Norway, May 17, 1897; s. Ernst A. and Mathilde (Laveness) H.; Cand.real., U. Oslo, 1920; Dr.phil., Berlin U., 1924; Dr.phil.h.c., Copenhagen (Denmark) U., 1950; Fildr.h.c., Stockholm (Sweden) U., 1960. Faculty Oslo U., 1925—, prof., dir. phys. chemistry dept., 1934-64. Decorated knight Order of St. Olav. Recipient Guinerus medal; Guldberg-Waage medal. Hon. fellow Chem. Soc. (London), Norwegian Chem. Soc. (past chmn.); mem. Norwegian Council for Scis. and Humanities, acads. scis. of Oslo, Trondheim, Stockholm, Uppsala, Gothenberg, Copenhagen. Author: Kristallchemie, 1934; also numerous articles. Norwegian editor Acta Chemicae Scandinavie, 1947-57. Research, numerous publs. on crystal and molecular structures using X-ray and electron diffraction methods and measurement of electric dipole moments, stereochemistry connected with 6-membered rings and conformational analysis, atomic arrangement in weak complexes formed by electron transfer from donor to acceptor molecules. Home: 10 Holsteinvein, Oslo 8, Norway.

HASSELMANN, C. Max, German physician; b. Frankfort am Main, 1897; ed. Inst. Tropical Medicine, Hamburg; M.D.; diploma in medicine and tropical hygiene; m. Margaret Kahlert, 1926; 1 son, Detlev Edgar Max. Med. asst., Frankfort, Wurzburg, Hamburg; practice medicine, asst. prof. dermatology and venereal diseases Coll. Medicine, Far Eastern U., Manila, P.I., 1926-41; practice medicine, instr. Pa. Med. Coll., St. John U., Shanghai, 1941-46; with U. Erlangen, 1947—, now dir. Clinic and Polyclinic. Cons. specialist WHO. Mem., corr., hon. mem. numerous German and fgn. assns. Research and numerous publs. on tropical illnesses, exzema, histopathology, leprosy, syphilis, serology, sexual problems. Address: 14 Hartmannstrasse, Erlangen, Germany.

HASSENFRATZ, Jean-Henri, French chemist, physicist; b. Paris, France, 1755; studied science in France, mining in Germany; dir. Lavoisier's lab., circa 1785; active Jacobin in Revolution; mem. French Commune; apptd. to reorganize mines and miners after the armistice; prof. Ecole Polytechnique. 1794-1824. Author: Cours revolutionarie d'administration militaire, 1794; Tableau de minéralogie, 1796; Cours de physique céleste, 1803-04; Sidérote chimique, 1812. Died 1827.

HASSID, Sami, architect, educator; b. Cairo, Egypt, Apr. 19, 1912; s. Joseph S. and Isabelle (Israel) H.; diploma in Architecture, Sch. Engring., Giza, 1932; B.A. with honours in Architecture, U. London, 1935; M.Arch., U. Cairo, 1943; Ph.D. (Fulbright grantee), Harvard, 1956; m. Juliette Mizrahi, June 29, 1941; children—Fred, Muriel. Came to U. S., 1957, naturalized, 1962. Tchr. tech. sch., Alexandria, Egypt, 1932-34; faculty U. Cairo, 1934-56, U. Ein-Shams, Cairo, 1957; faculty U. Cal., Berkeley, 1957—, prof. architecture, 1964—. Archtl. practice, Cairo, Alexandria, 1932-57, Berkeley, 1957—; commr. Cal. Bd. Archtl. Examiners, 1961—. Recipient 1st prize Al-Chams Competition, Cairo, 1947,

A.I.A. Hdqrs. competition, San Francisco, 1963; also other archtl. awards. Mem. A.I.A., Assn. Collegiate Schs. Architecture, Bldg. Research Inst. (recipient award). Author several books including: Architectural Construction Details, 1954; Development and Application of a System for Recording Critical Evaluations of Architectural Works, 1964; Architectural Education U. S. A., Our Changing Environment—How Does Architecture Prepare to Cope with Demands, 1967. Compiled annotated bibliographies on urban aesthetics and archtl. edn.; developed policies for archtl. research and grad. programs in architecture; studies on environmental design criteria and on systems of judgment of archtl. design, based on human judgment as a statistically quantifiable fact of nature; developed punch card system for recording critical evaluations of archtl. works; surveyed and appraised archtl. curricula in U. S. Home: 976 Oxford St., Berkeley, Cal. 94707.

HASSID, William Zev, biochemist; b. Jaffa, Palestine, Oct. 1, 1897; s. Mordecai and Esperanza (Hassid) H.; came to U. S., 1920, naturalized, 1926; A.B., U. Cal. at Berkeley, 1925, M.S., 1930, Ph.D., 1934; m. Lila B. Fenigston, Jan. 21, 1936. With U. Cal. at Berkeley, 1927—, successively asst. plant nutrition div., jr. chemist agrl. expt. sta., instr. plant nutrition, jr. chemist, asst. prof., asst. chemist, asso. prof., asso. chemist, 1927-47, prof. plant nutrition, also chemist, 1947-50, professor biochemistry and biochemist, 1950-65, prof. emeritus, biochemist agrl. sta., 1965—. Served as cpl. Brit. Army, 1918-20. Recipient Sugar Research award Nat. Acad. Scis., 1946; Guggenheim fellow, 1955, 62. Mem. Nat. Acad. Scis., Am. Soc. Plant Physiologists (Charles Reid Barnes hon. life mem. award 1964), Am. Soc. Biol. chemists, Am. Chem. Soc. (Hudson award div. carbohydrate chemistry), Chem. Soc. (Brit.), Sigma Xi. Research on structural carbohydrate chemistry; carbohydrate metabolism of plants. Home: 20 Northgate Av., Berkeley 8, Cal.

HÄSSIG, Alfred, Swiss physician; b. Zürich, Apr. 8, 1921; s. Alfred and Rosina (Basler) H.; M.D., U. Zürich, 1945; m. Paula Kaarina Siro, Aug. 8, 1953; children—Hans-Rudolf, Lena Elisabeth and Marietta Ruth (twins). Dir. central lab. Swiss Red Cross Blood Transfusion Service, 1955—; prof. immuno-pathology, blood transfusion, medico-legal serology, U. Bern, 1966. Recipient Marcel Benoist prize, 1962. Discovery of antigenic individual specificity of paraproteins in myeloma and macroglobulinemia Waldenström; research on therapeutic and prophylactic use of plasma fractions; incidence of hepatitis after transfusion of blood, plasma and fraction I of Cohn; use of blood donation to screen donor's health. Home: 32 Aeschiweg, 3066 Stettlen. Office: 10 Wankdorfstrasse, 3000 Bern, Switzerland.

HASSINGER, Hugo, Austrain geographer; b. Vienna, Austria, Nov. 8, 1877; prof., Basel, Switzerland, also Vienna. Pres. Geography Soc., Vienna. Author: Geographische Grundlagen der Geschichte, 1931; Osterreichische Anteil an der Erforschung der Erde, 1940. Research in anthropogeography, geomorphology, cultural geography; stressed importance of land forms for settlements. Died 1952.

HASSLER, Rolf, German physician; b. Berlin, Germany, Aug. 3, 1914; s. Rudolf and Elisabeth (Oehlmann) H.; Ph.D. in medicine; m. Carola Rickmers, 1940; children—Ulrike, Cordula, Thomas. Collaborator, KWI Research; asst., chief physician Neurol. Clinic, U. Freiburg; mem. sci. com. Max Planck Sci. Com.; dir. neuroanatomic dept. Max Planck Inst. Research. Mem. French Neurol. Soc. (hon.), Neurol. Soc. (hon.), Uruguay Neurosurg. Assn. Author: Extrapyramidalmotorische Syndrome und Erkrankungen; Anatomie des Thalamus und Wegungstorungen. Home: 21 Jägershäusleweg, Freiburg, Breisgau. Office: 47 Deutschordenstrasse, Frankfort am Main, Germany.

HASSNER, Alfred, chemist; b. Czernowitz, Rumania, Nov. 11, 1930; s. Siegmund and Mina (Koffler) H.; came to U. S., 1951, naturalized, 1957; student U. Tech., Vienna, Austria, 1949-51; M.S., U. Neb., 1954, Ph.D., 1956; m. Sidonia P. Schachter, Mar. 29, 1957; children—Lillian R., Ericka. Research chemist Splty. Products Co., Lincoln, Neb., 1952-54; research fellow Harvard, 1956-57; hon. research asso., 1966; faculty U. Colo., Boulder, 1957—, asso. prof. chemistry, 1963-67, prof. chemistry, 1967—; cons. Med. Sch., 1960-63. Monsanto fellow, 1955-56; NIH Spl. fellow, Zurich 1967. Mem. Am. Chem. Soc., Chem. Soc. London, German Chem. Soc., A.A.-A.S., Colo.-Wyo. Acad. Sci., Sigma Xi. Research and publs. on stereochemistry of organic nitrogen compounds, syntheses and reactions of steroids, novel heterocyclic compounds and potential drugs, sterochemistry addition reactions to olefins, synthesis of small ring compounds; discovered new methods of stereospecific introduction of nitrogen functions into organic molecules. Home: 340 18th St., Boulder, Colo. 80302.

HASTED, John Barrett, English physicist; b. Woodbridge, Eng., Feb. 17, 1921; s. John Ord Cobbold and Phyllis (Barrett) H.; student Winchester Coll., 1933-39; M.A. in Chemistry, New Coll. Oxford, 1939-41, D.Phil. Physics, 1948; m. Elizabeth Gregson, Dec. 29, 1948 (div. 1956); children—Anne, Belinda; m. 2d, Jocelyn Iris Meeanee Wynn-Harris, Nov. 1,

1958; 1 son, John-Andrew. Imperial Chem. Industries fellow U. Coll. London (Eng.), 1948-52, lectr. 1952-58, reader, 1958—. Cons. U.K. Atomic Energy Research Establishment, Harwell, Eng., 1965—, Inst. Battelle, Geneva, Switzerland, 1964—. Fellow Phys. Soc. (U.K.). Author: The Physics of Atomic Collisions, 1963. Research, publs. on establishment experimentally of dependence of electron capture by ions from atoms on adiabatic criterion; helped establish experimentally dielectric relaxation of water in microwave band, and apply this to non-destructive testing for moisture and to bound water study. Home: Riverbank House, Ladye Place, Ferry Lane, Shepperton, Middlesex, Eng. Office: Physics Dept. U. Coll. London, Gower St., London, W.C. 1, Eng.

HASTERLIK, Robert Joseph, physician; b. Chgo., Mar. 17, 1915; s. Henry and Antonia (Epstein) H.; S.B., U. Chgo., 1934; postgrad. Rush Med. Coll., 1934-38; M.D., U. Chgo., 1938. Faculty, U. Chgo., 1948—, prof. medicine, 1960—; dir. health div. Argonne Nat. Lab., Chgo., 1948-53, sr. scientist div. biol. and med. research, 1950-53; asso. dir. Argonne Cancer Research Hosp., U. Chgo., 1951-63. Mem. sci. com. 14 Nat. Council on Radiation Protection, 1955—, Ill. Legislative Commn. on Atomic Energy, 1957—, Internat. Cooperation Year Com. on Peaceful Uses Atomic Energy, 1965—. Diplomate Am. Bd. Internal Medicine. Fellow Acad. Policy Study; mem. Central Soc. for Clin. Research, Inst. Medicine, Health Physics Soc., A.A.A.S., A.C.P., Radiol. Soc. N.Am. Research and publs. on long term effects of bone seeking radioelements especially radium in man, human radiobiology. Home: 5801 S. Dorchester Av., Chgo. 60637.

HASTINGS, A(lbert) Baird, Am. biochemist; b. Dayton, Ky., Nov. 20, 1895; s. Otis Luther and Elizabeth (Henry) H.; B.S., U. Mich., 1917, Sc.D., 1941; Ph.D., Columbia, 1921; Sc.D., Harvard, 1945, Oxford (Eng.) U., 1952, Boston U., 1956, St. Louis U., 1965; m. Margaret Anne Johnson, May 13, 1918; 1 son, Alan Baird. Chemist, USPHS, 1917-21; staff Rockefeller Inst. Med. Research, 1921-26; faculty U. Chgo., 1926-35; Hamilton Kuhn prof. biol. chemistry, head dept. Harvard Med. Sch., 1935-38, prof. emeritus, 1959—; dir. Div. Biochemistry, Scripps Clinic and Research Found., La Jolla, Cal., 1959-62; head Lab. Metabolic Research, 1926-66; research asso. U. Cal. at San Diego, 1966—. Vis. prof. various fgn. univs.; cons. numerous govt. agys., colls., other orgns. Recipient Pres.'s Medal for Merit, OSRD, 1948, Distinguished Service award U. Chgo. Med. Alumni Assn. 1961, Banting Medal Am. Diabetes Assn., 1962, Award, Am. Coll. Physicians, 1964, Citation, USPHS, 1964, Distinguished Achievement award Modern Medicine, 1965, Citation for Service, Brookhaven Nat. Lab., 1965. Mem. Assn. Am. Physicians, Nat. Acad. Scis., A.A.A.S., Am. Acad. Arts and Scis., Am. Chem. Soc., Am. Soc. Biol. Chemists (pres. 1945-47), Am. Philos. Soc., Am. Physiol. Soc., Soc. Exptl. Biology and Medicine (pres. 1945-47), NRC, Harvey Soc., Western Assn. Physicians, Sigma Xi, Alpha Chi Sigma, Alpha Omega Alpha, several fgn. socs., others. Editor: Jour. Biol. Chem., 1941-54, 55-59; Am. Jour. Physiology, 1956-64; Jour. Applied Physiology, 1956-64; Endocrinology, 1963—; Handbook of Physiology, 1959—; Geriatrics, 1965—; others; adv. bd. Biochemical Preparations, 1945—; Comprehensive Biochemistry, 1957—, others. Publs. on devel. micromethod for measurement of acid-base balance; formation of bone salts; studies of tissue electrolytes, metabolism of glucose, modifying hormones and elements. Home: 5912 Bellevue St., La Jolla 92037. Office: P.O. Box 109, Dept. Neurosciences, School of Medicine, University of Cal., San Diego, La Jolla, Cal., 92037.

HASTINGS, Sir Charles, Brit. physician; b. Ludlow, Jan. 11, 1794; s. James Hastings; studied under 2 surgeons at Stourport; ed. Edinburgh U. 1815, M.D., 1818; m. eldest dau. of George Woodyatt, 1825; 1 son, G. W. Hastings; 2 daus. Apptd. physician to Worcester infirmary; leading practitioner in Worcestershire for many years. Knighted, 1850. Founded Provincial Med. and Surg. Assn., 1832, styled Brit. Med. Assn., from 1856, established its Jour., 1840. Hastings medal and prize in his honor awarded annually by Brit. Med. Assn. Author: Illustrations of Natural History of Worcestershire, 1834; A Treatise on Inflammation of the Mucous Membrane of the Lungs, 1820, also many memoirs in med. jours. Died July 30, 1866.

HASTINGS, Charles Sheldon, Am. physicist; b. Clinton, N.J., 1848; s. Panet Marshall and Jane (Sheldon) H.; Ph.B., Yale, 1870, Ph.D., 1873; m. Elizabeth Tracy Smith, June 28, 1878; 1 dau., Katherine. Instr. physics, Yale, 1871-73; later asso. prof. physics, Johns Hopkins; prof. physics. Sheffield Scientific Sch., Yale, from 1884. Fellow A.A.A.S.; mem. Nat. Acad. Scis. Author (with F. E. Beach) Text-Book on General Physics, 1899; Light—A Consideration of the More Familiar Phenomena of Optics, 1901. Research in phys. optics, astronomy; produced correcting lenses for transforming visual into photog. refractors; made improvements in spectroscope; designed telescopes. Died Feb. 1, 1932.

HASTINGS, J(ohn) Woodland, Am. biochemist; b. Salisbury, Md., Mar. 24, 1927; s. Vaughan A. and

Katherine (Stevens) H.; B.A., Swarthmore Coll., 1947; M.A., Princeton, 1950, Ph.D., 1951; m. Hanna Machlup, June 6, 1953; children—Jennifer, David, Laura, Karen. Faculty, Northwestern U., 1953-57, asst. prof., 1956-57; faculty U. Ill., Urbana, 1957-66, prof. biochemistry, 1963-66; instr. Marine Biol. Lab., Woods Hole, Mass., 1961-66, dir., 1962-66; prof. biology Harvard, 1966—; vis. lectr. biochemistry Sheffield (Eng.) U., 1961-62. Mem. panel on molecular biology NSF, 1963-66; mem. various coms. Nat. Acad. Scis.; mem. space biology subcom., space sci. steering com. NASA, 1966—. AEC fellow Johns Hopkins, 1951-53; John Simon Guggenheim Meml. Found. fellow, 1965-66. Mem. A.A.A.S., Am. Assn. U. Profs., Am. Soc. Biol. Chemists, Am. Soc. Zoologists, Biochem. Soc., Biophys. Soc., Soc. Am. Microbiologists, Soc. Gen. Physiologists (past pres.). Mng. editor Jour. Gen. Physiology, 1964—. Research, numerous publs. on molecular mechanism of bioluminescence and its biol. role in energy metabolism, phys. and chem. characterization of energy storage and excited state intermediates in luminescent reactions; quantum yields as related to reaction pathway; isolation and purification of enzymes and particulate fractions involved in bioluminescence: mechanisms of persistent daily rhythms; control of daily variations in concentrations of enzymes and substrates in rhythmic phenomena and role of macromolecular biosynthesis in biochem. mechanism. Home: 405 Commonwealth Av., Newton Centre, Mass. 02159.

HASTINGS, John Beazley, geologist; b. Liverpool, Eng., 1858; s. James Hastings; came to U. S., 1874, naturalized in 1880's; student Van der Naillen's Sch., San Francisco, 1877-79; m. Louise Ross Brodhead, 1884; 6 children. Cadel midshipman Pacific Mail Steamship Col., 1874-75; in shipping, 1876; surveyor's asst., 1876-79; with surveying party, Cal., 1879-80; miner U.B.H. mine, nr. Nevada City, Cal., 1880-81; with smelter, Hailey, Ida., 1881; surveyor, mining engr. U. S. Dept. Mineral Surveyor, Ketchum, Ida., 1881-90; mining engr., cons., Boise, Ida., 1891-96; visited mining dists. of Butt and Colo., 1892, Europe, 1892, B.C., 1896; became mgr. Center Star Mine, 1896; mng. dir. Booderham-Blackstock Syndicate, 1899-1900; mining engr., N.Y., 1901-03, Denver, from 1903. Fellow Geol. Soc. Am.; mem. Am. Inst. Mining Engrs. Helped place mining engring. on more sci. basis; contbd. to mining devel. of Far West. Died 1942.

HASTORF, Albert H.; Am. psychologist; b. N.Y.C., Nov. 26, 1920; s. Albert H. and Hilda (Menke) H.; A.B., Amherst Coll., 1942; M.A., Princeton, 1947, Ph.D., 1949; Doctor of Humane Letters, Amherst Coll., 1967; m. Barbara E. Reck, Oct. 4, 1943; children—Elizabeth C., Christine A. Instr. to prof. Dartmouth Coll., 1948-55, chmn. dept. psychology, 1955-61; prof., exec. head. dept. psychology Stanford U., Cal., 1961—. Served with USAAF, 1942-46. Center for Advanced Study fellow in behavioral scis., Palo Alto, Cal., 1954-55. Fellow Am. Psychol. Assn., A.A.U.P., Sigma Xi. Exptl. studies of the perceptual process, especially as an influence on social behaviour. Home: 571 Foothill Rd., Stanford, Cal.*

HASWELL, William Aitcheson, zoologist; b. Edinburgh, Scotland, Aug. 5, 1854; ed. Edinburgh Instn. and U.; m. Josephine Gordon Rich, 1891; 1 dau. Curator, Queensland (Australia) Mus., 1880; zool. cruise, tropical Queensland coasts, 1881; became demonstrator zoology, comparative anatomy and histology Sydney (Australia) U., 1882, Challis prof. biology, 1890-1917, dean faculty sci., 1913-17, emeritus prof., from 1917; acting curator Australian Mus., 1882. Fellow Royal Soc., 1897. (Clarke Meml. medal New S. Wales 1915); pres. Linnean Soc. New S. Wales, 1892-93. Author: (with T. Jeffrey Parker) A. Text-Book of Zoology, 1897; A Manual of Zoology, 1899. Died Jan. 24, 1925.

HATA, Sahachiro, Japanese bacteriologist; b. Tsumo, Japan, 1872; studied under Kitagato in Japan, under Wassermann in Berlin; worked with Paul Ehrlich, Frankfort/Main, Germany, 1908-10; mem. Katazato Bacteriological Research Inst.; prof. Keio U.; head dept. Keio Hosp., Tokyo, from 1911. Worked (with Ehrlich) on expts. leading to discovery of salvarsan, a medicine for syphilis; 1st to use salvarsan to treat rat-bite fever in Japan, 1912. Died 1938.

HATAI, Shinkishi, Japanese zoologist; b. Aomori Prefecture, Japan, 1876; student Tohoku Gakuin Coll.; grad. U. Chgo., 1903; Ph.D.; LL.D., Columbia U. Prof., Tohoku U., 1921-38, emeritus prof., 1938—; successively dir. Palau Tropical Life Lab., adminstr. Army; adviser Manila City Sci. Bur. Recipient Imperial Acad. prize for thesis, 1925. Author: Earthworm. Research on growth of shells, ecology of large snails from Palau Islands.

HATAKEYAMA, Hisanao, Japanese meteorologist; b. Niigata Prefecture, Japan, 1905; grad. Tokyo U., 1928; D.Sc. Mem. Tokyo Central Meteorol. Obs. Recipient Imperial prize, 1944. Mem. Meteorol. Soc. (dir.), Snow and Ice Research Soc. (dir.), Physics Soc. Research on meteorology, especially geomagnetics.

HATANO, Hiroyuki, Japanese chemist; b. Hyogo Prefecture, Japan, Sept. 27, 1924; s. Kenji Fujita and Matsue Hatano; grad. Faculty Sci., Kyoto (Japan) U., 1947, Dr.Sci., 1958; m. Yasuko Matsuda, May 25, 1954; children—Mika, Mary. Staff, Kyoto U. Faculty Sci., 1947-55, 58—, prof. chemistry, 1963—, chmn. dept. chemistry, 1965-66; instr. Kobe (Japan) U. Faculty Sci., 1955-58; research asso. Cornell U., Ithaca, N.Y., 1963-64. Mem. Am. Chem. Soc., Chem. Soc. Japan, Japanese Biochem. Soc., Japanese Biophys. Soc., Japanese Radiation Research Soc. Author: Automatic Amino Acid Analysis, 1963; also articles. Research on radiation effects on enzymes in vitor, electron spin resonance studies on biochems. (amino acids, sulfur compounds proteins), radiation chemistry of biochems. (amino acids, peptides, proteins), automatic liquid chromatography. Home: 944 1-Bancho; Takatsuki, Osaka, Japan. Office: Dept. Chemistry, Faculty Sci., Kyoto U., Oiwakecho Kitashirakawa, Kyoto, Japan.*

HATCH, Frederick Tasker, Am. biochemist; b. Boston, Aug. 27, 1924; s. Frederick Southard and Beatrice (Tasker) H.; B.A., Dartmouth, 1944; M.D., Harvard, 1948; Ph.D., Mass. Inst. Tech., 1960; m. Virginia Weeks, Mar. 3, 1946; children—Daniel F., Daphne A., Deborah J., Douglas E. Asso. in Medicine Med. Sch. Harvard, 1960-65; asst. physician, chief arteriosclerosis unit Mass. Gen. Hosp., 1960-65; biochemist Lawrence Radiation Lab. U. Cal., Livermore, 1965—. Mem. Arteriosclerosis Council, Am. Heart Assn., Am. Chem. Soc. Contbr. numerous articles to profl. jours. Research and publs. on structure, formation, function, methods of measurement of serum lipoproteins, lipid metabolism in atherosclerosis and other diseases, role of riboflavin coenzyme and vitamin B12 in enzymatic formation of methyl group of methionine, biochem. changes in stress, dietary treatment of atherosclerosis, hypertension and hyperlipemia, radiation effects on nucleic acid synthesis. Home: 3855 Yale Way. Office: Lawrence Radiation Lab., Livermore, Cal. 94550.*

HATCH, Lewis Frederic, Am. chemist; b. Puyallup, Wash., Aug. 21, 1912; s. Lewis Miles and Mary (Brayton) H.; B.S., State Coll. Wash., 1933; M.S., Purdue, 1934, Ph.D., 1937; m. Marie Jeanette Shultz, June 12, 1959; children—Carna Mary, Christine Ann. Research chemist Shell Devel. Co., Emeryville, Cal., 1937-40; mem. faculty U. Tex., Austin, 1940—, prof. chemistry, 1953-67, dean Sch. Sch., U. Tex., El Paso, 1967—; vis. prof. U. Marburg, Germany, summer 1966. Fulbright cons. Nat. Research Centre, Cairo, Egypt, 1960-61; cons. to chem., petrochem. industry. Fellow Am. Inst. Chemists, A.A.A.S., Tex. Acad. Sci.; mem. Am. Chem. Soc., Chem. Soc. London, Sigma Xi, Alpha Chi Sigma. Contributing editor Hydrocarbon Processing, published in 1950. Author: Introduction to Organic Chemistry (with R. J. Williams), 1948; (with G. W. Watt) The Science of Chemistry, 1949, 54; Organic Chemistry, 1955; Chemistry of Petrochemical Reactions, 1955; Higher Oxo Alcohols, 1957; Isopropyl Alcohol, 1961; Ethyl Alcohol, 1962; (with G. W. Watt and J. Lagowski) Chemistry, 1964; also numerous articles. Patentee. Research on influence of structure and substituents on reactivity of allylic halides, elucidation of mechanism for acid catalyzed cleavage of beta diketones by glycols and related compounds; discovered stereospecific rearrangement reaction between allylic halides and lithium aluminum hydride; discovered inhibitor used to stabilize butadiene and other dienes. Home: 524 Satellite Dr., El Paso, Tex. 79912.*

HATCH, Winslow Roper Hatch, Am. botanist; b. Lexington, Mass., Mar. 1, 1908; s. Roy Winthrop and Bertha (Roper) H.; A.B., Dartmouth, 1930; PM.D., Johns Hopkins, 1934; m. Dita (Keith), June 30, 1937; children—Robert Winslow, John Keith, Rosita Alvarado. NRC fellow Harvard, 1935-36; instr. Dartmouth, 1936, 39; faculty Wash. State Coll., 1939-55, prof., 1945—, head dept. botany, 1940-47, chmn. div. biol. scis., 1947-53; asso. dean Coll. Scis. and Arts, Wash. State U., 1953-55; dean Coll. Gen. Edn., Boston U., 1955-57, dir. Clearinghouse of Studies on Higher Edn., 1957-65; coordinator research div. higher edn. research U. S. Office of Edn., Washington, 1965—. Fellow A.A.A.S.; mem. Bot. Soc. Am., Am. Mycol. Soc., Phi Beta Kappa, Phi Kappa Phi. Author: Series on Inquiry; Quest for Quality; (with Howard) The Developing College; (with others) Search for Relevance; author, editor: New Dimensions in Higher Edn. also articles. Research on sexual phenomena in lower fungi; cytological studies of Allomyces arbusculus; Home: 6800 Churchill Rd., McLean, Va. 22101. Office: 400 Maryland Av. S.W., Washington 20202.*

HATCHETT, Charles, English chemist; b. London, Jan. 2, 1765; s. John and Elizabeth Hatchett; m. Elizabeth; research worker. Fellow Royal Soc., 1797; mem. French Acad. Scis., 1823, Lit. Club. Author: On the Spikenard of the Ancients, 1836; An Analysis of the Magnetical Pyrites, with remarks on Other Sulpherets of Iron, 1804; On an Artificial Substance which Possesses the Principal Characteristics of Tannin, 1805. Discovered element columbium (niobium), 1801; eponym of hatchettine (or hatchettite) and hatchettolite. Died Chelsea, Eng., Feb. 10, 1847.

HATFIELD, Efton Everett, Am. biologist; b. Mindenmines, Mo., Jan. 25, 1919; s. Thomas J. and Vina (Taylor) H.; B.S., U. Ark., 1942; M.S., Okla. State U., 1949; Ph.D., U. Ill., 1955; m. Ethelyn Broyles, Aug. 1, 1942; children—Stephen E., Douglas P., Mary L. Faculty, U. Ark., 1946-47, Okla. State U., 1947-49, Panhandle A. and M. Coll., 1949-51; faculty animal sci. dept. U. Ill., Urbana, 1951—, now prof. Mem. Am. Soc. Animal Sci., Am. Dairy Sci. Assn., Poultry Sci. Assn. Research, numerous publs. on nonprotein nitrogen sources in ruminant nutrition, antibiotics in nutrition, energy and mineral applications in animal prodn. Home: 1103 Westlawn St., Champaign, Ill. 61820. Office: U. Ill., Urbana, Ill. 61801.*

HATFIELD, Rufus Clay, Am. microbiologist; b. Nicholasville, Ky., Jan. 4, 1918; s. Rufus Clay and Lisa Maude (Young) H.; B.S., U. Dayton, 1941; M.A., U. Cal. at Los Angeles, 1947, Ph.D., 1950; m. Marilyn Jean Jestes, Aug. 30, 1947; children—Wendy Ann, Lisa Melinda Felicity, Brian Boyce Christian. With research div. Nat. Cash Register Co., 1941-43; faculty Cal. Poly. Inst., San Luis Obispo, 1949—, prof. microbiology, 1965—; chief facilities operations div. Assessment Labs., Ft. Detrick, Md., 1952-54; research with USPHS, WHO, Azul, Argentina, 1962-63. Mem. Internat. Union Biol. Scis., Sci. Research Soc. Am., Sigma Xi, Beta Beta Beta, Phi Sigma. Author: Practical Experiments in General Bacteriology, 1952, 58; Applied Microbiology, 1965; also articles. Research in epidemiology, chemically defined, synthetic media in evaluating microbial metabolism of antibiotic prodn. and assay. Home: 991 Wadsworth Av., Pismo Beach, Cal. 93449.*

HATHAWAY, Milicent Louise, nutrition specialist; b. North Tonawanda, N.Y., Sept. 12, 1898; d. Augustus J. and Kate (Smith) Hathaway; A.B., Wells Coll., 1920; M.A., U. Buffalo, 1925; Ph.D., U. Chgo., 1932. Faculty, U. Buffalo, 1925-29, U. Chgo., 1929-33, Coll. Medicine U. Ill., 1933-35, HE dept. U. Ill., Urbana 1935-37, Battle Creek Coll., 1937-38, Cornell U., 1938-46; nutrition specialist human nutrition research U. S. Dept. Agr., 1946-62; prof. home econs. Howard U., Washington, 1962-66. Recipient Borden award Am. Home Econ. Assn., 1947. Fellow A.A.A.S.; mem. Inst. Nutrition, Am. Chem. Soc., Am. Home Econ. Assn., N.Y. Acad. Scis., Phi Beta Kappa, Sigma Xi, Sigma Delta Epsilon. Publs. on research on sterols, critic acid metabolism, human metabolism of essential vitamins and minerals, relationship of heights and weights to age in children and adults. Home: 3383 S. Leisure World Blvd., Silver Spring, Md. 20906.*

HATIEGANU, Iuliu, Rumanian physician; b. 1884; univ. prof.; mem. Acad. Rumanian People's Republic. Author: (with others) Elementary Treatise on Medical Semiology and Pathology, 1937; Text-book on Medical Pathology, 1955. Founder Cluj sch. internal medicine; skilled diagnostician; investigated pathology of alimentary canal, cardio-vascular, excretory, nervous systems. Died 1959.

HATT, Philippe, French hydrographic engr.; b. Strasbourg, France, July 17, 1840; chief hydrographic engr. of navy. Mem. French Acad. Scis., Bur. Longitudes. Research on tides of French coast; geodetic studies in S. Vietnam. Died Guindales, France, Oct. 9, 1915.

HATT, Robert T(orrens), Am. mammalogist, museologist; b. Lafayette, Ind., July 17, 1902; s. William K. and Josie (Appleby) H.; student Purdue U., 1919-20; B.S., U. Mich., 1923; A.M., Columbia, 1925, Ph.D., 1932; m. Marcelle Boigneau, Mar. 29, 1929 (dec. Mar. 1951); children—Richard R., Peter K.; m. 2¹, Suzannah Beck, Jan. 10, 1953. Instr. biology N.Y. U., 1923-28; asst. curator mammals Am. Mus. Natural History, 1928-35; dir. Cranbrook Inst. Sci., Bloomfield Hills, Mich., 1935-67; research collaborator Mus. Zoology U. Mich., 1937—; cons. Ministry Edn., Pakistan, 1956. Recipient Founders medal Cranbrook Found., 1964. Fulbright research fellow, Iraq, 1952-53. Fellow A.A.A.S., Am. Mus. Natural History, Zool. Soc. (London); N.Y. Zool. Soc., Rochester Mus. Arts and Sci., mem. Mus. Assn. (Eng.), Assn. Sci. Mus. Dirs., Am. Soc. Naturalists, Am. Soc. Zoology, Soc. Systematic Zoology, Am. Soc. Mammalogists, Am. Assn. Museums. Author: (with others) Island Life in Lake Michigan, 1948. Research, publs. primarily in field of ecol., faunal, systematic studies mammals of Yucatan, Congo, Rhodesia, Iraq, Mich. Address: Cranbrook Inst. Sci., Bloomfield Hills, Mich. 48013.*

HATTECLYFFE, William, Brit. physician; b. Eng.; M.D.; physician, sec. to Edward IV; became original scholar King's Coll., Cambridge, 1440; apptd. physician to Henry IV, 1454; captured by Lancastrians, 1470; later master of requests, royal councillor; some of his med. prescriptions were preserved by Worsley. Died 1480.

HATZIKAKIDIS, Athanassios Demetrios, Greek chemist; b. Athens, Greece, Oct. 24, 1918; s. Demetrios Thomas and Antigone (Alavanos) H.; B.S., Athens U., 1941, D.Phys. Scis., 1948; Diploma, Greek Atomic Energy Commn., 1956; diploma Oak Ridge Inst. Nuclear Studies, 1957; m. Elisabeth Malcotsis, Dec. 20, 1961; 1 son, Demetrios. Dir. biochem. lab. Dispensary City of Athens, 1941-43; 1st asst. inorganic chemistry lab. U. Athens, 1943-45, bot. lab., 1945-49; head chem. dept. Hydrobiol. Inst., Nat. Acad. Athens, 1949—; head solar energy dept. Greek Atomic Energy Commn., 1961—; faculty various tech. schs., 1952-59. Decorated Silver Cross with Swords of Royal Order George I. Author several books, articles. Research on chem. oceanography, physics, bacteriology of waters, solar energy applications, history and philosophy sci. Home: 30 Atlantos Str. P. Phaliron, Athens. Office: 16 Possidonos Av., P. Phaliron Athens, Greece.*

HAUDRICOURT, Andre Georges, French anthropologist; b. Paris, Jan. 17, 1911; s. Evariste Maurice and Ernestine Suzanne (Remy) H.; Ingénieur Agronome, Institut agronomique, 1931, specialized in genetics and phytopathology, 1932; Certificat d'études supérieurs de Géography, 1933; Certificat d'Étude supérieures de Botanique, 1936; Diplomed Oceanian langs., 1945, École des langues Orientales, of Siamese, 1946, Laotian, 1947. Mem. Mission Ethnobotany, USSR, 1934-35; staff Centre National de la Recherche Scientifique, 1939-45, 50-65, dir. research, 1960-65; lectr. EPHE (Practical Sch. High Scholarship), Sorbonne, Paris, 1965—; mem. EFEO (French Sch. Far-East), Hanoi, 1948-49. Mem. sci. com. ORSTOM (Office Sci. Research in Overseas Ters.). Decorated Officier des Palmes académiques, 1963. Mem. Soc. Ling. de Paris, Soc. Bot. de Paris, Soc. asiatique. Author: (with L. Hedin) L'Homme et les plantes cultivées, 1945; (with Delamarre) L'Homme et la charrue, 1954; (with A. Juilland) Essai pour une histoire du phonetisme fran..ais, 1949; Le vocabulaire Bé, 1965; also articles. Research on influence of contrasted climate on human utilization of plants, influence of type of agr. on human mentality, influence of bilinguilism on phonetic evolution of langs.; how tones appeared in Far-Eastern langs.; new type of consonant (post-nasalized) in langs. of New Caledonia; Chinese origin of horse harness. Home: 47 rue d'Assas, Paris 6, 75, France.*

HAUER, Franz von, see von Hauer, Franz.

HAUG, Arne, Norwegian chemist; b. Horten, Norway, Sept. 15, 1926; s. Herman and Magnhild (Stang) H.; Siv.ing., Norwegian Tech. U., 1950, Dr.techn., 1965; m. Ragnhild Froslett, June 30, 1951; children—Harald, Marit. Chemist, Norwegian Inst. Seaweed Research, 1951—, dir., 1958—; postdoctoral fellow div. applied biology Nat. Research Council Can., Ottawa, Ont., 1955-56. Research and publs. on seaweed chemistry, chem. structure and chem. and phys. properties of alginic acid. Home: 29 Ronningsveien, Trondheim. Office: N.T.H., Trondheim, Norway.*

HAUG, (Gustave-) Émile, French geologist, paleontologist; b. Drusenheim, France, June 19, 1861; Dr.-es-Sc., Paris, 1891; prof. geology Paris Faculty Scis. Mem. French Acad. Scis. Author: Traité de géologie, 1908-11. First to distinguish clearly between continental platforms and their geosynical folds, in his theory of mountain range formation. Died Niederbronn, France, Aug. 28, 1927.

HAUGAARD, Niels, biochemist, pharmacologist; b. Copenhagen, Denmark, Feb. 25, 1920; s. Gotfred H. C. and Karen (Pedersen) H.; student U. Copenhagen, 1938-40; A.B. with honors Swarthmore Coll., 1942; Ph.D., U. Pa., 1949; m. Ella Elizabeth Shwartzman, Oct. 11, 1947; children—David, Lisa. Faculty U. Pa., Phila., 1949—, prof. pharmacology, 1965—. Guggenheim Found. fellow Carlsberg Lab., Copenhagen, 1953-54; Commonwealth Fund fellow U. Amsterdam, (Netherlands), 1965-66. Mem. Am. Soc. Biol. Chemists, Am. Soc. Pharmacology and Exptl. Therapeutics, Soc. Exptl. Biology and medicine. Sect editor Chem. Abstracts, 1960-65; editorial bd. Circulation Research, 1964—, Jour. Pharmacology and Exptl. Therapeutics, 1965—. Contbr. numerous articles to tech. jours. Research in intermediary metabolism, mechanism of hormone and drug action, enzyme chemistry, carbohydrate metabolism. Home: 129 Maple Av., Bala Cynwyd, Pa. 19004. Office: Dept. Pharmacology, Sch. Medicine, U. Pa., Phila. 19104.*

HAUGE, Steinar, Norwegian vet. scientist; b. Fredriksstad, Norway, Dec. 25, 1914; s. Engel Martin and Karoline Olava (Langgard) H.; grad. Norway's Vet. and Agrl. Coll., 1941; student bacteriology and virology, U. Cal.; m. Karen Elisa Johnson, 1947. Sci. asst. Norway's Vet. Coll., 1941-47, became prof., 1953; head dept. Vet. Inst., Oslo, 1952-53; made study tour to U. S. A., 1946-47, European countries, 1950. Mem. Radiation Adv. Council, Norway; Norwegian mem. Study Group on Food Irradiation; mem. radiation adv. council Norwegian Radium Hosp. Home: Oslo, Norway.

HAUGHTON, Geoffrey, immunologist; b. Yorkshire, Eng., June 10, 1932; s. Arthur and Dorothy (Reynolds) H.; B.Sc., Southampton U., 1952, Ph.D., 1958; m. Thelma Mary Smith, June 2, 1956; children—Keith Linsay, Adrian Peter, Jonathan Anthony; m. 2d, Solveig Kristina Torckell, Nov. 29, 1965. Sr. research fellow Microbiological Research Establishment, Porton, Eng., 1958-61, sr. sci. officer, 1961-64; vis. scientist Eleanor Roosevelt Research fellow dept. tumor biology Karolinska Inst., Stockholm, Sweden, 1964-65; asst. prof. dept. bacteriology, immunology Med. Sch. U. N.C., Chapel Hill, 1966—. Research on immunology of tissue transplantation, tumor-specific antigens of virus-induced tumors; purified and discovered mechanism of action of a blood-clotting enzyme produced only by certain pathogenic bacteria. Home: 1 Holloway Lane, Chapel Hill, N.C. 27514.*

HAUGHTON, Sidney Henry, geologist; b. London, Eng., May 7, 1888; s. Henry Charles and Alice (Aves) H.; B.A. with honors, Cambridge (Eng.) U., 1909; D.Sc., U. Cape Town (S. Africa), 1924, LL.D., 1948; D.Sc., U. Witwatersrand (S. Africa), 1962; D.Sc., U. Natol (So. Africa), 1967; m. Edith Hoal, Dec. 19, 1914; children—Edith Joan (Mrs. Rodney Maynard), Leslie Frank. Paleontologist, S. African Mus., Cape Town, 1911-14, asst. dir., 1914-20; sr. geologist S. African Geol. Survey, 1920-34, dir., 1934-48; cons. geologist S. African Atomic Energy Bd., 1948-54; inter-African sci. corr. for geology C.C.T.A., 1954-62; hon. dir. Bernard Price Inst. for Palaeontol. Research, Johannesburg, S. Africa, 1962—. Mem. Govt. Commn. on Indsl. Requirements, 1940-42, Govt. Commn. on Museums, 1945, Govt. Commn. on U. Finances, 1952. Recipient King George V Jubilee medal, 1924, Queen's Coronation medal, 1953, Gold medal Mountain Club S. Africa, 1932. Mem. Geol. Soc. London (Muarchison award 1932), Geol. Soc. S. Africa (past pres.), Geog. Soc. S. Africa (past pres.), Asso. Sci. and Tech. Socs. (past pres.), Royal Soc. S. Africa (past pres.), Geol. Soc. Am., Soc. géol. de Belgique, Soc. Econ. Geologists. Fellow Royal Soc., 1961. Author: Stratigraphy of Africa South of the Sahara, 1963; also numerous articles. Research on fossil invertebrates, reotiles, mammals, stratigraphy and econ. geology of So. Africa especially stratigraphic correlations in central and so. Africa. Home: 19 Mercury, 744 Schoeman, Pretoria. Office: Geol. Survey, Private Bag 112, Pretoria, S. Africa.*

HAUKSBEE, Francis (the elder), see Hawksbee, Francis (the elder).

HAUL, Robert Arnold Wilhelm, German phys. chemist; b. Hamburg, Germany, May 31, 1912; s. Karl and Erna (Behncke) H.; Dipl.-Ing., U. Danzig, 1937; Dr.Ing., U. Berlin, 1938; m. Liselotte Rahaus, Dec. 27, 1939; children—Sigrid, Rainer Andries, Cornelia. Faculties, Univs. Munich, Prague, Hamburg, 1939-49; prin. research officer Nat. Chem. Research Lab., S. African Council for Sci. and Indsl. Research, Pretoria, 1949-56; faculty U. Bonn (Germany), 1956-65, prof., dir. Inst. Phys. Chemistry, 1962-65; prof., dir. Inst. Phys. Chemistry, Hanover (Germany) Inst. Tech., 1965—, dean sci. faculty, 1966-67; mem. Sci. Council Nuclear Research Center, Jülich, 1956-67. Mem. German Chem. Soc., Bunsen Soc. for Phys. Chemistry, Faraday Soc., S. African Chem. Inst. Research, publs. in surface chemistry: physisorption, chemisorption; in solid state chemistry: decomposition of solids; on isotope exchange reaction for enrichment of stable isotopes. Address: 82 Stamme Strasse, Hanover, West Germany 3000.*

HAUPTSCHEIN, Murray, Am. chemist; b. N.Y.C., Mar. 15, 1923; s. William and Sylvia (Spandorfer) H.; B.S., Coll. City N.Y., 1943; Ph.D., Duke, 1950; m. Gloria D. Ogus, June 22, 1947; children—David F., Mark J. Research chemist Manhattan Project, S.A.M. Labs., Columbia U., 1943-45, Carbide and Carbon Chems. Corp., N.Y.C., 1945-46; research asso. Research Inst., Temple U., Phila., 1950-52, dir. organic research, 1952-55; with Pennsalt Chems. Corp., King of Prussia, Pa., 1955—, group leader, 1957-63, dir. organic research dept., 1963—. Mem. Am. Chem. Soc. (mem. nat. exec. com. div. fluorine chemistry 1966—, dir. Phila. sect. 1967—, 1st Phila. sect. award 1962), A.A.A.S., Phi Beta Kappa, Sigma Xi, Phi Lambda Upsilon. Research, publs., numerous patents on organic fluorine chemistry; discoverer, inventor new high performance products and processes, monomers, polymers, plastics, refrigerants, aerosols, water and oil repellent surface treatments for leather, paper, textiles, surfactants for fire fighting foams, manufacture plastics, dielectric fluids for aerospace applications, lubricants and mold release agts. Home: 513 Patricia Dr., Glenside, Pa. 19038. Office: 900 1st Av., King of Prussia, Pa. 19406.*

HAUROWITZ, Felix, biochemist; b. Prague, Czechoslovakia, Mar. 1, 1896; s. Rudolf and Emilie (Russ) H.; M.D., German U., Prague, 1922, D.Sc., 1923; m. Gina Perutz, June 23, 1925; children—Alice H. (Mrs. H. William Sievert), Martin O. Harwit. Came to U. S., 1947, naturalized, 1952. Faculty, German U., 1925, U. Istanbul, 1930-48; prof. chemistry Ind. U., Bloomington, 1948—, Distinguished Service prof., 1958—. Recipient Paul Ehrlich award and gold medal, 1960. Mem. German Acad. Scientists Leopoldina, Am. Soc. Biol. Chemists, Am. Assn. Immunologists, Am. Chem. Soc., Biochem. Soc. (London), Soc. Chimie Biologique (France), Soc. Exptl. Biology and Medicine. Author: Progress in Biological Chemistry, 1950; Biochemistry, an Introductory Textbook, 1955; Chemistry and Function of Proteins, 1963. Research, numerous publs. on template theory antibody formation, fetal human hemoglobin. Home: 417 S. Henderson St., Bloomington, Ind. 47403.*

HAURWITZ, Bernhard, meteorologist; b. Glogau, Germany, Aug. 14, 1905; s. Paul and Betty (Cohn) H.; student U. Breslau (now Wroclaw, Poland), 1923-24, U. Göttingen (Germany), 1924-25; Ph.D., U. Leipzig (Germany), 1927; m. Marion B. Wood, Jan. 15, 1961; 1 son, Frank D. Came to U. S., 1932, naturalized, 1946. Privat dozent U. Leipzig, 1931-32; research asso. Blue Hill Obs., Harvard, 1932-35; with Meteorol. Service Can., 1935-41; asso. prof. Mass. Inst. Tech., 1941-47; prof., chmn. meteorol. dept. N.Y. U., 1947-59; prof. meteorology U. Colo., Boulder, 1959-64; meteorologist Nat. Center for Atmospheric Research, Boulder, 1964—. Mem. Nat. Acad. Sci., Am. (Rossby award 1962), Royal meteorol. socs., Am. Geophys. Union, A.A.A.S., German Geophys. Soc. Author: Dynamic Meteorology, 1941; (with J. M. Austin) Climatology, 1944; also numerous articles. Research on theory of motions in atmosphere and ocean. Office: Nat. Center for Atmospheric Research, Boulder, Colo. 80302.*

HAURY, Emil Walter, Am. archaeologist; b. Newton, Kansas, May 2, 1904; s. Gustav A. and Clara K. (Ruth) H.; student Bethel Coll., Newton, Kan., 1923-25; A.B., U. of Ariz., 1927, A.M., 1928; Ph.D., Harvard, 1934; Doctor of Laws, University of New Mexico, 1959; m. Hulda E. Penner, June 7, 1928; children—Allan Gene, Loren Richard. Instr., U. of Ariz., 1928-29, research asst. in dendrochronology, 1929-30, assistant director, Gila Pueblo, Globe, Ariz., 1930-37, prof. anthropology, 1937—; Guggenheim fellow, 1949-50. Chmn. div. anthropology and psychology NRC, 1960-62; mem. Adv. Bd. Nat. Parks, Historic Sites, Bldgs. and Monuments, 1964—. Viking Fund medalist in anthropology, 1950; recipient Alumni Achievement award U. Ariz., 1957; Salgo-Noren Found. award for teaching excellence. Mem. Nat. Council Humanities, American Academy Arts and Sciences, National Speleological Society (honorary life mem.), Soc. for Am. Archaeology, Am. Anthrop. Association (pres. 1956), National Academy Sciences, Tree-Ring Society, A.A.A.S., Sigma Xi. Author publs. including: The Excavations of Los Muertos and Neighboring Ruins in the Salt River Valley, Southern Arizona, Peabody Museum Papers (Vol. XXIV, No. 1), 1945; (with others) The Stratigraphy and Archaeology of Ventana Cave, 1950. Study of Southwestern U. S. A. archaeology, Pueblo, Mogollon and Hohokam remains; early man. Home: 2749 E. 4th St., Box 4366, Tucson 85717.

HAUSCHILD, Fritz, German physician, chemist; b. Chemnitz, Germany, Dec. 8, 1908; s. Curt and Melanie (Gröber) H.; student U. Göttingen, 1928-30, U. Münich (Germany), 1930-32, U. Leipzig (Germany), 1932-35; m. Charlotte Richter, Nov. 30, 1935; 1 son, Dieter. Dir. Inst. Pharmacology and Toxicology, U. Leipzig, 1949—. Mem. Deutsche Akademie der Wissenschaften zu Berlin, Deutschen Pharmakologischen Gesellschaft (Nationalpreis 1957). Author: Pharmacologie und Einführung in die Toxikologie, 1956; (with Görisch) Einführung in die Pharmakologie und Arzneiverordnungslehre, 1963; also numerous articles. Research on aliphatic and aromatic amines, effect of methylene blue on methamoglobinamie, effect of methamphetamine, toxicologic problems; discovery of new medicinal drugs. Home: 26 Stieglitz-str. 7031 Leipzig, East Germany.*

HAUSCHKA, Theodore Spaeth, biologist; b. Reichenau, Austria, July 31, 1908; s. Hugo and Carola (Spaeth) H.; A.B., Princeton, 1930; M.S., U. Pa., 1941, Ph.D., 1943; m. Elsa Voorhees, Mar. 29, 1938; children—Stephen Denison, Peter Voorhees, Margaret Spaeth. Tchr. biology Chestnut Hill Acad., Phila., 1935-39; Harrison fellow zoology U. Pa., 1940-42, instr., 1942-43; biologist Lankenau Hosp. Research Inst., Phila., 1943-48, sr. mem., 1940-54; asso. dir. Marine Exptl. Sta., Truro, Mass. 1945-57; sr. mem. Inst. Cancer Research, Phila., 1949-54; dir. biol. research Roswell Park Meml. Inst., Buffalo, 1954—; research prof. biology State U. N.Y., Buffalo, 1954—. Cons. cancer research Lederle Lab., Am. Cyanamid Co., 1953-65; cons. in genetics Med. Found. Buffalo, 1962—. Fellow A.A.A.S.; mem. N.Y. Acad. Sci., Am. Soc. Zoologists, Am. Genetic Assn., Am. Soc. for Human Genetics, Soc. for Study Growth and Devel., Am. Naturalists Soc., Soc. for Exptl. Biology and Medicine, Am. Assn. Cancer Research (past pres.), Sigma Xi. Research, publs. on life cycles sporozoan parasites, morphogenesis marine hydroids, trypanosome endotoxins, naturally occurring growth inhibitors, mouse genetics, cell biology cancer tissue, tumor immunity, mammalian cytogenetics, immunogenetic aspects tissue transplantation. Home: 94 Lexington Av., Buffalo 14222.*

HAUSER, Charles Roy, Am. chemist; b. Cal., Mar. 8, 1900; s. Charles and Elizabeth (Rogan) H.; B.S., U. Fla., 1923, M.S., 1925; Ph.D., U. Ia., 1928; m. Madge L. Baltimore, June 30, 1929; children—Betty (Mrs. James E. Yourison), Frances (Mrs. Robert L. Grate), Charles F. Faculty, Duke, Durham, N.C., 1929—, prof. chemistry, 1946—, James B. Duke prof. chemistry, 1961—. Cons. Union Carbide Chems. Co., 1946-62; vis. lectr. Ohio State U., summer 1956. Recipient certificate of merit OSRD, 1947; Herty medal, 1962; Synthetic Organic Chem. Mfrs. Assn. medal, 1967. Mem. Nat Acad. Sci., Am. Chem. Soc. (award Fla. sect. 1957, nat. award 1962), A.A.A.S., Am. Assn. U. Profs, Phi Beta Kappa, Sigma Xi, Phi Lambda Upsilon, Gamma Sigma Epsilon. Contbr. chpts. to Organic Reactions, 1942, 54; Heterocyclic Compounds, 1961. Research, numerous publs. in organic chemistry involving mechanisms of organic reactions and syntheses of new organic compounds in fields of carbon-carbon condensations, cyclization, molecular rearrangements and elimination reactions.*

HAUSER, Gustav, German bacteriologist, anatomist; b. Germany, 1856; prof. med. pathology Erlangen (Germany) U.; 1st to isolate Proteus vulgaris (common cause of cystatis), circa 1885. Died 1935.

HAUSER, Philip M., Am. sociologist, demographer, urbanologist; b. Chgo., Sept. 27, 1909; s. Morris and Anna (Diamond) H.; A.A., Central YMCA Coll., 1927; Ph.B., U. Chgo., 1929, M.A., 1933, Ph.D., 1938; m. Zelda Barnett Abrams, Nov. 27, 1935; children—William Barry, Martha Ann. Faculty, Central YMCA Coll., Chgo., 1929-34; chief, labor inventory WPA, Washington, 1934-37; asst. chief statistician Nat. Unemployment Census, Washington, 1937-38; asst. chief statistician to dep. dir. U. S. Bur. Census, Washington, 1938-47, acting dir., 1949-50; faculty U. Chgo., 1933-38, prof. sociology, 1947-—. U. S. rep. Population Commn. UN, 1947-51; mem. council Nat. Inst. Child, Health and Human Devel., 1964-—. Recipient Eleanor Roosevelt Key award Roosevelt U. Mem. Population Assn. Am. (pres. 1950), Am. Statis. Assn. (pres. 1962), Am. Sociol. Research Assn. (pres. 1962), A.A.-A.S. (v.p. sect. K 1959), Am. Sociol. Assn. (v.p. 1955-—), Internat. Statis. Inst., Internat. Union for Sci. Study Population, Am. Philos. Soc., Social Sci. Research Council (dir.), Phi Beta Kappa, Kappa Delta, Author numerous books. Editor: Handbook for Social Research in Urban Areas (UNESCO); (with Leo Schnore) The Study of Urbanization, 1965. Contbr. to devel. census procedures and sample social survey methods; to devel. of demographic analysis and tng.; to orgn. of Demographic Research & Tng. Centers in U. S., S.E. Asia. Home: 5729 Kimbark Av., Chgo. 60637.*

HAUSHOFER, Karl, German geographer, polit. theorist; b. Munich, Germany, Aug. 27, 1869; followed successful mil. career to rank of maj. gen., until end of World War I; became prof. geography U. Munich, founder Inst. Geopolitics, 1924, dir., during 1930's. Author: Geopolitik des Pazifischen Ozenas, 1924; Bausteine zur Geopolitik, 1928; Weltpolitik von heute, 1934. Hitler greatly influenced by his theories as can be seen in Mein Kampf; recognized geog. position as important factor in world politics; believed Germany was expanding nation requiring Lebensraum (living or growing space) at the expense of declining, oceanic power of Gt. Britain; maintained that Eurasian "Heartland" was key to world domination. German attack on Russia and axis alliance, as well as other Nazi fgn. policy, may be interpreted in terms of his polit. theories. After World War II he was investigated as alleged war criminal, subsequently committed suicide, Munich, Mar. 10, 1946.

HAUSMANN, Johann Friedrich Ludwig, German mineralogist; b. Hanover, Germany, Feb. 22, 1782; ed. U. Göttingen (Germany); various positions Adminstrn. Mines, Clausthal, Germany, also Brunswick, Germany, 1803-06; explored Norway and Sweden, 1806-07; named insp. mines and salts Kingdom of Westphalia, 1809; tchr. mineralogy and mine exploitation U. Göttingen, from 1811. Mem. French Acad. Scis. Author: Krystallogische Beitraege, 1803; Norddeutsche Beitraege zur Berg- und Hüttenkunde, 1806-10; Reise durch Skandinavien, 5 vols., 1811-18; Handbuch der Mineralogie, 3 vols., 1813; Untersuchungen über die Formen der leblosen Natur, 1821; Umrisse nach der Natur, 1831; Ueber die Bildung des Harzgebirges, 1842; Beitraege zur metallurgischen Krystallkunde, 2 vols., 1850-52. Devised system of crystallography based on spherical trigonometry. Died Göttingen, Dec. 26, 1859.

HAUSNER, Henry Herman, metall. engr.; b. Vienna, Austria, June 1, 1901; s. Hans and Helene (Tritsch) H.; E.E., Technische Hochschule, Vienna, 1925; D.Eng., U. Vienna, 1938; m. Elizabeth Wallner, July 30, 1927 (dec.); m. 2d, Hedda M. John, Nov. 23, 1962 (dec.). Came to U. S., 1940, naturalized, 1946. With Elin A.G., Vienna, 1925-38, Elin Gluehlampenfabrik, Vienna, 1938-40, Am. Electro Metal Corp., Gen. Ceramics and Steatite Corp., Keasbey, N.J., 1940-45; cons. engr., 1945-—; with Sylvania Electric Products, Inc., Bayside, L.I., 1948-55; adj. prof. Bklyn. Poly. Inst., 1951-—; vis. prof. U. Cal. at Los Angeles, 1963-—. Recipient Powder Metall. Achievement award Stevens Inst. Tech., 1956. Fellow N.Y. Acad. Sci.; mem. Am. Inst. Mining, Metall. and Petroleum Engrs., Am., German socs. metals. Author: (with others) Powder Metallurgy, 1947, The Physics of Powder Metallurgy, 1951, Human Engineering, 1951; (with S. B. Roboff) Materials for Nuclear Power Reactors, 1955. Editor: Modern Materials, Vols. 1-4, 1958-64; Beryllium—Its Metallurgy and Properties, 1965. Editor: Internat. Jour. Powder Metallurgy, 1965-—. Contbr. numerous articles to profl. jours. Patentee in field. Research and development in powder metallurgy, nuclear metallurgy and space metallurgy. Home: 549 W. 123d St., N.Y.C. 10027. Office: 730 Fifth Av., N.Y.C. 10019.*

HAUSSER, Rolf, German physicist; b. Stuttgart, Germany, Sept. 30, 1932; s. Karl H. and Margarete (Eberhard) H.; student U. Tübingen (Germany), 1952-54; diploma Technische Hochschule Stuttgart, 1958, dr.rer.nat., 1964; m. Marie-Louise (Jetter), Aug. 28, 1965; 1 son, Lutz Karl Friedrich. Sci. asst. I. Physikalisches Institut, Technische Hochschule Stuttgart, 1959-64, leader study group on spin resonance, 1960-66, sr. asst., 1965-66, sr. adminstrv. adviser in fed. civil service, 1966-—. Lectr. on physics; indsl. adviser on radiation protection and application radioisotopes. Mem. Deutsche Physikalische Gesellschaft. Author: (with Gehrtsen-Kneser) Einfuehrung in der Experimentalphysik, 1966; (with Kohlrausch) Praktische Physik, 1967; also articles. Research on nuclear magnetic relaxation in liquids and solids; discovered spin-rotation interaction in water; electron resonance studies in electrolytic solutions; measurement nuclear rays; radiation protections; def. against effects of atomic weapons. Home: 38 Am. Illerdamm, Sonthofen. Office: 17 Berghoferstrasse, Sonthofen, Germany.*

HAUSZ, Walter, Am. electronic engr.; b. N.Y.C., May 17, 1917; s. Walter Otto and Marie (Kenz) H.; B.S., Columbia, 1937, M.S. in Elec. Engring., 1938; m. Linda Carolyn Miller, June 21, 1952; children—Jeffrey, Stephen, Leslie. With G. T. Southgate, Cons. Engr., 1938-39; with Gen. Electric Co., 1939-—, asst. to mgr. Electronics Lab., 1945-54, mgr. Advanced Electronics Center, 1954-56, mgr. engring. disciplines TEMPO, Santa Barbara, Cal., 1956-—. Recipient Coffin award Gen. Electric Co., 1952. Fellow I.E.E.E., A.A.A.S.; mem. Am. Inst. Aeros. and Astronautics, Operations Research Soc. Am. Contbg. author: Electron Optics, 1948; also articles. Research in radar, missile guidance, color TV, digital computers, solid state component applications. Home: 4520 Via Vistosa, Santa Barbara 93105. Office: P.O. Drawer QQ, Santa Barbara, Cal. 93102.*

HAUTEFEUILLE, Jean de, see de Hautefeuille.

HAUTEFEUILLE, Paul Gabriel, French chemist, mineralogist; b. Étampes, France, Dec. 2, 1836; grad. as engr. l'École centrale des artes et manufactures, 1858; M.D., 1864; Dr.-es-Scis., 1864. Répétiteur of mechanics, then indsl. chemistry, instr. course metallurgy École Centrale; adj. dir. chemistry lab. and master of confs. École normale supérieure; prof. mineralogy Faculty Scis. of Paris, after 1885; dir. mineral. chemistry lab. École des hautes études. Mem. French Acad. Scis. Author: Recherches sur l'acide perazotique, 1884; (with J. Chappuis) Recherches sur l'ozone, 1884; Henri Sainte-Claire-Deville, 1885. Research on mineral. chemistry; determined composition and synthesized many minerals, determined temperature at which they dissociate and temperature at which a mineral will crystallize in one crystal system or another, worked on oxides of nitrogen; mineral hautefeuillite named for him. Died 1902.

HAÜY, René Just, French mineralogist; b. St. Just, Picardy, France, Feb. 28, 1743; s. of weaver; studied at Navarre Coll., Paris. Tchr. Latin, Navarre Coll; ordained, then taught at Cardinal Lemoine Coll.; imprisoned during revolution of 1792, released soon after; apptd. dir. Sch. Mines, prof. physics Normal Sch., 1794; prof. mineralogy Mus. Natural History, Paris, 1802; granted pension by Napoleon; named canon of Notre Dame by Napoleon. Fellow Royal Soc., 1818; mem. French Acad. Sci. Author: Essai d'une théorie sur la structure des cristaux appliquée a plusieurs genres de substances cristallines, 1784; Traité de Minéralogie, 5 vols., 1801; Traité de Cristallographie, 3 vols., 1822. Founder of crystallography; formulated geometrical law of crystallization (named after him) after accidently dropping piece of calcareous spar and noting shape of fragments; also did important work on pyroelectricity. Died Paris, June 3, 1822.

HAUY, Valentin, French physician, educator; b. St.-Just, France, 1745; ed. Paris; founder inst. for teaching blind, Paris, 1784, later at St. Petersburg. Author: Essay on the Education of the Blind, 1786. Introduced system for edn. of blind, 1784; originated method of raised letters and embossed paper for teaching blind to read, 1786. Died 1822.

HAVA, Milos, Czechoslovakian physician, pharmacologist; b. Prague, Czechoslovakia, Oct. 15, 1927; s. Emanuel and Eta (Dorman) H.; M.D., Charles U., Prague, 1952, C.Sc. in Pharmacology, 1955; m. Maria Kovác, Sept. 5, 1951; 1 dau., Nadja. Asst., Inst. Pharmacology, Med. Sch., Charles U., 1952-55; sci. worker pharmacological dept. Czechoslovak Acad. Sci., 1955-58; chief dept. pharmacology Research Inst. for Natural Drugs, Prague, 1958-—. Guest lectr. 3d Med Inst. Pharmacology, Charles U., since 1965-—, cons. Czechoslovak Ministry Health. 1958-—. Recipient prize Czechoslovak Acad. Sci., 1956. Mem. Pharmacological Soc. J. E. Purkyne (mem. com.). Research and publs. on pharmacology of bacterial toxines, natural substances and steroid compounds; problems of neomycine toxicity, possibility of local application of androgenes in dental medicine. Home: 4 Uruguayska, Prague 2. Office: 8 U Electry, Prague 9, Czechoslovakia.*

HAVAS, Peter, physicist; b. Budapest, Hungary, Mar. 29, 1916; s. George G. and Irene (Harmos) H.; Absolutorium, Technische Hochschule, Vienna, Austria, 1938; Ph.D., Columbia, 1944; m. Helga Francis Höllering; children—Eva Catherine, Stephen Walter. Came to U. S., 1941, naturalized, 1948. Research asst. in mass spectroscopy U. Vienna, 1937-38; research fellow Institut de Physique Atomique, Lyon, France, 1938-41; lectr. in physics Columbia, N.Y.C., 1941-45; instr. physics Cornell U., Ithaca, N.Y., 1945-46; asst. prof. physics Lehigh U., Bethlehem, Pa., 1946-49, asso. prof., 1949-54, prof., 1954-65; mem. Inst. for Advanced Study, Princeton, N.J., 1953-54; prof. physics Temple U., Phila., 1965-—. Guggenheim fellow, 1953-54. Research on classical and quantum theories of radiation, theory of relativity, especially equations of motion of interacting elementary particles, foundation problems. Home: 240 Berkeley Rd., Glenside, Pa. 19038.*

HAVEL, Richard Joseph, Am. physician; b. Seattle, Feb. 20, 1925; s. Joseph and Anna (Fritz) H.; B.A., Reed Coll., 1946; M.S., U. Ore., 1949, M.D., 1949; m. Virginia Martin Johnson, June 28, 1947; children—Christopher Martin, Timothy Franklin, Peter Joseph, Julianne Virginia. Clin. asso. Nat. Heart Inst., Bethesda, Md., 1953-54, research asso., 1954-56; faculty U. Cal. Sch. Medicine, San Francisco, 1956-—, prof. medicine, 1964-—, asso. dir. Cardiovascular Research Inst., 1961-—. Mem. A.A.A.S. (Theobald Smith prize 1960), Am. Fedn. for Clin. Research (nat. councillor 1962-65, pres. 1965-66), Am. Heart Assn. (exec. com. council on arteriosclerosis 1960-63), Am. Soc. for Clin. Investigation, Western Soc. Clin. Research (v.p. 1965), Am. Physiol. Soc., Western Assn. Physicians. Research and numerous publs. on delineation of pathways of transport of fats in blood and mechanisms regulating these processes; defined variety of hyperlipemia resulting from deficiency of specific enzyme; elucidated role of sympathetic nervous system in control fat-mobilization from adipose tissue, influence of muscular exercise on fat-mobilization and utilization. Home: P.O. Box 719, Ross, Cal. 94957. Office: U. Cal. Med. Center, San Francisco 94122.*

HAVELOCK, Sir Thomas Henry, English mathematician; b. Newcastle, Eng., June 24, 1877; s. Michael Havelock; ed. Armstrong Coll., U. Durham, also St. John's Coll., Cambridge U.; hon. D.Sc., Hamburg, Germany, 1960; hon. D.C.L., U. Durham, 1958. Vice prin. Armstrong Coll., 1933-37; prof. math. King's Coll., Newcastle, Eng.; fellow St. John's Coll., Cambridge. Recipient Smith's prize; William Fronde Gold medal, 1956. Fellow Royal Soc., 1914. Research and publs. on math., geography, navigation. Home: 8, Westfield Dr., Grosforth, Newcastle, Eng.

HAVENER, William Henry, Am. physician; b. Portsmouth, Ohio, June 2, 1924; s. Gilbert Harry and Laura (Braunlin) H.; B.A., Wooster Coll., 1944; M.D., Western Res. U., 1948; M.S. in Ophthalmology, U. Mich., 1953; m. Phyllis Ann Johnson, Jan. 26, 1946; children—Michael, Mark, Ann, Gail, John, Amy, Neal. Instr. ophthalmology U. Mich., 1953-54; faculty Ohio State U., Columbus, 1954-—, prof., chmn. ophthalmology, 1959-61, prof., 1961-—. Mem. A.M.A., Ear, Eye, Nose and Throat Soc., Ohio Ophthalmol. Soc., Am. Acad. Ophthalmology and Otolaryngology. Author: Synopsis of Ophthalmology, 2d edn., 1963; (with William H. Saunders and Betty S. Bergerson) Nursing Care in Eye, Ear, Nose & Throat Disorder, 1964; Ocular Pharmacology, 1966; (with Sallie Goekner) Atlas of Diagnostic Technique and Treatment of Retinal Detachment, 1967. Research and numerous publs. on eye surgery. Home: 1859 Bedford Rd., Columbus, Ohio 43212.*

HAVENS, Walter Paul, Jr., Am. internist; b. Farmingdale, N.J., Dec. 20, 1911; s. Walter Paul and Jessie (Crouse) H.; B.A., Harvard, 1932, M.D., 1936; m. Ida Markle Hessenbruch, June 14, 1941; children—Florence Dreer, Timothy Markle, Michael Crouse, Peter Hessenbruch, John Paul. Faculty, Yale, 1946; faculty Jefferson Med. Coll., Phila., 1946-—, prof. medicine, 1957-—, mem. staff coll. hosp., 1946-—; chief sect. infectious disease Pa. Hosp., 1946-—. Mem. commns. Army Epidemiological Bd., 1944-—, commn. on viral infections, 1955-—; mem. expert adv. panel on virus diseases, also expert com. viral hepatitis WHO, 1952-—; cons. to various brs. of govt.; hon. Prof. Medicine U. Chile, Santiago, 1956. NRC fellow, 1941-42. Diplomate Am. Bd. Internal Medicine. Mem. A.M.A. Am. Assn. Study of Liver Diseases. Research and publs. on liver diseases and serologic techniques. Home: 139 Cheswold Lane, Haverford, Pa. 19041. Office: 1025 Walnut St., Phila. 19107.*

HAVERS, Clopton, English physician, anatomist; b. Eng., between 1650-60; s. Henry Havers; ed. Catharine Hall, Cambridge; M.D., Utrecht, 1685. Licentiate Royal Coll. Physicians, 1687; practiced in London, also spent time on anatomy. Fellow Royal Soc., 1686. Author: Osteologia Nova (gave 1st minute account of bone structure), 1691. Editor: Remmelini's Catoptrium Microscosmicum, entitled A Survey of the Microcosme, 1695. Haversian canals, glands, and Haversian system named after him; studies on bone structure and growth. Died Apr. 1702.

HAVIGHURST, Robert James, Am. social scientist; b. DePere, Wis., June 5, 1900; s. Freeman A. and Winifred (Weter) H.; A.B., Ohio Wesleyan U., 1921; LL.D., 1963; Ph.D., Ohio State U., 1924; D.Sc., Adelphi Coll., 1962; LL.D., Lincoln Coll., 1965; m. Edythe D. McNeely, June 21, 1930; children—Helen (Mrs. Marvin S. Berk), Ruth (Mrs. Samuel Neff), Dorothy (Mrs. Thomas Kucera), James, Walter. NRC fellow physics Harvard, 1924-26; asst. prof. chemistry Miami U., Oxford, O., 1927-28; asst. prof. physics, adviser Exptl. Coll., U. Wis., 1928-32; asso. prof. sci. edn. Ohio State U., 1932-34; asst. dir. for gen. edn. Gen. Edn. Bd., Rockefeller Found., N.Y.C., 1934-37, dir. for gen. edn., 1937-40; prof. edn., and human devel. U. Chgo., 1941-65, part-time 1965-—; prof. edn. U. Mo., Kansas City, part-time 1964-—; Fulbright prof. U. Canterbury, Christchurch, New Zealand, 1953-54, U. Buenos Aires (Argentina), 1961; co-dir.

Brazilian Govt. Center for Ednl. Research, UNESCO, 1956-59. Dir. survey Chgo. Pub. Schs., 1963-64. Mem. Gerontological Soc. Am. (past pres.), Am. Psychol. Assn. (past pres. div. on maturity and old age), Nat. Soc. for Study Edn. (dir. 1957-63, 64-67), A.A.A.S., Am. Ednl. Research Assn., Nat. Soc. for Study Edn., N.E.A. Author numerous books including: Human Development and Education, 1953; American Higher Education in the 1960's, 1960; La Sociedad y La Educacion en America Latina, 1962; The Public Schools of Chicago, 1964; The Educational Mission of the Church, 1965; (with Ruth Albrecht) Older People, 1953; (with Bernice Neugarten) American Indian and White Children, 1955, Society and Education, 1957; (with Robert DeHaan) Educating Gifted Children, 1957; (with others) Psychology of Character Development, 1960, Growing up in River City, 1962; (with J. Roberto Moreira), Society and Education in Brazil, 1965. Research and numerous publs. on atoms and electrons in crystals; child devel.; human behavior; human devel. at all ages in various environments, ednl. and social systems in large cities and met. areas; research on children, adolescents and adults according to social class structure of community. Home: 5844 Stony Island Av., Chgo. 60637.*

HAVILAND, Robert Paul, Am. electronic engr.; b. Cleve., Oct. 18, 1913; s. Irwin Edgar and Mary (Seeds) H.; A.A., Central Wesleyan Coll., 1933; B.S., Mo. Sch. Mines, 1939; m. Opal Evelyn Dorf, July 19, 1937; children—Kay (Mrs. William B. Freilich), Jean, Jon Robert. Engr., Schlumberger Well Surveying Corp., Houston, 1939-42; engr. Gen. Electric Co., Phila., 1947-—; mem. U. S. delegation to Internat. Telecommunications Union, 1959, Internat. Radio Consultative Com. Study Group IV, 1962. Fellow Am. Astron. Soc. (past dir.), Am. Inst. Aeros. and Astronautics; mem. I.E.E.E. (sr.), Brit. Interplanetary Soc., Franklin Inst., Eta Kappa Nu, Phi Kappa Phi. Author: Engineering Reliability and Long Life Design, 1964; (with C. M. House) Handbook of Satellites and Space Vehicles, 1965. Contbr. numerous articles to profl. jours. Initiated 1st U. S. studies on artificial earth satellites; project engr. 1st large two-stage rocket, 1st missile launching from Cape Canaveral; initial system work on Discoverer, Advent, other space vehicles. Home: 182 Clubhouse Rd., King of Prussia, Pa. 19406. Office: P.O. Box 855, Phila. 19101.*

HAWKES, H(erbert) E(dwin), Jr., Am. geologist; b. N.Y.C., Dec. 11, 1912; s. Herbert Edwin and Annette (Coit) H.; A.B., Dartmouth, 1934; postgrad. Columbia, 1935-36; Ph.D., in Geology, Mass. Inst. Tech., 1940; m. Evelyn Voorhees, Mar. 1, 1966; children—Alice, Samuel, Susan, Mary. Field party chief, Hans Lundberg, Toronto, Ont., Can., 1936-37; jr. geologist to geologist U. S. Geol. Survey, Washington, 1940-53; faculty Mass. Inst. Tech., 1953-57, U. Cal. at Berkeley, 1957-65; editor Geol. Soc. Am., N.Y.C., 1965-—. Mem. Soc. Econ. Geologists, Geochem. Soc. Author: (with J. S. Webb) Geochemistry in Mineral Exploration, 1962. Developer geochem. methods of exploration for minerals. Address: 4422 Macomb St. N.W., Washington 20016.*

HAWKINS, Caesar Henry, English surgeon; b. Bisley, Gloucestershire, Eng., Sept. 19, 1798; s. Rev. Edward Hawkins; ed. Christ's Hosp. and St. George's Hosp.; m. Miss Dolbel; m. 2d, Miss Ellen Rouse; no children. Surgeon, St. George's Hosp., 1829-61, cons. surgeon, 1861; Hunterian orator, 1849; sgt.-surgeon to Queen Victoria, 1862. Fellow Royal Soc., 1856; mem. Coll. Surgeons (pres. 1852, 61). His memoirs and lectrs. to med. jours. collected and pub. as The Hunterian Oration Presidential Addresses and Pathological and Surgical Writings, 1874. First successful practitioner of ovariotomy. Died July 20, 1884.

HAWKINS, Gerald Stanley, astronomer; b. Norfolk, Eng., Apr. 20, 1928; s. Frederick A. and Annie (Nichols) H.; B.Sc. in Physics with honors, Nottingham (Eng.) U., 1949; Ph.D., Manchester (Eng.) U., 1952, D.Sc., 1963; m. Dorothy Zoe Willacy-Barnes, July 15, 1955; children—Lisette Carole, Carina Geraldine. Came to U. S., 1954, naturalized, 1964. Electronic engr. Ferranti Bros., Manchester, 1952-54; research asso. Harvard Coll. Obs., 1954-—; faculty Boston U., 1957-—, prof. astronomy, 1964-—, chmn. dept., 1966-—, dir. Boston U. Obs., 1957-—; astronomer Smithsonian Astrophys. Obs., Cambridge, Mass., 1962-—, coordinator radiometeor program, 1965-—. Recipient Shell Faculty award Boston U., 1964, Exceptional Service award Smithsonian Instn., 1964, Arthur S. Flemming award D.C. Jr. C. of C., 1966. Fellow Meteoritical Soc. Am.; mem. Am. Astron. Soc., Archaeol. Inst. Am., Am. Geophys. Union, Internat. Astron. Union, A.A.A.S., Sigma Xi. Author: Splendor in the Sky, 1961; Meteors, Comets and Meteorites, 1964; (with J. B. White) Stonehenge Decoded, 1965; (with Wolfe, Battan, Flemming, Skornik) Earth and Space Science, 1966; also numerous articles. Showed that sporadic meteors came from two hot spots in sky; determined number of meteors and meteorites falling on earth; showed that Stonehenge was sun-moon obs. and computer for predicting eclipses. Home: 37 Maugus Hill, Wellesley Hills, Mass. 02181. Office: 60 Garden St., Cambridge, Mass. 02138.*

HAWKINS, Joseph Elmer, Jr., Am. physiologist; b. Waco, Tex., Mar. 4, 1914; s. Joseph Elmer and Maude (Schlenker) H.; A.B., Baylor U., 1933; postgrad. Brown U.; B.A., U. Oxford, 1937, M.A., 1966; Ph.D., Harvard, 1941; m. Jane Elizabeth Daddow, Aug. 24, 1939; children—Richard Spencer Daddow, Peter Douglas Huntington, James Marion Davis, William Alexander Parmley, Priscilla Ann. Instr., Harvard Med. Sch., 1941-45, spl. research asso. Psycho-Acoustic Lab., 1943-45; asst. prof. physiology Bowman Gray Sch. Medicine, Wake Forest Coll., 1945-46; research asso., head dept. neurophysiology Merck Inst. for Therapeutic Research, 1946-56; asso. prof. otolaryngology Med. Sch. N.Y. U., 1956-63; spl. fellow USPHS U. Göteborg, Sweden, 1961-63; prof. otorhinolaryngology, physiol. acoustics Med. Sch. U. Mich., Ann Arbor, 1963-—. Mem. sensory diseases study sect. NIH, 1958-62; mem. communicative disorders research tng. com. Nat. Inst. Neurol. Diseases and Blindness, 1965-—. Fellow Acoustical Soc. Am., A.A.A.S.; mem. Am. Physiol. Soc., Sigma Xi. Research, numerous publs. on damage to inner ear by intense sound and by streptomycin, related antibiotics and other drugs, anatomy and physiology of cochlear and vestibular endorgans. Home: 1081 Arlington Blvd. Office: Kresge Hearing Research Inst., 1301 E. Ann St., Ann Arbor, Mich. 48104.*

HAWKSBEE (or HAUKSBEE), Francis (the elder), English physicist; flourished 1700; pupil of Boyle. Famous instrument maker; Fellow Royal Soc., 1705; made original contbns. to knowledge of barometer and atmospheric refraction; observed that mercury shaken in glass vessel produces light, and that friction is cause, 1705; carried further observations by Gilbert and Boyle on electricity, inventing 1st glass elec. machine, 1706, and improving air-pump, 1709; experimented on transmission of sound in air and other media; determined relative weights of air and water; investigated elec. discharges in rarefied gases. Died circa 1713.

HAWKSWORTH, Frank Goode, Am. forest pathologist; b. Fresno, Cal., Apr. 30, 1926; s. William A. and Elsie E. (Goode) H.; B.S.F., U. Ida., 1949; M.F., Yale, 1952, Ph.D., 1958; m. Margaret Rosenberger, Feb. 25, 1956; children—David Lee, Mark Andrew, William Scott. With Bur. Plant Industry, U. S. Dept. Agr., Albuquerque, 1949-53; with U. S. Forest Service, Albuquerque, 1954-57, Ft. Collins, Colo., 1958-—, forest pathologist, 1958-—. Mem. Soc. Am. Foresters, Am. Phytopath. Soc., Ecol. Soc. Am., Save-the Dwarf Mistletoes League, Sigma Xi. Research and publs. on forest tree diseases of Rocky Mountains and Southwestern U. S. especially taxonomy, biology and control of dwarf mistletoes of Western conifers. Home: 1041 Meadowbrook Dr., Ft. Collins, Colo. 80521.*

HAWLEY, Amos Henry, Am. sociologist; b. St. Louis, Dec. 5, 1910; s. Amos H. and Margaret (Holtzclaw) H.; A.B., U. Cin., 1936; Ph.D., U. Mich., 1941; m. Gretchen Haller, Sept. 5, 1937; children—Steven A., Margie L. (Mrs. Craig McEwen), Susan E., Patrice A. Faculty, U. Mich., Ann Arbor, 1941-66, prof. sociology, 1951-66, dept. chmn., 1952-61; prof. U. N.C., Chapel Hill, 1966-—. Vis. prof. tech. asst. USOM, U. Philippines, 1953-54; dir. Census of Aruba, 1960; cons., Prime Minister's Office, Thailand, 1964-65. Fulbright Research scholar, Italy, 1959. Fellow Am. Sociol. Assn., A.A.A.S.; mem. Population Assn. Am., Internat. Union for Population Study, Regional Sci. Assn., Am. Assn. U. Profs. Author: Human Ecology, 1950; Changing Shape of Metropolitan America, 1956; (with R. Freedman, G. Lenski, H. Miner) Principles of Sociology, 1952; Demography and Public Administration, 1953; La Estructura de los Sistemas Sociales, 1966; also articles. Devel. theory of human ecology (man's adaptation to environment); research on causal significance of population size for social orgn. Home: 407 Brookside St., Chapel Hill, N.C. 27515.*

HAWLEY, Robert, English elec. engr.; b. Wallasey, Eng., July 23, 1936; s. William and Eva (Dawson) H.; B.Sc., Kings Coll., U. Durham, 1959, Ph.D., 1963; m. Valerie Clarke, Jan. 17, 1962. Staff, High Voltage lab. BICC Ltd., 1952-55; faculty King's Coll., U. Durham, 1955-61; with C. A. Parson & Co., Ltd., Newcastle upon Tyne, Eng., 1961-—, sect. leader research dept., 1963, in generator design dept., 1963-66, dep. chief generator engr., 1966-—. Mem. Instn. Elec. Engrs. (council 1965-—), I.E.E.E., Inst. Physics (asso.), Phys. Soc. (asso.). Author: (with A. Maitland) Vacuum as Insulator, 1967. Research and numerous publs. on elec. insulating properties and conduction and breakdown mechanisms in vacuum and in insulating liquids, detection and measurement of partial discharges in high voltage equipment, novel and conventional methods of elec. power generation. Home: 179 Benton Rd., Newcastle upon Tyne 7. Office: C. A. Parsons & Co. Ltd., Newcastle upon Tyne, 6, Eng.*

HAWLEY, Robert William, Am. chem. engr.; b. Trinidad, Colo., Apr. 23, 1921; s. Joseph William and Edith (Higby) H.; B.S. in Chem. Engring., U. Colo., 1943, M.S., 1947; m. Marcelle M. Easley, June 9, 1946; children—Stephen R., Thomas W., Susan J. Chem. engr., U. S. Naval Research Lab., 1943; compounder Gates Rubber Co., Denver, 1944; chem. engr. expt. sta., instr. chem. engring., U. Colo., 1944-47; scientist Los Alamos Sci. Lab., 1947-50; oil shale retorting Petroleum and Oil Shale Lab., U. S. Bur. Mines, Laramie, Wyo., 1950-52; head chem. engring. devel. Dow Chem. Co., Rocky Flats Plant, Denver, 1952-58, exec. v.p. staff, Midland, Mich., 1958-59, sales engr., Denver, Colorado, 1959-67, district manager, Kansas City, 1967-—. Registered engr., Colo. Mem. Am. Inst. Mining, Metall. and Petroleum Engrs., Am. Inst. Chem. Engrs., Am. Chem. Soc., Colo. Chem. Club. Research and publs. on heat transfers, solar energy utilization, high temperature oil shale retorting, atomic energy research and devel.; designed spl. equipment for high temperature and corrosive service. Home: 5101 W. 96th St., Shawnee Mission, Kan. 66207. Office: Commerce Tower, Kansas City, Mo. 64199.*

HAWLICZEK, Fritz, Austrian physicist; b. Vienna, Austria, Nov. 6, 1920; s. Johann and Marie (Vogel) H.; Dr.rer.nat., U. Vienna, 1941; m. Edith Lehr, July 19, 1952; children—Peter, Robert. With High Frequency Lab., GEMA, Berlin, Germany, 1943-45; tchr. math. and physics high sch., Steyr, Upper Austria, 1945-46; fellow Inst. for Radium Research and Nuclear Physics, Austrian Acad. Scis., 1946-55; hosp. physicist dept. radiation therapy Hosp. Lainz, Vienna, 1955-—. Sci. cons. Austrian Soc. for Atomic Energy, 1962-—, Inst. for Radiation Protection, 1962-—. Research and publs. on practical use of radioisotopes, med. use of isotopes and instrumentation, dosimetry of high energy radiation. Home: Friedrich Knauerg. 1-3 7/3 A1100, Vienna. Office: Dept. for Radiation Therapy, Hosp. Lainz A 1130 Vienna, 13, Austria.*

HAWORTH, Adrian Hardy, English botanist, entomologist; b. Hull, Eng., 1767; studied entomology, ornithology and botany; m. 3 times; children by each marriage. Articles to solicitor, renounced legal profession; settled at Cottingham, nr. Hull; lived in Little Chelsea, 1793-97, Cottingham, 1797-1817, Chelsea, 1817-33; founded Aurelian Soc., circa 1802, dissolved, 1806, Entomol. Soc. London, 1806 (later merged with Zool. Club of Linnean Soc.). Fellow Linnean Soc. Author: Botanical History of Rhus Toxicodendron, 1793; Observations on the genus Mesembryanthemum, 1794; Prodromus Lepidopterorum Britannicorum, 1802; 6th vol. of Andrew's Botanist's Repository, 1803; Lepidoptera Britannica, 3 parts, 1803, 10, 12, appendix, 1829; Synopsis Plantarum Succulentarum, 1812, supplement, 1819; Saxifragearum Enumeratio, 1821. Collected 1100 species and 300 varieties of lepidoptera, 40,000 insects and herbarium of 20,000 species (now incorporated with that of H. B. Fielding at Osford); cultivated succulent plants; helped form Hull bot. garden. Died Chelsea, Aug. 24, 1833.

HAWORTH, Leland John, Am. physicist; b. Flint, Mich., July 11, 1904; s. Paul Leland and Martha (Ackerman) H.; A.B., Ind. U., 1925, A.M., 1926, D.Sc., 1961; Ph.D., U. Wis., 1931, D.Sc., 1962; D.Sc., Bucknell U., 1961; D.E., Stevens Inst. Tech., 1961; D.C.L., Union Coll., 1964, Columbia U., 1965, U. Ill., 1965; LL.D., Rider Coll., 1964, L.I. Univ., 1965, Delaware State College, 1965; m. to Barbara Mottier, July 2, 1927 (dec. Feb. 1961); children—Barbara Jane, John Paul; m. 2d, Irene Benik, May 15, 1963. Tchr. high sch., Indpls., 1926-28; instr. physics U. Wis., 1930-37; Lalor fellow in phys. chemistry Mass. Inst. Tech., 1937-38, staff mem. Radiation Lab., 1941-46, group leader, 1942-43, div. head, 1943-46; asso. in physics U. Ill., 1938-39, asst. prof., 1939-44, prof., 1944-47; asst. dir. projects Brookhaven Nat. Lab., Upton, L.I., N.Y., 1947-48, dir., 1948-61; v.p. Asso. Univs., Inc., Upton, 1951-60, pres., 1960-61; mem. AEC, Washington, 1961-63; dir. NSF, Washington, 1963-—. Mem. Fed. Council for Sci. and Tech., 1963-—; mem. Interdepartmental Com. for Atmospheric Scis., 1963-68; cons. President's Sci. Adv. Com., 1964-—; ex-officio mem. Def. Sci. Bd., 1963-—; mem. President's Com. on Manpower, 1964-—; mem. Nat. Council on Marine Resources and Engring. Devel., 1966-—. Bd. dirs. Oak Ridge Inst. for Nuclear Studies, 1959-61. Recipient certificate of merit from Pres. for World War II research, 1948. Fellow Am. Nuclear Soc. (dir. 1955-60, pres. 1957-58), Am. Phys. Soc., N.Y. Acad. Scis.; mem. Am. Inst. Physics, Phi Beta Kappa, Sigma Xi, Gamma Alpha, Lambda Chi Alpha. Contbr. numerous articles to sci. jours. Research in crystal structure, X-ray diffraction, elec. contacts, acoustic instruments; contbr. to erection nuclear reactor, accelerator design. Home: 2000 S. Eads St., Arlington, Va. 22202. Office: 1800 G St. N.W., Washington 20550.*

HAWORTH, Sir (Walter) Norman, English chemist; b. Chorley, Lancashire, Eng., Mar. 19, 1883; s. Thomas Haworth; Ph.D., Göttingen (Germany) U., 1910; D.Sc., Manchester, 1911; m. Violet Chilton Dobbie, 1922; 2 sons. Prof., U. St. Andrews, Scotland, from 1912, Armstrong Coll., Newcastle, 1920; prof. U. Birmingham (Eng.), 1925-48; vice prin., 1947-48. Recipient (with Paul Karrer) Nobel prize, 1937. Fellow Royal Soc., 1928 (Davy medal 1934, Bakerian lectr. 1944, pres. 1947-48); mem. Chem. Soc. (pres. 1944-46, Longstaff medal 1933), Atomic Scientists Assn.; hon. mem. acads. scis. of Haarlem, Brussels, Munich, Vienna, Finland, India, Dublin, Swiss Chem. Soc. Author: The Constitution of Sugars, 1929, also articles. Synthesized vitamin C (suggested name ascorbic acid), 1933; important work on structure of carbohydrates (especially sugars); developed carbohydrate substitute for blood plasma; worked on separation of uranium isotopes by gaseous diffusion during World War II. Died Birmingham, Mar. 19, 1950.

HAWORTH, Robert Downs, English chemist; b. Cheadle, Cheshire, Eng., Mar. 15, 1898; s. John Thomas and Emily (Downs) H.; B.S., U. Manchester,

1919, M.S. (Mercer scholar), 1920, Ph.D. (Beyer fellow), 1922, D.Sc. (1851 Exhbn. scholar), 1927; B.A., U. Oxford, 1924; m. Dorothy Stocks, 1930; 1 dau., Elizabeth Downs. Firth prof. chemistry U. Sheffield (Eng.), 1939-63, emeritus prof., 1963—; Leverhulme Royal Soc. prof. chemistry U. Madras, 1963-64. Tilden lectr. Chem. Soc., 1942, Pedler lectr., 1961. Fellow Royal Soc., 1944 (Davy medal 1956), Royal Inst. Chemists. Research on gen. organic chemistry, chemistry of natural products. Home: 67 Tom Lane, Sheffield 10, Eng.*

HAWTHORNE, John Nigel, English biochemist; b. Dudley, Eng., Mar. 7, 1926; s. Ralph William and Alice (Baker) H.; B.Sc., U. Birmingham, 1946, Ph.D., 1949, D.Sc., 1963; m. Jennifer Browne, Aug. 21, 1954; children—Deborah, Barnie, Prudence. Research fellow Coll. Physicians and Surgeons, N.Y.C., 1951-52, U. Birmingham (Eng.) Med. Sch., 1952-54, Sorbonne, Paris, 1955; lectr. U. Birmingham Med. Sch., 1956-63, sr. lectr. med. biochemistry, 1963-66, reader biochemistry, 1966—. Mem. Biochem. Soc., Chem. Soc., Biophys. Soc., Internat. Soc. for Neurochemistry. Author: Questions of Science and Faith, 1960; (with G. B. Ansell) The Phospholipids, 1964; also numerous articles. Structural metabolic studies of phospholipids, particularly those containing inositol, their importance in excitable tissues such as brain and nerve; study of cell membrane and biol. transport. Home: 37, Grange Hill Rd., Birmingham 30, Eng.*

HAXO, Francis Theodore, Am. biologist; b. Grand Forks, N.D., Mar. 9, 1921; s. Henry Emile and Florence (Shull) H.; B.A., U. N.D., 1941; Ph.D., Stanford, 1947; m. Judith Morgan McLaughlin, Apr. 15, 1961; children—John F., Barbara, Philip Henry, Francis and Aileen. Research asst. Cal. Inst. Tech., 1946; research asso. Hopkins Marine Sta., 1946-47; faculty Johns Hopkins, Balt., 1947-52, asst. prof., 1949-52; faculty U. Cal. at LaJolla, 1952—, prof. biology, 1963—, chmn. marine biology dept., 1960-65. Vis. faculty U. Cal. at Berkeley, 1957, U. Wash., Seattle, 1963, Marine Biology Lab., Woods Hole, Mass., 1948-52. Fellow A.A.A.S.; mem. Am. Soc. Plant Physiologists, Bot. Soc. Am., Phycological Soc. Am., Soc. Gen. Physiologists, Western Soc. Naturalists, Phi Beta Kappa, Sigma Xi. Research and publs. on nature and distbn. of carotenoids and phycobilins in lower plants, function of accessory pigments and chlorophyll in photosynthesis. Home: 6381 Castejon Dr., La Jolla, Cal. 92037.*

HAXO, François Nicolas Benoit, French engr.; b. Lune Ville, France, 1774; ed. Paris, Coll. de Nevarre; mil. engr.; insp. gen. of fortifications during Restoration; dir. siege of citadel of Antwerp, 1882. Author: Mémorie pour le figure du terrain dans les cartes topographiques; Carte endiquant la circonscription des divers états de l'Europe en 1838. Noted for reconstrns. of Vauban's fortifications. Died 1838.

HAY, Oliver Perry, Am. vertebrate paleontologist; b. Saluda, Ind., May 22, 1846; s. Robert and Margaret (Crawford) H.; A.B., Eureka (Ill.) Coll., 1870, A.M., 1873; Yale, 1876-77; Ph.D., Ind. U., 1884; m. Mary Emily Howsmon, June 30, 1870; children—William Perry, Mrs. Mary Mennick, Frances Steele, Robert Howsmon. Prof. natural scis. Eureka Coll., 1870-72, Oskaloosa (Ia.) Coll., 1874-76; prof. biology and geology, Butler Coll., Indpls., 1879-92; asst. curator zoology, Field Mus. Natural History, Chgo., 1895-97; asst. and asso. curator vertebrate paleontology Am. Mus. Natural History, N.Y., 1901-07; engaged in pvt. investigations in vertebrate paleontology, 1907-11; research asso. Carnegie Instn., Washington, 1912-17, asso., from 1917, investigating history of Pleistocene vertebrata of N. America; asst. Geol. Survey of Ark., 1884-88, of Ind., 1891-94, 1911-12, of Ia., 1911-13. Asso. editor American Geologist, 1902-05. Author: Bibliography and Catalogue of the Fossil Vertebrata of North America, 1902; The Fossil Turtles of North America, 1908; Pleistocene Period in Indiana and its Vertebrates, 1912; Pleistocene Mammals of Iowa, 1914; Pleistocene and its Vertebrates, East Mississippi River, 1923; and of Middle Region of N.A., 1924; Pleistocene of Western Region, 1927; Second Bibliography and Catalogue of the Fossil Vertebrata of North America, Vol. I, 1929, Vol. II, 1930. Died Nov. 2, 1930.

HAYAISHI, Osamu, biochemist; b. Stockton, Cal., Jan. 8, 1920; s. Jitsuzo and Mitsuko (Uchida) H.; M.D., Osaka (Japan) U., 1942, Ph.D., 1949; m. Takiko, Nov. 10, 1946; 1 dau., Mariko. Research asso. dept. bacteriology Osaka U., 1942-46, asst. dept. bacteriology, 1946-55; William Waterman fellow enzyme chemistry U. Wis., 1949-50; spl. research fellow Nat. Inst. Arthritis and Metabolic Diseases, NIH, Bethesda, Md., 1951-52, chief sect. on toxicology, 1954-58; asst. prof. dept. microbiology Washington U. Sch. Medicine, St. Louis, 1952-54; prof., chmn. dept. med. chemistry Kyoto (Japan), 1958—. Recipient Matsunaga award, Japan Soc. Vitaminology, 1964; Asahi Cultural award, 1965; Japan Acad. Sci. award, 1967. Mem. Japan Soc. Biochemistry, Am. Soc. Biol. Chemists, Am. Chem. Soc. Editor: Archives Biochemistry, 1960—, Analytical Biochem., 1960—, Oxygenases, 1962; Biochim. Biophys. Acta, 1964—. Research and numerous publs. on metabolism of amino acids and nucleic acids, reaction mechanism of enzymes and regulation of enzyme activities; discovered oxygenases. Home: 23 Kitachanokicho, Shimogamo, Sakyo-ku, Kyoto, Japan.*

HAYAKAWA, Satio, Japanese physicist; b. Niihama, Japan, Oct. 16, 1923; s. Koichi and Michiko (Obata) H.; B.Sc., U. Tokyo (Japan), 1945, D.Sc., 1951; m. Michiko Fujimuro, Aug. 17, 1952; children—Yoichi, Hisao. Research ofcl. Meteorol. Research Inst., Tokyo, 1945-49; faculty Osaka (Japan) City U., 1949-54, asst. prof., 1950-54; prof. Kyoto (Japan) U., 1954-59, Research Inst. for Fundamental Physics, part-time 1959—; prof. physics Nagoya (Japan) U., 1959—; prof. Inst. Space and Aero. Sci., U. Tokyo, part-time 1964—. Mem. Phys. Soc. Japan, Soc. Terrestrial Magnetism and Electricity of Japan, Am. Phys. Soc. Author: Cosmic Rays, 1956; (with C. Hayashi) Nuclear Fusion, 1959; also numerous articles. Theoretical analyses of high energy cosmic ray phenomena; research on elementary particle theories, cosmic X-rays and electrons, theory of nuclear reactions. Home: RH-33, Nakata-Jutaku, Chikusa-ku, Nagoya, Japan.*

HAYASAKA, Ichiro, Japanese geologist, paleontologist; b. Miyagi Prefecture, Japan, 1891; grad. Tohoku U., 1915; D.Sc. Concurrently prof. Hokkaido U. and Kanazawa U.; later prof. sci. dept. Hokkaido U. Author: Study of History of Japanese Topography; The World of Fossils.

HAYASHI, Chushiro, Japanese astrophysicist; b. Kyoto, Japan, July 25, 1920; s. Seijiro and Ume (Kamoi) H.; B.Sc., U. Tokyo, 1942; D.Sc., Kyoto U., 1954; m. Yoshiko Hirai, May 25, 1951; 1 son, Nobuo. Research asso. U. Tokyo (Japan), 1942-46; research asso. Kyoto U., 1946-49, faculty, 1954—, prof. physics, 1957—; asst. prof. Naniwa U., Osaka, Japan, 1949-54. Recipient Nishina Meml. prize, 1963; Asahi Press prize, 1966. Mem. Phys. Soc. Japan, Astron. Soc. Japan, Am. Phys. Soc., Am. Astron. Soc. Research, publs. on origin of chem. elements and on stellar evolution, such as composition of primordial matter in expanding universe and stellar structure and evolution before and after stage of hydrogen burning. Home: 1 Mukaijima Yogorocho, Fushimiku, Kyoto. Office: Dept. Physics, Kyoto U., Kyoto, Japan.*

HAYASHI, Shizuo, Japanese physicist; b. Takasaki City, Japan, Mar. 23, 1922; s. Kaisuke and Hana (Kasuya) H.; grad. Tokyo Kyoiku U., 1945, D.Sc., 1961; m. Kiyono I'hori, Oct. 25, 1957; children—Hironao, Yurika. Tchr., Takasaki High Sch., 1946; prof. physics Gunma Coll. Edn., Maebashi, Japan, 1947-49; faculty Gunma U., Maebashi, 1949—, prof. physics, 1964—. Lectr., Kyoto (Japan) U., 1958—, Tokyo Kyoiku U., 1960—, Tokyo Met. U., 1964—. Mem. Phys. Soc. Japan, Chem. Soc. Japan, Soc. Polymer Sci. Japan. Author: Rheology Hand Book, 1965. Research and publs. on molecular theory of viscoelasticity of concentrated polymer solutions and amorphous polymeric substances especially molecular theoretical explanation of box type mech. spectrum and law that viscosity is proportional to 3.4th power of molecular weight. Home: 63 Tokiwacho, Takasaki City, Japan. Office: 1-14-7 Hiyoshicho, Maebashi City, Japan.*

HAYASHI, Tsuruichi, Japanese mathematician; b. Tokuschima, Japan, June 13, 1873; grad. Tokyo U.; prof., Tokyo Higher Normal Sch.; prof. Tohoku U. also largely responsible for establishing sci. sect. Compiler many math. textbooks. Research in wasan (Japanese math.). Died Matuje, Japan, Oct. 4, 1935.

HAYATA, Bunzo, Japanese botanist; b. Niigata Prefecture, Japan, 1874; grad. sci. sch. Tokyo U.; D.Sc.; visited Britain, 1909; apptd. prof. Tokyo U., 1922. Collected bot. specimens, Formosa, 1905-24; studied flora of Annam and Thailand, 1921-22; discovered over 1,200 new plant species. Died 1934.

HAYCOCK, Obed Crosby, Am. elec. engr.; b. Panguitch, Utah, Oct. 5, 1901; s. George A. and Elida (Crosby) H.; B.S. in Elec. Engring., U. Utah, 1925, postgrad., 1938; M.S. in Elec. Engring., Purdue U., 1931; postgrad. U. Mich., 1937; m. Mary Harding, Aug. 12, 1926 (dec.); children—Jean (Mrs. Carl Gardiner), Don H., Ralph H., Richard O., Lois (Mrs. Roland Porter); m. 2d, Ellen Smith, July 1, 1964. Asso. dir. Upper Air Research Lab., Salt Lake City, 1947-57, dir., 1957—; faculty U. Utah, Salt Lake City, 1926—, prof. elec. engring., 1947—. Recipient Outstanding Engr. award Utah Joint Engring. Council, 1962. Fellow I.E.E.E.; mem. Union Radio Scientifique Internationale, Sigma Xi, Tau Beta Pi. Research and numerous publs. on upper atmosphere using rockets and satellites. Home: 3390 Colemere Way, Salt Lake City 84109.*

HAYDEN, Ferdinand Vandiveer, Am. geologist; b. Westfield, Mass., Sept. 7, 1829; s. Asa and Melinda (Hawley) H.; A.B., Oberlin Coll., 1850; M.D., Albany Med. Coll., 1853; LL.D., U. Rochester, 1876, U. Pa., 1887; m. Emma Woodruff, Nov. 9, 1871. Went to Badlands of S.D. on collecting trip, 1853; 1st contbn. to geology was a vertical geol. sect. showing order of superposition of the strata; geologist on staff Lt. G. K. Warren of Topog. Engrs. in surveying expdn. of Yellowstone and Missouri rivers and Badlands of S.D., 1856-57; with F. B. Meek in explorations in Kan. Terr., 1858; explored Yellowstone and Missouri rivers with Capt. W. F. Raynolds, 1859-62; surgeon U. S. Army, 1861-65; prof. geology U. Pa., 1865-72; began survey of Neb. Territory in 1867 which laid foundation for U. S. Geol. Survey as it exists today; his work resulted in setting aside land for Yellowstone Nat. Park. Mem. Acad. Natural Scis. of Phila., Nat. Acad. Scis., Geol. Socs. of London and Edinburgh, Geologische Reichsanstalt of Vienna, Société Impériale of Moscow. Died Phila., Dec. 22, 1887.

HAYDEN, Sir Henry Hubert, geologist; b. 1869; s. James Hayden; ed. Hilton Coll., Natal, Trinity Coll., Dublin; B.A., B.A.I.; D.Sc. (hon.), Calcutta. Became mem. staff Geol. Survey India, 1895, later dir.; with Tirah Expeditionary Force, 1897-98, Tibet Frontier Commn., 1903-04; in service Amir of Afghanistan, 1907-08. Fellow Royal Soc., 1915, Geol. Soc. Author: (with S. G. Burrard) Sketch of the Geography and Geology of the Himalaya Mountains and Tibet; also papers. Died Aug. 1923.

HAYDEN, Horace H., Am. dentist, geologist; b. Windsor, Conn., Oct. 13, 1769; s. Thomas and Abigail (Parsons) H.; received D.D.S. as mem. Am. Soc. Dental Surgeons, 1840; M.D. (hon.), Med. Sch., U. Md., 1840; m. Marie Robinson, Feb. 23, 1805, 6 children including Handel. Licensed as dentist by Med. and Chirurgical Faculty of Md., 1810; 1st sec. Balt. Phys. Assn., 1818; pres. Md. Acad. Scis. and Lit. 1826; prin. founder Balt. Coll. Dental Surgery (world's 1st dental coll.), 1840, 1st pres., 1st prof. principles of dental sci., prof. dental physiology and pathology until 1844; influenced orgn. of Am. Soc. Dental Surgeons, N.Y.C., 1840; discovered nr. Balt. a form of chalazite (named Haydenite after him). Author: Geological Essays; or An Inquiry into Some of the Geological Phenomena to be Found in the Various Parts of America and Elsewhere, 1820. Died Balt., Jan. 26, 1844.

HAYEK, F. A. (von), economist; b. Vienna, Austria, May 8, 1899; s. August and Felicitas (von Juraschek) von Hayek; Dr.jur., U. Vienna, 1921, Dr.re.-pol., 1923; D.Sc. in Econs., U. London, 1941; Dr.jur.-hon.c., U. Rikkyo, Tokyo, Japan, 1964; children—Christine M. F., Lorenz J. H.; m. 2d, Helen Bitterlich, 1950. With Austrian Civil Service, 1921-26; dir. Austrian Inst. for Bus. Cycle Research, 1927-31; lectr. econs. and statistics U. Vienna, 1929-31; Tooke prof. econ. sci. and statistics U. London, London Sch. Econs., 1931-50; prof. social and moral sci. U. Chgo., 1950-62; prof. econs. U. Freiburg (Germany), 1962—. Fellow Brit. Acad., Royal Econ. Soc. (London), Econometric Soc.; mem. Am. Econ. Assn. Author: Prices and Production, 1931; Monetary Theory and the Trade Cycle, 1933; Monetary Nationalism and International Stability, 1937; Profits, Interest and Investment, 1939; The Pure Theory of Capital, 1941; The Road to Serfdom, 1944; Individualism and Economic Order, 1949; John Stuart Mill and Harriet Taylor, 1951; The Sensory Order, 1952; The Counter-Revolution of Science, 1952; The Constitution of Liberty, 1960; Studies in Philosophy, Politics and Economics, 1967; also numerous articles. Acting editor Economica, 1940-50. Research on relation between law, legislation and liberty. Home: 27 Urachstrasse, Freiburg i. Brg., W. Germany.*

HAYEM, Georges, French physician; b. Paris, Nov. 24, 1841; M.D., 1868; prof., Paris. Author: Recherches sur l'anatomie normale et pathologique du sang, 1878. Founder of modern chematology; research on diseases of stomach, composition of blood; discovered hematoblastes, 1877; introduced 1st accurate method for counting blood platelets, 1878; eponym of Hayem's corpuscle, Hayem solution. Died 1933.

HAYES, Augustus Allen, Am. chemist; b. Windsor, Vt., Feb. 28, 1806; s. Thomas Allen and Sophia (West) H.; attended med. sch. Dartmouth, M.D. (hon.), 1846; m. Henrietta Bridge Dana, July 13, 1836. Published account of the isolation of alkaloidal compound which he called sanguinaria, 1825; investigated certain chromium compounds, 1826-28; became dir. of large plant mfg. colors and other chems., Roxbury, Mass.; cons. chemist for several large dyeing, bleaching, gasmaking and smelting establishments in New Eng.; state assayer Mass.; devised methods for shortening the time needed in smelting iron and refining copper; work led to fundamental improvements in constrn. of furnaces and arrangement of steam boilers; conducted investigation of water supply, Charlestown, Mass., 1859-60; devised and used simple elec. method of detecting the limits of slight impurities in drinking water; 1st to suggest application of oxides of iron in refining of pig iron. Mem. Am. Acad. Arts and Scis., contbr. sci. papers to its Proceedings, also to Am. Jour. Sci. Died Brookline, Mass., June 21, 1882.

HAYES, Charles Amos, Jr., Am. mathematician; b. Winnipeg, Man., Can., Apr. 9, 1916; s. Charles Amos and Amy (Noblett) H.; came to U. S., 1923, naturalized, 1940; A.B. in Math., U. Cal. at Berkeley, 1937, M.A. in Math., 1938, Ph.D., 1942; m. Lola Thelma Valente, June 21, 1942; children—Rodney Charles, Laura Louise. Faculty, U. Cal. at Berkeley, 1946-47; faculty U. Cal. at Davis, 1947—, prof. math., 1959—, chmn. dept. math., 1959-64. Mem. Am. Math. Soc., Math. Assn. Am., Phi Beta Kappa, Sigma Xi, Pi Mu Epsilon. Author: Concepts of Real Analysis, 1964. Research and publs. on theory of differentiation and integration of set functions in abstract measure spaces, theory of constrn. of measures in abstract spaces, boundary value problem for elliptic systems of partial differential equations. Home: 302 11th St., Davis, Cal. 95616.*

HAYES, Edward Cary, Am. sociologist; b. Lewiston, Me., Feb. 10, 1868; s. Benjamin Francis (D.D.) and Arcy (Cary) H.; A.B., Bates Coll., Me., 1887; LL.D., 1927; studied Cobb Div. Sch., 1889-92, U. of Berlin, 1900-01; Ph.D., U. of Chicago 1902; LL.D., Grinnell, 1920; m. Annie Lee Bean, Oct. 23, 1895; children—Edward Bean, Robert Cary, Harmon Phillips. Ordained ministry, 1893; pastor Augusta, Me., 1893-96; dean Keuka Coll., N.Y., 1897-99; prof. economics and sociology, Miami U., Ohio, 1902-07; prof. sociology and head of dept., U. of Ill., 1907——. Advisory editor Am. Jour. Sociology; coöperating editor Jour. Applied Sociology. Pres. Ill. State Conf. Charities and Correction, 1910-11; mem. Ill. Bd. Commrs. Pub. Welfare, 1917-18; sec. social psychol. sect. of World's Congress of Science, St. Louis Expn., 1904. Taught in summer sessions, Harvard, Columbia, U. Chicago, U. Pa., U. Colo. Author: Introduction to the Study of Sociology, 1915; Sociology and Ethics, 1921. Influenced by Auguste Comte; had a unified view of society, and in his synthetic sociology tried to create a system from available knowledge of social activities. Died Urbana, Ill., Aug. 7, 1928.

HAYES, Francis Newton, Am. chemist; b. Shanghai, China, Apr. 25, 1924 (parents Am. citizens); s. L. Newton and Frances (Gray) H.; A.B., Yale, 1944; Ph.D., Northwestern U., 1949; m. Catherine Frances Hoffman, June 15, 1949; children—Geoffrey Newton, William Gray, Bradford Kent, Russell Scott. Chemist, Eastman Kodak, Co., Rochester, N.Y., 1944-45; teaching and research asst. Northwestern U., Evanston, Ill., 1945-48; asst. prof. chemistry Ill. Inst. Tech., Chgo., 1948-50; staff mem. Los Alamos Sci. Lab., 1950——, sect. leader molecular radiobiology, 1960——; cons. prof. chemistry U. N.M., Albuquerque, 1957——. Mem. Am. Chem. Soc., Chem. Soc. London, American Association Advancement Sci., Am. Inst. Chemists, Author: (with C. G. Bell, Jr.) Liquid Scintillation Counting, 1958. Research and numerous publs. on organic chemistry, synthesis and reaction mechanisms, radiation detectors using organic chems. that emit light; in biochemistry, nucleic acids and their information transfer mechanisms. Home: 119 Canyon Vista. Office: P.O. Box 1663, Los Alamos 87544.*

HAYES, Maurice Richard Joseph, physician; ed. Sacred Heart Coll., Mungret Coll., Limerick, Eire, also Cath. U., Dublin; m. Agnes Walsh, 1906; 2 sons, 2 daus. Became radiologist Mater Misericordiae Hosp., Dublin, 1907; apptd. dir. gen. Irish Free State Army Med. Service, 1922; prof. materia medica and therapeutics Univ. Coll., Dublin; cons. radiologist Nat. Maternity Hosp., Dublin; mem. Med. Registry Council, Irish Free State. Fellow Royal Coll. Surgeons Ireland. Research and publs. on X-ray diagnosis and treatment, internal hernia, treatment of lesions of perinaeum. Died Mar. 2, 1930.

HAYES, Robert Mayo, Am. information scientist; b. N.Y.C., Dec. 3, 1926; s. Dudley Lyman and Myra (Lane) H.; Ph.D., U. Cal. at Los Angeles, 1952; m. Alice Peters, Sept. 2, 1952; 1 son, Robert Dendron. Head applications group Electronics div. Nat. Cash Register Co., Hawthorne, Cal., 1953-55; head bus. systems dept. Magnavox Research Labs., West Los Angeles, Cal., 1955-59; v.p., sci. dir., pres. Advanced Information Systems Co., West Los Angeles, 1959-64; prof. U. Cal. at Los Angeles Sch. Library Service, 1964——, dir. Inst. Library Research, 1965——; cons. U. S. Navy, Nat. Library Medicine, Office Edn., State Library of Wash., John Wiley, Inc., N.Y.C. Mem. Soc. Indsl. and Applied Math., Assn. for Computing Machinery, Am. Math. Soc., Spl. Libraries Assn., Phi Beta Kappa, Sigma Xi, others. Author: (with Joseph Becker) Information Storage and Retrieval, 1963. Research and publs. on problems in orgn. of large files of ill-defined information. Home: 3943 Woodfield Dr., Sherman Oaks, Cal.*

HAYES, Samuel Perkins, Am. psychologist; b. Baldwinsville, N.Y., Dec. 17, 1874; s. M. D. L. and Mary Ellen (Perkins) H.; B.A., Amherst, 1896; B.D., Union Theol. Sem., 1902; M.A., Columbia, 1902; fellow Clark U., 1902-03; studied U. Berlin and Sorbonne, Paris, 1903-04; Ph.D., Cornell, 1906; studied Cambridge U., Eng., 1912; m. Agnes Hayes Stone, July 23, 1903; children—Lyman Stone, Mary Ellen, Samuel Perkins, Janet Card, Betsy Wanton. Prof. psychology Mt. Holyoke Coll., 1906-40; head dept. personnel Perkins Instn. and Mass. Sch. for Blind, 1940-55; cons. Am. Found. for Blind, 1932-54. Mem. Am. Psychol. Assn., Phi Beta Kappa, Sigma Xi. Author: Contributions to a Psychology of Blindness, 1941; numerous articles. Authority on color blindness and psychology of the blind; specialist in adminstrn. and interpretation of achievement and intelligence tests for blind children. Died Princeton, N.J., May 7, 1958.

HAYES, Samuel Perkins, Am. social scientist; b. South Hadley, Mass., Jan. 28, 1910; s. Samuel Perkins and Agnes (Stone) H.; A.B. (Moore fellow), Amherst Coll., 1931, LL.D., 1966; Ph.D., Yale, 1934; postgrad. U. Chgo., 1937-38; m. Alice Mary Cable, Mar. 25, 1937; children—Susan, Jonathan. Instr. Psychology Mt. Holyoke Coll., 1934-37; fellow Social Sci. Research Council, 1937-38; faculty social sci. Sarah Lawrence Coll., 1938-40; mem. sr. staff econs. and market research Young & Rubican, N.Y.C., 1940-42; with OPA, Office Lend Lease Adminstrn., Fgn. Econ. Adminstrn., Dept. State, ECA, Mut. Security Agy., 1942-45, 48-53; asso. dir. Marketing and Research service, Dun & Bradstreet, N.Y.C., 1945-48; dir. Found. for Research on Human Behavior, Ann Arbor, Mich., 1953-60, now trustee; prof. econs., dir. Center for Research on Econ. Devel., U. Mich., 1960-62; pres. Fgn. Policy Assn., N.Y.C., 1962——. Mem. Com. on Ednl. Interchange Policy, 1956-64. Fellow A.A.A.S., Am. Statis. Assn., Am. Psychol. Assn.; mem. Am. Econ. Assn., Soc. for Internat. Devel., Council Fgn. Relations, Phi Beta Kappa, Sigma Xi. Author: Measuring the Results of Development Projects, 1959; An Internat. Peace Corps, 1961; Evaluating Development Projects, 1966. Editor: (with Rensis Likert) Some Applications of Behavioral Research, 1957. Research and numerous publs. on price practices and competition in potash industry, devel. of short-cut methods of computation coefficient tetrachoric correlation and its standard error, promotion of application of social sci. concepts, methodology and findings in indsl. mgmt., adminstrn. of econ. devel. programs; pioneered nat. survey of voters attitudes and voting behavior. Home: 468 Riverside Dr., N.Y.C. 10027. Office: 345 E. 46th St., N.Y.C. 10017.*

HAYES, Walter, English math. instrument maker; flourished 1651-1692. Worked in London for William Leybourne and other English Mathematicians. Most skilled and well known math. instrument maker of that period.

HAYFLICK, Leonard, Am. microbiologist; b. Phila., May 20, 1928; s. Nathan Albert and Edna (Silbert) H.; B.A., U. Pa., 1951, M.S., 1953, Ph.D., 1956; m. Ruth Louise Heckler, Oct. 2, 1955; children—Joel, Deborah, Susan, Rachel, Ann. Research asst. bacteriology Sharpe & Dohme div. Merck & Co., 1951-52; asst. instr. med. microbiology U. Pa., 1955-56, asso. med. microbiology Med. Sch., 1958, now asst. prof. research medicine; James W. McLaughlin research fellow U. Tex., Galveston, 1956-58; asso. asst. dir. tissue culture Assn. U. Wis., 1961-64; asso. mem. Wistar Inst., Phila., 1958——. Vis. prof. U. P.R. Sch. Medicine, 1963——, U. W.I., 1965——; cons. WHO, Nat. Inst. Allergy and Infectious Diseases, NIH. Mem. Am. Soc. Microbiologists, A.A.A.S., Tissue Culture Assn., N.Y. Acad. Scis., Am. Soc. Cell Biology, Am. Soc. Exptl. Pathology, Soc. Exptl. Biology and Medicine, Gerontological Soc. Research and publs. on discovery of Eaton agent (cause of primary atypical pneumonia); 1st demonstration of cause of disease in man by a mycoplasma; research in tissue cultures and their use in prodn. of human virus vaccines; discoverer of finite lifetime of normal human cells in tissue culture and theory that limited growth potential of normal cells in culture is expression of aging at level of single cell. Home: 1510 Remington Rd., Phila. 19151. Office: Wistar Institute, 36th and Spruce Sts., Phila. 19104.*

HAYFORD, John Fillmore, Am. civil engr.; b. Rouse's Point, N.Y., May 19, 1868; s. Hiram and Mildred Alevia (Fillmore) H.; C.E., Cornell U., 1889; Sc.D. George Washington U., 1918; m. Lucy Stone, Oct. 11, 1894. Apptd. computer U. S. Coast and Geod. Survey, 1889; asst. astronomer to Internat. Boundary Commn. U. S. and Mexico, in charge of one of field parties, 1892-93; aid and later asst. U. S. Coast and Geodesic Survey, 1894-95; instr. civil engring., Cornell U., 1895-98; expert computer and geodesist, U. S. Coast and Geodesic Survey, 1898-99, insp. geodetic work and chief of computing div., 1900-09; dir. Coll. Engring., Northwestern U. from 1909. Mem. NACA, 1915-23; research asso. Carnegie Instn. of Washington. Author: Geodetic Astronomy, 1898. Established theory of isostasy. Died Mar. 10, 1925.

HAYGARTH, John, English physician; b. Garsdale, Eng., 1740; M.B., St. John's Coll., Cambridge, Eng., 1766; practiced medicine, Chester, Eng.; physician Chester Infirmary, 1767-98; practiced in Bath, Eng. Fellow Royal Soc., 1781, Royal Soc. Edinburgh. Author: Plan to Exterminate Small-pox and introduce General Inoculation, 1793; A Clinical History of Diseases, 1805-12; Synopsis pharmacopoeiae londinensis, 1810; also numerous articles. First to isolate fever patients, 1783; one of 1st to differentiate various fevers by their periods of incubation; 1st to insist on isolated sch. hosps. Died June 10, 1827.

HAYMAN, Walter Kurt, mathematician; b. Cologne, Germany, Jan. 6, 1926; s. Franz Samuel and Ruth Terese Haymann; B.A., Cambridge U., 1946, M.A., 1951, Sc.D., 1956; m. Margaret Riley Crann, Sept. 20, 1947; children—Daphne, Anne, Sheila. Lectr., King's Coll., Newcastle, 1947; lectr. Exeter U., 1947-53, reader, 1953-56; prof. pure math. Imperial Coll., U. London, 1956——; vis. lectr. Brown U., 1949-50, Stanford, 1950, Am. Math. Soc., 1961. Fellow Royal Soc., 1956; mem. London Math. Soc. (mem. council 1956-59, Berwick prizes 1955, 64), Cambridge Philos. Soc. Author: Multivalent Functions, 1958; Meromorphic Functions, 1964; also articles. Research on theory of functions of a complex variable. Home: 18 Campion Rd., London, S.W. 15., Eng.*

HAYMES, Robert C., Am. physicist; b. N.Y.C., July 3, 1931; s. Michael and Winifred (Konig) H.; B.A., N.Y.U., 1952, M.S., 1953, Ph.D., 1959; m. Jamie E. Buswell, Jan. 22, 1965. Asst. prof. physics N.Y.U., 1959-62; resident research appointee Jet Propulsion Lab. Cal. Inst. Tech., 1962-64; asst. prof. space sci. Rice U., 1964-66, asso. prof., 1966——. Cons., Transistor Spltys. Inc., 1959, Vitro Corp. Am., 1961-62, Oak Ridge Nat. Lab., 1966——. Mem. Am. Phys. Soc., Am. Geophys. Union, Am. Astron. Soc., Sigma Xi, Sigma Pi Sigma. Research and publs. on cosmic rays, gamma ray astronomy and planetary magnetism. Home: 10611 Candlewood Dr., Houston 77042.*

HAYNES, Elwood, Am. inventor; b. Portland, Ind., Oct. 14, 1857; s. Jacob March and Hilinda Sophia (Haines) H.; B.S., Worcester Poly. Inst., 1881; Johns Hopkins, 1884-85; m. Bertha Beatrice Lanterman, Oct. 21, 1887. Teacher sciences, Eastern Ind. Normal Sch., Portland, 1885-86; mgr. Portland Natural Gas & Oil Co., 1886-90; field supt. Ind. Natural Gas & Oil Co., 1890-1901; pres. Haynes Automobile Co., from 1898; pres. Haynes Stellite Co., 1912-20. Discovered tungsten chrome steel, 1881, alloy of chromium and nickel, 1897, alloy of cobalt and chromium, 1900; developed latter alloy for cutting instruments, 1910; designed and constructed a horseless carriage, 1893-94, which is the oldest American automobile in existence and is on exhibition at Smithsonian Instn.; first to introduce aluminum in automobile engine, 1895; invented and built rotary valve gas engine, 1903. Discovered alloys of cobalt, chromium, and tungsten, also alloys of cobalt, chromium and molybdenum, 1911-12; discovered "stainless steel," 1911 (patented 1919). Died Kokomo, Ind. Apr. 13, 1925.

HAYNES, Robert Hall, biophysicist; b. London, Ont., Can., Aug. 27, 1931; s. James Wilson and Lillian (Hall) H.; B.S., U. Western Ont., 1953, Ph.D., 1957; m. Nancy Joanne May, Sept. 1, 1954; children—Mark, Geoffrey, Paul. With physics dept. St. Bartholomew's Hosp. Med. Coll., U. London (Eng.), 1957-58; instr., asst. prof. biophysics U. Chgo., 1958-64; asso. prof. biophysics U. Cal., Berkeley, 1964——; mem. subcom. radiobiology Nat. Acad. Scis.-NRC. Brit. Empire Cancer Campaign Research fellow, 1957-58. Mem. Biophys. Soc., Radiation Research Soc., Genetics Soc. Research and publs. on anomalous flow properties of blood, mechanism of radiation actions on cells; developed an alpha particle microbeam for partial cell irradiation. Office: Donner Lab., U. Cal., Berkeley, Cal. 94720.*

HAYNES, William Clarence, bacteriologist; b. N.Y.C., Aug. 17, 1912; s. William Henry and Lucy Thirza Amy (Holthusen) H.; B.S., Cornell U., 1935, Ph.D., 1946; m. Dorothy Louise Starr, June 29, 1940; children—William Arthur, Dorothy Lorene. Chemist, Snider Packing Corp., Albion, N.Y., 1935; asst. in research N.Y. State Agrl. Expt. Sta., Geneva, 1935-41; asst. bacteriologist No. regional research lab. Agrl. Research Service, U. S. Dept. Agr., Peoria, Ill., 1941-42, 46-48, asso. bacteriologist, 1948-53, bacteriologist, 1953-58, prin. bacteriologist, 1953——. Fellow Am. Acad. Microbiology; mem: Am. Soc. for Microbiology (chmn. agrl. and indsl. div. 1967-68), Am. Inst. Biol. Scis., Internat. Assn. Microbiol. Socs. (chmn., sec. various coms.), Sigma Xi. Studies, publs. on various bacteria, particularly Pseudomonas, food poisoning, dextran for blood volume expansion, control of insects with bacteria. Home: 726 W. Ridge Rd., Peoria 61614. Office: 1815 N. University St., Peoria, Ill. 61604.*

HAYS, Isaac, Am. ophthalmologist, med. editor; b. Phila., July 5, 1796; s. Samuel and Richea (Gratz) H.; B.A., U. Pa., 1816, M.D., 1820; m. Sarah Minis, 1834, 4 children including Isaac Minis. Mem. staff Infirmary for Diseases of the Eye and Ear, Phila., 1822-27; surgeon Wills' Ophthalmic Hosp., Phila., 1834-54; fellow Coll. of Physicians, Phila., 1835-79; editor-in-chief Am. Jour. Med. Scis., 1827-79; established Med. News, 1843; published Monthly Abstract of Med. Science, 1874-79; a founder Franklin Inst., A.M.A.; pres. Acad. Natural Scis. of Phila., 1865-69. Pioneer in study of color blindness and astigmatism; invented spl. knife for cataract operations. Died Phila., Apr. 13, 1879.

HAYS, Kirby Lee, Am. insect ecologist; b. nr. Holly Pond, Ala., Aug. 11, 1928; s. Charles K. S. and Cora (Marsh) H.; B.S., Auburn U., 1948, M.S., 1954; Ph.D., U. Mich., 1958; m. Dean Styles, Dec. 21, 1958; 1 dau., Jane Elizabeth. Plant quarantine insp. U. S. Dept. Agr. Bur. Entomology and Plant Quarantine, 1948-50; research asst. U. Mich., 1954-57; faculty Auburn (Ala.) U., 1957——, prof. insect ecology, 1964——. Mem. Ecol. Soc. Am. (past treas.), Sigma Xi, Phi Kappa Phi, Phi Sigma, Gamma Sigma Delta. Research and publs. on effects of environment on populations of medically important insects, taxonomy of Tabanidae. Home: 1044 Terrace Acres Dr., Auburn, Ala. 36830.*

HAYWARD, C(harles) Lynn, Am. zoologist; b. Paris, Ida., July 16, 1903; s. William Gammon and Ellen (Neibaur) H.; B.S., Brigham Young U., 1927, M.S., 1931; Ph.D., U. Ill., 1941; m. Elizabeth Cook, Aug. 6, 1930; children—Margaret (Mrs. Bruce J. Taylor), Gerald Lynn. Tchr., Fielding High Sch., Paris, Ida., 1927-30; faculty Brigham Young U., Provo, Utah, 1930——, prof., 1954——, chmn. dept. zoology, 1958-62. Mem. A.A.A.S., Ecol. Soc. Am., Am. Soc. Mammalogists, Am. Inst. Biol. Scis., Am. Ornithologists Union. Author: A Laboratory Study of the Vertebrates, 1964. Research in ecology, birds and mammals especially in Western U. S. Home: 959 Cedar Av., Provo, Utah 84601.*

HAYWARD, George, Am. surgeon; b. Boston, Mar. 9, 1791; s. Lemuel Hayward; grad. Harvard, 1809; M.D., U. Pa., 1812. Physician to the almshouse, Boston; fellow Am. Acad. Arts and Scis., 1818; asst. surgeon Mass. Gen. Hosp., 1826; founder (with John Collins Warren and Enoch Hale) med. sch., Boston, 1830-38; lectr. Harvard Med. Sch., 1834, prof. principles of surgery, clin. surgery, 1835-49; surgeon-in-chief Mass. Gen. Hosp., 1838; pres. Mass. Med. Soc., 1835-55; fellow Harvard, 1852-63. Translator: Anatomie Generale (Bechat), 1822. Author: (1st Am. text of physiology) Outlines of Human Physiology, 1834. Performed 1st major operation (amputation) using ether, 1846; performed successful operation for vesico-vaginal fistula, 1839. Died Boston, Oct. 7, 1863.

HAYWOOD, Frederick Wardle, metallurgist; b. Kegworth N. Derby, Eng., Nov. 2, 1905; s. George William and Florence (Wardle) H.; D.L.C., Loughborough Coll., 1927; B.Sc., U. London (Eng.), 1926, Ph.D., 1932; m. Ethel Thornton, Dec. 26, 1932; children—John Thornton, Richard William. Lectr., Loughborough, Coll., 1927-34; sr. tech. staff Imperial Chem. Industries, Billingham, Eng., 1934-38; chief metallurgist Wild Barfield Electric Furnaces, Ltd., 1938-47, tech. dir., 1947—; tech. dir. Refractory Mouldings & Castings Ltd., 1951—, Watford Metal Fabrications Ltd., 1957—. External examiner metallurgy U. Surrey (Eng.), 1963—. Fellow Royal Inst. Chemistry, Inst. Metals; mem. Iron and Steel Inst., Brit. Ceramic Soc., Soc. Chem. Industry, Soc. for Analytical Chemistry, Am. Soc. Metals. Author: (with A. R. Wood) Metallurgical Analysis, 1951; also articles; contbg. author: Steels in Modern Industry, 1954. Research on methods of gaseous cementation, methods of prodn. of controlled atmosphere, metall. analysis and instrumental methods and means, vacuum fusion gas analyses, refractories; devel. new bonding methods. Home: 37, Oakroyd Av., Potters Bar, Herts. Office: Elecfurn Works, Watford By-Pass, Watford, Herts, Gt. Britain.*

HAZARD, John Beach, Am. physician; b. White Horse, Pa., Jan. 7, 1905; s. Frank Birdsall and Blanche Darrah (Stong) H.; B.S., U. Fla., 1924, M.S., 1925; M.D., Harvard, 1930; m. Etta Mae Holly, Sept. 3, 1931. Practice medicine, specializing in pathology, Cleve., 1946—; head dept. tissue pathology Cleve. Clinic Found., 1946—, chmn. div. pathology, 1957—, acting dir. div. research, 1966—; asso. pathology to clin. prof. pathology Western Res. U. Sch. Medicine, 1952—. Mem. study sect. USPHS, 1962-65. Mem. A.M.A., Am. Assn. Pathology and Bacteriology, Am. Soc. Clin. Pathology, Am. Thyroid Assn., Coll. Am. Pathologists, Internat. Acad. Pathology, A.A.-A.S., Am. Soc. Cytology, Royal Soc. Medicine, N.Y. Acad. Sci. Editor: The Thyroid, 1964. Research, numerous publs. on surg. pathology, thyroid disease, particularly thyroid neoplasia and thyroiditis. Home: County Line Rd., Box 171, Gates Mills, O. 44040. Office: 2020 E. 93d St., Cleve. 44106.*

HAZEL, Lanoy Nelson, biologist; b. Shannon, Tex., Apr. 23, 1911; s. Willis Martin and Mae (Skelton) H.; B.S., Tex. Technol. Coll., 1933; M.S., Tex. A. and M. Coll., 1938; Ph.D., Ia. State U., 1941; m. Frances Ethel Peterson, Feb. 17, 1940; children—Bonnie Lyn (Mrs. William Shoultz), Nancy Jo (Mrs. Allen Brelig), Robert Martin, Laurie Ann. Animal husbandman North Platte (Neb.) Agr. Expt. Sta., 1941-42; animal geneticist Western Sheep Breeding Lab., 1942-45; poultry geneticist Kimber Farms, Inc., Niles, Cal., 1946; prof. animal sci. dept. Ia. State U., Ames, 1947—. Recipient Morisson award Am. Soc. Animal Sci., 1960, Genetics award Am. Soc. Animal Sci., 1962. Curtiss Distinguished prof. Ia. State U., 1966. Mem. Am. Soc. Animal Sci., Genetics Soc. Am., Biometrics Soc., Am. Inst. Biol. Sci. Contbr. numerous articles to profl. jours. Research on devel. of selection indexes to maximize rates of genetic improvement, statis. methods for estimating heritabilities and genetic correlations, devel. of swine testing stas. Home: 2101 Country Club St., Ames, Ia. 50010.*

HAZELTINE, (Louis) Alan, Am. physicist, mathematician; b. Morristown, N.J., Aug. 7, 1886; s. Louis Rawson and Henrietta Maud (Ahern) H.; M.E., Stevens Inst. Tech., 1906, Sc.D., 1933; A.M., Columbia, 1938; m. Elizabeth Barrett, Jan. 16, 1931; children—Barrett, Patricia (Mrs. Alan Duhnkrack), Maud Denise (Mrs. Ansel Chaplin), Esther (Mrs. Jeffery Kramer), and Richard Deimel. Tester Gen. Electric Co., 1906-07; asst. Stevens Inst. Tech., 1907, successively instr., asst. prof., prof., prof. phys. mathematics, chmn. physics dept., 1943-44; cons. Hazeltine Corp., 1945—, dir., 1949-63. Member of the National Conference, 1922-24. Cons. radio engineer Washington Navy Yard, 1918-19; with OSRD, 1944-45. Fellow Am. Inst. E.E., Inst. Radio Engrs. (pres. 1936), A.A.-A.S.; mem. Math. Assn. Am., Radio Club of Am. (pres. 1946-47; recipient Armstrong Medal 1937), Tau Beta Pi. Author: Electrical Engineering, 1924. Inventor neutrodyne radio receiver; developed various kinds of radio apparatus; research involving operational calculus, new methods of presenting physics and math.; made 1st comml. receiver suited for gen. broadcasting reception. Address: 15 Tower Dr., Maplewood, N.J.

HAZEN, Harold Locke, engring. educator; b. Philo, Ill., Aug. 1, 1901; s. Wirt Mandeville and Elta (Brewer) H.; B.S., Mass. Inst. Tech., 1924, M.S., 1929, Sc.D., 1931; m. Katherine Pharis Salisbury, Sept. 5, 1928; children—Stanley Seamans, Martha (Mrs. William Liller), Nathan Lord, Anne (Mrs. John G. Bowen). With Mass. Inst. Tech., 1925—, prof., head dept. elec. engring., 1938-52, dean grad. sch., 1952-67, fgn. study adviser, 1967—; chief Div. 7 Nat. Defense Research Com., 1942-46; cons. engring. edn. Robert Coll. of Istanbul, Turkey, 1955, trustee, 1955, acting pres., 1961; cons. engring. edn. Am. U. of Beirut, 1957, Ministry of Edn., Iceland, 1958, U.N. Mission to U. Brasilia, 1962; mem. U. S. Naval Weapons Lab. Adv. Council, 1954-65. Recipient Levy medal of Franklin Inst., 1935, Lamme Gold medal, 1962. Fellow I.E.E.E., Am. Acad. Arts and Scis.; mem. Am. Soc. Engring. Edn., Franklin Inst., Sigma Xi, Tau Beta Pi, Eta Kappa Nu. Contbr. numerous articles in field to profl. jours. Research on automatic control, calculating machines, quantitative model of electric-power systems. Home: 81 Clark St., Belmont, Mass. 02178. Office: Mass. Inst. Tech., Cambridge, Mass. 02139.*

HAZEN, Henry Allen, Am. meteorologist; b. Serur, India (parents Am. citizens); grad. Dartmouth, 1871; student Thayer Sch. of Civil Engring. Asst. in drawing, Sheffield Sci. Sch., Yale, 1873-76, pvt. asst. to Prof. Elias Loomis in physics and meteorology, 1877-80. Joined U. S. Signal Service (now Weather Bur.), 1881, prof. meteorology, from 1891. Wrote numerous meteorol. papers. Devised sling psychrometer, 1884; established tables for reduction of barometric readings to sea-level; devised thermometer shelter; made five balloon ascensions (one to 16,000 feet) for meteorol. research. Died 1900.

HAZEN, Wayne Eskett, Am. physicist; b. Three Rivers, Mich., Feb. 8, 1914; s. Wirt M. and Elta (Brewer) H.; B.S., Mass. Inst. Tech., 1936; Ph.D., U. Cal. at Berkeley, 1941; m. Jean Mary Shearer, Aug. 19, 1939; children—Priscilla (Mrs. John Lillie), Gretchen (Mrs. Willard Crittenden), Virginia (Mrs. Andrei Weinert), Eric. Faculty, U. Cal. at Berkeley, 1941-47, asst. prof., 1945-47; faculty U. Mich., Ann Arbor, 1947—, prof. physics, 1957—. Guggenheim fellow Mass. Inst. Tech., 1947; Fulbright scholar Ecole Polytechnique, Paris, 1953-54; Guggenheim fellow Imperial Coll., London, 1954; Smith-Mundt prof. Am. U., Beirut, Lebanon, 1958-59. Fellow Am. Phys. Soc. Author: (with R. W. Pidd) Physics, 1965; also articles. Research on ionization by relativistic velocity particles, interactions of relativistic muons, cosmic-ray air showers. Home: 2117 Highland Rd., Ann Arbor, Mich. 48104.*

HAZLETON, Lloyd Walter, Am. pharmacologist; b. Chelan Falls, Wash., Feb. 4, 1911; s. William H. and Mabel (TenEyck) H.; B.S., U. Wash., 1933, M.S., 1937, Ph.D., 1939; m. Harriet L. Rosenzweig, June 25, 1939; children—H. Elaine (Mrs. Herbert Alfred Bolton), William L., Mary J. Instr. pharmacology Georgetown U. Med. Sch., 1939-40; asst. prof. George Washington U., 1940-45, asso. prof., 1945-46; profl. lectr. Georgetown U. Med. Sch., 1948-53; owner, dir. Hazleton Labs. (Name later changed to Hazleton Labs., Inc.), 1946-53, pres., dir., 1953—; pres., dir. Karloid Corp., Falls Church, Va.; director Hazleton-Carbia, Inc., Burtonsville, Md.; dir. Resources Research, Inc., Reston, Va.; v.p., dir. First Nat. Bank of Vienna (Va.). Vice-pres., trustee Fairfax (Va.) Hosp.; bd. dirs. The Epilepsy Found. Recipient Am. Pharmaceut. Assn. Ebert prize, 1946. Mem. Am. Soc. for Pharmacology and Exptl. Therapeutics, Soc. Exptl. Biology and Medicine, Am. Chem. Soc., Nat. Acad. Scis. (mem. food protection com.), Am. Indsl. Hygiene Assn., Soc. Toxicology. Contbr. chpt. on organic phosphates Vol. II Toxicology, 2d rev. edn. Industrial Hygiene and Toxicology, 1962. Contbr. numerous articles in field to sci. jours. language understandable to the lay researcher. Authority on research and testing necessary for the development of new products intended for human, agrl., vet., indsl., or mil. use. Home: 9120 Leesburg Pike, Vienna, Va. 22180. Office: P.O. Box 30, Falls Church, Va. 22046.*

HEACOCK, Ronald Arthur, chemist; b. London, Eng., Apr. 14, 1928; s. Arthur and Alice (Coombs) H.; B.Sc., U. London, 1949, Ph.D., 1952, D.Sc., 1965; m. Sheila Margaret McNulty, July 21, 1951; children—Susan Anne, Helen Jane, Anthony David. Postdoctoral research fellow div. pure chemistry NRC Can., Ottawa, Ont., 1954-56; sr. sci. officer Agrl. Research Council Wye Coll. U. London, Ashford, Kent, U.K., 1956-57; chief research biochemist psychiat. research unit Univ. Hosp., Saskatoon, Sask., Can., 1957-65; sr. research officer, head physiol. chemistry sect. Atlantic Regional Lab. NRC, Halifax, N.S., Can., 1965—; asso. prof. chemistry Faculty Grad. Studies, Dalhousie U. Mem. Chem. Soc. London, Royal Inst. Chemistry, Chem. Inst. Can. Contbr. numerous articles to profl. jours. Studies on devel. of new chromatographic procedures for analysis of indole compounds of possible physiol. importance, chemistry of aminochromes which may be of relevance to problems in psychiatry, chemistry of psychoactive compounds in gen. Home: 4 Hillwood Crescent, Clayton Park, Rockingham, N.S. Office: 1411 Oxford St., Halifax, N.S., Can.*

HEAD, Alan Kenneth, Australian physicist; b. Melbourne, Australia, Aug. 10, 1925; s. Rowland Henry and Elsie May (Burrell) H.; B.A., U. Melbourne, 1945, B.Sc., 1946; Ph.D., U. Bristol (Eng.), 1953; D.Sc., U. Melbourne, 1963; m. Gwenneth Nancy Barlow, Mar. 8, 1951. Research officer div. aeros. Commonwealth Sci. and Indsl. Research Orgn., Melbourne, 1947-50, sr. prin. research scientist div. tribophysics, 1957—; research officer Aero. Research Labs., Melbourne, 1950-51, 53-57; vis. prof. Brown U., Providence, 1961-62. Dep. chmn. structures and materials Australian Aero. Research Com., 1964—. Fellow Australian Inst. Physics, Inst. Physics (Eng.); mem. Internat. Congress on Fracture (exec. com. 1964—). Editorial bd. Internat. Jour. Fracture Mechanics, 1965—. Research, publs. on strength, fatigue and fracture of metals, dislocation theory, elasticity, electron diffraction, aspherical optical and radio systems, infra-red radiation; patentee celestial refrigerator. Home: 6 Duffryn Pl., Toorak, 3142 Victoria. Office: Commonwealth Sci. and Indsl. Research Orgn., Div. Tribophysics, U. Melbourne, Parkville, N.Z., Victoria, Australia.*

HEAD, Sir Henry, English neurologist; b. Stamford Hill, Stoke Newington, Eng., Aug. 4, 1861; s. Henry and Hester (Beck) H.; ed. U. Halle, Trinity Coll., Cambridge, 1880-84, German U. of Prague, Univ. Coll. Hosp., London; M.D., Cambridge, 1890, M.D., 1892; LL.D., Edinburgh U.; M.D. (hon.), Strassburg U.; m. Ruth Mayhew, 1904. House physician Univ. Coll. Hosp. and Victoria Park Hosp. for Diseases of Chest, 1892-94; clin. asst. County Mental Hosp., Rainhill, Liverpool, 1894; registrar London Hosp., physician, later cons. physician. Fellow Royal Coll. Physicians (Moxon medal 1897), Royal Soc., 1899. Author (with others) Studies in Neurology, 2 vols., 1920; Aphasia and Kindred Disorders of Speech, 2 vols., 1926. Distinguished between protopathic and epicritic cutaneous sensibility; also important work on aphasia and on functions of thalamus in feeling and emotion. Died Reading, Eng., Oct. 8, 1940.

HEAD, Richard Moore, Am. aero-physicist; b. Cedar Rapids, Ia., Oct. 24, 1919; s. Bernard Bryan and Lucille (Moore) H.; B.S., Cal. Inst. Tech., 1942; M.S. in Meteorology, 1942, M.S. in Aeros., 1943, Aero. Engr., 1943, Ph.D. cum laude, 1949; m. Laura Ellison, June 28, 1961; 1 son, Steven Murray. Research engr. flight test div. Lockheed Aircraft Corp., Burbank, Cal., 1943-44; faculty U. S. Naval Postgrad. Sch., Monterey, Cal., 1949-63, prof. aeros., 1951-63; asso. dir. Office Plans and Program Evaluation, NASA, Washington, 1963, dep. dir., 1964, acting dir. policy planning div., 1964-66, spl. asst. to dir. NASA Electronics Research Center, Cambridge, Mass., 1966-67, chief scientist, 1967—. Cons. to pvt. cos.; spl. lectr. Johns Hopkins, 1950-51. Douglas Aircraft scholar Cal. Inst. Tech., 1946-47. Asso. fellow Am. Inst. Aeros. and Astronautics; mem. A.A.A.S., Sigma Xi. Research, publs. on condensation phenomena in supersonic flows, plasma diagnostics, including spectrographic measurements, triggering mechanism of solar proton events; devel. of ultra-sensitive Schlieren optical system for hypersonic wind tunnels, interferometer and imageconverter camera system. Home: Jersey Lane, Manchester, Mass. 01944. Office: Code SC, NASA ERC, Cambridge, Mass. 02139.

HEADY, Earl Orel, Am. economist; b. Imperial, Neb., Jan. 25, 1916; s. Orel C. and Jessie L. (Banks) H.; B.Sc., U. Neb., 1939, M.Sc., 1940, D.Sc., 1960; Ph.D., Ia. State U., 1945; D.Sc., U. Uppsala (Sweden), 1965, Hungarian Acad. Sci., 1964; m. Marian R. Hoppert, Mar. 1, 1941; children—Marilyn (Mrs. Timothy Kling), Stephen, Barbara. Fieldman, FCA, Omaha, 1940; faculty Ia. State U., Ames, 1940—, prof. econs., 1949—, Curtiss Distinguished prof., 1956—, exec. dir. Center Econ. and Agrl. Devel., 1958—. Cons. to govt. agys., Ford Found., OECD, mem. White House Com. on Domestic Affairs, 1964-65; mem. adv. panel Dept. State, 1966—. Recipient Gamma Sigma Delta award for Outstanding Contbn. to Agr., 1962, Nat. award for outstanding and distinguished research on behalf Am. agr. Am. Assn. Agrl. Editors, 1965, also 5 awards for books pub. Fellow Am. Farm Econs. Assn. (Outstanding Research awards 1949, 52, 66), Econometric Soc., Am. Statis. Soc., A.A.A.S.; mem. Am. Econ. Assn., Internat. Agrl. Econ. Assn., Canadian Agrl. Econ. Soc. Author: Economics of Agricultural Production and Resource Use, 1952; Farm Management Economics, 1954; Agricultural Production Functions, 1960; Linear Programming Methods, 1958; Farm Records and Accounting, 1949; Agricultural Policy under Economic Development, 1962; Resource Structure and Demand in the Agricultural Industry, 1963; Roots of the Farm Problem, 1965; Policies and Problems of Developed Countries, 1966; Food, Agriculture and Public Policy, 1967; also articles. Developed and applied theories in solving econ. problems and devel. of agr. at levels of farm, region and nat. policy; devel. models for interdisciplinary research among economists and natural scientists in agrl. problems. Home: 919 Gaskill St., Ames, Ia. 50010.*

HEADY, Harold Franklin, Am. ecologist; b. Buhl, Ida., Mar. 29, 1916; s. Orah E. and Edith A. (Philbrick) H.; B.S., U. Ida., 1938; M.S., N.Y. State Coll. Forestry, 1940; postgrad. U. Minn., 1940-41; Ph.D., U. Neb, 1949; m. E. Eleanor Butler, June 12, 1940; children—Carol Marie, Kent Arthur. Range conservationist U. S. Soil Conservation Service, White Salmon, Wash., 1941; asst. prof. N.Y. State Coll. Forestry, Syracuse, N.Y., 1942; asst. prof. Mont. State U., Bozeman, 1942-47; asso. prof. Tex. A. and M. U., College Station, 1947-51; faculty U. Cal. at Berkeley, 1951—, prof. forestry, 1962—. Pasture cons. FAO,

Saudi Arabia, 1962-63. Fulbright Research scholar, East Africa, 1958-59, Australia, 1966; Guggenheim fellow, 1958-59. Mem. Ecol. Soc. Am., Am. Soc. Range Mgmt. Research and numerous publs. on ecol. relationships and methods sampling grasslands and deserts, influence of domestic and wild animals on vegetation, range mgmt. Home: 1864 Capistrano Av., Berkeley, Cal. 94707.*

HEAF, Frederick R. G., Brit. physician; b. Desborough, June 21, 1894; s. Julius and Alice (Beavon) H.; ed. Cambridge U., St. Thomas's Hosp., London; M.A., M.D.; m. Madeleine Stanmore, 1920; children—Peter, Frederic, Mary. Med. supt. Colindale Hosp., London, 1930; higher med. officer London County Council, 1936; Tb cons. Ministry Pub. Health, Office of Colonies and Council of Welsh Hosp., 1947; prof. U. Wales, 1949, prof. emeritus, 1960. Fellow Royal Coll. Physicians; mem. Royal Coll. Surgeons, Royal Geog. Soc., Gt. Britain, Turkey, Swedish, Salonika assns. against Tb. Author: Recent Advances in Tuberculosis; Symposium on Tuberculosis, also numerous articles. Home: 141 Lake Rd. W. Office: Welsh National School of Medicine, Cardiff, Wales, Gt. Britain.

HEALD, Frederick de Forest, Am. botanist; b. 1872; ed. U. S.D., U. Wis. Prof. biology Parsons Coll., Fairfield, Ia., 1897-1903; adj. prof. plant physiology U. Neb., 1903-05; asso. prof. botany, botanist Neb. Exptl. Sta., 1905-06, prof. agrl. botany, botanist, 1906-08; head Sch. Botany, U. Tex., 1908-12; pathologist Pa. Chestnut Tree Blight Commn., U. Pa., also agt. forest pathology U. S. Dept. Agr., 1915-17; head dept. plant pathology Wash. State Coll., Pullman, 1917—; collaborator Bur. Plant Industry, Washington, 1905-08, 15—, expert, 1909-10. Fellow A.A.A.S.; mem. Am. Phytopath. Soc., Am. Microscop. Soc., Bot. Soc. Am., Am. Forestry Soc. Asso. editor Phytopathology, 1911-16, 19-21, 31—. Author: Laboratory Manual in Elementary Biology, 1902; Symptoms of Disease in Plants, 1909; (with I. M. Lewis) Experiments in Plant Physiology; Manual of Plant Diseases, 1926; Introduction to Plant Pathology, 1937. Died Apr. 24, 1954.

HEALD, Henry Townely, Am. civil Engr.; b. Lincoln, Neb., Nov. 8, 1904; s. Frederick De Forest and Nellie (Townley) H.; B.S., State Coll., Wash., 1923; M.S., U. Ill., 1925; D.Eng., Rose Poly. Inst., 1942, Clarkson Coll. Tech., 1948; LL.D., Northwestern U., 1942, Rutgers U., 1952, Columbia, 1954. Hofstra Coll., 1955, Fairleigh Dickinson Coll., 1956, Princeton, U. Pitts.; D.C.L., N.Y. University, 1956; L.H.D. (hon.), Rollins College, 1953; D.Sc. (honorary), Pratt Institute, 1954, Newark College of Engring., 1954, Union Coll., 1956; m. Muriel Starcher, Aug. 4, 1928. Asst. engr. U. S. Bur., Reclamation, Pendleton, Ore., 1923-24; designer, bridge dept., I.C. R.R., Chgo., 1925-26; structural engr., bur. design, Bd. of Local Improvements, Chgo., 1926-27. Asst. prof. civil engring. Armour Inst. Tech., 1927-31, asso. prof., 1931-34, prof., 1934-40, asst. dean, 1931-33, dean of freshmen, 1933-34, dean of the Coll., 1934-38, pres., 1938-40; pres. Ill. Inst. Tech., Armour Research Found., and Inst. Gas Tech., 1940-52; chancellor N.Y.U., 1952-56; president and director The Ford Found., N.Y.C., 1956—. Regional adviser E.S.M.W.-T.P., U. S. Office Edn., 1940-45; adviser Bur. Naval Personnel, U.S.N., 1942-49; cons. War Manpower Commn., 1942-45; mem. Chem. Corps Adv. Bd., 1945-49; dep. chmn. com. on equipment and supplies Research and Development Bd., 1949-53; dir. Equitable Life Assurance Soc. of U. S. Stewart-Warner Corporation, Swift & Company, American Telephone & Telegraph Company. Recipient Washington award, 1952. Vice pres. for edn. Chgo. Assn. Commerce and Industry, 1946-49, dir., 1949-52; chmn. Mayor's Com. (to reform) Chgo. Bd. Edn., 1946; trustee John Crerar Library, 1945-52; commr. Nat. Commn. on Accrediting, 1950-56; chmn. N.Y. State Com. on Higher Edn., 1960. Mem. Council Fgn. Relations, New York Academy of Sciences, American Ordnance Assn., Am. Soc. Engring. Edn. (pres. 1942-43), A.A.A.S. (v.p. 1946-47), Am. Pub. Works Assn., Am. Soc. C.E., Am. Soc. M.E., Western Soc. Engrs. (pres. 1945-46), Tau Beta Pi, Sigma Tau, Phi Kappa Phi, Chi Epsilon, Theta Xi, Pi Tau Sigma (hon.), Phi Beta Kappa (hon.); hon. asso. A.I.A. Office: 477 Madison Av., N.Y.C. 22.

HEALD, Peter Joseph, Brit. biochemist; b. Alderley Edge, Cheshire, Eng., July 26, 1925; s. Thomas Hugh and Maud (Lofthouse) H.; grad. with 1st class honors in chemistry, U. Manchester, 1945, B.Sc., M.Sc., Ph.D., D.Sc.; m. Kathleen Homer, Sept. 22, 1951; children—Margaret Anne, Richard Peter, Thomas Jonathan Homer. Research fellow Faculty Tech., U. Manchester, 1945-46, demonstrator, 1946-48; sci. officer Rowett Research Inst., Bucksburn, Aberdeenshire, 1948-52; faculty Maudesley Hosp., London, Eng., 1952-61; vis. asso. prof. Mayo Meml. Clinic, 1959; head dept. animal biochemistry Twyford Labs., London, 1961-66; prof. biochemistry U. Strathclyde, Glasgow, Scotland, 1966—. Mem. Biochem. Soc., Soc. Endocrinology, Soc. Reprodn. and Fertility. Author: Phosphorus Metabolism of Brain, 1961. Editor: Biochem-Biophys. Acta. Research, publs. on digestive process in sheep, brain function particularly in relation to artificial epileptic type stimuli, control in reprodn. in fowl and mechanisms of hormone action. Office: Dept. Biochemistry, U. Strathclyde, Glasgow C-2, Scotland.*

HEALEY, John Edward, Jr., Am. anatomist; b. Bristol, Pa., Oct. 27, 1922; s. John Edward and Anne (Sharkey) H.; student U. P.R., 1940-42; B.S., St. Joseph's Coll., Phila., 1943; M.D., Jefferson Med. Coll., 1948; m. Kathryn Delores Byrnes, Apr. 17, 1948; children—John Edward III, Ellen, Mary, Kathryn, Michael, Daniel, Margaret, Edward, Joseph, Martin. Faculty, Jefferson Med. Coll., 1946-53; dir. research Price Diagnostic Clinic, Phila., 1953-57; dir. med. edn. St. Joseph's Hosp., Reading, Pa., 1957-58; asso. exptl. surgeon M.D. Anderson Hosp. and Tumor Inst., U. Tex., Houston, 1958—, asso. prof., 1959-65, prof. anatomy, 1965—. Recipient Schaeffer Anatomic League prize, 1945. Mem. Jefferson Soc. Clin. Investigation (pres. 1953), A.M.A., Am. Assn. Anatomists, Am. Soc. Artificial Internal Organs, So. Soc. Anatomists, Internat. Coll. Surgeons, N.Y. Acad. Scis., Am. Coll. Sports Medicine, Am. Geriatrics Soc., A.A.A.S. Research and publs. on resection for bronchial carcinoma; intrahepatic vascular anatomy, technique for resection; liver tumors; bowel anastomosis; surg. instruments. Home: 5075 Fieldwood St., Houston 77027.*

HEALY, William, physician, psychologist; b. Buckinghamshire, Eng., Jan. 20, 1869; s. William and Charlotte (Hearne) H.; brought to America in childhood; student Harvard Coll. and Med. Sch.; A.B., Harvard, 1899; M.D., Rush Red. Coll. (U. of Chicago), 1900; post-grad. work, Vienna, Berlin and London, 1906-07; m. Mary Sylvia Tenney, May 12, 1901; 1 son, Kent Tenney; m. 2d, Augusta F. Bronner, 1932. Physician, Wisconsin State Hospital, 1900-01; instructor gynecology, Northwestern Univ. Medical Sch., 1901-03; instr. and asso. prof. nervous and mental diseases, Chicago Policlinic, 1903-16; dir. Psychopathic Inst., Juvenile Court, Chicago, 1909-17; dir. Judge Baker Guidance Center, Boston, 1917-46; director emeritus since 1946; professor and research associate, Institute of Human Relations, Yale, 1929-32. Psychiatrist on American Law Institute. Youth-Justice Com., 1938-40. Lecturer Harvard U. and Boston U. Chmn. trustees Boston Psychopathic Hosp. Fellow Am. Acad. Arts and Sciences; mem. A.M.A., Am. Psychiatric Assn., Mass. Med. and Psychiatric socs., Am. Inst. Criminal Law and Criminology, Boston Soc. of Psychiatry and Neurology (pres.), Am. Neurol. Assn., Am. Ortho-psychiatric Assn. (pres.), Am. Psychopathol. Assn. (pres.); mem. St. Louis Med. Soc. (hon.) Ethical Culture Soc. Author: Case Studies of Mentally and Morally Abnormal Types (Harvard Univ. Press), 1912; The Individual Delinquent—A Textbook of Diagnosis and Prognosis, 1915; (with Mary Tenney Healy) Pathological Lying, Accusations, and Swindling, 1915; Honesty, 1915; Mental Conflicts and Misconduct, 1917; Judge Baker Foundation Case Studies (with A. F. Bronner), 1923; Delinquents and Criminals—Their Making and Unmaking (with same), 1926; Manual of Individual Mental Tests (with others), 1927; (with others) Reconstructing Behavior in Youth, 1929; (with others) The Structure and Meaning of Psychoanalysis, 1930; (with F. Alexander) Roots of Crime, 1935; (with A. F. Bronner) New Light on Delinquency and Its Treatment, 1936; Personality in Formation and Action (Salmon Memorial lectures), 1937; (with A. F. Bronner) Treatment and What Happened Afterward, 1940; (with B. S. Alper) Criminal Youth and the Borstal System, 1941. Pioneer in establishing guidance clinics in U. S. for problem children and young people, and in use of case study method in criminolcgical studies. Died Mar. 1962.

HEARD, Harry Gordon, Am. physicist; b. Reines, Tenn., Sept. 23, 1922; s. Pascal Harrison and Cliffie (Page) H.; B.S. in Elec. Engring. with honors, U. Cal. at Berkeley, 1950, M.S., 1951, postgrad. in Physics; m. Allison Norcross, Aug. 27, 1946; children—Pamela Suzanne, Todd Addison Crandall. Applied physicist Lawrence Radiation Lab., U. Cal. at Berkeley, 1950-59; chief engr. Levinathal Electronics, Palo Alto, 1959-61; v.p., tech. dir. Radiation at Stanford, Palo Alto, 1961-63, Energy Systems, Palo Alto, 1963-65; v.p., sec., dir. research hnu Systems div. Ohio Steel, Palo Alto, 1965—; cons. engr., Woodside, Cal., 1950—. Mem. I.E.E.E., Sci. Research Soc. Am. (pres. Sequoia br. 1966), Am. Phys. Soc., Optical Soc. Am. Author: (with C. Susskind, others) Electronics Encyclopedia, 1963; Laser Parameter Measurements, 1966; also numerous articles. Pioneered computer-controlled proton synchrotron, multiple beam control and target facility to obtain 6-fold increase in utility proton synchrotron, extended beam and constant energy pulsed particle accelerator, ion laser used in Ascension Island to earth-satellite communication; discovered 1st room temperature ultraviolet molecular gas laser. Home: 50 Skywood Way, Woodside, Cal. 94062. Office: 470 San Antonio Rd., Palo Alto, Cal. 94306.*

HEASLET, Maxwell Alfred, Am. mathematician; b. Bentonville, Ark., Feb. 17, 1907; s. Walter M. and Nannie (Austin) H.; B.A., U. Okla., 1927, M.A., 1929; Ph.D., Stanford, 1934; m. Helen Virginia Camp, June 15, 1935; children—Gary Austin, Jonathan Lee. Faculty, U. Okla., 1927-29, Stanford, 1929-34, 46-56; faculty San Jose State Coll., 1935-42; research scientist NASA, Moffett Field, Cal., 1942—; head theoretical br. Ames Research Center, 1947—. Mem. A.A.A.S., Am. Math. Assn., Am. Phys. Soc., Am. Geophys. Union, Sigma Xi. Author: (with J. V. Uspensky) Elementary Number Theory, 1939. Research and numerous publs. primarily in field of math. analysis of integral and differential equations. Home: 1501 Camino del Rio, Vero Beach, Fla. 32960. Office: Ames Research Center, Moffett Field, Cal.*

HEATH, Charles Joseph, English physician; b. Totnes, Devon, Eng., Dec. 25, 1856; s. John Heath; m. Agnes Fridzwede Wilson (dec. 1930); 2 daus. Former surgeon Throat Hosp.; asst. surgeon Central London Ear and Throat Hosp.; hon. cons. aural surgeon London County Ct. Mem. exec. com. Nat. Inst. for Deaf. Fellow Royal Coll. Surgeons (Eng.), Royal Soc. Medicine (mem. otol. and laryngol. sects.), Hunterian, Brit. Otolaryngol. socs. Devised mastoid operation for restoring hearing; designer over 100 surg. instruments for ear, throat, nose diseases; inventor chamberless wildfowling gun; defined principles necessary in designing army boots. Died July 13, 1934.

HEATH, Christopher, English surgeon; b. London, Mar. 13, 1835; ed. King's Coll. Sch., King's Coll. and Hosp., London. Became demonstrator anatomy Westminster Hosp., 1856, lectr. anatomy, asst. surgeon, 1862; named asst. surgeon, tchr. operative surgery U. Coll. Hosp., 1866, named Holme prof. clin. surgery, surgeon, 1875, emeritus prof. clin. surgery, cons. surgeon, 1900. Mem. Clin. Soc. London (pres. 1890-91), Coll. Surgeons (became pres. 1895). Author: A Course of Operative Surgery, 2d. edit., 1884; Manual of Minor Surgery and Bandaging for Use of House Surgeons, Dressers, and Junior Practitioners, 12th edit., 1901; Clinical Lectures on Surgical Subjects, 2d series, 1902; Practical Anatomy—a Manual of Dissections, 9th edit. 1902; Injuries and Disease of the Jaws, 4th edit. 1894; On the Treatment of Intrathoracic Aneurism by the Distal Ligature, 1871; The Student's Guide to Surgical Diagnosis, 2d edit. 1883. Died Aug. 8, 1905.

HEATH, Donald Albert, English pathologist; b. Henley-on-Thames, Oxfordshire, Eng., May 4, 1928; s. Edward and Florence (Holloway) H.; M.B., Ch.B., M.R.C.S., L.R.C.P., U. Sheffield, 1952, M.D., 1956, M.R.C.P., MC Path., 1963; Ph.D., U. Birmingham, 1959. Leverhulme research fellow Royal Coll. Physicians, 1954-56; lectr. pathology Birmingham U., 1956, sr. lectr. pathology, 1960-64, reader, 1964—; Rockefeller Travelling fellow in medicine Brit. Med. Research Council, Mayo Clinic, St. Paul, 1957-58. Mem. Royal Coll. Physicians, Thoracic Soc., Brit. Cardiac Soc., Path. Soc. Gt. Britain and Ireland. Author: (with P. C. Harris) The Human Pulmonary Circulation, 1962; (with C. A. Wagenvoort, J. E. Edwards) The Pathology of the Pulmonary Vasculature, 1964. Contbr. numerous articles to sci. jours. Demonstration of effects of diseases of arteries and lungs on clin. picture of patients with heart and lung disease and on surg. treatment of congenital heart disease; elucidation of relation between form and function of arteries; studies on microscopic structure of normal and path. human pulmonary circulation. Home: High Hall, Church Rd., Edgbaston, Birmingham. Office: Dept. Pathology, Birmingham U. Med. Sch., Eng.*

HEATH, Douglas Hamilton, Am. psychologist; b. Woodbury, N.J., Oct. 1, 1925; s. Russell M. and Eleanor (Conrow) H.; student Swarthmore Coll., 1943-44; B.A. summa cum laude, Amherst Coll., 1949; M.A., Harvard, 1952, Ph.D., 1954; postgrad. U. Mich., 1957-58; m. Harriet Elizabeth Frye, June 15, 1952; children—Russell, Wendilee, Anne Marie. Clin. psychology trainee, VA, 1950-54; psychologist Newton (Mass.) Counseling and Guidance Center, 1951-52; instr. Harvard, 1953-54; asst. prof. psychology Haverford (Pa.) Coll., 1954-59, asso., 1959-65, prof., 1965—, chmn. psychology, 1959-65, 67-68. Research cons. Acad. Religion and Mental Health, 1966—. NSF Faculty fellow, 1957; Fulbright traveling fellow, 1965-66. Mem. Am., Eastern psychol. assns., Soc. for Projective Techniques, Acad. Religion and Mental Health, Phi Beta Kappa, Sigma Xi (asso.). Author: Explorations of Maturity: Studies of Mature and Immature College Men, 1965. Research and publs. on determinants of maturing in coll. men, theory of mental health in principle religiocultural areas of world. Home: 21 Matlack Lane, Villanova, Pa. 19085. Office: Sharpless Bldg., Psychology Dept., Haverford Coll., Haverford, Pa. 19041.*

HEATH, Fred H(arvey), Am. chemist; b. Warner, N.H., Feb. 25, 1883; s. Benjamin Franklin and Julia Augusta (Wadleigh) H.; B.S., U. N.H., 1905; Ph.D., Yale, 1909; postgrad. U. Marburg (Germany), 1909; m. Winnifred A. Grant, Apr. 20, 1911 (dec. 1918); 1 son, Frank Harvey; m. 2d, Mrs. Ida M. Erickson, Sept. 7, 1921 (dec. 1928); m. 3d, Mrs. Errah Shannon Schindler, June 8, 1932. In chem. analysis of waters and water problem research N.D. State Biol. Sta., 1914-16; investigation on platinum ores of Washington, 1921-22. Instr. phys. chemistry Mass. Inst. Tech., 1909-10; instr. gen. chemistry Case Sch. of Applied Sci., 1910-11; instr. Wesleyan U. Conn., 1911-12; instr., asst. prof. U. N.D., 1912-17; asst. prof. U. Wash., 1917-23; prof. U. Fla., from 1923. Mem. Am. Chem. Soc. (pres. Fla. sect. 1943), Fla. Acad. Sci., Alpha Chi Sigma, Phi Lambda Upsilon, Gamma Sigma Epsilon. Author: Laboratory Manual of Quantitative Analysis, 1910, 1921-22, Laboratory Manual of General Chemistry (with W. H. Beisler),

1926, 1934, General Chemistry Text Book (with others), 1926, 2d edit., 1927; also chem. research articles in profl. jours. Invented selenium mustard gas, 1918; did spl. work in photography. Died Jan. 26, 1952.

HEATH, Russell La Verne, Am. physicist; b. Denver, June 13, 1926; s. Robert L. and Hazel (Gibbs) H.; B.S., Colo. State U., 1949; postgrad. Rutgers U., 1949-50; M.S. (AEC fellow), Vanderbilt U., 1951; m. Edna Neel, Aug. 4, 1949; children—Deborah Anne, Robert Louis. Staff, Oak Ridge Nat. Lab., 1951-53; sr. radiol. physicist Am. Cyanamid Co., 1953-54; group leader Phillips Petroleum Co., Nat. Reactor Testing Sta., 1954-56, physics sect. chief, 1956-66; with Ida. Nuclear Corp., 1966-—; asso. prof., collaborator Utah State U., 1965. Fellow Am. Phys. Soc.; mem. I.E.E.E., Am. Nuclear Soc. Contbg. author: Handbook of Physics and Chemistry, 1964. Research and numerous publs. on devel. of scintillation and solid state detectors for use in basic and applied gamma-ray spectroscopy, devel. multi-channel pulse-height analyzers, electronic equipment used by nuclear spectroscopists; developed computer techniques for analysis gamma-ray spectra. Home: 1620 Charlene St., Office: P.O. Box 1845, Idaho Falls, Ida. 83401.*

HEATH, Sir Thomas Little, English math. historian; b. Barnetly-le-wold, Eng., 1861; s. Samuel and Mary (Little) H.; ed. Clifton and Trinity Coll., Cambridge, Eng.; m. Ada Mary Thomas, 1914; 1 son, 1 dau. Fellow Royal Soc., 1912. Author: The Thirteen Books of Euclid's Elements, 1908; A History of Greek Mathematics, 1921; Greek Astronomy, 1932; others. Died 1940.

HEAVISIDE, Oliver, English math. physicist; b. Camdentown, Eng., May 18, 1850; s. Thomas and Rachel (West) H.; self-taught; Ph.D. (hon.) Göttingen (Germany), U. Telegrapher, Gt. No. Telegraph Co., 1870-74; independent studies, 1876-89; accepted Civil List Pension, 1896. Recipient Faraday medal Instn. Elec. Engrs. Fellow Royal Soc., 1891. Author: Electrical Papers, 1892; Electromagnetic Theory, 1893-94; also articles. Developed duplex telegraphy, 1873; research on theoretical electromagnetism, especially formulation of expansion theorem, also discovery of distortionless transmission by proper inductive loading, 1885-87; predicted existence of ionized air layer and its effect on radiotransmission (Kennelly-Heaviside layer), 1902. Died Torquay, Eng., Feb. 3, 1925.

HEBB, Donald Olding, Canadian psychologist; b. Chester, N.S., Can., July 22, 1904; s. Arthur Morrison and Clara (Olding) H.; B.A., Dalhousie U., 1925, LL.D., 1965; M.A., McGill U., Montreal, Que., Can., 1932; Ph.D., Harvard, 1936; D.Sc., U. Chgo., 1961, U. Waterloo, 1963; D.H.L., Northeastern U., 1963; D.Sc., York U., 1966; McMaster U., 1967; LL.D., Dalhousie U., 1965, Queen's University, 1965; m. Marion Isabel Clark, 1931 (dec. 1933); m. 2d, Elizabeth Nichols Donoran, 1937 (dec. 1962); children—Jane Nichols (Mrs. Ronald Larry Paul), Mary Ellen; m. 3d, Margaret Doreen Williamson, Aug. 31, 1966. Prof., McGill U., 1947-—. Chmn. Human Resources Research Adv. Com., Ottawa, Ont., 1951-52; chmn. asso. com. exptl. psychology NRC, Ottawa, 1956-62. Recipient Coronation medal Canadian Psychol. Assn., 1953, Warren medal Soc. Exptl. Psychologists, 1958. Claude Bernard medal U. Montreal Inst. Exptl. Medicine, 1966. Mem. Am. (past pres.), Canadian (past pres.) psychol. assns., A.A.A.S., Brit. Psychol. Soc., Societe Francaise Psychology, Sociedad Española de Psicologia, Am. Acad. Arts and Scis., Royal Soc. Can., Royal Soc. London, Sigma Xi. Author: Organization of Behavior, 1949; Textbook of Psychology, 1958, 66. Contbr. numerous articles profl. jours. Studies of visual perception, emotion, intelligence; effects of brain damage in man and animal; effects of sensory deprivation on young animal and adult human beings. Office: Dept. Psychology, McGill U., Montreal 2, Que., Can.*

HEBERDEN, William, English physician; b. London, Aug. 1710; s. Richard Heberden; B.A., St. John's Coll., Cambridge, Eng., 1728, became sr. fellow, 1749, M.D., 1739; m. Elizabeth Martin, 1752; m. 2d, dau. of William Wollaston, 1760; 9 children, including Thomas, William. Began practice, London, 1748; declined position of physician to Queen Charlotte, 1761; became Gulstonian lectr., 1749, Harveian orator, 1750; attended Johnson, Cowper, Warburton. Fellow Royal Coll. Physicians, Royal Soc., 1749; hon. mem. Royal Soc. Medicine (Paris). Author: Commentarii de morborum historia et curatione (Commentaries), pub. 1802, translated into English, 1803; also articles. First description of angina pectoris; described and differentiated chickenpox from smallpox, 1764; described bony nodules on fingers in osteoarthritis (Heberden's nodes), 1754. Died London, May 17, 1801.

HEBERER, Gerhard, German geneticist and anthropologist; b. Halle, Germany, Mar. 20, 1901; s. Richard H. and Hedwig (Quedenfeld) H.; univ. studies in zoology, genetics and anthropology; m. Gisela Fischer, Aug. 12, 1929; children—Gudrun, Sunhilt, Wolfran, Sigrun. Explored Indonesia, 1927-28; asst. zool. lab., Buitenzorg, Java, 1928; asst. Zool. Inst., U. Tübingen, 1929-35; prof. zoology U. Frankfurt U., 1935-36; prof. biology and anthropology U. Jena, 1938; prof. anthropology U. Göttingen, 1947-—. Recipient medal Nat. Sci. Soc. Jena. Mem. Zool. Soc., German Anthrop. Soc., Soc. German Natural Sci. and Physicians. Author: Rassengeschichtliche Forschungen im indogermanischen Heimatgeblet, 1943; Allgemeine Abstammungslehre, 1949; Die Inlandmalaien von Lombok und Sumbawa, 1950; Neuete Ergebnisse der menschlichen Abstammungslehre, 1951. Research on cytology, especially of chromosomes; studies on evolutionary theory; paleoanthropology of S.E. Asia. Address: 21 Riemannstrasse, Göttingen, Germany.

HÉBERT, Edmond, French geologist; b. Villefargeau, France, June 12, 1812; ed. École normale supérieure; aggregation, 1840, docteur ès scis., 1857; named prof. École normale supérieure, 1833; became prof. geology Paris Faculty Scis., 1857, dean, 1885. Mem. French Acad. Scis. Creator new method of geol. observation; 1st to show how facies may vary for same geol. period; revealed presence of Archean in Brittany; studied ancient riverbeds in Paris regions. Died Paris, Apr. 4, 1890.

HEBERT, Teddy Theodore, Am. plant pathologist; b. Lafayette, La., Nov. 24, 1914; s. Frank and Leocadie Hebert; B.S., U. Southwestern La., 1938; M.S., La. State U., 1940; Ph.D., U. N.C., 1946; m. Nell Heidelberg, Sept. 16, 1943; children—Stephen William, Teddy Thomas, Linda Nell, Michael James. Faculty, N.C. State U., Raleigh, 1946-—, prof. plant pathology, 1957-—. Fulbright lectr. U. Alexandria (Egypt), 1963. Mem. Am. Phytopath. Soc., N.C. Acad. Sci., Sigma Xi, Phi Kappa Phi, Alpha Zeta. Contbr. numerous articles to profl. jours. Developed method for controlling leaf spot diseases of peanuts; originated anaerobic treatment for control of smut diseases of wheat and barley; co-developer disease resistant varieties of wheat, oats and barley; introduced polyethylene glycol precipitation for purification of viruses. Home: 2703 Clark Av., Raleigh, N.C. 27607.*

HEBRA, Ferdinand Ritter von, see von Hebra, Ferdinand Ritter.

HECHT, Adolph, Am. botanist; b. Chgo., July 25, 1914; s. Mannassa and Bertha (Friedmann) H.; B.S., U. Chgo., 1936, M.S., 1937; Ph.D., Ind. U., 1942; m. Edith Goldstein, July 19, 1942; children—Anton L., Julia A. Field asst. No. Rocky Mountain Forest and Range Expt. Sta., U. S. Forest Service, 1937-38; instr. botany U. Chgo., 1946-47; faculty Wash. State U., Pullman, 1947-—, prof. botany, 1957-—, chmn. dept., 1955-—, chmn. genetics staff, 1963-65. Mem. Bot. Soc. Am., N.W. Sci. Assn. (past pres.), A.A.A.S. (pres. Pacific div. 1966-67), Am. Inst. Biol. Scis., Genetics Soc. Am., Am. Genetic Assn., Western Soc. Naturalists, Phi Beta Kappa, Sigma Xi. Contbg. author: Pollen Physiology and Fertilization, 1964. Research, publs. on evolutionary trends in Oenothera, subgenus Raimannia using cytogenetic analyses; developed procedures for inactivating self-incompatibility mechanisms in stylar tissues. Home: 306 Derby St., Pullman, Wash. 99163.*

HECHT, Hans H., cardiologist; b. Basel, Switzerland, Jan. 23, 1913; s. Hans P. and Hanna (Meinhold) H.; M.D., U. Berlin, 1936, U. Utah, 1946; M.S., U. Mich., 1942; m. Ilse Wagner, Nov. 8, 1937; children—Hannelore, F. Thomas, Susan B. Came to U. S., 1937, naturalized, 1941. Faculty, U. Utah, 1944-46, successively instr., asst. prof., asso. prof. medicine, 1944-58, prof. medicine, 1958-64, Viko prof. cardiology, 1957-64; prof. medicine and physiology U. Chgo., 1964-—, co-chmn. dept. cardiology, 1964-—, chmn. dept. medicine, 1966-—. Diplomate Am. Bd. Internal Medicine (mem. exam. bd. 1961-66), Nat. Bd. Med. Exam. Fellow A.C.P., Am. Coll. Chest Physicians, N.Y. Acad. Sci., Royal Soc. Medicine; mem. Am. (exec. com., sci. council 1953-57), Utah (pres. 1956), Ill., Chgo. heart assns., Western, (pres. 1953, Central socs. clin. research, Chgo. Inst. Medicine, Am. Physiol. Soc., Am. Soc. Clin. Investigation, Assn. Am. Physicians, Am. Fedn. Clin. Research, Western Assn. Physicians (councillor 1957-60, 63-65), A.M.A. Author monographs. Editorial bd.: Circulation, Diseases of the Chest, Excerpta Medica, Mallatti Cardiovascular. Contbr. articles tech. lit. Research on electrocardiography; cardiovascular diseases; circulation; physiology; internal medicine. Home: 1044 E. 49th St., Chgo. 60615.*

HECK, Nicholas Hunter, Am. hydrog. and geodetic engr.; b. Heckton Mills, Pa., Sept. 1, 1882; s. John Lewis and Mary Frances (Hays) H.; B.A., Lehigh, 1903, C.E., 1904, D.Sc., 1929; D.Sc., Fordham U., 1941. With U. S. Coast Survey, 1904-45; in charge wire drag parties, Atlantic Coast, 1906-16, comdr. schooner Mathlcess, 1917; lt. and lt. comdr. USNRF, New London, Conn., London, Eng., 1917-19, in charge location of submerged forest in Lake Washington, nr. Seattle, 1919-20; comdr. steamer Explorer, 1920-21; chief dir. geomagnetism and seismology Coast and Geodetic Survey, 1924-42, asst. to dir., 1942-45, ret. 1945. Recipient William Bowie medal, Am. Geophys. Union, 1942. Fellow A.A.A.S., Am. Geog. Soc., Wash. Acad. Sci., Am. Geophys. Union (chmn. 1935-38), Geol. Soc. Am; mem. Am. Soc. C.E. Seismol. Soc. Am. (pres. 1936-39), Soc. Am. Mil. Engrs., Internat. Seismol. Assn. (pres. 1936-45), Tau Beta Pi, Phi Beta Kappa, Sigma Xi. Author: Earthquakes; also govt. publs. concerning wire drag and sweep work of Coast and Geodetic Survey; compensation of magnetic compass; velocity of sound in sea water; radio acoustic method of determining position in hydrography; earthquake history of U. S. and articles relating to magnetism and seismology. Died Dec. 21, 1953.

HECKE, Erich, mathematician; b. Buck/Posen, Sept. 20, 1887; prof., Basel, Switzerland, also Göttingen, Hamburg (both Germany). Author: Vorlesungen über die Theorie der algebraischen Zahlen, 1923. Fundamental research in number theory. Died Copenhagen, Denmark, Feb. 13, 1947.

HECKEL, Édouard-Marie, French botanist; b. Toulon, France, Mar. 24, 1843; named prof. Faculty Scis., Marseille, France, 1875; became prof. Sch. Medicine, Marseille, 1877; visited Antilles, 1861, then Australia; became prof. natural history, Nancy, France, 1874. Mem. French Acad. Scis. Founder 1st colonial mus. in France, at Marseille; studied medicinal and oil-yielding plants from French colonies. Died Marseille, Jan. 20, 1916.

HECKHAUSEN, Heinz, German psychologist; b. Wuppertal, Germany, Mar. 24, 1926; s. Max and Hedwig (Steinhoff) H.; Diploma in Psychology, U. Münster, 1951, Dr. phil., 1954; m. Christa Kraneburg, Oct. 24, 1954; children—Jutta, Dorothee, Cordula, Felix. Asst., Psychol. Inst., U. Münster (Germany), 1953-63, docent, 1962-64; prof. psychology Ruhr U., Bochum, Germany, 1964-—, head adv. bd. Counseling Center, 1966-—. Mem. Deutsche Gesellschaft für Psychologie, Am. Psychol. Assn. (fgn. affiliate). Author: Hoffnung und Furcht in der Leistungsmotivation, 1963: The Anatomy of Achievement Motivation, 1967. Research and publs. on assessment of different tendencies in achievement motivation, analysis of achievement-motivated behavior and experience, devel. of achievement motivation, personality assessment by content analysis. Home: 7 Äskulapweg, Bochum, 463, Germany.*

HECKMANN, Otto Herman Leopold, German astronomer; b. Opladen, Germany, June 23, 1901; s. Max C. and Agnes (Gruter) H.; Dr.phil., U. Bonn (Germany), 1925, Dr.h.c., U. Marseille (France), 1966; m. Johanna Topfmeier, Feb. 2, 1926; children—Klaus H., Hildegard (Mrs. Lamersdorf), Ulrike. Asst., Bonn Obs., 1925-27; faculty Göttingen (Germany) Obs., 1927-35, asso. prof., 1935-41; prof. astronomy U. Hamburg (Germany), 1941-—, dir. Hamburg Obs., 1941-62; dir. European So. Obs., Hamburg, also Santiago de Chile, 1961-—. Recipient Watson medal Nat. Acad. Sci. Washington, 1963, Janssen medal Soc. Astron. de France, Bruce medal Astron. Soc. Pacific, San Francisco, 1964. Mem. Deutsche Akademie der Wissenschaften; Royal Astron. Soc., London; Royal Sci. Soc., Upsala; Royal Physiographic Soc., Lund. Author: Theorien der Kosmologie, 1942; also numerous articles. Determined precise star positions and proper motions, steller brightness in star clusters, orgn. large scale internat. collaboration, statis. dynamics of stellar systems, cosmological theories. Home: 12 von Anckeln Str., Hamburg, Bergedorf, West Germany.*

HECQUET, Philippe, French physician; b. Abbeville, France, Feb. 11, 1661; grad. medicine, Rheims, France, 1684; M.D., Paris, 1694. Became tchr. medicine, Paris, 1697; iatromathematician. Author: De la digestion des aliments . . . , 1712; Novus medicinae conspectus, 1722; also numerous treatises. Sometimes called French Hippocrates; advocated free use of lancet. Died Paris, Apr. 11, 1737.

HEDBERG, Hollis Dow, Am. geologist; b. Falun, Kan., May 29, 1903; s. Carl A. and Zada M. (Dow) H.; A.B., U. Kan., 1925; M.S., Cornell U., 1926; Ph.D., Stanford, 1937; m. Helen F. Murray, Nov. 8, 1932; children—Ronald M., James D., William H., Franklin A., Mary F. Asst. Kan. Geol. Survey, 1924-25; petrographer Lago Petroleum Corp., Venezuela, 1926-29; stratigrapher, dir. geol. lab. Mene Grande Oil Co., Venezuela, 1928-39, asst. chief geologist, 1939-46; chief geologist fgn. prodn. div. Gulf Oil Corp., 1946-51, exploration mgr., 1951-52, chief geologist, 1952-53, exploration coordinator, 1953-57, vice pres., exploration, 1957-64, exploration adviser, 1964-—. Professor geology Princeton U., 1959-—. Mem. Am. Com. Stratigraphic Nomenclature, 1946-56; pres. Internat. Com. Stratigraphic Terminology, 1952-—. Dir. Cushman Found. Foraminiferal Research, 1951-63; past chmn. com. Nat. Acad. Scis., now mem. lunar and planetary surfaces com. Space Sci. Bd.; exec. bur. Internat. Commn. on Stratigraphy, Decorated Medalla de Honor de la Instruccion Publica (Venezuela), 1941. Recip. Sidney Powers medal Am. Assn. Petroleum Geologists, 1963. Fellow Geol. Soc. Am. (pres. 1959-60); mem. Nat. Acad. Sci., Am. Assn. Petroleum Geologists (asso. editor 1937-—; pres. Eastern sect. 1948-49), A.A.A.S., Am. Petroleum Inst., Am. Geophys. Union, Paleontological Soc. Am. (v.p. 1952) Soc. Econ. Paleontological Minerology, Swiss. Am. geol. socs., Soc. Exptl. Geophys., Am. Inst. Mining and Metall. Engrs., Am. Geol. Inst. (pres. 1962), International Union of Geological Sciences (chairman U. S. nat. com. on geology 1965-66), Geol. Soc. Am. (asso. editor 1961-—), Geol. Soc. London (fgn.), Geological Society of London (foreign), Phi Beta Kappa, Sigma Xi. Research on micropaleontology; stratigraphy; sedimentary petrology; compaction of sediments; petroleum geology. Contbr. articles profl. jours. Home: 118 Library Pl., Princeton N.J. Office: Gulf Bldg., Pitts. 30.

HEDDLE, Douglas William Orr, Brit. physicist; b. Dundee, Scotland, Feb. 14, 1928; s. Eric W. M. and Winifred (Orr) H.; B.Sc., Reading U., 1948, Ph.D., 1952; m. Pauline Lovatt, Oct. 4, 1952; children—Andrew, Ian. Lectr. in physics U. Coll. London (Eng.), 1952-65; reader in physics U. York, Heslington, Eng., 1965—. Fellow Inst. Physics, Royal Astron. Soc. Research, publs. on use of Cherenkov radiation and scattering techniques in vacuum ultraviolet, stellar astronomy from rockets, electron-atom collisions. Home: Spring Barn, Heslington Lane, Heslington, York, Eng.*

HEDIN, Sven Anders, Swedish explorer; b. Stockholm, Feb. 19, 1865; s. Ludwig and Anna (Berhn) H.; ed. Stockholm, Uppsala, Berlin, Halle univs.; Ph.D., 1892. Explorations through Persia and Mesopotamia (1st white man to do so since Marco Polo), 1885-86, old silk-trade routes from Russia to Peking, 1893-97, Tibet, 1899-1902, 1905-1908, head Asiatic expdn. from China, 1927-33. Mem. French Acad. Scis., 1911. Author: Through Asia, 1898; Scientific Results of a Journey in Central Asia, 8 vols., 1904-08; Trans-Himalaya, 3 vols., 1909-13; From Pole to Pole, 1911; Southern Tibet, 9 vols., 1916-22; Bagdad, Babylon, Nineveh, 1917; Jerusalem, 1917; Mount Everest, 1922; The Grand Canyon of Arizona, 1925; Across the Gobi Desert, 1931; Jehol, City of Emperors, 1931; The Conquest of Tibet, 1934; Scientific Results of the Sino-Swedish Expedition, 35 vols., 1937-49; The Silk Road, 1938; The Flight of the Big Horse, 1938; The Wandering Lake, 1940; My Life as an Explorer, 7th edit., 1942; Great Men I Met, 2 vols., 1952. For about 50 years explored little known areas, particularly in Asia, amassing collections of meteorol., geol., zool., paleontol., and archeol. data; explored unknown sources of Brahmaputra, Indus and Sutlej rivers; added greatly to knowledge of central Asia, particularly Tibet. Died Stockholm, Nov. 26, 1952.

HEDVALL, Johan Arvid, Swedish chemist; b. Skara, Jan. 18, 1888; s. Johan August and Augusta (Nordstedt) H.; ed. univs. Uppsala, Lund, Göttingen; Ph.D. in sci.; Ph.D. honoris causa in sci. numerous fgn. univs.; Ph.D. honoris causa; D.Tech. honoris causa; m. Carin Carlsson, 1919; 1 son, Johan Anders. Asst., Inst. Meteorology and Chemistry, U. Uppsala, 1929-46; instr. chem. research Stora Kopporbergs Berglsag; prof. chemistry and tech. Sch. Orebro; asso. prof., full prof. applied chemistry, titular prof. chemistry for solid and silicate bodies, 1946-56; instr. research in constrn. material and archeology Poly. Sch., Göteborg; prof. emeritus Tech. Sch., Göteborg, 1956—. Vis. prof. Free U. Brussels, 1954, U. Istanbul, 1959; vis. lectr. all European countries, U. S., Mexico, Peru, Brazil, South Africa, Egypt, Israel. Recipient medal of merit U. Liege, Free U. Brussels, Gewerbeverein, Austria. Mem. Swedish Acad. Sci., Swedish Acad. Applied Sci., Acad. Sci. and Letters Göteborg, Acad. Physiography Lund, Spanish Soc. Physics (hon.); corr. mem. Acad. France, Acad. Finland, Acad. Göttingen and Mayence, Royal Soc. Sci. and Letters Norway, Royal Acad. Sci. (Denmark). Author: Reaktionsfahigkeit Fester Stoffe, 1938; Einführung in die Festkörperchemie, 1952; La Chimie des Solides, 1955; Chemie in Dienst der Archeologie Bautechnik Denkmalpflege, 1962; Memoires, 2 vols., 1955, 62. Address: Institute of Silicates, 5 Rue de Gibralter, Göteborg, Sweden.

HEDWIG, Johann, botanist; b. Kronstadt, Rumania, Dec. 8, 1730; student, Leipzig, Germany; practice medicine, Leipzig, also Chemnitz, became prof. pharmacology U. Leipzig, 1786, prof. botany, 1789. Mem. French Acad. Scis. Author: Fundamentum historiae naturalis muscorum frondosorum, 1782-83, 2 vols.; Stirpes cryptogamicae, 4 vols., 1787-93; Analytic Description and Designs of New and Doubtful Cryptogamus Plants, 4 vols., 1795; Observations on the True Parts of Generation in Mosses; Specieis Muscorum, frondosorum (posthumous), 1801. Founder muscology; 1st to interprete sex structures of mosses correctly; used spore-distributing mechanism of capsule for classification. Died Feb. 18, 1799.

HEEGAARD, Poul Esbern, Danish zoologist; b. Horsens, Denmark, Nov. 15, 1908; s. Aage Anker and Elna (Sorensen) H.; M.Sc., U. Copenhagen (Denmark), 1939, Dr.phil., 1947; M.Sc.h.c., Western Australia U., 1942; div.; children—Frederik, Susanne, Nina. Lectr., U. Western Australia, 1937-38; collaborater Danish Marine Research and Fisheries, 1939-52; prof., chmn. zoology U. Indonesia, 1952-59, Am. U., Beyrouth, Lebanon, 1959-60, U. Baghdad, 1960-61; research asso. Carlsberg Found., 1961-64; research asso. U. Copenhagen, 1964—. Lectr., UNESCO 1964—. Mem. Zoology Acad. (exec. council), Danish Naturhist. forening. Contbg. author: Phylogeny of the Arthropods, 1947; Larvae of Decapod Crustacea, Oceanic Penaeids, 1966. Research and publs. on relationship of crustacea to each other and other arthropods based on morphology, embryology and hist. devel. Home: Lykkevej 13, Charlottenlund, Denmark. Office: Zool. Mus., U. Copenhagen, Universitetsparkem 15, Copenhagen, Denmark.*

HEEGER, Alan Jay, Am. physicist; b. Sioux City, Ia., Jan. 22, 1936; s. Peter J. and Alice (Minkin) H.; B.S. with high distinction, U. Neb., 1957; Ph.D., U. Cal. at Berkeley, 1961; m. Ruthann Chudacoff, Aug. 11, 1957; children—Peter Scott, David Jerome. Research scientist Lockheed Missiles div. Lockheed Aircraft Corp., Palo Alto, Cal., 1958-59; faculty dept. physics U. Pa., Phila., 1962—, professor, 1967—. Mem. adv. com. Ann. Conf. on Magnetism and Magnetic Materials, 1966—; cons. Gen. Electric Co., 1966—, U. S. Naval Research Lab., 1966—. Alfred P. Sloan Found. fellow, 1963-66; John Simon Guggenheim fellow, 1968-69. Mem. Am. Phys. Soc. Research and publs. on magnetism especially magnetic resonance in magnetic solids, magnetism in metals and their alloys. Home: 134 Cornell Rd., Bala-Cynwyd, Pa. 19004. Office: Dept. Physics, U. Pa., Phila. 19104.*

HEER, Clifford V., Am. physicist; b. Archbold, O., May 31, 1920; s. Nelson V. and Minnie (Leu) H.; B.Sc. in Physics, Ohio State U., 1942, Ph.D., 1949; m. Esther Jean Leonard, Dec. 17, 1949; children—Barbara Jean, Deborah Ann, Daniel Nelson. Asst. physics Ohio State U., Columbus, 1946-48, instr., 1949-50, asst. prof., 1951-55, asso. prof., 1956-60, prof., 1960—. Cons., Ramo-Wooldridge Corp., Los Angeles, 1956-58, Space Tech. Lab., Los Angeles, 1959-64. Fellow Am. Phys. Soc. Research and publs on low temperature, solid state and laser physics; co-inventor of magnetic refrigerator, optical maser photon rate gyroscope. Home: 4174 Fairfax Dr., Columbus O. 43221.*

HEER, Hendrik van, alchemist; b. Tongres, Liège, circa 1570; M.D. Traveled in Germany, Italy, Spain, France and England; knew languages of all these countries and Latin, Greek, Hebrew; town physician, Liège, 1605; physician of Ernest Ferdinand, Elector of Cologne. Author: Spadacrene, hoc est Fons Spadonus, 1614; Observationes Medicae oppido rarae in Spa et Leodii animadversae cum medica mentis aliquot selectis, 1631. Criticised Van Helmont's book on Spa water. Died circa 1636.

HEER, Oswald, Swiss naturalist, biologist; b. Nieder-Utzwyl, Switzerland, Aug. 31, 1809; studied theology, Halle, Germany; became Protestant pastor, Nieder-Uzwil; founder, dir. bot. gardens, Zurich, 1835-83; prrof. botany and entomology U. Zurich, later at Polytechnicum. Author: Die Käfer der Schwiez, 1838-41; Die Insektenfauna der Tertiärgebilde von Oeningen und von Radoboj in Kroatien, 3 vols., 1847-53; Flora Tertiaria Helvetiae, 3 vols., 1857-58; Die Urwelt der Schweiz, 1865; Die fossile Flora der Polarländer, 7 vols., 1868-83. Pioneer in research on fossil plants; studied tertiary flora of Switzerland. Died Lausanne, Switzerland, Sept. 27, 1883.

HEEZEN, Bruce C., Am. oceanographer, geologist; b. Vinton, Ia., Apr. 11, 1924; s. Charles Christian and Esther (Shirding) H.; B.A., State U. Ia., 1948; M.A., Columbia, 1952, Ph.D., 1957. Geologist, expdn. leader Woods Hole (Mass.) Oceanographic Instn., 1948; Robert fellow Columbia, 1948-51, faculty, 1951-53, 54-56, 60—, asso. prof. geology, 1964—; research asst. Lamont Geol. Obs., 1953-54, sr. research scientist, 1958-60. Mem. spl. com. for oceanic research Internat. Indian Ocean Expdn. Working Group on Geology, Geophysics and Bathymetry, 1960—; mem. oceanwide survey panel com. on oceanography Nat. Acad. Scis.-NRC, 1961—. Recipient Henry Bryant Bigelow medal. Fellow A.A.A.S.; mem. Internat. Assn. Phys. Oceanography (gen. bathymetric chart oceans com. 1957—), Internat. Union Geodesy and Geophysics (com. for upper mantle project 1960-63), Am. Assn. Petroleum Geologists, Am. Geog. Soc., Am. Geophys. Union, Am. Soc. Limnology and Oceanography, Ia. Acad. Sci., Royal Astron. Soc., N.Y. Acad. Scis., Royal Astron. Soc., Seismol. Soc. Am., Soc. Exploration Geophysicists, Geochem. Soc., Sigma Xi. Author: (with Marie Tharp, Maurice Ewing) The Floors of the Oceans. Research, numerous publs. on turbidity currents and deep geostrophic currents, especially regarding deposition of deep sea sediments; discovered Mid-Oceanic Ridge and Rift; studies on submarine physiography. Home: 646 River Rd., Piermont, N.H. 10968. Office: Lamont Geol. Obs., Torrey Cliff, Palisades, N.Y.*

HEFFERLINE, Ralph Franklin, Am. psychologist; b. Muncie, Ind., Feb. 15, 1910; s. Samuel Thomas and Blanche (Cecil) H.; B.S., Columbia, 1941, M.A., 1942, Ph.D., 1947; m. Dorothy Halliday, Aug. 25, 1939. Faculty, Columbia, N.Y.C., 1948—, prof., 1967—, pre-med. adviser, 1949-54, chmn. dept. psychology, 1965—. Fellow Am. Psychol. Assn., A.A.A.S., N.Y. Acad. Scis.; mem. Psychonomic Soc., Eastern Psychol. Assn., Soc. for Gen. Systems Research, Soc. Psychophysiol. Research. Author: (with F. S. Perls and P. Goodman) Gestalt Therapy, 1950. Research on significance of patterns of muscular tension for psychoneurosis and anxiety, electromyographic study of very small responses as behavioral counterpart of mental processes. Home: 549 W. 123d St., N.Y.C. 10027.*

HEFLEY, Harold Martin, Am. biologist; b. Norman, Okla., Jan. 20, 1904; s. Harold M. and Annie Florence (Miller) H.; B.S., U. Okla., 1925, M.S., 1926, Ph.D., 1935; m. Vera Lue Corbin, Aug. 21, 1938. Bacteriologist, Tenn. Tb. Hosp., Memphis, 1954-55; faculty Panhandle State Coll., Goodwell, Okla., 1955—, prof., head dept. biology, 1962—; biologist summers Rocky Mountain Biol. Sta., Gulf Coast Research Lab. U. Okla., La. State U., U. Omaha. Fellow A.A.A.S., Tex., Okla. acads. scis.; mem. Ecol. Soc., Am. Forestry Assn., Sigma Xi, Alpha Epsilon Delta, Beta Beta Beta, Phi Sigma, others. Research in ecology of natural vegetation, parasitic relations in insects, conservation. Home: Box 147, Goodwell, Okla. 73939.*

HEFNER-ALTENECK, Friedrich, see von Hefner-Alteneck, Friedrich.

HEFTMANN, Erich, Am. biochemist; b. Vienna, Austria, Mar. 9, 1918; s. Salomon and Rosa (Seifert) H.; student U. Vienna Med. Sch., 1936-38; B.A., N.Y. U., 1942; postgrad. U. Md., 1943-44; Ph.D., U. Rochester, 1947; m. Lily Rubin, May 2, 1942; m. second, Brigitte Sander, March 14, 1968; children—Rex Walter, Lisa Valerie, Erica Sophie. Came to U. S., 1939, naturalized, 1945. Biochemist, USPHS, Boston, 1947-48, Nat. Cancer Inst., NIH, Bethesda, Md., 1948-50, Nat. Inst. Arthritis and Metabolic Diseases, 1950-63; biochemist Western utilization research and devel. div. U. S. Dept. Agr., Pasadena, Cal., 1963—; research fellow Cal. Inst. Tech., 1959, 61-64, research asso., 1964—; asso. prof. medicine U. So. Cal. Med. Sch., 1965—. Lectr. Georgetown U., 1958-59; scientist USPHS. Fellow A.A.A.S.; mem. Am. Soc. Biol. Chemists, Am. Chem. Soc., Soc. for Exptl. Biology and Medicine, Sigma Xi. Author: (with E. Mosettig) Biochemistry of Steroids, 1960; Chromatography, 1961, 67; numerous articles. Developed analytical methods for various steroids; research biosynthesis, metabolism and physiol. significance steroids in man, animals, plants and micro-organisms; developed rapid portable screening test for blood sugar. Office: Div. Biology, Cal. Inst. Tech., Pasadena, Cal. 91109.*

HEGAR, Alfred, German gynecologist; b. Darmstadt, Germany, Jan. 6, 1829; prof. Freiburg (Germany) U.; invented series of metal bougies or sounds of varying sizes for dilating cervix of uterus, 1872; described softening of lower segment of uterus in early stages of pregnancy (known as Hegar's sign), 1884. Died Freiburg, Aug. 6, 1914.

HEGEL, Georg Wilhelm, German philosopher, historian; b. Stuttgart, Germany, Aug. 27, 1770; s. George Ludwig and Maria Magdalena (Fromme) H.; Ph.D., Tübingen (Germany) U., 1790; theol. certificate, 1793; independently studied Gibbon, Montesquieu, Kant, 1793-96; m. Marie von Tucher, Sept. 16, 1811; 2 sons—Karl, Immanuel. Pvt. tutor, Bern, Switzerland, 1793-96; tutor Frankfort/Main, Germany, 1797-99, private docent, 1799-1800; lectr. U. Jena (Germany), 1801-04, asso. prof., 1805-07; editor Bamberger Zeitung, 1807-08; rector Aegidiengymnasium, Nürenberg, Germany, 1808-16; prof. U. Heidelberg (Germany), 1816-18; prof. philosophy, Berlin (Germany), U., 1818-31, rector, 1830-31. Decorated by Frederick William III, 1831. Author: Die Phänomenologie des Geistes (covers whole range of psychology; describes how human mind has risen from mere consciousness through self-consciousness, reason, spirit, religion to absolute knowledge), 1807; Wissenschaft der Logik, 1812-16; Enzyklopädie der philosophischen Wissenschaften im Grundriss (compendium of Hegelian system, divided into logic, nature, mind; these are subject among and within themselves to indefinitely circular process of dialectic, in which a positive thesis is opposed by an antithesis, thus producing a synthesis which again generates an antithesis), 1817; Grundlinien der Philosophie des Rechts (presents views on politics and history; holds that persons, not individuals, are subjects of rights, also that persons are all equal before law, but are required by it to be obedient without exception to the state; Hegel's state rests on instns. of family and guild and is characterized by strong centralization and respect for individual rights working in harmony; these views have had an impact on fascism and communism, as well as social sci.), 1821; Philosophy of History, 1831. A leading philosopher of 19th century Germany; attempted, as Aristotle had done, to comprehend universe with a systematic philosophy; grounded his system in faith; took Plato seriously; working in tradition of Immanuel Kant, he built on Kantian analysis of knowledge and reason, but was oriented toward history rather than phys. scis.; unlike Kant, he held that absolute knowledge is possible and his acceptance of fact made him less idealistic than Kant; saw human history as process of man's spiritual and moral progress; sought to develop more precise language to express ideas; study of Hegel has led to critical examination of religious history and dogma, also to studies noting growing importance of communism, as well as contbg. to rise of existentialism. Died Kupfergraben, Germany, Nov. 14, 1831.

HEGER, Paul, Belgian physiologist; b. Brussels, Belgium, 1846; became prof. physiology Free U. Belgium, 1873; named dir. Solvay Inst., 1895. Founder jour. Archives internationales de physiologie, 1904. Studied white corpuscles, vasomotors, nervous system, anthropology, psychiatry; one of 1st to use method of artificial circulation. Died 1925.

HEGETSCHWEILER, Johann Jakob, Swiss physician, botanist; b. Rifferswil, Switzerland, Dec. 14, 1789; practiced medicine, Stäfá, studied flora of Switzerland. Author: Helvetiens Flora, 1822; Flora der Schweiz, 1840. Died Zurich, Sept. 10, 1839.

HEGI, Gustav, botanist; b. Rickerbach, Switzerland, Nov. 13, 1876; prof., Munich, Germany; studied flora of Central Europe. Author: Illustrierte Flora von Mitteleuropa, 12 vols., 1906-29. Died Küsnacht, Apr. 21, 1932.

HEIBERG, Johan Ludwig, Danish mathematician, philologist; b. Aalbork, Denmark, 1854; hon. Ph.D., Oxford; prof. classical philology U. Copenhagen, 1896-1925; corr. mem. Brit. Acad. Edited much Greek literature, notably math. and med. works; attempted to restore Eudemus' original text, deleting Simplicius' additions; important research on Archimedes; with Menge did edition of complete writings of Euclid. Died 1928.

HEIDBREDER, Edna, Am. psychologist; b. Quincy, Ill., May 1, 1890; d. William Henry and Mathilda (Meyer) Heidbreder; A.B., Knox Coll., 1911, LL.D., 1964; A.M., U. Wis., 1918; Ph.D., Columbia U., 1924. With U. Minn., 1924-34, asso. prof., 1929-34; prof., chmn. dept. Wellesley Coll., 1934-55; vis. prof. U. Cal., Berkeley, 1950; mem. seminar staff Radcliffe Coll., 1955-61. Mem. Am., Eastern psychol. assns., A.A.A.S., N.Y. Acad. Scis., Phi Beta Kappa, Sigma Xi. Author: (with others) The Minnesota Mechanical Ability Tests, 1930; Seven Psychologies, 1933. Book rev. editor Jour. Abnormal and Social Psychology, 1934-47; asso. editor Psychol. Monographs, 1948-62. Publs. on studies in measurement of introversion, extraversion, and inferiority attitudes; forms of thought in problem solving; schools and systems in psychology. Home: 34 Cottage St., Wellesley, Mass. 02181.*

HEIDELBERGER, Michael, Am. chemist; b. New York, N.Y., Apr. 29, 1888; s. David and Fannie (Campe) H.; student Ethical Culture Schs., New York, 1900-05; B.S., Columbia U., 1908, A.M., 1910, Ph.D., 1911; student Federal Poly. Inst., Zurich, Switzerland, 1911-12; Dr. (hon.) Faculty of Medicine, U. Bordeaux, 1947, U. Oslo, 1956; Dr. Faculty Pharmacy, U. Paris, 1949; Faculty Philosophy, U. Upsala, Sweden, 1950; Faculty Pharmacy, U. Strasbourg, 1952; Dr. (hon.), Faculty of Scis., U. Aix-Marseille, 1959; Dr. (hon.), Faculty of Pharmacy, U. Nancy, 1960; D.Sc., Rutgers U., New Brunswick, N.J., 1961; m. Nina Tachau, June 29, 1916 (died July 1, 1946); 1 s., Charles; m. 2d, Charlotte Rosen, June 23, 1956. Asst. in qualitative analysis, supplementary term, Stevens Inst., Hoboken, N.J., 1909; successively chemist to Med. Service and chemist to Presbyn. Hosp., N.Y. City, 1928-56; asso. prof. medicine, Coll. Physicians and Surgeons, Columbia U., 1928-29, associate professor biochemistry, 1929-45, professor biochemistry, 1945-48, professor immuno-chemistry, 1948-56, prof. emeritus, 1956-—; vis. prof. immunochemistry Inst. of Microbiology Rutgers U., 1955-64; adj. prof. pathology N.Y.U. Sch. Medicine, 1964-—; cons. to Sec. War, 1941-46. Awarded A. Cressy Morrison prize (with F. E. Kendall), N.Y. Acad. Scis., 1929; Ehrlich silver medal, 1933; Albert Lasker Award, 1953; share in reward for cure of African sleeping sickness. Belgian Govt.; officer of Order of Leopold II, 1953; John Simon Guggenheim Memorial fellow, 1934, 36; Behring prize, 1954; Pasteur medal, Swedish Medical Society, 1960; Nat. Medal of Sci., 1967; N.Y. Acad. Medicine medal, 1968. Chairman research council N.Y.C. Pub. Health Research Inst., 1952-56; mem. biochem. panel Am. Inst. Biol. Scis. for Office Naval Research, 1950-52. Fellow Am. Acad. Microbiology; member Nat. Acad. Sciences, Am. Chem. Soc., A.A.A.S., Am. Philos. Soc., Soc. of Am. Biochemists, Soc. Am. Microbiologists, Am. Assn. Immunologists (pres. 1946-49), Harvey Soc. (pres. 1952-53), Royal Danish Acad. Sci. (corres.), Microgiol. Soc. France (hon.), Soc. de Biologie, French, Brit. socs. immunologists (hon.), Accademia Nazionale dei Lincei (Rome) (fgn. mem.), Soc. for Exptl. Biology and Medicine, Sigma Xi. United States delegate to the 50th Anniversary of death of Pasteur, Paris and Caen (France), 1946. Author textbooks. Research in organic chemistry, sodium-uranium compounds, volumetric determination of cerium, phthalones, cyclo-octatetraene, chemotherapy, alkaloids, oxyhemoblobin, chemistry of immune reactions complement; considered founder of immunochemistry. Home: 333 Central Park W., N.Y.C.

HEIDENHAIN, Rudolf Peter Heinrich, German physiologist; b. Marienwerder, Germany, 1834; ed. Königsberg, Halle, Berlin (all Germany); 1 son, Martin. Docent Halle, from 1857; prof. Breslau, Germany (now Wroclaw, Poland), from 1859; research with DuBois-Reymond. Author: Mechanische Leistung, Wärme entwicklung und Stoffumsatz bei der Muskeltätigkeit, 1864; Physiologie der Absonderungs Vorgänge, 1880. Described columnar or rodlike cells in renal tubules (Heidenhain's rods), 1861; formulated law of glandular secretion (secretion always involves change of structure in gland), 1866; described certain delomorphous and adelomorphous cells in glands of stomach (Heidenhain's cells), 1870; formulated secretion theory of renal function, 1874; introduced term zymogen (ferment generator), 1876; also worked on mechanics, metabolism, prodn. of heat in muscles. Died 1897.

HEIDT, Lawrence Joseph, Am. phys. chemist; b. Portage, Wis., Apr. 5, 1904; s. Frank and Barbara (Ehr) H.; B.A., U. Wis., 1927, M.S., 1928, Ph.D. (duPont fellow), 1930; m. Agnes Grace Kiley, June 16, 1933; children—Marianne (Mrs. Dean A. Ockerbloom), David, Barbara (Mrs. Paul J. Ryan). Postdoctoral research fellow Havard, 1930-35; faculty Mass. Inst. Tech., Cambridge, 1935-—; cons. various founds., industries, govt. agys.; vis. prof. various univs., 1959-63. Recipient Alfred P. Sloan Merit award Mass. Inst. Tech., 1956, Certificate of Appreciation U. S. Army, 1963. Guggenheim fellow, 1962-63. Fellow A.A.A.S., Am. Acad. Arts and Scis. (C. M. Warren Fund Research award 1936); mem. Am. Chem. Soc. (councillor 1963-—), N.Y. cad. Arts and Scis., Am. Assn. U. Profs., Sigma Xi, Phi Lambda Upsilon. Editor; (with R. S. Livingston, E. Rabinowitch, F. Daniels) Photochemistry Liquid and Solid States, 1960. Research, publs. on sugar reactions; fundamental studies of intra- and interspecies charge and energy transfer processes based on measurements of gross and quantum yields, visible and ultraviolet light absorption and emission spectra, reaction kinetics and properties of reaction intermediates; discovered in vitro cerium catalyzed solar energy conversion process for photochem. decomposition water into molecular hydrogen and oxygen, prodn. of ozone by light absorbed by peroxides in water, enhancement of photo-decomposition of gaseous ozone by water vapor, improvements in elec. insulating properties of solid and liquid organic materials by certain additives; inventor high intensity mercury arc lamp; colorless invert sugar solution for intravenous therapy. Home: 46 Bailey Rd., Arlington, Mass. 02174.*

HEILAND, Carl August, geophysicist; b. Hamburg, Germany, July 16, 1899; s. Carl Heinrich and Emilie (Gruetter) H.; grad. Wilhelm Gymnasium, Hamburg, 1917; Dr. rer. nat., U. Hamburg, 1923; student U. Heidelberg, 1922-23; came to U. S., 1925, naturalized, 1935; m. Peggy Johnston, May 24, 1947; children—Ann, John Thomas. In charge geophys. dept. Askania Werke, Berlin, 1924-25, Am. rep., 1925-26; prof. geophysics and head dept. Colo. Sch. of Mines, 1926-48; pres. Heiland Research Corp., Denver, pres. div. of Minneapolis-Honeywell Regulator Co. Collaborator in seismology, U. S. Coast & Geodetic Survey. Fellow Geol. Soc. Am.; mem. Am. Geophys. Union, Soc. Exploration Geophysicists, Am. Inst. Mining and Metall. Engrs., Seismol. Soc. Am., Am. Assn. Petroleum Geologists. Author: Geophysical Exploration, 1940; also many tech. articles. Used magnetic field scales in soil investigation, 1926; worked on photoelectric transmission time by gravity pendulum, seismic soil investigation; invented Geophon, 1930. Died Feb. 23, 1956.

HEILBRON, Sir Ian, Scottish chemist; b. Glasgow, Scotland, Nov. 6, 1886; s. David Heilbron; ed. Royal Tech. Coll., Glasgow; Ph.D., U. Leipzig; D.Sc., LL.D., Glasgow U.; LL.D., Edinburgh; m. Elda Marguerite Davis; 2 sons. Lectr., Royal Tech. Coll., 1909-14, prof. organic chemistry, 1919-20; Heath Harrison prof. organic chemistry Liverpool, 1920-23; prof. organic chemistry U. Manchester, 1933-35, Sir Samuel Hall prof. chemistry and dir. Chem. Labs., 1935-38; prof. organic chemistry, dir. Organic Chem. Labs., U. London, Imperial Coll. Scis. and Tech., 1938-49, prof. emeritus, 1949-59; dir. research Brewing Industry Research Found., 1949-58; mem. adv. council on sci. research and tech. devel. Ministry Supply, 1958-59. Fellow Royal Soc. (Davy medallist 1943, Royal medallist 1951), Royal Inst. Chemistry; mem. Chem. Soc. (London) (pres. 1948-50, Longstaff medallist 1939, Hugo Muller lectr. 1940, Pedler lectr. 1947), French Chem. Soc. (hon.), Royal Netherlands Acad. Scis. (fgn.). Recipient Priestly medal Am Chem. Soc., 1945. Editor-in-chief: Dictionary of Organic Compounds, 3 vols., 1937; chmn. editorial bd. Thorpe's Dictionary of Applied Chemistry, also numerous articles. Research on synthesis of naturally occurring organic compounds, vitamins A and D. Died Sept. 14, 1959.

HEILBRONN-WIKSTRöM, Edith, biochemist; b. Fürth, Germany, Jan. 9, 1925; d. Philipp and Else (Schwarz) Heilbronn; Fil.Kand., U. Stockholm (Sweden), 1952; fil.lic., U. Uppsala (Sweden), 1959, fil. dr., 1965, Docent, 1966; m. Bengt Axel Wilkström, Feb. 6, 1960 (dec. Oct. 1967); 1 son, Martin Axel. Staff, Wenner-Grens Inst., 1942-45; asst. Karolinska Institutet, 1950-51; with Research Inst. Nat. Def., Sundbyberg, Sweden, 1952-—, laborator, 1959-—, sect. leader, 1963-—. Mem. Svenska Kemistförbundet, Sveriges Biokemiska Förening, Sv. Fysiologföreningen. Research and publs. on biochemistry of enzymes, especially cholinesterases, chemistry and biochemistry of organophosphorus compounds reacting with esterases; biochemistry of antidotes to organophophorous compounds. Home: 30, Rörstrandsgatan, Stockholm, Sweden. Office: Research Inst. Nat. Def., Sundbyberg 4, Sweden.*

HEILM, Peter Jacob, Swedish chemist; b. Sunnerbo Harad, Sweden, 1746; ed. U. Uppsala (Sweden); became mem. staff Royal Mint, Stockholm, Sweden, 1782; named dir. chem. lab. Bur. Mines, Stockholm, 1794. Isolated metallic molybdenum, 1781. Died Stockholm, Oct. 7, 1813.

HEILMAN, Fordyce R(ussell), Am. bacteriologist, physician; b. Austin, Minn., June 16, 1905; s. Oliver Charles and Emma (Larson) H.; student N.D. Agrl. Coll., 1923-26; B.S., Northwestern U., 1930, M.D., 1931; fellow Mayo Foundn. Grad. Sch., U. of Minn., 1932-35, M.S., 1938, Ph.D. 1940; m. Dorothy Henderson, Aug. 10, 1934 (div. 1953). Intern St. Luke's Hosp., Chicago, 1930-31, resident physician, 1931-32; 1st asst. Mayo Foundn., 1935-39, cons. physician Mayo Clinic since 1939, head sect. on bacteriology since 1945; instr. in bacteriology Mayo Foundn. Grad. Sch., U. of Minn., 1939-41, asst. professor bacteriology, 1941-45, asso. prof., 1945-46, professor bacteriology since 1946. Mem. A.M.A., A.A.A.S., Am. Soc. Bacteriologists, Am. Assn. Cancer Research, Am. Soc. Clin. Pathologists, Soc. Exptl. Biology and Medicine, Am. Pub. Health Assn., N.Y. Academy Sciences, American Academy of Microbiology, Society for General Microbiology, Phi Rho Sigma, Sigma Xi, Alpha Omega Alpha. Contbr. to med. and sci. jours. Research in chemotherapy of infectious disease., med. bacteriology; produced crystalline procaine penicillin G. Home: 1516 3d Av. N.E. Office: Mayo Clinic, Rochester, Minn.

HEILMEYER, Ludwig, German physician; b. Munich, Germany, Mar. 6, 1899; s. Alexander and Barbara (Hötzel) H.; ed. U. Munich, U. Lena; M.D. honoris causa, univs. Santiago, Athens, Louvain, Frankfort on Main; m. Ingeborg von Mutius, 1947; children—Alexander, Renate, Lutz, Peter, Barbara, Sabine. Full prof. medicine, promotion to U. Munich, 1925; qualified U. Lena, 1928, asso. prof., 1936; full prof. U. Düsseldorf, 1945, U. Freiburg, Bresgau, 1946; now dir. Med. Clinic of Robert Koch Clinic for Tb Diseases. Recipient Fr. von Müller Commemorative medal, 1959, Robert Koch medal, 1960, Stratton Lecture medal, medal Med. Soc. Natural Scis. Mem. Royal Soc. London (corr., hon.), Leopoldina Acad. of German Natural Scis., Acad. Scis. Heidelberg, Swiss Soc. Hematology (hon.), Soc. Med. Scis. Lisbon, Soc. Medicine Finland, Soc. Internal Medicine Buenos Aires, Turkish Med. Soc. Author: Medizinische Spektrophotometrie, 1933 (English edit. 1939); Radioisotope in der Heilkunde, 1952; (with Gitter) Taschenbuch der Klinischen Funktionsprüfungen, 6th edit., 1960; Lehrbuch der Inneren Medizin, 1935, also numerous articles. Authority on hematology. Address: 100 Sonnhalde, Freiburg, Bresgau, Germany.

HEIM, Albert, Swiss geologist, seismologist; b. Zurich, Switzerland, Apr. 12, 1849; student natural scis., Zurich, Berlin, Germany; Ph.D., U. Bern (Switzerland), 1884, Oxford, Eng.; m. Marie Vöglin, 1875; children include Arnold. Prof., Zurich. Fellow Royal Soc., 1896; mem. French Acad. Scis. Author: Untersuchungen über den Mechanismus der Gebrigsbildung, 1878, 3 vols.; Handbuch der Gletcherkunde, 1885; Geologie der Schweiz, 2 vols., 1916-23, Founder gen. geol. collection, Poly. Sch., Zurich; studied mountain-forming processes in Alps. Died Aug. 3, 1937.

HEIM, Karl Ernst Wilhelm, German physician; b. Berlin, Germany, Nov. 2, 1906; s. Wilhelm and Emma Elfrieda (Hahn) H.; ed. U. Berlin, U. Innsbruck; M.D. Chief doctor of surgery; med. dir., prof. surgery, hon. prof. U. Berlin. Recipient Ernst von Bergmann plaque. Mem. Acad. for Promotion Medicine Berlin (pres.), German Soc. Blood Transfusion (treas.). Author: Einrichtung und Arbeitweise einer Blutbank; Die Operationen an der Brustdrüse; Aetiologie, Klinik und Prophylaxie der Anbiotischen Schäden in der Chirurgie; Radioaktive Isotopes in der Chirurgie. Home: 17 Lyckallee, Berlin 19. Office: 1 Augustenburger Platz, Berlin 65, Germany.

HEIM, Roger Jean, French botanist; b. Paris, Feb. 12, 1900; s. Henri and (Sauvion) H.; D.Sc., U. Paris, 1931; Ph.D. (hon.), Uppsala (Sweden) U.; m. Pauca Eftimin, Dec. 7, 1935; 1 son, Jean-Louis. Specialist cryptogamy French Nat. Mus. Natural History, 1927-—, now dir., chief cryptogamy lab.; dir. lab. for study mycology and tropical phytopathology Ecole Pratique des Hautes Etudes, 1941-—. Pres. botany commn. French Nat. Sci. Research Center; pres. Singer-Polignac Found., 1958-—. Decorated Legion of Honor, Palmes Acad. Mem. French Acad. Sci. (past pres.), Societé Botanique de France (past pres.), Societé Mycologique de France (past pres.), Academie d'Agriculture, Am. Philos. Soc., Royal Belgian Acad., Brit. Mycol. Soc. (hon.), Societé de Biologie (hon.). Author: (with R. G. Wasson) Les champignons hallucinogenes du Mexique, 1958; Les champignons toxiques et hallucinogenes, 1963. Dir., Revue de Mycologie; co-dir. Annales des Sciences Naturelles. Research on mycology, especially various varieties of hallucinogenic mushrooms of Mexico; tropical pathology; microscopical research on cell membranes; relations between fungi and insects; conservation of nature. Home: 11, rue des Médicis, Paris 6, France. Office: 57, rue Cuvier, Paris 5, France.

HEIM, Werner George, Am. biologist; b. Mülheim Ruhr, Germany, Apr. 4, 1929; s. Fred and Recha (Hirsch) H.; came to U. S., 1940, naturalized, 1946; B.A. in Zoology, U. Cal. at Los Angeles, 1950, M.A., 1952, Ph.D. (USPHS Research fellow Nat. Cancer Inst.), 1954; m. Julie I. Blumenthal, June 25, 1961; children—Susan Louise, David Lee. Jr. research zoologist U. Cal. at Los Angeles, 1954-56; instr. Brown U., Providence, 1956-57; faculty Wayne State U., Detroit, 1957-67, asso. prof. dept. biology, 1963-67, vice chmn. dept. biology, 1965-67; prof. biology Colo. Coll., Colorado Springs, Colo., 1967-—. Mem. Am. Soc. Zoologists, Soc. for Developmental Biology, A.A.A.S., Internat. Inst. Embryology, Sigma Xi.

Research and publs. on changes in blood serum proteins during devel. of chicken and rat, suppression of catalase activity by 3-aminotriazole, occurrence and role of rat slow alpha-two globulin. Home: 1010 Jupiter Dr., Colorado Springs, Colo. 80906.*

HEIMBECKER, Raymond Oliver, Canadian surgeon; b. Calgary, Alta., Can., Nov. 29, 1922; s. Harry Oliver and Dorothy (Turner) H.; B.A., U. Sask., 1945; M.D., U. Toronto, 1947, M.A., 1949, M.S., 1957; m. Kathleen Jensen, Nov. 18, 1950; children—Ray, Kathleen, Harry, Anita, Constance. Mem. surg. staff Toronto Gen. Hosp., U. Toronto, 1954—; research asso. Ont. Heart Found., 1954—; cardiovascular cons. Wellesley Hosp., Toronto, 1957—. Fellow A.C.S., Am. Royal Coll. Surgeons Can., Am. Assn. Thoracic Surgery; mem. Soc. U. Surgeons, Internat. Cardiovascular Soc., Soc. Vascular Surgery. Asso. editor Modern Medicine, 1965. Research and publs. on cardiac physiology, devel. of open heart surgery, artificial lung and heart control systems, artificial heart valves and valve transplants. Home: 75 Hillholme Rd., Toronto 7. Office: 170 St. George St., Toronto 2, Ont., Can.*

HEIMBERG, Murray, Am. pharmacologist, biochemist; b. Bklyn., Jan. 5, 1925; s. Gustav and Fanny (Geller) H.; B.S., Cornell U., 1948, M. Nutrional Sci., 1949; Ph.D., in Biochemistry, Duke, 1952; M.D., Vanderbilt U., 1959; m. 2d, Anna Langlois Knox, July 12, 1964; children—Richard G., Steven A.; stepchildren—Larry M. Knox, David S. Knox. Postdoctoral fellow biochemistry Washington U., St. Louis, 1952-54; research asso. physiology Vanderbilt U., Nashville, 1954-59, faculty, 1959—, professor pharmacology, 1967—; established investigator Am. Heart Assn., 1962—. Recipient Borden Undergrad. Research award in medicine Vanderbilt U., 1959. Mem. Am. Soc. For Pharmacology and Exptl. Therapeutics, Endocrine Soc., Am. Chem. Soc., A.A.A.S., Am. Oil Chemists Soc., Am. Soc. Biol. Chemists. Research and publs. on metabolism of lipids, effect of various hormones and drugs on lipid metabolism, abnormalities of lipid metabolism producing cardiovascular diseases. Home: 221 Leonard Av., Nashville 37205.*

HEIMBURGER, Robert Francis, Am. physician; b. Weishien, Shantung, China, July 12, 1917 (parents Am. citizens); s. LeRoy Francis and Louise (Corbett) H.; B.S., Drury Coll., 1939; M.S., Vanderbilt U., 1943; m. Elizabeth Whitaker Fletcher, Mar. 18, 1950; children—Paul Martin Fletcher (stepson), L. Corbett, Douglas C. Douglas Smith fellow U. Chgo., 1949; faculty Ind. U., Indpls., 1949—, prof. surgery, 1961—, dir. neurosurgery, 1949-65. Cons. to surgeon gen. USAF, 1958—. Mem. Soc. Neurol. Surgeons, Harvey Cushing Soc., So. Neurosurg. Soc., Am. Acad. Cerebral Palsy, Am. Inst. Ultrasonics in Medicine, Pan Pacific Surg. Soc., Internat. Soc. Stereoencephalotomy, Soc. Consultants to Armed Forces, Ind. Neuropsychiat. Assn., A.M.A. Research on spinal cord injuries and paraplegia, birth defects of nervous system, psychol. testing for brain damage, stereotopic surgery. Home: 4462 Central Av., Indpls., 46205.*

HEIM DE BALSAC, Raymond (Pierre Léon), French physician; b. Paris, France, July 5, 1903; s. Frédéric and Juliette Chavdiaguet (D'Anval) H. de B.; ed. Faculty Scis. and Medicine, Paris; M.D., M.A. in scis.; m. Colette Anjubault; children—Gerald, Renaud; m. 2d, Molly Boulanger, Dec. 21, 1944. Chief lab. Heart Clinic, Med. Sch., Paris instr. cardiovascular radiology, 1925-41; research on congenital heart defects and their surg. treatment, 1941-54; cardiologist Marie-Lannelorgue Surg. Center, Paris, 1954—. Mem. French Soc. Cardiology, Am. Heart Assn., Journees Medicales de France (v.p.). Author: Radiologie Clinique du Coeur et des Gros Vaisseaux, 1939; Traité des Cardiopathies Congenitales, 1954, also over 400 articles. Described idiopathic dilation of pulmonary artery (with C. Laubry and D. Routier), 1940; developed procedures for surg. treatment (open heart surgery) for cardiac patients.

HEINE, Heinrich Eduard, German mathematician; b. Mar. 15, 1821; student univs. Berlin and Göttingen, 1838-42. Lectr., U. Bonn, 1844, then prof. extraordinary; prof. math. U. Halle, 1856-81. Author: Handbuck der Kugelfunktionem, 1861. Co-developer of Heine-Borel theorem on point set theory; other work on spherical harmonics. Died Oct. 24, 1881.

HEINE, Jakob von, see von Heine, Jakob.

HEINEMAN, Robert Edwin, Am. physicist; b. Omaha, Oct. 8, 1926; s. Paul George and Annie (Lowe) H.; B.S. in Physics, U. Mich., 1948, B.S. in Math., 1948, M.S., 1949, Ph.D., 1953; m. Beverly Jane Hendrie, June 24, 1950; children—Robert Edwin, Virginia A., Thomas G. Research asst. Engring. Research Inst., U. Mich., 1951-53; research physicist Hanford Labs., Gen. Electric Co., Richland, Wash., 1953-57, mgr. reactor lattice physics, 1957-65; mgr. high temperature reactor physics Battelle N.W., Richland, 1965—. Mem. Am. Nuclear Soc., Am. Phys. Soc. Author: Reactors, 1961; Exponential and Critical Experiments, 1964. Research and publs. on electron distbn. in large cosmic ray cascades, effects of plutonium fuel in thermal reactors on operational characteristics, neutron and reactor physics. Home: 1948 Harris St. Office: Battelle-N.W., Richland, Wash. 99352.*

HEINEMANN, Edward Henry, Am. aero. engr.; b. Saginaw, Mich., March 14, 1908; s. Gustave Christian and Margaret (Schust) H.; student pub. schs. Mich. and Cal.; m. Zell Shewey, 1959; 1 dau. by previous marriage, Joan (Mrs. Allan Lamont). Engaged as designer, and project engr., also assistant chief engr. Moreland Aircraft Corp., Internat. Aircraft Corp., Northrop Aircraft Corp., 1931-36; chief engr. El Segundo (Cal.) div., Douglas Aircraft Co., Incorporated, 1936-58, corporate v.p. in charge of combat aircraft engineering, aircraft and missles, 1958-60; exec. v.p. Summers Gyroscope Co., 1960-62; v.p. European Sales, 1960; v.p. engring. and program devel. Gen. Dynamics Corp., N.Y.C., 1962—. Cons. to numerous govt. coms. Recipient Sylvanus Albert Reed award, 1952, Collier Trophy, 1953; So. Cal. Aviation, Man of Yr. award, 1954; Paul Tissandier diploma Fedn. Aeronautique Internationale, 1955. Fellow Am. Inst. Aeros. and Astronautics, Soc. Aero. Weight Engrs. (hon.), Royal Aero. Soc., Am. Astronautic Soc.; mem. Soc. Naval Architects and Marine Engrs., Am. Soc. Naval Engrs., Nat. Acad. Engring., Soc. Automotive Engrs., Tau Beta Pi (hon.). In charge design, development Navy BT, SB2D, TB2D, SBD Dauntless dive bombers, which sank more combatant tonnage than any weapon during World War II, Navy R3D-DC5 transport, A-20 Air Force Havoc attack bomber, DB-7 Boston attack bomber for British and French, B-26 Air Force Invader attack bomber, Navy AD Skyraider attack bomber series, A2D Skyshark attack bomber, F3D Skyknight night fighter, F4D Skyray interceptor D-558 Skystreak, D-558-2 Skyrocket research airplanes, first airplane to reach twice the speed of sound, A3D attack bomber, the largest jet bomber ever produced for Navy carriers, A4D Skyhawk attack bomber, the lightest and most efficient jet bomber yet developed. Home: 860 UN Plaza, N.Y.C. 10017. Office: 1 Rockefeller Plaza, N.Y.C.

HEINIVAARA, Olli Eino, Finnish physician; b. Helsinki, Finland, Feb. 23, 1923; s. Eino and Fanny (Mattsson) H.; M.D., U. Helsinki, 1947; m. Saara Tellervo Nieminen, Mar. 19, 1948; 1 dau., Kristiina. Community physician, Kuorevesi, Finland, 1948-52; with dept. internal medicine Central Hosp., Helsinki U., 1953-64, asst. prof. internal medicine, 1965—. Mem. Duodecim (dir.), European, Internat. socs. hematology, Internat., Scandinavian socs. internal medicine. Research, publs. on absorption of vitamin B12, disturbing effect of para-aminoslaicylic acid (PAS) in absorption of vitamin B12, pulmonary problems. Home: 8.B.35 Mannesik., Helsinki, Finland.*

HEINKEL, Klaus, German physician; b. Wittstock/Dosse, Germany, Feb. 4, 1921; s. Hans and Hedwig (Dumpig) H.; student J. Jena (Germany), 1941-44, Dr.med., Heidelberg (Germany) U., 1950; m. Ilse Dornauer, Oct. 7, 1950; children—Jutta Carola, Karen Marina Denise. Asst., Medizinische Poliklinik Würzburg (Germany), 1950-53; with Medizinische Universitätsklinik Erlangen (Germany), 1953-64, head physician, 1965-67, lectr., 1958—; med. dir. Clinic Internal Medicine, Sanatorium Frankenland, Bad Windsheim/Germany, 1965-66; med. dir. Med. Clin. Stuttgart (Bad Cannstatt, Germany), 1966—; lectr. U. Erlangen, 1958—. Mem. Deutsche Gesellschaft für Physiologische Chemie, Deutsche Gesellschaft für Verdauungs-und Stoffwechselkrankkeiten, Deutsche Gesellschaft für Photographie, Sekt. Medizin, Deutsche Gesellschaft für Innere Medizin, N.Y. Acad. Scis. Author: (with Schön) Ätiologie, Pathogenese, Klinik und Therapie der Pankreaserkrankungen, Schattauer, 1962. Research and publs. on technique of gastroenterological endoscopy; exptl. and clinic studies in pancreatic diseases; clin.-morphological diagnosis of chronic gastritis and intestinal diseases. Patentee sci. instruments. Home: 73 Wiesbadener Strasse. Office: Medizinische Klinik Stadtkrankenhaus, Stuttgart-Bad Cannstatt, Germany.*

HEINRICH, Hellmuth Carl, German biochemist, physician; b. Hamburg, Germany, May 28, 1928; s. Hellmuth and Elsa (van Stareren) H.; M.D., U. Hamburg, 1955, P.D. Med. Faculty, U. Hamburg, 1959—, prof. physiol. chemistry, 1966—, head dept. physiol. chemistry, 1967—, head research group clin. biochemistry and nuclear medicine, 1937—. Recipient Martini prize, 1966. Mem. Vereinigung Deutscher Wissenschaftler, Gesellschaft Deutscher Naturforscher und Ärzte. Author: European Symposium on Vitamin B12 and Intrinsic Factor, 1st, 1957, 2d, 1962. Research and numerous publs. on biochem. basis of diagnosis and therapy of vitamin B12 deficiencies, physiol. depot-preparations of vitamin B12, potent antagonists of vitamin B12; devel. and applications of large volume whole body radioactivity detectors in biomed. research and clin. diagnosis, absorption and whole body metabolism of labelled proteins and trace elements. Home: 57 Bökenkamp, Hamburg 52, Germany.*

HEINRICH, Helmut Gustav, aerodynamicist; b. Berlin, Germany, Aug. 5, 1910; s. Karl Hermann and Anna (Groth) H.; B.S., State Engring. Sch., Stettin, Germany, 1931; M.S., Technische Hochschule, Stuttgart, Germany, 1938, Dr.Eng. Sc., 1943; m. Luise W. Giller, Dec. 10, 1938; children—Hildegard-Crowley, Eva-Schutz, Klaus. Came to U. S., 1946, naturalized, 1954. Asst. prof. applied mechanics Technische Hochschule Stuttgart, 1938-40; head dept. aerodynamics research Inst. Graf Zeppehin, Stuttgart, 1940-46; tech. adviser Wright Air Devel. Center, Wright Field, O., 1946-56; prof. aero-space U. Minn., Mpls., 1956—. Cons. to pvt. cos. Fellow Am. Inst. Aeros. and Astronautics (asso., Thurman H. Bane award 1954); mem. Wissenschafthiche Gesellschaft für Luft-und Raumfahrt, Sigma Gamma Tau. Contbr. numerous articles to tech. jours. Pioneered sci. treatment parachute tech.; research in parachute stability, drag, stress analysis, opening dynamics, parachute functioning at supersonic speed; designed recovery system missiles and space craft; invented subsonic and supersonic guide surface parachutes, internal parachutes, automatic disreef systems. Home: 231 W. Elmwood Pl., Mpls. 55419.*

HEINRICHS, David Henry, geneticist; b. Melitopel, Russia, Apr. 29, 1914; s. Henry and Marie (Dick) H.; B.S., U. Sask., 1938, M.Sc., 1941; Ph.D., U. Minn., 1952; m. Ruth Malvina Thiessen, Feb. 6, 1943. children—Jennifer Louise, John David. With Canadian Dept. Agr., Swift Current, Sask., 1938—, head plant sci. sect., 1964—; dir. Agrl. Inst. Can., 1957-58. Mem. Sask. Inst. Agrologists, Agrl. Inst. Can., Canadian Soc. Agronomy, Genetics Soc. Can., Sigma Xi. Research and numerous publs. on field of alfalfa breeding; developed two creeping rooted varieties, Rambler and Roamer; recognized value of creeping character of alfalfa; made genetic material available across world; population genetics in alfalfa; methods of breeding and evaluating grasses in dry climate; agronomic studies related to growing forage crops; winter hardiness studies and ecol. studies related to persistance. Home: 1328 North Hill Dr. Office: Research Sta., Canadian Dept. Agr., Swift Current, Sask., Can.*

HEINROTH, Oskar, German ornithologist; b. Kastel/Mainz, Germany, Mar. 1, 1871; dir. Rossitten bird sanctuary; built aquarium of Berlin zoo; mem. expdn. to S. Sea, 1900-01. Author: Die Vögel Mitteleuropas, 4 vols., 1924-33. Died Berlin, May 31, 1945.

HEINS, Maurice Haskell, Am. mathematician; b. Boston, Nov. 19, 1915; s. Samuel and Rose (Golbert) H.; A.B., Harvard, 1937, A.M., 1939, Ph.D., 1940; m. Hadassah Sylvia Wagman, Aug. 25, 1940; children—Sulamith Hannah, Samuel David. Asst., Inst. for Advanced Study, Princeton, N.J., 1940-42; asst. prof. Ill. Inst. Tech., Chgo., 1942-44; mathematician Office Chief Ordnance, Washington, 1944-45; faculty Brown U., Providence, 1946-58, prof., 1947-58; prof. math. U. Ill., Urbana, 1958—; vis. prof. U. Cal. at Berkeley, 1963-64. Mem. Am. Math. Soc. (editor Proc. 1962-67, council 1962—), Am. Acad. Arts and Scis., Société Mathématique de France. Author: (with others) Analytic Functions, 1960; Classical Theory of Functions of a Complex Variable, 1962; Complex Function Theory; also numerous articles. Research on theory of analytic functions and Riemann surfaces. Home: 603 W. Illinois St., Urbana, Ill. 61801.*

HEINSHEIMER, George J., hydrologist; b. Vienna, Austria, Sept. 28, 1892; s. Maurice and Helen (Fischer) H.; Dr.jur., Dr.rer.pol., U. Innsbruck (Austria), 1921; m. Emily R. Hofbauer, July 30, 1917; children—George, Waltraud (Mrs. Helmut C. Hinrichs). With Agua y Energia Eléctrica, Buenos Aires, Argentina, Servicio de Hidrología Naval. Mem. Am. Geophys. Union, Glaciological Soc. Research, publs. in glaciology, hydrology and oceanography in Cordillero de San Juan, So. Patagonia, Antarctica. Home: 540 Avellaneda, Merlo-Province of Buenos Aires, Argentina. Office: 1544 Lavalle, Buenos Aires, Argentina.*

HEINTZ, Wilhelm Heinrich, German chemist; b. Berlin, 1807; prof., Halle, Germany; 1st to isolate bilirubin, 1851. Died 1880.

HEINTZEN, Paul Heinz, German pediatric cardiologist; b. Essen, Germany, May 8, 1925; s. Paul and Gertrud (Bahrenberg) H.; Staatsexamen, U. Bonn and Med. Acad., 1951; M.D., U. Dusseldorf (Germany), 1952; m. Heinke Thiessen, July 26, 1957; children—Matthias, Frauke, Christian, Andreas. Research fellow physiology U. Münster (Germany), 1954-59; pediatric cardiologist U. Kiel (Germany), 1954-59, faculty, 1959—, prof. pediatrics, 1965—, head dept. pediatric cardiology, 1966—. Mem. Assn. European Pediatric Cardiologists, German Soc. Pediatrics, German Soc. Circulation Research, German Soc. Biomed. Electronics. Author: Quantitative Phonokardiographie, 1960; Differentialdagnose von Krankheitssymptomen bei Kindern und Jugendlichen, 1963; also articles. Quantification of heart sound phenomena; methods for diacardiac sound transmission studies; devel. roentgen-densitometric methods for study heart function and circulation. Home: 112 Birkenweg, 2308 Preetz, Germany. Office: 15 Fröbelstr., 23 Kiel, Germany.*

HEINZ, Erich, German phys. biochemist; b. Essen, Germany, Jan. 10, 1912; s. Peter and Margarethe (Gubener) H.; student medicine U. Münster u. Kiel (Germany), 1934-39; grad. U. Kiel (Germany), 1939, Dr. med., 1941; m. Ursula M. Staak, Sept. 2, 1941; children—Bettina, Agnes, Peter. Asso. prof. biochemistry Tufts U., Boston, 1955-57; research asso. biol. chemistry Harvard, 1957, cons. biophys. lab. Med. Sch., 1956-58; research prof. physiology George Washington U., 1958; prof., chmn. Insts. Vegetative

Physiology, J. W. Goethe U., Frankfurt, Germany, 1959—; vis. investigator U. Uppsala (Sweden), 1956-58; vis. investigator NIH, Bethesda, Md., 1962-63. Mem. Gründungsausschuss Med. Hochschule Hanover (Germany), 1961—. Brit. Council fellow U. Liverpool (Eng.), 1949; USPHS Sr. Research fellow, 1958. Mem. Am. Soc. Biol. Chemists, Biophys. Soc. (charter), Gesellschaft für Physiol. Chem., Am. Heart Assn. Contbg. author: Handbuch der Zoologie, 1956. Research and numerous publs. on mechanism of acid secretion in stomach, kinetic and energetic aspects of active transport substances across biol. membranes; metabolism of calcium, albumin in blood plasma. Home: Ludwig-Rehn Strasse 14, Frankfurt, Germany.

HEINZ, Robert, German physician, pharmacologist; b. Wustegiersdorf, Germany, 1865. Dir., Inst. Pharmacology, Erlangen, Germany. Described certain refractile inclusion bodies seen in red blood cells in cases of poisoning, 1890 (inclusions also described by Paul Ehrlich, 1892), known as Heinz-Ehrlich bodies; research on dermatology and metallic colloids. Died 1924.

HEIRTZLER, James Ransom, Am. geophysicist; b. Baton Rouge, Sept. 16, 1925; s. William Ransom and Jimmie (Clark) H.; B.Sc. in Physics, La. State U., 1947, M.Sc., 1948; Ph.D., N.Y. U., 1953; m. Phyllis Virginia Trossen, Feb. 7, 1951; children—Fenton Ransom, Jason Dean. Asst. prof. Am. U. Beirut (Lebanon), 1953-56; sr. physicist Gen. Dynamics Corp., Rochester, N.Y., 1956-60; research scientist Lamont Geol. Obs., Columbia, Palisades, N.Y., 1960-64, sr. research scientist, 1964—. Mem. A.A.A.S., Am. Phys. Soc., Am. Assn. Physics Tchrs., Geol. Soc. Am., Am. Geophys. Union (chmn. rapid variation com. sect. on geomagnetism and aeronomy 1964—), Soc. Exploration Geophysicists. Research and publs. on interpretation of optical diffraction patterns resulting from interference light rays, atomic processes in Geiger counter, attentuation nuclear radiation by various materials, earth's magnetic field. Home: Upper Grandview St., Nyack, N.Y. 10960. Office: Lamont Geol. Obs., Palisades, N.Y. 10964.*

HEIS, Eduard, German astronomer; b. Cologne, Germany, Feb. 18, 1806; grad. Bonn, Germany, 1827; faculty math. and sci., Cologne, 1829-37, Aachen, Germany, 1837-52; prof. math. and astronomy, Münster, Germany, 1852—. Mem. Royal Astron. Soc. (fgn. asso.), Leopoldine Acad. (hon.), Sci. Soc. Brussels (hon.). Author math. textbooks including: Atlas coelestis. Research and publs. on zodiacal light shooting stars, variable stars, sunspots, Milky Way. Died 1877.

HEISE, George William, Am. electrochemist; b. Milw., June 27, 1888; s. Paul Edgar and Dora (Tyre) H.; B.S. in Chemistry, U. Wis., 1909, M.S., 1912; postgrad. U. Chgo.; D.Sc., Fenn Coll., 1960; m. Margaret Armstrong, Aug. 6, 1915; children—Margaret D. (Mrs. W. D. Stevens), Alice M. (dec.), George A. Instr., Grinnell Coll., 1909-10, DePaul U., 1910-11; with U. Wis., 1911-13; phys. chemist, chief sect, acting chief div. Bur. Sci., Manila, P.I., 1913-17; research electrochemist, engr. Nat. Carbon Co., Union Carbide Corp., 1919-53, dept. head Nat. Carbon Research Labs., Cleve., 1925-51, asso. dir. research, 1951-53, cons. in electrochemistry, 1953—; civilian NDRC, 1942-44. Cons. Office Naval Research, 1950-53 com. on undersea warfare Nat. Acad. Sci., 1953-65, chmn. panel on primary batteries, Recipient certificate of merit Cleve. Chem. Profession, 1953, citation of honor Ind. Tech. Inst., 1958. Fellow A.A.A.S.; mem. Electrochem. Soc. (past pres., hon. mem., Acheson medal 1954, medal Cleve. sect. 1960), Am. Chem. Soc., Am. Interprofl. Soc., Faraday Soc., Soc. Chem. Industry, Sigma Xi, Phi Lambda Upsilon, Alpha Chi Sigma. Author: (with A. S. Behrman) Philippine Water Supplies, 1918; also numerous articles. Patentee in field. Research on primary batteries, especially Leclanché, alkaline air depolarized systems, gas electrodes for fuel cells porous electrodes in electrosynthesis and electrowinning catalytic carbons. Address: 18550 Rivercliff Dr., Fairview Park, Cleve. 44126.*

HEISENBERG, Werner Karl, German physicist; b. Wurzburg, Germany, Dec. 5, 1901; s. August and Annie (Wecklein) H.; Dr. phil. U. Munich, 1923; Dr. phil. habil., U. Gottingen, 1924; m. Elisabeth Schumacher, Apr. 29, 1937; children—Wolfgang, Maria, Jochen, Martin, Barbara, Christine, Verena. Prof. theoretical physics U. Leipzig, 1927-41; dir. Kaiser-Wilhelm Inst. for Physics, prof. U. Berlin, 1941-45; dir. Max-Planck-Inst. for Physics prof. U. Gottingen, 1946-58; dir. Max-Planck-Inst. for Physics and Astrophysics, prof. U. Munich, 1958—. Pres. Alexander-von-Humboldt-Stiftung. Recipient Barnard medal Columbia, Matteucci medaille, Max-Planck medaille, Nobel prize for physics, 1932. Mem. Am. Philos. Soc. Author: Die physikalischen prinzipien der Quantentheorie, 1930; Wandlungen in den Grundlagen der Naturwissenschaft, 1935; Die Physik der Atomkerne, 1943; Vorträge über kosmische Strahlung, 2d edit., 1953; Das Naturbild der heutigen Physik, 1955; Physics and Philosophy, 1958. Noted for work on quantum theory; founded quantum mechanics, 1925; worked on atomic structure and Zeeman effect; evolved uncertainty principle (states impossibility of accurately determining simultaneously both position and velocity of a particle), 1927; suggested that laws of subatomic phenomena be stated in terms of observable properties such as intensity and frequency of radiation; current work on unified field theory. Home: 1 Rheinlandstrasse. Office: 6 Föhringer Ring, Munich, Germany.*

HEISER, Charles Bixler, Jr., Am. botanist; b. Cynthiana, Ind., Oct. 5, 1920; s. Charles Bixler and Inez (Metcalf) H.; A.B., Washington U., St. Louis, 1943, M.A., 1944; Ph.D., U. Cal. at Berkeley, 1947; m. Dorothy Gaebler, Aug. 19, 1944, children—Lynn Marie, Cynthia Ann, Charles Bixler III. Instr., Washington U., 1944-45; asso. botany U. Cal. at Davis, 1946-47; faculty Ind. U., Bloomington, 1947—, prof., 1957—. Guggenheim fellow, 1953; NSF Sr. Postdoctoral fellow, 1962. Mem. Am. Soc. Plant Taxonomists (pres. 1967), Bot. Soc. Am., Soc. Study Evolution (v.p. 1968), Am. Soc. Naturalists, A.A.A.S., Internat. Assn. for Plant Taxonomy, Phi Beta Kappa, Sigma Xi. Research and numerous publs. on systematics flowering plants, natural and artificial hybridization, origin cultivated plants. Home: 1018 Southdowns St., Bloomington, Ind. 47401.*

HEISING, Raymond Alphonsus, Am. engr., physicist; b. Albert Lea, Minn., Aug. 10, 1888; s. Charles and Anna A. (Fitzgerald) H.; E.E., U. N.D., 1912, D.Sc., 1947; M.S., U. Wis., 1914; m. Teresa A. Coneys, Nov. 25, 1920; children—William P., Charles R., Mary Ellen (Mrs. J. G. Ausman). Radio research engr. Western Electric Co., Inc., N.Y.C., 1914-25; radio research engr. Bell Telephone Labs., Inc., 1925-44, patent engr., 1945-53, ret., 1953; cons. engr., patent agt., Summit, N.J., 1953—. Recipient Modern Pioneer award N.A.M., 1940, Armstrong medal Radio Club Am., 1953. Fellow Inst. Radio Engrs. (past pres., Morris Liebmann Meml. prize 1921, Founder's medal 1957), A.A.A.S., Am. Inst. Elec. Engrs., Am. Phys. Soc. Trustee Armstrong Meml. Research Found. Author: Quartz Crystals for Electric Circuits, 1946. Contbr. articles to tech. jours. Patentee in field. Developed radio telephone transmitters for ships and planes, 1917-18; helped develop transatlantic long wave telephone circuit, 1922-29, short wave transoceanic circuits, 1925-29, ship-shore telephone circuit, 1929. Died Jan. 16, 1965.

HEISKANEN, W(eikko) A(leksanteri), Finnish geodesist, geophysicist; b. Kangaslampi, Finland, July 23, 1895; B.S., State U. Helsinki (Finland), 1917, M.S., 1919, Ph.D., 1924. Geodesist, Finnish Geodetic Inst., 1921-28; asst. prof. geodesy Helsinki U.,1926-49; prof. Finnish Inst. Tech., 1928-49, dean surveying dept., 1935-49; dir. Finnish Geodetic Inst., 1949—; project supr. mapping and charting research lab. Ohio State U., 1950—, dir. Inst. Geodesy, Photogrammetry, and Cartography, 1953—; dir. Isostatic Inst., Internat. Assn. Geodesy, 1936—; research asso. Cal. Inst. Tech., 1948. Decorated Order White Star (Estonia), St. Stephan (Hungary), White Rose (Finland); recipient grand award Finnish Govt., 1948, 49, 50, award Wihuri Found., 1953, Bowie medal Am. Geophys. Union, 1956. Mem. Geophys. Union, German Acad., Norwegian Acad., Dutch Geog. Soc., Finnish Acad., Finnish Geog. Soc. (hon., past pres.). Author: Columbus Geoid, 1956; The Earth and its Gravity, 1958. Modified isostatic-floating theory of G. B. Airy; originated hypothesis on thickness of earth's crust in Alps, Caucasus, Norway, Rocky Mountains, Ferghena Basin, Carpathians; developed isostatic reduction tables; studied gravimetric method, stellar astronomy. Home: 645 Beechwold Blvd., Columbus, O. Office: Geodeettinen Laitos, Bulevardi 40, Helsinki, Finland.

HEISTER, Lorenz, physician; b. Frankfort/Main, Germany, Sept. 19, 1683; ed. U. Giessen (Germany), Amsterdam, Leiden (both Netherlands); pupil of Ruysch and Boerhaave. Physician-gen. Dutch Mil. Hosp.; prof. anatomy and surgery, Altdorf, Switzerland, 1710-19, U. Helmstedt (Germany), from 1719. Author: Compendium anatomicum, 1717; Chirurgie, 1718; Anatomico Surgical Lexicon (a standard work), 1753. Founder modern German surgery; believed to have introduced spinal braces, 1710; described several tourniquets in use, 1718, folds of mucous membrane in cystic duct and in neck of gall bladder (Heister's valve), 1720, a sinus of external jugular vein (Heister's diverticulum), 1720; differentiated between direct and indirect inguinal hernias; opponent of Linnaeus. Died Helmstedt, Apr. 18, 1758.

HEITLER, Walter Heinrich, theoretical physicist; b. Karlsruhe, Germany, Jan. 2, 1904; s. Adolf and Ottilie (Rudolf) H.; student U. Karlsruhe, 1922-23, U. Berlin (Germany), 1923-24; Ph.D., U. Munich (Germany), 1926; D.Sc., U. Dublin (Ireland), 1954; m. Kathleen Winifred Nicholson, Mar. 21, 1942; 1 son, Eric. Rockefeller Research fellow, 1926-27; privatdozent sci. U. Göttingen (Germany), 1927-33; research fellow U. Bristol (Eng.), 1933-41; prof. Dublin Inst. for Advanced Studies, 1941-49, dir., 1946-49; prof. U. Zurich (Switzerland), 1949—. Recipient Max Planck medal, 1968. Fellow Royal Soc., 1948; mem. Royal Soc. Scis. Uppsala; hon. mem. Phys. Soc. Zurich, Royal Irish Acad., Phys. Soc. (Eng.), Swiss, Japanese phys. socs. Author: Quantum Theory of Radiation, 1936; Elementary Wave Mechanics, 1945; Man and Science, 1961. Research and numerous publs. on theory of chem. bond, theory of fast particles and cosmic rays, theory of nuclear forces, mesons, theory radiation damping. Home: 5 Am Guggenberg, Zurich 53, Switzerland.*

HEITZ, Emile, German geneticist; b. Strasbourg, Oct. 29, 1892; s. Paul and Mathilde (Schwalb) H.; ed. univs. Strasbourg, Basle, Munich, Heidelberg; Ph.D. in genetics; D. honoris causa, Faculty Scis., U. Cologne, U. Berlin, U. Hamburg; m. Elisabeth Staehelin, 1921; children—Roland, Thomas, Elisabeth, Sabastian. Asst. in Weihenstephan, Greifswald, Hamburg: prof. agrégé; asso. prof. Max Planck Soc., Basle; hon. prof. U. Tübingen. Recipient Schleiden medal Leopoldine Soc. Mem. Max Planck Soc., Swiss, German socs. botany, Swiss Soc. Genetics, Portuguese Biol. Soc. (hon.). Research and publs. on formation and structure of chromosomes in plants and animals, microscopy, electronic microscopes, structure of chloroplasts, genetic transformation of plants treated with x-rays. Home: 63 Sandweg, New-Allschwil. Office: Max Planck Society, University of Tübingen, Munich, Germany.

HEJDA, Stanislav, Czechoslovak nutritionist; b. Krochelavy, Czechoslovakia, Nov. 13, 1928; s. Vilem and Antonie (Svitakova) H.; M.D., Charles U., 1952; m. Zdenka Panenkova, Nov. 11, 1952; 1 son, Stanislava. With Inst. Human Nutrition, Prague, 1956—, head dept. for epidemiol. research; external lectr. Charles U., U. Brno. Mem. J. E. Purkyne Med. Assn. (Kose prize 1961, 65), Soc. for Rational Nutrition. Research, numerous publs. on nutritional status and dietary habits of different populations and profl. groups, nutritional disorders, nutritional status assessment, obesity. Home: 2520 Sporilov sidliste, Prague 4. Office: 800 Budejovicka, Prague 4, Czechoslovakia.*

HEJNY, Slavomil, Czechoslovak botanist; b. Lidman, Czechoslovakia, June 21, 1924; s. Josef and Ludmila (Dvoraková) H.; RNDr, Charles U., Prague, Czechoslovakia, 1948, C.Sc., 1956; m. Liselotta Teltscherová, June 12, 1952; children—Petr, Michal. Asst., Bot. Inst., Charles U., 1946-50; sci. worker Inst. Plant Prodn., 1953-56; sci. worker Bot. Inst., Czechoslovak Acad. Scis., Pruhonice, Czechoslovakia, 1956—, dir., 1962—. Lectr. synanthropical botany Charles U., 1963—. Mem. Czechoslovak Bot. Soc., Czechoslovak Limnological Soc., Commn. for Exploration of Danube, Internat. Assn. for Vegetation. Author: Okologische Charakteristik der Wasser-und Sumpfpflanzen in den slowakischen Tiefebenen, 1960; Eine Studie über die Ökologie der Echinochloa-Arten, 1957; also articles. Editor-in-chief Folia geobotanica et phytotaxonomica, 1965—. Research on Czechoslovak water and swamp plants and their communities, especially in rice fields and ponds; classified ecol. groups of Macrophytes; ecol. studies in weeds and xenophytes; floristical exploration of E. Slovakia and S. Bohemia. Home: 726 Sobínská, Praha 6-Ruzyne, Czechoslovakia. Office: Pruhonice nr. Prague, Czechoslovakia.

HEJTMANCIK, Milton R., Am. physician; b. Caldwell, Tex., Sept. 27, 1919; s. R. J. and Millie (Jurcak) H.; B.A., U. Tex., 1939, M.D., 1943; m. Myrtle Lou Erwin, Aug. 21, 1943; children—Kelly Erwin, Milton R., Peggy Lou. Practice medicine specializing in internal medicine, Galveston, Tex., 1949—; faculty dept. internal medicine Med. Br., U. Tex., 1949—, prof., 1965—, asst. dir. Heart Sta., 1949-65, dir., 1965—, dir. Heart Clinic, 1949—; cardiac cons. St. Mary's Hosp., 1951—, Galveston County Meml. Hosp., 1953—, USPHS Hosp., 1957—. Diplomate Am. Bd. Internal Medicine, Fellow A.C.P., Am. Coll. Chest Physicians, Am. Coll. Cardiology; mem. A.M.A., Am. Heart Assn. (fellow council clin. cardiology), Am. Fedn. Clin. Research, A.A.A.S., Phi Beta Kappa, Sigma Xi, Alpha Omega Alpha. Research and numerous publs. on clin. cardiology, especially electrocardiography, vectocardiography, arrhythmias of the heart, drug therapy of cardiac disorders. Home: 118 Marlin St., Galveston 77550.*

HEKTOEN, Ludvig, Am. pathologist; b. Westby, Wis. July 2, 1863; s. Peter P. and Olave Hektoen; A.B. Luther Coll., Ia., 1883, A.M., 1896; studied U. of Wis.; M.D., Coll. Phys. and Surg., Chicago, 1887; interne, Cook County Hosp., Chicago, 1887-89; studied Upsala, Prague, Berlin, 1890, 94, 95; M.D., ad eundem, Rush Med. College, 1896; hon. M.D., U. of Norway, 1911; Sc.D., U. of Mich., 1913, U. of Wis. 1916; U. of Chicago, 1940; LL.D., U. of Cincinnati, 1920. Western Reserve U., 1920, and Luther Coll., Decorah, Ia., 1936; m. Ellen Strandh, July 7, 1891; children—Aikyn (dec.), Josef Ludvig. Pathologist to Cook County (Ill.) Hosp., 1889-1903; lecturer pathology, Rush Med. Coll., 1890-92; physician to coroner's office, Chicago, 1890-94; prof. pathology, Coll. Phys. and Surg., Chicago, 1892-94; prof. morbid anatomy, Rush Med. Coll., 1895-98, prof. pathology, 1898-1933; prof. and head dept. pathology, U. of Chicago, 1901-32; dir. John McCormick Inst. for Infectious Diseases, 1902-39. Mem. Occupational Diseases Commission of Illinois, 1909-11. President Chicago Tumor Institute from 1938. Chmn. div. of med. sciences, Nat. Research Council, 1924-25, 1926-27, 1929-30; chmn. Nat. Research Council, 1936-38; mem. Nat. Advisory Health Council, U.S.P.H.S., 1934-38; exec. dir. Nat. Advisory Cancer Council, U.S.P.H.S., 1937-1944. Received Centennial Award, Wis. State Medical Society, 1941, Distinguished Service medal, American Medical Association, 1942. Gold-headed cane. Association American Pathologists and Bacteriologists,

1944; Howard Taylor Ricketts award, U. of Chicago, 1949. Mem. National Acad. Sciences, Assn. Am. Phys., A.M.A., Assn. Am. Pathologists and Bacteriologists (pres. 1901), Soc. Am. Bacteriologists (pres. 1929), Soc. Immunologists (pres. 1927), Inst. of Medicine of Chicago (pres. 1929), A.A.A.S. (v.p. 1909); hon. mem. Norwegian Med. Soc., Am. Soc. of Clinical Pathologists; mem. Norwegian Acad. of Sciences; honorary member Swedish Medical Society (Stockholm), Norwegian Pathological Society. Vienna Microbiologie Society. American Society of Bacteriologists, Academy of Medicine. Washington, College of American Pathologists (honorary member). Author: Post-mortem Technique, 1894; Introduction to Study of Infectious Diseases. Editor: Durck's Pathologic Histology; Collected Works of Christian Fenger: Contributions to Medical Science, by Howard Taylor Ricketts. Co-editor and contributor to American Text-Book of Pathology, 1902; wrote numerous articles on pathology, bacteriology and immunology. Co-compiler A Bibliography of Infantile Paralysis, 1789-1944, 1946. Editor Jour. Infectious Diseases, 1904-40, Archives of Pathology, 1926-49. Demonstrated that X-rays check or slow prodn. of antibodies, 1915; induced measles with injections of blood taken from measles patients, 1905; called attention to danger of isoagglutination in blood transfusions, 1907; produced exptl. bacillary cirrhosis of liver, 1906. Died Chgo., July 5, 1951.

HELBOK, Adolf, Austrian ethnographer; b. Hiltisau, Feb. 2, 1883; s. Adolf and Frieda (Wagenhauser) H.; ed. U. Innsbruck, U. Vienna; Ph.D. in ethnology; m. N. Flossmann, 1919. Instr. history, 1919; prof. U. Innsbruck, 1924, full prof. ethnography and natural history, 1941; affiliate prof. U. Berlin, 1934; full prof. U. Leipzig, also dir. numerous natural hist. and ethnographic instns., 1935. Mem. Austrian Soc. for Ethnographic Atlas (hon.), also numerous others, Deutsches Kulturwerk (pres. Austria). Collaborator: Deutscher Volkskundeatlas, 1929; Oesterreichscher Volkskundeatlas, 1953. Research and numerous publs. on ethnography of Germany and France. Address: Gotzens über Innsbruck, Austria.

HELBRONNER, Paul, French geodesist; b. Compiègne, France, Apr. 24, 1871; student École Polytechnique, Paris, 1882-94; Dr.-ès-sc. in Math., U. Paris, 1911; mem. Bur. des Longitudes, Paris, 1932-38, French Acad. Scis., 1927. Completed triangulation of an Alpine region; made accurate trigonometric liaison of area from Corsica to France. Died Paris, Oct. 18, 1938.

HELD, Richard Marx, Am. psychologist; b. N.Y.C., Oct. 10, 1922; s. Lawrence W. and Tessie (Klein) H.; B.A., Columbia U., 1943, B.S., 1944; M.A. in Psychology, Swarthmore Coll., 1948; Ph.D., in Exptl. Psychology, Harvard, 1952; m. Doris F. Bernays, June 29, 1951; children—Lucas D.B., Julia B., Andrew L.B. Research fellow Inst. Pub. Health, Harvard, 1952-53; faculty Brandeis U., 1953-63, prof., chmn. dept. psychology, 1961-62; prof. psychology Mass. Inst. Tech., 1963—; staff Inst. for Advanced Study, Princeton, N.J., 1955-56. Mem. exptl. psychology study sect. B, NIH, 1964—, chmn., 1966—; mem. com. on vision, NRC, 1965. Sr. Research fellow NSF, 1962-63. Fellow Am. Psychol. Assn., A.A.A.S., Am. Acad. Arts and Scis.; mem. Psychonomic Soc., Eastern Psychol. Assn. Co-editor, psychologische Forschung, 1966—; cons. editor Jour. Exptl. Psychology, Perspectives in Biology and Medicine. Research and publs. on mechanisms of sensorially guided behavior in higher mammals, devel. and maintenance of visuomotor coordination and aspects of perception of space. Home: 102 Appleton St., Cambridge, Mass. 02139.*

HELE-SHAW, Henry Selby, English engr.; b. Billericay, Essex, Eng., July 29, 1854; D.Eng. (hon.), Liverpool, 1931; D.Sc.; LL.D.; m. Ella Marion Rathbond; 1 dau. First prof. engring. Univ. Coll., Bristol, 1881-85; 1st prof. engring., Harrison chair Univ. Coll., Liverpool, 1885-1904; 1st prof. civil, mech. and elec. engring. Transvaal Tech. Inst., 1903-04, prin., 1904-05; emeritus prof. Liverpool U. Recipient Gold medal Internat. Inventions Exhbn., 1885, Gold metallist Inst. N.Am., 1899, Silver medallist Soc. Arts, 1908. Fellow Royal Soc., 1899, Soc. Engrs. (hon.), Royal Metall. Soc.; mem. Instn. Mech. Engrs. (hon. life, past pres.), Bristol Naturalists Soc. (hon.), Instn. Civil Engrs. (Watt gold medal, Telford premium), Instn. Chem. Engrs. (hon., Liverpool Engring. Soc. (pres. 1894-95), Royal Soc. South Africa (council 1905), Brit. Assn. (past pres. mech. sect.), Instn. Automobile Engrs. (past pres.), Assn. Engrs. in Charge (past pres.), Soc. Model and Exptl. Engrs. (past pres.). Contbr. numerous articles in math., phys. and engring. fields Inventor clutches, pumps, also other machines. Died Jan. 30. 1941.

HELFERICH, Burckhardt, German chemist; b. Greifswald, June 10, 1887; s. Heinrich and Natalie (Burckhardt) H.; ed. univs. Lausanne, Munich, Berlin; Ph.D. in chemistry; Ph.D. honoris causa in natural sci.; m. Hildegard Kohlrautz, 1922; children—Hans, Eva, Marianne, Hilde, Gisela. Instr., U. Berlin, 1922; full prof. chemistry U. Frankfort, 1922, U. Gesifswald, 1925, U. Leipzig, 1930, U. Bonn., 1947. Recipient Emil Fischer medal. Mem. Acad. Scis. Saxe, Leopoldina Soc. Halle, Soc. German Chemists, German Bunsen Soc., Am. Chem. Soc. Research and publs. on chemistry, organic chemistry, hydrates of carbon and fermentation. Home: 66 Talweg, Bonn 53. Office: 168 Meckenheimer Allee, Bonn 53, Germany.

HELFFERICH, Friedrich, Georg, chemist; b. Berlin, Germany, Aug. 1, 1922; s. Karl and Anna (von Siemens) H.; Chemiker-Vordiplom, U. Hamburg, 1949, diploma, 1952; Ph.D., U. Gottingen, 1955; m. Hana Maria Konecna, Feb. 24, 1961; children—Christiane, Cornelia, Stefanie. Came to U. S., 1956, naturalized, 1964. Research asst. Max Planck Institut für physikalische Chemie, Gottingen, Germany, 1951-56, vis. scientist, 1958; research asst. Mass. Inst. Tech., Cambridge, 1954, Cal. Inst. Tech., Pasadena, 1956-58; chemist Shell Devel. Co., Emeryville, Cal., 1958-65, supr., 1965—. Lectr., U. Cal. at Berkeley, 1961-62; chmn. Gordon Research Conf. on Ion Exchange, 1967; chmn. subcom. on terminology of ion exchange NRC, 1962—. Author: Ionenaustauscher, 1959; Ion Exchange, 1962; also numerous articles. Translator: General Chemistry (Linus Pauling), 1956; Kinetics and Mechanism (A. A. Frost, R. G. Pearson), 1964. Research in theory of bi-ionic membrane potentials, theory of ion-exchange kinetics, ligand exchange separation technique, kinetics of catalysis, computer programs for nonlinear diffusion problems, tracer-pulse technique for determining sorption isotherms and phase equilibria, gen. theory for multicomponent chromatography, chem. reaction kinetics. Home: 710 Wildcat Canyon Rd., Berkeley, Cal. 94708. Office: Shell Devel. Co., Emeryville, Cal. 94608.*

HELGASON, Sigurdur, mathematician, b. Akureyri, Iceland, Sept. 30, 1927; s. Helgi and Kara (Briem) H.; Cand. Phil., U. Iceland, 1946; Mag. Scient., U. Copenhagen (Denmark), 1952; Ph.D., Princeton, 1954; m. Artie Gianopulos, June 9, 1957; children—Thor, Anna. Instr., Mass. Inst. Tech., 1954-56; lectr. Princeton, 1956-57; asst. prof. U. Chgo., 1957-59; with Mass. Inst. Tech., 1959—, prof., 1965—. Vis. asst. prof. Columbia, 1959-60; mem. Inst. for Advanced Study, Princeton, 1964-66. Mem. Am. Math. Soc. Author: Differential Geometry and Symmetric Spaces, 1962. Research and publs. in geometric analysis; differential geometry; harmonic analysis on Lie groups and coset spaces. Home: 5 Benton Rd., Belmont, Mass. 02178.

HELIDOROS OF LARISSA, Greek mathematician; flourished after Ptolemy; Damianos, either his son or pupil. Work on perspective which was mainly a commentary on Euclid.

HELL, Maximilian, astronomer; b. Banská Stiavnica, Hungary, May 13, 1720; mem. Soc. Jesus; tchr., Löcse (Leutschau), Silesia, also Cluj, Rumania; dir. Vienna (Austria) Obs.; Mem. French Acad. Scis. Author: Ephemerides astronomicae, 1757; Wardehusii in Finnmarchia observationes, 1770. At invitation of King Christian VII of Denmark, observed transit of Venus, Lapland, June 1769 (pub. observations, 1770). Died Apr. 14, 1792.

HELLEGERS, André Eugène, obstetrician; b. Venlo, Netherlands, June 5, 1926; s. Anthony C. and Jane (Boland) H.; M.D., Belgian Central Jury, 1952; Dipl. Med. Aeron., Paris (France) U., 1952; m. Charlotte Fraser Lindsay Sanders, June 14, 1957; children—Paul Matthew, Caroline Ann, Désirée Eugenie, Renée Elizabeth. Josiah Macy Found. research fellow Yale, 1956-57; faculty Johns Hopkins, Balt., 1953-67, asso. prof. gynecology and obstetrics, 1962-67; prof. obstetrics and gynecology Georgetown U., 1967—; staff Johns Hopkins Hosp., 1953—, gynecologist-obstetrician, 1960—; sr. research scholar Josesph P. Kennedy, Jr. Meml. Found., 1961—; mem. Josiah Macy Found. Andean Expdn., Peru, 1958; cons. gynecologist-obstetrician Balt. City Hosps., 1960—. Cons., NIH, 1958—; dep. sec. gen. Papal Commn. on Population and Birth Control, 1966—; cons. Office of Sec., U. S. Dept. Health, Edn. and Welfare, 1964-65. Mem. Soc. for Gynecologic Investigation (pres. 1968, council 1964—). Research and numerous publs. in maternal, fetal and placental physiology. Home: 4805 Dorset Av., Chevy Chase, Md. Office: Georgetown U. Hosp., Washington 20007.

HELLEMS, Harper Keith, Am. physician; b. Sinksgrove, W.Va., Mar. 16, 1920; s. Harvey Kemy and Nilah (Eppling) H.; student U. Va., 1938-40, M.D., 1943; m. Anne Wheatley, Feb. 12, 1939; children—Harper Keith, Eric Wheatley, Harvey Kem. Med. research fellow Peter Bent Brigham Hosp., Harvard Med. Sch., 1946-48; practice medicine specializing in cardiology, Jersey City, 1960-65; faculty Wayne U. Coll. Medicine, 1950-60, prof. medicine, 1959-60; prof. medicine, dir. div. cardiovascular disease Seton Hall Coll. Medicine, Jersey City, 1960-65; prof., chmn. dept. medicine U. Miss. Sch. Medicine, Jackson, 1965—. Recipient Susan and Theodore Cummings Humanitarian award. Fellow A.C.P.; mem. Am. Coll. Cardiology, Am. Soc. for Clin. Investigation, Am. Fedn. for Clin. Research, Central Soc. for Clin. Research, Am. Physiol. Soc., Assn. U. Cardiologists, Council Clin. Cardiology, Am. Heart Assn. Editorial bd. Jour. Lab. and Clin. Medicine, 1960—. Contbr. articles to tech. jours. Devised technique for measuring pulmonary capillary pressure useful in evaluating patients with heart disease; research in hemodynamics particularly coronary circulation. Home: 5365 Suffolk Dr., Jackson, Miss.*

HELLER, Arnold, German pathologist, neurologist; b. 1840; discovered reactions for determining presence of phosphates and albumins in urine; described aortitis and aortic aneurysm (Doehle-Heller aortitis), established role of syphilis in their causation, 1899. Died 1913.

HELLER, Arnold Franz August, German chemist; b. Dec. 12, 1905; s. Franz and Elise (Sommerfeld) H.; Ph.D. in Chemistry, U. Berlin; m. Magdalene Dreist, 1932; children—Gert, Ingrid, Karin. With Prussian Inst. for Cleanliness of Water, Sky and Air, 1929; asst., 1935; sci. mem., 1939; sci. cons., prof., 1943; dir., prof., 1958. Mem. Commn. for Purity of Air; mem. Com. Experts on San. Air. Author: Luft in der Orts. und Landesplanung, 1953. Research and publs. on atmosphere and hygiene. Home: 119a Mörchingerstrasse, Berlin 37. Office: 1 Corrensplatz, Berlin 33, Germany.

HELLER, Carl G., Am. physiologist; b. Syracuse, Jan. 20, 1913; s. Ernst and Gertrude (Warmburg) H.; Ph.B., U. Wis., 1936, M.D., 1940, Ph.D., 1940; m. Meta Betz Ellis, July 23, 1962; children—Stephen C., Gary V. Fellow, Wayne State U., 1941-42, instr., 1942-44; faculty U. Ore. Med. Sch., 1944-57; head div. reproductive physiology Pacific N.W. Research Found., Seattle, 1958—. Recipient Ciba award in endocrinology, 1948; Squibb award Pacific Coast Fertility Soc., 1962; Oliver Bird lectr. Royal Coll. Physicians, London, 1965. Fellow A.C.P.; mem. A.M.A., Am. Soc. Clin. Investigation, Central, Western, N.W. socs. clin. research, Am. Physiol. Soc., A.A.A.S., N.Y. Acad. Scis., Am. Fedn. Clin. Research, Soc. Exptl. Biology and Medicine, Royal Soc. Medicine, numerous others. Research, numerous publs. on interrelationships of pituitary-gonad axis for men and women; developed assay methods for urinary gonadotropins, estrogens, testosterone hormones; pioneer in study of effects of progestins on women as possible oral contraceptive; potential male hormonal contraceptives; effects of radiation on human testis. Home: Route 2, Box 233, Poulsbo, Wash. 98370. Office: 732 Broadway, Seattle 98122.*

HELLER, Edward Lincoln, Am. mech. engr.; b. Newark, Feb. 12, 1912; s. Ira S. and Mary (Coyle) H.; B.S. in Mining Engring., Lehigh U., 1934; M.B.A. in Transp., Harvard, 1939; m. Irene M. Farrar, Apr. 27, 1946; children—Peter DeBois, Gregory Edward. Mining engr. Empire Zinc Co., Gilman, Colo., 1934-38; transp. engr. Standard Oil Co. Ohio, 1939-41; chief engr. MacLaughlin-Carr Assos., Indsl. Designers, N.Y.C., 1946-47; staff engr. Joint Congl. Com. on Atomic Energy, 1947-56; dir. Nuclear div. H. K. Ferguson Co., Cleve., 1956-59; tech. mgr. East Coast Office, Gulf Gen. Atomic, Incorporated, Washington, 1959—. Mem. Am. Nuclear Soc. (past v.p. Washington sect.), Am. Soc. M.E., Am. Nuclear Soc., Marine Tech. Soc. Contbr. articles to tech. jours. Devel. true sequential sampling tables for quality control application and application to major quality control insp. operations, process steam reactors. Home: 9515 Cable Dr., Kensington, Md. 20795. Office: 1025 Connecticut Av. N.W., Washington 20036.*

HELLER, Hàns, pharmacologist; b. Brünn, Czechoslovakia, Sept. 25, 1905; s. Joseph and Valerie (Schwarz) H.; student U. Vienna (Austria), 1927-28; Dr.rer.nat., U. Prague (Czechoslovakia), 1929; M.B.B. Ch., Emmanuel Coll., Cambridge (Eng.) U., 1938, M.D., 1948; m. Josephine Gertrude Libich, July 23, 1933; children—Katharine (Mrs. D. J. Jewell), Barbara (Mrs. Malcolm Faulk). Beit Meml. Research fellow U. Coll. Hosp., also dept. pharmacology U. Oxford (Eng.), 1938-41; faculty U. Bristol (Eng.), 1941—, prof., chmn. dept. pharmacology, 1949—, dean Faculty Medicine, 1966—. Hon. dir. group in neurosecretion Med. Research Council, 1965—. Mem. European Soc. for Comparative Endocrinology (pres.), Soc. for Endocrinology, German Physiol. and Pharmacological Soc. (corr.), Pakistan Pharmacological Soc. (hon.). Editor: The Neurohypophysis, 1957; (with R. B. Clark) Neurosecretion, 1963; (with U. S. von Euler) Comparative Endocrinology, 1963. Editor, Jour. Endocrinology, 1963—. Research and numerous publs. on control of water and salt content of body, evolution of pituitary gland. Home: 1 Ivywell Rd., Bristol, Eng.*

HELLER, Jack, Am. mathematician; b. Bklyn., Sept. 11, 1922; s. Michael and Lillian (Shohart) H.; B. Aero. Engring., Poly. Inst. Bklyn., 1944, M.Aero. Engring., 1946, Ph.D., 1950; m. Myra Minna Levine, Aug. 4, 1946; children—Glen, Cathy, Douglas, Adam. Instr., Newark Coll. Engring., 1946-50; research asso. Princeton, 1951-52; research scientist N.Y. U. Courant Inst. Math. Scis., 1953-61; faculty N.Y. U., N.Y.C., 1961—, prof. math., 1965—, dir. Heights Acad. Computing Facility, 1961—, dir. Inst. for Computer Research in Humanities, 1965—. Mem. Am. Math. Soc., Operations Research Soc., Soc. for Indsl. and Applied Math., Assn. for Computing machinery. Author index: (with Alice M. Pollin, Raquel Kersten) Revista de Filogia Espanola, 1964; also articles. Research in numerical analysis, computer scis., operations research and humanities data processing. Home: 354 Manor Rd., Englewood, N.J. 07635. Office: Math. Dept., N.Y. U., N.Y.C. 10453.*

HELLER, Jakob Rudolf Florian, German geologist; b. Nuremberg, July 12, 1905; s. Stefan and These

(Schiller) H.; ed. univs. Erlangen, Munich, Heidelberg; Ph.D. in natural sci.; m. Hedwig Bock, 1939; 1 son, Hartmut. Sci. collaborator Inst. Geology, South Halle; sci. asst. Inst. Geology, Giessen; curator, instr., affiliate prof. geology U. Heidelberg, Inst. Geology, U. Erlangen. Mem. Paleontol. Soc., German Soc. for Mammology, German Quarternary Union, Hugo-Obermaier Soc., Soc. Speleology. Author: Der Dritte Archaeopteryz-Fund aus den Solnhoefener Plattenkalken. Research and numerous publs. on Fossils, eocene mammals of Geiseltal near Halle, Saale. Home: 26 Rudrunstrasse, Nuremberg. Office: 5 Schlossgarten, Erlangen, Germany.

HELLER, Johann Florian, Austrian physiologist; b. Iglau, May 4, 1813; pupil of Liebig and Wöhler; prof. Vienna. Author: Die Harnconcretionen, 1860. Research on chemistry of urine; mem. New Vienna Sch.; devised ring test for albumen, 1844, caustic potash, test for sugar in urine, 1844; 1st to note retention of chlorides in pneumonic urine, 1847; inventor ureometer to measure specific gravity, 1848; introduced caustic potash test for blood in urine, 1858. Died 1871.

HELLER, John Herbert, Am. research scientist; b. N.Y.C., Nov. 28, 1921; s. Herbert Charles and Helen (Breschel) H.; B.A., Yale, 1942; M.D., Western Res. U., 1945; m. Mary Helen Strong, Oct. 5, 1946; children—John Christopher, Kathleen Karen, Susan Wick, Sandra Lee. Faculty, Yale, 1946-53; exec. dir. New Eng. Inst. for Med. Research, Ridgefield, Conn., 1954-—. Project dir. AEC, 1951-55; lectr. univs., 1955-56, 60-—; cons. USN, USAF, 1959, A.M.A., 1961-—; mem. various coms. NRC, NSF, 1961-62. Recipient Silver medal Belgian Govt.; Internat. Gold medal Soc. for Reticuloendothelial Research. James Hudson Brown fellow, 1948-49; Am. Heart Assn. fellow, 1949-51. Fellow N.Y. Acad. Scis., Am. Geriatrics Soc.; mem. Am. Astronautical Soc. (sr.), I.E.E.E., Biophys. Soc., Am. Chem. Soc., Am. Soc. for Clin. Investigation, Soc. for Nuclear Medicine, Am. Geophys. Union, A.A.A.S., Soc. for Exptl. Biology and Medicine, Ordnance Assn., Am. Physiol. Soc., Am. Fedn. for Clin. Research, Am. Heart Assn. (established investigator 1951-54), A.M.A., Reticuloendothelial Soc. (sec.-treas. 1954-64), Internat. Soc. RES (sec.-gen.), Sigma Xi. Author: Of Mice, Men and Molecules, 1960. Editor: Reticuloendothelial Structure and Function, 1960. Research, publs. in hypertension, enzyme kinetics, radiation biology; host def. system, shark behavior. Office: Box 308, Grove St., Ridgefield, Conn. 06877.*

HELLER, Laszlo, Hungarian engr.; b. Nagyvrad, Hungary, Aug. 6, 1907; s. Herman H. and Fanny (Weiss) H.; Dr.sc.-techn., Fed. Tech. U., Zürich, Switzerland, 1948; m. Ilona Kun, Dec. 4, 1945. Engr. praxis, 1931-36; cons. engr., 1936-48; tech. leader Thermoenergetical Design Bur., Budapest, Hungary, 1948-—; prof., chief dept. energetics Tech. U., Budapest, 1952-—. Mem. State Office Tech. Devel., 1962-—. Recipient Kossuth prize, 1951. Mem. Hungarian Acad. Scis. Contbr. articles to tech. jours. Pioneered field of econ. utilization of heat pump; elaboration of new schemes for better thermal efficiency of atomic power plants, waste heat recovery, use of gas turbine as indsl. heat and power generator; inventor System Heller (condensation using air). Home: 45. Bimb-ut, Budapest. Office: 3.Müegyetem-rkp., Budapest, Hungary.*

HELLER, Paul, physician; b. Komotau, Bohemia, Aug. 8, 1914; s. Alfred and Elsa (Hoenig) H.; M.D., U. Prague, 1938; m. Alice H. Florsheim, Aug. 3, 1946; children—Thomas Allen, Carol Elizabeth. Came to U. S., 1946, naturalized, 1948. Practice medicine specializing in hematology, Chgo., 1954-—; physician VA Westside Hosp., 1954-57, hematologist, dir. research, 1957-—; faculty Coll. Medicine, U. Ill., 1954-—, prof. medicine, 1963-—. Chmn. VA Hematology Research Com.; mem. hematology study sect. NIH. Mem. Am. Fedn. Clin. Research, Central Soc. Clin. Research, Soc. Exptl. Biology and Medicine, Am. Assn. Immunologists, Am., Internat. socs. hematology, A.C.P. Contbr. to Progress in Hematology, 1964. Research and numerous publs. on immunology and biochem. mechanisms in autoimmune disease, relationship of drug reactions to lupus erythematosus, cause and mechanisms of anemia in liver disease, enzyme abnormalities in megaloblastic anemia; discoverer several abnormal hemoglobins, pathogenetic mechanisms of certain hereditary anemias, models of autoimmune hemolytic anemias. Home: 376 Herrick Rd., Riverside, Ill. 60546. Office: 820 S. Damen St., Chgo. 60612.*

HELLER, Robert Leo, Am. geologist; b. Dubuque, Ia., Apr. 10, 1919; s. Edward W. and Olive (Bauck) H.; B.S., Ia. State U., 1942; M.A., U. Mo., 1943, Ph.D., 1950; m. Geraldine F. Hanson, Sept. 26, 1946; children—Roberta Lynn, Katherine Louise, Nancy Elizabeth. Geologist, U. S. Geol. Survey, 1943-44, Mo. Geol. Survey, summers 1947-48, 54, Minn. Geol. Survey, summer 1955; acting exec. dir. Am. Geol. Inst., summer 1960, dir. Teaching Resources Devel. Program, 1958-62; prof., head dept. geology U. Minn., Duluth, 1960-—, asst. to provost, 1965-—; dir. Earth Sci. Curriculum Project U. Colo., 1963-65. Earth sci. cons. McGraw-Hill Pub. Co., 1967-—; cons. U. S. Office Edn., 1966-—. Recipient Neil Miner award for outstanding contbn. in geol. edn., 1965. Mem. Geol. Soc. Am., Am. Assn. Petroleum Geologists, Paleontol. Soc., A.A.A.S., Polar Soc., Sigma Xi. Editor: Geology and Earth Sciences Sourcebook for Elementary and Secondary Schools, 1962. Research in earth sci. Home: 320 Morley Pkwy., Duluth, Minn. 55803.*

HELLER, Wilfried, chemist; b. Bad-Dürkheim, Germany, Dec. 13, 1903; s. Gustav and Lina (Strauss) H.; B.S. in Chemistry, U. Würtzburg (Germany), 1925; Ph.D., U. Berlin, 1931; m. Abbie H. Howe, Aug. 31, 1943. Came to U. S., 1938, naturalized, 1944. Research asso. Kaiser Wilhelm Institut, Berlin-Dahlem, Germany, 1931-33; instr., fellow U. Paris, France, 1933-38; lectr., research fellow U. Minn., Mpls., 1938-42; lectr. colloid chemistry, U. Chgo., also dir. colloid div. Rubber Research Project 1 (WPA), 1943-46; from asst. prof. to prof. Wayne State U., Detroit, 1946-—. Cons., Dow Chem. Co., Midland, Mich., 1951-—, Polymer, Sarnia, Ont., Can., 1957-—; chmn., Gt. Lakes Confs. Polymer and Colloid Sci., 1964-68. Recipient citation Eximium, U. Berlin, 1931, Liebig Fellowship award, 1931, Research award Wayne State U. chpt. Sigma Xi, 1959; NSF Sr. Postdoctoral fellow, 1965-66. Fellow Am. Phys. Soc., A.A.A.S.; mem. Am. Chem. Soc., Nat. Geog. Soc., Sigma Xi, Phi Lambda Upsilon. Contbr. to Ency. Britannica, 1944-—. Author: Propriétés Magnetooptiques des Solutions Colloidales, 1939; (with W. J. Pangonis, A. Jacobson) Tables of Light Scattering Functions for Spherical Particles, 1957; (with W. J. Pangonis) Angular Scattering Functions for Spherical Particles, 1960; (with A. F. Stevenson) Tables of Scattering Functions for Heterodisperse Systems, 1961. Important work includes research in various phases of colloid, polymer sci., biophys. chemistry, optics; invention of apparatus for determining molecular shapes from scattering in streaming solutions, new micro method for determining surface tension. Home: 12757 LaSalle Blvd., Huntington Woods, Mich. Office: Dept. Chemistry, Wayne State U., Detroit 48202.*

HELLMAN, Bo Olof Alexander, Swedish physician; b. Mariestad, Sweden, June 19, 1930; s. Erik Olof Alexander and Ingeborg (Hedberg) H.; M.L., Med. Faculty, U. Uppsala (Sweden), 1958, M.D., 1959; m. Birgit Inga Margareta Nilsson, Oct. 14, 1961; children—Björn Erik, Mats Gustav Ingemar, Jarl Olof Karl. Faculty, U. Uppsala, 1951-63, asst. prof., 1960-63; researcher in exptl. diabetes Swedish Med. Research Council, 1963-64; asso. prof. Karolinska Institutet, Stockholm, Sweden, 1964-66; prof., head histological dept. U. Umea, Sweden, 1966-—. Mem. Am. Diabetes Assn., European Assn. for Study Diabetes. Editor: (with S. E. Brolin and H. Knutson) The Structure and Metabolism of the Pancreatic Islets, 1964. Research and numerous publs. on pancreatic Islets of Langerhans in math. terms, evaluations of mechanisms of synthesis and secretion of insulin, relationship between obesity and diabetes. Home: 13 Murklevägen, Uppsala, Sweden. Office: Histological Dept., U. Umea, Umea 6, Sweden.*

HELLMAN, Milo, orthodontist; b. Jassy, Rumania, Mar. 26, 1872; s. Wolf and Fanny (Hellman) H.; came to U. S., 1888, naturalized, 1893; D.D.S., U. Pa., 1905; Sc.D., 1933; postgrad. Angle School of Orthodontia, N.Y. City, 1908; hon. Sc.D., Witwatersrand Univ., Johannesburg, South Africa, 1938; m. Helen Michelson, Nov. 30, 1905; children—Doris, Edith (Mrs. John L. Bull Jr.), Marion (Mrs. William T. Sandalls). Practiced dentistry, N.Y.C., 1905-08, orthodontia 1908-42, cons., from 1942; lecturer in orthodontia, Univ. of Pa. School of Dentistry, 1924-26, Harvard Dental Sch., 1927-28; prof. comparative dental anatomy, New York U. Coll. Dentistry, 1927-28, prof. orthodontia, 1928-29; prof. of dentistry, Sch. Dental and Oral Surgery, Columbia U., from 1932; research associate in phys. anthropology, Am. Museum Nat. History from 1917. Mem. exec. group of coms. for standardization of anthrop. techniques, Internat. Congress Anthropologists and Ethnologists, 1937-38; mem. S. African Expdn. of Am. Mus. of Nat. History, 1938; consultant Bur. of Med. Information, N.Y. Acad. of Medicine. Recipient Albert II. Ketcham Memorial award, 1939. Fellow N.Y. Acad. Science (v.p. 1932, 33), A.A.A.S., Am. Coll. of Dentists, Odontological Soc. Union of S. Africa; asso. fellow N.Y. Acad. Mem.; mem. Am. Dental Assn. (life), Internat. Assn. Dental Research (pres. N.Y. Sect. 1933), Am. Association Physical Anthropologists (life), American Assoc. Mammalogists, Ethnol. Soc., Am. Assn. Orthodontists (v.p. 1941). Author: The Dentition of Dryopithecus and the Origin of Man (with W. K. Gregory), 1926; Fossil Anthropoids of the Yale-Cambridge Expedition of 1935 (with others), 1938; also articles. Mem. editorial board Archives of Clinical and Oral Pathology. Editor American Orthodontist, 1910-12. Contributor of chapter "The Evidences of the Dentition on the Origin of Man" in Early Man, 1937; chapter "The Face in Its Developmental Career" in The Human Face, 1935; chapter "The Factors Influencing Occlusion" in Development of Occlusion, 1941. Died May 11, 1947.

HELLMANN, Kurt, pharmacologist; b. Nuremberg, Germany, May 12, 1922; s. Stefan and Senta (Levite) H.; D.Phil., Oxford U., 1953, B.M., B.Ch., 1958, D.M., 1964; m. Jane Rosamond Caseley, July 31, 1961. Mem. sci. staff Brit. Med. Research Council, 1951-56, Reckitt & Sons Ltd., 1960-62; head chemotherapy unit Imperial Cancer Research Fund, London, Eng., 1962-—. Lectr. various univs., 1960-—. Mem. Brit. Pharm. Soc., Physiol. Soc., Biochem. Soc. Research and publs. on autonomic nervous system physiology, skin transplantation, theories of origin of malformed infants, mechanism of action of thalidomide, new anticancer drugs. Home: Office: 44 Lincoln's Inn Fields, London W.C.2, Eng.*

HELLOT, Jean, French chemist; b. Paris, Nov. 20, 1685; insp.-gen. royal dye-works; mem. French Acad. Scis., 1735; Fellow Royal Soc., 1740. Author: De la fonte des mines et des fonderies, 1750-53; L'art de la teinture des laines . . . , 1772. Editor: Gazette de France, 1718-32. Research on bismuth, 1737, also on preparation techniques of sulphur, exploitation of mines; his analysis of zinc demonstrated that it is not composite mixture. Died Paris, Feb. 15, 1766.

HELLPACH, Willy, German psychologist; b. Öls, Silesia, Feb. 26, 1877; became prof., Karlsruhe, Germany, 1911, Heidelberg, Germany, 1926; named minister pub. instrn., Baden, 1922; pres. Baden, 1924-25. Author: Das Pathologische in der moderen Kunst, 1911; Das Problem der Industriearbeit, 1925; Elementares Lehrbuch der Socialpsychologie, 1933; Mensch und Volk der Gorssstadt, 1939; Klinische Psychologie, 1946; Kulturpsychologie, 1953; Der deutsche Charakter, 1954. Began his work in psychiatry; extended interest to anthropology and social psychology; studied reciprocal action between man and psychic reality and man's relationship to large scale bus. or indsl. concerns; attempted to synthesize sociology and psychology. Died Heidelberg, July 6, 1955.

HELLRIEGEL, Hermann, see Helriegel, Hermann.

HELLWARTH, Robert Willis, Am. physicist; b. Ann Arbor, Mich., Dec. 10, 1930; s. Arlen R. and Sarah (Townsend) H.; B.S. in Engring., Princeton, 1952; D.Phil., St. John's Coll., Oxford (Eng.) U., 1955; postgrad. (Hughes postdoctoral fellow) Cal. Inst. Tech., 1955-56; m. Abigail Gurfein, Sept. 20, 1957; children—Benjamin John, Margaret Eve, Thomas Abraham. Mem. tech. staff, senior scientist Hughes Research Labs., Malibu, Cal., 1956-—. Vis. lectr. in physics Cal. Inst. Tech., 1956-64, sr. research Fellow, 1966-—; vis. asso. prof. physics and elec. engring. U. Ill., Urbana, 1964-65. Mem. adv. council dept. elec. engring. Princeton, 1964-—. Mem. A.A.-A.S., Am. Phys. Soc., I.E.E.E. (asso. editor Jour. Quantum Electronics 1965-—), Sigma Xi. Research and publs. on microwave spectroscopy, masers, lasers, non-linear optics; inventor highpower pulsed output laser; co-discoverer stimulated Raman scattering. Home: 522 Avondale Ave., Los Angeles 90049. Office: Hughes Research Labs., Malibu, Cal. 90265.*

HELLWEGE, Karl Heinz, German physicist; b. Bremerhaven, Germany, Oct. 23, 1910; s. Klaus Johann Jakob and Henriette (Bullwinkel) H.; student U. Marburg (Germany), 1929, U. Munich (Germany), 1930, U. Kiel (Germany), 1931; Dr.phil., U. Göttingen (Germany), 1934, Dr.phil.habil., 1938; m. Anne Marie Röver, Oct. 27, 1939; children—Hans Henning, Helmut, Harro. With U. Göttingen, 1934-52, apl. prof., 1949-52; prof. tech. physics, dir. Inst. Tech. Physics, Tech. U. Darmstadt (Germany), 1952-—, dir. German Plastics Inst., 1953-—. Mem. Deutsche Physikalische Gesellschaft. Author: Einführung in die Physik der Atome, 1964; also articles. Editor: Landolt-Börnstein new series, 1961-—. Research on spectroscopy of ion crystals, solid state physics at low temperature, physics of macromolecular substances. Home: 6 Rehkopfweg, 61 Darmstadt, West Germany.*

HELLWIG, Martin Günter, German mathematician; b. Oberschöna, Freiberg, Feb. 9, 1926; s. Martin and Frida (Sohr) H.; Ph.D. in Natural Sci., U. Göttingen; m. Birgitta Oman, 1961; 1 dau., Annette. Qualified mathematician, 1952; prof. math. Tech. U. Berlin, 1952-58, affiliate prof. math., 1958; research inst., Fulbright fellow Inst. Math. Scis., N.Y.U., 1954-55. Author: Das Randwertproblem eines Linnearen Elliptisch, 1952; Verbiegbarkeit von Flachenstucken, 1955; Anfangswertprobleme bei Partiellen Differentialgleichungen, 1956; Partielle Differentialgleichungen, 1960. Home: 8 Reichskanzleplatz, 1 Berlin 19. Office: 34 Hardenbergstrasse, 1 Berlin 12, Germany.

HELMBERG VON WEITERSDORF, Gilbert Meinrad, mathematician; b. Vienna, Austria, June 2, 1928; s. Theodore and Ehrentrau (Lanner) H. von W.; Ph.D. sum auspiciis praesidentis rei publicae, U. Vienna, 1953; Habiltation, U. Mainz (Germany), 1960, U. Innsbruck (Austria), 1961; m. Thea Kittinger, Aug. 5, 1960; children—Arno, Wolfgang. Instr., U. Wash., Seattle, 1956-58; asst. prof. Tulane U., New Orleans, 1958-59; diätendozent, asst. U. Mainz, 1959-63; dozent U. Innsbruck, 1961-62; guest prof. U. Amsterdam (Netherlands), 1963-64; faculty Techn. U., Eindhoven, Netherlands, 1964-—, prof. math., 1966-—. Recipient Ring of Honor, Pres. Austrian Republic, 1953. Mem. Osterr. Math. Gesellschaft, Am. Math. Soc., Deutsche Math. Vereinigung, Wiskundig Genootschap Amsterdam. Author: Introduction to Spectral Theory in Hilbert Space, 1968; also articles. Research in abstract math. especially uniformly distributed sequences, topological groups, functional analysis, ergodic theory. Home: 5 Kalkofenweg, 6020 Innsbruck-Arzl, Austria. Office: 787 Vm. Montgomerylaan, Eindhoven, Netherlands.*

HELMCKE, Johann-Gerhard, German zoologist; b. Hanover, May 3, 1908; s. Gerhard and Paula (Müller) H.; Ph.D. in Zoology, U. Berlin; m. Magdalene Gdaniec, 1935; children—Conrad, Gerhard, Dietrich. Sci. asst. U. Greifswald, Mus. Physics, U. Berlin; asst. chief div. research in micromorphology; chief micromorphology div. Max Planck Inst.; full prof. biology and anthropology Tech. U. Berlin. Mem. German Soc. Electronic Microscopes, German Soc. Zoology, German Soc. Anthropology, German Soc. Physics and Medicine. Editor: Kükenthal-Krumbach, Handbuch der Zoologie. Research and publs. on electronic microscopes, electronic photometry, stereoscopes. Home: 7 Anzengruberstrasse, Berlin 44, Neukölln. Office: 16 Faradayweg, Berlin 33, Dahlem, Germany.

HELMER, Oscar Marvin, Am. biochemist; b. Portland, Ore., Mar. 13, 1900; s. Eric and Emely (Peterson) H.; B.S., Ore. State U., 1922; M.S., U. Ill., 1924; Ph.D., U. Chgo., 1927; m. Lois Ethel Steffy, July 10, 1928; 1 son, J. Eric. Faculty, Ind. U. Sch. Medicine, Indspls., 1943-—, asso. prof. exptl. medicine, biochemistry, 1950-—; research asso. clin. research Marion County Gen. Hosp., 1954-59, research adviser clin. biochemistry, hematology, 1960-65; mem. Confs. on Factors Regulating Blood Pressure Josiah Macey, Jr. Found. Diplomate Am. Bd. Clin. Chemistry. Fellow A.A.A.S.; mem. Am. Soc. Biol. Chemistry, Am. Chem. Soc., Central Soc. Clin. Research, Soc. Exptl. Biology and Medicine, Am. Heart Assn., Fedn. Clin. Research, N.Y. Acad. Sci., Scabbard and Blade, Sigma Xi, Gamma Alpha, Sigma Tau, Alpha Chi Sigma. Numerous publs. on discovery of angiotensin and its role in prodn. of hypertension; devel. of assay method for renin in plasma for diagnostic purposes. Home: 5015 N. Illinois St., Indpls. 46208. Office: Ind. U. Med. Center, 1100 W. Michigan St., Indpls. 46207.*

HELMHOLTZ, Hermann Ludwig Ferdinand von, German physiologist, physicist; b. Potsdam, Germany, Aug. 31, 1821; s. August Ferdinand Julius and Caroline (Penne) H.; studied medicine at Royal Friedrich-Wilhelm Institute for Medicine and Surgery, Berlin, 1838-42 (under Johannes Müller), M.D., 1842; m. Olga von Velten, 1849 (d. 1859); m. 2nd. Anna von Mohl, 1861; several children. Asst. surgeon, Charité Hosp., Berlin, 1842-43; Army-Surgeon, Potsdam, 1843-48; lectr. anatomy, Kunstakademie, Berlin, and asst. at Anatomical Museum, Berlin, 1848-49; prof. physiology, U. Königsberg, 1849-55; prof. physiology and anatomy, U. Bonn, 1855-58; prof. physiology, U. Heidelberg, 1858-71; prof. physics, U. Berlin, 1871-88; pres. Imperial Physico-Technical Institute (Reichsanstalt), Charlottenburg, 1888-94. Fellow, Royal Soc., 1860 (Copley Medal 1873); mem. Royal Soc. Edinburgh, and many other European learned societies. Author: Über die Erhaltung der Kraft, 1847; Handbuch der physiologischen Optik, 1856-66; Die Lehre von den Tonempfindungen als physiologische Grundlage für die Theorie der Musik, 1862; many scientific articles. Invented (independently) the opthalmoscope, 1851, and opthalmometer; described mechanisms of focusing and binocular vision in eye; investigated color vision and color blindness, reviving Young's three color theory of vision and expanding it (now known as Young-Helmholtz theory); explained accurately mechanism of bones of ear; advanced theories about functioning of cochlea; did important research on perception of tonal quality; founded fixed pitch theory of vowel tones; applied physiological acoustics to "subjective" area of music; is 1 of 3 men credited with discovery and explication of law of conservation of energy; work rendered possible quantitative consideration of processes of metabolism; showed that animal heat produced chiefly by contracting muscle and that acid formed in working muscle; opposed "vitalist" theories; measured speed of nerve impulse and reflex processes; developed Riemann's non-Euclidean geometry; indicated possibility of electromagnetic theory of light; studied electrical oscillations, 1869-71; determined velocity of propagation of electromagnetic induction, 1871; did research on theory of electrical double layer and on electrolysis; wrote on physical meaning of principle of least action and applied it to electrodynamics; studied vortex motion in liquids; denied doctrine of "innate ideas." Died Charlottenburg, Sept. 8, 1894.

HELMONT, Franciscus Mercurius van, Dutch physician, chemist; b. Vilvorde, Oct. 20, 1614; s. J. B. van Helmont; studied medicine (probably not at a university) and practiced as physician; joined band of gypsies to learn their language; published works of his father, 1648; went to Rome where his views on metempsychosis were noticed by Inquisition, 1662; visited Mannheim and Sulzbach where he worked with kabbalist Knorr von Rosenroth, 1663; friend of Leibnitz; invited by Electress of Brandenburg to Berlin. Author: Alphabeti vere naturalis hebraici brevissima delineatio quae simul methodum . . . quam surdi noti sunt sic informari porsunt, 1657; Paradoxal Discourses . . . concerning the Macrocosm and Microcosm, 1685; Opuscula Philosophia, 1690; others. Considered Hebrew to be the primitive language; thought Hebrew letters represent basic positions of organs of speech; published an instruction of deaf and dumb; described formation of ferrous sulphide; also of brass from copper and zinc; described volcanic sal ammoniac, weight increase in lead on oxidation; preparation of phosphorus from evaporated urine distilled with three parts of sand. Died Cologne/Spree, nr. Berlin, Germany, 1699.

HELMONT, Johannes Baptista van, physician, chemist; b. Brussels, Belgium, Jan. 12, 1579; s. Franciscus Mercurius and Marie (de Stassert) van H.; student arts, Louvain, Belgium, until 1594; attended Jesuit sch., Louvain, heard lectures on magic by Martin Del Rio; studied works of mystics Johann Tauler, Thomas à Kempis, then medicine and works of Hippocrates, Galen, Avicenna; greatly influenced by works of Paracelsus, but later rejected his more mystical views; refused to take any degrees; gave away books, devoted life to study of chemistry as true key to medicine; sent to Franciscan prison, Brussels, by Spanish Inquisition (due to his suggestion that relics of saints acted through magnetic property), 1634, released shortly thereafter, but confined to house arrest until 1636 or 38, formally acquitted posthumously, 1649. Author: De Magnetica Vulnerum Naturalis et Legitima Curatone (involved him in then current controversy on magnetic cure of wounds), 1621; Ortus Medicinae (collected works pub. by son, which were highly influential among iatrochemists), 1621, numerous Latin edits., until 1707, translated into English, 1662, German, 1683, French (parts only), 1670. Introduced concept of gas as wild spirit which could not be contained in vessels, also described gases originating from various sources; one of 1st to suggest use of balance in quantitative measurements; taught indestructibility of matter; emphasized that metals when dissolved in acids are not destroyed but can be recovered; explained that precipitation of one metal by another is not transmutation, however believed firmly in alchem. transmutation and gave account of own transmutation of mercury into gold using quarter grain of philosopher's stone; claimed to have universal solvent, alkahest; demonstrated acid to be responsible for digestion in stomach and alkali in duodenum; called father of biochemistry because studied and expressed vital phenomena in chem. terms; one of founders modern pathology which he based on study of external agents and anat. changes in disease; recognized role of inflammation, prodn. of pus; asso. kidney with dropsy and edema; recognized effect of silica dust in causing disease; rejected bleeding as method of healing; used graduated air thermoscope to measure temperature of body; determined specific gravities of metals; turned to creation account in Genesis for his doctrine of elements, water and air, stating that air or sky separates water above from that below firmament; rejected fire as element since it has no matter; rejected earth as matter since expts. seemed to show that it was formed of water (grew willow tree in weighed amount of earth adding only water, attributed increase in weight of wood at end of 5 years to water; this wood could be converted to ashes or earth, consequently earth could be formed from water); rejected Paracelsian 3 principles, salt, sulphur, mercury, as true elements, but accepted them on practical level as explanatory device; considered life to be combination of matter and force (archeus); thought origin of bodies are water and ferment (ferment is innate formative energy that attracts spirit of Archeus which in turn creates bodies after its own idea), also that specific ferments and archei in stomach and other organs brought about digestions and other physiol. changes; on larger scale archeus system was employed to explain changes in earth; explained disease as malfunction of archeus of stomach which sends acid ferments to various parts of body; vitalist who sought key to nature and life in chemistry and chem. analogies in contrast to formal logic and math. of Aristotelians. Died Brussels, Dec. 30, 1644.

HELMREICH, Ernst J. M., biochemist; b. Munich, Germany, July 1, 1922; s. Georg M. and Augusta (Hesselbach) H.; M.D., U. Erlangen (Germany), 1949; m. Rosemarie Hartmann, May 31, 1949; children—Irene K. M., Ilka A. E. Came to U. S., 1954, naturalized, 1962. Faculty, Munich Inst. Tech., 1953, U. Munich Sch. Medicine, 1954-57; vis. fellow Nat. Acad. Sci. dept. biol. chemistry Washington U. Sch. Medicine, St. Louis, 1954-56, faculty, 1957-—, asso. prof. biol. chemistry, 1962-—. Mem. Am. Soc. for Biol. Chemists, German Chem. Soc., German Soc. for Physiol. Chemistry. Contbg. author: Radioactive Isotopes in Physiology, Diagnostics and Therapy, 1961. Research and numerous publs. on rapid conversion of monosaccharides to pyruvate and acetyl-CoA in rat liver, reducing power for fatty acid synthesis, increase in muscle permeability following contraction, synthesis and secretion of antibody by isolated lymphnode cells, regulation of enzyme activity in living muscle, role of adenylic acid in activation of muscle glycogen phosphorylase, study of enzyme conformation with antibody. Home: 10 Deerfield Rd., Ladue St., St. Louis 63124.*

HELMS, Carl Wilbert, Am. biologist; b. New Brighton, Pa., May 26, 1933; s. Lloyd Alvin and Gladys (Leach) H.; B.A., U. Colo., 1955; A.M., Harvard, 1956, Ph.D., 1960; m. S. Ann Darby, Sept. 4, 1960; children—Katherine Carter, Margaret Porter. Research asst. Hathaway Sch. Conversation Edn., South Lincoln, Mass., 1957-60; faculty Bucknell U., Lewisburg, Pa., 1960-—, chmn. dept. biology, 1965-—, asso. prof. biology, 1965-—. Fellow A.A.A.S.; mem. Am. Inst. Biol. Scis., Ecol. Soc. Am., Am. Ornithol. Union, Soc. for Study Evolution, Wilson, Cooper ornithol. socs., N.Y. Acad. Sci., Am. Soc. Zoologists, Brit. Royal Australasian, Netherlands ornithologists unions, Northeastern Bird-Banding Assn., Am. Bot. Soc. Research and publs. in physiol. ecology, physiol. ecology of birds during ann. cycle, wild populations of birds, exptl. studies of caged individuals, metabolism, biochemistry, endocrinology, neuroanatomy. Home: 109 Faculty Ct., R.D. 1, Lewisburg, Pa. 17837.*

HELRIEGEL (Hellriegel), Hermann, German biologist; b. Mausitz, Germany, Oct. 21, 1831; ed. Tharandt; asst. to Adolph Stöckhardt, chem. lab. Acad., became dir. Exptl. Stat., Brandenburg, Germany, 1856; named dir. new exptl. sta., Bernburg, Germany, 1873; mem. French Acad. Scis., 1892, Soc. Agr. Author: Beiträge zu den naturwissenschaftlicher Grundlagen des Ackerbaues mit besonderer Rücksicht auf die agrikultur-chemischen Methoden der Sandkultur, 1883. Discovered (with Wilfarth) that leguminous plants fix atmospheric nitrogen by means of root nodules; proved rate of absorption of nitrogen is proportional to nodulation. Died Sept. 24, 1895.

HELSON, Harry, Am. psychologist; b. Chelsea, Mass., Nov. 9, 1898; s. William and Ida Helson; A.B., Bowdoin Coll., 1921; M.A., Harvard, 1922, Ph.D., 1924; m. Lida G. Anderson, Sept. 3, 1926; children—Henry Berge, Martha Alice (Mrs. William A. Wilson, Jr.). Instr., Cornell U., 1924-25, U. Ill., 1925-26; asst. prof. U. Kan., 1926-28; faculty Bryn Mawr Coll., 1928-49, prof., 1933-49; acting prof. Thomas Welton Stanford fellow Stanford, 1948-49; prof., chmn. dept. psychology Bklyn. Coll., 1949-51; prof. U. Tex., 1951-61; John C. Peterson Distinguished prof. Kan. State U., Manhattan, 1961-—; co-dir. antiaircraft fire-control project Foxbor Co., 1942-44; dir. USAF-U. Tex. Radiobiol. Lab., 1952-54. Hogg Found. research scholar, 1956-57. Fellow A.A.A.S., Soc. Exptl. Psychologists (Howard Crosby Warren medal 1959); mem. Am. (Distinguished Sci. contbn. award 1962), Midwestern, Southwestern psychol. assns., Am. Assn. U. Profs., Optical Soc. Am., Phi Beta Kappa, Sigma Xi. Author: The Psychology of Gestalt, 1926; Adaptation-Level Theory: an Experimental and Systematic Approach to Behavior, 1964; also numerous articles. Editor: Theoretical Foundations of Psychology, 1951; (with William Bevan) Contemporary Approaches to Psychology, 1967. Research on theory adaptation level in which all responses are referable to neutral or indifferent states which are weighted means of focal, background and residual sources of stimulation, principle of color conversion, reformulation of Weber-Fechner law, human factors in equipment design. Home: 1113 Hylton Heights Rd., Manhattan, Kan. 66502.*

HELVETIUS, Claude Adrien, philosopher; b. Paris, France, 1715; s. Jean Claude Adrien Helvetius; m. 1751; apptd. farmer gen., through influence of Queen Marie Leczinska, 1738, retired to devote himself to philos. studies. Author: De l'homme (influenced Bentham, posthumous publ. 1772); De l'esprit (condemned by the ch. and drs. of Sorbonne, it was publicly burned by the hangman, 1759), 1758. Using John Locke's theory that all knowledge originates in sensation, he concluded that minds of men are equal at birth and all differences of man are result of education; held to doctrine based on premise that self-interest is spring of all action; believed in existence of neurolymphatic capillaries. Died 1771.

HELVÉTIUS, Jean Claude Adrien, French physician; b. Paris, July 18, 1685; s. Jean Adrien Helvetius; grad. as physician, 1708; cured infant Louis XV of serious illness, 1719; named councillor of state, also physician to Queen. Fellow Royal Soc., 1755; mem. French Acad. Scis., Berlin, London acads. Author several treatises, including one on cures for prin. diseases, 1737. Died July 17, 1755.

HELVETIUS, Johann Friedrich, alchemist; b. Kothen, Duchy of Anhalt, 1625; student medicine. Went to Hague, Netherlands; became physician to Prince of Orange. Author: De Alchymia opuscula complura veterum philosophorum, 1650; Vitulus aureus . . . , 1667; Guildenes Kalb (among most famous accounts of transmutation), 1668. Died Hague, Netherlands Aug. 29, 1709.

HELWIG, Elson Bowman, Am. physician; b. Pierceton, Ind., Mar. 5, 1907; s. Llewellyn and Grace (Bowman) H.; B.S., Ind. U., 1930, M.D., 1932; m. Mildred Stoelting, Apr. 20, 1933; children—Alan Stoelting, Warren Bowman, Ann (Mrs. Thomas Gordon). Practice medicine specializing in pathology, Boston, 1936-39, St. Louis, 1939-46, Washington, 1946-—; sr. pathologist Armed Forces Inst. Pathology, 1946-47, chief dermatology, gastro-intestinal br., 1947-—, chief dept. pathology, 1955-—, asso. dir. for consultation, 1968-—; lecturer School of Medicine George Washington U., 1947-64, clin. prof. pathology, 1964-—; vis. prof. Sch. Medicine Temple U., 1958-—. Adviser, WHO. Recipient numerous civilian service awards. Diplomate Am. Bd. Pathology. Mem. A.M.A., Am. Acad. Dermatology, Assn. Mil. Dermatologists, Am. Soc. Dermatopathology (past pres.), Assn. Mil. Surgeons, Histochem. Soc., Coll. Am. Pathologists, Am. Soc. Clin. Pathologists, Internat. Acad. Pathology, Am. Assn. Pathologists and Bacteriologists. Research and numerous publs. primarily in field of pathology of skin and gastrointestinal tract. Home: 14 W. Maple St., Alexandria, Va. 22301. Office: Armed Forces Inst. Pathology, Washington 20305.*

HEMBERG, Torsten Nils Emil, Swedish plant physiologist; b. Stockholm, Sweden, Feb. 18, 1915; s. Sven Emil and Elin (Jakobsson) H.; Fil. cand., U. Stockholm, 1940, Fil. mag., 1941, Fil. licentiat in botany, 1943, Fil. dr., 1947; m. Elsie Margareta Hedström, July 4, 1942; children—Sven Anders Torsten, Elin Elsa Margareta. Amanuensis, Bot. Inst., U. Stockholm, 1940-46, tchr. in plant physiology, 1946-47, asst. prof. botany, 1947-52, prof. plant physiology, 1959—; lectr. botany Royal Pharm. Inst., Stockholm, 1952-59. Mem. bd. Royal Coll. Forestry, Stockholm. Mem. Scandinavian Soc. Plant Physiologists (v.p.), Swedish Bot. Soc. (v.p.). Author: (with Kurt Falck and Folke Fagerlind) Botanik, 1960, 64. Research, publs. on growth hormones in potato tuber; studies of occurrence of auxins and growth inhibitors in resting and non-resting plant organs (potato tubers and buds of Fraxinus) showing that rest period can be correlated with occurrence of acid inhibitor complex (B-inhibitor); studies of action of this inhibitor complex on respiration and on activity of amylases; studies of auxin balance in rooting hypocotyles of Phaseolus and in germinating kernels of maize. Home: 5 Ytterbystrand, Vaxholm, Sweden. Office: Botanical Inst., Lilla Frescati, Stockholm, 50, Sweden.*

HEMENWAY, Curtis Leland, Am. astrophysicist; b. Hope, Me., Sept. 11, 1920; s. Leland David and Clara (Hinckley) H.; A.B., Colby Coll., 1938; M.S., Rutgers U., 1946, Ph.D. in Physics, 1950; m. Vivian Bennett, July 29, 1948; children—Susan, David. Faculty, Rutgers U., 1943-44, 47-49; asst. prof. physics Union Coll., Schenectady, 1949-54, asso. prof., 1954-59, prof., 1959-64, research prof. physics, 1964—; dir. Dudley Obs., Albany, N.Y., 1956—; prof., chmn. astronomy and space sci. dept. State U. N.Y., Albany, 1964—. Mem. security bds., designee AEC, 1954—. Cons. Smithsonian Astrophys. Obs., 1957-58. Research fellow Harvard Coll. Obs., 1955-56. Fellow A.A.A.S., Am. Meteoritical Soc.; mem. Am. Phys. Soc., Am. Assn. Physics Tchrs., Am. Assn. U. Profs., Am. Astron. Soc., Internat. Astron. Union, Sigma Xi. Author: (with Richard W. Henry and Martin Caulton) Physical Electronics, 1962. Research and publs. on micrometeorites. Home: River Rd. Route 144, Selkirk, N.Y. 12158. Office: 100 Fuller Rd., Albany, N.Y. 12205.*

HEMHOLZ, August Carl, Am. physicist; b. Evanston, Ill., May 24, 1915; s. Henry Frederic and Isabel (Lindsay) H.; A.B., Harvard, 1936; student Cambridge (Eng.) U., 1936-37; Ph.D., U. Cal., Berkeley, 1940; m. Elizabeth Jane Little, July 30, 1938; children—Charlotte Ann (Mrs. Edward Merrick Chaffee, Jr.), George Lindsay, Frederic Vogel, Edith Little. Instr. physics U. Cal., Berkeley, 1940-43, asst. prof., 1943-48, asso. prof., 1948-51, prof., 1951—, chmn. dept. physics, 1955-62, research physicist, Lawrence Radiation Lab., 1940—. Guggenheim fellow CERN, Geneva, Switzerland, 1962-63. Fellow Am. Phys. Soc.; mem. A.A.A.S., Am. Assn. Physics Tchrs., Am. Assn. U. Profs., Am. Inst. Physics (governing bd. 1964-67), Phi Beta Kappa, Sigma Xi. Research in artificial radioactivity, nuclear isomerism, interaction of high energy gamma rays and neutrons, pion nucleon interactions. Home: 28 Crest Rd., Lafayette, Cal. 94549. Office: Dept. Physics, U. Cal., Berkeley, Cal. 94720.*

HEMMENDINGER, Arthur, Am. physicist; b. Bernardsville, N.J.; s. Max and Jeannette (Harris) H.; B.A., Cornell U., 1933; Ph.D., Cal. Inst. Tech., 1937; m. Margaret Elaine Ross, May 13, 1945; children—Ross, Anne, Dennis. Faculty, U. Okla., Norman, 1937-41; physicist U. S. Naval Ordnance Lab., Washington, 1941-44; staff mem. Los Alamos Sci. Lab., 1945-62, group leader, 1962—. Research in mass spectroscopy, underwater sound, nuclear reactions of light elements, neutron physics, cross-section measurements, nuclear weapons tech., radiation. Home: 1442 47th St. Office: Box 1663, Los Alamos 87544.*

HEMMER, Per Christian, Norwegian physicist; b. Arendal, Norway, May 14, 1933; s. Alv and Hilda (Moller) Kristiansen; Siv.Ing., Tech. U. Norway, 1956, Dr.Techn., 1959; m. Ellen Vorvik, Dec. 31, 1955; children—Hilde, Sara, Hallstein. Faculty, Tech. U. Norway, Trondheim, 1957-58, 61, 63—, asso. prof., 1967—; with Nordic Inst. Theoretical Atomic Physics, 1959-60, Rockefeller Inst., N.Y.C., 1961-62; research fellow dosent Inst. Theoretical Physics, Trondheim, 1963—. Mem. Royal Norwegian Acad. Scis. Research and publs. in foundation of Brownian motion, systematics of nuclear spectra, theory of condensation of gases, equation of state of fluids.*

HEMMERICH, Peter Erwin, German chemist; b. Frankfurt/Main, Dec. 30, 1929; s. Franz and Erna (Haurand) H.; Dr.phil.nat., U. Basel (Switzerland), 1957; m. Marianne Hirschfeld, Aug. 24, 1957; children—Stefan, Andreas, Christiane. Research asso. U. Basel, 1957-63, lectr., 1963—; research asso. U. Wis., Madison, 1964, vis. asso. prof., 1965; research asso. U. Mich., Ann Arbor, 1965; prof. biol. chemistry, U. Constance, 1967—. Mem. Swiss, German chem. socs. Contbr. articles in field. Research on structure and function of biol. metal complexes; action mechanism of metal enzymes and metal interaction of con-enzymes; free radicals in biology; redox-active metal-mercaptide clusters in proteins; synthesis of flavins; chemistry of univalent copper. Home: 100 Hoheneggstrasse, 775 Konstanz, West Germany.

HEMMETER, John Conrad, Am. physiologist; b. Balt., Apr. 25, 1864; s. John and Mathilde H.; grad. Royal Gymnasium, Wiesbaden, Germany; Balt. City Coll., 1881; M.D., U. Md., 1884, Sc.D., 1914; Ph.D., Johns Hopkins, 1890; LL.D., St. John's Coll., Annapolis, 1905; m. Helene Emilie Hilgenberg, Jan. 1893. Prof. physiology and clin. prof. medicine, and regent med. dept., U. Md., also dir. Physiol. Lab.; practice medicine limited to diseases of digestive organs. Asso. editor Archives for Digestive Diseases, Berlin, Internationale Beitr. z. Physiol. u. Pathol. d. Verdauung u.d. Stoffwechsels, Berlin, and Archives of Clinical Medicine. Pres. Am. sect. Internat. Assn. for History of Medicine; hon. pres. Am. Soc. Med. History. Author: Manual of Practical Physiology, 1912; Master Minds in Medicine—a History of Evolution of Ideas in Medicine, 1927. Research and publs. on physiol. history of medicine; inventor duodenal intubation: wrote 1st Am. treatise on diseases of digestive tube; pioneer in use of X-ray for diagnostic purposes. Died Feb. 25, 1931.

HEMPEL, Walter, German chemist; b. Pulsnitz, Germany, 1851; studied under Bunsen; prof., Dresden, Germany, 1879-1912. Author: Gas analytische Methoden. Extended Bunsen's lab. methods; devised pipette and burette named after him; determined atomic weight of cobalt; developed method for combustion of organic compounds subjected to high pressure; research on reactions in lead-chamber process, recovery of phosphorus, electrolytic prodn. of sodium carbonate. Died 1916.

HEMSLEY, William Botting, Brit. botanist; b. Easthoathley, Sussex, Dec. 29, 1843; privately ed.; entered Kew, 1860; LL.D. Former keeper herbarium and library Royal Botanic Gardens, Kew. Fellow Royal Soc., 1889, Linnean Soc.; hon. mem. Natural History Soc. (Mexico), Royal Hort. Soc. (Victoria medal honor), Royal Soc. New South Wales, New Zealand Inst.; corr. mem. German Bot. Soc. Author: Handbook of Hardy Trees, Shrubs and Herbaceous Plants, 1873, 77; Botany of the Challenger Expedition; Botany of Salvin and Godman's Biologia Centrali-Americana; Flora of China; Flora of Tibet, also numerous articles on geog. botany and insular floras. Died Oct. 7, 1924.

HENCH, Philip Showalter, Am. physician; b. Pitts., Feb. 28, 1896; s. Jacob Bixler and Clara John (Showalter) H.; A.B., Lafayette Coll., 1916, Sc.D., 1940; M.D., U. Pitts., 1920, Sc.D., 1951; post grad. study U. Freiburg and Ludwig-Maximilians-Universitat, Munich, 1928-29; M.S. in Internal Medicine, U. Minn., 1931; Sc.D., Lafayette Coll., 1940, Washington and Jefferson Coll., 1940, Western Res. U., Nat. U. Ireland, 1950, University of Pittsburgh, 1951; LL.D., Middlebury (Vt.) Coll., 1951; m. Mary Genevieve Kahler, July 14, 1927; children—Mary Showalter, Philip Kahler, Susan Kahler, John Bixler. Intern St. Francis Hosp., Pitts., 1920-21; with Mayo Found., Mayo Clinics, Grad. Sch., U. Minn., 1921—; cons., head sect. rheumatic diseases, Mayo Clinic, 1926—; prof. medicine Mayo Found. and Grad. School, University of Minnesota, 1947—. Recipient (with E. C. Kendall and Tadeus Reichstein) Nobel prize for physiology and medicine, 1950, numerous other awards. Fellow A.M.A., A.C.P. Chief editor Am. Rheumatism Reviews (Am. Rheumatism Assn.), 1932-48; asso. editor Annals of Rheumatic Diseases (London). Contbr. about 200 articles to med. jours. Research on rheumatoid arthritis; discovered that in certain instances its progress could be reversed; successfully used ACTH and cortisone, pituitary and adrenal cortex extracts as remedies for the disease; studied exact beneficial effects of jaundice and pregnancy on rheumatoid arthritis. Died Mar. 30, 1965.

HENCKE, Karl Ludwig, German astronomer; b. Driesen, Germany, 1793; discovered 5th and 6th minor planets Astrée and Hébé, 1845. Died Marienwerder, 1866.

HENCKEL (Henkel), Johann Friedrich, German physician, mineralogist; b. Merseburg, Germany, Aug. 11, 1679; physician, also dir. mines, Freiburg, Germany. Mem. Leopoldina Acad. Author: Flora saturnizans, die Verwandtschafft . . . , 1722; Pyritologia, oder Kiess-Historie . . . , 1725; Bethesda portuosa . . . , 1726; Giesshübelium redivivum . . . , 1729; D. Johann Friedrich Henkels . . . kleine mineralogische und chymische Schrifften . . . , 1744; Henckelius in mineralogia redivivus . . . , 1747; Described metallic arsenic obtained by sublimation in closed retorts and subliming vessels, also metallic zinc; recognized vitriol is compounded of metal with sulphuric acid; distinguished varieties of pyrites; wrote that sulphur has more affinity for copper than for iron. Died Freiburg, Jan. 27, 1744.

HENDERSON, Francis Glelland, Am. physician; b. Elwood, Ind., Oct. 30, 1917; s. F. G. and Zelda (Bitner) H.; B.S., Ind. U., 1939, M.D., 1944; m. Frances Pauline Graham, Sept. 23, 1939; children—Greg G., Shirley Jo. Practice medicine specializing in pharmacology, Indpls., 1944—; pharmacologist Eli Lilly & Co. Research Labs., Indpls., 1947-57, head dept. pharmacodynamics, 1957-63, research asso., 1963-65, head clin. physiology Lilly Labs. for Clin. Research, Marion County Gen. Hosp., 1965—. Mem. Soc. Pharmacology, Soc. Exptl. Biology, A.M.A., N.Y. Acad. Medicine, Am. Chem. Soc., Am. Heart Assn. Research and publs. on adrenergic-blocking agts.; cardiac glycosides; vertatrum alkaloids; senecio alkaloids; cardiovascular research. Home: 319 Ridgeview Dr., Indpls. 46219. Office: Lilly Labs. for Clin. Research, Marion County Gen. Hosp., Indpls. 46207.*

HENDERSON, George Gerald, Scottish chemist; b. Glasgow, Scotland, Jan. 30, 1862; s. George and Alexandrina (Kerr) H.; M.A., D.Sc., LL.D., U. Glasgow; postgrad. U. Leipzig; LL.D., St. Andrews; D.Sc., Belfast; m. Agnes Mackenzie, 1895. Lectr., demonstrator chemistry U. Glasgow, 1884-92, Regius prof. chemistry, 1919-37; lectr. chemistry Queen Margaret Coll., Glasgow, 1889-92; prof. chemistry Royal Tech. Coll., Glasgow, 1892-1919. Fellow Royal Soc., 1916, Inst. Chemistry (past pres.), Chem. Soc. (past pres.); mem. Soc. Chem. Industry (past pres.), Brit. Assn. (past pres. sect. B), Soc. Pub. Analysts (hon.). Author: (with M. A. Parker) An Introduction to Analytical Chemistry; Catalysis in Industrial Chemistry; Report on Chemical Industries of West of Scotland, also numerous articles on organic and inorganic chemistry. Died Sept. 28, 1942.

HENDERSON, George Hugh, Canadian physicist; b. Dec. 8, 1892; s. John A. Henderson; ed. Dalhousie U., Halifax, N.S., Can., Gonville and Caius Coll., Cambridge; m. Ruth Ross, 1929; 2 daus. Fellow NRC Can., 1918-19, also mem.; asst. prof. U. Sask., 1922-24; prof. physics Dalhousie U., from 1924; joint supt. Naval Research Establishment, Halifax, 1939-48, chief supt., from 1948; gov. N.S. Research Found. Fellow Royal Soc., 1942, Royal Soc. Can. Contbr. articles on radioactivity to sci. publs. Developed microphotometer for determining age of rocks. Died June 19, 1949.

HENDERSON, Lawrence Joseph, Am. biol. chemist; b. Lynn, Mass., June 3, 1878; s. Joseph and Mary Reed (Bosworth) H.; A.B., Harvard, 1898, M.D., 1902, Sc.D., 1932; postgrad. U. Strassburg, 1902-04; Sc.D., U. Cambridge (Eng.), 1934; Dr. Hon., U. Grenoble, 1934; m. Edith Lawrence Thayer, 1910; 1 son, Lawrence Joseph. Lectr. biol. chemistry Harvard, 1904-05, instr., 1905-10, asst. prof., 1910-19, prof., from 1919, founded lab. phys. chemistry in Med. Sch., 1920, organized fatigue lab. in Bus. Sch., 1927. Exchange prof. Harvard to U. Paris, 1921; Silliman lectr. Yale, 1928; Leyden lectr. U. Berlin, 1928; Mills lectr. U. Cal., 1931. Sr. fellow and chmn. Soc. Fellows (Harvard), 1933—. Fellow Am. Acad. Arts and Scis. Author: The Fitness of the Environment, 1913; The Order of Nature, 1917; Blood, 1928; Pareto's Sociology: A Physiologist's Interpretation, 1935. Investigated acid-base equilibria in blood plasma; stressed relation of phys. properties of matter to existence of life; influenced by Pareto's work on social equilibriums. Died Feb. 10, 1942.

HENDERSON, Malcolm Colby, Am. physicist; b. New Haven, Mar. 13, 1904; s. Yandell and Mary (Colby) H.; grad. Phillips Acad., 1921; Ph.B., Yale, 1925; Ph.D., Cambridge U., 1928; m. Katharine Gordon Linforth, July 28, 1933; children—Ian Yandell, Anthony Gordon. Sterling fellow Yale 1928-30, hon. research fellow, 1930-32, U. Cal. at Berkeley, 1932-35; instr. physics Princeton, 1935-40; asst. prof. Dartmouth Coll. Hanover, N.H., 1940-44, prof., 1944-46; physicist U. Cal. div. war research U. S. Navy Radio and Sound Lab., San Diego, 1941-45, asst. dir., 1945-46; physicist, G-2 AUS, Washington, 1946-49; dep. dir. intelligence AEC, Washington, 1949-53; tech. dir. atomic tests FCDA, Washington, 1953-54; research prof. physics, Cath. U. Am., Washington, 1954—, chmn. dept., 1965-66. Cons. NSF, Washington, 1957-59. Fellow, A.A.A.S., Phys. Soc., Acoustical Soc., mem. Philos. Soc. Washington (pres. 1964—), Soc. Cin. Home: 2900 29th St. N.W., Washington 20008.*

HENDERSON, Roland George, Am. geophysicist; b. Albia, Ia., Oct. 21, 1911; s. Nicholas and Ola (Johnson) H.; B.Ed., U. Wis. at Milw., 1935; M.S., Atlanta U., 1939; postgrad. U. Wis., Am. U.; m. Dora Mae Moseley, Apr. 4, 1938; children—Anita Jacquelyn, Roland George, Nicholas Robert. With U. S. Geol. Survey, Washington, 1942—, mathematician, 1951—. Cons. on computer applications, interpretation in gravity and magnetic exploration. Mem. Am. Geophys. Union, Soc. Exploration Geophysicists, Seismol. Soc. Am., European Assn. Exploration Geophysicists. Asso. editor Geophysics, 1967—. Research, publs. on use of analytical methods and electronic computers in interpretation of magnetic and gravity fields; identified with derivative methods and analytical continuation of fields. Home: 6014 2d St. N.W., Washington 20011. Office: U. S. Geol. Survey, Interior Bldg., Washington 20242.*

HENDERSON, Thomas, Scottish astronomer; b. Dundee, Scotland, Dec. 28, 1798; pupil in math. under Mr. Duncan, prin. Dundee Acad.; m. dau. of Alexander Adie, 1836; 1 dau. Sec. to Earl of Landerdale and Lord Jeffrey, 1819-31; astronomer royal, Cape of Good Hope, 1832-33; 1st Scottish astronomer royal, and prof. practical astronomy, Edinburgh, 1834-44. Fellow Royal Astron. Soc., Royal Soc.; mem. Astron. Instn. Edinburgh. Author: Edinburgh Observations, 1838-43, (edited by Plazzi Smyth) 1843-52; also papers. Observed Encke's and Biela's comets; observed transit of Mercury, 1832; measured parallax of Alpha Centuri, showing it in closest known star system, 1832, pub. 1839 (credit given to Bessel 1838). Died Edinburgh, Nov. 23, 1844.

HENDERSON, Velyien Ewart, Canadian pharmacist, toxicologist; b. Cobourg, Ont., Can., June 27, 1877; s. Joseph H.; student Upper Can. Coll.; B.A., U. Toronto, 1899, M.B., 1902, M.A., 1903; postgrad. univs. Prague, Marburg, 1903-04; m. Edith E. van der Smissen, 1911; 2 sons. Demonstrator physiology U. Pa., 1903-03; demonstrator physiology U. Toronto (Ont.), 1904-06, lectr. pharmacy and pharmacology, 1906-09, prof. pharmacology and toxicology from 1909. Fellow Royal Soc. Can. Author: Materia Medica and Pharmacy, 1932; Air Crew in Their Element, 1942; (with W. L. Chute) Materia Medica, 1943. Research in physiology and pharmacology; conducted animal expts. (with Lucas) on use of cyclopropane as anesthetic. Died Aug. 6, 1945.

HENDERSON, William, physician, b. Thurso, Scotland, Jan. 17, 1810; s. William Henderson; M.D., Edinburgh (Scotland) U., 1831; postgrad. Paris, France, Berlin, Germany, Vienna, Austria, 2 years. Physician, Fever Hosp., Edinburgh, 1832; pathologist Royal Infirmary until 1845; apptd. chair gen. pathology Edinburgh U., 1842, resigned 1869. Fellow Royal Coll. Physicians Edinburgh; mem. Medico Chirurg. Soc. Edinburgh. Author: An Inquiry into the Homeopathic practice of Medicine, 8 vols., 1845. Clinical studies, publs. on heart and larger blood vessels; described (independent of Robert Paterson) inclusion of bodies in molluscum contagiosum, 1841; also 1st to describe murmur of efflux in a case of sacculated aortic aneurism; 1st to demonstrate (as diagnostic sign of aortic regurgitation) that radial pulse followed that of heart by longer interval than usual. Died Apr. 1, 1872.

HENDERSON, Yandell, Am. physiologist; b. Louisville, Ky., Apr. 23, 1873; s. Isham and Sally Nielson (Yandell) H.; B.A., Yale, 1895, Ph.D., 1898; univs. Marburg, 1899, Munich, 1900; m. Mary Gardner Colby, Apr. 2, 1903; children—Malcolm Colby, Sylvia Yandell (Mrs. G. McL. Harper, Jr.). Instr. physiology Yale, 1900-03, asst. prof., 1903-11, prof., 1911-21, prof. applied physiology, 1921-38, prof. emeritus from 1938. Cons. physiologist, U. S. Bur. of Mines, 1913-25; chief of physiol. sect. U. S. war gas investigation, 1917-18; chmn. Med. Research Bd., Aviation Sect., Signal Corps, U. S. Army, 1917-18. Mem. A.M.A., Am. Physiol. Soc., Am. Pharm. Soc., Nat. Soc. Anesthetists, Physiol. Soc. (Gt. Britain), A.A.A.S., Nat. Acad. Scis., Am. Philos. Soc.; hon. mem. Am. Climatol. Assn., Assn. Physicians of Vienna, Austria, Coal Mining Inst. Am. Research on circulation, respiration, pharmacology of gases; demonstrated that poisonous effect of illuminating gas is caused by affinity of its carbon monoxide for hemoglobin, 1916. Died Feb. 18, 1944.

HENDLER, Edwin, Am. physiologist; b. Phila., Aug. 29, 1922; s. David and Elene (Kalman) H.; B.S., Pa. State U., 1943; M.S., U. Pa., 1956, Ph.D., 1959; m. May Snyder, May 13, 1945; children—Lynn Karen, Sandra Dee. With Aerospace Crew Equipment Lab., Naval Air Engring. Center, Phila., 1946—, supt. life scis. research div., 1955-63, mgr. life scis. research group, 1963—; research asso. physiology dept. U. Pa., Sch. Medicine, 1960—. Mem. Am. Physiol. Soc., Biophys. Soc., Aerospace Med. Assn., Biometric Soc., Inst. Environmental Scis., Sci. Research Soc. Am., A.A.A.S., Am. Assn. for Advancement Med. Instrumentation, Assn. Nat. Francaise d'Electronique Medicale et Biologique. Author: (with N. M. Burns and R. C. Chambers) Unusual Environments and Human Behavior, 1963; also articles. Proposed mechanism of cutaneous temperature sensation; research on effects of skin irradiation by infrared and microwave energy, physiologically evaluated artificial atmospheres proposed for spacecraft; developed and evaluated personal protective equipment and life-support systems; invented 1st naval aviators protective helmet. Home: 415 Brentwood Rd., Havertown, Pa. 19083. Office: Aerospace Crew Equipment Lab., Naval Air Engring. Center, Phila. 19112.*

HENDRICKS, Charles Durrell, Jr., Am. physicist; Lewiston, Utah, Dec. 5, 1926; s. Charles Durrell and Louise (McAlister) H.; B.S., Utah State U., 1949; M.S., U. Wis., 1951; Ph.D., U. Utah, 1955; m. Leah Funk, Mar. 4, 1948; children—Katherine, Martha Jane. Research asst. U. Utah, 1953-55; staff mem. Lincoln Lab., Mass. Inst. Tech., 1955-56; faculty U. Ill., Urbana, 1956—, prof. dept. elec. engring., 1961—, prof. nuclear engring., 1965—, also dir. charged particle research lab.; editor Blaisdell Pub. Co. Cons. indsl. firms. Mem. I.E.E.E., Am. Phys. Soc., Am. Assn. Physics Tchrs., Am. Inst. Aeros. and Astronautics, Sigma Xi, Phi Beta Kappa, Tau Beta Pi, Phi Kappa Phi, Eta Kappa Nu. Research on hypervelocity impact, artificial meteors, airborn radar, elec. propulsion of space vehicles, plasma physics, collision and coalescence of liquid drops, prodn. and control of small liquid and solid particles, electrohydrodynamics, mass spectroscopy of ionospheric D and E regions. Home: 403 Sunnycrest Ct., Urbana, Ill. 61801.*

HENDRICKS, Sterling B., Am. agrl. chemist; b. Elysian Fields, Tex., Apr. 13, 1902; s. James G. and Daisy (Gamblin) H.; B. Ch.E., U. Ark., 1922, LL.D., 1946; M.S., Kan. State Agrl. Coll., 1924; Ph.D., Cal. Inst. Tech., 1926; D.Sc., Kan. State U., 1963; m. Edith Ochiltree, Feb. 21, 1931; 1 dau., Martha. Research asso., Rockefeller Inst. Med. Research, N.Y.C., 1927-28; scientist U. S. Dept. Agr., 1928—; prin. chemist U. S. Bur. Plant Industry, 1945-48, chief chemist, 1948—. Recipient Hillebrand prize for work on crystal structure Am. Chem. Soc., 1938; sci. award for work on clays Washington Acad. Sci., 1940; Day medal Geol. Soc. Am., 1952; Pres.'s award for distinguished fed. service, 1958; recipient Rockefeller Public Service award, 1961; Hoblitzelle Nat. award Tex. Research Found., 1962. Fellow Am. Soc. Agronomy; mem. Nat. Acad. Scis., Am. Chem. Soc., Am. Mineral. Soc. (pres. 1954). Contbr. numerous articles on structure of matter, constn. of soils, biology of plants to various publs. Made 3d ascent of Mount McKinley, Alaska, 1942; research on structural aspects of organic and inorganic chemistry, chemistry and physics of crystal structure; phase rule; X-ray diffraction from solids; electron diffraction; plant physiology and nutriton; photoperiodism; determined causes of clay properties, fundamental controlling factor for plant growth and reprodn. Home: 1118 Dale Dr., Silver Spring, Md. Office: Plant Industry Station, Beltsville, Md.

HENDRICKSON, Adolph Alexander, Am. microbiologist; b. Holmen, Wis., Apr. 11, 1907; s. Samuel and Agatha (Pederson) H.; B.S., U. Wis., 1929, M.S., 1930, Ph.D., 1932; m. Wendell R. Barsness, Apr. 14, 1933; 1 son, Bruce Barsness. Wis. Research Fund fellow U. Wis., 1932-33; Albert Dickinson Co. fellow Ohio State U., 1933; chief bacteriologist Albert Dickinson Co., Chgo., 1933-57; dir. research, prodn. Nodogen Labs. Rudy-Patrick Seed div. W. R. Grace & Co., Princeton, Ill., 1957-66, research dir. Inoculant Labs., 1966—. Fellow A.A.A.S., Am. Acad. Microbiology; mem. Ill. Soc. Microbiology, Am. Soc. Microbiologists, Am. Soc. Agronomy, Soil Sci. Soc. Am., Internat. Soil Sci. Soc., N.Y., Ill. acads. scis., Sigma Xi. Research on mass prodn. of bacteria in liquid culture; use of antibiotics for control of plant diseases; strain variations of Rhizobia cultures; silica gels; physiology of bacteria; plant diseases; plant growth regulators; pre-inoculation of legume seeds. Home: 1206 S. Euclid Av. Office: P.O. Box 404, Princeton, Ill. 61356.*

HENDRICKSON, John Reese, Am. nuclear and chem. engr.; b. Seattle, Apr. 2, 1910; s. Harvey Stevenson and Cora Margaret (Reese) H.; B.S. in Chem. Engring., U. Wash., 1932, M.S., 1933; postgrad. U. Md., U. Pa.; m. Frances Gertrude Dunkin, Nov. 22, 1934; children—Frances Joanne (Mrs. Vernon Magness), John Reese, Evelyn Virginia (dec.), Harvey Joseph, Robert Lee. Asst. chemist Utah-Ida. Sugar Co., 1933-35; jr. chem. engr. Cal. and Hawaiian Sugar Refining Corp., 1936, Standard Oil of Cal., 1936-38; asst. chemist Colgate Palmolive C., 1938-42; supr. chem. research Lawrence Radiation Lab., 1942-43, Union Carbide Nuclear Corp., 1943-49; adj. prof. chemistry U. Tenn., 1947-49; chief research and devel. radiol. protection Nuclear Def. Lab., Edgewood, Md., 1949-56; radiol. health officer, cons. chem. and nuclear engr. Def. Electronics Plant, RA, Camden, N.J., 1956—. Mem. coms. U. S. A. Standards Inst., Am. Soc. Testing and Materials, I.E.E.E. Recipient Silver A Bomb Pin for chem. research for Manhattan Project. Mem. Am. Nuclear Soc., Am. Inst. Chem. Engrs. (nuclear engring. div.), Am. Ordnance Assn., Air Force Assn., I.E.E.E., Franklin Inst., Aerospace Systems Safety Soc. Author numerous articles. Inventor photoelectric reflectometer for grading brown sugar; developed vapor phase process for chlorination of uranium compounds, high efficiency distillation processes for separating new isomeric fluorocarbon mixtures, radiotracer techniques for measuring efficiency of filters for air purification; research chemist on 1st cyclotron built in 1942; project engr. for 1st hydrogen bomb test, 1951. Home: Lou-Mar Rd., Abingdon, Md. 21009. Office: RCA, Camden, N.J. 08102.*

HENDRIE, Joseph Mallam, Am. physicist; b. Janesville, Wis., Mar. 18, 1925; s. Joseph Munier and Margaret Prudence (Hocking) H.; B.S., Case Inst. Tech., 1950; Ph.D., Columbia, 1957; m. Elaine Kostell, July 9, 1949; children—Susan Debra, Barbara Ellen. Asst. physicist Brookhaven Nat. Lab., Upton, N.Y., 1955-57, asso. physicist, 1957-60, physicist, 1960—; chmn. steering com., project chief engr. high flux beam reactor design and constrn., 1958-65, asso. head engring. div., project mgr. pulsed fast reactor project, 1967—. Cons., Columbia Radiation Safety Com., 1962—; adv. com. reactor safeguards AEC, 1966—. Mem. Am. Phys. Soc., Am. Nuclear Soc., Am. Soc. M.E., Sigma Xi, Tau Beta Pi. Co-inventor high flux beam reactor; research and publs. on physics of nuclear reactors, engring. design of reactors, chem. physics of nitrogen dissociation process, structure of oxygen molecule. Home: 24 Thornhedge Rd., Bellport, N.Y. 11713. Office: Brookhaven Nat. Lab., Upton, N.Y. 11973.*

HENFREY, Arthur, Brit. botanist, physician; b. Aberdeen, Scotland, Nov. 1, 1819; ed. St. Bartholomew's Hosp., London; m. Elizabeth Anne Henfrey; 1 son, Henry William. Became lectr., med. sch. St. George's Hosp. 1847; named prof. botany King's Coll., London, 1853; examiner in natural history Royal Mil. Acad., Soc. Arts. Genus Henfreya named in his honor. Fellow Royal Soc., 1852, Linnean Soc.; mem. Coll. Surgeons. Author: (with A. Tulk) Anatomical Manipulations, 1844; Outlines of Structural and Physiological Botany, 3 vols., 1847; Elementary Course of Botany, 1857. Editor: (with Huxley) Scientific Memoirs, 1837; (with J. W. Griffith) Micrographic Dictionary, 1854; Anatomy of British Ferns (Francis), 1855. Died Turnham Green, Eng., Sept. 7, 1859.

HENGLEIN, Friedrich Arnim, German chemist; b. Cologne, Germany, May 23, 1926; s. Friedrich-August and Gertraude (Christ) H.; ed. Higher Tech. Sch., Karlsruhe, U. Mayence; Ph.D. in chem. engring., Ph.D. in natural scis.; m. Gudrun Fröhlich, 1961; children—Frank, Arwed. Asst. scientist Max Planck Inst. Chemistry, Mayence; physicist Beyer Color Mfg. Co.; instr. U. Cologne; dir. Hahn-Meitner Inst. Nuclear Research; prof. radiation chemistry Tech. U. Berlin. Mem. Mellon Inst., Pitts. Research and numerous publs. on spectrometry of earth, nuclear chemistry, kinetics, chemistry of polymers and radiations. Home: 84 Blienickerstrasse. Office: 100 Glienickerstrasse, Berlin-Wannsee, Germany.

HENIN, Françoise Gabrielle, Belgian physicist; b. Braine-le-Comte, Belgium, Mar. 19, 1934; d. Edouard Hubert and Julia (Delplace) H.; Licence en Chimie, U. Libre de Bruxelles, Belgium, 1955, Doctorate en Chimie, 1959; m. Jean Jeener, May 14, 1960. Fellow IRSIA, 1955-56, FNRS, 1956-65; chargé de cours asso. U. Libre de Bruxelles, 1965—. Research, publs. on irreversible processes using methods of statis. mechanics, classical electrodynamics. Home: 102, Av. de la Héronniere, Brussels, Belgium.*

HENIN, Stéphane Marie Victor, French agronomist; b. Paris, Sept. 22, 1910; s. Stefen and Suzanne (Coitteux) H.; Ingénieur, Grignon Agrl. Coll., 1929; Doctor-Engr., Sorbonne, 1939; m. Huguette de Saint-Genys, Jan. 29, 1934; children—Patrick, Christian, Jean-Francois, Pierre-Yves. Chef des travaux, soil lab., Institut Nat. de la Recherche Agronomique, Versailles, France, 1931-40, dir. soil lab., 1946-59, head dept. agronomy, 1965—; prof. Institut Nat. Agronomique, Paris, 1959-65. Prof. soil physics, chmn. tech. com. agronomy (ORSTOM). Mem. Acad. Agr. France. Author: (with G. Monnier, R. Gras, A. Feodoroff) Le profil cultural, 1960; (with S. Caillère) Mineralogie des argiles. Research, publs. on phys. properties of soil, including structure, stability, circulation of unsaturating water, water and organic matter balance, also on exptl. pedology, including characterization of clays, genesis and evolution of clays, minerals, devel. of soil profiles by exptl. methods. Home: 72, rue des Entrepreneurs, Paris. Office: Centre National de la Recherche Scientifique, Route de Saint-Cyr, 78-Versailles, France.*

HENISCH, Heinz Kurt, physicist; b. Neudek, Czechoslovakia, Apr. 21, 1922; s. Leo and Fanny (Soycher) H.; B.Sc., U. Reading (Eng.), 1942, Ph.D., 1949; m. Bridget Ann Wilsher, Feb. 6, 1960. Jr. sci. officer Royal Aircraft Establishment, Eng., 1942-46; lectr. physics U. Reading, 1948-62; vis. scientist Sylvania Electric Products, Bayside, N.Y., 1955-56, prof. applied physics Pa. State U., University Park, 1963—. Pub., Carnation Press, State College, Pa. Cons., Mining & Chem. Products, London, Eng., 1958-63, Polaroid Corp., Cambridge, Eng., 1963—. Fellow Inst. Physics, Am. Phys. Soc.; associate Royal Photog. Soc. Author: Metal Rectifiers, 1949; Rectifying Semiconductor Contacts, 1957; Electroluminescence, 1962; also numerous articles. Editor, Internat. Series Monographs on Semiconductors, 1957-67; joint editor-in-chief Materials Research Bull., 1966—. Research in field semiconducting and phosphor materials, particularly on properties of surfaces and contacts, crystal growth in gels. Patentee in field. Home: 346 W. Hillcrest Av., State College, Pa. 16801. Office: Research Bldg., University Park, Pa. 16802; also Carnation Press, P.O. Box 101, State College, Pa. 16801.

HENIZE, Karl Gordon, Am. astronomer; b. Cin., Oct. 17, 1926; s. Fred R. and Mabel (Redmon) H.; B.A., U. Va., 1947, M.A., 1948; Ph.D. (NSF fellow), U. Mich., 1954; m. Caroline R. Weber, June 27, 1953; children—Kurt Gordon, Marcia Lynn, Karen Skye. Astron. observer Lamont-Hussey Obs., Bloemfontein, Union South Africa, 1948-51; Carnegie postdoctoral fellow Mt. Wilson Obs., Pasadena, Cal., 1954-56; sr. astronomer Smithsonian Astrophys. Obs., Cambridge, Mass., 1956-59; faculty Northwestern U., Evanston, 1959—, prof. astronomy, 1964—, acting dir. Dearborn Obs., 1959-60; Am. Astron. Soc. vis. professorial lectr., 1958-64; guest investigator Mt. Stromlo Obs., Canberra, Australia, 1961-62; NASA astronaut, 1967—. Mem. adv. subcom. on astronomy NASA, 1965-68. Mem. Am., Royal astron. socs., Astron. Soc. Pacific, Internat. Astron. Union, Phi Beta Kappa, Sigma Xi. Contbr. articles to tech. jours. Spectroscopic survey of stars and nebulae in So. sky; research on distbn. of emission-line stars and nebulae in large and small Magellanic clouds, spectra and forms of planetary nebulae in So. Milky Way, So. S stars, spectra of peculiar emission-line stars, objective-prism ultraviolet spectroscopy of stars from Gemini and Apollo series of manned spaceflights. Home: 18630 Point Lookout Dr., Houston 77058. Office: Astronaut Office, NASA Manned Spacecraft Center, Houston 77058.

HENKEL, Johann Friedrich, see Henckel, Johann Friedrich.

HENLE, Friedrich Gustav Jacob, German pathologist, anatomist; b. Fürth, Bavaria, July 9, 1809; ed. univs. Heidelberg, Bonn; pupil of J. P. Müller; M.D., 1832; tchr., Zurich, Switzerland; became prof. pathology and anatomy, Heidelberg, 1844; prof. U. Göttingen (Germany), from 1852. Fellow Royal Soc., 1873. Author: Pathologische Untersuchungen, 1840; Allgemeine Anatomie, 1841; Handbuch der rationellen Pathologie, 2 vols., 1846-52; Handbuch der systematischen Anatomie des Menschen, 23 vols., 1855-73. Founder of histology; 1st to treat physiology and pathology as branches of one sci.; investigated anat. structure of hair, blood vessels, nails, nervous system, lacteal system; discovered epithelium, 1837, also looped portion of kidney tubule (Henle's loop); anticipated Pasteur in suggesting microorganisms cause disease, 1840. Died Göttingen, May 13, 1885.

HENLE, Gertrude Szpingier, virologist; b. Mannheim, Germany, Apr. 3, 1912; d. Theophil and Eleonore (Baumgart) Szpingier; M.D., U. Heidelberg, 1936; m. Werner Henle, Mar. 13, 1937. Came to

U. S., 1937, naturalized, 1943. Faculty, Sch. Medicine, U. Pa., Phila., 1937—, asso. prof. virology in pub. health and preventive medicine, 1947-64, prof. virology in pediatrics, 1965—; mem. research staff Children's Hosp., Phila., 1940—. Recipient Mead Johnson award Mead Johnson Co., 1949. Diplomate in virology Am. Acad. Microbiology. Mem. Soc. Bacteriology, Tissue Culture Assn. Research and numerous publs. in immunology and prevention of virus infections, hostvirus interactions, tumor viruses, interference phenomena, viral toxicity. Home: 533 Ott Rd., Bala-Cynwyd, Pa. 19004. Office: 1740 Bainbridge St., Phila. 19146.*

HENLE, Werner, virologist; b. Dortmund, Germany, Aug. 27, 1910; s. Adolf and Tina (Lang) H.; student U. Munich, 1928-29; M.D., U. Heidelberg, 1934; m. Gertrude Szpingier, Mar. 13, 1937. Came to U. S., 1936, naturalized, 1942. Faculty, Sch. Medicine, U. Pa., Phila., 1936—, prof. virology in pediatrics, 1947—, in pub. health and preventive medicine, 1947-64, dir. virus diagnostic lab., 1947-64; mem. research staff Children's Hosp., Phila., 1939—. Mem. virus and rickettsial study sect. div. research grants NIH, 1950-57; mem. com. viral infections Armed Forces Epidemiol. Bd., 1953—. Recipient Mead Johnson award Mead Johnson Co, 1949. Diplomate in virology and immunology Am. Acad. Microbiology. Mem. A.A.A.S., Soc. Exptl. Biology and Medicine, Soc. Bacteriology, Soc. Pediatric Research, Soc. Gen. Microbiology, Assn. Immunology, Pub. Health Assn., N.Y. Acad. Scis. Research and numerous publs. in field of immunology of virus infections, devel. of virus vaccines including field trials, diagnosis of virus infections, host-virus interactions, persistent viral infections, oncogenic viruses. Home: 533 Ott Rd., Bala Cynwyd, Pa. 19004. Office: 1740 Bainbridge St., Phila. 19146.*

HENLEIN, Peter (Hele), German inventor; b. Nuremberg, Germany, 1480; clockmaker, Nuremberg; constructed 1st pocket watch with spring driven steel wheels (Nuremberg egg), early 16th century; statue erected to his memory, Galshütte, 1903. Died 1542.

HENLEY, Ernest Mark, physicist; b. Frankfort, Germany, June 10, 1924; s. Fred S. and Josy (Dreyfuss) H.; came to U. S., 1939, naturalized, 1944; B.E.E., Coll. City N.Y., 1944; Ph.D., U. Cal. at Berkeley, 1952; m. Elaine Dimitman, Aug. 21, 1948; children—M. Bradford, Karen M. Physicist, Lawrence Radiation Lab., U. Cal. at Berkeley, 1950-51; research asso. physics dept. Stanford, 1951-52; lectr. physics Columbia, 1952-54; faculty U. Wash., Seattle, 1954—, now prof. physics. F. B. Jewett fellow, 1952-53, NSF Sr. fellow, 1958-59; Guggenheim fellow, 1967-68. Mem. Am. Phys. Soc., Am. Assn. Physics Tchrs., Sigma Xi. Author: (with W. Thirring) Elementary Quantum Field Theory, 1962. Research and numerous publs. on better understanding of nuclear reactions and high energy particle interactions. Office: Physics Dept., U. Wash., Seattle 98105.*

HENLEY, William, Brit. physicist, electrician; said to have worked as linen draper. Fellow Royal Soc., 1773; believed sharp point to be better condr. than knob; constructed needle electrometer, 1772; author (pamphlet) Experiments concerning different efficacy of pointed and blunted rods in securing buildings against the strike of lightning. Died circa 1779.

HENLEY, William Thomas, Brit. telegraphic engr.; b. Midhurst, Eng., circa 1813; began career as scientific instrument maker, circa 1838; employed by Wheatstone to make electric telegraph, also to assist in expts.; made instruments for 1st Electric Telegraph Co., 1846; formed Brit. and Irish Magnetic Telegraph Co., 1852. Recipient medal Exhbn., 1851. Inventor (with Forster) magnetic needle telegraph; made electric light apparatus, submarine cable. Died Dec. 13, 1882.

HENNEBERG, Georg Heinrich Hermann, German microbiologist; b. Charlottenburg (Berlin), Germany, Oct. 12, 1908; s. Wilhelm and Charlotte (Schwerin) H.; student U. Kiel (Germany), U. Giessen (Germany), U. Vienna (Austria); M.D., U. Kiel, 1935; m. Amalie Langer, Nov. 27, 1937. Staff, Hygiene Inst., U. Kiel, 1936-37; chief bacteriol. dept. Schering A.G., 1937-45; staff Robert Koch Inst., Berlin, 1945—, dir., 1949—, v.p. Fed. Health Office, 1960—; asso. prof. Free U., Berlin, 1949—. Named hon. dir. Kitasato Inst., Tokyo, Japan, 1964. Hon. mem. Greek, Austrian socs. microbiology and hygiene; mem. Deutsche Gesellschaft für Hygiene und Mikrobiologie, Berliner Mikrobiologische Gesellschaft, Berliner Medizinische Gesellschaft. Author: Einführung in die Untersuchungstechnik zur Penicillintherapie, 1947; Weg, Ziel und Grenzen der Streptomycintherapie, 1953; (with Köhler) Praktikum der Virusdiagnostik, 1961. Research on methods for testing resistance to antibiotics, hemagglutination inhibition test for Psittacosis, morphology of bacterial cultures, viremia after preventive vaccination, depression of antibodies after vaccination, influence of chemotherapy on immunity. Home: 15 Wachtelstr., Berlin FRG 1. Office: 20 Nordufer, Berlin FRG 1, Germany.*

HENNEGUY, Félix (Luis), French zoologist; b. Paris, Mar. 18, 1850; prof. comparative embryology Coll. de France, Paris. Mem. Acad. Medicine, Acad. Agr., French Acad. Scis. Author: Le corps vitellin de balbiani dans l'oeuf des vertébres; Leçons sur la cellule; Les insectes; Sur la parthenogenèse expérimental chez les amphibiens. Research in cytology, entomology, comparative anatomy. Died Jan. 16, 1928.

HENNESSY, Thomas Gerard, Am. physician; b. Grand Forks, N.D., Sept. 17, 1919; s. Daniel Joseph and Mary (Geraghty) H.; B.S., St. Johns U., 1939; B.S., U. N.D., 1941; M.D., U. Cal. at San Francisco, 1943; Ph.D., U. Cal. at Berkeley, 1950; m. Mary Catherine Harnett, Oct. 28, 1944; children—Patricia, Thomas Gerard, John, Mary, Margaret. Served from lt. (j.g.) to comdr. M.C., USN, 1943-54; research physician U. Cal. Sch. Medicine, Los Angeles. 1954—, asst. dir. lab. nuclear medicine and radiation biology, 1956—. Mem. Soc. Nuclear Medicine, Sigma Xi. Contbr. chpt. to Radiobiology at the Intra-Cellular Level, 1959. Research, publs. on iron metabolism using radioactive iron; established technique of studying radiation damage to bone marrow of mammals with radioactive iron; participated in 1st radioactive iron turnover studies in patients with blood diseases. Home: 457 El Medio, Pacific Palisades, Cal. 90272.*

HENNIG, Emil Hans Willi, German entomologist; b. Dürrhennersdorf, Apr. 20, 1913; s. Emil and Emma (Gross) H.; Ph.D. in Zoology, agrege, U. Leipzig; m. Irma Wehnert, 1939; children—Wolfgang, Bernd, Gerd. Head systematic entomology div., asso. dir. German Inst. Entomology, Berlin; prof. Tech. U. Berlin, 1961-63; div. head phylogenetic research Mus. Natural Sci., Stuttgart, 1963—. Mem. German Leopoldina Acad. Naturalists, German (Fabricius medal), Finnish (corr.) socs. entomology, German Soc. Zoology, Nat. Union Natural Sci. of Wurtemberg. Co-editor: Das Tierreich; Grundzüge einer Theorie der Phylogenetischen Systematik, 1950; Die Larvenformen der Dipteren, 1948-52. Home: 16 Denkendorferstrasse, Ludwigsburg-Pflugfelden. Office: 3 Arsenalplatz, Ludwigsburg, Germany.

HENNING, Edwin Georg Eugen, German geologist; b. Berlin, Germany, Apr. 27, 1882; s. Arthur and Eugenie (Schmädicke) H.; ed. U. Fribourg, Breslau, U. Berlin; Ph.D. in geology; m. Johanna Trendelenburg, 1913; children—Ingeborg, Waltraut, Sieglinde, Gerda. Dir. excavations in East Africa, 1909-11; prof. agrégé, 1912; affiliate prof., 1915; full prof. U. Tübingen, 1917; prof. emeritus, 1946. Recipient Silver medal of Leibniz, prize Acad. Scis. Mem., hon. mem. numerous sci. assns. Research and numerous publs., books on paleontology, geology of Württemberg, Germany and Africa. Address: 3 Nägelestrasse, Tübingen, Germany.

HENNION, George Felix, Am. chemist; b. South Bend, Ind., Aug. 23, 1910; s. Rene C. and Elodie (van de Walle) H.; B.S., U. Notre Dame, 1932, M.Sc., 1933, Ph.D., 1935; m. Alice C. Braunsdorf, Aug. 23, 1933; children—Mary Claire (Mrs. Jon Clauss), Alice (Mrs. Kenneth Robison), George Felix, Margaret (Mrs. George Filchak). Faculty, U. Notre Dame (Ind.), 1935—, dir. research in chemistry, 1942-45, Julius Arthur Nieuwland prof. chemistry, 1945—. Cons., Eli Lilly Co. Research Labs., Indpls., 1951—. Fellow A.A.A.S., Ind. Acad. Sci.; mem. Am. Chem. Soc., Sigma Xi. Contbr. numerous articles to tech. jours. Research and patents in acetylene chemistry, organoborn compounds, medicinal agts. Home: 1441 E. La Salle Av., South Bend, Ind. 46617. Office: U. Notre Dame, Notre Dame, Ind. 46556.*

HENNY, George Christian, Am. med. physicist; b. Cal., Feb. 22, 1899; s. David Christian and Julia (Wetzel) H.; A.B., Reed Coll., 1920; M.S., Cal. Inst. Tech., 1922; M.D., U. Ore., 1930; m. Harriet Elizabeth Gore, Aug. 23, 1926; children—Jeanette Thomas, David Christian. Faculty, Temple U., Phila., 1931—, prof., head dept. med. physics, 1945—. Diplomate Am. Bd. Health Physics, Am. Bd. Radiology. Fellow Am. Coll. Radiology; mem. Franklin Inst. (com. on sci. and arts), A.M.A. (Silver medal 1946). Author: (with Mona Spiegel-Adolf) X-ray Diffraction Studies in Biology and Medicine, 1947; also numerous articles. Research on cryogenics in treatment of tumors, other conditions; developed electrokymograph for rec. motions of heart. Home: 6714 Wissahickon Av., Phila. 19119.*

HENRICI, Arthur Trautwein, Am. bacteriologist; b. Economy, Pa., Mar. 31, 1889; s. Jacob Frederick and Viola (Irons) H.; M.D., U. Pitts., 1911; m. Blanche Ressler, Aug. 7, 1913; children—Carl Ressler, Ruth Elizabeth, Hazel Jean. Pathologist St. Francis Hosp., Pitts., 1912-13; instr. in pathology and bacteriology, U. Minn., 1913-16, asst. prof. bacteriology, 1916-20, asso. prof., 1920-25, prof. from 1925; Walker-Ames prof. U. Wash., 1941. Mem. Soc. Am. Bacteriologists (pres. 1939), Soc. Exptl. Biology and Medicine, Limnological Soc. Am., Mycol. Soc. of Am., Sigma Xi, Gamma Alpha, Alpha Omega Alpha. Author: Morphologic Variation and the Rate of Growth of Bacteria, 1928; Molds, Yeasts and Actinomycetes, 1930; The Biology of Bacteria, 1934, 2d edit., 1939. Contbr. to bacteriol. jours. Died Apr. 23, 1943.

HENRIJEAN, François, Belgian physiologist; b. Spa, Belgium, 1860; became prof. pharmacodynamics Liège (Belgium) Faculty Medicine, 1904; mem. Belgian Acad. Medicine (v.p.), Acad. Medicine, Paris. Author: L'antisépsie, 1887; Traité de pharmacodynamique; Le coeur et les médicaments cardiaques. Research on effect of drugs on living organisms, physiology of heart, cardiac medicines. Died 1932.

HENRIKSEN, Erle, Am. gynecologist; b. Fresno, Cal., Apr. 12, 1904; s. Louis and Ann (Andersen) H.; A.B., U. Nev., 1927; M.D., Johns Hopkins, 1931; m. Helen W. Wood, Oct. 18, 1942; children—Patrik, Helen Ann. Asso. clin. prof. U. So. Cal., Los Angeles, 1936-41, clin. prof. gynecology, 1947—, chmn. dept., 1947—; sr. gynecologist Hosp. Good Samaritan, Good Hope Med. Found., Los Angeles County Hosp.; sr. cons. Children's Hosp., Los Angeles, 1936—. Fellow A.C.S.; mem. A.M.A., Pacific Coast Surg. Soc., Pacific Coast Obstet. and Gynecol. Soc., Am. Gynecol. Soc. Research and publs. on dissemination of cervical and uterine cancer, female genital malignancies, ovarian tumors, infertility, premenstrual tension, stress incontinence, use of hormones in treatment of uterine cancer. Home: 1 Toluca Estates, Dr., North Hollywood, Cal. 91602. Office: 1136 W. 6th St., Los Angeles 90017.*

HENRION, Denis, French mathematician; prof., Paris; engr. to Prince of Orange. Author: Traité des globes et leur usage, 1618; Canon manuel des sinus, 1619; Cosmographie, 1626; Traité des logarithmes (1st French table of logarithms), 1626; also early French transl. of Euclid. Worked on perfecting Gunther's slide rule. Died Paris, circa 1640.

HENRY, James Paget, physiologist; b. Leipzig, Germany, July 11, 1914 (parents Am. citizens); s. Oscar John and Winifred (Paget) H.; M.A., Cambridge U., 1939, M.D., 1952; Ph.D., McGill U., 1955; m. Isabelle Johnston Jardine, Dec. 7, 1938; children—John, Peter, Mark. With physiology dept. U. So. Cal., 1943-47, vis. prof. physiology, 1962-65, prof., 1964—; with Wright Patterson AFB Aeromed. Lab., 1947-56; with Air Research Devel. Command, Brussels, Belgium, 1956-59; with NASA Manned Space Flight Program, 1959-62. Mem. Am. Inst. Biol. Scis. panel on behavioral biology for NASA. Recipient Tuttle award Aerospace Med. Assn., 1953, Jefferies award Inst. Aero. Scis., 1954. Fellow Aerospace Med. Assn.; mem. Am. Physiol. Soc., Internat. Acad. Astronautics. Author: Biomedical Aspects of Space Flight, 1966. Invented partial pressure high altitude suit; codiscoverer volume regulatory reflex connecting heart filling with kidney function; research and numerous publs. on animals in rocket flight and on chimpanzee studies preceding Mercury Flight, role of social stress in inducing high blood pressure in man and animals. Home: 8626 Skyline Dr., Los Angeles 90046.*

HENRY, Joseph, Am. physicist; b. Albany, N.Y., Dec. 17, 1797; s. William and Ann (Alexander) H.; received numerous hon. degrees including A.M., Union Coll., 1829; LL.D., S.C. Coll. (now U. S.C.), 1838, U. State N.Y., 1850, Harvard, 1851; m. Harriet Alexander, May 1830, 6 children. Conducted research in electromagnets, building most powerful of his day; set up telegraph one mile in length; discovered principle of induction, 1830 (however Faraday was 1st to publish and receive credit); invented electromagnetic motor, 1831; discovered principle of self-induction, 1832; laid basis for electromagnetic telegraph with his invention of elec. relay; prof. natural philosophy Coll. of N.J., (now Princeton), 1832-46; did research which anticipated some of the modern developments in science of electricity, 1838-42; invented low and high resistance galvonometers; 1st sec. and dir. Smithsonian Instn., 1846-78, (instn. as known today largely a result of his work); initiated system of basing weather forecasts on weather reports received by telegraph, 1850; investigated sun spots and solar radiation. Became mem. Light House Bd., 1852, pres., 1871-78; became mem. Am. Philos. Soc., 1835; an organizer A.A.A.S., pres. 1849; a founder Philos. Soc. Washington, 1871, pres. 1871-78; charter mem. Nat. Acad. Scis., v.p., 1866, pres., 1868-78; sci. unit of self-induction named Henry in his honor. Died Washington, D.C., May 13, 1878.

HENRY, Morris Henry, physician; b. London, Eng., July 26, 1835; s. Henry A. Henry; M.D., U. Vt., 1860; m. Elizabeth Hastings, 1872; m. 2d, Mrs. Harrison Everett Maynard, 1880; survived by 1 child. Came to N.Y.C., 1852; surgeon Northern. Dispensary, 1864; surgeon N.Y. Dispensary, 1869; surgeon-in-chief N.Y. State Emigrant Hosp., 1873-80; chief police surgeon of N.Y.C., 1872-84, organized ambulance service; founder, editor Am. Jour. Syphilography and Dermatology, 1870-75; published Am. edit. Skin Diseases: Their Description, Pathology, Diagnosis, and Treatment (W. T. Fox), 1871; invented numerous instruments including forceps, scissors for many purposes. Died May 19, 1895.

HENRY, Paul Pierre, French astronomer; b. Nancy, France, Aug. 21, 1848; astronomer aide Paris Obs., from 1864, adjunct astronomer, from 1876. With bro. Prosper, continued preparation of ecliptic atlas of sky begun by Chacornac; observed transit of Venus, 1882; improved process of astron. photography; discovered comet III, 1873, also 14 planetoids; improved photographic process, thus facilitating mapping of sky. Died Montrouge, France, 1905.

HENRY, Prosper-Mathieu, French astronomer; b. Nancy, France, Dec. 10, 1849; became aid in astronomy Paris Obs., 1864, adj. astronomer, 1876, titulaire astronomer, 1893. Contbr. to sci. publs. Discovered 7 planetoids; developed (with brother Paul Henry) astron. photography to high degree of perfection. Died Vanoise, 1903.

HENRY, Richard Joseph, Am. clin. chemist; b. Harrisburg, Pa., Mar. 4, 1918; s. John Rynard and Bessie (Ryan) H.; A.B., Gettysburg Coll., 1940; M.D., U. Pa., 1943; m. Maryon Margaret Dytche, June 10, 1943; children—Lauren Margaret, Janice Elizabeth. Investigator dept. bacteriology Sch. Medicine, U. Pa., 1942-46; biochemist, bacteriologist in bacteriologic warfare Camp Detrick, Md., 1945-48; instr. U. So. Cal. Sch. Medicine, Los Angeles, 1950, adj. prof. biochemistry, 1962—; dir. Bio-Sci. Labs., Los Angeles, 1948—. Fellow A.A.A.S., Am. Assn. Clin. Chemists (past pres.); mem. Am. Soc. Microbiology, Am. Chem. Soc. (sr.), Assn. Clin. Scientists, Soc. Exptl. Biology and Medicine, Biometric Soc., Western Soc. for Clin. Research, Am. Acad. Microbiology, Endocrine Soc., Phi Beta Kappa, Sigma Xi, Alpha Omega Alpha. Author: Clinical Chemistry: Principles and Technics, 1964. Research and numerous publs. on effects antibiotics on bacterial metabolism, clin. chemistry methodology. Home: 667 Ledo Way, Los Angeles 90049. Office: 7600 Tyrone Av., Van Nuys, Cal. 91405.*

HENRY, Thomas, Brit. chemist; b. Wrexham, Wales, Oct. 26, 1734; apprentice to apothecary, Wrexham; attended anat. lectures Oxford (Eng.) U.; m. Mary Kinsey, circa 1759; children—Thomas, William. Asst. to apothecary, Oxford, Eng.; began own bus., Knutsford, Eng., 1759; practiced as surgeon and apothecary, Manchester, Eng.; co-founder, lectr. on chemistry, bleaching, dyeing Manchester Coll. Arts and Scis. Fellow Royal Soc., 1775; mem. Am. Philos. Soc., Manchester Lit. and Philos. Soc. (sec. 1781, pres. 1807). Author: Experiments and Observations, 1773; also papers. Translator chem. essays of Lavoisier, 1776, 83; pub. Memoirs of Albert de Haller, 1783. Patentee process for preparing calcined magnesia; 1st to observe use of carbonic acid to plants. Died Manchester, June 18, 1816.

HENRY, Victor, French philologist; b. Alsace, France, Aug. 17, 1850; held appointments at Douai and Lille, then became prof. Sanskrit and comparative grammar U. Paris, 1889. Author: Précis de Grammaire camparée de l'anglais et de l'allemand, 1893; Précisim du Grec et du Latin, 1888; (with A. Bergaigne) Manuel pour étudier le Sanscrit Védique, 1890; Éléments de Sanscrit classique, 1902; Precis de grammaire Palie, 1904; Les Littérature de l'Inde: Sanscrit, Pali, Pracrit, 1904; La Magie dan l'Inde antique, 1904; Le Parisme, 1905; L'Agnistoma, 1906. Probably best known for comparative Greek-Latin and English-German grammar books; also important work on India and Indian languages; gave considerable attention to obscure languages such as Innok, Quichua, also of Greenland and local dialects. Died 1907.

HENRY, William, English chemist; b. Manchester, Eng., Dec. 12, 1775; ed. medicine, Edinburgh; practiced medicine, Manchester. Fellow Royal Soc., 1809 (Copley medal 1808). Discovered law which states amount of gas absorbed by a liquid is proportional to pressure (named after him). Died Pendlebury, Eng., Sept. 2, 1836.

HENRY OF BATE, (or Henry of Malines), Belgian astronomer; b. Malines, nr. Antwerp, Belgium, 1246; flourished in Liège, Belgium and Malines; canon, cantor Liège cathedral; in Orvieto, Italy, 1292. Author: Speculum divinorum et quorundam naturalium, 23 vols., 1280-1300 or 1310; Tabulae machlinienses, before 1281; Liber servi Dei de Machlinia super inquisitione et verificatione navitatis propriae, 1281. Translator (from French to Latin) astrological treatises of Ibn Ezra. One of most sci. astrologers of his day; influenced astrology up to 16th century; observed annular eclipse of sun, January 31, 1310. Died 1310 or after.

HENRY OF MALINES, see Henry of Bate.

HENRY OF MONDEVILLE, French surgeon; b. Normandy, circa 1260; ed. Montpellier, France, also Paris, under Theodoric Borgognoni in Italy; personal surgeon, also mil. surgeon in armies of Philip the Fair and Louis X, from 1301; lectr. anatomy and medicine, Montpellier, 1304. Author: Cyrurgia (1st treatise on surgery by a surgeon), 5 sects., Anatomy, General and Particular Treatment of Wounds, Ulcers, and Contusions, Special Surgical Pathology and Therapeutics, Fractures and Luxations, Antidotary, 1306-20. Introduced Theodoric's new methods of wound treatment into France. Died circa 1320.

HENRY THE NAVIGATOR (Henry of Portugal), Portuguese navigator, explorer; b. Oporto, Mar. 4, 1394; s. Joao I (King of Portugal) and Philippa. Knighted for distinguished service in battle over Ceuta, 1415, also created duke of Viseu and lord of Covilham, then gov. of Algarve, 1419; founder obs. and nautical sch., Algrave, 1419; enlisted skilled mariners for exploratory voyages which resulted in rediscovery of Porto Santo, 1418, also rediscovery of Madira, 1420; began colonization of Madeiras and tried to purchase the Caneries, 1424-25; sent out numerous voyages to Africa, then to Azores, 1427, 31, gold and slaves were brought back from Africa (resulted in exploratory voyages becoming popular). Noted particularly for his part in stimulating other European nations to begin explorations. Died nr. Cape St. Vincent, Nov. 13, 1460.

HENSCHEL, Austin Ferdinand, Am. physiologist; b. Princeton, Minn., Mar. 13, 1906; s. August and Mary (Senzel) H.; B.S., U. Minn., 1929, M.A., 1932, Ph.D., 1938; m. Helen Busch, Oct. 29, 1938. Asst. zoology U. Minn., 1930-34, asst. physiology, 1934-38, instr., 1938-41, asst. prof. physiol. hygiene, 1941-46, asso. prof., 1946-51; dir. environmental protection research div. Q.M.C., Research & Devel. Command, Natick, Mass., 1951-61; chief physiol. sect. div. occupational health USPHS, Cin., 1961—; vis. lectr. physiology Harvard Sch. Pub. Health, 1951-61; adj. asst. prof. environment health U. Cin., 1967—. Mem. Am. Physiol. Soc., A.A.A.S., Soc. Exptl. Biology and Medicine, Am. Indsl. Hygiene Assn., Sigma Xi. Author: (with others) Human Starvation, 2 vols., 1950. Research and publs. on human nutritional requirements, biology of starvation, effects of environmental temperatures on performance, ergonomics, occupational hazards. Home: 2696 Bonnie Dr., Cin. 45230. Office: 1014 Broadway, Cin. 45202.*

HENSEL, Herbert, physiologist; b. Prague, Czechoslovakia, Sept. 2, 1920; s. Walter and Olga (Pokorny) H.; student medicine univs. Tübingen, Strasbourg; M.D., U. Heidelberg, Germany; m. Rehanita Hein, Dec. 17, 1949; children—Michael, Cora. Mem. physiology depts. U. Heidelberg, 1946-48, 50-55, Royal Vet. Sch., Stockholm, Sweden, 1949-50; prof. physiology, dir. dept. physiology U. Marburg/Lahn, Germany, 1955—, dean faculty medicine, 1964-65, rector, 1965-66; Walker-Ames vis. prof. U. Wash., Seattle, 1960. Recipient Adolph Fick prize Med. Soc. Würzberg, 1954; Feldberg prize U. London, 1961. Mem. Royal Soc. Medicine London, Acad. Scis. Finnland, Internat. Soc. Biometeorology. Author: (with Precht, Christophersen) Temperatur und Leben, 1955; (with Golenhofen, Hildebrandt) Durchblutungsmessung Mit Warmeleitelementen, 1963; Allgemeine Sinnesphysiologie. Hautsinne, Geschmack, Geruch, 1966. Research, publs. on sensory physiology, thermoregulation, circulation; discovery of excitation mechanism of temperature receptors; 1st records of single afferent signals in human nerve fibres; inventor Fluvograph for rec. local blood flow in various organs. Home: 48 Sandweg, Marburg/Lahn, Germany.*

HENSEL, Kurt W. S., German mathematician; b. Königsberg, Germany (now Kaliningrad, USSR), Dec. 29, 1861; prof., Marburg, Germany. Author: Theorie der algebraischen Zahlen, 1908; Zaalentheorie, 1913. Creator theory in which numbers are represented by power series. Died Marburg, Jan. 6, 1941.

HENSELER, Heinz, German agronomist; b. Euskirchen, May 7, 1885; s. Josef and Elisabeth (Breuer) H.; Ph.D. in Agronomy, Agrl. U., Halle, Salle; m. Marion Nentwig, 1921; 1 son, Heinz Arno. Asst., Inst. Meteorology and Botany, 1900; asst. Inst. Animal Breeding and Dairy Farming, U. Halle, Salle, 1910, dir., 1913-14; prof. animal breeding, dir. Inst. Animal Breeding and Lab. Bacteriological Chemistry Dairy Farming, U. Göttingen, 1916-20; instr. gen. agronomy and agrl. econs. U. Munich, Technische Hockschule, Munich, 1920-54, emeritus prof., 1951; lectr. gen. biology Technische Hochschule, Munich, 1954—. Recipient Dr. Paul-Parey prize, 1913, Max von Eyth Silver medal, 1925, Max von Eyth Bronze medal, 1931, Silver medal of agr. of France, 1935. Mem. German Soc. Agrl. Sci., German Soc. Animal Breeding, Profl. Soc. Animal Breeding and Agr. (dir.). Author: Untersuchungen über den Einfluss der Ernahrung auf die Morphologische und Physiologische Gestaltung des Tierkörpers, 1913; Die Mendelsche Lehre und ihre Bedeutung für die Praktische Tierzucht, 1912, 21; Aus der Tierzucht von Australien und Neuseeland, 1928; Die Anwendung der Genetik und Rasseverbesserung auf Milchkühe und Schafe; Die Grundlagen der Vererbung und die Zucht des Hundes, 1935, also numerous articles. Home: 10/6 Schellingstrasse, Munich 8. Office: Munich Technische Hochschule, Munich, Germany.

HENSEN, Victor, German physiologist, pathologist; b. Kiel, Germany, Feb. 10, 1835; prof., Kiel, 1871-1911; research in embryology; studied anatomy and physiology of sense organs; began research on and introduced word plankton; eponym of cells of Hensen and Hensen's duct (both in ear). Died Apr. 5, 1924.

HENSHAW, Henry Wetherbee, Am. ornithologist, ethnologist; b. Cambridge, Mass., Mar. 3, 1850; s. William and Sarah (Holden) H.; ed. Cambridge pub. schs. Naturalist, Wheeler Survey for exploration of West, 1872-79; adminstrv. work Bur. of Ethnology, 1879-93, part of which time was editor Am. Anthropologist; adminstrv. biologist, Biol. Survey, U. S. Dept. Agr., from 1905, chief of Biol. Survey, from 1910. Fellow Author: Report on Ornithology of Nev., Utah, Calif., Colo., N.Mex., and Ariz., 1875; Animal Carvings from the Mounds of the Mississippi Valley, 1883; Birds of the Hawaiian Islands, 1902. Died Aug. 1, 1930.

HENSHAW, Paul Stewart, Am. biophysicist; b. Salt Fork, Okla., Aug. 23, 1902; s. Carter Lee and Elizabeth Ann (Stewart) H.; A.B., Southwestern Coll., Winfield, Kan., 1925; M.S., U. Wis., 1927, Ph.D., 1930; m. Christine Thygeson; children—Robert Eugene, Elizabeth Joanne. Exec. dir. Nat. Com. Maternal Health, Inc., N.Y.C., 1954-56; biophysicist div. biology, medicine AEC, Washington, 1956-66; dir. Avery Postgrad. Inst., Insts. For Achievement of Human Potential, Phila., 1966—. Mem. A.A.A.S., Radiation Research Soc., Am. Acad. Polit. and Social Sci. Author: Adaptive Human Fertility, 1955. Editor: Peaceful Uses of Atomic Energy, Proceedings 2d Internat. Conf., Geneva, 1958; Isotopes in Medicine, Conf. Proceeding AEC, 1959, Late Irradiation Effects: Inventory of Capability, 1965. Research and numerous publs. on radiobiology, exptl. embryology, cancer, human fertility, level of living, systems scis., neurol. orgn. Home: 201 W. Evergreen Av., Phila. 19118. Office: 8801 Stenton Av., Phila. 19118.*

HENSHAW, Samuel, Am. entomologist; b. 1852; A.M. (hon.), Harvard, 1903; asst. entomologist, librarian, mus. comparative zoology Harvard, curator, 1904-12, dir., 1912-27, emeritus, from 1927, dir. univ. mus., 1918-27, emeritus, from 1927. Mem. Soc. Zoologists, Entomol. Soc., Am. Acad. Author: List of the Coleoptera of America North of Mexico, 1885; Report on the Gypsy Moth of Massachusettes, 1892. Died 1941.

HENSLOW, John Stevens, English botanist; b. Rochester, Eng., Feb. 6, 1795; s. John Prentis Henslow; grad. St. John's Coll., Cambridge, 1818, M.A., 1821; m. Harriet Jenyns, 1823; children—Leonard, George, Frances (Mrs. J. D. Hooker), Anne (Mrs. Barnard), Louisa. Prof. mineralogy Cambridge U., 1822-27, prof. botany, 1827-61; became vicar of Hitcham, Eng., 1839; asst. to W. J. Hooker, Kew; examiner in botany, mem. senate London U. Fellow Linnean Soc.; founder (with Adam Sedgwick) Cambridge Philos. Soc., 1821. Author: A Catalogue of British Plants, 1829, 2d edit., 1835; Principles of Descriptive and Physiological Botany, 1825; Letters to the Farmers of Suffolk (on sci. agr.), 1843; Dictionary of Botanical Terms, 1857; also part of Flora of Suffolk (E. Skepper), 1860. Tchr. of Charles Darwin, whom he recommended as naturalist to Beagle; made geol. expdns. (with Sedgwick) to Isle of Wight, Isle of Man; discovered phosphatic nodules in Suffolk Crag, 1842; leader Brit. Assn. discussion on Origin of Species, 1861. Died May 16, 1861.

HENSON, James Bond, Am. veterinarian; b. Colorado City, Tex., Nov. 13, 1933; s. John Lee and Beatrice (Porter) H.; B.S., Tex. A. and M. U., 1956, D.V.M., 1958, M.S., 1959; Ph.D., Wash. State U., 1964; m. Evie Leone Callihan, Mar. 28, 1956; children—Sarah Ruth, Benjamin Lee, James Dodd. Asst. veterinarian Tex. Agrl. Expt. Sta., Angleton, 1958-60; faculty dept. vet. pathology Wash. State U., Pullman, 1960—, asso. prof. vet. pathology, 1965—. Recipient award for distinguished teaching in vet. medicine, 1964, Research Career Devel. award, 1965. Diplomate Am. Coll. Vet. Pathologists. Mem. Am. Soc. Exptl. Pathologists, Am. Vet. Med. Assn., Conf. For Research Workers in Animal Diseases. Research and numerous publs. on etiology of infectious keratitis of cattle and characterization of agt., pathogenesis of hypergammaglobulinemia in mink, immunological mechanisms in animals, equine infectious anemia, comparative pathology, ultra structure of renal disease. Home: Box 505C.S., Pullman, Wash. 99163.*

HENTGES, James Franklin, Jr., Am. animal nutritionist; b. Perry, Okla., Feb. 6, 1925; s. James Franklin and Edna (Golliver) H.; B.S., Okla. State U., 1948; M.S., U. Wis., 1950, Ph.D., 1952; m. Iris Lavaun McGill, Mar. 10, 1946; children—Douglas, Eric James, Kurt William. Instr., U. Wis., 1951-52; faculty U. Fla., Gainesville, 1952—, prof. animal sci., 1966—, dir. U. Fla.-AID contract, Cuba, 1957. Cons. U. Central Venezuela, 1961. Mem. Am. Inst. Nutrition, Am. Soc. Animal Sci., Am. Dairy Sci. Assn., Sigma Xi, Phi Kappa Phi, Alpha Zeta, Phi Eta Sigma, Gamma Sigma Delta. Research, numerous publs. on comparative physiology of Brahman and Hereford cattle, use of indsl. by-products for cattle nutrition, effect of texture of diet, feed additives on feedlot cattle, diet formulation by computers. Home: 1915 N.W. 5th Av., Gainesville, Fla. 32601.*

HENTSCHEL, Hans Emil, German geochemist; b. Leipzig, Oct. 14, 1898; s. Emil and Helene (Hendler) H.; Ph.D. in Geochemistry, U. Leipzig; agrégé; m. Dorothea Cuneua, 1932. Asst. supt., prof. agrégé mineralogy and petrography U. Leipzig; titular prof., mandate prof. petrography U. Mayence; dir. depts. petrography and geochemistry, study of deposits Office Pedology of Hesse, Wiesbaden. Mem. German Mineralogy Soc., German Geol. Soc., Assn. Geology, Soc. Geochemistry. Research and publs. on petrography, rock formations, mountain crystals, volcanic mineralogy. Home: 2 Fontanestrasse, Wiesbaden. Office: Johannes Gutenberg University, 21 Saarstrasse, Mainz, Germany.

HENWOOD, William Jory, English mineralogist; b. Perron Wharf, Eng., Jan. 16, 1805; s. John Henwood; clerk to Messrs. Fox & Co., Perron Wharf; assay-master, supr. tin, duchy of Cronwall, 1832-39; placed in charge Congo-Soco mines, Brazil, 1843; studied metals of Kumaon and Gurhwal for govt. of India, 1855; ret., Penzance, Eng., 1858. Recipient Telford medal Instn. Civil Engr., 1837. Fellow Geol. Soc. (Murchison medal 1874), Royal Soc., 1840; pres. Royal Instn. Cornwall, 1859. Contbr. papers to jours. Eponym of Henwoodite, a hydrous phosphate of aluminum and copper. Died Aug. 5, 1875.

HENYEY, Louis G(eorge), Am. astronomer; b. McKees Rocks, Pa., Feb. 3, 1910; s. Belá and Mary (Floszmann) H.; B.S., Case Inst. Tech., 1932, M.S.,

1933; Ph.D., U. Chgo., 1937; m. Elizabeth Rose Belak, Apr. 29, 1934; children—Thomas Louis, Francis Stephen, Elizabeth Maryrose. From instr. to asst. prof. Yerkes Observatory, U. Chgo., 1937-47; from asst. to asso. prof. U. Cal., Berkeley, 1947—, now prof., also dir. Computer Center, 1956-60. chmn. astronomy dept., 1959-64; research asso. Princeton, 1951-52; prin. investigator optical project NDRC. Guggenheim fellow, 1940-41. Mem. Am., Royal astron. socs., Astron. Soc. Pacific (pres.), Sigma Xi. Study of interstellar matter; illumination of reflection nebulae; stellar atmosphere; stellar evolution; electronic computation. Home: 1020 Contra Costa Dr., El Cerrito, Cal. 94532. Office: U. Cal., Berkeley, Cal. 94720.

HENZE, Henry Rudolf, Am. chemist; b. New Haven, Jan. 11, 1896; s. Henry and Wilhelmina (Tamm) H.; Ph.B. magna cum laude, Sheffield Sci. Sch., 1918; Ph.D., Yale, 1921; m. Elizabeth Sledge, Aug. 24, 1933; 1 son, Henry Rudolf. Head dept. pharm. chemistry med. br. U. Tex., Galveston, 1921-27, prof. chemistry, Austin, 1927—, grad. prof., 1949—, research prof., 1945-46, chmn. dept. chemistry and chem. engring., 1929-39. Chem. cons. to pharm. cos., 1932-62. Recipient L. E. Scarbrough Found. award for excellence in teaching, 1956, Alpha Epsilon Delta Outstanding Tchr. award 1961-62. Fellow A.A.A.S.; mem. Am. Chem. Soc. (Southwestern Regional award for service to chemistry 1953), Tex. Acad. Sci., Beta Phi Sigma (hon.), Alpha Epsilon Delta, Kappa Xi. Asso. editor Jour. Organic Chemistry. Research and numerous publs. on synthesis of anti-convulsant drugs, and heterocyclic nitrogen compounds, calculations of numbers of isometric aliphatic organic compounds. Home: 309 Moore Blvd., Austin, Tex. 78705.*

HENZL, Milan, Czechoslovakian physician; b. Bratislava, Slovakia, Feb. 2, 1928; s. Joseph and Anna (Vymetalová) H.; M.D., U. Olomouc, Moravia, Czechoslovakia, 1951; C.Sc., Charles U., Prague, Czechoslovakia, 1961; m. Vera Sjnohová, Sept. 4, 1954; children—Helen, Renata, David. Mem. research unit gynecic endocrinology Inst. Care of Mother and Child, Prague-Podolí, Czechoslovakia, 1955—. Mem. Czechoslovak Soc. Obstetrics And Gynecology. Author: (with Horsky, Presl, Valenta) Pneumopelvigraphy in Gynecology, 1961. Research and numerous publs. on diagnostic methods in developmental anomalies of female internal genitalia, mechanism of action of hemostasis of hormonal steroids. Home: 5 Pod Kláudiánkou, Office: 157 Marxovo nábr., Prague 4, Czechoslovakia.*

HEPBURN, Joseph Samuel, Am. chemist; b. Phila., Aug. 25, 1885; s. Samuel Martin and Mary M. W. (Danenhower) H.; A.B., Central High Sch., 1903, A.M., 1908; B.S. in Chemistry, M.S., U. Pa., Ph.D., Columbia, 1913; D.Chem., Hahnemann Med. Coll., 1946, M.D., 1934. With Food Research Lab. Bur. Chemistry, U. S. Dept. Agr., 1907-12, 13-20, asst. chemist, 1908-20; fellow biol. chemistry Columbia, 1912-13; with Hahnemann Med. Coll. and Hosp., Phila., 1920—, prof. chemistry, 1939-50, emeritus prof. biol. chemistry, 1950—, research asso. gastro-enterology, 1937-50; archives researcher Franklin Inst., 1954—. Recipient Edward Longstreth medal of merit, 1911; Franklin Inst. certificate of merit, 1921, certificate appreciation Atwater Kent Mus., 1958. Diplomate Am. Bd. Clin. Chemistry. Mem. Am. (abstractor), Pa. chem. socs., Am. Soc. Biol. Chemists, Am. Coll. Gastroenterology, Am. Inst. Homeopathy, Acad. Natural Scis. Phila., Physiol. Soc. Phila., Am. Assn. U. Profs., Am. Assn. History Medicine, Franklin Inst., Phi Beta Pi, Sigma Xi, others. Author: (with William A. Pearson) Physiological and Clinical Chemistry, 1925, 1938. Research on influence of time and temperature, i.e., cold storage, chill room, household refrigerator, room, upon chem., bacteriol., and histol. changes in perishable animal foodstuffs; chemistry and physiology of insectivorous plants, especially pitcher plants of Am. and East Indies; temperature of human digestive tract, and influence thereon of hot and cold foods, externally applied sources of heat and cold; action of drugs in small doses upon man; studies on history of sci., medicine in U. S., and on history of Phila. Home: 1360 Marlborough St., Phila. 19125. Office: 20th St. and Benjamin Franklin Pkwy., Phila. 19103.*

HEPLER, David Skeen, Am. electronics engr.; b. Greensboro, N.C., July 24, 1927; s. William L. and Verla (Skeen) H.; B.S. in Elec. Engring., N.C. State Coll., 1950; M.S., U. Md., 1956; m. Carolyn Lou Brown, June 25, 1950; children—Elizabeth Jean, Ann Carolyn. Staff, U. S. Naval Research Lab., Washington, 1950-58; electronics engr. NASA, Goddard Space Flight Center, Greenbelt, Md., 1958—. Tech. cons. to Canadian, Brit., and French space programs, European Space Research Orgn. Mem. I.E.E.E. Designed R-f systems for sci. spacecraft; designed 1st R-f command system used in satellite. Home: 313 Onondaga Dr., Washington 20021. Office: Goddard Space Flight Center, Greenbelt, Md. 20771.*

HEPP, Joseph Andrew, Am. physician; b. Pitts., Mar. 10, 1900; s. Adolph G. and Clara (Snyder) H.; B.S., U. Pitts., 1923, M.D., 1925; m. Eleanor Rachel Keller, June 30, 1937; children—Joseph Andrew, Richard Stephen. Faculty, U. Pitts., 1929—, clin. prof. obstetrics and gynecology, 1959—, chmn. dept. gynecology, 1949-59; sr. gynecologists Magee Womens Hosp., Pitts., 1950—, St. Francis Gen. Hosp., Pitts., 1950—. Cons. to hosps., USPHS. Fellow Am. Cancer Soc., A.M.A., A.C.S., Am. Coll. Obstetricians and Gynecologists (founding); mem. Internat. Corr. Soc. Obstetricians and Gynecologists. Research and numerous publs. on clin. gynecology, endocrine gynecology, cancer of female reproductive tract. Home: 7417 Richland Manor Dr., Pitts. 15208.*

HEPPNER, James Paul, Am. geophysicist; b. Winona, Minn., Aug. 9, 1927; s. Diederick John and Alice (Gengnagel) H.; student U. Neb., 1944-45; B.Physics, U. Minn., 1948; M.S., Cal. Inst. Tech., 1950, Ph.D., 1954; m. Edith L. Summa, July 23, 1955; children—Alex J., Annet S., Peter W. Research asso. U. Alaska, College, 1950-52; physicist Naval Research Lab., Washington, 1954-58; asst. head fields and particles br. Goddard Space Flight Center, NASA, Greenbelt, Md., 1959-63, head fields and plasmas br., 1963—. Mem. Am. Geophys. Union, Am. Inst. Aeros. and Astronautics, A.A.A.S., Soc. Exploration Geophysicists, Internat. Union Radio Sci., Sigma Xi. Research and publs. in magnetic field measurements from satellites; 1st measurements revealing magnetic tail of earth and magnetosphere boundary; studies of aurora, ionosphere, geomagnetic disturbances; rocket-borne expts. for studies of airglow, ion composition, magnetic fields. Home: 47 Shaw Av., Silver Spring, Md. 20904. Office: Goddard Space Flight Center, Greenbelt, Md. 20904.*

HEPTING, George Henry, Am. plant pathologist; b. Bklyn., Sept. 1, 1907; s. George and Lena (Schuler) H.; B.S., Cornell U., 1929, Ph.D., 1933; m. Anna Johnson Love, Mar. 17, 1934; children—George Carleton, John Bartram. Instr., Cornell U., 1929-31; U. S. Dept. Agr., 1931-62, chief forest disease research Southeastern Forest Expt. Sta., Asheville, N.C., 1953-62; prin. research scientist U. S. Forest Service, Asheville, N.C., 1962—. Recipient Soc. Am. Foresters Biol. Research award, 1963, U. S. Dept. Agr. Superior Service award, 1954. Fellow Am. Phytopathol. Soc., Soc. Am. Foresters (chmn. Appalachian sect. 1943); mem. So. (pres. 1946), internat. shade tree confs., Forest History Soc., N.C. Forestry Assn., So. Appalachian Mineral Soc. Asso. editor Phytopathology, 1937-39, Jour. Forestry, 1950-52, Forest Sci., 1959-65. Contbr. numerous articles to sci. jours. Patentee field wilt-resistant mimosa trees, method of stimulating gum flow in pines by a fungus; developed the INTREDIS electronic system of literature retrieval for forest pathology; determined cause and control of many forest and shade tree diseases, including pioneer work on air pollution effects on forest trees. Home: 11 Maplewood Rd., Asheville, 28804. Office: Post Office Bldg., Box 2570, Asheville, N.C. 28802.*

HEPTINSTALL, Robert Hodgson, physician; b. Keswick, Eng., July 22, 1920; s. James Alfred and Mabel (Sander) H.; M.B., King's Coll., London, Eng., 1944, M.D., 1948; m. Ann Enraght Porter, Jan. 25, 1949; children—Bridget Ann, Gillian Margaret, Jonathan Richard, James Nigel, Caroline Jane, and Christopher Edward Heptinstall. Began career as registrar Wright-Fleming Inst. London, 1947, pathology dept. St. Mary's Hosp., London, 1947-49; jr. lectr. St. Mary's Hosp., London, 1949-51, sr. lectr. pathology, 1951-60; vis. prof. pathology, Washington U. Sch. Medicine, St. Louis, 1960-62; prof. pathology Johns Hopkins Sch. Medicine, 1967—, acting chmn. dept. pathology, 1966—; acting pathologist in chief Johns Hopkins Hosp., Balt., 1966—. Mem. pathology study sect. NIH, 1963—. Mem. Path. Soc. Gt. Britain, Assn. Am. Pathologists, Internat. Acad. Pathology, Renal Assn. Author: Pathology of the Kidney, 1966; also articles. Research in exptl. arterial and kidney diseases. Home: 104 Longwood Rd., Balt. 21210. Office: Johns Hopkins Hosp., Balt. 21205.*

HERACLIDES OF PONTOS, (or Heraclides Ponticus), Greek astronomer; b. Heraclea, Pontos (now Bander Eregli, east of Istanbul), circa 388 B.C.; studied under Speusippos, Plato, Aristotle, in Plato's Acad.; author numerous works on astronomy and geometry, few of which survive; suggested that rotation of earth causes movement of heavenly bodies, also that Venus and Mercury revolve around sun. Died Athens, circa 315 B.C.

HERACLIDES OF TARENTUM, physician; flourished circa 75 B.C.; ed. in Herophiliean sch.; probably worked in Alexandria; most important empirical physician of antiquity; probably dissected human bodies; praised by Galen for objectivity and tech. skill; author Symposium, at least 13 other works on pharmacology, therapeutics, dietetics, other subjects, some preserved in fragments in Galen.

HERACLITOS OF EPHESOS, natural philosopher; b. Ephesos, nr. Miletos, circa 540 B.C.; influenced sci. thought of his time; believed change the only certainty of the universe and fire its main principle; held that all men have a universal soul; known as weeping philosopher for his pessimistic view of life. Died circa 475 B.C.

HERAPATH, John, English physicist, mathematician; b. Bristol, Eng., May 30, 1790; m., 1815; 1 son, Edwin John; opened math. acad., Bristol, Eng., 1815; became math. tutor, Cranford, Eng., 1820; moved to Kensington, Eng., 1832; became prop., mgr. Ry. Mag., 1836, later Herapath's Ry. and Comml. Jour. (a weekly). Author: Mathematical Physics, 2 vols., 1847; also papers. Offered a controversial determination of principle of gravity to Royal Soc., 1820; produced a table of temperatures disputed by Tredgold; corrected math. works of Brougham. Died Catford Bridge, Lewisham, London, Eng., Feb. 24, 1868.

HERAPATH, William B., Brit. toxicologist, physician; b. 1820; s. William Herapath; M.D. Fellow Royal Soc., 1859; author several med. treatises; discovered polarization of light by plates of iodo-quinine, 1852. Died Oct. 12, 1868.

HERB, Raymond G., Am. physicist; b. Navarino, Wis., Jan. 22, 1908; s. Joseph and Annie (Stadler) H.; Ph.B., U. Wis., 1931, Ph.D., 1935; hon. degrees U. Sao Paulo, Brazil, U. Basel, Switzerland; m. Anne Williamson, Dec. 26, 1945; children—Stephen, Rebecca, Sara, Emily, William. With U. Wis., Madison, 1936—, prof. physics 1945—, Charles Mendenhall prof. physics, 1960—. Fellow Am. Phys. Soc.; mem. Nat. Acad. Sci. Developed high voltage electrostatic accelerators, conducted research in nuclear physics, developed Getter-ion vacuum pumps. Home: Route 2, Gammon Rd., Madison, Wis.*

HERBART, Johann Friedrich, German philosopher; b. Oldenburg, May 4, 1776; ed. univs. Jena, Bremen, Göttingen; Ph.D., 1809; pvt. tutor to sons of gov. of Interlaken; became docent, Göttingen, 1809; prof. philosophy, Königsberg, 1809-33, Göttingen, 1833-41. Author: De platonici systematis fundamento, 1805; Hauptpunkte der metaphysik, 1808; Lehrbuch zur Physiologie, 1816; Psychologie als Wissenschaft, 2 vols., 1824-25; Allgemeine Metaphysik, 2 vols., 1828-29; Umriss pädagogisher Vorlesungen, 1835. Father of sci. pedagogy; influenced exptl. psychologists, particularly Fechner and Wundt; 1st to emphasize dependence of sci. edn. on psychology; believed psychology a science grounded on experience, metaphysics, math; developed idea of threshold of consciousness; thus sometimes held responsible for introducing doctrine of unconsciousness; his educational ideas very influential. Died Göttingen, Aug. 14, 1841.

HERBER, Elmer C., Am. parasitologist; b. New Tripoli, Pa., Jan. 26, 1900; s. Alfred James and Amanda (Sieger) H.; A.B., Ursinus Coll., 1925; M.A., U. Pa., 1929; Sc.D., Johns Hopkins, 1941; m. Verna Weiss, June 15, 1929; 1 son, Charles J. Faculty, Dickinson Coll., 1929—, prof., 1950—, head dept. biology, 1955-65; parasitologist Stream Control Commn., Mich., 1940-41; hon. collaborator Smithsonian Instn., 1957—. Recipient Darbaker prize in microbiology, 1954, 61. Fellow A.A.A.S.; mem. Pa. Acad. Sci. (pres. 1954), Am. Soc. Parasitology, Am. Soc. Trop. Medicine, Am. Soc. Biol. Tchrs., Am. Soc. Zoology, Sigma Xi. Author: Correspondence Between Spencer F. Baird and Louis Agassiz—Two Pioneer American Naturalists, 1963. Research and publs. on discovery of life cycles of certain trematodes and studies on their ecology and physiology; contbns. to basic history of Smithsonian Instn. Home: 416 W. South St., Carlisle, Pa. 17013.*

HERBERSTEIN, Baron Siegmund (Sigismund) von, see von Herberstein, Baron Siegmund (Sigismund).

HERBERT, Edward, Am. biochemist; b. Hartford, Conn., Jan. 28, 1926; s. Nathan and Celia (Katz) H.; B.S., U. Conn., 1948; Ph.D., U. Pa., 1952; m. Phyllis Sydney Torgan, June 8, 1946; 1 son, Edward Aaron. Postdoctoral fellow U. Wis., 1953-55; instr. Mass. Inst. Tech. 1955-57, asst. prof. biology, 1957-61, asso. prof. biology, 1961-63; asso. prof. chemistry U. Ore., 1963—. Cons. editor Addison Wesley Pub. Co.; head exam. com. Advanced Placement in Biology. Recipient USPHS Career award. Mem. Am. Soc. Biol. Chemists. Research and publs. on relationship between structure of complex biol. molecules and their function; role of nucleic acids in living cell. Home: 3190 Emerald St., Eugene, Ore. 97405.*

HERBERT, Victor Daniel, Am. physician; b. N.Y.C., Feb. 22, 1927; s. Allan Charles and Rosaline (Margolis) H.; B.S., Columbia, 1948, M.D., 1952; m. Jacqueline Lubin, June 19, 1953; children—Robert, Steven, Kathy. Faculty, Albert Einstein Coll. Medicine, 1955-57; sr. research fellow hematology Montefiore Hosp., Bronx, N.Y., 1957-58; research asst. Mt. Sinai Hosp., 1958-59, asso. dir. hematology, 1964—, prof. medicine Mt. Sinai Sch. Medicine, 1966—; faculty Harvard Med. Sch., 1959-64, asso., 1960-63, asst. prof., 1963-64; asso. clin. prof. medicine Columbia Coll. Phys. and Surgs., 1964—. Med. and sci. cons. on nutritional anemias WHO, 1964—. Mem. Am. Chem. Soc., Am. Soc. Hematologists, Am. Phys. Soc., Royal Soc. Medicine (London), Harvey Soc., N.Y. Acad. Medicine, Am. Fedn. Clin. Research, Am. Inst. Nutrition, Am. Soc. Clin. Nutrition, Fed. Am. Soc. Exptl. Biology, Am. Soc. Clin. Investigation, Soc. Exptl. Biol. Medicine. Author: The Megaloblastic Anemias, 1959. Research and numerous publs. on nature, diagnosis and treatment of nutritional anemias; developed diagnostic assays for deficiency folate and vitamin B12; demonstrated nature human serum folate activity, minimal daily adult requirement for folate and vitamin B12, transport nature human B12 binding protein, suppression of hematopoiesis by alcohol, increased purine metabolism in patients with megaloblastic anemia, coated charcoal technic for assay vitamins, minerals, hormones and their binders. Home: 12 York Dr., Great Neck, N.Y. 11021. Office: Mt. Sinai Hosp., N.Y.C. 10029.*

HERBERT, William, Brit. naturalist; b. Jan. 12, 1778; s. Henry and Elizabeth Alicia Maria (Wyndham) H.; B.A., Exeter Coll., Oxford Eng., 1798; M.A., Merton Coll., 1802; B.C.L., D.C.L., 1808; B.D., 1840; m. Letitia Emily Dorothea (Allen), May 17, 1806; 4 children including Henry. Ordained, 1914; dean Manchester, Eng., 1840-47; mem. Parliament Hampshire, 1806, Crickdale, 1811; probably also practiced law. Author: Amaryllidacae, 1837; (ed. J. Lindley 1847) Crocuses; also History of the Species of Crocus; contbr. articles to jours. Editor: Musae Etonenses (poems), 1795; also assisted in editions of White's Selborne, 1833, 37. Research and publs. on bulbous plants which he cultivated, hybridization; fern (genus Herbertia) named after him by Sweet; studied linguistics. Died London, May 28, 1847.

HERBERTSON, Andrew John, Brit. geographer; b. 1865; M.A., Oxford (Eng.) U.; Ph.D., Freiburg im Breisgau, Germany; named reader in geography U. Oxford, 1905, prof., 1910; mem. Royal Commn. on Canals and Inland Waterways, 1906-10. Hon. sec. Geog. Assn.; pres. geog. sect. Brit. Assn., 1910. Co-author: Man and His Work, 1899. Author: Outlines of Physiography; Commercial Geography; Distribution of Rainfall, 1901; Descriptive Geographies, 1902-06; Natural Regions of the World, 1905; A Handbook of Geography. Editor: Oxford Geographies, Oxford Wall Maps, Geography Tchr. Co-editor: Atlas of Meteorology, 1899; Oxford Survey of the British Empire, 6 vols., 1914. Contbr. to encys., sci. jours. Died July 31, 1914.

HERBIG, George H(oward), Am. astronomer; b. Wheeling, W.Va., Jan. 2, 1920; A.B., U. Cal. at Los Angeles, 1943; Ph.D. (Kellogg fellow), U. Cal., 1948; m. 1943; 4 children. Asst., Lick Obs., U. Cal., Mt. Hamilton, 1944-46, jr. astronomer, 1948-50, asst. astronomer, 1950-55, asso. astronomer, 1955—; vis. prof. Yerkes Obs., U. Chgo., 1959. NRC fellow Mt. Wilson Obs., Carnegie Instn., Palomar Obs., Yerkes Obs., U. Chgo., McDonald Obs., Chgo.-Tex., 1948-49. Mem. Astron. Soc. (Warner prize 1955), Internat. Astron. Union, Nat. Acad. Scis. Research on stellar spectroscopy. Office: Lick Obs., U. Cal. Santa Cruz, Cal. 95060.*

HERBST, Walter Alexander, Am. chemist; b. West Hazleton, Pa., July 14, 1910; s. Walter H. and Katie (Smith) H.; B.S., Pa. State Coll., 1931, M.S., 1932, Ph.D., 1935; m. Catherine Pearl Strong, Jan. 1, 1931; children—Walter W., Joan M. (Mrs. Donald Hankins), Joyce A. (Mrs. Samuel Furiness). Chemist, Farm Bur. Oil Co., Indpls., 1935-36; research chemist Esso Research & Engring. Co., Linden, N.J., 1936—. Mem. Soc. Automotive Engring., Am. Soc. M.E., Am. Chem. Soc., Sigma Psi. Patentee in field. Research and publs. on relationship between chem. structure and temperature coefficient of viscosity of organic liquids, efficiency studies of packed distillation column, gasoline quality and performance, fluidized solids systems. Home: 93 Clark Pl., Union, N.J. 07083. Office: Esso Research & Engring. Co., P.O. Box 51, Linden, N.J. 07036.*

HERBUT, Peter Andrew, pathologist; b. Edson, Alta., Can., July 6, 1912; s. Andrew and Ahafia (Smetana) H.; student U. Alta., 1930-35; M.D., McGill U., 1937; m. Margaret Fetsko, Feb. 16, 1940; children—Linda (Mrs. S. Walter Foulkrod III), Paula. Practice medicine specializing in pathology, Phila., 1939—; faculty Jefferson Med. Coll. and Hosp., 1939—, prof., 1948—, head dept., 1948-66, pres., 1966—, dir. labs., 1952-66, cons., 1966—; chief dept. pathology Methodist Hosp., 1952—. Cons. VA Hosp., Phila., 1956-66. Recipient Ward Burdick award Am. Soc. Clin. Pathologists, 1950. Fellow Coll. Am. Pathologists, A.C.P.; mem. Am. Assn. Cancer Research, Am. Soc. Pathologists and Bacteriologists, Am. Soc. Clin. Pathologists, Am. Soc. Microbiology, Soc. Am. Bacteriologists, A.M.A., Am. Soc. Cytology, A.A.A.S., Am. Med. Writers Assn., World, Pan Am. med. assns., Internat. Acad. Pathology, Am. Soc. Exptl. Pathology, Assn. Am. Med. Colls., Internat. Acad. Cytology, Royal Soc. Health. Author: Surgical Pathology, 1948; Urological Pathology, 1952; Gynecological and Obstetrical Pathology, 1953; Pathology, 1955. Studies, and numerous publs. on elucidation of etiology of cancer; protective mechanisms against cancer; chemotherapy of cancer. Home: 1024 Great Springs Rd., Rosemont, Pa. 19010. Office: 1025 Walnut St., Phila. 19107.*

HERDER, Johann Gottfried von, see von Herder, Johann Gottfried.

HERDMAN, Sir William Abbott, Scottish marine naturalist; b. Edinburgh, Scotland, Sept. 8, 1858; s. Robert and Emma (Abbott) H.; grad. Edinburgh U., 1879; several hon. degrees; m. Sarah Douglas, 1882; 2 daus.; m. 2d, Jane Brandeth Holt; 1 son, 1 dau. Asst. to Charles Wyville Thomson, sec. of Challenger, 1879; demonstrator in zoology U. Edinburgh, 1880; 1st Derby prof. natural history U. Liverpool (Eng.), from 1881, organized marine research lab., 1891; founded George Herdman chair geology, 1916, chair oceanography, 1919 (held chair 1 year). Founder Liverpool Marine Biology Com., 1885; co-founder govt. mus. Douglas, Eng. Fellow Royal Soc., 1892 (chmn. grain pests com.); mem. Brit. Assn. (gen. sec. 1903-19); Linnean Soc. (pres. 1904-08). Author: The Founders of Oceanography, 1923. Contbr. numerous articles on fauna of Liverpool Bay and Irish Sea. Leading authority on tunicata (group of marine organisms); instrumental in establishing fish hatchery at Piel Island in Barrow Strait, 1901; investigated pearl oyster fisheries of Gulf of Manaar, Ceylon, 1901; interested in co-ordinating fishing industry with sci. research. Died July 21, 1924.

HERFELD, Hans Ernst, German chemist; b. Wuppertal, Germany May 26, 1907; s. Ernst and Antonie (Zisenis) H.; student German Tanner-Sch., 1926-27, Munich Inst. Tech., 1927-30; Dipl.Ing., Berlin Inst. Tech., 1931, Dr.Ing., 1932, Dr.Ing.habil., 1957, prof., 1967; m. Margarete Witt, April 4, 1934; children—Hanspeter, Dieter, Klaus. Mem. staff, tchr., vice dir. German Leather Inst., Freiberg, 1933-57; dir. with testing and research dept., West German Tanner Sch., Reutlingen, 1957—. Cons. leather industry. Recipient Assn. Tanning Chemistry and Tech. award, 1964. Mem. Assn. Tanning Chemistry and Tech., Soc. German Chemists. Author: Grundlagen der Lederherstellung, 1950; Qualitätsbeurteilung von Leder, Lederaustauschstoffen und Lederbehandlungsmitteln, 1942, 50; also numerous articles. Research in chemistry and tech. of leather prodn.; devel. of analysis methods and standards of leather. Home: 198 Ringelbachstrasse. Office: 9 Erwin-Seiz-Strasse, Reutlingen/Württemberg, West Germany.*

HERGENROTHER, Rudolf Clemens, electronics engr.; b. Chemnitz, Germany, Sept. 5, 1903; s. George Ewald and Anna Alvina (Beyer) H.; came to U. S., 1907, naturalized, 1913; A.B., Cornell U., 1925; M.S., Pa. State U., 1928; Ph.D., Cal. Inst. Tech., 1931; m. Sara Kathryn Meck, Aug. 4, 1928; children—Rudolf Meck, Karl Meck, George Robert. Engr., Westinghouse Lamp Co., Bloomfield, N.J., 1925-26; research fellow physics Washington U., St. Louis, 1932-34; physicist Farnsworth Television Corp., Germantown, Pa., 1934-35; electronics engr. Hazeltine Electronics Corp., Little Neck, N.Y., 1935-45; electronics engr. Raytheon Co., Lexington, Mass., 1945-47, lab. mgr., 1947-62, staff cons., Waltham, Mass., 1962—. Fellow I.E.E.E.; mem. Am. Phys. Soc. Patentee in field. Research and publs. in x-ray crystal structure analysis, photo-electricity, ionization of gases, TV and devel. of cathode ray, electron, storage, and microwave tubes. Home: 14 Holden Rd., West Newton, Mass. 02165. Office: Raytheon Co., Waltham, Mass. 02154.*

HERGET, Paul, Am. astronomer; b. Cin., Jan. 30, 1908; s. Conrad Fred and Clara (Brueckner) H.; A.B., U. Cin., 1931, M.A., 1933, Ph.D., 1935; m. Harriet Louise Smith, July 27, 1935; 1 dau., Marilyn Jean. Faculty, Cin. Obs., U. Cin., 1931—, prof. astronomy, 1943—, Distinguished Service prof., 1965—, dir. obs., 1943—; dir. Minor Planet Center, Internat. Astron. Union, Cin., since 1947—. Cons. to govt. projects; mem. adv. panel NSF, 1954-57, 65—. Recipient James Craig Watson Gold medal, Nat. Acad. Scis., 1965, William Howard Taft medal U. Cin. Alumni Assn., 1965; named Engr. of Year, Tech. and Sci. Socs. Council, 1957. Mem. Am. Astron. Soc., A.A.A.S., Am. Assn. U. Profs., Engring. Soc. Cin., Phi Beta Kappa, Sigma Xi. Author: The Computation of Orbits, 1948. Research and numerous publs. on computation of orbits of planets, comets, satellites, space vehicles; use of electronic computers in large scale sci. problems. Home: 1332 Ault View St., Cin. 45208.*

HERIGONE, Pierre, French mathematician; b. Paris, flourished 1634; author Cursus mathematicus, 6 vols. (Supplementum or vol. 6 contains oldest printed record of Fermat's method of maximus and minimis), 1634-44; also devel. own system of math. notation.

HERING, Constantine, homeopathic physician; b. Oschatz, Saxony, Jan. 1, 1800; s. Christian Karl and Christiane (Kreutzberg) H.; attended Surg. Acad. of Dresden, U. Leipzig; diploma in medicine, surgery, obstetrics U. Würzburg, 1826; m. Charlotte Kemper, 1829; m. 2d, Marianne Hussmann; m. 3d, Theresa Buckheim, 1845; children—Carl, Rudolph. Tchr. math. Blochmann Inst., 1826; delegated to go to S.Am. to make researches in zoology in Surinam, 1827; practiced medicine in Paramaribo; came to U. S., 1833; an organizer N.Am. Acad. Homeopathic Healing Art (1st sch. homeopathic therapeutics in world), Allentown, Pa., 1835, pres. and prin. instr., 1836-42; presided at 1st meeting of Am. Inst. Homeopathy, 1844; founded (with Jacob Jeanes and Walter Williamson) Homeopathic Med. Coll. of Pa., 1848, prof. insts. homeopathy, practice medicine, 1864-67; formed Hahnemann Med. Coll. of Phila., 1867 (joined with Homeopathic Med. Coll., 1869), dean, 1867-71, prof. insts. and materia medica, 1867-69, 70-71, prof. insts. and practice, 1869-70, prof. emeritus, 1876-80. An editor N.Am. Homeopathic Jour., 1851-53, Homeopathic News, 1854-56, Am. Homeopathic Materia Medica, 1867-71. Author: The Homeopathist or Domestic Physician, 1835; Analytical Therapeutics, 1875, Guiding Symptoms, 10 vols., 1878-91. Died Phila., July 23, 1880.

HERING, Erich Martin Richard, German entomologist; b. Heinersdorf, Nov. 10, 1893; s. Richard and Amalie (Boedewig) H.; ed. U. Berlin, U. Kalinigrad; Ph.D. in entomology; m. Xenia von Stryk, 1930; children—Vera, Tamara. Founder, organizer, 1913-14; curator, prof. Zool. Mus., U. Berlin, 1921-57. Recipient Fabricius medal German Soc. Entomology, Royal Belgian medal for sci. studies. Mem. Internat. Entomol. Congress (hon., corr.), Royal Soc. Entomology Belgium, Entomol. Soc. France, Brit. Soc. Entomology, World Acad. Scis., Internat. Union Biol. Sci. Address: 21 Reichensteiner Weg, Dahlem, Germany.

HERING, Ewald, German physiologist, psychologist; b. Alt-Gersdorf, Saxony, Apr. 5, 1834; med. student, Leipzig, from 1853; studied under E. H. Weber, Fechner, O. Funke, P. J. Carus. Practiced medicine, Leipzig, from 1860; became clin. asst. to Ernst. Wagner, 1860; lectr. physiology U. Leipzig, 1862-65, apptd. prof., 1895; prof. physiology and medico-physics Josephs Acad., Vienna, 1865-70; prof. physiology U. Prague, 1870-95. Author: Beiträge zur Physiologie, 1861-64; Die Lehre vom binokularen Sehen, 1868; Über das Gedachtnis als eine allgemeine Funktion der organisierten Materie, 1870; Zur Lehre vom Lightsinne, 1878; Der Raumsinn und der Bewegung des Auges, 1879; Grundzüge der Lehre vom Lichtsinne, 1905. Early researcher in visual space perception; favored nativistic theory of space perception in opposition to empiristic theory of von Helmholtz; made extensive studies on color, postulating important theory of color vision and inventing apparatus to demonstrate color phenomena; influenced phenomenological tradition which later became important part of Gestalt psychology. Died Leipzig, Jan. 26, 1918.

HERING, Heinrich E., physiologist; b. Germany, 1866; worked under Knoll, Prague, before 1913; prof. path. physiology, Cologne, Germany, 1913-34. Described carotid sinus reflex, 1923. Died 1948.

HERLIN, Melvin Arnold, Am. physicist; b. Salt Lake City, Apr. 25, 1923; s. M. Arnold and Leone E. (Hooper) H.; B.S., U. Utah, 1943; Ph.D., Mass. Inst. Tech., 1948; m. Eugenia Tingey, Mar. 20, 1943; children—Lynette, Melvin A. With Mass. Inst. Tech., 1943—, asst. prof. physics, 1949-55, staff mem., group leader Lincoln Lab., 1955-63, asso. div. head, 1963—. Research in microwave techniques gaseous electronics, low temperature physics, radar, reentry physics. Home: Stonehedge, Lincoln, R.F.D. 1, Mass. 01773. Office: Lincoln Lab., Mass. Inst. Tech., Lexington, Mass. 02173.*

HERLY, Louis, Am. surgeon; b. Vienna, Ill., Aug. 11, 1881; s. Leopold and Regina (Popper) H.; student Coll. City N.Y., 1894-97, Columbia Coll., 1897-99; M.D., Columbia, 1903; m. Queenie Brown, Dec. 15, 1928; children—Irene Winifred (Mrs. Robert C. Wilson), Lillian Isabelle. Gen. surgery at St. Marks, Lincoln, Community, Lenox Hill, Flower, Sydenham, Flower-Fifth Av. hosps., N.Y.C., from 1908; staff mem. Fordham Med. Coll. Hosp. for Women from 1911; cancer research, Inst. for Cancer Research Columbia U., Meml. Hosp., N.Y.C., from 1920; collaborated in cancer work at Radium Hemmet, Stockholm, and Radium Inst., Copenhagen, 1923; asso. in cancer research, Columbia U., from 1946. Mem. A.M.A., Am. Cancer Research Assn., Internat. Cancer Congress. Wrote articles including: A Critical Investigation of the Freund-Kaminer Reaction, 1920; Adenomyoma of the Uterus; Relation to Malignancy, 1924; Experimental Production of Tumor in White Rat, 1926. Studies in Selective Differentiation of Tissues by Means of Filtered Ultra-violet Light, 1944, A Simple Diagnostic Blood Test for Cancer, 1947. Early investigator of tar cancers in animals; discovered new method of diagnosing malignancy of breast tumors by means of filtered ultra-violet light; produced sarcoma in mice by means of sterile and cell free ascitic fluid; exhibited blood test for early cancer at 4th Internat. Congress of Cancer Research, St. Louis, 1947. Died July 14, 1952.

HERMAN, Lloyd George, bacteriologist; b. New Hamburg, Ont., Can., Feb. 23, 1911; s. John and Lilly (Berlet) H.; B.S.A., U. Toronto, 1934; Ph.D., McGill U., Montreal, Que., Can., 1948; m. Jean Wilson Macpherson, June 14, 1941; children—John Edward, Margaret Jean. Came to U. S., 1948, naturalized, 1958. Bacteriologist, sanitarian Wilson & Co., Chgo., 1948-53; bacteriologist Burroughs Wellcome & Co., Tuckahoe, N.Y., 1953-56, D.C. Labs., Washington, 1956-58; bacteriologist, chief sanitation sect. NIH, Bethesda, Md., 1958—. Cons. on hosp. sanitation, 1960—. Mem. Am. Soc. Microbiology (v.p. Washington sect. 1965—), Inst. Food Technologists, N.Y. Acad. Scis., Soc. Indsl. Microbiology, Internat. Soc. Chemotherapy. Author: (with Gray) Antimicrobial Agents Annual, 1960; also articles on bacteriology and sanitation. Research on oyster bed sanitation, dairy products, frozen and dehydrated foods, meat products, mold control, Rocky Mountain spotted fever, and tularemia vector survey, staphylococcus isolation by egg yolk, flavobacteria isolation, air microbiology; invented process for microbial preservation. Home: 4613 Highland St. Office: Nat. Inst. Health, Bethesda, Md. 20014.*

HERMANN, Edward Robert, Am. environmental engr.; b. Newport, Ky., Oct. 9, 1920; s. Joseph G. and Beatrice (Beyland) H.; B.S., U. Ky., 1942, C.E., 1953; S.M., Mass. Inst. Tech., 1949; Ph.D., U. Tex., 1957; m. Eleanor M. Hill, June 10, 1946; children—Mary E., Carolyn B., Catherine J., Georgia A., Joseph F., Michael E., John E. Jr. engr. TVA, Knoxville, 1942; faculty U. Ky., 1943-44; partner Hermann Cons. Engrs., 1946-47; teaching asst. Mass. Inst. Tech., 1947-49; engr. AEC, Los Alamos, 1949-54; research engr. U. Tex., 1954-57; indsl. hygienist, health engr. Humble Oil & Refining Co., 1957-62; cons., 1942-57, 63——; asso. prof. environmental health engring. Northwestern U., Evanston, Ill., 1963-——; mem. subcom. on waste mgmt. in space Nat. Acad. Scis.-NRC Space Sci. Bd., 1966-——. Recipient Mich. Soc. award for outstanding publ. in Am. Indsl. Hygiene Assn. Jour., 1964. Fellow Am. Pub. Health Assn., Am. Soc. C.E.; mem. A.A.A.S., Am. Indsl. Hygiene Assn. (sect. pres.), Am. Soc. Engring. Edn., Am. Water Works Assn. (cited for research 1960), Assn. Mil. Engrs., Nat. Soc. Profl. Engrs. (past chpt. pres.), Water Pollution Control Fedn., (Eddy medal for outstanding research 1959), Sigma Xi. Research on wastewater treatment processes-indsl. and domestic, wastewater reclamation nuclear waste disposal, radio. health practices, indsl. hygiene of petroleum refining and petrochem. operations, audiometry in relation to noise control and hearing conservation. Home: 627 Dartmouth Pl., Evanston, Ill. 60201.*

HERMANN, Jacob, Swiss mathematician; b. Basel, Switzerland, July 16, 1678; studied under Bernoulli; prof. math. U. Padua, Italy, 1707-13, Frankfort, Germany, 1713-24, St. Petersburg, Russia, 1724-31; became prof. philosophy, Basel, Switzerland, 1731; mem. St. Petersburg Acad. Scis. Author: (in Latin) Treatise on the Forces and Movements of Solids and Fluid Bodies, 1715. Followed the Leibnizian-Wolffian school of philosophy and mathematics. One of 1st to use infinitesimal calculus which he applied to geometric problems. Died July 11, 1733.

HERMANN, Jean, French physician, naturalist; b. Barr, France, Dec. 31, 1738; Doctorat en Médecine, U. Strasbourg (France), 1763; became prof. Strasbourg Sch. Pub. Health, 1784; prof. natural history École du Bas-Rhine, from 1796. Mem. French Acad. Scis. Author: Tabula affinitalum animalium (on theory that animal species are interrelated in devel.), 1783. Died Oct. 4, 1800.

HERMANN, Paul, botanist; b. Halle, Germany, 1646; practiced medicine in E. Indies, about 8 years; became prof. botany, Leiden, Netherlands, 1679. Author: West Indian Herbal, 1670-77, pub. by Linnaeus, 1747; Catalogue of the Botanic Garden of Leyden, 1687; Paradisus Batavus, 1698. A leading botanist of his time; divided plants into those with uncovered seeds and those in capsules. Died 1695.

HERMANN THE DALMATIAN (or Hermann the Slav), astrologer, philosopher; flourished mid-12th century; studied at Chartres or Paris; pupil of Theodoric of Chartres; lived in Spain, 1138-42, in Languedoc, 1143. Translator (from Arabic to Latin): Pronostica, or Liber sextus astronomie (Sahl ibn Bishr), 1138; Kitab al-madkhal ila 'ilm ahkam al-nujum (Abu Ma'shar), circa 1140; an Arabic transl. of Ptolemy's Planisphaerium, with commentary (Maslama ibn Ahmad al-Majriti). Author: De essentiis (on astronomy and geography), 1143; Liber ymbrium, on meteorology; other works on arithmetic, geometry, astrolabe.

HERMANN THE GERMAN (or Hermannus Alemannus, Germanicus Teutonicus), translator; flourished Toledo, Spain, 1240, 56; in service Manfred (king of Naples 1258-66); bishop of Astorga, León, Spain, 1272. Translator (from Arabic into Latin): middle commentary on Aristotle's Ethics (Averroes) 1240; middle commentary on Aristotle's Rhetoric and Ethics (Averroes) 1256; Summa quorundam Alexandrinorum (pertaining to the Ethics), 1243-44; commentary on the rhetoric (al-Farabi); also translated psalter into Castilian. Died 1272.

HERMANNUS CONTRACTUS (or Hermann the Lame of Reichenau), German mathematician, astronomer; b. Altshausen, July 18, 1013; s. Wolferad II (Count of Altshausen) and Hiltrudis; Entered sch. of Benedictine abbey at Reichenau, 1020, later became monk, also tchr. at abbey sch. Author: Chronicon (world history from birth of Christ to 1054, which was continued by his pupil Bertold to 1066); De utilitatibus astrolabii; De mensura astrolabii (also ascribed to Pope Silvester II); De Mense lunari; De divisione; De conflictu rithmimachiae; De figura quadrilatera; also compiled a martyrology, and wrote poem in several meters on 8 principle vices, also theoretical treatise on music stating his own system of notation with letters. Died Reichenau, Sept. 24, 1054.

HERMES TRISMEGISTUS (Hermes the Thrice Great), identified with god Hermes by Greeks and with Mercurius by Romans; thought to be same as Egyptian Thoth, divinity of wisdom and scribe of Egyptian pantheon; large Greek literature written circa 100-300 A.D. ascribed to Hermes Trismegistus; most important works the Asclepius (supposedly describing Egyptian religion) and collection of 15 dialogues titled Corpus Hermeticum (includes Pimander which contains account of creation of world similar to Genesis); works reflect popular Greek philosophy, Platonism, Stoicism, Jewish and Persian influences of late antiquity; emphasis on astrology, astral influences, occult sciences, talismans, secret virtues of plants and stones; and universal sympathetic magic; works presented in context of mysticism and religion; proved highly popular and influential as approach to reality at a time when traditional religion and education seemed moribund; assertion that there was truth in ancient wisdom; general belief (accepted by Church Fathers Lactantius and St. Augustine) that Hermes had been a real person who

lived about same time as Moses and had thus been aware of Biblical truths; Hermes often portrayed as Gentile prophet who had foreseen coming of Christianity; similarity of some Hermetic books with Mosaic books of Bible and also to Platonic views gave added weight to collection among neo-Platonists and Christians in late antiquity; Hermetic works also influential among 12th century neo-Platonists such as Hugh of St. Victor; Corpus Hermeticum translated into Latin prior to works of Plato by Marsilio Ficino (who equated Hermes Trismegistus with Zoroaster) at request of Cosimo de'Medici, 1463; work of Florentine neo-Platonists helped place a Hermetic stamp on Renaissance philosophy of nature thereby promoting natural magic, alchemy and astrology as legitimate truths in nature and supernature; proof of late composition of Corpus Hermeticum given by Isaac Casaubon, 1614; his work influential in gradual decline of interest in Hermetic texts among 17th century scholars.

HERMITE, Charles, French mathematician; b. Lorraine, France, Dec. 24, 1822; s. Ferdinand and Madeleine (Lallemand) H.; ed. École Polytechnique; m. Louise Bertrand, 1848. Became lectr. École Normale, 1862; apptd. examiner École Polytechnique, 1863, prof. analysis, 1869; prof. higher algebra Sorbonne, 1869-97. Fellow Royal Soc., 1873; mem. French Acad. Scis., 1892. Author: Sur la theorie des fonctions elliptiques, 1863; Sur l'equation de V. degré, 1866; Cours d'analyse de l'École Polytechnique, 1873. Studied theory of algebraic forms, gen. theory of equations, elliptic functions; contbd. law of reciprocity named after him, also 1st skew invariant to calculus of invariants; 1st to solve 5th degree equation; proved that base of natural logarithms (e) is not an algebraic number. Died Paris, Jan. 14, 1901.

HERMS, William Brodbeck, Am. entomologist; b. Portsmouth, O., Sept. 22, 1876; s. Carl Julius Herman and Rosa Emma (Brodbeck) H.; B.Sc., Baldwin-Wallace Coll., 1902, D.Sc., 1935; postgrad. Western Reserve U., 1905; Harvard, 1907-08; M.A., Ohio State U., 1906; m. Lillie (Carrie) Magly, June 14, 1902; children—William Magly, Herbert Parker, George Walter. Head of Sch. of Commerce, prof. theory and practice of domestic commerce Baldwin-Wallace Coll., 1902-05, instr. biology, 1904-05; teaching fellow zoology Ohio State U., Columbus, 1905-06, instr. exptl. zoology and invertebrate zoology, summer sessions, 1907-08; acting head dept. zoology Ohio Wesleyan U., 1906-07; Edward Austin fellow in zoology Harvard, 1907-08; asst. prof. parasitology U. Cal. at Berkeley, 1908-15, asso. prof., 1915-20, prof., 1920-46, emeritus, 1946, head div. entomology and parasitology, 1923-46; Officer in charge malaria investigations Cal. Bd. Health, 1910-13, cons. entomologist and parasitologist, from 1913, made malaria-mosquito survey Cal. Del. to 7th Internat. Zool. Congress, Boston, 1907, Internat. Hygiene Exhibit, Dresden, Germany, 1911, Pan Pacific Food Conservation Congress, Honolulu, 1924, 4th Internat. Congress of Entomology, Ithaca, N.Y., 1928; investigated coconut pests of Fanning and Washington Islands, 1924. Fellow A.A.A.S., Cal. Acad. Sci., Entomol. Soc. Am. (pres. 1941); hon. mem. Nat. Malaria Com. Helminthological Soc. Washington, Am. Assn. Econ. Entomologists (pres. 1928), Pacific Coast Entomol. Soc. (chmn. 1925-26), Am., Western socs. naturalists, Am. Soc. Parasitologists (v.p. 1936), Am. Soc. Tropical Medicine, Ohio Acad. Sci., Sigma Xi, Alpha Zeta, Phi Sigma, Delta Omega. Author: Malaria, Cause and Control, 1913; Laboratory Guide to the Study of Parasitology, 1913; Textbook in Medical and Veterinary Entomology, 1915-23, 4th edit., 1949; Mosquito Control (with H. F. Gray), 2d edit., 1944. Died May 9, 1949.

HERNANDEZ, Teme Paul, Am. horticulturist; b. Duson, La., May 15, 1919; s. Saul and Birtha (Lagneaux) H.; B.S. La. State U., 1940, M.S., 1942; Ph.D., U. Wis., 1948; m. Lauris Tate, Mar. 28, 1946; children—Patricia Ann (Mrs. John Iry Quebadeaux), Sandra Ann, Teme Paul. Supt., Sweet Potato Research Center, Chase, La., 1949-57; faculty La. State U., Baton Rouge, 1957-——, prof. horticulture, 1958-——. Mem. Am. Soc. Genetics, A.A.A.S., Am. Soc. Hort. Sci., Phi Kappa Phi, Gamma Sigma Delta, Alpha Zeta. Contbr. numerous articles to profl. jours. Co-developer several varieties of vegetables, sweet potato and tomato. Home: 625 Kimbro St., Baton Rouge 70808.*

HERNANDEZ, Thomas, Am. pharmacologist; b. Lafayette, La., May 17, 1914; s. Saul and Bertha (Lagneaux) H.; A.B., La. State U. 1936, M.S., 1938, M.D., 1947; Ph.D., State U. Ia., 1942; m. Dorothy M. Shaw, Apr. 11, 1943; children—Cary A., Jeanne, James T., Donna. Insp. explosives, chemist Day and Zimmerman, Inc., Burlington, Ia., 1942-43; faculty La. State U. Sch. Medicine, 1948-51, 53-——, prof. pharmacology, 1953-——, chmn. dept., 1960-——; sr. asst. NIH, USPHS, Atlanta, 1951-53. Cons. clin. chemistry lab. Mercy Hosp., New Orleans, 1956-——; mem. malaria commn. Armed Forces Epidemiological Bd. 1956-——. Mem. Am. Physiol. Soc., Am. Soc. for Exptl. Biology and Medicine, A.A.A.S., Am. Soc. for Pharmacology and Exptl. Therapeutics, Sigma Xi. Author: (with R. A. Coulson) Biochemistry of the Alligator, 1964; also articles. Research on biochemistry and physiology of fluid and electrolytes, renal function, human testing antimalarial drugs, carbohydrate and amino acid metabolism, nitrogen metabolism. Home: 2316 Odin St., New Orleans 70122.*

HERO (Heron) OF ALEXANDRIA, Greek mathematician, inventor; dates uncertain, probably flourished 1st century A.D.; believed to have lived in Alexandria. Author: Pneumatica; Belopoeica; Mechanica; Geometria; Geodaesia; Mensurae; Metrica; On the Dioptra; commentary on Euclid's Elements. Inventor devices operated by steam, water or compressed air, including fountain, fire engine pump, siphons, steam engine in which steam revolved a hollow sphere; solved quadratic equations arithmetically; gave algebraic solutions to 1st and 2d degree equations; approximated sq. and cube roots; gave formula for area of triangle as function of its sides; demonstrated in optics that angle of incidence equals angle of reflection.

HERODOTOS, Greek historian; b. Halicarnassos, Caria, circa 484 B.C.; s. Lyxes and Rhaeo (or Dryo); called father of history; traveled throughout Europe, Asia, Africa; lived in Samos, then Athens during ascendency of Pericles; colonist to Thurii, circa 444. Author: The Persian Wars (IX Books). His history contains much sci. and ethnographic lore, including 1st-hand description of prehistoric Macedonian lake village, description of eclipse dated spring of 480 B.C. before Thermopolae (none occurred then), discussion of Nile River; deduced from presence of fossilized sea shells that lower Egypt was once covered by sea. Died Thurii, 425 B.C.

HEROLD, Edward William, Am. physicist; b. N.Y.C., Oct. 15, 1907; s. Carl Frederick and Marie (Wollersheim) H.; B.Sc., U. Va., 1930; M.Sc., Poly. Inst. Bklyn., 1942, D.Sc., (hon.), 1961; m. Alexandra Dacis, Aug. 4, 1931; 1 dau., Linda M. (Mrs. Robert C. Johnson). Tech. asst. Bell Telephone Labs., N.Y.C., 1924-26; radio engr. E. T. Cunningham, Inc., N.Y.C., 1926-29; research physicist RCA, Harrison, N.J., 1930-42, physicist, sect. head RCA Labs., Princeton, N.J., 1942-51, dir. electronic research lab., 1951-59, corporate staff RCA, 1965-——; v.p. research Varian Assos., Palo Alto, Cal., 1959-64. Chmn. adv. group on electron devices Dept. Def., 1966-67; mem. adv. council elec. engring. dept. Princeton, 1965-——. Fellow I.E.E.E.; mem. Am. Phys. Soc., Am. Inst. Aeros. and Astronautics, Soc. Photog. Scientists and Engrs., Assn. Ednl. Data Systems, Phi Beta Kappa, Sigma Xi. Research, patents and publs. on various types electron tubes especially frequency converters and amplifiers; negative resistance, electron beams, fluctuation noise phenomenon, semiconductor physics, junction transistor; dir. devel. shadow-mask color TV picture tube. Home: 332 Riverside Dr. Office: RCA, Princeton, N.J. 08540.*

HEROMIDES (or Herundes), Greek mathematician. Alluded to by al-Nairizi as defining a line as a magnitude extended one way (circa 910).

HEROPHILOS, Greek physician, anatomist; b. Chalcedon (nr. modern Istanbul); flourished 300 B.C.; pupil of Praxagoras of Cos; practiced in Alexandria; author treatise in 3 vols. on anatomy and eyes, textbook for midwives. Considered father of sci. anatomy; 1st to dissect human body in public; compared structure of human body with that of animals; improved technique and terminology of anatomy; distinguished tendons from nerves; recognized brain as center of nervous system; classified nerves as sensory or motor; studied spinal cord; described ventricles of brain, liver, spleen, eye, alimentary canal; introduced names retina, duodenum; distinguished arteries from veins; 1st to count pulse; studied reproductive organs; described ovaries and tubes connecting them to uterus; observed and named prostate gland.

HEROULT, Paul Louis Toussaint, French chemist; b. Thury-Harcourt, France, Apr. 10, 1863; ed. École des Mines, Paris; inventor (independently of C. H. Hall in Am.) Héroult process for producing aluminum by electrolysis of alumina in cryolite, 1886; inventor Heroult electric furnace for prodn. of electric steel. Died May 9, 1914.

HERRATH, Ernst Albert Gustav von, German histologist; b. Krefeld, June 7, 1907; s. Karl and Amalie (Steinfartz) H.; M.D., u. Cologne; m. Helmtrud Lentz; children—Ulrich, Dietrich, Angelika, Michael, Matthias. Doctorate, U. Cologne, 1934, qualified for anatomy; agrégé U. Berlin, 1938; dir. Inst. Histology, U. Fribourg (Switzerland), 1938-41; asso. prof. U. Giessen, 1941; full prof. dir. Inst. Anatomy, Free U. Berlin, 1949. Mem. Deutscher Hochschulausschuss für Leibesubengen. Author: Bau und Funktion der Normalen Milz, 1958; Atlas der Normalen Histologie und Mikroskopischen Anatomie des Menschen. Research and publs. on comparative histology of rat. Home: 5 Schellendorffstrasse, 1 Berlin 33. Office: 15 Königin-Luise-Strasse, 1 Berlin 33, Germany.

HERRE, Karl Wolf, German zoologist, anatomist; b. Halle, Salle, May 3, 1909; s. Karl and Ida (Taatz) H.; ed. U. Halle, U. Graz; Ph.D. in natural sci.; m. Ilse Rabes, 1941; children—Renate, Ursula. Asst. Inst. Breeding; instr. zoology and comparative anatomy U. Halle, 1936; asst. dir. Mus. Natural History, Braunschweig, 1942; with Zool. Inst. and Mus., U. Kiel, 1945, dir. Inst. Vet. Sci., 1947, full prof. zoology, anatomy and vet. physiology, 1950. Author: Die Urodelen des Braunkohle des Geiseltales, 1935; Das Ren als Haustier, 1950; Abstammung und Entwicklung der Haustiere, 1958; Der Rasse- und Artbegriff, 1961. Home: 24 Goethestrasse. Office: New University, Institute of Veterinary Science, 40-60 Olshausenstrasse, Kiel, Germany.

HERRELL, Wallace Edgar, Am. physician; b. Marshall, Va., Oct. 1, 1909; s. Bennett F. and Bessie (Ballard) H.; M.D., U. Va., 1933; m. Margaret Harwick, Jan. 18, 1936; children—Steven, John, Sarah (Mrs. James B. Foust). Faculty, Mayo Clinic, Rochester, Minn., 1934-53, prof. medicine Mayo Found. Grad. Sch., U. Minn., 1952-53; cons. in medicine Lexington (Ky.) Clinic, 1953-67; prof. clin. medicine U. Ky., Lexington, 1961-67; editor in chief Med. Digest, Internal Medicine Digest, Clin. Medicine, Mid. Digest Publs., Northfield, Ill., 1964——. Mem. panel on infectious diseases of U. S. Pharmacopeia, 1950-——. Fellow A.M.A., A.C.P., A.A.A.S.; mem. Am. Soc. Clin. Investigation, Am. Therapeutic Soc., Am. Fedn. Clin. Research, Am. Med. Writers Assn., Am. Soc. Microbiology, Sigma Xi, others. Contbr. chpts. to books, numerous articles to profl. jours. Research on antibiotics and infectious disease. Home: 296 Crestwood Village. Office: 445 Central Av., Northfield, Ill. 60093.*

HERRERA, Juan Bautista, Spanish mathematician, architect; b. Asturias, Spain, 1530; student sci. U. Brussels (Belgium); architect to Philip II, architect in chief, 1569. Mem. Madrid Acad. Math. (founder). Translator some works of Alberti. In charge constrn. of Escurial Palace; built bridge at Segovia, Spain; verified maps of India and Am.; disseminated Arabic math. Died Madrid, Spain, 1597.

HERRERA, Teofilo S., Mexican biologist; b. Mexico City, Feb. 24, 1924; s. Teofilo Herrera Ostos and Victoria Suarez; grad. Faculty Scis. U. Mexico, 1948, Dr. Biology, 1964; student U. Wis., 1953, Mexican Nat. Poly. Inst., 1954. Staff, faculty Inst. Biology, U. Mexico, Mexico City, 1944——, prof. Faculty Scis., 1952-——, sec., 1961-65, cons. Bot. Garden, 1959——. Technician, Infan Labs., 1954-56. Mem. Mexican Mycol. Soc. (past pres.) Mycol. Soc. Am., Bot. Soc. Mexico, Mexican Soc. Microbiology, Internat. Soc. Plant Taxonomy, Natural History Mexican Soc. Research, publs. on taxonomy and ecology of fungi, particularly Gasteromycetes, edibile fungi, hallucinogenic mushrooms, cultivation of higher fungi, fermentation microbiology and biochemistry, autotrophic bacteria, taxonomic and ecol. botany. Home: 50 Benjamin Franklin, Mexico D.F. 18. Office: Inst. Biologia, Ciudad Universitaria, Mexico 20, D.F., Mexico.*

HERRERO, C. Mariano, Spanish engr.; b. Madrid, Spain, Mar. 30, 1918; s. Roque and Carmen (Campanero) H.; ed. Poly. Sch., Madrid, Stanford; Ph.D. in engring. and electronics; m. Angela Herrero, 1944; children—Jose Carlos, Mariano, Carmen, Paloma, Angel, Dolores. Prof. electronics, dir. Battelle Inst.; mem. engring. council. Mem. I.E.E.E., Iron and Steel Inst., Engrs. of Armaments, Sigma Xi. Author: Resonance Phenomena in Timer Varying Circuits, 1954; Electronica Racional, 7 vols. Home: 5 Paseo de Moret, Madrid 8. Office: 18 Valle Hermoso, Madrid, Spain.

HERRESHOFF, J(ohn) B(rown) Francis, Am. chemist; b. Bristol, R.I., Feb. 7, 1850; s. Charles Frederick and Julia Ann (Lewis) H.; student Brown U., class of 1870, Ph.B., by spl. vote, 1905, hon. A.M., 1890, Sc.D., 1909; m. Grace Eugenia Dyer, Feb. 9, 1876; m. 2d, Emily Duval Lee, Oct. 25, 1882; m. 3d, Carrie Ridley Enslow, June 9, 1919 (died 1924); m. 4th, Irma Grey, Ridley, Apr. 14, 1925. Prof. analytical chemistry Brown U., 1869-72; supt. Laurel Hill Chem. Works, L.I., 1876; v.p. and trustee Nichols Copper Co.; dir. Granby Consol. Mining, Smelting & Power Co.; hon. v.p. Gen. Chem. Co. Recipient Perkin medal (1st time awarded in America) for work in chem. and metall. industries. Invented process for mfg. sulphuric acid. Died Jan. 30, 1932.

HERRESHOFF, James Brown, Am. inventor; b. Bristol, R.I., Mar. 18, 1834; s. Charles Frederick and Julia Ann (Lewis) H.; student Brown U., 1853-56; m. Jane Brown, 1875; children—James Brown, Charles Frederick, William Stuart, Jane Brown, Ann Frances. Supt. Rumford (R.I.) Chem. Works, 1858-63; mfr. fish oil, Prudence Island, R.I., 1863-69. Inventor first naphtha driven motorcycle of internal combustion type in America (1870); coil boiler; fin-keel for sailing yachts; mercurial antifouling paint; sliding-seat for rowboats; thread-tension regulator for sewing machines; apparatus for measuring specific heat of gases; sounding apparatus. Died Dec. 5, 1930.

HERRICK, Charles Judson, Am. anatomist, neurologist; b. Mpls., Oct. 6, 1868; M.S., Denison U., 1895; B.S., U. Cin., 1891; Ph.D., Columbia U., 1900. Instr. natural sci. Granville Acad., 1891-92; prof., Ottawa, Kan., 1892-93; instr. biology Denison U., 1893-96, asst. prof., 1897-98, prof. zoology, 1898-1907; neurologist U. Chgo., 1907-34, emeritus prof. research, 1934-37, emeritus prof., 1937-——. Asso. comparative neurology Path. Inst. State Lunacy Commn., N.Y., 1897-1901. Recipient Cartwright prize, 1899, 1905. Mem. Internat. Assn. Acads. (mem. brain commn.), Path. Acad., Am. Soc. Naturalists, Am. Assn. Anatomists, Am. Soc. Zoologists, Am. Physiol. Soc. Mng. editor Jour. Comparative Neurology, 1894-1927, editor, 1894-1948. Research on nerve components of vertebrates, anatomy of brain of amphibians and

fishes, evolution of cerebral cortex. Office: 236 Morningside Dr. S.E., Grand Rapids 6, Mich.

HERRICK, Edward Claudius, Am. entomologist; b. New Haven, Conn., Feb. 24, 1811; s. Claudius and Hannah (Pierpont) H.; A.M. (hon.), Yale, 1838. Advanced theory of periodic occurrence of large number of meteors, 1837; contbr. to Am. Jour. of Sci., 1837-62; published A Brief Preliminary Account of the Hessian Fly and Its Parasites, 1841; 1st to find and describe parasites of eggs of spring canker-worm moth. librarian Yale, 1843-58. Died New Haven, June 11, 1862.

HERRICK, Francis Hobart, Am. biologist; b. Woodstock, Vt., Nov. 19, 1858; s. Marcellus Aurelius and Hannah Andrews (Putnam) H.; A.B., Dartmouth, 1881; Ph.D., Johns Hopkins, 1888; Sc.D., Western U. of Pa., 1897, Western Reserve U., 1936; m. Josephine Herkomer, Eng., June 24, 1897; children—Agnes Elizabeth (Mrs. Hans Platenius), Francis Herkomer. Instr. biology, 1888-91, prof., 1891-1929, prof. emeritus, 1929, Adelbert Coll., Western Reserve U. Fellow A.A.A.S., Am. Ornithologists' Union. Trustee Cleveland Mus. of Natural History, from 1920, v.p., from 1928. Author: Audubon the Naturalist, 2 vols., 1917; The American Eagle on the Shores of Lake Erie, 1925, 27; The American Eagle—A Study in Natural and Civil History, 1934; Wild Birds at Home, 1935. Research in devel. and morphology Crustacea, habits and instincts wild birds, origin and devel. instincts. Died Cleve., Sept. 11, 1940.

HERRICK, Glenn W., Am. entomologist; b. Otto, N.Y., Jan. 5, 1870; B.S.A., Cornell U., 1896; postgrad. Harvard, 1897; m. 1898; 3 children. Tchr. pub. schs., 1888-90; instr. Nat. Sci. Camp, N.Y., 1895-96; prof. biology Miss. A. and M. Coll., 1897-1908, vice dir. expt. sta., 1906-08; faculty econ. entomology Cornell U., 1909—, prof., 1912-35, emeritus prof., 1935—; entomologist Expt. Sta., 1809-35. Fellow Entomol. Soc. (pres. Assn. Econ. Entomologists 1915), Am. Geog. Soc. Contbg. editor Ency. Britannica Yearbook. Research on Miss. mosquitoes, Thysanoptera, Coccidae, pecan insect pests, fruit tree leaf roller, red clover pests, apple pests, external parasites of domestic fowls, malaria. Home: 219 Kevin Pl., Ithaca, N.Y.

HERRICK, James Bryan, Am. physician; b. Oak Park, Ill., Aug. 11, 1861; s. O. W. and Dora E. (Kettlestrings) H.; A.B., U. Mich., 1882, A.M. (hon.), 1907, LL.D., 1932; Sc.D. (hon.), U. Chgo., 1938, Northwestern U., 1940; M.D., Rush Med. Coll. (U. Chgo.), 1888; m. Zellah P. Davies, 1889; children—Helen Powers (Mrs. George H. Gilbert, Jr.), John Origen. Intern Cook County Hosp., Chgo., 1888-89, instr. medicine Rush Med. Coll., 1890-94, adj. prof., 1894-1900, prof., 1900-27; attending physician Presbyn. Hosp., Chgo., 1895-1945. Mem. Assn. Am. Physicians A.M.A. (Distinguished Service medal 1939), hon. mem. N.Y. Acad. Medicine, Cardiac Soc. of Great Britain and Ireland. Author: A Short History of Cardiology, 1942; Memories of Eighty Years, 1949; also articles. Described clin. features observed in cases involving sudden obstruction of coronary arteries. Died Mar. 7, 1954.

HERRICK, Samuel, Am. astronomer; b. Madison County, Va., May 29, 1911; s. Samuel and Fanny (Field) H.; B.A., Williams Coll., 1932, Sc.D., 1962; Ph.D., U. Cal. at Berkeley, 1936; m. Betulia Toro, June 15, 1934; children—Henry, Nike, Rufus. Faculty, U. Cal. at Los Angeles, 1937-——, prof. astronomy, 1952-62, prof. astronomy and engring., 1962-——, chmn. dept. astronomy, 1943-45, 46-51; mathematician Nat. Bur. Standards, 1948-49. Cons. indsl. firms; Hunsaker prof. Mass. Inst. Tech., 1961-62, Minta Martin lectr. Inst. Aero Space Sci., 1962. Guggenheim Meml. fellow, 1945-46, 52-53. Fellow Brit. Inst. Navigation, Royal Astron. Soc., Brit. Interplanetary Soc., Am. Astronautical Soc.; mem. Internat. Astron. Union, Am. Astron. Soc., Astron. Soc. Pacific, History Sci. Soc., Am. Inst. Navigation (past pres.), Am. Rocket Soc. (past dir.), Am. Inst. Aeros. and Astronautics (past dir.), Internat. Acad. Astronautics, Phi Beta Kappa. Research and numerous publs. in celestial mechanics, gravitational astronomy, astrodynamics, space navigation, guidance and control. Home: 13500 Mulholland Dr., Beverly Hills, Cal. 90210. Office: U. Cal., Los Angeles 90024.*

HERRIN, Eugene Thornton, Am. geologist, educator; b. Dallas, Nov. 19, 1929; s. Eugene Thornton and Dorothy (Johnston) H.; B.S., So. Methodist U., 1951, M.S., 1953; Ph.D., Harvard, 1958; m. Barbara Bradshaw, Dec. 20, 1953; children—Kathleen, Laura. With So. Methodist U., Dallas, 1955-——, prof. geology, 1964-——, dir. Dallas Seismol. Obs., 1958-——, cons. prof. S.W. Center for Advanced Studies, 1964-——. Recipient Grove Karl Gilbert award, 1964. Carnegie Inst. Washington fellow, 1964. Mem. Am. Geophys. Union, Dallas Geophys. Soc., Geol. Soc. Am., Seismol. Soc. Am., Seismology of Am. Geophys. Union (v.p. 1964). Research and numerous publs. in seismology, especially determination of seismic travel-times and location of earthquakes and explosion. Home: 5555 Winston Ct., Dallas 75220.*

HERRING, William Conyers, Am. physicist; b. Scotia, N.Y., Nov. 15, 1914; s. William Conyers and Mary (Joy) H.; A.B., U. Kan., 1933; postgrad. Cal. Inst. Tech., 1933-34; Ph.D., Princeton, 1937; m. Louise C.

Preusch, Nov. 30, 1946; children—Lois Mary, Alan John, Brian Charles, Gordon Robert. NRC fellow Mass. Inst. Tech., 1937-39; instr. math., research asso. Princeton, 1939-40; instr. U. Mo., 1940-41; sci. staff div. war research Columbia, 1941-45; prof. applied math. U. Tex., 1946; tech. staff Bell Telephone Labs., Murray Hill, N.J., 1946-——; mem. Inst. for Advanced Study, 1952-53. Mem. NRC, 1961-63. Recipient Army-Navy Certificate Appreciation, 1947. Fellow Am. Phys. Soc. (Oliver E. Buckley Solid State Physics prize 1959, Am. Acad. Arts and Scis.; mem. A.A.A.S. Contbg. author: Magnetism, 1966. Research and numerous publs. in solid-state theoretical physics, including quantum mechanic electrons in solids, elec. conduction in semiconductors, sintering and surface phenomena and magnetism; introduced orthogonatized plane wave method of calculating behavior of electron waves in crystals. Home: 3 Hawthorne Pl., Summit, N.J. 07901. Office: Bell Telephone Labs., Murray Hill, N.J. 07971.*

HERRISANT, François-David, French anatomist; b. Rouen, France, Sept. 29, 1714; regent Faculty Medicine, Paris. Fellow Royal Soc., 1750; mem. French Acad. Scis. Author: Sur la structure des cartilages des côtes de l'homme et du cheval, 1748; Sur la conformation de l'émail des dents, 1758. Research on respiratory organs, vocal chords, formation of enamel on teeth. Died Paris, Nov. 12, 1774.

HERRLICH, Albert, German physician; b. Munich, Germany, Mar. 4, 1902; s. Heinrich and Anna (Steininger) H.; M.D., U. Munich, 1930, docent, 1945; m. Helma Saemmer, Dec. 18, 1939; children—Peter, Gisela. Faculty U. Munich, 1950-——, prof. tropical medicine, 1964-——, dean faculty, 1966-——; dir. Inst. Tropical Medicine, 1952-——. Recipient Bavarian Medal of Honor, 1964. Mem. German Soc. for Tropical Medicine (exec. bd.). Author: Handbuch der Schutzeinpfungen, 1965; Die Pocken, 1967. Research on pox diseases, smallpox, vaccination, methods to prevent postvaccinal neural complications. Home: 19 Zamboninistrasse, Munich 19, Germany.*

HERRLINGER, Robert Bernard Charles, med. historian; b. Antwerp, Belgium, Apr. 24, 1914; s. Carl and Elise (Hattich) H.; ed. U. Heidelberg, U. Lena; M.D., agrégé, Ph.D.; m. Gertrud Jochheim, 1939; children—Margot, Jörg-Dieter, Axel, Regine. Prof. agrégé of anatomy, 1944; 1st lectr. history of medicine, Ratisbon, 1949; prof. history of medicine U. Wurzburg, 1958; titular prof. history of medicine U. Kiel, 1962. Mem. Internat. Acad. History of Medicine. Author: (with H. Voss) Taschenbuch der Anatomie, 3 vols., 12th edit., 1963; Volcher Coiter, 1534-72, 1952; Die Nobelpreisträger der Medizin, 1963. Editor communications Inst. History Medicine of Wurzburg, 1957-62, Kiel, 1963-——. Died Feb. 8, 1968.

HERRMANN, Guenter Friedrich, German chemist; b. Greiz, Germany, Nov. 29, 1925; s. Ernst Walter and Anna (Schmidt) H.; Dr.rer.nat., U. Mainz, 1956; m. Ellen Emig, Dec. 23, 1954; 1 dau., Anne. Research and teaching asst. Inst. Inorganic and Nuclear Chemistry, U. Mainz (Germany), 1955-66, group leader, 1966-——, asso. prof. nuclear chemistry, 1967-——. Research and numerous publs. in nuclear fission, decay of radioactive isotopes, application of radioactivity in chemistry. Home: 1 Martin-Kirchner Strasse, Mainz-Bretzenheim 65. Office: Inst. for Inorganic and Nuclear Chemistry, University of Mainz, Mainz 65, Germany.*

HERRMANN, Karl, German food chemist; b. Leipzig, Germany, Feb. 12, 1919; s. Willy and Anna H. (Fink) H.; student U. Leipzig, 1941-46; Pharmac. examen, U. Jena (Germany), 1946; food chemist examen U. Halle/Saale (Germany), 1948, Dr.rer.nat., 1948, Habil., 1953; m. Maria H. Schulze, Oct. 14, 1950; 1 dau., Gloria. Asst., U. Halle/Saale, 1946-54, docent, 1954-58; chief food labor, Hannover, Germany, 1959-65; chief food labor, Stuttgart, Germany, 1965-——. Author: Obst, Obstdauerwaren und Obsterzeugnisse, 1966. Research and numerous publs. on plant phenols in fruits and vegetables; revs. of chemistry and tech. of fruit products. Home: 24 Drosselweg, Fellbach, Western Germany. Office: 184 Nürnbergerstrasse, Stuttgart, Western Germany.*

HERRMANN, Roy George, Am. pharmacologist; b. Middle, Ia., Oct. 22, 1920; s. C. H. and Johanna (Jeck) H.; S.B. in Pharmacy, State U. Ia., 1942; M.S., 1944; Ph.D., U. Chgo., 1948; m. Elaine R. Wegert, Jan. 9, 1947; children—Suzanne E., Roy W. Research chemist Chem. Warfare Agts., U. Chgo., 1944-48; Life Ins. Med. research fellow, 1948-49; head biochem. sect. William S. Merrell & Co., 1949-51; research pharmacologist Lilly Research Lab., Eli Lilly & Co., Indpls., 1951-58, research asso., 1968-——. Mem. Am. Soc. Pharmacology and Exptl. Therapeutics, Fedn. Am. Socs. for Exptl. Biology, Microcirculatory Soc. Research and publs. on heart attacks, blood coagulation in blood vessels, blood lipids. Home: 7304 N. Audubon Rd., Indpls. 46250. Office: Lilly Research Labs., Indpls. 46206.*

HERSCH, Joseph, Swiss mathematician; b. Geneva, Switzerland, June 12, 1925; s. Liebmann Peissach and Liba (Lichtenbaum) H.; diplomas mathematician Fed. Inst. Tech., Zürich, Switzerland, 1948, Dr.sc.math., 1953; m. Eda Goldstein, Mar. 25, 1950; children—

Roger-David, Marc, Daniel. With Fed. Inst. Tech., Zürich, 1948-54, 55—, prof., 1962—; fellow Centre National de la Recherche Scientifique, Paris, 1954-55; research worker Battelle Meml. Inst., Geneva, 1957-62. Recipient Kern prize Fed. Inst. Tech., 1953. Mem. Swiss Math. Soc., Société Mathématique de France. Research and publs. on conformal and quasi-conformal mapping (method of extremal length); applied math., including finite differences, cell functions; math. physics, including variational principles, method of one-dimensional aux. problems, isoperimetric inequalities for membrane and Stekloff eigenvalues, generalized symmetry properties. Home: 497 Lindauerstrasse, Nürensdorf, Zürich, 8303, Switzerland.*

HERSCHBACH, Dudley Robert, Am. chemist; b. San Jose, Cal., June 18, 1932; s. Robert Dudley and Dorothy Edith (Beer) H.; B.S. in Math., Stanford, 1954, M.S. in Chemistry, 1955; A.M. in Physics, Harvard, 1956, Ph.D. in Chem. Physics, 1958; m. Georgene Botyos, Dec. 26, 1964; children—Lisa, Brenda. Jr. fellow Harvard, 1957-59; faculty U. Cal. Berkeley, 1959-63, asso. prof. chemistry, 1961-63; prof. chemistry, Harvard, 1963—; Phillips lectr. Haverford Coll., 1962; Falk-Plaut lectr. Columbia, 1963; vis. prof. Göttingen (Germany) U., summer 1963. Alfred P. Sloan fellow, 1959-63; Harvard lectr. Yale, 1964; Debye lectr. Cornell, 1966. Fellow Am. Phys. Soc., Nat. Acad. Scis., Am. Acad. Arts and Scis.; mem. Am. Chem. Soc. (Pure Chemistry prize 1965), A.A.A.S., Phi Beta Kappa, Sigma Xi. Research on molecular dynamics of chem. reactions and other collision processes.

HERSCHEL, Alexander Stewart, Brit. astronomer; b. Feldhausen, South Africa, Feb. 5, 1836; s. Sir John Frederick William; B.A., Trinity Coll., Cambridge, Eng., 1859, M.A., 1877; student meteorology Royal Sch. Mines, London, Eng., 1861; D.C.L., Durham U. (Eng.), 1866. Began meteorol. observations Royal Sch. Mines; prof. physics Glasgow, Scotland, 1866-71; also Durham Coll., Newcastle, Eng., 1871-86; observed solar eclipse, Spain, 1905. Fellow Royal Soc., 1884, Royal Astron. Soc.; mem. Phys. Soc. London, Soc. Arts. Contbr. meterol. articles to sci. jours.; also made annual reports on meteors observed and progress of meteoric sci. to Brit. Assn., 1862-81. Research on summation reduction; noted for precision in recording meteor paths in relation to stars; formed (with R. P. Greg) extensive catalogues of radiant points of meteor streams; determined radiant point of Nov. Leonids (led Schiaparelli to deduce identity of their orbit with that of Temple's Comet of 1866); experimented with photography. Died Slough, Eng., June 18, 1907.

HERSCHEL, Caroline Lucretia, Brit. astronomer; b. Hanover, Germany, Mar. 16, 1750; d. Issac and Anna Ilse (Moritzen) Herschel; unmarried. Studied violin, trained as concert singer; worked as dressmaker, 1767; joined her bro., William at Bath, Eng., 1772; then continued singing studies, sang in oratorios at Bath, Bristol and elsewhere, until 1782, also studied arithmetic and astronomy under her bro., then assisted him in astron. studies throughout his life, read his instruments and copied his observations, eventually executed extensive calculations necessary to his work, also edited his papers and prepared his catalogues for publ.; began independent studies of the heavens, 1782; given salary as brother's asst., from George III, 1787; after brother's death, 1822, returned to Hanover and catalogued his observations. Recipient Gold medal Astron. Soc., Gold medal in sci. from King of Prussia. Hon. mem. Royal Astron. Soc., Royal Irish Acad. Author: Index to Flamsteed's Observations of the Fixed Stars (gave reference to every star in Brit. catalogue), 1798; Reduction and Arrangement in the Form of a Catalogue in Zones of all the Star Clusters and Nebulae observed by Sir William Herschel, 1828. First important woman astronomer; discovered 3 nebulae, 1783, also 8 comets (5 of them with undisputed priority), 1786-97. Died Hanover, Jan. 9, 1848.

HERSCHEL, Sir John Frederick William, Brit. astronomer; b. Slough, Eng., Mar. 7, 1792; s. Sir William H. and Mary Baldwin (Pitt) H.; ed. St. John's Coll., Cambridge, circa 1809-13; studied law, Lincoln's Inn, 1814; M.A., 1816; D.C.L., Oxford U., 1839; m. Margaret Brodie, Stewart, Mar. 3, 1829; children—William James, Alexander Stewart, John, 9 daus. Fellow, St. John's Coll., 1813; founder (with G. Peacock and C. Babbage) Analytical Soc. Cambridge, 1813, writer, pub. Trans.; active in found. Royal Astron. Soc., 1st fgn. sec., pres., 1827-32, Gold medal, Lalande prize, 1825, medal, 1936; traveled on continent with Babbage, 1821-22; conducted stellar observations from Cape of Good Hope, South Africa, 1834-38, instituted system of nat. edn. for colony; lord rector Marischal Coll., Aberdeen, 1842; mem. Royal Commn. examining curricula at Oxford and Cambridge, 1850; master of mint, 1850-55, ret. to Collingwood, 1855. Mem. Royal Commn. on Standards, 1838-43. Fellow Royal Soc., 1813 (sec., Copley medal 1821, 47, Royal medal 1833, 36, 40), 1813; mem. Brit. Assn., Académie des sciences (fgn. asso.), also numerous others in Europe and U. S. Author: On the Study of Natural Philosophy, 1830; Results of Astronomical Observations made during the years 1834-8 at the Cape of Good Hope, 1847; Outlines of Astronomy, 1849; Familiar Lectures on Scientific Subjects; Collected Addresses, also articles to Ency. Brit. Reexamined and revised binary stars observed by his father, 1821-23, reviewed nebulae and star-clusters of northern hemisphere, 1825-33; pioneer in chem. analysis of solar spectrum, 1819; discovered 525 new nebulae; one of 1st to approach dynamical problems presented by star-clusters; made barometrical determination of height of Mt. Etna, 1924; a founder of southern sideral astronomy; studied nebulae and star-clusters of southern hemisphere; revised nomenclature of southern stars; experimented on solar radiation, 1826, made 1st satisfactory measures of direct solar radiation, 1836; studied structure of solar floccules, 1864; made one of earliest studies of influence of earth's orbital eccentricity upon climate, 1830; studied optics; contbd. to knowledge of terrestrial magnetism, photography, color photography; 1st to describe prints as positive and negative; revitalized math. sci. in Eng. (with Babbage and Peacock); wrote on differential calculus, finite differences, summation of series, calculus of operations. Died Collingwood, Hawkhurst, Kent, May 11, 1871.

HERSCHEL, Sir William (originally Frederich Wilhelm), Brit. astronomer; b. Hanover, Germany (then possession of George II of Eng.), Nov. 15, 1738; s. Isaac and Anne Ilse (Moritzen) H.; hon. doctorate U. Edinburgh, U. Glasgow; m. Mary Pitt Baldwin, May 8, 1788; 1 son, John Frederick William. Musician, Hanovarian army; deserted army because of poor health and Seven Years War, smuggled himself into Eng., 1757; engaged by Earl of Darlington to train band of Durham milita, 1760; organist, condr., composer, music tchr. at Doncaster, circa 1761, Halifax, 1765, Bath, 1766-82; became interested in astronomy; began to build telescopes and other instruments, 1773; erected his 1st large reflector behind his house, 1775; returned to Hanover to pick up his sister Caroline, brought her to Eng., 1772; she devoted her life to his work; received royal summons to bring his instruments to London for royal inspection, given formal pardon for his desertion from Hanovarian army by George III, apptd. ct. astronomer, 1782, pension he received for this enabled him to stop teaching music; continued making and selling telescopes for revenue, until 1788; received grants from George III to build and maintain larger and better instruments. Fellow Royal Soc., 1781 (Copley medal 1781); mem. French Acad. Scis. (fgn.), also numerous others. Made finest and most powerful telescopic equipment of his time; discovered planet Uranus, 1781, 1st new planet to be discovered in historic times, doubled extent of known solar system; discovered 2 Uranian satellites, 1787; discovered and named 2 new satellites of Saturn, 1789; timed Saturn's period of rotation and showed that its rings also rotate; discovered over 800 double stars, showed that some of these at least are binary or phys. double-stars which rotate about each other in accord with Newton's law of gravity, thus supplied 1st direct evidence of validity of law of universal gravitation outside our solar system; 1st systematic reporter on variable stars; virtual founder of sideral sci.; 1st to represent our solar system as tiny speck in vast universe of stars; 1st to suggest that sun itself moves and calculated that its motion is toward a point in constellation Hercules; catalogued and studied some 2500 nebulae and/or star clusters; initiated classification of these and examined their distbn. relative to Milky Way; a founder of study of stellar chromatics; tested temperature of various portions of sun's spectrum, found that highest temperature occurred in area beyond that of visible light (area now called infrared); pioneer in stellar photometry; author numerous papers and sci. memoirs pub. largely in Philos. Trans. of Royal Soc., Memoirs of Royal astron. Soc. Died Slough, Buckinghamshire, Eng., Aug. 25, 1822.

HERSEY, John Brackett, Am. geophysicist; b. Wolfeboro, N.H., Aug. 20, 1913; s. Fred Edgar and Alice (Brown) H.; A.B., Princeton, 1934, A.M., 1935; Ph.D., Lehigh U., 1943; m. Sally Magowan, Jan. 7, 1946; children—Cyrus Brackett, Joslyn Helen. With U. S. Coast and Geodetic Survey, 1935-36; with Phillips Petroleum Co., 1936-39; faculty Lehigh U., 1941; physicist Naval Ordnance Lab., 1941-44; geophysicist Woods Hole (Mass.) Oceanographic Inst., 1946-66; prof. oceanography Mass. Inst. Tech., 1959-66; with Office Naval Research, Washington, 1966—. Recipient Fleming medal Am. Inst. Geonomy and Natural Resources, 1958. Mem. Am. Geophys. Union, Geol. Soc. Am., Acoustical Soc. Am., Soc. Exploration Geophysicists. Editor: Deep Sea Photography, 1967. Contbr. chpts. to The Sea, Ideas and Observations, 1960, 63. Research and numerous publs. primarily in field of resonance in sound scattering by fishes, sound transmission, seismic reflection in ocean structure, thickness and phys. properties of earth's crust. Home: 5232 Westpath Way, Washington 20016. Office: Code 102-05, Office Naval Research, Navy Dept., Washington 20360.*

HERSEY, Mayo Dyer, Am. engr.-physicist; b. Pawtuxet Neck, R.I., Aug. 30, 1886; s. George Milbank and Alice (Budlong) H.; B.A., Colo. Coll., 1907; B.S., Mass. Inst. Tech., 1910; M.A., Olivet Coll., 1910; postgrad. Harvard, 1915-16; m. Frances Lester Warner, June 24, 1922. Mem. staff Mass. Inst. Tech., 1909-10, asso. prof., 1920-22; research asso. mech. engring., cons., 1942-47; physicist U. S. Bur. Standards, Washington, 1910-20, 26-31, U. S. Bur. Mines, Pitts., 1922-26; mech. engr. Vacuum Oil Co., Paulsboro, N.J., 1931-33; faculty Brown U., Providence, 1934-36, vis. prof. engring., 1957—; engr. Kingsbury Machine Works, Phila., 1936-39; research dir. Morgan Constrn. Co., Worcester, Mass., 1939-41; mech. engr. U. S. Naval Engring. Expt. Sta., Annapolis, Md., 1947-57. Lectr. Harvard, 1936-40, 42-43; cons., project dir. NRC, 1941. Recipient Levy medal Franklin Inst., 1936; Certificate Appreciation, U. S. Sec. of War, 1945; Fellow Am. Soc. M.E. (Mayo D. Hersey award 1965); mem. Am. Soc. Lubrication Engrs. (hon.), Soc. Rheology, Am. Soc. Naval Engrs., Washington Acad. Scis., Sigma Xi, Tau Beta Pi, others. Author: Theory of Lubrication, 1936, 38; Theory and Research in Lubrication, 1966. Research and publs. on methods of testing aeronautic instruments; devel. of theory of errors of phys. measurements with reference to economy of time; lubrication research includes 1st application of dimensional analysis to bearings, and 1st measurement of viscosity of oils at high pressure. Home: Providence.*

HERSH, A(mos) H(enry), Am. biologist; b. Lancaster, Pa., Nov. 2, 1891; s. George and Margaret (Rudy) H.; A.B., Franklin and Marshall Coll., 1914, A.M., 1915, Sc.D., 1946; postgrad. Princeton, 1915-16; Ph.D., U. Ill. (fellow in zoology, 1919-22), 1922; m. Roselle Karrer, Aug. 19, 1922; children—Charles Karrer, Robert Tweed. Instr. zoology Kan. State Coll., 1916-18; instr. biology Marquette U., 1918-19; instr. zoology U. Mich., 1922-23; instr. biology Western Res. U., 1923-25, asst. prof., 1925-36, asso. prof., 1936-46, prof. biology from 1946. Fellow Ohio Acad. Sci., A.A.A.S.; mem. Am. Soc. Zoologists, Am. Soc. Naturalists, Genetics Soc. of Am., Am. Soc. Human Genetics, Soc. for Study Devel. and Growth; Am. Assn. U. Profs., Gamma Alpha, Sigma Xi, Phi Kappa Sigma. Contbr. articles on genetics and relative growth to sci. jours.; Blakiston's New Gould Medical Dictionary, 1949. Died Aug. 28, 1955.

HERSHEY, Alfred Day, Am. biologist; b. Owosso, Mich., Dec. 4, 1908; s. Robert Day and Alma (Wilber) H.; B.S., Mich. State Coll., 1930, Ph.D., 1934; m. Harriet Davidson, Nov. 15, 1945; 1 son, Peter Manning. Faculty Washington U., 1934-50; staff dept. genetics Carnegie Instn. of Washington, Cold Spring Harbor, N.Y., 1950—, dir., 1962—. Recipient Lasker award Am. Pub. Health Assn., 1958. Mem. Am. Acad. Arts and Scis., Am. Soc. Microbiology, Nat. Acad. Scis. Expts. (with Martha Chase) confirmed earlier indications that material basis of heredity resides in nucleic acids, suggesting that nucleic acid structure provides genetic code for organisms, 1952. Home: Cold Spring Harbor. Office: Carnegie Instn., Cold Spring Harbor, L.I., N.Y.

HERSHEY, Falls Bacon, Am. physician; b. Chgo., Aug. 16, 1918; s. Charles Owens and Emma (Eby) H.; B.S. magna cum laude, U. Ill., 1939; M.D., Harvard, 1943; m. Julia K. Elder, Oct. 15, 1955. Research asso. dept. biology Mass. Inst. Tech., 1949-50; faculty Washington U., St. Louis, 1953-64, asso. prof. clin. surgery, 1959-60; chmn. div. surgery, dir. surg. research Michael Reese Hosp. and Med. Center, Chgo., prof. surgery Chgo. Med. Sch., 1964-66. Diplomate Am. Bd. Surgery. Mem. A.C.S., Western Surg. Soc., Soc. U. Surgeons, Internat. Cardiovascular Soc., Soc. Cryobiology, Sigma Xi, Alpha Omega Alpha, others. Author: Atlas of Vascular Surgery, 1963. Research and publs. on irradiated yeast cells, arteriography; histochemistry of human skin, breast cancer. Home: 11 Wydown Terrace, Clayton, Mo. 63105. Office: 18 S. Kingshighway, St. Louis 63108.*

HERSHEY, H(oward) Garland, Am. geologist; b. Quarryville, Pa., Oct. 1, 1905; s. Howard Risser and Pearl (Reinhart) H.; A.B., Johns Hopkins, 1929, Ph.D., 1936; m. Erna Madelyn Eybs, Oct. 24, 1931; children—Howard Garland, Timothy J. Geologist, B. & O. R.R., 1930; staff Md. Geol. Survey, 1931-36, Ia. Geol. Survey, 1936-39; asst. geologist State of Ia., 1939-43, asso. state geologist, 1944-47, state geologist, dir., 1947—; dist. geologist U. S. Geol. Survey, 1944-55. Mem. various state gov. coms.; mem. steering com. Internat. Hydrological Decade Com., 1964—; mem. bd. consultants Ia. Hydraulics Research Inst., 1965—. Fellow Geol. Soc. Am. (past councilor); mem. Ia. Natural Resources Council (chmn. 1949—), Midwestern States Flood Control Conf. (past chmn.), Nat. Water Conservation Conf. (v.p. 1953—), Am. Assn. State Geologists (past pres., past historian), Soc. Econ. Geologists, Soil Conservation Soc. Am., A.A.A.S., Am. Geophys. Union, Am. Water Works Assn., Ia. Engring. Soc. (past pres.), Ia. Acad. Sci. (past pres.), others. Author reoorts, maps, articles in tech., sci. jours. Home: 329 Beldon Av. Office: 16 W. Jefferson St., Iowa City, 52240.*

HERSHEY, J(ohn) Willard, Am. chemist; b. Gettysburg, Pa., Feb. 6, 1876; s. Abraham and Hosie (Eyster) H.; B.S., Gettysburg Coll., 1907, M.S., 1910; postgrad. Harvard, 1907-08, Johns Hopkins, 1910-11; Ph.D., U. Chgo., 1924; m. Effie Bowman, Aug. 24, 1916; 1 son, Ardys Willard. Prof. chemistry, Bridgewater (Va.) Coll., 1908-10, Defiance (O.) Coll., 1911-18, McPherson (Kan.) Coll. from 1918. Fellow A.A.-A.S., Internat. Coll. Anesthetists (life); mem. Am. Chem. Soc. (councilor 1930-32; pres. Wichita sect. 1937-39), Am. Inst. Chemists, Kan. Acad. Science (pres. 1933-34), Sigma Xi, Phi Beta Kappa. Author: A Laboratory Guide to Study of Qualitative Analysis, 1927; (with L. A. Enberg) Laboratory Manual for General Chemistry, 1933; The Book of Diamonds, 1940;

also research papers on synthetic diamonds and components of the atmosphere and synthetic gases in relation to animal life. Died Sept. 27, 1943.

HERSHEY, Solomon George, Am. physician; b. N.Y.C., June 23, 1914; s. Mark and Cecil (Brunner) H.; B.S., Coll. City N.Y. 1934; M.D., N.Y. U., 1939; m. Lenore Oppenheimer, Dec. 21, 1941; 1 dau., Jane. Faculty, N.Y. U., 1944-66, prof. anesthesiology, 1964-66; prof. anesthesiology Albert Einstein Coll. Medicine, 1966——; mem. staffs Bellevue, Beth Israel, Univ., Bronx Municipal, Lincoln, Einstein hosps. Vis. prof. anesthesiology U. Cal., San Francisco, 1963. Mem. Shock com. NRC-Nat. Acad. Scis. Mem. Am., N.Y. State (pres. 1958) socs. anesthesiologists, Am. Coll. Anesthesiology (bd. govs. 1952-58, chmn. 1957-58), N.Y. Acad. Medicine (chmn. sect. on anesthesiology 1965-66), N.Y. Acad. Scis., Assn. U. Anesthesiologists, Soc. Pharmacology and Exptl. Biology, Soc. Exptl. Biology and Medicine, A.A.A.S., Phi Beta Kappa, Sigma Xi, Alpha Omega Alpha. Author: Shock, 1964. Editor Jour. Anesthesiology, 1963——. Research on physiology of peripheral circulation; effects of anesthetics and various drugs on microcirculation in normal and various disease states; extensive research in field of shock; tech. and clin. studies related to anesthetic considerations of various surg. problems. Home: 750 Ladd Rd., N.Y.C. 10471. Office: 550 1st Av., N.Y.C. 10016.*

HERSHKOVITZ, Philip, Am. zoologist; b. Pitts., Oct. 12, 1909; s. Aba and Bertha (Halpern) H.; B.S., U. Mich., 1938, M.S., 1940; m. Anne-Marie Dode, Sept. 15, 1945; children—Francine, Michael Dode, Mark Alan. Asst. dept. zoology U. Pitts., 1930-31; undergrad. asst. zoology U. Mich., 1931-32, grad. asst. zoology, 1939-41; Walter Rathbone Bacon travelling scholar Smithsonian Inst. Zool. Expdn., Colombia, 1941-43, 1946-47; asst. curator mammals Field Mus. Natural History, 1947-53, asso. curator, 1954, curator, 1956-62, zool. expdn. Suriname, 1961-62, research curator, 1962——; zool. expdn., Colombia, 1948-52. Cons. govtl., pvt. philantropical and med. agys. for identifications of tropical mammalian hosts and vectors of viruses. Mem. Am. Soc. Mammalogists, Soc. for Study Evolution, Soc. Systematic Zoology, Systematic Assn., Biol. Soc. Washington, Internat. Primatol. Soc., Assn. Trop. Biology. Author: Evolution of Neotropical Cricetine Rodents, 1962; Catalog of Living Whales 1965. Contbr. articles to Ency. Brit., Ency. Americana, others. Publs. on studies of origin, evolution, classification, zoogeography and life histories of tropical mammals; stabilization of zool. nomenclature; discovery of new forms of mammals, birds, reptiles, amphibians, ectoparasites. Home: South Holland, Ill. 60473. Office: Field Mus. Natural History, Chgo. 60605.*

HERSKOVITS, Melville Jean, Am. anthropologist; b. Bellefontaine, O., Sept. 10, 1895; s. Herman and Henrietta (Hart) H.; Ph.B., U. Chgo., 1920; A.M., Columbia, 1921, Ph.D., 1923; Sc.D., Sarah Lawrence Coll., 1962; m. Frances S. Shapiro, July 12, 1924; 1 dau., Jean Frances. Fellow in anthropology Bd. Biol. Scis., NRC, 1923-26; lectr. anthropology Columbia, 1924-27, Howard U., 1925; asst. prof. anthropology Northwestern U., 1927-30, asso. prof., chmn. dept., 1931-35, prof. since 1935, prof. African affairs, 1960; dir. Program of African Studies since 1951, hon. prof. anthropology Fac. de Filosofia, Bahia, Brazil; Guggenheim Meml. fellow, 1937-38; field research in Dutch Guiana, 1928, 29, W. Africa, 1931, Haiti, 1934, Trinidad, 1939, Brazil, 1941-42, Sub Saharan, Africa, 1953, 57; vis. prof. grad. sch. U. Ill., 1948-49. Chmn. com. on African anthropol.; chmn. com. on internat. relations in anthropology, div. anthropology and psychology NRC, 1942-50; chmn. com. Negro studies, Am. Council Learned Socs., 1939-50. Mem. permanent council Internat. Anthrop. Congress; v.p., mem. exec. com. Internat. Union Anthropol. and Ethnol. Scis. Decorated officer Order of Honor and Merit, Haiti; officer Order of Orange-Nassau, The Netherlands. Hon. fellow Royal Netherlands Geog. Assn. Royal Anthrop. Inst.; fellow Nat. Acad. Scis., A.A.A.S. (v.p. 1934), Soc. Research Child Devel. Am. Anthrop. Assn. (pres. Central sect. 1939, exec. bd. 1947, editor Am. Anthropologist, 1949-52); African Studies Assn. (pres. 1957-58); mem. Am. Assn. Phys. Anthropologists, Am. Folklore Soc. (pres. 1945), Society des Africanistes de Paris, Internat. African Inst. (exec. council), Council Fgn. Relations. Author: The American Negro, A Study in Racial Crossing, 1928; Anthropometry of the American Negro, 1930; Outline of Dahomean Religious Belief (with Frances S. Herskovits), 1933; Rebel Destiny, Among the Bush Negroes of Dutch Guiana (with same), 1934; Suriname Folklore (with same), 1936; Life in a Haitian Valley, 1937; Dahomey, 1938; Acculturation, 1938; The Economic Life of Primitive Peoples, 1940; The Myth of the Negro Past, 1941; Trinidad Village (with Frances S. Herskovits), 1947; Man and His Works, 1948; Economic Anthropology, 1952; Franz Boos, The Science of Man in the Making, 1953; Cultural Anthropology, 1955; Dahomean Narrative, a Cross-Cultural Analysis (with Frances S. Herskovits), 1958; The Human Factor in Changing Africa, 1962; Economic Transitio in Africa (with M. Harwitz), 1964; Selected Papers in Afro-American Studies, 1966; Cultural Dynamics, 1964; (with others) The Influence of Culture on Visual Perception, 1966. Research on anthropology, ethnology of Negro communities in U. S., Africa, elsewhere. Died Feb. 25, 1963.

HERTER, Christian A., Am. physician; b. Glenville, Conn., Sept. 3, 1865; s. Christian and Mary (Giles) H.; med. degree Coll. Phys. and Surg. (Columbia), 1885; m. Susan Dows, Dec. 1886. Worked under William Welch at Johns Hopkins, also in Europe; settled in N.Y., 1888; vis. physician N.Y.C. Hosp., 1894-1904; named prof. path. chemistry Bellevue Hosp. Med. Coll., 1898; prof. pharmacology and therapeutics Columbia, from 1903. Mem. Am. Soc. Biol. Chemists (charter). Author: The Diagnosis of Diseases of the Nervous System, 1892; Diseases of the Cranial Nerves. Founder, editor Jour. Biol. Chemistry. Described diarrheal children's disease (possibly sprue), 1908; founder lectureships at Johns Hopkins, Bellevue. Died Dec. 5, 1910.

HERTER, Gustav Adolf Wilhelm Konrad, German zoologist; b. Berlin, Germany, Dec. 12, 1891; s. Ernst and Elisabeth (Wiebe) H.; ed. U. Fribourg, U. Berlin; Ph.D. in zoology; m. Margarethe Fasquel, 1922. Asst., Inst. Zoology, U. Göttingen, U. Berlin, 1921-26; agrégé, Berlin, 1924; asso. prof., 1930; affiliate prof., 1939; dir. Inst. Zoology, U. Humboldt, U. Berlin, 1945-51; titular prof., dir. Inst. Zoology, Free U. Berlin, 1952-59, prof. emeritus, 1959. Mem. German Zool. Soc., German Soc. Sci. of Mammals. Author: Die Physiologie und Ökologie der Hirdnieen. 1936-37; Die Biologie der Europäischen Lgel, 1952; Der Temperaturisnn der Saügetiere, 1952; Der Temperatursinn der Insekten, 1953; Die Fischdressuren und Ihre Sinnesphysiologischen Grundlagen, 1953. Home: 5 Wrangelstrasse, Berlin-Steglitz. Office: Zoological Institute, University of Fribourg, 34 Grunewaldstrasse, Berlin-Steglitz II, Germany.

HERTIG, Arthur Tremain, Am. physician; b. Mpls., May 12, 1904; s. Charles Marshall and Florence (Long) H.; B.S., U. Minn., 1928; M.D., Harvard, 1930; m. Linda Woodworth, Dec. 22, 1932; children—Helen Learned (Mrs. Thomas Goodrich Craig), Andrew Woodworth. With Kala Azar field studies Rockefeller Found., Peking, China, 1925-26; faculty Harvard Med. Sch., 1931——, Shattuck prof., 1952——; staff Boston Lying-in Hosp., 1931——, pathologist, 1939-52, cons. pathologist, 1952——; staff pathologist Free Hosp. for Women, Brookline, Mass., 1938-52, cons. pathologist, 1952——. Cons. to hosps., Armed Forces Inst. Pathology. Recipient Outstanding Achievement award U. Minn., 1951. Diplomate Am. Bd. Obstetrics and Gynecologists, Am. Bd. Pathology (trustee 1959——). Mem. A.A.A.S., Am. Gynecol. Soc. (award 1949), Am. Assn. Anatomists, A.M.A., Am. Assn. Pathologists and Bacteriologists. Contbg. author: Progress in Gynecology, 1950; Anderson's Textbook of Pathology, 1957; Atlas of Tumor Pathology, 1956. Research and numerous publs. on normal and abnormal early human embryos, significant stages in devel. of potential tumor-like conditions of human placenta and afterbirth, sequential stages in formation of cancers of uterus, formation of corpus luteum and maturation of oocytes in human ovaries. Home: 21 Everett Av., Winchester, Mass. 01890. Office: 25 Shattuck St., Boston 02115.*

HERTWIG, Oscar, German embryologist, anatomist; b. Friedberg, Hessen, Apr. 21, 1849; ed. Jena, Zurich, Bonn; became lectr. anatomy U. Jena (Germany), 1875, prof. anatomy, 1881; prof. anatomy U. Berlin, 1888-1921; founder, dir. Anat. Inst. Berlin, 1888-1921. Author: Lehrbuch der Entwicklungsgeschichte, 1886; Die Zelle und die Gewebe, 2 vols., 1893-98; Handbuch der vergleichen und experimenten Entwicklungslehre, 3 vols., 1901-06; Das Werden der Organismen, 1916. Aware of cell div. in developing fetus; realized significance of cell nucleus for heredity; proved (with Fol) that only one sperm normally enters to fertilize egg, 1879. Died Berlin, Oct. 25, 1922.

HERTWIG, Richard Carl Wilhelm Theodor von, see von Hertwig, Richard Carl Wilhelm Theodor.

HERTY, Charles Holmes, Am. chemist; b. Milledgeville, Ga., Dec. 4, 1867; s. Bernard R. and Louisa T. H.; student Ga. Mil. and Agrl. Coll., 1880-84; Ph.B., U. Ga., 1886; Ph.D., Johns Hopkins, 1890; at univs. of Berlin and Zürich, 1899, 1900; hon. Ch.D., U. Pitts., 1917; D.Sc., Colgate U., 1918, Oglethorpe U., 1934, U. Fla., 1937; LL.D., U. Ga., 1928, U. N.C., 1933; m. Sophie Schaller, Dec. 23, 1895 (dec. 1929); children—Charles Holmes, Frank Bernard, Sophie Dorothea. Asst. chemist Ga. State Expt. Sta., 1890-91; instr. U. Ga., 1891-94, adj. prof. chemistry, 1894-1902; collaborator Bur. of Forestry, U. S. Dept. Agr., 1901-02, expert, 1902-04; with Chattanooga Pottery Co., 1904-05; prof. chemistry U. N.C., 1905-16; editor Jour. Industrial and Engring. Chemistry, 1917-21; pres. Synthetic Organic Chemical Mfrs.' Assn., 1921-26; adviser to Chem. Found., Inc., 1926-28; became indsl. cons., N.Y.C., 1928; dir. div. pulp and paper research Ga. State Dept. Forestry and Geol. Development, 1932-33; dir. Pulp and Paper Lab. of Industrial Com. of Savannah, Ga. Fellow A.A.A.S., Chem. Soc. (London). Contbd. to theory of chem. compounds; did early work in double halides of lead, antimony, platinum and mercury; developed efficient method of gathering turpentine. Died July 27, 1938.

HERTY, Charles Holmes, Jr., Am. metallurgist; b. Athens, Ga., Oct. 6, 1896; s. Charles H. and Sophie (Schaller) H.; B.S., U. N.C., 1918; M.S., Mass. Inst. Tech., 1921, D.Sc., 1924; m. Kathleen Malloy, Nov. 13, 1929; children—Dorothea, Charles H. III, Kathleen, Timothy. Research asso., Sch. of Engring. Practice, Mass. Inst. Tech., at Lackawanna plant of Bethlehem Steel Co., 1924-26; in charge ferrous metall. research U. S. Bur. of Mines, 1926-31; dir. research Metall. Advisory Bd., Pitts., 1931-34; research engr. Bethlehem Steel Co., 1934-42, asst. to v.p. in charge of operations, from 1942. Mem. Am. Iron and Steel Inst., Am. Inst. Mining and Metall. Engrs. (Hunt medalist; Howe lectr.), Am. Soc. for Metals (Campbel lectr., Saveur Award), Am. Chem. Soc., Nat. Acad. Scis., Phi Beta Kappa, Sigma Xi, Alpha Chi Sigma. Contbr. papers to tech. lit. Specialized in phys. chemistry of steel making, open hearth steel processing. Died Jan. 17, 1953.

HERTZ, Sir Arthur Frederick (name changed to Hurst 1916), English physician; b. Eng., 1879; grad. Magdalen Coll., Oxford, 1901; B.M., Guy's Hosp. (Oxford U.), 1904. Became asst. physician in charge neurol. dept. Guy's Hosp., 1907, physician, 1918-39. Pioneer in clin. sci.; built up team of colleagues at New Lodge Clinic, Windsor Forest; specialized in alimentary tract; 1st to describe dumping syndrome which may occur after gastrectomy and is characterized by palpitations, hot flushes, cold perspiration, 1913. Died 1944.

HERTZ, Carl Hellmuth, physicist; b. Berlin, Germany, Oct. 15, 1920; s. Gustav Ludwig and Ellen (Dihlmann) H.; Abitur, Schule Schloss Salem, Germany, 1939; Ph.D., U. Lund (Sweden), 1955; m. Birgit Nordbring, Feb. 22, 1953; children—Hans Martin, Thomas. Asst. nuclear physics group, dept. physics U. Lund, 1949-55, asst. prof., dir. biophysics group, 1958-63; research physicist Siemens Reiniger Werke, Erlangen, Germany, 1956-57; prof., dir. dept. elec. measurements Lund Inst. Tech., 1963——. Mem. Fedn. Med. Electronics and Biol. Engring. Research, publs. on heart diagnosis by ultrasound echo method, gen. med. physics, geoelectric effect and related topics in plants, new methods for rec. of fast elec. signals, gas discharges (corona), hygrometer. Home: 8 Skolbanks vagen, Lund, Sweden.*

HERTZ, Gustav Ludwig, German physicist; b. Hamburg, Germany, July 22, 1887; s. Gustav and Auguste (Arning) H.; grad. U. Munich, also U. Berlin, 1911; m. Ellen Dihlmann, 1919 (d. 1941); children—Hellmuth, Johannes; m. 2d, Charlotte Jollasse, 1943. Research asst. Physics Inst., Berlin U., 1913-20; with physics lab. Phillips Incadescent Lamp Factory, Eindhoven, Germany, 1920-25; resident prof., dir. Physics Inst., U. Halle (Germany), 1925-28; dir. Physics Inst., Charlottenburg Technol. U., 1928-35; dir. Research Lab., Siemans Co., 1935-45; research in USSR, 1945-54; prof., dir. Physics Inst., Leipzig, Germany, 1955-61. Recipient (with J. Franck) Nobel prize in physics, 1925; Max Planck medal German Phys. Soc. Mem. German Acad. Scis. Berlin, Gottingen Acad. Scis. (corr.), Hungarian Acad. Scis. (hon.), Czechoslovakian Acad. Sci., USSR Acad. Sci. (fgn.). Research and publs. (with Franck) on effect of impact of electrons on atom showing connection between spectral lines and energy states of atoms; measurement of ionization potentials of atoms; separation of isotopes by diffusion cascade.

HERTZ, Hans Georg, astronomer; b. Göttingen, Germany, Aug. 8, 1915; s. Paul and Helen (Markiel) H.; student U. Hamburg, (Germany), 1934-37; Ph.D., Yale, 1941. Came to U. S., 1937, naturalized, 1943. Sterling Research fellow Yale, 1941-42, instr., 1942-43; jr. astronomer U. S. Naval Obs., Washington, 1943-44, asst. astronomer, 1944-47, asso. astronomer, 1947-58; astronomer U. S. Army Map Service, Washington, 1958-60; astronomer Goddard Space Flight Center, NASA, Greenbelt, Md., 1960——. Mem. Am. Astron. Soc., Internat. Astron. Union, Am. Math. Soc., Am. Geophys. Union, Math. Assn. Am., Am. Assn. Computing Machinery, A.A.A.S., Sigma Xi. Research and publs. on celestial mechanics, orbits of Trojan asteroids, satellite orbits, reduction of occultations. Home: 2112 Florida Av. N.W., Washington 20008. Office: Goddard Space Flight Center, Greenbelt, Md. 20771.*

HERTZ, Heinrich Rudolf, German physicist; b. Hamburg, Germany, Feb. 22, 1857; student Dresden (Germany) U., 1 semester; at Munich until 1878; studied under H. von Helmholtz, U. Berlin, doctorate magna cum laude, 1880; asst. to von Helmholtz, 1880-32; at Kiel, Germany, 1 semester; prof. exptl. physics Inst. Tech. Karlsruhe, Germany, probably 1884-1889; succeeded Clausius as prof. U. Bonn, 1889-94. Author: Electric Waves, 1890, English edit., 1893; Principles of Mechanics, 1894, English edit., 1899. Research on discharge of electricity in rarefied gases, Maxwell's electromagnetic theory of light; 1st to observe electromagnetic waves (Hertzian, or radio waves), circa 1886, also showed they can be refracted, reflected and polarized in manner of light, measured their velocity and length. Died Bonn, Jan. 1, 1894.

HERTZLER, Arthur Emanuel, Am. surgeon; b. West Point, Ia., July 26, 1870; s. Daniel and Hannah M. (Krehbiel) H.; B.S., Southwestern Coll., Winfield, Kan., 1897; M.D., Northwestern U., 1894; postgrad. Berlin, 1899-1901; Ph.D., Ill. Wesleyan U., 1902; LL.D., Washburn Coll.; LL.D., Southwestern Coll., 1939; Sc.D., Boston U.; 1939; m. Myrtle Arnold, May 1, 1894; children—Agnes H., Helen L., Margaret L.; m. 2d, Irene A. Koeneke, June 8, 1945; Tchr. pathology, 1902-07, surgeon and gynecologist Univ. Med. Coll., Kansas City, Mo., 1907-09; prof. surgery, U. of Kan., from 1909; founder, 1902, and pres. Agnes Hertzler Meml. Hosp. Mem. A.M.A., Western Surg., Assn., Am. Micros. Soc. (pres. 1911), Assn. Am. Anatomists, Am. Acad. Medicine. Author: Treatise on Tumors, 1912; Surgical Operations Under Local Anaesthesia, 1913; Principles of Abdominal Surgery, 1918; Diseases of the Peritoneum, 1919; Case Histories in Surgery, 1919; Diseases of the Thyroid Gland, 1922; Treatise on Minor Surgery (with V. E. Chesky), 1925; Pathology of the Surgical Diseases of Bone, 1930; Pathology of Genitourinary Diseases, 1930; Pathology of the Diseases of the Skin, 1931; Pathology of the Female Generative Organs, 1931; Surgical Pathology of the Diseases of the Mammary Gland; Surgical Pathology of the Diseases of the Peritoneum; Surgery of General Practice (with Chesky), 1934; Surgical Pathology of the Gastro Intestinal Tract, 1936; Surgical Pathology of the Thyroid Gland, 1936; Surgical Pathology of Diseases of the Neck, 1937; The Horse and Buggy Doctor, 1938; Surgical Pathology of Diseases of the Mouth and Jaws, 1938; The Doctor and His Patient, 1940; Study of the Diseases of the Thyroid Gland, 1941. Grounds of an Old Surgeon's Faith, 1944; Ventures in Science of a Country Surgeon, 1944; Always the Child, 1944. Research in diseases of peritoneum, from 1894, also work on local anaesthesia, diseases of thyroid gland. Died Sept. 12, 1946.

HERTZSPRUNG, Ejnar, astronomer; b. Frederiksberg, Denmark, Oct. 8, 1873; s. Severin and Henrietta (Frost) H.; ed. chem. engring., Copenhagen, Denmark, 1898; ed. Leipzig, Germany, 1901; became asst. prof., Gottingen, Germany, 1909; observer, asst. dir., later dir. Leyden (Netherlands) Obs., 1914-44; prof. U. Leyden, 1920——. Recipient Gold medal Royal Astronomy Soc., 1929, Bruce Gold medal, Cal., 1937. Mem. Danish Acad., Dutch Acad., Royal Astronomy Soc., Am. Acad. Arts and Scis., Am. Astronomy Soc. Research on double stars and colors, especially Pleides and Cepheides; occulation vacillatings; distance of small Magellan clouds; estimation of size and luminosity of universe. Home: Villavej 6, Tollose, Denmark.

HERWEG, (August) Julius, German physicist; b. Helmstedt, Germany, Oct. 17, 1879; Ph.D., Würzburg, Germany, 1905. Became lectr. physics U. Greifswald (Germany), 1907; prof., 1913; mil. service, 1914-18; became lectr. U. Halle (Germany), 1917; named asso. prof. high frequency physics and founds. of physics and photography Tech. U., Hanover, Germany, 1923, also lectr. on theoretical physics; founder Inst. for High Frequency Physics (became Inst. for High Frequency Physics and Photography 1929), 1924; named prof., 1929, emeritus, 1935. Research on high frequency physics, photography, telegraphy, telephony. Died Bad Reichenhall, Germany, Jan. 26, 1936.

HERXHEIMER, Karl, German dermatologist; b. Weisbaden, Germany, 1861; prof. dermatology and syphilis U. Frankfort (Germany). Author: Die Behandlung der Krankheiten des Behaartenkopfes; Die Teerbehandlung von Hautkrankheiten. Described exacerbation in syphilitic symptoms and lesions sometimes following adminstrn. of antisyphilitic drugs (Herxheimer's reaction, or Jarisch-Herxheimer reaction), 1902. Died 1944.

HERZ, Werner, chemist, educator; b. Stuttgart, Germany, Feb. 12, 1921; s. Alfred and Hedwig (Loewenstein) H.; brought to U. S., 1937, naturalized, 1944; B.A., U. Colo., 1943, M.A., 1945, Ph.D., 1947; m. Marcia Lucile King, Feb. 22, 1945; children—Michael John, Patrick Werner, Monica Lucile, Andrea Lauren. Instr. math. U. Colo., 1946-47; Am. Cyanamid fellow U. Ill., 1947-49; with Fla. State U., Tallahassee, 1949——, prof., 1959——; mem. chemistry panel Cancer Chemotherapy Nat. Service Center, 1959-62, NSF, 1961-64; cons. Nat. Cancer Inst., 1962-65; mem. cancer chemotherapy study sect. NIH, 1962——. Mem. Am. Chem. Soc. (councilor Fla. sect. 1960——), Chem. Soc. London, Phi Beta Kappa, Sigma Xi, Sigma Pi Sigma, Pi Mu Epsilon, Phi Lambda Upsilon. Author: The Shape of Molecules, 1963. Editorial bd. Jour. Organic Chemistry, 1962-63, sr. editor, 1963——. Research and numerous publs. on isolation and structure determination of plant products with emphasis on possible applications to chemotaxonomy, structure synthesis and transformations of terpenoid substances; studies of molecular rearrangements in chemistry. Home: 314 Saratoga Dr., Tallahassee 32303.*

HERZBERG, Frederick Irving, Am. psychologist; b. Lynn, Mass., Apr. 18, 1923; s. Lewis and Gertrude (Copleman) H.; B.Social Service, Coll. City N.Y., 1946; M.S., U. Pitts., 1948, Ph.D., 1950; M.P.H., U. Pitts., 1951; m. Shirley Bedell, June 1, 1944; 1 son, Mark Allen. Lectr. psychology U. Pitts., 1946-48; prin. personnel adminstr. City of Richmond, Va., 1948-50; research dir. Psychol. Service Pitts., 1951-57; prof., chmn. dept. psychology Western Res. U., 1957——. Mgmt. cons. to industry, govt., pvt. orgns. Mem. Am. Psychol. Assn., Am. Assn. U. Profs., Sigma Xi. Author: Job Attitudes: Research and Opinion, 1957; The Motivation to Work, 1959; Work and the Nature of Man, 1966; also numerous articles. Originated motivation-hygiene theory, M-H theory of illness. Home: 3729 Meadowbrook Blvd., University Heights, O. 44118. Office: Dept. Psychology, Western Res. U., Cleve. 44118.*

HERZBERG, Gerhard, physicist; b. Hamburg, Germany, Dec. 25, 1904; s. Albin and Ella (Biber) H.; Dr.Ing., Darmstadt (Germany) Inst. Tech., 1928; LL.D., U. Sask., U. Toronto, Dalhousie U., U. Alta.; D.Sc., McMaster U., Nat. U. Ireland, Oxford U., U. B.C., University of Chicago, University Stockholm; m. Luise Hedwig Oettinger, Dec. 29, 1929; children—Paul Albin, Agnes Margaret. Research staff U. Bristol (Eng.), 1929-30; faculty Darmstadt Inst. Tech., 1930-35, U. Sask., 1935-45, Yerkes Obs., U. Chgo., 1945-48; mem. staff NRC, Ottawa, Ont., Can., 1948——, dir. div. pure physics, 1955——. Recipient Frederic Ives medal Optical Soc. Am., 1964. Fellow Royal Soc. Can. (v.p. 1964-66, pres. 1966-67), Royal Soc., 1951 (Bakerian lectr. 1960), Indian Acad. Scis. (hon.); academician Pontifical Acad. Scis.; mem. Canadian Assn. Physicists (pres. 1956-57), Internat. Union Pure and Applied Physics (v.p. 1957-63), Hungarian Acad. Scis. (hon.), Am. Acad. Arts and Scis. (hon.), Royal Soc. Can. (Henry Marshall Tory medal 1953). Author: Atomic Spectra and Atomic Structure, 1944; Molecular Spectra and Molecular Structure I. Spectra of Diatomic Molecules, 1939, II. Infrared and Raman Spectra of Polyatomic Molecules, 1945; III. Electronic Spectra and Electronic Structure of Polyatomic Molecules, 1966. Research and publs. on atomic and molecular spectroscopy; determined (with assos.) structures of large number of diatomic and polyatomic molecules, including structures of many free radicals; also successfully applied spectroscopic studies to identification of certain molecules in planetary atmospheres, comets and interstellar space. Home: 190 Lakeway Dr. Office: 100 Sussex Dr., Ottawa 7, Ont., Can.*

HERZBERGER, Maximillian Jakob, math. physicist; b. Charlottenburg, Germany, Mar. 7, 1899; s. Leopold and Sonja Herzberger; Ph.D., U. Berlin, 1923; m. Edith Kaufmann, May 31, 1925; children—Ruth (Mrs. Roy A. Rosenberg), Ursula (Mrs. P. Bellugi), Hans George. Came to U. S., 1935, naturalized, 1940. With Leitz, Wetzlar, 1923-27; math. asst. to dir., Zeiss, Germany, 1927-34; docent optics U. Delft (Netherlands), 1934; sect. dir. geometric optical research Eastman Kodak, Rochester, N.Y., 1935-65; prof. Fed. Inst. Tech., Zürich, Switzerland, 1965——. Recipient Cressy Morrison prize N.Y. Acad. Scis., 1945. Fellow A.A.A.S., Optical Soc. Am. (Ives medal 1962); mem. Am. Math. Soc., Bavarian Acad. Scis., German Optical Soc., Swiss Com. Optics, Swiss Phys. Soc., Sigma Xi. Author: Strahlenoptik, 1932; also numerous articles and patents. Asso. editor: Jour. Optical Soc., 1947-59. Research in geometrical optics, theory of microscopic vision, design of optical instruments, glass dispersion, optical lenses, photography, gen. field theory. Address: 42 Feldeggstrasse, Zürich 8, Switzerland.*

HERZENBERG, Arvid, physicist; b. Vienna, Austria, Apr. 16, 1925; s. Harry and Wilhelmine (Pfeiffer) H.; B.Sc., U. Manchester, 1949, Ph.D., 1952, D.Sc., 1964; m. Marjorie Swift, Nov. 30, 1949; children—Catherine, Anne, Stephen. Faculty, U. Manchester (Eng.), 1952——, reader in theoretical physics, 1965——. Visitor, Inst. Theoretical Physics, Copenhagen, Denmark, 1956-57; cons. Westinghouse Electric Corp., Pitts., 1966. Mem. Brit., Am. phys. socs., Brit. Biophys. Soc. Research and publs. on theory of forces between alpha particles, proof of existence of self-maintaining dynamo motions in homogeneous fluids, contbns. to theory of unstable negative ions with applications to electron-molecule and ion-molecule scattering, contbr. to theory of excited states of biopolymers and x-ray analysis. Home: Wall End, Coppice Lane, Disley, Cheshire, Eng. Office: Theoretical Physics Dept., University of Manchester, Manchester, Eng.*

HERZFELD, Karl Ferdinand, physicist; b. Vienna, Austria, Feb. 24, 1892; s. Charles August (M.D.) and Camilla (Herzog) H.; student Schotten Gymnasium, Vienna, 1902-10, U. of Vienna, 1910-12, U. of Zurich, 1912-13, U. of Göttingen, 1913-14; Ph.D., U. of Vienna, 1914; D.Sc. honoris causa, Loyola Coll. (Baltimore), 1932. Marquette U., 1933; D.Sc., U. Md., 1956, Manhattan Coll., 1959; hon. degrees Fordham U., 1960, Inst. of Technology, Stuttgart, 1962, Cath. U. of Am., 1963; m. Regina Flannery, 1938. Privatdocent Univ. Munich, 1920, a. o. prof., 1923; Speyer guest prof., Johns Hopkins, 1926, prof. physics, 1926-36; head of dept. of physics, Catholic U. of America, since 1936. Served as 1st lt., Austrian Army, 1914-18. Awarded Secchi medal, Georgetown University, 1938; USN Meritorious Pub. Service award, 1964. Fellow Am. Phys. Soc., A.A.A.S.; mem. Washington Academy Science, Washington Philos. Soc., Bavarian Physical Soc., Catholic Assn. Internat. Peace, Am. Academy Arts and Scis., National Acad. Scis., Phi Beta Kappa, Sigma Xi. Awarded Mendel medal, 1931. Received Certificate of Exceptional Service to Navy Ordnance Development, 1946. K.C. Author: Kinetische Theorie der Waerme, 1925; Absorption and Dispersion of Ultrasonic Waves (with T. A. Litovitz), 1959. Research on thermodynamics; solid state; kinetic theory of gases; ultrasonics; electronic structure of molecules; reaction velocities; interior ballistics; semiconductors. Home: 3726 Connecticut Av. N.W. Address: Cath. U. Am., Washington 17.

HERZOG, Edgar Friedrich, German psychologist; b. Berlin, Germany, Dec. 24, 1891; s. Gotthold and E. (Poppe) H.; ed. U. Leipzig, U. Berlin; Ph.D. in psychology; m. J. Duerck, 1943; 1 son, Valentin. Prof., dir. Internat. Sch. Modern Expts., dismissed for polit. reasons, 1934; instr. psychotherapy German Inst. Psychol. Research and Psychotherapy, Berlin, 1938; instr., monitor Inst. Psychol. Research and Psychotherapy, Munich, 1946; practice psychotherapeutic medicine; dir. studies. Mem. German Soc. Psychotherapy and Psychology, Internat. Soc. Psychotherapy, Kant Soc. Germany. Co-editor: Erziehungshilfe; Erziehungsschwierigkeiten im Schulalter, 1939-42; Persönlichkeitsprobleme des Lehrers in der Erziehung, 1953; Psyche und Tod Wandlung des Todesbildes im Mythos und in Träumen Heutiger Menschen, 1960. Address: 22 Bruhildenstrasse, Munich, Germany.*

HERZOG, Emil Rudolph, astronomer; b. Basel, Switzerland, Mar. 18, 1917; s. Emil and Bertha (Goetschin) H.; Ph.D., U. Basel, 1947; m. Gertrud M. Steiner, Nov. 21, 1946; 1 son, Adrian D. Came to U. S., 1956, naturalized, 1964. Tchr. math. Institut Athenaeum, Basel, 1947-49, 51-56; sr. research asst. Cal. Inst. Tech., Pasadena, 1949-51, 56——; vis. asso. prof. astronomy U. So. Cal., Los Angeles, 1963——. Mem. Astron. Soc. Pacific, Am. Astron. Soc., A.A.-A.S., Swiss Math. Soc., Naturforschende Gesellschaft Basel, Sigma Xi. Author: (with F. Zwicky) Catalogue of Galaxies and of Clusters of Galaxies, vols. 1-4, 1961-68; also articles. Research on photometry of galaxies, statistics of galaxies, novae and supernovae, morphological astronomy, astronomic-geodetic methods, spherical and positional astronomy, celestial mechanics, theory of numbers. Home: 426 Heather Heights, Monrovia, Cal. 91016. Office: 1201 E. California Blvd., Pasadena, Cal. 91109.*

HERZOG, Henry Walter, Swiss physician; b. Zurich, Switzerland, Mar. 26, 1920; s. Paul Adolph and Frieda K. (Hotz) H.; M.D., U. Zurich, 1950; student Med. Sch., U. Zurich also U. Geneva (Switzerland), 1939-47; postgrad. U. Davos, U. Paris (France), U. Zurich, U. Basel (Switzerland); m. Dorothy Elizabeth Christ, Oct. 7, 1950. Registrar, Med. U. Clinic Basel, 1955-60, dir. dept. chest diseases, 1960——; faculty U. Basel, 1959——, prof. medicine, 1963——, dir. cardiopulmonary lab., 1953——. Recipient Jubilee award Swiss Soc. Radiology, 1951. Mem. Am., Brit thoracic socs., Schweiz. Gesellschaft der Lungen- und Tuberkuloseärzte, Société Francaise de Pathologie Respiratoire, N.Y. Acad. Scis., Schweiz. Gesellschaft für Innere Medizin, Deutsche Gesellschaft für Lungenfunktionsforschung. Editor: Respiration (Internat. Rev. Thoracic Diseases), 1963——. Research and publs. on new method for treatment of Tb by long lasting intravenous infusion of several tuberculostatics in high dosage, surg. method of treatment of tracheobronchial collapse in pulmonary obstructive emphysema, method for study of distbn. of bronchial flow resistance along bronchial system, theoretical basis and clin. performance of artificial ventilation of lungs with pressure triggered respirators. Home: 18 Passwangstrasse. Office: Med. U. Clinic, Bürgerspital, Basel, Switzerland.*

HERZOG, Leonard Frederick, II, Am. geochemist; b. Syracuse, N.Y., June 16, 1926; s. Leonard Frederick and Alma Elizabeth (Jillson) H.; certificate civil engring. Ore. State Coll., 1946; B.S. in Geology, Cal. Inst. Tech., 1948; Ph.D. in Nuclear Geophysics, Mass. Inst. Tech., 1952; postgrad. Harvard, 1950-52; m. Elisabeth Jean Moffitt, June 18, 1955; children—Leonard Frederick, III, Heather Elisabeth, Clayton Ethan. Vis. scientist dept. terrestrial magnetism Carnegie Inst. Washington, 1950-52; dir. Nuclear Geophysics Lab., Mass. Inst. Tech., 1952-56; faculty Pa. State U., State College, 1956——, asso. prof. geophysics and geochemistry, 1960——; founder Nuclide Analysis Asso., 1954; chmn. bd., pres. Nuclide Corp., State College, Medford, Mass., 1961——. Mem. com. on isotope measurements standards AEC, 1962——. Recipient Free Enterprise Assn. award, 1964. Mem. Geol. Soc. Am., Am. Soc. C.E., Am. Chem. Soc., Am. Inst. M.E., Am. Soc. for Testing and Materials, Am. Phys. Soc., Am. Vacuum Soc., Am. Geophys. Union, Meteoritical Soc., Geochem. Soc. Contbr. articles to tech. jours. Pioneered applications of mass spectroscopy to earth and solar system, rubidium-strontium age determination using isotope dilution and ion exchange columns, chem. analysis for trace constituents by isotope dilution or spark-source mass spectrography; developed techniques and new analytical instrument component; research on meteorites and determination of planetary atmosphere composition by rocket-borne spectrometers, stable isotope tracers in biochemistry especially human metabolism, beryllium dating of sediments, negative ion prodn. by thermionic process; Home: R.D. 1, Box 456, Boalsburg, Pa. 16827. Office: 642 S. College Av., State College, Pa. 16801.*

HERZOG, Richard Franz Karl, physicist; b. Vienna, Austria, Mar. 13, 1911; s. Karl and Aloisia (Hickel) H.; Ph.D. in Physics and Astronomy, U. Vienna, 1933. Came to U. S., 1953, naturalized, 1958. Tit. ao prof. physics, U. Vienna, 1950-53; chief systems research br. Photochemistry Lab., Geophysics Research Directorate, U. S. Air Force Cambridge (Mass.) Research Center, 1953-58; chief scientist space sci. operations GCA Corp., Bedford, Mass., 1958——. Contbr. numerous articles to tech. jours. Devel. and application of ion optics for mass spectrometers, especially for analysis of solid materials. Home: 34 Whipple Rd., Lexington, Mass. 02173. Office: GCA Corp., Bedford, Mass. 01730.*

HESCHL, Richard, Austrian anatomist; b. Wellsdorf, 1824; prof. path. anatomy U. Vienna; also prof., Graz, Austria. Author: Compendium die allgemeinen und speziellen pathologischen Anatomie, 1855. Described transverse temporal convolutions of brain (Heschl's gyrus), 1855; suggested word porencephaly to describe a type of brain cyst, 1859. Died Vienna, May 26, 1881.

HESEHOUS, Nikolaus, Russian chemist; b. Russia, 1845. Prof. Tech. Inst. St. Petersburg. Conducted experiments on the absorption of hydrogen by palladium. Died 1919.

HESLER, Lexemuel Ray, Am. botanist; b. Veedersburg, Ind., Feb. 20, 1888; s. Clinton F. and Laura Iris (Youngblood) H.; A.B., Wabash Coll., 1911, LL.D., 1953; Ph.D. (U. Indsl. fellow), Cornell U., 1914; m. Esther Collins, July 2, 1914. Faculty, Cornell U., 1912-19; faculty U. Tenn., Knoxville, 1919-58, dean, prof. emeritus, 1958—. Mem. Am. Phytopath. Soc., Mycol. Soc. Am., Tenn. Acad. Sci. (pres. 1931). Author: Mushrooms of the Great Smokies, 1960; (with H. H. Whetzel) Manual of Fruit Diseases, 1917; (with A. H. Smith) N.A. Species of Hygrophorus, 1962, N.A. Species of Crepidotus, 1965. Research and publs. on plant pathology and mycology, including numerous studies on agarics of Southeastern U. S., involving discovery and publ. numerous new species. Home: 3600 Timberlake Rd., Knoxville, Tenn. 37920.*

HESLOP-HARRISON, John, botanist; b. Middlesborough, Yorkshire, Eng., Feb. 10, 1920; s. John William and Christian Watson (Henderson) H.-H.; B.Sc., U. Durham (Eng.), 1941, M.Sc., 1947; Ph.D., U. Belfast (Ireland), 1949; D.Sc., U. Durham, 1959; m. Yolande Massey, Sept. 23, 1950; 1 son, John Seymour. Radio officer Ministry Supply, 1941-42; lectr. U. Durham, 1945-46; lectr. botany Queen's U., Belfast, 1946-50, prof., head dept. botany, 1954-60; faculty U. Coll., U. London (Eng.), 1950-54, reader, 1953-54; Mason prof. botany, head dept. botany U. Birmingham (Eng.), 1960-67; prof. botany U. Wis., Madison, 1967—; Brittingham vis. prof. U. Wis., Madison, 1965. Fellow Linnean Soc., Royal Soc., Edinburgh, Inst. Biology; mem. Royal Irish Acad. Author: New Concepts on Flowering Plant Taxonomy, 1953; also numerous articles. Editor: Annals of Botany, 1961-67. Research on growth substances in relation to sex expression in plants, cytoplasmic organelles and wall formation during meiosis and microsporogenesis, role of endoplasmic reticulum in determining pollen wall pattern, photoperiodic and thermoperiodic effects on pollen fertility, mechanisms of apomixis, photoperiodic control of incidence of apomixis, ultrastructure of chloroplast, especially ontogeny of lamellar system, effects of analogues on plastid structure, behavior of plastids in inheritance; genecology especially technique; theory of taxonomy. Home: 1824 Rowley Av., Madison, Wis.*

HESS, Alfred Fabian, Am. physician; b. N.Y.C., Oct. 19, 1875; s. Selmar and Josephine (Solomon) H.; A.B., Harvard, 1897; M.D., Coll. Phys. and Surg. (Columbia), 1901; post-grad. work in Prague, Vienna and Berlin; m. Sara Straus, Oct. 12, 1904; children—Eleanor Straus, Margaret Straus. Practiced in New York, from 1901; prof. clin. pediatrics, Univ. and Bellevue Hosp. Med. Coll. Awarded John Scott medal, 1927, by Franklin Soc. for devising a method of producing a vitamin factor in food by ultra-violet light. Author: Scurvy, Past and Present, 1920; Rickets, Including Osteomalacia and Tetany, 1927. Showed that hemorrhagic tendency in scurvy is due to fragility of capillaries, rather than to abnormal composition of blood. Died Dec. 7, 1933.

HESS, Arthur, Am. anatomist; b. N.Y.C., Feb. 19, 1927; s. David and Anna (Kruger) H.; B.S., U. Ark., 1946, M.S., 1947; Ph.D., Univ. Coll. London, 1949; D.Sc., U. London, 1959; m. Gloria Joy Tomsen, Dec. 25, 1953; children—Douglas Thomas, Elisa Tilda. Faculty, Wash. U. Sch. Medicine, 1951-61, asst. prof. anatomy, 1954-61; asso. prof. physiology U. Utah Coll. Medicine, 1961-67; prof., chmn. dept. anatomy Rutgers Med. Sch., 1967—. Mem. Am. Assn. Anatomists. Research and publs. on fine structure of nerve and muscle-electron microscopy, histochemistry, histology. Home: P.O. Box 93, Franklin Park, N.J. 08823. Office: Dept. Anatomy, Rutgers Med. Sch., New Brunswick, N.J. 08903.*

HESS, Eckhard Heinrich, psychologist; b. Bochum, Germany, Sept. 27, 1916; s. Heinrich Porter and Wilhelmina (Salewski) H.; came to U. S., 1927, naturalized, 1943; A.B., Blue Ridge Coll., 1941; M.A., Johns Hopkins, 1947, Ph.D., 1948, m. Dorothea Burghard-Nawiasky, Sept. 29, 1942. Faculty, U. Chgo., 1948—, prof., 1959—, chmn. dept. psychology, 1963—, dir. W. C. Allee Lab. Animal Behavior, 1960—. Vis. prof. Swarthmore Coll., 1957, U. Cal. at Berkeley, 1958, Stanford, 1965; cons., dir. Perception Lab., Interpub., N.Y.C., 1960—; fellow Center Advanced Studies Behavioral Scis., Palo Alto, Cal., 1955-56. Fellow A.A.A.S.; mem. Am. Psychol. Assn., Internat. Brain Research Orgn., Sigma Xi. Author: (with others) New Directions in Psychology. Research and publs. on critical period in early experience, pupillary changes as indicator of mental activity or interest, animal and human perception. Home: 1151 E. 56th St., Chgo. 60637.*

HESS, Eugene Lyle, Am. molecular biologist; b. Superior, Wis., May 14, 1914; s. Lyle Sidney and Olga (Otteson) H.; B.Ch., U. Minn., 1938; M.S., U. Wis., 1942, Ph.D., 1947; m. Lila A. Huhtala, Dec. 20, 1941; children—Gretchen, Jennifer. Research asso. Rheumatic Fever Research Inst., Northwestern U., Chgo., 1948-57; sr. scientist Worcester Found. for Exptl. Biology, Shrewsbury, Mass., 1957-65; affiliate prof. chemistry Clark U., Worcester, Mass., 1962-65; program dir. metabolic biology NSF, Washington, 1965-66, sect. head molecular biology, 1966—. Mem. Am. Chem. Soc., Am. Soc. Biol. Chemists, A.A.A.S., N.Y. Acad. Sci., Biophys. Soc., Sigma Xi. Research, publs. on biochem. and biophys. properties of blood plasmaproteins and cellular components of streptococci and small lymphocytes; viscosity, electrophoresis, ultracentrifugation, light scattering, interactions of macromolecules.*

HESS, Germain Henri, chemist; b. Geneva, Switzerland, Aug. 7, 1802; student of Berzelius, 1828; became prof. chemistry, St. Petersburg, Russia. Founder of thermochemistry; formulated Hess's law that net heat evolved or absorbed in chem. reactions is constant and not dependent on number of stages, 1840, later shown a result of law of conservation of energy. Died St. Petersburg, Russia, Nov. 20, 1850.

HESS, Harry Hammond, Am. geologist; b. N.Y.C., May 24, 1906; s. Julian S. and Elizabeth (Engel) H.; B.S., Yale, 1927; Ph.D., Princeton, 1932; m. Annette Burns, Aug. 15, 1934; children—George Burns, Frank Deming Mather. Geologist Loangwa Concessions, Ltd., N. Rhodesia, 1928-29; geologist gravity measuring cruises U. S. submarines S-48 and Barracuda, 1931, 36; faculty mem. Princeton since 1934, now professor of geology, curator of mineralogy; exchange prof. U. Cape Town, 1949-50. Chmn. Space Science Bd., 1962. Fellow Geol. Soc. Am. (pres. 1962), Mineral Soc. Am. (pres. 1954-55); mem. Nat. Acad. Sci., Mineral Soc. London, Geol. Soc. S. Africa, Soc. Econ. Geologists, Am. Geophys. Union (pres. sect. geodesy 1951-53, pres. sect. tectonophysics 1956—), NRC (chmn. division of earth sciences, chmn. committee on disposal radioactive waste 1955—). Author articles in science journals. Discoverer greatest depth in oceans, 1945. Research on submarine geology; gravity anomalies island arcs; rock forming minerals; petrology of periodotites. Home: 150 Fitzrandolph Rd., Princeton, N.J.

HESS, Sidney M., biochem. pharmacologist; b. New Brunswick, N.J., Aug. 24, 1916; s. Abraham Hess and Fannie Lipshitz; B.S., U. Ill., 1940; M.S. in Biochemistry, Georgetown U., 1950, Ph.D., 1952; m. Doris Rush, June 7, 1941; children—Arthur, Marjorie. With FDA, 1948-54, Nat. Heart Inst., NIH, 1954-61; director of Biochem. Pharmacology Labs., Squibb Inst. for Med. Research, New Brunswick, 1961—. Hon. prof. dept. biochemistry and microbiology Rutgers U. Coll. Agr. and Environmental Sci., 1962—. Mem. A.A.A.S., Am. Chem. Soc., Am. Soc. Pharmacology and Exptl. Therapeutics, Soc. Exptl. Biology and Medicine, N.Y. Acad. Scis. Research and publs. on absorption, disposition, metabolism, excretion and mode of action of pharmacologic substances. Office: Squibb Inst. for Medical Research, Georges Rd., New Brunswick, N.J. 08903.*

HESS, Victor Francis, physicist; b. Waldstein, Austria, June 24, 1883; s. Vincens and Seraphine (Grossbauer) H.; student U. of Graz, 1901-05, U. of Vienna, 1905-08; Ph.D., 1906; Sc.D., Fordham U., 1946, U. Chgo., 1956; m. Mary Bertha Warner (Breisky), Sept. 6, 1920 (dec. 1955); m. 2d, Elizabeth M. Hoenke, 1955. Prof. physics, university Vienna, Graz and Innsbruck, Austria, 1919-38; chief physicist U. S. Radium Corp., New York and Orange, New Jersey, 1921-23; prof. physics, Fordham U., 1938-56, professor emeritus, 1956—. Recipient (with C. D. Anderson) Nobel prize in physics for discovery cosmic rays, 1936. Fellow Am. Physical Soc., Sigma Xi; member Pontifical Academy of Science Rome, Academy of Sci. in Vienna (life), Am. Geophys. Union. Author: Conductivity of the Atmosphere and Its Causes, 1928; Ionization Balance of the Atmosphere, 1933; Cosmic Rays and Their Biological Effects, 1949. Extensive research in cosmic rays atmospheric electricity, radioactivity and refractive indices of mixtures of liquids; collaborated with R. W. Lawson in determining the number of alpha-particles expelled by a gram of radium. Died Dec. 17, 1964.

HESS, Walter Cohen, Am. biochemist; b. Phila., July 19, 1899; s. Edward and Theresa (Solis-Cohen) H.; B.S., U. Pa., 1920; Ph.D., George Washington U., 1930; m. Jeanette Samuel, June 11, 1922; 1 son, Walter. With E. I. Du Pont de Nemours & Co., 1920-26; chemist USPHS, 1926-31; faculty Georgetown U., Washington, 1931—, prof., chmn. dept. biochemistry, 1945-64, asst. dean med. and Dental Schs., 1957, asso. dean, 1960-63, acting dean Med. Sch., 1963, prof. emeritus, 1964—, grants and contracts adminstr. 1964-66, asst. v.p. for planning, 1966-68; president of Dunbarton College, 1968—. Cons. to USN Grad. Sch., Washington, 1945—; chmn. biochem. sect. Dental Nat. Bd., 1960—. Recipient 175th Anniversary Gold medal for distinguished service Georgetown U., 1965. Mem. Soc. for Exptl. Biology and Medicine (past chmn. Washington sect., counsellor 1958—), Internat. Assn. Dental Research (past chmn. Washington sect., counsellor 1960—), Am. Chem. Soc., Am. Soc. Biol. Chemists. Internat. Assn., Dental Research, Washington Acad. Sci., Phi Beta Kappa, Sigma Xi, Omicron Kappa Upsilon. Author: Biochemistry for Dental Students, 1959. Research and numerous publs. on biochemistry proteins, their structure and metabolism, sulfer containing amino acids; pioneered biochemistry of teeth. Home: 3607 Chesapeake St., N.W., Washington 20008. Office: Dunbarton College, Washington 20008.*

HESS, Walter Rudolf, Swiss physiologist; b. Frauenfeld, Switzerland, Mar. 17, 1881; s. Clemen and Gertrud (Fischer Saxon) H.; student medicine univs. Lausanne, Berne, Berlin, Kiel and Zurich, 1900-05; M.D., U. Zurich, 1906; Dr. honoris causa in philosophy, U. Berne, 1933, in medicine U. Geneva, 1944, in science, McGill U., Montreal, Can., 1953, in medicine, Freiburg, Germany, 1960; married to Luise Sandmeyer, 1909; children—Gertrud, Rudolf. Asst. physician and ophthalmologist, 1905-08; oculist, 1906-12; asst. physiology, Zurich, Switzerland and Bonn, Germany, 1913-17; prof. physiology, dir. Physiol. Inst., U. Zurich, 1917-51, ret., 1951. Awarded Marcel Benoist prize (Swiss), 1933; Ludwig Medal, German Soc. for Circulation Research, 1938; Nobel prize for medicine and physiology (with Prof. Egas Moniz), 1949. Mem. or hon. mem. numerous Swiss and fgn. learned socs. Author books, including: Die Regulienung von Blutkreilauf und Atmung, 1932; Beiträge zur Physiologie des Hirnstammes, I, II, 1932, 1938; Die funktionelle Organisation des vegetativen Nervensystemes, 1948; Das Zwichenhirn, 1949, 2d edit., 1953; Diencephalon, 1954; The Functional Organization of the Diencephalon Hypothalamus and Thalamus, 1956; The Biologie of Mind, 1964. Formulated principles of automatic nerve function and of sympathetic and parasympathetic nerve impulses; systematically investigated effects of elec. stimulation on body function at hundreds of different points of midbrain; made topog. analysis of stimulation areas in relation to their function. Home: 6 Via Gabbio, 6612 Ascona, Switzerland.

HESS, Wilmot Norton, Am. physicist; b. Oberlin, O., Oct. 16, 1926; s. Walter Norton and Rachel (Metcalf) H.; B.S. in Elec. Engring., Columbia, 1946; M.S. in Physics, Oberlin Coll., 1949; Ph.D. in Physics, U. Cal. at Berkeley, 1954; m. Winifred Esther Lowdermilk, June 16, 1950; children—Walter Craig, Alison Lee, Carl Ernest. Instr., Oberlin Coll., 1947-48; staff Lawrence Radiation Lab., U. Cal. at Berkeley and Livermore, 1954-58, plowshare div. leader, Livermore, 1959-61; chief theoretical div. Goddard Space Flight Center, NASA, Greenbelt, Md., 1961-67, dir. sci. and applications Manned Spacecraft Center, NASA, Houston, Texas, since 1967—. Fellow Am. Geophys. Union, Am. Phys. Soc.; mem. Sigma Xi. Editor: Introduction to Space Science, 1965; asso. editor Jour. Geophys. Research, 1961—, Jour. Atmospheric Sci., 1962—, Space Sci. Revs., 1962—, Rev. Geophysics, 1963—. Research and publs. on utilization nuclear explosions for indsl. applications, origin terrestrial Van Allen radiation belt, measurement neutron flux in space; developed quantitative theory neutron decay origin inner belt and diffusion protons in outer belt in from solar wind. Home: 1010 Shorewood Dr., Seabrook, Tex. 77586. Office: Manned Spacecraft Center, Houston.*

HESSABY, Mahmoud, Iranian physicist; b. Tehran, Iran., Feb. 25, 1903; s. Abbas and Goharshad Hessaby; B.A., Am. U., Beirut, Lebanon, 1924; Civil Engr., French Engring. Coll., Beirut, 1923; Elec. Engr., E.-S.E., Paris, France, 1925; Ph.D., Sorbonne, Paris, 1927; m. Sedigheh Haeri, Sept. 6, 1949; children—Iraj, Anousheh. Civil engr. Pub. Works Adminstrn., Beirut, Damascus, Syria, 1923-24; elec. engr. French R.R.'s, Paris, 1925-26; civil engr. Pub. Works Ministry, Tehran, 1928-29; prof. physics Nat. Tchrs. Coll., Tehran, 1929-34; prof. physics Faculty Sci., U. Tehran, 1934—, dean Faculty Engring., 1934-36, Faculty Sci., 1942-57; minister edn. Iran, 1950-51; mem. Iranian Senate, 1949-61. Decorated Crown Decoration (Iran); commandeur de la Legion d'Honneur (France). Mem. Am. Phys. Soc., Iranian Acad., Soc. for Ancient Iranian Culture, Iran-Am. Soc. Author: Notre Voie, 1935; Iranian Proper Names, 1940; Physical Optics, 1960; Electromagnetic Theory, 1966; also articles. Proposed theory of infinitely extended elementary particles, modification of Newton's law of gravitation and Coulomb's law of electric field. Home: 8 Charrah Hessaby (Tajrish), Tehran, Iran.

HESSE, Ludwig Otto, German mathematician; b. Königsberg, Germany (now Kaliningrad, USSR), Apr. 22, 1811; pupil of Jacobi, Königsberg; named prof. U. Halle (Germany), 1855, Munich Polytechnic, 1868; also prof., Heidelberg. Author: Vorlesungen über analytische Geometrie, 1861; Analytische Geometrie der Ebene, 1865. Applied newly developed theory of algebraic forms to analytic geometry of curves and surfaces; eponym of Hessian covariant; made classic studies of curves of 3d and 4th orders. Died Munich, Aug. 4, 1874.

HESSELBACH, Franz Caspar, German surgeon, anatomist; b. Hammelburg, Germany, Jan. 27, 1759; ed. Würzburg, Germany; prof. anatomy and surgery, Würzburg. Author: Vollstaendige Anleitung zur zergliederungskunde des menschlichen Koerpers; also anat.-surg. treatise on origin of hernias. Described complete femoral hernia (Hesselbach's femoral hernia), 1798; described Hesselbach's triangle, bounded by rectus abdominis muscle, medial half of inguinal ligament, deep inferior epigastric vessels, 1806; distinguished external from internal inguinal hernias, 1810. Died July 23, 1816.

HESSELTINE, Henry Close, Am. physician; b. Promise City, Ia., May 10, 1901; s. Henry Elmer and Arminda Alice (Close) H.; B.S., U. Ia., 1923, M.D., 1925, M.S., 1928; m. Grayce Chapman, Dec. 19, 1925; children—Glen C., Marvin H. Instr., U. Ia., 1927-30; faculty U. Chgo., 1931——, prof., sec. dept. obstetrics and gynecology, 1948-58, Mary Campau Ryersen prof., 1958-66, prof. emeritus, 1966——. Commr., Joint Commn. on Accreditation of Hosps.; v.p., chmn. Ill. Med. Service. Recipient Ann. prize Central Assn. Obstetricians and Gynecologists, 1932, 33. Mem. A.M.A., (past chmn. sect. on obstetrics and gynecology), Ill. Med. Soc. (past pres.), Chgo. Gynecol. Soc. (past pres.), Am. Gynecol. Soc., Am. Assn. Obstetricians and Gynecologists, Am. Coll. Obstetricians and Gynecologists. Research, numerous publs. on microbiologic disorders of vagina; invented umbilical cord clamp; discovered fungicide for vaginal mycosis; devised one step operation for radical vulvectomy. Home: 5807 S. Dorchester Av., Chgo. 60637. Office: 1836 W. 87th St., Chgo. 60620.*

HESSER, Carl Magnus, Swedish physiologist, physician; b. Stockholm, Sweden, Nov. 23, 1919; s. Carl A. H. and Thyra (Almquist) H.; M.D., Karolinska Inst. Sch. Medicine, 1945, Ph.D. in Physiology, 1949; m. Rose A. M. Barkel, June 10, 1948; children—Helene, Ulf, Christian. Instr., research asst., asst. prof. dept. physiology Karolinska Inst. Faculty Medicine, Stockholm, 1940-52, head dept. naval medicine, 1952——; asso. prof. naval medicine Swedish Nat. Research Com. on Aviation and Naval Medicine, Swedish Med. Research Council, 1963-66, prof., 1966——. Mem. Physiol. Soc. Stockholm, Royal Swedish Soc. Naval Scis. Research and numerous publs. on chem. and nervous control of respiration, pulmonary gas exchange, muscular exercise, breath-holding, body functions at high atmospheric pressure. Home: 46 Artillerigatan, Stockholm, Sweden.*

HESSLAND, Ivar Rudolf, Swedish geologist; b. Grebbestad, Sweden, Apr. 4, 1914; s. Ludvig and Hilma (Alexandersson) H.; D.Sc., Uppsala (Sweden) U., 1943; m. May Ingegerd Hedlund, June 6, 1942; children—Karin, Ingegerd, Görel, Lillemor. Asst. prof. Uppsala U., 1943-50; prof. La. State U., Baton Rouge, 1951-52; prof. Stockholm (Sweden), 1953——, dir. Geol. Inst., 1953——, dean, 1959-60. Decorated knight Order of Pole Star (Sweden). Mem. Royal Swedish Acad. Scis. Author: Hard Rock Geology of Scandinavia (wall map), 1967; (with others) Treatise on Invertebrate Paleontology, 1961; also articles. Founder, Stockholm Contributions in Geology, 1957 editor, 1957——. Research on re-entering of marine shell-bearing faunas to W. Sweden after last Ice-Age, invertebrate faunas and their environment in Tertiary, Cretaceous, Ordovician, and Cambrian deposits; developed marine geophys. research on Swedish shelf; sedimentological investigations. Home: Movägen 9, Enebyberg, Sweden. Office: Geologiska Inst., Kungstensgat. 45, Stockholm, Sweden.*

HESSTVEDT, Eigil, Norwegian meteorologist; b. Oslo, Norway, Sept. 16, 1920; s. Knut Scharnhorst and Asgerd Gunfrid (Ekerholt) H.; cand.real, U. Oslo, 1947, Dr.philos., 1961; m. Mimi Johnson, June 29, 1946; children—Goril, Ola. Coll. tchr., Oslo, 1948-49; states meteorologist, Oslo, 1949-61; asso. prof. U. Stockholm (Sweden), 1961 62; asso. prof., U. Oslo, 1962-65; prof. meteorology, 1966——. Mem. Det Norske Videnskaps-Akademi Oslo, Norsk Geofysisk Forening. Research and publs. on supercooling of water, mother-of-pearl clouds, noctilucent clouds, atmospheric composition in stratosphere and mesosphere, especially oxygen and hydrogen compounds. Home: Borgestadveien 16, Oslo 8, Norway.*

HESTON, Walter Enoch, Am. biologist; b. Lucas, Ia., Aug. 23, 1909; s. George L. and Rosanna (Schnebly) H.; B.S., Ia. State U., 1932; M.S., Mich. State U., 1934, Ph.D., 1936, LL.D., 1958; m. Vivian Mable Janney, June 5, 1937; children—David James, Donald Walter, Thomas Janney. Prof. biology, head dept. McMurry Coll., 1936-38; Nat. Cancer Inst. Research fellow Roscoe B. Jackson Meml. Lab., 1938-40; staff Nat. Cancer Inst., NIH, Bethesda, Md., 1940—, chief Lab. Biology, 1962——, mem. genetics study sec. NIH, 1958. Recipient Mich. State U. Centennial award, 1956; Alessandro Pascoli prize Perugia U., 1965. Fellow A.A.A.S.; mem. Am. Assn. for Cancer Research (past dir.), Genetic Assn. (past pres.), Genetics Soc., Soc. for Exptl. Biology and Medicine, Soc. for Human Genetics (past treas.), Am. Soc. Naturalists, Am. Soc. Exptl. Pathologists, Societa Italiana di Cancerologia (fgn. corr.). Contbg. author: Biology of the Laboratory Mouse, 1941; Mammary Tumors in Mice, 1945; Genetics and Cancer, 1959; Methodology in Mammalian Genetics, 1963. Research and numerous publs. on causation of cancer. Home: 2121 P St., N.W., Washington 20037. Office: Nat. Cancer Inst., NIH, Bethesda, Md. 20014.*

HETÉNYI, Miklós, engr.; b. Debrecen, Hungary, Nov. 5, 1906; s. Géza and Etel (Jakab) H.; Eng. Dipl., U. Tech. Scis., Budapest, Hungary, 1931, Hon. D., 1965; Ph.D., U. Mich., 1936; m. Jeanie G. Ritchie, Nov. 27, 1941; children—Agnes Jeanie, John Gilchrist. Came to U. S., 1934, naturalized, 1943.

Staff, Office of Danube Bridge Design, 1930, 32-34; engr. Hungarian State Rys., 1934; Jeremiah Smith fellow U. Ill., also U. Mich., 1934-36; lectr. Grad. Sch., Horace H. Rackham fellow U. Mich., 1936-37; research engr. Westinghouse Electric Corp., 1937-46; prof. theoretical and applied mechanics Northwestern U., 1946-50, Walter P. Murphy prof. engring. sci., 1950-62; prof. engring. mechanics and structural engring. Stanford, 1962——, chmn., div. engring. mechanics, 1965——. Chmn., U. S. Nat. Com. for Theoretical and Applied Mechanics 1962——. Recipient Hollan prize Hungarian Soc. Engrs., 1935; cited as Distinguished Alumni, U. Mich., 1953. Mem. Soc. for Exptl. Stress Analysis (hon., past pres.), Internat. Union Theoretical and Applied Mechanics (U. S. del. 1957——, mem. assembly 1958——). Author Beams on Elastic Foundation, 1946; Handbook of Experimental Stress Analysis, 1950. Research and numerous publs. on theory of structures, theory elasticity and strength of material; devel. of 3-dimensional photoelastic method of stress analysis. Home: 460 Woodside Dr., Woodside, Cal. 94062. Office: Div. Engring. Mechanics, Stanford, Stanford, Cal. 94305.*

HETTCHE, Hans Otto, German hygienist, chemist; b. Frankfurt/Main, Germany, June 8, 1902; s. Adam and Wilhelmine (Grosch) H.; student univs. Frankfurt, Marburg, Freiburg; Dr. phil., U. Munich, 1926; Dr. med., U. Greitswald, 1932; m. Margarete Adam, July 27, 1937; children—Hildegard, Hans, Helga, Helmut, Hermann. Asst., Inst. Hygiene, U. Konigsberg, 1932-35; faculty U. Munich, 1935-46, U. Hamburg, 1951-63; prof. air hygiene U. Bochum, also dir. Landesoenstalt fur Immissions-und Bodnnutzungs Schultz, Land Nordheim Westfallen, Essen, West Germany, 1963——. Recipient Golden medal Verein Deutscher Ingenieure. Mem. WHO, Bundesgsundheitstrat, others. Author: (with others) Nahrboden u. Farben in Bakteriologie, 1935; Atiologic, Pathogenese, Prophylaxe der Struma, 1954. Mem. editorial staff Stadtehygiene, Internat. Jour. Air and Water Pollution, London, Eng. Research and publs. on bacteriological diagnostic, disinfection and sterilization control of water pollution; cause of Graves disease; control of air pollution. Home: 10 Am Buchenhain, 43 Essen-Heidhausen, West Germany. Office: 160 Est strasse, 43 Essen, West Germany.*

HETTNER, Alfred, German geographer; b. Dresden, Aug. 8, 1859; s. Hermann Hettner; ed. Halle, Bonn, Strasbourg. Prof., U. Heidelberg; explored Andes of Colombia, 1882-84, Andes of Peru and Bolivia, 1888-90, East Asia and India, 1913-14, also Europe and Africa. Author: Grundzüge der Länderkunde, 2 vols., 1907-24; Der Gang der Kultur über die Erde, 1923; Die Geographie, ihre Geschichte, ihr Wesen und ihre Methoden, 1927; Der Kilmate der Erde, 1930; Vergleichende Länderkunde, 4 vols., 1933-35. Stressed relationships among different geog. phenomena and causation in relationships; pioneer in use of modern sci. methods in systematic geography. Died Heidelberg, Germany, Aug. 31, 1941.

HEUBEL, Joseph, French chemist; b. Strasbourg, France, Feb. 27, 1920; s. Henri and Anne (Douvier) H.; Licencié ès Scis., Faculty Scis. Strasbourg, 1942; Docteur ès Scis., Faculty Scis. Paris, 1948; m. Annie Cuniasse, May 24, 1952; children—Pierre-Henri, Veronique, Vincent. Became asst. Faculty Scis. Paris, 1943, chef de travaux, 1953; apptd. maitre de conférences, Faculty Scis., Lille (France) 1957, titular prof., chair mineral chemistry, 1962——, also vice dean. Mem. Chem. Soc. France (pres. Lille sect. 1967——). Research and publs. on derivatives of sulfuric anhydrides (synthesis, reaction mechanisms), reactions of nitrite salts, chemistry of non-aqueous solvents. Home: 234 Bd. Victor Hugo, Lille, France.*

HEUBNER, Johann Otto Leonhard, German pediatrician; b. Mühltroff, Jan. 21, 1843; prof. Leipzig, Berlin, Dresden (all Germany). Author: Lehrbuch der Kinderheilkunde, 2 vols., 1902-06. Accurately described inflammation of intima of cerebral vessels due to syphilis (Heubner's disease), 1874; credited with isolating meningococci from cerebrospinal fluid, 1896; determined (with Max Rubner) caloric requirements of infants and introduced caloric feeding, 1897. Died Dresden, Oct. 17, 1926.

HEUDE, Pierre, zoologist; b. Fougères, France, June 25, 1836; became mem. Soc. Jesus, 1856; ordained, 1867; sent as missionary to China, 1868; founder mus., Zi-ka-wei, nr. Shanghai, China; studied botany and zoology; made series of studies of mammals in China, Japan, Indochina, Philippines, Guam. Died Zi-ka-wei, Jan. 2, 1902.

HEURALT, Hendrik van, see van Heuralt, Hendrik.

HEURNIUS, Johannes (Jan van Heurne), Dutch physician; b. Utrecht, Netherlands, Jan. 25, 1543; ed. U. Louvain, Paris and Padua, M.D. 1571. Practised medicine, Utrecht, 1573; became physician of Prince of Egmont; prof. medicine, U. Leyden, 1581-1601; six times rector of U.; physician to Maurice of Nassau. Author: De Peste Liber; De Febribus Liber. Wrote commentary on Hippocrates, pub. 1609; dealt with special diseases and history of medicine; wrote in defense of alchemy; 1st to give practical demonstrations of anatomy at Leyden. Died Aug. 11, 1601.

HEUSER, Carlos, surgeon; b. Argentina, 1879; credited with introducing iodized oil for contrast medi-

um in X-ray diagnosis of uterine cavity, 1921. Died 1934.

HEUSLER, Karl Friedrich, Swiss chemist; b. Basel, Switzerland, Apr. 27, 1923; s. Friedrich and Elisabeth (Brenner) H.; student Realgymnasium Basel, 1933-42; Ph.D., U. Basel, 1949; postgrad. Harvard, 1949-50; m. Regula Vest, June 14, 1949; children—Andreas, Lucas, Sibylle. With CIBA Ltd., Basel, 1951-63; asst. dir. Woodward Research Inst., Basel, 1963-——. Recipient Ruzicka prize in chemistry Fed. Inst. Tech. Zurich, 1965. Mem. Am., Swiss (sec. 1966-68, prize 1967) chem. socs., Chem. Soc. London. Research, publs. on synthetic studies in steroid field, especially corticoid hormones such as aldosterone, mechanisms of radical reactions and antibiotic synthesis (Cephalosporin C). Home: 40, Im tiefen Boden, 4000 Basel, Switzerland.*

HEUSNER, Ludwig, surgeon; b. Barmeu, Germany, 1846; performed 1st successful operation to close perforated gastric ulcer, in pvt. house, 1892. Died 1916.

HEVELIUS, Johannes (or Hevel, Höwelcke), German astronomer; b. Danzig, Germany, Jan. 28, 1611; s. Abraham and Cordelia (Hecker) H.; ed. Leiden, Netherlands, 1630-31; m. Katharina Rebeschke, 1635 (dec. 1662); m. 2d, Catherina Elisabetha Koopman, 1663; 4 children. Traveled in Eng. and France, 1631-34; settled in Danzig, 1634; admitted to brewer's guild, 1636; elected alderman, 1641; admitted to city council, 1651; built astron. obs., 1641; obs. and library destroyed by fire, 1679; granted pensions by Louis XIV and John III, Sobiesky. Fellow Royal Soc., 1664. Author: Selenographia, 1647; Prodromus cometicus, 1665; Cometographia (1st book on comets exclusively), 1668; Machina coelestis, Part I, 1673, II, 1679; Annus climactericus, 1685; Catalogus stellarum fixarum, 1687; Firmamentum Sobiescianum, 1690; Prodromus astronomiae, 1690. Last great astronomer to make important observations without a telescope; observed eclipses of sun and moon, 1639; made detailed studies of moon, 1642-44; published 1st detailed atlas of lunar surface, thus often called founder of lunar topography; made observations of sunspots, 1642-45; distinguished phases of Mercury, 1644, also studied those of Saturn; undertook review of fixed stars, 1647; began intensive observations of comets, 1652; discovered 4 new comets, 1652, 61, 72, 77; held that comets move in parabolic paths around the sun; observed great comet of Dec. 1680; discovered moon's libration in longitude, 1657; observed transit of Mercury, 1661; observed and named variable star Mira; delineated 7 new constellations; prepared star catalog of about 1500 stars at least as accurate as that of Tycho Brahe. Died Danzig, Jan. 28, 1687.

HEVESY, George Charles de, chemist; b. Budapest, Hungary, Aug. 1, 1885; s. Louis and Baronesse Eugenie (Schosberger) de H.; Ph.D. Freiburg, 1908; D.Sc. (hon.) Cape Town U., 1929, Gandh, 1950; D.Phil. (hon.), U. Uppsala, 1945, U. Freiburg, 1949, U. Copenhagen, 1950, Liege, 1959, London, 1960; M.D. (hon.), Sao Paulo U., Rio de Janeiro U., Turin, 1957, Freiburg, 1958; Jur.D. (hon.), Burlington, Vt.; hon. degree University of Cambridge, England, 1964; m. to Pia Riis, September 24, 1924, children—Jenny, George Louis, Ingrid, Pia. Asst., Tech. High Sch. of Zurich, Switzerland, 1908-10; hon. research fellow U. Manchester, Eng., 1912-14; asso. Inst. Theoretical Physics, U. Copenhagen, Denmark, since 1920, asso. Inst. for Research in Organic Chemistry since 1943; prof. in physical chem., U. of Freiburg, 1926-35; Franqui prof. U. Gandh, 1949-50. Recipient Canizzaro prize, Academy of Sciences, Rome, Italy, 1929; Nobel prize, Stockholm, 1943; Copley medal, Royal Soc., 1949, Faraday medal, 1951; Baily medal London Coll. Physicians, 1952; Silvanus Thompson medal, Brit. Inst. Radiology, 1956; Atoms-for-Peace Award, 1958; Niels Bohr medal, 1961; Chgo. U. Rosenberg medal, 1961. Hon. fellow German Bunsen Soc. and Physiol. Soc., Chem. Soc. London, Japan and Helsingfors, Royal Instn., London; fellow German Order Pour le mèrite; mem. Acad. of Scis. Copenhagen, Gotenburg, Stockholm, Brüssells and Rome, Halle and Heidelberg, Am. Acad. Arts and Sciences, Acad. Sci. India Acad. Pontefic, Fellow, Royal Soc., 1939 Soc. Nuclear Medicine, Med. Soc. Sweden. Author: Das Element Hafnium, 1927; Quantitative Analysis by X Rays, 1932; Praktikum der Rontgenspektroskopischen Analyse, 1932; Radioactive Indicators, 1948; also many papers. With D. Coster, discovered new element, hafnium, 1923; calculated relative abundance of chemical elements in universe; discovered use of radioactive tracers in study of chemical processes, especially in living organisms; research using lead and phosphorus as tracers, 1934——; discovered dynamic state of chemical change in the body; research on separation of isotopes by physical means. Died Freiburg, Germany, July 5, 1966.

HEWES, Leslie, Am. geographer; b. nr. Guthrie, Okla., Feb. 25, 1906; s. Willis and Pearl E. (Gifford) H.; B.A., U. Okla., 1928; Ph.D., U. Cal. at Berkeley, 1940; m. Elma Graham Beary, June 14, 1933; children—Carolyn Louise (Mrs. Daniel John Toft), Robert Willis. Faculty, U. Okla., 1932-45, asso. prof., 1943-45; prof. geography U. Neb., Lincoln, 1945——, chmn. dept., 1946——; Fulbright prof. U. Vienna (Austria) 1958-59. Mem. nat. adv. com. on geography Fulbright Program, 1955-58; mem. com. on geography advice Office of Naval Research, 1957-58. Mem. Assn. Am. Geographers (past mem. council, editorial bd. Annals

1958-62, cited for meritorious contbn. in geography 1965), Internat. Geographic Union, Phi Beta Kappa (pres. Okla. 1945, Neb. 1959-60), Sigma Xi (pres. Neb. 1952-53, counselor 1962—). Research and publs. on Okla. Ozarks as Land of Cherokees, delimination No. wet prairie U. S. and effects its reclamation through drainage, character early woodland and prairie settlement in Central Ia., patterns and causes of wheat failure in Neb. and Central Gt. Plains, role of suitcase farming in Central Gt. Plains, conservation res. and other indicators marginality in dry farming in Central Gt. Plains. Home: 3022 S. 27th St., Lincoln, Neb. 68502.*

HEWETT, Donnel Foster, Am. geologist; b. Irwin, Pa., June 24, 1881; s. George Claude and Hetty Barclay (Foster) H.; student Georgia School of Technology, Atlanta, Ga., 1895-97; Metall. Engr., Lehigh University, 1902; D.Sc. (honorary), 1942; graduate student Yale, 1909-11, Ph.D., in Geology, 1924; m. Mary Amelia Hamilton, Jan. 14, 1909. Asst. in dept. of metallurgy, Lehigh U., 1902-03; mining practice, Pittsburgh, Pa., 1903-09; geologist U. S. Geol. Survey since 1911, in charge section of metalliferous deposits 1935-44, also strategic mineral investigations, 1939-44, research geologist, 1963; research asso. Stanford Univ., Stanford, Cal. Recipient D.S.M., Dept. Interior; Penrose medal Soc. Economic Geologists, 1956; award American Acad. of Achievement, 1965. Mem. Geological Soc. Am. (council 1931-33; v.p. 1935, 45, Penrose medal 1964), Nat. Acad. of Scis., Am. Chem. Soc., Mineral. Soc. Am., Am. Inst. of Mining and Metall. Engrs., Am. Assn. Advancement Sci., American Acad. Arts and Sciences, the Society of Economic Geologists (president 1936), Sigma Xi, Phi Beta Kappa. Author: Anticlines of the Bighorn Basin, Wyo. (with C. T. Lupton), 1916; Geology, Oil and Coal Resources in the Oregon Basin, Meeteetse, and Grass Creek Basin Quadrangles, Wyo., 1926; Geology and Ore Deposits of the Goodspring Quad., Nevada, 1931; Geology and Ore Deposits of the Ivanpah Quadrangle, Nev.-Calif., 1956; also numerous smaller sci. reports on mineralogy and ore deposits. Research on structural geology; economic geology; geology and mineral deposits of the Mojave Desert; metalliferous mineral deposits; mineralogy and genesis of manganese deposits. Home: 360 Everett Av., Palo Alto, Cal. Office: 345 Middlefield, Menlo Park, Cal.

HEWITT, Eric John, chemist; b. Yoxford, Eng., Dec. 4, 1906; s. John and Bessie (Shoulding) H.; came to U. S., 1914, naturalized, 1922; A.B., Amherst Coll., 1927; A.M., Columbia, 1928, Ph.D., 1932; m. Jean White, May 21, 1947; children—Gordon John, Geoffrey Lee. Research asso. Columbia Coll. Phys. and Surg., N.Y.C., 1932-37; research dir. Nat. Oil Prodn. Co., Newark, 1937-45; vice president of Evans Research & Devel. Corp., N.Y.C., 1945—; instr. Hunter Coll., 1943-45. Mem. Soc. Chem. Industry (exec. com. 1964-67), Chemists Club N.Y., Am. Chem. Soc., Am. Inst. Chemists, Inst. Food Technologists, Textile Inst., Sigma Xi, A.A.A.S., Phi Lambda Upsilon, Alpha Chi Sigma. Research and publs. on organic chemistry, enzymes, vitamins, foods, food flavors, biosynthesis. Home: 75 Midland Av., Tarrytown, N.Y. 10591. Office: 250 E. 43d St., N.Y.C. 10017.*

HEWITT, Eric John, English plant physiologist; biochemist; b. London, Eng., Feb. 27, 1919; s. Harry Edward and Blanche (DuRoveray) H.; B.Sc. with 1st class honors, Kings Coll., U. London, 1939, Diploma Edn., 1940; Ph.D., U. Bristol (Eng.), 1947, D.Sc., 1967; m. Eluned Williams, Aug. 26, 1943; 1 son, David Richard. Chemist, Royal Ordnance Factory, 1940-42; plant physiologist Long Ashton (Bristol, Eng.) Research Sta., U. Bristol, 1942—, head plant physiology, 1947—, hon. lectr., 1949—. Adviser on nutrition Nigerian Inst. for Oil Palm Research, 1959—; U.K. dep. rep. trace elements coop. research project FAO, 1959—. Fellow Inst. Biology; mem. Biochem. Soc., Soc. for Exptl. Biology, Assn. Applied Biologists, Phytochem. Group, Photobiology Group, Am. Soc. Plant Physiologists, Japanese Soc. Plant Physiologists, Société Française de Physiologie Végétale. Author: Sand and Water Culture Methods Used in Study Plant Nutrition, 1952; also numerous articles. Exposition and devel. modern methods for study plant nutrition especially trace element nutrition in culture media; research on symptomology of mineral disorders in plants and roles of trace elements especially molybdenum, induced synthesis of enzyme proteins in plants and biochem. processes of nitrate and nitrite reduction. Office: Long Ashton Research Sta., Bristol, U.K.

HEWITT, John Theodore, English chemist; b. Windsor, Eng., Oct. 12, 1868; s. John and Alice Mary Hewitt; student Hartley Instn. (now Univ. Coll.), Southampton, Eng., 1881-84, Royal Coll. Sci., 1884-87; M.A., Cambridge U., 1890; Ph.D., Heidelberg, Germany; D.Sc., London, 1893. Became jr. demonstrator Cambridge U., 1891; in charge lab. Dept. Explosives Supply, 1916-19; prof. chemistry Queen Mary Coll., London U.; cons. chemist; hon. fellow Imperial Coll. Sci.; examiner various univs. Fellow Royal Soc., 1910. Author: Dyestuffs Derived from Pyridine, Quinine, Acridine and Xanthene, 1922; Chemistry of Wine, 1928; also papers. Died 1954.

HEWITT, Peter Cooper, inventor; b. N.Y.C., 1861; s. Abram Stevens and Sarah Amelia (Cooper) H.; hon. Sc.D., Columbia, 1903, Rutgers Coll., 1916; m. Lucy Work; m. 2d, Maryon J. Bruguiere. Dir. N.Y. & Greenwood Lake Ry., Cooper, Hewitt & Co., Midvale Water Co., Hexagon Realty Co., Ringwood Co., Hewitt Realty Co., Lehigh & Oxford Mining Co. First v.p. Naval Consulting Bd., from 1915. Trustee Cooper Union for Advancement of Science and Art. Invented mercury-vapor electric lamp, 1900, mercury-vapor rectifier; discovered fundamental principle of vacuum-tube amplifier (used in radio); experimented with hydro-airplanes and helicopters; (with F. B. Crocker) built a helicopter, 1918. Died Neuilly-sur-Seine, France, Aug. 25, 1921.

HEWITT, William Francis, Jr., Am. biologist; b. Chgo., May 25, 1914; s. William Francis and Ada Alice (Monroe) H.; A.B., Princeton, 1935; M.Sc., U. Chgo., 1937, Ph.D., 1942; m. Dorothy de Moure, June 14, 1941; children—Alice de Moure (Mrs. Robert Seyerle), Meredith Monroe (Mrs. Lee Repassy), Rosalie Godfriaux (Mrs. David R. Farnham), Willa Frances. Member of the faculty U. Wichita, 1942-43; faculty Coll. Osteo. Physicians and Surgeons, Los Angeles, 1943-46, Des Moines, 1952-57, 60-62; lit. scientist Smith Kline & French Labs., Phila., 1946-48; faculty Howard U., 1948-52; with Mead Johnson & Co., Evansville, Ind., 1957-59; pharmacologist Plough, Inc., Memphis, 1959-60; prof. pharmacology Coll. Pharmacy, Ohio No. U., Ada., 1962-65; chief editorial sect. Norwich Pharmacal Co., N.Y., 1965—. Exec. sec. Com. Health and Research, Washington, 1949-51; lectr., cons. to colls., pharm. firms. Fellow A.A.A.S.; mem. Am. Osteo. Hist. Soc., Am. Soc. Zoologists, Am. Med. Writers Assn., Brit. Soc. for History of Sci., Soc. Exptl. Biology and Medicine, Am. Inst. Biol. Scis., Soc. Social Responsibility in Sci. (v.p., councillor), Sigma Xi, Rho Chi, others. Research and publs. on endocrinology and changes in pregnancy, drug effects on blood flow, toxicology; history of sci. and medicine, organ studies. Home: 4 Ft. Hill Park, Oxford, N.Y. 13830. Office: Box 191, Norwich, N.Y. 13815.*

HEWITT, William Lane, physician; b. Hebron, Neb., Nov. 25, 1916; s. William Thomas and Iva Lee (Lane) H.; B.A., U. Cal. at Berkeley, 1938, M.D., 1942. Instr. medicine Boston U. Sch. Medicine, 1946-53; asso. prof. medicine U. Cal. at Los Angeles, 1953-58, prof., 1958—; acting chief medicine Harbor Hosp., Torrance, Cal., 1965-66. Fellow A.A.A.S.; mem. A.M.A., A.C.P., N.Y. Acad. Medicine. Research on infectious diseases, chemotherapy, cardiovascular diseases, pharmacology, antibiotic agts. Home: 11771 Montana St., Los Angeles 90049.*

HEWSON, E(dgar) Wendell, Am. meteorologist; b. Amherst, N.S., Can., July 12, 1910; s. Edgar Ellis and Helen (Bell) H.; B.A., Mt. Allison U., 1932; M.A., Dalhousie U., 1933; M.A., U. Toronto, 1935; D.I.C., Imperial Coll. Sci. and Tech., London, 1937; Ph.D., U. London, 1937; m. Julia Elizabeth O'Brien, Aug. 17, 1935; children—David Garnet, Barbara Elizabeth (Mrs. Douglas M. Foley). Came to U. S., 1948, naturalized, 1956. With Meteorol. Service of Can., Toronto, 1938-48; dir. diffusion project Mass. Inst. Tech., Round Hill Field Sta., South Dartmouth, Mass., 1948-53; prof. meteorology U. Mich., Ann Arbor, 1953—. Cons. to industry, govt. agys. Fellow Royal Meteorol. Soc. (Buchan prize 1939), Royal Soc. Can.; mem. Am. Meteorol. Soc., Am. Phys. Soc. Author: (with R. W. Longley) Meteorology, Theoretical and Applied, 1944; also articles. Research in application of meteorology to solution of air pollution problems, research on aeroallergens. Home: 10 Harvard Pl., Ann Arbor, Mich. 48104.*

HEWSON, William, Brit. surgeon, anatomist; b. Hexham, Eng., Nov. 14, 1739; s. William Hewson; apprenticed to father; pupil of Mr. Jambert, Newcastle, Eng.; m. Mary Stevenson, 1770; 2 children; partner Dr. William Hunter, 1762-71; began lecturing, 1772. Fellow Royal Soc., 1770 (Copley medal 1769). Author: An Experimental Inquiry into the Properties of the Blood, 1771; also works edited by Sydenham Soc., 1846. Contbr. papers to sci. jours. Discovered that fibrinogen (which he called coagulable lymph) is essential to blood clotting, 1768; theorized that white corpuscles derive from thymus and lymph glands and are transformed into red corpuscles in spleen, 1774; studied forms of red corpuscles in various animals. Fatally wounded himself while dissecting, May 1, 1774.

HEXTER, William Michael, Am. biologist; b. Canton, O., Aug. 10, 1927; s. Milton J. and Jean (Bonda) H.; A.A., U. Cal. at Los Angeles, 1947; B.A., U. Cal. at Berkeley, 1949, M.A., 1951, Ph.D., 1953; m. Rachel Ringler, Sept. 9, 1950; children—James, Marla, Karen. Faculty, Amherst (Mass.) Coll., 1953—, prof. biology, 1966—. Fellow A.A.A.S.; mem. Am. Inst. Biol. Scis., Am. Assn. U. Profs., Genetics Soc. Am., Am. Soc. Zoologists, Sigma Xi, Phi Beta Kappa. Research and publs. genetic fine structure, nature gene in Drosophila melanogaster; discovered 1st case of gene conversion in Drosophila. Home: 40 Orchard St., Amherst, Mass. 01002.*

HEY, William, English surgeon; b. Leeds, Eng., Aug. 23, 1736; s. Richard and Mary (Simpson) H.; surgeon's apprentice, Leeds; student St. George's Hosp., London, 1757-59; m. Alice Banks, 1761; 11 children. Promoted found. Leeds Infirmary, 1767; sr. surgeon, 1773-1812; mayor of Leeds, 1787-8, 1901-02. Fellow Royal Soc., 1775; mem. Leeds Lit. and Philos. Soc. (pres. 1783). Author: Observations on the Blood, 1779; Practical Observations in Surgery, 1803; several papers. Devised operation named after him for partial amputation of foot, also spl. saw for certain skull fractures; pioneer in work on internal derangements of knee; 1st to describe infantile hernia; friend of Priestly. Died Mar. 23, 1819.

HEYCOCK, Charles Thomas, Brit. metallurgist; b. Aug. 21, 1858; s. Frederick Heycock; ed. King's Coll., Cambridge; M.A.; m. Caroline Sadler; 1 son, 2 daus. Lectr., asst. tutor King's Coll., Cambridge, elected fellow, 1895; prime warden Goldsmiths' Co., 1922-23; Fellow Royal Soc., 1895 (Davy medal 1920); mem. Brit. Assn. (pres. sect. B 1920). Author (with F. H. Neville) numerous papers on alloys. Died June 3, 1931.

HEYDECKER, Walter, biologist; b. Fürth, Germany, June 5, 1918; s. Hermann and Nelly (Röthler) H.; B.E. in Horticulture, U. Reading (Eng.), 1941; M.Sc. U. Nottingham (Eng.), 1954, Ph.D., 1959; nat. diploma in horticulture with honors, 1948; m. Betty E. G. Cox, Oct. 23, 1943; children—Jane Caroline, Deborah, Judith, David, Benjamin. Foreman math. sect., tech. officer Chase Continnouscloke Manufacture, 1947-51; asst. lectr. horticulture U. Coll. S.W. Eng., 1947-51; lectr. horticulture U. Nottingham, Longhborough, Eng., 1951—. Mem. Internat. Seed Testing Assn. (chmn. vigor test com.), Inst. Biologists, Soc. for Exptl. Biology Assn. Applied Biology, Hort. Edn. Assn. Research and numerous publs. on seed germination, establishment of seedlings in soil (especially vegetables); effect of atmospheric gases, soil composition, water and air requirements for germination; relationship between germination in lab. and capacity for field establishment. Home: 31 Park Rd., Longhborough, Leic., Eng.*

HEYDEN, Francis Joseph, Am. astronomer; b. Buffalo, May 3, 1907; s. Frederick John and Clara Elizabeth (Drescher) H.; M.A. in Philosophy, Woodstock Coll., 1931, S.T.L., 1938; M.A. Harvard, 1942, Ph.D., 1944; D.Sc., Georgetown U., 1964. Chief astronomer Manila Obs., 1931-34; prof. astronomy Georgetown U., Washington, 1945—, dir. radio-TV activities, 1946—. Mem. Internat. Astron. Union, Royal Astron. Soc. Britain and Can., Internat. Union Geodesy and Geophysics, Washington Acad. Sci., Phi Beta Kappa, Research and publs. on photometry of stars and spectroscopy of sun, moon and planets. Home: Georgetown U., Washington 20007. Office: Georgetown Coll. Obs., Washington 20007.*

HEYDON, Sir Christopher, English astrologer; s. Sir William Heydon; m. Mirabel Rivel; several sons, including Sir John; m. 2d, Anne Dodge (dec. 1642); 4 daus., 1 son. Elected mem. English Parliament, 1588. Author: A Defence of Judiciall Astrologie, in Answer to a Treatise Lately Published by M. John Chamber . . . , 1603; Recital of the Caelestiall Apparitions of This Present Trygon (unpublished); An Astrological Discourse with Mathematical Demonstrations, Proving the Powerful and Harmonical Influence of the Planets and Fixed Stars upon Elementary Bodies . . . , 1603, printed 1650. Died 1623.

HEYERDAHL, Thor, Norwegian ethnologist, explorer; b. Larvik, Norway, Oct. 6, 1914; s. Thor and Alison (Lyng) H.; Realartium, Larvik Coll., 1933; grad. student U. Oslo; field study, Polynesia, B.C.; library study U. S., Can., Germany, Norway; Ph.D. (hon.), Oslo U., 1961; m. Liv Coucheron Torp, Dec. 24, 1936; children—Thor, Bjorn; m. 2d, Yvonne Dedekam-Simonsen, Mar. 7, 1949; children—Anette, Marian, Elisabeth. Ethnol. collection and research primitive man, his habits, Polynesia and British Columbia, 1937-40; sci. expdns. Pacific Islands to test theory that inhabitants of these islands partly originated in prehistoric S.A.; to conduct expt., replica of prehistoric Inca balsa-wood raft was fashioned, crew 6 Norwegian scientists, tech. experts, went on-board off Callao, Peru, drifted westward until safely landed on Polynesian Atoll Raroia, Tuamotu Archipelago (direct oversea drift of 4300 miles); meteorol., hydrographic, zool. research carried out during expdn., also tests for Am., Brit. war depts., 1947; prod. authentic film, Kon-Tiki, 1951; leader, organizer Norwegian archaeol. expedition Galapagos, 1953; research Andes region, 1954; prod. film, Galapagos, 1955; leader, organizer Norwegian archaeological expedition Easter Island and the East Pacific, 1955-56. Mem., lectr. Internat. Congress Americanists, Cambridge, 1952, Sao Paulo, 1954, San Jose, 1958, Vienna 1960, Barcelona, Madrid, Sevilla, 1964, Internat. Congress Anthropology and Ethnology, Paris, 1960, Moscow, 1964, Internat. Pacific Scientific Congress, Honolulu, 1961. Founder, bd. mem. Kon-Tiki Mus. Oslo. Recipient Retzius medal Royal Swedish Anthrop. and Geog. Soc., 1950, Vega medal, 1962, Mungo Park medal Royal Scottish Geog. Soc., 1951, U. S. Camera achievement award, 1951; apptd. commander Order St. Olav (Norway), 1951; documentary award Acad. Motion Picture Arts and Scis., 1951; Prix Bonaparte-Wyse, Société de Geographie Paris, 1951; Elish Kane gold medal Geog. Soc. Phila., 1952, Lomonosov medal Moscow U., 1962, Royal Gold medal Royal Geog. Soc., 1964. Fellow N.Y. Acad. Scis.; mem. Belgian, Brazilian, Peruvian, Russian, Swedish Geog. Soc. (hon.), Norwegian Acad. Sci., Norwegian Geog. Soc. (hon.). Author: Paa Jakt Efter Paradiset (Oslo), 1938; Kon-Tiki (Am. edit.), 1950; American Indians in the Pacific: The Theory Behind the Kon-Tiki Expedition, 1952; Archaeological Evidence of pre-Spanish Visits to the Galapagos Islands, 1956; Aku-Aku, The Secret of Easter Island, (Am. edit.) 1958; Reports of the Norwegian Archaeological Expedition to Easter Island and the East Pacific: Vo. 1, Archaeology of Easter Island, 1961, Vol. 2, Miscellaneous Papers, 1965; Sea Routes to Polynesia, 1968; also articles sci. and popular mags. Address: Kon-Tiki Mus. Oslo, Norway.*

HEYFELDER, Johann Ferdinand Martin, surgeon; b. Germany, 1798; author treatise on resections and amputations, 1854, other works; introduced ethyl chloride in anesthesia, 1848. Died 1869.

HEYL, Paul R(enno), Am. physicist; b. Phila., June 30, 1872; B.S., U. Pa., 1894; Ph.D., 1899. Instr., Clinton Inst., Ft. Plain, N.Y., 1896-97; tchr. high schs., Pa., 1898-1910; physicist Comml. Research Co., 1910-20; staff Nat. Bur. Standards, 1920-42. Tyndale fellow, 1894-96; Harrison fellow, 1897-98; recipient Boyden premium Franklin Inst., 1907; Potts medal, 1943; Magellan medal Am. Philos. Soc., 1922. Mem. A.A.A.S., Phys. Soc. Research on gravitation anisotropy in crystals; absolute determination of gravity at Washington; theory of light on hypothesis of 4th dimension; redetermination of gravitation constant. Home: 2800 Ontario Rd. N.W., Washington.

HEYMANN, Hans, biochemist; b. Cologne, Germany, Oct. 31, 1915; s. Bernhard H. and Johanna (Ransohoff) H.; student U. Munich (Germany), 1934-38; M.A., Harvard, 1939, Ph.D., 1941; m. Jeanne Dupree Roquemore, Aug. 16, 1952; children—Hans, Paul C., Andrew D., Brooks T., Mary R., Maia. Came to U. S., 1938, naturalized, 1944. Asst., Rockefeller Found. grantee, Harvard, 1941-46, faculty, 1942-46, research asso., 1949-53; asst. prof. chemistry U. Ore., 1946-49; sr. chemist CIBA Pharm. Co., Summit, N.J., 1953-64, head microbial biochemistry, 1964—. Mem. Am. Chem. Soc. (chmn. biochem. discussion group N.J. sect. 1966), Am. Soc. Biol. Chemists. Research and publs. on aromatic synthesis, chemotherapeutics, reaction mechanisms, steroids, enzymes, microbial cell walls, structure, function, biosynthesis. Home: R.D. 1, Far Hills, N.J. 07931. Office: Morris Av., Summit, N.J. 07901.*

HEYMANN, Walter, physician; b. Brussels, Belgium, Nov. 22, 1901; s. Gustav and Selma (Kaufman) H.; student U. Bonn (Germany) Med. Sch., 1918-23, U. Leipzig (Germany), 1918-19, U. Munich, (Germany), 1919-21; M.D., U. Kiel (Germany), 1923; m. Marion Oberdorfer, Aug. 28, 1939; 1 son, Peter W. Came to U. S., 1933, naturalized, 1939. Faculty, Western Res. U., Cleve., 1933—, prof. pediatrics, 1964—. Mem. A.M.A., Am. Acad. Pediatrics, Am. Pediatric Soc., Soc. for Exptl. Biology and Medicine, Soc. Exptl. Pathology, A.A.A.S., Sigma Xi. Research and publs. on carbohydrate metabolism in infants and children, mode action vitamin D, kidney diseases infants and children. Home: 3060 Woodbury Rd., Cleve. 44120.*

HEYMANS, Corneille J(ean) F(rançois), Belgian physiologist; b. Ghent, Belgium, Mar. 28, 1892; s. Jean-François and Marie-Henriette (Henning) H.; M.D., University of Ghent, 1920; post grad. work, College of France (Paris), Universities of Lausanne and Vienna, University College (London), Western Res. Univ. Med. School; M.D. hon. University of Utrecht, 1938, Univ. of Louvain, 1940, M.D. (hon.), University of Montevideo, 1948; m. Bertha May, Jan. 18, 1921; children—Marie-Henriette, Pierre, Jean, Berthe. Lecturer in pharmacology U. of Ghent, 1923-30, prof. pharmacology since 1930; advanced fellow of Belgian-Am. Ednl. Foundation, New York and Brussels, 1927-28; Herter lecturer, New York U., 1934; Dunham Memorial lecturer, Harvard Univ. Med. Sch., 1937; Hanna Foundation lecturer, Western Res. Med. Sch., 1937; Greensfelder Memorial Foundation lecturer, Chicago, 1937; Purser Memorial Foundation lecturer, Trinity Coll., Dublin, 1939; also lecturer in other large univs. throughout world. Head of med. dept. Belgian Relief Com., 1940-44. Awarded Nobel prize for physiology and medicine (for studies on physiology and pharmacology of respiration), 1938; prizes from Royal Acad. Medicine (Belgium), Pontifica Academia Scientarium, Academie médecine de Paris, Institut de France, Acad. of Science (Bologna), U. of Berne, and others. Mem. Royal Acad. of Medicine of Belgium, Pontificia Academia Scientarium, Physiol. Soc. of Gt. Britain, Soc. for Exptl. Biology and Medicine, Société belge de biologie, Nederlandsche Vereeniging voor Physiologie en Pharmacologie, Vlaamsche Chemische Vereening, Biol. Soc. Paris, French Soc. Endocrinology, Biol. Soc. Barcelona, Biol. Soc. Argentina, Societa Italiana biologia sperimentale, N.Y. Acad. Medicine, Am. Med. Assn., Acad. Medicine of Buenos Aires, Med. Soc. Argentine, Pharmacol. Soc. Argentine, Soc. Biol. Montevideo, Alpha Omega Alpha. Author: Le Sinus Carotidien, 1933. Contbr. articles on physiol. and pharmacol. problems to scientific publs. Showed importance of sinus and aortic mechanisms in regulation of respiration and blood pressure, 1933; developed new physiol. techniques, including cross circulation, a means of passing blood from one animal into organs of another. Home: 142 Zoutelaan, Knokke-Zoute. Office: J. F. Heymans Institute for Pharmacology and Therapeutics, University of Ghent, 3 Albert Baertsoenkaai, Ghent, Belgium.

HEYNE, Elmer George, Am. agronomist; b. Wisner, Neb., Apr. 4, 1912; s. Henry A. and Elise (Von Segren) H.; B.Sc., U. Neb., 1935; M.Sc., Kan. State U., 1938; Ph.D., U. Minn., 1952; m. Marjorie O. Francis, June 10, 1938; children—George, Kathryn, Robert, David. Agronomist, Soil Conservation Service, U. S. Dept. Agr., San Antonio, 1935-36, Manhattan, Kan., 1938-46; grad. asst. Kan. State U., 1936-38, prof., 1946-50, 51—. Fellow Am. Soc. Agronomy (dir. 1956-58); mem. Phytopath. Soc., Genetics Soc., Sigma Xi, Phi Kappa Phi, Gamma Sigma Delta. Contbr. numerous articles to sci. jours. Devel. improved varieties of corn, oats, wheat; inheritance studies in corn, wheat, oats, barley on drought resistance, resistance to diseases and quality. Home: 918 Ratone St., Manhattan, Kan. 66502.*

HEYNS, Ockert Stephanus, S. African obstetrician, gyencologist; b. Paarl, S. Africa, Nov. 27, 1906; s. Johannes Ockardus and Mildred (Botha) H.; M.S., D.Sc., U. Cape Town (S.Africa), London U., Queen's U., Belfast, Ireland, U. Edinburgh (Scotland), U. Manchester (Eng.), U. Witwatersrand, S. Africa; m. Yvonne Veronica Parsons, Aug. 13, 1936; children—John, Anthony Julian. Faculty, Med. Sch., Johannesburg, S. Africa, 1939—, prof., head dept. obstetrics and gynecology, 1947—, dean Faculty Medicine, 1956-58. Recipient Havena prize S. African Acad. Fellow Royal Coll. Obstetricians and Gynecologists, Internat. Coll. Surgeons (internat. ho. of dels.); mem. S. African Acad. Sci. and Arts, Brit., S. African gynecol. socs. Author: Abdominal Decompression, 1963; also articles. Research on female pelvis, uterine action, devel. method of abdominal decompression used to relieve pain in labor, backache and to improve fetal blood flow. Home: Hermanus, Cape Province, S. Africa. Office: Med. Sch., Hospital St., Johannesburg, S. Africa.*

HEYROVSKY, Jaroslav, Czechoslovakian physicochemist; b. Prague, Czechoslovakia, Dec. 20, 1890; s. Leopold and Clara (Hanl) H.; Ph.Dr., U. Prague, 1918; D.Sc., U. Coll., London (Eng.) U., 1921; Dr.-ChemSci. (h.c.), U. Warsaw (Poland), 1950; Dr.rer.-nat. (h.c.), U. Dresden (Germany), 1955, U. Aix-Marselles (France), 1959, U. Paris (France), 1960, Charles U., Prague, 1965; Dr.med.h.c., U. Vienna (Austria), 1965; Dr.phil.nat.(h.c.), J. W. Goethe, U. Frankfurt/Main, Germany, 1966; m. Mary Koránová, Feb. 22, 1926; children—Jitka (Mrs. Charles Cerny), Michael. Faculty sci. Charles U., 1919-54, prof. phys. chemistry, 1926-54, dir. Polarographic Inst. (Czechoslovak Acad. Scis., 1950, 52-63; academician Czechoslovak Acad. Scis., 1952. Carnegie Vis. prof. U. Cal. at Berkeley, 1933. Recipient Nobel prize in chemistry, 1959; state prize 1st grade, Prague, 1951; Gold medal Czechoslov. Acad. Scis., 1962. Hon. mem. Am. Acad. Arts and Scis., Indian, Polish, Warsaw, Danish, Hungarian acads. scis., Chem. Soc. London, others; fgn. mem. Royal Soc. Author: Use of Polarographic Method in Practical Chemistry, 1933; Polarographie, 1941; Polarographisches Praktikum, 1948; (with P. Zuman) Introduction to Polarography, 1950; (with J. Kuta) Principles of Polarography, 1962; (with R. Kalvoda, J. Forejt) Oscillographic Polarography, 1953; (with Kalvoda) Oscillographic Polarography with Alternating Current, 1960; also numerous articles. Research on electrochemistry; invention, devel. polarographic and oscillo-polarographic methods. Died Mar. 27, 1967.

HEYTESBURY, William of, Brit. logician, mathematician; flourished 1340; became fellow Merton Coll., Oxford, 1330, 38-39; possibly became original fellow Queen's Coll., Oxford U., 1340, probably became chancellor Oxford U., 1371. Author: Regulae solvendi sophismata; Probationes conclusionum; Sophismata. Analyzed and defined instananeous velocity; defined uniform acceleration and stated that uniform acceleration is equivalent to uniform motion in which velocity is equal to instantaneous velocity possessed by uniformly accelerating body at middle instant of time in respect to space traversed in a given time (Merton Mean Speed theorem).

HEYTING, Arend, Dutch mathematician; b. Amsterdam, Netherlands, May 9, 1898; s. Johannes and Clarissa (Kok) H.; student U. Amsterdam, 1916-22, dr.'s degree, 1925. Lector, U. Amsterdam, 1938-48, prof. math., 1948—. Mem. Koninklijke Nederlandse Akademie van Wetenschappen, Orde van de Nederlandse Leeuw, Assn. for Symbolic Logic. Author: Mathematische Grundlagenforschung, 1934; Intuitionism, an Introduction, 1956; Axiomatic Projective Geometry, 1963; also articles. Elucidated principles of intuitionistic math., according to ideas of L. E. J. Brouwer; reconstrn. of math. in agreement with these principles. Home: 1 Prinses Margrietstraat, Castricum, Netherlands. Office: 121 Nieuwe Achtergracht, Amsterdam, Netherlands.*

HIÄRNE, (Hjarne), Urban, Swedish physician, chemist, mineralogist; b. Skworitz, Sweden, 1641; s. Erlandus Jonae Hiärne; student Uppsala (Sweden) U., London; M.D., U. Angers (France), 1673; postgrad. in chemistry, France, 1673-74; returned to Sweden, 1764; practiced medicine, Stockholm; chmn. Collegium Medicum; v.p. Bd. Mines; head Laboratorium Chemicum. Named Archiater (highest honor for Swedish physicians). Fellow Royal Soc., 1669. Author: De obstructione lacteorum vasorum; Xylobalsamus artificialis; De duplici formicarum sale. Produced experimentally substitutes for wine and oil, also materials to protect iron from rust, corpses from decay; discovered balsam said to protect wood from rot for many years; investigated formic acid; 1st Swedish physician to study hydrotherapy. Died 1724.

HIATT, Howard H., Am. physician; b. Patchogue, N.Y., July 22, 1925; s. Alexander and Dorothy (Askinas) H.; M.D. cum laude, Harvard, 1948; m. Doris Bieringer, Nov. 29, 1947; children—Jonathan, Deborah, Frederick. Faculty, Harvard Med. Sch., Boston, 1955—, Herrman L. Blumgart prof. medicine, 1963—; mem. staff Beth Israel Hosp., Boston, 1955—, physician-in-chief dept. medicine, 1963—. American Cancer Soc. scholar, 1958-59; Lederle fellow, 1959-62; Commonwealth Fund Traveling fellow, 1960-61. Diplomate Am. Bd. Internal Medicine, Nat. Bd. Med. Examiners. Member American Soc. Biol. Chemists, Am. Soc. for Clin. Investigation, Assn. Profs. Medicine, Assn. Am. Physicians, Am. Acad. Arts and Scis., Am. Fedn. for Clin. Research, A.A.A.S., N.Y. Acad. Scis., Soc. for Exptl. Biology and Medicine, Am. Assn. for Cancer Research, Mass. Med. Soc., Alpha Omega Alpha. Studies of mineral metabolism and parathyroid function in man, intermediary carbohydrate metabolism in animals and man, control mechanisms in normal and neoplastic mammalian tissues. Home: 22 Hyslop Rd., Brookline, Mass. 02146. Office: 330 Brookline Av., Boston 02215.*

HIATT, Robert Worth, Am. zoologist; b. San Jose, Cal., Dec. 23, 1913; s. Elwood B. and Berneice (Bane) H.; A.B., San Jose State Coll., 1936; Ph.D., U. Cal. at Berkeley, 1941; m. Elizabeth A. Matthews, July 17, 1938; children—Judith L. (Mrs. Roderick Tuttle), Gerald A., William R. Faculty, Mont. State Coll., 1941-43; faculty U. Hawaii, Honolulu, 1943—, sr. prof. zoology, 1954—, chmn. dept. zoology and entomology, 1946-55, dean Grad. Sch., dir. research, 1955-63, v.p. acad. affairs, 1963—. Head upland game bird investigation Mont. Fish and Game Commn., 1941-43; dir. Eniwetok Marine Biol. Lab., AEC, 1952—; biologist Bikini Sci. Resurvey, 1947; mem. biology panel Office Naval Research, 1950-56, sci. liaison officer, U. S. embassy, London, Eng., 1957-58; cons. marine biology divisional com. for biol. and med. scis. NSF, 1956—, mem. adv. council office instl. programs, 1959-65; mem. com., on marine biology NRC-Nat. Acad. Scis., 1958—; mem. adv. com. facilities and spl. programs div. biol. and med. scis. NSF, 1965—; mem. panel experts fisheries div. FAO, UNESCO, 1962—. Fellow A.A.A.S.; mem. Am. Fisheries Soc., Am. Ornithol. Union, Am. Soc. Ichthyologists and Herpetologists, Am. Soc. Limnology and Oceanography, Am. Zool. Soc., Ecol. Soc. Am., Wildlife Soc., Sigma Xi. Author: Directory of Hydrobiological Laboratories and Personnel in North America, 1954; World Directory of Hydrobiological and Fisheries Institutions, 1963; also numerous articles. Research on upland game birds, marine biology, fate of radioactive fission on marine organisms of Pacific, European marine stations. Home: P.O. Box 234, Waimanalo, Hawaii 96795. Office: 2444 Dole St., Honolulu 96822.*

HIBBARD, Angus Smith, Am. elec. engr.; b. Milw., Feb. 7, 1860; s. William B. and Adaline H.; ed. Racine Coll.; D.Sc., Carleton Coll., Northfield, Minn., 1939; m. Lucile Ray, Dec. 4, 1884; 1 dau., Janet (Mrs. Janet H. Henneberry). Began in railroading; later sec. to gen. supt. Northwestern Telegraph Co.; studied telephony; supt. Wis. Telephone Co., 1881-86; first gen. supt. Am. Telephone & Telegraph Co., 1886-93, inaugurating their long-distance lines; gen. mgr. Chgo. Telephone Co., 1893-1911, later also 2d v.p.; adviser to exec. dept. Am. Telephone & Telegraph Co., N.Y., 1911-16. Mem. Am. Inst. E.E. Invented and patented many improved devices for use with telephone. Designer of Blue Bell sign of telephone, used throughout world. Died Oct. 21, 1945.

HIBBARD, Claude William, Am. vertebrate paleontologist; b. Toronto, Kan., Mar. 21, 1905; s. Charles E. and Evie (Johnson) H.; B.A., U. Kan., 1933, M.A., 1934; Ph.D., U. Mich., 1941; m. Faye Louise Ganfield, Sept. 8, 1934; 1 dau., Katherine Hibbard (Mrs. John Phillip Mull). Wildlife technician Nat. Park Service, Ky., 1934-35; with U. Kan., 1935-37, 38-46, curator, asst. prof., 1941-46; faculty U. Mich., Ann Arbor, 1946—, prof. geology, 1953—, curator fossil vertebrates, 1953—. Fellow Geol. Soc. Am.; mem. Kan. (past pres.) Mich. (past pres.) acads. sci., Mich. Geol. Soc. (past pres.), Soc. Vertebrate Paleontology (past pres.), Soc. Mammalogists, Soc. Ichthyologists and Herpetologists, Am. Assn. Petroleum Geologists, Paleontol. Soc. Research, numerous publs. on extinct vertebrates of plains of N.Am., interpretations of their paleocology and late cenozoic climate; discovery nearly complete Late Cenzoic faunal sequence. Home: 1709 Morton Av., Ann Arbor, Mich. 48104.*

HIBBEN, Frank Cummings, Am. anthropologist; b. Lakewood, O., Dec. 5, 1910; s. Fred M. and Lucy (West) H.; A.B., Princeton, 1933; M.S., U. N.M., 1934; Ph.D., Harvard, 1940; m. Eleanor Brown, June 6, 1936; children—Nora (Mrs. William Liddell), Margaret (Mrs. Tom Bahti), Patrick (dec.). Archaeologist, Ohio State Mus., 1928-29; asst. archaeology and sociology Cleve. Mus., 1930-33; faculty U. N.M., Albuquerque, 1934—, prof., 1952—, dir. Mus. Anthropology, 1959—. Lectr., Archeol. Inst., 1961—; del. Geneva Conf., Dept. State, 1957; chmn. Albuquerque Zoo Bd., 1960—, N.M. Dept. Game and Fish Commn., 1961—. Ford Found. grantee, 1956; NSF grantee, 1961. Author: The Lost Americans, 1946; Prehistoric Man in Europe, 1958; Digging Up America, 1960; also articles. Contbns. include excavations

Sandia cave, delineation of Paleo Indian cultures of Am. S.W.; compilation of N.Am. archaeology translated into major langs. of world; outline of Paleolithic sequences in Africa. Home: 3005 Campus N.E., Albuquerque 87106.*

HIBBERT, Harold, chemist; b. Manchester, Eng., 1877; B.S., Victoria, Eng., 1897, M.S., 1900, D.Sc., 1911; Ph.D., Leipzig, Germany, 1906. Instr. chemistry Wales, 1899-1904; research chemist I. E. DuPont, 1910-14; fellow Mellon Inst., 1914-16; cons., research chemist, 1916-19; asst. prof. chemistry Yale, 1919-21, asso. prof., 1921-25; prof. indsl. and cellulose chemistry McGill U., from 1925. Fellow Inst. Chemistry, London; mem. Chem. Soc., Inst. Textile Research, Chemists Club, Canadian chem. Assn., Textile Inst. Eng., London Soc. Chem. Industry. Research on explosives, tautomerism, cellulose, artificial silk, acetylene derivatives, pharm. products, aliphatic chemistry, pulp and paper, lignin. Died 1945.

HIBBS, John W., Am. animal nutritionist; b. Cleve., June 2, 1917; s. Edwin G. and Stella (Thompson) H.; B.Sc. in Agr., Ohio State U., 1940, M.S., 1941, Ph.D., 1947; m. Marie Maxwell, May 7, 1942; children—David William, Samuel Edwin. Staff, Ohio Agr. Research and Devel. Center, 1943, prof., dept. dairy sci., 1953—; faculty Ohio State U., 1948—, staff Inst. Nutrition and food Tech., 1948—. Recipient Am. Feed Mfg. award for achievement in nutrition and physiol. research, 1962. Fellow A.A.A.S. Mem. Am. Dairy Sci. Assn. (Borden award 1952), Am. Soc. Animal Sci., Am. Inst. Nutrition. Research, numerous publs. on early rumen devel. dairy calves, use vitamin D. in prevention milk fever, thyroidal effects roughages in dairy cattle, neonatal anemia in calves. Home: 185 Cherry Lane. Office: Ohio Agr. Research and Devel. Center, Wooster, O. 44691.*

HIBBS, Russell A., Am. surgeon; b. Birdsville, Ky., Sept. 1, 1869; student Vanderbilt; M.D., U. Louisville, 1890. Practiced medicine, Tex.; intern Polyclinic Hosp., N.Y., 1893-94; resident surgeon N.Y. Orthopedic Dispensary and Hosp., 1894-99, chief surgeon, 1899-1932; founder N.J. Orthopedic Hosp., 1904; prof. orthopedic surgery Coll. Physicians and Surgeons, N.Y.C., from 1918. Used surgery for relief of Tb of joints; devised operation for stabilizing treatment of flail knee, 1907, spinal fusion operation, 1911, also operation to lengthen Achilles tendon; applied fusion technique to Tb of hip, 1926. Died Sept. 16, 1932.

HICETAS, astronomer; b. possibly in Syracuse, Sicily; flourished 370 B.C.; studied eclipses; said to have believed earth rotates on axis; sun, fixed stars remain at rest; acknowledged by Copernicus for providing 1st suggestion for his system.

HICKEY, Joseph James, Am. biologist; b. N.Y.C., Apr. 16, 1907; s. James B. and Sarah (Mooney) H.; B.S., N.Y.U., 1930; M.S., U. Wis., 1943; postgrad. U. Chgo.; Ph.D., U. Mich, 1946; m. Margaret Brooks, June 20, 1942; 1 dau., Susan. With Wis. Soil Conservation Commn., 1941-43, Toxicity Lab., U. Chgo., 1943-44, Mus. Zoology, U. Mich., 1944-46; Guggenheim fellow, 1946-47; faculty dept. wildlife ecology U. Wis., Madison, 1948—, prof., 1958—. Fellow Am. Ornithol. Union (gov.), Linnaean Soc. N.Y. (past pres.); mem. Wildlife Soc., Wilson, Cooper ornithol socs., Ecol. Soc. Am., Nature Conservancy, Wis. Acad. Arts, Sci. and Letters, Wis. Soc. for Ornithology (past pres.). Author: A Guide to Bird Watching, 1943. Editor: Jour. Wildlife Mgmt., 1956-59. Research and numerous publs. on mortality rates of birds, population studies of birds, effect of pesticides on wildlife. Home: 5517 Dorsett Dr., Madison, Wis. 53711.*

HICKMAN, C. J., botanist; b. Eng., Apr. 1, 1914; s. Walter James and Winifred (Sansom) H.; B.Sc. with honors, U. Birmingham, Eng., 1934, M.Sc., 1936, Ph.D., 1946; m. Eveline Chapman, Sept. 5, 1940; children—Michael, Patricia Anne. Asst. adv. mycologist U. Bristol, Eng., 1936-38; plant pathology research officer Agrl. Research Council Eng., 1938-44; faculty U. Birmingham, 1944-60; prof. botany, head dept. U. Western Ont., London, Can., 1960—. Author: (with M. Abercombie, M. L. Johnson) Dictionary of Biology, 1951, 66; also articles. Mem. numerous mycol., plant pathol. socs. in Eng., N.Am. Research on biology and control of fungi causing plant diseases. Home: 624 Santa Monica Rd., London, Ont., Can.*

HICKMAN, Henry Hill, English physician and surgeon; b. 1800. Began practice in Ludlow, Eng., circa 1820; moved to Paris, France, 1828 because his ideas were coldly received in Eng.; however, his request that his ideas be presented to the Acad. of Medicine was denied and Hickman returned to Eng., 1829. Author: A Letter on Suspended Animation, 1824. Rendered animals unconscious by exclusion of air and substitution of carbon dioxide; proved painless operations could be performed with good results, 1824. Died 1830.

HICKMAN, James Blake, Am. chemist; b. Charleston, W.Va., Nov. 29, 1921; s. James Howard and Bessie (Barnsgrove) H.; B.S. in Chemistry, W.Va. U., 1942, M.S., 1943; Ph.D., Pa. State U., 1950; m. Martha Louise Hornor, June 25, 1948. Chemist, Carbide and Carbon Chems. Corp. div. Union Carbide & Carbon Co., South Charleston, W.Va., 1943-45; faculty W.Va. U., Morgantown, 1946—, prof. chemistry, 1962—. Fellow A.A.A.S. (rep. W.Va. Acad. Sci. 1959—), Chem. Soc. (London), mem. Am. Chem. Soc. (past chmn. No. W.Va. sect.), W.Va. Acad. Sci. (sec. 1963—), Photog. Soc. Am., Phi Beta Kappa (pres. W.Va. 1962), Sigma Xi, Phi Lambda Upsilon, Alpha Epsilon Delta, Pi Mu Epsilon, Delta Phi Alpha. Research and publs. on theory of binary liquid mixtures, structure peroxychromic acid, philosophy of sci. Home: 307 Maple Av., Morgantown, W.Va. 26505.*

HICKMAN, Kenneth Claude Devereux, chemist; b. London, Eng., Feb. 4, 1896; s. Claude and Blanche (Edwards) H.; Ph.D. in Photog. Chemistry, Royal Coll. Sci., U. London, 1925; m. Eleanor Whiting, Aug. 19, 1939; children—Elizabeth Blanche (Mrs. Terry Schilling), Bryan Devereux, Margaret Anne. Came to U. S., 1925, naturalized, 1943. Chemist, Eastman Kodak Co., Rochester, N.Y., 1925-39; v.p., dir. research Distillation Products, Inc., 1939-48; cons. chemist, Rochester, 1948—; v.p., tech. dir. Aquastills, Inc., Rochester, 1956—; research prof., dir. distillation research lab. Rochester Inst. Tech., 1960—. Cons. to various indsl. cos. Recipient Best Sci. Apparatus award Rev. Sci. Instruments, 1930, 31, Wetherill medal Franklin Inst., 1950. Fellow Chem. Soc. London, Royal Photog Soc. (Williamson award 1923, life), N.Y. Acad. Scis. (life); mem. Am. Chem. Soc., Am. Vacuum Soc. (life). Research and numerous publs. and patents in photography and recovery of silver, vacuum tech., chem. processing, vitamin oils, distillation, pump fluid and saline water conversion. Home: 136 Pelham Rd., Rochester 14610. Office: 50 W. Main St., P.O. Box 3421, Rochester, N.Y. 14614.*

HICKOX, George Harold, Am. civil engr.; b. Spokane, Wash., Mar. 1, 1903; s. Ryal M. and Mary (Doak) H.; B.E., U. Ia., 1925, M.S., 1926; Ph.D., U. Cal. at Berkeley, 1939; m. Doris E. Miller, June 11, 1925; children—Mary L. (Mrs. Stewart W. McCormick), Ethel J. (Mrs. Roy D. Greenwood), Paul G. Hydraulic engr., dir. hydraulic lab. TVA, Norris, 1935-48, cons., 1948-54; asso. dir. engring. Expt. Sta., U. Tenn., 1948-54; program dir. engring. scis. NSF, Washington, 1954-56; dir. research U. S. Army Engr. Research and Devel. Labs., Ft. Belvoir, Va., 1956-62; staff asst. (research) Boeing Co., Seattle, 1962—. Mem. Am. Geophys. Union, Am. Soc. C.E. (Collingwood prize 1934, James R. Croes medal 1945, James Laurie prize 1949), Am. Soc. M.E., Am. Soc. Engring. Edn., A.A.A.S., Soc. Am. Mil. Engrs., Wash. Acad. Sci., Sigma Xi, Tau Beta Pi. Author: (with M. P. O'Brien) Applied Fluid Mechanics, 1937. Contbr. sects. on hydraulic models to Handbook of Applied Hydraulics, 1942, Colliers Ency., 1951. Research and publs. primarily in field of models, hydraulic friction; developed, refined model techniques particularly for open channel flow and high dams. Home: 9310 Allwood Ct., Alexandria, Va. 22309.*

HICKS, Henry, English geologist, physician, b. St. David's, Eng., May 26, 1837; s. Thomas Hicks; ed. Guy's Hosp.; M.D., St. Andrews U.; m. Mary Richardson, 1864. Practiced medicine specializing in mental diseases; physician, St. David's, 1862-71, later at Hendon, Middlesex; Recipient Bigsby gold medal, 1883. Fellow Royal Soc., 1885; mem. Geol. Soc. (pres. 1896-98), Geologists Assn. (pres. 1882-84). Author: On the Lower Lingula Flags of St. David's, Pembrokeshire, 1863; Report on Further Researches in the Lingula Flags of South Wales, 1865; Descriptions of New Fossils from the Cambrian Rocks, 1869, 71, 72; On the Discovery of a Hyaena Den near Laugharne, Carmarthenshire, 1869; On the Pre-Cambrian (Dimetian and Pebidian) Rocks of St. David's, 1877; On the Pre-Cambrian Rocks of West and Central Rossshire, 1880; On the Discovery of some Remains of Plants at the Base of Denbighshire Grits, 1881; On some Researches in Bone Caves in North Wales, 1886; The Geology of North Wales with New Map, 1888; On the Discovery of Mammoth and Other Remains in Endsleigh Street, London, 1892; On the Morte Slates of North Devon and West Somerset, 1896-97. Died Nov. 18, 1899.

HICKS, Samuel Pendleton, Am. physician; b. Bryn Athyn, Pa., Nov. 21, 1913; s. Curtis Kepler and Luelle (Pendleton) H.; A.B., U. Pa., 1936, M.D., 1940; m. Mary Louise Trimmer, May 31, 1941; children—John Trimmer, Michael Gregory. Chief pathologist U. S. Naval Med. Center, Bethesda, Md., 1946-47; asso. prof. pathology Harvard Med. Sch., 1948-62; pathologist New Eng. Deaconess Hosp., Boston, 1948-62; prof. pathology U. Mich. Med. Center, Ann Arbor, 1962—. Cons. NIH, United Cerebral Palsy, Nat. Acad. Sci.-NRC. Recipient Max Weinstein award for outstanding research in cerebral palsy, 1951. Mem. Am. Chem. Soc., A.M.A., Am. Soc. for Exptl. Pathology, Genetics Soc. Am., Am. Soc. Human Genetics, Am. Acad. Neurology. Author: (with Shields Warren) Neuropathology, 1950; also numerous publs. in exptl. pathology of devel., radiation, and nervous system. Home: 1112 Meadowbrook St., Ann Arbor, Mich. 48104.*

HIDA, Kinzo, Japanese physicist; b. Osaka, Japan, Jan. 31, 1922; s. Tatsuzo and Teru (Segawa) H.; B.Sc., Osaka Imperial U., 1944; D.Sc., Kyoto (Japan) U., 1956; m. Atuko Hamamoto, Nov. 2, 1948; children—Takeo, Yasuo. Asst., Kyoto U., 1945-46; faculty U. Osaka Prefecture, Sakai, Japan, 1949—, prof. physics, 1965—; research fellow Cal. Inst. Tech., Pasadena, 1959; vis. asso. prof. U. Cin., 1963-64; lectr. Naniwa Sr. High Sch., Osaka, 1947-49. Mem. Am. Inst. Aeros. and Astronautics, Phys. Soc. Japan, Japan Soc. for Aeros. and Space Scis. Research and publs. on hydro- and aero-dynamics especially high speed flow of compressible fluid with or without magnetic field, flow with detached shock wave in front of blunt body. Home: 462-6 Minami-Noda, Sakai, Osaka, 588, Japan.*

HIDAKA, Koji, Japanese oceanophysicist; b. Miyagi Prefecture, Japan, 1903; grad. Tokyo U., 1940; D.Sc. Began as engr. Kobe Marine Meteorol. Obs.; became prof. Tokyo U., 1942; guest prof. Washington U., St. Louis, Tex. Agrl. and Tech. U. Mem. Japan Oceanographical Soc. (pres.). Author: Science of Sea; also numerous articles. Research on oceanography, oceanophysics, including waves, depth of ocean currents.

HIDE, Raymond, geophysicist; b. Doncaster, Eng., May 17, 1929; s. Stephen and Rose (Cartlidge) H.; B.Sc. with 1st class honors, Manchester U., 1950; Ph.D., Cambridge U., 1953; m. Phyllis Ann Licence, Mar. 29, 1958; children—Julia Ann, Stephen James, Kathryn Margaret Rosanne. Research asso. in astrophysics U. Chgo., 1953-54; sr. research fellow Atomic Energy Research Establishment, Eng., 1954-57; lectr. physics U. Durham, Eng., 1957-61; prof. geophysics and physics Mass. Inst. Tech., Cambridge, 1961—. Fellow Am. Acad. Arts and Scis.; mem. Am. Geophys. Union, Am. Meteorol. Soc., Sigma Xi. Research and publs. in exptl. and theoretical geophys. fluid dynamics, including planetary atmospheres with spl. interest in major planets, magnetohydrodynamics, planetary magnetic fields, hydrodynamics of rotating fluids.*

HIESEY, William McKinley, Am. plant biologist; b. Denver, Aug. 21, 1903; s. Alexander and Caroline (Imobersteg) Heusi; B.S., U. Cal. at Berkeley, 1929, Ph.D., 1940; m. Louise Sophia Boogaert, Dec. 22, 1935; children—Ralph William, Elaine Louise, With Carnegie Instn. of Washington, 1926—, staff mem. dept. plant biology, Stanford, Cal., 1939—; prof. Stanford, 1951—. Recipient Pope medal Cranbrook Inst. Sci. 1949. Mem. Am. Soc. Plant Physiologists, Bot. Soc. Am., Am., Western socs. naturalists, Cal. Bot. Soc., A.A.A.S., Am. Inst. Biol. Scis., Soc. for Study Evolution, Am. Soc. Plant Taxonomists, Soc. for Developmental Biology, Phi Beta Kappa, Phi Sigma, Alpha Zeta. Author: (with Jens Clausen and David D. Keck) Experimental Studies on the Nature of Species, 1940, vol. II, 1945, vol. III, 1948, vol. IV, 1958; also articles. Research on relationships between plants, degrees of cytogenetic relationship between species, analysis of species composition, transplant responses of ecol. races of same or closely related species in contrasting environments, studies on comparative physiology of ecol. races and species from contrasting environments. Home: 455 Marion Av., Palo Alto, Cal. 94301. Office: Carnegie Instn. of Washington, Stanford, Cal. 94305.*

HIGASHI, Akira, Japanese physicist; b. Tokyo, Japan, Jan. 20, 1922; s. Arata and Kane (Monma) H.; B.S., Hokkaido U., 1945, D.Sc., 1951; m. Margarett Akiko Shiga, May 1, 1954; children—Mariko, Emiko. Faculty, Hokkaido U., Sapporo, Japan, 1950—, prof. applied physics, 1964—. Contract scientist U. S. Army Snow, Ice and Permafrost Research Lab., Wilmette, Ill., 1955-58; project leader Hokkaido U. Alaskan Glacier Expdn., I, 1960, II, 1964. Mem. Phys. Soc. Japan, Japan Soc. Applied Physics, Am. Geophys. Union, Glaciological Soc., Arctic Inst. N.Am. Author: (in Japanese) Artificial Rain Making, 1954; Glaciers, 1967; also articles. Research on mechanism of frost heaving of ground, desiccation cracking of soil in relation to formation mechanism of patterned ground in arctic area, solid state study on mech. properties of ice single crystal, formation mechanism of large single crystals of ice at temperate glaciers. Home: 426 Teine-Miyanosawa, Sapporo-shi, Hokkaido, Japan.*

HIGASHI, Ken-ichi, Japanese chemist; b. Saitama Prefecture, Japan, 1905; grad. Tokyo U., 1930; D.Sc. Past prof. Japanese Army Arty. and Engring. Sch.; past instr. Tokyo U.; became prof. Hokkaido U., 1944. Recipient prize for effect of solvent on bi-polar efficiency, Japan Sci. Assn. Author: Science of Air Defence; Theoretic Organic Chemistry; Foundation of Structure of Molecule. Research on structure of molecules, including bi-polar efficiency.

HIGBEE, Edward, Am. geographer, agronomist; b. N.Y.C., Dec. 29, 1910; s. R. B. and Celia (Walker) H.; B.A., U. Wis., 1932, M.A., 1938; Ph.D., Johns Hopkins, 1949. Soil conservationist, sr. agronomist U. S. Dept. Agr., 1939-46; asst. prof. Johns Hopkins, 1947-48; vis. asso. prof. Yale, 1950; faculty Clark U. Grad. Sch. Geography, 1951-57; prof. U. Del., 1957-62, U. R.I., Kingston, 1962—. Dir. research 20th Century Fund study of Am.'s Pub. Environment, 1964—; cons. Brookings Instn. Urban Policy Confs., 1964—; mem. Pres.'s Group on Domestic Affairs, 1964—. Mem. Am. Geog. Soc., A.A.A.S., Soil Sci. Soc., Am., Am. Soc. Agronomy, Sigma Xi. Author: The American Oasis—The Land and Its Uses, 1957; American Agriculture—Geography, Resources, Conservation, 1958; The Squeeze—Cities Without Space, 1960; Farms and Farmers in an Urban Age, 1963; also articles. Research on land use and land use policies, rural and urban. Home: Mooresfield, Farm, Kingston, R.I.*

HIGGINS, Bryan, chemist, physician; b. County Sligo, Ireland, perhaps 1737; M.D., Leiden (Netherlands) U.; m. Jane Welland; founder sch. chemistry, Soho, Eng., 1774; disputed with Priestley over priority of air expts., 1775; invited by Czarina Catherine to Russia, circa 1785; asst. in improvement of Muscovado sugar and rum in Jamaica, 1797-99. Author: A Philosophical Essay concerning Light, 1776; Experiments and Observations relating to Acetous Acid, Fixable Air . . . , 1786; Minutes of the Society for Philosophical Experiments and Conversations, 1795. Patentee inexpensive, durable cement, 1779; expressed perceptive observations and views, but had little success as experimentalist. Died 1820.

HIGGINS, Edwin Stanley, Am. biochemist; b. L.I., N.Y., Mar. 12, 1925; s. John Thomas and Nettie (Stein) H.; B.A. cum laude, Alfred U., 1952; Ph.D., Syracuse U., 1956; m. Barbara Jean Wilson, June 20, 1958; children—Elizabeth Jean, James Edwin, Glenn David. Research fellow State U., Syracuse, 1952-56; faculty Med. Coll. Va., Richmond, 1956—, professor biochemistry, 1968—; vis. scientist Univ. Center in Va., Richmond, 1960—. Sigma Xi lectr., 1963. Mem. A.A.A.S., Am. Chem. Soc., Soc. for Exptl. Biology and Medicine, Am. Soc. for Cell Biology, Va. Acad. Sci., Sigma Xi. Editorial bd. Va. Jour. Sci., 1963—; collaborating editor Enzymologia, 1966—. Research and publs. in enzyme chemistry, metabolism of mold fungi, alcohol metabolism, effects ethanol on metabolism of brain.

HIGGINS, Huntly Gordon, Australian physicist, chemist; b. Perth, Western Australia, Jan. 8, 1917; s. Harold Darwin Conway and Cora (Moore) H.; B.Sc. with honors (Hackett scholar), 1939; D. Applied Sci., Melbourne U., 1962; m. Irena Taube, Jan. 8, 1941; children—Peter, Barbara, Matthew. Geologist, Island Exploration Co., New Guinea, 1938; mining geologist, lectr., Kalgoorlie, Western Australia, 1940; meteorologist Royal Australian A.F., 1941-45; with C.S.I.R.O., Melbourne, Australia, 1945—, now sr. prin. research scientist, div. forest products, head, chemistry sect., 1960—. Fellow Inst. Physics, Australian Inst. Physics, Royal Australian Chem. Inst.; mem. Appita (exec. com.), T.A.P.P.I. (paper physics com.), Australian Pugwash Com. (chmn.), Internat. Assn. Sci. Paper Makers (v.p.). Research, publs. on structure and properties of paper. Home: 59 Fellows St., Kew, Victoria, Australia. Office: 69 Yarra Bank Rd., South Melbourne, Victoria, Australia.*

HIGGINS, William, Irish chemist; b. Sligo County, Ireland, 1763; ed. Pembroke Coll., Oxford; chemist Apothecaries' Co. Ireland, 1791-95; apptd. chemist, librarian Royal Dublin Soc., 1795; declined fellowship Royal Soc. Author: A Comparative View of the Phologistic and Antiphlogistic Theories with Inductions (foreshadowed law of multiple proportions, adopted independently by Dalton 1802, and also foreshadowed theory of valency bonds), 1789; Experiments and Observations on the Atomic Theory and Electrical Phenomena, 8 vols., 1814; also articles. Speculated on chem. combination of particles, but did not try to find relative weights of particles; work unknown to Dalton. Died 1825.

HIGGINSON, John, pathologist; b. Belfast, No. Ireland, Oct. 16, 1922; s. William and Elma (Rogers) H.; B.A., U. Dublin (Ireland), 1945, M.B., B.Ch., B.A.O., 1946; M.D., Royal Coll. Physicians, London, Eng., 1949; m. Nan McKee, Nov. 26, 1949; children—Jacqueline, Wendy Ann. Asst. in pathology and bacteriology U. Glasgow, 1947-49; pathologist South African Inst. for Med. Research and Baragwanath Hosp., Johannesburg, 1950-58; Am. Cancer Soc. prof. pathology U. Kan. Med. Center, Kansas City, 1961-66; dir. Internat. Agy. for Research on Cancer, Lyon, France, 1966—. Chmn. geog. pathology com. Internat. Union Against Cancer, 1962—; cons. Armed Forces Inst. Pathology, 1961, WHO, 1958—. Mem. Am. Assn. Pathologists and Bacteriologists, Soc. Pathology and Bacteriology, Am. Assn. for Cancer Research, Sigma Xi. Research in geog. pathology of disease, especially cancer, liver, heart disease; organized cancer surveys on degenerative diseases in primitive communities to establish etiological factors. Office: 16 Av. Maréchal Foch, 69-Lyon (6 éme), France.*

HIGHMAN, Benjamin, pathologist; b. Russia, July 13, 1909; s. Max and Mary (Landis) H.; brought to U. S., 1912, naturalized 1929; A.A., Crane Jr. Coll., 1927; B.S. with honors, U. Ill. Coll. Med., 1930, M.S., 1932, M.D., 1933; m. Helen Wienshienk, May 7, 1939; children—Barbara, Lawrence Marshall. Instr. pathology U. Ill. Coll. Medicine, 1936; with NIH, USPHS, Bethesda, Md., 1941—, med. dir., 1954—, chief sect. path. anatomy Nat. Inst. Arthritis and Metabolic Diseases, 1960—, liaison officer to Armed Forces Inst. Pathology, chief radiopathology div., 1965—. Diplomate Am. Bd. Pathology. Founding Fellow Coll. Am. Pathologists; mem. A.M.A., Am. Assn. Pathologists and Bacteriologists, Am. Soc. Exptl. Pathology, Assn. Mil. Surgeons U. S., Biol. Stain Commn., Histochem. Soc., Internat. Acad. 13, 1909; s. Max and Mary (Landis) H.; brought Pathologists, Maryland Society of Pathologists, Washington Society Pathologists (pres. 1968), Electron Microscope Soc. Am., Soc. Exptl. Biology and Medicine. Research and publs. on calcified epitheliomas and mixed tumors, effects of halogenated hydrocarbons, aerosols, catecholamines, environmental stresses, parasitic diseases, bacterial endocarditis. Home: 5202 W. Cedar Lane. Office: NIH, Bethesda, Md. 20014.*

HIGHMORE, Nathaniel, English physician, anatomist; b. Fordingbridge, Eng., 1613; s. Nathaniel Highmore; ed. Trinity Coll., Oxford; M.D., 1642; practiced medicine, Sherburne, Eng. Author: Corporis humani disquisito anatomica, 1651; History of Generation, 1651. Supported notion of indestructible atoms in human frame (similar to Buffon's organic molecules); accurately described cavity in superior maxillary bone (antrum of Highmore), median partition of testes (Highmore's body). Died Sherburne, Mar. 21, 1685.

HIGHTON, Henry, English inventor; b. Leicester, Eng., 1816; s. Henry Highton; B.A., Queen's Coll., Oxford, 1837, M.A., 1840; asst. master Rugby, 1841-59; prin. Cheltenham Coll., 1859-62. Author: (paper) Telegraphy without Insulation, 1872. Developed uninsulated telegraphy as cheap means of internat. communication; patentee artificial bldg. and paving stone. Died Dec. 23, 1874.

HIGHTOWER, Nicholas Carr, Am. physiologist; b. Nashville, Sept. 26, 1918; s. Nicholas Carr and Iva Catherine (Baxter); B.S., U. No. Tex., 1941; M.D., U. Tex., 1944; M.S., U. Minn., 1949, Ph.D., 1952; m. Don Ann Watson, Sept. 18, 1941; children—Nikki Ann, James Edward, Sandra Sue, Cynthia Kay. Fellow, Mayo Found., 1945-49, U. Minn., 1949-51; staff sect. physiology Mayo Clinic, 1951-52; dir. clin. research Scott and White Clinic, Temple, Tex., 1952—. Cons. gastroenterology Temple VA Hosp., 1952—; lectr. physiology U. Tex. Med. Br. 1953—. Recipient Outstanding Achievement award U. Minn., 1964; Outstanding Alumni award N. Tex. U., 1965. Mem. A.M.A. (chmn. sect. pathology and physiology 1966), Tex. (past chmn. sect. gastroenterology), So. (Seale Harris award 1960) med. assns., Am. Physiol. Soc., Am. Gastroent. Assn., Central Soc. for Clin. Research. Research and numerous publs. in gastrointestinal physiology and clin. gastroenterology. Home: 3208 W. Av. T. Office: 2401 S. 31st St., Temple, Tex. 76501.*

HIGINBOTHAM, William A(lfred), Am. physicist; b. Bridgeport, Conn., Oct. 25, 1910; A.B., Williams Coll., Williamstown, Mass., 1932; student, Cornell Univ., 1932-40; Ph.D. (hon.), Williams Coll., 1963, D.Sc., 1963; m. Julie Ann Burritt, July 9, 1949; children—Julie Eileen, Robin Ann, William. Radar research, Radiation Lab., Mass. Inst. Tech., 1941-43; Manhattan Project, Los Alamos, N.M., 1943-45, head electronics group, 1944-45; chmn. Fedn. of Am. Scientists, Washington, 1946, 50, exec. sec., 1947, vice-chmn., 1948, 51; asso. head, electronics div., Brookhaven Nat. Lab., 1947-51, head instrumentation div., 1951—. Fellow Am. Phys. Soc., Inst. Elec. and Electronic Engrs., Am. Nuclear Soc., A.A.A.S.; mem. Federation of American Scientists (chmn. 1946, 50, 65). Invented Higinbotham scaler circuit; research on thermionic emission, radar, electronics, nuclear instrumentation. Home: 11 N. Howell's Pt. Road., Bellport, N.Y. 11713. Office: Brookhaven National Laboratory, Upton, N.Y. 11973.*

HIGMAN, Graham, English mathematician. b. Louth, Eng., Jan. 19, 1917; s. Joseph and Ethel (Ellis) H.; ed. Balliol Coll., Oxford (Eng.) U., 1934-40; m. Ivah May Treleaven, 1941; children—James, Joseph, Simon, Andrew, Rebecca, Roger. With Meteorol. Office, 1940-46; lectr. U. Manchester (Eng.), 1946-55; reader math. Oxford U., 1955-60, sr. research fellow Balliol Coll., 1958-60, Waynflete prof. pure math. Oxford U., also fellow Magdalen Coll., 1960—. Fellow Royal Soc., 1958. Research and publs. on algebra, especially theory of groups. Home: 64 Sandfield Rd., Oxford, Eng.*

HIGUCHI, Jiro, Japanese chemist; b. Tokyo, Japan, Jan. 23, 1928; s. Shigeji and Sei (Takahira) H.; B.Tech., Tokyo Inst. Tech., 1951; D.Sc., U. Tokyo, 1957; m. Akiko Oka, Mar. 15, 1960; children—Yasuko, Tsuneharu. Research fellow Tokyo Inst. Tech., 1957-65; vis. scholar, tech. staff Bell Telephone Labs., Murray Hill, N.J., 1961-63; asso. prof. Yokohama (Japan) Nat. U. 1965—; vis. fellow Mellon Inst., Pitts., 1966-67. Research, publs. on electronic structures of molecules, especially free radicals, fine structures and hyperfine structures of free radicals. Home: 5-27 3-Chome, Nakameguro, Meguro-ku, Tokyo, Japan. Office: Dept. Chemistry, Yokohama Nat. U., 702 Ohkamachi, Minami-ku, Yokohama, Japan.*

HIGUCHI, Kentaro, Japanese dermatologist; b. Japan, May 14, 1907; s. Yasuto and Yasuyo Higuchi; student Med. Sch., Kyushu U., 1929-34, Dr. degree, 1937; m. Michie Higuchi, Feb. 20, 1934. Prof. dermatology Djakarta Med. Sch., Indonesia, 1943-46, Kurume Med. Sch., Japan, 1947-49; prof. dermatology, dir. dept. dermatology, Kyushu U., Fukuoka, Japan, 1950—, also dir. hosp. Mem. Japanese Dermatol. Assn., others. Research, numerous publs. on dermatology, med. mycology, syphilogy, leprology, cosmetic field. Home: 979 Hamamatsucho (Maedashi), Fukuoka. Office: 1.276 Katakasu, Fukuoka, Japan.*

HIGUCHI, Takeru, Am. pharm. chemist; b. Los Altos, Cal., Jan. 1, 1918; s. Iyekichi and Chiye (Shiki) H.; student San Jose State Coll., 1935-38; A.B. in Chemistry with honors, U. Cal. at Berkeley, 1939; Ph.D. in Phys. and Organic Chemistry, U. Wis., 1943; D.Sc., U. Mich., 1967; m. Aya Toki, Jan. 1, 1944; children—Kenji W., Junji H., Chie Susan, Peter T. Research asso. U. Wis., 1943-44; research chemist U. Akron (O.) Office Rubber Res., 1944-47; faculty U. Wis. Sch. Pharmacy, Madison, 1947-67, prof., 1954-64, Edward Kremers prof. pharm. chemistry, 1964-67; Regents Distinguished prof. chemistry and pharmacy U. Kan., Lawrence, 1967—. External examiner pharmaceutics U. Singapore, 1962-65; cons. div. research grants NIH, 1965—. Recipient Sturmer Lecture award Rho Chi Soc., 1956. Mem. Japanese Pharm. Soc. (hon.), Am. Pharm. Assn. (life, past chmn. sci. sect., Ebert awards 1951, 52, 54, Justin Powers award 1964, Research Achievement award 1962), Am. Chem. Soc. (past chmn. Wis. sect.), Am. Oil Chemists Soc., Chem. Soc. London, Wis. Acad. Sci., Acad. Pharm. Scis. (pres. 1965—). Author: (with Einar Brochmann-Hanssen) Pharmaceutical Analysis, 1961; also numerous articles. Developed applications of phys. chem. principles to pharm. system. Home: 2811 Schwarz Rd., Lawrence, Kan. 66044.*

HIKI, Yoshisato, Japanese physicist; b. Kanagawa Prefecture, Japan, 1893; grad. Tokyo U., 1921; M.D. Became lectr. Tokyo U., 1929; prof. Nippon U., 1933—, apptd. head Hosp., 1947, head univ. med. dept., 1950—. Mem. Japan Sci. Council. Author: Tuberculosis and Allergy. Research on relations between Tb and allergy, salivary glands, beteroplasty of human cancer.

HIKOSAKA, Tadayoshi, Japanese physicist; b. Toyohashi, Japan, Dec. 25, 1902; s. Teiji and Hiro (Sakakura) H.; B.Sc., Tohoku Imperial U., 1926, Ph.D., 1950; m. Kyo Atooda, Dec. 27, 1931; children—Tai, Hiromi (Mrs. Hideyuki Mizutani), Michichika, Naomichi, Akiko, Kiyoko, Masamichi. Prof., Yamaguchi Koto Gakko, 1939-45; prof. Coll. Engring. of Ryojun, Port Arthur, Manchuria, 1945-50; prof. physics Iwate U., Morioka, 1950-51; prof. physics Niigata (Japan) U., 1951-67, senator, 1964-67; dir. Niigata Airglow Obs., 1957-67. Mem. Phys. Soc. Japan, Am. Geophys. Union. Author: (in Chinese) Physics, 1948; also articles. Developed theory of shell structure of atomic nuclei, 1935; discovered type of low-latitudinal, invisible aurora. Home: 21 Asahigaoka, Tagajyo-mati, Miyagi-ken. Office: Tohoku-Gakuin U., Tagajyo-mati, Miyagi-ken, Japan.*

HILBERT, David, German mathematician; b. Königsberg, Germany, Jan. 23, 1862; prof., Göttingen, Germany, 1895-1943. Fellow Royal Soc., 1928. Author: Grundiagen der Geometrie, 1899; (with R. Courant) Methoden der mathematischen Physik, 1924; (with P. Bernays) Grundlagen der Mathematik, 2 vols., 1934-39. Produced theorem of polynomial ideal with finite basis, during early 1890's; perfected analysis of foundations of geometry, 1898-99, thus beginning modern postulational method and foreshadowing abstract math. in gen.; investigated number theory and theory of relative fields; gave existential proof of Waring's conjecture in theory of numbers, 1909; developed Hilbert space in work on integral equations; worked on proving consistency of math. and math. analysis. Died Göttingen, Feb. 14, 1943.

HILD, Walther Johannes, anatomist; b. Wesel, Germany, Nov. 3, 1919; s. Peter and Sophie (Kirchner) H.; student U. Freiburg, Germany, 1941-43; M.D., U. Kiel, 1949; m. Ursula Marianne Schuster, Feb. 12, 1955; children—Susanne Barbara, Peter George. Came to U. S., 1954, naturalized, 1959. Asst. dept. anatomy U. Kiel, 1949-54; Rockefeller Found. fellow U. S., 1952-53; with U. Tex. Med. br., 1954—, prof. anatomy, 1966—. Mem. Am. Assn. Anatomists, Tex. Acad. Sci., Anatomische Gesellschaft, Endocrinologische Gesellschaft. Author: Das Neuron, 1959. Co-editor Ergebnisse der Anatomie und Entwicklungsgeschichte, 1965—. Research and numerous publs. on neurosecretion leading to recognition of physiol. significance of hypothalamic neurosecretion in mammals, in vitro cultivation of central nervous tissue in connection with electrophysiol. expts. involving single neurons and neuroglia cells, discovery of specific electrophysiol. properties of neuroglia cells. Home: 126 Tuna Av., Galveston, Tex. 77550.*

HILDEBRAND, Joel Henry, Am. chemist; b. Camden, N.J., Nov. 16, 1881; s. Howard Ovid and Sarah (Swartz) H.; B.S., U. Pa., 1903, Ph.D., 1906, Sc.D., 1939; LL.D., U. Cal., Berkeley 1954; postgrad. U. Berlin; m. Emily Josephine Alexander, Dec. 17, 1908; children—Louise (Mrs. Ferdind Klein), Alexander, Milton, Roger. Faculty U. Pa., 1907-13; faculty U. Cal. at Berkeley, 1913-52, prof. chemistry emeritus, 1952—. Lectr. various univs. Recipient King's medal Brit. Govt., 1948; William Proctor prize Sci. Research Soc. Am., 1962; Joseph Priestley award Dickinson Coll., 1965. Hon. fellow Royal Soc. Edinburgh; mem. Nat. Acad. Sci. (mem. council), Am. Philos. Soc., Am. Chem. Soc. (William H. Nichols medal 1939, Edn. award 1952, Willard Gibbs medal 1953, James Flack Norris award 1961, Priestley medal 1962, pres. 1955), Am. Inst. Chemists (hon.), Faraday Soc. (hon. life). Author: Science in the Making, 1956; (with R. E. Powell) Principles of Chemistry, 7th edit., 1964; Is

Intelligence Important, 1963; (with R. L. Scott) Solubility of Nonelectrolytes 24, 36, 1950; Regular Solutions, 1962. Theory of regular solutions; research on intense absorption of ultraviolet light by chem. complexes; structure of simple liquids; diffusion in liquids; use of helium-oxygen mixtures in deep diving. Home: 500 Coventry Rd., Berkeley, Cal. 94707.*

HILDEBRAND, Otto, surgeon; b. Berne, Switzerland, Nov. 15, 1858; prof., Basel, Switzerland, also Berlin, Germany; author textbook on gen. surgery and surg.-topog. anatomy; rep. of period of surgery in which anatomy was held to be primary basis. Died Berlin, Oct. 18, 1927.

HILDEBRAND, Roger Henry, Am. physicist; b. Berkeley, Cal., May 1, 1922; s. Joel Henry and Emily (Alexander) H.; A.B. in Chemistry, U. Cal. at Berkeley, 1947, Ph.D., 1951; m. Jane Roby Beedle, May 28, 1944; children—Peter Henry, Alice Louise, Kathryn Jane, Daniel Milton. Physicist, Radiation Lab., U. Cal. at Berkeley, 1942-51, Tenn.-Eastman Corp., Oak Ridge Nat. Lab., 1945; asst. prof. U. Chgo., 1952-55, asso. prof., 1955-60, prof. physics Enrico Fermi Inst. for Nuclear Studies, 1960—, dir., 1965-68; asso. lab. dir. for high energy physics, Argonne (Ill.) Nat. Lab., 1958-64, acting dir. high energy physics div., 1959-64. Chmn. sci. policy com. Stanford (Cal.) Linear Accelerator Center, 1962-66. Recipient Llewellyn John and Harriet Manchester Quantrell award for excellence in undergrad. teaching U. Chgo., 1960; John S. Guggenheim fellowship, 1968-69. Fellow Am. Phys. Soc.; mem. Midwestern Univs. Research Assn. (dir. 1956-59, 62-66), Phi Beta Kappa, Sigma Xi. Demonstrated that neutral pion is psuedoscaler (with B. J. Moyer), principle of conservation of isotopic spin in pion production, developed liquid hydrogen bubble chamber (with D. E. Nagle), made 1st observation of muon capture by free protons, discovered positrons in primary cosmic radiation (with J. A. DeShong and P. Meyer). Home: 5722 S. Kimbark, Chgo. 60637.*

HILDEBRANDT, Albert Christian, Am. plant pathologist; b. State College, Pa., Apr. 10, 1916; s. Albert F. and Emma (Kucera) H.; B.S., Pa. State U., 1939, M.S., 1941; Ph.D., U. Wis., 1945; m. Lillian A. Goggins, May 8, 1954. Fellow, U. Wis., Madison, 1945-49, asst. prof. plant pathology, 1949-46, asso. prof., 1956-60, prof., 1960—. Fellow A.A.A.S.; mem. Am. Phytopath. Soc., Bot. Soc. Am., Tissue Culture Assn., Sigma Xi, Gamma Alpha, Phi Sigma. Research and numerous publs. on bacterial plant diseases, diseases of ornamental plants, insect galls, plant viruses, plant tissue cultures. Home: 1309 Wingra Dr., Madison, Wis. 53715.*

HILDEGARD, Saint (Hildegardis de Pinguia), German physician; b. Böckelheim/Nahe, Germany, 1098; d. Hildebert and Mechtildis; placed with recluse Jutta, circa 1016; became Benedictine nun; founder Rupertsberg convent, 1147; author mystic works, including Scivias, 1141-50, Liber divinorum operum simplicis hominis, 1163-70, also 1st med. writing in Germany, sci. works based on Benedictine and popular traditions: Physica (ency. of natural history), Causae et curae (compendium of pathology and therapeutics). Died Sept. 17, 1179.

HILDEMANN, William Henry, Am. immunogeneticist; b. Los Angeles, Nov. 18, 1927; s. William A. and Elizabeth (Schulz) H.; B.A., U. So. Cal., 1950, M.Sc., 1951; Ph.D., Cal. Inst. Tech., 1956; m. Dorothy Vera Wegner, June 11, 1954; children—Lynn Mary, Lori Ann. USPHS fellow Univ. Coll., London, Eng., 1956-57; faculty U. Cal. at Los Angeles Sch. Medicine, 1957—, prof. immunogenetics, 1966—. Mem. allergy and immunology study sect. NIH, 1963—. NSF Sr. fellow Jackson Lab., Bar Harbor, Me., 1963-64; recipient Lederle Med. Faculty award, 1965—. Mem. Genetics Soc. Am., Am. Assn. Immunologists, Am. Soc. for Microbiology, A.A.A.S., Phi Beta Kappa, Sigma Xi. Editorial bd. Transplantation, 1960—. Research and publs. in immunogenetics and tissue transplantation immunology, comparative studies ontogeny and phylogeny of immunologic responsiveness, transplantation disease mechanisms, immuno-competence of small lymphocytes, immunogenetics of lower vertebrates. Home: 2555 Amherst Av., Los Angeles 90064.*

HILDING, Anderson Cornelius, Am. physician; b. Tacoma, June 29, 1892; s. Gustav Adolph and Anna Lavinia (Tilderquist) Anderson; student U. Wash., 1910-13; B.S., U. Minn., 1915, B.M., 1918, M.D., 1919, M.A., 1922, Ph.D., 1929; D.Sc., Gustavus Adolphus Coll. 1957; m. Inez M. Melander, Jan. 1, 1929; children—David, Cecile (Mrs. Richard E. Swenson), Wendell, Stephen, Jean (Mrs. John Ellstrom). Practice medicine specializing in ophthalmology and otorhinolaryngology, Duluth, Minn., 1921-54; cons., surg. practice in ophthalmology and otology, Duluth, 1954-64; clin. prof. otolaryngology U. Minn. Med. Sch., 1944-60, prof. emeritus, 1960—; vis. research prof. Columbia, 1951-52; dir. research lab. St. Luke's Hosp., Duluth, 1954—. Hill Found. research grantee, 1954-62, NIH grantee, 1962—. Mem. A.C.S., Am. Laryngol. Assn. (v.p. 1948, 60, pres. 1963-64, Casselberry prize, 1934, council 1964—), Am. Acad. Ophthalmology and Otolaryngology (1st v.p. 1963, award for originality 1943, Wherry meml. lectr. 1960), Am. Triological Soc. (council 1953), Collegium Oto-Rhino-Laryngologicum Amicitiae Sacrum (Shambaugh prize 1963), A.A.A.S., A.M.A., Am. Otol. Soc., Assn. for Research in Ophthalmology, Am. Broncho-Esophagological Assn., Pan. Am. Assn. Ophthalmology, N.Y. Acad. Scis., Royal Physiographical Soc. of Lund (Sweden), Sigma Xi, others. Contbr. chpts. to books, numerous articles to profl. jours. Home: 421 36th Av. E., Duluth 55804. Office: Research Lab., St. Luke's Hosp., Duluth, Minn. 55805.*

HILDRETH, Eugene Augustus, Am. internist; b. St. Paul, Mar. 3, 1924; s. Eugena A. IV and Lila K. (Clator) H.; B.S., Washington and Jefferson Coll., 1943; M.D., U. Va., 1947; m. Dorothy Anne Myers, Mar. 26, 1946; children—Jeffrey Reed, William Myers, Anne Sarver, Katherine Clator. USPHS research fellow U. Pa. Hosp., Phila., 1949-51, staff U., chief resident medicine, 1953-54, mem. staff U. Pa., Phila., 1954—. chief allergy and immunology div., dept. medicine, 1954—, asso. prof. medicine, 1960—, asso. dean sch. medicine, 1965—; cons. to VA, 1955—. Cons. U. Taiwan (Formosa), 1951-53; mem. drug panel Nat. Acad. Sci., 1966—. John and Mary R. Markle Found. scholar, 1958-63. Diplomate Am. Bd. Internal Medicine. Fellow A.C.P., Am. Acad. Allergy, Coll. Physicians Phila.; mem. A.A.A.S., Peripatetic Soc., Fedn. Am. Soc. for Exptl. Biology, Am. Clin. and Climatol. Assn., Am. Acad. Allergy, N.Y. Acad. Scis., Internat. Soc. Internal Medicine, Am. Heart Assn., others. Author: (with D. M. Hildreth) Therapy with Low Fat Diets, 1952. Editorial bd. Annals of Internal Medicine, 1960—. Research and publs. on effect of dietary fats on cholesterol content, antibody inhibitory effects of corticosteroid drugs, transplant immunology, use of percutaneous kidney biopsy in diagnosis of kidney disease, renovascular hypertension. Home: 631 Winsford Rd., Bryn Mawr, Pa. 19010. Office: U. Pa. Hosp., 3600 Spruce St., Phila. 19104.*

HILDRETH, Samuel Prescott, Am. physician, historian; b. Methuen, Mass., Sept. 30, 1783; s. Samuel and Abigail (Bodwell) H.; studied medicine with father, then with Dr. Thomas Kittredge, Andover, Mass.; M.D., Med. Soc. of Mass., 1805; m. Rhoda Cook, Aug. 19, 1807, 6 children. Practiced medicine, Hampstead, N.H., 1805, Marietta, O., 1806-61; pres. 3d Med. Conv. of Ohio. Recorded discoveries of curative effect of malaria on epilepsy and value of charcoal and yeast in malignant fevers; wrote med. papers on epidemics, especially great epidemic of 1822-23; as naturalist contbd. articles to Am. Jour. Science, 1826-33, wrote one of earliest papers of presence of petroleum in salt springs; wrote reports of meteorol. observations pub. Smithsonian Contributions to Knowledge, Vol. XVI, 1870. Died Marietta, July 24, 1863.

HILE, Ralph, Am. biologist; b. Plainville, Ind., Mar. 18, 1904; s. John A. and Agnes (Bryn) H.; A.B. summa cum laude, Ind. Central Coll., 1924, LL.D., 1961; Ph.D., Ind. U., 1930; m. Lula Mathews, Jan. 27, 1934. Asst., Bur. Fisheries, 1930-31, asst. aquatic biologist, 1931-37; asso. aquatic biologist U. S. Fish and Wildlife Service, 1937-47, aquatic biologist, 1947-49, supervisory biologist fishery research, 1949-58; asst. lab. dir. Bur. Comml. Fisheries, Ann Arbor, Mich., 1958-61, sr. scientist, 1961-64, tech. editor div. biol. research, 1965—. Research asso. zoology U. Mich., 1948—; councilor Great Lakes Research Inst., 1949-53. Recipient Distinguished Service citation Dept. Interior, 1962. Fellow Am. Inst. Fishery Research Biologists; mem. A.A.A.S., Am. Fisheries Soc., Am. Soc. Ichthyology and Herpetology, Am. Soc. Limnology and Oceanography, Am. Soc. Zoologists, Biometrics Soc., Internat. Limnology Assn., Ind. Acad. Sci., Mich., Wis. acads. sci., arts and letters. Research and numerous publs. on extent and factors fluctuations growth and success reprodn. in stocks fish, population fluctuations in Gt. Lakes, methods treatment comml. fishery statistics. Home: 1307 W. Madison St., Ann Arbor 48103. Office: P.O. Box 640, Ann Arbor, Mich. 48107.*

HILGARD, Ernest Ropiequet, Am. psychologist; b. Belleville, Ill., July 25, 1904; s. George E. and Laura (Ropiequet) H.; B.S., U. Ill., 1924; Ph.D., Yale, 1930; Sc.D., Kenyon Coll., 1964; m. Josephine E. Rohrs, Sept. 19, 1931; children—Henry Rohrs, Elizabeth (Mrs. Jerald W. Jecker). Faculty, Stanford, 1933—, prof. psychology and edn., 1938—, exec. head dept. psychology, 1942-51, dean grad. div., 1951-55; pres., dir. Ann. Revs., Inc., Palo Alto, Cal. Recipient Warren medal in exptl. psychology, 1940. Fellow Brit. Psychol. Assn.; mem. Am. Psychol. Assn. (past pres.), Nat. Acad. Scis., Nat. Acad. Edn., Am. Acad. Arts and Scis., Soc. Exptl. Psychologists. Author: (with Donald G. Marquis) Conditioning and Learning, 1940; Theories of Learning, 1948, rev., 1956, 66; Introduction to Psychology, 1953, rev., 1957, 62, 67; Hypnotic Susceptibility, 1965. Studies of animal and human conditioned responses, and in man, their integration with vol. behavior; human motivation and unconscious processes, in part through exptl. studies of hypnosis. Home: 1129 Hamilton Av., Palo Alto, Cal. 94301. Office: 582 Alvarado Row, Stanford, Cal. 95305.*

HILGARD, Eugene Woldemar, geologist; b. Zweibrucken, Bavaria, Jan. 5, 1833; brought to U. S., 1836; doctorate, Heidelberg, Germany; also studied in Zurich, Switzerland, Freiberg, Germany; hon. degrees Columbia, univs. Miss., Mich., Cal. Became asst. state geologist, Miss., 1856; apptd. chemist Smithsonian Instn., Washington, 1858; conducted the work on cotton prodn. for 10th decennial census, 1880; mem. faculties univs. Miss., Mich., Cal., after 1865. Author: Geology of the Mississippi Delta, 1870; Soils, their Formation, Properties, Composition, and Relation to Climate and Plant Growth in the Humid and Arid Regions, 1906. Introduced sci. methods of cultivating cotton; his geol. work laid found. for later studies of Gulf Coast plain. Died Jan. 8, 1916.

HILL, Albert Vickery, Australian plant pathologist; b. Melbourne, Australia, Apr. 8, 1903; s. Albert V. and Ada (Pollard) H.; B.Agr.Sc., Melbourne U., 1926, M.Agr. Sci., 1936; m. Rosina Lorinda Roe, Apr. 4, 1931; children—Helen (Mrs. Donald Woolley), June (Mrs. Andrew Straw), Patricia (Mrs. Max Grant), Bruce Vickery, Ian Vickery. With Australian Tobacco Investigation, Canberra, 1929-38; research scientist Commonwealth Sci. and Indsl. Research Orgn., Canberra, 1938-66, prin. research scientist, 1949—; cons. Greek govt., 1961, Centre de Cooperation pour les Recherches Scientifiques Relatives au Tabac, 1962. Mem. Australian Inst. Agrl. Sci. Research, publs. on occurrence, devel. and control of all tobacco diseases in Australia and particularly of Peronospora disease. Home: 68 Dominion Circuit. Office: P.O. Box 109, Canberra, A.C.T., Australia.*

HILL, Archibald Vivian, English physiologist; b. Bristol, England, Sept. 26, 1886; s. Jonathan and Ada Priscilla (Rumney) H.; attended Blundell's Sch., Tiverton; M.A., Sc.D., Cambridge (Trinity), Hon. D.Sc., Oxford, Manchester, Pa., Bristol, Algiers Columbia, Liege, Johns Hopkins, University of Brazil, also Rockefeller Institute, Rochester, N.Y.; LL.D., Edinburgh, Belfast; M.D. (honorary), Louvain, Brussels, Toulouse; m. Margaret Neville Keynes, June 18, 1913; children—Mary Eglantyne (Mrs. Humphreys), David Keynes, Janet Rumney (Mrs. Humphrey), Maurice Neville. Prof., physiology, Manchester U., 1920-23; prof. physiology Univ. Coll., London, 1923-25; Foulerton research prof., Royal Soc., 1926-51, ret. (sec., 1935-45, fgn. sec., 1946). Chmn. exec. com., Nat. Physical Lab., 1940-45; com. of award, Commonwealth Fund Fellowships, 1934-39; pres. Soc. Protection of Sci. and Learning. Mem. of Parliament, 1940-45. Sci. adviser, govt. of India, 1943-44. Recipient (with O. Meyerhof) Nobel prize Physiology and Medicine, 1922; Medal of Freedom with Silver Palm (U. S.) for sci. research, World War II. Fellow Royal Soc., 1918; mem. Marine Biol. Assn. United Kingdom (pres. 1955-60), Soc. Vis. Scientists (pres.) Nat. Academy if Sciences, American Philos. Society, Am. Acad. of Arts and Scis., and others. Author: Muscular Activity, 1925; Living Machinery, 1927; Muscular Movement in Man, 1927; The Ethical Dilemma of Science, 1960; Trails and Trials in Physiology, 1965. Contbr. sci. publs. Research yielding quantitative results on oxygen consumption during muscular action; demonstrated that oxygen is required only for recovery phase of activity, not for actual contraction. Home: 16 Bishopswood Rd., Highgate, London, N. 6. Address: Physiology Dept., University Coll., Gower St., London, WC 1, Eng.

HILL, Sir Arthur William, botanist; b. Oct. 11, 1875; s. D. Hill; ed. King's Coll., Cambridge; M.A., Sc.D.; D.Sc., Adelaide. Named fellow Kings' Coll., Cambridge, 1901, dean, 1902, hon. fellow, 1932; univ. lectr. botany, 1905-07; asst. dir. Royal Botanic Gardens, Kew, 1907-22, dir., from 1922; traveled in Iceland, 1900, Andes of Bolivia and Peru, 1903, W.I., 1911, 24, W. Africa, 1921, Australia, New Zealand, Java, Malaya, Ceylon, 1927-28, S. and E. Africa, 1920-31, India, 1937. Recipient Veitch Gold medal Royal Hort. Soc., 1937. Fellow Royal Soc., 1920, Linnean Soc.; mem. Royal Soc. New Zealand (hon.), N.Y. Acad. Scis. (hon.). Contbr. papers to sci. jours. Died Nov. 3, 1941.

HILL, Austin Bradford, English med. statistician; b. London, Eng., July 8, 1897; s. Leonard Erskine and Janet (Alexander) H.; B.Sc., U. London, 1922, Ph.D., 1926, D.Sc., 1929; D.Sc. (hon.) Oxford (Eng.) U., 1963; M.D. (hon.), U. Edinburgh (Scotland), 1968; m. Florence Maud Salmon, July 31, 1923; children—Phyllis Rosemary, Ian David, Peter Anthony. Research worker Indsl. Health Research Bd. and Med. Research Council Gt. Britain, 1923-33; reader epidemiology and vital statistics and epidemiology, 1945-61, dean, 1955-57, prof. emeritus, 1961—; hon. dir. statist. research unit Med. Research Council, 1945-61. Civil cons. in med. statistics Royal Air Force, 1945—, Royal Navy, 1958—. Hon. fellow Faculty Medicine, U. Chile, 1959—; fellow U. Coll. London, 1955—; recipient Gold medal Royal Inst. Pub. Health and Hygiene, 1961, Soc. Apothecaries, 1959; Heberden Soc. medal, 1965. Fellow Royal Soc., 1954, Royal Statis. Soc. (Gold medal 1953), Internat. Statis. Inst.; hon. fellow Royal Coll. Physicians, Inst. Actuaries, Am. Pub. Health Assn., Royal Soc. Medicine (Jenner medal 1965), Soc. Occupational Medicine, Soc. Social Medicine, Soc. Med. Officers of Health. Author: Principles of Medical Statistics, 1937, 8th edit., 1966; Statistical Methods in Clinical and Preventive Medicine, 1962. Research and numerous publs. on epidemiology of smoking and lung cancer, application of statistics to new drug assessment, indsl. diseases. Home: Green Acres, Little Kingshill, Gt. Missenden, Buckinghamshire, Eng.*

HILL, D. C., Canadian nutritional biochemist; b. Guelph, Ont., Can., Jan. 7, 1914; s. Sanford Pierce and Phoebe (Calvert) H.; B.S.A., U. Toronto, 1937, M.S.A., 1939; Ph.D., U. Minn., 1943; m. Kathryn Eilene Mosier, Aug. 20, 1942; children—Douglas Alan, Kathryn Carol. Mem. faculty Ont. Agrl. Coll., U. Guelph, 1937-—, prof. nutrition, 1964-—. Mem. Nutrition Soc. Can., Chem. Inst. Can., Biometrics Soc. Research and numerous publs. on mode of action of antibiotics as growth stimulants, protein and amino acids requirements of birds, factors influencing amino acids levels in blood plasma. Home: 83 Mary St., Guelph, Ont., Can.*

HILL, Eben Clayton, Am. roentgenologist; b. Balt., Oct. 9, 1882; s. Charles Ebenezer and Kate Watts (Clayton) H.; A.B., Johns Hopkins, 1903, M.D., 1907; research student U. Freiburg (Germany), 1904, 05; grad. Army Med. Sch., 1909; m. Carolyn Sherwin Bailey, Oct. 14, 1936. Asst. in anatomy, Johns Hopkins U. Med. Sch., 1907-08; practiced, Balt., 1907-08, Poughkeepsie, N.Y., 1913-20; pathologist and radiologist Vassar Hosp. and Dispensary, 1911-13, roentgenologist, 1912-20; instr. Johns Hopkins, 1920-21, asso. in Roentgenol. anatomy, 1921-22, then lectr. roentgenology. Fellow A.A.A.S.; fellow and life mem. A.C.P., A.M.A. Author: Cross Roads of the Mind, 1939. Contbr. research X-ray technic for studying collateral circulation, sacroiliac injuries and effects of rays on cellular life. Proved the necessity of massive doses of diphtheria antitoxin in laryngeal and other serious cases of diphtheria, 1909; proved the importance of carriers in the spread of diphtheria, and the relative unimportance of disinfection and fumigation, 1910; showed that salvarsan, even in frequent dosages, is not specific in action, and is not the complete curative drug, 1912; completed anat.-surg. studies of sacroiliac joint, 1937; invented radiopaque injection method. Died June 15, 1940.

HILL, Edward Lee, Am. physicist; b. Hartford, Ark., Nov. 3, 1904; s. Robert Lee and Louise (McKnight) H.; B.S. in Elec. Engring., U. Minn., 1925, Ph.D., 1928; m. Irene Ellison, June 8, 1928. Fellow NRC, Harvard, 1928-30; faculty physics U. Minn., Mpls., 1930-37, asso. prof., 1937-46, prof., 1946-62, prof. physics and math., 1962-—. Research physicist Tech. Inst., Leningrad, USSR, 1934-35. Mem. Am. Phys. Soc., Am. Math. Soc., Am. Geophys. Union, I.E.E.E., A.A.A.S., Sigma Xi, Tau Beta Pi. Editor: Phys. Rev., 1950. Research on quantum mechanics, theory of relativity, atmospheric physics. Home: 1921 East River Terrace, Mpls. 55414.*

HILL, Fredric William, Am. nutritionist; b. Erie, Pa., Sept. 2, 1918; s. Vaino A. and Mary (Holmstrom) H.; B.S., Pa. State U., 1939, M.S., 1940; Ph.D., Cornell U., 1944; m. Charlotte H. Gummoe, Apr. 1, 1944; children—Linda Charlotte, James Fredric, Dana Edwin. Head nutrition research sect. Western Condensing Co., Appleton, Wis., 1944-48; faculty Cornell U., 1948-59, prof. animal nutrition and poultry husbandry, 1953-59; prof., chmn. poultry husbandry U. Cal. at Davis and Berkeley, 1959-65, prof., chm. nutrition, Davis, 1965-—, acting asso. dean family and consumer scis., 1965-66. Mem. various subcoms. NRC; commr. Cal. Poultry Improvement Commn., 1959-65; cons. Eli Lilly & Co., 1959-66. Recipient Nutrition award Am. Feed Mfrs. Assn., 1958; Newman Internat. Research award Brit. Poultry Assn., 1959. Danforth Found. fellow, 1938; Guggenheim Found. fellow, 1966. Fellow A.A.A.S.; mem. Am. Chem. Soc., Am. Inst. Nutrition, Poultry Sci. Assn. (Poultry Sci. Research prize 1957, Borden award 1961), World Poultry Sci. Assn., Am. Soc. for Animal Sci., Soc. for Exptl. Biology and Medicine, Am. Inst. Biol. Scis., Nutrition Soc. Gt. Britain, Sigma Xi, Phi Eta Sigma, Phi Kappa Phi, Gamma Sigma Delta, Delta Theta Sigma, Gamma Alpha. Home: 643 Miller Dr., Davis, Cal. 95616.*

HILL, George Andrews, Am. astronomer; b. Elizabeth, N.J., Apr. 11, 1858; s. late Rev. I. N. and Annie M. H.; ed. Columbia Coll.; married; children—George Cooper, Edgar Montgomery (dec.). An astronomer U. S. Naval Obs., from 1893. Spl. work in line of fundamental determination of star positions, determination of longitude by wireless, astronomical constants, variation of latitude. Died Aug. 29, 1927.

HILL, George William, Am. astronomer, mathematician; b. N.Y., Mar. 3, 1838; s. John William and Catherine (Smith) H.; A.B., Rutgers, 1859, A.M., 1862; (hon. Ph.D., Rutgers, 1872; Sc.D., U. of Cambridge, Eng., 1892; LL.D., Columbia, 1894, Princeton, 1896); unmarried. Became asst. in office of The Am. Ephemeris and Nautical Almanac, 1861; lectr. in celestial mechanics, Columbia U., 1898-1901. Researches in connection with the lunar theory secured him gold medal of Royal Astron. Soc., London, 1887; recipient Damoiscan prize, Paris Acad. Sciences, 1898. Asso. fellow Am. Acad. Arts and Sciences; Fellow Royal Soc., 1902. Author: Collected Mathematical Works, vol. 1, 1905. Known for work in celestial mechanics and math.; used infinite determinants for 1st time in analyzing motion of moon's perigree; developed theory of motion of Jupiter and Saturn, theory on effect of moon's motion on the planets. Died Apr. 16, 1914.

HILL, James Peter, embryologist; b. Fifeshire, Scotland, Feb. 21, 1873; ed. Royal Coll. Sci., London; D.Sc., fellow, U. Edinburgh; hon. Sc.D., Dublin, Belfast; m. dau. of John Steele, 1900; 2 daus. Became demonstrator biology U. Sidney, 1892, lectr. embryology, 1904; Jodrell prof. zoology U. London, 1906-21, prof. embryology, 1921-28, emeritus 1938. Recipient Mueller medal Australian Assn. Advancement Sci., 1906, Linnean Gold medal, 1930, Darwin medal, 1940. Fellow Royal Soc., 1913 (Croonian lectr. 1929), mem. numerous sci. socs. Research, publs. on zool., including embryology of Monotremata, Marsupialia, Eutherian mammals. Died May 24, 1954.

HILL, Jerald Everett, Am. physicist; b. Cooper Twp., Mich., Nov. 12, 1907; s. Frank E. and Mary (Shaw) H.; B.S., Western Mich. U., 1929; M.A., U. Mich., 1930; Ph.D., U. Rochester, 1940; m. Dorothy Madelyn Coffeen, Aug. 10, 1938; children—Stephen Coffeen, Douglas Spencer. Instr., Kalamazoo Coll., 1930-37; Westinghouse Research fellow Westinghouse Research Labs., Pitts., 1940-42, research physicist, 1942-43, sect. head, 1943-46; sr. physicist Oak Ridge Nat. Labs., 1946-48; phys. scientist Rand Corp., Santa Monica, Cal., 1948-—. Mem. Am. Phys. Soc. Research, publs. on nuclear reaction kinetics in light elements, slowing down neutrons in water, water-aluminum mixtures and in graphite, effects nuclear weapons, microwave components, theory lunar cratering; inventor microwave components. Home: 14613 Bestor Blvd., Pacific Palisades, Cal. Office: 1700 Main St., Santa Monica, Cal. 90406.*

HILL, John Benjamin, Am. physician; b. N.Y.C., Mar. 8, 1923; s. Maurice Charles and Dorothy (Arkush) H.; student Brown U., 1940-42, U. Ill., 1943-44; B.S., U. Wis., 1945; Ph.D., Columbia, 1950, M.D., 1952; m. Elaine T. MacKenzie, Aug. 21, 1950; children—Christopher M., Charles C., Alison, Cynthia R., Susan P. Instr. pharmacology Columbia, N.Y.C., 1949-50; faculty U. N.C., Chapel Hill, 1952-—, asso. prof. pharmacology, 1961-—. Cons., Technicon Instruments Corp., 1961-64, B-C Remedy Co., Durham, 1963-—. Mem. A.A.A.S., Am. Soc. for Pharmacology and Exptl. Therapeutics. Research and publs. on absorption insulin to glass, microinsulin assay, automated methods for glucose, magnesium, calcium and phenylalanine, alpha and beta anomers glucose in blood mutarotase. Home: 800 Houston Rd., Chapel Hill, N.C. 27514.*

HILL, Julian Werner, Am. chemist; b. St. Louis, Sept. 4, 1904; s. Werner Kamlah and Pearl (Samesreuther) H.; B.S., Washington U., 1924; Ph.D., Mass. Inst. Tech., 1928; LL.D., Juniata Coll., 1966; Sc.D., Kenyon Coll., Lebanon Valley Coll., 1966; m. Mary Louisa Butcher, July 23, 1932; children—Louisa Custer (Mrs. Richard K. Spottswood), Joseph John, Jefferson Borden. With E. I. du Pont de Nemours & Co., Inc., Wilmington, Del., 1928-—, dir., Crystal Trust, 1964-—. Mem. Am. Chem. Soc., A.A.A.S., Franklin Inst. (com. sci. and arts). Research and publs. on organic polymers, including work leading to discovery of nylon, large ring compounds and synthetic musks, cellulose chemistry, fiber research and gen. organic chemistry. Home: 1106 Greenhill Av., Wilmington 19805. Office: DuPont Bldg., Wilmington, Del. 19898.*

HILL, Kenneth Robson, English pathologist; b. Washington, Eng., Apr. 20, 1911; s. Frederick and Lydia (Robson) H.; B.Sc., Kings Coll., London, Eng., 1932; M.B.B.S., Westminster Hosp. Med. Sch., 1937; M.D., London U., 1945; m. Elsie Wade, Sept. 2, 1938; children—David John, Katherine Lydia. Prof. pathology U. W.I., 1949-56, Royal Free Hosp., London, 1956-—. Mem. expert panel on treponematous WHO, 1950-67; med. adv. com. Overseas Devel. Ministry, 1961-—, Brit. Vol. Programme, 1965-—. Fellow Coll. Pathology; mem. Brit. Med. Assn., Coll. Physicians (London), Coll. Surgeons (Eng.). Author: (with Kodjat and Sardadi) Atlas of Franboesia, 1950. Research and numerous publs. on tropical medicine, yaws, liver diseases, leprosy, arterial disease. Home: 12 Aldenham Av., Radlett, Herts., Eng. Office: Royal Free Hospital, London, Eng.*

HILL, Sir Leonard Erskine, Brit. physiologist; b. June 6, 1866; s. George Birkbeck Norman Hill; M.B. (fellow), U. Coll., London, Eng.; LL.D., Aberdeen; m. Janet Alexander, 1891; 4 sons, 2 daus. Dir. research St. John Clinic and Inst. of Phys. Health; dir. research dept. applied physiology Nat. Inst. Med. Research Mt. Vernon, Hampstead; prof. physiology London Hosp. Knighted, 1930. Pres., Assn. San. Inspectors; mem. Naval, Army med. adv. bds. Fellow Royal Soc., 1900; hon. asso. Royal Inst. Brit. Architects. Author: The Physiology and Pathology of the Cerebral Circulation, 1896; Manual of Physiology for Beginners; Sunshine and Open Air; (with A. Campbell) Health and Environment; (with Mark Clement) Common Colds, 1929; Science of Ventilation and Open Air Treatment; also contbr. articles to jours. Editor: Recent and Future Advances of Physiology. Research and publs. on blood circulation, cerebral anemia, caisson disease, diver's palsy; inventor: Katathermometer (for measuring cooling power of air). Died Mar. 30, 1952.

HILL, Mason Lowell, Am. geologist; b. Pomona, Cal., Jan. 17, 1904; s. James R. and Laura (Ogden) H.; B.A., Pomona Coll., 1926; postgrad. U. Cal. at Berkeley; M.A., Claremont Coll., 1929; Ph.D., U. Wis., 1932; m. Katharine Maple, June 12, 1932; children—James N., Robert H., Laurence O., John M., Thomas K. Geologist, Shell Oil Co., 1927-37; geologist, exploration mgr. Internat. div. Atlantic Richfield Co., Los Angeles, 1937-—. Chmn., U. S. Nat. Com. on Geology, 1963-65; mem. com. on geologic hazards Cal. Resources Agy., 1964-—. Fellow Geol. Soc. Am. (chmn. Cordilleran sect. 1954-55, councilor 1957-59); mem. Am. Assn. Petroleum Geologists (pres. Pacific sect. 1955-56, pres. 1961-62). Contbr. articles to tech. jours. Application of geology to petroleum discoveries; research on stratigraphy, structure, faulting, seismicity, fault classifications. Home: 1114 Summit Dr., Whittier, Cal. 90601. Office: 555 S. Flower St., Los Angeles 90054.*

HILL, Maurice Neville, English geophysicist; b. Cambridge, Eng., May 19, 1919; s. Archibald Vivian and Margaret (Keynes) H.; B.A., King's Coll., Cambridge, 1947, M.A., Ph.D., 1950; m. Philippa Pass, Feb. 12, 1944; children—Mark Oliver, Julia Margaret, Alison Maynard, James David, Griselda Katharine. Faculty, Cambridge U., 1949-—, reader in marine geophysics, 1965-—, dir. studies in natural sci. King's Coll., 1961-65. Recipient Charles Chree medal Phys. Soc., 1963. Fellow Royal Soc., 1962, Royal Astron. Soc. (council 1961-64), Geol. Soc.; mem. Marine Biol. Assn. U.K. (council 1957-—). Editor: The Sea, 3 vols. Publs. on geol. and geophys. structure of sea floor, earth as a planet. Died Jan. 11, 1966.

HILL, Micaiah John Muller, Brit. mathematician; b. Berhampore, Bengal, Feb. 22, 1856; s. Samuel John Hill; ed. Univ. Coll., London, (hon. fellow) St. Peter's Coll., Cambridge, (scholar) U. London; M.A., LL.D., Sc.D.; m. Minnie Grace Tarbottom, 1892 (dec. 1920); 2 sons, 1 dau. Prof. math. Mason Coll., Birmingham, 1880-84; prof. pure math. Univ. Coll., London, 1884-1907, acting head dept. pure math., 1923-24; Astor prof. math. U. London, 1907-23, vice chancellor, 1909-11, emeritus prof. math. Fellow Royal Soc., 1894; mem. Math. Assn. (pres. 1926-27) Author: Fifth and Sixth Books of Euclid's Elements; The Theory of Proportion, also articles on spherical vortex. Died Jan. 11, 1929.

HILL, Nathaniel Peter, Am. metallurgist; b. Montgomery, N.Y., Feb. 18, 1832; s. Nathaniel P. and Matilda (Crawford) H.; grad. Brown U., 1856; m. Alice Hale, July 1860, 2 daus., 1 son. Prof. chemistry Brown U., 1860-64; mem. commn. of Mass. and R.I. mfrs. to investigate mineral deposits in Colo., 1864; studied metallurgy in Europe, 1865-66, 66-67; organized Boston & Colo. Smelting Co. (introduced smelting process increasing gold ore yield), 1867; developed refining process for separating precious metals from copper, 1873; mem. U. S. Senate from Colo., 1879-85; mem. Internat. Monetary Commn. (studying question of internat. metal currency), 1891; del. Bimetallic Conf., 1893. Died Denver, May 22, 1900.

HILL, Reuben L(orenzo), Am. chemist; b. Ogden, Utah, Mar. 24, 1888; s. George Richard and Elizabeth Nancy (Burch) H.; student Brigham Young U., 1908-11; B.S., Utah State Agrl. Coll., 1912; Ph.D., Cornell U., 1915; m. Mary Theresa Snow, Oct. 11, 1911; children—Reuben Lorenzo, Cornelia, (Mrs. Mac Novak), Richard Snow, Theresa Marie (Mrs. Donald Ashdown), Wesley Sherwin, Alwyn Spencer, Edward Eyring, Carl David. Instr. in biochemistry Cornell U., 1914-16; biochemist Bur. of Chemistry, U. S. Dept. Agr., Washington, 1916; biochemist Md. Agr. Expt. Sta., 1916-18; head dept. chemistry Utah State Coll. of Agr., Logan, from 1919, human nutritionist, expt. sta., 1919-41. Mem. Am. Chem. Soc., Am. Assn. U. Profs., A.A.A.S., Utah Acad. Arts, Sci. and Letters, Res. Officers Assn., Sigma Xi. Research and publs. on milk secretion, soft curd milk; patentee original equipment used in Hill Curd Test. Died Jan. 22, 1953.

HILL, Robert Dickson, physicist; b. nr. Me.-bourne, Australia, July 3, 1913; s. Robert W. and May (Dickson) H.; B.S., U. Melbourne, 1935, D.Sc., 1946; postgrad. U. Cambridge, U. Ill.; m. Judith A. Fowler, Dec. 22, 1942; 1 son, Robert B. Came to U. S., 1947, naturalized 1953. Research officer Air Ministry, London, Eng., 1939-40; lectr. U. Melbourne, physicist Radar Research Lab., 1941-45, sr. lectr. 1945-47; prof. physics U. Ill., Urbana, 1948-—. With Gen. Research Corp., Santa Barbara, Cal., 1965-—. Recipient Exhbn. of 1851 award, 1937, Fulbright Sr. Research award, 1960. Guggenheim fellow, 1961. Fellow Am. Phys. Soc. Author: (with L. H. Martin) Manual of Vacuum Practice, 1946; Tracking Down Particles, 1963. Research and publs. on elementary particle physics and high energy interactions. Home: 612 Alston Rd., Santa Barbara 93103. Office: Gen. Research Corp., Santa Barbara, Cal. 93105.*

HILL, Robert Gardiner, English surgeon; b. Louth, Eng., Feb. 26, 1811; s. Robert Hill; apprenticed to surgeon, Louth; student, Grainger's, Guy's, St. Thomas's hosps.; house surgeon Lincoln lunatic asylum, 1835-40; joint-propr. Eastgate House asylum, 1840-63; major of Lincoln, 1852; propr. Earl's Ct. Houst, Old Brompton, 1863-78. Mem. Royal Coll. Surgeons. Author: A Concise History of the entire Abolition of Medical Restraint in the Treatment of the Insane, and of the success of the Non-Restraint System, 1857; Lunacy, its Past and Present . . . , 1870; also articles. Originated non-restraint system in treatment of mentally ill. Died May 30, 1878.

HILL, Robert L., Am. biochemist; b. Kansas City, Mo., June 8, 1928; s. William Alfred and Geneva (Scurlock) H.; student Kansas City Jr. Coll., 1945-47; A.B., U. Kan., 1949, M.A., 1951, Ph.D., 1954; m. Helen A. Roote, June 26, 1948; children—Terry, Amy, Geneva, Rebecca. Asst. instr. biochemistry U. Kan., 1949-51, instr., 1951-53; faculty U. Utah, 1956-61, asso. research prof. biochemistry, 1960-61; faculty Duke, Durham, N.C., 1961—, prof. biochemistry, 1965—. Cons. biochem. tng. com. NIH, 1966—, chmn., 1967—; mem. biochemistry com. Nat. Bd. Med. Examiners, 1966—. Mem. Am. Soc. Biol. Chemists, A.A.A.S., Am. Assn. U. Profs. Research and publs. on proteins, especially antibodies, hemoglobin, fumarase and lactose synthetase, structure-function relationships among these proteins. Home: 2510 Perkins Rd., Durham, N.C. 27705.*

HILL, Robert M., Am. biochemist; b. Carthage, Ill., Oct. 27, 1894; s. William Kuhns and Katharine (Griffith) H.; B.S., Carthage Coll., 1915; M.S., U. Ill., 1921, Ph.D., 1923; m. Katherine Stanley, Aug. 12, 1929; children—Peter Stanley, Susan Applegate, Muriel McClaughry (Mrs. John Reed Cronin), Katherine Llewellyn. Asst. prof. biochemistry Loyola U. Med. Sch., Chgo., 1923-25; prof. biochemistry U. Colo. Sch. Medicine, Denver, 1925-63; head biochem. dept. Mercy Inst. Biomed. Research, Inc., Denver, 1960—, also dir., trustee. Mem. Soc. Exptl. Biology and Medicine, Am. Chem. Soc., Am. Assn. Clin. Chemists (past pres.), Am. Soc. Biol. Chemists. Research, numerous publs. on deficiencies of carbohydrates in gastric juices, metabolism of manganese and its role in nerve functions. Home: 6611 Richtofen Pkwy., Denver 80220. Office: 2920 E. 16th Av., Denver 80206.*

HILL, Robert Thomas, Am. geologist; b. Nashville, Aug. 11, 1858; B.S., Cornell, 1886; LL.D., Baylor, 1920; m. Justina Robinson, Dec. 28, 1887. Asst. paleontologist Smithsonian Instn., 1885; geologist U. S. Geol. Survey, 1885-1904; asso. geologist Ark. Geol. Survey, 1888-90; prof. geology U. Tex., 1890-91; geologist in charge U. S. investigation underground water and arid regions, 1891-92; cooperator with Prof. A. Agassiz in W. Indian and Central Am. explorations, 1895-1905; contract work Geol. Survey of So. Cal., U. S. Geol. Survey, 1911-17; lectr. various colls. and univs.; explored for first time great canons of the Rio Grande; spl. commr., Martinique eruptions for Nat. Geog. Soc.; cons. practice, from 1918. Fellow Am. Geol. Soc., A.A.A.S. Died July 28, 1941.

HILL, Robert Towner, Am. endocrinologist; b. Brookings, S.D., June 6, 1905; s. Henry Douglas and Lora (Towner) Hill; B.Sc., S.D. State U., 1928; M.Sc., Kan. State Coll., 1929; Ph.D., State U. Ia., 1932; D.Sc., S.D. State U., 1965; m. Vivian F. Kay Klinefelter, July 26, 1929; children—Dolores Vivian (Mrs. Robert F. Smith, Jr.), Richard William. NRC fellow U. Wis., 1932-33; staff Nat. Inst. Med. Research, London, Eng., 1933-34; instr. Yale Med. Sch., 1934-37; faculty Ind. U. Med. Sch., 1937-49, asso. prof. anatomy, 1940-48, prof., 1948-49; prof. U. Miami Med. Sch., Coral Gables, Fla., 1949-56; vis. staff mem. Nat. Inst. Med. Research, London, 1956, Coll. de France, Paris, 1956-57; exec. sec. endocrinology study sect. NIH, Bethesda, Md., 1957-61, dep. chief spl. program, 1961-64, 66—; chief human reprodn. program WHO, Geneva, Switzerland, 1964-66. Mem. Endocrine Soc., Am. Assn. Anatomists, N.Y. Acad. Scis., Soc. for Study Fertility, Internat. Planned Parenthood Fedn. (research com. 1965—), Soc. for Exptl. Biology and Medicine, Sigma Xi, Nu Sigma Nu. Author: Anatomy of Head and Neck, 1946; also numerous articles. Patentee condiment server, fungicide. Research on physiology of reprodn.; initiated WHO activities in family planning. Home: 4977 Battery Lane. Office: NIH, Bethesda, Md. 20014.*

HILL, Rodney, English mathematician; b. Leeds, Eng., June 11, 1921; s. Harold Harrison and Lena (Clark) H.; B.A., Cambridge (Eng.) U., 1942, M.A., 1945, Ph.D., 1948, Sc.D., 1959; m. Jeanne Kathlyn Wickens, June 1, 1946; 1 dau., Caroline Elizabeth. Head solid mechanics sect. Brit. Iron and Steel Research Assn., 1949; research fellow dept. theoretical mechanics U. Bristol (Eng.), 1950-52, reader, 1953; prof. applied math. U. Nottingham (Eng.), 1953-62, professorial research fellow, 1962-63; Berkeley Bye-fellow Gonville and Caius Coll., Cambridge, 1963—. Fellow Royal Soc., 1961, Inst. Math. Author: Mathematical Theory of Plasticity, 1950; Principles of Dynamics, 1964; also numerous articles. Editor-in-chief Jour. Mechanics and Physics of Solids, 1952—. Theoretical investigations of elastic and plastic deformation of metals especially shaping processes, mech. testing, structural instability, composite and polycrystalline materials. Office: Dept. Applied Math., Silver St., Cambridge, Eng.*

HILL, Roscoe Earle, Am. entomologist; b. Atkinson, Neb., Jan. 16, 1911; s. Leon R. and Leita (Mohrman) H.; student Neb. Wesleyan U., 1930-32; B.S., U. Neb., 1934, M.S., 1936; Ph.D., Ia. State U., 1946; m. C. Norene Johnston, Nov. 12, 1938; children—Kelvin J., Kenton R., Douglas D., Gary L. Dist. supr. U. S. Dept. Agr., 1938-40; faculty U. Neb., Lincoln, 1940—, chmn. dept. entomology, 1950-66, prof. entomology, 1957—. Fellow A.A.A.S.; mem. Entomol. Soc. Am. (past chmn. N.Central States br.), Ecol. Soc. Am., Am. Inst. Biol. Sci., Soc. Systematic Zoology. Research and publs. primarily in field of biology, ecology, control insects affecting field crops. Home: 848 Carlos Dr., Lincoln, Neb. 68505.*

HILL, Sir Rowland, English inventor; b. Kidderminster, Eng., Dec. 3, 1795; s. Thomas Wright and Sarah Lea; student at father's Hill Top Sch., Birmingham, Eng., from 1803; continued studies self-taught and under his father in astronomy and surveying; hon. D.C.L., U. Oxford (Eng.), 1864. Became tchr. Hill Top Sch., at 12 years of age; managed father's affairs from 1811; built and founded sch. governed by students; mem. assn. for colonizing South Australia, 1833; sec. South Australian Commn., 1835; described adhesive postage stamp, 1837; made exhaustive calculations and proposed uniform postage (accepted by govt., 1839); employed in treasury to oversee his postage plan, until 1842; dir., then chmn. Brighton Ry., 1843-46; sec. to post-master gen., 1846-54; sec. gen., 1854-64. Knighted, 1860. Fellow Royal Soc., 1857. Inventor: penny postage, rotatory printing press used to fix stamps (delayed 35 years because treasury would not permit stamps to be machine fixed), instrument to measure time in connection with astron. observations. Died Hampstead, Eng., Aug. 26, 1879.

HILL, Terrell Leslie, Am. chemist; b. Oakland, Cal., Dec. 19, 1917; s. George Leslie and Ollie (Moreland) H.; A.B., U. Cal. at Berkeley, 1939, Ph.D., 1942; m. Laura Etta Hill, Sept. 23, 1942; children—Julie Lisbeth, Carolyn Jo, Ernest Evan. Instr., Western Res. U., 1942-44; research asso. Manhattan Project, Berkeley, Cal., 1944-45; postdoctoral fellow U. Rochester, 1945-46, asst. prof., 1946-49; research chemist Naval Med. Research Inst., Bethesda, Md., 1949-57; prof. chemistry U. Ore., Eugene, 1957-67; prof. chemistry U. Cal. at Santa Cruz, 1967—. Vis. prof. Howard U., 1951-56, Yale, 1952-53, Cambridge (Eng.) U., 1960-61, Weizmann Inst., 1964. Recipient Arthur S. Flemming award U. S. Govt., 1955; Distinguished Civilian Service award U. S. Navy, 1956. Guggenheim fellow, 1952-53; Alfred P. Sloan fellow, 1958-62. Mem. Am. Chem. Soc., Am. Phys. Soc., A.A.A.S., Nat. Acad. Scis., Phi Beta Kappa, Sigma Xi. Author: Statistical Mechanics, 1956; Statistical Thermodynamics, 1960; Thermodynamics of Small Systems, 1963, 64; Matter and Equilibrium, 1966; Thermodynamics for Chemists and Biologists, 1968. Research and numerous publs. on theory of phys. adsorption, surface tension, protein solutions, Donnan membrane equilibrium, phase transitions, contractility, imperfect gases, small system thermodynamics, steady state statis. thermodynamics. Home: 1220 Laurent St., Santa Cruz, Cal. 95060.*

HILL, Thomas George, Brit. botanist; b. London, Feb. 13, 1876; s. Henry William Hill; ed. Royal Coll. Sci.; D.Sc.; lectr. biology and botany U. London, named reader, 1912, prof. plant physiology, 1929-45; became asst. prof. botany Univ. Coll., London, 1910, head dept. botany, dir. bot. labs., 1929. Asso. Royal Coll. Sci.; mem. Brit. Assn. (pres. botany sect. centenary meeting 1931). Author: (with P. Haas) Chemistry of Plant Products, 1913; The Essentials of Illustration, 1915; also papers on anatomy, ecology and physiology of plants. Died June 25, 1954.

HILLARY, Sir Edmund Percival, New Zealand explorer; b. Auckland, New Zealand, July 20, 1919; s. Percival A. and Gertrude (Clark) H.; ed. Auckland Grammar Sch.; m. Louise Rose, 1953; 1 son; 2 daus. Apiarist, 1936-43; navigator Catalina flying boats, 1944-45; mem. New Zealand Gawhal Expdn., 1951, Brit. Everest Reconnaissance, 1951, Brit. Cho Oyu Expdn., 1952, Everest Expdn., 1953; leader New Zealand Alpine Expdn. to Barun Valley, 1954. Recipient Hubbard medal, U. S., 1954; Star of Nepal 1st Class; U. S. Gold Cullum Geog. medal, 1954; Founder's Gold medal Royal Geog. Soc., 1953; Polar medal, 1958. Author: High Adventure; (with George Lowe) East of Everest, 1956; (with Sir Vivian Fuchs) The Crossing of the Antarctica, 1958; No Latitude for Error, 1961; (with Desmond Doig) High in the Thin Cold Air, 1963; School House in the Clouds, 1965. Leader, New Zealand Transantarctic Expdn. which was 1st to cross entire continent, 1958. Home: 278A Remuera Rd., Auckland S.E.2., New Zealand.

HILLARY, William, English physician; b. circa 1700; M.D., U. Leyden (Netherlands), 1722. Practiced medicine at Ripon, Eng.; at Bath, Eng. from 1734; in Barbadoes Is. from 1752; returned Eng., practiced in London from 1758. Author: Rational and Mechanical Essay on the Small-Pox, 1735; Observations on the Changes of the Air and the Concomitant Epidemical Diseases in Barbadoes with a Treatise on the Bilious Remittent Fever . . . , 1759; An Enquiry into the . . . Medicinal Virtues of Spaw Water near Bath, 1743; the Nature, Properties, and Laws of Motion of Fire, 1759; The Means of Improving Medical Knowledge, 1761. Research included systematic observations on climate and prevalent diseases. Died London, Apr. 22, 1763.

HILLE, (Carl) Einar, Am. mathematician; b. N.Y.C., June 28, 1894; s. Carl (August) and Edla (Ekman) H.; ed. U. Stockholm, 1911-16, 1917-18 doctor's degree); student Harvard, 1920-21, univs. Copenhagen and Göttingen, 1926-27; A.M. (hon.), Yale, 1933; m. Kirsti Ore, Aug. 10, 1937; children—Harald, Bertil. Instr., Stockholm U., 1915-16, 1919-20; Swedish Civil Serv., 1918-21; instr. in math. Harvard, 1921-22, Princeton U., 1922-23; asst. prof. math. Princeton U., 1923-30, asso. prof., 1930-33; prof. math., Yale, 1933-62, emeritus and sr. research asso., 1962—, dir. grad. studies in math., 1938-62; Fulbright lectr. U. Nancy (France), 1952, Sorbonne, 1953, U. Mainz (Germany), 1956; vis. prof. U. Stockholm and U. Upsala, 1949, 62, Tata Inst., Bombay, 1963, U. Mainz, 1956. Fellow Am. Acad. Arts and Scis.; mem. Nat. Acad. Scis., Am. Math. Soc. (pres. 1947-48), Conn. Acad. Arts and Sciences, Royal Physiog. Soc. of Lund, Royal Swedish Acad. Sci. Guggenheim fellow, 1952-53. Author: Functional Analysis and Semi-Groups, 1948; Analytic Function Theory, 1959, rev. 1962; Analysis, 1964. Editor Annals of Math., 1929-33. Asso. editor Transactions Am. Math. Soc., 1923-37, editor 1937-43. Work in differential equations, Dirichlet series, LaPlace transforms, Fourier series and transforms, Cauchy's problem, analytical theory of semi-groups. Address: 210 Edwards St., New Haven 11.

HILLEBOE, Herman Ertresvaag, Am. physician; b. Westhope, N.D., Jan. 8, 1906; s. Peter S. and Inga (Jacobson) H.; B.S., U. Minn., 1927, M.B., 1929; M.D., Minn. Med. Sch., 1931; M.P.H., Johns Hopkins, 1935; m. Alida C. Champean, Sept. 28, 1929; children—Joyce Elaine (Mrs. Kiaer), Theresa Ann (Mrs. McUmber), Herman Ertresvaag. Practice gen. medicine, rural Minn., 1929-30; Rockefeller Found. fellow Inst. Child Welfare, 1932-33, Johns Hopkins Sch. Hygiene and Pub. Health, Balt., 1934-35, chief Tb and crippled children's programs Minn. Bd. Control, 1935-39; chief med. unit Minn. Div. Social Welfare, 1939-41; commd. sr. asst. surgeon USPHS, 1939, advanced through grades to asst. surgeon gen., 1947; chief Tb control officer U. S. Coast Guard, 1942-44; chief div. Tb control, 1944-46; asso. chief Bur. State Services, 1946-47; ret., 1959; commr. N.Y. State Dept. Health, Albany, 1947-63; DeLamar prof. pub. health practice, head div. pub. health practice Columbia Sch. Pub. Health and Adminstrv. Medicine, N.Y.C., 1963—. Diplomate Am. Bd. Preventive Medicine and Pub. Health. Mem. N.Y. State Pub. Health Council, 1963—. Mem. Assn. State and Territorial Health Officers (past pres.), Am. Pub. Health Assn. (past pres.). A.A.A.S., Am. Epidemiological Soc., Am. Statistical Soc. Research and publs. on pub. health and adminstrv. medicine; initiated practical method mass screening for Tb. Home: 4901 Hudson Pkwy., Riverdale, N.Y. 10471. Office: Sch. Pub. Health and Adminstrv. Medicine, Columbia, N.Y.C. 10032.*

HILLEBRAND, William Francis, geochemist; b. Honolulu, Dec. 12, 1853; s. William and Anna (Post) H.; attended Cornell, 1870-72; studied 6 yrs. in Germany; Ph.D., U. of Heidelberg, 1875; studied at U. of Strasburg, and Mining Acad., Freiberg, Saxony; m. Martha May Westcott, Sept. 6, 1881. Assayer, Leadville, Colo., 1879-80; chemist U. S. Geol. Survey, 1880-1908; chief chemist Bur. of Standards, Washington, from 1908; prof. gen. chemistry and physics, Nat. Coll. of Pharmacy, 1892-1910. Author: Some Principles and Methods of Rock Analysis (2 edits.), 1900, 1902; Methods of Silicate and Carbonate Analysis, 1907, 10, 19. Determined age of radioactive rocks from measurements of percentage of lead. Died Feb. 7, 1925.

HILLEMAN, Maurice Ralph, Am. virologist; b. Miles City, Mont., Aug. 30, 1919; s. Robert A. and Edith (Matson) H.; B.S., Mont. State Coll., 1941; Ph.D., U. Chgo., 1944; m. Thelma L. Mason, Dec. 31, 1943 (dec. Nov. 1962); children—Jeryl Lynn, Kirsten Jeanne; m. 2d, Lorraine Witmer, Aug. 3, 1963. Asst. bacteriologist U. Chgo., 1942-44; research asso. virus labs. E. R. Squibb & Sons, New Brunswick, N.J., 1944-47, chief virus dept. 1947-48; med. bacteriologist, asst. chief virus and Rickettsial diseases Army Med. Service Grad. Sch., Walter Reed Army Med. Center, 1948-56, chief respiratory diseases, 1956-58; dir. virus and cell biology research Merck Inst. for Therapeutic Research Merck & Co., Inc., West Point, Pa. 1958-66, exec. dir., 1966—; vis. lectr. Rutgers U., 1947. Vis. investigator Hosp. Rockefeller Inst. for Med. Research, 1951; vis. prof. U. Md., 1953-57; cons. Surgeon Gen. U. S. Army, 1958-63; mem. expert adv. panel virus diseases WHO, 1952—; U. S. Armed Forces Epidemiology Bd., 1955-58; mem. study sect. microbiology and immunology Grants-in-Aid Program, USPHS, 1953-61, mem. spl. cons. panel for respiratory and related viruses, 1960-64, mem. nat. cancer inst. Primate Study Group, 1964—. Recipient Howard Taylor Ricketts prize, 1945; Arthur S. Flemming award, 1956. Fellow Am. Pub. Health Assn., American Association for Advancement of Sci., Am. Acad. Microbiology; mem. Am. Soc. Microbiology, Soc. Exptl. Biology and Medicine, Am. Assn. Immunologists, N.Y. Acad. Scis., Am. Assn. Cancer Research, Infectious Disease Soc., Internat. Assn. Microbiol. Socs. Editorial bd. Internat. Jour. Cancer, 1964—, Institute for Sci. Information, 1968. Research and publs. on field virology and cell biology, with spl. reference to psittacosis, influenza, adenovirus, poliomyelitis, measles, rubella, mycoplasma, other agts.; co-discoverer adenoviruses, Asian influenza and cancer virus SV40; devel. and proof of efficacy of virus vaccines. Home: 4107 Fields Dr., Lafayette Hill, Pa. 19444. Office: Merck & Co., Inc., West Point, Pa. 19486.*

HILLEMANN, Howard Herbert, Am. zoologist; b. Illmo, Mo., Apr. 4, 1910; s. Gustav Frederick and Anna (Uelsmann) H.; B.A., Marquette U., 1933; M.A., U. Wis., 1939, Ph.D., 1942; m. Emma Hansen, June 30, 1936; children—Ellen Kay, Lois Ann (Mrs. David Hadley). Faculty, U. Wis., 1939-42, Gonzaga U., 1945-46; faculty Ore. State U., Corvallis, 1946—, prof. developmental anatomy, 1956—, dir. NSF summer insts. for coll. biology profs., 1958-60. USPHS Research grantee, 1953-56, 64-66; Nat. Acad. Sci. grantee, Pallanza, Italy, 1960. Fellow A.A.A.S., Internat. Coll. Applied Nutrition; mem. N.W. Sci. Assn., Ore. Acad. Sci., Am. Assn. U. Profs., Am. Soc. Zoologists, Ore. State Employees Assn., Sigma Xi, Phi Kappa Phi, Phi Sigma. Author: The Vertebrate Organism, 1960. Research and numerous publs. in vertebrate anatomy, devel., embryology, placenta, reprodn. Home: 712 N. 26th St., Corvallis, Ore. 97330.*

HILLER, Johannes-Erich, German mineralogist; b. Berlin, German, Nov. 14, 1911; s. Bruno and Annemarie (Richter) H.; ed. univs. Kiel, Zurich, Berlin; Ph.D. in mineralogy; m. Ingeborg Boecker, 1944; children—Christian, Andreas, Mathias. Instr., U. Berlin; prof. mineralogy Higher Tech. Sch., Stuttgart. Mem. German Com. Mineralogy, Com. German Chemists, Mineral. Soc. Am. Author: Grundiss der Kristalchemie; Die Mineralischen Rohstoffe. Home: 1A Alte Weinsteige. Office: 5 Werdweg, Stuttgart 1, Germany.

HILLER, Lejaren Arthur, Jr., Am. chemist; b. N.Y.C., Feb. 23, 1924; s. Lejaren A. and Sara (Plummer) H.; B.A., Princeton, 1944, M.A., 1946, Ph.D., 1947; M.Mus., U. Ill., 1958; m. Elizabeth Halsey, Apr. 18, 1945; children—Amanda Kate, David Halsey. Research chemist E. I. du Pont de Nemours & Co., Inc., Waynesboro, Va., 1947-52; research asso., asst. prof. chemistry U. Ill., Urbana, 1952-55, asst. prof., 1958-61, asso. prof. music, 1961-66, prof. music, 1966—, dir. exptl. music studio, 1961—. Mem. Am. Assn. U. Profs., A.S.C.A.P. Author: (with L. M. Isaacson) Experimental Music, 1959; Principles of Chemistry (with R. H. Herber), 1960; also articles. Research in cellulose and high polymar chemistry and dye chemistry, thermodynamics, reaction kinetics, statis. mechanics, applications electronic computers and electronic to music and acoustics, applications information theory. Home: R.R. 3, Hudson Acres, Urbana, Ill. 61803.*

HILLER, Wilhelm Gustav Paul, German geophysicist; b. Altdorf, Böblingen, Feb. 2, 1899; s. Wilhelm and Berta (Wanner) H.; ed. Technische Hochschule, Stuttgart, U. Tübingen; Ph.D. in engring.; m. Tilly Streich, 1928; children—Marianne, Lore, Karl. Asst., 1923-34; dir. seismol. service of Stuttgart, 1934-61; prof. geophysics Technische Hochschule, Stuttgart, 1962. Mem. German Geophys. Soc., Am. Geophys. Union, Seismol. Soc. Am., Léopoldina (Halle, Salle). Coeditor: Geophysikalische Monographien. Research and numerous publs. on seismology and geophysics. Home: 14 Stalinweg. Office: 44 Richard Wagnerstrasse, Stuttgart, Germany.

HILLHOUSE, William, Brit. botanist; b. Bedford, England, Dec. 17, 1850; s. John Paton Hillhouse; M.A., M.Sc., Trinity Coll., Cambridge. Asst. master Bedford Modern Sch., 1867-77; asst. curator Univ. Herbarium, 1878-82; lectr. botany in Univ., also Girton and Newnham colls.; prof. botany Mason Coll. (later U. Birmingham), 1882. Fellow Linnean Soc.; mem. Birmingham, Bedfordshire (co-founder) natural history socs., Birmingham Bot. and Hort. Soc. (chmn., dir. Bot. Gardens), Midland Reafforesting Assn. (chmn. council), Assn. Internat. Botanists, Birmingham Bot. Hort. Soc. (hon. sec. 1892-1905). Author: (with Strasburger) Practical Botany, 1886, 6th edit., 1907. Co-founder, co-editor: Cambridge Rev., 1879; co-editor Midland Naturalist, 1887-94, also numerous articles. Died Jan. 27, 1910.

HILLIER, James, physicist; b. Brantford, Ont., Can., Aug. 22, 1915; s. James and Ethel Anne (Cooke) H.; B.A., U. Toronto (Ont., Can.), 1937, M.A., 1938, Ph.D., 1941; m. Florence Marjory Bell, Oct. 24, 1936; children—James Robert, William Wynship. Came to U. S., 1940, naturalized, 1945. Research physicist fundamental electron microscope research RCA, Princeton, N.J., 1940-53, adminstrv. engr., 1954-55, chief engr. comml. electronics products, 1955-57, gen. mgr. labs., 1957-58, v.p., 1958—; dir. research dept. Melpar, Inc., 1953-54. Mem. indsl. adv. com. NASA, 1962-64; pres. Indsl. Reactor Labs., 1964-65; mem. commerce tech. adv. bd. U. S. Dept. Commerce, 1964-68; chmn. adv. council to dept. elec. engring. Princeton, 1965—. Recipient Albert Lasker award Am. Pub. Health Assn., 1960. Fellow A.A.A.S., Am. Phys. Soc., I.E.E.E.; mem. Indsl. Research Inst. (governing bd. 1962-65, president 1963-64), National Academy of Engineering, Electron Microscope Society of Am. (past pres.), Am. Inst. Physics (governing bd. 1962-65), N.A.M. (research com.), Am. Mgmt. Assn., Sigma Xi, Eta Kappa Nu. Author: (with others) Electron Optics and Electron Microscope, 1945. Contbr. numerous articles to tech. publs. Designer, builder (with Albert Prebus) 1st successful high-resolution electron microscope in western hemisphere, 1939-40. Patentee electron diffraction, ultra-thin sectioning, viral and bacteriological techniques. Home: 22 Arr Office: RCA Labs., Princeton, N.J. 08540.*

HILLION, Pierre Théodore, French physicist; b. Saint Brieuc, France, Jan. 31, 1926; s. Pierre Alexandre and Olive (Marion) H.; Baccalauréat A. Collège Cordeliers, 1942; Baccalauréat Math., Collège Stanislas, 1945; Philo., Ecole Superieure d'Electricité, 1949, ingenieur, 1949; Licence, Faculté des Sciences, Paris, France, 1955, Doctorat in Scis., 1957; m. Jane Garde, July 9, 1955; children—Catherine, Pierre, Joelle, Hervè. Engr. electronic dept. Société Le Materiel Electrique S-W, Paris, 1950-55; engr. atomic group Section Technique de l'Armée, Paris, 1955-60; chief dept. physics-math., study and research center of mil. atomics Fort d'Aubervilliers, Paris, 1960-64; chief of sect. Laboratoire Central de l' Armement, Ft. Montrouge, Paris, 1964—. Electronic prof. Arts et Metiers, Lyon, France, 1952-53; research asso. l'Institut Henri Poincaré, Paris, 1957-66, asso. centre democrite de physique mathematique, 1966—. Recipient chevalier du mérite for research and invention. Mem. Société Francaise des Electriciens, Société d'Encouragement for Recherche et l'invention, Syndicat de la Presse Scientifique, Société Mathematique de France. Research, publs. on relativistic hydrodynamics in frame of interpretation of quantum mechanics; a model of relativisticorotator, of elementary particles considered as a fluid droplet; application of group theory in high energy physics; computation methods for penetration of neutral particles and charged particles in matter. Home: 12 Av. du Centre, Le Pecq, Yvelines 78, France. Office: Fort de Montrouge, Arcueil, Val-de-Marne 94, France.*

HILLS, Edwin Sherbon, Australian geologist; b. Melbourne, Australia, Aug. 31, 1906; s. Edwin Sherbon and Eva (Cameron-Toe) H.; B.Sc., U. Melbourne, 1937, M.Sc., 1928; Ph.D., Diploma, Imperial Coll., Royal Coll. Sci., U. London (Eng.), 1931; D.Sc., U. Melbourne, 1938; D.Sc., U. Durham (Eng.), 1960; m. Claire Doris Fox, Aug. 26, 1932; children—Elizabeth (Mrs. John L. Loder), David, Richard. Faculty, U. Melbourne, 1932—, prof. geology and mineralogy, 1944-62, dept. vice chancellor, research prof. geology, 1962—. Chmn., Australia-UNESCO Com. for Natural Scis., 1965—. Fellow Imperial Coll. Sci. and Tech., Royal Soc., 1954; mem. Australian Acad. Sci., Geol. Soc. (London); hon. mem. Inst. Geographers Australia. Author: Physiography of Victoria, 1941; Outlines of Structural Geology, 1940; Elements of Structural Geology, 1963; Arid Lands, 1966; also numerous articles. Research on Australian fossil fishes, geomorphology of arid and humid areas of Australia; discovered structure Devonian volcanic complexes; described major fracture-pattern of Australia and related to world patterns of geologic structures. Home: 25 Barry Kew, E 4, Victoria, Australia.*

HILLS, Frederick Jackson, Am. agronomist; b. Oakland, Cal., Feb. 27, 1919; s. Frederick B. and Almah (Hollenback) H.; B.S., U. Cal. at Berkeley, 1941; M.S., U. Cal. at Davis, 1951, Ph.D., 1961; m. Juanita Alberta Wood, June 23, 1943; children—Donald Jackson, Christine Eleth, Cynthia Anne, Diane Lee. With Spreckels Sugar Co., Cal., 1941-51; faculty U. Cal. at Davis, 1951—, extension agronomist, 1951—. Mem. Am. Soc. Sugar Beet Technologists, Am. Assn. Agronomy, Am. Assn. Phytopathologists, Sigma Xi, Phi Kappa Phi. Research, numerous publs. on aspects of sugar beets, including effects of plant population on prodn., devel. of methods to eliminate hand labor in establishment of stands, effects and control of aphid-borne viruses, fertilizer mgmt. Home: 848 Oeste Dr., Davis, Cal. 95616.*

HILLS, Orin Ancil, Am. entomologist; b. Pomona, Cal., Oct. 21, 1903; s. Percy V. and Nellie (Wixon) H.; B.S., Ore. State U., 1927, M.S., 1929; m. Mary Nadine Caves, Aug. 25, 1927; children—Edward O., Shirley M. (Mrs. Richard C. Kuhn). With U. S. Dept. Agr., 1929—, investigations leader vegetable and sugarbeet vector investigations, Mesa, Ariz., 1964—. Mem. Entomol. Soc. Am., Am. Soc. Sugar Beet Technologists. Contbr. numerous articles to sci. jours. Determination of insects causing low germination of sugarbeet seed and devel. controls for these insects; determination of effect of time of infestations of insect vectors of virus diseases on sugarbeet seed crop and methods of preventions; devel. controls for insects affecting lettuce and cantaloupes. Home: 906 Broadmor Dr., Tempe, Ariz. 85281. Office: P.O. Box 858, Mesa, Ariz. 85201.*

HILLYER, John Carpenter, Am. chemist; b. Winona, Minn., Sept. 8, 1908; s. George J. and Katherine (Carpenter) H.; B.A., Carleton Coll., 1929; Ph.D., Yale, 1932; m. Dorothy D. Dalton, Aug. 18, 1934; 1 son, John Carpenter, II. Instr., Carleton Coll., 1932-37; with Phillips Petroleum Co., Bartlesville, Okla., 1937—, sect. chief, 1940-60, br. mgr. research div., 1960—. Fellow Am. Inst. Chemists, A.A.A.S.; mem. Am. Chem. Soc., Phi Beta Kappa, Sigma Xi, Alpha Chi Sigma. Contbr. articles to tech. jours., Ency. Chemistry, 1957, 66. Patentee in field. Developed catalysts for dehydrogenation, process for dehydrogenation butane to butadiene, butadiene analytical methods and specifications, butadiene reactions, ammonium nitrate coating and other fertilizer devels., vinyl fluoride synthesis, use organic fluids for nuclear reactor cooling, uranium ore refining, furfural decomposition reactions. Home: 1920 Southview St., Bartlesville 74003. Office: Phillips Petroleum Co., Bartlesville, Okla. 74004.*

HILMBAUER, Karl Franz, Austrian biologist; b. Jan. 20, 1930; s. Karl and Anna Hilmbauer; Ph.D. in Philosophy, U. Vienna; m. Gusti Aichinger, 1954; children—Werner, Gabriele, Karl. Biologist; technician for preservation plants. Mem. Study Commn. for Preservation Plants, Vienna. Mem. Soc. Zoology and Botany (Vienna). Author: Obstbauschädlings Fibel; Unkraut Fibel. Research and publs. on cells of Euglenaccae, Trachelomonas. Home: 21 Gärtnergasse, Korneuburg. Office: 6 Dr. Karl Luegerring, Vienna I, Austria.

HILTNER, W(illiam) A(lbert), Am. astronomer; b. Continental, O., Aug. 27, 1914; s. John Nicholas and Ida Lavina (Schafer) H.; B.S., U. Toledo, 1937; M.S., U. Mich., 1938, Ph.D., 1942; m. Ruth Moyer Kreider, Aug. 12, 1939; children—Phyllis Anne, Kathryn Jo, William Albert, Stephen Kreider. Faculty, U. Chgo., 1943—, prof. astronomy, 1955—, dir. Yerkes Obs., 1963—. Asso. Univs. for Research Astronomy, 1959—. NRC fellow, 1942-43. Mem. Astron. Soc. Pacific, Am. Astron. Soc. (councilor 1962-65), A.A.A.S. Coauthor: Photometric Atlas of Stellar Spectra, 1946. Editor: Astronomical Techniques, 1962. Research in photoelectric photometry; electronic image intensification, stellar spectroscopy; interstellar polarization of starlight. Home: Parkhurst Pl. Office: Yerkes Observatory, Williams Bay, Wis. 53191.*

HILTON, James Garrett, Am. physician; b. Washington, Apr. 7, 1920; s. James C. and Harriett (Williams) H.; B.A., Princeton, 1942; M.D., Columbia, 1945. Research asso. Columbia Coll. Phys. & Surg., 1948-51, asso. clin. prof. medicine, 1965; fellow N.Y. Heart Assn., Goldwater Meml. Hosp., N.Y.C., 1949-51, dir. lab. for renal physiology, 1950-51; with St. Luke's Hosp., N.Y.C., 1951—, chief renal-endocrine physiology sect., 1952—, chief Endocrine Clinic, 1955—, asso. attending physician, 1958—; cons. VA Hosps., Orange, N.J., Vallhalla, N.Y., 1958—; career scientist Health Research Council City N.Y., 1960—. Mem. Am. Soc. Clin. Research, Am. Fedn. Clin. Research, Am. Physiol. Soc., Endocrine Soc., Internat. Soc. Nephrology, N.Y. Acad. Medicine, A.C.P., Harvey Soc., Sigma Xi. Author: Ideas, Inertia and Achievement, 1960. Contbr. numerous articles to med. jours. Investigated mechanism of action of mercurial diuretics; devised Hilton Pouch, research technique for investigation of adrenal gland physiology; research to perfect artificial kidney. Office: 421 W. 113th St., N.Y.C. 10025.*

HILTON, James Gorton, Am. pharmacologist; b. Balt., Sept. 21, 1923; s. George Edward and Ethel (Schaefer) H.; B.S., Va. Poly. Inst., 1947; M.S., U. Tenn., 1952, Ph.D., 1954; m. Elizabeth Lindsay, Sept. 21, 1946; children—James Lindsay, William Edward. Research asso. Pharmacology U. Va., 1948-50; faculty U. Tenn., 1950-53, U. Miss., 1953-58, Marquette U., 1959-61; faculty U. Tex. Med. Br., Galveston, 1961—, prof. pharmacology, 1963—; lectr. various univs. Fulbright lectr. Nat. U. San Agustin, Arequipa, Peru, 1959-60; cons. council on drugs for new and nonofcl., 1963—; mem. pharmacology and endocrinology fellowships rev. panel NIH, 1964—. Fellow A.A.A.S., Am. Coll. Clin. Pharmacology and Chemotherapy; mem. Am. Heart Assn., Am. Physiol. Soc., Am. Soc. Pharmacology and Exptl. Therapeutics, Soc. Exptl. Biology and Medicine, A.M.A. (affiliate), Latin Am. Soc. Pharmacology, Sigma Xi. Research and publs. on factors responsible for regulation and maintenance of blood pressure, nerve transmission and its effects upon blood pressure. Home: 2626 Gerol Ct., Galveston, Tex. 77550.*

HILTON, John, English surgeon; b. Castle Hedingham, Eng., 1804; ed. Chelmsford; entered Guy's Hosp., 1824; demonstrator anatomy Guy's Hosp., surgeon, 1849-70, made dissections of human body reproduced in wax by Joseph Towne for anat. mus.; prof. human anatomy and surgery Coll. Surgeons, 1860-62, pres., 1867; mem. Royal Coll. Physicians. Author: Notes on Some of the Developmental and Functional Relations of certain Portions of the Cranium helected from Hilton's Lectures on Anatomy By F. W. Povy, 1855; On Rest and Pain (classic work in surgery), 1863; also clin. lectures in Guy's Hosp. Reports, and Hunterian Oration, 1867. Performed (with G. Bird) 1st operation for relief of internal strangulation of small intestine, 1847; advocated complete rest in cases of surg. disorders. Died Sept. 14, 1878.

HILTON, Peter John, mathematician; b. London, Eng., Apr. 7, 1923; s. Mortimer and Elizabeth (Freedman) H.; M.A., Oxford (Eng.) U., 1948, D.Phil., 1950; Ph.D., Cambridge (Eng.) U., 1952; m. Margaret Mostyn, Sept. 14, 1949; children—Nicholas, Timothy. Lectr., Manchester (Eng.) U., 1948-52, sr. lectr., 1956-58; lectr. Cambridge U., 1952-55; Mason prof. pure math. Birmingham (Eng.) U., 1958-62; prof. math. Cornell U., Ithaca, N.Y., 1962—; guest prof. Eidgenössiche Technische Hochschule, Zurich, Switzerland, 1966-67. Co-chmn., Cambridge Conf. on Sch. Math., 1965. Mem. Math. Assn. Am., Am., Belgium (hon.), London math. socs., Cambridge Philos. Soc. Author: Homotopy Theory, 1953; (with S. Wylie) Homology Theory, 1960; Homotopy Theory and Duality, 1966; also numerous publs. Editor: Ergebnisse der Mathematik, 1964—, Ill. Jour. Math., 1962—, Topics in Modern Topology, 1968. Research in algebraic topology, homological algebra and category theory, problems of math. edn. Home: 132 N. Sunset Dr., Ithaca, N.Y. 14850.*

HIMELICK, Eugene Bryson, Am. plant pathologist; b. Summitville, Ind., Feb. 11, 1926; s. Virgil B. and Madalene (Bryson) H.; B.S., Ball State U., 1950; M.S., Purdue U., 1952; Ph.D., U. Ill., 1959; m. Elizabeth Ann Oyler, June 21, 1951; children—David Eugene, Kirk Joseph, Douglas Neal. With Ill. Natural History Survey, Urbana, 1952—, plant pathologist, 1965—. Mem. Internat. Shade Tree Conf. (chmn. edn. and research com. Midwestern chpt. 1959—, Am. Phytopath. Soc., Ill. Tech. Forestry Assn., Ill. Acad. Sci., Am. Forestry Assn., Gamma Sigma Delta, Sigma Xi. Research and publs. on vascular wilt diseases, leaf diseases, various physiol. problems of forest and shade trees, supplemental methods for controlling Dutch elm and oak wilt disease. Home: 601 Burkwood Ct. E. Office: Natural Resources Bldg., Urbana, Ill. 61801.*

HIMMELBLAU, David Mautner, Am. chem. engr.; b. Chgo., Aug. 29, 1923; s. David and Roh (Mautner) H.; B.S., Mass. Inst. Tech., 1947; M.B.A., Northwestern U., 1950; M.S., U. Wash., 1956, Ph.D., 1957; m. Betty Hartman, Sept. 1, 1948; children—Andrew, Margaret Ann. Cost engr. Internat. Harvester Co., Chgo., 1947-48; cost analyst Simpson Logging Co., Seattle, 1952-53; mgr. Excell Battery Co., Seattle, 1955-56; faculty U. Tex., Austin, 1957—, prof. chem. engring., 1964—. Pres., Univ. Fed. Credit Union, 1964—. Mem. Am. Inst. Chem. Engrs., Am. Chem. Soc., Am. Math. Soc. Author: Basic Calculations in Chemical Engineering, 1962, 2d edit., 1967; (with K. B. Bischoff) Chemical Process Analysis, 1967; also articles. Research on diffusion gas molecules dissolved in water leading to revised concept liquid state matter, methods analysis and decomposition systems large size. Home: 4609 Ridge Oak Dr., Austin, Tex. 78731.*

HINCHLEY, John William, Brit. chem. engr.; b. Grantham, Eng., Jan. 21, 1872; ed. Imperial Coll. Sci. and Tech., became Nat. scholar, 1892; m. Edith Mary Mason. Successively became asso. Royal Sch. Mines, 1896, mgr. Nat. Colour Photography Co., 1897, asst. to Oscar Guttman, chem. engr., 1899, tech. head New Siamese Mint, Bangkok, 1903, lectr. chem. engring. Battersea Poly., 1909; became lectr. Imperial Coll. Sci. and Tech., 1911, prof. chem. engring., 1926; cons. chem. engr. to several firms. Fellow Inst. Chemistry, Chem. Soc.; mem. Instn. Chem. Engrs. (hon. sec.), Royal Sch. Mines (asso.). Author: Chemical Engineering; also articles. Died Aug. 13, 1931.

HIND, John Russell, English astronomer; b. Nottingham, Eng., May 12, 1823; s. John Hind; LL.D., Glasgow, 1882; m., 1846; 6 children; asst. to civil engr., London; became mem. staff magnetic and meteorol. dept. Royal Obs., Greenwich, Eng., 1840; dir. obs. founded by George Bishop in Regent's Park, 1844-95. Fellow Royal Soc., 1863; mem. Royal Astron. Soc. (pres. 1880-81), French Acad. Scis., 1851. Author: The Solar System, 1852; An Introduction to Astronomy, 1852; The Comets, 1852; The Illustrated London Astronomy, 1853; also articles. Supt. Nautical Almanack, 1953-91. Discovered 10 asteroids, 2 comets, a nebula in Jaurus, also several variable stars, including temporary apparition of May 1848. Died Twickenham, Eng., Dec. 23, 1895.

HINDENBURG, Karl Friedrich, German mathematician; b. 1741. Considered founder of modern theory of combinations and permutations; originated combinatorial sch. Died 1808.

HINDLE, Edward, Brit. zoologist; b. Sheffield, Eng., Mar. 21, 1886; s. Edward James and Sarah Elizabeth (Dewar) H.; A.R.C.S., Royal Coll. Sci., 1906; B.A., Magdalene Coll., Cambridge, U., 1912, M.A., 1919, Sc.D., 1925; Ph.D., U. Cal. at Berkley, 1910. Research asso. Liverpool (Eng.) Sch. Tropical Medicine, 1907-08; Kingsley Bye fellow Magdalene Coll., Cambridge, Eng., 1913-19; prof. biology Sch. Medicine, Cairo, Egypt. 1919-24; Milner Research fellow London Sch. Tropical Medicine, 1924-25; mem. Kala Azar Commn., Royal Soc. to N. China, 1925-27; Beit Research fellow in tropical medicine, 1928-33; Regius prof. zoology U. Glasgow (Scotland), 1935-43; sci. dir. Zool. Soc. London, 1944-51. Recipient Croix Civique (1st class), Beligium, 1931; Rhodesian Gold medal, 1912; Medaille Geoffrey St. Hilaire (En or), France, 1951. Fellow Royal Soc., 1942, Inst. Biology (hon., founder, pres.), Royal Geog. Soc. (hon. sec., v.p. 1951—), Brit. Assn. for Advancement Sci. (past gen. sec.), Linnean Soc. London, Univs. Fedn. for Animal Welfare (pres. 1944—), Zool. Soc. Glasgow (past pres.), Royal Philos. Soc. Glasgow (past pres.). Author: Flies and Disease—Bloodsucking Flies, 1914; also numerous articles. Described cytology of artificial parthenogenesis in sea urchin, life-cycles of various flagellates transmitted by flies; co-discoverer carrier of Chinese Kala Azar; prepared 1st vaccine against yellow fever; established lab. strain of yellow fever; described nature of feline distemper and prepared vaccine against infection; studied spirochaetes. Home: 51 Warwick Av., London W.9., Eng.*

HINDMAN, James Victor, Australian radio astronomer; b. Sydney, Australia, Aug. 26, 1919; s. Henry James and Lily (Druyve) H.; ed. Sydney Tech. Coll.; m. Anita Dulcie Cale, Apr. 22, 1943; children—Allen James, Lynne Anita, Christine Anne, Katherine Jane. With Amalgamated Wireless, Australia, 1936-42; mem. staff radio physics div. Commonwealth Sci. and Indsl. Research Orgn., Australia, 1942-66; officer-in-charge Siding Spring Obs., Coonabarabran, NSW, Australia, 1966—. Fellow Royal Astron. Soc.; grad. mem. Australian Inst. Physics. Research, publs. on wartime radar, solar radio, confirmation of original detection of hydrogen line, first detection (with F. J. Kerr) of H-line radiation from Magellanic Clouds. Address: Siding Springs Obs., Coonarbarabran NSW, Australia.*

HINE, Charles Henri, II, Am. physician; b. Toledo, Mar. 31, 1916; s. Charles Henry and Grace (Gibson) H.; B.A., St. Norbert Coll., 1937; M.A., U. Wis., 1942, Ph.D., 1942, M.D., 1943; m. Betty Dixon, Aug. 11, 1945; children—Holly Elizabeth, Charles Henri, III. Wis. Alumni research fellow U. Wis., 1938-42; chief toxicology sect. U. S. Naval Med. Research Inst., 1946-47; acting chief, head toxicologists U. S. Naval Def. Lab., San Francisco Naval Shipyards, 1947; practice medicine specializing in occupational medicine and toxicology, San Francisco, 1947—; faculty U. Cal. Sch. Medicine, San Francisco, 1947—, clin. prof. toxicology and occupational medicine, 1964—; toxicologist, coroner City and County of San Francisco, 1950—; pres., dir. Hine Labs., Inc., San Francisco, 1961—; toxicologist Cal. Dept. Pub. Health, 1955—. Fellow Am. Acad. Forensic Scis.; mem. Cal. Acad. Preventive Medicine, Am. Acad. Occupational Medicine, Indsl. Med. Assn., Am. Indsl. Hygiene Assn., A.M.A., Am. Soc. for Pharmacology and Exptl. Therapeutics, Soc. Toxicology. Contbg. author: Industrial Hygiene and Toxicology, 1962. Research and numerous publs. on detection, determination of mechanism of action, treatment and prevention of diseases arising from drugs, toxic chems. and noxious environment. Home: 60 King Av., Piedmont, Cal. 94611. Office: Sch. Medicine, U. Cal., 3d and Parnassus Sts., San Francisco 94122.*

HINE, Frederick Roy, Am. psychiatrist; b. Cin., Nov. 16, 1925; s. Fred Ludwig and Alma (Waltamath) H.; B.S., Yale, 1946, M.D., 1949; m. Corinne Lucille Kopp, Sept. 16, 1948; 1 son, Frederick Clifford. Fellow in psychiatry Tulane U. Sch. Medicine, 1950-53, instr., 1952-59; staff psychiatrist S.E. La. Hosp., Mandeville, 1953-55, dir. tng. and research, 1955-58, clin. dir., 1958-59; faculty Duke Med. Center, Durham, 1959—, asso. prof. psychiatry, 1963—. Mem. Am. Psychiat. Assn., Am. Group Psychotherapy Assn. Research and publs. on tranquilizing drugs, nature psychiat. concepts., importance of conflict in psychodynamic theory. Home: 2317 Prince St., Durham, N.C. 27707.*

HINE, Fumio, Japanese chemist; b. Osaka, Japan, Nov. 23, 1927; s. Asakichi and Koma (Kitaguchi) H.; student dept. elec. engring. Miyakojima Tech. Coll., Osaka; Ph.D. in Electrochem. Engring., Kyoto U., 1960; m. Teruyo Nakabo, Feb. 8, 1949; children—Shiro, Takako. With Kyoto U., 1948—, prof. Uji Campus, 1965—. Fulbright fellow Western Res. U., Cleve., 1960-61. Mem. Electrochem. Soc., Am. Inst. Chem. Engrs., Comité International de Thermodynamique et de Cinétique electrochimiques, Sigma Xi. Recipient award and prize for achievement in amalgam type chlorine cell KINKI Chem. Industry Assn., Osaka, 1959. Numerous publs. on chem. engring. studies of brine purification system, design amalgam decomposer; design graphite anode for chlorine cell; electrolysis of hydrochloric acid with oxygen depolarized cathode. Home: 13-22 Hagoromo-1, Hamadera, Osaka, Japan. Office: Kyoto U. at Uji, Uji, Kyoto, Japan.*

HINE, Maynard Kiplinger, Am. dentist; b. Waterloo, Ind., Aug. 25, 1907; s. Clyde L. and Delia (Kiplinger) H.; D.D.S., U. Ill., 1930, M.S., 1932; m. Harriett Foulke, Apr. 30, 1932; children—Maynard Kiplinger, Judith F. (Mrs. John Hyde), William C. Research staff U. Rochester, 1934-36; faculty U. Ill., 1936-44; prof. Sch. Dentistry, head dept. oral histopathology and periodontics Ind. U., Bloomington, 1944—, dean Sch. Dentistry, 1945—. Exec. bd. Ind. Bd. Health, 1948-63, chmn., 1959-60; mem. nat. adv. dental Research council U. S. Dept. Health, Edn. and Welfare, 1966—; mem. adv. council on dentistry VA Med. Care Program, 1964—. Named Periodontist of Year, Tufts U. Sch. Dental Medicine, 1957; recipient Distinguished Dental Alumnus of Year award U. Ill. Alumni Assn., 1962. Diplomate Am. Bd. Endodontics (founder), Am. Bd. Periodontology (chmn. 1962). Mem. Am. Assn. Dental Schs., Internat. Assn. for Dental Research (past pres.), Am. Assn. Dental Editors (past pres.), Am. Assn. Endodontists (past pres.), Indpls. Dist. Dental Soc. (past pres.), Ind. Dental Assn. (past pres.), Am. Acad. Periodontology (past pres.), Am. Dental Assn. (past pres.), A.A.A.S. (past v.p.), Sigma Xi, Omicron Kappa Upsilon. Research and publs. on periodontology. Home: 4580 N. Meridian St., Indpls. Office: Sch. Dentistry, Ind. U., Indpls. 46202.*

HINES, Colin Oswald, physicist; b. Toronto, Ont., Can., June 4, 1927; s. Oswald Otis and Winnifred (Mills) H.; B.A., U. Toronto, 1949, M.A., 1950; Ph.D., U. Cambridge (Eng.), 1953; m. Bernice Eleanor Bishop, May 29, 1948; children—David Bruce, Michael Andrew, Margot Lynn, Karen Louise. Sci. officer Def. Research Bd. Can., 1950-51, 53-62, supt. Radio Physics Lab., 1958-60, head Theoretical Studies Group, 1960-62; prof. aeronomy U. Chgo., 1962—. Mem. asso. com. on radio sci. Canadian NRC, 1961-63. Recipient Napier Shaw Meml. prize Royal Meteorol. Soc. London, 1962. Mem. Internat. Sci. Radio Union (exec. com. U. S. commn. IV 1962-65, mem. U. S. nat. com. 1963—, vice chmn. internat. commn. III 1963—), Canadian Assn. Physicists, Am. Geophys. Union. Editor: (with I. Paghis, T. R. Hartz, J. Fejer) Physics of the Earth's Upper Atmosphere, 1965; Contbr. numerous articles to tech. jours. Analysis of electromagnetic wave propagation in ionized media, radio reflections from meteor trails, motions upper atmosphere, hydromagnetic motions in earth's magnetosphere, sun-earth relations, theory geomagnetic and auroral storms.*

HINES, Marion, Am. neuroanatomist; b. Carthage, Mo., June 11, 1889; d. Frank Bristow and Laura (Saunderson) H.; student Drury Coll., 1909-11; A.B., Smith Coll., 1913, Sc.D., 1943; Ph.D., U. Chgo., 1917; Sc.D., Emory U., 1965. Faculty, U. Chgo., 1917-25; faculty Johns Hopkins, 1925-47, vis. prof., 1965—; faculty Emory U., 1947-65, dir. postgrad. tng. grant neuro anatomy, 1959-65. Recipient Distinguished Service award Drury Coll., 1952. Fellow A.A.A.S.; mem. Am. Assn. Anatomists, Am. Physiol. Soc., Am. Neurol. Assn., Assn. for Research in Nervous and Mental Diseases. Studies on growth of cerebral cortex in man, anatomy of cerebral cortex in so-called lower forms, innervation of skeletal muscle, control of muscle activity by central nervous system. Home: 1514 Berwick Rd., Ruxton, Md. 21204. Office: Dept. Anatomy, Johns Hopkins, Balt. 21205.*

HINGSON, Robert Andrew, Am. physician; b. Anniston, Ala., Apr. 13, 1913; s. Robert A. and Elloree (Haynes) H.; A.B., U. Ala., 1935, postgrad. Med. Sch.; M.D., Emory U., 1938; H.H.D., Monrovia (Liberia) Coll., 1962; LL.D., William Jewell Coll., 1963, Pa. Baptist Coll., 1963; Litt. D., Hardin Simmons U., 1965; m. Gussie Dickson, Mar. 2, 1940; chidren—Dickson James, Roberta Ann, Andrew Tobian, Ralph Waldo, Luke Lockhart. Commd. asst. surgeon USPHS, 1942, advanced through grades to med. dir., 1963; asso. prof., anesthesiologist dept. obstetrics Johns Hopkins, 1948-51; faculty anesthesiology Royal Coll. Surgeons, Eng., 1949-66; prof. anesthesiology Western Res. U., also dir. anesthesiology U. Hosp., Cleve., 1951—. Dir. operations Bro.'s Bro., Liberia, 1962, dir. service missions to Africa, S. Pacific, Middle East, Latin Am., 1954-66; sr. med. cons. Amigos de Honduras, 1965. Named one of Am.'s Ten Outstanding Young Men, U. S. Jr. C. of C., 1947; decorated knight grand comdr. Humane Order African Redemption (Liberia). Fellow Internat. Coll. Surgeons (dir. 1959—); mem. Am. Soc. Anesthesiologists, Am. Coll. Anesthesiologists, Internat. Anesthesiology Research Soc. (v.p. 1948—), A.M.A., Ohio, Cleve. med. assns., Cleve. Acad. Medicine, World Fedn. Socs. of Anesthesiologists (pres. Anes. Edn. and Relief Found.), Phi Kappa Alpha. Author: (with Clifford B. Lull) Control of Pain in Childbirth, 1943; (with George Pitkin, James L. Southworth) Pitkin's Conduction Anesthesia, 1946; (with Louis M. Hellman) Anesthesia for Obstetrics, 1956; also numerous articles. Developed technique continuous caudal and peridural analgesia, hypospray for clin. use, jet injector for mass inoculation, dilution technique for smallpox vaccination by jet inoculation; invented portable anesthesia gas machine and resuscitator. Home: 2889 N. Park Blvd., Cleve. 44118.

HINKS, Arthur Robert, English astronomer, geographer; b. London, May 26, 1873; s. Robert and Mary (Hayward) H.; ed. Trinity Coll., Cambridge U., 1892-94; m. Lily Mary Packman, 1899; children—Roger, David. Second asst. obs. Cambridge, 1895, chief asst., 1903-13, demonstrator practical astronomy, 1895-1913, Gresham lectr. astronomy, 1913-41. Lectr. surveying and cartography Royal Geog. Soc., 1908-13, asst. sec., 1913-15, sec. Royal Astron. Soc., 1909-13, v.p., 1913. Fellow Royal Soc., 1913. Author: Map Projections, 1912; Maps and Survey, 1913. Major contbn. in study of map projections and calculating oblique Mercators; interest in surveying by aerial photography brought this into realm of Royal Geog. Soc. Died Royston, Hertfordshire, Apr. 18, 1945.

HINMAN, Richard Leslie, Am. chemist; b. Utica, N.Y., Mar. 8, 1927; s. Harold William and Mildred (Durr) H.; B.A., Columbia, 1949; Ph.D., U. Ill., 1952; m. Rosalind V. S. Ellam, Sept. 23, 1967. Merck postdoctoral fellow Cambridge (Eng.) U., 1952-53; instr. organic chemistry State U. Ia., Iowa City, 1953-57; research chemist Union Carbide Research Inst., Tarrytown, N.Y., 1958-61, group leader exploratory organic chemistry, 1961-65, dir. chem. research and devel. pharm. div. Union Carbide Corp., 1965-67, mgr. pharm. tech., 1967—. Mem. Am. Chem. Soc., Chem. Soc. (London), A.A.A.S. Author: (with Robert E. Buckles) Study Questions for Organic Chemistry, 1957; also articles. Research on free radical chemistry of oxidation, chlorination and other indsl. processes; discovered mechanism of destruction of natural plant growth hormone, indoleacetic acid. Home: Lounsbury Rd., Ridgefield, Conn. 06877. Office: P.O. Box 278, Tarrytown, N.Y.*

HINNERS, Scott William, Am. nutritionist; b. Metropolis, Ill., Jan. 2, 1913; s. William John and Elizabeth (Wadeking) H.; B.S. in Agr., U. Ill., 1934, Ph.D. in Animal Sci., Nutrition and Physiology, 1958; M.S. in Poultry Sci., Purdue U., 1942; postgrad. So. Ill. U.; m. Mary Jane Harris, Mar. 20, 1937; children—Jane Ann (Mrs. Helmer R. Engh, Jr.), Karen Sue, Scott William. With Purdue U., 1936-42, Ind. Poultry Assn., 1942-44; mgr. Fairview Poultry Farms, 1944-51; faculty So. Ill. U., Carbondale, 1951-55, 58—, prof. animal industries, 1966—; asst. animal sci. U. Ill., 1955-57. Recipient Merit award Poultry and Egg Nat. Bd., 1961. Mem. A.A.A.S., Am. Inst. Biol. Sci., Poultry Sci. Assn., Worlds Poultry Sci. Assn., Ill. Egg Council (sec., Outstanding Service award 1963). Contbr. numerous articles to profl. jours. Developer biol. assay for evaluation of protein supplements for chick growth, interrelation of fat and protein in laying hens, egg quality at producer, retail and consumer level; co-inventor sound signal to measure albumen height of eggs. Home: 507 S. Dixon St., Carbondale, Ill. 62901.*

HINRICHS, Gustavus Detlef, chemist; b. Lunden, Holstein, Denmark, Dec. 2, 1836; s. Johan Detlef and Caroline C. E. (Andersen) H.; ed. Poly. Sch. and U. Copenhagen (Denmark); m. Auguste S. F. Springer, Apr. 1861 (dec. 1865); m. 2d, Anna C. M. Springer, July 1867 (dec. 1910); son, Carl Gustav. Prof. phys. sci. State U. of Ia., 25 years; prof. chemistry Coll. Pharmacy, St. Louis, 1889-1903; prof. chemistry, med. dept. St. Louis U., 1903-07. Originated graded courses in lab. work, mng. classes of several hundred students, about 1870; founded, 1875, and sustained, 1st state weather service in U. S.; did practical scientific work for U. S. and state authorities. Hon. and corr. mem. many scientific socs. in Austria, Eng., France, Germany and U. S. Author numerous books including: The Proximate Constituents of the Chemical Elements (32 plates), 1904; The Amana Meteorites, 1905; La Matière est Une, Paris, 1906. Contbr. articles to Comptes Rendus, Acad. Scis. Paris, Moniteur Scientifique, Paris, Trans. Acads. Scis. of Vienna, Paris, Proc. Am. Philos. Soc., others. His writings were mainly devoted to math. demonstration of unity of matter by quantitatively determined phys., chem. and crystallographical properties of all known chem. compounds. Died Feb. 14, 1923.

HINRICHS, Marie Agnes, Am. physiologist; b. Chgo., Sept. 22, 1892; d. Fred and Anne (Link) Hinrichs; A.B., Lake Forest Coll., 1917; Ph.D., U. Chgo., 1923, M.D., Rush Med. Coll., 1934. Instr., Vassar Coll., 1920-21; asst. physiology Nela Research Lab., 1922-24; research asso. U. Chgo., 1926-34; faculty So. Ill. U., 1935-49, prof., 1938-49, head dept. physiology, 1935-49; asso. prof. U. Ill., 1949-53; dir. bur. health service Chgo. Pub. Schs., 1953-58; lectr. health edn. Roosevelt U., 1958-65. Med. cons. A.M.A., 1960—. NRC fellow, 1924-26; cited by Alumni Lake Forest Coll., 1957. Mem. A.M.A., Ill., Chgo. med. socs., Am., Ill. pub. health assns., Nat. Ret. Tchrs. Assn., Am. Sch. Health Assn. (Distinguished Service award 1962), Ill. Nutrition Assn., Soc. Exptl. Biology and Medicine, Am. Physiol. Soc., Fedn. Am. Socs. for Exptl. Biology, Am. Ednl. Research Assn., Am. Heart Assn., Phi Beta Kappa, Sigma Xi. Research and numerous publs. on physiol. effects radiation on lower animals, physiol. effects alkaloids, studies in student health. Home: 344 E. Quincy St., Riverside, Ill. 60546. Office: 535 N. Dearborn St., Chgo. 60610.*

HINSBERG, Karl, German biochemist; b. Rombach, Lorraine, July 29, 1894; s. Robert and Ida (Emmel) H.; ed. U. Berlin, U. Fribourg; Ph.D. in biochemistry; m. Frieda Hinsberg, 1923; children—Christel, Helga. Full prof. physiology and biochemistry; with Med. Clinic, Fribourg; dir. lab. Med. Clinic, Cologne; with chemistry dept. Inst. Pathology, Berlin, Inst. Physiology and Biochemistry, Dusseldorf; prof. emeritus, 1963—. Author: (with Lang) Medizinische Chemie); (with Merten) Chemische Bestimmungs-Methoden in Klinischen Laboratorien; Das Geschwulstproblem. Home: 28 Solenanderstrasse. Office: Institute for Physiological Chemistry, Medical Academy, Dusseldorf, Germany.

HINSDALE, Wilbert B., Am. physician; b. Wadsworth, O., May 25, 1851; s. Albert and Clarinda (Eyles) H.; B.S., Hiram (O.) Coll., 1875, M.S., 1878; M.D., Cleveland Homoe. Med. Coll., 1887; hon. A.M., Hiram Coll., 1897, U. Mich., 1934; m. Estella Stone, Nov. 25, 1875 (dec.); 1 son, Albert Euclid (dec.). Prof. internal and clin. medicine, homoe. dept. U. Mich., Ann Arbor, 1895-1922; dean homoe. med. dept., 1895-1922, also med. dir. of coll. hosp.; custodian Mich. archaeology, Univ. Museums, 1922-29; asso. in charge of div. of the Great Lakes, Mus. of Anthropology, U. Mich. from 1929. Pres. Mich. Acad. of Science, Arts and Letters, 1931. Mem. Am. Assn. Archaeologists. Author: Primitive Man in Michigan; The Indians of Washtenaw County; The First People of Michigan; Archaeological Atlas of Michigan; Distribution of the Aboriginal Population of Michigan; Perforated Indian Crania in Michigan (with E. F. Greenman), 1936. Died July 25, 1944.

HINSELMANN, Hans, German gynecologist; b. Neumünster, Aug. 6, 1884; practiced medicine, specializing in gynecology, Hamburg-Alton, Germany; dir. Hamburg Women's Clinic; introduced examination of vagina and cervix by colposcope. Died Apr. 18, 1959.

HINSEY, Joseph Clarence, Am. anatomist; b. Ottumwa, Ia., Apr. 29, 1901; s. Joseph Edgar and Sarah Belle (Majors) H.; student Iowa Wesleyan College, 1918-20; B.S., Northwestern U., 1922, M.S., 1923, D.Sc., 1951; Ph.D., Washington U. Sch. of Medicine, St. Louis, Mo., 1927; Doctor of Science (honorary), Union College, 1955, Iowa Wesleyan College, Mount Pleasant, Ia., 1958; married Sarah Lillian Callen, June 18, 1926; children—Elaine (Mrs. Donald P. Reynolds), Joseph. Assistant in zoology Northwestern University, 1921-23; instructor in biology Western Reserve University, 1923-24; asst. in neuro-anatomy, Washington U., 1924-27, asst. prof., 1927-28; asst. prof. in neuro-anatomy, Northwestern U. Med. Sch., 1928-29, asso. prof., 1929-30; prof. of anatomy, Stanford U., 1930-36; prof. of physiology and head dept., Cornell U. Med. Sch., New York, 1936-39; prof. of anatomy and head dept., 1939-53, dean, 1942-53; dir. N.Y. Hosp.-Cornell Med. Center, 1953—; prof. neuroanatomy Cornell U. Med. Coll., 1956—. Chairman of China Medical Board of N.Y., 1956—. Mem. President's Commn. on Med. Needs of Nation, 1952. Mem. bd. mgrs. Meml. Hosp., N.Y. City. Trustee Sloan-Kettering Institute; trustee, China Med. Bd.; mem. bd. Lahey Clinic Found., 1965—, Inst. for Muscle Disease, Inc., 1965—. Recipient Merit award Nu Sigma Nu, 1956; Abraham Flexner award for service to medical education Association American Medical Colleges, 1958; Medal award New York Medical College, 1960; distinguished service award N.C. Coll. Medicine, 1964. Fellow A.A.A.S. (vice pres. 1950), N.Y. Acad. of Sci. (v.p. 1949), Assn. Am. Med. Colls. (pres. 1949; chmn. exec. council 1946-54), California Academy of Science; member Am. Assn. Anatomists, Am. Physiol. Soc., Am. Neurological Soc., Soc. Exptl. Biology and Med., Harvey Soc., N.Y. Acad. Medicine, Calif. Acad. Medicine, S.A.R., Pilgrims U. S., Newcomen Soc., Saint Andrews Society, Phi Delta Theta, Nu Sigma Nu, Phi Beta Kappa, Alpha Omega Alpha, Sigma Xi. Contbr. to med. jours. Home: 156 Brewster Rd., Scarsdale, N.Y. Office: 525 E. 68th St., N.Y.C. 10021.*

HINSHAW, J. Raymond, Am. physician; b. Butler, Okla., Oct. 8, 1923; s. J. R. and Lucille Iris (Whitenack) H.; B.A., U. Okla., 1943, M.D., 1946; D.Phil., Oxford U., 1951. Faculty, U. Rochester (N.Y.), 1955—, prof. surgery, 1967—; chief surgery Rochester Gen. Hosp., 1966—; cons. VA Hosp., Batavia, N.Y., also Highland, Rochester State, Genesee hosps. (all Rochester). Rhodes scholar Oxford U. Mem. Soc. U. Surgeons, Am. Soc. Exptl. Pathology, A.M.A., Central Surg. Soc., Med. Soc. State of N.Y., Internat. Soc. Surgery, Phi Beta Kappa. Research, numerous publs. on burns and wound healing. Home: 748 Quaker Rd., Scottsville, N.Y. 14546.*

HINSHAW, Randall, Am. economist; b. La Grange, Ill., May 9, 1915; s. Virgil G. and Evelyn (Piltz) H.; A.B., Occidental Coll., 1937, M.A., 1939; Ph.D., Princeton, 1944; m. Pearl Electa Stevens, June 19, 1949; children—Frederic Randall, Robert Louis, Elisabeth Mary. Spl. adviser internat. financial policy U. S. Mission to OEEC, USRO, Paris, 1952-57; vis. research fellow Council Fgn. Relations, N.Y.C., 1959-60; prof. econs. Claremont (Cal.) Grad. Sch., 1960—, chmn. dept., 1967—; vis. prof. Yale, 1957-58, Oberlin Coll., 1958-59, U. So. Cal., 1963-64, Johns Hopkins, 1965-67. U. S. del. UN Spl. Com. on East-West Payments, Geneva, 1955-57; U. S. expert OEEC spl. com. central banking experts which drafted European Monetary Agreement, 1955. Mem. Am. Econ. Assn., Econometric Soc., Phi Beta Kappa. Author: The European Community and American Trade: A Study in Atlantic Economics and Policy, 1964. Editor, contbr. Monetary Reform and the Price of Gold, 1967. Research on price elasticity of demand for U. S. imports. Home: 755 W. 8th St., Claremont, Cal. 91711.*

HINSHELWOOD, Sir Cyril (Norman), English chemist; London, Eng., June 19, 1897; s. Norman Macmillan and Ethel (Smith) H.; M.A., Balliol Coll., Oxford U., 1924, D.Sc., 1947, D.C.L., 1960; D.Sc., univs. Dublin, 1936, London, 1947, Leeds, 1952, Sheffield, 1954, Cambridge, 1955, Bristol, 1956, Hull, 1957, Wales, 1958, Ottawa, 1961, Southampton, 1964. Fellow, tutor Trinity Coll., Oxford, 1921-37; Dr. Lees prof. chem. Oxford, 1937-64; sr. research fellow Imperial Coll., London, 1964—; fellow Exeter Coll., 1937—. Chmn. Fuel Research Bd., 1950-55; mem. adv. council sci. policy Brit. Govt., 1953-56; hon. adv. sci. com. Natl. Gallery. Created Knight, 1948; recipient Lavoisier medal French Chem. Soc., Guldberg medal Oslo; Davy and Royal Medalist Royal Soc.; Longstaff and Faraday medalist Chem. Soc.; Nobel prize for chemistry (with N. N. Semenov) 1956; Grand Officer Order of Merit (Italy), 1956; Avogadro medal Accad. dei XL, Rome, 1956. Fellow Royal Soc., 1929 (past pres.); mem. Nat. Acad. of Sciences, Pontifical Academy. Chemical Society (president 1946-48); honorary member French, Belgian, Swiss, Italian chem. socs., Spanish Society Physics and Chemistry; fgn. mem. Accademia dei XL Rome, U.S.S.R. Academy of Scis., Accademia Nazionale dei Lincei (Rome), Real Academia de Ciencias Madrid, A.A.A.S. Author: The Kinetics of Chemical change in Gaseous Systems, 1926; Kinetics of Chemical Change, 1946; Chemical Kinetics of the Bacterial Cell, 1946; The Structure of Physical Chemistry, 1951; also papers in sci. jours. Authority on chem. kinetics; elucidated important aspects of hydrogen-oxygen reaction in formation of water which led to better understanding of inorganic and organic reactions in gen.; discovered certain cellular kinetic phenomena, as bacterial adaptation. Died 1968.

HINTENBERGER, Heinrich, physicist; b. Ober Grafendorf, Austria, Feb. 7, 1910; s. Heinrich and Betty (Gotsbachner) H.; Ph.D. in Physics, U. Vienna; m. Theresa Zippelt, 1942; 1 son, Heinrich-Karl. Asst. Inst. Physics, U. Vienna, with Chemistry Lab., 1935-38; with Lab. Physics Research Siemenswerke, Berlin, 1938-43, Kaiser-Wilhelm Inst. Chemistry, Berlin, 1944-48, Theodor-Kocher Inst., U. Bern, 1949-51; instr. U. Bern; with Max Planck Inst. Chemistry, 1952-55; instr. U. Mayence, 1952-55; affiliate prof., 1956-57; sci. collaborator Max Planck Soc., 1956-57; dir. div. earth spectroscopy Max Planck Inst. Chemistry of Mayence, 1958—. Mem. German Soc. Physics, Am. Phys. Soc. Author: Methoden und Unwendungen der Massenspektrsocopie, 1953; Nuclear Masses and Their Determination, 1957. Research and numerous publs. on earthspectroscopy, nuclear physics, geology, isotopes, meteorites. Address: 23 Saarstrasse, Mainz, Germany.

HINTEREGGER, Hans Erich, Am. physicist; b. Waidhofen, Austria, Sept. 3, 1919; s. Johann and Marie (Hrusehka) H.; Dipl. Ing., Inst. Tech., Vienna, 1944, Dr.techn., 1947; m. Erna E. Waldmuller, May 16, 1942; children—Hans F., George F., Elisabeth E., Peter H., Bernhard G., Maria E. Came to U. S., 1951, naturalized, 1956. Sr. scientist, chief solar ultraviolet br. Air Force Cambridge Research Labs., Bedford, Mass., 1962—; mem. space sci. steering com. NASA. Recipient Meritorious Civilian Service award, 1961; Air Force Exceptional Service award, 1966. Fellow Optical Soc. Am.; mem. Am. Phys. Soc., Am. Geophys. Union, Internat. Union Geodesy and Geophysics, Internat. Astron. Union. Research and numerous publs. in solar physics, ionospheric physics and planetary aeronomy, quantitative measurement of absolute intensity and variation of solar radiation in range between 1 and 1300 Angstroms and its effects on earth's upper atmosphere. Home: 140 Newtonville Av., Newton, Mass. 02158. Office: Air Force Cambridge Research Lab., L. G. Hanscom Field, Bedford, Mass. 01730.*

HINTON, Sir Christopher (Lord Hinton of Bankside), English engr.; b. Tisbury, England, May 12, 1901; s. Fredrick Henry and Kate (Christopher) H.; M.A., Trinity Coll., Cambridge U., 1926; D.Eng. (honorary), Liverpool U., 1955; D.Sc. (hon.), London U., 1956, Oxford U., 1957, Cambridge University, 1960, Southampton University, 1962; LL.D., Edinburgh University, 1958; married Lillian Boyer, January 19, 1931; 1 daughter, Susan Mary. Chief engr. I.C.I. (Alkali) Ltd., 1931-40; dep. dir. gen. Royal Filling Factories, 1942-46; head indsl. div. Brit. Atomic Energy Orgn., 1946-57; chairman of the Central Electricity Generating Bd., 1958-64; special adviser to the minister of transport on coordination of transport, 1965—. Hon. asso. Manchester Coll. Tech.; hon. fellow Trinity College of Cambridge University, 1957. Created Knight Bachelor, 1951; Knight Comdr., Order of the British Empire, 1957; recipient 1st class honors in mech. scis., 1925, John Wimbolt prize, 1926, 2d Yeats prize, 1926, Cambridge University; created baron, 1965. Fellow Royal Society, 1954, Royal Society Arts, Institute of Fuel, British Institute Management; mem. Instn. Gas Engrs. (hon.), Inst. Chem. Engrs., Inst. of Electrical Engineers, Institute of Metals (hon.), Institute of Mech. Engrs., Inst. Civil Engrs. Designed and directed Brit. nuclear indsl. establishments, notably large nuclear power plant at Calder Hall. Home: Tiverton Lodge, Dulwich, Common, London S.E. 21, Eng.

HINTON, Howard Everest, entomologist; b. San Luis Potosi, Mexico, Aug. 24, 1912; s. George Boole and Emily (Wattley) H.; B.Sc., U. Cal. at Berkeley, 1934; Ph.D., U. Cambridge (Eng.), 1939, Sc.D., 1957; m. Margaret Rose Clark, Dec. 18, 1937; children—Charlotte Boole (Mrs. Terry Goodhill), James Sebastian, Geoffery Everest, Teresa Ann. Jr. curator U. Mus. Zoology, Cambridge 1937-39; asst. keeper Brit. Mus. Natural History, 1939-49; faculty U. Bristol (Eng.), 1949—, prof. entomology, 1964—. Hon. cons. entomologist Ministry Agr., 1947-49. Fellow Royal Soc., 1961, Australian Acad. Scis. (sr.), Inst. Biology, Royal Entomol. Soc. (past v.p.); mem. Soc. Brit. Entomology (past pres.). Research and numerous publs. on taxonomy and classification of beetles and lepidoptera, biology and classification of insect pupae; discovered plastron respiration in pupae and eggs of insects, suspended animation in insects; research on significance of pupal stage. Home: 16 Victoria Walk, Bristol 6, Eng.*

HINTON, James, English surgeon; b. Reading, Eng., 1822; s. John H. Hinton; diploma St. Bartholomew's Hosp., 1847; m. Margaret Haddon, 1852; surgeon, passenger ship to China; med. officer, ship to Jamaica; returned to Eng., 1850; began practice medicine in London, 1850. Author: Man and his Dwelling-Place, 1858; Life in Nature, 1862; The Mystery of Pain, 1866; Thoughts on Health, 1871; Atlas of Diseases of the Membrana Tympani, 1874; Questions of Aural Surgery, 1874; The Place of the Physician, 1874. Co-editor Yearbook of Medicine, 1863. Most skillful aural surgeon of times; performed 1st operation for mastoiditis in Eng., 1868. Died 1875.

HINZ, Carl Frederick, Jr., Am. physician; b. Cleve., Apr. 9, 1927; s. Carl Frederick and Marie (Jones) H.; B.S., Western Res. U., 1948, M.D., 1951; m. Joan Herndon, June 5, 1953; children—Elizabeth, Richard, Catherine, Gretchen. Faculty dept. medicine Western Res. U. Sch. Medicine, Cleve., 1953-67, asst. prof., 1961-67, research asso. div. research in med. edn., 1964-67; prof., asso. dean U. Conn. Sch. Medicine, 1967——. Markle scholar, 1959-64. Mem. Am. Soc. Clin. Investigation, Am. Assn. Immunologists, Am. Soc. Hematology, Central Soc. Clin. Research, Am. Fedn. Clin. Research, Ohio Med. Assn., Acad. Medicine Cleve. Research in immunology, hemolytic anemias, complement and autoimmunity. Developer methods for measuring clin. performance of physicians. Home: 11 Highwood Dr., Avon, Conn. 06001. Office: U. Conn. Sch. Medicine, Hartford Plaza, Hartford 06105.*

HIPPARCHOS, Greek astronomer, mathematician; b. Nicaea, Bithynia (now Iznik, Turkey), circa 190 B.C.; made observations mainly from Rhodes. Author: Geography, On Things Borne Down by their Weight (both lost), a commentary on Aratus and Eudoxus. Earliest known systematic astronomer, greatest astronomer of antiquity; introduced notion of epicycles and eccentrics; calculated inclination of ecliptic; discovered precession of equinoxes, eccentricity of sun's apparent orbit, inequalities in moon's orbit; 1st to classify stars according to brightness; made 1st known comprehensive map of sky, listing more than 850 stars; determined parallax of moon; made improved measurements of distances and sizes of moon and sun; invented stereographic projection; measured length of year to 6 minutes; 1st to use latitude and longitude as designations for place; laid founds. of trigonometry with constrn. of table of chords (trigonometric sines); believed to have made observations and conclusions upon which Ptolemy based his geocentric theory. Died 125 B.C.

HIPPASOS, mathematician; flourished 500 B.C.; added to doctrine of Pythagoras shortly after his death; constructed circumscribed sphere of dodecahedron; developed theory of harmonic mean; added double octave and 5th beyond octave to 3 constant intervals (octave, 5th, 4th); discussed theory of mus. ratios with mus. Lasos; became leader of sci. group of Pythagoreans, known as Mathematikoi. Expelled for deviating from Pythagoras' teachings and died in shipwreck.

HIPPIAS OF ELIS, mathematician; flourished 2d half of 5th century B.C.; tchr., orator throughout Greece; employed on state bus. by city of Elis; depicted in Plato's Hippias Major, Hippias Minor; author numerous works, few known by name; discovered quadratrix, the 2d curve recognized by Greek geometers.

HIPPOCRATES, Greek physician; b. Cos (Aegean Island), circa 460 B.C.; s. Heraclides; reputedly ed. by his father and Herodicus, also to have visited Egypt early in life and studied med. works attributed to Imhotep; probably married; children are said to include Thessalus, Dracon. Allegedly taught at various places including Athens; founder sch. of medicine at Cos; author of some 60 works called Hippocratic Collection (possible that most or all of collection was written by his followers). Regarded as father of medicine; rejected superstitions about diseases and based his med. practice on observation and study; held that disease is purely phys. phenomenon (said to have even attributed epilepsy to natural rather than demonic causes), and thus attempted to dissociate medicine from superstition; held that disease is caused by imbalance of vital body fluids (humors), and considered imcomplete indigestion to be cause of such imbalance; considered the body an organism and held that it must be treated as a whole; accurately described disease symptoms; recognized and described pneumonia, left earliest known description of puerperal fever, and gave first known description of epilepsy in children; held that physician should interfere as little as possible with natural healing processes, but stated that desperate diseases need desperate remedies; emphasized importance of noting individual differences in patients concerning their ability to cope with diseases and tolerate therapeutic measures; placed emphasis on diet, fresh air, medicinal water, and gymnastics for better health; believed in moderation of diet, and in cleanliness and rest for sick or wounded; developed improved diagnostic techniques; developed theory of human procreation and pre-natal devel.; his primary importance was to later generations who considered him the ideal physician, and his high ethical and med. standards are still commemorated in the Hippocratic Oath which is ascribed to him. Died Larissa, Thessaly, circa 377 B.C.

HIPPOCRATES OF CHIOS, Greek mathematician; b. Chios, Greece, circa 470 B.C.; began career as mcht., but lost money, circa 430 B.C.; became geometer, Athens; said to have been 1st math. tchr. to accept money for teaching. Author: Elements of Geometry (1st book of its kind, anticipates much of Euclid). Concerned with squaring circle; reduced problem of doubling cube to finding 2 mean proportionals; demonstrated proof by contradiction. Died circa 400 B.C.

HIPPON (or Hipponax), philosopher; probably from Samos; flourished in Athens, Greece, 2d third of 5th century B.C.; explained physiol. and path. facts as processes of desiccation or humidification; derived water as principle of all things from his observations of animal semen; considered semen basis of soul, located in brain; noted devel. of human body from embryo to maturity.

HIRAI, Tokuzo, Japanese virologist; b. Osaka, Japan, Nov. 19, 1910; s. Masahichi and Tatsu (Uesugi) H.; grad. Kyoto U. Faculty Agr., 1938, Dr.Sci. in Agr., 1956; m. Yuriko Kato, Mar. 19, 1939; children—Atsushi, Masashi, Kiyoshi, Tamiko. Plant pathologist Imperial Agrl. Exptl. Sta., Tokyo, Japan, 1938-45; prin. plant pathologist Nat. Tohoku Agrl. Expt. Sta., Morioka, Japan, 1946-54; prof. Faculty Agr., Nagoya (Japan) U., 1955——. Mem. Japanese (councilor), Am. phytopath. socs., Japanese Virus Soc. (councilor), Scandinavian Plant Physiologists Soc. Author: Plant Virology, 1959; Plant Pathology, 1962, rev. 1967; also numerous articles. Discovered 2 antibiotics which inhibit plant virus multiplication in plant tissues. Home: 5-55 Nishisatocho, Nagoya, Aichi, Japan.*

HIRAKI, Kiyoshi, Japanese physician; b. Okayama City, Japan, Dec. 30, 1910; s. Kazuta and Hisayo Hiraki; M.D., Okayama Med. Coll., 1934; m. Kumiko Himei, May 10, 1935; children—Masako (Mrs. Hiroshi Fujiwara), Yoshio, Kenjiro, Shunkichi. Faculty, Okayama Med. Coll., 1942-52, asst. prof., 1946-52; prof. medicine Okayama U. Med. Sch., 1952——. Recipient Exhibit prizes VI, VIII, Internat. Congress Hematology, 1956, 60. Mem. N.Y. Acad. Scis., Japan Soc. Internal Medicine (mem. com.), Japanese Cancer Assn., Japan Hematological Soc. Author: (with H. Sanada) Manual for Daily Hematological Practice, 1964; Agricultural Drug Poisoning, 1966. Research and numerous publs. on structure of blood vessels of bone marrow, clin. applications of bone marrow tissue culture for diagnosis of leukemias, physiopathology and treatment of agrl. drug poisoning, role of virus in malignant tumors of man and animals, cancer chemotherapy. Home: 3-chome Uchida honmachi, Okayama City, Japan.*

HIRASAWA, Kō, Japanese anatomist; b. Niigata Prefecture, Japan, 1900; grad. Kyoto U., 1924; M.D.; studied in Germany, Switzerland. Asst. prof. Kyoto U.; asst. prof. Niigata Med. Coll., became prof., 1930; named prof. Kyoto U., 1946. Recipient Japan Acad. Prize, 1951. Author: The Highest Center of Cerebrum; Progress of Medicine. Research on central nervous system.

HIRASE, Sakugoro, Japanese botanist; b. Fukui, 1856; painter with botany sect., sci. sch. Tokyo U., from 1888; later lived in Kyoto. Recipient prize Imperial Acad., 1912. Studied history of gingko tree; 1st to discover moving spermatozoon of phanerogam, 1896. Died 1925.

HIRATA, Morizo, Japanese physicist; b. Hiroshima, Japan, 1906; grad. Tokyo U., 1928; D.Sc. Began as asst. prof. Tokyo U., prof., 1932——. Invented flat-headed harpoon (replaced old-type harpoons in all whaling vessels); research on destruction various solid bodies, elec. method of whaling, underwater franjectory, rebound of pointed head rifle shot from surface of water.

HIRATA, Yukimasa, Japanese physician; b. Gifu Prefecture, Japan, May 5, 1925; s. Jiro and Katuko (Hirose) H. degree Kyushu (Japan) U., 1948, Spl. Degree, 1953; m. Fumi Yamamoto, Nov. 27, 1949; children—Michi, Nobuko. Sr. instr. internal medicine Faculty Medicine, Kyushu U., Fukuoska, Japan, 1956-60, 62——; research fellow Inst. Exptl. Pathology, Jewish Hosp., St. Louis, 1960-62. Mem. Japan Diabetic Soc. (mem. council 1958——, Japan Endocrinological Soc., Japanese Soc. Internal Medicine, Am. Diabetes Assn. Author: Diabetes Mellitus, 1960 (with Goroo Kusunoki); also numerous articles. Performed diabetes survey, Kyushu, Japan, 1957——; discovered diabetes in obese mice produced by goldthio-glucose, strong effect of xylitol on insulin secretion in dogs; research on relation between insulin antibody and blood vessel lesions. Home: 3-30, Yanagochi 2, Fukuoka City, Japan.*

HIRAYAMA, Seiji, Japanese astronomer; b. Sendai, Japan, 1874; grad. sci. sch. Tokyo U.; became prof. Tokyo U., 1919; mem. Imperial Acad. Research on calendar and astron. dynamics, latitudes and comets, star evolution; made. statis. surveys of asteroids. Died 1943.

HIRN, Gustav Adolphe, French physicist; b. Alsace, France, 1815, pub. research on mech. equivalent of heat, 1858. Investigated speed of streaming gas, 1839, ventilators, 1845; developed method for studying heat-motors; built curve and torsion dynamometer, also one-cylinder steam engine, exhaust gas warmer; constructed overheater (hyperthermogenerator), 1855; developed mech. heat theory, 1862, (with M. Hallaker) a practical thermodynamics, 1882-83; theoretical and applied research on current law and gas pressure as function of temperature, 1886; sought to determine mech. equivalent of heat. Died 1890.

HIROHATA, Ryozo, Japanese biochemist; b. Onomichi, Japan, Dec. 1, 1893; s. Kushiro and Kane (Takagaki) H.; student Okayama Med. Coll., 1913-17; grad. Kyoto Imperial U. Med. Sch., 1922, Dr.Med. Sci., 1928; m. Katsuko Utsumi, Mar. 23, 1922; children—Emiko (Mrs. Yasukazu Abe), Nobuko (Mrs. Kiyoshi Kuramoto), Tatsuo, Tomio. Research asso. Inst. for Physiol. Chemistry, U. Leipzig (Germany), 1931-32, Chem. Inst., U. Zürich, 1933; prof. biochemistry Govt. Med. Coll. (later Taipei Imperial U.), Formosa, 1923-43; prof. med. chemistry Kyushu U. Med. Sch., 1943-57, emeritus, 1957——; prof., chief lab. for protein chemistry Yamaguchi Med. Sch., Ube, Japan, 1957-63; prof. biochemistry Dailchi Coll. Pharm. Scis., 1963-66; prof., dir. div. food and nutrition Nakamura Gakuen Coll., Fukuoka, Japan, 1966——. Adj. prof. biochemistry Kumamoto Med. Sch., 1947; adj. prof. nutrition Women's Coll. Fukuoka Prefecture, 1954-55. Mem. Am., Japanese chem. socs., Japanese Biochem. Soc. (hon.), Japanese Soc. Food Nutrition, Japanese Vitamin Soc., Fukuoka Med. Assn., Soc. Mass Spectroscopy Japan, Japanese Soc. Biophysics. Co-author: Diagnosis and Treatment of Malnutrition, 1946; Standard Practical Biochemistry, 1953; Handbook of Biochemistry, 1953; Neurochemistry, 1954; Series of Biochemistry, 1958; also numerous articles. Research on mugiline (protamine of mugil jap), including clarification of amino acid sequence; Japanese foods, including chem. constituents, absorption rate, biol. value of their proteins, increasing nutrition value of soy-bean protein; rabbit kidney demethylase, including its specificity, activity mechanism, and its co-enzyme; hydrolysis velocity of dipeptides by means of acids; discovered steric hindrance cause by some amino acids in peptides using many kinds of dipeptides synthesized as substrates. Home: 14-3 Honnancho, Takamiya, Fukuoka, Japan.*

HIRONO, Takuzo, Japanese seismologist; b. Hyogo Prefecture, Japan, Mar. 24, 1912; s. Masazo and Koai (Okamoto) H.; student Tokyo (Japan) U., 1932-35, D.Sc., 1951; m. Toshiko Murauchi, May 14, 1944; children—Mayumi, Hideki. Staff seismol. sect. Central Meteorol. Obs., 1935-52; chief seismol. lab. Meteorol. Research Inst., Tokyo, 1952-57, 65——; chief seismol. sect. Japan Meteorol. Agy., Tokyo, 1957-65. Mem. Seismol. Soc. Japan, Volcanic Soc. Japan, Am. Geophys. Union, Seismol. Soc. Am. Author: (with Kiyoo Wadati) Earthquakes and Tsunami, 1955; also articles. Discovered cause of ground subsidence common in Japan; mathematically elucidated character elastic waves generated in half-space elastic solid by forces applied on surface. Home: 1894 Shimo-Jujo, Kita-ku, Tokyo. Office: 35 Koenji Kita-4, Suginami-ku, Tokyo, Japan.*

HIROSE, Hideo, Japanese astronomer; b. Himeji, Japan, 1909; grad. Tokyo U., 1932. Past chief technician photog. sect. Tokyo Astron. Obs.; prof. Tokyo U.; observed solar eclipse, Reibun Island, Japan, 1948. Recipient prize for re-discovery of Daniel Comet, Japan Astronomy Acad.; Research Fund grantee. Author: The Sun and the Moon; also articles. Research on comets, variable stars; photog. observations of comets and minor planets.

HIROSE, Teruo, surgeon; b. Tokyo, Japan, Jan. 20, 1926; s. Yohei and Seiko (Ogushi) H.; B.S., Tokyo Coll., 1944; M.D., Chiba U., 1948, Ph.D., 1958; m. Shigeko Masuda, Dec. 18, 1955; 1 son, George. Resident thoracic surgery Hahnemann Hosp., Phila., 1955-56; practiced medicine, specializing in gen. and thoracic surgery, Bronx, N.Y., 1965——; asst. prof. surgery Chiba U. Hosp., 1958; faculty Sch. Medicine Chiba U., 1958-63, N.Y. Med. Coll., 1963-65; attending surgeon St. Barnabas Hosp., 1965——, dir. cardiovascular lab., dir. exptl. surg. lab. dept. thoracic surgery, 1966; chief vascular surgery Union Hosp., 1966——. Recipient Bronze medal for exhibit A.M.A., 1965, 1st prize for exhibit N.Y. State Med. Assn., 1966. Fellow A.C.S., Am. Coll. Chest Physicians, Am. Coll. Cardiology; mem. Am. Assn. Thoracic Surgery, Soc. Thoracic Surgeons. Contbr. chpt. to Bleeding Problem in Surgery, 1966. Research and numerous publs. on single pass low prime-heart lung machine, revascularization of posterior myocardium with gastro epiploic arterial implant, reconstruction of aortic valve with autologous tissue, low cardiac pressure suction machine, influence of ethylene oxide gas to hemolysis of blood during heart lung bypass. Address: 5830 Tyndall Av., Bronx, N.Y. 10471.*

HIROTO, Ikuichiro, Japanese otolaryngologist; b. Yonago-City, Japan, June 21, 1918; s. Setsuzo and Taka (Watanabe) H.; grad. Kyoto (Japan) U., 1942, Dr.Med.Sc., 1950; m. Kiyoe Nakashima, July 2, 1942; children—Masako, Junko, Michiko. With dept. otolaryngology Kyoto U., 1947-60, asst. prof., 1960; prof. otolaryngology Kurume U., Fukuoka, Japan, 1960——. Mem. Otorhinolaryngol. Soc. Japan, Japan Soc. Logopedics and Phoniatrics, Japan Soc. Head and Neck Surgery. Author: (with T. Goto, T. Kitamura) Oto-Rhino-Laryngology, 1960; Minor Otolaryngology, 1964; also numerous articles. Research on function of laryngeal muscles during phonation of Japanese speech sounds, mechanism of phonation in normal and path. condition were classified using X-ray cinematography, electromyography, measurement subglottal air pressure. Home: 6-459 Sasayama-cho, Kurume, Fukuoka, Japan.*

HIRSCH, Edwin Frederick, Am. physician; b. Milw., Aug. 18, 1886; s. John Fred and M. Louisa (Weiland) H.; A.B., Northwestern U., 1910; A.M., U. Ill., 1911; Ph.D., U. Chgo., 1914; M.D., Rush Med. Coll., 1916; D.Sc., Morningside Coll., 1955; m. Helen Kotas, Mar. 19, 1949; children—Catharine H. (Mrs. Alex Ayala), Helen S. (Mrs. W. P. Kent), Jean H. (Mrs. Robert Priest). Dir., Henry Baird Farill Lab., St. Luke's Hosp., Chgo., 1919-59, dir. emeritus, 1959-—; faculty dept. pathology U. Chgo. Sch. Medicine, 1913-51, asso. prof., 1944-51, asso. prof. emeritus, 1951-—. Recipient George H. Coleman award Inst. Medicine Chgo., 1960. Mem. Am. Assn. Pathology and Bacteriology, A.M.A., Am. Soc. Clin. Investigation, A.C.P., Am. Coll. Pathology, Am. Soc. Clin. Pathology, Central Soc. Clin. Research. Author: Pathology in Surgery, 1953; Biography of Frank Billings, 1966; also numerous articles. Research on role of lipids in evolution of atherosclerosis, innervation of heart in vertebrates, innervation lungs and effects lung reimplantation. Home: 5830 Stony Island Av., Chgo. 60637. Office: 1753 Congress Pkwy., Chgo.*

HIRSCH, Gottwalt Christian, German paleontologist; b. Magdebourg, Nov. 14, 1888; s. Max and Julie (Diesterweg) H.; ed. univs. Tübingen, Naples, Halle, Salle; Ph.D. in natural sci.; m. Anna Marie Jaekel, 1934; 4 children. Reader, instr. U. Tübingen; full prof. U. Utrecht, 1920-45; prof. cytology U. Göttingen, 1945. Mem. Acad. Scis. Leopoldina, zool. and cytological socs. Germany, Eng., U. S. Author: Physiology and Comparative Anatomy; General Cytology; Cytology and the Secret. Home: 24 von Ossietskystrasse, 34 Göttingen. Office: 28 Berlinerstrasse, Göttingen, Germany.

HIRSCH, H., German physician; s. V. and E. (Haselbach) H.; Ph.D., U. Bonn, 1948, M.D., 1949; m. Irene Winkhaus, Apr. 8, 1952; 3 children. Prof. physiology U. Cologne (Germany), 1962-—, dir. Inst. Normal and Path. Physiology. Research on survivial, revival, oxygen consumption, blood flow and electrophysiology of brain, microcirculation. Home: 13 Theodor-Körner-Strasse, 5038 Rodenkirchen. Office: 39 Robert-Koch-Strasse, 5 Cologne, West Germany.*

HIRSCH, James Gerald, Am. physician; b. St. Louis, Oct. 31, 1922; s. Mack J. and Henrietta B. (Schiffman) H.; B.S., Yale, 1942; M.D., Columbia, 1946; m. Marjorie Manne, June 6, 1943; children —Ann I., Henry J. NRC fellow Rockefeller Inst., N.Y.C., 1950-52, faculty medicine and microbiology, 1952-—, prof., mem. inst., 1960-—. Mem. Harvey Soc.; Am. Assn. Immunologists, Soc. Am. Bacteriologists, Am. Acad. Microbiology, Soc. Exptl. Biology and Medicine, A.A.A.S., Am. Soc. Clin. Investigation, Assn. Am. Physicians, Alpha Omega Alpha. Author: (with R. J. Dubos) Bacterial and Mycotic Infection of Man, 1965. Research and publs. in cell biology, immunology especially mechanisms, role of white blood cell. Home: 18 The Sail, East Islip, N.Y. 11730. Office: Rockefeller U., 66th St. and York Av., N.Y.C. 10021.*

HIRSCH, Jerry, Am. psychologist; b. N.Y.C., Sept. 20, 1922; s. Samuel M. and Millie (Barnett) H.; student Johns Hopkins, 1938-40; B.A., U. Cal., 1952, Ph.D., 1955; m. Marjorie J. Barrie, Aug. 29, 1950; 1 son, Wesley M. NSF fellow U. Cal., Berkeley, 1955-57; asst. prof. psychology Columbia, 1956-60; NIH fellow Center For Advanced Study in Behavioral Scis., Stanford, Cal., 1960-61; asso. prof. U. Ill., Urbana, 1960-63, prof., 1963-—. Mem. com. Genetic behavior Social Sci. Research Council, 1962-65; mem. behavioral scis. tng. com. Nat. Inst. Gen. Med. Scis. NIH, 1966-—. Recipient Aux. Research award Social Sci. Research Council, 1962. Mem. A.A.A.S., Am. Assn. U. Profs., Animal Behavior Soc., Am. Genetics Assn., Am. Inst. Biol. Scis., Am. Psychol. Assn., Am. Soc. Naturalists, Ecol. Soc., Genetics Soc. Am., Psychonomic Soc., Soc. Study Evolution. Contbr. to Roots of Behavior, 1962; Expanding Goals of Genetics in Psychiatry, 1962. Research, publs. on heredity in behavioral scis., analysis of racial differences and behavior, chromosome effects in tropistic behavior. Home: 2012 Zuppke Circle, Urbana, Ill. 61801.*

HIRSCHBERG, Joseph Gustav, Am. physicist; b. Chgo., Apr. 13, 1921; s. Joseph Gustav and Lillian (Kahn) H.; A.B., Dartmouth, 1943; M.S., U. Wis., 1951, Ph.D., 1952; m. Ginette Henriette Tetard, Apr. 26, 1947; children—Dorothy Jean, Joseph Gerald, Anne Marie, Lynn Susan. Fulbright scholar Ecole Normale Superieure, Paris, France, 1952-53; research asso. U. Wis., Madison, 1953-58; research physicist Princeton, 1958-65; prof. physics, chmn. dept. U. Miami, Coral Gables, Fla., 1965-—. Cons., AEC Princeton Plasma Physics Lab., 1965-—; exchange prof. U. Paris, 1964. Fellow Am. Phys. Soc. (plasma physics div.), Optical Soc. Am.; mem. Phi Beta Kappa, Sigma Xi, Sigma Pi Sigma. Contbr. to McGraw Hill Handbook of Physics, 1958, Methods of Experimental Physics, vol. 5, 1963. Research and publs. in spectrum of ionised helium; devel. methods of applying the photomultiplier to Fabry Perot interferometer; co-discoverer of absorption line of telluric sodium in light of sun nr. horizon; devel. multichannel Fabry Perot interferometer; applications of spectroscopy to plasma physics. Home: 1046 Alfonso Av., Coral Gables, Fla. 33146.*

HIRSCHBERG, Julius, German ophthalmologist; b. Potsdam, Germany, Sept. 18, 1843; asst. to von Graefe; prof. ophthalmology U. Berlin. Author: Der Elektromagnet in der Augenheilkunde, 1885; Wörterbuch der Augenheilkunde, 1887; Hilfswörterbuch zu Aritophanes, 190 1898; Geschichte der Augenheilkunde im Altertum, 1899; Die arabischen Lehrbücher der Augenheilkunde, 1905. Founder jour. Zentralblatt für praktische Augenheilkunde. Credited with introducing electromagnet in ophthalmology, 1885. Died Berlin, Feb. 17, 1925.

HIRSCHBOECK, John Stephen, Am. physician; b. Milw., Mar. 25, 1910; s. Stephen H. and Katherine (Heiser) H.; B.S., Marquette U., 1931, M.D., 1937, M.S., 1941; m. Rosemary Louise Bach, May 22, 1943; children—Paula, John Jarl, Lisbeth, Katherine, Laura Marie. Faculty, Marquette U. Sch. Medicine, Milw., 1938-—, dean Sch. Medicine, 1947-65, v.p., 1965-66; coordinator The Wis. Regional Med. Program, 1966-—. Cons. hematology Milw. County Hosp., VA Hosp., Wood, Wis. Diplomate Am. Bd. Internal Medicine. Mem. Internat. Soc. Hematology, A.M.A., A.C.P., A.A.A.S., Soc. for Exptl. Biology and Medicine, Central Soc. for Clin. Research, Milw. Acad. Medicine, Alpha Omega Alpha. Research and publs. on erythrocyte physiology and blood coagulation. Home: 3948 N. Harcourt Pl., Milw. 53211. Office: 110 E. Wisconsin Av., Milw. 53202.*

HIRSCHFELD, Jan Karl-Erik, serologist; b. Wroclaw, Poland, Feb. 5, 1932; s. Robert and Ida (Koritzinsky) H.; M.D., Karoline Inst., 1958, Ph.D., 1961; m. Anne-Marie Skantze, July 5, 1959; 1 son, Erik Robert. Asst. lectr. bacteriol. dept. Karoline Inst., Stockholm, Sweden, 1954-60; lab. doctor State Bacteriol. Lab., Stockholm, 1958-60; lab. dr. State Inst. for Blood Group Serology, Stockholm, 1960-62, asso. prof., 1962-67; asso. prof. medicine and med. genetics U. Wis., Madison, 1967-—. Cons., LKB-Industries, 1960-—. NIH Internat. fellow dept. genetics, U. Wis., Madison, 1964-65; recipient Jean Julliard prize, 1964. Hon. mem. Deutsche Gesellschaft für Gerichtliche und Soz. Medizin, Deutsche Gesellschaft für Humangenetik und Anthropologie. Research and publs. on different ways of interpreting serological data; discovered genetically controlled serum protein systems in man and other animals demonstrable by serological methods. Address: U. Wis., Madison, Wis.*

HIRSCHFELD, Ludwik, physician; b. Switzerland, 1884; isolated and described Bacterium paratyphosum C, 1919. Died 1954.

HIRSCHFELD, Magnus, German physician; b. Kolberg, Germany, May 14, 1868; practiced medicine, Berlin; applied term 3d sex to homosexuals, advocated tolerance toward them. Died Nice, France, May 15, 1935.

HIRSCHFELDER, Joseph Oakland, Am. chemist, physicist; b. Balt., May 27, 1911; s. Arthur Douglass and May Rosalie (Straus) H.; student U. Minn., 1927-29; B.S., Yale, 1931; Ph.D., Princeton, 1936, postgrad.; m. Elizabeth Stafford Sokolnikoff, Mar. 7, 1953. Research asso. Wis. Alumni Research Found., 1937-40; faculty U. Wis., Madison, 1940-—, prof., 1946-—, Homer Adkins prof. chemistry, 1962-—, dir. Naval Research Lab., 1946-59, dir. Theoretical Chemistry Lab., 1959-62, dir. Theoretical Chemistry Inst., 1962-—. Cons. NDRC on interior ballistics guns and rockets, group leader Geophys. Lab., 1942-43, group leader theoretical physics and ordnance Los Alamos Atomic Bomb Lab., 1943-46; head theoretical physics div. Naval Ordnance Test Sta., Inyokern and Pasadena, 1945-46; chief phenomenologist Bikini Atomic Bomb Test, 1946; cons. Army and Navy ordnance AEC; chmn. phys. chemistry com. NRC, 1958-61; mem. policy adv. bd. Argonne Nat. Lab., 1962-67. Recipient Egerton gold medal Combustion Inst., 1966. Fellow Am. Phys. Soc., Am. Acad. Arts and Scis., Faraday Soc., Phys. Soc. (London, Eng.); mem. Am. Chem. Soc. (chmn. div. phys. chemistry 1959-61, Peter Debye award 1966), Nat. Acad. Scis., Japanese Phys. Soc., Royal Norwegian Soc. (fgn.), Am. Soc. M.E. (hon. life), Sigma Xi, Gamma Alpha, Phi Lambda Upsilon. Author: Molecular Theory of Gases and Liquids, 1954, 64; Theory of Perturbations, 1964; also articles. Chmn. bd. editors The Effects of Atomic Weapons, 1949. Editor: Intermolecular Forces, 1967. Research on aerodynamics, absolute reaction rates, molecular theory of gases and liquids, gas imperfections, statistical mechanics, scattering gamma radiation, theory of detonations and flame propagation, theoretical chemistry, radioactive fallout, math. determination of chem. and phys. properties of materials. Home: Thorstrand Rd., Madison, Wis. 53705.*

HIRSCHHORN, Isidor Solomon, chemist; b. Prague, Czechoslovakia, Oct. 25, 1915; s. Ephraim and Amelia (Kornfeld) H.; came to U. S., 1922, naturalized, 1932; A.B., Montclair State Coll., 1936; postgrad. N.Y. U.; M.A., Columbia, 1940; postgrad. U. Wis., (Gen. Electric fellow) Union Coll., Schenectady; m. Ellen S. Stein, Apr. 2, 1947; children—Susan, Robert. Supr. tng. VHF electronics U. S. Air Force Tech. Tng. Command, 1942-44; chem. engr. Hercules Powder Co., 1945; instr. chemistry Drew U., 1945-49; tech. dir. New Process Metals, Inc., 1949, gen. mgr., 1950-58; v.p. Ronson Metals Corp., Newark, 1959-—. Mem. Am. Chem. Soc., Electrochem. Soc., Am. Inst. Mining, Metall. and Petroleum Engrs., Chemists Club, Am. Soc. for Metals. Contbg. author: Ency. Electrochemistry, 1964. Research on rare earth and thorium metals and alloys, double-base rocket propellants. Home: 56 Greenwood Av., West Orange, N.J. 07052. Office: 55 Manufacturers Pl., Newark, 07102.*

HIRSCHHORN, Kurt, physician; b. Vienna, Austria, May 18, 1926; s. Emanuel and Helen (Mayberger) H.; came to U. S., 1940, naturalized, 1944; student U. Pitts., 1944; B.A., N.Y. U., 1950, M.D., 1954, M.S., 1958; m. Rochelle Reibman, Dec. 20, 1952; children—Melanie D., Lisa R., Joel N. USPHS clin. trainee, instr. N.Y.U. Sch. Medicine, N.Y.C., 1956-57, asst. prof., 1958-64, Am. Heart Assn. Advanced Research fellow, 1958-60, established investigator, 1960-65, asso. prof., 1964-66; John Berquist, Population Council fellow Inst. Med. Genetics, Upsala, Sweden, 1957-58; seminar asso. Columbia, 1958-—; asst. vis. physician Bellevue Hosp., N.Y.C., 1956-64, asso. vis. physician, 1964-66; asst. attending physician N.Y. U. Hosp., 1946-64, asso., 1964-66; prof. pediatrics, chief div. med. genetics Mt. Sinai Sch. Medicine, N.Y.C., 1966-—, vis. physician Mt. Sinai Hosp., 1966-—; Cohen professor of genetics Mt. Sinai School of Medicine, 1968-—. Career scientist N.Y.C. Health Research Council, 1965-—. Fellow A.A.A.S.; mem. Am. Soc. Human Genetics (dir., pres.-elect), Genetics Soc. Am., Am. Heart Assn., Am. Soc. for Clin. Investigation, Am. Assn. Immunologists, Harvey Soc. Mem. editorial bd. Cytogenetics, Humangenetik, Jour. Clin. and Exptl. Immunology. Research and numerous publs. in cytogenetics; described 1st case of mosaicism in hermaphrodite, new chromosomal abnormalities and normal morphologic variants; in immunology, elucidated physiology of cultured human lymphocyte including immunoglubulin prodn. and role of lysosomes, mode of inheritance of hypercholesterolemia and hyperlipemia. Home: 29 Washington Sq., N.Y.C. 10011.*

HIRSCHMAN, Isidore Isaac, Jr., Am. mathematician; b. Washington, Nov. 22, 1922; s. Isidore Isaac and Margaret (Ostrander) H.; A.B., Harvard, 1942, Ph.D., 1947; Sc.M., Brown U., 1943; m. Miriam Diamant, Sept. 18, 1943; children—Lynette, Helen K., Stephanie. Faculty math. Washington U., St. Louis, 1960-67. Guggenheim fellow, 1952-53. Mem. Am. Math. Soc., Math. Assn. Am. Author: (with D. V. Widder) The Convolution Transform, 1955; Infinite Series, 1962. Research and numerous publs. on harmonic analysis, variation diminishing integral transformations, quasi-analytic functions, Fourier series, Toeplitz operators. Home: 6933 Columbia Pl., University City, Mo. 63130. Office: Washington University, St. Louis 63130.*

HIRSCHMANN, Johannes Kurt, German physician; b. Plauen, Vogtl., Mar. 7, 1910; M.D., U. Leipzig; m. Gertrude Ende; children—Wolf-Dietrich, Siegrun, Gisela, Ingeborg. Sci. asst., med. specialist neurology and psychiatry; chief med. psychotherapist, dir. Neurol. Clinic, Tübingen; founder, dir. Clinic and Polyclinic of Neurology, U. Tübingen. Mem. german Soc. Neurology, Com. Criminal Biology, Soc. Psychiatry and Neurology. Research and numerous publs. on neurology, psychiatry, criminology, psychoneurosis. Home: 16 Bohnenbergerstrasse. Office: 20 Liebermeierstrasse, Tübingen 74, Germany.

HIRSCHOWITZ, Basil Isaac, physician, b. Bethal, South Africa, May 29, 1925; s. Morris and Dorothy (Drieband) H.; B.Sc., Witwatersrand U., South Africa, 1943, M.B., B.Ch., 1947, M.D., 1954; mem. Royal Coll. Physicians, London, 1952; m. Barbara Louise Burns, July 8, 1958; children—David E., Karen, Edward A. Came to U. S., 1953, naturalized, 1961. House officer in medicine and surgery Johannesburg Gen. Hosp., South Africa, 1948-50; registrar Central Middlesex Hosp., London, 1951-53; instr. internal medicine U. Mich., Ann Arbor, 1953-55, asst. prof., 1955-56; asst. prof. Sch. Medicine, Temple U., Phila., 1957-59; asso. prof., dir. div. gastroenterology Med. Center of U. Ala., Birmingham, 1959-64, prof. medicine, asso. prof. physiology, 1964-—. Fellow A.A.A.S., Royal Coll. Physicians Edinburgh, American College of Physicians; member Physiological Society, Fedn. Clin. Research, Med. Research Soc. Gt. Britain, Soc. Exptl. Biology and Medicine, Am. Gastroent. Assn., N.Y. Acad. Scis., Brit., South African, Ala. med. assns., Sigma Xi., Alpha Omega Alpha. Research and numerous publs. on diseases of gastrointestinal tract, physiology of gastric secretion relating to pepsinogen secretion and electrolyte secretion, mechanisms of vagal nerve stimulation, endoscopy of gastrointestinal tract with fiber optic instruments, including photography using these; inventor fiber-optic gastroduodenoscope.*

HIRSCHSPRING, Harald, pediatrician; b. 1830; prof., Copenhagen, Denmark; described Hirschsprung's disease of large intestine in children, 1888. Died 1916.

HIRST, Sir Edmund Langley, organic chemist; b. Preston, Eng., July 21, 1898; s. Sim and Elizabeth (Langley) H.; M.A., St. Andrews U., Scotland, 1918, B.Sc., 1919, Ph.D., 1921, LL.D., 1951; D.Sc., U. Birmingham (Eng.), 1929, LL.D., 1962; M.Sc., U. Manchester (Eng.), 1947; LL.D., U. Aberdeen (Scotland), 1960, U. Strathclyde (Scotland), 1965; Sc.D., U. Dublin (Eire), 1959; m. Kathleen Jennie Harrison, Feb. 12, 1949. Asst. lectr. Manchester U., 1923-24, prof., 1944-47; lectr. Armstrong Coll., U. Durham (Eng.), 1924-26; faculty Birmingham U., 1927-36, reader, 1934-36; prof. Bristol U., 1936-44; prof. Edinburgh (Scotland), 1947-—. Chmn. dept. sci. and indsl. research Chem. Research Bd., 1950-55. Fellow

Royal Soc., 1934, Chem. Soc. (past pres.), Royal Soc. Edinburgh (Scotland), 1947-68. Chmn. dept. sci. and indsl. research Chem. Research Bd., 1950-55. Knighted, 1964. Fellow Royal Soc., 1934, Chem. Soc. (past pres.), Royal Soc. Edinburgh (past pres.), Royal Inst. Chemistry; mem. Soc. Chem. Industry, Biochem. Soc., Swiss, Polish (hon.), chem. socs., Brit. Assn. for Advancement Sci. Research and numerous publs. on molecular structure of carbohydrates especially ring structures of simple sugars, chemistry vitamin-C, chemistry polysaccharides, polysaccharides of grasses and fodders. Office: Chemistry Dept., U. Edinburgh, Edinburgh 9, Scotland.*

HIRST, George K(eble), Am. virologist; b. Eau Claire, Wis., Mar. 2, 1909; B.S. Yale, 1930; M.D., 1933; m. 1937; nine children. Intern, New Haven Hosp., Conn., 1933-34; med., Univ. Hosps., Cleveland, O., 1934-35; asst. res., 1935-36; Rockefeller Inst. Hosp., 1936-40; mem., Rockefeller Found., 1940-46; chief, dept. virology, Public Health Research Institute, 1946-—, dir., 1956; prof., microbiology, N.Y.U. College Medicine, 1957-—. Mem., Nat. Acad. Scis., Microbiological Soc., Harvey Soc., Assn. Immunology. Research on virus diseases; influenza, streptococci, rheumatic fever. Office: Public Health Research Institute, 445 First Ave., New York, N.Y. 10016.

HIRSZFELD, Ludwik, Polish hematologist; b. Warsaw, Poland, Aug. 5, 1884; prof. Faculty Medicine, Warsaw, Poland (imprisoned during German occupation); organized faculty Medicine, Wroclaw, Poland, also inst. immunology and exptl. therapy named after him, 1945; mem. Faculty Lublin (Poland). Publs. on serological differentiation in man, incompatibility of mother and child; demonstrated (with Emil von Dungern) that blood groups are inherited according to Mendelian laws, pub. 1910. Died 1954.

HIRT, Robert Charles, Am. chemist; b. Warren, O., July 2, 1919; s. Theodore Samuel and Bessie (Bailey) H.; B.A., Coll. of Wooster, 1940; Ph.D., Brown U., 1947; m. Claire Anita Bernier, Feb. 6, 1943; 1 son, Theodore Charles. Asst. scientist Manhattan Project, U. Chgo., 1942-44; asso. scientist Manhattan Project, Los Alamos, 1944-45; group leader ultraviolet spectroscopy Am. Cyanamid Co., Stamford, Conn., 1947-—. Mem. Optical Soc. Am., Am. Phys. Soc., Am. Chem. Soc., Soc. for Applied Spectroscopy, Am. Soc. for Testing and Materials (com. E-13 1952-—). Research in ultraviolet spectroscopy, fluorescence, solar radiation, photochemistry. Home: 44 Barrett Av., Stamford 06905. Office: 1937 W. Main St., Stamford, Conn. 06904.*

HIRUMI, Hiroyuki, biologist; b. Tokyo, Japan, Sept. 19, 1932; s. Koichiro and Taga (Watanabe) H.; B.S., Shizuoka U., Japan, 1957; M.D., Wakayama Med. Coll., Japan, 1965; m. Kazuko Tajima, Nov. 30, 1959; children—Eri, Atsusi. Asst. in cytology Nat. Inst. Genetics, Japan, 1958-59; asst. in anatomy Wakayama Med. Coll., 1959-62, electron microscopist, 1964-66; research asso. Boyce Thompson Inst., Yonkers, N.Y., 1962-64, 66-—. Mem. Genetic Soc. Japan, Anat. Soc. Japan, N.Y. Soc. Electron Microscopists. Research on cultivating leafhopper-borne virus vector tissues and nematode tissues in vitro; studies on plant pathogenic viruses in their insect vector bodies and cultured cells, insect and nematode ultrastructures by means of electron microscopy. Home: 255 S. Broadway, Hastings on Hudson, N.Y. 10706. Office: 1086 N. Broadway, Yonkers, N.Y. 10701.*

HIRVONEN, Leo Leopold, Finnish physiologist; b. Savonlinna, Finland, Nov. 15, 1924; s. Antti Juhana and Helmi (Jarvelä) H.; Med. candidate examination U. Helsinki (Finland), 1947, D.Med. Sc., 1955; Med. Licentiate examination U. Turku (Finland), 1950; m. Hellin Selma Sucksdorff, Aug. 6, 1950; children—Kari Juhani, Tarja Anneli, Vesa Vilhelm. Jr. asst. dept. physiology U. Helsinki, 1953-55; staff U. Turku, 1955-—, asso. prof. physiology, 1957-—, head cardiorespiratory research unit, 1959-—. Mem. Nordisk Förening för Fysiologi. Author: Electrographic Studies of the Isolated Rabbit Auricle, 1955; (with T. Peltonen), Experimental Studies on Fetal and Neonatal Circulation, 1965; also numerous articles. Research on clin. physiology, cardiovascular studies, especially fetal and neonatal period. Home: Kaarina, Finland. Office: 10 Kiinamyllynkatu, Turku 3, Finland.*

HIRVONEN, Reino Antero, Finnish geodesist; b. Savonlinna, Finland, Nov. 14, 1908; s. Simo and Wendla (Paavilainen) H.; M.Sc., U. Helsinki (Finland), 1931, Ph.D., 1934; m. Linnea Mansnerus, May 29, 1932; children—Esko, Eero. Sinikka (Mrs. Juhani Lindgren), Lauri, Laura. State geodesist Geodetic Inst., Helsinki, 1932-38; faculty Tech. U., Helsinki, 1932-—, prof. geodesy, 1950-—; vis. prof. Ohio State U., Columbus, 1951. Named comdr. Order Finland's Lion, 1965. Mem. Academia Scientiarum Fennica. Author: The Continental Undulation of the Geoid, 1934; Textbook of Adjustments in Geodesy, 1966. Research and numerous publs. on gravity measurements by pendulums, computation of geoid, new theories of gravimetric and 3-dimensional geodesy, observations of solar eclipses for geodetic purposes, theory adjustments with applications to photogrammetry. Home: Tiilimaki 2, Helsinki 33, Finland.*

HIRZEBRUCH, Friedrich Ernst Peter, German mathematician; b. Hamm, Westphalia, Oct. 17, 1927; s. Fritz and Martha (Holtschmit) H.; ed. U. Munster, Fed. Poly. Sch., Zurich; Ph.D. in sci.; m. Ingeborg Spitzley, 1952; children—Ulrike, Barbara, Michael. Sci. asst. U. Erlangen, 1950-52; with Inst. for Advanced Study, Princeton, N.J., 1952-54; instr. U. Munster, 1954-55; asst. prof. Princeton, 1955; prof. U. Bonn, 1956. Mem. German, Am., French math. socs. Author: Neue Topologische Methoden in der Algebraischen Geometrie, 1956. Co-editor: Mathematische Annalen; Jour. für die Mathematik; Topology. Home: 7 Endenicher Allee. Office: 10 Wegelerstrasse, 53 Bonn, Germany.

HIS, Wilhelm, anatomist; b. Basel, Switzerland, July 9, 1831; at least 1 son, Wilhelm; prof., Basel, 1857-72, Leipzig, Germany, 1872-1904. Author: Anatomie menschlicher Embryonen, 1880; Zur Geschichte des menschlichen Rückenmarks und der Nervenwurzeln, 1886. A founder of human embryology; introduced microtome into microscopic research, 1865, thus making possible cutting of serial sects.; introduced concept of developmental mechanics to explain formation of organs and structures, 1874; composed a list including, neuroblast, spongioblast, other Latin anat. terms, adopted by German Anat. Soc., Basel, 1895; gave 1st description of fetal thyroglossal duct, 1901. Died Leipzig, May 1, 1904.

HIS, Wilhelm, Jr., physician, anatomist; b. Basel, Switzerland, Dec. 29, 1863; s. Wilhelm His; prof. internal medicine U. Berlin, 1907-26, also dir. 1st med. clinic. Author: Die Tätigkeit des embryonalen Herzens, in series Beiträge zur Pathologie des Kreislaufs, 1893. Studied impulse conduction system of heart; described His's bundle of nerve fibers connecting auricles with ventricles, 1893. Died Nov. 10, 1934.

HISAW, Frederick Lee, Am. zoologist; b. Newtonia, Mo., Aug. 23, 1891; s. Frederick Louis and Anna (Cummins) H.; A.B., U. Mo., 1914, B.S., 1915, A.M., 1916, LL.D., 1938; Ph.D., U. Wis., 1924; m. Minnie Nancy Reynolds, Sept. 1, 1917; children—Lois Lee (Mrs. David T. Crockett Jr.), Frederick Lee II. Asst. in zoology U. Mo., 1913-16; asso. prof. zoology U. Miss., 1916-17; asst. prof. zoology Kan. Coll., 1919-24; faculty zoology U. Wis., 1924-35, prof., 1929-35; prof. zoology Harvard, 1935-53, Fisher prof. natural history, 1953-62, emeritus, 1962-—. Bd. sci. dirs. Yerkes Labs. Primate Biology, 1942-62. Recipient Anniversary award Am. Gynecol. Soc., 1952. Fellow Harvard Soc. Fellows; mem. Endocrine Soc. (medal 1956), Soc. Zoologists (pres. 1937), Am. Acad. Arts and Scis., Am. Philos. Soc., Nat. Acad. Scis. Research and numerous publs. in comparative physiology of reprodn. in vertebrate animals, including endocrines regulating growth and devel. of reproductive organs, estrous and menstrual cycles, ovulation, placentation, gestation, parturition, physiol. evolution of viviparous conditions. Office: Biol. Labs., Harvard, Cambridge, Mass. 02138.*

HISER, Homer Wendell, Am. atmospheric physicist; b. Ava, Ill., Nov. 21, 1924; s. Clinton O. and Augusta (Bishop) H.; A.B., Washington U., St. Louis, 1948; B.S., U. Ill., 1951, M.S., 1954; postgrad. U. Chgo., 1951-52; m. Wanda Jean Leach, June 5, 1953. Research asso. Radar Meterol. Research, Ill. Water Survey, U. Ill., 1950-55; prof., head radar meterol. research U. Miami, Coral Gables, Fla., 1955-—. Mem. Am. Meteorol. Soc. (com. on radar meteorology 1956-65), Am. Geophys. Union (past mem. com. on precipitation), I.E.E.E. (sr.). Author: (with W. L. Freseman) Radar Meteorology, 1959; also numerous articles. Research on atmospheric physics, radar meteorology, microwave propagation. Home: 4705 University Dr., Coral Gables, Fla. 33146.*

HISINGER, Wilhelm, Swedish geologist, mineralogist; b. Ridoarhytan, Sweden, Dec. 22, 1766. Author: Lethea Suecica, seu petrificata Sueciae, 1837-40; other works. Research on electrolysis of salts; found amount of decomposition proportional to affinity of compound components and to surface contact with conductor; discovered (with Berzelius) cerium. Died Skinnskatteberg, Sweden, June 28, 1852.

HISKES, John Robert, Am. physicist; b. Chgo., May 30, 1928; s. John and Alice (Boonstra) H.; A.B., U. Cal., Berkeley, 1951, M.A., 1952, Ph.D., 1960; m. Dolores Grant, Sept. 1, 1951; children—Robin, John Grant. Physicist, Cal. Research & Devel. Co., Livermore, 1952-54, Lawrence Radiation Lab., Livermore, 1954-55, 60-—, Berkeley, Cal., 1955-60, Argonne Nat. Lab., 1958. Visitor, Culham Lab., Abingdon, Berkshire, Eng., 1963-64; cons. Gen. Elec. Co., Santa Barbara, Cal., 1961-62. Mem. Am. Phys. Soc. Theoretical physics research in atomic and molecular physics with application to controlled fusion research; particle accelerator; nuclear fission calculations. Home: 792 Adams Av. Office: P.O. Box 808, Livermore, Cal. 94550.*

HISKEY, Richard G., Am. chemist; b. Emporia, Kan., May 21, 1929; s. Clifford and Ethyl Mary (Grant) H.; A.B., Kan. State Coll., 1951; M.A., Kan. State U., 1953; Ph.D., Wayne State U., 1955; m. Joan T. Crooke, June 13, 1953; children—Kathleen, Elizabeth, Timothy Grant, Charles Richard, Jonathan. Research asso. Poly. Inst. Bklyn., 1955-58; faculty U. N.C., Chapel Hill, 1958-—, prof. organic chemistry, 1966-—. Mem. Am. Chem. Soc., Chem. Soc. London, N.Y. Acad. Sci., Elisha Mitchell Sci. Soc., Sigma Xi, Phi Lambda Upsilon. Research, publs. on synthetic peptides, mechanism of enzyme action, role of sulfur in biol. systems, organic sulfur chemistry, azomethine chemistry. Home: 521 Lakeshore Lane, Chapel Hill, N.C.*

HISS, Philip Hanson, Jr., Am. bacteriologist; b. Balt., Sept. 17, 1868; A.B., Johns Hopkins, 1891; M.D., Columbia, 1895; asst. bacteriologist Columbia, 1895-99, instr. hygiene and bacteriology, 1899-1903, named adj. prof. bacteriology, 1903; asst. bacteriologist Dept. Health N.Y., 1896-99; prof. hygiene Woman's Med. Coll., N.Y. Mem. Pub. Health Assn., N.Y. Path. Soc., Harvey Soc., other profl. socs. Research on bacteriology of dysentery; improved techniques in differentiation of typhoid and colon bacilli; developed method of isolating typhoid bacilli; studied relation of serum-globulin and diphtheritis antitoxin, also differentiation of pneumococcus and streptococcus, capsule staining methods. Died 1913.

HITCHCOCK, Charles Baker, Am. geographer; b. Boston, Mar. 16, 1906; s. John and Esther Mary (Baker) H.; A.B., Harvard Univ., 1928; M.A., Columbia, 1933; D.Sc. (honorary), Temple University, 1954; m. Agnes Murchie, Dec. 3, 1931; children—Gail, Suzanne, Esther Lee; married second, Anita Kincaid, January 19, 1957. Joined Am. Geog. Soc., N.Y. City, 1930, chief dept. Hispanic Am. research and assistant director, 1943-48, acting dir., 1949, exec. sec., dir., 1953-—, chmn. adv. com. Am. cartography, 1948-—, research and Development Bd., 1949; mem. adv. com. Am. Geography Pan Am. Inst. Geog. and History; mem. nat. atlas com. Nat. Research Council; mem. adv. com. census atlases Bur. Census; expdns. Am. Mus. Natural History, So. Venezuela, 1929, Phelps Venezuela 1947-49, 51-53, U. S. del. to Internat. Geog. Congress, Amsterdam, 1938, Rio de Janeiro, 1956, London, England, 1964. United States delegate Pan American Inst. Geography and History, Lima, 1941, Caracas, 1946, Buenos Aires, 1948, Santiago, Chile, 1950, Dominican Republic, 1952, Mexico, 1955, Ecuador, 1959, Stockholm, Sweden, 1960; alternate U. S. delegate to the Commn. Cartography, 1957-—. Mem. Assn. Am. Geographers (treas., 1948-49), Arctic Institute, N.Y. Acad. Scis., Soc. Geog. de Lima (corr. mem.), Soc. Venezolana de Ciencias Naturales (honorary), Scottish Geographical Society (honorary), A.A.A.S., Geol. Soc. Am. Editor Map of Hispanic America, 1938-47. Contbr. articles tech. publs. Directed making of 107 page Millionth Map of Hispanic Am., 1938-45; mapped parts of Venezuela; as dir. Geog. Soc., helped expand its research to include natural resources, population, disease and land use. Home: R.F.D. 1, Pound Ridge, N.Y. Office: American Geographical Society, Broadway at 156th St., N.Y.C. 10032.*

HITCHCOCK, Charles Henry, Am. geologist; b. Amherst, Mass., Aug. 23, 1836; s. Edward and Orra (White) H.; A.B., Amherst Coll., 1856, M.A., 1859, LL.D., 1896; student Yale Div. Sch. and Andover Theol. Sem.; Royal Sch. Mines, London; Ph.D., Lafayette, 1870; m. Martha Bliss Barrows, June 19, 1862. Lectr. zoology, Amherst Coll., 1858-64; non-resident prof. geology and mineralogy Lafayette Coll., 1866-70; prof. geology and mineralogy Dartmouth, 1868-1908, emeritus prof., from 1908; later lived in Honolulu. Asst. state geologist, Vt., 1857-61; state geologist, Me., 1861-62, N.H., 1868-78; prof. geology, Va. Agrl. and Mech. Coll., 1880; prof. natural history, Williams Coll., 1881; lectr. geology Mt. Holyoke Coll., 1870-96. Headed expdn. occupying Mt. Washington, N.H., in winter of 1870-71, the 1st high mountain observatory in the U. S. Author: Elementary Geology (with Edward Hitchcock), 1861; Mt. Washington in Winter, 1871; Report on Geology of New Hampshire (3 vols.), 1873-78 (state publ.); Geological Map of the United States, 1881; Hawaii and Its Volcanoes, 1909. Compiled several geol. maps of U. S.; research in ichnology, geology of crystalline schists, glacial geology. Died Nov. 5, 1919.

HITCHCOCK, Claude Raymond, Am. physician; b. Mpls., Oct. 14, 1917; s. Ralph Carleton and Lucy (Morris) H.; B.A., U. Minn., 1940, B.S., Med. Sch., 1943, M.B., 1944, M.D., 1945; Ph.D. in Surgery, 1954; m. Wilma Ruth Baker, Apr. 23, 1949; children—Jeffry, Claudia. Faculty U. Minn. Med. Sch., 1954-—, prof. surgery, 1962-—, dir. Cancer Detection Research Center, 1952-55; chief surgery Hennepin County Gen. Hosp., Mpls., 1955-—; pres. Mpls. Med. Research Found., 1957-—. Recipient So. Minn. Med. Assn. award for Sci. merit, 1964. Diplomate Pan Am. Med. Assn. Mem. Minn. (dir. 1963-—), Hennepin County (dir. 1963-—) cancer socs., Minn., Hennepin County med. socs., A.M.A., Central, Minn., Mpls. (dir. 1963-—) surg. socs., Soc. Head and Neck Surgeons, Soc. for Exptl. Biology and Medicine, N.Y. Acad. Scis., Soc. for Cryobiology. Contbg. author: Treatment of Cancer and Allied Disease, 1957; Progress in Clinical Cancer, 1965; also numerous articles. Research on cancer detection and screening techniques, renal homotransplantation immunosuppressive techniques, cancer chemotherapy, effects thiazides and potassium on gastrointestinal tract; developed preservation methods for kidney during transplantation, hyperbaric oxygenation methods for medicine and surgery; invented suction-injection device for gastrointestinal tract, automatic bed positioner. Home: 6616 W. Shore Dr., Edina,

Mpls. 55424. Office: Hennepin County Gen. Hosp., Mpls. 55415.*

HITCHCOCK, Edward, Am. geologist; b. Deerfield, Mass., May 24, 1793; s. Justin and Mercy (Hoyt) H.; grad. Yale Theol. Sem., 1820; m. Orra White, 1821, 6 children including Charles Henry, Edward. Prin., Deerfield Acad., 1815-19; pastor Congregational Ch., Conway, Mass., 1821-25; prof. chemistry and natural history Amherst (Mass.) Coll., 1825-45, pres., 1845-54, prof. theology and geology, 1854-64; made 1st complete geol. survey of Mass., 1830; state geologist Vt., 1857-61. Author: Geology of Connecticut Valley, 1823; Elementary Geology, 1840; Fossil Footsteps, 1848; Religion of Geology, 1851; Illustrations of Surface Geology, 1857. Research includes fossil footsteps in new red sandstone of Connecticut River Valley, glacial drift, gneiss formation; formulated doctrine of distortion and alteration of fragmental constituents of sediments. Died Amherst, Feb. 27, 1864.

HITCHINGS, George Herbert, Am. chemotherapist; b. Hoquiam, Wash., Apr. 18, 1905; s. George Herbert and Lillian (Matthews) H.; B.S. cum laude, U. Wash., 1927, M.S., 1928; Ph.D., Harvard, 1933; m. Beverly Reimer, June 24, 1933; children—Laramie (Mrs. R. C. Brown), Thomas. Instr., tutor Harvard, 1932-36, research fellow, 1934-36, asso., 1936-39; sr. instr. Western Res. U., 1939-42; with Wellcome & Co., Tuckahoe, N.Y., 1942—, asso. research dir., 1955-63, research dir. chemotherapy div., 1963-66, v.p. in charge research, 1966—. Cons. cancer chemotherapy study sect. USPHS, 1959-60. Mem. Am. Assn. for Cancer Research (past dir.), Am., Westchester (past chmn.) chem. socs., Chem. Soc. (laondon), Am. Soc. Biol. Chemistry, A.A.A.S., Soc. for Exptl. Biology and Medicine. Contbr. numerous articles to sci. jours. Invented chemotherapeutic agts., including pyrimethamine for malaria, mercaptopurine for leukemia, diaveridine for coccidiosis, azathioprine for organ transplantation and auto-immune disease, trimethoprim for bacterial diseases, allopurinol for gout and hyperuricemia, also drugs based on rationale of specific biochem. modifications. Home: 50 Primrose Av., Yonkers, N.Y. 10710. Office: 1 Scarsdale Rd., Tuckahoe, N.Y. 10707.*

HITCHOCK, Fred A., Am. physiologist; b. Akron, O., Oct. 30, 1889; s. George E. and Florence (Tucker) H.; Ph.B., U. Akron, 1912; M.S., Ohio State U., 1923, Ph.D., 1927; postgrad. U. Chgo., Cambridge (Eng.) U.; m. Mary Alice Rines, Apr. 17, 1917. Instr., Columbus (O.) High Sch., 1919-23; faculty dept. physiology Ohio State U., Columbus, 1923—, prof., 1940-60, prof. emeritus, 1960—; vis. prof. physiology Pahlevi U., Shiraz, Iran, 1963-64; civilian OSRD, 1941-46; dir. research CAA and comml. airlines, 1949-56. Cons., Avco Corp., 1959-63; dir. ednl. activities Am. Inst. Biol. Scis., 1961-62. Recipient Arnold D. Tuttle award Aero. Med. Assn., 1955, Distinguished Service award Ohio State U., 1964. Fellow A.A.A.S., Am. Aero. Assn., Aero-Space Med. Soc.; mem. Am. Phys. Soc. (past mem. council), Sociedade Interpanetaria Brasileira (life mem. sci. council), Am. Physiol. Soc., Am. Inst. Nutrition, Am. Zool. Soc., N.Y. Acad. Scis., Soc. for Exptl. Biology and Medicine. Co-translator: La Pression Barometrique (Paul Bert), 1943. Research and numerous publs. on effects of explosive decompression and need for pressurized air transport, effects of use of safety belts. Home: 133 Amazon Pl., Columbus, O. 43214.*

HITTMAIR, Otto, Austrian physicist; b. Innsbruck, Austria, Mar. 16, 1924; s. Rudolf and Margarete (Schumacher) H.; Ph.D., U. Innsbruck, Austria, 1949; m. Anna Rauch, Dec. 3, 1956; children—Maria Christine, Elisabeth, Georg, Margarete. Physicist, AEC of Argentina, Buenos Aires, 1957; physicist Atomic Inst. Austrian U., 1958-60; lectr. U. Innsbruck, 1953-58; lectr. U. Vienna (Austria), 1958; lectr. Tech. U. Vienna, 1959-60, asso. prof., 1960-63, prof., 1963—. Mem. Austrian Acad. Sci., Austrian Phys. Assn., Chem.-Phys. Assn. (pres. 1967-68). Author: (with S. T. Butler) Nuclear Stripping Reactions, 1957; also numerous articles. Research on nuclear levels by analysis of angular distbns. in nuclear reactions. Home: 58b Gallitzinstrasse, A-1160, Vienna. Office: 13 Karlsplatz, A-1040, Vienna, Austria.*

HITTORF, Johann Wilhelm, German physicist, chemist; b. Prussia, Mar. 27, 1824; pupil of Plücker; prof. chemistry, physics U. Münster (Germany); ret., 1889. Mem. French Acad. Scis., 1900. Author: Über die Elecktrizitätsleitung der Gase, 1869; Über die Wanderung der Ionen während der Elektrolyse, 2 vols., 1903-04. Pioneer in electrochem. research; studied migration of ions during electrolysis; gave 1st suggestion of transport numbers, 1853; anticipated Crookes' report of 1878 with research on cathode rays, including study of deflection of rays by magnet; research on elec. phenomena in rarefied gases; eponym for Hittorf tube. Died Münster, Nov. 28, 1914.

HITZIG, Eduard, physiologist, neurologist; b. Germany, 1838. Author: (paper) Ueber die elektrische Erregbarkeit des Grosshirns, 1870. Demonstrated (with Fritsch) that stimulation of different specific regions of cerebral cortex activates different muscle groups, 1870. Died 1907.

HIYAMA, Yoshio, Japanese fishery scientist; b. Tokyo, Japan, 1909; grad. Tokyo U., 1943; D.Agr. Faculty, Tokyo U., 1947—, now prof. Tech. expert Investigation Commn. for Bikini Atom Bomb Test, 1954. Recipient Japan Agr. Soc. Prize for thesis. Mem. Japan Marine Products Soc., Internat. Sci. and Cultural Assn. Author: Outline of Fishery Science; Thesis on Experimental Ecology of Fishes. Research on fishes, including taxology, histology and ecology.

HJARNE, Urban, see Hiärne, Urban.

HJELMQVIST, Sven, geologist; b. Lund, Sweden, Jan. 23, 1908; s. Theodor and Elisabeth (Feuk) H.; Ph.D., U. Lund, 1935; m. Anna-Lisa Bergqvist, June 9, 1946; children—Ulf, Karin, Eva. Faculty, U. Lund, 1934—, prof. mineralogy and petrology, 1951—, head dept. geology, 1965—; geologist Geol. Survey of Sweden, 1936-42, state geologist, 1942-51. Mem. Swedish Sci. Research Council, 1962. Fellow Mineral. Soc. Am., Royal Physiographic Soc. of Lund. Author: Zur Geologie d. sudschwedischen Grundgebirges, 1934; The Titaniferous Iron-ore Deposit of Taberg, 1950; On the Occurrence of Ignimbrite in the Pre-Cambrian, 1956; also articles. Research on Archaean rocks and ores of Sweden; history of pre-Cambrian period. Home: 20 Studentgatan, Lund, Sweden.*

HJELT, Edvard Immanuel, Finnish chemist; b. Wichtis, Finland, June 28, 1855; s. Otto E. A. Hjelt; Ph.D., U. Helsinki (Finland), 1879; LL.D., U. Aberdeen, 1906; m. Ida Ostroem, 1878; 2 sons, 5 daus. Became docent chemistry U. Helsinki, 1880, prof. chemistry, 1882-1907, rector, 1899-1907, vice-chancellor. Mem. Finnish Soc. Scis., Royal Swedish Acad. Scis. Author: Principles of Organic Chemistry; Aus J. Berelius und G. Magnus Briefwechsel; Berelius, Liebig, Dumas, ihre Stellung zur Radikaltheorie; Geschichte d. organ. Chemie; other biog. works. Continued (with Brühl and Aschan) Ausführliches Lehrbuch der Chemie von Roscoe-Schorlemmer. Died July 2, 1921.

HJELT, Lars Holger Edvard, Finnish pediatric pathologist; b. Forssa, Finland, Oct. 31, 1917; s. John Edvard and Lucie (Henning) H.; M.D., U. Helsinki, 1945, docent, 1956; m. Sarkka Kirsti Maija Kyllikki, Apr. 4, 1944. Asst. physician Inst. Pathology, Helsinki U., 1947-52, prof. gen. pathology, 1957-58; staff U. Children's Hosp., Helsinki, 1947-55, chief pathologist, 1959—; pediatrician U. Women's Clinic, Helsinki, 1955-57. Cons. pathologist Cancer Detection Clinic, Helsinki, 1957, chief path. lab., 1963—. Gen. sec. Found. for Pediatric Research, Helsinki, 1963—. Mem. Finnish Med. Assn., Finnish Soc. Pediatrics, Finnish Soc. Pathology. Research and numerous publs. on kidney diseases of childhood especially congenital nephrotic syndrome. Home: 17 Viitasuontie. Office: 11 Stenbackstreet, Helsinki, Finland.*

HJORT, Axel Magnus, Am. physician; b. Chgo., Nov. 28, 1889; s. Axel F. and Mary Charlotte (Mellin) H.; A.B., U. Ill., 1914, M.S., 1915; Ph.D., Yale, 1918, M.D., 1921; m. Justine Caroline Brockett, Aug. 6, 1919. Research biochemist Parke Davis & Co., Detroit, 1923-26; prof. pharmacology Dartmouth Med. Sch., 1926-29; civilian chief med. research U. S. Army Chem. Warfare Sect., 1929-31; chief physiology, pharmacology Burroughs Wellcome & Co. Labs., 1931-41; practice medicine specializing in internal medicine, Scarsdale, N.Y., 1941-55; ret., 1955. Fellow A.C.P.; mem. Am. Soc. Biol. Chemists, Sigma Xi, Alpha Chi Sigma, Nu Sigma Nu. Research and numerous publs. on contbns. to discovery of parathyroid hormone, circulatory drugs, hypnotics, endocrinology and allergy. Home: 3341 S.W. 20th St., Ft. Lauderdale, Fla. 33312.*

HJORT, Johan, Norwegian biologist; b. Christiania (now Oslo), Norway, Feb. 18, 1869; s. Johan Storm Aubert and Johanne Elisabeth (Falsen) H.; Ph.D., U. Munich (Germany), 1902; Sc.D., U. Naples (Italy); hon. doctorates univs. Cambridge, London, Harvard; m. Constance Gran. Dir. fisheries Norwegian govt., 1900-17; prof. U. Oslo, 1921-39; pres. Council for Internat. Study Sea, from 1938. Fellow Royal Soc.; mem. Norwegian, French acads. scis., Am. Philos. Soc. Author: (with John Murray) The Depths of the Oceans, 1912; The Unity of Science, 1921; The Emperor's New Clothes, 1931; The Human Value of Biology, 1938. Research on fish migration by means of marked cod, also on embryology and life history of marine animals. Died Oct. 7, 1948.

HLADYSHEVSKY, Evhen Ivanovich, Russia chemist; b. Rekynec, Livi, Ukranian SSR, Apr. 14, 1924; s. Ivan Vasylovich and Francisca (Sarkhman) H., Chemist, U. Lviv, 1947, candidate chem. scis., 1953; m. Tatiana Salash, Oct. 5, 1951; children—Nadia, Roman. Mem. faculty U. Lviv, 1947—, asst. prof. chemistry, 1954—, dean chem. dept., 1964—. Mem. Mendeleev Chem. Soc. Author numerous articles. Research on crystal chemistry of intermetallic compounds; discovery of about 350 intermetallic compounds. Home: 8 Sholokhov. Office: 6 Lomonosov, Lviv, Ukrainian SSR.*

HLAVACEK, Vladimir, Czechoslovakian otolaryngologist; b. Briza, Czechoslovakia, July 12, 1898; s. Vaclav and Marie (Maskova) H.; grad. Faculty of Medicine, Prague, Czechoslovakia, 1922, asst. prof., 1933, ordinarius, 1952; m. Hildegarde Petriková, Oct. 16, 1929; children—Alena, Darla, Vladimira. With surg. clinic, 1923-24, asst. otolaryn. clinic, 1925-33; dir. otolaryn. dept. Czech Policlinic, 1934-39; chief otolaryn. dept. Hosp. Prague XII, 1939-52; dir. otolaryn. clinic, faculty of hygiene U. Charles IV, 1952—, asst. prof., 1933-52, asso. prof., 1952—. Fellow Internat. Coll. Surgeons; mem. French Otolaryngol. and Broncho esophageal Soc. (corr.), Collegium Otolaryngolicum Amicitiae Sacrum, Czechoslovak Otolaryn. Soc., Czecheslovak Allergological Soc., German Soc. Immunology and Allergy. Author: Allergy in otolaryngology, 1933; Konstitution in upper airways, 1940; Inflammation of the internal ear, 1948; (with Vl. Chládek) Otosclerosis, 1958. Research, publs. in allergy on original findings in histology of membranes of upper airways concerning role of mastcells and eosinophils and their mut. relations; pioneer allergy in nose and paranasal cavities. Home: 14 Francouzska, Prague 2. Office: 50 Srobarová, Prague 10, Czechoslovakia.*

HLAWKA, Edmund, Austrian mathematician; b. Bruck an der Mur, Nov. 5, 1916; s. Leopold and Melanie (Ronnert) H.; Dr. phil., U. Vienna, 1938; m. Rosa Reiterer, Dec. 22, 1943. Docent, Vienna Inst. Tech., 1945-48; docent U. Vienna, 1945-48, prof., head dept., 1948—, dean philosophy faculty, 1955-56; faculty Inst. Advanced Studies, Princeton, N.J., 1959—. Recipient Austrian Art and Sci. medal, 1964; Dannie Heinemann prize, Göttingen, Germany, 1963. Mem. Austrian Acad. Scis., Leopoldina. Research, numerous publs. on geometry of numbers, uniform distbn. model, applications on statistics, numerical analysis, kinetic theory of gases, asymptotic expansions of solutions of differential equations and multiple integrals; rhythmic sequences on compact groups. Home: 8 Khünplatz, Vienna 1040. Office: 4 Strudlhofgasse, Vienna, Austria 1090.*

HLYNKA, Isydore, chemist; b. Ternopil, Ukraine, Feb. 17, 1909; s. Hryhory and Katerina (Krywaniuk) H.; B.S., U. Alta., 1935, M.S., 1937; Ph.D., Cal. Inst. Tech., 1939; m. Olga Lysyk, Sept. 25, 1939; children—L. Denis, Myron H. Agrl. scientist dairy chemistry, grain research lab. Bd. Grain Commn., Can. Dept. Agr., Winnipeg, Man., 1947—, asst. dir. Mem. Am. Assn. Cereal Chemists (Brabender award 1966). Chemical Inst. Can., Brit. Soc. Rheology. Editor: Wheat: Chemistry and Technology, 1964; editorial bd. Chemistry in Canada, 1949-54; Cereal Chemistry, 1955-58. Research and numerous publs. on role of chem. constituents in devel. of flavors and off-flavors especially in cheddar cheese; introduced concepts of three-dimensional network structure of wheat proteins; helped lay found. for better use and better understanding of phys. properties of dough and gluten in relation to protein quality in bread wheat. Home: 121 Oakview Av. Office: Grain Exchange Bldg., Winnipeg, Man., Can.*

HOADLEY, Leigh, Am. biologist; b. Northampton, Mass., Aug. 14, 1895; s. Alfred Henry and Grace (Leigh) H.; A.B., U. Mich., 1921; Ph.D., U. Chgo., 1923; A.M., Harvard, 1942; m. Harriet Leigh Warner, Sept. 12, 1923; children—Alfred Warner, Nancy Leigh (Mrs. Richard Norton Fryberger). Asst. in zoology U. Mich., 1916-17, 19-21, U. Chgo., 1921-23; instr. zoology St. Xavier Coll., 1923; NRC fellow in biology, 1923-25; gen. edn. bd. fellow biology, Bruxelles, Berlin-Dahlem, Freiburg, 1925-26; asst. prof. zoology Brown U., 1926-27; faculty zoology Harvard, 1927—, prof., 1930-62, emeritus, 1962—; instr. embryology Marine Biol. Lab., Woods Hole, Mass., 1928-38; Harvard Exchange prof. at Sorbonne (Paris), 1939. Mem. A.A.A.S., Am. Soc. Zoologists, Am. Assn. Anatomists, Am. Soc. Naturalists, Am. Acad. Arts and Scis. (corr. sec. 1937-40), Soc. Developmental Biology, others. Contbr. articles to sci. jours. Sci. contbns. relating to cytology of egg and early embryo; physiology of fertilization and early devel.; developmental responses to alterations in composition of environment of developing embryo and to various constant temperatures of environment; developmental capacities of isolated embryonic structures; interrelations of histogenesis and morphogenesis; neuro-embryology; theoretical embryology and importance of morphochoresis as a process in establishment of pattern at all stages and at all levels, especially as agt. in origin and elaboration of functional diversity within individual. Home: 14 Scott St., Cambridge, Mass. 02138.*

HOAG, Warren George, Am. veterinarian; b. Roosevelt, L.I., N.Y., Nov. 8, 1919; s. George Watson and Mae (Oertel) H.; D.V.M., Cornell U., 1944; M.P.H., Harvard, 1957; m. Euleta Ross Vincent, June 17, 1945; children—Michael George, Donna Sharon, Linda May, Mark Joseph, Matthew William and Martin Douglas (twins). Practice vet. medicine, 1944-45, Mt. Upton, N.Y., 1945-48; field veterinarian Cornell U., 1948-50; prof. animal pathology Va. Poly. Inst., Blacksburg, 1953-57; staff Jackson Lab., Bar Harbor, Me., 1957-67, asst. dir., 1959-67, sr. staff scientist, 1964-67; prof., dir. Center for Lab. Animal Resources, Michigan State University, East Lansing, 1967—. Commended by Brit. Govt., 1951. Fellow Royal Soc. Health (Gt. Britain); mem. Am. Soc. Microbiology, Am. Soc. Exptl. Pathology, Animal Care Panel (exec. bd. 1965—), Am., Me. vet. med. assns., A.A.A.S., N.Y. Acad. Scis., Conf. Animal Disease Research Workers, Assn. for Applied Gnotobiotics, Am. Assn. for Accreditation Lab. Animal Care (council on accreditation 1965—), Sigma Xi. Research and numerous publs. on infectious diseases common to man and animal, control and prevention of leptospirosis, Salmonellosis, Brucellosis, Pseudomonas infection. Home: 5115 W. St. Joseph St., Lansing, Mich. 48917.*

HOAGLAND, Dennis Robert, Am. agriculturist, botanist; b. Golden, Colo., 1884; A.B., Stanford, 1907; A.M., U. Wis., 1913; m. Jessie A. Smiley, 1920; 3 sons. Asst. chemist, lab. of M. E. Jaffa, U. Cal., 1908-10, asst. prof. agrl. chemistry, 1913-22, asso. prof. plant nutrition, 1922, prof., 1927-49, faculty research lectr., 1942; with FDA, U. S. Dept. Agr., 1910-12. Cons., collaborator Salinity Lab., Riverside, Cal., Animal Nutrition Lab., Cornell; Prather lectr. Harvard, 1942. Recipient Stephen Hales award Am. Soc. Plant Physiologists, 1929. Mem. Nat. Acad. Sci., Western Soc. Soil Sci. (pres. 1924), Bot. Soc. Am., Western Soc. Naturalists, A.A.A.S. (prize 1940). Cons. editor Soil Sci., Am. Jour. Botany, Plant Physiology. Research on inorganic nutrition of plants, soil solution, giant kelp, absorption of nutrients by plants, movement of solutes in plants potassium fixation by soils; devised Hoagland's culture solution for use in studies; established that plant absorption of mineral nutrients is a metabolic process. Died 1949.

HOAGLAND, Hudson, Am. physiologist; b. Rockaway, N.J., Dec. 5, 1899; s. Mahlon L. and Ella (Baylis) H.; A.B., Columbia, 1921; M.S., Mass. Inst. Tech., 1924; Ph.D., Harvard, 1927; Sc.D., Colby Coll., 1945, Wesleyan U., Conn., 1959, Clark U., 1962, Bates Coll., 1965, Boston U., Worcester Poly. Inst., 1966; m. Anna Plummer, June 9, 1920; children—Mahlon, Ann (Mrs. Bruce Crawford), Peter, Joan (Mrs. Burton Humphrey). Affiliate research prof. biology, physiology in psychiatry Boston U., 1944—; exec. dir. Worcester Found. Exptl. Biology, Shrewsbury, Mass., 1944—. Recipient Modern Medicine award for distinguished achievement, 1965; Humanist of Year award Am. Humanist Assn., 1965. Mem. A.A.A.S. (dir.), Nat. Assn. Mental Health, Am. Acad. Arts and Scis. (past pres.). Author: Pacemakers in Relation to Aspects of Behavior, 1935. Editor: Hormones, Brain Function and Behavior, 1957; (with R. Burhoe) Evolution and Man's Progress, 1962; (with others) Symbols and Society, 1964. Research and publs. on sensory nerve messages, elec. brain waves and their chem. determinants, chem. basis of our time sense, stresses and responses of adrenal cortex, biochem. aspects of schizophrenia. Home: Deerfoot Rd., Southboro, Mass. 01772. Office: 222 Maple Av., Shrewsbury, Mass. 01545.*

HOAGLAND, Mahlon Bush, Am. biochemist; b. Boston, Oct. 5, 1921; s. Hudson and Anna (Plummer) H.; student Harvard, 1941-43, M.D., 1948; m. Olley V. Jones, Jan. 10, 1961; children—Judith, Susan, Mahlon Bush, Robin. Staff dept. medicine Mass. Gen. Hosp., Boston, 1948-51; researcher, Copenhagen, Denmark, 1951-52, Cambridge, Eng., 1957-58; with Harvard Med. Sch., 1952-67, asso. prof. bacteriology and immunology, 1960-67; prof., chmn. dept. biochemistry Dartmouth Med. Sch., 1967—. Mem. biochem. study sect. NIH, 1961-64; Am. Cancer Soc. scholar in cancer research, 1953-58. Fellow Am. Acad. Arts and Scis.; mem. Am. Soc. Biol. Chemists. Research, publs. on mechanism of cancer-producing effects of beryllium, mechanism of liver regeneration, control of growth; discovered mechanism of amino acid activation in Protein synthesis, transfer ribonucleic acid. Home: Thetford, Vt. 05074. Office: Dartmouth Med. Sch., Hanover, N.H.*

HOAGLAND, Robert John, Am. physician; b. N.Y.C., 1909; s. William and Clara (Bunsen) H.; A.B., U. Mich., 1929; M.D., Cornell U., 1933; m. Thurma Pritchard Justus, 1932; children—Robert Dawson, William Justus, Cary. Commd. 1st lt. U. S. Army, 1937, advanced through grades to col., 1950; med. cons. to Surgeon Gen. in Europe, 1953, Far East, 1961-64; ret., 1967; prof. medicine Sch. Medicine Emory U., Atlanta, 1967—. Recipient medallion Surgeon Gen. U. S. Army for outstanding profl. achievement, 1967. Mem. A.M.A., A.C.P., Am. Therapeutic Soc. (v.p.). Author: Infectious Mononucleosis, 1967. Research and numerous publs. in clin. manifestations, incubation period and mode of transmission of infectious mononucleosis; discovered pharmacologic treatment of heat stroke and that coma due to many causes, including brain injury, could be ended with Ritalin; improved treatment of delirium tremens with great reduction of mortality. Home: 3680 Peachtree Rd. N.E., Atlanta 30319. Office: 69 Butler St. S.E., Atlanta 30303.*

HOAK, Richard Daugherty, Am. chem. engr.; b. Lancaster, Pa., Aug. 13, 1905; s. Frank B(aldwin) and Ella M. (Daugherty) H.; S.B., Mass. Inst. Tech., 1928, S.M., 1929; Ph.D., U. Pitts., 1948; m. Orpha Grace Hostetter, Feb. 24, 1940. Asst. dir. Chem. Engring. Practice Sch., Buffalo sta. Mass. Inst. Tech., 1929-31; with Mead Corp., Chillicothe, O., 1931-32; partner Hoak & Andes, Lancaster, Pa., 1932-34; chemist and bacteriologist City of Lancaster, 1934-36; dist. engr. Pa. Dept. Health, Harrisburg, 1936-40; sr. fellow Mellon Inst., Pitts., 1940—. Mem. adv. com. on water data U. S. Geol. Survey, 1965—; mem. Nat. Tech. Task Com. on Indsl. Wastes, 1950—, chmn., 1958. Fellow A.A.A.S., Am. Inst. Chemists, Am. Geog. Soc.; mem. Am. Inst. Chem. Engrs. (past chmn.), Am. Chem. Soc. (past chmn. exec. com., councillor 1962—, mem. adv. bd. Indsl. and Engring. Chemistry, also Chem. and Engring. News 1951-54), Pa. Chem. Soc. (bd. govs. 1946—, sec. treas. 1960—), Water Pollution Control Fedn. (standards methods com., 1947—, chmn., 1965—), Am. Soc. for Testing and Materials (Max Hecht award 1965), Am. Water Works Assn. (past trustee Western Pa. sect.), Air Pollution Control Assn., Pa. Water Works Operations Assn. (Freeburn Cup), N.Y. Acad. Scis., Sigma Xi. Patentee; Research and numerous publs. on water resources, design of settling basins, identification organic compounds in water. Home: R.D. 1, Franklin Rd., Mars, Pa. 16046. Office: 4400 5th Av., Pitts. 15213.*

HOAR, Thomas Percy, English metallurgist; b. Rochester, Eng., Nov. 16, 1907; s. George Percy and Lily (Hollings) H.; B.A., U. Cambridge, 1929, M.A., 1932, Ph.D., 1933, Sc.D., 1958; m. Joan Bailey, Mar. 1, 1935; 1 dau., Sally. Investigator, Internat. Tin Research and Devel. Council, 1933-39; research and devel. Ministry of Supply, Admiralty, also Ministry Aircraft Prodn., 1936-46; cons. to industry, 1938—; faculty U. Cambridge, 1946—, reader chem. metallurgy, 1965—. Dir. Metals Research Ltd. Fellow Royal Instn. Chemistry, Instn. Metallurgists; mem. Inst. Metal Finishing (past pres., Hothersall medal 1962), Internat. Com. for Electrochem. Thermodynamics and Kinetics (past pres.), Faraday Soc., Electrochem. Soc., Soc. Chem. Industry, Inst. Metals, Iron and Steel Inst., Chem. Soc. Contbg. author: Modern Aspects of Electrochemistry, 1959; also numerous articles. Research on mechanism of corrosion, stress-corrosion of metals, mechanism of passivity of metals, methods of protection of metals, theory of electrode processes on metals. Home: 26 Storey's Way, Cambridge, Eng.*

HOAR, William Stewart, Canadian zoologist; b. Moncton, N.B., Can., Aug. 31, 1913; s. George W. and Nina (Steeves) H.; B.A., U. N.B., 1934, D.Sc., 1965; M.A., U. Western Ont., 1936; Ph.D., Boston U., 1939; m. Margaret MacKenzie, Aug. 13, 1941; children—Stewart George, David Innes, Kenzie Margaret, Melanie Frances. Asst. prof. biology U. N.B., 1939-42, prof. zoology, 1943-45; research asso. U. Toronto, 1942-43; prof. zoology and fisheries U. B.C., Vancouver, 1945-64, prof., head dept. zoology, 1964—; research scientist Fisheries Research Bd. Can., 1935-47. Fellow Royal Soc. Can., Canadian, U. S. socs. zoology and physiology. Author: General and Comparative Physiology, 1966. Research and publs. on behavior, physiology of fishes, particularly migration, hormonal control of reprodn. Home: 4173 W. 16th Av., Vancouver, B.C., Can.*

HOARE, Cecil Arthur, protozoologist; b. Middelburg, Holland, Mar. 6, 1892; s. Arthur Stovell and Aimé (Charlet) H.; B.Sc., U. St. Petersburg (Russia), 1917; D.Sc., U. London, 1927; m. Marie Leserson, May 6, 1931. Fellow Petrograd (Russia) U., 1917-20; lectr. Mil. Med. Acad., 1918-20; head protozool. dept. Wellcome Labs. Tropical Medicine, 1923-57; Wellcome Research fellow Wellcome Hist. Med. Mus., London, 1957—; staff Uganda Med. Service, Trypanosomiasis Research Inst., 1927-29. Mem. expert panels London U., WHO, Geneva. Recipient Gaspar Vianna medal Brazilian Acad. Scis., 1962; Patrick Manson prize, 1963. Fellow Royal Soc., 1950, Inst. Biology; hon. mem. Soc. Protozoologists, Brit. Parasitological Soc.; fgn. mem. French, Belgian sci. socs. Author: Handbook of Medical Protozoology, 1949; also numerous articles. Research on parasitic protozoa; worked out life cycles of various blood parasites; systematics and host-parasite relations of human and animal trypanosomes; revised classification of coccidia; studied bionomics of copozoic ciliates and host-parasite relations of dysentery amoeba; developed concept of biol. races in parasitic protozoa. Home: 77, Sutton Ct. Rd., London, W.4, Office: Euston Rd., London, N.W.1., Eng.*

HOARE, Edward Wallis, Brit. veterinarian; b. Oct. 9, 1863; s. William Jesse Hoare; ed. New Vet. Coll., Edinburgh; m. Emily Helen, 1899; 1 son, 5 daus. Lectr. vet. sci. Univ. Coll., Cork, Ireland; mem. bd. examiners Royal Coll. Vet. Surgeons, also fellow; external examiner in vet. jurisprudence, toxicology and san. law U. Liverpool; practiced vet. surgery. Hon. mem. Am. Vet. Med. Assn. Author: A System of Veterinary Medicine, 2 vols., 1916; Veterinary Therapeutics, 3d edit., 1916; also papers. Editor Veterinary News. Died Nov. 26, 1920.

HOBBES, Thomas, English philosopher; b. Westport, Eng., Apr. 5, 1588; s. Thomas H.; B.A., Magdalen College, Oxford, 1608. Tutored William Cavendish, eldest son of 1st Earl of Devonshire, for 20 years; toured France, Germany, Italy with Cavendish, from 1610; spent leisure studying classical literature; became traveling tutor to son of Sir Gervase Clifton, Paris, 1629; ret. to Eng. to tutor Cavendish's son, 1632; made 2 more trips to continent meeting many eminent writers and philosophers; ret. to Eng., 1637; began writing treatises embodying his philosophy; ret. to Paris when his political writings brought him into disfavor in Eng., 1640; taught exiled Charles, Prince of Wales, circa 1646; offended churchmen and intellectuals in France and fled to Eng., 1651; lived with Cavendish, 1653; after Charles II came to power in Eng., Hobbes became popular and spent rest of his life as professional writer. Author: De Cive, 1642; Human Nature, 1650; Leviathan, 1651; De Homine, 1658; De Corpore Politico, 1680; Quadratura Circuli, 1889; others. A convinced materialist; rejected supernatural explanations for all phenomena; accepted explanations in terms of finite, natural forces; rejected inductive methods and accepted deductive reasoning; considered by some to be one of founders of empirical psychology; claimed as one of founders of analytic philosophy and scientific sociology; impressed by Galileo's theory of motion (that it is natural state of bodies), he applied it to question of sensation; concluded that sense organs were agitated by motions, and without motion there could be no sensation; explained dreams, passions, imagination, language and perception in terms of modes of motion; explained human action through concept of "efficient causes"; did not employ theological and ethical interpretations in his examination of man and society; presented sensationalist interpretation of the nature of the state and of man; Hobbesian social contract theory is based on fear of violence and anarchy in absence of strong central authority; said men are basically antisocial; maintained constitutional limitations on power of sovereign are impossible; his views influential in development of doctrine of economic determinism and movements for scientific legislation; his concept of conatus also influential in mechanics. Died Hardwick, Eng., Dec. 4, 1679.

HOBBS, Henry Edwin, English ophthalmic surgeon; b. London, Apr. 2, 1910; s. Henry and Florence (Saunders) H.; student London U., London Hosp. Med. Coll., 1932-39; m. Jean Kennedy, Apr. 11, 1942; children—Jennifer, Bridget, Charlotte. House surgeon London Hosp., 1939-40, Oxford (Eng.) Eye Hosp., 1942-42; ophthalmic specialist RAF, 1943-46; chief clin. asst., registrar, hon. cons. Moorfields Eye Hosp., 1947-56; ophthalmic surgeon Royal Free Hosp., 1948—, Nat. Hosps. for Nervous Diseases, 1946—, Homes of St. Giles, 1954—; research asst. Inst. Ophthalmology, 1949-56; cons. ophthalmic surgeon, London, 1946—; faculty ophthalmology U. London, 1947—. Decorated Comdr. Mil. and Hospitaller Order St. Lazarus of Jerusalem, Officer Most Venerable Order of Hosp. St. John of Jerusalem; recipient Sr. County award, 1934, Buxton Anatomy prize, 19 , Andrew Clark Medicine/Pathology prize, 1937. Fellow Royal Soc. Medicine; (hon. sec. ophthalmology); mem. Ophthal. Soc. U.K., Internat. Congress Ophthalmology, Royal Colls. Physicians and Surgeons, Brit. Orthoptic Bd., Oxford Congress, Med. Soc. London. Author: Modern Ophthalmology (with A. Sorsby), 1961; Principles of Ophthalmology, 1965; also articles; contbg. author: The Place of Surgery in Loss of Vision, Surgical Aspects of Medicine, 1959; Internat. Ophthalmology Clinics, 1963. Research on value of cortisone in treatment of ocular disease, assessment of visual defects in diagnosis of intracranial lesions, angiography in diagnosis of vascular lesions resulting in proptosis; demonstrated safer surgery for treatment of cataract in diabetes, toxic ocular effects of antimalarial drugs. Home: 1, Wells Rise, St. John's Wood, London, N.W.8, Eng. Office: 46, Wimpole St., London, W.1, Eng.*

HOBBS, William Herbert, Am. geologist; b. Worcester, Mass., July 2, 1864; s. Horace and Mary Paine (Parker) H.; S.B., Worcester Poly. Inst., 1883, D. Engring., 1929; fellow geology Johns Hopkins, 1887-88. A.M., Ph.D., 1888; postgrad. Heidelberg, 1888-89; LL.D., U. Mich., 1939; m. Sara K. Sale, June 23, 1896; 1 dau., Winifred (Mrs. J. N. Lincoln). Curator Geol. Mus., 1889-90. asst. prof. mineralogy and metallurgy U. Wis., 1890-99, prof. mineralogy and petrology, 1899-1906; prof. geology and dir. Geol. Lab., U. Mich., 1906-34, prof. emeritus from 1934. With U. S. Geol. Survey, 1886-1906; U. S. asst. geologist, 1896; exchange prof. Technische Hoogeschool, Delft, 1921-22; Greenland expdns., U. Mich., 1926-31, hon. dir. 1932-33. Vice-pres. Internat. Glacier Com., 1930-36; adv. mem. OSS, 1941-45. Fellow A.A.A.S. (v.p. 1932). Geol. Soc. Am. (1st v.p. 1922); mem. Am. Philos. Soc. (mem. council 1930), Assn. Am. Geographers (1st v.p. 1917; pres. 1936), Mich. Acad. Sci. (pres. 1917). Decorated chevalier Legion of Honor (France). Author: Earthquakes, 1907 (German transl. 1910); Characteristics of Existing Glaciers, 1911; Earth Features and Their Meaning, 1912; The World War and Its Consequences (with Introd. by Theodore Roosevelt), 1919; Leonard Wood Administrator, Soldier and Citizen, 1920; Earth Evolution and Its Facial Expression, 1921; Cruises Along By-ways of the Pacific, 1923; The Glacial Anti-cyclones, 1926; Exploring About the North Pole of the Winds, 1930; Peary, 1936; Explorers of the Antarctic, 1941; Fortress Islands of the Pacific, 1945; Glacial Studies of the Pleistocene on N. Am., 1947; (autobiography) An Explorer-Scientist's Pilgrimage 1952. Designer project for combined open-cut and tunnel sea-level ship-canal across Tehuantenec Isthmus in Mexico, sea level shipcanal across Honduras, 1952. Died Jan. 1, 1953.

HOBHOUSE, Leonard Trelawney, English sociologist, philosopher; b. Ive, Cornwall, Sept. 8, 1864; s. Reginald and Caroline (Trelawney) H.; ed. Corpus Christi Coll., Oxford, 1883-87, fellow Merton Coll., 1887; hon. D.Litt., Durham, 1913, LL.D., St. Andrews, 1919; m. Nora Hadwenm, 1891; 1 son, 2 daus. Became tutor Corpus Christi Coll., 1890, fellow, 1894; with Manchester Guardian, 1897-1902; mem. staff Free Trade Union, 1903-05, editor Sociol. Review, 1906-07; with Tribune, 1906-17; prof. sociology London U., 1907-29. Assisted in forming Sociol. Soc., 1903. Author: The Labor Movement, 1893; The Theory of Knowledge, 1896; Mind in Evolution, 1901; Democracy and Reaction, 1904; (with J. L. Hammond) Memoir of Lord Hobhouse, 1905; Morals in Evolution, 1906; Social Evolution and Political Theory, 1911; Liberalism, 1911; Development and Purpose, 1913; (with G.

C. Wheeler, M. Ginsberg) The Material Culture and Social Institutions of the Simpler People, 1915; The Metaphysical Theory of the State, 1918; The Rational Good, 1921; The Elements of Social Justice, 1922; Social Development, 1924. Used anthrop. and psychol. data in study of society; attempted to show biol. and spiritual evolution of man were simultaneous developments. Died Normandy, France, June 21, 1929.

HOBSON, Ernest William, Brit. mathematician; b. Derby, Oct. 27, 1856; ed. Christ's Coll., Cambridge, Sc.D.; LL.D., St. Andrews, Aberdeen; D.Sc., Oxford, Dublin, Sheffield, Manchester Univs; m. 1882. Examiner math. and natural philosophy U. London, 1895-1900; Stokes lectr. U. Cambridge, 1903-10, Sadleirian prof. pure math., 1910-31, emeritus prof., 1931-33. Gifford lectr., Aberdeen, 1921, 22. Fellow Royal Soc., 1893 (Royal medal for math. investigations 1907), Royal Astron. Soc.; mem. Brit. Assn. (pres. math. and phys. sect. 1910), Cambridge Philos. Soc. (pres. 1906-08), London Math. Soc. (pres. 1900-02, De Morgan medal 1920), Royal Irish Acad. (hon.), Acad. Scis. Bologna (corr.), Carol Leopold Akademie Halle (fgn.). Author: Treatise on Plane Trigonometry, 1891; Treatise on the Theory of Functions of a Real Variable, 1907; Squaring the Circle, 1913; The Domain of Natural Science (Gifford lecture), 1923; The Theory of Spherical and Ellipsoidal Harmonics, 1931, also articles. Showed that all Euclidean constructions can be carried out by use of compasses alone; gave simplified proof of impossibility of squaring the circle; work important for introducing modern theory of functions in Eng. Died Apr. 18, 1933.

HOBSON, John Atkinson, English social economist; b. Derby, Eng., July 6, 1858; s. William and Josephine (Atkinson) H.; ed. Lincoln Coll. Oxford (Eng.) U.; m. Florence Edgar, 1885; 2 children. Tchr. classics, schs. at Faversham and Exeter, Eng., 1880-87, English lit., econs. Delegacy Extension U. and London Soc. for Extension U. Teaching; lectr., traverler U. S., Can., S. Africa. Mem. Bryce Com. (prepared plan for a league nations); founder (with others) Union Dem. Control. Author: (with A. F. Mummery) The Physiology of Industry: Being an Exposure of Certain Fallacies in Existing Theories of Economics, 1889; Problems of Poverty, 1891; The Evolution of Modern Capitalism, 1894; The Problem of the Unemployed, 1896; John Ruskin, Social Reformer, 1898; The Economics of Distribution, 1900; The War in South Africa, 1900; The Psychology of Jingoism, 1901; The Social Problem, 1901; Imperialism, 1902; Canada Today, 1906; The Industrial System, 1909; The Science of Wealth, 1911; Gold, Prices and Wages, 1913; Work and Wealth, 1914; Towards International Government, 1915; The New Protectionism, 1916; Democracy after the War, 1917; Taxation in the New State, 1919; Richard Cobden, International Man, 1919; The Economics of Unemployment, 1922; Free Thought in the Social Sciences, 1926; Wealth and Life, 1929; Rationalization and Unemployment, 1930; Confessions of an Economic Heretic, 1938. Contbr. articles jours. of social opinion, 1899-1920. Approached econs. basically as humanist-sociologist, rather than pure econ. thinker, sought to integrate realms of ethics, politics, econs., criticizing much existing theory as class-biased; his theory of underconsumption related econ. health to consumption-capital investment equilibrium; saw prepondrance of wealth and power held by minority in capitalist soc. as interfering with attainment of such equilibrium; argued for recognition of effects of vested interest on behavior; proposed steeply graduated taxation, extension social services, nationalization of monopolies; as opponent of Boer (S. Africa) war, wrote maj. study on econ. basis imperialism; contbd. to enlargement of scope of econs. and to greater reliance on realistic investigation. Died Hampstead, Eng., Apr. 1, 1940.

HOCH, Michael, metall. engr.; b. Budapest, Hungary, Feb. 9, 1923; s. Joseph and Jolanda (Herzfeld) H.; Chemist, Swiss Fed. Inst. Tech., Zurich, 1945, D.Sc., 1947; m. Claire Hoch, July 19, 1956. Came to U. S., 1950, naturalized, 1955. Research asso. U. Zurich, 1947-50; research asso. Ohio State U., Columbus, 1951-56; faculty U. Cin., 1956—, prof. material sci., 1963—. Cons. to labs., indsl. firms. Fellow Am. Ceramic Assn.; mem. Am. Phys. Soc., Am. Chem. Soc., German Ceramic Soc., Am. Soc. for Metals, Am. Inst. Mining, Metall. and Petroleum Engrs. Research and publs. on solid refractory materials at high temperatures, thermal conductivity, specific heat, emittance in temperature range 2000-6000° F, phase relationship in refractory systems, vaporization mechanisms, thermodynamic properties, non stoichiometry in refractory systems. Home: 2920 Scioto St., Cin. 45219.*

HOCHE, Alfred E., psychiatrist; b. Wildenhain, Saxony, 1865; student U. Berlin; M.D., Heidelberg, Germany, 1888; asst. various clinics; prof. psychiatry and neuropathology, Freiburg. Research on organic diseases of nervous system. Died 1945.

HOCHENEGG, Julius, see von Hochenegg, Julius.

HOCHREIN, Max, German physician; b. Nuremberg, Aug. 2, 1897; s. Johann and Anna (Edelhäuser) H.; ed. univs. Erlangen, Heidelberg, Munich; M.D.; m. Irene Schleicher. Asst., Physiology Inst., Munich, 1922-24, Univ. Clinic, Cologne, 1924-26; asst., prof., agrégé internal medicine Univ. Clinic, Leipzig, 1926-28, chief physician, 1929-39; prof. Harvard, 1928-29; asso. prof. internal medicine, 1932; full prof., dir. Univ. Polyclinic, 1939; 1st dir. Univ. Inst. Medicine of Labor, 1940-45; chief physician Med. Clinic, municipal Hosp., Ludwigshafen, Rhine, 1948-63; with Inst. Specialized Medicine, 1963. Mem. Am. Coll. Angiology, German Assn. Internal Medicine, Deutsche Gesellschaft für Kreislaufforschung, German Soc. for Rheumatology, Deutsche Gesellschaft für Verkehrsmedizin, Deutsche Gesellschaft für Verdauungs- und Stoffwechselkrankheiten. Author: Koronarkreislauf, 1930; Der Myorardinfarkt, 3 edits.; (with I. Schleicher) Rheumatische Erkankungen, 2d edit.; Herz Kreislauferkrankungen, 1959, also works on preventive, clin., gerontological medicine. Address: 7 Saarbrückestrasse, Ludwigshafen, Rhine, Germany.

HOCHSTADT, Harry, mathematician; b. Vienna, Austria, Sept. 7, 1925; s. Samuel and Amalie (Dorn) H.; brought to U. S., 1938, naturalized, 1944; B.Ch.E., Cooper Union, 1949; M.S., N.Y. U., 1950, Ph.D., 1956; m. Pearl Ruth Schwartzberg, Mar. 29, 1953; children—Julia Phyllis, Jesse Frederick. Research engr. W. L. Maxson Corp., N.Y.C., 1951-57; research asso. N.Y. U. Courant Inst. Math Sci., 1959-61; faculty Poly. Inst. Bklyn., 1957—, prof., 1961—, head dept. math., 1963—; Mem. Am., London, Indian math. socs., Math. Assn. Am., Soc. Indsl. and Applied Math., Am. Assn. U. Profs., Sigma Xi, Tau Beta Pi. Author: Special Functions of Mathematical Physics, 1961; Differential Equations, 1964. Editor SIAM Applied Math. series. Research and publs. on area of ordinary and partial differential equations, especially problems in phys. optics, such as diffraction, non-linear resonance and wave propogation in periodic structures. Home: 126 Joralemon St., Bklyn. 11201. Office: 333 Jay St., Bklyn. 11201.*

HOCHSTETTER, Ferdinand von, see von Hochstetter, Ferdinand.

HOCKING, George Macdonald, pharmacognosist; b. Newquay, Cornwall, Eng., Mar. 21, 1908; s. Humphrey Scoble and Tamar Elizabeth (Giles) H.; student Reed Coll., 1924-26; B.S. in Pharmacology, U. Wash., 1931; M.S., U. Fla., 1931-32, Ph.D., 1942; m. Betty Susan Willis, Oct. 27, 1933; 1 dau., Barbara Jean (Mrs. Dean Theodore Anderson). Came to U. S., 1924, naturalized, 1935. With George Washington U., 1933-35; with various comml. firms, 1935-37; faculty Ohio No. U., 1937-40; chief pharmacognosist S. B. Penick & Co., 1942-46; faculty U. Buffalo, 1946-48, U. N.M., 1948-51; prof. pharmacognos. Auburn (Ala.) U., 1951—. Mem. exec. com., chmn. subcom. on pharmacognosy Com. on Nat. Formulary, 1947-60; adviser FAO to Pakistan Govt., 1951; mem. Internat. Commn. for Plant Raw Materials, 1965—, subcom. on med. plants Pacific Sci. Assn. Hon. mem. Commisao de Estudos de Plantas Brasileiras, Medicinais e Toxicas (Sao Pulo, Brazil); mem. Am. Pharm. Assn., Am. Soc. Pharmacognosy, Am. Inst. Hist. Pharmacy, Internat. Assn. Plant Taxonomy, Fédn. Internat. Pharmacognosy and Economic Botany, 1955; also numerous articles. Abstracter books, jours. Research on morphology and histology of Mentha piperita, Mentha spicata, Mentha cardiaca, Liatris root. Home: 1021 S. Gay St., Auburn, Ala. 36830.*

HODES, Horace Louis, Am. physician; b. Phila., Dec. 21, 1907; s. Morris and Anna (Jacobson) H.; A.B., U. Pa., 1927, M.D., 1931; m. Anne E. Reber, June 10, 1931; children—Ruth (Mrs. Steven Rabin), David. Dir. children's out-patient dept. Johns Hopkins, 1935-36, asso. prof. pediatrics, 1946-49; asst. pathology and bacteriology Rockefeller Inst., 1936-38; officer in charge virus lab. Naval Med. Research Unit, Guam, 1944-45; cons. Sec. of War, 1940-42; clin. prof. pediatrics. Columbia Coll. Phys. and Surg., 1949—; pediatrician in chief Mt. Sinai Hosp., N.Y.C., 1949—; prof., chmn. pediatrics Mt. Sinai Med. Sch., 1964—. Recipient Mead Johnson award for research in pediatrics Acad. Pediatrics, 1946. Mem. Am. Pediatric Soc., Soc. for Pediatric Research (past pres.), N.Y. Acad. Medicine, Soc. for Exptl. Biology and Medicine, Am. Acad. Pediatrics. Author: (with Philip M. Stimson) Common Contagious Diseases, 1956; also numerous articles. Demonstrated that mosquito larvae could be infected with encephalitis virus and transmit infection as adults; developed ultraviolet-killed effective rabies vaccine; discovered antiviral substance against herpes virus. Home: 41 Sutton Crest, Manhasset, N.Y. 11030. Office: Mt. Sinai Hosp., N.Y.C. 10029.*

HODGDON, Albion Reed, Am. plant taxonomist; b. Boothbay, Me., Nov. 1, 1909; s. Lewis Percival and Laura (Hodgdon) H.; B.S., U. N.H., 1930, M.S., 1932; Ph.D., Harvard, 1936; m. Audrey Mackown, Aug. 11, 1940; children—Alan Lewis, Anthony Jason, Ariel Josephine. Faculty, U. N.H., Durham, 1936—, head botany sect. biology, 1941-47, head dept. botany also Botany Agr. Expt. Sta., 1947—, prof. botany, 1948—. Fellow A.A.A.S.; mem. Am. Soc. Plant Taxonomists, Ecol. Soc. Am., New Eng. Bot. Club (editor Rhodora 1962—), Soc. for Study Evolution, Sigma Xi. Author: (with Frederic Steele) Woody Plants of New Hampshire, 1958; also numerous articles on higher plant flora of N.H., islands and coastal areas Bay of Fundy in Eastern Can., specis of blackberries in Northeastern Am. Home: R.F.D., Box 97, Nottingham, East Barrington, N.H., 03825. Office: Nesmith Hall, U. N.H., Durham, N.H. 03624.*

HODGE, Paul Van, Jr., Am. geophysicist; b. Loco, Okla., Dec. 1, 1917; s. Paul Van and Mae (Stewart) H.; student Ranger Jr. Coll., 1936, So. Meth. U., 1937, Ind. Inst. Tech., 1939-40; m. Hazel Robinson, Mar. 29, 1959; 1 son, Milton Morris. With Geophys. Service Internat., 1945—, mgr. Far East Marine area, Sydney, Australia, 1964—; one of founders Tex. Instruments, 1953, mgr. marine operations, 1962—. Mem. Soc. Exptl. Geophysicists, Am. Geophys. Union, European Assn. Econ. Geophysicists, Australian Petroleum Exploration Assn. Discovered unknown maj. salt dome, oil fields in La., Tex., Cal., Okla., Africa, Persian Gulf, Australia; inventor Diesel engine automatic lube pressure protector, underwater towed gravity meter array, underway deep ocean sensor transport. Home: 94 Harbour Tw's, North Sydney. Office: 120 Christie St., St. Leonard's, Sydney, Australia.*

HODGE, Philip Gibson, Jr., Am. mech. engr.; b. New Haven, Conn., Nov. 9, 1920; s. Philip Gibson and Muriel (Miller) H.; A.B., Antioch Coll., 1943; Ph.D., Brown U., 1949; m. Thea Drell, Jan. 3, 1943; children—Susan E., Philip T., Elizabeth M. Research asst., asso. Brown U., 1947-49; faculty U. Cal., Los Angeles, 1949-53, Poly. Inst. Bklyn., 1953-57; prof. mechanics Ill. Inst. Tech., Chgo., 1957—. Mem. basic research adv. com. NRC-Nat. Acad. Scis., 1963—; mem. reactor engring. rev. com. Argonne Nat. Lab., 1964—. Mem. Am. Assn. U. Profs., Math. Assn. Am., Am. Soc. M.E. Author: (with W. Prager) Plastic Analysis of Solids, 1951; (with J. N. Goodier) Elasticity-Plasticity, 1958; Plastic Analysis of Structures, 1959; Limit Analysis of Rotationally Symmetric Plates and Shells, 1963. Research and numerous publs. primarily in field of plasticity and plastic analysis of structures. Home: 5628 S. Harper Av., Chgo. 60637.*

HODGE, Sir William Vallance Douglas, Brit. mathematician; b. Edinburgh, Scotland, June 17, 1903; s. Archibald James and Janet (Vallance) H.; M.A., Edinburgh U., 1923; B.A., Cambridge (Eng.) U., 1925, M.A., 1930, Sc.D., 1950; LL.D., Edinburgh U., 1958; D.Sc., U. Bristol (Eng.), 1957, U. Sheffield (Eng.), 1960, U. Leicester (Eng.), 1959, U. Wales, 1961, U. Liverpool (Eng.), 1961, U. Exeter (Eng.), 1961; m. Kathleen Anne Cameron, July 27, 1929; children—Michael Robert, Gillian Janet. Lectr., U. Bristol, 1926-31; vis. fellow Princeton, 1931-32; fellow St. John's Coll., Cambridge U., 1930-33, faculty Cambridge U., 1933—, prof. astronomy and geometry, 1936—, fellow Pembroke Coll., 1935-58, master, 1958—. Vis. lectr. Harvard, 1950. Fellow Royal Soc., 1938 (past sec., Royal medal 1957); mem. London Math. Soc. (pres. 1947-49, Berwick prize 1952, De Morgan medal 1959), Cambridge Philos. Soc. (past pres.), Internat. Congress Mathematicians (pres. 1958), Internat. Math. Union, (past v.p.); fgn. mem. U. S. Acad. Sci., Am. Philos. Soc., Acad. Arts and Scis., Royal Danish Acad., Akademie der Wissenshaften in Göttingen. Author: Theory and Application of Harmonic Integrals, 1965; Methods of Algebraic Geometry (with D. Pedoe), 3 vols. Research on algebraic geometry especially function theoretic and topological properties of algebraic varieties over the complex field, properties of complex manifolds. Home: Master's Lodge, Pembroke Coll., Cambridge, Eng.*

HODGES, Clarence Vernard, Am. surgeon; b. Terry, S.D., Nov. 11, 1914; B.S., Ia. State Coll., 1937; M.D., U. Chgo., 1940. Research fellow in surgery U. Chgo. Clinics, 1940-41, intern, 1941-42, research in urol. surgery, 1945-46; instr. Hopkins Hosp., 1947-48; asso. prof. urology U. Ore. Med. Sch., 1948-58, prof. surgery, 1958—, head div. urology, 1948—. Research on prostatic and bladder carcinoma, lower nephron nephrosis; used stilbestrol in treatment of metastatic carcinoma of prostate (with C. Huggins), 1941. Office: University of Oregon Medical School, Portland, Ore. 97201.*

HODGES, Nathaniel, English physician; b. Kensingston, Eng., Sept. 14, 1629; s. Thomas Hodges; student Trinity Coll., Cambridge U.; B.A., Oxford U., 1651, M.A., 1654, M.D., 1659. Practised medicine in London; became poor and imprisoned for debt; Harveian orator, 1683. Mem. Royal Coll. Physicians, 1659, fellow, 1672 (censor 1682). Author: Voindiciae Medicinae et Medicorum, an Apology for the Profession and Professors of Physic, 1966; Pestis nuperae apud Populum Londinensem grassantis narratio historica, 1672. Wrote good description of symptoms of the plague and results of treatments; probably only writer to describe pericarditis in a case of the plague. Died Ludgate (debtor's) Prison, June 10, 1688.

HODGES, Paul Chesley, Am. physician; b. Anderson, Ind., Jan. 6, 1893; s. Fred Jenner and Josephine (Chesley) H.; B.S., U. Wis., 1919, Ph.D., 1924; M.D., Washington U., St. Louis, 1918; m. Merle Rhea, Aug. 10, 1918 (dec.); children—Lorna (Mrs. Guillermo Santin), Josephine (Mrs. Phil Rogers), Patricia (dec.). Paul Chesley, Jean. Instr., Harvard Med. Sch. Shanghai, 1915-16; faculty Peking (China) Union Med. Coll., 1919-27, prof. radiology, 1926-27; faculty U. Chgo., 1927-58, prof., 1928-58, prof. emeritus, 1958—; vis. prof. radiology U. Taiwan, 1960-63, U. Fla., Gainesville, 1964—. Decorated Cravat of Order Brilliant Star; recipient Grubbe medal Chgo. Med. Soc., 1964. Mem. A.M.A., Radiol. Soc. N.Am., Am. Roentgen Ray Soc. (Caldwell medal 1953 past pres.), Sigma Xi, Alpha Omega Alpha. Author: (with D. B. Phemister, Alex Brunschwig) Diseases of Bone, 1936;

Life and Times of Emil H. Grubbe, 1964; also numerous articles. Dept. editor Postgrad. Medicine, 1958—. Measurement of pelvis and fetus, photoelectric timing of X-ray exposures, protection against radiation hazards. Home: 2306 S.W. 13th St., Gainesville, Fla. 32601.*

HODGES, Robert Edgar, Am. physician; b. Marshalltown, Ia., July 30, 1922; s. Wayne H. and Blanche E. (McDowell) H.; B.A., U. Ia., 1944, M.D., 1947, M.S., 1949; m. Norma Lee Stempel, June 8, 1946; children—Jeannette Louise, Robert William, Karl Wayne, James Wolter. Faculty, U. Ia., 1952—, prof. internal medicine, 1964—, dir. metabolic ward, 1952—, chmn. publs. com. Coll. Medicine, 1962—. Mem. numerous profl. coms., cons. govt. agys.; cons. in nutrition to various Far East nations; chmn. com. on nutrition Am. Heart Assn.; asso. editor Nutrition Revs., 1959—, Am. Jour. Clin. Nutrition, 1962—; bd. editors Jour. Lab. and Clin. Medicine, 1962—. Diplomate Am. Bd. Internal Medicine. Fellow A.C.P.; mem. A.M.A., Endocrine Soc., Am. Fedn. Clin. Research, Am. Pub. Health Assn., Soc. Exptl. Biology and Medicine, Am. Rheumatism Assn., Central Soc. Clin. Research, Am. Soc. Clin. Nutrition (pres. 1966-67), A.A.A.S., Nutrition Soc. London, Sigma Xi, others. Contbr. numerous articles to profl. jours. Research, numerous publs. on diet and heart disease; induced vitamin deficiencies in man. Home: 365 Ellis Av., Iowa City 52240.*

HODGKIN, Alan Lloyd, English biophysicist; b. Feb. 5, 1914; s. G. L. and M. F. (Wilson) H.; student Trinity Coll., Cambridge, fellow, 1936—; M.D. (hon.), univs. Berne, Louvain; D.Sc., U. Sheffield, 1963; m. Marion de Kay Rous, 1944; 1 son, 3 daus. Sci. officer radar Air Ministry, also Ministry Aircraft Prodn., 1939-45; lectr., then asst. dir. research Cambridge, 1945-52; Foulerton research prof. Royal Soc., 1952—; mem. Med. Research Council, 1959-65. Recipient Baly medal, 1955, Nobel prize for medicine and physiology (with A. F. Huxley, J. C. Eccles), 1963. Fellow Royal Soc., 1948 (Royal medal 1958, Copley medal 1965); mem. Physiol. Soc. (fgn. sec. 1960—), Am. Acad. Arts and Scis. (fgn. hon.), Royal Danish Acad. Scis., Leopoldina Acad. Author: Conduction of the Nervous Impulse, 1963; also sci. papers on nature of nervous conduction. Devised (with Fielding Huxley) system of math. equations describing nerve impulse; worked with giant nerve fibers of squid, proving that electricity was direct causal agt. of impulse propagation.*

HODGKIN, Dorothy Crowfoot, chemist; b. Cairo, Egypt, 1910; student Somerville Coll., Oxford, Eng., 1928-31, Cambridge U., 1932-34; m. Thomas L. Hodgkin, 1937. mem. faculty Oxford U., 1934—. Recipient Nobel Prize in chemistry, 1964. Fellow Royal Soc., 1947. Determined structure of vitamin B12, cholesterol iodide, and penicillin using X-ray crystallographic analysis. Home: 20c Bradmore Rd., Oxford, Eng.*

HODGKINSON, Eaton, English mathematician; b. Anderton, Eng., 1789; m. Catherine Johns; m. 2d, dau. of Henry Holditch; farmer, then pawnbroker in Salford, Manchester, Eng., 1811; began Sci. research, Manchester; royal commr. on use of iron in ry. structures, 1847-49; named prof. mech. principles of engring. Univ. Coll., London, 1847. Fellow Royal Soc., 1841 (Royal medal), Geol. Soc., Royal Irish Acad.; mem. Manchester Lit. and Philos. Soc. (pres. 1848-50), Inst. Civil Engrs. (hon.). Author: Experimental Researches on the Strength and Other Properties of Cast Iron, 1846. Research in strength of materials; developed concept of mech. principle by which position of neutral line in sect. of rupture or fracture can be determined; offered theoretical expositions of neutral line, 1830, leading to expts. to determine strongest beam and resulting in discovery of Hodgkinson's beam. Died 1861.

HODGSON, Corrin Haley, Am. physician; b. Fergus Falls, Minn., June 7, 1908; s. Fred E. and Anastasia (Haley) H.; B.S., M.B., U. Minn., 1931, M.D., 1932, M.S. in Medicine, 1945; m. Florence M. Pitman, Aug. 6, 1932; children—C. John, Stephen F., Clague P. Pvt. practice medicine, Fergus Falls, 1934-37; fellow in internal medicine Mayo Found., Rochester, Minn., 1937-39; practice medicine specializing in internal medicine, Lima, Peru, S.Am., 1939-44; cons. internal medicine Mayo Clinic, 1944-66; prof. clin. medicine Mayo Grad. Sch. Med. U. Minn., 1964-66; Med. dir. Minn. Mining & Mfg. Co., St. Paul, 1966—; hon. prof. University of San Marcos, Lima, Peru. Trustee Minn. Med. Found., 1959—, pres., 1963-64. Mem. A.M.A., A.C.P., Am. Coll. Chest Physicians, Minn. Med. Assn., Zumbro Valley Med. Soc. (past pres.), Am. Thoracic Soc., Minn. Soc. Internal Medicine, Sigma Xi. Publs. on studies of chest diseases, including Tb, tumors, other granulomatous and vascular diseases, typhoid fever, Carrion's disease. Home: 49 E. Pleasant Lake Rd., North Oaks, Minn. 55110. Office: 3M Center, St. Paul 55101.*

HODGSON, James, English mathematician; b. Eng., 1672; m. niece of John Flamsteed; several children; tchr.; apptd. master math. sch. Christ's Hosp., 1709; Fellow Royal Soc., 1703 (council 1733). Author: Doctrine of Fluxians founded on Sir Isaac Newton's Method, 1736; other works. Co-editor: Atlas coelestis (Flamsteed). Died Eng., June 25, 1755.

HODGSON, Joseph, English surgeon; b. Penrith, Eng., 1788; apprentice to dr., Birmingham, Eng.; student St. Bartholomew's Hosp., London, Eng.; diploma Coll. Surgeons, 1811; m. Practiced medicine, London, then Birmingham, 1818-49; surgeon Gen. Dispensary, then Gen. Hosp., 1848-49; founder (with others) Birmingham Infirmary, 1824; examiner surgery London U. and Coll. Surgeons, from 1849. Recipient Jacksonian prize, 1811. Fellow Royal Soc., 1831; mem. Coll. Surgeons (mem. council, pres. 1864), Medico-Chirurg. Soc. (pres. 1851). Author: Diseases of the Arteries and Veins (provided excellent illustrations of aortic valoular endocarditis; 1st description aneurysmal dilation of aortic arch or Hodgson's Disease), 1815. Editor, contbr. London Med. Review. Research, publs. on aneurysmal disease, diseases arteries and veins, also wounds; accomplished lithotomist. Died Feb. 7, 1869.

HODIERNA, Giovanni Battista, astronomer; b. Ragusa, Sicily, Apr. 15, 1597; archpriest, Palermo, Sicily; mathematician to Duke of Palma. Author: Universae facultatis directorium physico-mathematicum (on astrological hypotheses), 1629; Archimede redevivo (on static), 1644; L'occhio della mosca, 1644; Nunzio della terra, 1644; Thaumantiae miraculum, 1652; De systemate orbis cometici, 1656; Medicaeorum epemerides (credited as 1st book with observations on eclipses of Jupiter's satellites), 1656; Menologiae Jovis compendium (astron. tables on satellites of Jupiter), 1656; De admirandis phasibus in sole et luna visis, 1656; Protei coelestis vertigines (on appearances of Saturn). Discovered motions of satellites of Jupiter; verified positions of fixed stars; research on prism, insects, eye of fly; recognized function of queen bee. Died Palma, Apr. 6, 1660.

HODR, Jaroslav, Czech. gynecologist; b. Prague, Czechoslovakia, Feb. 8, 1921; s. Anton and Rose (Hodr) H.; M.D., Charles U., 1950, candidat sci., 1959; m. Anna Novak, Mar. 3, 1956; 1 son, Zbynek. Mem. dept. obstetrics and gynecology, also dept. surgery Dist. Hosp., Vysoke Myto, Bohemia, 1950-52; mem. staff Inst. for Care of Mother and Child, Prague, 1952—, asst. prof., head research unit, 1955—. Mem. Med. Soc. Czechoslovakia. Research, numerous publs. on metabolic changes in mother and foetus related to uterine activity, influencing of protracted uterine activity by insulin, oxytocin, and methyloxytocin, changes of enzyme activity in parturients and healthy and hypoxic fetusses, functional tests in pregnant women with threatening premature delivery. Home: 816 Nabrezi K. Marxe, Prague 4. Office: 157 Nabrezi K. Marxe, Prague 4, Czechoslovakia.*

HOEFER, Paul Frederick Adam, Am. physician; b. Munich, Germany, Nov. 21, 1903; s. Paul Adam and Sophie (Gutkin) H.; Ph.D., U. Berlin (Germany), 1927; M.D., U. Wuerzburg (Germany), 1928; Maria Kuehl, Sept. 27, 1932. Came to U. S., 1934, naturalized, 1939. Asst. physiology U. Berlin, 1929-32; asst. neurology U. Heidelberg (Germany), 1932-33; instr. Tufts U., 1934-36; research fellow, asst. neurology Harvard, 1936-39; faculty Columbia Coll. Phys. and Surg., 1939-62—, prof. neurology, 1962—; exchange prof. Free U. Berlin, 1953-54. Cons. Walter Reed Hosp., Washington, 1946-50. Mem. A.M.A., Am. Neurology Assn., Am. Physiol. Soc., Am. Electroencephalograph Soc., Royal Soc. Medicine Eng., Am. Psychiat. Assn., Am. Acad. Neurology, Assn. for Research in Nervous and Mental Disease. Research, numerous publs. on electrophysiology, nerve, muscle, and brain, modern research instrumentation, motor disorders central nervous system, neurol. disorders, studies Myasthenia Gravis. Home: 355 E. 72d St., N.Y.C. 10021.*

HOEFER, Wolfgang, physician; b. Freising, Bavaria, 1614; M.D., U. Ingolstadt, 1653; practiced medicine, Straubing, also Linz, Austria, and Raab; ct. physician, Vienna. Author: Hercules medicus, 1657. Gave early description of myxedema, 1657. Died 1681.

HOEG, Donald Francis, Am. chemist; b. Bklyn., Aug. 2, 1931; s. Harry H. and Charlotte (Bourke) H.; B.S. summa cum laude, St. John's Coll., 1953; Ph.D. in Chemistry (Armour Research Found. fellow, Ta Ping Lin scholar) Ill. Inst. Tech., 1957; m. Patricia C. Fogarty, Aug. 30, 1952; children—Thomas E., Robert F., Donald J., Mary Beth. Research fellow Armour Research Found., 1953-54; research chemist W. R. Grace and Co., 1956-58, sr. research chemist, 1958-61; group leader addition polymers Borg-Warner Corp. Roy C. Ingersoll Research Center, Des Plaines, Ill., 1961-64, mgr. polymer chemistry, 1964-66, asso. dir. chemistry research, 1966—. Mem. Am. Chem. Soc., (chmn. Chgo. sect. polymer group 1966—), A.A.A.S., Research Soc. Am., Sigma Xi, Phi Lambda Upsilon. Contbg. author: Stereochemistry of Macromolecules, 1967. Research, publs. and numerous patents in organic, organometallic and polymer chemistry, chemistry of ozone, catalysis of olefin polymerizations, elucidation of Ziegler-Natra catalysts, comml. devel. polypropylene in U. S.; a pioneer in use of carbon-14 isotope effects on organic reaction rates to elucidate mechanisms; discoveries in synthesis of polymer structures; discovered several organometallic compounds, including carbenoid alpha-haloalkyllithium reagents. Home: 313 S. Elmhurst Av., Mt. Prospect, Ill. 60057. Office: Roy C. Ingersoll. Research Center, Des Plaines, Ill. 60018.*

HOEG, Ove Arbo, Norwegian botanist; b. Larvik, Norway, Nov. 25, 1898; s. T. Arbo and Sigrid (Bugge) H.; Dr.philos., U. Oslo, 1942; m. Hjordis Holm, Sept. 6, 1962. Became curator Paleontol. Mus., U. Oslo, 1924; head bot. dept. Mus. Trondheim; prof. botany U. Oslo, 1947-67; dir. Birbal Sahni Inst. Paleobotany, Lucknow, India, 1951-53. Research in Devonian flora and other early land plants, fossil algae, Arctic fossil floras. Home: Asryggen 4, Nordstrandshogda, Norway. Office: P.O. Box 1068, Univ., Blindern, Oslo 5, Norway.*

HOEHN, Willard Max, Am. chemist; b. Mt. Olive, Ill., Apr. 27, 1909; s. William J. and Linda (Opp) H.; Asso. B.S., Blackburn Coll., 1930; B.S., U. Ill., 1932; Ph.D., 1936; m. Jeanette Elizabeth Ross, July 4, 1936; children—Margaret Linda (Mrs. John W. Heiser), Marian Caroline (Mrs. Garth A. Hull), Elizabeth Jean. Instr., Mayo Found., U. Minn., 1936-39, research asso., 1936-39, sr. chemist George A. Breon and Co., Kansas City, Mo., 1939-42, asst. dir., 1942-44, lab. dir., 1944-48; asso. prof. U. Kansas City (Mo.), 1948-51; sr. investigator G. D. Searle & Co., Skokie, Ill., 1951-57, head spl. syntheses, 1958—. Instr. Kahler Sch. Nursing, Rochester, Minn., 1936-39. Mem. A.A.A.S., Am. Swiss chem. socs., Chem. Soc., Chgo. Chemists Club. Research, numerous publs. on elucidation structure of adrenalcortical hormones, conversion cholic acid to deoxycholic acid for conversion to cortical hormones, preparation analgesics, antispasmodics, diuretics and hormones, diterpenic acids. Home: 1539 Walnut Av., Wilmette, Ill. 60091. Office: P.O. Box 5110, Chgo. 60680.*

HOEKSTRA, Justin Bernard, Am. pharmacologist; b. Grand Rapids, Mich., July 1, 1922; s. Peter and Wilhelmina (Zaagman) H.; A.B., Calvin Coll., 1943; Ph.D., U. Ill., 1947; m. Joyce H Larson, Aug. 23, 1963. Sr. research scientist in pharmacology Bristol Labs. div. Bristol Myers Co., Syracuse, N.Y., 1947—. Mem. A.A.A.S., Am. Soc. for Pharmacology and Exptl. Therapeutics, N.Y. Acad. Sci. Research, publs. in normal and abnormal functions colon and action drugs on colon dogs and man, basic actions phenyltoloxamine, ambucetamide, amino-pentamide, ambutonium bromide, synthetic penicillins, teracycline, kanamycin. Home: 204 Ridge Crest Rd., DeWitt, N.Y. 13214. Office: Bristol Labs., Syracuse, N.Y. 13201.*

HOELKER, Rudolf Franc, mathematician; b. Halle, Westphalia, Germany, Mar. 16, 1912; s. Cornelius and Christina (Ochsenfarth) H.; D.Sc., U. Muenster, Germany, 1942; m. Anneliese M. Wibben, Apr. 27, 1948; 1 dau., Melanie Cornelia. Came to U. S., 1945, naturalized, 1955. Chief flight mechanics sect. Army Ballistic Missile Agy., Huntsville, Ala., 1951-58, dep. dir. aeroballistics div., chief future projects br., 1958-60; dep. dir. aeroballistics div., chief future projects br. Marshall Space Flight Center, NASA, Huntsville, 1960-63, chief astrodynamics and guidance theory div., 1963-65, sr. scientist guidance lab. Electronic Research Center, Cambridge, Mass., 1965—. Recipient Astronautics Engr. Achievement award, 1959. Mem. Am. Inst. Aeros. and Astronautics, Soc. Indsl. and Applied Math., Math. Assn. Am., Hermann Oberth Soc. Germany. Devel. flight mechanics, control and guidance theory for guided missiles and space vehicles; research on space orbits and earth-moon trajectories. Home: 29 Brook St., Wellesley, Mass. 02181. Office: NASA/Electronic Research Center, 575 Tech. Sq., Cambridge, Mass. 02139.*

HOELTKER, Georg, German anthropologist; b. Ahaus, May 22, 1895; s. Bernhard and Johanna (Nabers) H.; ed. Faculty Ethnology, U. Vienna, U. Berlin; Dr. phil. Prof., St. Gabriel Sem., Vienna, 1925-36, Troppen Inst., Basle, 1942-52; prof. ethnology U. Friburg, Switzerland, 1948-54, St. Augustin Sem., Bonn, 1960—. Mem. sci. mission to New Guinea, 1936-39. Mem. Anthropol. Soc. Victoria (Australia), Anthropos Inst. Author: Die Familie bei den Azteken; Das Sündenbewusstein bei den Azteken; Einiges über Steinkeulenköpfe und Steinbeile in Neuginea; Steinerne Ackerbaugeräte; Leichenbrand und anderes vom Unteren Ramu. Co-editor: Anthropos, 1926—. Address: Arnold-Janssenstrasse, St. Augustin, Bonn, Germany.

HOENE-WRONSKI, Jozef Maria, see Wronski, Jozef Maria.

HOERMANN, Helmut Walter Ludwig Willibald, chemist; b. Innsbruck, Austria, May 24, 1926; s. Ekkehard and Hildegard (Reder) H.; dr., U. Innsbruck, 1951; m. Christel Beige, July 15, 1957; children—Reinhold, Ingrid. Asst., Max Planck Inst. Protein and Leather Research, Regensburg, 1952-56, Munich, Germany, 1956—; faculty U. Munich, 1966—. Mem. Soc. German Chemists. Research, publs. in protein chemistry: C-terminal amino acids, electrophoresis of enzymes, collagen, sugar chemistry: color reactions for quantitative estimations; chromotography: devel chromatographic separations on polyamide. Home: 49 Nymphenburgerstrasse. Office: 46 Schillerstrasse, 8 Munich, West Germany.*

HOERNES, Rudolf, Austrian geologist, seismologist; b. 1850; s. Moritz Hoernes; studied with Eduard Suess, U. Vienna (Austria), Ph.D., 1874; prof. geology U. Graz (Austria). Author: Erdbeben-Studien; Erdbebenkunde, 1893; also a criticism of Rudolf Faulb's earthquake theories, reports of Syrian earthquakes and Greek earthquakes of 1902, 04. Classified earthquakes as rock-fall, volcanic, tectonic. Died 1912.

HOEVEN, Jan van der, see van der Hoeven, Jan.

HOFER, Gustav, Austrian physician; b. Vienna, Austria; s. Karl and Olga (von Scharmitzer) H.; ed. U. Vienna; D. honoris causa, Faculty Medicine, U. Thessalonica; m. Helene Stoll Cirheimb, 1944. Agrégé, 1920; prof. U. Vienna, 1931; prof., dir. Otology, Rhinology and Laryngology Clinic, U. Graz, 1932. Mem. Austrian Laryngology-Rhinology Soc. (hon.), German Soc. Otologists, Rhinologists and Laryngologists, Internat. Coll. Otolaryngologicum, Sci. Assn. Physicians Styria (hon. pres.), Austrian Soc. Otology, Rhinology and Laryngology (pres.). Research and numerous publs. on otology, rhinology, laryngology, bacteriology and neurology. Address: 8 Teichhof, Mariatrost, Graz, Austria.

HÖFER, Hans, seismologist, geologist; b. Ellbogen, Austria, May 17, 1843; prof. geology U. Leoben (Austria). Author: Das Erdöl, 6 vols., 1909-25. First to suggest earthquakes may originate in two different foci. Died Vienna, Feb. 9, 1924.

HOFER, Helmut, German zoologist; b. Weisskirchen, Moravia, Oct. 22, 1912; s. Ewalt and Ferdinanda Carola N. Hofer; Ph.D. in Zoology, U. Vienna; m. Hertha Komos, 1944; children—Wolfgang, Thomas, Klaus. Dept. head Mus. Zoology, Dresden; instr. U. Vienna; prof. U. Giessen; with dept. primates Max Planck Inst. Research on Mental Diseases, Frankfort. Mem. numerous sci. socs. Author: Vergleichende Anatomie des Schädels; Palaeo-Neurologie des Primatenhirnes; Circumventrikuläre Organe; Primatologie; Bibliotheca Primatologica; Folia Primatologica. Editor: (with A. H. Schultz and D. Starck) Handbuck der Primatologie. Home: 47 Ludwigstrasse, Giessen. Office: Institut für Hirnforschung, Deutschordenstrasse 46, Frankfort on Main, Germany.

HOFF, Clarence Clayton, Am. biologist, educator; b. Fortuna, N.D., July 6, 1908; s. Clarence Harrison and Jennie (Glinz) H.; A.B., Bradley Coll., 1930; M.S., U. Ill., 1939, Ph.D., 1941; m. Oris Lillian Foster, Aug. 27, 1930; children—Audrey Dawn, Marcia Beth (Mrs. James Morris MacDougall). Grad. asst. zoology dept. U. Ill., 1939-41; head biology dept. Quincy Coll., 1941-46; asst. prof. zoology and parasitology Colo. State Coll., 1946-47; faculty U. N.M., Albuquerque, 1947—, prof. biology dept., 1955—; research asso. Am. Mus. Natural History, N.Y.C., 1955—. Fellow A.A.A.S.; mem. N.M. Acad. Sci. (pres. 1950-51), Soc. Systematic Zoology, Ecol. Soc. Am., Am. Soc. Parasitologists, Etomol. Soc. Am., Phi Beta Kappa, Sigma Xi, Phi Sigma, Phi Kappa Phi, numerous others. Research, publs. on ecology, distbn., taxonomy of various invertebrates. Home: 308 Fontana Pl. N.E., Albuquerque 87108.*

HOFF, Ebbe Curtis, Am. neurophysiologist; b. Rexford, Kan., Aug. 12, 1906; s. Hans and May (Knudson) H.; B.Sc., U. Wash., 1928; B.A., U. Oxford (Eng.), 1930, Ph.D., 1932, M.A., 1936, B.M., B.Ch., 1941, M.D., 1953; m. Phebe Margaret Flather, June 2, 1934; children—Phebe May (Mrs. Leigh Van Valen), David Christensen. Sterling fellow Yale, 1932-33, Coxe Fellow, 1933-34, staff, 1934-36, 41-43; faculty Med. Coll. Va., Richmond, 1943—, prof. dept. psychiat. div. psychiat. research, 1963—, dean Sch. Grad. Studies, 1957-66.; med. dir. div. alcohol studies and rehab. Va. Health Dept., Richmond, 1948—, mem. administrv. council, 1956-66. Cons. to govt. agys. Mem. A.M.A., Am. Physiol. Soc., Am. Soc. Biol. Psychiatry, Am. Coll. Neuropsychopharmacology. Author: (with Leon J. Greenbaum) A Bibliographical Sourcebook of Compressed Air, Diving and Submarine Medicine, vol. I, 1948, vol. II, 1954, vol. III, 1966; also numerous articles. Research on cerebral regulation autonomic function and its relation to psychosomatic mechanisms and behavior, etiologies, treatment and evaluation alcoholism. Home: 117 Gaymont Rd., Richmond, Va. 23229.*

HOFF, Hebbel Edward, Am. physiologist; b. Urbana, Ill., Dec. 2, 1907; s. Hans Jacob and May (Knudson) H.; B.S., U. Wash., 1928; B.A., Oxford (Eng.) U., U., 1930, D.Phil. (Rhodes scholar), 1932; M.D., Harvard, 1936; m. Helen Curtis Sullivan, Aug. 21, 1936; children—Johanna (Mrs. Jerry Chastain), Victoria (Mrs. Ralph Worsham). Instr., Yale, 1932-34, asst. prof. physiology, 1936-39, asso. prof., 1939-42; Ware fellow Harvard Med. Sch., 1934-36; Joseph Morley Drake prof. physiology McGill U., Montreal, Que., Can., 1942-48; Benjamin Franklin Hambelton prof. physiology Baylor U. Coll. Medicine, Houston, 1948—. Mem. Am. Physiol. Soc. Research, numerous publs. on physiology cardiovascular system, nature electrocardiogram, location, function respiratory center. Home: 78 Patti Lynn St., Houston 77024.*

HOFF, Jacobus Hendricus Van't, see Van't Hoff, Jacobus Hendricus.

HOFF, Karl Ernst Adolf von, see von Hoff, Karl Ernst Adolf.

HOFF, Nicholas John, Am. aero. engr.; b. Magyarovar, Hungary, Jan. 3, 1906; s. Miklos and Lenke (Meller) H.; diploma in engring. Fed. Poly. Inst., Zurich, Switzerland, 1928; Ph.D., Stanford, 1942; m. Vivian Church, July 20, 1940. Came to U. S., 1939, naturalized, 1944. Airplane designer Manfred Weiss Aeroplane and Motor Works, Budapest, 1929-39; research asst. Stanford U., 1939-40; prof., head dept. aero. engring. and applied mechanics Poly. Inst. Bklyn., 1940-57; prof., head dept. aeros. and astronautics Stanford, 1957—. Mem. research adv. com. on aircraft structures NASA, 1948-67, research and tech. adv. council, 1968—; pres. 11th Internat. Congress Applied Mechanics; sci. adv. bd. to chief of staff USAF, 1948-51, 60-62. Recipient medal Soc. Engrs. and Architects of Sweden, 1949; medal U. Liege, Belgium, 1956. Fellow Am. Inst. Aeros. and Astronautics, Royal Aero. Soc., Am. Soc. M.E.; mem. Am. Soc. C.E., Soc. Exptl. Stress Analysis, Am. Soc. Engring. Edn., Soc. Engring. Sci., Nat. Acad. Engring., Nat. Com. for Theoretical and Applied Mechanics (chmn. 1957-60), Internat. Union for Theoretical and Applied Mechanics (exec. com. 1960—), Author: The Analysis of Structures, 1956; also articles. Research in calculation of loads under which various structural elements collapse in consequence of compressive loads, strength and deformation of thin-walled shells used in aircraft and missile structures subjected to internal pressure or to concentrated loads; developed founds of behavior of structures at high temperatures, particularly when material exhibits creep. Home: 782 Esplanada Way, Stanford, Cal. 94305.*

HOFFA, Albert, surgeon; b. Richmond, S. Africa, Mar. 31, 1859; prof., Würzburg, Germany. Author: Lehrbuch der orthopädischen Chirurgie, 1891. Founder Zietschrift für orthopädische Chirurgie (1st orthopedic jour.). Argued fatal lesions in anthrax are caused by poisonous action of toxins formed by bacilli, rather than by blocking of minute blood vessels, or abstraction of oxygen from blood by bacilli; introduced Hoffa-Lorenz operation for reduction of congenital dislocation of hip, 1890; described traumatic lipoma of knee joint (Hoffa's disease). Died Cologne, Germany, Dec. 31, 1907.

HOFFER, Abram, Canadian psychiatrist; b. Hoffer, Sask., Can., Nov. 11, 1917; s. Israel and Clara (Schwartz) H.; B.S., U. Sask., 1938, M.S., 1940; Ph.D., U. Minn., 1944; M.D., U. Toronto, 1949; m. Rose Beatrice Feb. 14, 1942; children—Bill, John, Miriam. Dir. psychiat. research Dept. Pub. Health Sask., 1950-67; asso. prof. psychiatry U. Sask. Coll. Medicine, 1955-67; pvt. practice psychiatry, Saskatoon, Sask., 1967—. Vice pres. Am. Schizophrenia Found. Fellow Am. Psychiat. Assn.; mem. Am. Coll. Psychopharmacology. Author: Chemical Basis Clinical Psychiatry, 1960; Niacin Therapy in Psychiatry, 1962; How to Live With Schizophrenia, 1966. Editor: Jour. Schizophrenia, 1966—. Research in adrenochrome hypothesis of schizophrenia, psychedelics: concept and therapy, niacin therapy. Home: 1027 University Dr. Office: 800 Spadina Crescent E., Saskatoon, Sask., Can.*

HOFFLEIT, Ellen Dorrit, Am. astronomer; b. Florence, Ala., Mar. 12, 1907; d. Fred and Kate (Sanio) Hoffleit; A.B., Radcliffe Coll., 1928, M.A., 1932, Ph.D., 1938. Mem. research staff Harvard Obs., Cambridge, Mass., 1929-43, 48-56, astronomer, 1948-56; mathematician Ballistic Research Labs., Aberdeen Proving Ground, Md., 1943-48, expert, 1948-61; lectr. Wellesley (Mass.) Coll., 1955-56; research asso. Yale Obs., New Haven, 1956—; dir. Maria Mitchell Obs., Nantucket, Mass., 1957—. Recipient certificate of appreciation War Dept., 1946; Achievement award Grad. Soc. Radcliffe Coll., 1964. Fellow A.A.A.S.; mem. Am. Astron. Soc., Internat. Astron. Union, Am. Geophys. Union, Meteoritical Soc. (editor 1957), Am. Assn. Variable Star Observers, Phi Beta Kappa, Sigma Xi. Author: Some Firsts in Astronomical Photography, 1950; Bright Star Catalogue, 1964; also numerous articles. Discovered 1000 new variable stars and studied their modes of variation; discovered relation between velocities of meteors and their changes in brightness; research in stellar spectra. Home: 149 Elm St., New Haven 06511; also 3 Vestal St., Nantucket, Mass. 02554. Office: Maria Mitchell Observatory, Nantucket, Mass. 02554; also Yale U. Observatory, New Haven 06520.*

HOFFMAN, Alan Jerome, Am. mathematician; b. N.Y.C., May 30, 1924; s. Jesse and Muriel (Schrager) H.; A.B., Columbia, 1947, Ph.D., 1950; m. Esther Walker, May 30, 1947; children—Eleanor, Elizabeth. Mem. Inst. for Advanced Study, 1950-51; mathematician Nat. Bur. Standards, Washington, 1951-56; sci. liaison officer Office Naval Research, London, Eng., 1956-57; cons. Gen. Electric Co., N.Y.C., 1957-61; research staff mem. IBM Research Center, Yorktown Heights, N.Y., 1961—; adj. prof. math. City U. N.Y., 1965—, Yeshiva U., 1963—, adj. prof. engring. Columbia, 1964,67, Israel Inst. Tech., 1965. Mem. Am. Math. Soc., Math. Assn. Am. Research, publs. on math. studies in combinational math., linear inequalities and programming matrix theory. Home: 32 Old Country Rd., New Rochelle, N.Y. Office: P.O. Box 218, Yorktown Heights, N.Y.*

HOFFMAN, Charles John, Am. chemist; b. Cleve., July 9, 1918; s. John Fred and Charlotte (Froehlich) H.; B.S., Case Inst. Tech., 1942, M.S., 1948; M.S., U. Ill., 1948, Ph.D., 1951; m. Ruth Helen Burton, Sept. 17, 1955; 1 dau., Lisa. Research chemist Diamond Alkali Co., Painesville, O., 1942-43; chemistry asst. Case Inst., Cleve., 1947, U. Ill., Urbana, 1948-50; cryogenist, staff mem. Los Alamos Sci. Lab., 1951-53; chemist Merrill Co. div. Arthur D. Little, Inc., San Francisco, 1953-55, Lawrence Radiation Lab., Livermore, Cal., 1955-58; staff scientist Lockheed Missiles & Space Co., Palo Alto, Cal., 1958—. Mem. Am. Chem. Soc., Am. Phys. Soc., Chem. Soc. (London, Eng.). Contbr. articles to tech. jours. Pioneered use an interpretation of nuclear magnetic resonance for elucidation molecular structure. Home: 1665 Edmonton Av., Sunnyvale, Cal. 94086. Office: 3251 Hanover St., Palo Alto, Cal. 94304.*

HOFFMAN, Joseph Gilbert, Am. biophysicist; b. Buffalo, Aug. 19, 1909; s. Joseph and Helene (Seyler) H.; A.B. with honors, Cornell U., 1935, Ph.D., 1939; m. Ruth A. Buckland, Aug. 17, 1940; children—Joseph H., Paul G. Research asst. physics Cornell U., 1935-39; staff Roswell Park Meml. Inst., 1939-46, dir. cancer research, 1946-54, now cons. cancer research; physicist Carnegie Inst. of Washington, 1940-42; physicist Nat. Bur. Standards, 1942-44; staff sci. Los Alamos Sci. Labs., 1944-46, cons., 1946—; research prof. Sch. Medicine, U. Buffalo, 1947—, prof. biophysics, 1954—, prof. physics, 1957—. Recipient awards, Naval Ordnance, 1943, Manhattan Dist., 1945, OSRD, 1945. Fellow Am. Phys. Soc.; mem. Am. Assn. Cancer Research, A.A.A.S., Soc. Exptl. Biology and Medicine, Soc. Exptl. Biology and Medicine of Western N.Y., Austin Flint Soc. Med. Research, N.Y. Acad. Scis., Sigma Xi, Phi Kappa Phi. Author: Size and Growth of Tissue Cells, 1953; Life and Death of Cells, 1957. Contbr.: Acute Radiation Syndrome, Ann. Internal Medicine, Vol. 36, 1952. Research on electronic circuit design, neutron interreactions with matter, transplantable mouse tumors, quantitative measures of tissue growth, radiation reaction of tissues, genetics. Home: 195 Crescent Av. Office: 3435 Main St., Buffalo 14.

HOFFMAN, Murray Mitchell, Am. dental surgeon; b. N.Y.C., Jan. 20, 1910; s. Richard and Bella (Bellanca) H.; B.S., U. Ill., 1935, D.D.S., 1937, M.S., 1939; m. Rhea Esther Kaplan, July 29, 1935; children—Howard, Daniel R. Research asst. U. Ill., 1933-41; practice dentistry, specializing in oral maxillofacial surgery, Chgo., 1946—; dir. health program St. Gregory's Parish, 1947-50; dir. sect. oral maxillofacial surgery N.W. Hosp., Chgo., 1955—. Mem. Am. Dental Soc., Internat. Soc. Dental Research, Am. Soc. Oral Surgery, A.A.A.S., Ill. Acad. Sci., Sigma Xi. Research, numerous publs. on growth, devel. calcium metabolism, pharmacology analgetics, ataractics, anti-inflammatory enzymes, ultra high speed equipment in oral surgery. Home: 8555 Keeler Av., Skokie, Ill. 60076.*

HOFFMAN, Roger Alan, Am. biologist, educator; b. Willimantic, Conn., Feb. 23, 1924; s. Albert J. and Vera (Jewett) H.; B.S., U. Conn., 1950; M.S., Purdue U., 1952, Ph.D., 1956; m. Elizabeth Pawlikowski, Jan. 14, 1946; children—Christine E., Patricia A., Roger Alan II. Biologist, U. S. Edgewood Arsenal, Md., 1956-61, 62-65; instr. dept. biology Colgate U., 1961-62; asso. prof. biology Colgate U., Hamilton, N.Y., 1965—. Mem. Am. Soc. Physiology, Am. Soc. Zoologists, A.A.A.S., Am. Inst. Biol. Sci., Soc. Mammologists, Md. Acad. Sci., Sigma Xi. Editor; (with P. F. Robinson and H. Magalhaes) The Golden Hamster: Its Biology and Use in Medical Research, 1968. Research, publs. in areas of temperature regulation and hibernation, endocrinology of reprodn., pineal-pituitary interrelationships, neuroendocrinology; co-developed technique of pinealectomy. Home: Box 84, Spring St., Hamilton, N.Y. 13346.*

HOFFMAN, Samuel Kurtz, Am. propulsion engr.; b. Williamsport, Pa., Apr. 15, 1902; s. George and Louise (Bruner) H.; B.S., Pa. State U., 1925, M.E., 1945; m. Genevieve Wieland, June 18, 1932; children—Jean (Mrs. Edward Harker), Susan (Mrs. Robert Laughlin), Louise (Mrs. Ronald Rosequist), John. Sales engr. Reliance Elec. & Engring. Co., 1925-27; design engr. Fairchild, Farmingdale, L.I., 1927-28, Lycoming Mfg. Co., 1928-29, Allison div. Gen. Motors Corp., 1929-30; engr. Gen. Motors Research Lab., 1930-32; project engr. Lycoming div. Aviation Corp., 1932-34, asst. chief engr., 1934-36, chief engr., 1936-45; prof. aero. engring. Pa. State U., 1945-49; chief propulsion sect. aerophysics N.Am. Aviation, Inc., 1949-54, mgr. propulsion center, 1954-55, v.p., 1957-67, gen. mgr. Rocketdyne div. N.Am. Aviation, Inc., Canoga Park, Cal., 1955-57, v.p., gen. mgr., 1957-59, div. pres., 1960-67; v.p. aerospace and systems group, pres. Rocketdyne div. N.Am. Rockwell Corp., 1967—. Bd. dirs. Missile, Space and Range Pioneers, Inc. Recipient Goddard Meml. trophy Missiles and Rockets Mag., 1959; ARS Goddard award, 1959; co-recipient Louis W. Hill Transp. award Inst. Aero. Scis., 1960; Spirit of St. Louis award Am. Soc. M.E., 1962, Distinguished Alumnus award Pa. State U., 1963; Am. Rocket Soc. Propulsion award, 1962. Fellow Am. Inst. Aeros. and Astronautics (nat. dir., pres. So. Cal. sect. 1953), Inst. Aero. Scis., Royal Aero. Soc.; mem. Am. Soc. M.E., Soc. Automotive Engrs. (v.p. 1943), Am. Mgmt. Assn., Am. Ordnance Assn., Nat. Space Club (gov.), Internat. Astronautical Fedn. (corr.), Sigma Tau, Theta Xi. Leader in devel. U. S. rocket engines, ballistic missiles, space booster vehicles. Home: 5432 Wilbur Av., Tarzana, Cal. 91356. Office: 6633 Canoga Av., Canoga Park, Cal. 91304.*

HOFFMAN, William Samuel, Am. physician; b. Balt., July 5, 1899; s. Louis B. AND Lena (Miller) H.; A.B., Johns Hopkins, 1918, Ph.D., 1922; M.D., U. Chgo., 1929; m. Miriam Berliner, July 26, 1928; children—Paul Arthur, Nancy Regina (Mrs. Howard LeVant). With Johns Hopkins Med. Sch., 1922-27, instr., 1925-27; NRC fellow in medicine Rush Med. Coll., 1931-32; prof. physiol. chemistry Chgo. Med. Sch., 1932-44, asso. prof. medicine, 1942-44; acting dir. Hektoen Inst. for Med. Research, Cook County Hosp., Chgo., 1944-45, dir. biochem. research, 1946-53, cons. biochemistry, 1953—; med. dir. Sidney Hillman Health Centre of Chgo., 1953—; cons. biochemistry, metabolism VA Hosp., Hines, Ill., 1954—; cons. pesticide research Chgo. Bd. Health, 1962—. Mem. A.M.A., Soc. Exptl. Biology and Medicine, Am. Soc. Biol. Chemists, A.C.P., Am. Coll. Clin. Pharmacology and Chemotherapy, Am. Soc. Clin. Pathologists, Am. Pub. Health Assn., Am., Greater Chgo. (trustee), diabetes assns., Chgo. Soc. Internal Medicine, Inst. Medicine Chgo. Author: Photelometric Clinical Chemistry, 1941; The Biochemistry of Clinical Medicine, 1954, 3d ed., 1964; also numerous articles. Pioneer in devel. photo-electric methods for analysis blood and urine; research in protein nutrition, kidney function, water and electrolyte metabolism, gout and pesticides. Home: 6101 N. Sheridan Rd., Chgo. 60626. Office: 333 S. Ashland Blvd., Chgo. 60607.*

HOFFMANN, Banesh, theoretical physicist; b. Richmond, Surrey, Eng., Sept. 6, 1906; s. Maurice and Leah (Brozel) H.; B.A. with 1st class honors, Oxford (Eng.) U., 1929; Ph.D., Princeton, 1932; m. Doris Marjorie Goodday, July 10, 1938; children—Laurence David, Deborah Ann. Came to U. S., 1929, naturalized 1940. Research asso. U. Rochester, 1932-35; mem. Inst. for Advanced Study, Princeton, 1935-37, 47-48; from instr. to asso. prof. math. Queens Coll. City U. N.Y., 1937-52, prof., 1953—. Mem. exec. com. Conf. on Methods in Philosophy and Scis., 1952—, chmn., 1952-53; cons. on tests Westinghouse Sci. Talent Search, 1944—. Recipient Distinguished Tchr. award Queens Coll. Alumni Assn., 1963; 1st prize Gravity Research Found., 1964. Fellow Am. Phys. Soc.; mem. Am. Math. Soc., Sigma Xi. Author: The Strange Story of the Quantum, 1947; The Tyranny of Testing, 1962; About Vectors, 1966; also numerous research and expository articles. Editorial bd. Jour. Math. and Mechanics, 1958-67. Contbns. to projective theory of relativity, showed (with Einstein and Infeld) relation between motion and field equations of gen. relativity. Home: 43-17 169th St., Flushing, N.Y. 11358.*

HOFFMANN, Conrad Edmund, Am. microbiologist; b. Lawrence, Kan., Apr. 15, 1920; s. Conrad and Louise (Bischoff) H.; B.S., Cornell U., 1942, M.S., 1943; Ph.D., Western Res. U., 1952; m. Margaret Elizabeth Daniels, Dec. 20, 1942; children—Margaret Louise (Mrs. David C. Reichard), Frances Lee, John Richard. Research asso. Owens-Ill. Glass Co., Toledo, 1946-47; research technician nutrition, physiology Lederle Labs., Pearl River, N.Y., 1947-49; teaching asst. microbiology dept. Western Res. U., 1949-52; with E. I. du Pont de Nemours & Co., Wilmington, Del., 1952—, research mgr. microbiology sect., 1961—. Mem. Am. Soc. Microbiology, N.Y. Acad. Sci., Sigma Xi. Research, publs. on vitamins, growth factors; devel. assay system for Vitamin B12; studies on isolation and characterization of protogen (a-lipoic acid), metabolism of DNA, effect of herbicides on photosynthesis; devel. anti-influenza compound amantadine HC1, Symmetrel R for prevention of human influenza. Home: 159 Welsh Tract Rd. Office: Stine Lab., E. I. du Pont de Nemours & Co., Newark, Del. 19711.*

HOFFMANN, Erich, German dermatologist; b. Witzmitz, Pomerania, Apr. 25, 1868; Ph.D., U. Berlin (Germany). Pvt. docent U. Berlin, 1904-09; prof. U. Halle (Germany), from 1910; prof. venereal disease, also clinic dir. U. Bonn (Germany), prof. emeritus, from 1934. Mem. numerous med. socs. Author: Frühzeit Venensyphilis; (with F. Schaudinn) Entdeckungen der Erregers der Syphilis, 1905; Lehrbuch über Behandlungen der Haut und Geschlechtskrankheiten, 1917; Begründrisse der Früherkenn und Heiligen der Syphilis; Eisophylaxie, nur innen Gerichtete Schutz und Heilkraft dr Haut; Herkunft der Syphilis aus den Tropischen Tramboesie als deren Tochterkrankheiten; Enstehung dr Tabes und Paralyse; Intuition und Genie; Syphilis und Genie; Neomakrobiotik; Naso und Onkogenetik. Discovered (with Schaudinn) spirochaeta pallida bacillus as cause of syphillis; research on pathology and therapy of syphillitic diseases.

HOFFMANN, Friedrich, German physician; b. Halle, Germany, Feb. 19, 1660; s. Friedrich Hoffmann; studied chemistry, Erfurt, Germany; M.D., Jena, Germany, 1681; became physician Principality of Halberstadt, 1688; named to 1st chair medicine, new univ. at Halle, 1693, also formed statutes for new med. faculty; physician to king of Prussia, 1708-12. Fellow Royal Soc., 1720. Author: Medicina rationalis systematica, 9 vols., 1718-40; Medicina consultatoria, 12 vols., 1721-39. Mem. Iatrophys. sch. medicine; experimented with mineral waters and introduced them into therapeutic practice; developed compound spirit of ether (Hoffmann's anodyne), many other new drugs; gave early description of German measles, appendicitis; recognized regulatory role of nervous system; studied pediatrics and meteorology. Died Halle, Nov. 12, 1742.

HOFFMANN, Hermann, German botanist; b. Rödelsheim, Germany, 1819. Author: Principles of Plant Climatology, 1857; The Development of Mushroom Spores, 1860; Icones analyticae fungorum, 1861-65; Index fungorum, 1863; On Adjustment to Thermatic Conditions, 1876. Died Giessen, Germany, 1891.

HOFFMANN, Horst Rudolf Karl, German physicist; b. Breslau, Germany (now Wroclaw, Poland), Nov. 5, 1932; s. Albert and Charlotte (Guderjahn) H.; student U. Tubingen, 1952-53; Diplomphysiker, U. Munich, 1959, Ph.D., 1961, Habilitation, 1965, privatdozent, 1965; m. Karin Glufke, Aug. 14, 1957; children—Birgit, Kirsten, Dirk, Lars. Asst., U. Munich, 1961-65, asst. prof. physics, 1965-66, sci. adviser, 1966—; dozent Oskar-von-Miller-Polytechnikum, Munich, Germany, 1959-65. Vis. asso. prof. Cal. Inst. Tech. Research, publs. on expts. on very thin magnetic fields, concerning saturation magnetization and structure dependend properties; theory of fine-structure of magnetization distbn. (so called magnetization ripple); stray-field problems in micromagnetism. Home: 14 Hans-Leipelstrasse, 8 Munich 23, Germany. Office: Schellingstr. 2-8, 8 Munich 13, Germany.*

HOFFMANN, Johann, neurologist; b. Germany, 1857; described hereditary familial spinal muscular atrophy affecting forearms, hands, legs, feet (Hoffmann's atrophy, Hoffmann-Werdnig syndrome), circa 1889. Died 1919.

HOFFMANN, Johann Moritz, German physician; b. Altorf, Bavaria, Oct. 6, 1653; s. Moritz Hoffmann; studied Latin and Greek at Herspruck in Franconia, medicine at Altorf and Frankfort an der Oder; M.D., Altorf, 1675. Spent 2 years in Italy, returned to studies at Altorf, 1674; extraordinary prof. anatomy and chemistry U. Altorf, 1677, ordinary prof., 1682, also lectr. botany, resigned chair anatomy, 1709, practiced medicine; physician of princes of house of Anspach, resided at court, 1713-27. Mem. Leopoldine Acad. (dir. 1721). Author: Laboratorium novum, chemicum apertum medicinae cultoribus, 1683, also chem. articles in Acta of Leopoldine Acad. Wrote chiefly on anatomy and physiology; 1st prof. chemistry at Altorf. Died Anspach, Oct. 31, 1727.

HOFFMANN, Moritz, anatomist; b. probably Altdorf, Switzerland; 1622; at least 1 son, Johann Moritz; prof. surgery, Altdorf; credited with discovering excretory duct of pancreas in fowl, 1642, possibly leading to Wirsung's discovery of duct in man, 1642. Died 1698.

HOFFMANN, Tibor Andrews, Hungarian physicist; b. Budapest, Mar. 6, 1922; s. Mikel and Lenke (Krausz) H.; M.A., U. Budapest, 1946, Ph.D., 1954, Dr. Physics, 1956; m. Hedda Koncz, May 5, 1959; children—Clara J., Andrews P., Judith Takats. Asst., Tech. U. Budapest, 1945-52; head theoretical dept. Research Inst. for Telecommunication, Budapest, 1952-65; head computer center of Chem. Industries, Budapest, 1965—; sci. adviser Tungsram Works, Budapest, 1949-63; chief scientist Central Research Inst. for Physics, Budapest, 1955—. Recipient Rezso Schmied Meml. award, 1950; Silver Grade, Order of Work, 1966. Mem. Eotvos Lorand Phys. Soc., Hungarian Biophys. Soc. Author: What Atomic Physics Tells Us About the Theory of Metals, 1953; also articles. Research in basic quantum mechanics, applied quantum chemistry, electronic structure of DNA, decontaminating height of reactor stacks, linear programming, theory of melting, theory of solids, optical maser. Home: 36 Maros, Budapest XII. Office: 1/c Erzsebet Kiralyne, Budapest XIV, Hungary.*

HOFFMANN-BERLING, Hartmut, German biologist; b. Danzig, Germany, Apr. 7, 1920; s. Walter Kurt and Elisabeth (Berling) H.-B.; student univs. Königsberg, Freiburg; Dr.med., U. Bonn, 1944; Dr.rer.nat., U. Heidelberg, 1955; m. Liselotte Frick, July 22, 1944; children—Eberhard, Manfred. With Inst. Virus Research, Heidelberg, 1948-54; asst. Max Planck Inst. Med. Research, Heidelberg, 1954—, dir., 1966—; unscheduled prof. microbiology U. Frankfurt/Main, 1961-66; prof. molecular biology, U. Heidelberg, 1966—. Mem. German Soc. Biol. Chemistry, European Molecular Biology Orgn., (Am.) Soc. for Study Development and Growth, German Soc. Phys. Biology. Numerous publs. on detection and characterization of contractile proteins in non-muscular cells; study of small DNA and RNA containing E. coli phages, including their mode of replication. Address: 13 Tischbeinstrasse, Heidelberg, West Germany.*

HOFFMEISTER, Donald F(rederick), Am. biologist; b. San Bernardino, Cal., Mar. 21, 1916; s. Percival George and Julia Bell (Hillgartner) H.; A.B., U. Cal. at Berkeley, 1938, M.A., 1940, Ph.D., 1944; m. Helen E. Kaatz, Aug. 11, 1938; children—James Ronald, Robert George. Research, curatorial asst. Mus. Vertebrate Zoology, U. Cal. at Berkeley, 1941-44, teaching asst. zoology at univ., 1943-44; asso. curator modern vertebrates Mus. Natural History, U. Kan., 1944-46, asst. prof. zoology, 1944-46; dir. Mus. Natural History, U. Ill. 1946—, faculty univ., 1946—, prof. zoology, 1959—. Fellow A.A.A.S.; mem. Am. Soc. Mammalogists (sec. 1946-52, v.p. 1961-64, pres. 1964-66), Midwest Museum's Conf. (exec. v.p. 1962-63, pres. 1963-64), Am. Assn. Museums, Australian Mammal Soc., Southwestern Naturalists, Soc. Vertebrate Paleontology, Ill. Acad. Scis. Author: Mammals, 1955; Mammals, 1963; Fieldbook of Illinois Mammals, 1957; Zoo Animals, 1967; also articles, reports. Research on taxonomic, evolutionary and phylogenetic studies of mammals; distbn., ecology and taxonomy of mammals of southwestern U. S., particularly Ariz. and Ill. Home: 1505 W. Charles St., Champaign, Ill. 61822. Office: Museum Natural History, University of Ill., Urbana, Ill. 61801.*

HOFFMEISTER, F. Stanley, surgeon; b. Prague, Czechoslovakia, Nov. 7, 1914; s. Ferdinand and Kamila (Steklova) H.; M.D., Charles U., Prague, 1939. Came to U. S., 1947, naturalized, 1954. Surg. fellow Johns Hopkins, 1952-53, instr., 1953-54; asso. chief Roswell Park Meml. Inst., Buffalo, 1954-56, chief dept. reconstructive and head and neck surgery, 1956-67; asso. Sch. Medicine U. Buffalo, 1957-58, asst. prof. surgery, 1958-67; asso. clin. prof. Albany (N.Y.) Med. Coll., 1967—. Diplomate Am. Bd. Plastic Surgery. Mem. A.M.A., A.C.S., Am. Assn. Cancer Research, Soc. Plastic and Reconstructive Surgery, Royal Soc. Medicine, Soc. Head and Neck Surgeons, Council on Research Plastic Surgery, Canadian Soc. Plastic Surgery, Am. Assn. Plastic Surgeons. Research, publs. on congenital deformities, circulation in flaps, burn treatment, surg. mgmt. head and neck cancer. Address: 1465 Western Av., Albany, N.Y. 12203.*

HOFMAN, Klaus Heinrich, biochemist; b. Karlsruhe, Germany, Feb. 21, 1911; s. Heinz Kaeppele and Marianne (Bally) K.; Ph.D., Fed. Inst. Tech., Zurich, Switzerland, 1936; m. Frances Finn, Feb. 26, 1965. Came to U. S., 1938, naturalized, 1952. Fellow, Rockefeller Inst., N.Y.C., 1938-40, Cornell Med. Coll., N.H.C., 1940-42; sci. guest Ciba Pharm., Inc., Summit, N.J., 1942-44; faculty U. Pitts., 1944—, research prof., 19—64, chmn. biochemistry dept., 1952-64, Commonwealth prof., dir. Protein Research Lab., 1964—. Recipient Borden award, 1963; Myrtle Wreath award Hadassah Found., 1965. Mem. Nat. Acad. Sci., Am. Chem. Soc. (Pitts. award 1962), Am. Assn. Biol. Chemists. Author: Fatty Acid Metabolism in Micro-organisms, 1963; also numerous articles. Research on pituitary hormones, specificity proteolytic enzymes, chem. studies on cyclopropane fatty acids, lipid metabolism in micro-organisms, melanocyte-stimulating hormones. Home: 1467 Mohican Dr., Pitts. 15228.*

HOFMANN, Albert, Swiss chemist; b. Baden, Switzerland, Jan. 11, 1906; s. Adolf and Elisabeth (Schenk) H.; Dr.phil., U. Zürich, 1929; Dr.pharm. h.c., Royal Pharm. Inst. Sweden, 1966; children—Dietrich, Andreas, Garrielle, Beatrix. Research chemist, lab. pharm. chemistry Sandoz Ltd., Basel, 1929—, dir. research, dept. natural products, 1956—. Fellow World Acad. Art and Sci. Author: Die Mutterkornalkaloide, 1964; also numerous articles. Research on cardioactive glycosides of squill, alkaloids of ergot, partial synthesis of ergonovine, synthesis of ergotamine, of lysergic acid diethylamide, rauwolfia alkaloids, magic plants of Mexico. Home: 11 Oberwilerstrasse, 4103 Bottmingen. Office: Sandoz A.-G., 4002 Basel, Switzerland.*

HOFMANN, August Wilhelm von, see von Hofmann, August Wilhelm.

HOFMANN, Frederick Gustave, Am. endocrinologist; b. Detroit, May 25, 1923; s. Gustave A. and Florence (Heinsman) H.; A.B., U. Mich., 1943; Ph.D., Harvard, 1952; m. Adele Otis Dellenbaugh, July 28, 1956; children—Peter F., Anne G. Fellow NRC, Harvard, 1952-53; faculty Columbia, N.Y.C., 1953—, asso. prof. pharmacology Coll. Phys. and Surg., 1961—. Markle scholar in med. scis., 1955-60. Mem. Endocrine Soc., Am. Soc. Pharmacology and Exptl. Therapeutics, Am. Soc. Zoologists, Harvey Soc. Editor: (with W. S. Root) Physiological Pharmacology, 10 vols., 1963—; editor in chief Endocrinology, 1963-67. Research, publs. on characterization of cofactor requirements of guinea pig adrenal steroid hydroxylases, comparative studies of steroid hydroxylases of adrenal cortex and testis, nature of steroid C-17 hydroxylase deficit in adrenal cortex of rat and mouse. Home: 85 Mayflower Dr., Tenafly, N.J. 07670.*

HOFMANN, Fritz, chemist; b. Kölleda, Thuringia, Nov. 2, 1866; prof., Breslau (now Wroclaw, Poland); discovered artificial rubber (Buna). Died Hanover, Germany, Oct. 29, 1956.

HOFMANN, Gustav Ludwig, Austrian psychiatrist; b. Vienna, Austria, Aug. 11, 1924; s. Gustav and Ludovika (Lukasek) H.; Dr.med.univ., U. Vienna, 1951; m. Gertraud Hiessmanseder, June 26, 1954; children—Margarete, Gustav, Katharina. Staff, U. Vienna, 1951—, head biochem. Lab. psychiat. dept., 1960-67, head detoxication center, 1967—, venia legendi, 1964—, lectr., 1964—; lectr. Wiener Kath. Akademi, 1965—. Recipient Innitzer award, 1963. Mem. Gesellschaft der Arzte Wien, Verein für Psych. u. Neurologie Wien, Gesellschaft Österr. Nervenarzte und Psychiater, Österr. Biochem. Gesellschaft, Wr. Kath. Akademia, Coll. Int. Neuropsychopharmacologium, Wr. Arbeitskrels fur Tiefenpsychologie. Author: Experimentelle Grundlage der multifaktoriellen Genese der Schizophrenie, 1963; also numerous articles. Research on organic basis of mental disorders especially schizophrenia, treatment of drug intoxication. Home: 6 Breitenfeldergasse, Wien 8, Austria.*

HOFMANN, Karl Andreas, German chemist; b. Ansbach, Apr. 2, 1870; prof. Tech. U. Charlottenburg (Berlin). Author: Die radioaktiven Stoffe, 1902; Lehrbuch der anorganischen Chemie, 1918. Research on radioactivity, catalysis, other areas of inorganic chemistry. Died Berlin, Oct. 15, 1940.

HOFMANN, Karl Heinrich, mathematician; b. Heilbronn, Germany, Oct. 3, 1932; s. Wilhelm and Auguste (Rau) H.; Dr. Rer. Nat., U. Tübingen, Germany, 1958, Habilitation, 1962; m. Isolde Rösler, May 11, 1963. Faculty, U. Tübingen, Germany, 1958-60, 62-63, dozent, 1962-63; faculty Tulane U., 1960-61, 63—, prof., 1965—. Mem. Deutsche Mathematiker Vereinigung, Am. Math. Soc. Author: (with Paul S. Mostert) Elements of Compact Semigroups, 1966. Research, publs. on structure theory of locally compact groups and compact semigroups, application to related structures, topol. rings. Office: Dept. Math., Tulane U., New Orleans 70118.*

HOFMANN, Ulrich, German chemist; b. Munich, Germany, Jan. 22, 1903; s. K. A. and Emma (Burger) H.; ed. Technische Hochschule, Berlin; Ph.D. in chemistry; m. Renate Schiebeler, 1935; children—Peter, Gisela, Fritz, Franz, Klaus. Habilitation, Berlin, 1931; prof. chemistry U. Rostock, 1937, Technische Hochschule, Vienna, 1942, Technische Hochschule, Darmstadt, 1951, U. Heidelberg. Mem. Acad. Scis. Heidelberg, Leopoldina Acad. Naturalists (Halle). Author: Hoffmann-Rüdorf Anorganische Chemie, 17 edits. Research and publs. on carbon-graphite connections, clay minerals and collagens. Home: 28 Turnerstrasse. Office: Institute of Inorganic Chemistry, 2 Tiergartenstrasse, Heidelberg, Germany.

HOFMEIER, Kurt Fritz Robert, German physician; b. Königsberg, Germany, Sept. 9, 1896; s. Fritz and Paula (Ramm) H.; ed. univs. Marbourg, Fribourg, Breslau, Wurtzburg; M.D.; m. Edith Breitschuh, 1929; children—Rosemarie, Hans Melchior, Friedrich Alexander. Intern, Wurzburg, Frankfort on Main, Leipzig; pediatrician, Berlin; dir. Kaiser August Victoria Clinic, 1938-41; dir. Pediatric Clinic, U. Strasbourg, 1941-45; instr. U. Tübingen. Mem. German Soc. Internal Medicine, Germany Soc. Pediatricians. Author: Die Bedeutung der Erbanlagen für der Kinderheilkunde, 1937; Das Biologische Anrecht des Kindes, 1952; Die Theraphie der Epidemischen Kinderlähmung, 1948. Address: 7211 Neukirch über Rottweill, Germany.

HOFMEISTER, Wilhelm Friedrich Benedict, German botanist; b. Leipzig, May 24, 1824; self-educated. With father's pub. and bookselling bus., 1841-63; prof. botany Heidelberg, 1863; dir. bot. garden; prof. Tübingen, from 1872. Mem. French Acad. Scis., 1865. Author: Die Entstehung des Embryos der Phanerogamen, 1849; Vergleichende Untersuchungen der Keimung, Entfaltung und Fruchtbildung höherer Kryptogamen, 1851. Foremost German botanist of his time; one of 1st to describe ovule and devel. of embryo from fertilized ovum; proved embryo forms from an ovum, thus refuting Schleidan's pollen tube theory; discovered alternating cycle of mosses and ferns and their relationship to angiosperms and gymnosperms (had important effect on plant classification). Died Lindenau, nr. Leipzig, Jan. 12, 1877.

HOFSTADTER, Robert, Am. physicist; b. N.Y.C., Feb. 5, 1915; s. Louis and Henrietta (Koenigsberg) H.; B.S. magna cum laude (Kenyon prize), Coll. City of N.Y., 1935; M.A. (Procter fellow), Princeton, 1938, Ph.D., 1938; LL.D., City U. N.Y., 1961; D.Sc., Gustavus Adolphus College; Laurea Honoris Causa, Univ. of Padua, 1964; m. Nancy Givan, May 9, 1942; children—Douglas Richard, Laura James, Mary Hinda. Coffin fellow Gen. Electric Co., 1935-36; Harrison fellow University of Pennsylvania, 1939; instr. physics College City of N.Y., 1941; physicist Norden Lab. Corp., 1943-46; asst. prof. physics Princeton, 1946-50; asso. prof. physics Stanford, 1950-54, prof., 1954—, Guggenheim Fellow, Ford Found., Geneva, Switzerland, 1958-59; Cal. Sci. of year, 1959; Recipient (with R. L. Moessbauer) Nobel prize in physics, 1961; Townsend Harris medal College City of New York, 1961. Fellow of the American Physical Society, Physical Society London; mem. Italian Phys. Soc., Nat. Acad. Scis., Am. Assn. U. Profs., Phi Beta Kappa, Sigma Xi. Editor: Investigations in Physics, 1952. Asso. editor Phys. Review, 1951-53; mem. editorial bd. Review Sci. Instruments, 1953-55, Reviews of Modern Physics, 1958-61. Discovered and described form factors in nucleons; discovered that proton and neutron are considerably complex, (with R. Herman) crystal trapping states, crystal conductor factors and scintillation counters; research on infrared spectra, photoconductivity, high voltage electrostatic generator, servomechanisms, Compton effect at high energies, electromagnetic showers, electron scattering at high energies, nuclear charge distbns. Home: 639 Mirada Av., Stanford, Cal.*

HOFSTEE, Evert Willem, Dutch sociologist; b. Stedum, Netherlands, Oct. 15, 1909; s. Hendrik and Carolina H. (Venderbosch) H.; candidate degree cum laude, U. Amsterdam, 1931, doctoral degree cum laude, 1933, Ph.D. cum laude, 1937; m. Caroline Henriette Venderbosch, May 7, 1941; children—Rudolf H., Christine R. G., Caroline H. Tchr. secondary edn., 1933-37; head social research dept. Provincial Devel. Bd., Province of Groningen, Netherlands, 1938-46; lectr. sociology U. Groningen, 1943-46; prof. sociology, head dept. Agrl. U. Wageningen (Netherlands), 1946—, mem. bd., 1960—. Adviser state agys. for reclamation and colonization of Zuiderzee-polders, 1943—; mem. Govt. Bd. for Town and Country Planning 1956—, Bd. InterU. Inst. for Social Research, Central Bd. Statistics, Council for Sci. Policy, Council for Edn. Decorated knight Order of Dutch Lion. Mem. Royal Netherlands Acad. Scis. (pres. 1959—, social sci. council), European Soc. for Rural Sociology (pres. 1957—), Am. Sociol. Assn., Am. Rural Sociol. Soc., European Soc. for Rural Sociology, Nederlandse Sociologische Vereniging. Author, and co-author numerous books, numerous articles. Research on socio-cultural aspects of migration, birth rate, rural economy, problems related to colonization, land-div. Home: 2a Generaal Foulkesweg, Wageningen, Netherlands.*

HOFSTETTER, Henry W., Am. optometrist; b. Windsor, O., Sept. 10, 1914; s. Kaspar and Augusta (Kresin) H.; student Western Res. U., 1931-33, Kent State U., 1933; B.S., Ohio State U., 1939, M.S., 1940, Ph.D., 1942; D. Optometric Sci., Los Angeles Coll. Optometry, 1954; m. Frances Jane Elder, July 5, 1941; children—Ann Kresin, Susan Claire. Elementary sch. tchr. Middlefield (O.) Pub. Sch., 1933-36; faculty Ohio State U., 1942-48, asso., prof., 1947-48; dean Los Angeles Coll. Optometry, 1948-52; prof. Ind. U., Bloomington, 1952—. Mem. Nat. Adv. Council on Edn. for Health Professions, 1964—; mem. com. on vision Armed Forces NRC, 1961—. Mem. Optical Soc. Am., Am. Optometric Assn., Am. Acad. Optometry, Assn. Schs. and Colls. (past pres.), Am. Optometric Assn. (trustee 1962—). Author: Optometry, 1948; Industrial Vision, 1956; (with M. Schapero, David Cline) Dictionary of Visual Science, 1960; also numerous articles. Research on graphical analysis of relationship between accommodation and convergence of eyes and application to clin. techniques, accommodation fatigue and age-amplitude relationships, heredity in astigmatism, role of stereopsis in vehicle control. Home: 1107 Southdowns Dr., Bloomington, Ind. 47401*

HOGAN, Albert Garland, Am. biochemist; b. Maryville, Mo., Dec. 31, 1884; A.B., U. Mo., 1907, B.S., 1909, A.M., 1912; Ph.D., Yale, 1914; m. 1920; 3 children. Instr. chemistry Mo. State Normal Sch., Maryville, 1909-15; asst. physiol. chemist Sheffield Sci. Sch., Yale, 1913-14; chemist Kan. State Coll., 1914-17, asst. prof., 1917-19; prof. physiology chemistry Sch. Medicine, Ala., 1919-20; animal nutritionist U. Mo., 1920-55, chmn. dept. agrl. chemistry, 1922-55. Recipient Mead Johnson & Co. award, 1944, Morrison award, 1951, Borden award, 1955, Osborne and Mendel award, 1956. Mem. Soc. Animal Products, A.A.A.S., Soc. Biol. Chemists, Chem. Soc., Soc. for Exptl. Biology. Isolated (with E. M. Parrott) folic acid; studied vitamins, proteins, minerals. Deceased.

HOGARTH, Cyril Alfred, English physicist; b. London, Eng., Jan. 22, 1924; s. Alfred and Florence (Farrow) H.; B.Sc. with honors, Queen Mary Coll., U. London, 1944, Ph.D., 1948; m. Audrey Jones, Sept. 4, 1951; children—Celia, Yvonne, Adrian. Faculty, Reading (Eng.) U., 1949-51; sr. sci. officer Royal Radar Establishment, 1951-58; prof., head physics dept. Brumel U., London, 1958—. Cons. physicist to pvt. cos., govt. depts., 1958—. Fellow Inst. Physics (chmn. membership com. 1965—). Author: (with J. Blitz) Techniques of Non-Destructive Testing, 1960; Materials Used in Semiconductor Devices, 1965; also articles. Research in solid state physics, especially semiconductors, semiconductor device physics. Home: Shepherds Hey, Orchehill Av., Gerrards Cross, Bucks., Eng. Office: Physics Dept., Brunel U., London, W.3., Eng.*

HOGARTH, David George, English archeologist; b. Barton-on-Humber, Lincolnshire, May 23, 1862; s. George and Jane Elizabeth (Uppleby) H.; Magdalen Coll., Oxford; D.Litt., Oxford, 1918; Litt.D., Cambridge U., 1924; m. Laura Violet Uppleby, 1894; 1 son. Excavated at Paphos, Cyprus, 1888, Der el-Behri, Alexandria, Fayum, Egypt, 1894-96; dir. Brit. Sch., Athens, 1897-1900; excavated at Phylokopi, Melos, Naukratis, Egypt, 1899, 1903, Dictaean Cave, Knossos, 1900, Zakro, Crete, 1901, Artemisium, Ephesus, 1904-05, Carchemish, 1911; keeper Ashmolean Mus., Oxford, 1908-27. Fellow Royal Geog. Soc. (pres. 1925-27), Soc. Antiquaries; mem. Hellenic Soc. (v.p.). Author: Devia Cypria, 1890; Modern and Ancient Roads in Eastern Asia Minor, 1892; A Wandering Scholar in the Levant, 1896; Philip and Alexander of Macedon, 1897; Authority and Archaeology, 1899; The Nearer East, 1902; The Penetration of Arabia, 1904; The Archaic Artemisia of Ephesus, 1908; Ionia and the East, 1909; Accidents of an Antiquary's Life, 1910; The Ancient East, 1914; Carchemish I, 1914; The Balkans, 1915; Hittite Seals, 1920; Arabia, 1922; The Wandering Scholar, 1925; Kings of the Hittites, 1926; C. M. Doughty, A Memoir, 1928. Asso. with many of most important excavations in East (with Evans at Knossos, Naukratis, Ephesus); primary contbn. is work on Hittite civilization; wrote sects. on Hittites in Cambridge Ancient History. Died Nov. 6, 1927.

HOGBEN, Lancelot, Brit. physiologist, statistician; b. Southsea, Hants, Eng., Dec. 9, 1895; s. Thomas and Margaret (Prescott) H.; M.A., (Sr. scholar), Trinity Coll., Cambridge, Eng., 1915; S.Sc., U. London (Eng.), 1922; D.Sc. (hon.) U. Wales, 1963; M.Sc. (hon.), U. Birmingham (Eng.), 1941, LL.D., 1963; m. Enid Charles, 1918; (div. 1957); m. 2d, Sarah Jane Evans, Oct., 1957. Mackinnon student Royal Soc., 1923; lectr. zoology Imperial Coll. Sci., 1919-22; asst. dir. animal breeding research dept., 1923; lectr. exptl. physiology U. Edinburgh (Scotland), 1923-25; asst. prof. zoology McGill U., Montreal, Que., Can., 1925-27; prof. U. Cape Town (South Africa), 1927-30; prof. social biology U. London, 1930-37; Regius prof. natural history U. Aberdeen (Scotland), 1937-41; Mason prof. zoology Birmingham U., 1941-47, prof. med. statistics, 1947-61, hon. fellow linguistics, 1961-63; vice chancellor U. Guyana, Georgetown, Brit. Guiana, 1963-65. Vis. prof. U. Wis., 1940-41; dep. dir. med. statistics Brit. War Office, 1942-46. Recipient Frank Smart prize Trinity Coll., 1915, 1942. Fellow Royal Soc., 1936 (Croonian lectr. 1942), Royal Soc. Edinburgh (Keith prize, gold medal 1936); mem. Soc. Exptl. Biology (hon., 1st sec. 1963). Author: Comparative Physiology, 1925; The Comparative Physiology of Internal Secretion, 1926; Nature and Nurture, 1933; Introduction to Mathematical Genetics, 1941; Statistical Theory, 1957; also numerous articles. Asso. editor Brit. Jour. Exptl. Biology, 1923-25. Research on comparative physiology of ductless glands especially metamorphosis, chromatic function and ovarian cycle, math. genetics, med. statistics. Home: Tregeiroiog, Denbighshire, North Wales.*

HÖGFELDT, Erik, Swedish chemist; b. Dalarö, Sweden, July 2, 1924; s. Einar and Karin (Ljunggren) H.; studentexamen, U. Stockholm (Sweden), 1943, Bromma, fil kand., 1951, fil lic., 1953, fil. dr. docent, 1964; m. Birgit Elisabeth Holgersson, Aug. 18, 1953; children—Elsa Karin Birgitta, Nils Erik, Lars Einar. Research asso. U. Stockholm, 1949-51; research asso. Royal Inst. Tech., Stockholm, 1951-56, 59-65, asso. prof., 1965—; research asso. Brookhaven Nat. Lab., Upton, L.I., N.Y., 1956-58; vis. scientist chemistry dept. Mass. Inst. Tech., Cambridge, 1958-59; vis. prof. Middle E. Tech. U., Ankara, Turkey, 1965. Cons. Swedish Atomic Energy Co., 1960-64; cons. com. on ion exchange terminology Nat. Acad. Sci., 1961-66. Mem. Swedish Chem. Soc. Author: (with L. G. Sillén, G. Lundgren) Stöhiometri och Jámvihtslara, 1951; Fysikalisk kemi, 1963; also articles. Research on thermodynamics of solutions, ion exchange equilibria, amine extraction, hydration in concentrated electrolytes, gas adsorption on heterogeneous surfaces. Home: 28 Kämperägen, Sollentuna 4, Sweden. Office: Dept. Inorganic Chemistry, Royal Inst. Tech., Stockholm, 70, Sweden.*

HOGG, Benjamin Gregory, Canadian physicist; b. Winnipeg, Man., Can., July 26, 1924; s. David C. and Harriet (Gregory) H.; B.Sc. with honors, U. Man., Winnipeg, 1946; M.A., Wesleyan, U., Conn., 1947; Ph.D., McMaster U., Ont., 1953; m. Emilie Shipel, June 1, 1949; children—Kristine, David. Research scientist Def. Research Bd., 1949-51; faculty Royal Mil. Coll., Kingston, Ont., 1953-57, asso. prof., 1954-57; faculty U. Man., 1957—, prof. physics, 1965—. Mem. Am. Phys. Soc., Canadian Assn. Physics (past dir.), Am. Assn. Physics Tchrs., Sigma Xi. Research, publs. on determination of atomic masses, positron annihilation. Home: 1587 Wolseley Av., Winnipeg, Man., Can.*

HOGG, Helen Battles Sawyer, astronomer; b. Lowell, Mass., Aug. 1, 1905; d. Edward Everett and Carrie (Sprague) Sawyer) A.B., Mt. Holyoke Coll., 1926, D.Sc., 1958; A.M., Radcliffe Coll., 1928, Ph.D., 1931; D.Sc., U. Waterloo, 1962; m. Frank Scott Hogg, Sept. 6, 1930; children—Sarah Longley (Mrs. Sarah MacDonald), David Edward, James Scott. Instr., Smith Coll., 1927; instr. Mt. Holyoke Coll., 1930-31, asst. prof., acting chmn. dept. astronomy, 1940-41; with Dominion Astrophys. Obs., B.C., 1931-34; faculty U. Toronto, Ont., Can., 1936—, prof. astronomy, 1957—. Dir., Bell Telephone Co. of Can. Program dir. for astronomy NSF, 1955-56. Recipient citation Mt. Holyoke Coll., 1952; Rittenhouse medal Rittenhouse Astron. Soc., 1967; Can. Centennial medal, 1967. Fellow Royal Soc. Can. (pres. sect. 1960-61); mem. Am. Astron. Soc. (council 1965-68; Annie J. Cannon prize 1950), Am. Assn. Variable Star Observers (pres. 1939-41), Royal Astron. Soc. Can. (pres. 1957-59; Service medal 1967), Royal Can. Inst. (pres. 1964-65), Internat. Astron. Union, Fedn. Ont. Naturalists, Phi Beta Kappa, Sigma Xi. Astronomy columnist Toronto Daily Star, 1951—. Contbr. numerous articles to profl. jours., newspapers. Discovered new variable stars in globular star clusters; determined their periods of light variation from photographs taken with reflecting telescopes. Home: 98 Richmond St. Office: David Dunlap Obs., Richmond Hill, Ont., Can.*

HÖGLER, Franz, Austrian physician; b. Stazern, Nov. 2, 1893; s. Franz and Maria (Ostermann) H.; M.D., U. Vienna; m. Gertrude Neubauer, 1964; children—Michael, Eva. Physician, dept. chief numerous hosps., until 1947; prof. internal medicine U. Vienna, 1939—. Mem. Soc. Internal Medicine. Author 2 books on diabetes, also numerous articles. Home: 14 Haubenbiglstrasse, Vienna XIX. Office: 10 Rooseveltplatz, Vienna IX, Austria.

HOHENBERG, Fritz, Austrian mathematician; b. Graz, Austria, Jan. 4, 1907; s. Ferdinand and Luise Hohenberg; ed. U. Vienna, Higher Tech. Sch., Vienna; Ph.D. in math.; m. Elisabeth Töpfl, 1940; children—

Reinhard, Günter, Harald. Prof., 1934-39; affiliate prof. Higher Tech. Sch., Vienna, 1939-47; prof. Higher Tech. Sch., Graz, 1947—; pres. Inst. for Study Geometry; constrn. engr., mathematician. Mem. Austrian, German assns. mathematicians. Author: Konstruktive Geometrie in der Technik, also articles on geometry. Home: Felix Dahnplatz 7/11. Office: 24 Kopernikusgasse, Graz, Austria.

HOHENEMSER, Kurt Heinrich, mech. engr.; b. Berlin, Germany, Jan. 3, 1906; s. Richard and Alice (Salt) H.; Dr. Ing., Technische Hochschule Darmstadt, 1929; m. Katharina Dietrich, Feb. 11, 1933; children—Christoph, Veronica Sutherland. Came to U. S., 1947, naturalized, 1953. Sr. scientist McDonnell Aircraft Co., St. Louis, 1947-65; affiliate prof. applied mechanics Washington U., St. Louis, 1957-65, prof. aerospace engring., 1965—. Recipient Grover E. Bell award, 1957. Fellow Am. Helicopter Soc. (Alexander Klemin award 1964), Am. Inst. Aeros. and Astronautics. Author: (with W. Prager) Dynamik der Stabwerke, 1932; (with R. V. Mises) Fluglehre, 1936. Contbr. numerous articles to profl. jours. Contbns. to theory of vibrations, mechanics of continua, helicopter flight mechanics, gas dynamics, helicopter tech. Home: 2421 Remington Lane, St. Louis 63144.*

HOHENHEIM, see Paracelsus, Theophrastus Bombastus von Hohenheim.

HOHENNER, Werner Wilhelm, physicist; b. Braunschweig, Germany, Oct. 25, 1907; s. Heinrich and Emilie (Oberlaender) H.; diploma Tech. U. Munich, 1932; Ph.D., Tech. U. Darmstadt, 1944; m. Felizia Schoenemann, May 21, 1960; children—Rosmarie, Paul O., Frederickson, Erika, Paul Douglas, Harro, Wolfgang. Came to U. S., 1947, naturalized, 1955. Research engr. Schenck, Maschinenfabrik, Darmstadt, Germany, 1932-34; dir. Inst. for Physics, German Research Inst., 1934-45; consultant Mauser Waffenfabriken, Oberndorf, 1945-47; cons. U. S. Naval Air Missile Test Center, Point Mugu, Cal., 1947-54; head specialist br. U. S. Naval Air Devel. Center, Johnsville, Pa., 1954-56; asst. to dir. Spl. Project Officer, Bur. Weapons, Washington, 1956-57; cons. Fleet Ball System; adv. scientist Westinghouse Electric Corp. Aerospace Div., Balt., 1957—. Research, numerous publs. in gas dynamics, servo mechanisms, internal and external ballistics, Fleet Ballistic System (Polaris); devel. of figures of merit of various weapon systems; devel. weapons system analysis, advanced weapons systems. Home: 349 S. St. Johns Lane, Ellicott, Md. 21043. Office: P.O. Box 746, Balt. 21203.*

HOHMANN, Georg, German orthopedist; b. Eisenach, Germany, Feb. 28, 1880; s. Louis Hohmann; ed. univs. Jena, Würzburg, Munich, Berlin; Dr. honoris causa univs. Würzburg, Munich, Frankfort/Main, Gressen (all Germany); m. Ursula Neumann, 1927; 8 children. Asst., later head dr. Munich Orthopedic Clinic, until 1910; practiced orthopedics, Munich, from 1910; mil. dr., 1914-18; apptd. pvt. docent U. Munich, 1918, dir. Orthopedic Clinic, rector univ. from 1946; prof. U. Frankfort/Main from 1923, med. dir. Orthopedic Clinic, 1930-46, rector univ., 1945-46. Served as pres. German Welfare Soc. for Cripples, and Bavaria, Germany March of Dimes. Mem. orthopedic socs. Austria, Italy, Scandinavian countries, and France. Author: Fuss und Bein, 1920; Orthopädische Technik, 1941; Orthopäd, 1949; Gymnastik; Hand und Arm, 1949. Established (with others) orthopedics as ind. sci.

HöHN, Otto (E(mil), physiologist, ornithologist; b. Basel, Switzerland, Mar. 14, 1919; s. Otto Frederick and Paula (Eiche) H.; B.Sc. with honors, U. London, 1939, M.B., B.S., 1943, M.Sc., 1946, Ph.D., 1951, postgrad. Worcester Found. for Exptl. Biology, also Clark U.; m. Barbara Constance Moncrieff Wood, Mar. 31, 1944; children—Howard Falcon, Peter Angun. From demonstrator to asst. lectr. physiology dept. Guy's Hosp. Med. Sch., U. London (Eng.), 1943-47; faculty physiology dept. U. Alta., Edmonton, Can., 1947—, now asso. prof. Fellow Arctic Inst. N.Am., Zool. Soc. London; mem. Brit., Canadian phys. socs., Soc. for Endocrinology, Brit. Med. Assn., Canadian, Am. socs. zoologists, Brit., Am. ornithologists unions. Author: The Phalaropes, 1965; Hormones in Man and Animals, 1966. contbg. author: Biology and Comparative Physiology of Birds, 1961; also articles. Research on role hormones in mammals and birds, ornithology Eng., Arctic, subarctic Can., Western Alaska, recorded animal names in N.Am. Indian and Eskimo dialects. Home: 11511 78th Av., Edmonton, Alta., Can.*

HöHNK, Johann Willy Georg, German microbiologist; b. Bremen, Sept. 18, 1899; s. Wilhelm and Catharine (Gehrels) H.; ed. univs. Hambourg, Marbourg, Frankfort, Wis.; Ph.D. in microbiology; m. Irmgard Fischer, 1932; children—Klaus, Dierk. Founder, curator, acting dir., del. dir. Mem. German Sci. Commn. Maritime Exploration, German Soc. Microbiology and Hygiene, German Soc. Naturalists and Physicians, German Soc. Botany. Editor: Veröffentlichungen des Instituts für Meeresforschung, also numerous articles on aquatic mycology. Home: 5 Virchowstrasse. Office: 12 Am Handelshafen, Bremerhaven, Germany.

HOIJER, Harry, Am. anthropologist; b. Chgo., Sept. 6, 1904; s. John Oscar and Agnes (Peterson) H.; A.B., U. Chgo., 1927, A.M., 1929, Ph.D., 1931; m. Dorothy Addis Jared, June 7, 1927; children—Charlotte (Mrs. P. Bert Teeples), Peter, Susan (Mrs. Stephen Fisher). Instr., U. Chgo., 1931-40; faculty U. Cal. at Los Angeles, 1940—, prof. anthropology, 1948—, chmn. dept. anthropology, 1942-43, 48-52, 34th Ann. Faculty Research lectr., 1958. Fellow Center for Advanced Study in Behavioral Scis., 1959-60. Fellow Am. Anthrop. Assn. (past pres.); mem. Linguistic Soc. Am. (past pres.), Sigma Xi. Author: (with others) Linguistic Structures of Native America, 1946; (with Ralph L. Beals) An Introduction to Anthropology, 1953, 59, 65; (with others) Studies in Athapaskan Languages, 1963; also articles, monographs. Editor: Language in Culture, 1954. Research on Am. Indian langs. Alaska, N.W. Can., Ore.-Cal. coast, Ariz.-N.M. area. Home: 191 Beloit Av., Los Angeles, 90049.*

HOIJTINK, Gerrit Jan, Dutch chemist; b. Alkmaar, Jan. 14, 1925; s. Gerrit Hendrik and Maaike Elisabeth (Rijkers) H.; Ph.D. in Natural Sci., Free U. Amsterdam; m. Theodora J. Kramer, 1952; children—Gerrit Hendrik, Frans, Gerrit Jan, Maaike J. Elisabeth. Reader physics and theoretical chemistry Free U. Amsterdam, 1953-57, prof., 1957-60; prof. U. Amsterdam, 1960—. Named laureate, recipient Crismer-Penny prize Chemistry Soc. Belgium, 1963. Mem. Royal Netherlands Assn. Chemistry. Author: Recueil de Travaux Chimiques aux Pays-Bas, 1952-58; Molecular Physics, 1958-63; Chemical Physics. Home: 100 Linneeushof. Office: 125 Nieuwe Prinsengracht, Amsterdam, Netherlands.

HöILAND, Einar, Norwegian math. physicist; b. Lista, Norway, Apr. 5, 1907; s. Andreas and Aline (Jolle) H.; M.Sc., Ph.D. in Math., U. Oslo; m. Randi Höiland, Feb. 11, 1946; 1 son, Klaus. Asst. math. physics Carnegie Found., 1935-45; researcher Center Def. Research, Norway, 1946-47; affiliate prof. aerodynamics and hydrodynamics U. Oslo, 1947-54, full prof., 1954; vis. prof. dept. meteorology U. Cal. at Los Angeles, 1950-51; dir. Inst. Climatic and Meteorologic Research, Oslo, 1951-61. Mem. Videnskaps Akademiet, Norsk Geofysisk Forening, Norsk Fysisk Selskap. Author: On the Interpretation and Application of the Circulation Theorems of V. Bjerknes, 1939; On Horizontal Motion in a Rotating Fluid, 1950; Fluid Flow over a Corrugated Bed, 1951; Two-Dimensional Perturbation of a Flow with Constant Shear of Stratified Fluid, 1953; Discussion of a Hyperbolic Equation Relating to Inertia and Gravitational Fluid Oscillations, 1962. Research on stability of motion of a homogeneous and a stratified fluid; qualitative treatment of uprolling of hydrodynamic discontinuity surfaces; stability and motion of a rotating atmosphere. Home: 75 A Bygdö Alle, 1 Oslo 2. Office: P.O. Box 1054, Blindern, Oslo 3, Norway.*

HOINKES, Herfried Carl, Austrian meteorologist, geophysicist; b. Bielitz, Mar. 9, 1916; s. Carl and Hildegard (Schmidt) H.; Ph.D. in Meteorology and Geophysics, U. Innsbruck; m. Gertrude Riedmayr, 1943; children—Christian, Georg. Asst., Meteorol. and Geophys. Inst., U. Innsbruck, 1940, prof. agrégé, 1949, asso. prof., 1956, pres. Meteorol. and Geophys. Inst., full prof., 1958; mem. Am. Antarctic Expdn., 1956-58. Pres., Commn. Snow and Ice. Recipient medal of honor U. Innsbruck, 1958, Rüpell medal Geog. Soc. Frankfort, 1961. Mem. Internat. Assn. Hydrological Sci., Austrian Alpine Soc., Austrian, Am. meteorol. socs., Glaciological Soc. (Cambridge), Leopoldina Acad. Research and numerous publs. on meteorology and glaciology. Home: 8 Salurnerstrasse. Office: 41 Schöpfstrasse, Innsbruck, Austria.

HOISINGTON, Laurence Earl, Am. physicist; b. Orange, Cal., Feb. 14, 1915; s. Morris Earl and Bessie (Henderson) H.; student Mesa Jr. Coll., 1932-34; B.A., U. Colo., 1936; Ph.D., U. Wis., 1941; m. Frances Eunice Rumsey, Apr. 26, 1943; children—Charles Morris, Hugh Evans, Elizabeth Ann. Head extrapolator sect. Naval Ordnance Lab., Washington, 1940-42; mem. mine warfare operations research group Office of Chief of Naval Operations, Washington, 1942-46, head mine warfare tech. sect., 1946—. Mem. Am. Phys. Soc., Operations Research Soc. Am. Research, numerous publs. on determination of optimum operational characteristics of sea mines and mine countermeasures equipment; devel. optimum tactics and strategy for their use; direction USN mine warfare research, devel. programs; devel. analysis of U. S. mine warfare plans; guidance and analysis of mine warfare operations research studies; direction of U. S. cooperation in mine warfare with NATO, and other fgn. countries and commands. Home: 10017 Frederick Av., Kensington, Md. 20795. Office: OPNAV, Op-323T, Navy Dept., Washington 20350.*

HOITINK, Arnoldus Wilhelm Jan Hendrik, Dutch physician; b. Winterswijk, Feb. 15, 1905; s. Arnold W. and J. C. (de Rode) H.; M.D., U. Utrecht; m. Maria Bleeker; children—Arnold Rein, Marjoleine Maria, Johanna Else. Sci. asst. Lab. Physiol. Chemistry, U. Utrecht, practice medicine, 1930-41; subdir., chief hygiene and physiology dept. Inst. Preventive Medicine, Leyden, 1941-49; studied hygiene and aero. medicine, Eng., 1946; mem. sci. mission to U. S. and Can., 1947; commandant, dir. Netherlands Mil. Sch. for Hygiene and Preventive Medicine, 1948; prof. physiology U. Indonesia, dir. Lab. Physiology, Bogor, 1949-52; dir. Lab. Physiology, prof. physiology Free U. Amsterdam. Mem. numerous sci. assn. Author: Zijn er kunstmatige sera met een bijzonder gunstige werking bij bloedverlies?, 1934; Reformatie van de Universiteit?, 1945; Vitamine C en Arbeid, 1946; Verantwoording der Wetenschap, 1950, also numerous articles. Home: Hoickenrode, Zelhem. Office: Physiology Lab., Van der Boechorststraat 7, Amsterdam, Netherlands.

HOKIN, Lowell Edward, Am. biochemist; b. Chgo., Sept. 20, 1924; s. Oscar E. and Helen (Manfield) H.; student U. Chgo., 1942-43, Dartmouth, 1943-44, U. Ill. Sch. Medicine, 1946-47; M.D., U. Louisville, 1948; Ph.D., U. Sheffield (Eng.), 1952; m. Mabel Neaverson, Dec. 1, 1952; children—Linda Ann, Catherine Esther, Samuel Arthur. Postdoctoral fellow dept. biochemistry McGill U., 1952-54, faculty, 1954-57, asst. prof., 1955-57; faculty U. Wis., Madison, 1957—, prof. physiol. chemistry, 1961—. Recipient Career Research award NIH, 1962—. Mem. Am. Soc. Biol. Chemists, Biochem. Soc. (U.K.), A.A.A.S., Am. Chem. Soc. Contbr. numerous articles to tech. jours., chpts. to numerous books on phospholipids, biol. transport, pancreas. Research on phospholipids in cell membranes. Home: 1234 Sherman Av., Madison, Wis. 53703.*

HOL, Jacoba Brigitta Louisa, Dutch geographer; b. Antwerp, Sept. 21, 1886; d. Richard and Maria Theresia (Koene) Hol; Ph.D. in Sci., U. Utrecht. Chief of works Geog. Inst., U. Utrecht; instr. Higher Catholic Studies, Tilburg, 1914-46; prof. geog. physics and geomorphology, dir. Inst. Geography, Utrecht, 1946-58. Mem. Royal Netherlands Soc. Geography (hon.), Belgian Soc. Geographic Study, Geog. Soc. Netherlands, Collegium Studiosorum Veritas, Royal Belgian Soc. Geography (corr.), Geol. Soc. Belgium, Adelbert Soc., Thijm-Genootschapl Utrechts. Author: Beiträge zur Hydrographie der Ardennen; Geomorphologie van Nederland, also numerous articles. Address: 6 Damiaanberg, Meerssen, Limburg, Netherlands.

HOLBOROW, Eric John, Brit. immunologist; b. Shrewton, Wiltshire, U.K., Mar. 30, 1918; s. Edward Ratcliffe and Marian (Crutchley) H.; M.B., B.Chir., M.A., Cambridge (Eng.) U., 1942, M.D., 1953; m. Cicely Mary Foister, Mar. 31, 1943; children—Jonathan, Paul, Margaret Mary. Bacteriologist, Rheumatism Research Unit, Taplow, U.K., 1948-53, cons. bacteriologist, 1953-58; sci. staff rheumatism research unit Med. Research Council, Canadian Red Cross Meml. Hosp., Taplow, 1958—. Mem. Coll. Pathologists Brit. Soc. for Immunology (gen. sec. 1962—), Heberden Soc., Am. Assn. Immunologists. Author: (with L. E. Glynn) Autoimmunity and Disease, 1965; also articles. Research on autoimmunity in relation to disease especially rheumatic group of diseases, exptl. prodn. of autoimmunity, spontaneous autoimmune disease in mice, and modes of suppression. Home: 23 Hampden Hill, Holtspur, Beaconsfield, Bucks, U.K. Office: Canadian Red Cross Meml. Hosp., Taplow, Maidenhead, Berks, U.K.*

HOLBROOK, John Edwards, Am. zoologist; b. Beaufort, S.C., Dec. 30, 1794; s. Silas and Mary (Edwards) H.; grad. Brown U., 1815; M.D., U. Pa., 1818; m. Harriott Pinckney Rutledge, May 1827. A founder Med. Coll. of S.C., 1824, prof. anatomy, 1824-54; med. officer Confederate Army, head S.C. Examining Bd. of Surgeons, 1861-65; specialized in study of Am. reptiles and fishes; considered most important Am. zoologist of his time. Author: American Herpetology: or a Description of Reptiles Inhabiting the United States, 1842; Ichthyology of South Carolina, 1855, 2d edit., 1860. Died Norfolk, Mass., Sept. 8, 1871.

HOLCIK, Juraj, Czechoslovakian ichthyologist; b. Trnava, Czechoslovakia, Oct. 18, 1934; s. Juraj and Edita (Furdíková) H.; student biol. faculty Comenian U., Bratislava, Czechoslovakia, 1953-56; RNDr., Charles U., Prague, Czechoslovakia, 1958, Candid. Sci., 1966; m. Kristína Kovácová, July 15, 1965; 1 son, Martin. Custos, Regional Mus., Trnava, Czechoslovakia, 1958-60; asst. engr., custos, Lab. Fisheries, Agr. Acad., Bratislava, 1960-62; sci. asst. dept. zoology Charles U., 1965-66; head hydrobiol. expdn. Czechoslovak Acad. Scis. in Cuba, 1966; custos, sci. worker, dept. zoology Slovak Nat. Mus., Bratislava, 1966—. Cons. Fishery Research Inst., Vodnany, Czechoslovakia, 1966—. Recipient financial award, 1963. Mem. Slovak (sec. ichthyology sect.), Czechoslovak zool. socs., Internat. Soc. Ichthyology and Hydrobiology, Am. Soc. Ichthyologists and Herpetologists. Contbr. articles to profl. jours. Discovered (with J. J. Duyvené de Wit) geog. meridional variability of fishes, 1964; described lampreys and fish, also many new hybrids of Acheilognathinae (with J. J. Duyvené de Wit); determined devel. of fish fauna in new reservoirs and its mgmt.; studied taxonomy of several fishes. Home: 3 Pazického. Office: 2 Vajanského, Bratislava, Czechoslovakia.*

HOLCOMB, Amasa, Am. telescope maker; b. Southwick, Mass., June 18, 1787; s. Elijah and Lucy (Holcomb) H.; m. Gillett Kendall, 1808; m. 2d, Maria Holcomb, circa 1861. Began mfg. telescopes, circa 1825, his telescopes described as being same quality as more expensive European makes; recommended by com. on science and arts of Franklin Inst. (Phila.) for award and medal from John Scott Legacy Fund, 1835. Died Southwick, Feb. 27, 1875.

HOLDEN, Abe Noel, Am. metallurgist; b. Wisconsin Rapids, Wis., Aug. 1, 1921; s. Everett B. and Daisy B. (Brower) H.; B.S. in Chem. Engring., U. Wis., 1943, M.S. in Metallurgy, 1947; m. Joyce A. Hungerford, June 23, 1945; children—Karen Ann, Beverly Lynn, Pamela Gale, Noel Paul, Lee Adrian. Metallurgist, Research Lab., Gen. Electric Co., Schenectady, 1947-50, Knolls Atomic Power Lab., 1950-54, mgr. metallurgy, 1954-56, mgr. metallurgy and ceramics Vallecitos Atomic Lab., Pleasanton, Cal., 1956-67; mgr. fuel and materials control Westinghouse Astronuclear Lab., 1967—. U. S. del. 3d Internat. Conf. on Peaceful Uses of Atomic Energy, Geneva, 1964. Mem. Am. Inst. Metall. Engrs. (chmn. nuclear metallurgy symposium 1966, Rossiter W. Raymond award 1953), Am. Soc. Metals, Am. Ceramic Soc., Am. Electrochem. Soc., Am. Soc. Testing and Materials. Author: Physical Metallurgy of Uranium, 1958; Dispersion Fuel Elements, 1968; also articles. Research on phys. metallurgy, nuclear power plant materials. Home: 2225 Country Club Dr., Pitts. 15241. Office: P.O. Box 10864, Pitts. 15236.*

HOLDEN, Edward Singleton, Am. astronomer; b. St. Louis, Nov. 5, 1846; s. Edward and Sarah Frances (Singleton) H.; B.S., Washington U., 1866, A.M., 1879; grad. U. S. Mil. Acad., 1870; LL.D., U. of Wis., 1886, Columbia, 1887; Sc.D., U. of the Pacific, 1896; Litt.D., Fordham U., 1910; m. Mary Chauvenet, May 8, 1871. Lt. engrs., U. S. Army, 1870-73; prof. mathematics, U.S.N., 1873-81; dir. Washburn Obs., Wis., 1881-85; pres. U. of Calif.; 1885-88; dir. Lick Obs., 1888-98; librarian U. S. Mil. Acad., from 1901. Mem. Nat. Acad. Sciences. Author: Bastian System of Fortification, 1872; Index Catalogue of Nebulae, 1877; Life of Sir William Herschel, 1881; Writings of Sir William Herschel, 1881; Astronomy (with Simon Newcomb), 1887; Hand-book of the Lick Observatory, 1888; Briefer Astronomy (with Simon Newcomb), 1892; Mogul Emperors of Hindustan, 1895; Mountain Observatories, 1896; Memorials of W. C. and G. P. Bond, 1897; Pacific Coast Earthquakes, 1898; Earth and Sky, 1898; Our Country's Flag, 1898; Primer of Heraldry, 1898; Elementary Astronomy, 1899; Family of the Sun, 1899; Essays in Astronomy, 1900; Stories of the Great Astronomers; Real Things in Nature, 1903; The Sciences, 1903; Galileo, 1905. Editor: Publications Washburn Observatory, 4 vols., 1881-85; Publications Lick Observatory, 3 vols., 1888-94; Centennial U. S. Military Academy, 1902 (2 vols.). Best known as astron. writer, especially on phys. features of planets and nebulae; also made notable contbns. to astron. bibliography. Died Mar. 16, 1914.

HOLDEN, Frederick Clark, Am. obstetrician, gynecologist; b. Tremont, Me., Nov. 4, 1868; s. Simeon A. and Hannah A. (Verrill) H.; M.D., N.Y. U., 1892, D.P.H., 1941; m. Rachel Maud Wilson, Nov. 17, 1897; 1 son, W. Wilson. Asst. attending surgeon, St. Mary's Hosp., Bklyn., 1895-98; asst. obstetrician and gynecologist Williamsburg Hosp., 1902-06, Meth. Hosp., 1906-08; attending obstetrician Meth. Hosp., 1908-14; chief obstetrician and gynecologist, Greenpoint Hosp., 1914-19; asst. prof. gynecology and obstetrics L.I. Coll. Hosp., 1914-19; dir. dept. gynecology Bellevue Hosp., 1919-34; prof. obstetrics and gynecology N.Y. Univ. Med. Coll., 1919-32; dir. gynecology Jersey City Med. Centre, from 1940; attending obstetrician and gynecologist, French Hosp.; cons. numerous other hosps. Diplomate Am. Bd. Obstetrics and Gynecology. Fellow A.C.S.; mem. Am. Gynecol. Soc. (pres.), A.M.A., Am. Soc. for Control of Cancer, N.Y. Acad. Medicine, N.Y. Obstet. Soc. (past pres.), Alpha Kappa Kappa, Alpha Omega Alpha. Died Aug. 27, 1944.

HOLDEN, Roy Jay, Am. geologist; b. Sheboygan Falls, Wis., Oct. 21, 1870; s. Harvey J. and Sarah Diana (Danforth) H.; B.S., U. Wis., 1900, Ph.D., 1915; m. Elizabeth Virginia Evans, June 29, 1915; children—Sarah Virginia, Elizabeth Flora, Roy Jay. Teacher country schs., 1892-95, high sch., Sheboygan Falls, 1895-97; sci. tchr. high sch., Beloit, Wis., 1900-02; with Va. Poly. Inst. from 1905, successively asso. in geology and mineralogy, asso. prof. and prof. geology; also cons. comml. geologist. Fellow A.A.A.S., Geol. Soc. Am.; mem. Am. Inst. Mining and Metall. Engrs., Am. Assn. Petroleum Geologists, Soc. Econ. Geologists, Va. Acad. Sci., Phi Kappa Phi, Sigma Xi. Contbr. various articles and reports on geol. resources in Wis., Va. Determined geol. structure of Valley Coal Field of Va., adding 100 sq. miles to previously known coal-bearing terr.; located 1st gas well in Va. Died Dec. 16, 1945.

HOLDEN, William Douglas, Am. physician; b. Pittsfield, Mass., Aug. 25, 1912; s. Harry Douglas and Katherine (MacInnis) H.; A.B., Cornell U., 1934, M.D., 1937; m. Janet Cobb, Dec. 28, 1936; children—John, Frank, Katherine. Faculty, Western Res. U. Sch. Med., 1946—, Oliver H. Payne prof. surgery, 1950—, dir. surgery U. Hosps. of Cleve., 1950—. Mem. Am. Surg. Assn., A.C.S., Soc. U. Surgeons, Soc. Vascular Surgery, Soc. Exptl. Biology and Medicine, Am. Bd. Surgeons. Author: Acute Peripheral Arterial Occlusion, 1952. Studies, publs. on metabolism after injury, transplantation immunity, shock, clotting disorders, clin. surgery. Home: 2195 Demington Dr., Cleveland Heights, O. 44106. Office: 2065 Adelbert Rd., Cleve. 44106.*

HOLDER, Douglas William, English aero. engr.; b. London, Eng., Apr. 14, 1923; s. William Arthur and Anne (Daniel) H.; B.Sc., Imperial Coll. Sci. and Tech., 1943, diploma Imperial Coll., 1944, Ph.D., 1951, D.Sc., 1955; M.A., Oxford U., 1961; m. Barbara Woods, July 4, 1946; children—Jane Anne, Sarah Ellen. Asst., Aircraft and Armament Exptl. Establishment, Boscombe Down, Eng., 1943; staff Nat. Phys. Lab., Teddington, Eng., 1944-61, sr. prin. sci. officer, 1953-57, dep. chief sci. officer, 1957-61; head dept. engring. sci. Oxford (Eng.) U., 1961—, prof. engring. sci., 1961—, fellow Brasenose Coll., 1961—. Mem. various coms. Aero. Research Council, 1946—; mem. adv. council on sci. research Army Dept., 1964—; vis. prof. U. Mich., 1964-65. Bd. govs. Oxford Coll. Tech. Fellow Royal Soc., 1962; mem. Royal Aero. Soc. (chmn. aerodynamics com. 1964—), Instn. Civil Engrs., Am. Inst. Aeros. and Astronautics. Author: (with others) Modern Developments in Fluid Dynamics, 1953; (with R. C. Parkhurst) Wind tunnel Techniques, 1952; (with R. J. North) Schlieren Methods, 1962; also numerous articles. Responsible for providing facilities of Nat. Phys. Lab. for research on gasdynamics and design of aircraft and missiles, for providing new labs. for research on all brs. of engring. at Oxford. Home: Grazeley House, Iffley Turn, Oxford, Eng.*

HÖLDER, Helmut, German paleontologist; b. Stuttgart, Germany, Jan. 18, 1915; s. Karl and Helene (Lorberg) H.; Ph.D. in Natural Sci., U. Tübingen; m. Erna Werner, 1944; children—Irmela, Isolde, Dorothee. Sci. asst., chief curator Inst. of Mus. Geology and Paleontology, U. Tübingen; prof. Inst. Geology and Paleontology, U. Munster. Mem. German Soc. Paleontology. Author: Geologie und Paläontologie; Orbis Academicus, 1960, also articles on cephalopod fossils and Jurassic stratigraphy. Home: 11 Kaiser-Wilhelm Circle. Office: 3 Pferdegasse, Munster, Germany.

HÖLDER, Ludwig Otto, German mathematician, b. 1859. Specialized in group theory; devel. group theorems important in study of structure of any finite discrete group; inaugurated modern theory of divergent series through his method of summation by arithmetic means. Died 1937.

HOLINGER, Paul Henry, Am. physician; b. Chgo., Mar. 13, 1906; s. Jacques and Cora (Lange) H.; B.S., U. Chgo., 1928; M.S., Northwestern U., 1930, M.D., 1933; m. Julia Drake, June 26, 1940; children—Lauren Drake, William Jacques, Paul Campbell, Richard Lange. Faculty, U. Ill. Coll. Medicine, Chgo., prof. otolaryngology, attending laryngologist, bronchoesophalogists Research and Edn. Hosp., Ill. Ear and Eye Infirmary; staff Presbyn.-St. Luke's Hosp., Children's Meml. Hosp. Mem. Am. Laryngological Assn. (past pres., Casselberry award 1946, James Newcomb award 1962, deRoaldes award 1966), Inst. Medicine Chgo. (past chmn. bd. govs.), Am. Cancer Soc. (past pres. Ill. div.), A.M.A. (chmn. interspity. com. 1966—), A.C.S. (regent 1965—). Research, numerous publs. on laryngology and bronchoesophagology; developed 1st color still and motion pictures taken of interior of bronchial tubes and esophagus. Home: 1500 N. Lake Shore Dr., Chgo. 60610. Office: 700 N. Michigan Av., Chgo. 60611.

HOLL, Frederick John, Am. biologist; b. Buffalo, Mar. 22, 1898; s. Fred and Augusta (Lembke) H.; B.S., U. Buffalo, 1922; M.S., U. Wis., 1926; Ph.D., Duke, 1928. Faculty, U. Buffalo, 1922-60, asso. prof. biology, 1937-60. Mem. Phi Beta Kappa, Sigma Xi. Research, publs. on effect high frequency radiation on Trichennele spiralis, ecology parasites; described new species trematodes and nematodes. Home: 69 E. Winspear Av., Buffalo 14214.*

HOLLAENDER, Alexander, biophysicist; b. Samter, Germany, Dec. 19, 1898; s. Heymann and Doris (Rotholz) H.; came to U. S., 1921, naturalized, 1927; A.B., U. Wis., 1929, M.A., 1930, Ph.D., 1931; m. Henrietta Wahlert, Oct. 10, 1925. Asst. phys. chemistry dept. U. Wis., Madison, 1929-31, NRC fellow biol. scis., 1931-33; investigator Rockefeller Found., N.Y.C., 1934; investigator in charge radiation work NRC Project, Madison, 1934-37; biophysicist Washington Biophysics Inst., NIH, USPHS, 1937-50, OSRD, USN, 1940-45; dir. biology div. Oak Ridge Nat. Lab., 1946-66, sr. research adviser, 1967—; prof. radiation biology U. Tenn., 1957-66, prof. biomed. scis. U. Tenn. Oak Ridge Grad. Sch. Biomed. Scis., 1966—. Fellow A.A.A.S., Am. Acad. Arts and Scis., Tenn. Acad. Scis.; mem. Am. Phys. Soc., Am. Soc. Microbiology, Am. Soc. Naturalists, Biophys. Soc., Genetics Soc. Am., Nat. Acad. Scis., Radiation Research Soc. Author: Radiation Biology, 3 vols., 1956; Radiation Protection and Recovery, 1960. Publs. on studies of biolog. effects of ultraviolet and x-ray radiation, mechanisms of radiation damage, recovery, protection. Home: 48 Outer Dr., Office: P.O. Box Y, Oak Ridge Nat. Lab., Oak Ridge 37830.*

HOLLAND, Charles Thurstan, Brit. radiologist; b. 1863; s. William Thomas Holland; ed. Univ. Coll. Hosp., London; LL.D., Liverpool; m. Lilian Fergusson; 1 son. Hon. radiologist Royal So. Hosp., Liverpool, 1896-1904, Royal Infirmary, Liverpool, 1904-23, Royal Liverpool Children's Infirmary, 1907-32; radiologist 1st Western Gen. Hosp., 1914-19; hon. cons. radiologist King Edward VII Welsh Nat. Meml. Assn., Liverpool Royal Infirmary, Shropshire Orthopaedic Hosp. Fellow Royal Coll. Surgeons, Am. Electrotherapeutic Assn. (hon.), Royal Photog. Soc.; fellow or mem. Liverpool Med. and Lit. Soc. (pres.), Roentgen Soc. London (pres.), Brit. Med. Assn. (pres. electrotherapeutic sect. Liverpool 1912), Royal Soc. Medicine (pres.), International Congress Medicine (v.p. radiology sect.), Congress Radiology and Physio-therapy (pres. radiology sect. 1922), First Internat. Congress Radiology (pres. 1925), Brit. Inst. Radiology (pres.), Brit. Orthopaedic Assn. Contbr. to Med. Ann., 1914-30. Developed technique for detection of bullets and shell fragments in body (during World War I); an early follower of Roentgen in work on X-ray. Died Jan. 16, 1941.

HOLLAND, Heinrich Dieter, geochemist; b. Mannheim, Germany, May 27, 1927; s. Otto and Jeannette (Liebrecht) H.; B.A., Princeton, 1946; M.S., Columbia, 1948, Ph.D., 1952; m. Alice Tilghman Pusey, June 22, 1953; children—Henry L., Anne L., John P., Matthew T. Faculty dept. geology Princeton, 1950—, prof., 1966—, dir. summer studies, 1962-66. Mem. Geol. Soc. Am., Geochem. Soc. (councilor 1964—), Mineral. Soc., Am. Geophys. Union, Am. Assn. U. Profs., Sigma Xi. Contbr. articles to tech. jours. Developed consistent model for chem. evolution atmosphere and ocean water; showed how mineral assemblages of ore deposits can be used to define their condition of formation; research on solubility carbonate minerals and trace element composition calcite and aragonite. Home: 118 Washington St., Rocky Hill, N.J. 08533. Office: Dept. Geology, Princeton, Princeton, N.J. 08540.*

HOLLAND, James Frederick, Am. physician; b. Morristown, N.J., May 16, 1925; s. Albert H. and Mary (Layer) H.; A.B., Princeton, 1944; M.D., Columbia U., 1947; m. Jimmie Coker, July 7, 1956; children—Diane, Steven, Mary, Sally, Peter, David. Staff, Nat. Cancer Inst., Bethesda, Md., 1953-54; chief medicine Roswell Park Meml. Inst., Buffalo, 1956—; asso. research prof. medicine State U. N.Y., Buffalo, 1962—; dir. Cancer Clin. Research Center, Roswell Park Meml. Inst., 1963—. Mem. Acute Leukemia Group B, 1955—, Eastern Solid Tumor Group, 1956—; chmn. Acute Leukemia Coop. Group B, 1963—; mem. clin. panel Cancer Chemotherapy Nat. Service Center, 1955-58. Mem. Am. Assn. for Cancer Research, A.A.-A.S., Am. Fedn. for Clin. Research, Am. Soc. Hematology, Am. Soc. Clin. Investigation. Editor: (with Hreshchyshyn) Choriocarcinoma, 1967. Research, numerous publs. on chem. control cancer, chemotherapy of leukemia and related diseases, design and conduct clin. investigation on nature of illness caused by cancer. Home: 137 Depew Av., Buffalo 14214. Office: Roswell Park Meml. Inst., 666 Elm St., Buffalo 14203.*

HOLLAND, John Philip, inventor; b. Liscanor, County Clare, Ireland, Feb. 29, 1840; s. John and Mary (Scanlon) H.; m. Margaret Foley, Jan. 17, 1887, 4 children. Taught sch., Ireland, 1858-72; came to U. S., 1873, taught sch., Paterson, N.J.; offered design of submarine to U. S. Navy, 1875 (offer rejected); constructed his 1st submarine with financial backing of revolutionary Fenian Soc. (sank on 1st trial, 1878); launched Fenian Ram, 1881 (proved impractical); contracted to build submarine Plunger for U. S. Navy, 1895, but his designs were radically altered; launched his own submarine the Holland, 1898 (1st submarine equipped to move underwater by electric power and on surface by gasoline engine; 1 of 1st designed to dive by inclining its axis); sold Holland, with 6 sister ships, to U. S. Navy, 1900; also built submarines for Russia, Japan, Gt. Britain; invented respirator for escape from disabled submarines, 1904. Died Newark, N.J., Aug. 12, 1914.

HOLLAND, Joshua Zalman, Am. meteorologist; b. Chgo., June 23, 1921; s. Samuel H. and Lillie (Perlman) H.; B.S., U. Chgo., 1941, certificate meteorology, 1942; Ph.D., U. Wash., 1968. m. Anabel Schreiber, Apr. 27, 1947; children—Joseph, Susannah, Sarah. Meteorologist in charge U. S. Weather Bur., Oak Ridge, 1948-53, meteorologist, Washington, 1953-55; tech. adviser to asst. gen. mgr. for research and devel. U. S. AEC, Washington, 1956-57, meteorologist, 1957-59, chief fallout studies br., 1959—, exec. sec. adv. com. on reactor safeguards, 1956-57; faculty U. Chgo., 1942, N.Y. U., 1943. Recipient U. S. Dept. Commerce Silver medal, 1957; U. S. AEC Superior Service award, 1963. Mem. Am. Meteorol. Soc., Am. Geophys. Union, A.A.A.S. Research, publs. on climatology of Oak Ridge, methodology for determining turbulence and small-scale atmospheric transport in hilly terrain, formula for cumputing height of rise of hot stack gases, nomogram for computing height of rise of hot stack gases, nomogram for computing gamma radiation dose from an airborne cloud radioactivity, characteristics, transport and deposition of radioactive fallout, detailed structure of atmospheric turbulent eddies. Home: 5609 Glenwood Rd., Bethesda, Md. 20034. Office: Div. Biology and Medicine, U. S. AEC, Washington 20545.*

HOLLAND, Leslie, English physicist; b. London, Eng., June 14, 1921; s. Charles Arthur and Kate (Mines) H.; student Acton Tech. Coll., 1935-38, 41-42; D. honoris causa, Rouen. m. Doris Elsie Rider, May 20, 1944; children—John R., Jennifer S., Peter R. Head vacuum deposition lab. Edward High Vacuum Ltd., 1944-62, head surface physics lab., 1962-64; dir. research central research lab. Edwards High Vacuum Internat. Ltd., Crowley, Eng., 1964—; asso. reader Brunel U., 1963—. Recipient Silver medal Plastics Inst., 1955. Fellow Inst. Physics (chmn. vacuum physics group 1965—), Soc. Glass Tech.; mem.

Am. Vacuum Soc., Am. Optical Soc. Author: Vacuum Deposition of Thin Films, 1956; The Properties of Glass Surfaces, 1964; Thin Film Microelectronics, 1965; also articles. Editor high vacuum series Chapman & Hall Ltd., 1964—. Research, patents on growth and phys. properties of thin films, high vacuum subjects, surface phys. studies, especially in vacuo; applications in devel. of optical and electronic devices. Home: Hazelwood, Balcombe Rd., Crawley. Office: Central Research Lab., Edward High Vacuum Internat., Ltd., Manor Royal Crawley, Sussex, Eng.*

HOLLAND, Robert Francis, Am. dairy scientist, educator; b. Holley, N.Y., Sept. 21, 1908; s. Robert B. and Mary (Parker) H.; B.S., Cornell U., 1936, M.S., 1938, Ph.D., 1940; m. Ruth McCargo, Aug. 10, 1930; children—Robert G., Daniel M., Deborah (Mrs. Charles Stewart), James S. Agrl. chemist Co-op. G.L.F. (now AGWAY), Ithaca, N.Y., 1941-45; faculty Cornell U., Ithaca, N.Y., 1945—, prof. dairy sci., 1945-55, head, food sci. dept., 1956—. Cons. Agway, Corning Glass Co., Thatcher Glass Mfg. Co., Allied Chem. Co. Recipient Am. Agrl. award of merit, 1956. Mem. Inst. Food Technologists, Am. Dairy Sci. Assn., Am. Chem. Soc., N.Y. State Assn. Milk and Food Sanitarians (pres. 1965-66), Sigma Xi, Phi Kappa Phi. Developed system for ultra-high temperature pasteurization of milk; publs. on microbiology of new milk packaging methods, chemistry of detergents and automatic cleaning systems; formulations for low-fat fluid dairy products. Home: 71 Old Main St., Trumansburg, N.Y. 14886. Office: Cornell U., Ithaca, N.Y. 14850.*

HOLLAND, Sir Thomas Henry, Brit. geologist; b. Helston, Eng., Nov. 22, 1868; s. John and Grace Treloar (Roberts) H.; ed. Normal Sch. Sci., Royal Sch. Mines; hon. doctorated, Calcutta, Melbourne, Witwatersrand, Manchester, Glasgow, St. Andrews, Edinburgh, Aberdeen, Queens univs.; m. Frances Maud Chapman, 1896; 1 son, 1 dau.; m. 2d, Helen Ethleen Verrall, 1946. Asst. supt. Geol. Survey India, 1890-1903, dir., 1903-09; prof. Manchester U., 1909-18; pres. Indian Indsl. Commn., 1916; Indian Munitions Bd., 1917; mem. Gov.-Gen.'s Council, India, 1920-21; rector Imperial Coll. Sci. and Tech., 1922-29; prin., vice-chancellor Edinburgh U., 1929-44. Fellow Royal Soc. London, 1904, v.p., 1924-25; fellow or mem. Brit. Assn. (sect. pres.), Inst. Mining Engrs. (pres. 1915-16), Royal Soc. Arts (chmn. 1925-27), Inst. Mining and Metallurgy (pres. 1925-27), Petroleum Technologists (pres. 1925-27), Brit. Assn. (pres. 1929), Royal Soc. Edinburgh (v.-p. 1932-35), Geol. Soc. London (pres. 1933-36), Mineral. Soc. (pres. 1933-36). Author: The Mineral Sanction, an aid to International Security. Discovered hypersthene rock formations in India; added to knowledge of mineralogy, petrology, geology of India; studied mountain landslips in Himalayas; wrote on elaeolite-syenites and mica deposits. Died May 15, 1947.

HOLLANDE, Charles Augustin, French physician; b. Chambery, Jan. 29, 1891; s. Dieudonné and Mathilde (Duinez) H.; ed. Faculty Chamberg, U. Grenoble, U. Lyons; Ph.D. in sci., Ph.D. in pharmacy; M.D.; m. Aline Huquet, 1900; children—Armand, André. Prof., Faculty Pharmacy, U. Nancy, U. Strasbourg, U. Montpellier, now hon. prof. Author: Cytologie; Fonction Hepathique chez l'Insecte; Hérmatologie Insecte; Origine de la Vie. Address: 7 Avenue of School of Agriculture, Montpellier, Herault, France.

HOLLANDER, Bernard, physician; b. Vienna, Austria, 1864; ed. King's College, London; various continental Us., M.D.; m. Louisa Vogel, 1906. Asst. to Prof. Krafft-Ebing, Vienna; moved to London, 1883; asst. to Sir David Ferrier, King's College Hosp.; physician, Brit. Hosp. for Mental Disorders and Brain Diseases. Founder, Ethnological Soc. (pres. 1904-29); pres., Société Internationale de Philologie, Sciences et Beaux-Arts; mem., Royal College Surgeons; lic., Royal College Physicians; corr. mem., Royal Acad. Medicine, Madrid. Author: Die psychischen Thätigkeiten des Gehirns, 1900; The Mental Functions of the Brain, 1901; Scientific Phrenology, 1902; Crime and Responsibility, 1907; Psycho-Therapeutics of Insanity, 1908; The Unknown Life and Works of Dr. Francis Joseph Gall, 1909; Hypnotism and Suggestion, and Mental Symptoms of Brain Disease, 1910; Change of Life in Man, 1910; Eugenics and Marriage, and the First Signs of Insanity, their Prevention and Treatment, 1912; The Insanity of Genius, 1913; Nervous Disorders of Men, their Causes, Effects and Treatment, 1916; Abnormal Children, 1916; In Search of the Soul and the Mechanism of Human Thought, Emotion and Conduct, 1920; Psychology of Misconduct, Vice and Crime, 1922; Methods and Uses of Hypnosis and Self-Hypnosis, 1928; Brain, Mind and the External Signs of Intelligence, 1931; Seeing Ourselves in the Light of Modern Psychology, 1931; Old Age Deferred, 1933. Studied mind and character and their deviations from the normal; research on cerebral localization. Died Feb. 6, 1934.

HOLLANDER, Jack Marvin, Am. nuclear chemist; b. Youngstown, O., Apr. 13, 1927; s. Isadore M. and Adele (Feuer) H.; B.S., Ohio State U., 1948; Ph.D., U. Cal. at Berkeley, 1951; m. Margie J. Schnarr, Aug. 26, 1948; children—Judith, Jeffrey, Allan. Instr. chemistry U. Cal., Berkeley, 1951-53; sr. staff mem. Lawrence Radiation Lab., Berkeley, 1953—. J. S. Guggenheim fellow, 1958, 65. Mem. Am. Phys. Soc. Author: (with Lederer, Perlman) Table of Isotopes, 1967. Research, numerous publs. on nuclear energy levels and nuclear structure, compilations of radioisotope and nuclear energy level data. Home: 61 Edgecroft Rd., Berkeley, Cal. 94707.*

HOLLANDER, Joseph Lee, Am. physician; b. Ferguson, Mo., Mar. 8, 1910; s. Charles Samuel and Elsa (Windstosser) H.; A.B., Cornell U., 1932; M.D., U. Pa., 1935; m. Olive Wright McKinney, July 2, 1936; children—Elizabeth Jean (Mrs. John Duncan McCallum), Susan Lee. Physician, Pa. Hosp., 1937-46, dir. Arthritis Clinic, 1937-46; demonstrator medicine Jefferson Med. Coll., 1939-46; practice medicine, specializing in internal medicine, rheumatology, Phila., 1946—; faculty U. Pa., 1946—, prof. medicine, 1962—; cons. rheumatology Childrens' Hosp., Naval Hosp. Fellow A.C.P., Am. Rheumatism Assn. (pres. 1961-62), numerous others. Editor-in-chief Arthritis, 7th edn., 1966. Contbr. numerous articles to sci. jours. Pioneered in study of joint diseases, particularly in findings of joint fluid, diagnosis of gout, intra-articular steroid therapy; pathogenesis of rheumatoid arthritis, and controlled climate chamber on effect of climate on arthritis. Home: 45 University Mews, 45th and Spruce Sts., Phila. 19104. Office: University Hosp., 3400 Spruce St., Phila. 19104.*

HOLLANDER, Nina, Am. biochemist; b. N.Y.C., July 28, 1927; B.A., Hunter Coll., 1948; M.D., N.Y. U., 1951; m. Vincent Paul Hollander. Instr. chemistry Hunter Coll., 1951; research asso. S.W. Found. for Edn. and Research, San Antonio, 1954-55; USPHS fellow, 1956-58; fellow in cancer research dept. med. U. Va., 1958, research asso. dept. med., 1960, asst. prof. biochemistry, 1961; asst. sect. leader dept. biochemistry Research Inst. Skeletomuscular Diseases, Hosp. Joint Diseases, N.Y.C., 1963—. Mem. Endocrine Soc. Contbr. numerous articles to med. jours. Home: 500 E. 83d St., N.Y.C. 10028. Office: Research Inst. for Skeletomuscular Diseases, Hosp. Joint Diseases, 1919 Madison Av., N.Y.C. 10035.*

HOLLANDER, Vincent Paul, Am. biochemist; b. N.Y.C., June 18, 1917; s. Sidney and Elizabeth (Labs) H.; B.S., U. Chgo., 1941, M.S., Ph.D. in Biochemistry, 1944; M.D., Northwestern U., 1947; m. Nina Labounsky, Aug. 2, 1952; children—Lucille, Marie, Vincent Paul. Faculty, U. Va., 1952-63, Am. Cancer Soc. prof. internal medicine, 1960-63; dir. research Inst. for Skeletomuscular Diseases, Hosp. for Joint Disease and Med. Center, N.Y.C., 1963—; attending physician in internal medicine, Hosp. for Joint Diseases, 1964—. Spl. cons. to endocrine evaluation cancer chemotherapy Nat. Service Center, NIH, 1957—, mem. endocrine study sect. div. research grants, 1964—. Fellow A.C.P., N.Y. Acad. Sci.; mem. Am. Fedn. Clin. Research, Am. Soc. for Cancer Research, A.A.A.S., Am. Chem. Soc., Soc. for Exptl. Biology and Medicine, Am. Soc. Biol. Chemists, Endocrine Soc., Sigma Xi. Research, numerous publs. in adrenalectomy in treatment breast cancer, enzymes, endocrinology. Home: 500 E. 83d St., N.Y.C. 10028. Office: 1919 Madison Av., N.Y.C. 10035.*

HOLLANDUS, Isaac and Johann Isaac, alchemists. Most authorities regard Isaac as father, Johann Isaac as son, but may have been same person. Dates assigned them vary from 16th to early 18th centuries; natives of either Stolk or Stolkwijk in Krimpenaarwaard. Authorship: (under name Magistri Joannis Isaaci Hollandi) Opera Mineralia, 1600; De Tribus Ordinibus Elixiris and Lapidis Theoria, 1608, Opera Mineralia et Vegetabilia, 1616, Opus Saturni, 1604, Des hocherleuchteten, 1659, Das Dritte Theil des Mineral-Wercks, 1665, Dess Weit und Breit Berühmten Johannis Isaaci Hollandi Geheimer und Biss dato Verborgen Gehaltener Trefflicher Tractat, 1667; (under name Isaaci Hollandi) Tractatus de Lapide Philosophico . . . , 1669, Sonst auch Flandri Genannt . . . , 1714; Sammlung Unterschiedlicher Bewährter Chymischer Schriften . . . , 1667, Fragmentum de Lapide Philosorum, 1959, Fragmenta, 1647; Tractatus Johannis Isaaci Hollandi de Krina Quo Modo per Spiritum Ejus Omnes Tincturae Sint Extra Hendae, 1661; Tractatus Isaaci Hollandi de Salibus and Oleis Metallorum, 1723. Works explain forms of the Philosopher's Stone and medicinal virtues, transmutation of lead, mercury or silver into gold, and alchem. recipes (salt preparation); mention potassium sulphate and what might have been calcium chloride and sodium phosphate and 3 prin. ingredients, mercury, sulphur, salt.

HOLLDACK, Klaus, German physician; b. Stuttgart, Germany, May 3, 1912; s. Hans and Lehene (Ehlers) H.; student univs. Rostock, Leipzig; state exam. U. Leipzig, 1937, Dr.med., 1938; m. Karin Stefanie Bruns, Dec. 11, 1957; children—Bettina (Mrs. Frank Bernatzky), Claudia (Mrs. Hase), Friedericke, Johanna, Catharina. Became prof. U. Heidelberg, 1956, prof. Free U. Berlin, 1958—; med. dir. Berlin-Neukölln Municipal Hosp., 1957—. Mem. Internat. Heart Assn. (mem. research com.), Laennec Soc., German Soc. Internal Medicine, Berlin Med. Soc. Author: Lehrbuch der Auskultation and Perkussion, 6th edit., 1967; (with D. Wolf) Atlas und kurzgefasstes Lehrbuch der Phonokardiographie, 3d edit., 1967, Herzschallfibel, 3d edit., 1966; also numerous articles. Research on physical fundamentals of heart auscultation and phonocardiography. Address: 56 Rudowerstrasse, 1 Berlin 47, West Germany.*

HOLLE, Gottfried Johannes, German pathologist; b. Chemnitz, Germany, Aug. 13, 1912; s. Johannes Alfred and Marie (Richter) H.; ed. at univs. Leipzig, Wien, Innsbruck, Munich, Konigsberg; m. Maria Inez Renate Weinreich, Dec. 23, 1940; children—Bernd, Jörg, Ulf, Jens, Albrecht, Thomas, Petra, Gottfried, Gisbert. Asst., Inst. Pathology, Chemnitz, 1937-38, U. Leipzig, 1939-45; dozent U. Leipzig, 1945-49; dir. Inst. Pathology, Gera, 1949-53; prof. pathology U. Greifswald, 1953-59, U. Leipzig, 1959—. Mem. Soc. Pathologists and Anatomists East Germany (chmn. 1962-66), Soc. German Pathologists, European Soc. Pathology (mem. council). Author: Textbook of General Pathology, 1967; also numerous articles. Research on arteriosclerosis, pathology of kidney, liver and gall passages, electronmicroscope and histochemistry. Home: Abtuaundorferstr. 60, 7024, Leipzig, East Germany.*

HOLLEMAN, Arnold Frederik, Dutch chemist; b. Oisterwyk, Netherlands, Aug. 28, 1859; Chem.D., U. Leyden, 1887; ed. U. Heidelberg, U. Munich; LL.D., U. St. Andrews, Scotland; D.Sc., U. Leeds, U. Caen. Asst. to prof. van't Hoff, 1887; dir. Agrl. Lab., Groningen, 1889; prof. chemistry U. Groningen, 1893-1905; prof. U. Amsterdam, 1905-24. Mem. French Acad. Scis. (corr.), Royal Acad. Scis. Amsterdam, Royal Acad. Scis. Edinburgh (hon.), Chem. Soc. France, Royal Inst. London, Netherlands Chem. Soc., Chem. Soc. Warsaw. Author: Die direkte Einführung von Substituenten in den Benrolkern, also articles. Published well-known chemistry textbook. Died Bloemendaal, Netherlands, Aug. 11, 1953.

HOLLENDER, Marc Hale, Am. psychiatrist; b. Chgo., Dec. 19, 1916; s. Abraham R. and Anna (Winsberg) H.; student Loyola U., Chgo., 1934-35, Northwestern U., 1935-37; B.S., U. Ill., 1939, M.D., 1941; m. Betty Jane Schultz, July 7, 1943; children—Mary Jo David Albert. Faculty, U. Ill. Coll. Medicine, 1946-56; faculty State U. N.Y. Upstate Med. Center, Syracuse, 1956-66, prof., chmn. psychiatry, 1956-66; prof. U. Pa., Phila., 1966—. Dir. Syracuse Psychiat. Hosp., 1957-64; staff mem. Inst. for Psychoanalysis, Chgo., 1953-56. Fellow Am. Psychiat. Assn.; mem. Am. Psychoanalytic Assn., Am. Psychosomatic Soc. Author: The Psychology of Medical Practice, 1958, The Practice of Psychoanalytic Psychotherapy, 1965. Contbr. articles to profl. jours. Home: 508 Waldron Terrace, Merion Station, Pa. 19066. Office: Gates Pavillion, Hosp. of U. Pa., Phila. 19104.*

HOLLERITH, Herman, Am. inventor; b. Buffalo, Feb. 29, 1860; s. George and Franciska (Brunn) H.; grad. Columbia U. Sch. Mines, 1879; m. Lucia Beverly Talcott, Sept. 15, 1890; children—Lucia, Nannie, Virginia, Herman, Richard, Charles. Asst. to statistician William Petit, Columbia, 1880; instr. mech. engring. Mass. Inst. Tech., 1882; went to St. Louis, 1883; with Patent Office, Washington, 1884-90; invented tabulating machine which worked on principle of punched holes in non-conducing material (counting took place as electric current passed through holes); his tabulating machine used for Census of 1890; read paper concerning his invention before Berne session Internat. Statis. Inst., 1895; founded Tabulating Machine Co., N.Y., 1896, merged with 2 other cos. to become Computing-Tabulating-Recording Co., 1911, later became IBM. Died Washington, Nov. 17, 1929.

HOLLEY, Howard Lamar, Am. physician; b. Marion, Ala., July 14, 1914; s. Warren Alton and Lula (Fretwell) H.; B.S., U. Sc., 1935; M.D., Med. Coll. S.C., 1941; m. Martha Holcomb, Sept. 7, 1946; children—Dan, Nancy, Warren, Howard, Jane. Faculty, Ala. Med. Coll., 1947—, prof. internal medicine, 1959—, dir. div. rheumatology, 1950—. Arthritis tng. grants com. Nat. Inst. Arthritis, chmn., 1964-65. Recipient Seale Harris Research award, 1962. Fellow A.C.P. (gov. Ala. 1966—); mem. So. Soc. Clin. Investigation, Am. Rheumatism Assn., A.M.A., Am. Fedn. Clin. Research, Med. Assn. State Ala., Ala. Acad. Sci., So. Med. Assn., Sigma Xi. Author: (with Allen E. Hussar) Antibiotics and Antibiotic Therapy, A Clinical Manual, 1954; (with others) Potassium Metabolism in Health and Disease, 1955. Contbr. numerous articles to med. jours. Research in clin. characteristics of rheumatic diseases that reflect some of biochem. and physiol. alterations in path. process. Home: 4016 Old Leeds Circle, Birmingham 35213. Office: 1919 7th Av., S., Birmingham, Ala. 35233.*

HOLLING, Crawford Stanley, biologist; b. Theresa, N.Y., Dec. 6, 1930; s. Stanley Arnold and Claudia (Guichard) H.; B.A., U. Toronto, 1952, M.A., 1954; Ph.D., U. B.C., 1957; m. Beverley Joan Rowley, Mar. 21, 1953; children—Christopher, Nancy. Scientist, Can. Dept. Forestry, Sault Ste. Marie, Ont., 1952-64, Victoria, B.C., 1965-67; vis. scientist Bur. Comml. Fisheries, Honolulu, 1964; vis. prof. U. Cal., Berkeley, 1965; prof. U. B.C., Vancouver, 1967—. Cons. Ford Found., 1967—. Mem. Ecol. Soc. Am. (George Mercer award 1966), Canadian Entomol. Soc., Brit. Ecol. Soc., A.A.A.S. Author: Systems Ecology, 1966. Research on exptl. and math. modelling of complex population processes. Home: 5789 College Highroad, Vancouver 8, B.C., Can.*

HOLLINGSHEAD, August deBelmont, Am. sociologist; b. Lyman, Wyo., Apr. 15, 1907; s. William Thomas and Daisy (Rollins) H.; A.B. cum laude, U. Cal. at

Berkeley, 1931, M.A., 1933; Ph.D., U. Neb., 1935; M.A. (hon.), Yale, 1952; m. Carol Evelyn Dempsey, Nov. 4, 1931; children—Ann Marie (Mrs. Gary English Hanna), Ellen Mae (Mrs. Russell Wade Steele). Faculty, State U. Ia., 1935, U. Ala., 1935-36, Ind. U., 1936-41, 45-47; faculty Yale, 1947—, William Graham Sumner prof., 1963—, chmn. dept. sociology, 1959-65. Cons. Nat. Resources Planning Bd., 1937-41; mem. nat. adv. bd. Planned Parenthood Assn., 1954-57, research adv. bd. Nat. Assn. for Retarded Children, 1959-65; cons. Surgeon Gen. USPHS, 1950-54, 60-65, 65—. Recipient McIver award Am. Sociol. Assn., 1960. Social Sci. Research Council fellow, 1941-42; sr. Fulbright Research scholar, Eng., 1957-58. Mem. Am. Assn. U. Profs. (pres. Yale chpt. 1955-59), Sigma Xi, Alpha Kappa Delta (pres. united chpts. 1950-52). Author: Principles of Human Ecology, 1938; Elmtown's Youth, 1949; (with F. C. Redlich), Social Class and Mental Illness, 1958; (with L. H. Rogler) Trapped: Families and Schizophrenia, 1965 (with R. S. Duff) Sickness and Society, 1968. Home: Enoch Dr., Woodbridge, Conn. 06525. Office: Yale U., New Haven 06520.*

HOLLISTER, Leo Edward, Am. physician; b. Cin., Dec. 3, 1920; s. William Baker and Ruth (Appling) H.; B.S., U. Cin., 1941, M.D., 1943. Chief med. services VA Hosp., Palo Alto, Cal., 1953-60, asso. chief staff, 1960—; asst. clin. prof. medicine Stanford Sch. Medicine, 1953—. Exec. com. VA Coop. Studies in Psychiatry, 1955—; cons. NIH, 1960—, A.M.A. Council on Drugs, 1964—, Cal. State Dept. Mental Hygiene, 1965—. Recipient VA Meritorious Service award, 1960; William S. Middleton award, 1966. Diplomate Am. Bd. Internal Medicine. Mem. A.C.P., N.Y. Acad. Scis., Soc. Pharmacology and Exptl. Therapeutics, Am. Fedn. Clin. Research, Coll. Internat. Neuropsychopharmacologicum, Am. Therapeutics Soc., Am. Coll. Clin. Pharmacology and Therapeutics. Research, publs. on delineation of clin. usefulness of variety of psychotherapeutic drugs including antipsychotic, tranquilizing, antidepressant and psychotomimetic drugs, description and elucidation of complications of these drugs; devel. methodology of such agts. Home: 3237 Benton St., Santa Clara, Cal. 95051. Office: VA Hosp., Palo Alto, Cal. 94304.*

HOLLISTER, Ned, Am. zoologist; b. Delavan, Wis., Nov. 26, 1876; s. Kinner Newcomb and Margaret Frances (Tilden) H.; pvt. studies in zoology, 1896-1901; m. Mabel Pfrimmer, Apr. 15, 1908. Field work in vertebrate zoology, U. S. Biol. Survey, in Tex., N.M., Alaska, B.C., Wash., Ore., Cal., Utah, Nev., La. and Ariz., 1902-09; asst. curator of mammals U. S. Nat. Mus., 1910-16; mem. Canadian Alpine Club expdn. to explore Mt. Robson region of B.C. and Alta., 1911; mem. Smithsonian-Harvard expdn. to Altai Mountains, Siberia and Mongolia, 1912; supt. Nat. Zool. Park, Smithsonian Instn., Washington, from 1916. Author: The Birds of Wisconsin, 1903; A Systematic Synopsis of Muskrats, 1911; Mammals of the Philippine Islands, 1912; Mammals of Alpine Club Expedition to Mount Robson, 1913; Philippine Land Mammals in the U. S. National Museum, 1913; A Systematic Account of the Grasshopper Mice, 1914; A Systematic Account of the Prairie-dogs, 1916; East African Mammals in the U. S. National Museum, vol. 1, 1918, vol. 2, 1919. Died Nov. 3, 1924.

HOLLMANN, Wildor, German physician; b. Menden, Sauerland, Jan. 30, 1925; s. Albert and Henriette (Bomnüter) H.; M.D., U. Cologne; m. Ingeborg Cüsters, 1954; children—Helmut, Ulrike. State exam., 1953; promotion, 1954; asst., 1954-55; sci. asst., 1954-58; founder, dir. Inst. for Research on Circulation and Athletic Medicine, 1958—. Mem. internat. research com. UNESCO. Recipient medal of honor Pan-Am. Congress for Sportive Medicine. Mem. German Assn. Athletic Medicine (1st pres.). Author: (monographs) Der Arbeits- und Trainingseinfluss auf Kreislauf und Atmung; Höchst- und Dauerleistungsfähigkeit des Sportiers, also numerous articles. Home: 9 Gelagweg, Brüggen, North Rhine. Office: Lindenburg, Cologne, Germany.

HOLLO, Janos, Hungarian chem. engr.; b. Szentes, Hungary, Aug. 20, 1919; s. Gyula and Margit (Mandl) H.; grad. U. Tech. Scis., Budapest, 1941; Dr.'s degree in engring., U. Tech. Sci., Budapest, 1947; m. Hermina Milch, May 2, 1944; m. 2d, Vera Novak, Dec. 27, 1956; children—Andrew, Michael, Dorottya. With various chem. plants, 1941-46; engr. brewery, Budapest, 1946-48; works mgr. Inc. Breweries, Budapest, 1948-52; dir. Inst. Agr. Chem. Tech., U. Tech. Sci., Budapest, 1952—, prof., 1956—, dean Faculty for Chem. Engring., 1956-57, 63—. Vice pres. Commn. Internationale des Industries Agricoles, 1959—; tech. adviser UNO; pres. ISO Coma Cercals and Pulses. Named Outstanding Worker, Hungarian Food Industry, 1948, 63; commandeur l'Ordre du mérite pour la recherche et l'invention, 1962; recipient award Commn. internationale des Industries Agricoles, 1963. Mem. Hungarian Acad. Scis., Hungarian Sci. Assn. Food Industry (chief sec. 1948—), Internat. Soc. for Fat Research (pres. 1964-66), Deutsche Gesellschaft für Fettwissenschaft, Am. Assn. Cereal Chemists, European Fedn. Chem. Engrs. Food Industries (dir.). Author: Malting and Brewing, 3 vols., 1952, 63; Food Industries, 2 vols., 1952, 54, 57; Automatization in the Food Industry; Biochemical Engineering, 1967; also monographs, numerous articles. Research in biochem. engring., including chemistry and physico-chemistry of starch, fine mechanism of amylolytic and starch synthesizing enzymes, fermentation, molecular distillation and drying of biol. sensitive materials; developed econ. prodn. of fibre-free fodder, new methods in treatment of process-waters and sewages. Home: 9, Guyon Richard, Budapest, Hungary.*

HOLLOMON, John Herbert, Am. metallurgist; b. Norfolk, Va., Mar. 12, 1919; s. John Herbert and Pearl H. (Twiford) H.; B.S., Mass. Inst. Tech., 1940, D.Sc., 1946; m. Miss Wheeler, Aug. 12, 1941; children—Jonathan Bradford, James Martin, Duncan Twiford, Elizabeth Wheeler. With Gen. Electric Co., 1946-62, gen. mgr., gen. engring. lab., Schenectady, 1960-62; asst. sec. for sci. and tech. U. S. Dept. Commerce, 1962-67; pres. U. Okla., 1968—. adj. prof. Rensselaer Poly. Inst., 1950-62. Cons. Pres.'s Sci. Adv. Council, 1962—; mem. Fed. Council Sci. and Tech., 1962—. Recipient Alfred Noble award, 1947. Fellow A.A.A.S.; mem. Am. Phys. Soc., Am. Soc. Metals, Inst. Mining, Metall. and Petroleum Engrs. (Raymond award 1946), N.Y. Acad. Sci., Brit. Inst. Metals (Rosenhain medal 1958). Research, numerous publs. on corrosion of metals, plastic deformation and fracturing of metals, phase transformations in solids and related subjects, phys. metallurgy, ferrous metallurgy, phys. chemistry, sci. and engring. policy, tech. and econ. devel., engring. edn., related subjects. Address: 712 W. Lindsey St., Norman, Okla. 73069.*

HOLLWICH, Fritz, German ophthalmologist; b. Munich, Bavaria, Germany, July 13, 1909; s. Friedrich and Rosa (Wastian) H.; grad. U. Munich, 1935. Faculty, U. Munich, 1948-56; apl. prof. U. Frankfort (Germany), 1956-58; dir. Eye-Clinic, U. Jena (Germany), 1958-64; dir. Eye Clinic, U. Muenster (Westphalia, Germany), 1964—. Mem. Instituto Barraquer (hon.), Acad. Scis. N.Y., Akademie der Naturforscher und Arzte Leopoldina, Mitglied des wissenschaftlichen Beirates der Deutschen Gesellschaft für Plastische und Wiederherstellungschirurgie, Mitglied des wissenschaftlichen Beirates der Deutschen Neurovegetativen Gesellschaft, ophthal. socs. of Austria, France, Germany, Italy, Switzerland. Author: Einführung in die Augenheilkunde, 1966; Schielen (Strabismus), 1960; Ophthalmologen-Verzeichnis-Bio- und Bibliographie, 1964; Lähmungsschielen (Strabismus paralyticus), 1966; also numerous articles. Research on influence of light via eye on organism, eye and gen. diseases. Home: 49, Sentruper Höhe, Muenster. Office: 15, Westring, Muenster, Westphalia 44, Germany.*

HOLLYER, Robert Nelson, Jr., Am. physicist; b. Detroit, Sept. 4, 1919; s. Robert Nelson and Jane (Wixon) H.; B.S. in Physics, Wayne U., 1942; M.S., U. Mich., 1948, Ph.D., 1953; m. Phyllis E. Priest, Sept. 6, 1946; children—Phyllis Jane, Thomas Robert. Sr. staff Applied Physics Lab., Johns Hopkins, 1953-56; sr. physicist Research Labs., Gen. Motors Corp., Warren, Mich., 1956—, asst. dept. head physics dept., 1963—. Mem. Am. Phys. Soc., Philos. Soc. Washington, A.A.A.S., Sigma Xi, Sigma Pi Sigma. Research, publs. on compressible boundary layer, shock tube performance and expts.; pioneered observation and study luminous shock waves. Home: 3628 Middlebury Lane, Birmingham, Mich. 48010. Office: 12 Mile and Mound Rds., Warren, Mich. 48090.*

HOLMAN, Ralph Theodore, Am. biochemist; b. Mpls., Mar. 4, 1918; s. Alfred Theodore and May (Nilson) H.; A.A., Bethel Jr. Coll., 1937; B.S., U. Minn., 1939; M.S., Rutgers U., 1941; Ph.D., U. Minn., 1944; m. Karla Calais, Mar. 26, 1943; 1 son, Nils Teodor Calais. Instr., U. Minn., 1944-46, asso. prof., 1951—, prof., 1958-59, prof. physiol. chemistry Hormel Inst., 1956—; asso. prof. Tex. A. and M. U., 1948-51. NRC fellow Med. Nobel Inst., Stockholm, Sweden, 1946-47; Am. Scandinavian Found. fellow U. Uppsala (Sweden), 1947; spl. fellow NIH, U. Gothenburg (Sweden), 1962. Mem. Am. Chem. Soc., Am. Soc. Biol. Chemists, Am. Inst. Nutrition (Borden award 1966), Am. Oil Chemistry Soc., Soc. for Exptl. Biology and Medicine, Sigma Xi. Editor: (with W. O. Lundber and T. Malkin) Progress in the Chemistry of Fats and other Lipids, vols. 1-6, 1951-63, vols. 7-9, 1963—. Research, numerous publs. on spectrophotometric studies fat oxidation, isolation and characterization lipoxidase, displacement chromatography lipids, biochem. characterization fatty acid deficiency; established nutritional requirements essential fatty acids; research on metabolism polyunsaturated fatty acids, near-infrared spectra lipids, mass spectrometry lipids; developed methods for lipid analysis. Home: 1403 2d Av. S.W., Austin, Minn. 55912.*

HOLMBERG, Allan Richard, Am. anthropologist; b. Renville, Minn., Oct. 15, 1909; s. Axel Rudolph and Anna Catherine (Carlson) H.; A.B., U. Minn., 1935; student U. Chgo., 1937-38; Ph.D. (Sterling fellow 1945-46), Yale, 1947; m. Laura Mary Hines, May 7, 1945; children—Anna Katherine, David Hines, Eric Allan. Fellow Social Sci. Research Council, 1940-41; econ. analyst Am. embassy, La Paz, Bolivia, 1942; field technician Rubber Devel. Corp., Bolivia, 1942-45; cultural anthropologist Inst. Social Anthropology, Smithsonian Instn., 1946-48; asst. prof. anthropology Cornell U., Ithaca, N.Y., 1948-51, asso. prof., then prof., 1951—, chmn. dept., 1962-67, dir. Cornell-Peru Project; prof. anthropology U. San Marcos, Lima, Peru, 1946-48; anthropol. field trips to Mexico, Peru, Bolivia. Fellow Am. Anthropol. Assn.; mem. Soc. Applied Anthropology, Sigma Xi. Author: Nomads of the Long Bow, 1950; co-author: Social Change in Latin America Today, 1960. Contbr. articles profl. publs. Research on highland communities of S.Am.; application of method of participant intervention as new technique for community devel. in S.Am. Died Oct. 13, 1966.

HOLMBERG, Bengt Rudolf, Swedish physicist; b. Lund, Sweden, Jan. 12, 1919; s. N. Rudolf and Hanna (Bengtsson) H.; M.S., U. Lund, 1941, licentiate, 1945, D.Sc., 1948; m. Elsa Gertrud Sundgren, May 14, 1955; children—M. Tuve R., Elna M.B. Docent U. Lund, 1949-56; scientist Swedish Research Inst. Nat. Def., Stockholm, 1956-58; faculty Chalmers U. Tech., Gothenburg, Sweden, 1958—, prof. mechanics, 1960—, dean faculty engring. physics, 1962—. Mem. Swedish Nat. Com. Mechanics, Am. Math. Soc., Am. Phys. Soc. Author: Lecture Books in Rational Mechanics and Hydrodynamics; Research on neutron-proton and electron scattering problems, determination of scattering potential from asymptotic phases, quantum-mech. 3 body-problem, nuclear structure, theory, hydrodynamical implosion problems. Home: 53 Blidvädersgatan, Göteborg H, Sweden.*

HOLMBERG, Erik Bertil, Swedish astronomer; b. Skillingaryd, Sweden, Nov. 13, 1908; s. Per Malcolm and Anna (Nilsson) H.; Dr.phil., Lund (Sweden) U., 1938; m. Märta Asdahl, Mar. 29, 1947; 1 dau., Asa Berenike. Faculty, Lund U. Obs., 1938-59, asso. prof., 1951-59; prof. astronomy Uppsala (Sweden) U. 1959—, dir. Obs., 1959—. Fellow Center for Advanced Studies, Wesleyan U., Middletown, Conn., 1963; guest investigator Mt. Wilson Obs., also Palomar Obs., Pasadena, Cal., 1940-41, 47, 55. Mem. Royal Physiographic Soc. Lund, Royal Soc. Scis. Uppsala, Royal Acad. Scis. Stockholm, Am., Swedish (chmn. 1964—) astron. socs. Research and publs. on investigations of galaxies, especially photometry of galaxies, double galaxies, distances, dimensions, luminosities, masses, space motions, and evolution of galaxies. Address: Astron. Obs., Uppsala, Sweden.*

HOLMBOE, Carl Fredrik, Norwegian engr.; b. Oslo, Norway, Aug. 25, 1862; s. Jens Anton and Marie (Aarreberg) H.; grad. Technische Hochschule, Berlin, Germany, 1902; D.Sc., 1924; m. Clara Ragnhild Willemsen, 1905; 3 children. Planning and research engr., Calvert & Co., Göteborg, Sweden, from 1905; cons. engr., Oslo, Norway, 1910-15, for Lord Leverhulme, Port Sunlight, Eng., 1915-21; mng. dir., de Nordiske Fabriker De-No-Fa A/S, Oslo, Norway, 1921—. Bd. dirs. Norsk Teknisk Mus. Recipient Fridtjof Nansen prize. Mem. Norwegian Acad. Sci. (v.p. sect. math. and natural sci. 1934, pres. 1935), Royal Inst. Gt. Britain. Author: Die Heissdampp Schiffsmaschinen en Norsk Innsats: Forskning og Teknikk, 1947; Ingenior ser sig Tilbake, 1948. Contbr. articles to sci. jours. Developed new process for catalytic hydrogenation of oils. Home: Brantenborgveien 3, Stemdal, V. Aker, Norway.

HOLMES, Arthur, Brit. geologist; b. Hebburn/Tyne, Eng., Jan. 14, 1890; s. David and Emily (Dickinson) H.; ed. Imperial Coll. Sci. and Tech., U. London (Eng.); B.Sc., 1909; D.Sc., 1917; m. Margaret Howe, 1914 (dec. 1938); 1 son; m. 2d, Doris Livesy Reynolds, June 1939. Geologist with expdn. to Mozambique, 1911-12; demonstrator in geology Imperial Coll., 1912-20; chief geologist Yomah Oil Co., Burma, 1920-23; prof. geology U. Durham (Eng.), 1924-43, dir. sci. lab., 1940-43; regius prof. geology, mineralogy U. Edinburgh (Scotland), dir. Grant Inst. Geology, Edinburgh, from 1943. Exchange prof. U. Basel (Switzerland), 1931; Lowell lectr. Boston, 1932, mem. U. S. NRC com. on age of earth. Recipient Murchison medal Geol. Soc. London, Vetlesen prize, 1964. Fellow Royal Socs. (London, Cornwall, Yorkshire, Edinburgh), Mineral Soc., Royal Geog. Soc., Mineral. Soc. Am.; asso. Royal Coll. Sci.; mem. geol. socs. Am., Belgium (corr.), Royal Irish Acad., Am. Acad. Arts (hon.). Author: The Age of the Earth, 1912; The Nomenclature of Petrology, 1920; Petrographic Methods and Calculations, 1927; Principles of Physical Geology, 1944; Radioactivity and Geological Time; Petrology of the Bufumbia Volcanoes, 1937; Principles of Physical Geology, 1944. Contbr. pamphlets on geology. Investigated ages of rocks by measuring their radio-active constituents; proposed 1st quantitative time scale for geology, 1913. Died Sept. 20, 1965.

HOLMES, Francis Oliver, Am. virologist; b. Cambridge, Mass., Nov. 26, 1899; s. Frank William and Fannie Elizabeth (Greenleaf) H.; B.S., Mass. Inst. Tech., 1921; Sc.D., Johns Hopkins, 1925; m. Ruth Deem, Aug. 30, 1924; 1 son, Francis William. Asso. mem. Rockefeller Inst. Med. Research, Princeton, N.J, 1932-48; asso. prof. Rockefeller U., N.Y.C., 1948-65; vis. prof. dept. botany U. Ill., Urbana, 1966-67. Vis. investigator U. P.R. Agrl. Expt. Sta., 1944, 46, 48, 58, 59, U. Hawaii, 1946; cons. virology UN FAO, Philippines, 1961-62, India, 1964. Fellow Am., Indian phytopath. socs. Author: Handbook of Phytopathogenic Viruses, 1939; The Filterable Viruses, 1948. Research, numerous publs. on inheritance of resistances to viral diseases, cure of several viral diseases in vegetatively propagated plants, nectar secretion in plants and responses of honey bees; developed methods for quantitative study of viruses in plants by counts of primary lesions on inoculated leaves; introduced classification of filterable viruses, using Latin binomials. Home: Craney Hill Rd., Henniker, N.H. 03242.

HOLMES, Harry Nicholls, Am. chemist; b. Lawrence County, Pa., July 10, 1879; s. John Pattison and Eliza (Nicholls) H.; B.S., Westminster Coll., New Wilmington, Pa., 1899, M.S., 1907; LL.D., 1941; Ph.D., Johns Hopkins, 1907; m. Mary V. Shiveley, July 15, 1909; children—Charles Shiveley, Richard Remsen. Prof. chem. Earlham Coll., Richmond, Ind., 1907-14; Oberlin Coll., 1914-45. Mem. Nat. Research Council, 1923-29 (chmn. sub-com. on chem. of colloids 1919-25); cons. NDRC, 1942. Recipient Kendall Award in colloid chemistry, 1954, James Flack Norris Award in teaching, 1955; Oberlin Alumni medal, 1945; gold medal American Institute of Chemists, 1951; Westminster Alumni medal, 1957. Fellow A.A.A.S., American Inst. Chemistry; mem. Am. Chem. Soc. (councilor at large, 1926-29, 1930-33, 1938-41, pres. elect, 1941, pres., 1942), Gamma Alpha, Sigma Xi, Alpha Chi Sigma, Phi Lambda Upsilon; hon. mem. Met. Chem. Inst. of South Africa, 1942. Author: Outline of Qualitative Analysis, 1908-45; Laboratory Manual of General Chemistry, 1909, 30, 37, 49; General Chem., 1921, 30, 36, 41, 49; Lab. Manual of Colloid Chemistry, 1921, 28, 34; Bibliography Colloid Chemistry, 1923; Elements of Chemistry (with Louis W. Mattern), 1927; Introductory Coll. Chem. 1925, 31, 39, 46, 51; Out of the Test Tube, 1934, 37, 41, 43, 57; Have You Had Your Vitamins, 1938; Strategic Materials and National Defense, 1942; also articles giving results of original chem. research. Known for work in colloidal chemistry; isolated (with Ruth E. Corbet) crystalline vitamin A concentration, 1937; isolated (with others) an alcohol from bone marrow which increases white blood corpuscles; introduced lab. technique of chronatography in U. S.; used vitamin C to treat shock in World War I, also in treatment allergies. Died July 1, 1958.

HOLMES, James, Am. chemist; b. Balt., Apr. 29, 1916; s. James B. and Rosalie (Rutherfoord) H.; B.A. in Chemistry, U. Va., 1936; M.S., Ga. Sch. Tech., 1938; Ph.D., Rice U., 1941; m. Helena T. Johnston, June 24, 1944; children—James T., Julia G., Rosalie R. Staff, NDRC project Princeton, 1941-42, Johns Hopkins, 1941-44, chief investigator, 1944; chemist Rohm & Haas Co., Phila., 1945; group leader Houdry Process Co., Linwood, Pa., 1946-48; research chemist E. I. DuPont de Nemours & Co., Wilmington, Del., 1949—. Mem. Am. Chem. Soc., Sigma Xi. Contbr. articles to tech. jours. Characterization catalysts and catalysts supports by X-ray diffraction, nitrogen adsorption, chemisorption and mercury porosimeter evaluation, high polymers by X-ray diffraction; alteration treatments gas mask charcoals. Home: 1806 Shipley Rd. Office: Plastics Dept., E. I. DuPont de Nemours & Co., Wilmington, Del. 19803.*

HOLMES, John Richard, Am. physicist, educator; b. Chula Vista, Cal., Sept. 24, 1917; s. Robert and Mary Elizabeth (Burns) H.; A.B. in Physics, U. Cal. at Berkeley, 1938, M.A., 1941, Ph.D., 1942; m. Louise Murphy, 1951; children—Susan Diana, Ronald John, Sandra Kathleen. With radiation lab. U. Cal. at Berkeley, 1942-45; faculty physics U. So. Cal., Los Angeles, 1945-63, prof., 1954-63, chairman of department of physics, 1956-62; prof., chmn. physics dept. U. Hawaii, Honolulu, 1963—; Fulbright lectr. U. Madrid, Spain, 1962-63. Cons. Autonetics Corp., Anaheim, Cal. Douglas Aircraft, Santa Monica, Cal., Electro-Optical Systems, Pasadena, Cal.; lectr. Edwards AFB, Loyola U., Los Angeles. Fellow Am. Phys. Soc., Optical Soc. Am.; mem. A.A.A.S. Research in spectroscopy and optics, especially isotopic shift in spectra, forbidden lines, lasers. Home: 41-543 Kalanianaole Hy., Waimanalo, Hawaii. Office: 315 PSB, U. Hawaii, Honolulu.*

HOLMES, Major Edward, Am. ceramic engr.; b. La Grange, Ky., Jan. 8, 1882; s. Jesse Munroe and Laura N. (Maddox) H.; A.B., Ind. U., 1908; M.S., Cornell U., 1910, Ph.D., 1920; m. Florence Juanita Garr, Apr. 27, 1922. Tchr. and high sch. prin., Kempton, Ind., 1901-06; devel. engr. Nat. Carbon Co., 1910-19; mgr. chemistry dept. and acting gen. mgr., Nat. Lime Assn., 1919-22; devel. engr. U. S. Gypsum Co., 1922-23, Dolomite, Inc., 1923-26; prof. and head dept. ceramic engring. and dir. Mo. Clay Testing and Research Lab., U. Mo., 1926-32; dean State Coll. of Ceramics, Alfred U., from 1932. Fellow Am. Ceramic Soc. (v.p. 1942-43); mem. Am. Soc. for Testing Materials, N.Y. Ceramic Industries Assn. (sec.-treas.), Sigma Xi, Alpha Chi Sigma, Contbr. manuals and scientific articles. Invented and improved electric dry cells, quick setting lime plaster and stable dolomite clinker and refractories of various kinds. Died May 2, 1946.

HOLMES, Oliver Wendell, Am. physician, author; b. Cambridge, Mass., Aug. 29, 1809; s. Rev. Abiel and Sarah (Wendell) H.; grad. Harvard, 1829, M.D., 1836; hon. degrees, Edinburgh, Harvard, Cambridge (Eng.) U, 1887, Oxford (Eng.) U., 1887; m. Amelia Lee Jackson, June 15, 1840; children—Amelia (Mrs. Turner Sargent), Edward Jackson, Chief Justice Oliver Wendell. Wrote poem Old Ironsides which caused preservation of U.S.S. Constitution, 1830; winner Boylston prize for med. essay, 1836; prof. anatomy Dartmouth, 1838-40; Parkman prof. anatomy and physiology Harvard Med. Sch., 1847-82, dean, 1847-53, prof. emeritus, 1882-94. Elected to Hall of Fame for Great Americans, 1910. Established Atlantic Monthly, 1857, contbd. essays noted for humor and common sense later published as The Autocrat of the Breakfast-table, 1858, The Professor at the Breakfast-table, 1860, The Poet at the Breakfast-table, 1872, Pages from an Old Volume of Life, 1883, Over the Teacups, 1891; mem. Saturday Club (also included James Russell Lowell, Louis Agassiz, William Wadsworth Longfellow, Ralph Waldo Emerson, others); poems The Chambered Nautilus, The Deacon's Masterpiece or, The Wonderful One-Hoss Shay included in Poems, 1836; other works include Songs in Many Keys, 1862; Songs of Many Seasons, 1875; The Iron Gate and Other Poems, 1880; Before the Curfew and Other Poems, 1887. Author: novels including Elsie Venner, 1861; The Guardian Angel, 1867; A Mortal Antipathy, 1885 (novels were anti-Calvinist, dealt with med. problems, were ahead of their time in understanding of psychology); sci. writings include Lectures on Homeopathy and Its Kindred Delusions, 1842; The Contagiousness of Puerperal Fever, 1843; Currents and Counter Currents in Medical Science, 1861; Border Lines in Some Provinces of Medical Science, 1862; Medical Essays (reissue), 1883. Promoted exploratory surgery as means of diagnosis; 1st to recognize contagiousness of puerperal fever, 1843, recommended aseptic methods to prevent it; introduced aseptic techniques in obstetrics and surgery; devised term anesthesia to describe process of using ether to induce unconsciousness. Died Boston, Oct. 7, 1894.

HOLMES, Ralph Jerome, Am. geologist; b. N.Y.C., Oct. 31, 1906; s. Charles Jerome and Blanche (Holley) H.; B.S., Columbia, 1933, Ph.D., 1946; m. Jane W. Hegeman, June 12, 1954. Faculty, Columbia, N.Y.C., 1936—, prof. geology, 1962—. Sci. adviser Gemological Inst. Am., 1946—; geol. adviser Venezuelan Ministry Edn., 1963; participant various geol. missions UN, Dept. State, 1953, 56-58. Fellow Mineral. Soc. Am. (sec., councillor 1966—), Geol. Soc. Am.; mem. Soc. Econ. Geologists, Am. Inst. Mineral. and Metall. Engrs., A.A.A.S., Mineral. Soc. Gt. Britain, Am. Crystallographic Assn., Assn. Geology Tchrs., Soc. Francaise de Minéralogie et de Cristallographie, Geochem. Soc., Am. Geophys. Union, Rochester Acad. Sci. (hon.). Author: Reference Clay Localities of Europe, 1950; Mineral Deposits of Somalia, 1954. Field and lab. investigations of minerals to clarify mineral relations and mode of origin; x-ray studies of sulphide and arsenide mineral groups; studies of gem materials to determine methods of distinguishing between real, synthetic, and imitation gems; investigation of distbn. and econ. aspects of mineral deposits of Iran and Somalia; systematic study of turquoise deposits of U. S. Home: 1237 Old Nassau Rd., Jamesburg, N.Y. 08831. Office: Dept. Geology, Columbia U., N.Y.C. 10027.*

HOLMES, Robert Lewis, Brit. anatomist; b. Wakefield, U.K., Feb. 3, 1926; s. Robert Smith and Kathleen (Bedford) H.; B.Sc., U. Leeds (Eng.), 1947, M.Sc., 1948, M.B., Ch.B., 1951, Ph.D., 1957; D.Sc., U. Birmingham (Eng.), 1965. House surgeon Leeds Gen. Infirmary, 1951; lectr. anatomy U. Leeds, 1952-58, prof., 1965—; faculty U. Birmingham (Eng.), 1958-65, reader neuroanatomy, 1964-65. Fellow Zool. Soc. London, Royal Micros. Soc.; mem. Anat. Soc. Gt. Britain and Ireland, Physiol. Soc., Soc. for Exptl. Biology, Soc. for Endocrinology. Author: Living Tissues: An Introduction to Functional Histology, 1965; also numerous articles. Research on structure and function of mammalian nervous and endocrine systems especially pituitary gland; electron microscopy of mammalian neurosecretory system. Home: Boyle Hall, West Ardsley, nr. Wakefield, Yorkshire, Office: Sch. Medicine, Leeds 2, Yorkshire, U.K.*

HOLMES, Samuel Jackson, Am. zoologist; b. Henry, Ill., Mar. 7, 1868; s. Joseph and Avis Folger (Taber) H.; B.S., U. Cal., 1893, M.S., 1894; Ph.D., U. Chgo., 1897; D.Sc., U. Mich., 1948; m. Cecelia Warfield Skinner, 1909; 5 children. Instr. zoology U. Mich., 1899-1905; asst. prof. U. Wis., 1905-12; asso. prof. U. Cal. at Berkeley, 1912-17, prof., 1917-39, emeritus prof., 1939-64. Instr. Marine Biology Lab., Woods Hole. Mem. A.A.A.S., Soc. Exptl. Biology, Am. Soc. Naturalists (pres. 1931), Am. Acad. Arts and Scis., Eugenics Soc. Am. (pres.), Am. Soc. Zoologists (pres. 1927), Am. Genetics Assn. Author: The Evolution of Animal Intelligence, 1911; Studies in Animal Behavior, 1916; Studies in Evolution and Eugenics, 1923; A Bibliography of Eugenics, 1924; Life and Evolution, 1926; The Eugenic Predicament, 1933; The Negro's Struggle for Survival, 1937; Life and Morale, 1948. Research on animal behavior, Crustacea taxonomy, eugenics and population, evolution, heredity, comparative psychology. Died 1964.

HOLMES, Thomas Hall, III, Am. physician; b. Goldsboro, N.C., Sept. 20, 1918; s. Thomas Hall and Elizabeth (Stephenson) H.; A.B., U. N.C., 1939; M.D., Cornell U., 1943; m. Janet Lawrence, Dec. 29, 1942; children—Thomas Stephenson, Janet, Eleanor Scott, Elizabeth Lawrence. Faculty, U. Wash. Sch. Medicine, Seattle, 1949—, prof. psychiatry, 1958—; staff Univ., King County, Seattle VA hosps.; cons. Firland Sanatorium, Bremerton Naval Hosp. Fellow Am. Psychiat. Assn. (Hofheimer prize 1953); mem. A.A.A.S., Am. Fedn. Clin. Research, A.M.A., Am. Psychosomatic Soc., Am. Sociol. Soc., Assn. Am. Med. Colls., Am. Coll. Psychiatrists, No. Pacific Soc. Neurology and Psychiatry, Am. Thoracic Soc., Assn. Research Nervous and Mental Diseases. Author: (with H. Goodell, S. Wolf, H. G. Wolff) The Nose: An Experimental Study of Reactions within the Nose of Human Subjects During Varying Life Experiences, 1950; also numerous articles. Research on relationship psychol. and sociol. phenomena to natural history disease. Office: Dept. Psychiatry, U. Wash., Seattle 98105.*

HOLMES, William Henry, Am. anthropologist, archaeologist; b. Harrison County, O., Dec. 1, 1846; s. Joseph and Mary (Heberling) H.; grad. McNeely Normal Coll., 1870, D.Sc., George Washington U., 1918; m. Kate Clifton Osgood, Oct. 1883 (died 1925); children—Osgood, William Heberling. Normal sch. teacher, 1871-72; asst., 1872-80, geologist, 1880-89, U. S. Geol. Survey; curator dept. aboriginal pottery, U. S. Nat. Museum, 1882-93; archaeologist Bur. Am. Ethnology, in charge of explorations, 1889-98; curator anthropology, Field Mus. of Natural History, and prof. anthropic geology, U. of Chicago, 1894-97; head curator dept. anthropology, 1898-1902; curator prehistoric archaeology, 1903, and Nat. Gallery of Art, 1907, U. S. Nat. Mus.; chief Bur. Am. Ethnology, Oct. 1902-09. Head curator, anthropology, U. S. Nat. Mus., 1910-20; curator, Nat. Gallery of Art, 1910-20; dir., 1920—. U. S. del. 1st Pan-Am. Scientific Congress, Santiago, Chile, 1908-09; chmn. organizing com. Internat. Congress of Americanists, 1915; chmn. section of anthropology, 2d Pan-American Congress, 1915. Chmn. mng. com. School of Am. Archaeology; pres. Washington Acad. Sciences, 1917-18. Chmn. com. on anthropology of Nat. Research Council, 1917. Pottery of the Ancient Pubelos, 1886; Archaeological Studies Among the Ancient Cities of Mexico, 1895; Stone Implements of the Potomac-Chesapeake Tidewater Province, 1897; Hand book of Aboriginal American Antiquities, 1918. Authority on primitive art and artifacts of N.Am. Died Apr. 20, 1933.

HOLMGREN, Alarik Frithiof, Swedish physiologist; b. 1831; research on ophthalmology, especially color blindness; demonstrated retinal action current, 1865; introduced various colored wool skeins for testing color vision, 1874; advocated gymnastics done without apparatus. Died 1897.

HOLMGREN, Albert, physicist; b. Sweden, 1824; prof., Marienberg, Lund, Sweden; research on influence of temperature on magnetism. Died 1905.

HOLMGREN, Harry D., Am. physicist; b. Mpls., Apr. 21, 1928; s. Harry W. and Myrtle (Dahl) H.; B.S., U. Minn., 1949, M.S., 1950, Ph.D., 1954; m. Phyllis Cleveland, Oct. 6, 1949; children—Diane, Bruce, Cheryl, Cynthia. Physicist, U. S. Naval Research Lab., Washington, 1954-61; faculty U. Md., College Park, 1961—, prof. physics, astronomy, dir. Cyclotron Project, 1965—. Recipient Edward O. Hulburt award U. S. Naval Research Lab.; Arthur S. Flemming award Jr. C. of C. Washington. Mem. Am. Phys. Soc., Phi Beta Kappa, Sigma Psi. Research, numerous publs. on exptl. nuclear physics, basic structure of nucleus of atoms and processes by which nuclear particles interact with atomic nucleus, theoretical studies of nuclear interaction processes, astrophys. studies related to prodn. of stellar energy, nuclear instrumentation, accelerators for nuclear research, effects of nuclear weapons. Home: 4681 Leslie Av., Washington 20031. Office: U. Md., College Park, Md. 20740.*

HÖLN, Helmut, German physicist; b. Mannheim, Feb. 10, 1903; s. Theodor and Mathilde (Haeberlein) H.; Ph.D. in physics; m. Guda von Wolff; children—Annette, Brigitte, Cordelia. Prof., U. Munich, 1926, Technische Hochschule, Stuttgart, 1933; asso. prof. U. Erlangen, 1940; full prof. U. Freiburg, 1949. Mem. German Leopoldina Acad. Physics. Author: Mecanique du Champ des Corpuscules Élémentaires; Problème d'Intensité du Spectre Lumineux; Théorie de la Dispersion des Rayons Ronttgen; Théorie Géneral de la Relativité. Home: 117 Zasiusstrasse, Freiburg 1. Office: 3 Hermann-Herderstrasse, Freiburg 1, Breslau, Germany.

HOLPER, Jacob Charles, Am. microbiologist; b. Bosworth, Mo., May 20, 1924; s. Amil George and Lou E. (Kemble) H.; B.S. in Bacteriology, Kan. U., 1949, M.A. in Virology, 1951; Ph.D., U. Mich., 1954; m. Patricia A. Johnson, Dec. 26, 1946; children—Ann Holper, David Holper. Virologist, Abbott Labs., North Chicago, Ill., 1954-57, sect. head, 1957-59, head infectious disease research, 1959-64, dir. infectious disease research div., 1964—. Fellow Am. Acad. Microbiology; mem. Soc. for Exptl. Biology and Medicine, Tissue Culture Assn., Am. Soc. for Microbiology. Research, publs. on virus propagation, immunology, tissue culture growth, infectious diseases, respiratory viruses, antibiotics, viral induced cancer, animal health problems. Home: 2206 Hawthorne Lane, Waukegan, Ill. 60064. Office: 14th and Sheridan Rd., North Chicago, Ill. 60064.

HOLSER, William Thomas, Am. mineralogist; b. Bakersfield, Cal., July 4, 1920; s. Courtney Talbot and Margaret (Fraser) H.; B.Sc., Cal. Inst. Tech., 1942, M.Sc., 1946; Ph.D., Columbia, 1948; m. Mary Ann Harris, Dec. 23, 1954; children—Thomas Dana, Alec Stuart, Margaret Ann. Lectr. geology Columbia, 1946-48; asst. prof. geology Cornell U., 1948-54; geologist U. S. Geol. Survey, 1948-54; research geochemist Battelle Meml. Inst., Columbus, O., 1954-55; research asso. Inst. Geophysics U. Cal. Los Angeles, 1955-58; sr. research asso. Chevron Oil Field Research

Co., La Habra, Cal., 1958—. Vis. prof. dept. geology and mineralogy U. Mich. 1962. Fellow Mineral. Soc. Am., Geol. Soc. Am., Am. Geophys. Union; mem. Geochem. Soc., Am. Crystallographic Assn. Editor Am. Mineralogist, 1966—; vol. editor Internat. Tables X-ray Crystallography, 1964-66. Studies, publs. contbg. to understanding geochem. distbn. in space and time of such elements as bromine, chlorine, sulfur, oxygen, beryllium, iron; theoretical analysis of problems in crystallography; generalized symmetry, twinning, polymorphism, electromagnetic adsorption. Home: 1041 Citrus Dr. Office: Box 446, La Habra, Cal. 90631.*

HOLST, Axel, German physician; b. 1861; produced scurvy (which he cured with cabbage) in guinea pigs, thus suggesting their value in study of antiscorbutic substances, circa 1907. Died 1931.

HOLSTIUS, Elvin Albert, Am. pharm. chemist; b. Woonsocket, R.I., Feb. 9, 1921; s. Albert R. and Mary (Elpe) H.; B.S., U. R.I. Coll. Pharmacy, 1943; M.S., Purdue U., 1949, Ph.D., 1950; m. Eleanor C. Andeen, Mar. 28, 1942; children—Faith E. (Mrs. Daniel Carlson), Mark E. Research pharmacist Burrough Wellcome, Tuckahoe, N.Y., 1945-47; sr. chemist Merck & Co., Inc., Rahway, N.J., 1950-53; asso. prof. U. Kansas City (Mo.), 1953-56, dir. pharm. research, 1954-56; asst. head tech. prodn. Geigy Pharms. div. Geigy Chem. Corp., Ardsley, N.Y., 1956-60, dir. pharm. devel. Geigy Research div., 1960-67; tech. dir. Endo Labs., Inc., Garden City, N.Y., 1967—. Mem. A.A.A.S., Am. Pharm. Assn., Am. Chem. Soc., Sigma Xi, Kappa Psi, Rho Chi. Patentee in field. Research in pharm. and vet. dosage forms, medicinal chemistry. Home: 15 Primrose Av. W., White Plains, N.Y. 10607. Office: Endo Labs., Inc., 1000 Stewart Av., Garden City, N.Y.*

HOLT, Ethan Cleddy, Amer. agronomist; b. Brilliant, Ala., Feb. 6, 1921; s. Austin F. and Anna (Ingle) H. B.S., Auburn U., 1943; M.S., Purdue U., 1948, Ph.D., 1950; m. Jean Jordan, June 18, 1944; children—Janet Marie, Robert A. Faculty, Tex. Agrl. Expt. Sta., College Station, 1948-57; prof. soil and crop scis. Tex. A. and M. U., College Station, 1957—. Recipient Merit award Tex. Turfgrass Assn., 1955. Fellow Am. Soc. Agronomy (v.p., chmn. crops So. br. 1966); mem. A.A.A.S., Crop Sci. Soc. Am., Sigma Xi, Gamma Sigma Delta. Contributed to basic knowledge of method of reproduction and mechanisms of reprodn. in Paspalum dilatatum and Pennisetum ciliare; developed improved forage grasses, agronomic prodn. and mgmt. practices for pastures in Southwestern U. S., also weed control practices. Home: 1110 Ashburn St., College Station, Tex. 77840.*

HOLT, L(uther) Emmett, Am. pediatrician; b. Webster, N.Y., Mar. 4, 1855; s. Horace and Sabrah (Curtice) H.; A.B. U. Rochester, 1875, A.M., 1878; M.D., College of Physicians and Surgeons (Columbia U.), 1880; (LL.D., Rochester, 1902; Sc.D., Columbia U., 1904, Brown U., 1914); m. Linda Mairs, Apr. 26, 1886. Prof. diseases of children, N.Y. Polyclinic, 1890-1901; clin. prof. diseases of children Coll. Phys. and Surg., from 1901; physician in chief Babies Hosp. Mem. bd. dirs. and sec. Rockefeller Inst. Med. Research; trustee U. Rochester. Author: Care and Feeding of Children, 1894, 1902, 18; Diseases of Infancy and Childhood, 1896, 1902, 18. Introduced (with W. H. Howell) term heparin, 1918; influential in separating pediatrics from gen. medicine. Died Jan. 14, 1924.

HOLT, Luther Emmett, Jr., Am. physician; b. N.Y.C., Mar. 20, 1895; s. Luther Emmett and Linda (Mairs) H.; A.B., Harvard, 1916; M.D., Johns Hopkins, 1920; m. Olivia Cauldwell, June 17, 1921; children—Neil Maclean, Arnold Rich, Linda M. (Mrs. Peter H. Holz). Acad. pediatrics Johns Hopkins Hosp., 1922-44, N.Y. U., N.Y.C., 1944—; prof., chmn. dept. pediatrics N.Y. U., 1944-60, prof. pediatrics, 1944—. Mem. protein adv. group WHO, 1957-62; mem. expert com. protein requirements FAO, 1955; mem. expert com. protein requirements WHO, FAO, 1963. Recipient Modern Medicine award, 1959; Mannerheim medal (Finland), 1948; medal Charles U. (Prague), 1946; Order White Lion (Czechoslovakia), 1946. Mem. Am. Soc. Biol. Chemists, Am. Inst. Nutrition (Osborne-Mendel award 1964), Am. Soc. Clin. Nutrition, Assn. Am. Physicians, Am. Soc. Clin. Investigation, Harvey Soc., Am. Pediatric Soc. (Howland award 1966), Soc. Pediatric Research, Am. Acad. Pediatrics (Borden award 1955), N.Y. Acad. Medicine (Harlow Brooks medal 1957). Author: Care and Feeding of Children, 1926; (with R. McIntosh) Diseases of Infancy and Childhood, 1933, Pediatrics, 1953; (with R. L. Duffus) Pioneer of a Children's Century, 1940; The Good Housekeeping Book of Infant and Child Care, 1956. Studies, publs. on mineral, fat, protein and amino acid metabolism, nutritional requirements of children, diarrhea, allergy. Home: 415 E. 52d St., N.Y.C. 10022. Office: 550 1st Av., N.Y.C. 10016.*

HOLT, Robert Rutherford, Am. clin. psychologist; b. Jacksonville, Fla., Dec. 27, 1917; s. Walter John and Grace (Hilditch) Watson; B.A. with highest honors, Princeton, 1939; M.A., Harvard, 1941, Ph.D., 1944; m. Louisa C. Pinkham, Feb. 1944 (div. Feb. 1952); children—Dorothy, Catherine; m. 2d, Crusa Adelman, Dec. 27, 1957 (dec. Aug. 1959); m. 3d, Joan Esterowitz, Aug. 2, 1963; children—Daniel, Michael. Study dir. div. program surveys Bur. Agrl. Econs., Dept. Agr., Washington, 1944-46; clin. psychologist VA Hosp., Topeka, 1946-49; with Menninger Found., Topeka, 1947-53, sr. psychologist, 1948-53, dir. psychol. staff, 1951-53; faculty N.Y. U., 1953—, prof. psychology, 1958—, dir. Research Center for Mental Health, 1953-63, co-dir., 1963—; clin. asst. prof. U. Kan., 1946-50. Mem. rev. panel NIH Mental Health Fellowships, 1963-65. Recipient Research Career award Nat. Inst. Mental Health, 1962—; fellow Center for Advanced Study in the Behavioral Scis., 1960-61. Fellow Am. Psychol. Assn. (past div. pres.), A.A.A.S.; mem. Council on Bibliography (pres. 1965—), N.Y. State Psychol. Assn., Psychonomic Soc., Fedn. Am. Scientists, Soc. for Psychol. Study Social Issues, Phi Beta Kappa, Sigma Xi. Author: (with others) Developments in the Rorschach Technique, vol. I, 1954; (with Luborsky) Personality Patterns of Psychiatrists, 1958; also numerous articles, book revs. Editor: Motives and Thought, 1967; Rapaport's Diagnostic Psychological Testing, 1968. Developed clin. methods assessing personality for selection physicians for psychiat. tng., related logical analysis inferential process clin. prediction, operational technique measuring manifestation central psychoanalytic concept, primary process thinking and application to study thought disordered by perceptual isolation, history and clarification psychoanalytic theory. Home: 4 Washington Pl., N.Y.C. 10003.*

HOLTE, Per Gunnar, Swedish physicist; b. Linköping, Sweden, Nov. 19, 1920; s. Oskar and Naemi C. (Lundquist) H.; Ph.D. in Physics, U. Uppsala (Sweden), 1951; m. Inga Westerfors, May 18, 1948; children—Göran, Jan. Asst. prof. theoretical physics U. Uppsala, 1951-53; head theoretical physics sect. AB Atomenergi, Stockholm, Sweden, 1953-57, head dept. for reactor physics, 1957-61, dir. research, 1961—. Mem. Swedish Nat. Com. for Physics, 1960—, Swedish Atomic Research Council, 1964—. Research, publs. on theory of slowing down of neutrons. Home: 16, Sommar V., Nyköping. Office: Studsvik, Nyköping, Sweden.*

HOLTEDAHL, Olaf, Norwegian geologist; b. Oslo, Norway, June 24, 1885; s. Arne Holtedahl and Mathilde (Madsen) H.; Cand.real. U. Oslo, 1909, Dr.-Philos., 1913; Dr.h.c., U. Stockholm, 1960; m. Tora Gurstad, Sept. 19, 1912; children—Aase (Mrs. Odd Gulliksen), Hans, Sonni (Mrs. Lars Andersgaard). Staff, U. Oslo, 1910-55, prof. geology, 1920-55, emeritus, 1955—; state geologist, 1916-20; dean Sci. Faculty, U. Oslo, 1946-47. Decorated comdr. Norwegian St. Olav Order; ridder Swedish Nordstjerne order; recipient Andrée medal Swedish Geog. Soc., 1927; Wollaston medal Geol. Soc. London, 1951; v. Buch medal Deutsche Geologische Gesellschaft, 1955; Gunnerus medal Kgl. Vidnsk. Selskab, Trondheim, Norway, 1955. Mem. Norwegian Acad. Sci. and Arts (pres. 1956, v.p. 1957), Norwegian Geog. Soc. (pres. 1939-45), Norwegian Geol. Soc. (pres. 1915, 32); corr. fgn. or hon. mem. sci. socs. in Denmark, Finland, Sweden, Eng., Scotland, Germany, Belgium, Austria, U. S. Editor report on sci. results Norwegian Expdn. Nov. Zemlya 1921, 3 vols., 1924-30, report sci. research Norwegian Antarctic Expdns. 1927-28, 3 vols., 1935-58. Research, publs. on geology of Norway (including Spitsbergen, Bear Island), Novaya Zemlya, W. Antarctica, paleontology of Arctic Can. Home: 16 Torgny Segerstedts vei. Office: Institutt for geologi, Blindern, Oslo 3, Norway.*

HOLTER, Heinz, biologist; b. Leonding, Austria, June 5, 1904; s. A. and Albine (Fischer) H.; Ph.D., U. Vienna, 1928; M.D. (hon.), Ghent (Belgium), 1957; Ph.D. (hon.), U. Copenhagen (Denmark), 1960; m. Karen Teisen, June 27, 1942; children—Susanne (Mrs. Jens Haagen Hansen), Peter. Asst., Chemistry Inst., Vienna (Austria) U., 1928-30; staff Carlsberg Lab., Copenhagen, 1936—, chief subdept. cytochemistry, 1942-56, chief physiol. dept., 1956—; lectr. cytology Copenhagen U., 1964—. Decorated Order Dannebrog, 1964. Mem. Royal Danish Acad. Scis. and Letters, Kungl. Vetenskaps Societen Uppsala, Kungl. Fysiografiska Sällskapet Lund. Research and numerous publs. in organic chemistry, enzyme chemistry, histochemistry, cytochemistry, cell physiology. Address: 10 Gamle Carlsbergvej, Copenhagen Valby, Denmark.*

HOLTFRETER, Johannes Friedrich Karl, biologist; b. Richtenberg, Germany, Sept. 1, 1901; s. Johannes and Sabine (Peters) H.; Ph.D., U. Freiburg, 1924; postgrad. Marine Biol. Stas., Naples and Heligoland, U. Greifswald, Germany; m. Hiroko Ban, Mar. 12, 1959. Came to U. S., 1946, naturalized, 1953. Faculty, Kaiser-Wilhelm Inst. Biology, Berlin-Dahlem, Germany, 1928-33, Zool. Inst., U. Munich, Germany, 1933-38; Rockefeller and Guggenheim fellow McGill U., Montreal, Can., 1942-46; faculty U. Rochester (N.Y.), 1946—, prof. zoology, 1948—, Tracy N. Harris prof., 1965. Mem. internat. bd. control Hubrecht Lab., Utrecht, Holland, 1934—; lectr. U. S. fgn. countries. Recipient hon. plaque Tokyo Soc. Med. Scis., 1962, U. Montreal, 1959. Rockefeller fellow, 1936-37. Fellow Zool. Soc. India, Am. Acad. Arts and Scis., A.A.A.S., John Simon Guggenheim Meml. Found.; mem. Am. Assn. Anatomists, Indian Soc. Zoologists, Inst. Internat. d'Embryologie, Society for the Study of Development and Growth, National Academy of Scientists, Deutsche Akademie der Naturforscher Leopoldina, Assn. Venezolana para el avance de La Ciencia, Swedish Acad. Scis. Uppsala. Editorial bd. Jour. Embryology and Exptl. Morphology, London. Research, publs. on phenomena of devel. in Amphibia; induction, orgn., structure, differentiation and movement of cells; cell inclusions. Home: 29 Knolltop Dr., Rochester, N.Y. 14610.*

HOLTHUIS, Lipke Bijdeley, systematic zoologist; b. Probolinggo, E. Java, Indonesia, Apr. 21, 1921; s. Bernard Jan and Neeltje (bij de Ley) H.; D.Sc., U. Leiden (Netherlands), 1946. Staff, Rijksmuseum van Natuurlijke Historie, Leiden, 1941—, curator, 1950-60, sr. curator Crustacea, 1960—; research asso. Allan Hancock Found., Los Angeles, 1947-48; zoologist NRC, Washington, 1952-53, 59-60; systematic zoologist Crustacea, Smithsonian Instn., Washington, 1960. Mem. Internat. Commn. on Zool. Nomenclature, 1953—, v.p., 1964—, acting pres., 1965—; adviser on crustacea Fla. Bd. Conservation, 1963—. Mem. Zool. Soc. London (corr.), Carcinological Soc. Japan (hon.). Research, numerous publs. on systematic study of Decapod and Stomatopod crustacea of world especially shrimps, Isopod crustacea of Netherlands. Home: 290 Boerhaavelaan, Leiden. Office: 2 Raamsteeg, Leiden, Netherlands.*

HOLTMAN, Darlington Frank, Am. bacteriologist; b. Randolph, Kan., Oct. 12, 1903; s. Frank O. and Adelia (Hall) H.; A.B., U. Kan., 1927; M.A., U. Tenn., 1930; Ph.D., Ohio State U., 1937; m. Harriet Hitchcock, Sept. 4, 1937; children—Virginia Ann (Mrs. Walter K. Thigpen), Mary Jean. With U. Tenn., Knoxville, 1927-32, instr., 1929-32, prof. bacteriology, head dept., 1943—; asst. bacteriology Ohio State U., 1932-36, instr., 1936-42; asst. prof. bacteriology and hygiene Western Res. U., 1942-43; cons. TVA, 1944-48, Oak Ridge Nat. Lab., 1954-58; chief microbial ecology br. Chem. Corps Biol. Labs., Ft. Detrick, Md., 1949-50; dir. Tenn. Civil Def. for Biol. Warfare, 1950—. Diplomate Am. Bd. Microbiology. Fellow Am. Acad. for Microbiology, A.A.A.S.; mem. Am. Soc. for Microbiology, Soc. for Biology and Exptl. Medicine, Radiation Soc., Sigma Xi. Author (with P. W. Allen) Food Bacteria Manual, 1931; (with others) Microbes That Help or Destroy Us, 1941. Studies, publs. on heterophile antigens of bacteria; environment in susceptibility to poliomyelitis; antibiotics, including co-discovery of anti-fungal antibiotic Tennecetin; diseases of poultry, role of proteins, carbohydrates and lipids in disease. Home: 390 West End Lane, Knoxville, Tenn. 37919.*

HOLTSMARK, Johan Peter, Norwegian physicist; b. Oslo, Norway, Feb. 13, 1894; s. Gabriel and Margrethe (Weisse) H.; student U. Würzburg, 1912-14, U. Leipzig, 1916-17, U. Göttingen, 1917-18, U. London (Kings College) 1919-20; U. Oslo, 1911-12; Dr.philos., U. Oslo, 1918; m. Astrid Gundersen, Apr. 12, 1954; children—Signe Margrethe Vennemoe, Anne Elisabeth Lindvik, Bent. Asst. physics U. Leipsig, U. Göttingen, U. Oslo, 1921-23; prof. of physics, Tech. U., Trondheim, Norway, 1923-42; prof. of physics, U. Oslo, 1942-64; emeritus, 1964. Acoustical cons., 1930-64; mem. Acad. Scis., Oslo, Royl Soc. Scis., Trondheim. Author: Physics; also articles. Research on spectroscopy, atomic physics, acoustics, esp. stochastic properties of reverberation. Home: Askliveien 5, Blommenholm, Norway. Office: U. Oslo, Oslo 3, Norway.*

HOLTTUM, Richard Eric, Brit. botanist; b. Cambridge, Eng., July 20, 1895; s. Richard and Florence (Bradley) H.; M.A., Sc.D., Cambridge U.; m. Ursula Massey, 1927; children—Elizabeth, Catherine. Dir., Botanic Garden, Singapore, 1925-49; prof. U. Singapore, 1949-54, later hon. prof.; now hon. collaborator for Malaysian flowers Royal Botanic Gardens of Kew. Mem. Royal Netherlands Assn. Botany, Linnaean Soc., Brit. Pteridological Soc., Am. Fern Soc. Author: Plant Life in Malaya; Orchids of Malaya; Ferns of Malaya; Gardening in the Lowlands of Malaya; Pteridophyta in Flora Malesiana. Home: 80 Mortlake Rd. Office: Royal Botanic Gardens, Kew, Richmond upon Thames, Eng.

HOLTZ, Peter Wilhelm Joseph, German physician; b. Stolberg, Rhone, Feb. 6, 1902; s. Karl and Barbara (Dreuw) H.; ed. univs. Bonn, Heidelberg, Wurzburg, Fribourg, Munich; M.D.; m. Dorothea Schümann, 1939. Agrégé of pharmacology and toxicology U. Greifswald, 1936; prof. physiol. chemistry U. Rostock, 1938, dir. Inst. Physiol. Chemistry, 1945, full prof. pharmacology and toxicology, dir. Inst. Pharmacology; full prof. pharmacology and toxicology, dir. Inst. Pharmacology U. Frankfort on Main. Recipient Schmiedeberg Plakette. Mem. Léopoldina Acad., Acad. Scis. Berlin, German Soc. Pharmacology, German Soc. Physiol. Chemistry, German Soc. Naturalists and Physicians. Author: Dopadecarboxylase; Noradrenaltin; Physiologie und Pharmakologie des Vegetativen Nervensystems. Co-editor: Naunyn-Schmiedebergs Archiv. exp. Pathologie und Pharmakologie. Home: 21 Grosse Fischerstrasse. Office: Pharmacology Institute, University of Frankfort on Main, 14 Ludwig Rehnstrasse, Frankfort on Main, Germany.

HOLTZ, Wilhelm, German physicist; b. Saatel, Germany, 1836; became docent Greifswald (Germany) Inst., 1881, prof. physics, 1884. Author (in German): The Theory and Installation of Lightning Rods, 1878;

The Danger of Lightning and Its Causes. Inventor one of 1st machines to produce electricity at high potential, 1865. Died Greifswald, 1913.

HOLTZMANN, Oliver Vincent, Am. plant pathologist; b. Highmore, S.D., June 26, 1922; s. Alphonse J. and Mary Lona (St. Pierre) H.; B.S., Colo. State U., 1950, M.S., 1952; Ph.D., Wash. State U., 1955; m. Cecelia Catherine Lucas, June 7, 1957; children—Fredrick John, Kathryn Marie, Nicholas Edward, Joseph Albert, Eleanor Ann. Asst. research prof. plant pathology N.C. State Coll., Raleigh, 1956-57; asst. prof. plant pathology U. Hawaii, Honolulu, 1957-66, asso. prof., 1966—. Mem. Am. Phytopath. Soc., Soc. Nematologists, Hawaiian Acad. Sci., Hawaiian Bot. Soc., Sigma Xi. Research, publs. in pathogenicity, biology and control of plant parasitic nematodes, control of diseases of tropical fruits and nuts, factors influencing the resistance of plants to nematode diseases. Home: 1235 Manu-Mele, Kailua, Hawaii 96734. Office: Dept. Plant Pathology, U. Hawaii, Honolulu 96822.*

HOLVECK (or Holweck), Fernand, French physicist; b. Paris, France, 1890; ed. École de physique et chimie, Paris. Dir. research Centre National de la Researche Scientifique, during World War II; among 1st resistance fighters against German occupation. Studied X-rays with long wave lengths; invented molecular pump of high precision, also (with Lejay) Holveck-LeJay apparatus for determining relative intensity of gravity; established continuity between ultraviolet and X-rays, 1920. Arrested and killed by Nazis, Paris, 1941.

HOLYOKE, Edward Augustus, Am. physician; b. Marblehead, Mass., Aug. 1, 1728; s. Edward and Margaret (Appleton) H.; grad. Harvard, 1746, M.D. (hon.), 1783, LL.D., 1815; m. Judith Pickman, June 1755; m. 2d, Mary Viall, Nov. 22, 1758; 12 children. Practiced medicine, Salem, Mass., 1749-1821; in charge of smallpox hosp. in Salem during epidemic, 1777; taught medicine privately; a founder Mass. Med. Soc., pres., 1782-84, 86-87; a founder Am. Acad. Arts and Scis., pres., 1814-20. Pioneer in smallpox vaccination in U. S. Died Salem, Mar. 31, 1829.

HOLZER, Helmut, German biochemist; b. Nuremberg, June 14, 1921; s. Emil and Emma (Steinmeyer) H.; Ph.D. in Natural Sci., U. Munich; m. Erika Vogel, 1951. Instr. biochemistry U. Munich, U. Hamburg; prof. biochemistry, dir. Biochemistry Inst., U. Freiburg, Breslau. Mem. numerous European and U. S. sci. socs. Research and over 150 publs. on biochemistry.

HOLZER, Robert Edward, Am. geophysicist, educator; b. Portland, Ore., Nov. 21, 1906; s. Charles Emil and Emma Elizabeth (Schlegel) H.; A.B., Reed Coll., 1926; M.A., U. Cal. at Berkeley, 1928, Ph.D., 1930; m. Wilma Edith Botts, Aug. 6, 1931; children—Roberta, William, Thomas. Instr. physics U. Cal. at Berkeley, 1930-31, 34-35, prof. geophysics U. Cal. at Los Angeles, 1947—; NRC fellow in physics U. Chgo., 1931-33; asst. prof. Fenn Coll., 1933-34; instr. physics U. N.M., 1935-37, asst. prof., 1937-39, asso. prof., 1939-43, prof., 1943-46, dir. pre-meteorology tng. program, 1943, asst. dir. research projects, 1943-46; prof. physics, head dept. Pomona Coll., 1946-47. Coordinator, N.M. Council Nat. Def., 1941; war research OSRD, 1941-45. Fellow Am. Phys. Soc.; mem. A.A.A.S., Am. Meteorol. Soc., Am. Geophys. Union, Am. Assn. U. Profs., Sigma Xi. Contbr. articles on lightning, thunderstorm structure, extremely low frequency natural electromagnetic spectrum to profl. jours. Home: 1514 Veteran Av., Los Angeles 90024.

HOLZER, Wolfgang, Austrian physician, engr.; b. Krems, Austria, Apr. 20, 1906; s. Valentin and Anna (Feichtinger) H.; ed. Poly. Sch. Berlin, U. Vienna; engring. diploma; Ph.D. in engring.; M.D.; m. Josefine Höger, 1950. High tension elec. engr.; physiologist; prof. psychiatry and neurology; neurologist. Author: Development of Rheocardiograph. Research and numerous publs. on phys. medicine, diagnosis of therapy, neurology, therapy by short waves and rheocardiography. Address: 17 Herrengasse, Graz, Austria.

HOLZKNECHT, Guido, Austrian radiologist; b. Vienna, Dec. 3, 1872; prof., Vienna. Author: Röntgenologie, 1920; Röntgentherapie, 1924; also classic work on diagnosis of stomach tumors, 1906. Research in X-ray diagnosis and therapy. Died Oct. 30, 1931.

HOLZMANN, Max, Swiss physician; b. Zurich, Switzerland, Mar. 31, 1899; s. Maurice and Anna (Lerch) H.; ed. U. Zurich, U. Lausanne; M.D. m. Lise Rüegg, 1931; children—Annelise, Rosemarie, Marianne, Regula. Prof. agrégé; titular prof. cardiology, affiliate Med. Clinic, U. Zurich. Mem. Swiss Soc. Cardiology (co-founder), also mem. numerous cardiological socs. Author: Klinische Elektrokardiographie; Die Rhythmusstorungen des Herzens; Die Erkrankungen des Herzens und der Gefässe, also numerous articles on cardiology. Home: 2 Schlossbergstrasse, Zollikon. Office: 56 Bahnhofstrasse, Zurich, Switzerland.

HOMANS, John, Am. surgeon; b. Boston, Dec. 26, 1836; grad. Harvard, 1858; M.D., 1862; postgrad. Vienna, Austria, Paris, France. Surg. intern Mass. Gen. Hosp., house surgeon, early Civil War; asst. surgeon USN, 1862; asst. surgeon U. S. Army, St. James Hosp., New Orleans, surgeon-in-chief 1st div., 19th Army Corps, med. insp. staff Gen. Sheridan, to 1866; practice medicine Boston, from 1866; surgeon Boston Dispensary, Children's Hosp., Carnedy Hosp., from 1868, St. Margaret's Hosp.; cons. Carnedy Hosp., from 1880; instr. in diagnosis and treatment of ovarian tumors Harvard Med. Sch., from 1881. In his honor, John Homans professorship of Surgery, Harvard, established 1906. Mem. Am. Surg. Assn. One of foremost operators on ovarian tumors in New Eng. (performed 601 ovariotomies, 1872-1901, 1st used carbolic spray); (with S. J. Crowe, H. Cushing) demonstrated that excision of pituitary gland results in atrophy of genitalia; one of 1st to operate for abscess of appendix; began abdominal hysterectomies, 1881. Died Feb. 7, 1903.

HOMBERG, Wilhelm, chemist, naturalist; b. Batavia, Java, Jan. 8, 1652; grad. medicine, Wittenberg, Germany. Practice medicine, Paris, 1682-85, 91—, Rome, Italy, 1685-90; dir. Lab. French Acad. Scis.; prof. chemistry; 1st physician to Duke of Orleans, 1694. Mem. French Acad. Scis. Author: Essais de Chemie, 1702; Diverses expériences du phosphoro, 1692. Discovered boric acid, 1702; one of 1st to describe green color produced by copper in fire; research on salt crystallization, sulphur of antimony, phosphorous; proportions in chem. combinations. Died Paris, Nov. 13, 1715.

HOMBURGER, Freddy, exptl. pathologist; b. St. Gall, Switzerland, Feb. 8, 1916; s. Ludwig and Cecile (Gaille) H.; student U. Vienna (Austria) 1936-37; M.D., U. Geneva (Switzerland), 1941; m. Regina Thürlimann, Nov. 8, 1939. Came to U. S., 1941, naturalized, 1952. Research fellow Yale, 1941-42, Thorndike Meml. Lab., Harvard, 1943-45; research clinician, chief div. clin. investigation Sloan-Kettering Inst. for Cancer Research, N.Y.C., 1945-48; faculty Tufts U. Sch. Medicine, 1948-57; pres., dir. Bio-Research Inst., Inc., Cambridge, Mass., 1957—, also pres., treas. Bio-Research Consultants, Inc., 1957—. Chmn. adv. com. on Am. students U. Geneva Med. Sch., 1950-57, mem. adv. bd. Lachaise Found., 1965—. Fellow A.A.A.S., N.Y. Acad. Scis.; mem. Am. Assn. Cancer Research, Am. Fedn. Clin. Research, Am. Geriatrics Soc., A.M.A., Am. Med. Writers Assn., Am. Soc. Exptl. Pathology, Endocrine Soc., Gerontol. Soc., N.Y. Acad. Medicine, Soc. Exptl. Biology and Medicine, Soc. Toxicology, Swiss Med. Assn., Teratology Soc., New Eng. Arthritis and Rheumatism Soc., Sigma Xi, others. Author: The Medical Care of the Aged and Chronically Ill, 1955, rev. edit. (with C. D. Bonner) The Medical Care and Rehabilitation of the aged and Chronically Ill, 1964; The Biologic Basis of Cancer Management, 1957. Editor: The Physiopathology of Cancer, 1953, 2d edit., 1958; vols. 1-3 Symposia on Research Advances Applied to Medical Practice, 1955-58, vols. 1-9, Progress in Experimental Tumor Research, 1960-67. Research and publs. in oncology; endocrinology; exptl. pathology. Home: 759 High St., Dedham, Mass. 02026. Office: 9 Commercial Av., Cambridge, Mass. 02141.*

HOME, Sir Everard, English physician; b. Hull, Eng., May 6, 1756; s. Robert Boyne and Mary (Hutchinson) H.; student surgery with John Hunter, London, Eng.; m. Jane Tunstall, 1792; 2 sons, James, William Archibald, 4 daus. Asso. John Hunter in exptl. work for 20 years; surgeon St. George's Hosp., 1793-1827, Chelsea Hosp., from 1821; prof. anatomy, surgery Coll. Surgeons, 1804-13, 21; surgeon to king, from 1808; Hunterian orator, 1822. Recipient Gold medal Lyceum Medicum Londenense, 1788. Fellow Royal Soc., 1787; mem. Coll. Surgeons (ct. examiners 1809, master 1813, pres. 1821), French Acad. Scis., 1813. Author: A Dissertation on the Properties of Pus. Contbr. treatises on structure of urethra, ulcers of leg, cancer and diseases of prostate; custodian Hunterian Collection (after Hunter's death, pub. about 116 papers, not determined whether based on Hunter's or Home's work). Died Aug. 31, 1832.

HOME, Francis, Scottish physician; b. Scotland, 1719; M.D., U. Edinburgh (Scotland), 1750; student Leiden (Netherlands) U. Prof. materia medica Edinburgh U., 1768-98, prof. insts. medicine, agr.; surgeon of dragoons, Seven Years' War. Author: Principa Medicinae, 1758; An Enquiry into the Nature, Cause and Cure of the Croup (contains 1st clear systematic study of diphtheria), 1765; Experiments on Bleaching (advocated use of dilute sulphuric acid instead of sour or buttermilk in souring of linen). Vaccinated children with material from measles, obtaining some immunity; 1st observer to discover that yeast ferments sugar in diabetic urine. Died 1813.

HOMEYER, Eugen Ferdinand von, see von Homeyer, Eugen Ferdinand.

HOMMA, Hans, Austrian pathologist; b. Vienna, June 9, 1894; s. Franz and Marie (Rumler-Aichenwehr) H.; Ph.D. in Pathology, U. Vienna; m. Johanna Böttner, 1930. Mem. council dirs. Inst. Pathology, Am. U., Beirut; mem. Inst. Bacteriology and Histology, Gynecology Clinic Vienna, 1936-44, Inst. Pathology, Municipal Polyclinic Vienna, 1940-45, Inst. Pathology, Salzburg, 1949. Vice pres. council, dir. Health Service of Salzburg. Mem. Internat. Paracelsus Assn. (v.p.). Author: Das Formproblem in der Biologie, also numerous works on gynecol. histopathology and serological determination of problems of paternity. Home: 4 Vierthalerstrasse. Office: 54 Müllner-Hauptstrasse, Salzburg, Austria.

HON, Edward Harry Gee, physician; b. Canton, China, Jan. 12, 1917; s. Harry Gee and Cecilia Wong (See) H.; student Sydney Tech. Coll., 1942-44; B.S., Union Coll., 1950; M.D., Coll. Med. Evangelists, 1950; m. Audrey Quay, June 15, 1948; children—Shirley June, Robert William, Edward David. Came to U. S., 1945, naturalized, 1959. Practice medicine, specializing in obstetrics, gynecology, New Haven, 1954-60, 64—, Los Angeles, 1960-64; faculty Yale, 1954-60, 64—, asso. prof. obstetrics, gynecology, 1964—; vis. prof. U. Sydney, 1959; prof. Loma Linda U., 1960-64; asso. obstetrician, gynecologist Yale New Haven Hosp., 1956-60; sr. attending physician Los Angeles County Gen. Hosp., 1960-64. Markle scholar in med. sci. Yale, 1955-60; recipient AID Humanitarian of Year award, 1960. Diplomate Am. Bd. Obstetrics and Gynecology. Mem. Am. Coll. Obstetrics and Gynecology, A.M.A., Internat. Fedn. Gynecology and Obstetrics, Soc. For Gynecologic Investigation, Soc. For Study Sterility, Sigma Xi, Alpha Omega Alpha. Author: A Manual of Pregnancy Testing, 1961. Publs. on application of biophys. techniques to study of human fetus; analysis of fetal heart rate patterns; continuous monitoring during labor. Office: 333 Cedar St., New Haven 06511.*

HONDA, Kotaro, Japanese physicist; b. Aichi Prefecture, Japan, 1890; grad. physics Tokyo (Japan) U., 1897; Dr. Sci., 1908. Became prof. Tohoku U., 1911, dir. Metal Research Inst., 1922, pres., 1931; named pres. Tokyo U. Sci., 1948. Recipient Imperial Acad. prize, 1916, Cultural Order, 1937. Developed K.S. steel (strongest steel then known), 1917, produced still stronger K.S. steel, 1919. Died Feb. 12, 1954.

HONDA, Minoru, Japanese astronomer; b. Shimane Prefecture, 1913; studied astronomy by himself. Technician, Kurashiki Astron. Obs., 1925—, now dir. Recipient prize for discovery of new comet, Astron. Soc. Japan. Mem. Japan Meteorol. Soc. Discovered comet, 1940, 41, 47 (now known as Honda Comet), discovered new comet in Perseus Constellation, 1948.

HONDIUS (or De Hondt), Jodocus (or Joos, Josse), engraver, cartographer; b. Wacken, Flanders, 1563; s Olivier and Petronella (van Havertuyn) de H.; ed. in Ghent; m. Collette van der Keere, Apr. 11, 1587; 13 children, incl. Jodocus, Hendrik. Supposedly began engraving own compositions on copper, ivory at age 8; apprentice painter; employed by Alessandro Farnese, Duke of Parma, Gov. Netherlands; came to Eng. circa 1584; typefounder, engraver, globe and math. instrument maker, London; moved to Amsterdam, Netherlands, circa 1594. Pub. new edit. Mercator's Atlas, 1606. Publs. on constrn., use of globes, 1597, calligraphy, 1594. Engraved some of earliest maps Eng.; constructed largest celestial, terrestrial globes thus known; illustrated Drake's, Cavendish's voyages; also engraved portraits of notables, John Speed's work. Died Amsterdam, Feb. 10, 1611.

HONECK, Henry Charles, Am. physicist; b. Batavia, N.Y., Oct. 4, 1930; s. Henry and Sigrid (Petersen) H.; B.S., Rensselaer Poly. Inst., 1952; postgrad. Oak Ridge Sch. Reactor Tech.; Sc.D., Mass. Inst. Tech., 1959; m. Mary Buxton, Feb. 5, 1966; children—Lorraine, Christine. Nuclear engr. Pratt & Whiting Aircraft, Hartford, Conn., 1952-56; nuclear engr. Brookhaven Nat. Lab., Upton, N.Y., 1959-65; physicist U. S. AEC, Washington, 1965-66, E. I. du Pont Savannah River Lab., Aiken, S.C., 1966—. Cons. reactor theory, 1960-65. Mem. Am. Nuclear Soc. (exec. com. math. and computation div. 1963-68). Research, publs. in nuclear reactor theory and applied maths. including study neutron transport theory, neutron thermalization, cross sect. evaluation and processing, numerical analysis and application of digital computers to reactor physics. Home: 118 Vivion Dr. Office: E. I. du Pont, Savannah River Lab., Aiken, S.C. 29801.

HONESS, Arthur P., Am. geologist; b. Berea, O., 1886; s. Pharoah and Anna (Riddles) H.; student Tri-State Coll.; B.S., Oberlin Coll., 1914; M.S., Princeton, 1917, D.Sc., 1924; m. Ethel Wortley; a dau., Mary Ann. Tchr. pub. schs. Ind., 1909-10; joined faculty as instr. Pa. State Coll., 1917, became prof. mineralogy, 1931. Author: The Nature, Origin and Interpretation of Etch Figures on Crystals, 1927; also articles. Research on descriptive mineralogy, igneous rocks, including their genesis, alteration and replacement phenomena, also mica peroditite dike of Western Pa., Pa. bentonite; discovered relationship between porosity and mineral composition in Pa. oil sands. Died 1942.

HONEYMAN, Merton Seymour, Am. geneticist; b. Hartford, Conn., Sept. 27, 1925; s. Edward L. and Dorothy (Cohen) H.; B.A., U. Conn., 1950, M.S., 1951; Ph.D., Ohio State U., 1954; m. Harriet Shirley Chatzek, June 15, 1952; children—Elisa, Michael, Jason. USPHS postdoctoral fellow Yale, 1954-57; geneticist Nat. Cancer Inst., 1957-59; geneticist, dir. Conn. Twin Registry, Conn. State Dept. Health, Hartford, 1959—; dir. planning and program analysis Office Mental Retardation, Conn. Dept. Health, 1966—. Adj. asst. prof. genetics U. Hartford, 1960—; lectr. pub. health Yale Sch. Medicine, 1962—. Mem. Am. Soc. Human Genetics, Am.

Genetics Assn., Am. Pub. Health Assn., Am. Heart Assn., Am. Assn. Mental Deficiency, Sigma Xi. Research, publs. in human genetics chronic diseases, twin family studies related to coronary heart disease, twin epidemiology. Home: 9 Alderwood Dr., West Hartford, Conn. 06117. Office: 79 Elm St., Hartford, Conn. 06115.*

HONG, Suk Ki, Korean physiologist; b. Kyunggi Do, Korea, Oct. 16, 1928; s. Sung Cho and Kwan R. (Lew) H.; M.D., Severance Union Med. Coll., Seoul, Korea, 1949; Ph.D., U. Rochester, 1956; m. Kyung Im Min, June 27, 1959; children—Bum Suk, Dae Suk, Kwi Suk. Faculty, U. Buffalo, 1956-59, vis. asso. prof. physiology, 1965; faculty Yonsei U., Seoul, 1959—, prof. physiology, 1966—. Recipient Samil Cultural award in Natural Scis., 1963. Mem. Korean Physiol. Soc. Author: (with Pyung Hee Lee) Textbook of Physiology, 1966; also articles. Research in physiology of kidney, including mechanism by which renal tubule secretes acidic compounds, effects of lowered body temperature on renal functions, mechanics of breathing, physiology of Korean women divers; especially their respiration and body temperature regulation. Home: 94-13 Sukyo Dong, Mapo Ku, Seoul, Korea.*

HONIG, John Gerhart, operations research scientist; b. Vienna, Austria, Oct. 30, 1923; s. Walter and Gertrude (Weiss) H.; came to U. S., 1940, naturalized, 1943; A.B., Drew U., 1947; M.S. in Phys. Chemistry, U. Mich., 1948; Ph.D., Georgetown U., 1956; m. Erna Appenzeller, Sept. 18, 1949; children—Gary Walter, Judy Gail. Phys. chemist Nat. Inst. Cleaning and Dyeing, 1948-51, Naval Research Lab., 1951-56; staff operations evaluation group Office Chief Naval Operations, 1956-59; asso. project leader Weapon System Evaluation Group, Office Sec. Def., Washington, 1959-62; chief naval warfare tech. Honeywell, Inc., Washington, 1962-66; with Weapons Evaluation and Control Bur., U. S. Arms Control and Disarmament Agy., Washington, 1966-67, Office of Chief Staff, U. S. Army, 1967—. Vice president Mil. Operations Research Symposia, 1964—. Mgmt. cons. to govt. agys. Fellow A.A.A.S.; mem. Nat. Security Indsl. Assn. (chmn. tech. com. antisubmarine warfare adv. com. 1964-66), Washington Operations Research Council (pres.), Washington Acad. Sci., Operations Research Soc. Am., Inst. Mgmt. Sci. Research, publs. on nature large molecules in anhydrous systems, weapon system analyses; developed basic methodologies, means to measure non-Newtonian systems in anhydrous environment. Home: 7701 Glenmore Spring Way, Bethesda, Md. 20034. Office: Office of Chief of Staff, Dept. Army, Washington 20310.*

HONIGBERG, Bronislaw M(ark), biologist; b. Warsaw, Poland, May 14, 1920; s. Zachary Z. and Mary R. (Laks) H.; A.B., U. Cal. at Berkeley, 1953, M.A., 1946, Ph.D., 1950; m. Rhoda L. Springer, Feb. 7, 1948; children—Paul Mark, Martin Philip. Came to U. S., 1941, naturalized, 1948. Abraham Rosenberg research fellow zoology U. Cal. at Berkeley, 1949-50; faculty U. Mass., Amherst, 1950—, prof. zoology, 1961. Research asso. pathobiology Johns Hopkins Sch. Hygiene and Pub. Health, Balt., 1958-59; USPHS spl. research fellow, vis. scientist Lab. Parasitic Diseases, Nat. Inst. Allergy and Infectious Diseases, NIH, Bethesda, Md., 1965-66. Fellow A.A.A.S., N.Y. Acad. Scis.; mem. Soc. Protozoologists (pres. 1965-66, chmn. U. S. delegation Internat. Commn. Protozoologists 1965—), Am. Micros. Soc. (past pres., asso. editor Trans. 1966—), Am. Soc. Parasitologists, Am. Soc. Zoologists, Soc. Systematic Zoologists, Soc. for Cryobiology, Biol. Stain Commn., Tissue Culture Assn., Sigma Xi, Phi Beta Kappa, Phi Kappa Phi. Contbg. author: Manual of Determinative Bacteriology, 1957; Chemical Zoology, 1967; Immunity to Parasitic Animals, 1968; Biology of Termites, 1968. Editorial bd. Jour. Protozoology, 1959—. Research, publs. on systematics and evolution of protozoan, especially of non-pigmented (zoomastigophorean) flagellates, cytology, physiology, biochemistry and immunology of parasitic flagellates, especially of trichomonads; cytological, cytochem., biochem. and immunological studies of mechanisms by which trichomonad parasites injure their hosts. Home: 95 Red Gate Lane, Amherst, Mass. 01002.*

HONJO, Susumu, Japanese geologist, electron microscopist; b. Osaka, Japan, Mar. 6, 1933; s. Sadajiro and Kimi (Otsuki) H.; M.S., Hokkaido U., Sapporo, Japan, Ph.D.; m. Kazuko Watanabe, Oct. 17, 1958. Research asst. U. Kan., 1959-60; research asso. Hokkaido U., 1961-62, 64-66, asso. prof., 1966—; research asso. Princeton, 1962-64; dir. electron microscopic lab. dept. geology Hokkaido U. Mem. Geol. Soc. Japan, Geol. Soc. Am., Sigma Xi. Author: (with A. G. Fischer, R. E. Garrison) Electron Micrographs of Limestone and their Nannofossils; also articles. Discovered various techniques of applying regular emission electron microscopy geologic and paleontol. study; introduced scanning electron microscopy for study planktonic foraminfera. Home: 6, 3-chome, Izumicho, Makomanai, Sapporo, Japan.*

HONKANEN, Erkki Juhani, Finnish chemist; b. Riihimaki, Finland, Jan. 23, 1930; s. J. N. and L. M. (Roschin) H.; M.Sc., Finland's Inst. Tech., 1954, Dr. Tech., 1961; m. Rauni Hellevi Aalto, July 10, 1954; children—Sakari Johannes, Tiina Johanna. Asst., Finland's Inst. Tech., 1954-56; chemist pharm. mfrs. Laaketehdas Orion Oy, Finland, 1956-59, 63—; chemist lab. found. chem. research Biochem. Inst., Helsinki, Finland, 1959—. Mem. Suomalaisten Kemistein Seura, Soc. Biochem. Biophys. et Microbiol. Fenn. Research, publs. on synthetic and analytical organic chemistry, synthetic pharm. chemistry, flavor chemistry, mass spectrometry. Home: 1 A 13, Vanha Viertotie, Helsinki. Office: 56 B, Kalevankatu, Helsinki, Finland.*

HONTI, Géza, Hungarian physician; b. Szeged, Hungary, Dec. 16, 1923; s. Alexander and Maria (Malatinsky) H.; diplom U. Szeged Med. Faculty, 1948; specialization, U. Pécs (Hungary), 1952; m. Catherine Hertelendy, Sept. 10, 1949; children—George, Catherine. Staff, Pharmacological Inst., Szeged U., 1946-49; staff Hospital Hódmezővásárhely, 1949-51; staff II Clinic Medicine, Pécs U., 1951-53; with Policlinic of Kecskemét, 1953—; head biol. sect. Agrl. Research Inst., Kecskemét, 1966—. Cons., Social Home, Kecskemét; tchr. Nursing Sch., Kecskemét. Mem. Hungarian Biol. Soc., Hungarian Assn. Medicine, Hungarian Microbiol. Soc., Sci. Documentation Center. Editor, Biologiai Szemle. Research and publs. on difference of local and general effect of histamine, role of histamine in regulation of human body, role of histamine and histamines in defence mechanism of human body; studies (with Putnoky) on relationship between histamine and tumors; described effect of regenerating medicine (Hisactin). Home: 1 Mátyási. Office: 7 Komszomotter, Kecskemét, Hungary.*

HOOBLER, Sibley W., Am. physician; b. N.Y.C., Apr. 30, 1911; s. B. Raymond and Madge (Sibley) H.; B.A., Princeton, 1933; Sc.D., Johns Hopkins, 1937, M.D., 1938; m. Katherine Taylor, Mar. 16, 1940; children—Raymond T., Patricia Ann. Practice medicine, specializing in internal medicine, Ann Arbor, Mich., 1945—; faculty U. Mich., 1945—, prof. internal medicine, 1959—; dir. hypertension unit U. Mich. Hosp., 1947—. Mem. Am. Soc. Clin. Investigation, Am. Physiol. Soc., Am. Heart Assn., A.C.P. Author: Hypertensive Disease: Diagnosis and Treatment, 1959; also numerous articles. Research on effects of blocking autonomic innervation in man by drugs, various aspects of high blood pressure in man and in exptl. animal. Address: U. Mich. Med. Center, Ann Arbor, Mich. 48104.*

HOOD, Bertil, Swedish physician; b. Malmo, Sweden, Feb. 4, 1917; s. Bert and Anna (Nielsen) H.; M.D., U. Lund (Sweden), 1943; m. Birgitta Gullander, Nov. 28, 1942; children—Tomas, Anne, Daphne. With dept. internal medicine and pediatrics U. Lund, also Hosp. of Falun, 1943-49; staff dept. I, Sahlgren's Hosp., Gothenburg, Sweden, 1949—; fellow physiology N.Y. U. Coll. Medicine, 1946-47. Research, numerous publs. on kidney physiology, treatment hypertension and chronic pyelonephritis, hyperlipiodic disorders enzyme research in human fat and human liver. Home: 7 Snöskategatan. Office: Sahlgrens Hosp., Gothenburg, Sweden.*

HOOD, Donald Wilber, Am. chem. oceanographer; b. New Castle, Pa., July 12, 1918; s. Charles E. and Ida (Blews) H.; student Westminster Coll., 1936-38; B.S., Pa. State U., 1940; M.S., Okla. State U., 1942; Ph.D., Tex. A. and M. U., 1950; m. Betty Ellen Jackson, Nov. 22, 1945; children—Rebecca Jean, Barbara Joan, Susan Marie. Research fellow biochemistry and nutrition Tex. A. and M. Coll., 1950-54, faculty, 1954-65, prof., 1960-65; dir. Inst. Marine Sci., U. Alaska, College, 1965—. Chmn. bd. dirs. Indsl. Waste Disposal Corp., Houston, 1957—. Cons. pvt. cos.; mem. study sect. USPHS, 1963-67. Recipient Faculty Achievement award for research Tex. A. and M. Coll., 1963; NSF Sr. Postdoctoral fellow, 1963-67. Fellow Am. Inst. Chemists; mem. Am. Chem. Soc. (regional lectr. 1963), A.A.A.S., Pollution Control Fedn., Am. Geophys. Union, Am. Soc. Limnology and Oceanography, Sigma Xi, Phi Lambda Upsilon, Phi Kappa Phi. Contbg. author: Annual Review of Oceanography and Marine Biology, 1963; Earth Science Ency., 1966; also numerous articles. Patentee solvent extraction process for saline water conversion. Research in chemistry oceans, elucidation carbon dioxide system in oceans and atmosphere, organic chemistry sea water, trace metal chemistry, waste disposal in marine environment, saline water conversion. Home: Box 567, College, Alaska 99735.*

HOOD, John, surveyor, inventor; b. Moyle, Donegal, Ireland, 1720. Author: Tables of Difference of Latitude and Departure for Navigators, Land Surveyors . . . , 1772. Described diurnal variation of magnetic needle and its correction; inventor Hood's compass theodolite (forerunner of modern theodolites); credited with work on invention of Hadley's quadrant. Died 1783.

HOOD, Thomas, English mathematician; flourished 1582-98; s. Thomas Hood; B.A., Trinity Coll., Cambridge (Eng.) U., 1577 or 78, M.A., 1581, Ph.D. in Physics, 1585. Math. lectr. to City London, Eng., 1588-92; later pvt. tchr. Author: A Copie of the Speache . . . (argument favoring application of math. to other subjects), 1588; Elements of Geometrie, 1590; The Use of the Celestial Globe . . . , 1590; The Use of Jacobs Staffe, 1590; Touching the Use of the Crosse Staffe, 1590; The Use of Both Globes Celestiall and Terrestriale, 1592; The Marriners Guide . . . , 1596; Elements of Arithmeticke, 1596; The Making and Use of the Geometricall Instrument Called the Sector, 1598. Studied nav., designed charts and instruments.

HÖÖK, Olle, Swedish physician; b. Skinnskatteberg, Sweden, Oct. 30, 1918; s. Robert and Elsa (Arnberg) H.; med.lic., Karolinska Institut, 1944, M.D., 1958; m. Kerstin Lundvall, Jan. 31, 1943; children—Eva, Peter, Annika, Lars. Asst. chief physician dept. neurology Serafimerlasarettet, Karolinska Institutet, Stockholm, Sweden, 1957-63; head dept. neurol. rehab. Karolinska Hosp., 1963-66; prof. med. rehab. Sahlgren Hosp., U. Gothenburg (Sweden), 1966—; asst. prof. neurology Karolinska Instituet, 1959-66. Cons., Nat. Swedish Bd. Civil Aviation, 1960—; mem. com. for tech. devices for disabled Nat. Swedish Bd. Health, 1962—; mem. med. sci. council Nat. Bd. Health, 1967—. Mem. Swedish Neurol. Soc. (past chmn.), Royal Acad. Arts and Scis. Uppsala (corr.). Author: Study on Subarachnoid Haemorrhage and Intracranial Arterial and Arteriovenous Aneurysms, 1958; also numerous articles. Research on vascular anomalies of the brain and spinal cord; tech. inventions for neurologically disabled people. Home: 7, Änggardsplatsen. Office: Sahlgrenska Sjukhuset, Göteborg SV, Sweden.*

HOOKE, Robert, English natural philosopher; b. Freshwater, Isle of Wight, July 17, 1635; s. John Hooke; for short period pupil of artist Sir Peter Lely; studied for five years at Westminster School under Dr. Busby; entered Christ Church, Oxford U., as Servitor, 1653; asst. to Thomas Willis in chem. and nat. philosophy, 1655; asst. to Robert Boyle (largely responsible for constructing Boyle's air pump); first curator of experiments for Royal Soc., 1662-1703 (office made perpetual, 1665); M.A., Oxford U., 1663; Fellow Royal Soc., 1663 (sec. 1677-83); fund for annual lectrs. established by Sir John Cutler for Hooke's benefit, 1664 (Cutlerian Lectures); Gresham prof. geometry, 1665; became surveyor after Great London Fire of 1666; prepared plan for rebuilding London and designed Montague House, Bethleham Hosp., Coll. Physicians; created doctor of physic at Doctors' Commons by warrant from Archbishop Tillotson, 1691. Author: Micrographia, 1665; Lectures de Potentia restitutiva, 1678; Lampas, or Descriptions of some Mechanical Improvement of Lamps and Waterpoises, 1677; Lectiones Cutlerianae, 1679; Posthumous Works, 1705; Philosophical Experiments and Observations, 1726; others; altogether some 35 published books and papers on wide variety of topics. Used freezing and boiling points of water as fixed points on the thermometric scale (as van Helmont had suggested earlier); described iridescent colors of thin films and transparent plates; explained them as manifestations of optical interference; held light to be vibratory motion in a homogeneous medium; explained heat as a form of motion of the small parts of body; suggested that force of gravity could be measured by motion of pendulum, 1666; measured vibrations of 200 foot long pendulum in steeple of St. Paul's Cathedral, London; attempted to show center of gravity of earth and moon is point describing an ellipse around sun, 1666; applied circular pendulum to watches; used spiral spring to regulate balance of watches; invented anchor escapement for clocks; invented many other instruments; among 1st to use microscope to examine minute objects; discovered honeycomb cavities (which he called cells) in thin sections of cork; discovered Hooke's Law of Elasticity, 1660; carried out experiments on capillary action, 1680; 1st to describe thermal expansion as general property of matter; conceived air to be particulate in nature with particles separated from one another at relatively great distances; collected different airs in bladders; carried out experiments on combustion and respiration with air pump; like earlier iatrochemists insisted that active part of air required for combustion was "aerial niter," 1665; showed that a dog could live with its lungs motionless if blown up with a pair of bellows, 1663; one of 1st to discuss crystal structure as build up of spherical particles, 1665; made 1st Gregorian telescope; suggested that Jupiter rotates, 1664; gave detailed sketches of Mars (helped in determination of its rate of rotation in 19th century); discovered 5th star in Orion, 1664 (rediscovered, 1826); attempted unsuccessfully to determine parallax of fixed star with telescope, 1669; described diffraction of light, 1672, 1675 (anticipated by Grimaldi); stated a law of inverse squares for planetary motions in his Cometa, 1678; in letters to Newton, 1679, presented conjectures and queries on paths of projectiles which induced Newton to return to his own investigations in mechanics; involved in controversies with Newton, Hevelius and Oldenburg; his complaints led Newton to suppress publication of his Opticks until after Hooke's death; Hooke also asserted that Newton had not given him enough credit in Principia Mathematica for his statement of inverse square law (which in fact only Newton proved for elliptical paths). Died London, March 3, 1703.

HOOKE, Robert, Am. mathematician; b. Chattanooga, Apr. 8, 1918; s. Malcolm King and Lucy (Annis) H.; B.A., U. N.C., 1938, M.A., 1939; A.M., Princeton, 1940, Ph.D., 1942; m. Annis Marie Hines, June 26, 1941; children—William Hines, John Allen. Faculty, N.C. State U., 1941-46, asst. prof., 1943-46; vis. asst. prof. Duke, 1944; asso. prof.

math. U. South, Sewanee, Tenn., 1946-51; staff operations evaluation group U. S. Dept. Navy, Washington, 1951-52; research asso. Princeton, 1952-55; staff Westinghouse Research Labs., Pitts., 1954—, mgr. statistics sect. 1956-63, mgr. math., 1963—. Fellow Am. Statis. Assn., A.A.A.S.; mem. Operations Research Soc. Am., Inst. Math. Statistics, Soc. for Indsl. and Applied Math. Author: Introduction to Scientific Inference, 1963; (with Fox, Garbuny) Science of Science, 1964; (with D. H. Shaffer), Math and Aftermath, 1965; also articles. Research on modern algebra, theory topological groups, math. statistics, numerical analysis, process control; inventor device for optimizing control. Home: 2630 McCrady Rd. Office: Westinghouse Research and Devel. Center, Pitts. 15235.*

HOOKER, Arthur Lee, Am. plant pathologist; b. Lodi, Wis., Oct. 12, 1924; s. Robert Lee and Dora (Leuth) H.; B.S., U. Wis., 1948, M.S., 1949, Ph.D., 1952; m. Ellen Margaret Zimmerman, July 5, 1950; children—David Lee, Margaret Ann. With U. Wis., 1948-52, project asso. plant pathology, 1951-52, asst. prof. plant pathology, 1954-58; asst. prof. botany and plant pathology Ia. State U., 1952-54; plant pathologist U. S. Dept. Agr., 1954-58; faculty U. Ill., Urbana, 1958—, prof. plant pathology and genetics, 1963—; cons. to seed industry, 1958—. John Simon Guggenheim Meml. Found. fellow, 1964-65. Mem. Am. Phytopath. Soc., Crop Sci. Soc. Am., Genetics Soc. Am., A.A.A.S. Studies, publs. on genetics and physiology of plant parasite interactions, variation in plant pathogenic fungi, sources and types of disease resistance in plants, and devel. of disease resistant crop plants. Home: 4 Illini Circle, Urbana, Ill. 61801.*

HOOKER, Sir Joseph Dalton, English botanist; b. Halesworth, Suffolk, Eng., June 30, 1817; s. Sir William Jackson Hooker; ed. U. Glasgow; M.D., D.C.L., LL.D.; m. Frances Harriet Henslow, 1851 (dec. 1874); m. 2d, Hyacinth Symonds Jardine, 1876; 6 sons, 2 daus. Surgeon, naturalist in H.M.S. Erebus in Antarctic Expdn. under Sir James Ross, 1839-43; as naturalist visited Himalaya Mountains, Eastern Bengal, Khasia Mountains, 1847-51, Syria and Palestine, 1860, Morocco and Greater Atlas, 1871, Rocky Mountains and Cal., 1877; asst. dir. Royal Gardens, Kew, 1855-65, dir., 1865-85. Fellow Royal Soc., 1847 (pres. 1872-77), Linnaean Soc.; mem. French Acad. Scis., 1866. Author: Botany of the Antarctic Expedition; Handbook of the New Zealand Flora; Himalayan Journal; Student's British Flora; (with J. Ball) The Rhododendrons of the Sikkim Himalaya, Morocco, and the Great Atlas; The Flora of British India; (with George Bentham) Genera Plantarum, 7 vols., 1862-63 (gave important system of classification). Authority on Antarctic flora; collaborated with Darwin, influential in persuading Darwin to publ. On the Origin of Species. Died Dec. 10, 1911.

HOOKER, Sir William Jackson, English botanist; b. Norwich, Eng., July 6, 1785; s. Joseph Hooker; ed. privately; LL.D., Glasgow, Scotland; D.S.L., Oxford, Eng.; m. Maria Turner, 1915; children—Joseph Dalton, 2 daus. Studied botany in Scotland, 1806, Iceland, 1809, France, Switzerland, No. Italy, 1914; settled in Halesworth, Suffolk, Eng., 1915-20; became Regius prof. botany U. Glasgow, 1820; dir. Royal Gardens, Kew, Eng., 1841-65; founder (with J. S. Henslow 1st mus. of econ. botany at Kew Gardens, 1847. Fellow Linnean Soc., Royal Soc., 1812, Wernerian Soc. Edinburgh, French Acad. Scis. (corr. mem.), Author numerous books on botany including: British Jungermanniae, 1816; Flora Scotica, 1821; Icones Plantarum, 10 vols., 1827-54; The British Flora, 2 vols., 1830-36; Genera Filicum, 1842; Species filicum, 5 vols., 1846-64; Synopsis filicum, 1868. Research on ferns; discovered species of moss new to Eng., 1805; sent plants from Kew to various parts of the Empire. Died Kew Aug. 12, 1865.

HOOKER, Worthington, Am. physician; b. Springfield, Mass., Mar. 3, 1806; s. John and Sarah (Dwight) H.; grad. Yale, 1825; M.D., Harvard, 1829; m. Mary Ingersoll, Sept. 30, 1830; m. 2d, Henrietta Edwards, Jan. 31, 1855. Prof. theory and practice of medicine Med. Instn. of Yale Coll., 1852-67; a dir. Conn. Hosp. Soc.; v.p. A.M.A., 1864. Author: Physician and Patient, 1849; Homeopathy, 1852; Human Physiology for Colleges and Schools, 1854; Rational Therapeutics, 1857. Died New Haven, Conn., Nov. 6, 1867.

HOOLEY, Joseph Gilbert, Canadian chemist; b. Vancouver, B.C., Can., Sept. 26, 1914; s. Joseph Stringfellow and Cecelia (Frisby) H.; B.A., U. B.C., 1934, M.A., 1936; Ph.D., Mass. Inst. Tech., 1939; m. Agnes Schroeder, Sept. 16, 1939. Research chemist Corning (N.Y.) Glass Works, 1939-42; prof. chemistry U. B.C., Vancouver, 1942—, dept. chmn., 1949-55; vis. scientist Canadian Atomic Energy Plant, Chalk River, Ont., 1947-48. Fellow Chem. Inst. Can.; mem. Am. Chem. Soc., Faraday Soc. London. Research on hydrogen bond behavior at low temperatures, U.V. transmitting and fluorescent glass compositions, mechanism of intercalation of graphite; pioneer all glass recording vacuum balance. Home: 4769 W. 7th St., Vancouver 8, B.C., Can.*

HOOPER, Leslie James, Am. hydraulic engr.; b. Essex, Mass., Feb. 15, 1903; s. William J. and Ethel (Cruze) H.; B.S. in Mech. Engring., Worcester Poly. Inst., 1924, M.E., 1928, D.Sc., 1964; m. Edith C. Stockwell, May 24, 1930; children—Donald Leslie, William Robert, Lee Monroe, Neal Cruze. Hydraulic test engr. Canadian and Gen. Finance Co., Sao Paulo, Brazil, 1924-27; with Worcester Poly. Inst., 1928—, prof. hydraulic engring., dir. Alden Hydraulic Lab., 1952—. Freeman fellow, 1934-36. Fellow Am. Soc. C.E., Am. Soc. M.E. (Jr. award 1937); mem. Am. Soc. Engring. Edn., Congress on Large Dams, Instrument Soc. Am., Boston Soc. Civil Engring., Internat. Assn. for Hydraulic Research, Am. Water Works Assn., Sigma Xi, Tau Beta Pi. Author: (with C. M. Allen) Piezometer Measurement, 1938; Salt Velocity Measurements of Low Velocities in Pipes, 1940; American Hydraulic Laboratory Practice, 1936; Representative Hydraulic Laboratories in the United States and Canada, 1938; also articles. Research in hydraulic measurements, gen. heat pollution problem. Home: 31 Highland Av., Holden, Mass. 01520. Office: Alden Research Labs., Worcester Poly. Inst., Worcester, Mass. 01609.*

HOOTON, Earnest Albert, Am. anthropologist; b. Clemansville, Wis., Nov. 20, 1887; s. William and Margaret Elizabeth (Newton) H.; B.A., Lawrence Coll., Appleton, Wis., 1907, Sc.D., 1933; M.A. University of Wisconsin, 1908, Ph.D., 1911; also LL.D. (honorary), 1954; Rhodes Scholar at Oxford U., 1910-13, diploma in anthropology, 1912, B.Litt., 1913; m. Mary Beidler Camp, June 3, 1915; children—Jay Camp, William Newton, Emma Beidler. Instr. anthropology, 1913, asst. prof., 1921, asso. prof., 1927, prof. since 1930, Harvard. Asst. curator somatology, Peabody Museum, 1913-14, curator since 1914. Fellow A.A.A.S. (v.p. sect. H, 1923-24), Royal Anthropol. Inst., Am. Acad. Arts and Sciences; mem. Am. Anthropol. Assn., Am. Assn. Phys. Anthropol., Am. Genetic Assn., Am. Philos. Soc., Am. Soc. Naturalists, Nat. Acad. Sci., Phi Beta Kappa; hon. fellow Am. Acad. Dental Science. Author: Ancient Inhabitants of the Canary Islands, 1925; The Indians of Pecos, 1930; Up from the Ape, 1931; rev. ed., 1946; Apes, Men and Morons, 1937; Crime and the Man, The American Criminal, Vol. I, Twilight of Man, 1939; Why Men Behave Like Apes and Vice Versa, 1940; Man's Poor Relations, 1942; "Young Man, You are Normal," 1945; also papers on physical anthropology. Investigations of criminals in Mass.; generations of Harvard graduates; adult Irish males; Canary Islanders; New Mexico Indians. Died May 3, 1954.

HOOVER, Charles Franklin, Am. physician; b. Miamisburg, O., Aug. 2, 1865; s. Abel and Clara Elizabeth (Hoff) H.; Ohio Wesleyan U., 1882-85; A.B., Harvard, 1887, M.D., 1892; univs. Vienna and Strassburg, 1890-94; m. Katherine Fraser, Aug. 9, 1900. Practiced in Cleve., from 1894; tchr. physical diagnosis and visiting physician, Cleveland City Hosp., 1894-1907; prof. medicine, Med. Coll. Western Reserve U., from 1907. A prominent med. cons., especially in diagnosis pulmonary, cardiac and hepatic diseases. Visiting physician Lakeside Hosp. Died June 15, 1927.

HOOVER, C(harles) R(uglas), Am. chemist; b. Oskaloosa, Ia., Sept. 30, 1885; s. Hiram Alonzo and Edith Adaline (Crane) H.; Ph.B., Penn Coll., Oskaloosa, Ia., 1906; B.S., Haverford, 1907, M.A., 1908; Ph.D., Harvard, 1915; m. Anna Mary Johnson, Sept. 7, 1912; children—Albert Charles, John Crane. Prof. chemistry, Penn Coll., 1909-10; Austin fellow in chemistry Harvard, 1910-12, Carnegie research fellow in chemistry, 1912-13; asso. prof. chemistry, Syracuse U., 1913-15; asso. prof. chemistry Wesleyan U., Middletown, Conn., 1915-18, prof. from 1918, v.p., 1926-27. Gas chemist research div. C.W.S., U. S. Army, 1918; cons. chemist same, 1917-19; cons. chemist, State Water Commn. and State Commn. Fisheries and Game, Conn., from 1928. Consultant Nat. Defense Research Com. Mem. Conn. State Board of Registration for Engineers; mem. State Flood Control and Water Policy Commn. Mem. of committees of Nat. Research Council and Assn. Harvard Chemists; fellow Am. Inst. Chemists, A.A.A.S.; mem. Am. Chem. Soc. (councillor, also pres. Conn. Valley sect.), New England Sewage Works Assn., Soc. Chem. Industry, Am. Public Health Assn., N.E.A., Sigma Xi, Phi Beta Kappa, Alpha Chi Sigma. Inventor of gas absorbent and gas detector, circa 1917; worked on determination atomic weights, analysis of gases, lab. constrn. and equipment. Author: Laboratory Construction and Equipment; contbr. on chem. topics. Died June 8, 1942.

HOOVER, John Irvin, Am. physicist; b. Altoona, Pa., Sept. 11, 1911; s. George C. and Hulda Ann (Sweeney) H.; A.B., Am. U., 1935; postgrad. George Washington U., 1940-41; m. Mary Ann Hodgson, May 18, 1940; children—John Blake, Marjorie Jane, Mary Anne. Staff, U. S. Naval Research Lab., Washington, 1939-47, 47—; staff Monsanto Chem. Co., Oak Ridge, 1947. Recipient Distinguished Civilian Service award USN, 1947, Meritorious award, 1957. Mem. A.A.A.S., Research Soc. Am., Am. Phys. Soc., Washington Acad. Sci., Washington Philos. Soc. Author: (with Abelson, Rosen) Liquid Thermal Diffusion, 1951; also articles. Research in energy conversion, liquid thermal diffusion method isotope separation, isotope separation in molten salts, gamma ray polarization, neutron detecting phosphors, Measurement nuclear constants, nuclear instrumentation, radiation monitoring systems, activation analysis as applied to cloud physics and oceanography. Home: 5313 Briley Pl., Washington 20016. Office: U. S. Naval Research Lab., Washington 20390.*

HOOVER, John Russel Eugene, Am. chemist; b. New Enterprise, Pa., Jan. 3, 1925; s. Jason E. and Hattie (Miller) H.; B.S., Juniata Coll., 1947; M.S., U. Pa., 1949, Ph.D., 1953; m. Janet Louise Gochnour, Mar. 11, 1945; children—Carol Lynn, John Lawrence, Karen Ann. Sr. scientist Wyeth Inst. Applied Biochemistry, Am. Home Product Corp., 1951-53; research asso. cancer chemotherapy program U. Pa., 1953-56; with Smith Kline & French Labs., Phila., 1956—, group leader, 1959-61, medicinal chemistry sect. head, 1961—. Mem. A.A.A.S., Am. Chem. Soc., N.Y. Acad. Scis., Am. Inst. Chemists, Sigma Xi, Alpha Chi Sigma. Patentee in field. Research, publs. in medicinal chemistry related to microbial, viral and cancer chemotherapy, heterocycles, cage compounds, antibiotics. Home: 624 Crescent Av., Glenside, Pa. 19038. Office: 1500 Spring Garden St., Phila. 19101.*

HOPE, James, English physician; b. Stockfort, Cheshire, Eng., Feb. 23, 1801; s. Thomas Hope; student medicine Edinburgh (Scotland) U.; diploma Royal Coll. Surgeons, 1826; licentiate Coll. Physicians, 1828; m. Anne Fulton, Mar. 10, 1831; 1 son, Theodore C. Established pvt. dispensary, 1829; physician to Marylebone Infirmary, from 1831; lectr. diseases of chest, 1832; asst. physician, lectr. St. George's Hosp., from 1834, full physician, from 1839; lectr. Aldersgate St. Sch. Medicine; house physician, house surgeon Royal Infirmary. Fellow London, Royal colls. physicians, Royal Soc., 1832; mem. Royal Med. Soc. Edinburgh (pres.). Author: A Treatise on the Diseases of the Heart and Great Vessels, 1831, revised 3d edit., 1839; Principles and Illustrations of Morbid Anatomy, 1834. Contbg. author: Cyclopaedia of Practical Medicine, 1833-35. Contbr. articles to sci. publs. Helped prove value of asculsation; many observations on phys. signs cardiac disease; expts. prodn. heart sounds; Hope's murmur named for him. Died 1841.

HOPE, John, Scottish botanist; b. Edinburgh, Scotland, May 10, 1725; s. Robert Hope; ed. Edinburgh U., European med. schs.; M.D., U. Glasgow (Scotland), 1750; m. Juliana Stevenson; 4 sons (including Thomas Charles), 1 dau. Prof. botany, materia medica Edinburgh U., from 1761, regius prof. medicine, botany, from 1768; King's botanist for Scotland; supt. Royal Garden Edinburgh; physician to Edinburgh Royal Infirmary, to 1786. Genus Hope named for him by Linnaeus. Fellow Royal Soc., 1767; mem. Edinburgh Coll. Physicians (pres. 1786). Editor: Genera Animalium (Linnaeus), 1781; pub. Alston's lectures on materia medica, 1770. Erected monument to Linnaeus in Edinburgh Bot. Gardens, also arranged plants according to Linnean system. Died 1786.

HOPE, Thomas Charles, Scottish chemist; b. Edinburgh, Scotland, 1766; s. John and Juliana (Stevenson) H.; grad. Edinburgh U., 1787. Prof. chemistry Glasgow (Scotland) U., 1787-89, asst. prof. medicine, 1789-95; joint prof. chemistry (with Joseph Black) Edinburgh U., 1795-99, sole prof. (on Black's death), from 1799, taught for 50 years, founder chem. prize, 1828. Fellow Royal Soc., 1810. Author: Tentamen Inaugurale quaedam de Plantarum Motibus et Vita, complectens . . . , 1787. First chemist in Eng. to teach Lavoisier's views on combustion (learned from him personally); (with Crawford) 1st to distinguish between baryta and strontia; proved strontian contained peculiar earth; estimated maximum density of water. Died Edinburgh, June 13, 1844.

HOPF, Harry Arthur, management engr.; b. London, Eng., Apr. 3, 1882; s. Charles and Franziska (Grote) H.; came to U. S., 1898; naturalized, 1903; prep. edn. in Germany, 1892-98; B.C.S., New York U., 1906, M.C.S., 1914, M.B.A., 1922; post-grad. student, Columbia, 1927-28; hon. M.S., Bryant Coll., Providence, R.I., 1937; Dr. Engring., Rensselaer Polytechnic Institute, 1942; m. Flora Paine, 1908; children—Elliott Arthur, Gordon Allen; m. 2d, Rita Hilborn, 1926. Underwriting and organization work Germania (now Guardian) Life Ins. Co., New York, 1902-14; organization and office planning Phoenix Mutual Life Ins. Co., Hartford, Conn., 1914-17, E. I. duPont de Nemours & Co., Wilmington, Del., 1917-18; professional practice as management engr., New York, 1918-19; orgn. counsel Federal Reserve Bank of New York, 1919-22; management engr. at H. A. Hopf & Co. since 1922; pres. Hopf Inst. of Management, Ossining, N.Y.; adviser to President Taft's Commn. on Economy and Efficiency, 1909; lecturer in office management, New York U., 1919-25, Columbia, 1931-33, adjunct prof. management Grad. Sch. Bus. Adminstrn., New York U., 1947-49. Mem. engrs. adv. com. Div. of Contract Distribution, Office of Production Management, 1941. Licensed professional engineer, N.Y.; chmn. bd. trustees Bard Coll., Annandale-on-Hudson, 1948-49. Del. 1st Internat. Management Congress, Prague, 1925, 5th Congress, Amsterdam, 1932, 6th Congress, London (chairman U. S. del.), 1935, 7th Congress, New York (chmn. organizing com.), 1938. Fellow Inst. of Industrial Administration (London). Mem. Acoustical Soc. America, A.A.A.S., Am. Management Assn. (v.p. 1929), Am. Marketing Soc., Am. Soc. Mech. Engrs., N.Y. State Soc. Professional Engrs., Am. Soc. for Pub. Adminstrn., Am. Statis. Assn., Assn. of Cons. Management Engrs. (charter mem.; pres. 1933-35; dir. 1935-36), Inst. of Management of Am. Management Assn. (charter fellow, pres. 1929), Econometric Soc.,

Nat. Bureau Economic Research, Nat. Office Management Assn. (hon. fellow; president 1932-34; director 1934-36), Personnel Research Federation. Special Libraries Association, Comité Nat. de l'Organisation Française (Paris), Comité Nat. Belge d'Organisation Scientifique (Brussels), Internat. Com. Sci. Management (dep. pres., 1935-38, hon. mem. council, 1938-49), Internat. Indsl. Relations Inst. (The Hague), Masaryk Acad. Work (Praha), Nat. Management Council of the U.S.A. (founder; chmn. 1932-35), Soc. for Advancement of Management (advisory council, v.p. and mem. operating council 1937-39); fellow Royal Econ. Soc., Royal Soc. of Arts (both London). Recipient: gold medal by Internat. Com. of Scientific Management, 1938; Taylor Key award for distinguished services in management Soc. Advancement Management, 1947; Knight Royal Order of North Star by direction of King Gustaf VI of Sweden, 1947. Author of numerous articles and brochures on management and organization; contributing editor, The Spectator. Pioneer in applying scientific management to office; studied office procedures analysis, clerical work standardization; production control and job analysis. Died Ossining, N.Y., June 3, 1949.

HOPF, Heinz, mathematician; b. Breslau, Germany (now Wroclaw, Poland), Nov. 19, 1894; s. Wilhelm and Elisabeth (Kirchner) H.; ed. univs. Breslau, Heidelberg, Berlin, Göttingen; D.phil.; D. honoris causa, Princeton, Freiburg, Manchester, Paris, Brussels, Lausanne; m. Anna Marie von Mickwitz, 1928. Privatdozent U. Berlin, 1926-31; prof. ETH, Zurich, 1931-65, prof. emeritus, 1965—. Hon. mem. various acads. and math. socs. Address: Alte Landstrasse 37, Zollikon (ZH), Switzerland.

HOPFF, Heinrich, chemist; b. Kaiserslautern, Germany, Oct. 19, 1896; s. Arthur and Rosa (Fischer) H.; doctor, U. Munich, 1921; m. Ida Klein, June 6, 1926; children—Wolfgang, Lore, Ervin Kovats. Superior high sch, Kaiserslautern, 1907-16; with U. Munich, 1916-21; research chemist BASF, Ludwigshaven, Germany, 1921-52; prof. organic tech. U. Mainz (Germany), 1945-51, Fed. Tech. High Sch. Zürich, Switzerland, 1952-67; Cons. to pvt. cos. Recipient Lavoisier medal Societé Chimique de France; named hon. citizen U. Mainz, 1964. Mem. Deutsche Chemische Gesellschaft, Schweiz. Chem. Gesellschaft, Am. Chem. Soc. Author: (with A. Müller, Fritz Wenger) Die Polyamide, 1954; Introduction in Inorganic and Organic Chemistry, 16th edit., 1962; also numerous articles. Research, numerous patents on dyestuff and plastics; devel. new intermediates. Home: 5, Raenkestreet Kuesnacht-ZH, Switzerland.*

HOPFIELD, John Joseph, Am. physicist; b. Chgo., July 15, 1933; s. John Joseph and Helen (Staff) H.; A.B., Swarthmore Coll., 1954; Ph.D., Cornell U., 1958; m. Cornelia Fuller, June 30, 1954; children—Alison, Jessica. Mem. tech. staff Bell Telephone Labs., 1958-60; vis. research physicist Ecole Normale Superieure, Paris, France, 1960-61; asst. prof., then asso. prof. physics U. Cal. at Berkeley, 1961-64; prof. physics Princeton, 1964—. Mem. Am. Phys. Soc., Phi Beta Kappa, Sigma Xi. Research on quantum theory of solids, relation between electronic structure and dielectric response function (optical properties) of solids. Home: 183 Hartley Av., Princeton, N.J. 08540.*

HOPKINS, B. Smith, Am. chemist; b. Owosso, Mich., Sept. 1, 1873; s. Loren Hopkins and Clara Sibley (Norgate) H.; A.B., Albion (Mich.) Coll., 1896, A.M., 1897, Sc.D., 1926; studied Columbia, 1900-01; Ph.D., Johns Hopkins, 1906; LL.D., Carroll Coll., 1940; m. Maude Childs, June 25, 1901; children—Harvey Childs, B. Smith; m. 2d, May L. Whitsitt, Dec. 17, 1942, Began with pub. schs., Menominee, Mich., 1897; prin. high school. 1898-1900, supt. schs., 1901-04; prof. chemistry Neb. Wesleyan U., 1906-09, Carroll Coll., Waukesha, Wis., 1909-12; with U. Ill., from 1912, prof. inorganic chemistry, from 1923, dir. sci. Gen. Studies Div., 1941-42, dir. chemistry teaching Army Specialized Tng. Program, 1942-43; spl. summer lectr. Northwestern U., 1910-11, Western Reserve U., 1929, emeritus, 1941. Fellow A.A.A.S.; mem. Am. Chem. Soc., Am. Electrochem. Soc., Ill. Acad. Science (pres. 1933-34), Central Assn. Science and Mathematics Tchrs., Am. Philos. Soc., Sigma Xi, Phi Beta Kappa, Phi Lambda Upsilon, Alpha Chi Sigma, Alpha Tau Omega. Author: Exercises in Chemistry (with W. A. Noyes), 1917, 19; Chemistry of the Rarer Elements, 1923; Laboratory Exercises in General Chemistry (with H. A. Neville), 1925, 26, 31; General Chemistry for Colleges, 1930 (rev. 1937, 42); Essentials of College Chemistry, 1932; with J. C. Bailar, Jr. 1945; Laboratory Exercises in General Chemistry (with M. J. Copley), 1937, (with M. J. Copley and F. B. Schirmer, Jr., 1942); (with T. Moeller and F. B. Schirmer, Jr., 1946). Co-author: Chemistry and You, 1939, 44, 49; chapters in Chemistry of Less Familiar Elements, 1940. With colleagues discovered element 61, named it illinium, 1926; worked mainly with rare earth group of chem. elements; responsible for many improvements in tech. of methods of purification and for determination of accurate atomic weights; made many contbns. to chemistry of little known elements including cesium, beryllium, selenium, tellurium; contbd. largely to devels. in ednl. field of chemistry. Died Aug. 27, 1952.

HOPKINS, Clarence Yardley, Canadian chemist; b. Kinmount, Ont., Can., June 20, 1903; s. Harry and Mabel (Soward) H.; B.A., Queen's U., 1924, M.A., 1926; postgrad. Northwestern U.; Ph.D., N.Y. U., 1929; m. Kathleen Kerr, June 4, 1928; 1 son, Edward. Instr., U. Toronto (Ont.), 1929-30; research chemist NRC, Ottawa, Ont., 1930—, head fats and oils sect. div. pure chemistry, 1952—. Lectr., Carleton U., Ottawa, evenings 1945-49. Fellow Chem. Soc. London, Chem. Inst. Can.; mem. Am. Chem. Soc., Am. Oil Chemists Soc., Internat. Soc. for Fat Research. Contbg. author: Progress in the Chemistry of Fats, vol. 8, 1965. Research, publs. on discovery and identification of 12 new fatty acids in seed oils, including isomers of linoleic acid; discovered 1st goitrogenic sulphur compound in plants, 1938; developed ofcl. method of analysis for starch in cereals, 1934. Home: 180 Carleton Rd., Ottawa 2. Office: NRC, Ottawa, Ont., Can.*

HOPKINS, Cyril George, Am. agrl. chemist; b. nr. Chatfield, Minn., July 22, 1866; s. George Edwin and Caroline (Cudney) H.; B.S., S.D. Agrl. Coll., 1890; M.S., Cornell, 1894, Ph.D., 1898; studied agrl. chemistry at Göttingen, 1899-1900; m. Emma Matilda Stelter, May 11, 1893. Asst. in chemistry Agrl. Coll. and Expt. Sta., Brookings, S.D., 1890-92, 1893-94; Cornell, 1892-93; acting prof. pharmacy S.D. Agrl. Coll., 1893-94; chemist Ill. Agrl. Expt. Sta., 1894-1900, prof. agronomy and chief in agronomy and chemistry, from 1900, vice-dir., from 1903; dir. agr. So. Settlement and Devel. Orgn., 1913-14. Author: Soil Fertility and Permanent Agriculture, 1910; The Story of the Soil, 1911; The Farm That Won't Wear Out, 1913; also bulls. on soil investigations and papers on chem. and agrl. subjects. Invented Hopkins' condenser, Hopkins' distilling tube, 1898, Hopkins' limestone tester, 1917. Died Oct. 6, 1919.

HOPKINS, Sir Frederick Gowland, English biochemist; b. Eastbourne, Eng., June 30, 1861; student Guy's Hosp.; Ph.D., London, Eng., 1894. Joined faculty Guy's Hosp., London, 1894; joined faculty chemistry, Cambridge, Eng., 1898, became reader chem. physiology, 1902, prof. biochemistry, 1914. Recipient Royal medal, 1918, Copley medal, 1926, Order of Merit, 1935, (with Eijkman) Nobel prize in physiology and medicine, 1929. Fellow Royal Soc., 1905 (pres. 1930-35). Isolated tryptophane, 1901; discovered essential amino acids, accessory food factors (vitamins), 1906-07; explained (with Walter M. Fletcher) connection between lactic acid and muscular contraction, 1907; isolated glutathione from living tissue, 1921. Died Cambridge, May 16, 1947.

HOPKINS, Harold Hoffman, Am. botanist; b. Bison, Kan., May 1, 1918; s. Arthur J. and Eunice (Peevey) H.; B.A., Ft. Hays Kan. State Coll., 1940, M.S., 1941; Ph.D., U. Neb., 1950; m. Pauline Felter, Dec. 22, 1943; children—David, Norman, Nona, Marcia, Jolene. Soil conservationist U. S. Dept. Agr., 1941, 45; asst. prof. Ft. Hays Kan. State Coll., 1946, 49-57; asso. prof. St. Cloud (Minn.) State Coll., 1957-59, prof., chmn. dept. biology, 1959—. Cons. AID, India, 1966. Mem. A.A.A.S., Ecol. Soc. Am. Research, publs. on ecology of grassland utilization. Home: 818 9th Av., St. Cloud, Minn. 56301.*

HOPKINS, Nevil Monroe, Am. engr., inventor; b. Portland, Me., Sept. 15, 1873; s. Francis Nevil and Frances Anna (Monroe) H.; B.S., Columbian (now George Washington) U.; 1899, M.S., 1900, Ph.D., 1902; grad. student, Harvard, 1901; m. Katherine Guy, Jan. 5, 1897; children—Anne Dorsey (Mrs. James W. Allison), Frances Monroe (Mrs. Horace W. Peaslee); m. 2d, Raymonde Briggs, June 22, 1932. Instr. chemistry, Columbian Univ., 1899-1902; asst. prof. chemistry George Washington U., from 1902; professional engr. lecturer, Coll. of Engring., N.Y. U., from 1934; mem. faculty Institute for Industrial Progress; mem. Munroe, Hall & Hopkins, cons. engrs. Electrician, Gen. Electric Co., Schenectady, N.Y.; editorial rep. Electrical World and Engineer, N.Y.; trustee and in charge div. elec. engring., Inst. of Industrial Research; v.p. and elec. engr. Electric Tachometer Co. Elec. engr. Navy Dept. in charge power plant design and constrn. at all navy yards and stas., 1905-08; expert engr. U. S. Office of Public Roads, Washington, from 1909; tech. adviser, design sect. of gun div. Bur. of Ordnance, from 1917; cons. engr. numerous companies; dir. research Burnot Fireproofing Products; pres. New-Mix Products, Inc., Internat. Tube Co. Recipient, John Scott medal, Franklin Inst., 1900; George Washington U. Alumni Award for notable achievement in science, 1942. Fellow A.A.A.S., American Institute Mech. Engrs., mem. Am. Chem. Soc., Am. Electrochem. Soc., Am. Soc. Testing Materials, Am. Soc. M.E., Am. Inst. Chem. Engrs., Soc. Am. Mil. Engineers, Army Ordnance Assn., Inst. of Social Sciences, Author: Model Engines and Boats, 1898; Twentieth Century Magic, 1904; Experimental Electro-Chemistry, 1905; The Strange Cases of Mason Brant, 1916; The Racoon Lake Mystery; Over the Threshold of War; The Outlook for Research and Invention; The Inventor and His Workshop; The Horrors in the Grew Mystery; also articles in scientific and engring. jours., and short stories in mags. Inventor of electric and mech. devices, instruments for high temperature measurements, etc.; inventor and developer submersible battle cruiser and long range naval and antiaircraft guns, high explosive antiaircraft shells and battleship wrecking bombs; blast meter and system for U. S. Army for measuring force of high explosives in the field; electro chronograph for ballistic measurements; inventor Synchronous Electric Registration and Voting System, and Home Registration Voting Stations making possible mass voting by radio and newspaper announcement; automatic radio-electric survey system showing the number of radio receiving sets tuned in to any particular broadcasting station wave length, at any time; inactivators for destroying time bombs; designer of torpedo and magnetic-mine protection equipment for freighters at sea; also super rocket guns and rocket missiles. Died Mar. 26, 1945.

HOPKINS, Prynce Charles, Am. psychologist; b. Oakland, Cal., Mar. 5, 1885; s. Charles H. and Mary (Booth) H.; Ph.B., Yale, 1906; M.A., Columbia, 1910; Ph.D., U. London, 1927; m. Eileen Thomas, 1920; 1 dau., Eileen (Mrs. Ames), m. 2d, Fay Cartledge, 1933; children—Jennifer (Mrs. Jorgen Hansen), David C. Prin., Boy Land Exptl. Sch., Santa Barbara, Cal., 1912-18; founder, supr. Chateau de Bures Sch. for Boys, France, 1925-35; hon. lectr. psychology U. Coll., U. London, 1927-40; lectr. psychology Claremont (Cal.) Grad. Sch., 1941-46; ret. Fellow Am. Psychol. Assn., Brit. Psychol. Soc., Royal Geog. Soc.; asso. mem. Brit. Psychoanalytic Assn.; mem. Am. Sociol. Soc. Author: Psychology of Social Movements, 1938; Aids to Successful Study, 1941; A Westerner Looks East, 1951; Orientation, Socialization and Individualization, 1963; Social Psychology of Religious Experience, 1962; World Invisible, 1963. Research on unconscious motivation of individual and social behavior, various Occidental and Oriental norms especially proclivities to violence and proliferation, values and beliefs, personality-maturation from instinct-dominated autism, through realistic egotism, conformism. Address: 1920 Garden St., Santa Barbara, Cal. 93101.*

HOPKINS, Reginald Haydn, Brit. biochemist; b. Birmingham, England, Feb. 13, 1891; s. Albert and Jane (Rosser) H.; Ph.D. in Biochemistry, U. Birmingham; m. Phyllis Harrison, 1953; 1 son, Patrick Adrian. Asso. chem. analyst Pub. Health Com of Warwyck, 1911-17; instr. breweries and biochemistry Heriot Watt Coll., Edinburgh, 1920-31; prof. breweries and biochemistry U. Birmingham, 1931-56. Hon. prof. Higher Sch. Fermentation, Ghent, 1947. Fellow Royal Inst. Chemistry; mem. numerous Brit. sci. socs. Coauthor: Biochemistry Applied to Malting and Brewing, 1937, 48, also numerous articles. Address: Portarlington Rd., Curzon Ct., Bournemouth, Eng.

HOPKINS, Sewell Hepburn, Am. zoologist; b. Gloucester, Va., Mar. 24, 1906; s. Nicholas Snowden and Selina (Hepburn) H.; B.S., William and Mary Coll., 1927; postgrad. Johns Hopkins; M.A., U. Ill., 1929, Ph.D., 1933; m. Pauline Harriet Cole, 1929; children—Thomas Johns, Nicholas Arthur. Instr., Danville Jr. Coll., 1933-35; jr. parasitologist Bur. Animal Industry, U. S. Dept. Agr., 1934; faculty Tex. A. and M. U., College Station, 1935—, prof. biology, 1947—. Recipient Faculty Distinguished Achievement award for research Tex. A. and M. U., 1957. Mem. A.A.A.S., Am. Inst. Biol. Scis., Micros. Soc. Am., Soc. Parasitologists, Nat. Shellfisheries Assn., Soc. Limnologists and Oceanographers, Ecol. Soc., Soc. Systematic Zoology, Phi Beta Kappa, Sigma Xi, Phi Kappa Phi. Research, numerous publs. on systematics and life cycles of parasites, crabs, oysters, predators and parasites of oysters, fishes, boring sponges. Home: 709 Garden Acres, Bryan, Tex. 77801.*

HOPKINS, Terence Kilbourne, Am. sociologist; b. New Rochelle, N.Y., Nov. 20, 1928; s. Frank Warren and Eleanor (Matthews) H.; student Oberlin Coll., 1947-49; A.B. cum laude, N.Y. U., 1952; Ph.D. in Sociology, Columbia, 1959; Faculty, Columbia, N.Y.C., 1958—, asso. prof. sociology, 1964—, research asst. Bur. Applied Social Research, 1952-57, research asso., 1958—, exec. sec. Columbia Council for Research in Social Scis., 1963—, staff Inst. African Studies, 1964—, asso. E. African Inst. Social Research, 1961-62, 63. Cons. to numerous orgns. Recipient award joint com. Social Sci. Research Council and Am. Council Learned Socs., 1961-62; Fulbright award, 1961-62; award Columbia Council for Research in Social Scis., 1963. Fellow Social Sci. Research Council; mem. Am. Sociol. Assn., Eastern Sociol. Soc., African Studies Assn., Am. Assn. U. Profs., Econ. History Assn., Acad. Polit. Sci. Author: (with Herbert H. Hyman, Charles R. Wright) Applications of Methods of Evaluation, 1962; The Exercise of Influence in Small Groups, 1964; also articles, revs., monographs. Developed theory distbn. relative influence among mems. small group, methodology for assessing effectiveness social action programs, theory and method in comparative study of polit. econs. Home: 438 W. 116th St., N.Y.C. 10027.*

HOPKINS, Thomas Cramer, Am. geologist; b. Center County, Pa., May 4, 1861; s. Isaac Cramer and Mary Ann (Glenn) H.; B.S., De Pauw U., 1881; M.S., 1890; A.M., Leland Stanford Jr. U., 1892; Ph.D., U. Chgo., 1900; D.Sc., Colgate U., 1923; m. Edistina Farrow, Jan. 8, 1890 (dec. May 1907); m. 2d, Elizabeth G. Hendrix, Mar. 31, 1909. Prin. high sch., Rising Sun, Ind., 1887-88; instr. in chemistry De Pauw U., 1888-89; asst. geologist Ark. Geol. Survey, 1889-92, Ind. Geol. Survey, 1895, 96, 1901; prof. geology Pa. State Coll., 1896-99, Syracuse U., 1900-32. Mem. Geol. Soc. Am., Sigma Xi, Phi Beta Kappa, Phi Kappa Phi. Author: Marble and Other Limestones, 1893; Brownstones of Pennsylvania, 1896; Clays and Clay Industries of Pennsylvania, 1898; Elements of Physical Geography (text-book), 1908; Laboratory Manual on Physical Geography, 1909; also numerous papers on geol. subjects pub. in tech. jours. and in various state reports. Died Apr. 3, 1935.

HOPKINS, William, English mathematician, geologist; b. Kingston, Eng., Feb. 2, 1793; s. William Hopkins; grad. Cambridge U., 1827, M.A., 1830; m. Miss Braithwaite; m. 2d, Caroline Boys; 1 son, 3 daus. Pvt. tutor Cambridge U., after 1827; apptd. syndic for bldy. Fitzwilliam Mus., 1835, 37. Recipient Wollaston medal, 1850; prize established in his honor by Cambridge Philos. Soc. Fellow Royal Soc., 1837. Pres. Geol. Soc., 1851, Brit. Assn., 1853. Author: Elements of Trigonometry, 1833; Abstract of a Memoir on Physical Geology, 1836; Researches in Physical Geology, 1839, 40; Theoretical Investigations of Motion of Glaciers, 1842; Transport of Erratic Blocks, 1844. Applied math. methods to phys. theories of geology; studied melting temperatures of substances under pressure, concluding that conducting power of strata or melting temperature increases with depth in earth; used astron. phenomena of precession of equinoxes to test whether earth's interior is solid or molten. Died Oct. 13, 1866.

HOPKINSON, Bertram, English engr., physicist; b. Birmingham, Eng., Jan. 11, 1874; s. John and Evelyn (Oldenbourg) H.; degree Trinity Coll., Cambridge, Eng., 1895; m. Mariana Siemens, 1903; 7 daus. Trained for bar; took up engring., 1898; became prof. mechanism and applied mechanics, Cambridge, Eng., 1903, professorial fellow Kings Coll., 1914; commd. in Royal Engrs., 1914. Mem. Adv. Com. to Govt. on Sci. Problems of War; served in dept. mil. aeros. in charge offensive armament of aircraft. Fellow Royal Soc., 1910 (sec.). Research on endurance of metals under stress, internal combustion engines, gas explosions, detonation pressure of high explosives, magnetic properties of iron and its alloys, airplanes; his work led to improved defensive apparatus against mines and torpedoes for ships. Died London, Aug. 26, 1918.

HOPKINSON, John, English engr.; b. Manchester, Eng., July 27, 1849; s. Alderman Hopkinson; ed. Owens Coll., Manchester; Trinity Coll., Cambridge (Eng.) U. (1st in math. exam., 1871); D.Sc., U. London (Eng.); m. Evelyn Oldenburg, 1873; children—Jack, Alice. Engr., dir. lighthouse sect. Chance Bros. Optical Plant, 1872-78; adv. engr. London, from 1878; prof. electro-technics King's Coll., London U., from 1881. Judge electricity display Paris, France, 1881. Mem. Civil, Mech., Elec. (twice pres.) engring. instns., Royal Soc., 1878 (medal for research on properties of iron, 1890), Athenaum. Author: Group Flashing Lights, 1874; On Electric Lighting, 1879; Some Points in Electrical Lighting, 1883. Recommended new group flashing procedures for lighthouses; research principles of synchromotors, electromagnetic occurrences in dynamo, properties of dielectrics which involved magnetism (interference principle established in form a basic part of theory of dielectrics); patentee more than 40 inventions. Died 1898.

HOPMANN, Josef, astronomer; b. Berlin, Dec. 22, 1890; s. Franz and Klara (Müller-Vanvolxem) H.; Ph.D., Classic Sch.; m. Maria Horster, 1920; children—Franz Josef, Wilhelm, Hansmichel. Agrégé, 1920; with U. Bonn, 1923; titular prof., dir. Obs. of Leipzig, 1930, Obs. of Vienna, 1951. Mem. Internat. Astron. Union, Acad. Scis. Leipzig, Vatican, Vienna. Author numerous works on astronomy. Address: 17 Türkenschanzstrasse, 1180 Vienna, Austria.*

HOPPE, Gunnar, Swedish phys. geographer; b. Skällvik, Sweden, Dec. 24, 1914; s. Carl and Stina (Björkman) H.; Fil.mag., U. Uppsala (Sweden), 1938, Fil.lic., 1942, Fil.Dr., 1945, Docent, 1945; prof. geography U. Stockholm (Sweden), 1954-55, prof. phys. geography, 1955—, prorector, 1966—. Mem. Swedish Royal Acad. Scis. Research, publs., books on glacial morphology, glaciology, Pleistocene geology, air photo interpretation. Home: Virebergsvägen 2, Solna, Sweden. Office: Drottningatan 120, Stockholm Va, Sweden.*

HOPPER, Arthur Frederick, Am. biologist; b. Plainfield, N.J., Sept. 7, 1917; s. Arthur Frederick and Catherine (Hoenig) H.; A.B., Princeton, 1938; M.S., Yale, 1942; Ph.D., Northwestern U., 1948; m. Amy Patricia Hull, Dec. 28, 1940; children—Arthur III, Geoffrey, Gregory, Christopher. Faculty, Northwestern U., 1948, Wayne State U., 1948-49; research asso. Detroit Inst. Cancer Research, 1948-49; faculty Rutgers U., New Brunswick, N.J., 1949—, prof., 1966—. Scientist, Oceanographic R/V Vema, Columbia, 1955-57; research collaborator Brookhaven Nat. Lab., 1963—. Mem. Am. Soc. Zoologists, A.A.A.S., Soc. Study Growth and Devel., Sigma Xi. Research, numerous publs. primarily in field of hormonal control of fin regeneration in fish, effect of nutrition on cell physiology and recovery from radiation damage in mammals, effect of X-rays and neutrons on intestinal damage in mammals. Home: Mine Mount Rd., Bernardsville, N.J. 07924. Office: Nelson Biol. Lab., New Brunswick, N.J. 08903.*

HOPPER, Rex D., Am. sociologist; b. Nashville, Ind., July 4, 1898; s. Martin Jackson and Estelle (Taggart) H.; A.B., Butler U., 1922, A.M., 1925; Ph.D., U. Tex., 1943; m. Ida Tobin, Sept. 15, 1924 (div.); children—Rex D., John M.; m. 2d, Janice Harris, June 10, 1961. Treas., mem. faculty Colegio Internacional Asuncion, Paraguay, 1926-31; from tutor to asst. prof. dept. sociology Inst. Latin Am. Studies, U. Tex., 1931-46; faculty Bklyn. Coll., City U. N.Y., 1941-66, prof., chmn. dept. sociology and anthropology; sr. staff scientist Spl. Operations Research Office, Am. U., Washington, 1964-66. Vis. prof. Nat. U. Mex., summers 1943-46, Mex. City Coll., 1950; Smith-Mundt prof. Law Sch., Nat. U. Paraguay, 1957; Fulbright prof., dept. sociology Nat. U. Buenos Aires, Argentina, 1959. Mem. Am. Sociol. Assn. (mem. council, chmn. com. internat. co-operation), Eastern Sociol. Soc., Am. Polit. Assn., A.A.A.S., Soc. Study Social Problems, Soc. Sci. Study Religion, Soc. Internat. Devel., Sigma Xi, Alpha Kappa Delta, Phi Kappa Phi, Sigma Delta Pi. Author: (with Feliks Gross, Samuel Koenig) Sociology: A Book of Readings, 1953; (with Gross) A Century of Revolution, 1958. Studies, publs. on social change and social movements; devel. of revolutionary model; effects of automation on social structure. Home: 824 25th St. N.W., Washington 20037. Office: 5010 Wisconsin Av. N.W., Washington 20016. Died June 17, 1966.*

HOPPE-SEYLER, Ernst Felix Emmanuel, German physiol. chemist; b. Freiburg, Germany, Dec. 26, 1825; med. degree, Berlin, Germany, 1850. Prosector, Greifswald, 1854-56; asst. under Virchow, Path. Inst., Berlin, 1856-64; prof. applied chemistry Tubingen, 1861-72; prof. physiol. chemistry, Strassburg, 1872-95, (all Germany). Author: Handbuch der Physiologisch und Pathologisch Chemischen Analyse, 1858; Physiologische Chemie, 4 vols., 1877-81. Founded Zeitschrift fur Physiologische Chemie (1st sci. jour. specializing in biochemistry). Research on physics of percussion, 1854; classified percussion notes according to phys. characteristics; discovered and named hemoglobin, 1862; demonstrated hemoglobin to be carrier of oxygen, also affinity of hemoglobin for carbon monoxide; discovered absorption spectra of blood; described bubbles in blood caused by caisson disease; known as leading physiol. chemist of his time. Died Aug. 10, 1895.

HOPPMANN, William Henry II, Am. mech. engr.; b. Charleston, S.C., Sept. 23, 1908; s. William Henry and Marie (Dean) H.; B.S., Coll. Charleston, 1929; M.A., George Washington U., 1935; Ph.D., Columbia, 1947; m. Gladys A. Stearley, Sept. 16, 1932; children—Colette (Mrs. Edward Dowling), William Henry III. Prof. mech. engring. Johns Hopkins, 1947-57; prof. mechanics Rensselaer Poly. Inst., Troy, N.Y., 1957—; prof. of engring. U. S.C., 1967—. Fellow Am. Soc. M.E.; mem. Soc. Rheology, Am. Assn. U. Profs., Sigma Xi, Pi Tau Sigma. Contbr. numerous articles to profl. jours. Invented, designed, developed impact machines for research on materials; developed analytic methods for studying resistance of structures to explosive loads, theory and expts. for research on dynamic loading of bars and beams, mech. shock and vibration of structures, flow of Newtonian and non-Newtonian liquids. Home: 521 Nottingham Rd., Columbia, S.C. 29210.*

HOPWOOD, Frank Lloyd, English physicist; b. 1884; s. William and Elizabeth Hopwood; ed. Royal Coll. Sci., also U. Coll., London; Hon. M.A., Cambridge; D.Sc., London; m. Helen Sproxton, 1909; 1 son. Vice dean, prof. physics St. Bartholomew's Hosp. Med. Coll., U. London; physicist then cons. physicist to St. Bartholomew's Hosp., London; prof. emeritus U. London, 1949—. Hone. sec. Brit. Com. for Radiol. Units, 1925-50. Fellow Inst. Physics, Royal Coll. Sci. (asso.), L'Associazione Italiana di Radio-Biologia (hon. asso.), Brit. Inst. Radiology (hon.), Röntgen Soc. (hon.), Faculty Radiologists (hon.). Contbr. articles to tech. jours. Invented method for underwater signalling, sci. instruments. Died May 2, 1954.

HOPWOOD, M(ortimer) Lloyd, biochemist; b. Jamaica, W.I., Jan. 10, 1918; s. Alexander I. and Winifred (Parke) H.; B.S., N.Y. U., 1940; M.S., Colo. State U., 1952, Ph.D., 1960; m. Grace M. Duncan, Feb. 1, 1941; children—Duncan A., Christina M., Control chemist Flintkote Co., East Rutherford, N.J., 1940-43; biochemist U. S. Naval Med. Research Inst., Bethesda, Md., 1946-47; Marcuse fellow chemistry U. Colo., 1947-48; faculty Colo. State U., Ft. Collins, 1948—, prof. chemistry and physiology, 1963—, head endocrine sect., 1960—. Cons. AEC, 1963—, G. D. Searle & Co., 1965—, Mattox & Moore, 1964—. Mem. Endocrine Soc., Am. Physiol. Soc., Am. Chem. Soc., Soc. for Exptl. Biology and Medicine, A.A.A.S., Soc. for Study Fertility, N.Y. Acad. Sci., Am. Dairy Sci. Assn., Am. Soc. Animal Sci., Sigma Xi. Research, publs. in reprodn. including metabolism of hormones in domestic animals, metabolism of spermatozoa, accessory sex organ function, hemodynamics and vascular metabolism of swine arteries, irradiation of bovine testicles and evaluation of influence of sperm prodn., physiology and biochemistry of antithyroid drugs, regulation of estrus cycle with hormones. Home: 1510 S. College Av., Ft. Collins, Colo. 80521.*

HORACKOVA, Eva Heyrovska, Czechoslovakian psychologist; b. Prague, Czechoslovakia, Feb. 28, 1923; d. Leopold and Siebertova (Heyrovska) H.; Ph.D., Charles U., Prague, 1950, C.Sc., 1963; m. Vladimir Horacek, Apr. 12, 1951; children—Michal, Katherine. Clin. psychologist Mental Hosp., Prague-Bohnice, 1955-56; research psychologist Inst. Human Nutrition, Prague, 1956—. Mem. Czechoslovak Acad. Sci., Psychol. Soc. Research, numerous publs. in psychopharmacology, psychology of personality and interpersonal communication, co-discoverer meaning of one-sided relationship for personality diagnosis, co-inventor ambulant reactometric apparatus. Home: 7 Platnerska, Pague 1. Office: 800 Budejovicka, Prague 4, Czechoslovakia.*

HORAK, Zdenek Frantisek, Czechoslovakian physicist; b. Prague, Czechoslovakia, Oct. 6, 1898; s. Frantisek and Milada (Hrdlicka) H.; RNDr., Prague U., 1923; D.Sc., Sorbonne, Paris, France, 1957; m. Bozena Vávra, July 22, 1930. With Tech. U. Prague, 1920—, prof., 1945—, head dept. physics, 1952—, mem. sci. council, 1960—. Mem. Czechoslovak Acad. Scis. (mem. sci. collegium 1962—), Société Française de Physique, Internat. Acad. Astronautics (mem space relativity com.). Author: Practical Physics, 1947; Molecular and Atomic Physics, 1955; (with Krupka) Technical Physics, 1960; (with Kucera) Tensors in Electrotechnics and in Physics, 1963; also numerous articles. Research on concept of non holonomic spaces; new phenomena at impact of rough bodies; new method of calculus of observations; new relativistic theory of electromagnetism, gravitation and inertia. Home: 2 Vietnamská, Prague 6, Czechoslovakia.*

HORANYI, Janos, Hungarian physician, surg. pathologist; b. Acsa, Hungary, Oct. 5, 1899; s. St. Hofhauser and A. (Kommer) H.; Diplom.med. doctor Budapest (Hungary) Med. U., 1922, Diplom. Surgery, 1926; m. Éva Havel, Jan. 20, 1945; children—Janos, Christina. With Budapest Med. U., 1922—, clin. chief surgeon 2d surg. clinic, 1946—, pvt. prof. surg. pathology, 1939—, mandatory lectr., 1957-64. Research, numerous publs. on mut. effects digestive juices, pathology of bronchi and peribonchial lymph glands, forms of pneumogranuloma, histodevelopmental disorders of bronchi. Home: 48 Baross. Office: 23 Barossu. 48, Budapest, Hungary.*

HORDER, Baron Thomas Jeeves, English physician; b. Shaftesbury, Eng., Jan. 7, 1871; s. Albert Horder; B.Sc. in Physiology, U. London, (Eng.) 1893, M.B. (1st class honors), 1898, M.D., 1899 (honors, gold medals medicine, midwifery, forensic med.), 1899.; also ed. Corr. Coll. U. London, St. Bartholomew's Hosp.; D.C. L. (hon.), Dunelm.; M.D. (hon.), univs. Melbourne, Adelaide, Australia; m. Geraldine Rose Doggett, 1902; 1 son, Thomas Mervyn, 2 daus. Joined St. Bartholomew's Hosp., London as demonstrator biology, 1895, subsequently advanced to sr. cons. physician and gov. of hosp., from 1936; cons. physician Cancer Hosp., Fulham, Royal Orthopedic and Royal Northern hosps., hosps. at Bury St. Edmunds, Swindon, Stratford, Leatherhead, Beckenham, Finchley (all Eng.) chmn. adv. commn. Mt. Vernon Hosp., Eng.; pvt. practice medicine, London; physician in ordinary to Prince of Wales (later King Edward VIII); from 1923, extra physician King George VI, Queen Elizabeth II; hon. cons. physician Ministry Pensions, from 1939. Mem. adv. com., also chmn. Brit. Empire Cancer Campaign; adviser Minister of Food; chmn. adv. com. on med. questions recruitment ministries Labor and Nat. Service, Smoke Abatement Soc., Empire Rheumatism Council, 1936-43; pres. Indsl. Health Edn. Soc., Food Edn. Soc. mem. Noise Abatement League, Mobile Physiotherapy, Cremation Soc., Eugenics Soc. Fellow Royal Coll. Physicians; mem., pres. Harveyian Soc. London, Fellowship Medicine, Med. Soc. London, Internat. Fedn. Phys. Medicine, Brit. Assn. Author: Clinical Pathology in Practice, 1910; Cerebro-Spinal Fever, 1915; Medical Notes, 1921; Essentials of Medical Diagnosis (considered authoritative work), 1929; Health and a Day, 1937; Obscurantism: 50 Years of Medicine. Maintained that personal relationship of doctor, patient is most important factor in diagnosis and treatment, also that pub. health was affected by both phys. and social environments. Advocated cremation, eugenics, birth control; opposed legalized euthanasia; specialist radiology, cancer, orthopedics, rheumatism. Died Aug. 13, 1955.

HORECKER, Bernard L(eonard), Am. biochemist; b. Chicago, Oct. 31, 1914; s. Paul and Bessie (Bornstein) H.; B.S., U. Chicago, 1936, Ph.D., 1939; m. Frances Goldstein, July 12, 1936; children—Doris, Marilyn, Linda. Research asso. chemistry U. Chicago, 1939-40; examiner U. S. Civil Service, 1940-41; biochemist U.S.P.H.S., Nat. Insts. Health, Bethesda, Md., 1941-59, with biochemistry study sect., 1956-59, chief lab. of biochemistry and metabolism Nat. Inst. Arthritis and Metabolic Disease, 1956-59; professorial lectr. enzyme chemical George Washington U., 1950-57; guest research-worker Pasteur Inst., Paris, France, 1957-58; prof. microbiology, chmn. dept., N.Y. University, College of Medicine, 1959-63; professor of molecular biology, chairman department Albert Einstein College Medicine, 1963—; vis. prof. biochem. U. Cal., 1954; vis. lectr. U. Ill., 1956; Ciba lectr. Rutgers U., 1962; Phillips lectr. Haverford College, 1965; vis. prof. biochemistry and molecular biology Cornell U., 1965. Research Career Award Com., Nat. Inst. Gen. Med. Scis., 1966—. Recipient 1952 Paul Lewis Labs. award in enzyme chemistry; Rockefeller Pub. Service award, 1957; Hillebrand prize, Am. Chem. Soc., 1954; award in biol. Scis., Washington Acad. of Science, 1954; Fulbright Travel award, 1963. Commonwealth Fund fellow, 1967. Fellow A.A.A.S., Am. Acad. Arts and Scis.; mem. Japanese Biochemical Society (honorary member), Am. Chem. Soc., Biochem. Soc. (Eng.), Swiss Biochem. Soc. (hon.), Am. Soc. Biol. Chemists. chmn. editorial committee 1962-63), Soc. Gen. Microbiology, Nat. Acad. Scis., Harvey Society (honorary), Phi Beta Kappa, Sigma Xi. Author: Pentose Metabolism in Bacteria, published 1961; articles sci. publs. Editor: Biochemical and Biophysical Communications, 1959—, Archives of Biochemistry and Biophysics, 1960—. Research on characterization and isolation of respiratory enzymes; carbohydrate metabolism; spectrophotometry of hemoglobin and derivatives; enzymology. Home: 340 E. 64th St., N.Y.C. 10021. Office: Albert Einstein College of Medicine, Eastchester Rd. and Morris Park Av., N.Y.C.

HORGAN, Stephen Henry, Am. inventor; b. Norfolk, Va., Feb. 2, 1854; ed. Cork, Ireland, also Coll. St. Francis Xavier, N.Y.; became photographer N.Y. Daily Graphic, 1874; named art dir. N.Y. Herald, 1893. Honored on 50th anniversary of perfection of half-tone process, Eng., 1930. Author: Three Color Process Work, 1902; Horgan's Half-tone and Photo-Mechanical Processes, 1912. Inventor half-tone engraving process, used in N.Y. Daily Graphic, 1880, adapted for high speed press and used by N.Y. Tribune; developed half-tone process using glass screen made of fine parallel lines; contbd. to tech. of transmission of color pictures by wire. Died Orange, N.J., Aug. 30, 1941.

HÖRHAMMER, Ludwig, German physician; b. Freising, June 12, 1907; Ph.D., M.D., U. Munich; m. Elisabeth Seidl, 1935; children—Hans-Peter, Rolf, Elke. Promotion summa cum laude, 1932, qualified, 1940; asso. prof., dir. Inst. Pharmacology, 1960, full prof., 1960. Recipient medal of honor U. Santa Maria Rio Grande do Sul Faculty Medicine, 1962. Mem. Uniao Farmaceutica (Brazil) (Hon.), Soc. Pharmacy and Chemistry (Sao Paulo) (hon.). Author: Die Pharmakognostische Teeanalyse, 1939; Teeanalyse; also 150 articles. Home: 19 Fasanenstrasse, Munich, Obermenzing. Office: 29 Karlstrasse, 8 Munich 2, Germany.

HORI, Jun-ichi, Japanese physicist; b. Kyoto, Japan, Oct. 6, 1926; s. Takeo and Shizu (Tomonaga) H.; B.Sc., Hokkaido U., 1950, D.Sc., 1959; m. Yoko Ozazaki, Mar. 29, 1952; children—Akiko, Hisashi. Staff, Hokkaido U., Sapporo, Japan, 1950—, prof. physics dept. physics, 1966—. Mem. Phys. Soc. Japan. Author: (with Takeo Hori) Optics, 1953; Spectral Properties of Disordered Chains and Lattices, 1968; also articles. Research on theories of information and prediction, irreversible thermodynamics and solid-state physics, lattice dynamics, especially theory of disordered lattices; established theory, phase theory, which can explain and predict various characteristic spectral properties of disordered systems. Home: 765, 8-jo Higashi 5-chome, Kotoni-8-Ken, Sapporo, Japan.*

HORI, Takeo, Japanese physicist; b. Toyama Prefecture, Dec. 7, 1899; grad. Tokyo U., 1923; D.Sc. Instr., 3d Higher Sch., asst. Phys. and Chem. Research Inst., 1925; research student physics in Europe and U. S., 1926-28; prof. Ryojun (Port Arthur) Tech. Coll., Manchuria; prof. Hokkaido U., also Low Temperature Sci. Inst., 1935—. Author various works, including: Molecular Spectrum and Isotopes; The Spectroscopical Study of Atomic Nucleus. Research on phys. optics; inventor phase-difference microscope.

HORIBA, Shinkichi, Japanese chemist; b. Kyoto Prefecture, 1886; grad. Kyoto U., 1910; D.Sc. Asst. prof. Kyoto U., later prof., dir. sci. dept., ret., 1947, now emeritus prof.; studied physics and chemistry in Europe and U. S.; prof. Kyoto Inst. Tech.; dir. engring. dept. Doshisha U.; pres. Naniwa U., 1952. Recipient awards for study of corrosion-proof paint for ship bottoms, paint for wooden boats, paint without mercury and copper for ship bottoms. Mem. Japan Acad., Imperial Acad. (Imperial prize for article). Research on methods to analyze speed of chem. reactions.

HORIGUCHI, Yoshiki, Japanese geophysicist; b. Gifu Prefecture, Japan, 1885; grad. Tokyo U., 1911, Dr. Sci. m. Masue. Dir., Kobe Weather Sta., then studied meterology in Europe and U. S.; returned to Japan, apptd. successively, engr. Central Meterol. Obs., army engr., dir. meteorol. div. Japanese Mil. Govt. Team, Malay, dir. Marine Meterol. Obs. Recipient Japan Acad. prize, 1929. Author: Change in Construction of Typhoon; Researches in the Typhoons of the Philippines; Treatise on Typhoons in the Far East.

HORIUCHI, Juro, Japanese chemist; b. Hokkaido, Japan, 1901; grad. Sci. Sch., Tokyo U., 1925. Prof. chemistry Himeji Higher Sch., 1929; studied in Europe; prof. Hokkaido U., 1935, also dir. Catalyzer Research Inst.; mem. Physico-Chem. Research Inst. Recipient prize for research in velocity of chem. reactions Imperial Acad., 1940. Author: Heavy Water and Isotope Chemistry, also other books. Authority on quantum dynamics.

HORIUCHI, Kazuya, Japanese physician; b. Tokyo, Japan, Oct. 1, 1912; s. Yajiro and Aiko Horiuchi; grad. Keio Gijuku U., M.D., 1937, Dr.Med. Sci., 1944; m. Mariko Tohyama, Dec. 4, 1941; children—Yataro, Masaya Horiuchi. Physician, St. Lukes Internat. Med. Center, Tokyo, 1937-41; asst. dept. hygiene Keio U. Sch. Medicine, 1941-44, instr., 1944-45, prof. hygiene, 1945-49; prof. preventive medicine and publ. health, chmn. dept. Osaka (Japan) City U. Med. Sch., 1949—. Cons. Ministry Health and Welfare, Ministry Labor, Osaka City; tech. adv. mem. WHO, ILO; mem. Permanent Commn. Occupational Health. Fellow Japan Hygiene Soc. (dir.), Japan Assn. Indsl. Medicine (dir.), Japan Assn. Pub. Health (dir.), Japan Assn. Air Pollution (dir.); mem. Internat. Assn. Occupational Health. Author: Industrial Medicine; Preventive Medicine and Public Health; also numerous articles. Research on prevention occupational diseases especially those caused by heavy metals; determination of maximum allowable concentrations of heavy metals allowable for humans and animals. Home: Kamo 2 2-13, Takaishi-Cho, Osaka, Japan.*

HORN, George Henry, Am. physician, entomologist; b. Phila., Apr. 7, 1840; s. Philip Henry and Frances Isabella (Brock) H.; M.D., U. Pa., 1861. mem. Entomol. Soc. of Phila., 1860, pres., 1866; vice dir. Acad. of Natural Scis., 1876-83, dir. entomol. sect., 1883-97; prof. entomology U. Pa., 1889; prin. contbr. to study and classification of Coleoptera, responsible for naming and describing more than 1,550 species; pres. Am. Entomol. Soc., 1883-97; hon. mem. Entomol. Soc. France. Author papers: "Description of New North American Coleoptera in the Cabinet of the Entomological Society of Philadelphia," 1860; (with John L. LeConte) The Classification of the Coleoptera of North America, 1883. Died Beesley's Point, N.J., Nov. 24, 1897.

HORN, Herwarth Max Germund, German hygienist, microbiologist; b. Lommatzsch, Nov. 17, 1924; s. Wilhelm J. F. and Walpurga (Anderle) H.; student U. Greifswald, 1946-47; M.D., Humboldt U., Berlin, 1950, Dr.med.habil., 1966; children—Christian, Marie-Louise. Faculty, Inst. Med. Microbiology and Epidemiology, Humboldt U., Berlin, 1954-65, head dept. epidemiology, 1961-63, dep. dir., 1963-65; dir. Inst. Hygiene, Med. Acad., Erfurt, 1965—; faculty U. Jena, 1966—. Med. officer for environmental sanitation Ministry Health, 1957-65; med. councillor, 1965—. Mem. German Soc. Hygiene. Author: (with J. Grober and F. Oberdoerster) Gesundheitstaschenbuch für die warmen Länder, 1960; Zum Problem der biologischen Prüfung der Sterilisation, 1966; also articles. Research on sterilization; analysis of routes of infection; prophylaxis of tropical infections; pub. health legislation; biographical research on Robert Koch. Home: 38 Goethestrasse. Office: 6 Predigerstrasse, 50 Erfurt, East Germany.*

HORN, John L(eonard), Am. psychologist; b. St. Joseph, Mo., Sept. 7, 1929; s. John Leonard and Nellie (Weldon) H.; B.A., U. Denver, 1956; postgrad. U. Melbourne (Australia); A.M., U. Ill., 1961, Ph.D., 1965; m. Bonnie Colleen Hoskins, July 31, 1955; children—John Leonard, James Bryan, Julie Lynn, Jennifer Lee. Research asst. U. Melbourne, 1957; staff Lab. Personality Assessment, U. Ill., 1958-61, asst., asso. prof. dept. psychology U. Denver, 1961—; asst. dir. Inst. Personality and Ability Testing-Western, Boulder, Colo., 1962—. Fulbright fellow, 1956-58; NIH fellow, 1958-61. Mem. Am. Assn. U. Profs., Am. Psychol. Assn., Psychometric Soc., Soc. Multivariate Exptl. Psychologists, Phi Beta Kappa, Phi Kappa Phi, Sigma Xi. Author: articles. Devel. statis.-math. solutions for gen. problems encountered in attempts to identify characteristics personality; improved statis.-math. techniques required to assess change in personality over short periods of time; refined description ability, motive and temperament characteristics; demonstrated 2 maj. components intelligence have different patterns change over life span from adolescence to old age. Home: 196 S. Corona St., Denver 80209.*

HORN, Robert Chisholm, Jr., Am. physician; b. Allentown, Pa., July 7, 1913; s. Robert Chisholm and Zelie (Soléliac) H.; B.S., Muhlenberg Coll., 1933; M.D., Yale, 1937; m. Dorothy Louise App, Jan. 1, 1940; children—Robert C. III, Thomas L., Ethel M. Surg. pathologist, pathologist Hosp. U. Pa., 1942-55, faculty Sch. Medicine, 1942-55, prof. surg. pathology, 1955, dir. tumor clinic, 1942-55; chmn. dept. labs. Henry Ford Hosp., Detroit, 1955—. Cons. Armed Forces Inst. Pathology, Washington, 1950—; chmn. Nat. Com. for Careers in Med. Tech., 1960—. Trustee Detroit Inst. Cancer Research, pres., 1965—. Recipient Achievement awards Muhlenberg Coll., Geisinger Meml. Hosps., 1965. Mem. A.M.A. (past chmn. sect. pathology and physiology), Coll. Am. Pathologists (speaker assembly 1964—), Am. Assn. Pathologists and Bacteriologists, Am. Soc. Clin. Pathology, Internat. Acad. Pathologists, Mich. Soc. Pathologists, Am. Thyroid Assn. (pres. 1965), N.Y. Acad. scis. Contbr. Research, numerous publs. in diseases thyroid, endocrine glands, diseases gastrointestinal tract, cancer. Home: 90 N. Edgewood Dr., Grosse Pointe, Mich. 48236. Office: Henry Ford Hosp., 2799 W. Grand Blvd., Detroit 48202.*

HORNADAY, William Temple, Am. zoologist; b. Plainfield, Ind., Dec. 1, 1854; s. William and Martha (Varner) H.; ed. Ia. State Coll.; studied zoology and museology in U. S. and Europe; Sc.D., U. Pittsburgh, 1906; A.M., Yale 1917; Ph.M., Ia. State Coll., 1923; m. Josephine Chamberlain, Sept. 11, 1879; 1 dau., Helen Ross (Mrs. George T. Fielding). As collecting zoologist visited Cuba, Fla., the W.I., S. America, India, Ceylon, the Malay Peninsula and Borneo, 1875-79; chief taxidermist U. S. Nat. Mus., 1882-90; in real estate business, Buffalo, N.Y., 1890-96; dir. New York Zool. Park, 1896-1926 (retired). Gold medalist Republic of France; British Royal Soc. for Protection of Birds, Royal Zool. Soc. of Antwerp; New York Zool. Soc.; Inter-Nat. Congress for Study and Protection of Birds; silver medalist Société Nationale d'Acclimation of France. Active in promoting game preserves and new laws for the protection of wild life generally; took initiative in creation of Mont. Nat. Bison Range, Wichita Nat. Bison Range, and Elk River Game Preserve (Mont.); the Bayne law to prohibit the sale of native game; and new tariff law to prohibit all importations of wild birds' plumage into U. S. for millinery purposes (1913); organized Permanent Wild Life Protection Fund, 1913-14. Author: Two Years in the Jungle, 1885; The Extermination of the American Bison, 1887; Taxidermy and Zoölogica Collecting, 1892; The American Natural History, 1904; Camp-Fires in the Canadian Rockies, 1906; Camp-Fires on Desert and Lava, 1908; Wild Life Conservation in Theory and Practice, 1914; Minds and Manners of Wild Animals, 1922; Tales from Nature's Wonderlands, 1924; A Wild-Animal Round-Up, 1925; Wild Animal Interviews, 1928; Thirty Years War for Wild Life, 1931. Died Stamford, Conn., Mar. 6, 1937.

HORNBLOWER, Jonathan Carter, English engr.; b. Chacewater, Eng., July 5, 1753; s. Jonathan Hornblower; worked (with father and 3 brothers) for Watt in building engines. Author: Description of a Machine for Communicating Motion at a Distance, 1786; also several articles. Inventor double beat valve; patentee engine which represented 1st attempt to use steam expansively and anticipated principle of compound engine, 1781, but remained undeveloped because it infringed on Watt's steam engine patent, later rediscovered by Woolf. Died Mar. 1815.

HORNE, Frank Robert, English botanist; b. London, June 9, 1904; s. Thomas and Jessie Binnie (Baxter) H.; M.A., N.D.A., Cambridge U.; N.D.D. honoris causa; m. Marjorie Bannister, 1929; children—Rosemary Baines, Jennifer Baines, Thomas Christopher Bannister, Diana Silvester. Dir. botany dept. Seale-Hayne Agr. Coll., 1927-44; dir. Nat. Inst. Agrl. Botany, 1944—. Pres., European Productivity Agy., OECD, 1954-59, Com. on Herbage Varieties and Internat. Seed Certification, 1939-59, World Seed Year Meeting, 1962; mem. European Barley Com.; mem. nat. tech. com. FAO. Mem. Brit. Grassland Soc. (pres.), Brit. Ecol. Soc., Internat. Commn. for Nomenclature Cultivated Plants, Seed Research Assn. Sweden (hon.). Author: Ryegrass Strains Indigenous to South-Western England, 1946; Winter Cauliflower; History and Breeding in the South West, 1954. Home: Hill Farm, Lolworth, Cambridge. Office: Nat. Institute of Agricultural Botany, Huntingdon Rd., Cambridge, Eng.

HORNE, Ralph Albert, Am. chemist; b. Haverhill, Mass., Mar. 10, 1929; s. Ralph L. and Flora T. (Kelly) H.; S.B., Mass. Inst. Tech., 1950; M.S., U. Vt., 1952; M.A., Boston U., 1953; Ph.D. (Higgins fellow), Columbia, 1955; postgrad. Sch. Mus. of Fine Arts. Research asso. Brookhaven Nat. Lab., Upton, L.I., N.Y., 1953-55, dept. chemistry Lab. for Nuclear Sci., Mass. Inst. Tech., 1955-57; sr. scientist RCA, Needham, Mass., 1957-59, Joseph Kaye & Co., Cambridge, Mass., 1959-60; chief scientist Lab. for Phys. Chemistry and Exptl. Oceanography, Arthur D. Little, Inc., Cambridge, Mass., 1960—. Mem. Am. Chem. Soc., Electrochem. Soc., Am. Geophys. Union, A.A.A.S., History of Sci. Soc., Chem. Soc. London, Faraday Soc., Sigma Xi. Research, publs. on effects of temperature, pressure, electrolyte type and concentration, a second solvent, and interfaces on structure of liquid water and dependence of transport processes. Home: Currier House, Currier Rd., Deerfield, N.H. Office: 15 Acorn Park, Cambridge, Mass. 02140.*

HORNE, Robert William, biologist; b. Ascot, Eng., Jan. 21, 1923; s. Alfred E. and Alice M. (Leaver) H.; M.A., U. Cambridge (Eng.), 1958, Sc.D., 1968; m. Doris M. Bedingfield, Mar. 18, 1946; children—Corinne Patricia, Pamela Jane. Asst. in research Cavendish Lab., 1947-53, asst. dir. E.M. sect., 1958-62; head electron microscope unit Inst. Animal Physiology, Babraham, Cambridge, Eng., 1962-68; head electron microscope lab. John Innes Inst., Norwich, Eng., 1968—. Fellow Royal Micros. Soc. (v.p. 1964—), Royal Soc. Arts. Author: (with K. C. L. Smith) Electron Microscopy; Methods in Virology; Quantitative Electron Microscopy; also articles. Research on application of electron microscope to biol. structure at macromolecular level especially viruses; devel. (with S. Brenner) negative staining technique for electron microscope. Home: 9 Birkdale, Sunningdale, Norwich. Office: John Innes Inst., Colney Lane, Norwich, Eng.*

HORNEMANN, Friedrich Konrad, explorer; b. Hildesheim, Germany, Sept. 1772; ed. Göttingen, Germany; explorer for African Assn., London, from 1796; traveled across N. Africa disguised as mameluke trading with the Fezzan. Author: Tagebuch einer Reise von Cairo nach Murzuck (journal), 1802. First modern European to traverse the Sahara and to establish exact location of Hausa country. Died Nupe, Africa, Apr. 7, 1800.

HORNER, Johann Friedrich, Swiss ophthalmologist; b. Zurich, Switzerland, Mar. 27, 1831; prof., Zurich; described oculopupillary syndrome (Horner's syndrome), 1869; indicated eye symptoms in some gen. diseases; introduced Lister's antisepsis into ophthalmology. Died Dec. 20, 1886.

HORNER, Leopold, German chemist; b. Kehl, Rhine, Aug. 24, 1911; s. Leopold and Maria (Winkelmaier) H.; ed. U. Heidelberg, U. Munich; Ph.D. in natural philosophy; Ph.D., agrege; m. Gerda Marie Krause, 1945; children—Christoph, Michael, Irene. Asst., instr., affiliate prof., asso. prof., full prof. Mem. Assn. German Chemists; Am. Chem. Soc. Research and over 150 publs. on organic chemistry. Home: 17 Alfred Mumbächerstrasse. Office: University of Mainz, Mainz, Rhine, Germany.

HORNER, Seward Ellis, Am. geologist; b. Abilene, Kan., Apr. 13, 1906; s. Charles and Elma (Biggart) H.; B.S. in Geology, Kan. State Coll., 1933; m. Leona Zoe Tibbetts, Aug. 23, 1940; 4 children. Employed to locate water for Kan. Emergency Relief Com. for Ogden S. Jones; joined WPA for Kan., 1935; became sr. geologist Kan. Hwy. Comm., 1937, chief geologist, 1944—; chief geologist U. S. Pub. Rds. Adminstrn. on Alcan Hwy., 1942-44; part time geol. cons. U. S. Geol. Survey, 1946—. Mem. landslide and hwy. subdrain com. Hwy. Research Bd., NRC. Mem. Kan. Geol. Soc., Flint Hills Geol. Soc., Geol. Soc. Am., Soc. Econ. Geologists, Am. Assn. Petroleum Geologists, Kan. Engring. Soc., Kan. Acad. Sci., Sigma Xi, Sigma Gamma Epsilon. Developed geol. system of rock classification for excavation; studied flood on Bear Creek bridge found., 1951; electro-chem. stablzn. of clay. Died July 8, 1954.

HORNER, William Edmonds, Am. anatomist; b. Warrenton, Va., June 3, 1793; s. William and Mary (Edmonds) H.; grad. U. Pa., 1814; m. Elizabeth Welsh, Oct. 26, 1820, 10 children. Adj. prof. anatomy U. Pa., 1819, later prof., dean med. dept., 1822-52; a founder St. Joseph's Hosp. Author: The American Dissector, 1819; A Treatise on Pathological Anatomy for the Use of Dissectors, 1823; Treatise on Special and General Anatomy, 2 vols., 1826; Treatise on Pathological Anatomy (1st path. text pub. in America), 1829. Described for 1st time tensor tarsi, spl. muscle connected with lachrymal apparatus (muscle of Horner), 1824; described incisor teeth marked by grooves because of enamel deficiency (Horner's teeth), 1829. Died Phila., Mar. 13, 1853.

HORNER, William George, English mathematician; b. nr. Bristol, Eng., 1786; s. William Horner; ed. Kingswood Sch. nr. Bristol; several children, including William. Became asst. master Kingswood Sch., 1802, head master, 1806; founder sch. at Grosvenor Place, Bath, Eng., 1809, tchr. until his death. Author: (pamphlet) Natural Magic, 1832; also articles. Discovered solutions to numerical equations of any degree by continuous approximation which was previously known only to Chinese (Horner's method). Died Grosvenor Place, Sept. 22, 1837.

HORNEY, Karen, psychiatrist; b. Hamburg, Germany, Sept. 16, 1885; dau. Berndt and Clotilde (Von Ronzelen) Danielson; M.D., Univ. Berlin, 1913; m. Oscar Horney, Oct. 1909; children—Brigitte (Mrs. K. Tschetwerikoff), Marianne (Mrs. W. von Eckardt), Renate (Mrs. F. Crevenna). Came to U. S., 1932, naturalized, 1938. Instr. Inst. for Psychoanalysis, Berlin, 1920-32; asso. dir., Chicago Inst. for Psychoanalysis, 1932-34; lecturer New Sch. for Social Research, New York City, since 1935; dean Am. Inst. for Psychoanalysis since 1941. Mem. Assn. for Advancement of Psychoanalysis, Am. Psychiatric Assn. Author: The Neurotic Personality of Our Time, 1936; New Ways in Psychoanalysis, 1939; Self-Analysis, 1942; Out Inner Conflicts, 1945; Neurosis and Human Growth, 1950. Most important contbns. were emphasis on cultural and social factors in personality devel. and growth of neurosis which differed from Freudian emphasis on biol. factors, and research in area of feminine psychology; described anxiety as failure of neurotic defense mechanisms in helping individual to cope with world. Died N.Y.C., Dec. 1952.

HORNICH, Hans, Austrian mathematician; b. Vienna, Austria, Aug. 28, 1906; s. Rudolf and Anna (Pracher) H.; Dr.phil., U. Vienna, 1929; m. Michaela Rabenlechner, Dec. 5, 1936; 1 son, Richard. Privatdozent U. Vienna, 1933-48; prof. Tech. U., Graz, Austria, 1948-57; faculty Tech. U. Vienna, 1957—, dean faculty natural scis., 1961-62; vis. prof. Cath. U. Am., Washington, 1965; hon. prof. U. Vienna, 1966—. Mem. Austrian Acad. Scis. (corr.), Austrian Math. Soc., Am., German math. socs. Author: Lehrbuch der Funktionentheorie, 1950; Existenzprobleme bei lin. part. Diff. Gleich. erster Ordn., 1960; also numerous articles. Solution of mixed boundary value problem of potential theory; research on integrals of first order on transcendental Riemann surfaces, existence theorem of linear part differential equations, uniform analytical functions, ordinary differential equations of high order. Home: 14 Würthgasse Vienna A1190. Office: Karlsplatz 13, Vienna A1040, Austria.*

HORNIG, Donald Frederick, Am. chemist; b. Milw., Wis., Mar. 17, 1920; s. Chester Arthur and Emma (Knuth) H.; B.S., Harvard University, 1940, Ph.D., 1943; LL.D., Temple U., 1964; D.H.L., Yeshiva U., 1965; D.Sc., Notre Dame, 1965, U. Md., 1965, Rensselaer Poly. Inst., 1965, Ripon Coll., 1966; m. Lilli Schwenk, July 17, 1943; children—Joanna, Ellen, Christopher, Leslie. Research asso. Woods Hole (Mass.) Oceanographic Instn., 1943-44; scientist, group leader Los Alamos Lab., N.M., 1944-46; asst. prof. chemistry Brown U., 1946-49, asso. prof., 1949-51, prof., 1951-57, dir. Metcalf Research Lab., 1949-57, asso. dean grad. asch., 1952-53, acting dean, 1953-54; vis. prof. Princeton, 1957, prof. chemistry, 1957-64, chmn. dept., 1958-64, Donner prof. sci., 1959-66; spl. asst. sci. and tech. to Pres., 1964—; dir. Office Sci. and Tech., 1964—; chmn. Fed. Council Sci. and Tech., 1964—. Board directors W. A. Benjamin, Incorporated. Member President's Science Adv. Com., 1960—, chmn. 1964—; chmn. Project Metcalf Office Naval Research, 1951-52. Mem. bd. overseers Harvard Coll., 1964—. Guggenheim fellow, 1954-55; Fulbright fellow, 1954-55. Fellow Am. Phys. Soc., Am. Acad. Arts and Scis.; mem. Nat. Acad. Scis., Am. Chem. Soc., A.A.A.S., Faraday Soc. (London), Sigma Xi. Author articles sci. jours. Research in theoretical chemistry; shock and detonation waves; spectra of crystals; molecular spectroscopy. Home: 2810 Brandywine St. N.W., Washington 20008. Office: White House, Washington.

HORNING, Donald Oury, Am. mech. engr.; b. Lewiston, Ida., Nov. 26, 1910; s. Charles Ernest and Josephine (Oury) H.; B.A., U. Neb., 1934, M.E., 1951; m. Margaret E. Walker, Sept. 1, 1938; children—Virginia Louise (Mrs. John Templeton), Martha Margaret (Mrs. Ralph Suess), George Walker. Service engr. Standard Oil Co., 1938-42; test engr. Consol.-Vultee Aircraft Co., 1942-46; faculty U. Cal., 1946—, research engr., lectr. mech. engring., Richmond, 1954—, supervising engr. Vehicle Tech. Facility, 1966—. Mem. San Francisco Bay Area Engring. Council (chmn. 1966-67), Am. Soc. M.E., Cal. Soc. Profl. Engrs., Solar Energy Soc., Am. Inst. Aeros. and Astronautics, Marine Tech. Soc. Contbr. numerous articles to profl. jours. Participant in devel. 1st supersonic low density wind tunnel; developed high intensity exotic arc image furnace for solar simulation and high temperature research; patentee device for reduction of sewage affluents, deep hole stress meter. Home: 620 San Fernando Av., Berkeley, Cal. 94707.*

HORNING, Evan Charles, Am. chemist; b. Phila., June 6, 1916; s. Samuel and Mary (Schnader) H.; B.S., U. Pa., 1937; Ph.D., U. Ill., 1940; m. Marjorie Groothuis, Sept. 25, 1941. Instr., Bryn Mawr Coll., 1940-41; instr. U. Mich., 1941-43, research asso., 1943-45; with U. Pa., 1945-50, asso. prof., 1947-50; chief Lab. of Chem. of Natural Products, NIH, Bethesda, Md., 1950-61; prof. chemistry Baylor U. Coll. Medicine, 1961—, chmn. dept. biochemistry, 1962-66, dir. Inst. for Lipid Research, 1966—. DuPont fellow, 1939-40; Rohm & Haas fellow, 1940; Guggenheim fellow, 1958. Fellow N.Y. Acad. Sci.; mem. Am., Swiss chem. socs., Biochem. Soc. (London), Am. Geriatrics Soc., A.A.A.S., Tex. Acad. Sci. Author: Organic Syntheses, vol. III, 1955; Effects of Drugs on Synthesis and Mobilization of Lipids, 1963; also numerous articles. Devel. gas. phase analytical biochem. methods for study of steroids, lipids, urinary acids, glucuronides and other compounds of biol. significance; studies in gas chromatography and gas chromatography-mass spectrometry, and applications in studies of atherosclerosis and related metabolic problems; isolation and structural studies of naturally occurring compounds. Home: 11610 Starwood Dr., Houston 77024.*

HORNO LIRIA, Ricardo, Spanish physician; b. Zaragoza, Mar. 8, 1910; s. Ricardo and Luisa (Liria) Horno; ed. U. Zaragoza, U. Madrid; M.D.; m. Maria-Isable Gonzalez, 1936; children—Maria-Marta, Miguel, Alfonso, Ricardo. Head maternity services of state; head service of gynecology S.O.E., Spanish Red Cross; dir. med. jour. Clínica y Laboratorio. Mem. Royal Acad. Medicine (Silver medal, prize), Soc. Obstetrics and Gynecology Region of Ebre, Soc. Med. Writers and Artists Madrid, Internat. Assn. Physicians, La Cadiera Soc., Spanish Gynecol. Soc. Author: Puericultura Ante-Natal; Hidatidosis Génito-Urinaria; Cancer del Aparato Genital Femenino; La Crisis de la Natalidad en Nuestra Epoca. Address: 31 Calle Calvo Sotelo, Zaragoza, Spain.

HORNSBY, Thomas, English astronomer; b. Oxford, Eng., Aug. 28, 1733; s. Thomas Hornsby; grad. Corpus Christi Coll., Oxford, Eng., 1749, B.A., 1753, M.A., 1757, D.D. by diploma, 1785; children—Thomas, George. Became Savilian prof. astronomy Oxford, 1763, 1st Radcliffe observer, 1772, Sedleian prof., 1782, Radcliffe librarian, 1783. Fellow Royal Soc., 1763. Editor: Astronomical Observations (Bradley) 1798. Deduced solar parallax almost identical to modern results from observations of transits of Venus, 1761, 69; studied eclipses of sun. Died Oxford, Apr. 11, 1810.

HORNYKIEWYTSCH, Théophil, physician; b. Luczyne, Austria, Aug. 25, 1919; s. Théophil and Anna (von Sas-Jovorsky) H.; M.D., U. Vienna; m. Hannelore Böhmer, Aug. 5, 1948; children—Theo, Georg. Asst., U. Vienna; with U. Marburg; div. head radiology and radiography Med. Clinic, U. Giessen. Mem. German Soc. Radiography, German Soc. for Biophysics, German Soc. for Research on Light, German Soc. for Digestion Troubles and Metabolism. Author: Intravenose Cholangiographie, 1956, also over 100 monographs, books on radiology. Editor: Internistische Praxis, Tagliche Praxis. Home: 59 Goethestrasse. Office: 32 b. Klinikstrasse, Giessen, Germany.

HOROVITZ, Zola Phillip, Am. pharmacologist; b. Pitts., Oct. 12, 1934; s. Reuben and Jean (Liff) H.; B.S. in Pharmacy, U. Pitts., 1955, M.S. in Pharmacology, 1958, Ph.D., 1960; m. Marlene C. Davis, Aug. 24, 1958; children—Bonna Lynn, Reid Alan. Research investigator VA Research Labs. in Neuropsychiatry, Pitts., 1959-60; sr. research neuropharmacologist Squibb Inst. Med. Research, New Brunswick, N.J., 1960-66, research supr. neuropsychopharmacology, 1966-67, director dept. of pharmacology, since 1967—. Recipient A. E. Bennett Neuropsychiatry award for original investigation in biol. psychiatry Soc. Biol. Psychiatry, 1965. Fellow A.A.A.S., Am. Found. for Pharm. Edn.; mem. Am. Soc. Pharmacology and Exptl. Therapeutics, N.Y., N.J. acads. scis., Am. Pharm. Assn., Sigma Xi, Rho Chi. Contbr. articles to tech. jours. Developed techniques for testing central nervous system activity; localized site action anti-depressant compounds; discovered anti-depressant activity thiazesim. Home: 5 Stearns Rd., East Brunswick, N.J. 08816. Office: Squibb Inst. Med. Research, New Brunswick, N.J. 08903.*

HOROWITZ, Norman Harold, Am. biologist; b. Pitts., Mar. 19, 1915; s. Joseph and Jeannette (Miller) H.; B.S., U. Pitts., 1936; Ph.D., Cal. Inst. Tech., 1939; m. Pearl Shykin, June 16, 1939; children—Joel Lawrence, Elizabeth Anne. NRC fellow Stanford, 1939-40; research fellow Cal. Inst. Tech., 1940-42; sr. research fellow, 1946-47, faculty, 1947—, prof. biology, 1953—; research asso. Stanford, 1942-46; chief biosci. sect. Jet Propulsion Lab., Pasadena, 1965—. Fulbright dellow, Guggenheim fellow Laboratoire de Génétique, U. Paris (France), 1954-55. Mem. Am. Soc. Biol. Chemists, Genetics Soc. Am., A.A.A.S., Am. Assn. U. Profs., Sigma Xi. Research, numerous publs. on biochem. nature inheritance, formation enzymes by genes, evolution biochem. mechanisms, pathways amino acid synthesis in living cells, regulation enzyme synthesis, strategy and tactics search for extra-terrestrial life.

HORREBOW, Peder, Danish astronomer; b. Jutland, 1679; at least 1 son, Christoffer; became prof. astronomy, Copenhagen, Denmark, 1710. Author: Copernicus triumphans, sive de parallaxi orbis annui, 1727; Clavis astronomiae (contains more exact solar parallax, favors theory of Descartes). Died 1764.

HORROCKS, Donald Leonard, Am. chemist; b. Dearborn, Mich., July 14, 1929; s. Wilfrid and Deena (Hendrick) H.; B.A., Reed Coll., 1951; Ph.D., Ia. State U., 1955; m. Margaret Annette Powell, July 30, 1955; children—Andrea Jean, Cynthia Kay. Research asst. Ames Lab. AEC, 1951-55; research chemist Nat. Carbon Research Lab., Cleve., 1955-56; asst. chemist Argonne (Ill.) Nat. Lab., 1956-58, asso. chemist, 1958—, chmn. Internat. Symposium on Organic Scintillators, 1966; lectr. St. Procopius Coll., 1958-60. Mem. Am. Chem. Soc., Sigma Xi. Contbr. chpt. to Analytic Chemistry, 1966. Studies on nuclear chemistry with spl. interest in fission process and transuranic elements, properties of certain organic compounds which convert other forms of energy into light; proved correlation between structure of compounds and scintillation efficiency. Home: 312 Hazelwood Dr., Naperville, Ill. 60540. Office: 9700 S. Cass Av., Argonne, Ill. 60439.*

HORROCKS (or HORROX), Jeremiah, English astronomer; b. Liverpool, Eng., 1617 or 19; s. William Horrocks; student Emmanuel Coll., Cambridge, 1632-35; acquainted with William Crabtree, from 1636; began tidal observations, 1640. Marble scroll inscribed in his memory by Dean Stanley, Westminster Abbey, 1875. Author: Venus in sole visa, 1662; Opera posthuma, 1672, 78. Predicted and made 1st observation of transit of Venus across sun, 1639; detected long inequality of Jupiter and Saturn; probably identified solar attraction with terrestrial gravity; developed theory of lunar motions and determined elliptical orbit used by Newton; made most accurate estimate of his day of distance between earth and sun. Died Jan. 3, 1641.

HORROCKS, John Edwin, Am. psychologist; b. Cohoes, N.Y., Aug. 7, 1913; s. Samuel Edwin and Ada (Wooster) H.; B.A., N.Y. State Coll. for Tchrs., 1937; M.A., Syracuse U., 1942, Ph.D., 1945; m. Jane Kehrer, Aug. 27, 1962. Tchrs., N.Y. State Pub. Schs., 1937-43; instr. Syracuse U., 1944-45; faculty Ohio State U., 1945—, prof. psychology, 1952—, chmn. Inst. for Child Devel. and Family Life, 1954-58; social research scientist USAF, 1951, mem. adv. com. Personnel Research Lab., 1966—. Editorial adviser in devel. psychology Houghton Mlifflin Co., 1966—. Mem. Am., Ohio (past pres.) psychol. assns., Soc. for Research in Child Devel., Royal Soc. Medicine (Eng.), N.Y. Acad. Scis., Internat. Council Psychologists. Author: Psychology of Adolescence, 1951, 62; (with Pressey, Robinson) Psychology in Education, 1959; Assessment of Behavior, 1965; also numerous articles. Editor: Internat. Series in Behavioral Scis., 1962—; Jour. Genetic Psychology, 1966—; Genetic Psychology Monographs, 1966—. Research on self-concept, cognitive behavior, psychol. needs, human factors, group behavior. Home: 190 E. North Broadway, Columbus, O. 43214.*

HORROX, Jeremiah, see Horrocks, Jeremiah.

HORSEFIELD, Thomas, Am. naturalist, physician; b. Bethlehem, Pa., May 12, 1773; s. Timothy and Juliana Sarah (Parsons) H.; M.D., U. Pa., 1798. Surgeon, Dutch Colonial Army, Java, 1801; joined Brit. East India Co., 1811; went to London with collections of Java flora, 1819; curator East India Co. Museum, 1820-59. Author: An Experimental Dissertation on Rhus Vernix, Rhus Radicans and Rhus Glabrum (pioneer contbn. to study of poison ivy and sumac in exptl. pharmacology), 1798; Plantae Javanicae Rariores, 5 vols., 1838-52. Died July 24, 1859.

HORSFALL, Frank Lappin, Jr., Am. virologist; b. Seattle, Dec. 14, 1906; s. Frank Lappin and Jessie (Ludden) H.; B.A., U. Wash., 1927; M.D., C.M. (Banting fellow), McGill U., 1932, D.Sc. (hon.), 1963; Ph.D. (hon.), Uppsala U., 1961; LL.D., U. Alta., 1963; m. Norma Campagnari, July 1, 1937; children—Frank Lappin III, Susan Ludden, Mary Elizabeth. Mem. staff Rockefeller Inst., N.Y.C., 1934-37, 41-60, Rockefeller Found., N.Y.C., 1937-41; pres., dir., trustee Sloan-Kettering Inst. Cancer Research, N.Y.C., 1960—, dir. research Meml. Sloan-Kettering Cancer Center, also Meml. Hosp. Cancer and Allied diseases, 1965—; prof. medicine Cornell 1960—, dir., prof. microbiology Sloan-Kettering div. Grad. Sch. Med. Scis., 1960—. Chmn. research council, bd. dirs. Pub. Health Research Inst. City N.Y., 1956—; mem. Health Research Council City N.Y., 1958—, commn. health services, 1959—; mem. coms. Nat. Found., N.Y., 1959—; mem. Pres.'s Commn. on Heart Disease, Cancer and Stroke, 1964-65; mem. numerous coms., also cons. to various brs. of govt. Recipient Holmes Gold medal McGill U., 1932, Casgrain and Carbonneau award, 1942; Eli Lilly award, 1937; John F. Lewis prize Am. Philos. Soc., 1959; others. Fellow Montreal Medic-Chirurg. Soc.; mem. Am. Assn. Cancer Research, A.A.A.S., Am. Assn. Immunologists (councillor 1962-66, pres. 1967-68), Am. Assn. Pathologists and Bacteriologists, A.M.A., Am. Soc. Clin. Investigation (v.p. 1943-44), Am. Soc. Microbiology, Assn. Am. Physicians, Nat. Acad. Scis., N.Y. Acad. Medicine, Royal Soc. Medicine, London (fgn.), Soc. Exptl. Biology and Medicine, Harvey Soc. (pres. 1956-57), Sigma Xi, Alpha Omega Alpha, others. Co-editor Viral and Rickettsial Infections of Man, 1959, 65; asso. editor Virology, 1954-60, Jour. Exptl. Medicine, 1963—; mem. editorial bd. World-Wide Abstracts of Gen. Medicine, 1958—, others. Research, publs. on viral infections of human beings, chem. inhibition of viruses, incitants of cancer, nature of cancerous change in cells. Home: 410 E. 68th St., N.Y.C. 10021. Office: Sloan-Kettering Inst. for Cancer Research, N.Y.C. 10021.*

HORSFALL, James Gordon, Am. plant pathologist; b. Mt. Grove, Mo., Jan. 9, 1905; s. Frank and Margaret Atwood (Vaulx) H.; B.S., U. Ark., 1925; Ph.D., Cornell, 1929; D.Sc., University of Vermont, 1958; Doctor honoris causa, Turin University, Italy, 1964; m. to Sue Belle Overton, June 30, 1927; children—Margaret Eleanor, Anne Vaulx. Asst. plant pathology Cornell, 1925-28, instr., 1928-29; research asso., asst. prof. plant pathology N.Y. State Agr. Exptl. Sta., Geneva, 1929-36, research chief, prof., 1936-39; chief dept. plant pathology and botany Conn. Agr. Exptl. Sta., 1939-50, dir. since 1948; lectr. Yale since 1950. Chmn. executive com. Chem.-Biol. Coordination Center, Nat. Research Council, 1952-57, mem. plant protection com., 1942-44; U. S. del. Atoms for Peace Conf., Geneva 1958; v.p. 2d Internat. Congress Crop Protection, London, Eng., 1949; trustee Biological Abstracts, 1952-58; mem. Nat. Adv. Commn. Food and Fiber, 1965-67; adv. com. 1st Internat. Congress Plant Pathology. Life member India Internat. centre, New Delhi. Distinguished Alumnus, U. Ark., 1951. Mem. Am. Phytopathology Soc. (pres. 1950), A.A.A.S. (council 1947-48), Bot. Soc. Am., American Academy of Arts and Sciences, Indian Phytopathological Soc., Assn. Applied Biologists of England, National Acad. Scis., Soc. Indsl. Microbiologists (pres. 1954), Accademia Nazionale di Agricoltura (Italy) (corresponding associate), Societa Phytoatria Italy (honorary member), Sigma Xi. Author: Fungicides and their action, 1945; Principles of Fungicidal Action, 1956; also articles sci. jours. Contbr. Ency. of Chem. Technology, Am. Peoples Ency. Co-editor: Plant Pathology, An Advanced Treatise, 3 vols., 1960. Edit. com. Ann. Rev. Plant Physiology, 1957-61; editor Ann. Rev. Phytopathology, 1962—. Research on chemotherapy of plant diseases; root rot diseases; vascular diseases of plants; abnormal plant physiology; nutrition in plant diseases; organic fungicides; mechanisms of fungicidal action. Home: 49 Woodstock Rd., Hamden, Conn. 06517. Office: Conn. Agrl. Expt. Sta., 123 Huntington St., New Haven 06504.

HORSFALL, William Robert, Am. entomologist; b. Mountain Grove, Mo., Jan. 11, 1908; s. Frank and Margaret (Vaulx) H.; B.S., U. Ark., 1928; M.S., Kan. State U., 1929; Ph.D., Cornell U., 1933; m. Annie Laurie Ellis, Sept. 7, 1930. Prof. biology Ark. A. and M. Coll., 1933-37; asst. prof. entomology S.D. State Coll., 1937-38; faculty U. Ark., 1938-47, asso. prof., 1945-47; faculty U. Ill., Urbana, 1947—, prof., 1955—. Entomol. cons. TVA, 1964—, WHO, 1966. Cited by Soc. Zoology and Botany Finland, 1964. Mem. A.A.A.S., Entomol. Soc. Am., Ill. Acad. Sci., Sigma Xi, Phi Kappa Phi. Author: Mosquitoes: Their Bionomics, 1955; Medical Entomology, 1962; also numerous articles. Research on life history and habits of floodwater mosquitoes, fundamentals of devel. of mosquitoes, census procedures for floodwater mosquitoes, techniques for colonizing mosquitoes, techniques for inducing hatching of eggs of floodwater mosquitoes. Home: 503 W. Vermont St., Urbana, Ill. 61801.*

HORSFORD, Eben Norton, Am. chemist; b. Moscow, N.Y., July 27, 1818; s. Jerediah and Charity (Norton) H.; grad. Rensselaer Poly. Inst., 1838; A.M., Harvard, 1847; studied analytical chemistry with Liebig, Giessen, Germany, 1844-46; m. Mary L'Hommedieu Gardiner, 1847, 4 daus.; m. 2d, Phoebe Dayton Gardiner, 1857, 1 dau. Prof. math. and natural scis. Albany (N.Y.) Female Acad., 1840-44; taught chemistry, research Lawrence Scientific Sch. (now part of Harvard), Cambridge, Mass., 1847-63; pres. bd. visitors Wellesley Coll.; early mem. Am. Chem. Soc. Author: The Theory and Art of Breadmaking, 1861; also many articles in sci. jours. Research in food preservation (condensed milk, baking powder), restoration of phosphates to bread flour, gen. food chemistry. Died Cambridge, Jan. 1, 1893.

HORSLEY, Sir Victor Alexander Haden, English surgeon; b. Kinsington, Eng., 1857; s. J. C. Horsley; ed. Univ. Coll. Hosp.; B.S.; M.D., Halle; LL.D., D.C.L., Montreal U., McGill U.; m. Eldred Bramwell, 1887; 2 sons, 1 dau. Prof., supt. Brown Instn., 1884-90; Fullerian prof. Royal Instn., 1891-93; prof. pathology Univ. Coll., 1893-96; surgeon Nat. Hosp. for Paralysis and Epilepsy, 1886; emeritus prof. clin. surgery, cons. surgeon Univ. Coll. Hosp., 1906-16. Fellow Royal Soc. (Royal medallist), Royal Coll. Surgeons; mem. Sci. Soc. Sweden, Acad. Med. Paris (fgn. asso.), Acad. Sci. Berlin (fgn. asso.), also others. Recipient Cameron gold medal, Fothergill gold medal, 1st medallist Lannelongue Internat. Prize in Surgery, 1911. Author: Functions of the Marginal Convolutions, 1884; (with others) Experiments on the Functions of the Cerebral Cortex, 1888; Alcohol and the Human Body, 1907; also numerous articles on nervous system. A founder neurosurgery in Eng.; produced artificial myxoedema in monkey by excision of thyroid gland, 1884; operated on tumors of spine, 1887; perfected methods of craniotomy; showed (with Gotch) that active mammalian brain produces elec. currents which they recorded with string galvonometer, 1891. Died July 16, 1916.

HÖRSTADIUS, Sven Otto, Swedish zoologist; b. 1898; Ph.D., U. Stockholm, 1930; hon. doctorates from U. Paris, U. Cambridge. Reader zoology U. Stockholm, 1928-32, asso. prof., 1932-42; head dept. developmental physiology and genetics Wenner Gren Inst. Exptl. Biology, 1938-42; prof. zoology Uppsala U., 1942—; research marine biol. stations. Decorated comdr. No. Star (Sweden). Recipient Albert Brachet award Belgian Royal Acad. Sci., 1938. Fellow Royal Swedish Acad. Sci., Royal Soc. Sci. Uppsala, Royal Danish Academy of Science, Royal Physiography Society of Lund Pontifical Academy; member International Congress Ornithology (gen. secretary, 1950), Internat. Union Biol. Scis. (pres. 1953-58), Internat. Council Sci. Unions (past pres.), Swedish Ornithol. Soc. (pres. 1947-68), Internat. Council Bird Preservation (pres. 1960—); fgn. mem. Royal Soc. London, Zool. Soc. London; hon. mem. Belgian Royal Zool. Soc., Brit. Ornithologists Union, Biology Soc. Paris, Linnean Soc. London, Royal Inst. Gt. Britain. Author: Über die Determination des Keimes bei Echinodermen, 1928; Über die Determination im Verlaufe der Eiachse bei Seeigelm, 1935; The Neural Crest, 1950. Research on embryology of sea urchins, including testing potentialities for differentiation of different parts of egg by micro-operations and treatment with chem. substances to reveal change of devel. in different directions, analysis of neural crest in amphibians, descriptive embryology, bird photography and popular ornithology. Home: 6B Mellanvägen, Uppsala, Sweden. Office: Zool. Inst., Uppsala, Sweden.*

HORSTMANN, August Friedrich, German chemist; b. Mannheim, Germany, 1842; ed. Heidelberg, Zürich, Bonn; hon. prof. theoretical chemistry Heidelberg. Founder of phys. chemistry; 1st to work in thermodynamics of chem. reactions; applied kinetic theory to chem. equilibrium and dissociation of gases, 1868; research on nature of solution, heats of formation, vapor densities of ammonium sulfide, ammonium chloride, hydrogen sulfide and their deviations from law of Gay-Lussac; stated gas equation PV=RT. Died 1929.

HORSTMANN, Dorothy Millicent, Am. physician; b. Spokane, Wash., July 2, 1911; d. Henry John and Anna (Hunold) H.; A.B., U. Cal. at Berkeley, 1936, M.D., 1940; D.Sc. (hon.), Smith Coll., 1961; D.M.S. (hon.), Women's Med. Coll., 1963. Commonwealth Fund fellow sect. preventive medicine Yale, 1942-43, faculty, 1943-44, 45-47, 48—, prof. epidemiology and pediatrics, 1961—; instr. medicine U. Cal. at San Francisco, 1944-45; NIH fellow Nat. Inst. for Med. Research, London, 1947-48. Mem. vaccine devel. com. Nat. Inst. Allergy and Infectious Disease, 1964-67. Recipient award for contemporary leadership in poliomyelitis research, 1953. Fellow A.C.P., Am. Acad. Pediatrics (hon. asso.); mem. Am. Soc. for Clin. Investigation, Am. Epidemiological Soc., Assn. Tchrs. Preventive Medicine, Am. Pediatric Soc., Infectious Diseases Soc. Am. Editor: Virology and Epidemiology, 1962. Contbr. numerous articles to jours., textbooks. Research in infectious diseases, clin. epidemiology, virology, poliomyelitis, rubella, virus, others. Home: 11 Autumn St., New Haven 06511.*

HORSTMANN, Ernst Hermann Erich, German zoologist, physician; b. Pirmasens, Mar. 2, 1909; s. Wilhelm and Bertha (Heimüller) H.; Ph.D. in zoology; M.D.; m. Ingeborg Recker, 1936; children—Ute, Peter, Ulrich, Gisela. Prof. agrégé U. Heidelberg, 1939; asso. prof. U. Kiel, 1955; full prof. U. Hamburg, 1959. Mem. German Anat. Soc. (council), German Zool. Soc. (council). Author: Die Haut. Handbuch der Mikroskopischen Anatomie, 1957; Elektronenmikroskopie der Nervensystems und der Menschlichen Spermiogenese; Motorik der Lymphgefässe. Home: 42 Klotzemoor, Hamburg. Office: Anatomic Institute, University of Hamburg, 52 Martinistrasse, Hamburg-Eppendorf, Germany.

HORT, Edward Collett, Brit. physician; b. 1868; ed. Emmanuel Coll., Cambridge, Guy's Hosp.; m. Ethel Augusta Gordon, 1896; 2 daus. Research in pvt. lab. Lister Inst.; cons. physician. Mem. adv. council for med. research Nat. Health Ins. Commn. Fellow Royal Coll. Physicians (Edinburgh), Royal Soc. Medicine, Royal Acad. Medicine (Ireland); mem. Path. Soc. Gt. Britain and Ireland, Assn. Physicians. Author: Rational Immunization in the Treatment of Pulmonary Tuberculosis, 1909; Fever, Its Causes and Treatment, 1912; also articles. Research on method to diagnose cancer by blood exam., new method to treat gastric and duodenal ulcer with serum and dry protein. Died Oct. 15, 1922.

HORTOLOMEI, Nicolae, Rumanian surgeon; b. 1886; mem. Acad. Rumanian People's Republic. Author: Surgical Treatment of Hyperthyreoses, 1954; (with others) Surgery, 4 vols., 1955-58. Research in gastric surgery, dynamics of urinary passages, also surg. treatment of mitral stenosis, blue disease, hyperthyroidism. Died 1961.

HORTON, Charles Abell, chemist; b. Buffalo, Mar. 31, 1918; s. Harvey S. and Mildred (Abell) H.; A.B., Cornell U., 1941; Ph.D., U. Mich., 1949; m. Elsa Carmen Hart, June 27, 1947; children—Nancy Helen, John Charles. Analytical chemist Mead Johnson, Evansville, Ind., 1941-42, Bell Aircraft, Buffalo, 1942-44, U. Rochester, 1944-45, U. Mich., 1945-49; analytical chemist Union Carbide, Oak Ridge, 1949—; sr. officer IAEA, Vienna, Austria, 1966-68. Fellow A.A.A.S., Am. Inst. Chemists; mem. Am. Chem. Soc., Am. Nuclear Soc., Soc. Analytical Chemists (Eng.), Sigma Xi, Alpha Chi Sigma. Contbg. author: Fluorine Chemistry, 1954; Advances in Analytical Chemistry and Instrumentation, 1960; Treatise on Analytical Chemistry, II, 1962; also articles. Research on analytical chemistry of fluorine, uranium, and their compounds; determination of other elements in their compounds. Home: 20B/8 Leystrasse, Vienna 1200. Office: 11 Kaerntnerring, Vienna 1010, Austria.*

HORTON, Claude Wendell, Am. physicist; b. Cherryvale, Kan., Sept. 23, 1915; s. Roy Wesley and Marie (Terwilleger) H.; B.A. with honors Rice Inst., 1935, M.A., 1936; Ph.D., U. Tex., 1948; m. Louise C. Walthall, Nov. 23, 1938; children—Claude Wendell, Margaret Elaine (Mrs. Landess Morefield). Fellow in physics Rice Inst., Houston, 1935-36; with Shell Oil Co., Houston, 1936-37, 38-43; asst. in math. Princeton, 1937-38; research asso. Underwater Sound Lab., Harvard, 1938-43; prof. physics U. Tex., Austin, 1943—, chmn. dept., 1956-62, prof. geology, 1965—. Mem. corp. Woods Hole Oceanographic Instn., 1966—. Fellow Acoustical Soc. Am.; mem. Am. Phys. Soc., Soc. Exploration Geophysicists, Am. Geophys. Union. Research, numerous publs. in geophysics, underwater acoustics, diffraction of waves; patentee. Home: 3213 Cherry Lane, Austin, Tex. 78703.*

HORTON, Derek, chemist; b. Birmingham, Eng., Aug. 31, 1932; s. George Frederick and Eva (Mead) H.; B.Sc. with honors, U. Birmingham, 1954, Ph.D., 1957; m. June Margaret d'Adrian Sculfer, Dec. 21, 1957; children—Martin, Veronica, Fiona. Head sci. dept. Sebright Sch., Worcestershire, Eng., 1957-59; with Ohio State U., Columbus, 1959—, asso. prof. chemistry, 1965—. Fellow of the Royal Institute of Chemistry; mem. Chemical Society (London), A.A.A.S., Am. Chem. Soc. (past program chmn. div. carbohydrate chemistry), Sigma Xi, Phi Lambda Upsilon. Editor; Carbohydrate Research, 1965—. Research, numerous publs. on organic chemistry of carbohydrates and their derivatives, chem. synthesis of biologically important compounds, biochemistry. Home: 3456 Clearview Av., Columbus, O. 43221.*

HORTON, Frank, Brit. physicist; b. Aug. 20, 1878; s. A. Horton; M.Sc., Mason Univ. Coll., Birmingham; ed. St. John's Coll., Cambridge, Sc.D. D.Sc., London; m. J. M. Vera Fulton, 1911; 1 dau.; m. 2d, Ann Catherine Davies, 1939. Fellow St. John's Coll., Cambridge, 1905-13, lectr. Cavendish Lab., 1905-14; prof. physics U. London, 1914-46, dean Faculty Sci., 1930-34, vice chancellor, 1939-45, emeritus prof., 1946-57. Fellow Royal Soc., 1923. Co-editor: Monographs on Physics. A History of the Cavendish Laboratory, 1871-1910. Research and publs. on exptl. elasticity and electricity. Died Oct. 31, 1957.

HORTULAIN (or Orthulain), alchemist; b. France. Author: Practica vera alchimica, 1358; Practica chimica. Wrote on preparation of alcohol, aqua regia, nitric acid, also on formula for magic elixir, table of emeralds used in Middle Ages.

HORVATH, I. István, Hungarian biochemist; b. Gálosfa, Hungary, June 6, 1920; s. István and Hermin (Györfi) H.; Ph.D. in Chemistry, Physics and Minerology, Pázmány Péter U., Budapest, Hungary, 1944; candidate biol. scis. Med. U., Budapest, 1964; m. Éva Fehér; children—Zsolt László, Klára László, István T. Asst. prof. dept. pharmacology, Budapest, 1945-48; chief dept. microbiology Research Inst. for Pharm. Chemistry, Budapest, 1950—. Mem. Hungarian Microbial Soc. (sec.). Research, publs. on antibiotics and microbiol. oxidation of steroids, molecular biology, especially regulation of valine and isoleucine biosynthesis. Home: 33 Naphegy utca, Budapest I. Office: 47-49 Szabadságharcosok, Budapest IV, Hungary.*

HORVATH, Milan, Czechoslovakian physiologist; b. Krenovice, nr. Brno, Czechoslovakia, Dec. 8, 1921; s. Otakar and Rosalie (Kynerova) H.; D.nat.sci., Charles U., 1947, C.Sc., 1965; m. Cecilie Michalova, Dec. 9, 1950 (dec.); children—Michaela, Pavel. Faculty dept. physiology Med. Faculty, Charles U., Prague, Czechoslovakia, 1946-50, asso. prof. gen. hygiene (physiology), 1965—; chief dept. physiology of higher nervous activity Inst. Indsl. Hygiene and Occupational Diseases, Prague, 1951—. Chmn. research team Ministry Health Research Council, 1955—. Recipient award Ministry Health Sci. Council, 1963. Mem. Internat. Brain Research Orgn., CINS-Internat. Assn. Psychiatry (councillor), Internat. Assn. Occupational Health (sec. subcom. for study higher nervous functions), Czechoslovak Med. Soc., Czechoslovak Psychol. Soc., Pavlovian Soc. N.Am., French Ergonomic Soc. Co-author text books, including Occupational Physiology, 1956, 58. Editor: Activitas Nervosa Superior, 1959—; Symposium Higher Nervous Functions and Occupational Health, 1966; co-editor Psychopharmacological Methods—a Symposium, 1963; editorial bd. Conditional Reflex. Research, publs. on relative neurotropic potency of indsl. chem. substances and drugs, their mechanism of action, search for preventive and therapeutic measures, analysis of functional overstrain of central nervous system and/or psychoemotional stress caused by modern working and living conditions, introduction systematic environmental studies of higher nervous functions in occupational health. Home: 14, Volynska, Prague 10 Vrsovice. Office: 48, Srobarova, Prague 10 Vinohrady, Czechoslovakia.*

HORVATH, William John, Am. physicist; b. N.Y.C., Sept. 13, 1917; s. John and Anna (Horvath) H.; B.S., Coll. City N.Y., 1936; M.S., N.Y. U., 1938, Ph.D., 1940; m. Rebecca Sue Badger, Feb. 23, 1963; 1 dau., Susan Grace. Instr., Poly. Inst., Bklyn., 1940; physicist Bur. Ordnance, U. S. Navy, 1940-43, analyst Operations Research Group, 1943-45; sect. head Operations Evaluation Group, Chief Naval Operations, Washington, 1945-47, dept. dir., 1947-49; mem. rev. bd. weapons systems evaluation group Joint Chiefs Staff, Washington, 1949-52; cons. Sylvania Electric & Philco Corp., 1952-55; sect. head Biomed. Elec. Airborne Instrument Lab., Mineola, N.Y., 1955-58; research physicist Mental Health Research Inst., U. Mich., Ann Arbor, 1958—. Cons. to govt. agys.; mem. health services research study sect. NIH, 1962—. Recipient Naval Ordnance Devel. award, 1945, Presdl. certificate merit, 1947. Mem. Phys. Soc., Operations Research Soc., I.E.E.E., Biophys. Soc., Am. Pub. Health Assn. Contbr. articles to tech. jours. Research on application math. methods to analysis submarine and antisubmarine operations, evaluation large scale mil. systems through systems analysis; devel. automatic systems med. screening; math. studies social, psychol. and biol. phenomena. Home: 2451 Trenton Ct., Ann Arbor, Mich. 48105.*

HORWITT, Max Kenneth, Am. biochemist; b. N.Y.C., Mar. 21, 1908; s. Harry and Bessie (Kenetz) H.; B.A., Dartmouth, 1930; Ph.D., Yale, 1935; m. Frances Levine, Sept. 3, 1933; children—Ruth Ann (Mrs. Donald H. Singer), Mary Louise (Mrs. Samuel Goldman). Asst. biochemist Yale Sch. Medicine, 1932-35, Yale fellow, 1935-37; dir. biochemistry Research Lab., Elgin (Ill.) State Hosp., 1937-59, dir. Mendel Research Lab., 1960-68; faculty U. Ill. Coll. Medicine, 1942—, prof. biochemistry, 1962—; acting dir. div. research services Ill. Dept. Mental Health, Chgo., 1967-68; prof. dept. biochemistry St. Louis U. Sch. Medicine, 1968—. Mem. various coms. NRC, NIH; mem. FAO/WHO expert group on vitamin requirements, 1965—. Diplomate Am. Bd. Clin. Chemistry, Am. Bd. Nutrition, Fellow A.A.A.S., Gerontological Society, New York Acad. Scis.; mem. Am. Soc. Clin. Nutrition (Osborne and Mendel award 1961), Soc. for Exptl. Biology and Medicine, Assn. Vitamin Chemists, Am. Chem. Soc., Inst. Food Tech., Society for Biological Psychiatry, Sigma Xi. Contbg. author: Modern Nutrition, 1964; The Vitamins, 1954; Hawk's Physiological Chemistry, 1965; also numerous articles. Established significance controlling nutrition and metabolism in mental health research; determined currently accepted human requirements for thiamine, riboflavin, niacin-trypotophan and tocopherol; evaluation nutritional requirements aged; research on biology schizophrenia, effect dietary fats in altering metabolism brain and other tissues. Home: 750 S. State St. Office: Mendel Research Lab., Elgin State Hosp., Elgin, Ill., 60120.*

HORWITZ, Nahmin, Am. physicist, educator; b. Duluth, Minn., Oct. 28, 1927; s. Aaron B. and Bertha (Fineman) H.; B.S., Western Res. U., 1949; M.S., U. Minn., 1951, Ph.D. (NSF fellow), 1954; m. Leah June Gressel, Jan. 29, 1949; children—David Jeffrey, Susan Beth, Ann Elissa, Amy Ruth. Physicist, Lawrence Radiation Lab., Berkeley, Cal., 1954-59; faculty Syracuse (N.Y.) U., 1959—, prof. physics, 1965—. Vis. physicist Brookhaven Nat. Lab., summers 1962, 63, 64; research asso. Rutherford High Energy Lab., Chilton, Eng., 1965-66. Mem. Am. Phys. Soc., Am. Assn. Physics Tchrs. Research, publs. in exptl. elementary particle physics. Home: 920 Westcott St., Syracuse, N.Y. 13210.*

HOSACK, David, Am. physician; b. N.Y.C., Aug. 31, 1769; s. Alexander and Jane (Arden) H.; attended Columbia, 1786; grad. Princeton, 1789; m. Catharine Warner, 1792; m. 2d, Mary Eddy, 1797; m. 3d, Magdalena Coster; 10 children including Alexander Eddy. Began med. practice, Alexandria, Va., 1791; prof. botany Columbia, 1795-1811, prof. materia medica, 1795-1811, prof. materia medica Coll. Phys. and Surg., 1807-08, prof. theory and practice of physic, 1811; attending surgeon at Burr-Hamilton duel, 1804; an incorporator Am. Acad. Fine Arts, 1808; a founder Rutgers Med. Coll., pres. until 1830; a founder Bellevue Hosp., N.Y.C., 1820; a founder N.Y. Hist. Soc., pres., 1820-28; established Am. Med. and Philos. Register, editor, 1810-14. Author: A System of Practical Nosology, 1819. First American to litigate femoral artery for aneurysm, 1808. Died N.Y.C., Dec. 22, 1835.

HOSEMANN, Hans, German gynecologist; b. Rostock, Sept. 29, 1913; s. Gerhard and Anna-Dorothea (Kobert) H.; ed. univs. Freiburg, Rostock, Königsberg, Exeter; qualified M.D.; m. Ursuta Schameitat; children—Peter, Christine, Annette, Axel. Asst., U. Freiburg, 1937; asst. U. Göttingen, 1938, prof., 1948, physician Gynecology Clinic, 1948-62. Mem. German Assn. for Gynecology, also numerous others. Author: Die Grundlagen der Statistischen Methoden für Mediziner und Biologen, 1948; Geburtshilfe und Gynaekologie; Biologie und Pathologie des Weibes, 1949; Schmerzlinderung mit Trichloratylen, 1952; Normale and Abnorme Schwangerschaftsdauer in Seitzamreich; Vaterschatsgutachten für die Gerichtliche Praxis, 1956, also over 200 articles. Address: 14 Hermann Löns-Strasse, Emden, Germany.

HOSEMANN, Rolf, German physicist; b. Rostock, Germany, Apr. 20, 1912; s. Gerhard and Dorothea (Kobert) H.; Dr.Phil.Nat., U. Freiburg (Germany), 1936; Dr.Phil. Nat. Habil., Technische Hochschule Stuttgart, Germany, 1941; m. Ursula Hedwig Siebold, Apr. 17, 1938; children—Gerhard, Friedrich Wilhelm, Klaus Günther, Hans Jochen. Asst. to profs. univs. Freiburg, Tübingen, Berlin, 1931-34, 36-37, 50-60; prof. physics Technische Universität Berlin, 1954—; head dept. Fritz Haber Institut, Max Planck Gesellschaft, Berlin, Germany, 1957—. Cons. AEG-Telefunken, Berlin, European Research Asso., Brussels, Belgium. Mem. Bunsen Gesellschaft, Kolloid Gessellschaft, Deutsche Phys. Gesellschaft. Author: (with S. N. Bagchi) Direct analysis of Diffraction by Matter, 1962. Co-editor: Die Makromolekulare Chemie, Kolloid Zeitschrift; hon. adviser Jour. Materials Sci. Research, publs. on radioactivity of samarium, small angle scattering of solids, liquids, highpolymers; structure of paracrystals (a new theory); tertiary structure of protein of quantasomes; new x-ray tubes with microfocus and with high currents; a self-powered dosimer for x-rays and neutrons. Home: 26A Schmorlemerallee. Office: 4 Faradayweg, 1 Berlin 33, Germany.*

HOSHINO, Sadao, Japanese physicist; b. Tokyo, Japan, Sept. 7, 1926; s. Shuichi and Eiko (Godai) H.; B.Sc., Osaka (Japan) U., 1948, D.Sc., 1958; m. Toki Kusaka, Nov. 14, 1954; children—Moriyuki, Hiroshi, Takashi. Research asst. Osaka U., 1948; instr. Tokyo Inst. Tech., 1949-56, asst. prof., 1959-60; research asso. Pa. State U., 1956-59; vis. physicist Brookhaven Nat. Lab., Upton, N.Y., 1958-59; asso. prof. physics Tokyo U., 1960-66, prof., 1967—. Mem. Phys. Soc. Japan, Am. Phys. Soc., Cryst. Soc. Japan (gen. sec. 1965-66). Author: Neutron Diffraction, 1961; also articles. Research on phase transition of crystals, structure analysis of some ferroelectric substances, high intensity X-ray tube and automatic neutron diffractometer, neutron diffraction study ferroelectric and antiferromagnetic substances. Home: 41 Kamaya-cho, Hodogaya-ku, Yokohama, Japan. Office: Inst. Solid State Physics, Tokyo U., 7-Roppongi, Tokyo, Japan.*

HOSHINO, Toshio, Japanese organic chemist; b. Niigata Prefecture, Japan, 1899; grad. Tohoku U., 1924; postgrad. U. Munich. Former research worker Phys. and Chem. Research Inst.; asst. prof. Tokyo Tech. Coll., now prof. Mem. Japan Chem. Soc. (v.p.). Research on synthesis of synthetic fiber.

HOSKING, Eric, English ornithologist; b. London, Oct. 2, 1909; s. Albert and Margaret (Steggall) H.; ed. Stationer's Co.'s Sch., London; m. Dorothy Sleigh, 1939; children—Margaret, Robin, David. Illustrator 700 books of natural history of world; editor photographs New Naturalists, Brit. Birds; made expdns. to Cote Donana, So. Spain, 1956, 57, Bulgaria, 1960, Hungary, 1961, Jordan, 1963; head expdn. to Cazorla Valley, Spain, 1959, 65, 67. Hon. fellow Royal Photog. Soc.; mem. Brit. Ornithologists Union, Brit. Trust for Ornithology, Zool. Soc., Royal Soc. for Protection Birds. Author: The Art of Bird Photography, 1944; Birds of the Day, 1944; Birds of the Night, 1945; Masterpieces of Bird Photography, 1947; Birds in Action, 1949; Bird Photography as a Hobby, 1961; Nesting Birds, 1967. Address: 20 Crouch Hall Rd., London N.8, Eng.

HOSLER, Charles Luther, Jr., Am. meteorologist; b. Honey Brook, Pa., June 3, 1924; s. Charles Luther and Miriam (Stauffer) H.; student Bucknell U., 1943-44, Mass. Inst. Tech., 1944-45; B.S., Pa. State U., 1947, M.S., 1948, Ph.D., 1951; m. Gladys Irene Cheesbrough, June 23, 1948 (div. 1966); children—Sharon, Lynn, David, Peter. Hydrographer, Pa. Dept. Forests and Waters, 1949-59; faculty Pa. State U., University Park, 1948—, prof. meteorology, 1958—, head meterology dept., 1960-65, dean Coll. Earth and Mineral Scis., 1965—. Cons. to pvt. cos., law firms. Mem. A.A.A.S., Am. Meteorol. Soc., Am. Geophys. Union. Research, publs. on ice formation in clouds and phys. consequences; developed techniques increasing precipitation. Home: 102 Clearview St., State College, Pa. 16801. Office: Dieke Bldg., University Park, Pa. 16802.*

HOSOYA, Eikichi, Japanese pharmacologist; b. Maebashi, Japan, Nov. 11, 1910; s. Tsugio- and Fiji (Aramaki) H.; M.D., Keio U., 1935, D.Med. Sc., 1952; m. Kimi Sakai, Oct. 24, 1942; 1 son, Hideo. Faculty dept. pharmacology Keio U. Sch. Medicine, Shinjukuku Tokyo, Japan, 1935—, prof. pharmacology, 1963—, chmn. dept., 1965—; instr. U. Mich. Med. Sch., 1953-54. Mem. expert com. on dependence-producing drugs WHO, 1965-66. Recipient Kitasato award for studies on conjugation of morphine with glucuronic acid, 1960. Mem. Japanese Soc. for Pharmacology (trustee 1966—), Japanese Soc. Neuropsychiatry, Japanese Soc. Pharmacognosy, Japanese Soc. Neurochemistry, Japanese Soc. Biophysics. Author: Analgesics, 1963; also numerous articles. Research on metabolism of morphine in living bodies, prevention of re-addiction to morphine and heroin; developed new method for detection very small amount morphine in urine; screening of dependence-liability of drugs with small animals. Home: 35-3 Jingumae 6-Chome. Shibuyaku, Tokyo, Japan.*

HOSOYA, Seigo, Japanese bacteriologist; b. Tokyo, 1894; grad. Tokyo U., 1919, M.D., 1919, Igaku Hakushi, 1926. Joined Govt. Inst. for Infectious Diseases, 1919; apptd. asst. prof. Tokyo U., 1930; became prof. Taihoku (Taipeh) U., 1936; named prof. Tokyo U., 1937; ret., 1955; chief dir. Tokyo Microscope Acad. Author publs. including: Prevention of Diphtheria; also articles in bacteriology, especially anaerobic bacteria; Discovered Trichomycin. Died 1957.

HOSOYA, Sukeaki, Japanese physicist; b. Kobe, Japan, Dec. 5, 1924; s. Sukemitsu and Sanae (Hirota) H.; B.Sc., Inst. for Physics, U. Tokyo (Japan), 1947, D.Sc., 1962; Ph.D. (Brit. Council scholar), U. Wales, 1958; m. Yoko Tsunashima, May 10, 1959; children—Sukeharu, Keiko. Asst. prof. First Higher Sch., Tokyo, 1947-49; asst. U. Tokyo, Coll. Gen. Edn., 1950-58, asst. prof. Inst. for Solid State Physics, 1958—. Mem. Phys. Soc. Japan, Crystallographic Soc. Japan, Japan Soc. Applied Physics, Materials Sci. Soc. Japan. Author several books on physics; also articles. Determined electronic states in various inorganic substances using precise measurements of X-ray intensity diffracted from powder crystals; X-ray structure analyses of heterocyclic compounds such as thianthrene dioxides, its tetraoxide and related substances; discovered rule about shapes of these molecules, some being folded, others planar. Home: 10-2, 315 Higashi-Ohizumi, Nerima-ku, Tokyo, Japan.*

HOSPERS, Jan, Dutch geophysicist; b. Veendam, Netherlands, June 19, 1925; s. Johannes and Marchiena (Slim) H.; B.S., U. Groningen, 1948; M.S., U. Utrecht, 1952; Ph.D., U. Cambridge, 1953; m. Kerstin Barbro Helga Tunback, June 21, 1954; children—Michiel, Martin, Peter. Geophysicist, Royal Dutch Shell Group, Turkey, Venezuela, Nigeria, 1953-62, research geophysicist Royal Dutch Shell Exploration and Prodn. Lab., 1962-63; lectr. U. Utrecht, 1962-63; prof. geophysics U. Alta., Edmonton, Can., 1963-65; prof. solid earth geophysics U. Amsterdam, 1965—, head div. geophysics, dir. Geol. Inst., 1965—. Fellow Geol. Soc., Royal Astron. Soc.; mem. European Assn. Exploration Geophysics, Soc. Exploration Geophysicists, Am. Geophys. Union. Publs. on research in palaeomagnetism particularly occurrence of reversals of earth's magnetic field and problems of continental drift; gravimetry, particularly deeper structure of earth's crust in mountain ranges and deltas.*

HOTCHKISS, Benjamin Berkeley, Am. inventor, mfr.; b. Watertown, Conn., Oct. 1, 1826; s. Asahel A. and Althea (Guernsey) H.; m. Maria Bissell, May 27, 1850. Inventor (with his brother) new form of cannon projectile, demonstrated at Washington Navy Yard, 1855; made gift of some projectiles to liberal govt. of Mexico, 1859; furnished several hundred projectiles to Japanese, 1860, succeeded in getting small order from U. S., 1860; founder factory, N.Y.C., when Civil War produced demand for projectiles; patentee practical machine gun, 1872; inventor magazine rifle, 1875 (rifle used by U. S. Army, then by U. S. Navy); organizer Hotchkiss & Co., with hdqrs. in U. S., factories in Eng., Germany, Austria, Russia and Italy, 1882; other inventions include an explosive shell and packing for projectiles, percussion detonator, rapid fire revolver. Died Paris, France, Feb. 14, 1885.

HOTCHKISS, Henry Thomas, Am. chem. engr.; b. Bklyn., Apr. 15, 1900; s. Henry Thomas and Alice (Muns) H.; Chem.E., Poly. Inst. Bklyn., 1923; M.S. (Bache research fellow), N.Y. U., 1925; m. Lilian Van Winkle Ball, Oct. 10, 1928; children—Phyllis B. (Mrs. Fred G. Schwartz), Alice V. (Mrs. W. Lee Hidy). Faculty, N.Y. U. Washington Sq. Coll., 1925-27, Pratt Inst., 1934-40, Poly. Inst. Bklyn., 1940-42; with Tex. Gulf Sulphur Co., 1927-32, Feedwaters, Inc., 1932-34; supervising chemist Village of Larchmont, 1935-42; materials engr. Asiatic Petroleum Corp., N.Y.C., 1942-62. Cons. Village of Ossining, New Rochelle Water Co., Westchester County Dept. Pub. Works, Village of Larchmont. Mem. Am. Chem. Soc. Patentee in field. Research, publs. on synthetic drugs for lowering blood pressure, recovery of sulfur from pyrites, improved water filtration and purification. Address: 35 Coolidge St., Larchmont, N.Y. 10538.*

HOTCHKISS, Rollin D(ouglas), Am. biochemist; b. South Britain, Conn., Sept. 8, 1911; s. Charles Leverett and Eva (Platt) H.; B.S., Yale, 1932, Ph.D. (Loomis fellow), 1935, Sc.D., 1962; m. Shirley Dawson, June 24, 1933 (div. 1967); children—Paul, Cynthia; m. 2d, Magda Gabor, May 19, 1967. Mem. staff of The Rockefeller Univ., 1935—. mem., professor The Rockefeller Univ., N.Y.C., 1955—; Rockefeller Found. fellow Carlsberg Lab., Copenhagen, 1937-38, vis. prof. Mass. Inst. Tech., 1957-58; Dyer lectr. NIH, 1962. Recipient Commercial Solvents award in antibiotics, 1954. Member American Society Cell Biology (councilor 1962-64), Harvey Soc. (pres. 1958-59), L.I. Biol. Assn. (dir.), Am. Soc. Biol. Chemists, Am. Acad. Arts and Scis., Nat. Acad. of Science, Genetics Soc. America, American Chem. Soc., A.A.A.S., Am. Soc. Naturalists, N.Y. Acad. Sci., Sigma Xi. Contbr. articles tech., sci. publs. Research on immunochemistry of bacterial polysaccharides, 1935-37, protein chemistry, 1937-39, (with R. J. Dubos) devel. and purification, chem. study of antibiotics gramicidin and tyrocidine, 1939-43, bacterial metabolism and physiology, peptide synthesis, 1944-47, genetic biochemistry of deoxyribonucleic acids, 1947—; developed genetic transformation bacteria to drug resistance and sensitivity, 1951. Home: 342 E. 67th St., N.Y.C. 10021. Office: The Rockefeller U., 66th St. and York Av., N.Y.C. 10021.*

HOTEL, Hoyt Clarke, Am. chem. engr.; b. Salem, Ind., Jan. 15, 1903; s. Louis Weaver and Myrtle (Clarke) H.; A.B., Ind. U., 1922; M.S., Mass. Inst. Tech., 1924; m. Nellie Louise Rich, June 11, 1929; children—Lois (Mrs. William M. Wood, Jr.), Hoyt Clarke, Barbara Ellen (Mrs. Richard M. Willis), Elizabeth Leah (Mrs. John Rutledge Chapin). Faculty, Mass. Inst. Tech., 1927—, prof. fuel engring., 1941-65, Carbon P. Dubbs prof. chem. engring., 1965—, dir. fuels research lab., 1934—; cons. on fuels and combustion. Chmn., U. S. Com. on Flame Radiation, 1952—; adv. panel bldg. research div. Bur. Standards, 1963—; v.p. Combustion Inst, 1952-64. Recipient U. S. medal for merit, 1945; King's medal (Gt. Britain), 1945; William H. Walker award Am. Inst. Chem. Engrs.; Egerton Gold medal Combustion Inst., 1960; Melchett medal Inst. Fuel (Britain), 1960; Max Jakob award Am. Soc. M.E.,-Am. Inst. Chem. Engrs., 1966; Founders award Am. Inst. Chem. Engrs., 1967. Mem. Nat. Acad. Scis., Am. Acad. Arts and Sci., Am. Chem. Soc., Am. Inst. Chem. Engrs., Am. Soc. M.E., Phi Beta Kappa, Sigma Xi, Tau Beta Pi, Phi Gamma Delta. Author: (with G. C. Williams, C. N. Satterfield) Thermodynamic Charts for Combustion Processes, 1949; (with A. F. Sarafim) Radiative Transfer, 1967; also articles. Research on radiant heat transfer from water vapor and carbon dioxide, mechanism of combustion of solid carbon, indsl. furnace design, pulverized fuel combustion, two-color optical pyrometry of flames, solar energy utilization, diffusion flame structure, gaseous combustion in well-stirred reactor, jet mixing, allowance for temperature gradients in gas radiation, radiative scatter. Home: 27 Cambridge St., Winchester, Mass. 01890. Office: Mass. Inst. Tech., Cambridge, Mass. 02139.*

HOTELLING, Harold, Am. statistician, economist; b. Fulda, Minn., Sept. 29, 1895; s. Clair Alberta and Susy (Rawson) H.; A.B., U. Wash., 1919, M.Sc., 1921; postgrad. U. Chgo., LL.D., 1955; Ph.D., Princeton, 1921, postgrad.; Sc.D., U. Rochester, 1963; m. Floy Tracy, Dec. 27, 1920 (dec. Oct. 1932); children—Eric Bell, Muriel (Mrs. Glenn L. Burrows); m. 2d, Susanna P. Edmondson, June 14, 1934; children—George Alfred, William Edmondson, Edward Rawson, Harold Addison, James Maynard. Instr. math Princeton, 1922-24; staff Food Research Inst., Stanford, 1924-27, asso. prof. math., 1927-31; prof. econs. Columbia, 1931-46; prof. statistics U. N.C., Chapel Hill, 1946—, Kenan prof. statistics, 1961—. Adviser univs., govt. agys., pvt. cos. Recipient medal Free U. Brussels (Belgium), 1951. Fellow A.A.A.S. (past sects. chmn.), Am. Econ. Assn. (Distinguished fellow 1965); mem. Social Sci. Research Council (past dir.), Elisha Mitchell Sci. Soc. (past pres.), Indian Statis. Congress (past pres.), Am. Math. Soc., Econometric Soc. (past pres.), Am. Statis. Assn. (past v.p.), Inst. Math. Statistics (past pres.), Internat. Statis. Inst., Royal Statis. Soc. Research, numerous publs. in pure maths., topology and differential geometry, matrices and comutation, math. econs., taxation, welfare, gen. multi-commodity problems, application calculus of variations to mineral econs., theory statistics, multivariate analysis, canonical correlations, use nonnormal distbns., rank correlations. Home: P.O. Box 62, Chapel Hill, N.C. 27514.*

HOTTEL, Hoyt Clarke, Am. engineer; b. Salem, Ind., Jan. 15, 1903; s. Louis Weaver and Myrtle (Clarke) H.; A.B., Ind. U., 1922; M.S., Mass. Inst. Tech., 1924; m. Nellie L. Rich, June 11, 1929; children—Lois, Hoyt C., Barbara E., Elizabeth L. Mem. faculty Mass. Inst. Tech. since 1927, prof. fuel engring. 1941-66, Carbon P. Dubbs prof. chem. engring., since 1966—, dir. fuels research lab. since 1934; cons. fuels and combustion. Sect. chief, on fire warfare Nat. Def. Research Com., 1942-45; chmn. U. S. Committee on Flame Radiation, National Academy-National Research Council Com. on Fire Research, AFSWP Thermal Panel, 1949-58; v.p. Combustion Inst., 1952-64. Decorated Medal for Merit; King's Medal (Brit.); recipient Wm. H. Walker award Am. Inst. Chem. Engrs.; Egerton Gold medal. Combustion Inst., 1960; Melchett medal, Inst. of Fuel (Britain), 1960; Jakob award Am. Soc. M E.-Am. Inst. Chem. Engrs., 1966. Member American Chemical Society, American Institute Chemical Engineers, American Soc. M.E., Am. Acad. Arts and Scis., National Academy of Sciences, also member of Soc. Sigma Xi, Phi Beta Kappa. Author: Thermodynamic Charts for Combustion Processes (with G. W. Williams and C. N. Satterfield), 1949; also sects. in handbooks, and tech. articles on combustion, radiant heat transmission, pyrometry. Research on optical methods of temperature measurement; solar energy utilization; combustion mechanisms; combustion in turbines and ramjets. Home: 27 Cambridge St., Winchester, Mass. Office: Mass. Inst. of Tech., Cambridge 39, Mass.

HOUBEN, Joseph, German organic chemist; b. Waldfeucht, Germany, Oct. 27, 1875; prof., Berlin, Germany. Author: (with Weyl) Die Methoden der organischen Chemie. Prolific synthesist; applied Gattermann's aldehyde synthesis to nitrites. Died 1940.

HOUBOLT, John Cornelius, Am. physicist; b. Altoona, Ia., Apr. 10, 1919; s. John H. and Hendreika (Van Ingen) H.; B.S., U. Ill., 1940, M.S., 1942; Ph.D., Swiss Fed. Inst. Tech., Zurich, 1958; m. Mary Morris, June 14, 1949; children—Mary Cornelia, Joanna, Julie. Bridge engr. I.C. R.R., 1940; city engr., Waukegan, Ill., 1941; aero. research scientist NASA, Hampton, Va., 1942-49; asso. chief dynamic loads div. NACA-NASA, 1949-62, chief theoretical mechanics div. NASA, 1962-63; exec. v.p., dir. Aero. Research Asso. Princeton Inc., (N.J.), 1963—. Instr. grad. extension div. U. Va., 1944—, Va. Poly. Inst., 1958—; exchange scientist Royal Aircraft Establishment, Eng., 1949. Recipient Rockefeller Pub. Service award, 1956; NASA Exceptional Sci. Achievement award, 1963. Asso. fellow Am. Inst. Aeros. and Astronautics. Asso. editor Jour. Spacecraft and Rockets. Research, numerous reports in aeros., aeroelasticity, structures, atmosphere turbulence, space flight and landing. Home: 105 Elm Rd. Office: 50 Washington Rd., Princeton, N.J. 08540.*

HOUCK, John Candee, Am. biochemist; b. N.Y.C., Feb. 19, 1931; s. John Walter and Marjorie (Candee) H.; B.A., Columbia, 1953; M.Sc. with honours, U. Western Ont., Can., 1955, Ph.D., 1956; m. Elizabeth Stuart, July 3, 1953; children—Leslie, Mary, John, Eric. Sr. Postdoctoral fellow Surg. Biochem. Lab., Georgetown U. Med. Sch., 1956-57, dir. surg. biochem. research lab., asst. prof. dept. biochemistry, 1957-59, asso. prof., 1964—; dir. biochem. research lab. Children's Hosp., Washington, 1959—, dir. Core Labs., Chem. Research Center, 1965—. Recipient Morrison award N.Y. Acad. Sci., 1962, Am. Dermatol. Assn. Research award, 1965. Mem. Am. Soc. Biol. Chemists, A.A.A.S., Am. Chem. Soc., Soc. for Exptl. Biology and Medicine, Soc. for Investigative Dermatol., So. Soc. for Clin. Research. Research, numerous publs. on chemistry inflammation and tissue necrosis, determination effects drugs and hormones on chemistry skin; discovered enzymatic destruction skin collagen, process inflammation. Home: 8621 Redwood Dr., Vienna, Va. Office: Biochem. Research Lab., Children Hosp., Washington.*

HOUDRY, Eugene J., engr.; b. Domont, France, 1892; came to U. S., naturalized, 1941. Developed catalytic methods for prodn. of motor fuels from coal, lignite, etc.; then invited to U. S. by Vacuum Oil Co., to use methods for comml. prodn. of gasoline; became pres. Houdry Process Corp., 1931. Started comml. prodn. of gasoline (using catalytic method of cracking heavy oil), 1936; later research on producing aviation gasoline from naphtha, also catalytic prodn. of synthetic rubber.

HOUEL, Jules, French mathematician; b. Thaon, France, Apr. 7, 1823; prof. Bordeaux (France) Faculty Sci.; author works on analysis and logarithms; spread non-Euclidan geometry throughout France; brought attention to works of Bolyai and Lobatschefsky. Died Caen, France, June 14, 1886.

HOUGEN, Jon Torger, Am. chem. physicist; b. Sheboygan, Wis., Oct. 23, 1936; s. Edward Thomas and Mildred (Dulmes) H.; student Oberlin Coll., 1952-54; B.S., U. Wis., 1956; A.M., Harvard, 1958, Ph.D., 1960. Postdoctoral fellow NRC Can., Ottawa, Ont., 1960-62, staff mem., 1962-66; staff mem. Nat. Bur. Standards, Washington, 1967—. Mem. Am. Phys. Soc. Research in quantum mech. devels. of theory which aid in our understanding of nature of molecular energy levels. Office: Nat. Bur. Standards, Washington 20234.*

HOUGH, George Washington, Am. astronomer; b. Montgomery Co., N.Y., Oct. 24, 1836; s. William and Magdalene (Selmser) H.; grad. Union Coll., 1856 (A.M., LL.D.); m. Emma C. Shear, 1870. Asst. astronomer Cin. Obs., 1859-60; astronomer and dir. Dudley Obs., Albany, N.Y., 1860-74; dir. Dearborn Obs. and prof. astronomy U. Chgo., 1879-87; prof. astronomy Northwestern U., dir. Dearborn Obs., until 1909. Discovered more than 600 new stars, measured numerous double stars, and made a systematic study of the planet Jupiter; invented many instruments pertaining to astronomy, meteorology and physics, including machine for mapping and cataloging stars, recording and printing barometer (1865), meteorograph, automatic anemometer. Died 1909.

HOUGH, Paul Van Campen, Am. physicist; b. Ellwood City, Pa., May 21, 1925; s. John Healey and Margaret (Southerton) H.; B.A., Swarthmore Coll., 1945, Ph.D., Cornell U., 1950; m. Barbara Raymond, Oct. 24, 1945; children—David Southerton, Judith Anne. Faculty, U. Mich., Ann Arbor, 1949-53, asso. prof., 1957-59; Guggenheim fellow Cern, Geneva, Switzerland, 1959-60; physicist Brookhaven Nat. Labs., Upton, L.I., 1961—. Fellow Am. Phys. Soc.; mem. Phi Beta Kappa, Sigma Xi. Research on nuclear structure, analysis of particle interactions in nuclear emulsion and bubble chamber photographs. Home: 3 Henhawk Lane, Huntington, N.Y. Office: Brookhaven Nat. Labs., Upton, L.I., N.Y. 11973.*

HOUGHTON, Henry G(arrett), Am. meteorologist; b. N.Y.C., Feb. 2, 1905; s. Henry Garrett and Ivy Estelle (Smith) H.; B.S., Drexel Inst. Tech., 1926; S.M., Mass. Inst. Tech., 1927; D.Sc., Drexel Inst. Tech., 1947; m. Dorothy H. Jenness, July 10, 1933. With Bell Telephone Co. Pa., Phila., 1927-28; with Mass. Inst. Tech., 1928—, prof. meteorology, head dept., 1945—; chmn. bd. U. Corp. for Atmospheric Research, Boulder, Colo., 1959-62. Recipient Robert M. Losey award Inst. Aero. Scis., 1940. Mem. Am. Meteorol. Soc. (Charles F. Brooks award 1958, past pres.), Am. Geophys. Union (pres. sect. on meteorology 1964-68), Royal Meteorol. Soc., A.A.A.S., Am. Acad. Arts and Scis. Contbg. author: Listen to Leaders in Science, 1965. Research, publs. on transmission light thru fog, measurements droplet size and number in natural fog and clouds, devel. method for artificial dissipation natural fog, processes by which rain and snow are formed, ways in which solar radiation is distributed over earth. Home: 29 Edmunds Rd., Wellesley Hills, Mass. 02181. Office: Mass. Inst. Tech., Cambridge, Mass. 02139.*

HOUGHTON, John Theodore, Brit. physicist; b. Dyserth, Wales, Dec. 30, 1931; s. Sydney M. and Miriam (Yarwood) H.; B.A., Oxford U., 1951, D.Phil., 1955; m. Margaret E. Broughton, Apr. 7, 1962; children—Janet, Peter. Research fellow Royal Aircraft Establishment, Farnborough, Eng., 1954-57; lectr. in atmospheric physics Oxford U., 1958-62, reader, 1962—, fellow Jesus Coll., 1960—. Recipient Darton prize Royal Meteorol. Soc., 1954, Buchan prize, 1966. Mem. Inst. Physics, Am. Meteorol. Soc., Am. Geophys. Union. Author: (with S. D. Smith) Infrared Physics, 1966; also articles. Research on measurements and calculation of radiative transfer in atmosphere; devel. of spectroscopic techniques for remote sounding of atmospheric structure from satellites. Home: 1 Begbroke Lane, Oxford, Eng.*

HOULI, Jacques, Brazilian physician; b. Rio de Janeiro, Brazil, Apr. 7, 1925; s. Samuel and Sarah (Barki) H.; M.D., U. Brazil, 1947; m. Sylvia Pederneiras, Oct. 6, 1949; children—Leonor, Sarita Ester. With U. Brazil, Rio de Janeiro, 1947—, prof. internal medicine Sch. Medicine and Surgery, 1966—, prof. Postgrad. Sch. Medicine, 1952—. Cons. rheumatologist Previdence Unit Govt. Brazil, 1950—. Bd. dirs. Med. jour. Arquivos Brasileiros Medicina. Fellow Cornell U., Harvard; Internat. fellow A.C.P., 1954; vis. fellow Brit. Council and French Govt., 1961, 65; recipient Miguel Couto award, 1954, 55. Fellow Brazilian Coll. Surgeons; mem. Sociedade Brasileira de Reumatologia, A.C.P., Ligue Internationale Contre le Rhumatisme. Author: Rheumatoid Arthritis, 1954; Ankylosing Spondylitis, 1954; Synovial Fluid, 1955; Osteo-Arthritis of the Knee, 1957; Manual of Rheumatology, 1960; Lessons on Rheumatic Diseases, 1961; Laboratory in Rheumatic Diseases, 1964; also numerous articles. Research on rheumatic diseases in tropical and sub-tropical countries, including determination of incidence rate in Brazil of rheumatoid arthritis; bone marrow studies; synovial inflamation of osteoarthritis of knee previously considered only as degenerative joint condition; imprint of synovial tissue; studies on contbn. of synoviocytes to elaboration of mucopolysaccharides (hyaluronic acid); treatment of blood effusions in hemarthroses with steroids; blood distbns. of people with rheumatic diseases; biopsy studies on synovial tissues. Home: 131 Apt. 102 Rua

Miguel Lemos, ZC-07, Rio de Janeiro, Brazil. Office: 775 Rua Mariz e Barros, Rio de Janeiro, Brazil.*

HOULLEVIGUE, (Aimé-Charles-) Louis, French physicist; b. Honfleur, France, 1863; student École Normale, 1882-85; docteur ès scis., 1895; prof. physics Faculty of Marseilles (France). Author: De l'influence de l'aimantation sur les phénomènes thermo-électriques, 1895; L'évolution des sciences, 1908; Éléments d'électricité industrielle, 1910; Leçons d'électricité appliquée, 1910; La matière vie du globe et la science moderne, 1929. Research on nervous influence and cathodic emanations, indsl. electricity, magnetic phenomena. Died Marseilles, 1944.

HOUSE, Earl Lawrence, Am. anatomist; b. Hornell, N.Y., June 18, 1914; s. Claude E. and Bertha (Lawrence) H.; B.S., Hamilton Coll., 1936; M.A., Cornell U., 1938, Ph.D., 1941; m. Grace Ruth Moore, Feb. 6, 1943; children—Stephen L., David A. Faculty, Emory U., 1941-44, asst. prof., 1944-45; faculty N.Y. Med. Coll., N.Y.C., 1946-—, prof. anatomy, 1961-—. Fellow A.A.A.S.; mem. Am. Assn. Anatomists, Am. Soc. Zoologists, N.Y. Acad. Scis., A.M.A. (asso.), Assn. Am. Med. Colls., Sigma Xi, Phi Chi. Author: (with B. Pansky) A Functional Approach to Neuroanatomy, 1960; Review of Gross Anatomy, 1964; also articles. Research on embryology pituitary gland cattle, anatomy hyoid apparatus and muscles pharynx rat, blood in normal and aging hamster, alloxan diabetes, transplantation hamster pancreas, extraction insulin-like factor from thymus, description granular reticular cells in thymus. Home: 167-10 Crocheron Av., Flushing, N.Y. 11358. Office: Fifth Av. at 106th St., N.Y.C. 10029.*

HOUSE, Homer Doliver, Am. botanist; b. Oneida, N.Y., July 21, 1878; s. Doliver E. and Alice J. (Petrie) H. B.S., Syracuse U., 1902; M.S., Columbia, 1903, Ph.D., 1905; m. Erma N. H. Hotaling, Dec. 21, 1908. Prof. botany and bacteriology, Clemson Coll., S.C., 1906-07; asso. dir. and lecturer botany and dendrology, Biltmore (N.C.) Forest Sch., 1908-13; asst. state botanist, N.Y., 1913-14, state botanist from 1914. Mem. Torrey Bot. Club, Am. Bot. Soc., Am. Mycolo. Soc., Am. Fern Soc. Author: North American Species of Ipomoca, 1908; Wild Flowers of New York, 1923; Annotated List of Ferns and Flowering Plants of New York State, 1924; Wild Flowers, 1935; also profl. reports, articles. Died Dec. 21, 1949.

HOUSE, Robert Ernest, Am. physician; b. Dallas, Aug. 3, 1875; s. John Ford and Marguerite Janie (Harper) H.; student U. Tex., 1896-98; M.D., Tulane U., 1899; postgrad. New Orleans Polyclinic, also in Chgo., N.Y.; m. Mary Alma Orr, Feb. 28, 1900; children—John Ford, Samuel David. House surgeon City Hosp., Dallas, 1897-98; practiced at Ferris, Tex., 1899-—, specializing in obstetrics. Originated the Florence-Rosser method in obstetrics; introduced the use of scopolamin to determine guilt or innocence of an individual charged with crime; also for the diagnosis and treatment of insanity by determining the cause of delusions. Died July 15, 1930.

HOUSE, Verl Lee, Am. biologist; b. Wellsville, Mo., Apr. 10, 1919; s. Ralph and Annie (Tipton) H.; A.B. in Zoology, U. Cal. at Berkeley, 1941, M.A., 1948, Ph.D. in Genetics, 1950; m. Marilyn Florence Grace, Sept. 11, 1948; children—Jeffrey Lee, Robert Garth, Verl Tipton. Quail mgmt. researcher Cal. Forest and Range Exptl. Sta., 1938-39; fisheries researcher U. S. Bur. Reclamation, 1941-42; asso. zoology U. Cal. at Davis, 1948-49; faculty Johns Hopkins, Balt., 1950-54, asst. prof., 1953-54; prof. biology, chmn. dept. woman's div. Va. Poly. Inst., Radford, 1954-58; asso. prof. zoology and entomology Ohio State U., Columbus, 1958-67, prof. genetics, 1967-—, asso. dir. fellowships Nat Acad. Scis.-NRC, Washington, 1966-67. Cons. biology Indian Summer Inst. in Biology, U. S. AID, 1965-—. Soc. Sigma Xi grantee, 1951-52; NSF Grantee in genetics, 1954-58; Ohio State U. Devel. Fund grantee for research in genetics, 1960-66. Fellow Ohio Acad. Sci.; mem. Soc. Am. Zoologists, Am. Soc. Human Genetics, Am. Genetics Soc., Phi Beta Kappa, Sigma Xi, Alpha Zeta, Xi Sigma Pi. Research, publs. on developmental genetics in Drosophila. Home: 378 Walhalla Dr., Columbus, O. 43202.*

HOUSEHOLDER, Alston Scott, Am. mathematician; b. Rockford, Ill., May 5, 1904; s. Earl and Mary (Scott) H.; B.S., Northwestern U., 1925; M.A., Cornell U., 1927; Ph.D., U. Chgo., 1937; D. Natural Sci. (hon.), Munich Tech. Inst., 1965; m. Eleanor Belle Noonan, Mar. 3, 1926; children—Jaclin (Mrs. Charles E. Christian), John A. Faculty, Washburn Coll., 1930-37, asst. prof., 1931-37; Rockefeller Found. fellow U. Chgo., 1937-39, faculty, 1939-44, asst. prof., 1941-44; sr. research psychophysiologist applied psychology panel NDRC Brown U., 1944-45; math. cons. Naval Research Lab., Washington, 1945-46; sr. mathematician Oak Ridge Nat. Lab., 1946-—. Intr., N. Ill. Coll. Optometry, 1940; prof. math. U. Tenn., 1964-—. Fellow A.A.A.S.; mem. Am., Math. Soc., Biometric Soc., Soc. for Indsl. and Applied Math. (past pres.), Assn. Computing Machines (past pres.), Math. Assn. (past v.p.). Author: (with H. D. Landahl) Mathematical Biophysics of the Central Nervous System, 1945; Principles of Numerical Analysis, 1953; The Theory of Matrices in Numerical Analysis, 1964; also numerous articles. Research in math. biology and psychology, numerical analysis. Home: 116 Cahill Lane. Office: P.O. Box X, Oak Ridge 37830.*

HOUSSAY, Bernardo Alberto, Argentinian physiologist; b. Buenos Aires, Argentina, Apr. 10, 1887; s. Alberto and Clara (Lafont) H.; Pharmacist, 1904; M.D., Buenos Aires, 1911; Dr. Honoris causa in medicine, universities Paris, 1935, Montreal, 1946, Lyon, 1946, Geneva, 1946, Asunción, 1943, Catholic (Chile), 1942, Montevideo, 1948, Brussels, Catholique of Louvain, 1949, Strasbourg, 1949, also Düsseldorf, Montpellier, also Alger; in sciences, Harvard, 1936, Sao Paulo, 1936, Oxford, 1947, Mexico, Toronto, Columbia, New York, Cambridge (England) University, 1961; L.H.D. (hon.), Georgetown U., Washington; LL.D. (hon.), Glasgow; m. Maria Angelica Catán, Dec. 22, 1920; children—Alberto Bernardo, Héctor Emílio José, Rául Horacio. Prof. physiology Vet. Faculty, Buenos Aires, 1910-19, Faculty Medical Scis., 1919-43, 45-46, 55-57; prof. honoris causas Faculty Medicine of La Habana; hon. prof. faculties of med. univs. of Montevideo, Santiago, Bogota, Lima, Brazil, Bahia, Porto Alegre, Minas Gerais, San Carlos, de Guatemala, Veterinary Sch., Buenos Aires and Lima; faculty sci., Lima; Hitchcock prof. U. Cal., 1948; now dir. Inst. Biology and Exptl. Medicine, was research prof. Physiology Faculty Med. Scis., Buenos Aires. Recipient Nat. Award Scis., Buenos Aires, 1923; Charles Mickle Fellowship, Toronto, 1945; Banting medal Am. Diabetes Association, 1946; research award Am. Pharm. Mfrs. Assn., 1947; Baly medal Royal Coll. Physicians, London, 1947; Nobel Prize for Physiology and Medicine, (with C. F. and G. T. Cori), 1947; James Cook medal, Sidney, 1948. Fgn. asso. mem. Nat. Acad. Scis. U. S. A., Royal Soc. London, Am. Philos. Soc., Swedish Acad. Scis., Acad. Medicine Paris, Acad. Sciences, Paris, Deutsche Akademie für Naturforschung, Royal Acad. Medicine Belgium, Academia Nazionale dei Lincei (Italy), Academia Inst. Egypt, Ciencias Exactas, Fisicas y Naturales, Lima; hon. mem. academies of medicine of Rio de Janeiro, Madrid, Mexico, N.Y., Lombardia, Bogota; hon. mem. Am. Physiol. Soc., A.A.-A.S., Royal Soc. Edinburgh, Harvey Soc. N.Y., Mus. de la Plata Argentina, Acad. Scis. Cordoba, Acad. Medicine Washington; hon. mem. numerous sci. socs.; pres. Argentine Soc. Biology; past pres. Argentine Assn. Advancement Sci., Nat. Acad. Medicine Buenos Aires, Internat. Union Philosophy of Scis. Author: Human Physiology, 1951; also numerous sci. papers. Extensive research in endocrinology, physiology and pharmacology; investigated relationship of hypophysis gland to carbohydrate metabolism, especially with regard to diabetes, 1923-37; demonstrated that a hormone secreted by the pituitary prevented metabolism of sugar and that injections of pituitary extract induced symptoms of diabetes; studies in physiology of circulation, digestion and nervous system, also snake and spider toxins. Home: Viamonte 2790. Laboratory: Obligado 2490, Buenos Aires, Argentina.

HOUSTON, William Vermillion, Am. physicist; b. Mt. Gilead, O., Jan. 19, 1900; s. William and Lena (Vermillion) H.; B.A., B.S. in Edn., Ohio State U., 1920, Ph.D., 1925, D.Sc., 1950; S.M., U. Chgo., 1922; LL.D., U. Cal., 1956; m. Mildred H. White, June 6, 1924; 1 dau., Harriet Anne (Mrs. T. H. Coley). Faculty, Cal. Inst. Tech., 1925-46; dir. spl. studies Div. War Research, Columbia, N.Y.C., 1941-45 (on leave from Cal. Inst. Tech.); faculty Rice U., Houston, 1946-—, hon. chancellor, prof. physics, 1961-—. Recipient medal of Merit, USN, 1947. Guggenheim fellow, Munich, Leipzig, 1927-28. Mem. Am. Phys. Soc. (pres. 1961), Am. Philos. Soc., Nat. Acad. Scis. Author: Principles of Mathematical Physics, 1934; Principles of Quantum Mechanics, 1951; Dover, 1959. Detailed studies of spectrum of atomic hydrogen; spectroscopic measurement of e/m; application of ware mechanics to elec. resistance; analysis of vibrations in crystals; behavior of superconductors moving in magnetic fields. Home: 8615 Chatsworth Dr., Houston 77024.*

HOUTMAN, Johannes Paulus Willem, chemist, chem. engr.; b. Salatiga, Indonesia, Dec. 16, 1917; s. Johannes Paulus Willem and Jeannette M.A. (van Aken) H.; Eng. degree cum laude, Technische Hogeschool Delft (Netherlands), 1940; m. Catharina Alida Kok, June 26, 1948; children—Annette Elisabeth, Johannes Paulus Willem (dec. Apr. 1967), Catharina Aleida, Dirk Gerard. Asst., Technische Hogeschool Delft, 1940-44, prof. nuclear chemistry, dir. reactor instituut, 1959-—; staff Kon./Shell Lab., Amsterdam, Netherlands, 1944-48; prof. chem. tech. and chem. engring. Faculty Technische Scis. Bandung (Indonesia, 1948-55; staff Dutch State Mines at KEMA Arnhem, 1955-59; head chem. and metall. dept. Reactor Centrum Nederland, Petten, 1960-62. Mem. Royal Netherlands Chem. Soc., Royal Dutch Inst. Engrs., Lab. Invest. Objects Art. Research, publs. on heterogeneous catalysis, distillation theory, drying oils activation, radiochemistry, activation analysis, recoil chemistry, tech. applications of radiation and isotopes, applications to works of art. Home: 134 Meermanstraat, Delft, Netherlands.*

HOUTZ, Sara Jane, Am. kinesiologist, exercise physiologist; b. Sioux City, Ia., Sept. 28, 1913; d. Howard Kline and Mabel (Bach) Houtz; certificate Neb. State Tchrs. Coll., 1934, Cook County Grad. Sch., 1950, Children's Rehab. Inst., 1950; B.S., U. Colo., 1939; phys. therapy Walter Reed Hosp., U. S. Army, 1942; M.S., Med. Coll. Va., 1949. Tchr. phys. edn. high sch., Grand Island, Neb., 1940-42; faculty Med. Coll. Va., 1945-55, Research and Edn. Hosp., U. Ill., 1951-55; cons. research and edn. Detroit Orthopaedic Clinic, 1955-—. Guest lectr. univs. Fellow Am. Med. Writers Assn.; mem. Am. Phys. Therapy Assn., A.A.A.S., Nat. Soc. for Med. Research, Detroit Physiol. Soc. Author: (with Robert A. Groff) Manual of Diagnosis and Management of Peripheral Nerve Injuries, 1945. Abstract editor Exerpta Medica, 1956; asst. editor Jour. Am. Phys. Therapy Assn., 1953-55, editor-in-chief, 1956-58. Publs. on electromyographical studies of normal, disabled persons and influence of drug and bracing therapy. Home: 9595 Mettetal St., Detroit 48227. Office: 5447 Woodward St., Detroit 48202.*

HOUZEAU, Auguste, French chemist; b. Elbeuf, France, Mar. 3, 1829; dir. Lower Seine Agrl. Research Lab.; prof. agrl. chemistry Rouen (France) Research Lab.; mem. French Acad. Scis. Research on ozone, oxygenized water, fertilizer titration; inventor ozonoscopic paper. Died Rouen, Feb. 17, 1911.

HOUZEAU DE LEHAIE, Jean Charles, Belgian astronomer; b. Mons, Belgium, Oct. 7, 1820. Astronomer at Brussels Obs., Belgium, from 1843, apptd. dir. 1876; went to U. S., 1857; became dir. Negro jour. Tribune; returned to Brussels, 1876. Author: Études sur les facultés mentales des animaux, comparées a celle de l'homme, 1872; Uranometrie générale, 1878; Vademecum, de l'astronomie, 1882; Bibliographie générale de l'astronomie, 1882; also pub. star catalogue. Died Brussels, July 12, 1888.

HOVANITZ, William, Am. biologist; b. Chgo., Nov. 6, 1915; s. Julius and Appolonia (Volauvshek) H.; B.S., U. Cal. at Berkeley, 1938; Ph.D., Cal. Inst. Tech., 1943; m. Barbara Jean Bjorkman, Aug. 14, 1949; children—Eric, Christine, Karl. NRC fellow Rockefeller Lab., Villavicencio, Colombia, 1944-45, Tallahassee, 1945; faculty U. Mich. Sch. Pub. Health, 1945; asst. biologist lab. vertebrate biology U. Mich., 1946-47; asst. prof. biology Wayne U., Detroit, 1948-49; vis. asst. prof. U. Cal. at Santa Barbara, 1949; prof. biology U. San Francisco, 1949-55; entomologist Cal. Inst. Tech., 1955-59; prof. zoology Cal. State Coll., Los Angeles, 1955-—; geneticist Cal. Arboretum Found., Arcadia, 1955-66; dir. Lepidoptera Research Found., 1965-—. Fellow Cal. Acad. Scis., A.A.A.S.; mem. Ecol. Soc. Am., Entomol. Soc. Am., Western Soc. Naturalists, Am. Soc. Zoologists, Soc. for Study Evolution. Author: Textbook of Genetics, 1953; also numerous articles. Research on geog. distbn., genetics, ecology and population studies butterflies, chromosome structure, genetics of color characteristics of plants and animals, relationships of insects to disease transmission in humans; field research in arctic of N. Am. and Europe, mountains of N. and S.Am., Europe and Asia. Address: 1160 W. Orange Av., Arcadia, Cal. 91006.*

HOVEY, Charles Mason, Am. horticulturist; b. Cambridge, Mass., Oct. 16, 1810; s. Phineas and Sarah (Stone) H.; m. Anna Chaponil, Dec. 25, 1835. Established nursery, Cambridge, 1832; founder editor, Mag. of Horticulture, Botany, and All Useful Disconeries and Improvements in Rural Affairs, 1835-68; mem. Am. Pomol. Soc.; pres. Mass. Hort. Soc., 1863-66. Author: Fruits of America, 3 vols., 1847-56. Developed Hovey strawberry (1st variety of fruit to be developed by planned breeding in N.Am.), 1834. Died Sept. 2, 1887.

HOVI, Väinö, Finnish physicist; b. Muola, Finland, Sept. 29, 1913; s. Arndt and Elli (Lahti) H.; M.A., U. Helsinki (Finland), 1938, Ph.D., 1948; m. Saara Ekman, Apr. 20, 1940; children—Mikko, Marja, Matleena. Tchr. physics, math. and chemistry Kallion Yhteiskoulu, Helsinki, 1941-47; asst. physics U. Helsinki, 1945-53, docent physics, 1950-—; asst. physics Inst. Tech., Helsinki, 1945-46; prof. physics U. Turku (Finland), 1953-—, head Wihuri Phys. Lab., 1957-—; vis. fellow U. Ill., Urbana, 1957. Mem. adv. bd. Phys. kondens. Materie, Switzerland, 1962-—; mem. commn. 1 Internat. Inst. Refrigeration, France, 1962-—. Recipient Internat. prize in physics Wihuri Found., 1961. Mem. Phys. Soc. Finland, Am. Phys. Soc., Finnish Acad. Scis. (ordinary), Phys. Soc. Japan. Asso. editor Acta Metallurgica, U. S., 1953-—. Research and numerous publs. on properties and theory of alkali halide solid solutions; X-ray studies of transition in ammonium halides, NMR in ionic crystals at low temperatures, color centers in alkali halides, order-disorder phenomena in binary alloys; organized and established low temperature physics in Finland, 1957. Home: 7 b A 7 Eerikink, Turku, Finland.*

HOVLAND, Carl I., Am. psychologist; b. Chicago, June 12, 1912; s. Ole C. and Augusta (Anderson) H.; A.B., Northwestern U., 1932, A.M., 1933; Ph.D., Yale, 1936; m. Gertrude Raddaty, June 4, 1938 (dec. Aug. 1960); children—David, Katharine. Instr. dept. of psychology and researcher Inst. of Human Relations, Yale, 1936-37, asst. prof., 1937-42, dir. grad. studies, 1941, asso. prof. (in absentia), 1942; chief psychologist, dir. of exptl. studies, information and edn. div., Office Chief of Staff, War Dept., 1942-45; prof. psychology, chmn. of dept. and dir. lab. of psychology Yale, 1945-51, Sterling prof. psychology, 1947-—; cons. Bell Telephone Labs., Inc., Am. Telephone & Telegraph Co. Cons. U. S. Dept. of Justice, 1940, Sec. of War, 1942, Social Sci. Div., Rockefeller Found., Research Development Bd., 1947—; mem. sci. adv. bd. Office Chief of Staff Air Forces. Rep. Nat. Research Council, exec. com., 1954—; mem. exec. com. Inst. of Human Relations; dir. Social Sci. Research Council; dir. Vocational counselling service; member psychobiology panel Nat. Sci. Found.; mem. advisory committee Ford Foundation Center for Advanced Study; also consultant Fund for Adult Education trustee Russell Sage Found.; dir. Hamden Hall Country Day Sch.; mem. adv. com. Human Resources Research Inst., Maxwell Field, 1948-51, Cross-Cultural Research Group; mem. self-study vis. com. Stanford U. Recipient Distinguished Scientific Contribution award, American Psychological Association, 1957, 58, Warren medal Soc. Exptl. Psychologists, 1961. Fellow Am. Psychological Association (director; pres. div. gen. psychology 1953-54; council), A.A.A.S.; member Am. Acad. Arts and Scis., Society of Experimental Psychologists, Eastern Psychol. Assn. (pres. 1950), National Academy of Science, American Philosophical Society, Connecticut State Psychol. Soc. (pres. 1946-47), Sigma Xi, Phi Beta Kappa. Co-author: Experiments on Mass Communication, 1949; Communication and Persuasion, 1954; co-author, editor: Order of Presentation in Persuasion 1957. Co-author, editor: Personality and Persuasibility; Attitude Organization and Change, 1960. Contributor to mags. and books. Coop. editor, Personnel Psychology; mem. editorial bd. Ann. Revs. of Psychology 1948-51; cons. editor Human Relations Psychol. Rev., Audio-Visual Communications Rev. Research on psychology of communication. Died Apr. 16, 1961.

HOVORKA, Frank, chemist; b. Cernikovice, Bohemia, Aug. 5, 1897; s. Frank and Anna (Pavlova) H.; B.A., State Coll. Ia., 1922; M.S., U. Ill., 1923, Ph.D., 1925; m. Sophie Paul Nickel, June 12, 1926. Faculty, Western Res. U., Cleve., 1925—, dir. chem. labs., 1942-58, dir. Navy Ultrasonic and Electrochemistry Lab., 1948-64, chmn. dept. chemistry, 1950-58, 62-64, Hurlbut prof. chemistry, 1954—; research asso. Argonne Nat. Lab., 1952-58, bd. dirs. Asso. Midwest Univs., 1958-62. Recipient Nat. Coll. Chemistry Tchr. award Mfg. Chemists Assn., 1963; Alumni Achievement award State Coll. Ia., 1964. Fellow A.A.A.S., Chem. Soc. London; mem. Am. Chem. Soc., Am. Assn. U. Profs., Faraday Soc., Soc. Chem. Industry (Eng.), Am. Electrochem. Soc. Contbr. chpts. to books, numerous articles to profl. jours. Research on electrolytic solutions, electrodes, hydrocarbon chains, deposition of metals, molten binary alloys, application of emission spectroscopy to med. problems. Home: 2593 Exeter Rd., Cleveland Heights, O. 44118. Office: Western Res. U., Cleve. 44106.*

HOWARD, Alma (Mrs. Michael Ebert), biologist; b. Montreal, Que., Can., Oct. 23, 1913; d. E. Edwin and Evalyn (Peverley) Howard; B.Sc., McGill U., Montreal, 1934, Ph.D., 1938; m. 2d, Michael Ebert, Jan. 17, 1958; m. P. W. Rolleston, June 10, 1939 (dec. 1947); children—Francis S., Patrick H. Cytologist, radio-therapeutic research unit Med. Research Council, Hammersmith Hosp., London, 1948-55; biologist research unit in radiobiology Brit. Empire Cancer Campaign, Mt. Vernon Hosp., Northwood, Eng., 1955-62; head radiobiology lab., dep. dir. research Paterson Labs., Christie Hosp. and Holt Radium Inst., Manchester, Eng., 1962—. Editor series: (with M. Ebert) Current Topics in Radiation Research, 1965—. Research, publs. on determination of time in life cycle in which cells synthesize DNA, measurement of effects of agts. which modify killing of cells by radiation. Office: Paterson Labs., Christie Hosp. and Holt Radium Inst., Manchester 20, Eng.*

HOWARD, Ephraim Manasseh, Am. applied scientist; b. Detroit, Aug. 15, 1921; s. Julius A. and Anna (Horvitz) H.; B.Sc. (Detroit Bd. Edn. scholar) Wayne State U. 1946; M.Sc. in Engring., U. Mich., 1947; Ph.D. in Applied Science, U. Cal. at Davis-Livermore, 1967; m. Joyce Louise Holzberg, Aug. 15, 1942; children—Sandy Gail, Fran Renee, Brian David. Aero. research scientist Lewis Flight Propulsion Lab., NASA, Cleve., 1947-50; cons. engr. E. M. Howard & Assos., Cleve., 1950-51; br. mgr. Aro, Inc., Tullahoma, Tenn., 1951-54; sr. project engr. Allison div. Gen. Motors Corp., Indpls., 1954-56; sr. staff engr. Marquardt Aircraft Co., Van Nuys, Ca., 1956-58; engring. mgr. Aerojet Gen. Corp., Azusa, Cal., 1958—. Recipient D.A.R. award, 1940, Am. Legion award, 1940, Am. Helicopter Soc. award, 1946. Asso. fellow Am. Inst. Aeros. and Astronautics, Inst. Aero. Sci. (chmn. indpls. sect. 1955-56, mem. nat. council 1961-62; mem. Am. Rocket Soc. Research, publs., devel. on turbojet and ramjet engines; developed methods flight simulation testing to design and devel. advanced propulsion systems. Home: 2030 E. Santa Clara St., Santa Ana, Cal. 92705. Office: 1100 W. Hollyvale St., Azusa, Cal. 91702.*

HOWARD, Harvey James, Am. ophthalmologist; b. Churchville, N.Y., Jan. 30, 1880; s. Charles William and Mary Jessie (Williamson) H.; A.B., U. Mich., 1904; M.D., U. Pa., 1908, A.M., Harvard, 1917; Oph.D., U. Colo., 1918; m. Maude Irene Strobel, June 25, 1910 (dec. May, 1948); children—Margaret Strobel (Jackson), James Howell, Martha Williamson Blake; m. 2d, Alice Tilson Eastes, Aug. 24, 1948. Head dept. ophthalmology U. Med. Sch., Canton, China. 1910-13; ophthalmologist Canton Christian Coll., 1912-15; ophthalmic surgeon Canton Hosp., 1912-15; fellow China Med. Bd. of Rockefeller Found. at Harvard, 1916-18, at Vienna U., 1923-24; ophthalmic asst. Harvard Post-Grad. Med. Sch., and Mass. Charitable Eye and Ear Infirmary, 1917-18; prof., head of dept. ophthalmology, Peking Union Medical Coll., 1918-27; med. adviser to Dept. Aeros., Chinese Govt., 1920-23 ophthalmologist to boy emperor of China, 1921-23; prof. ophthalmology Washington Univ. Sch. Medicine, 1927-33. Med. dir. Mo. Com. for the Blind, 1931-48. Certificate of Am. Bd. of Ophthalmic Examiners. Fellow A.C.S.; mem. A.M.A., Am. Ophthal. Soc., Am. Acad. Ophthalmology and Otolaryngology, China Med. Assn., So. Med. Assn., Alpha Omega Alpha, Nu Sigma Nu. Author: Ten Weeks with Chinese Bandits; also numerous articles pertaining to ophthalmology. Devised depth perception test for aviators for U. S. Army, Navy and Dept. Commerce. Died Nov. 6, 1956.

HOWARD, John Eager, Am. physician; b. nr. Balt., Aug. 27, 1902; s. John Duvall and Mary Greenwood (Smith) H.; A.B., Princeton, 1924; M.D., Johns Hopkins, 1928; m. Lucy Iglehart, June 30, 1928; children—John Eager, William James, Lucy Anne Calhoun. Practice medicine, specializing in internal medicine, Balt., 1934—; mem. staff Johns Hopkins, Union Meml. hosps.; faculty Johns Hopkins 1937—, prof. medicine, 1960—. Recipient Modern Medicine award for Distinguished Achievement, 1964. Jacques Loeb fellow, 1932-33; John D. Archbold fellow, 1933-34. Mem. Assn. Am. Physicians, Endocrine Soc. (pres. 1960-61), Am. Diabetes Assn., Am. Clin. and Climatol. Assn., A.M.A., Balt. Med. Soc., Med. and Chirurg. Faculty Md., Am., Md. (pres. 1961-62) socs. internal medicine. Research, publs. on disorders of endocrine or metabolic nature. Home: "Waverley", R.F.D. 1, Lutherville, Md. 21093. Office: care Johns Hopkins Hosp., Balt. 21205.*

HOWARD, John Nelson, Am. physicist; b. Phila., Feb. 27, 1921; s. Harold and Ann Louise (Vandegrift) H.; B.Sc., U. Fla., 1943; M.Sc., Ohio State U., 1949, Ph.D., 1954; m. Irene Russell Rogers, Sept. 7, 1950; children—Martha Louise, Katherine Ann, Rogers Vandegrift, John James. Asst. chemist dept. soils U. Fla., 1943-44; asso. physicist Spectroscopy Lab., NACA, 1944-46; research asso. dept. physics Ohio State U., 1948-54; staff Air Force Cambridge Research Labs., Bedford, Mass., 1954—, chief optical physics lab., 1960-64, chief scientist, 1964—. Fellow Optical Soc. Am. (editor Applied Optics 1960—); mem. Am. Phys. Soc., Am. Meteorol. Soc., Soc. for Applied Spectroscopy, Phi Beta Kappa. Research, numerous publs. on transmission infrared radiation through gases, molecular spectroscopy simple molecules. Home: 7 Norman Rd., Newton Highlands, Mass. 02161. Office: Air Force Cambridge Research Labs., Bedford, Mass. 01730.*

HOWARD, Leland Ossian, Am. entomologist; b. Rockford, Ill., June 11, 1857; s. Ossian Gregory and Lucy Dunham (Thurber) H.; B.S., Cornell, 1877, M.S., 1883; Ph.D., Georgetown U., 1896; M.D. (hon.), George Washington U., 1911; LL.D., U. Pitts., 1911, U. Cal., 1929; Sc.D., U. Toronto, 1920, Rutgers U., 1930; m. Marie T. Clifton, Apr. 28, 1886; children—Lucy Thurber, Candace Leland (Mrs. Edward De Mille Payne), Janet Moore. Asst. entomologist Bur. of Entomology, U. S. Dept. Agr., 1878-94, chief of bur., 1894-1927, prin. entomologist, 1927-31, ret. Hon. curator dept. of insects, U. S. Nat. Museum, from 1895; cons. entomologist USPHS, from 1904, sr. entomologist, 1919. Mem. com. on agr. Nat. Council Defense, 1917; chmn. sub-com. on med. entomology NRC, 1917. Permanent sec. A.A.S.S., 1898-1920; pres. Assn. of Econ. Entomologists, 1894. Biol. Soc. Washington, 1897-98, Washington Acad. Scis., 1916, A.A.A.S., 1920; v.p. Internat. Congress of Agr., Paris, 1923; hon. pres. Internat. Conf. of Phytopathologists and Econ. Entomologists, Holland, 1923; chmn. Pan-Pacific Food Conservation Congress, Honolulu, 1924; pres. sect. econ. zoology Internat. Congress of Zoölogy, Budapest, 1927; pres. Internat. Congress Entomology, Ithaca, N.Y., 1929. Fellow Am. Acad. Arts and Scis.; mem. Am. Philos. Soc., Nat. Acad. Sciences; hon. mem. many fgn. sci. socs. Decorated officer Legion of Honor, officer Order of Agrl. Merit (both France); medalist Holland Soc. of New York, 1924; 2d Capper award, 1931. Author: Mosquitoes—How They Live, Etc., 1901; The Insect Book, 1901; The House-Fly—Disease Carrier, 1911; monograph, Mosquitoes of North America, Carnegie Instn., 1912-17; History of Applied Entomology (Smithsonian Inst.). 1930; The Insect Menace, 1931; Fighting the Insects—The Story of an Entomologist, 1933. Specialized in econ. aspects of entomology, with particular reference to parasitic and disease breeding insects. Died May 1, 1950.

HOWARD, Luke, English meteorologist; b. London, Nov. 28, 1772; s. Robert Howard; m. Mariabella Eliot, 1796; children—Robert, John Eliot. Apprenticed to Olive Sims, chemist; began bus. as chemist, London, 1793; partner (with William Allen) wholesale and retail chem. firm, 1796-1903, also head mfg. dept., Plaistou, Eng.; moved to Tottenham, Eng., 1812. Fellow Royal Soc., 1821. Author: Climate of London, 1818-20; Essay on the Modifications of Clouds, 1832; Seven Lectures on Meteorology, 1837; Papers on Meteorology, 1854; also articles. Editor The Yorkshireman, 1833-37. A founder meteorology; began meteorol. register, 1806; applied method of Linnaeus to classification of clouds, 1802; corresponded with Goethe and John Dalton. Died Mar. 21, 1864.

HOWARD, Robert Adrian, Am. physicist; b. Los Angeles, Feb. 23, 1913; s. Robert and Mary (Taylor) H.; B.S., Cal. Inst. Tech., 1934, M.S., 1935; Ph.D., Washington U., St. Louis, 1938; m. Jane Elizabeth Morgens, June 2, 1939; children—Eileen (Mrs. Ernest Manes), Brian, Donald, Kathleen. Research physicist Carter Oil Co., Tulsa, 1938-42; staff mem. Radiation Lab., Mass. Inst. Tech., 1943-45; research physicist Hydrodynamics Lab., Cal. Inst. Tech., 1946-47; faculty U. Okla., Norman, 1947—, prof. physics, 1955—, dir. Nuclear Reactor Lab., 1958-60; mem. tech. staff Thompson-Ramo-Wooldridge Corp., 1955-57; dir. Instituto Central de Física, Universidad de Concepción, Chile, 1962-64. Mem. Am. Phys. Soc., Am. Assn. Physics Tchrs. Author: Nuclear Physics, 1963; also articles. Research on movement oil in water-wet sands, oil accumulation in reservoirs, flow oil through reservoir rocks, interactions, particles very high energy. Patentee devices for measuring power and frequency microwaves. Home: 115 E. Farmer St., Norman, Okla. 73069.*

HOWARD, Robert Stearns, Am. ecologist; b. Akron, O., June 24, 1921; s. Harry Claude and Fanchon (Seeds) H.; B.S., U. Chgo., 1947; M.S., U. Miami, 1949; Ph.D., Northwestern U., 1952. Instr., U. Pa., Phila., 1952-53; asst. prof. U. Del., Newark, 1953-61; asso. prof. biology Ursinus Coll., Collegeville, Pa., 1961—. Mem. A.A.A.S., Am. Entomol. Soc., Am. Inst. Biol. Scis., Am. Soc. Zoologists, Animal Behavior Soc., Atlantic Estuarine Research Soc., Ecol. Soc. Am., Entomol. Soc. Am., Soc. Natural History Del. (past pres.), Soc. Systematic Zoologists, Entomol. Soc. Pa., Sigma Xi, Beta Beta Beta. Contbg. author: Comparative Psychology; Research in Animal Behavior, 1964; also articles. Research on ecology and disbn. intertidal insects; discovered several new species. Home: 1st Av., Collegeville, Pa. 19426.*

HOWARD, Walter Egner, Am. vertebrate ecologist, ethologist; b. Davis, Cal., Apr. 9, 1917; s. Walter Lafayette and Maybelle (Cooper) H.; A.B., U. Cal. at Berkeley and Davis, 1939; M.S., U. Mich., 1941, Ph.D., 1947; m. Elizabeth Ann Kendall, June 21, 1940; children—Thomas Kendall, Kathryn Spencer, John Casey. faculty dept. zoology U. Cal. at Davis, 1947-54, Agrl. Field Stas., 1954-63, vertebrate ecologist dept. animal physiology, 1964—. Mem. sub.-com. on vertebrates Agr. Bd., Nat. Acad. Scis.-NRC, 1958-60. Fulbright Research fellow New Zealand, Australia, 1957; New Zealand Dept. Sci. and Indsl. Research animal ecology div. grantee, 1962-63; New Zealand Forest Service Travel grantee, 1962-63. Mem. Am. Soc. Mammals, Wildlife Soc., Ecol. Soc. Am., Animal Behavior Soc., Soc. for Exptl. Analysis Behavior, Brit. Ecol. Soc., Western Soc. Naturalists, Am. Inst. Biol. Scis., Am. Assn. Advancement Science, Wildlife Disease Assn., Sigma Xi, Phi Kappa Sigma. Research, numerous publs. on ecology, behavior, population dynamics, mgmt., olfactory-taste acuities, starch-gel electrophoresis blood serum, sound perception, antifertility agts., intestinal bacteria of rodents, other wild vertebrates. Home: 24 College Park, Davis, Cal. 95616.*

HOWARD-FLANDERS, Paul, biophysicist; b. Bristol, Eng., June 30, 1919; s. Richard Leonard and Millicent (Franks) H.-F.; B.S.C., London U. (Eng.), 1940, Ph.D. in Physics, 1956; M.A., Yale, 1964; m. June Daphne Caine, Sept. 23, 1950; children—Rob Stewart, Mark Richard. Physicist, Radiotherapeutic Research unit med. research council Hammersmith Hosp., London, 1946-53; sci. staff exptl. radiopathylology research, 1953-57, 58-59; lectr. physics Donner Lab., U. Cal. at Berkeley, 1957-58; faculty Yale Sch. Medicine, 1959—, prof. radiobiology and molecular biophysics, 1966—. Mem. research adv. council Am. Cancer Soc., 1963—. Mem. Am. Soc. for Microbiology, Genetics Soc., Physics Soc., Soc. for Cellular Biology. Contbr. numerous articles to sci. jours. Elucidation of genetic repair mechanisms, genetic recombination, bacterial genetics and nucleic acid research, sensitization cells to X-rays by nitric oxide and organic nitroxides. Home: 173 Martin Lane, Orange, Conn. 06477. Office: 333 Cedar St., New Haven 06510.*

HOWE, Calderon, Am. microbiologist; b. Washington, Mar. 8, 1916; s. Walter Bruce and Mary (Carlisle) H.; B.A., Yale, 1938; M.D., Harvard, 1942; m. Sarah Ann Drury, Feb. 2, 1944; children—Calderon Carlisle, Sarah Bruce, Ann Mandeville, John Robins. Markle scholar in med. sci. Columbia, 1952-57, faculty, 1958—, prof. microbiology, 1962—, acting chmn. dept., 1964-65; asso. attending microbiologist Presbyn. Hosp., N.Y.C. Mem. Am. Assn. Immunologists, Am. Soc. Tropical Medicine and Hygiene, Mass. Med. Soc., Am. Soc. for Microbiology. Research, numerous articles on immunology and virology, clin. aspects infectious disease. Home: 4940 Arlington Av., N.Y.C. 10471.*

HOWE, Clifton Dexter, Am. physician; b. Morrisville, Vt., July 29, 1914; s. Carlton Dexter and Alice (Durfee) H.; B.S., U. Vt., 1936, M.D., 1939; m. Eleanor Self, Oct. 1, 1945; children—Nancy, Alice, Carlton, Sally, Mary. Internist, chief gen. med. service, head dept. medicine U. Tex. M. D. Anderson Hosp. and Tumor Inst., Houston, 1953—, prof. medicine, 1965—; prof. medicine div. continuing edn. U. Tex. Grad. Sch. Biomed. Scis., Houston, 1963—. Chmn. sect. on solid tumors in adults S.W. Cancer Chemotherapy Study Group, Nat. Cancer Inst., 1959-62, prin. investigator adult sect., 1962——. Fellow A.C.P.; mem. Am. Assn. for Cancer Research, Am. Heart Assn., Houston Acad. Medicine, Alpha Omega Alpha. Contbr. numerous articles to med. jours., chpts. to books. Research on anti-tumor agts. Home: 1902 Sunset Blvd., Houston 77005. Office: 6723 Bertner Av., Houston 77025.*

HOWE, Elias, Am. inventor; b. Spencer, Mass., July 9, 1819; s. Elias and Polly (Bemis) H.; m. Elizabeth J. Ames, Mar. 3, 1841; married a 2d time. Invented sewing machine that equated speed of 5 of swiftest hand sewers, making 250 stitches per minute, 1844-45, granted patent, 1846; made and marketed a number of sewing machines, N.Y.C.; organized Howe Machine Co., Bridgeport, Conn., 1865; won Gold medal for Howe Machine at Paris Exhbn., 1867. Died Bklyn., Oct. 3, 1867.

HOWE, Henry Marion, Am. metallurgist; b. Boston, Mar. 2, 1848; s. Samuel G. and Julia (Ward) H.; ed. Harvard, Mass. Inst. Tech.; m. Fanny Gay, Apr. 9, 1874. Became supt. Bessemer steel works, Joliet, Ill., 1872; designed and built works Orford Nickel & Copper Co., Capelton, Can., also Bergen Point, N.J., 1880-82; cons. metallurgist, Boston, also lectr. Mass. Inst. Tech., 1883-97; became prof. metallurgy, Columbia U., 1897. Named juror class mining and metall. process Paris Exhbn., 1889; became pres. jury on mines and mining World's Columbian Exhibit, 1893. Recipient Bessemer medal, 1895; gold medal Verein zur Beforderung der gewerbfleisses, 1895; Elliot Cresson Gold medal Franklin Inst. Phila., 1895. Mem. Am. Inst. Mining Engrs. (v.p. 1879-81, became pres. 1893), Am. Philos. Soc., Am. Acad. Arts and Scis. Author: Copper Smelting, 1885; Metallurgy of Steel, 1890; Metallurgical Laboratory Notes, 1902; Iron, Steel and other Alloys, 1903. Improved metall. processes; developed indsl. applications. Died Bedford Hills, N.Y., May 14, 1922.

HOWE, Herbert Alonzo, Am. astronomer; b. Brockport, N.Y., Nov. 22, 1858; s. Alonzo J. and Julia M. (Osgood) H.; A.B., U. Chgo., 1875; A.M., U. Cin., 1877; Sc.D., Boston U., 1884; LL.D., U. Denver, 1910, Colo. Coll., 1913; m. Fannie McClurg Shattuck, Dec. 23, 1884; children—Julian Osgood, Hubert Shattuck, Warren Francis, Ernest Joseph. Student and asst. Cin. Obs., 1875-80; prof. astronomy, U. Denver, from 1880; dean Coll. Liberal Arts and dir. Chamberlin Obs., University Park, Denver, from 1891. Author: A Study of the Sky, 1896; Elements of Descriptive Astronomy, revised, 1909. Designed an impersonal micrometer for equatorial telescopes. Died Nov. 2, 1926.

HOWE, James Lewis, Am. chemist; b. Newburyport, Mass., Aug. 4, 1859; s. Francis A. and Mary F. (Lewis) H.; A.B., Amherst, 1880; A.M., Ph.D., Göttingen, 1882; hon. M.D., Hosp. Coll. of Medicine, Louisville, 1886; Sc.D., Washington and Lee U., 1946; m. Henrietta Leavenworth Marvine, Dec. 27, 1883 (dec. Oct. 1943); children—Guendolen, Frances Ray (dec.), James Lewis. Prof. chemistry, Central U., 1883-86; scientist and lectr. Polytechnic Soc. Ky., 1886-94; prof. chemistry and head dept. chemistry, Washington and Lee U., 1894-1938, emeritus prof. of chemistry and univ. historian, from 1938; also dean Sch. Applied Sci., 1921-32. Mem. spl. com. on platinum NRC, 1917. Fellow A.A.A.S. (gen. sec., 1895, v.p. chem. sect., 1900); mem. Am. Chem. Soc., Va. Acad. Sci. (pres. 1924), Nat. Inst. Social Science, Soc. Chem. Industry, Washington Acad. Scis., Deutsche Chemische Gesellschaft, Chem. Soc. (London); Phi Beta Kappa, Omicron Delta Kappa, Delta Kappa Epsilon. Author: A Bibliography of the Metals of the Platinum Group, 1897, and continuing updated edits., 1919, 1947, 1953; Inorganic Chemistry According to the Periodic Law (with Francis Preston Venable), 1898, revised as Inorganic Chemistry for Schs. and Colleges, 3d edit., 1921; also numerous papers. Died Dec. 20, 1955.

HOWE, Percy Rogers, Am. dental researcher; b. Providence, Sept. 30, 1864; s. James Albert and Elizabeth Rachel (Rogers) H.; A.B., Bates Coll., 1887, Sc.D., 1927; D.D.S., Phila. Dental Coll., 1890; LL.D., Harvard, 1941; m. Rose Alma Hilton, Dec. 21, 1891 (died 1942); children—James Albert, John Farwell; m. 2d, Ruth Loring White, 1943. Began practice at Auburn, Me., 1890; moved to Lewiston, 1891, to Boston, 1898; chief of research labs., Forsyth Dental Infirmary for Children, from 1915, dir. from 1927; asso. prof. dental research, Harvard, 1915-25, Thomas Alexander Forsyth prof. dental sci., instr. in pathology Harvard Med. Sch., 1925-40. Recipient Am. Dental Assn. award, 1945. Fellow dental surg., Royal Coll. Surgeons (Eng.); hon. fellow Internat. College Dentists, 1947. Mem. Am. Dental Assn. (pres. 1928-29), Am. Acad. Arts and Sciences, Am. Acad. Dental Sci., Hist. Science Soc., New Eng. Pediatrics Soc., Norwegian Dental Soc., Acad. Internacional de odontologia, Fedn. Dentaire Internat., Sociedad Cubana de Odontologia Infantil, Sigma Xi, Phi Beta Kappa, Delta Sigma Delta. Research in nutrition; originator of silver reduction treatment for infected dentine and septic roots, used extensively in the Army, World War; isolated group of bacteria from dental caries. Died Feb. 28, 1950.

HOWE, Robert Hsi Lin, engr.; b. Swatow, China, Jan. 2, 1922; s. Zulin and Afia (Lin) H.; B.S., Meth. U., Soochow, China 1943; B.C.E., St. John's

U., Shanghai, China, 1945; M.S., Cornell U., 1949, M.C.E., 1950; Ph.D., Purdue U., 1955; m. Jean Ma, Dec. 23, 1953; children—David J., Roberta C., Albert G. Came to U. S., 1948, naturalized, 1962. With Meth. U., 1947, lectr. chem. engring., physics, 1945-47; scholar Meth. Ch., Cornell U., 1948-50; fellow Ind. Dept. Conservation, Lafayette, 1950-52; san. engr. Eli Lilly & Co., Lafayette, 1952-55, project engr. 1955-63, sr. san. engr., 1963-65, research scientist, 1966——; Fulbright-Hays lectr. prof. Istanbul Teknik U., 1965-66, prof.-advisor, San. Engring. chair, 1966——; prof. lectr. Istanbul U., 1966, Milan Poly. U.; spl. profl. lectr. Middle East Tech. U., 1966. Mem. Am. Chem. Soc., Nat. Soc. Profl. Engrs., Royal Soc. Health, Am. Soc. Photogrammetry, Ind. Acad. Sci., Inst. Advanced Sanitation Research. Author: Applied Chemistry for Wastes Treatment and Water Purification, vol. 1, 1966. Contbr. numerous articles to sci. jours. Developed economical method for discovery and evaluation of water resources, a biochem. method for degradation of cyanides, methods for wastes, and water treatment; conducted pollution survey along the Bosphorus Strait. Home: 106 Drury Lane, West Lafayette, Ind. 47906. Office: Lilly Rd., Lafayette, Ind. 47902.*

HOWE, William, English botanist; b. London, Eng., 1620; ed. Merchant Taylor's Sch., from 1632; commoner St. John's Coll., Oxford (Eng.) U., B.A., 1641, M.A., 1644; student medicine. Served in King's army, promoted to troop command; returned to med. practice, London. Author: Phytologia Britannica . . . , 1650; Matthiae de Lobel Stirpium illustrationes . . . , 1655. Pub. 1st works devoted exclusively to Brit. plants. Died London, Aug. 31, 1656.

HOWELL, Benjamin Franklin, Am. geophysicist; b. Princeton, N.J., June 12, 1917; s. Benjamin Franklin and Claire Homan (Mead) H.; A.B., Princeton, 1939; M.S., Cal. Inst. Tech., 1942, Ph.D., 1949; m. Constance M. Benson, June 30, 1943; children—Barbara C., Bonnie A., James B. Research engr. div. war research U. Cal., 1942-45; geophysicist United Geophys. Co., 1946-49; prof. geophysics Pa. State U., 1949——; chief seismologist Vibratech Engrs., Inc., Hazelton, Pa., 1953—. Mem. exec. com. div. earth scis. NRC, 1965-67. Mem. Soc. Exploration Geophysics, Seismol. Soc. Am. (dir. 1959-66, past pres.), Am. Geophys. Union (sec. seismol. and tectonophysics sects. 1956-64), Geol. Soc. Am., Pa. Acad. Sci. Author: Introduction to Geophysics, 1959; also articles. Research on theory of transmission and absorption of seismic waves, structure of Earth's crust. Home: 308 W. Prospect Av., State College, Pa. 10801. Office: Mineral Scis. Bldg., University Park, Pa. 16802.*

HOWELL, Thomas Jefferson, Am. botanist; b. nr. Pisgah, Mo., Oct. 9, 1842; s. Benjamin and Elizabeth (Matthews) H.; self-educated; m. Effie Hudson McIlwane, 1892. Farmer, stock breeder; began to organize herbarium, 1877; made specimens from his collections into sets and distributed to bot. centers of Eastern U. S. and Europe, 1885-95. Author: A list of flowering plants of Ore., 1881; A Catalogue of the Known Plants (Phaenogamia and Pteridophyta) of Oregon, Washington, and Idaho, 1887; Flora of Northwest America, 1897-1903. Discovered weeping spruce, also over 50 new species. Died Portland, Ore., Dec. 3, 1912.

HOWELL, Thomas Raymond, Am. zoologist; b. New Orleans, June 17, 1924; s. Walter Lyall and Frances (Raymond) H.; B.S., La. State U., 1946; M.A., U. Cal. at Berkeley, 1949; Ph.D., 1951; m. Trudi Hanna Gubler, Sept. 1, 1959; children—Thomas Raymond, Yvonne H., Heidi F. Faculty, U. Cal. at Los Angeles, 1949——, prof. zoology, 1965——. Mem. Am. Ornithologists Union (1st v.p. 1966——), Cooper (pres. 1965-67), Wilson ornithol. socs., Soc. for Study Evolution, A.A.A.S., Am. Soc. Mammalogists, Am. Inst. Biol. Scis., Sigma Xi. Research, publs. on speciation, systematics, distbrn., physiol. ecology, behavior of birds. Home: 15450 Milldale Dr., Los Angeles 90024.*

HOWELL, Trevor Henry, English physician; b. Barnsley, Eng., Oct. 6, 1908; s. Trevor and Florence (Davis) H.; student Bradfield Coll., 1922-26, St. John's Coll., Cambridge, 1927-28, St. Bartholomew's Hosp., 1929-34; m. Margaret Greig Bannochie, Oct. 16, 1937; children—Anthea, Isobel, Christopher, Peter. House physician St. Olave's Hosp., London, Eng., 1934-35; med. tutor Brit. Postgrad. Med. Sch., 1935-37; pvt. practice medicine, Worthing, 1938-39; dep. physician, surgeon Royal Hosp., Chelsea, Eng., 1939-44; physician Queen's Hosp., Croydon, Eng., 1952——, Geriatric Research unit St. John's Hosp., London, 1946——; vis. prof. geriatrics Alexandria U., Egypt, 1961; lectr. problems old age St. Bartholomew's Hosp., 1948-64; chmn. mgrs. Westmoor Half Way House for Aged Sick, London, 1950——. Recipient Willard Thompson Gold medal Am. Geriatrics Soc., 1967. Founder Brit. Geriatrics Soc. Author: Old Age, 2d edit., 1950; Chronic Bronchitis, 1952; Our Advancing Years, 1953; Student's Guide to Geriatrics, 1963. Research, publs. on physiol. changes in old age, morbid anatomy of old age, half-way houses, description of rupture of right ventricle. Home: 10 Ockley Rd., London, Eng.*

HOWELL, Wallace Egbert, Am. meteorologist; b. Central Valley, N.Y., Sept. 14, 1914; s. James Cox and Alice (Egbert) H.; A.B., Harvard, 1936; M.S.,

Mass. Inst. Tech., 1941, Sc.D., 1948; m. Christine Gallagher, Dec. 30, 1942; children—Stephen Barnard, Jeremy, Holly Catherine, Jane Christine, James Jessup. Chief meteorologist Mid-Continent Airlines, Kansas City, Mo., 1937-39; asst. regional forecast U. S. Weather Bur., Boston, 1939-41, weather officer, 1941-45; research asst. fellow Blue Hill Obs., Harvard, 1946-51; pres. W. E. Howell Assos., Inc., Lexington, Mass., 1951-67; prin. sci. exec. EGQG, Incorporated, 1967——. Mem. subcom. on icing problems NACA, 1948-51, cons. adv. com. on weather control, 1954-57; pres. Mt. Washington Obs., 1954——. Mem. Am. Meteorol. Soc., Am. Geophys. Union, No. New Eng. Acad. Sci. Asso. editor Jour. Applied Meteorology, 1962——. Contbr. articles to profl. jours. Established role of condensation in determining composition of newly formed clouds; proposed dynamic effects of cloud seeding that are basis of hurricane-modification expts.; research in theory and practice in reduction of cloud seeding to useful applications, especially precipitation stimulation in tropical and temperate climates. Home: 35 Moon Hill Rd. Office: EGQG, Inc., Crosby Rd., Bedford, Mass. 01730.*

HOWELL, William Henry, Am. physiologist; b. Balt., Feb. 20, 1860; s. George Henry and Virginia Teresa H.; A.B., Johns Hopkins, 1881, Ph.D., 1884; hon. M.D., U. Mich., 1890; LL.D., Trinity, 1901, U. Mich., 1912, Washington U., 1915, U. Edinburgh, 1923; Sc.D., Yale, 1911; m. Anne Janet Tucker, June 15, 1887; children—Janet Tucker (Mrs. A. H. Clark), Roger, Charlotte Teresa (Mrs. Edward O. Hulburt). Asso. prof. physiology Johns Hopkins, 1888-89; prof. physiology and histology U. Mich., 1889-92; asso. prof. physiology Harvard, 1892-93; prof. physiology Johns Hopkins U., 1893-1931, dean med. faculty, 1899-1911, asst. dir. Sch. Hygiene, 1917-26, dir., 1926-31, emeritus from 1931. Chmn. NRC, 1932-33. Mem. Nat. Acad. Scis., Am. Philos. Soc., etc.; hon. mem. English Physiol. Soc. Author: Text-book of Physiology, 1905. Editor An American Text-book of physiology, 1896. Described certain small round or oval bodies observed in some blood cells (known as Howell's bodies), 1890; (with Luther Emmett Holt) isolated an anti-coagulant, termed it heparin, 1918; formulated Howell method of calculating blood clotting time. Died Feb. 6, 1945.

HOWELLS, John Gwilym, Brit. psychiatrist; b. Amlwich, U.K., June 24, 1918; s. Richard David and Mary (Hughes) H.; M.D., U. London, 1950, M.B.B.S., 1943, Diploma Psychol. Medicine, 1947; postgrad. Göttingen (Germany) U.; m. Ola Margaret Harrison, Dec. 11, 1943; children—David John Barry, Richard Keith, Cheryll Mary, Roger Bruce. Registrar, Inst. Psychiatry and Inst. Neurology, London U., 1947-49; cons. psychiatrist Inst. Family Psychiatry, Ipswich, U.K., 1949——; vis. prof. U. Neb., 1962. Adviser, WHO, intermittently, 1961——. Recipient Warneford medal, 1939. WHO fellow, U.S.A., 1961. Distinguished fellow Am. Psychiat. Assn.; mem. Royal Soc. Medicine, Royal Medico-Psychol. Assn., World Psychiat. Assn. Author: Theory and Practice of Family Psychiatry, 1968; (with J. R. Lickorish) Family Relations Indicator, 1962; Family Psychiatry, 1963; also articles. Editor: Modern Perspectives in Psychiatry Series, 1965——. Originated family psychiatry, a system of psychiat. practice using family group as unit, vector therapy a system for repattern-emotional forces, experiential psychiatry, explanation of psychopathology and its resolution by analysis of real life experiences; research in parent-child separation and problem families. Home: 89 Bucklesham Rd. Office: 23 Henley Rd., Ipswich, Suffolk, Eng.*

HOWELLS, William White, Am. anthropologist; b. N.Y.C., Nov. 27, 1908; s. John Mead and Abby Macdougall (White) H.; grad. St. Paul's Sch., 1926; S.B., Harvard University, 1930, Ph.D., 1934; m. Muriel Gurdon Seabury, June 15, 1929; children—Muriel Gurdon (Mrs. Richard E. Metz), and William Dean II. Engaged as teaching fellow anthropology Harvard Univ., 1930; research asso. phys. anthropology Am. Mus. Natural History, 1932-43; asst. prof. anthropology U. Wis., 1939-46, asso. prof. 1946-48, prof. 1948-54, prof. integrated liberal studies, 1948-54; prof. anthropology Harvard, 1954——, also curator somatology, faculty Peabody Mus., 1955——; asso. Lowell House, 1955; excavation of Gallen Priory, Harvard Irish Survey, 1935. Recipient Viking Fund medal in phys. anthropology, 1954. Fellow Am. Acad. Arts and Scis., Am. Anthropol. Assn. (pres. 1951), A.A.-A.S.; mem. Nat. Acad. Scis., Am. Assn. Phys. Anthropologists, Society of American Archaeology, Am. Soc. Human Genetics, Am. Soc. Naturalists, Am. Eugenics Soc. Inst. Human Paleontology; corr. mem. Geog. Soc. Lisbon, Anthrop. Soc. Paris, Anthrop. Soc. Vienna. Author: Mankind So Far, 1944; The Heathens, 1948; Back of History, 1954; Mankind in the Making, 1959, rev. edit., 1967. Editor: Early Man in the Far East, 1949; Ideas on Human Evolution, 1962; Am. Jour. Phys. Anthropology, 1949-54; asso. editor Human Biology, 1955——. Contbr. articles sci. and gen. publs. Developed new descriptive methods for comparing human size and form variation with predictions of genetic theory. Home: Kittery Point, Me. 03905. Office: Peabody Mus., Cambridge, Mass. 02138.*

HOWERTON, Hugh King, Am. physicist; b. Edina, Mo., Aug. 21, 1921; s. Hugh King and Vera (Ray) H.; B.A., N.E. Mo. State Coll., 1941; M.A., George Washington U., 1947; m. Doris Batta, Mar. 21, 1943;

children—Hugh King, Barbara Lee, Ronald Earl. Indsl. Research fellow Central Sci. Co., Chgo., 1941-42; chief atmospheric simulation sect. U. S. Ordnance Lab., White Oak, Md., 1943-47; with Am. Instrument Co., Silver Spring, Md., 1947—, chief product devel., 1956-62, asst. to pres., 1962—. Cons. to U. S. govt., industry, schs; adviser Nat. Acad. Scis.-NRC, 1962—. Mem. N.Y. Acad. Scis., Am. Phys. Soc., Am. Chem. Soc., Soc. for Applied Spectroscopy, Am. Soc. for Testing and Materials (acting chmn. subcom. on molecular fluorescence 1962), Optical Soc. Am. Contbr. articles to tech. jours. Patentee spectrofluorescence measuring instruments, diaphragm pump protective system. Research on methods and apparatus for measuring small quantities water in gases, calibration hygrometers at low temperatures, prodn. neutrons, strain in materials at elevated temperatures, determination trace amounts materials by fluorescence and phosphorescence, rec. small optical absorption in turbid media. Home: 9203 Whitney St., Silver Spring 20901. Office: 8030 Georgia Av., Silver Spring, Md. 20901.

HOWES, Edward Lee, Am. surgeon; b. New Haven, June 16, 1903; s. Albert E. and Caroline K. (Nesbit) H.; B.S., Yale, 1925, M.D., 1928; Diplomate (Guggenheim fellow) U. Strasbourg, 1932; M.S., Columbia, 1933, Sc.D., 1934; m. Ruth Larash, June 24, 1929; children—Edward Lee, Caroline (Mrs. Harry Weinroth). Faculty, Columbia, 1932-36, asso. clin. prof. Coll. Phys. and Surg., 1941—, asso. prof. clin. surgery, 1950—, prof. surgery Med. Sch., 1960—; prof., head dept. surgery Howard U., 1936-41, emeritus, 1941. Dir. Surg. Research and Basic Sci. Labs., Hartford Hosp., 1960—, cons. VA Hosp., Newington, Conn., 1965—. Recipient medal Institute Pasteur, 1960. Fulbright fellow, 1959. Diplomate Am. Bd. Surgery. Fellow N.Y. Acad. Scis.; mem. Soc. U. Surgeons, Soc. for Proc. Exptl. Biology and Medicine, Societe Internationale de Chirurgie, A.C.S., A.M.A., New Eng. Surg. Soc., Am. Chem. Soc., others. Research, publs on healing and regeneration of tissues. Home: 15 Birch Rd., West Simsbury, Conn. 06092. Office: 80 Seymour St., Hartford, Conn. 06115.*

HOWES, George Bond, English zoologist; b. London, Eng., Sept. 7, 1853; s. Thomas Johnson Howes; ed. pvt. sch.; D.Sc. Victoria; LL.D., St. Andrews. Asst. to Prof. Huxley biol. div. Royal Sch. Mines, 1874; demonstrator biology Normal Sch. Sci., Royal Sch. Mines, 1881, asst. prof., 1885; lectr. comparative anatomy St. George's Hosp. Med. Sch.; prof. zoology Royal Coll. Sci., London, 1895-1905. Fellow Royal Soc., 1897; mem. Zool. Soc. (former v.p., mem. council), Linnean Soc. London (former hon. zool. sec.), Anat. Soc. (past hon. treas.), Malacological Soc. London (past pres.), Brit. Assn. (past pres. sect. D), N.Y. Acad. Scis. (corr.); hon. mem. New Zealand Inst., Royal Soc. Victoria. Author: Atlas of Practical Elementary Biology (now revised as Atlas of Elementary Zootomy); also numerous articles on vertebrate morphology. Died Feb. 4, 1905.

HOWES, James Raymond, physiologist; b. Norfolk, Eng., Mar. 17, 1924; s. James Richard and Victoria (Branch) H.; B.Sc., U. London, 1949; N.D.A., Edinburgh U., 1949; M.Sc., McGill U., 1951; Ph.D., U. Fla., 1960; m. Grace Mary Murchison, May 22, 1951; children—Richard Ian, Robert Scott. Came to U. S., 1956, naturalized, 1962. With Imperial Coll. Tropical Agr., Trinidad, W.I., 1951-56, asso. prof. animal physiology, 1954-56; faculty Auburn (Ala.) U., 1960-67, asso. prof., physiologist, 1963-67; prof., physiologist Tex. A. and M. U., 1967—. Brit. Ministry of Agr. scholar McGill U., 1950; Royal Soc. and Nuffield Found. travelling fellow, East and South Africa, 1954; Internat. Cooperation Adminstrn. Travelling fellow, U.S.A., 1955. Mem. N.Y. Acad. Sci., Soc. Exptl. Biology and Medicine, Am. Chem. Soc., Am. Inst. Nutrition, Am. Assn. Poultry Sci., Am. Soc. Animal Sci., Am. Zool. Soc., Internat. Soc. Biometeorology. Contbr. numerous articles to sci. jours. Contbns. to study of ability of some animals to withstand heat stress better than others; studies of effect of high temperatures on calcium metabolism; application of telemetry to small animal physiology; devel. of composting as a method of utilizing animal waste products. Home: 3710 Sunnybrook Lane, Bryan, Tex. 77801. Office: P.O. Box 46, College Station, Tex. 77840.*

HOWLAND, John, Am. pediatrician; b. N.Y.C., Feb. 3, 1873; s. Henry Elias and Sarah Louise (Miller) H.; A.B., Yale, 1894; M.D., Univ. Med. Coll., N.Y. U., 1897; m. Susan M. Sanford, Oct. 12, 1903. Began practice, N.Y.C., 1897; physician Willard Parker and Riverside and St. Vincent's hosps., pathologist Foundling Hosp., others; prof. pediatrics, Washington U., St. Louis, 1911-12; prof. pediatrics Johns Hopkins, and pediatrician in chief Johns Hopkins Hosp., from 1912. Research on nutritional disorders of infancy, chem. aspects of disease; established 1st pediatric clinic in U. S. at Johns Hopkins. Died June 20, 1926.

HOXIE, Charles A., Am. inventor; b. Constable, N.Y., circa 1867; s. John Clark and Victoria Adelaide (Welch) H.; ed. pub. schs.; m. Jeannie Helena Benalloch, Oct. 4, 1892; children—Earle Alfred, Lilla May, Merril Clark, Florence Benallock (Mrs. John H. Blackwood). With Gen. Electric Co., 1912-32. Known as father of sound pictures; inventor 1st process for changing sound into light and recording it on transparent film; introduced system of recording radio code on continuous strip of paper, thus permitting increased speed of nearly 100 per cent in message transmission and reception; developer photophone. Died 1941.

HOYLE, Fred, English astrophysicist; b. Yorkshire, Eng., June 24, 1915; s. Benjamin and Mabel (Pickard) H.; M.A., Emmanuel Coll., Cambridge U., 1939; m. Barbara Clark, Dec. 28, 1939; children—Geoffrey, Elizabeth Jeanne (Mrs. Richard A. Lowndes). Fellow St. John's Coll., Cambridge U., 1939—, univ. lectr. math., 1945-58, Plumian prof. astronomy and exptl. philosophy, 1958—; mem. staff Mt. Wilson, also Palomar observatories, 1956—. Mayhew prizeman in math. tripos, 1936, Smith's prizeman, 1938. Fellow Royal Soc., 1957, Royal Astron. Soc.; mem. Am. Acad. Arts and Scis. Author: Some Recent Researches in Solar Physics, 1949; The Nature of the Universe, 1950; A Decade of Decision, 1953; Frontiers of Astronomy, 1955; (novel) The Black Cloud, 1957; (novel) Ossian's Ride, 1959; (play) Rockets in Ursa Major, 1962; (TV play) A for Andromeda, 1962; (novel) Fifth Planet, 1963; Of Men and Galaxies, 1964; Galaxies, Nuclei and Quasars, 1965; (novel) October 1st Is too Late, 1966; Galaxies, Nuclei and Quasars, 1966; also papers. Research on devel. of stars (giants, white dwarfs), age of stars; helped formulate several major theories in cosmology. Address: St. John's Coll., Cambridge Univ., Cambridge, Eng.

HOYT, Samuel L(eslie), Am. metallurgist; b. Mpls., May 29, 1888; s. Alfonzo O. and Huldah L. (Hunt) H.; E.M., U. Minn., 1909; Ph.D., Columbia, 1914; postgrad. Columbia, Royal Inst. Tech., Berlin, Germany; Sc.D., S.D. Sch. Mines and Tech., 1953; m. Edyth Armstrong, Nov. 1, 1946. Head dept. metallurgy, U. Minn., 1913-19; research metallurgist Gen. Electric Co., 1919-30; research metallurgist, head dept. A.O. Smith Corp., Milw., 1930-39; tech. adviser Battelle Meml. Inst., Columbus, O., 1939-53; metall. cons., 1953-59; staff rep. research and engring. div. U. S. Dept. Def. to NATO, Frankfurt, West Germany, 1959-60; metall. cons., Berkeley, Cal., 1960—. Mem. sci. mission ALSOS, Germany, 1945; cons. govt. agys., indsl. firms. Recipient Outstanding Achievement award U. Minn., 1950. Mem. Am. Inst. Mining Engrs., Am. Soc. Metals, A.A.A.S., Inst. of Metals, (Eng.), Iron and Steel Inst. (Eng.), Am. Welding Soc., Sigma Xi. Author: Principles of Metallography, 1920; Metals and Common Alloys, 1920; Metal Data, 1950; also numerous articles. Developed cemented tungsten carbide carboloy; inventor, devel. Smith alloys number 10, superior elec. heating alloy; research on problems brittle fracture, welding metallurgy, steel procurement and use, stainless steel, steel fabrication. Home; 88 Purdue Av., Berkeley 94708. Office: 2827 7th St., Berkeley, Cal. 94710.*

HOYT, William Dana, Am. biologist; b. Rome, Ga., Apr. 16, 1880; s. William Dearing and Florence West (Stevens) H.; A.B., U. Ga., 1901; M.S., 1904; Ph.D., Johns Hopkins, 1909; postgrad. U. Heidelberg, 1909-10; research work, Naples, 1910; m. Margaret Howard Yeaton, Dec. 27, 1910 (dec. Sept. 1943); children—William Dana, Southgate Yeaton, Robert Stephens. Tutor in biology, U. of Ga., 1901-04; fellow Johns Hopkins, 1908-09; scientific asst. U. S. Bur. Fisheries, Beaufort, N.C., 1902-09; Adam T. Bruce fellow Johns Hopkins, 1909-10; instr. in botany Rutgers Coll., 1910-12; fellow Johns Hopkins 1912-15; asso. prof. biology Washington and Lee U., 1915-20, prof. from 1920; instr. in botany, Marine Biol. Lab., Woods Hole, Mass., 1917-19. Chmn. adv. council State Parks and Forests, 1930-33. Fellow A.A.A.S.; mem. Bot. Soc. Am., Am. Forestry Assn., Am. Genetic Assn., Am. Eugenic Soc., Am. Soc. Naturalists, Assn. Research Human Heredity, Va. Acad. Sci., Phi Beta Kappa. Author: Marine Algae of Beaufort, N.C., and Adjacent Regions, 1920. Research and publs. on life habits and physiology of algae; proved alternation of generations in algae. Died Sept. 24, 1945.

HRDLICKA, Ales, anthropologist; b. Humpolec, Bohemia, Mar. 29, 1869; s. Maxmilian and Karolina H.; M.D., N.Y. Eclectic Coll., 1892, New York Homoe. Coll., 1894; Md. Allopathic State Bd., 1894; hon. Sc.D., Prague U., 1920, Brno U., 1929; investigator among insane and other defective classes, N.Y. State Service, 1894-99; asso. in anthropology, N.Y. State Pathol. Inst., 1896-99; studied in Paris U. and Anthrop. Sch., first half 1896; married 1896. Tour over European prisons, insane asylums and museums, 1896; in charge phys. anthropology of Hyde expdns. for Am. Mus. Natural History, 1899-1903; asst. curator in charge div. phys. anthropology, 1903-10, curator since 1910, U. S. Nat. Museum. Anthrop. expdns. to many countries throughout period since 1898. Author exhibits phys. anthropology and prehistoric Am. pathology, San Diego Expn., 1915-16. Asso. editor Am. Naturalist, 1901-08; founder, and editor Am. Jour. Phys. Anthrop. since 1918. Sec. gen. XIX Internat. Congress Americanists, 1915, sec. Sect. 1, anthropology, 2d Pan-Am. Scientific Congress, 1915-16; sec. com. on anthropology, Nat. Research Council, 1917-18; etc. Fellow Am. Acad. Arts and Sciences, A.A.A.S. (life); mem. Assn. Am. Anatomists, Am. Anthrop. Assn. (pres. 1925-26), Nat. Acad. Sciences, Am. Philos. Soc., Washington Acad. Sciences (pres. 1928-29), Archaeol. Inst. America, Am. Assn. Physical Anthropology (pres. 1928-32; founder and life mem.); hon. and corr. member various foreign acads. and socs. Huxley medal lecturer, London, 1927. Author: Physical Anthropology, 1914; Anthropometry, 1920; Old Americans, 1925; Skeletal remains of early man, 1930; Children who Run on All Fours, 1931; Alaska Diary, 1926-31, 1943, numerous others. Did valuable work on origins and evolution of man; known for his work on theory that Am. Indians migrated from Asia. Died Washington, Sept. 5, 1943.

HRDY, Ivan, Czechoslovakian entomologist; b. Olomouc, Czechoslovakia, Oct. 21, 1928; s. Theodor and Anna (Reissova) H.; student Palacky U., 1948-50; R.N.Dr., Charles U., Prague, 1952, postgrad.; m. Jitka Truxova, May 22, 1954; children—Judita, Ivan. Mem. Inst. Entomology, Czechoslovakian Acad. Scis., Prague, 1954—, head dept. insect toxicology, 1964—; Mem. Czechoslovakian Entomol. Soc., Czechoslovakian Zool. Soc., L'Union internationale pour l'etude des Insectes Sociaux, European Assn. Editors of Biol. Periodicals. Editor: The Ontogeny of Insects, 1960; Actaentomologica bohemoslovaca. Research, publs. on effects of insecticides, ecol. aspects of chem. control of pests, social insects. Home: 4 Brezinova, Zbraslav n. Vlatavou II, Czechoslovakia. Office: 7 Vinicna, Prague 2, Czechoslovakia.*

HRISTOV, Vladimir Kirilov, Bulgarian geodesist; b. Sofia, Bulgaria, Dec. 18, 1902; s. Kiril and Nevena (Palasheva) H.; Ph.D., Leipzig (Germany), 1925; m. Jordanka Stefanova, Sept. 16, 1928; children—Nevena Vladimirova (Mrs. Kosta Petrov Nikolov). Head astron. sect. Mil. Geog. Inst., 1925-48; prof. high geodesy High Inst. Civil Engring., Sofia, Bulgaria, 1948—; acad. dir. Central Lab. Geodesy, Bulgarian Acad. Scis., 1958—. Decorated Red banner of Labour; Peoples' Republic of Bulgaria I degree; Dimitrov Prize Laureate II degree; Honoured worker Culture. Mem. Union Sci. Workers. Author: Gauss' and geog. co-ordinates on Krassowsky's ellipsoid, 1955; Gauss-Krüger coordinates on the Ellipsoid of Rotation, 1957; General Theory of the Coordinates Applied in Geodesy, 1959; Foundations of the Theory of Probability and Mathematical Statistics and the Method of the Least Squares, 1961; Extention of the Compensation by the method of the Least Squares, 1966. Research on theory of all coordinates applied in geodesy, especially Gauss-Krüger coordinates; developed further theory of compensation of observations, including correlated observations. Home: 71 Boyanski vazhod, Sofia, Bulgaria.*

HRUBAN, Zdenek, physician; b. Prerov, Czechoslovakia, June 15, 1921; s. Jaroslav and Aloisie (Rieger) H.; cand.med. U. Rostock (Germany), 1943; MUC, Charles U., Praha, Czechoslovakia, 1948; M.D., U. Chgo., 1956, Ph.D., 1963; m. Jarmila Stanek, Aug. 27, 1955; children—Paul, Ralph, Diana. Came to U. S., 1951, naturalized, 1957. USPHS postdoctoral fellow, 1957-60; Am. Cancer Soc. Clin. fellow, 1960-64; faculty U. Chgo., 1960—, asso. prof. dept. pathology, 1967—. Recipient Lederle Med. Faculty award, 1963-67. Mem. Am. Soc. for Cell Biology, Am. Soc. for Exptl. Pathology, Electron Microscope Soc. Am., Am. Assn. for Cancer Research. Research, publs. on fine structure of cellular injury, enzyme localization in microbodies, lysosome formation by focal degradation. Home: 5427 S. Greenwood Av., Chgo. 60615.*

HRUBY, Karl, Austrian oculist; b. Steinwan/Krain, Oct. 20, 1912; s. Karl and Theresia (Zdrahal) H.; M.D U. Vienna; m. Elfriede Schindler, July 11, 1937; children—Johanna, Karl, Michael. Med. asst. ophthalmology clinic U. Vienna, 1937-40, asst. to Prof. Lindner, 1945-55, agrégé, 1946, titular prof., dir. ophthalmology clinic, 1955; asst. Ophthalmology Clinic, Prague, 1940-45. Mem. Royal Soc. Medicine (London), Instituto Barraquer (Barcelona), West Bengal Ophthal. Soc., Van Swieten Soc. (Vienna), Austrian Soc. Ophthalmologists. Author: Spaitlampenmikroskopie der hint. Augenabschnittes, 1950; Der Augenarzt, 1960; Die bedrohlichen Erkrankungen und Verletzungen des Auges, 1962. Home: Berggasse 15, Vienna IX. Office: Ophthalmology Clinic, University of Vienna, Vienna I, Austria.

HRUZA, Zdenek, physiologist; b. Prague, Czechoslovakia, Oct. 3, 1926; s. Antonin and Zdenka (Kutscherova) H.; M.D., Charles U., Prague, 1950; Ph.D., Acad. Scis., Prague, 1955, D.Sc., 1966; m. Judit Ilkovics, Feb. 11, 1951; children—Eva, Jiri. Instr. dept. exptl. pathology Med. Sch., Prague, 1950-52; asst. scientist Inst. Human Nutrition, Prague, 1952-56; chief lab. for exptl. gerontology Inst. Physiology, Acad. Sci., Prague, 1957-66; research prof. dept. pathology N.Y. U. Med. Sch., N.Y.C., 1966—. Fellow Am. Gerontol. Soc., Czechoslovakian Gerontol. Soc., Physiol. Soc. Czechoslovakia, Am. Physiol. Soc.; mem. Gertontologic Soc. Prague (sec.). Author: Science About Aging, 1966; also numerous articles. Editor: Ceskoslovenska Fyziologie, 1964-66; regional editor Exptl. Gerontology, London, Eng., 1963—. Research on aging process, use of properties of connective tissue for estimation of biol. age, mechanism of adaption to repeated stress; discovered decrease in catabolism of fats with aging. Office: 550 1st Av., N.Y. U. Med. Sch., N.Y.C. 10016.*

HRYNKIEWICZ, Andrzej Zygmunt, Polish physicist; b. Wilno, Poland, May 29, 1925; s. Józef M. and Stanislawa (Paszkowska) H.; student Nicolaus Copernicus U., Torun, Poland, 1945-46; Ph.D., Jagellonian U., Cracow, Poland, 1950; m. Halina Cholewa, Apr. 16, 1966. Faculty, Jagellonian U., 1946-61, prof.

chair nuclear physics, 1961; head nuclear magnetic resonance lab. Inst. Nuclear Physics, Cracow, Poland, 1954-61, head gamma ray spectroscopy lab., 1961——, vice dir., 1962——; vice dir. Joint Inst. for Nuclear Research, Dubna, USSR, 1966——. Recipient Polonia Restituta; 3 awards Govt. Council for Use Atomic Energy. Mem. Polish, Am. phys. socs. Research, numerous articles on new method of relaxation-time measurements in nuclear magnetic resonance, magnetic moment measurements for short-lived nuclear states; research on nuclear magnetic resonance; nuclear spectroscopy and Mössbauer effect. Home: 4, Pl. Wolnosci. Office: Inst. Nuclear Physics, Cracow, Poland.*

HSIA, David Yi-Yung, pediatrician; b. Shanghai, China, Aug. 22, 1925; s. Ching-lin and Waitsung (New) H.; came to U. S., 1939, naturalized, 1961; A.B. magna cum laude, Haverford Coll., 1944; M.D. cum laude, Harvard, 1948; m. Hsio-Hsuan Shih, July 29, 1949; children—David, Judith, Lisa, Peter. Research fellow Harvard Med. Sch., 1951-55, instr., 1953-56; hon. research asst. Galton Lab., U. Coll., London, Eng., 1956-57; faculty Northwestern U. Med. Sch., Chgo., 1957——; prof. pediatrics, 1960——; mem. staff, dir. div. clin. biochemistry Children's Meml. Hosp., Chgo., 1957——; dir. research and genetic clinic, 1958——; mem. staffs Evanston (Ill.) Hosp., Cook County Hosp. Chmn. med. adv. com. Chgo. chpt. Cystic Fibrosis Research Found., 1963——. Recipient E. Mead Johnson award, 1965. Diplomate Nat. Bd. Med. Examiners. Am. Bd. Pediatrics. Mem. Am. Chem. Soc., Am. Assn. Clin. Chemists, Am. Genetic Soc., Am. Soc. Human Genetics, A.A.A.S., Soc. Exptl. Biology and Medicine, Am. Fedn. Clin. Research, Central Soc. Clin. Research, Am. Acad. Pediatrics, Midwestern Soc. Pediatric Research, Soc. Pediatric Research, Am. Pediatric Soc., Phi Beta Kappa. Research, publs. on human biochem. genetics. Home: 2752 Bennett St., Evanston, Ill. 60201. Office: 707 Fullerton, Chgo. 60614.*

HSIAO, Chu (Chou Ting Wang), Chinese physician, botanist; flourished 1390; s. Hung-Wu, 1st Ming emperor; created a bot. and exptl. garden on his estates nr. Kai-Feng-Fu, Honan, China, 1382-1400; was banished later; restored to honor by his brother the 3d emperor, Yung Lo. Received the title of Chou Wang (Prince Chou), 1378; received the posthumous name of Chou Ting Wang. Author: Chiu Huang Pen Ts'ao (Famine Herbal or Herbal to Relieve Famine), 1378. P'u-Chi Fang (a treatise on medicine), compiled circa 1378. Described and illustrated 414 species of plants including many new species; acclimatized wild plants for use of food in famine times. Died 1425.

HSIAO, Sidney Chihti, biologist; b. Wuchang, China, Oct. 24, 1905; s. Langsing and Shi (Dunn) Hsiao; B.A., Shanghai U., 1928; M.A., Yenching U., 1933; Ph.D., Harvard, 1938; m. Erica Karawina, June 21, 1938. Came to U. S., 1935, naturalized, 1951. Research asst. Harvard, 1938-39, China fellow, 1939-41; faculty Huanchung U., Wuchang, China, 1941-46, prof. biology, 1942-46, dean Coll. Sci., 1941-43; Seessel fellow Yale, 1946-48; vis. asst. prof. biology N.Y. U., 1948-49; faculty U. Hawaii, Honolulu, 1949——, prof. zoology, 1958——; vis. prof. zoology Nat. Taiwan U., Republic China, 1955-56. Dir. Summer Inst. in Radiation Biology, 1959-63; Fulbright Lectureship grantee, Nat. Taiwan U., 1962-63; vis. scholar Academia Sinica, 1965. Fellow A.A.A.S., N.Y. Acad. Sci.; mem. Am. Soc. Zoologists, Am. Soc. Limnologists and Oceanographers, Soc. for Study Devel. and Growth, Hawaiian Acad. Sci., Sigma Xi, Phi Kappa Phi. Research, publs. on Limnology Erh Hai, radiocalcium uptake in sea urchin, alkaline phosphatase in echinoids, effect ionizing radiation in early devel. sea urchin. Home: 3529 Akaka Pl., Honolulu 96822.*

HSU, Francis Lang-Kwang, anthropologist; b. Chuang-ho, China, Oct. 28, 1911; s. Chung-ting and Lee (Shih) H.; B.A., U. Shanghai, 1933; Ph.D., U. London, 1940; m. Vera Hi-nan Tung, Apr. 26, 1943; children—Eileen Yi-nan, Penalope Se-hwa. Faculty, Nat. Yunnan U., Kunming, China, 1941-44, Columbia, 1944-45, Cornell U., 1945-47; faculty Northwestern U., Evanston, Ill., 1947——, prof., 1955——, chmn. dept. anthropology, 1957——; cons. govt. agys., hosps. Wenner-gren Found. fellow, 1944-45, 49-50, 55-57, 64-65; Social Sci. Research Council fellow, 1949-50; Rockefeller Found. fellow, 1955-57; Carnegie Found. fellow, 1964-65; Govt. India fellow, 1955-57. Fellow Am. Anthrop. Assn.; mem. Am. Assn. U. Profs. Author: Under the Ancestors' Shadow, 1948; Religion, Science and Human Crises, 1952; Americans and Chinese, 1953; Psychological Anthropology, 1961; Clan, Caste and Club, 1963; Study of Literate Civilizations, 1967. Research, numerous publs. patterns of culture and personality of Chinese, Hindus, Japanese and White Americans. Home: 310 Wesley Av., Evanston, Ill. 60202.*

HSU, Hsi Fan, parasitologist; b. Huangyen, Chekiang, China, Mar. 9, 1906; s. Fan Chin and Shih (Wang) H.; B.S., Amoy U., China, 1929; D.Sc. U Neuchatel, Switzerland, 1935; M.D., U. Philippines, 1948; m. Shu Ying Li, Apr. 18, 1954. Came to U. S., 1954, naturalized, 1963. Chief lab. parasitology Nat. Inst. Health, Nanking, China, 1948-49; with Nat. Taiwan U., 1949-54, chmn. dept. pre-medicine, 1949-50, prof., head dept. zoology, 1949-54; with U. Ia., 1954——, research prof. preventive medicine, environ-

mental health, 1961——. Mem. Soc. Exptl. Biology and Medicine, Am. Assn. Immunology, Am. Soc. Tropical Medicine and Hygiene, Am. Soc. Parasitology. Publs. on discovery of zoophilic strain Schistosoma japonicum in Taiwan, parasites; characterization 4 geog. strains Schistosoma japonicum, successful immunization of rhesus monkeys against Schistosoma japonicum with cercariae of zoophilic strain or x-irradiated cercariae human strain. Home: 1512 Derwen Dr., Iowa City, Ia. 52240.*

HSU, Konrad Chang, chemist; b. Taichow, Kiangsu, China, Aug. 28, 1901; s. George Chien and Wu (Yu) H.; B.S., St. John's U., Shanghai, 1921; M.A., Columbia, 1923, Ph.D., 1924; m. Katharine D. Hawley, Jan. 31, 1951; children—Victoria Ruffner, Theodora Du, Alicia Wohl, Adelina, Konrad T., Lydia Yu. Came to U. S., 1950, naturalized, 1962. Faculty, Gt. China U., 1924-26; with Chinese Govt., 1926-28; chemist, gen. mgr. Chan Hwa & Co., 1926-46; v.p. Sino Hawaiian Corp., 1946-49; faculty Columbia, N.Y.C., 1954——, asso. prof. microbiology, 1965——; vis. prof. U. Rome, U. Bonn, also vis. scientist Inst. Gustave Roussey, Villejuif, France, 1966, Sydney (Australia) U., 1967. Mem. Am. Assn. Immunologists, Am. Soc. Microbiology, Harvey Soc., N.Y. Acad. Scis., A.A.A.S., Sigma Xi. Research, publs. on diseases resulting from immune reactions using fluorescein and ferritin labeled antibodies and antigens; immunochem. investigation; pathogenesis of some sequelae of streptococcal disease. Home: 24 Schreiber St., Tappan, N.Y. 10983.*

HSU, Shu Ying, parasitologist; b. Foochow, Fukien, China, Aug. 16, 1920; s. Chiao Ping and Hsinfu (Chen) Li; B.S., Biology, Nat. Peiping Normal U., Peiping, China, 1940; Ph.D. in Hygiene, State U. Ia., 1957; m. Hsi Fan Hsü, Apr. 18, 1954. Came to U. S., 1952, naturalized, 1961. Instr. parasitology dept. bacteriology Mich. State Coll., East Lansing, Mich., 1952-53; research asso. dept. tropical pub. health Harvard Sch. Pub. Health, Boston, 1953-54; faculty U. Ia., Iowa City, 1957——, research asso prof. preventive medicine and environmental health U. Ia., Iowa City, 1963——. Recipient Taiwan Provincial Sci. award, 1953. Mem. Soc. for Exptl. Biology and Medicine, Am. Assn. Immunologists, Am. Soc. Tropical Medicine and Hygiene, Am. Soc. Parasitologists. Research, publs. on discovery of zoophilic strain Schistosoma japonicum in Taiwan, characterization 4 geog. strain Schistosoma japonicum, successful immunization rhesus monkeys against Schistosoma japonicum with cercariae of zoophilic strain or X-irradiated cercariae human strain. Home: 1512 Derwen Dr., Iowa City 52240.*

HSU, T(ao) C(hiuh), biologist, educator; b. Shaohsing, Chekiang, China, Apr. 17, 1917; B.S., Chekiang U., 1941; Ph.D., U. Tex., 1951. Came to U. S., 1948, naturalized, 1953. Asst. prof. U. Tex. Med. Br., 1953-55; asso. biologist U. Tex. M.D. Anderson Hosp. and Tumor Inst., Houston, 1955-60, biologist, 1960-65, biologist, prof. biology, 1965——. Vis. prof. Baylor U. Sch. Medicine, 1966——; cons. NIH, 1963——. Mem. A.A.A.S., Am. Soc. for Cell Biology, Genetics Soc. Am., Am. Soc. for Exptl. Pathology, Am. Assn. for Cancer Research, Tissue Culture Assn. Research, numerous publs. on chromosomes, their morphology, structure, physiology and role in evolution. Home: 3036 Albans St., Houston 77005.*

HU, Funan (Mrs. Lan An Hsu), dermatologist; b. Shanghai, China, Sept. 13, 1919; d. Tun-fu and Kwei-Shing (Hwa) Hu; M.D., Nat. Med. Coll. Shanghai, 1942; m. Lan An Hsu, Dec. 24, 1944. Came to U. S., 1950, naturalized, 1955. Practice medicine, specializing in dermatology, Detroit, 1953-65, Beaverton, Ore., 1965——; asso. dept. dermatology Henry Ford Hosp., 1953-65; scientist Ore. Regional Primate Research Center, 1965——; prof. dermatology Med. Sch. U. Ore., Portland, 1965——. Diplomate Am. Bd. Dermatology and Syphilology. Mem. Am. Acad. Dermatology, A.A.A.S., Am. Assn. Cancer Research, Am. Soc. Cell Biology, Am. Soc. Dermatopathology, N.Y. Acad. Scis., Soc. Investigative Dermatology, Tissue Culture Assn. Editor: (with M. Montagna) Advances in Biology of Skin, 1967. Research, numerous publs. on cytology, pigment cell biology, melanogenesis and pigmentary disorders. Home: 4175 S.W. Crestwood Dr., Portland, Ore. 97225. Office: 505 N.W. 185th Av., Beaverton, Ore. 97005.*

HU, Stephen Moi Kee, Am. entomologist; b. Honolulu, May 16, 1903; s. Akana Tung and Tam (She) H.; B.S., Cornell U., 1928, M.S., 1929; Sc.D., Johns Hopkins, 1931; m. Mildred Li, Apr. 3, 1934; children—Richard C. C., Patricia (Mrs. Herbert Chew), Rosemary (Mrs. Kwang Ping Hsu). Head dept. entomology, dir. malariology Nat. Health Adminstrn., Nanking, China, 1931-33; head dept. entomol. Henry Lester Inst. Med. Research, Shanghai, China, 1933-46; entomologist, health div. UNRRA, 1946-47; chief bur. mosquito control Honolulu Dept. Health, 1948-55; entomologist div. virus, rickettsial diseases 406th Med. Gen. Lab., Zama, Japan, 1955-57; head dept. entomology U. S. Naval Med. Research Unit 2, Taipei, Taiwan, 1957-60; malaria specialist Pub. Health Div., Internat. Co-Op. Adminstrn., 1960——, acting chief of health div. U. S. AID, Labore, West Pakistan, 1962-64. Rockefeller Found. fellow, 1931-32. Mem. Am., Royal socs. tropical medicine and hygiene, Am. Mosquito Control Assn., Pub. Health Assn., Sigma Xi.

Research, publs. on dog filaria-mosquito relationship in U. S., malaria vectors of China, encephalitis vectors in Japan and Taiwan. Home: 207 Beechwood Dr., Ellicott City, Md. 21043. Office: U. S. AID, Philippines, METC, APO San Francisco 96528.*

HUANG, Hsing Tsung, chemist; b. Malacca, Malaya, Sept. 9, 1921; s. Y. Y. and W. (Yu) H.; B.S., U. Hong Kong, 1941; D.Phil. Oxford (Eng.) U., 1947; m. Rita Quan, Dec. 17, 1949; children—Pamela, Terence. Naturalized U. S. citizen, 1957. Research fellow U. Rochester (N.Y.), 1947-48, Cal. Inst. Tech., Pasadena, 1948-51; research biochemist Rohm & Haas Co., Phila., 1951-55; research biochemist Charles Pfizer & Co., Bklyn., 1955-61, group supr., Groton, Conn., 1961-64; asso. dir. biochem. Bioferm div. Internat. Minerals & Chem. Corp., Wasco, Cal., 1964-68, dir. biol. scis. research and devel. div., Libertyville, Ill., 1968——. Mem. Am. Soc. Biol. Chemists, Am. Chem. Soc., A.A.A.S. Research, publs., patents in organic chemistry and fermentation processes; improvement processes to make antibiotics and amino acids; discovered new method to produce new semi-synthetic penicillins. Home: 1211 Stratford Rd., Deerfield, Ill. 60015. Office: Internat. Minerals & Chem. Corp., Growth Science Center, Libertyville, Ill. 60048.*

HUANG, Kee Chang, pharmacologist; b. Canton, China, July 22, 1917; s. Chon Yue and Mary S. F. (Lee) H.; B.S., Dr. Sun Yet San U., Canton, 1940, M.D., 1940; Ph.D., Columbia, 1953; m. Shoushan Chang, Feb. 16, 1947; children—Kou-chu, Anna K., Karen T. Came to U. S., 1949, naturalized, 1962. Research fellow NIH of China, 1940-46; instr. Nat. Shanghai Med. Sch. China, 1946-49; research asso. U. Louisville, 1953-56, faculty, 1956——, prof. pharmacology, 1963——. Mem. Am. Soc. Physiology, Am. Soc. Pharmacology and Therapeutics, Soc. Exptl. Biology and Medicine, Coll. Clin. Pharmacology and Therapeutics, Royal Soc. Medicine (London, Eng.). Research, publs. on renal physiology and pharmacology. Home: 154 Forest Dr., Jeffersonville, Ind. 47130. Office: 511 S. Floyd St., Louisville 40202.*

HUANG, Su-Shu, astrophysicist; b. Changshu, Kiangsu, China, Apr. 26, 1915; s. Zin Zee and Hoo (Cheng) H.; B.S., Nat. Chekiang U., 1937; M.S., Nat. Tsing-Hua U., 1943; Ph.D., U. Chgo., 1949. Came to U. S. 1947, naturalized, 1959. Instr., U. Chgo., 1949-51; Guggenheim fellow, 1951-52; research astronomer U. Cal. at Berkeley, 1952-59; physicist Goddard Space Flight Center, NASA, Greenbelt, Md., 1959-60, 61-64; mem. Inst. for Advanced Study, Princeton, N.J., 1960-61; prof. Cath. U. Am., Washington, 1964-65; prof. astro-physics Northwestern U., Evanston, Ill., 1965——. Mem. Am. Astron. Soc., Internat. Astron. Union, Astron. Soc. Pacific, A.A.A.S., Sigma Xi. Research, numerous publs. in atomic physics and astrophysics, especially theory stellar atmosphere, stellar spectroscopy, binary stars, bioastronomy. Home: 1725 Orrington Av. Office: Lindheimer Astron. Research Center, Northwestern U., Evanston, Ill. 60201.*

HUARTE DE SAN JUAN, Juan, Spanish physician, psychologist; b. Saint-Jean-Pied-Port, Navarre, circa 1530; grad. in medicine U. Huesca; practiced medicine, Madrid; distinguished himself during plague at Baeza, 1566. Author: Exémen de ingenios para las ciencias, 1575. His research on physical constitution is regarded as early psychol. study which attempts to connect physiology and psychology. Died 1592.

HUBAYS BEN AL-HASAN AL-A'SAM, see al-Hasan al-A'sam.

HUBBARD, Bert Earl, Am. mathematician; b. Cameron, Ill., Aug. 6, 1928; s. Everett S. and Lula (Avise) H.; B.S., Western Ill. U., 1949; M.S., State U. Ia., 1952; Ph.D., U. Md., 1960; m. Doris Mae Blecker, Dec. 27, 1952; children—David, Bari-Lynn. Mathematician, U. S. Naval Ordnance Lab., 1955-61, cons. research mathematician, 1961-65; research asst. prof. Inst. for Fluid Dynamics and Applied Math. U. Md., College Park, 1961-64, research asso. prof., 1964——. Mem. Am. Math. Soc., Soc. For Indsl. and Applied Math. (dir. vis. lectureship program), Sigma Xi, Phi Kappa Phi. Research, publs. on numerical analysis, especially the approximation of solutions of partial differential equations of math. physics. Home: 3517 Duke St., College Park, Md. 20740.*

HUBBARD, Edward Leonard, Am. physicist; b. Phoenix, July 7, 1921; s. Edward Robert and Sue (Leonard) H.; B.A., U. Cal. at Los Angeles, 1943, M.A., 1948, Ph.D., 1951; m. Bonnie Eula Cushman, Oct. 4, 1952; children—Paul Edward, Glenda Kay, Ruth Susan, Alison Ann. Asst. physics U. Cal., Los Angeles, 1943-44, technician Cyclotron Lab. 1948-51; physicist U. S. Naval Ordnance Lab., Washington, 1944-45; physicist Lawrence Radiation Lab., Berkeley, Cal., 1951-68, Nat. Accelerator Lab., Oakbrook, Ill., 1968——. Mem. Am. Phys. Soc., Sigma Xi. Research, publs. on particle accelerators, penetration of atomic particles through matter, nuclear reactions with heavy ions. Home: 622 Newton Av., Glen Ellyn, Ill. 60137. Office: Nat. Accelerator Lab., 1301 W. 22d St., Oakbrook, Ill. 60521.*

HUBBARD, Harvey Hart, Am. physicist; b. Swanton, Vt., June 17, 1921; s. Horace Waite and Elbie

(Hart) H.; B.S., U. Vt., 1942; m. Sadie Margaret Miller, Dec. 6, 1947; children—Thomas W., Susan H., Pamela L., Walter R. Research engr. NACA-Langley, Hampton, Va., 1945-58, head atmospheric and acoustics br., 1958, head acoustics br. NASA, 1959—; mem. com. on hearing and bioacoustics NRC. Fellow Acoustic Soc. Am.; mem. Am. Phys. Soc., Am. Inst. E.E., Soc. Automotive Engrs. Contbg. author: Tech. Aspects of Sound, 1956; Handbook of Shock and Vib Control, 1961; Agardograph on Aircraft Noise, 1963. Contbr. numerous articles to profl. jours. Research on aircraft and spacecraft noise problems, including jet noise generation, propeller and rotor noise, sonic boom generation, propagation and effects, sonic fatigue, low frequency noise effects in spacecraft design. Home: 23 Elm Av., Newport News, Va. 23601. Office: NASA, Langley Sta., Hampton, Va. 23365.*

HUBBARD, Joseph Stillman, Am. astronomer; b. New Haven, Conn., Sept. 7, 1823; s. Ezra Stiles and Eliza (Church) H.; grad. Yale, 1843. m. Sarah E. L. Handy (dec. 1860), Apr. 27, 1848; one child (dec. 1856). Asst. to Sears C. Walker at High Sch. Obs., Phila., 1844-45; prof. math. U. S. Naval Acad., stationed at U. S. Naval Obs., 1845-63. Especially interested in question of parallax of Alpha Lyrae; contbr. numerous articles to Astron. Jour.; calculated orbit of comet of 1843; confirmed identity of Neptune, and determined its orbit with great precision. Died Aug. 16, 1863.

HUBBARD, Oliver Payson, Am. geologist; b. Pomfret, Conn., Mar. 1809; s. Stephen and Zeruiah (Grosvenor) H.; grad. Yale, 1828; M.D., S. C. Med. Coll., 1837; LL.D., Hamilton, 1861; m. Faith Wadsworth Stilliman; children—Harriet, Henrietta, Grosvenor. Prof., 1836-66; lecturer, 1866-71, chemistry, pharmacy, mineralogy and geology, afterward prof. chemistry and pharmacy, 1871-83 (emeritus), Dartmouth. Mem. N.H. legislature, 1863-64. Aided Charles Goodyear in research leading to process of vulcanizing rubber. Died New York, N.Y., 1900.

HUBBERT, M(arion) King, Am. geologist, geophysicist; b. San Saba, Tex., Oct. 5, 1903; s. William Bee and Cora Virginia (Lee) H.; student Weatherford Coll., 1921-23; B.S., U. Chgo., 1926, M.S., 1927, Ph.D., 1937; m. Miriam Graddy Berry, Nov. 11, 1938. With Shell Oil Co., Houston, 1943-55, asso. dir. research, 1945-51, chief cons. gen. geology, 1951-55, cons. gen. geology Shell Devel. Co., Houston, 1955-63; research geophysicist U. S. Geol. Survey, Washington, 1964—; vis. prof. geology, geophysics Stanford, 1961-63, prof., 1964-68; vis. prof. geography Johns Hopkins, 1968. Chmn. div. earth scis. Nat. Acad. Scis.-NRC, 1963-65; lectr. various univs., socs. Mem. Am. Acad. Arts and Scis., Nat. Acad. Scis., Geol. Soc. Am. (Arthur L. Day medal 1954, pres. 1962), Am. Assn. Petroleum Geologists, Soc. Exploration Geophysicists (hon. life), Am. Geophys. Union, Soc. Petroleum Engrs. Am. Inst. Mining, Metall. Engrs., A.A.A.S. Author: Theory of Ground-Water Motion, 1940; Energy Resources, 1962. Research, numerous publs. on underground fluids, especially water through porous rocks and migration and entrapment of oil and gas, structural geology, especially phys. theory of faulting, mineral resources, philosophy of sci. and edn. Home: 5208 Westwood Dr., Washington 20016. Office: U. S. Geol. Survey, Washington 20242.*

HUBBLE, Edwin Powell, Am. astronomer; b. Marshfield, Mo., Nov. 20, 1889; s. John Powell and Virginia Lee (James) H.; B.Sc., U. Chgo., 1910, Ph.D., 1917; Rhodes scholar from Ill. at Oxford (Eng.) U., 1910-13, B.A. in Jurisprudence, 1912; hon. D.Sc., Oxford, 1934, Princeton, 1936, Brussels, 1937; LL.D., Occidental, 1936; U. Cal., 1949; m. Grace Burke, Feb. 26, 1924. Admitted to Ky. bar, 1913; research work, Yerkes Obs., U. Chgo., 1914-17; astronomer on staff Mt. Wilson Obs., Pasadena, Cal., from 1919; chmn. research com., Mt. Wilson and Palomar Observatories; Hitchcock lectr. U. Cal., 1948; Silliman lectr. Yale, 1935; Rhodes Meml. lectr. Oxford, 1936. Chief ballistician and dir. of Supersonic Wind Tunnels Lab., Ballistic Research Lab., U. S. War Dept., 1942-46. Recipient Barnard medal for sci. service, 1935, Bruce medal, 1938, Franklin medal, 1939, Royal Astron. Soc. medal, 1940, Medal for Merit, 1946. Fellow Royal Astron. Soc., Eng.; mem. Astron. Soc., Astron. Soc. of Pacific, Nat. Acad. Scis., Am. Philos. Soc., Sigma Xi; hon. fellow, Queen's Coll., Oxford, 1948; membre de l'Institut, Academie de France. Author: The Realm of Nebulae, 1936; The Observational Approach to Cosmology, 1937; The Nature of Science. His indication of existence of extragalactic nebulae initiated the study of the universe beyond our galaxy; classified galaxies according to shape, discussed their evolution; discovered that radial velocities of receding galaxies are proportional to their distance (Hubble's law), 1929, which led to an estimation of size of universe. Died San Marino, Cal., Sept. 28, 1953.

HUBBS, Clark, Am. zoologist; b. Ann Arbor, Mich., Mar. 15, 1921; s. Carl Leavitt and Laura (Clark) H.; A.B., U. Mich., 1942; Ph.D. Stanford, 1950; m. Catherine Vickery Symons, Sept. 10, 1949; children—Laura Ellen, John Clark, Ann Frances. Faculty, U. Tex., Austin, 1949—, prof. zoology, 1963—. Mem. Am. Soc. Ichthyologists and Herpetologists, Am. Fisheries Soc., Am. Soc. Naturalists, A.A.A.S., Soc. for Study Evolution, Soc. Systematic Zoologists, Nat. Audubon Soc., Tex. Acad. Sci., Southwestern Assn. Naturalists (pres. 1966—), Ecol. Soc. Am., N.Y. Zool. Soc., Am. Soc. Limnology and Oceanography, Am. Soc. Zoologists, Genetics Soc. Am. Contbr. numerous articles, book revs., abstracts to tech. jours., encys. Exptl. analyses evolutionary relationships fishes; racial variation fishes which adapt fish to environment. Home: 5719 Marilyn Dr., Austin, Tex. 78731.*

HUBEL, David Hunter, neurophysiologist; b. Windsor, Ont., Can., Feb. 27, 1926 (parents Am. citizens); s. Jesse Hervey and Elsie (Hunter) H.; B.Sc., McGill U., 1947, M.D., 1951; M.A. (hon.) Harvard, 1962; m. Shirley Ruth Izzard, June 20, 1953; children—Carl Andrew, Eric David, Paul Matthew. Neurophysiologist, Walter Reed Army Inst. Research, Washington, 1957-58; sr. fellow neurol. scis. group Johns Hopkins Hosp., Balt., 1958-59; with Harvard Med. Sch., Boston, 1959—, prof. neurophysiology, 1965-67, prof. physiology, chmn. dept., 1967—. Fellow Am. Acad. Arts and Scis.; mem. Am. Physiol. Soc. Asso. editor Jour. Neurophysiology, 1965—. Research, publs. on devel. electrodes and instrumentation for brain research, function eye and visual part of brain in higher mammals, especially cats and monkeys, response of brain cells to retinal stimulation with lights, shapes and colors, handling of information originating from eye by cerebral cortex, effects depriving animals of vision or distorting visual input. Home: 59 Carlton Rd., Waban, Mass. 02168. Office: Dept. Physiology, Harvard Med. Sch., 25 Shattuck St., Boston 02115.*

HUBER, Bruno, botanist; b. Hall, Tirol, Austria, Aug. 19, 1899; s. Rudolf and Maria (von Wildauer) H.; Ph.D., U. Vienna; D. honoris causa, Vienna Agrl. Inst.; m. Lucie Gerber, Aug. 14, 1928; children—Waltraud, Klaus, Peter, Rolf. Asst. at Vienna, 1920-25, Greifswalt, 1925-27, Fribourg, 1927-32, to Prof. Darmstadt, 1932-34, Dresde Tharandt 1934-46; dir. Botanic Inst. Silviculture, Munich, 1946—. Mem. Austrian Acad. Scis., German Acad. Natural Sci. (Leopoldina), German Inst. Archaeology. Author: Wärmehaushalt der Pfanzen, 1935; Saftströme der Pflanzen, 1956; Grundzuge der Pflanzenanatomie, 1961, also over 220 articles. Home: Königinstrasse 69/11, Munich 22. Office: Amalienstrasse 52, Munich 13, Germany.

HUBER, Daniel, mathematician; b. 1768; prof. in Basel, Switzerland; developed (independently of Gauss) the method of least quadrature; died 1829.

HUBER, François, Swiss naturalist; b. Geneva, Switzerland, July 2, 1750; s. Jean Huber; student of Saussure, U. Geneva; m. Mlle. Lullin; 1 son, Pierre; spent life in village nr. Paris; although blind, carried out research with aid of wife and an asst.; mem. French Acad. Scis., 1813. Author: Nouvelles observations sur les abeilles, 1792; Memoire sur l'influence de l'air et des diverses substances gazeusis dans la germination des differentes plantes, 1801. Research on bees revealed functions of antennae, expulsion of drones, fertilization of queen bee in flight, other aspects of bee life. Died Pregny, nr. Geneva, Dec. 21, 1831.

HUBER, Gerd Friedrich, German psychiatrist; b. Echterdingen, Germany, Dec. 3, 1921; s. Friedrich and Elsbeth (Hartmann) H.; M.D., U. Heidelberg, 1948. With psychiat. and neurol. clinic U. Heidelberg, 1949, 51-62, faculty, 1957-62, prof., 1962; head physician neurol. clinic, sci. adviser, prof., U. Bonn, 1962 . Mem. Deutsche Gesellschaft für Psychiatric und Nervenheilkunde, Neuroradiologische Arbeitsgemeinschaft, Vereinigung Deutscher Neuropathologen, others. Research, publs. on endogenous psychoses, chronic schizophrenia, delusion. Home: 17 Am. alten Forsthaus, Bonn-Röttgen, Germany.*

HUBER, Otto, Swiss physicist; b. Zürich, Switzerland, Aug. 13, 1916; s. Jakob and Elisabeth Huber; Dr. rer. nat., Fed. High Sch. Tech., 1944; m. Elisabeth Stoller, Apr. 3, 1954; children—Kathrin, Franziska, Markus, Susanna. Private docent Fed. High Sch. Tech., 1951, titular prof., 1952; prof., dir. Inst. Physics, U. Fribourg, Switzerland, 1953—. Mem. phys. socs. Switzerland, Italy, U. S. Research on nuclear reactions, nuclear spectroscopy (to determine and interpret levels of the nucleus); contbd. to surveillance of radioactivity in Switzerland. Home: 8 Beustweg, Zürich, CH-8032, Switzerland. Office: Perolles, Fribourg, CH-1700, Switzerland.*

HUBER, Paul, Austrian surgeon; b. Hall, Tirol, Austria, May 25, 1901; s. Rudolf and Maria (von Wildauer) H.; M.D., U. Vienna; m. Gabrielle Rehwald, Oct. 17, 1932; children—Heinrich, Brigitte, Ulrike, Christopher. Titular prof. U. Vienna, 1945; full prof., dir. Surg. Clinic, U. Innsbruck, 1956. Mem. Internat., German socs. surgeons, Internat. Coll. Surgery, Austrian Soc. Surgery and Traumatology, Austrian Soc. Anesthesiology, Doctors Soc. Vienna. Research and numerous publs. on path. anatomy, surgery. Home: Fischerstrasse 26. Office: Anichstrasse 35, Innsbruck, Austria.

HUBER, Paul, Swiss physicist; b. Rekingen (AG), Switzerland, Oct. 1, 1910; s. Jakob and Elisabeth (Huber) H.; student Tchrs. Coll., Wettingen, Switzerland, 1926-30; Dr.sc.nat., Swiss Inst. Tech., Zurich, 1937; m. Margrit Kellenberger, Feb. 28, 1942; children—Regula, Franziska, Ines. U. lectr. Swiss Inst. Tech., 1941; prof. Kantonales Technikum Winterthur, 1941-42; prof., dir. physics dept. U. Basel (Switzerland), 1942—. Vice pres. Swiss Nat. Sci. Found., 1966—. Mem. Swiss Acad. Natural Scis. (pres.), Am. Phys. Soc., Swiss Acad. Natural Scis. Author: (with P. Frauenfelder) Einführung in die Physik, vol. I, 1951, vol. II, 1958; (with H. B. Willard, L. C. Biedenharn, E. Baumgartner) Fast Neutron Physics, 1963; also numerous articles. Research on neutron physics, nuclear structure, sources of polarized protons and deutrons. Home: 13, Hungerbachweg, Riehen/Basel, Switzerland.*

HUBER, Walter Anton, Swiss zoologist; b. Olten, Switzerland, Apr. 1, 1917; s. Adolf and Maria (Steiner) H.; Dr.phil.nat., U. Bern (Switzerland), 1946; Diplôme d'Etudes supérieures, U. Strasbourg (France), 1949; m. Dora Lilli Roth, Oct. 7, 1949; children—Marianne Elisabeth, Dorothée, Christine. Asst., Zoologisches Inst., U. Bern, 1945-47; research staff Centre Nat. de la Recherche Scientifique, Lab. Embryology, U. Strasbourg, 1947-50; curator Naturhistorisches Mus., Bern, 1951-63, dir., 1964—; asso. prof. U. Bern, 1965—. Expert museology Africa UNESCO, 1964—. Mem. Schweizerische Zoologische Gesellschaft, Schweizerische Naturforschende Gesellschaft, Schweizerische Entomologische Gesellschaft. Author: Das Alpenmurmeltier, 1965; also articles. Research on topogenesis of chicken head; biometrical studies of mammalian skulls, mammalian tails; electron microscope studies of peritrophic membranes of insects; genital cycles of roe deer and chamois. Home: 65 Brunnadernstrasse. Office: 15 Bernastrasse, Bern, Switzerland.*

HUBERT, Lester Ferris, Am. meteorologist; b. Granada, Minn., Oct. 27, 1913; s. Paul Charles and Viola (Hoffman) H.; B.S., U. Chgo., 1947, M.S., 1948; m. Dorothy Green, Nov. 5, 1938; children—Loralee (Mrs. James Woodward), Linda Katherine. Research meteorologist U. S. Weather Bur., 1948-59, head synoptic br. Meteorol. Satellite Lab., Suitland, Md., 1959—. Fellow Washington Acad. Scis.; mem. Am. Meteorol. Soc. Research, publs. on structure and motion hurricanes, interpretation meteorol. satellite data. Home: 4704 Mangum Rd., College Park, Md. 20740. Office: Meteorol. Satellite Lab., Environmental Scis. Services Adminstrn., Washington 20233.*

HÜBNER, Jacob, German entomologist; b. Augsburg, Germany, June 20, 1761. Author: Sammlung europäischer Schmetterlinge, 1805-24; Geschichte europai scher Schmetterlinge, 1806-18; Sammlung exotischer Schmetterlinge, 1806-24; Tentamen determinationis digestionis atque denominationis singularum . . . , 1806; Verzeichniss bekannter Schmetterlinge, 1816. First great lepidopterist; gave sound definitions and classification of genera, named many species of Lepidoptera. Died Sept. 13, 1826.

HUBRECHT, Ambrosius Arnold Wilhelm, Dutch zoologist; b. Rotterdam, Netherlands, 1853; prof., Utrecht, Netherlands. Author: Die Phylogenese des Amnions, 1895; Early Otogenic Phenomena in Mammals, 1908. Research in history of devel. of vertebrates. Died 1915.

HUCHARD, Henri, French physician; b. Auxon, France, 1844; certified physician, 1828. Author: (with Axenfield) treatise on neuroses, 1883; Traité clinique des maladies du coeur et des vaisseaux, 1893; Consultations médicales, 1900-06. Founder Jour. des Practiciens, 1896. Research in heart and blood diseases, therapeutics. Died Clamart, France, 1910.

HUCKEL, Walter Karl Friedrich, German chemist; b. Charlottenburg, Feb. 18, 1895; s. Armand and Marie (Maier) H.; Ph.D. in Chemistry, U. Göttingen; D. honoris causa, U. Rennes; m. Hildegard Schimpf, Dec. 28, 1935; children—Dietlind, Ulrich, Konrad. Instr., 1923; asso. prof. at Fribourg/Br., 1927; full prof. chemistry at Wroclaw (Breslau), 1935; full prof. pharm. chemistry U. Tübingen, 1948. Recipient Lavoisier medal, Stas medal. Mem. Heidelberg, Göttingen, Helsinki acads. sci., Acad. Leopoldina Naturalists (Halle), Soc. German Chemists, Soc. Pharmacists, Pharm. History Soc., Soc. Finnish Chemists (hon.), Chem. Soc. Belgium. Author: Theoretische Grundlagen organischer Chemie; Anorganische Strukturchemie. Home: Mohlstrasse 78. Office: Wilhelmstrasse 27, Tübingen, Germany.

HUDDART, Joseph, English hydrographer; b. Cumberland, Eng., Jan. 11, 1741; m., 1762; 5 sons; went to sea for fish-curing bus.; entered service E. India Co., 1771; apptd. comdr. ship Royal Adm., 1778; made 4 voyages to East; ret., 1788. Fellow Royal Soc., 1791. Gave 1st reliable description of color blindness (in letter to Joseph Priestley), 1777; charted Sumatra and Indian coast from Bombay to Godavery; surveyed Hebrides, 1788-91; inventor improved method of rope manufacture. Died London, Aug. 19, 1816.

HUDDE, Jan, Dutch mathematician; b. Amsterdam, 1628; pupil of van Schooten; burgomaster of Amsterdam. Author: De reductione aequationum et de maximus et minimus (letter), 1713. Improved (with René François de Sluse) theory of maxima and minima, also Descartes' and Fermat's methods of drawing tangents; 1st to use 3 variables in analytic geometry; developed method for finding equal roots; 1st to use letter as symbol for negative quantity. Died Amsterdam, Apr. 16, 1704.

HUDLICKA, Olga, Czechoslovak. physiologist; b. Prelouc, Czechoslovakia, July 11, 1926; s. Ing. Jaroslav and Marie (Babackova) H.; M.D., Charles U., Prague, 1950; Ph.D., Inst. Physiology, Prague, 1955; m. Andrej Klein, June 24, 1950; children—Olga, Pavel. With Inst. Exptl. Pathology, Prague, 1948-49, Inst. Physiology, Czechoslovak Acad. Scis., Prague, 1950; Vis. prof. Karolinska Inst., Stockholm, Sweden, 1960, Duke Med. Center, Durham, N.C., 1964. Mem. Czechoslovak Med. Soc. J. E. Purkyne, Czechoslovak Physiol. Soc. (sec. 1960—). Author: Denervated Muscle, 1962; Regulation of Skeletal Muscle Circulation, 1967; also articles. Research on nervous regulation of blood flow in skeletal muscle, relations between muscle metabolism and blood flow, vasomotor changes during muscle atrophy. Home: 50 Cukrovarnicka, Prague 6. Office: 1083 Budejovicka, Prague 4, Czechoslovakia.*

HUDSON, Claude Silbert, Am. chemist; b. Atlanta, Jan. 26, 1881; s, William James and Maude Celestia (Wilson) H.; B.S., Princeton, 1901, M.S., 1902, Ph.D., 1907; D.Sc. (hon.), 1947. Research asst. in phys. chemistry, Mass. Inst. Tech., 1903-04; instr. physics, Princeton, 1904-05, U. Ill., 1905-07; assistant physicist, U. S. Geol. Survey, 1907-08; assistant chemist, Bureau of Chemistry, U. S. Dept. Agr. and later chief of Carbohydrate Laboratory, 1908-19; chemist Bureau of Standards, U. S. Dept. Commerce, 1923-28; professor chemistry, U.S.P.H.S., 1928-51; mem. exec. com., editor Advance in Carbohydrate Chemistry. Councillor Internat. Union of Chemistry, 1930. Mem. Am. Com. on Organic Chem. Nomenclature, 1912. Mem. Am. Chem. Soc. (Nichols medal, 1916; Gibbs medal, 1929; Hillebrand prize, 1930; Richards medal, 1940; Borden award, 1941; chmn. organic div. of same soc. 1933), Nat. Acad. Sciences, German Acad. Natural Scientists (Halle). Asso. editor: Journal of Am. Chem. Soc., 1938. Author: Collected Papers of C. S. Hudson, 2 volumes, 1946, 1948. Recipient Cresson medal, Franklin Institute, 1942, Grand Prize by Sugar Research Found., Inc., 1950, First Fed. Security Agency Distinguished Service Award, 1950, Distinguished Service medal, U.S.P.H.S., 1951. Died Dec. 27, 1952.

HUDSON, Donald Edwin, Am. physicist; b. Butte, Mont., July 17, 1921; s. Edwin Emory and Mae (Page) H.; B.Physics, U. Minn., 1942; Ph.D., Cornell U., 1950; m. Hazel Gruner, Mar. 21, 1944; children—Howard, Sharon, Toren, Darrel. Jr. scientist Los Alamos Sci. Lab., 1944-46; asst. Cornell U., Ithaca, N.Y., 1946-49; instr. Princeton, 1949-51; faculty Ia. State U., Ames, 1951-64, asso. prof., 1957-64; prof., chmn. physics dept. Cal. State Coll., Los Angeles, 1964—. Mem. Am. Assn. Physics Tchrs., Phi Beta Kappa, Sigma Xi. Research, publs. on drift velocities of electrons in gases, high energy showers in cosmic rays, grant air showers in cosmic rays, cohesion of rare-earth elements, cohesion of intermetallic compounds, properties of charge carriers in boron, imperfection photo-conductivity in diamond, work function of tungsten crystal planes, elec. conductivity of diamond. Home: 1565 Washburn Rd., Pasadena, Cal. 91105.*

HUDSON, Edward, dentist; b. County Wexford, Ireland, Oct. 1772; s. Capt. Henry Edward and Jane (de Tracey) H.; studied dentistry under Dr. Hudson at Trinity Coll., Dublin, Ireland; m. Maria Bridget Bryne, Apr. 1804; m. 2d, Maria Elizabeth Becker; m. 3d, Marie Mackie; 8 children. Became involved with Thomas More and Robert Emmet (leaders of Irish Revolution) while at Trinity Coll.; imprisoned by British in Ft. George, Scotland, 1798-1802; exiled to Holland at conclusion of Treaty of Amiens, 1802; came to Am., settled in Phila., 1803; practiced dentistry, Phila., 1810-33; one of 1st to perform operation removing dental pulp and filling root of tooth with gold foil. Died Jan. 3, 1833.

HUDSON, G(eorge) Donald, geographer; b. Osaka, Japan, Apr. 26, 1897 (parents Am. citizens); s. George Gary and Delia A. (Herndon) H.; Ph.B., U. Chgo., 1925, M.A., 1926, Ph.D., 1934; m. Nellie Ruckelshausen, Mar. 20, 1926; children—Donald G., James W., Glen G. Dir. Middle Sch., Am. U., Beirut, Lebanon, 1926-29; acting dean Blackburn Coll., Carlinville, Ill., 1929-30; asso. prof. Wis. State Coll., Whitewater, 1933-34; chief land planning div. TVA, Knoxville, Tenn., 1934-39; prof. Northwestern U., Evanston, Ill. 1939-51, chmn. dept. geography, 1945-51; prof. U. Wash., Seattle, 1951-67, chmn. dept. geography, 1951-63. Mem. Assn. Am. Geographers (past pres.), Chgo. Geog. Soc. (past pres.), Sigma Xi. Contbg. author: Land Classification in the United States, 1941. Geog. editor: Ency. Brit. World Atlas, 1942-66. Publs. on devel. of field research techniques, primarily in classification of land resources and land uses; practical application of geog. research to land use planning; devel. curricula in geog. edn. and profl. tng. on grad. level. Home: 6024 Princeton Av. N.E., Seattle 98115.*

HUDSON, George, English anatomist; b. Edenfield, Eng., Aug. 10, 1924; s. George and Edith (Bennett) H.; B.Sc. with Honors in Physiology, Victoria U., Manchester, Eng., 1946, M.B., Ch.B., 1949, M.Sc., 1962; M.D., Bristol (Eng.) U., 1959, D.Sc., 1966; m. Mary Patricia Hibbert, Apr. 14, 1955; 1 dau., Elisabeth Helen. House surgeon Manchester Royal Infirmary, 1949-50; faculty Bristol U., 1950-51, 53—, reader anatomy, 1963—, pre-clin. tutor, 1962-64, preclin. dean, 1964—; vis. asst. prof. U. Minn., 1959-60. Mem. Anat. Soc. Gt. Britain and Ireland, Brit. Soc. for Hematology, Royal Micros. Soc., Brit. Med. Assn., Chartered Soc. Physiotherapy (examiner). Contbg. author: Bone Marrow Reactions, 1966; also articles. Research on hemopoiesis including estimates of total vol. of bone marrow in guinea-pig, cat and human foetus; quantitative studies on mobilzn. of marrow res. of eosinophil leucocytes in fgn. protein response and its inhibition by cortical hormones and ACTH; electron microscope appearances of granules of eosinophil leucocytes modified by electron stains; demonstrated (with J. M. Yoffey) lymphocytes passing through wall of marrow sinusoids. Home: 14 Cotham Lawn Rd., Bristol 6, Eng.*

HUDSON, N(oel) Paul, Am. virologist; b. Lincoln, Ill., Jan. 9, 1895; s. George Gary and Delia (Herndon) H.; A.B., Millikin U., 1917, Sc.D. (hon.), 1951; Ph.D. in Bacteriology, U. Chgo., 1923; M.D. with honors, Harvard, 1925; Sc.D. (hon.), Ohio State U., 1965; m. Emily Madlin, June 11, 1925; children—Robert Paul, Stephen Perry. Faculty, U. Chgo., 1922-24, 30-35; researcher Rockefeller Found., Nigeria, W.Africa, also Rockefeller Inst., 1927-30; faculty Ohio State U., Columbus, 1935—, dean Grad. Sch., 1946-55, asst. dean Coll. Medicine, 1956-61, prof. emeritus, microbiology, 1961—. Cons. Surgeon Gen. U. S. Army, 1943-45, So. region Edn. Bd., 1952-55. Fellow A.A.A.S., Am. Pub. Health Assn., Ohio Acad. Sci., Am. Acad. Microbiology (charter); mem. Am. Soc. for Microbiology, Am. Soc. Tropical Medicine and Hygiene (pres. 1942-43), Assn. Grad. Schs. (pres. 1954-55), Ohio Acad. Med. History (pres. 1963-64), Phi Beta Kappa, Sigma Xi, Alpha Omega Alpha. Editor, contbr. Ohio State University College of Medicine History, 1934-58, 1961. Research, publs. on viral agents of yellow fever, polio, vaccinia, influenza, rat-bite fever, fetal susceptibility to viruses and bacteria. Home: Wesley Manor, Jacksonville, Fla. 32223.*

HUDSON, Robert Francis, English chemist; b. Lincoln, Eng., Dec. 15, 1922; s. John Frederick and Ethel (Oldfield) H.; B.Sc., Royal Coll. Sci., 1943, Ph.D., D.I.C., 1945; m. Monica Ashton Stray, Aug. 3, 1945; children—John Martin Edward, Sarah Elizabeth, Mary Alexandra. Asst. lectr. Royal Coll. Scis., 1945-47; lectr. Queen Mary Coll. London, 1947-59; group dir. Cyanamid European Research Inst., Geneva, Switzerland, 1960-66; prof. organic chemistry U. Kent, Canterbury, Eng., 1966—; cons. Wolsey Ltd., Leicester, 1945-50. Mem. Royal Inst. Chemistry, Chem. Soc. London (council 1967). Author: Wool—Its Physics and Chemistry, 1954; Structure and Mechanism in Organophosphorus Chemistry, 1965; also numerous articles. Investigations of chem. reactivity, in particular, alkylation, acylation, phosphorylation and elimination; theoretical interpretation of organic reactions; application of phys. and theoretical principles to reactions of organophosphorus compounds. Home: 32 Puckle Lane, Canterbury, Kent, Eng.*

HUDSON, William, English botanist; b. Kendal, Eng., circa 1730; apprentice to apothecary, London; resident sub-librarian Brit. Mus., 1757-58. Genus Hudsonia named in his honor. Fellow Royal Soc., 1761; original mem. Linnean Soc. Author: Flora Anglica (established Linnaean botany in Eng.), 1762; Praefectus horte, 1765-71. Adapted (more accurately than John Hill) Linnaean nomenclature to plants described by Ray; discovered Trochus terrestris. Died May 23, 1793.

HUDSON, William Henry, naturalist, ornithologist; b. Quilmes, Argentina, 1841; s. Daniel and Catherine (Kemble) H.; m. Emily Wingrave, 1876. Spent youth in Argentina; went to Eng., 1869, naturalized, 1900; writer natural history of S. Am. Author: The Purple Land that England Lost, 1885; (with N. P. L. Sclater) Argentine Ornithology, 2 vols., 1888-89; The Naturalist in La Plata, 1892; Birds in a Village, 1893; British Birds, 1895; Hampshire Days, 1903; Green Mansions, 1904; Afoot in England, 1909; A Shepherd's Life, 1910; Adventures Among Birds, 1913; Far Away and Long Ago, 1918; The Book of a Naturalist, 1919; A Little Boy Lost, 1920; An Aldthorn, 1920; A Hind in Richmond Park, 1922. A bird sanctuary containing Epstein's Rima decoration was erected in his memory in Hyde Park, London, 1925. Died 1922.

HUDSPETH, Allen Sherrill, Am. physician; b. Statesville, N.C., Nov. 15, 1929; s. Nelson L. and Hattie E. (Brandon) H.; student Wake Forest Coll., 1947-49; M.D., Bowman Gray Sch. Medicine, 1953, postgrad.; m. Beth Schaufelberger, Sept. 11, 1954; children—Dudley, Grant, Deborah, Mark, Glenn. Asst. surgery Bowman Gray Sch. Medicine, Winston Salem, N.C., 1956-61, research fellow, 1962-63; faculty, 1963—, asst. prof., 1964—; practice surgery, Winston Salem, 1963—; staff N.C. Bapt. Forsyth Meml. hosps., Winston-Salem. NIH Spl. research fellow, 1962-63. Fellow A.C.S.; mem. N.C. State, Forsyth County med. socs., Am. Assn. for Thoracic Surgery, So. Thoracic Surg. Assn. Publs. on new procedures in thoracic and cardiovascular surgery, care and mgmt. patients in shock and post-operatively, studies pulmonary embolism and infarction. Home: 411 N. Stratford Rd., N., Winston-Salem 27104. Office: 300 S. Hawthorne Rd., Winston-Salem, N.C. 27103.*

HUDSPETH, Emmett Leroy, Am. physicist; b. Denton, Tex, Dec. 3, 1916; s. Junia Evans and Ethel (Burns) H.; student Arlington State Coll., 1933-34; B.A., Rice U., 1937, M.A., 1938, Ph.D., 1940; m. Mary Alice Barnes, Dec. 2, 1944; children—John, Philip, Anne, Paul. Fellow in physics Rice U., 1937-40; fellow Bartol Research Found., Swarthmore, Pa., 1940-41, asst. dir., 1946-50; staff mem. Radiation Lab., Mass. Inst. Tech., 1941-45; prof. physics, dir. Nuclear Physics Lab., U. Texas, Austin, 1950—, mem. sci. adv. com. Radiobiol. Lab., U. Tex. and USAF, 1954-58. Cons., USN, 1943-45; adv. Sec. War, 1942; sci. adv. com. USAF, 1954-58. Bd. dirs. Tex. Nuclear Corp., Austin, Nuclear-Chgo. Corp. Fellow Am. Phys. Soc., A.A.A.S.; mem. Phi Beta Kappa, Sigma Xi. Contbr. articles on nuclear physics in tech. jours. Patentee, radar. Home: 6104 Janey Dr., Austin, Tex. 78731.*

HUEBER, Eduard Franzjosef, Austrian physician; b. Sept. 9, 1907; s. Eduard and Margareth (Mayr-Millesi) H.; Med. Dr., Vienna (Austria) Sch. Medicine, 1934; m. Marie Kalat, Mar. 21, 1939; children—Pamela, Edward Franz, William. Research fellow Physiol. Lab., Cambridge, Eng., 1934; research asst. Pharmacological Lab., Vienna, 1935-37; staff 1 Med. dept. Vienna Sch. Medicine, 1937—, now sr. cons. physician; postgrad. cardiac unit Mass. Gen. Hosp., Boston, 1946-52; asso. prof. internal medicine Vienna Sch. Medicine, 1958—. Mem. Soc. Internal Medicine, Internat. Soc. for Prophylactic Medicine and Social Hygiene (pres.). Research, numerous publs. on internal medicine, especially heart and pulmonary diseases, their pathogenesis, symptomatology and their treatment. Home: 8 Löwelstrasse, Vienna 1010. Office: 23 Spitalgasse, 1 Med. Univ. Klinik, Vienna 1090, Austria.*

HUEBNER, Robert Joseph, Am. virologist; b. Cin., Feb. 23, 1914; s. Joseph F. and Philomena (Brickner) H.; student Xavier U., 1932-35; student U. Cin., 1937-38, LL.D., 1965; M.D., St. Louis U., 1942; m. Grace Berdine Hoffmann, Sept. 29, 1939; children—Elizabeth, F. Katherine, Geraldine, R. James, Virginia, Roberta Sue, Edward, Mary Louise, R. Daniel. With NIH, 1944—, chief Lab. Infectious Diseases, Nat. Inst. Allergy and Infectious Diseases, Bethesda, Md., 1956—. Cons. orgns.; lectr. various colls.; vis. lectr. microbiology Harvard, 1957—. Recipient Bailey K. Ashford award Am. Soc. Tropical Medicine, 1949; certificate of merit, Am. Vet. Med. Assn., 1950; citation Variety Children's Hosp., 1960; Pasteur medal Pasteur Institute (Paris, France), 1965. Fellow Am. Pub. Health Assn., A.M.A. (certificate of Merit 1947), N.Y. Acad. Scis.; mem. Nat. Acad. Scis., A.A.A.S., Am. Assn. Immunologists, Am. Epidemiological Soc., Am. Thoracic Soc., Soc. Exptl. Biology and Medicine, Washington Acad. Scis. (award in biol. scis. 1949), Am. Acad. Microbiology, Am. Assn. Cancer Research, Sigma Xi, Alpha Omega Alpha. Studies, publs. on etiology and epidemiology of several viral and rickettsial diseases; transmission of animal cancer viruses. Home: Hidden Hills Farm, Ijamsville, Md. 21754. Office: Dept. Health, Edn. and Welfare, Pub. Health Service, NIH, Bethesda, Md. 20014.*

HUEBSCHMAN, Eugene Carl, Am. elec. engr.; b. Evanston, Ind., Oct. 31, 1919; s. Louis L. and Alvine (Englebrecht) H.; B.S., Concordia Coll., 1941; M.S., Purdue U., 1946; postgrad. U. Chgo.; Ph.D. in Physics, U. Tex., 1957; m. Edna Arledt, Mar. 7, 1948; children—Donald, Michael, Ruth, JoAnna. Prof., Concordia Coll., River Forest, Ill., 1943-46; mgr. analysis sect. Inertial Guidance Test Facility, Holloman AFB, N.M., 1957-59; tech. adviser Weapons Guidance Lab., Wright-Patterson AFB, O., 1959-60; mgr. advanced analysis advanced planning PanAm World Airways, Cape Kennedy, Fla., 1960-63; chief scientist Teledyne, Inc., Alexandria, Va., 1965—; now with U. Tenn. Space Inst., Tullahoma. Chmn. grad. physics dept. Brevard Engrng. Coll., Melbourne, Fla., 1960—; tchr. grad. physics, math. elec. engrng. Purdle U., 1942-44, U. Tex., 1948-56, U. N.M., 1956-59, San Diego State Coll., 1960, U. Cal. at San Diego 1960. Named Outstanding Inventor, Wright (Cal.) Air Devel. Center, 1960. Mem. Am. Phys. Soc., Am. Inst. Aeros. and Astronautics, Luth. Acad. for Scholarship, Am. Astronautical Soc., Marine Tech. Soc., Sigma Psi, Beta Sigma Psi. Patentee, magnetic torsion accelerometer, solid state accelerometer, continuous flow pump, hall accelerometer, light converter. Home: 1100 Bragg Circle, Tullahoma, Tenn.*

HUECK, Alexander Friedrich, anatomist; b. Reval, Germany, Dec. 1802; studied in Berlin, Munich, Göttingen, Paris; M.D., U. Dorpat (Estonia), 1826; became prosector, anat. inst. U. Dorpat, 1830, prof. anatomy, 1833. Author: Lehrbuch der anatomie des menschen, 1833-35; Ueber das Studium der Anatomie, 1833; Gerüste der Anatomie; Eine Uebersicht der vorzüglichsten Theile des menschlichen Körpers, 1833; De cranüs Estonum commentatio anthropologica, 1838; Des Sehen, seinen äusseren Process nach entwickelt, 1830; Die apendrehung des Auges, 1838; Die Bewegung der Krystallinse, 1840. Research in anatomy and physiology, paleontology, archeology; described pectinate ligament of iris, 1828. Died July 28, 1842.

HUEPER, Wilhelm Carl, physician; b. Schwerin, Germany, Nov. 4, 1894; s. Wilhelm E. and Karoline (Dornemann) H.; student U. Marburg, 1913-14,

U. Rostock, 1914, 18-19; M.D., U. Kiel, 1920; m. Martha Sennhenn, July 18, 1924; 1 son, Klaus Wilhelm. Came to U. S., 1923, naturalized, 1929. Asst. dir. Warner Inst. Therapeutic Research, N.Y.C., 1938-48; chief environmental cancer sect. Nat. Cancer Inst., Bethesda, Md., 1948-64. Cons. U. S. Dept. Labor, 1954-58; chmn. com. cancer prevention Unio Internationale Contra Cancrum, 1954-66. Recipient Anne Frankel Rosenthal Meml. award, 1959; UN award for outstanding research in causes of cancer, 1962; Golden Plate award Acad. Achievement, 1962; Modern Medicine award, 1963; Humanitarian award Nat. Health Assn., 1965. Diplomate Am. Bd. Pathology, Am. Bd. Preventive Medicine. Author: Occupational Tumors and Allied Diseases, 1942; Berufskrebse, 1964; (with Walter Conway) Chemical Carcinogens and Cancers, 1965; Occupational and Environmental Cancers of Respiratory System, 1966; also numerous articles. Research in occupational, indsl., environmental causes of cancer in man and animals, causes and causative mechanisms of arteriosclerosis, pathology of diseases in man and animals induced by macromolecular substances of natural and synthetic origin. Home: 4558 McGregor Blvd., Ft. Myers, Fla. 33901.*

HUEPPE, Ferdinand, German bacteriologist; b. 1852; prof. German U., Prague, Czechoslovakia, from 1889; research and publs. on bacteriology, vegetarianism, alcoholism, hygiene, phys. tng.; studied use of related but benign bacteria in inoculation against pathogenic bacteria; demonstrated that certain bacteria were polymorphic bacteria rather than specific pathogens; his research on infectious diseases led to simplification of disinfection. Died 1938.

HUET, Pierre-Daniel, French mathematician, anatomist; b. Caen, France, Feb. 8, 1630; given Jesuit edn., Caen; tutor to dauphin (son of Louis XIV), 1670; became bishop of Soissons, 1685, of Avranches, 1691; ret. to Jesuits in Paris, 1701. Mem. French Acad. Author: Traité de la traduction, 1661; Demonstratio evangelica, 1679; Origines de Caen, 1706; Histoire de la commerce et de la navigation des anciens, 1710; Censura philosophiae Cartesianae; Faiblesse de l'esprit humain; Essai sur l'origine des romans. Editor Commentaria in Sacram Scripturam (manuscript of Origin discovered on visit to Swedish Ct. in 1652), 1668. Noted as a mathematician, sceptic, Hellenist, Hebraist; performed interesting anat. dissections of eye. Died Paris, Jan. 26, 1721.

HUFELAND, Christoph Wilhelm, German physician; b. Langensalza, Germany, Aug. 12, 1762; ed., Jena and Göttingen, Germany; physician, Weimar, Germany, 1783-93; prof. pathology, Jena, and Berlin, Germany; physician of German classicists in Weimar. Mem. French Acad. Scis. Author book on long life. Research on smallpox vaccination. Died Berlin, Aug. 25, 1836.

HUFF, Clay G., Am. parasitologist; b. Cory, Ind., Sept. 10, 1900; s. Howard and Estella (Coble) H.; A.B., Southwestern Coll., Kan., 1924; Sc.D., Johns Hopkins, 1927; m. Florence May Clark, Sept. 1, 1927; children—Eskin, Elaine (Mrs. Jay Samstag). Asso. prof. U. Ga., 1927-28; NRC fellow Harvard Med. Sch., 1928-30; faculty U. Chgo., 1930-47, prof. parasitology, 1941-47; head parasitology dept. Naval Med. Research Inst., Bethesda, Md., 1947—. Recipient Navy Distinguished Civilian Service award, 1958, Dept. Def. Distinguished Civilian Service award, 1958. Mem. Nat. Malaria Soc. (past v.p.), Am. Soc. Parasitologists (past pres.), Am. Soc. Tropical Medicine and Hygiene (past pres.), Am. Acad. Tropical Medicine (Theobald Smith award 1947), A.A.A.S., Am. Soc. Protozoologists, Am. Soc. Parasitologists, Tissue Culture Assn., Am. Acad. Microbiology. Author: (with Hegner, Root, Augustine) Parasitology, 1938; A Manual of Medical Parasitology, 1943; also numerous articles. Discovered inheritance susceptibility mosquitoes to malarial parasites; described first malarial parasite with tissue stages, 1st complete pre-erythrocytic cycle malarial parasite; discovered better culture methods for tissue stages malaria. Home: 6510 River Rd., Bethesda 20034. Office: Naval Med. Research Inst., Bethesda, Md. 20014.*

HUFF, Jesse William, Am. biochemist; b. Westmoreland County, Pa., Dec. 8, 1916; s. Robert F. and Ethel (Peters) H.; B.S. in Chemistry, U. Pitts., 1940; Ph.D. in Biochemistry, Duke, 1945; m. Catherine Bedsworth, Feb. 19, 1944; children—Karen, Larry, Debra, David, Randy. Nutrition Found. fellow Duke Med. Sch., 1942-46, instr. 1945-46, biochemist OSRD, 1944-46; biochemist dept. biochemistry Sharp & Dohme, Glenolden, Pa., 1946-49, asst. dir. biochem. research, 1949-50, dir. biochem. research, West Point, Pa., 1950-58; dir. biochemistry Merck Inst. Therapeutic Research, Rahway, N.J., 1959—. Mem. Am. Chem. Soc., Am. Fedn. Biol. Chemists, N.Y. Acad. Sci. Research, numerous publs. on metabolism vitamins, isolation and identification unknown biol. compounds, metabolism lipids and nucleic acids, nutrition exptl. animals, exptl. atherosclerosis and heart disease. Home: 466 Channing Av., Westfield, N.J. 07090. Office: Merck & Co., Rahway, N.J. 07065.*

HUFF, Rex Lamar, Am. physician; b. Bremen, Ind., June 2, 1918; s. Mason Wade and Eva (Shearer) H.; B.S., Purdue U., 1941; M.D., Ind. U., 1944; m. Marie Evelyn Shearer, June 9, 1940; children—Cynthia Ann, Elizabeth Eleanor, Shelley Diane. Instr. pharmacology Ind. U. Sch. Medicine, Indpls., 1943-45; Permanente Found. fellow med. physics Donner Lab., U. Cal. at Berkeley, 1947-48, AEC fellow, 1948-50, research asso., lectr. div. med. physics, co-dir. health physics program Radiation Lab., Donner Lab., 1950-52; faculty U. Wash. Sch. Medicine, Seattle, 1952-59, asso. prof. medicine, 1956-59; dir. radioisotope unit VA Hosp., 1952-59; practice medicine specializing in internal medicine, Elizabethtown, Ky., 1959—. Fellow A.C.P., N.Y. Acad. Sciences; mem. Soc. Nuclear Medicine (charter), Soc. for Clin. Investigation, Soc. for Exptl. Biology, Endocrine Soc. Contbr. articles to jours., numerous chpts. in books, revs. Research on metabolism of iron, original ferrokinetics studies with radioiron, effects changes in barometric pressure on red cell prodn., effects of present natural ionizing radiation on cell growth, effect of decreasing amount, high energy radiation effects in mammals, protective mechanism for high energy ionizing radiation, analogue simulations and solutions of hemodynamic and other clin. problems. Address: 211 N. Mulberry St., Elizabethtown, Ky. 42701.*

HUFFAKER, Carl Barton, Am. entomologist, ecologist; b. Monticello, Ky., Sept. 30, 1914; s. DeWitt Talmadge and Elizabeth (Wray) H.; B.A., U. Tenn., 1938, M.S., 1939; postgrad. U. N.C.; Ph.D., Ohio State U., 1942; m. Saralyn Knight, June 16, 1936; children—Ronald Wray, Harry Knight, Carolyn Sue (Mrs. John Noack), Thomas Keith. Asst. entomologist U. Del., 1941-43; entomologist health and sanitation div. Inst. Inter-Am. Affairs, Colombia, Haiti, 1943-46; staff expt. sta. U. Cal. at Berkeley, 1946—, entomologist, 1953-63, prof. entomology, entomologist, 1963—. John Simon Guggenheim Found. fellow, 1963-64; recipient NIH awards, 1957-67. Mem. Entomol. Soc. Am., Ecol. Soc. Am., Pacific Coast Entomol. Soc., A.A.A.S., Am. Inst. Biol. Scis., Japanese Soc. Population Ecology, Internat. Biol. Program. Research, numerous publs. on role of enemies, intraspecific competition and phys. conditions on population fluctuation and mean density, population balance; predator-prey host-parasite systems, practical biol. control of Klamath weed, olive scale and cyclamen mite. Home: 3540 Springhill Rd., Lafayette, Cal. 94549. Office: 1050 San Pablo Av., Albany, Cal. 94706.*

HUFFMAN, Eugene Harvey, Am. chemist; b. Lawrenceburg, Ind., Dec. 5, 1905; s. Harvey Russell and Atta (Smashey) H.; A.B., U. Colo., 1927; M.S., U. Wash., 1929; Ph.D., U. Ill., 1937; m. Elizabeth Allampress Brecher, July 13, 1961. Instr., U. Wis., 1937-38; asst. prof. Coll. Puget Sound., 1938-40; faculty Kan. State Coll., 1940-41; chemist U. S. Bur. Mines, 1941-43; chemist, sr. staff Lawrence Radiation Lab., U. Cal. at Berkeley, 1943—. Mem. Am. Chem. Soc., A.A.A.S., Sigma Xi, Phi Beta Kappa, Phi Kappa Phi, Alpha Chi Sigma, Phi Lambda Upsilon. Editorial bd. Analytical Chemistry of the Manhattan Project, 1950. Research, publs. on inorganic complex compounds, thermodynamic properties of compounds metall. interest, anion exchange separations, chelation-extraction separations. Patentee zirconium and hafnium separation process, rare earth separation by anion exchange. Home: 70 Panoramic Way, Berkeley, Cal. 94704.*

HUFFMAN, Max Niel, Am. chemist; b. Norwood, Mo., Apr. 1, 1911; s. Francis Marion and Jessie (Shepard) H.; A.B., U. Mo., 1937; Ph.D., St. Louis U., 1941; NRC fellow in med. scis., Columbia, 1941-42; m. Edna Taborn, Dec. 8, 1939; children—John Chavis, Thomas Niel, William Francis. Head sect. medicinal chemistry Okla. Med. Research Found., 1950-57; prof. research biochemistry U. Okla. Sch. Medicine, 1950-63; dir. research Lasdon Found. Research Inst. Chemotherapy, 1957-64; distinguished lectr. Colo. Coll., 1958-64; prof. physiology, pharmacology Creighton U. Sch. Medicine, Omaha, 1964—. Fellow A.A.A.S.; mem. Am. Soc. Biol. Chemists, Am. Chem. Soc., Soc. Exptl. Biology and Medicine, Endocrine Soc., Phi Beta Kappa, Sigma Xi, Delta Epsilon, Alpha Sigma Nu. Contbr. numerous articles to profl. jours. Isolated (with E. A. Doisy) ovarian hormone in human; opened 16-keto steroid field synthetically, synthesized many natural estrogens, some of which are on world drug market. Home: 9187 Boyd St., Omaha 68134. Office: 629 N. 27th St., Omaha 68131.*

HÜFNER, Carl Gustav, see von Hüfner, Carl Gustav.

HUG, Otto F. J., physician, radiobiologist; b. Marktzeuln, July 26, 1913; s. Alfons and Hedwig (Bertele) H.; M.D., U. Frankfort; m. Else Stock, May 15, 1954; children—Dorothea, Sabine, Eugen. Asst., Path. Inst., U. Berlin; asst. Max-Planck Inst. Biophysics, Frankfort, 1948-56; prof. biology, 1956-57; sr. scientist IAEA, 1957-58; prof. radiobiology, dir. Inst. Radiology and Biology, U. Munich, 1959—, dir. Inst. for Research for Protection against Radiation, Neuherberg-Munich, 1959. Mem. Soc. Pathologists, Röntgen Soc., Radiation Research Soc. Research and numerous publs. on pathology, supersonic waves, electron microscopy, radiobiology, biophysics. Address: Bavariaring 19, Munich, Germany.

HUGGETT, Richard William, Jr., Am. physicist; b. Frazee, Minn., July 2, 1930; s. Richard William and Alice (Hammer) H.; B.A., Concordia Coll., 1951; M.S., Ind. U., 1953, Ph.D., 1957; m. Ethel Caroline Furbay, Aug. 16, 1952; children—Laurie, James, Heather, Ariana. Faculty, La. State U., Baton Rouge, 1957—, prof. physics, 1965—, dir. ultra high energy physics research group, 1960—. Mem. Am. Phys. Soc., A.A.A.S., Am. Assn. U. Profs. Research, publs. on existence and properties Theta Meson, properties ultra high energy nuclear interactions. Home: 955 Verdun Dr., Baton Rouge 70810.*

HUGGHINS, Ernest Jay, Am. zoologist; b. Bryan, Tex., Dec. 25, 1920; s. Ernest Clarence and Dua (Harris) H.; B.A., Baylor U., 1943; M.S., Tex. A. and M. U., 1949; postgrad. U. Mich. Biol. Sta., Marine Biol. Lab., Woods Hole, Mass.; Ph.D. (NIH Research fellow), U. Ill., 1952; m. Mildred K. Shields, Aug. 12, 1952; children—Susan Jane, Arley Jay, Mildred Kay. Faculty, S.D. State U., Brookings, 1952—, prof. zoology, 1961—, asso. zoologist Agrl. Expt. Sta., 1954-60; vis. prof. U. Okla. Biol. Sta., 1960. Cons. 1st Internat. Congress Parasitology, Office Naval Research, Rome, Italy, 1964. La. State U. InterAm. fellow to Caribbean and S.Am., 1963. Fellow A.A.A.S.; mem. Am. Soc. Parasitologists (rep. to council from Midwestern conf. parasitologists 1964—), Helminthological Soc. Washington, Micros. Soc., Am. Soc. Zoologists, Am. Inst. Biol. Scis., Wildlife Soc., Wildlife Disease Assn., Am. Soc. Mammalogists, S.D. Acad. Sci. (past pres.), Sigma Xi, Phi Kappa Phi, Gamma Sigma Delta. Research, publs. on life history and ecol. relationship Hysteromorpha triloba, parasites fishes in S.D., parasites wildlife, pesticide residues in wildlife. Home: 1034 6th Av., Brookings, S.D. 57006.*

HUGGINS, Charles Brenton, surgeon; b. Halifax, N.S., Can., Sept. 22, 1901; s. Charles Edward and Bessie (Spencer) H.; B.A., Acadia U., 1920, D.Sc., 1946; M.D., Harvard, 1924; M.Sc., Yale, 1947; D.Sc., Washington U., 1950, Leeds (Eng.) U., 1953, Turin (Italy) U., 1957, Trinity Coll., 1965; m. Margaret Wellman, July 10, 1927; children—Charles Edward, Emily Wellman (Mrs. Max Fine). Came to U. S., 1920, naturalized, 1933. Faculty, U. Chgo., 1927—, William B. Ogden Distinguished Service prof., 1961—, dir. Ben May Lab. for Cancer Research, 1951—. Macewen lectr. U. Glasgow, Scotland, 1958. Recipient gold medals for research A.M.A., 1936, 40; research award Am. Urol. Assn., 1948; Francis Amory award, 1948; Am. Cancer Soc. award, 1953; Bertner award M.D. Anderson Hosp., 1953; award Am. Pharm. Mfrs. Assn., 1953; Gold medal Am. Assn. Genito Urinary Surgeons, 1955; Borden award Assn. Am. Med. Colls., 1955; Comfort Crookshank award Middlesex Hosp., London, 1957; Cameron prize Edinburgh U., 1958; City of Hope award, 1958; Oscar B. Hunter award Am. Therapeutic Soc., 1962; Ferdinand Valentine award N.Y. Acad. Medicine, 1962; Albert Lasker award for clin. research, 1963; Gold medal Rudolf Virchow Med. Soc., 1964; Passano Found. award, 1965; Nobel prize for physiology and medicine, 1966. Fellow Am., Royal (Edinburgh, Eng. Walker prize 1955-60) colls. surgeons; mem. Nat. Acad. Sci. (Meyer award 1943), Am. Assn. Cancer Research (pres. 1947-48), Royal Soc. Medicine London, Am. Philos. Soc., Alpha Omega Alpha. Research, publs. on concept of hormonal-dependence of cancers of animals, man; bilateral adrenalectomy in man; protein sulfide interchange; transforming principle in uroepithelium; studies of bone formation, bone marrow, serum enzymes. Home: 5759 Kenwood Av., Chgo. 60637.

HUGGINS, Clyde Griffin, Am. biochemist; b. Watertown, Tenn., June 21, 1922; s. Charles T. and Mattie (Murphy) H.; B.S., Middle Tenn. State Coll., 1945; B.S. in Pharmacy, U. Miss., 1948, M.S., 1950; Ph.D., Tulane U., 1954; m. Evelyn Cummins, Aug. 10, 1944; children—Lydy, Peter, Lisa. Instr. pharmacy U. Miss., 1948-51, asst. prof. physiol. chemistry, Sch. Medicine, 1954-55; faculty U. Kan. Med. Sch., 1955-61, asso. prof. pharmacology, 1960-61; asso. prof. biochemistry Tulane U. Sch. Medicine, 1961-66; prof., 1966—. Cons. VA Hosp., New Orleans, 1964—. Recipient Mellon award for excellence in teaching Tulane U., 1965, Owl Club award, 1966. Mem. Am. Soc. Biol. Chemists, Am. Soc. for Pharmacology and Exptl. Therapeutics, Assn. Am. Med. Colls., A.A.A.S., Sigma Xi. Research, publs. on biochemistry phosphoinositide complex in kidney and other secretory organs; chemistry, physiology and biochemistry vasopressor polypeptide, use smooth muscle and biol. response for measuring enzymatic formation and destruction active polypeptides. Home: 221 Walnut St., New Orleans, 70118.*

HUGGINS, Maurice Loyal, Am. chemist; b. Berkeley, Cal., Sept. 19, 1897; s. Amos Williamson and Mary Abigail (Hackley) H.; A.B., B.S., U. Cal. at Berkeley, 1919, M.S., 1920, Ph.D., 1922; m. Dorothy Bates Gettell, June 12, 1928; children—Robert Alan, David Glenn. Research fellow Harvard, 1922-23, Cal. Inst. Tech., 1923-25; instr. Stanford, 1925-26, asst. prof., 1926-33, sr. research scientist Stanford Research Institute, Menlo Park, California, 1959-67, manager of physical and inorganic chemistry, 1959-61, consultant, 1967—; asso. Johns Hopkins, 1933-36; research chemist Eastman Kodak Co., Rochester, N.Y., 1936-58. Fulbright lectr., vis. prof. Osaka and Kyoto Univs. (Japan), 1955-56. Recipient Morrison prize N.Y. Acad. Scis., 1941; Frank Forrest award Am. Ceramic Soc., 1956. Johnstone scholar Johns Hopkins, 1932-33. Fellow A.A.A.S., Am. Phys. Soc. (past chmn. div. high polymer physics), Am. Soc. X-ray and Electron Diffraction (past pres.); mem. Am.

Chem. Soc. (past chmn. Rochester chpt.), Am. Crystallographic Assn., Research Soc. Am., Sigma Xi. Asso. editor Chem. Physics Jour., 1941-43; mem. adv. bd. Polymer Sci. Jour., 1946-56; editorial bd. Catalysis Jour., 1962—. Patentee in field. Research, publs. on theories of structure of atoms, molecules, crystals, glasses, proteins, high polymers, theories of thermodynamic and other properties of high polymers and their solutions. Address: 135 Northridge Lane, Woodside, Cal. 94062.*

HUGGINS, Robert Alan, Am. metallurgist, educator; b. Stanford, Cal., Mar. 26, 1929; s. Maurice L. and Dorothy (Gettell) H.; A.B. cum laude, Amherst Coll., 1950; M.S., Mass. Inst. Tech., 1952, Sc.D., 1954; m. Eleanor J. Mitchell, July 7, 1951; children—Alan H., John M., Mark W. Instr. dept. metallurgy Mass. Inst. Tech., 1953-54; faculty Stanford, 1954—, prof. dept. materials sci., 1962—, dir. Center for Materials Research, 1961—, acting exec. head dept. metall. engring., 1957-58. NSF fellow, Gottingen, 1965-66. Mem. Am. Soc. for Metals, Am. Welding Soc., Am. Inst. Mining, Metall. and Petroleum Engrs. (Robert Lansing Hardy Gold medal 1957), Am. Phys. Soc., Sigma Xi. Asso. editor Materials Research Bull., 1966—. Research, numerous publs. on materials sci. related to imperfections in crystals, solid state kinetics, mech. behavior, magnetic behavior of solids. Home: 824 San Francisco Ct., Stanford, Cal. 94035.*

HUGGINS, Sir William, English astronomer; b. London, Eng., Feb. 7, 1824; s. William Thomas Huggins; ed. City of London Sch., pvt. tchrs.; LL.D., Cambridge, 1870; D.C.L., Oxford, 1871; m. Margaret Lindsay Murray, 1875. Built obs. nr. London, 1855, began study of phys. constitution of stars, planets, comets and nebulae. Recipient Copley medal, Rumford medal, several prizes from French Acad. Scis. Fellow Royal Soc., 1865 (pres. 1900-05); mem. Royal Astron. Soc. (pres. 1876-78, 2 medals), Brit. Assn. (pres. 1891-92), Micros. Soc. Author: Lines of the Spectra of Some of the Fixed Stars, 1863; Spectrum Analysis, 1866; (with wife) An Atlas of Representative Stellar Spectra, 1900; The Royal Society or Science in the State. Pioneer in devel. of spectroscopic photography; showed that elements existing on earth are also in stars, 1863; proved bright nebulae consist of luminous gas, 1864; 1st to study spectrum of a nova, 1866, showed it to be enveloped by hydrogen thus confirming structure of stars similar to that of sun; studies on spectra of comets showed them to be composed of glowing carbon compounds; devised methods for photographing spectra, 1875; determined relative velocity of star Sirius by noting shift of a line towards red end of spectrum, 1868; made similar observations on other stars to determine their relative velocities; also photographed spectra of moon, planets; research on presence of calcium in sun, 1897. Died London, May 10, 1910.

HUGH OF CITTA DI CASTELLO (or Ugo de Castello, Hugo de Civitate Castellis), Italian astrologer; flourished circa 1337; ed. St. Honore (Dominican convent), Paris, France. Became Dominican (Umbrian) monk. Author: Commentary on the Sphere of Sacrobosco, 1337; De diebus criticis secondum astrologos, 1358 (or possibly 1338).

HUGH OF LUCCA, (Borgognoni, Ugo), Italian physician, surgeon; b. Lucca, Italy, 3d quarter 12th century; 4 sons, including Theodoric. Accompanied Bolognese crusaders to Syria and Egypt where he served as physician; attended siege of Damietta conducted by Jean de Brienne; founder surg. sch., Bologna, circa 1214. Simplified treatment of dislocations, fractures and wounds; studied sublimation of arsenic. Died circa 1252-58.

HUGH OF SANTALLA, Spanish astrologer, chemist; b. probably Santalla, Spain; worked under patronage of Michel, bishop of Tarazona from 1119-51. Translator (from Arabic into Latin): De nativitatibus, Liber yubrium (both Mashallah); treatise on geomancy (earliest Latin work on geomancy); 2 treatises on spatulamancy; Liber Apollonii de principalibus rerum causis; Centiloquium (Ptolemy); Emerald Table (alchem. text, earliest Latin version); Al-Buruni's commentary on al-Farghani's astronomy.

HUGHES, Arthur Llewelyn, physicist; b. Liverpool, Eng., Dec. 18, 1883; s. John and Elizabeth (Parry) H.; B.S., Liverpool U., 1906, M.S., 1908, D.Sc., 1912; B.A., Cambridge U., 1910; m. Jessie Alice Paterson Smyth, July 15, 1919; 3 children. Naturalized, 1936. Asst. prof. physics Rice Inst., Houston, 1913-19; research prof. physics Queen's U., Kingston, Ont., Can., 1913-19; prof. physics Washington U., St. Louis, 1923-53, ret., 1953; asst. dir. Manhattan Dist. Project, Los Alamos, 1943-44. Mem. Am. Phys. Soc. (councillor 1924-28, 41), A.A.A.S. (sec. sect. B 1929-32, v.p., chmn. 1940). Author: Photoelectricity, 1913; (with L. A. Dubridge) Photoelectric Phenomena, 1932. Research on beta-ray spectroscopy, photoelectricity, nuclear physics, electron scattering, ionization in gases, energies of photoelectrons. Home: 245 Union Blvd., St. Louis 63108.

HUGHES, Daniel Richard, Am. mathematician; b. Cin., Aug. 7, 1927; s. Donald M. and Dorothy (Conger) H.; B.Sc., U. Md., 1950, M.A., 1952; Ph.D., U. Wis., 1955; m. Carol May Rodgers, June 27, 1953; children—Monica, Stephen, Sarah, Christa, John, Matthew. Instr., U. Wis., 1955; NSF Postdoctoral fellow, 1955-56; asst. prof. Ohio State U., 1956-58; research asso. U. Chgo., 1958-60; faculty, U. Mich., Ann Arbor, 1960-64, asso. prof. math., 1961-64; reader Westfield Coll., U. London (Eng.), 1964-67, prof., 1967—; prof. U. London, 1962-63, U. Rome, 1963-64. Mem. London, Am. math. socs. Research, publs. on combinatorial algebra and geometry, especially finite combinatorial structures and geometries with their asso. algebraic structures such as collineation or automorphism groups. Home: 7, Aldenham Av., Radlett, Herts., Eng. Office: Westfield Coll., Kidderpore Av., London N.W.3, Eng.*

HUGHES, David Edward, electrician, inventor; b. London, Eng., May 16, 1831; son of David Hughes; m. Anna Chadbourne. Came with parents to U. S., 1838; ed. St. Joseph's Coll., Bardstown, Ky.; taught music and natural philosophy; became interested in telegraphic experimentation; settled in Bowling Green, Ky., 1853; supported his researches by giving music lessons; worked on improved methods of telegraphic printing, sold his still-uncompleted printing device to Comml. Printing Telegraph Co. for $100,000, 1855, patented invention, 1856; started working for Am. Telegraph Co., 1856; went to Eng. in unsuccessful attempt to introduce his methods there, 1857-60; had his system adopted by France, 1861, also by all other major European countries, by 1869; lived in London, 1877-1900. Recipient Albert medal, Soc. Arts. Fellow Royal Soc., 1880 (v.p., Gold medal 1885); mem. Soc. Telegraph Engrs. (pres. 1886), Royal Instn. (mgr. 1889, v.p. 1891). Considered to be inventor of microphone, 1878; invented induction balance, 1883; did work in aerial telegraphy, anticipating some of work done independently by Marconi and Lodge (never published results). Died London, Jan. 22, 1900.

HUGHES, David Edward, Brit. microbiologist; b. London, Eng., Apr. 10, 1915; s. Herbert Edward and Ada Millicent (Jellis) H.; B.Sc., Sheffield U., 1948, Ph.D., 1953; M.A., Oxford (Eng.) U., 1960, D.Sc., 1966; m. Ivy Day, Aug. 23, 1936 (dec. Apr. 1941); children—Brenda J. (Mrs. K. Taylor), Dennise (Mrs. L. Biggins), Mary E. (Mrs. A. Starkey); m. 2d, Joyce G. (Dent) Sharp, Apr. 17, 1943; children—Stephen G., Richard E. Staff Med. Research Council, 1935-64, sci. staff unit for cell metabolism Oxford U., 1954-64, hon. dir. group for microbial structure and function, 1965—; prof. microbiology U. Wales, Cardiff, U.K., 1964—; vis. prof. Dartmouth Med. Sch., N.H., 1961-62. Mem. biol. engring. panel Sci. Research Council, 1965—. Fellow Inst. Biology; mem. Soc. Gen. Microbiology (mem. profl. com. 1966—), Biochem. Soc., Biol. Engring. Soc. Contbg. author: Respirometrie Biochemie des Hororgans. Research, numerous publs. on microbial nutrition and enzymology; isolated enzymes such as bacterial glutaminase, nicotinamide deamidase, nicotinic hydroxylating system (1st isolate of an aromatic oxidase); bacterial disintegration; proposed (with W. L. Nyborg) gen. theory for biol. action of ultrasound; biochemistry of inner ear; action of ultrasound intreating Menieres disease, vitamin B absorption in animals, microbial structure and function with reference to membranous organelles. Home: 60 Beulah Rd., Cardiff, Glamorgan, U.K.*

HUGHES, Edward Charles, Am. physician; b. Syracuse, N.Y., Feb. 13, 1901; s. Arthur G. and Marie (Heindorf) H.; B.S., Syracuse U., 1922, M.D., 1924; m. E. Marion Peters, June 16, 1927; children—Jeanne (Mrs. Russell Greenhalgh), Edward Charles. Practice medicine specializing in obstetrics and gynecology, Syracuse, 1930—; prof., chmn. dept. obstetrics State U. N.Y., Upstate Med. Center, Syracuse, 1944-61, prof. obstetrics and gynecology, 1944—. Diplomate Bd. Obstetrics and Gynecology. Mem. Am. Coll. Obstetricians and Gynecologists (past pres.), Internat. Coll. Surgeons (past pres. obstetric and gynecologic sect.), Am. Assn. Obstetrics and Gynecology, Gynecological Society, Medical Society of State N.Y. (president 1968-69). Research and publs. on carbohydrate metabolism of endometrium of uterus, growth traits of cancer of uterus experimentally produced in animals, endocrine relationships to normal and abnormal pregnancy. Home: Peck Hill Rd., Jamesville, N.Y. 13078. Office: 325 University Av., Syracuse, N.Y. 13210.*

HUGHES, Griffith, English naturalist; b. Towyn, Eng., 1707; s. Edward Hughes; grad. St. Johns Coll., Oxford, Eng., 1729, B.A., M.A., 1748. Fellow Royal Soc., 1748. Author: Natural History of Barbados, 1750; also articles. First to use term, yellow fever, 1750. Died 1779.

HUGHES, John Russell, Am. neurophysiologist; b. DuBois, Pa., Dec. 19, 1928; s. John Henry and Alice (Cooper) H.; A.B. summa cum laude, Franklin and Marshall Coll., 1950; B.A. with honors, Oxford (Eng.) U., 1952, M.A. with honors, 1955; Ph.D., Harvard, 1954; m. Mary Ann Dick, June 14, 1958; children—John Russell (dec.), Christopher Alan, Thomas, Cheryl. Neurophysiologist, NIH, 1954-56; dir. electroencephalograph dept. Meyer Hosp., State U., N.Y., 1956-63; dir. div. electroencephalographology Northwestern U. Med. Center, 1967—, professor neurophysiology, 1964—; staff Wesley Meml., Passavant Meml. hosps. (both Chgo.). Cons. to hosps. Mem. Am. (treas. 1965—), Eastern (sec.-treas. 1961-64) electroencephalograph socs., A.A.A.S. (lectr.), Am. Physiol. Soc., Am. Epilepsy Soc., Am. Acad. Neurology, Sigma Xi (lectr.). Author: Functional Organization of the Diencephalon, 1957. Research, numerous publs. on coding in central nervous system, new theory on neural mechanisms in olfaction, electroencephalograph clin. correlations in different types of epilepsy, organic aspects in juvenile delinquency. Home: 720 Roslyn Terrace, Evanston, Ill. 60201. Office: 303 E. Chicago Av., Chgo. 60611.*

HUGHES, Raymond Hargett, Am. physicist, educator; b. Walla Walla, Wash., June 1, 1927; s. Clifford R. and Frances (Hargett) H.; A.B., Whitman Coll., 1949; M.S., U. Wis., 1951, Ph.D., 1954; m. Olive Jane Wipson, Feb. 8, 1952; children—Diane Frances, Marshall Raymond, Clayton Wipson, Randall Clifford. Faculty dept. physics U. Ark., Fayetteville, 1954—, prof., 1965—. Mem. Am. Phys. Soc., Phi Beta Kappa, Sigma Xi. Research, publs. in atomic structure and spectra, atomic collisions, atomic radiative processes. Home: Route 4, Fayetteville, Ark. 72701.*

HUGHES, Vernon Willard, Am. physicist; b. Kankakee, Ill., May 28, 1921; s. Willard Vernon and Jean (Parr) H.; A.B., Columbia, 1941, Ph.D., 1948; M.S., Cal. Inst. Tech., 1942; M.S. (hon.), Yale, 1960; m. Inge Michelson, Sept. 22, 1950; children—Gareth Albert, Emlyn Willard. Research asso. radiation lab. Mass. Inst. Tech., 1942-46; lectr., instr. physics Columbia, 1949-52, asso. prof., 1958-59; asst. prof. physics U. Pa., 1952-54; faculty Yale, 1954—, prof., 1960—, asso. chmn. physics dept., 1960-61, chmn., 1961—. Cons. Inst. for Space Studies NASA, Oak Ridge Nat. Lab.; physics survey com. Nat. Acad. Sci., 1963—. Fellow Am. Phys. Soc., A.A.A.S. Author: Handbuch der Physik; Atomic and Molecular Beam Spectroscopy, 1959; Gravitation and Relativity, 1963; also numerous papers. Home: 117 Marvel St., New Haven.*

HUGHES, William Frank, Am. mech. engr.; b. Ash, N.C., Oct. 20, 1930; s. Olan T. and Elma (Frink) H.; B.S., Carnegie Inst. Tech., 1952, M.S., 1953, Ph.D., 1955; m. Jane Thomas, June 27, 1959; children—Christopher T., Eric Olin. NSF postdoctoral fellow Cambridge (Eng.) U., 1957-58; faculty Carnegie Inst. Tech., Pitts., 1955—, prof. mech. and elec. engring., 1966—, co-ordinator space scis. program, 1963—. Fulbright lectr. U. Sydney (Australia), 1963. Mem. Am. Soc. M.E., Am. Phys. Soc., Soc. Automotive Engrs., Sigma Xi, Tau Beta Pi. Author: (with F. J. Young), Electromagnetodynamics of Fluids, 1966; also articles. Research in magneto-fluid mechanics, lubrication and friction, space scis., fluid mechanics. Home: 295 Constitution Dr., Pitts. 15236.*

HUGHES-SCHRADER, Sally, Am. zoologist; b. Hubbard, Ore., Jan. 25, 1895; d. Evan Peris and Ellen (Blackburn) Hughes; B.S., Grinnell Coll., 1917; Ph.D., Columbia, 1924; m. Franz Schrader, Nov. 1, 1920. Instr. zoology Grinnell Coll., 1917-19; lectr. Barnard Coll., Columbia, 1919-21; demonstrator biology Bryn Mawr Coll., 1922-24, instr., 1924-30; sci. faculty Sarah Lawrence Coll., 1930-40; research guest Columbia, N.Y.C., 1941-47, research asso. in cytology, 1947-59; guest investigator Duke, 1959-62, vis. prof., 1962-65, emeritus, 1965—. Sarah Barliner fellow Am. Assn. U. Women, 1929-30; Hargitt Research fellow Duke, 1961-62. Mem. Am. Soc. Zoologists, Genetics Soc. Am., Soc. for Study Evolution, Am. Acad. Arts and Scis., Phi Beta Kappa, Sigma Xi. Editorial bds. Chromosoma, 1962—, Biol. Bull., 1962—. Research, publs. in field of comparative cytology of insects, structure and behavior of chromosomes and evolution of chromosomal systems. Address: Dept. Zoology, Duke U., Durham, N.C. 27706.*

HUGI, Franz Josef, Swiss naturalist; b. Grenchen, Switzerland, Jan. 23, 1796; became prof. physics Solothurn (Switzerland) U., 1823, prof. natural history, 1835-37. Author: Die Erde als Organismus, 1841; Über das Wesen der Gletscher, 1842; Die Gletscher und der erratischen Blöcke, 1843. Developed a theory of glaciers; research on light, movement of ocean. Died Solothurn, Mar. 25, 1855.

HUGUIER, Pierre Charles, French surgeon; b. Sézanne, France, 1804; M.D., Faculty Medicine, Paris, 1834; aggregate physician, 1835; intern, Paris, 1828; surgeon Beaujon Hosp.; prof. anatomy Sch. Fine Arts, Paris; mem. Acad. Medicine. Author: Diagnostic différentiel des maladies du coude, 1842; Mémorie sur les maladies de la Glande vulvo-vaginale et les divers appareils sécréteurs de la vulve, 1846. Described Huguier's canal for chorda tympani nerve in temporal bone, 1834; gave 1st description of esthiomene (chronic ulceration of vulva), 1849. Died 1874.

HUHNER, Max, surgeon; b. Berlin, Germany, June 30, 1873; s. Edward and Minna (Jakmuss) H.; came to U. S., 1876; ed. Coll. City N.Y.; M.D., Coll. Phys. and Surg., Columbia, 1893; studied in Europe; m. May Levy (dec. Dec. 11, 1936); 1 dau., Minna H. Schulz. Attending genito-urinary surgeon Bellevue Hosp.; chief of clinic genito-urinary dept., Mt. Sinai Hosp. Dispensary. Diplomate Am. Bd. Urology. Fellow A.M.A., Am. Urol. Assn., N.Y. Acad. Medicine; mem. Soc. Med. Jurisprudence. Author: Sterility in the Male and Female and Its Treatment, 1913; Disorders of the Sexual Function, 1916 (Spanish transl. 1920); The Diagnosis and Treatment of Sexual Disorders in the Male and Female including Sterility and Impotence, 1937; also numerous articles in med. jours. and reference works. Devised accepted test for sterility, known as the Huhner Test. Died Nov. 8, 1947.

HUIDOBRO, Fernando, Chilean pharmacologist; b. Santiago, Chile, Nov. 27, 1910; s. Pedro and Luisa (Toro) H.; M.D., U. Chile Med. Sch., 1935; m. Maria Isabel Toro, Nov. 11, 1945; children—M. Isabel, Juan

Pablo, Cristian, Macarena, Magdalena, Margarita. Faculty, dept. physiology Cath. U. Med. Sch., Santiago, 1930-39, 42-—, prof. pharmacology, head dept., 1948-—; prof. physiology U. Cochabamba (Bolivia), 1940; dir. Cath. U. Med. Sch., 1955-59, dean, 1963-64. Grantee for study in U. S., 1940-41, Latin Am., 1941. Mem. Med. Soc. Santiago, Biol. Soc., Chile, Latin Am. Assn. Physiol. Scis. Research, numerous publs. on cardiovascular action epinephrine its surrogates and similar compounds, effects diverse drugs on neuromuscular junction, mice chronical intoxication with morphine and its consequences, devel. tolerance and phys. dependence. Home: 2591 Bustos, St., Santiago, Chile.*

HUISGEN, Rolf, German chemist; b. Gerolstein, Eifel, Germany, June 13, 1920; s. Edmund and Maria (Flink) H.; student U. Bonn (Germany), 1939, U. München (Germany), 1939-43; Dr.rer.nat., U. Munich (Germany), 1943, habilitation, 1947; m. Trudl Schneiderhan, July 25, 1945; children—Birge, Helga. Asso. prof. organic chemistry U. Tübingen (Germany), 1949-52; prof. chemistry U. Munich, 1952-—, dir. Inst. für Organische Chemie, 1952-—. Recipient Liebig medal Gesellschaft Deutscher Chemiker, 1961; Medaille Lavoisier, Société Chimique de France, 1965. Mem. Bayer. Akademie der Wissenschaften, Am. Acad. Arts and Scis., Deutsche Akademie Leopoldina. Research, numerous publs. on strychnine alkaloids, naphthalene substitution and ring closures, aromatic radical phenylation, nitrosoacylamines, medium-sized rings, benzyne chemistry, cyclo-additions, amine N-oxide rearrangements, valence tautomerism in unsaturated compounds. Home: Kaulbachstrasse 10, 8 Munich 22, Germany. Office: U. of Munich, Karlstrasse 23, 8 Munich 2, Germany.*

HUISMAN, Alphonse, Belgian physician; b. Anvers, June 4, 1880; s. Henri and August Rosalie Huisman; M.D., Free U. Brussels; m. Marguerite Weyerman, Feb. 4, 1908; children—Louise, Georgette. Physician in Brussels hosps.; chief service Anderlecht Hosp.; publs. asst. at Saint-Gilles and Schaerbeek; med. controller; med. adviser; prof. Ecole Centrale de Puériculture; med. insp.; physician Intercommunity Anti-Venereal Dispensary. Author: Nouveau traitement du paludisme; Méthodes de coloration du sang; La Crémation; Sur la tuberculose. Address: 136, avenue Eugène-Demolder, Brussels 3, Belgium.

HUIZENGA, John Robert, Am. nuclear chemist; b. Fulton, Ill., Apr. 21, 1921; s. Harry M. and Josie (Brandts) H.; A.B., Calvin Coll., 1944; Ph.D., U. Ill., 1949; m. Dorothy J. Koeze, Feb. 1, 1946; children—Linda, Jann, Robert, Joel. Lab. supr. Tenn. Eastman Co., Oak Ridge, 1944-46; instr. Calvin Coll., Grand Rapids, Mich., 1946-47; asso. sci. Argonne (Ill.) Nat. Lab., 1949-58, sr. sci., 1958-67; prof. U. Rochester (N.Y.), 1967-—; profl. lectr. U. Chgo., 1963-—. Fulbright fellow, Netherlands, 1954-55; Guggenheim fellow, vis. prof. U. Paris, 1964-65. Recipient Orlando Lawrence Meml. award A.E.C., 1966. Fellow Am. Phys. Soc.; mem. Am. Chem. Soc., Phi Beta Kappa, Sigma Xi, Phi Kappa Phi. Contbr. chpts. to books, numerous articles to profl. jours. Research in nuclear properties of heavy elements, nuclear reaction mechanisms, aspects of nuclear fission process, determination of abundances of trace elements in meteorites; joint discoverer, identification of several nuclides in periodic chart including elements Einsteinium and Fermium. Home: 51 Huntington Meadow, Rochester, N.Y. 14625.*

HUKOVIC, Seid, Yugoslavian pharmacologist; b. Sarajevo, Yugoslavia, Sept. 1, 1925; s. Ibrahim and Semsa (Kundalic) H.; M.D., Med. Faculty, U. Zagreb (Yugoslavia), 1950; Sci.D., Med. Faculty, U. Sarajevo, 1958; m. Ilduza Bubic, Mar. 23, 1962; 1 son, Nedim. Staff, Hosps. Med. Faculty, U. Sarajevo, 1951-53, U. Milan (Italy), 1957-58, U. Oxford (Eng.), 1959-60, U. Mainz (Germany), 1962, U. Cairo (Egypt), 1963; faculty U. Sarajevo, 1953-—, prof. pharmacology, 1962-—. Mem. Acad. Sci. and Arts of Bosnia and Herzegovina, Brit. Physiol. Soc., Brit. Pharmacol. Soc., N.Y. Acad. Sci., Yugoslav Physiol. Soc., German Pharmacological Soc. Research, numerous publs. on isolation of heart, lungs, urinary bladder, vas deferens, blood vessels, with nerves, action and different drugs on effect nerve stimulation. Home: 35/4 R. Jankovic, Sarajevo, Yugoslavia.*

HUKUHARA, Takesi, Japanese physiologist; b. Niigata, Japan, Apr. 3, 1904; s. Nisiro and Hideko Hukuhara; Ph.D., Niigata U. Med. Sch., 1928; m. Yayoi Homma, June 20, 1929; children—Takehiko, Tosihiko. Faculty dept. physiology Niigata U. Med. Sch., 1928-38, asst. prof., 1933-38; prof. physiology Nat. U. Med. Sch., Peking, China, 1938-45; prof. physiology Tottori U. Med. Sch., 1948-54; prof. physiology Okayam (Japan) U. Med. Sch., 1954-—. Mem. Physiol. Soc. Japan, Japan Soc. Smooth Muscle Research. Author: The Experimental Physiology, 1950; The Motility of Digestive Tract, 1953; also numerous articles. Cinematographed normal movements of gastro-intestinal tract through abdominal window; discovered intrinsic intestinal reflexes, criteria used to analyse function of intramural ganglion cells, specific muscular tissue endowed with inherent rhythmicity on valves of heart; postulated new hypothesis on vagus-respiratory reflex. Home: 305 Uchida, Okayama, Japan.

HULETT, James Edward, Jr., Am. sociologist; b. Hattiesburg, Miss., June 30, 1909; s. James E. and Anna P. (Schneider) H.; A.B., Miss. Coll., 1929; A.M. U. Wis., 1935, Ph.D., 1939; m. Jo Day, Sept. 3, 1937; 1 son, David Todd. Sr. research asst. WPA, Washington, 1934-35; asso. social psychologist U. S. Dept. Agr., Washington, 1939-40; clin. psychologist OSS, Washington, 1944-45; faculty U. Ill., Urbana, 1940-—, prof. sociology, 1957-—, chmn. dept. sociology and anthropology, 1953-59, acting head of department sociology, 1968-—. Fellow Am. Psychol. Assn., Am. Sociol. Assn., A.A.A.S.; mem. Midwest Sociol. Soc. Editor: (with Ross Stagner) Problems in Social Psychology, 1951; editorial cons. Sociol. Quar., 1959-62. Contbr. articles, book revs. to profl. lit. Research on pub. opinion, social movements, polygamous family structure; contbr. to theory of social roles and statuses; originator theory of human communication based on symbolic interactionist social psychology of George Herbert Mead. Home: 105 W. Pennsylvania Av., Urbana, Ill. 61801.*

HULL, Albert Wallace, Am. physicist; b. Southington, Conn., Apr. 19, 1880; s. Lewis Caleb and Frances Reynolds (Hinman) H.; A.B., Yale, 1905, Ph.D., 1909; Sc.D., 1947; Sc.D., Union U., 1930; Middlebury Coll., 1931; Eng.D., Worcester Poly. Inst., 1944; m. Mary Shore Walker, June 14, 1911; children—Robert Wallace, Harriet. Instr. physics Worcester Poly. Inst., 1909-11, asst. prof., 1911-13; research physicist Gen. Electric Co., Schenectady, 1914-50, asst. dir. research lab., 1928-50, cons. 1950-66. Recipient Potts medal for work on x-ray crystal analysis Franklin Inst., 1923. Fellow Am. Phys. Soc. (pres. 1942), I.R.E. (Morris Liebman prize for work on vacuum tubes 1930, medal of honor 1958); mem. Nat. Acad. Scis., Phi Beta Kappa, Sigma Xi. Contbr. articles to profl. jours. Inventor magnetron; used powder method to perform x-ray crystal analysis; developed new vacuum tubes. Died 1966.

HULL, Alvin C., Jr., Am. range scientist; b. Whitney, Ida., Mar. 25, 1909; s. Alvin C. and Ella (Maughan) H.; B.S., Utah State U., 1936, Ph.D., 1959; M.S., Brigham Young U., 1940; m. Mayme Laird, June 10, 1936; children—Nancy Ann (Mrs. Dale Manning), Susan Lee (Mrs. Carl Burstedt), James Laird, Mary Kay. With U. S. Forest Service, 1936-54, range research Ogden, Utah, 1936-42, Boise, Ida., 1942-48, Ft. Collins, Colo., 1948-51, Washington, 1951-54; range seeding, fertilization research U. S. Agrl. Research Service, Logan, Utah, 1954-—. Range cons. U. S. Dept. Agr., Egypt, Israel, 1952-53, Peru (N.C.) State Coll., 1958. Mem. No. Utah Safety Council, American Grassland Council, Utah Historical Association, Am. Soc. Range Management, Utah Acad. Arts and Scis., Sigma Xi. Research, numerous publs. on tests of adaptability of hundreds of species and all practicable methods of seeding western range lands; determined best species and methods of seeding ranges, especially arid lands; developed techniques which reduce costs and insure higher percentage of success; determined basic plant and environmental reactions which cause success or failure of range seedings. Home: 341 N. 4th E. Office: Crops Research Lab., Utah State U., Logan, Utah 84321.*

HULL, Clark Leonard, Am. psychologist; b. Akron, N.Y., May 24, 1884; s. Leander G. and Florence L. (Trask) H.; grad. Alma (Mich.) Academy, 1905; student Alma College 2 years; A.B., U. of Mich., 1913; Ph.D., U. of Wis., 1918; m. Bertha E. Iutzi, 1911; children—Ruth Trask, Richard Hazard. Prin. pub. sch., Sickels, Mich., 1909-11; acting prof. psychology, Eastern Ky. State Normal Sch., Richmond, Ky., 1913-14; with U. of Wis. as instr. psychology, 1916-20, asst. prof., 1920-22, asso. prof. and dir. lab., 1922-25, prof., 1925-29; prof. psychology, Inst. Human Relations (Yale), 1929-47, Sterling professor of psychology, from 1947. Fellow A.A.A.S.; member Am. Psychol. Assn. (council 1931-33, president 1935-36), Nat. Acad. Sciences, Am. Acad. Arts and Sciences, Sigma Xi. Author: The Evolution of Concepts, 1920; Influence of Tobacco Smoking on Mental and Motor Efficiency, 1924; Aptitude Testing, 1928; Hypnosis and Suggestibility—An Experimental Approach, 1933; (with Hovland, Ross, Hall, Perkins and Fitch) Mathematico-Deductive Theory of Rote Learning, 1940; Principles of Behavior, 1943; Essentials of Behavior, 1951; A Behavior System, 1952. Contbr. to psychol. jours. Leader in aptitude testing; experimented with hypnosis and suggestibility; sought to further understanding of learning process; applied hypothetical-deductive method to psychology; influenced devel. of psychology in U. S. Died New Haven, May 10, 1952.

HULL, Derek, Brit. phys. metallurgist; b. Blackburn, U.K., Aug. 8, 1931; s. William and Nellie (Hayes) H.; B.Sc. with 1st class honors in Metallurgy, U. Wales, 1953, Ph.D., 1956, D.Sc., 1966; m. Pauline Scott, Aug. 5, 1953; children—Andrew, Sian, Karen, Beverly. Staff, Harwell, Clarendon labs. metallurgy div. Atomic Energy Research Establishment, U. Oxford (Eng.), 1956-60, sr. sci. officer, sect. leader, 1958-60; faculty U. Liverpool (Eng.), 1960-—, prof. phys. metallurgy, 1964-66, Henry Bell Wortley prof. metallurgy, 1966-—. Mem. basic properties and ferous metals coms. Inter Service Metall. Research Council, 1961-66. Fellow Instn. Metallurgists; mem. Inst. Metals (pres. local sect. 1967-68), Iron and Steel Inst. Author: Introduction to Dislocations, 1965; also articles. Research on phys. metallurgy including crystallography of martensitic transformation, radiation damage, low temperature mech. properties, creep, brittle fracture and fatigue, yield and fracture of body centred cubic metals. Home: Lanyork IV, Church Meadow, Heswall, Cheshire, Eng. Office: Dept. Metallurgy, U. Liverpool, Liverpool 3, Eng.*

HULL, Edward, Brit. geologist; b. Antrim, Ireland, May 21, 1829; s. Rev. John Dowson H.; M.A., LL.D., Trinity College, Dublin; m. d. of C. T. Cooke, 1857 2 sons, 4 daus. Apptd. to Geological Survey of U.K., 1850 Scot. District Surveyor, 1867; Ireland dir., 1869; prof. geology, Royal College of Sci., until 1890. Mem., Royal Commission on Coal Reserves. Fellow Royal Soc., 1867; Royal Geological Soc., Ireland (pres. 1873); Victoria Inst. (sec. 1900). Author: A Treatise on Building and Ornamental Stones of Great Britain and Foreign Countries, 1872; Contributions to the Physical History of the British Isles, 1882; Mount Seir, Sinai, and Western Palestine, 1885; Memoir on the Physical Geology of Arabic Petraea and Palestine, 1886; Physiology (Physical Geography), 1888; Physical Geology and Geography of Ireland, 2d. ed., 1891; Volcanoes Past and Present, 1892; Our Coal Resources at the Close of the Nineteenth Century, 1897; The Coal Fields of Great Britain, 5th ed., 1905; Reminiscences of a Strenuous Life, 1910; Monograph on the Sub-oceanic Physiography of the North Atlantic Ocean, 1912. Conducted scientific expedition through Arabia and Palestine, 1883-84; investigations of geology of British Isles, especially coal fields. Died Oct. 19, 1917.

HULL, Frank Montgomery, Am. biologist; b. Coahoma, Miss., Nov. 3, 1901; s. Dabney Herndon and Anna Baldwin (Montgomery) H.; B.Sc., Miss. State U., 1922; M.Sc., Ohio State U., 1925; Ph.D., Harvard, 1937; m. Mary Marguerite Chappell, Feb. 4, 1926; children—Martha Cecil, Frank Montgomery, Clovis Malcolm Sillers. Instr., State Coll. Wash., 1925-26; asst. prof., acting state entomologist N.M. State Coll., 1926-27; entomologist Expt. Sta., Tex. A. and M. Coll., 1926-27; prof. biology U. Miss., University, 1930-—, head dept., 1930-52; research asso. Am. Mus. Natural History, 1953-62; hon. research asso. Smithsonian Instn., 1960-—. Mem. Am. Soc. Zoologists, Entomol. Soc. Am., Entomol. Soc. Washington, Systematic Zoologists, Brazilian Entomol. Soc., Am. Assn. U. Profs., Sigma Xi. Author: Robber Flies of the World, vols. I and II, 1962; also numerous articles. Research on insects in Cuba, Panama, Mexico, Europe, S. Africa, Australia, U. S., regeneration in planarians, econ. aspects, insects, Diptera. Home: Country Club Rd., Oxford, Miss. 38655. Office: P.O. Box 413, University, Miss. 38677.*

HULL, Robert William, Am. biologist; b. LaCrosse, Ind., Apr. 3, 1924; s. Frank Hollis and Adah (Bell) H.; B.S., U. Ill., 1949, M.S., 1950, Ph.D., 1953; m. Grace Marie Houf, Feb. 25, 1945; children—Bradford Kenyon, Sandra Gail, Penna Lee. Research asst. U. Ill., 1951-53; research asst. Ill. Natural History Survey, Urbana, 1950-51; faculty dept. biol. sci. Northwestern U., 1953-63, asso. prof., 1961-63; prof., chmn. dept. biol. sci. Fla. State U., Tallahassee, 1963-—. Mem. Soc. Protozoologists (sec.), Am. Soc. Zoologists, Soc. Gen. Microbiology (Eng.), Soc. Parasitologists, Am. Microscopic Soc., Am. Soc. Naturalists. Author: (with others) Experiments in Biology, 1957; Laboratory Exercises in Biology, 1959. Research, publs. field biochemistry and physiology of feeding mechanisms in Suctarian Protozoa; metabolism and physiology of bird malaria with particular regard to host blood changes during infection with malaria organisms; devel. of test techniques for assay of carcinogenic hydrocarbons in air pollutants, particulary tobacco smoke. Home: 3310 North Shore Circle, Tallahassee 32303.*

HULLIGER, Fritz, Swiss physicist; b. Horgen, Switzerland, Feb. 4, 1929; s. Fritz and Emma (Suter) H.; Diploma in Exptl. Physics, Fed. Inst. Tech., Zürich, Switzerland, 1953, Dr. Degree, 1959; m. Elfriede Menga Luessi, Feb. 2, 1957; children—Fritz, Maja, Oskar. Coworker of Prof. Dr. G. Busch, Laboratorium fuer Festkoerperphysik ETH, Zürich, 1954-61; research scientist Cyanamid European Research Inst., Geneva, Switzerland, 1961-—. Mem. Swiss Phys. Soc. Research, publs. on structural chemistry of semiconductors, effect of electronic configuration of constituent atoms on structure, electric and magnetic properties of transition element compounds; synthesized new compounds, including sulvanite, cobaltite, arsenopyrite and ternary rare-earth chalcogenide phases. Home: 17 Schoerenweg, CH 8713, Uerikon, Zürich, Switzerland. Office: 91 Route de la Capite, Cologny, CH 1223 Geneva, Switzerland.*

HULSE, Frederick Seymour, Am. anthropologist; b. N.Y.C., Feb. 11, 1906; s. Hiram Richard and Frances (Seymour) H.; student Williams Coll., 1923-25; A.B., Harvard, 1927, M.A., 1928, Ph.D., 1934; m. Leonie Robinson Mills, Aug. 27, 1934; children—Richard, Christopher. Research fellow Harvard, Cuba, 1928-29, Seville, Spain, 1930, U. Hawaii, 1931, Japan, 1931-32, NRC, Cal., 1934; instr. U. Hawaii, 1935-36, U. Wash., 1936-37; archaeol. dir. W.P.A. Ga., 1938-42; asst. prof. Colgate U., 1946-48; asst. prof. U. Wash., 1948-49, asso. prof., 1949-58; prof. U. Ariz., Tucson, 1958-—; field research, U. S., fgn. countries, 1952-—. Mem. Am. Assn. Phys. Anthropologists (pres. 1967-—), Am. Anthrop. Assn., Am. Soc. Human Genetics, Soc. Study Human Biology. Author: The Human Species, 1963; also articles. Editor Am. Jour. Phys. Anthropology, 1963—. Demonstrated type and degree of bodily change in offspring of Japanese migrants to U. S., that exogamy among Swiss leads to hybrid vigor, that sexual selection in Japan leads to lighter skin color in upper classes. Home: 2200 E. River Rd., Tucson 85718.*

HULSIUS, Levinus, mathematician; b. Ghent, Belgium; flourished Nuremburg, Germany, circa 1600. Author: De Quadrante Geometrico (usually attributed to Cornelius de Judeis, who only did the drawings), 1594. Studied geography, math., metals, proportional compass.

HULSIZER, Robert Inslee, Jr., Am. physicist; b. East Orange, N.J., Nov. 25, 1919; s. Robert Inslee and Dorothy Joy (Price) H.; B.S., Bates Coll., 1940; M.A., Wesleyan U., Middletown, Conn., 1942; Ph.D., Mass. Inst. Tech., 1948; m. Bernice Lord, June 21, 1941 (div. 1965); children—Stephen, Ann, Deborah, Cynthia, m. second, to Carol Ascher, May 27, 1967. Staff mem. Radiation Lab., Mass. Inst. Tech., 1942-46, grad. research nature cosmic rays, 1946-49; research cosmic rays U. Ill., 1949-51, devel. computer control naval air def. situation, Control Systems Lab., 1951-57, research low temperature physics, 1957-60; research elementary particle physics and computer aids to data analysis, 1960-64; prof. physics, dir. Sci. Teaching Center, Mass. Inst. Tech., 1964—. Cons. Office Naval Research, 1957-59; mem. steering com. Phys. Sci. Study Com., 1959-64; mem. Commn. Coll. Physics, 1960-66, chmn. panel selection fellowships in physics NSF, 1962-66. Fellow Am. Phys. Soc.; mem. Am. Assn. Physics Tchrs. (Outstanding Teaching citation 1964), Am. Assn. U. Profs. Editor vol. 22, Electronic Time Measurements of Radiation Laboratory Technical Series, 1946. Research on elementary particle physics, vibrations of unsymmetrical crystal bars, search for electrons in cosmic rays from outer space; determination of radioactive decay scheme of cobalt at 0.003 degree kelvin; measurement of reactions of K-mesons with protons, including elastic, charge exchange, strange particle resonances. Home: 31 Maple St., Lexington, Mass. 02173. Office: Physics Dept., Mass. Inst. Tech., Cambridge, Mass. 02139.*

HULST, Hendrik Christoffel van de, see Van de Hulst, Hendrik Christoffel.

HULTÉN, Eric Oskar Gunnar, Swedish botanist; b. Halla, Sweden, Mar. 18, 1894; s. August Johan and Agnes (Forsberg) H.; student U. Stockholm (Sweden), 1913-31; D.Sc., U. Lund (Sweden), 1937; D.Sc., U. Montreal (Que., Can.), 1959; m. Signe Elisabet Vougt, Jan. 21, 1920; children—Karl Gunnar Pontus, Maj Anita. Curator herbarium U. Lund, 1931-44; prof., dir. bot. dept. Swedish Mus. Natural History, Stockholm, 1945-61; mem. bot. expdns. to Kamchatka, E. Siberia, 1920-22, Kola Peninsula, 1927, Mexico, 1931-32, Alaska, 1932, 60-65. Mem. Royal Swedish Acad. Sci., Linnean Soc. London (fgn. mem.), Acad. Sci. Phila. (corr.). Author: Flora of the Aleutian Islands, 1937; Flora of Alaska and Yukon, 1941; Atlas of the Distribution of Plants in North West Europe, 1950; The Amphi-Atlantic Plants, 1958; The Circumpolar Plants, 1961; also numerous articles. Research on plant geography and taxonomy of arctic and boreal bells. Home: Odinvägen 21, Djursholm, Sweden. Office: Naturhistoriska Riksmus., Stockholm, Sweden.*

HULTGREN, Herbert Nils, Am. physician; b. Santa Rosa, Cal., Aug. 29, 1917; s. Adolf William and Hilda (Hakanson) H.; jr. certificate Santa Rosa Jr. Coll., 1937; A.B., Stanford, 1939, M.D., 1943; m. Barbara Brooke, Aug. 7, 1948; children—Peter, Bruce, John. Fellow cardiology Thorndike Meml. Lab., Boston, 1947-48; faculty Stanford Med. Sch., Palo Alto, Cal., 1948—, asso. prof. medicine, dir. cardiology div., 1958—. Markle scholar med. sci., 1952. Mem. Am., Soc. Clin. Investigation, Western Soc. Clin. Research, Am. Physiol. Soc., Assn. U. Cardiologist, Western Assn. Physicians. Research, numerous publs. on congenital heart disease, phenocardiography, hemodynamics, vavular heart disease, coronary artery disease, high altitude physiology, high altitude pulmonary edema in Peruvian Andes. Home: 827 San Francisco Ct., Stanford, Cal. 94305. Office: 300 Pasteur Dr., Palo Alto, Cal. 94304.*

HULTGREN, Ralph Raymond, Am. metallurgist; b. Spokane, Wash., Sept. 7, 1905; s. Charles August and Augusta (Benson) H.; B.S., U. Cal., 1928; M.S., U. Utah, 1929; Ph.D., Cal. Inst. Tech., 1933; m. Lesta Teresa Wood, June 11, 1933; children—Neilen Wood, Glen Owen, Eric Carl. Research fellow NRC, Mass. Inst. Tech., 1933-35; advanced from instr. to asst. prof. metallurgy Harvard, 1935-41; with U. Cal. at Berkeley, 1941—, asst., then asso. prof., 1941-48, prof. metallurgy, 1948—, asst. dean Coll. Engring., 1958-61, chmn. dept. mineral tech., 1961-65. Chairman, United States Calorimetry Conference, 1965-66. Member of the Mechanics Institute of San Francisco (trustee), Am. Inst. Mining and Metall. Engrs., Phi Beta Kappa, Sigma Xi. Author: Fundamentals of Physical Metallurgy; Selected Values of Thermodynamic Properties of Metals and Alloys, 1963. Research on theory of chem. bonding, crystal structure, thermodynamics of metallic state. Home: 1501 LeRoy Av., Berkeley, Cal. 94708.*

HULTH, Anders Gustav, Swedish orthopaedic surgeon; b. Uppsala, Sweden, Dec. 7, 1916; s. Johan Marcus and Sina (Johansson) H.; M.D., Sch. Medicine, U. Uppsala, 1944, Ph.D. in Medicine, 1956; m. Svea Thorborg Ohlsén, July 3, 1943; children—Kerstin, Lars, Ingrid, Eva. Asso. prof. surgery U. Hosp., Uppsala, 1956-63; asst. prof. orthopaedic U. Hosp. Gothenburg, 1963-65; surgeon in chief dept. orthopaedics Gen. Hosp., U. Lund, Malmö, Sweden, 1965—. Mem Scandinavian Assn. Orthopaedics, A.A.A.S. Research, publs. on vascularity of head of femoral bone and effect of proteolytic enzymes on cartilage growth. Office: Gen. Hosp., Malmö, Sweden.

HULTHÉN, Lamek, Swedish theoretical physicist; b. Östra Ljungby, Sweden, Dec. 14, 1909; s. Per Martensson and Anna (Lindblad) H.; student U. Stockholm (Sweden), 1928-38, Ph.D.; m. Berta Maria Eriksson, May 30, 1936; children—Eva (Mrs. Per Wejke), Gyno (Mrs. Göran Linde), Marit (Mrs. Björn Askert), Marten. Reader in mechanics, math. physics U. Stockholm, 1938-39, U. Lund (Sweden), 1940-49; prof. applied math. Royal Inst. Tech., 1949-54, prof. math. physics, from 1954, dean physics dept., 1954-58. Mem. Swedish Natural Sci. Research Council, 1955-61, Swedish Council for Nuclear Research, 1965—, Nobel com. for physics, 1966—; chmn. Swedish Space Research Com., 1959-62, 64-67, sci. and tech. working group Commission Pré-paratoire Européenne de Recherches Spatiales, 1961-64, Swedish Nat. Com. for Physics, 1966—; Swedish del. to ESRO, 1964—, to CERN, 1966—; mem. bd. NORDITA, 1965—. Fellow Phys. Soc. London; mem. Swedish Acad. Engring. Scis., Royal Swedish Acad. Scis. Author: Über das Austauschproblem eines Kristalles, 1938; (with M. Sugawara) Handbuch der Physik, 1957. Research, publs. on antiferromagnetism, 1935-38, nuclear forces and neutron-proton problem, 1939—(Hulthén potential and wave function, 1940); contbns. to theory of continuous spectra, including variational problem for continuous spectrum of Schrödinger Equation, 1944, reformulated, 1948; operations research.*

HULTQVIST, Bengt Karl Gustaf, Swedish physicist; b. Hemmesjö, Sweden, Aug. 21, 1927; s. Erik Bernhard and Elsa (Carlsson) H.; Fil.Kand, U. Stockholm, 1951, Fil.Mag., 1952, Fil.Lic., 1954, Fil.Dr., 1956; m. Gurli Anita Gustafsson, Apr. 4, 1953; children—Anders B. E., Hans B. H., Anna-Karin. Research physicist Inst. Radiophysics, Stockholm, Sweden, 1951-56; dir. Kiruna (Sweden) Geophys. Obs., 1956—. Decorated knight Order of North Star. Mem. Am. Geophys. Union, Internat. Acad. Astronautics. Author: Introduktion till Geokosmofysiken, 1967; also numerous articles. Research on natural radioactivity and physics of upper atmosphere of earth with spl. regard to disturbance phenomena in auroral latitudes. Home: 2 Grönstensvägen, 98100 Kiruna. Office: Kiruna Geophys. Obs., 98101 Kiruna, Sweden.*

HULUBEI, Horia, Rumanian physician; b. 1896; prof. Bucharest (Rumania) U.; mem. Acad. Socialist Republic Rumania, numerous fgn. sci. acads. and socs. Author: Contribution à l'étude de la diffusion qu antique des rayons X, 1934; other works on optical spectroscopy, spectroscopy of X-rays, high energy physics, elementary particles, physics of nuclear reactors.

HUMBERT, (Marie) Georges, French mathematician; b. Paris, Jan. 7, 1859; at least 1 son, Pierre; chief mining engr., Paris; prof. analysis École polytechnique; prof. math. Coll. de France; mem. French Acad. Scis. Investigated Kummer's surfaces; discovered some properties of surfaces; gave geometrical representation of continuous fraction, 1916. Died Paris, Jan. 22, 1921.

HUMBOLDT, Baron Friedrich Wilhelm Heinrich Alexander von (von Humboldt), German naturalist; b. Berlin, Sept. 14, 1769; s. Prussia's chamberlain and Maria (von Colomb) H.; studied finance U. Frankfort/Oder; matriculated Gottingen, 1789; studied commerce and languages, Hamburg, geology, Freiberg, anatomy, Jena; unmarried. Apptd. assessor of mines, Berlin, 1792; served Prussian King as diplomat, after fall of Napoleon; received title of councillor, Berlin, 1829. Made several sci. excursions in Europe, from 1790 (Eng., 1790, Vienna, 1792, 97); made geol. and bot. survey, Switzerland and Italy, 1795; travelled to South and Central Am., 1799-1804; explored course of Orinoco (covered 1725 miles, showed communication between water systems of Orinoco and Amazon), 1800; also extensively explored previously unvisited areas of South Am. Mem. French Inst., 1810, also all principle acads. of world. Fellow Royal Soc., 1815. Author: Voyage de Humboldt et Bonpland, 23 vols., 1805-34; On the Geographical Distribution of Plants according to the Temperature and Altitude, 1817; Critical Examination of the Geography of the New Continent, 1835-38; Central Asia: Researches on the Chains of Mountains and the Comparative Climatology, 1843; Kosmos: Entwurfeiner physischen Weltbeschreibung (most important work, gives conception and description of entire phys. world), 1845. Known especially for excursions in Central and South Am., which formed the basis for sci. founds. of phys. geography and meteorology; made observations which led to discovery of periodicity of meteor showers and investigated fertilizing qualities of guano; established use of isotherms; studied origin and course of tropical storms, increase in magnetic intensity from equator toward poles, and volcanos; made pioneer investigations on relationship between geog. environment and plant distribution. Died Berlin, May 6, 1859.

HUME, David, Scottish philosopher, polit. economist, historian; b. Edinburgh, Apr. 26, 1711; s. Joseph and Catherine (Falconer) H.; matriculated U. Edinburgh, 1723; studied law several years. With mcht. firm, Bristol; gave up law and commerce, went to France to study philosophy, 1734; returned to Ninewells, Eng., 1737; became pvt. tutor after being unable to get professorship at Edinburgh, 1745-47; sec. to Gen. St. Clair in expdn. intended to operate against Can., 1746; apptd. judge-adv. by St. Clair, accompanied him on mil. embassy to Vienna and Turin, 1748; returned to Ninewells, 1749; librarian Advocates Library, Edinburgh, 1752; accompanied Lord Hertford to Paris, 1763; undersec. to Gen. Conway, 1767; settled in Edinburgh, 1767. Author: Treatise of Human Nature, 1737; Essays, 1741-42; Political Discourses, 1751; History of England, 1753; Inquiry concerning the Principles of Morals, 1754; Four Dissertations; The Natural History of Religion, of the Passions, of Tragedy, of the Standard of Taste, 1757; Dialogues concerning Natural Religion, 1779. Received fame during his lifetime as result of his writings on history; wrote one of 1st hist. works in English which emphasized sociol. and cultural aspects of nat. life and presented comprehensive and correlated survey of hist. facts; one of major Brit. empiricist philosophers. Died Edinburgh, Aug. 25, 1776.

HUME, William Fraser, geologist; b. 1867; s. George Hume; ed. College Galliard, Lausanne, Switzerland, Royal Coll. Sci., Royal Sch. Mines, London; D.Sc.; m. Ethel Gladys Williams; 1 son, 1 dau. Mem. teaching staff Royal Coll. Sci., 1890, later sr. demonstrator; examiner in geology Royal Indian Engring. Coll.; mem. Geol. Survey of Egypt, 1897, later supt. and dir., tech. counsellor, until 1940. Fellow Royal Geol. Soc., Geol. Soc. (Lycell medallist 1919), Royal Soc. Edinburgh; mem. Royal Sch. Mines (asso.), Royal Coll. Sci. (asso.), Geol. Soc. France (fgn. asso.), Royal Geol. Soc. Egypt (pres. 1926-40), Inst. Egypt (pres. 1928-29). Author: (with Barron) Topography and Geology of Eastern Desert of Egypt, 1903; Topography and Geology of Southeastern Sinai, 1906; Petroleum: Its Occurrence and Origin, 1910; Geology of Egypt, vol. 1, Surface Features of Egypt and Their Determining Causes, vol. 2, The Fundamental Pre-Cambrian Rocks of Egypt and the Sudan, 1934; Terrestrial Theories, 1947; also numerous articles. Died Feb. 23, 1949.

HUME-ROTHERY, William, English metallurgist; b. Worcester Park, Surrey, Eng., May 15, 1899; s. Joseph Hume and Ellen Maria (Carter) H.-R.; M.A., Magdalen Coll., Oxford, Eng., 1926, D.Sc., 1935; PhD., Royal Sch. Mines, London, 1926; D.Met. (hon.), Sheffield U., 1966; D.Sc., U. Manchester; m. Elizabeth Alice Fea, Mar. 28, 1931; 1 dau., Jennifer Ann (Mrs. Anthony David Moss). Armourers and Brasiers' Co. Research fellow Oxford U., 1929, Royal Soc. Warren Research fellow, 1932-42, 43-55, faculty, 1938-66, George Kelley reader metallurgy, 1955-57, Isaac Wolfson prof., 1958-66; prof. emeritus, 1966—; hon. lectr. Sheffield (Eng.) U., 1945—. Named mem. Order Brit. Empire, 1951; recipient Sir George Beilby Meml. award, 1934; Am. Inst. Mining and Metall. Engrs. ann. lectr., 1946; Francis J. Clamer medal Franklin Inst., 1949; Roozeboom Gold medal Royal Netherlands Acad., 1950; Luigi Losana prize medal, Italy, 1955. Fellow Imperial Coll. Sci. and Tech. London, Royal Soc., 1937, Inst. Physics, Phys. Soc., Instn. Metallurgists; hon. mem. Inst. Metals (Platinum medal 1949), Am. Soc. for Metals (life), Société Française de Metallurgie; mem. Iron and Steel Inst. Author several books including: The Metallic State, 1932; The Structure of Metals and Alloys, 1936; Atomic Theory for Students of Metallurgy, 1946; Electrons, Atoms, Metals and Alloys, 1948; (with J. W. Christian, W. B. Pearson) Metallurgical Equilibrium Diagrams, 1952; Elements of Structural Metallurgy, 1961; The Structures of Alloys of Iron: An Elementary Introduction, 1966; also numerous articles. Research on structures of metals and alloys, effects different elements on melting and freezing points of copper, silver, and iron alloys, lattice distortion produced by one metal when dissolved in another; demonstrated that if atomic diameters of 2 kinds of atoms differ by more than 15%, they do not fit together easily on a common lattice, that the solid solubility limits are related to electron concentrations in some alloys of copper, silver, and gold, that electron concentration effects control structures of some transition metal alloys. Home: Cherry Orchard, Abberbury Rd. Iffley, Oxford, Eng.*

HUMES, Arthur G(rover), Am. zoologist, educator; b. Seekonk, Mass., Jan. 22, 1916; s. Edwin Judson and Agnes (Gillis) H.; A.B., Brown U., 1937; M.S., La. State U., 1939; Ph.D., U. Ill., 1941. Instr. biology U. Buffalo, 1941-42; asst. prof. zoology U. Conn., 1946-47; prof. Boston U., 1947—. Guggenheim fellow, 1954-55. Mem. Am. Acad. Arts and Scis., A.A.-A.S., Am. Soc. Zoology, Am. Soc. Parasitologists, Am. Micros. Soc., Soc. Systematic Zoologists, Phi Beta Kappa, Sigma Xi. Contbr. numerous articles to profl.

jours. on taxonomy, morphology, life histories of parasitic and free-living copepods. Home: 149 Cabot St., Newton, Mass. 02158. Office: 2 Cummington St., Boston 02215.*

HUMM, Douglas George, physiologist; b. St. Anne de Bellvue, Que., Can., May 10, 1917; s. George Herbert and Mable (Charles) H.; B.S., Yale, 1939; Ph.D., Stanford, 1948; m. Jane Harrison, Mar. 29, 1947; children—Alan Douglas, Sandra Margaret. Came to U. S., 1917, naturalized, 1959. Asst. prof. U. N.M., 1945-51; faculty U. N.C., Chapel Hill, 1951—, prof. zoology, 1961—. NIH fellow, 1964. Mem. Am. Soc. Zoologists, Soc. Gen. Physiologists, Soc. Developmental Biology, A.A.A.S. Research, publs. on tissue culture and embryological transplantation, tyrosinase levels in developing salamander, determination nitrogen by micromethods, analysis amino acids in salamander blood by manometric methods, cellular physiology, intermediary metabolism, embryological transplantation, biochem. embryology. Home: Smith Level Rd., Chapel Hill, N.C. 27514.*

HUMM, Harold Judson, Am. marine biologist; b. Lorain, O., Feb. 26, 1912; s. William Alfred and Iza B. (Shepherd) H.; B.S., U. Miami, 1934; M.A., Duke, 1942, Ph.D., 1945; m. Olga Squires Minor, June 24, 1936; children—Sandra Susan (Mrs. Craig T. Smith), Roger Brian, Marilyn Jeanette. Resident investigator Marine Lab., Duke, Beaufort, N.C., 1942-45, asst. dir., 1945-48, dir., 1948-49, asso. prof. botany, Durham, N.C., 1954-65; prof. biology Queens Coll., Charlotte, N.C., 1965-67; prof. marine sci. U. S. Fla., 1967—. Dir. Oceanographic Inst., prof. botany Fla. State U., 1949-54. Indsl. specialist WPB, 1942-44; Jacques Loeb asso. marine biology Rockefeller Inst., N.Y.C., 1959-60. Mem. Assn. Southeastern Biologists (pres. 1967-68), Internat., Am. phycological socs., Am. Inst. Biol. Scis., Am. Soc. Limnology and Oceanography, Phi Beta Kappa, Sigma Xi. Research, numerous publs. on taxonomy, distbn., ecology marine algae of Southeastern coast N.Am., agar and other polysaccharides of red algae, marine bacteriology. Patentee new kind of agar. Office: U. South Fla., St. Petersburg, Fla. 33701.*

HUMMEL, Alfred, German engr.; b. Heilbronn, Sept. 7, 1891; s. Gottlob and Ida (Langenbacher) H.; student engring. tech. schs., Karlsruhe, Stuttgart; m. Margarete Stehle, Mar. 28, 1925; children—Evamaria, Christof. Asst. at tech. sch.; div. head research instr.; dir. inst. for materials control; asso. prof. U. Berlin; prof. tech. sch., Aix-la-Chapelle, dir. Inst. for Research on Constrn. Recipient Emil-Morsch medal. Mem. Tech. Assn. (hon.). Author publs. on tech. of constrn. materials. Home: Im Brockenfeld II. Office: Technische Hochschule, Tempelgraben, Aachen, Germany.

HUMMEL, Katharine Pattee, Am. biologist; b. St. Paul, Oct. 2, 1904; d. John Adolph and Adeline (Pattee) H.; B.A., Carleton Coll., 1926; M.A., U. Minn., 1927; Ph.D., Cornell U., 1934. Instr., Carleton Coll., 1927-28, 29-32, Mt. Holyoke Coll., 1935-36, Cornell U., 1936-37; faculty Douglass Coll., Rutgers U., 1939-43, asst. prof., 1941-43, acting head dept. zoology, 1942-43; staff scientist Jackson Lab., Bar Harbor, Me., 1943—, 1943-62, sr. staff scientist, 1962—. Mem. Am. Assn. for Cancer Research, Am. Assn. Anatomists, Am. Soc. Zoologists, Growth Soc. Contbg. author: Biology of the Laboratory Mouse, 1966. Research, publs. on mammary tumor agt. and breast cancer in mice, developmental abnormalities in mice, morphology, causation and inheritance, gross and microscopic anatomy mice, dystrophic calcification heart muscle in mice. Home: Cleftstone Rd. Office: Jackson Lab., Bar Harbor, Me. 04609.*

HUMPHREY, Edward William, Am. surgeon; b. Fargo, N.D., Dec. 6, 1926; s. Edward W. and Minnie (Ramstad) H.; B.A., U. Minn., 1948, M.D., 1951, Ph.D. in Physiology, 1959; m. Noreen Sander, Sept. 23, 1950; children—Katherine Lisa, Joan Karen. Faculty, U. Minn., Mpls., 1958—, prof. dept. surgery, 1965—; mem. staff VA Hosp., Mpls., 1958—, chief, surg. service, 1962—. Mem. A.C.S., Minn. Surg. Soc., Am. Physiol. Soc., Am. Soc. Cell Biology, Soc. for Exptl. Biology and Medicine, Sigma Xi, Alpha Omega Alpha. Research, publs. in field cancer: chemotherapy of cancer; electrolyte flux in various tissues. Home: 9734 Russell Circle, Mpls. 55431. Office: VA Hosp., 48th Av. and 54th St., Mpls. 55417.*

HUMPHREY, George, psychologist; b. Boughton, Kent, England, July 17, 1889; s. Edmund and Emily (Maddex) H.; M.A., Ph.D., Oxford U.; m. 2d, Berta Wolpert, Mar. 26, 1956; 1 dau., Anne. Dir. humanities, prof. ancient history St. Francis Xavier U., N.S., Can.; asst. prof. psychology Wesleyan U.; prof. psychology Queen's U., Can.; prof. exptl. psychology, dir. Inst. Exptl. Psychology, Oxford U. Fellow Royal Soc. Can.; mem. Am. (hon.), Canadian psychol. assns., Brit. Psychol. Soc. Author: Nature of Learning; Directed Thinking; Social Psychology through Experiment; Thinking; Chemistry of Our Thinking. Address: 52 Sherlock Close, Cambridge, Eng.

HUMPHREY, George Frederick, Australian biochemist; b. Sydney, Australia, June 28, 1920; s. Harry and Priscilla (Pope) H.; M.Sc., Sydney U., 1940, Ph.D., 1949; m. Beverly Arline Franklin, Nov. 3, 1945; 1 dau., Sonia Denise. Asst. research officer fisheries sect. Commonwealth Council for Sci. and Indsl. Research, 1942; faculty Sydney U., 1942-56, sr. lectr. biochemistry, 1949-56; chief div. fisheries and oceanography Commonwealth Sci. and Indsl. Research Orgn., Sydney, 1956—. Pres., Sci. Com. Oceanic Research, 1961-64. Mem. Biochem. Soc. (London), Australian and New Zealand Assn. for Advancement Sci. Research, publs. on biochemistry of oysters, metabolism of unicellular plants and animals. Home: 79 Lansdowne Crescent, Oatley, Sydney 2223. Office: Box 21, Cronulla, Sydney, Australia.*

HUMPHREY, George Murray, English surgeon; b. Sudbury, Eng., July 18, 1820; s. William Wood Humphrey; became student St. Bartholomew's Hosp., London, 1839; M.B., 1852, M.D., 1859. Dep. prof. anatomy, Cambridge, Eng., 1847-66, prof. human anatomy, 1866-83, prof. human anatomy and surgery, 1883—. Author: A Treatise on the Human Skeleton including the Joints, 1858; Observations on the Limbs of Vertebrate Animals, 1860; Observations in Myology, 1872; also others. Founder, Jour. Anatomy and Physiology, 1867. Died 1896.

HUMPHREY, John Herbert, English immunologist; b. West Byfleet, Eng., Dec. 16, 1915; s. Herbert Alfred and Mary (Hornblow) H.; B.A., Cambridge (Eng.) U., 1937, M.B., 1940, M.D., 1947; m. Janet Rumney Hill, Nov. 11, 1939; children—Simon John, Nicholas Keynes, Sarah Maynard, Andrea Vanessa, Charlotte Miranda. Jenner Research student Lister Inst. Preventive Medicine, 1940-41; asst. pathologist Central Middlesex Hosp., 1942-46; staff Med. Research Council, London, 1946—; head div. immunology Nat. Inst. for Med. Research, 1957—, dep. dir., 1962—. Mem. expert com. on biol. standardization WHO, 1956—, mem. expert com. on immunology, 1963—. Fellow Royal Soc., 1963; mem. Am. Assn. Immunologists, Brit. Soc. for Immunology, Physiol. Soc. Biochem. Soc., Path. Soc., Soc. for Gen. Microbiology. Author: (with R. G. White) Immunology for Students of Medicine, 1963; also numerous articles. Asso. editor Immunology, 1957—; co-editor Advances in Immunology, 1960—. Research on mechanisms of action, biosynthesis, stimulation of antibodies. Home: 12 Hampstead Sq., London N.W.3. Office: Nat. Inst. for Med. Research, Mill Hill, London N.W. 7., Eng.*

HUMPHREY, Philip Strong, Am. ornithologist; b. Hibbing, Minn., Feb. 26, 1926; s. Watts Sherman and Katharine (Strong) H.; B.A. cum laude, Amherst Coll., 1949; M.S., U. Mich., 1951, Ph.D., 1955; m. Mary Louise Countryman, Jan. 1, 1946; children—Margaret H., Stephen S. Research asso. U. Mich., 1955-57; asst. curator ornithology, asst. prof. zoology Peabody Mus. Natural History, Yale, 1957-62; curator birds U. S. Nat. Mus., Smithsonian Instn., 1962-65; chmn. dept. vertebrate zoology, 1965-67; dir. Mus. Natural History, chmn. dept. zoology U. Kan., 1967—; research asso. Fla. State Mus., 1961; research prof. U. Md., 1964. John Simon Guggenheim Meml. fellow, 1960-61. Mem. Am. Ornithologists' Union, Nat. Audubon Soc., Am. Soc. Zoologists, Soc. Systematic Zoologists, Cooper Ornithol. Soc., Wilson Ornithol. Soc., Internat. Council for Bird Preservation, Conn. Acad. Scis., Assn. for Tropical Biology. Research and publs. in ecology and distribution of New World tropical birds. Home: 612 Louisiana St., Lawrence, Kan. 66044.

HUMPHREY, Robert Regester, Am. range ecologist; b. Palo Alto, Cal., June 14, 1904; s. Harry Baker and Olive (Mealey) H.; student George Washington U., 1924-26; B.A., U. Minn., 1928, M.A., 1930, Ph.D., 1933; m. Roberta January, Sept. 12, 1929; children—Shirley (Mrs. Carl Schoof), Lois (Mrs. Robert Angus), Alan B., Beth P. Range ecologist U. S. Forest Service, 1933-35; range conservationist U. S. Soil Conservation Service, 1936-48; faculty U. Ariz., Tucson, 1948—, prof. range mgmt., 1955-66, emeritus, 1966—. Cons. on range surveys, Mexico, Brazil. Recipient Distinguished Service award Ariz. Water Resources Com., 1967. Mem. A.A.A.S., Ecol. Soc. Am., Am. Inst. Biol. Scis., Am. Soc. Range Mgmt. Author: Range Ecology, 1962; also numerous articles. Developed site analysis range condition concept; research on fire as ecol. factor in vegetation types of S.W. U. S. Home: Route 8, Box 106 A, 9211 E. Rosewood, Tucson, Ariz. 85710.*

HUMPHREY, Rufus Richard, Am. anatomist; b. Manistee County, Mich., Apr. 12, 1892; s. Ellis and Helena (Lumley) H.; A.B., Eastern Mich. U., 1917; A.M., Cornell U., 1920, Ph.D., 1923. Instr. biology Berea Coll., 1916-17; instr. histology Cornell U., 1919-23; with U. Buffalo, 1923—, prof. anatomy, 1924-57, emeritus prof., 1957—; vis. prof. anatomy Ind. U., 1957-58, research scholar in zoology, 1957—. Mem. Am. Assn. Anatomists, Am. Soc. Zoology, Am. Soc. Human Genetics, Am. Genetic Assn., Soc. for Study Devel. and Growth, A.A.A.S., Internat. Inst. Embryology. Studies, publs. on interstitial cells and other features of testis in urodeles, primordial germ cells of amphibia, sex reversal in ambystomid salamanders, proof of female heterogamety in these urodeles by mating reversed females, studies on polyploidy in salamanders; pioneer to obtain progeny from triploid and tetraploid vertebrates; studies of several mutations in Mexican axolotl; discovery of maternal effects of three mutant genes; tumors of testis in Mexican axolotl.

HUMPHREYS, Curtis Judson, Am. physicist, govt. ofcl.; b. Alliance, O., Feb. 17, 1898; s. James and Olive E. (Conser) H.; B.S., Ohio Wesleyan U., 1918; M.S., U. Ky., 1921; Ph.D., U. Mich., 1928; m. Jeanetta Mae Raum, Dec. 25, 1922; children—Richard Raum, Jean Carolyn (Mrs. George Blakeslee), Kathryn Faye (Mrs. Joel Harold Sundquist), Jamie Louise. Instr. physics U. Ky., 1919-21, Western Res. U., 1922-27, Mass. Inst. Tech., 1934-35; research asst. U. Mich., 1921-22; physicist Nat. Bur. Standards, Washington, 1928-33, 35-44, chief radiometry sect., 1944-51, chief infrared spectroscopy sect., Corona, Cal., 1951-53; head infrared spectroscopy div. U. S. Naval Ordnance Lab., Corona, 1953-55, head phys. sci. dept., 1955-57, head research dept., 1957—. Mem. com. on line spectra elements NRC, 1948—. Recipient award for distinguished achievement in sci. U. S. Navy, 1963, award for participation in Manhattan Project Dept. Def., 1945, Meritorious Civilian Service award Dept. Commerce, 1953. Fellow Am. Phys. Soc., Optical Soc. Am.; mem. Washington Acad. Scis., Philos. Soc. Washington, Internat. Astron. Union (commn. 14), Internat. Council for Sci. Unions (corr. mem. triple commn. for spectroscopy), Phi Beta Kappa, Sigma Xi. Author: Wave-length Standards in the Infrared (with D. H. Rank, K. N. Rao), 1965. Contbr. sects. to encys. Placed uranium in proper place in periodic series; discovered 6th series atomic hydrogen; extended internat. system wavelength standards to 3.5 microns; invented sapphire spectrum tube for microwave excitation. Home: 3693 Yosemite Way, Riverside, Cal. 92506. Office: U. S. Naval Ordnance Lab., Corona, Cal. 91720.*

HUMPHREYS, Lloyd Girton, Am. research psychologist; b. Lorane, Ore., Dec. 12, 1913; s. John P. and Gertrude (Stephenson) H.; B.S., U. Ore., 1935; A.M. Ind. U., 1936; Ph.D., Stanford, 1938; m. Dorothy Jane Windes, Dec. 28, 1937; children—John Daniel, Michael Stephenson, Margaret Anne, Susan Jeanne. Faculty, Northwestern U., 1939-45 (on leave 1941-45), U. Wash., 1946-48, Stanford, 1948-52 (on leave 1951-52); tech. dir. Personnel Lab. USAF, San Antonio, 1951-57; prof. psychology U. Ill., Urbana, 1957—, head, dept. psychology, 1959—. Recipient Civilian award for Exceptional Service USAF, 1958. NRC fellow, Yale, 1938-39; Carnegie fellow, Columbia, 1941-42. Mem. Am. (div. pres. 1957-58, 60-61), Midwestern psychol. assns., Psychometric Soc. (pres. 1961-62), Psychonomic Soc. (chmn. governing bd. 1962-63), A.A.A.S. (v.p. 1963), Soc. Exptl. Psychologists, Phi Beta Kappa, Sigma Xi, Delta Upsilon, Beta Gamma Sigma, Phi Delta Kappa. Research, publs. in field. Home: Rural Route 1, White Heath, Ill. 61884.*

HUMPHREYS, Richard Franklin, Am. physicist; b. Greenville, O., May 16, 1911; s. Robert Thomas and Tunia (Cunningham) H.; B.A., DePauw U., 1933, D.Sc., 1963; M.A., Syracuse U., 1936; Ph.D., Yale, 1939. Physics faculty Syracuse U., 1939-40; faculty Yale, 1940-49, asso. prof. physics, 1947-49; physicist in underwater sound Bur. Ships, USN, 1944-45; v.p., physics chmn. Armour Research Found. (now Ill. Inst. Tech. Research Inst.), Chgo., 1949-61; pres. Cooper Union, N.Y.C., 1961—. Mem. Am. Phys. Soc., A.A.A.S., Newcomen Soc., Sigma Xi. Author: (with Robert Beringer) First Principles of Atomic Physics, 1950. Research, publs. in molecular spectroscopy, ultrasonics, cyclotron in nuclear particle reactions, nuclear reactor research, research adminstrn. Home: 70 E. 10th St., N.Y.C. 10003. Office: Cooper Union, Cooper Sq., N.Y.C. 10003.*

HUMPHREYS, William Jackson, Am. physicist; b. Gap Mills, Monroe County, W.Va., Feb. 3, 1862; s. Andrew Jackson and Eliza Ann (Eads) H.; A.B., Washington and Lee U., 1886, C.E., 1888, Sc.D. (hon.), 1942; grad. schs. Physics and Chemistry, U. Va., 1889; fellow in physics Johns Hopkins, 1895-96, Ph.D., 1897; m. Margaret Gertrude Antrim, Jan. 11, 1908. Prof. physics and math. Miller Sch., Va., 1889-93; prof. physics and chemistry Washington Coll., Md., 1893-94; instr. physics U. Va., 1897-1905; prof. meteorol. physics U. S. Weather Bur., 1905-35, ret., 1935, collaborator, from 1936; prof. meteorol. physics, George Washington Univ., 1911-34, emeritus, from 1934. Dir. Research Sta., Mt. Weather, Va., 1905-08; mem. U. S. Naval Observatory's eclipse expdn. to Sumatra, 1901; sec. Physics of the Electron, Internat. Congress of Arts and Sciences, St. Louis, 1904. Fellow Am. Phys. Soc., Optical Soc. Am. Astron. Soc. America, Seismol. Soc. Am., A.A.A.S. (gen. sec. 1924-28); mem. Am. Math. Soc., Am. Meteorol. Soc. (pres. 1928-29), Am. Acad. Arts and Sciences, Am. Philos. Soc., Am. Geophys. Union (chmn. 1932-35), Franklin Inst., Washington Acad. Scis. (pres. 1922), Phi Beta Kappa, Sigma Xi; corr. mem. Meteorol. Soc. of Hungary, State Russian Geog. Soc. Author: Physics of the Air, 1920, 2d edit., 1929, 3d edit., 1940; Weather Proverbs and Paradoxes, 1923, 2d edit., 1934; Rain Making and Other Weather Vagaries, 1926; Fogs and Clouds, 1926; Snow Crystals (with W. A. Bentley), 1931; Weather Rambles, 1937; Ways of the Weather, 1942; Fogs, Clouds and Aviation, 1943. Of Me (autobiography), 1947; also articles. Spl. editor (meteorology) Webster's New Internat. Dictionary, 2d edit.; asso. editor Jour. of Franklin Inst.; editor Monthly Weather Rev., 1931-35. Died Nov. 10, 1949.

HUNAIN IBN ISHAQ AL-IBADI (Joannitius), physician; b. Hira, Mesopotamia, 809; one son, Ishaq (also a translator). Court Physician at Bagdad. Author: Quaestiones medicinae; Ten Dissertations on the Eye. Made over 200 translations from Greek into Syriac and Arabic; translated works of Aristotle, Plato, Dioscorides, Hippocrates, Galen; created technical terms and scientific language of the Arabs. Died 873.

HUNAUD, François-Joseph, see Hunauld, François-Joseph.

HUNAULD (or Hunaud), François-Joseph, French anatomist; b. Chateaubriant, France, Feb. 24, 1701; named prof. anatomy and surgery Royal Sch. Medicine, 1730; Fellow Royal Soc., 1733; mem. French Acad. Scis. Author: Chirurgien médecin, 1726; Sur les maladies des os, 1728. Research on bones of human skull; described early case of hydrocephalus. Died Paris, Dec. 15, 1742.

HUNDLEY, James Manson, Am. physician; b. Summitville, Ind., Apr. 17, 1915; s. Frank Martin and Mabelle (Johnson) H.; B.S., Ind. U., 1937, M.D., 1940; m. Grethel Ann Hendricks, Apr. 19, 1940; children—Phyllis Jean (Mrs. Christopher Keenan), Marjorie Ann, June Marie, Joyce Eileen, Frank Martin. Commd. asst. surgeon USPHS, 1941, advanced through grades to asst. surgeon gen., 1961; with NIH, Bethesda, Md., 1943-60, nutrition research chief nutrition sect., 1948-50, chief Lab. Biochemistry and Nutrition, 1950-53, chief lab. research Nat. Inst. Arthritis and Metabolic Diseases, 1953-56, spl. asst. for internat. affairs, NIH, 1958-60; asst. surgeon gen. for plans Office of Surgeon Gen., Washington, 1960-63, asst. surgeon gen. for operations, 1963-66; ret., 1966; dir. Inst. Med. Sci., Presbyn. Med. Center, San Francisco, 1966—. Vice chmn. Surgeon Gen.'s Adv. Com. on Smoking and Health, 1962-64. Recipient Distinguished Service medal USPHS, 1964; Fleming award Washington C. of C., 1954; McLandon Research fellow, 1939-40. Diplomate Am. Bd. Nutrition. Fellow Am. Pub. Health Assn.; mem. A.M.A., Am. Inst. Nutrition, A.A.A.S. (charter), Am. Soc. Clin. Nutrition, N.Y., Washington acads. sci., Sigma Xi, Alpha Omega Alpha. Contbr. numerous articles to tech. jours, chpts. to textbooks. Research on vitamin and amino acid requirements and function, human nutrition, interrelations of diet and health, obesity, diet and heart disease. Home: 8 Inverness Dr., San Rafael, Cal. Office: Inst. Med. Sci., Presbyn. Med. Center, San Francisco 94115.*

HUNECK, Siegfried Robert, German chemist; b. Floh (Thuringia), Sept. 9, 1928; s. Emil and Frieda (Vogt) H.; Dipl.-Chem., U. Jena, 1956, Dr.rer.nat., 1959; habilitation Tech. U. Dresden, 1964; m. Ruth Göhler, June 10, 1960; children—Reinhard, Rolf. Asst., U. Jena, 1956-61; chief asst. Inst. Plant Chemistry, Dresden Inst. Tech., 1961-65, sci. asso., 1965—. Research, many articles on chemistry of triterpenes, photochemistry, chemotaxonomy and structural elucidation of lichen substances and constituents of mosses. Home: 77 W.-Pieck-Strasse, Freital-V. Office: Institute of Plant Chemistry, Tharandt, East Germany.*

HUNERMANN, Theodor, otorhinolaryngologist; b. Mainz, Germany, Oct. 20, 1893; s. Rudolf H. and Carolina (Hiltermann) H.; ed. univs. Fribourg, Giessen, Bonn; M.D.; m. Hete Devin; children—Walter, Gabriele, Theo, Hete. Asst., Inst. Pathology, Fribourg, 1919, Surg. Clinic, U. Fribourg, surg. clinics univs. Vienna, Innsbruck, clinic of otorhinolaryngology univs. Berlin, Düsseldorf; agrégé, 1930; asso. prof. at Düsseldorf, 1936; prof. at large, physician in chief otorhinolaryngol. sect. Marienhospital, Düsseldorf, 1937—. Mem. German Soc. Otorhinolaryngol. Physicians (hon.), Internat. Bronchoesophagological Soc. (v.p.), French, Austrian otorhinolaryngol. socs., Greek Otoneuro-ophthal. Soc., Thessaloniki Soc. Author: Operationen am Ohr; Verletzungen in Gebiet von Hals, Nase, Ohren; Früherkennung des Kehlkopfkrebses in Prophylaktische Med., also numerous articles. Address: Reichstrasse 51, Düsseldorf, Germany.

HUNKINS, Kenneth Leland, Am. oceanographer; b. Lake Placid, N.Y., Mar. 3, 1928; s. Harlan Kenneth and Florence (Olsen) H.; B.Sc., Yale, 1950; M.Sc., Stanford, 1956, Ph.D., 1960; m. Julia Bruce Bontjes, May 28, 1960; children—Sarah Bruce, Elizabeth Ann. With Lamont Geol. Obs., Palisades, N.Y., 1957—, research scientist, 1960-64, sr. research asso., 1964—. Mem. Am. Geophys. Union, Acoustical Soc. Am., A.A.A.S., Sigma Xi. Research, publs. on geophys. exploration Arctic Ocean from drifting ice stas., ocean currents, bottom topography, sediments, underwater acoustics and ocean waves; mapped Arctic Ocean floor. Home: 23 Hickory Hill Rd., Tappan, N.Y. 10983. Office: Lamont Geol. Obs., Palisades, N.Y. 10964.*

HUNSAKER, Jerome Clarke, Am. engr., educator; b. Creston, Ia., Aug. 26, 1886; s. Walter Jerome and Alma (Clarke) H.; B.S., U. S. Naval Acad., 1908; M.S., Mass. Inst. Tech., 1912, Sc.D., 1916; m. Alice Porter Avery, June 24, 1911; children—Sara Swope, Jerome Clarke, Peter (dec.), Alice. Officer to comdr. USN, 1909-26; asst. v.p. Bell Telephone Labs., 1926-28; v.p. Goodyear-Zellelin Corp., 1928-33; prof., head dept. mech. engring. Mass. Inst. Tech., 1933-47, head dept. aero. engring., 1933-51. Mem. Nat. adv. Com. for Aeros., 1938-58, chmn., 1941-57. Decorated Navy Cross, D.S.M. Recipient Guggenheim medal, 1933; Franklin medal, 1942; Presdl. medal for Merit, 1946; Legion of Honor, 1949; Wright Trophy, 1951; Langley medal, 1955; Gold medal Royal Aeros. Soc. (Gt. Britain), 1957; USN award Distinguished Pub. Service, 1958; Presdl. citation for service to NASA, 1958. Fellow Am. Phys. Soc., A.A.A.S., Am. Soc. Naval Architects and Marine Engrs., Nat. Acad. Scis., Sigma Xi. Author: (with B. G. Rightmire) Engineering Applications of Fluid Mechanics, 1947; Aeronautics at the Mid-Century, 1952. Research in aeronautic design and communications. Home: 10 Louisburg Sq., Boston, 02108. Office: Mass. Inst. Tech., Aeros. 33-207, Cambridge, Mass. 02139.

HUNSBERGER, Isaac Moyer, Am. organic chemist; b. Quakertown, Pa., Aug. 3, 1921; s. Amos Franklin and Eliza (Moyer) H.; B.S., Lehigh U., 1943, M.S., 1946, Ph.D., 1948; m. Elizabeth Rita Ochnich, Mar. 19, 1944; children—Donald Moyer, Elizabeth Anne, Gretchen, Mark, Carol, Luke, Heidi. Postdoctoral fellow U. Ill., 1948-49; faculty Antioch Coll., 1949-55; asso. prof. Fordham U., 1955-60; faculty U. Mass., Amherst, 1960—, head dept. chemistry, 1960-61, dean Coll. Arts and Scis., research prof. chemistry, 1961—. Mem. com. modern methods of handling chem. information NRC, 1959-65. Mem. Am. Chem. Soc. (chmn. subcom. chem. notations of organic div. 1964-66), A.A.A.S., Am. Assn. U. Profs., Phi Beta Kappa, others. Author: (with others) Survey of Chemical Notation Systems, 1964. Editor sect. on condensed carbocyclic compounds Chem. Abstracts, 1960-64. Research, publs. on structure of aromatic hydrocarbons by use of infrared and nuclear magnetic resonance spectra of hydrogen-bonded derivatives; chemistry and reactions of sydnones, sulfoxides, N-oxides and N-nitrosoimines. Home: 346 N. Pleasant St., Amherst, Mass. 01002.*

HUNSCHER, Helen Alvina, Am. nutritionist, educator; b. Gates Mills, O., Aug. 5, 1904; d. Ernest Henry and Cora (Knapp) H.; A.B., Ohio State U., 1925; Ph.D., U. Chgo., 1932; m. Howard Paul Wilkinson, Dec. 23, 1939; children—Jean (Mrs. Mark Edwin Goebel), James Howard. Fellow nutrition research Merrill-Palmer Sch., Detroit, 1926-27; Laura Spelman Rockefeller fellow, 1927-28, NRC fellow, 1929-30; asst. nutrition research Merrill-Palmer Sch. and Children's Hosp. Mich., 1928-29; instr. home econs. U. Chgo., 1930-31; asso. research Children's Fund of Mich., Detroit, 1931-37; profl. lectr. nutrition Wayne U., 1931-37; faculty Western Res. U., 1937—, prof., chmn. dept. nutrition, 1963—. Recipient Marjorie Hulsizer Copher award, 1958. Mem. Am. Bd. Nutrition, Am. Diet. Assn. (pres. 1947-48), Am. Home Econs. Assn., Am. Inst. Nutrition, Am. Chem. Soc., Pub. Health Assn., Soc. Exptl. Biology and Medicine, Soc. Research in Child Devel., A.A.A.S., Am. Assn. Maternal and Child Health (exec. bd. Ohio chpt.), Phi Beta Kappa, Sigma Xi, others. Editorial bd. Heinz Handbook of Nutrition, 2d edit., 1965. Research, studies, publs. on vitamin content and chemistry of human milk, mineral metabolism of women during pregnancy and lactation, child nutrition, child growth and devel. Home: 35 Old Mill Rd., Gates Mills, O. 44040.*

HUNT, Charles Butler, Am. geologist; b. West Point, N.Y., Aug. 9, 1906; s. Irvin Leland and Annie (Butler) H.; A.B., Colgate U., 1928; postgrad. Yale; m. Alice Parker, Oct. 23, 1930; children—Eugene Parker, Anne B. (Mrs. John T. MacDonald). Geologist, U. S. Geol. Survey, Mt. Taylor, N.M., Henry Mountains, Utah, 1930-42, asst. chief mil. geology unit, 1943, chief, 1944-45, regional geologist, Salt Lake City, 1946-47, chief gen. geology br., 1948-53, research, Death Valley, Cal., 1956-60; exec. dir. Am. Geol. Inst., Washington, 1954-55; prof. Isaiah Bowman dept. geography Johns Hopkins, 1961—. Recipient U. S. Dept. Interior Distinguished Service award, 1961. Mem. Geol. Soc. Am., Am. Assn. Petroleum Geologists, Nat. Assn. Geology Tchrs., Phi Beta Kappa. Author: Geology of the Mount Taylor Volcanic Field, 1938; Geology and Geography of the Henry Mountains Area, Utah, 1953; Geology of Death Valley, California, 1965; Physiography of the United States, 1966; others; also articles. Office: 1 University Pkwy. E., Balt. 21218.*

HUNT, Chester Leigh, sociologist; b. Duluth, Minn., July 24, 1912; s. Ray Erwin and Ada (Bash) H.; A.B., Neb. Wesleyan U., 1934; M.A., Washington U., St. Louis, 1937; Ph.D., U. Neb., 1948; m. Maxine Cole, Aug. 28, 1942; children—Joanne, Leigh Rae. Asst. prof. Neb. State Tchrs. Coll., Chadron, 1941-42; prof. sociology Western Mich. U., Kalamazoo, 1948—, chmn. Asian studies com., 1964—, asso. dir. Center for Sociol. Research, 1956—; acting head dept. sociology U. Philippines, 1952-54; prof. sociology Silliman U., Dumaguete City, Philippines, 1961-62. Fulbright lectr., 1941-42, 52-54. Mem. Philippine, Am., Mich. (past pres.) sociol. socs., Soc. for Study Social Problems, Asian Soc., Soc. for Internat. Devel., Alpha Kappa Delta. Author: (with others) Sociology in Philosophy Setting, 1954, rev., 1963; (with Paul Horton) Sociology, 1964; Social Foundation of Community Development, 1964; Social Aspects of Economic Development, 1966; also articles. Editor: Philippine Sociol. Rev., 1953-54. Research on application sociol. prins. to study change in developing areas, delineation role perception in race relations; designed community survey procedure to measure alcoholism. Home: 2248 Crest Dr., Kalamazoo 49001.*

HUNT, Edward Eyre, Am. anthropologist, educator; b. Washington, Mar. 9, 1922; s. Edward Eyre and Virginia (Fox) H.; A.B., Harvard, 1942, A.M., 1949, Ph.D., 1951, postgrad. Sch. Pub. Health; m. Vilma Maxine Rose Dalton-Webb, Apr. 13, 1952; children—Margaret Rose, William Webb, Louise Cassilly, Catherine McCord. With USAAF, 1945-46, Peabody Mus. (Harvard) Expdn. to Yap, Micronesia, 1947-48; faculty Harvard, 1947, 49-50, 51-55, lectr., 1955-60, asst. prof., 1960-66; staff anthropologist Forsyth Dental Center, 1951-66; research fellow Peabody Mus., 1955-66, research asso., 1966—; prof. anthropology City Univ. N.Y., 1966—. Fulbright vis. lectr. U. Melbourne, 1956-57; mem. study sect. in applied physiology NIH, 1965—. Fellow A.A.A.S., Am. Anthrop. Assn., N.Y. Acad. Scis.; mem. Am. Soc. Human Genetics, Am. Eugenics Soc., Am. Assn. Phys. Anthropologists. Author: (with C. S. Coon) The Living Races of Man, 1965. Co-editor: Anthropology A to Z, 1963. Research, publs. on phys. growth of children including obesity, physique and body composition, x-ray studies of tooth formation and resorption, growth of head and teeth in Australian aborigine children, genetic control of human devel., population genetics, palm and finger prints of Micronesians. Home: 17 Buxton Lane, Riverside, Conn. 06878. Office: Grad. Center, City U. N.Y., 33 W. 42d St. N.Y.C. 10036.*

HUNT, Franklin Livingston, Am. physicist; b. Manchester, N.H., Sept. 3, 1883; s. Charles Franklin and Flora N. (Webber) H.; B.S., Mass. Inst. Tech., 1909, Ph.D., 1919; A.M., Harvard, 1913; postgrad. Berlin, Paris, Cambridge, Eng.; m. Dorothy Helen Allen, Jan. 18, 1926 (dec. Feb. 1963); children—Barbara Standish (Mrs. David W. Dodge), Allen Standish. Faculty, Mass. Inst. Tech., 1909-13, 15-17, Harvard, 1915-16; physicist Bur. Standards, Washington, 1917-27, rep. in Europe, 1919-20; asst. mgr. engring. and research Victor Talking Machine Co., Camden, N.J., 1927-29; lectr. George Washington U., 1925-27; mem. tech. staff Bell Telephone Labs., N.Y.C., 1929-48. Mem. sub com. on aeros. Nat. Adv. Commn. for Aeros., Washington, 1921-24. Fellow Am. Phys. Soc., Acoustical Soc., A.A.A.S. Researcher in aeros., electrolytic conduction aqueous solutions, sound rec. and reprodn., x-rays, telephony. Home: 67 Robinson Pkwy., Burlington, Vt. 05401.*

HUNT, Graham Robert, phys. chemist; b. Deniliquin, Australia, Dec. 2, 1930; s. Robert David and May (Levett) H.; B.Sc., Sydney U., 1951, B.Sc. with honors, 1952, M.Sc., 1954, Ph.D., 1958; m. Jeanette M. Chaston, Aug. 24, 1959; children—Gillean, Lysette, Bronwyn. Teaching fellow Sydney U., 1957-58; research officer Commonwealth Sci. and Indsl. Research Orgn., 1959; research asso. Tufts U., 1959-62; research asso. Mass. Inst. Tech., 1962-63; spectroscopist Air Force Cambridge Research Labs., Hanscom Field, Bedford, Mass., 1964—; lectr. U. Vt., 1963. Cons. Polaroid Corp. Recipient citations Dept. Air Force, 1964, 65, 66. Mem. Am. Geophys. Union, Royal Australian Chem. Inst., Optical Soc. Am. Research, publs., patents in field of spectroscopy, electronic, molecular spectra-normal coordinate analyses techniques; design of spectrometers; application of spectroscopic techniques to determination of composition of lunar, planetary surfaces and atmospheres, telescopic observation, instrumentation; inventor spectrum matching technique, thermal enhancement methods for remote sensing. Home: 28 Thornberry Rd., Winchester, Mass. 01890. Office: Air Force Cambridge Research Labs., Hanscom Field, Bedford, Mass. 01730.*

HUNT, Harry, mathematician; flourished Eng., 1673-1713; studied constrn. of instruments under Robert Hooke, also pvt. asst.; became operator Royal Soc., 1676, named keeper Library and Repository, 1696; read meteorol. instruments at Gresham House. Built marine barometers, weather-glass for Samuel Pepys, instruments for meteorol. studies.

HUNT, Howard Francis, Am. psychologist; b. Morgantown, W.Va., May 29, 1918; s. Harrison Randall and Jane (Fisher) H.; B.A. with high honors, Mich. State U., 1940; Ph.D., U. Minn., 1943; m. Ida Altman, Aug. 16, 1941; children—Carol Ann, William Harrison, Steven Charles, John Howard. Instr. psychology U. Minn., 1943-44; asst. prof. psychology Stanford, 1946-48; faculty U. Chgo., 1948-62, prof., 1954-62, chmn. dept., 1955-62; chief psychiat. research N.Y. State Psychiat. Inst., N.Y.C., 1962—; prof. psychology, med. psychology Columbia Coll. Physicians and Surgeons, 1962—. Cons. VA, 1948—, Nat. Inst. Mental Health, 1956—. Fellow Center for Advanced Study in Behavioral Scis., 1959-60. Diplomate Am. Bd. Examiners in Prof. Psychology. Fellow Am. Psychol. Assn., N.Y. Acad. Scis., A.A.A.S.; mem. Eastern Psychol. Assn., Am. Psychopath. Assn., Sigma Xi, Phi Kappa Phi. Author: Hunt-Minnesota Test for Organic Brain Damage, 1943. Editor: Jour. Abnormal Psychology. Publs. on method of intellectual loss measurement after brain injury, effects of lesions on emotional behavior and learning in animals, effects of drugs. Home: Langdon Av., Ardsley Park, Irvington-on-Hudson, N.Y. 10533. Office: 722 W. 168th St., N.Y.C. 10032.*

HUNT, James Ramsay, Am. neurologist; b. Phila., 1874; s. William R. and Eva (Ramsay) H.; M.D., U. Pa., 1893, hon. Sc.D., 1931; studied Vienna, Berlin and Paris; m. Alice St. John Nolan, Sept. 26, 1908;

children—James Ramsay, Alice St. John. Asso. prof. nervous diseases, Columbia, 1910-15; clin. prof. nervous diseases, Columbia, 1924; prof. clin. neurology, from 1931; dir. neuropsychiatric division, New York Neurol. Institute; cons. neurologist Psychiatric Inst., the Babies', N.Y. Eye and Ear and Lenox Hill hosps., and Letchworth Village for Mental Defectives; formerly cons. psychiatrist to Lying-In Hosp., New York Hospital and Randall's Island institutions. Consulting neuropathologist to Craig Colony for epileptics. Contbr. numerous articles in field to profl. jours. Described syndrome asso. with disease of geniculate ganglion of 7th nerve, marked by herpes, ear ache, sometimes facial paralysis (Hunt's syndrome). Died July 22, 1937.

HUNT, J(oseph) McV(icker), Am. psychologist; b. Scottsbluff, Neb., Mar. 19, 1906; s. R. Sanford and Carrie Pearl McVicker (Loughborough) H.; A.B., U. Neb., 1929, M.A., 1930; Ph.D., Cornell U., 1933; Sc.D., (hon.), Brown University, 1958; Sc.D., University of Neb., 1967; m. Esther Dahms, Dec. 25, 1929; children—Judith Ann, Carol Jean. With U. Neb., 1929-31; vis. instr. Cornell U., 1935, asst., 1931-33; NRC fellow N.Y. Psychiat. Inst., Columbia, 1933-34, Worcester State Hosp., Clark U., 1934-35; research asso. St. Elizabeth Hosp., Washington, 1936; faculty Brown U., 1936-46, asso. prof., 1944-46; research cons. Inst. Welfare Research, Community Service Soc. N.Y., 1944-46, dir., 1946-51; prof. psychology U. Ill., Urbana, 1951—, coordinator tng. in clin. psychology, 1951-62; lectr. Columbia, 1948-50; adj. prof. N.Y. U., 1950-51; dir. Coordination Center, Nat. Lab. of Early Childhood Education, since 1967—. Member committee on anthropology and psychology NRC, 1953-56, mem. disaster com., 1956-58; mem. tng. com. Nat. Inst. Mental Health, 1959-64, mem. social problems com., 1965—, chmn., 1966—. Recipient award for excellence in research Am. Personnel and Guidance Assn., 1950, 60, Research Career award Nat. Inst. Mental Health, 1962—. Diplomate Am. Bd. Profl. Psychology. Fellow A.A.A.S., Am. Psychol. Assn. (past pres.); mem. Am. Sociol. Assn., Am. Statis. Assn., Am. Acad. Psychotherapists, Eastern Psychol. Assn. (past pres.), Soc. for Research in Child Devel. Author: (with L. S. Kogan) Measuring Results in Social Casework: a Manual on Judging Movement, 1950; (with Kogan, M. Blenkner) Testing Results in Social Casework: A Field-Test of Movement Scale, 1950; (with Kogan, P. Bartelme) A Follow-up Study of the Results of Social Casework, 1953; Intelligence and Experience, 1961; also numerous articles. Editor: Personality and the Behavior Disorders, 2 vols., 1944; Jour. Abnormal and Social Psychology, 1950-55. Research on psychol. deficit in psychoses and neurol. disorders, effects infantile experience on adult behavior, change in clients with social casework and psychotherapy, role of early experience in devel. of intelligence and maturation, ordinal scales of psychol. devel. in human infants. Home: 1807 Pleasant St., Urbana, Ill. 61801.*

HUNT, John Meacham, Am. geochemist; b. Cleve., Dec. 1, 1918; s. Raymond E. and Marguerite (Meacham) H.; A.B., Western Res. U., 1941; M.S., Pa. State U., 1943, Ph.D., 1946; m. Doris Press, Apr. 26, 1947; children—Randall Kieth, Lawrence Lee. Instr., Pa. State U., 1946-47; research chemist Jersey Prodn. Research Co., 1948-55, head geochem. research, 1956-63; chmn. dept. chemistry Woods Hole (Mass.) Oceanographic Instn., 1964—. Lectr. for continuing edn. com. Am. Assn. Petroleum Geologists, asso. editor Bulletin. Mem. Am. Chem. Soc., Geochem. Soc., Geol. Soc. Am., Am. Geol. Inst., Am. Geophys. Union, Am. Assn. Petroleum Geologists, A.A.A.S. Research, publs., patents in fields of origin, migration, accumulation of petroleum; devel. geochem. techniques for exploring for oil; nature and distbrn. of organic matter in marine environment. Home: P.O. Box 112. Office: Woods Hole Oceanographic Instn., Woods Hole, Mass. 02543.*

HUNT, John Nigel, English mathematician; b. Hillingdon, Eng., July 18, 1928; s. Cyril and Mabel (White) H.; B.Sc., Imperial Coll., London, Eng., 1950, Ph.D., 1952; m. Lydia Dixon, July 25, 1956; children—Penelope Ann, Lucinda Rosemary, Miranda Jane. Sci. officer Hydraulics Research Sta., Wallingford, Berkshire, Eng., 1952-55; sr. sci. officer A.W.-R.E., Aldermaston, Berkshire, 1955-58; asso. prof. math. Ga. Inst. Tech., 1958-59; faculty Imperial Coll., 1959-64, reader phys. oceanography, 1961-64; prof. applied math. Reading (Eng.) U., 1964—. Cons. in hydrodynamics, indsl. and engring. applications, 1958—. Fellow Royal Astron. Soc., Royal Meteorol. Soc. Author: Incompressible Fluid Dynamics, 1964; also articles. Research on gravity waves, tides, vortex flows, hydrodynamic instability, sediment transport. Home: 49 Northcourt Av., Reading, Berkshire, Eng.*

HUNT, Reid, Am. pharmacologist; b. Martinsville, O., Apr. 20, 1870; s. Milton L. and Sarah E. (Wright) H.; A.B., Johns Hopkins, 1891, Ph.D., 1896; student U. of Bonn, Germany; M.D., U. of Md., 1896, Sc.D., 1925; student Ehrlich's Institut, Frankfort, 1902-04; m. Mary Lillie Taylor, Dec. 12, 1908. Tutor in physiology, Coll. Phys. and Surg. (Columbia), 1896-98; asso. and asso. prof. pharmacology, Johns Hopkins, 1898-1903; chief of div. and prof. pharmacology USPHS, 1904-13; prof. pharmacology, Med. Dept. Harvard, 1913-36; visiting prof. Peking-Union Med. Coll., 1923. Cons. pharmacologist Mass. State Dept. Pub. Health; mem. advisory bd. Hygienic Lab., USPHS. Mem. Permanent Standards Com. of League of Nations. Fellow Am. Acad. Arts and Sciences; mem. Nat. Acad. Sciences, Assn. Am. Physicians, Am. Physiol. Soc., Am. Soc. Pharmacology and Exptl. Therapeutics, A.M.A. (ex-chmn. council on pharmacy and chemistry), Am. Chem. Soc. (chmn. Northeastern sect. 2 yrs.), Leopold-Carol Akademie, Deutsch Pharmacologie Gesellschaft, Phi Beta Kappa, Alpha Omega Alpha. Pres. U. S. Pharmacopeia, 1920-30. Contbr. Am. and European med. jours. and govt. publs. Joint-Author: Non-Alkaloidal Organic Poisons, in Vol. II of Peterson, Haines and Webster's Text-Book of Legal Medicine and Toxicology, 1923; Studies in Experimental Alcoholism, 1907; Studies on the Thyroid (with A. Seidell), 1909; Effects of Various Diets Upon Resistance to Poisons, 1910; Effects of Derivatives of Choline and Analogous Compounds, 1911. Contbr. to Heffter's Handbuch der Exper. Pharmakologie, 1923. Demonstrated that thyroid activity was proportional to iodine content, also demonstrated for 1st time presence of thyroid hormone in human blood. Died Mar. 10, 1948.

HUNT, Robert, English sci. writer, physician, physicist; b. Plymouth Dock (now Devonport), Eng., Sept. 6, 1807; studied with surgeon, anatomy under Joshua Brookes. Practiced medicine with physician, for more than 5 years; head med. dispensary London, Eng., 4 years; practiced Cornwall, Eng.; with firm chem. mfrs., London; keeper mining records, 1845-78; prof. exptl. physics Sch. Mines. Mem. Coal Commn., 1866. Fellow Royal Soc., 1854; mem. Royal Cornwall Poly. Soc. (pres. 1859). Author: Popular Treatise of the Art of Photography, 1841 (1st in English); Elementary Physics, 1851; Mineral Statistics, 1855-84; Dictionary of Arts; British Mining, 1884. Discovered that photosulphate of iron could be used as developing agent, chem. rays of solar spectrum accelerate germination of seeds. Died Oct. 17, 1887.

HUNT, Thomas Sterry, Am. chemist, geologist; b. Norwich, Conn., Sept. 5, 1826; s. Peleg and Jane (Sterry) H.; attended Yale; m. 1877. Chemist mineralogist Geol. Survey of Can., 1847-52; prof. chemistry Laval U., Que., 1856-62, Can., McGill U., Montreal, Can., 1862-68; prof. geology Mass. Inst. Tech., 1872-78. An organizer, sec. 1st Geol. Congress; v.p. Geol. congresses, Paris, 1878, Bologna, 1881, London, 1888. Fellow Royal Soc., 1859, Royal Soc. Can. (a founder, 1st pres. 1884), A.A.A.S. (pres. 1870); mem. Am. Inst. Mining Engrs. (pres. 1877), Am. Chem. Soc. Author: Chemical and Geological Essays, 1875, 78; Coal and Iron in Southern Ohio . . . , 1881; Mineral Physiology and Physiography, 1886; A New Basis for Chemistry: A Chemical Philosophy, 1887; Systematic Mineralogy, 1891. Invented green ink with which Am. paper money is printed, 1859; research on nitrogen compounds; first to define organic chemistry as the chemistry of carbon and its compounds; his studies of polymerism of mineral species opened new field for mineralogy; his researches into chem. and mineral composition of rocks were probably more extensive than those of any of his contemporaries; explained for 1st time the relations of gypsum and dolomites; made 1st systematic geol. classification of stratiform crystalline rocks. Died N.Y.C., Feb. 12, 1892.

HUNT, William Alvin, Am. psychologist, educator; b. Hartford, Conn., Nov. 10, 1903; s. Alvin A. and Mabel (Hodges) H.; A.B. summa cum laude, Dartmouth, 1928; A.M., Harvard, 1929, Ph.D., 1931; m. Edna Reeve Bossen, June 15, 1929 (dec. Apr. 1959); m. 2d, Diana Bengston Theobald, Dec. 19, 1960; 1 dau., Margt. Faculty, Harvard, 1930-31, Dartmouth, 1931-33, Conn. Coll. for Women, 1933-37, Wheaton Coll., 1938-45; with N.Y. State Psychiat. Inst. and Hosp., 1937, Neuropsychiat. Inst. of Harford Retreat, 1937-38; prof. psychology Northwestern U., Evanston, Ill., 1946—, chmn. dept., 1951—. Cons. in clin. psychology to govt. agys. including Nat. Security Agy., 1962—, U. S. Office Edn., 1965—. Fellow Am. Psychol. Assn. (pres. div. clin. psychology 1954), Soc. Exptl. Psychologists; mem. A.A.A.S., Am. Psychopath. Assn., Phi Beta Kappa, Sigma Xi. Author: (with Carney Landis) The Startle Pattern, 1939; The Clinical Psychologist, 1956. Editor: Psychology in Schs.; asso. editor Jour. Clin. Psychology. Research, numerous publs. on psychophysiology, psychopathology, psychiat. screening and clin. judgment. Home: 2815 Sheridan Rd., Evanston, Ill. 60201.*

HUNTEN, Donald Mount, physicist; b. Montreal, Que., Can., Mar. 1, 1925; s. Kenneth William and Winnifred (Mount) H.; B.Sc., U. Western Ont., 1946; Ph.D., McGill U., 1950; m. Isobel Ann Rubenstein, Dec. 28, 1949; children—Keith Atherton, Mark Ross. Research asso. to prof. physics U. Sask., Can., 1950-63; physicist Kitt Peak Nat. Obs., Tucson, 1963—; Cons. NASA, 1964—. Editorial bd. Jour. Geophys. Research, 1967—. Mem. Internat. Union Geodesy Geophysics, Internat. Astron. Union, Canadian Assn. Physicists (editor 1961-63), Am. Phys. Soc., Am. Astron. Soc., Am. Geophys. Union. Author: Introduction to Electronics, 1964. Research, publs. on phys. mechanisms in aurora and related properties of upper atmosphere; twilight emissions from upper atmosphere; atmospheres of inner planets. Office: Box 4130, Tucson 85717.

HUNTER, Albert Sinclair, Am. soil scientist, educator; b. Worthington, Ind., Oct. 21, 1908; s. Sinclair and Effie A. (Carpenter) H.; B.S. in Chemistry, Utah State U., 1938; M.S. in Chemistry, Wash. State U., 1940; Ph.D. in Soils, Rutgers U., 1943; m. Mildred P. Bowers, Sept. 5, 1947; 1 dau., Alice Jane. With Spl. Emergency Guayule Project, Salinas, Cal., 1943-45, U. S. Plant, Soil and Nutrition Lab., Ithaca, N.Y., 1945-48, U. S. Natural Rubber Research Sta., Salinas, 1948-49; sr. soil scientist U. S. Dept. Agr., Corvallis, Ore., 1949-57; prof. soils Ore. State U., 1949-57; prof. soil tech. Pa. State U., University Park, 1957—; soil fertility specialist IRI Research Inst., Recife, Brazil, 1965-66 (on leave). Fellow A.A.-A.S.; mem. Am. Soc. Agronomy, Soil Sci. Soc. Am., Internat. Soil Sci. Soc., Soil Conservation Soc. Am. Research, publs. on soil chemistry, fertilizer needs of crops, irrigation, interactions of lime and fertilizer, mech. equipment for field plot research on crops. Home: 1424 Harris St., State College, Pa. 16801. Office: Tyson Bldg., Pa. State U., University Park, Pa. 16802.*

HUNTER, Francis Robert, physiologist; b. N.Y.C., Oct. 3, 1912; s. George William and E. Isabel (Jobbins) H.; B.S., Cal. Inst. Tech., 1933; M.A., Wesleyan U., 1934; Ph.D., Princeton, 1937; m. Alice Schwarze, Sept. 11, 1954; children—Debora Alice, Linda Cecilia. Fellow, Princeton, 1936-37, asst. in biology, 1934-36; instr. zoology R.I. State Coll., 1937-42; asst. prof. zool. sci. Okla. U., 1942-47; asso. prof. Fla. State U., 1947-50, prof., head dept. physiology, 1950-54; research asso. U. Ill. Coll. Medicine, 1954-56; prof. physiology So. Ill. U., 1956-57; prof., head dept. biology U. Andes, Bogota, Colombia, 1957-65; prof., head dept. biology Centro Exptl. de Estudios Superiores, Barquisimeto, Venezuela, 1966—. Fellow A.A.A.S.; mem. Soc. Zoology, Soc. Exptl. Biology, Physiol. Soc., Am. Inst. Biol. Scis., Soc. Gen. Physiology, Sigma Xi. Author: Biology in Our Lives, 1949; College Zoology, 1949; also numerous articles. Research on cell permeability particularly concept of facilitated diffusion of non-electrolytes and anions in erythrocytes. Home: Urbanizacion Las Acacias, Quinta Lupe, Barquisimeto, Venezuela.*

HUNTER, George William, Am. biologist; b. Mamaroneck, N.Y., Apr. 7, 1873; s. George William and Emma Louise (Cartwright) H.; A.B., Williams, 1895, A.M., 1896; fellow in zoölogy, U. Chgo., 1896-99; post-grad. work and lecturer in methods of teaching nature study and biology, N.Y. Univ., 1907-14, Ph.D., 1918; m. Emily Isabel Jobbins, June 19, 1899; children—George William III, Cartwright, Francis Robert. Teacher of biology at Hyde Park High Sch., Chicago, 1898-99; teacher biology DeWitt Clinton High Sch., New York City, 1899-1906, head dept. biology, 1906-19; prof. biology, Carleton Coll., 1919-20; prof. biology, Knox Coll., 1920-26; adjunct prof. biology, Pomona Colls., 1926-29; lecturer in methods in science, Claremont Coll., from 1930. Asst. Marine Biol. Lab., Woods Hole, Mass., summers, 1900-10. Fellow A.A.A.S.; mem. N.E.A., American Zoölogical Society, Nat. Assn. for Research in Science Teaching, National Society for Study of Edn., Phi Delta Kappa. Author: Elements of Biology, 1907; Essentials of Biology, 1911, 23; A Civic Biology, 1914; Laboratory Problems in Civic Biology, 1916; History of Y.M.C.A. War Work in the Washington District, 1919; Laboratory Manual in Biology, 1923; New Civic Biology, 1926; Teachers Manual, 1927; New Laboratory Problems in Civic Biology, 1927; Problems in Biology, (4th edition, revised, 1940); Pupil's Workbook for Problems in Biology, 1932; Teacher's Manual for Problems in Biology, 1932; The Teaching of Science at the Junior and Senior High School Levels, 1934; Life Science. A Social Biology, 1941. Co-Author: Laboratory Manual of Biology, 1903; Civic Science in the Home, 1921; Civic Science in the Community, 1922; Civic Science in Home and Community, 1923; Civic Science Manual, 1924; Problems in General Science, (4th edit., rev., 1944); Teacher's Manual and Key for Problems in General Science, 1931; Readings in Science, 1931; Workbook in General Science, 1932; A Testing Program in General Science, 1933; A Testing Program in Biology, 1933; The March of Science Series (My Own Science Problems; Science in Our Social Life; Science in Our World of Progress), 1935; Biology, The Story of Living Things, 1937; Work book and testing program for life science, 1941-42; Doorways to Science, 1947; Work book, Doorways to Science, 1947. Contbr. numerous educational and scientific articles. Died Feb. 4, 1948.

HUNTER, George William, III, Am. parasitologist; b. N.Y.C., Jan. 27, 1902; s. George William and E. Isabel (Jobbins) H.; B.S., Knox Coll., 1923; M.S., U. Ill., 1924, Ph.D., 1927, certificate in Tropical and Mil. Medicine, 1944; m. Adelaide Louise White, July 11, 1941; 1 dau., Gay. Prof., Affiliated Units Grad. Sch., Med. Sch. Baylor U., 1952-55; prof. U. Fla., Coll. Medicine, Gainesville, 1956—; prof. La. State U. Sch. Medicine, 1961-63. Exec. officer Commn. of Schistosomiasis, U. S. Army, Philippines, 1945; dep. blood adminstr. Far Tast Command, 1950-51; pres., chmn. exec. com. Biol. Abstracts, 1938-42, trustee, 1938-47; trustee Rocky Mountain Biol. Lab., 1953-55, 62—, pres., 1954-55. Recipient Knox Coll. Alumni Achievement award, 1954. Fellow A.A.A.S., Am., Royal socs. tropical medicine and hygiene, Am. Pub. Health Assn.; mem. Am., Japan socs. parasitologists, Entomol. Soc. Am., Am. Soc. Zoologists, Assn. Southeastern Biologists, Boars, Phi Beta Kappa, Sigma Xi, Delta Sigma Rho, Phi Eta. Author: (with

George W. Hunter, H. E. Walter) Biology—The Story of Living Things, 1937; Manual of College Biology, 1938; (with T. T. Mackie, C. B. Worth) A Manual of Tropical Medicine, 1945; (with F. R. Hunter) College Zoology, 1949; (with W. W. Frye, J. C. Swartzwelder) A Manual of Tropical Medicine, 1960, 4th edit., 1966. Studies, numerous publs. on schistosomiasis in Japan; charge whole blood for Armed Forces during Korean War. Home: 1401 Nashua Circle, Sun City Center, Fla. 33570.*

HUNTER, Gordon Denis, English biochemist; b. Chelsea, London, Eng., Apr. 29, 1927; s. George Williamson and Kathleen (Maclaughlan) H.; B.Sc., A.R.C.S., Imperial Coll., London U., 1947, Ph.D., D.I.C., 1950, D.Sc., 1961; m. Ellen Joy Topping, July 16, 1952; children—Philip Antony, Stephen George, Nicholas James. Radiochemist dept. biochemistry Nat. Inst. for Med. Research, Mill Hill, Eng., 1950-52, Glaxo Labs. Ltd., Fermentation Research div., Stoke Poges, Bucks, Eng., 1952-54; biochemist dept. phys. chemistry Chester Beatty Research Inst., Inst. for Cancer Research, Fulham, London, 1954-61; head dept. biochemistry A.R.C. Inst. for Research on Animal Diseases, Mewbury, Berks., Eng., 1961—. Mem. Biochem. Soc., Soc. Gen. Microbiology, Vet. Research Club. Biochem. editor Internat. Abstracts Biol. Scis., 1961—. Contbr. numerous articles to sci. jours. Discovery some pathways for biosynthesis of streptomycin, devel. new methods for chem. degradation of fatty acids and steroids, assessment of part played by cytoplasmic membrane complex in synthesis of proteins by bacteria; analysis of biochem. changes in scrapie brain and investigations of nature of scrapie agt. Home: 127 Andover Rd., Newbury. Office: A.R.C. I.R.A.D., Compton, Newbury, Berks, Eng.*

HUNTER, James de Graaff, topographer; b. Chester, Sept. 11, 1881; s. James and Sarah Jane (Pierrepont) H.; ed. Pembroke Coll., Cambridge; M.A., Sc.D.; m. Gwendoline Maud Henville Davis, Apr. 25, 1929; 1 dau., Susan Jane. Sec. to Lord Kelvin, 1904-05; asst. Nat. Phys. Lab. of Gt. Britain, 1905-07; math. cons. for topog. mosaic of East Indies, 1907-36; dir. geodetic service in land survey of East Indies, 1929-32; dir. war research work in land survey of East Indies, 1943-47. Fellow Royal Soc., 1935; mem. Internat. Assn. Geodesy (pres. 1954-57, hon. pres. 1954—). Author: Topographic Mosaic of the East Indies, also numerous articles on geodesy. Address: 7 Beaconsfield Rd., Mosman, New South Wales, Australia.

HUNTER, James George, Brit. chemist; b. Glasgow, Scotland, Dec. 9, 1915; s. John and Jessie (Morrison) H.; B.Sc., Ph.D., U. Glasgow; m. Willaimina Donald, Aug. 12, 1950; children—Anne, John, Ross. Agronomical chemist West of Scotland Agrl. Coll., 1938-46; head plant physiology dept. Macaulay Inst. for Soil Research, Aberdeen, 1946-52; officer commandant Trelawney Tobacco Research Sta., Southern Rhodesia, 1952-55; head research dept. for study of soil, dir. adj. Levington Research Sta., Ipswich, 1955—. Fellow Royal Inst. Chemistry; mem. Brit. Soil Sci. Soc., Soc. Applied Biology, Inst. Biology. Research and numerous publs. on toxic elements and chem. methods of analysis. Address: Levington Research Station, Ipswich, Suffolk, Eng.

HUNTER, John, Scottish anatomist, surgeon; b. Lanarkshire, Scotland, Feb. 13, 1728; s. John and Agnes (Paul) H.; ed. Chelsea (Eng.) Hosp. (pupil of William Cheselden), St. Bartholomew's Hosp. (pupil of Pott); student St. Mary's Hall, Oxford (Eng.) U., 1755-56; m. Anne Home, July, 1771; children—John, Agnes (Mrs. James Campbell). Cabinet-maker with bro.-in-law, Glasgow, Scotland; asst. in dissection with bro. William, London, Eng., from 1748; house surgeon St. George's Hosp., from 1756, surgeon, from 1768; with Belleisle expdn., 1761; with army in Portugal, 1762; began practice medicine London, 1763; lectr. surgery, from 1773 (pupils included Edward Jenner, Astley Cooper, Abernathy); surgeon extraordinary to George III, 1776; surgeon gen., from 1790. Croonian lectr., 1776-82; built mus. 1784-85. Fellow Royal Soc., 1767 (Copley medal 1787). Author: A Treatise on the Natural History of the Human Teeth, 1771; A Treatise on the Venereal Disease, 1786; Observations on certain parts of the Animal Economy, 1786. Treatise on the Blood, Inflammation, and Gunshot Wounds, 1794; Observations and Reflections on Geology, 1859; Memoranda on Vegetation, 1860. Contbr. numerous papers to profl. jours. First to study teeth in sci. manner (classified them into molars, bicuspids, cuspids, incisors, 1778); advocated use of braces for dental corrections; introduced artificial feeding by means of flexible stomach tube; investigations include work concerning course of olfactory nerves, descent of testes in fetus, placental circulation, coagulation of blood, recovery of apparently drowned people, formation of pus, function of lymphatics, structure of whales, bees, deer's antlers, digestion in hibernating lizards and snakes; 1st to ligate femoral artery for aneuryism, 1785. Died Oct. 16, 1793.

HUNTER, Lloyd Phillip, Am. physicist, educator; b. Wooster, O., Feb. 11, 1916; s. R. Hayes and Cora (Work) H.; B.A., Coll. Wooster, 1939; B.S., Mass. Inst. Tech., 1939; M.S., Carnegie Inst. Tech., 1940, D.Sc., 1942; m. Esther Kinch, Jan. 9, 1943; children—Joyce, John, Ross, Phillip, Gregory. Research physicist Westinghouse Electric Corp., Pitts., 1942-51; with IBM, Poughkeepsie, N.Y., 1951-63, resident mgr. research lab., 1958-60, dir. component engring., 1960-61; prof. elec. engring. U. Rochester (N.Y.), 1963—. Vis. prof. elec. engring. Stanford, 1962-63; cons. IBM, 1964—. Fellow Am. Phys. Soc., I.E.E.E.; mem. Soaring Soc. Am. Author: Introduction to Semiconductor Phenomena and Devices, 1966. Editor: Handbook of Semiconductors, 2d edit., 1962; asso. editor Solid State Design Mag., 1960—; editorial bd. Solid State Electronics Mag., 1960—. Patentee solid state electronics. Home: 10 Schoolhouse Lane, Rochester, N.Y. 14618.*

HUNTER, Oscar Benwood, Jr., Am. physician; b. Washington, Oct. 27, 1915; s. Oscar Benwood and Sidney (Pearson) H.; B.S., Cath. U., Washington, 1936; M.D., Georgetown U., 1940; m. Anne Battaile, Dec. 27, 1941; children—Anne Hay (Mrs. James P. Ganley), Sidney (Mrs. John J. Hastings), Margaret, Patricia, William, Michael, Ellen. Practice medicine specializing in pathology, Washington, dir. Oscar B. Hunter Meml. Lab., 1946—; dir., dept. pathology Drs. Hosp., Sibley Meml. Lab., Loudoun County Hosp., Leesburg, Va., Montgomery Gen. Hosp., Olney, Md.; profl. lectr. forensic pathology Georgetown U., adj. prof. clin. pathology Am. U. Recipient Distinguished Unit Citation, Armed Forces Inst. Pathology; Distinguished Service award D.C., 1958. Diplomate Am. Bd. Pathology. Fellow Am. Coll. Pathologists (pres., bd. govs.), Am. Soc. Clin. Pathologists; mem. Am. Assn. Blood Banks (pres. 1958), A.M.A., Southeastern Surg. Conf., Am. Fedn. Clin. Research, Internat. Soc. Hematology, Nat. Med. Vets. Soc. (past pres.), Internat. Soc. Clin. Pathology, Soc. Nuclear Medicine, Hippocrates-Galen Soc. (sec.-treas. 1964) Internat. Med. Club (treas.), Am. Heart Assn., So. Med. Assn. (pres.), others. Research, publs. on blood diseases, including diagnosis, pathogenesis, treatment of leukemia, anemia, erythroblastosis; uses of radioisotopes; cancer treatment. Home: 6408 Garnett Dr., Kenwood, Chevy Chase, Md. 20015. Office: 915 19th St. N.W., Washington 20006.*

HUNTER, Richard Sewall, Am. psychophysicist; b. Washington, Oct. 25, 1909; s. Herbert Coleman and Isabel (Sewall) H.; A.B., George Washington U., 1937; m. Elizabeth C. Landman, June 22, 1940; children—Philip S., Paul L. Physicist, Nat. Bur. Standards, Washington, 1927-46; chief engr. Gardner Lab., Inc., Bethesda, Md., 1946-52; pres. Hunter Assos. Lab., Inc., Fairfax, Va., 1952—. Cons. in colorimetry and glossimetry, 1952—. Recipient Armin J. Bruning award for pioneer work in color measurement of paints Fedn. Socs. for Paint Tech., 1962. Fellow Optical Soc. Am.; mem. Am. Soc. for Testing and Materials (award of merit 1961), T.A.P.P.I., Inst. Food Technologists, Inter-Soc. Color Council, Washington Acad. Sci. Research, publs. on devel. of method of photoelectric tristimulus colorimetry, instruments to measure reflectance, color, and gloss; originated concepts of different aspects of gloss. Home: 1703 Briar Ridge Rd., McLean, Va. 22101. Office: 9529 Lee Hwy., Fairfax, Va. 22030.*

HUNTER, Samuel John, entomologist; b. Ireland, Nov. 11, 1866; s. James and Rebecca (Davison) H.; came to U. S. in infancy; A.B., A.M., U. Kan., 1893; grad. student, Cornell U., 1896; investigator Marine Biol. Lab., 1901-02; m. Lida W. Campbell, June 16, 1897 (died June 1929); 1 dau., Geneva (Mrs. Edwin J. Simmons). Prin. Columbus (Kan.) High Sch., 1890-91, Atchison County High Sch., 1893-96; absent on leave, 1894, to visit zool. laboratories of Europe; asst. prof. entomology U. Kan., 1896-99, asso. prof., 1899, asso. prof. comparative zoology and entomology, 1901, head dept. entomology, 1902, prof., 1906, curator entomol. collections, 1909. State entomologist; mem. Kan. State Entomol. Commn., 1907, 24; collaborator Federal Bur. Entomology; mem. Nat. Com. on Unification of State and Federal Hort. Inspection Legislation. Fellow A.A.A.S.; mem. Kan. Acad. Science, Phi Beta Kappa, Sigma Xi. Author: Elementary Studies in Insect Life, 1902; An Account of Kansas Coccidae and Their Hosts, 1903; Morphology of Artificial Parthenogenesis, 1904; Insect Parthenogenesis, 1906; Problems in Parthenogenetic Parasites, Pathogenic Parasites, 1911. Chmn. Science Bull. editorial com., 1916-24. Died July 10, 1946.

HUNTER, Walter King, Scottish physician; b. Glasgow, Scotland, 1867; s. William Hunter; ed. Glasgow U., King's Coll., London; M.D.; D.Sc.; LL.D.; Former med. officer Royal Hosp. for Sick Children; lectr. practice medicine Queen Margaret Coll., Glasgow U., Muirhead prof. medicine, 1911-34, then emeritus prof.; cons. physician Glasgow Royal Mental Hosp., Glasgow Royal Infirmary. Fellow Royal Faculty Physicians and Surgeons Glasgow; mem. Royal Medico Chirurg. Soc. Glasgow (past pres.), Assn. Physicians. Author: (articles) Recent Advances in Haematology, 1911; Acute Degenerative Changes in the Nervous System, as Illustrated by Snake-venom Poisoning, 1910; Certain Chronic Glandular Enlargements, 1912, also numerous others. Died Nov. 7, 1947.

HUNTER, Walter Samuel, Am. psychologist; b. Decatur, Ill., Mar. 22, 1889; s. George and Ida (Weakley) H.; student Polytechnic Coll., Ft. Worth, Tex., 1905-08; A.B., U. of Tex., 1910; Ph.D., University of Chicago, 1912; honorary M.A., Brown University, 1913; married Katharine Pratt, January 1, 1913 (died Apr. 1, 1915); 1 dau., Thayer; m. 2d, Alda Barber, Aug. 1, 1917; 1 dau., Helen Barbara. Instructor philosophy, 1912-14, adjunct prof. psychology, 1914-16, U. of Texas; prof. psychology, U. of Kansas, 1916-25; G. Stanley Hall, prof. genetic psychology, Clark U., 1925-36; prof. psychology and dir. psychol. lab. Brown University, from 1936. Expert consultant to U. S. Sec. War. 1941-42; chief, applied psychology panel, N.D.R.C., 1943-45, mem. 1945-46; mem. Moe sub-com. for Bush Report, 1944-45; cons. Research and Development Bd., from 1947; mem. Undersea Warfare Committee N.R.C., 1946-51. Visiting prof. Tulane U., 1915, State U. of Ia., 1920, U. of Chicago, 1923, 30, U. of Calif. at Los Angeles, 1926, Northwestern U., 1927, Harvard, 1927-29 and 1932, U. of Minn., 1936. Awarded Presidential Medal for Merit, 1948. Fellow A.A.A.S. (vice pres. and chmn. Sect. I, 1932); mem. Am. Psychol. Assn. (council 1921-23, pres. 1931), Eastern Psychol. Assn. (pres. 1941), Soc. Exptl. Psychology, Internat. Congress Psychology (exec. sec. 9th congress, internat. com., 1929—), Am. Acad. Arts and Sci., 1933-47, Nat. Acad. Scis., Am. Philos. Soc., Phi Beta Kappa, Sigma Xi. Mem. Social Science Research Council, 1943-45; member div. anthropology and psychology, National Research Council, 1926-29, 1933-36 and 1939-42, chmn. same division, 1936-38. Member Research Board for Nat. Security, 1945, Sci. Advisory Group, Headquarters A.A.F., 1946-47. Author: General Psychology, 1919, 1923; Human Behavior, 1928. Asso. editor Psychology Bull, 1916-24, Jour. Animal Behavior, 1914-20, Jour. Comparative Psychology, 1921-25, Jour. Genetic Psychology, 1925-34, Genetic Psychology Monographs, 1926-34, Am. Jour. Psychol., 1940—; editor Behavior Monographs, 1922, Comparative Psychol. Monographs, 1922-27, Psychol. Index, 1925-36, Psychol. Abstracts, 1926-46. Contbr. on exptl. and theoretical psychology to scientific jours. Pioneer in use of documentation and objective methods in psychology; important work on delayed reaction, learning problems and retinal streaming; advocated use of term, anthroponomy, for sci. research. Died Providence, Aug. 3, 1954.

HUNTER, William, English anatomist, obstetrician; b. Long Calderwood, East Kilbride, Lanarkshire, Eng., May 23, 1718; s. John and Agnes (Paul) H.; student St. George's Hosp.; M.D., Glasgow, Scotland, 1750. Asst. dissector to Dr. James Douglas, 1675-1742; assisted by his bro. John Hunter, 1748-59; became surgeon accoucheur to Middlesex, Eng., 1748, Brit. Lying-in Hosp., 1749; named physician extraordinary to Queen Charlotte, 1764; became 1st prof. anatomy Royal Acad., 1768. Fellow Royal Soc., 1767, Soc. Antiquaries; mem. Med. Soc. (became pres. 1781), Royal Med. Soc. Paris (fgn. asso.), French Acad. Scis., 1782. Author: Medical Observations and Inquiries, 1757; Medical Commentaries, 1762-64; Anatomical Description of Human Gravid Uterus, 1774; also numerous articles. First to describe retroversion of uterus, 1771; discovered separate nature of fetal and maternal circulations, 1774. Died London, Mar. 30, 1783.

HUNTINGTON, Edward Vermilye, Am. mathematician; b. Clinton, N.Y., Apr. 26, 1874; s. Chester and Katharine Hazard (Smith) H.; A.B., Harvard, 1895, A.M., 1897; Ph.D., U. Strassburg. Germany, 1901; hon. Sc.D., U. San Marcos, Lima, Peru, 1925; m. Susie Edwards Van Volkenburgh, July 6, 1909. Instr. mathematics, Harvard, 1895-97, Williams Coll., 1897-99; in Europe, 1899-1901; instr. mathematics, 1901-05, asst. prof., 1905-15, asso. prof., 1915-19, prof. mechanics, 1919-41, emeritus since 1941, Harvard U. Western Exchange prof. from Harvard to Beloit, Carleton, and Knox Colls., 1925. Consultant, Nat. Defense Research Com. since 1942. Editor Annals of Mathematics, 1902-11. Fellow Am. Acad. Arts and Sciences (chmn. com. of pub., 1914-19), A.A.A.S. (vice pres. and chmn. section A, 1926), Am. Inst. of Math. Statistics; mem. Am. Philos. Soc., Am. Math. Soc. (v.p. 1924; rep. on NRC 1923-26), Math. Assn. America (pres. 1918), Am. Statistical Assn., Am. Standards Assn. (mem. sectional com.; chmn. subcommittee on math. symbols, 1928), Assn. for Symbolic Logic, Am. Acad. Polit. Sci., Phi Beta Kappa. Author: Four-Place Tables of Logarithms and Trigonometric Functions, arranged for decimal division of the degree, 1907; Monograph IV on "The Fundamental Propositions of Algebra," in work entitled "Mathematical Monographs" (edited by J. W. A. Young), 1911; The Continuum and Other Types of Serial Order, 1917; Handbook of Mathematics for Engineers, 1918; chapters I and IV in "Handbook of Mathematical Statistics" (edited by H. L. Rietz), 1924; Survey of Methods of Apportionment in Congress (Senate Doc. No. 304), 1940. Contbr. to math. journals. Devised the method of apportioning representatives in Congress which became law Nov. 1941. Best known for studies on systems of postulates, forming bases of elementary math. theories. Died Nov. 25, 1952.

HUNTINGTON, Ellsworth, Am. geographer; b. Galesburg, Ill., Sept. 16, 1876; s. Henry Strong and Mary Lawrence (Herbert) H.; B.A., Beloit (Wis.) Coll., 1897, D.Sc.; Harvard, 1901-03, M.A., 1902, non-resident fellow, 1906-07; Ph.D., Yale, 1909; D.Litt. Clark; m. Rachel Slocum Brewer, Dec. 22, 1917; children—Charles Ellsworth, Anna Slocum. George Herbert (dec.). Asst. to pres., instr. Euphrates Coll., Harput, Turkey, 1897-1901; explored cañons of Euphrates River, 1901, and was awarded Gill Memorial by the Royal Geog. Soc. of London; research assistant Carnegie Institution, Washington, D.C., and member Pumpelly expedition to Russian Turkestan, 1903-04 (spent 1½ years in Turkestan and Persia); member Barrett expedition to Chinese Turkestan (spent 1 1/2 years in India, China and Siberia, 1905-06) and was awarded Maunoir Medal by Geographic Society of Paris, and club medal by Harvard Travelers Club. Instructor in geography. 1907-10; assistant prof., 1910-15, and research asso., Yale, 1917-45, emeritus since 1945. Made expdn. Syrian Desert, Palestine, and Asia Minor, 8 mos., 1909, as rep. of Yale U., and spl. corr. Harper's Magazine. Research asso. of Carnegie Instn., Washington, for climatic investigations in U. S., Mex. and Central America, 1910-13. Asso. editor Geog. Rev., Econ. Geography, Ecology and Social Philosophy. Chmn. com. on atmosphere and man, Nat. Research Council. Attended Pan-Pacific Scientific Congress, Australia, 1923. Fellow Geol. Soc. America, A.A.-A.S., Am. Acad. Arts and Sciences; mem. Assn. Am. Geographers (pres. 1923), Ecol. Soc. America (pres. 1917), Population Assn. of America (dir.), Am. Eugenics Soc. (dir., treas., pres. 1934-38), Nat. Research Council (geol. and geog. div. 1919-23, 1935; biol. and agrl. div., 1921-24); hon. mem. and medalist Geog. Soc., Phila (Award of Merit, Council Geog. Teachers, 1943). Author: Explorations in Turkestan, 1905; The Pulse of Asia, 1907; Asia—A Geography Reader, 1912; Palestine, and Its Transformation, 1911; The Climatic Factor, 1914; Civilization and Climate, 1915; World Power and Evolution, 1919; Red Man's Continent, 1919; Principles of Human Geography (with S. W. Cushing), 1920; Business Geography (with F. E. Williams), 1922; Climatic Changes (with S. S. Visher), 1922; Earth and Sun, 1923; The Character of Races, 1924; Modern Business Geography (with S. W. Cushing), 1924; West of the Pacific, 1925; Quaternary Climates, 1925; The Pulse of Progress, 1926; The Builders of America (with L. F. Whitney), 1927; The Human Habitat, 1927; Living Geography (with F. M. McMurry and C. F. Benson), 1932; Economic and Social Geography (with F. E. Williams and S. van Valkenberg), 1933; After Three Centuries (with Martha Ragsdale), 1934; Europe (with S. van Valkenberg), 1935; Tomorrow's Children—The Goal of Eugenics, 1935; Season of Birth, 1938; Principles of Economic Geography, 1940; Mainsprings of Civilization, 1945. Well-known for research on influence of climate and climatic changes on human civilization; developed theory of rhythmic climactic changes, theory of direct causal relationship of solar and terrestrial changes; postulated that solar storms are conducive to intellectual progress; related studies on influences of selective migration and of intermarriage of certain types of people. Died New Haven, Oct. 17, 1947.

HUNTINGTON, Hillard Bell, Am. physicist; b. Wilkes-Barre, Pa., Dec. 10, 1910; s. Frederick L. and Gertrude (Bell) H.; B.A., Princeton, 1932, M.A., 1933, Ph.D., 1941; m. Ruth Smedley Wheeler, June 24, 1939; children—Frederic Wright, Hillard Griswold, David C. Teaching asst. U. Pa., 1941; physics instr. Washington U., 1941-42; staff mem. Radiation Lab., Mass. Inst. Tech., 1942-46; asst. prof. Rensselaer Poly. Inst., 1946-48, asso. prof., 1948-50, prof., 1950-—, chmn. dept. physics, 1961-—, chmn. div. solid state physics, 1965. Fellow Am. Phys. Soc.; mem. Fedn. Am. Scientists, Sigma Xi. Author: Elastic Constants of Crystals. Research on theory of defects in crystals, ultrasonic measurements of elastic constants; mass motion in solids induced by high electric currents and thermal gradients. Home: 219 Pinewoods Av., Troy, N.Y. 12180.*

HUNTINGTON, Robert Watkinson, Jr., Am. physician; b. Hartford, Conn., July 2, 1907; s. Robert Watkinson and Constance (Willard) H.; B.A., Yale, 1928, M.D., 1933; m. Katherine Upchurch, Mar. 21, 1936; children—Robert W., Ann B. (Mrs. Alan Wohl Heldman), Edith W., Deborah L. Powers fellow New Haven Hosp., 1933-34; Catlin fellow St. Louis Children's Hosp., 1935-38; instr. Cornell Med. Coll., 1938-41; faculty U. So. Cal., Los Angeles, 1946-—, clin. prof., 1965-—; pathologist Kern County Hosp., Bakersfield, Cal., 1950-—, Kern County Coroners Office, 1950-—. Mem. Coll. Am. Pathologists, Am. Soc. Clin. Pathologists, Acad. Forensic Scis., Am. Soc. for Microbiology, Soc. for Exptl. Biology and Medicine. Research, publs. on pathologic anatomy and racial distbn. of Gravi Coccidioidomycosis, coccidioidal menningitis, cocciioidal penuemonias and bronchiectasis, tumors brain, germinal tumors in young children, forensic pathology. Home: 901 Pershing St., Bakersfield 93304. Office: Kern County Gen. Hosp., Bakersfield, Cal. 93305.*

HUNTSMAN, Archibald Gowanlock, Canadian biologist; b. Tintern, Ont., Nov. 23, 1883; s. Lution Erotas and Elizabeth (Gowanlock) H.; student St. Catharines Collegiate Institute; B.A., U. Toronto, 1905, M.B., 1907, M.D., 1933; m. Florence Marie Stirling, Sept. 11, 1908; children—M. Elinor (wife of Dr. Colin A. Mawson), Edith G. (wife of Dr. Cosmo Marchant), H. Elizabeth. Asst. biology U. Toronto, 1905-07, lectr. biology, 1907-16, asso. prof. marine biology, 1917-27, prof. marine biology, 1927-54; curator Atlantic Biol. Sta., Fisheries Research Bd. of Can. (formerly Biol. Bd. of Can.), 1911-13, 16-19, dir., 1919-34; dir. Atlantic Exptl. Sta. for Fisheries, 1924-28, editor, 1934-48, cons. dir., 1934-53. Vice pres. Bermuda Biol. Sta., 1935-54, trustee, 1926-54; trustee Woods Hole Oceanographic Instn., 1931-48. Recipient Flavelle medal Royal Soc. Can., 1952. Mem. A.A.-A.S., Ecol. Soc., Soc. Zoologists, Soc. Limnologists and Oceanographers, Fisheries Soc. (pres. 1937), Royal Soc. Can. (pres. 1938), Royal Canadian Inst., Am. Museum Natural History, Inst. on Religion in an Age of Sci.; fgn. hon. mem. Am. Acad. Arts and Scis., Marine Biology Assn. (Great Britain). Author: Life and the Universe, 1959. Contbr. The Light and the Flame, 1956, and articles various jours. Research on herring, salmon, Ascidians, Deeapoda, Amphipoda, plankton, Marine fishes. Homes: 217 Indian Rd., Toronto, Ont., and St. Andrews, N.B., Can.

HUNZIKER, Juan Héctor, Argentinian plant geneticist; b. Buenos Aires, Argentina, Aug. 26, 1925; s. Juan Santiago and Manuela (Landolt) H.; Ingeniero Agrónomo, U. Buenos Aires, 1949; M.S., U. Cal. at Berkeley 1953, Ph.D., 1959; m. Homayoun Taj Mortazavi, July 27, 1955; children—Marcelo Dario, Isabel Julia, Eduardo Armando. Botanist, Bot. Inst., Ministry Agr., Argentina, 1947-58; botanist Nat. Inst. Agrl. Tech., Buenos Aires, 1958-60; prof. genetics Sch. Exact and Natural Scis., U. Buenos Aires, 1960-—. Mem. career for research workers Nat. Council for Sci. and Applied Research of Argentina, 1962-—; research asso. geneticist U. Cal., 1963. Guggenheim Found. fellow, 1957-59. Mem. Bot. Soc. Am., Soc. for Study Evolution, Torrey Bot. Club, Internat. Soc. Plant Morphology, Internat. Assn. Plant. Taxonomy. Publs., research on evolution of species of grasses and gymnosperms; origin by natural hybridization of useful willows in Argentina; chromosomal and protein differentiation within a self-pollinated wheat grass. Home: 253 Muniz, Buenos Aires, Argentina.*

HUPPERT, Milton, Am. microbiologist; b. Jersey City, Dec. 23, 1919; s. Martin and Annie (Teitelbaum) H.; B.S., Coll. City N.Y., 1940; postgrad. Va. Poly. Inst.; Ph.D., Columbia, 1955; m. Roslynd P. Froelich, May 2, 1946; children—Jeffrey E., Andrew R. Lab. asst. Med. Mycology Diagnostic Lab., Columbia U. Phys. and Surg., N.Y.C., 1946-48; research mycologist Scripps Clinic, La Jolla, Cal., 1948-50; instr. dept. med. microbiology U. N.C. Sch. Medicine, Chapel Hill, N.Y.C., 1950-55; chief Mycology Research Lab., VA Hosp., San Fernando, Cal., 1955-—; sec., dir. Central Lab., VA Armed Forces Coccidioidomycosis Study Group, 1956-—. Chmn. infectious disease com. Los Angeles County Tb and Health Assn., 1962-63; VA liaison rep. bacteriology and mycology A study sect. NIH. Diplomate Am. Bd. Microbiology. Mem. Am. Soc. for Microbiology, Am. Thoracic Soc., Internat. Soc. for Human and Animal Mycology, N.Y. Acad. Scis., A.A.A.S. Research, publs. on lab. techniques for lab. diagnosis human fungus infections; devel. chems. for use chemotherapeutic agts. in human fungus infections. Home: 5900 Buffalo Av., Van Nuys, Cal. 91401. Office: 13000 Sayre St., San Fernando, Cal. 91342.*

HURD, Charles De Witt, Am. chemist; b. Utica, N.Y., May 7, 1897; s. Charles John and Susan (Perlet) H.; B.S., Syracuse U., 1918, Sc.D., 1943; postgrad. U. Minn.; Ph.D., Princeton, 1921; m. Mary Ormsby Nelson, Aug. 29, 1921; 1 son, Richard Nelson. With Thomas A. Edison, West Orange, N.J., 1917; instr. U. Ill., Urbana, 1921-24; faculty Northwestern U., Evanston, Ill., 1924-—, prof. chemistry, 1933-49, Morrison prof. 1949-51, Clare Hamilton Hall Research prof. organic chemistry, 1951-65, Hall prof. emeritus, 1965-—. Lectr., U. Chgo., 1928-39, U. So. Cal., 1936, Stanford, 1940, U. Buffalo, 1946. Fellow A.A.A.S.; mem. Am. Chem. Soc. (past chmn. Chgo. sect., Midwest award 1958), NRC. (past chmn. com. on nomenclature). Author: The Pyrolysis of Carbon Compounds, 1929; contbg. author Organic Chemistry, 1938; also numerous articles. Co founder Jour. Organic Chemistry; cons. all chem. definitions Merriam-Webster New Internat. Dictionary, 3d edit. Research on mode pyrolysis organic compounds, attachment hydrocarbon groups to carbohydrates, heterocyclic series, chemistry ketene and diketene, chemistry acylais, mechanisms rearrangement allyl ethers phenols, N-alkylanilines, also hydroxamic acids, synthesis polypeptides, mechanism furfural formation from pentoses, browning carbohydrate substances by interaction with amino acids or proteins; demonstrated formation benzene, naphthalene and polycyclic arenes from diverse open-chain or heterocyclic compounds 800-850° Centigrade; pioneered theory quasi 6-membered ring in transition state chem. reactions. Home: 2649 Lawndale Av., Evanston, Ill. 60201.*

HURLBUT, Cornelius Searle, Jr., Am. mineralogist; b. Springfield, Mass., June 30, 1906; s. Cornelius Searle and Marion (Adams) H.; A.B., Antioch Coll., 1929; M.A., Harvard, 1932, Ph.D., 1933; m. Anne Dawson, June 18, 1932 (dec. 1954); children—Cornelius Searle, Patricia Ann (Mrs. John L. Williams), Marcus Dawson; m. 2d, Margaret Richards Carver, Oct. 27, 1956. Faculty, Harvard, 1933-—, prof. mineralogy, 1954-—. Mem. Mineral. Soc. Am. (past pres.), Geol. Soc. Am., Soc. Econ. Geologist, Mineral. Soc. Gt. Britain, Geol. Soc. S. Africa, Author: (with H. Wenden) The Changing Science of Mineralogy, 1964; also articles, chpts. in books. Research in descriptive mineralogy and crystallography. Home: 53 Woodbine Rd., Belmont, Mass. 02178. Office: Geol. Mus., Oxford St., Cambridge, Mass. 02138.*

HURLEY, Desmond Eugene, New Zealand biologist; b. Hawera, New Zealand, Mar. 6, 1928; s. Eugene and Mary Agnes (Dwyer) H.; B.Sc., Victoria U. Coll., Wellington, New Zealand, 1948, M.Sc. with honors, 1950, Ph.D., 1953; m. Marina Greaves, July 14, 1962; children—Jane Melanie, Dominic Andrew. Nuffield Research asst. Portobello Marine Biol. Sta., New Zealand, 1952-54; Fulbright scholar Allan Hancock Found., U. So. Cal., 1954; marine biologist New Zealand Oceanographic Inst., Wellington, 1955-—; Nuffield Research fellow Brit. Mus. Natural History, 1962-63. Fellow Zool. Soc.; mem. Royal Soc. New Zealand, Brit., New Zealand ecol. socs., New Zealand Marine Scis. Soc., Soc. Systematic Zoology, Marine Biol. Assn. Research, publs. on systematics of marine and terrestrial animals, especially Crustacea Isopoda and Amphipoda, ecology of terrestrial Amphipoda, Benthic ecology in New Zealand waters. Home: 27 Clark St., Khandallah, Wellington N.5. Office: New Zealand Oceanographic Inst. Dept. Sci. and Indsl. Research, P.O. Box 8009, Wellington, New Zealand.*

HURLEY, Patrick Mason, geologist; b. Hong Kong, China, Jan. 12, 1912; s. Frederick Charles Mason and Anne (Peacock) H.; B.A., B.A.Sc., U. B.C., 1934; Ph.D., Mass. Inst. Tech., 1940; m. Margaret Macurda, Aug. 9, 1941; children—David M., Peter M., Pamela M. Came to U. S., 1937, naturalized, 1943. Royal Soc. Can. fellow Mass. Inst. Tech., 1939-41; mining engr., geologist gold mines, B.C., Can., 1934-37; research asso. Nat. Def. Research Com., 1942-45, U. Wis., 1945-46; faculty Mass. Inst. Tech., Cambridge, 1946-—, prof. geology, dept. exec. officer, 1953-—. Cons. to pvt. cos., govt. agys. Fellow Am. Acad. Arts and Scis., Geol. Soc. Am., Am. Geophys. Union, A.A.A.S.; mem. Soc. Econ. Geologists, Geochem. Soc., Am. Inst. Mining Engrs. Author: How Old is the Earth, 1959; also numerous articles. Editor: Advances in Earth Science, 1966. Research on application nuclear physics to earth sci., geochronology, isotope geochemistry applied to origin crustal materials. Home: 36 Oakmount Circle, Lexington, Mass. 02173. Office: Mass. Inst. Tech., Cambridge, Mass. 02139.*

HURMUZESCU, Dragomir, Rumanian physicist; b. 1865; Author: Nouvelle determination du rapport V entre les unités électrostatiques et électromagnétiques, 1896; Sur les modifications mécaniques, physiques et chimiques qu 'prouvent les diffrents corps par l'aimantation, 1898. Contbd. to knowledge of electricity, diffraction, endosmosis, physics of X-rays; built electroscope, bearing his name, with dielectrine (used by H. Becquerel in initial research on radioactivity and by Curies) as insulation. Died 1954.

HURON, Roger Jules, physician, statistician; b. Toulouse, Aug. 20, 1913; s. Louis and Marie (Abizzanda) H.; ed. univs. Toulouse, Bordeaux, Paris; Dr. es sc., M.D.; m. Cecile Zwilling, Nov. 27, 1936; children—Elisabeth, Helene, Jacqueline, Marguerite, Paul-Louis. Agrégé, prof. lycées of Roanne, Rodez, Toulouse, Faculty Sci., Toulouse. Recipient Gold medal Faculty Medicine. Mem. French Acad. Sci. (prize 1959-61.) Author: Probabilités, Erreurs; Statistique mathématique; Les Méthodes en génétique générale et humaine. Home: 9, rue des Pyrénées. Office: Faculty of Sciences, route de Narbonne, Toulouse, France.

HURST, Sir Arthur Frederick, see Hertz, Sir Arthur Frederick.

HURST, Clarence Thomas, Am. zoologist, archaeologist; b. Kingston, Ky., June 20, 1895; s. Alexander Lusk and Margaret Katherine (Folkerts; H.; B.Pd., Colo. State Normal Sch., Gunnison, 1920, M.Pd., 1922; A.B., Western State Coll. of Colo., 1923, A.M., 1923; Ph.D., U. of Calif., 1926; m. Blanche Hendricks, June 1, 1919. Pub. sch. teacher, Colo., 1916-21; teaching fellow in zoology, U. of Calif., 1923-24, 1925-26, tech. asst., 1924-25; asst. prof. of zoology, Mills Coll., 1926-28; prof. zoology, Western State Coll. of Colo., from 1928, head dept., from 1928 and dean Grad. Sch. from 1930; chmn. div. of natural science and math., from 1937; dir. Museum of Archeol., Western State Coll., from 1935; director field expeditions in archeology, Museum of Archeology, Western State Coll., from 1939; research asso. in zoology, U. Cal., 1933-34. Fellow A.A.A.S.; mem. Am. Assn. of Univ. Profs., Am. Soc. of Zoologists, Colo.-Wyo. Acad. Sci. (pres. 1940-41). N.E.A., American Association Museums, Am. Museum of Natural History, Soc. for Am. Archeology, Phi Sigma, Beta Beta Beta (mem. editorial bd.), Kappa Delta Pi, Sigma Xi. Author: Hegner's Invertebrate Nomenclature—A Dictionary of Zoology, 1934; Colorado's Old-Timers; The Indians Back to 25,000 Years Ago, 1946. Contbr. numerous articles on biology and anthropology to Am. and European publs. Editor of Southwestern Lore (jour. Colo. Archeol. Soc.). Died Jan. 17, 1949.

HURST, Donald Geoffrey, physicist; b. St. Austell, Eng., Mar. 19, 1911; s. George Leopold and Sarah (Inns) H.; B.Sc., McGill U., 1933, M.Sc., 1934, Ph.D., 1936; m. Margaret Christina McCuaig, Dec. 23, 1939; children—Dorothy June, David Alan. 1851 Exhbn. scholar Radiation Lab., U. Cal., 1936-37, Cavendish Lab., U. Cambridge (Eng.), 1937-39; research officer NRC, Ottawa, Can., 1939-44, Montreal, Que., Can., 1944-45, Chalk River, 1949-52; with Atomic Energy of Can., Ltd., Chalk River, 1952-—, dir. reactor research div., 1960-—, on leave as dir. div. nuclear power and reactors Internat. Atomic Energy Agy., Vienna, Austria, 1965-67. Fellow Royal Soc. Can., Am. Nuclear Soc.; mem. Can. Assn. Physicists. Research, publs. in optical properties of solids, structure of solids, liquids and gases by neutron diffraction, nuclear transmutations and nuclear properties, physics of nuclear reactors, basic devel. of power reactors. Address: Atomic Energy of Can. Ltd., Chalk River, Ont., Can.*

HURST, George Sam, Am. physicist; b. Pineville, Ky., Oct. 13, 1927; s. James H. and Myrtle (Wright) H.; A.B., Berea Coll., 1947; M.S., U. Ky., 1948; Ph.D., U. Tenn., 1959; m. Betty Partin, July 3, 1948; children—Karen Louise, Donald Edward. Staff, Oak Ridge Nat. Lab., 1948-66, sect. chief radiation physics, 1956-66; vis. research prof. Fla. State U., 1962-63; asso. prof. physics U. Tenn., Knoxville, 1963-66; prof. physics U. Ky., Lexington, 1966—. Mem. Am. Phys. Soc., Radiation Research Soc., Health Phys. Soc. Contbr. numerous articles to tech. jours. Developed methods radiation dosimetry; research in ionization and excitation gases by fast particles, electron diffusion and capture in gases. Office: Dept. Physics, U. Ky., Lexington, Ky.*

HURT, Wesley Robert, Am. anthropologist; b. Albuquerque, Sept. 20, 1917; s. Wesley Rosecrans and Amy (Passmore) H.; B.A., U. N.M., 1938, M.A., 1942; Ph.D., U. Mich., 1952; m. Mary Catherine Darden, Feb. 28, 1948; children—Steven, Rosalind, Teresa. Dir. Mus., prof. anthropology U. S.D., 1949-63; dir. Mus., prof. anthropology Ind. U., Bloomington, 1963—; vis. prof. anthropology U. Cal., Berkeley, 1961. Fellow Am. Anthrop. Assn.; mem. Soc. Am. Archaeology, Ind. Acad. Sci., Plains Archaeol. Conf., Midwest Mus. Assn., Sigma Xi. Contbr. articles to sci. jours. Directed excavations of numerous archaeol. sites, S.D.; directed archaeol. excavations for mus. N.M.; directed archaeol. project Ala. Mus. Natural History; directed archaeol. excavations in Minas Gerais, Brazil, Parana, Brazil, Santa Catarina, Brazil. Home: 120 Concord Rd., Bloomington, Ind. 47401.*

HURUHATA, Masaaki, Japanese astronomer; b. Nagano, Japan, Sept. 18, 1912; s. Ginsaku and Tome (Hiraide) H.; grad. astron. dept. Tokyo (Japan) U., 1938, Dr.Sci., 1955; m. Kei Yoshida, Apr. 29, 1943; 1 dau., Kuniko. Asst., Harvard Coll. Obs., Mass., 1938-41; faculty, Tokyo U., 1942—, prof. astronomy, 1957—; sr. specialist East-West Center, U. Hawaii, 1963-64; Dir. World Data Center, IGY, 1957—, mem. Japanese nat. com., 1956—. Mem. Internat. Astron. Union (v.p. commissional 1964—). Research, publs. on astronomy and upper atmosphere physics (airglow). Address: Tokyo Astron. Obs., Mitaka, Tokyo, Japan.*

HURVICH, Leo Maurice, Am. psychologist; b. Malden, Mass., Sept. 11, 1910; s. Julius S. and Celia (Chikinsky) H.; B.A., Harvard, 1932, M.A., 1934, Ph.D., 1936; m. Dorothea A. Jameson, Oct. 23, 1948. Asst. instr. Harvard, 1936-40, research asst. Grad. Sch. Bus. Adminstrn., 1940-47; research psychologist Eastman-Kodak Co., Rochester, N.Y., 1947-57; prof. psychology, chmn. dept. Washington Sq. Coll., N.Y. U., 1957-62; prof. psychology U. Pa., Phila., 1962—. Fellow A.A.A.S., Optical Soc. Am., N.Y. Acad. Sci.; mem. Am. Psychol. Assn., Soc. Exptl. Psychologists, Psychonomic Soc., Biophys. Soc., Phi Beta Kappa, Sigma Xi. Author: (with D. Jameson) The Perception of Brightness and Darkness, 1966; translator (with Jameson) Outlines of a Theory of the Light Sense (E. Hering), 1964. Contbr. articles to tech. jours. Research, publs. in sensory mechanisms, vision, taste, smell, quantification opponent colors theory. Home: 286 St. James Pl., Phila. 19106.*

HURWITZ, Adolph, Swiss mathematician; b. Switzerland, 1859; prof. U. Zurich; constructed extensive arithmetics of ordinary quaternions; extended and simplified method of differential operators; worked with theory of numbers and negative roots. Died 1919.

HURWITZ, Alfred, Am. surgeon; b. Boston, Apr. 12, 1909; s. Max and Rose (Goldsmith) H.; A.B., Harvards 1929; M.D., Johns Hopkins, 1933; m. Dorothy Solomon, Apr. 12, 1938; children—Richard L., Susan R., Tobey E. Practice medicine specializing in thoracic surgery, Boston, 1938-42, 45-46; faculty Harvard Med. Sch., 1937-46, Yale, 1946-55, State U. N.Y. Coll. Medicine, 1955-63; clin. prof. surgery U. Miami (Fla.) Sch. Medicine, 1965—, dir. surg. services Mt. Sinai Hosp., Miami Beach, Fla., 1965—; Decorated Bronze Star, Presdl. citation. Kirstein fellow, 1941. Diplomate Am. Bd. Surgery, Am. Bd. Thoracic Surgery. Fellow A.C.S. (editorial bd. surgery 1960—); mem. Am. Thoracic Surg. Assn., A.M.A., New Eng. Surg. Soc., Am. Surg. Assn. Sigma Xi, others. Author: (with George Degenshein) Milestones of Modern Surgery, 1958; (with Philip Cooper) Craft of Surgery, 1964. Research, publs. on surg. treatment of various conditions, use of Vitamin K1, designing of gastrostomy tube, thoracic trocar, esophageal clamp. Home: 1130 N.E. 100 St., Miami Shores, Fla. 33138. Office: 4300 Alton Rd., Miami Beach, Fla. 33140.*

HURWITZ, Henry, Jr., Am. physicist; b. N.Y.C., Dec. 25, 1918; s. Henry and Ruth (Sapinsky) H.; B.A., Cornell U., 1938; M.A., Harvard, 1939, Ph.D. in Physics, 1941; m. Jean Klein, June 29, 1944 (div. June 1949); 1 son, Harry I., m. 2d, Alma Rosenbaum, Apr. 15, 1951; children—Robin Elaine, Julia Lea, Wayne Mark. Instr. physics Cornell U., Ithaca, N.Y., 1941-44; research asso. Los Alamos Sci. Lab., U. Cal., 1944-46; research asso., supr. theoretical physics, cons. physicist advanced naval reactor physics activity Knolls Atomic Power Lab., Gen. Electric Co., Schenectady, 1946-56, mgr. nucleonics and radiation sect. Research Lab., 1957—. Recipient Ernest Orlando Lawrence Meml. award AEC, 1961. Fellow Am. Phys. Soc., Am. Nuclear Soc., A.A.A.S., N.Y. Acad. Scis. Exec. editor plasma physics, accelerators, thermonuclear research Jour. Nuclear Energy. Contbr. articles on reactor theory, nuclear engring., plasma physics to profl. publs. Patentee nuclear reactors, energy conversion, miscellaneous devices. Home: 827 Jamaica Rd., Schenectady 12309. Office: P.O. Box 1088, Schenectady 12301.*

HUSKEY, Harry Douglas, Am. automatic computer scientist; b. Whittier, N.C., Jan. 19, 1916; s. Cornelius and Myrtle (Cunningham) H.; B.S., U. Ida., 1937; post-grad. Ohio U.; M.S., Ohio State U., 1940, Ph.D., 1943; m. Velma E. Roeth, Jan. 2, 1939; children—Carolyn (Mrs. Joe Dickinson), Roxanne (Mrs. Richard Dwyer), Harry Douglas, Linda Louise. Math. instr. U. Pa., 1943-46; prin. sci. officer Nat. Phys. Labs., Eng., 1947; mathematician, Inst. for Numerical Analysis, Nat. Bur. Standards, 1948-54; prof. math., elec. engring. U. Cal. at Berkeley, 1954—; vis. prof. various insts., univs. Research adviser System Devel. Corp., Santa Monica, Cal., 1962—; dir. URS Corp., Burlingame, Cal. Fellow I.E.E.E. (editor-in-chief computer group 1965—), A.A.A.S. (mem. council); mem. Assn. for Computing Machines (past pres.), Am. Math. Soc., Math. Assn. Am., Simulation Councils, Author: (with G. A. Korn) Computer Handbook, 1962; also numerous articles. Design and constrn. automatic computers; work on on-line systems, man-machine communication. Home: 2655 Buena Vista, Berkeley, Cal. 94708.*

HUSSERL, Edmund, German philosopher; b. Prossnitz, Moravia, Apr. 8, 1859; ed. Gymnasium Olmütz; studied astronomy, math., physics, philosophy Leipzig, Berlin, Wien univs.; Ph.D., U. Vienna, 1886; studied philosophy under Franz Brentano. Lectr. philosophy U. Halle, 1887-1901; prof. philosophy U. Gottingen, 1901-16, then at Freiburg, im Breisgan, 1916-28. Author: Logische unersuchungen, 1900-01; Ideen zu einer reinen Phänomenolögie, 1913-52; Carksianische Meditationen, 1931; also others. Discovered phenomenological method as new philos. approach to describe and define genuine essence of conscious data, circa 1896; stated that intentionality (the necessary relation of every conscious act to meaningful object) is universal law of awareness, and that intuition is source of content and value of all knowledge; his method and theories had wide influence in philosophy, especially in Germany (Max Scheler, Martin Heidegger) and France (Satre, Maurice Merleau-Ponty), also on psychology (Ludwig Binswanger, F. Buytendijk). Died Freiburg, Apr. 27, 1938.

HUSSEY, Keith Morgan, Am. geologist; b. Rock Island, Ill., Dec. 2, 1908; s. Ernest Samuel and Diana Hill (Stow) H.; A.B., Augustana Coll., 1936; M.S., La. State U., 1939, Ph.D., 1940; m. Lillian Alberta Papping, Dec. 26, 1937; children—Michael Keith, Patricia Ann. Instr., U. Houston, 1940-42; asst. prof., then asso. prof. Okla. U., 1945-49; faculty Ia. State U., 1949—, prof. geology, 1954—, head dept., 1961—; spl. research micropaleontology of Gulf Coast, geomorphology and stratigraphy of Colo., Wyo., Ia. and No. Alaska; vis. geoscientist lectr. Am. Geol. Inst., 1960-62. Fellow Geol. Soc. Am.; mem. Am. Assn. Petroleum Geologists, Ia. Acad. Sci., Arctic Inst. N. Am., A.A.A.S., Am. Assn. U. Profs., Wyo. Geol. Assn., Nat. Assn. Geology Tchrs. (pres. central sect. 1958), Sigma Xi, Phi Kappa Phi. Contbr. profl. jours. Co-editor: Dictionary of Geological Terms, 1961. Described major Middle Eocene foraminiferal assemblage from La.; co-determined cause of orientation of No. Alaska's oriented lakes; determined assn. of arctic ground patterns to terrain elements and their use in photog. interpretation of arctic terrains. Home: 1910 Meadowlane Av., Ames, Ia. 50010.*

HUSSEY, William Joseph, Am. astronomer; b. Mendon, O., Aug. 10, 1862; s. John Milton and Mary Catherine (Severns) Hussey; B.S., Univ. of Michigan, 1889; Sc.D., Brown U., 1912; m. Ethel Fountain, June 27, 1895; children—Roland Fountain, Allis Fountain; m. 2d, Mary McNeal Reed, Sept. 1, 1917. Instr. in mathematics, U. of Mich., 1889-91; acting dir. Detroit Obs., 1891-92; asst. prof. astronomy, 1892-93, asso. prof., 1893-94, prof., 1894-96, Leland Stanford Jr. U.; astronomer in Lick Obs., 1896-1905; prof. astronomy and dir. obs., U. of Mich., from 1905; also prof. astronomia y geodesia, U. of La Plata, Argentina and dir. Observatorio Nacional de La Plata, La Plata, Argentina, Sept. 1, 1911-17. Expert on observatory sites in Southern Calif., Ariz. and Australia, to com. on observatories of Carnegie Instn., 1903; in charge of Lick Obs. eclipse expdn. to Egypt, 1905, of La Plata eclipse expedition to Brazil, 1912; discoverer of 1,650 double stars; awarded Lalande prize of French Acad., 1906, for double star discoveries and investigations. Foreign asso. Royal Astron. Soc.; hon. mem. Sociedad Astronomica de Mexico. Author: Logarithmic and Other Mathematical Tables, 1891, 1895; Mathematical Theories of Planetary Motions, 1892; Micrometrical Observations of the Double Stars Discovered at Pulkowa (Vol. V. Lick Observatory publ.), 1901. Known for micrometrical measurements of comets, satellites and double stars, also for discoveries of double stars. Died London, Eng., Oct. 28, 1926.

HUSTIN, Albert, Belgian surgeon; b. Ethe, Belgium, July 15, 1882; s. Joseph and Emile (China) H.; M.D., U. Brussels (Belgium), 1906; m. Mathilde Houyoux, 1910; 3 children. Asst. surgeon hosps., Brussels, also asst. in superior instruction U. Brussels, from 1913; chief Surg. Service, from 1923; prof. surgery U. Brussels, from 1925, also dir. Superior Inst. Phys. Edn. Pres., Office Medico-Legal, Ministry Pub. Health. Mem. Belgian Surg. Soc. (pres.), French Assn. Surgery, Internat. Soc. Surgery, Belgian Royal Acad. Medicine (corr.), French Acad. Surgery (fgn. corr.). Discovered sodium citrate's anticoagulant properties and advocated its use in transfusions, 1914. Home: 194 Avenue Winston Churchill, Brussels, Belgium.

HUSTON, Mervyn James, Canadian pharmacologist, educator; b. Ashcroft, B.C., Can., Sept. 4, 1912; s. William Mervyn and Irene (Gray) H.; B.Sc., U. Alta. (Can.), 1937, M.Sc., 1941; Ph.D., U. Wash., 1944; m. Helen Margaret McBryan, Dec. 18, 1938; children—Bryan Mervyn, Dorna Helen. With U. Alta., Edmonton, 1939—, dean faculty pharmacy, 1955—. Mem. Canadian Conf. Pharm. Faculties (chmn. 1948), Canadian Found. for Advancement Pharmacy (dir. 1960), Canadian Drug Adv. Com., Chem. Inst. Can., Am., Canadian (mem. council 1961, pres. 1968), Pharm. assns., Sigma Xi, Phi Sigma, Rho Chi. Author: Textbook of Pharmaceutical Arithmetic, 1959; Test and Improve Your Scientific Word Power, 1960; The Great Canadian Lover—Musson, 1964. Editor Canadian Jour. Pharm. Scis. Research, publs. on effect of drugs on metabolism of tissues, fluid distbn. in molluscs. Home: 11562 80th Av., Edmonton, Alta., Can.*

HUSTON, Paul Eger, Am. physician; b. Delphos, O., Aug. 18, 1903; s. William A. and Anna (Eger) H.; B.S., Purdue U., 1926; postgrad. Harvard, 1926-30; Ph.D., 1937; M.D., Yale, 1939; m. Margaret A. Flinn, Feb. 22, 1932; children—John William, David. Instr., Jr. Coll. New Haven, 1935-39, Smith Sch. Social Work, 1938; staff Psychopath. Hosp., Iowa City, 1940—, prof. psychiatry, 1950—, head dept., 1955—, dir., 1956—; dir. Ia. Mental Health Authority, 1956—. Cons. VA, 1947—; mem. Gov.'s Com. on Mental Health, 1956—, Gov.'s Com. on Corrections. Mem. A.M.A., Am. Psychiat. Assn. (mem. conf. on grad. edn.), Group for Advancement Psychiatry, Ia. Psychiat. Soc., Am. Psychopath. Assn., Sigma Xi. Contbr. numerous articles in schizophrenia, depression, exptl. psychopathology to tech. jours. Home: 223 Lucon Dr. Office: 500 Newton Rd., Iowa City 52241.*

HUTCHENS, John Oliver, Am. physiologist; b. Noblesville, Ind., Nov. 8, 1914; s. Bernayse E. and Della (Moore) H.; A.B., Butler U., 1936; Ph.D., Johns Hopkins, 1939; m. Eleanore Mae Mothersill, June 3, 1939; children—Margaret Ann (Mrs. Gray Holbrook), Judith Marie, Helen Louise. NRC fellow biochemistry Harvard, 1939-40; Johnston scholar Johns Hopkins, 1940-41; faculty U. Chgo., 1941—, prof. physiology, 1948—. Sci. liaison officer Office Naval Research, London, Eng., 1954-55. Mem. Am. Physiol. Soc., Soc. Gen. Physiology, Biochem. Soc. (Brit.), A.A.A.S. Research, publs. on application 2d law of thermodynamics to living organisms, entropies of amino acids and proteins. Home: 5633 S. Drexel Av., Chgo. 60637.*

HUTCHERSON, William Robert, Am. mathematician; b. Glasgow, Ky., Dec. 20, 1898; s. John Cy and Ida Elizabeth (Lyon) H.; A.B., U. Ky., 1922, A.M., 1924; Ph.D., Cornell U., 1931; m. Sally May Grainger, Aug. 28, 1924; children—Sarah Ann (Mrs. Morgan Eugene Wing), Martha Lou (Mrs. Eugene Culpepper Cochran), William Robert. Prof., U. Fla., Gainesville, 1949—. Mem. Am. Math. Soc., Math. Assn. Am., Am. Assn. U. Profs., Sigma Xi. Author: (with Dekker) Introduction to Physical Science, 1947. Asso. editor: Ky. Acad. Sci. Bull., 1939-41. Research, publs. on involution in algebraic geometry. Home: Route 3, Box 210, Gainesville, Fla. 32601.

HUTCHINGS, Richard Henry, Am. physician; b. Clinton, Ga., Aug. 28, 1869; s. Richard H. and Cornelia (Greaves) H.; student U. Ga., 1 year: M.D., Bellevue Hospital Medical College (New York Univ.), 1891; D.Sc., Colgate Univ., 1940; m. Lillie Beall Compton, Sept. 6, 1893; children—Richard Henry, Charles Wyatt, Dorothy (Mrs. R. N. Alberts). Entered New York State Hospital Service, 1892; served as asst. physician to 1903, med. supt., 1903-19, St. Lawrence State Hosp., Ogdensburg, N.Y.; med. supt. Utica (N.Y.) State Hosp., 1919-39. Prof. clin. psychiatry (emeritus), Syracuse U. Mem. A.M.A., Am. Psychiatric Assn. (pres. 1938), Alpha Omega Alpha. Author: A Psychiatric Word Book, 7th edit., 1943. Editor of Psychiatric Quarterly. Contbr. numerous articles on care and treatment of nervous and mental diseases. Investigated epidemic of typhoid faver, 1903, traced its origin to infected ice, a source of contagion not previously recognized; made study of the care of insane in State of Ga. Died Oct. 28, 1947.

HUTCHINS, Ross Elliott, Am. entomologist; b. Ruby, Mont., Apr. 30, 1906; s. Elliott John and Helen (Pierce) H.; B.S., Mont. State Coll., 1929; M.A.,

Miss. State U., 1931; Ph.D., Ia. State Coll., 1935; m. Annie Laurie McClanahan, June 6, 1932. Faculty, Miss. State U., State College, 1930—, prof. entomology, 1945; exec. officer Miss. Plant Bd., State College, 1951—. Mem. Am. Entomol. Soc. Author: numerous books including: Insects, Hunters and Trappers, 1957; Insect Builders and Craftsman, 1958; Wild Ways, 1959; Strange Plants, 1960; This is a Leaf, 1961; This is a Flower, 1962; This is a Tree, 1963; The Amazing Seeds, 1965; Insects, 1966; Caddis Insects, 1966; Plants without Leaves, 1966; The Travels of Monarch X, 1966; The Ant Realm, 1967; also articles. Research on insect control and insect biology. Home: 502 N. Montgomery St., Starkville, Miss. 39762. Office: Drawer EH, Miss. State U., State College, Miss. 39762.*

HUTCHINSON, Arthur, English mineralogist; b. Woodside, Eng., July 6, 1866; s. George and Deborah (Richardson) H.; student Christ's Coll., Cambridge (Eng.) U., 1884-88; student of Emil Fischer, W. K. Roentgen, 1889-91; Ph.D.; m. Evaline Shipley, 1901; 2 sons, 1 dau. Research (with M. M. Pattison Muil), 1888-89; demonstrator chemistry Gonville and Caius Coll., Cambridge U., 1891, named fellow Pembroke Coll., coll. lectr. in natural scis., 1892, asst. tutor, 1901-26, master, 1928-37; became demonstrator in mineralogy, 1895, lectr. in crystallography, 1923, prof. mineralogy, 1926. Hon. fellow Christ's Coll. Fellow Royal Soc., 1922. Discovered stokesite, 1899; measured refractive indices of stibnite and proved its orthorhombic symmetry; devised inverted goniometer to determine optical and crystallographic nature of small crystals, also protractor for constrn. of stereographic and gnomonic projections; research on gas masks. Died Cambridge, Dec. 12, 1937.

HUTCHINSON, Arthur Cyril William, oral surgeon; b. Nottingham, Eng., July 26, 1894; s. Edward William Roberts and Annie Lowater (Goddard) H.; B.D.S., L.D.S., U. Manchester, Eng., 1916, M.D.S., 1929; D.D.S., U. Witwatersrand, S. Africa, 1934; D.D.S. (hon.), U. Edinburgh (Scotland), 1968; m. Dorothy Mary Orme, Aug. 22, 1918. Demonstrator in operative dental surgery Dental Hosp. of Manchester, 1913-23, hon. dental surgeon, 1920-33; hon. dental surgeon Oldham Royal Infirmary, 1913-23; lectr. dental prosthetics, dental mechanics U. Manchester, 1925-33; lectr. dental bd. U.K., 1931; dean Edinburgh Dental Hosp. and Sch., 1934-51; prof. dental surgery, dir. dental studies U. Edinburgh, 1951-58; vis. prof. diagnosis Northwestern U. Dental Sch., 1958—. Mem. Gen. Dental Council U.K., 1950-59; cons. Royal Navy, 1940—; assessor U. Grants Com., 1958—; external examiner to Manchester, Liverpool, Leeds, Glasgow, Durham univs. Fellow Royal Soc. Edinburgh, Royal Coll. Surgeons Edinburgh, Royal Society of Medicine, London. Mem. American, British dental associations, Odonto-Chirurgical Soc. Scotland, Internat. Dental Fedn., Chgo. sect. Internat. Soc. Dental Research, Brit. Soc. Dental Radiologists, Royal Soc. Health, London Royal Soc. Arts, Nutrition Soc. Author: (with F. C. Thompson) Some Observations on Cast and Swaged Plates, 1932; Dental and Oral X-Ray Diagnosis, 1954. Research, publs. on structure of human dental enamel as shown by electron microscope and by microradiography, early and late caries of human dental enamel as shown by electron microscope and microradiography. Home: 12 Glencairn Crescent, Edinburgh 12, Scotland. Office: Northwestern U. Dental Sch., 311 E. Chicago Av., Chgo. 60611.*

HUTCHINSON, Charles Angevine, Jr., Am. chem. engineer; b. Boulder, Colo., June 27, 1923; s. Charles Angevine H.; B.S. Colorado, 1944, M.S., 1949; m. 1947; four children. Jr. engr. Atlantic Refining Co., 1949-51; admin. asst., research, 1951-54; sr. research engr., 1954-56; research supvr., production research, 1956—. Mem., Nat. Acad. Scis., Assn. Petrol. Geology. Research on petroleum production; fluid flow; surface chemistry. Office: Atlantic Refining Co., Box 1981, Corpus Christi, Texas 01035.

HUTCHINSON, Franklin, Am. biophysicist; b. Bklyn., Feb. 29, 1920; s. Franklin and Marjorie (Rollhaus) H.; B.S., Mass. Inst. Tech., 1942; Ph.D., Yale, 1948; m. Edith Arnold Pringle, Sept. 16, 1944; children—Bruce, Franklin IV, Alexander, Mary Candace. With Mass. Inst. Tech. Lab., 1942-45; instr. physics and radiology Yale, 1948-51, asst. prof. physics, 1951-56, asso. prof. biophysics, 1956-60, prof. biophysics, 1960. Cons. radiol. physics Yale-New Haven Hosp., 1948—, Hartford (Conn.) Hosp., 1950—. Dir. Raycon, Inc., South Windsor, Conn. Guggenheim fellow U. London, 1963-64. Mem. Biophys. Soc., Radiation Research Soc., Am. Phys. Soc., A.A.A.S. Studies, numerous publs. on the molecular basis for action of ionizing radiations on living cells, new methods for determining structure of biol. molecules. Home: 862 Grassy Hill Rd., Orange, Conn. 06477.*

HUTCHINSON, John, British botanist; b. Apr. 7, 1884; s. Michael and Annie (Wylie) H.; LL.D., U. London; m. Lilian Florence Cook, Oct. 29, 1910; children—Violet, Joan, Alan, Nora, John. Botanist, mus. curator Royal Botanic Gardens of Kew, Surrey. Fellow Royal Soc., 1947. Author: Families of Flowering Plants; A Botanist in Southern Africa; British Wild Flowers; Genera of Flowering Plants; Story of Plants. Home: Heather Bank, Lightwater. Office: Herbarium, Kew, Eng.

HUTCHINSON, Sir Jonathan, English surgeon; b. Selby, Eng., July 23, 1828; s. Jonathan and Elizabeth (Massey) H.; ed. St. Bartholomew's Hosp., London; M.D.; LL.D.: 4 sons, 4 daus. Hunterian prof. Royal Coll. Surgeons; emeritus prof. surgery London Hosp. Coll.; cons. surgeon. Mem. Royal Commn. on Smallpox Hosps., 1884, Royal Commn. on Vaccination, 1890-96. Fellow Royal Coll. Surgeons (pres. 1889-90), Royal Soc., 1882. Author: The Centuries; Rare Diseases of the Skin; The Pedigree of Disease; Syphilis, a Manual; Illustrations of Clinical Surgery; Syphilitic Affections of the Eye and Ear; Lesser Atlas of Clinical Illustrations; Fish-eating and Leprosy, also articles. Described numerous rare diseases and disease characteristics named after him. Died June 23, 1923.

HUTCHINSON, Sir Joseph Burtt, English plant geneticist; b. Burton Latimer, Northants, Eng., Mar. 21, 1902; s. Edmund and Lydia Mary (Davy) H.: B.A., St. John's Coll., Cambridge U., 1923, M.A., 1931, Sc.D., 1949; m. Martha Leonora Johnson, July 9, 1930; children—Helga Leonora, Dennis Procter. With Empire Cotton Growing Corp., 1926-33, 37-57, dir. Research Sta., Namulonge, Uganda, 1949-57; geneticist, botanist Inst. Plant Industry, Indore, India, 1933-37; Drapers prof. agr., Cambridge U., 1957—, professorial fellow St. John's Coll. Fellow Royal Soc., 1951, Linnean Soc.; mem. Brit. Assn. for Advancement of Sci. (pres. 1965-66). Author: The Genetics of Gossypium, 1947; Application of Genetics to Cotton Improvement, 1959; Essays on Crop Plant Evolution, 1965. Research on genetics and evolutionary history of world's cottons, world agr. with reference to food supply problems, effect of govt. policies on productivity. Home: Huntingfield, Huntingdon Rd., Cambridge, Eng.*

HUTCHISON, Dorris Jeannette, Am. microbiologist; b. Carrsville, Ky., Oct. 31, 1918; d. John W. and Maud (Short) H.; B.S., Western Ky. State Coll., 1940; M.S., U. Ky., 1943; Ph.D., Rutgers U., 1949. Instr., Russell Sage Coll., 1942-44, Vassar Coll., 1944-46; staff Rugers U., 1946-49, research asso., 1948-49; instr. Wellesley Coll., 1949-51; staff Sloan Kettering Inst., N.Y.C., 1951-56, asso. mem., 1960—, acting chief div. chemotherapy, 1965-66, div. chief drug resistance, 1967—; faculty Sloan Kettering Inst. div. Cornell U. Med. Coll., New York City, 1952—, asso. prof. microbiology, 1958—. Faculty fellow Vassar Coll., 1946; USPHS fellow, 1951-53; Philippe Found. fellow, Paris, 1959. Fellow N.Y. Acad. Sci., Am. Acad. Microbiology (charter), N.Y. Acad. Medicine (asso.); mem. A.A.A.S., Am. Assn. Cancer Research, Harvey Soc., Genetics Soc. Am., Am. Inst. Nutrition, Am. Soc. for Microbiology, (past pres. N.Y. br.), Soc. for Cryobiology, Am. Genetic Assn. Research, numerous publs. on antibiotics and chems. effective in treatment of Tb and leukemia, reports on mechanisms explaining how leukemic cells become resistant to treatment; search for more effective antileukemia drugs. Home: Southgate Apts., Bronxville, N.Y. 10708. Office: 145 Boston Post Rd., Rye, N.Y. 10580.*

HUTCHISON, John, English physician; b. Newcastle, Eng., 1811; ed. London U.; asst. prof. Hosp. for Consumption, Brompton, Eng. Author: The Spirometer and Stethoscope and Scale Balance, their use in discriminating diseases of the chest and their value in life-offices, 1852. Inventor early spirometer for measuring air inhaled and exhaled by lungs, 1846. Died Fiji, Sandwich Islands, 1861.

HUTCHISON, (William) Kenneth, chem. engr.; b. Assam, India, Oct. 30, 1903; s. William and Barbara (McCormack) H.; B.A., Corpus Christi Coll., Oxford (Eng.) U., 1925, B.Sc., 1926, M.A., 1960; m. Dorothea Marion Eva Bluett, Sept. 14, 1929; 1 dau., Ann (Mrs. Christopher Buckley). Staff, Gas, Light & Coke Co., 1926-48, controller by-products, 1945-48, mng. dir., 1947; asst. dir. hydrogen prodn. Air Ministry, 1940-42, dir., 1942-43, dir. compressed gasses, 1943-45; chmn. S. Eastern Gas Bd., 1948-59; dep. chmn. Gas Council, London, 1960-66, cons., 1967—; chmn. Internat. Mgmt. and Engring. Group Ltd., 1967—; dir. Newton Chambers & Co. Ltd. Created knight, 1962; decorated comdr. Brit. Empire, 1954. Fellow Royal Soc.; mem. Instn. Chem. Engrs. (Moulton medal 1942), Instn. Gas Engrs. (hon.), (H. E. Jones Gold medal 1938, Birmingham medal 1966). Research, publs. on mechanism of gas reactions, prodn. and purification of gas; devel. of cheaper methods of gas prodn. by gasification of oil and introduction of natural gas into Brit. gas industry as liquid or as discovered in N. Sea. Home: 2, Arlington Rd., Twickenham, Middlesex, U.K. Office: 7/8 Savile Row, London W. 1, Eng.*

HUTCHISON, Miller Reese, Am. inventor, engr.; b. Montrose, Ala., Aug. 6, 1876; s. William Peter and Tracie (Magruder) H.; student Marion Mil. Inst., 1889-91, Spring Hill Coll., 1891-92, University Mil. Inst., Mobile, 1892-95; B.S. in E.E., Ala. Poly. Inst., 1897, E.E., 1913; Ph.D., Spring Hill Coll., 1914; attended Ala. Med. Coll.; m. Martha J. Pomeroy, May 31, 1901. Chief elec. engr. U. S. Light House Establishment, 7th and 8th dists., during Spanish-Am. War, engaged in laying submarine mines and cables, Gulf Harbors; established Hutchison Laboratory, New York, 1899; invented and marketed many elec. and mech. appliances among which were "Acousticon" for the deaf, "Dictograph," "Klaxon Horn"; Presented with spl. gold medal by Queen of Eng. for exceptional merit in the field of invention, 1902; gold medals, St. Louis Expn., 1904 for Acousticon and commercially operated wireless telephone. Became associated with Thomas A. Edison in spl. work on storage batteries, 1910, apptd. chief engr. Edison Lab. and all affiliated Edison interests, chief engr. to and personal rep. of Thomas A. Edison, 1913, adv. mgr. Edison Storage Battery Co., 1912-17; engr. adviser Thomas A. Edison, 1917-18; formed Miller Reese Hutchison, Inc., 1917, to act as sole distributors Edison Storage Batteries for all govt. purposes all nations, of which became pres.; sold rights back to Edison Co., 1918, to devote entire time to govt. service for period of war; propr. Hutchison Lab. Mem. Internat. Elec. Congress, St. Louis, 1904, Internat. Engring. Congress, San Francisco, 1915; mem. Naval Consulting Bd. Mem. Am. Acad. Polit. and Social Science, A.A.A.S., Am. Inst. E.E., Am. Inst. Radio Engrs., Am. Soc. M.E., Am. Soc. Naval Engrs., Nat. Inst. Social Sciences, Navy League U. S., New York Elec. Soc., Soc. Automotive Engrs., Nat. Geog. Soc., Soc. Am. Mil. Engrs., U. S. Naval Inst., Optical Soc. Am., Accademia Internazionale di lettre e Scienze (Napoli), Royal Soc. for Encouragement of Arts, Manufacture and Commerce (London). Died Feb. 16, 1944.

HUTCHISON, Victor Hobbs, Am. biologist; b. Blakely, Ga., June 15, 1931; s. Joseph Victor and Veva (Hobbs) H.; B.S., N. Ga. Coll., 1952; Ph.D., Duke, 1959, M.A., 1956; m. Theresa Dokos, Aug. 14, 1952; children—Victoria Ann, John Christopher, David Michael, Kenneth Hobbs. Instr., Duke, 1957-58, faculty fellow, So. Fellowship Fund fellow, 1958-59; faculty U. R.I., Kingston, 1959—, professor biology, 1968—, dir. Inst. Environmental Biology, 1966—. Research prof. Universidad de Los Andes, Bogota, Colombia, 1965-66; prin. investigator Nat. Geog. Soc.-U. R.I. Herpetological Expdn. to Colombia, 1964-65. John S. Guggenheim Meml. fellow, 1964-66. Mem. A.A.A.S., Am. Inst. Biol. Sci., Am. Soc. Icthyologists and Herpetologists, Am. Soc. Mammalogists, Am. Soc. Zoologists, Ecol. Soc. Am., Herpetologists League, Assn. Latin Am. Herpetologists, Sigma Xi, Phi Sigma. Author: (with George C. West) Laboratory Manual in Animal Biology, 1963, 2d edit., 1967; also articles. Taxonomic studies on amphibians and reptiles, zoogeography; research on animal-alga symbiosis, heat tolerances of lower vertebrates, effects of daylength on metabolism and temperature tolerance of lower vertebrates, ecology of amphibians and reptiles; discovered that pythons are able to become facultative endotherms during brooding of eggs. Home: 41 Highland Av., Wakefield, R.I. 02879. Office: Dept. Zoology, U. R.I., Kingston, R.I. 02881.*

HUTCHISSON, Elmer, Am. physicist; b. Cleve., Dec. 29, 1902; s. Harry and Anna Bertha (Merrick) H.; B.S., Case Inst. Tech., 1923, D.Sc., 1957; M.S. (Charles A. Coffin fellow), Mass. Inst. Tech., 1924; Ph.D., U. Minn., 1926; postgrad. U. Berlin (Germany); D.Sc., Washington Coll., Chestertown, Md., 1958; m. Rose Valasek, Sept. 14, 1925. Faculty, U. Pitts., 1926-44, prof. physics, 1937-44, head dept., 1938-44; dean faculty Case Inst. Tech., Cleve., 1945-55, acting pres., 1950-52, dean grad. sch., 1955-57; dir. Am. Inst. Physics, N.Y.C., 1957-64, dir. emeritus, 1965—, chmn. editors, 1945-54, Tech. aide NDRC, 1941-44; mem. com. on sci. abstracting UNESCO, Paris, 1949, London, 1950, U. S. Nat. Commn., 1958-63; v.p. Internat. Abstracting Bd. Internat. Council Sci. Unions, 1953-65; vis. scholar Stanford University, 1965—; mem. documentation commn. Internat. Union Pure and Applied Physics, 1950-57, Indsl. adv. com. AEC, 1947-57; chmn. adv. com. documentation Nat. Acad. Scis.-NRC, 1960-65; mem. U. S. nat. com. Internat. Documentation Fedn., 1961—, chmn., 1963-65; mem. sci. information council NSF, 1958-63. Recipient Presdl. certificate merit, 1948; Achievement gold medal U. Minn., 1951; Meritorious Service award Case Alumni Assn., 1956. Fellow Am. Phys. Soc.; mem. Optical Soc. Am., Am. Assn. Physics Tchrs. (Distinguished Service citation 1957), Acoustical Soc. Am., A.A.A.S., Sigma Xi, Tau Beta Pi, Gamma Alpha. Author: (with others) Outline of Atomic Physics, 1933; (with O. Blackwood) Laboratory Manual of Physics, 1933; (with Osgood, Ruark) Atoms Radiations and Nuclei, rev., 1965. Editor: Jour. Applied Physics, 1937-53. Research, publs. in molecular studies, including energy calculations, intensity of band structure, and specific heat calculations. Home: The Sequoias, 501 Portola Rd., Portola Valley, Cal. 94025.*

HUTER, Carl, German physician; b. Heinde, 1861. Author: Menschenkenntnis durch Körper-Lebens-Seelen- und Gesichtsausdruckskunde, 5 vols., 1904. Editor book on constitution, typology, and physiognomics. Died Dresden, Germany, 1912.

HUTNER, Seymour Herbert, Am. microbiologist; s. Julius and Fannie (Zuckerman) H.; B.S., Coll. City N.Y., 1932; Ph.D., Cornell U., 1937; m. Margarita Silva, Aug. 18, 1956; 1 son, Reed A. Research asso. Mass. Inst. Tech., 1935-36; technician div. labs. and research N.Y. State Health Dept., 1938-41; staff mem. Haskins Labs., N.Y.C., 1941-65, asso. research dir., 1964—; vis. prof. Inst. Microbiology U. Brazil, Rio De Janeiro, 1963-64; adj. prof. biology Fordham U., 1964. Bus. mgr. Jour. Phycology, 1964—. Fellow Am. Acad. Microbiology; mem. Soc. Protozoologists, Am. Soc. for Microbiology, Soc. Gen. Microbiology, N.Y. Acad. Sci. Editor: (with Andre Lwoff) Biochemistry and Physiology of Protozoa, vol. 2, 1953; Biochemistry and Physiology of Protozoa, vol. 3, 1964.

Editorial bd. Jour. Protozoology, 1953—. Research, numerous publs. on chelating agts. in microbial nutrition, hydroponic cultivation higher plants, protozool. assays for vitamins in blood. Home: 142 W. End Av., N.Y.C. 10023. Office: 305 E. 43d St., N.Y.C. 10017.*

HUTT, Frederick Bruce, geneticist; b. Guelph, Ont., Can., Aug. 20, 1897; s. Howard Laing and Annie (Pook) H.; B.S.A., Ont. Agr. Coll. (now U. Guelph), U. Toronto, Ont., 1923; M.S., U. Wis., 1925; M.A., U. Man. (Can.), 1927; Ph.D., U. Edinburgh (Scotland), 1929, D.Sc., 1939; D.Sc. h.c. U. Brno (Czechoslovakia), 1965; m. Alice Jean Bacon, June 30, 1930; children—Frederick Bruce, Robert Bacon, Margaret Ann. Lectr., U. Man., 1923-27; faculty U. Minn., 1928-34; faculty Cornell U., Ithaca, N.Y., 1929—, prof. poultry husbandry and animal genetics, 1934-40, prof. zoology, 1940-44, prof. animal genetics, 1940-65, emeritus prof., 1965—, head dept. poultry husbandry, 1934-40, chmn. dept. zoology, 1939-44; Vis. prof. N.C. State Coll., 1956; vis. scholar Va. Poly. Inst., 1966; guest prof. genetics Ore. State U., 1966; George Scott Robertson Meml. lectr. Queens U., Belfast, N. Ireland, 1957. Recipient Newman Trust Internat. Poultry award, 1961. Fellow A.A.A.S., Poultry Sci. Assn. (Research prize 1929, Borden award 1946, past pres.); mem. Genetics Soc. Am., Am. Soc. Zoologists, Am. Genetic Assn., Am. Soc. Naturalists, Am. Soc. Human Genetics, Genetics Soc. Can., Sigma Xi. Author: Genetics of the Fowl, 1949; Genetic Resistance to Disease in Domestic Animals, 1958; Animal Genetics, 1964; also numerous articles. Editorial bd. Jour. Heredity, 1957—. Research on heredity in domestic animals, breeding fowls resistant to leucosis, other diseases, mutations, linkage, hereditary defects. Home: 107 Woodcrest Terrace, Ithaca, N.Y. 14850.*

HUTT, Michael Stewart Rees, pathologist; b. Shrewsbury, Eng., Oct. 1, 1922; s. Arthur Cyril and Dorothy (Peck) H.; M.B.B.S., U. London, 1945, M.D., 1949; m. Elizabeth Newton-Jones, June 4, 1946; children—Caroline, Jane, Rosamunde, Mark. Cons. pathologist St. Thomas' Hosp., 1958-62; prof. pathology Makerere U. Med. Sch., Kampala, Uganda, 1962—, U. East Africa, 1962—; cons. pathologist New Mulago Hosp., Kampala, 1962—. Mem. Royal Soc. Medicine, Renal Soc., Brit. Soc. Haematology, Assn. Clin. Pathologists. Research, numerous publs. on geog. distbn. of diseases, particularly cancer and heart disease in Africa. Address: Makerere Univ. College, P.O. Box 2072, Kampala, Uganda.*

HUTT, Sidney John, English psychologist; b. Bilston, Eng., July 29, 1935; s. William Arthur and Irene (Johnson) H.; B.A. with honors, Manchester U., 1956; M.A., Oxford U., 1965; m. Corinne Mack, Aug. 10, 1957; 1 son, Simon John. Research psychologist Inst. Aviation Medicine, Farnbrough, 1956-60; psychologist St. John's Hosp., Aylesbury, 1960-61; sr. research psychologist Park Hosp. for Children, Oxford, 1961-65; research fellow St. Catherines Coll., Oxford, 1965—. Hon. research fellow Warneford and Park hosps. Asso. mem. Brit. Psychol. Soc. Research, publs. on application of techniques derived from study of animal behavior in wild to study of human behavior; proposed physiol. theory for explanation of infantile autism. Home: The Gables, Newington, Warborough, Oxford, Eng.*

HUTTON, Charles, English mathematician; b. Newcastle-on-Tyne, Eng., Aug. 14, 1732; trained as tchr.; LL.D., Edinburgh, 1779; m. twice; 2 daus., including Mrs. Henry Vignoles; 1 son, George Henry. Hewer, pit at Long Benton, Eng.; opened math. sch., Newcastle, 1760; prof. math. Woolwich Acad., 1773-1807. Recipient Copley medal, 1778. Fellow Royal Soc., 1774 (fgn. sec. 1779). Author: Principles of Bridges, 1772; Mathematical Tables, 1785; also papers. Editor Ladies Diary, 1773-1818; abridged Philos. Trans., 1809. Prepared map of Newcastle, 1770; computed mean density of earth, 1778; studied and expressed in math. terms initial velocity and air resistance of shells. Died Jan. 27, 1823.

HUTTON, Frederick Wollaston, biologist; b. Gate Burton, Lincolnshire, Eng., Nov. 16, 1836; s. H. F. Hutton; ed. Naval Acad., Gosport; m. Annie G. Montgomerie, 1863. In army, 1855-66, emigrated to New Zealand, 1866; asst. geologist New Zealand Geol. Survey, 1871; curator Otago Mus., 1873; prof. natural sci. Otago U., 1877-1905; prof. biology U. New Zealand, 1880-93; curator Museum, Christchurch, New Zealand. Fellow Royal Soc., 1892, Geol. Soc.; mem. New Zealand Inst. (pres.). Author: Darwinism and Lamarckism, 1899; The Lesson of Evolution, 1902; Animals of New Zealand, 1904, also numerous articles. Died Oct. 29, 1905.

HUTTON, James, Scottish geologist; b. Edinburgh, Scotland, June 3, 1726; s. William Hutton; ed. univs. Edinburgh, Paris; M.D., U. Leiden (Netherlands), 1749. With James Davie produced salammoniac from coal soot; settled Edinburgh, 1768. Mem. Edinburgh Philos. Soc., Royal Soc. Edinburgh (contbr. to Transactions). Author: Theory of the Earth, 1795 (verified by visits to Glen, Tilt, Galloway, Arran, Isle of Man); Theory of Rain; Dissertations, 1792; Investigations of Principles of Knowledge, 1794. Joint-editor: Essays on Philosophical Subjects (Adam Smith), 1795. Originator modern theory of formation of earth's crust, uniformitarian theory of geology (sometimes considered father of geology); his theory required much greater age for earth than previously held; studies on rainfall (observed that amount of moisture air could hold rose with temperature, predicted necessary conditions for rainfall). Died Edinburgh, Mar. 26, 1797.

HUUS, Torben, Danish physicist; b. Copenhagen, Denmark, Dec. 5, 1919; s. Ove Raahauge and Else (Clemmensen) H.; Mag.Scient., U. Copenhagen, 1943, Dr.Phil., 1956; Research asso. Niels Bohr Inst., U. Copenhagen, 1943-50, 52-60, amanuensis, 1956-60, prof. physics, 1960—; research asso. Cal. Inst. Tech., Pasadena, 1951. Mem. Royal Danish Acad. Scis. Research, publs. on exptl. nuclear physics especially nuclear structure using Coulomb excitation. Home: 11 Vordingborggade, Copenhagen 2100, Denmark.*

HUXFORD, Walter Scott, Am. physicist; b. Neligh, Neb., Dec. 15, 1892; s. Herbert C. and Cora (Scott) H.; B.A., Doane Coll., 1917; M.S., U. Neb., 1924; Ph.D., U. Mich., 1928; m. Mary Bertha Whalen, Aug. 28, 1917; children—Sara May (Mrs. George Ball), Barbara Jane (Mrs. Joseph Nicol), Charles K., Mary Patricia. Engaged in elec. communications work, 1917-19; high sch. instr. physics, 1919-21; instr. later prof. physics Doane Coll., Crete, Neb., 1923-26 and 1928-29; physicist dept. of engring. research U. Mich., 1930-32; asst. prof., later asso. prof. physics Northwestern U., Evanston, Ill., 1932-39, prof. from 1939; dir. research, Nat. Defense Research Council, 1943-45. Recipient Army and Navy certificate of merit, 1947. Fellow Am. Phys. Soc.; mem. Optical Soc. Am., Ill. Acad. Sci., Sigma Xi. Research and publs. in gaseous electronics and infrared communication systems, methods of optical communications. Died Feb. 12, 1958.

HUXLEY, Andrew Fielding, English physiologist; b. London, Eng., Nov. 22, 1917; s. Leonard and Rosalind (Bruce) H.; B.A., Cambridge (Eng.), 1938, M.A., 1941; M.D., U. Saar, 1964; D.Sc., U. Sheffield (Eng.), 1964; m. Jocelyn Richenda Gammell Pease, July 5, 1947; children—Janet Rachel, Stewart Leonard, Camilla Rosalind, Eleanor Bruce, Henrietta Catherine, Clare Marjory Pease. Research staff Anti-Aircraft Command, 1940-42, Admiralty, 1942-45; fellow Trinity Coll., Cambridge, 1941-60, dir. studies, 1952-60; demonstrator Cambridge U., 1946-50, asst. dir. research, 1951-59, reader exptl. biophysics, 1959-60; Jodrell prof. physiology U. Coll. London, 1960—. Recipient (with A. L. Hodgkin, J. C. Eccles) Nobel prize for medicine and physiology, 1963. Fellow Royal Soc., 1955; mem. Physiol. Soc., Biophys. Soc., Brit. Biophys. Soc. Editor: Jour. Physiology, 1950-57, chmn. bd. Publs. on analysis of nerve conduction (with Hodgkin), physiology of striated muscle, devel. of interference microscope and ultramicrotome. Home: Manor Field, Grantchester, Cambridge, Eng. Office: Dept. Physiology, U. Coll. London, Gower St., London W.C.1., Eng.*

HUXLEY, Hugh Esmor, English molecular biologist; b. Birkenhead, Eng., Feb. 25, 1924; s. Thomas Hugh and Olwen (Roberts) H.; B.A., Christ's Coll., Cambridge (Eng.) U., 1948, M.A., 1950, Ph.D., 1952, Sc.D., 1964. Research student molecular biology unit Med. Research Council, Cavendish Lab., Cambridge, 1948-52, sci. staff, 1955-56; external staff dept. Med. Research Council biophysics U. Coll., London, 1956-61; sci. staff Med. Research Council Lab. Molecular Biology, 1962—. Commonwealth fellow dept. biology Mass. Inst. Tech., Boston, 1952-54; fellow Christ's Coll., Cambridge U., 1955-56, fellow King's Coll., 1961-67, fellow Churchill Coll., 1967—. Decorated mem. Order Brit. Empire. Recipient Feldberg prize, 1963; Hardy prize, 1965. Fellow Royal Soc., 1960; mem. Physiol. Soc., Brit. Biophys. Soc., European Molecular Biology Orgn., Am. Acad. Arts and Scis. (hon. fgn. mem.). Editor: Progress in Biophysics and Molecular Biology, 1960-66; editorial bd. Jour. Cell Biology, 1959-63, Jour. Molecular Biology, 1962—, Jour. Cell Sci., 1966—. Research, publs. on ultrastructures of striated muscles, especially by electron-microscopy and X-ray diffraction leading to sliding filament theory of contraction (with Jean Hanson; simultaneously proposed by A. F. Huxley and R. Niedergerde); studies on electron microscopy of viruses, ribosomes and other nucleic-acid containing structures. Home: Binsted, Herschel Rd. Office: Med. Research Council Lab. Molecular Biology, Hills Rd., Cambridge, Eng.*

HUXLEY, Sir Julian (Sorell), English biologist; b. Eng., June 22, 1887; s. of Leonard Huxley; scholar Eton and Balliol Coll., Oxford U. (awarded Newdigate Prize for poetry, 1908); m. Marie Juliette Baillot, 1919; 2 sons. Lecturer in zoology, Balliol Coll., 1910-12; research asso. (travelling in Germany), Rice Inst., Houston, Tex., 1912-13, asst. prof., 1913-16; fellow and sr. demonstrator zoology, New Coll., Oxford, 1919-25; mem. Oxford expdn. to Spitsbergen, 1921; prof. zoology, King's Coll. (London), 1925-27, hon. lecturer, 1927-35; Fullerian prof. physiology, Royal Inst., Eng., 1926-29; gen. supervisor, biol. films, Gaumont Brit. Instructional, Ltd., 1933-36, Zool. Film Prodns., Ltd., 1937-42; sec. Zool. Soc. of London, advisory editor Zoo Mag., 1935-42; Beatty lectr. McGill U., Montreal, 1956. Created Knight, 1958. Fellow Royal Soc., 1938 (Darwin medal 1956). Former officer several profl. assns. Mem. commn. on Higher Education in West Africa, 1944. Mem. com. Nat. Parks, United Kingdom, 1945-46. Exec. sec. United Nations Educational and Cultural Organization Prep. Commission, 1946; director general UNESCO, 1947-48, v.p. Commn. for Sci. and Cultural History of Mankind, mem. Jordan expedition, 1963. Author or editor, some with others, over 40 books 1911—, including The Individual in the Animal Kingdom, 1911; Religion without Revelation, 1927, rev. 1957; Animal Biology (with J. B. S. Haldane), 1927; The Sci. of Life (with H. G. and G .P. Wells), 1929; What Dare I Think?, 1931; The Elements of Experimental Embryology (with G. R. de Beer), 1934; Scientific Research and Social Needs, 1934; If I Were Dictator, 1934; T. H. Huxley's Diary on the Rattlesnake (Ed.), 1935; We Europeans, 1936; The Living Thoughts of Darwin (with J. Fisher), 1939; The Uniqueness of Man, 1941; Evolution, the Modern Synthesis, 1942; Evolutionary Ethics (Romanes lecture, Oxford, 1943); Man in the Modern World, 1947; Evolution and Ethics, 1947; Heredity, East and West, 1949; Evolution in Action (ed.), 1953; The Evolutionary Process (ed.), 1953; The Kingdom of Beasts (with W. Suschitzky), 1956; Secrets of Life, 1957; Biological Aspects of Cancer, 1957; New Wine in New Bottles (essay), 1957; The Study of Evolution, 1959; The Humanist Frame (editor), 1961; Conservation of Wild Life in Central and East Africa, 1961; Essays of a Humanist, 1964; (with H. B. Kettlewell) Darwin and His World, 1965. Biol. editor Ency. British, 14th edit.; editorial bd. New Naturalist series, 1944—. Authority on evolution; applied principles of sci. knowledge to polit. and social problems; formulated pragmatic ethical theory based on principle of natural selection. Home: 31 Pond St., Hampstead, London N.W. 3, Eng.

HUXLEY, Thomas Henry, English biologist; b. Ealing, Middlesex, Eng., May 4, 1825; s. George and Rachel (Withers) H.; studied medicine Charing Cross Hosp.; med. degree, 1845; m. H. A. Heathorn, 1855. Became surgeon in H.M.S. Rattlesnake which did survey work in Torres Straits, 1846; lectr. Sch. Mines, 1854, held chair natural history 31 years; naturalist Geol. Survey, 1955; mem. numerous royal commns., held many pub. positions, 1862-64. Fellow Royal Soc., 1851 (council 1851, sec. 1871-80, pres. 1881-85, Royal medal 1851). Author: On the Causes of the Phenomena of Organic Nature, 1863; Zoological Evidences as to Man's Place in Nature, 1863; Elementary Lessons in Physiology, 1866; Lay Sermons and Essays and Reviews, 1870; Manual of Comparative Anatomy of Vertebrated Animals, 1871; The Crayfish, 1880; Science and Culture, 1881; Evolution and Ethics, 1893; Collected Essays, 1894. Did extensive research on tropical animals while with Rattlesnake; postulated class hydrozoa to add to Cuvier's natural classification system; demonstrated existence of previously unrecognized inner sheath of hairs (known as Huxley's layer); named phylum coelenterata; disproved theory that origin of skull was vertebrate, 1858; coined phrase describing protoplasm as phys. basis of life; had great influence on English nat. elementary edn.; exerted great influence on pub. opinion, did much to popularize Darwin's theory of evolution. Died Eastbourne, Sussex, June 29, 1895.

HUYGENS (or Huyghens), Christiaan, Dutch mathematician, astronomer, physical scientist; b. The Hague, Netherlands, Apr. 14, 1629; s. Constantijn H.; studied law and math. at Leyden and Breda; hon. LL.D. U. Angers, 1655. Traveled widely on Continent and in England; accompanied mission of Henry, Count of Nassau, to Denmark, 1649; lived in Paris 1661-81; returned to Netherlands, 1681. Fellow Royal Soc., 1663; mem. French Acad. Scis., 1666. Author: Theoremata de quadratura hyperboles ellipsis et circuli, 1651; De circuli magnitudine inventa, 1653; De Saturni luna observatio novis, 1656; Systema Saturnium 1659; Horologium Oscillatorium, 1673; Traité de la lumière (written 1678, pub. 1690, with supplement Discours sur la cause de la pesanteur); Cosmotheoros, 1698. Preferred Descartes' vortices to Newton's attractions and absolutes; presented mechanical model of gravity; discovered laws of centrifugal force; invented pendulum clock, 1655; a manometer, 1661; maintained light is wave motion in the aether, opposing Newton's corpuscular theory, 1690. Died The Hague, June 8, 1695.

HUYSER, Earl Stanley, Am. chemist; b. Holland, Mich., May 27, 1929; s. Stanley Quirinus and Gertrude (Westra) H.; A.B., Hope Coll., 1951; Ph.D., U. Chgo., 1954; m. Barbara L. Van Kolken, June 26, 1952; children—Nancy, Thomas, David, Gretchen. Research asso. U. Chgo., 1954-55; postdoctoral fellow Columbia, N.Y.C., 1956-57; chemist Dow Chem. Co., Midland, Mich., 1957-59, cons., 1959—; faculty U. Kan., Lawrence, 1959—, prof., 1966—; vis. prof. organic chemistry U. Groningen (Netherlands), 1964-65. Mem. Am. Chem. Soc., Chem. Soc. London, Kan. Acad. Scis., Sigma Xi, Phi Lambda Upsilon. Research, publs. on free radical reactions in solvent, kinetics of radical chain reactions, organo-phosphorous compounds. Home: 1821 W. 21st St., Lawrence, Kan. 66044.*

HUZARD, Jean-Baptiste, French veterinarian; b. Paris, Nov. 3, 1755; ed. Alfort Sch.; at least 1 son, Jean-Baptiste; insp. vet. schs.; prof. Alfort Vet. Sci. Sch., 1773-92; mem. French Acad. Scis., Acad. Medicine, Soc. Agr. Author: Sur les soins à donner aux cheveaux, 1794; Sur les maladies qui affectent les vaches laitières, 1795. Research on bovine and epizootic diseases; developed project to introduce merino sheep into France. Died Dec. 1, 1815.

HWANG, Kao, pharmacologist; b. Changsha, Hunan, China, Apr. 3, 1916; s. Hung Hsi and Tsung-su (Hsu) H; came to U. S., 1945, naturalized, 1956; M.D.,

Hsiang Ya Med. Coll., Kweiyang, China, 1940; M.S., U. Ill., 1947, Ph.D. in Physiology, 1953; m. Sheila Ning-tso Chen, July 17, 1948; children—Leila, Miriam, Catherine, Jason Kao. Faculty, Hsiang Ya Med. Coll., 1940-45; instr. Nat. Tsing Hua U., 1945; fellow Mayo Found., 1945-46; faculty U. Ill. Med. Sch., 1946-61, asst. prof. dept. clin. sci., 1958-61; sr. pharmacologist Abbott Labs., North Chicago, Ill., 1950-57, group leader, 1957-61, sect. head dept. pharmacology, 1961—. Boxer Rebellion Indemnity Fund scholar Tsing Hua U., 1944; recipient Spl. Certificate of Merit, Abbott Labs., 1955. Fellow Am. Coll. Clin. Pharmacology and Chemotherapy; mem. A.A.A.S., Soc. for Exptl. Biology and Medicine, Am. Soc. for Pharmacology and Exptl. Therapeutics, Ill. Soc. Med. Research, Am. Med. Writers Assn., Am. Soc. for Microbiology, Sigma Xi. Contbg. author: Diseases of the Digestive System, 1953. Research, publs. on physiology esophagus, mechanisms of laxative action of certain poorly absorbed salts, anticholinergics as inhibitors of gastric secretions, evaluation of new antibiotics; discovered nerves and their function in relations to upper end of esophagus; introduced hexocyclium methylsulfate for peptic ulcer; developed new diuretic-antihypertensive, methylclothiazide. Home: 1338 Chestnut St., Waukegan, Ill. 60085. Office: 1400 Sheridan Rd., North Chicago, Ill. 60064.*

HYATT, Alpheus, Am. biologist, paleontologist; b. Washington, Apr. 5, 1838; s. Alpheus and Harriet R. (King) H.; academic edn. Yale; grad. Lawrence Sci. Sch., Harvard, S.B., 1862; LL.D., Brown, 1898; m. Andella Beebe, Jan. 7, 1867; 3 children, including Anna. Custodian, Boston Soc. Natural History, 1870-81, curator, 1881-1902; established marine lab., Annisquam, Mass. (later moved to Woods Hole, Mass.), 1879; prof. zoology and paleontology Mass. Inst. Tech., 1870-88, Boston U., 1887-1902; asst. for paleontology Cambridge Mus. Comparative anatomy, 1886, paleontologist U. S. Geol. Survey, 1889. Mem. Am. Soc. Naturalists (founder, 1st pres.), Nat. Acad. Scis., Am. Acad. Arts and Scis.; fgn. mem. Geol. Soc. London. Author: Genesis of the Arietidae, 1889; Phylogeny of an Acquired Characteristic, 1894; Larval Theory of the Origin of Cellular Tissue, 1884; Revision of North American Porifera, 1875; Observations on Polyzoa, 1866. Founder, editor American Naturalist, 1867-71. Founded new sch. of invertebrate paleontology; discovered law of acceleration in evolution of Cephalopoda and mech. causes of their evolution; contbd. greatly to increase of knowledge about forms and methods of devel. primitive organisms; gathered collection of fossil cephalopods at Harvard Mus. Comparative Zoology that stimulated much research in mechanisms of evolution. Died Cambridge, Mass., Jan. 15, 1902.

HYATT, John Wesley, Am. inventor; b. Starkey, N.Y., Nov. 28, 1837; s. John Wesley and Anne (Gleason) H.; ed. Eddytown Sem.; m. Anna E. Taft, July 21, 1869. Recipient Perkin medal Soc. Chem. Industry, 1914. Invented and patented a knife-sharpener, 1861, method of making dominoes and checkers, 1869; discovered method of dissolving pyroxylin under pressure and with his brother I. Smith H., invented "celluloid"; established mfg. at Newark, N.J.; began mfg. of sch. slates, 1875; invented Hyatt billiard ball, both material and machinery; invented water purifying system, 1881; about 1892 invented Hyatt Roller Bearing and organized Hyatt Roller Bearing Co., Harrison, N.J.; invented, 1900, lockstitch sewing machine, with 50 needles, for sewing belting; also invented machine for squeezing juice from sugar cane, method of solidifying Am. hard woods, from which bowling balls, golf heads, mallets, etc., are made. Died May 10, 1920.

HYDE, Alvin Seymour, Am. physician, physiologist; b. Bklyn., Jan. 21, 1928; s. Reuben M. and Sophie (Lossef) H.; B.S., U. Ark., 1950; Ph.D., Tulane U., 1953, M.D., 1957; m. Harriet J. Bartlett, Sept. 8, 1951; children—Marshall B., Hillary Ann. Research fellow biophysics Tulane U., 1953-58; project engr. Aerospace Med. Research Labs., Wright-Patterson AFB, O., 1958-59, chief acceleration sect., 1960-63, sr. investigator, 1963-64, chief environmental medicine div., 1965-68; asst. prof. dept. preventive medicine Ohio State U., Columbus, 1960-68; dir. research Bionetics Research Labs., 1968—. Mem. biodynamics com. adv. group for aero. research and devel. NATO, 1960-65; mem. com. on hearing bioacoustics and biodynamics Nat. Acad. Sci.-NRC, 1965—. Mem. Am. Physiol. Soc., Aerospace Med. Soc., Am. Inst. Aero. and Astronautics. Editorial bd. Aerospace Medicine, 1965—. Research, publs. on absorption and distbn. iron in mammals, biol. effects microwave radiations, physiologic performance effects prolonged acceleration, interactions environmental stresses. Home: 301 G St. S.W., Washington 20024. Office: Bionetics Research Labs., Primate Research Center, 5510 Nicholson Lane, Kensington, Md. 20795.*

HYDE, Earl K., nuclear chemist; b. Rossburn, Man., Can. Aug. 9, 1920 (presents Am. citizens); s. Howard Earl and Evelyn (Black) H.; student Carleton Coll., 1938-39; B.S., U. Chgo., 1941, Ph.D., 1946; m. Ethel Jean Babbitt, Jan. 1, 1949; children—Carol, Wendy, Charles, Howard. Jr. chemist inorganic materials project U. Chgo., 1942-44, Manhattan Dist. Plutonium project, 1944-46; chemist Argonne (Ill.) Nat. Lab., 1946-49; mem. sr. staff nuclear chemistry div. Lawrence Radiation Lab., U. Cal., Berkeley, 1950—. Mem. Am. Chem. Soc., Am. Phys. Soc. Author: (with I. Perlman, G. T. Seaborg) The Nuclear Properties of the Heavy Elements, vol. I, 1964, Systematics of Nuclear Structure and Radioactivity, vol. II, 1964, Detailed Radioactivity Properties, vol. III, 1964, Fission Phenomena, 1964. Research, publs. on chemistry of borohydride compounds; chemical properties of plutonium and transuranium elements; nuclear and chem. properties of francium and astatine; synthesis, identification and study of decay schemes of isotopes of elements above lead; phenomena asso. with high energy nuclear reactions; study of nuclear isomerism. Office: Lawrence Radiation Lab., U. Cal., Berkeley, Cal. 94720.*

HYDE, Roscoe Raymond, Am. immunologist; b. Cory, Ind., Mar. 23, 1884; s. John Andrew and Mary Ann (Michaelree) H.; A.B., Ind. State Tchrs. Coll., 1908; A.B. and A.M., Ind. U., 1909; Ph.D., Columbia, 1913; m. Elsie A. Coss, Sept. 18, 1910; children—Dr. Gertrude Martina, Dr. Margaret Irene Moore, Edith Raymond. Asst. prof., later prof. and head of dept. zoology and physiology, Ind. Tchrs. Coll., 1909-19; lectr. in pathology, Terre Haute Veterinary Coll., 1912-19; fellow, Johns Hopkins, 1918-19, asso., 1919-22, asso. prof. immunology, 1922-28, asso. prof. filterable viruses, and head dept., 1928-32, prof. immunology and dir. labs. filterable viruses and immunology from 1932; vis. prof. U. Chgo., 1930; mng. editor Am. Jour. Hygiene, 1927-32. Fellow A.A.A.S., Am. Soc. Zoölogists, Am. Soc. Immunologists, Am. Soc. Geneticists, Am. Soc. of Naturalists, Nat. Geog. Society, Sigma Xi, Delta Omega. Contbr. to scientific publs. Died Sept. 15, 1943.

HYDE, Walter Lewis, Am. optical engr.; b. Mpls., May 30, 1919; s. Walter Lloyd and Edith (Drake) H.; S.B., Harvard, 1941, A.M., 1943, Ph.D., 1949; m. Elizabeth Sanford, Aug. 14, 1941; children—Lee, Lewis, Benjamin, Elizabeth, Rebecca. With Polaroid Corp., 1943-46, Baird Assos., 1947-50, Office Naval Research, 1950-53; asst. dir. research Am. Optical Co., Southbridge, Mass., 1953-60, dir. devel., Pitts., 1960-63, cons., 1963—; prof. optics, U. Rochester, 1963—, dir. Inst. Optics; provost Univ. Heights Center, N.Y. U., Bronx. Sec., treas. Internat. Commn. Optics, 1965—; cons. Am. Inst. Physics. Mem. Am. Phys. Soc., Optical Soc. Am. (dir.-at-large), Royal Micros. Soc., Phys. Soc. (London, Eng.), Sigma Xi. Inventor automobile rear-view periscope using cylindrical lenses and mirrors, airplane periscope, polarizing microscope, sensitive infrared receivers, interferometers using polarized light. Home: 2195 Andrews Av., Bronx, N.Y. 10453.*

HYLAND, K(erwin) E(llsworth), Am. zoologist; b. York, Pa., Apr. 7, 1924; s. Kerwin E. and Mabel (Fisher) H.; B.S., Pa. State U., 1947; M.S., Tulane U., 1949; Ph.D., Duke, 1953; m. Jean E. Scammon, May 7, 1966; children—John, Jeanne, Janet, Jeffrey. Sci. instr. Christchurch (Va.) Sch., 1951-53; faculty U. R.I., Kingston, 1953—, prof. zoology, 1966—, chmn. dept. 1966—; vis. prof. U. Lovanium, Léopoldville, Congo, 1964. Fulbright research scholar Institut de Médecine Tropicale Prince Léopold, Antwerp, Belgium, 1960-61. Mem. Am. Soc. Parasitologists, A.A.A.S., Entomol. Soc. Am., Royal Entomol. Soc. Belgium, Sigma Xi, Phi Kappa Phi, Phi Sigma. Research, publs. on host-parasite relations of chigger mites on amphibian host, systematics and life cycles of nasal mites infesting birds and mammals. Home: 967 Kingstown Rd., Peace Dale, R.I. 02879. Office: Dept. Zoology, Univ. of Rhode Island, Kingston, R.I. 02881.*

HYLLERAAS, Egil Andersen, Norwegian physicist; b. Engerdal, Norway, May 15, 1898; s. Ole and Inger (Romoen) A.; B.A., 1924; Ph.D., 1933; student theoretical physics (Rockefeller scholar) U. Göttingen, 1926-28; m. Magda Cathinka Christiansen, 1926. Asst., Inst. Physics, Oslo, 1922-24; Univ. scholar, 1926-31; prof. theoretical physics U. Oslo, 1937—. Mem. Oslo Sci. Acad., Uppsala Sci. Soc., Uppsala, Copenhagen sci. acads. Author: Fra Fysikkens Verden, 1947; Matematisk and Teoretisk Fysik, 1950. Made theoretical investigations and calculations of atoms, molecules, crystalgrating's energy; pub. sci. treatises on exptl. roentgenographical investigations of crystal structures. Died Oct. 28, 1965.

HYMAN, Herbert Hiram, Am. sociologist; b. N.Y.C., Mar. 3, 1918; s. David Elihu, and Gisella (Mautner) H.; A.B., Columbia, 1939, A.M., 1940, Ph.D., 1942; m. Helen Raphael Kandel, Sept. 30, 1945; children—Lisa Davis, David Kandel, Alex Raphael. Social sci. analyst U. S. Dept. Agr., Washington, 1942, OWI, 1942-44; dir. field surveys morale div. U. S. Strategic Bombing Survey, Germany, 1944-45, cons., 1945-46; research asso. Nat. Opinion Research Center, U. Chgo., 1947-57; asst. prof. Bklyn. Coll., 1946-47; vis. prof. U. Cal. at Berkeley, 1950, U. Oslo (Norway), 1950-51, U. Ankara (Turkey), 1957-58; faculty Columbia, 1951—, prof. sociology, 1956—, chmn. dept., 1965—. Program dir. UN Research Inst. for Social Devel., Geneva, Switzerland, 1964-65. Fulbright award, 1950; Guggenheim award, 1961; Julian Woodward Meml. award Am. Assn. Pub. Opinion Research, 1956. Mem. Am. Assn. for Pub. Opinion Research (past pres.), Am. Sociol. Assn. (past chmn. methodology sect.), Soc. for Psychol. Study Social Issues (past mem. exec. council). Author: Psychology of Status, 1942; Pre-Election Polls, 1948; (with others) Interviewing in Social Research, 1954; Survey Design and Analysis, 1955; Political Socialization, 1959; Applications of Methods of Evaluation, 1962; also articles, other books. Developed and systematized reference group and concept of polit. socialization; research on field of sample survey methods. Home: 38 Woodside Av., Westport, Conn. 06880. Office: Columbia, N.Y.C. 10027.*

HYMAN, Herbert Hyman, Am. chemist; b. N.Y.C., Sept. 27, 1919; s. Nathaniel and Estelle (Machinist) H.; B.S., Coll. City N.Y., 1938; M.S., Poly. Inst. Bklyn., 1941; Ph.D., Ill. Inst. Tech., 1960; m. Ruth Dixler, Dec. 28, 1943; children—Mark Nathaniel, David Steven. With Argonne (Ill.) Nat. Lab., 1944—, sr. chemist, 1948—. Mem. Am. Chem. Soc., Am. Nuclear Soc., A.A.A.S., N.Y. Acad. Scis., Fedn. Am. Scientists, Am. Soc. for Technion, Research Soc. Am. (past br. pres.), Sigma Xi, Phi Lambda Upsilon. Editor: Process Chemistry, 1956, 58, 61; Noble-Gas Compounds, 1963. Research, publs. on non-aqueous solution; noble-gas compounds, plutonium processing. Home: 1347 E. Park Pl., Chgo. 60637. Office: 9700 S. Cass Av., Argonne, Ill. 60439.*

HYMAN, Libbie Henrietta, Am. zoologist; b. Des Moines, Ia., Dec. 6, 1888; d. Joseph and Sabina (Neumann) Hyman; S.B., U. Chgo., 1910, Ph.D., 1915, Sc.D., 1941; Sc.D., Goucher College, 1958, Coe Coll., 1959. Research appt. U. of Chicago, 1916-31, research on physiology and morphology of lower invertebrates; research appointment (hon.) Am. Mus. Natural History, N.Y.C., 1937—. Recipient Gold Medal, Linnean Soc. London, 1960. Mem. Am. Soc. Zoologists, Am. Micros. Soc., Am. Soc. Limnology and Oceanography, National Academy of Sciences, American Society of Naturalists, Phi Beta Kappa, Sigma Xi. Author: A Laboratory Manual for Elementary Zoology, 1919; A Laboratory Manual for Comparative Vertebrate Anatomy, 1922; Comparative Vertebrate Anatomy, 1942; The Invertebrates, 5 vols. 1940-59; also articles sci. jours. Recipient Elliott Gold Medal, 1951. Research in invertebrate zoology, oxygen consumption in relation to oxygen tension; physiol. studies in planaria, also metabolic gradiants of vertebrate embryos. Office: American Museum of Natural History, N.Y.C. 10024.*

HYNE, James Bissett, chemist; b. Dundee, Angus, Scotland, Nov. 23, 1929; s. William Simpson and Winnifred (Bissett) H.; B.Sc., St. Andrews U., Dundee, 1951, Ph.D., 1954; m. Ada Leah Jacobson, Sept. 3, 1958. Fellow, NRC Can., Ottawa, Ont., 1954-56; instr. Yale, 1956-59; asst. prof. Dartmouth, Hanover, N.H., 1959-60; faculty U. Alta. (Can.), Calgary, 1960—, prof., 1965—, head dept. chemistry, 1960-66, dean Faculty Grad. Studies, 1966—; research dir. Alta Sulphur Research, Ltd., Calgary, 1964—. Cons. to oil and gas industry, 1962—, Canadian Def. Research Bd., 1964—. Fellow Chem. Inst. Can. (sec. organic sect. 1965—); mem. Am. Chem. Soc., Faraday Soc., Sigma Xi. Research, publs. on intermolecular forces in solutions, pressure effects on rates reactions in solution, phys. chem. studies by nuclear magnetic resonance, mechanism chem. reactions biol. significance, chemistry and tech. sulphur and sulphur compounds, analogue computer techniques. Home: 312 Superior Av., Calgary S.W., Alta., Can.*

HYNEK, Joseph Allen, Am. astronomer; b. Chgo., May 1, 1910; s. Josef and Bertha (Waska) H.; B.S., U. Chgo., 1931, Ph.D., 1935; m. Miriam Curtis, May 31, 1942; children—Scott, Roxane, Joel, Paul, Ross. Dir. McMillin Obs., Ohio State U., Columbus, 1946-55; asso. dir. Smithsonian Astrophys. Obs., Cambridge, Mass., 1956-60; dir. Dearborn Obs., Northwestern U., Evanston, Ill., 1961—, chmn. dept. astronomy, 1960—, dir. Lindheimer Astron. Research Center, 1965—; in charge UFO investigations, Project Bluebook, Wright-Patterson AFB, Dayton, O. Fellow Royal Astron. Soc.; mem. Am. Astron. Soc., Internat. Astron. Union. Author: Astrophysics, 1951; Challenge of the Universe, 1962; also numerous articles. Study of double star, with spl. application to stellar evolution. Home: 2623 Ridge Av., Evanston, Ill. 60201.*

HYNIE, Sixtus, Czechoslovakian pharmacologist; b. Prague, Czechoslovakia, Mar. 15, 1933; s. Rudolf and Marie (Dolezalová) H.; M.D., Faculty of Gen. Medicine, Charles U., Prague, 1958, C.Sc., 1963; m. Hana Kochmanová, July 17, 1958; children—Lucie, Kristina. Asst. dept. pharmacology Faculty Gen. Medicine, Charles U., 1958-61, asst. prof., 1961—; research asso. Lab. Chem. Pharmacology, NIH, Bethesda, Md., 1964-65. Research and publs. on investigation of the role of the sympathetic nervous system in mobilisation of energy fuel by glycogenolysis and lipolysis; spl. interest has been given to the receptor theory and the mechanism of the action of catecholamines. Home: 3. Námestí 14. rijna, Prague 5. Office: 4. Albertov, Prague 2, Czechoslovakia.*

HYPATIA, mathematician, philosopher; b. circa 370; d. Theon of Alexandria; prof. Platonic philosophy, Alexandria; author commentaries on Diophantus, Apollonius, Ptolemy; only notable woman scholar of antiquity; last of Alexandrian sch. of math. Stoned to death for pagan beliefs by adherents to Christianity, 415.

HYPSICLES OF ALEXANDRIA, astronomer, mathematician; b. Alexandria, Egypt; flourished 180 B.C.; disciple of Isidorus; probably lived in Alexandria. Author 14th Book of Euclid's Elements (8 propositions on regular cosahedrons and dodecahedrons); Anaphorai (on rising stars); also writings in Pythagorean tradition on polygonal numbers and on harmony of spheres. Calculated times of rising and setting of zodiac signs in Babylonian manner (Greek math. could not yet solve this problem).

HYRTL, Joseph, anatomist; b. Eisenstadt, Hungary, Dec. 7, 1810; ed. U. Vienna (Austria); became prosector U. Vienna, 1833, prof. anatomy, 1845-74; apptd. prof. anatomy, Prague, Czechoslovakia, 1837. Marble statue erected in his honor, U. Vienna. Author: Lehrbuch der Anatomle des Menschen mit Rücksich auf physiologische Begründung und praktische Anwendung, 1846; Handbuch der topographischen Anatomie, 2 vols., 1847; Vergleichende Angiologie, 1850; Onomatologia anatomica, 1880. Research in comparative anatomy of fishes, constrn. of ear and testicles, angiology; demonstrated that entrance to coronary arteries is not covered by semilunar valves during systole, also that blood does not enter these arteries during diastole, 1854. Died Perchtoldsdorf, nr. Vienna, July 17, 1894.

HYSLOP, James Hervey, Am. psychologist; b. Xenia, O., Aug. 18, 1854; s. Robert and Mary Ann (Boyle) H.; A.B., U. of Wooster, 1877; U. of Leipzig, 1882-84; Ph.D., Johns Hopkins, 1887; (LL.D., Wooster, 1902); m. Mary Fry Hall, Oct. 1, 1891. Instr. philosophy, Lake Forest U., Ill., 1880-82 and 1884-85, Smith Coll., 1885-86, Bucknell U., Pa., 1888-89; tutor philosophy, ethics and psychology, 1889-91; instr. ethics, 1891-95, prof. logic and ethics, 1895-1902, Columbia; organizer, and sec. Am. Inst. for Scientific Research, 1903——. Editor Proceedings and Journal Am. Society for Psychical Research. Author: Elements of Logic, 1892; Ethics of Hume, 1893; Elements of Ethics, 1895; Democracy, 1899; Syllabus of Psychology, 1899; Logic and Argument, 1899; Problems of Philosophy, 1905; Science and a Future Life, 1905; Enigmas of Psychical Research, 1906; Borderland of Psychical Research, 1906; Psychical Research and the Resurrection, 1908; Psychic Research and Survival, 1913; Life After Death, 1918; Contact with the Other World, 1919. One of 1st Am. psychologists to relate psychology to psychic research. Died Montclair, N.J., June 17, 1920.

HYTTEN, Frank Eyvind, physiologist; b. Hobart, Tasmania, Australia, July 8, 1923; s. Torleiv and Margaret (Compton) H.; M.B., B.S., Sydney (Australia), 1946, M.D., 1962; Ph.D., Aberdeen U. (Scotland), 1954; m. Catherine Amy Hudson, Mar. 10, 1949; children—Patricia Anne, Margaret Kari, Peter Leif. Walter and Eliza Hall Travelling Research scholar, 1949-51; research fellow U. Aberdeen, 1951-54; sci. staff Med. Research Council, Newcastle upon Tyne, Eng., 1954——. Lectr. dept. physiology U. Aberdeen, part-time 1959-65; hon. mem. staff U. Newcastle upon Tyne, 1959-65. Mem. Blair Bell Research Soc. (chmn. 1965——), Nutrition Soc. (mem. council 1964-66). Author: (with I. Leitch) The Physiology of Human Pregnancy, 1964; also articles, chpts. in books. Research on composition of human milk, relationship of breast feeding to infant nutrition, maternal body changes during pregnancy. Home: 9, King Johnis Ct., Darras Hall, Ponteland, Northumberland, Eng. Office: Reprodn. and Growth Research Unit, Med. Research Unit, Princess Mary Maternity Hosp., Newcastle upon Tyne, 2, Eng.*

I

IACOB, Caius, Rumanian mathematician; b. 1912; prof. Bucharest (Rumania) U.; mem. Acad. Socialist Republic Rumania, several fgn. sci. socs. Author: Introduction mathématique à la mécanique des fluides, 1959. Developed theory of jets at high sub-sonic speeds; drew up methods for exact and approximative direct or hodographical methods of calculation.

IAMBLICHOS, mathematician, philosopher; b. Chalcis, Coele Syria, probably circa 250; pupil of Porphyry in Rome or Sicily, also of Anatolios; lived and founded sch. in Syria. Author: Protrepticus (extracts from earlier writers), 3 treatises on math. (authorship of one disputed), forming with a 5th work, part of an ency. of Pythagoreanism; Reply of Abammon to Porphyry's Letter to Anebo (defense of ritualistic magic, also known as De mysteriis); lost writings include excerpts preserved in Stobaeos, a work used by Macrobius and Julian, an exposition of Chaldaean theology, commentaries on Plato and Aristotle, quoted by Proclos. Contbd. to Neoplatonic system of philosophy, but placed greater stress on magic than his predecessors; emphasized number mysticism. Died Syria, circa 330.

IAMPIETRO, P. F., Am. physiologist; b. Middleboro, Mass., Jan. 5, 1925; s. Leonard and Maria (Repoli) I.; B.S., U. Mass., 1949; M.A., U. Mass., 1951; Ph.D., U. Rochester, 1954; m. Josephine Marie Fava, Nov. 27, 1954; children—Nancy Ellen, Michael, Pat. Research fellow U. Rochester, 1951-54; physiologist U. S. Army Research and Engrng. Center, Natick, Mass., 1954-58, chief environmental physiology sect., 1958-60, acting chief physiology br., 1959-60; chief Physiology Lab., FAA, Oklahoma City, 1960——; professor research physiology U. Okla. Med. Sch., Oklahoma City, 1967——, adj. asso. prof. zoology U. Okla., Norman, 1962——. Cons. NASA, 1964; panelist Bd. U. S. Civil Service Examiners, 1965——. Mem. Am. Physiol. Soc., Soc. for Exptl. Biology and Medicine, Am. Polar Soc., Sigma Xi, Kappa Sigma. Research, numerous publs. on temperature regulation and factors which influence ability man and animals to regulate temperature, regulation body fluids and electrolytes. Home: 1103 S. Ponca Av., Norman, Okla. 73069. Office: P.O. Box 25082, Oklahoma City 73125.*

IANITSKII, Ivan Vitalievich (also known as Janickis, J., Ianistskis, Ioans Vitalievich), Soviet phys. chemist; b. Aug. 3, 1906; ed. Dresden Higher Tech. Sch., 1930; with Kaunas U., 1933-51; faculty Kaunas Poly., 1951——. Research in electrochemistry, tech. of silicates, chemistry of sulphur, selenium. Mem. Lithuanian Acad. Sci.

IANSHIN, Aleksandr Leonidovich, Soviet geologist; b. Mar. 28, 1911; joined Moscow Geologic Survey Inst., 1932; mem. staff Geologic Inst., USSR Acad. Sci., 1936——, in charge tectonic sect., 1956——. Recipient Karpinskii prize, 1953. Research on Western Kazakhstan, So. Urals; participated in drawing tectonic maps of USSR 1952, 56.

IBANEZ DE IBERO, Carlos, geodesist; b. Barcelona, Spain, Apr. 14, 1825; dir. Geographic and Statistic Inst. Europe; pres. Internat. Commn. on Weights and Measures, 1872-91. Mem. Acad. Scis. Madrid, French Acad. Scis. Author: Tableau géographique et statistique de l'Espagne, 1888. Recognized geodetic junction of N. Africa to Europe. Died Nice, France, Jan. 29, 1891.

IBERS, James Arthur, Am. chemist; b. Los Angeles, June 9, 1930; s. Max Charles and Esther (Imerman) I.; B.S., Cal. Inst. Tech., 1951, Ph.D., 1954; m. Joyce Audrey Henderson, June 10, 1951; children—Jill Tina, Arthur Alan. NSF postdoctoral fellow Commonwealth Sci. and Indsl. Research Orgn., Melbourne, Australia, 1954-55; chemist Shell Devel. Co., Emeryville, Cal., 1955-61; chemist Brookhaven Nat. Lab., Upton, N.Y., 1962-64; prof. dept. chemistry Northwestern U., Evanston, Ill., 1965——. Sec., Nat. Com. for Crystallography, 1965——. Mem. Am. Chem. Soc., Am. Phys. Soc., Am. Crystallographic Assn. Research, numerous publs. on nature of bonding between atoms in solid state, rare-gas compounds, nature of bonding of molecular oxygen to metals in synthetic molecular oxygen carriers. Home: 2657 Orrington Av., Evanston, Ill. 60201.*

IBN ABU USAYBI'A (Muwaffaq al-din Abu al-'Abbas Ahmad ben al-Qasim) (Abu al-'Abbas Ahmad ben al-Qasim), Arabic physician; b. Arabia, 1203-04; staff hosp. in Cairo; in service to amir of Sarhad, Syria; studied botany with Ibn al-Suri, Lebanon. Author: Book of the Sources of Information on the Classes of Physicians (information on history of Arabic medicine). Died 1270.

IBN AL-AKFANI (Abu'Abd Allah ibn Alakfani) (Sams al-din ibn Ala-kfani) (Muhammad ibn al-Akfani), Iraqi physician, naturalist; b. Sinjar, Jazirah; worked in Cairo, Egypt. Author: Discovery of the Defect in the Conditions of the Eye, 3 books; Domestic Medicine for Use in the Physician's Absence; Book on Bloodletting; Advice on the Buying of Slaves (work of anthrop. interest which applies sci. of physiognomy to buying of slaves); On Precious Stones; The Direction of the Discoverer to the Highest Questions (on 60 Sciences). Died circa 1349-49.

IBN AL-'AWWAM (Abu Zakariya Yahya ibn Muhammad ibn Ahmad ibn al-'Awwam al-Ishbili), Spanish Muslim agriculturist; flourished end of 12th century, Seville, Spain. Author: The Book of Agriculture (most important medieval work on agriculture; observations on soils and manure, grafting, numerous plants).

IBN AL-BANNA' (Abu al-'Abbas Ahmad ben Muhammad ben 'Utman al-Azdi), Arabic mathematician; b. Marrakesh, French Morocco, circa 1256; lived in Morrocco; became sufi; student Marraksh, Fas, French Morocco. Author: Introduction to Euclid; Four Discourses on Calculation; Resume of the Operations of Arithmetic; also numerous treatises on geometry, arithmetic, algebra, astronomy, astrolabe, astrology, calendar. Improved methods using fractions; rule of double false position; 1st to use term almanac. Died Marrakesh, 1321.

IBN AL-BATRIG (Abu Zakariya Yahya ibn al-Batrig), Arabic translator; flourished beginning 9th century; s. Abu Yahya al-Batrig; translated into Arabic, Secreta secretorum (he attributed it to Aristotle); Timaeus (Plato); De theriaca (Galen); De coelo, De mundo, De anima, Meteorologica (all Aristotle).

IBN AL-BAYTAR (Abu Muhammad 'Abd Allah ben Ahmad ibn al-Baytar) (Diya al-din al-Malaqi) (Ibn al-Baytar Diya al-din al-Malaqi) (Abdullah ibn Ahmad ibn al-Baytar), botanist, pharmacist; b. Málaga, Spain, 1197; studied under al-Nabata, botanist, Seville, Spain. First collected plants with al-Nabati; left Spain to travel in N. Africa, 1219, in Bulgaria, 1220; traveled through Constantine, Tunis, Tripoli, Barca; entered service of sultan of Egypt, al-Kamil, after 1224; went to Damascus, Syria, 1237; lived in Cairo, Egypt, after 138. Author: Book of the Collections of Simple Medicines; The Sufficient Book on Simple Medicaments. Considered greatest botanist and pharmacologist in Islam. Died Damascus, 1248.

IBN AL-DURAIHIM (Taj al-din 'Ali ibn Muhammad ibn 'Abd al-Aziz ibn al-Duraihim al-Thalabi al-Shafi'i al-Mawsili), zoologist; b. Musul, Dec. 1312; s. Muhammad ibn 'abd al-Aziz ibn al-Duraihim; traveled to Damascus and Cairo; went to Aleppo, 1348, then apptd. prof. in mosque of Banu Umayya, Damascus; moved to Cairo, 1359. Wrote treatise including some 250 miniatures painted on gold background, representing animals with great accuracy, circa 1354; treatise is written in Eastern script, gives description of man including liquids and secretions of body, also descriptions of domestic and wild animals, birds. Died on diplomatic mission to Abyssinia, at Qus, on Upper Nile, Dec. 1360.

IBN AL-FARRAKHAN (Umar ibn al-Farrakhan), astrologer; flourished 1st half of 9th century; s. Abu Hafs Umar; a son, Muhammad. Author: of Nativities and Question (De nutritatibus et interrogationibus, 1503, 1525 edit. translated by Solomon the Jew); his works compiled by Frederick II. His astrological writings influenced sci. thought in medieval Europe.

IBN AL-HA'IM, see al-Ha'im.

IBN AL-HATIB (Abu 'Abd Allah ibn al-Hatib) (Lisan al-din ibn al-Hatib) (Muhammad ibn al-Hatib), Arabic historian, biographer, geographer, physician; b. Loja, Spain, 1313. Prime minister Kingdom of Granada, Spain. Wrote several histories of caliphs of Orient, Spain and Africa, history of Granada; works on his travels; treatises on medicine, Black Plague of 1348, health in different seasons, fetal devel. Died 1374.

IBN AL-KHATIB, Spanish physician; b. Loja nr. Granada, Spain, Nov. 15, 1313; ed., Granada; became high ofcl. at Nasri Ct. Wrote one of earliest accounts of black plague, occurring in Granada, 1348; work on gen. and spl. pathology, including eye disease, fever, surgery, pediatrics, aphrodisiacs, sex, cosmetics; also other works on medicine, history, travel. Considered possibility that disease is contagious. Died circa 1374-75.

IBN AL-LITH (Muhammad ibn al-Lith) (Abu al-Jud Muhammad ibn al-Lith) (Ibn al-Lit), Arabic mathematician; b. flourished 10th-11th century; contemporary of al-Biruni. Wrote treatise on inscription of regular hectagon in a circle; also manuscript in Khedivial library, Cairo, Egypt. Solved problem of trisection of angle, Albirunic problems using intersecting conics; classified equations and their reduction to conic sections; studied regular heptagon and enneagon.

IBN AL-LUBUDI (Ibn Muhammad ibn 'Adan al-Sahib Najm al-din ibn al-Lubudi) (Abu Zakariya Yahya ibn al-Lubudi) (Abu Zakariya Ahmad ibn al-Lubudi), physician, mathematician, astronomer, philosopher; b. Halab, Syria, 1210-11; studied medicine under al-Dakhwar, Damascus, Syria. Wazir to al-Mansur Ibrahim (ruler of Hims, 1239-45); after his death joined service of Al-Salih Najm al-din Ayyab (ruler of Egypt 1240-49); apptd. govt. insp., Alexandria; later held similar post in Syria. Author several med. works, treatises on rheumatism, questions of Hanoin ibn Ishaq, aphorisms of Hippocrates. math., physiology; also commentary on Qanun (by Aricenna). Died after 1267.

IBN AL-MADJA (Shihab al-din Abu al-'Abbas Ahmad ibn Rajab ibn Tibugha ibn al-Majdi), Egyptian astronomer, mathematician; flourished Egypt, circa 1358-59. Author: Choice Words Concerning the Determination of Time and the Discovery of the New Moon as Soon as it Appears (explains use of sine quadrant); Unveiling of the Truth Concerning the Calculation of Degrees and Minutes (discusses sexagesimal fractions); also astron. observations and tables, treatises on use of quadrants. Died Jan. 27, 1447.

IBN AL-MUNDHIR (Abu Bakr ibn Badr ibn al-Mund-hir al-Baitar), Muslim veterinarian; flourished 1298-1340, during reign of Mamluk sultans who ruled Egypt and Syria, circa 1293-40; s. Badr al-din; probably employed by Sultan. Author: Treatise on Hippology and Hippiatry (probably most complete medieval book on the subject).

IBN AL-NAFIS (Ibn al-Nafis al-Qarsi al-Misri al-Safi'i) ('Ala al-din ibn al-Nafis) (Abu al-Hazm ibn al-Nafis), Egyptian or Syrian physician; b. circa 1205; studied under Ibn al-Dakhwar, Damascus, Syria. Author: Commentary on the Anatomy of Ibn Sina; also treatises on eye disease, diet; commentaries on med. writings of Hippocrates, Hunain ibn Ishaq. Described pulmonary transit of blood, and for this reason is sometimes considered a forerunner of Harvey. Died 1288.

IBN AL-QUFF (Ibn al-Quff al-Masihi al-Karaki Amin al-Dawla) (Abu al-Faraq Ya'qub ben Ishaq ibn al-Quff), Arabic physician; b. Karak, Moab, 1232-33; studied under Ibn abi Usaibia. Author: Collection Concerning the Conservation of Health and the Prevention of Illness; Book of the Foundation of the Art of Surgery; also commentaries of Aphorisms (by Hippocrates), works of Ibn Sina. Died Damascus, Syria, 1286.

IBN AL-QUNFUDH (Abu al-Abbas Ahmad ibn al-Hasan ibn al-Khatib al-Qustantini), Algerian historian, mathematician, astronomer; s. Al-Hasan (or Husain); student in Spain, Morocco, Tunis. Became qadi, Constantine, Algeria. Author numerous books including: History of the Banu Hafs; Obituaries of Learned Men Arranged Year by Year to 1404; Biography of Abu Madyan Shuaib ibn al-Hasan al-Tilimsani; Help to the Student for the Determination of the Positions of Planets (commentary on Book of Ease, Tables of the Wandering Stars by Ibn al-Banna'); Commentary on the Talkhis of ibn al-Banna'; Commentary on the Astrological Poem (Arjuza) of Ibn abi-L-Rijal, written 1372. Used many algebraic symbols. Died 1407-08.

IBN AL-RAQQAM (Abu Abdallah ibn Ibrahim ibn al-Raqqam al-Awsi al-Mursi), Spanish Muslim physician, mathematician, astronomer; practiced medicine, Granada, Spain. Author: Discussion of Sundials; also articles on sci. instruments; astron. tables for Andalusia. Died May 27, 1315.

IBN AL-SA'ATI (Ridwan ibn Muhammad ibn 'Ali Fakhr al-din ibn al-Sa'ati) (Fakhr al-din ibn al-Sa'ati), physician; b. Damascus, Syria; s. Muhammad ibn 'Ali ibn Rustam al-Khura-sani al-Sa'ati; flourished in Damascus; joined service of Ayyubid princes al-Fa'iz Ibrahin and al-Mu'azzam 'Isa sons of Saphadin ruler of Egypt and Damascus (ruled until 1218); al-Mu'azzam ruled Damascus until 1227. Author commentaries on Ibn Sina's Qanun, also supplement on his treatise on gripes. Repaired and improved clock built by his father; wrote book explaining its constn. and use (most important source of information on Early Muslim clocks), 1203.

IBN AL-SAMH (Abu al-Qasim Asbag ben Muhammad ibn al-Samb), Arabic mathematician; b. 979; flourished at Granada, Spain; author treatises on arithmetic, geometry, also astron. tables; De cuemo peude ell ome tazer una lámina a cada planeta segund lo mostró el sabio Abulcacim Abnacam, which appears in Los libros del Saber, is probably an extract from his astron. tables. Died 1035.

IBN AL-SHATIR, see al-Shatir.

IBN AL-SURI (Mansur ibn Abu Fadllibn 'Ali Rashid) (Abu Mansur), Arabic botanist, physician; b. Sur (Tyre), Lebanon, circa 1177-78; studied medicine under 'Abd al-Latif, Damascus, Syria; worked with hosp. in Jerusalem; apptd. chief of physicians; established himself in Damascus. Author: al-adwiya al mufrada (treatise on medicine). Studied flora of Syria. Died circa 1242.

IBN AL-TAYYIB, (Abu al-Farag 'Abd Allah ibn al-Tayyib), Arabic physician at hosp. founded by 'Adud al-dawla, Bagdad; author several commentaries on Greek med. treatises; translator several works, including pseudo-aristotlelian De plantis. Died 1043-44.

IBN AL-TILMID (Ibn al-Tilmid Amin al-Dawla) (Abu al-Hasan Hibat Allah ibn Sa'id ibn al-Tilmid), Arabic physician; b. circa 1073; travelled in Iran; physician to Caliph al-Mugtafi, Bagdad; many eminent physicians were his disciples. Author: An Antidotary (supplanted the work of Saburibn Sahl); Treatise on Bloodletting; A Collection of Observations; also numerous other treatises. Died 1165.

IBN AL-WAFID, see Abenguefit.

IBN AL-WAHSIYA AL-KALDANI AL-NABATI, see al-Wahsiya.

IBN BAJJA, astronomer, philosopher; b. Saragossa, Spain, 1106; lived in Granada, Saragossa, Fez; Author: Kitab; Tadbir al-mutawahhid (guide of the solitary); Tardiya (poem on hunting); Risalat al-wada' (farewell letter to a friend leaving Spain for Egypt); treatise on materia medica (quoted by Ibn al-Baitar). Leader in movement to revise Ptolemy through systematic research; his criticisms of Ptolemy's assumptions prepared way for Ibn Tufail and al-Bitruji; influenced thought of Averroes, Albert the Great; interested in geometry and music. Died (possibly poisoned) Fez, 1138/39.

IBN BASA (Abu 'Ali al-Hasan ibn Muhammad ibn Basa), Spanish Muslim mathematician, astronomer; a son, Abu Ja'far Ahmad ibn al-Hasan; chief time computer in gt. mosque of Granada; converted to Islam from Spanish Jewish family. Studied calculus and astronomy; improved and simplified Saphea of Al-Zarqali so that it could serve any latitude with a single tablet; built astron. instruments, including sundials, astrolabes. Died Granada, 1316.

IBN BASA (Ahmad ibn Basa), architect; b. Seville, Spain, 12th century; ct. architect to 1st Muwahhid Abd al-Mumwl (ruled 1130-63), also to Abu Yaqubysuuf al-Mansur (ruled 1163-84); designed Mosque of Seville and decorated its minaret, the Giralda.

IBN BATTUTA (Abu Abd Allah Mohammed ibn Abd Allah), Muslim geographer; b. Tangier, Morocco, 1304; left Morocco for pilgrimage to Mecca, 1327; returned after 15 years having traveled through N. Africa, Asia, So. Russia; sent to China, visited Spain, Sudan. Author: Travels, pub. in English translation, 1829, also French translation of his work pub. with original test, 4. vols., 1853-59. Died Fez, Morocco, 1377.

IBN EZRA, philosopher, translator; b. Toledo, Spain, circa 1089; traveled extensively, to Rome, 1140, Salerno, 1141, Lucca, 1144, 48, Mantua, 1145, Verona, 1146, Bziers, 156, London, 1158, Narbonne, 1160. Translator: 3 treatises on grammar, 1140, 2 treatises on astrology (Mashallah) before 1148, Commentary on al-Khwarizmi's tables (al-Biriuni), 1160 (all from Arabic to Hebrew) Wrote books on math., astrology, calendar, astrolabe; writings show interest in magic squares, mystical properties of numbers; explained decimal system of numeration; astrol. treatises exerted considerable influence on both Jews and Christians; fame due to biblical commentaries (wrote only few short treatises, but expounded neoplatonic views in commentaries); one of 1st to translate Muslim writings into Hebrew; through travels helped propagate among Jews of Christian Europe, rationalistic, sci. points of view which had been developed in Spain by Muslims, Jews on basis of Greco-Muslim knowledge; inspiration Robert Browning's Rabbi Ben Ezra. Died probably Calahorra, Spain, 1164.

IBN GULGUL (Abu Da'ud Sulayman ben Hassan), Arabic physician; flourished Cordoba, Spain, circa 976-1009; physician to Caliph al-Hisam II (ruled 976-1009); wrote history of Spanish scholars and physicians of period; commentary on Dioscorides, circa 982.

IBN JAMI Abu-l-Makarim Hibatallah ibn Zain ibn al-Hasan), Judeo-Egyptian physician; b. Fustat, flourished end of 12th century; a son, Abu Tahir Isma'il. Wrote medical treatises in Arabic, including Direction for the improvement of souls and bodies (general treatise on medicine, completed by son); commentary on 5th book of Qanun; description of Alexandria and its climate.

IBN KAMMUNA ('Izz al-Dawla sa'd ben Mansur), Arabic philosopher, physician; converted to Islam, 1280; flourished in Egypt; author: The New Philosopher (on logic); On the Immortality of the Soul; A Commentary on Ibn Sina's Logical Treatise; A Commentary on the Philosophical Treatise Kitab al-talwihat of al-Suhrawardi; treatises on ophthalmology, alchemy; Remarks on the Tokhis al-muhassal; also treatise on comparison of Judaism, Christianity, and Islam. Died 1277-78.

IBN-KHALDUN, 'Abd-Al-Rahman, Arabic social philosopher, historian; b. Tunis, May 27, 1332; received traditional Arab-Muslim education in Tunis; apptd. aide under Marinid sultan Abu Inan (Faris I), Fez, 1352; imprisoned on suspicion of lack of integrity, 1356-58; served rulers at Fez, Granada, Bougie, also twice at Tlemcen (services terminated because of enmity of top ministers or seizure of throne in some cases); employed by sultan of Tunis, 1378; lectr., tchr., Cairo; grand qadi of Malikite rite for Cairo, from 1384 (reformed adminstrn. of justice in Egypt), from 1384, work interrupted by grief over death of family, also by several temporary removals from office; sent by Egyptians to Damascus in connection with expdn. against Tamerlane, 1400, captured by Tamerlane, but permitted to return to Egypt. Author: Kitab al-'Ibar wa-Diwan al-Mubtada' w-al- Khabar fi Ayyam al-'Arab w-al 'Ajam w-al Barbar (a universal history, noted for its Muqaddimal, or preface). A great forerunner of sociology and philosophies of history; analyzed influences from climate to civilization on character; traced social devel. of nomads, town dwellers, Arabic life in gen.; explained rise and fall of states by waxing and waning of group solidarity (Asabiya); considered greatest Arab historian. Died Cairo, Egypt, Mar. 19, 1406.

IBN KHURDABDBIH (Abu al-Qasim 'Ubayd Allah ben Abd Allah), Iranian geographer; b. circa 825; flourished in al-Jibal, Media, later in Samarra, Iraq. Author: Book of Roads and Provinces, (important source of hist. topography; also contains abridged narratives of journeys to various countries.), circa 846. Died circa 912.

IBN MAGID (Ahmad) (Sihab al-din); navigator; b. Gulfar, circa 1430; came from family of Arabian seas pilots; called lion of sea; served as Vasco da Gama's pilot. Author numerous writings including: Book of Instructions on the Principles of Navigation and the Rules (summary of nautical sci. known at end of Middle Ages), 1489-90; Summary of Nautical Principles. Described local winds, routes and latitudes of Indian Ocean.

IBN MASAWAIH (Mesuë Major, Mesuë the Elder, Abu Zakariya Yuhanna ibn Masawaih), physician, translator; b. Jundishapur, 777; studied under Jibril ibn Bakhtyashu, Bagdad; tchr. of Hunain ibn Ishaq; Caliph al Mu'tasim supplied him with apes for dissection, circa 836. Author numerous anat. and med. writings including: Disorder of the Eye (earliest systematic treatise on ophthalmology extant in Arabic). Translated Greek medical works into Syrian. Died Samarrah, 857.

IBN MATQAH (Judah ben Solomon Ha-Kohen), mathematician, astronomer, philosopher; b. Toledo, Spain, circa 1219; in corr. with Theodore of Antioch, 1237; attended Imperial Ct. in Toscana, 1247. Compiled encyclopaedic treatise in Arabic, 1247, later translated it into Hebrew under title The Search for Wisdom (discussed Aristotelian logic, physics and metaphysics, Bibl. commentary, math.).

IBN RUSAYD (Abu Abd Allah Muhammad ben Umar) (Muhibb aldin), (al-Sabti) (al-Fihri) (al-Andalusi); Arabic historian, geographer; b. Centa, 1259; lived in Granada, Spain; wrote on Spain and Africa: The Two Voyages (information on natural history, lit. history, geography). Died Fas, 1321.

IBN RUSHD, see Averroës.

IBN SINA, see Avicenna.

IBN THABIT, Sindh, Muslim physician, mathematician, astronomer; b. Harrán, Mesopotamia, 860; physician to 3 successive caliphs, 908-40; lived in Bagdad; became Muslim in later life; adminstr. to Bagdad Hosps. Author various math. and astron. works. Attempted to raise sci. standards for medicine; enforced rule that med. practitioners must be examined and given a diploma, 931-32.

IBN TUFAIL (Abu Bakr), Arabic physician, philosopher; b. Guadix, Spain, early 12th century; practiced medicine, Granada, Spain; adminstr. in Granada, then Centa, Tanger; later physician to Sultan Abu Ya'Qub Yusuf (ruled 1163-84). Author: Secrets of Illuminative Philosophy (on Muslim history and philosophy, also gives natural classicification of sciences, discussion on spontaneous generation); The Living Son of He Who Watches; Hayy ben Yakzhan (philos. novel translated by Pococke); Philosophus autodidactus translated into French by Léon Gauthier, 1900. Died circa 1185-86.

IBN VELI IBN HAMZA, Ali, see Ali ibn Veli ibn Hamza.

IBN WADIH AL-YA'QUBI, see al-Ya'qubi.

IBN YUNUS (Abu al-Hasan 'ali ben abi Sa'id 'Abd al-Rahman ben Ahmad) (al-Safadi) (al-Misri), Egyptian astronomer; b. Cairo, Egypt, circa 950; worked in obs. founded by Fatimid caliph al-Hakim (ruled 996-1020). Author: The Hakemite Tables (astron. and math. tables based on 200 years of Arabian observations). Solved problems in spherical trigonometry; established secular acceleration of moon's mean motion; improved theory of obliqueness and ecliptiqueness, also theory of precession of equinoxes. Died Cairo, 1009.

IBN YUNUS (Kamal al-din) (Abu al-Fath) (Abu Imran) (Musa ben Yunus ben Muhammad ibn Man'a), Arabic mathematician; b. Mawsul, 1156; lived in Bagdad; taught al-Abhari Kaves; Frederick II submitted various questions to him, including one in which he was asked to find a square equal to a given circular segment. Died Mawsul, 1242.

IBN YUSUF AL-QIFTI, see al-Qifti.

IBN ZUHR (Avenzoar) (Abu Marwan 'Abd al-Malik ibn Abu al-'Ala' Zuhr), physician; b. Seville, Spain, circa 1091; s. Alguazor Albuleizor; a son, Abu Bakr Muhammad ibn 'Abd al-Malik, 1 dau., ed. by father; traveled in N. Africa; served under Semorarudes; after their defeat by Unitarians, became wazir, physician to 'Abd al-Mu'min (1st Muwahhid); tchr. of Averroes. Author: Book of the iqtisad concerning the Reformation of Souls and Bodies, 1121-22 (summary of therapeutics and hygiene); Book of Simplification concerning Therapeutics and Diet (study path. conditions); Book of Foodstuffs; also 3 others now lost. Attempted to keep medicine and chemistry united. Died Seville, circa 1161-62.

IBRAHIM AL-NAZZAM, see al-Nazzam.

IBRAHIM BEN HABIB BEN SULAYMAN BEN SAMURA IBN GUNDAB AL-FAZARI (Abu Ishaq), Iranian astronomer; 1st Muslim to construct astrolabe. Died circa 777.

IBRAHIM IBN SINAN (Abu Ishaq ben Tabit ben Qurra), Muslim mathematician, astronomer, physician; b. Baltan, Mesopotamia, 908; s. Abu Sa'id Sinan. Author: (in translation) Abhandlung über die Ausmessung der Parabel; also commentaries on Appolonius, Almagest; wrote on conics, dialing, elementary geometry. Developed simplest math. method before invention of integral calculus. Died 946.

IBSEN, Heman Lauritz, Am. geneticist; b. Chgo., Sept. 16, 1886; s. Oluf August Martin and Gemaliah (Larsen) I.; B.S. in Agr., U. Wis., 1912; M.S. in Genetics, 1913, Ph.D., 1916; m. Elma Ruth Stewart, Dec. 22, 1927; 1 dau., Jane Ruth. Asst. in genetics U. Wis., 1913-17, in zoölogy, 1917-19; prof. genetics Kan. State Coll., Manhattan, from 1919. Fellow

A.A.A.S.; mem. Am. Soc. Animal Prodn., Am. Dairy Sci. Assn., Am. Soc. Zoologists, Am. Soc. Naturalists, Am. Genetic Assn., Sigma Xi, Phi Kappa Phi, Gamma Sigma Delta, Alpha Zeta. Specialized in research on inheritance and physiology of reproduction in guinea pigs, rabbits, cattle, and rats. Died Manhattan, Kansas, Jan. 29, 1955.

ICHIYE, Takashi, phys. oceanographer; b. Kobe, Japan, Oct. 1, 1921; s. Mankichi and Ume (Yumoto) I.; B.S., U. Tokyo, 1944, D.Sc., 1953; m. Chiyoko Nagao, Oct. 8, 1952; children—Toshiko, Keiko. Research oceanographer Kobe Marine Obs., 1946-54; asso. chief oceanography sect. Japan Meteorol. Agy., 1954-57; vis. scientist Woods Hole Mass. Oceanographic Instn., 1957-58; research asso. Oceanographic Instn., Fla. State U., 1958-59, asst. prof., 1959-63; research scientist Lamont Geologic Obs., Palisades, N.Y., 1963-64; sr. research associate 1965—. Mem. Am. Geophys. Union, Am. Meteorol. Soc., Japan Oceanographical Soc., Japan Meteorol. Soc., Sigma Xi. Author: Oceanography, 1955; also numerous articles. Research on dynamics ocean circulation by maths., hydrography Pacific Ocean, dynamics tsunamis and storm surges, diffusion and turbulence in ocean, rotating tank expts. on circulation Gulf Mexico, Antarctic Ocean. Home: 5 Skye Pl., Spring Velley, N.Y. 10977. Office: Lamont Geologic Obs., Palisades, N.Y. 10964.*

IDDINGS, Joseph Paxson, Am. geologist; b. Balt., Jan. 21, 1857; s. William Penn and Almira (Gillet) I.; Ph.B., in engring. course. Sheffield Sci. Sch., Yale, 1877, postgrad. in chemistry and mineralogy, 1877-78, in geology and assaying Columbia Sch. of Mines, 1878-79, in microscopic petrography, Heidelberg, 1879-80; Sc.D., Yale, 1907. Asst. geologist U. S. Geol. Survey, 1880-88, geologist, 1888-92, and from 1895; asso. prof. petrology U. Chgo., 1892-95, prof. 1895-1908; Silliman Lectr. Yale, 1914; hon. asso. in petrology, U. S. Nat. Museum, from 1917. Author: Rock Minerals, 1906; Igneous Rocks, 1909, Vol. II, 1913; The Problem of Volcanism, 1914. Joint author: Geology of the Yellowstone National Park, 1899; Quantitative Classification of Igneous Rocks, 1903. Translated and abridged H. Rosenbusch's Microscopical Physiography of the Rock-Making Minerals, 1898. Died Sept. 8, 1920.

IDRISI (Edrisi Mohammed) (Abu 'Abd Allah Muhammad ibn Muhammad ibn 'Abd Allah ibn Idris, al-Sharif, al-Hasani, al-Qurtubi, al-Sqali), geographer; b. Ceuta, Spain, 1099; s. Spanish Arabic parents; student, Cordova, Spain; lived in Sicily under patronage of King Roger II. Author: Roger's Book (description of world); Pleasure of Men and Delight of Souls (geo. ency. compiled for William I), circa 1161, divided earth into 7 climatic zones; recognized sphericity of earth; also treatise on botany and materia medica. Died 1166.

IDYLL, Clarence Purvis, biologist; b. Edmonton, Alta., Can., Feb. 10, 1916; s. Albert Charles and Annabelle (Purvis) I.; B.A., U. B.C., 1938, M.A., 1940; Ph.D., U. Wash., 1951; m. Marion Janet Daniels, June 28, 1941; children—Marilyn Judith (Mrs. Richard Dana Hamly), Janice Leah, Jacqueline Margaret. Came to U. S., 1941, naturalized, 1951. Tchr. math., B.C., 1940-41; biologist Internat. Pacific Salmon Fisheries Com., 1941-48; faculty Inst. Marine Sci., U. Miami (Fla.), 1948—, chmn. div. fishery scis., 1953—; prof. biology, 1956—; chmn. Gulf and Caribbean Fisheries Inst., Miami, 1954—. Mem. Am. Soc. Fishery Research Biologists, Am. Fisheries Soc., Am. Soc. Ichthyologists and Herpetologists, Am. Inst. Biol. Scis., Sigma Xi. Author: Abyss—The Deep Sea, 1964; also numerous articles. Research on life histories salmon, mullet, shrimp, conservation of natural resources. Home: 716 Tibidabo Av., Coral Gables, Fla. 33143. Office: 1 Richenbacker Caseway, Miami, Fla. 33149.*

IDZKOWSKY, Henry Joseph, Am. biologist; b. Pitts., Mar. 19, 1908; s. Harry Francis and Mary Driesch (Loebach) I.; B.S., U. Pitts., 1932, M.S., 1933, Ph.D., 1936; m. Velva Seyler, Aug. 30, 1936; children—Betty (Mrs. George G. Hazlett), Gretchen. Buhl Found. research fellow U. Pitts., 1936-37; faculty St. Francis Coll., Loretto, Pa., 1937-45; faculty U. Pitts. at Johnstown, Pa., 1945—, prof. biology, head dept.; vis. faculty Mercy Hosp., Johnstown. Mem. A.A.A.S., Am. Inst. Biol. Scis., Internat. Oceanographic Found., Pa. Acad. Sci., Am. Littoral Soc., Found. for Study of Cycles. Research, publs. on embryological devel. and behavior, endocrine relationships between adrenal gland and gonads, conservation. Home: 1324 Christopher St., Johnstown, Pa. 15905.*

IERUSALIMSKII, Nikolai Dmitrievich, Russian microbiologist; b. 1901; grad. Moscow State U., 1931; with Chemico-Pharm. Inst., Moscow, 1930-35; with Inst. Microbiology, USSR Acad. Scis., 1935—, became dep. dir., 1950; dep. sect. chief Sci. Research Lab. on Indsl. Fermentation, 1935-38; prof. Moscow State U., 1954—. Mem. USSR Acad. Scis. (corr.). Author: Structure of Bacterium, 1940; Microbiology of Cellulose, 1953; The Laws Governing the Growth and Development of Microorganisms, 1959; A Method of Stream Culturing Organisms and Opportunities for Using It, 1962; coauthor: The Relations of Baccillus megaterium to the Conditions Prevailing in the Medium in the Life Cycle Transition Process, 1960; A Change in Various Physiological Requirements of Yeast as a Result of Adaptation to Streptomycin, 1963. Research on nutrition related to devel. microorganisms. Home: 1-aya Cheremushkinskaya 4/34, Moscow, USSR.

IGLESIAS, Rigoberto, Chilean physician; b. Lebu, Chile, Nov. 22, 1911; s. Antonio and Januaria (Bastias) I.; student Sch. Medicine, U. Concepcion, Chile, 1934; M.D., U. Chile, 1938. With Inst. Exptl. Medicine, Nat. Health Service, Santiago, Chile, 1938—, sub-dir., 1957-60, dir., 1960—; research asso. Alton Ochsner Med. Found., Tulane U., New Orleans, 1948-49; vis. asso. Sloan-Kettering Inst. for Cancer Research, N.Y.C., 1955-56; head Chilean Cancer Research Coop. Center, WHO, 1964—. Chilean rep. adv. panel Ciba Found., Eng., 1964. WHO fellow cancer research labs. Europe, 1962, Mexico, U. S. A., 1964. Mem. Am. Assn. Cancer Research, N.Y. Acad. Scis., Royal Soc. Medicine, Peruvian (hon.), Chilean (past pres.) socs. endocrinology, Chilean Soc. Biology, Chilean Soc. Cancer (pres. 1967), Chilean Soc. Pathology. Research, numerous publs. on abdominal fibroids, transplantable tumors of various endocrine glands and effect of steroids upon them. Home: Tenderini 26. Office: 849 Avenida Irarrázaval, Santiago, Chile.*

IGO, George Jerome, Am. physicist; b. Greeley, Colo., Sept. 2, 1925; s. Henry Jerome and Ida (Danielsen) I.; A.B., Harvard., 1949; M.S., U. Cal. at Berkeley, 1951, Ph.D., 1953; m. Nancy Tebow, May 15, 1953; children—Saffron Igo, Peter Alexander. Research asso. Sloane Physics Lab., Yale, 1953-54; research asso. exptl. physics Brookhaven Nat. Lab., Upton, L.I., N.Y., 1954-56; faculty Stanford, 1956-58, acting asst. prof., 1957-58; guest prof. U. Heidelberg Inst. for Theoretical Physics, also Max Planck Inst. for Nuclear Physics, Heidelberg, Germany, 1958-59; staff Lawrence Radiation Lab., U.Cal. at Berkeley, 1959-63; dir. Cyclotron Inst., Tex. A. and M. U., College Station, 1963-64; vis. staff mem. Los Alamos Sci. Lab., 1964—. Fulbright Travel fellow, Germany, 1958-59. Mem. Am. Phys. Soc. Research, numerous publs. in exptl. nuclear physics, especially study particles in medium energy range. Home: Box 233, Route I, Santa Fe 8750l. Office: P. Division, Los Alamos Sci. Lab., Los Alamos 87544.*

IIDA, Shuichi, Japanese physicist; b. Kobe, Japan Jan. 30, 1926; s. Shunzo and Sono (Uyeda) I.; B.Sc., U. Tokyo, 1947, D.Sc., 1948; m. Kyoko Matsuoka, Apr. 29, 1955; children—Mariko, Junko. Asst., Faculty Sci., U. Tokyo (Japan), 1952-58, asso. prof. physics, 1958—. Mem. Phys. Soc. Japan, Am. Inst. Physics, Japan Inst. Metals, Crystallographic Soc. Japan, Biophys. Soc. Japan, Japan Soc. Powder and Powder Metallurgy (research contbn. prize 1964). Author: (with M. Kawakami), Ferrites and Their Application, 1957; Magnetic Properties of Crystals, 1957; Ferrites, 1959; Magnetic Structure, 1960; Sample Preparations, 1966; Magnetic Measurements, 1967. Research, numerous publs. on origin of induced anisotropy and its relaxation by ionic and electronic migration for ferrites, rare earth iron garnets, switching mechanism of ferrites, very high pressure synthesis of magnetic oxides. Home: 431-4, Funabashi-cho, Setagaya-ku, Tokyo, Japan.*

IIJIMA, Isao, Japanese zoologist; b. Hamanatsu, Shizuoka Prefecture, Japan, 1861; grad. zool. sect. Tokyo U., 1882; postgrad., Germany; D.Sc., 1891; apptd. prof. Tokyo U.; later dir. Marine Exptl. Sta., Misaki, Japan; mem. Imperial Acad. Author: An Outline of Zoology, 1918; standard works on porifera. Founder Japanese parasitology; contbd. to Japanese ornithology. Died 1921.

IIMORI, Satoyasu, Japanese chemist; b. Ishikawa Prefecture, Japan, Oct. 19, 1885; grad. Sci. Sch., Tokyo U., 1910; student mineral chemistry Cambridge U., also Oxford U.; doctorate, 1916; eldest dau., Kenshin Haru (Mrs. Ojima). Became prof. 1st Higher Sch., 1915; later head Iimori Lab., mem. Physico-Chem. Research Inst., now emeritus researcher. Recipient Asahi Shimbun cultural prize Imperial Acad., 1944, Sakurai prize Japan Chemistry Assn. Author books including: Analytical Chemistry. Research on soil analysis, luminous phenomena of minerals, radioactive minerals in pegmatite, spl. procelain.

IKE, Nobutaka, Am. polit. scientist; b. Seattle, June 6, 1916; s. Yasuji and Tsuya (Tanaka) I.; B.A., U. Wash., 1940, postgrad.; Ph.D., Johns Hopkins, 1949; m. Tai Inui, Aug. 23, 1942; children—Linda Y., Brian Y. Instr., Japanese, U. S. Naval Tng. Sch., U. Colo., 1942-46; Lectr., Charles Pack Fellow Johns Hopkins, 1948-49, curator Japanese Collections, 1949-58, asso. prof. polit. sci., 1955-63, prof., 1963—, exec. head dept. polit. sci., 1963-64. Recipient Demobilization award Social Sci. Research Council, 1946-48; Ford Found. fellow, 1953-55; Rockefeller Found. fellow, 1964-65. Mem. Phi Beta Kappa, Pi Sigma Alpha. Author: The Beginnings of Political Democracy in Japan, 1950; Japanese Politics, 1957; Japan's Decision for War, 1967; also articles. Asso. editor Far Eastern Quar., 1950-55. Research on voting behavior in Japan, relationship of polit. participation and party support to urbanization. Home: 621 Alvarado Row, Stanford, Cal. 94305.*

IKEDA, Tetsuro, Japanese astronomer; b. Shimane Prefecture, Japan, 1894; grad. Kyoto U., 1922; also D.Sc. Past cons. Tokyo Astron. Obs., also Central Meteorol. Obs.; became dir. Latitude Obs., Mizusawa, Japan, 1943. Research on effect of weather in observation of latitude, upper air-current, precision clock for astron. observation.

IKEGAMI, Hideo, Japanese physicist; b. Tokyo, Japan, Nov. 21, 1932; s. Kichijuro and Yukiko (Takayam) I.; B.S., U. Tokyo, 1955, M.D. (research scholar), 1959, D.Phys., 1963; m. Masako Yokoi, May 1, 1958; children—Takashi, Hitoshi. Research staff Elec. Communication Labs., 1959-62; research fellow, asso. prof. Inst. Plasma Physics, Nagoya (Japan) U., 1963—; guest research asso. Stanford, 1964-66. Mem. Phys. Soc. Japan. Research, publs. on microwave radiation from plasmas; invented resonance probe method to measure electron density and temperature of ionosphere. Office: Inst. Plasma Physics, Nagoya U., Nagoya, Japan.*

IKEHARA, M(orio), Japanese chemist; b. Tokyo, Japan, Jan. 1, 1923; s. Minami and Masako (Kobayashi) I.; grad. Faculty Pharm. Scis., Sch. Medicine, Tokyo U., 1947, Ph.D., 1954; m. Eiko Kimura, Nov. 11, 1952; children—Hiroshi, Susumu. Asst., Faculty Pharm. Scis., Tokyo U., 1951-54; asso. prof. Faculty pharm. Scis., Sch. Medicine, Hokkaido, Sapporo, Japan, 1954-66, prof., 1966—. Recipient award Pharm. Soc. Japan, 1955. Mem. Pharm. Soc. Japan, Biochem. Soc. Japan. Research and publs. on chemistry of isoquinoline, chem. synthesis of nuclosides and nucliotides, interaction of ATP analog and actomyosin, chem. and enzymatic synthesis of polynuclotides, mechanism of action of antibiotic nucleosides. Home: N-27, W-9, Sapporo, Japan.*

IKENO, Seiichiro (or Seiitiro), Japanese botanist; b. Edo, Japan, 1866; s. Tamegoro Ikeno; student Tokyo Sch. Fgn. Langs., Meiji English Sch.; D.Sc., Tokyo U.; postgrad., Europe, 1906; became prof. botany Tokyo U., 1891; recipient prize Imperial Acad.; hon. mem. French Acad. Scis. Author: Recherches sur les spermatozoïdes de cycas, 1896; Mémoires sur la formation des spores dans les taphrina, sur la spermatogénèse de marchantia ploymorpha; treatise in Japanese on philogeny of plants. A founder of modern genetics; discovered spermatozoon of cycad and their pollination methods. Died 1943.

ILIEV, Ljubomir Georgiev, Bulgarian mathematician; b. Tirnovo, Bulgaria, Apr. 20, 1913; s. Georgi Iliev and Katya (Haritonova) I.; student Sofia U., 1932-36, D.Sc., 1938; m. Maria Tarakchieva, Dec. 25, 1950; 1 son, Georgi Iliev. Staff, Sofia U., 1941—, prof. math., 1952—, vice rector, 1951-60; sci. sec. gen. Bulgarian Acad. Sci., 1961—, dir. Math. Inst., 1963—, academician, 1967—. Recipient award for sci., 1948; Dimitrov prize laureate, 1951; order Narodna Republica Bulgaria, 1963. Mem. Union Sci. Workers in Bulgaria, Balcan Math. Union (v.p.). Author: Analytische Nichtfortsetzbrakeit und Überkonvergenz, 1960; also articles. Research on theory of univalent functions, analytical uncontinuability of series, distbn. of values of polynomials and entire functions as well as theory of spl. functions. Home: B.2 Iztok Dist., Sofia, Bulgaria.*

ILIIN, Boris Vladimirovich, Russian physicist; b. Mar. 31, 1888; ed. Moscow U., 1911; mem. staff Moscow Vet. Inst., 1912-24; with Moscow U., 1918—. Author: Molecular Forces and Their Electric Properties, 1929; The Nature of the Forces of Adsorption, 1952. Research on molecular and adsorption forces; devel. (with B. B. Tarasov, V. K. Semenchenko) electric theory adsorption forces for polar substances.

ILINSKI, Michael Aleksandrovich, Russian chemist; b. Russia, Nov. 1, 1856; s. Aleksander; specialist in chem. tech. of synthetic dyes; discovered a still useful theory of valence, 1887, also division of valence, existence of free radicals, solvatation of ions. Died Nov. 18, 1941.

ILLIES, Henning Jürgen, German geologist; b. Hamburg, Mar. 14, 1924; s. Konrad and Katherina (Brachmann) I.; Dr. es sc. nat. in Geology, U. Halle, U. Hamburg; m. Gisela Kohlrausch, May 3, 1952. Asst., prof. U. Hamburg, 1948-53; prof. U. Freiburg, 1954-55, U. Valdivia (Chile), 1956-57; dir. Geol. Inst., Ecole Technique Supérieure, Karlsruhe, 1958—. Mem. Assn. German Geologists, Geologists Assn., Quartär, Soc. Econ. Paleontologists and Mineralogists. Author 34 books and articles on geology in N.W. Germany, Rhenish region, S.Am. Home: Von Beckstrasse 8. Office: Kaiserstrasse 12, Karlsruhe, Germany.

ILLIES, Kurt, German marine engr.; b. Hamburg, Germany, Nov. 18, 1906; Dipl.Ing., Tech. U. Munich; Dr.Ing., Tech. U. Braumsberg, 1940; m. Irmgard Kissel, Aug. 25, 1935; children—1 son, 2 daus. Practical experience on shipbd., 1930-35; constrn. engr. Blolun and Voss, Hamburg, chief constrn. engr., 1940-50; faculty U. Hanover; prof. marine engring., U. Hamburg; with Tech. U. Hanover. Mem. German Soc. Naval Architects and Marine Engrs. (chmn., Silberne medal). Author: Shipboilers, vols. 1, 2, 3. Research, publs. on marine engring. Home: 20 Babendickstr. Hamberg-Blanhenese, Germany. Office: 15 Calhmstr., Hanover Tech. U., Hanover, Germany.*

ILYUSHIN, Aleksey Antonovich, Russian mech. engr.; b. Jan. 20, 1911; grad. Physico-Math. Faculty,

Moscow U., 1934. Prof., Moscow U., 1938—; asso., head strength of materials dept. Inst. Mechanics, USSR Acad. Sci., 1943—, dir., 1953-60. Recipient Stalin prize, 1948. Mem. USSR Acad. Sci. (Corr.). Author: Elastic and Plastic Deformations, 1948; Plasticity: Principles of General Mathematical Theory, 1963. Research and publs. on theory of elastic and plastic deformations, theory of stability of plates and casings beyond limits of elasticity, 1942-48, developer simulation theory for pressure-working of metals, 1951-52, established isotropy hypothesis of gen. theory of plasticity, 1953-54; designer pile driver to test materials and model structures at high rates of deformation. Address: Moscow University, Leninskie gory, Moscow, USSR.

ILYUTCHENOK, Rostislav Yulianovich, Russian pharmacologist; b. Minsk, USSR, June 12, 1929; s. Yulian Vikentievich and Alexandra (Lavrusevich) I.; D.Med. Sci., U. Moscow, 1963; m. Lidia Vasilievna Devoino, June 6, 1954; 1 dau., Inna. Asst. prof. worker Chem.-Pharm. Inst., Moscow, 1957-60; head Lab. Pharmacology, Inst. Exptl. Medicine, Novosibirsk, 1960-63, head Lab. Neurophysiology and Pharmacology of Behavior dept. gen. and ecol. physiology Inst. Cytology and Genetics, Siberian br. Acad. Scis. USSR, 1963-67; prof. pharmacology Novosibirsk U., 1965—; head lab. neurophysiology and pharmacology of behavior Inst. Physiology, Siberian Br. Acad. Scis. USSR, 1967—. Mem. All-Union Soc. Pharmacologists, All-Union Soc. Physiologists. Author: Neurohumoral Mechanisms of the Brain Stem Reticular Formation, 1965. Contbr. numerous articles to profl. jours. Established neurochem. mechanisms of reticular brain formation, heterochemism of reticular neurons and cholinergicity of cortex neuron of ascending reticular formation, demonstrated cholinergic mechanism of emotional fear reaction. Address: 44/10 Morskoy Prospect. Novosibirsk, USSR.*

IMAHORI, Kazutomo, Japanese biophysicist; b. Osaka, Japan, June 1, 1920; s. Tomichi and Kazue. (Masuda) I.; D.Sc., U. Tokyo (Japan), 1952; m. Aiko Kazaoka, Nov. 16, 1947; children—Tomoko, Tadasu. Faculty, U. Tokyo, 1944—, prof. biophysics, 1961—. Mem. Biophys. Soc. Japan (pres. 1963-64). Author: Optical Rotation, 1960; Principle of Physical Chemistry, 1964. Editor: Jour. Biochemistry, Tokyo, 1962—. Research, publs. primarily on conformation and function of biopolymers; conformational study of homo and copolypeptides, structure and function of several enzymes and some works on protein and nucleic acid biosynthesis are included. Home: 24-25 2-chome Kakinokizaka Meguroku, Tokyo. Office: 865 Komabacho Meguroku, Tokyo, Japan.*

IMAI, Isao, Japanese physicist; b. Dairen City, Manchuria, 1914; grad. Tokyo U., 1936; D.Sc., 1943. Became instr. sci. dept. Tokyo U., 1938, later asst. prof.; mem. Aero. Research Inst. Recipient Imperial award Japan Acad., 1959, Asahi Cultural Prize, 1950. Research on theory of high speed liquid dynamics and compressed liquid; explained inconsistency in liquid dynamics; dir. research on wings of jet planes in Am.

IMAI, Tamaki, Japanese pathologist; b. Saiki City, Japan, Feb. 24, 1908; s. Komakichi and Ryu (Izawa) I.; D.Medicine, Kyushu U. Faculty Medicine, Kyushu U., Fukuoka, Japan, 1932, Igakuhakushi, 1939; m. Atuko Izawa, Nov. 12, 1937; children—Hiroshi, Kazuko. Faculty, Kyushu U., 1941—, prof. pathology, 1945—, dir. Cancer Research Inst., Faculty Medicine, 1961-66. Mem Japanese Cancer Assn. (exec. com. 1956—, pres. 1965). Author: Pathology, 1963; Cancer, 1963; also numerous articles. Presented new histological classification human cancer based on analysis of mode of growth; histogenesis and history of human stomach cancer, mode of elaboration of Bitter virus in mouse mammary carcinoma. Home: 7-3 Igakubu-Kita, Fukuoka, Japan.*

IMAMURA, Akitsune, Japanese seismologist; b. Kagoshima Prefecture, 1870; grad. Tokyo U.; postgrad., Europe, Am.; became prof. Tokyo U.; mem. Imperial Acad. Author: Theoretical and Applied Seismology. Accurately predicted Kanto earthquake of 1923. Died 1947.

IMAMURA, Chisho, Japanese mathematician; flourished 17th century; pupil of Mori Kambei; known (with Yoshida Koyu, Takahora Kisshu), as one of Sanshi, or 3 honorable scholars. Author: Jugai-roku (1st to assign a value, that of 0.51, to spherical vol. of unit diameter), 1639; Inki-Sanka (math. rules in verse), 1640. Gave 3.162 as value of pi; suggested surface of sphere equals circumference squared, divided by 4.

IMANISHI, Kinji, Japanese biologist; b. Kyoto, Japan, 1902; grad. Kyoto U., 1928; later D.Sc. Instr. sci. dept; dir. Seihoku Research Inst., Kyoto U.; mgr. Kyoto U. expdn. to Gt. Karakorum Mountains, 1955. Author: Logic of Society of Living Things; Society before Mankind. Research and publs. on animal soc., life zones, aquatic life in mountain torrents of Japan.

IMBERT, Fleury, French physician; b. Lyons, France, 1793; Docteur en médicine, Paris, France; m. widow of Gall; physician at la Charité and l'Hotel-Dieu of Lyons. Author: Nécessite d'une théorie en médicine, 1832; Voyage phrénologique à la Grandechartreuse, 1833; Traité des maladies des femmes. First to perform subcutaneous symphysiectomy; research on diseases of women. Died 1851.

IMBO, Giuseppe, Italian geophysicist; b. Procida, Naples, Dec. 6, 1899; s. Alfonso and Luisa Scotta (Lavina) I.; Ph.D. in Physics, U. Naples; m. Anna Grimaldi, 1932. Asst., Vesuvius Obs., later curator, now dir.; geophysicist, geophysicist-in-chief Central Office Meteorology and Geophysics; full prof. geophysics U. Naples. Mem. Nat. Assn. Geophysics (pres.), Italian Geophys. Commn. (pres.), dei Lincei Acad., Nat. Soc. Scis., Letters and Arts, Pontaniana Acad. Naples (sec. gen.). Author numerous articles on terrestrial physics, vulcanology, volcanic physics. Address: via Largo San Marcellino 10, Naples, Italy.

IMBRIE, John, Am. geologist; b. Penn Yan, N.Y., July 4, 1925; s. Charles Kisselman and Margaret (Fleming) I.; student Coe Coll., 1942-43; B.A., Princeton, 1948; M.S., Yale, 1950, Ph.D., 1951; m. Barbara A. Zeller, Oct. 11, 1947; children—Katherine Palmer, John Zeller. Asst. prof. geology U. Kan., 1951-52; asst. prof. geology Columbia, 1952-54, asso. prof., 1955-60; prof., 1960—, exec. com. dept. geology, 1959—, chmn. dept., 1965—; Geologist, Kan. Geol. Survey, 1952—; research asso. Am. Mus. Natural History, N.Y.C., 1953—. Faculty fellow NSF, project dir., study evolution Paleozoic biofacies. Fellow Geol. Soc. Am.; mem. Soc. Econ. Mineralogists and Paleontologists (sec.-treas. 1959-60), Am. Assn. Petrol. Geologists, Paleontol. Soc., Soc. Systematic Zoologists, Phi Beta Kappa, Sigma Xi. Home: 111 Highwood Av., Leonia, N.J. Office: Columbia University, N.Y.C. 27.*

IMHOF, Maximus von, see von Imhof, Maximus.

IMHOTEP, Egyptian physician, architect, astrologer; flourished circa 2980-2950 B.C.; 1st physician known by own name; physician, adviser to Zoser; architect of step pyramid, Sakkara; astrologer; prime minister; deified as Egyptian god of medicine and worshiped as son of Prah, god of Memphis, during Saite Period, 500 B.C.

IMOTO, Minoru, Japanese chemist; b. Kyoto, Japan, July 11, 1908; grad. Osaka U., 1932; later D.Eng. Asst. prof. Osaka U.; became prof. Osaka Municipal U., 1949. Recipient Japan Chem. Soc. Prize, 1952. Author: Synthetic Fiber; On Organic Electron; Chemistry of Synthetic High Molecule.

IMPERATI, Luigi, Italian surgeon; b. Pietra Montec.no, Italy, Mar. 9, 1909; s. Giovanni and Maria P. (Borreca) I.; Degree in Medicine, Naples (Italy) U., 1932; m. Candida Petrone, Dec. 27, 1949; children—Gabriella, Paola, Giovanna. Asst. prof. surgery Naples U., 1934-50; asso. prof. Sassari (Sardinia) U., 1950-63, Rome U., 1957-65; surgeon in chief Hosp. Foggia (Italy), 1964—; dir. surg. path. clinic Sassari U., 1950-56. Decorated Italian Republic Commendator. Mem. Soc. Italiana di Chirurgia, Soc. Internat. de Chirurgie, Royal Soc. Medicine, Internat. Congress on Hydatidosis (gen. sec. Rome 1960). Author books, numerous articles. Research on shock role of acetylcholine in pathogenesis of shock, personal method in surg. treatment of hydatidosis, surgery of cancer of colon and rectum, vascular researches. Home: 67 via Lanciani, Rome, Italy. Office: Gen. Hosp., Dept. Surgery, Foggia, Italy.*

IMPERATO, Ferrante, Italian pharmacist; b. circa 1550; pharmacist, Naples, Italy; founder bot. garden, Naples; corresponded with Aldrovandi, Bauhin, Clusius, Mattioli; translator works into Italian. Author: Dell'historica naturale libri XXVIII (catalogue of plants, animals, fossils, minerals, includes discussions of litharge, haematite, shapes of gem crystals), 1599; De fossilibus (opposed current beliefs about fossils), 1610; Discorso sopra le matazioni dei pasesi, 1672. Died circa 1631.

IMSHENETSKY, Aleksandr Aleksandrovich, Russian microbiologist b. Jan. 9, 1905; grad. Voronezh U., 1926; D. Biol. Sci., 1939. Asso., Inst. Microbiology, USSR Acad. Sci., 1930-40, head dept., 1941-49, dir., 1949—; instr. microbiology Leningrad Chemicotechnol. Inst. Pub. Catering Engring., 1932-34; with Moscow Higher Tech. Fish Inst. (formerly fish Industry Faculty, Timiryazev Agr. Acad.), 1935-37. Mem. USSR Acad. Sci., All-Union Microbiol. Soc. (pres.). Author: Variability and Selection of Microorganisms, 1951; The Microbiology of Cellulose, 1953; Man's Invisible Helpers, 1963. Chief editor Microbiology; co-editor Microbiology sect Large Med. Ency., 2d edit.; mem. editorial council Antibiotics. Research and publs. on structure, history and individual devel., biology of nitrifying bacteria, variability and physiology of microorganisms, variability and selection of fungoid and bacterial yeasts, biol. study of cellulose-destroying bacteria, found differential nuclear structures in group of bacteria; formulated principles of selection of active strains; established that speed of processes caused by thermophils is much greater than with mesophilic bacteria, his findings used in production of antibiotics, enzymes, alcohol. Address: Inst. Microbiology, USSR Acad. Sci., Leninsky prospect 33, Moscow, USSR.

IMSTER, Harry Frederick, Am. aero. engr.; b. Kennebec, S.D., Apr. 16, 1925; s. Frederick and Anna (Hefflefinger) I.; B.S. in Aero. Engring., U. Mich., 1946; M.S. in Aeros., Cal. Inst. Tech., 1947, profl. degree in aeronautics, 1948; m. Margery Dodson, Dec. 30, 1958; children—Eleanor, Harry Frederick. Design engr. Applied Physics Lab., Johns Hopkins, Silver Springs, Md., 1948-50, Aro, Inc., St. Louise, 1950-51; sr. systems integration engr. Project Gemini, McDonnell Aircraft Corp., St. Louis, 1951—. Named in Congl. Record for contbns. to project Gemini, 1962. Mem. Am. Inst. Aeros. and Astronautics (sec. St. Louis sect. 1965—), Sigma Xi, Tau Beta Pi. Research, publs. in fields fluid mechanics and heat transfer; design and devel. Talos and Quail missiles, Phantom fighter aircraft, Gemini spacecraft. Home: 6325 Waterman St., St. Louis 63130. Office: P.O. Box 516, St. Louis 63130.*

INABA, Yanosuke, Japanese engr.; b. Ibaraki, Japan, 1910; grad. Tokyo U., 1934; D.Eng., 1950. Became asst. prof., mem. Inst. for Aero. Research, Tokyo U., 1942; prof. engring. Hokkaido U.; became chief Central Research Inst., Toyo High Pressure Industry Co., 1952. Author: Computation Tables of Lubricating Oil Viscosity and Humidity. Research and publs. in oil chemistry and high molecule chemistry, synthetic fiber oil chemistry.

INADA, Ryukichi, Japanese bacteriologist; b. Aichi Prefecture, Japan, 1874; ed. Tokyo U.; studied in Germany until 1903; became asst. prof., Fukuoka U.; prof. U. Tokyo; physician to Imperial household: mem. Imperial Acad. Discovered (with R. Hoki, Y. Ido, H. Ito, R. Kaneko) that Leptospira icterocaimorrhagiae causes spirochetal jaundice, that Leptospira icterohaemorrhagiae causes Weil's disease, also developed serum for leptospiral jaundice, reported 1916. Died 1950.

INCH, William Rodger, Canadian radiobiologist, b. Port Hope, Ont., Can., Feb. 21, 1928, s. Hector Walker and Elva (Tinney) I.; B.Sc., Queen's U., 1950; Ph.D., U. Western Ont., 1954; m. Isobel June Terry, June 26, 1954; children—Leanne, Carol, Nicole. With Brit. Empire Cancer Campaign research unit, London, Eng., 1956-57, London (Can.) Clinic of Ont. Cancer Found., 1957-64; faculty U. Western Ont., London, 1957—, asso. prof. 1963—; with Lawrence Radiation Lab., Berkeley, Cal., 1965. Mem. Canadian Assn. Med. Physicists, Biophysics Soc., Am. Assn. Cancer Research, Radiation Research Soc. Research, publs. on changes in tissue components in growth of malignant tumors; developed treatment modalities combining ionizing radiation and drugs for treatment of solid tumors. Home: 78 Lonsdale Dr., London, Ont., Can.*

INFANTELLINA, F., Italian physiologist; b. Palermo, Italy, Apr. 18, 1916; s. Simone and Rosa (Lo Manto) I.; M.D., U. Palermo, 1941. Head dept. human physiology U. Catania, Italy, 1958-63; head dept. human physiology U. Bologna, Italy, 1963—. Mem. N.Y. Acad. Sci., Acad. Sci. Med. Palermo, Acad. Gioenia Catania, Soc. Italian Physiology, Soc. Chem. Biology Paris, Soc. Italian Biology, Soc. Med. Chir. Catania. Research, numerous publs. on functional relationships between cerebellum and cerebral cortex, technique for preparation of isolated cerebellar cortex, labyrinthine and cerebellar epilepsy, studies on claustrum. Home: 17, Via San Donato, Bologna, Italy.*

INFELD, Leopold, Polish physicist; b. Kraków, Poland, Feb. 8, 1898; s. Salomon and Ernestyna (Kahane) I.; Ph.D., Jagiellonian U. (Kraków), 1921; m. Helen Schlauch, Apr. 12, 1939; children—Eryk, Joan. Rockefeller Found. fellow Cambridge, Eng., 1933-35; physicist Inst. Advanced Study, Princeton, 1936-38; lectr. to prof. physics Toronto U., Can., 1938-50; prof. theoretical physics U. Warsaw, Poland, 1950-68, dir. Inst. Theoretical Physics 1951-68. Mem. Pugwash continuing Com. Fellow Am. Phys. Soc.; mem. Polish Acad. Scis. (presiding officer 1955-68). Author: (with A. Einstein) The Evolution of Physics, 1938; Quest, 1941; Whom the Gods Love, 1948; Albert Einstein, 1950; (with J. Plebanski) Motion and Relativity, 1960; Sketches from the Past, 1965; also numerous articles. Research in classical, relativistic, and quantum field theory. Died Jan. 15, 1968.

INGALLS, Theodore Hunt, Am. epidemiologist; b. Utica, N.Y., Apr. 8, 1908; s. Frederic Charles and Edith (Ingersoll) H.; A.B., Hamilton Coll., 1929, Sc.D., 1955; M.D., Harvard, 1933; m. Mary Parker Smith, Sept. 17, 1933; children—Mary Tooke, Theodore Smith, Elizabeth Robinson (Mrs. Thomas B. Jansen). Faculty, Harvard, 1939-58, asso. prof. epidemiology, 1951-58; prof. preventive medicine and epidemiology U. Pa. Sch. Medicine, Phila., 1958-67; research prof. preventive medicine (epidemiology) Boston U. Sch. Medicine, dir. Epidemiology Study Center, Framingham (Mass.) Union Hosp., 1967—; dir. Henry Phipps Inst. for Research in Community Disease, 1960-65. Fellow A.M.A., Am. Acad. Pediatrics (chmn. com. on congenital malformations 1961-65); mem. Am. Acad. Neurology, Am. Epidemiol. Soc. (pres. 1967-68), Soc. Pediatric Research, Am. Soc. Human Genetics, Am. Genetic Assn., Am. Pub. Health Assn. Asso. editor progress sect. Am. Jour. Med. Scis., 1952-67; editorial bd. Archives of Environmental Medicine, 1960—. Research, numerous publs. on causation of congenital malformations, epidemiol. analysis of non-infectious diseases of childhood, lead poisoning. Home: 52 Oaks Rd. Office: Epidemiology Study Center, 113 Lincoln St., Framingham, Mass. 01701.

INGALS, E(phraim) Fletcher, Am. physician; b. Lee Centre, Ill., Sept. 29, 1848; s. Charles Francis and Sarah (Hawkins) I.; M.D., Rush Med. Coll., Chgo., 1871; A.M., U. Chgo., 1879; m. Lucy S. Ingals, Sept. 5, 1876. Asst. prof. materia medica Rush Med. Coll., 1871-73, lectr. diseases of chest and physical diagnosis, 1874-83, prof. laryngology, 1883-90, laryngology and practice medicine, 1890-93, laryngology and diseases of chest, 1893-98, diseases of chest, throat and nose, and comptroller, from 1898; prof. diseases throat and chest, Northwestern Woman's Med. Sch., 1879-98; prof. laryngology and rhinology Chgo. Polyclinic, from 1890; professorial lectr. medicine U. Chgo., from 1901. Chmn. sect. laryngology, Pan-Am. Med. Congress, 1893. Author: Diseases of the Chest, Throat and Nose, 4th edit., 1900. Introduced partial excision of nasal septum for relieving deviation, 1882. Died Apr. 29, 1918.

INGARDEN, Roman Stanislaw, Polish physicist; b. Zakopane, Poland, Oct. 1, 1920; s. Roman Witold and Maria (Pol) I.; ed. U. Lwów; Jagellonian U., Carcow, M.A., 1946; Ph.D., Warsaw U., 1949; m. Regina Urbanowicz, May 5, 1956; sons: Krzysztof; Jacek Janusz. Asst., U. Wroclaw, 1945; adjoint, then dep. prof. theoretical physics, U. Wroclaw, 1949; extraordinary prof., U. Wroclaw, 1954; prof., U. Wroclaw, 1964—; head, Group of Geometrical Optics, Inst. Mathematics, Polish Acad. Scis., 1950-55; head, Low Temperature Dept., Inst. of Physics, Polish Acad. Scis., 1956-60; head, Lab. of Theoretical Physics, Low Temperature Dept. Inst. Physics, Polish Acad. Scis., 1960—. Sr. research asso., Physico-Technical Inst., Acad. Sci., Ukrainian SSR., Kharkov, 1954-55; Ford Found. Fellow, Courant Inst. Math. Sci., N.Y. U., New York, 1961-62; vis. prof., U. Rochester, N.Y., 1962. Recipient Cross of Merit, 1954; Medal of 10th Anniversary of Polish People's Republic, 1954; Cavalier Cross of Polonia Restituta, 1964; Sci. award, Polish Acad. Sci., 1663. Mem., Polish Physical Soc. (Sci. award, 1964); Polish Math. Soc.; Soc. of Sci. and Letters, Wroclaw. Author articles and review articles in theoretical physics. Investigated generalization of Rubinowicz boundary wave method in theory of diffraction; solution of problem of imbedding of Finsler space in Minkowski space and its application to theory of electron microscope; new formulation of information theory and statistical thermodynamics (with K. Urbanik); definition of temperatures of high orders and their application to non-equilibrium thermodynamics, especially to lasers and masers and coherence of light. Home: 57 Kochanowskiego, Wroclaw, Poland. Office: 36 Cybulskiego, Wroclaw, Poland.*

INGBAR, Sidney Harold, Am. physician; b. Denver, Feb. 12, 1925; s. David Harry and Belle (Friedland) I.; student U. Cal., Los Angeles, 1941-43; M.D. magna cum laude (fellow), Harvard, 1947; m. Mary Lee Gimbel Mack, May 28, 1950; children—David Harry, Eric Edward, Jonathan Clarence. Practice medicine, specializing in endocrinology, Boston, 1949—; asso. in medicine Med. Sch. Harvard, 1955-58, asst. prof. medicine, 1958-62, asso. prof., 1962—; program dir. Harvard Clin. Research Center, Boston City Hosp., 1962—, physician-in-charge Outpatient Endocrine Clinic, 1963—; asso. dir. Thorndike Meml. Lab., 1963—; cons. Newton-Wellesley Hosp., 1957—, Mass. Soldiers Home, Chelsea, 1964—. Mem. Surgeon Gen.'s Adv. Com. on Gen. Medicine, chmn. subcom. on endocrinology and metabolism, 1963—. Recipient Maimonides award Greater Boston Med. Soc., 1947. Mem. Assn. Am. Physicians, Am. Soc. for Clin. Research, Endocrine Soc. (Earnest Oppenheimer award 1965), Am. Thyroid Assn., Am. Physiol. Soc., Soc. Exptl. Biology and Medicine, Sigma Xi, Alpha Omega Alpha. Editorial bd. Endocrinology, 1957—. Research, numerous publs. on thyroid physiology and disease, mechanisms of thyroid hormone synthesis and especially their relation to energy metabolism of the thyroid, effects of iodine on the thyroid, transport of thyroid hormone in blood, peripheral metabolism of thyroid hormone. Home: 41 Montvale Rd., Newton Centre, Mass. 02159. Office: 818 Harrison Av., Boston 02118.*

INGELMAN-SUNDBERG, Axel Gustaf Isidor, Swedish gynecologist; b. Uppsala, Sweden, Dec. 22, 1910; s. Isidor and Maria (Ingelman) Sundberg; M.L., Uppsala U., 1938; M.D., Caroline Inst., 1947; m. Mirjam Furuhjelm, Mar. 10, 1946; children—Björn, Henrik, Catharina, Magnus. Asst. gen. chemistry Med. Faculty, U. Uppsala, 1929, asst. histology, 1930-32; faculty Caroline Inst., Stockholm, 1947—, Royal prof. obstetrics and gynecology, 1959—; chmn. univ. dept. obstetrics and gynecology Sabbatsberg Hosp., Stockholm, 1959—. Named hon. prof. U. Montevideo, 1962. Author: Rectal Injuries Following Radium Treatment of Cancer of the Cervix Uteri, 1947. Research and publs. on ovarian and pituitary function in hypovitaminosis E, sperm antiagglutinic factor in women, etiology and treatment of stress incontinence, uterine motility in vivo and in vitro, new methods for operation on fistulas in irradiated tissue, influence of prostaglandins on tubal motility. Home: 1A Fjalarstigen, Djursholm, Sweden. Office: Sabbatsberg Hospital, Stockholm 6, Sweden.*

INGELSTAM, Erik Paul Gerhard, Swedish physicist; b. Tidaholm, Sweden, July 15, 1909; s. Carl O. and Ester (Svensson) Carlsson; Ph.D., U. Uppsala (Sweden), 1937; Dr ès sc. honoris causa, Besançon, France, 1962; m. Margit Andersson, Apr. 4, 1936; children—Lars, Bengt, Rolf, Eva (Mrs. Leif Strand), Hans. Acting prof. physics Chalmers Inst. Tech., Göteborg, Sweden, 1937-40; faculty Royal Inst. Tech., Stockholm, Sweden, 1943—, prof. physics, 1956—; dir. Inst. Optical Research, Stockholm, 1949—; vis. prof. U. Rochester (N.Y.), 1960. Vice pres. Internat. Commn. Optics, 1953-59, pres., 1959-62. Fellow Optical Soc. Am.; mem. Swedish Nat. Com. Physics, Japan Soc. for Applied Physics (hon.), Royal Swedish Acad. Engring. Scis., Royal Swedish Acad. Scis. Research, numerous publs. on precision measurements and apparatus design in X-ray spectroscopy, phase contrast and interferometry; research and design of apparatus for analysis of optical imagery using optical transfer function applied to optical lenses, photography and radiology. Home: Bromma Kyrkväg 461 A, Bromma, Sweden. Office: Royal Inst. Tech., Stockholm 70, Sweden.*

INGENHOUSZ, Jan, botanist, physician; b. Breda, Holland, Dec. 8, 1730; ed. Louvain U., Belgium, U. Leyden, Holland. Traveled to Eng., where he learned of smallpox inoculation, 1764; inoculated Royal House, Vienna; became physician to Empress Maria Theresa, 1772; traveled in Europe, then settled in Eng.; mem. Aulic Council, Vienna, Austria; Fellow Royal Soc., 1769. Author: Experiments on Vegetables, discovering their Power of Purifying the Air, 1779; Essay on Food Plants; Experiments and Observations on Various Physical Subjects. Demonstrated that green parts of plants absorb carbon dioxide during day and give it off at night, while giving off oxygen in day, also that nongreen parts always give off carbon dioxide; devised method for comparing heat conductivities. Died Wiltshire, Eng., Sept. 7, 1799.

INGENIEROS, José, Argentinian psychiatrist, sociologist, philosopher; b. Buenos Aires, Argentina, Apr. 24, 1877; M.D., U. Buenos Aires, 1900; became dir. psychopathic wards Buenos Aires Hosps., 1901; apptd. prof. exptl. psychology U. Buenos Aires, 1904; founder Buenos Aires Inst. Criminology, 1907. Author: Psicopatología en el arte, 1902; Simulación de la locura, 1903; El hombre mediocre, 1913; Princípios de psicologia biológica, 1913; La evolución de la ideas argentinas, 2 vols., 1918-20. Editorial dir. Archivios de psiquatría y criminología, 1902-13, Revista de filosofía, 1915-25. Leading exponent of positivism in Argentina; influenced by Comte and Spencer; applied principles of positivism to criminology, sociology and edn.; worked in biol. psychology. Died Buenos Aires, Nov. 1, 1925.

INGERSLEV, Fritz Halfdan Bent, Danish physicist; b. Aarhus, July 6, 1912; s. Christian and Helga Victorine (Raeder) I.; Dr. es sc. in Elec. Engring., Royal Tech. U., Copenhagen; m. Else Ingeborg Margarethe Heiberg, May 29, 1949; children—Ib, Olaf, Dan, Dorrit. Prof., Tech. U., Copenhagen, 1954. Sec., Acoustics Commn. Mem. Am. Acoustical Soc., Internat. Union Pure and Applied Physics, Acad. Tech. Scis., Acoustical Soc. Denmark, Royal Soc. Physicists (London). Author: Acoustics in Modern Building Practice, 1952; Measurement of Linear and Non-linear Distortion in Electrodynamic Loudspeakers, 1954. Home: Vilvordevej 22, Charlottenlund, Copenhagen. Office: Royal Technical University, Lyngbye, Denmark.

INGERSOLL, Alfred Cajori, Am. environmental health engr.; b. Madison, Wis., June 8, 1920; s. Leonard R. and Helen (Flint) I.; B.S., U. Wis., 1942, M.S. in Civil Engring., 1948, Ph.D., 1950; m. Elizabeth R. McNamara, Feb. 22, 1946; 1 son, John Thomas. Mem. tech. staff lab. Linde Air Products Co., Tonawanda, N.Y., 1942-46; faculty U. Wis., 1946-49, project asst. engring. expt. sta., 1949-50; faculty Cal. Inst. Tech., 1950-60; guest prof. applied mechanics Bengal Engring. Coll., U. Calcutta (India), 1954-55; dean Sch. Engring., U. So. Cal., 1960—; cons. U. S. Naval Ordnance Test Sta., 1958. Res. sr. san. engr. USPHS. Fellow Am. Soc. C.E. (Rudolph Hering medal 1957, Daniel W. Mead prize 1942); mem. Am. Soc. Engring. Edn., Engring. Colls. Administrv. Council (v.p. West sects. and brs. 1966—), Am. Assn. U. Profs., Nat. Soc. Profl. Engr., Sigma Xi, Chi Epsilon (pres. 1956-58), Tau Beta Pi, Phi Kappa Phi, Pi Epsilon Tau, Phi Eta Sigma. Author: (with R. L. Daugherty) Fluid Mechanics with Engineering Applications, 5th edit., 1954; (with L. R. Ingersoll and O. J. Zobel) Heat Conduction with Applicatioons in Engineering and Geology, 1954. Research and publs. on understanding of sedimentation in settling tanks, as for water purification and sewage treatment, and separating chambers as used in the separation of oil from refinery waste waters; recently a model study of the resuspension of flocculent solids arising from currents in settling tanks. Home: 1400 Cresthaven Dr., Pasadena, Cal. 91105. Office: Univ. Southern Calif., Los Angeles 90007.*

INGERSOLL, Arthur William, Am. organic chemist; b. nr. Burr, Neb., Nov. 8, 1894; s. Arthur Merritt and Amelia (Wahrer) I.; B.S., U. Neb., 1917, M.S., 1918, Ph.D., U. Ill., 1922; m. Valentine Minford, June 10, 1922; 1 son, Arthur William. Mem. faculty Vanderbilt U., 1922—, prof., 1938-63, prof. emeritus, 1963—; chmn. div. natural sci., 1945-52. Mem. Am. Chem. Soc., Sigma Xi, Phi Beta Kappa, Phi Lambda Upsilon, Alpha Chi Sigma, Alpha Zeta, Gamma Alpha. Author and editor: (with Roger Adams and others) Organic Reactions, Vol. II, 1944. Research and publs. on devel. new methods and agts. for researching racemic forms of compounds into their active components, especially concerning derivatives of alcohols, amines, amino acids and terpenoid classes; topics of gen. organic-biochem. and sterochem. interest. Home: 2305 Elliston Pl., Nashville 37203.*

INGERSOLL, Leonard Rose, Am. physicist; b. N.Y.C., June 1, 1880; s. Hiram Day and Mary Augusta (Rose) I.; B.S., Colo. Coll., 1902; Ph.D., U. Wis., 1905; m. Barbara Ethel Smeigh, June 19, 1907 (dec. Apr. 1917); children—Barbara M., Hugh D.; m. 2d, Helen (Flint) Wallace, Aug. 2, 1918; 1 son, Alfred C. Instr. physics U. Wis., 1905-08, asst. prof., 1908-10, asso. prof., 1910-25, prof., from 1925. Mem. Smithsonian Expdn. to Mt. Wilson, Cal. for solar investigations, 1905, 06, 09. Fellow A.A.A.S., Am. Phys. Soc.; mem. Wis. Acad. Sci., Arts and Letters, Phi Beta Kappa, Sigma Xi. Author; (with O. J. Zobel) An Introduction to the Mathematical Theory of Heat Conduction, 1913; (with M. J. Martin) Experiments in Physics, 1942; (with O. J. Zobel and A. C. Ingersoll), Heat Conduction with Engineering and Geological Applications, 1948. Research and publs. on subjects connected with electro-magnetic theory of light; invented glarimeter for measuring gloss of paper (adopted as standard govt. test, 1925). Died Apr. 25, 1958.

INGERSON, Earl, Am. geochemist; b. Barstow, Tex., Oct. 28, 1906; s. Fred Percy and Mamie (Carson) I.; A.B., Simmons Coll., 1928, M.A., 1931; Ph.D., Yale, 1934; m. Martha Anna Duncan, June 5, 1930; children—Mary Zoe (dec.), Earl. With Carnegie Instn., 1935-47, petrologist, 1943-47; geologist, chief br. geochemistry and petroleum U. S. Geol. Survey, Washington, 1947-58; prof. geology U. Tex., Austin, 1958—, asso. dean Grad. Sch., 1961-64. Chmn. various coms. NRC, NSF; sec. Internat. Commn. Geochemistry, 1960-63. Recipient Distinguished Service award U. S. Dept. Interior, 1959. Fellow Mineral. Soc. Am., Geophys. Union, Geol. Soc. Am. (Day medal 1955), Geochem. Soc. Am. (hon.); mem. Am. Chem. Soc., Geochem. Soc. (past pres.), Internat. Union Pure and Applied Chemistry (past v.p., geochem. com.), Internat. Assn. Geochemistry and Cosmochemistry (pres. 1965-68), Soc. Econ. Geologists, Mineral. Soc. Gr. Britain and Ireland, Geochem. Soc. France, Mineral. Soc. France, Mineral. Soc. Japan, German Mineral. Soc., Geochem. Soc. Japan Geol. Soc. Brazil, Mexican Geol. Soc., Sigma Xi. Author: (with E. B. Knopf) Structural Petrology, 1938; (with George W. Morey) Pneumatolytic and Hydrothermal Alteration and Synthesis of Silicates, 1937; also numerous articles. Research on synthesis minerals, mineral systems in presence water, applications liquid inclusions in minerals, origin and history igneous, sedimentary and metamorphic rocks, isotope abundances, geol. thermometry. Home: 3402 Mt. Bonnell St., Austin, Tex. 78731.*

INGHEN, Marsilius van, see van Inghen, Marsilius.

INGHIRAMI, Francesco, Italian archeologist; b. Volterra, Italy, 1772; librarian. Author: Monumenti etruschi (most complete description of antiquities of Etruria), 10 vols., 1820-27; Galleria Omerica, 3 vols., 1827-38; Letters on Etruscan Erudition, 1828; Storia della Toscana, 16 vols., 1841-45. Died Badia, Italy, May 17, 1846.

INGHIRAMI, Giovanni, Italian astronomer; b. Volterra, Italy, Apr. 16, 1779; s. Niccolo and Lidia (Venuti) I.; student St. Michael's Coll., Volterra; tchr. math. Volterra, Florence, Italy; dir., obs., Florence; mem. Acad. Astronomy, Acad. Geography London; author works on physics, hydraulics, astronomy, math.; also topog. atlas of Tuscany and sect. of astron. atlas for Berlin Acad. Scis. Died Aug. 15, 1851.

INGHRAM, Mark Gordon, Am. physicist; b. Livingston, Mont., Nov. 13, 1919; s. Mark Gordon and Luella (McNay) I.; A.B., Olivet Coll., 1939; Ph.D., U. Chgo., 1947; m. Evelyn Mae Dyckman, May 12, 1946; children—Cheryl Ann, Mark Gordon III. Physicist Manhattan Project, N.Y.C., 1942-45; sr. physicist Argonne Nat. Lab., Lemont, Ill., 1945-49; faculty physics U. Chgo., 1947—, prof., 1954—, chmn. dept., 1959—, asso. dean div. phys. scis., 1964—. Fellow Am. Phys. Soc.; mem. Nat. Acad. Scis. (Smith medal 1957). Research in nuclear physics, chem. physics, geophysics; use of mass spectrometric techniques. Home: 1534 E. 59th St., Chgo. 60637.

INGLE, Dwight Joyce, Am. physiologist; b. Kendrick, Ida., Sept. 4, 1907; s. David J. and Mattie (Self) I.; B.S., U. Ida., 1929, M.S., 1931, D.Sc. (hon.), 1962; Ph.D., U. Minn., 1941; m. Geneva McGarvey, Oct. 25, 1930; children—David, Ann, Jane. Mayo Found. fellow, 1934-38; George S. Cox med. research fellow U. Pa., 1938-41; Upjohn fellow Upjohn Research div., 1941-43, sr. research sci. Upjohn Co., 1943-53; prof. physiology Ben May Lab., U. Chgo., 1953—; prof., chmn. dept. physiology, U. of Chgo., 1959—. Mem. adv. com. Am. Cancer Society; mem. com. Nat. Found.; mem. panel USPHS. Nat. Foundation. Recipient Roche-Organon award Laurentian Hormone Conf., W. E. Upjohn prize; Outstanding

Achievement award U. Minn., 1964. Fellow Am. Acad. Arts and Scis., N.Y. Diabetes Assn.; mem. Am. Physiol. Soc., Soc. Exptl. Biology and Medicine (pres. 1965-67), A.A.A.S., Am. Diabetes Assn., Endocrine Soc. (pres. 1959-60; Koch award 1963), Nat. Acad. Scis., Sigma Xi, Alpha Omega Alpha, Phi Beta Kappa. Author: Physiological & Therapeutic Effects of Corticotropin (ACTH) and Cortisone, 1953; Principle of Research in Biology and Medicine, 1958; I Went to See the Elephant, 1962. Editor: Perspectives in Biology and Medicine. First to demonstrate biologic activity of cortisone, also the compensatory atrophy of adrenal cortex in response to corticosteroid excess; developed concept of permissive actions of hormones; studies of liver regeneration and factors affecting survival of liverless animals; studies on role of adrenal cortex, and muscle work. Home: 5514 Woodlawn Av., Chgo. 60637.*

INGLIS, Sir Claude Cavendish, Brit. civil engr.; b. Dublin, Ireland, Mar. 3, 1883; s. John Malcolm and Caroline (Johnston) I.; B.A., B.Engring., Trinity Coll., Dublin, 1905; M.Engring., 1949; m. Vera St. John Blood, Sept. 25, 1912; 1 son, Brian St. John. Engr., Indian Service Engrs., 1906-15; irrigation and hydraulic research Bombay (India) Presidency, 1916-39, Govt. of India, 1939-45; dir. hydrautics Research Sta., Gt. Britain Dept. Sci. Research, 1947-58; cons., Henfield, Sussex, Eng., 1958—. Named Companion of Indian Empire, 1938; knighted, 1945; recipient Ewing medal, 1958. Fellow Royal Soc., 1953; mem. Inst. Civil Engrs. (past mem. council), Am. Soc. C.E., Instn. Engrs. India. Author: The Behavior and Control of Rivers and Canals with the Aid of Models, 2 vols., 1947; also numerous articles. Research on river and estuary problems, floods, irrigation, ecology of crops, rainfall, water measurements. Address: 10 S. View, Henfield, Sussex, Eng.*

INGLIS, David Rittenhouse, Am. physicist; b. Detroit, Oct. 10, 1905; s. William and Carolyn (Rittenhouse) I.; A.B., Amherst Coll., 1928; D.Sc., U. Mich., 1931; D.Sc., Amherst Coll., 1963; m. Dorothy Rosalind Kerr, Mar. 26, 1934; 1 son, John Lockwood. Faculty Ohio State U., 1931-34; research asso. U. Leipzig and Eidgenosche Technische Hoscschule, Zurich, 1932-33; asst. prof. U. Pitts., 1934-37; asst. prof. Princeton, 1937-38; asso. Johns Hopkins, 1938-41, asso. prof., 1941-49; sr. physicist Argonne (Ill.) Nat. Lab., 1949—; professorial lectr. U. Chgo., 1965—. Physicist, OSRD, 1942, Aberdeen Proving Ground, 1943, Los Alamos Sci. Lab., 1943-46; vis. prof. U. Cal. at Berkeley, 1955-56, U. Grenoble, 1964. Fellow Am. Phys. Soc.; mem. Fedn. Am. Scis. (chmn. 1959-60), Phi Beta Kappa, Sigma Xi, Alpha Delta Phi. Bd. editors Revs. Modern Physics, 1959-62, Nuclear Physics, 1962-64, Bull. of Atomic Scis., 1956—. Contbr. numerous articles on polit. adjustment to sci. advance, promotion of nuclear test ban to profl. jours., polit. mags. Discovered nuclear spin-orbit coupling, 1936, intermediate coupling in nuclei, 1951, nucleon participation in collective rotation, 1953, speculations on origin of geomagnetism, clarification of nuclear scattering and gamma ray patterns. Home: 3840 Ellington Av., Western Springs, Ill. 60558. Office: Argonne Nat. Lab., Argonne, Ill. 60440.*

INGOLD, Sir Christopher, English chemist; b. London, Eng., Oct. 28, 1893; s. William Kelk and Harriet (Newcomb) I.; student U. Southampton (Eng.), 1911-13; B.Sc., Imperial Coll., U. London, 1913, M.Sc., 1919, D.Sc., 1923; D.Sc. (hon.), U. Oslo (Norway), U. Leeds (Eng.), U. Sheffield (Eng.), U. Southampton, Oxford (Eng.), U. Bologna (Italy), U. Paris (France), U. Montpellier (France); m. Edith Hilda Usherwood, July 11, 1923; children—Sylvia Ray (Mrs. Stanley William Owen Ivermee), Keith Usherwood, Dilys Rosemary (Mrs. Lionel Hope Jones). Chemist, Cassel Cyanide Co., Glasgow, Scotland, 1918-19; lectr. chemistry Imperial Coll., 1920-24; prof. organic chemistry U. Leeds (Eng.), 1924-30; prof. chemistry U. Coll., London, 1930-61, emeritus prof., spl. lectr., 1961—; vis. lectr. Stanford, 1932, Cornell U., 1950, Notre Dame U., 1957, U. Minn., 1957. Recipient Meldola medal Royal Inst. Chemistry, 1922. James Flack Norris award for phys. organic chemistry, 1965; NSF fellow Vanderbilt U., 1964; named Knight Batchelor, 1958. Fellow Royal Soc., 1924 (Royal medals 1946, 52); hon. mem. Royal Acad. Scis., (Spain), Council for Higher Sci. Spain, Weizmann Inst. (Israel), N.Y. Acad. Scis., Am. Acad. Arts and Sci.; mem. Chem. Soc. (Longstaff medal 1951), Faraday Soc. (medal 1962), Soc. Chem. Industry, Inst. Chemistry, Royal Irish Acad. Author: Structure and Mechanism in Organic Chemistry, 1953; Substitution at Other than Carbon, 1959; also numerous articles. Research on theories of organic and phys.-organic chemistry; structure of aromatic compounds. Home: 12 Handel Close, Edgware, Middlesex, Eng. Office: Univ. Coll., Gower St., London W.C.1, Eng.*

INGOLS, Robert S(malley), Am. biochemist; b. Newark, Mar. 5, 1911; s. George A. and Nellie (Smalley) I.; B.Sc. in Biology, Bucknell U., 1931; M.A. in Biochemistry, Columbia, 1934; Ph.D. in Sanitation, Rutgers U., 1939; m. Dorothy Ohlson, Nov. 4, 1939; children—Marcia R., Cynthia A., George A. Research asst. Rutgers U., 1935-41; chemist sewage treatment plant, Hackensack, N.J., 1941-43; research fellow Fla. Citrus Commn., 1943-44; instr. Sch. Pub. Health, U. Mich., 1944-47; faculty Ga. Inst. Tech., 1947—, prof., head dept. applied biology, 1957-60, dir. Sch. Applied Biology, 1960-65, prof. 1965—. Chmn., Joint Com. on Uniformity Methods for Water Exam., research participant Oak Ridge Inst. Nuclear Studies, 1950. Fulbright lectr. Istituto Politecnico di Milano, Italy, 1956-57. Mem. A.A.A.S., Am. Chem. Soc., Am. Water Works Assn., Water Pollution Control Fedn., Sigma Xi. Mem. editorial bd. 12th edit. Standard Methods for the examination of Water and Wastewater, 1957. Research and publs. on the mechanism of biol. processes of waste water treatment; the mechanism of chlorine disinfection and its control in water treatment; biol. surveys in stream pollution problems; methods of analysis of waste water; and toxicity of waste water. Home: 2973 Margaret Mitchell Ct., Atlanta 30327.*

INGRAHAM, Hollis Steadman, Am. physician; b. Brookline, Mass., Mar. 10, 1908; s. Alward and Grace (Steadman) I.; A.B., Harvard, 1930, M.D., 1933, M.P.H., 1935; D.Sc., Union U., Albany, N.Y., 1964; m. Helena Johnson, June 1, 1931; children—Priscilla (Mrs. Albert Pultz), Irad, Sylvia (Mrs. Harry Goetzmann), Mark. Epidemiologist, N.Y. State Dept. Health, 1935-38, dist. health officer, Kingston, N.Y., 1938-46, dir. epidemiology and communicable disease control, 1946-48, dep. commr., 1948-53, 1st dep. commr., 1953-63, commr. health, Albany, N.Y., 1963—. Prof. community health Albany Med. Coll., 1963—; adj. asso. prof. pub. health practice Columbia, 1963—; pres. bd. dirs., Health Research, Inc.; mem. vis. com. Johns Hopkins Sch. Hygiene and Pub. Health, 1965—. Recipient medal U. S. Am. Typhus Commn., 1944. Mem. Am. Assn. Pub. Health Physicians, Am. Epidemiological Soc., Am. Sch. Health Assn., Am. Bd. Preventive Medicine and Pub. Health, A.M.A., Am., N.Y. State pub. health assns., Soc. Am. State N.Y., N.Y. State Acad. Preventive Medicine, Capitol Dist. Soc. for Pub. Adminstrn. Contbr. articles on epidemiology to tech. jours. Home: 291 McCormack Rd., Slingerlands, N.Y. 12159. Office: 84 Holland Av., Albany, N.Y. 12208.*

INGRAHAM, John Lyman, Am. microbiologist; b. Berkeley, Cal., Sept. 22, 1924; s. Dean Clement and Velma (Lewis) I.; B.S. in Chemistry, U. Cal. at Berkeley, 1947, Ph.D. in Microbiology, 1951; m. Marjorie F. Mitchell, June 30, 1950; children—Catherine Ann, Thomas Mitchell. Research scientist E. I. du Pont de Nemours & Co., Inc., 1951-56; chemist Western Regional Lab., U. S. Dept. Agr., 1956-58; faculty U. Cal. at Davis, 1956—, prof. bacteriology, 1956—. Guggenheim fellow, 1965-66. Mem. A.A.A.S., Soc. for Gen. Microbiology, Am. Soc. for Microbiology, Sigma Xi. Research, publs. on growth temperature response of Psychophilie bacteria; determined pathway of formation of n-propyl iso-amyl, iso-butyl, active amyl and n-amyl alcohols by yeasts; isolated cold sensitive mutants of bacteria and determined metabolic basis of their lesions. Home: 540 W. 8th St., Davis, Cal. 95616.*

INGRAM, Richard Ernest, Irish mathematician, geophysicist; b. Belfast, Ireland, July 27, 1916; s. John and Edith (Kelly) I.; B.Sc., U. Coll., Dublin (Ireland), 1938, M.Sc., 1939; postgrad. Milltown Park, Dublin; Ph.D., John's Hopkins, 1948. Entered Soc. Jesus, 1933, ordained priest Roman Catholic Ch., 1944; vis. fellow Cal. Inst. Tech., 1948-49; dir. Seismic Obs., Rathfarnham Castle, 1949-63; asso. prof. U. Coll. Dublin, 1966—. Fellow Royal Astronautical Soc.; mem. Royal Irish Acad., Am. Math. Soc., Seismol. Soc. Am. Editor: (with H. Halberstam) Collected Works of Sir W. R. Hamilton, III, 1967. Research, publs. on characters of representations in group theory, focal mechanism of earthquakes, and interpretation of recorded movements. Home: 35 Lower Leeson St., Dublin 2, Ireland.*

INGRAM, Vernon Martin, biochemist; b. Breslau, Germany, May 19, 1924; B.S. in Pure Math., Chemistry, Zoology, U. London, 1943, B.S. in Chemistry 1st class, 1945, Ph.D., 1949; D.Sc., 1961. Mem. sci. staff Med. Research Council unit for molecular biology Cavendish Lab., Cambridge, Eng., 1952-58; faculty Mass. Inst. Tech., Cambridge, 1958—, prof. biochemistry, 1961—; Jesup lectr. Columbia, 1962; Harvey Soc. lectr., 1965. Mem. Am. Chem. Soc., Genetical Soc., Chem. Soc., Biochem. Soc., Am. Soc. Biol. Chemists, Am. Acad. Arts and Scis. Research hemoglobin synthesis during devel., structure and function of s-RNA. Home: 45 Bellevue Av., Cambridge, Mass. 02138.*

INGRASSIA, Giovanni Filippo, anatomist, physician; b. Palermo, Sicily, 1510; ed. U. Padua; prof. anatomy, practical medicine, U. Naples (Italy) practiced medicine, Naples; apptd. 1st physician of Sicily by King Philip II of Spain, 1563. Author: De tumoribus praeter naturam, 1553. Discovered stirrup-bone of middle ear, seminal vesicles; differentiated between measles and scarlet fever, 1556, also between chicken pox and scarlatina; described lesser wings (Ingrassia's wings) of sphenoid bone, circa 1568; distinguished in work during epidemic of 1575. Died Palermo, 1580.

INGVAR, David H(enschen), Swedish clin. neurophysiologist; b. Lund, Sweden, Feb. 3, 1924; s. Sven and Ingegerd (Henschen) I.; M.D., U. Lund, 1950; m. Elisabet Ulfsparre, Sept. 14, 1946; children—Christian, Malin, Martin, Anna. Research fellow Montreal (Que., Can.) Neurol. Inst., 1951-53; asst. Nobel Inst. for Neurophysiology, 1954-56; asso. prof. clin. neurophysiology U. Lund, 1956—; head Clin. Neurophysiol. Lab. U. Hosp., Lund, 1956—. Cons. physician for EEG to hosps. in So. Sweden. Mem. Internat. Brain Research Orgn. Editor: Regional Cerebral Blood Flow (with N. Lassen), 1965. Research and publs. on color vision, EEG, epilepsy and cerebral circulation; developed (with Lassen) a method for measurement of carebral blood flow using isotopes. Home: Sunna, Dalbyvägen, Lund, Sweden.*

INIGUEZ ALMECH, José Maria, Spanish mathematician; b. Calatayud, July 28, 1897; s. Francisco and Maria (del Pilar) I. A.; Ph.D., U. Madrid; m. Maria de la Cinta, June 28, 1929. Aux. prof. U. Madrid, 1919-22; prof. theoretical mechanics U. Zaragoza, 1922—. Mem. Higher Council Sci. Research; pres. math. tng. center Garcia de Galdeano, Zaragoza. Mem. Acad. Exact, Physico-chem. and Natural Scis. Zaragoza, Royal Acad. Exact, Phys. and Natural Scis. Madrid. Author: Curso de matemáticas para estudiantes de física, química e ingeniería; Mecánica cuántica; Operadores lineales en los espacios métricos; Mecánica téorica. Address: calle Bolonia, 4-3°, Zaragoza, Spain.

INKERI, Kustaa Adolf, Finnish mathematician; b. Laitila, Finland, Nov. 12, 1908; s. Kustaa Edvard and Hilma (Ala-Kauppi) I.; mag.phil., U. Turku (Finland), 1936, Ph.D., 1946; m. Fanny Ragnhild Vänni, Dec. 3, 1933; children—Pirkka Kalevi, Liisa Sinikka (Mrs. Pekka Vainio), Heikki Kustaa, Seppo Ilkka, Jukka Pekka. Sr. tchr. high sch., Forssa, also Turku, 1938-45; faculty U. Turku, 1945—, prof. math., 1950—. Mem. Finnish Acad. Scis. Decorated K.SL, VR.3, m.k., VR.4. Research, publs. on theory of numbers and algebra. Home: 5 A 33 Kurjenkaivonk, Turku, Finland.*

INMAN, James, English mathematician; b. Yorkshire, Eng., 1776; s. Richard Inman; B.A., St. John's Coll., Cambridge, Eng., 1800, M.A., 1805, D.D., 1820; m. Mary Williams. Astronomer with Flinders in Investigator and Porpoise, 1903-04; prof. math. Royal Naval Coll., Portsmouth, Eng., 1808-39; prin., sch. naval architecture, 1810. Author: Navigation and Nautical Astronomy for British Seamen, 1821, 3d edit., (with intro. of new trigonometrical function, addition of logarithms of halfversine, or haversine, to tables, simplification of practical solution of spherical triangles), 1835; Introduction to Naval Gunnery, 1828. Died Feb. 2, 1859.

INMAN, Verne Thomson, Am. physician, educator; b. San Jose, Cal., Nov. 6, 1905; s. Jesse Jay and Lois Elizabeth (Headen) I.; B.A., U. Cal. at Berkeley, 1928, M.A., 1929, Ph.D., 1934; M.D., U. Cal. at Berkeley, 1928, M.A., 1929, Ph.D., 1934; M.D., U. Cal. at San Francisco, 1932; m. Irene Patricia Cootey, May 9, 1930; children—Robert Anthony, Verne Thomas, Richard Headen. Asst. in anatomy U. Cal., Berkeley, 1928-29, instr. anatomy, 1932-36; fellow in orthopaedic surgery San Francisco Hosp., 1939-40; faculty U. Cal. at San Francisco, 1940—, prof. orthopaedic surgery Sch. Medicine, 1952—, chmn. dept., 1957—, dir. Biomechanics Lab., 1957—. Mem. coms. Nat. Acad. Sci.-NRC, Nat. Found., Office Vocational Rehab., USPHS; Fulbright lectr. U. Assiut, UAR, 1964-65. Diplomate Am. Bd. Orthopaedic Surgery. Mem. Am., Western orthopaedic assns., A.M.A., Am. Acad. Orthopaedic Surgeons, Orthopaedic Research and Edn. Found., Phi Beta Kappa, Alpha Omega Alpha, Phi Sigma. Asso. editor Jour. Bone and Joint Surgery, 1948-53. Research, numerous publs. on human locomotion, prosthetic and orthotic design, lower extremity amputation. Home: 212 Edgewood St., San Francisco 94117.*

INNES, Robert Thornton Axton, astronomer; b. Edinburgh, Scotland, Nov. 10, 1861; D.Sc. (hon.), Leyden, Netherlands; m.; 3 sons; sec. Royal Obs., Cape, 1896-1903; dir. meteorology, Transvaal, 1903-11; Union astronomer, 1911-27; chmn. Innes Film Projection Co., Ltd. Author: Double Stars; Discovery of the nearest star, Proxima Centauri; Proof of oscillations in length of the day. Died Mar. 13, 1933.

INO, Tadataka, Japanese geographer; b. Sawara, Chiba Prefecture, Japan; 1742; s. Sadatsune Jimbo (or Saburozaemon); adopted by Ino family; began study of calendar under Yoshitoki Takahashi, Edo, circa 1792; commd. by govt. to survey Hokurikudo, east coast of Hokkaido, other parts of Japan, 1800-18; made surveys with own instruments and presented findings often as accurate as those today. Author: Udai Yochi Zenzu (altas). Died 1818.

INOKUTI, Mitio, physicist; b. Tokyo, Japan, July 6, 1933; s. Haruhisa and Takako (Kure) I.; B.Sc., U. Tokyo, 1956, M.S., 1958, Ph.D., 1962; m. Makiko Omori, Mar. 12, 1960; 1 dau., Mika. Instr. math. physics U. Tokyo, 1960-65; research asso. Northwestern U., 1962-63; resident research asso. Argonne (Ill.) Nat. Lab., 1963-65, asso. physicist, 1965—. Fellow Phys. Soc. London; mem. Am. Phys. Soc., Phys. Soc. Japan. Author: (with others) Introduction to Radiation Chemistry (in Japanese), 1961. Research, publs. in effects of ionizing radiation on molecular substances as basis for fundamental radiation physics, chemistry and biology, aspects of absorption of ionizing radiation by molecules, effects of radiation on high-polymeric substances. Home:

45 W. 59th St., Westmont, Ill. 60559. Office: 9700 S. Cass Av., Argonne, Ill. 60439.*

INSLEY, Herbert, Am. mineralogist; b. Nanuet, N.Y., May 16, 1893; s. Earle and Annie (Hutton) I.; B.S., Hamilton Coll., 1914; Ph.D., Johns Hopkins, 1919; Sc.D. (hon.), Hamilton Coll., 1956; m. Margarette Hiteshew, Oct. 8, 1921; children—Robert H., Herbert H. Asst. physicist Nat. Bur. Standards, 1917-18, petrographer, 1922-44, asst. chief mineral products div., 1946-47, chief, 1947-53; ret., 1953; now cons. mineralogist; asst. geologist U. S. Geol. Survey, 1919-21; petrographer U. S. Bur. Mines, 1921-22; head dept. earth sci. Pa. State Coll., 1944-45. Trustee Edward Orton Jr. Ceramic Found., 1947-53. Fellow Geol. Soc. Am., Mineral. Soc. Am., Am. Ceramic Soc. (v.p. 1959, Albert Victor Bleininger award 1963; Ross Coffin Purdy award); mem. Optical Soc. of America, Geochem. Soc., Washington Academy of Sci., Phi Beta Kappa, Sigma Xi. Author: Phase Diagrams for Ceramists (with F. P. Hall), 1947; Microspy of Ceramics and Cements (with V. D. Frechette), 1955. Research and publs. on the petrography of mine dusts; the constn., microstructure and thermal chemistry of ceramics, cements and artificial silicates; the synthesis of inorganic crystals; phase equilibrium diagrams of inorganic materials; phase relations of the fluorides of the alkalies, beryllium, uranium, thorium, zirconium. Address: 5219 Farrington Rd., Washington 20016.*

INUI, Tetsuro, Japanese physicist; b. Tokyo, Japan, 1905; grad. physics sect. sci. dept. Tokyo U., 1929. Instr. 1st Higher Sch., later of Tokyo U.; became prof. sci. and engring. dept. Keijo (Seoul) U., 1941; prof. engring. dept., 1944—. Author: Applied Differential Equation and Sphere Function; Cylinder Function and Geometrical Function. Research on quantum property of matter, applied math.

INUISHI, Yoshio, Japanese physicist; b. Kyoto-fu, Japan, Feb. 2, 1921; s. Totaro and Sute (Hitomi) I.; B.S., Faculty Engring., Osaka U., 1944, Ph.D., 1957; m. Kazuko Funakoshi, May 5, 1948; children—Michiko, Nobuko, Masahide. Faculty elec. engring. dept. Osaka (Japan) U., 1948-55, 56—, prof. electro physics, 1961—; Fulbright vis. fellow Lab. for Insulation Research, Mass. Inst. Tech., 1955-56. Mem. Am., Japan phys. socs., Japan Inst. Elec. Engrs. Author: (with T. Tanaka) Electric Materials. Editorial bd. Jour. Inst. Elec. Engrs. Japan, 1958-62, Jour. Applied Physics Japan (Oyo Butsuri), 1965—. Research, publs. on mechanism of conduction and breakdown in ionic crystals, measurements of carrier mobility in dielectric liquid, behaviour of electrons at higher electric field in semi-conductors, interaction of electrons with acoustic waves in piezo-electronic semiconductors. Home: 7-2, 7 chome Habikino-shi, Osaka-fu, Japan.*

IOFFE, Abram Feodorovich, Russian physicist; b. USSR, Oct. 17, 1880; dir. Physico-Technol. Inst., Acad. Sci. Research on properties of crystal; dir. studies on durability and wear resistance of metals.

IOFFE, Vladimir Ilich, Russian microbiologist, immunologist; b. 1898; grad. Med. Faculty, Kazan U., 1921; D.Med. Sci. Asso., Leningrad Inst. Exptl. Med., 1923-41; head dept. med. microbiology Inst. Exptl. Medicine, USSR Acad. Med. Sci., Leningrad, 1946—. Mem. USSR Acad. Med. Sci. (corr.). Co-author Immunological Study of Malignant Tumors, 1951. Co-editor Microbiology sect. Large Med. Ency., 2d edit.; mem. editorial council Problems of Virology. Research and publs. on immunity and immunology of malignant tumors, serological variability of intestinal bacteria, intestinal and children's droplet infections; analyzed serological characteristics of hemolytic streptococci, determined their prescriptive structure; differentiated typhoid and dysentery bacilli by fermentative (intestinal group) and catalase (dysentery group) indices, devised method to compare size of bacterial foci and course of its devel. and extinction in children's infections, suggested method for early diagnosis of pertussis during incubation period. Address: Inst. Exptl. Medicine, USSR Acad. Med. Sci., Leningrad, USSR.

IONESCU, Sisesti Gheorghe, Rumanian agronomist; b. 1885; mem. Acad. Socialist Republic Rumania; organizer, dir. Inst. for Agronomical Research. Author: Maize Growing, 1955; Text-Book on Agriculture, 2 vols., 1957. Research on soil fertility in Rumania; created new wheat species; improved methods of plant culture in Rumania.

IONESCU, Theodor V., Rumanian physicist; b. 1899; prof. Bucharest (Rumania) U.; mem. Acad. Socialist Republic Rumania. Author: Electricity, 1957. Research on plasma in magnetic field; built 1st plasma oscillator, 1936; discovered vibrations of negative labile oxygen ions, also of molecular hydrogen; set off multiple gyro-magnetic effect and foretold its existence in ionosphere, 1940.

IONESCU, Thoma, Rumanian surgeon; b. 1860; hon. mem. Rumanian Acad. Author: Anatomie topographique du duodénum et hernies duodénales, 1899; Interventions chirurgicales dans les affections non cancereuses de l'estomac, 1905; La sympathique cervico-thoracique, 1923. Collaborator: French Treatise on Anatomy. Introduced aseptic practices into Rumania. Died 1926.

IONESCU-SOLOMON, Irina, Rumanian chemist; b. Barlad, Rumania, Oct. 12, 1916; d. Iacob Gheorghe and Ada (Ferhat) Solomon; B.Pharmacy, Faculty of Pharmacy, Bucharest, Rumania, 1939; m. Stefan Ionescu, Dec. 31, 1960. Sci. researcher Inst. for State Control of Drugs and Pharm. Research, Bucharest, 1939—, leader Lab. for Chem. Control of Drugs, 1957—. Mem. Rumanian Union Med. Scis. Co-author: Rumanian Pharmacopoeia, vol. VII, 1957, vol. VIII, 1965. Research, publs. on methods for qualitative and quantitative determinations involving volumetric titrations (oxido-reduction, nonaqueous media, complexometry) and phys. determinations such as photocolorimetry for various inorganic and organic substances. Home: 158 Bd. 1 Mai. Office: 48 Av. Sanatescu, Bucharest, Rumania.*

IPATIEFF, Vladimir Nikolaevich, chemist; b. Moscow, Russia, Nov. 9, 1867; s. Nicolay and Anna (Glyky) I.; student Arty. Acad., St. Petersburg, 1889-92; Ch.D., U. St. Petersburg, 1907; hon. Dr., univs. Munich, Strasbourg, Sofia, Northwestern U.; m. Barbara Ermakoff, July 26, 1892; children—Nicolay, Vladimir, Anna, Dimitry. Came to U. S., 1931. Prof. chemistry Arty. Acad., St. Petersburg, 1898-1906, U. St. Petersburg, 1906-15, Northwestern U., 1931-35, prof. emeritus from 1935; dir. chemical research, Universal Oil Products Co. In charge chem. work Russian govt., World War I, 1914-18; founder Leningrad. Inst. High Pressures, 1927, Ipatieff Catalytic High Pressure Lab. at Northwestern U., 1940. Decorated Comdr. Legion d'Honneur (France). Received Berthelot gold medal from French Chem. Soc., 1928, Lavoisier medal, 1939; silver medal from King Boris of Bulgaria, 1939; Palmes d'officier from French Acad. of Sci., 1939; Willard Gibbs medal Am. Chem. Assn., 1940; Fawcett honor award for work on aviation gasoline, 1943. Mem. Am. Chem. Soc.; hon. mem. Deutsche Chemische Gesellschaft (Berlin), Acad. of Göttingen (Germany), Acad. of Sci. St. Petersburg, Nat. Acad. Sci. (U. S.). Author: Text Book Organic Chemistry, 1903-30; Aluminum Oxide as Catalyst, 1929; Catalytic Reactions at High Pressures and Temperatures, 1936; The Life of One Chemist (memoirs), 1946. Research, publs. and patents in field of high pressure catalytic reactions (especially in reference to petroleum refining, hydrocarbon, artificial rubber synthesis); developed polymerization process of mfg. high octane gasoline, method of making olefins from alcohols. Died Chgo., Nov. 29, 1952.

IPPEN, Arthur T(homas), engr., b. London, England, July 28, 1907; s. Peter Joseph and Augusta (Hechelmann) I.; diplom-ingenieur, Tech. Univ., Aachen, Germany, 1931; S.M., California Institute of Technology, 1935, Ph.D., 1936; Hon. Doctorate, U. Toulouse (France), 1962; Dr. Ing. honori's causa, Tech. U., Karlsruhe, Germany, 1967; m. Elizabeth Wagenplatz, Dec. 25, 1937 (dec.); children—Erich Peter, Karin Ann; married 2d, Ruth M. Calvert, April 10, 1955. Came to United States, 1932. Asst. geodesy, Aachen, 1932; research, teaching asst. hydraulics, Calif. Inst. of Tech., research engr., instr., 1936-38; instr., Lehigh U., 1938-39, asst., prof. civil engring., charge of hydraulic lab. 1939-45; asso. prof. hydraulics, Mass. Inst. of Tech., 1945-48, prof., 1948—, Ford professor of engineering, 1965—, also director of the hydrodynamics lab.; registered prof. engineer, Massachusetts; consultant various companies. Recipient Vincent Bendix Research award Am. Soc. Engring. Edn., 1963. Fellow Am. Acad. Arts and Scis., Am. Soc. C.E., American Geophys. Union, National Acad. Engring.; mem. Am. Soc. M.E. Boston Soc. Civil Engrs. (past pres.), Japan Society of Civil Engineers (honorary), Am. Geophys. U., Am. Soc. for Engring. Edn., Internat. Assn. Hydraulic Research (past pres., hon. mem.), Sigma Xi. Author tech. articles. Research on sediment motion; waves; density currents; fluid mechanics of open channels; flow resistance. Home: 49 Pequossette Rd., Belmont, Mass. 02178. Office: Hydrodynamics Lab., Civil Engring. Dept., Mass. Inst. Tech., Cambridge, Mass. 02139.

IPPEN, Hellmut Karl Gerd, German physician, chemist; b. Stettin, Germany, Mar. 15, 1925; s. Herbert R. H. and Elisabeth (Brauss) I.; student U. Greifswald; Dipl.Chem., U. Hamburg, 1951, Dr.med., 1954; m. Margot Petersen, Aug. 18, 1949; children—Michael, Jean-Raffael, Dagmar. Chemist, C. Feldten Nachfolger, Hamburg, Germany, 1946-52; sci. asst. Med. Acad., Düsseldorf, Germany, 1955-62; chief physician U. Dermatol. Clinic, Düsseldorf, 1962—. Mem. Soc. German Chemists, German socs. dermatology, indsl. medicine, cosmetic chemists, European Soc. Pharm. Toxicology, others. Author: Lichtschutz und Lichtschaden, 1957; Hautphysiologie, 1965; Allergologie, 1966; Glucose, 1966; also numerous articles. Research on light protection, light dermatoses, porphyrias, skin allergy, dermatol. pharmacology. Home: 23 Am Massenberger Kamp. Office: 5 Moorenstrasse, 4 Düsseldorf, West Germany.*

IPPOLITO, Felice, Italian engr.; b. Naples, Nov. 16, 1915; s. Girolamo and Angelica (Giulani) I.; m. Annemaria Perusini, 1952; children—Angelica, Susanna. Prof. geology and Applied Geology, U. Naples, 1958. Vice pres. cons. com. Euratom. Author: L'Italia et l'Energia Nucleare; Fabbisogno energetico e Energia Nucleare; Energia, ricerca scientifica e piano di sviluppo; Saggi e studi di Geologia, also publs. on geology, tech. and applied geology, mineral lodes, uranium prospecting in Italy, problems of nuclear policy. Home: via E. Ximenes 12. Office: via Belisario 15, Rome, Italy.

IRANOV, Leonid Aleksandrovich, Russian biologist; b. Feb. 24, 1871; grad. Moscow U., 1895. Prof., Institute of Forestry, 1904-41; head, photosynthesis lab., USSR Acad. Scis. Institute of Plant Physiology, 1938-47; lab. chief, USSR Acad. of Scis. Institute of Forests, since 1944. Corr. mem., USSR Acad. Scis., 1922. Author: Physiology of Plants, 1936; General Course on Systematics of Plants, 1937; Anatomy of Plants, 1939; Light and Moisture in the Life of Our Wood Strains, 1946. Studied influence of light and moisture on wood; established an original method of studying photosynthesis and designed new devices for it; developed theoretical basis of tapping conifers; anatomy of wood strains; systematics of simple plants, investigation of processes of fermentation and respiration and transformation of phosphorus in plants. Died Apr. 12, 1962.

IRGON, Joseph, phys. chemist; b. Polonnoe, Russia, Dec. 30, 1919; s. Joseph and Ida (Galperin) I.; came to U. S., 1922, naturalized, 1942; B.S. in Chem. Engring, Northeastern U., 1943; Ph.D. in Phys. Chemistry, Mass. Inst. Tech., 1948; m. Thelma Pugach, Apr. 11, 1948; children—Deborah Lisa, Judith Martha, Adam Edward. Research asso. Chem. Warfare Service Devel. Lab., Mass. Inst. Tech., Cambridge, Mass., 1943-45; project leader Central Labs. div. Gen. Foods Corp., Hoboken, N.J., 1948-52; chief theoretical analysis dept. Reaction Motors, Inc., Denville, N.J., 1952-56; v.p. Fulton-Irgon Corp., Dover, N.J., 1956-60, Hydro-Space Tech., Inc., West Caldwell, N.J., 160-62; pres. Proteus, Inc., Mountain Lakes, N.J., 1962—. Cons. to industry and govt. undersea techs., 1961—. Charles A. Coffin fellow, 1947-48. Mem. Marine Tech. Soc. (charter), Am. Inst. Aeros. and Astronautics, Am. Chem. Soc., Nat. Security Indsl. Assn. Contbr. articles to tech. jours. Developed rapid theoretical evaluation methods for rocket propellant systems; introduced statis. tools in research new materials for deep ocean environments; invented high-energy solid rocket propellants, hydrogen gas generator-operated emergency buoyancy system for deep diving submarines. Home: 144 Emmans Rd., Flanders, N.J. 07836. Office: Box 72, Denville, N.J. 07834.*

IRINO, Shozo, Japanese physician; b. Kochi Prefecture, Japan, Jan. 20, 1928; s. M. and N. (Yamasaki) I.; M.D., Okayama (Japan) U. Med. Sch., 1953; m. Etsu Yamasaki, May 3, 1954; children—Noriko, Satoko. Research fellow Okayama U. Med. Sch., 1954-59, lectr. dept. internal medicine, 1960—; research fellow Institut de Recherches sur Les Maladies du Sang, Faculté de Medécine, U. Paris, 1961-62. Recipient Hayashibara prize Okayama U. Med. Sch. for cancer research, 1963. Mem. Internat. Histamine Club, Japanese Cancer Assn., Soc. Internal Medicine, Soc. Hematology, Coll. Angiology. Research, numerous publs. on role of histamine in inflammatory reactions, role of virus in chem. carcinogenesis, role of virus in rad. leukemogenesis clin. application of lymphography. Home: 1-4-17 Kobashi, Okayama, Japan.*

IRION, Arthur Lloyd, Am. psychologist; b. Springfield, Mo., May 14, 1918; s. Theophil William Henry and Edith Grace (Ham) I.; B.A., U. Mo., 1939; M.A., State U. Ia., 1941, Ph.D., 1947; m. Isabelle Virginia Cox, 1944; children—John, Millard, Janet. Instr. to asso. prof. U. Ill., 1947-51; prof., chmn. dept. psychology Tulane U., 1951—; cons. editor Jour. Exptl. Psychology, Perceptual and Motor Skills. Mem. Am., Midwestern psychol. assns., So. Soc. Philosophy and Psychology, A.A.A.S., Phi Beta Kappa, Sigma Xi. Author: (with John McGeoch) The Psychology of Human Learning, 1952. Research and publs. on human verbal and motor learning and memory with particular reference to the phenomenon of reminiscence and set factors in retention. Home: 1315 Broadway, New Orleans 70118.*

IRSIGLER, F. J., neurosurgeon; b. Zwinelag, Austria, Sept. 9, 1903; s. Johann and Therese (Jany) I.; M.D., German U., Prague, 1927; M.B., Ch.B., U. Pretoria (S. Africa), 1953, M.D., 1954; m. Alice Emma Hanssen, Sept. 2, 1936; children—Volker Franz, Gernot Bruno, Burkhard, Gudrun Therese, Uta Maria. Asst. surg. clinic U. Erlangen (Germany), 1931-36; sr. surgeon Red. Cross Hosp., Hanover, Bayreuth, Neustettin (all Germany), 1937-39, Charité Hosp., Berlin, 1939-45; sci. mem. Brain Research Inst., Berlin-Buch, 1939-45; fgn. asst. neurosurg. clinic, Zürich, Switzerland, 1948-51; sr. neurosurgeon, Pretoria, Durban, S. Africa, 1955-59; med. practice, Krugersdorp, S. Africa, 1959—. Mem. Med. Assn. S. Africa. Author: The Neurosurgical Approach to Intracranial Infections, 1961; also contbns. to handbooks. Research, numerous publs. in exptl. biology, clin. medicine, neurology and neurosurgery. Home: 68 Paul Kruger Dr. Office: Med. Center, Ockerse St., Krugersdorp, Transvaal, South Africa.*

IRVIN, Joseph Logan, Am. biochemist; b. Jacksonville, Fla., Nov. 24, 1913; s. Joseph Logan and Eva (Hawkins) I.; student Georgia Southwestern Coll., 1930-32; B.S., U. S.C., 1934; Ph.D., U. Pa., 1938; m. Elinor Moore, Dec. 26, 1941. Asst. instr. U. Pa., 1936-38; instr. Wayne U., 1938-41; instr. Johns Hopkins Sch. Medicine, 1941-43, asst. prof., 1943-50; faculty U. N.C., Chapel Hill, 1950—, prof., chmn. dept. biochemistry, 1957—. Civilian staff OSRD,

1943-44; cons. health research facilities br. div. research facilities and resources USPHS, 1965. Mem. Am. (mem. research planning and policy com. 1964), N.C. (chmn. research com. 1962-64, chmn. bd. of dirs. 1968—) heart assns., Am. Soc. Biol. Chemists, Am. Chem. Soc., Soc. Exptl. Biology and Medicine, Am. Assn. Cancer Research. Guggenheim fellow, 1956. Contbr. numerous articles to tech. jours. Isolated and identified nitrogenous compounds in muscle; synthesized octopine; research on secretion bile acids in bile, phys. chemistry antimalarial drugs, phys. chemistry interaction small molecules with proteins and nucleic acids, metabolism cell nuclei, mechanisms regulation biosynthesis nucleic acids and proteins, biochem. control cell div., mode action antitumor drugs and antimetabolites. Home: Route 1, Box 347, Chapel Hill, N.C. 27514.*

IRVINE, Sir James Colquhoun, Scottish chemist; b. Glasgow, Scotland, May 9, 1877; ed. Royal Tech. Coll. Glasgow; Ph.D., U. Leipzig (Germany); D.Sc., U. St. Andrews (Scotland), (hon.), Cambridge (Eng.) U., U. Pa., Yale; LL.D., U. Glasgow, U. Aberdeen (Scotland), others; D.C.L. (hon.), Oxford U., U. Durham (both Eng.); m. Mabel Violet Williams, 1905; 1 son (dec.), 2 daus. Prof. sci. U. St. Andrews, prin., vice-chancellor, from 1921. Vis. lectr. various U. S. univs., 1926, 29, 31; chmn. several coms. on higher edn., Scottish Univs. Entrance Bd., 1920-44, Forest-Products Research Bd., 1927-39, Viceroy's Com. on Indian Inst. Sci., 1936; mem. Prime Minister's Com. on Tng. Biologists, 1931. Recipient Longstaff medal Chem. Soc. London. Fellow Royal Soc., 1918 (Davy medal), Royal Soc. Edinburgh (Gunning Victoria Jubilee prize); hon. mem. Am. Chem. Soc. (Willard Gibbs medal), Franklin Inst. (Elliott Cresson medal), Am. Philos. Soc. Contbr. numerous articles to sci. publs. Research on chemistry of sugars. Died June 12, 1952.

IRVINE, John Withers, Jr., Am. chemist; b. Marshall, Mo., July 15, 1913; s. John Withers and Nadine (Young) I.; A.B., Mo. Valley Coll., 1934, Sc.D. (hon.), 1952; Ph.D., Mass. Inst. Tech., 1939; Sc.D. (hon.), U. Gent (Belgium), 1963; m. Fredna Tweedt, Aug. 14, 1941; children—Mary Jane, Kathryn, Janne Elizabeth. Mem. faculty Mass. Inst. Tech., 1937—, prof. chemistry, 1958—. Sci. liaison officer Office of Naval Research, London, Eng., 1957-58; sr. chemist Clinton Labs. Monsanto Chem. Co., Oak Ridge, Tenn., 1946. Mem. Am. Chem. Soc., Am. Acad. Arts and Sci., Sigma Xi, Sigma Nu. Contbr. articles tech. jours. Research on radioactivity; separation methods in radiochemistry; organic scintillators; prodn. of radionuclides; anion exchange; solvent extraction. Office: Mass. Inst. Tech., Cambridge, Mass. 02139.

IRVINE, William Michael, Am. astrophysicist; b. Los Angeles, Aug. 31, 1936; s. S. Rodman and Mary (Dailey) I.; B.A. summa cum laude, Pomona Coll., 1957; M.A. in Physics, Harvard, 1958, Ph.D. (NSF fellow), 1961; m. Susan Wynn Ross, June 10, 1959; children—Douglas Ross, Kenneth Dwight, Peter Rodman. NATO fellow in astronomy Leiden (Holland) U., 1961-62; physicist Smithsonian Astrophys. Obs., Cambridge, Mass., 1962-66; research fellow Harvard Coll. Obs., Harvard, 1962-66, lectr., 1964-66, investigator, NASA grantee, 1964—; head astronomy program U. Mass., Amherst, head Four Coll. Astronomy Dept., asso. prof., 1966—. Mem. Royal, Am., Astron. Soc., Am. Geophys. Union, Am. Phys. Soc., Optical Soc. Am., Phi Beta Kappa. Research, publs. in gen. relativity and cosmology, condition under which galaxies may form in universe, light scattering and radiative transfer, photoelectric observations of planets. Home: 54 Jeffrey Lane, Amherst, Mass. 01002.*

IRVING, James Tutin, physiologist; b. New Zealand, May 3, 1902; s. William and Florence (Tutin) I.; A.B., Caius Coll., Cambridge (Eng.) U., 1923, Ph.D., 1927, M.D., 1931; M.A., Beit Meml. Research fellow, Trinity Coll., Oxford (Eng.) U., 1927; A.M., Harvard, 1961; m. Janet M. O'Connor, June 23, 1937. Came to U. S., 1959. Lectr. physiology Bristol (Eng.) U., 1931-34, Leeds (Eng.) U., 1934-36; head physiology dept. Rowett Research Inst., Aberdeen, Scotland, 1936-40; faculty, Cape Town (S. Africa) U., 1940-53, fellow, 1947-53; prof. exptl. odontology, dir. dental research unit, U. Witwatersrand, Johannesburg, S. Africa, 1954-59; prof. anatomy Harvard Sch. Dental Medicine and Forsyth Dental Center, 1959-61, prof. physiology, 1961—; vis. prof. U. Ill., 1947, U. Pa., 1951, U. Cal. at San Francisco, 1956. Sr. med. officer Civilian Protective Service, Cape Town, 1941-45. Recipient medal for non-mil. war services, S. Africa, 1948. Fellow Odontological Soc. S. Africa; mem. Am. Physiol. Soc., N.Y. Acad. Scis. (hon. life), Nutrition Soc. So. Africa (hon. life), Internat. Soc. Dental Research, British Physiol. Soc., British Biochem. Soc., British Nutrition Soc., British Bone and Tooth Society, Omicron Kappa Upsilon (hon. member), Sigma Xi. Author: Calcium Metabolism, 1957. Editor: Archives of Oral Biology. Research and publs. on bone and tooth formation, especially the effects of nutritional factors. Home: Rockmarge, Prides Crossing, Mass. 01965. Office: Forsyth Dental Center, 140 Fenway, Boston 02115.*

IRVING, Laurence, Am. biologist; b. Boston, May 3, 1895; Wm. Nathaniel and Esther (Messenger) I.; A.B., Bowdoin Coll., 1916, Charles Carroll Everett grad. scholar, 1916-17, Sc.D., 1959; A.M., Harvard, 1917; Ph.D., Stanford, 1924; M.D. (hon.), U. Oslo, 1956; m. Florence A. Binsley; children—Susan, William Nathaniel, Laurence, Alan. Mem. faculty Wm. Warren Sch., Menlo, Cal., 1921-23, Stanford, 1924-27; NRC fellow U. of Frankfurt-am-Main, 1925-27; faculty, U. Toronto, 1927-37; prof. biology Swarthmore Coll., 1937-49, chmn. dept. of zoology, also dir. Edward Martin Biol. Lab., 1937-49; lectr. physiology, U. Pa., 1937-49; chief physiology sect. Arctic Health Research Center, USPHS, Anchorage, 1949-62; prof. zoophysiology Lab. Zoophysiology, U. Alaska, 1962—, dir. Inst. Arctic Biology, 1964—. Sci. dir. Arctic Research Lab., Pt. Barrow, Alaska, 1948; George Cyril Graves lectr. physiology, U. Ind., 1948; hon. research asso. Woods Hole Oceanographic Inst., 1956—, Smithsonian Instn., 1958—. Fellow Arctic Inst. of N.A. (gov. 1963—, chmn. 1964-65), A.A.A.S., Am., Can. (hon.) physiol. socs., Am. Soc. Zoologists, Am. Soc. Naturalists, mem. Norwegian Acad. Sci. and Letters, Marine Biol. Lab. (life), Delta Kappa Epsilon, Phi Beta Kappa, Sigma Xi. Research and publs. on physiol. reactions of diving animals and in gen. for endurance of asphyxia; maintenance of warmth in arctic animals; use of cold extremities and surfaces for preservation of warmth and physiol. implications of heterothermus operation; orgn. of bird migrations for arctic nesting. Home: P.O. Box 5070, College, Alaska 99735.*

IRWIN, George Rankin, Am. physicist; b. El Paso, Tex., Feb. 26, 1907; s. William Rankin and Mary (Ross) I.; A.B., Knox Coll., 1930; M.S., U. Ill., 1933, Ph.D., 1937; m. Georgia Shearer Irwin, June 10, 1933; children—Joseph Ross, Mary Susan (Mrs. Charles A. Gillett), Sarah Belle (Mrs. Darald Lofgran), John Shearer. Asso. prof. physics Knox Coll., Galesburg, Ill., 1935-36; fellow physics U. Ill., 1936-37; physicist U. S. Naval Research Lab., Washington, 1937-67; U. prof. Lehigh U., Bethlehem, Pa., 1967—; vis. prof. Ford Found., U. Ill., 1961. Recipient Navy Distinguished Civilian Service award, 1947; Knox Coll. Alumni Achievement award, 1949. Mem. Am. Phys. Soc., Am. Soc. for Testing and Materials (Dudley medal 1960), Washington Philos. Soc., Washington Acad. Sci. Contbr. articles to tech. jours. Developed lab. penetration ballistics techniques, new armor materials; introduced fracture mechanics method. Home: 7306 Edmonston Av., College Park, Md. 20740. Office: Lehigh U., Bethlehem, Pa. 18015.*

IRWIN, John Henry Barrows, astronomer; b. Princeton, N.J., July 7, 1909; s. Frank and Mary (Barrows) I.; B.S., U. Cal. at Berkeley, 1933, Ph.D., 1946; m. Ruth Catherwood, July 19, 1936; children—Esther (Mrs. Gareld I. Gilman), Paul Manning, Nancy Barrows (Mrs. Franz Haaf), Alan William. Jr. astronomer U. S. Naval Obs., Washington, 1937-39; research asst., asso. Cal. Inst. Tech., Pasadena, Cal., 1941-45; asst. prof. U. Pa., 1946-48; asso. prof. Ind. U., 1948-51, prof., 1951-64; prof. U. So. Cal., staff asso. Carnegie Inst. Washington, La Serena, Chile, 1964—; vis. prof. U. Cal., Los Angeles, 1961. Lick fellow, 1941; Guggenheim fellow, 1962; Fulbright prof., Chile, 1959. Mem. Am. (council 1957-60), Royal astron. socs., A.A.A.S., Internat. Astron. Union, Astron. Soc. Pacific, Sigma Xi. Research, numerous publs. theory of eclipsing binaries, photoelec. photometry of variable stars especially cepheid variables, site investigations for large telescopes. Address: Casilla 61-D, La Serena, Chile.*

IRWIN, Malcolm Robert, Am. biologist; b. Artesian, S.D., Mar. 2, 1897; s. Joseph Speer and Mary T. (McCollum) I.; B.S., Ia. State Coll., 1920, M.S., 1925, Ph.D., 1928; m. Margaret Chandler House, July 5, 1929; children—Joe Robert, Harriet Anne. NRC fellow Harvard, 1928-29; with Rockefeller Inst. for Med. Research, 1929-30; faculty U. Wis., Madison, 1930—, prof. genetics, 1939—, chmn. dept., 1951-65; coop. agt. U. S. Dept. Agr., 1930-56. Recipient Alumni Merit award Ia. State U., 1960; H. V. Nathusius medaille Deutsche Gesellschaft Zuchtungskunde E.V., 1965. Mem. Am. Soc. Naturalists (past treas.), Nat. Acad. Scis. (Elliot medal 1938), Genetics Soc. Am. (past pres.), Am. Assn. Immunologists, A.A.A.S., Am. Soc. Animal Sci. (Morrison award 1962), Am. Soc. Zoologists, Soc. for Study Evolution, Soc. for Exptl. Biology and Medicine, Swedish Royal Acad. Agr. (fgn. mem.), Sigma Xi, Phi Kappa Phi. Contbr. numerous articles to tech. jours., chpts. to books. Research on role heredity in resistance to bacterial infection in lab. animals, role leucocytes and antibodies in resistance and susceptibility rabbits and cattle to Brucella abortus; demonstrated and genetically analyzed cellular antigens that differentiate various species pigeons and doves that hybridize, and differentiate individuality among various species domesticated animals. Home: 1220 Dartmouth Rd., Madison, Wis. 53705.*

IRWIN, Richard Leslie, Am. pharmacologist; b. nr. Fullerton, Neb., Sept. 3, 1917; s. Lewis Richardson and Eva (Graves) I.; student U. Neb., 1937-40; B.A., U. Denver, 1949; Ph.D., U. Colo., 1953; m. Lorella Elizabeth Garton, May 25, 1943; children—Cynthia, Martha, Richard. With Nat. Inst. Neurol. Diseases and Blindness, NIH, Bethesda, Md., 1953—, head sect. pharmacology, 1953—. Mem. Am. Soc. Pharmacology and Exptl. Therapeutics, Sigma Xi. Contbr. articles to profl. jours. Discovered anticholinesterase and muscle effect of lycoramine alkaloids, calcium deprivation contractures of slow striated muscle. Home: 105 Upton St., Rockville, Md. 20850. Office: Nat. Inst. Neurol. Diseases and Blindness, NIH, Bethesda, Md. 20014.*

ISAAC BEN JOSEPH BEN ISRAEL OF TOLEDO (Isaac Israeli the Younger), astronomer; flourished 1310; s. Joseph; one son, Joseph ben Isaac Israeli. Lived in Toledo, Spain. Author: Gate of Heaven; Gate of Space; Foundation of the World, 1310. Represents height of mediaeval Jewish astronomy; works include complete solution of right-angled spherical triangles; Ptolemaic views as regised by al-Bitruji; ideas of sun and moon; Jewish chronology and calandar with new moons as calculated by Maimonides; chronologies of various nations. Died circa 1330.

ISAAC IBN SID HA-HAZZAN, Jewish translator; flourished circa 1263-1277; employed by Alfonso X (el Sabio); made observations of eclipses, 1263-66; prepared (with Yudah ben Moses) Alphonsine Tables (replaced those of al-Zorgali), 1272; translated al-Battani and several anonymous astron. works; possibly built astron. instruments.

ISAAC JUDAEUS (or Isaac Israëli, the Elder, Abu Ya'Kub Ishak ibn Sulaiman al-Isra'ili), Egyptian physician; b. Egypt, circa 830-832; pupil of Ishak ben Amram, in Kairuwan; physician under several Egyptian rulers; named physician to Fatamid, al-Mahdi, 908. Author: De elementis, De Febribus, De urina (all influenced Western medieval medicine); De diaetis universalibus et particularibus; Omnia opera Ysaac; Pantegni; The Guide of Physicians. Important rep. of Arabian sch. medicine; noted as oculist; credited with writing 1st separately printed book on diet. Died circa 932.

ISAACS, Alick, Brit. virologist; b. Glasgow, Scotland, July 17, 1921; s. Louis and Rosine Isaacs; M.D. with honors, Glasgow U., 1954, M.B.CH.B., 1944; M.D. (hon.) Cath. U. Louvain (Belgium), 1962; m. Susanna Foss, May 6, 1949; children—David, Stephen, Harriet Jane. Rockefeller Travelling fellow Walter and Eliza Hall Inst. Med. Research, Melbourne, Australia, 1948-50; sci. staff Nat. Inst. for Med. Research, London, Eng., 1950—, head lab. for research on interferon, 1964—; faculty U. London, 1964—. Fellow Royal Soc., Inst. Biology; mem. Soc. for Gen. Microbiology, European Molecular Biology Orgn. Research, numerous publs. on virology, viral interference, interferon; discovered (with J. Lindenmann) interferon. Died Jan. 26, 1967.

ISAACS, John Dove III, Am. oceanographer; b. Spokane, Wash., Mar. 28, 1913; s. John Dove Jr. and Constance (Ashley) I.; student Ore. State U., 1930-41; B.S., U. Cal. at Berkeley, 1944, postgrad. San Diego; m. Mary Carol Zander, Nov. 11, 1938; children—Ann Katherine, Caroline Marie, Jon Berkeley, Kenneth Howard. Faculty, Scripps Inst. Oceanography U. Cal., San Diego, 1948—, prof. oceanography, dir. marine life research, 1957—; research engr. U. Cal., Berkeley, 1948. Mem. Am. Geophys. Union, Am. Soc. Limnology and Oceanography, World Acad. Art and Sci., Western Soc. Naturalists, Found. for Ocean Research, Sigma Xi. Research, numerous articles on motion of waves, organisms and sediments, marine resources, mil. oceanography, climatol. history, larval fish, planetary food potential, desalination and world water problems. Home: El Camino Real and La Orilla, Rancho Santa Fe, Cal. 92067. Office: U. Cal., San Diego, La Jolla, Cal. 92037.*

ISAACSON, Robert Lee, Am. psychologist; b. Detroit, Sept. 26, 1928; s. Emil Alfred and Evelyn (Johnson) I.; A.B., U. Mich., 1950, M.A., 1954, Ph.D., 1958; m. Susan Doherty, Dec. 29, 1956; children—Gunnar, Lars, Mary Ingrid, Mary Christina. Faculty, U. Mich., Ann Arbor, 1956—, prof. dept. psychology, 1967—. Mem. Am., Mich. psychol. assns., N.Y. Acad. Scis., A.A.A.S., Psychonomic Soc., Sigma Xi, Phi Kappa Phi, Phi Sigma. Author: (with M. L. Hutt and M. L. Blum) Psychology: The Science of Behavior, 1965, Psychology: The Science of Interpersonal Behavior, 1966. Editor: Basic Readings in Neuropsychology, 1965. Research, publs. on ways structures of oldest parts of brain (limbic system) interact with each other and with neocortex to control and regulate behavior, effects of brain lesions on behavior and developing nervous system. Home: 810 Oxford St., Ann Arbor, Mich. 48104.*

ISABELLE, Didier Bernard, French physicist; b. Paris, France, Aug. 9, 1934; s. Bernard Etienne and Jacqueline (Beurdeley) I.; Baccalaureat ès sciences, U. Paris, 1951, License ès Sciences, 1957, Docteur ès Sciences, Physique Nucléaire, 1961; m. Rosine Peycelon, July 22, 1958; children—Nathalie, Valérie. Research asso. Laboratoire de l'Accélérateur Lineaire, Orsay, France, 1957-61; staff Nat. Bur. Standards, Washington, 1963-64; maitre de recherches Laboratoire de l'Accelerateur Linéaire, 1964-65; faculty U. Clermont-Ferrand (France), 1964—; maitre de conférences Faculté des Sciences de Clermont-Ferrand, 1965—. Mem. Société Francaise de Physique, Societa Italiana di Fisica, Am. Phys. Soc. Research, publs. on electron scattering on nuclei, electron secondary emission at high energy and its application to current monitors for electron accelerators. Home: 10 Av. R. Bergougnan, 63-Clermont-Ferrand, France.*

ISAEV, Vasili Isaevich, microbiologist; b. Russia, 1854; s. Isaev; physician at Cromstadt Mariae Hosp.;

discovered immunity against cholera; confirmed immunity of Mechnikov theory. Died June 12, 1911.

ISAGULYANTS, Vache Ivanovich, Russian organic chemist; b. May 1, 1893; grad. Plekhanov Inst. Nat. Econs., Moscow, 1921. With research lab. Karpov Chem. Pharm. Plant, 1917-23, 25-35, Plekhanov Inst. Nat. Econs., 1923-30, Chem. Inst., Armenian Acad. Sci., 1942-48, Yerevan U., 1942-46; with Moscow Petroleum Inst., 1933-36, prof., 1936—. Mem. Armenian Acad. Sci. Author: A New Method of Producing Magnesium Organic Compounds and Its Application, Vol. 1, 1932, Vol. 2, 1935; Synthetic Aromatic Substances, 1946; Ionites and Their Use in Catalytic Synthesis; co-author: The Chemical Processing of Shale Gasoline, 1960. Research and publs. on chemistry of fine organic synthesis and chemistry of petroleum and acetylene hydrocarbons. Address: Moscow Petroleum Inst., Leninsky prospect 65, Moscow, USSR.

ISARD, Walter, Am. regional scientist; b. Phila., Apr. 19, 1919; A.B., Temple U., 1939; M.A., Harvard, 1941, Ph.D., 1943; postgrad. U. Chgo., 1941-42; m. Caroline Berliner, July 3, 1943; children—Peter, Susan, Toni, Michael, Scott A., Roberta J. Mem. faculty Harvard, 1949-53, Mass. Inst. Tech., 1953-56; prof. econs. U. Pa., 1956—, also chmn. dept. regional sci.; vis. prof. regional sci. Yale, 1960-61; founder, exec. sec. Peace Research Soc. (Internat.), 1965—, co-editor Papers, 1963—. Cons. TVA, 1951-52, Ford Found., 1955; cons. Resources for Future, Inc. Mem. Regional Sci. Assn. (pres. 1956, hon. chmn. 1957—), Regional Sci. Research Institute), Am. Econ. Assn., Am. Sociol. Soc., Econometric Soc., Assn. Am. Geographers. Author: Atomic Power: An Economic and Social Analysis, 1952; Location and Space-Economy, 1956; Municipal Costs and Revenues, 1957; Methods of Regional Analysis, 1960. Home: 3218 Garrett Rd., Drexel Hill, Pa. 19026. Office: W-45 Wharton School, U. Pa., Phila.*

ISBELL, Horace Smith, Am. chemist; b. Denver, Nov. 13, 1898; s. Harvey G. and Mary E. (White) I.; B.S., U. Denver, 1920, M.S., 1923; Ph.D. (USPHS fellow), U. Md., 1926; m. May Davidson, June 26, 1930. Asst. chemist Am. Smelting & Refining Co., Pueblo, Colo., 1920-21, Bur. Animal Industry, Dept. Agr., Washington, 1923-25; research chemist, chief organic chemistry sect. Nat. Bur. Standards, Washington, 1927—. Recipient meritorious award Dept. Commerce, 1950; Hillebrand award Washington sect. Am. Chem. Soc., 1951; Distinguished Alumni award U. Denver, 1953; Hudson Honor award, div. carbohydrate chemistry Am. Chem. Soc., 1954. Mem. Washington, N.Y. acads. sci., Am. Chem. Soc. (Hillebrand award 1951, Hudson award 1954). Patentee in field of sugars, sugar derivatives, tritium-labeled carbohydrates; research and publs. on the importance of ring shape or conformation with respect to the reactions and properties of organic compounds; first to establish that reaction rates are influenced by the angular positions of groups attached to a 6-membered pyranose ring. Home: 4704 Blagden Av., Washington 20011. Office: West Bldg., Nat. Bur. Standards, Washington 20025.*

ISBELL, John Rolfe, Am. mathematician; b. Portland, Ore., Oct. 27, 1930; s. Henry Wyatt and Dana (Martin) I.; B.S., U. Chgo., 1951; Ph.D., Princeton, 1954; m. Joan Gilbreath, Aug. 1, 1960; children—Margaret McCulloch, John Claiborne, Brecht Wyatt. Research instr. Tulane U., 1954; NSF fellow Inst. for Advanced Study, Princeton, N.J., 1956-57, mem., 1963-64; faculty U. Wash., Seattle, 1957-65, prof., 1962-65; prof. Case Inst. Tech., Cleve., 1965-67, Case Western Res. U., 1967—. Asso. editor Soc. Indsl. Applied Math. Jour. Applied Math., 1967—. Mem. Am. Math. Soc. Author: Uniform Spaces, 1964; also numerous articles. Developed basic relations between different notions of uniform dimension; basic relations between invariants of cofinal type; normal completion theory of categories. Home: 2501 Wellington Rd., Cleveland Heights, O. 44118.*

ISCH, Louis François, French physician; b. Montauban, Nov. 8, 1918; s. Louis and Suzanne (Piepenbring) I.; M.D., U. Strasbourg; m. Cecile Treussard, Oct. 7, 1949; 1 dau., Evelyne. Former extern hosps. of Strasbourg, Paris; former intern Hosp. of Strasbourg; former head neurology clinic Faculty Medicine, Strasbourg, asst., later physicians of hosps.; chief electromyography services hosp. center U. Strasbourg; chief consultations. Mem. French Soc. Neurology. Research and numerous publs. on clin. neurology and electromyography. Home: 14, rue Fischart. Office: Lab. of Electromyography, Hospital Center, University of Strasbourg, Strasbourg, France.

ISELIN, Beat Martin, Swiss chemist; b. Basel, Switzerland, Mar. 26, 1917; s. Martin and Elisabeth (Goetzinger) I.; Ph.D., U. Basel, 1944; m. Dorothe Grauer, July 24, 1945; children—Felix, Christine, Dominik. Research asst. U. Basel, 1944-46; research asso. Banting Inst., Toronto, Ont., Can., 1946-48, Cal. Inst. Tech., Pasadena, 1948-49; research chemist Squibb Inst. Med. Research, New Brunswick, N.J., 1949-51; research chemist Ciba Ltd., Basel, 1951—. Mem. Am., Swiss chem. socs., Swiss Soc. Biochemistry. Research and publs. on natural products including carbohydrates, steroids and enzymes; synthesis of peptide hormones. Home: 59 Ruetiring Riehen 4125, Switzerland. Office: 141 Klybeckstrasse, Basel, 4000, Switzerland.*

ISELIN, Columbus O'Donnell, Am. oceanographer; b. New Rochelle, New York, Sept. 25, 1904; s. Lewis and Marie (de Neufville) I.; A.B., Harvard, 1926, A.M., 1928; m. Eleanor E. Lapsley, 1929; children—Eleanor E., Columbus O'Donnell, Marie de Neufville, Victoria D., Thomas H. Asst. curator of oceanography, Museum Comparative Zoölogy, Cambridge, Mass. since 1929; asso. prof. physical oceanography, Harvard U., since 1939; dir. Woods Hole Oceanographic Instn. 1940-50, senior physical oceanographer, 1950-58, Henry Bryant oceanographer, 1958—; professor oceanography Mass. Inst. of Technology, Cambridge, 1959—. Mem. Nat. Acad. Scis., N.Y. Acad. Scis., American Acad. Arts and Sciences, Am. Geophysical Union. Contbr. articles on oceanography to various periodicals. Research on distbn. temperature and salinity, oceanic circulation, current variability, under water acoustics; focus on oceanography of Gulf Stream. Home: Vineyard Haven, Mass. Office: Woods Hole, Mass.

ISELIN, Marc, French surgeon; b. Paris, France, Feb. 15, 1898; s. Armand and Thérèse (Engel) I.; ed. Paris Faculty Medicine; m. Germaine Leewitz, Nov. 14, 1929; children—François, Jerôme, Brigitte Marty-Lavauzelle. Rockefeller fellow Johns Hopkins Hosp., 1927-28; became mem. staff Charity Hosp., Berlin, 1929; chief surg. dept. Nanterre Hosp., 1941-54. Mem. Acad. Surgery. Author: Chirurgie de la main, edited by Manon, 4 edits., 1928-55; Traité de chirurgie de la main, 1967; also articles; (with others) Atlas de chirurgie de la main, 1958. Address: 1, Auguste Vacquerie, Paris 16, France 75.*

ISELY, Frederick B., Am. biologist; b. Fairview, Kan., June 20, 1873; s. C. H. and Elise (Dubach) I.; B.S., Fairmount Coll., Wichita, Kan., 1899; M.S., U. Chgo., 1909; Sc.D., Trinity U., 1946; m. Mary E. Nickerson, May 8, 1901; children—Marion Frances, Harold Nickerson, Ralph Dubach, Frederick B. Prin. Central Sch., Hiawatha, 1899-1901; tchr. biology, high sch., Wichita, 1901-06; State Prep. Sch., Tonkawa, Okla., 1906-12; prof. biology, Central Coll., Fayette, Mo., 1912-20; dean and prof. biology Culver-Stockton Coll., Canton, Mo., 1920-22, Tex. Woman's Coll., 1922-31; prof. biology Trinity U., from 1931. Fellow A.A.A.S., Entomol. Soc. Amer., Okla. (sec. 1909-12), Tex. (pres. 1937-38) acads. sci.; mem. Am. Soc. Zoölogists, Ecol. Soc. Am., Phi Sigma. Research and publs. on fresh water mussels and ecology of orthoptera. Died Dec. 30, 1947.

ISENBURGER, Herbert R(udolf), indsl. radiologist; b. Frankfort/Main, Germany, July 22, 1900; student schs. of Frankfort/Main and Berlin, Germany; m. Anne Landsman, July 11, 1930. Came to U. S., 1925, naturalized, 1932. Asso. Dr. Ancel St. John, 1927; pres. St. John X-Ray Service, Inc., 1933; owner St. John X-Ray Lab. Califon, N.J., 1946—. Mem. gov.'s adv. com. on radiation protection, 1954-56. Mem. Health Phys. Soc., Am. Soc. M.E., Am. Soc. Testing Materials, Am. Soc. for Metals, Soc. for Nondestructive Testing, Am. Crystallographic Assn., Am. Nuclear Soc. Author: (with Ansel St. John) Industrial Radiology, 1943; also bibliographies on indsl. radiology, X-ray stress analysis and filmbadge monitoring; contbr. to Ency. Chem. Tech. Pioneer in field of indsl. radiology; expert in X-ray diffraction and X-ray stress analysis; developed film monitoring for civil def. Address: Califon, N.J. 07830.*

ISHAQ IBN HUNAIN (Abu Ya'qub Ishaq ibn Hunain ibn Ishaq al-Ibadi), Arabic translator, physician, mathematician; s. Hunain ibn Ishaq. Translated works of Menelaos, Ptolemy ("Almagest"), Euclid, Aristotle, Archimedes, Autolycos, Hypsicles; most important is Elements of Euclid as revised by Thabit ibn Qurra; credited with translation of several works of Galen. Died Baghdad, circa 910.

ISHCHENKO, Ivan Nikolaevich, Russian surgeon; b. Pustovarovka (now Kiev Oblast), 1891; grad. Med. Faculty, Kiev U., 1917; D.Med. Sci. Surgeon, Kiev Mil. Clin. Hosp., 1918-38; asso. various Kiev research insts., 1933-41; head chair surgery 2d Kiev Med. Inst., 1937-41; prof., 1937—; maj. gen. Med. Service, 1943—; chief surgery Kiev Mil. Dist., 1944-54; head chair Faculty Surgery, Kiev Med. Inst., 1944—. Decorated Order of Lenin. Mem. Ukrainian Acad. Sci. (corr.), Ukrainian Soc. Surgeons (bd. mem.). Mem. editorial council Clin. Medicine. Research and numerous publs. on clin. and exptl. surgery of organs of abdominal cavity, traumatic injuries of nervous system, shock, anesthesiology, transplantation of organs and tissues. Address: Kiev Med. Inst., b. Tarasa Shevchenko 13, Kiev, Ukraine SSR, USSR.

ISHIBASHI, Massayoshi, Japanese chemist; b. Chiba, Japan, 1896; grad. Kyoto U., 1921; later Sc.D. Became prof., dean physics dept. Kyoto U., 1936; also dir. Oceanic Chemistry Research Inst. Mem. Japan internal com. bd. UNESCO; mem. Japan Acad. Sci. Council. Recipient Acad. Sci. Promotion subsidy, Japan Chemistry Council Sakurai Prize. Mem. Japan Acad. Sci. Council. Research and publs. on analytical chemistry and oceanic chemistry.

ISHIDA, Eiichiro, Japanese ethnologist, educator; b. Osaka, Japan, June 30, 1903; s. Hachiya and Yuko (Ishida) I.; student Vienna U., 1937-39; D.Litt., U. Tokyo, 1961; m. Fusako Okamura, Apr. 9, 1936; 1 dau., Sahoko (Mrs. Tomohiro Takayama). Prof., Hosei (Japan) U., 1948-51; prof. U. Tokyo, 1951-63; prof. Tohoku U., 1963-67; pres. Tama U. Art, Tokyo, 1968—. dir. U. Tokyo Sci. Expdn. to Andes, 1958. Mem. Japanese Soc. Ethnology (chief dir. 1960-62). Author: Kappa Legend, 1948, 66; Notes on Cultural Anthropology, 1955, 66; Mother of Momotaro, 1955, 66; Introduction to Cultural Anthropology, 1959, 66; Maya Civilization, 1967; also numerous articles. Editor: Japanese Jour. Ethnology, 1959-61. Established and promoted ednl. and research instns. in cultural anthropology in Japan; contbd. to theoretical devel. in anthropology. Home: 2-9-12, Asagaya Minami, Suginami-ku, Tokyo, Japan.*

ISHIDA, Nakao, Japanese bacteriologist, virologist; b. Naoetsu, Japan, Mar. 6, 1923; s. Kenji and Katsu (Morita) I.; M.D., Tohoku U., 1946, doctorate in med. sci., 1951; m. Harue Suzuki, Mar. 27, 1949; children—Kazuko, Toshio, Kenneth Norio. Staff, Tohoku U. Sch. Medicine, Sendai, Japan, 1948—, prof. bacteriology and virology, chmn. dept., 1960—. Japanese rep. com. WHO, 1965—. Fulbright fellow Mich. U. Sch. Pub. Health, 1954-56. Mem. Japan Bacteriological Soc., Japan Virus Soc., Japan Cancer Assn., Japan Soc. Chemotherapy, N.Y. Acad. Scis. Author: (with Hinuma) Pathogenic Microbiology, 1955; (with Higashi) Virology (Virus Gaku), 1964; Antivirus Substance, 1966; also numerous articles. Research on antibiotics produced by Streptomyces, discoveries include Myxoviromycin (antivirus antibiotic), 1957, Shinkomycins, 1965, neocarzinostatin, 1960, antitumor antibiotic; isolated Sendai virus, 1953; studied virus multiplication and regulation. Home: 5-40 Tsunogoro-1-cho, Sendai, Japan.*

ISHIGURO, Takeo, Japanese pharmacologist; b. Niigata, Japan, 1904; grad. Tokyo U., 1929; later Ph.D. Prof. med. dept. Kyoto U. Recipient Acad. Sci. Promotion subsidy, Edn. Ministry subsidy, Japan Acad. Sci. Assn. Prize. Mem. Japan Pharmacy Council, Japan Chemistry Council, Japan Physiol. Chemistry Research Assn., Japan Chemistry Research Soc., Kyoto Pharmacist Soc. (chmn.). Author: Organic Drug-Making Chemistry; Organic Pharmaceutical Chemistry. Research and publs. on polyethylene glycol and derivative, chemotherapeutic medicines.

ISHIHARA, Jun, Japanese physicist; b. Tokyo, 1881; grad. Tokyo U.; postgrad., Germany; D.Sc.; prof. Tohoku U., until 1921; interpreter during Einstein's visit to Japan; recipient Onshisho (Imperial prize) for research in relativity, gravitation, quantum theory, 1919; Author: Scientific World Image; Relativity. Died 1947.

ISHIHARA, Shinobu, Japanese ophthalmologist; b. Tokyo, Japan, 1879; ed. Tokyo U.; studied in Germany, 1912-14; instr. Mil. Med. Coll.; became prof. Tokyo U., 1922; named med. officer Ministry of Communications Hosp., 1940; dir. Maebashi Med. Coll., 1943-46. Author: Brief Study of Ophthalmology; Cause of Trachoma; Test Diagram of Daltonism. Developed Ishihara test for color vision by placing figures or lines of colored dots on a field of dots of another color, published 1917. Deceased.

ISHIKAWA, Haruji, Japanese physicist; b. Iwata, Shizuoka, Japan, Apr. 13, 1915; s. Sahei and Mura (Hanai) I.; B.Sc., Tokyo U., 1941; D.Sc., Geophys. Inst., Kyoto (Japan) U., 1961; m. Tamiko Watanabe, Jan. 21, 1945; children—Manabu, Izumi, Osamu. Physicist, Govt. Mech. Lab., Ministry of Trade and Industry, 1941-51; faculty Research Inst. Atmospherics, Nagoya U., Toyokawa, Japan, 1951—, prof. physics, 1961—, dir. Research Inst. of Atmospherics, 1968. Cons. Central Radio Propagation Lab., Nat. Bur. Standards, Boulder, Colo., 1961-63; sec. gen. Japanese arrangement com. 4th Internat. Conf. on Atmospheric Electricity, 1965—. Mem. Soc. Terrestrial Magnetism and Electricity of Japan (Tanakadate award 1957), Meteorol. Soc. Japan. Author: Nature of Lightning Discharges as Origins of Atmospherics, 1961; also articles. Discovered predominance of beta type stepped leader over alpha type in lightning discharge to ground in Japan Island; prediction of complex streamer mechanism in lightning discharge in thunder-cloud; verified return stroke contbn. to ELF sferic radiation in lightning discharge to ground. Home: 2-5 Hatcho Dori, Toyohashi, Aichi. Office: Research Inst. Atmospherics, Nagoya U., Toyokawa, Aichi, Japan.*

ISHIKAWA, Yoshikazu, Japanese physicist; b. Yokohama, Japan, Oct. 24, 1930; s. Bunju and Kôko (Inomata) I.; B.S., U. Tokyo, 1953, D.Sc., 1958; m. Hiroko Kobayashi, May 17, 1959; 1 dau., Marie. Faculty, U. Tokyo (Japan), 1958—, asst. prof. Inst. for Solid Physics, 1959—; prof. Grenoble (France), 1962-64. Research, publs. on neutron diffraction study of antiferromagnetic Fe-Cr., Fe-Mn alloys; discovered ferrimagnetism of ilmenite-hematites system and clarified its origin; discovered mechanism of reverse thermoremanent magnetism of FeTiO3-Fe2O3; established method to investigate magnetically dilute system. Home: RG 302, 1029 Shimoishihara, Chôfu, Tokyo. Office: Inst. for Solid State Physics, U. Tokyo, Roppongi, Azabu, Minato-ku, Tokyo, Japan.*

ISHIZAKI, Tatsushi, Japanese physician, parasitologist; b. Tochigi, Japan, Mar. 8, 1915; s. Takaji and

Satoko I.; M.D., Tokyo (Japan) U., 1939, Ph.D., 1951; Diploma Pub. Health, Singapore U., 1959; m. Yuriko Sase, Nov. 5, 1946; children—Terumi, Michiharu. With U. Tokyo Sch. Medicine, 1939, sr. asst. dept. phys. therapy and medicine, 1946-55; chief 2d div. dept. parasitology Nat. Inst. Health, Japan, 1953-67, chief dept., 1967—; lectr. clin. allergy U. Tokyo Sch. Medicine, 1956—. Panelist Japan-U.S.A. Coop. Study Air Pollution, 1965—. Mem. Japanese Soc. Allergy, Japanese Soc. Parasitology, Japanese Soc. Tropical Medicine, Japanese Soc. Internal Medicine and Chest Diseases, Am. Soc. Tropical Medicine and Hygiene, Am. Soc. Parasitologists. Research, numerous publs. on skin tests for various antigens; standardization of criteria of positive and negative response and basic phenomena of skin reaction from allergical point of view; analysis of onset of asthma attacks especially related to air pollution, weather; analysis of onset of symptoms of ancylostomiasis from allergical view point. Home: 34-1, Itabashi 3-chome, Itabashi-ku, Tokyo. Office: Nat. Inst. Health, 10-35 Kamiosaki-2-chome, Shinagawa-ku, Tokyo, Japan.*

ISHLINSKY, Aleksandr Yulevich, Russian mathematician; b. Aug. 6, 1913; grad. Moscow U., 1935, postgrad., until 1938; D.Physico-Math. Sci., 1943. Head chair Moscow U., 1935-48, prof., head chair, 1955—; dir. Inst. Math., Ukrainian Acad. Sci., 1948-55, head dept. gen. mechanics, 1956—; prof. Kiev U., 1948-55. Mem. USSR, Ukrainian acads. sci. Author: The Mechanics of Special Gyroscopic Systems, 1952; The Dynamics of Ground Masses, 1954; The General Theory of Plasticity with Linear Consolidation, 1954; Essays on the History of Technology, 1955; The Theory of the Horizon-Compass, 1956; The Theory of the Gyroscopic Pendulum, 1957; The Theory of a Double Gyroscopic Gyro-Vertical, 1957. Research in general mechanics, elasticity, oscillations; developed theory of gyroscopic devices; studied behavior of complex gyroscopic systems on movable base; gave theoretical basis for space gyroscope. Address: Moscow University, Leninskie gory, Moscow, USSR.

ISIDOR, Pierre, French pathologist; b. Paris, France, May 8, 1904; s. Isidor and Isidor (Kolp) I.; Faculty of Medicine, Paris; m. Jacqueline Weber, Mar. 3, 1959. Histology asst. Paris Faculty of Medicine, 1926-30, path. anatomy asst., 1931-38, head lab., 1931-38; dir. pathology lab. Hosp. Center, St. Germain en Laye, France, 1937—. Mem. Assn. for Study of Cancer, Soc. Obstetrics and Gynaecology, Royal Belgian Soc. Obstetrics and Gynaecology, Anatom. Soc. Paris, Assn. Anatomists, Soc. Obstetricians and Gynaecologists. Author: Precis de Techniques Histologiques a l'usage des Laboratoires Hospitaliers, 1934. Research, publs. on histology, pathology of endocrine glands, genitals, neurovegetative system, cancer cytology. Home: 1 Tintoret Asnieres, Seine, France. Office: Centre Hospitalier, Saint-Germain en Laye 78, France.*

ISIDORE OF MILETOS (or Isidorus the Elder), Greek architect, mathematician; employed (with Anthemius of Tralles) by emperor Justinian as architect for St. Sophia, Constantinople, 532-37; possibility that 15th book of Euclid's Elements came from Isadore's sch.

ISIDORE OF SEVILLE, Spanish scholar; b. Seville, Spain, circa 560; became Archbishop of Seville, 609. Author: Chronica (history down to the times); Historia Gothorum; De natura rerum; Differentiae, 2 vols.; Quaestiones in vetus Testamentum; Etymologiae (or Origines, ency. now in 20 vols.), pub. between 622 and 633. Influential in early Middle Ages by means of his ency., which borrowed from Greeks, especially Pliny, and covered 7 liberal arts, natural history, geography, medicine, prodigies, foods, drinks, other topics; often credulous. Died Seville, Apr. 4, 636.

ISMA'IL AL-JURJANI, see al-Jurjani.

ISOMURA, Kichitoku (or Kihe), Japanese mathematician; studied math. under Yoshitane Takahara; apptd. castellan (rank of magistrate) Nihonmatsu Castle; helped lay water pipes to Castle from Mt. Adachi Taro. Author: Sampo Ketsugisho (textbook). Influenced Takakazu Seki; gained followers later known as students of Isomura sch.; developed theory of calculating cubic vol. Died 1710.

ISRAEL, Hans, German geophysicist; b. Herschdorf/Thuringia, Germany, Apr. 7, 1902; s. Karl and Antonie (Schmiedeknecht) I.; student univs. Marburg/Lahn, Munich; Dr.phil., 1926; m. Lieselotte Schaefer, Sept. 26, 1934; children—Gerhard W., Christian Karl. Sci. asso. William G. Kerckhoff Inst., Bad Nauheim, Germany, 1933-36, (Meteorol. Obs., Potsdam, Germany, 1936-45; with Württemberg-Hohenzollern weather service, 1946-53; docent U. Tübingen, Germany, 1947-49, unscheduled prof., 1949-53; dir. Meteorol. Obs., Aachen, Germany, 1953-63; faculty Aachen Inst. Tech., 1953—, prof., 1963—. Mem. German Geophys. Soc., German, Am. meteorol. socs., Am. Geophys. Union, UGGI, numerous others. Author: Das Klima von Bad Nauheim, 1937; Das Gewitter, 1948; Radioaktivität, 1940; Atmosphärische Elektrizität, 2 vols., 1957, 60; (with A. Krebs) Kernstrahlung in der Geophysik, 1961; Probleme der Gewitterforschung, 1964/66; also numerous articles. Research on atmospheric electricity, radioactivity, bioclimatology, radiometeorology, aerosols. Address: 186 Turmstrasse, Aachen, West Germany.*

ISRAEL, Lucien, French physician; b. Paris, France, Apr. 14, 1926; s. Jacques and Alice (Molho) I.; M.D., Paris Faculty Medicine, 1956; m. Germaine Bach, Jan. 31, 1949; children—Danièle, Dominique, Guillaume. Head clinic Paris Faculty Medicine, from 1957; prof. agrégé, chair pneumo-phtesiology Laënnec Hosp., Paris, from 1962, dir., from 1964; dir. in Paris hosps., 1964—. Mem. Med. Soc. Paris Hosps., French Soc. Respiratory Pathology and Tb. Research, publs. on mucoviscidosis in chronic bronchitis in adults, simultaneous polychemotherapy with 7 drugs in lung cancer, phytohemaglutinism in man as hematological protector against chemotherapy, immunotherapy of cancer in man. Home: 85 Blvd. Pasteur, Paris 15. Office: rue de Sèvres, Paris 7, France.*

ISRAELS, Martin Cyril Gordon, English physician; b. Manchester, Eng., Sept. 24, 1906; s. Herman and Jeanette (Gordon) L.; M.Sc., U. Manchester (Eng.), 1928, M.B., 1932, M.D., 1935; m. Ivy D. Livesley, Sept. 25, 1935; 1 dau., Jane. Lady Tata Research fellow, 1933; Foulerton Royal Soc. fellow, 1938; faculty U. Manchester, 1946—, reader clin. haematology, 1962—, dir. clin. hematology dept., 1962—, dean postgrad. med. studies, 1964—; cons. physician Manchester Royal Infirmary, 1948—. Fellow Royal Coll. Physicians; mem. Assn. Physicians, Brit. Hematological Soc., Med. Research Soc. Author: Diagnosis and Treatment of Blood Diseases, 1964; Atlas of Bone Marrow Pathology, 1966; also numerous articles. Research on diagnosis, treatment and abnormalities in diseases of blood. Home: Beech Holme, Stamford Rd., Bowdon, Altrincham, Cheshire, Eng. Office: Dept. Clin. Hematology, Royal Infirmary, Manchester, 13, Eng.*

ISSELBACHER, Kurt Julius, physician; b. Wirges, Germany, Sept. 12, 1925; s. Albert and Flori (Strauss) I.; came to U. S., 1936, naturalized, 1945; A.B., Harvard, 1946, M.D. cum laude, 1950; m. Rhoda Solin, June 22, 1955; children—Lisa, Karen, Jody, Eric. Investigator, NIH, Arthritis and Metabolism Inst., 1953-56; chief gastrointestinal unit, prof. medicine Mass. Gen. Hosp., Harvard Med. Sch., Boston, 1956—. Cons. NIH. Editorial bds. Jour. Clin. Investigation, 1963—, Gastroenterology, 1962—, Medicine, 1964—. Research, numerous publs. on genetic and metabolic abnormalities; discovered cause of galactosemia as 1st definitely proven disease due to hereditary enzyme defect; elucidated mechanism of intestinal fat absorption and causes of fatty liver. Home: 20 Nobscot Rd., Newton Centre, Mass. 02159. Office: Mass. Gen. Hosp., Boston 02114.*

ISSELS, Josef, German physician; b. München-Gladbach, Germany, Nov. 21, 1907; s. Joseph and Adelheid (Heinen) I.; ed. univs. Bonn, Munich, Vienna, Rostock, Düsseldorf, Wurzburg; M.D.; m. 2d, Ilse-Maria Klos, Nov. 6, 1959; children—Ruthild, Rolf, Helmut. Asst. various hosps., 5 years; practiced medicine, Mönchen-Gladbach, 1938-51; founder spl. clinic for tumors and chronic illnesses with biol. therapy for malignant tumors, 1951. Mem. Deutsche medizinische Gesellschaft für Herdforschung und Herdbekämpfung, German Soc. for Research, Soc. German Naturalists and Physicians Internat. Med. Soc. for Circulatory and Infectious Diseases, Free Internat. Acad. Author: Grundlagen und Richtlinien für eine interne Krebstherapie, 1953; Therapeutische Richlinien bei inoperablen malignen Tumoren, 1953; Ergebnisse und Erkenntnisse nach vierjähriger klinisch-interner Therapie bei inkurablen Krebskranken, 1956; Könne wurzelbehandelte Zähne Krebs erzeugen?, 1956. Home: Rueppweg 8. Office: Ringbergstrasse 30, Tottach-Egern/Tegernsee, Germany.

ISTIFAN BEN BASIL, Greek translator; flourished 9th century; translated into Arabic 9 works of Galen, also Dioskyrides, 1st version of works of Oreibasios.

ISTRATI, Constantin I., Rumanian chemist, physician; b. 1850; mem. Rumanian Acad.; investigated salt, petroleum, amber, ozokerite in Rumania; discovered new class of dyestuffs (Franceines); helped establish sci. nomenclature in Rumanian lang. Died 1918.

ITARD, Jean-E.-Marie-Gaspard, French physician; b. Oraison, France, 1774; became surgeon Val-de-Grace, Paris, 1796; physician Inst. Deaf-Mutes, from 1800. Mem. Acad. Medicine. Author: Traité des malades deloreille et de l'audition; Mutisme produit par lésion des facultés intellectuelles; Treatise on the diseases of the ear, 2 vols., 1821. Pioneer of modern otology; perfected methods of exploring ear and naso-pharnyx; invented eustachian catheter (Itard's catheter), also developed precise technique of catheterization of Eustachian tube; established criteria for puncture of ear drum. Died Paris, July 5, 1838.

ITO, Eiji, Japanese biochemist; b. Sapporo, Hokkaido, Japan, Sept. 9, 1925; s. Mitsuji and Eiko (Ito) I.; grad. Imperial U. Hokkaido, 1944, B.A. in Chemistry, 1947; Ph.D. in Biochemistry, U. Hokkaido, Sapporo, 1958; m. Reiko Miura, May 5, 1955; children—Satoshi, Yako, Akiko. Staff, U. Hokkaido, Sapporo, 1949-58, 60—, prof. chemistry, 1964—; research fellow Washington U., St. Louis, 1958-60. Mem. Chem. Soc. Japan, Biochem. Soc. Japan, Japanese Soc. Bacteriology. Research and publs. on biochemistry and pharmacology of bacterial cell wall, especially mechanism of antibiotic action of oxamycin and penicillin on enzymatic level; discovered enzyme system synthesizing macromolecule of cell wall. Home: North 7, West 11, Sapporo, Hokkaido, Japan.*

ITO, Han-ya, Japanese economist; b. Kyoto, Japan, 1894; grad. jr. course Tokyo Higher Comml. Sch.; later Ph.D. in Econs. Asst. prof. Tokyo U. Commerce, became prof., 1934; became chief econs. dept. Hitotsubashi U., 1951. Mem. Japanese Acad. Author: Principle of Public Finance; System and Growth of Basic Principles of Taxes; Outline of Social Policy; Essays in Public Finance. Pioneered studies in public finance and established firm base of its methodology.

ITO, Junkichi, Japanese physicist; b. Hyogo, Japan, 1914; D.Sc., Osaka U., 1936; asst. prof., later prof. Osaka U. Author: Artificial Transformation and Artificial Radiation; Atomic Physics. Research on artificial radiation, super microwaves.

ITO, Keisuke (or Kiyotami Ito), Japanese botanist; b. Nagoya, 1803; s. Gendo Nishiyama; studied medicine; later studied botany together with Siebold; entered service Ministry Edn., 1870; apptd. prof. Kyoto U., 1881. Author: Explanatory Diagrams of Japanese Plants. Pioneer in botany. Died 1901.

ITO, Seiya, Japanese agriculturist; b. Niigata, Japan, Aug. 7, 1883; grad. Tohoku U., 1908; Ph.D. in Agr., 1919; studies in Eng., U.S.A., Germany, France.; m. Toshihiko Hara. Became asst. prof. Tohoku U., 1909; dir. Bot. Gardens, Hokkaido U., became dean agr. dept., 1941, pres., 1945, also 1949. Recipient Asahi Newspaper Sci. prize, 1952, Japan Agr. Sci. Prize, 1935. Mem. Japanese Acad. Sci. Author: Rice Plant Fever; Japanese Bacteria; A History of Plant Pathology. Research on plant pathology and bacteriology, wheat mosaic disease.

ITO, Teiichi, Japanese geologist; b. Osaka, Japan, 1898; grad. Tokyo U., 1923; Sc.D., 1943. Instr., then asst. prof. Tokyo U. Recipient Imperial Acad. Prize. Mem. Acad. Sci. Research Congress, Japanese Acad. Scis. Author: The Mineral Resources of Japan, I, II; Minerals in Japan: Illustrated, I-IV; X-ray Studies on Polymorphism. Research on mineralogy, crystallography.

ITO, Tokunosuke, Japanese physicist; b. 1894; grad. Tokyo U., 1918; later Sc.D. Engr., Central Meteorol. Obs.; lectr., later prof. Kyushu U.; became prof. physics Kyushu U., 1934, also dean Physics Faculty. Author: Studies on Soft Elastic Body; Ocean Current and Tidal Current. Research and numerous publs. on applied math., artificial rain.

ITO, Toshio, Japanese anatomist; b. Aichi, Japan, Aug. 5, 1904; s. Yukichi and Tane (Ishiguro) I.; M.D., Keio U. Sch. Medicine, 1930, D.M.S., 1936; m. Mariko Kondo, Mar. 17, 1941; children—Seita, Hiroko. Asst., Keio U. Sch. Medicine, 1930-32, instr., 1932-34, asst. prof., 1934-41; prof. Tokyo Women's Med. Coll., 1941-48; prof. Gunma U. Sch. Medicine, 1948—, dean, 1961-63. Mem. Japanese Soc. Anatomy (dir. 1964—). Research, numerous publs. on cytological analysis of structure of human and animal sweat glands revealing presence of two types of secretory cells in eccrine gland and of a transitional portion between secretory portion and excretory duct; cytological and comparative histological studies on livers of vertebrates and discovery of fat-storing cell in their sinusoid wall. Home: 1-11-14 Minami-cho, Kokubunji, Japan. Office: 3-39-22 Showa-machi, Maebashi, Japan.*

ITTELSON, William H(oward), Am. psychologist; b. N.Y.C., May 4, 1920; s. Ralph B. and Julia (San) I.; B.A., Columbia, 1941, B.S. in Elec. Engring., 1942; M.S., Princeton, 1948, Ph.D. in Psychology, 1950; m. Martha Lane, Feb. 16, 1946; 1 son, Lane. Staff, Naval Research Lab., 1942-43; faculty Princeton, 1947-55; asso. prof. psychology Bklyn. Coll., 1955-61, prof., chmn. dept., 1961—; acting exec. officer grad. program psychology City U. N.Y., 1963; cons. VA, 1952—; Fulbright lectr., Japan, 1961-62. Author: The Ames Demonstrations in Perception, 1952; Visual Space Perception, 1960; (with S. B. Kutash) Perceptual Changes in Psychopathology, 1961. Research and publs. in visual space perception; the relation between perception and psychopathology; the environmental psychology of the psychiatric ward. Home: 1585 Westervelt Av., Baldwin, N.Y. Office: Bklyn. Coll., Bklyn. 10.*

ITTNER, Martin Hill, chemist; b. Berlin Heights, O., May 2, 1870; s. Conrad Smithman and Sarah Content (Hill) I.; B.Ph., Washington U., 1892, B.Sc., 1894, LL.D., 1938; A.M., Harvard, 1895, Ph.D., 1896; D.Sc., Colgate U., 1930; m. Emilie A. Younglof, Nov. 20, 1900 (dec. Dec. 1933); children—Irving Hill, Lois Elizabeth (Mrs. Eldon Bisbee Sullivan); m. 2d, Hildegard Hirsche, July 21, 1934; 1 son, Robert Austen. Became pvt. asst. to Dr. Wolcott Gibbs, Newport, R.I., 1896; chief chemist Colgate & Co., N.Y.C. and Jersey City, 1896-1928 Colgate-Palmolive-Peet Co. from 1928; also served as chem. engr. and dir. of research of both cos. U. S. del. to First World Chem. Engring. Congress, London,

1936. Recipient Modern Pioneer award N.A.M., 1940; Perkin medal, Soc. Chem. Industry, 1942. Mem. Am. Inst. Chem. Engrs. (pres. 1936, 37), Am. Chem. Soc. (chmn. nat. com. on indsl. alcohol 1920-42), Soc. Chem. Industry (England), Am. Inst. Chemists, Am. Oil Chemists Soc. Franklin Inst., Assn. Harvard Chemists. Developed and patented methods for hydrogenation of fatty oils, methods of making soap from petroleum, hydrocarbons, glycerol. Died Apr. 22, 1945.

ITTNER, William Butts, III, Am. physicist; b. St. Louis, Sept. 28, 1923; s. William Butts and Mignon (Morrow) I.; A.B., Washington U., St. Louis, 1947, M.A., 1949, Ph.D., 1953; m. Mary Phyllis Cady, Oct. 14, 1950; 1 dau., Stacy Phillips. Electronic design engr. McDonnell Aircraft Co., 1949-50; research physicist IBM, Poughkeepsie, N.Y., 1953-60, program mgr., 1963-67, engring. lab. mgr., Kingston, N.Y., 1967——. Cons. Wilson Meml. Hosp., Binghamton, N.Y., 1954-55. Mem. Am. Phys. Soc., I.E.E.E., Phi Beta Kappa, Sigma Xi, Phi Mu Epsilon. Research, publs. in med. physics, cryogenic electronics, high-speed solid state electronic memory technologies and systems. Home: 98 Ridgeview Rd., Poughkeepsie 12601. Office: IBM Lab., Neighborhood Rd., Kingston, N.Y.

IUNUSOV, Adkham Iunusovich, Russian physiologist; b. Jan. 14, 1910; grad. Central Asian Sericultural Inst., 1932, Tashkent Med. Inst., 1943. Mem. staff Tashkent Med. Inst., 1934-43, in charge physiology, 1949-52; mem. staff Tashkent Agrl. Inst., 1945-49; acad. sec. Uzbek Acad. Sci., 1953-56, dir. Physiology Lab., Regional Medicine Inst., 1956——. Mem. Uzbek Acad. Sci. Research in high temperature effects on man, including food and water intake under high temperatures; water-sodium metabolism.

IUNUSOV, Sabir Yunusovich, Soviet organic chemist; b. Nov. 11, 1909; ed. Central Asian U., 1935. Dir. chemistry of alkaloids lab. Chemistry Inst., Uzbek Acad. Scis., 1943——, prof., 1952——; dir. Chemistry Inst., 1950-52, v.p., 1952——. Mem. USSR Acad. Sci. (corr.). Research and publs. on alkaloids.

IURIEV, Vasilii Iakovlevich, Russian plant geneticist; b. Feb. 22, 1879; ed. Novo-Aleksandriisk Inst. Agr. and Forestry, 1905; staff Kharkov Selection Sta., 1909——, dir., 1944——; prof. Kharkov Agrl. Inst., 1937——; dir. Ukrainian Acad. Sci. Genetics and Selection Inst., 1946——. Recipient Stalin Prize, 1947, 2d gold medal Hammer and Sickle, 1959; named Honored Scientist Ukraine, 1958. Mem. Ukrainian Acad. Agrl. Scis. (hon.), Ukrainian Acad. Sci., All-Union (Lenin) Acad. Agrl. Scis. Author various books on selection. Developed several new varieties of rye, millet, corn, winter and spring wheat.

IURKEVICH, Ivan Danilovich, Russian forester, geobotanist; b. July 5, 1902; ed. Byelorussian Agrl. Acad., 1930; mem. staff Byelorussian Forest Economy Inst., Gomel, 1930-53; staff Forest Inst., Byelorussian Acad. Sci., 1953-56, dir., 1954——, dep. acad. sec., 1956—— also in charge sect. geobotany Biology Inst. Recipient Stalin prize, 1951. Research on natural and artificial reforestation of native variety of rubber plant.

IUSKOYETS, Moisei Kallinikovich, Russian vet. physician; b. Aug. 15, 1898; ed. Moscow Vet. Inst., 1925; veterinarian various state collective farms; vet. physician All-Union Inst. for Exptl. Vet. Medicine, 1935-41, 52-57; mem. Sci. Control Inst., USSR Ministry Agr., 1943-52; acad. sec. livestock and vet. medicine sect. Byelorussian Inst. Agrl. Scis. Acad., 1957——. Mem. Byelorussian Acad. Sci., Byelorussian Acad. Agrl. Scis. Research on control diseases in animals contagious to man, including Tb, brucellosis.

IUSUPOVA, Saradzhan Mikhailovna, Russian geochemist; b. May 18, 1910; ed. Uzbek U., Samarkand, 1935; mem. staff Dadjik U., Stalinabad, 1948——, prof., 1949——. Mem. Tadjik Acad. Sci. Author: Colloido-chemical Properties of Clays of Uzbekistan, 1941; Mineralogical Characteristics of Loesses of Central Asia, 1951. Her research has been on geochemistry of celestite; colloidal minerals, especially clays; mineral springs of Tadjikistan.

IVADY, Gyula, Hungarian physician; b. Pétervására, Hungary, Dec. 9, 1914; s. Ignac and Margarete (Zámori) I.; grad. U. Med. Sch., Szeged, Hungary, 1940; m. Pénzes Margarete, May 22, 1945; children—Réka, Péter, Pál. Staff, Munkács, asylum for destitute children, 1941-43, Physiol. Inst. U. in Kolozsvar, Cluj, Hungary, 1943-45, dept. pediatrics Med. Sch. U. Szeged, 1945-61, dept. pediatrics Dist. Hosp., Meissen, Germany, 1961-64; physician Paul Heim's Hosp. for Children, Budapest, 1964——. Recipient die Medaille für ausgezeichnete Leistungen, Germany, 1963. Mem. Soc. Hungarian Pediatrics (mem. pharm. com. pub. health), Sächsisch-thüringische med. wiss. Gesellschaft für Kinderheilkunde, Deutsche Gesellschaft für klin. Medizin. Author: Basis of Pharmacotherapy in Pediatrics, 1963; also articles. Invented Pentamidin treatment in Pneumocystis carinii pneumonie, treatment of rose-hips in infancy enteritis; elaborated new methods of dosing drugs in infancy and childhood; research on origin of dermatitis Leineri, antidiuretic effects of cerebrospinal fluid; elaborated new probe of albumin in urine with bromwater. Home: 57. Arpád fejedelem, Budapest. Office: 86. Üllöi, Budapest, Hungary.*

IVANENKO, Dmitriy Dmitrevich, Russian physicist; b. July 29, 1904; grad. Leningrad U., 1927. Prof. univs., insts. in Leningrad, Kharkov, Tomsk, Sverdlovsk, Kiev, Moscow, 1930——; prof. Moscow U., 1942——; asso. Inst. History of Natural Sci. and Engring., USSR Acad. Sci., 1949-61. Recipient Nobel prize in physics (with I. Ye. Tamm), 1958. Author: Classical Field Theory, 1951; Quantum Field Theory, 1952. Hypothesized structure of atomic nucleus as being composed of protons and neutrons, 1932; developer (with I. Ye. Tamm) principles of theory of specific nuclear forces, 1934-36, developer (with I. Ya. Pomeranchuk and A. A. Sokolov) theory prediction of "synchrotron" electromagnetic radiation emitted by electrons accelerated to high energies in betatron and synchroton accelerators, 1944-48; proposed theory of parallel transfer of electron spinor wave functions permitting generalization of Dirac quantum equation in presence of gravity (with V. Fock), 1929; developed hypothesis of expansion of earth; suggested new linear matrix geometry, non linear spinor equations. Address: Moscow University, Leninskie gory, Moscow-234, USSR.

IVANIC, Stevan, Yugoslavian hygienist; b. Mali Mokri Lug, Yugoslavia, Dec. 25, 1884; s. Zoran and Marija (Petrovic) I.; ed. Med. Sch., Vienna; student bacteriology, immunology, serology, Vienna, Hamburg, Berlin; received M.D.; m. Andjelka Popovic, 1912. With Hygienic Service, 1920——; dir. Central Inst. Hygiene, Belgrade, Yugoslavia. Decorated Order Yugoslavian Crown; French Silver medal. Mem. Soc. for Pub. Health (pres.), Health Coop. Assn. (v.p.), Yugoslavian League Against Tb (treas.), Yugoslavian Culture Soc., Yugoslavian and Serbian Med. Assn., Yugoslavian Microbiol. Soc. Author: Textbooks of Hygiene for High School, 4 vols.; Le service hygiènique dans le Royaume des Serbes, Croates et Slovenes; Le service d'hygiène dans Yougoslavie; also numerous sci. and popular articles, pamphlets. on pub. health, epidemiology, dietetics.

IVANOFF, Alexandre, physicist, oceanographer; b. Poltava, Russia, July 22, 1917; s. Alexandre and Mary (Kosmatcheff) I.; Ingénieur l'Ecole de Supérieure de Physique et de Chimie de Paris (France), 1939; Docteur ès Sciences, U. Paris, 1946. Prof. Ecole Supérieure de Physique et de Chimie de Paris, 1947——; asso. dir. Lab. Applied Physics, Nat. Mus. Natural History, 1950-61; prof. physics and oceanography, U. Paris, 1961——. Mem. Société Française de Physique. Research, numerous publs. on optics of sea, underwater photography, optical properties of sea waters, underwater daylight. Office: 10 rue Vauquelin, Paris 5e, France.

IVANOV, Artemii Vasilevich, Russian zoologist; b. 1906; grad. Leningrad U. 1930. Asst., Pacific Ocean Inst. Fisheries and Oceanography, Vladivostok, USSR, 1930-32; sr. asso. State. Hydrological Inst., Leningrad, 1932-34; instr. Leningrad U., 1934——. Recipient Lenin Prize, 1961. Mem. German Acad. Natural Scis. in Galle. Author: Pogonophora, 1960.

IVANOV, Dimitre Ivanov, Bulgarian chemist; b. Makotzevo, Sofia, Bulgaria, Oct. 13, 1894; s. Ivan Stoyanov and Maria (Ivanova) Petrov; student U. Lyons (France), 1914-15; dipl. chemist, U. Sofia, 1920; chem. engr. U. Nancy (France), 1922, Dr.Sci., 1923; laureate Inst. de France, 1932; m. Maria Christova Trinkarova, Oct. 16, 1921; children—Emil, Lydia Ivanova (Mrs. Kyoseva). Faculty dept. chemistry U. Sofia, 1920——, prof. chemistry, 1929——. Decorated for civic service; for civic service St. Alexander IV; Corona d'Italia III, 1940. Mem. Bulgarian Acad. Scis., Union Scientists in Bulgaria, Chemische Gesellschaft, Société Chimique de France. Author: Textbook of Organic Chemistry, 1942; (with C. Ivanov) Laboratory Practice of Organic Chemistry, 1946; also numerous articles. Research on organometallic compounds (magnesium and lithium), essential oils, including rose oil, lavender oil, mentha piperita oil; synthesized polyfunctional organometallic compounds (Ivanov reagts. which give substituted beta-hydroxy-acids with aldehydes and cetones (Ivanov reaction). Home: 18 Janko Sakasov. Office: Institut Organic Chemistry, Acad. Sci., Sofia, Bulgaria.*

IVANOV, Leonid Aleksandrovich, Russian botanist; b. Feb. 24, 1877; ed. Moscow U., 1895; prof. Forest Inst. (now Kirov Inst. Forest Tech.), 1904-41, head forest trees physiology and ecology lab., 1944——; in charge lab. photosynthesis Inst. Plant Physiology, USSR Acad. Sci., 1938-47. Mem. USSR Acad. Sci. (corr.). Author: Plant Physiology, 1936; General Course in the Systematization of Plants, 1937; Plant Anatomy, 1939; Light and Moisture in the Life of our Trees, 1946.

IVANOV, Vadim Nikolaevich, Russian physician; b. USSR, Apr. 30, 1892; grad. Med. Faculty, Kiev (USSR) U. (now Kiev Bogomolets State Med. Inst.), 1916. From intern to lectr. Kiev Bogomolets State Med. Inst., 1918-34; head, chair fac. therapy 2d State Kiev State Med. Inst., 1933-41; head, chair therapy Kiev State Med. Inst., 1944-50, head, chair hosp. therapy clinic, 1951-57; head dept. clin. physiology Bogomolets Inst. Physiology USSR Acad. Sci. 1953——; head, chair fac. therapy clinic Kiev Bogomolets State Med. Inst., 1958——. Mem. USSR Acad. Med. Sci., Ukrainian Acad. Sci., All-Union, Ukrainian (dep. chmn.) socs. oncologists; Ukrainian Soc. Therapeutists, All-Union Sci. Soc. Therapeutists (hon.). Author: The Motions of the Empty Stomach in Healthy Persons, 1926; Rejection of the Contents of the Duodenum in to the Stomach Outside the Digestive Process, 1926; Clinical Forms of Gastric Cancer, 1932; Acidity of the Gastric Contents and the Secretions in Cases of Gastroduodenal Ulcers, 1935; Diverticulae of the Stomach, 1947; The Diagnosis of Lung Cancer, 1949; The Diagnostical and Clinical Value of Tomofluorography of the Lungs, 1950; Forty Years of Progress in Internal Medicine in the Ukrainian SSR, 1957. Works deal with clin. physiology and pathology and diseases of digestive organs. Address: Institut fiziologii, Shevchenko 18, Kiev, Ukrainian SSR, USSR.

IVANOV-SMOLENSKY, Anatoliy Georgievich, Russian psychiatrist, pathophysiologist; b. St. Petersburg, May 16, 1895; grad. Petrograd Mil. Med. Acad., 1917; D.Med. Sci., 1921. With Petrograd Psychoneurol. Inst. and Psychiat. Clinic, Petrograd. Mil. Med. Acad., 1919-21; founder (with Bekhterev) USSR Psychiat. Inst., 1920; asso. Mil. Med. Acad., 1921-25; founder USSR's 1st chair physiology and pathology of higher nervous activity at Gertsen Pedagogical Inst., 1924; with Physiol. Inst., USSR Acad. Med. Sci., head chair physiology and pathology of nervous activity Leningrad Pedagogical Inst., 1925-31; founder, head psychiat. clinic Pavlov's Lab., 1931-45; head Moscow dept. Pavlov Inst. Evolutionary Physiology and Pathology of Higher Nervous Activity, 1945-50; dep. dir. Inst. Higher Nervous Activity, USSR Acad. Med. Sci., 1950-52, dir., 1952-57. Recipient Pavlov prize, 1941, Stalin prize, 1950. Mem. USSR Acad. Med. Sci. Author: Methods of Studying Conditioned Reflexes in Man, 1928, 2d edit., 1933; Natural Science and the Science of Human Behavior, 1929; Basic Problems in the Pathophysiology of Human Higher Nervous Activity, 1933; Basic Problems in the Therapy of Schizophrenia, 1941; Outline Pathophysiology of the Central Nervous System, 1949, 2d edit., 1952; The Teachings of I. P. Pavlov and Pathophysiology, 1952; An Objective Study of the Functioning and Interaction of the Signal Systems of the Brain, 1963. Mem. editorial council S. S. Korsakov Jour. Neuropathology and Psychiatry. Research and over 200 publs. (including monographs) on pathophysiology, therapy of nervous disturbances caused by contusions and concussions, 1941-45, psychiatry, physiology and pathology of higher nervous activity in animals and man. Address: Journal of Neuropathology and Psychiatry, ulitsa imeni 1905 goda 8, Moscow, D-22, USSR.

IVANOVICS, György, Hungarian med. microbiologist; b. Budapest, Hungary, June 11, 1904; s. Lajos and Irma (Lacza) I.; M.D., U. Budapest, 1928; Privat Docent, U. Szeged (Hungary), 1936, D.Med. Scis. 1951; m. Sári Kompaszky, Aug. 3, 1933; 1 dau., Georgiana (Mrs. László Sáfrány). Vol. dept. pathology U. Budapest, 1927-28; faculty U. Szeged, 1929-34, 35——, prof. gen. path. bacteriology, 1940-50, prof. microbiology, 1950——, dir. Inst. Microbiology, 1950——, dean Med. Faculty, 1948-49. Vis. lectr. dept. bacteriology Harvard Med. Sch., Boston, 1960; Macfarlane prof. exptl. medicine Glasgow (Scotland) Royal Infirmary, 1965-66. Rockefeller fellow Johns Hopkins Sch. Hygiene, Balt., 1934-35. Recipient Kossuth prize, 1948, 52; Semmelweis medal, 1964. Mem. Hungarian Acad. Scis., Soc. Gen. Microbiology (Gt. Britain), Soc. Hungarian Microbiologists, Soc. Hungarian Physiologists. Author: Chemotherapie der bakteriellen Infektionen, 1944; Viral and Rickettsial Diseases of Men, 1953; (with Z. Alfödi, K. Rauss), Medical Microbiology, 1960; also numerous articles. Research on infection and immunity to anthrax, biology of B. anthracis, chemotherapeutics of sulfanilamides and related compounds, viruses; discovered D-glutamic acid polypetide, mode of action of sodium salicylate, megacin. Home: 3 Bodogh Janos U, Szeged, Hungary.*

IVANYI, John, Hungarian physician; b. Békéssámson, Hungary, Dec. 12, 1924; s. John and Elisabeth (Csete) I.; M.D., U. Szeged (Hungary), 1949; m. Ella Tóth, July 1, 1950; children—John, Adalbert, Tibor. Asst. physician I. Clinic Medicine, Szeged, 1949-55; head physician Town Hosp., Jászberény, Hungary, 1955-60; head physician dept. medicine, dept. for infectious diseases County Hosp., Guyla, Hungary, 1960——; primarius internal medicine County of Békés, 1966——, leader drs. postgrad. course, 1966——; cons. dietetic physician, 1960——; head Diabetic Ambulance, 1960——. Mem. Nat. Sci. Assn., European Diabetic Assn., Hungarian Int. Assn., Hungarian Phys. Assn. for Infectious Disease. Research, numerous publs. on diabetes and its care, liver and stomach diseases, various clin. pharmacological problems. Home: 36 Béke sugárut, Gyula. Office: County Hosp.:1 Semmelweis Str., Gyula, Hungary.*

IVERSEN, Johannes, Danish geologist; b. Sönderborg, Dec. 27, 1904; s. Hans and Anna (Asmussen) I.; Ph.D., U. Copenhagen; D.honoris causa, U. Upsala; m. Aase F. Enf, 1939; 1 child, Mette. Asst. geologist U. Copenhagen, 1931, geologist, 1938, prin. geologist, instr., 1942. Mem. Soc. Phytogeographica (Sweden), Soc. Zoologica Botanica Fennica, Royal Acad. Sci. and Letters, Finnish Soc. for Flora and

Fauna, Bot. Soc. Netherlands. Author: Biologische Pflanzentypen als Hilfsmittel in der Vegetationsforschung, 1936; Landnam i Danmarks Stenalder, 1936; Textbook of Pollen Analysis, 1950. Home: Rörskaersvej 10, Copenhagen (Gen A ofte). Office: Radhusvej 36, Charlottenlund, Denmark.

IVERSEN, Olav Hilmar, Norwegian pathologist; b. Horten, Norway, Mar. 24, 1923; s. Hilmar and Agnes (Andersen) I.; student medicine U. Copenhagen (Denmark), 1945-48, U. Bergen (Norway), 1948-51; M.D., U. Oslo (Norway), 1962; m. Ella Bredvei, July 17, 1949 (dec. June 1958); children—Jon, Tore (dec. July 1966), Sven, Anne; m. 2d, Ulla Marianne Ising, Feb. 20, 1960; children—Lena, Anders; Gen. practice medicine, 1954-56; research fellow Inst. Gen. and exptl. Pathology, U. Oslo, 1957-59, faculty, 1959—, prof., head inst., 1964—; head dept. surg. pathology State Hosp. Norway, Oslo, 1964—. Eleanor Roosevelt Cancer Research fellow, Institut de Anatomie Pathologique, Liège, Belgium, 1963. Mem. Internat. Acad. Pathology, Brit. Assn. Cancer Research. Author: (with A. Evensen) Experimental Skin Carcinogenesis in Mice, 1962; (with R. Bjerknes) Kinectics of Epidermal Reaction to Carcinogens—A Cybernetic Model Analysis, 1963; also articles. Research on early effects of carcinogens and other skin irritants on epidermal cell population in hairless mice; described tetrazolium test, changes in epidermal population kinetics introduced by carcinogens; interpretation of cell population kinetics in epidermis using math. model and electron computers; research on prins. normal growth control of epidermis. Home: 66 Hvikveien, Bekkestua, Norway. Office: Rikshospitalet, Oslo 1, Norway.*

IVES, Frederic Eugene, Am. inventor; b. Litchfield, Conn., Feb. 17, 1856; s. Hubert Leverit and Ellen A. (Beach) I.; ed. pub. schs., Litchfield, Norfolk, Newtown, Conn.; m. Mary Elizabeth Olmstead, June 15, 1879 (dec. 1904); 1 son, Herbert Eugene; m. 2d, Mrs. Margaret Campbell Cutting, Nov. 15, 1913 (dec. 1928). In charge of photography lab. Cornell U., 1874-78; realized first practically successful process of orthochromatic photography, and invented 1st practically successful process of half-tone photo-engraving, 1878; invented the half-tone photo-engraving process still used, 1886; (inscribed gold testimonial Internat. Photo-Engravers' Assn., 1911; gold medals United Typothetae of America, Printing House Craftsman, Poor Richard Club, 1926); began expts. in color photography on the trichromatic principle, 1878, culminating in the three-color printing process in typographic press, and in Kromskop, Tripak, Hicrom and 1931 Polychrome processes; and a successful process for moving pictures in natural colors; invented a single-objective variety of binocular microscope. Recipient Cresson gold medal, Franklin Inst., Phila.; spl. gold medal, Photog. Soc. Phila., Progress medal Royal Photog. Soc. for work in color photography; Science medal, Royal Photo. Society, London, and Scott Legacy medal, Franklin Inst., for The Parallax Stereogram; Rumford medal, Am. Acad. Arts and Scis. for inventions in color photography and photoengraving; other medals by scientific socs. for various inventions and discoveries. Author: The Autobiography of an Amateur Inventor, 1928. Died Phila., May 27, 1937.

IVES, Herbert Eugene, Am. physicist; b. Phila., July 31, 1882; s. Frederic Eugene and Mary Elizabeth (Olmstead) I.; B.S., U. Pa., 1905; Ph.D., Johns Hopkins, 1908; hon. Sc.D., Dartmouth and Yale, 1928, Pa., 1929; m. Mabel Agnes Lorenz, Nov. 14, 1908; children—Ronald Lorenz, Barbara Olmstead (Mrs. Charles Beyer), Kenneth Holbrook. Asso. with Ives Kromskop Co., Phila., 1898-1901; physicist Bur. of Standards, Washington, 1908-09, Nat. Electric Lamp Assn., Cleve., 1909-12, United Gas Improvement Co., Phila., 1912-18, Bell Telephone Labs., N.Y., 1919-47. De Forest lectr. Yale, 1928; Lowell lects., Boston, 1932; Thomas Young orator Phys. Soc., London, 1933; Traill-Taylor meml. lectr. Royal Photog. Soc., 1933. Sect. head NDRC, 1941-46. Fellow A.A.A.S. (v.p. Sect. B, 1938), Am. Inst. E.E.; mem. Am. Philos. Soc., Am. Phys. Soc., Optical Soc. Am. (pres. 1924-25, asso. editor jour.), Illuminating Engring. Soc. (v.p. 1911-12), Am. Astron. Soc., Franklin Inst., Am. Numis. Soc. (pres. 1942-47), Nat. Acad. Scis. Phys. Soc. of London, Phi Beta Kappa, Sigma Xi; corr. mem. British Illuminating Engring. Soc. Recipient medals Franklin Inst. for diffraction color photography, artificial daylight and studies of Welsbach mantle; John Scott medal and award, 1927, for electric telephotography and television; medal of the Optical Society for distinguished work in optics, 1937, U. S. Medal for Merit, 1948, Rumford medal Am. Acad. Arts and Sci., 1951. Author: Airplane Photography, 1920; also articles in sci. jours., Ency. Brit. Inventor apparatus for testing visual acuity, various photometric instruments, illuminating devices, means for producing artificial daylight, relief pictures, electrical photo-engraving, apparatus for transmission of pictures over telephone lines, 1924; in charge of exptl. and devel. work culminating in first demonstration of television by wire and radio, 1927, television in color, 1929, 2-way television in conjunction with telephone, 1931; developed scientific trichromatic palette for artists' use; gave 1st exptl. demonstration of change of rate of moving atomic clocks (crucial expt. for theories of Larmor, Lorentz and Einstein's theory of relativity). Died Nov. 13, 1953.

IVES, James Edmund, physicist; b. London, Eng., Sept. 19, 1865; s. James Thomas Bostock and Mary Collins (Johns) I.; student U. Pa., 1888-89, Harvard, 1894, U. Cambridge (Eng.), 1896; Ph.D., Clark U., Mass., 1901; m. Georgiana Luvanne Stone, June 25, 1903; 1 dau., Elizabeth Laura (Mrs. Ives Lowe). Asst. curator Acad. Natural Scis., Phila., 1887-93; instr. physics Drexel Inst., 1893-1901, lectr., research asso. in physics, 1912-21; instr. physics, U. Cin., 1901-03, asst. prof. physics, 1905-09, asso. prof., 1909-12; scientific expert with the DeForest Wireless Telegraph Co., N.Y., 1903-05; physicist USPHS with Office of Industrial Hygiene and Sanitation, 1921-31, sr. physicist, 1931-36. Recipient silver medal, St. Louis Expn., 1904, for work in wireless telegraphy. Fellow Am. Acad. Arts and Scis.; mem. Acad. Natural Science Phila., Am. Phys. Soc., Illuminating Engring. Soc., Optical Soc. Am., Washington Acad. Scis. Author: An Annotated List of Experiments in Physics, 1912; also many papers in sci. jours. and rev. Made 1st broad study of distn. sunlight and ultraviolet radiation (with W. A. Gills). Died Jan. 2, 1943.

IVORY, Sir James, Brit. mathematician; b. Dundee, Scotland, Feb. 17, 1765; s. James Ivory; ed. St. Andrews U., Edinburgh (Scotland) U. Prof. math. Royal Mil. Coll., Marlow, Eng., 1805-19. Fellow Royal Soc., 1815 (Copley medal 1814, Royal medal 1826, 39); mem. French Acad. Scis., 1820, Royal Acad. Berlin, Royal Soc. Göttingen (Germany). Author: Theory of Astronomical Refractions, 1839; also 15 papers in Philos. Trans. Completely resolved problem of attractions for all classes of homogeneous ellipsoids with Ivory theorem, 1809; skill in applying infinitesimal calculus to phys. investigations equaled that of Laplace, Lagrange, Legendre; determination of orbit of comets; theory of astron. refractions. Died Sept. 21, 1842.

IVY, Andrew Conway, Am. physiologist; b. Farmington, Mo., Feb. 25, 1893; s. Henry McPherson and Cynthia (Smith) I.; A.B., Cape Girardeau (Mo.) State Tchrs. Coll., 1913; B.S., U. Chgo., 1916, M.S., 1917, Ph.D., 1918; M.D., Rush Med. Coll., 1922; LL.D., Loyola U., 1950; D.Sc., U. Neb., 1947, Grinnell Coll., 1947, Boston U., 1948; D.Sch., Hastings Coll., 1951, Coe Coll. (Ia.), 1952; m. Emma A. Kohman, Dec. 24, 1919; children—John Henry, William Harvey, Andrew C., Horace Kohman, Robert Emerson. Faculty, U. Chgo., 1916-19, 23-25, Loyola U., 1919-23, Northwestern U., 1925-46; head dept. clin. sci. U. Ill., 1946-62, v.p., 1946-53, Distinguished prof. emeritus, 1962—; research prof. biochemistry Roosevelt U., 1962—. Exec. dir. Nat. Adv. Cancer Council, 1944-51; cons., witness Nuremberg Trials, 1946-47; pres. Internat. Com. for Prevention Alcoholism, 1951-63; co-founder, pres. Nat. Conf. Educators to eliminate discrimination in higher edn., 1954. Recipient Carpenter award Am. Roentgen Ray and Radium Soc., 1946; Distinguished Service award Am. Congress Phys. Medicine, 1947; Distinguished Service award Nat. Soc. Med. Research, 1953; Pope Leo XIII award, 1953; Eloy Alfaro award, 1953. Mem. A.C.P., Am. Soc. Internal Medicine, Am. Physiol. Soc., Am. Inst. Nutrition, Am. Gastroenterology Assn. Author: (with M. I. Grossman, Wm. H. Bachrach) Peptic Ulcer, 1950. Research, numerous publs. on hormone causing gall bladder evacuation, gastric secretion, bile secretion, gallstones, peptic ulcer, cardiac pain, body balance of cholesterol, minimal water requirements, desalination of sea water, anti-flash burn cream, anti-cancer substance in body tissues, poison gases, resuscitation from drowning and co-asphyxia, artificial respiration, cancer producing properties of overheated fats. Home: 8158 S. Merrill St., Chgo. 60617. Office: 178 W. Randolph, Chgo. 60601.*

IVY, Robert Henry, plastic surgeon; b. Southport, Eng., May 21, 1881; s. Robert Sutcliffe and Annie Edith (Cryer) I.; D.D.S., U. Pa., 1902, M.D., 1907; D.Sc., 1954; m. Norma C. Crossland, June 19, 1912; children—Cynthia Thompson, Robert Henry, Eleanor Anne, Peter Cryer. Practiced medicine in Phila., 1907-15, Milw., 1915-17; prof. maxillo-facial surgery, U. Pa., 1919—, prof. plastic surgery, 1944-51, emeritus, 1951—; chief cleft palate sect. Pa. Health Dept., 1949-65. Mem. com. NRC 1940-46; cons. office surgeon gen. U. S. Army; mem. bd. hon. consultants, Army Med. Library. Recipient certificate of appreciation, War Dept., 1946, U. Pa. Alumni Award of Merit, 1946. Fellow A.C.S.; mem. Am. Soc. Plastic and Reconstructive Surgery (hon.), A.M.A., Am. Surg. Assn., Am. Dental Soc., Soc. Consultants to Armed Forces, Am. Assn. Plastic Surgeons, Loyal Legion; hon. mem. several fgn. profl. assns., Deutsche Akad. der Naturforscher. Author: Applied Anatomy and Oral Surgery, 1911; Applied Immunology (with B. A. Thomas), 1916; Essentials of Oral Surgery (with V. P. Blair), 1923; Fractures of the Jaws (with L. Curtis), 1931; A Link With the Past (autobiography), 1962. Editor-emeritus Plastic and Reconstructive Surgery. Home: 104 Dalton Rd., Paoli, Pa. 19301.*

IWANCIOW, Bernard Louis, Am. phys. chemist; b. Milw., Aug. 8, 1922; s. Michael and Mary (Kurowski) I.; B.S., U. Wis., 1947, Ph.D., 1950; m. Kathleen Gibbons, Aug. 17, 1946; children—Michael George, Stephen Franklin, Richard Bernard. Group leader ballistics Rohm & Hass Co., Redstone Arsenal, Huntsville, Ala., 1950-54, sr. chemist, 1954-60; sect.

chief ballistics research United Tech. Center, Sunnyvale, Cal., 1960——. Mem. Am. Inst. Aeros. and Astronautics. Research, publs. in interior ballistics rockets. Home: 1188 E. Vanderbilt Ct., Sunnyvale 94-87. Office: 1050 E. Arques St., Sunnyvale, Cal. 94088.*

IWATSUKI, K., Japanese anesthesiologist; b. Takahama, Aichi Prefecture, Japan, Nov. 13, 1913; s. Kinzaburo and Koma (Kamiya) I.; M.D., Tokyo U., 1940, Ph.D., 1946; m. Sadako Nakamura, Nov. 3, 1939; children—Naofumi, Kazuhiko, Noriyuki. Chief surgery Chigasaki (Japan) City Hosp., 1945-50; asst. prof. surgery, anesthesia Shinshu U. Sch. Medicine, Matsumoto, Japan, 1950-58; mem. dept. anesthesiology Boston City Hosp., Boston, 1954; prof. anesthesiology Tohoku U. Sch. Medicine, 1958——, chief dept. anesthesiology Tohoku U. Hosp., Sendai, Japan, 1958——. Mem. Japan Soc. Anesthesiology (pres. 1960), Japan Soc. Surgery, Japan Soc. Thoracic Surgery. Author: (with N. Hoshiko) Introduction to Modern Anesthesiology, 1952; Basic Science in Anesthesiology, 1956; Anesthesiology, 1959; (with H. Yamamura) Text Book of Anesthesiology, 1963; (with T. Saito) Halothane Anesthesia, 1963; also numerous articles. Editor: Japanese Jour. of Anesthesiology, Far East Jour. Anesthesia. Introduction of modern anesthesia and its devel. in Japan; research of muscle relaxants in anesthesiology. Home: 175 Kita-8-Bancho, Sendai. Office: 85 Kita-4-Bancho, Sendai, Japan.*

IYA, K(rishnaswami) K(illara), Indian microbiologist; b. Bombay, India, June 26, 1920; s. K. L. N. and Saradabai Iya; I.D.D. with honors, Nat. Dairy Research Inst., Bangalore, India, 1939; B.Sc., St. Xavier's Coll., Bombay, 1942; A.I.I.Sc., Indian Inst. Sci., Bangalore, 1945; Ph.D., U. Wis., 1948; m. Sharada Krishnaswami, Apr. 28, 1944; children—Arun Kilara, Gurunath Kilara, Sandhya. With Nat. Dairy Research Inst., Bangalore, Karnal, India, 1948-65; dairy devel. adviser Ministry Food and Agr., Govt. India, New Delhi, 1965-66; dep. dir. gen. animal scis. Indian Council Agrl. Research, New Delhi, 1966——. Chmn., Bd. Dairy Edn., Govt. India, 1960-65; mem. Nat. Dairy Devel. Bd., 1965-66; mem. expert panel on milk quality FAO, UN, 1963-65. Mem. Indian Dairy Sci. Assn. (pres.), Assn. Food Technologists (pres.). Author: Handbook on Manufacture of Western Dairy Products in India. Research, publs. on use of Triphenyl Tetrazolium bromide for dye reduction tests of milk, dried starter cultures for Dahi making in India, test for detecting mixtures of cow milk with buffalo milk, fistulation of 4th stomach of calf for tapping rennet from living calves, mfr. of bacterial rennet for use in cheese making. Home: C-11-32, Tilak Marg, New Delhi. Office: Krishi Bhavan, Dr. Rajendra Prasad Rd., New Delhi, India.*

IYANAGA, Shokichi, Japanese mathematician; b. Tokyo, Japan, Apr. 2, 1906; D.Sc., Tokyo U., 1929; m. Sumiko Kikuchi. Became asst. prof. sci. dept. Tokyo U., 1935, prof., 1942; also dir. Japan Math. Acad. Mem. Internat. Math. Union (exec. mem.), Japan Sci. Council. Author: Theory of Free Groups; The Fundamental Conception of Modern Mathematics. Research and publs. on phase mathematics, basic theory math., theory of integral numbers.

IZOTOV, Aleksandr Aleksandrovich, Russian geodesist; b. 1907; grad. Moscow Inst. Geodetic, Aerial Survey and Cartography Engring., 1932. Instr., Moscow Inst. Geodetic, Aerial Survey and Cartography Engring., 1932-46, dep. dir., 1946-48, prof., 1951—; dir. lab. higher geodesy Central Research Inst. Geodesy, Aerial Survey and Cartography, 1939——. Recipient Stalin prize, 1952. Mem. All-Union Astron. and Geodetic Soc. (v.p. 1960——). Author: Triangulation Accuracy Evaluation, 1936; Theory of Sequential Solution of Conventional Equation Systems, 1937; New Original Geodetic Dates of the USSR, 1948; The Shape and Dimensions of the Earth According to Present Data, 1950; The Theory of Determining Original Geodetic Dates, 1953; Geodetic Methods of Studying Movements of the Earth's Crust, 1963. Research on original geodetic dates of USSR; his definition of earth's shape and dimensions forms basis of Soviet geodetic works (Krasovsky's ellipsoid). Address: Moscow Inst. Geodetic Aerial Survey and Cartography Engring., Gorokhovsky p. 4, Moscow, USSR.

IZRAELSON, Zigfrid Isidorovich, Russian hygienist; b. 1893; grad. Med. Faculty, Yurev U., 1916; D. Med. Sci., 1940. Sr. asso. Moscow Inst. Labor Protection, 1929-44; asst., later lectr. dept. hygiene 1st Moscow Med. Inst., 1932-44; prof., head chair labor hygiene Sechenov 1st Moscow Med. Inst., 1944-66. Mem. Commn. on Maximum Permissible Concentrations of Poisons in Atmosphere, Commn. on Air Pollution, Commn. for Combatting Silicosis. Mem. All-Union Hygiene Soc. (bd. mem.). Author: Labor Hygiene in the Basic Chemical Industry, 1940; co-author, editor: The Toxicology of Rare Metals, 1963, 67. Mem. editorial bd. Labor Hygiene and Occupational Diseases. Research and numerous publs. on labor hygiene in chem. industry, regulation of toxic substances in atmosphere in factory bldgs., toxicology of new metals and alloys. Address: Sechenov 1st Moscow Med. Inst., B. Pirogovskaya ulitsa 2-6, Moscow, USSR.

J

JAAG, Otto, Swiss biologist; b. Beringen, Switzerland, Apr. 29, 1900; s. Otto and Elisabeth (Selwyn) J.; Dr.rer.nat., U. Geneva, 1929; Dr.h.c. for tech. scis., Inst. Tech. Stuttgart, 1959; m. Elisabeth Schenk, Oct. 10, 1940; children—Dorothe, Tobias, Regina, Elisabeth, Christopher. Faculty, Swiss Fed. Inst. Tech., Zurich, 1929—, prof. biology, 1963—. Dir. Swiss Fed. Inst. Water Supply, Sewage Purification and Water Pollution Control, 1952—. Mem. Linnean Soc. London, Finnish Soc. Fauna and Flora, Swiss Soc. Natural Scis., Swiss Assn. for Water and Air Protection, Internat. Assn. for Water Pollution Research, others. Author: Coccomyxa Schmidle, 1933; Untersuchungen über die Vegetation und Biologie der Algen des nackten Gesteins in den Alpen, im Jura und im schweizerischen Mittelland, 1945; also numerous articles. Research in algology, plant physiology, plant pathology, limnology, waste water treatment. Home: 37 Keltenstrasse, 8044 Zurich, Switzerland.*

JAAP, Robert George, geneticist, animal breeder; b. Thedford, Ont., Can., Oct. 27, 1905; s. George Rowand and Sarah Ellen (Close) J.; B.S., U. Sask., 1927, B.S.A., 1929; M.S., U. Wis., 1930, Ph.D., 1933; m. Edna F. Jaap, July 3, 1934. Came to U. S., 1929, naturalized, 1936. Research asst. U. Wis., 1929-33; hatchery mgr. Swift & Co., Waterloo, Ia., 1933-34; prof. genetics and animal breeding Okla. State U., 1934-46, Ohio State U., Columbus, 1946—. Mem. panel on animal breeding and climatology FAO, 1962—. Fulbright research scholar, Edinburgh, Scotland, 1951-52. Fellow Ohio Acad. Sci., A.A.A.S.; mem. Poultry Sci. Assn., Worlds Poultry Sci. Assn., Genetics Soc. Am., Sigma Xi, Gamma Sigma Delta (Outstanding Profl. award 1966). Research, numerous publs. on genetics of plumage coloration in ducks, turkeys and chickens, genetics of body conformation in turkeys and chickens, use of estrogens in fattening, developer breeding and selection methods for improvement growth and egg prodn. in chickens, discoverer role of bursa of fabricins in chickens. Home: 4435 Shields Pl., Columbus, O. 43214.*

JABIR IBN HAIYAN (Geber), Islamic alchemist; physician; flourished in Kufa, circa 776. Alleged author: Book of Royalty; Book of the Kingdom; Little Book of the Balances; Book of Mercy; Book of Concentration; Book of Eastern Mercury; total of some 500 treatises attributed to him, but generally agreed that most date from 9th-12th centuries. Works allegorical and mystical in content and show influence of Pythagorean and neo-Platonic thought; described animistic universe; developed sulphur-mercury theory of metals. Known to Latin West in middle ages as Geber; works appearing in Latin under this name probably date from early 14th century and include Summa perfectionis magisterii; De investigatione perfectionis; De inventione veritatis; others. Latin works systematic; describe properties of metals and chemical processes in considerable detail; are among most advanced texts in field for period; works translated into German, English, and French; frequently printed 16th-18th centuries.

JABLOCHKOV (or JABLOCHKOFF), Paul, physicist; b. Saratof, Russia, 1847; ed. sch. engring., St. Petersburg, mil. sch. electrotechnics; chief engr. telegraph service Moskow-Kursk R.R.; resided chiefly in Paris, France; inventor arc lamp (Jablochkov light), 1877, which he used in London, then as theatrical device for opera, Paris, 1878; inventor devices for study of electricity. Died 1894.

JABLONSKI, Alexander, Polish physicist; b. Woskresenowka, Ukraine, Feb. 26, 1898; s. Wladyslaw and Maria (Bilinska) J.; Ph.D., U. Warsaw, 1930, docent exptl. physics, 1934; m. Wiktoria Gutowska, Nov. 14, 1922; children—Halina Borecka, Danuta Frackowiak. Asst., Warsaw U., 1925-38, Vilno U., 1938; lectr. Polish Sch. Medicine, Edinburgh, Eng., 1943-45; adj. Warsaw U., 1945; prof. physics Nicholas Copernicus U., Torun, Poland, 1946—. Mem. Polish Phys. Soc., Polish Acad. Scis. Publs. on studies of photoluminescence, pressure broadening of spectral lines, Franck-Condon principle. Home: 38 Moniuszki, Torun, Poland.*

JABONERO, Vicente, Spanish anatomist; b. Plasencia, Spain, July 12, 1914; s. Marcos and Maria (Sanchez) J.; M.D., U. Madrid; m. Antonia Avila, Feb. 24, 1940; children—Rosa-Maria, Mari-Paz, Mari-Luz, Nicolas. Prof. histology Med. Faculty, U. Valladolid, sect. morphopathology C.S.I.C., 1942-43, prosector, 1943-46, sec. anatomy, 1946-59; with Cajal Inst., Madrid, 1959-61, sect. autonomic nervous system, 1961—; with Mil. Path. Hosp. Gomez Ulla, Madrid, 1961—; dir. Mil. Hosp., Oviedo, Spain, 1961—. Mem. Anat. Soc. Spain, Soc. Histopathology (Paris), Anatomische Gesellschaft, Soc. Int. Neurobiol. comparée, Societé Internat. Syst. Neurovégétatif. Research, publs. on anat. orgn. of vegetative periphery, synapses of autonomic nervous system, vegetative innervation of organs. Home: 6 Hnos M.Pidal, Oviedo. Office: Dir. Hospital Militar, Oviedo, Spain.*

JABOULAY, Mathieu, French surgeon; b. St.-Genis-Laval, France, 1860; doctoral thesis accepted, also permitted to teach anatomy, 1886; became surgeon of hosps., 1892; named prof., clinic of surgery, 1902. Author: Chirurgie des centres nerveux, des viscères et des membres, 1901; Chirurgie du grand sympathique et du corps thyroïde, 1901; Myxosporidies et tumeurs de l'homme, 1903; Leçons de clinique chirurgicale professées à l'Hôtel-Dieu de Lyon en 1902-1903, 1904; Recherches sur les tumeurs épithéliales, 1909. Developed techniques for difficult operations, especially surgery of sympathetic nerve; 1st to attempt resection in treatment of Basedow's malady; cut rami communicantes to relieve pelvic pain, 1899; performed cervical sympathectomy for exophthalmic goiter, 1900. Died Melun, France, 1913.

JACCARINO, Vincent, Am. physicist; b. Bklyn., May 12, 1924; s. Louis and Eva (Golde) J.; B.S., Bklyn. Coll., 1948; Ph.D., Mass. Inst. Tech., 1952; m. Lies Nijssen, July 3, 1965; children—Gina Ellen, Judith Carol. Research asso. Mass. Inst. Tech., 1952-54; mem. tech. staff Bell Telephone Labs., 1954-63, head solid state physics research dept., 1963-66; prof. physics U. Cal., Santa Barbara, 1966—. Fellow Am. Phys. Soc. Contbr. to Treatise On Magnetism, 1965. Research, numerous publs. on electromagnetic moments of nuclei and electronic properties of solids; discovered 1st nuclear magnetic octupole moment.

JACCHIA, Luigi Giuseppe, astronomer; b. Trieste, Italy, June 4, 1910; s. Giuseppe Ulderico and Beatrice (Prandina) J.; Ph.D., U. Bologna (Italy), 1932. Came to U. S., 1939, naturalized, 1944. Faculty, U. Bologna, 1928-38; research asso. Harvard Obs., Cambridge, Mass., 1939—, lectr. astronomy, 1962—; research asso. elec. engring. Mass. Inst. Tech., 1949-53; physicist Smithsonian Astrophys. Obs., 1956—. Mem. Internat. Astron. Union, Internat. Union Geodesy and Geophysics, Am. Astron. Soc., Sigma Xi. Author: Le Stelle Variabili, 1933; (with L. Campbell) The Story of Variable Stars, 1941. Contbr. numerous articles to profl. jours. Research photog. meteors; pioneer progressive fragmentation meteors, upper atmosphere variations and their relation to solar and geomagnetic activity. Home: 6 Washington Av., Cambridge 02140. Office: Smithsonian Astrophys. Obs., 60 Garden St., Cambridge, Mass. 02138.*

JACCOUD, François-Sigismond, physician; b. Geneva, Switzerland, 1830; studied medicine in Paris (France); degree in medicine, 1859; joined the Faculty of Medicine, 1862; named prof. internal pathology, 1876; held a chair of clin. medicine Faculty of Medicine. Mem. Acad. Medicine (perpetual sec.), De l'organisation des Facultés de médecine en Allemagne. Author: Leçons cliniques, 1867, 72, 85; Traité de pathologie interne, 1869-72; Du froid comme cause de penumonie, 1887. Directed the editing of Nouveau dictionnaire de médecine et de chirurgie pratiques. Studied causes of albuminuria, aortic dilation; fever in tuberculous meningitis of adults named Jaccoud's dissociated fever after him. Died Paris, 1913.

JACKMAN, Raymond Joseph, Am. physician; b. Emmetsburg, Ia., May 16, 1906; s. Joseph and Mary (Jennings) J.; M.D., State U. Ia., 1930; M.S. (Mayo Found. fellow), U. Minn., 1938; m. Lois Hovenden, Jan. 18, 1934; children—Colette (Mrs. Walter E. Miller), Roger J., Steven J. Practice medicine, specializing in proctology, Rochester, Minn., 1935—; faculty Mayo Grad. Sch. U. Minn., 1938—, head dept. proctology, 1954—, prof. proctology, 1961—; pres. Minn. Bd. Health, 1963—. Mem. Am. Proctologic Soc. (v.p.), Sigma Xi. Author: Lesions of the Lower Bowel, 1958. Contbr. numerous articles to profl. jours. Research, publs. on diseases of rectum and colon. Home: 913 9th Av. S.W. Office: Mayo Clinic, Rochester, Minn. 55902.*

JACKSON, Benjamin Daydon, Brit. botanist; b. London, Eng., Apr. 3, 1846; ed. pvt. schs.; Ph.D. (hon.), Upsala (Sweden) U.; m. Jane Hunt. Sec. to departmental com. Her Majesty's Treasury on Bot. Work, 1900-01; curator Linnean Collections, from 1926. Mem. Linnean Soc. (sec. 1880-1902, gen. sec. 1902-26). Author: Life of John Gerard, 1877; Life of Dr. William Turner, 1878; Guide to the Literature of Botany, 1881; Vegetable Technology, 1882; Index Kewensis, 1893-95, (with Thomas Durand), supplement, 1901-06; Glossary of Botanic Terms, 1900, 3d edit., 1916; Darwiniana, 1910; Index to the Linnean Herbarium, 1912; Catalogue of Linnean Specimens of Zoology, 1913; Notes on a Catalogue of the Linnean Herbarium, 1922; Linnaeus: the Story of his Life, 1923. Editor: Pryor's Flora of Herts, 1887; New Genera and Species of Cyperaceae (C. B. Clarke), 1908; Illustrations of Cyperaceae (C. B. Clarke), 1909. Contbd. Life of George Bentham to English Men of Science. Died Oct. 12, 1927.

JACKSON, Charles Ian, Brit. geographer; b. Keighley, Eng., Feb. 11, 1935; s. Harry S. and Nellie (Crabtree) J.; B.A., London (Eng.) Sch. Econs., London U., 1956; M.Sc., McGill U., Montreal, Que., Can., 1959, Ph.D., 1961; m. Margaret Cochrane Storrie, July 10, 1963. Meteorologist Canadian Govt. expdns. to Ellesmere Island, 1957-58, 61, 66; lectr. geography London Sch. Econs., 1959—. Recipient Evan Durbin prize Inst. Econ. Affairs, 1966. Fellow Royal Geog. Soc., Royal Meteorol. Soc. (Darton prize 1963); mem. Hakluyt Soc. (council 1967—), Inst. Brit. Geographers, Geog. Assn., Am. Geophys. Union, Arctic Inst. N.Am. Research and publs. on arctic meteorology especially at Lake Hazen, Ellesmere Island, econ. aspects of climate and water supply in U.K., hist. geography of Arctic. Home: 50 Berryhill, London S.E.9., U.K.*

JACKSON, Charles Loring, Am. chemist; b. Boston, Apr. 4, 1847; s. Patrick Tracy and Susan Mary (Loring) J.; A.B., Harvard, 1867, A.M., 1870; studied under Bunsen in Heidelberg and under Hofmann in Berlin, 1873-75. Asst. in chemistry, Harvard, 1868-71, asst. prof., 1871-81, prof., 1881-99, Erving prof. chemistry, 1899-1911, emeritus, from 1911. Mem. Am., German chem. socs., A.A.A.S., Brit. Assn. for Advancement Sci., Nat. Acad. Scis. Contbr. numerous articles profl. jours. Contbns. include 1st new organic compound in Harvard lab. (parabrombenzyl bromide), significant work on exptl. preparation of halogen substitution products of aromatic hydrocarbons and later quinones. Died 1935.

JACKSON, Charles Thomas, Am. chemist; b. Plymouth, Mass., June 21, 1805; s. Charles and Lucy (Cotton) J.; M.D., Harvard, 1829; studied at Sorbonne and École des Mines, Paris, France, 1829; m. Susan Bridge, Feb. 27, 1834, 5 children. Became interested in mineralogy after visiting and collecting mineral specimens in Nova Scotia, circa 1828; travelled through Europe studying geology, mineralogy and meeting leading med. men, 1829-32; practiced medicine, Boston, 1832-36; mem. survey team in Me., Mass., 1837-39; became state geologist of Me., 1837; surveyor of R.I., 1839-40; state genologist, N.H., 1840, surveyor of N.H., 1841-44; a U. S. geologist in survey of Lake Superior region, 1847-48. Contbr. articles to Am. Jour. of Science, 1828-29. Author: A Manual of Etherization, Containing Directions for the Employment of Ether, Chloroform and other Anaesthetic Agents, 1861. Involved in controversy with Samuel F. B. Morse over invention of telegraph, claimed to have made 1st working model, 1840; claimed to have 1st developed surg. anesthesia, 1846; claimed priority in discovery of guncotton. Died Somerville, Mass., Aug. 28, 1880.

JACKSON, Chevalier, Am. laryngologist; b. Pitts., Nov. 4, 1865; s. William Stanford and Katharine Ann (Morange) J.; student Western U. Pa., 1879-83; M.D., Jefferson Med. Coll., Phila., 1886; m. Alice Bennett White, July 19, 1899. Hon. prof. brocheosophagology Temple U., Phila. Mem. A.M.A., Am. Laryngol. Assn., Am. Laryngol., Rhinol. and Otol. Soc., Pan Am. Med. Assn., A.C.S., Am. Bronchoscopic Soc., Am. Therapeutic Soc., Am. Assn. Thoracic Surgery, Am. Acad. Ophthalmology and Otolaryngology, Assn. des Médicins de Langue Francaise de l'Amerique du Nord, hon. fellow Royal Soc. Medicine, Nat. Acad. Medicine Mexico, Swedish Med. Soc., Nat. Acad. Medicine Brazil, Scottish, French, Italian, Polish, Rumanian, Belgian and Spanish otolaryngol. socs. Decorated Officer Legion of Honor (France); Chevalier Order of Leopold (Belgium), Comdr. Order of Crown of Italy; Cross of Brazil. Author: Peroral Endoscopy and Laryngeal Surgery, 1914; Bronchoscopy, Esophagoscopy and Gastroscopy, 1934; Foreign Body in Air and Food Passages, 1934. Editor: (also contbr.) The Nose, Throat and Ear and Their Diseases, 1929; Bronchoscopy and Esophagoscopy. Contbr. Systems of Surgery and Medicine. Developed method of removal of fgn. bodies from lungs by insertion of tubes through the mouth, also contbd. to development of laryngeal surgery; perfected bronchoscope for removal fgn. bodies from bronchial tube, 1915, esophagoscope for exam. esophagus, 1932. Died Aug. 16, 1958.

JACKSON, Clarence Martin, Am. anatomist; b. What Cheer, Ia., Apr. 12, 1875; s. John Calvin and Adeline (Hartman) J.; B.S., U. Mo., 1898, M.S., 1899, M.D., 1900, LL.D., 1923; postgrad. U. Chgo., 1900-01, U. Leipzig, 1903-04; U. Berlin, 1904; m. Helen Clarahan, June 1898; children—Margaret, Helen, Dorothy, Mary. Faculty, U. Mo., 1899-13, asst. prof. anatomy and histology, 1900-02, prof., 1902-13, dean faculty of medicine, 1909-13; prof., dir. dept. anatomy U. Minn., Mpls., 1913-41, emeritus from 1941. Chmn. med. div. NRC, Washington, 1923-24. Fellow A.A.A.S.; mem. Am. Assn. Anatomists (pres. 1922-24), Anatomische Gesellschaft, A.M.A., Phi Beta Kappa, Sigma Xi, Alpha Omega Alpha. Editor of Morris's Human Anatomy. Author: Effects of Inanition and Malnutrition upon Growth and Structure; also articles. Died Jan. 17, 1947.

JACKSON, Daniel Dana, Am. chem. engr.; b. Gloucester, Mass., Aug. 1, 1870; s. Daniel and Lucy Agnes (Langsford) J.; B.S., Mass. Inst. Tech., 1893; postgrad. Harvard, 1896-97; M.S., N.Y. U., 1908; Sc.D., U. Pitts., 1924; m. Ella Howard Phillips, Nov. 26, 1902; children—Daniel Dana, Elizabeth Purdy. Chemist, Boston Water Works, 1893; biologist, Mass. Bd. of Health, also lectr. Mass. Inst. Tech., 1895-97; chief chemist Bklyn. Water Supply Dept., 1897-1904; dir. labs. N.Y. dept. water supply, gas and electricity, 1904-12; lectr. san. engring. and bacteriology Columbia U., N.Y.C., 1911-13, asst. prof. chem. engring., 1913-17, asso. prof., 1917-18, prof. and exec. officer, dept. chem. engring., from 1918; v.p. Leavitt-Jackson Engring. Co., 1912-20; tech. mgr. Permutut Co., 1912-17; dir. Chem. Treatment Co. Mem. advisory com. on training camps, office of sec. of war, dean Sch. of Mil. Photography, U. S. Signal Corps, prof. Sch. of Explosives, Ordnance Dept., World War I. Mem. adv. com. N.Y. Health Dept. Cons. for cities and companies on water and drainage, factory processes. Died Sept. 1, 1941.

JACKSON, Derek Ainslie, physicist; b. London, Eng., June 23, 1906; s. Charles James and Ada (Williams) J.; M.A., Cambridge U., 1927, D.Sc., Oxford (Eng.) U., 1936; m. Pamela Mitford, Dec. 30, 1936; m. 2d, Janetta Kee, Oct. 10, 1951; 1 dau., Rose Janetta; m. 3d, Marie-Christine Reille, Mar. 8, 1968. Research in spectroscopy Oxford U., 1927-39, univ. lectr., 1934, prof. spectroscopy, 1945-48; vis. prof. Ohio State U., 1950; hon. asst. Dunsink Obs., 1952-53; research prof. Lab. A-Cotton, Centre National de la Recherche Scientifique, Bellevue, France, 1953——. Fellow Royal Soc., 1947. Contbr. articles to profl. jours. Made 1st quantitative determination of magnetic moment of nucleus, 1928; research on measurements of nuclear moments by spectroscopic methods; developed (with H. Kuhn) methods of increasing spectroscopic resolving power, their application to investigation of properties of nucleus. Home: 12 Chemin de Roseneck, Lausanne, Switzerland; also 19 rue Auguste Vacquerie, Paris XVI, France. Office: Laboratoire A-Cotton, Centre National de la Recherche Scientifique, Bellevue, Seine-et-Oise, France.*

JACKSON, Douglas, English dental surgeon; b. Oldham, Lancashire, Eng., May 3, 1918; s. Horace and Margaret (Sumner) J.; B.D.S., Victoria U., Manchester, Eng., 1942; m. Marjorie Sara Kenyon, Apr. 15, 1944; children—David, Katherine, Barbara. House surgeon Manchester Dental Hosp., 1942; resident house surgeon Manchester Royal Infirmary, 1943; lectr. Manchester U., 1947-49; pub. dental service, 1950-51; lectr. Leeds U., 1951-64, prof., 1965——. Mem. Brit. Dental Assn., European Orgn. for Research on Fluorine and Dental Caries Prevention; Brit. Soc. for Study Orthodontics, Internat. Assn. for Dental Research. Research, numerous publs. on metabolism of fluorides in animals and man, epidemiology of dental disease, clin. trials, new theories of aetiology of caries and periodontal disease, relationship of these diseases to autoimmunity and aging process. Home: 5, Croft Park, Menston, Nr. Ilkley, Yorks, Eng. Office: Dental Sch. and Hosp., Blundell St., Leeds, Eng.*

JACKSON, Dudley Pennington, Am. physician; b. Roanoke, Va., Apr. 1, 1924; s. Waddie Pennington and Bessie (Gills) J.; student Randolph-Macon Coll., 1941-43; M.D., Johns Hopkins, 1947; m. Patricia Custer, May 15, 1948. Faculty, Johns Hopkins Sch. Medicine, Balt., 1955——, asso. prof. medicine, 1960——; staff Johns Hopkins Hosp., Balt., 1955; cons. physician Balt. City Hosps., 1963——. John and Mary R. Markle scholar physiol. chemistry Johns Hopkins Sch. Medicine, 1954-59. Mem. Am. Soc. for Clin. Investigation, Am. Fedn. for Clin. Research, Am. Physiol. Soc., Am. Soc. Hematology, Soc. for Exptl. Biology and Medicine, A.A.A.S., A.M.A. Fellow A.C.P.; Contbr. articles to tech. jours. Research in hemorrhagic diseases, blood coagulation and blood platelets, mechanisms abnormal bleeding in various disease states. Home: Trotter Rd., Clarksville, Md. 21029. Office: Johns Hopkins Hosp., Balt. 21205.*

JACKSON, Dugald Caleb, Am. elec. engr.; b. Kennett Square, Pa., Feb. 13, 1865; s. Josiah and Mary Detweiler (Price) J.; B.S. in Civil Engring., Pa. State Coll., 1885, C.E., 1888; postgrad. Cornell U., 1885-87; D.Sc., Columbia, 1932; D.Eng., Northeastern U., 1938; m. Mabel A. Foss, Sept. 24, 1889; children—Catharine Emma (Mrs. Philip L. Alger), Dugald Caleb. With Western Engring. Co., Lincoln, Neb., 1887-89, Sprague Electric Ry. and Motor Co., 1889; chief engr. central dist. Edison Gen. Electric Co. 1890; cons. engr. 1891-1931, sr. mem. firm D. C. and Wm. B. Jackson 1902-18, Jackson & Moreland, 1919-30; prof. elec. engring. U. Wis., 1891-1907; prof. elec. engring. and in charge dept. Mass. Inst. Tech., Cambridge, 1907-35, emeritus, 1935-51; lectr. Inst. E.E. of Japan, Iwadare Found., 1935; lectured on engring. edn. to univs. in China, 1936. Lt. col. engrs. in France, chief engr. Power Bd., AEF; chief Engring. Estimates Bd. to Estimate War Damage in Allied Countries, Am. Peace Commn., 1918-19. Mem. internat. jury, Chgo. Expn., 1893, Buffalo Expn., 1901. Chmn. Research Com. on Indsl. Illumination, 1926-36; U. S. Govt. del. to World Engring. Congress, Tokyo, 1929; chmn. div. engring. and industrial research NRC, 1930-33. Decorated Chevalier Legion of Honor (France); recipient Lamme medal of Soc. Promotion Engring. Edn. for tech. teaching, 1931. Fellow Am. Inst. E.E. (Edison medal 1938), Am. Arts and Scis. (pres. 1937-39), A.A.A.S. (chmn. sect. M. 1932), Am. Phys. Soc., Am. Soc. M.E.; mem. Am. Philos. Soc., Am. Soc. C.E., Instn. E.E. (London), Société Francaise des Electriciens (Paris), Am. Soc. for Engring. Edn. (pres. 1905-06), Am. Inst. Cons. Engrs. (pres. 1938-40), Sigma Xi, Tau Beta Pi. Author: Text-book on Electro-Magnetism and Construction of Dynamos, 1893; Electricity and Magnetism, 1895; Alternating Currents and Alternating Current Machinery (joint author, 1896, new edit. 1913; An Elementary Book on Electricity and Magnetism and Their Applications (joint author), 1902, rewritten, 1919; Street Railway Fares, Their Relation to Length of Haul and Cost of Service (joint author), 1917; Engineering's Part in the Development of Civilization, 1939; Present Status and Trends of Engineering Education in the United States, 1939; Present-Day Salaries of Engineering Schools in the United States and Canada (with D. C. Jackson, Jr.), 1947; also numerous papers. Patentee in fields of electric rotating machinery and transformers, electric motors and instruments, electric motor starting devices, telephone equipment, train lighting. Died July 1, 1951.

JACKSON, Dunham, Am. mathematician; b. Bridgewater, Mass., July 24, 1888; s. William Dunham and Mary Vose (Morse) J.; A.B., Harvard, 1908, A.M., 1909; postgrad. univs. of Göttingen and Bonn, 1909-11; Ph.D., Göttingen, 1911; m. Harriet Spratt Hulley, June 20, 1918; children—Anne Hulley, Mary Eloise. Instr. math. Harvard, 1911-16, asst. prof., 1916-19; prof. math. U. Minn., Mpls., from 1919. Fellow Am. Acad. Arts and Scis., Am. Phys. Soc., A.A.A.S. (sec. 1941-44, Sect. A); mem. Am. Math. Soc. (v.p. 1921), Math. Assn. Am. (pres. 1926), Nat. Acad. Sci., Phi Beta Kappa, Sigma Xi. Author: The Theory of Approximation, 1930; Fourier Series and Orthogonal Polynomials, 1941. Editor Trans. Am. Math. Soc., 1926-31. Contbr. to math. jours. Died Nov. 6, 1946.

JACKSON, Francis Charles, Am. surgeon; b. Rutherford, N.J., Sept. 2, 1917; s. Frank Emil and Margaret C. (Kuhn) J.; B.A., Yale, 1939; M.D., U. Va., 1943; m. Joan Gloria Mortenson, Sept. 1, 1949; children—Geoffrey P., Bradford M., Gregory C., Donna E. Mem. faculty sch. medicine Cornell U., 1943-50; faculty U. Pitts., 1953——, prof. surgery, 1965——; chief surgeon, cons. Arabian Am. Oil Co., Dhahran, Saudi Arabia, 1951; asst. chief surg. service VA Center, Togus, Me., 1952; chief surg. service, dir. gen. surg. residency program VA Hosp., Pitts., 1952——; mem. cons. staff Presbyn.-U. Hosp., Pitts., 1959——. Cons., Bur. Health Services, USPHS; mem. com. on emergency med. services Nat. Acad. Sci.-NRC. Diplomate Nat. Bd. Med. Examiners, Am. Bd. Surgery. Fellow A.C.S. (chmn. subcom. on disaster surgery); Mem. A.M.A., Am. Assn. for Surgery of Trauma, Central Surg. Assn., Soc. Surgery of Alimentary Tract, others. Research and publs. on diagnostic angiography; portal hypertension, portacaval shunt, disaster care, emergency med. services, wound healing. Address: VA Hosp., University Dr. C, Pitts. 15240.*

JACKSON, Frederick George, explorer; b. Alcester, Eng., Mar. 6, 1860; s. George Frederick Jackson; ed. Denstone Coll., Edinburgh (Scotland) U.; m. Marguerite Wigan Hernu Fisher, 1929. Traveled in Australian deserts; made famous journey across Gt. Tundra and Lapland in midwinter; comdr. Jackson-Harmsworth Polar Expdn. to Franz Josef Land; crossed Africa from Beira to Banana Point, 1925-26; traveled across Mashonaland, Matabeleland, No. Rhodesia and up Lake Tanganyika; trekked with carriers across Urundi and Ruanda countries, visiting volcanic Virunga Mountains and ascending Mt. Sabiuyo; crossed Congo Forest from Lake Kivu to Lualaba River; followed length of Congo River to sea; visited sources of Nile, Zambesi, Congo rivers. Recipient Gold medal Paris Geog. Soc., medal Royal Humane Soc. Gt. Britain. Author: The Great Frozen Land, 1895; A Thousand Days in the Arctic, 1899; The Lure of Unknown Lands, 1935; contbd. papers and articles to socs., publs. Died Mar. 13, 1938.

JACKSON, George Gee, Am. physician; b. Provo, Utah, Oct. 5, 1920; s. Elvon L. and Adelia May (Gee) J.; B.A., Brigham Young U., 1942; M.D., U. Utah, 1945; m. Amy Smith Cox, Sept. 4, 1943; children—Janet, Sandra, Christopher George, Amy Adelia, John Gee. Practice medicine, specializing in infectious diseases, Chgo., 1953——; faculty Coll. Medicine U. Ill., 1953——, prof. medicine, 1959——; mem. Commn. Acute Respiratory Diseases, Commn. on Influenza, Armed Forces Epidemiological Bd.; mem. bd. vaccine devel. Nat. Inst. Allergy and Infectious Diseases. Mem. A.A.A.S., Am. Assn. Immunologists, A.C.P., Am. Epidemiological Soc., Am. Fedn. Clin. Research, Central Soc. Clin. Research, Am. Soc. Clin. Investigation, Infectious Disease Soc. Am., Assn. Am. Physicians. Contbr. numerous articles to profl. jours. Research, publs. on usefulness and mechanisms of action of antimicrobial drugs, including chemotherapeutic antibiotics and antiviral agts., causes and pathogenesis of common respiratory infections resulted from researches; demonstrated specific immunity to the common cold. Home: 315 N. Lincoln St., Hinsdale, Ill. 60521. Office: 840 S. Wood St., Chgo. 60612.*

JACKSON, Hall, Am. surgeon; b. Hampton, N.H., Nov. 11, 1739; s. Dr. Clement and Sarah (Leavitt) J.; studied medicine under father in Portsmouth, N.H., also in London hosps.; m. Molly (Dalling) Wentworth, Dec. 1, 1765, 2 children. One of earliest Am. physicians to perform cataract-couching operation of eye; administered inoculations during smallpox epidemic, Boston, 1764; an organizer smallpox hosp. on Henzell's Island, nr. Portsmouth after return from Boston; cared for wounded of Battle of Bunker Hill, 1775; surgeon in Pierce Long's regt. at capture Ft. Ticonderoga. Mem. Mass. (hon.), N.H. (charter) med. socs. Died Sept. 28, 1797.

JACKSON, Harold, Brit. pharmacologist; b. Ashton-on-Mersey, Gt. Britain, Mar. 31, 1912; s. Clarence Boyd and Alice (Collier) J.; B.Sc. with honors in Chemistry, U. Manchester (Eng.), 1934, M.Sc., 1935, Ph.D., 1937, M.B., Ch.B. with honors, 1945, D.Sc., 1964; m. Mary Kathleen Finch, July 23, 1938; children—Celia Marion, Alan Howard, Roger Harold, Nigel Colin. With U. Manchester, 1937-42, Imperial Chem. Industries fellow, 1946-47, lectr. bacteriology, 1947-48; head exptl. chemotherapy Christie Hosp., Manchester, 1940——; mem. external sci. staff Med. Research Council London, 1958——; hon. reader exptl. chemotherapy U. Manchester, 1966——, head unit reproductive pharmacology, 1965——. Recipient Oliver Bird medal and prize for research into pharmacology of reprodn., London, 1960. Mem. Pharmacological Soc., Soc. for Study Fertility. Author: Antifertility Substances in Male and Female, 1966; also numerous articles. Research on devel. of anticancer drugs and their fate in body; discovered various chem. agts. capable of damaging male sex cells at various stages of their devel. without interference with sexual activity. Home: 25 Barcheston Rd., Cheadle, Cheshire, Gt. Britain. Office: Peveril Mount, 92 Alexandra Rd., S., Manchester 16, Gt. Britain.*

JACKSON, Sir Herbert, English chemist; b. London, Eng., Mar. 17, 1863; s. Samuel and Clenentina (Grant) J.; ed. King's Coll., London, 1879-83; m. Amy Collister, 1900. From asst. prof. to Daniell prof. chemistry King's Coll., London, 1902-18, fellow, 1907; dir. research Brit. Sci. Instrument Research Assn., 1918-33, later cons.; emeritus prof. chemistry, mem. senate U. London. Gov. Imperial Coll. Sci. Fellow Royal Soc., 1917, Inst. Chemistry (pres. 1918-21), Inst. Physics, Chem. Soc., Royal Micros. Soc. (hon.); mem. Röntgen Soc. (pres. 1901-03), Brit. Inst. Radiology (hon.). Expts. in 1890's on excitation of phosphorescence by means of discharge tubes anticipated Röntgen's discovery of x-rays; used concave cathode and inclined anti-cathode to make focus tube which became prototype of later x-ray tubes; studied weathering of stone, action of soaps and solvents in laundry work; research on coloring agents in glasses, also during World War I, determined formulae for number of different lab., heat-resisting and optical glasses. Died Hampstead, Eng., Dec. 10, 1936.

JACKSON, Jabez North, Am. surgeon; b. Labaddie, Mo., Oct. 6, 1868; s. John Wesley (M.D.) and Jennie Clark (North) J.; A.B., Central Coll., Fayette, Mo., 1889, A.M., 1890; M.D., Univ. Med. Coll., Kansas City, 1891; D.Sc., Park Coll., 1926; LL.D., U. Mo., 1927; post-grad. N.Y. Polyclinic, 1891; m. Virlea Wayland, Oct. 12, 1899; children—Virginia, Margaret. Demonstrator anatomy Univ. Med. Coll., Kansas City, Mo., 1891-95, prof., 1895-98, prof. surg. anatomy and adj. prof. surgery, 1898-1900, prof. principles and practice of surgery and clin. surgery, 1900-11, also trustee, 1891-1911, and pres.; surgeon Kansas City Gen. Hosp., Research Hospital, Trinity Hosp.; dir. of health, Kansas City, Mo., from 1923. Mem. A.M.A. (pres. 1927), Am., Western (pres. 1913) surg. assns., others. Described Jackson's veil (thin membrane sometimes observed covering 1st part of large intestine), 1909. Died Mar. 8, 1935.

JACKSON, James, Am. physician; b. Newburyport, Mass., Oct. 3, 1777; s. Johnathan and Hannah (Tracy) J.; A.B., Harvard, 1796, A.M., 1799, M.B., 1802, M.D., 1809; studied anatomy, London, Eng., 1799; m. Elizabeth Cabot, Oct. 3, 1801; m. 2d, Sarah Cabot; 9 children, including James. Learned to vaccinate in Eng., became 1st in Am. to apply scientific investigation to vaccination; apptd. physician to Boston Dispensary, 1802; a founder Harvard Med. Sch., 1810, Hersey prof. theory and practice of physics, 1813. Author: On the Theory and Practice of Physics, 1825; Letters to a Young Physician, 1855; Another Letter to a Young Physician, 1861. First to describe peripheral neuritis caused by alcoholism. Died Aug. 27, 1867.

JACKSON, John, Brit. astronomer; b. Feb. 11, 1887; s. Matthew and Jeanie (Millar) J.; ed. Glasgow (Scotland) U., Trinity Coll., Cambridge (Eng.) U.; M.A., D.Sc.; m. Mary Beatrice Marshall, 1920. Chief asst. Royal Obs., Greenwich, Eng., 1914-33; His Majesty's astronomer Cape of Good Hope, 1933-50. Fellow Royal Soc., 1938, Royal Soc. South Africa (pres. 1949), Royal Astron. Soc. (hon. sec. 1923-29, pres. 1953-55, gold medal 1952). Author: (with Dr. Knox-Shaw, W. H. Robinson) Reduction of Hornsby's Observations at Oxford, 1774-1798. Editor: The Observatory, 1919-27. Contbr. articles on dynamical and fundamental astronomy to Monthly Notices of Royal Astron. Soc. Died Dec. 9, 1958.

JACKSON, John David, physicist, educator; b. London, Ont., Can., Jan. 19, 1925; s. Walter David and Lillian (Ferguson) J.; B.Sc., U. Western Ont., 1946; Ph.D. in Physics, Mass. Inst. Tech., 1949; m. Marilyn Barbara Cook, June 26, 1949; children—Ian, Nancy, Maureen, Mark. Came to U. S., 1957. Research asso. Mass. Inst. Tech., 1949; faculty McGill U., Montreal, Que., Can., 1950-57; asso. prof. physics U. Ill., Urbana, 1957-58, prof., 1958——. Cons., Space Tech. Labs., Los Angeles, 1958-60, Argonne (Ill.) Nat. Lab., 1961——. Guggenheim fellow Princeton, 1956-57; Ford Found. fellow European Orgn. for Nuclear Research, Geneva, Switzerland, 1963-64. Fellow Am. Phys. Soc.; mem. Italian Phys. Soc., Am. Assn. U. Profs. Author: Physics of Elementary Particles, 1958; Classical Electrodynamics, 1962, Mathematics for Quantum Mechanics, 1962; contbg. author: Dispersion Relations, 1960; Elementary Particle Physics, 1962. Contbr. articles to profl. jours. Research in theoretical nuclear and high energy physics. Office: Dept. Physics, U. Ill., Urbana, Ill. 61803.*

JACKSON, John Harry, Am. metallurgist; b. Troy, N.Y., Apr. 19, 1915; s. Myron J. and Irene (Yates)

J.; Ch.E., Rensselaer Poly. Inst., 1936; m. Marian Anderson, Oct. 12, 1940; children—Elizabeth I., Anne V. (Mrs. John E. Hohmann), John H. Welding and metall. engr. Caterpillar Tractor, Peoria, Ill., 1936-41; sect. editor Chem. Abstracts, Columbus, O., 1950-59; research scientist, mgr. metall. research Battelle Meml. Inst., Columbus, O., 1941-61; gen. dir. metall. research div. Reynolds Metals Co., Richmond, Va., 1961—, exec. v.p., 1966—. Cons. to govt. agys.; chmn. materials adv. bd. Nat. Acad. Scis. Com. to U. S. Dept. Def. on Composites, 1963-66, Engring. Found., 1964—. Mem. Am. Inst. Mining, Metall. and Petroleum Engrs. (dir.), Metall. Soc., (pres. 1967—), Am. Soc. for Metals (past chmn. Columbus), Am. Soc. for Testing and Materials, Am. Ordnance Assn., Am. Ordnance Assn., Am. Foundry Soc., Am. Soc. M.E., Am. Inst. Aeros. and Astronautics, Nat. Acad. Scis. (mem. materials adv. bd. 1965—), Engrs. Club Richmond, Sigma Xi. Research, publs. and patentee in phys. metallurgy; devel. alloys for elec. applications, alloys based on metals aluminum, beryllium, magnesium, titanium, iron, nickel and refractory metals. Home: 8201 Diane Lane, Richmond 23227. Office: Reynolds Metals Co., 4th and Canal Sts., Richmond, Va. 23218.*

JACKSON, John Hughlings, English neurologist; b. Providence Green, Eng., Apr. 4, 1835; s. Samuel and Sarah (Hughlings) J.; student York Med. and Surg. Sch., London U.; M.D., St. Andrews U., 1860; m. Elizabeth Dade Jackson, 1865. House surgeon, dispensary, York, 1856-59; named physician to new Nat. Hosp. for Nervous Diseases, Queen Sq., 1862. Croonian, Lumleian lectr. Hughlings Jackson lecture created in his honor at Neurol. Soc. London, 1897. Licentiate Apothecaries Soc. Fellow Royal Soc., 1878, Royal Coll. Physicians; mem. Royal Coll. Surgeons. Gave classic description of hemiplegic or focal epilepsy (Jacksonian epilepsy), 1863; 1st to demonstrate importance of ophthalmoscopy in study of diseases of nervous system, 1864; corresponded localized epileptiform movements to lesions of motor area of cerebral cortex, 1875; developed concept of control of voluntary movements by cortex. Died London, Oct. 7, 1911.

JACKSON, L. W. R., Am. forester; b. Lookout, Wis., Feb. 21, 1900; s. Edward and Clara (Larson) J.; B.S., U. Minn., 1926, M.S., 1927; Ph.D. (Harrison fellow), U. Pa., 1932; m. Jessie Measel, July 1, 1939. In charge Phila. office U. S. Div. Forest Pathology, 1932-40, Athens (Ga.) office, 1940-46; prof. forestry U. Ga., Athens, 1946—, mem. grad. faculty, 1947—. Fellow Ga. Acad. Sci., A.A.A.S.; mem. Ecol. Soc. Am., Am. Phytopathol. Soc. (emeritus), Soc. Am. Foresters, Sigma Xi, Xi Sigma Pi (nat. pres. 1956), Gamma Sigma Delta, Phi Kappa Phi. Research and numerous publs. on heritability of tracheid length and specific gravity of So. pines, heritability of fibril angle in wood of So. pines, effect of thinning of So. pines on stimulation on growth of basal area and specific gravity with height in stem; discovered 1st virus disease of forest tree. Home: 480 Milledge Terrace, Athens, Ga. 30601.*

JACKSON, Ray Dean, Am. physicist; b. Shoshone, Ida., Sept. 28, 1929; s. Oran A. and Joyce (Johnson) J.; student Ida. State Coll., 1949-50; B.S., Utah State U., 1956; M.S., Ia. State U., 1957; Ph.D., Colo. State U., 1960; m. Donna Mae Jenson, Aug. 8, 1952 (div. Oct. 1965); children—Linda Kaye, Joyce, John Ray. Soil scientist Agrl. Research Service, U. S. Dept. Agr., Ft. Collins, Colo., 1957-60, research physicist U. S. Water Conservation Lab., Phoenix, 1960—. Recipient Letter commendation Sec. of Agr. Orville Freeman, 1965. Fellow A.A.A.S.; mem. Am. Soc. Agronomy, Soil Sci. Soc. Am. (asso. editor Proc. 1966—), Sigma Xi. Research and publs. on behavior and flow of water in fine capillaries, water vapor adsorption and flow in soils, heat transfer in porous materials; co-inventor desert survival still. Office: 4331 E. Broadway, Phoenix 85040.*

JACKSON, Raymond Carl, Am. botanist, cytogeneticist; b. Medora, Ind., May 7, 1928; s. Thornton C. and Flossie (Booker) J.; A.B., Ind. U., 1952, A.M., 1953; Ph.D., Purdue U., 1955; m. Thelma June Snyder, Oct. 24, 1947; children—Jeffrey Wayne, Rebecca June. Instr. biology U. N.M., Albuquerque, 1955-57; asst. prof., 1957-58; faculty U. Kan., Lawrence, 1958—, prof. botany, 1964—. Mem. Am. Inst. Biol. Sci., Am. Soc. Plant Taxonomists, A.A.A.S., Internat. Assn. for Plant Taxonomy, Soc. for Study Evolution, Bot. Soc. Am., Am. Soc. Naturalists. Contbr. articles to tech. jours. Research on natural hybridization and introgression in sunflowers, cytogenetic analyses and chromosomal evolution in various genera Compositae, cytogenetic studies of Haplopappus gracilis. Home: 2614 Orchard Lane, Lawrence, Kan. 66044.*

JACKSON, Richard Willet, Am. biochemist; b. Camp Point, Ill., June 14, 1901; B.S., Eureka Coll., 1922; M.S., U. Ill., 1923, Ph.D., 1925; m. Dorothy Vernon Gee, Nov. 26, 1925; children—Marianna, William T. Asst. prof. biochemistry U. Louisville Med. Sch., 1925-27; Seessel fellow, NRC fellow Yale, 1927-31, asst. prof. physiol. chemistry, 1931-33; asso. prof. biochemistry Cornell U. Med. Coll., N.Y.C., 1933-39; chief protein div. Eastern Regional Research Lab., U. S. Dept. Agr., Phila., 1939-47, chief fermentation Lab., No. Regional Research Lab., Peoria, Ill., 1947—. Trustee Eureka Coll. Recipient Superior Service award U. S. Dept. Agr., 1952, Distinguished Service award with gold medal, 1957, Centennial citation Eureka Coll., 1955. Mem. Am. Chem. Soc., Am. Inst. Nutrition, Am. Soc. Microbiology, Am. Soc. Biol. Chemists (past sec.), Harvey Soc., A.A.A.S., Ill. State Hist. Soc., Sigma Xi, Pi Kappa Delta. Author: (with L. L. Wallen, F. H. Stodola) Type Reactions in Fermentation Chemistry, 1959; also numerous articles. Patentee in field. Discovered that methionine is indispensable in diet; research on synthesis and metabolic relationships of essential amino acids, methionine, cystine and tryptophane and their derivatives; devel. artificial fibers and bristles from casein; absorption fat-soluble vitamins and methods for microbiol. prodn. several vitamins and other products. Home: 1319 N. Institute Pl., Peoria 61606. Office: 1815 N. University St., Peoria, Ill. 61604.*

JACKSON, Robert Lawrence, Am. pediatrician; b. Clare, Mich., Nov. 30, 1909; s. Lawrence W. and Josephine (Cour) J.; B.S., U. Notre Dame, 1930; M.D., U. Mich., 1934; m. Sara Elizabeth Soisson, Sept. 6, 1937; children—Ann (Mrs. Thomas Allen), Mary Jo, Sarah, Katherine, Margaret, Martha, Robert. Faculty U. Ia., 1936-54, prof., 1951-54; prof., chmn. dept. pediatrics U. Mo., 1954—; vis. prof. Am. U., Beirut, Lebanon, 1961-63. Mem. Am. Acad. Pediatrics, Nat. Acad. Sci. Food and Nutrition Bd., A.M.A., Am., Mo. heart assns., Am. Diabetes Assn., Mo. Diabetic Assn., Sigma Xi, Alpha Omega Alpha. Research, publs. on defects of metabolism, endocrinology of diabetes. Home: 1103 Stewart Rd., Columbia, Mo. 65201.*

JACKSON, Robert Tracy, Am. paleontologist; b. Dorchester, Mass., July 13, 1861; s. Dr. John Barnard Swett and Emily Jane (Andrews) J.; S.B., Lawrence Scientific Sch., Harvard, 1884, Sc.D., 1889; m. Fanny Esther Roberts, June 27, 1889; children—Esther, Dorothy Quincy (Mrs. John E. Bastille). Asst. prof. paleontology Harvard, 1899-1909; asso. in paleontology, Museum of Comparative Zoology, Harvard, 1911-16, curator of fossil echinoderms, 1928-39. Fellow Am. Acad. Arts and Scis., A.A.A.S.; mem. Am. Soc. Naturalists, Geol. Soc. Am. Author: Phyolgeny of the Pelecypoda, 1890; Studies of Melonites Multiporus (with T. A. Jaggar, Jr.), 1896; Studies of Palaechinoidea, 1896; Localized Stages in Development in Plants and Animals, 1899; Phylogeny of the Echini, 1912; Studies of Arbacia, 1927; Palaeozoic Echini of Belgium, 1929; The Status of Bothriocidaris, 1929. Died Oct. 24, 1948.

JACKSON, T. Burr, Am. electronics engr.; b. Penn Yan, N.Y., Jan. 4, 1919; s. Carlton T. and Jennie (Burr) J.; B.S. in Elec. Engring., Tri-State Coll., 1941; m. Vera C. McMillin, Dec. 18, 1943; 1 dau., Joanne Carol. Electronic engr. instrumentation Eastman Kodak Co., Rochester, N.Y., 1940-42, U. S. Naval Research Lab., Washington, 1942-48; electronic scientist, head telemetry group Nat. Bur. Standards, Washington, 1948-52; supervisory electronic engr. head instrumentation div. Naval Weapons Center, Corona (Cal.) Labs., 1952—. Mem. I.E.E.E. (sr.), A.A.A.S., Inter-Range Instrumentation Group (mem. telemetry working group since 19———). Contbr. numerous articles to profl. jours., trade mags., govt. publs. Designer, developer indicator, display equipment for mil. radar systems, telemetry transmitter, receiver and data reduction systems, components for Navy guided missiles. Home: 6864 De-Anza Av., Riverside, Cal. 92506. Office: Naval Weapons Center, Corona Labs., Corona, Cal. 91720.*

JACKSON, Wilfrid J(ames), physicist; b. Jacksonville, N.S., Can., May 11, 1900; s. William G. L. and Helen E. (Musgrave) J.; student Sydney (N.S.) Acad., 1914-17; B.A., Dalhousie U., 1921, A.M., 1923; A.M., Princeton, 1926, Ph.D., 1927; m. Mabel E. Mott, June 21, 1932; children—Charles Wilfrid, Marilyn Mott. Came to U. S., 1923, naturalized, 1939. Instr. physics Princeton U., 1923-25; James W. Queen fellow, physics King's Coll., Dalhousie U., 1925-26, research asst., 1926-27, interim lectr. math., physics, 1927-28; instr. physics Rutgers U., New Brunswick, N.J., 1928-29, asst. prof., 1929-43; asso. prof. Douglass Coll., Rutgers State U., 1943-46, prof., chmn. dept. physics, from 1946; vis. lectr. Mass. Inst. Tech. ESMDT, 1941-42, radar sch., 1942-44, vis. asso. prof. radar sch., 1944-46; chief scientist N.Y. br. office, Office of Naval Research, 1950. Fellow A.A.A.S.; mem. Am. Phys. Soc., Am. Inst. Physics, Am. Assn. Physics Teachers, N.B. Sci. Soc., Sigma Xi. Contbr. articles in physics jours. Died Mar. 13, 1959.

JACKSON, William Arthur Douglas, geographer; b. Toronto, Ont., Can., Nov. 25, 1923; s. Joseph Walter and Sarah (Palmer) J.; B.A., Victoria Coll. U. Toronto, 1946, M.A., 1949; Ph.D., U. Md., 1953; student U. Wis., 1948-49, Russian Inst., Columbia U., 1952-53. Lectr., asst. prof. U. Ia., 1953-55; asst. prof. U. Wash., 1955-58; fellow Russian Research Center Harvard U., 1958; asso. prof. U. Wash., 1958-60, prof., 1960—, asst. dir. Far Eastern and Russian Inst., 1963-66. Mem. Am. Geograph. Soc., Am., Canadian assns. geographers, A.A.U.P., Am. Assn. Advancement Slavic Studies. Author: Philadelphia Waterfront Industry: Industrial Land and Its Potentials on the Delaware River, 1955; The Soviet Union, 1962; The Russo-Chinese Borderlands, 1963; Politics and Geographic Relationships, 1964. Contbr. articles in field to sci. jours. Research, elaboration of geographic aspects underlying certain aspects of Soviet agricultural policy and crop patterns, particularly in relation to the Khrushchev virgin and idle land program; elaboration of the basic trends in contemporary political geography; assessment of geographic factors underlying the Sino-Soviet boundary dispute. Home: 7307 56th Av. N.E., Seattle 98115.*

JACKSON, OF BURNLEY, Baron (Sir Willis), English elec. engr.; b. Burnley, Eng., Oct. 25, 1904; s. Herbert and Annie (Hiley) J.; B.Sc., Manchester (Eng.) U., 1925. D.Sc., 1936; D.Phil., Oxford (Eng.) U., 1936; Dr.Sc. Tech., U. Zurich (Switzerland), 1955; D.Sc., U. Bristol (Eng.), 1957, U. Dublin, 1962, City U. London, 1966; D.Tech., U. Bradford, 1966; D.Eng., Sheffield (Eng.) U., 1958; LL.D., U. Aberdeen (Scotland), 1959; m. Mary Elizabeth Boswall, Dec. 28, 1938; children—Anne Boswall, Ruth Lesley, Prof. elec. engring. Manchester U., 1938-46; prof. Imperial Coll., London, Eng., 1946-53, 61—, prorector, 1967—; dir. research and edn. Asso. Elec. Industries Ltd., Manchester, 1953-61. Mem. Council on Sci. Policy, 1961—; chmn. Com. on Manpower Resources for Sci. and Tech., 1963—, TV Adv. Com., 1963—. Fellow Royal Soc., 1953; Inst. Physics; mem. Inst. Elec. Engrs. (past pres.), Brit. Assn. for Comml. and Indsl. Edn. (pres. 1962—), Instn. Mech. Engr., Brit. Assn. Advancement Sci. (pres. 1966—). Author: High Frequency Transmission Lines, 1946; Communication Theory, 1953; Insulation of Electrical Equipment, 1953; also numerous articles. Research on materials and techniques used at ultra-high frequencies. Home: 25 Manor Rd., Cheam, Surrey, Eng. Office: Imperial Coll., London S.W. 7, Eng.*

JACNIEWITZ, Jean, see Jaskiewicz, Jean.

JACOB, Arthur, ophthalmologist; b. Knockfin, Ireland, June 1790; s. John and Grace (Alley) J.; studied medicine with father, also at Steerien's Hosp., Dublin; M.D., U. Edinburgh (Scotland), 1814; m. Sarah Carroll, Sept. 21, 1824; 5 sons. Became demonstrator anatomy Trinity Coll., Dublin, 1819; prof. anatomy, Dublin U., 1826-69. Founder (with C. Benson and others) City of Dublin Hosp., 1832; founder (with others) Med. Benevolent Fund Soc. Pres. Irish Coll. Surgeons; founder Irish Med. Assn. Author: A Treatise on the Inflammation of the Eyeball, 1849; On Cataract and the Operation for its Removal by Absorption, 1851. Contbr. articles to Cyclo. of Anatomy, Cyclo. of Practical Medicine, Philos. Trans. (with H. Maunsell) Dublin Med. Press, 1839, editor, 1839-59. Described Jacob's ulcer of eyelids and other parts of face, 1827; discovered layer of rods and cones in retina (Jacob's membrane), 1816; revived cataract operation through cornea with curved needle. Died Sept. 21, 1874.

JACOB, François, French biologist; b. Nancy, France, June 17, 1920; s. Simon and Therese (Franck) J.; M.D., Faculty of Medicine, Paris, 1947; D.Sc., Faculty of Scis., Paris, 1954; D.Sc. (hon.), U. Chgo., 1965; m. Lysiane Bloch, Nov. 27, 1947. children—Pierre, Laurent, Odile, Henri. Asst., Pasteur Inst., from 1950, head, dept. cellular genetics, 1960—; prof. cellular genetics, Coll. of France, 1964—. Recipient Charles Leopold Mayer prize, 1962; Nobel prize in physiology and medicine (with A. Lwoff and J. Monod), 1965. Research on genetics of bacterial cells and viruses; contbd. to mechanisms of information transfer (messenger RNA) and genetic basis of regulatory circuits. Home: 20 rue Guynemer, Paris 6, France. Office: Pasteur Inst., Paris 15, France.*

JACOB, Gerhard, physicist; b. Hanover, Germany, Nov. 5, 1930; s. Walter and Hertha (Rave) J.; B.Sc. in Physics, Fed. U. Rio Grande do Sul, 1952, B.Sc. in Math., 1953, Ph.D., 1964; m. Thereza Christina de Azevedo, July 5, 1956; children—Angela, Monica, Magda. Faculty, U. Rio Grande do Sul, Porto Alegre, Brazil, 1953-56, 58—, prof. theoretical physics, dep. dir. Inst. Physics, 1958—, mem. research council, 1964—; prof. Cath. U. Rio Grande do Sul, 1955; research asso. Inst. Atomic Energy, Sao Paulo, 1956-58, U. Heidelberg, 1958-59, U. Copenhagen, 1961-62; vis. prof. U. Heidelberg, 1962-63. Mem. Brazilian NRC, 1965—. Mem. Brazilian Soc. for Advancement Sci., Brazilian Acad. Sci., Brazilian Phys. Soc. Editor: (with others) Many-Body Problems and Other Selected Topics in Theoretical Physics, 1966. Research, publs. on structure of atomic nuclei, quantum electrodynamics. Home: 52 Travessa Fonte da Saude, Porto Alegre, R.S., Brazil.*

JACOB, Joseph Jean Christian, pharmacologist; b. Liège, Belgium, Sept. 19, 1918; s. Pierre Chretien and Marie (Jamsin) J.; M.D., U. Liège, 1943; m. Yvonne Mulier, July 31, 1943; children—Perrine, Hélène, Lise, Philippe. Chargé de recherches F.N.R.S., Liège, 1943-46; asst. Pasteur Inst., Paris, France, 1946-47, became dir. lab., 1948, dir. service, 1965. Asst., Lab. of Pharmacodynamics, Gand, Belgium, 1946-47; cons. O.M.S. Recipient Prix de la Ville de Paris, 1963. Mem. French Soc. Therapeutics and Pharmacodynamics, French Soc. Psychopharmacology, French Soc. Physiology, Brit. Pharmacological Soc., European Soc. for Study of Toxicity of Medicines, Internat. Soc. Biochem. Pharmacology. Research and publs. on mechanisms of action of drugs in relation to cardiovascular and mental diseases, pharmacology of autonomic nervous system, functions of biogenic amines in the brain, pharmacology of learning and of pain, animal models for study of psychotomimetic drugs, side effects and toxicity of drugs. Home: 121 Ave. Mozart, Paris 16. Office: 28 rue du Dr. Roux, Paris 15, France.*

JACOB, Louis, physicist; b. Riga, Russia, May 1, 1904; s. David and Rachel (Mofsovitz) J.; B.Sc., Royal Coll. Sci., Dublin, Ireland, 1926; M.Sc., U. Coll. Dublin, 1927, Ph.D., 1941, D.Sc., 1959; m. Adeline Elkinson, July 24, 1957; 1 dau., Ruth. Sci. staff, research labs. Gen. Electric Co., Wembley, Eng., 1928-45; I.C.I. fellow U. Manchester, 1945-47; lectr. U. Liverpool, 1947-56; sr. lectr. U. Strathclyde, Glasgow, Scotland, 1956—. Fellow Inst. Physics; asso. mem. Instn. Elec. Engrs. Author: High Voltage, 1934; Electron Optics, 1951; also articles. Discovered elec. oscillations in low energy electron beams; invented method for detecting single atoms and molecules; patentee electron optical transfer from anode to cathode; research on fission of virus by electron bombardment; studies in high voltage, gas discharges, thermionics, electron optics, soft X-ray spectra, semiconductors. Home: 178 Wilton St., Glasgow N.W., Scotland.*

JACOB, P., French physician; b. Plainfaing, France, Apr. 11, 1907; s. Jean Nicolas and Mathilde (Lhuillier) J.; student Faculté Médecine Nancy (France), 1925-32; Doctorat en Médecine et spécialités; m. Antoinette Georgette Tailly, Feb. 2, 1931; children—Jean-Pierre, Marguerite (Mme. Terrad), François, Denis, Michel, Irene (Mme. Pfirsch). Asst., hosps. of Nancy; head. radiology dept. Regional Center for Fight Against Cancer, Nancy, 1936-51; lectr. Faculty Medicine Nancy, 1937-51; head electroradiology dept., dir. Regional Center for Fight Against Cancer, Caen, France, 1952—. Mem. Soc. Med. Electro-radiology of France. Contbr. papers to med. publs. Research on pathology of cancer, biol. effect of ionized radiations, radiodiagnostics, radiotherapy, orthokinetic radiotherapy. Home: 4 August Lechesne, Caens 14. Office: 2 Georges Clemenceau, Caen 14, France.*

JACOB, Stanley W., Am. surgeon; b. Phila., Jan. 7, 1924; s. Abraham and Belle (Shulman) J.; B.A., Ohio State U., 1945, M.D. cum laude, 1948; m. Beverly Blomquist, Feb. 29, 1964; children—Stephen, Jeffrey, Darren, Robert. Instr. surgery Harvard Med. Sch., 1958-59; faculty U. Ore. Med. Sch., 1959—, asso. prof. surgery, 1965—. Kemper Found. Research scholar A.C.S., 1957-60; Markle scholar, 1960-65. Recipient Outstanding N.W. Scientist Gov.'s award, 1965. Mem. A.C.S., Soc. U. Surgeons, Portland Surg. Soc., A.M.A., Phi Beta Kappa, Alpha Omega Alpha. Author: (with Clarice Francone) Structure and Function in Man, 1965, Laboratory Guide for Structure and Function in Man, 1966. Studies, numerous publs. on preservation of whole organs prior to transplantation; original description of membrane penetrant actions, primary pharmacology and clin. usefulness of dimethyl sulfoxide. Home: 19117 S.W. Old River Rd., Oswego, Ore. 97034.*

JACOB, T. Mathai, Indian chemist; b. Kerala, India, June 6, 1927; s. Mathai Ipe and Annamma (John) J.; B.Sc., U. Madras (India), 1949, M.A., 1951, Ph.D., 1957; m. Rose Thomas, July 7, 1952; children—Anita, Maya, Matthew, Teresa. Lectr., St. Joseph's Coll., Bangalore, India, 1951-52; NRC fellow Indian Inst. Scis., Bangalore, 1957-59; research asso. U. Toronto (Ont., Can.), 1960, Stevens Inst. Tech., Hoboken, N.J., 1961; project asso. Inst. for enzyme Research, U. Wis., Madison, 1962-63, asst. prof., 1964-66; with Dept. Biochemistry, Indian Inst. Sci., Bangalore, 1967—. Recipient GUHA Research medal Indian Inst. Sci., 1957. Mem. A.A.A.S. Contbr. articles to tech. jours. Synthesized short DNA molecules of known base sequence used to prepare enzymatically long RNA molecules of known sequence; these RHA molecules were used to synthesize specific polypeptides; established nature of genetic code. Address: Dept. Biochemistry, Indian Inst. Sci., Bangalore 12, India.*

JACOB, William Stephen, Brit. astronomer; b. Somerset, Eng., Nov. 19, 1813; s. Stephen Long Jacob; entered E. India Co.'s Coll., Addiscombe, 1828; mil. edn. at Chatham, Eng.; m. Elizabeth Coates; 6 sons, 2 daus. Served with Bombay engrs.; dir. Madras (India) Obs., 1849-59. Fellow Royal Astron. Soc. Author: Subsidiary Catalogue of 1,440 Stars selected from the British Association Catalogue, 1848-52; Plurality of Worlds, 1855; Results of magnetical observations at Madras (1846-50), 1854; also pub. Singapore meteorol. observations from 1841-45, 1850, Dodabetta observations from 1851-55, 1857. Catalogued 244 double stars observed from Poona, India; reobserved and corrected 317 stars from Brit. Assn. Catalogue; discovered triplicity of Scorpii, 1847; observed transparency of Saturn's dusky ring, 1852. Died Poona, India, Aug. 16, 1862.

JACOB BEN MAHIR IBN TIBBON, mathematician, astronomer, translator; b. Marseilles, France, circa 1236. Flourished in Montpellier; chiefly known as astronomer; quoted by some leading 16th century astronomers, Copernicus, Erasmus, Reinhold, Christ of Clavius. Original thinker as well as translator; 1 of Jews who most contributed to bringing Arab high culture to Christian West; translated: De sphaera mota of Autolykos, 1273; Elements, 1255, and Data, 1272, of Euclid; Menelaos' Sphaerica, 1273; Usage of the Celestial Sphere of Qusta 'b Luga, 1276; Configuration of the World by Ibn-al-Haytam, 1271; Usage of the Astrolabe of al-Zargali, 1263; Compendium of the Organon, 1289, and Commentary on the Zoology of Aristotle by Ibn Rusd; also wrote describing quadrant which he invented; compiled astronomical tables for Montpellier (Al-manach perpetuum Prophati). Died circa 1304.

JACOBAEUS, Hans Christian, surgeon; b. Skarhult, Sweden, 1879; prof. Karolinska Inst.; invented thoracoscope by using cystoscope to examine endopleural space and other cavities of body, reported 1910; devised method of pneumonolysis by using galvano cautery to divide pleural adhesions, 1915. Died 1937.

JACOBAEUS, Oliger (or Holger Jacobi), Danish naturalist; b. Aarhus, Denmark, July 6, 1650; ed. univs. in Germany, France, Italy; M.D., Leyden, Holland; son was Jacques Jacobaeus. Prof. history, geography, medicine in Denmark; counsellor of justice, asst. judge Supreme Ct., Copenhagen, Denmark; mem. French Acad. Scis., 1699. Author: Observationes de ranis et lacertis, 1676; Compendium institutionum medicarum, 1688-92; Museum regium, 1695. Died Copenhagen, June 18, 1701.

JACOBI, Abraham, physician; b. Hartum, Westphalia, May 6, 1830; studied univs. of Greifswald, Göttingen; M.D., Bonn, 1851; LL.D., U. Mich., 1898, Columbia, 1900, Yale, 1905, Harvard, 1906, Jefferson, 1913; m. Mary C. Putnam, 1873 (dec. 1906). Identified with German revolutionary movement; in detention, Berlin and Cologne, 1851-53; started med. practice in N.Y.C., 1853; prof. diseases of children N.Y. Med. Coll., 1860-64, N.Y.U., 1865-70; prof. diseases of children Coll. Phys. and Surg., Columbia, 1870-1902, prof. emeritus, from 1902. Pres. N.Y. State Med. Soc., 1882, N.Y. Acad. Medicine, 1885-89, Am. Pediatric Soc., 1891, 1906, Assn. Am. Physicians, 1896, Am. Climatol. Soc., 1896, A.M.A., 1912-13. Author: Dentition and Its Derangements, 1862; The Raising and Education of Abandoned Children in Europe, 1870; Infant Diet, 1872, 1875; Diphtheria, 1876; Treatise on Diphtheria, 1880; Pathology of the Thymus Gland, 1889; Therapeutics of Infancy and Childhood, 1896, 1898, 1903; Intestinal Diseases; Collectanea Jacobi, 8 vols., 1909. Co-editor Am. Jour. Obstetrics and Diseases of Women and Children, 1868-71. Died July 10, 1919.

JACOBI, Joannes, Spanish physician; b. perhaps at Lérida/Segre, Spain; possibly ed. U. Lleida; Chancellor U. Montpellier (France), 1364-84; cons. physician to several popes and kings; translator works from Arabic into Catalan. Author: Tractatus de pestilentia; Secretarium practicae medicinae; Tractatus de calcaulis in vesica, probably written 1373. Died 1384.

JACOBI, Karl Gustav Jacob, German mathematician; b. Potsdam, Germany, Dec. 10, 1804; Ph.D., Berlin, Germany, 1825; became asso. prof. math., Königsberg, Germany, 1827, prof., 1829-42; royal pensioner in Berlin, 1842-51. Fellow Royal Soc., 1833; mem. French Acad. Scis., 1830. Author: Fundamenta nova theoriae functionum ellipticarum, 1829; Canon arithmeticus, 1839; Mathem. Werke, 3 vols., 1846-71; Ges. Werke, 8 vols., 1881-91. Developed (independently of abel) theory of elliptical functions; an early founder of theory of determinants; invented functional determinant (the Jacobian); introduced theory of last multiplier; advanced theory of configurations of rotating liquid masses; research on differential equations, calculus of variations, number theory, analytical mechanics and dynamics; expert in devising math. algorithms. Died Berlin, Germany, Feb. 18, 1851.

JACOBI, Moritz Hermann von, see von Jacobi, Moritz Hermann.

JACOBS, Aletta H., Dutch physician; b. 1849; suffragist leader; first woman physician in Holland; opened the first birth control clinic in the world, Amsterdam, Netherlands. Died 1929.

JACOBS, Israel Samson, Am. physicist; b. Buffalo, July 20, 1925; s. Elias Rex and Ida (Feinberg) J.; B.S., U. Mich., 1947; S.M., U. Chgo., 1951, Ph.D., 1953; m. Judith Booth, Sept. 4, 1950; children—Eve, Michael. Teaching asst. U. Chgo., 1949-50; physicist Gen. Electric Research and Devel. Center, Schenectady, 1954—; vis. scientist Lab. of Electrostatics and Physics of Metals, Grenoble, France, 1965-66. Organizing com. Internat. Congress on Magnetism, 1962-67, exec. chmn., 1966-67; adv. bd. Conf. on Magnetism and Magnetic Materials, 1959—, publs. chmn., 1962-64. Fellow Am. Phys. Soc. Research and publs. demonstrated pressure-induced shifts in optical coloring of X-ray colored alkali halides; new processes in magnetization behavior of fine ferro-magnetic particles including permanent magnet behavior and thermally unstable magnetic behavior; magnetization processes in very high magnetic fields for antiferromagnetics and ferrimagnetics. Home: 3060 Rosendale Rd., Schenectady 12309. Office: Gen. Electric Research and Devel. Center, Box 8, Schenectady 12301.*

JACOBS, Konrad, German mathematician; b. Rostock, Aug. 24, 1928; s. Werner and Waldtraut (Eilers) J.; ed. U. Munich, U. Hamburg; Ph.D., 1954, agrégé, 1957; m. Annemarie Kreppel, Feb. 28, 1956; children—Ursula, Susanne, David, Dorothee. Sci. asst., U. Munich, 1956-58; instr. statistics U. Göttingen, 1958-59, titular prof., dir. Inst. Statistics, 1959—. Mem. Union German Mathematicians, Am. Soc. Math. Author: Neuere Methoden und Ergebnisse der Ergodentheorie; Lectures on Ergodic Theory. Home: Merkelstrasse 53. Office: Bunsenstrasse 3, Göttingen, Germany.

JACOBS, Leon, Am. med. microbiologist; b. Bklyn., Mar. 26, 1915; s. Samuel and Evelyn (Rosenthal) J.; B.A., Bklyn. Coll., 1935; M.A., George Washington U., 1938, Ph.D., 1947; m. Eva Eisenberg, Nov. 26, 1946; children—Jonathan H., Alice E., Abby M. Zoologist, NIH, 1937-43, 46-58, chief, lab. parasitic diseases, 1959-64, acting sci. dir. Inst. Allergy and Infectious Diseases, 1964-65, sci. dir. Div. Biol. Standards, 1966-67; dep. asst. sec. for science U. S. Dept. of Health, Education and Welfare, Washington, 1967—. Recipient Arthur S. Flemming award, U. S. Jr. C. of C., 1954; citation Washington Acad. Sci., 1955; Alumnus award Bklyn. Coll., 1955; Henry B. Ward medal Am. Soc. Parasitologists, 1963, Distinguished Service Medal USPHS, 1966. Mem. A.A.A.S., Am. Soc. Parasitologists, Assn. Immunologists, Am. Soc. Tropical Medicine and Hygiene, Soc. Protozoologists, Sigma Xi. Research and numerous publs. in cultivation of parasitic protozoa, immunology of toxoplasmosis; proved toxoplasma gondii to be cause of ocular disease. Home: 3705 Morrison St. N.W., Washington 20015. Office: Dept. Health, Edn. and Welfare, Washington 20201.*

JACOBS, Melville, Am. anthropologist; b. N.Y.C., July 3, 1902; s. Alexander and Rose (Blau) J.; A.B., Coll. City N.Y., 1922; A.M., Columbia U., 1923, Ph.D., 1931; m. Elizabeth L. Derr, Jan. 3, 1931; Faculty dept. anthropology U. Wash., Seattle, 1928—, prof., 1951—; vis. prof. U. Wis., Milw., 1961-62. Mem. Am. Anthropol. Assn., Am. Ethnol. Soc., Linguistic Soc. Am., Am. Folklore Soc. (past pres), Am. Council Learned Socs. (del 1966—), Sigma Xi. Author: A Sketch of Northern Sahaptin Grammar, 1931; Northwest Shatpin Texts, 2 vols., 1934, 37; Coos Texts, 2 vols., 1939, 40; Kalapuya Texts, 1945; (with B. J. Stern) General Anthropology, 1947; Clackamas Chinook Texts, 2 vols., 1958, 59; (with Elizabeth D. Jacobs) Nehalem Tillamook Tales, 1959; The Content and Style of an Oral Literature, 1959; The People are Coming Soon, 1960; Pattern in Cultural Anthropology, 1964; also articles. Descriptive data on vanishing langs., folklores, ethnography Am. Indian peoples N.W. U. S.; devel. theory and methods structural-functional analysis folklores, humor systems. Home: 2016 N.E. Ravenna Blvd., Seattle 98105.*

JACOBS, Merkel Henry, Am. physiologist; b. Harrisburg, Pa., Dec. 6, 1884; A.B., 1905, Ph.D., 1908, U. Pennsylvania; studied Berlin; m. 1912; 1 child. Instr., zoology, U. Pennsylvania, 1909-13; asst. prof., 1913-23; prof. general physiology, 1923-55, emeritus, 1955—; instr., Marine Biology Lab., Woods Hole, 1921-29; assoc. dir., 1925-26, dir., 1927-37, trustee, 1938—. Mem. Nat. Acad. Scis., AAAS, Soc. Zoologists (vice-pres., 1928, pres. 1938); Soc. Gen. Physiol; Internat. Soc. Cell Biol.; Soc. Experimental Biol.; Am. Chem. Soc.; Am. Philos. Soc. Investigated effects of high temperature, carbon dioxide and desiccation on organisms; physiological differences of species; erythrocyte's osmotic properties; cell permeability and diffusion; hemolysis. Address: Dept. Physiology, University of Pennsylvania School of Medicine, Philadelphia, Pennsylvania 19104.

JACOBS, Patricia Ann, Brit. biologist; b. London, Eng., Oct. 8, 1934; d. Cyril and Sadie (Jones) Jacobs; B.Sc. with honors, St. Andrews (Scotland) U., 1956, D.Sc., 1966. Research asst. Mt. Holyoke Coll., Mass., 1956-57; sci. staff Med. Research Council, Edinburgh, Scotland, 1957—. Spl. lectr. dept. medicine U. Edinburgh, 1966. Mem. Genetical Soc. Research, publs. on chromosomes of man especially individuals with abnormal sex chromosome constns., surveys of various populations, described 1st abnormal set chromosome constn. in man, contribution of Y chromosome to abnormal behavior. Home: 80 Falcon Ct., Edinburgh 10. Office: Med. Research Council, Western Gen. Hosp., Edinburgh 4, Scotland.*

JACOBS, Sydney, Am. physician; b. New Orleans, Nov. 23, 1907; s. Wolf Nathan and Hannah (Abramson) J.; B.S., Tulane U., 1928, M.D., 1930; m. Sadie Helen Frumin, June 12, 1938; children—Alan Norman, Jerome Mark, Myron Hillel, Joel Frumin. Practice medicine specializing in diseases of chest; faculty Tulane U., New Orleans, 1932—, prof. clin. medicine, 1961—; staff Touro Infirmary, New Orleans, 1932—, now chief medicine. Cons. to hosps. Recipient So. Tb. Conf. award, 1965. Diplomate Am. Bd. Internal Medicine, Am. Bd. Pulmonary Diseases. Fellow A.M.A., A.C.P.; mem. Am. Coll. Chest Physicians, Am. Thoracic Soc., Am., So., Nat. (pres. 1966-67) tb. assns., Am. Heart Assn., Nat. Rehab. Assn. Research and numerous publs. on diseases of chest especially Tb. Home: 3704 Octavia St., New Orleans 70125. Office: Touro Infirmary, 1400 Foucher St., New Orleans 70115.*

JACOBS, Walter Abraham, Am. chemist; b. New York, N.Y., Dec. 24, 1883; A.B., 1904, M.A. 1905, Columbia U.; Ph.D., Berlin, 1907; m. 1908; two children. Fellow, Rockefeller U., 1907-08; asst., chemistry, 1908-10; assoc., 1910-12; assoc. mem. chemotherapy, 1912-23; mem., 1923-49; emeritus mem., 1949—. Mem., Nat. Acad. Scis.; Am. Chem. Soc.; AAAS; Soc. Biol. Chem.; Harvey Soc.; Soc. Pharmacology. Research on organic compounds; amino and nucleic acids; structure of alkaloids; chemotherapy; cardiac glycosides and sapogenins. Home: 2548 Laurel Pass Road, Los Angeles, Calif. 90046.

JACOBS, William P., Am. biologist; b. Boston, May 25, 1919; s. Vincent Henry and Elizabeth (Kennedy) J.; B.A., Harvard, 1942, Ph.D., 1946; m. Jane Shaw, Mar. 12, 1949; children—Mark, Anne. Research asso. biology Harvard, 1946-47; faculty Princeton, 1948—, prof. biology, 1962—. Fellow Harvard Soc. Fellows, 1947-48; Sheldon Travelling Fellow, 1945-46; Lalor Fellow at Woods Hole, Mass., 1950-51; NSF fellow, 1956-57, 1962; Guggenheim fellow, 1967. Recipient Morrison prize, 1951. Mem. Soc. Study Development and Growth (pres. 1960-61, Am. Soc. Plant Physiologists, Japanese Soc. Plant Physiology, Botan. Soc. Am., Internat. Soc. Plant Morphologists, Internat. Phycological Soc., British Soc. Exptl. Biology, A.A.A.S. Author (with C. E. LaMotte): Regulation in Plants by Hormones, 1964. Contbr. articles in field to sci. jours. Research on the role of plant growth hormones in controlling the regeneration and differentiation of the vascular cells that transport materials in higher plants; the control of longevity of plant organs, and the relation between hormone movement and biol. polarity. Home: 72 Western Way, Princeton, N.J. 08540.*

JACOBS, Woodrow C(ooper), Am. oceanographer, meteorologist; b. Pasadena, Cal., Sept. 11, 1908; s. William Rozel and Mabelle (Cooper) J.; student Va. Mil. Inst., 1926-27; A.B., U. Cal. at Los Angeles, 1930, Ph.D., 1948; M.S., U. So. Cal., 1934; m. Dorothy Cecelia Quinn, June 15, 1933; 1 dau. Marilyn Rozel (Mrs. Wilbur M. Ott). With U.S. Weather Bur., San Diego, Sandberg, Cal., 1931-36, 46-48; with Scripps Instn. Oceanography, 1936-41; chief civilian meteorologist hdqrs. USAAF, 1942-46; with Air Force Air Weather Service, 1948-60; phys. scientist Library of Congress, 1960-61; dir. Nat. Oceanographic Data Center, Washington, 1961-67; dir. Environmental Data Service, Environmental Sci. Services Adminstrn., Silver Spring, Md., 1967—; also lectr. Mem. adv. com. Smithsonian Instn., 1962—; adv. council Mission Bay Research Found. San Diego, 1964—; adv. panel air-sea research program U. S. Dept. Commerce, 1964—; U. S. mem. oceanographic com. UNESCO, 1962—, chmn. group World Meteorol. Orgn., 1965—; U. S. del. com. Internat. Indian Ocean Expdn., 1964, also U. S. observer Internat. Council for Exploration of Sea, Copenhagen, 1964. Recipient Certificate of Appreciation, USAAF, 1946. Mem. Am. Geophys. Union (council 1961), Am. Meteorol. Soc. (asso. editor publs. 1946—, council 1961-64), Royal Meteorol. Soc. London, Washington Acad. Scis., Oceonographical Soc. Japan, Soc. Lymnology and Oceanography, Marine Tech. Soc. Author: Meteorological Results of Curise VII of the Carnegie, 1943; Wartime Developments in Applied Climatology, 1947; The Energy Exchange Between Sea and Atmosphere, 1951; Arctic Meteorology (with Sverre Petterssen), 1956; Meteorological Satellites, 1962. Research, publs. on periodic ceiling height variations, atmospheric nucleation, variations in sea level, minimum temperature forecasting, energy exchanges between ocean and atmosphere, synoptic climatology for mil. planning. Home: 6309 Bradley Blvd., Bethesda, Md. 20034. Office: 8060 13th St., Silver Spring, Md. 20910.*

JACOBSEN, Carlyle, Am. psychologist; b. Minneapolis, Minn., Jan. 17, 1902; s. Christian and Ane (Hanson) J.; student Hamline University, St. Paul, Minnesota, 1921-22; A.B., University of Minn., 1924, Ph.D., 1928; m. Marion Myer, Dec. 17, 1927 (dec.); 1 dau. Patricia; m. 2d, Ellen Townley Cook, August 12, 1958. Instructor in psychology U. of Minn., 1924-28; prof. of medical psychology Washington U., St. Louis, Mo., 1938-46, asst. dean Sch. of Medicine, 1942-46; dean Grad. Coll., State U. of Iowa, 1946-47, exec. dean, div. health sciences and services, 1947-50; exec. dean for med. edn. State U. of N.Y., 1955-57; pres. Med. Center, Upstate Med. Center of State U. of N.Y., 1957—. Exec. v. chmn. Med. Sch. Grants Adv. Com., Ford Found., 1956-57; mem. USPHS Nat. Advisory Council, 1956-61. Awarded Howard Crosby Warren medal for expl. psychology, 1938. Mem. Am. Psychol. Assn., Assn. for Research in Nervous and Mental Disease, Nat. Inst. Gen. Med. Scis. Council, Sigma Xi. Made studies of apes to demonstrate that heart works as hard in everyday actions as in extreme muscular exertion, and that emotional stimulii produce such phys. responses as sweating and blood pressure changes. Home: 8 1/2 Ledyard Av., Cazenovia, N.Y. Office: 766 Irving Av., Syracuse, N.Y. 13210.

JACOBSEN, Erik Sophus Alexander, Danish pharmacologist; b. Copenhagen, Denmark, July 10, 1903; s. Julius Eskild and Alpha (Brammer) J.; M.D., U. Copenhagen, 1928, Ph.D., 1933; m. Henny Gyrithe Andreasen, Sept. 23, 1928; children—Inge Abigael (Mrs. Per Da), Ulla Merete (Mrs. Bent Rune). With Hosp. Service Copenhagen, 1928-33; asst. biochemist Inst. U. Copenhagen, 1930-32; dir. research Medicinalco, 1934-60, Dumex, 1960-62; prof. pharmacology Royal Danish Sch. Pharmacy, Copenhagen 1962—. Decorated Ridder af Dannebrog; recipient Holger Petersen's prize, 1951; MEFA prize, 1962; Thorvald Madsen prize, 1967. Mem. Danish Acad. Tech. Scis., Scandinavian (hon. sec.), Danish, Brit. pharmacological socs., Am. Soc. Pharmacology and Exptl. Therapeutics. Asc. Medica Soc. de Med. Legal y Toxicol. (hon.), Royal Physiographic Soc. Lund Sweden). Author several books including: Cellernes Aanding, 1936; Hormonerne, 1941; Omgang med Alkohol, 1941; also numerous articles. Research on phosphatases, discovery and devel. of disulfiram, benactyzine, devel. psychopharmacological methods. Home: 32 Rahbeks Alle, Copenhagen V, Denmark.*

JACOBSEN, Niels Kingo, Danish geographer; b. Gladsakse, Denmark, June 13, 1921; s. Chr. K. and A. (Schaumburg) J.; cand. mag., U. Copenhagen, 1947; D.Sc., U. Copenhagen, 1964. Staff mem., Skalling-Laboratoriet, 1948-64, dir. 1964—; staff mem., De Danske Vade- og Marskundersogelser, 1953-64, dir., 1964—; staff mem., U. Copenhagen, Geographical Institute, 1958-64, prof. 1964—. Ed., Meddelelser om Gronland, 1947-59, mem. ed. board, 1965; ed., Geografisk Tidsskrift, Folia Geographica Danica, Kulturgeografiske Skrifter, Atlas of Denmark, Meddelelser fra Skalling-Laboratoriet, since 1956. Research in soil geography; arctic regions; land reclamation, especially salt marshes. Address: U. Copenhagen, Frue Plads, Copenhagen, Denmark.

JACOBSEN, Theodor Siegumfeldt, astronomer; b. Nyborg, Denmark, Feb. 6, 1901; s. Hans Christian and Ellen (Siegumfeldt) J.; came to U. S., 1917, naturalized, 1926; A.B., Stanford, 1923; Ph.D. (Lick Obs. fellow) U. Cal. at Berkeley, 1926; Computer, Mt. Wilson Obs., Carnegie Instn., 1922; asst. Lick Obs., Cal., 1926-28; asst. prof. astronomy and math. U. Wash., Seattle, 1928-41, asso. prof., 1941-52, prof., 1952—; acting asst. prof. Stanford, 1937; guest investigator Mt. Wilson, also Palomar Obs., 1953. Mem. Am. Astron. Soc., Assn. Variable Star Observers, Astron. Soc. Pacific, Royal Astron. Soc. London, Royal Astron. Soc. Can., Internat. Astron. Union. Contbr. articles to tech. jours. Research on radial velocity determination, radial velocities cepheid variable stars, green flash at sunset, surface moon. Home: 6205 17th Av. N.E., Seattle 98115.*

JACOBSON, Edmund, Am. physician, physiologist; b. Chgo., Apr. 22, 1888; s. Morris and Frances (Blum) J.; A.B., Northwestern U., 1908; M.A., Harvard, 1909, Ph.D., 1910; M.D., U. Chgo., 1915; LL.D., George Williams Coll., 1962; m. Elizabeth Ruth Silberman, Dec. 1925; children—Ruth Frances (Mrs. Engelbert Ludovicus Grommers), Edmund, Nancy Elizabeth (Mrs. Stanley Engelsberg). Practice medicine specializing in internal medicine, psychiatry, Chgo., N.Y.C., 1917—; research asso., asst. prof. physiology U. Chgo., 1926-36; dir. Lab. Clin. Physiology, Chgo., 1936—; pres. Nat. Found. for Progressive Relaxation, Chgo., 1954—; hon. mem. faculty U. Stuttgart, 1968. Fellow A.C.P., A.A.A.S., Internat. Coll. Angiology; Mem. St. Louis Med. Soc. (hon. life); A.M.A., Am. Physicians Soc. for Physiol. Tension Control (past pres.), Internat. Soc. Internal Medicine, Chgo. Med. Soc., Acad. Psychochem. Medicine. Author several books including Progressive Relaxation, 1929, rev., 1938; You Must Relax, 1934. You Can Sleep Well, 1938; How to Relax and Have Your Baby, 1959; Tension Control for Businessmen, 1963; Anxiety and Tension Control, 1964; Biology of Emotions, Tension in Medicine, 1967; also numerous articles. Research in physiology, psychology, internal medicine, measurements mental activities; pioneered measurement muscular tension in intact man, animals, isolated muscles with amplifier galvonometer assemblies or neurovoltometer; introduced elec. rectification and integration into physiology. Home: 5532 S. Shore Dr., Chgo. 60637. Office: 55 E. Washington St., Chgo. 60602.*

JACOBSON, Eugene Donald, Am. physiologist; b. Bridgeport, Conn., Feb. 19, 1930; s. Morris David and Mary (Mendelsohn) J.; student Ohio Wesleyan U., 1947-48; B.A., Wesleyan U., Middletown, Conn., 1951; M.D., U. Vt., 1955; M.S., State U. N.Y., 1960; m. Joyce Elma Bravender, Dec. 19, 1953; children—Laura Ellen, Susan Ruth, Morris David, Daniel Frederick, Miriam Louise. Faculty dept. physiology Sch. Medicine, U. Cal. at Los Angeles, 1964—, asso. prof., 1965-66; Prof., chmn. dept. physiology U. Okla. Med. Center., 1966—. Mem. Am. Physiol. Soc., Western Soc. for Clin. Research, Soc. for Exptl. Biology and Medicine, Am. Gastroent. Assn. Research and numerous publs. on relationship between blood flow and secretion in stomach. Home: 4620 N.W. 31st St., Oklahoma City 73112. Office: Dept. of Physiology, U. Okla. Med. Center, 800 N.E. 13th St., Oklahoma City, Oklahoma 73104.

JACOBSON, Harold Gordon, Am. physician; b. Cin., Oct. 12, 1912; s. Samuel and Regina (Ditman) J.; B.S., U. Cin., 1934, B.M., 1936, M.D., 1937; m. Ruth Enenstein, Aug. 10, 1941; children—Richard J., Arthur J. Instr. radiology Yale Sch. Medicine, 1942; staff VA Hosp., Bronx, N.Y., 1946-53, chief radiology services, 1950-53, cons., 1958—; dir. dept. radiology Hosp. for Spl. Surgery, N.Y.C., 1953-55; faculty N.Y. U. Coll. Medicine, 1953—, prof. clin. radiology, 1959—; practice medicine specializing in radiology, Galveston, Tex., 1941-42, New Haven, 1942, N.Y.C., 1946—; chief div. diagnostic radiol. services Montefiore Hosp. and Med. Center, N.Y.C., 1955—; prof. radiology Albert Einstein Coll. Medicine, N.Y.C., 1964—. Fellow Am. Coll. Radiology (bd. chancellors 1962—); mem. Radiol. Soc. N.Am. (pres. 1966—), A.M.A., N.Y. State, N.Y. County med. socs., Am. Roentgen Ray Soc., Alpha Omega Alpha. Author numerous books including: (with others) Cardiac Calcifications, 1963; (with Davidoff, Zimmerman) Neuroradiology Workshop, vol. III, Intracranial Lesions Other than Meningiomas, 1967; (with Clarence J. Schein, Wilhelm Z. Stern) The Common Bile Duct: Operative Cholangiography, Biliary Endoscopy and Choledocholithotomy, 1966; also numerous articles, books chpts. Research in anat. status and pathologic abnormalities of vaterian segment, diagnosis and significance of intracardiac calcifications, ant. and physiol. studies lower and upper esophagus, neuroradiology, diseases of skeletal system. Home: 3240 Henry Hudson Pkwy., N.Y.C. 10467. Office: 111 E. 210th St., Bronx, N.Y. 10467.*

JACOBSON, Leon Orris, Am. physician; b. Sims, N.D., Dec. 16, 1911; s. John and R. Patrine (Johnson) J.; B.S., N.D. State Coll., 1935; M.D., U. Chicago, 1939; m. Elizabeth Benton, Mar. 18, 1938; children—Eric Paul, Judith Ann. Intern, U. Chicago, 1939-40, asst. resident medicine, 1940-41, asst. in medicine, 1941-42, instr., 1942-45, asst. prof., 1945-48, asso. dean, div. biol. scis., 1945-51, asso. prof., 1948-51, prof. medicine, 1951—, chmn. dept. medicine, 1961-65, dean division of the biological sciences, 1965—, head hematology sect. U. Chicago Clinics, 1941-62, mem. Inst. Radiobiology and Biophysics, 1949-54; asso. director health Plutonium project Manhattan Dist., 1943-45, dir. health, 1945-46; dir. Argonne Cancer Research Hospital, Univ. Chicago, 1951—. Dir. Packard Instrument Company, Inc. United States rep. First and Second UN Conf. Peaceful Uses Atomic Energy, Geneva, 1955, 58; U. S. rep. WHO conf., Research Radiation Injury, Geneva, 1959. Consultant biology division Argonne Nat. Lab., mem. adv. com. on isotope distbn. AEC, 1952-56; mem. nat. adv. com. on radiation, USPHS, 1961, mem. com. radiation studies, cons. hematology study sect. USPHS; mem. com. cancer diagnosis and therapy NRC, 1949-55; mem. bd. sci. counselors Nat. Cancer Inst., 1963—; lectr. Internat. Soc. Hematology and Internat. Congress Radiology, Eng., France, Norway, Sweden, 1950, Fifth Internat. Cancer Congress, Paris, 1950, Internat. Soc. Hematology, Argentina, 1952, Paris, 1954, others. Recipient Janeway Medal, 1953; Robert Roesler de Villers award, The Lukemia Society, Borden Award med. scis., Association American Medical Colleges, 1962, Modern Med. and Am. Nuclear Soc. awards, 1963; Kennecott lecturer, 1963. Fellow Am. Coll. Physicians; member American Society Clin. Investigation, Assn. Am. Physicians, Soc. Exptl. Biology and Medicine, Central Soc. Clin. Research, Am. Assn. Cancer Research, Internat. Soc. Hematology, A.M.A., National Academy Science, Central Clin. Research Club, Am. Assn. Advancement Sci., Radiation Research Soc., Am. Soc. Exptl. Pathology, Sigma Xi. Author chpts. on specialized items, various med. books; also articles med. jours. Research on radiobiology; radiation injury; effects of radioisotopes and chemotherapeutic agents on neoplastic diseases of the blood forming tissues; hematology. Home: 1222 E. 56th St., Chgo. 37. Office: 950 E. 59th St., Chgo., Ill. 60637.

JACOBSON, Ludwig Levin, Danish surgeon, anatomist; b. Copenhagen, Denmark, Jan. 10, 1783; studied under Cuvier in Paris, 1811-13; served with French army before battle of Leipzig; specialized in comparative anatomy; mem. French Acad. Scis., 1833; produced numerous writings. Described rudimentary canal above vomeronasal cartilage opening in side of nasal septum (organ of Jacobson), 1809, tympanic, or Jacobson's, plexus, 1818, tympanic, or Jacobson's, nerve, 1818. Died Copenhagen, Aug. 29, 1843.

JACOBSON, Martin, Am. chemist; b. N.Y.C., Nov. 11, 1919; s. Harry and Mollie (Bauch) J.; B.S., Coll. City N.Y., 1940; postgrad. George Washington U., 1945-48; m. Nettie Jamnitsky, June 14, 1942; 1 dau., Barbara Diane. Research asst. bacteriology and toxicology NIH, USPHS, Bethesda, Md., 1941-42; research chemist U. S. Dept Agr., entomology research div., Beltsville, Md., 1942—; asso. in chemistry George Washington U., 1947-57. Recipient Dept. Agr. Merit award, 1962, 65, 67, 68; Hillebrand award Chem. Soc. Washington, 1964. Mem. Am. Chem. Soc., Entomol. Soc. Am., Washington Acad. Sci., A.A.A.S., N.Y. Acad. Scis., Am. Oil Chem. Soc. Author: Insect Sex Attractants, 1965; also numerous articles. Research on isolation and identification insecticides and insect attractants from natural products; 1st synthesis of natural insect sex attractant. Home: 8504 16th St., Silver Spring, Md. 20910. Office: Agr. Research Center, Beltsville, Md. 20705.*

JACOBSON, Nathan, mathematician; b. Warsaw, Poland, Sept. 8, 1910; s. Charles and Pauline Ida (Rosenberg) J.; brought to U. S., 1917, naturalized, 1921; A.B., U. of Ala., 1930; Ph.D., Princeton, 1934, Procter fellow, 1934-35; Nat. Research fellow U. of Chicago, 1936-37; m. Florence Dorfman, Aug. 25, 1942; children—Michael Sidney, Pauline Ida. Asst. in mathematics Inst. for Advanced Study, Princeton, 1933-34; lectr. Bryn Mawr Coll., 1935-36; instr. U. of N.C., 1937-38, asst. prof., 1938-40, asso. prof., 1941-42; vis. asso. prof. Johns Hopkins, 1940-41, asso. prof., 1943-47; asso. ground sch. instr. Navy Pre-Flight Sch., 1942-43; asso. prof. Yale, 1947-48, prof. since 1949, now Henry Ford II prof. of math., 1964—; visiting prof. U. Chgo., summer 1947; Fulbright research grant for France, 1951-52, Guggenheim Meml. fellow, 1951-52. Member American Academy of Arts and Sciences, American (councilor 1943-48, editor bull. 1948-54, v.p., 1956-58), French, Japan math. socs., Nat. Acad. Scis., Phi Beta Kappa, Sigma Xi. Author: Theory of Rings, 1943; Lectures in Abstract Algebra, 3 vols., 1953-64; Structure of Rings, 1956; Lie Algebras, 1962. Study of topological algebra; semi-linear transformations; non-associative algebra; structure theory of rings; Galois theory. Home: 2 Prospect Ct., Hamden 14, Conn.

JACOBSON, Norman Leonard, Am. agriculturist; b. Eau Claire, Wis., Sept. 11, 1918; s. Frank R. and Elma (Baker) J.; B.S., U.Wis., 1940; M.S., Ia. State U., 1941, Ph.D., 1947; m. Gertrude A. Neff, Aug. 24, 1943; children—Gary, Judy. Faculty Ia. State U., Ames, 1947-—, prof., 1953-—, Charles F. Curtiss Distinguished prof. agr., 1963-—. Mem. subcoms. Nat. Acad. Sci.-NRC. Recipient Am. Feed Mfrs. Award in dairy nutrition, 1955; Borden award in dairy prodn. Borden Co. Found., Inc., 1960; Moorman Travel award Moorman Mfr. Co., 1966. Fellow A.A.A.S.; mem. Am. Dairy Sci. Assn. (past dir.), Am. Inst. Nutrition, Am. Soc. Animal Sci., Sigma Xi, Phi Kappa Phi, Gamma Sigma Delta. Editor: (with others) Physiology of Digestion in the Ruminant, 1965. Contbr. numerous articles to tech. jours. Research on effect dietary lipids on blood lipids calves, effects feeding antibiotics to cattle, carbohydrate absorption and metabolism in the calf, effect diet on anat. and functional devel. ruminant stomach, efficiency feed utilization by dairy cows, etiology ruminant bloat; codiscoverer, patentee prevention bloat by a combination antibiotics. Home: 2004 Country Club Blvd., Ames, Ia. 50010.*

JACOBSON, Paul, chemist. b. 1859; became prof., Heidelberg, Germany, 1891; sec. German Chem. Soc. Author: (with Victor Meyer) Lehrbuch der organischen Chemie. Became editor Berichte, 1896; revised and edited edits. of Beilstein. Research on phenyl mercaptans, reduction of azo compounds, rearrangement of hydrazo compounds. Died 1923.

JACOBUS, David Dinkel, Am. mech. engr.; b. Jersey City, Feb. 16, 1900; s. David Schenck and Laura (Dinkel) J.: M.E., Stevens Inst. Tech., 1921; Sc.D., Mass. Inst. Tech., 1930; m. Margaret Penman, Feb. 24, 1926; children—David Penman, John Henry; m. 2d, Elinor Hughes, July 14, 1957. With Stone & Webster Engring. Corp., 1922-26; research asso. Mass. Inst. Tech., 1927-30; asst. prof. Stevens Inst. Tech., 1931-40, prof. chemistry, 1946-48; dir. research Keuffel & Esser Co., 1936-42; staff Radiation Lab., Mass. Inst. Tech., 1942-46; head mech. engring. div. Brookhaven Nat. Lab., Upton, N.Y., 1949-57, sr. mech. engr., 1966-—; research fellow in mech. engring. Harvard, 1957-66. Recipient certificate War and Navy Depts., 1947. Mem. Am. Phys. Soc., Am. Soc. M.E., Contbg. author: Radar Scanner Engineering, vol. 126. Design mech. components for radar scanners, graphite nuclear reactor, proton accelerator, electron accelerator. Home: Bellport, N.Y. 11713. Office: Brookhaven Nat. Lab., Upton, N.Y. 11973.*

JACOBY, Harold, Am. astronomer; b. N.Y., Mar. 4, 1865; s. Max and Eve M. (Jackson) J.; A.B., Columbia, 1885, Ph.D., 1895; m. Annie Maclear, Dec. 28, 1890; children—Maclear, Eve (Mrs. Edward Terhune Van de Water). Asst. and instr. geodesy and practical astronomy Columbia, 1888-92, instr. astronomy, 1892-94, adj. prof., 1894-1904, prof., 1904-30, acting dir. Obs., 1903-06, dir., 1906-30. Asst. astronomer U. S. eclipse expdn. to W. Africa in U. S. S. Pensacola, 1889-90. Civilian instr. navigation Submarine Officers' Material Sch., Pelham Bay Naval Training Sta., 1918. Author: Practical Talks by an Astronomer, 1891; Astronomy, a Popular Handbook, 1913; Navigation, 1917. Died 1932.

JACOBY, Henry Sylvester Am. civil engr.; b. Springtown, Pa., Apr. 8, 1857; s. Peter Landis and Barbara (Shelly) J.; C.E., Lehigh U., 1877, E.D., 1941; m. Laura Louise Saylor, May 18, 1880; children—John Vincent, Hurlbut Smith, Freeman Steel. Chief draughtsman U. S. Engr. Office, Memphis, 1879-85; instr. civil engring. Lehigh U., 1886-90; prof. bridge engring Cornell, 1890-1922. Fellow A.A.A.S. (v.p. Sect. D., 1901); hon. mem. Am. Soc. C.E.; mem. Soc. Promotion Engring. Edn. (pres. 1915-16), Am. Ry. Engring. Assn., Washington Acad. Scis., Tau Beta Pi, Sigma Xi. Author: Notes and Problems in Descriptive Geometry, 1892; Outlines of Descriptive Geometry (3 parts), 1895-96-97; TextBook on Plain Lettering, 1897; Text-Book on Roofs and Bridges (4 parts), 1890-98 (with Mansfield Merriman); Structural Details, or Elements of Design in Timber Framing, 1909; Foundations of Bridges and Buildings (with Roland P. Davis), 1914; Timber Design and Construction (with Roland P. Davis), 1929; also numerous papers tech. jours. Died Aug. 1, 1955.

JACOPI, Giulio, Italian archaeologist; b. Trieste, Italy, Sept. 7, 1898; s. Giuseppe and Luciana (Anzulovich) J.; ed. U. Graz, U. Vienna, grad. U. Rome, 1920; post-grad. degree, Scuola Archeologica Italiana, Rome and Athens, 1923; m. Marica Montesanto, Apr. 19, 1919; one dau., Livia Irene. Dir., Monuments and Excavations for Italian Islands of Aegean, 1924-33; head, Ital. arch. expdn. Anatolia, 1935-38; sec., Ital. Inst. Ancient History, 1939-42; Supt. of antiquities; Emilia, Romagna, Reggio, Cababria, 1942-54; supt. antiquities, Rome, 1954-—; docent archaeology Us. Rome and Bologna, from 1933; prof. archaeology, Bologna, 1944-45, Messina, 1946-56; Lecce, 1961-—. Cavalier, Order of Malta, 1937; commendatore della Corona d'Italia, 1942; grande ufficiale al Merito della Republica, 1962; commendatore dell' Orange-Nassau, 1962. Founder, Inst. Storico-Archeologico a Rodi (1st. sec., 1927-33); mem., Inst. of Etruscan Studies; German Archaeological Inst. Author: Clara Rhodos, (vol. I-III); Scavi di Afrodisiade; L'antro di Tiberio a Sperlonga; over 100 pub. articles. Directed important excavations in Rhodes, Anatolia, Caria, Cababria and Ostia; discovered Greek statuary, including what may be original Laocoon group, 1957; dir. exploration of ancient city of Aphrodisia, 1935-38. Address: 72 Via di Monte Verde, Rome, Italy.

JACOX, Harold William, Am. radiologist; b. Detroit, Aug. 31, 1904; s. William Garrett and Cora (McGraw) J.; B.S., U. Mich., 1926, M.D., 1928; m. Lois Kimball, June 21, 1928; children—Judith (Mrs. Norman Peachey), Elizabeth (Mrs. James Warner). With U. Mich., 1929-36, asst. prof. dept. radiology, 1934-36, in charge radiation therapy, 1931-36; dir. dept. radiation, phys. therapy Western Pa. Hosp., Pitts., 1936-47; chief radiotherapy div. Radiol. Services Presbyn. Hosp., N.Y.C., 1947-—; prof. radiology Coll. Physicians and Surgeons Columbia U., 1947-—; cons. radiotherapist to Surgeon Gen. U. S. Army and VA; cons. radiotherapist numerous metropolitan hosps. Mem. Am. Roentgen Ray Soc. (past v.p.), Am. Radium Soc., Radiol. Soc. N.Am. Editor: Medical Year Book of Radiology, 1951-65. Research, publs. in neoplasms, radiation fractures; pulmonary sulcus tumors; diseases of the reticulo-endothelium. Home: 77 Glenwood Rd., Tenafly, N.J. 07670. Office: 622 W. 168th St., N.Y.C. 10032.*

JACOX, Marilyn Esther, Am. chemist; b. Utica, N.Y., Apr. 26, 1929; d. Grant Burlingame and Mary (Dunn) Jacox; B.A., Syracuse U., 1951; Ph.D., Cornell U., 1956. Postdoctoral research asso. U. N.C., Chapel Hill, 1956-58; fellow fundamental research Mellon Inst., Pitts., 1958-62; phys. chemist Nat. Bur. Standards, Washington, 1962-—. Mem. Am. Chem. Soc., Am. Phys. Soc., Sigma Xi. Contbr. articles to profl. jours. Research on prodn. of highly reactive, unstable chem. species in inert environment at very low temperatures for observation of their infrared and ultraviolet absorption spectra for information on geometry and chem. bonding of such species. Home: 10203 Kindly Ct., Gaithersburg, Md. 20760. Office: U. S. Dept. of Commerce, Nat. Bureau of Standards, Washington 20234.*

JACQUELAIN, Victor-Auguste, chemist; b. Goro, Italy, 1802; studied medicine, Paris; asst. to Gay-Lussac at Sorbonne; asst. to J. B. Dumas at L'Ecole Centrale, wrote many memoirs while there, 1832-73; showed allotropic transformation of diamonds from graphite. Died Romaneche-Thorins, 1885.

JACQUEMAIN, René, French chemist; b. Ferrières, France, Apr. 7, 1898; s. Charles and Marie (Brunet) J.; Bachelier, 1916; Chem. engr., 1922; doctorate Besançon (France) U., 1929; D.Sc., 1943; m. Jeanne Baillet, Apr. 23, 1924; 1 dau., Françoise. Became asst. (tutor) Faculty Sci. Besançon, 1922, head instr., 1947, prof. indsl. chemistry, 1952, head chem. inst., 1952, dean, 1958-—, also prof. med. sch., 1947; apptd. lectr. Faculty Pharmacy, Nancy, France, 1948. Mem. French Chem. Soc. Research and publs. on local anesthetics, organic syntheses, distillation of wood. Home: 2, rue Felix Vielle, Besançon (25), France.*

JACQUEMIER, Jean Marie, obstetrician; b. Tutegny, France, 1806; author of Recercé de anatomie et de physiologie sur le developpement des etres organises, 1837; Manuel de Obstetrique, 1845; Manuel de accouchements, 1841; Developpement de l'oeuf humain, 1851. Described bluish or violet-colored spot on vaginal mucous membrane just below the urethral opening which appears in early pregnancy (Jacquemier's sign), 1838. Died 1879.

JACQUEMONT, Victor, explorer; b. Paris, France, Aug. 8, 1801; s. Frederic Jacquemont; student Imperial Sch., until 1816; then studied chemistry with Thenard, also botany, natural history; went to Am. and Haiti, 1826; sent by Jardin du Poi on expdn. to India, 1828; explored Himalayas, Tartar, China, Kashmir, Tibet, 1830. Author: Voyage dans l'Inde, 4 vols., 1828-32. Collected information on geology, geography, weather, plant and animal life of regions he explored. Died Bombay, India, Dec. 7, 1832.

JACQUES, William Henry, Am. engr., naval architect; b. Phila., Dec. 24, 1848; ed. pub. schs., Newark, N.J.; grad. U. S. Naval Acad., 1867; remained in service until 1887; resigned, 1887, to inaugurate mfr. of heavy ordnance and armor at works of Bethlehem Iron Co.; served as asst. in U. S. Coast and Geod. Survey. Mem. internat. jury on marine transp. and war material, Chicago Expn., 1893. Mem. bd. visitors U. S. Naval Acad., 1905. Author: The Establishment of Steel Gun Factories in the United States, 1884. Introduced into U. S. the fluid compression and hydraulic forging of heavy masses of steel; invented reforging process. Died Nov. 23, 1916.

JACQUESSON, Raymond Paul Pierre, French physicist; b. Vernouvilliers, France, Apr. 22, 1900; s. Paul and Aline (Guyot) J.; student Ecole Normale Supérieure de St.-Cloud, 1919-21; Agrégé de Faculté des Sciences, Poitiers, France, 1927, Dr.es sc., 1943; m. France Guillon, Aug. 3, 1925; children—Pierre, Jean-Claude, Eveline (Mrs. Mathe). Prof., Ecole primaire supérieure de Poitiers, 1924-28, Lycée de Nantes, 1928-45, Faculté des Sciences de Poitiers, 1946-—; sci. staff Commissariat à l'Energie Atomique. Decorated commandeur l'Ordre des Palmes académiques; chevalier la Légion d'Honneur. Mem. Groupement français d'Edudes de Contraintes (hon. pres.), Société français de Physique, Sté français Métallurgie, Group. français de Rhéologie. Research, publs. on structural changes of metal according to mech. or heat treatment, influence of variations of structures on phys. and mech. properties especially spontaneous variations of damping capacity of metals, activation of structural changes by mechanical energy, disperted phases in metals. Home: 73 rue Ranaudot, Poitiers, France.*

JACQUET, Lucien, French dermatologist; b. Sauviat, France, 1869; named physician of hosps., 1896; dermatologist St.-Antoine Hosp., Paris; a dir. Atlas iconographique des affections cutanées St.-Louis Hosp. Mus. Author: (with Brocq) Précis élémentaire de dermatologie, 1898; (with Besnier and Brocq) Pratique dermatologique, 1900-04. Showed influence of food and alcohol on skin diseases; studied eczema in children, also syphilitic lesions; described dermatitis (gluteal erythema or Jacquet's disease) in diaper area of babies, 1889. Died Royan, France, 1914.

JACQUEZ, John Alfred, physiologist, biomathematician; b. Pfastt, Alsace, France, June 26, 1922; s. Albert and Victorine (Oestermann) J.; came to U. S., 1929, naturalized, 1937; m. Marianne Reibel, Mar. 20, 1948; children—Albert, Nicholas, Geoffrey, Philip. Staff, Sloan-Kettering Inst., N.Y.C., 1947-53, 56-62, asso. mem., 1960-62; vis. investigator Rockefeller Inst. for Med. Research, 1947-48. faculty Cornell U., 1952-53, asst. prof., 1956-58; fellow in medicine, mem. Center for Cancer and Allied Diseases, N.Y.C., 1956-63; asso. prof. physiology Med. Sch., U. Mich., Ann Arbor, 1962-—, asso. prof. biostatistics Sch. Pub. Health, 1962-—. Member of the scientific advisory committee The Ore. Regional Primate Research Center, Beaverton, 1963-—. Mem. Am. Assn. for Cancer Research, A.A.A.S., The Harvey Soc., Am. Physiol. Soc., Biophys. Soc., Assn. for Computing Machinery, Soc. for Indsl. and Applied Math. Contbr. numerous articles to tech. jours. Research on active transport amino acids, permeability and transport cancer chemotherapeutic agts., mathematical modeling biol. systems, digital computations on such math. models.

JACQUIER, François, French physicist; b. Vitry-le-Francois, France, June 7, 1711; ed. Rome, Italy; became friar Minor, 1717; named tchr. scriptures Coll. de propagande, 1733; became prof. exptl. physics Collegio Romano, 1746, prof. math., 1773; named tutor to infante Don Fernando, Parma, Italy, 1763. Fellow Royal Soc., 1741; mem. French Acad. Scis., 1743, other European sci. socs. Author: (with Leseur) Isaaci Newtoni philosophiae naturalis principia mathematica, 1739-42; Elementi de perspettiva secondo i principi di Taylor, 1755; Eléments de calcul intégral (then the best work on integral calculus), 1768; Trattato intorno la sphera, 1775; Elogio academico del celebre mathematico, 1786. First to determine apogee of moon, 1742. Died Paris, July 3, 1788.

JACQUIN, Nicolas-Joseph (Baron), botanist; b. Leyden, Netherlands, Feb. 16, 1727; ed. Antwerp, Belgium, Leycle, M.D. Leyden, Paris; studied under Bernard de Jussieu. Apptd. imperial botanist by Francis I., 1752; made bot. expdn. to Am., 1754-59; became councillor for mines, prof. chemistry, mining and metallurgy at Schemnitz; prof. chemistry and botany U. Vienna Fellow Royal Soc., 1788; mem. French Acad. Scis., 1804. Author: Hortus botanicus Vindobonensis, 3 vols., 1770-76; Florae Austriacae (flora of Austria), 5 vols., 1773-74. also illustrated books on Viennese gardens; retired 1797. Created baron, 1806. Discovered 60 new plants; studied plants in Caribbean islands, 1760; introduced Am. plants and animals into gardens of Schönbrunn palace; genus Jacquinie named in his honor by Linnaeus; defended Joseph Black's fixed air against Meyer's "acidum pingue." Died Vienna, Oct. 24, 1817.

JACQUINOT, Pierre, French physicist; b. Frouard, France, Jan. 18, 1910; s. Georges A. and Eugenie (Vicq) J.; student U. Nancy (France), 1927-32; agrégation de physique, 1932; Ph.D., U. Paris, 1937; m. Françoise, Touchot, June 3, 1937; children—Cecile (Mme. Loubignac), Jean, Jacques-François, Denis. Research fellow Centre Nat. de la Recherche Scientifique, Paris, 1933-42, dir. gen., 1962-—, also dir. Aimé Cotton Lab.; prof. U. Clermont-Ferrand (France), 1942-46, U. Paris, 1946-—. Mem. French Acad. Scis., French Soc. Physics (pres. 1958-59), Am. Phys. Soc., Optical Soc. Am. Contbr. papers to jours. Research on optical diffraction, new methods of optical spectroscopy, including interference spectroscopy by means of interferometers, Fourier transformation; atomic spectroscopy (Zeeman effect, forbidden lines, nuclear effects). Home: 16, rue Pierre Curie. Office: 15, quai Anatole France, Paris, France.*

JAEGER, Alphons Otto, chemist; b. Bergzabern, Palatine, Germany, Oct. 14, 1886; s. Philip and Scholastica Adolphine Wilhemine (Stoecker) J.; student Inst. Tech. of Friedberg (Germany), 1906-07, U. Zurich (Switzerland), 1907, U. Basel (Switzerland), 1911, B.S., M.A., Ph.D., 1913; m. Hedwig Maria Wuermell, Sept. 9, 1920; children—Carl Heinz, Raymond Alphons, Marian Scholastica, Lucia Constancia. Came to U. S. 1923, naturalized, 1929. Research chemist Badische Anilin & Soda Fabrik, Germany, 1914-23. group leader research dept. Nat. Aniline & Chem. Co., Buffalo, 1923-25; cons. chemist, 1925-26; tech. dir. The Selden Co., Bridgeville, Pa., 1926-29, v.p., 1929-34; gen. mgr., research dir. Selden div. Am. Cyanamid & Chem. Corp., 1934-38, chmn. devel. com. Am. Cyanamid Co., N.Y.C., from 1938, dir. gen. tech. div. Mem. Am. Assn. Textile Chemists and Colorists, Am. Chem. Soc., Am. Soc. for Testing Materials, Internat. Soc. Leather Trades Chemists, Chem. Soc. (Eng.). Awarded numerous patents relating to apparatus and catalysts for Selden and Monsanto contact sulfuric acid processes, and with catalytic processes, apparatus, and catalysts for organic oxidations, oil cracking, catalysts, ammonia and Co oxidation catalysts, etc., dye intermediates and dyes, numerous organic acids, wetting agents and other textile, leather and paper chemicals, paint, lacquer, varnishes, terpenes, Talloil. Contbr. tech. publs. Died July 21, 1953.

JAEGER, Robert Gottfried, German radiologist; b. Berlin, May 6, 1893; s. Wilhelm and Johanna (Griessbauer) J.; ed. univs. Munich, Marburg, Berlin; Ph.D.; m. Mila Fresenius, Mar. 30, 1936. Prof., Gutenberg U., Mayence; supporting cons. for govt., mem. Inst. Tech. Physics of Reich; officer IAEA; liaison officer WHO, Geneva; mem. group experts Internat. Bur. Weights and Measures, Paris; mem. Standardization Commn. for Radiology, Euratom, Brussels, Internat. commn. on Radiology. Mem. German Physicists Soc., German Rontgen Soc., Asso. Brit. Hosp. Physicians. Author: Physikalisch-Technische Daten für Dosimetrie und Strahlenschutz, 1959, also numerous publs. on radiation measurement, standard dosimetry, bases for protection against radiation, quasi-adiabatic calorimeter for measure of dose in rad, 1963. Home: Otto-Weissstrasse 10, Bad Nauheim. Office: Klin. Strahleninst., University of Mainz, Langenbeckstrasse 1, Germany.

JAENICKE, Rainer, German biophysicist; b. Frankfort/Main, Germany, Oct. 30, 1930; s. Johannes and Erna (Buttermilch) J.; student U. Marburg/Lahn; Dipl. Chem., U. Frankfort/Main, 1954, Dr.phil.nat., 1957; m. Agathe Calvelli-Adorno, June 13, 1956; children—Susanna, Johannes, Alexander. Asst., U. Frankfort/Main, 1954-63, docent phys. chemistry, chief asst. Inst. Phys. Biochemistry, 1964—. Research and publs. on dielectric analogue of streaming birefringence; heat aggregation and denaturation of proteins; molecular structure of proteins in solution, especially quaternary structure of globular proteins. Address: 38 Im Heidenfeld, 6 Frankfort/Main 21, West Germany.*

JAENSCH, Erich R., German psychologist; b. Breslau, Germany, Feb. 26, 1883; ed. under Müller; doctorate Göttingen, Germany, 1908; prof., Marburg, Germany, from 1913. Author: Zur Analyse der Gesichtswahrnehmungen, 1909; Ueber die Wahrnemung des Raumes, 1911; Ueber den Aufbau der Wahrnehmungswelt, 1923; Die Eidetik, 1925; Grundformen menschlichen Seins, 1929; Uber den Aufbau des Bewusstseins, 2 vols., 1930; Zur Eidetik und Integrationstypologie, 1941. Studied sight and space perception; analyzed eidetics; tried to establish a closer relation between psychology and philosophy. Died Jan. 12, Marburg, Germany, 1940.

JAFFE, Max, German biochemist; b. Grünberg, Germany, 1841; prof. pharmacology, Breslau, also Königsberg, Germany; discovered urobilin in urine, 1866, reported 1868; isolated indican from urine, 1877; credited with discovering some pathogenic bacteria. Died 1911.

JAFFE, Rudolf, physician; b. Berlin, Germany, Oct. 4, 1885; s. Benno and Helen (Salomon) J.; ed. univs. Freiburg, Berlin, Munich; M.D.; M.D. honoris causa, U. Caracas, U. Frankfort; prof. honoris causa, U. de los Andes en Merida-Venezuela; m. Emilie Fellner, Apr. 2, 1912; children—Helmuth, Werner, Erwin, Ilse. Med. asst. in chief Inst. Pathology, Frankfort, 1917-25; dir. Inst. Pathology, Moabit-Berlin, 1925-36; titular prof. anat. pathology Central U. Caracas (Venezuela), 1936—. Mem. Assn. Profs. Anat. Pathology (hon. life pres.), German Assn. Prof. Anat. Pathology, also hon. mem. various Venezuelan med. socs. Research and numerous publs. on cirrhosis of liver, appendicitis, tropical diseases, auto-allergy of heart. Home: La Florida, av. Los Mangos, Quinta descanso, Caracas. Office: Institute of Anatomical Pathology, Central University of Caracas, Caracas, Venezuela.

JAFFE, Sigmund, Am. chemist; b. New Haven, Mar. 1, 1921; s. Morris and Rose (Blosveren) J.; A.B. with high distinction in Chemistry, Wesleyan U., Middletown, Conn., 1949; Ph.D., Ia. State U., 1953; m. Elaine Leventhal, Aug. 25, 1946; children—Matthew Lee, Paul Jonathan. Researcher, Ames (Ia.) Lab., 1949-53, labs. Air Reduction Corp., 1953-58; prof. chemistry Cal. State Coll. at Los Angeles, 1958—, chmn. dept., 1958-64; researcher Jet Propulsion Lab., Pasadena, Cal., 1960-64. NIH fellow Wiezmann Inst. Sci., Israel, 1964-65. Mem. Am. Chem. Soc., Phi Beta Kappa, Sigma Si, Phi Lambda Upsilon. Research and publs. on rare earths, carbides, metal and high temperature inorganic reactions, solid propellant fuel systems, photo-chemistry and gas phase kinetics; recently research has revealed the nature of several oxygen atom reactions which are of specific interest in the study of the photo-chemistry of urban air pollution. Home: 326 E. Loma Alta Dr., Altadena, Cal. Office: Dept. Chemistry, Cal. State Coll., Los Angeles 90032.*

JAFFIOL, C., French physician; b. Anduze, France, July 10, 1933; s. M. and O. (Valette) J.; ed. Faculté de Médecine, Montpellier, France; m. July 28, 1958; children—Catherine Jaffiol, Françoise. Prof. agrégé Faculté de Médecine. Mem. French Endocrinology Soc., French Diabetology Soc. Research and numerous publs. on multiple myeloma (origin of disprotidose), 1962; thyroid (isotopic techniques); automation of glycerine recording. Home: 4 rue des Tourterelles, Montpellier 30, France.*

JAGENDORF, Andre T., Am. plant biochemist; b. N.Y.C., Oct. 21, 1926; s. Moritz Adolf and Sophie (Sokolsky) J.; B.A., Cornell U., 1948; Ph.D., Yale, 1951; m. Jean Elizabeth Whitenack, June 12, 1952; children—Suzanne E., Judith C., Daniel Z. S. Fellow, U. Cal. at Los Angeles, 1951-53; faculty Johns Hopkins, Balt., 1953—, prof. plant biochemistry, 1965—. Named Outstanding Young Scientist award Md. Acad. Sci., 1961; Weizmann fellow, 1962; recipient Kettering Research award, 1963. Fellow A.A.A.S.; mem. Am. Soc. Plant Physiologists, Am. Soc. Biol. Chemists, Am. Soc. Cell Biology, Soc. Gen. Physiologists, Phi Beta Kappa, Sigma Xi. Research, numerous publs. on mechanism photosynthesis, especially conversion from light energy to phosphate bond energy, pathway electron transport in isolated chlorplasts, existence of obligate coupling between electron transport and phosphorylation; discovered existence of high energy condition in illuminated chloroplasts with phosphorylation occurring in dark in second step, existence of hydrogen ion migration during course electron transport, formation ATP without electron transport using artificial changes hydrogen ion concentration. Home: 3629 Courtleigh Dr., Randallstown, Md. 21133. Office: Johns Hopkins, Charles and 34th St., Balt. 21218.*

JÄGER, Emilie, mineralogist; b. Atzelsdorf, (Austria, Jan. 4, 1926; Michael and Maria (Frey) J.; Dr. Degree, U. Vienna, 1952. With U. Bern (Switzerland), 1952—, prof. mineralogy, 1965—. Research and publs. on age determinations of minerals and rocks from the Alps. Home: 13 Finkenrain, Bern, Switzerland.*

JÄGER, Gustav, German zoologist; b. Bürg am Kocher, Germany, June 23, 1832; ed. Urach, Stuttgart, Tübingen, Germany, Vienna, Austria; founder, dir. zool. garden, Vienna, until 1866; prof., Stuttgart. Author: Die Darwinsche Theorie, 1869; Deutschlands Tierwelt nach irhen Standorten, 2 vols., 1874; Handwortenbuch der Zoologie, Anthropologie und Ethnologie, 8 vols., 1880-1900; Die Normalkleidung (advocated hygienic clothing named after him), 1880. Helped spread Darwinian thought; anticipated Weismann's theory in an account of reprodn. of protoplasm, 1876. Died Stuttgart, May 23, 1916.

JAGER, Otto Arnold, physician; b. Brussels, Aug. 21, 1900; s. Emil and Emilie Jager; ed. univs. Marburg, Kiel, Frankfort, Munich; M.D.; m. Selke Eschholz, July 20, 1950; children—Nora, Benvenuto, Jorinde, Angela. Pediatrician, 1929-45; head central adminstrn. for pub. health in Germany, 1945-48; personal physician to Shah of Iran, 1950-53; counselor WHO, 1953-62; dir. dept. tng. centers German Found. for Underdeveloped Countries, 1962—. Author: Miniatures éthiopiennes, 1957; Ethiopia, 1961; Problèms de la santé publique dans les pays en voie de développement, 1962; La santé publique dans les pay en voie de développement. Address: Reiherverder, 1 Berlin, 27, Germany.

JAGER, Rolf, German biologist; b. Jena, Germany, Sept. 27, 1905; s. Carl and Margarete Jager; ed. univs. Erlangen, Munich, Frankfort; Ph.D. in natural sci.; m. Maria Jager, July 14, 1950; children—Werner, Bernhard, Walpurga, Lorenz. Mem. Research Inst. Colloids, U. Frankfort, 1939, dir. research, 1948, instr., 1951, venia legendi agrégé, 1959. Mem. Colloid Soc., Soc. German Chemists, Senckenbergische Naturforschende Gesellschaft. Author: Wissenschaftliche Forschungsberichte; also publs. on diseases of epidermis and colloido-biol. principles of pathogenesis of pneumoconioses. Address: Immanuel-Kantstrasse 10, 638 Bad Hombourg, Germany.

JÄGERSTEN, Karl Gustav (Gösta) Magnus, Swedish zoologist; b. Öglunda, Sweden, Dec. 3, 1903; Dr.philos., Uppsala (Sweden) U., 1935; m. Eva Aminoff, Dec. 24, 1936; children—Bo Gunnar, Ulla Birgitta (Mrs. Dic Aronson). Prof. zoology U. Stockholm (Sweden), 1945-47; prof. zoology U. Uppsala, 1947—, dir. Zool. Inst.; dir. Klubban's Biol. Sta., Fiskebäckskil, Sweden. Mem. Swedish Acad. Sci. Author: Life in the Sea, 1959. Research on devel., morphology, ecology in annelids, myzostomids, entoprocts, pogonophors; phylogeny of Metazoa; developed Bilaterogastraea theory, 1955, 59. Home: 12, Norbyvägen, Uppsala, Sweden.*

JAGGAR, Thomas Augustus, Am. geologist. b. Phila., Jan. 24, 1871; s. Thomas Augustus and Anna Louisa (Lawrence) J.; B.A., Harvard, 1893, A.M., 1894, Ph.D., 1897; D.Sc., Dartmouth, 1938; LL.D., U. Hawaii, 1945; postgrad. Munich U., 1894, Heidelberg U., 1895; m. Helen Kline, Apr. 15, 1903; children—Kline Eliza Bowne; m. 2d Isabel P. Maydwell, Sept. 17, 1917. Instr. geology Mass. Inst. Tech., 1904-17, head of dept., 1906-12. Asst. geologist U. S. Geol. Survey, 1898-1904, in charge of work in S.D., Ariz. and Mass.; conducted volcano expdns. to Martinique, 1902, Vesuvius, 1906, Aleutian Islands, 1907, Hawaii and Japan, 1909, Costa Rica, 1910, Sakurajima, Japan, 1914, New Zealand, 1920; investigated Tokyo Earthquake, 1923, Alaskan volcanoes, 1927; leader of Nat. Geog. Soc. Pavlof Expdn., 1928; directed Aleutian expdns., 1929, 31, 32; geologist U. S. Naval Obs. Eclipse Expdn. to Niuafoou, Tonga, 1930; established volcano expt. sta., Hawaii, 1911; volcanologist in charge Hawaiian Volcano Obs., U. S. Weather Bur., 1919-24, U. S. Geol. Survey, 1924-35, chief of sect. of volcanology, 1926-35, with stations in Hawaii, Cal. and Alaska; volcanologist U. S. Nat. Park Service, 1935-40; later research asso. in geophysics, U. Hawaii. Recipient Burr prize for devel. amphibian vehicles, Nat. Geog. Soc., 1945. Dir. expeditions through research fellows of Hawaiian Volcano Research Assn., in Hawaii, also in Chile, 1929-30. Fellow Am. Acad. Arts and Scis. Washington Acad. Scis. (non-resident v.p., 1918). Author: Volcanology, 1931; Volcanoes Declare War, 1945; Origin and Development of Craters, 1947; Union through the ages, 1948; Steam Blast Volcanic Eruptions, 1949; Abrasion Hardness, 1950; My Experiments with Volcanoes, 1952. Contbr. to memoirs of Geol. Soc. Am., reports of U. S. Geol. Survey, to scientific jours., reports and bulls. of Hawaiian Volcano Obs.; also editor of Volcano Letter. Died Jan. 17, 1953.

JAGODZINSKI, Ernst Heinz, German mineralogist; b. Aschersleben, Apr. 20, 1916; s. Ernst and Sophie (Papajewski) J.; Ph.D., U. Göttingen; m. Margarethe Brandenburg, May 30, 1942; children—Wolfgang, Erika, Ingeborg. Asst., instr., head dept. Max Planck Inst. Silicate Research, Wurzburg; prof. U. Wurzburg; full prof. Technische Hochschule, Karlsruhe, U. Munich. Mem. German Physicists Soc., German Soc. Mineralogists, German Soc. Metallography, Max Planck Soc. Author: Handbook of Physics, also publs. on disorder in and structure of crystals. Home: Muhlfeldstrasse 23, 8033 Planegg. Office: Institute for Crystalography and Mineralogy, Luisenstrasse 37, 8 Munich, Germany.

JAHANGIR (or Jehangir, Jehan-Geer, Djehanguire, Djahanguir, Djahan Ghyr, Dschehangir, Prince Selim), emperor of Hindustan 1605-27; b. 1569; s. Akbar the Great; rebelled against father and carried on long wars in Deccan; patron of art, lit., architecture; emperor when British first came to India. Author: Tuzuk-i-Jehangiri (memoirs, translated by Rodgers and Beveridge 1909-14). His memoirs include extensive information of flora and fauna of India, astronomy, chem. tech., epidemics, biographies of scientists of his time. Died 1627.

JAHN, Alfred, Polish geographer; b. Lwów, Poland, Apr. 22, 1915; s. Fryderyk and Helena (Jaworska) J.; M.Sc., U. Lwów, 1937, D.Sc., 1939; m. Maria Szaynowska, Aug. 4, 1940; children—Jadwiga, Maria. Docent geography Maria Curie Sklodowska U., Lublin, Poland, 1945-49; docent U. Poznan (Poland), 1946. prof. U. Wroclaw, 1949—, prorector, 1959-62, rector, 1962—; sci. expdns. Greenland, 1937, Spitsbergen, 1957, 58, Alaska, Iceland, 1960. Decorated Gold Cross of Merit, Commandrery Cross of Order Poland's Revival; recipient medal U. Liège (Belgium), 1959. Mem. Internat. Geog. Union, Commn. on Evolution of Slopes (co-pres. 1964—, Periglacial Com., Internat. Assn. Univs. Author several books including: Lublin Plateau, 1956; Alaska, 1966; also numerous articles. Exploration of Polar region especially relating to frost activity/permafront, patterned ground, tundra, study of erosion, evolution of slopes, origin of loess; study of ice age glaciation of Poland; originated concept of denudational balance of slopes. Home: 24, Pugeta, Wroclaw, Poland.*

JAHN, Else, Austrian biologist; b. Klagenfurt, Aug. 28, 1913; d. Franz and Else (Supersperg) Jahn; Ph.D., U. Vienna. Staff chemistry asst., 1938-39; collaborator U. Vienna, 1939-40, dir. asst. services Inst. Sylviculture, Sch. Agr., 1940-45, now instr., full prof.; cons. scientist for forestry protection of Tyrol, Innsbruck, 1946-54; dir. entomol. services Fed. Exptl. Forestry Sta., Vienna, Innsbruck, 1954—. Mem. German Soc. Applied Entomology, German Soc. Agr., Austrian Soc. Pure and Applied Biophysics. Author: Insektenvisen; Heimatklange; also numerous articles on entomology and zoology. Home: Fasangartenstrasse 5-7/11 8. Office: Federal Exptl. Forestry Station, Obertiroler Garten, Vienna XIII, Austria.

JAHN, Otto, German archeologist; b. Kiel, Germany, June 16, 1813; prof. in Greifswald, Leipzig, also Bonn (all Germany); lit. and music historian. Author: Mozart, 4 vols., 1856-59; Beschreibung der Vasensammlung Ludwigs I. von Bayern, 1845; Griechischer Bilderchroniken, 1873; Persius, 1843. Applied critical methods to classical archeology. Died Göttingen, Germany, Sept. 9, 1869.

JAHNS, Richard Henry, Am. geologist; b. Los Angeles, Mar. 10, 1915; s. Alfred H. and Cecelia (Schnackenbeck) J.; B.S., Cal. Inst. Tech., 1935, Ph.D., 1943; M.S., (teaching fellow) Northwestern U., 1937; m. Frances M. Hodapp, Sept. 5, 1936; children—Alfred, Jeannette. Mem. faculty Cal. Inst. Tech., 1937-60; prof. geology Pa. State U., 1960-65, chmn. division earth scis., 1960-61, dean Coll. Mineral Industries, 1962-65; prof. geology, dean Sch. Earth Scis., Stanford, 1965—. Mem. U. S. Geol. Survey, 1937—, sr. geologist, 1949—. Mem. A.A.A.S., Am. Assn. Petrol. Geol., Am. Geophys. Union, Am. Inst. Mining and Metall. Engrs., Geol. Society Soc. Am., Assn. Engring. Geologists, Am. Inst. Profl. Geologists, Am. Assn. U. Profs., Mineral. Soc. Am., Soc. Econ. Geologists, Soc. Vertebrate Paleontologists. Geochem Soc., Sigma Xi (nat. lectr. 1965). Author: Hand Speciman Petrology. Editor: Geology of Southern California. Assistant editor of Economic Geology. Research and publs. on economic, glacial and structural geology, and petrology; also advanced knowledge of landslides and other elements of ground instability as affect, and are influenced by, works of man; developed theories for the genesis of pegmatites, other grantitic rocks, and related mineral deposits. Home: 2312 Branner Dr., Menlo Park, Cal. 94026. Office: Stanford U., Stanford, Cal. 94305.*

JAIRAJPURI, M. Shamim, Indian nematologist; b. Azamgarh, India, Apr. 8, 1942; s. M. Moazzam and Qudsia J.; B.Sc., Aligarh Muslim U., 1959, M.Sc., 1961, Ph.D., 1964; m. Durdana S. (Durdana Faroogr), Dec. 19, 1965, 1 daughter, Sabiha Qudsi. Research fellow dept. zoology Aligarh Muslim U., Aligarh, India, 1961-64, lectr. nematology, 1964—; research officer, prin. investigator I.C.A.R. Cotton Nematode Research Scheme, 1966-67. Mem. Helminthological Soc. Washington, Soc. European Nematologists, Sci. and Technol. Soc. Kanpur. Research and publs. on ecology, biology and pathogenecity of nematodes; discovered and described large number of nematodes found asso. with plants. Home: Ashraf Manzil, Amir Nishan, Aligarh, U.P., India. Office: Section of Nematology, Dept. Zoology, Aligarh Muslim U., Aligarh, U.P., India.

JAKIMOVSKI, Ammon, Israeli mathematician; b. Jerusalem, Israel, June 11, 1925; s. Baruch and Bella (Rappoport) J.; student Hebrew U., Jerusalem, 1944-47, M.Sc., 1952, Ph.D., 1955; m. Lea Herenzon, Feb. 2, 1948; children—Joram, Esther, Dov. Vis. lectr. Tel-Aviv (Israel) U., 1956, sr. lectr., 1962-63, asso. prof., 1964; research fellow Hebrew U., 1957-63; sr. lectr. Technion (Israel Inst. Tech.), Haifa, 1963-64, asso. prof. math., 1965—. Mem. Israel Math. Union (president, 1966), Am. Math. Soc., London Math. Soc. Research and publs. on summability, Tauberian theorems, inversion of integral transforms, spl. methods of approximation. Home: 21 Dizengovf, Tel-Aviv, Israel.

JAKOBI, H., zoologist, physiologist; b. Erlangen, Germany, Apr. 20, 1928; s. Joseph and Katharina (Dotterweich) J.; Dr.rer.nat., U. Erlangen-Nuernberg (Germany), 1951; Dr.habil., U. Parana (Brazil), 1962; m. Ivette Zanello-Dra, Dec. 18, 1952; children—Hans Hyperides, Paul Robert, Heinz Roland, Rolanda Maria. Vol. asst. Zoology Inst., U. Erlangen-Nuernberg, 1951; researcher U. Parana, 1952-56, prof. animal physiology zool. dept., 1961—; researcher-in-chief NRC, Brazil, 1957-60. Mem. Sociedad Paranaense de Ciencias Naturais (pres. council), Sociedade Brasileira para o Progresso da Ciencia (mem. council), Union Internationale pour l'Etude des Insectes Sociaux. Cons. editor Ency. Sci. and Tech. Research, publs. on systematics and ecology of syncarida and Harpacticoidea, exptl. ecology of sand layer inhabiting microfauna, catalase in Crustacea, solubility of wild bee wax, thiocyanate and formation of malign tumor in higher digestive tube, oxygen consumption of interstitial fauna. Home: 2180 Al.Da.Izabel, Curitiba, Paraná, Brazil.*

JAKOBOVITS, Antal, Hungarian pathologist, gynecologist; b. Torokkanizsa, Hungary, Feb. 27, 1925; s. Dezso and Piroska (Forgach) J.; M.D., U. Szeged, 1951; m. Rozalia Molnar, Mar. 24, 1951; children—Kynga, Akos. Mem. faculty dept. pathology U. Szeged, 1951-53, dept. obstetrics and gynecology, 1953—; mem. dept. obstetrics and gynecology Humboldt U., Berlin, 1958-59, U. Hamburg, 1964-65. Fellow Hungarian Soc. Pathology, Hungarian Soc. Obstetrics and Gynecology, Hungarian Soc. Endocrinology. Author: Endokrinologie des Ovars. Physiologie, Pathologie und Klinik, 1965. Studies, publs. on endocrinology of ovary; androgenic effect of male fetuses on the mother. Home: Jozsef Attila sut. 91, Szeged. Office: Semmelweis u.1, Szeged, Hungary.*

JAKOBY, William Berhard, biochemist; b. Breslau, Germany, Nov. 17, 1928; s. Leo and Sally (Hauser) J.; came to U. S., 1937, naturalized, 1946; B.S., Bklyn. Coll., 1950; Ph.D., Yale, 1954; m. Ruth Kerr, June 24, 1956; children—Michael, Robert. USPHS fellow depts. pharmacology and biochemistry N.Y. U. Sch. Medicine, 1953-55; sr. asst. scientist USPHS, NIH, Bethesda, Md., 1955-57, sr. investigator Nat. Inst. Arthritis & Metabolic Diseases, 1957—, chief sect. enzymes and cellular metabolism, faculty grad. sch., 1958—. Mem. Am. Soc. Biol. Chemists, Soc. Am. Microbiology, Am. Chem. Soc. Contbr. articles to profl. jours. Contbns. to methodology of enzyme purification, including a gen. method of enzyme crystallization; discovery, isolation and characterization of a variety of enzymes dealing with the synthesis of nicotinic acid and its derivatives, with amino acid synthesis and degradation, and with bacterial metabolism; contbns. to enzyme topography and kinetics.

JAKOSKY, John Jay, Am. cons. engr.; b. Vinita, Okla., Jan. 20, 1896; s. John Lewis and Jennie (Meyerhart) J.; B.S. in M.E., U. Kan., 1920, M.E., 1926; student Carnegie Inst. of tech., 1923-24; B.S. in E.E., U. Pitts., 1925; Sc.D., U. Ariz., 1933; m. Katharine Fulkerson, Jan. 22, 1923; 1 son, John Jay. Asst. research engr., U. S. Bur. of Mines, 1920-25; with Western Precipitation Co., 1925-27, Southwestern Engring Co., 1927-29; dir. research Electroblacks, Inc., 1929-38; pres., tech. dir., 1945—; pres. Internat. Geophysics, Inc., 1929—; pres. Trija Co., also Trija Mfg. Co. of Cal., 1945—; cons. research engr., 1945—; pres. Electrophysics Corp., 1960—. Dir., U. Kan. Research Found., dean Sch. of Engring. and Architecture, U. of Kan., 1940-43; faculty U. So. Cal., 1943-45. Registered profl. mech., elec. and petroleum engr., Mem. Soc. Promotion of Engring. Edn., Am. Inst. Mining Engrs., Soc. Exploration Geophysicists (v.p. 1945, pres. 1946), A.A.A.S., Am. Assn. Petroleum Geologists, Delta Upsilon, Sigma Xi, Tau Beta Pi, others. Author: Exploration Geophysics, 1941. Home: 1042 W. Bay Av., Newport Beach, Cal. Office: 3017 W. Coast Hwy., Newport Beach, Cal.*

JAKOWSKA, Sophie (Mrs. C. L. Jeannopoulos), biologist; b. Warsaw, Poland, Feb. 12, 1922; d. Josef and Maria (Swiergocka) Jakowski; certificate U. Rome (Italy), 1941; M.S., Fordham U., 1945, Ph.D., 1947; m. C. L. Jeannopoulos, June 11, 1941; children—Peter, John, Marie-Helene. Came to U. S., 1942, naturalized, 1947. Faculty, Coll. Mt. St. Vincent, N.Y.C., 1948-58, asso. prof. biology, 1954-58; asst. mem. Sloan-Kettering Inst., N.Y.C., 1947-48; research collaborator dept. biology Brookhaven Nat. Lab., L.I., N.Y., 1953-60; research asso. N.Y. Aquarium, N.Y. Zool. Soc., 1947-66, Beth Israel Hosp. Dept. Labs., N.Y.C., 1958—; asst. v.p. med. affairs Nat. Cystic Fibrosis Research Found., 1961-62; research asso. Inst. for Crippled and Disabled, N.Y.C., 1962; head dept. Pathology Food and Drug Research Labs., Inc., Maspeth, N.Y.C., 1963—. Cons. Inst. Marine Biology, Autonomous U. Santo Domingo, Dominican Republic, 1962—, Animal Med. Center, N.Y.C., 1964—. Fellow N.Y. Acad. Scis., N.Y. Zool. Soc., A.A.A.S.; mem. Am. Micros. Soc., Am. Assn. Cancer Research, Soc. Protozoologists, Am. Soc. Zoologists, Radiation Research Soc., Société Zoologique de France, Soc. Zoologorumi Ploniae, Am. Soc. Ichthyologists and Herpetologists, Am. Assn. Anatomists, Harvey Soc., Internat. Acad. Pathology, Am. Soc. Exptl. Pathology, Am. Assn. Neuropathology, Am. Inst. Biol. Scis., Animal Care Panel, Royal Soc. Medicine, Latin Am. Assn. Ichthyologists and Herpetologists, Bermuda Biol. Sta. for Research, Assn. Island Marine Labs. Carribean, Intersci. Liaison Assn. (founder, exec. dir.). Contbr. numerous articles to tech. jours. Discovered protozoan parasites in electric eels, protozoa in nerve cells of angler fish, antibiotics from marine sponges; research in comparative hematology and pathology, radiation effects toxicology and exptl. surgery in newts. Home: 27 W. 96th St., N.Y.C. 10025. Office: 58th St. and Maurice Av., Maspeth, N.Y.C. 10078.*

JAKUBOWSKI, Janusz Lech, Polish elec. engr.; b. Warsaw, Poland, Sept. 12, 1905; s. Wladylaw and Wiktoria (Handzelewicz) J.; E.E., Tech. U. Warsaw, 1931, Dr.Tech.Sci., 1935, Dr.Habil., 1938; m. Hanna Wiszniewska, June 21, 1931; 1 dau., Krystyna (Mrs. Wieslaw Chtokowski); m. 2d, Zofia Wysocka, Dec. 11, 1948. Faculty, Tech. U. Warsaw, 1929-52, prof. high voltage engring., 1945-52, dean elec. faculty, 1945-52; dir. State Research Elec. Inst., Warsaw, 1945-56; pres. Com. on Electrification of Poland, 1957-61; dir. Center Polish Acad. Scis., Paris, 1959-61; UNESCO project mgr. Ecole Nationale Polytechnique d'Alger, 1967—. Mem. presidium Polish Acad. Sci., 1952-67. Decorated officer Polonia Restituta, 1951, comdr., 1954; Etandard de Travail, 1966; State prize for sci., Corr. mem. Acad. Sci. Toulouse, 1950. Author: New Method of High-Voltage Measurement, 1935; Measurement of Distorted High Voltages, 1937; Actual Problems of Distorted High Voltage Engineering, 1939; High Voltage Engineering, 1951; Traveling Waves in High Voltage Systems, 1962; Over-Voltages in High Voltage Systems, 1968; also numerous articles. Research on high voltage measuring methods, theory of traveling waves, lightning protection of elec. systems, methods for long-term planification of electrification. Home: 9 Iganska, Warsaw. Office: Politechnika, Warsaw, Poland.*

JALABERT (or JALLABERT), Jean, Swiss physicist; b. Geneva, Switzerland, July 26, 1712; named prof. exptl. physics, Geneva, 1737, succeeded Cramer as head depts. math. and philosophy, 1752; became mem. Council Two Hundred, 1746, later Council of State, syndic of Geneva. Fellow Royal Soc., 1740; mem. French Acad. Scis., acads. Berlin, London, Bologna, Montpellier, Lyons. Author: De libertate humana, 1734; Expériences sur l'électricité, avec quelques conjectures sur ses causes et ses effects, 1748. Research on static electricity, also on electric shock as treatment for paralysis. Died Geneva, Mar. 11, 1768.

JAMES, Arthur Montague, English chemist; b. Poole, Dorset, Eng., Feb. 27, 1923; s. Montague John and Hilda (Legg) J.; B.A., Oxford U., 1944, B.Sc., 1945, M.A., 1947, D.Phil., 1947, D.Sc., 1964; m. Mary Patricia Lord, Apr. 3, 1954; children—Frank Arthur, Charles Robert. Research chemist Glaxo Labs. Ltd., 1947-49; sr. lectr. phys. chemistry Chelsea Coll. Sci. and Tech., 1949-58, reader, 1958-64; reader phys. chemistry Queen Elizabeth Coll., U. London (Eng.), 1964—. Fellow Royal Inst. Chemistry (assessor in phys. chemistry). Author: Practical Physical Chemistry, 2nd edit., 1967. Contbr. articles to profl. jours. Research on electrochemistry of bacterial surface with view to identifying nature of surface components and correlating elec. properties with known biol. properties, study of lipid content of antibiotic resistant and sensitive strains of Gram-positive cocci, and phys. study of model particles. Home: 50 Brodrick Rd., London S.W.17, Eng.*

JAMES, Charles, chemist; b. Earls Barton, nr. Northampton, Eng., Apr. 27, 1880; s. William and Mary Diana (Shatford) J.; grad. Inst. of Chemistry, London, 1904, fellow, 1907; hon. D.Sc., U. of N.H., 1928; m. Marion Elizabeth Templeton, 1915. Chemist, New Cransley Iron & Steel Co., Kettering, Eng.; came to U. S. 1906, naturalized citizen, 1920; asst. prof. inorganic chemistry, New Hampshire Coll., 1906-12, prof., from 1912. Recipient Ramsay silver medal, 1901; Nicholas medal, 1911. Discovered rare earth, lutecia (credited to Urbain who pub. his discovery first); expert on separation of rare earths; 1 separation method now known by his name was by fractional crystallization of double magnesium nitrates; prepard many salts of uncommon composition, including sebacates, pyromucates, tungstates, cacodylates; effected separation of element 61 (lost priority of discovery to Hopkins, who pub. results while James' work was still being checked by X-ray). Died Boston, Dec. 10, 1928.

JAMES, Edwin, Am. botanist, geologist; b. Weybridge, Vt., Aug. 27, 1797; s. Daniel and Mary (Emmes) J.; grad. Middlebury Coll., 1816; studied medicine, botany and geology, Albany, 1816-19; m. Clarissa Rogers, Apr. 5, 1827, 1 child. Became botanist, 1820; geologist and surgeon of expdn. commanded by Maj. Stephen H. Long to explore country between Mississippi River and Rocky Mountains; with two companions reached Pike's Peak (1st white men to accomplish the feat), July 14, 1820; asst. surgeon in U. S. Army, 1823-29; resigned, 1833; asso. editor Temperance Herald and Journal, 1836; agt. for Potawatmie Indians at Old Council Bluffs (Ia.), 1837-38; engaged in agrl. pursuits, nr. Burlington, Ia., from 1836; abolitionist, mem. Underground Railroad. Author: Expedition to the Rocky Mountains, 1819-20, published 1823; The Narrative of John Tanner, 1830; translated New Testament in Ojibway (Indian) Language, 1883. Earliest bot. explorer of Rocky Mountains; his name originally given to what is now known as Pike's Peak. Died Rock Spring, Ia., Oct. 28, 1861.

JAMES, Harold Lloyd, geologist; b. Nanaimo, B.C., Can., June 11, 1912; s. Evan and Blodwen (Davies) J.; came to U. S., 1923, naturalized, 1933; student Western Wash. Coll., 1933; B.S., Wash. State U., 1938; student U. Wash., 1938-40; Ph.D., Princeton, 1945; m. Ruth Graybeal, Feb. 13, 1936; children—David E., Robert C. L., Hugh L., Herbert T. Field asst. U. S. Geol. Survey, 1938-40, geologist, 1941-61, 64—; vis. lectr. Northwestern U., 1952-53; prof. geology U. Minn., 1961-64. Mem. Nat. Acad. Scis., Mineral. Soc. Am. (council 1964—), Soc. Econ. Geologists (council 1962-65), Geol. Soc. Am. (council 1959-62), Geochem. Soc., Phi Beta Kappa, Sigma Xi, Phi Kappa Phi. Research and publs. on origin and structure of iron-rich rocks and iron ores of the Lake Superior region; use of oxygen isotopes to determine temperatures of rock formation. Home: 1414 17th, Washington 20036. Office: U. S. Geol. Survey, Washington 20006.*

JAMES, Hubert M(axwell), Am. physicist; b. Clarksburg, W.Va., Mar. 10, 1908; s. Ernest Wilbur and Edna Virginia (Maxwell) J.; A.B., Randolph-Macon Coll., 1928, D.Sc. (hon.); A.M., Harvard, 1930, Ph.D., 1934; m. Madeline Roxane Fitzpatrick, Aug. 29, 1932; 1 dau., Martha Virginia. Mem. faculty Purdue U., 1936—, prof., 1944—, head physics dept., 1959-66; Guggenheim Meml. Foundn. fellow, Oslo, Norway, 1939-40; researcher Mass. Inst. Tech. Radiation Lab., 1941-46; tech. editor Radiation Lab. Series. Fellow Am. Phys. Soc.; mem. Am. Assn. Physics Tchrs., Am. Assn. U. Profs., Phi Beta Kappa, Sigma Xi, Chi Beta Phi, Sigma Pi Sigma, Kappa Alpha. Methodist. Author (with another): Advances in Colloid Science, Vol. II, 1946. Editor and co-author: Theory of Servomechanisms, 1947. Collaborated (with A. S. Coolidge) in earliest quantum-mech. calculations on the hydrogen molecule, with sufficient accuracy to provide an ind. test of the validity of wave mechanics; made early contbns. to Quantum theory of other atoms and molecules; collaborated (with E. Guth) in developing the network theory of ideal rubber-like materials; contbd. to theoretical understanding of disordered crystals and of molecular crystals. Home: 316 Forest Hill Dr., West Lafayette, Ind. 47906.*

JAMES, Joseph Hidy, Am. chemist; b. Jeffersonville, O., Nov. 3, 1868; s. John A. and Mary J. (Hidy) J.; B.S., Buchtel Coll., 1894; postgrad. Columbia, 1897, U. Pa., 1898-99; Ph.D., 1899; m. Edith Mallison, Nov. 28, 1899. children—Mary Alice, Virginia, Josephine. Asst. chemistry and physics, Buchtel Coll., Akron, O., 1894-97; chief chemist Lake Superior Power Co., Sault Ste. Marie, 1899-1902; asst. prof. textile chemistry, Clemson (S.C.) Coll., 1902-05; asst. prof. tech. chemistry Carnegie Inst. Tech., Pitts., 1905-06, asst. prof. chem. practice in charge dept. of chemistry, 1906-07, asso. prof., 1907-08, prof. chemistry from 1908. Mem. Am. Chem. Soc., Am. Inst. Chem. Engrs. Research and patents on acetylene storage and on oxidation products of petroleum. Died Feb. 12, 1948.

JAMES, Preston Everett, Am. geographer; b. Brookline, Mass., Feb. 14, 1899; s. Frank Everett and Gertrude (Woodworth) J.; A.B., Harvard, 1920, A.M., 1921; Ph.D., Clark U., 1923; D.Sc., Eastern Mich. U., 1967; m. Dorothy Tenney Upham, Apr. 3, 1922 (divorced); m. 2d Eileen Woodbury Bowles, July 23, 1943; 1 son, Everett Woodbury. Teaching assistant in geography, Harvard, 1919-21, Radcliffe College, 1920; instr. Clark U., 1921-23; faculty U. Mich., 1923-45; prof. geography, Syracuse U., 1945—, chmn. dept., 1951-68, Maxwell prof. geography, 1964—; geog. cons. Conselho Nacional de Geografia (Brazil) 1949-50; Fulbright prof U. Edinburgh, 1957. Mem. NRC; active U. S. del. to various consultations in geography; chief U. S. del. mem. U. S. delegation to Internat. Geog. Congress, Rio de Janeiro, 1956, London, Eng., 1964. Recipient George Morgan Ward medal Rollins Coll., 1965, Pan-Am. med Pan-Am. Inst. Geography and History, 1965; Livingstone Centenary medal Am. Geog. Soc., 1967. Mem. Assn. Am. Geographers (pres. 1951, hon. pres. 1966), Am. Meteorol. Soc., A.A.A.S., Nat. Council Geog. Edn. (Distinguished Writing award 1963, Distinguished Service award 1964), Nat. Council for Social Studies, Council for Latin Am. Affairs (pres. 1957), Phi Kappa Phi, Sigma Xi. Author: An Outline of Geography, 1935; Latin America, 1942, rev. edit. 1959; A Geography of Man, 1949, rev. edit. 1966; series grade-school geographies (with Gertrude Whipple), 1947-55; (with Nelda Davis) The Wide World. A Geography, 1959, rev. edit., 1967; One World Divided, 1964; An Introduction to Latin America, pub. 1964. Editor: American Geography, Inventory and Prospect (with C. F. Jones), 1954; New Viewpoints in Geography (Yearbook of the Nat. Council for Social Studies), 1959; also Brazil-Geography in Handbook of Latin Am. Studies, 1935-59. Home: 220 Standish Dr., Syracuse, N.Y. 13224.*

JAMES, Robert, Brit. physician; b. Kinurston, Scotland; 1703; s. Edward James; B.A., St. John's Coll., Oxford, Eng., 1726; M.D., Cambridge (Eng.) U., 1728; 1 son, Pinkstan. Practiced medicine, Sheffield, Lichfield, Birmingham, finally in London. Extralicentiate Coll. Physicians, London. Author: Medical Dictionary (pub. with help of Dr. Johnson), 3 vols., 1743-45; Practise of Physic; On Canine Madness; Dissertations on Fevers; Vindication of the Fever Powder, 1778. Used febrifuge powder composed of lime phosphate and calcines antimony. Died London, Mar. 23, 1776.

JAMES, Robert Clarke, Am. mathematician; b. Bloomington, Ind., July 30, 1918; s. Glenn and Inez (Clarke) J.; B.A., U. Cal., Los Angeles, 1940; Ph.D., Cal. Inst. Tech., 1947; m. Edith Maria Peterson, Oct. 28, 1945; children—Judith Marie, Linda Inez, David Vernon, Robert Glenn. Benjamin Peirce instr. math. Harvard, 1946-47; with U. Cal., Berkeley, 1947-51, asst. prof., 1949-51; asso. prof. Haverford Coll., 1951-57; prof., chmn. dept. Harvey Mudd Coll., 1957-67; prof. State U. N.Y., 1967-68; prof., chmn. dept. Claremont (Cal.) Grad. Sch. and Univ. Center, 1968—; mem. Inst. for Advanced Study, Princeton, N.J., 1962-63. Mem. Am. Math. Soc., Math. assn. Am. (gov.), Soc. Indsl. and Applied Math., Fedn. Am. Scientists, A.A.U.P., Phi Beta Kappa, Sigma Xi. Author: University Mathematics, 1963, Advanced Calculus, 1966. Studies, publs. structural properties of Banach spaces, including devel. of an example of a non-reflexive Banach space isometric to its second conjugate and of characterizations of weak compactness, e.g. a closed convex subset is weakly compact if and only if each continuous linear functional attains its supremum on the subset. Office: Claremont Grad. Sch., Claremont, Cal. 91711.*

JAMES, Thomas Chalkley, Am. physician; b. Phila., Aug. 31, 1766; s. Abel and Rebecca (Chalkley) J.; grad. U. State of Pa., 1787, M.D., 1811; studied under Dr. John Hunter, also at Story Street Lying In Hosp. under Drs. Osborne and John Clark, London, Eng.; m. Hannah Morris, 1802. Ship's surgeon on voyage to Cape of Good Hope, 1788-90; returned to Phila., 1793; gave series of lectures on obstetrics, U. Pa., 1810; physician Pa. Hosp., 1807-32, became obstetrician, 1810; frequent lectr., read papers before Phila. Coll. Surgeons; editor Eclectic Repertory, 11 years. One of the 1st male physicians to deliver children. Died July 5, 1835.

JAMES, Thomas N., Am. physician; b. Amory, Miss., Oct. 24, 1925; B.S., Tulane U., 1946, M.D., 1949; m. Gleaves E. Tynes, June 22, 1948; children—Mark, Terry, Peter. Cardiologist, Ochsner Clinic, asst. prof. clin. medicine Tulane U., 1955-59; chmn. sect. on cardiovascular research Henry Ford Hosp., Detroit, 1959-68; prof. medicine, prof. pathology U. Ala. Med. Center, Birmingham, 1968—. Diplomate Am. Bd. Internal Medicine. Fellow A.C.P., Am. Coll. Cardiology (Mich. gov.), Council on Clin. Cardiology of Am. Heart Assn. Author: Anatomy of the Coronary Arteries, 1961; (with J. W. Keyes) Etiology of Myocardial Infarction, 1963. Research, numerous publs. on structure and function of heart, coronary arteries and conduction system, clin. problems of arrhythmias, myocardial infarction, sudden death, enlarged heart. Home: 4561 Sharpsburg Dr., Birmingham, Ala.*

JAMES, Thomas Potts, Am. botanist; b. Radnor, Pa., Sept. 1, 1803; s. Isaac and Henrietta (Potts) J.; m. Isabella Batchelder, 1851. In wholesale drug bus. with brother in Phila. from 1831 for 35 years; prof., examiner Phila. Coll. Pharmacy, many years; moved to Cambridge, Mass., devoted rest of life to study of mosses, 1866; conferred with W. Ph. Schimper, Europe, 1878, compared Am. and Old World species mosses; became recognized as foremost Am. specialist on mosses, began collaboration with Charles Leo Lesquereux on Manual of North American Mosses, died before its completion in 1884; contbr. to sci. mags including Proc. Acad. of Natural Scis. of Phila. Mem. Am. Philos. Soc., A.A.A.S., Am. Pomol. Soc. (a founder). Died Feb. 22, 1882.

JAMES, William, Am. psychologist, philosopher; b. New York, Jan. 11, 1842; s. Rev. Henry J. and Mary R. (Walsh) J.; student Lawrence Scientific Sch. (Harvard), 1861-63; M.D., Harvard, 1869; Ph.D., and Litt.D., Padua, 1893; LL.D., Princeton, 1896, Edinburgh, 1902, Harvard 1903; D.Sc., Oxford, 1908; Litt.D., Durham, 1908; D.Sc., Geneva, 1909; m. Alice H. Gibbens, 1878. Instr. physiology, 1872-76, anatomy, 1873-76, asst. prof. physiology, 1876-80, philosophy, 1880-85, prof. same, 1885-89, prof. psychology, 1889-97, philosophy, 1897-1907, emeritus prof. 1907, Harvard. Gifford lecturer on natural religion, U. of Edinburgh, 1899-1901; Hibbert lecturer on philosophy, Oxford, 1908. Author: Principles of Psychology (2 vols.), 1890; Psychology—Briefer Course, 1892; The Will to Believe, and Other Essays in Popular Philosophy, 1897; Talks to Teachers on Psychology and to Students on Life's Ideals, 1898; Human Immortality—Two Supposed Objections to the Doctrine, 1899; The Varieties of Religious Experiences, 1902; Pragmatism—A New Name for Some Old Ways of Thinking, 1907; A Pluralistic Universe, 1908; The Meaning of Truth, 1909. Considered to be America's most influential psychologist; responsible for establishing psychology on a physiological basis in America; initiated first psychological laboratory in America, at Harvard, 1876; led the way for functional psychology in America. Known for James-Lange theory of emotion (anticipates behavioral psychology); as a philosopher was known as a leader of pragmatism. Died Chocorua, N.H., Aug. 26, 1910.

JAMES OF VENICE, translator; flourished circa 1128-36. Translator: (from Greek to Latin) Topics; Prior & Posterior Analytics; Sophistici elenchi, including his commentary, 1128 (all parts of the Aristotelian Organon called The New Logic).

JAMESON, David Lee, Am. biologist; b. Ranger, Tex., June 3, 1927; s. Aubrey Murl and Gertrude (Harwood) J.; B.S., So. Meth. U., 1948; M.A., U. Tex., 1949, Ph.D., 1952; m. Marianne Mayo, June 11, 1949; children—Roy Alan, David Laurence, Robert Carey, Carol Lee. Asst. prof. biology Pacific U., Forest Grove, Ore., 1952-53; faculty U. Ore., Eugene, 1953-57, asst. prof. biology, 1954-56; faculty San Diego (Cal.) State Coll., 1956-67, prof. biology, 1963-67; faculty U. Houston, 1967—. Mem. Am. Soc. Ichthyologists and Herpetologists, Soc. for Study of Evolution, Genetics Soc. Am., Am. Soc. Naturalists, Ecol. Soc. Am., Sigma Xi. Mng. editor Copeia, Jour. Am. Soc. Ichthyologists and Herpetologists, 1960-64. Publs. on research on population genetics of amphibians, evolution of behavior studies. Office: U. Houston (Tex.).*

JAMESON, Horatio Gates, Am. physician; b. York, Pa., 1778; s. David and Elizabeth (Davis) J.; studied medicine under father; M.D., U. Md., 1813; m. Catherine Shevell, Aug. 3, 1797; m. 2d, Hannah (Fearson) Ely, 1852; 7 children. Practiced medicine, Somerset County, Pa., 1795-1810; moved to Balt., 1810; physician to fed. troops, 1812; surgeon Balt. City Jail, 1814-35; cons. physician Balt. Bd. Health, 1821-35; founder Washington Med. Coll., Balt., 1827, obtained univ. charter, built hosp. by 1835, forced to close schs. because of financial difficulties, 1851; prof. surgery Med. Coll. of Ohio, Cin., 1835-36. Author: Treatise on Epidemic Cholera, 1855; published accounts of his unusual operations in leading med. jours. including Am. Med. Recorder. Used animal ligature (his best known contbn. to surgery); credited with 1st recorded excision of superior maxilla, 1820. Died N.Y.C., Aug. 26, 1855.

JAMESON, Robert, Scottish mineralogist, geologist; b. Leith, Scotland, July 11, 1774; ed. Edinburgh U.; studied under Werner at U. Freiburg (Germany), 2 years. Became asst. to surgeon, Leith, 1796; named Regius prof. natural history, keeper univ. mus., Edinburgh, 1804. Founder Wernerian Natural History Soc., 1808. Fellow Royal Soc., 1826. Author: Mineralogy of Shetland Islands and of Arran, 1798; System of Mineralogy, 3 vols., 1804-08; Manual of Minerals and Mountain Rocks, 1821; Elements of Mineralogy, 1840; also articles in encys. Founder Edinburgh Philos. Jour., 1819. First great exponent of Werner's geol. tenets in Britain; later accepted Hutton's views. Died Apr. 19, 1854.

JAMIESON, John Calhoun, Am. geophysicist; b. St. Joseph, Mo., Jan. 5, 1924; s. William Thomas and Glessie (McPike) J.; S.B., U. Chgo., 1947, S.M., 1951, Ph.D., 1952; m. Ruth Virginia Lamb, Mar. 26, 1949; children—William Thomas, Virginia Anne, John Booth. Mem. faculty U. Chgo., 1953—, prof. geophysics 1965—; with Cal. Research Corp., summer 1953, Stanford Research Inst., summers 1959-63, U. S. Geol. Survey, 1958—; cons. Los Alamos Sci. Lab., 1964—. NSF fellow, 1953. Mem. Geol. Soc. Am., Am. Geophys. Union, Seismol. Soc. Am., A.A.A.S., Geochem. Soc., Sigma Xi. Research and publs. on devel. techniques for study matter under conditions of earth's far interior; devel. theories concerning constitution and properties of earth's interior; devel. and use of ultra high pressure apparatus; study of matter using explosive waves. Address: Dept. Geophys. Scis., U. Chgo., Chgo. 60637.*

JAMIN, Jules-Célestin, French physicist; b. Termes, France, May 30, 1818; ed. L'école Normale Supérieure; tchr. physics, Caen, France, Coll. Bourbon (now Lycée Condorcet), Louis Le Grand; prof. Polytechnic Sch., 1852-81; prof. Faculty Scis. Paris, from 1863; dir. phys. lab. Sorbonne, 1868-86; became perpetual sec. French Acad. Scis., 1884. Author: Réflexion métallique (doctoral thesis); Constitution des aimants (memoirs); Cours de physique de l'Ecole polytechnique (on interferences, life of light, electro-magnetic currents), 1858-61, reissued under collaboration of Bouty, 1878-87. Invented brilliant candle to study reflection of metal surface; discovered effect of polarization of light by lenses; measured index of refraction of many substances; studied velocity of light; constructed model of artificial magnet; invented interferometer to make interference fringes by method originated by Brewster; also studied botany and geology. Died Paris, Feb. 12, 1886.

JAMISON, Frank Stover, Am. horticulturist; b. Bellwood, Pa., Feb. 12, 1903; s. Isaac McKinney and Mary (Stover) J.; B.S., Pa. State Coll., 1924; M.S., Ia. State Coll., 1925; Ph.D., Cornell U., 1934; m. Ellen Van Arnam, Apr. 1, 1933. Faculty Tex. A. and M. Coll., 1925-30, Cornell U., to 1934; horticulturist Fla. Agr. Expt. Sta., U. Fla., from 1934, extension vegetable crops specialist, 1948—, chmn. dept. vegetable crops, 1956—; asst. econ. commr. U. S. Dept. State, Paris, France, 1952-53. Cons. to Phelps-Dodge, 1954-59, Monsanto, 1964-65. Recipient Distinguished Service award Fla. Seedmen, 1951, award merit, 1954, Research award Fla. Fruit and Vegetable Assn., 1960; named Hon. State Farmer Future Farmers Am., 1963. Fellow Am. Soc. Hort. Sci. (past pres., past chmn. bd.), A.A.A.S. (council 1964—); hon. mem. Fla. State Hort. Soc. (past v.p.), Am. Inst. Biol. Sci. Contbr. articles to tech. jours., state exptl. sta. and extension publs. Introduced superior methods using fertilizer, new crops, methods handling for delivery quality products. Home: 709 N.W. 22d St., Gainesville, Fla. 32601.*

JAMISON, Vernon Carnahan, Am. soil physicist; b. Preston, Ida., July 14, 1907; s. William Henry and Mary (Jamison) J.; B.S., Utah State U., 1936; Ph.D., Cornell U., 1941; m. Ruth Woodward, May 27, 1936; children—Sherwin W., Bonnie Fay (Mrs. Paul Sampson), Max W., Bruce W., Karl W. Faculty Utah State U., 1935-36, Cornell U., 1937-41; soil chemist, physicist Citrus Expt. Sta., Lake Alfred, Fla., 1941-47; soil physicist Agrl. Res. Service, Auburn, Ala., 1947-55; U. Mo., Columbia, 1955—, staff N. Central Watershed Hydrology Res. Center, 1962—. Mem. Am. Soc. Agronomy, Soil Sci. Soc. Am., Am. Geophys. Union, Am. Geog. Soc., Fla. Hort. Soc., Fla. Soil Sci. Soc., Sigma Xi, Phi Kappa Phi. Contbr. numerous articles to profl. jours. Expressed quantitative relationships between soil hardness, moisture content and porosity; characterized moisture flow patterns in claypan and loessal soils. Home: 218 E. Parkway. Office: Watershed Hydrology Research Center, U. S. Dept. Agr., Columbia, Mo.*

JANCIK, Erich, Czechoslovakian physician; b. Rymarov, Czechoslovakia, Feb. 16, 1911; s. Jan and Irma (Spitzerova) J.; M.D., Charles U., Prague, 1935, D.Sc., 1962; m. Marcela Máková, Dec. 21, 1963; children—Hana Wurmová, Eva Smutná, Tomás Londzin. Hosp. and santoria practice, 1935-38; head Tb dept. Hosp. Zlín, 1939-49; dir. Tb Sanatorium Prosecnice nr. Prague, 1949-53; head Tb dept. U. Hosp., Brno, Czechoslovakia, 1954-55; head clin. dept., dep. dir. Tb Research Inst., Prague, 1956-63; head Tb dept. U. J. E. Purkyne, Czechoslovakia, 1963—, faculty, 1963—; head chest diseases dept. U. Hosp., Brno, Jihlavaská, Czechoslovakia, 1963—; faculty Postgrad. Med. Sch., Prague, 1955—. Recipient médaille Robert Koch, Weimar, 1964; médaille de bronze Academie de Médecine, Paris, 1964. Mem. Czechoslovak Pneumophtisiological Soc. (past pres. sci. and post grad. com., pres. 1961—), Internat. Union Tb (past sec. therapy com.). Author: Treatment of Empyema,

1956; Antimicrobial Tb Treatment, 1954; also numerous articles. Research on orgn. of control, prevention, treatment of Tb, epidemiologic studies in chronic bronchitis. Address: 100 Jihlavská, Brno, Czechoslovakia.*

JANET, Paul, French physicist; b. Paris, Jan. 10, 1863; s. Paul Janet; became asst., then prof. P.C.N. Paris, 1901; prof. Sorbonne; dir. Laboratoire Central, also of École Supérieure d'Electricité. Mem. French Acad. Scis., 1919; Author: Problèmes et exercices d'electricité générale, 1921; Lecons d'électro-technique générale, 1925-28; Premiers principes d'électricité industrielle, 1929. Died Malakoff, France, Feb. 21, 1937.

JANET, Pierre, French psychologist; b. Paris, May 30, 1859; ed. U. Paris, Doctorate, 1889, M.D., 1893; student of Charcot at Paris. Taught philosophy lycée, Lehavre; became dir. psychology lab. (by invitation of Charcot), Salpetrière, 1890; apptd. prof. U. Paris, 1895, then at Coll. France, 1902-36; lectr. Harvard, 1906; hon. pres. XIth Internat. Congress of Psychology, 1937. Mem. Acad. Moral Scis. Author: L'état mental des hystériques, 1893; Les Névroses 1909; Les méditations psychologiques, 1919; De l'angoisse à l'extase, 1927; also others. Known especially for initiating movement to bring clin. and academic psychology together in hope of finding 1 coherent, intelligible set of concepts; extensive research on mental pathology hysteria and attempt to order all known data on subject, use of hypnosis, also studied obsessions, amnesia, neuroses and related subjects; 1st to describe psychasthenia; founder of automatic psychology. Died Feb. 24, 1947.

JANETSCHEK, Heinz Georg Maria, German biologist; b. Blüdenz/Voralberg, Aug. 3, 1913; s. Carl and Maria Olga (Muller) H.; ed. U. Vienna, U. Innsbruck; Ph.D.; m. Anne Elizabeth Hofmann, June 15, 1941; children—Gloria, Gunter. Asst., prof. Stella Matutina Coll., St. Blasien, also various schs., Munich, 1938-45; asst. Inst. Zoology, U. Innsbruck, 1945-—, agrégé in zoology, permanent instr., 1947, asso. prof., 1951, full prof., dir. Inst. Zoology, 1963-—. Sci. dir. research group in Mt. Everest area, 1961; made zool studies in Antarctic, 1961-62, So. Alps of New Zealand, 1962. Mem. German Zool. Soc., German Entomology Soc., Entomol. Soc. Munich. Author: Tierische Sussessionen auf hochalpinem Neuland, 1949; also articles on zoology, ecology and zoogeography of regions of Alps, Himalayas and Antarctica, taxonomy of insecta machilida. Home: Leopoldstrasse 9. Office: Institute of Zoology, Universitstrasse 4, Innsbruck, Austria.

JANEWAY, Theodore Caldwell, Am. physician; b. N.Y., Nov. 2, 1872; s. Edward Gamaliel and Frances Strong (Rogers) J.; Ph.B., Yale, 1892; M.D., Coll. Phys. and Surg., Columbia, 1895; hon. A.M., Yale, 1912; Sc.D., Washington U., 1915; m. Eleanor C. Alderson, Sept. 27, 1898. Instr., then lectr. on med. diagnosis N.Y. U., 1898-1906; asso. in medicine, Coll. Phys. and Surg., Columbia, 1907-09, prof. practice of medicine, 1909-14; prof. medicine Johns Hopkins U. and physician-in-chief Johns Hopkins Hosp., Balt., from 1914. Sec. Russell Sage Inst. Pathology, N.Y., from 1907; mem. editorial bd. Archives of Internal Medicine, from 1908; mem. bd. scientific dirs. Rockefeller Inst. for Med. Research, from 1911. Author: The Clinical Study of Blood-Pressure, 1904. Introduced clinico-pathol. conf. at City Hosp. (widely copied); began use of routine blood pressure check. Died Dec. 27, 1917.

JANICK, Jules, Am. horticulturist; b. N.Y.C., Mar. 16, 1931; s. Carl and Frieda (Tullman) J.; B.S., Cornell U., 1951; M.S., Purdue U., 1952, Ph.D., 1954; m. Shirley Reisner, June 15, 1952; children—Peter Aaron, Robin Helen. Faculty, Purdue U., Lafayette, Ind., 1954-—, prof. horticulture, 1963-—; research fellow U. Coll., London, Eng., 1963; horticulturist Rural Universidade do Estado de Minas Gerais, Brazil, 1963-65. Mem. Genetics Soc. Am., Bot. Soc. Am., Am. Soc. Hort. Sci., Am. Pomol. Soc., Sigma Xi. Author: Horticultural Science, 1963. Research on genetics of hort. plants, breeding of hort. crops. Home: 106 Black Hawk Lane, West Lafayette, Ind. 47906.*

JANICKI, Jozef, biochemist; b. Bremen, Germany, Apr. 21, 1904; s. Antoni and Elzbieta (Sadzinska) J.; M.Phil. Chemistry, U. Poznan, 1929, Dr.Phil., 1930; m. Felicja Ast, June 28, 1931; children—Andrzej, Kryzysztof. Faculty, U. Poznan, 1927-—, asst. prof., 1930-—; with Agrl. Acad., Poznan, 1951-—, prof., 1957-—, dean faculty agr. and food tech., 1962-—. Mem. Sci. Bd., Ministry Food Industries, Inst. Leather Industries; organizer internat. congresses, 1959, 65. Recipient awards Ministry Higher Edn., 1962, 67; award for sci. achievement City and Dist. of Poznan, 1963. Mem. Polish Biochem. Soc., Polish Chem. Soc., Am. Inst. Nutrition, Polish Copernicus Soc. Naturalists. Books, research, articles on amylases action in cereal grains, biosynthesis of Vitamin B12; standardization of vitamin estimation methods in cereals; use of waste products in feed preparations. Home: 2 Slaska. Office: 48 Mazowiecka, Poznan, Poland.*

JANIK, Jerzy Antoni, Polish physicist; b. Lwow, Poland, Apr. 30, 1927; s. Alfred and Irena (Budzynowska) J.; grad. Jagiellonian U., Krakow, Poland, 1944-48; Ph.D., 1950; m. Janina Machaczka, Apr. 15, 1952; children—Barbara, Joanna, Malgorzata. With Jagiellonian U., 1947-—, prof. physics, 1960-—, head chair structure research, 1963-—; head lab. for structure research Inst. for Nuclear Physics, Krakow, 1966-—. Mem. Polish Phys. Soc. Research and publs. on atomic internal motions in crystals, using neutron inelastic scattering. Home 8 ul. Marka, Krakow. Office: Inst. Nuclear Physics, Krakow 23, Poland.*

JANIS, Irving Lester, Am. psychologist; b. Buffalo, May 26, 1918; s. M. Martin and Etta (Goldstein) J.; B.S., U. Chgo., 1939, postgrad., 1939-40; Ph.D. in Psychology, Columbia, 1947; m. Marjorie Graham, Sept. 4, 1939; children—Cathy, Charlotte Janis. Sr. social sci. analyst, spl. war policies unit Dept. Justice, 1941-43; research asso., research fellow Social Sci. Research Council, 1945-47; faculty Yale, 1947-—, prof. psychology, 1960-—. Cons. adv. panel for sociology and social psychology div. social scis. NSF, 1965-66; research cons. RAND Corp., 1948-—. Ford Found. grantee, 1956. Fulbright Research scholar, U. Oslo (Norway), 1957-58; Social Sci. Research Council Faculty Research fellow, 1961-62, 66-67; Yale Sr. Faculty fellow Tavistock Clinic and Inst., London, Eng., 1961-62; recipient Hofheimer prize Am. Psychiat. Assn., 1959. Fellow Am. Psychol. Assn. (mem. com. on psychology in nat. and internat. affairs 1963-66, chmn. 1965-66), A.A.A.S. (rep. Am. Psychol. Assn. on council 1966-—; Socio-Psychol. prize 1967). Author: Air War and Emotional Stress: Psychological Studies of Bombing and Civilian Defense, 1951; (with C. I. Hovland, H. H. Kelley) Communication and Persuasion: Psychological Studies of Opinion Change, 1953; Psychological Stress, 1958; (with others) Personality and Persuasibility, 1959; also articles, chpts. in books. Asso. editor Sociometry, 1955-58; cons. editor Jour. Abnormal and Social Psychology, 1955-65, Jour. Exptl. Social Psychology, 1965-—. Research in personality and social psychology, attitudes and decisions, stress, emotional behavior. Home: Race Brook Rd., Woodbridge, Conn. 06525. Office: Dept. Psychology, Yale, 333 Cedar St., New Haven 06510.*

JANKOVIC, Branislav Dragoljuub, Yugoslavian physician, immunologist; b. Golubac, Serbia, Yugoslavia, Feb. 22, 1920; s. Dragoljub I. and Ljubica (Pavlovic); M.D., U. Belgrade (Yugoslavia), 1947, Dipl. Bacteriology and Immunology, 1951, Ph.D., 1954; m. Ema Stojimirovic, July 14, 1948 (div. 1955); children—Maja, Milos. Med. faculty, U. Belgrade, 1948-—, prof. microbiology, Faculty Pharmacy, Microbiol. Inst., 1962-—, head Microbiol. Inst., 1956-—; head dept. developmental biology and genetics Inst. for Biol. Research, Belgrade, 1962-—; head immunology unit, 1962-—. Recipient Nat. award for sci., 1957, 63, Medal of Merit with silver rays, 1965. Mem. Serbian Med. Assn., Serbian Soc. Biology, Yugoslav Soc. Physiology, Am. Assn. Immunologists. Author: Fundamentals of Blood Transfusion, 1951; Fundamentals of Immunohaematology, 1954; Textbook of Microbiology, 1957; also numerous articles. Research on exptl. auto-immune diseases, role of bursa of Fabricus in immune reactions of birds, activity of antibrain antibodies on various brain regions; co-discoverer function of thymus gland in immunity.*

JANKOVIC, Zlatko, Yugoslavian physicist; b. Varazdin, Yugoslavia, Sept. 27, 1916; s. Ivan and Zlata (Dezelic) J.; Ph.D., U. Zagreb, 1950; m. Nada Marijan, Aug. 13, 1949. Mem. faculty sci., U. Zagreb, 1939-—, prof. applied math., 1959-—, head applied math. dept. 1960-—, dean Faculty Sci., 1962-64. Mem. Inst. Ruoter Boskovic, 1951-—; mem. sci. council Yugoslav Fed. Nuclear Energy Com., 1960-65. Corr. mem. Yugoslav Acad. Sci.; mem. other sci. socs. Author: Teorijska mehanika, 1963; also articles. Research on theoretical mechanics, especially axiomatic found. and parallel devel. of classical mechanics and spl. theory of relativity; theory of nuclear reactions; spl. functions. Home: 50 Prilaz JA, Zagreb, Yugoslavia.*

JANKOWIAK, Józef, Polish physician; b. Hamborn, Jan. 20, 1904; s. Waclaw and Antonina (Sloma) J.; student U. Poznan, Poland, 1923-29; m. Helena Kirsch, Mar. 19, 1930; Chief phys. medicine dept. Social Ins., Poznan, 1929-52; dir. State Inst. Balneoclimatology, Poznan, 1952-—; lectr. phys. medicine Med. Acad., Poznan, 1960-—. Chmn., State Council for Health Resorts Problems, 1961-—; state cons. for balneoclimatology, 1963-—. Decorated State Order Odrodzenié; medal French Fedn. Balneology and Climatology. Hon. mem. Italian, Austrian, Czechoslovakian socs. balneology; mem. Polish Soc. Balneoclimatology (chmn.). Author: Outline of Physiotherapy, 1934; Physical Therapy, 1946; also numerous articles. Chief editor: Clinical Balneology, 1962, Balneologia Polska, 1958-—, Wiadomosci Uzdrowiskowe, 1956-—; editorial bd. Zeitschrift für Rehabilitation, 1962-—, Physicalische Therapie Berlin, 1967-—. Research on influence of balneological and peat-mud stimuli on body, influence of physicotherapeut stimuli, especially effect of ultrasounds on body. Home: 24 Nowowiejskiego, Poznan, Poland.*

JANNACCONE, Amedeo, Italian agronomist; b. Avellino, Jan. 20, 1910; s. Giovanni and Luisa (Matarazzo) J. Mem. staff univ. profs., 1954; full prof. agronomy and herbaceous cultures U. Catania, also former dean Faculty Agr. Mem. supervisory council Publs. Inst. Recipient medal of merit for teaching, culture and art. Author numerous publs. on agronomy, physiology, biology, problems of meridional agr. Address: via Firenze 225, Catania, Italy.

JANNASCH, Holger Windekilde, microbiologist; b. Holzminden, Germany, May 23, 1927; s. Hans W. and Heidi (Schmieder) J.; student U. Munich, U. Göttingen (both Germany), 1949-55; Ph.D., U. Göttingen, 1955; m. Friederun Goldschmidt, Jan. 14, 1956; 1 son, Hans W. Rockefeller fellow Scripps Inst. Oceanography, 1957-58; research asso. U. Wis., Madison, 1958-59. research asst. Max Planck Soc., 1957-60; asst. prof. U. Göttingen, 1960-63, pvt. docent, 1963; sr. scientist Woods Hole (Mass.) Oceanographic Inst., 1963-—. Research and publs. on marine and freshwater microbiology, growth and physiology of bacteria in continuous culture, denitrification. Office: Woods Hole Oceanographic Instn., Woods Hole, Mass. 02543.*

JANNE D'OTHEE, Ecuyer Henry Marie Jean Ferdinand Joseph, Belgian physicist; b. Liège, Nov. 11, 1884; s. Charles and Theodosie (d'Othee) J.; Ph.D. in Physics and Math., U. Liège; m. Louise Bourguignon, Oct. 7, 1920; children—Claire, Monique, André, Charles, Véronique, Paul. Asst. lectr., instr., full prof. Faculty Sci., U. Liège, now prof. emeritus. Mem. Royal Soc. Sci. Liège, Sci. Soc. Brussels, Math. Soc. Belgium, Physics Soc. Belgium, Belgian Soc. Logic and Sci. Philosophy. Author: Sur la variation des latitudes; Extension des theoremes de Cosserat; De la portee de l'explication dans la physique contemporaine; Les lois de conservation de la physique, leur combinaisons et leur eventuelle interindependance. Address: III, rue Louvrex, Liège, Belgium.

JANOWITZ, Morris, Am. sociologist; b. Paterson, N.J., Oct. 22, 1919; s. Samuel and Rose (Myers) J.; A.B., N.Y. U., 1941; Ph.D., U. Chgo., 1948; m. Gayle Schulenberger, Dec. 22, 1951; children—Rebecca, Naomi. Faculty, U. Chgo., 1946-51, 61-—, prof., 1961-—, dir. Center for Social Orgn. Studies, 1961-—; faculty U. Mich., 1951-61; Fulbright prof. U. Frankfurt (Germany), 1954-55; cons. to numerous govt. agys. Center For Advanced Study Behavioral Scis. fellow, 1958-59. Mem. Am. Sociol. Assn., Soc. For Study Social Problems, Am. Assn. Pub. Opinion Research. Author: (with Bruno Bettelheim) The Dynamics of Prejudice, 1950; The Community Press in an Urban Setting, 1952; Campaign Pressures and Democratic Consent: An Interpretation of the 1952 Election, 1956, rev., 1964; Public Administration and the Public, 1958; Sociology and the Military Establishment, 1959, rev., 1965; The Professional Soldier, 1960; (with Robert D. Vinter) The Comparative Study of Juvenile Correctional Institutions: A Research Report, 1961; A Study of the Military Retired Pay System and Certain Related Subjects: A Report to the Committee on Armed Services, United States Senate, 1961; The Military in the Political Development of New Nations, 1964; (with Bruno Bettelheim) Social Change and Prejudice, 1964; The New Military: Changing Patterns of Organization, 1964. Editor: (with Bernard Berelson) Reader in Public Opinion and Communication, 1950, rev., 1952; (with Heinz Eulau and Samuel Eldersveld) Political Behavior, 1956. Psychological Warfare: A Case Book, 1958; Community Political Systems: International Yearbook of Political Behavior Research, 1961, Heritage of sociology series. Publs. on basic studies of Am. mil. profession; research on factors asso. with prejudice and inter group hostility, polit. process; investigation of organizational processes. Home: 1357 E. 55th Pl., Chgo. 60637.*

JANSEN, Laurens, chem. physicist; b. Terborg, Netherlands, Mar. 25, 1923; s. Laurens and Everdiena (van Drumpt) J.; Cand.Ex. (cum laude), U. Utrecht (Netherlands), 1947, Doct.Ex. (cum laude), 1950; D.Sc. (cum laude), U. Leyden (Netherlands), 1955; m. Liselotte Nothnagel, Aug. 1, 1950; children—Laurens Erwin, Inga-Lisa. Faculty, Inst. Molecular Physics, U. Md., College Park, 1952-58, asso. prof., 1955-58; vis. prof. U. Uppsala, (Sweden), also Fed. Inst. Tech., Zürich, Switzerland, 1958-59; research scientist Batelle Meml. Inst., Geneva, Switzerland, 1959-—; asso. dir. Battelle Inst.-Advanced Studies Center, Geneva, 1966-—. Recipient Poor Richard Club award, Franklin Soc., 1956. Mem. Swiss Phys. Soc. Author: (with M. H. Boon) Finite Group Theory and Applications, 1967; also articles. Research on dielectric theory of compressed gases, 3-atom and 3-ion interactions in crystals, crystal stability theory. Home: 20 Im Lägi Küsnacht, 8700, Switzerland. Office: 7 Route de Drize, Carduge, Geneva 1227, Switzerland.*

JANSEN, Zacharias, Dutch optician; b. Le Haye, 1580; credited with inventing (with his father) compound microscope, circa 1590; also with inventing (independently of H. Lippershey and J. M. Adriaenszoon) telescope, built in model form, 1609. Died Amsterdam, Netherlands, circa 1638.

JANSKY, Jan, Czechoslovakian physician, psychiatrist; b. 1873. Classified human blood into 4 groups, 1907, which he numbered I to IV; Jansky's groupings correspond to Landsteiner O, A, B and AB groupings. Died 1921.

JANSKY, Karl, Am. radio engineer; b. Norman, Oklahoma, Oct. 22, 1905; ed. U. Wisconsin. Began as engr. for Bell Telephone Laboratories, 1928. Discovered that source of weak radio static was constellation of Sagittarius, 1932; (this discovery led to development of radio astronomy); unit of radio-wave emission strength called Jansky in his honor. Died Red Bank, New Jersey, Feb. 14, 1950.

JANSSEN, Carl Luplau, Danish astronomer; b. Frederiksberg, Nov. 3, 1889; s. Ernek and Gerda (Ramloss) J.; ed. univs. Lund, Copenhagen, Sorbonne; Ph.D.; m. Aase Mollerup, Mar. 25, 1939; 1 dau., Anna. Librarian, Univ. Library Copenhagen; dir. Urania Obs., Copenhagen; head Municipal Obs., Copenhagen Mem. Internat. Astron. Union. Research and publs. on astronomy. Address: Dr. Olgasvej 25, Copenhagen F., Denmark.

JANSSEN, Paul Adriaan Jan, Belgian physician, pharmacologist; b. Turnhout, Sept. 12, 1926; s. Constant and Margareta (Fleerackers) J.; ed. univs. Namur, Louvain, Gand; M.D., agrege in teaching of pharmacology; m. Theodora Arts, July 1, 1957; children—Graziella, Herwig, Yasmine, Paul. Dir. Research Lab. Dr. C. Janssen, Beerse, Belgium. Author: Over de Pharmacologie van een reeks Propylaminen, 1956; Synthetic Analgesics. Part I: Diphenylpropylamines, 1960; The Action of Neuroleptic Drugs. Home: Antwerpsesteenweg 31, Vosselaar/Turnhout. Office: Research Lab. Dr. C. Janssen, Beerse, Belgium.

JANSSEN, Pierre Jules César, French astronomer; b. Paris, Feb. 22, 1824; ed. U. Paris; made expdns. to determine magnetic equator in Peru, 1857-58, to study telluric rays in solar spectrum in Italy, 1861-62, 64, to make magnetic and topog. studies in Azores, 1867; made various solar eclipse expdns.; became prof. gen. sci., sch. architecture U. Paris, 1865; named dir. Meudon Obs., Paris, 1876; founder, dir. obs. on Mont Blanc, from 1893. Fellow Royal Soc., 1875 French Acad. Scis., 1873. Author: Atlas des photographies solaires, 1904. Used observation of solar spectrum made from Monte Blanc in 1893 to prove that atmospheric oxygen causes strong oxygen absorption lines; made spectroscopic observations of planets and sun; 1st to use photography regularly to study sun; discovered method for observing solar prominences in daylight. Died Meudon, Dec. 23, 1907.

JANTSCHEW, Wassil Georgiev, Bulgarian oncologist, gastroentologist; b. Haskovo, Bulgaria, Mar. 20, 1918; s. Georgi and Maria (Wassileva) J.; Dr.vet. med., Sofia, Univ., 1942, Dr. medicine, 1947; m. Margarita Ilieva, Feb. 28, 1943; 1 son, Givko Wassilev. Asst. physician Vet. Med. Faculty, Sofia, Bulgaria, 1942-44, sci. worker Inst. Oncology, 1947-50; staff dept. internal medicine and gastroenterology Postgrad. Med. Inst., Sofia, 1950-—, sr. asst. physician, 1956-—. Cons. Sci. Research Inst. Tb., Sofia. Mem. Gastroenterology Assn. Bulgaria, Union Sci. Workers Bulgaria. Author: Rectoromanoscopy, 1958; The Procain (Novocain) in the Contemporary Therapy, 1963; (with Tashev) Diseases of the Stomach, Intestine and Peritoneum, 1963; (with others) Created Morbid States, 1965. Research, publs., inventor on tonus and motoric function of colon and rectum, pathomorphology and pathophysiology of digestive tract, cancerous and rare tumors. Home: 18 Benkowski, Sofia, Bulgaria.*

JANUARY, Lewis Edward, Am. physician; b. Haswell, Colo., Nov. 14, 1910; s. Frank Puleng, and Estella Lela (Miller) J.; B.A., Colo. Coll., 1933; M.D., U. Colo., 1937; D.Sc. (hon.) Colo. Coll., 1966; m. Virginia Eloise Taylor, Sept. 13, 1941; children—Alan Frank, Craig Taylor. Faculty, U. Ia., Iowa City, 1946-—, prof. cardiology, 1953-—; staff U. Hosps., Iowa City, Des Moines VA hosps. Recipient Ia. Heart Assn. medal, 1955. Fellow A.C.P., Council on Clin. Cardiology (past chmn.); mem. Am. Heart Assn. (pres. 1966-67, dir. 1955-—, exec. com. 1961-—), Central Soc. Clin. Research, Am. Fedn. Clin. Research, Am. Soc. Internal Medicine, Pan Am. Med. Soc., Assn. Univ. Cardiologists. Contbg. author: Cyclopedia of Medicine, 1952; Current Therapy, 1952, 53, 54, 59, 62, 63, 66; also numerous articles. Research on cardiovascular disease and electrocardiography; co-developer treatment diabetes insipidus by pitressin tannate in oil. Home: 425 Lexington Av., Iowa City 52240.*

JAPHET IBN ABU AL-HASAN AL-BARQAMANI AL-ISRA'ILI AL-ISKANDARI, see al-Barqamani.

JAPP, Francis Robert, chemist; b. Dundee, Scotland, Feb. 8, 1848; s. James Japp; ed. U. St. Andrews; student law Edinburgh (Scotland) U.; student chemistry under Bunsen and Kekule, Bonn, Heidelberg univs. (both Germany); M.A., LL.D.; m. Elizabeth (Tegetmeyer), 1879; 2 daus. Research asst., later lectr. chemistry Normal Sch. Sci., Kensington, Eng., 1878-90; prof. chemistry Aberdeen (Scotland) U., from 1890. Fellow Royal Soc., 1885; mem. Chem. Soc. (fgn. sec. 1885-91, Longstaff medallist 1891, v.p. 1895-99, 1915-18), Brit. Assn. (pres. chmn. sec. 1898), Inst. Chemistry of Gr. Britain and Ireland (v.p. 1901-04). Author: (with Sir Edward Frankland) Inorganic Chemistry, 1884; also contr. articles to Journal Chemical Society, from 1879. Research on benzil, benzoin and phenanthraquinone. Died Aug. 1, 1925.

JAQUES, Louis Barker, Canadian pharmacologist; b. Toronto, Ont., Can., July 10, 1911; s. Robert Herbert and Ann Bella (Shepherd) J.; B.A., U. Toronto, 1933, M.A., 1935, Ph.D., 1941; m. Helen Evelyn Delane, May 15, 1937; 1 dau., Mary (Mrs. Stephen S. Hall). Faculty, U. Toronto, 1934-46; faculty U. Sask., Saskatoon, 1946-—, prof. physiology, head physiology, pharmacology, 1946-—. Ofcl. Canadian rep. Council and Gen. Assembly Internat. Union Physiol. Scis., Leyden, 1962. Fellow N.Y. Acad. Scis., Royal Soc. Arts, Royal Soc. Can., Internat. Soc. Hematology; mem. Am. Physiol. Soc., Am. Soc. Pharmacology and Exptl. Therapeutics. Author: The Prayer Book Companion, 1963; Anticoagulant Therapy: Pharmacological Principles, 1965; also numerous articles. Research on metabolism and action of anticoagulant and related drugs basic to use of these compounds. Home: 682 University Dr., Saskatoon, Sask., Can.*

JARCHO, Julius, gynecologist, obstetrician; b. Ukraine, Russia, Oct. 15, 1882; ed. Berlin, Vienna, Prague, Budapest; M.D., Columbia U., 1904; hon. Ph.D., Hebrew U., Jerusalem, 1955; m. 1st (d. 1955); m. 2d. Fanny; children: Saul, Leonard W.; Mrs. Daniel Rose. Attending gynecologist, obstetrician, Sydenham Hosp., N.Y., 1920-48; dir., obstetrics and gynecology, 1948-50; consultant, 1950. Mem. Board of Governors, Hebrew U., Jerusalem; founder and vice-pres., Am. Jewish Physicians Committee; Fellow, Am. College Surgeons; Internat. College Surgeons, N.Y. Acad. Medicine; World Medical Assn.; diplomate, Am. Board of Obstetrics and Gynecology. Author: Gynecological Röntgenology, 1931; The Pelvis in Obstetrics, 1933; Postures and Practices Among Primitive Peoples, 1934; over 30 pub. articles. Designed equipment for use in gynecology and obstetrics including: manometer (called Jarcho's pressometer) to measure pressure in hysterosalpingography, pyelography; instrument to diagnose sterility, 1930; tenaculum forceps; obstetric table, gynecological x-ray table; studied tumors, malformations of uterus. Died New York, N.Y., May 14, 1963.

JARJAVAY, Jean François, French physician; b. Si Savignac-les-Églises, France, 1815; ed. Paris; became intern, 1841; surgeon, anatomist; prof. anatomy and clin. surgery, Paris. Author: Aponévrose du Périnée, principalement chez la femme, 1846; De l'influence des efforts sur les maladies chirurgicales, 1847; Pouce et la luxation du dernier os de ce doigt, 1849; Sur les opérations applicables aux Corps fibreux de l'uterus, 1849; Sur l'ensemble des généralités relatives aux fractures articulaires, 1850; Traité d'anatomie chirurgicale, 2 vols., 1852; Recherches sur l'urètre de l'homme, 1856; Du mode de reductionet des maintien des fragments, dans la fracture de l'extremité inférieure du radius, 1866; Sur la luxation du Tendon de la longue portion du biceps humeral, 1867. Sur la fracture de la cloison des fosses nasales, 1867. Described a branch of sphincter urethrae membranaceae, known as depressor urethrae muscle, or Jarjavay's muscle, 1856. Died Paris, 1868.

JÄRNEFELT, Gustaf Juhana, Finnish astronomer; b. Helsinki, May 6, 1901; s. Tauno and Ebba (Schauman) J.; Ph.D., U. Helsinki; m. Irma von Rosenkampff, Sept. 22, 1928; children—Johan, Margareta, Carl Gustaf, Wilhelm. Asst., Astron. Obs., U. Helsinki, 1923-41; asst., Helsinki Obs., 1927-45, full prof. astronomy, dir., 1945-—. Mem. Finnish Soc. Sci., Finnish Acad. Sci. Author: Das Einkorperproblem in dem sich ausdehnenden Raume der Einstein de Sitter'schen Welt, 1942; Visual Observations of Satellites in Finland, 1957-64; On a Finite Approximation of the Linear Harmonic Oscillator, 1958. Research and publs. on relativity, finite geometries. Home: 3 Ohjaajantie, Helsinki 40. Office: Astron. Obs., Helsinki 13, Finland.*

JARRETT, Alan Hamer, Brit. physicist; b. Manchester, Eng., Feb. 2, 1925; s. Walter George and Julie (Hamer) J.; B.Sc., Leeds U., 1946; M.Sc., Manchester U., 1949; Ph.D., St. Andrews U., 1952; m. Hester Margaret Kinnear, Sept. 16, 1959; 1 dau., Fiona Margaret. Lectr. astronomy St. Andrews U., 1953-59; lectr. astronomy Queen's U. Belfast, No. Ireland, 1959-64, sr. lectr. applied math. dept., 1964-—. Fellow Royal Astron. Soc.; mem. Inst. Physics. Research, publs. on solar corona and investigations of earth's upper atmosphere. Home: 35, Deramore Dr., Belfast, No. Ireland.*

JARRETT, Arthur, English physician; b. Portsmouth, Eng., Nov. 26, 1919; s. Charles James and Ruth (Clarke) J.; M.B.Ch.B., U. Birmingham, 1943; M.R.C.P., Edinburgh, 1952, F.R.C.P., 1960, M.C. Path., 1963; D.Sc., U. London, 1965; m. Lily Allport, Sept. 23, 1943; 1 son, Peter James. Dermatol. registrar Bradford, Edinburgh, Cambridge, 1947-52; faculty U. London, 1952-—, reader in dermatol. histopathology, 1959-—, dir. dept. Hon. cons. U. Coll. Hosp. Fellow Royal Soc. Medicine, Royal Micros. Soc., Brit. Assn. Dermatology; mem. Soc. Exptl. Biology, Soc. Cosmetic Chemists. Author: Science of the Skin, 1964; (with R. I. C. Spearman) Histochemistry of the Skin: Psoriasis, 1965; (with R. I. C. Spearman, P. A. Riley) Functional Introduction to Dermatology, 1966; also articles. Research on hormonal effects on surface sebum, fluorescence microscopy of normal and path. skin, keratinization of human and animal epidermis, histochem. studies of enzymes in human and animal skin, melanocytes and their function. Home: 22 Michleham Down, Woodside Park, London N.12, Eng.*

JARRETT, Howard Starke, Am. physicist; b. Charleston, W.Va., Oct. 24, 1927; s. Howard S. and Frances (Baer) J.; B.S., Rensselaer Poly. Inst., 1947, M.S., 1948; Ph.D., Mass. Inst. Tech., 1951; m. Ann Wyllie, Apr. 7, 1951; children—William H., David S., Andrew W., Sara Jane. Research physicist E. I. duPont de Nemours & Co., Wilmington, Del., 1951-55, research supr., 1955-—; mem. steering com., adv. com. Ann. Conf. on Magnetism and Magnetic Materials, 1966-—, also program chmn. 11th Ann. Conf. Mem. Am. Phys. Soc. Contbr. chpt. to Solid State Physics, 1963. Editor: (with C. W. Haas) Magnetic Materials Digest, 1966. Investigated paramagnetic resonance in organic free radicals, transition metal chelates and ferromagnetic materials; studies on magnetic phase transitions and magnetic spin configurations in ferromagnetic materials. Home: 805 Sycamore Lane, Wilmington 19807. Office: Central Research Dept., E. I. duPont & Co., Wilmington, Del. 19898.*

JARRICOT, Jean, French physician; b. Saint Genis Laval, Rhone, France, July 14, 1877; Diplome d'etudes superieures; diplome de physiologie generale et comparee; M.D., Lyons (France) U.; lab. dir. Med. Sch., Lyons U.; mem. staff for med. research in pharmacodynamics at Saint Rambert l'Ile Barbe, Rhone. Decorated Officer, Legion d'honneur, Officer de l'Instruction publique, Officier du Nicham (Tunis), Laureat de l'Academie francaise, 1960. Mem. Rhone Homeopathy Soc. (hon. pres.), Barcelona Medico-Homeopathic Soc. (hon.), Hahnemanian Inst. Brazil (asso.). Author: Pendule et Medicine, 1949; Radiesthesie, 1958. Research in radiesthesia or dowsing, with negative results in paranormal phenomena. Died Nov. 13, 1962.

JARS, Antoine Gabriel, French metallurgist, mineralogist; b. Lyons, France, Oct. 1, 1732; ed. sch. of Ponts and Chausses; mining engr.; prospected for coal in eastern France; visited Holland, Germany, Scandinavia, 1766; mem. French Acad. Scis., 1768. Author: Voyages métallurgiques, 3 vols., 1774-77. Studied (with Trupaitie) French mines, also (with Geullt) fgn. mines; publicized in France the Brit. method for refining aluminum; developed process for copper refining which consumed less wood, 1775. Died Clermont Ferrand, France, Aug. 23, 1769.

JARVIK, Murray Elias, Am. psychopharmacologist; b. N.Y.C., June 1, 1923; s. Jacob and Minnie (Haas) J.; B.S., Coll. City N.Y., 1944; M.A., U. Cal. at Berkeley, 1945, M.D., 1951, Ph.D., 1952; m. Lissy Feingold, Dec. 19, 1954; children—Laurence, Jeffrey. Research technician phys. chemistry Rockefeller Inst., 1943-44; asst. exptl. psychology U. Cal., Los Angeles, 1944-45, asst. physiol. psychology, Berkeley, 1945-46; research asso. comparative and physiol. psychology Yerkes Lab., Fla., 1951-53; lectr. physiol. psychology Columbia, 1953-55; research asso. Mt. Sinai Hosp., 1953-55; vis. asst. prof. physiol. psychology U. Cal., Berkeley, 1955; research asso. psychopharmacology L.I. Biol. Assn., N.Y.C., 1955-56; adj. asst. prof. physiol. psychology grad. div. N.Y. U., 1957; faculty Albert Einstein Coll. Medicine, Yeshiva U., N.Y.C., 1956-—, asso. prof. pharmacology, 1960-—. Mem. N.Y. State Med. Assn., Am. Acad. Neurology, Am. Coll. Neuropsychopharmacology, Eastern Psychiatric Assn., Am. Psychol. Assn., Am. Soc. Pharmacology and Exptl. Therapeutics, A.A.A.S., N.Y. Acad. Sci., Fedn. Am. Soc. Exptl. Biology, Psychonomics Soc., A.M.A., Collegium Internationale Neuro-Psychopharmacologium, Phi Beta Kappa, Sigma Xi. Publs. on studies of effects of drugs upon learning and retention, neurophysiol. basis of learning, drugs and central nervous system, psychopharmacology, implantation of arterial catheters. Home: 2727 Palisades Av., N.Y.C. 10463.*

JÄRVINEN, Pentti Antero, Finnish obstetrician, gynecologist; b. Helsinki, Finland, Aug. 22, 1923; s. Kustaa Esa and Ingrid Maria (Koskinen) M.B., U. Helsinki, 1946, Licenciate Medicin, 1949, Dr.med. and Chir., 1953; m. Marjatta Margit Sattuma, Jan. 10, 1948; children—Riitta Marjatta, Olli Antero, Jarkko Juhani, Kirsi Inkeri. Asst. surgeon II Womens Clinic, Helsinki U., 1953-56, asst. chief surgeon, 1959-65, asst. surgeon radiol. dept., 1957-58; chief surgeon, prof. Womens Clinic Oulu (Finland) U., 1965-—, vice dean med. faculty, 1966-—. Mem. Internat. Fertility Assn., Internat. Fedn. Gynecology and Obstetrics, Soc. for Study Fertility. Contbg. author: Sterility, 1964. Research, numerous publs. on ectopic pregnancy, endocrinology of severe vomiting of pregnancy, physiology of pregnant uterine musculature. Home: 14 Lektoranta, Oulu, Finland.*

JARVIS, Deming, Am. inventor; b. probably Boston, circa 1790; s. John and Hannah (Seabury) J.; m. Anna Smith Stutson, 1815; 2 children—John, James. Bought (with Amos Binney and Daniel Hasting) Boston Crown Glass Co., Cambridge, Mass., 1817, granted charter to manufacture flint and crown glass; constructed exptl. furnace in which he compounded litharge or red lead which enabled co. to compete with English glass cos.; broke with former assos. and organized Boston and Sandwich Glass Co., Sandwich, Mass., 1826, produced apothecary and chem. supplies in addition to glass products, tableware, chandeliers, mantle lamps; reproduced certain shades of English glass by using barytes earth in his mixture; compiled directions for bldg. kilns, 1828; organized Cape Cod

Glass Co., Sandwich, 1858. Author: Reminiscences of Glass Making, 1854. Died Boston, Apr. 15, 1869.

JARVIS, Edward, Am. physician; b. Concord, Mass., Jan. 9, 1803; s. Francis and Milicent (Hosmer) J.; grad. Harvard, 1826; studied physiology and anatomy under Dr. Josiah Bartlett; attended Mass. Med. Sch. (now Harvard Med. Sch.); M.D., U. Vt., 1830; m. Almira Hunt, Jan. 9, 1834, 2 children. Taught sch., Concord, 1827. practiced medicine, Northfield, Mass., 1830, Louisville, Ky., 1837-43; became interested in vital statistics due to Lemuel Shattuck's influence; contbr. articles to Louisville Med. Jour.; returned to Dorchester, Mass., opened house for treatment of insane, 1843; mem. commn. that studied condition of insane and idiots in Mass., 1854. Author: (pamphlets) Practical Physiology, 1847, Primary Physiology, 1848; also many articles, speeches, other pamphlets. Suggested improved methods in census procedure. Died Dorchester, Oct. 31, 1884.

JARVIS, Roger George, physicist; b. Hugglescote, England, Apr. 26, 1928; s. Thomas Wilfrid and Dorothy (Fowkes) J.; B.A., Oxford (Eng.) U., 1949, M.A., 1952, D.Ph., 1953; m. Ruth Anne Wilson, Nov. 13, 1954; children—Alan Richard, Hugh William. Nat. Research Council of Can. fellowship nuclear physics Chalk River, Ont., 1953-54; Imperial Chem. Industries Research fellowship nuclear physics Liverpool (Eng.) U., 1955; advanced projects Gen. Elec. Co. of England Atomic Power div., Erith. Eng., 1956; sr. research officer reactor physics Atomic Energy of Can., Ltd., Chalk River, Ont., 1956—. Mem. Am. Nuclear Soc., Canadian Assn. Physicists. Contbr. numerous articles in field to profl. jours. Expts. low energies in nuclear reactions of light nuclei; measurements of neutron scattering at low and high energies; theory of thermal neutron spectra in reactors. Home: 38 Parkdale Av., Deep River, Ont., Can. Office: Atomic Energy of Canada Ltd., Chalk River, Ont., Can.*

JARVIS, William Chapman, Am. laryngologist; b. Fortress Monroe, Va., May 13, 1855; s. Nathan Sturges Jarvis; M.D., U. Md., 1875; attended Johns Hopkins. Practiced medicine, N.Y.C., 1877; worked as asst. in Prof. Frank H. Bosworth's Nose and Throat Service, Bellevue Hosp., N.Y.C.; prof. U. City of N.Y. (now N.Y.U.), 1886-93, prof. emeritus, 1893. Known for innovations in diagnosis and treatment of nasal and laryngeal diseases such as use of local anesthetic; contbr. papers to periodical lit., 1880-92; invented snare used to remove polyps from nose and throat (Jarvis' snare), 1881. Died West Point, N.Y., July 30, 1895.

JASINSKI, Wladyslaw Kazimierz, Polish physician; b. Warsaw, Poland, Jan. 22, 1916; s. Wladyslaw and Zofia (Pawlik) J.; M.D., Jagiellonian U., Cracow, Poland, 1946; m. Janina Groyecka, May 3, 1941; children—Wlaydslaw, Margaret, With Inst. Oncology, Radium Inst., Warsaw, 1937-41; chief radiotherapist, Gliwice, Poland, 1947-57; chief radiotherapist, Warsaw, 1957—, dir., 1960—; with dept. radiology Gen. Hosp., Warsaw, 1941-45, Cracow U. Sch. Medicine, 1945-47; prof. radiotherapy Postgrad. Med. Sch., Warsaw, 1959—. Mem. Polish Acad. Sci. Author: Physical Bases of Radiotherapy of Cancer, 1950; Clinical Use of Radioactive Isotopes, 1965, contbr. to Handbook of Oncology, 1955, 65. Editor: The Use of Radioactive Calcium Ca 47, 1966. Research in radiotherapy of cancer, clin. research on mineral metabolism, organic cancer. Home: 34-A Tamka. Office: 15 Wawelska, Warsaw, Poland.*

JASKIEWICZ, Jean, Polish physician; b. Lvov, Poland, July 6, 1749; physician Faculty Medicine Vienna (Austria); became prof. chemistry, botany, natural history U. Cracow (Poland), 1780; pres. Cracow Coll. Physicians; mem. French Acad. Scis., 1781. Author: De l'atmosphère, 1785. Founder 1st chem. lab. in Cracow; research on composition of atmosphere; drew up a mineralogical chart. Died Pinczow, Poland, Nov. 14, 1809.

JASKOSKI, Benedict Jacob, Am. biologist; b. Velva, N.D., July 25, 1915; s. Leo Martin and Anna (Kokott) J.; A.B., Jamestown Coll., 1939; M.S., U. Notre Dame, 1942, Ph.D., 1950, D.Sc., 1964; hon. degree Ill. Coll. Podiatry, 1964; m. Josephine Dilso, Apr. 7, 1956; 1 son, Joseph Michael. Asst. prof. Creighton U., Omaha, 1950-54; prof. biology Loyola U., Chgo., 1954—. Spl. cons. World Book Ency., 1963—; cons. research Am. Podiatry Assn., 1960—. NIH grantee 1957-64. Fellow A.A.A.S.; mem. Sigma Xi. Contbr. articles to tech. jours. Research in parasitology especially parasitic round worms, parasites zoo animals, dermatophytic fungi. Home: 6336 Magnolia St., Chgo. 60626.*

JASMIN, Gaetan, Canadian physician; b. Montreal, Que., Can., Nov. 24, 1924; s. Horace and Antoinette (Piquette) J.; B.A., Coll. St. Laurent, Montreal, 1945; M.D., U. Montreal, 1951; Ph.D., 1956; m. Suzanne DuPont, Oct. 4, 1952; children—Eve, Luc, Pierre. Faculty U. Montreal, 1951—, asso. prof. dept. pathology Faculté de Médecine, 1957—. Bd. dirs. Revue Canadienne de Biologie, 1965—; med. research asso. Med. Research Council Can., 1957—. Mem. Endocrine Soc., Soc. Exptl. Biology, Histochem. Soc., Canadian Rheumatism Assn., Am. Physiol. Soc., N.H. Acad. Scis., Assn. Canadienne Francaise pour l'avancement des Scis. Editor: Methods and Achievements in Experimental Pathology, 1966. Research and numerous publs. on muscle diseases, exptl. carcinogenesis. Home: 189 Glengarry, Mt. Royal. Office: Dept. Pathology, Faculty of Medicine, U. Montreal, 2900 Blvd., Mt. Royal, Montreal, Que., Can.*

JASPER, Herbert Henry, neurophysiologist; b. LaGrande, Ore., July 27, 1906; s. Franklin Merrill and Lina (Dupertuis) J.; B.A., Reed Coll., 1927; M.A., U. Ore., 1927; Ph.D., U. Ia., 1931; D.Sci., U. Paris (France), 1935; M.D.C.M. McGill U., 1943; m. Margaret M. Goldie, Aug. 17, 1940; children—Stephen F., Joan G. Instr. U. Ore., 1927-29, U. Ia., 1929-31; NRC fellow U. Paris, 1931-33; asst. prof. Brown U., 1933-38; with McGill U., 1938-64, prof. exptl. neurology, 1946-64; prof. neurophysiology U. Montreal (Que., Can.), 1965—; editor Internat. Jour. Electroencephalography and Clin. Neurophysiology, 1946. Fellow Royal Soc. Can.; mem. Am. Electroencephalographic Soc. (pres., 1950), Internat. Brain Research Orgn. (exec. sec., 1961-63), Am., Canadian physiol. socs., Am. Neurol. Assn., Am. Acad. Neurology, Canadian Neurol. Soc., Assn. for Research in Nervous and Mental Disease. Author: (with W. Penfield) Epilepsy and the Functional Anatomy of the Human Brain, 1953. Numerous publs. as pioneer in studies of the elec. activity of the brain and the devel. of electroencephalography and clin. neurophysiology; discovered new principles for the central integration of brain function regulating consciousness and attention; established elec. diagnostic techniques for brain diseases, including microelectrode techniques for recording solitary brain cell activity. Home: 804 Upper Lansdowne Av., Westmount, Que. Office: Dept. Physiology, U. Montreal, Montreal, Can.*

JASTRAM, Philip Sheldon, Am. physicist; b. Providence, Feb. 28, 1920; s. Edward Perkins and Laura (Whitney) J.; S.B., Harvard, 1943; Ph.D., U. Mich., 1948; m. Frances Falconer Grant, June 10, 1947; children—John Carter, Mary Whitney. Research asso. Radio Research Lab., Cambridge, Mass., 1943-45; instr. U. Mich., 1948-49, research physicist Bur. Ordnance project, 1949; asst. prof. physics Washington U., St. Louis, 1949-54; faculty Ohio State U., Columbus, 1955—, prof., 1964—. Cons. U. S. AID, 1964, 65. OEEC fellow, Copenhagen, Denmark, 1961. Fellow Am. Phys. Soc.; mem. I.E.E.E., A.A.A.S., Fedn. Am. Scientists, Phi Beta Kappa, Sigma Xi. Contbr. articles to tech. jours. Research in nuclear structure and nuclear spectroscopy, angular and polarization correlation measurements, nuclear orientation at low temperatures. Home: 115 W. Royal Forest, Columbus, O. 43214.*

JASTROW, Joseph, psychologist; b. Warsaw, Poland, Jan. 30, 1863; s. Marcus and Bertha (Wolfsohn) J.; A.B., U. Pa., 1882, A.M., 1885; fellow in psychology, Johns Hopkins, 1885-86, Ph.D., 1886; LL.D., Wittenberg, 1928; m. Rachel Szold, Aug. 2, 1888 (dec. 1926); 1 son, Benno B. (adopted). Prof. psychology U. Wis., 1888-1927. In charge psychol. sect. Chgo. Expn., 1893; lectr. Columbia U., 1910, New Sch. for Social Research, N.Y.C., 1927-33. Pres. Am. Psychol. Assn., 1900. Author: Time Relations of Mental Phenomena, 1890; Epitomes of Three Sciences (with others), 1890; Fact and Fable in Psychology, 1900; The Subconscious, 1906; The Qualities of Men, 1910; Character and Temperament, 1915; The Psychology of Conviction, 1918; Keeping Mentally Fit, 1928; Piloting Your Life, 1930; Effective Thinking, 1931; The House that Freud Built, 1932; Wish and Wisdom, 1934; Sanity First, 1935; The Betrayal of Intelligence, 1938. Editor and contbr.: The Story of Human Error, 1936. Contbr. on psychol. subjects in scientific journals and mags. Conducted daily syndicated articles under title Keeping Mentally Fit, 1928-32. Did much to establish psychology as a new science, and to popularize sci. psychology; dealt with psychology of occult; did original work in psychophysics, important work on method of determining differential limen; studied abnormal psychology, devel., analysis, correlation of mental processes. Died Jan. 8, 1944.

JASTROW, Robert, Am. atmospheric physicist; b. N.Y.C., Sept. 7, 1925; s. Abraham and Marie (Greenfield) J.; B.A., Columbia Coll., 1944; M.A., Columbia U., 1945, Ph.D., 1948; postgrad. (postdoctoral fellow) Leyden U., 1948-49; m. Ruth Witenberg, 1967. Mem., Princeton Inst. for Advanced Study, 1949-50, 1935; faculty U. Cal. at Berkeley, 1950-53; asst. prof. Yale, 1953-54; cons. nuclear physics U. S. Naval Research Lab., Washington, 1958-62; head theoretical div. Goddard Space Flight Center NASA, 1958-62, chmn. lunar exploration com., 1959-60, mem., 1960-63; dir. Goddard Inst. Space Studies, 1961—; adj. prof. geophysics Columbia, 1961—, dir. summer inst. space physics, 1962—. Fellow Am. Geophys. Union (v.p. planetary sci. sect. 1961-63), Royal Astron. Soc.; mem. Am. Phys. Soc., Internat. Acad. Astronautics, A.A.A.S. (commn. on sci. edn.). Recipient, medal of Excellence, Columbia, 1962, Arthur S. Fleming award, 1965. Author: Exploration of Space, 1960; Red Giants and White Dwarfs: The Evolution of Stars, Planets and Life, 1967. Editor, Jour. Atmospheric Sci., 1962—; coeditor: The Origin of the Solar System, 1963; editorial bd., Sci. Mag., 1962—. Study of nuclear and atmospheric physics; physics of moon and terrestrial planets. Home: 22 Riverside Dr., N.Y.C. 10023. Office: 2880 Broadway, N.Y.C. 10025.*

JASTRZEBSKI, Zbigniew Damazy, chemist; b. Warsaw, Poland, Sept. 15, 1910; s. Toefil and Apolonia (Niesiobedzka) J.; Dipl. Ing., Tech. U. of Warsaw, 1935, D.Sc., 1961; m. Irena Siedlecka, Apr. 17, 1948; children—Christopher Z., George W. Came to U. S., 1953, naturalized 1958. Sr. asst. lectr. research engr. Road Research Inst., Warsaw Polytech., 1935-39; dir. research labs., Kabul, Afganistan, 1939-43; prof., head dept. chemistry and chem. engring. Polish U. Coll., also vice prin., London, Eng., 1946-53; prof. chem. engring. Lafayette Coll., Easton, Pa., 1953—, head dept., 1961—. Mem. Am. Inst. Chem. Engrs., Am. Soc. Engring. Edn., Nat. Assn. Corrosion Engrs., Am. Assn. U. Profs., Instn. Chem. Engrs. Gt. Britain. Author: Nature and Properties of Engineering Materials, 1959. Research on rheological behavior concentrated suspensions and electrochem. Studies of corrosion phenomena. Home: 218 Burke St., Easton, Pa. 18042.*

JAUCH, Josef Maria, physicist; b. Lucerne, Switzerland, Sept. 20, 1914; s. Josef Alois and Emma (Conti) J.; M.S., Fed. Inst. Tech., 1938; Ph.D., U. Minn., 1940; m. Tonia Hegland, Jan. 1, 1940; children—Karl Martin, Eldri-Therese (Mrs. Pierre Marechal), Aletha Lucia. Came to U. S., 1938, naturalized, 1945. Asst. prof. Princeton, 1942-45; with Bell Telephone, 1945-46; asso. prof. U. Ia., 1946-50, prof., 1950-59; with Office Naval Research, London, Eng., 1959-60, cons., 1961—; prof., dir. inst. theoretical physics U. Geneva (Switzerland), 1960—. Vis. scientist CERN, 1960—; mem. Research Council Switzerland, 1963—; cons. Batelle Inst., Geneva, 1966—. Author: (with F. Rohrlich) Theory of Photons and Electrons, 1956; Foundations of Quantum Mechanics, 1968; also papers. Home: 22 Rue des Charmilles, Geneva, Switzerland.*

JAUGEON, Jacques, French mechanician; b. 1646; mem. French Acad. Scis., 1699. Author: Description de l'art de l'imprimerie. Research on silk prodn., bookbinding; pub. collection of alphabets known then. Died Paris, Dec. 30, 1724.

JAUHO, Pekka Antti Olavi, Finnish physicist; b. Oulu, Finland, Apr. 27, 1923; s. Antti Arvid and Sylvi Matilda (Pajari) J.; M.Sc., U. Helsinki, 1948, Ph.D., 1950; m. Lahja Kyllikki Hakala, Apr. 1, 1948; children—Antti-Pekka, Veli-Matti. Faculty Tech. U. Helsinki 1953—, prof. physics, 1957—. Mem. Finnish Phys. Soc., Finnish Acad. Scis., Finnish Acad. Tech. Scis., Am. Nuclear Soc. Author: Atomi-ja ydinfysikka, 1962; also articles. Research on determination of nuclear forces using scattering expts., thermodynamics, neutron slowing down and thermalization. Home: 6 L. Menninkäisentie, Tapiola, Finland.*

JAUMOTTE, Andre Louis, Belgian physicist; b. Jambes, Dec. 8, 1919; s. Jules and Marie (Braibant) J.; ed. in civil, mech. and elec. engring. U. Brussels; m. Valentine Demoulin, May 14, 1946; 1 dau., Irene. Dir., Inst. Applied Mechanics and Electronics, Aero. Inst. Recipient August Sacred prize Royal Acad. Belgium. Mem. Inst. Mech. Engrs., Royal Belgian Soc. Engrs. and Industrialists (laureate), Assn. Engrs. U. Brussels. Author: La propulsion par fusees; Rocket Propulsion; also numerous publs. on thermodynamics, aerodynamics, turbomachines. Home: 33, avenue Jeanne. Office: 50, avenue Franklin-Roosevelt, Brussels 5, Belgium.

JAUNCEY, George Eric MacDonnell, physicist; b. Adelaide, Australia, Sept. 21, 1888; s. George and Agnes Binnie (Davis) J.; B.S., U. Adelaide, 1910, D.Sc., 1922; M.S., Lehigh U., 1916. m. Ethel Sarah Turner, Jan. 16, 1913; 1 dau., Molly Horsfall. Came to U. S. 1914; asst. prof. physics, Washington U., St. Louis, 1921-24, asso. prof., 1924-30, prof. from 1930. Fellow Am. Phys. Soc., A.A.A.S.; mem. Sigma Xi, Phi Beta Kappa. Author: Modern Physics, 1932, rev., 1937; also numerous articles on scattering of X-rays. Co-author: M.K.S. Units, 1940. Asso. editor Phys. Rev., 1926-28. Made spl. researches in theory of Compton effect in X-rays and diffuse scattering of X-rays by crystals. Died May 19, 1947.

JAVAL, Louis Emile, French ophthalmologist; b. Paris, France, May 5, 1839; grad. Paris Med. Faculty, 1868; postgrad. under Graebe, Berlin, Germany. Founder ophthal. lab. Sorbonne, 1878, dir. to 1900. Mem. French Acad. Medicine. Author: Applications a la Theorie de la Vision, 1868; L'Hygiene des Ecoles Primaires et des Ecoles Maternelles, 1884; Entre Aveugles, 1903; Manuel theorique et practique du strabisme. Translator: (with Klein) Physiologic Optic (Helmholtz). One of foremost authorities on visual hygiene; introduced astignometer for diagnosis of astigmatism, 1876; (with H. Schiotz) invented ophthalmometer for determining refractive power of eye, 1881; conceived binocular optometer; devised method of eye exercises in treatment of strabismus; 1st to emphasize importance of orthopedic tng. in various types of muscular imbalance; discovered form of eye movements in reading; work on edn. of blind (pleaded for adoption of Esperanto as auxiliary lang. of aid to blind, also suggested simplification of Esperanto). Died Paris, Jan. 20, 1907.

JAVERT, Carl Theodore, Am. physician, surgeon; b. Depew, N.Y., Nov. 12, 1907; s. Eric Gustav and Selma (Asplund) J.; M.D., U. Buffalo, 1932; m. Nancy Viscariello, June 4, 1954; children—Yvonne

Marshall, Carl Tyrrell. Asso. prof. obstetrics, gynecology Cornell U., New York City, 1946-56; clin. prof. Coll. Phys. and Surg., Columbia, 1956——; pathologist N.Y. Lying-in Hosp., 1947-56; dir. obstetrics-gynecology Womans Hosp., N.Y.C., 1956-63; practice medicine specializing in obstetrics and gynecology, N.Y.C., 1940——. Pres. Well Born Found., N.Y.C., 1958——. Fellow A.C.S.; mem. Coll. Obstetricians and Gynecologists, N.Y. Obstet. Soc. (past pres.), Am., N.Y. (pres. 1965——) gynecol. socs. Author: Spontaneous and Habitual Abortion, 1957; also numerous articles. Research on X-ray pelvimetry, endometiousis theory benign metastasis, cancer ovary, cervix and endometrium, habitual abortion, erythroblastosis, Rh and HR factor. Home: 20 Stellar Pl., Pelham Manor, N.Y. Office: 580 Park Av., N.Y.C. 10021.*

JAVID, Manucher J., neurol. surgeon; b. Tehran, Iran, Jan. 12, 1921; s. Asdolah and Touba (Ahdiyyeh) J.; came to U. S., 1944, naturalized, 1957; M.D., U. Ill., 1946; m. Lida Emma Fabbri, Oct. 19, 1951; children—Roxane Maria, Daria Deanna, Claudia Marguerite, Jeffrey James. Fellow neuropathology Ill. Neuropsychiat. Inst., Chgo., 1948-49; fellow in neurosurgery Lahey Clinic, 1949-50; teaching fellow in surgery Harvard Med. Sch., 1952; faculty U. Wis. Med. Sch., 1953——, prof. neurol. surgery, 1962——, chmn. div. neurol. surgery, 1963——. Mem. A.M.A., A.C.S., Harvey Cushing Soc., Central Neurosurg. Soc., A.A.A.S., Soc. Pan Am. Med. Soc., Am. Bd. Neurol. Surgery, Sigma Xi. Contbr. articles to profl. jours. Introduced clin. use of hypertonic solutions of urea for reduction of intracranial and intraocular pressure. Home: 4750 Lafayette St., Madison, Wis. 53705.*

JAVILLIER, Jean-Maurice, French chemist, biologist; b. Nevers, France, Feb. 5, 1875; became asst. Pasteur Inst., 1908; adjt. prof. Paris Faculty Scis., 1919, Conservatory Arts and Scis., 1931. Mem. French Acad. Scis., 1936 (pres. 1951), Acad. Medicine, Acad. Agr. Research and publs. on phosphorus, magnesium, catalytic elements, vitamins, enzymes; specialist in agrl. chemistry. Died Paris, 1955.

JAWETZ, Ernest, microbiologist; b. Vienna, Austria, June 9, 1916; s. Karl and Angela (Goldhammer) J.; came to U. S., 1939, naturalized, 1944; student U. Vienna, 1934-38; M.S., U. N.H., 1940; Ph.D., U. Cal. at San Francisco, 1942. M.D., Stanford, 1945; m. Mary Jean Morse, Oct. 29, 1954; children—Katherine, Steven, Michael, Ann. Sr. asst. surgeon USPHS, 1946-48; faculty U. Cal. Med. Center, San Francisco, 1948——, prof. microbiology, lectr. medicine and pediatrics, 1953——, chmn. dept. microbiology, 1962——; cons. VA, 1952——; A. Wright lectr., London, Eng., 1952; McArthur lectr., Edinburgh, Scotland, 1952; lectr. U. London, 1962. Fellow N.Y. Acad. Scis.; mem. Am. Fedn. Clin. Research, Western Soc. Clin. Research (pres. 1961), Am. Soc. Clin. Investigation, Western Assn. Physicians. Author: Review of Medical Microbiology, 1968. Mem. various editorial bds. med. jours. Research and publs. on infectious diseases, chemotherapy, virology, immunology. Home: 19 Cushing Dr., Mill Valley, Cal. 94941. Office: Med. Center, San Francisco 94122.*

JAYARAMAN, Aiyasami, chem. physicist; b. Madras, India, Dec. 5, 1926; s. Pichu Aiyasami and Mangalam (Ramiah) J.; B.Sc., U. Madras, 1946, M.Sc., 1954, Ph.D., 1960; m. Kamala Ramayya, May 21, 1945; children—Geetha, Chitra. Came to U. S., 1960. Faculty Raman Research Inst., Bangalore, India, 1949-60; asst. research geophysicist U. Cal. at Los Angeles, 1960-63. mem. tech. staff Bell Telephone Labs., Murray Hill, N.J., 1963——. Fellow Indian Acad. Scis. (treas. 1954-60); mem. Am. Phys. Soc. Editor Current Sci., 1957-58. Contbns. to high pressure solid state research pertaining to phase transformations, elec., electronic and magnetic properties of solids; also contbns. on structure, spectroscopic and optical properties of precious and semiprecious crystals. Home: 161 Ryder Way, Murray Hill 07974. Office: care Bell Telephone Labs., Murray Hill, N.J. 07974.*

JAYME, Georg, German chemist; b. Obermodau-Hesse, Apr. 10, 1899; s. Balthasar and Margarete (Adelberger) J.; engring. diploma Technische Hochschule, Darmstadt; m. Roseth Hjordis, Dec. 25, 1930; children—Erik, Dagmar, Astrid. Instrn. asst., Darmstadt; dir. research Verein, Glanzstoff Fabriken A. G., Berlin, Canadian Internat. Paper Co., Hawkesbury; full prof., dir. Chem. Inst. Cellulose, Technische Hochschule, Darmstadt. Recipient Alexander-Mitscherlich medal, 1942, Valentin-Hottenroth medal, 1961. Mem. German Soc. Chemists, Am. Chem. Soc., T.A.P. P.I., Assn. Cellulose Chemistry, German Soc. Wood Research, Internat. Union Pure and Applied Chemistry. Author: Zellstoff, naturl. und kunstl. Faserstoff, 1939; Zellstoff und der Papierfabrikation, 1940; Zellstoff, 1939-46, 1953; Festschrift zum 50 jahr. Bestehen des Instituts für Cellulosechemie, 1958, also over 450 articles. Home: Am Weidenborn 7. Office: Alexanderstrasse 24, Darmstadt, Germany.

JEAN DE LINIERES, French astronomer, mathematician; b. Diocese of Amiens, France. Author: Canones super tabulas magnas, 1320; Canones tabularum astronomie, 1322; Theorica planetarum (textbook), 1335. Accepted Ptolemaic views of eccentrics and epicycles and rejected homocentric spheres; doubted motion of fixed stars and of apogees of planets; explained difference between theoretical astronomy and observational astronomy; described universal astrolabe and other instruments; explained use of tables; determined distance of declination for 1332; gave tables for sines at each half-degree, with radius (sinus totus, sinus perfectus) at 60 degrees, also tables for tangents at every degree, with radius at 12; based his catalogue of stars on his observations at vernal equinox of 1350. Died circa 1350-55.

JEAN DE MEURS (or John of Murs), French mathematician, astrologer; b. Diocese of Lisieux, Normandy; with Sorbonne, 1320-39, or longer. Author: Canones tabule tabularum, 1321; Opus quadripartium numerorum sive de mensurandi ratione (important in history of math. teaching and diffusion, also most clearly anticipated idea of decimal fractions until Stevin's work of 1585), 1343; (with Firmin de Beauval) work on calendar reform that prefigured Gregorian reform, 1344; Arithmetica communis; De moventibus et motis (treatise on fundamental principle of Aristotelian dynamics); De ponderibus et metallis (on properties of floating bodies and determination of specific weights). Died after 1350.

JEANES, Allene Rosalind, Am. chemist; b. Waco, Tex., July 19, 1906; d. Lonnie E. and Viola (Herring) J.; B.A., Baylor U., 1928; M.A., U. Cal. at Berkeley, 1929; Ph.D., U. Ill., 1938. Tchr. high sch., Laredo, Tex., 1930; head sci. dept. Athens (Ala.) Coll., 1930-35; instr. chemistry U. Ill., 1936-37; Corn Industries Research Found. fellow NIH, USPHS, 1938-40; research chemist No. Regional Research Lab., U. S. Dept. Agr., Peoria, Ill., 1941——. Recipient Distinguished Service award U. S. Dept. Agr., 1953, Fed. Woman's award U. S. Civil Service Commn., 1962. Mem. Am. Chem. Soc. (Garvan medal 1956), Sigma Xi, Iota Sigma Pi. Contbr. articles to profl. jours. Research on prodn., properties and structure of dextrans and other microbial polysaccharides. Home: 511 W. Stone St., Peoria 61606. Office: 1815 N. University St., Peoria, Ill. 61604.*

JEANLOZ, Roger William, biochemist, educator; b. Berne, Switzerland, Nov. 3, 1917; s. William M. and Rose (Poisat) J.; Baccalaureate, Coll. Geneva (Switzerland), 1936; Chem.E., U. Geneva, 1941, D.Sc., 1943; A.M. (hon.), Harvard, 1961; m. Dorothea A. H. de Passavant, Dec. 20, 1945; children—Claude-Andre, Raymond Francois, Danielle Renee, Sylvie Anne. Came to U. S., 1947, naturalized, 1953. Research asso. U. Geneva, 1943-45, U. Basel, 1945-46; asst. U. Montreal, 1946-47; sr. research fellow NIH, 1947-48; sr. scientist Worcester Found. Exptl. Biology, 1948-51; asso. biochemist Mass. Gen. Hosp., Boston, 1951-61, biochemist, 1961——; research asso. Harvard Med. Sch., 1951-57, asso. organic chemistry, 1957-60, asst. prof. biol. chemistry, 1960-61, asso. prof., 1961——. Mem. study sect. physiol. chemistry div. research grants NIH, 1964——. Recipient medal Societe de Chimie Biologique de France, 1960, U. Liege, 1964. Mem. Am. Soc. Biol. Chemists, Am., Swiss chem. socs., Chem. Soc. (London), Am. Rheumatism Assn. Author: (with Balazs) The Amino Sugars, 4 vols., 1965. Contbr. over 175 articles to profl. jours. Research on chemistry and biochemistry of amino sugars, glycoproteins, glycolipids, chem. structure of bacterial cell walls. Home: 42 Ruthven Rd., Newton, Mass. 02158. Office: Mass. Gen. Hospital, Boston 02114.*

JEANS, Sir James Hopwood, English math. physicist; b. Sept. 11, 1877; s. W. T. Jeans; ed. Trinity Coll., Cambridge; hon. degrees from numerous univs.; m. Charlotte Tiffany Mitchell, 1907; 1 dau.; m. 2d, Susi Hock, 1935; 2 sons, 1 dau. Fellow, Cambridge U., 1901, hon. fellow, 1942; univ. lectr. math., 1904; prof. applied math. Princeton, 1905-09; Stokes lectr. applied math. Cambridge U., 1910-12; research asso. Mt. Wilson Obs., 1923-44; prof. astronomy Royal Instn. Mem. adv. council Dept. Sci. and Indsl. Research, 1924-29, 34-39; mem. Royal Commn. for 1851 Exhbn., also chmn. sci. com. Recipient Franklin medal Franklin Inst., 1931, Mukerjee medal Indian Assn. for Cultivation Sci., 1937, Calcutta medal Royal Asiatic Soc. Bengal, 1938. Fellow Royal Soc., 1906 (sec. 1919-29, Royal medal 1919), Inst. Physics (hon.).; mem. Royal Astron. Soc. (pres. 1925-27, Gold medal 1922), Brit. Assn. for Advancement Sci. (pres. 1934), Indian Sci. Congress Assn. Author: The Dynamical Theory of Gases, 1904; Theoretical Mechanics, 1906; The Mathematical Theory of Electricity and Magnetism, 1908; Radiation and the Quantum-Theory, 1914; Problems of Cosmogony and Stellar Dynamics, 1919; Atomicity and Quanta, 1926; Astronomy and Cosmogony, 1928; Eos, or the Wider Aspects of Cosmogony, 1928; The Universe around Us, 1929; The Mysterious Universe, 1930; The Stars in their Courses, 1931; The New Background of Science, 1938; Through Space and Time, 1934; Science and Music, 1937; Introduction to the Kinetic Theory of Gases, 1940; Physics and Philosophy, 1942, others. Investigated stellar dynamics; a propounder fission theory of birth of binary and multiple star systems; developed theory about formation of spiral nebulae; elaborated on Chamberlin's catastrophic or tidal theory of planetary origin, 1917 (held that planets were formed from solar material withdrawn from sun by gravitational attraction of nearby passing star); postulated that throughout universe matter is constantly being formed at very slow rate; studied behavior of rapidly spinning bodies, examined their methods of breaking up under stress of centrifugal force; pointed out that kinetic energy of all stars is approximately equal since small stars move more rapidly than large ones; calculated temperature in degrees corresponding to uranium radiation; known for his philos. interpretations of modern sci.; held that laws of sci. are imprecise, that relations of cause and effect devolve from relative viewpoint which is individual and subjective; popular interpreter of difficult sci. concepts including quantum theory and theory of relativity. Died Dorking, Surrey, Sept. 16, 1946.

JEANS, Philip Charles, Am. pediatrician; b. Hillsboro, O., Jan. 3, 1883; s. Frank Hibben and Anna Mary (Stafford) J.; A.B., U. Kan., 1904; M.D. Johns Hopkins, 1909; m. Grace Whittier Cushing, Dec. 22, 1914; 1 son, Robert Philip. Asst. instr., asso. prof. pediatrics Washington U., St. Louis, 1913-24; prof. and head pediatrics U. Ia., 1924-52, prof. emeritus, 1952. Midwest cons. on nutrition USPHS. Mem. com. Revision U. S. Pharmacopeiea, 1951; mem. Steering com. and vice-chmn. food and nutrition bd. NRC; mem. sci. adv. com. Nutrition Found., Inc.; sent to S.E. Europe by Red Cross for rehab. work with children, 1921-22, to Hawaii by USPHS for lectures on nutrition, 1950, to C.Am. by WHO as nutrition cons., 1952. Fellow A.M.A. (mem. council on foods and nutrition from 1930, chmn. 1952), Am. Pediatric Soc. (pres. 1950-51); mem. Soc. for Exptl. Biology and Medicine, Am. Inst. Nutrition, Soc. for Research in Child Devel., Soc. for Pediatric Research, Nu Sigma Nu, Alpha Omega Alpha, Sigma Xi. Author: (with J. V. Cooke) Prepubescent Syphilis, 1930; (with W. Rand) Essentials of Pediatrics, 1934 (rev. edit. with W. Rand and F. Blake, 1946); Infant Nutrition (with W. M. Marriott) 1941 (rev. edit. 1947). Contbr. to many tech. articles. Died Panama City, Oct. 22, 1952.

JEANSELME, Edouard, French physician; b. Paris, 1858; physician, instr. hosps; prof., clinic of skin diseases and syphilis, Faculty Paris, St. Louis Hosp.; mem. Acad. Medicine. Author: Traité de la syphilis, 1931. Contbd. to knowledge of leprosy and treatment of syphilis; described extra-auricular, or Jeanselme's nodules; described gout in Byzantine court on basis of local histories and patristic writings. Died 1935.

JEFFCOTT, Henry Homan, Irish engr.; b. County Donegal, Ireland; s. William Jeffcott; B.A., B.A.I., Sc.D., Dublin (Ireland) U.; m. Louise E. Howard Lang. Asst. engr. Siemens Bros. and Co., Ltd., Woolwich, Stafford (both Eng.); with shops of W. G. Armstrong, Whitworth & Co., Ltd., Manchester, Eng.; head meteorology dept. Nat. Phys. Lab., 1905-10; prof. engring. Royal Coll. Scis. for Ireland, 1910-22, dean faculty, 1914-22. Sec., Water Power Resources of Ireland Sub-com., 1918-21; mem. bd. studies in civil, mech. engring. U. London, from 1922; examiner in mech. engring. U. Belfast (Ireland), Nat. U. Ireland. Mem. Instn. Civil Engrs. (sec. from 1922), Instn. Mech. Engrs., Royal Irish Acad. Contbr. papers on whirling shafts, vibration of beams, elastic deformation, screw threads, elec. transmission lines, surge chambers, hydro-electric investigations, hydraulic pipe lines, heat engines cycles, surveying instruments. Patentee direct reading tacheometer, others. Died June 29, 1937.

JEFFERSON, Sidney, English physicist, engr.; b. Bolton, Eng., Sept. 27, 1907; s. Sidney Roberts and Annie (Jefferson) J.; grad. Imperial Coll. Sci., London, 1931; m. Nellie Boyce, Mar. 25, 1939; children—Jill Mary, Clive Stuart. Employed in comml. research and devel.; joined Air Ministry, 1936, worked under Robert W. Watson-Watt, designed 1st operational radar receivers; leader Tele-communications Research Establishment Group on Indsl. Applications of Wartime Electronic Techniques, 1947-51; with U.K.A.E.A. Indsl. Applications of Isotopes, 1951-55; became head Tech. Irradiation Group (for exploration of direct effects of massive radiation sources), 1956, head radiation br. for expansion of work of the group, 1960. Author: Radioisotopes; Massive Radiation Techniques, 1964. Pioneer in devel. of radar; developed indsl. application of electronic and isotopic techniques; inventor radar anti-jamming circuits, electronic tension guages, check weighing machines; holder patents relating to leak detection by radioisotopes, radiation-induced ionization in cold cathode tubes, designs of gamma radiation processing equipment. Home: 2 Appleford Dr., Abingdon, Berks. Office: Wantage Research Lab. (AERE), Wantage, Berks., Eng.*

JEFFERSON, Thomas, Am. polit. theorist, zoologist, 3d Pres. U. S.; b. "Old Shadwell," Goochland (now Albemarle County), Va., Apr. 13, 1743; s. Peter and Jane (Randolph) J.; attended Coll. William and Mary, 1760-62; studied law under George Wythe; m. Martha Wayles Skelton, Jan. 1, 1772; 6 children (only Martha and Marie attained maturity). County lt. Albemarle County, 1770, county surveyor, 1773; admitted to bar, 1776; mem. Va. Ho. of Burgesses, 1769-75; mem. com. which created Va. Com. of Correspondence; introduced (with others) resolution for a fast day in Va. in sympathy with Boston Port Bill (resolution resulted in dissolution of Ho. of Burgesses); wrote A Summary View of the Rights of British America, 1774 (not adopted by Va. Ho.); mem. Continental Congress,

1775-76; mem. com. of 5 to draw up Declaration of Independence, personally wrote declaration (with minor changes by John Adams and Benjamin Franklin and by Congress as finally adopted), signed declaration, 1776; mem. Va. Ho. of Dels., 1776-79; gov. Va., 1779-81, struck blow at vested privilege by initiating abolition of primogeniture (achieved in 1785) and entail; originated bill to establish freedom of religion and opinion (passed in 1786); urged public sch. and library system; a resolution calling for inquiry into his mil. conduct as gov. was found groundless by Ho. of Dels. and resolutions of thanks were adopted; went into semi-retirement and finished his scientific work Notes on the State of Virginia (privately printed 1785); again mem. Continental Congress, 1783-84, drafted com. report urging adoption of dollar as unit of money system based on decimal notation, 1784; drafted resolution known as Ordinance of 1784, providing for temporary govt. of western territory; named as commr. to help carry out his formula for negotiating treaties of commerce based on universal reciprocity; succeeded Benjamin Franklin as minister to France, 1785-89; 1st U. S. sec. of state under new constn., 1790-93, chief architect of policy of neutrality; became leader of Anti-Federalist forces (Republicans); resigned, retired to Monticello, 1793; vice pres. U. S., 1796-1801; wrote Ky. Resolutions in answer to Alien and Sedition Acts which grew out of Am.-French trouble of the time, 1798; Pres. U. S., 1801-09; his 1st administrn. marked by La. Purchase, 1803; sent Lewis and Clark to explore new territory; his 2d adminstrn. beset with troubles stemming from English-French wars on the Continent; maintained Am. neutrality largely through econ. measures such as Non-Importation Act, 1806, Embargo Act, 1807; forced by econ. distress to ease embargo through Non-Intercourse Act, 1809; retired to Monticello for remainder of his life, 1809; an architect of renown, partly planned City of Washington (D.C.), designed and built Monticello; a prin. founder U. Va., mem. 1st bd. visitors, a rector (1819-26), also conceived univ.'s distinctive architecture and ednl. perspective, personally compiled several thousand titles in all academic fields as basis for its library; pres. Am. Philos. Soc., 1797-1815; maintained scientific interests which led him into studies and writings on paleontology, ethnology, geography and botany; writings include: Manual of Parliamentary Practice, 1801, thereafter used in U. S. Senate. Regarded as 1st great shaper of Am. democracy based on individual liberties, people's capabilities and checks on fed. power; believed the solution to democracy's shortcomings was greater democracy. Died Monticello, Albemarle County, July 4, 1826.

JEFFERY, Geoffrey M(arron), Am. parasitologist; b. Dundee, N.Y., May 13, 1919; s. Joseph Ewart and Augusta (Knapp) J.; A.B., Hobart Coll., 1940; M.A., Syracuse U.; 1942; Sc.D., Johns Hopkins, 1944; M.P.H., Yale, 1961; m. Jane Wicker, Aug. 16, 1941; children—Janet (Mrs. Robert B. Harrison), Thomas W., Sarah V., Susan E. From sr. asst. scientist to scientist malaria research Lab. Tropical Diseases NIH, USPHS, Milledgeville, Ga., 1948-54; from scientist to sr. scientist Lab. Tropical Diseases, Columbia, S.C., 1954-60, scientist dir., Columbia, 1960-63; asst. chief lab. parasite chemotherapy Nat. Inst. Allergy and Infectious Diseases, Bethesda, Md.; 1963-66, acting chief, 1966-67, chief, 1967—; mem. expert panel on malaria WHO, 1963—, cons. drug. resistant malaria, W.Africa, 1965; mem. Central Am. Malaria Assessment Mission AID, 1964, asso. mem. Commn. on Malaria Armed Forces Epidemic Board, 1965-67, full member, since 1967—. Recipient Jefferson awards S.C. Acad. Sci., 1952, 56, 60, Bailey K. Ashford award Am. Soc. Tropical Medicine and Hygiene, 1959. Mem. Am., Royal socs. tropical medicine and hygiene, Am. Soc. Parasitologists, A.A.A.S., Assn. Southeastern Biologists, Sigma Xi. Contbr. numerous articles to profl. jours. Research, publs. on epidemiology, chemotherapy, immunology and pathology of malaria; other parasitic infections in man. Home: 6516 Willmett Rd., Bethesda 20034. Office: Lab. Parasite Chemotherapy, NIAID, NIH, Bethesda, Md. 20014.*

JEFFERY, George Barker, Brit. mathematician; b. May 9, 1891; s. George and Elizabeth (McDonald) J.; student King's Coll.; B.Sc., U. London (Eng.), 1911, M.A., 1914, D.Sc., 1921; m. Elizabeth Schofield, 1915; 1 son, 2 daus. Asst. in applied math. U. Coll., London, from 1912, univ. reader in math., from 1921; Astor prof. math., 1924-45, pro-provost in Bangor, Wales, 1939-44, dir. Inst. Edn., from 1945, also mem. senate; prof. math. King's Coll., London, 1922-24. Fellow Royal Soc., 1926 (v.p. 1938-40); mem. London Math. Soc. (pres. 1935-37), Math. Assn. (pres. 1947), others. Contbr. papers to profl. publs. Wide range of work including theory of relativity, hydronamics, elasticity, also theory of relevant differential equations; much work concerned with exact solutions in fields where such solutions are rare and some form of approximation is usually accepted. Died Apr. 27, 1957.

JEFFERY, James William, English crystallographer; b. Maidstone, Eng., Aug. 14, 1911; s. Samuel Webster and Mabel (Dorman) J.; B.A., Clare Coll., Cambridge, U., 1934; Ph.D., Birkbeck Coll., London U., 1950; m. Nora Mary Jarvis, Dec. 15, 1934; children—Jill (Mrs. George Bingham), David, Stephen. Tchr. math. Gillingham County Sch., 1934-35; tchr. math. and sci. Birkenhead Inst., 1935-46; staff Birbeck Coll., London, 1946—, reader crystallography, 1964—. Cons. on problems of identification by X-ray diffraction, 1950—. Fellow Inst. Physics; mem. Mineral. Soc. Contbg. author: International Tables of X-ray Crystallography, vol. III, 1962; Chemistry of Cements, 1964. Research, publs. on application of X-ray diffraction to cement chemistry, elucidation of atomic arrangement in cement compounds, determination of phases present in hydrated cement; devel. methods and techniques for obtaining positions of atoms in crystals of all kinds. Home: 7 Pearfield Rd., London, S.E. 23, Eng.*

JEFFREY, George Alan, crystallographer; b. Cardiff, Eng., July 29, 1915; s. George F. and Beatrice (Hand) J.; B.Sc., U. Birmingham (Eng.), 1936, Ph.D., 1939, D.Sc, 1953; m. Maureen Ward, Sept. 5, 1942; children—Susan M., Paul D. Came to U. S., 1953, naturalized, 1962. Crystallographer, Brit. Rubber Producers Assn., 1939-45; lectr. U. Leeds (Eng.), 1945-53; prof. chemistry and physics, dir. crystallography lab. U. Pitts., 1953—, prof. crystallography, 1965—; sec. U. S. Nat. Com. Crystallography, 1956-58, chmn., 1965-66. Research in crystallographic computing; crystal structure analysis of organic and inorganic compounds. Home: 200 Mary Ann Dr., Glenshaw, Pa. 15116.

JEFFREY, John, botanist; b. Perthshire, Scotland, Nov. 14, 1826; gardener at Edinburgh (Scotland) Royal Botanic Garden; sent to Am. N.W. to collect plant specimens, 1851-53; traveled through Can., stopping at Hudson Bay posts, then went to western U. S.; discovered and introduced to Gt. Britain Dodecatheon jeffreyi (shooting star), Pinus jeffreyii, Pinus balfouriana (fox-tail pine); introduced red-flowered gooseberry and Western hemlock. Probably died in N.M., 1853.

JEFFREYS, Harold, English astronomer, geophysicist; b. Fatfield, Eng., Apr. 22, 1891; s. Robert Hall and Elizabeth (Sharpe) J.; B.Sc., Armstrong Coll., Newcastle, Eng., 1910, D.Sc., 1917; B.A., St. John's Coll., Cambridge (Eng.) U., 1913, M.A., 1917; D.C.L., U. Durham (Eng.); LL.D., U. Liverpool, 1953; Sc.D., U. Dublin (Ireland), 1956; m. Bertha Swirles, Sept. 6, 1940. Meteorol. officer, 1917-22; faculty Cambridge (Eng.) U., 1922-58, Plumian prof. astronomy, 1946-58. Recipient Victoria medal Royal Geog. Soc.; Bowie medal Am. Geophys. Union; Guy medal Royal Statis. Soc. Fellow Royal Soc., 1925 (Copley medal, Royal medal); mem. Royal Astron. Soc. (Gold medal 1937, Vetlesen prize 1962), Geol. Soc. (Wollaston medal, Murchison medal), Seismol. Soc. Am., Nat. Acad. Scis. Author: The Earth, 1924; Scientific Inference, 1957; Theory of Probability; (with B. Jeffreys), Methods of Mathematical Physics; 1946; Earthquakes and Mountains, 1950; also numerous articles. Concerned with origin of solar system and theory of variation of latitude; developed J. Jeans' theory of tidal evolution of solar system and theory of variation of latitude; developed J. Jean's theory of tidal evolution of solar system, estimating its age at a few thousand million years; also calculated that moon took 4000 million years to reach its present position; One of 1st to investigate effect of radioactivity on cooling of earth; calculated accurate travel times of seismic waves; made meteorol. studies of wind dynamics. Home: 160 Huntingdon Rd., Cambridge, Eng. Office: St. John's Coll., Cambridge, Eng.*

JEFFREYS, Julius, Brit. inventor, surgeon; b. Kent, Eng., 1801; s. R. Jeffreys; ed. Edinburgh, Scotland, London, Eng.; apptd. to med. establishment, Bengal, India, 1822; became staff surgeon, Cawnpore, India, 1824; returned to Eng., 1835. Fellow Geog. Geol. Soc., Royal Soc., 1840; Mem. Med. and Chirurg. Soc. Author: The Construction and Use of the Respirator, 1836; Statics of the Chest, 1845; The Atmospheric Treatment of the Chest, 1845; Remarks on Climate and Affections of the Throat and Lungs, 1849. Made meteorol. observations in India which led to his recommendation that hill stations be formed as health resorts; invented respirator, 1836. Died May 13, 1877.

JEFFRIES, Carson Dunning, Am. physicist; b. Lake Charles, La., Mar. 20, 1922; s. Charles William and Yancey (Dunning) J.; B.S., La. State U., 1943; Ph.D., Stanford, 1951; m. Elizabeth Dyer, Sept. 15, 1945; children—Andrew, Patricia. Research asso. Radio Research Lab., Harvard, 1943-45; research asst. Stanford U., 1946-50; instr. Physikalisches Institut der Universität, Zurich, Switzerland, 1951; faculty U. Cal. at Berkeley, 1952—, prof., 1963—, dir. AEC, Office Naval Research research projects in solid state physics, 1953—. Sr. Postdoctoral fellow NSF, Oxford (Eng.) U., 1958, Harvard, 1965-66; Fulbright prof., France, 1959. Mem. Am. Phys. Soc. Author: Dynamic Nuclear Orientation, 1963; also numerous articles, revs. Research in nuclear magnetic resonance, nuclear moments, microwave paramagnetic resonance and relaxation, low temperature physics, hyperfine interactions, orientation of nuclei, polarized proton targets; originator of microwave dynamic polarization scheme for polarizing protons and of nuclear spin refrigerator for cooling spins to ultra low temperatures.*

JEFFRIES, John, Am. physician, balloonist; b. Boston, Feb. 5, 1745; s. David and Sarah (Jaffrey) J.; grad. Harvard, 1763; M.D., Marischal Coll., U. Aberdeen (Scotland), 1769; m. Sarah Rhoads, 1770; m. 2d, Hannah Hunt, Sept. 8, 1787; 14 children. Asst. surgeon on Brit. naval vessel, 1771-74; surgeon maj. with Brit. troops, 1775-79; surgeon gen. of forces in N.S.; surgeon gen. Am. forces at Charleston, S.C., 1780; 1st to attempt to gather scientific data of free air; made 2 ascents, one over London, 1784, other over English Channel for scientific purposes, Jan. 1785, flying from Dover to Forest of Guines, Ardes, France (1st crossing of English Channel by air); gave 1st public lecture on anatomy in New Eng., 1789. Author: Narrative of Two Aerial Voyages, 1786. Died Boston, Sept. 16, 1819.

JEFFRIES, Zay, Am. metall. engr.; b. Willow Lake, S.D., Apr. 22, 1888; s. Johnston and Florence (Sutton) J.; B.S. in Mining Engring., S.D. State Sch. Mines, 1910, Met.E., 1914, D.Engring. (hon.), 1930; D.Sc., Harvard, 1918; D.Sc. (hon.), Case Inst. Tech., 1937, Clarkson Coll. Tech., 1950; m. Frances Schrader, Dec. 27, 1911; children—Elizabeth, Marian. Faculty Case Inst. Tech., 1911-17, mem. corp., trustee; cons. Nat. Lamp works Gen. Electric Co., 1914-25, incandescent lamp dept., 1925-36, tech. dir. lamp dept., 1936-45, v.p. chemistry dept., 1945-49; dir. research Aluminum Castings Co. and successor, Aluminum Mfrs., Inc., 1916-20. Cons. metall. lab. Manhattan Dist.; mem. tech. adv. panel on materials Dept. Def., chmn., 1954-59; dir.-gen. World Metall. Congress, 1951, 57; vice chmn. War Metallurgy Com.; chmn. metals conservation and substitution group, adv. bd. on metals and minerals NRC-Nat. Acad. Scis. Hon. chmn. bd. trustees Battelle Meml. Inst. Recipient 1st Annual Power Metallurgy medal Stevens Inst. Tech., 1945, Francis J. Clamer silver medal Franklin Inst., 1945, John Fritz gold medal, 1946. Fellow Am. Phys. Soc., Am. Inst. E.E.; mem. A.A.A.S., Nat. Acad. Scis., Am. Philos. Soc., Am. Chem. Soc., Japan Inst. Metals, Am. Inst. Mining, Metall. and Petroleum Engrs. (James Douglas medalist 1927), Am. Soc. for Metals (Sauvuer award 1935, gold medal 1943, Distinguished service award 1948, past pres.). Author: (with R. S. Archer) The Science of Metals, 1924; (with J. D. Edwards and F. C. Frary) The Aluminum Industry, 2 vols., 1930; (with T. W. Frech) Business Ideals, Principles and Policies. Research in metallography; grain growth in metals; effect of grain size and temperature deformation on mechanical properties of metals. Died May 21, 1965.*

JEFIMENKO, Oleg Dimitri, physicist; b. Kharkov, USSR, Oct. 14, 1922; s. Dimitri N. and Galina J.; Vordiplom, U. Gottingen (Germany), 1950; B.A., Lewis and Clark Coll., 1952; M.A., U. Ore., 1954, Ph.D., 1956; m. Valentina, Nov. 14, 1945. Came to U. S., 1951. Faculty, U. Ore., 1952-56; faculty W.Va. U., Morgantown, 1956—, professor, 1967—. Recipient Sigma Xi award, 1956. Mem. Am. Phys. Soc., Am. Assn. Physics Tchrs., Sigma Xi. Author: Electricity and Magnetism, 1966. Studies on satellite bands in spectra of gaseous mixtures, effects of earth magnetic field on motion of artificial satellites; devel. of method for observing electric field lines of current-carrying conductors, thin shell atomic model. Office: Physics Dept., W.Va. U., Morgantown, W.Va. 26506.*

JEHLE, Herbert, physicist; b. Stuttgart, Germany, Mar. 5, 1907; s. Julius and Maria (Gminder) von J.; Dipl. Ing., Inst. Tech., Stuttgart, 1929; Dr.Ing., U. Berlin (Germany), 1933; postgrad. Cambridge (Eng.) U., 1933-34; m. Dietlinde von Kuenssberg, May 30, 1952; children—Eberhard v. k., Dietrich v. k. Came to U. S., 1941, naturalized, 1955. Editorial asst. Jahrbuch Fortschritte der Mathematik, 1935-36; research asst. Southampton U. Coll., 1937-38; research asso. U. Brussels, 1938-40; instr. Harvard, 1942-46; mem. Inst. for Advanced Study, Princeton, 1947-48; asst. prof. U. Pa., 1947-49; asso. prof. U. Neb., 1949-59; research asso. Cal. Inst. Tech., 1956-57; prof. physics George Washington U., 1959—. NIH fellow, 1965-66. Fellow Am. Phys. Soc.; mem. Fellowship Reconciliation (internat. com.), Soc. Social Responsibility in Sci., Fedn. Am. Scientists. Research in spinors and relativistic wave equations, statis. mechanics, intermolecular forces and biol. specificity, replication of biol. macromolecules. designed projection magnifier for semiblind under NRC and Franklin Inst., 1946-47. Home: 1208 Sherwood Rd., Charlottesville, Va. 22901. Office: Physics Dept., George Washington U., Washington 20006.*

JELINEK, Karl, Austrian meteorologist; b. Brünn, Austria, Oct. 23, 1822; prof. in Prague (Czechoslovakia), and Vienna (Austria); dir. Central Inst. for Meteorology, Vienna; founder Austrian Meteorol. Soc. Died Vienna, Oct. 19, 1876.

JELKS, John Lemuel, Am. surgeon; b. Bells, Tenn., July 5, 1870; s. Lemuel Marshall and Nannie (Lane) J.; student U. Ark.; M.D., Memphis Hosp. Med. Coll., 1892; m. Minnie Rollwage, Oct. 14, 1903; m. 2d, Louise Whitmire Speegle, Jan. 25, 1940. Practiced, Memphis, from 1893, specializing in enteroproctology. Fellow Am. Proctologic Soc. (pres.), A.C.S., Nat. Gastro-Enterol. Assn.; mem. A.M.A., Southern Med. Assn., Tenn. Acad. Scis., A.A.A.S., Internat. Soc. Gastro-Enterology, Am. Med. Editors and Authors. Introduced theory of intestinal parasitic infection as the etiology of pellagra; contbr. chpt. on Dysentery in Hirschman's Diseases of the Rectum, Sigmoid and

Colon; chpt. on intestinal protozoa in man, The Cyclopedia of Medicine. Died 1945.

JELLIFFE, Derrick Brian, pediatrician; b. Rochester, U.K., Jan. 20, 1921; s. Reginald Eric and Dorothy (Hebb) J.; student Middlesex Hosp. Med. Sch., 1938-43; M.D., U. London (Eng.), 1945, Diploma in Child Health, 1945, Diploma in Tropical Medicine and Hygiene, 1947; m. Patricia Macready, June 13, 1942. Staff, Sudan Med. Service, 1949; lectr. pediatrics U. Coll., Ibadan, Nigeria, 1949-52; sr. lectr. U. Coll., Jamaica, W.I., 1952-54; WHO prof. pediatrics U. Calcutta (India), 1954-56; vis. prof. tropical medicine Tulane U. Med. Sch., New Orleans, 1956-59; prof. pediatrics Makerere Med. Sch., Kampala, Uganda, 1959-66; prof. community nutrition, dir. Caribbean Food and Nutrition Inst., U. W.I., Jamaica, 1966—. Mem. panel on nutrition WHO, 1962—. Fellow Am. Pub. Health Assn. (hon.), Am. Acad. Pediatrics (hon.), Royal Coll. Physicians; mem. Am. Soc. Tropical Medicine, Royal Soc. Tropical Medicine and Health. Author: Infant Nutrition in the Tropics, 1955; Diseases of Children in Tropics, 1958; The Assessment of the Nutrition of the Community, 1966; also numerous articles. Devel. of methods for nutritional assessment of children in developing tropical countries. Home and office: P.O. 140, Mona P.O., Jamaica.*

JELLIFFE, Smith Ely Am. neurologist; b. N.Y.C., Oct. 27, 1866; s. William Munson and Susan E. (Kitchell) J.; A.B., Brooklyn Poly. Inst., 1898; M.D., Coll. Physicians and Surgeons (Columbia), 1889, Ph.D., 1899, A.M., 1900; post-grad. work, Europe, at various times since 1890; m. Helena Dewey Leeming, Dec. 20, 1894; children—Sylvia Canfield, Winifred, Smith Ely, Wm. Leeming (dec.), Helena; m. 2d, Bee Dobson, Dec. 20, 1917. Visiting neurologist, City Hosp., 1903-13; clin. prof. mental diseases, Fordham U., 1907-12; instr. materia medica and therapeutics, 1903-07, prof. pharmacognosy and tech. microscopy, 1897-1907, Coll. of Pharmacy, Columbia; adjunct prof. diseases of the mind and nervous system, Post-Grad. Hosp. and Med. Sch., 1911-17; cons. neurologist, Manhattan State Hosp., Kings. Park Hospital. Editor Med. News, N.Y., 1900-05; asso. editor New York Med. Journal 1905-09; mng. editor Journal of Nervous and Mental Disease, 1902-45, editor (with W. A. White) Nervous and Mental Monograph Series, 1907-45, Psychoanalytic Review, 1913-45. Mem. Am. Neurol. Assn. (pres. 1929-30), N.Y. Neurol. Soc. (pres.). N.Y. Psychiat. Society (pres.), A.M.A., Am. Psychiatric Assn., Am. Psychopathol. Society (pres.), Am. Psychoanalytic Soc. (pres.), Phila. Neurol. Soc., N.Y. Acad. Med.; corr. mem. Neurol. Soc. Paris, Acad. Medicine of Brazil. Author: Essentials of Vegetable Pharmacognosy, 1895; Morphology and Histology of Plants, 1899; also "Nervous Diseases" in Butler's Diagnostics, 1902; Outlines of Pharmacognosy, 1904. Reviser: Butler's Materia Medica, 1902; Shaw on Nervous Diseases, 1903. Editor and Translator: Dubois' Psychoneuroses, 1905; Grasset-Demi Fous, 1907; Payot, Education of Will; Paranoia; the Wassermann Serum Reaction in Psychiatry; Dejerine, Psychoneuroses and Psychotherapy; Rank, Myth of the Birth of the Hero; Eppinger and Hess, Vagotonia; Silberer, Problems of Mysticism and Its Symbolism; W. Hess, Vegetative Nervous System and Psyche; also Hysteria, Tetany, Migraine, in Osler's System of Medicine, etc. Co-editor: Ency. Americana, 1904, 18; (with Dr. W. A. White) Modern Treatment of Nervous and Mental Diseases, 2 vols.; Diseases of the Nervous System, 7 edits.; Technique of Psychoanalysis; Psychoanalysis and the Drama; Respiratory Disorders in Encephalitis; Oculogyric Crises and Encephalitis. Contbr. to medical press. Testified in famous acquittal case of Harry K. Thaw; pioneer in psychoanalysis in U. S.; one of outstanding leaders in psychiatric thought; freely disseminated knowledge of Sigmund Freud's work; contbd. to concept of psychosomatic medicine. Died Sept. 25, 1945.

JELLINEK, Elvin M(orton), Am. biometrician; b. N.Y.C., Aug. 15, 1890; s. Ervin Marcell and Rose (Jacobsen) J.; student U. Berlin, 1908-11, U. of Grenoble, 1911; M.Edn., U. Leipzig, 1914, Sc.D., 1935; m. Thelma Pierce. Oct. 18, 1935; 1 dau., Ruth (Mrs. Charles L. Surrey). Biometric consultant, library and field research, 1914-19; biometrician elder plant research, Sierra Leone, West Africa, 1920-25; biometrician, later asst. dir. of research, United Fruit Co., Tela, Honduras, 1925-30; chief biometrician, later asso. dir. of research, Memorial Foundation for Neuro-Endocrine Research, Worcester St. Hosp., 1931-39, dir. study of effects of alcohol from 1939; asso. prof. applied physiology, Yale U., from 1941, dir. Sch. of Alcohol Studies since 1943; vice-chmn. scientific com. Research Council or Problems of Alcohol from 1941; chmn. survey com. on tests of intoxication since 1941; asso. editor Quarterly Jour. of Studies on Alcohol since 1941. Fellow Am. Acad. Arts and Sciences, A.A.A.S.; mem. Am. Statis. Assn., Am. Psychopathol. Assn. Author: (with H. W. Haggard) Alcohol Explored, 1942. Editor: The Effect of Alcohol on the Individual, 1942. Contbr. to scientific jours. Pioneered concept of alcoholism as a disease and as pub. health problem; his investigations (with assos.) gave basis for diagnosis of potential alcoholic (classic estimate of alcoholism rates, spectrum of alcoholic phases described). Died Palo Alto, Cal., Oct. 22, 1963.

JELLINEK, Harry, Hungarian physician; b. Rimaszombat, Hungary, Nov. 25, 1924; s. Henrik and Aranka (Lendvay) J.; Spl. exam. pathohistology, Med. Sch. Budapest (Hungary), 1954, candidate med. sci., 1959, d.med.scis., 1965; m. Genovéva Süle, Mar. 14, 1951; 1 dau., Kinga. Staff, Med. U. Budapest, 1949—, lectr., 1956—, vice dean, 1962-66; cons. chief physician path. dep. Kállai Éva-Street Hosp., 1952—. Named honored physician, 1951; recipient Jendrassik medal, 1965. Mem. Hungarian Path. Soc. (mem. bd.), Internat. Acad. Pathology, Deutsche Gesellschaft für Pathologie. Author: Practice in Pathology, 1949; Practice in Histopathology, 1949; also numerous articles. Research in vessels pathology, including vessel wall necrosis, regeneration of great vessels, vessel wall changes and differentiation of these changes from arteriosclerosis. Home: 10, Csörsz, Budapest, Hungary XI.*

JELLINEK, Karl W. K., physico-chemist; b. Vienna, Austria, Nov. 5, 1882; s. M. J. and Emma (Steiner) J.; ed. univs. Vienna, Leipzig (Germany), Berlin; Ph.D., U. Göttingen (Germany), 1905; m. Melitta Baden, Nov. 24, 1913; children—Elisabeth Charlotte, Hans Helmut. Asst. prof. univs. Göttingen, Berlin, 1904-06; chemist Bayer & Co., Elberfeld, Germany, 1907-08; asst. prof. Kruger Technische Hochschule, 1908-12; privatdocent for phys. chemistry Technische Hochschule, Danzig, Germany, from 1910; docent for analytical chemistry, from 1916, also prof. Author: Grundzuge der physiken Chemie der Hydrosulfit, 1911; Anorganisches organische, und technische Chemie der Hydros, 1912; Nernstsche Warmetheorien und seine Beziehung zur Zuantentheorie, 1912; Weltather und Relativitie Theorie, 1922; Lehrbuch der physikalishen Chemie, 1928; Elemente der Wellen mechanik, 1950; numerous basic texts on phys. chemistry. Research in phys., inorganic and analytical chemistry.

JELLINGER, Kurt, Austrian neuropathologist; b. Vienna, May 28, 1931; s. Alois and Rosa (Proksch) J.; M.D. summa cum laude, U. Vienna, 1956; m. Elisabeth Manoschek, Aug. 17, 1960. Mem. staff Neurol. Inst., Med. Faculty, U. Vienna, 1957—, mem. staff Psychiat.-Neurol. Clinic, 1958—, asst. prof., 1960—; mem. staff Vienna Alcoholics' Hosp., 1963—, asso. prof., 1967—. Recipient Th. Theodor Körner prize, 1962; Hoechst AG prize, 1965; Kardinal Innitzer prize, 1966. Mem. Austrian Commn. Neuropathology (sec.-gen. 1964—), Vienna Med. Soc., Austrian Soc. Neurologists and Neuroanatomists, Soc. German Neuropathologists and Neuroanatomists; corr. mem. Comm. Neuropathology World Fedn. Neurology, Soc. Austrian Pathologists; fgn. mem. Neurol. Soc. France; ord. mem. Internat. Med. Soc. Paraplegia. Author: Zur Orthologie und Pathologie der Rückenmarksdurchblutung, 1966. Publs. on neuropathology of virus encephalitides, demyelinating diseases, poisons; effects of coma, trauma, vascular and traumatic disorders of spinal cord, postvaccinal complications on brain. Address: 18 Kenyongasse, Vienna VII. Austria.*

JELLISON, William Livingston, Am. parasitologist; b. LaSalle, Mont., Feb. 28, 1906; s. Walter Fremont and Hester (Hay) J.; B.S., Mont. State Coll., Bozeman, 1929; M.S., U. Minn., 1934, Ph.D., 1940; postgrad. Stanford, Johns Hopkins, Harvard, Duke; m. Gretchen Loretta Gayhart, Aug. 13, 1932; 1 son, William Henry. With USPHS, 1929-31, 31-62, parasitologist Rocky Mountain Lab., Hamilton, Mont., 1931-62, ret., 1962, now cons. Parasitologist Pan-Am. Zoonosis Center, Azul, Argentina, 1961; research scientist Harvard, 1961, U. Okla., 1964-65. Recipient U. S. Typhus Commn. medal, 1945. Mem. Am. Soc. Parasitologists, Am. Soc. Mammalogists, Am. Microscopic Soc., Wildlife Soc., Canadian Field Naturalists, Arctic Inst. N.Am., Sigma Xi, others. Research and publs. on bacterial, rickettsial, fungal diseases and their arthropod transmission to man. Home: 504 S. 3d St. Hamilton, Mont. 59840.

JELSTRUP, Hans Saverin, Norwegian astronomer; b. Oslo, Norway, Feb. 8, 1893; s. Arne and Gudrun (Logn) J.; M.Sc., Oslo U., 1918; diploma Meudon Obs., Paris, France, 1921; m. Alice Constance Gandois, 1924; Asst., Norwegian Meteorol. Inst., 1914-16; prof. math. and astronomy Naval Acad., 1918-19; astronomer Norwegian Geog. Soc., 1922—. Pres. Norwegian Geod. Commn., 1946—. Mem. L'Association Internationale de Géodésie (v.p. 1946—), Norsk Astronomisk Selskap (v.p.), Norsk Matematisk Forening (sec.), Det Norske Videnskapsakademi (Oslo), Société Astronomique (France), Astronomische Gesselschaft, Norse Polarklub, Norsk Kartografisk Forening, Den Norske Gramalings Kommission. Author: Mesures de Longeur par Ondes stationaires électromagnétiques, 1927; Determination of Astronomical Longitude, Latitude and Azimuths, 1929, 31; numerous treatises in French on observations of eclipses of sun. Editor: Norwegian Ofcl. Almanac, 1942—. Address: Norges Geogr. Oppmal, St. Olavs Gate 32, Oslo, Norway.

JEN, Chih Kung, physicist, educator; b. Shansi, China, Aug. 15, 1906; s. Shiang Ting and Shih (Tsui) J.; student Tsing Hua Coll., Peking, China, 1920-26; S.B., Mass. Inst. Tech., 1928; S.M., U. Pa., 1929; Ph.D. Harvard, 1931; m. Paocheng Tao, July 27, 1937; children—May (Mrs. George Koo), Linda, Phyllis, Erica. Came to U. S., 1946, naturalized, 1955. Asst. physics Harvard, 1930-32, instr., 1932-33, research lectr. electronics, 1946-50; prof. physics Nat. Shantung U., China, 1933-34; prof. physics, elec. engring. Nat. Tsing Hua U., Peking, 1934-37; prof. physics, dir. radio inst. Nat. Tsing Hua U., Kunming, China, 1937-45; physicist, prin. staff mem. Applied Physics Lab., Johns Hopkins U., Silver Spring, Md., 1950—, vice chmn. research center, 1958—, Parsons vis. prof., 1966-67. Fellow Am. Phys. Soc., Academia Sinica, Taiwan; mem. Washington Acad. Scis., Philos. Soc. Washington. Research and publs. on ionosphere (1929), electronic tubes, especially high frequency oscillator, theoretical aspects of negative hydrogen ion, microwave spectroscopy. Home: 10203 Lariston Lane, Silver Spring, Md. 20903.*

JENCKS, William Platt, Am. biochemist; b. Bar Harbor, Me., Aug. 15, 1927; s. Gardner and Elinor (Melcher) J.; student Harvard, 1944-47, M.D., 1951; m. Miriam Ehrlich, June 4, 1950; children—Helen Esther, David Alan. Faculty Brandeis U., Waltham, Mass., 1957—, prof. biochemistry, 1963—. Recipient Am. Chem. Soc. award in biol. chemistry Eli Lilly Co., 1963. Mem. Am. Soc. Biol. Chemists, Am. Chem. Soc., A.A.A.S., Alpha Omega Alpha. Contbr. numerous articles to tech. jours. Research on mechanism enzyme catalysis and mechanism non-enzymic reactions and their catalysis in aqueous solution, nature interaction between molecules in water responsible for properties proteins and nucleic acids. Home: 11 Revere St., Lexington, Mass. Office: Grad. Dept. Biochemistry, Brandeis U., South St., Waltham, Mass. 02154.*

JENDEN, Donald James, pharmacologist; b. Horsham, Sussex, Eng., Sept. 1, 1926; s. William Herbert and Kathleen Mary (Harris) J.; B.Sc. with 1st class honors, U. London, 1947, M.B., B.S. (U. Gold medal), 1950; m. Jean Ickeringill, Nov. 18, 1950; children—Patricia Mary, Peter D., Beverly J. Naturalized U. S., 1958. Faculty, U. London, 1947-50, U. Cal. at San Francisco, 1950-53; faculty U. Cal. at Los Angeles, 1953—, prof. pharmacology, 1960—. Cons. neuropharmacology VA Hosp., Long Beach, Cal. Mem. Am. Soc. Exptl. Pharmacology and Therapeutics, Am. Physiol. Soc., A.A.A.S., Physiol. Soc. London, Brit. Med. Assn. Research and numerous publs. in chem. and biochem. pharmacology, physiology and pharmacology of muscle, cholinergic mechanisms, math. biology. Home: 3814 S. Castlerock Rd., Malibu, Cal. 90265. Office: Dept. Pharmacology, U. Cal. Los Angeles, Los Angeles 90024.*

JENEY, Endre (Andrew), Hungarian microbiologist; b. Radnót, Hungary, Aug. 26, 1891; s. Eugen and Adele (Nagy) J.; ed. Med. Sch. Kolozsvár and Budapest, 1909-14, med. diplom, Kolozsvár, 1914; m. Magdalen Gindele, July 5, 1932; children—Andrew, Enikö-Maria (Mrs. John Hankiss). Asst., U. Kolozsvár, 1913-18, asst. inst. gen. pathology and therapy, 1918-26; Rockefeller Found. fellow, 1923-25, 28; faculty Szeged U., 1926-28; faculty U. Debrecen, Hungary, 1934—, prof. hygiene Inst. Hygiend, 1958—, dean Med. Faculty, 1942-43. Decorated Red Cross, War decorations; recipient Kossuth prize laureate, 1963, Kivalo orvos, 1961, Munkaerdemrend, 1955. Mem. Internat. Soc. Chemotherapy, Duodecin Soc., Hungarian socs. of Hygiene, Microbiology, Parasitology, Pharmacology, Physiology, Biology, Hydrology, Biochemistry, Oncology. Author: numerous textbooks, latest is: Bacterial Metabolism and Chemotherapy. Research, publs on bacterial variability, antagonism, metabiosis, drug-resistance bacteriophagy. Home: 78. Nagyerdei Blvd., Debrecen 12, Hungary.*

JENKIN, (Henry Charles) Fleeming, elec. engr.; b. Dungeness, Eng., Mar. 25, 1833; s. Charles and Henrietta (Jackson) J.; student Edinburgh (Scotland) Acad.; M.A., Genoa (Italy) U., 1850; m. Anne Austin, Feb. 21, 1859. Apprentice Fairbairn works, Manchester, Eng.; with ry. survey, Switzerland, 1855; with Liddill & Jordon in marine telegraphy, before 1859; with R. S. Newall and Co., makers of submarine telegraph cables; prof. engring. Univ. Coll., London, 1865-68, Edinburgh U., 1868-85. Fellow Royal Soc., 1865. Author: Electricity and Magnetism, 1873. Research (with Lord Kelvin) on insulation and resistance of submarine telegraphy cables; inventor telpherage, 1882; worked on resistance of guttapercha. Died June 12, 1885.

JENKINS, Charles Francis, Am. physicist, inventor; b. nr. Dayton, O., Aug. 22, 1867; s. Amasa Milton and Mary Ann (Thomas) J.; prep. edn., high sch., Fountain City, Ind., and Spiceland (Ind.) Acad.; student Earlham Coll.; spl. lectures, Johns Hopkins; Sc.D., Earlham Coll., Richmond, Ind., 1928; m. Grace Love, Jan. 30, 1902. Pres., Jenkins Laboratories; research v.p. Jenkins Television Corp.; dir. Park Savings Bank, Federal-Am. Bank & Trust Co. Medalist, Franklin Inst. and City of Philadelphia. Author: Picture Ribbons, 1896; Animated Pictures, 1898; Motion Picture Handbook, 1908; Vision by Radio, Radio Photographs, 1925; Visual Radio and Television, 1928. Inventor, patentee motion picture projector having intermittent movement, 1895; spiral-wound waxed-paper box; inventor braking device for airplanes, altimeter, conical paper drinking cup, one of 1st automobile self-starters, devices in radiophotography, TV, radiomovies (granted 400 domestic and fgn. patents in this field). Died June 6, 1934.

JENKINS, George Neil, English dental physiologist; b. Wallasey, Eng., Oct. 23, 1914; s. David and Mary

Jenkins; student U. Liverpool (Eng.), U. Cambridge; m. Olive Harvey, Mar. 31, 1945; children—Oliver, Hilbre. Research worker biochemistry Cambridge U., 1936-39; lectr. physiology St. Bartholomew's Hosp. Med. Coll., 1939-45; faculty King's Coll., Newcastle, Eng., 1945-63, reader physiology to dental students; prof. oral physiology U. Newcastle, 1963—. Mem. Nutrition Soc., Physiology Soc., Biochemistry Soc., Internat. Assn. Dental Research. Author: Physiology of the Mouth, 1954; also numerous articles. Research on mode of action of fluoride in dental caries, significance of refinement of food in dental caries, chemistry of dental plaque and its mode of formation. Home: 4 Jesmond Dene Terrace, Newcastle-upon Tyne. 2, Eng.*

JENKINS, Glenn L., Am. pharmacologist; b. Sparta, Wis., Mar. 25, 1898; s. Thomas and Laura (Rathbun) J.; B.S., U. Wis., 1922, M.S., 1923, Ph.D., 1926; LL.D., U. Toronto, 1963; m. Serena Elizabeth Forberg, June 29, 1926; children—Serena (Mrs. George Willard Ford), Thomas N., Glenn L. II, Carol Ruth. Instr. U. Wis., 1923-27; prof. pharmacal chemistry U. Md., 1927-36, U. Minn., 1936-41; prof. medicinal chemistry, dean Purdue U. Sch. Pharmacy and Pharmacal Sci., 1941-66; emeritus dean, 1966—. Recipient Ebert medal for Research, Remington medal for Leadership. Mem. Am. Assn. Colls. Pharmacy, Am. Pharm. Assn., Ind. State Bd. Health, Am. Chem. Soc., Am. Council Pharmacal Edn., A.A.A.S., Am. Soc. Hosp. Pharmacy, Ind. Acad. Sci. Author (with A. G. DuMez): Quantitative Pharmacal Chemistry, 1931, (with G. J. Sperandio, Clifton Latiolais): Clinical Pharmacy, 1965. Contbr. numerous articles in field to profl. jours. Inventor calcium methionate against milk fever; inventor methanamine mandelate as urinary antiseptic; development analytical methods for assay of drugs. Home: 328 Forest Hill Dr., West Lafayette, Ind. 47906. Office: Purdue U., Lafayette, Ind. 47907.*

JENKINS, Herbert Theodore, Am. civil engr.; b. Detroit, July 30, 1902; s. Herbert T. and Imogene Belmont (Stone) J.; B.S. in C.E., U. Mich., 1930, M.S.E., 1932; m. Mary Elisabeth Gill, Dec. 26, 1925; children—Peter VanWyck, Paul Andrew. Pvt. practice civil engring., Ann Arbor and Detroit, 1932-35; mem. faculty, Cornell, 1935-57; cartographer U. S. Geol. Survey, 1950; cons. engr. Bogema, Gifft, & Jenkins, Ithaca, 1948-57; mem. faculty U. Mich., Ann Arbor, prof. civil engring., chmn. dept. engring. graphics, 1957—. Mem. Am. Soc. C.E., Am. Soc. Engring. Edn., Am. Assn. Univ. Profs., Seal and Serpent, Chi Epsilon, Tau Beta Pi. Author: Soil Mechanics Laboratory Manual, Physical Properties of Soils, 1947. Research and publs. in developing methods of plotting math. and graphical data by a digital computer and its accessories. Home: 1217 Island Dr. Office: Dept. Engring. Graphics, 412 W. Engring. Bldg., Ann Arbor, Mich. 48104.*

JENKINS, James Jerome, Am. psychologist; b. St. Louis, July 29, 1923; s. Joe E. E. and Frances (Reynolds) J.; B.S., U. Chgo., 1944; A.B., William Jewell Coll., 1947; M.A., U. Minn., 1948, Ph.D., 1950; m. Geraldine Lucille Schoech, Aug. 11, 1946; children—Richard Mark, Robert Noel, Christopher James, Lynn Lucille. Faculty, U. Minn., Mpls., 1950—, prof., 1959—, dir. research Human Learning Center, 1965—; com. mem. Social Sci. Research Council, 1954-60; cons. Aphasia Clinic VA Hosp., 1957—; fellow Center Advanced Study in Behavioral Scis., 1958-59, 64-65. Mem. Am. Psychol. Assn., A.A.A.S., Am. Assn. U. Profs. Author: (with Donald G. Paterson) Studies in Individual Differences: The Search For Intelligence, 1961; (with David Palermo) Word Association Norms Grade School Through College, 1964; (with others) Aphasia in Adults, 1964. Contbr. numerous articles to profl. jours. Research, publs. on psychology of lang. Home: 2252 Folwell St., St. Paul 55108. Office: Center for Research in Human Learning, 400 Ford Hall, U. Minn., Mpls. 55455.*

JENKINS, Samuel Forest, Jr., Am. plant pathologist; b. nr. Oxford, N.C., June 15, 1930; s. Samuel Forest and Fannie (Smith) J.; student Elon Coll., 1948-49; B.S., N.C. State U., 1958, M.S., 1960, Ph.D., 1962; m. Anne Rebecca Oakes, Jan. 4, 1951; children—John Forest, Kay Frances, Steven Paul. Grad. research asst. plant pathology N.C. State U., Raleigh, 1958-61, asst. prof. plant pathology, 1965—; asst. plant pathologist Coastal Plain Expt. Sta., U. Ga., 1961-65. Mem. Am. Phytopath. Soc., Sigma Xi. Contbr. numerous articles to profl. jours. Studies on effect of various atmospheric gases on plant pathogenic organisms; research on serology to detect plant pathogenic organisms, vegetable and tobacco diseases; described new plant disease causing organism on cucurbits, new pathogenic races of a fungus on cucurbits; developed new technique for detecting fungi in soil. Home: 5026 Melbourne Rd., Raleigh 27606. Office: P.O. Box 5397, Raleigh, N.C. 27607.*

JENKINS, William Robert, Am. nematologist; b. Hertford, N.C., Sept. 12, 1927; s. William Herman and Dorothy (Perrow) J.; A.A., Norfolk div. Coll. William and Mary, 1949; B.S., Coll. William and Mary, 1950; M.S., U. Va., 1951; Ph.D., U. Md., 1954; m. Mary Frances Earhart, Aug. 18, 1951; children—William Brian, Robert Edward, Mary Ellen. Faculty U. Md., 1951-59, asst. prof., 1955-59; faculty Rutgers U., 1960—, research prof., 1963—, head nematology sect. Dept. Entomology and Economic Zoology, 1960—. Mem. Am. Inst. Biol. Sci. (mem. governing bd.), Soc. Nematologists, Am. Phytopathol. Soc., Helminthological Soc. Wash., A.A.U.P. Author: (with J. N. Sasser) Fundamentals and Recent Advances, 1960; (with D. P. Taylor) Plant Nematology, 1967. Contbr. numerous articles in field to sci. jours. Research, publs. on the role of plant parasitic nematodes in causing plant damage and their control. Home: R.D. 3, Box 285, Flemington, N.J. 08822. Office: Dept. Entomology and Economic Zoology, Rutgers U., New Brunswick, N.J. 08903.*

JENKS, Albert Ernest, Am. anthropologist; b. Ionia, Mich., Nov. 28, 1869; s. Stillman Leek and Sophia Parnell (Keeny) J.; B.S., Kalamazoo Coll., 1896, Sc.D., 1924; B.S., U. Chgo., 1897; Ph.D., U. Wis., 1899; m. Maud Huntley, Oct. 22, 1901 (dec. June 1950); 1 son, Clifford Huntley (dec.). Econ. editor Am. Thresherman, 1900-01; ethnologist Bur. Am. Ethnology, Washington, 1901-02; asst. chief Bur. Non-Christian Tribes, Manila, P.I., 1902; chief Ethnol. Survey P.I., 1903-05; chief ethnol. dept., Philippine exhbn. of St. Louis Expn., 1904; asst. prof. sociology U. Minn., 1906-07, prof. anthropology, 1907-38, chmn. dept. sociology and anthropology, 1915-18, dir. Americanization tng. course, 1918-23, chmn. dept. anthropology, 1918-38, dir. archeol. research fund, 1929-38, emeritus from 1938. Chmn. div. anthropology and psychology NRC, 1923-24; conducted investigations U. S. desert Indians, 1913-14, African desert Berber, Kabyl, Arab, 1914, southern and eastern European emigration at its sources, 1914, Ojibwa Indian-White ethnic amalgamation, 1915, prehistoric culture sites, in Europe, 1925, prehistoric Mimbres culture, N.M., 1928, 29, 31, cultures northern Africa and Europe, 1930; researches in prehistory of Minn. and Dakotas, 1932-38. Fellow A.A.A.S. (v.p.); mem. Am. Anthrop. Assn., Am. Genetic Assn., Sigma Xi. Author: The Childhood of Jishib, the Ojibwa, 1900; The Wild Rice Gatherers of the Upper Lakes, 1901; The Bontoc Igorot, 1905; Ba-long-long, the Igorot Boy, 1907; Present Conditions and Future Needs of the Science of Anthropology (with W. H. R. Rivers and S. G. Morley), 1913; Indian-White Amalgamation, 1916; Chart of Prehistoric Man and Culture, 1927; Pleistocene Man in Minnesota, a Fossil Homo Sapiens, 1936; Minnesota's Browns Valley Man and Asso. Burial Artifacts, 1937; Sauk Valley Skeleton (with L. A. Wilford), 1938; also articles. Died June 6, 1953.

JENKS, George Frederick, Am. geographer; b. Oneonta, N.Y., July 9, 1916; s. J. Frederick and Mae (Snook) J.; B.S. in Edn., State Tchrs. Coll., Albany, N.Y., 1941; M.A. in Geography, Syracuse U., 1947, Ph.D., 1950; m. Madelyn B. Ronnie, Dec. 4, 1943; children—Kiane Kathryn, Kathleen Mae, Debra Sue. Faculty, U. Kan., 1949—, prof. dept. geography, 1960—. Mem. Am. Assn. U. Profs., Assn. Am. Geographers (citation for meritorious contbn. to geography 1952), A.A.A.S., Am. Congress Surveying and Mapping, Sigma Xi. Author: Atlas of Kansas 1952; also articles. Research on map making procedures, map symbolization, data processing for mapping. Home: 1932 Hillview Rd., Lawrence, Kan. 66044.*

JENNER, Edward, English physician; b. Berkeley, Gloucestershire, May 17, 1749; s. Stephen J. and Miss (Head) J.; med. apprentice to Dr. Ludlow of Sodbury; studied under John Hunter, London, St. George's Hosp., London, 1870-72; M.D., U. St. Andrews, 1792; M.D. (hon.), Oxford U., 1813; m. Catharine Kingscote, Mar. 6, 1788; several children. Employed by Sir J. Banks to prepare biol. specimens procured from Pacific by Capt. Cook, 1771; offered post of naturalist for Cook's 2d voyage, declined; practiced medicine, Berkeley, 1773-1823; made frequent visits to London to promote smallpox vaccinations, from 1798; spent much time in obtaining and distbg. lymph needed for vaccine, 1798-1800; head Royal Jennerian Instn. (for promotion vaccinations), 1803, replaced with govt.-supported Nat. Vaccine Establishment, 1808; elected corr. med. and surg. sect. 1st class French Acad. Scis., 1808, fgn. asso., 1811; awarded grants by Parliament, 1803, 06. Recipient Gold medal from med. officers Brit. Fleet, 1801. Fellow Royal Soc., 1789; mem. Med. and Chirurg. Soc. (founder). Author: (most important work) An inquiry into the causes and effects of the variolae vaccinae, 1798. Discovered that cowpox lymph matter could be safely used as innoculation to prevent smallpox, 1st vaccination performed May 14, 1796; experimented on various vehicles which could be used for transmission of cowpox lymph, circa 1801; dedicated himself to promotion of vaccination against smallpox in many parts of world; monuments erected of him in Eng., France, Italy; Bavaria honored Jenner's discovery by making smallpox vaccination compulsory, 1803. Died Berkeley, Jan. 24, 1823.

JENNER, Sir William, English physician; b. Chatham, Eng., Jan. 30, 1815; s. John Jenner; ed. London U.; D.C.L., LL.D.; m. Adela Adey, 1858; a son, Walter. Prof. path. anatomy Univ. Coll.; physician-in-ordinary to Queen Victoria, Prince of Wales (later King Edward VII). Fellow Royal Soc., 1864, Royal Coll. Physicians (pres. 1881-88). Author: On the Identity and Non-identity of Typhoid Fever, 1850; Lectures on Fever and Diphtheria, 1893. Distinguished typhoid fever from typhus, 1849; clearly described emphysema of lungs, 1857. Died Bishop's Waltham, Eng., Dec. 11, 1898.

JENNESS, Robert, Am. biochemist; b. Rochester, N.H., Sept. 21, 1917; s. Myron I. and Ruth (Libby) J.; B.S., U. N.H., 1938; M.S., U. Vt., 1940; Ph.D., Minn., 1944; m. Katherine Ward, Aug. 30, 1940; children—Douglas F., Malcolm I., David R. With Vt. Agr. Expt. Sta., 1938-40; faculty U. Minn., St. Paul, 1940—, prof. biochemistry, 1953—. Fulbright scholar, Netherlands, 1961-62. Fellow A.A.A.S.; mem. Am. Chem. Soc. (Borden award in dairy chemistry 1953); Am. Soc. Biol. Chemists, Am. Dairy Sci. Assn., Am. Soc. Mammalogists, Sigma Xi, Phi Kappa Phi, Phi Lambda Upsilon, Alpha Zeta. Author: (with Patton) Principles of Dairy Chemistry, 1959; also numerous articles. Research on structure and properties milk proteins in relations to species and genetic variation, immunology denaturation, indsl. utilization. Home: 942 Oak Ridge Av., St. Paul 55112.*

JENNINGS, Herbert Spencer, Am. naturalist; b. Tonica, Ill., Apr. 8, 1868; s. Dr. George N. and Olive Taft (Jenks) J.; B.S., U. of Mich., 1893; A.M., Harvard, 1895, Ph.D., 1896; studied Jena (Germany), 1896-97, LL.D., Clark University, 1909, University of Pa., 1940, University of California, 1943, S.D., University of Mich., 1918, U. of Pa., 1933, Oberlin Coll., 1933, University of Chicago, 1941; m. Mary Louise Burridge, 1898; 1 son, Burridge; m. 2d, Lulu Plant Jennings, 1939. Assistant professor botany, Texas Agricultural College, 1889-90; professor botany, Montana State Agrl. College, 1897-98; instr. zoology, Dartmouth Coll., 1898-99; asst. prof. zoology, U. of Mich., 1900-03; asst. prof. zoology, U. of Pa., 1903-05; prof. exptl. zoology, 1906-10, Henry Walters prof. zoology and dir. zool. lab., 1910-38, emeritus prof. since 1938, Johns Hopkins Univ.; research associate U. of Calif. at Los Angeles since 1939; visiting prof., Keio University, Tokyo, 1931-32, George Eastman visiting prof. and fellow of Balliol College, Oxford Univ., England, 1935-36; visiting prof. U. of Calif. at Los Angeles, 1939; Terry lecturer Yale, 1933; Vanuxem lecturer, Princeton, 1934; Leidy lecturer, University of Pa., 1940; Patten lecturer, University of Indiana, 1943; hon. fellow, Stanford, 1941. Specialist in research work on physiology of micro-organisms, animal behavior, and genetics. Director U. S. Fish Commn. Biol. Survey of the Great Lakes, 1901; trustee Marine Biol. Lab., Woods Hole, Mass.; biometrician, U. S. Food Administration, 1917-18; mem. Nat. Research Council, 1922-25. Asso. editor Jour. of Experimental Zoology, of Genetics, and of Biol. Bulletin. Pres. Am. Zool. Society, 1908-09, Am. Society Naturalists, 1910-11; fellow Am. Acad. Arts and Sciences, A.A.A.S.; hon. fellow Royal Micros. Soc. Great Britain; mem. Nat. Acad. Sciences, American Philosophical Soc., Philadelphia Academy Natural Sciences; Royal Society of Edinburgh; corr. member of Russian Academy of Science, Société de Biologie de Paris. Author: Anatomy of the Cat (with Jacob Reighard), 1901; Behavior of Lower Organisms, 1906; Life and Death, Heredity and Evolution in Unicellular Organisms, 1919; Contributions to the Study of the Lower Organisms, 1919; Prometheus—or Biology and the Advancement of Man, 1925; The Biological Basis of Human Nature, 1930; Genetics of the Protozoa, 1929; The Universe and Life, 1933; Genetics, 1935; Genetic Variations in Relation to Evolution, 1935. Contbr. of numerous papers in zool. and physiol. jours. Investigated heredity and variation of micro-organisms; conducted biol. survey of Gt. Lakes for U. S. Fish Commn., 1902; made study of protozoa psychologically (behavior). Died Apr. 14, 1947.

JENNISON, Marshall W(alker), Am. microbiologist; b. Portland, Me., May 27, 1905; s. William Walker and Harriet (Marshall) J.; B.S., Mass. Inst. Tech., 1927, Ph.D., 1932; m. Cynthia M. Lamb, Aug. 15, 1929 (dec. June 1950); children—Cynthia M. (Mrs. Richard Harnett), Margaret W. (Mrs. Edward Benton); m. 2d, Margaret Kerfoot, Jan. 3, 1952. Faculty, Mass. Inst. Tech., 1927-46; prof. bacteriology Syracuse U., 1946—, chmn. dept. bacteriology and botany, 1956—; vis. lectr. Harvard Sch. Pub. Health, 1945. Fellow Am. Acad. Microbiology, Am. Pub. Health Assn., A.A.A.S.; mem. Am. Inst. Biol. Scis., Am. Soc. Microbiology, Mycol. Soc. Am., Soc. Indsl. Microbiology, Sigma Xi, Kappa Sigma. Editorial bd. Jour. Bacteriology, 1931-51, Applied Microbiology, 1959-62. Research and publs. in droplet prodn. in sneezing, physiology wood-rotting fungi, bacterial population cycles and statis. evaluation errors in counting bacteria. Home: 307 Standish Dr., Syracuse, N.Y. 13224.*

JENNISON, Roger Clifton, English physicist; b. Grimsby, Eng., Dec. 18, 1922; s. George Robert and Elsie (Clifton) J.; student Hull Tech. Coll., 1941-42; B.Sc. with honors, U. Manchester (Eng.), 1950, Ph.D., 1954; m. Jean Gordon Gray, Nov. 7, 1952; children—Michael Alexander Clifton, Heather Clifton, Timothy Martin Clifton. Mackinnon Research student Royal Soc., 1952-55; lectr. in radio astronomy Manchester U., 1955-61, sr. lectr., 1961-66, prof. phys. electronics, dir. Electronic Labs., U. Kent, Canterbury, Eng., 1966—. Fellow Royal Astron. Soc., Inst. Physics, Instn. Elec. Engrs. (past pres.), Instn. Electronics. Author: Fourier Transforms and Convolutions for the Experimentalist, 1961; Introduction to Radio Astronomy, 1966; also articles. Numerous discoveries in radio astronomy including 1st double radio source, limb brightening of Cassiopeiai and Cassiopeia spur, various techniques in radio astronomy including 1st phase sensitive stellar interferometry and synthesis, techniques for space research, low penetrating power of interplanetary dust and asso. discoveries, theoretical contbns. to phenomenon of rotation. Home: Wildwood, Nackington, Canterbury, Kent., Eng.*

JENSEN, Aage Johannes Christian, Danish marine biologist; b. Ronnebak, Denmark, Feb. 9, 1898; s. Ludvig and Dagmar (Thiesen) J.; cand.mag., U. Copenhagen (Denmark), 1924, dr.phil., 1933; m. Ellen Marie Petersen, Nov. 27, 1925; children—Leif Farregaard, Ide Cathrine Farregaard (Mrs. Aage Hegen), Helle Farregaard (Mrs. Gunnar Larsen). Asst., Internat. Commn. for Exploration of Sea, 1919-21; staff Danish Sea and Fishery Investigations, Charlottenlund, since 1919—, state biologist, since 1959—; leader permanent exhbn. Fishery Ministry, Fisken og Havet, 1953. Recipient Ridder of Dannebrog, 1952. Author several books including: Periodic Fluctuations in the Size of Various Stocks of Fish and Their Causes, 1933; also numerous articles. Research on biology of comml. fishes, Norway lobster, influence of currents upon climate; proved influence of hydrographical factor on fluctuations of many comml. fisheries. Home: Jagersborgalle 1A, Charlottenlund, Denmark; also Enemarket, Kongsgaarde pvt. Knebel, Denmark. Office: Charlottenlund Castle, Charlottenlund, Denmark.*

JENSEN, Arne, Danish mathematician; b. Vordingborg, Denmark, Feb. 16, 1920; s. Alfred M. and Anna (Klenke) J.; Actuary, Copenhagen U., 1944, Sc.D., 1954; m. Anne M. Bundgaard, Oct. 29, 1960; 4 children. Asst., Serum Inst., Copenhagen, 1944-45; actuary Copenhagen Telephone Co., 1945-63; lectr. Copenhagen U., 1945-51, asst. prof., 1951-63; faculty Tech. U. Denmark, Lyngby, Denmark, 1963—. Mem. organizing com. Internat. Congresses on Teletraffic and Adminstrn.; co-founder Danish Folk U., 1945. Mem. Internat. Assn. for Statistics in Phys. Scis. (mem. council), Union Internat. pour l'Etude Scientific de la Population, Danish Operations Research Soc. (chmn.). Author: An Elucidation of A. K. Erlang's Statistical Works through the Theory of Stochastic Processes, 1948; Moe's Principle, 1950; Stormflod og Digebygning med saerligt henblik pa Sondre Limfjordstange, 1958; Population Statistics, 1960; also articles on traffic theory, decision theory, risk theory. Home: 181 Bolbrovej, Horsholm 2970, Denmark. Office: Instituttet for Matematisk Statistik og Operationsanalyse, Denmarks Tekniske Hojskole, Lyngby, 88 14 33 Bygning 305, Denmark.*

JENSEN, Cyril D(ewey), Am. civil engr.; b. Brownton, Minn., Aug. 17, 1898; s. Peder B. and Anna Marie (Thompson) J.; S.B., U. Minn., 1921, C.E., 1931; S.M., Lehigh U., 1929; m. Bessie Evans, June 26, 1923; children—Grace (Mrs. James Randolph), Thomas Evans. Surveyor, field engr., county and state hwys., hydroelectric surveys, Minn., Wis., 1921-25; faculty Lehigh U., 1925-42, prof. engring., chmn. div. soil mechanics, 1946-58; dir. research projects U. S. Army, 1946-50, Pa. Dept. Hwys., 1950-51; welding engr., Lehigh Structural Steel Co. 1930, U. S. Naval Engring. Expt. Sta., Annapolis, Md., 1942-46, Lockheed Aircraft Corp., summers 1952, 53, dir. bur. materials research and testing. Pa. Dept. Hwys., Harrisburg, 1958—. Registered profl. engr. Pa. Fellow Am. Soc. C.E.; mem. Am. Welding Soc., Am. Soc. for Engring. Edn., Sigma Xi, Tau Beta Pi, Theta Tau. Author: Summary of Recent Advances in Underwater Cutting, 1944; Manual for Underwater Welding, 1945. Patentee and discoverer (with another) that underwater arc welding was possible; during World War II further developed underwater arc welding for USN; also patented and developed a new underwater cutting process known as the arc-oxygen process, which is the present standard method used by the Navy and industry. Home: 117 Carol St., New Cumberland, Pa. 17070. Office: Pa. Dept. of Highways, 1118 State St., Harrisburg, Pa. 17120.*

JENSEN, Elwood Vernon, Am. biochemist; b. Fargo, N.D., Jan. 13, 1920; s. Eli A. and Vera (Morris) J.; A.B., Wittenberg Coll., 1940, Sc.D., 1963; Ph.D., U. Chgo., 1944; m. Mary Collette, June 17, 1941; children—Karen Collette, Thomas Eli. Mem. faculty U. Chgo., 1943—, prof. Ben May Lab., 1960-63, Am. Cancer Soc.-Charles Hayden Found. Research prof., Ben May Lab., also dept. physiology, 1963—. Vis. prof. Max Planck Institut für Biochemie, Munich, Germany, 1958; cons. Upjohn Co., USPHS; mem. Ill. Clin. Chemistry and Blood Bank Adv. Bd., 1966—. Guggenheim Found. fellow Eidgenössische Technische Hochschule, Zurich, 1946-47. Mem. Am. Chem. Soc., Am. Soc. Biol. Chemists, Am. Assn. Cancer Research, Endocrine Soc., A.A.A.S. Research and numerous publs. on sulfhydryldisulfide interchange in proteins, mechanism of estrogen action-estrogen receptors in target tissues, free radical addition of carbon tetrahalides to olefins, electrophilic fluorination with perchloryl fluoride, novel organophosphorus chemistry. Home: 5650 Dorchester Av., Chgo. 60637.*

JENSEN, Hans Laurits, Danish microbiologist; b. Sigerslevvester, Denmark, June 27, 1898; s. Anders and Anna (Christensen) J.; B.Sc.Agr., Royal Vet. and Agr. Coll., Copenhagen, 1920, M.Sc., Agr., 1923, dr. agronomiae, 1941; Agriculturae et artis silvaticae doctor honoris causa, U. Helsinki (Finland), 1958; m. Helene Zink, Nov. 12, 1926; children—Anna Elisabeth (Mrs. Hugo Nielsen), Margarethe (Mrs. Nils Eric Boesgaard), John Peter. Research asst. Statens Planteavls Laboratorium, 1923-27; Rockefeller scholar Rothamsted Expt. Sta., Britain, 1927-29; bacteriologist Linnean Soc. N.S.W., Sydney, Australia, 1929-47; head dept. bacteriology Statens Planteavls-Laboratorium, Lyngby, Denmark, 1947—, dir., 1961—; external examiner U. Copenhagen (Denmark), 1962—. Decorated Knight Order of Dannebrog (Denmark). Mem. Acad. Tech. Scis., Microbiol. Soc. Denmark, Soc. for Applied Microbiology (Eng., hon.), Am. Soc. Microbiology (hon.). Research, numerous publs. on fungi and actinomycetes in soil, taxonomic relationships of actinomycetes, mycobacteria and cornyneform bacteria, effect, microorganisms in decomposition of organic matter in soil, importance of free-living and symbiotic nitrogen-fixing bacteria in nitrogen balance of nature, role of microorganisms in biol. detoxification of chem. pesticides and herbicides. Home and office: 24 Lottenborgvej, Lyngby, Denmark.*

JENSEN, Henning Höjgaard, Danish physicist; b. Blidstrup, Denmark, Nov. 21, 1918; s. Knud Höjgaard and Nanni (Bruhn) J.; cand.mag., U. Copenhagen, 1942; m. Arendse Möller, Apr. 3, 1943; children—Jens, Sven, Bodil, Ellen. Prof., Tech. U., Copenhagen, 1952-60; prof. U. Copenhagen, 1960—. Dir. NORDITA. Mem. exec. com. Danish Atomic Energy Commn., since 1960—; mem. Danish Tech. Research Council, since 1967—; Danish rep. sci. com. NATO, 1962-64. Recipient Gold medal U. Copenhagen, 1944. Mem. Danish Acad. Tech. Scis. (v.p., 1965), mem. Danish Assn. for Advancement Sci. (sec. 1953-60, mem. bd. directors, 1960); Internat. Standardization Orgn. (chmn. tech. com. on units 1957—). Author: Mechanics of Deformable Bodies, 1955; also articles. Research on thermodynamics, acoustics, solid state physics. Home: 18A Rödtjörnevej, Vanlöse, Copenhagen, Denmark.*

JENSEN, Henry, Danish geophysicist; b. Copenhagen, Denmark, Oct. 7, 1915; s. Hans Peter and Valborg (Hansen) J.; Ph.D. in Geodesy, U. Copenhagen, 1938; Dr.phil. in Seismology, 1961; m. Gudrun Valborg Nielsen, June 23, 1939; children—Hanne (Mrs. Hans Christian Eisen), Elin, Iben, Erik. With Royal Danish Geodetic Inst., 1938-65, head seismic dept., 1953-65; prof. geophysics U. Copenhagen, 1965—. Pres., European Seismol. Commn., 1966—. Decorated knight Order of Dannebrog, Medal of Def. Mem. Internat. Union Geodesy and Geophysics (mem. nat. com. 1963—), Acad. Tech. Scis. Author: Earthquakes, 1963; also articles. One of 1st to use matrix-symbolism in adjustment computations; studies on and new formulas for astron. corrections to precise levelings; methods for determination of direction of approach of microseisms (method which works without tacit assumptions as to nature of waves). Home: Raunstrupvej 13, 272o Vanlose, Denmark. Office: Inst. Geophysics, U. Copenhagen, Copenhagen, Denmark.*

JENSEN, J. Hans D., German physicist; b. Hamburg, Germany, June 25, 1907; s. Karl and Helene (Ohm) J.; student univs. Hamburg, Freiburg; Dr.rer.nat., U. Hamburg, 1932. Docent, U. Hamburg, 1937-41; prof. theoretical physics Hanover Inst. Tech., 1941-48; prof. U. Heidelberg, 1949—; hon. prof. U. Hamburg, 1947—; guest prof. U. Wis., Madison, 1951, Inst. Advanced Study, Princeton, N.J., 1952, U. Cal. at Berkeley, 1952, Cal. Inst. Tech., Pasadena, 1953, Ind. U., Bloomington, 1953, U. Minn., Mpls., 1956, U. Cal. at La Jolla, 1961. Recipient (with Maria Goeppert Mayer and Eugene P. Wigner) Nobel prize physics, 1963. Mem. Heidelberg Acad. Scis., Max Planck Soc. Author: (with M. Goeppert-Mayer) Elementary Theory of Nuclear Shell Structure, 1955; also numerous articles in handbooks, profl. lit. Editor (with O. Haxel) Zeitschrift für Physik, 1955—. Explained characteristics of atomic nuclei using nuclear shell model, according to which nucleons move in definite orbits of shells. Address: 16 Philosophenweg, Heidelberg, West Germany.*

JENSEN, J. L. W. V., Scandanavian mathematician; b. 1859; produced important and detailed discussion of factorials, 1884; proved that Dirichlet series has an abscissa of convergence, thus laying found. for modern work in the series. Died 1925.

JENSEN, James Herbert, Am. plant pathologist; b. Madison, Neb., June 16, 1906; s. Jens and Eda (Hansen) J.; B.S., U. Neb., 1928, M.A., 1930; postgrad. Columbia; Ph.D., U. Wis., 1935; m. Lucille Irene Christopher, Nov. 2, 1931; children—James Michael, Karen (Mrs. James A. Bailey), Stephen Christopher, Roger Anthony. Asst. plant pathologist Rockefeller Inst. for Med. Research, Princeton, N.J., 1932-35; asst. prof. pathology U. Neb., 1937-44; prof., head plant pathology N.C. State Coll., 1945-53; chief biology br. U. S. AEC, Washington, 1948-49; prof. botany, provost Ia. State U., 1953-61; prof. botany, pres. Ore. State U., Corvallis, 1961—. Chmn. subcom. waste disposal and decontamination Nat. Radiation Protection Com., 1948-57; pres. Midwest U., Argonne, Ill., 1958-59, bd. dirs., 1958-64; mem. agrl. research planning com. U. S. Dept. Agr., 1964-66; mem. sci. adv. com. Boyce Thompson Inst., Yonkers, N.Y., 1958—; chmn. pesticides residues com. Nat. Acad. Scis., 1964-65; mem. Nat. Commn. on Accrediting, 1965-68. Mem. Nat. Assn. State U. and Land Grant Colls. (pres. elect 1965), Am. Phytopath. Soc. (past pres.), Sigma Xi, Gamma Sigma Delta, Alpha Zeta. Contbn. to understanding of origin and isolation of strains of tobacco mosaic virus, methods of disease control in sugarcane, potatoes, beans, grapefruit, papayas. Home: 3520 Hayes St., Corvallis, Ore. 97330.*

JENSEN, Kai Arne, Danish chemist; b. Copenhagen, Denmark, Mar. 27, 1908; Mag.Scient., U. Copenhagen, 1932, Dr.Phil., 1937; m. Ida Reichardt, Aug. 3, 1932; 3 daus. Faculty, U. Copenhagen, 1943—, prof., head chem. lab. II, 1950—. Mem. Internat. Union Pure and Applied Chemistry (chmn. commn. on nomenclature inorganic chemistry), Royal Danish Acad. Scis., Danish Acad. Tech. Scis. Recipient Julius Thomsen medal. Research numerous publs. on coordination chemistry and organic chemistry. Author textbooks on gen. and organic chemistry. Home: 64 Bogehoj, Copenhagen, Hellerup, Denmark.*

JENSEN, Mead LeRoy, Am. geologist; b. Salt Lake City, June 11, 1925; s. Joseph Mead and Gertrude (Grayson) J.; B.S., U. Utah, 1948; Ph.D., Mass. Inst. Tech., 1951; m. LouDeen Davis, Oct. 10, 1947; children—Joseph, Pamela, Patricia, Janice, Robert. Faculty, Yale, 1951-65; prof., dir. lab. isotope geology U. Utah, Salt Lake City, 1965—; pres., mng. dir. Geoventures, Inc., 1952—. Sr. vis. scientist Australian Acad. Sci., 1963. Fellow Geol. Soc. Am.; mem. A.A.A.S., Am. Geophys. Union, Soc. Econ. Geology, Soc. Exptl. Geophysicists, Mineral. Soc. Am. Author: Biogeochemistry of Sulfur Isotopes, 1952. Research, numerous publs. on origin of mineral deposits by application of stable isotopic abundances to genesis of sulfur and oxygen bearing minerals, role of bacteria in forming mineral deposits. Home: 2327 Berkeley St., Salt Lake City 84109.*

JENSEN, William August, Am. botanist; b. Chgo., Aug. 22, 1927; s. William McKinley and Gertrude (Hild) J.; Ph.B., U. Chgo., 1948, M.S., 1950, Ph.D., 1953; m. Joan Nancy Sell, June 20, 1948; children—Scott William, Christina Cathrine. NIH fellow Carlsberg Lab., Copenhagen, Denmark, 1952-53, Cal. Inst. Tech., Pasadena, Cal., 1953-55; NSF fellow U. Brussels (Belgium), 1955-56; asst. prof. dept. biology U. Va., Charlottesville, 1956-57; faculty dept. botany U. Cal. at Berkeley, 1957—, prof., 1963—, asso. dean Coll. Letters and Sci., 1963-66. Recipient Distinguished Teaching award U. Cal. at Berkeley, 1960, N.Y. Botanic Garden award for bot. research, 1964. Mem. Soc. for Study Devel. and Growth (past sec.), Bot. Soc. Am. (program dir. 1963—), A.A.A.S., Am. Inst. Biol. Scis., Bot. Soc. Am., Am. Soc. Plant Physiologists, Soc. for Developmental Biology, Am. Soc. Cell Biology, Histochem. Soc. Author: Botanical Histochemistry, 1962; The Plant Cell, 1964; The Ultrastructure of the Cell (with R. Park), 1967; also articles. Devel. bot. histochem. procedures and application of these procedures to problems of early cell devel. in plants especially root tips and embryos; research on distbn. of nucleic acids in plant tissues especially embryos and cell wall devel. Home: 1419 Campus Dr., Berkeley, Cal. 94708.*

JENTOFT, Ralph E., Jr., Am. chemist; b. Tacoma, Nov. 30, 1918; s. Ralph E. and Helena (Nielsen) J.; student Coll. Puget Sound, 1937-38; B.S., U. Wash., 1941, Ph.D., 1952; m. Betty Louise Eshelman, Apr. 17, 1954; children—Elisabeth Dale, Rolf Erik. Chemist, U. S. Fish and Wildlife Service, P.I., 1947-48; research asso. in oceanography Office Naval Research, Seattle, 1949-52; sr. research asso. Chevron Research Co., Richmond, Cal., 1952—. Mem. Am. Chem. Soc., Am. Geophys. Union, A.A.A.S., Geochem. Soc., Sigma Xi, Phi Lambda Upsilon. Publs. on devel. method for analysis for potassium in sea water; determined concentration ratio for potassium in sea water; phase studies and thermodynamic measurements for hydrocarbon systems; devel. improved apparatus for separation and purification of hydrocarbon substances; methods of separation and analysis for trace substances in petroleum and petrochem. mixtures. Home: 880 Penny Royal Lane, San Rafael, Cal. 94903. Office: 576 Standard Av., Richmond, Cal. 94802.*

JEPSEN, Glenn Lowell, Am. geologist; b. Lead, S.D., Mar. 4, 1903; s. Victor Theodore and Kittie Elizabeth (Gallup) J.; student U. Mich., 1922-23, S.D. State Sch. Mines, Rapid City, 1923-25; B.S., Princeton, 1927, Ph.D., 1930; m. Janet E. Mayo, June 14, 1934 (div. Dec. 1953); 1 dau., Katherine Alice. Instr., S.D. State Sch. Mines, 1924-25; faculty Princeton, 1930—, Sinclair prof. vertebrate paleontology, 1946—, dir. Scott Fund Expdns., also curator vertebrate paleontology, 1935—, now dir. Princeton Natural History Museum. Recipient Addison Emery Verrill Medal, Yale, 1962. Mem. A.A.A.S., Am. Soc. Mammalogists, Geol. Soc. Am. (councilor 1951-54), Paleontol. Soc. Am. (v.p. 1955-56), Soc. Vertebrate Paleontol. (pres. 1944-45), Am. Philos. Soc., Am. Assn. U. Profs., Geol. Soc. N.J. (pres. 1959-61, dir. 1961-63), Phi Beta Kappa, Sigma Xi. Research and publs. on the geologic history of several groups of mammals such as rodents, chalicothers, edentates, and certain primates by discovering their oldest-known fossil remains. Home: 144 Patton Av., Princeton, N.J.*

JERLOV, Nils Gunnar, oceanographer; b. Bosjökloster, Sweden, Oct. 12, 1909; s. David and Hilma (Henriksson) J.; M.A., Lund U., 1932, Ph.D., 1939; m. Elvi Johanna Maria Galeen, Feb. 19, 1949. First sci. asst. Svenska hydrografisk-biologiska Kommissionen, Göteberg, Sweden, 1939-48; sci. head hydrographical dept. Royal Bd. Fisheries, Göteborg, 1948-57; faculty Göteborg U., 1953-65, asso. prof. oceanography, 1957-65; prof. phys. oceanography, Copenhagen (Denmark) U., 1965—. Partner Swedish Deep-Sea expdn., 1947-48, internat. expdn. Adriatic Sea, 1955, Discovery II, 1959; leader Dana expdn. Sargasso Sea, 1966; diver (with Piccard) in bathyscaphe Mediterranean, 1957. Recipient Albatross medal, 1948; Bathscaphe medal; named knight Order North Star, 1960. Mem. Internat. Assn. Phys. Oceanography (chmn. com. on radiant energy in sea), Goteborgs Kungl. Vetenskaps-och Vitterhetssamhälle. Research, publs. on influence of chem. combination on X-ray emission spectra, thermodynamical problems, underwater light problems, particles distbn. in Atlantic, Indian and Pacific Ocean, equatorial currents in Indian and Pacific Ocean. Home: 18 Hartmannsvej, Kgs Lyngby, Denmark. Office: 83 Solvgade, Kobenhavn, Denmark.*

JERRARD, George Birch, Brit. mathematician; s. Joseph Jerrard; B.A., Trinity Coll., Dublin, 1827. Author: Mathematical Researches (contbd. to solution of gen. quintic equation), 8 vols., 1832-35; Essay on the Resolution of Equations, 1858. Died Norfolk, Eng., 23, 1863.

JERSILD, Arthur Thomas, Am. psychologist; b. Elk Horn, Ia., Nov. 12, 1902; s. Thomas Nielsen and Anne (Bille) J.; A.B., U. Neb., 1924, LL.D., 1962; Ph.D., Columbia, 1927; m. Catherine Livingston Thomas, Feb. 22, 1930; children—Catherine Anne (dec.), Alice Jane, Johanna (dec.), John. Mem. faculty U. Neb., 1922-24, Barnard Coll., 1927-29, U. Wis., 1929-30; faculty Tchrs. Coll., Columbia, 1930—, prof. edn., 1938—, prof. psychology, edn. emeritus; cons. psychologist CBS, 1948-55; vis. expert civil affairs div., reorientation br. Dept. of Army, Japan, 1948-49. Recipient outstanding research citation Am. Ednl. Research Assn., 1936. Dept. Japan, 1948-49. Fellow Research Child Devel. Soc., Am. Psychol. Assn.; mem. Phi Beta Kappa, Sigma Xi, Phi Delta Kappa. Editor: Child Development Monographs. Author several books on child psychology including: In Search of Self, 1952; Child Psychology, 6th edit., 1968; Psychology of Adolescence, 2d edit., 1963; (with another) Education for Self Understanding and others. Research and publs. on learning, childrens fears, interests, social devel., and the psychology of the self.*

JERUSALEM, Christoph, anatomist, histologist; b. Schreiberhan, Germany, Feb. 27, 1922; s. Hanns David and Katherina (Uhlig) J.; student medicine U. Tübingen (Germany), 1945-50, Dr.Med., 1953, Priv.-Doz., 1961; m. Ruth Sutorius, May 17, 1953. With U. Tübingen, 1951-63, sci. asst. anthropology and anatomy, 1952-63; sci. chief officer R.K. U., Nijmegen (Netherlands), 1963-66, chief Inst. Cytohistology, 1966—, prof. cytohistology; with Gondi-Shapoor U., Ahwaz, Persia, 1962. Mem. Anat. Gesellschaft Nederl. Research, publs. on induction of cell proliferation by mitosis and amitosis during compensatory hypertrophy of kidney and liver, daily rhythm of nuclear size and DNA content in liver cells, immunologic phenomena in exptl. malaria infection, dependency of state of immunity on time of parasitemia; discovered different immunity to malaria toxins and malaria parasites and possibility of active immunization to malaria parasites. Home: 190, Vassenlaan, Nijmegen, Netherlands.*

JERVIS-SMITH, Frederick J., Brit. inventor; b. Taunton, Eng., Apr. 2, 1848; s. F. J. Smith; B.A., Pembroke Coll., Oxford (Eng.) U., M.A., Trinity Coll.; M.A. (hon.), Adelaide (Australia) U.; m. Annie Eyton Taylor; 1 son. Vicar, St. John's Ch., Taunton, 1884-86; patron livings of St. John's, St. Andrew's chs., Taunton; univ. lectr. mechanics; Millard lectr. in exptl. mechanics, engring. Trinity Coll., Oxford U. Mem. com. on explosions Home Office, 1895-96; Oxford U. rep. to Tercentenary of Torricelli, Faenza, Italy, 1908. Recipient medal for dynamometer French Exhbn.; silver medal for integrator Inventions Exhbns.; medal for sav. life Humane Soc. Fellow Royal Soc., 1894. Contbr. numerous sci. papers. Inventor several forms of dynamometers, integrators, chronographs, tramchronograph for measuring speed of flying bullets, also inductoscript; pioneer wireless telegraphy (analyzed 1st installation in Oxford). Died Aug. 23, 1911.

JESSERER, Hans, Austrian physician; b. Vienna, Austria, Oct. 5, 1914; s. Johann and Marie (Hanzl) J.; D.medicinae universae, U. Vienna, 1938, Dr.med.-habil, 1943; m. Herta Kanzler, Dec. 14, 1940. Mem. I. med. Clinic, U. Vienna, 1938-62, attendant dir., 1963-64, prof. internal medicine, 1959—. dir. II med. dept. Kaiser-Franz-Joseph-Spital, Vienna, 1965—. Mem. Gesellschaft der Ärtze in Wien, Arbeitsgemeinschaft für Erkrankungen des Bewegungsapparates (chmn. 1965—). Author: Tetanie, 1958; Osteoporose, 1963; Atlas der Knochen- und Gelenkkrankheiten, 1963; Histologischer Atlas der Knochen und Knochenmarkbiopsie, 1965; also numerous articles. Research on diseases of parathyroid glands, etiological differentiation of types of tetany, pharmacology of vitamin D, differential diagnosis of bone and joint diseases, tissue damages correlated to glucocorticoid medication. Home: 10 Albertgasse, 1080 Vienna. Office: II.Medizinische Abteilung des Kaiser-Franz-Joseph-Spitals 3, Kundratstrasse, 1100 Vienna, Austria.*

JESSIOT, Ernest-Paulin-Joseph, French mathematician; b. Marseille, France, 1865; ed. École Normale Supérieure; prof. Faculty Scis. Lyons (France), also at Sorbonne; dir. École Normale Superieure; mem. French Acad. Scis., 1943. Author: Leçons de géométrie moderne; Cours de mathématiques générales; Géométrie supérieure; other textbooks. Research on group theory and wave propagation; noted as tchr. of many well-known mathematicians. Died LaBauche, France, 1952.

JESSOR, Richard, Am. psychologist; b. Bklyn., Nov. 24, 1924; s. Thomas and Clara (Merken) J.; B.A., Yale, 1946; M.A., Columbia, 1947; Ph.D., Ohio State U., 1951; m. Shirley Louise Glasser, Sept. 13, 1948; children—Kim, Tom. Asst. prof. U. Colo., Boulder, 1951-56, asso. prof., 1956-61, prof. psychology, from 1961, also dir. research program Social and Personal Problem Behavior, Inst. Behavioral Sci. Cons., Fed. Correctional Instn., Englewood, Colo. Mem. Am. Psychol. Assn., Soc. Psychol. Study of Social Issues, Soc. Study of Social Problems. Author: Society, Personality, and Deviant Behavior, 1968. Contbr. articles to jours., also joint-editor Cognition, Personality, and Clinical Psychology, 1967; editorial cons. Sociometry, Psychological Monographs. Studies on phenomenological view of personality, role of expectations in human behavior, relations of sociocultural and personality determinants to deviant behavior. Home: 595 Euclid, Boulder, Colo. 80302.*

JETER, Wayburn Stewart, Am. microbiologist; b. Cooper, Tex., Feb. 16, 1926; s. Joseph Plato and Beulah (Stewart) S.; B.S., U. Okla., 1948, M.S., 1949; Ph.D., U. Wis., 1950; m. Margaret Ann McDonald, May 30, 1947; children—Randall Mark, Monette Ann, Marcus Kent. Faculty U. Ia., 1950—, asso. prof., 1958-63; prof. microbiology U. Ariz., Tucson, 1963—. Mem. Am. Acad. Microbiology, Am. Assn. Immunologists, A.A.A.S., Am. Soc. for Microbiology, Soc. for Exptl. Biology and Medicine, Sigma Xi. Diplomate Am. Bd. Microbiology. Contbr. articles to tech. jours. Research on factors contbg. to resistance animal body to infectious disease, mechanism allergy to chem. substances, transplantation tissues from one individual to another of same species. Home: 4834 E. Glenn St., Tucson 85716.*

JEVONS, Frederic Raphael, biochemist; b. Vienna, Austria, Sept. 19, 1929; s. Fritz and Hedwig (Scharf) Bettelheim; came to Eng., 1939, naturalized, 1949; student Langley Sch., 1940-46; B.A., King's Coll., Cambridge, 1950, M.A., 1953, Ph.D., 1953; D.Sc., U. Manchester, 1966; m. Dita Bradel, Dec. 21, 1956; children—Colin Peter, Norman Thomas. Fellow dept. biochemistry U. Seattle, 1953-54; fellow King's Coll., Cambridge, Eng., 1953-59; demonstrator biochemistry U. Cambridge, 1956-59; lectr. biol. chemistry U. Manchester, 1959-66, sr. lectr., 1966, prof. liberal studies in sci., 1966—. Mem. chemistry com. Cotton, Silk and Man-Made Fibres Research Assn., 1961—; dir. Langley Sch., 1967—. Mem. Biochem. Soc., Brit. Soc. History of Sci. Author: The Biochemical Approach to Life, 1964, Italian edit., 1965, Spanish Japanese edits., 1968. Research, numerous publs. on chem. changes in clotting of fibrinogen and activation of chymotrypsinogen, structure of mucoproteins and their enzymatic degradation, history of sci. ideas of Boerhaave, Harvey, Paracelsus, Plotinus. Home: Langley, The Lowes, Bowdon, Altrincham, Cheshire, Eng. Office: Dept. Liberal Studies in Sci. U. Manchester, Manchester 13, Eng.*

JEVONS, William Stanley, English economist; b. Liverpool, Eng., Sept. 1, 1835; s. Thomas and Mary Anne (Roscoe) J.; ed. U. Coll., London, A.M., 1862; hon. doctorate, U. Edinburgh, 1875; m. Harriet Ann Taylor, Dec. 19, 1867. Assayer at mint, Sydney, Australia, 1854-59; tutor, lectr., later prof. Owens Coll., Manchester, Eng., 1863-76; prof. U. Coll., 1876-80, also examiner in logic, mental and moral philosophy. Fellow Royal Soc., 1876. Author: The Coal Question, 1865; Lessons in Logic, 1870; Theory of Political Economy, 1871; The Principles of Science, 1874; Money and the Mechanism of Exchange, 1875; A Primer on Political Economy, 1878; Studies in Deductive Logic, 1880; The State in Relation to Labour, 1882; Investigations in Currency and Finance, 1884; The Principles of Economics, 1905; also articles. Analyzed the dependence of value upon utility; pioneer proponent of marginal theory of value; furthered use of math. and statistics in econ. science; speculated on relationship between comml. crises and sun-spots; developed theories in logic on probability, induction and their relation. Died nr. Hastings, Eng., Aug. 13, 1882.

JEWETT, Frank Baldwin, Am. electrical engr.; b. Pasadena, Calif., Sept. 5, 1879; s. Stanley P. and Phebe (Mead) J.; A.B., Throop Poly. Inst. (now Calif. Inst. Tech.), 1898; Ph.D., U. of Chicago, 1902; D.Sc., New York U., Dartmouth, 1925, Columbia Univ. and Univ. of Wis., 1927, Rutgers University, 1928, U. of Chicago, 1929, Harvard, 1936, Univ. of Pa., 1940, Boston U., 1944; Dr. Eng., Case School of Applied Science, 1928; LL.D., Miami U., 1932, Rockford College, 1939; Norwich Univ., 1944, Yale, 1946; married Fannie C. Frisbie, Dec. 28, 1905; children—Harrison Leach, Frank Baldwin. Research assistant to Professor A. A. Michelson, Univ. of Chicago, 1901-02; instructor physics and electrical engineering, Mass. Institute of Technology, 1902-04; transmission engr. Am. Telephone & Telegraph Co., 1904-12; asst. chief engr., 1912-16, chief engr., 1916, vice-pres., 1922, Western Electric Co.; v.p. Am. Telephone & Telegraph Co., in charge development and research 1925-44; pres. Bell Telephone Laboratories, Inc., 1925-Oct. 1, 1940, chmn. bd. 1940-44. Maj. Signal Corps, U.S.R., 1917; lt. col. Signal Corps, A.U.S., Dec. 1, 1917; was advisory mem. Spl. Submarine Bd. of the Navy and mem. State Dept. Spl. Com. on Cables. Vice-chmn. Engring. Foundation, 1919-25; chmn. Div. of Engring. and Industrial Research, Nat. Research Council, 1923-27, now mem. com. on scientific aids to learning; mem. President Roosevelt's Science Advisory Board, 1933-35; mem. Nat. Defense Research Committee of Office of Scientific Research and Development; mem. coordination and equipment division, Signal Corps; consultant to Chief of Ordnance. President, Nat. Academy of Sciences, since 1939. Pres. and trustee, New York Museum of Science and Industry. Life mem. Mass. Inst. Tech. Corp.; pres. M.I.T. Alumni Association, 1939-40. Trustee Princeton U., Carnegie Instn. of Washington, Woods Hole Oceanographic Inst. Tabor Academy, Carnegie Inst. of Tech. Fellow Inst. Radio Engrs., A.A.A.S., Am. Physical Soc., Acoustical Soc. of Am., Acad. Arts and Sciences; mem. Am. Inst. Electric Engrs. (pres. 1922-23), Inst. Elec. Engrs. (Brit.), Am. Soc. for Engring. Edn., Am. Philos. Soc.; member (hon.) N.Y. Electrical Society, Sigma Xi, Tau Beta Pi. Awarded Edison medal, 1928; Faraday medal, 1935; Franklin medal, 1936; Washington award, 1938; John Fritz medal, 1939; Medal for Merit, 1946. Author brochures, articles and pub. addresses on physical and elec. subjects. Best known for advancements in fields of transcontinental and trans-oceanic telegraphy, aircraft communications, radio and TV. Died Nov. 18, 1949.

JEWETT, Frank Fanning, Am. chemist; b. Newton Corner, Mass., Jan. 8, 1844; s. Charles and Lucy Adams (Tracy) J.; A.B., Yale, 1870; A.M., 1873; univs. of Göttingen and Berlin, Germany; m. Frances Gulick, July 30, 1880. Tchr. Norwich Free Acad., later pvt. asst. to Dr. Wolcott Gibbs, of Harvard; prof. chemistry, Imperial U. Japan, 1877-80; prof. chemistry and mineralogy Oberlin Coll., 1880-12, prof. emeritus, 1912. Trustee Oberlin Missionary Home Assn. Mem. Am. Chem. Soc., A.A.A.S., Deutsche Chemische Gesellschaft, Alpha Delta Phi. Author: Tables for Qualitative Chemical Analysis, 1883; Laboratory Manual of Inorganic Chemistry, 1885; Laboratory Manual in General Chemistry, 1885; (with Frances Jewett) The Chemical Department of Oberlin College, from 1883-1912, 1922. Closely identified with devel. aluminum industry; established prestige of chemistry dept. Oberlin Coll.; one of his pupils (Charles M. Hall) discovered electrolytic method of separating aluminum from its various compounds. Died July 1, 1926.

JEWKES, John, English economist; b. Lancashire, Eng., June 29, 1902; s. John and Fanny (Cope) J.; M. Commerce, U. Manchester (Eng.), 1922; M.A., Oxford (Eng.) U.; m. Frances Sylvia Butterworth, Aug. 14, 1929; 1 dau., Ann (Mrs. Bryan Clarke). Rockefeller Found. fellow, 1929-30; prof. social econs. Manchester U., 1936-46, prof. polit. economy, 1946-48; prof. econ. orgn. Oxford U., 1948—, fellow Merton Coll., 1948—; vis. prof. U. Chgo., 1953-54, Princeton, 1961. Mem. Royal Commn. on Doctors Remuneration, 1957. Decorated Comdr. Order of Brit. Empire. Fellow Royal Econ. Soc. Author: Ordeal by Planning, 1948; Sources of Invention (with D. Sawers, R. Stilleman), 1958; Public and Private Enterprise, 1965; also articles. Study of history of invention and part played by sci. in econ. growth. Home: Entwood, Red Copse Lane, Boars Hill, Oxford, Eng.*

JEZKOVA, Zdenka, Czechoslovakian bacteriologist, immunologist; b. Vimperk, Czechoslovakia, Mar. 9, 1921; d. Emanuel and Marie Jezková (Polanková) J.; Phil.D. in Pharmacy, Brno, Czechoslovakia, 1961. Head bacteriological control biol. products Transfusion Sta., Banská Bystrica and Kosice, Czechoslovakia, 1952-54; substitute Inst. Hematology and Blood Transfusion, Prague, Czechoslovakia, 1955; head bacteriological and immunological dept. Inst. Hematology and Blood Transfusion, Prague, 1955—. Mem. Med. Soc. J.E. Purkyne (award diabetic sect. 1963). Research and numerous publs. on bacteriological control of biol. preparations, evaluation of medium, microbes growth curves in transfusion preparations, immunologic studies tissue antibodies at different diseases. Home: 15 Leger, Prague. Office: 1 U nemocnice, Prague, Czechoslovakia.*

JILEK, Lubor, Czechoslovakian physiologist; b. Prague, Czechoslovakia, Jan. 8, 1926; s. Frant. and Ludmila (Hemerova) J.; M.D., Charles U., Prague, 1951, Ph.D., 1958, D.Sc., 1965; m. Helena Civínová, Apr. 17, 1951; 1 son, Ondrej. Med. faculty, Charles U., Prague, 1951—, prof. med. physiology, 1966—, vice dir. Inst., 1959—, vice dean Med. Faculty, 1961—; vis. scientist Galesburg (Ill. State Research Hosp., 1964. Mem. Czechoslovac Physiol. Soc. (mem. com. 1958—). Author: Stagnant Hypoxia and Anoxia of the Brain, 1966; also numerous articles. Research on metabolical functional and structural devel. of nervous system during ontogeny, reaction and adaptation of nervous tissue to hypoxia and anoxia in course of post-natal life, reversibility and prognosis of anoxic changes of brain, influence of different agts. on anoxia. Home: 6 Havanska, Prague 7, Czechoslovakia.*

JIMENO GIL, Emilio, Spanish chemist; b. Calatayud, Mar. 21, 1886; s. Antonio and Elena (Gil) J.; ed. U. Zaragoza, U. Madrid; Ph.D. in chemistry; Dr. Eng. honoris causa, U. Hanover; m. Irene Esteve, Dec. 24, 1921. Prof. inorganic chemistry U. Oviedo, U. Barcelona, U. Madrid. Recipient Francisco Franco prize, 1941. Fellow Royal Soc. Edinburgh (hon.); mem. Nat. Assn. Chemists Spain (hon.), Royal Acad. Sci. Madrid, Inst. Solder, Assn. Argentine Chemists (corr.). Author: Metalografía, 1922; Química general, 1941; El problema de la corosión metalica, 1947; Ciencia y sociedad, 1952; Metalurgía general, 1955. Address: calle Marques de Urquijo 34, Madrid, Spain.

JINKS, John Leonard, English geneticist; b. Stoke-on-Trent, Eng., Oct. 21, 1929; s. Jack and Beatrice (Holtom) J.; B.Sc. with honors, in Botany, U. Birmingham (Eng.), 1950, Ph.D. in Genetics, 1952, D.Sc., 1964; m. Diana Mary Williams, Aug. 3, 1954; children—Peter John, Jane Catherine. Sci. officer Agrl. Research Council unit biometrical genetics, 1953-56, sr. sci. officer, 1956-60, prin. sci. officer, 1960-65; faculty U. Birmingham, 1956—, prof. genetics, head dept., 1965—. Mem. Genetical Soc. Gt. Britain, Soc. Gen. Microbiology, Mycol. Soc. Author: Extrachromosomal Inheritance, 1964; also numerous articles. Devel. analytical techniques for studying genetic difference of econ. importance and applying them to problem of breeding improved varieties of plants; research on animal behaviour, variation in microorganisms especially fungi; inheritance other than by nuclear gene mechanism. Home: 81 Witherford Way, Birmingham, Warwickshire, Eng.*

JINNAI, D(ennosuke), Japanese surgeon; b. Fukuoka, Japan, Nov. 8, 1912; s. Dempachi and Mineko (Kitahara) J.; M.D., Kyushu U., Fukuoka, 1936, Ph.D., 1944; m. Sachiko Matsuo, May 1, 1938; children—Motoko (Mrs. Takahiro Oka), Shinnosuke, Kônosuke. Asst., Kyushu U. Hosp., 1942-47; prof. surgery Okayama (Japan) U. Med. Sch., 1948-62; prof. surgery Osaka (Japan) U. Med. Sch., 1962—; dir. surg. dept. Osaka U. Hosp., 1962—. Fellow Internat. Congress Neurol. Surgery; mem. Internat. Coll. Surgeons (Asian fedn. sec.), Internat. Symposia Stereoencephalotomy. Author: (with T. Ogata, F. Kosaka) Pre- and postoperative Care and Complications, 1963; Pre-and Postoperative Care in Neurosurgery, 1966; also numerous articles. Research on epilepsy; advocated allergic genesis of epilepsy; invented stereotaxic destruction of Forel-H-field as treatment for epilepsy, surg. technique of extended radical operation for gastric cancer. Home: 108 Hon-machi 6, Toyonaka, Osaka. Office: 1, Dojimahama-dori 3, Fukushima, Osaka, Japan.*

JIRASEK, Jan Evangelista, physician; b. Pardubice, Bohemia, July 26, 1929; s. Jaroslav and Marie (Kuncova) J.; M.D., Charles U., 1953, C.Sc., 1960; m. Vera Polakova, Feb. 24, 1954; 1 dau., Jana. Asst., Inst. Histology and Embryology, Med. Faculty, Hradec Kral, 1950-51, Inst. Embryology, Med. Faculty, Prague, Czechoslovakia, 1953-60; chief Lab. Embryology and Histochemistry, Inst. for Care Mother and Child, Prague, 1960—; faculty, cons. Clinic of Obstetrics and Gynecology, Postgrad. Med. Sch., 1960—. Mem. Czechoslovakian Med. Soc., Anatomical Soc., Pathol. Anatomical Soc., Gynecol. Soc. Author: Vogelwelt am Wasser, 1963. Contbr. numerous articles to med. jours. Research on histogenesis of human gonads and diagnosis of intersexual malformations in human beings; described localization of various enzymes, eunuchoid form of testicular feminization, research on histochem. analysis of human embryos and their decidual membranes. Home: 1851 V. Rybnickách. Office: 157 Nabr k. Marxe, Prague, Czechoslovakia.*

JIRGENSONS, Bruno, chemist; b. Adazhi, Latvia, May 16, 1904; s. Fridrichs and Anna (Menca) J.; Mag. chem. University of Latvia, 1926, Dr. chem., 1933, Dr. chem. habil., 1934; m. Anna Resnais, June 1, 1927; children—Juris Bruno, Andris. Faculty U. Latvia, Riga, 1926-41, prof., 1940-41; asso. prof. United Nations Relief & Rehab. Adminstrn. U. Munich, Germany, 1946-47; Imperial Chem. Industries Research fellow Manchester U., (Eng.) 1947-49; prof. chemistry Tex. Lutheran Coll., 1949-52; asso. biochemist U. Tex. M.D. Anderson Hosp. and Tumor Inst., 1952-57, biochemist, 1957—; prof. biochemistry U. Tex. Grad. Sch. Biomed. Scis., 1964—, chief sect. protein structure; mem. editorial bd. internat. jour. Die Makromolekulare Chemie. Mem. Am. Chem. Soc., Am. Soc. Biol. Chemists. Author (with M. E. Straumanis) Colloid Chemistry, 1954, Organic Colloids, 1958, Natural Organic Macromolecules, 1962. Research, numerous publs. on polymers, proteins (contbr. to elucidation of immunoglobulins, Bence-Jones proteins, myeloma globulins); co-discoverer Lutz-Jirgensons rule of amino acid configuration. Home: 6022 Fordham St., Houston 77005.*

JIROUT, Jan., Czechoslovakian physician; b. Praha, Czechoslovakia, June 3, 1912; s. Jan and Emilie (Balcarova) J.; grad. Charles U., Prague, 1937, D. Med. Scis., 1959; m. Milena Pokorná, Jan. 1, 1946; children—Milena, Jan. Staff Neurol. Clinic, Charles U., Prague, 1937—, head roentgen dept., 1942—, prof. neurology, 1958—. Mem. Neurol. Soc., Roentgen Soc., Med. Soc. J. E. Purkyn. Author: Encefalografie, 1948; Obecná Neuroradiologie, 1953; Speciální neuroradiologie, 1956; Pnévmomielografija, 1964. Neuroradiologie, 1966; Pneumomyelography, 1967; also articles. Research on therapy of multiple sclerosis, pneumographic studies of spinal canal, spinal cord, and meninges, (Roentgen studies of the spine and its movements. Home: Praha 2, Na Moráni 7, Czechoslovakia. Office: Neurologic Clinic, Praha 2, Katerinská, 30, Czechoslovakia.*

JIROVEC, Otto, biologist, Czechoslovakian parasitologist; b. Prague, Czechoslovakia, Jan. 31, 1907; s. Frantisek and Otylie (Merhautová) I.; D.Sc., Faculty of Natural Scis., Charles U., Prague, 1929, D. honoris causa Med. Faculty in Berlin, 1960, Faculty Scis. in Clermont-Gerrand, France, 1963; m. Marie Spacková, Apr. 24, 1933; children—Jana, Jirí, Pavel. Asst., Zool. Inst., Charles U., 1930-39, asst. prof. parasitology Faculty Natural Scis., 1947—; chief parasitological lab. Institut Hygiene, Prague, 1940-46. Recipient State prize, 1953. Mem. Czechoslovak Acad. Scis., Parasitological Soc. in Czechoslovakia (pres. 1958—), Polske towarzystwo parazytologiczné (hon.), Parasitolog. Gesellschaft der DDR (hon.), Proctol. Soc. U. S. A. (hon.). Author: Parasitologie für Ärzte, 1960. Protozoologie, 1953; Life under the Microscope, 1956; Zoologická technika, 1942, 47, 52; also numerous articles. Research on various molds and parasites in man; discovered pneumocystis carinii parasitic agt. from interstitial plasmacellular pneumonia in man. Home: Praha-Hodkovicky, Nad údolím 14. Czechoslovakia. Office: Zool. Institut, Praha 2, Vinicná 7, Czechoslovakia.

JOACHIM, Georges (or Rheticus), astronomer; b. Feldkirchen, Austria, 1514; tchr. math., Wittenberg; follower, asst. to Copernicus, from 1539; Author: Narratio de libris revolutionum Copernici, 1540-51; Orationes de astronomia, 1542; Ephemeris ex fundamentis copernici, 1550; Opus palatinum de triangulis (trigonometric tables with sines calculated to 10 places, pub. by follower Valentin Othon), 1596; Thesaurus mathematicus (tables calculated to 15 decimal places, pub. by Pitiscus), 1613. Attacked Ptolemaic system. Died Kassa, Hungary, 1576.

JOACHIMSTAHL, Ferdinand, German mathematician; b. Goldberg, Silesia, May 9, 1818; pupil of Jacobi; Ph.D., Berlin, 1842; became tchr. Realschule, Berlin, 1842; names pvt. docent, philos. faculty Berlin U., 1846; apptd. prof. math., Halle, Germany, 1856, Breslau (now Wroclaw, Poland), 1858. Contbr. articles to jours. Research in analytic geometry, calculus, theory of functions; developed theorem on lines of curvature of a surface. Died Breslau, Apr. 5, 1861.

JOANNES ACTUARIOS, Byzantine physician; s. Zacharias; flourished Constantinople, circa 1328-41; court physician under Andronicos III Palaeologus in Constantinople. Author: Methodus medendi, Treatise on urine (most elaborate Greek treatise on urine, includes physiology, diagnosis, aetiology, prognosis), 7 parts; De actionibus et affectibus spiritus animalis huisque nutritione. (important in history of pneumatism and psychopathology). Gave early description of pin worm; also probably 1st description of paroyxsmal hemoglobinuria.

JOANNES DE JANUA, see John of Genoa.

JOANNES DE SANCTO AMANDO, see John of Saint Amand.

JOANNES PHYSICUS, see John of Glogau.

JOBERT DE LAMBALLE, Antoine Joseph, French surgeon; b. Matignon, France, Dec. 17, 1799; ed. Paris; became surgeon, Paris, 1829, aggregate prof., 1830; named cons. surgeon to king, 1831; apptd. surgeon to emperor, 1852; named prof., faculty medicine, 1854. Mem. Acad. Medicine, French Acad. Scis., 1856. Author: Sur les hemorroides, 1828; Traité theorique et practique des maladies chirurgicales du canal intestinal, 1829; Traité des plaies d'armes à feu, 1833; Études sur le système nerveux, 2 vols., 1838; Recherches sur les appareils electriques des poissons, 1858; Traité de chirurgie plastique, 2 vols., 1849; De la reunion en chirurgie, 2 vols., 1864. First to use ether in France, 1846; contbd. to therapeutics of uterine diseases; inventor cytoplasty; developed method of intestinal invagination still used; credited with radical cure of fistula of vesicovagina by autoplasty (electroplasty). Died Paris, Apr. 25, 1867.

JOBLOT, Louis, French bacteriologist; b. Bar-le-Duc, France, 1645; became asst. prof. math., bacteriology, geometry; perspective Royal Acad. Painting and Sculpture, Paris, 1680, prof., 1699-1721; invented and built microscopes with which he examined animalcules and infusoria; opposed theory of spontaneous generation; showed that infusoria are absent from infusions that have been boiled and kept from air. Died 1723.

JOCHIMS, Johannes Christian, German physician; b. Hamburg, Sept. 9, 1899; s. Johannes and Bertha (Schmidt) J.; ed. univs. Hamburg, Friburg, Munich, Berlin; M.D.; m. Elisabeth Menck, Oct. 6, 1926; children—Johanns, Karsten, Reimer, Wiebke, Silke. With Kaiser Wilhelm Inst. Phys. Chemistry, Hamburg, 1923, Inst. Pathology, Friburg, 1925-26, Children's Clinic, Kiel, 1927-42; instr. Children's Clinic, Lübeck, 1930, prof., 1936, chief physician, 1942—. Cons., Assn. for Protection Milk Products. Mem. German Soc. Children's Medicine. Author: Darmbrand, 1949; Praxis der antiboit. Behandlung im Kindesalter, 1951; Bedeutung der Nahgrungsfette in der Padiatrie, 1961. Home: Elsasserstrasse 41. Office: Kronsforderallee, 71-73, Lübeck, Germany.

JOERG, Eduard, German physician; b. Leipzig, Germany, 1808; M.D., 1852. Author: Darstellung des Nachthéilinger Einfluss es des Tropentilimas, 1851; Anweisung die Tropenrankheiten, 1854; Die gaenzliche Underdruchung der asiatischen Cholera, 1855. Described pulmonary atelectasis in infants and showed its importance in determining whether or not air has entered the lungs of a dead newborn. Died 1878.

JOERG, W(olfgang) L(ouis) G(ottfried), Am. geographer; b. Brooklyn, N.Y., Feb. 6, 1885; s. Oswald and Denise (Coulin) J.; grad. Poly. Prep. Sch. of Brooklyn, 1899, Thomas Gymnasium, Leipzig, Germany, 1904; student U. of Leipzig, 1904, Columbia, 1904-06, U. of Göttingen, Germany, 1906-11; m. Hannah Heaton, Nov. 14, 1911; children—Oswald Heaton, Norton Coulin. Mem. scientific staff Am. Geog. Soc. 1911-37—asst. editor of Bull., 1911-15, asso. editor Geog. Review, 1915-20, editor research series, 1920-25, research editor, 1925-37; apptd. chief Div. of Maps and Charts, The Nat. Archives, 1937; member Federal Bd. of Surveys and Maps, 1937-42; mem. div. of geology and geography Nat. Research Council, 1924-27, and of exec. com., 1931-36, vice chmn., 1933-36; sec. U. S. nat. com. Internat. Geog. Union, 1931-37, chmn., 1937-39; mem. advisory com. U. S. Geog. Bd., 1931-34, Com. on Mapping Services of the Federal Govt., 1934; mem. U. S. Bd. on Geog. Names, 1937-47 (chmn. exec. com. 1938-47, com. on Antarctic names, 1943-47, adv. mem. since 1947; cons. prof. hist. geog., Univ. of Maryland, since 1947. Fellow A.A.A.S.; mem. Assn. Am. Geographers (v.p. 1928; pres. 1937), Am. Geophys. Union, Am. Soc. of Photogrammetry, Nat. Congress on Surveying and Mapping (charter mem. 1941), Arctic Inst. of North American (charter mem. 1948), Soc. Am. Archivists, Am. Assn. for State and local history; hon. mem. Geog. Soc. Neuchatel (Switzerland). Unitarian. Club: Cosmos (Washington, D.C.). Author: Recent Geographical Work in Europe, 1922; Brief History of Polar Explorations since the Introduction of Flying, 1930; Work of the Byrd Antarctic Expedition of 1928 to 1930, 1930; and numerous papers on cartography, the regional geography of N. America and land utilization. Editor: Problems of Polar Research, 1928; Pioneer Settlement 1932; (with W. A. Mackintosh) Canadian Frontiers of Settlement, 9 vols. since 1934; contbg. editor Geographical Review since 1937. Died Jan. 7, 1952.

JOEST, Ernst, German bacteriologist; b. Wallefeld, Germany, 1873; became prof. at Dresden (Germany), 1904, at Leipzig (Germany), 1923. Author: Handbuch der Speziellen pathologischen Anatomie der Haustiere, 5 vols., 1919-29. Died Leipzig, 1926.

JOFFE, Abram Feodorovich, Ukrainian physicist; b. Romny, Ukraine, Oct. 29, 1880; student Leningrad (Russia) Tech. Inst.; Ph.D., U. Munich (Germany); Dr. honoris causa univs. Graz (Austria), Bucharest (Rumania). Joined Polytech. Inst. Leningrad, as asst., 1906; prof. physics from 1913. Also dean phys.-mech. faculty; dir. Phys.-Tech. Inst. State X-ray and Radiology Inst., Leningrad, 1918-51; dir. phys.-Tech. Inst., USSR, 1951-54; dir. Inst. Semi-Conductors, USSR, from 1954; dir. Phys.-Agronomical Inst. from 1932; mem. USSR Acad. Sci. (v.p. 1942). Recipient Stalin prize, 1942, 3 orders Lenin, Hero Socialist Labor award. hon. fgn. mem. Am. Acad. Sci., Indian, Brit. French, Chinese Phys. Socs.; fgn. mem. Acad. Lincei, Rome, Italy; mem. (corr.) Göttingen, Berlin (Germany) Acads. Sci. Author: Elementary Photo-Electric Effect; Magnetic Field of Cathode Rays, 1913; Lectures on Molecular Physics, 1923; Physics of Crystals, 1929; Semi-Conductors in Physics and Engineering, 1940; My Life and Work, 1933; Physics of Semi-Conductors, 1955; Semi-Conductor Thermo-Elements, 1958. Research on mech. properties crystals di-electrics and semi-conductors; important findings elasticity crystals; proved existence of magnetic field of cathode rays. Died Romny, Ukraine, USSR, Oct. 14, 1960.

JOFFE, Joseph, chem. engr.; b. Moscow, USSR, Oct. 14, 1909; s. Lew Moses and Sophie (Joffe) J.; came to U. S., 1921, naturalized, 1926; A.B., Columbia, 1929, B.S., Engring. Sch., 1930, M.A., 1931, Ph.D., 1933; m. Bertha Pashkovsky, June 20, 1931; children—Robert, Paul, Richard. Faculty, Newark Coll. Engring., 1932—, prof. chem. engring., 1940—, chmn. dept. chem. engring., 1963—, dir. research, 1961-63; chem. engr. Esso Research & Engring. Co., Florham Park, N.J., summers 1950—. Mem. Am. Inst. Chem. Engrs., Am. Chem. Soc., Am. Soc. for Engring. Edn., N.J. Edn. Assn. Editor (translation): Palatnik Landau, Phase Equilibria in Multicomponent Systems, 1964. Research, publs. on phys. and thermodynamic behavior gases and gas mixtures.*

JOFFROY, Alex, French physician, neurologist; b. Stainville, France, 1844; became physician of hosps., 1879, instr., 1880; named prof. Clinic for Mental Diseases as mem. Paris Faculty, 1893; mem. Acad. Medicine. Asst. dir. Archives de physiologie normale et pathologique, then Archives de médecine expérimentale et d'anatomie pathologique. Described some nervous and muscular symptoms named after him; discovered (with Jean Martin Charcot) lesions of spinal cord in muscular atrophy, atrophy of anterior horns of spinal cord in poliomyelitis, 1869; wrote on dangers of alcohol; opposed Fournier's ideas on relation between syphilis and gen. paralysis. Died 1909.

JOHANNES AKTUARIOS, see Joannes Actuarios.

JOHANNES DE MIRFELD, see John of Mirfeld.

JOHANNES HISPALENSIS (or Juan de Sevilla, Avendeut), translator; flourished 12th century; apparently translated from Arabic into Castilian in connection with Dominicus Gundisalvus who translated the Castilian into Latin; attributed to them are one arithmetic work, 12 astron. or astrological works, 1 med. work, 7 philos works, including Epistula Aristotelis ad Alexandrum de conservatione corporis humani (extract from Secreta secretorum), astron. and astrological writings of Masallah, al-Fargani, al-Hayyat, al-Balhi, al-Kindi, al-Tabari, al-Battani, Tabit ben Qurra, al-Qabisi, philos. works of al-Kindi, Qusta ben Luqa, al-Farabi, Inb Sina, Ibn Gabirol, al-Gassali, possibly al-Khowarizmi's algebra.

JOHANNSEN, Albert, Am. geologist; b. Belle Plaine, Ia., Dec. 3, 1871; B.S., U. Ill., 1894, U. Utah, 1894; Ph.D., Johns Hopkins 1903; m. 1904; 2 children. Geology asst. Md. Geol. Survey, 1901-03; geologist U. S. Geol. Survey, 1903-25, acting chief petroleum sect., 1907-10; asst. prof. petroleum U. Chgo., from 1909. Petrografo, Geology Inst. Mexico, 1919—. Mem. Geol. Soc., Nat. Acad. Sci. Mexico. Author: Determination of Rockforming Minerals, 1908; Manual of Petrographic Methods, 1914; A Descriptive Petrography of the Igneous Rocks, 4 vols., 1931-38. Address: Box 566, Winter Park, Fla.

JOHANNSEN, Wilhelm Ludwig, Danish botanist; b. Copenhagen, Denmark, Feb. 3, 1857; ed. Helsinki, Finland; prof., Copenhagen. Author: Über Erblichkeit in Populationen und in reinen Linien, 1903. Pioneer in modern exptl. genetics; introduced words genotypes, phänotypes, other simplified terminology into genetics; developed pure-line theory after expts. with Princess beans. Died Copenhagen, Nov. 11, 1927.

JOHANOVSKY, Jirí, Czechoslovakian microbiologist, immunologist; b. Prague, Czechoslovakia, Feb. 27, 1925; s. Karel and Marta (Krupicková) J.; MUDr., Charles U., Prague, 1950; m. Dagmar Vondrová, May 24, 1952; children—Dagmar, Ludmila. Staff, Inst. Biology, Czechoslovak Acad. Scis., 1951-59; research dir. Inst. Sera and Vaccines, 1959-65; dir. Research Inst. Immunology, Prague, Czechoslovakia, 1966—; lectr. immunology Med. Faculty, Prague, 1960—. Mem. expert panel on biol. standardization WHO, 1964—; chmn. research com. for immunology Czechoslovak Ministry Health, 1960—. Mem. Brit. Soc. for Immunology, Royal Soc. Medicine (London), Czechoslovak Med. Soc. J. E. Purkyne. Research, publs. on pathogenesis and immunology of staphylococcal infections, mechanism of delayed (cellular) hypersensitivity, especially tissue culture and systemic pyrogenic reaction, role of mediator substances; pub. health aspects in vaccine research. Home: 2498 Sporilov II, Praja 4. Office: 108 W. Pieck Praha 10, Czechoslovakia.*

JOHANSEN, Hans Christian, ornithologist, zoogeographer; b. Riga, Russia, Dec. 2, 1897; s. Jens Christian and Ingeborg (Laudrup) H.; grad. U. Tomak, Siberia, 1921; Ph.D., U. Munich, 1924; m. Agnia Tabokova, May 5, 1919; 1 dau., Ljudmila (Mrs. Jorgen Svane). Prof., U. Tomak, 1925-37, German U. Riga, 1938-39, U. Konigsberg, 1940-44; ornithologist Zool. Mus., Copenhagen, 1945—. Decorated Ridder of Danebrog, 1961. Mem. numerous ornithol. socs. Author: Der Baikalsee, 1925; Strana Zverel, 1935; Revision and Origin of the Arctic Birdfauna, 1956; also numerous articles. Zoogeograph. research on birds of N.Am., S.Am., Siberia, Arctic. Home: 8 Skovsvinget, Lyngby, Denmark. Office: Zoological Museum, 15 Universperken, Copenhagen, Denmark.*

JOHANSEN, Kjell, physiologist; b. Oslo, Norway, Sept. 30, 1932; s. Ivar and Karen (Johannessen) J.; D.Phil., U. Oslo, 1962; m. Hermana A. Knudtzon, Mar. 18, 1958; children—Tonje, Kjetil. Research fellow U. Oslo, 1955-58, 61-62; postdoctoral research U. Wash., Seattle, 1959; research physiologist Arctic Aeromed. Lab., Fairbanks, Alaska, 1960; faculty U. Wash., 1961—, asso. prof. zoology, 1965—. Established investigator Am. Heart Assn., 1965. Fulbright grantee, 1959-60. Mem. Scandinavian Soc. Biochem. Physiology, Am. Physiol. Soc. Research, numerous publs. on comparative physiology of heart and circulation. Home: 6530 51st Av. N.E., Seattle 98115.*

JOHANSEN, Sophus Hauberg, Danish anesthesiologist; b. Copenhagen, Denmark, Aug. 23, 1921; s. Andreas and Margaret (Hauberg) J.; ed. Finnish Med. Sch., U. Copenhagen, Finsen Inst., Gentofte Hosp., Copenhagen. Jr. asso. in anesthesia Peter Bent Brigham Hosp., Boston, 1958-59; instr. anesthesia Mass. Gen. Hosp., Boston, 1959-60, vis. lectr., 1965; chief, anesthesia dept. II, Gentofte Hosp., 1960—; apptd. instr. Anesthesiology Center of Copenhagen, 1961. Mem. Danish, Scandinavian anesthesiological socs. Editor: Barbiturate Poisoning and Tetanus. Research and publs. on immunizing agts., their interaction and effect on respiratory system in animals, including man, also on respiratory response to analgesics. Home: 3 Vemmetofte Alle, Gentofte, Denmark. Office: Dept. of Anesthesia, Gentofte Hospital, Copenhagen, Denmark.*

JOHANSSON, Arne Semb, Norwegian zoologist; b. Oslo, Norway, Sept. 20, 1919; s. Fredrik Wilhelm and Astrid (Semb) J.; student U. Oslo, 1939-47, Dr.Phil., 1958; postgrad. Cornell U., 1948-49, U. Colo., 1954-55; m. Eva Borgen, Dec. 27, 1946; children—Pal, Tom. Faculty, U. Oslo, 1948—, prof., 1959—, head dept. zoology, 1963—. Norwegian del. Nordic Council for Terrestrial Ecology, 1962—. Recipient King's medal for Courage (Eng.); King's Gold medal (Norway). Mem. Norwegian Acad. Sci. Research, publs. on insect endocrinology and population studies of Norwegian lemming. Home: 4 Blindernv., Oslo, Norway.*

JOHANSSON, Bengt Wilhelm, Swedish physician; b. Lund, Sweden, Feb. 28, 1930; s. Nils V. and Ulla (Karlsson) J.; M.D., U. Lund, 1957; m. Ulla Margareta Petersson, June 5, 1954; 1 dau., Ulla Birgitta. Asso. prof. Heart Lab. Gen. Hosp., Malmö, Sweden, 1967—. Mem. Swedish Cardiological Soc. Author: Complete Heart Block, 1966; also numerous articles. Research on cardiology, especially disorders of heart beat, hibernation, hypothermia. Home: 12 Ferievägen, Lund, Sweden. Office: Heart Lab., Gen. Hosp., Malmö, Sweden.*

JOHANSSON, Ingebrigt, Norwegian mathematician; b. Narvik, Norway, Oct. 24, 1904; s. Isak and Gjertrud (Kletten) J.; student U. Oslo (Norway), 1923-28, Dr.Phil., 1931; m. Gidske Schult, Apr. 8, 1941; children—Per, Jan. Scholar, U. Oslo, 1931-36; faculty U. Oslo, 1936—, prof. math., 1942—, dir. Inst. Math., 1948-66. Mem. Nordic Com. for Modernization Sch. Math., 1960—. Mem. Norwegian Acad. Research, publs. on surface topology containing proof that every subgroup of a free group is free, minimal calculus, a reduced calculus for intuitionistic logic; studies on Dehn's lemma for a singular disc; geometric representation of quaternions. Home: 4 Blindernveien, Oslo 3, Norway.*

JOHANSSON, Ivar Karl, Swedish animal geneticist; b. Vislanda, Sweden, Nov. 7, 1891; s. J. August and Eva Johansson; B.Sc.Agr., Alnarp Agr. Coll., 1918; M.S., U. Wis., 1926, Ph.D., 1930; Dr.agr. h.c. U. Kiel (Germany), 1951; Agr. Dr. h.c., U. Uppsala (Sweden), 1953; Dr. Agr. et Forestry, U. Helsinki (Finland), 1958; m. Ingrid Bolin, June 18, 1936 (dec. 1937); 1 dau., Ingrid. Tchr. animal husbandry Hvilan Agr. Sch., Sweden, 1918-23; acting prof. animal husbandry Ultuna Agr. Coll., 1926-32; asst. prof. genetics U. Wis., 1932-33; prof. animal genetics Agr. Coll. Sweden, Uppsala, 1933-58, asst. dean, 1950-55; vis. prof. dairy sci. U. Ill., 1958-59, Ia. State U., 1959. Cons. in animal genetics Swedish Fur Breeders Assn., 1960—. Mem. Royal Swedish Acad. Agr. and Forestry (Nilsson-Ehle gold medal in genetics 1966, hon.), Norwegian Acad. Sci., Royal Soc. Sci. Uppsala. Author several books on animal prodn. including: (with Jan. Rendel) Genetics and Animal Breeding, 1967; also numerous articles. Co-editor: Handbuch der Tierzuchtung, vols. I, 1958, vol. II, 19—, vol. III. Research on inheritance of qualitative and quantitative traits in cattle, sheep, pigs, foxes and mink, re-productive physiology of foxes and mink. Home: 36 Artillerigatan, Uppsala, Sweden.*

JOHANSSON, Karl Richard, Am. microbiologist; b. Bay City, Mich., June 28, 1920; s. Karl G. and Elizabeth (Sherwood) J.; B.S., U. Wis., 1942, M.S., 1946, Ph.D., 1948; m. Dorothy Alice Heilig, May 27, 1943; children—Sandra Lee, Peter Gustaf, Steven Herb. Instr. dairy microbiology U. Cal., Davis, 1948-49; faculty U. Minn. Med. Sch., 1949-59, asso. prof. bacteriology, 1955-59; exec. sec. virology and rickettsiology study sect. NIH, Bethesda, Md., 1959-61, 65—; asso. prof. environmental health engrng. Cal. Inst. Tech., 1961-63; chief research grants Nat. Inst. Neurol. Diseases and Blindness, NIH, 1963-65. Cons. Gen. Mills, Mpls., 1952-54, Honeywell Corp., Mpls., 1957-59. Mem. Am. Soc. Microbiology, Am. Acad. Microbiology, Am. Inst. Nutrition, Soc. Gen. Microbiology, A.A.A.S., Soc. Exptl. Biology and Medicine, Sigma Xi. Author: (with Walter Krueger) Principles of Microbiology, 2d edit., 1959. Research, publs. on interrelationships and ecology of microorganisms; effects of antibiotics. Home: 10305 Dickens Av. Office: NIH, Bethesda, Md. 20014.*

JOHN, Fritz, mathematician; b. Berlin, Germany, June 14, 1910; s. Hermann and Hedwig (Buergel) Jacobsohn; Ph.D., Goettingen (Germany) University, 1933; student Cambridge (Eng.) University, 1934-35; m. Charlotte Woellmer; children—Thomas Franklin, Charles Frederic. Came to U. S., 1935, naturalized, 1941. Asst., then asso. prof. U. Ky., 1935-42; mathematician Aberdeen Proving Grounds, 1942-45; prof. math N.Y. U., 1946—; dir. Research Inst. Numerical Analysis, Nat. Bur. Standards, 1950-51; spl. research applied math., math. analysis. Rockefeller fellow, 1935, 42; Fulbright lectr., Goettingen U., 1935; Guggenheim travel grantee, 1963. Mem. Nat. Acad. Scis., Am. Math. Soc., A.A.A.S., Math. Assn. Am., Sigma Xi. Author: Plane Waves and Spherical Means, 1955; (with L. Bers and M. Schechter), Partial Differential Equations, 1964; (with R. Courant) Introduction to Calculus and Analysis, 1965. Study of geometry; analysis; partial differential equations; surface waves in liquids; ballistics; non-linear elasticity. Office: New York Univ., 251 Mercer St., Courant Inst., N.Y.C. 10012.

JOHN, Walter, Jr., Am. physicist; b. Newkirk, Okla., Feb. 16, 1924; s. Walter and Carrie (Hollingsworth) J.; B.S., Cal. Inst. Tech., 1950; Ph.D., U. Cal. at Berkeley, 1955; m. Carol S. Salin, Jan. 20, 1954; children—Kenneth Elliot, Laura Ann, Claudia Susan, Leslie, Ellen. Instr., U. Ill., Urbana, 1955-58; physicist Lawrence Radiation Lab., Livermore, Cal., 1958—. Tchr., U. Cal. Extension at Livermore, 1959. Fellow Am. Phys. Soc. Research, publs. on nuclear reactions, fission, prodn. mesons, photoneutrons, polarization photoneutrons, gamma ray spectroscopy with bent crystal spectrometer. Home: 21 Gary Way, Alamo, Cal. 94507. Office: Lawrence Radiation Lab., Livermore, Cal. 94551.*

JOHN OF ERZINJAN, (or John Bluss, John the Blue), astronomer; b. Erzinjan, Armenia, circa 1271, or earlier; ed. monastery of Mt. Sepuh; traveled across Armenia and Syria to Jerusalem; tchr. monastery of Dsordsor. Author astron. treatise, 1284 (later rewritten in verse), an Armenian grammar modeled on Dionysios Thraz's Greek grammar, a compilation of explanations of earlier Armenian grammarians, 1293, a collection of religious and moral precepts, 2 hagiological discourses made at Sepuh, 1288; also completed Nerses the Graceful's commentary on Gospel of St. Matthew; translated St. Thomas Aquinas' treatise on sacraments from Latin. Died Mt. Sepuh, 1326.

JOHN OF GADDESDEN, English physician; b. Little Gaddesden, 1280; A.B., Merton Coll., Oxford, 1300; A.M., 1303, M.D., 1309, also theol. degree; ordained as priest; apptd. to stall in St. Paul's London, 1342; practiced medicine, Oxford, also London; 1st Englishman apptd. ct. physician to an English monarch (Edward II). Author: Rosa medicinae (med. treatise in 5 parts), circa 1314. Died 1361.

JOHN OF GENOA (or Joannes de Janua), astronomer; possibly surgeon and physician to Clement VI, 1348; compiled tables for computation of eclipses. Author: Canones eclipsium, 1332; Investigation eclipsis solis anno Christi, 1337.

JOHN OF GLOGAU (or Joannes Physicus), physician; b. circa 1300-10; became canon of Glogau on the Oder, Silesia, 1354, archdeacon, 1377, dean, 1379; participant in struggle for bishopric of Breslau (vacant 1376-82). Author: Causae et signa pestilentiae et summa remedia contra ipsam (on causes of plague, proper diet), circa 1371-73. Died 1377 or after.

JOHN OF GLOGAU, mathematician, astronomer; b. Glogau, Silesia (now Poland), 19445; of German origin; ed. U. Cracow (Poland), became instr., prof., 1470; author works on dialectic and doctrine of Aristotle. Died Cracow, 1507.

JOHN OF HOLLYWOOD, mathematician, astronomer; b. Halifax, Yorkshire, Eng., circa 1200; ed. Oxford (Eng.) U. Tchr. math., astronomy Paris, France, from circa 1230. Author: Tractatus de sphaera (paraphrase of Ptolemy's Almagest), pub. 1472; Algorismus (study of arithmetic), pub. 1488; De anni ratione (treatise on calendar). First to use astron. writings of Arabs; books were influential and widely read. Died Paris, 1256.

JOHN OF JANDUN, French natural philosopher; b. Reims diocese, France, 1280; ed. Paris; became master Coll. Navarre, Paris, 1316; elected canon of Senlis by John XXII, 1316; apptd. bishop of Ferrara by Emperor Ludwig of Bavaria, 1328. Author: Quaestiones super libros physicorum (on motion, celestial bodies, elements, growth and change, qualities such as dryness and cold, metals, stones, meteorology, soul, passions, plants, animals, also discussions of Aristotle's Physics, and his zool. treatises, De coelo et mundo, De generatione, 2d book of De anima, Albert the Great's De nutrimen to et nutribili, Ibn Rushd's Colliget); Quaestio disputata super libros physicorum; Quaestiones super libros De coelo et mundo, printed 1501; Expositio et quaestiones super ibrum De substanita orbis (commentary on Ibn Rushd's commentary of De coelo et mundo), printed 1481; preface to edit. of Pietro d'Abano's commentary on Problemata. Died 1328.

JOHN OF LONDON (possibly John Peckham, Archbishop of Canterbury), English astronomer, mathematician; b. London, circa 1246; received instrn. in langs., math., optics, under patronage of Roger Bacon; lectr., Oxford, Eng., circa 1210-13, still lectr., 1252; sent by Bacon to explain his works to Pope Clement IV, Rome, 1267; said to have been received by Bacon into Order of St. Francis; tchr. of John of Garland; credited with authorship De trigonio circinoque analogico, also De speculis comburentibus; expounded new Aristotle.

JOHN OF MIRFELD (or Johannes de Mirfeld), English physician; b. Eng.; ed. Oxford; studied under London practitioner; became canon regular of St. Austin, Priory of St. Bartholomew, Smithfield, London. Author: Breviarium Bartholomei (med. work); Florarium Bartholomei (theol. treatise with med. chpt.). Drew on classical and medieval sources in his works, chiefly important for showing state of English med. knowledge of his times. Died 1407.

JOHN OF MURS, see Jean de Meurs.

JOHN OF SAINT AMAND (or Joannes de Sancto Amando), physician; b. Tournai, Hainut, Belgium, or St.-Anand in Pévèle, nr. Valenciennes; flourished in Paris. Author: Exposito sive additup super antidotarium Nicolai (treatise on gen. therapeutics and action of drugs, antidotary of Nicholas of Salerno); Revocativum memoriae (med. compendium). His works provided main source in West for Arabacized Greek medicine; aware to some extent of exptl. method. Died circa 1300.

JOHN OF SAINT GILES, physician; b. nr. St. Albans, Eng.; flourished 1230; ed. Oxford (Eng.) U., Paris, Montpellier (both France); lectr. medicine, Montpellier; became 1st physician to Philip Augustus, king of France, circa 1209; lectr. medicine, philosophy, theology U. Paris; presented Hopital de St. Jacques to Dominicans; lectured against Albigenses, Toulouse, France, 1233-35; invited by Grosseteste to Eng.; head Dominican schs., Oxford; became chancellor of Lincoln, 1239, archdeacon of Oxford, circa 1239, royal councillor, 1239; attended Grosseteste, also Richard de Clare, Earl of Gloucester. Author: Experimenta Joannis de S. Aegidio.

JOHN OF SALISBURY, polit. theorist; b. Salisbury, Eng., circa 1115; student of Abelard and other important scholars of Paris and Chartres, France, from circa 1136. Apptd. sec. to Theobold, Archbishop of Canterbury, circa 1150, visited Rome and other Continental centers on diplomatic missions for the see, continued as sec. to Theobold's successor, Thomas à Becket; took part in long dispute between Becket and King Henry II; withdrew to France, became sec. to Bishop of Chartres, 1176-80, active in council of Lateran, 1179. Author: Policraticus (described as earliest elaborate medieval treatise on politics), 1159; Metalogicus (study of logic); Entheticus (poem); Historia Pontificialis; also letters and lives of Becket and Saint Anslem. Conceived of the state and its components as comparable to human body, with the church (as the body's soul) having primacy; held that law and temporal rulers must serve God's justice and equity. Died nr. Chartres, Oct. 25, 1180.

JOHN OF SAXONY (or Danck, Dancko, Joannes de Madenburch or Magdeburg), astronomer; flourished circa 1323-61; disciple of John de Linieres in Paris; Author: Tempus est mensura motus (canons on Alphonsine tables), 1327. Introduced (with others) Alphonsine tables in Paris and other Western European centers of learning; included rules of eclipses in canons.

JOHN OF SEVILLE, translator; florished Toledo, Spain, circa 1135-53; probably s. of David. Translator (from Arabic to Latin): Liber alghoarismi de practica arismetrice; Commentary on Ptolemy's Centiloquium (Ahmad ibn Yusuf ibn al-Daya), circa 1130-36; Kitab fi harakat al-samawiya wa jawami' al-nujum (al-Farhhani), 1134-35; Albumasar de magnis conjunctionibus et annorum revolutionibus ac eorum profectionibus (al-Kindt), 1489; De nativitibus et interrogationibus (Umar ibn al-Farrukhan), 1503; Centiloquium (al-Battani); De imaginibus astronomicus (Thabit ibn Qurra); Kitab al-madkhal ila sina 'at ahkam al-nujum, De conjunctionibus planetarum in duodecim signis (both al-Qabisi); De astrolabio (Maslama ibn Ahmad al-Majriti). Deeply influenced devel. of scholastic philosophy through his transls. of philos. works.

JOHN OF YPRES, see Yperman, Jan.

JOHN THE CANON, see Canonicus, Joannes.

JOHNE, (Heinrich) Albert, German bacteriologist, veterinarian; b. Dresden, Germany, Dec. 10, 1839; studied under Cohnheim; prof., Dresden; contbd. to lit. of actinomycosis and trichinosis; discovered causative organism (Johne'b bacillus) of pseudo-Tb in cattle, 1895; developed method for staining bacterial capsules; instrumental in intro. of meat inspection. Died Kleinsedlitz/Pirna, Germany, Dec. 5, 1910.

JOHNS, Harold Elford, biophysicist; b. Chengtu, West China, July 4, 1915; s. Alfred E. and Myrtle (Madge) J.; B.A., McMaster U., 1936; M.A., U. Toronto, 1937, Ph.D., 1939; LL.D., U. Sask., 1959; m. Alice Sybil Hawkins, June 15, 1940; children—Gwyneth (Mrs. Clive Greenstock), Claire, Marilyn. Faculty, U. Sask., 1945-56; physicist Sask. Cancer Commn., 1945-56; head physics div. Ont. Cancer Inst., Toronto, 1956—; prof. physics dept. U. Toronto, 1956—, prof. med. biophysics, 1958—, head dept. med. biophysics, 1962—; mem. Internat. Commn. on Radiol. Units, 1952-66. Recipient Canadian Assn. Physicists medal, 1965. Charles Mickle fellow, 1966. Mem. Am. Radium Soc., Brit. Inst. Radiology, Canadian Assn. Physicists, Canadian Assn. Med. Physicists, Biophys. Soc., Radiation Research Soc., Royal Soc. Author: The Physics of Radiation Therapy, 1953; Physics of Radiology, 1961. Contbr. numerous articles to profl. jours. Research in application of physics to diagnostic and therapeutic radiology, creation of first Cobalt 60 unit for cancer therapy, measurement phys. parameters for many types of radiations used in radiation therapy, effects of ultraviolet radiation on DNA components and basic investigations of photochemistry of pyrimidines. Home: 24 Anderson Av. Office: 500 Sherbourne St., Toronto, Ont., Can.*

JOHNS, Martin Wesley, physicist; b. Chengtu, China, Mar. 23, 1913; s. Alfred and Myrtle (Madge) J.; B.A., McMaster U., Ont., Can., 1932, M.A., 1934; Ph.D., U. Toronto (Ont.), 1938; m. Margaret Hilborn, July 15, 1939; children—Robert, Beth, Kenneth, Kathryn. Prof. physics Brandon Coll., 1937-46; research physicist Atomic Energy Can., Chalk River, Ont., 1946-47; faculty McMaster U., Hamilton, Ont., 1947—, prof., 1953—, chmn. dept. physics, 1961—; vis. prof. Clarendon Lab., Oxford, Eng., 1959-60. Fellow Royal Soc. Can.; mem. Canadian Assn. Physicists (past registrar), Am. Phys. Soc. Research, publs. on decay mechanisms of radioactive nuclei using magnetic spectrometers, scintillation spectrometers, coincidence circuits. Home: 116 Sterling St., Hamilton, Ont., Can.*

JOHNS, Richard James, Am. physician, educator; b. Pendleton, Ore., Aug. 19, 1925; s. James Shanard and Pearl (McKenna) J.; B.S., U. Ore., 1947; M.D., Johns Hopkins, 1948; m. Carol Greacen Johnson, June 27, 1953; children—James Ashmore, Richard Clark, Robert Shanard. Physician, Johns Hopkins Hosp., Balt., 1956—; with Johns Hopkins, 1951—, prof. medicine, 1966—, asst. dean, 1962-66, prin. profl. staff applied physics lab., 1967—. Chmn. med. bd. Myasthenia Gravis Found. Mem. Am. Soc. Clin. Investigation, Assn. Am. Physicians, Sigma Xi, Alpha Omega Alpha. Studies, publs. on neuromuscular function in man, clin. investigation of myasthenia gravis. Home: 203 E. Highfield Rd., Balt. 21218. Office: Johns Hopkins Hosp., Balt. 21205.*

JOHNS, William Davis, Jr., Am. minerologist, geochemist; b. Waynesburg, Pa., Nov. 2, 1925; s. William Davis and Beatrice (VanKirk) J.; B.A., Coll. Wooster, 1947; M.A., U. Ill., 1951, Ph.D., 1952; m. Mariana Paull, Aug. 28, 1948; children—Sydney Ann, Susan Helen, David William, Amy Matilda. Spl. research asst. petrology Engring. Expt. Sta., U. Ill., 1949-52; faculty U. Ill., 1952-55; faculty Washington U., St. Louis, 1955—, prof. earth scis., 1964—, chmn. dept., 1962—. Fulbright sr. research scholar U. Goettingen (Germany), 1959-60. Fellow Geol. Soc. Am., Mineral. Soc. Am.; mem. Mineral. Soc. Great Britain and Ireland, Mineral. Soc. Can., Deutsches Mineralogisches Gesellschaft, Geochem. Soc. Research and publs. on crystal chemistry of silicate minerals; interactions of organic molecules and clay mineral constituents in sediments; reactions between seawater and marine muds, important in affecting the composition of sea water. Home: 29 Rosemont Av., Webster Groves, Mo. 63119. Office: Dept. Earth Scis., Washington, St. Louis 63130.*

JOHNSEN, Kjeli, physicist; b. Meland, Norway, June 11, 1921; s. Georg and Borhhild (Hagen) J.; Ph.D. in Civil Engring. and Tech. Sci., Tech. U. Norway; m. Aase Jordal, Dec. 29, 1945; children—Arnlaug, Kjetil, Ottar. Asst., Tech. U. Norway, 1947-48, tech. prof. theoretical electronics, 1957-59; research asst. Christian Inst. Bergen, 1947-48; physicist European Orgn. for Nuclear Research, Cern-Geneva, 1948-52; head physicist Cern, 1959—. Mem. Norwegian Acad. Tech. Sci., Norwegian Soc. Physicists. Author: Heavy Beam Loading in Linear Accelerator; Effects of Non-Linearities on the Phase-Transition, 1956; High Trapping Efficiency in Synchrotons with Phase Lock, 1961; Features of the Next Generation of Proton Accelerators. Home: En Marnex, Commugny, Vaud. Office: Cern, Geneva, Switzerland.

JOHNSEN, Russell Harold, Am. chemist, educator; b. Chgo., Aug. 5, 1922; s. Harold Gunnar and Irene (Gaul) J.; B.S., U. Chgo., 1947; Ph.D., U. Wis., 1951; m. Dorothy Ruth Pehta, Jan. 20, 1948; children—Peter B., Margaret A. Research chemist Ninol Labs., Chgo., 1947-48; teaching asst. U. Wis., 1948-51; faculty Fla. State U., Tallahassee, 1951—, prof., 1961—, chmn. dept. phys. sci., 1951—, sr. scientist Electron Van de Graaf program, 1958—; cons. editor W. A. Benjamin, Inc., N.Y.C., 1961—. Mem. A.A.A.S., Am. Chem. Soc., Am. Phys. Soc., Chem. Soc. (London), Radiation Research Soc., Am. Assn. U. Profs., Sigma Xi. Author: (with E. M. Grunwald) Atoms, Molecules and Chemical Change, 1960, 2d edit., 1965. Research, publs. on chem. effects of ionizing radiation, with emphasis on primary reaction steps, reaction mechanisms of gaseous hydrocarbons and condensed phase alcohols. Home: 1425 Devil's Dip, Tallahassee 32303.*

JOHNSON, A. W., ornithologist; b. Lake County, Cal., July 14, 1894; s. Alfred William and Emily (Renmitz) J.; m. Elsie Whicaelow, Sept. 10, 1920; children—Dorothy Marion (Mrs. Nichols) Jean Audrey (Mrs. Wight), William Bryan. With Agua Santa Nitrate & Ry. Co., 1912-30, gen. mgr., 1924-30; exec. head office Consorio Sabioncello, 1930-31; partner Katz Johnson & Co. Ltd., 1932-46; bd. dirs. Katz Johnson S.A.C., Frio-Lux S.A.I., Santiago, S.P.L. Hayskrim S.A. Corr. fellow, Am., Brit., Argentinian ornithologists unions; mem. Cooper Ornithol. Soc. Cal. Author: Les Aves de Chile, vol. I, 1946, vol. II, 1951, supplements, 1957, 64 (with R. A. Philippi, J. D. Goodall); The Birds of Chile and Adjacent Regions of Argentina, Bolivia and Peru, vol. I, 1965; also articles in field. Research on birds of Chile. Home: 2454 Pedro de Villagra, Santiago. Office: 1920 General McKenna, Santiago, Chile.*

JOHNSON, Alan Julian, Am. physician; b. Washington, Mar. 19, 1919; s. Edmund C. and Aimée L. (Weil) J.; B.A., Dartmouth, 1940; M.A., U. Wis., 1942; M.D., L.I. Coll. Medicine, 1945; m. Ruth Waters, Aug. 16, 1952; children—Eleanor, Amy, David. Faculty, N.Y. U. Sch. Medicine, N.Y.C., 1948—, asso. prof. medicine, 1961—; mem. staffs Bellevue, VA, U. hosps., N.Y.C. Asso. research dir. blood program ARC, 1961—, dir. research lab. Eastern div., 1961—; mem. NIH coms. Mem. Internat., Am. socs. haematology, Internat. Soc. Blood Transfusion, Am. Soc. Clin. Investigation, Am. Soc. Exptl. Pathology, Am. Physiol. Soc., Soc. for Study Blood, Soc. Exptl. Biology and Medicine, Am. Fedn. Clin. Research, Am. Coll. Clin. Pharmacology, Harvey Soc., N.Y. Acad. Medicine, N.Y. Acad. Sci., A.A.A.S. Research, numerous publs. on fibrinolysis, blood coagulation, fibrinolytic agts. and inhibitors, thromboembolic disease; isolation, purification, function and kinetics of proteolytic enzymes and blood coagulation components. Home: 127 W. 12th St., N.Y.C. 10011.*

JOHNSON, Alexander, mathematician; M.A., Trinity Coll., Dublin, Ireland; D.C.L. (hon.), Bishop's Coll.; LL.D., U. N.B. (Can.). Prof. math. and natural philosophy McGill U., Montreal, Can., from 1857, emeritus prof. pure math., vice-prin. emeritus, fellow, dean Faculty of Arts. McGill U. rep. to tercentenary festival U. Dublin. Fellow Royal Soc. Can. (pres. 1905-06). Secured (through Brit. Assn. and Royal Soc. Can.) establishment of tidal and hydro. surveys for Can. Died Feb. 10, 1913.

JOHNSON, Arthur Newhall, Am. civil engr.; b. Lynn, Mass., Nov. 11, 1870; s. David Newhall and Amanda Malvina (Richardson) J.; S.B. in civ. engring., Lawrence Scientific Sch. (Harvard), 1894; hon. Dr. Engring., U. of Md., 1924; m. May Louise Ash, Sept. 12, 1900. Instr. descriptive geometry, Harvard, 1895-96; asst. engr. Calumet and Hecla Mine, Calumet, Mich., 1896-97; asst. engr. Mass. Highway Commn., 1897-98; state highway engr. of Md., 1898-1905; chief engr. U. S. Office of Pub. Roads, Washington, 1905; state highway engr. of Ill., 1906-14; with Bur. Municipal Research of New York, 1914-16; consulting highway engr., Portland Cement Assn., Chicago, 1916-20; dean Coll. of Engring., U. of Md., 1920-36, emeritus. Chmn. Highway Research Bd., Nat. Research Council, 1923-26; del. to Pan Am. Road Congress, Buenos Aires, 1925. Received Bartlett award for outstanding contribution to highway progress, 1933. Authority on hwy. constrn. and design; called father of hard-surface road system of Md. Died July 10, 1940.

JOHNSON, Bruce, Australian biologist; b. Melbourne, Australia, Sept. 5, 1925; s. Frank Norman and Marion (Taylor) J.; B.Sc. Agr. with 1st class honors, U. Sydney, 1952; Ph.D., U. London, 1955; m. Bridget Agatha Meade Waldo, Aug. 6, 1952; 1 son, Alban. Lectr., sr. lectr., reader in entomology Waite Agrl. Research Inst., U. Adelaide, South Australia, 1955-66; prof. zoology U. Tasmania, Hobart, Australia, 1966—. Nuffield fellow, 1963. Mem. Royal Soc. Tasmania, Entomol. Soc. Australia. Research, publs. on biology and physiology of aphids, electron microscope studies of neurosecretory system. Home: 42 Maning Av., Sandy Bay, Hobart, Tasmania, Australia.*

JOHNSON, Bruce Connor, biochemist; b. Regina, Sask., Can., Apr. 28, 1911; s. Wilfred Connor and Edna (Young) J.; B.A., McMaster U., 1933, M.A., 1934; Ph.D., U. Wis., 1940; m. Elizabeth Marie Peterson, Sept. 7, 1940; children—Bruce Connor II, Peter Y., Stephen P., Lisa C., Christina M. Chemist, Canadian Canners Ltd., Hamilton, Ont., 1934-37; grad. asst. U. Wis., 1937-40; post-doctoral fellow U. Ill., 1940-42; research biochemist Golden State Co., San Francisco, 1942-43; faculty U. Ill., 1943-65, prof. biochemistry, 1951-65; prof. biochemistry, chmn. dept. U. Okla. Coll. Medicine, Oklahoma City, 1965—; head biochem. sect. Okla. Med. Research Found., Oklahoma City, 1965—. Recipient Nutrition Council $1000 award, 1960. Mem. Am. Soc. Biol. Chemistry, Am. Inst. Nutrition, Biochem. Soc., Am. Chem. Soc., A.A.A.S., N.Y. Acad. Sci., Soc. Exptl. Biology and Medicine. Author: Methods of Vitamin Determination, 1948; also numerous articles. Research field vitamin requirements of new born calves, lambs, pigs; first discovery of B-vitamin deficiencies in these species; metabolism of vitamins with discovery of a new metabolite of niacin; studies of function of vitamin B12, A, K and E; role of vitamin B12 in mammalian enzyme reactions; role of vitamin K in prothrombin biosynthesis; nutrition and enzyme induction.*

JOHNSON, Charles Willison, Am. entomologist; b. Morris Plains, N.J., Oct. 26, 1863; s. Albert Fletcher and Sarah Elizabeth (Willison) J.; ed. pub. and pvt.

schs., Morristown, N.J.; m. Carrie W. Ford, Jan. 14, 1897 (died 1931). Moved to St. Augustine, Fla., 1880, continuing studies in natural history, and making large collection of insects, mollusca and fossils. Curator Mus. of Wagner Free Inst. of Science, Phila., 1888-1903; curator Boston Soc. Natural History since Mar. 1903. Especially interested in study of mollusca and diptera, and contbr. to biol. jours. of papers relating to these spltìes.; asso. editor and mgr. The Nautilus, from 1890. Fellow Am. Acad. Arts and Sciences, Entomol. Soc. Am. (pres. 1924), A.A.A.S.; mem. Phila. Acad. Natural Sciences, Entomol. Soc. Washington, Malacol. Soc. of London. Died May 8, 1932.

JOHNSON, Charles Yothers, Am. physicist; b. Washington, Aug. 16, 1920; s. Theodore Carlton and Florence (Yothers) J.; B.E.E., U. Va., 1942; postgrad. U. Md.; m. Virginia Frances Focht, May 7, 1945; children—Bradish Focht, Philemon Wye, Quinley Yothers. With U. S. Naval Research Lab., Washington, 1942—, head air and ion composition sect., 1954-58, head aeronomy sect., 1958—, chmn. research rocket scheduling com. Naval Ordnance Missile Test Facility, 1961-67; cons. NASA. Mem. A.A.A.S., Am. Phys. Soc., Sci. Research Soc. Am., Am. Geophys. Union, I.E.E.E. Pioneered investigation ionosphere ion composition with ion mass spectrometers; discovered nitric oxide ion. Home: 3909 Malcolm Ct., Annandale, Va. 22003. Office: U. S. Naval Research Lab., Washington 20390.*

JOHNSON, Clarence Leonard, Am. aircraft engr.; b. Ishpeming, Mich., Feb. 27, 1910; s. Peter and Christine (Anderson) J.; B.S. (Sheehan fellow 1932), U. Mich., 1932, M.S., 1933, D. Eng. (hon.), 1964; D.Sc., U. So. Cal., 1964; LL.D., U. of Cal. at Los Angeles, 1965; m. Althea Louise Young, Aug. 31, 1936. With Lockheed Aircraft Corp., Burbank, Cal., 1933—, chief research engr., 1938-52, chief engr., 1952-56, v.p. charge research and devel., 1956-58, v.p. advanced devel. projects, 1958—. Recipient Lawrence Sperry award Inst. Aero. Scis., 1937, Sylvanus A. Reed award, 1956, 66; Wright Bros. medal Soc. Automotive Engrs., 1941; named aviation man of year Airlines Activities Com.; Collier award for design of air frame, 1959; Distinguished Alumnus award in engring., U. Mich., 1953; Gen. Hap Arnold gold medal award Vets. Fgn. Wars, 1960; Collier trophy Nat. Aero. Assn., Look Mag., 1964, Theodore Von Karman award Air Force Assn., 1963; Presdl. Medal of Freedom, 1964, Nat. Medal of Sci., 1966, Thomas D. White Nat. Def. award USAF Acad., 1966; hon. mem. Aerospace Med. Assn., 1963. Fellow Inst. Aero. Scis. (chmn. W. Coast sect. 1946-47, nat. v.p. 1948-53), Nat. Acad. Engring., Nat. Acad. Scis., Soc. Automotive Engrs., Sigma Xi. Research on power plants; aeronautical design; boundary layer control. Home: 16801 Oak View Dr., Encino, Cal. Office: 2555 Hollywood Way, Burbank, Cal.

JOHNSON, Curtis Carl, Am. elec. engr.; b. Long Beach, Cal., Nov. 7, 1932; s. Carl G. and Katharine (Parsons) J.; B.S., Cal. Inst. Tech., 1954, M.S., 1955; Ph.D., Stanford, 1958; children—Steven C., Susanne C., James C., D. Eric. Mem. tech. staff Gen. Electric Co., Palo Alto, Cal., 1955-58, Hughes Research Labs., Malibu, Cal., 1958-61; asso. prof. U. Utah, Salt Lake City, 1961-67; NIH fellow U. Wash., Seattle, 1967-68, asso. prof. elec. engring., asst. dir. bioengring., 1968—. Mem. I.E.E.E., Sigma Xi. Author: Field and Wave Electrodynamics, 1965. Research in devel. microwave devices for radar and communications, application of field equations to understanding of wave propagation effects in materials such as plasma, biol. substances. Office: Dept. Physiology and Biophysics, U. Wash., Seattle 98105.*

JOHNSON, D(avid) Gale, Am. agrl. economist; b. Vinton, Ia, July 10, 1916; s. Albert D. and Myra Jane (Reed) J.; B.S., Ia. State Coll., 1938; M.S., U. Wis., 1939; postgrad. U. Chgo.; Ph.D., Ia. State U., 1945; m. Helen Wallace, Aug. 10, 1938; children—David Wallace, Kay Ann. Asst. prof. agrl. econs. Ia. State Coll., 1941-44; with U. Chgo., 1944—, prof. econs., 1954—, asso. dean div. social scis., 1957-60, dean div. social scis., 1960—. Cons. RAND Corp., 1953—, Nat. Adv. Commn. on Food and Fiber, 1965—; v.p. bd. trustees Nat. Opinion Research Center, Chgo., 1960-62, pres., 1962—. Mem. Am. Farm Econ. Assn. (past pres.). Author: Forward Prices for Agriculture, 1947; Trade and Agriculture, 1950; (with Gustafson) Grain Yields and the American Food Supply, 1952; also numerous articles. Research on effects of uncertainty on agrl. prodn. decisions and income, prodn. function analysis, effect factor supply conditions on agrl. supply response. Home: 5617 S. Kenwood Av., Chgo. 60637.*

JOHNSON, Donald Bruce, Am. polit. scientist; b. Cloquet, Minn., Mar. 4, 1921; s. Oscar C. and Hulda (Holm) J.; B.A., U. Minn., 1943, M.A., 1948; Ph.D., U. Ill., 1952; m. Eleanor M. Thomas, Feb. 28, 1945; children—Donald Bruce, Kristen Eleanor, Brian Thomas. Faculty, Duluth Jr. Coll., 1946-48, U. Ill., 1948-51; faculty U. Ia., Iowa City, 1951—, prof. polit. sci., 1962—; vis. prof. Emory U., 1955, San Francisco State Coll., 1962, U. Me., 1966. Nat. asso. dir. Citizenship Clearing House, N.Y.C., 1959-60; regional cons. U. S. Commn. on Civil Rights, 1961—; examiner-cons. N.Central Assn. Colls. and Secondary Schs., 1963—. Mem. Am. Polit. Sci. Assn., Am. Assn. U. Profs., Midwest Conf. Polit. Scientists, Am. Civil Liberties Union. Author: The Republican Party and Wendell Willkie, 1960; (with Kirk Porter) National Party Platforms, 1957, 61, 65; (with Jack Walker) Dynamics of the American Presidency, 1964; also articles. Studies in Am. polit. parties, Am. instns., and Am. presidency. Home: 336 Beldon Av., Iowa City, Ia. 52240.*

JOHNSON, Dorothy Durfee Montgomery, Am. physicist; b. Rochester, N.Y., May 3, 1909; d. Charles G. and Abigail (LeRoy) Durfee; A.B., Vassar Coll., 1930; M.S., Yale, 1932, Ph.D.., 1951; m. Carol Gray Montgomery, June 1, 1933 (dec. Dec. 1950); children—Charles Gray, Katherine A.; m. 2d, Thomas H. Johnson, Sept. 1, 1965. Research fellow Bartol Found. of Franklin Inst., Swarthmore, Pa., 1933-40; research asso. Yale, 1940-51; mem. staff Mass. Inst. Tech. Radiation Lab., 1941-45; asso. editor radiation lab. series Office Publs., 1945-46; prof. physics, chmn. dept. Hollins Coll., 1952-66; film rev. editor Am. Jour. Physics, 1962-66; phys. sci. study com., mem. staff Ednl. Services, Inc., 1958-59, cons., 1959-62. Fellow Am. Phys. Soc., A.A.A.S.; mem. Am. Assn. Physics Tchrs. Research, numerous publs. on nature and origin of cosmic radiation, microwaves. Home: Denmark, Me. 04022.*

JOHNSON, Douglas Wilson, Am. geologist; b. Parkersburg, W.Va., Nov. 30, 1878; s. Isaac H. and Jennie A. (Wilson) J.; Denison U., 1896-98, D.Sc., 1932; B.S., U. N.M., 1901; Ph.D., Columbia, 1903, D.Sc., 1929; Docteur, honoris causa U. Grenoble (France), 1924, Nancy, France, 1932, Montpelier, France, 1933; m. Alice Adkins, Aug. 11, 1903; m. 2d, Edith Sanford Caldwell, 1943. Asst., U. S. Geol. Survey, 1901, 03, 04, 05; instr. in geology, 1903-05, Mass. Inst. Tech., asst. prof., 1905-07; asst. prof. physiography, Harvard, 1906-12; asso. prof. physiography, Columbia U., N.Y.C., 1912-19, prof. from 1919, exec. officer, dept. of geology, from 1937. Asst. N.J. Geol. Survey, 1911; dir. Shaler Meml. Expdn., 1911-12; geog. adviser U. S. Dept. State, 1919-20; cons. physiographer to Canadian govt. in Labrador boundary dispute, 1926; chief of div. boundary geography, Am. Commn. to Negotiate Peace, Paris, 1918-19; exchange prof. to France, in engring. and applied sci., 1923-24; pres. sect. phys. geography of Internat. Geog. Congress, Paris, 1931. Chmn. nat. com. U. S. Internat. Geographic Union, 1933-37; pres. Internat. Terrace Commn., 1934-38. Recipient numerous awards including Janssen medal Paris Geog. Soc., 1920; A. Cressy Morrison prize N.Y. Acad. Sci., 1924, 30; Cullum medal Am. Geog. Soc., 1935; chevalier Legion of Honor (France). Mem. Nat. Acad. Sci., Geol. Soc. Am. (pres. 1942), Am. Acad. Arts and Scis., Am. Mus. Natural History N.Y. Acad. Scis., A.A.A.S., Am. Philos. Soc., Assn. Am. Geographers, Am. Geog. Soc., NRC, Sigma Xi, Phi Beta Kappa; hon. mem. numerous fgn. learned socs. Author: Lettre d'un Américain à un Allemand, 1916; Topography and Strategy in the War, 1917; Peril of Prussianism, 1917; My German Correspondence, 1917; Shore Processes and Shoreline Development, 1919; Battlefields of the World War, 1921; The New England-Acadian Shoreline, 1925; Paysages Américains et Problèmes Géographiques, 1927; Stream Sculpture on the Atlantic Slope, 1931; The Origin of Submarine Canyons, 1939; The Origin of the Carolina Bays; also numerous scientific bulls., papers. Editor of Jour. Geomorphology. Pioneer work on concept of coastal erosion. Died Feb. 24, 1944.

JOHNSON, Duncan Starr, Am. botanist; b. Cromwell, Conn., July 21, 1867; s. Edward Tracy and Lucy Emma (Starr) J.; B.S., Wesleyan U., Conn., 1892, D.Sc., 1932; student, Yale, 1887; Ph.D., Johns Hopkins, 1897, U. Munich, 1901; m. Mary E. G. Lentz, June 22, 1904; children—George Duncan, David Starr. Asst. in botany Johns Hopkins U., Balt., 1898-99, asso., 1899-1901, asso. prof., 1901-06, prof., from 1906, dir. bot. garden, from 1913. In charge of botany Cold Spring Harbor, L.I., N.Y., 1896-1900, cryptogamic botany, 1902-11; botanical exploration and investigation, Jamaica, B.W.I., 6 times to 1932; investigator, Carnegie Instn., Washington, 1912, 15; in charge of inspection for disease of grain fields of Atlantic States for Bureau of Plant Industry, 1918; v.p. Mt. Desert Biol. Lab., from 1921; lectr. Mt. Lake Biol. Sta., U. Va., summer 1935. Dir. 6th Bot. Expdn. of Johns Hopkins U. to West Indian tropics, and Guatemala, 1932, and of 7th Expdn. of Bot. Dept. to Jamaica and Central America, summer 1936. Mem. NRC, 1927-1933, chmn. division biology and agr., 1931-32. Fellow A.A.A.S. (v.p. sect. G, 1912); mem. Bot. Soc. Am. (sec. 1907-09). Author: The Relation of Plants to Tide Levels (with H. H. York), 1915; The Fruit of Opuntia fulgida and Proliferation in Fruits of the Cactaceae, 1918; Littoral Vegetation on Mt. Desert Island (with A. F. Skutch), 1928. Died Balt., Feb. 16, 1937.

JOHNSON, Elbe Herbert, Am. physicist, educator; b. Traverse City, Mich., Aug. 9, 1887; s. Frank H. and Ella (Brown) J.; B.A., Olivet Coll., 1911, M.A. Physics, 1913; postgrad. U. Wis., Columbia Tchrs. Coll.; Ph.D., U. Chgo., 1926; Sc.D., Kenyon Coll., 1955; m. Bertha B. Beckwith, July 26, 1911; children—Raymond Herbert, Barbara Beckwith. Student asst. in physics Olivet Coll., 1909-11; asst. instr. physics U. Wis., 1911-14; teaching fellow U. Chgo., 1925-26; faculty physics Kenyon Coll., Gambier, O., 1914—, prof., head dept. physics, 1918-55, Henry G. Dalton prof. physics emeritus, 1955—; lectr. Gt. Lakes Naval Tng. Sta., 1918, Ind. U., summer 1929. Mem. Source Research Council, Washington, 1930—, Ohio Pub. Expenditures Council, 1943—. Recipient Copernican award Kosciuszko Found., 1944. Fellow A.A.A.S., Am. Phys. Soc. (co-organizer, charter mem. Ohio sect.), Ohio Acad. Sci. (v.p. physics-astronomy div. 1928-29, 41-42), History of Sci. Soc. (mem. found., patron, nat. v.p. 1947-49); mem. Illuminating Engring. Soc., Am. Assn. U. Profs., Central Ohio Physics Club, Ohio Physics Assn. (past pres.), Assn. Physics Tchrs., Central Assn. Sci. and Math. Tchrs., Knox County Hist. Soc. (charter, life, trustee, exec. bd.), Sigma Xi, Sigma Pi Sigma. Author: Laboratory Physics, 1923; (with others) Johann Kepler, 1931; General Physics Laboratory Guide, 1941; also articles. Died Dec. 15, 1967.

JOHNSON, Ernest Frederick (Jr.), Am. chem. engr., educator; b. Jamestown, N.Y., Apr. 4, 1918; s. Ernest Frederick and Esther Marie (Engstrom) J.; B.S., Lehigh U., 1940; Ph.D., U. Pa., 1949; m. Marjorie Ruth McMullin, July 15, 1944; children—David Shaffer, Carolyn L., Arthur B., Melissa. With Barrett div. Allied Chem. Corp., Phila., 1940-46; faculty Princeton, 1948—, prof. chem. engring., 1959—, asso. plasma physics lab., 1955—, asso. dean faculty, 1962-66; cons. petroleum, chem., engring. firms, 1949—. Mem. trustees com. Nat. Radio Astronomy Obs.; mem. adv. bd. Indsl. and Engring. Chemistry. Trustee Asso. Univs., Inc., 1962—, chmn. bd., 1965-67. Fellow A.A.A.S.; mem. Am. Chem. Soc., Am. Soc. Engring. Edn., Am. Inst. Chem. Engrs., Sigma Xi, Tau Beta Pi, Phi Eta Sigma, Alpha Phi Omega. Author: Automatic Process Control, published 1967. Research and publs. in the measurement and correlation of the thermodynamic and transport properties of pure fluids and solutions; in the behavior of process control systems and design procedures therefor; and in problems related to the devel. of thermonuclear power devices. Home: 90 Galbreath Dr., Princeton, N.J. 08540.*

JOHNSON, Francis Severin, Am. physicist; b. Omak, Wash., July 20, 1918; s. Ralston Severin and Elizabeth (Gruenes) J.; B.Sc. with honors in Physics, U. Alta., 1940; postgrad. U. Cal. at Berkeley; M.A. in Physics and Meteorology, U. Cal. at Los Angeles, 1942, Ph.D. in Meteorology, 1958; m. Maurine Marie Green, Sept. 12, 1943; 1 dau., Sharan Kaye. Head high atmosphere research sect. U. S. Naval Research Lab., Washington, 1946-55; mgr. space physics research Lockheed Missiles & Space Co., Palo Alto, Cal., 1955-62; head atmospheric and space scis. div. S.W. Center for Advanced Studies, Dallas, 1962-64, dir. earth and planetary scis. lab., 1964—. Mem. various coms., cons. NASA, Am. Inst. Aeros. and Astronautics, NSF, Nat. Acad. Scis., others. Fellow Am. Geophys. Union, A.A.A.S., Am. Inst. Aeros. and Astronautics (Space Sci. award 1966, asso.); mem. Am. Astron. Soc., Am. Phys. Soc., Am. Meteorol. Soc. (profl.), I.E.E.E. (sr.), Internat. Union Geodesy and Geophysics Internat. Sci. Radio Union, Research Soc. Am., Sigma Xi. Author: Satellite Environment Handbook, 1961. Research on solar ultraviolet spectrum, solar energy received at earth, hydrogen geocorona surrounding earth, upper atmosphere of earth. Home: 13619 Sprucewood Dr., Dallas 75240. Office: P.O. Box 30365, Dallas 75230.*

JOHNSON, Frank Harris, Am. cell physiologist, microbiologist; b. Raleigh, N.C., July 31, 1908; s. A. R. D. and Mary Victoria (Harris) J.; A.B., Princeton, 1931, Ph.D., 1936; M.A., Duke, 1932; student N.C. State Coll., 1933, Vanderbilt U., 1934; m. Mary Frances McGhee, June 11, 1933; children—Virginia Lane, Mary Frances, Charlotte Elizabeth. Teaching fellow Vanderbilt U.; grad. research fellow Woods Hole Oceanographic Inst.; Eli Lilly & Co. research fellow biology, Princeton, 1936-37, faculty, 1937—, prof., 1956—; vis. prof. U. Utah, 1950-51. Program dir. NSF, 1952-53, cons., 1953-56; com. photobiology NRC, 1952-64; adv. panel NASA. Rockefeller Found. fellow, 1939; Guggenheim fellow, 1944-46, 50-51; recipient award (with others), A.A.A.S., 1942. Fellow Am. Acad. Microbiology; mem. Am. Soc. Zoology, Am. Physiol. Soc., Soc. Am. Bacteriology, Soc. Exptl. Biology and Medicine, Soc. Gen. Physiology, A.A.A.S., Sigma Xi, Phi Beta Kappa. Author: (with H. Eyring, M. J. Polissar) The Kinetic Basis of Molecular Biology, 1954. Editor: The Luminescence of Biological Systems, 1955; The Influence of Temperature on Biological Systems, 1957; (with Y. Haneda) Bioluminescence in Progress, 1966. Editorial bd. Am. Jour. Physiology, Jour. Applied Physiology, 1958-64; fgn. editor Canadian Jour. Microbiology, 1953-57; hon. editor Jour. Photobiology and Photochemistry, 1961—. Discoverer and partial elucidation of a fundamental relationship in the biol. effects of temperature, hydrostatic pressure and narcotics mediated through inter-related influences on the activity and reversible denaturation of enzymes; and (with others) purification and partial elucidation of about half of the total number of extracted bioluminescence systems. Home: 590 Lake Dr., Princeton, N.J. 08540.*

JOHNSON, Franklin Paradise, Am. anatomist, urologist; b. Hannibal, Mo., Jan. 7, 1888; s. Horace William and Lillie May (Paradise) J.; A.B., U. Mo., 1908; A.M., Harvard, 1910, Ph.D., 1912; postgrad. U. Freiburg (Germany), summer, 1911; M.D., Johns

Hopkins, 1920; m. Juliette Omohundro, Sept. 4, 1923; children—Lillian Paradise, Virginia Martin, Louise Carter. Austin teaching fellow Harvard Med. Sch., 1908-10, instr. histology and embryology, 1910-12; asst. prof. anatomy U. Mo., 1912; asso. prof., 1913, prof., 1919-20; practice medicine, specializing in urology, Portland, Ore., from 1923; asst. clin. prof. urology, U. Ore., from 1929. Diplomate Am. Bd. Urology. Fellow A.C.S., Am. Urol. Assn.; mem. A.M.A., Am. Assn. Anatomists, Sigma Xi, Phi Beta Pi. Research and publs. on development of the digestive tract, embryology and histology of the liver, urethra, urol. subjects; collaborator Young's Practice of Urology, Nelson's Loose Leaf Surgery, Morris' Anatomy, 9th and 10th edits. Died Feb. 12, 1943.

JOHNSON, Sir George, Brit. physician; b. Gaudhurst, Eng., Nov. 29, 1818; s. George and Mercy (Corke) J.; apprenticed in medicine to uncle, 1837; M.D., King's Coll., London, 1844; m. Charlotte Elizabeth White, 1850; 5 children. Became asst. physician King's Coll. Hosp., 1847, physician, 1856; prof. materia medica and therapeutics, 1857-63, prof. medicine, 1863-76, prof. clin. medicine, cons. physician, from 1886; named physician extraordinary to Queen Victoria, 1889. Fellow Royal Soc., 1872, Royal Coll. Physicians (Gulstonian lectr. 1852, materia medica lectr. 1853, Lumieian lectr. 1877, Harveian orator 1882, v.p. 1887). Author: Diseases of the Kidney, their Pathology, Diagnosis, and Treatment, 1852; Lectures on Bright's Disease, 1873; The Pathology of the Contracted Granular Kidney, 1896; also papers in Brit. Med. Jour. Discovered hypertrophy of small arteries in Bright's disease; reintroduced picric acid test for albumen, picric acid and potash test for sugar. Died June 3, 1896.

JOHNSON, H. Herbert, Am. zoologist; b. Macon, Ga., Aug. 29, 1898; s. H. Herbert and Wilhelmina (Wheeler) J.; B.S., Mercer U., 1918; A.M., Columbia U., 1920, Ph.D., 1930; m. Selina Tetzlaff, June 1, 1927; children—Jaqueline (Mrs. Donald Horvath), Frank Wheeler. Field asst. U. S. Bur. Entomology, 1919; lab. asst. Columbia, 1919-22; field investigator China Med. Bd., 1924; instr. Soochow U., 1925; instr. Hunter Coll., 1926-27; faculty U. City N.Y., 1925—, prof. zoology, 1948-65, emeritus, 1965—. Dir. Sci. Mus., Maria Mitchell Assn., 1959-65; treas. Bergen Community Mus., Leonia, N.J., 1965—. Fellow A.A.-A.S., N.Y. Acad. Scis.; mem. Am. Soc. Zoologists. Author: Centrioles and Other Cytoplasmic Components of Gryllidae, 1930. Discovered self-perpetuating centrioles in male germ cells; nature of protoplasm, vertebrate embryology. Home: 24 Hawthorne Terrace, Leonia, N.J. 07605.*

JOHNSON, H(arold) Daintree, English surgeon; b. London, May 26, 1910; s. Stanley and Edith (Heather) J.; student Christ Coll., Cambridge U., 1919-22, M.A., M.B., B.Ch., 1936, M.D., 1964, M.Ch., 1967; student St. Thomas's Hosp., London, 1922-36; m. Margaret Dixon, May 27, 1944; children—David, Piers (dec.). First asst. London Hosp., 1942-43; faculty Postgrad. Med. Sch., London, 1946—, sr. lectr. surgery, 1964—, surgeon Royal Free Hosp., London, 1947—; examiner U. London, 1960-64, Leverhulm scholar surg. research Royal Coll. Surgeons Eng., 1948, mem. ct. examiners, 1960-64. Fellow Assn. Surgeons Gt. Britain (sr.), Royal Soc. Medicine, Surg. Research Soc. (corr.); sr. mem. Brit. Soc. Gastroenterology. Author: Surgical Aspects of Medicine, 1959; The Cardia and Hiatus Hernia, 1968; also articles. Classified gastric ulcers and correlated types with secretion patterns and blood groups; research on function of cardia, relation of physiol. observations with laws of physics. Home: 5 Holly Terrace, London N.6., Eng.*

JOHNSON, Harold David, Am. physiologist; b. Verona, Mo., Feb. 28, 1924; s. John E. and Docia (Smith) J.; B.S., Drury Coll., 1949; M.S., U. Mo., 1952, Ph.D., 1956; m. Lonetta, P. Gallion, Jan. 21, 1949; children—Michael, Deborrah, Tina, Leah. Pharm. rep. Kendall Drug Co., 1949-51; faculty U. Mo., Columbia, 1954—, prof. environmental physiology dairy dept., 1965—. Mem. A.A.A.S., Am. Physiol. Soc., Soc. Animal Production, Dairy Sci. Assn., Internat. Soc. Biometerology. Research, publs. on effects of environmental factors on physiology domestic and other lab. animals, physiol. mechanisms of adaptability to hot and cold environments. Home: Shepard Hills, Route 1, Columbia, Mo. 65202.*

JOHNSON, Harry Gordon, economist; b. Toronto, Ont., Can., May 26, 1923; s. Henry Herbert and Lily (Muat) J.; B.A., U. Toronto, 1943; B.A., Cambridge (Eng.) U., 1946, M.A., 1951; M.A., U. Toronto, 1947; M.A., U. Manchester (Eng.), 1960; M.A., Harvard, 1949, Ph.D., 1958; LL.D., St. Francis Xavier U., 1965, U. Windsor, 1966; m. Elizabeth Scott Serson, May 28, 1948; children—Steven Ragnar, Karen Eve. Lectr., Cambridge U., 1950-56, fellow King's Coll., 1950-56; prof. econ. theory Manchester U., 1956-59; prof. econs. U. Chgo., 1959—; prof. econs. London (Eng.) Sch. Econs., 1966—. Mem. Rev. Com. on Balance Payments Statistics, 1963-65, Ad Hoc Com. on Basic Research and Nat. Goals, 1964-65. Fellow Am. Acad. Arts and Scis.; mem. Am. Econ. Assn. (exec. Cee 1964-66), Royal Econ. Assn., A.A.A.S., Canadian Polit. Sci. Assn. (pres. 1965-66). Author: International Trade and Economic Growth, 1958; Money, Trade and Economic Growth, 1962; The Canadian Quandary, 1963; The World Economy at the Crossroads, 1965; Economic Policies Toward the Less Developed Countries, 1966; also numerous articles. Research on theory effects econ. growth on internat. trade, theory tariff structure and its effects on internat. trade.*

JOHNSON, Helgi, geologist; b. Akureyri, Iceland, Feb. 3, 1904; s. Gisli and Gudrun Helga (Finnsdottir) J.; B.S., U. Man., 1926, A.M., U. Toronto, 1929, Ph.D., 1934; m. Helen Mary Eliza Hunter, Sept. 12, 1933. Came to U. S., 1929, naturalized, 1948. Research asst. Royal Ont. Mus. Paleontology, Toronto, 1926-29; faculty Rutgers U., 1929—, prof. geology, 1944—, dir. Mus. Geology and Paleontology since 1944, dir. Bur. Mineral Research, 1944—, chmn. dept. geology, 1945—, mem. joint com. N.J. Hwy-Rutgers, 1947—. Geologist for various govt. brs. U. S., Can., Newfoundland; sec., dir. Viking Minerals, Inc.; Mineral Resources, Inc., Trenton, N.J. Joint Research Fellow Geol. Society, Am., Paleontol. Soc.; mem. Am. Assn. Petroleum Geologists, Am. Inst. Mining and Metall. Engrs., Am. Geophys. Union, Icelandic Nat. League, Yellowstone-Bighorn Research Assn., Inc. (exec. dir.), Sigma Xi. Contbr. articles in field to sci. publs. Study of Pleistocene geology western Newfoundland; paleontology and paleozoic stratigraphy of Newfoundland and northeastern N. Am.; geol. studies in Iceland and Greenland, snow and ice studies. Home: Highwood, Easton Av., Somerset, N.J. 08873. Office: Geology dept., Rutgers U., New Brunswick, N.J.

JOHNSON, Hildegard Binder, geographer; b. Berlin, Germany, Aug. 20, 1908; d. Albert Wilhelm and Emma (Gartenschlager) Binder; M.A., Ph.D., U. Berlin, 1933; m. Palmer O. Johnson, Aug. 20, 1936 (dec.); children—Gisela Charlotte, Karin Luise. Came to U. S., 1936, naturalized, 1944. Tchr. high sch., Eng., 1934; instr. Mills Coll., 1935; faculty geography Macalester Coll., 1947—, chmn. dept., since 1955; vis. prof. univs. Ga., Minn., Cal., Wash. Pres., Minn. Council for Social Studies, 1958-60; mem. Minn. Gov.'s Com. for Atomic Devel. Mem. Am., Canadian assns. geographers, Am. Geog. Soc., Minn. Acad. Sci., Nat. Council for Geog. Edn., Minn. Hist. Soc., African Studies Assn. Author: Carta Marina 1525, 1963; also articles. Research on pioneer settlement in Am. Middle West, influence of watersheds on culture groups, truth in hist. maps, significance of original site selection for devel. in Africa. Home: 3312 Edmund Blvd., Mpls. 55406.*

JOHNSON, Irving Stanley, Am. exptl. biologist; b. Grand Junction, Colo., June 30, 1925; s. Walter Glenn and Frances (Tuttle) J.; student Westminster Coll., 1943-44, Cornell U., 1944, Harvard, 1945; A.B., Washburn Municipal U., Topeka, 1948; postgrad. Duke, Northwestern U.; Ph.D., U. Kans., 1953; m. Alwyn Neville Ginther, Jan. 29, 1949; children—Rebecca Lyn, Bryan Glenn, Kirsten Shawn, Kevin Bruce. Bacteriologist, Eli Lilly & Co., Indpls., 1953-58, sr. bacteriologist, 1958-60, research asso., 1960-63, asst. dir., 1963-66, dir. biol. research, 1966—. Mem. Am. Assn. for Cancer Research, Kans., N.Y. acad. scis., Am. Soc. for Cell Biology, A.A.A.S., Internat. Soc. for Chemotherapy, Soc. for Exptl. Biology and Medicine, Sigma Xi, Phi Sigma. Contbr. to profl. textbooks. Research, publs. on cancer chemotherapy; developed tissue culture techniques, viral chemotherapy, virol oncology. Home: 5329 Hedgerow Dr., Indpls. 46226. Office: 740 S. Alabama St., Indpls. 46206.*

JOHNSON, James McIntosh, Am. chemist; b. Newberry, S.C., Aug. 15, 1883; s. William and Mary Eugenia (Kibler) J.; B.S., Newberry Coll., 1902, A.M., 1903; Ph.D., Johns Hopkins, 1908; m. Mabel Rebecca Earnshaw, Mar. 15, 1912; children—William Mercier, Mabel Eleanor, Phoebe Rebecca, Frances Lillian. Chemist, Johns Hopkins Hosp., 1908-09; chemist Bur. Chemistry, U. S. Dept. of Agriculture, 1910-18; research chemist Hygienic Lab. (now NIH), USPHS, Washington, from 1918. Mem. Am. Chem. Soc. Co-discoverer method of making sulpharsphenamine. Contbr. to scientific jours. Died Mar. 2, 1953.

JOHNSON, John Raven, Am. chemist; b. Chgo., Aug. 9, 1900; s. John William and Johanna (Raven) J.; B.S., U. Ill., 1919, M.S., 1920, Ph.D., 1922; student (Am. Field Service fellow), Coll. de France, Paris 1922-24; m. Hope Anderson, Aug. 24, 1929; children—Keith Raven, Leonard Williams. Instr. U. Ill., 1924-27; faculty Cornell U., 1927—, Todd prof. chemistry, 1952-65, prof. emeritus, 1965—. Sci. liaison officer London Mission OSRD, 1942-43; mem. sect. NDRC, 1940-42, 44-45; mem. div. NRC, 1938-41, 44-47. Cons. E. I. duPont de Nemours & Co., 1937—. Fellow N.Y. Acad. Scis., Chem. Soc. London; mem. Nat. Acad. Scis., Am. Chem. Soc., Alpha Chi Sigma, Sigma Xi, Gamma Alpha. Contbr. to sci. publs. Research on organic boron compounds; furan derivatives; explosives; biosynthesis of isoprene derivatives; mechanism of reactions; structure of gliotoxin. Home: Deer Valley Rd., Townshend, Vt. 05353.

JOHNSON, Jotham, Am. archaeologist; b. Newark, Oct. 21, 1905; s. Jotham Clarke and Edith Jennette (Compson) J.; A.B., Princeton, 1926; fellow Am. Sch. Classical Studies, Athens, 1927-28; Ph.D. (Harrison fellow), U. Pa., 1931; m. Elsa Diederich, June 7, 1930 (div.); m. 2d., Sarah Jane Coates, Aug. 16, 1941; 1 son, Jotham. Field dir. U. Pa. Museum expdn., Minturnae, Italy, 1931-34; with U. Pittsburgh, 1937-46; faculty N.Y. U., 1946—, prof., 1953—, chmn. dept. classics, Washington Sq. Coll., 1948—, head, all-u. dept., 1958—; editor Classical Weekly, 1938-42; archaeol. editor Classical Jour., 1946-50; editor Archaeol. Newsletter, 1946-56, Archaeology, 1948-51; editorial cons. New Century Classical Handbook, 1962; Norton lectr. Archaeol. Inst. Am., 1950-51; research scholar under Fulbright Act, Rome, Italy, 1952-53. Mem. Archaeol. Inst. Am. (pres. 1961-64), Am. Assn. of U. Profs., A.A.A.S., Am. Philol. Assn., Classical Assn. Atlantic States, Council Old World Archaeology (trustee), Vergilian Soc. Am., Phi Kappa Sigma. Research and publs. on ancient time-reckoning; calendar revision; Italic fortifications (established criteria for relative dating of polygonal and ashlar city walls, 4th-1st centuries B.C.); alphabet reform. Died Feb. 8, 1967.

JOHNSON, Karl McKibben, Am. physician; b. Terre Haute, Ind., May 14, 1929; s. Marion A. and Laura (McKibben) J.; A.B., Oberlin Coll., 1951; M.S., U. Rochester, 1955, M.D., 1956; m. Patricia Ann Webb, Mar. 21, 1964. Sr. asst. surgeon USPHS, Nat. Inst. Allergy and Infectious Diseases, Bethesda, Md., 1958-60, staff mem., 1960—, chief virus diseases sect. Middle Am. research unit, 1962-64, dir. MARU, Balboa Heights, C.Z., 1964—. Decorated Order of Condor, Govt. of Bolivia, 1964. Mem. Am. Epidemiological Soc., Am. Soc. Microbiology, Soc. for Exptl. Biology and Medicine, Am. Soc. Tropical Medicine and Hygiene. Research, publs. on isolation and characterization of viruses pathogenic for man, myxoviruses, rhinoviruses, epidemiology of respiratory viral infections, isolation, characterization and epidemiology of tropical viruses, Machupo viruses, etiologic agt. of hemorrhagic fever, epidemiology of arthropod-transmitted viruses diseases in tropical Am. Office: Box 2011, Balboa Heights, C.Z.*

JOHNSON, Kenneth Alan, Am. physicist; b. Duluth, Minn., Mar. 26, 1931; s. Fred Henry and Edith Victoria (Nystrom) J.; B.S., Ill. Inst. Tech., 1952; Ph.D., Harvard, 1955; m. Gladys Diaz de los Arcos, June 8, 1954. Research fellow, lectr. Harvard, 1955-57; NSF fellow Inst. for Theoretical Physics, Copenhagen, Denmark, 1957-58; faculty Mass. Inst. Tech., Cambridge, 1958—, prof., 1965—; vis. asso. prof. Princeton, 1964. Sloan Research fellow, 1961-63. Mem. Am. Phys. Soc. Author: (with D. B. Lichtenberg, J. Schwinger, S. Weinberg) Lectures on Particles and Field Theory, 1965; also articles. Research on quantum theory of fields in particular quantum electrodynamics.*

JOHNSON, Laverne Clarence, Am. psychologist; b. Blue Springs, Ala., Jan. 7, 1925; s. William C. and Carrie (Shehane) J.; B.A., Stanford, 1949, M.A., 1952, Ph.D., 1954; m. Margaret Jane Woolen, June 28, 1952; children—Susan Louise, Carol Ruth, David Michael. Instr., Washington U. Sch. Medicine, St. Louis, 1954-58, asst. prof. med. psychology, 1958-60; dir. psychology service Malcolm Bliss Mental Health Center, St. Louis, 1957-60; head psychophysiology div. USN Med. Neuropsychiat. Research Unit, San Diego, 1960—; lectr. psychology San Diego State Coll., 1961—. Diplomate Am. Bd. Examiners in Profl. Psychology. Fellow Am. Psychol. Assn.; mem. San Diego County Psychol. Soc. (pres. 1966—), Am. Western electroencephalographic socs., A.A.A.S., Soc. for Psychophysiol. Research, Assn. for Psychophysiol. Study Sleep, Western Psychol. Assn. Research, publs. on relation brain wave activity to psychophysiology and behavior. Home: 8409 Abbotshill Rd., San Diego 92123. Office: USN Med. Neuropsychiat. Research Unit, San Diego 92152.*

JOHNSON, LeRoy Peter Vernon, Canadian geneticist; b. Stavely, Alta., Can., July 28, 1905; s. Lars Frederick and Mary (Johnson) J.; B.S., U. Alta., 1931, M.S., 1933; Ph.D., State U. Wash., 1935; m. Reta May Delta Hewitt, Mar. 5, 1938; children—Robert Frederick, Elaine Marilyn, Linda Elise. Forest geneticist NRC, Ottawa, Ont., 1938-46; faculty U. Fla., 1946-47, State U. Wash., 1947-48; prof. genetics U. Alta., Edmonton, 1948—. With FAO, Syria, 1953; Nuffield Found. fellow, Eng., 1959-60; with Ford Found., Philippines and Pakistan, 1965-67. Mem. Genetics Soc. Can., A.A.A.S. Author: An Introduction to Applied Biometrics, 1950; In The Time of the Thetans, 1960; Biometrical Methods, 1963; also numerous articles. Research in genetics and biometrics, breeding of new varieties of forage and cereal crops and forest trees. Home: 11735 91st Av., Edmonton, Alta., Can.*

JOHNSON, Manuel John, astronomer; b. Macao, China, May 23, 1805; s. John William Johnson; B.A., Oxford, 1839, M.A., 1842; m. dau. of Dr. Ogle; several children. Entered St. Helena arty., 1821, became a.d.c. to Gen. Walker; in charge St. Helena Obs.; keeper Radcliffe Obs., 1839. Astron. prize founded in his honor at Oxford, 1862. Fellow Royal Soc., 1856; mem. Royal Astron. Soc. (pres. 1857-58, Gold medal). Author: A Catalogue of 606 Principal Fixed Stars in the Southern Hemisphere (only source besides Madras catalogues for exact places of fixed stars in regions beyond reach of European observatories), 1835; Radcliffe Observations. Observed solar eclipse, 1832; made observations and measurements with large heliometer; used elec. transit recorder, 1858. Died Feb. 28, 1859.

JOHNSON, Martin Wiggo, Am. marine biologist; b. Chandler, S.D., Sept. 30, 1893; s. Christian Hans and Julia M. (Hansen) J.; B.S. cum laude, U. Wash., 1924, M.S., 1930, Ph.D., 1931; m. Lelia T. Clutter, Apr. 16, 1924; children—Byron M., Phyllis T. Curator, Puget Sound Biol. Sta. (Friday Harbor Labs.), U. Wash., 1924-29, research asso., 1933-34, prof., summers 1960-62; biologist Internat. Passamaquoddy Fisheries Commn., 1932-33; faculty Scripps Instn. Oceanography U. Cal., La Jolla, 1934—, biologist div. War Research, 1942-46, prof., 1949-62, prof. emeritus, research biologist, 1962—. Mem. staff Operation Crossroads USN, 1946; del. NRC, U. Cal. 7th Pacific Sci. Congress, New Zealand, 1949, 8th Congress, P.I., 1953; prof. Marine Biol. Lab., Woods Hole, Mass., summer 1956; mem. UNESCO adv. com. Biol. Center, India, 1962-63. Recipient certificate of commendation for outstanding research World War II USN, 1947; Agassiz Gold medal for contbns. to field oceanography Nat. Acad. Sci., 1959. Fellow Cal. Acad. Sci., San Diego Soc. Naturalists, A.A.A.S.; mem. Soc. Limnology and Oceanography (past sect. pres.), Western Soc. Naturalists (past pres.), Am. Micros. Soc., Soc. Systematic Zoology, Sigma Xi. Author: (with H. U. Sverdrup, R. H. Fleming) The Oceans, 1942. Contbr. numerous articles to profl. jours. Pioneer gen. biol. nature sonic deep scattering layer; research biol. underwater sounds of importance to sonar. Home: 2524 Ellentown Rd. Office: Scripp Inst. Oceanography, La Jolla, Cal. 92037.*

JOHNSON, Montgomery Hunt, Am. physicist; b. Utica, N.Y., Nov. 21, 1907; s. Montgomery Hunt and Frances (Munger) J.; A.B., Harvard, 1929, M.A., 1931, Ph.D., 1932; m. Gwyneth Mary Johnson, Apr. 6, 1946; children—Sarah Langhorne, Gwyneth Mary. Prof., N.Y. U., 1932-41; physicist Sperry Gyroscope Co., Garden City, L.I., N.Y., 1941-42, Mass. Inst. Tech. Radiation Lab., Cambridge, 1942-45, Naval Research Lab., Washington, 1945-52, U. Cal. Radiation Lab., Livermore, 1952-54, Lockheed Missile Systems Div., Van Nuys, Cal., 1954-55; physicist, v.p., treas. Systems Research Corp., Burbank, Cal., 1955-56; physicist, chief scientist research lab., aeronutronic div. Ford Motor Co., Newport Beach, Cal., 1956-63, dir. physics Philco Research Labs., 1964-66, dir. devel. planning Aeronutronic div. Philco Ford, 1966—. Cons. U. Cal. Radiation Lab., 1960—; faculty applied sci. dept. U. Cal. at Davis, 1964—, dir. Aeronutronic Systems, Inc., Glendale, Cal. Fellow Am. Phys. Soc.; mem. Research Soc. Am., Am. Rocket Soc., Phi Beta Kappa, Sigma Xi. Contbr. articles to profl. jours. Research on quantum theory; nuclear and atomic physics; absorbent materials; electrodynamics. Home: 19002 E. Dodge Ave., Santa Ana, Cal. Office: Ford Rd., Newport Beach, Cal.

JOHNSON, Noah R., Am. nuclear chemist; b. Kingsport, Tenn., Oct. 15, 1928; s. Noah R. and Alma (Lambert) J.; B.S., E. Tenn. State U., 1950; Ph.D., Fla. State U., 1956; m. Rosemary McElroy, Aug. 19, 1950; children—Kurt McElroy, Gregory Alan, Gwendolyn Rae. Nuclear chemist Oak Ridge Nat. Lab., 1956-62, 63—; with Inst. Theoretical Physics, Copenhagen, Denmark and Danish Atomic Energy Commn., Riso, 1962-63. Guggenheim fellow, 1962-63; Fulbright Research scholar, 1962-63. Mem. Am. Phys. Soc., Research Soc. Am., Sigma Pi Sigma. Author: (with Eugene Eichler, G. Davis O'Kelley) Nuclear Chemistry, 1963. Contbr. articles to profl. jours. Developer exptl. techniques for measurements nuclear properties by scintillation spectrometry; research in nuclear structure studies from disintegration schemes of radioactive nuclei and from nuclear reactions. Home: 997 W. Outer Dr., Oak Ridge 37831. Office: P.O. Box X, Oak Ridge 37830.*

JOHNSON, Peter Dexter, Am. phys. chemist; b. Norwich, Conn., July 1, 1921; s. Philip Adams and Edith (Dexter) J.; S.B., Harvard, 1942; M.A., U. N.C., 1948, Ph.D., 1949; m. Jessie Lois Jones, Oct. 3, 1943; children—Peter Dexter, Carol Osborne, William Todd. Supr. ballistic testing Hercules Powder Co., Radford, Va., 1942-43; research asso. Gen. Electric Co. Research & Devel. Center, Schenectady, 1949—. Vis. asso. prof. Cornell U., Ithaca, N.Y., 1958-59. Fellow Am. Phys. Soc.; mem. Am. Chem. Soc., Optical Soc. Am., A.A.A.S., Sigma Xi. Research on luminescence leading to fundamental understanding of mechanisms of optical absorption and emission in inorganic phosphors; new equipment designs for far ultraviolet spectroscopy; research and devel. new types of arc-discharge lamps. Patentee new light sources and radiation-measuring devices. Home: 1100 Merlin Dr., Schenectady 12309. Office: care Gen. Electric Research & Devel. Center, Schenectady 12301.*

JOHNSON, Philip Carl, Am. physician; b. White Plains, N.Y., Nov. 12, 1924; s. Philip Carl and Josephine (Losee) J.; B.S., U. Mich., 1948, M.D., 1949; m. Virginia Snyder, July 1, 1950; children—Philip Carl III, Christopher Carl. Faculty, U. Mich., 1952-54, asst. chief radioisotope lab. U. Mich. Hosp., 1954; chief radioisotope service, asst. chief profl. service for research VA Hosp., Oklahoma City, 1955-60; asso. prof. medicine U. Okla., 1958-60; dir. radioisotope Lab., Meth. Hosp., Houston, 1960—; asso. prof. medicine Baylor U., Houston, 1960-67, prof. medicine, 1967—. Mem. A.M.A., A.A.A.S., Central Soc. for Clin. Research, Nuclear Medicine Soc., A.C.P., Sigma Xi. Research, numerous publs. on use radioisotopes in clin. medicine, thyroid, cardiac function, drugs. Home: 12122 Rip Van Winkle St., Houston 77024. Office: 6515 Bertner Av., Houston 77025.*

JOHNSON, Ralph Gordon, Am. biologist; b. Oak Park, Ill., July 5, 1927; s. Ralph Emil and Edna (John) J.; B.A., N. Central Coll., 1950; Ph.D., U. Chgo., 1954; m. Helen Francis Tobin, June 3, 1949; children—David Scott, Gordon Tobin, Maria Louise. Faculty, U. Chgo., 1953—, asso. prof., 1963—; adj. prof. paleontology U. Pacific, 1963—; research asso. Field Mus. Natural History, 1963—. Mem. Am. Soc. Zoologists, Soc. For Study Evolution, Am. Soc. Limnology and Oceanography, Soc. Vertebrate Paleontologists. Editor: Evolution, 1966—. Research in fossil and modern marine communities; developed methods for analysis modern and fossil marine communities; described variations phys. environment and effects upon marine organisms. Home: 509 Blair St., Park Forest, Ill. 60466. Office: Dept. Geophys. Scis., U. Chgo., Chgo. 60637.*

JOHNSON, Ralph M., Am. biochemist, educator; b. Ririe, Ida., Apr. 9, 1918; s. Ralph M. and Gertrude (Marler) J.; B.S., Ida. State U., 1940; M.S., U. Wis., 1947, Ph.D., 1948; m. Genevieve Porter, Aug. 8, 1940; children—Karen, Christian, Wilford. Research asso., asst. to sci. dir. Detroit Inst. Cancer Research, 1948-59; research prof., dir. research labs. Inst. Nutrition and Food Tech., Ohio State U., Columbus, 1959-64, dir. Inst., 1963—, dean coll. Biol. Scis., 1966—. Mem. Am. Soc. Biol. Chemists, Am. Inst. Nutrition, Phi Kappa Phi, Sigma Xi, Gamma Sigma Delta. Research, numerous publs. on biochemistry of cell division, role of dietary fats in health and disease. Home: 4103 Fairfax Dr., Columbus, O. 43212.*

JOHNSON, Robert Eugene, Am. physiologist; b. Conrad, Mont., Apr. 8, 1911; s. Arthur D. and Florence (Disbrow) J.; B.S., U. Wash., 1931; B.A. (Rhodes scholar), Oxford U., 1934, Ph.D., 1935; M.D., Harvard, 1941; m. Margaret Hunter, Jan. 11, 1935; children—Thomas Arthur, Charles William, Katherine (dec.). Jr. demonstrator biochemistry Oxford U., 1934-35; research asst. to asst. prof. indsl. physiology Harvard Fatigue Lab., 1935-46; dir. U. S. Army Med. Nutrition Lab., Chgo., 1946-49; prof. physiology U. Ill., Urbana, 1949—, head, dept., 1949-60, dir. honors program, 1959-67. NSF postdoctoral fellow, 1957-58, Guggenheim fellow, 1964-65. Mem. Am. Physiol. Soc., Am. Soc. Clin. Investigation. Author: (with Consolazio, Marek) Metabolic Methods, 1953; (with Consolazio, Pecora) Physiological Measurements, 1963. Research, publs. on metabolic and nutritional effects of stress, as exercise, starvation, and exposure to heat, cold, and altitude. Home: 804 W. Green St., Urbana, Ill. 61801.*

JOHNSON, Robert Joseph, Am. anatomist; b. Toppenish, Wash., Feb. 8, 1915; s. Joseph Oliver and Minnie M. (Godfrey) J.; student Ia. State Coll., 1932-37; M.D., U. Ia., 1943; m. Dorothy Etsinger, Dec. 23, 1941; children—Lynn S., Patricia E., Nora J. Mem. faculty U. Wash. Sch. Medicine, 1946-57, U. Wash., 1951-57, U. W.Va., 1957-63; prof., chmn. dept. anatomy Grad. Sch. Medicine, U. Pa., 1963—. Basic sci. examiner anatomy Wash. State, 1948-57; coroner Monongalia County, W.Va., 1961-63. Pres. W.Va. div. Am. Cancer Soc., 1961-63. Mem. Canadian, Am. (exec. com. 1960-64), assns. anatomists, Am. Acad. Forensic Scis. Research and publs. on the sympathetic nerves in the neck and pelvis, on the circulation of blood in the fetus and newborn, on the distbn. of the phrenic nerves within the diaphragm, and on the embryologic explanation of anal anomalies. Home: 1234 Wyngate Rd., Wynnewood, Pa. 19096. Office: U. Pennsylvania, Phila. 19104.*

JOHNSON, Roy Melvin, Am. microbiologist; b. Chgo., Sept. 8, 1926; s. Carl Henning, and Ebba D. (Helstrom) J.; A.B., U. Chgo., 1949, M.S., 1951; Ph.D., U. N.M., 1955; m. Betty Lou Schutt, Sept. 6, 1952; children—Renee Marcia, Roberta Maureen, Rhonda Margo, Regan Michelle. Instr. biology Ariz. State U., Tempe, 1952-53; faculty, 1955—, prof. microbiology, 1965—; scientist 8th Cruise, Indian Ocean Expdn., 1964. NIH postdoctoral fellow U. Ind., 1963-64. Mem. Am. Soc. Microbiology, A.A.A.S., Am. Assn. U. Profs., Ariz. Acad. Sci., Sigma Xi. Research, publs. on physiology and microculture sporulating bacteria, effect antibiotics and bacterial virus, physiology and taxonomy marine bacteria and viruses. Home: 1826 Aster St., Tempe, Ariz. 85281.*

JOHNSON, Samuel William, Am. agrl. chemist; b. Kingsboro, N.Y., July 3, 1830; s. Abner A. and Annah W. (Gilbert) J.; studied at Yale (now Sheffield) Scientific School and univs. of Leipzig and Munich. Prof. analytical and agrl. chemistry, 1856-74; prof. theoretical and agrl. chemistry Sheffield Scientific Sch., 1874-95. Dir. Conn. Agrl. Expt. Sta., 1877-1899. Pres. Am. Chem. Soc., 1878; chmn. sub-section of chemistry, A.A.A.S., 1875; mem. Nat. Acad. Sciences. Author: Essays on Peat Muck and Commercial Manures, 1859; Peat and Its Uses as Fertilizer and Fuel, 1866; How Crops Grow, 1868, 1890; How Crops Feed, 1870. Devised and improved apparatus for analysis of fertilizers. Died 1909.

JOHNSON, Sidney, Am. chemist; b. Connersville, Ind., Mar. 20, 1925; s. Herbert Halston and Kate (Worrell) J.; student St. Lawrence U., 1943-44, Rensselaer Poly. Inst., 1944, N.C. U., 1944, Cornell U., 1944-46; B.S., Purdue U., 1948, Ph.D., 1953; m. Roxana Marie Fye, July 27, 1947; children—Andrew Jay, Sherry Lou, Judith Ann, Daniel Dee, Thomas Lee. Chemist, Metal Hydrides Inc., Beverly, Mass., 1953, group leader chem. research labs., 1953-58, asst. dir. research and devel. labs., 1958-62; chief, br. physics and surface chemistry Office of Saline Water, U. S. Dept. Interior, Washington, 1962-65, chief, div. applied sci., 1965—. Cons. hydrides for propellant and nuclear applications. Mem. Am. Chem. Soc., Electrochem. Soc., A.A.A.S., Am. Inst. Aeros. and Astronautics. Research, publs. on phosphorus pentafluoride, hydrides. Home: 1001 Potomac Lane, Alexandria, Va. 22308. Office: Office of Saline Water, U. S. Dept. Interior, Washington 20240.*

JOHNSON, Sture Archie Mansfield, Am. physician; b. Morgan, Ore., Apr. 24, 1907; s. Per and Olga (Miller) J.; Ph.C., N. Pacific Coll., 1928; B.A., U. Ore., 1934, M.D., 1938; m. Geneva Frances Beane, June 7, 1936. Asst. prof. U. Mich., 1944-46; prof., head dept. dermatology U. Wis. Med. Sch., Madison, 1946—; chief cons. dermatology VA Hosp., Madison, 1961—. Fellow A.C.P.; mem. Am. Dermatol. Assn., Soc. Investigative Dermatology (v.p.), Acad. Dermatologists (dir. 1947-51, 64—), Wis. Dermatol. Soc. (past pres.). Splty. editorial bd. Post Grad. Medicine, 1960—. Research, numerous publs. on fungus diseases, collagen diseases, tryptophanemetabolism in various diseases, amino acids on surface skin, syphilis. Home: 313 New Castle Way, Madison, Wis. 53704.*

JOHNSON, Thomas, Brit. botanist; b. Selby, Eng., 1604/05; B.Physic, Oxford (Eng.) U., 1642, M.D., 1643. Apothecary in London by 1626; in bus. with physic garden, Snow Hill, London, 1629; joined Royalists during Civil war. Mem. Apothecaries Co. Author: Iter plantarum investigationis ergo susceptum a decem sociis in agrum cantianum (1st English plant catalogue), 1629; The Herball . . . gathered by John Ferarde . . . very much enlarged and amended by Thomas Johnson, citizen and apothecary of London, 1633; Mercurius botanicus, sive plantarum gratia suscepti itineris . . . , 1634; Mercurii Bot. pars altera, 1641. Described new plant species. Killed in siege of Basing Castle, Sept. 28, 1644.

JOHNSON, Thomas, Brit. botanist; b. Barton-on-Humber, 1863; ed. Elmfield Coll., York, Eng.; Royal Coll. Sci., London; D.Sc., London; m. Bessie Stratton Rowe; 2 sons, 1 dau. Demonstrator botany Coll., Sci., Dublin, 1885-90, prof. botany, 1890-1928; keeper, founder bot. div. Nat. Mus., Dublin, 1891-1928; 1st dir. seed-testing, plant disease stas. Dept. Agr., 1900-10. Fellow Linnaean Soc.; mem. Brit. Assn. for Advancement Sci. (life), Royal Irish Acad. (v.p.), Royal Dublin Soc. Author: The Inometer: a new form of Food Chart; also papers on plant pathology, parasitic plants, seaweeds, seed-testing, Irish fossil plants. Died Sept. 9, 1954.

JOHNSON, Thomas Hope, Am. physicist; b. nr. Coldwater, Mich., Sept. 12, 1899; s. Henry E. and Anna (Darling) J.; B.A., Amherst Coll., 1920, M.S. (hon.), 1940; Ph.D., Yale, 1926; m. Anna Griffin, Jan. 24, 1930; m. 2d, Dorothy D. Montgomery, Sept. 1, 1965. Asst. dir. Bartol Research Found., Swarthmore, Pa., 1927-42; asso. dir. Ballistic Research Lab., Aberdeen, Md., 1942-47; chmn. physics dept. Brookhaven Nat. Lab., Upton, N.Y., 1947-52; dir. research U. S. AEC, Washington, 1952-57; v.-p., mgr. research Raytheon Co., Waltham, Mass., 1957-65, cons., 1965—. Recipient Medal for Merit with Citation, 1947; Outstanding Service medal, U. S. AEC, 1957. Study of geomagnetic effects; molecular beams; mesons; diffraction of atoms; high pressure cloud chambers; application of microwaves for measuring movement of projectiles in the bore. Home: 44 Chestnut St., Boston 02108. Office: Raytheon Co., Waltham, Mass. 02154.

JOHNSON, Thomas Nick, Am. neuroanatomist; b. Davenport, Ia., Aug. 20, 1923; s. Nick Thomas and Martha (Lekos) J.; B.S., St. Ambrose Coll., 1944; M.S., Mich. State Coll., 1949; Ph.D. in Neuroanatomy, U. Mich., 1953; m. Despina Mary Gerachis, June 26, 1955; children—Debbie Martha, Madeline Helen, Donna Mary. Faculty, Mich. State Coll., 1946-54; faculty George Washington U. Sch. Medicine, Washington, 1954—, asso. prof. neuroanatomy, 1959—; spl. trainee dept. anatomy Med. Center, Nat. Inst. Neurol. Diseases and Blindness, Los Angeles, 1958-59. Spl. cons. Nat. Inst. Mental Health, 1962—, St. Elizabeth's Hosp., Washington, 1962—. Mem. A.A.A.S., Assn. Anatomists, Am. Acad. Neurology (asso.). Research, publs. on comparative neurology mammalian midbrain and forebrain, exptl. neurology extrapyramidal motor systems. Home: 12610 Laurie Dr., Silver Spring, Md. 20904. Office: 1335 H St. N.W., Washington 20005.*

JOHNSON, Treat Baldwin, Am. chemist; b. Bethany, Conn., Mar. 29, 1875; s. Dwight Lauren and Harriet (Baldwin) J.; Ph.B., Sheffield Scientific Sch. (Yale), 1898; Ph.D., Yale, 1901; m. E. Estelle Amerman, June 29, 1904. Instr. in organic chemistry, Sterling Lab., Yale, 1901-08, asst. prof., 1908-16, prof. 1917-25, Sterling professor chemistry, 1925-43; prof. emeritus in 1943. President Bethany Library Association, Inc.; director Bethwood Research Lab., Bethany, Conn. Mem. bd. trustees, Yale-in-China. Member Society Biol. Chemists, American Biological Society, Am.

Chem. Soc., Am. Assn. Univ. Profs., Nat. Acad. Sciences, Conn. Acad. Arts and Sciences, Nat. Geog. Soc., Am. Inst. Chemists (pres. 1926-27), Nat. Research Council, Deutsche Chemische Gesellschaft, Acad. Science (Halle, Germany), Sigma Xi, Alpha Chi Sigma. Mem. ednl. advisory com. N.B.C. Radio Corp. Research, publs. on biochemistry, organic chemistry, prodn. synthetic organic compounds including investigations of chemistry of vital cells, bacteria and tubercle bacilli (discovered that nucleus of tubercle bacillus has many elements giving life to human body and that germ is protected by waxy armor, 1925). Died July 28, 1947.

JOHNSON, Victor, Am. physiologist; b. Chgo., Jan. 19, 1901; s. Eric and Carrie (Pearson) J.; Ph.B., U. Chgo., 1926, Ph.D., 1930, M.D., 1939; D.Sc., Rockford Coll., 1954; m. Maria Bacca, Jan. 12, 1962; 1 son, Victor Raymond. Mem. faculty U. Chgo., 1929-47; Dir. Mayo Grad. Sch. Medicine, Rochester, Minn., 1947-66, prof. physiology, U. Minn., 1947-66; cons. Edn. and Tng. Div., Office Surgeon Gen., AUS, 1958-66. Dir. Am. Med. Edn. Found., 1950-61; adv. council Nat. Fund for Med. Edn., 1950-62; mem. sci. adv. council Inst. Advancement Medical Communication, 1960-66; nat. civilian cons. to surgeon gen. USAF, 1961-66. Mem. Am. Assn. U. Profs., A.A.A.S., A.M.A., A.C.P., Am. Physiol. Soc., Phi Beta Kappa, Sigma Xi, Alpha Omega Alpha (dir. Honor Med. Soc. 1959-66, pres.). Author: (with another) Elements of Electrocardiographic Interpretation, 1944; Machinery of the Body, 1941, 5th edit., 1961. Contbr. sci. articles to Am. Jour. Physiology. Research in physiology of blood and circulation. Home: Costa de la Calma, Mallorca, Spain.*

JOHNSON, Victor Joseph, Am. mech. engr.; b. Lyons, Neb., Apr. 30, 1916; s. Oliver V. and Anne J. (Lawless) J.; B.S. in Mech. Engring., Tri-State Coll., Angola, Ind., 1939; postgrad. U. Md., 1952, U. Colo., George Washington U. Design engr. Bucyrus Erie Co. South Milwaukee, Wis., 1939-41; indsl. engr., structural designer Panama Canal, C.Z., Washington, 1941-46; head facilities engring. br. Naval Research Lab., Washington, 1946-50; with Nat. Bur. Standards, 1950—, dir. Cryogenic Data Center, Inst. for Materials Research, Boulder, Colo., 1958—. Vis., lectr. U. Colo., 1955-56; cons. to pvt. cos., 1954—. Recipient Dept. Commerce Gold medal for exceptional service, 1953. Mem. Am. Soc. Heating, Refrigerating and Air Conditioning Engrs. (Distinguished Service award 1966), A.A.A.S., Sci. Research Soc. Am., Am. Soc. M.E., Am. Inst. Aeros. and Astronautics, Sigma Xi, Tau Sigma Eta, Pi Tau Sigma. Editor, contbg. author: Compendium of Properties of Materials at Low Temperatures, 4 vols., 1960. Research, publs. on refrigeration and cryogenics, purification of gases, heat exchanger design, liquefaction cycle efficiencies, ortho-para conversion of hydrogen, standards reference data on properties of fluids and structural solids at low temperature, design and devel. of world wide data center and tech. information retrieval system. Home: 1380 55th St. Office: Nat. Bur. Standards, Boulder, Colo. 80302.*

JOHNSON, Virgil Allen, Am. agronomist; b. Lincoln, Neb., June 28, 1921; s. Oscar J. and Fairy (Johnson) J.; B.S., U. Neb., 1948, Ph.D., 1952; m. Betty Ann Tisthammer, July 29, 1943; children—Karen (Mrs. Ronald Eakes), Leslie, Reed, Scott. Faculty, U. Neb., Lincoln, 1952—, prof. agronomy, 1954—, research agronomist U. S. Dept. agr., 1954—, regional leader Hard Red Winter Wheat Research. Sec., Hard Red Winter Wheat Improvement Com.; mem. Nat. Wheat Improvement Com.; mem. exec. com. Hard Winter Wheat Quality Adv. Council. Recipient Ofcl. Commendation Bd. Regents U. Neb., 1966; award for outstanding agrl. achievement Neb. Certified Seed Producers, 1966. Mem. Am. Soc. Agronomy, Sigma Xi, Gamma Sigma Delta, Alpha Zeta. Co-developer nine improved varieties of hard red winter wheat; discovered genes for male fertility restoration in common wheat which are necessary for hybrid wheat; research on inheritance and physiology of high protein in wheat. Home: 3849 Dudley St., Lincoln, Neb. 68503.*

JOHNSON, Virginia Eshelman, Am. psychologist; b. Springfield, Mo., Feb. 11, 1925; d. Harry Hershel and Edna (Evans) Eshelman; student Drury Coll., 1940-42, Mo. U., 1944-47; m. George Johnson, June 13, 1950 (div. Sept. 1956); children—Scott Forstall, Lisa Evans. Adminstrv. asst. St. Louis Daily Record, 1947-50; marketing research staff CBS, sta. KMOX, St. Louis, 1950-51; research staff Washington U. Sch. Medicine, St. Louis, 1957-64; research asso. Reproductive Biology Research Found., St. Louis, 1964—. Fellow Soc. for Sci. Study Sex. Author: (with Dr. William H. Masters) Human Sexual Response, 1966; also articles. Assisted (with Dr. Masters) in defining physiology of human sexual response using methods of direct measurement; devel. therapy technique of altering problems of sexual effectiveness by combining behavioral therapy with physiologic orientation. Home: 50 Salem Estates, Ladue, Mo. 63124. Office: 4910 Forest Park, St. Louis 63108.*

JOHNSON, Walter Curtis, Am. engr., educator; b. Weikert, Pa., Jan. 6, 1913; s. David C. and Mary (Ely) J.; B.S., Pa. State Coll., 1934, E.E., 1942; m. Carolyn Shirk, Sept. 1, 1934; children—Walter Curtis, William Stanford, David Edward. Engr., Gen. Electric Co., Schenectady, 1934-37; Mem. faculty, Princeton, 1937—, Arthur LeGrand Doty prof. engring., 1963—, chmn. dept., 1950-65; engring. cons. various cos. Fellow I.E.E.E. (Nat. award 1939); mem. Am. Soc. Engring. Edn. (Western Electric award 1967), Sigma Xi. Author: Mathematical and Physical Principles of Engineering Analysis, 1944. Transmission Lines and Networks, 1950; (with P. R. Clement) Electrical Engineering Science, 1960. Patentee in field of telemetering systems, electronic switching, and spl.-purpose analog computers. Research and publs. in analog computation, analysis of linear systems, wave guides, theory of magnetic amplifiers. Home: 20 McCosh Circle, Princeton, N.J. 08540.*

JOHNSON, William, Australian geologist; b. Perth, Western Australia, Feb. 26, 1920; s. John William and Clara (Stevens) J.; B.S. with honors, U. Western Australia, 1942; m. Patricia Maria O'Sullivan, Apr. 14, 1945; children—Christine R., John W., Geoffrey A., Michael P., David P., Andrew C. Geologist, Geol. Survey Western Australia, Perth, 1946-50; geologist in charge Met. Water Sewerage and Drainage Bd., Sydney, 1950-55; geologist Geol. Survey S. Australia, Adelaide, 1955-57, sr. geologist, 1957-63; cons. geologist, partner Fitzpatrick, Johnson & Assos., Adelaide, 1963-67; mng. dir. W. Johnson & Assos. Pty., Ltd., Adelaide, 1967—. Mem. Geol. Soc. Australia, Australian Inst. Mining and Metallurgy, Royal Soc. S. Australia, Royal Soc. Western Australia, Am. Assn. Petroleum Geologists. Research, publs. on occurrence of oil, coal, ground water, uranium, other metallic minerals. Home: 5 Horn Ct., Walkerville, South Australia. Office: 323 Wakefield St., Adelaide, South Australia.*

JOHNSON, William Howard, Am. agrl. engr.; b. Sidney, O., Sept. 3, 1922; s. Russell Earl and Dollie (Gamble) J.; B.S., Ohio State U., 1948, M.S., 1953; Ph.D., Mich. State U., 1960; m. Wyoma Jean Swift, Oct. 2, 1943; children—Lawrence Alan, Cheri Ellen, Dana Sue. Faculty Ohio Agrl. Expt. Sta., Wooster, O., 1948-64; faculty Ohio Agrl. Research and Devel. Center, Wooster, 1964—, prof., asso. chmn. dept. agrl. engring., 1959—; part-time prof. Ohio State U., 1964—. Cons. farm equipment cos. Mem. Am. Soc. Agrl. Engrs., Sigma Xi, Tau Beta Pi. Author: (with B. J. Lamp) Principles, Equipment and Systems for Corn Harvesting, 1966; also articles. Research on soil-plant-machine relationships, harvesting, design for soiltillers, planters, harvesters. Home: 935 Fenwick Pl., Wooster, O. 44691.*

JOHNSON, William Summer, Am. chemist; b. New Rochelle, N.Y., Feb. 24, 1913; s. Roy Wilder and Josephine (Summer) J.; B.A., Amherst Coll., 1936; M.A., Harvard, 1938, Ph.D., 1940; D.Sc., Amherst Coll., 1956; m. Barbara Allen, Dec. 27, 1940. Mem. faculty Amherst Coll., 1936-37; faculty U. Wis., 1940-60, Homer Adkins prof. chemistry, 1954-60; prof., exec. head dept. chemistry Stanford, 1960—; vis. prof. Harvard, 1954-55. Mem. chem. adv. panel NSF, 1952-56; sec. organic sect. Internat. Congress Pure and Applied Chemistry, 1951. Recipient medal Synthetic Organic Chem. Mfrs. Assn., 1958. Fellow London Chem. Soc.; mem. Am. Acad. Arts and Scis., Swiss, Am. chem. socs., Nat. Acad. Scis., Phi Beta Kappa, Sigma Xi. Bd. editors Organic Syntheses, Vol. 34, 1954; Jour. Am. Chem. Soc., 1956-65, Jour. of Organic Chemistry, 1954-56. Research and publs. on total synthesis of natural products, particularly steroids (for example equilenin, estrone, testosterone, others); also on the cyclization of polyofenic systems, and devel. of new stereoselective approaches to polycyclic substances. Home; 191 Meadowood Dr., Portola Valley, Cal. 94206. Office: Dept. Chemistry, Stanford U., Stanford, Cal. 94305.*

JOHNSON, William Woolsey, Am. mathematician; b. Owego, N.Y., June 23, 1841; s. Charles Frederick and Sarah Dwight (Woolsey) J.; A.B., Yale U., 1862, A.M., 1868; LL.D., St. John's Coll., Md., 1915; m. Susannah Leverett Batcheller, Aug. 12, 1869 (died 1916). In U. S. Nautical Almanac Office, Cambridge, Mass., 1862-64; asst. prof. mathematics, U. S. Naval Acad., at Newport, R.I., and Annapolis, 1864-70; prof. mathematics, Kenyon Coll., Ohio, 1870-72, St. John's Coll., Md., 1872-81, U. S. Naval Acad., from 1881; prof. mathematics in the Navy, from 1913. Author: Differential Calculus (Rice and Johnson), 1879; An Elementary Treatise on the Integral Calculus, 1881; Curve Tracing in Cartesian Co-ordinates, 1884; The Theory of Errors and Method of Least Squares, 1890; Treatise on Differential Equations, 1889; Theoretical Mechanics, 1901; Treatise on Differential Calculus, 1904; Differential Equations (math. monographs), 1906; Treatise on Integral Calculus, 1907; An Elementary Treatise on the Differential Calculus, 1908. One of best known expository mathematicians of his time, chiefly due to his numerous contbns. to math. lit. Died May 14, 1927.

JOHNSTON, Alexander Keith, Brit. geographer; b. Kirkhill, Dec. 28, 1804; s. Andrew and Isabel (Keith) J.; ed. U. Edinburgh, LL.D., 1865; m. Margaret Johnston, Aug. 3, 1837; 11 children, including Alexander Keith. Engraver early in career; apptd. geographer in ordinary to Queen at Edinburgh, 1840; toured Germany, 1842, Palestine, 1863. Recipient medal London Internat. Exhbn., 1851. Fellow Royal Soc., 1810, Royal Geog. Soc. (Victoria medal 1871), Royal Geol. Soc.; mem. Royal Soc. Edinburgh, Edinburgh Geol. Soc., Scottish Meteorol. Soc. (founder), also corr. mem. various other geog. socs. Author: National Atlas of Historical, Commercial and Political Geography, 1843; The Physical Atlas of Natural Phenomenon . . . , 1848; Atlas of Physical Geography, 1852. The Royal Atlas of Modern Geography, 1861 (most accurate atlas of that time). First to produce globe illustrative of phys. geography. Died Ben Rhydding, Yorkshire, July 9, 1871.

JOHNSTON, George Ben, Am. surgeon; b. Tazewell, Va., July 25, 1853; s. John Warfield and Nicketti (Floyd) J.; student U. of Va., 1870-75; M.D., Univ. Med. Coll. (New York U.), 1876; LL.D., St. Francis Xavier Coll., 1897; m. Helen Rutherfoord, 1892. Prof. surgery, Med. Coll. of Va.; surgeon Memorial Hosp., Richmond, to Johnston-Willis Sanatorium, and to Abingdon (Va.) Hosp. Fellow Internat. Surg. Soc., American Surg. Assn. (pres. 1904-05). First surgeon in Va. to give pub. demonstration of Listerism; performed 1st kidney removal operation reported in southern U. S. Died Dec. 20, 1916.

JOHNSTON, Harold S(ledge), chemist; b. Woodstock, Ga., Oct. 11, 1920; s. Smith L. and Florine (Dial) J.; A.B., Emory U., 1941, D.Sc., 1965; Ph.D., Cal. Inst. Tech., 1947; m. Mary Ella Stay, Dec. 29, 1948; children—Shirley Louise, Linda Marie, David Finley. Research asst. NDRC, Cal. Inst. Tech., 1942-45; faculty Stanford, 1947-56, asso. prof., 1954-56; asso. prof. Cal. Inst. Tech., 1956-57; prof. chemistry U. Cal. at Berkeley, 1957—. Sloan fellow, 1955-59; Guggenheim fellow, Brussels, Belgium, 1961; NATO vis. prof., Rome, 1964. Mem. Nat. Acad. Scis., Am. Chem. Soc. (Gold medal Cal. sect. 1956), Am. Phys. Soc., A.A.A.S. Author: Gas Phase Reaction Rate Theory, 1966; also numerous articles. Research on gas-phase reaction rates; photochemistry; unimolecular reactions; kinetic isotope effects. Home: 132 Highland Blvd., Berkeley, Cal. 94708.*

JOHNSTON, Sir Harry (Hamilton), Brit. explorer; b. London, June 12, 1858; s. John Brookes and Esther Laetitia (Hamilton) J.; ed. King's Coll., London; student Royal Acad. Arts, 1876-80; D.Sc., Cambridge U.; m. Winifred Irby; traveled in N. Africa, 1879-80; explored Portuguese W. Africa, River Congo, 1882-83; comdr. sci. expdn. Royal Soc. to Mt. Kilimanjaro, 1884; vice consul, Cameroons 1885; acting consul Niger Coast Protectorate, 1887; consul for province of Mozambique, 1888; expdn. to Lakes Nyasa and Tanganyika (founding of Brit. Central Africa Protectorate), 1889; consul-gen. Regency of Tunis, 1897-99; spl. commr., comdr.-in-chief, consul-gen. Uganda Protectorate, 1899-1901. Recipient Gold medals Zool. Soc., Royal Geog. Soc., Royal Scottish Geog. Soc., Silver medal Zool. Soc., 1896. Hon. mem. Italian Geog. Soc., Royal Irish zool. socs. Author: River Congo, 1884; Kilimanjaro, 1885; History of a Slave, 1889; Life of Livingstone, 1891; British Central Africa, 1897; A History of the Colonisation of Africa by Alien Races, 1899-1913; The Uganda Protectorate, 1902; British Mammals, 1903; The Nile Quest (History of the Exploration of the Nile), 1903; Liberia, 1906; George Grenfell and the Congo, 1908; Pioneers in West Africa, Canada, India, Australasia, Tropical America, and South Africa, 6 vols., 1911-13; Comparative Study of the Bantu and Semi-Bantu Languages, Vol. I, 1919, Vol. II, 1922; also blue-books and reports on W., Central, N. and E. Africa, 1888-1901. Died July 31, 1927.

JOHNSTON, Herrick Lee, Am. chemist; b. North Jackson, O., Mar. 29, 1898; s. Edgar Francis and Aelaide Sarah (Simpson) J.; A.B., Muskingum Coll., 1922; B.S., Wooster Coll., 1922, D.Sc., 1943; Ph.D., U. of Calif., 1928; m. Margaret Gardner Vanderbilt, June 14, 1923; children—William Vanderbilt, Margaret Louise, Robert Edgar. Instr. Shiloh (O.) High Sch., 1916-17; grad. asst. chemistry, Ohio State U., 1923-24; teaching fellow chemistry, U. of Calif., 1925-28, instr., 1928-29; asst. prof. chemistry, Ohio State U., 1929-33, asso. prof., 1933-38, prof., 1938-54; fellow John Simmons Guggenheim Mem. Foundation, U. of Goettingen, Ger., 1933; research engr. Gen. Electric Research Labs., Schenectady, N.Y., 1937; dir. Manhattan Project Research, Ohio State U., 1942-46; dir. Cryogenic Lab., Ohio State U., also Rocket Motor Lab., 1946-51; 1942-54; president Herrick L. Johnston, Inc., Johnston Research and Development Laboratories, 1954—. Fellow American Physical Soc., A.A.A.S.; mem. Internat. Inst. Refrigeration (tech. com. since 1948), Internat. Union Pure and Applied Physics (com. low temperature physics 1948-51, internat. com. thermo-dynamics and electro chemistry), Am. Chem. Soc. (past nat. sec. and nat. chmn., div. phys. and inorganic chemistry; past chmn. divisional officers group; past chmn. Columbus sect.), N.A.M., NRC (mem. com. deuterium research, 1934-37; com. thermodynamic constants, since 1940, com. on natural and phys. constants, com. on phys. chem. since 1948), Am. Soc. Mech. Engrs. (com. properties of gases), Ohio Post War Planning comm. (sci. com.), Am. Soc. Refrigeration Engrs. (com. on Low Temperature Research), Am. Rocket Soc. Author: Titanium and Its Compound. Asso. editor, Jour. Chem. Physics, 1935-38; contbr. technical articles to sci. publications. Asso. with Giauque in work on low temperatures in which less than a degree from absolute zero was reached; also studies on high temperature, calorimetry and vapor pressures, molecular spectra, isotope identification and separation, high pressure thermodynamics, transport properties of gases, photochemistry. Home: 177 Brevoort Rd. Office: 1992 Kenny Rd., Columbus, O.

JOHNSTON, John Black, Am. neurologist; b. Belle Center, O., Oct. 3, 1868; s. Robert H. and Hannah M. (Clyde) J.; Ph.B., U. Mich., 1893, Ph.D., 1899, Sc.D. (hon.), 1933; m. Juliet Morton Butler, Aug. 29, 1899; children—Stanwood (dec.), Norris. Asst. and instr. zoölogy, U. Mich., 1893-99; asst. prof., W.Va. U., 1899-1900, prof. zoölogy, 1900-07; asst. prof., U. Minn., 1907-08, asso. prof., 1908-09, prof. comparative neurology, from 1909, sec. faculty of medicine, 1910-13, editor-in-chief research publs., 1911-14, dean Coll. Science, Lit. and Arts, 1914-37. Charter mem. Mich. Acad. Science; fellow A.A.A.S.; mem. Am. Soc. of Zoölogists, Assn. Am. Anatomists, Am. Soc. Naturalists, Minn. Neurol. Soc. (charter mem.). Mem. editorial bd. Journal of Comparative Neurology, 1908-32; chmn. com. on ednl. testing Am. Council on Edn. Author: The Nervous System of Vertebrates, 1906; The Liberal College in Changing Society, 1930; Education for Democracy, 1934; Scholarship and Democracy, 1937; also numerous articles on neurology and edn. Died Nov. 19, 1947.

JOHNSTON, Thomas Baillie, Brit. anatomist; b. July 1883; s. A. Johnston; ed. George Watson's Coll., Edinburgh, Scotland; M.B., Ch.B. with 1st class honors, Edinburgh U., 1906, M.D. (Gold medal), 1937. Demonstrator, lectr. anatomy Edinburgh U., 1907-14; lectr. U. Coll., London, Eng., 1914-19; prof. Guy's Hosp. Med. Sch., U. London, 1919-48, supt. hosp., 1937-48; group officer sector X, E.M.S., 1939-46. Author: Medical Applied Anatomy, 1915; (with Lewis Beesly) Manual of Surgical Anatomy; Synopsis of Regional Anatomy, 8th edit., 1957; also articles. Editor, Gray's Anatomy, 1928-58. Died Oct. 8, 1960.

JOHNSTON, Thomas Harvey, zoologist; b. Dec. 9, 1881; s. Thomas Johnston; M.A., D.Sc., U. Sydney (Australia); m. Alice Pearce Petersham; 1 son, 1 dau. Lectr. zoology and physiology Sydney Tech. Coll., 1907-08; asst. govt. microbiologist, Sydney, 1909-11; lectr. in charge biology dept. U. Brisbane, 1911-19, prof. biology, 1919-22; prof. zoology U. Adelaide, from 1922, hon. prof. botany, 1928-34. Chmn. Queensland Govt. Sci. Travelling Commn. on prickly pear control in fgn. lands, 1912-14; sci. controller Commonwealth Govt. Prickly Pear Control Investigations (with labs. in Queensland, New S. Wales, Argentina, Fla., Tex.), 1920-23; hon. dir. S. Australian Mus., 1927-30; chief biologist Brit., Australian and New Zealand Antarctic Research Expdn., 1929-30, 30-31. Recipient Syme Research prize and medal U. Melbourne, 1913; Polar medal, 1934; Verco medal for research Royal Soc. S. Australia. Fellow Australian and New Zealand Assn. for Advancement Sci. (Mueller medal 1939); pres. Royal Soc. Queensland, 1915, Royal Soc. S. Australia, 1931-32; mem. Australian NRC. Author: Control of Prickly Pear Pest; Economic Entomology; also papers on Australian parasitology. Died Aug. 30, 1951.

JOHNSTONE, Donald B., Am. microbiologist; b. Newport, R.I., July 25, 1919; s. Fred W. and Reba (Boyes) J.; B.S., U. R.I., 1942; M.S., Rutgers U., 1943, Ph.D., 1948. m. Helen Beardslee, Aug. 20, 1949; children—Elizabeth, Margaret, Fred. Bacteriologist, Woods Hole Oceanographic Inst., 1942-43, 46; microbiologist Bikini Atomic Bomb Tests in Pacific, 1946; faculty U. Vt., Burlington, 1948—, prof. microbiology, chmn. agr. biochemistry, 1958. Fellow Am. Acad. Microbiology, Am. Inst. Biol. Scis., Soc. Am. Microbiology, A.A.A.S. Contbr. articles to tech. jours. Discovered copper bacteria in ocean, streptomycin producing culture in Bikini, various antibiotics from plants and bacteria; research on bacterial slimes, growth in heavy water, ecology Lake Champlain. Home: Box 2192, South Burlington, Vt. 05401. Office: U. Vt., Burlington, Vt., 05401.*

JOHNSTONE, Paul Nugent, Am. surgeon; b. Kansas City, Mo., Aug. 17, 1900; s. Paul Alexander and Bertha (Nugent) J.; A.B., U. Mo., 1922; postgrad. (Dennison fellow) Johns Hopkins, 1923, M.D., 1926; m. Cecilia Alta Taylor, Dec. 20, 1919; children—Emma Jean (Mrs. William Dean Ray), Paul Nugent. Faculty, U. Ga., 1923, U. Kan., 1928-48; prof. vertebrate anatomy U. Mo., Kansas City, 1964—; practice medicine specializing in gen. surgery, Kansas City, Mo., 1928—; staff Surgeon Research Hosp. and Med. Center, Kansas City, 1932—; vis. staff St. Joseph Hosp., Trinity Lutheran Hosp., 1932—. Pres., Med. Research Found. Mo., 1962—. Prin. investigator John A. Hartford Found., 1959—. Recipient Selective Service medal U. S. Congress, 1945. Fellow A.M.A., A.C.S.; mem. Internat. Coll. Surgeons, So. Soc. Anatomists, So. Soc. Anatomists, Mo. State Surg. Soc., Southwestern Surg. Congress, Mo. State Med. Assn., Jackson County Med. Soc. (award merit 1948), Kansas City Southwest Clin. Soc., Henry Ford Hosp. Med. Soc., Johns Hopkins Med. Soc., N.Y. Acad. Scis., Sigma Xi. Research, publs. on gross anatomy conductive system heart, physiol. anatomy embryonic heart, intestinal obstruction, acute renal failure. Home: 117 W. 69th Terrace, Kansas City, 64113. Office: Bryant Bldg., Kansas City, Mo. 64106.*

JOHNSTON-LAVIS, Henry James, physician, seismologist; b. 1856; ed. Univ. Coll., London; practiced medicine among Brit. and Am. residents, Naples, Italy, 1879-94; became prof. volcanology U. Naples, 1892; sec. com. on volcanic phenomena of Vesuvius, Brit. Assn., 1885-90. Author: Monographs of the Earthquakes of Ischia, 1885. Surveyed Vesuvius, 1880-88, pub. map in 6 sheets on scale 1:10,000, 1891; determined small depth of focus in volcanic earthquakes. Died 1914.

JOINER, Jasper Newton, Am. horticulturist; b. Winter Garden, Fla., June 1, 1921; s. Jasper N. and Neva (Tanner) J.; student N.C. State Coll., 1939-41; B.S.A., U. Fla., 1947, M.Agr., 1955; Ph.D., Ohio State U., 1958; m. Corinne Terrell, Jan. 23, 1944; children—James Terrell, Harry Jasper, Jan Corinne. Owner, operator florist and nursery bus., Clermont, Fla., 1947-50; asst. agr. editor Agr. Expt. Sta. and Extension Service, U. Fla., Gainesville, 1950-53, horticulture specialist, 1953-57, prof. ornamental horticulture, 1957—. Mem. Am. Hort. Soc., Sigma Xi, Gamma Sigma Delta. Research, publs. on plant nutrition, interactions of nutrition with photoperiod, temperature and growth regulator responses; use of radioisotopes in physiol. and biochem. studies propagation and photoperiodism in plants. Home: 715 N.W. 40th Terrace, Gainesville, Fla. 32601.*

JOINVILLE, Jean de, see de Joinville.

JOKL, Ernst, physician; b. Breslau, Germany, Aug. 3, 1907; s. Hans and Rosa (Oelsner) J.; M.Ed., U. Breslau, 1928, M.D., 1930; M.B., B.Ch., U. Witwatersrand, Johannesburg, S. Africa, 1936; m. Erica Lestmann, June 2, 1933; children—Marion (Mrs. John Ball), Peter. Came to U. S., 1950, naturalized, 1956. Chmn. research inst. sports medicine U. Breslau, 1932-33; chmn. dept. phys. edn. U. Stellenbosch, also Witwatersrand Tech. Coll., 1936-47; chief med. research officer Union Dept. Edn., S. Africa, 1946-50; prof., dir. research labs. sportsmedicine and phys. edn. U. Ky., Lexington, 1950—. Pres., exec. com. Internat. Council Sport and Phys. Edn., UNESCO, 1956—. Recipient Buckston Browne Brit. Empire medal for med. research Harveian Soc. London, 1943; S. African medal for war services, 1945; Nat. medal award for phys. fitness leadership U. S. C. of C., 1964; Distinguished Prof. award U. Ky., 1966. Fellow Am. Coll. Cardiology, Am. Coll. Sportsmedicine (founder), A.A.H.P.E.R. Author: Textbook Aviation Medicine, 1940; Medical Aspects of Boxing, 1940; Training and Efficiency, 1941; Syncope in Athletics, 1945; also numerous articles, monographs. Founder sci. sportsmedicine; demonstrated inhibitory effect sustained phy. activity on aging; clarification of path. bases of sudden death asso. with exercise; classification of clin. sportsmedicine, theory of role of exercise in rehab. Home: 340 Kingsway St., Lexington, Ky. 40502.*

JOKLIK, Wolfgang Karl, biochemist; b. Vienna, Austria, Nov. 16, 1926; s. Karl F. and Helene (Giessl) J.; B.Sc. with 1st class honors, U. Sydney, Australia, 1948, M.Sc., 1949; D.Phil. (Australian Nat. U. scholar), U. Oxford, Eng., 1952; m. Judith Vivien Nicholas, Apr. 9, 1955; children—Richard G., Vivien H. Australian Nat. U. research fellow Copenhagen, Denmark, 1953, Canberra, Australia, 1954-56, fellow, 1957-62; asso. prof. cell biology Albert Einstein Coll. Medicine, Bronx, N.Y., 1962-65, prof. cell biology, 1965-68, Siegfried Ullmann prof. biochem. virology, 1966-68; chmn. dept. microbiology and immunology Duke U. Med. Center, Durham, N.C., 1968—. Mem. Am. Soc. Cell. Biology, Am. Soc. Microbiology, Am. Soc. Biol. Chemists, Am. Soc. Immunology. Research, publs. in viruses, function of animal cells infected with viruses. Address: Duke U. Med. Center, Durham, N.C.*

JOLIBOIS, (Médard) Pierre, French chemist; b. Paris, May 23, 1884; prof. gen. and applied chemistry École Nationale Supérieure des Mines; mem. French Acad. Scis., 1944. Author: Recherches sur le phosphose et les phosphures metalliques, 1910. Research on reduction of iron oxides by carbon monoxide, also on electrolysis, electrophoresis. Died Paris, Feb. 18, 1954.

JOLIOT-CURIE, Frédéric, French physicist; b. Paris, France, Mar. 19, 1900; s. Henri and Emilie (Roederer) Joliot; ed. Lycée Lakanal, Bourg-la-Reine; Ingénieur, Ecole de physique et de chimie industrielles de la Ville U. de Paris, 1923; Licencié (B.A.) in Sciences, University of Paris, 1927, Dr. Sciences, 1930; m. Iréne Curie, Oct. 9, 1926; children—Hélene, Pierre. Asst., Faculty of Sciences, Univ. of Paris, 1932; maitre de Recherches, Caisse Nationale des Sciences, 1933; lecturer at the Sorbonne, 1935; prof., Coll. of France, 1937; dir. of lab. of atomic synthesis, Ivry, 1937; vice pres. Haut Comité des Recherches Scientifiques, 1938; mem. of commn. on atomic weights Union Internationale de la Chimie, 1938; dir. Centre National de la Recherche Scientifique, 1944; high commr. for atomic energy, 1946-50; prof. science University of Paris, also director Laboratoire Curie de l'Institut du Radium, 1956—. Mem. com. Nat. Inst. Nuclear Scis. and Techniques. Became mem. Internat. Commn. on Standards for Radium, 1934, sec., 1936. Pres. Société de Chimie-physique de France, 1936-38. Comité de Gestion des Tables Annuelles de Constantes (com. of management of annual tables of constants), 1936; sec.-gen. Natl. Union of Intellectuals, 1945. Decorated Grand Croix de la Legion d'Honneur; recipient (with Irene Joliet-Curie) Nobel prize for chemistry, 1935. Mem. Acad. of Medicine, Acad. of Sciences, Société Royale des Sciences de Liége, Acad. Royale des Sciences d'Amsterdam, Acad. des Sciences du Denmark, and many other scientific associations; president du Conseil Mondial de la Paix. Author: Le Noyau des Atomes (reports to Congrès de Leningrad), 1933; (with Iréne Joliot-Curie) Les Rayonnements des Atomes sous l'Action des Rayons Alpha (reports to Congrès de Physique Solvay), 1933; Le Neutron, Le Positron et la Radioactivité Artificielle; (brochures, in collaboration with Iréne Joliot-Curie) L'existence du neutron; l'électron positif; Radioactivité Artificielle; La Constitution de la Matiere et la Radioactivité Artificielle (Scientific Encyclopedia, Library of Teaching Technique), 1937; Prix Staline International. With Iréne Joliet-Curie, made numerous researches on structure of atom, dematerialization of electrons and inverse transformation; combined radiol. with biochem. research in synthesis of hormones and thyroid substances containing radioactively labeled elements; most important discovery was that of artificial radioactivity; produced an artificial radioactive substance by bombarding boron with fast alpha particles, 1933; (with Halban and Kowarski) proved experimentally that neutron emission takes place in nuclear fission, 1939. Died Aug. 14, 1958.

JOLIOT-CURIE, Iréne, French nuclear scientist; b. Paris, France, Sept. 12, 1897; d. Pierre and Marie (Sklodowska) Curie; Licentiate in Physics and Math. U. of Paris, 1920, Sc.D., 1925; Dr. Honoris Causa, Edinburgh U., 1939, U. of Oslo, 1946, U. Cracow, 1951, es sciences physiques, Univ. de Sofia, 1948; m. Frédéric Joliot, Oct. 9, 1926; children—Hélene, Pierre. Served as hosp. nurse in radiography during World War I. Asst. to Marie Curie at Inst. of Radium of Paris, 1918, head of studies, 1932; head of research Caisse Nationale de la Recherche Scientifique, 1935; under-sec. of state for sci. research, June-Sept. 1936; lecturer, Faculty of Sciences of Univ. of Paris, 1937, prof. without chair, 1937; commr. atomic energy, prof. Faculty of Sciences, and dir. Laboratoire Curie, 1946. Mem. Internat. Commn. of Standards for Radium, 1934; mem. Com. of Management of Office of Astrophysical Research, 1936. Awarded Mateucci Medal by Italian Soc. of Sciences, 1932, Henri Wilde prize, 1933, and Marquet prize, 1934, by Acad. of Sciences of Paris, Nobel Prize for Chemistry, 1935 (all with Frédéric Joliot-Curie), French Medal of Recognition, 1918, Medal of Soc. of Civil Engrs., 1937, Chevalier of Legion of Honor, 1935, Officer, 1939. Fgn. corresponding mem. Akademja Nauk Technieczynch w. Warszawie, Acad. Royale de Medecine de Belgique; cor. mem. U.S.S.R. Acad. Scis., 1947, de l'academie des sciences de Berlin; mem. Am. Phys. Soc.; mem. (hon.) de l'Institute Grant Ducal du Luxembourg, 1947; de la Société Shimique des Indes, 1948, mem. (hon.), Edinburgh U. Women's Assn., 1947. Mem. sci. sect. Upper Council of French Radiodiffusion, 1936; mem. outside sect. Upper Council of Scientific Research, 1938. Author: Les Radioéléments Naturels: Propriétés; Preparation et Dosage, 1946. With husband, Frédéric, produced artificial radioactive elements by bombarding boron with alpha rays, 1933; found that a material containing hydrogen, if exposed to what they considered gamma rays, would emit protons; gave 1st chem. proof of artificial transmutation, also proof of the capture of alpha particle; 1st to prepare positron emitters. Died Paris, Mar. 17, 1956.

JOLLÈS, Pierre Felix, biochemist; b. Vienna, Austria, Oct. 9, 1927 (citizen Liechtenstein); s. Henri and Erica (Weinberg) J.; Baccalaureat, 1945; Licence ès Sciences, 1947, Ingenieur-Chimiste, 1949, Ecole Superieure de Chimie Industrielle, Lyons, France; postgrad. biochemistry, ETH, Zurich, Switzerland, 1949-50; Docteur-Ingenieur, Faculty Scis., Paris, 1952, Docteur ès Sciences Physiques, 1964; m. Jacqueline Thaureaux, Apr. 5, 1957; children—Béatrice, Anne, Marie-Hélène. With Faculty Scis., Paris, 1950—, staff Lab. Bio-Chemistry, 1952—; staff Nat. Center Sci. Research, Paris, 1952—, maitre, 1960-64, dir. research, 1964—. Mem. commn. genetics immunology and molecular pathology Nat. Inst. Health and Med. Research, Paris, 1964—. Mem. French Biochem. Soc. (mem. council 1967—), Biochem. Soc. (London), Swiss Biochem. Soc., French Chem. Soc., Swiss Soc. Cellular and Molecular Biology, N.Y. Acad. Scis. Research and numerous publs. on chem. structure of hen egg-white lysozyme, 1961, chem. structure and comparative biochemistry of lysozymes of different origins, K-caseins, K-caseino-glycopeptides, action of rennin during milk-clotting process, biochemistry of waxes D, active peptido-glycolipids of Mycobacteria (Freund's adjuvant effect), transplantation antigens. Home: 2 Rue Gerbillon, Paris 6, France.*

JOLLIE, Malcolm Thomas, Am. biologist; b. Lakewood, O., July 11, 1919; s. Oliver Clinton and Ada Clara (Meek) J.; B.S., Western Res. U., 1941; M.S., U. Colo., 1943; Ph.D. (NSF fellow), Stanford, 1954; m. Alma Blanche Sanders, June 5, 1950; children—Malcolm R., Susan A. Faculty, N.M. State Tchrs. Coll., 1945-47, U. Ida., 1947-56, U. Pitts., 1956-65; prof. biology No. Ill. U., DeKalb, 1965—. NSF Faculty fellow, Germany, 1964-65. Mem. A.A.A.S., Am. Soc. Zoologists, Evolution Soc., Sigma Xi. Author: Chordate Morphology, 1962; also articles. Research on head skeleton from fish to mammals, systematics order falconiformes. Home: Nottingham Woods, Elburn, Ill. 60119. Office: Biol. Scis., No. Ill. U., De Kalb, Ill. 60115.*

JOLLY, Frederic, German physician; b. Hiedelberg, Germany, 1844; s. Philipp von Jolly; ed. Munich, Göttingen (both Germany). Became pvt. docent, Würzburg, Germany, 1871; named dir. clinic of nervous diseases, prof. U. Strasbourg (now France), 1873; became prof., dir. clinic of diseases of brain, 1890. Author: Bericht über die Irrenabteilung des Juliusspitals, 1873; Untersuchung über den elektrischen Leitungswiderstand des menschlichen Körpers, 1884; Ueber Irrtum und Irrsinn, 1893. Editor Archiv für Psychiatrie und Nervenkrankheiten, after 1890. Research on cervical sclerosis, hallucinations, delirium, malfunction of memory, treatment of insane in Ecosse, hosps. for epileptics. Died Berlin, Germany, 1904.

JOLLY, John, Irish geologist; b. Kings County, Ireland, 1857; s. J. P. and Julia Jolly; ed. Trinity Coll., Dublin, Ireland; became prof. mineralogy and geology Dublin U., 1897; fellow Royal Soc., 1892. Inventor: photometer (based on paraffin wax slabs separated by tinfoil sheet), used to measure illumination, 1888; calculated age of earth by measuring sodium content of oceans, 1889; formulated theory of thermal cycles (based on radioactive elements in earth's crust); proved pleochroic haloes are result of alphaparticles shot out from minute crystals of zicron and allanite (led to method of estimating age of rocks); pioneer work on cooling of earth with respect to theory of isostasy and radioactive heating of crust; developed (with Stevenson) Dublin Method of radium extraction, 1914; research on radium treatment of cancer, color photography. Died Dec. 8, 1933.

JOLLY, Justin, French cytologist; b. Melun, France, Aug. 6, 1870; prof. histology Coll. de France, 1925-40; mem. French Acad. Scis., 1939, Acad. Medicine. Author: Traité d'hématologie, 1928. Research on morphology of blood and blood-forming tissues, mammal embryology; discovered small oval structures which appear in blood after splenectomy (Howell-Jolly bodies). Died Paris, Feb. 1, 1953.

JOLLY, Philipp Johann Gustav von, see von Jolly, Philipp Johann Gustav.

JOLLY, William Adam, physiologist; b. Edinburgh, Scotland; s. James Jolly; ed. George Watson's Coll. and U., Edinburgh; M.B., Ch.B., D.Sc., LL.D., Edinburgh; m. Aimée Murray; 2 sons, 1 dau. Asst. to prof. physiology Edinburgh U., 1904, lectr. exptl. physiology, 1909; Carnegie research fellow, research student Leyden (Netherlands) U., 1908; prof. physiology Univ., Cape Town, S. Africa.; dean med. faculty, pres. Brit. Med. Assn. Cape Western Div. Research and publs. on electrocardiagram, elec. response of eyeball to light, thyroid gland and reflexes. Died July 1939.

JOLY, A. B., Brazilian botanist; b. Itatiba, Brasil, Nov. 15, 1924; s. Alfredo Alves and Maria Carolina (Brandao) J.; Bacharel Ciec. Nat., Faculty Fil. Cienc. Letras, U. Sao Paulo (Brazil), 1945, Espec.Bot., 1946, Dr.Sc., 1950, Livre Docente Bot., 1957; m. Mathilde Grabher, Feb. 5, 1947; children—Ana Maria, Maria Cristina, Carlos Alfredo. Faculty, U. Sao Paulo, 1945—, asso. prof. botany, 1962—, counselor research funds Oceanographic Instn., 1960-64, Inst. Marine Biology, 1965—, subs. dir. dept. botany, 1961—; guest prof. Marine Biol. Sta., Puerto Deseado and Inst. Marine Biology, Mar del Plata, Argentina, 1962, Inst. Marine Biology, U. Ceará, 1964, Tax., Bot., U. Ceará, 1966. Counselor research funds State Bot. Inst., 1961. Rockefeller fellow, Ann Arbor, Mich., also Marine Biol. Lab., Woods Hole, Mass., 1951-52; NRC Brazil research grantee; Fundacao de Amparo à Pesauisa do estado de Sao Paulo research grantee. Mem. Internat. Assn. for Plant Taxonomy and Nomenclature, Am., Internat. phycological socs., Sociedad Brasileira para o Progresso da Ciencia, Sociedad Botanica do Brasil, Sociedade Argentina de Botanica. Author: Botanica, Introaducao à Taxonomia Vegetal, 1966; Generos de algas marinhas da costa Atlantica Latino Americana, 1966; also articles. Research on orgn., ecology, distbn., reprodn. of algae along coast of Brazil. Home: 190 Hungria, Sao Paulo, S.P., Brasil.*

JOLY, Charles Jasper, Irish astronomer; b. Tullamore, Ireland, June 27, 1864; s. John Swift and Elizabeth (Slator) J.; ed. Trinity Coll., Dublin, Ireland, Berlin (Germany) U.; M.A.; m. Jessie Meade, 1897. Royal astronomer Ireland; Andrews prof. astronomy U. Dublin, from 1897. Trustee Nat. Library Ireland; visitor, Sci. and Art Mus., Dublin. Fellow Trinity Coll., Dublin. Fellow Royal Soc., 1904, Royal Astron. Soc.; mem. Royal Irish Acad. (sec. 1902 contbd. papers), Internat. Assn. for Promoting Study of Quaternions and Allied Systems of Math. (pres.). Author: A Manual of Quaternions. Editor: Elements of Quaternions (William Rowan Hamilton); Theory of Light (Preston), 3d edit. Died Jan. 4, 1906.

JOLY, Henry, French thoracic surgeon; b. Tours, France, July 5, 1905; s. Robert and Marie (Frouin) J.; student univs. Tours, Paris; m. Marcelle Plouvier, July 7, 1949; children—Anne, Françoise, Catherine. Surgeon, sanatorial sta., Passy, France, 1936—. Recipient French Pub. Health Service award, 1958; medals French Med. Nat. Acad., 1936, 64. Mem. French Thoracic Surg. Soc. (pres. 1957-58). Author: Surgical Treatment of Pulmonary Tuberculosis, 1947; also numerous articles. Research on surg. collapse, surg. pulmonary resects., surg. treatment of pulmonary Tb in children. Address: Station sanatoriale de Passy, Haute-Savoie, France.*

JOLY, John, Brit. physicist, geologist; s. J. P. and Julia (de Lusi) J.; ed. Trinity Coll. (fellow), Dublin, Ireland; B.A.I., M.A., D.Sc.; LL.D., U. Mich.; Sc.D. (hon.), Cambridge (Eng.) U., Nat. U. Ireland. demonstrator civil engring., 1882-91; demonstrator exptl. physics Trinity Coll., Dublin, 1893, prof. geology, mineralogy from 1897. Warden, Alexandra Coll. for Higher Edn. of Women; sr. commr. Irish Lights; sci. adviser Dr. Steevens' Hosp., Dublin; mem. Brit. ednl. mission to U. S., 1918. Fellow Royal Soc., 1892 (Royal medal 1910), Geol. Soc. (Murchison medal 1923); mem. Royal Dublin Soc. (pres., Boyle medal 1911), Brit. Assn. (pres. geol. sect. 1908), Acad. Sci. Russia (hon.). Author: On the Specific Heats of Gases at Constant Volume; On a Method of Photography in Natural Colours, 1896; Radio-activity and Geology; The Local Application of Radium in Therapeutics, 1914; The Birth-time of the World and other Scientific Essays, 1915; Synchronous Signalling in Navigation; Reminiscences and Anticipations, 1920; Radioactivity and the Surface History of the Earth, 1924; The Surface History of the Earth, 1925, 2d. edit., 1930. Editor: (with others) Philos. Mag., from 1901. Contbns. to procs. of sci. socs. Devised diffusion phytometer, meldometer, steam calorimeter; uniform radiation method for use in cancer treatment; color photography pioneer; research on crust formation of earth. Died Dec. 8, 1933.

JOLY, Maurice Henri, French biophysicist; b. Paris, France, Apr. 6, 1911; s. Maurice and Aimee (Nivert) J.; Licence ès Sciences, Faculté des Sciences de Paris, 1935, Dr. ès scis., 1946; m. Madeleine Rollet, July 20, 1946; 1 dau., Monique. Research asst. dept. phys. chemistry Institut Pasteur, Paris, 1945-50; research master Nat. Center Sci. Research, 1950-60, research dir. dept. biophysics Pasteur Inst., Paris, 1960—. Mem. French Group Rheology (pres.), Société de Chimie Physique, Société de Chimie Biologique, Faraday Soc., Kolloid Gesellschaft. Author: A Physico-Chemical Approach to the Denaturation of Proteins, 1965; also numerous articles. Devel. surface rheology, theory of surface viscosity and phase transition in monolayers, structural changes induced by flows, study of rheoturbidity, molecular aspect of solgel transition, interactions between macromolecules in aqueous phases, structural basis of protein denaturation, physico-chem. study artificial nucleoproteins. Home: 55 Rue Lacordaire. Office: 28 Rue du Dr. Roux, Paris 15, France 75.*

JOLY, Nicolas, French physiologist; b. Toul, France, July 11, 1812; prof. zoology, anatomy, comparative physiology Faculty Toulouse (France); mem. French Acad. Scis., 1875. Author: Considérations sur la vie physique et ses principales manifestations; Principles d'ostéologie comparée; Recherches ser les vers à soie et leurs maladies, 1858; Recherches sur l'origine, le génération et la fructification de la levure de bière, 1861; Études sur l'embryogénie de l'axolotl du Mexique (in Études de phychologie comparée 1876). Research on heterogeny; defended theory of spontaneous generation; his biol. research gathered support for evolutionary theory. Died Toulouse, Oct. 17, 1885.

JOLY, Sebastiana, Brazilian microbiologist; b. Rio das Pedras, Brazil, Oct. 30, 1920; d. Octavio and Antonia (Leite) J.; doctor degree Escola Superior de Agricultura Luiz de Queiroz, U. Sao Paulo, Piracicaba, Brazil, 1954. Asst. prof. Instituto Zimotécnico, Piracicaba, 1952—. Rockefeller Found. fellow U. Pisa (Italy), 1956-57. Research, publs. on soil fungi (keratinophilic group), pathogenic fungi (Dermatophytes) and yeasts, actinomycetes (antibiotic and biol. properties), bacteria (family Enterobacteriaceae), water, soil and sugar-cane microbiology. Home: 1785, Gov. P. de Toledo. Office: C. postal 56, E.S.A.L.Q., Piracicaba, Sao Paulo, Brazil.*

JOLYCLERC, Nicolas, French naturalist; prof. natural history Oise Central Sch. Author: Système sexuel des vegataux de Linné (1st French transl. of Linnaeus), 1798; Cours de minéralogie, 1802. Wrote on biol. and bot. terms, also on mineralogy. Died Paris, 1817.

JOMARD, Edmé François, French egyptologist; b. Versailles, France, Nov. 27, 1777; ed. École Polytechnique; mem. French sci. commn. to Egypt, 1798, sec., 1802; became curator-dir. Royal Library, 1828, founder Geog. Soc. Paris, 1821; helped found Institut des Egyptiens. Author: Description de l'Egypte, 1809-13; Remarques sur les rapports de l'Ethiopie et l'-Egypte, 1822; Voyage à l'Oasis de Syouah, 1823; also many articles. Stimulated interest of scientists and gen. readers in study of Egypt and other ancient cultures; gave impetus to attempts at deciphering hieroglyphs. Died Paris, Sept. 23, 1862.

JONAS, August Frederick, Am. surgeon; b. Arlington, Wis., June 12, 1858; s. August Otto and Fredericka (Gundlach) J.; ed. pub. schs., Madison, Wis.; M.D., Bennett Med. Coll., Chicago, 1877; Ludwig Maximilian U., Munich, 1884; post-grad., Vienna, Berlin, Paris; LL.D., U. of Neb., 1928; m. Jessica Stebbins, Nov. 16, 1907; children—August F., Mary Elizabeth, Carl Stebbins. Began practice of medicine, 1877; became prof. surgery, med. dept., U. Neb., 1892, emeritus; later surgeon to Neb. Methodist and Douglas County hosps.; chief surgeon U.P. R.R., div. surgeon C.&N.W. Ry., asst. surgeon C.,St.P.,M&O. R.R.; retired from practice, 1928. Produced a modification in operative method for invertebrate and relapsed cases of talipes-equino-varus, 1899. Died Nov. 13, 1934.

JONASCH, Erich Walter, Austrian physician; b. Vienna, Austria, Sept. 10, 1922; s. Adolf and Josefa (Lautner) J.; M.D., U. Vienna, 1950; Staff, Unfallkrankenhaus Wien, Vienna, since 1957—, 1st asst., since 1964—. Mem. Austrian Traumatologic Soc. (founding sec. since 1965—). Author: Das Kniegelenk, 1964; Unfallchirurgische Operationen, 1965; also articles. Research on traumatologic surgery. Home: Vienna, 20. Webergasse 2, Accident Hosp., Austria.*

JONASSEN, Christen Tonnes, Am. sociologist; b. Lodshavn, Norway, Sept. 5, 1912 (parents Am. citizens); s. Tonnes Omar and Sigrid (Tonnessen) J.; A.B., Bklyn. Coll., 1937; A.M., N.Y. U., 1941, Ph.D., 1947; m. Lillian Alice Dolan, Dec. 25, 1938; 1 son, Eric Demarest. Faculty Ohio State U., Columbus, 1947—, prof. sociology, 1960—; dir. research projects NRC, 1951-54, Kellogg Found., 1954-59, U. S. Office Edn., 1964-65, Dir. Internat. House, Columbus, 1949; cons. bus. and govt., 1954, 55, 66; Fulbright prof. Inst. Social Research and U. Oslo, 1962-63. Fellow Am. Sociol. Assn.; mem. Am. Assn. U. Profs., Am. Acad. Polit. and Social Sci., Am. Statis. Assn., Ohio Valley Sociol. Soc. (editor Ohio Valley Sociologist 1953-57). Author: The Norwegians in Bay Ridge, 1947; The Shopping Center versus Downtown, 1955; (with Sherwood H. Peres) Interrelationships of Dimensions of Community Systems, 1960; also articles. Sociol. study of Norwegian immigrants in Bklyn.; established systematic relationships between 83 community variables. Home: 276 W. Kenworth Rd., Columbus, O. 43214.*

JONEK, Józef Jan, Polish physician; b. Kolonowski, Poland, Mar. 8, 1928; s. Wiktor and Anastazja (Dreja) J.; physician Med. Acad., Zabrze, Rokitnica, Poland, 1955, M.D., 1957; m. Teresa Bobik, July 4, 1954. Staff, dept. histology and embryology Silesian Acad. Medicine, 1952—, head dept., 1961—, asst. Clinic of Gynecology and Maternity, 1955—; sci. worker Inst. Medicine, Zabrze, 1956—; faculty Acad. Medicine, Katowice, Poland, 1955—, prof., 1967—. Mem. Polish socs. of anatomists, physiologists, gynecologists, endocrinologists, histochemists and cytochemists. Collaborator: (with T. Pawlikowski) Histology, 1962. Research, publs. on influence of oestrogens on genital organs, genital tract of mammals, influence of ionizing radiation on living organisms, behavior of enzymes in tissues in different path. and physiol. conditions, influence of different toxic agts. on living organisms. Address: 19 K.Marx, Zabrze-Rokitnica, Poland.*

JONES, Alfred Russell, Am. chemist; b. Chgo., Oct. 23, 1920; s. Alfred and Stella (Blain) J.; B.S., U. Chgo., 1942; Ph.D., U. Wis., 1948; m. Edna Wanda Belford, June 3, 1945; children—Janara Jean, Alfred Russell, Glynis Ann. Jr. chemist Gen. Electric Co. Mass., 1942; with U. S. Naval Research Lab., Washington, 1942-44; chemist Oak Ridge Nat. Lab., 1948-53, sr. chemist, 1953—; lectr. Panjab U. Sci. adviser AEC Pakistan, 1963-64; vis. scientist Hahn-Meitner Inst., Berlin, 1964-65. Mem. Am. Chem. Soc., A.A.A.S. Research, publs. primarily in field of compounds containing radioactive carbon, radiation chemistry of organic compounds, energy transfer in solid and liquid systems. Home: 7217 Wellswood Lane, Knoxville, Tenn. 37919. Office: P.O. Box X, Oak Ridge 37831.*

JONES, Alister Vallance, physicist; b. Christchurch, New Zealand, Feb. 4, 1924; s. Frederick Edmund and Nellie (Vallance) J.; B.Sc., U. New Zealand, 1945, M.Sc., 1946; Ph.D., Cambridge (Eng.) U., 1950; m. Catherine Ferguson, Dec. 1, 1951; children—Elizabeth Marie, Catriona Anne, Frederick. Fellow, NRC Ottawa, Ont., Can., 1949-51; faculty U. Sask., Saskatoon, Can., 1953-68, prof. physics, 1963-68; with radio and elec. engring. div. NRC of Can., Ottawa, Ont., 1968—. Fellow Royal Soc. Can.; mem. Canadian Assn. Physicists. Editor: Physics in Can., 1963-66. Research, numerous publs. on auroral spectrum; discovered (with A. W. Harrison) 1.58 u twilight oxygen emission, Ca2 emission in twilight spectrum. Home: 105 Bottomley Av. N., Saskatoon, Sask., Can.*

JONES, Arthur Letcher, chemist; b. New Canton, Va., May 2, 1914; s. Plummer Flippen and Lottie (Pitts) J.; B.S., Hampden-Sydney Coll., 1936; postgrad. Cornell U.; Ph.D., U. Va., 1943; m. Ellen Jane Vidal, Sept. 21, 1946; children—Catherine Vidal, Nancy Ellen, Adelaide Elizabeth, Diana Pitts. Instr. chemistry, math. Hampden-Sydney Coll., 1936-39; instr. chemistry Cornell U., 1943-46; with Standard Oil Co. (O.), Cleve., 1946—, coordinator long range planning research and devel., 1964—. Recipient Aerospace award, 1963. Mem. Am. Petroleum Inst. (com. analyt. research), Am. Chem. Soc. (social chmn. petroleum div. 1960), Cleve. Astronom. Soc. (pres.), Anti-Tb League Cleve. (trustee), Ohio, N.Y. acads.

sci., Cleve. Assn. Research Dirs., A.A.A.S. Author: Technique of Organic Chemistry, vol. III, part 1, 1956; Thermal Diffusion of Organic Liquids, 1956. Research, publs. on organic analytical reagents and discovery of Tiron, a new colorimetricreagent, magnetic susceptibility, separations of hydrocarbons, liquid thermal diffusion, phys. techniques in petroleum research, fuel cells, radio satellite tracking. Inventor, co-inventor several different types of thermal diffusion apparatus. Home: 32340 Woodsdale Lane, Solon, O. 44139. Office: Midland Bldg., Prospect Av., Cleve. 44115.*

JONES, Arthur Wynne, Am. parasitologist; b. Washington, Dec. 10, 1913; s. Allan D. and Marguerite (Bullard) J.; B.A., U. Va., 1934, LL.B., 1936, M.S., 1941, Ph.D., 1943; m. Ellen Sherburne, Nov. 27, 1938; children—Sherburne (Mrs. Larry Webb Barber), Anthony Bullard, Jonathan Allan, Mary Ellen. Faculty, Southwestern U., 1943-45; faculty U. Tenn., Knoxville, 1945—, prof. zoology, 1961—. Mem. Am. Soc. Parasitologists, Am. Micros. Soc., Soc. Study Evolution, Tenn. Acad. Sci., Sigma Xi. Author: (with J. M. Carpenter) Microtechnique, 1954, rev., 1960, 67; Introduction to Parasitology, 1967; also numerous articles. Research in effects of radiation on parasites, especially tapeworms; introduced mouse bile duct tapeworm as lab. animal. Home: 5517 Pinellas Dr., Knoxville, Tenn. 37919.*

JONES, Sir Bennett Melvill, English aero. engr.; b. Birkenhead, Eng., Jan. 28, 1887; s. Benedict and Henrietta (Melvill) J.; student Emmanuel Coll., Cambridge (Eng.) U., 1906-09; m. Dorothy Laxton Jotham, Oct. 25, 1916; children—Margaret (Mrs. Edmund Wright), Warren, Geoffrey. With Woolwich Arsenal, London, Eng., 1909-11, Nat. Phys. Lab., Teddington, Eng., 1911-13, Messrs. Armstrong Whitworth, London, 1913-14, Royal Aircraft Establishment, 1914-16, Armament Exptl. Sta., Orfordness, Eng., 1916-18; prof. aero. engring. Cambridge U., 1919-52; cons. Royal Aircraft Establishment, 1952-60. Named comdr. Brit. Empire, Knight Bachelor, 1942. Recipient, Air Force Cross, 1918. Fellow Royal Soc., 1939; hon. fellow Royal Aero. Soc., Am. Inst. Aeros. and Astronautics, Canadian Aeros. and Space Inst. Author: (with J. C. Griffiths) Aerial Surveying by Rapids Methods, 1925; also articles; contbg. author: Aerodynamic Theory, 1935. Wind tunnel and in-flight research on aircraft efficiency and control especially concerning stalling. Home: Emmanuel Coll., Cambridge, Eng.*

JONES, Brynmor, Brit. chemist; b. Wrexham, North Wales, Sept. 25, 1903; s. William Edward and Hannah Jane (Roberts) J.; ed. Univ. Coll. North Wales, St. John's Coll., Cambridge; B.Sc., Ph.D., D.Sc. Prof. chemistry U. Sheffield; G. F. Grant prof. chemistry Univ. Coll., U. Hull; dean Faculty Sci., U. Hull. Research and publs. on organic and physical chemistry. Home: 4 Hull Rd., Cottingham, East Yorkshire. Office: University of Hull, Hull, Eng.*

JONES, Burton Wadsworth, Am. mathematician; b. Redwood Falls, Minn., Oct. 1, 1902; s. Arthur Julius and Ethel (Rounds) J.; A.B., Grinnell Coll., 1923; A.M., Harvard, 1924; Ph.D., U. Chgo., 1928; m. Marian Grace Snelling, Sept. 10, 1932; children—Marian Louise, Christopher, Phyllis Wadsworth (Mrs. Richard C. Miller). Faculty, Western Res. U., 1924-26, U. Chgo., 1928-29, Cornell U., 1930-48; prof. math. U. Colo., Boulder, 1948—. Mem. Inst. for Advanced Study, Princeton, N.J., 1939-40; math. regional specialist AID, Central Am., 1964-65. NRC fellow, 1929-30; Research Corp. grantee, 1947-48. Mem. Am. Math. Soc., Math. Assn. Am. (asso. sec. 1946-48), Nat. Council Tchrs. Math. (dir.). Author: Elementary Concepts of Mathematics, 1947; Arithmetic Theory of Quadratic Forms, 1950; The Theory of Numbers, 1955; Modular Arithmetic, 1964; also articles. Study of matrices; theory of numbers; quadratic forms. Home: 901 Cascade St., Boulder Colo. 80302.

JONES, Butler Alfonso, Am. sociologist; b. Birmingham, Ala., July 22, 1916; s. Jackson C. and Nettie (Butler) J.; A.B., Morehouse Coll., 1937; A.M., Atlanta U., 1938; Ph.D., N.Y. U., 1955; m. Lillian E. Webster, Dec. 27, 1939. Tchr., Atlanta U. Lab. Schs., 1938-42; prof. social scis. Talladega Coll., 1943-52; asso. prof., prof. sociology Ohio Wesleyan U., Delaware, 1952—; vis. prof. Oberlin Coll., 1962-63, Hamline U., 1966-67. Mem. Am. Sociol. Assn., Soc. for Study Social Problems, Ohio Valley Sociol. Soc., Assn. for Study Negro Life and History, Soc. for Applied Anthropology. Research on effectiveness of law as an instrument for creating and directing social change along a predetermined path, comparison of conditions obtaining before intervention of law with those existing after legal intervention, cases taken primarily from sch. segregation litigation in U. S. Home: 18 Westgate Dr., Delaware, O. 43015.*

JONES, Calvin, Am. physician; b. Great Barrington, Mass., Apr. 2, 1775; s. Ebeneezer and Susannah (Blackmer) J.; m. Temperance Williams, 1819. Practiced medicine, Great Barrington, 1792-95, Smithfield, N.C., 1795, Raleigh, N.C., 1803-32; a founder N.C. Med. Soc., 1799; co-owner newspaper Star, Raleigh, 1808-15. One of 1st physicians in state to urge use of inoculation against smallpox. Died at estate Pontine, nr. Bolivar, Tenn., Sept. 20, 1846.

JONES, David Charles Lloyd, Am. physiologist; b. Oakland, Cal., Jan. 13, 1923; s. David Stuart Lloyd and Ethel Beatrice (Ostrom) J.; A.B., U. Cal. at Berkeley 1943, M.A., 1948, Ph.D., 1954; m. Betty Yvonne Smith, Dec. 29, 1943; children—Marsha Lynn, David Christopher Lloyd, Daniel Charles Lloyd. Physiologist, U. S. Naval Radiol. Def. Lab., San Francisco, 1948—, now head of physiology-psychology branch. Member Am. Physiol. Soc., Radiation Research Soc., Gerontological Soc., A.A.A.S., Sigma Xi. Research, publs. on physiology irradiated animal from exposure to death, establishing interrelationships between radiobiology and gerontology. Home: 1870 Oakdell Dr., Menlo Park. Cal. 94025. Office: U.S.N.R.D.L., San Francisco 94135.*

JONES, Douglas Samuel, Brit. mathematician; b. Corby, Northants, Eng., Jan. 10, 1922; s. Jesse Dewis and Bessie (Streather) J.; M.A., U. Oxford, 1947; D.Sc., U. Manchester (Eng.), 1957. m. Ivy Styles, Sept. 23, 1950; children—Helen Elizabeth, Philip Andrew. Faculty, U. Manchester, 1948-56, prof. math. U. Keele, 1956-64; prof. U. Dundee, Scotland, 1965—. Decorated Order Brit. Empire. Fellow Inst. Math. and its Applications, Royal Soc. Edinburgh. Author: Electrical and Mechanical Oscillations, 1960; Theory of Electromagnetism, 1964; Generalized Functions, 1966; also articles. Editor: Jour. Inst. Math. and its Applications, Quar. Jour. Mechanics and Applied Math. Solution of problems in wave propagation and devel. math. methods. Office: Math. Dept., U. Dundee, Scotland.*

JONES, E(dmund) Ruffin, Jr., Am. zoologist, educator; b. Charlottesville, Va., Oct. 1, 1905; s. Edmund Ruffin and Jane (Dabney) J.; B.A., B.S., U. Va., 1927, M.A., 1928, Ph.D., 1930; m. Helen Purdum Bell, June 11, 1936; children—Helen Bell (Mrs. William Edward Kerby), Frances Dabney (Mrs. Donald Moore Giles). Lectr. biology Dalhousie U., Halifax, N.S., Can., 1930-31; faculty Norfolk div. Coll. William and Mary, 1931-46, prof., 1937-46, chmn. div. natural sci., 1940-46, dir. Evening Coll., adult edn., 1942-44; faculty U. Fla., Gainesville, 1946—, prof., 1947—, dir. preprofl. counseling, 1956—, asst. dean Grad. Sch., 1961-64, asst. dean coll. arts and scis., 1964—. Mem. tng. com. Nat. Inst. Dental Research, 1962-66, chmn., 1966-67. Fellow A.A.A.S., (pres. conf. 1962); mem. Am. Soc. Zoologists, Assn. So. Biologists (pres., 1963-64), Soc. Systematic Zoology, Am. Microscopic Soc., Corp. Marine Biol. Lab., Am. Assn. U. Profs., Va., Fla. (pres., 1959) acads. sci., Phi Beta Kappa, Sigma Xi, Phi Sigma, Alpha Epsilon Delta. Research, publs. on evolutionary relationships among lower invertebrates, with descriptions of a number of new species. Home: Route 3, Box 200 B1, Gainesville, Fla. 32601.*

JONES, Emrys, Brit. geographer; b. Aberdare, South Wales, Aug. 17, 1920; s. Samuel Garfield and Annie (Williams) J.; B.Sc. with 1st class honors, U. Coll. Wales, 1941, M.Sc., 1945, Ph.D. (fellow), 1947; m. Iona Vivien Hughes, Aug. 7, 1948; children—Catrin Prydderch, Rhianon Elen. Faculty, U. Coll., London, Eng., 1947-50, Queen's U., Belfast, Ireland, 1950-58; reader social geography London Sch. Econs., 1958-60, prof. geography, 1960—. Cons. on planning, 1963—; UN cons. on urbanization to Govt. of Venezuela, 1963. Fellow Royal Geog. Soc.; mem. Inst. Brit. Geographers. Editor: Belfast in Its Region, 1952; Social Geography of Belfast, 1961; Human Geography, 1965; Towns and Cities, 1966. Research, publs. on social groups in towns, methodology, growth and change in city. Home: 4, Apple Orchard, Hemel Hempstead, Eng. Office: Houghton St., London W.C.1, Eng.*

JONES, E(rnest) Lester, Am. hydrographic and geodetic engr.; b. E. Orange, N.J., Apr. 14, 1876; s. Charles Hopkins and Ada (Lester) J.; mem. class of 1898, Princeton, A.M., 1919; m. Virginia Brent Fox, Sept. 28, 1897; children—Mrs. Elizabeth Brent Barker, Cecil Lester. Business, research and secretarial work 10 yrs.; U. S. dep. commr. fisheries, 1913-15; apptd. supt. U. S. Coast and Geod. Survey, 1915, title changed to dir., 1919. Commr. Internat. Boundary between U. S. and Can. (Alaska and Can.); mem. first Aerial Coastal Patrol Commn.; mem. Fed. Bd. Surveys and Maps, Fed. Personnel Bd. Decorated by govts. of Italy and France. Author: Alaska Investigations, 1914; Hypsometry, 1915; Elements of Chart Making, 1915; Neglected Waters of the Pacific, 1916; Safeguard the Gateways of Alaska, 1917; Aerial Surveying, 1919; Earthquake Investigation in United States, 1925; Tide and Current Investigations of the U.S.C. and G.S., 1926 (all govt. publs.); Surveying from the Air, 1922; The Evolution of the Nautical Chart, 1924; Science and the Earthquake Perils, 1926. Died Apr. 9, 1929.

JONES, E(va) Elizabeth, Am. comparative pathologist; b. Ottawa, Kan., Sept. 12, 1898; d. Edward Archbold and Mary (Redmond) Jones; A.B., Radcliffe, Coll., 1920, Ph.D., 1930; M.A., U. Me., 1924. Research asst. Sta. for Exptl. Evolution, Cold Spring Harbor, 1920-22; instr. U. Me., 1924-25; fellow Radcliffe Coll., Harvard, 1925-26, research asst., 1926-27, fellow, 1927-28, research fellow dept. 1930-34; mem. faculty Wellesley Coll., 1934—, prof. zoology, 1949-64, prof. zoology emeritus, 1964—, Lewis Atterbury Stimson prof. zoology, 1961, chmn. dept. zoology and physiology, 1949-55; instr. Simmons Coll., 1936-37. Research fellow Fearing Research Lab., Free Hosp. for Women Brookline, Mass., 1934-38; spl. cancer research fellow Nat. Cancer Inst., 1947-48, 55-56; research asso. Children's Cancer Research Found., 1964—. Recipient Harbison award Wellesley Coll., 1956-57. Fellow A.A.A.S.; mem. N.Y. Acad. Sci., Soc. Study Devel. and Growth, Am. Assn. Cancer Research, Am. Soc. Exptl. Pathology, Phi Beta Kappa, Sigma Xi. Research and publs. on coccidiosis, a parasitic disease in birds; described epidemic tremor, a virus diseases of chickens; exptl. work on cancer of mice; described spontaneous diabetes mellitus in mice. Home: 1120 Beacon St., Brookline, Mass. 02146. Office: Jimmy Fund Bldg., 35 Binney St., Boston 02115.*

JONES, (Alfred) Ernest, psychologist; ed. Cardiff, London, Munich, Paris, Vienna univs.; M.D.; B.S.; hon. D.Sc., U. Wales, D.P.H., Cambridge; m. Katherine Jokl, 1919; 2 sons, 1 dau. Mem. staff hosps., London, then prof. psychiatry U. Toronto, also dir. Ontario Clinic for Nervous Diseases; spent 2 years on Continent in research work, then returned to Eng., 1913, confining work to med. psychology. Hon. cons. Graylingwell Mental Hosp.; dir., cons. physician London Clinic for Psycho-Analysis; founder-editor Internat. Jour. Psychoanalysis; hon. pres. Internat. Psycho-Analytic Assn., Am. Psychoanalytic Assn., Brit. Psycho-Analytic Soc., Inst. Psychoanalysis. Fellow Royal Coll. Physicians, Royal Soc. Medicine, Royal Soc. Arts, Royal Anthrop. Inst., (hon.) Royal Medico-Psychol. Assn., Am. Psychiatric Assn., Am. Psychopath. Assn., also others. Author: Sigmund Freud, Life and Work, 3 vols., 1953-57; Free Associations, 1960, also others, including numerous monographs on neurology, psychology, anthropology; editor Social Aspects of Psychoanalysis, 1924. Introduced study of psychoanalysis into Eng. and U. S. Died Feb. 11, 1958.

JONES, Frank Culver, Am. physicist; b. Ft. Worth, July 30, 1932; s. Kenneth Hugh and Nancy (Culver) J.; B.A., Rice U., 1954; M.S., U. Chgo., 1955, Ph.D. in Physics, 1961; m. Ardythe Eunice Grube, Aug. 27, 1955; children—Cheryl Kay., Timothy Kenneth. Instr., Princeton, 1960-63; resident research asso. Nat. Acad. Sci.-NRC, Lab. for Theoretical Studies, Goddard Space Flight Center, Greenbelt, Md., 1963-65, staff Flight Center, 1963—. Mem. Am. Phys. Soc., Am. Assn. Physics Tchrs., A.A.A.S. Research, publs. on gamma radiation in cosmic rays, cosmic-ray electrons, their prodn. in space via high-energy collisions cosmic-ray protons and interstellar hydrogen gas, electrons interactions with starlight in inverse Compton effect. Home: 416 Hillsboro Dr., Silver Spring, Md. 20902. Office: Lab. for Theoretical Studies, Goddard Space Flight Center, Greenbelt, Md. 20771.*

JONES, Frank Morton, Am. entomologist; b. Wilmington, Del., Jan. 13, 1869; ed. Franklin Inst., certificate, 1921; hon. D.Sc., U. Del., 1931; fellow Entomological Soc., Royal Entomological Soc. London; mem. Ecology Soc., Soc. Natural History Del. (pres. 1928-43, emeritus, 1944). Research on psychidae, lepidoptera of Martha's Vineyard island, insectivorous plants and their insect associates, also adaptive coloration of insects.

JONES, Galen Everts, Am. microbiologist; b. Milw., Sept. 9, 1928; s. Galen and Grace (Everts) J.; A.B., Dartmouth, 1950; M.A. Williams Coll., 1952; Ph.D., Rutgers U., 1956; m. Edith Agnes Boehme, July 17, 1954; children—Galen Randolph, Gwenith Grace, Christopher Thomas, With U. Cal. SIO, La Jolla, Cal., 1955-63, asst. research microbiologist, 1957-63; asso. prof. biology Boston U., 1963-66; prof. microbiology, dir. N.H. Estuarine Lab., U. N.H., Durham, 1966—. Fellow A.A.A.S.; mem. Am. Inst. Biol. Sci., Soc. Gen. Microbiology, Geochem. Soc., Am. Soc. Microbiology, Am. Soc. Limnology and Oceanography. Research, publs. on fractionation of stable isotopes of sulfur by microorganisms, enumeration and ecol. relationships of marine bacteria, suppression of bacterial growth in seawater, effect of heavy metal content of seawater on bacteria. Home: Becky Branch Rd. at Crest, Southern Pines, N.C.*

JONES, George William, Am. mathematician; b. East Corinth, Me., Oct. 14, 1837; s. George William and Cordelia (Allen) J.; A.B., Yale, 1859, A.M., 1862. m. Caroline Tuttle Barber, Aug. 11, 1862. Tchr. math. Russell's Mil. Sch., New Haven, 1859-62, Del. Lit. Inst., Franklin, N.Y., 1862-68; prof. math. Ia. State Coll., 1868-74; asst. prof. Cornell U., Ithaca, N.Y., asso. prof., prof. math., 1877-1907. Author: Treatise on Algebra (joint author), 1882; Treatise on Trigonometry (joint author), 1881; Logarithmic Tables, 1889; Drill-book in Algebra, 1892; Drill-book in Trigonometry, 1896; Five-place Logarithms, 1896; Four-place Logarithms, 1896; Some Proofs in Elementary Geometry, 1904. Died 1911.

JONES, Grinnell, Am. chemist; b. Des Moines, Jan. 14, 1884; s. Richard and Carrie Holmes (Grinnell) J.; S.B., Vanderbilt U., 1903, S.M., 1905; A.M., Harvard, 1905, Ph.D., 1908; m. Genevieve Lupton, Aug. 18, 1910; 1 son, Grinnell. Instr. chemistry, U. Ill., 1908-12; instr. chemistry, Harvard, Cambridge, Mass., 1912-16, asst. prof., 1916-21, asso. prof. 1921-34, prof. from 1934. Chief chemist, U. S. Tariff Commn., 1917-19, cons. chemist, 1919-26. Bd. Mem. Am. Chem. Soc., Am. Acad. Arts and Scis., Am. Assn. of Textile Chemists and Colorists, Am. Inst.

Chem. Engrs., Phi Beta Kappa, Sigma Xi, Alpha Chi Sigma, Phi Lambda Upsilon. Developed fire-resistant paint, method of purifying salt water. Died June 23, 1947.

JONES, Hardin Blair, Am. biologist; b. Los Angeles, June 11, 1914; s. Henry Hardin and Maude (Blair) J.; A.B., U. Cal. at Los Angeles, 1937; M.A., U. Cal. at Berkeley, 1939, Ph.D., 1944; m. Helen Mary Cook, Mar. 17, 1940; children—Carolyn Frances, Hardin Cook, Nancy Helen, Mark Blair. Faculty, U. Cal. at Berkeley, 1946—, prof. med. physics and physiology, 1954—, chmn. group in biophysics and med. physics, 1956-62, 64—, asst. dir. Donner Lab., 1948—. Mem. Am. Physiol. Soc., Gerontology Soc., Eugenics Soc., Radiation Research Soc., A.A.A.S., Sigma Xi. Research, numerous publs. on blood flow through tissues, relationship to aging; effects of environmental factors on rate of aging; established concept of proportionality instead of threshold as guide to estimation of radiation effect; inventor chromic phosphate, other radioactive colloids, demographic methods. Home: 2816 Oak Knoll Terrace, Berkeley, Cal. 94705.*

JONES, Sir Harold Spencer, English astronomer; b. Kensington, London, Eng., Mar. 29, 1890; s. H. C. Jones; B.A., Jesus Coll., Cambridge, 1911, M.A., 1914, elected fellow, 1914, Sc.D., 1925, hon. fellow, 1933; D.Sc., univs. Paris, Brussels, Delhi, Durham, Oxford, Calcutta; Ph.D. (hon.), U. Copenhagen; LL.D., U. Glasgow; m. Gladys Mary Owers, 1918; 2 sons. Chief asst. Royal Obs., Greenwich, 1913-23; editor, 1914-23; His Majesty's astronomer, Cape of Good Hope, 1923-33; asst. dir. inspection optical supplies Ministry Munitions, World War I; astronomer royal, 1933-55, ret., 1955. Recipient gold medal Royal Astron. Soc., 1943, Janssen medal Soc. Astronomy France, 1945, gold medal Brit. Horological Inst., 1947, Bruce gold medal Astron. Soc. Pacific, 1948, gold medal Stoke-on-Trent Assn. Eng., 1950, Lorimer medal Astron. Soc. Edinburgh, 1953, Rittenhouse medal Rittenhouse Astron. Soc., 1955, Joykissen Mookerjee gold medal Indian Assn. Cultivation Sci., 1957. Fellow Royal Soc., 1930 (Royal medal 1943), Royal Soc. Edinburgh (hon.); mem. numerous assns. Author: General Astronomy, 1922; Worlds without End, 1935; Life on Other Worlds, 1940; A Picture of the Universe, 1947. Translator: Atlas of the Sky (Vincent de Callatay), also articles. Worked on analysis of accumulated meridian observations; his work in astron. photography contbd. greatly to knowledge of stellar distances in southern sky; most famous for his redetermination (1941) of solar parallax (and thereby astron. unit) from observations made of close approach of asteroid Eros, 1931; conducted important research on motion of earth's poles and mass of moon; also studied irregularity of earth's rotation, was responsible for moving site of Royal Obs. from Greenwich to Herstmonceux Castle, Sussex; instrumental in introduction of ephemeris time in fundamental ephemerides. Died Nov. 3, 1960.

JONES, Harry Clary, Am. chemist; b. New London, Md., Nov. 11, 1865; s. William and Joanna C. J.; A.B., Johns Hopkins, 1889, fellow, 1891-92, Ph.D., 1892; univs. of Leipzig, Amsterdam and Stockholm, 1892-94; m. Harriet Brooks, May 22, 1902. Instr. physical chemistry Johns Hopkins, 1895-98, asso., 1898-1900, asso. prof., 1900-04, prof., from 1904. Asso. editor Journal de Chimie Physique, Journal of Franklin Inst., and Zeitschrift für physikalische Chemie. Langstreth medalist, Franklin Inst., 1913. Author: Freezing Point, Boiling Point, and Conductivity Methods, 1897; The Modern Theory of Solutions (Harper's Scientific Series), 1898. Translator: Blitz's Practical Methods for Determining Molecular Weights, 1899; The Theory of Electrolytic Dissociation, 1900; Outlines of Electro-chemistry, 1902; Elements of Physical Chemistry, 1902 (translated into Russian, 1911, and Italian, 1912); Principles of Inorganic Chemistry, 1903; Elements of Inorganic Chemistry, 1903; The Electrical Nature of Matter and Radioactivity, 1906; Hydrates in Aqueous Solutions, 1907; Conductivity and Viscosity in Mixed Solvents, 1907; The Absorption Spectra of Solutions, 1909; Introduction to Physical Chemistry, 1910; The Absorption Spectra of Solutions, 1910, 11; Electrical Conductivity of Salts and Organic Acids, 1912; A New Era in Chemistry, 1913; Absorption Spectra Studied by Radiomicrometer, 1913; Freezing Points, Conductivities and Viscosities of Solutions of Salts in Mixed Solvents, 1913; The Absorption Spectra of Solutions; The Conductivities, Dissociations and Viscosities of Solutions of Electrolites in Aqueous, Nonaqueous and Mixed Solvents, 1915; Conductivities and Viscosities in Pure and in Mixed Solvents; Radiometric Measurements of the Formation Constants of Indicators, 1915. Died Mar. 19, 1916.

JONES, Herbert L., Am. elec. engr.; b. Copperton, N.M., Dec. 2, 1904; s. Herbert M. and Myra E. (Lukens) J.; B.A., U. Ore., 1926; M.A., Ore State Agrl. Coll., 1934, Ph.D., 1936; m. Margaret Zimmerman, Mar. 23, 1928; children—Margaret Elizabeth, Herbert Edmund. Telephone engr. Pacific Tel. & Tel., 1926-29; radio telephone engr. Bell Telephone Labs., 1929-32; instr. Ore. State Coll., 1935-36, instr. U. N.M., 1936-46; prof. elec. engring. Okla. State U., 1946—, dir. Tornado Lab., 1947—. Recipient NSF, grant, 1959. Profl. engr., N.M., Okla. Mem. Inst. Radio Engrs., Am. Inst. E.E., Nat. Soc. Profl. Engrs., Am. Meteorol. Soc., Am. Geophys. Union, Internat. Platform Assn., Phi Beta Kappa, Sigma Xi, Sigma Phi Epsilon, Phi Kappa Phi, Sigma Tau, Pi Mu Epsilon, Kappa Mu Epsilon, Eta Kappa Nu. Author: The Identification of Lightning Discharges by Speric Characteristics, 1958; The Tornado Pulse Generator: A Reality, 1960; The Tornado Pulse Generator as a Criteria for the Definition of the Severe Storm, 1963; A Review of the Frequency Spectrum of Cloud-to-cloud and Cloud-to-Ground Lightening, 1967. Discoverer of tornado pulse generator. Home: 823 Blakely St., Stillwater, Okla. 74074.*

JONES, J. Knox, Jr., Am. zoologist; b. Lincoln, Neb., Mar. 16, 1929; s. J. Knox and Virginia (Bowen) J.; B.Sc., U. Neb., 1951; M.A., U. Kan., 1953, Ph.D., 1962; m. Janet Helen Glock, Sept. 12, 1953; children—Amy, Sarah, Laura. Staff, Commn. on Hemorrhagic Fever, U. S. Army, 1954-55; faculty dept. zoology U. Kan., Lawrence, 1959—, professor zoology, 1968—, asst. curator Mus. Natural History, 1959-65, asso. curator, 1965—, asst. dir., 1962—; field and research asso. Neb. State Mus., 1957—. Mem. Am. Soc. Mammalogists (mng. editor Jour. Mammology), Soc. Systematic Zoology, Soc. for Study Evolution, Biol. Soc. Washington, Southwestern Soc. Naturalists, Sigma Xi. Author: Distribution and Taxonomy of Mammals of Nebraska, 1964; also numerous articles. Editor: (with Sydney Anderson) Mammals of the World, a Synopsis of Recent Families, 1967. Research on systematics and biogeography of temperate and tropical N.Am. mammals and mammals of Northeastern Asia, especially rodents and bats. Home: 914 Wellington Rd., Lawrence, Kan. 66044.*

JONES, Jack Colvard, Am. entomologist; b. Birmingham, Ala., Nov. 25, 1920; s. Felix Henry and Ethel (Colvard) J.; B.S., Auburn U., 1942, M.S., 1947; Ph.D., Ia. State U., 1950; m. Elizabeth Deese Jones, June 1, 1946; children—Robin, Megan, Brian. Med. entomologist NIH, Bethesda, Md., 1950-58; faculty U. Md., College Park, 1958—, prof. entomology, 1964—. Guggenheim award, 1965; NIH Career Devel. award, 1964—. Mem. Entomol. Soc. Am., Am. Physiol. Soc., Am. Mosquito Control Assn., Entomol. Soc. Washington. Research, numerous publs. on types and functions insect blood cells, insect heart, anatomy and physiology mosquitos. Home: 10234 Carrol Pl., Kensington, Md. 20795. Office: Dept. Entomology, U. Md., College Park, Md. 20740.*

JONES, John, Am. surgeon; b. Jamaica, L.I., N.Y., 1729; s. Dr. Evan and Mary (Stephenson) J.; grad. U. Rheims, 1751. Prof. surgery and obstetrics King's Coll. (now Columbia), 1767-76; founder (with Dr. S. Bard), attending physician N.Y. Hosp., 1771; credited as an organizer Med. Dept. of Continental Army; attending physician Pa. Hosp., 1780; pres. Humane Soc.; attended Pres. Washington in Phila., 1790; 1st v.p. Coll. Physicians of Phila., 1787; Author: Plain, Concise, Practical Remarks on the Treatment of Wounds and Fractures, Designed for the Use of Young Military Surgeons of America (1st surg. textbook in Am. colonies), 1775; translated Diseases Incident to Armies (VanSwieten). Known for his ability in lithotomy operations. Died Phila., June 23, 1791.

JONES, John Kenyon Netherton, chemist; b. Birmingham, Eng., Jan. 28, 1912; s. George Edward and Florence (Goodchild) N.; B.Sc. with 1st class honors, Birmingham U., 1933, Ph.D., 1936, D.Sc., 1947; m. Marjorie Ingles Noon, July 29, 1937; children—Stephanie (Mrs. William Buckeridge), Stephen Howard, Jonathan Ingles. Faculty, U. Bristol, Eng., 1936-45, 45-48, U. Manchester, Eng., 1945-48; Chown research prof. Queen's U., Kingston, Ont., Can., 1953—. Bd. dirs. Ont. Research Found. Fellow Royal Soc., 1957, Royal Soc. Can. Chem. Inst. Can.; asso. Royal Inst. Chemists. Research, numerous publs. on chemistry of sugars, biol. origin and function in man and plants. Home: Treasure Island, Box 31, Route 1, Kingston, Ont., Can.*

JONES, John Willis, Am. physicist; b. Riverside, Cal., Apr. 7, 1930; s. M. D. and Ora L. (Smith) J.; B.S., U. Wash., 1952; m. Barbara E. Burns, May 20, 1955; children—Christopher, Kevin, Margaret. Physicist, E.I. duPont de Nemour & Co., 1954-55, physicist rocket propellants, 1957-60; project engr. guided missiles Picatinny Arsenal Ordnance Corps, U. S. Army, 1955-57; mgr. solid rocket propellant structure integrity dept. Grand Central Rocket Co., 1960-61; mgr. solid rocket propellant structural integrity dept. Lockheed Propulsion Co., Redlands, Cal., 1961-65, advanced program mgr., 1965—; cons., 1953-54. Mem. Rocket Soc. Research in phys. behavior of rocket propellants, explosion phenomena, earthquake seismology, gun ballistics. Home: 1610 Margarita Dr. Office: P.O. Box 1, Redlands, Cal. 92373.*

JONES, Lawrence William, physicist; b. Evanston, Ill., Nov. 16, 1925; s. Charles Herbert and Fern (Storm) J.; B.S., Northwestern U., 1948, M.S., 1949; Ph.D., U. Cal. at Berkeley, 1952; m. Ruth Reavley Drummond, June 24, 1950; children—Douglas Warren, Carol Anne, Ellen Louise. Research asst. U. Cal. Radiation Lab., 1950-52; prof. physics U. Mich., Ann Arbor, 1952—. Vis. physicist Lawrence Radiation Lab., Berkeley, 1959, cons., 1964—; vis. scientist CERN, Geneva, Switzerland, 1961-62, 65; vis. physicist Brookhaven Nat. Lab., Upton. N.Y., 1963—; cons. Argonne (Ill.) Nat. Lab., 1963—. Guggenheim fellow, 1965. Mem. Am. Phys. Soc. Research, numerous publs. on photonuclear reactions, constrn. and operation of 1st F.F.A.G. particle accelerator, colliding beams devices and expts. novel types of spark chambers, cosmic ray, devel. 1st luminescent chambers used in physics research, 1st systematic expts. in high energy elastic scattering of pions and in dipion prodn. Home: 447 Hilldale Dr., Ann Arbor, Mich. 48105.*

JONES, Leslie McLaury, Am. aeronomist; b. Yonkers, N.Y., Dec. 22, 1917; s. Harold Morgan and Emma (McLaury) J.; B.S., U. Mich., 1940; m. Jean Jessop, Dec. 2, 1938 (dec. 1950); children—Thomas M., David F., John R.; m. 2d, Anne Denison Hough, Jan. 10, 1959. Sound engr. Woodall Industries, Detroit, 1940-41; prodn. engr. Western Elec., Kearey, N.J., 1941-42; instrument engr. Physicists Research Co., Ann Arbor, Mich., 1942-46; with U. Mich., 1947—, dir. high altitude engring. lab., 1949—, prof. dept. aerospace engring., 1963—, mem. com. on extension to standard atmosphere. Mem. com. on upper atmosphere rocket research Space Sci. Bd., Nat. Acad. Sci. Mem. Am. Inst. Aeros. and Astronautics, Am. Geophys. Union, Sigma Xi. Research on upper atmosphere using rockets and satellites, especially using small rockets; invented falling sphere method and asso. accelerometer. Home: 3595 E. Huron River Dr. Office: Dept. Aerospace Engring. U. Mich., Ann Arbor, Mich.

JONES, Lewis Ralph, Am. botanist; b. Brandon, Wis., Dec. 5, 1864; s. David and Lucy (Knapp) J.; student Ripon Coll., 1883-86; Ph.B., U. of Mich., 1889, grad. studies, 1901, 1904, Ph.D., 1904; hon. Sc.D., U. of Vt., 1910, U. of Cambridge (Eng.), 1930, U. of Wis., 1936; LL.D., U. of Mich., 1935; has also carried out investigations in lab. of Bur. of Plant Industry, Washington, under direction of Dr. Erwin F. Smith, 1899, and in Europe, 1904; m. May I. Bennett, June 24, 1890; m. 2d, Anna M. Clark, July 27, 1929. Prof. botany, U. of Vt. and botanist Vermont Expt. Sta., 1889-1910; prof. plant pathology, U. of Wis., 1910-35, emeritus prof. since 1935. Collaborator of Bureau of Plant Industry, U. S. Dept. Agriculture; ex-sec. bd. park commrs., Burlington, Vt.; ex-sec. Vt. Bot. Club; ex-pres. Vt. Forestry Assn.; fellow A.A.A.S.; ex-pres. Bot. Soc. America; mem. Nat. Acad. of Sciences, N.E. Bot. Club, Wis. Acad. Arts and Sciences, Am. Soc. Naturalists, Am. Philos. Soc.; corr. Assn. of Applied Biologists, England, Société de Pathologie Végétale et d'Entomologie Agricole de France; hon. mem. Vereinigung für Angewandte Botanik, Japanese Phytopathol. Soc.; ex-chmn. div. of biology and agr. Nat. Research Council; ex-pres. Am. Phytopathol. Soc. Joint author of "Flora of Vermont," and author bot. reports, bulls. and contbns. to scientific jours. Editor of Phytopathology; Am. Jour. Botany; bacteriol. terms, Webster's New Internat. Dictionary. First thorough research in mechanisms of infection in bacterial diseases of plants; developed cabbage resistant to yellows, also Wis. soil-temperature tank (thermostatically controlled equipment for plant growing. Died Mar. 31, 1945.

JONES, Lynn William, Am. physicist, engr.; b. Sparta, Wis., May 2, 1902; s. Fred Hubert and Jennie (McPeak) J.; A.B., Ripon Coll., 1925; postgrad. (fellow) Washington U., St. Louis, Wis. U.; A.M., U. So. Cal., 1931; m. Helen Edith Frable, Aug. 29, 1928; children—Lynn Winfield, Cecelia Ivaine (Mrs. Richard C. White). Prof. physics, engring. U. Redlands, 1927-43; sr. elec. engr. Kaiser Steel Co., 1943-46; supt. Basic Magnesium Corp. Nev., 1946-47; liaison engr. Stone & Webster Co., 1947-48; chief elec. design Frederick R. Harris Co., 1948-51; head facilities engring. research and devel. Hughes Aircraft Co., 1951-55; sr. staff engr. Thompson-Ramo-Wooldridge Systems Group, Los Angeles, 1955—; cons. engr., Redlands, Cal., 1955—. Fellow A.A.A.S.; mem. Eugene Field Soc. (honors), Soc. Mech. Engring., Research Soc. Am., I.E.E.E., Computers Soc. Contbr. articles to tech. jours. on electronics, missiles, high frequency phenomena. Died Feb. 10, 1967.

JONES, Marcus Eugene, Am. botanist, geologist; b. Jefferson, O., Apr. 25, 1852; s. Publius Virgilius and Lavinia (Burton) J.; A.B., Ia. (now Grinnell) Coll., 1875, A.M., 1878; m. Anna E. Richardson, Feb. 18, 1880; children—Mabel Anna, Howard Marcus, Mildred Lavinia. Tutor, Ia. Coll., 1876-77; prin. Lemars (Iowa) Acad., 1877; prof. natural science, Colorado Coll., 1879; same at Salt Lake City, 1880-81; prin. Jones High Sch., Salt Lake City, 1884-86; spl. expert, U. S. Treasury Dept., 1889; geologist Rio Grande Western R.R., 1890-93; spl. field agt., U. S. Dept. Agr., 1894-95; botanist, geologist and mining expert. Pres. Utah Acad. Sciences. Hon. curator of botany, Pomona Coll. Author: Excursion Botanique, 1879; Ferns of the West, 1883; Salt Lake City, 1889; Some Phases of Mining in Utah; Utah supplement to Tarr and McMurry's Geography, 1902; Contbns. to Western Botany (18 parts), 1879-1934. Exploring in Mexico, 1924. Made extensive collection of plants of western U. S. (Colo., Utah, Great Basin), discovered number of new species. Died June 3, 1934.

JONES, Mark Wallon, Am. physicist; b. Wheeling, W.Va., Dec. 20, 1916; s. John William and Virginia (Thomas) J.; A.B., U. W.Va., 1938; M.S., U. Fla., 1948, Ph.D., 1953. Observer dept. terrestrial magnetism Carnegie Instn. Washington, 1939-47; fellow, sr. physicist Geophys. Inst., U. Alaska, 1948-50; sr. physicist Ordnance Engring. Corp., Rockville, Md., 1953-55; pres., chief physicist Geo-Sci., Inc., Alamogordo, N.M., 1955——. Pres., N.M. Research Inst., Alamogordo, 1958——. Recipient Naval Ordnance Devel. award, 1945. Mem. I.E.E.E. (chmn. local sect. 1959), Am. Assn. on Emeriti (v.p. 1962——), A.A.A.S., Am. Geophys. Union, Soc. Exploration Geophysicists, Sigma Xi. Research, publs. in theoretical geophysics, upper atmospheric and ionospheric physics, biophysics and biol. effects radiation. Home: 1610 Juniper Dr. Office: P.O. Box 1128, Alamogordo, N.M. 88310.*

JONES, Reginald Victor, Brit. physicist; b. London, Eng., Sept. 29, 1911; s. Harold Victor and Alice Margaret (May) J.; M.A., D.Phil., Wadham Coll., also Balliol Coll., Oxford (Eng.) U., 1934; m. Vera Margaret Cain, Mar. 21, 1940; children—Susan Primrose (Mrs. John Parente), Robert Bruce, Rosemary Anne. Sci. officer Air Ministry, 1936, Admiralty, 1938-39, Air Staff, 1939; asst. dir. intelligence, 1941-46, dir. intelligence, 1946; prof. natural philosophy U. Aberdeen (Scotland), 1946——; dir. sci. intelligence Ministry Def., 1952-53. Chmn. electronics research council Ministry Tech., 1964——, Safety in Mines Research Bd., 1956-60; mem. adv. council BBC, 1964——; mem. adv. council on calibration and measurement Ministry Tech., 1967——; vis. bd. Nat. Phys. Lab., 1967——. Named comdr. Order Brit. Empire, companion Order of Bath; recipient U. S. medal of Freedom with silver palm, U. S. medal for merit; B.O.I.M.A. prize, 1934; Duddell medal, 1960; Parsons medal, 1967; hon. fellow Wadham Coll., Oxford U. Fellow Royal Soc. London (chmn. Paul fund com. 1960——), Royal Soc. Edinburgh, Royal Astron. Soc., Inst. Physics, Crabtree Found.; mem. Am. Geophys. Union. Research, numerous publs. on infra-red detection, instrument design, crystal growth, def. sci. Home: 8 Queen's Terrace, Aberdeen, Scotland.*

JONES, Reuben Gardner, Am. chemist; b. Cedar City, Utah, Nov. 17, 1913; s. Henry L. and Artamesia (Gardner) J.; student Brigham Young U., 1934-35; B.S., U. Utah, 1937, M.A., 1938; Ph.D., Ia. State Coll., 1941; m. Josephine Jackson, Sept. 10, 1940; children—Reuben J., Mary Margaret, Jeanette, Elizabeth. Research chemist Eli Lilly & Co., Indpls., 1941, 43-46, head organic research dept., 1946-55, dir. organic chemistry div., 1955-63, research adviser, 1963——; research chemist OSRD, Ia. State Coll., 1942-43; faculty Ind. U. Sch. Medicine, 1960-63. Mem. Am. Chem. Soc., A.A.A.S., N.Y. Acad. Scis. Research, numerous publs. on chemistry penicillins, anti-thyroid drugs, histamine-like compounds, alkaloids, antibiotics, vitamins, organometallic compounds; co-developer penicillin V. Home: 4625 Laurel Circle, Indpls. 46226. Office: Eli Lilly & Co., Indpls. 46206.*

JONES, Richard Uriah, Am. chemist; b. Ottawa, Minn., Jan. 25, 1877; s. William R. and Mary (Hughes) J.; B.A., Macalester Coll., St. Paul, Minn., 1901, hon. D.Sc., 1926; student U. of Minn., 1901-02, U. of Chicago, summers 1908, 09; M.A., U. of Wis., 1916; m. Mary Helen Smith, Aug. 18, 1909; children—Richard Herbert, Donald Caldwell, George William. Prof. physics and chemistry Macalester Coll., 1903, prof. chemistry, from 1907, dean of coll., 1917-36. Specialized in researches in organic chemistry, particularly sugars, starches and volatile oils. Author: The Scientific Eye of Faith. Died July 9, 1941.

JONES, Robert Clark, Am. physicist; b. Toledo, June 30, 1916; s. William Clark and Thora (Klagstad) J.; A.B., Harvard, 1938, A.M., 1939, Ph.D., 1941; m. Lois Kathryn Bertholf, June 15, 1938. Tech. staff Bell Telephone Labs., Murray Hill, N.J., 1941-44; sr. physicist Polaroid Corp., Cambridge, Mass., 1944——. Mem. U. S. Nat. Com. for Optics, 1959-64. Fellow Oytical Soc. Am. (dir. at large 1962-64, Adolph Lomb medal 1944), Acoustical Soc. Am., Soc. Photog. Scientists and Engrs. (past editorial v.p.), Am. Acad. Arts and Scis.; mem. Assn. for Computing Machinery, Archaeol. Inst. Am. Research, numerous publs. on behavior of optical systems containing polarizers and retarders, detecting ability and information capacity of detectors for visible and infrared radiation, resolution and granularity of photog. films. Home: 1716 Cambridge St., Cambridge 02138. Office: 730 Main St., Cambridge, Mass. 02139.*

JONES, Roy Winfield, Am. zoologist; b. Shawnee, Okla., Sept. 16, 1905; s. William Winfield and Grace (McCreery) J.; A.B., Oklahoma City U., 1927; M.S., Kan. State U., 1928; Ph.D., U. Okla., 1937; m. Maurine King, Sept. 1, 1928; children—Neil Winfield, Marian K. (Mrs. James Lee West). Tchr., Bolton High Sch., Alexandria, La., 1928-29; faculty Central State Coll., Edmond, Okla., 1929-47, prof. biology, head dept., 1935-39, dean coll., acting dean men, 1939-47; prof. zoology, chmn. biol. scis. Okla. State U., Stillwater, 1947-51, prof. zoology, head dept., 1951——. Chmn. com. on improvement of instrn. in sci. Okla. Curriculum Improvement Commn., 1957-68. Spl. faculty fellow NIH, 1963. Mem. A.A.A.S., Am. Inst. Biol. Scis., Am. Soc. Zoology, Am. Microbiol. Soc., Am. Soc. Exptl. Biology and Medicine, N.E.A., Southwestern Assn. Naturalists, Okla. Acad. Sci., Nat., Okla. sci. tchrs. assns., Nat. Biology Tchrs. Assn., Sigma Xi, Phi Kappa Phi, Phi Sigma. Author: Laboratory Exercises in Physiology, 1932; Biological Science Notebook, 1952, 54; also articles. Editor: The Human Element in College and University Administration, 1949. Research on resistance to parasitism due to previous infection, rate of cell div. to organ differentiation, fish embryology, use of fish embryos in screening chems. which affect cell div. and embryonic differentiation, culture of fish cells. Home: 2030 W. Admiral Rd., Stillwater, Okla. 74074.*

JONES, Stephen Barr, Am. geographer; b. Seattle, Feb. 23, 1903; s. Richard Saxe and Margaret (Barr) J.; B.S. magna cum laude, U. Wash., 1924; M.A., Harvard, 1939, Ph.D., 1934; m. Marjorie Jean Sadler, Sept. 2, 1933; 1 son, Douglas Barr. Instr., Ore. Coll. Edn., Monmouth, 1932-35; faculty U. Hawaii, Honolulu, 1935-42, asso. prof., 1941-42; sr. divisional asst. U. S. Dept. State, 1942-43. faculty Yale, 1943——, prof. geography, 1948——. Vice chmn. div. geology and geography NRC, 1948-51. Fellow Am. Geog. Soc. (hon.); mem. Assn. Am. Geographers (past chmn. honors com., hon. pres. 1959), Am. Meteorological Society. Author: Boundary-Making, 1945; (with M. J. Murphy) Geography and World Affairs, 1950; also articles. Contbg. editor Geog. Rev. Research on theory of polit. geography especially role of movement and circulation, nature and functioning of polit. boundaries. Home: 46 Morse St., Hamden, Conn. 06517. Office: Geography Dept., Yale, New Haven 06520.*

JONES, Thomas Oswell, Am. chemist; b. Oshkosh, Wis., May 13, 1908; s. Hugh Edwards and Jane (Davies) J.; B.E., Wis. State U., 1930; M.Philosophy, U. Wis., 1934, Ph.D. in Inorganic Chemistry, 1937; m. Phyllis Elizabeth Jackson, Aug. 19, 1950; children—Elizabeth Carol, Phyllis Jane. Faculty, Haverford Coll., 1937-56; acting head Office Sci. Information, NSF, Washington, 1956-58, program dir. Antarctic programs, 1958-61, head Office Antarctic Programs, 1961-65, div. dir. environmental scis., 1965——. Asst. to sect. chief Metall. Lab., Chgo., 1944-45, sect. chief information div., 1945-46; vis. prof. chemistry U. Wis., 1954-55. Recipient Order "Al Merito" Chilean govt., 1963; Outstanding award NSF, 1963. Mem. Am. Chem. Soc., Am. Geophys. Union, A.A.A.S., Sigma Xi, Phi Beta Kappa. Statis. calculations, publs. on values of and atomic ratios between hydrogen and oxygen; application isotope tracer techniques to chemistry. Home: 7504 Holiday Terrace, Bethesda, Md. 20034. Office: 1800 G St., N.W. Washington 20550.*

JONES, Thomas Rupert, English geologist; b. London, Oct. 1, 1819; ed. Taunton, Ilminster; m. Mary Harris; 2 sons, 2 daus.; m. 2d, Charlotte Ashburnham Archer; 2 sons, 3 daus. Prof. geology Royal Mil. and Staff Colls., Sandhurst; ret., 1880. Fellow Royal Soc., 1844, Geol. Soc. (Lyell medal); hon. mem. many Brit., fgn. sci. socs. Collected and described old stone implements, fossil foraminifera and entomostraca. Died Apr. 13, 1911.

JONES, Thomas Wharton, physician, oculist; b. St. Andrew's, Scotland, 1808; prof. ophthalmic medicine U. Coll., London, Eng. Fellow Royal Soc., 1840. Author: Treatise on Ophthalmic Medicine and Surgery; discovered the ameboid movement of leukocytes, 1853. Died 1891.

JONES, Walter, Am. physiol. chemist; b. Balt., Apr. 28, 1865; s. Levin and Zeanette J. (Bohen) J.; A.B., Johns Hopkins, 1888, Ph.D., 1891; m. Grace C. Clarke, Sept. 1, 1891; 1 dau., Marion E. (Mrs. Gilbert A. Jarman). Prof. chemistry, Wittenberg Coll., Springfield, O., 1891-92; prof. analyt. chemistry, Purdue U., 1892-95; asst., asso. prof. Johns Hopkins, 1895-1908, and prof. of physiol. chemistry, 1908-27, prof. emeritus, 1927. Author: Nucleic Acids, 1921. Research on nucleic acids, melanins; described a-menthyl-quinoline as constituent in secretion of anal glands of M.Mephitica. Died Feb. 28, 1935.

JONES, William Richard, Am. engr.; b. Hazleton, Pa., Feb. 23, 1839; m. Harriet Lloyd, Apr. 14, 1861, 4 children. Journeyman machinist, 1853; with Cambria Iron Co., Johnston, Pa., 1859; gen. supt. Edgar Thomson Steel Co., Braddock, Pa., 1875-89; cons. engr. Carnegie, Phipps & Co., 1888-89. First American to be invited to see Krupp steel works at Essen; erected Carnegie Bessemer plants, A through G series blast furnaces and rolling mills. Inventor and patentee numerous devices connected with prodn. steel, most important was Jones mixer (mixed molten iron from blast furnaces for converter), 1889. Died Pitts., Sept. 28, 1889.

JONES, Winston William, Am. horticulturist; b. Eclectic, Ala., Jan. 18, 1910; s. Fred F. and Ida (McManus) J.; B.S., Auburn U., 1931; M.S., Purdue U., 1933; Ph.D., U. Chgo., 1936; m. Gladys L. Davidson, June 17, 1933; children—Carol Ann (Mrs. Arden Adams), Sharon Jean (Mrs. Gary Saffer). Plant physiologist U. Hawaii, Honolulu, 1936-42; horticulturist, head dept. U. Ariz., Tucson, 1942-46; prof. horticulture, horticulturist U. Cal. at Riverside, 1946——. Recipient Cal. Citrograph Research award, 1964. Mem. Am. Soc. Hort. Sci., A.A.A.S., Am. Inst. Biol. Sci., Am. Soc. Plant Physiologists, Internat. Hort. Soc. Contbr. numerous articles to tech. jours., chpts. in books. Research on mineral nutrition of fruit trees, especially citrus. Home: 1152 Blazewood St., Riverside, Cal. 92507.*

JONES, (Frederic) Wood, Brit. anatomist, anthropologist; b. London, Jan. 23, 1879; D.Sc., M.B., B.S.; m. Gertrude Clunies. Demonstrator anatomy London, St. Thomas's hosps.; lectr. anatomy U. Manchester; med. officer, Far East; anthropologist Egyptian Govt.; mem. Archaeol. Survey Nubia; Arris and Gale lectr. Royal Coll. Surgeons, 1914-16, 19, Sir William H. Collins prof. human and comparative anatomy, 1945-52, hon. curator Hunterian collection human and comparative anatomy, from 1952; prof. anatomy London Sch. Medicine for Women; prof. anatomy U. Adelaide, 1919-26; Rockefeller prof. phys. anthropology U. Hawaii, Honolulu, 1927-30; prof. anatomy U. Melbourne, 1930-37, Manchester U., 1938-45; temporary dir. anatomy Peiping Union Med. Coll., 1932-33; examiner anatomy London U., Conjoint B. Royal Colls. Physicians and Surgeons, Soc. Apothecaries. Licentiate Royal Coll. Physicians. Fellow Royal Soc., 1925, Royal Australian Coll. Surgeons, Zool. Soc.; Author: Coral and Atolls: various articles on corals and coral islands; Arboreal Man; The Principles of Anatomy as seen in the Hand; The Mammals of South Australia; The Matrix of the Mind; Man's Place among the Mammals, 1929; Sea Birds Simplified, 1934; Structure and Function as seen in the Foot, 1944; Hallmarks of Mankind, 1948; Trends of Life, 1953; also articles on anatomy. Research on corals as living things interacting with their environment; discoverer several new marsupial species, also new invertebrate species. Died Sept. 29, 1954.

JONGKEES, Leonard Barend Willem, Dutch surgeon; b. Gronigen, Netherlands, June 5, 1912; s. Willem Johan Adriaan and Emily (Spruyt) J.; grad. U. Utrecht (Holland), 1938; m. Jacoba Christina van Pernis, June 10, 1944; children—Laura Sophie, Therese Dorine, Jan Arie. Ear nose and throat surgeon Zwolle, also dir. Emergency Hosp., 1943-45; head clinic Ear Nose and Throat, U. Hosp., Utrecht, 1945-50; prof. ear nose and throat, head ear nose and throat dept. U. Amsterdam, 1950——. Decorated Knight Dutch Lion; recipient Silver medal U. Helsinki. Hon. or corr. mem. Irish, Danish, Swiss, Peruvian, Buenos Aires, Littoral ear nose and throat socs.; mem. Royal Cancer Inst. Amsterdam (past pres.), Collegium ORL Amicitiae Sacrum (gen. sec. 1960——), Société Internationale contre l'otospongiose (past pres.), Leopoldina Akademie, Royal Netherlands Acad. Author: (with Oogheelkunde H. Dekking) Keel neus Oorheelkunde, 1965; Slechthoren, 1966; also numerous articles. Editorial bd. Annals ORL, Practica ORL Basel, Acta Otolaryngologica Stockholm, Excerpta Medica St. XI. Research on physiology and diseases of organ of equilibrium, including methods to examine its function (cupulometry, torsion swing, parallel swing); surg. treatment of paralysis offacial nerve, methods of studying psychology of deafness. Home: 157 Koninginneweg, Amsterdam, O.Z. Office: Ear Nose Throat Dept., Wilhelmina Gasthuis, Amsterdam OW, Holland.*

JONKE, Albert Aloysius, Am. chem. engr.; b. Cleve.; Nov. 27, 1920; s. Joseph and Mary (Staltzer) J.; B.-Chem. Engring., Fenn Coll., 1943; M.S., Ill. Inst. Tech., 1956; m. Virginia C. Donahue, Apr. 26, 1947; children—Karen M., Kenneth D., David B., Timothy A., Thomas A., Kathleen A. Chem. engr. Allied Chem. Corp. div. Baker & Adamson, Marcus Hook, Pa., 1943-47; staff Argonne (Ill.) Nat. Lab., 1947——, sr. engr., 1962——, sect. head, 1964——. Mem. adv. bd. indsl. and engring. Chemistry Mag., 1964——. Mem. Am. Chem. Soc. (sec. div. indsl. and engring. chemistry 1963, exec. com. 1964-65), Am. Inst. Chem. Engrs., Am. Nuclear Soc., Research Soc. Am. Contbr. articles to tech. jours. Developed processes for refining uranium ore, prodn. uranium compounds, recovering uranium and plutonium from nuclear reactor fuels, treatment radioactive waste products; invented chem. processes involving solven extraction separations, fluidization methods, fluidized-bed apparatus. Home: 264 Highview Av., Elmhurst, Ill. 60126. Office: 9700 S. Cass Av., Argonne, Ill. 60440.*

JONKER, Charles Christiaan, Dutch physicist; b. Wormerveer, Netherlands, July 22, 1908; s. Jan Hendrik and Sijtje (van Bemmel) J.; doctorandus in physics, U. Utrecht (Netherlands), 1933; Ph.D., U. Leiden (Netherlands), 1937; m. Anna Hendrika de Waal Malefijt, Mar. 29, 1961. Tchr. physics and maths. secondary schs., 1934-46; faculty Free U., Amsterdam, Netherlands, 1946——, prof. theoretical physics, 1951——, vice dir. Phys. Lab., 1951-65, dir., 1965——. Pres. nuclear physics group Found. Fundamental Research Matter, Utrecht, 1954——; Pres. governing bd. Instituut voor Kernfysisch Onderzock, Amsterdam, 1960-67. Mem. Dutch Phys. Soc., Netherlands Orgn. for Pure Sci. Research (mem. governing bd. 1967——), Am. Phys. Soc. Author: Intensitätsverhältnisse im Cäsium Spektrum, 1937. also articles. Research on elastic and inelestic scattering of fast neutrons, nuclear models for neutrongamma reactions and muon capture. Home: 30II Watteaustraat, Amsterdam, Netherlands.*

JONKER, Fredrik P., Dutch botanist; b. Nov. 10, 1912; s. Jannes and J. (Schuurman) J.; Ph.D., U. Utrecht; m. A. M. E. Verhoef, Sept. 4, 1947. Asst., Mus. and Herbarium, U. Utrecht, also prof. botany. Mem. numerous assns. Research and publs. on botany, phytogeography, paleobotany, palynology. Home: Byronstraat 16. Office: Lange Nieuwstraat 106, Utrecht, Netherlands.

JONSSON, Bjarni, mathematician; b. Draghals, Iceland, Feb. 15, 1920; s. Jon and Steinunn (Bjarnadottir) Petursson; B.A., U. Cal. at Berkeley, 1943, Ph.D., 1946; m. Amy Sprague, Dec. 16, 1950 (div. 1967); children—Eric M., Meryl S. Came to U. S., 1941, naturalized, 1963. Faculty, Brown U., 1946-56, asst. prof., 1948-56; vis. prof. U. Iceland, 1954-55; vis. asso. prof., U. Cal., Berkeley, 1955-56, vis. prof., research mathematician, 1962-63; faculty U. Minn., 1956-66, asso. prof., 1956-59, prof., 1959-66; distinguished prof. Vanderbilt U., Nashville, 1966—. Mem. Am. Math. Assn., Assn. for Symbolic Logic, Am., Icelandic math. socs., Icelandic Acad. Sci. Research, publs. in lattice theory, universal algebra, foundations of algebra, group theory. Home: 5810 Vineridge Dr., Nashville 37205.*

JOOSSENS, Jozef Victor, Belgian physician; b. Borgerhout, Belgium, Feb. 12, 1915; s. August and Josephine (Mast) J.; M.D., U. Leuven (Belgium), 1939; m. Agnes Buys, Sept. 12, 1942; children—André, Filip, Jacky, Luc. Staff, Clinic Internal Medicine, Leuven, 1940-46; head internal medicine St. Luc Clinic, Ekeren-Antwerp, Belgium, 1948-54; faculty U. Leuven, 1954—, prof. medicine, 1958—, head dept. clin. chemistry, 1954—; head dept. cardiology St. Raphael U. Clinic, Leuven, 1955—; dir. heart center Sch. Pub. Health, Leuven, 1963—, head dept. epidemiology, 1965—. Decorated officer Order Leopold II. Mem. Société Belge de Cardiologie (vice chmn. 1966—), Société Belge De Chimie clinique (vice chmn. 1960—), Am. Assn. Clin. chemists, Royal Flemish Acad. Medicine Belgium (corr.), Société Francaise de Cardiologie (corr.). Research, numerous publs. on atherosclerosis and epidemiology, statis. analysis of mortality data; formulated genetic hypothesis on influence of X chromosome on mortality. Home: 11 Kerselarenweg, Herent, Belgium. Office: 35 Capucijnenvoer, Leuven, Belgium.*

JORDAN, Alexis, French botanist; b. Lyons, France, 1814. Author: Observations sur plusieurs plantes rares, 1846; De l'origine des diverses variétés ou espèces d' arbres cultivés, 1853; Icones ad floram Europae, 1866; Notes sur diverses espèces de plusierus plantes nouvelles. Founder Jordian sch. which studied species as part of a group and believed species varied only in small details, in contrast to Linnaeian emphasis of species: studied mutation, rare plants of France; undertook methodical breeding. Died 1897.

JORDAN, David Starr, Am. zoologist; b. Gainesville, N.Y., Jan. 19, 1851; s. Hiram and Huldah Lake (Hawley) J.; M.S., Cornell, 1872; M.D., Ind. Med. Coll., 1875; Ph.D., Butler U., 1878; LL.D., Cornell, 1886, Johns Hopkins, 1902, Ill. College, 1903, Ind. U., 1909, U. of Calif., 1913, Western Reserve U., 1915; m. Susan Bowen, Mar. 10, 1875 (died 1885); m. 2d, Jessie L. Knight, Aug. 10, 1887; children—Edith, Harold, Thora (dec.), Knight, Barbara (dec.), Eric (dec.). Instr. botany, Cornell, 1871-72; prof. natural history Lombard U., 1872-83; prin. Appleton (Wis.) Collegiate Inst., 1873-74; teacher Indianapolis High Sch., 1874-75; prof. biology, Butler U., 1875-79; prof. zoology, 1879-85, pres., 1885-91, Ind. U.; pres. Stanford, 1891-1913, chancellor, 1913-16, emeritus. Cooperating asst. to U. S. Fish Commn., 1877-91, 1894-1909; also U. S. commr. in charge of fur seal and salmon investigations; internat. commr. of fisheries, 1908-10. Chief dir. World Peace Foundation, 1910-14; pres. World's Peace Congress, 1915; v.p. Am. Peace Soc. Pres. A.A.A.S., 1909-10, N.E.A., 1915. Author: Manual of Vertebrate Animals of Northern United States, 1876-1929; Fishes of North and Middle America, 4 vols. (with B. W. Evermann), 1896; Care and Culture of Men, 1896; Footnotes to Evolution, 1898; The Story of Matka, 1898; The Strength of Being Clean, 1900; Standeth God Within the Shadow, 1900; Animal Life (with V. L. Kellogg), 1900; The Philosophy of Hope, 1902; The Blood of the Nation, 1902; Animal Forms (with others), 1903; Food and Game Fishes of North America (with B. W. Evermann), 1905; A Guide to the Study of Fishes, 1905; the Call of the Twentieth Century, 1905; The Human Harvest, 1907; Evolution and Animal Life (with V. L. Kellogg), 1907; Fishes, 1907; The Stability of Truth, 1909; Unseen Empire, 1912; War's Aftermath (with Harvey E. Jordan), 1912; Eric's Book of Beasts (children), 1913; War and Waste, 1914; World Peace and the College Man, 1914; War and the Breed, 1915; Ways to Lasting Peace, 1915; The Genera of Fishes, 1918-20, 4 parts; Democracy and World Relations, 1918; Fossil Fishes of Southern California, 1919-26, 9 parts; Autobiography (2 vols.); Classification of Fishes, 1922; (with K. D. Cather) High Lights of Geography, North America, 1925, Europe, 1925; The Trend of the American University, 1929. Research on marine industries of Pacific coast; a leading Am. authority on ichthyology. Died Sept. 19, 1931.

JORDAN, Denis Oswald, chemist; b. London, Eng., Sept. 23, 1914; s. Walter William and Rosa (Waters) J.; B.Sc., Sir. John Cass Coll., U. London, 1936, M.Sc., 1938, Ph.D., 1944, D.Sc., 1953; D.Sc. (hon.), U. Adelaide (Australia), 1954; m. Margery Gauge, Dec. 30, 1939; children—Susan Margaret, Patricia Ann. Research asst. dept. sci. and indsl. research U. London, 1933-39; faculty U. Nottingham (Eng.), 1939-53, reader, 1952-53; Angas prof. chemistry U. Adelaide, 1954—, dean Faculty Sci. Pres., Australian Inst. Nuclear Sci. and Engring., 1959-60, 62-63. Commonwealth Fund fellow, Princeton, 1948-49. Mem. Chem. Soc. London, Faraday Soc., Royal Australian Chem. Inst. Author: Chemistry of Nucleic Acids, 1960; (with M. Patridge) Structure of Molecules, 1950; also numerous articles. Devel. ideas on structure of nucleic acids, hydrogen bond formation between bases, denaturation and renaturation of nucleic acids, conformation of polyelectrolytes in solution at low concentrations; binding of dyes and ions to deoxyribonucleic acid. Home: 33 Craighill Rd., St. Georges, S. Australia. Office: Dept. Phys. and Inorganic Chemistry, U. Adelaide, S. Australia.*

JORDAN, Edward Conrad, educator; b. Edmonton, Can., Dec. 31, 1910; s. Conrad Edward and Erna Elizabeth (Penk) J.; B.Sc., U. Alberta, 1934, M.Sc., 1936; Ph.D. (Battelle Meml. fellow), Ohio State U., 1940; m. Mary Helen Walker, Sept. 3, 1941; children—Robert E., David W., Thomas C. Came to U. S., 1937, naturalized, 1950. Control operator, engr. radio sta. CKUA, Edmonton, 1928-35; jr. elec. engr. Internat. Nickel Co., Sudbury, Ont., 1936-37; instr. elec. engring. Worcester Poly. Inst., 1940-41; instr. asst. prof. elec. engring. Ohio State U., 1941-45; asso. prof. elec. engring. U. Ill., 1945-47, prof., 1947—, head dept., 1954—. Fellow Inst. Radio Engrs.; mem. Am. Inst. E.E., Am. Soc. Engring. Edn., Internat. Sci. Radio Union (mem. U. S. nat. com.), Sigma Xi, Eta Kappa Nu, Tau Beta Pi. Author: (with others), Fundamentals of Radio, 1942, Fundamentals of Radio and Electronics, 2d edit., 1958, Foundations of Future Electronics, 1961; Electromagnetic Waves and Radiating Systems, 1950. Editor: Electromagnetic Theory and Antennas, 2 vols., 1963. Research on radio detection finding and antennas. Home: 415 W. Indiana Av., Urbana, Ill.

JORDAN, Edward Daniel, Am. physicist; b. Bridgeport, Conn., Mar. 14, 1931; s. Edward James and Jessie (Palsak) J.; B.S. in Engring. Physics, Fairfield Coll., 1953; M.S. in Physics, N.Y. U., 1955; Ph.D. in Nuclear Engring., U. Md., 1965; m. Margaret Ann Moran, July 20, 1957; children—Christopher, Kathleen, Daniel, David. Physicist, Naval Research Lab., Washington, 1952, Sylvania Research Lab., N.C., 1953; teaching fellow N.Y. U., N.Y.C., 1953-55, instr., 1955; guest scientist Brookhaven Nat. Lab., Upton, N.Y., 1955; reactor physicist Foster Wheeler Corp., N.Y.C., 1955-57, U. S. AEC, Washington, 1957-59; chmn. div. nuclear sci. and engring., prof. Catholic U., Washington, 1959—. Named Inventor of Month, Sci. Digest Mag., 1964. Mem. Am. Nuclear Soc. (units standard com. 1959—), I.E.E.E., Am. Assn. for Engring. Edn., Am. Assn. U. Profs., Sigma Xi. Research, publs. on nuclear reactors and reactor physics, other aspects of nuclear sci. and engring. Home: 2915 Rittenhouse St., N.W., Washington.*

JORDAN, Edwin Oakes, Am. bacteriologist; b. Thomaston, Me., July 28, 1866; s. J. L. and E. D. (Bugbee) J.; S.B., Mass. Inst. Tech., 1888; Ph.D., Clark U., 1892; student Pasteur Inst., Paris, 1896; Sc.D., U. Cincinnati, 1920; m. Elsie Fay Pratt, June 16, 1893; children—Henry Donaldson, Edwin Pratt, Lucia Elisabeth. Chief asst. biologist Mass. State Bd. of Health, 1888-90; lecturer on biology, Mass. Inst. Tech., 1889-90; fellow in morphology Clark U., 1890-92; asso. in anatomy U. Chgo., 1892-93, instr., 1893-95, asst. prof. bacteriology, 1895-1900, asso. prof., 1900-07, prof., 1907-33, and chmn. dept. hygiene and bacteriology, 1914-33. Chief of serum div. and trustee McCormick Memorial Inst. for Infectious Diseases, Chicago, Ill. Editor of Journal of Preventive Medicine; joint editor (with L. Hektoen and W. H. Taliaferro) Jour. Infectious Diseases. Mem. Internat. Health Bd. of Rockefeller Foundation, Feb. 1920-27; mem. bd. scientific dirs., Internat. Health Div. of Rockefeller Foundation, 1930-33; mem. Med. Fellowship Bd. of NRC. Author: General Bacteriology, 1908; Food Poisoning, 1917, 2d edit., 1931; A Pioneer in Public Health—W. T. Sedgwick (with G. C. Whipple and C. E. A. Winslow), 1924; Epidemic Influenza, 1927; (with I. S. Falk) The Newer Knowledge of Bacteriology and Immunology, 1928. Research and publs. in bacteriology, especially in connection with food poisoning, typhoid fever, influenza; did 1st extensive investigations of self-purification of streams. Died Sept. 2, 1936.

JORDAN, Ernst Pascual, German physicist; b. Hanover, Oct. 18, 1902; s. Ernst Pascual and Eva (Fischer) J.; ed. Technische Hochschule, Hanover, U. Göttingen; Ph.D., 1924; m. Hertha Stahn, Apr. 24, 1930; children—Pascual, Michael. Faculty U. Göttingen, 1927-28; asso. prof. U. Rostock, 1928-35; prof. 1935-44; prof. U. Berlin, 1944-51, U. Hamburg, 1951—. Recipient Max Planck medal, 1942, Gauss medal, 1955. Mem. Acad. Scis. and Letters Mayence. Author: (with Max Born) Elementare Quantenmechanik, 1932; Anschauliche Quantentheorie, 1936; Schwerkraft und Weltall, 1952; Der Naturwissenschaftler vor der religiösen Frage, 1963; Die Expansion der Erde, 1966; also numerous articles. Founder (with Born and Heisenberg) of quantum mechanics and (with Pauli, O. Klein, and Wigner) of quantum electrodynamics; theoretical work in biophysics, cosmology, general relativity, geophysics, gravitational problems, pure mathematics. Home: Isestrasse 123. Office: Jüngiusstrasse 9, Hamburg, Germany.*

JORDAN, Frank Craig, Am. astronomer; b. Cordova, Ill., Sept. 24, 1865; s. John Henry and Louisiana (Craig) J.; B.Ph., Marietta Coll., 1889, M.A., 1892; Ph.D., U. of Chicago, 1914; Sc.D., Marietta (Ohio) Coll., 1929; m. Cora A. Ross, June 22, 1893; 1 son, Frank Warren (dec.); m. 2d, Mrs. Harriet C. Roy, Nov. 25, 1909; 1 son, John William. Instr. astronomy and mathematics, Marietta Coll., 1889-1900, high sch., Portland, Ore., 1900-02, at Colorado Springs, Colo., 1902-05; fellow Yerkes Obs., Williams Bay, Wis., 1905-08; with Allegheny Obs., from 1908, asst. dir., 1920-30, dir., from 1930; asst. prof. astronomy U. of Pittsburgh, 1910-19, prof., from 1919. Specialized in photometry; determined and published light curves of many short period variable stars. Died Feb. 15, 1941.

JORDAN, George Lyman, Jr., Am. surgeon; b. Kinston, N.C., July 10, 1921; s. George Lyman and Sally (Herndon) J.; B.S., U. N.C., 1942; M.D., U. Pa., M.S. in surgery, Tulane U., 1949; m. Florence Henszey, June 23, 1945; children—George Lyman III, Florence Elizabeth, Amy Henszey, Jacob Henszey. Fellow in surgery Tulane U., 1947-49, Mayo Found., 1949-52; faculty Baylor U. Coll. Medicine, 1952—, prof., 1964—; chief surgery VA Hosp., Houston, 1955-60; dep. chief surgery Ben Taub Gen. Hosp., Houston, 1961—. Sr. cons. surgery Nat. Inst. Gen. Med. Scis., Bethesda, Md., 1966. Recipient Lederle Med. Faculty award. Mem. Harris County Med. Soc., Tex., Am. med. assns., Houston, Tex. surg. socs., A.C.S., Soc. U. Surgeons, Soc. Vascular Surgery, Houston, Am. gastroent. assns., Am., Western surg. assns., Southwestern Surg. Congress, So. Soc. Clin. Investigation, N.Y. Acad. Scis., Internat. Cardiovascular Soc., Am. Heart Assn., Soc. Exptl. Biology and Medicine, Soc. Surgery Alimentary Tract, Am. Soc. Exptl. Pathology, Am. Assn. Cancer Research, Am. Heart Assn., Phi Beta Kappa, Sigma Xi. Author: (with J. M. Howard) Surgical Diseases of the Pancreas, 1960. Research on aortic homografts and prostheses; homotransplantation; postgastrectomy syndromes; exptl. prodn. of atherosclerosis. Home: 1748 North Blvd., Houston 77006. Office: 1200 Moursund Av., Houston 77025.

JORDAN, Joseph, chemist; b. Timisoara, Rumania, June 29, 1919; s. Victor and Maria (Purjusz) J.; M.S., Hebrew U., Jerusalem, Israel, 1942, Ph.D., 1945; m. Colina L. Fischer, Nov. 26, 1952; children—Saskia, Sharon, Naomi, Adlai. Came to U. S., 1950, naturalized, 1959. Faculty, Hebrew U., 1945-49; research fellow Harvard, 1950; faculty U. Minn., 1951-54; faculty Pa. State U., University Park, 1954—, prof. chemistry, 1960—; Frontiers of Chemistry lectr. Wayne State U., 1958; vis. prof. U. Cal., Berkeley, 1959, Swiss Fed. Inst. Tech., Zurich, 1961-62, Cornell U., 1965. Fellow A.A.A.S.; mem. Am. Chem. Soc., Chem. Soc. (London), Assn. Harvard Chemists, Commn. on Electrochemistry, Internat. Union Pure and Applied Chemistry, Comite Internat. de Thermodynamique et de Cinetique Electrochimiques, Sigma Xi, Phi Lambda Upsilon. Contbr. numerous articles to profl. jours. Developed convection electrodes, hydrodynamic voltammetry, thermometric titrations, mixed carriers for gas chromatography; originated direct injection enthalpimetry. Home: 1007 Glenn Circle N., State College, Pa. 16801. Office: Whitmore Lab., Pa. State U., University Park, Pa. 16802.*

JORDAN, Pierre Aloïs, Swiss chemist; b. Vevey, Switzerland, Feb. 27, 1921; s. Aloïs and Angeline (Dupont) J.; Dipl. as chem.eng., Swiss Fed. Inst. Tech., Zurich, 1944, Ph.D., 1951; m. Fernande Novello, Aug. 6, 1949. Head radiochem. lab. Physics Inst., Swiss Fed. Inst. Tech., 1952-58, head isotope lab. Organic Chemistry Inst., 1958—; affiliate Argonne (Ill.) Nat. Lab., 1962. Research, publs. on tracer chemistry, hot atom chemistry and radioactivity measurement. Home: 41 Carmenstr., Zurich, Switzerland.*

JORDANUS NEMORARIUS (or Jordanus Teutonicus, Jordanus Saxo or de Saxonia, Jordanus de Nemore), mathematician, physicist; b. Borgentreich, Westphalia, flourished circa 1200. Professed in Dominican order, Paris, 1220; became 2d gen. of Dominicans, 1222-37; preached in Paris and Bologna; died at sea while returning from Holy Land, 1237; under his leadership Dominican order increased considerably. Author: Elementa super demonstationem ponderis; Demonstratio de algorismo; Arithmetica decem libris demonstrata; Tractatus de numeris datis; De triangulis; Planisphaerium. Developed Aristotle's, Nicomachos' and Boethius' ideas; devised new concept of gravity; developed axiom that that which can lift a given weight to a given height can lift a weight k times heavier to a height k times smaller; proposed basic idea of statical moment and applied it to study of lever and inclined plane; wrote algebraic work containing rules and problems leading to linear and quadratic equations; wrote on math. astronomy; used letters to denote magnitudes; developed new propositions in theory of numbers; 1st to ennunciate principles of theory of stereographic projection; projected sphere on plane tangent to North Pole; had considerable influence on teaching of sci. of his age. Died 1237.

JORDEN, Edward, English physician, chemist; b. High Halden, Kent, Eng., 1569; younger son of gentleman of good family; studied at Hart Hall, Oxford U.; M.A. Peterhouse College, Cambridge U., 1586; traveled on continent and took M.D. at Padua, circa 1591; m. daughter of Mr. Jordan. Numbered among his friends Andreas Libavius and Henry Briggs; had successful London practice and acquired confidence of King James I. Became Licentiate, 1595, and then Fellow, 1597, of Royal College Physicians; after some years moved to Bath, Eng. Author: A Briefe Discourse of a Disease called the Suffocation of the Mother, 1603; A Discourse of Natural Bathes and Mineral Waters (5 eds. 1631-73). Rejected supernatural as valid form of scientific explanation; employed by James I to examine young girl called a witch, rejected suggestion of demonic possession and attributed attack to natural causes, 1603; rejected sympathetic action as satisfactory mode of scientific explanation; explained heat as form of motion of small parts of a body, but was unable to determine any increase in temperature of water at cataracts of Rhine River; described analysis of mineral waters in great detail utilizing crystallization procedures for identification of salts; used scarlet dye to distinguish between "salts" (bases) and acids (perhaps 1st to describe this color test), 1631; accepted growth of metals in the earth from metallic seeds through "fermentation" process which made commonly accepted concept of internal terrestrial fire unnecessary; thought similar earthly fermentation processes to be cause of hot springs and medicinal spas. Died Bath, Eng., Jan. 7, 1632.

JORGENSEN, Christian Klixbull, chemist; b. Aalborg, Denmark, Apr. 18, 1931; s. Sven K. and Ingrid (Sorensen) J.; Dr.Phil., U. Copenhagen, 1957; m. Micheline Prouvez, June 26, 1957; 1 son, Philippe. Faculty chemistry dept. Tech. U., Copenhagen, Denmark, 1953-58; with NATO, 1959-60; dir. group for theoretical inorganic chemistry Cyanamid European Research Inst., Geneva, Switzerland, 1961—. Mem. Royal Acad. Scis. (Copenhagen). Author: Energy Levels of Complexes and Gaseous Ions, 1957; Absorption Spectra and Chemical Bonding in Complexes, 1962; Orbitals in Atoms and Molecules, 1962; Inorganic Complexes, 1963. Research, numerous publs. on explaining colors and absorption spectra of complexes of five transition groups, understanding chem. bonding and group-theoretical properties of energy levels of chromophores, established nephelauxetic series of central atoms and optical electronegativities, classified categorical propositions in formal logics. Home: 51 Route de Frontenex, Geneva 1200. Office: 91 Route de la Capite, Cologny 1223, Switzerland.*

JÖRGENSEN, Gerhard, German geneticist; b. Heide, Germany, Nov. 20, 1924; s. Georg and Minna (Voss) J.; M.D., U. Kiel, 1950; m. Waltraut Matzat, June 26, 1948. Asst. at various clinics, 1949-55; with U. Med. Clinic, Kiel, 1955-59; with Inst. Human Genetics, U. Göttingen, 1960—, lectr., 1963—. Mem. Deutsche Gesellschaft für innere Medizin, 1954, Deutsche Gesellschaft für Humangenetik und Anthropologie, Gesellschaft für Versuchstierkunde. Author: Untersuchungen zur Genetik der Sarkoidose, 1965; also numerous articles. Research on human clin. genetics, serogenetics, pharmacogenetics. Home: 3 Ludwig-Beck, Göttingen, Germany.*

JORGENSEN, Theodore Prey, Am. physicist; b. Longridge, Conn., Nov. 13, 1905; s. Theodore and Annie (Prey) J.; A.B., U. Neb., 1928, M.A., 1930; Ph.D., Harvard, 1935; m. Helene Hansen, Sept. 13, 1931 (dec. Sept. 1959); 1 dau., Joanna (Mrs. Peter C. Kaestner); m. 2d, Dorothy A. Goebel, June 18, 1960. Tutor, instr. Harvard, 1935-36; asst. prof. Clark U., 1936-38; staff Manhattan Project, Los Alamos, 1943-46; faculty U. Neb., Lincoln, 1938—, prof., 1950—, chmn. dept. physics, 1950-52. Fellow Am. Phys. Soc.; mem. Am. Physics Tchrs. Assn. Research, publs. in molecular spectra Lithium hydride; nuclear and atomic research using Cockscroft Walton accelerator. Home: 3455 L St., Lincoln, Neb. 68510.*

JORGENSON, Dale Weldeau, Am. economist; b. Bozeman, Mont., May 7, 1933; s. Emmett B. and Jewell (Torkelson) J.; B.A., Reed Coll., 1955; A.M., Harvard, 1957, Ph.D., 1959; m. Margaret Irma Jefraim, Aug. 17, 1957. Faculty, U. Cal. at Berkeley, 1959—, prof. econs., 1963—; Ford research prof. econs. U. Chgo., 1962-63. Fellow Econometric Soc., Am. Statis. Assn. (asso. editor Jour. 1962-64); mem. Am. Econ. Assn., Royal Econ. Soc. Author: (with J. J. McCall, R. Radner) Optimal Replacement Policy, 1967; also articles. Am. editor Rev. Econ. Studies, 1964-67; editor Am. Econ. Rev., 1967—. Research on optimal policies for replacement stochastically failing equipment, theory and econometrics of investment behavior, theory of devel. of a dual economy. Home: 1900 Yosemite Rd., Berkeley, Cal. 94707.*

JOSE, Paul Douglas, Am. astronomer; b. Irwin, Pa., Feb. 10, 1906; s. John F. and Stella (Douglas) J.; B.A., Washington and Jefferson Coll., 1929; M.A., Syracuse U., 1932; postgrad. Ohio Wesleyan U.; Ph.D., U. Mich., 1939; m. Helen May Porter, June 17, 1937; children—Alice R. (Mrs. William F. Hack), Martha H. (Mrs. Harold D. Heitmann), Stella Ann, Frances Jane. Prof. math. N.M. Western Coll., Silver City, 1937-42; asso. prof. astronomy, asso. dir. Steward Obs., U. Ariz., Tucson, 1942-46; asso. prof. math. Washington and Jefferson Coll., Washington, Pa., 1946; resident astronomer, research asso. McDonald Obs., U. Chgo., and U. Tex., Ft. Davis, Tex., 1946-50; with Ballistic Research Lab., Aberdeen, Md., 1950-53; staff Holloman AFB, N.M., 1953-63, research physicist Office Research Analyses, 1963-68. Mem. Am. Astron. Soc., Astron. Soc. Pacific. Contbr. articles to tech. jours. Demonstrated period 178.8 year in sun's motion induced by planets corresponds to 178.6 years in sun spot cycle. Home: Fort Davis, Tex. 79734.*

JOSEPH, Daniel D., Am. aero. engr.; b. Chgo., Mar. 26, 1929; s. Samuel and Mary (Simon) J.; M.A., U. Chgo., 1950; B.S. in Mech. Engring., Ill. Inst. Tech., 1959, M.S., 1960, Ph.D., 1963; m. Ellen Broida, Dec. 18, 1948; children—Karen, Michael, Charles. Asst. prof. mech. engring. Ill. Inst. Tech., 1962-63; faculty U. Minn., 1963—, asso. prof. fluid mechanics dept. aeros. and engring. mechanics, 1965—. Mem. Am. Phys. Soc., Am. Soc. M.E. (sponsor in hydrodynamics 1966—). Contbr. articles to sci. jours. Contbns. to math. theory of hydrodynamic stability; methods for determining quantitative conditions for change of one fluid motion into another, e.g. laminar to turbulent flow. Home: 3156 Shorewood Dr., St. Paul 55112. Office: U. Minn., Mpls. 55455.*

JOSEPH, Norman Ross, Am. chemist; b. Bradford, Pa., Mar. 27, 1908; B.Sc., Harvard, 1928, Ph.D. in Med. Sci., 1935. Fellow pharmacology U. Pa., 1935-38, chemistry U. Cin., 1939-40; biochemist Joseph E. Seagram & Sons, Inc., 1940-44; research chemist U. Ill., 1944-46, asst. prof. chemistry, 1946-57, asso. prof., 1957—. Mem. Chem. Soc., Soc. Biol. Chemistry. Research in phys. chemistry of amino acids and proteins, cardiac physiology, ionic calcium, electrolytes of body fluids, physiology of articular structures, connective tissue electrolytes; (with Starr Rawson, Schroeder) introduced ballistocardiograph, device for recording stroke volume of heart, 1939. Office: Dept. Chemistry, U. Ill., Chgo. 60612.*

JOSKE, Richard Alexander, Australian physician; b. Melbourne, Australia, Oct. 6, 1925; s. Esmond Shirley and Molly (Roberts) J.; M.B., B.S., Trinity Coll., U. Melbourne, 1948, M.D., 1952; m. Enid Jocelyn Prudence Apperly, Mar. 5, 1952; children—Michael Alexander, William Apperly, David John, Alexander Richard. Med. officer Royal Melbourne Hosp., 1948-50; with Walter and Eliza Hall Inst., 1951-55; Nuffield Dominion fellow U. Coll. Hosp., London, Eng., 1955-56; Commonwealth Fund Advanced fellow Mass. Gen. Hosp., 1956-57; Adolph Basser Research fellow Royal Australasian Coll. Physicians, U. Western Australia, Perth, 1957-62, reader exptl. medicine, 1962—. Fellow Royal Australasian Coll. Physicians; mem. Australian, Brit. med. assns., Royal Soc. Medicine, Australian and New Zealand Assn. for Advancement Sci., Gastroent. Soc. Australia. Research, publs. on clin. gastroenterology, including small bowel disease, mechanism of anemia in renal disease, use of thiazide diuretics in bromide intoxication; developed concept of autoimmune liver disease, gastric biopsy and gastritis, classification of pancreatitis, biochem. tests in clin. gastroenterology; discovered malabsorption caused by anoxia and acute bacterial infection. Home: 14 Walter St., Claremont, Western Australia. Office: Sch. Medicine, Victoria Sq., Perth, Western Australia.*

JOSSELYN, Irene Milliken, Am. psychiatrist; b. La-Grange, Ill., Jan. 19, 1904; d. Orris J. and Hattie (Fagersten) Milliken; B.A., Rockford Coll., 1925; M. in Social Sci., Smith Coll., 1926; M.D., U. Chgo., 1934; postgrad. Chgo. Inst. Psychoanalysis. m. Livingston Josselyn, Aug. 29, 1931 (dec. Apr. 1957); 1 dau., Helen (Mrs. J. R. Scott); m. 2d, Eugene F. Englehard, Oct. 27, 1960. Faculty, U. Ill. Coll. Medicine, 1938-39, U. Chgo. Sch. Social Service Adminstrn., 1948-49; tng. analyst So. Cal. Psychoanalytic Inst., 1962—, chmn. child analytic com., 1963—; faculty Ariz. State U. Sch. Social Work, Phoenix, 1963—; clin. prof. child psychiatry U. So. Cal., Los Angeles, 1964—; cons. North Shore Mental Health Clinic (now called Irene M. Josselyn Clinic) Winnetka, Ill., 1955—, VA Adminstrn. Hosp., Phoenix, 1961—. Mem. profl. adv. com. div. mental health Ariz. Dept. Health, 1961-66, gov.'s adv. com., 1964-66, also cons. Diplomate Am. Bd. Psychiatry and Neurology. Mem. A.M.A., Am. Psychoanalytic Assn., Am. Psychiat. Assn., Am. Acad. Child Psychiatry, Internat. Psycho-Analytic Assn., others. Author: Psychosocial Development of Children, 1948; The Adolescent and His World, 1952; The Happy Child, 1955. Editor: Jour. Am. Acad. Child Psychiatry, 1960-65. Research, publs. on adolescent problems, working mothers, unwed mothers, role of parents. Home: 702 W. Monte Vista Rd., Phoenix, 85007.*

JOSSELYN, John, English botanist; b. Essex, Eng.; s. Thomas and Theordora (Cook) J.; sailed to Boston, 1638, visited John Cotton and John Winthrop; remained in Me., until October 1639; returned to Eng.; in Me., 1663-71; returned again to Eng. to pub. material on Am. plants and animals. Author: New England's Rareties Discovered in Birds, Beasts, Fishes, Serpents and Plants of That Country, 1672; An Account of Two Voyages to New England, 1674.

JOST, Alfred, French biologist; b. Strasbourg, France, July 27, 1916; s. Marcel and Jeanne (Mertz) J.; D.Sc., Agrégé ès Sciences Naturelles, 1947; M.D. honoris causa, U. Basel (Switzerland), 1958; m. Christiane Oguse, July 13, 1940; children—Madeleine, Jean-Louis, Francois. Sub-dir. Lab. Gen. Physiology, Mus. National d'Histoire Naturelle, 1945-49; faculty Faculty Scis., U. Paris (France), 1949—, prof. biology, 1954—, head dept. comparative physiology, 1956—. Mem. Comité Nat. de la Recherche Scientifique, 1959, Comité Consultafif des Universités, 1958. Named Chevalier de la Légion d'Honneur, 1964; recipient Prix Jaffe de l'Institut de France, 1964; Grand Prix Scientifique de la Ville de Paris, 1965. Mem. Société de Biologie; hon. mem. Harvey Soc., Sociedad Argentina de Biologia, Sociedad de Biologia de Montevideo. Research, numerous publs. on role fetal hormones in devel. young, intra-uterine surgery on fetus in animals, roles fetal pituitary, adrenals, thyroid, gonads.*

JOST, Hans, German chemist; b. Cologne, Germany, Mar. 14, 1894; s. Johannes and Catherine (Heil) J.; ed. univs. Bonn, Cologne; m. Catherine Kreutzwald, 1927. Asst. Freiburg U., 1922-24; prof., dir. dept. clin. chemistry Frankfort/Main (Germany) U., 1924-33; prof. Frankfort U., 1934; prof., inst.dir. Innsbruck (Austria) U., 1939-45; prof. Cologna U., from 1945. Research on intermediary metabolism, glycolysis, other areas of biochemistry. Address: 5 Meister-Eckehart Str., Köln-Lindenthal, Germany.

JOST, Hans Hubert, German physician; b. Cologne, Germany Mar. 14, 1894; s. Johannes and Katharina (Heil) J.; ed. U. Bonn, U. Cologne; M.D.; m. Katha Kreutzwald, July 17, 1927. Asst., Med. Clinic, Cologne, Inst. Physiol. Chemistry, U. Freiburg, 1921-22; asst., instr, asso. prof. Inst. Physiol. Chemistry, U. Frankfort, 1924-38; dir. Inst. Physiol. Chemistry, U. Innsbruck, 1939-45. Research and publs. on intermediate phase of muscular metabolism. Address: Meister-Ekkehartstrasse 5, Cologne-Lindenthal, Germany.

JOST, Res Wilhelm, Swiss physicist; b. Bern, Switzerland, Jan. 10, 1918; s. Wilhelm and Hermine (Spycher) J.; Diplom für Höheres Lehramt, U. Bern, 1943; Dr. Phil. II, U. Zurich (Switzerland), 1946; m. Hilde Fleischer, July 25, 1949; children—Res, Beat, Inge. With Inst. for theoretical Physics, Copenhagen, Denmark, 1946; asst. to Prof. W. Pauli, Swiss Fed. Inst. Tech., Zurich, 1946-49, prof., 1955—; with Inst. Advanced Studies, Princeton, N.J., 1949-55. Research and publs. in math. physics. Home: 32 Rebhaldenstrasse, Unterengstringen, 8103. Office: 60 Hochstrasse, Zurich, 8044, Switzerland.*

JOST, (Friedrich) Wilhelm, German chemist; b. Friedberg, Germany, June 15, 1903; Dr. sc. nat., U. Halle (Germany), 1927. Pvt. asst. to Dr. Bodenstein, U. Berlin, 1926-29; asst. phys. chemistry Tech. U. Hanover (Germany), from 1929, also lectr., asso. prof., from 1935; Rockefeller research fellow Mass. Inst. Tech., Cambridge, 1932-33; asso. prof. applied phys. chemistry U. Leipzig (Germany), 1937-43; prof. phys. chemistry U. Marburg (Germany), 1943-51, Tech. U. Darmstadt (Germany), 1951-53, U. Göttingen (Germany), 1953—. Author: Leitfähigkeit v. Kristallen, 1934; Diffusion und chemische Reaktion 'n festen Stoffen, 1937; Explosions und Verbrennungsvorgänge in Gasen, 1939; Diffusion in Solids, Liquids and Gases, 1952. Home: Bürgestrasse 50, Göttingen, Germany.

JOUBERT, Jules-François, French physicist; b. Tours, France, 1834; ed. École Normale; tchr. physics lycees of Tours, Niort, Pointiers, Montpellier, then prof. physics Coll. Rollin, U. Paris, 1874-88 (all France); became insp. Academy, 1888, insp. gen., 1893. Author: Sur la phosphorescence du phosphore, 1874; (memoir) Sur la theorie des machines magneto-electriques, 1881; (with Mascart) Lecons sur l'electricité et le magnetisme, 1882; Traité élementaire d'electricité, 1888; also articles. Worked (with Pasteur) on fermentation of urine, germ theory, 1876-78, also on spontaneous generation, phosphorescent conditions; research on alternating currents. Died Paris, 1910.

JOUBERT, Laurent, French physician; b. Valence, France, 1529; bachelor's degree, Montpellier, France, 1551, M.D., 1558; also studied at univs. Paris, Turin, Padua, Ferrara, Bologna; at least 1 son, Isaac; instr., Montpellier, became prof., 1566, chancellor of faculty, 1574; physician to Henry III, also King of Navarre. Author: Histoire entière des poissons, 1558; Erreurs populaires au fait de la médecine et régime de santé, 1578; Seconde partie des erreurs populaires, 1579; Pharmacopée, 1579; La grande chirurgie de Guy de Chauliac, restituée par Joubert, 1598; Paradoxes (attack on Galen's system, by attributing medicative power to laws of nature and to course of reaction). Died Lombers, France, 1583.

JOUBIN, Louis-Marie-Adolphe, French naturalist; b. Epinal/Paris, France, Feb. 27, 1861; practiced medicine, then became interested in natural scis.; entered lab. of Lacaze-Duthiers; adminstr., maritime labs., Banyuls-sur-Mer, also Roscoff; named head lectr. Faculty Rennes (France), 1888; became prof. zoology Mus. Natural History, 1903; tchr. marine biology, oceanographic inst. founded by prince of Monaco; mem. various sci. expdns. Mem. French Acad. Scis., 1921. Author: La vie dans les océans, 1912; Le fond de la mer, 1920; Les métamorphoses des animaux marins. Collaborator, Larousse du XXe siècle. Died Paris, Apr. 24, 1935.

JOULE, James Prescott, English physicist; b. Salford, Eng., Dec. 24, 1818; s. Benjamin and Alice (Prescott) J.; studied under Dalton, Manchester (Eng.) U., 1835; hon. LL.D., Dublin, 1857; hon. D.C.L., Oxford, 1860; m. Amelia Grimes, Aug. 18, 1847; 1 son,

1 dau. Research in pvt. lab. Recipient Royal Medal, 1852, Copley medal, 1860. Fellow Royal Soc., 1850; mem. French Acad. Scis., Manchester literary and Philos. Soc. (pres. 1860). Writings include his collected sci. papers, 2 vols., 1885-87; also contbr. numerous articles to jours, especially to Philosophical Transactions. Devised formula governing development of heat by electric current; described attempt to measure electric current in units, 1840; determined phys. constant (Joule's equivalent or "J") which is equal to mech. equivalent of heat; 1st to describe magnetostriction, also to calculate velocity of gas molecule; experimented (with Sir William Thompson) on caloric effects of fluids in movement, 1853; stated law that energy of gas is function of its temperature and that molecular heat of a solid compound is equal to the sum of atomic heats of its components; investigated thermo-dynamic properties of solids and suggested improved measuring apparatus for electric currents; research on. mech. equivalent to heat provided basis for theory of conservation of energy; demonstrated Joule-Thomson effect is fall in temperature when gas expands without doing external work, also Joule effect is heating caused by electric current flowing through a resistance. Died Cheshire, Eng., Oct. 11, 1889.

JOUQUET, (Jacques-Charles) Émile, French physicist; b. Bessèges, France, Jan. 5, 1871; gen. insp. mines; prof. mechanics École des mines, École polytechnique; mem. French Acad. Scis., 1930. Author: Mécanique des explosifs, 1917. Research on wave diffusion, movement of fluids, explosives; fundamental work on hydrodynamic theory of detonation. Died Montpellier, France, Apr. 2, 1943.

JOURDONAIS, Leonard Francis, Am. physician; b. Havre, Mont., July 27, 1904; s. Lucien A. and Camille (Wyrn) J.; B.A., U. Mont., 1926; M.A., Northwestern U., 1932, M.D., 1933; m. Ruth Anderson, Sept. 1, 1927. Faculty, Northwestern U. Med. Sch., Evanston, 1928-—, prof. medicine, 1964-—; attending physician Evanston Hosp., 1938-—, chmn. dept. medicine, 1956. Recipient Service award Northwestern U. Alumni Assn., 1958. Mem. A.C.P., Ill., Chgo. med. socs., Am. Diabetes Assn., A.M.A., Chgo. Soc. Internal Medicine, Am. Soc. Endocrinology, Assn. Tchrs. Preventive Medicine. Research, publs. on mechanism of pneumonia, ulcerative colitis, mechanims of insulin resistance. Home: 425 Grove St., Evanston 60201. Office: 2650 Ridge Av., Evanston, Ill. 60201.*

JOUSSET DE BELLESME, Georges-Louis-Marie-Felicien, French physiologist, pisiculturist. entomologist; b. Paris, 1839; M.D.; asst. to Claude Bernard; tchr. physiology Sch. Medicine, Nantes, France, 1875-82; dir. Paris aquarium, 1882-91; organized fisheries of Ottoman Empire, 1893. Author: Recherches expérimentales sur les fonctions du balancier chez les insectes dipteres; Acclimation du saumon de Californie dans le bassin de la Seine, 1892; Le sang et la rate dans l'alimentation des alevins, 1893; La pisiculture practique, 6 vols., 1895-1900. Demonstrated necessity of breeding fish; acclimatized rainbow trout and Cal. salmon to tributaries of Seine. Died Brussels, Belgium, Apr. 27, 1925.

JOUVET, Michel Valentin, French neurophysiologist; b. Lons-le-Saunier, France, Nov. 16, 1925; s. Andre and Suzanne (Camus) J.; M.D., Lyon Sch. Medicine, 1956; m. Daniele Mounier, May 20, 1961. Research in neurophysiology U. Cal. at Los Angeles, 1954-55; in charge research Centre Nat. de la Recherche Scientifique, 1957-61; prof. exptl. medicine U. Lyon, 1961-—. Recipient Bing prize Swiss Acad. Sci., 1966. Publs. on contbns. to electro-physiology of conditioned learning; mechanisms of paradoxical sleep and role of lower brainstem in its prodn. Home: 36 L. Thevenet, Lyon 4, France. Office: 8 Rockefeller, Lyon, France.*

JOY, Alfred Harrison, Am. astronomer; b. Greenville, Ill., Sept. 23, 1882; M.A. Oberlin College, 1904; Ph.B., 1903, hon. D.Sc., 1945, Greenville College; m. 1919; 2 chidren. Thaw fellow, Princeton U., 1910-11; instr. Am. U., Beirut, 1904-10; prof. astronomy and dir. observatory, 1910-16; instr. astronomy, Yerkes Observatory, Chicago, 1914-15; astronomer, Mt. Wilson Observatory, 1915-48; sec., 1920-48; assoc., 1950-52; lectr., Calif. Inst. Tech., 1949-52; mem. Lick Observatory eclipse expdn., Egypt, 1905; ret. Mem., Nat. Acad. Scis.; Internat. Astron. Union; AAAS; Am. Astron. Soc. (vice-pres. 1946, pres. 1949); hon. mem. Royal Astron. Soc., Canada; assoc. fellow, Royal Astron. Soc. Studied stellar luminosities and distances; stellar parallax; spectra of variable stars. Address: Mt. Wilson Observatory, Pasadena, Calif. 91106.

JOYCE, Thomas Athol, English anthropologist, ethnologist; b. 1878; s. T. Heath Joyce; M.A., Hertford Coll., Oxford (Eng.) U. With Brit. Mus., from 1902, dep. keeper Dept. Ethnography, 1921, sub-keeper, 1932, ret. 1938. Mem. Royal Anthrop. Inst. (hon. sec. 1903-13, v.p. 1913-17, 23-25, pres. 1931-33), Hakluyt Soc. (hon. sec. 1923), Brit. Assn. (pres. sect. H 1934). Author: Les Bushongo, 1910; Handbook to the Ethnographical Collections in the British Museum, 1910; South American Archaeology, 1912; Mexican Archaeology, 1914; Central American Archaeology, 1916; Maya and Mexican Art, 1927. Contbr. papers on anthropology and archaeology to sci. jours. Led archeol. expdns. to Brit. Honduras, 1926, 27, 29, 31. Died Jan. 3, 1942.

JOYEUX, Bernard, French pathologist; b. Langres, July 29, 1899; s. Jules and Agnes (Godard) J.; ed. univs. Lyon, Nancy, Paris; M.D.; m. Jeanne Barrat, Nov. 20, 1943; children—Pierre-Alexis, Claude-Michele. Agrégé, Faculty Medicine, Indochina, 1925-55; hosp. physician, Vientiane, 1925; head labs. bacteriology, Hue, 1927; head path. anatomy labs. Inst. Pasteur, Hanoi, 1927-30; with Indochina Cancer Inst., 1927-46; instr., later prof. histology and path. anatomy U. Hanoi, 1930-39; asst., titular prof. U. Saigon, 1946-55; holder chair path. anatomy Ecole Nationale de Med., Grenoble. Dir., Anticancer Soc. Grenoble. Recipient silver medal of epidemics. Author: Le cancer de la vulve; Le cancer en Indochine; Les cancer des voies digestives au Vietnamn. Address: 5, rue Pierre-Loti, Grenoble (Isere), France.

JOYLIFFE, George, English physician; b. East Stower, Dorsetshire, Eng., 1621; s. John Joyliffe; B.A., Pembroke Coll., Oxford, 1640, M.A., 1643; M.D., Clare Hall, Cambridge, 1652; lectr. on vasa lymphatica; Fellow Royal Coll. Physicians; discovered lymph ducts, pub. by Francis Glisson, 1654. Died Nov. 11, 1658.

JUAN, Veichow C., Chinese geologist; b. Peking, China, May 23, 1912; s. Chingchen and Liho Juan; B.S., Peking U., 1935; Ph.D., U. Chgo., 1945; m. H. Yolanda Sun, May 22, 1946; children—Gordon, Miranda. Asso. geologist Nat. Geol. Survey China, 1935-41, U. S. Geol. Survey, 1943-45; prof. geology Peiyang U., Tietsin, China, 1946-47, Peking U., Peiping, China, 1947-49; prof. geology Nat. Taiwan Taipei, 1950-—, head dept. geology, 1952-62, dean Coll. Sci., 1954-62; dir. gen. Academia Sinica, 1962-64; nat. research prof. Chinese Nat. Sci. Council, 1960-—. Recipient Essey prize Peiping Acad. Sci., 1935; China Found. scholar, 1941-43; Tsinghua U. fellow for investigation sci. edn. in European Countries and U. S., 1956-57. Mem. Geol. Soc. China (past pres.), Academia Sinica (mem. council 1960-—), Internat. Union Geodesy and Geophysics (chmn. Chinese nat. com. 1960-—), Am. Assn. Petroleum Geologists, Am. Geophys. Union, A.A.A.S., Internat. Union Geol. Scis., Internat. Assn. Sedimentology, Sigma Xi. Author: Mineral Resources of China, 1946; Thirteen Lectures on Geology, 1952; Physiography and Geology of Taiwan, 1955; also articles. Estimated potentiality of mineral resources; discovered basic glassy rock in Taiwan named Taiwanite; research on rock fragments as end-member in classification of sediments, tertiary geol. history of Taiwan island as eugeosyncline with oil-producing miogeosynclinal sediments. Home: 6 Lane 16, Wenchow St., Hoping E., Rd. Sect. 1, Taipei, Taiwan.*

JUAN Y SANTACILLA, Jorge, Spanish geodesist; b. Novelda, Spain, Jan. 5, 1713; mem. Condamine's expdn. to Peru, 1735; dir. Madrid Coll.; mem. Acad. Madrid, French Acad. Scis., 1746. Author: Relación histórica del viaje a la America meridional, 1748; Examen marítimo-teórico-práctico, 1761. Worked to determine demarcation line in S.Am.; developed method for measuring mountain heights with barometer; research on new methods of naval constrn. Died Madrid, June 21, 1773.

JUCKER, Ernst, chemist; b. Tomsk, Russia, Nov. 4, 1918; s. Ernst Emil and Sonja (Kuks) J.; Ph.D., U. Zurich (Switzerland), 1945; m. Angela Grunauer, Oct. 20, 1966; children—Rolf, Peter. Research asst., chem. inst. U. Zürich, until 1947; with chem. research labs. Sandoz Ltd., Basel, Switzerland, 1947-49, personal asst. to pres., 1949-58, dep. dir. for synthetic drug research and patents, 1958-—. Mem. Swiss, German, Am. chem. socs., German Pharm. Soc., N.Y. Acad. Sci., Swiss Soc. Allergy, AIPPI. Author: (with P. Karrer) Carotinoide, 1948, English transl., 1950. Editor: Arbeiten aus dem Gebiet der Naturstoffchemie, Festschrift A. Stoll, 1957; Progress in Drug Research, ann. vols., 1959-—. Research and publs. on carotenoids, synthetic and natural alkaloids, synthetic drugs, antispasmodics, analgesics, drugs for Parkinson's disease, oral diuretics, psychotropic drugs. Home: Steinweg 28, CH-4107 Ettingen, Switzerland. Office: Sandoz Ltd., Lichstrasse 35, Ch-4002 Basel, Switzerland.*

JUDAH BEN MOSES (Judah ben Moses ha-Kohen), Spanish physician, astronomer, translator; flourished Toledo, Spain, last half of 13th century; employed to translate Arabic into Spanish by Alfonso X. One of main authors of Alfonsine Tables, 1272. Translator: Treatise on the Sphere (by Qusta ibn Luga), 1259; Kitab al-bari (astrological treatise by Ibn abi-l-Rijal), 1256; Kitab al-Kawakib (Book of Stars, by Abd al-Rahman al-Suft), 1256; A Lapidary by Abolays (dealt with 360 stones one for every day of the year, possibly by Abu-l-Aysh).

JUDAH BEN SOLOMON AL-HARIZI, see al-Harizi.

JUDAH BEN SOLOMON HA-KOHEN, see Ibn Matqah.

JUDAY, Chancey, Am. zoologist; b. nr. Millersburg, Ind., May 5, 1871; s. Baltzer and Elizabeth (Heltzel) J.; A.B., Ind. U., 1896, A.M., 1897, LL.D., 1933; m. Magdalen Evans, Sept. 6, 1910; children—Chancey Evans, Mary, Richard Evans. Tchr. sci., high sch., Evansville, Ind., 1898-1900; biologist Wis. Geol. and Natural History Survey, 1900-01; acting prof. biology U. Colo., 1903-04; instr. in zoölogy U. Cal., 1904-05 biologist Wis. Geology and Natural History Survey, 1905-31; lectr. zoology U. Wis., Madison, 1908-31; prof. limnology from 1931; asst. U. S. Bur. Fisheries, summers, from 1907; dir. Trout Lake Limnological Lab. from 1925. Mem. biol. and agrl. div. NRC, 1940-43; asso. editor, Ecological Monographs, 1940-43. Fellow A.A.A.S.; mem. Am. Soc. Zoologists, Am. Soc. Naturalists, Am. Micros. Society (pres. 1923), Ecol. Soc. Am., (pres. 1927), Internat. Limnol. Soc., Wis. Acad. Scis. (pres. 1937-39), Am. Limnol. Soc. (pres. 1935-36), Phi Beta Kappa, Sigma Xi. Author: Dissolved Gases of Wisconsin Lakes, 1911; Hydrography and Morphometry of Wisconsin Lakes, 1914; Plankton of Wisconsin Lakes, 1922—all with E. A. Birge; also papers dealing with the physics, chemistry and biology of lakes. Died Mar. 1944.

JUDD, Charles Hubbard, psychologist; b. Bareilly, Brit. India, Feb. 20, 1873; s. Charles Wesley and Sarah (Hubbard) J.; came to U. S., 1879; A.B., Wesleyan U., 1894; Ph.D., U. Leipzig, 1896; A.M., Yale, 1907; LL.D., Miami, 1909, Wesleyan, 1913, U. Ia., 1923, Colo. Coll., 1923; D.Sc., U. Louisville, 1937; m. Ella Le Compte, Aug. 23, 1898 (dec. 1935); 1 dau., Dorothy; m. 2d, May Diehl, Aug. 28, 1937. Instr. philosophy, Wesleyan U., 1896-98; prof. psychology N.Y. U., 1898-1901; prof. psychology and pedagogy U. Cin., 1901-02; instr. psychology Yale, 1902-04, asst. prof., 1904-07, prof. psychol. and dir. Psychol. Lab., 1907-09; prof., head dept. edn. U. Chgo., 1909-38, emeritus from 1938, chmn. dept. psychology, 1920-25; mem. science com. of Nat. Resources Planning Board, 1937-40; Inglis lectr. Harvard, 1928; Sir John Adams lectr. U. Cal., Los Angeles, 1942; cons. Nat. Youth Administrn., 1938-40, War Dept., 1942-43; chmn. Am. Council on Edn., 1929-30; mem. Social Sci. Research Council, 1930-40; mem. Internat. Inquiry on Sch. and Univs. Exams., 1935, Nat. Adv. Com. on Edn., 1937. Fellow A.A.-A.S.; mem. Am. Psychol. Assn. (pres. 1909), Nat. Soc. Coll. Tchrs. of Edn. (pres. 1911, 15), North Central Assn. Colls. and Secondary Schs. (pres. 1923), N.E.A. (policies commn. 1935-37). Author: Genetic Psychology for Teachers, 1903; (transl.) Outline of Psychology, by W. Wundt (from German), 3d edit., 1907; Psychology, General Introduction, 1907, 2d revised edit., 1917; Psychology Laboratory Course, 1907; Psychology Laboratory Equipment and Methods, 1907; Psychology of High School Subjects, 1915; Measuring the Work of the Public Schools, 1916; Introduction to the Scientific Study of Education, 1918; Evolution of a Democratic School System, 1918; Silent Reading, 1923; Psychology of Social Institutions, 1926; Psychology of Secondary Education, 1927; Psychological Analysis of the Fundamentals of Arithmetic, 1927; The Unique Character of American Secondary Education, 1928; Problems of Education in the United States, 1933; Education and Social Progress, 1934; Education as Cultivation of the Higher Mental Processes, 1936; Preparation of School Personnel, 1938; Educational Psychology, 1939; (with J. D. Russell) The American Educational System, 1940; (with W. C. Reavis) The Teacher and Educational Administration, 1942; contbr. to scientific and ednl. jours. Editor: Monograph Supplements of Psychol. Rev., Studies from Yale Psychol. Lab., 1905-09, Elementary Sch. Jours., Sch. Rev., Lessons in Community and Nat. Life, 1917-18. Research in ednl. problems, including nature and devel. of reading, variation and nature of visual perceptions, number ideas and their devel.; influential in altering Am. edn. (which he felt did not adequately deal with contemporary problems). Died July 18, 1946.

JUDD, Edward Starr, Am. surgeon; b. Rochester, Minn., Oct. 16, 1911; s. Edward Starr and Helen (Berkman) J.; grad. Deerfield Acad., 1928; A.B., Dartmouth, 1932; M.D., Rush Med. Coll., 1936; M.S., U. Minn., 1940; m. Virginia Helm, June 24, 1936; children—Thomas Helm, Jill Starr. Mem. surg. staff Mayo Clinic, Rochester, 1942-—; prof. surgery Mayo Grad. Sch. U. Minn., 1957-—; trustee Mayo Found. Recipient Geisinger Med. Center Distinguished Alumnus award, 1965. Investigations, publs. in clin. surgery, gastroenterology, thyroidology. Home: 1913 7th St. S.W., Rochester 55901. Office: Mayo Clinic, Rochester, Minn. 55902.*

JUDD, John Wesley, English geologist; b. Portsmouth, Eng., Feb. 18, 1840; ed. Royal Sch. Mines; LL.D.; m. Jeannie Frances Jeyes, 1878; 1 son, 1 dau. With Geol. Survey Eng. and Wales, 1867-70; insp. schs., 1871; prof. geology, 1876-1905; dean Royal Coll. Sci., London, Eng., 1895-1905; emeritus prof. geology Imperial Coll. Sci. and Tech., 1913. Fellow Royal Soc., 1877, Geol. Soc. (pres. 1887-88). Author: Geology of Rutland, 1875; Volcanoes: What they are and what they Teach, 1878; The Student's Lyell, 1896, 1911; The Coming of Evolution, 1910; contbr. sci. memoirs. Known for brilliant demonstrations of relationships of apparently different strata in Eng. and on continent. Died Mar. 3, 1916.

JUDD, William Wallace, Canadian entomologist, educator; b. Windsor, N.S., Can., Oct. 22, 1915; s. W. Wallace and Ruth (Alley) J.; B.A., McMaster U., 1938; M.A., U. Western Ont., 1940; Ph.D., U. Toronto, 1946; m. Kathryn Elizabeth Baker, May 18,

1946; children—Kenneth W., Donald W., Ruth E., Dorothy G. Lectr., asst. prof. zoology McMaster U., 1946-50; faculty dept. zoology U. Western Ont., London, 1950—, prof., 1963—. Mem. Entomol. Soc. Am., Entomol. Soc. Can., Fedn. Ont. Naturalists. Editor: (with J. M. Speirs) A Naturalist's Guide to Ontario, 1964. Research, numerous publs. on insect anatomy and histology, aquatic insects, insect galls, ecology of bogs, ecology of reptiles. Home: 50 Hunt Club Dr., London, Ont., Can.*

JUDET, Jean, French orthopedist; b. Paris, 1905; s. Henri and Berthe (Froment) J.; M.D., Faculty Medicine, Paris; m. Jeanne Mettas, May 7, 1937; children—Henri, Berthe, Claudine, Francine. Became intern, Paris, 1928, clin. physician, 1936; head of orthopedics Hosp. for Sick Children, from 1948. Mem. Conseil de l'Ordre des médecins. Mem. Acad. Surgery, Internat. Orthopedic Soc. Latin Am., orthopedic socs. Spain, Portugal, Germany, Italy, Belgium. Employed (with R. L. Judet) acrylic in constrn. of artificial femoral head in plastic surgery of hip joint, 1950. Address: 6, Square Jouvenet, Paris, France.

JUDSON, Sheldon, Am. geologist; b. Utica, N.Y., Oct. 18, 1918; s. Salmon Sheldon and Dorothy (Eurich) J.; A.B., Princeton, 1940; M.A., Harvard, 1946, Ph.D., 1948; m. Anne Perrin Galpin, Feb. 13, 1943; children—Stephanie Dean, Anne Perrin, Lucy Sheldon. Mem. faculty U. Wis., 1948-54; faculty Princeton, 1954—, Knox Taylor prof. geology, 1964—; dir. Princeton Coop. Sch. Program, 1964—; pres. Princeton Jr. Museum, 1964-68. Bd. dirs. Belgian Am. Ednl. Found. Faculty fellow Fund Advancement Edn., 1954-55; Guggenheim fellow, 1960-61, 66-67; Fulbright fellow, 1960-61. Fellow A.A.A.S., Geol. Soc. Am.; mem. Arctic Inst., Sigma Xi. Research and publs. on application of geology to the study of man's history. Home: 18 Aiken Av., Princeton, N.J. 08540.*

JUGENHEIMER, Robert William, Am. agronomist; b. Scout County, Ia., Nov. 6, 1904; s. William and Frances (Linder) J.; B.S., Ia. State U., 1934, M.S., 1936, Ph.D., 1940; m. Mabel C. Hobert, June 17, 1933; children—Robert William, Donald W. Agronomist, U. S. Dept. Agr., Ia. State U., Ames, 1928-38, U. S. Dept. Agr., Kan. State U., Manhattan, 1938-44; research dir. Pfister Asso. Growers, El Paso, Ill., 1944-45; prof. plant genetics U. Ill., Urbana, 1945—, asst. coordinator internat. cooperation programs, 1956-61, asst. dean, asst. dir., 1959-61, dir. internat. devel. projects, 1961—; research adminstr. U.P. (India) Agrl. U., 1968—. Cons. to fgn., U. S. govt. agys. Fellow Am. Soc. Agronomy, A.A.A.S., Nat. Geog. Soc.; mem. Am. Assn. U. Profs., Agr. Research Inst., Nat. Acad. Sci. (agrl. bd. 1959—), Ill. Acad. Sci. Author: Hybrid Maize Breeding and Seed Production, 1958; also numerous articles. Developed comprehensive corn breeding, genetics and prodn. research in Ill., Kan., other states, countries; developed many dent and popcorn lines and hybrids; studied resistance to diseases, insects and drought. Home: 1307 W. University St., Champaign, Ill. 61820.*

JUGLAS, Jean-Jacques, French geographer; b. Bergerac, June 10, 1904; s. Fernand and Marie (Lacoste) J.; Agrege in history and geography, U. Bordeaux; m. Marguerite Lacotte, Aug. 11, 1927; 1 dau., Francoise. Prof. provincial lycee, Lycee Louis-le-Grand, Paris; prof. Ecole Normale Superieure; mem. constituent assemblies, 1945-46, Nat. Assembly, 1946-56; pres. Commn. Overseas Tys., 1946-55; minister of overseas France, 1955; dir. gen., later pres. Office Sci. Research and Tech Overseas; prof. Nat. Conservatory Arts and Trades, Inst. Higher Studies Overseas, 1962; pres. Sci. Research and Tropical Tech. Com., 1963. Mem. Com. on Geography, Acad. Sci. for Overseas. Author: Traité de géographie économique, also articles on orgn. of tropical research. Home: 137, rue de la Tour, Paris 16. Office: 292, rue Saint-Martin, Paris, France.

JUHLIN, Lennart, Swedish physician; b. Stockholm, Sweden, Aug. 27, 1926; s. Bertil A. and Ruth (Lagerman) J.; M.D., U. Uppsala (Sweden), 1952, Docent, 1956; m. Ingrid Schück, July 5, 1951; children—Claes B., Ann Marie. Faculty, U. Uppsala, 1948—, asso. prof. dermatology, 1964—; research asso. dept. dermatology U. Pa., Phila., 1960-61, asso. prof., 1964-65. Mem. Swedish Drs. Assn., Swedish Dermatol. Soc., Swedish Allergological Soc., N.Y. Acad. Scis., Soc. Reticuloendothelial Systems. Research, publs. on connective tissue barrier in inflammation, phagocytic activity of reticulo-endothelial system, vascular reactions in skin disorders and physiopathology of basophil leukocytes; worked out methods for staining and estimation of histamine. Home: 30B Döbelnsgatan, Uppsala, Sweden.*

JUILFS, Johannes, German physicist; b. Hanover, Dec. 15, 1911; s. Arthur and Elfriede (Welz) J.; Ph.D., U. Berlin; m. Hannelore Juilfs, July 9, 1960. Asst. to Prof. Laue, 1938; teaching asst. at Rostock, 1940-45, at Helmstedt, 1948-52; established textile research at Krefeld, 1950-52; prof. Hanover Poly. Sch. Theoretical Physics, 1955—. Mem. Internat. Coll. Textile Scis., Assn. German Physicists, Union German Engrs. Author: Physik der gegen art; also numerous works on wear and tear of fibers, phys. structure and polymerization of textiles. Home: Humboldtweg 2, Ahlem. Office: Technische Hockschule, Hanover, Germany.

JUKES, Joseph Beete, Brit. geologist; b. Birmingham, Eng., Oct. 1811; s. John and Sophia Jukes; pupil of Sedgwick; B.A., St. John's Coll., Cambridge, 1836; m. dau. of J. Meredith, 1849; geol. surveyor of Newfoundland, 1839-40; naturalist H.M.S. Fly, survey of N.E. Australian coast, 1842-46; employed in N.Wales; dir. Irish survey, 1850-69; named mem. Royal Commn. on Coalfields, 1866. Fellow Royal Soc., 1853. Author: Excursions in and about Newfoundland, 1842; Narrative of the Voyage of H.M.S. Fly, 1847; Sketch of the Physical Structure of Australia, 1850; The Student's Manual of Geology, 1857; also papers. Played important role in establishing Huttonian doctrine that valleys have been excavated by running water and that most surface features of earth were formed through action of rain and rivers. Died July 29, 1869.

JUKES, Thomas Hughes, biol. chemist; b. Hastings, Eng., Aug. 25, 1906; s. Edward Hughes and Ann Mary (Barton) J.; B.S.A., U. Toronto, 1930, Ph.D., 1933; NRC fellow U. Cal. at Berkeley, 1933-34; m. Marguerite Esposito, July 2, 1942; children—Kenneth Hughes, Caroline Elizabeth, Dorothy Mavis. Came to U. S., 1925, naturalized, 1939. Instr., asst. prof. U. Cal. at Davis, 1934-42; with pharm. div. Lederle Labs., 1942-45, dir. nutrition and physiology research sect. Research Division, Am. Cyanamid Co., Pearl River, N.Y., 1945-58; vis. sr. research fellow Princeton, 1962-63; prof. in residence Donner Lab. Med. Physics, Univ. of Cal. at Berkeley, 1963—. Recipient Borden award Poultry Sci. Assn., 1947. Fellow Am. Soc. Animal Sci.; mem. Am. Soc. Biol. Chemists, Am. Inst. Nutrition, Soc. Exptl. Biology and Medicine, Am. Chem. Society, Trustees for Conservation, Biophys. Soc., Sigma Xi, Delta Tau Delta. Author: B Vitamins for Blood Formation, 1952; Antibiotics in Nutrition, 1955; Molecules and Evolution, 1966. Research and publs. on vitamins of the B complex; growth-promoting effects of antibiotics; amino acid code; molecular evolution; effects of folic acid antagonists on expt. cancer and leukemia. Home: 170 Arlington Av., Berkeley, Cal. 94707.*

JUKES-BROWNE, Alfred John, English geologist; b. Penn Fields nr. Wolverhampton, Eng., Apr. 1851; s. A. H. and C. A. (Jukes) Browne; B.A., St. John's Coll., Cambridge (Eng.) U.; m. Emma Jessie Smith, 1881; 1 dau. Appt. to staff Geol. Survey, 1874, mapped parts of Suffolk, Cambridge, Rutland, Lincoln, until 1883, examined and partly resurveyed cretaceous dists. in Herts, Bedford, Bucks, Oxford, Berkshire, Wiltshire, Dorset, Devon, for purpose of preparing monograph on Brit. cretaceous rocks; surveyed geology of Barbados, 1888-89 and (with J. B. Harrison) prepared papers on that island, ret. 1902. Fellow Royal Soc., 1909, Geog. Soc. Author: Student's Handbook of Physical Geology, 2 edits.; Student's Handbook of Historical Geology, 1886; The Building of the British Isles, 3d edit., 1911; Student's Handbook of Stratigraphical Geology, 2d edit., 1912; The Cretaceous Rocks of Britain, 3 vols. Contbr. papers to sci. jours. Died Aug. 16, 1914.

JULESZ, M., Hungarian physician; b. Kiskunfélegyháza, Hungary, Dec. 5, 1904; s. Leó and Rose (Haas) J.; grad. U. Budapest (Hungary), 1927, M.D.-Sci., 1956; m. Elisabeth Winkler, May 10, 1936. Faculty, I.Med.Clin.U. Budapest, 1944-58, prof. without chair, 1956-58; prof. medicine U. Szeged (Hungary) II Med. Clinic, 1959—. Recipient prize Med. Soc. Budapest, 1936, 48; Esterházy prize, 1940, 41; Physician meritus, 1961; Kossuth, prize, 1962; Golden Decor.f.Work, 1964. Mem. Royal Soc. Medicine (affiliate London), Hungarian Endocrine Soc. (pres.). Author: Pathological and Diagnostic Neuroendocrinology, 1957; (with Holló) Die Therapie der endokr. Krankheit und ihre theor. Grundl., 1966; (with Kovács) Rolle des Endokrinsystems in Allergie, 1959; also numerous articles. Demonstrated that ketosis increases function of adenohypophysis especially prodn. of follicle-stimulating hormone; demonstrated first presence of androgenic ketosteroids in human skin and hairs and proved they are synthesized in skin; described several endocrine diseases. Home: 6. Somogyi, Szeged, Hungary.*

JULIA, Marc, French chemist; b. Paris, France, Oct. 23, 1922; s. Gaston and Marianne (Chausson) J.; ed. École Normale Supèrieure, Paris; Imperial College of Sci. and Technology, London; m. Elisabeth de Vitry, Oct. 8, 1951; children: Bernard, Marianne, Pierre. Assoc. prof., École Polytechnique, Paris; assoc., then prof., organic chem., U. Paris; in charge of medicinal chem. dept., Institute Pasteur, Paris. Research in organic chemistry, drugs and biochemistry. Address: Institute Pasteur, Paris, France.

JULIA FONTELLE, Jean Sebastien Eugène (or De Fontelle), chemist; b. Narbonne, France, 1790; studied under Fourcroy at Berthollet; went to Barcelona, Spain, to study plague, 1820; became chief physician gen. hosp., Catalonia, Spain, 1823. Author: Recherches sur l'air marécageux (prize Acad. Lyons, France), 1823; Manuel de chimie médicale, 1824; Manuel de physique amusant, 1826; Guide pour les recherches et observation microscopiques, 1836; Histoire naturelle des "Fables" de la Fontaine, 1841. Died 1842.

JULIAN, Percy Lavon, Am. chemist; b. Montgomery, Ala., Apr. 11, 1899; s. James S. and Elizabeth Lena (Adams) J.; A.B., DePauw U., 1920, D.Sc., 1947; A.M., Harvard, 1923; Ph.D., U. of Vienna, 1931; D.Sc., Fisk U., 1947, W.Va. State College, 1948, Northeastern U., 1948, Morgan Coll., 1950, DePauw U., 1947, Northwestern University, 1951, Howard University Washington, 1951, Lincoln U., 1954, Roosevelt U., 1961, Va. State Coll., 1962, Morehouse Coll., 1963, Oberlin Coll., 1964; m. Anna Johnson, Dec. 24, 1935; children—Percy Lavon, Faith Roselle. Instr. chemistry Fisk U., 1920-22; Austin fellow chemistry, Harvard, 1922-23, research fellow biophysics, 1923-24, George and Martha Derby scholar chemistry, 1924-25, univ. scholar, 1925-26; prof. chemistry W.Va. State Coll., 1926-27; asso. prof. chemistry, acting head dept., Howard U., 1927-29; Gen. Edn. Bd. fellow, U. of Vienna, 1929-31; prof., head dept. chemistry, Howard U., 1931-32; research fellow, teacher organic chemistry DePauw U., 1932-36; dir. research Soya Products Div., The Glidden Co., 1936-45, dir. research, mgr. fine chemicals, 1945-53; pres. Julian Labs., Inc., Laboratorios Julian de Mexico, S.A., 1953-64, Julian Research Institute Julian Assos., Incorporated, 1964—. Trustee Fisk U., Nashville, Howard U., Washington, So. Union Coll., Wadley, Alabama, Roosevelt University; director Chgo. Theol. Sem., Chicago, Ill. Recipient Phi Beta Kappa Key, DePauw U., 1920, Sigma Xi Key, Northwestern, 1945; Spingarn Medal, 1947; Distinguished Service Award for 1949-50, The Phi Beta Kappa Assn., Chicago, 1949; Chicagoan of the Year Award from Chicago Sun-Times, 1950, Honor Scroll award Am. Inst. Chemists, 1964, also others. Fellow Chem. Soc. of London, N.Y. Acad. Sci., Am. Inst. Chemists; mem. Phi Beta Kappa Assos. Contbr. articles to profl. jours. Holder patents. Perfected a soya protein for use in coating paper; synthesized physostigmine (used in treatment of glaucoma) and prostigmin; research on indoles, amino-acids and pregnenolone, an anti-fatigue drug. Address: Julian Research Institute, 9352-58 W. Grand Av., Franklin Park, Ill.

JULIANO, Bienvenido Ochoa, Philippine chemist; b. Los Baños, Laguna, Philippines, Aug. 15, 1936; s. Jose Buencamino and Teodora (Ochoa) J.; B.S. in Agr. magna cum laude, U. Philippines, 1955; M.Sc., Ohio State U., 1958, Ph.D., 1959; m. Linda Alvarez, Apr. 10, 1965; 1 son, Bienvenido Jose. Asst. instr. U. Philippines, 1955-56, affiliate mem. grad. faculty, 1961—; staff Ohio State U., 1956-57, NSF fellow, 1958; C. F. Kettering Research Found. fellow, 1958-59; jr. chemist Philippine Refining Co., Manila, 1959-61, head lab. and devel. sects., 1961; asso. chemist Internat. Rice Research Inst., Los Baños, 1961—, acting head dept. chemistry, 1966—; prof. chemistry Far Eastern U., 1959-61. Named one of Ten Outstanding Young Men in Philippines, 1964. Mem. Am. Assn. Cereal Chemists, Inst. Food Technologists, Chem. Soc. Philippines, Soc. for Advancement Research, Sigma Xi, Gamma Sigma Delta, Phi Kappa Phi, Phi Lambda Upsilon. Research, publs. on amylose content of rice grain, physicochem. properties of rice starch and protein; established existence of rice protein in form discrete particles. Home: Lopez Av., Office: Internat. Rice Research Inst., Los Baños, Laguna, Philippines.*

JULIEN, Alexis Anastay, Am. geologist; b. N.Y.C., Feb. 13, 1840; s. Pierre Denis and Magdalen (Cantine) J.; A.B., Union Coll., 1859, A.M., 1864; hon. Ph.D., N.Y. U., 1881; m. Annie Walker Nevius, June 1, 1882. Resident chemist, guano island of Sombrero, 1860-64; studied geology and natural history there; collected birds and land shells and made meteorol. observations on the island for Smithsonian Instn.; asst. analytical chemistry Sch. of Mines, Columbia, 1865-85, instr. microscopy, microbiology, 1885-97, curator geology, 1897-1909. Connected with Mich. Geol. Survey, 1872, N.C. Geol. Survey, 1875-78; reported on bldg. stones of N.Y.C., 10th U. S. Census. Died May 7, 1919.

JUNCEDA, Juan, Spanish physician; b. Oviedo, Spain, Mar. 22, 1924; s. Eladio and Carmen (Avello) J.; student U. Oviedo, 1942; B.A., U. Santiago de Compostela, 1943, sobresaliente, 1949; M.D., U. Madrid (Spain), 1955; m. Maria Isabel Moreno, Nov. 5, 1954; children—Juan, Maria Isabel, Carmen, Eladio, Pablo. Asst., ophthal. clinic, N.Y.C., 1953-54, Hopital de Baviers, Liege, Belgium, 1955-56; ophthalmologist Nat. Pub. Health, 1958—; ophthalmic surgeon Nat. Social Security, 1962—; chief ophthalmic service Gen. Hosp. Asturias, Oviedo, 1963—; Hon. cons. Nat. Orgn. for Blind, 1951—. Recipient award ophthalmology Gen. Direction of Health, 1951, 52. Mem. Royal and Nat. Acad. Medicine, Ophthal. Soc. Madrid. Research, publs. on pathology and mgmt. of glaucoma in man; invented radiol. technique for diagnoses of ocular fgn. bodies. Home: Principado 7, Oviedo, Spain.*

JUNCKEN, Johann Helfrich, see Jungken, Johann Helfrich.

JUNCKER, Johann, German physician; b. Lehndorf, nr. Giessen, Germany, 1679-83; student, Giessen, Marburg, Halle (all Germany); M.D., Halle, Germany, 1717; tchr., Halle, 1701-02; practiced medicine; became physician to Royal Pedagogy and Orphanage, Halle, 1716; prof. medicine U. Halle. Author: Conspectus chemiae theoretico-practicae . . . , 1730-38; others. Helped spread Stahl's theories of phlogiston and animistic medicine; pub. textbook of chemistry which covered phlogiston theory, also definition, purposes, divs. of chemistry, symbols, instruments; originator of polyclinics in Germany. Died Halle, Oct. 25, 1759.

JUNG, Adolphe Michel, French surgeon; b. Strasbourg, Dec. 17, 1902; s. Adolphe and Emilie (Rich) J.; M.D., U. Strasbourg; m. Marlise Schertzer, Apr. 27, 1934; children—Pierre-Michel, Jean-Daniel, Catherine, Frank-Martin. Extern, Paris Hosps., 1926; intern Strasbourg Hosps., 1927; asst. to Nat. Sci. Treasury, 1931-34; head surg. clinic, 1932-35; agrégé in gen. surgery, 1939; hosp. surgeon, 1945; prof. surg. pathology, 1954. Research and publs. on gen., endocrine and osseous surgery. Address: 2, allee de la Robertsau, Strasbourg (bas-Rhin), France.

JUNG, Carl Gustav, Swiss psychologist, psychiatrist; b. Kesswyl (Thurgau), Switzerland, July 26, 1875; s. Paul and Emilie (Preiswerk) J.; student U. Basel, 1895-1900; M.D., Zurich, 1902; postgrad. U. Paris, 1902-03; hon. degrees include D.Sc., Harvard, Oxford univs.; m. Emma Rauschenbach, Feb. 26, 1903; children—Agathe, Gret, Franz, Marianne, Helene. Intern, 1900-04, 2d physician, Psychiatric Clinic, U. Zurich, 1905-09; lectr. psychiatry, U. Zurich, 1905-13; lectr. psychology Fed. Tech. U., Zurich, 1933-40; prof. psychology Basel U., 1943; pvt. practice medicine specializing in psychotherapy, Küsnacht-Zurich, Switzerland, from 1909. Founder Freud Soc., Zurich, later First Internat. Psychoanalytical Congress, 1908. Fellow Royal Soc. Medicine London; mem. Swiss Acad. Med. Sci., Swiss Soc. Psychiatry, Swiss Soc. Practical Psychology, others. Author: Psychologie der Dementia Praecox, 1906; Versuch einer Darstellung der psychoanalytischen Theorie, 1912; Collected Papers on Analytical Psychology, 1916; Studies in Word Association, 1918; Psychological Types, 1923; Psychology of the Unconscious, 1921; Two Essays on Analytical Psychology, 1928; (with Wilhelm) The Secret of the Golden Flower, 1931; Modern Man in Search of a Soul, 1933; Psychology and Religion, 1937; The Integration of the Personality, 1940; Paracelsica, 1942; Psychologie und Alchemie, 1944; Essays on Contemporary Events, 1947; Zur Psychologie der Uebertragung, 1946; Symbolik des Geistes, 1948; Gestaltungen des Unbewussten, 1950; Aion, 1951; Ueber Synchronizit, 1952; Antwort auf Hiob, 1952; Von den Wurzeln des Bewusstseins, 1953; Mysteriums Coniunctionis, 3 vols., 1955-57; Collected Works. Worked with Freud, but later broke with him to become founder of analytical psychology; based his psychology on psychic totality and energism; conceptualized 2 dimensions in unconscious: personal (repressed events in individual's life), and collective; introduced concepts of introvert and extrovert, anima/animus, and sychronicity (coincidence of events which are not causally related having similar meanings); felt that achievement of harmony between conscious and unconscious is an important life-long task; stressed connection between physiol. and psychol. aspects of disease; numerous studies of occult, yoga, witchcraft, and alchemy. Died Küsnacht, June 6, 1961.

JUNG, Frederic Theodore, Am. physiologist; b. Sheboygan, Wis., Jan. 15, 1898; s. Jacob and Louisa Mathilda (Winter) J.; B.S., U. Wis., 1919; Ph.D., U. Chgo., 1925; M.D., Northwestern U., 1932; m. Ruth Elvira Westlund, Sept. 8, 1928; 1 son, Paul. Faculty, Northwestern U. Med. Sch., Chgo., 1926-46, asso. prof., 1945-46; dept. editor, abstracts and translations Jour. A.M.A., Chgo., 1956-63, cons., 1963-—. Mem. A.M.A., Am. Physiol. Soc., I.E.E.E., Inst. Medicine Chgo., others. Author: (with E. C. Earle) Anatomy and Physiology, 1953; also articles. Research on function of parathyroid glands, postoperative treatment of gastrectomy, growth of rodent teeth, gall bladder and bile ducts, relation of nightblindness to vitamin A deficiency; 1st systematic study of breast-knots in boys in early adolescence; demonstration of centripetal drift as frequent fallacy in therapeutic research. Home: 521 Ridge Av., Evanston, Ill. 60202. Office: 535 N. Dearborn St., Chgo. 60610.*

JUNG, Glenn Harold, Am. oceanographer; b. Lyons, Kan., Oct. 11, 1924; s. Walter Benjamin and Elsie (Coppock) J.; student Colo. State Coll. Edn., 1941-43; B.S., Mass. Inst. Tech., 1949, M.S., 1952; Ph.D., Tex. A. and M. Coll., 1955; m. Jean McLean Clements, Aug. 28, 1948; children—Lynn Stockton, Lawrence Ervin, Margaret Leslie, Richard Brooks, Kenneth Dale. Faculty, Tex. A. and M. Coll., 1952-57; research oceanographer Gulf Cons., 1957-58; faculty U. S. Naval Postgrad. Sch., Monterey, Cal., 1958-—, prof., 1965. Cons., Tex. Co., 1958-60, Pacific Oceanographic Group Can., 1963, Data Dynamics, 1965, Naval Oceanographic Office, 1965-—. Fulbright scholar, U. Oslo, Norway, 1954-55. Mem. Am. Meteorol. Soc., Sigma Xi. Contbr. to Ency. of Oceanography, 1966. Described heat transported by ocean currents, especially in Atlantic Ocean; investigated ocean-atmospheric interactions; developed oceanographic forecasting relations, especially for upper ocean thermal structure. Home: Rio Vista and Segundo Drs., Carmel, Cal. 93921. Office: Naval Postgrad. Sch., Monterey, Cal. 93940.*

JUNG, Joachim (or Junge), German naturalist, physician; b. Lubeck, Oct. 22, 1587; student, Rostock, Germany; M.A. in Philosophy U. Giessen (Germany); M.D., Padua, Italy, 1618. Prof. math U. Giessen, 1608-14, Rostock, 1924-25; prof. medicine, Helmstadt, 1625; physician, Brunswick, Wolfenbuttel; became prof. math. U. Rostock, 1626; rector Hamburg Gymnasium, 1628-29; founder Societas Ereunetica, 1622. Author: Geometrica empirica, 1627; Logica Hamburgensis, 1638; De principis corporum naturalim, 1642; Doxoscopiae physicae minores, 1662; Isagoge phytoscopica, 1678; Opuscula botannicaphysica, 1747. An early champion of atomic theory; opponent of scholastic philosophy; anticipated Descartes in philosophy, emphasizing importance of math. discipline for sound philos. thinking; developed method for plant classification based on genus and species, thus foreshadowing Linnaeus; anticipated later views on chem. affinity; recognized that air plays a role in combustion; gave modern definition of a saturated solution. Died Hamburg, Germany, Sept. 23, 1657.

JUNGERMAN, John Albert, Am. physicist; b. Modesto, Cal., Dec. 28, 1921; s. Albert Augustus and Freda (Durst) J.; A.B., U. Cal., at Berkeley, 1943, Ph.D., 1949; m. Nancy Lee Kidwell, Oct. 23, 1948; children—Mark, Eric, Roger, Anne. AEC Postdoctoral fellow Cornell Inst. Nuclear Studies, 1949-50, Radiation Lab., U. Cal. at Berkeley, 1950-51; faculty U. Cal. at Davis, 1951-—, prof. physics, 1960-—, sr. investigator AEC, 1956-—, dir. Crocker Nuclear Lab., 1966-—; research staff Centre d'Etudes Nucleaires de Saclay, France, 1966-67. Fellow Am. Phys. Soc., Am. Assn. U. Profs., Sigma Xi. Research, publs. on beta ray spectroscopy, nuclear systems at intermediate energies; supervised constrn. of 2 spiral ridged cyclotrons. Home: 712 Elmwood Dr., Davis, Cal. 95616.*

JUNGERMAN, L., botanist; b. Leipzig, Germany, 1572; brother of Gottfried Jungerman; prof. botany, Altdorf, Switzerland; author (with M. and H. Hoffman) a flora of Germany. Died 1653.

JUNGFLEISCH, Emile-Clement, French chemist; b. Paris, Dec. 21, 1839; D.Sc. in Physics, Ecole supérieure de pharmacie de Paris, 1868, in Chemistry, 1869; student of Berthelot; intern in pharmacy, 1864-68; tchr. organic chemistry Ecole supérieure de pharmacie, Coll. de France, 1869, 74, 76, became prof., 1877; named prof. gen. chemistry Conservatoire National des Arts et Metiers, 1890; prof. Sorbonne, from 1908. Mem. French Acad. Scis., 1909 (Jecker prize 1872), also Acad. Medicine. Author: (with Berthelot) Traité de chimie organique, 2 vols., 2d edit., 1881; Manipulations de chimie, 1886; Notice sur E.M. Peligot, 1891; La production de la gutta-percha, 1892; La pharmacie et les marques de fabrique, 1894. Worked (with Berthelot) on law of partition, 1872, (with Lecoq de Bosibaudran) on isolation of gallium and indium in quantity; research on levulose, inverted sugar, reciprocal transformation of optical changes of even bodies; showed relation between number of substituting groups and melting points in studies of phys. properties of chlorine derivatives of benzene; showed that different forms of tartaric acid are transformed into each other when heated in water, also that an equilibrium exists; converted ethylene into succinic acid; showed camphoric acids exist in 4 isomeric forms; resolved inactive lactic and malic acids into their active forms. Died Paris, Apr. 24, 1916.

JUNGHUHN, Franz Wilhelm, botanist; b. Mansfeld, Holland, Oct. 26, 1809; ed. Halle, Berlin (both Germany); surgeon Prussian army; san. officer with French in Algeria; settled in Java; in Holland, 1849-55. Author: Java, seine Gestalt, Pflanzendecke, und sein innerer Bau, 4 vols., 1850-54; Die Bättalander auf Sumatra, 1847. Research on geology, geography and botany of Java; collected fossils plants, later described by other naturalists. Died Lembang, Java, Apr. 24, 1864.

JUNGKEN, Johann Helfrich (or Junken, Juncken, Jüngken), physician; b. Caldern, Hesse, Dec. 19, 1648; M.D., Heidelberg, Germany, 1671; physician to various notabilities, also physicus ordinarius, Frankfurt, Germany. Author: Chymia experimentalis . . . (on mercury preparations, analysis of human blood, artificial gems, other topics), 1681; Corpus pharmaceuticochymico-medicum . . . , 1697. Died Frankfurt, Jan. 5, 1726.

JUNGMANN, Antonin, physician; b. Hudlice, Bohemia, May 19, 1775; s. Tomas Simon and Katerina Jungmann; M.D., 1805; m. Marie Hola, 1812; 1 son, Adolf. Prof. gynecology, Prague, Czechoslovakia, 1811-50; became rector Charles U., Prague, 1838. Author books on obstetrics and veterinary diseases, also many articles. Founder 1st gynecol. clinic in middle Europe, 1842; founder modern Czech med. terminology. Died Prague, Apr. 10, 1854.

JUNKEN, Johann Helfrich, see Jungken, Johann Helfrich.

JUNKERS, Hugo, German inventor; b. Rheydt, Germany, Feb. 3, 1859; founder Junkers Co., 1895; became prof. École technique supérieure, Aix-la-Chapelle (Aachen, Germany), 1897; founder Junkers Aircraft works, 1919; began internal air service in Germany, 1921, later merged with Deutsche Luft Hanza. Patentee all-wing airplane which accommodated 2 engines, goods and passengers all within wings, 1910; built 1st Junkers all-metal airplane of steel, 1915, 1st all-light alloy airplane, 1916; developed G.24, G.38; constructed a calorimeter. Died Munich, Germany, Feb. 3, 1935.

JUNOD, Victor Theodore, physician; b. Bonvillars, Switzerland, 1809; M.D., Paris, 1833; recipient Montyon prize for large-scale application of principle of ventilator by means of pneumatic apparatus (hemospasic method) Institut, 1836; grand prize in medicine and surgery, 1870. Author: Traité théorique et pratique de l'hemospasie, 1875. Inventor apparatus for compressed air baths, also a substitute for water in hydrotherapy; studied effects of rarefied and compressed air on body; inventor Junod's boot which fit hermetically on leg or arm and could be pumped free of air. Died London, 1882.

JUNQUEIRA, Luiz Carlos, Brazilian biologist; b. Sao Paulo, Brazil, Aug. 5, 1920; s. Ozorio and Amelia (Uchoa) J.; M.D., U. Sao Paulo, 1947, Docente, 1949; m. Luiza Salles, May 10, 1956; children—Luiz Carlos, Vera, Luiz Roberto. Faculty, U. Sao Paulo Med. Sch., 1946-—, prof. biology, 1957-—; research asso. U. Chgo. Med. Sch., 1948. Vice pres. Brazilian Inst. for Edn., Sci. and Culture, 1960-64. Named prof. honoris causa U. San Marcos, Lima, Peru. Mem. Brazilian Acad. Scis. Research, publs. on cell physiology especially cell secretion, protein synthesis, comparative biology of secretion, pancreas and salivary glands. Home: 316 Rua Rio de Janeiro, Sao Paulo, Brazil.*

JÜPTNER VON JONSTORFF, Baron Hans, Austrian chem. technologist; b. Vienna, Austria, May 22, 1853; ed. Technische Hochschule, Vienna, 1870-74. Chief chemist and engr., exploration eastern Alps; prof. Technische Hochschule, Vienna, after 1902, dean faculty physico-chem. sci., 1905-08, rector, 1910-11. Author: Grundzüge der Siderologie, 3 vols., 1900-02; Lehrbuch der chemischen Technologie der Energien, 3 vols., 1905-08; several other works on applied chemistry, physics. Research on furnace tech., metallurgy, energetics. Died 1941.

JURAND, Artur, biologist; b. Krakow, Poland, Mar. 30, 1914; s. Marek Bieberstein and Emilia (Kaufler) J.; B.Sc. in Chemistry, Jagiellonian U., Krakow, 1937, B.Sc., in Biology, 1952, Ph.D. in Drug Chemistry, 1947, D.Sc. in Biology, 1954; m. Jadwiga Reiner, Jan. 15, 1946; 1 dau., Maria K. Faculty, Jagiellonian U., 1947-57, sr. lectr., 1952-57; lectr., research fellow Inst. Animal Genetics, Edinburgh, Scotland, 1957-—. Mem. Royal Phys. Soc. Edinburgh, Soc. Exptl. Biology, Soc. Developmental Biology. Research and publs. on congenital malformations induced by drugs, electron microscopy of limb buds in chick and mouse embryos; ultrastructure cytology in Paramecium and other protozoa. Home: 46 Ladysmith Rd., Edinburgh, Scotland.

JURIN, James, physician; b. London, Eng., 1684; s. John Jurin; student Christ's Hosp. and Trinity Coll., Cambridge (Eng.) U.; B.A. (fellow), 1706, M.A., 1709, M.D., 1716; 1 son, James. Tutor to Mordecai Carey, Christ's Hosp., 1708-09; master Newcastle/Tyne grammar sch., 1709-15; practiced medicine, London; physician Guy's Hosp., 1725-32; attending physician to Sir Robert Walpole, during his last illness. Fellow Royal Soc., 1717 (sec. 1721-27); mem. Royal Coll. Physicians (censor, council, 1748-49, pres. 1750). Author: An Account of the Success of Inoculating for Smallpox in Great Britain, 1724; Dissertationes Physico-Mathematical, 1732; The Minute Mathematician, 1735. Advocated inoculation for smallpox; attempted to make physiology an exact science; advanced law for rising height of fluid in capillary tube as function or its diameter, 1718; determined specific gravity of blood serum (1,030) and blood (1,053), 1719; formulated Jurin's law; supporter of Newtonian theory; studied theory of sight and principles of dynamics. Died London, Mar. 29, 1750.

JURKOVIC, Vilo, Czechoslovakian physician; b. Stará Bystrica, Czechoslovakia, Mar. 12, 1915; s. Josef and Ester (Bistricová) J.; M.D., Med. Sch., Brno, Czechoslovakia, 1939; Sc.D., Acad. Sci., Prague, Czechoslovakia, 1965; m. Irena Langerová, Aug. 23, 1945; children—Natasa, Josef. Staff, Bulovka Hosp., Prague, 1945-50, instr. medicine, 1948-50; physician, dep. chief Center Mil. Hosp., Prague, 1950-51; faculty Med. Sch., Charles U., Hràdec Kràlové, Czechoslovakia, 1951-—, prof. medicine, 1961-—, physician in chief IId dept. medicine, 1955-—, head dept. medicine, 1961-—, dir. Research Lab. Radiation Disease, 1958-—. Vis. prof. medicine, Strasbourg, France, 1966-67. Recipient Purkyne medal, 1964. Author: Textbook of Medicine, 1956, 60; Ventricular Tachycardia and Some Circulatory Changes in Radiation Diseases (with summary in English), 1965; also numerous articles. Research on origin of irregular heart action, cardiovascular changes during acute stage of radiation disease, some nervous mechanisms in disordered heart beat. Home: 1 Ambrozova, Hradec Králové, Czechoslovakia.*

JURSA, Adolph Simon, Am. physicist; b. Westerly, R.I., Sept. 17, 1923; s. Simon John and Mary (Tamulis) J.; B.S. in Physics, U. R.I., 1949; M.S., Ind. U., 1951. Physicist, Air Force Cambridge Research Lab., Bedford, Mass., 1951-—, supervisory physicist, 1959-—. Mem. Optical Soc. Am., Sci. Research Soc. Am. Contbr. articles to tech. jours. Research, publs. on molecular spectroscopy, physics of atmosphere. Home: 8 Snow Dr., Westford, Mass. 01886. Office: Air Force Cambridge Research Lab., L. G. Hanscom Field, Bedford, Mass. 01730.*

JUS, Andrzej, Polish physician; b. Lwow, Poland, Oct. 16, 1914; s. Ludwik and Estelle (Kober) J.; M.D., U. Lwow, 1939; Med.Fac., U. Wroclaw, 1946; Prof. Med.Fac., Acad. Medicine, Warsaw, 1953; m. Karolina Fryst, Aug. 1, 1941. Asst. in psychiat. dept. U. Lwow, 1939-41, U. Wroclaw, 1945-49; vice dir. Psychoneurol. Inst. Pruszkow, 1950-53; prof. psychiatry U. Lodz, 1953-55; vice dir. Psychoneurol. Inst., 1955-59; prof. psychiatry Warsaw Acad. Medicine, 1959—. Expert on mental health for WHO, 1961. Corr. fellow Am. Psychiat. Assn., Med.-psychol. Soc.; mem. Coll. Internat. Psychopharmacology. Author: (with K. Jus) Elektroencefalografia, 1953; Problems of Contemporary Psychopathology, 1957; Psychopharmacological Therapy, 1960. Numerous publs. on application of methods of modern neurophysiology to psycopath. phenomena, psychopharmacology, clin. studies on schizophrenia and epilepsy and sleep and dreams in mental illness. Home: 7, Al. Wyzwolenia, Warsaw, Poland.*

JUS, Karolina Fryst, physician; b. Vienna, Austria, Dec. 31, 1914; d. Juliusz and Dorota (Bannet) Fryst; LL.D., Cracow (Poland) U., 1937; M.D., U. Wroclaw (Poland), 1950; m. Andrzej Jus, Aug. 1, 1941. Asst. neurol. dept. U. Wroclaw, 1946-49; chief EEG dept. Psychoneurol. Inst., Pruszkow, Poland, 1950-53, Neurol. Clinic, U. Lodz (Poland), 1953-56; chief dept. EEG Clin. Neuro-physiol. Acad. Medicine, Warsaw, Poland, 1953-66, faculty, 1955—, prof. psychiatry, 1966—; tchr. Inst. Postgrad. Tng., Warsaw, 1953—. Recipient awards Acad. Medicine, Lodz, 1955, Acad. Medicine, Warsaw, 1965. Fellow Am. Psychiat. Assn. (corr.); mem. Polish Soc. for EEG and Clin. Neurophysiology (founder, pres. 1961—), Internat. Fedn. EEG (council 1961—). Author: (with A. Jus) Elektroencefalografia 1953; Elekroencefalografia Klinicsna, 1967; also numerous articles. Research on neurophysiol. methods in condition, electrophysiol. analysis of subthreshold stimuli, memory disturbances, sleep and dreaming in psychiat. patients. Home: 7 Al. Wyzwolenia, Warsaw, Poland.*

JUSSIEU, Adrien de, French botanist; b. Paris, France, Dec. 23, 1797; s. Antoine Laurent de Jussieu; M.D., U. Paris. Prof. botany Mus. Natural History, from 1826, later dir.; prof. vegetable organography Faculté des Sciences, U. Paris, from 1845. Recipient 1st prize Competition of 1714. Mem. French Acad. Scis., 1831 (pres. 1853), French Agrl. Soc. Author: Rutacées, 1825; Méliacées, 1830; Malpighiacées, 1843; Embryons monocotyledonés, 1844; Tiges des Lianes, 1845; Cours elementaire de botanique, 1842-44. Died Paris, June 29, 1853.

JUSSIEU, Antoine-Laurent de, French botanist; b. Lyons, France, Apr. 12, 1748; student botany under Bernard de Jussieu; 1 child, Adrian. Prof. botany at Jardin du Roi; dir. mus.; became asst. botanist at Royal Acad. Scis., 1773, apptd. resident, 1786; became titular mem. dept. botany and plant physics 1st class of Nat. Inst., 1795; employed to reorganize bot. sch. of Jardin des Plantes, Paris (France). Fellow Royal Soc., 1829. Author: Exposition d'un nouvel ordre des plantes, 1774; Genera plantarum secundum ordines naturales disposita, 1789; Principes de la methode naturelle des vegetaux, 1824. Expanded the natural system of plant classification formulated by his uncle (Bernard); his system influenced work of Candolle and Cuvier, and became basis of modern botanical classification. Died Paris, Sept. 17, 1863.

JUSSIEU, Bernard de, French botanist; b. Lyons, France, Aug. 17, 1699; s. Christophe de Jussieu; M.D., Montpellier, France, 1720. Accompanied brother Antoine on bot. expdn. to Spain, 1716; apptd. prof. Jardin du Roi, Paris, 1722; named supt. royal garden by Louis XV, Trianon, France, 1758. Honored by Linnaeus in name of genus Jussieua, 1737. Mem. French Acad. Scis., 1725. Author: Histoires des plantes des environs de Paris (rev. edit. of Tournefort's work), 1725; Ordines naturales in Ludovici XIV horto Trianonensi dispositis. Made 1st attempt at natural classification of plants, according to his understanding of the embryo (Jussieu system). Died Paris, Nov. 6, 1777.

JUSSIEU, Joseph de, French botanist; b. Lyon, France, Sept. 3, 1704; s. Christophe de J.; M.D. degree; served as botanist on expdn. to Peru, 1735; collected valuable information on natural history until 1771 when he returned to Paris. Mem., French Acad. Scis., 1742. Introduced various ornamental plants to Europe; notable for introduction of garden heliotrope. Died Paris, France, Apr. 11, 1779.

JUST, Ernest Everett, Am. zoologist; b. Charleston, S.C., Aug. 14, 1883; s. Charles Frazier and Mary (Mathews) J.; A.B., Dartmouth, 1907; Ph.D., U. Chgo., 1916; m. Ethel Highwarden, June 26, 1912. With Howard U., Washington, from 1907, prof. zoölogy, from 1912. Author: (with others) General Cytology, 1924; The Biology of the Cell Surface, 1939; Basic Methods for Experiments on Eggs of Marine Animals, 1939. Contbr. papers on physiology of devel., including results of research on fertilization, artificial partheonogenesis, cell division. Asso. editor Physiol. Zoölogy (Chicago) and The Biol. Bull. (Woods Hole, Mass.), Jour. Morphology. Died Oct. 27, 1941.

JUSTI, Eduard Wilhelm Leonhard, physicist; b. Hong Kong, May 30, 1904; s. Karl H. and Maria (Kuelz) J.; student U. Kiel, U. Berlin; 1922-27; Dr.phil., U. Marburg (Germany), 1929; Dr.phil.habil., U. Berlin (Germany), 1937; m. Ursula Gaede, Mar. 16, 1935; children—Carola (Mrs. Eugenio Andrés Puente); Irene. Staff Physikalisch-Technische Reichsanstalt, Berlin, 1929-44; prof. applied physics U. Posen, 1944-45; prof. tech. physics U. Braunschweig (W. Germany), 1946—, pres., 1953-54. Chmn., German Joint Program Direct Energy Conversion, 1960—; cons. physicist to German elec. firms. Decorated Comdr. Royal Swedish N. Star Order. Mem. Acad. Scis. and Lit. (past pres.), Deutsche Physikal. Gesellschaft, Am. Electrochem. Soc., Royal Swedish Acad. Engring. Sci., Royal Swedish Acad. Sci. Gothenburg. Author: Conductivity and Conduction Mechanism of Solids, 1948; Specific Heat, Enthalpy, Entropie, Dissociation of Gases, 1938; (with A. Winsel) High Drain Hydrogen Diffusion Electrodes, 1959, Fuel Cells Cold Combustion, 1962; Conduction Mechanism and Energy Conversion in Solids, 1966; also numerous articles. Research and numerous patents on theoretical thermodynamics, conductivity mechanism of pure metals and semiconductors, fuel cells for ambient temperature and pressure; discovery of semiconductivity of intermetallic compounds, new superconductors with high transition temperature, Home: 5 Brockenblick, 33 Braunschweig, Germany.*

JUSTICE, Oren Lloyd, Am. botanist; b. Pikeville, Ky., Nov. 11, 1907; s. William B. and Amanda (Maynard) J.; B.S., Ohio U., 1932, M.A., 1937; Ph.D., Cornell U., 1940; m. Ellen Kytta, Mar. 9, 1935; children—Roger Lee, Keith Oren. Asst. botany Cornell U., 1936-40; state seed analyst, dir. seed div. Ala. Dept. Agr., Montgomery, 1940-44; botanist U. S. Dept. Agr., Beltsville, Md., 1944-62, research botanist, leader seed quality investigations, 1962—; lectr. internat. seed tng. courses, Cambridge, Eng., 1954, Campinas, Brazil, 1964. Cons. seed sci., tech., Brazil, 1964. Mem. Botan. Soc. Am., Soc. Economic Botany, Washington Botan. Soc., Internat. Seed Testing Assn. (pres. 1962-65), Assn. Ofcl. Seed Analysts, Am. Inst. Biol. Sci. Author: (with others) Manual for Testing Seeds, 1952. Publs. on studies of seed dormancy, devel. methods for determination of seed quality. Home: 901 Copley Lane, Silver Spring, Md. 20904. Office: Plant Industry Sta., Beltsville, Md. 20705.*

JUSTUSSON, William Matt, Am. metallurgist; b. Bessemer, Mich., Jan. 10, 1930; s. William Mathew and Lydia (Ruhinen) J.; B.S., Mich. Tech. U., 1951; M.S., U. Wis., 1952; m. Nadine Ada Mattson, Dec. 18, 1948; children—Jerald William, Daniel Matt, Chris Lee, Cynthia Colleen, Lisa Kay, Teresa Lynn. With atomic power div. Westinghouse Electric Corp., Pitts., 1952-53, Gen. Dynamics Corp., Fort Worth, 1953-54; sci. research staff Ford Motor Co., Dearborn, Mich., 1954—, supr., 1962—. Mem. Am. Soc. Metals, Am. Inst. Mining and Metall. Engrs. Research, publs. on mechanisms to strengthen steels by thermomech. treatments, devel. iron-aluminum alloys, discovered fatigue-resistant steel, two uranium-base nuclear fuel alloys, vibration casting of thin castings; patentee in field. Home: 30146 Woodbrook Ct., Farmington, Mich. 48024. Office: 20500 Oakwood Blvd., Dearborn, Mich. 48011.*

K

KAADA, Birger Rygg, Norwegian physiologist; b. Stvanger, July 11, 1918; s. Theodor and Anna Kaada; M.D., U. Oslo; m. Ingrid Vedholm, July 14, 1945; children—Arvid, Geir, Ragnar. Research asst., Yale, 1947-48; asst. Neurol. Inst., Montreal, Que., Can., 1949; instr. neurophysiology U. Oslo, 1949 53, prof., 1959—; cons. electroencephalography Oslo Hosp. Recipient Monrad-Krohn prize, 1954. Mem. Norwegian Soc. Encephalography, Norwegian Soc. Biochemistry and Physiology (pres.), Scandinavian Soc. Physiology (pres.), Norwegian Soc. Neurology, Norwegian Acad. Sci., Sigma Xi. Author: Somatomotor, Autonomic and Electrocorticographic Responses to Electrical Stimulation of Rhinenephalic and Other Structures in Primates, Cat and Dog, 1951. Research on nervous control of motor visceral functions, emotions, sleep and waking, brain lesions. Home: Frits Kiaers vei, 17, Oslo 3. Office: Neurophysiological Lab., University of Oslo, Karl Johans gate, 47, Oslo, Norway.

KABACHNIK, Martin Izrailevich, Russian organic chemist; b. Sept. 9, 1908; grad. Moscow Higher Chemicotech. Sch., 1931. Asso., Inst. Organic Chemistry, 1939-54, Inst. Elemental-Organic Compounds, USSR Acad. Sci., 1954—. Decorated Order of Lenin; recipient Stalin prize, 1946. Mem. USSR Acad. Sci. Co-author: Dual Reaction Capacity and Tautomerism, 1955; author: Some Problems of Tautomerism, 1956; New Means for Practical Application of Elemental-Organic Compounds, 1956. Mem. editorial bd. News of USSR Acad. Sci. Research and publs. on tautomerism of organic compounds, chemistry of phosphororganic compounds, synthesis of phosphoro-organic insecticides. Address: Inst. Elemental-Organic Compounds, USSR Acad. Sci., Leninsky prospect 31, Moscow, USSR.

KABANOV, Nikolay Matveevich, Russian hydrobiologist, algologist, hydrochemist; b. Moscow, 1902; grad. biol. dept. Physico-Math. Faculty, Moscow U., 1926; Cand. Biol. Sci., 1938. Asso., 'Oka Biol. Sta., Murom, later Nizhniy Novgorod, 1924-31; with Nizhniy Novgorod Limnological Inst., 1931-32; dep. sci. dir. Gorky Hydrological Inst., 1932-34, asso. Volga dept., 1934-36; hydrobiol. asst. Gorky Civil Engring. Inst., 1932-33; lectr. hydrobiology and hydrology Gorky U., 1940-41; sr. asso. Moscow Inst. Gen. and Communal Hygiene, USSR Acad. Med. Sci., 1946—. Co-author: A Reference Manual of the Water Resources of the USSR. IV. The Central Volga Area, part 1, 1935, III. The Upper Volga and the Oka Basins, parts 1-2, 1936. Research and publs. on morphology and systematics of lower plants; discovered new type of biocenosis-microscopic community of psammonae, 1924-26, 1st to describe this community (has green coloration due to presence of algae); developer methods for studying bioflux of rivers and for making quantitative estimates of net seton; established ratio between biomass among various higher water plants and percentage of dry matter in them. Address: Inst. Gen. and Communal Hygiene, Pogodinskaya ulitsa 10, Moscow, USSR.

KABARA, Jon Joseph, Am. pharmacologist; b. Chgo., Nov. 26, 1926; s. John S. and Mary (Wielgus) K.; B.S., St. Marys Coll. (Minn.), 1948; M.S., U. Miami (Fla.), 1950; Ph.D., U. Chgo., 1959; m. Ginger Christie, Dec. 18, 1948; children—Christie Ann, Mary Katherine, Sheila Jon, Patricia Lee. Research asst. U. Chgo., 1953-57; faculty U. Detroit, 1957—, prof., 1965—, dir. biochemistry research, 1957—. Curriculum coordinator Joe Berg Found. High Sch. Program, 1958-60; dir. NSF Summer Inst. High Sch. Students, 1959; dir. life scis. Council on United Research in Devel. and Aging, 1965. Recipient Educator of Year award Alpha Epsilon Delta, 1964; citation of merit Mich. Acad. Sci., Arts and Letters, 1966. Damon Runyon Cancer fellow U. Miami, 1949. Mem. Am. Chem. Soc., N.Y. Acad. Sci., Assn. Analytical Chemists, Brit. Biochem. Soc., A.A.A.S., Am. Soc. Clin. Pathologists, Am. Assn. Clin. Chemists, Am. Oil Chemists Soc., Soc. For Exptl. Biology and Medicine, Sigma Xi, Sigma Pi. Contbr. numerous articles to profl. jours. Instrumental in evaluating and establishing critique and instrumentation in lipid chemistry; studies in brain metabolism; helped establish new techniques in radiochemistry; isolated active prin. in cobra venom. Home: 23336 Thorncliffe St., Southfield, Mich. 48075. Office: 4001 W. McNichols Rd., Detroit 48221.*

KABAT, Elvin Abraham, Am. immunochemist; b. N.Y.C., Sept. 1, 1914; s. Harris and Doreen (Otis) K.; B.S., Coll. City N.Y., 1932; M.A., Columbia, 1934, Ph.D., 1937; m. Sally Lennick, Nov. 28, 1942; children—Jonathan, Geoffrey, David. Lab. asst. immunochemistry Presbyn. Hosp., N.Y.C., 1933-37, research asso. biochemistry Columbia, Presbyn. Hosp., 1941-46, mem. sci. staff Div. War Research, Columbia, 1943-45, asst. prof. bacteriology, 1946-48, asso. prof., 1948-52, prof. microbiology, 1952—, microbiologist med. service Neurol. Inst., Presbyn. Hosp., 1956—. Mem. panel on plasma NRC, 1953—, ad hoc panel and com. on tissue transplantation, 1957-61; mem. med. adv. bd. Nat. Multiple Sclerosis Soc., 1953—, research rev. panel, 1956-59; mem. sci. and ednl. council Am. Found. for Allergic Diseases, 1954—, biochemistry adv. panel Office Naval Research, 1957-62, sci. adv. com. Inst. Microbiology, Rutgers U., 1962—; lectr., vis. prof. various univs.; mem. expert adv. panel on immunology WHO, 1965—. Recipient Eli Lilly award in bacteriology and immunology, 1949; Golden Hope Chest award Nat. Multiple Sclerosis Soc., 1962. Fellow A.A.A.S., N.Y. Acad. Scis.; mem. Am. Assn. Immunologists (pres. 1965-66), Am. Soc. Biol. Chemists, Am. Soc. for Microbiology, Internat. Assn. Allergists, Societe Francaise d'Allergie (hon.), Biochem. Soc. (Eng.), Chem. Soc. (Eng.), Am. Neurol. Assn. (asso.), Assn. for Research in Nervous and Mental Diseases (asso.), Harvey Soc., Am. Assn. U. Profs., Association de Microbiologists de Langue Francaise (asso.), Nat. Acad. Scis., Am. Acad. Arts and Scis., Phi Beta Kappa, Sigma Xi, Alpha Omega Alpha (hon.). Author: (with M. M. Mayer) Experimental Immunochemistry, 1948, 2d edit., 1961; Blood Group Substances, Their Chemistry and Immunochemistry, 1956; Structural Concepts with Immunology and Immuno chemistry, 1968; also articles. Editorial bd. Transplantation Bull., 1957-60, Jour. Immunology, 1961. Research on methods for measuring antigens and antibodies, analysis of gamma globulin in cerebrospinal fluid, size of combining site on an antibody molecule which reacts with antigen. Home: 630 W. 168th St., N.Y.C. 10032.*

KABATA, Zbigniew, parasitologist; b. Jeremicze, Poland, Mar. 17, 1924; s. Piotr and Helena (Wojciechowicz) K.; B.Sc. with 1st class honors, Aberdeen (Scotland) U., 1955, Ph.D., 1959, D.Sc., 1966; m. Mary Ann Montgomery, Sept. 8, 1953; children—Marta Maria, Andrzej Jozef. With Scottish Dept. Agr. and Fisheries, Aberdeen, 1955-67, prin. sci. officer, head parasitology dept.; parasitologist Fisheries Research Bd. of Can., Biol. Sta., Nanaimo, B.C., 1967—. Mem. Inst. Biology, Brit. Soc. Parasitology, Marine Biol. Assn. U.K. Translator from Russian: Parasitology of Fishes (Dogiel), 1961; (with J. M. Shewan) Marine Microbiology (Kriss), 1963; General Parasitology (Dogiel), 1964. Research, publs. on applications of parasitology to fisheries research, effects of parasites on stocks of marine fishes, use of parasites as labels, parasitic Copepoda of marine fishes. Home: Bon-Accord, 260 Cilaire Dr. Office: Biol. Sta., Nanaimo, B.C., Can.*

KABURAGI, Masaki, Japanese astronomer; b. Kanazawa, Japan, July 24, 1902; grad. Tokyo U.; received Sc.D.; became asst. prof. Tokyo U., 1935, prof., 1946—. Author: Construction of the Universe; Applied Astronomy. Research on Milky Way.

KAC, Mark, mathematician; b. Krzemieniec, Poland, Aug. 3, 1914; s. Bencion and Chana (Rojchel) K.; magister of Philosophy, U. Lwow (Poland), 1935, Ph.D., 1937; m. Katherine Elizabeth Mayberry, Apr. 4, 1942; children—Michael Benedict, Deborah Katherine. Came to U. S., 1938, naturalized, 1943. Teaching asst. U. Lwow, 1935-37; jr. actuary Phoenix Co., Lwow, 1937-38; fellow Parnas Found., Johns Hopkins, 1938-39; faculty Cornell U., 1939-61, Andrew D. White prof.-at-large, 1965—; mem. Inst. Advanced Study, Princeton, N.J., 1951-52; prof. Rockefeller Inst., N.Y.C., 1961—; H. A. Lorentz vis. prof. U. Leiden, The Netherlands, 1963. Guggenheim fellow, 1946-47; recipient Chauvenet prize Math. Assn. Am., 1950, 68. Mem. Am. Acad. Arts and Scis., Am. Math. Soc., Math. Assn. Am., Nat. Acad. Scis., Inst. Math. Statistics, Sigma Xi. Research and publs. primarily on probability theory and its applications in various brs. of math. and physics. Home: 6 Rectory Lane, Scarsdale, N.Y. 10583. Office: Rockefeller U. N.Y.C. 10021.*

KACHARAVA, Iván Vissariónovich, Russian geologist; b. Sept. 5, 1894; ed. Tbilisi U., 1924; prof. Tbilisi U., 1942—; dir. sect. on paleontology Geology Inst., Georgian Acad. Sci. 1941—. Author: The Rachinsko-Lechkhumsky Basin and Adjacent Areas in the Paleogene Period, 1944; Nummulites in Some Areas of Eastern Georgia, 1948; Paleogene of the Kartalina Depression and its Adjacent Areas, 1955; Oligocene Fauna of the Akhaltsikhe Area, 1960. Research in paleontology and stratigraphy; described various groups of fossils; studied stratigraphy of Palogene deposits in Georgia.

KACZKA, Edward Anthony, Am. chemist; b. North Little Rock, Ark., July 20, 1914; s. S. T. and Mary (Chudy) K.; B.S., Miss. State U., 1937; Ph.D., U. N.C., 1943; m. Nancy Lynch, June 7, 1947; children—Mark and Leslie Ann (twins), Richard, James, Mary Elizabeth. Instr. chemistry Miss. State U., 1937-39; research fellow Merck, Sharp & Dohme Research Labs., Rahway, N.J., 1943—. Mem. Am. Chem. Soc., A.A.A.S. Research and publs. on isolation and evaluation of natural tannins from trees and other woody plants of Southeastern U. S.; structural elucidation of penicillin and rhizopterin; isolation and characterization of antibiotic, novobiocin; isolation, characterization and synthesis of antitumor agt., Hadacidin. Home: 94 Clark Pl., Union, N.J. 07083. Office: Merck, Sharp & Dohme Research Labs., Rahway, N.J. 07065.*

KACZKOWSKI, Zbigniew Ignacy, Polish engr.; b. Wloszczowa, Poland, Mar. 26, 1932; s. Bronislaw and Antonina (Lao) K.; B.Electronic Engring., Tech. U., Warsaw, Poland, 1954, M.Sci.Engring., 1956; D.Tech. Scis., Polish Acad. Scis. Inst. Basic Tech. Problems, 1964; m. Danuta Gorecka, Mar. 3, 1956; children—Beata, Anita. Staff, Polish Acad. Scis., Warsaw, 1953-67, adj. Inst. Basic Tech. Problems, 1958-66, Inst. Electron Tech., 1966—, vice dir. inst., 1966—, chief piezomagnetic lab., 1967—, mem. sci. council, 1966-67; ind. sci. research, 1967—. Vis. research staff Inst. Acoustics, Soviet Acad. Scis., Moscow, also physics dept. U. Moscow, 1962-63, U. Sheffield (Eng.), 1965-66; lectr. Tech. U., Warsaw, 1966—. Research, publs. on desacommodation in magnetics; design and studies in miniaturized magnetic components, magnetic measurement circuits and methods; research in magnetistrictive ferrites, magnetistrictive and piezomagnetic temperature properties of aluminum—iron and other alloys, magnetic histeresis of piezomagnetic coefficients; research in Ni-Span C alloys; discovered shifting of extrema of piezomagnetic coefficients as function of polarization on temperature characteristics in cobalt containing ferrites. Home: Warszawa, Dlugosza Nr 8 m.7, Poland.*

KADANOFF, Dimiter Dimitrov, Bulgarian anatomist, anthropologist; b. Shumen, Bulgaria, Apr. 12, 1900; s. Dimiter and Wassilka (Dugmedjiewa) K.; M.D., U. Würzburg (Germany), 1925; m. Bozana Iwanova Pandowa, Dec. 29, 1960. Asst., anat. inst. Med. Faculty U. Würzburg, 1925-29; prosector Anat. Inst. Würzburg, 1929-33; asso. prof. anatomy Med. Faculty Sofia (Bulgaria), 1938-42, prof., 1942-65; prof., anthropologist Bulgarian Acad. Scis., Sofia, 1965—. Recipient Nat. prize, 1959, Dimitrov prize. Mem. Bulgarian Acad. Scis. (corr), Academia Leopoldina Halle, Bulgarian Soc. Anatomista and Histologists (pres. 1963). Author: Handbook of Anatomy, 2 vols., 1953, 55, 64; (monograph) Morphologie der Rezeptoren, 1957, 63; also articles. Research in innervation of hair, spl. innervation of pharynx; exptl. morphological investigations on degeneration and regeneration of nerve fibers; new findings on nerve fibers and endings in epithelium of several organs of man, roof of mouth, nasal passages, larynx. Home: 40, Boul. D. Blagoev, Sofia 6, Bulgaria.*

KADEN, Heinrich, German physicist; b. Bremen, Jan. 1, 1902; s. Gustav and Lina Kaden; Ph.D. in Physics, U. Gottingen; m. Herta Borowski, Oct. 4, 1930; children—Heinrich, Editha, Ulrich. With Central labs. S. A. Siemens & Halske, Berlin, 1930, Munich, 1947—; prof. agrégé, 1952; prof. Technische Hochschule, Munich, 1958. Mem. Phys. Soc., Gesellschaft für angewandte Math. und Mechan., Deutscher Alpenverein, Verein Deutscher. Author: Impulse und Schaltvorgänge in der Nachrichtentechnik, 1957; Wirbelströme und Schirmung in der Nachrichtentechnik, 1959; Taschenbuch der Hochfrequenztechnik, 1962. Research and numerous publs. on electromagnetic field, transmission of signals, telecommunications. Home: Bastian-Schmid-Platz 2, Munich-Solln. Office: Siemens & Halske Gesellschaft, Hofmannstrasse 51, Munich 25, Germany.

KADOMTSEV, Boris Borisovich, Russian physicist; mem. USSR Acad. Scis. (corr.). Author: Convective Instability of a Plasma Column, 1959; Plasma Equilibrium in Helical Symmetry, 1959; Stabilization of plasma with the Aid of Heterogenous Magnetic Fields, 1959; Instability of an Electric Cloud in a Magnetron, 1959; Magnetic Stability of Plasma in a Magnetic Field, 1962. Research on properties of plasma, plasma physics. Office: USSR Acad. Scis., Leinskii Prospekt 14, Moscow.

KADOTA, Hajime, Japanese microbiologist; b. Kyoto, Japan, Dec. 21, 1920; s. Jirokichi and Eiko (Akiyama) K.; grad. Hokkaido Imperial U., 1944; Ph.D. in Agr., Kyoto U., 1956; m. Momoko Konoshima, May 14, 1947; children—Yutaka, Osamu. Faculty, Kyoto U., 1949—, prof. Research Inst. for Food Sci., 1957—. Mem. Japanese Soc. Sci. Fisheries (councilor 1957—, Biology prize 1957), Agrl. Chem. Soc. Japan (councilor 1961—). Contbg. author: Marine Microbiology, 1963; Marine Boring and Fouling Organisms, 1959. Research, publs. on ecology, biochem. roles of marine bacteria, structure and function of bacterial spore coats. Home: 54 Higashimachi, Tojiin, Kita-ku, Kyoto. Office: Research Inst. for Food Sci., Kyoto U., Kyoto, Japan.*

KAEISER, Margaret, Am. botanist; b. McAlester, Okla., June 20, 1912; d. Henry Shanklin and Florence (Clayton) Kaeiser; B.S., U. Okla., 1934, M.S., 1936; Summer scholar, U. Wyo., 1935; Ph.D. (sr. botany fellow), U. Ill., 1940. Faculty, U. Ill., 1937-39, Chatham Coll., Pitts., 1940-41, Cedar Crest Coll., Allentown, Pa., 1941-44, St. Joseph Coll., West Hartford, Conn., 1944-47; faculty So. Ill. U., Carbondale, 1947—, asso. prof., 1950—. Dir. Regional Nat. Mus. Palisades Interstate Park Commn., Bear Mountain, N.Y., 1944; wood technologist Central States Forestry Exptl. Sta., 1956; vis. prof. biology Central Mo. State Coll., Warrensburg, 1956. Recipient grants Sigma Xi-Research Engring. Soc. Am., 1951, Ill. Acad. Sci., 1960, NSF, 1960-62. Fellow A.A.A.S.; mem. Torrey Bot. Club, Bot. Soc. Am., Internat. Soc. Plant Morphology, Internat. Assn. Wood Anatomists, Phi Beta Kappa, Sigma Xi, Phi Sigma, Sigma Delta Epsilon. Research, numerous publs. on anatomy of wood. Home: 1217 Hill St., Carbondale, Ill. 62901.*

KAESBERG, Paul Joseph, Am. biophysicist; b. Engers, Germany, Sept. 26, 1923; s. Peter Ernst and Gertrud (Mueller) K.; came to U. S., 1926, naturalized, 1933; B.S., U. Wis., 1945, Ph.D., 1949; m. Marian Hanneman, June 13, 1953; children—Paul, James, Peter. Faculty, U. Wis., Madison, 1949—, prof. biophysics, 1961—. Research Career award USPHS, 1962. Mem. Biophys. Soc., Am. Soc. Biol. Chemists, Am. Soc. Microbiology. Research, numerous publs. on phys. and chem. properties of viruses; demonstrated that many viruses have an icosahedral shape. Home: 5002 Bayfield Terrace, Madison, Wis. 53705.*

KAESER, Heinrich Ernst, Swiss physician; b. Aarau, Switzerland, Mar. 25, 1924; s. Otto and Marie (Lüthy) K.; student Med. Sch. Geneva, Switzerland, 1944-46, U. Basel (Switzerland), 1946-47; U. Zürich (Switzerland), 1947-50; m. Rosmarie Lüscher, Sept. 29, 1954; 1 son, Urs. Staff, Psychiat. Clinic Zürich, 1953-54; staff Neurol. Clinic and Policlinic Basel, 1954—; staff Mayo Clinic, Rochester, Minn., 1959-61, faculty U. Basel, 1962—, prof. neurology, head Neurol. Clinic, 1965—. Recipient Robert Bing prize, 1962. Mem. Swiss Soc. Neurology, French EEG Soc., World Fedn. Neurology. Author: Die Experimentelle diphtherische Polyneuritis, 1962; also articles. Research in electromyography and neuromuscular diseases; described new disease entities including scapuloperoneal muscular atrophy, hereditary myoclonus of musculus mentalis, paralysis of cricothyroid muscle. Home: 38 38 Bosenhaldenweg 4125 Riehen/BS, Switzerland.*

KAESTNER, Alfred Karl Ludwig Heinrich, German zoologist; b. Leipzig, Germany, May 17, 1901; s. Karl Ludwig and Hedwig (Muller) K.; Ph.D., U. Leipzig, 1929; m. Gertrud Engel, Oct. 4, 1930. Curator, Dohrn entomol. collection Naturkundemuseum, Stettin, Germany, 1930-45; curator arachnological dept. Zoolog. Museums Berlin (Germany), 1946-49; prof. specialized zoology Humboldt U. Berlin, 1949-51; prof. specialized zoology, 1st dir. Bavarian natural sci. collection U. Munich, 1957-66. Hon. fellow Academia Zoology Agra (India); mem. Deutsche Akademie der Wissenschaften Berlin, Academie Leopoldina Halle. Author: Lehrbuch der Speziellen Zoologie in 5 parts, 1954-67; also articles. Research on comparative anatomy, embryology and taxonomy of arachnids. Home: 225b Grünwalder, 8 Munich 9, Germany.*

KAFKA, Victor, physician, chemist; b. Karlsbad, Czechoslovakia, Oct. 12, 1881; s. Theodor Kafka; grad. U. Prague, 1906; postgrad., Munich, Hamburg; m. 2d, Elisabeth Holtappels. Became clk. Clinic and Hosp. Friedrichsberg, Hamburg, Germany, 1917, physician, 1920-33. Mem. Internat. Neurol. Congress, Bern, Switzerland, 1931. Mem. Soc. Psychiatry and Neurology, Hamburg, Kalloidchemischen Gesellschaft. Author: Taschenbuch der praktischen Untersuchungsmethoden der Körperflüssigkeiten, 3d edit., 1938; Die Zerebrospinalflüssigkeit, 1931; Sexualpathology, 1932; many other publs. on serology, psychiatry, internal secretions. Research on paraffin reactions of liquor cerebrospinalis, also albumin regulation, haemolysin reaction, normonastix reaction. Died May 5, 1955.

KAGAN, Benjamin Milton, Am. pediatric researcher; b. Washington, Pa., July 18, 1913; s. Alexander and Sarah K.; A.B. from Washington and Jefferson Coll., 1933; M.D., Johns Hopkins, 1937; m. Katherine Hamburger, June 2, 1940; children—Christopher, Robert. Instr. pediatrics Med. Coll. Va., Richmond, 1940-42, St. Phillip's Sch. Pub. Health Nursing, Richmond, 1940-42; asst. in pediatrics Columbia Coll. Physicians and Surgeons, 1946; clin. asst. prof. pediatrics U. Ill., 1947-53, clin. asso. prof., 1953; asso. prof. pediatrics Northwestern U., 1953-55, prof., 1955; faculty U. Cal. at Los Angeles, 1955—, prof.-in-residence, 1965—; chmn. dept. pediatrics Michael Reese Hosp., 1946-55; dir. chmn. dept. pediatrics Cedars of Lebanon Hosp., Los Angeles, 1955—, cons. dept. medicine, 1961—. Mem. numerous med. adv. bds., 1947—. Diplomate Am. Bd. Pediatrics. Fellow Am. Acad. Pediatrics, A.C.P., Am. Coll. Chest Physicians; mem. Am. Pediatric Soc., Soc. Pediatric Research, Western Soc. Pediatric Research, Fedn. Am. Socs. Exptl. Biology, Am. Inst. Nutrition, Soc. Exptl. Biology and Medicine, A.A.A.S., Am. Soc. Microbiology, Am. Soc. Infectious Disease, Am. Soc. Clin. Nutrition, Central, Western socs. clin. research, Nat. Assn. Standard Med. Vocabulary (cons.), A.M.A., Western Assn. Physicians, Phi Beta Kappa, Sigma Xi. Editorial bd. Cal. Medicine, 1963-68, Jour. Pediatric Surgery, 1966—; cons. editor Jour. Pediatric Ophthalmology, 1963-67. Studies, publs. on serum proteins, antibiotics, clin. use and pharmacology, nephrotic syndrome cystic fibrosis of pancreas, and infectious diseases, conditions of fetus, premature and neonate. Home: 5005 Finley Av., Los Angeles 90027. Office: Cedars-Sinai Med. Center, Los Angeles 90029.*

KAGAN, Jerome, Am. psychologist; b. Newark, Feb. 25, 1929; s. Joseph and Myrtle (Liebermann) K.; B.S., Rutgers U., 1950; Ph.D., Yale, 1954; M.A. (hon.), Harvard, 1964; m. Cele N. Katzman, June 20, 1951; 1 dau., Janet I. Faculty, Ohio State U., 1954-55; chmn. dept. psychology Fels Research Inst., Yellow Springs, O., 1957-64; prof. dept. social relations Harvard, Cambridge, Mass., 1964—. Mem. pediatrics staff Mass. Gen. Hosp., 1965—; cons. NIH, 1966—. Mem. Am. Psychol. Assn. (pres. div. developmental psychology 1967—, Hofheimer prize 1963), A.A.A.S., Soc. Research Child Devel. Author: (with H. Moss) Birth to Maturity, 1962; (with P. H. Mussen, J. J. Conger) Child Development and Personality, 1963; (with G. S. Lesser) Contemporary Issues in Apperception Methods, 1961. Research on long term stability of human psychol. characteristics, thinking in young children, mental devel. in infant, inventor test to assess aspect of mental functioning in child. Home: 135 Goden St., Belmont, Mass. 02178.*

KAGEYAMA, Daitetsu (Daisuke Ikkan-do), Japanese physician; b. Inaba, Japan, circa 1800. Practiced medicine at Hoki Province, later at Osaka; learned Dutch vaccination method from Kanebumi Imai; author books on uses of vaccine. Helped vaccinate Osaka populace during smallpox epidemic; vaccinated about 40,000 people, circa 1850-68. Died c. 1870.

KAHANE, Jean Pierre, French mathematician; b. Paris, Dec. 11, 1926; s. Ernest and Marcelle (Wurtz) K.; Agrégé, École Normale Supérieure, 1949, Docteur, 1953; m. Agnes Kaczander, July 11, 1951; children—Geneviève, François, Catherine. From lectr. to prof. Faculté des Sciences, Montpellier, France 1954-61; prof. Faculté des Sciences, U. Paris, 1961-65; prof. Faculté des Sciences, Orsay, France, 1965—. Author: (with R. Salem) Ensembles parfaits et series trigonométriques, 1963; also articles. Research on approximation of functions by differential linear combinations; algebraic study of Fourier transformers, role of perfect groups in harmonic analysis, uncertain Fourier series. Home: 11 rue du Val-de-Grace, Paris, France. Office: Faculté des Sciences, 91-Orsay, France.*

KAHLBAUM, Georg Wilhelm August, phys. chemist; b. Berlin, Germany, Apr. 8, 1853; ed. Heidelberg, Germany, Berlin, Strasbourg, France; prof. chemistry and physics, Basel, Switzerland. Author: Monographien zur Geschichte der Chemie, 8 vols., 1897-1904; also biographies of Wöhler, Faraday, Berzelius, Liebig. Invented mercury pump; distilled metals in a vacuum; developed boiling point methods and conditions for securing accurate results in phys. chemistry. Died Basel, Switzerland, Aug. 28, 1905.

KAHLBAUM, Karl Ludwig, German psychiatrist; b. Driesen, Dec. 12, 1828; author Gruppierung der psychischen Krankheiten, 1863; Die Katatonie, 1874; most important contbn. was his description of catatonia and suggestion to correct approach toward understanding it; also described hebephrenia, and attempted to simplify classification of mental diseases. Died Gorlitz, Apr. 15, 1899.

KAHLENBERG, Louis Albert, Am. chemist; b. Two Rivers, Wis., Jan. 27, 1870; s. Albert and Bertha (Albrecht) K.; B.S., U. Wis., 1892, M.Sc., 1893; Ph.D., summa cum laude, U. Leipzig, 1895; m. Lillian Belle

Heald (B.L., U. of Wis., 1893), July 21, 1896; children—Hester (Mrs. Jas. R. Davidson), Herman Heald, Eilhard (dec.). Instr., 1895-97, asst. prof. phys. chemistry, U. Wis., Madison, 1897-1900, prof., 1900-07, prof. chemistry, from 1907, chmn. chem. dept. and dir. course in chemistry, 1908-19. Fellow A.A.A.S. (v.p. Sect. C, 1907-08); mem. Am. Chem. Soc., Am. Electro-chem. Soc. (pres. 1930-31), Am. Soc. Plant Pathologists, Washington Acad. Scis., Wis. Acad. Scis., Arts and Letters (pres. 1906-09), Phi Beta Kappa. Engaged in study of the celluloses, the keratins, separation of crystalloids by dialysis, potentiometric titrations, gas electrodes, and nature of metallic state, activation of gases by metals; exptl. studies of elemental carbon and phosphorus; with Dr. Edward H. Ochsner, boric acid treatment of blood poisoning, also use of colloidal gold in cases of malignancy, and the introduction of the use of dichloracetic acid in med. practice. Author: Laboratory Exercises in General Chemistry, 1907, 09; Outlines of Chemistry, 1909, 15; Qualitative Chemical Analysis (with J. H. Walton), 1911; Chemistry and Its Relations to Daily Life (with E. B. Hart), 1914. Asso. editor Jour. Phys. Chemistry, Jour. de Chimie Physique. Inventor of Equisetene, skin suture material. Died Mar. 18, 1941.

KAHLSON, Georg Sigfrid, physiologist; b. Högfors, Finland, Sept. 23, 1901; s. Gustaf Parker and Edith (Lindqvist) K.; B.M., Helsingfors U., 1923; M.D., U. Munich, 1928; hon. doctorial degree U. Oslo, 1946; LL.D., St. Andrews (Scotland) U., 1963; m. Louise Mayer, July 24, 1930. Prof. physiology, dir. Inst. Physiology, U. Lund (Sweden), 1938—; staff Nat. Inst. for Med. Research, London, 1938. Mem. Swedish Med. Research Council, 1946-52; prin. adviser aviation physiology Swedish Air Force, 1963—. Rockefeller fellow, 1937-38. Recipient Royal Danish medal, 1946, King's medal, 1948. Mem. Royal Swedish, Royal Danish acad. scis., Royal Swedish Acad. Mil. Sci., Am. Physiol. Soc. (past chmn. European editorial com. for Physiol. Revs.). Author: Minerva och Bulldoggen, 1944; Ett urval anföranden om medicinsk forskning och utbildning, 1955; also numerous articles. Research on physiology of histamine. Home: 9 Rabygaban, Lund, Sweden.*

KAHN, Bernd, chemist; b. Pforzheim, Germany, Aug. 16, 1928; s. Eric and Alice (Mayer) K.; came to U. S., 1938, naturalized, 1943; B.S., Newark Coll. Engring., 1950; M.S., Vanderbilt U., 1952; Ph.D., Mass. Inst. Tech., 1961; m. Gail Pressman, Aug. 6, 1961; children—Jennifer, Elizabeth. Radiochemist, Oak Ridge Nat. Lab., 1951-54; with USPHS, 1954—, chief nuclear engring. lab., Nat. Center for Radiol. Health, Cin., 1967—. Mem. Am. Chem. Soc., Am. Phys. Soc., Sigma Xi. Research, publs. on transfer of radionuclides through environment, chem. behavior of radionuclides, measurement of radionuclides. Home: 2621 Briarcliffe Av., Cin. 45212. Office: 4676 Columbia Pkwy., Cin. 45226.*

KAHN, Reuben Leon, immunologist, serologist; b. Kovno, Lithuania, July 26, 1887; s. Lazarus and Lottie (Wolpert) K.; came to U. S., 1899, naturalized, 1905; B.S., Valparaiso (Ind.) U., 1909, LL.D., 1943; M.S., Yale, 1913; D.Sc., N.Y. U., 1916; Sc.D., Inst. Divi Thomae, Cin., 1954; M.D., Nat. U. Greece, 1963; Ph.D., Far East U., Manila, 1964; m. Dina Hope Weinstein, May 31, 1917; children—Lyra Justine (Mrs. Frank S. Morgan), David Curry. Faculty, U. Mich., Ann Arbor, 1928—, dir. lab. U. Hosp., 1928-48, prof. serology, 1951-57, prof. emeritus, research cons. dermatology, 1957—. Recipient Gold medal Phi Lambda Kappa, 1937; Bronze medallion 25th Anniversary Kahn Reaction, 1948. Mem. A.A.A.S. (11th Ann. award 1933), Am. Assn. Immunologists, Am. Soc. Microbiology, Soc. Exptl. Biology and Medicine, Radiation Research Soc., Med. Soc. Study Venereal Disease (London). Author: Serum Diagnosis of Syphilis by Precipitation, 1925; The Kahn Test, 1928; Tissue Immunity, 1936; Serology and Syphilis Control, 1942; Serology with Lipid Antigen, 1950; An Introduction to Universal Serologic Reaction, 1950. Contbr. numerous articles to profl. jours. Developed Kahn test for syphilis; tissue immunity; universal serologic reaction. Home: 2205-2 Hubbard St., Ann Arbor, Mich. 48105.

KAHN, Robert Louis, Am. social psychologist; b. Detroit, Mar. 28, 1918; s. George Arthur and Mabel (Sufinsky) K.; B.A., U. Mich., 1939, M.A., 1940, Ph.D., 1952; m. Beatrice Goldstein, Aug. 25, 1940; children—Judith, Marcia, Janet. Tchr. secondary schs., Detroit, 1940-42; analyst to acting chief, field div. U. S. Bur. Census, Washington, 1942-48; faculty U. Mich., Ann Arbor, 1948—, program dir. Survey Research Center, 1952—, prof. psychology, 1960—. Vis. prof. Mass. Inst. Tech., 1965-66. Fellow Am. Psychol. Assn., Am. Statis. Assn.; mem. Am. Assn. U. Profs., Sigma Xi, Phi Beta Kappa. Author: (with C. Cannell) The Dynamics of Interviewing, 1957; (with A. S. Tannenbaum) Participation in Union Locals, 1958; (with D. M. Wolfe and others) Organizational Stress, 1964; (with Daniel Katz) The Social Psychology of Organizations, 1966. Editor (with E. Boulding): Power and Conflict, 1964. Contbns. in devel. of organizational theory, especially factors determining the effectiveness of large scale orgns.; effects of social environment on phys. and mental health; research methods for social sci., especially techniques of data collection. Home: 2211 Avalon Pl., Ann Arbor, Mich. 48104.*

KAHN, Robert Phillip, Am. plant pathologist; b. Chgo., Apr. 20, 1924; s. Charles S. and Edith (Kramer) K.; student Northwestern U., 1941-42; B.A. with honors in Botany, U. Ill., 1948, Ph.D. in Plant Pathology, 1951; m. Judith Aronson, Dec. 21, 1947; children—Charles D., James R., Andrew B., Jeffrey A. Plant pathologist Chem. Crops, Ft. Detrick, Frederick, Md., 1952-55, supervisory plant pathologist, 1955-57; plant pathologist, plant quarantine div. Agrl. Research Service, U. S. Dept. Agr., U. S. Plant Introduction Sta., Glenn Dale, Md., 1957—. Mem. Am. Phytopath. Soc., Am. Inst. Biol. Scis., A.A.A.S., Washington Bot. Soc. Research, publs. on infection process of fungi and plant viruses; devel. methods for detecting obscure pathogens especially viruses in vegetatively propagated temperate and tropical crops, serological methods for plant viruses, purification and characterization of plant viruses, plant quarantine pathology. Home: 14104 Flint Rock Terrace, Rockville, Md. 20853. Office: U. S. Plant Introduction Station, Glenn Dale, Md. 20769.*

KAHN, Samuel George, Am. nutritionist; b. Belleville, N.J., May 20, 1929; s. Maxwell M. and Edith (Feldstein) K.; B.S., Ill. Wesleyan U., 1951; M.S., U. Ill., 1953, M.S., 1954, Ph.D., 1955; m. Irma M. Berger, July 17, 1960; children—Kenneth B., Edith Suzanne. Research asst. U. Ill., Urbana, 1952-55; sr. research scientist Squibb Inst. Med. Research, New Brunswick, N.J., 1956-58, asst. research supr., 1958-67, research mgr., 1967—; hon. prof. Rutgers U., 1967—. Mem. Am. Physiol. Soc., Soc. Exptl. Biology and Medicine, Poultry Sci. Assn., Am. Inst. Nutrition, Am. Chem. Soc., A.A.A.S., N.Y. Acad. Scis., Sigma Xi. Research, publs. in atherosclerosis, vitamin metabolism, obesity, use antibiotics in animal feeds for promotion of improved body growth. Home: 57 Beacon Hill Dr., Metuchen, N.J. 08840. Office: Squibb Inst. Med. Research, New Brunswick, N.J. 08903.*

KAHRSTEDT, Albrecht, German astronomer; b. Neisse, Aug. 24, 1897; Ph.D., U. Berlin. Asst., observer, chief observer Research Inst. Astronomy; asst. dir. Acad. Sci. Berlin, 1945—; dir. Babelsberg Obs., 1955—, ret., 1961. Mem. Astronomy Soc., Internat. Astron. Soc. Collaborator: Kleine Planeten; Berliner Astronom. internat. Research on astron. computing, astrometry. Address: Potsdamer Strasse 43, Berlin 45, Germany.

KAILA, Armi Kaarina, Finnish agrl. chemist; b. Helsinki, Finland, Sept. 25, 1920; d. Kustavi and Marja (Makkonen) K.; Candidate, U. Helsinki, 1944, Dr.Agr., 1948. Asst., Agrl. Research Centre, Tikkurila, 1944-51; faculty U. Helsinki, 1948—, acting prof. agrl. chemistry, 1962—. Mem. Sci. Agrl. Soc. Finland, Societas Biochemica, Biophysica et Microbiologica Fenniae, NJF. Research, publs. on soil phosphorus (fractions, fixation, moblzn.), potassium and nitrogen, humification of organic matter, peat soils. Home: Runeberginkatu 29 B 31, Helsinki 10, Finland.*

KAINDL, Fritz, Austrian physician; b. Stockerau, Austria, Apr. 19, 1922; s. Adolf and Josefine (Muhic) K.; student U. Vienna; m. Elisabeth Charlotte Unterberger, July 19, 1947; children—Elisabeth Friederiek, Hans Georg. Staff, Pharmacological Inst., Vienna, 1948, Hosp. for Lung Diseases, 1949, Neurol. Clinic, Vienna, 1949, Carolinska Inst., Stockholm, Sweden, 1950; asst. 2d med. clinic U. Clinic, Vienna, 1946-53; faculty U. Vienna, 1957—, prof. medicine, 1964—. Author: (with K. Polzer, F. Schuhfried) Rheography, 1959; (with E. Mannheimer, L. Fleger-Swarz, B. Thurnher) Lymphangiographi, 1960; also numerous articles, chpts. in books. Research on cardioangiologic questions; organizer first cardiac alarm instn. in Vienna. Home: 19 Rathaustreet, Vienna T., Austria.*

KAISER, Elmer Robert, Am. mech. engr.; b. New Holstein, Wis., Aug. 29, 1909; s. Henry and Aramenta (Hanssen) K.; B.S., M.S. in Mech. Engring., U. Wis., 1934; m. Ruby H. Anglin, June 6, 1936. Fuel engr. Battelle Meml. Inst., Columbus, O., 1935-44; asso. dir. research Bituminous Coal Research, Inc., Columbus, Pitts., 1944-54; dir. research Am. Soc. Heating and Air Conditioning Engrs., Cleve., 1954-57; project dir., sr. research scientist N.Y. U. Sch. Engring., N.Y.C., 1957—; cons. engr. fuels, combustion, air pollution, incineration, Bronx, N.Y., 1957—. Mem. Am. Soc. M.E., Air Pollution Control Assn., Am. Pub. Works Assn., A.A.A.S. Contbr. numerous articles to profl. jours. Developed furnaces for smokeless burning automobile bodies, copper wire, other refuse; analyses refuse components; research on combustion process in incinerators. Home: 7 Kaateskill Pl., Scarsdale, N.Y. 10584. Office: W. 177th St. and Cedar Av., Bronx, N.Y. 10453.*

KAISER, Hans, German chemist, pharmacist, dietician; b. Villingen, Sept. 24, 1890; s. Karl and Emma (Mayer) K.; D.Eng., Carlsruhe Tech. Sch.; m. Emmy Schnepf, Feb. 16, 1921; children—Rosemarie Stubler, Christa Schubnell. Pharm. and chem. studies; chem. engr., dietetics, pharmacy, chem. toxicology work. Mem. bd. Inst. Pharm. Research, Munich. Recipient Sertürner medal, Carl-Wilhelm-Scheele plaque, Hans-Krauth plaque. Mem. Arbeitsgeneinschaft Deutscher Krankenhausapotheker (hon. pres.), Internat. Assn. for History Pharmaceutics (hon.), Assn. German Pharmacists, Wurtenburg Assn. German Pharmacists, German Assn. for Research Medicinal Plants. Author 7 books and over 200 articles. Address: Am Bopserweg 10, Stuttgart-S., Germany.

KAISER, Wolfram Karl, German physician; b. Rheinberg, Germany, Aug. 20, 1923; s. Karl Maria and Hanna (Kühler) K.; Dr.med., U. Halle, 1951; m. Sigrid Postal, Mar. 28, 1964. Docent, U. Halle, 1961-67, prof. internal medicine, 1967—. Recipient Purkine medal, 1963. Mem. German Soc. Internal Medicine, German Soc. Clin. Medicine, German Endocrine Soc., U. Halle Med.-Sci. Soc. (sec.). Author: Aktuelle Probleme der Innerliche Medizin I, 1964, II, 1965; also numerous articles. Research in endocrinology; calcium metabolism, bone metabolism; Kaiser-Ponsold edetic acid test; studied parathyroid function. Address: 3 Heinrich-Heine-Strasse, Halle, East Germany.*

KAISHEV, Rostislav Atanasov, phys. chemist; b. St. Petersburg, Russia, Feb. 16, 1908; s. Atanas Petkov and Anastasia (Hadjimarinova) K. grad. U. Sofia (Bulgaria), 1930; m. Milka Chakalova, July 2, 1939; 1 dau., Anastasia. Dir. engring. Technische Hochschule, Breslau, 1932; asst. phys. chemistry dept. U. Sofia, from 1933, pvt. asst. prof., 1941-44, asst. prof., 1944-47, prof. head dept., from 1947; mem. Inst. Physics and Tech., Kharkov, USSR, 1935, Inst. Phys. Chemistry, Munich, Germany, 1937-38; mem. Bulgarian Acad. Sci., 1947 (dep. dir. Phys. Inst., 1957, v.p. 1961). Mem. (corr.) German Acad. Sci., Berlin, Czechoslovakian Acad. Sci., Union Sci. Workers. Recipient sci. awards, 1947, Dimitrov prize Laureate, 1950. Research, publs. on contemporary molecular-kinetic theory crystal growth (with Prof. Stransky), influence fgn. substances and substrata adsorption nuclei upon nucleation new phases of crystallization, and mechanism electro-crystallization. Home: 98 Vitosha, Sofia, Bulgaria.

KAISIN, Felix-Joseph, Belgian geologist; b. Heverlée, July 8, 1907; s. Felix and Madeleine (Parfait) K.; License ès Sciences geologiques, U. Louvain; m. Marie-Claire Fouarge, Sept. 8, 1941; children—Françoise, Anne-Michèle, Chantal, Thérèse, Jean-Marie. Asst., Coal Mus., 1932, instr., dir., 1936; prof. U. Louvain, 1939. Author: Géologie du bassin de Charleroi; Géologie du génie civil, 1948. Home: 325 Chaussée de Malines, Sterrebeek. Office: 6, rue Saint-Michel, Louvain, Belgium.

KAJAK, Zdzistaw, Polish hydrobiologist; b. Warsaw, Poland, July 14, 1929; s. Jan and Janina Kajak; magister U. Warsaw, 1951, doctorate, 1960; m. Anna Ostrihanska, Mar. 21, 1955; children—Piotr, Danuta. With inst. of ecology Polish Acad. Scis., 1953—, head lab. of hydrobiology, 1954-63, head dept. hydrobiology, 1964—. Convener PF sect. Polish com. IBP, 1964—. Mem. Polish Hydrobiol. Soc., Polish Soc. Naturalists. Research and publs. on numbers of organisms and their prodn. in fresh water environments. Home: Szwolezerow 3m16, Warszawa 36. Office: Dept. of Hydrobiology, Inst. of Ecology, Nowy Swiat 72, Warszawa, Poland.*

KAJI, Aritsune, Japanese chemist; b. Osaka City, Japan, Feb. 8, 1925; s. Satoru and Matsuyo (Okazaki) K.; B.Sc., Kyoto U., 1947, D.Sc., 1961; m. Chiyo Havashi, Oct. 3, 1951; children—Minoru, Yuko. With Nipoon Soda Co., Ltd., 1947-65, chief Inst. Agrl. and Medicinal Chems., 1961-65; prof. synthetic organic chemistry Kyoto (Japan) U., 1965—. Mem. Chem. Soc. Japan, Pharm. Soc. Japan, Agrl. Chem. Soc. Japan, Physico-chem. Soc. Japan (council), Soc. Organic Synthetic Chemistry Japan. Patentee in field. Research, publs. on new reaction mechanisms of thermal decomposition of aryl thiocarbamoylthiocarbonates and electrophilic thiocyanation of aromatic compounds, inventor novel compounds. Home: 1, Senyuji-Sannaicho, Higashiyama-ku, Kyoto, Japan.*

KAJIURA, Mutsuo, Japanese ophthalmologist; b. Ehime-Ken, Japan, Dec. 6, 1913; s. Kamajiro and Nobu (Ikeda) K.; grad. Okayama U., 1938, Dr.Med., 1942; m. Kazuko Matsumoto, Oct. 1, 1942; children—Takuichi, Ikuko, Yasuko. Staff, Okayama U. Sch. Medicine, 1938-48, asst. prof., 1943-48; asst. prof. Tokyo Medico-Dental Coll., 1948-51; prof. dept. ophthalmology Fukushima (Japan) Med. Coll., 1951—. Mem. Soc. Ophthalmologists Fukushima-Ken (pres. 1951—), Internat. Fedn. Ophthalmol. Socs., Japanese Ophthalmol. Soc., Japanese Physiol. Soc. Author books including: (with others), Modern Ophthalmology, 1961; (with others) Manuel of Ophthalmological Technics, 1961; also numerous articles. Research on ophthalmic optics, especially contact lens, slit lamp microscopy, astigmatism and accomodation of human eye, eye strain. Home: 2-10, Roppon-matsu, Nodamachi, Fukushima-shi, Japan.*

KAKU, Michitaka, Japanese physician; b. Takawa City, Fukuoka, Japan, Mar. 25, 1904; s. Hirokichi and Hajo (Funaji) K.; M.D., Tokyo (Japan) Imperial U., 1928, Ph.D., 1935; m. Kimiko Kikuya, Nov. 8, 1933; children—Ryutaro, Kotaro. Research asso. Tokyo U., 1933-38, asst. prof., 1938-44; prof., chmn. dept. obstetrics and gynecology Kumamoto (Japan), 1944—, dir. Kumamoto U. Hosp., 1964-65. Mem. Sci. Council Japan, 1960-66. Mem. Japanese Soc. Obstetrics and Gynecology (past pres., mem. bd.), Japanese Soc. Allergy (pres. 1966—), Japanese Soc. Cancer Therapy (mem. bd. 1963—), Japanese Soc. Clin. Cytology (mem. bd. 1961—), Japanese Soc. Cancer (mem.

council 1960—), Japanese Soc. Nephrology, Japanese Soc. Sterility (mem. bd. 1960—), Japanese Soc. Chemotherapy (mem. council 1965—). Author: Textbook of Obstetrics, 1953; Case Studies in Obstetrics and Gynecology, I, II, 1957; also numerous articles. Research on toxemia of pregnancy with advocation of theory of allergic etiology, uterine cancer, sterility, obesity. Home: 3-10-4 Oe-Machi, Kumamoto, Japan.*

KALCKAR, Herman M(oritz), biochemist; b. Copenhagen, Denmark, Mar. 26, 1908; s. Ludvig and Bertha (Melchior) K.; Ph.D. in Med. Research, U. Copenhagen, 1933; M.A. (hon.), Harvard U.; D.Sc., Washington U., St. Louis; m. Barbara Evelyn Wright. Apr. 20, 1951; children—Sonja Evelyn, Nina Marie, Niels Wright. Came to U. S., 1953, naturalized, 1956. Sci. asst. Inst. Med. Physiology, U. Copenhagen, 1934-37, instr. physiology, 1934-37, asst. prof., 1937; Rockefeller research fellow Cal. Inst. Tech., Hopkin Marine Sta. of Stanford and Washington U. Sch. Medicine, 1939; research asso. Pub. Health Research Inst., N.Y.C., 1943-46; asso. prof. physiology U. Copenhagen, 1946-49, research prof., dir. Inst. for Cytophysiology, 1949-54; vis. scientist Nat. Insts. Health, Bethesda, Md., 1953-56, chief sect. metabolic ezymes, 1956-58; prof. biology and biochemistry, dept. biology and McCollum-Pratt Inst., Johns Hopkins, 1958-61; prof. biol. chemistry Harvard Med. Sch., 1961—; chief biochem. research lab. Mass. Gen. Hosp., Boston, 1961—. Recipient Presdl. award Internat. Poliomyelities Congress, 1956, Medaille Internat., Societe de Chimie Biolgique, Paris, 1957. Fellow Am. Acad. Arts and Scis.; mem. Soc. Biol. Chemists, Harvey Soc. (hon.), Kongelige Danske Videnskabernes Selskab (fgn.), Nat. Acad. Scis. Author numerous articles. Research on biochemical genetics in man and microorganisms with particular reference to galactose metabolism. Office: Mass. Gen. Hosp., Boston 02114.

KALESNIK, Stanislav Vikentevich, Russian geographer, glaciologist; b. Jan. 23, 1901; grad. Leningrad U., 1929; D.Geog. Sci., 1937. Prof., Leningrad U., 1938—. Mem. USSR Acad. Sci. (corr.), USSR (pres.), All-Union (learned sec. 1940-52, v.p. 1952—) geog. socs. Author: Mountain and Glacial Areas of the USSR, 1937; General Glaciology, 1939; Principles of General Geography, 1947; Short Course in General Geography, 1957. Research and numerous publs. on gen. glaciology, theory of phys. and gen. geography, characteristics of regressive phase and evolution of glaciers. Address: Leningrad University, Universitetskaya n. 7-9, Leningrad, USSR.

KALISKI, Sylvester, Polish tech. physicist; b. Torun, Poland, Dec. 12, 1925; s. Wincenty and Waleria (Szynalewska) K.; dr.phil., Tech. U. Warsaw (Poland), 1954, dr.sci., 1957; m. Irena Jankowicz, July 23, 1950; 1 son, Wojciech. Staff, Institut Basic Techn. Problem dept. vibration Polish Acad. Scis., Warsaw, 1954—, chief sci. group, 1954—; chair mechanic and tech. physics Techn. U. in Warsaw, 1957—. Recipient Nat. prize I degree in tech. scis., Poland, 1964. Author books, numerous articles. Editor, Proc. Vibration Problems, 1959—. Mem. Polish Acad. Sci. (corr.). Research on math. physics, applied mechanics, acoustics, solid state physics, electron-phonon amplifiers. Home: 17 Einstein. Office: 21 Swietokrzyska, Warsaw, Poland.*

KALISS, Nathan, Am. biologist; b. N.Y.C., Aug. 1, 1907; s. Philip and Anna (Bruslavsky) K.; B.S., City Coll. N.Y., 1929; M.A., Columbia, 1931, Ph.D., 1938; m. Rebecca Bess Weiss, Nov. 5, 1928; children—Anthony M., Jeffrey D., William E. Research asso. Cornell U., 1940-43; med. statistician, chief presentation sect. VA, 1946-47; fellow Am. Assn. Cancer Research, Jackson Lab., Bar Harbor, Me., 1947-50, research asso., 1950-57, asst. dir. research, 1958-62, staff sci., 1957-59, sr. staff sci., 1959—. Guggenheim Found. fellow, 1956-57; vis. fellow Sloan-Kettering Inst., 1948. Fellow A.A.A.S., N.Y. Acad. Sci.; mem. Am. Assn. Cancer Research, Am. Soc. Naturalists, Am. Soc. Zoologists, Soc. Exptl. Biology and Medicine, Soc. for Study Devel. and Growth, Am. Assn. Immunologists. Mng. editor Transplantation Bull., 1953-62, editor, 1963—. Research, numerous publs. on tumor immunity, clarification of basis for immunol. enhancement of tissue grafts (normal and cancerous) showing that anti-graft antibodies may ensure survival of homograft rather than its rejection. Home: Seely Rd. Office: Jackson Lab., Bar Harbor, Me. 04609.*

KALKWARF, Donald Riley, Am. chemist; b. Portland, Ore., Aug. 17, 1924; s. Harry O. and Irene (Joslin) K.; B.A., Reed Coll., 1947; Ph.D., Northwestern U., 1951; m. Carol Louise Rider, Aug. 29, 1949; children—Kristi Ann, Timothy Onnen, Heidi Johanna, Trina Louise. Research chemist Hanford Labs. Gen. Electric Co., Richland, Wash., 1951-53, sr. research scientist, 1953-62, research specialist, 1962-66; mgr. radiation chemistry Pacific Northwest Labs., Battelle Meml. Inst., Richland, 1966—. Mem. Am. Chem. Soc., Chem. Soc. London, Sigma Xi. Contbr. articles to sci. jours. Devel. spectrophotometric methods of chem. analysis, chem. methods for dosimetry of ionizing radiation, evaluation of rate constants for radiation-induced chemical reactions; discovery chems. protective agts. to prevent radiation damage, long-lived free radicals in biochem. systems. Home: 1201 Birch Av. Office: Battelle Meml. Inst., Pacific Northwest Labs., 325 Bldg., 300 Area, Richland, Wash. 99352.*

KALLE, Kurt Albert Edmund Theodor, German chemist, oceanographer; b. Hamburg, May 8, 1898; s. Ludwig and Ella (Overbeck) K.; Dr. es sci. nat., U. Hamburg; m. Hildegard John, Oct. 20, 1933; children—Gerda, Helmut. With Inst. for Study Cancer, Hamburg-Eppendorf, 1925-29, Office Chem. Investigation, Hamburg, 1929-30, German Commn. for Exploration of Sea, 1930-35, Hamburg Marine Obs., 1933-45, Hydrographic Inst., Hamburg, 1945-63; instr. oceanography U. Hamburg, 1945-53; prof. U. Hamburg, 1956. Mem. German Soc. Chemists, German Soc. Geophysics, Soc. German Physicists and Physicians. Author: Der Stoffhaushalt des Meeres, 1943; Allgemeine Meereskunde, 1957. Research on chemistry and physics of ocean. Address: Rothenbaumchaussee 2, Hamburg 13, Germany.

KÄLLÉN, Bengt, Swedish embryologist; b. Kristianstad, Sweden, June 1, 1929; s. Anders Olof Yngve and Karin (Redin) K.; M.D., Lund U., 1958; m. Ingegerd Mörck, June 14, 1951; children—Anders, Ragnar, Rune, Barbro. Docent, U. Lund (Sweden), 1952—, prof. embryology, head dept., 1965—. Hon. research asst. dept. anatomy U. Coll., London, Eng., 1952-53; Rockefeller fellow Washington U. St. Louis, 1953-57. Mem. World Fedn. Neurology (commn. comparative neuroanatomy 1960) Internat. Inst. Embryology. Research, publs. on neuroembryology, cytogenetics, immunology. Home: 26 Galjevangsvagen. Office: 7 Biskopsgatan, Lund, Sweden.*

KALLIO, Kalle Emil, Finnish physician; b. Rautalampi, Finland, May 23, 1899; s. Kalle and Hilma Lydia (Pietilainen) K.; licensed physician U. Helsinki (Finland), 1927, D. Medicinae et Chirurgiae, qualified surgeon, 1932, orthopaedic surgeon, 1943; m. Aili Korckmann, June 22, 1928; 1 son, Kaarlo Erkki. Chief, Indsl. Hosp. Enso, 1934-40; asso. Orthopaedic Hosp. Invalid Found., Helsinki, 1940-50; chief dept. orthopaedics and traumatology U. Helsinki, 1950—; dir. Toolo Accident Hosp., Helsinki, 1960—. Decorated Cross of Liberty, Commdr. Lion of Finland. Mem. Finnish (past pres.), Scandinavian (past pres.), orthopaedic assns., Finnish Surg. Assn. (past pres.); hon. mem. Internat. Acad. Surgeons, Société Francaise d'Orthopedie, Deutsche Orthopadische Gesellschaft, Polish, Philippine orthopaedic assns., l'Academie de Chirurgie, Paris, German Acad. Leopoldina, Sociedad Lationo Americana do Ortopedia. Research, numerous publs. on war amputees from Finnish-Russian War, joint surgery, use of skin as surg. material in reconstructive surgery. Address: 47 A Runebergink, Helsinki 26, Finland.*

KALLOS, Paul, physician; b. Budapest, Hungary, Dec. 14, 1902; s. Armin and Rosa (Schenk) K.; M.D., Royal Elisabeth U., Pècs, Hungary, 1930; m. Liselotte Deffner, Nov. 16, 1932; children—Annalisa, Daniel. Research asso., head lab. Tb. Sanitarium, Baron Fr. v. Korányi, Budapest, 1924-29; research asso., clin. asst. pediatric clinic U. Leipzig (Germany), 1929-30; cons. internist Dermatologic Clinic, Nuremberg, Germany, 1930-32; staff Tomarkin Found., Locarno, Switzerland, 1933, Tb. Research Laboratorium, Davos, Switzerland, 1933; research asso., cons. allergist Acad. Hosp., Uppsala, Sweden, 1934-37; head immunological research lab. Wenner-Gren Inst., U. Stockholm, 1937-44; head Margareta and Eric Whilborg Lab. for Research in Allergy, Helsingborg, Sweden, 1946—; cons. allergist City Hosp., Helsingborg, 1951—. Recipient Von Pirquet Gold medal Am. Coll. Allergists, 1950. Hon. fellow Soc. Française d'Allergie, Soc. Argentina Allergists, Brit. Soc. Immunologists, Am. Coll. Allergists; mem. Internat. Assn. Allergology (founder), Collegium International Allergologicum (founder, sec.), Royal Soc. Medicine (London) (founder mem. sect. clin. immunology). Joint editor-in-chief Internat. Archives Allergy, 1939—; editor Progress in Allergy, 1939—; editor series Monographs in Allergy, 1965. Research and publs. on basic and clin. aspects of allergy, especially asthma. Address: Sundstorget 3, Helsingborg, Sweden.*

KALM, Peter, Finnish botanist, entomologist, ornithologist: b. Angermanie, Finland; student theology, Abo, Finland. Sent to N.Am. by French Acad. Scis. to study and collect plants, 1747-51; became prof. U. Abo, 1752; founder bot. garden, Abo. Mem. French Acad. Scis. Genus of mountain laurel, Kalmia, named in his honor. Author: En Resa Norra Amerika, 3 vols., 1753-1761, repub. as Travels in North America (editor A. B. Benson), 1937; also works on European natural history. Wrote 1st account of insects, plantations and agr. in N.Am. Died 1779.

KALMAN, Sumner M., Am. pharmacologist; b. Boston, Nov. 14, 1918; s. Max Manuel and Bessie (Richmond) K.; A.B., Harvard, 1940; M.D., Stanford, 1950; m. Anneliese Friedsam, Oct. 19, 1952; 1 dau., Susan S. Fellow dept. pharmacology and therapeutics Stanford, 1951-53; USPHS Nat. Cancer Inst. fellow Carlsberg Lab., Copenhagen, Denmark, 1953-54; faculty Stanford Sch. Medicine, 1954—, prof. pharmacology, 1967—. Chmn. pharmacy com. Palo Alto-Stanford Hosp., 1966—. Mem. Am. Soc. for Pharmacology and Exptl. Therapeutics, Soc. Gen. Physiologists, Am. Soc. Biol. Chemists, San Francisco Fedn. for Clin. Research (mem. council). Author: (with Avram Goldstein, Lewis Aronow) Principles of Drug Action; also articles. Research on mechanism of action of estrogens and their influence on growth and function of female reproductive tract, mechanisms for control of enzyme activities and growth in bacterial and mammalian cells. Home: 2299 Tasso St., Palo Alto, Cal. 94301.*

KALMUS, Hans, geneticist, physiologist; b. Prague, Czechoslovakia; Nov. 1, 1906; s. Ernst and Elsa (John) K.; Sc.D. in Zoology, German U., Prague, 1928, M.D., 1935; m. Anna Rosenberg, Feb. 11, 1931; children—Peter, George, Elsa. Faculty, German U., Prague, 1929-38; faculty U. Coll., London, Eng., 1939-49, lectr., 1946-49; asst. prof. McGill U., Montreal, Que., Can., 1949-50; reader biology U. London, Eng., 1950-67, prof., 1967—; vis. lectr. Hebrew U., Jerusalem, Israel, 1954, IFE U., Nigeria, 1966. Expert human genetics AEC, Mexico City, 1961. Recipient Dreyfus prize for genetics U. Sao Paulo (Brazil), 1953. Fellow Inst. Biology; mem. Soc. for Study Animal Behavior, (pres. 1965—), Assn. Brit. Biologists (treas.), Internat. Soc. for Biol. Rhythms (founding). Author: Paramecium, 1929; Genetics, 1948; Simple Experiments with Insects, 1948; Variation and Heredity, 1948; (with I. S. Hubbard) The Chemical Senses, 1960; Genetics of Color Vision, 1965; also numerous articles. Editor: Regulations and Control, 1966. Research on behavior of invertebrates as controlled by their sense organs, genetics of color blindness, taste blindness, tone deafness, Circadian rhythms, math. models of biol. systems. Home: 13 Milton Rd., Harpenden, Herts., Eng. Office: U. Coll., Gower St., London, W.C.1, Eng.*

KALMUS, Herbert Thomas, Am. metallurgist, physicist; b. Chelsea, Mass., Nov. 9, 1881; s. Benjamin G. and Ada Isabella (Gurney) K.; B.S., Mass. Inst. Tech., 1904; Ph.D., U. of Zurich, Switzerland, 1906; E.D., Northeastern U., 1951; m. Natalie Mabelle Dunfee, July 23, 1902 (div. 1921); m. 2d Eleanore King, Sept. 6, 1949. Prin. U. Sch., San Francisco, 1904-05, graduate fellow, Massachusetts Institute Technology, studying in Europe, 1905-06; research associate, 1906-07, instructor, 1907-10, M.I.T.; asst. prof. physics, 1910-12, prof. electrochemistry and metallurgy, 1913-15, Queen's U., Kingston, Ont.; dir. Research Lab. of Electrochemistry and Metallurgy, Canadian Govt., 1913-15; pres. Kalmus, Comstock & Westcott, Inc., 1912-25, Am. Protein Corp., Boston, 1920-25; v.p. The Exolon Co., Buffalo and Thorold, Ont., 1914-19, pres. 1919-23; pres., gen. mgr. Technicolor Corp. (motion pictures in natural color), 1915-59, pres. subsidiaries, 1956-59; pres., gen. mgr. Technicolor, Inc., 1922-59; chmn. Technicolor, Ltd., London, 1935-59, Dir. Stanford Research Inst., from 1953. Hon. fellowship British Kinematograph Soc. London, Eng., 1951; mem. Am. Inst. Chem. Engrs., Am. Physical Soc., Soc. Motion Picture Engrs. (N.Y.), A.A.A.S., Acad. Motion Picture Arts and Scis. Contbr. articles to tech. jours. Developed technicolor (with others); research in elec. conductivity of fused salts, electromotive forces in human body, cobalt, alundum and carborundum. Died July 11, 1963.

KALNITSKY, George, Am. biochemist; b. Bklyn., Oct. 22, 1917; s. Harry and Fannie (Kalnitsky) K.; B.A., Bklyn. Coll., 1939; Ph.D., Ia. State U., 1943; m. Sylvia Esther Duitch, Dec. 25, 1940; children—Katherine, Carol, Robert. Research asso., instr. U. Chgo., 1944-45; faculty U. Ia., Iowa City, 1946—, prof. biochemistry, 1957—. U. Ia. travelling fellow, Oxford U., 1956-57; Guggenheim fellow Weizmann Inst., Israel, 1965-66. Mem. Am. Chem. Soc. (sec.-treas. Ia. 1948-49, chmn. 1950), Am. Soc. Biol. Chemists, Soc. Exptl. Biology and Medicine, A.A.A.S., Sigma Xi, Alpha Chi Sigma, Phi Lambda Upsilon. Research, numerous publs. in fields of isolation and purification of enzymes involved in intermediary metabolism of carbohydrates, lipids and proteins; mechanism of action of inhibitors, isolation, purification and determination of properties of mold proteolitic enzymes, mechanism of action of enzymes. Home: 10 Lakeview Dr., Rural Route 1, Iowa City, Ia. 52240.*

KALOVOULOS, John, Greek agronomist; b. Smyrna, Aug. 29, 1923; s. Michael and Angeliki (Papadaki) K.; Ph.D. in agronomy, U. Athens; license in chemistry; m. Ionna Alexiadou, Jan. 9, 1955; children—Angelos, Loukia. Prof. applied sci. of soil U. Salonika. Mem. Internat. Soc. Soil Sci. Author: The Italian Soil Conditioner Flotal, 1955; The Aging of Hydrogen Clay Suspensions, 1960; Characteristics of Some Alluvial Soils of Northern Greece, 1962; Potassium Uptake by Cation Exchange Resin Amberlite IR-120 from Fithian Illite, 1963; Determination of the Total Cation Exchange Capacity of Triphenylmethylphosphonium Bromide, 1964. Home: Od. Marasli 20. Office: Aristotelian University, Thessaloniki, Greece.

KALOW, Werner, pharmacologist; b. Cottbus, Germany, Feb. 15, 1917; s. Johannes B. and Maria (Heyde) K.; student Greifswald U., 1935-36, 37-38, Graz U., 1936-37, Gottingen U., 1939-40; M.D., Konigsberg U., 1941; m. Brigitte D. von Gaza, Dec. 21, 1946; children—Peter B., Barbara I. Faculty, U. Toronto, Ont., Can., 1951—, prof., chmn. dept. pharmacology, 1966—; dir. biol. research C. H. Boehringer Sohn, Ingelheim, W. Germany, 1965-66. Mem. Pharmacol. Soc. Can. (past pres.), Am. Soc. Pharmacology and Exptl. Therapeutics, Canadian Soc. Forensic Scis., Canadian Physiol. Soc., N.Y. Acad. Scis., Am. Soc. Human Genetics, Genetics Soc. Can., Internat. Soc. Biol. Pharmacology. Research, numerous publs. on heredity and response to drugs. Home: 361 Blythwood Rd., Toronto 12, Ont., Can.

KALTER, Seymour Sanford, Am. virologist; b. N.Y.C., Mar. 19, 1918; s. Aaron Henry and Jessie (Schulman) K.; B.S., St. Joseph's Coll., 1940; M.A., U. Kan., 1943; postgrad. U. Pa.; Ph.D., Syracuse U., 1947; m. Gloria Vivian Verstein, Mar. 3, 1946; children—Susan Patty, Steven Paul, Debra Ilene. Asst. instr. U. Kan., 1940-43; research asst., Harrison scholar U. Pa., 1943-45; faculty State U. N.Y., 1945-56, asso. prof. microbiology, 1953-56; chief virus diagnostic methodology unit Communicable Disease Center, Atlanta, 1956-60; chief virology sect. Sch. Aerospace Medicine, Brooks AFB, Tex., 1960-63; chmn. dept. microbiology S. W. Found. for Research and Edn., San Antonio, 1963-66, dir. div. microbiology and infectious diseases, 1966—. Cons. Virus Lab., Neurol. Inst., U. Cologne (Germany), 1961—, Pan Am. San. Bur., 1959—; lectr. dept. virology and epidemiology Baylor U. Med. Sch., Houston, 1963—. Diplomate Am. Bd. Microbiology. Fellow A.A.A.S.; mem. Am. Acad. Microbiology, N.Y. Acad. Sci., Soc. for Exptl. Biology and Medicine, Soc. Am. Immunologists, Research Soc. Am., Sigma Xi. Author: Procedures for Routine Laboratory Diagnosis of Virus and Rickettsial Diseases, 1963; also articles. Research on microbiology of baboon and its use as an exptl. animal, virus diagnosis and virus tumors. Home: 1418 Haskin St., San Antonio 78209. Office: P.O. Box 2296, San Antonio 78206.*

KALUGIN, Pavel Ivanovich, Russian geologist, hydrologist; b. USSR, Jan. 27, 1904; grad. Don Poly. Inst., Novocherkassk, 1930. Field work Central Asia, 1930-46; dep. dir. Ukraine Geol. Research Inst., Borislav-Lvov, 1946-52; dir. Inst. Geology, Turkmen Acad. Sci., 1952-57, became mem., 1954, acad. sec., 1956. Author: A Brief Outline of the Gaurdak Oil Deposits, 1934; The Diagonal Rifts of Central Kopet-Dag, 1946; The Main Tectonic Features of Kopet-Dag, 1956; The Development of the Carpathian Flysch Basin During the Cretaceous Period, 1956. Research in gas deposits, oil geology, hydrogeology, regional geology in Turkemenia, Northeastern Iran, Western Ukraine. Office: Akademiya rauk Turkmen SSR, Ashkhabad, Turkmen SSR, USSR.

KALVODA, Jaroslav, chemist; b. Bratislava, Czechoslovakia, Mar. 26, 1929; s. Jaroslav and Maria (Koubek) K.; Dipl.Ing.Chem., E.T.H., Zurich, Switzerland, 1952, Dr.Sc.Techn., 1955; m. Maria Ménard, Dec. 14, 1956; 1 dau., Vera. Postdoctoral fellow, sci. coworker E.T.H., 1955-58; research chemist pharm. dept. CIBA Ltd., Basle, Switzerland, 1958—; exchange chemist Ciba Pharm. Co., Summit, N.J., 1966. Mem. Am., Swiss chem. socs. Research, publs. on stereochem. correlation of steroids, sesqui- and diterpenes and morphin alkaloids; partial synthesis of steroidal hormones, intramolecular free radical reactions. Home: 51 Leimenstrasse, Basel. Office: Klybeck Str., Basel, Switzerland.*

KALYUZHNY, Denis Nikolaevich, Russian hygienist; b. Kalyuzhny, Lebedinsky Uezd, Kharkov Guberniya, 1900; grad. sch. for med. orderlies, 1918; grad. Kharkov Med. Inst., 1926, postgrad. dept. gen. hygiene, 1931-33; D.Med. Sci., 1947. Asst. dept. gen. hygiene Kharkov Med. Inst., 1931-33; jr. asso. Ukrainian Central Inst. Communal Hygiene, Kharkov, 1932-39, sr. assoc., 1940-41; instr. hygiene Kharkov Mil. Med. Sch., 1937-41; head dept. for hygiene of atmospheric air, dep. dir. Ukrainian Inst. Communal Hygiene, 1946-55; head chair hygiene Kiev Postgrad. Med. Inst., 1948-56; prof., 1948—; head chair communal hygiene Kiev Med. Inst., 1956-60; dir. Ukrainian Research Inst. Communal Hygiene, Kiev, 1956 —. Hygiene cons. to design, archtl. and engring. orgns. Mem. USSR Acad. Med. Sci. (corr.), Ukrainian Soc. Hygienists (chmn.). Author: Material on the History of Hygiene and Public Health in the Ukrainian SSR, 1959. Research and publs. on gen., communal and mil. hygiene, hygiene of atmosphere, effect of pollution on health, incidence of upper respiratory tract ailments. Address: Ukrainian Research Inst. Communal Hygiene, ulitsa Kirova 7, Kiev, Ukrainian SSR, USSR.

KAMAL AL-DIN, see Ibn Yunus.

KAMARINSKY, Abram Moiseevich, Russian agrobotanist; b. Warsaw, Poland, 1900; student natural sci. dept. Physico-Math. Faculty, Dnepropetrovsk U., 1919-23. With All-Union Research Inst. Humid Subtropics, 1934-38; with All-Union Research Inst. Tea and Subtropical Crops, 1938—, head editorial and publs. dept., 1944—. Author: Citrus Plants, 1935; Cocoa, 1937; The Grapefruit, 1938; The Protection of Citrus Crops against Frost, 1940; The Olive, 1941; The Avocado, 1946; Subtropical Crops, 1960. Research and publs. on flora of Ukrainian forest-steppe zone, subtropical and tropical plants. Address: All-Union Research Inst. Tea and Subtropical Crops, Anasauli, Gruz. SSR, USSR.

KAMBARA, Shu, Japanese chemist; b. Tokyo, Japan, Sept. 23, 1906; grad. Waseda U., 1930; m. Kiyoko Isomura. After grad. became asst. prof. Tokyo Coll. Engring., became prof., 1944. Lectr. on rubber univs. and research insts., U. S., 1952. Research and publs. in natural and synthetic rubber; developed theory on use of sulphur in synthetic rubber.

KAMEDA, Haruo, Japanese physician; b. Utsunomiya, Japan, Aug. 7, 1925; s. Tadashi and Toh (Morito) K.; M.D., U. Tokyo, 1957; m. Kazuko Kikkawa, Nov. 5, 1955; children—Keiko, Yumiko. Instr., Faculty of Medicine, U. Tokyo (Japan), 1964—. Mem. Japanese Soc. Medicine, Japanese Soc. Gastroenterology. Author: Clinical Examinations, 1964; also numerous articles. Research in hepatic circulation and liver diseases, cin. application of infrared absorption spectrum for gallstone analysis, gallstone disease in Japan. Home: 28 Udagawa-cho, Shibuyaku, Tokyo, Japan.*

KAMEFUCHI, Susumu, Japanese theoretical physicist; b. Ishikawa-ken, Japan, Sept. 13, 1927; s. Shizuka and Chiyo (Ishido) K.; M.Sc., Nagoya U., 1950, Ph.D., 1954; m. Seija Lahja Tuulikki Lyytikäinen, July 27, 1962. Asst. physics dept. Nagoya U., 1955-60; research fellow Niels Bohr Inst., U. Copenhagen (Denmark), 1956-58, Imperial Coll., U. London (Eng.), 1958-63; asso. prof. physics dept. Tokyo (Japan) U. Edn., 1963—. Sr. sci. staff mem. Internat. Centre for Theoretical Physics, IAEA, Trieste, Italy, 1964-65. Mem. Japan, Am. phys. socs. Research, publs. on quantum field theory and elementary particle physics; derived with H. Umezawa spectral representation of 2-point vacuum expectation values; obtained renormalizability condition of interactions; made detailed studies on para-field theory. Home: 8-6-403 Hamamidaira, Chigasaki, Kanagawaken, Japan. Office: 3-29-1 Otsuka, Bunkyo-ku, Tokyo, Japan.*

KAMEL, Raafat Wasef, Egyptian physicist; b. Cairo, Egypt; b. June 10, 1926; s. Kamel and Kokab (Anton) K.; B.Sc. with honors, U. Cairo, 1946, M.Sc., 1950, Ph.D., 1954, D.Sc., 1966; m. Kamilia Naguib, July 20, 1952; children—Wafaa R., Osama R. Faculty, U. Cairo, 1946—, asst. prof. physics, 1961—; research fellow Bristol (Eng.) U., 1958-59, Lorand Eotvos U., Budapest, Hungary, 1965-66. Lectr., Mil. Engring. Coll., Cairo, 1964—. Recipient Egyptian State Prize for physics research, 1963; Egyptian medal for Sci. and Arts, 1963. Research, publs. on mapping of annealing spectra of cadium, silver and cadium-silver alloys, transformation in copper and gold-copper alloys anelastically and by stead creep methods, quenching and micro-creep relaxation for values of Em and Ef for vacancies in gold and cobalt; introduced internal friction as tool for study of lattice defeats in solids, concept of dissociated vacancy. Home: 7 Gameat El Kahira St., Giza, Cairo, Egypt.*

KAMEN, Martin David, biochemist; b. Toronto, Ont., Can., Aug. 27, 1913; s. Harry and Goldie (Achber) K.; came to U. S., 1913, naturalized, 1938; B.S. cum laude, U. Chgo., 1933, Ph.D. in Phys. Chemistry, 1936; m. Beka Doherty, Mar. 16, 1949 (dec. Nov. 1963); 1 son, David Martin. Prof. biochemistry U. Cal. at San Diego, 1960—, acting dean grad. studies, 1965-66. Group leader Manhattan Project, 1941-44. Mem. Nat. Acad. Scis., Am. Chem. Soc. (counsellor 1963; award applications nuclear chemistry 1963), Radiation Research Soc., Am. Acad. Arts and Scis., Phi Beta Kappa, Sigma Xi. Mem. editorial bd. Jour. Biol. Chemistry, 1959-64, Ann. Revs. Biochemistry, 1963—. Research and publs. on codiscovery of long-lived carbon; assignment of mass to long-lived sulfur isotope; nitrogen fixation in photo-synthetic bacteria; cytochrome proteins in anaerobes and in photosynthetic apparatus; pioneered in application of isotope tracer methods to biology. Home: 1104 Pearl St., La Jolla, Cal. 92037.*

KAMENOV, Evgeni Gantchev, Bulgarian economist; b. Sofia, Bulgaria, Dec. 29, 1908; s. Gancho Kamenov and Raina (Kostova) Jotkovski; student Sofia U. Law Faculty, 1927-31; m. Bogdana Pavlova Vinarova, June 8, 1939; children—Roumiana Miko (Mrs. Marinova), Meglena Evgenieva. Practice law, Sofia, 1934-44; staff Sofia U., 1940—, prof. Law and Econs. Faculty, 1949—, head polit. economy dept., 1952—; dir. Econs. Inst., Bulgarian Acad. Sci., 1952-54; legal adviser, sec. gen. Ministry of Interior, 1944-47; dep. minister Ministry of Fgn. Affairs, 1948-51; ambassador plenipotentiary, France, 1954-60. Decorated Order of 9th of Sept.—1st Degree; Decoration of Civil Service—III degree; Order of Cyril and Metody; badge of honor City Council—I degree. Mem. Bulgarian Acad. Sci. (sec. sect. philosophy, econs. and law 1961—, mem. presidium), Bulgarian Soc. Sociology, Internat. Law Assn. (pres. Bulgarian br.), Union Lawyers in Bulgaria, Union Sci. Workers. Author numerous books including Common Market: Causes and Consequences, 1964; Certain Problems concerning the Economic Relations between Socialist and Capitalist Countries, 1965; Bulgaria's Policy of Peaceful Coexistence and the Development of its Economic Relations with Capitalist Countries, 1966; also numerous articles. Pioneered research on influence of natural, social and historic agents on Bulgaria's econ. devel; studied early period of econ. devel. of peoples' democracies, influence of Soviet assistance in Bulgaria's socialist devel., econ. interrelations of socialist and capitalist countries, peaceful co-existence, Common Market from Marxist's point of view. Home: 12 Aksakov St., Sofia, Bulgaria.*

KAMERLINGH ONNES, Heike, Dutch physicist; b. Gröningen, Netherlands, Sept. 21, 1853; ed. U. Gröningen, 1870; doctorate, 1879; studied under Bunsen and Kirchhoff, Heidelberg, Germany; became prof. exptl. physics U. Leiden (Netherlands), 1882, founder cryogenic lab., 1894, also dir. Recipient Rumford medal Royal Soc., 1912, Nobel prize in physics, 1913. Mem. French Acad. Scis., 1920. Author: Over bet bereiiten van Temperaturen Belangrisiit Beneden het Kookpunt van Helium, 1910; Experiments at the Cryogenic Laboratory at Leyden, 1911; Liquefied Helium: The Last of thee Permanent Gases, 1908; General Theory of the Fluids, 1881. Research on isothermics of gases at low temperatures, properties of metals, especially nickle, and manganese iron alloys at near zero temperatures, optical-magnetic and magneto-optical properties of matter; cooled liquid helium to within one degree above absolute zero; produced liquid helium for 1st time, 1908; discovered certain metals, including lead and mercury, have no elec. resistance at liquid helium temperatures (he called this phenomenon superconductivity) and proved superconductivity could be eliminated by imposing large magnetic field on metal, 1911; research in superconductivity which led to discovery of helium II, also important in devel. of theory of elec. conduction in solids. Died Feb. 21, 1926.

KAMIN, Henry, biochemist; b. Warsaw, Poland, Oct. 24, 1920; s. Benjamin and Paula (Mirkowicz) K.; came to U. S., 1926, naturalized, 1932; B.S. in Chemistry, Coll. City N.Y., 1940; Ph.D., in Biochemistry, Duke, 1948; m. Dorothy Lee Lingle, Oct. 30, 1943. USPHS Postdoctoral fellow in biochemistry Duke, 1948-50, faculty, 1950—, prof. biochemistry, 1965—; prin. scientist VA Hosp., Durham, N.C., 1953—. VA liaison rep. biochemistry study sect. NIH, 1962-65; chmn. basic sci. program review com. VA., 1962-66. Fellow A.A.A.S.; mem. Am. Soc. Biol. Chemists, Am. Chem. Soc., Sigma Xi. Research, publs. on action oxidation enzymes; demonstrated existence key enzyme in microsomal electron transport pathway, electron flow through flavin enzyme, enzyme-substrate and enzyme-inhibitor complexes. Home: 2417 Perkins Rd., Durham, N.C. 27706.*

KAMINS, Robert Martin, Am. educator; b. Chgo., Mar. 10, 1918; s. Philip Eugene and Lena (Silverman) K.; student U. Cal. at Los Angeles, 1936-37; A.B., U. Chgo., 1940, M.A. (Walgreen fellow), 1948, Ph.D., 1950; m. Shirley Ritter, Oct. 14, 1940; children—Diane Eve (Mrs. Richard Quinn), Melissa. Research asst. dept. econs. U. Chgo., 1939-41, Fedn. Tax Adminstrs., Pub. Adminstrn. Clearing House, Chgo., 1941-42; asso. economist Fed. Bd. Investigation and Research, Washington, 1942; faculty U. Ill., 1946-47; faculty U. Hawaii, Honolulu, 1947—, prof. econs., 1963—, dean acad. devel., 1963—; vis. prof. Inst. for Social Studies; The Hague, 1953. Dir. research Hawaii Constl. Conv., 1950; adminstrv. asst. U. S. Senator Oren Long, 1959-61. Ford Found. grantee, 1963. Mem. Am., Western, Hawaii econ. assns., Am. Assn. U. Profs. (nat. council 1956-58), Nat. Tax Assn., Am. Soc. Pub. Adminstrn., Am. Assn. UN (pres. Hawaii chpt. 1952). Author: The Tax System of Hawaii, 1952; also articles. In state finance, devised system of income tax credits for reducing regressivity of sales taxes; in state legislation devised means of introducing proposals for new laws (short-form bills).*

KAMIYA, Kisaku, Japanese physician; b. Anjo, Aichi, Japan, Nov. 8, 1914; s. Sakajiro and Jitsu (Kato) K.; M.D., Nagoya U., 1941, D.M.S., 1949; m. Sadako Kiyohara, Feb. 26, 1944; children—Yasuko (Mrs. Yoshikazu Suzuki), Takashi, Ken. Asst. prof. surgery Tokushima U. Sch. Medicine, 1951-54; chief surgeon Nagoya First Red Cross Hosp., 1955-56; asst. prof. surgery Nagoya U. Sch. Medicine, 1956—. Mem. Japanese Surg. Soc. (councilor 1957—), Japanese Assn. Thoracic Surgery, Internat., Japanese colls. angiology. Author: Surgical Treatment of Vascular Emergency, 1963. Research, numerous publs. on etiology, treatment of phlebothrombosis, pathology of arteritis on relationship among Buerger's disease, Takayasu's disease, giant cell arteritis; made proposition of positive anticoagulant therapy for prophylaxis of thrombosis after operation on blood vessels and clarified its technique. Home: 1-3 Nekogahora-dori, Chikusa-ku, Nagoya. Office: 65 Tsurumai-cho, Showa-ku, Nagoya, Japan.*

KAMIYA, Noburô, Japanese biologist; b. Tokyo, Japan, July 23, 1913; s. Ichiro and Tami (Ukai) K.; grad. in botany, U. Tokyo, 1936; D.Sc., 1948; postgrad. exptl. cytology U. Giessen (Germany), cell physiology U. Pa., 1939-42; m. Miyeko Mayeda, July 3, 1946; children—Ritsu, Tohru. Lectr., U. Tokyo, 1943-49; prof. Osaka (Japan) U., 1949—; vis. prof. Princeton, 1962-63. Fellow A.A.A.S.; mem. Zool. Bot. Gesellschaft Wien (hon.), Japanese Soc. Plant Physiologists, Bot. Soc. Japan, Biophys. Soc. Japan, Japan Soc. for Cell Biology. Author: Protoplasmic Streaming, 1959; also numerous articles. Editorial or adv. bd. Protoplasma, 1956—, Plant and Cell Physiology, 1959—, Protoplasmatologia, 1966—, Biorheology, 1962—, Cellular Chemistry, 1966—, Japanese Jour. Botany, 1967—. Physiol. and biophys. analysis of protoplasmic streaming; cell surgeries on giant algal cells; measurement and analysis of water permeability of giant algal cells by transcellular osmosis. Home: 81 Sanjocho, Ashiya, Hyogo-Ken, Japan. Office: Dept. Biology, Faculty Sci., Osaka U., Toyonaka, Osaka, Japan.*

KAMMERER, Frederic, Am. surgeon; b. N.Y. City, Feb. 4, 1856; s. Joseph (M.D.) and Léonie (von Weisseneck) K.; M.D., U. of Freiburg, 1880; m. Ida Knapp, Mar. 23, 1907. Returned to U. S. from Europe, 1886, and since practiced in N.Y. City; prof. clin. surgery, Cornell U. Med. Sch., New York, 1898-1909, Coll. Phys. and Surg. (Columbia), 1909—; attending surgeon, Lenox Hill Hosp.; consulting surgeon, St. Francis Hosp. Fellow Am. Coll. Surgeons. Developed "right rectus" incision for appendicitis, 1897; successful ligation of 1st portion of subclavian artery for anuerysm, 1899; among 1st to demonstrate thyroidectomy in U. S.; influenced development of aseptic surgery in U. S. Died 1928.

KÄMMERER, Hermann, chemist; b. Milan, July 25, 1911; s. Adam and Klara (Braun) K.; Dr. ès Sci. Nat., U. Fribourg; m. Marie Antoinette Pfaff, Mar. 26, 1940; children—Hermann, Evemarie, Hans-Georg, Martin. Sci. asst.; prof. agrégé, sci. cons., prof. at large; prof. U. Mainz, 1947—. Mem. German Soc. Chemists. Research and publs. on macromolecular chemistry. Home: Friedrich von Pfeifferweg 5. Office: Institute for Organic Chemistry, University of Mainz, Mainz-Rhein, Germany.

KAMMERER, Paul, Austrian geneticist; b. Vienna, Austria, Aug. 17, 1880. Author: Das Gesetz der Serie, 1919; Allgemeine Biologie, 1920. Studied adaptation, also hereditary characteristics of amphibians; determined genetic structure of toad, 1909. Died Puchberg/Schneeberg, Sept. 23, 1926.

KAMMINGA, Christiaan Egbert, Dutch biochemist; b. Arnhem, Jan. 20, 1914; s. Egbert H. and N. (Woudt) K.; Ph.D. in Chemistry, U. Amsterdam; m. C. Noomen, Oct. 1, 1943; children—Harmke, Alex, Els. Asst., Physiol. Chemistry Labs., U. Amsterdam, 1940-46; collaborator Elsevier Publ. Cy, Amsterdam, 1946-48; research asst. Wilhelmina Hosp., Amsterdam, 1948-52; prof. biochemistry Baghdad Med. Coll., 1952-53; dir. research starch plant, 1954-55; dir. Found. Clin. Chemistry Labs., lab. chief Rotterdam Hosp., 1955—. Mem. various Dutch sci. assns. Contbr. articles to chem. publs. Home: av. Concordia 34b. Office: Glashaven 68, Rotterdam, Netherlands.

KAMNEV, Ivan Emelyanovich, Russian cytologist; Kapoostin, USSR, June 30, 1902; s. Emelyan Titovich and Yevdokeya Ilyinichna (Soina) K.; grad. Leningrad Guertzen Inst., 1929; postgrad. student Leningrad U., 1934; D. Biol. Scis.; m. Raisa Victorovra Pevgova, 1934; 1 son, Victor. With Bekhterev Inst. Brain Study, All-Union Inst. Exptl. Medicine, Leningrad. 1934-49; mem. faculty Rostov U., also Rostov Med. Inst., 1949—, now prof. biology. Author works on cytology and physiology of a cell. Home: 127, Pushkinsky St., Rostov on Don. Office: 38, Nakhichevansky Lane, Rostov on Don, USSR.*

KAMPELMACHER, E. H., Dutch microbiologist; b. Vienna, Austria, May 6, 1920; s. Karl and Rifka (Heissmann) K.; Ph.D., U. Utrecht, 1954; m. Carola Hoflich, Dec. 20, 1953; children—Michael, Raphael. With Dutch Council Applied Scis., 1950-52; gen. practice vet. medicine, Breukelen, Netherlands, 1952-54; microbiologist lab. for zoonoses Nat. Inst. Pub. Health, Utrecht, The Netherlands, 1954—, head lab., 1959—. Cons. WHO, 1960—; pres. bd. Dutch Vet. Jour., 1962—; mem. Dutch Health Council, 1962—. Fulbright research fellow, Atlanta, 1958-59. Mem. World Assn. Vet. Food Hygienists (past sec.-treas., award 1965), Dutch, Brit. microbiol. assns., World Royal Dutch (bd. 1968—) vet. assns. Research, numerous publs. sterility in cattle; salmonellosis, food poisoning problems, listeriosis, trichinosis, anisakiasis, antibiotics, problems in vet. pub. health, especially food hygiene. Home: 6, Korte Boslaan, Bilthoven, The Netherlands. Office: 1, Sterrebos, Utrecht, The Netherlands.*

KÄMPFER, Engelbert, German botanist; b. Lemgo, Lippe, Sept. 16, 1651; studied medicine and langs. at Danzig, Thorn, Cracow; M.D., U. Leyden. Accompanied Swedish amb. to Persia, 1683; surgeon Dutch East India Co.; visited Bengal, Java, Sumatra, Japan, spent 2 years in Japan, 1692-94, returned to Europe, 1693; also visited India and Siam. Author: Amoenitates Exoticae, 1712 (results of research in Persia); History of Japan and Siam, 2 vols., 1727; Icones selectae plantarum, 1791. Died Lemgo, Nov. 2, 1716.

KAMPP, Aage, Danish geographer; b. Mön. s. Mads, Sept. 27, 1906; s. Peter and Eline Marie (Dideriksen) K.; Ph.D., U. Copenhagen; m. Ebba Kampp, Apr. 12, 1933; children—Jorgen, Mogens, Bodil. Prof., St. George Coll., 1928; lectr. Tchrs. U., 1950; prof. U. Bonn, 1960-61; agrégé U. Copenhagen, 1961. Mem. Royal, Swedish socs. geography, Internat. Union Geog. Assns. Author: Faeroerne, 1950; Agro-Geographical Investigations of Denmark, 1959; Tal i Billder, 1961; Cartography, 1963, also articles. Home: Ceciliavej 50, Copenhagen Valby. Office: Emdrupvej 101, Copenhagen NV, Denmark.

KAMRAN, Helvadji, Turkish engr.; b. Diyarbékir, Turkey, 1891; ed. High Sch. for Electricity, Turkey, France; m. 1913. Engr. elec. plant, Damas, Turkey; prof. Sch. for Bridges and Rds., Istanbul, Electro-Mech. Industry, Istanbul; engr. Turkish Elec. Soc.; dir. indsl. affairs Municipality of Smyrna; dir. Sch. Arts et Métiers, Ankara. Author: Industrial Electricity, 3 vols.; Internal Combustion Engines; Experimental Physics, 2 vols., also articles. Designed device for automatically calculating trajectory of naval shells, anti-submarine mines, balustre compass for mgf. geometric spirals.

KÄMTZ, Ludwig Friedrich, German physicist, meteorologist; b. Treptow on the Rega, Jan. 11, 1801. Prof. in Halle, Dorpat, St. Petersburg. Author: Lehrbuch der Meteorologie, 3 vols., 1831-36; Repertorium für Meteorologie, 3 vols., 1859-64. Died St. Petersburg, Dec. 20, 1867.

KAMYNIN, Leonid Ivanovich, Russian mathematician; b. Krasnodar, USSR, May 31, 1923; s. Ivan Ivanovich and Mathilda (Tomasyk) grad. in Mech. Math., Lomondsov's State U., Moscow, USSR, 1949, postgrad., 1952; m. Vera Nicholaevna Mashennikov, Aug. 12, 1952; 1 son, Vitaly. Faculty, dept. of math. analysis and mech. math. Lomonosov U., 1952—, prof., 1955—. Mem. Moscow Math. Soc. Research, publs. on direct methods with pheline solution for parabolical equalization in class of fast developing functions; discovered comprehensive conditions using formed tepid potentiality theory; obtained accurate conditions for solution of number of regional problems for parabolical equalization with explosive coefficients arising in B-10 physics, contractive heat conductibility and hydrotech. Home: House 23/56, Bldg. 1, Garibaldi Dist., Moscow, USSR.*

KANAMORI, Junjiro, Japanese physicist; b. Osaka, Japan, Mar. 7, 1930; s. Kenji and Fumie (Sekine) K.; B.S., Osaka U., 1953, D.Sc., 1957; m. Sachiko Kuno, Nov. 1, 1961; 1 son, Yoshio. Faculty, Osaka U., 1957—, prof. dept. physics, 1965—; research asso. inst. study of metals U. Chgo., 1958-60, exchange prof., asso. prof. Faculte des Sciences, U. Paris (France), 1964. Recipient (with others) Asahi cultural prize. Mem. Phys. Soc. Japan, Am. Phys. Soc. Research, publs. in theoretical solid state physics, especially theory of magnetism in solids. Home: 485-17 Mitsugarasu, Nara, Nara, Japan. Office: Dept. Physics, Osaka U., Osaka, Toyonaka, Japan.*

KANASH, Sergey Stepanovich, Russian phytologist, selectionist; b. 1896; grad. Central Asian U., Tashkent, 1923. Asso., Central Selection Sta., All-Union Research Inst. Cotton-Growing, Tashkent, 1923—; dir. Central Selection Sta., All-Union Research Inst., Tashkent, 1941-47. Decorated Order of Lenin; recipient Stalin prize, 1941. Mem. Uzbekistan Acad. Sci., Lenin All-Union Acad. Agrl. Sci. Author: Methods and Techniques of Selection Work, 1933; Cotton, 1960; co-author: The Results of 25 Years of Work of the Central Selection Station at the All-Union Research Institute of Cotton, 1948. Research and publs. on genetics, selection and seed growing of cotton; developer varieties of wilt-resistant cotton, 1941-47. Address: Uzbekistan Acad. Sci., ulitsa Kuybysheva 15, Tashkent, Uzbekistan SSR, USSR.

KANAVEL, Allen Buckner, Am. surgeon; b. Sedgwick, Kan., Sept. 2, 1874; s. George W. and Mary A. (Paugh) K.; Ph.B., Northwestern U., 1896, M.D., cum laude, 1899, hon. D.S., 1924; postgrad., Vienna, Austria, 1899; m. Olive Rosenkranz, Oct. 1907; children—Patricia, David, Richard. Instr. clin. surgery Northwestern U. Med. Sch., 1901, asst. prof. surgery, 1908-17, asso. prof., 1917-19, prof. surgery, from 1919; attending surgeon Wesley and Passavant hosps. (both Chgo.). Fellow Am. Surg. Assn., A.C.S. (regent, pres., 1931-32), Western Surg. Assn., Soc. Clinical Surgery. Author: Infections of the Hand, 1912; numerous other publs. on subjects including surgery of intestines, transplantation of free flaps of fat, injuries of spinal cord, brain tumors, inflammations and abscesses, surg. anatomy of trigeminal nerve, spastic paralysis. Editor of Surgery, Gynecology and Obstetrics. Died Cal., May 27, 1938.

KANE, Elisha Kent, Am. surgeon, explorer; b. Phila., Feb. 3, 1820; s. John Kintzing and Jane (Leiper) K.; attended U. Va., 1838-39; grad. Med. Dept., U. Pa., 1842. Asst. surgeon U. S. Navy in ship Brandywine, 1843; served as surgeon in China, 1844, African coast, 1846, Mexico, 1848, in Mediterranean, 1849; attached to U. S. Coast Survey, served as surgeon in Advance; sailed for Arctic in Grinnell Expdn. sent by U. S. Govt. in search of English expdn. under Sir. Franklin (which had been lost since 1845), 1850-51; sailed for No. seas in command 2d Grinnell Expdn., 1853-55; abandoned ship, hiked to Upernauik, May 1855, arrived, Aug. 1855, returned home, Oct. 1855; brig. entered unknown waters now called Kane Basin. Recipient Gold medal Royal Geog. Soc. London, 1856, Paris Geog. Soc., 1858. Author: The U. S. Grinnell Expedition in Search of Sir John Franklin, 1853; Arctic Explorations: The Second Grinnell Expedition in Search of Sir John Franklin in the Years 1853, 1854, 1855, pub. 1856. Died Havana, Cuba, Feb. 16, 1857.

KANE, Rajaram Purushottam, Indian physicist; b. Damoh, India, Nov. 12, 1926; s. Purushottan Lakshman and Kamala (Bedekar) K.; I.Sc., 1942, B.Sc., Agra U., 1944; M.Sc., Banaras U., 1946; Ph.D., Bombay (India) U., 1953; m. Sindhu S. Karve, May 5, 1958; children—Vijay, Neelima. Fulbright Smith-Mundt scholar U. Chgo., 1953-54; became reader Phys. Research Lab., Ahmedabad, India, 1956-62, now asso. prof. Research and publs. on long-term and short-term variations of cosmic ray intensity and their relationship with solar and geophys. phenomena. Home: 2/23 Satbhai Ki Goth, Lakshmiganj, Lashkar, Gwalior, India. Office: Phys. Research Lab., Ahmedabad, India.*

KANE, Sir Robert John, Irish chemist; b. Dublin, Ireland, Sept. 24, 1809; s. John Kane; student medicine, Dublin, Trinity Coll.; LL.D., U. Dublin; m. Katherine Baily. Prof. natural philosophy, Dublin, 1834-45; named dir. Mus. Irish Industry, 1845; became vice chancellor Royal U. Ireland, 1880. Fellow King and Queens Coll. Physicians; mem. Royal Irish Acad. (Gold medal 1843). Recipient Graves award, 1830. Fellow Royal Soc., 1849 (medal 1841). Author: Elements of Chemistry, 1841. Advocated radical theory; before Liebig, suggested alcohol, ether and muriatic ether were compounds of a specific radical; studied structure of ammonia compounds; wrote chem. history of archil and litmus. Died Dublin, Feb. 16, 1890.

KANEDA, Takashi, Japanese chemist; b. Tokyo, Japan, Jan. 1, 1922; s. Sohji and Kuniyo (Fujimoto) K.; B.S., Tokyo U., 1943, M.S., 1945, Ph.D., 1953; m. Masako Kobayashi, Oct. 17, 1950; 1 son, Kazuhiko. Asst. dept. fisheries chemistry Faculty Agr., Tokyo U., 1945-47, lectr., 1955—, mem. staff Tokai Regional Fisheries Research Lab., Tokyo, 1947-65, chief oil and fat div., 1955-65; prof. food chemistry dept. Faculty Agr., Tohoku U., Sendai, Japan, 1965—. Recipient Edn. Ministers prize. Mem. Japanese Soc. Sci. Fish. (prize), Eiyo to Shokuryo, Soc. Biochemistry, Japanese Oil Chem. Soc., Agrl. Chem. Soc. Japan. Author books, numerous articles; research on nutritive value of marine animal oils; toxicity of autoxidized oils, effects of seaweeds on cholesterol metabolism in rats, effects of mushrooms on cholesterol metabolism in rats. Home: 39-1 Sanjo-Dori, Sendai, Japan.*

KANEKO, Ziro, Japanese psychiatrist; b. Nishinomiya, Japan, Feb. 12, 1915; s. Rokuro and Kaeko (Sekitani) K.; M.D., Osaka (Japan) U., 1938, Dr. Med. Sci., 1949; m. Mitsuko Fujimura, May 26, 1946; children—Hiroshi, Masami. Faculty, Nara Med. Coll., 1949-56, prof., 1953-56; prof. psychiatry Osaka U. Med. Sch., 1956—, dir. U. Hosp., 1957—. Mem. Japanese Soc. Psychiatry and Neurology, Japanese Psychosomatic Soc., Japanese Geriatric Soc. Author: Psychiatric Aspects of Senility, 1956; Psychology of Patients, 1958; also numerous articles. Research on neuropsychiatry of senile patients and investigation of ultrasonic blood-rheograph for estimation of cerebral circulation, psychophysiol. studies on neuromuscular diseases and narcolepsy. Home: 57 Tsutsuiiga, Momoyama, Fushimiku, Kyoto, Japan. Office: 3-Chome, Dojimahamadori, Fukushimaku, Osaka, Japan.*

KANGAS, Esko Mauno, Finnish forest entomologist; b. Virrat, Finland, Jan. 7, 1906; s. Aapo and Lempi (Vaali) K.; Dr. es sc.; children—Railikki, Ilpo, Tero, Leimo. Became asst. sylviculture U. Helsinki, 1928; named dist. forester Evo Dist., 1932; became asst. Helsinki Inst. Forestry Research, 1933. entomologist, 1943; agrégé, 1948; prof. forestry, zoology, agr. U. Helsinki, 1954—. Recipient Cajander medal Finnish Forestry Soc. Mem. Finnish Entomology Soc. (pres.), Finnish Acad. Sci. Research and publs. on entomology, including Symphyta, Coleoptera, Hymentoptera; forest entomology including orientation of scolytidea and other insects living in tree bark; sylviculture. Home: Pihlajatie 49B, Helsinki; Finland. Office: Viikki, Pihlajamäki-Helsinki, Finland.

KANGRO, Carl Walther Nicolai, chemist; b. Riga, Aug. 13, 1889; s. Carl and Helen (Neumann) K.; Ph.D., U. Leipzig; m. Aenne Hermanns, Oct. 10, 1941; children—Hans, Claus, Erna, Brigitte. Univ. asst., 1914-23; lab. head Osram-Conzern, 1923-25; editor Gmelin, 1925-27; chief engr. Technische Hochschule, Brunswick, 1927-39; titular prof., 1930; asso. prof., dir. Metall.-Chem. Inst., 1936-40; ret., 1956. Mem. German Bunsen Soc., German Soc. Chemists, German Soc. Chem. Equipment. Author: Adiabatische Kalorimetrie; Eisen (III)-osyd; Konzentierte wässrige Lösungen; Flüssigkheitsaccumulator; Gewinnung von Schwermetallen mit Hilfe von Chlorgas. Research on electrochemistry, phys. chemistry, chem. metallurgy. Address: Wolfenbüttelerstrasse 9, Braunschweig, Germany.

KANITZ, Felix Philipp, Austrian archaeologist; b. Budapest, Aug. 2, 1829; studied history, archaeology, ethnography at univs. in Austria, Germany, France. Author: Die römischen Funde in Serbien, 1861; Serbiens byzantinis che Monumente, 1862; Serbien historisch-ethnographische Reisestudien, 1868; Donau-Bulgarien und das Balkan, 3 vols., 1875-79; 3 Katechismus der Ornamentik, 1877; Reisein Südserbien und Nordbulgarien, 1878; Römische Studien aus Serbien, 1892; Das Königreich Serbian und das Serbenvolk, 3 vols., 1904-14. Traveled through the Balkans to study Roman and Byzantine remains, then became interested in history and culture of Balkans, and contbd. to knowledge of successive civilizations in these lands. Died Vienna, Austria, Jan. 1, 1904.

KANKELEIT, Egbert, physicist; b. Hamburg, Germany, Apr. 16, 1929; diplomphysiker Technische Hochschule, Munich, Germany, 1956, Ph.D., 1961; m. Irene Anastasia Filia, July 15, 1956; children—Alexandre Claudia, Katharina Isabella. Research asst. Laboratorium fur Technische Physik, Munich, 1961-62; research fellow phys. dept. Cal. Inst. Tech., Pasadena, 1962-64, sr. research fellow, 1964-65, vis. asst. prof., 1965—. Research in nuclear and atomic physics, especially Mossbauer effect. Home: 404 S. Mentor St., Pasadena, Cal.*

KANTATOULIS, Athanase, Greek cardiologist; b. Paris, July 17, 1899; s. Constantine and Vassiliki (Christaki) K.; M.D., U. Lyon; m. Helen Kapsambelis, Apr. 30, 1940. Chief, Lab. Path. Anatomy, Broussais Hosp.; dir. 1st cardiol. center, Athens; prof. Mil. Med. Sch., Athens; physician, chief cardiology service Evangelismos Hosp. Mem. French Cardiology Soc., Internat. Com. Cardiol. and Angiological News, Paris Med. Writers Club. Author: Cardiologie; Hygiène du cardiaque; Le secret d'Hippocrate. Inventor electrosphygmomanometer and tele-electrocardiograph. Address: 1, place d'Egypte, Athens, Greece.

KANTEMIR, Izzet, Turkish physician; b. Istanbul, Aug. 5, 1903; s. Ismail Musa and Umuy (Kadioglu)

K.; student Istanbul U. Med. Sch., 1920-25; m. Safiye Kantemir, Dec. 31, 1938; 1 dau., Iris (Mrs. Kemal Alper). Regional physician, Thrace, 1926-30; staff German Hosp., Istanbul, 1930-33; specialist internal diseases Berlin U. Charite, 1933-35, Numune Hosp., Ankara, Turkey, 1936-42; faculty Med. Sch., Ankara, 1946—, prof. pharmacology, 1954—. Recipient Decoration Hamdi Suat, Aknar Conf. Mem. Turkish Assn. for Cancer Research and Control (gen. sec.). Author: (with Safiye Kantemir) Diseases and Hygiene of Hot Countries; Prescription Books; Pharmacology; Drug Therapy; also articles. Discovered fungicide Hexachlorbenzen causes Porphyria, that plant Juglans Regia has spasmodic effect; tobacco consumption in Turkey. Home: 18 Kennedy, Ankara, Turkey.*

KANTHACK, Alfred Antunes, pathologist; b. Brazil, Mar. 4, 1865; s. Emilio Kanthack; ed. Hamburg, Lüneburg, Gütersloh (all Germany), Liverpool (Eng.) Coll., Liverpool Univ. Coll., St. Bartholomew's Hosp., St. John's Coll., Cambridge (Eng.) U., Berlin; M.A. (hon.), Cambridge U., 1897; M.D.; m. Lucie Henstock. Mem. Leprosy Commn., 1890-01; John Lucas Walker student, 1891-92; med. tutor Liverpool Univ. Coll., 1892-93; lectr. pathology and bacteriology St. Bartholomew's Hosp., 1893-97; prof. physiology and bacteriology Bedford Coll., 1895-96; dep. prof. pathology Cambridge U., 1896-97, prof. pathology, from 1897. Fellow Royal Coll. Physicians, Royal Coll. Surgeons (Jacksonian prize 1895). Author: Leprosy in India, 1892; Manual of Practical Morbid Anatomy, 1894; Practical Bacteriology, 1895; also articles. Died Dec. 21, 1898.

KANTOROVICH, Leonid Vitaljevich, Russian mathematician; b. Jan. 19, 1912; grad. Leningrad U., 1930; D.Physico-Math. Sci., 1935. Instr., Leningrad Inst. Indsl. Constrn. Engring., 1930-39; instr. Leningrad U., 1932-34, prof., 1934—; vice-dir. Inst. of Math., Siberian Dept., USSR Acad. Sci. Recipient State prize, 1949. Mem. USSR Acad. Sci. Author: Mathematical Methods of Organizing and Planning Production, 1939; Functional Analysis and Applied Mathematics, 1948; Economic Calculation and the Best Use of Resources, 1959; co-author: Functional Analysis in Semi-Ordered Spaces, 1950, Approximated Methods of Higher Analysis, 1952; Functional Analysis in Normed Spaces, 1959. Research and publs. on theory of functions of real variable, approximated methods of analysis, functional analysis, problems in use of high-speed electronic computers, math. for planned econ. analysis. Address: Inst. of Math., Novosibirsk, USSR.*

KANTROWITZ, Adrian, Am. surgeon; b. N.Y.C., Oct. 4, 1918; s. Bernard Abraham and Rose (Esserman) K.; A.B., N.Y.U., 1940; M.D., L.I. Coll. Medicine, 1943; grad. student physiology Western Res. U., 1950; m. Jean Rosensaft, Nov. 25, 1948; children—Niki, Lisa, Allen. Gen. rotating intern Jewish Hosp. of Bklyn., 1944; asst. resident and resident surgery Mt. Sinai Hosp., N.Y.C., 1947; asst. resident Montefiore Hosp., N.Y.C., 1948, asst. resident pathology, 1949, fellow cardiovascular research group, 1949, chief resident surgery, 1950, adj. surg. service, 1951-55; USPHS fellow cardiovascular research, dept. physiology Western Res. U., 1951-52, teaching fellow physiology, 1951-52; asst. vis. surgeon Flower Hosp., Bird S. Coler Hosp., Met. Hosp., Morrisania Hosp., N.Y.C., 1952-55; instr. surgery N.Y. Med. Coll., 1952-55; cons. surgeon Good Samaritan Hosp., Suffern, N.Y., 1954-55; asst. prof. surgery State U. N.Y., Coll. Medicine, 1955-56, asso. prof. surgery, 1957-64, prof., 1964—; dir. cardiovascular surgery Maimonides Med. Center, Bklyn., 1955-64, dir. surgery, 1964—. Recipient H. L. Moses prize to Montefiore Alumnus for outstanding research accomplishment, 1949; 1st prize sci. exhibit Conv. N.Y. State Med. Soc., 1952; Gold Plate Award Am. Acad. Achievement, 1966; Max Berg award for outstanding achievement in prolonging human life, 1966; Theodore and Susan B. Cummings humanitarian award Am. Coll. Cardiology, 1967. Diplomate Am. Bd. Surgery. Fellow N.Y. Acad. Sci., A.C.S.; mem. Internat. Soc. Angiology, Am. Soc. Artificial Internal Organs (pres. elect 1967-68), N.Y. County Med. Soc., Harvey Soc., N.Y. Soc. Thoracic Surgery, N.Y. Soc. Cardiovascular Surgery, Am. Heart Assn., Am. Physiol. Soc., Am. Coll. Cardiology, Am. Coll. Chest Physicians, Bklyn. Thoracic Surgery Soc. (pres. 1967-68), Pan Am. Med. Assn. Author articles profl. jours. Pub. pioneer motion pictures taken inside living heart, 1950; contbd. to development of pump-oxygenators for human heart surgery; pioneered devel. mech. artificial hearts. Home: 546 E. 17th St., Bklyn. Office: 4802 10th Av., Bklyn. 11219.

KANTROWITZ, Arthur Robert, Am. physicist; b. N.Y.C., Oct. 20, 1913; s. Bernard A. and Rose E. (Esserman) K.; B.S., Columbia, 1934, M.A., 1935, Ph.D., 1947; m. Rosalind Joseph, Sept. 12, 1943; children—Barbara Ann, Lore Ellen, Andrea Ruth. Chief gas dynamics sect. NACA, Langley Field, Va., 1937-46; prof. aero. engring. and engring. physics Cornell U., Ithaca, N.Y., 1946-58; dir. Avco-Everett Research Lab., Everett, Mass., 1955—, v.p. dir. Avco Corp., N.Y.C., 1956—; vis. Inst. prof. Mass. Inst. Tech., 1957—. Mem. adv. council det. aero. engring. Princeton, 1959—. Fulbright scholar Cambridge U., also U. Manchester (Eng.), 1954; Gugenheim fellow, 1954. Fellow Am. Inst. Aeros. and Astronautics; mem. Am. Phys. Soc., Internat. Acad. Astronautics, Am. Acad. Arts and Scis., Nat. Acad. Sci. Research, publs. on phys. gas dynamics, re-entry heating problems of intercontinental ballistic missiles; pioneered application of shock tube to high temperature gas problems, devel. of MHD electric power generation. Home: 25 Spring Valley Rd., Arlington, Mass. Office: 2385 Revere Beach Pkwy., Everett, Mass. 02149.*

KANWAR, J. S., Indian soil chemist; b. Khera Kalmot Hosiarpur, Punjab, India, Dec. 10, 1922; s. Fateh Singh and Risal Devi; B.Sc., Pb. Agrl. Coll., Lyallpur, India, 1944; M.Sc., Govt. Agrl. Coll., Ludhiana, India, 1950; Ph.D. (research fellow), Waite Agrl. Research Inst., Adelaide, Australia; also post doctorate studies U.S.A., USSR; m. Vidya Devi, Feb. 8, 1951; children—Neena, Kiran, Jatinder, Nayindar, Sweety. Mem. faculty Govt. Agrl. Coll., Ludhiana, 1954-62, prof., 1962; also chemist Punjab govt., 1957-62; dir. research Punjab Agrl. U., Ludhiana, 1963-66; dep. dir. gen. soils, agronomy, engring. Indian Council Agrl. Research, New Delhi, 1966—. Fellow Indian Chemist, Sigma Xi.; mem. Indian Soc. Soil Sci. (v.p. 1962-64), Internat. Soc. Soil Sci. (v.p. 1964—). Author: (with S. L. Chopra) Practical Agricultural Chemistry, 1960. Research and publs. on soil fertility, soil salinity, alkalinity, soil structure, irrigation of water, micronutrients and sulphur. Home: k4, Green Park Extension, New Delhi, India. Office: Indian Council Agrl. Research, New Delhi, India.*

KANZIG, Werner, Swiss physicist; b. Zurich, Switzerland, Feb. 7, 1922; s. Hans and Julie (Haude) K.; M.S., Swiss Fed. Inst. Tech., 1947, Ph.D., 1951; m. Erika L. Huber, Aug. 13, 1948; children—Verena, Christian Hans, Henry Werner, David Frederick. Asst., Swiss Fed. Inst. Tech., Zurich, 1948-51, research asso., 1951-53, prof. physics, 1961—, chmn. dept., 1966-68; research asst. prof. U. Ill., 1953-55; staff mem. Gen. Electric Research Lab., Schenectady, 1955-61. Mem. NRC of Switzerland, 1963—; cons. IBM Lab., Zurich, 1962-65. Recipient Silver medal Swiss Fed. Inst. Tech., 1951. Fellow Am. Phys. Soc.; mem. Swiss Phys. Soc. Author: Ferroelectrics and Antiferroelectrics, 1957. Editorial bd. Helvetica Physica Acta, 1962-68; co-editor Physics of Condensed Matter, 1962—. Research and discoveries in field of ferroelectricity, color centers. Home: Hettlerstrasse 5, 8104 Weiningen ZH, Switzerland.*

KAO, Frederick Fengtien, physiologist; b. Peking, China, Jan. 29, 1919; s. Po Wu and Hsi (Tai) K.; B.S., Yenching U., Peiping, 1945; M.D., W.China Union U., Chengtu, 1947; M.S., Northwestern U., 1950, Ph.D., 1952; m. Edith Chung Ying Ling, July 5, 1949; 1 son, John Jien. Came to U. S., 1948, naturalized, 1958. Faculty State U. N.Y., 1952—, prof. physiology, 1965—; vis. prof. Nat. Taiwan U., Taipei, 1956-57. Vis. scientist U. Lab. Physiology, Oxford, Eng., 1959; vis. prof. Inst. Clin. Physiology, Göttingen, Germany, 1964-65. Recipient Morse medal W.China Union U., 1947. Mem. Am. Physiol. Soc., Harvey Soc., Soc. Exptl. Biology and Medicine A.A.A.S., N.Y. Acad. Scis., Am. Assn. U. Profs. Author: An Introduction to Respiratory Physiology, 1957; Cerebrospinal Fluid and the Regulation of Ventilation, 1965. Research, numerous publs. on nervous, chem. control breathing and heart at rest and during muscular activity, effects anesthesia on heart and respiration. Home: 1004 Bryant Av., New Hyde Park, N.Y. 11040. Office: 450 Clarkson Av., Bklyn. 11203.*

KAO, Shih-Kung, atmospheric scientist; b. Foochow, China, Mar. 9, 1918; s. T. C. and Kang-Yi (Cheng) K.; B.S., Hua U., 1939; M.A., U. Cal. at Los Angeles, 1948, Ph.D., 1952; m. Yasuko Kao, Apr. 1, 1959. Came to U. S., 1947, naturalized, 1965. Lectr., mem. research staff Aero. Research Inst., Nat. Tsing Hua U., China, 1939-46; univ. fellow U. Cal. at Los Angeles, 1948-52, vis. asst. prof., 1956-57, research sci., 1958-60; research asso. Johns Hopkins, 1953-55; prof. meteorology U. Utah, Salt Lake City, 1960—; vis. prof. U. Cal., summer 1964. Vis. scientist Nat. Center for Atmospheric Research, 1963; councilor Univ. Corp. for Atmospheric Research; dir. Atmospheric Turbulence and Diffusion Project, 1961—. Mem. Royal Meteorol. Soc. Gr. Britain, A.A.A.S., Am. Meteorol. Soc., Am. Geophys. Union, Sigma Xi, Phi Kappa Phi. Research, publs. on atmospheric turbulence and diffusion, theoretical and exptl. research on characteristics of turbulence and diffusion of radioactive debris in atmosphere, wave motion in rotating fluids, atmospheric circulation. Home: 2567 Blaine Av., Salt Lake City 84108.*

KAPANY, Narinder S., physicist; b. Moga, India, Oct. 12, 1926; s. Sundar Singh and Kindan (Kaur) K.; B.Sc., D.A.V. Coll., Dehra Dun, India, 1948; D.I.C., Imperial Coll. Sci. and Tech., London, 1952; Ph.D., Imperial Coll. Sci. U. London, 1954; m. Satinder Kaur, Feb. 4, 1954; children—Rajinder S., Kiren K. Supr. ordnance factory, Rajpur, Dehra Dun, India, 1949-51; lens designer Barr & Stroud Optical Co., Anniesland, Glasgow, Scotland, 1952; research asso. Inst. Optics U. Rochester (N.Y.), 1955-57; supr. optics sect. Armour Research Found., Chgo., 1957-61; pres., dir. research Optics Tech. Belmont, Cal., 1961—; dir. Optics Tech., Inc., Societe Belge d'Optique et d'Instruments de Precision, Ghent, Belgium. Cons. Bausch & Lomb Optical Co., Argus Camera Co., 1955-57, Johns Hopkins, 1956-57. Fellow Optical Soc. Am., Am., London (Eng.) phys. socs.; mem. Am. Inst. Physics, I.E.E.E. (sr.), A.AA.S., Sigma Xi. Research, articles on geometrical and phys-optics, fiber optics, asphevics, interference microscopy, image evaluation, photoelectronics. Home: 2126 Greenways Dr., Woodside, Cal. Office: 248 Harbor Blvd., Belmont, Cal.*

KAPIL, Ravinder Nath, botanist; b. Multan (now Pakistan), Feb. 26, 1929; s. Pran Nath and Sobhagyawati (Mody) K.; B.Sc., D.A.V. Coll., Jullundur, India, 1948; M.Sc., Birla Coll. Sci., Pilani, 1952; Ph.D., U. Delhi (India), 1959; m. Nirmal Arora, Feb. 23, 1959; 1 dau. Staff, U. Delhi, 1952-58; sr. research fellow CSIR, 1958-59, lectr., 1959-61; sr. research fellow Nat. Inst. Scis. India, U. Delhi, 1961-63, reader, 1963—. Sr. research fellow NISI, 1961; research fellow UNESCO, 1966. Fellow Indian Bot. Soc.; mem. Internat. Soc. Plant Morphologists, Indian Assn. Biol. Scis. Author: (with P. Maheswari) Fifty Years of Science in India—Progress of Botany, 1963. Research, publs. on morphology and embryology of various families of flowering plants, demonstrated amply decisive importance of embryology in solving systematic controversies. Home: 59/8 W.E.A., New Delhi-5. Office: Dept. Biology, U. Delhi, Delhi-7, India.*

KAPITSA, Petr Leonidovich (Kapitza), Russian physicist; b. Kronstadt, July 8, 1894; grad. Elec. Engring. Faculty, Petrograd. Poly. Inst., 1918; D.Phys. and Math. Sci.; Ph.D. (fellow), Trinity Coll., Cambridge U., 1926; hon. dr. degrees Paris, Algiers, Norwegian univs. Asso., Petrograd. Phys. and Tech. Inst., 1918-21; with Cavendish Lab., Cambridge, Eng., 1921-30; dir. Mond Lab., Cambridge U., 1930-34; dir. Inst. Physics Problems, USSR Acad. Scis., 1935-46, 1955—, prof. Moscow Physico-Tech. Inst., 1955—. Decorated Order of Lenin; recipient State prize, 1941, 43, Faraday medal, 1942, Franklin medal, 1944, Lomonosov Gold medal, 1960, Niels Bohr Gold medal, 1964, Rutherford medal, 1966, Kamerlingh Onnes medal, 1968. Fellow Royal Soc., 1929, Brit. Inst. Physics; mem. USSR Acad. Sci. (Presidium mem. 1960—); hon. mem. Brit. Inst. Metals, Franklin Inst., Danish, U. S. Nat., Indian, Irish acads. sci., also others. Author: The Heat Transfer and Super-Fluidity of Helium II, 1941; Research into the Mechanism of Heat Transfer in Helium II, 1941; High-Power Electronics, 1962; Collected Papers, 3 vols., 1964-67. Chief editor Jour. Exptl. and Theoretical Physics; mem. editorial bd. Sci. Events of USSR Acad. Sci. Research and publs. on electron inertia and properties of radioactive radiation; developer equipment for work with very powerful magnetic fields, 1924, high-prodn. hydrogen liquefier, 1932; designer, constructor original apparatus for liquefying large quantities of helium, 1934, 38, built apparatus for producing large quantities of liquid oxygen, 1939; studied nature of ball lightning. Address: Inst. Physics Problems, Vorobevskoe sh. 2, Moscow, USSR.

KAPLAN, Albert Sydney, Am. microbiologist; b. Phila., Nov. 29, 1917; s. Harry and Rose (Koussevitsky) K.; B.S., Phila. Coll. Pharmacy and Sci., 1941; M.S., Pa. State Coll., 1949; Ph.D., Yale, 1952; m. Tamar Ben-Porat, Feb. 14, 1959; 1 dau., Nira, and one son, Daniel. Civilian chemist with the United States Army, Olean, N.Y., 1941-43, Phila., 1946-47; research asst. preventive medicine Yale, 1949, instr., 1954-55; research asso. bacteriology U. Ill., 1955-58; head, dept. microbiology research labs. Albert Einstein Med. Center, Phila., 1958—. Research prof. Temple U. Sch. Medicine, 1964—. Nat. Found. for Infantile Paralysis fellow Pasteur Inst., Paris, 1952-53, Yale, 1953-54. Recipient Career Devel. award Nat. Inst. Allergy and Infectious Diseases, 1964. Fellow A.A.A.S.; mem. Am. Assn. Immunologists, Am. Soc. Microbiology, N.Y. Acad. Scis., Sigma Xi. Asso. editor Virology, 1966, Jour. Virology. Research publs. on biochem. and biol. aspects of virus synthesis, particularly virulent and tumor inducing DNA-containing viruses. Home: 418 S. Sterling Rd., Elkins Park, Pa. 19117. Office: York and Tabor Rds., Phila. 19141.*

KAPLAN, Henry S(eymour), Am. radiologist; b. Chgo., Apr. 24, 1918; s. Nathan M. and Sarah (Brilliant) K.; B.S., U. Chgo., 1938; M.D., Rush Med. Coll., 1940; M.S. in Radiology, U. Minn. (Nat. Cancer Inst. tng. fellow), 1944; m. Leah Hope Lebeson, June 21, 1942; children—Ann Sharon, Paul Allen. Mem. faculty Yale, 1944-47; radiologist Nat. Cancer Inst., USPHS, 1947-48, mem. council, 1959-63; prof. radiology, exec. head dept., sch. medicine Stanford, 1948—, dir. biophysics lab., 1957-64. Mem. com. radiology NRC, 1950-56; mem. panel on pathological effects radiation Nat. Acad. Sci.-NRC; mem. adv. com. biology, Oak Ridge Nat. Lab.; mem. subcom. Commn. Research Internat. Union Against Cancer, chmn., 1957-58; Commonwealth Fund fellow, vis. scientist NIH, 1954-55. Decorated Legion of Honor (France). Mem. Am. Soc. Exptl. Pathology, Assn. U. Radiologists (pres. 1954-55), Radiol. Soc. N.A., Am. Coll. Radiology, A.A.A.S., Soc. Exptl. Biology and Medicine, Radiation Research Soc. (pres. 1956-57), Western Soc. Clin. Research, Am. Assn. Cancer Research (v.p. 1965-66, bd. dirs. Am., Internat. clubs therapeutic radiologists, Am. Soc. Biol. Chemists. Author: Congenital Heart Disease: An Illustrated Diagnostic Approach (2d edit, with S. J. Robinson, H. L. Abrams, 1965); Angiocardiographic Interpretation in Cogenital Heart Diseases (with H. L. Abrams), 1955. Mem. Editorial adv. bd. Cancer; mem. editorial com. Ann. Rev. Nuclear Medicine. Research and publs. on exptl. contbns. to the mechanism of induction of leukemia and cancer, especially the discovery that radiation-induced leukemia in mice is actually due to a latent virus; improvement of the results of radiotherapy for cancer, particularly Hodgkins disease; exptl. contbns. to the molecular mechanisms responsible for the lethal effects of radiation on cells. Home: 631 Cabrillo Av., Stanford, Cal.*

KAPLAN, Joseph, physicist; b. Tapolcza, Hungary, Sept. 8, 1902; s. Henry and Rosa (Lowy) K.; brought to U. S., 1910; B.S., Johns Hopkins, 1924, A.M., 1926, Ph.D., 1927; D.Sc., U. Norte Dame, 1957, Carleton Coll., 1957; L.H.D., Hebrew Union Coll., Yeshiva U., U. Judaism; m. Katherine E. Feraud, June 24, 1933. Research fellow, Princeton, 1927; faculty U. Cal. at Los Angeles, 1928—, prof., 1940—; chmn. dept. of physics and meteorology, 1938-44, dir. Inst. of Geophysics, 1946-47; v.p. Internat. Geodesy and Geophysics, 1960-63, pres. 1963-67; representative com., exec. bd. Internat. Council Sci. Unions, 1960—. Dir. Microdot, Inc. Mem. internat conf. sci. advancement new states, also gov. Weizmann Inst., Israel, 1960; chmn. acad. council Am. Friends Hebrew U. Mem. bd. govs. Hebrew U. Jerusalem, Hebrew Union Coll.-Jewish Inst. Religion; chmn. U. S. nat. com. IGY; dir. Reiss-Davis Clinic Child Guidance; chmn. sci. adv. group Office Aerospace Research. Decorated for exceptional civilian service, War Dept., 1947, Air Force, 1960; recipient Hodgkins medal Smithsonian Inst., 1965. Fellow Am. Phys. Soc., Inst. Aero. Scis.; mem. Am. Rocket Soc., Nat. Acad. Scis., Am. Astron. Soc., Am. Astron. Pacific, Soc. Research on Meteorites, Optical Soc. Am., Internat. Union Geodesy and Geophysics (pres.), Am. Meteorol. Soc. Tau Beta Pi, Tau Delta Phi, Scabbard and Blade. Author: (with another) Physics and Medicine of the Upper Atmosphere, 1951; Across the Space Frontier, 1951. An originator of the sci. of aeronomy; research and publs. primarily on clarification of results obtained from direct observations made in space. Home: 1565 Kelton Av., Los Angeles 90024.*

KAPLAN, Lewis David, Am. meteorologist; b. N.Y.C., June 21, 1917; s. Joel and Annie (Elstein) K.; B.S., Bklyn. Coll., 1939; S.M., U. Chgo., 1947, Ph.D., 1951; m. Lillian Vond Epstein, Feb. 1, 1942; 1 dau., Rebecca Miriam. Meteorologist, U. S. Weather Bur., Washington, 1940-54; research asso. Mass. Inst. Tech., Cambridge, 1957-61; prof. atmospheric physics U. Nev., Reno, 1961-64; chief planetary studies sect. Jet Propulsion Lab., Cal. Inst. Tech., Pasadena, 1961-62, staff sci., 1962—. Mem. Inst. Advanced Study, Princeton, N.J., 1954-56; guest Imperial Coll., London, 1956-67; vis. prof. Oxford U., 1964-65; vis. prof. U. Oslo, 1965, Mass. Inst. Tech., 1968. Fellow Am. Meteorol. Soc.; mem. Royal Meteorol. Soc., Internat. Astron. Union, Sigma Xi, others. Editor: (with S. S. Penner) Quantitative Spectroscopy and Applications in Space Science, 1963. Asso. editor: jour. Quantitative Spectroscopy, 1964—; Icarus, 1967—. Discovered water vapor, carbon monoxide and low surface pressure on Mars, hydrogen halides on Venus; originator method for indirect atmospheric sounding from satellites; discovered multiplication principle in molecular band absorption; infrared expt. on Mariner II Venus mission. Home: 1202 Avoca Av., Pasadena 91105. Office: 4800 Oak Grove Dr., Pasadena, Cal. 91103.*

KAPLAN, Morton A., Am. polit. scientist; b. Phila., May 9, 1921; s. Lewis J. and Anthea (Ginsburg) K.; B.S., Temple U., 1943; Ph.D., Columbia, 1951. Staff mem. Brookings Instn., 1954-55; faculty U. Chgo., 1956—, chmn. com. internat. relations, 1959—, dir. Ford Workshop Programs Internat. Relations, 1961—, prof. polit. sci., 1965—; vis. asso. prof. Yale, 1961-62. Acting dir. Harris Meml. Found. Internat. Relations, 1957-58; lectr. Army War Coll., Command and Gen. Staff Sch.; cons. Com. Econ. Devel., Hudson Inst. Center Internat. Studies fellow Princeton, 1952-53, 58-61; Center For Advanced Study in Behavioral Scis. fellow, 1955-56; Carnegie Travelling fellow, 1959-60. Mem. Am., Internat. polit. sci. assns., Inst. Strategic Studies (London). Author: (with Reitzel, Coblenz) United States Foreign Policy 1945-1955, 1956; System and Process in International Politics, 1957; Some Problems of Strategic Analysis in International Politics, 1959; The Strategy of Limited Retaliation, 1959; The Communist Coup in Czechoslovakia, 1960; United States Foreign Policy in a Revolutionary Age, 1961; (with Katzenbach) Political Foundations of International Law, 1961. Editor, contbr.: The Revolution in World Politics, 1962. Originated, developed comparative systems theory of internat. politics, applications of internat. systems theory to internat. law; supervised devel. of computer models of internat. systems theory; contbns. to mil. strategy, including limited strategic retaliation and war of resolution. Office: 1126 E. 59th St., Chgo. 60637.*

KAPLAN, Oscar Joel, Am. psychologist; b. N.Y.C., Oct. 21, 1915; s. Philip and Rebecca (Uttef) K.; A.B., U. Cal. at Los Angeles, 1937, M.A., 1938; Ph.D., U. Cal. at Berkeley, 1940; m. Rose Zankan, Dec. 28, 1942; children—Stephen Paul, Robert Malcolm, David T. A. Faculty, U. Cal., Berkeley, 1940-41, U. Ida., 1941-46; prof. psychology San Diego State Coll., 1946—, dir. Center for Survey Research, 1948—. Cons. govt. agys.; mem. coms. on aging; vis. prof. pub. health U. Cal., Los Angeles, 1965-66. Mem. Am. Psychol. Assn. (pres. div. maturity and old age 1954-55), Gerontological Soc., A.A.A.S., Am. Assn. Pub. Opinion Research, Western Gerontological Soc. (pres. 1956-57), Sigma Xi. Author: Mental Disorders in Later Life, 1956, also articles. Editor-in-chief Gerontologist, 1960-66; mem. internat. bd. editors Gerontology and Geriatrics, 1958—. Contbns. in field of gerontology, including co-authorship of psychol. abilities scale for seniles. Home: 5409 Hewlett St., San Diego 92115.*

KAPLAN, Reinhard Walter, German biologist; b. Glauchau, Germany, Aug. 30, 1912; s. Josef Franz and Lina (Strauss) K.; student U. Innsbruck (Austria); Dr.phil., U. Leipzig, 1937; m. Charlotte Röschke, Oct. 1, 1938; children—Rosemarie (Mrs. Ulrich Winkler), Hans-Gerhard. With Kaiser Wilhelm Inst. Plant Breeding Research, Müncheberg (Germany), 1937-45; head mutation research lab. Max Planck Inst. Plant Breeding Research, Voldagsen, Germany, 1946-53; faculty U. Marburg (Germany), 1952-53; research asso. Columbia U., N.Y.C., 1953-55; asso. prof. U. Frankfurt/Main (Germany), 1955-63, prof. microbiology, 1963—. Mem. German Assn. Biologists (pres. 1958-62), German Soc. Hygiene and Microbiology (v.p. 1965-67). Research and numerous publs. on mutation induced by X-rays in plants; discovery of similarity between mutation process in bacteria and higher organisms; discovered that ultra violet and X-ray induced mutation in bacteria and virus is initiated by absorption of radiation primarily in DNA. Address: 5 Teplitz-Schönauer Strasse, 6 Frankfurt/Main, West Germany.*

KAPLAN, Samuel, physician; b. Johannesburg, South Africa, Mar. 28, 1922; s. Aaron L. and Emma (Kahan) K.; M.D., U. Witwatersrand, Johannesburg, 1944; m. Molly Eileen McKenzie, Oct. 17, 1952. Came to U. S., 1950, naturalized, 1958. Lectr., U. Witwatersrand, 1946-49; registrar in cardiology Postgrad. Med. Sch. London (Eng.), 1949-50; faculty U. Cin. Coll. Medicine, 1950—, prof. pediatrics, 1966—; dir. div. cardiology Children's Hosp. of Cin., 1954—; attending physician Gen. Hosp., Cin., 1954—. Mem. Am. Fedn. for Clin. Research, Soc. Pediatric Research, Internat. Cardiovascular Soc., Am. Assn. Artificial Internal Organs, Am. Pediatric Soc., Am. Acad. Pediatrics. Research, numerous publs. in physiology of total extracorporeal circulation, polarography, physiology and natural history of congenital heart disease. Home: 2829 Victoria Av., Cin. 45208. Office: Children's Hosp., Cin. 45229.*

KAPLAN, Sylvan Julian, Am. behavioral scientist; b. San Antonio, Aug. 10, 1919; s. Morris T. and Betty (Katz) K.; B.A., U. Tex., 1940, M.A., 1946; Ph.D., Stanford, 1950; m. Barbara Elizabeth Williams, Jan. 24, 1959; children—David J., Jonathon L. Instr., Stanford, 1948, Yale, 1949; dir. radiobiology lab. USAF-U. Tex., 1950-54; prof., head dept. psychology Tex. Technol. Coll., Lubbock, 1954-62; chief field selection, dept. dir. div. selection U. S. Peace Corps., 1962-64; chmn. dept. behavioral scis. Armed Forces Radiobiology Research Inst., Def. Atomic Support Agy., Bethesda, Md., 1964-67; asso. dean div. research Va. Poly. Inst., Blacksburg, 1968—; practice psychology, Lubbock, 1954-62, Washington, Bethesda, Md., 1962—; professorial lectr. dept. psychology George Washington U., 1963—. Mem. Am. Psychol. Assn., Radiation Research Soc., Sigma Xi. Editor, co-author: The Marine Corps Reserve: A History, 1966; Toward the Marine Corps University, 1975, 1967; also articles. Research on effects of ionizing radiation on behavior and physiology of rodent and infra-human primate, on prenatal organisms, massive radiation doses from atomic bomb and nuclear reactors, social behavior of man, clin. and group behavior. Home: 130 Stonegate Dr., Blacksburg, Va. 24060.*

KAPLAN, William, Am. mycologist; b. N.Y.C., Apr. 27, 1922; s. Harry and Lena (Jaffe) K.; B.S., Cornell U., 1943, D.V.M., 1946; M.P.H. U. Minn., 1951; student Coll. City N.Y., 19——; m. Margaret Kaprielian, June 1, 1952. Veterinarian, clinician UNRRA, 1946-47; practice vet. medicine, Maysville, Mo., 1947; asst. dist. supr. U. S. Dept. Agr., Mexico, 1947-50; vet. mycologist USPHS Communicable Disease Center, Atlanta, 1951—. Mem. Am. Vet. Med. Assn., Am. Pub. Health Assn., Am. Soc. Microbiology, Assn. Mil. Surgeons, Am. Coll. Lab. Animal Medicine, Internat. Soc. Human and Animal Mycology. Editorial bd. Am. Jour. Vet. Research, 1963—, Sabouraudia, Jour. Internat. Soc. Human and Animal Mycology. Research, numerous publs. on ecology, epidemiology and pathogenesis of Q-Fever and equine encephalitis, animal ringworm and its prevention of transmission to man, epidemiology and ecology of mycotic disease; developed reagents for rapid diagnosis of systemic and subcutaneous mycotic diseases. Home: 1646 Rainier Falls Dr., Atlanta 30329. Office: Communicable Disease Center, Atlanta 30333.*

KAPLANSKY, Irving, mathematician; b. Toronto, Ont., Can., Mar. 22, 1917; s. Samuel and Anna (Zuckerman) K.; B.A., U. Toronto, 1938, M.A., 1939; Ph.D., Harvard, 1941; m. Rachelle Brenner, Mar. 16, 1951; children—Steven, Daniel, Lucille. Came to U. S., 1940, naturalized, 1955. Instr., Harvard, 1941-44; research mathematician Columbia, 1944-45; faculty U. Chgo., 1945—, prof. math., 1955—, chmn. dept., 1962—. Guggenheim fellow, 1948-49. Mem. exec. com., div. math. NRC, 1959-62. Mem. Am. Math. Soc., Math. Assn. Am., Nat. Acad. Scis. Author: Infinite Abelian Groups, 1952; An Introduction to Differential Algebra, 1957; also numerous articles. Research in fields algebra and functional analysis. Home: 5825 S. Dorchester Av., Chgo. 60637.*

KAPLANSKY, Samuil Yakovlevich, Russian biochemist; b. 1897; grad. Med. Faculty, Moscow U., 1921; D.Med. Sci., 1936. Asst. prosector dept. gen. pathology Moscow U., 1921-25, asst., head biochem. lab., 1925-29, lectr. mineral metabolism, 1929-31; biochemist Inst. Skin Tb, 1925-29; head biochem. dept. Biol. Research Inst., Timiryazev Communist Acad., Moscow, 1931-35; dep. head dept. metabolism All-Union Inst. Exptl. Medicine, 1935-36, dep. dir. for sci. work, 1939—, prof., 1938—; head chair biochemistry 2d Moscow Med. Inst., 1939-51; head physiochemistry dept. Inst. Biol. and Med. Chemistry, USSR Acad. Med. Sci., 1944—. Mem. Moscow Soc. Physiologists, Biochemists and Pharmacologists (bd. Mem.). Author: Biochemistry of the Skin, 1931; Mineral Metabolism, 1938. Co-editor Chemistry sect., author various articles Large Med. Ency., 2d edit.; mem. editorial council Problems of Med. Chemistry, Problems of Nutrition; mem. editorial bd. Bull. Exptl. Biology and Medicine. Research and numerous publs. on extracts, lungs, muscles, other tissues, synthesis of amino acids, skin chemistry, metabolism. Address: Inst. Biol. and Med. Chemistry, USSR Acad. Med. Sci., Pogodinskaya ulitsa 10, Moscow, USSR.

KAPLON, Morton Fischel, Am. physicist; b. Phila., Feb. 11, 1921; s. Myer and Ida (Abramson) K.; B.S., Lehigh U., 1941, M.S., 1947; Ph.D., U. Rochester, 1951; m. Anita Joanne Harle, June 16, 1946; children—Keith Victor, Bryna Myra, Andrea Joanne. Research asso. to prof. physics U. Rochester, N.Y., 1951—, NSF postdoctoral fellow, 1959, chmn. dept. physics, astronomy, 1964—, asso. dean Coll. Arts, Scis., 1963-65. Fellow Am. Phys. Soc.; mem. Italian Phys. Soc., A.A.A.S., Am. Assn. U. Profs., Am. Geophys. Union. Research in elementary particles, high energy nuclear physics, cosmic radiation, gamma ray astronomy. Home: 404 Warren Av., Rochester, N.Y. 14618.*

KAPOOR, Lachaman Das, Indian botanist; b. Muzaffarabad, India, Sept. 27, 1916; s. Bhagwand Das and Janki (Devi) K.; F.Sc., S.P. Coll., 1935; B.Sc., Forman Christian Coll., 1937; M.Sc. in Agrl. Botany, Banaras Hindu U., 1940; Ph.D., Sch. Pharmacy, London, Eng., 1958; m. Prakash Nagpal, Feb. 22, 1943; children—Suman (Mrs. Harish Chander Virmani), Anil, Ashish. Technician botany dept. Drug Research Lab., 1940-42; forest exploitation officer minor forest products forest dept. Jammu and Kashmir Govt., botanist Drug Research Lab., 1944-55; with Regional Research Lab. (formerly Drug Research Lab., 1958-64; asst. dir. Nat. Bot. Garden, Lucknow, India, 1964—, officer in charge field research sta. Fellow Indian Acad. Scis., Linnean Soc., Am. Soc. Pharmacognosy; mem. Inst. Biology (London). Author: (with R. N. Chopra, I. C. Chopra, K. L. Handa) Indigenous Drug of India, 1958; also numerous articles. Research on medicinal and aromatic plants, including their survey, ecology, botany, agronomy, physiology, anatomy and preparation for market; introduced some exotic medicinal and aromatic plants to India for cultivation. Home: C-409 Mahanagar. Office: Nat. Botanic Gardens, Lucknow, U.P., India.*

KAPOSI, Moritz (originally Moriz Kohn), dermatologist; b. Kaposuar, Hungary, 1837; studied under Hébra; became prof., Vienna, Austria, 1875; succeeded Hebra, founder new sch., at dermatologic Clinic, 1879. Mem. French Acad. Medicine (corr.). Author: Handatlas der Hautkrankheiten für Studirende und Artze, 1898; (with Ferdinand Ritter von Hebra) On Diseases of the Skin, including Exanthemata, 5 vols., 1866-80; Pathologie et traitment des maladies de la peau, 1891; Pathologie und Therapie der Hautkrankheiten in Vorlesungen für praktische Artze und Studirende, 1894; also numerous articles. Described multiple bluish nodules in skin marked by hemorrhages (Kaposi's sarcoma), 1872, xeroderma pigmentosum, disease characterized by brown spots and ulcers of skin (Kaposi's disease), 1882, lymphodermia perniciosa characterized by enlargement of lymph nodes, 1885, impetigo herpetiformis characterized by appearance of groups of pustules in pregnant women, 1887. Died Vienna, 1902.

KAPP, Gisbert, elec. engr.; b. Mauer, nr. Vienna, Austria, 1842; ed. Poly. Sch., Zurich, Switzerland; D.Eng. (hon.), Karlsruhe, Dresden (both Germany); m. Therese Mary Krall, 1884; 2 sons. Engr. in Eng. and other countries, from 1876; mgr., works of Messrs. Crompton and Co., Chelmsford, Eng., 1882-84; became cons. engr., Westminster, Eng., 1884; became lectr. elec. engring., Charlottenburg, Germany, 1894; prof. elec. engring. U. Birmingham (Eng.), 1904-18. Recipient Telford medal, 1886-88. Pres. German Assn. Elec. Engrs., Instn. Elec. Engrs., engring. sect. Brit. Assn. Author: Electric Transmission of Power; Transformers; Dynamo Machines; Construction of Electric Machinery; Alternating Currents; Principles of Electrical Engineering and their Application; also papers. Editor Elektrotechnische Zeitschrift, 11 years. Developed a dynamo; drew transformer diagram, circa 1890; designed transformers and power stas.; devised methods of elect. testing; Kapp Line (unit of magnetic force) named after him; applied math. to elect. engring. Died Aug. 10, 1922.

KAPPELER, Moritz Anton, Swiss geologist; b. Willisau, Switzerland, June 9, 1685; ed. Lucerne Mil. Acad.; practice medicine Lucerne, Switzerland. Research on composition of Pilatus mountains laying founds. for devel. geol. sci. in Switzerland; invented a seismograph; discovered new method for drawing more precise geol. charts; began a crystallography based on geometry. Died Beromünster, Switzerland, Sept. 16, 1769.

KAPPER, Oskar Gustavovich, Russian silviculturist, dendrologist; b. 1888; grad. St. Petersburg Forestry Inst., 1913. Asst. forester, instr. Khrenovskoy Forestry Sch., Voronezh Guberniya, 1913-16; sr. asst. dept. dendrology Forestry Faculty, Voronezh Agrl. Inst., 1916-24, lectr., 1924-30, also co-founder forestry corr. dept.; prof., head chair silviculture and dendrology Voronezh Forestry Mgmt. Inst., 1930—. Decorated Order of Lenin. Author: A Study of the Ecological Types of Arboreous Species, 1946. Research and publs. on biology and ecology of arboreous species, vegetative reproduction of arboreous species in relation to environment, planting time and felling seasons. Address: Voronezh Forestry Mgmt. Inst., ulitsa Timiryazeva 34, Voronezh, RSFSR, USSR.

KAPPERT, Hans, German biologist; b. Munster, Aug. 24, 1890; s. Wilhelm and Catharina (Schoenefeld) K.; Ph.D., Paulinum U.; D.honoris causa; m. Martha Rother, Jan. 25, 1921. Asst., Kaiser Wilhelm Inst.; agrege Berlin Sch. Agr.; full prof. genetics at Berlin; hon. prof. U. Munster. Mem. Acad. Naturalists and Physicians. Author: Grundlagen der Gärtner Pflanzenzüchtung; Vererbungswissensschaftliche Grundlagen der Züchtung; Handbuch der Pflanzenzüchtung; Zeitschrift für Pflanzenzüchtung. Home: Langenstrasse 11. Office: Botanical Institute, University of Munster, Munster, Germany.

KAPPLER, Eugen, German physicist; b. Schönberg, Germany, Apr. 1, 1905; s. Johannes and Marie (Hartmann) K.; Ph.D., U. Munich; m. Margarete Kühne, Mar. 12, 1936; children—Christa, Ulrich, Lieselotte. Asst. in high sch.; instr., asso. prof., full prof. U. Munster, also dir. Phys. Inst. Mem. Union German Physics Socs. Author: Brownsche Molekularbewegung; Schwankungserscheinungen; Materialprüfung; Fest-Koperphysik. Home: St. Mauritz, Erlen-Allee, Munster. Office: Physical Institute, Schlossplatz 7, Munster, Germany.

KAPPUS, Adolf Ludwig, microbiologist; b. Offenbach, Main, Germany, July 29, 1900; s. Martin and Minna (Weirich) K.; student U. Frankfurt (Germany), 1919, U. Frieborg (Germany), 1920-21, U. Dusseldorf (Germany), 1922; M.D., U. Munich (Germany), 1925; postgrad. U. Berlin (Germany), 1927, degree in Pub. Health, 1933; degree in Tropical Medicine, U. Hamburg (Germany), 1932. 1 dau., Lola. Prof., chmn. dept. microbiology Marquette U. Sch. Medicine, Milw. Office: 561 N. 15th St., Milw. 53233.*

KAPTEYN, Jacobus Cornelius, Dutch astronomer; b. Barneveld, Holland, Jan. 19, 1851; prof. astronomy U. Groningen, (Netherlands), 1878-1921. Fellow Royal Soc., 1919; mem. French Acad. Scis. Measured and plotted stars in Gill's plates of So. hemisphere, 1904, and discovered 2 streams of stars moving in opposite directions in plane of Milky Way; originated idea that shape of galaxy could be estimated by counting stars in random areas of sky (thus originating statis. astronomy), 1906; used studies of motion to determine stellar distances previously impossible to measure; discovered star which has 2d most rapid motion (Kapteyn's star); developed model of Milky Way (Kapteyn Universe). Died Amsterdam, Holland, June 18, 1922.

KAPTSOV, Nikolai Aleksandrovich, Russian physicist; b. Feb. 3, 1883; ed. Moscow U., 1904; researcher lab. of P. N. Lebedev, 1866-1912; staff All-Union Electrotech. Inst., 1920-28; with electric plants, 1928-33; with Moscow U., 1923—, prof., 1931—. Author: Electric Phenomena in Gases and in Vacuum, 1950.

KAPUR, J. N., Indian mathematician; b. Delhi, India, Sept. 7, 1923; s. L. N. and Sukho (Devi) K.; B.A. with honors, Hindu Coll., Delhi U., 1942, M.A., 1944, Ph.D., 1958; m. Chandra Vati, Feb. 4, 1945; children—Uma, Nina, Sanjay. Head math. dept. Hindu Coll., Delhi U., 1948-59, Inst. Post-grad. Studies, 1959-61; prof., head math. dept. Indian Inst. Tech., Kanpur, India, 1961—. Sec. congresses Theoretical and Applied Mechanics, India, 1961-64; organizing sec. numerous math. summer schs., 1958-64. Fellow Indian Acad. Scis., Nat. Acad. Scis., Inst. Math. and its Applications, U.K.; mem. Bharat Ganit Parisad (v.p.), Indian Soc. Theoretical and Applied Mechanics (v.p.), Kanpur Math. Soc. (pres.). Author: (with Shanti Narayan) Vector Calculus, 4th edit., 1966; Mathematical Statistics (with H. C. Saxena), 4th edit. 1966; Dynamics (with J. D. Gupta), 1966; Essays on Mathematics Education, 1966; The Spirit of Mathematics, 1966. Research, publs. on gen. fluid dynamics, internal ballistics, gen. theories of composite and moderated charges, operations research and probability theory. Home: 462, Indian Inst. Tech. Campus, Kanpur, U.P., India.*

KAPUSCINSKI, Witold Juliusz, Polish physician; b. Halle, Germany, Apr. 5, 1910; s. Witold and Janina (Psarska) K.; student dept. medicine U. Poznan (Poland), 1927-33, M.D., 1936; m. Jadwiga Korwin-Piotrowska, July 16, 1938; children—Salomea, Witold. Asst., U. Poznan, 1933-39; asst. Rotschild Inst., Paris, 1936; head surgeon Eye Hosp., Warszawa, Poland, 1941-44; prof. ophthalmology U. Wroclaw (Poland), 1946—; dir. dept. ophthalmology Med. Acad., Wroclaw, 1950—, prof. ophthalmology, 1961—; hon. oculist U. Hosp. Calixto Garcia, Havana Cuba, 1961—. Decorated Polonia Restituta Order. Mem. Polish (v.p. 1964—), French (rep. for Poland), Cuban ophthal. socs, Sci. Soc. in Wroclaw, World Union Writers-Physicians (mem. gen. council Polish dept.). Contbg. author: Handbook of Ophthalmological Therapy, 1957; Author: Ophthalmia Sympatica, 1955; Bonne Chance Cuba, 1965; Immunological Eye Diseases, 1967; Pillars of Hercules, 1968; also numerous articles. Devel. of stress therapy in ophthalmology, diagnostic and inflammatory mechanism in uveitis, physiopathology of rebounding pulse in retina, new surg. procedures in glaucoma, echogramm diagnostics in ophthalmology. Home: 10 Mickiewicza, Wroclaw 9, Wroclaw, Poland.*

KAPUSTINSKII, Anatolii Fedorovich, Russian chemist; b. Dec. 29, 1906; staff Inst. Applied Mineralogy (now All-Union Inst. Mineral Resources) until 1941; staff Inst. Gen. and Inorganic Chemistry, USSR Acad. Sci.; 1941—; prof. Gorkii U., 1934-37, Moscow Steel Inst., 1937-41, Kazan U., 1941-43, Moscow U., 1945-49; prof. Moscow Chemico-Technol. Inst., 1943—. Corr. mem. USSR Acad. Sci. Research and numerous publs. on phys. chemistry; originated hypothesis on isothermicity and metallization of inner nucleus of Earth.

KARAEV, Abdulla Ismail Oglu, Russian physiologist; b. Baku, USSR, 1910; s. Ismail Abdulla Oglu and Miasma Chanum (Djafarova) K.; ed. Azerbaijan Nat. U., 1930, Azerbaijan Med. Inst., 1936; married 1940; children—Naila Abdulla Kyri (Mrs. Alman Ismailov), Tomrida Abdulla Cyri (Mrs. Buniat Sardarov), Nuschabe Abdulla. Decorated Order of Lenin, Order Red Banner (2), Order Red Star. Author: Textbook of the Physiology of Man and Animals, 1934-45; Interrceptors and Metabolism, 1957; Physiology of the Phagacytose, 1960; Reticular Formation and Vegetative Nervous System, 1965; also numerous articles. Advanced theory about bioelectric phenomena, mechanism of physiol. action of the medicina naphtalan oil, decrement conduct subliminal excitation, interoceptio exchanged reflexes. Home: 8 Miasnikova Bar. Office: Sector Physiology, Acad. Bakyo, Azberbaijan USSR.*

KARANDEEV, Konstantin Borisovich, Russian elec. engr.; b. USSR, July 18, 1907; grad. Leningrad Poly. Inst., 1930. With Electro-Phys. Inst., Leningrad, 1929-35; with Leningrad Inst. Signal Engring., 1937-42, All-Union Sci. Research Inst. Metrology, 1937-42; became prof. Lvov (Poland) Poly. Inst., 1944; later with Inst. Machine Studies and Automation, Ukrainian Acad. Scis.; became dir. Inst. Automation and Electrometry, Siberian br. USSR Acad. Scis., 1957—. Named Honored Scientist of Ukrainian SSR, 1954. Mem. Ukrainian Acad. Scis. (corr.), USSR Acad. Scis. (corr.). Author: Methods of Electrical Measurement, 1952; Bridge Methods of Measurement, 1953; Semiconductor Rectifiers in Measuring Techniques, 1954; Direct Current Galvanometers, 1957; Combined Cascade Capacitor Circuit, 1962; coauthor: Automatic Large-Scale Control of Radio Components, 1963. Developed exact methods for elec. measurement, methods for measuring low and high currents and theory of bridge methods; studies on semiconductor rectifiers, geophys. apparatus, telemetry, computers used in measurement. Office: Inst. Automation and Electrometry, Siberian Br. of USSR Acad. Scis., Novosibirsk, Siberia.

KARANOV, Sary Karanovich, Russian ophthalmologist; b. Mar. 16, 1909; grad. Tashkent Med. Inst., 1936. With Tashkent Med. Inst., Ashkhabad, 1948—, prof., 1953—; dep. dir. Turkmen Trachoma Inst., 1949—. Recipient medal for valiant labor, 1961. Mem. Turkmenia Acad. Sci. Author: Glaucoma and Struggle against it in Turmenistan, 1957. Co-editor ophthalmology sect. Bolshaya Meditsinskaya entsikiopediya; editor, mem. Jour. Ophthalmology. Research on epidemic conjuntivitis, trachoma, glaucoma. Home: Ashkhabad, Turkm.SSR, Novaya 10, Office: Turkmensky Meditsinsky Inst., USSR.

KARAPETOFF, Vladimir, elec. engr.; b. St. Petersburg, Russia, Jan. 8, 1876; s. Nikita Ivanovitch and Anna Joakimovna (Ivanova) K.; C.E., Inst. of Ways of Communication, St. Petersburg, 1897, M.M.E., 1902; Technische Hochschule, Darmstadt, Germany, 1899-1900; hon. Mus. Doc., N.Y. Coll. Music, 1934; D.Sc., Poly. Inst. Bklyn., 1937; m. Frances Lulu Gillmor, Aug. 2, 1904 (dec. 1931); m. 2d, Rosalie Margaret Cobb, Nov. 25, 1936. Cons. engr. for Russian Govt., and instr. elec. engring. and hydraulics in 3 colls., St. Petersburg, 1897-1902; spl. engring. apprentice with Westinghouse Electric & Mfg. Co., East Pittsburgh, Pa., 1903-04; prof. elec. engring. Cornell U., 1904-39, emeritus from 1939; vis. prof. for grad. students, Poly. Inst. Bklyn., 1930-32; lectr. Stevens Inst. Tech., 1939-40. Cons. engr. and patent expert to various enterprises; cons. U. S. Bd. Econ. Welfare, 1942-43, Bethlehem. Steel Co., from 1944; mem. adv. bd. U. S. Naval Acad., 1916; chmn. com. on physics Conf. on Elec. Insulation of NRC, 1928-35. Recipient Montefiore prize, 1923, and Elliot Cresson gold medal, Franklin Inst., 1927, both for kinematic models of elec. machinery. Fellow Am. Inst. E.E., A.A.A.S., Am. Phys. Soc.; mem. U. S. Naval Inst., Am. Assn. U. Profs. (charter), Am. Math. Soc., Math. Assn. Am., Franklin Inst., Sigma Xi.; hon. mem. Eta Kappa Nu, Tau Beta Pi, Phi Mu Alpha (Sinfonia). Author: Ueber Mehrphasige Stromsysteme, 1900; Resistance of Ships to Propulsion (in Russian), 1st part 1902, 2d part 1911; Experimental Electrical Engineering, 2 vols., 1908; The Electric Circuit, 1910; The Magnetic Circuit, 1911; Engineering Applications of Higher Mathematics, part I, 1911; parts II to V, 1916; also numerous articles. Research editor Elec. World, 1917-26. Inventor, patentee several elec. devices, also five-stringed cello. Died Jan. 11, 1948; buried Ithaca, N.Y.

KARAPETYAN, Sahak Karapetovitch, Armenian physiologist; b. Armenia, USSR, May 16, 1906; s. Karapet Martirossovitch Karapetian; grad. All-Union Zoo-Vet. Inst., Yerevan, USSR, 1933; Dr.Biol. Scis., 1961. Became v.p. Armenian br. USSR Acad. Scis., 1943; chief Lab. Physiology Reprodn. and Stimulation Reproductive Function of Agrl. Animals; head chair physiology Armenian Pedagogical Inst.; became dir. Armenian Inst. Animal Breeding, Yerevan, 1954. Mem. Armenian Soc. Physiologists (chmn.), Internat. Assn. Poultry Breeding, Internat. Soc. Physiologists, All-Union Soc. Physiology and Central Council. Author several books, monographs numerous articles. Research on neuroendocrine mechanism of action of artifical illumination on reproductive function and higher nervous activity of animals; neurohumoral regulation of reprodn. in agrl. animals and birds; effects of ionizing radiation and ultraviolet on reproductive function of birds in ontogenesis; established role of large hemispheres of brain in reprodn., connection between neurosecretory function of hypothalamo-hypophysial system and sexual activity. Office: Inst. Physiology, Armenian Acad. Scis., Orbeli str. 3, Yerevan 28, USSR- Armenia.*

KARAS, Karl Maria, German mechn. engr.; b. Graz/Steiermark, Nov. 24, 1890; s. Karl Friedrich and Henriette (Schiman) K.; Ph.D. in Tech. Engring., U. Graz; m. Julie Pawlik-Koschany, Feb. 7, 1926; m. 2d, Waltraut Welzl, Mar. 28, 1959; children—Gotlinde, Meinhard, Waltraut. Mgr. research dept. Skoda Factories, Pilsen, 1915-19; asst. Deutsche Technische Hochschule, Brno, 1919-25, prof. agrégé, 1925-35; prof. agrege, Prague, until 1945; prof. agrégé at Darmstadt, 1949-59, prof. emeritus, 1959—. Mem. Assn. German Engrs., Gesellschaft für angewandte Math. und Mechan., German Acad. Prague (math. and phys. scientists). Author: Die Kritischen Drehzahlen wichtiger Rotoformen; Fachgebiet Hydraulik 1939-45; Fiat-Reviews; Medizin und Naturforschung in Deutschland. Home: C. Mierendorffstrasse 38, Darmstadt-Eberstad. Office: Hochschulstrasse, Darmstadt, Germany.

KARAVAEV, Nikolay Mikhaylovich, Russian chemist; b. 1890; grad. Moscow High Tech. Sch.; hon. Dr. Tech. Sci., 1940. Instr., Moscow Higher Tech. Sch., 1920-30; with All-Union Thermal Engring. Inst., 1924-32, Moscow Chemicotech. Inst., 1925-32, Inst. Combustible Minerals, USSR Acad. Sci., 1939-51; with Moscow Inst. Chem. Machine-Bldg., 1946—, prof., 1949—. Mem. USSR Acad. Sci. (corr.). Author: The Properties and Quality of the Coals of the USSR, Vol. 1, 1939; Lignites, 1957; co-author: Machines and Apparatus of the Coal-Tar Chemical Industry, Vol. 1, 1955. Research and numerous publs. on chemistry and fuel tech., semicoking of Siberian coals under indsl. conditions, derivation, chemistry and tech. of solid fuels and their products; hydrogenation of solid fuels and resins, 1933-36; origin and classification of fossil fuel. Address: Moscow Inst. Chem. Machine-Bldg., ulitsa Karla Marksa 21-4, Moscow, USSR.

KARAVANOV, Georgiy Grigorevich, Russian surgeon; b. 1899; grad. Kharkov Med. Inst., 1924; D.Med. Sci., 1939. Intern, asst. dept. faculty surgery Kharkov Med. Inst., 1930-35, lectr., 1935-41; asso. Ukrainian Inst. Exptl. Med., 1930-41; chief surgeon hosps., Lvov Oblast, 1944-47; dep. dir. for sci. work Lvov Blood Transfusion Inst., 1944-48; head chair faculty surgery Lvov Med. Inst., 1944—, also chmn. problems commn. for study and elimination goiter in Carpathian endemic focus. Mem. learned med. council Ukrainian Ministry Health. Mem. All-Union (bd. mem.), Ukrainian (bd. mem.), Lvov (bd. mem.) socs. surgeons. Author over 120 works including: Blood Transfusion, 1934; Data on Free Spinal Grafts, 1940. Address: Lvov Med. Inst., Pekarskaya ulitsa 71, Lvov, Ukrainian SSR, USSR.

KARCZMAR, Alexander George, pharmacologist; b. Warsaw, Poland, May 9, 1918; s. Stanislaus and Helena (Billauer) K.; student J. Pilsudski U., 1934-39; M.A., Columbia, 1941, Ph.D., 1947; m. Marion Hope Allen, Jan. 18, 1946; children—Gregory Stanislaus, Christopher Anthony. Came to U. S., 1940, naturalized, 1946. Research asst. N.Y. U., 1942-44; research asst. Amherst Coll., 1944-45; teaching fellow Columbia, 1945-46; lectr. Charleston Coll., 1946; asst. to asso. prof. dept. pharmacology Georgetown U. Med. Center, 1946-54; pharmacologist Sterling-Winthrop Research Inst., 1954-56; profl. lectr. Albany Med. Coll., 1954-56; prof., chmn. dept. pharmacology and therapeutics Stritch Sch. Medicine Loyola U., 1956——; sr. co-dir. Inst. Study of Mind, Drugs and Behavior, 1964——. Research cons. Hines VA Hosp.; cons. Ill. State Psychiat. Inst.; prin. investigator USPHS, 1950——, State of Ill. Psychiat. Tng. and Research Fund, 1956——. Charter fellow Am. Coll. Neuropsychopharmacology (council mem. 1965——); mem. Am. Soc. Pharmacology and Exptl. Therapeutics, others. Author: (with T. Koppanyi) Experimental Pharmacodynamics, 1955. Contbr. numerous articles to profl. jours. Editorial bd., sect. editor Internat. Ency. Pharmacology and Therapeutics; editorial bd. Internat. Jour. Neuropharmacology. Co-developer therapeutic agts. used in myasthenia gravis and in coronary disease; studies cholinergic drugs, synapic transmission, biol. bases of behavior and of differences in psychol. profiles. Home: 327 N. Euclid Av., Chgo. 60302.*

KARCZMARZ, Kazimierz, Polish botanist; b. Lubartów, Poland, May 7, 1933; s. Michal and Franciszka (Bodzak) K.; Magister diploma in Botany, Marie-Curie-Sklodowska U., Lublin, Poland, 1957, dr.'s degree in Botany and Bryology 1962. Staff, Inst. Systematic and Plant Geography, Marie-Curie Sklodowska U., 1957——, prospective docent, 1967——, adjunct, 1963——, faculty univ. dept. botany, since 1957——, dir. bryological lab. and herbarium mosses, since 1962——. Mem. Polski Towarzystwo Botaniczne, Internat. Bur. for Plant Taxonomy and Nomenclature (corr.), Internat. Assn. for Plant Taxonomy. Research, publs. on vascular plants of Southeastern Poland, mosses of Armenia, Arctic Regions and Middle Europe, Cretaceous floras of Poland algae, history of botany. Home: 23, Grazyna Str., Lublin, Poland.*

KARE, Morley Richard, physiologist; b. Winnipeg, Man., Can., Mar. 7, 1922; s. Isaac and Rose (Cohn) K.; B.S., U. Man., 1943; M.S., U. B.C., 1948; Ph.D., Cornell U., 1952; m. Carol Abramson, Sept. 5, 1951; children—Susan, Jordin. From asst. prof. to prof. physiology Cornell U., 1952-61; prof. physiology N.C. State U., Raleigh, 1961-67; prof. physiology, dir. Monell Chem. Senses Center, U. Pa., 1967——. Recipient Borden award, 1962. Fellow A.A.A.S.; mem. Am., Canadian physiol. socs., Soc. Exptl. Biology and Medicine, Animal Behavior Soc., Sigma Xi, Phi Kappa Phi. Author: (with B. P. Halpern) Physiological and Behavioral Aspects of Taste, 1961; (with E. E. Bernard) Biological Prototypes and Synthetic Systems, 1962; (with O. Maller) Chemical Senses and Nutrition, 1967. Contbr. numerous articles to profl. jours. Described taste world of variety of species of animals including birds; reported on various physiol. and nutritional factors that modify the sense of taste. Address: Monell Chem. Senses Center, U. Pa., Lippincott Bldg., 25th and Locust, Phila. 19104.*

KARELITZ, Samuel, physician; b. Chomsk, Russia, June 15, 1905; s. A. Samuel and Sara (Sporofsky) K.; M.D., Yale, 1923; m. Ethel Rosin; children—Susan (Mrs. William Ginsberg), Judith. Practice medicine specializing in pediatrics, N.Y.C., 1927-55, New Fairfield, Conn., 1951——, New Hyde Park, N.Y., 1955——; dir. pediatrics L.I. Jewish Hosp., New Hyde Park, 1964——; pediatrician-in-chief Queens Hosp. Center; asso. clin. prof. Columbia Coll. Phys. and Surg., 1934-55; clin. prof. pediatrics State U. N.Y. Downstate Med. Center, Bklyn., 1954——. Cons. Internat. Social Service, 1957——. Recipient Jacobi medal Alumni Assn. Mt. Sinai Hosp., N.Y.C., 1956, Medal of City, Aix-en-Provence, France, 1960; Lauer award Child Welfare League Am., 1961; Naomi Lehman Meml. award 1967. Mem. Am. Acad. Pediatrics, Am. Pediatric Soc., A.M.A., Soc. for Pediatric Research. Author: When Your Child Is Ill, 1954; also numerous articles. Introduced continuous intravenous fluid therapy into pediatrics, study of infant cries to determine brain damage; pioneered in child adoption; studied measles prophylaxis and measles vaccines. Home: 145-15 Bayside Av., Flushing, N.Y. 11354. Office: 270-05 76th Av., New Hyde Park, N.Y. 11040.*

KARGIN, Valentine Alexejevitch, Russian chemist; b. Dnepropetrovsk, USSR, Jan. 23, 1907; s. Alexei K. and Louise (Monretin) K.; D.Sc., Moscow (USSR) U., 1936; m. Kalerija Velitchko, Dec. 9, 1933; 1 dau., Olga. Research worker Karpov Physico-Chem. Inst., 1924-37 chief lab., 1937——; prof. Moscow U., 1953——, chief chair high molecular compounds, 1956——. Mem. State Com. for Sci. and Technics USSR, 1962. Recipient State prizes, 1943, 47, 50, Lenin prize, 1962, Order Lenin, 1954, 61, 66. Mem. USSR Acad. Scis. (vice sec. dept. gen. and tech. chemistry 1953——, pres. sci. polymer council 1963——). Author: (with G. L. Slonimsky) Brief Essays on Physical Chemistry of Polymers, 1960; also numerous articles. Elucidated role of chemosorptional processes in colloid systems behavior; devel. theory of mech. properties of amorphous polymers in wide temperature range; discovered super molecular structure in amorphous polymers, polymerization in complex and activated monomer state; research on role of polymer structures in polymer synthesis. Home: kw4 7 Gaidara, Moscow 64. Office: USSR Acad. Scis., Leninsky prospect 14, Moscow, USSR.*

KARIM, Syed Mujtaba, Indian physicist; b. India, July 10, 1908; s. Syed Noorul Hassan and Rokaiya; M.Sc., Muslim U., Aligarh, India, 1932, Ph.D., 1937; B.Sc. with honors (Sir W. M. Ejaz Rasool scholar, univ. medal), 1931; Ph.D.(V. Massay scholar), U. Toronto (Can.) 1951; m. Maimoona Khatoon, Oct. 11, 1935; children—Saad, Masood, Safar, Farooq, Hamid. Demonstrator in physics Aligarh Muslim U., 1932-38; phys. asst. Indsl. Research Bur., Alipore, Calcutta, India, 1932-42; research officer Sci. and Indsl. Research India, 1943-48; faculty U. Dacca, 1948-52; faculty U. Karachi (Pakistan), 1952——, dean faculty sci., 1956-59, prof. head physics dept., 1962——. Mem. Internat. Sci. Meeting, 1954, World Symposium Applied Solar Energy, 1955, 2d Internat. Conf. Atomic Energy, 1958. Fellow Physics Soc. (London, Eng.); mem. Applied Solar Energy Soc. U. S. A., Pakistan Assn. Advancement Science. Research, publs., patents on vibrational analysis of triatomic molecules, second viscosity, nuclear physics, theories of hearing. Home: B II C, 7/1a. Nazimabad, Karachi-18. Office: Physics dept. U. Karachi, U. Rd., Karachi-32, Pakistan.*

KARKHANAVALA, Minocher Dadabboy, Indian chemist; b. Bombay, India, Nov. 29, 1926; s. Dadabhoy and Manijeh (Vazifdar) K.; B.Sc., Wilson Coll., 1947, M.Sc., 1949; M.S., N.Y. State Coll. Ceramics, 1950; Ph.D., Pa. State U., 1952; m. Silloo Dinshaw, Dec. 29, 1952; children—Khurshed Aramaiti. Sci. officer chem. div. Atomic Energy Establishment, 1953——; head solid state studies sect., chemistry div. Bhabha Atomic Research Center, Bombay, 1957——. Fellow Indian Chem. Soc.; mem. Sigma Xi. Research and publs. in solid state chemistry, reactor materials, particularly mechanism of oxidation of uranium oxides. Home: 239 Tardeo Rd., Bombay, India.*

KARKHI, Abu Bakr Muhammad ibn al-Hasan (Husain) al-Hasib al-, Arab mathematician; flourished, circa 1020. Lived in Baghdad; among greatest Muslim mathematicians; pub. most scholarly Arabic works on algebra. Author: Fakhri (algebra based on Diophantos, includes work on complete solution quadratic equations, addition and subtraction of radicals, summation of series); Kafi fil Hisab (arithmetic of his time, derived from Hindu includes work rule of quarter squares, approximations for roots, mensuration of plane figures), circa 1010-16. Did not use numerical system of India, instead wrote numerals as words. Died circa 1019-29.

KARL, Georges Edouard, chemist; b. Ségonzac, France, June 11, 1886; s. Albert and Irma (Damesmme) K.; Ph.D. in Phys. Sci., U. Geneva; m. Marguerite Descamps, Sept. 25, 1920; 1 dau., Simone. With research labs. Haen Soc., Seelze-Hanover, 1910-14; dir. S.A. Aarau Franco-Swiss, 1918-20; mfg. dir. Hoffmann-La Roche Altstatten and Bale Co., 1921-22; tech. dir. Hoffmann-La Roche Paris, Produits Roche, chems. and pharms., 1923-48. Mem. Biolchem. Soc. London, Paris Soc. Indsl. Chemistry, Chem. Soc. France, Paris Soc. Biochemistry, Am., German chem. socs., Geneva Nat. Inst. Author: Les amines biologiques. Research on biochemistry, pharmacology, radioisotopes. Address: 15, chemin de Roches, Geneva, Switzerland.

KARLE, Jerome, Am. physicist; b. Bklyn., June 18, 1918; s. Louis and Sadie (Kun) Karfunkle; B.S., Coll. City N.Y., 1937; M.A., Harvard, 1938; M.S., U. Mich., 1942, Ph.D., 1944; m. Isabella Helen Lugoski, June 4, 1942; children—Louise Isabella, Jean Marianne, Madeleine Diane. Chair sci., head lab. for structure matter Naval Research Lab., Washington, 1944——; asso. prof., prof. math., physics U. Md., 1952——; guest prof. U. Kiel, Germany, 1960. Recipient Research Soc. Am. award in pure sci., 1959. Fellow Am. Phys. Soc., Washington Acad. Scis.; mem. Am. Chem. Soc., Am. Math. Soc., Am. Crystallographic Assn., A.A.A.S., Contbr. numerous articles to profl. jours. Developed methods for analyzing basic structure of materials in terms of arrangements of their atoms by means of electron, x-ray and neutron diffraction techniques, sector-microphotometer method and diffraction spectroscopy in electron diffraction, background theory and procedures for solving the phase problem directly from diffracted intensities in x-ray diffraction. Home: 6304 Lakeview Dr., Falls Church, Va. 22041. Office: Naval Research Lab., Washington 20390.*

KÄRLING, John S., Am. biologist; b. Austin, Tex. Aug. 2, 1889; s. Theodore L. D. and Mary Ruth (Jacobson) K.; student Tex. Wesleyan Coll., 1914-15; A.B., U. Tex. A.M., 1920; student U. Mexico City, 1920, Oxford U. (Eng.), 1923; Ph.D., (fellow) Columbia, 1925; m. Page Burwell Johnston, Aug. 10, 1940; 1 dau. Sayre Christian. Dean, Texas Wesleyan Coll., 1919-20; faculty U. Tex., 1920, Columbia, 1924-48; head dept. biol. scis. Purdue U., 1948——, dir. Ross Biol. Res., 1948-59, John Wright distinguished prof. 1959-65, now prof. emeritus; Fulbright-Hays fellow, New Zealand, 1965-66; Sir C. V. Roman vis. prof. U. Madras (India); marine mycologist Internat. Indian Ocean Expdn., research fellow, 1963. Fulbright grantee, 1965-66; also recipient numerous research fellows. Fellow A.A.A.S.; mem. Bot., Soc. Am. (sec., mem. edit. bd. 1945-50, v.p. 1951), Am. Soc. Indsl. Microbiol. (dir. 1950-52), Am. Biol. Soc. (sec.), Torrey Bot. Soc. (pres. 1941; historian), Ind. Acad. Sci., Mycol. Soc. Am., Am. Phytopath. Soc., Brit. Mycol. Soc., Am. Inst. Biol. Scis., Sigma Xi. Author: The Plasmodiophorales, 1942; The Simple Holocarpie Biflagellate Phycomycetes, 1942; Synchytrium, 1963. Cons. editor: Botanical Revs. Home: 1219 Tuckahoe Lane, Lafayette, Ind. 47906.*

KARLSON, Alfred Gustav, Am. microbiologist; b. Virginia, Minn., Apr. 26, 1910; s. Knute John and Pauline (Johnson) K.; B.S., Ia. State U., 1934, D.V.M., 1935, M.S., 1938; Ph.D., U. Minn., 1942; m. Janice Ruth Stillians, June 24, 1938; children—Alfred Lennart, Karl John, Kathy Jean, Trudy Ann, Julie Kay. Instr. vet. bacteriology Ia. State U., 1935-38; fellow Mayo Found., Rochester, Minn., 1938-39, cons. exptl. medicine, 1946-53; instr. vet. research U. Minn., 1939-41; cons. med. microbiology Mayo Clinic, Rochester, 1953——; prof. comparative pathology Mayo Med. Grad. Sch., 1962——. Recipient Ia. State U. Alumni Achievement award, 1965. Diplomate Am. Bd. Microbiology. Mem. Conf. Research Workers in Animal Diseases (sec.), Am. Coll. Vet. Pathologists (pres. 1949), Nat. Tb Assn. (dir.), Minn. Thoracic Soc. (pres. 1964-65. Sect. editor Biol. Abstracts, 1940; editorial bd. Am. Jour. Vet. Research, 1960——. Home: 428 16th Av. S.W. Office: 200 1st St. S.W., Rochester, Minn. 55901.*

KARLSON, Peter, German biochemist; b. Berlin, Germany, Oct. 11, 1918; s. Friedrich Gustaf and Ludmilla (Perli) K.; Diplomchemiker, U. Berlin, 1940, Dr.rer.nat., 1942; m. Lieselotte Poschmann, Feb. 20, 1945; children—Jurgen Peter Ilse Katharina, Martin Gunther. Staffs, U. Tübingen, 1942-56, U. Munich, 1956-63; prof. physiol. chemistry U. Marburg/Lahn (West Germany), 1964——, dir. physiol. Chemistry Inst., Philipps-U., Marburg/Lahn, 1964——. Mem. Gesellschaft Deutscher Chemiker, Gesellschaft Deutscher Naturforscher u.Arzte, Gesellschaft biologische Chemie. Author: Kurzes Lehrbuch der Biochemie, 1961, 5th edit. 1966; 8 translations English: Introduction to Modern Biochemistry, 1963, 65. Research, publs. on 1st insect hormone to be isolated: ecdysone, moulting hormone of insects; gen. theory that hormones act by activating certain genes. Home: 32 Schulstrasse. Office: 1-2 Deutschhausstrasse, 355 Marburg/Lahn, West Germany.*

KARMAN, Theodore von, see von Karman, Theodore.

KARMARSCH Karl, Austrian technologist; b. Vienna, Oct. 17, 1803; student Poly. Inst., Vienna, 1819-23; became 1st dir. Poly. Sch., Vienna, 1830. Mem. Inst. Arts and Crafts Hanover (Austria) (editor Pub. Rev.). Author: Grundriss der mechanische Technologie, 2 vols., 1837-41; (with Heeren) Technisches Wörterbuch, 3 vols., 1843-44; Geschichte der Technologie seit Mittelalters der Mitte der 18 Jahrhundert, 1872; also articles. Invented engraving machine which copies medals and other reliefs; originated new system of technology. Died Hannover, Germany, Sept. 24, 1879.

KARMEN, Arthur, physician; b. N.Y.C., Feb. 25, 1930; s. Hyman M. and Bella (Schneiderman) K.; M.D., N.Y. U., 1954; m. Marcia B. Benham, Jan. 12, 1955; children—Peter Barry, Carol Lynn, Richard Alan. Clin. asso. Nat. Heart Inst., Bethesda, Md., 1956-58, research staff Lab. Tech. Devel., 1958-63; asso. prof. radiology and radiol. sci. John Hopkins, Balt., 1963——. Recipient Alfred Sloan award for cancer research, 1955. Mem. Am. Chem. Soc., Instrument Soc. Am., A.A.A.S., Alpha Omega Alpha. Research, publs. on measurement of serum transaminases, methods and instruments for microanalyses of biol. materials by gas chromatography, lipid metabolism and diagnostic uses of radioisotopes. Home: 2308 W. Rogers Av., Balt. 21209. Office: 615 N. Wolfe St., Balt. 21205.*

KARPINSKY, Alexander Petrovich, Russian geologist; b. Bogolowsk, Russia, Dec. 26, 1846; s. Peter Karpinsky; civil Mining Engr., Mining Inst., St. Petersburg, Russia, 1866; 4 daus., 1 son. Became geologist, prospector, So. Urals, 1867, coal prospector, Ufa, Orenburg, 1879; joined faculty Mining Inst., St. Petersburg, 1868, named prof. geology and petrology, 1896; field work for Russian Mining Dept.; became imperial dir. mining research, 1885-1916; dir. Russian Geol. Com. Mem. Mineral. Soc. (became dir. 1899), Russian Acad. Scis. (pres. 1817-1936). Author: Outline of the Geological History of European Russia, 1883-94. Research in paleontology, petrology, mineralogy; prepared 1st geol. map of European Russia, 1892; studied the Urals, genus Volboztella. Died No. Urals, July 15, 1936.

KARPLUS, Robert, physicist; b. Vienna, Austria, Feb. 23, 1927; s. Hans and Isabella (Goldstern) K.; came to U. S., 1938, naturalized, 1943; B.S., Harvard, 1945, M.A., 1946, Ph.D., 1948; m. Elizabeth Frazier, Dec. 27, 1948; children—Beverly, Margaret, Richard, Barbara, Andrew, David, Peter. Asst. prof. physics Harvard, 1950-54; faculty U. Cal., Berkeley, 1954——, prof., 1958——, dir. Sci. Curriculum Improvement Study, 1961——. Research in analytical chemistry; quantum field theory of elementary particles; geomagnetism; magnetohydrodynamics; microwave spectroscopy; molecular structure; field theories. Home: 57 Overhill Rd., Orinda, Cal. 94563.

KARPLUS, Walter Joseph, elec. engr.; b. Vienna, Austria, Apr. 23, 1927; s. Robert and Garda (Schiff) K.; B.E.E., Cornell U., 1949; M.S., U. Cal. at Berkeley, 1951; Ph.D., U. Cal. at Los Angeles, 1954. Came to U. S., 1938, naturalized, 1944. Chief field party Sun Oil Co., 1949-50; elec. engr. Internat. Geophysics, 1951-52; research engr. Hughes Aircraft Co., 1955-56; faculty U. Cal. at Los Angeles, 1954-

—, prof. engring., 1955—, head Engring. Dept. Computer Lab., 1962—. Cons. to NATO, 1961-62. Fulbright Research scholar Sorbonne, Paris, France, 1961-62. Mem. I.E.E.E., Am. Soc. for Engring. Edn., Am. Inst. Elec. Engrs., Internat. Assn. for Analog Computation, Internat. Fedn. Automatic Control (mem. simulation council 1956—, Sigma Xi, Eta Kappa Nu. Author: Analog Simulation: Solution of Field Problems, 1958; (with W. Soroka) Analog Methods: Computations and Simulation, 1959; (with R. Tomovic) High Speed Analog Computers, 1962; On-line Computing, 1967. Address: P.O. Box 24673, Los Angeles 90024.*

KARPOV, Sergey Petrovich, Russian microbiologist; b. 1903; grad. Med. Faculty, Tomsk U., 1927; D.Med. Sci. Asso., Tomsk Bacteriological Inst., 1929-39; prof., head chair microbiology Tomsk Med. Inst., 1937—; sci. dir. Tomsk Research Inst. Vaccines and Sera, 1939—. Mem. USSR Acad. Med. Sci. (corr.), All-Russian Soc. Epidemiologists, Microbiologists and Infectionists (bd. mem., co-founder, chmn. Tomsk dept.). Author: Water as the Means of Spreading Tularemia: A New Factor in Its Epidemiology, 1936; The Results of Studies of Natural Foci Infections, 1957; co-author: Bacterial Dysentery, 1954. Co-editor Microbiology sect. Large Med. Ency., 2d edit; mem. editorial council Problems of Virology, Jour. Microbiology, Epidemiology and Immunobiology. Research and numerous publs. on role of water in epidemiology of tularemia, epidemiology and prophylaxis of spring-summer tick-borne encephalitis; discoverer hemolytic properties of Brucellae; showed toxin formation of Corynebacterium diphtheriae in presence of heavy metal salts, proved some ticks, mosquitoes and sandflies aid in natural circulation of Pasteurella tularensis, demonstrated feasibility of phage prophylaxis in typhoid. Address: Tomsk Med. Inst., Moskovsky trakt 2, Tomsk, USSR.

KARRAS, Hans, German physicist; b. Halberstadt, German, Feb. 16, 1930; s. Josef and Erika (Lucas) K.; Dipl.-Phys., U. Jena, 1955, Dr.rer.nat., 1961; m. Annerose Richarda Pfeifer, Feb. 7, 1953; 1 son Hans-Dieter. Scientist, VEB Carl Zeiss, Jena, 1955-64, lab. mgr., 1964—. Fellow Phys. Soc. London; mem. Leopoldina, East German Phys. Soc. Mem. adv. bd. jour. Kristall und Technik. Research, publs. in solid state physics; investigations of nature of point defects in alkaline earth fluoride crystals and alkali halide crystals, spectroscopic properties of activated laser crystals, crystal growth, low temperature physics. Home: 29 Herder-Strasse. Office: 1 Carl-Zeiss-Strasse, Jena 69, East Germany.*

KARRER, Enoch, Am. physicist; b. Rich Hill, Mo., May 23, 1887; s. Frank Xavier and Theresa (Braun) K.; A.B., U. Wash., 1911, A.M., 1912; Ph.D., Johns Hopkins, 1914; m. Ethel Walther, Aug. 2, 1919; children—Enoch, Aurora, Ethelda, Rathe. Research asst. United Gas Improvement Co., Phila., 1914-18; chief searchlight sect. U. S., Bur. of Standards, 1919-21; with Gen. Electric Co.'s research lab., 1921-26; research asso. Cushing Lab. for Exptl. Medicine, Western Reserve U., Cleve., 1923-31; research physicist B. F. Goodrich Co., 1926-31; cons. research engr., 1931-36; economics and govt. research, Washington, 1934-35; research biophysics Smithsonian Instn., 1935; tech. cons. Am. Instrument Co., 1936; sr. physicist, U. S. Dept. of Agr., from 1936, at Southern Regional Research Lab., New Orleans, from 1941. Recipient James A. Moore prize in physics, Seattle, 1913, Longstreet medal (co-winner), Franklin Inst., 1918. Mem. Washington Philos. Soc., Sigma Xi, Phi Beta Kappa, and several tech. socs. Patentee in field. Contbr. papers to tech. jours. Died Mar. 27, 1946.

KARRER, Karl, Austrian physician; b. Kufstein, Austria, June 12, 1923; s. Karl and Rosa (Simmaier) K.; M.D., U. Innsbruck, 1951. With Inst. Pathology, U. Vienna (Austria), 1951-53; with Austrian Cancer Research Inst., 1953—, head for exptl. pathology, head dept. statistics, 1958—. Research, numerous publs. on chemotherapy in surgery for human cancer, epidemiology and statistics of cancer disease in Austria. Address: 8a Borschkegasse, Vienna 1090, Austria.*

KARRER, Paul, chemist; b. Moscow, Russia, Apr. 21, 1889; s. Paul and Julie (Lerch) K.; Dr.Phil., U. Zurich, 1911; numerous hon. degrees, European univs.; m. Helene Froelich, Apr. 14, 1914; children—Jürg, Heinz. Chemist, Georg Speyer-Haus, Frankfurt, Germany, 1912-18; faculty U. Zurich, 1911-12, 18-59, prof., head dept. chemistry, 1919-59. Recipient Nobel Prize for chemistry, 1937 (with W. N. Haworth); Marcel Benoist prize, 1923; Cannizzaro prize, 1935. Mem. Nat. Acad. Scis., numerous other sci. socs., acads. Author: Lehrbuch der organischen Chemie, 1927; Carotinoide, 1948. Pioneer work on carotenoids, flavins and vitamins A and B; determined structure of and synthesized vitamin B2; also synthesized vitamins A and E. Home: 30 Spyristeig, Zurich, Switzerland.

KARSTEN, Gustave, German physicist; b. Berlin, 1820. Author: Theories of Mechanics, 1849-53. Reformed weights and measures (led to unification of German measurement system); research on dissolution of salt by water. Died Kiel, Germany, 1900.

KARTULIS, Stephanos, Greek physician; b. Greece, 1852. 1st to isolate Entamoeba histolytica from tropical abscess of liver (led to recognition of amebae as cause dysentery in man). Died 1920.

KARZON, David Theodore, Am. pediatrician, virologist; b. N.Y.C., July 8, 1920; B.S., Ohio State U., 1940, M.S., 1941; M.D., Johns Hopkins, 1944; m. Allaire Urban, May 18, 1946; children—David Theodore, Elizabeth U. Faculty, State U. N.Y. at Buffalo Sch. Medicine, 1952—, prof. pediatrics, 1963—; asso. attending physician Children's Hosp., Buffalo, 1952-64, attending physician, 1964—. Cons. USPHS, 1959—, NIH, 1963—; mem. spl. adv. com. on immunization practice to Surgeon-Gen., 1964—. Lowell M. Palmer Sr. fellow, 1953-56; Markle scholar, 1956-61; recipient Research Career award NIH, 1962—. Diplomate Am. Bd. Pediatrics, Am. Bd. Microbiology. Mem. Am. Acad. Pediatrics, Soc. for Pediatric Research, Am. Acad. Microbiologists, Am. Soc. for Microbiology, Am. Epidemiological Soc., N.Y. Acad. Sci., Am. Assn. Immunologists, Soc. for Exptl. Biology and Medicine, Phi Beta Kappa. Asso. editor Am. Jour. Epidemiology, 1966—. Research, numerous publs. on biol. and immunological aspects animal viruses; devel. and evaluation viral vaccines. Home: 79 Chapin Pkwy., Buffalo, 14209.*

KASANEN, Alpo Antero, Finnish physician; b. Helsinki, Finland, Oct. 16, 1927; s. Alpo and Sylvi (Jarvi) K.; Med.lic., U. Turku (Finland), 1953, Med.doctor, 1958; m. Terttu Totto, May 3, 1951; 1 son, Eero Olavi. Asso. chief dept. internal medicine U. Turku, 1958-65, asst. prof. internal medicine, 1965—, chief dept. artificial kidney, 1962—; chief doctor lab. Testi, Turku, 1963—. Mem. Internat. Soc. Nephrology, Nordisk forening for intern Medicin, Duodecim. Research, numerous publs. on nephrotoxis effect of analgetics, chronic pyelonephritis, studies of causes of heart infarction and significance of psychosomatic reasons for some gen. diseases. Home: 20.C.60. Itainenkatu, Turku, Finland.*

KASATKIN, Nikolay Ivanovich, Russian physiologist; b. 1903; grad. Gertsen Pedagogical Inst., Leningrad, 1929, Moscow Med. Inst.; D.Biol. Sci., 1947. Asso., Central Research Inst. for Mother and Child Care, Moscow, 1931-41; head lab. higher nervous activity of child Inst. Pediatrics, USSR Acad. Med. Sci., 1944-54; sr. asso. Sechenov Inst. Evolutionary Physiology, USSR Acad. Sci., 1954—. Mem. USSR Acad. Med. Sci. (corr.). Author: The Development of Auditory and Visual Conditioned Reflexes and their Differentiation in Infants, 1935; Early Conditioned Reflexes in the Ontogenesis of Man, 1948; The Food Center and Non-Food Conditioned Reflexes in Infants, 1951; An Outline of the Development of Higher Nervous Activity in Infants, 1951; The Early Ontogenesis of Conditioned Reflexes in Man, 1953; The Appearance of Early Conditioned Reflexes in the Child, 1955. Mem. editorial bd. Pavlov Jour. Higher Nervous Activity. Research and publs. on physiology of higher nervous activity of child, early ontogenesis of conditioned reflex activity; established experimentally sequential origin of functional cortical activity in child as appearance of 1st conditioned reflex links (refuted viewpoint that cortex functionally inactive in 1st 3-4 months of infant's life); proved there is regular sequence of temporary links within different analyzers; developer new methods to study conditioned reflexes in sucklings. Address: Sechenov Inst. Evolutionary Physiology and Biochemistry, USSR Acad. Sci., Thorez prospect 52, Leningrad K-223, USSR.

KASDON, Solomon Charles, Am. physician; b. N.Y.C., Dec. 18, 1912; s. David C. and Sarah (Mirkin) K.; B.S., Yale, 1933, M.S., 1934, M.D., 1938; m. Muriel C. Cohen, Dec. 23, 1943; children—David L., Madeline A., Louisa B. Research fellow Yale Sch. Medicine, 1938-39; faculty Tufts U. Sch. Medicine, Boston, 1940—, asso. prof., 1958—; chief staff Booth Meml. Hosp., Boston, 1953—; chief obstetrics Waltham (Mass.) Hosp., 1961—; practice medicine specializing in obstetrics and gynecology, Boston, Waltham, Newton, 1947—. Pres., Med. Research Found., Boston, 1951—. Mem. A.M.A., Mass. Med. Soc., Am. Bd. Obstetrics and Gynecology, A.C.S., Am. Coll. Obstetricians and Gynecologists, Am. Assn. Cancer Research, Endocrine Soc. Author: (with Bamford) Atlas of In Situ Cytology, 1963; also numerous articles. Research in cancer enzymology, cytology, clin. surg. research in obstetrics and gynecology. Address: 127 Bay State Rd., Boston 02215.*

KASE, Masao, Japanese physician; b. Tokyo, Japan, Jan. 6, 1914; s. Ryuji and Sae (Saito) K.; B.M., Tokyo U., 1941, M.D., 1949; m. Kikue En-ya, July 19, 1918; children—Nanae, Midori, Masaki. Physician, Tokyo U. Hosp. 1946-53; chief 3d dept. internal medicine Kanto Communication Hosp., Tokyo, 1953—; lectr. Tokyo U. Sch. Medicine. Mem. Japanese Soc. Neurology, Japanese Soc. Psychiatry and Neurology, Japanese Soc. Internal Medicine, Japanese Circulation Soc. Author: Diseases of Muscle, 1966; with (T. Tsubaki et al) Clinical Neurology, 1966; also numerous articles. Research in clin. neurology, cerebral vascular diseases; contbd. to understanding pathogenesis of cerebral hemorrhage. Home: 14-11, 3-chome, Hongo, Bunkyo-ku, Tokyo. Office: 22-9 5-chome Higashi-Gotanda, Shinagawa-ku, Tokyo, Japan.*

KASEMIR Heinz W., physicist; b. Tilsit, Germany, Aug. 30, 1913; s. Edwin S. and Erma (Lunbinger) K.; B.S., Tech. Coll., Danzig, 1935; M.S., Tech. Coll., Munich, 1938; Ph.D., Tech. Coll., Aachen, 1954; m. Poldi Tischer, Mar. 25, 1946; children—Wolfram, Imogen, Astride, Verena. With Telegraph & Elec. Co., Grafelting German Radio Research Inst., U. S. Army Electronics Research and Devel. Lab., Ft. Monmouth, N.J., ESSA, Atomic Phys. Chem. Lab., Boulder, Colo., Research, publs. on current flow theory in atmospheric electricity, matching effect in current measurements, atmospheric electric current radiosound, 2 component field mill for field measurement from an airplane. Home: 22 S. Ponderosa Dr., Boulder, Colo. 80302.*

KASHKAY, Mir-Ali Seid-Aliogly, Russian petrographer, geochemist; b. 1907. Acad. sec. dept. geology and chem. sci. Azerbaijan Acad. Sci., 1947-59, acad. sec., 1959-62. Mem. Azerbaijan Acad. Sci. Author: Geology of Azerbaijan: Petrography, 1952; Metallogeny of Azerbaijan, 1962. Research and publs. on bedrock, ultrabasic rocks and mineral resources of Azerbaijan, petrography and ore deposits of Azerbaijan, No. Caucasus and Western Siberia; formed new theory of formation of iron pyrites mineralizations in Azerbaijan; classified pyrite deposits on basis of their genetic connection with acid magma derivatives. Address: Azerbaijan Acad. Sci., Baku, Azerbaijan SSR, USSR.

KASHKIN, Pavel Nikolaevich, Russian microbiologist; b. Soligalich, Kostroma Guberniya, 1902; grad. Leningrad Med. Inst., 1926; postgrad. dept. microbiology 1st Leningrad Med. Inst., from 1926; Cand. Med. Sci., 1935; D.Med. Sci., 1937. Asst. dept. microbiology 1st Leningrad Med. Inst., until 1935, head chair gen. and tech. microbiology Chemico-Pharm. Faculty, 1935-36; head chair microbiology Leningrad Pharm. Inst., head bacteriology dept. Leningrad Inst. Epidemiology and Microbiology, learned specialist microbiology lab. USSR Acad. Sci., head mycology and bacteriology dept. Leningrad Skin and Venereal Diseases Inst., 1936-60; prof., 1938—; head chair microbiology Leningrad Postgrad. Med. Inst. Presidium mem. learned med. council RSFSR Ministry Health; mem. permanent group experts on antibiotics WHO. Recipient Stalin prize, 1951. Mem. Leningrad. Soc. Microbiologists (bd. mem.), Leningrad Soc. Dermatologists (bd. mem.). Author over 200 works including Guide to Research on Dermatophytes, 1931; Variation of Dermatophytes, 1938; Microbiology, 1938; Dermatomycosis Causative Organisms, Laboratory Diagnosis, Epidemiology, 1950; Practical Use of Antibiotics, 1952; Candidiasis—Causative Organisms, Clinical Aspects and Epidemiology, 1958; co-author: Epidemiology, 1941; Coccidioidomycosis, 1960. Address: Leningrad Postgrad. Med. Inst., ulitsa Saltykova-Shchedrina 41, Leningrad, USSR.

KASNER, Edward, Am. mathematician; b. N.Y., Apr. 2, 1878; s. Bernard and Fanny (Ritterman) K.; B.S., Coll. City N.Y., 1896; A.M., Columbia, 1897, Ph.D., 1899, Sc.D., 1954; postgrad. U. Göttingen, 1900. Tutor math. Columbia U., N.Y.C., 1900-05, instr., 1905-06, adj. prof., 1906-10, prof. from 1910, Adrain prof. from 1937; staff of trans. and bulletins, Am. Math. Soc., Revue Semestrielle des Mathématiques (Amsterdam). Speaker Internat. Congress of Arts and Sciences, St. Louis, 1904, Harvard, 1936; mem. com. math. NRC.; del. Internat. Math. congresses, Bologna and Zurich. Mem. Nat. Acad. Scis., Am. Math. Soc. (v.p. 1908), A.A.A.S. (v.p. and chmn. Section Mathematics and Astronomy, 1909), Circolo Matematico di Palermo, Société Mathématique de France, Phi Beta Kappa. Author: Invariants of the Inversion Group, 1900; Present Problems of Geometry, 1905; Differential-Geometric Aspects of Dynamics (Princeton Colloquium Lectures), 1913; Conformal Geometry, 1915; Einstein's Theory of Gravitation, 1920; Polygenic Functions, 1927; New Names in Mathematics, 1937; Isothermal Families (in Revista and Pastor vols.), 1943; (with James Newman) Mathematics and the Imagination, 1940. Editor: Scripta Mathematica. Coined the words googol and googolplex; did studies in relativity, invariants, divergent series, conformal and differential geometry; made study of trajectories with aid of analytic and contact transformations. Died Jan. 7, 1955.

KASS, Edward Harold, Am. physician; b. N.Y.C., Dec. 20, 1917; s. Hyman A. and Anna (Selvansky) K.; A.B., U. Ky., 1939, M.S., 1941; Ph.D., U. Wis., 1943; M.D., U. Cal. at San Francisco, 1945; M.A. (hon.), Harvard, 1958; Sc.D. (hon.), U. Ky., 1963; m. Fae Golden, Nov. 27, 1943; children—Robert, James, Nancy. Cons. in medicine specializing in infectious diseases, Boston; asso. vis. physician Harvard Med. Services, Boston City Hosp., 1955—; asst. prof. medicine Harvard, 1955-58, asso. prof. bacteriology and immunology, 1958—; asst. physician Thorndike Meml. Lab., Boston, 1958—; asso. dir. Mallory Inst. Pathology, Boston, 1958-63; dir. Channing Lab., dir. dept. med. bacteriology Boston City Hosp., 1963—. Mem. sci. adv. bd. Nat. Kidney Found., com. on urology NRC, adv. com. Iran Found. Recipient Borden Undergrad. award U. Cal., 1947. Mem. Infectious Diseases Soc. Am. (sec.-treas. 1963—), Am. Acad. Arts and Scis., Assn. Am. Physicians, Am. Soc. for Clin. Investigation, Am. Assn. Immunologists, A.C.P., Coll. Am. Pathologists, Am. Pub. Health Assn., Phi Beta Kappa, Sigma Xi, Alpha Omega Alpha. Editor: Progress in Pyelonephritis, 1965; (with E. L. Quinn) Biology of Pyelonephritis, 1960; editorial bd. Jour. Clin. Investigation and Antimicrobial Agts. and Chemotherapy. Research, publs. on mechanisms of resistance to infection, prevention of chronic diseases, prematurity. Home: 37 Barberry Rd., Lexington, Mass. 02173. Office: Channing Lab., Boston City Hosp., Boston 02118.*

KASSABIAN, Mihran Krikor, physician; b. Cesaria, Asia Minor, Turkey, Aug. 25, 1868; ed. Argeus (Am. missionary) High Sch., Cesaria, taught there, 1887-90; came to U. S., 1894; M.D., Medico-Chirurg. Coll., Phila., 1898; married. Enlisted 1898, in hosp. med. corps U. S. A.; instr. electro-therapeutics and skiagrapher, Medico-Chirurg. Coll. and Hosp., 1899-1902; dir. Roentgen Ray Lab., lectr. Roentgen ray Phila. Gen. Hosp., from 1903. Del. to Internat. Congress, Physio-therapy and Internat. Congress, Radiology and Ionization, Liège, Belgium, 1905, to Am. Congress for Tb, 1902. Mem. Am. Electro-Therapeutic Assn. (v.p.). Author: Electro-Therapeutics and Roentgen Rays with chapters on Radium and Phototherapy, 1907. Died 1910.

KASSIRSKY, Iosif Abramovich, Russian internist, hematologist; b. Fergana, 1898; grad. Med. Faculty, Saratov U., 1921; D. Med. Sci. Asst., Med. Faculty, Central Asian U. (now Tashkent Med. Inst.), 1925-26, sr. asst., lectr., 1926-32, prof., head chair faculty therapy, 1932-34; sci. dir. Semashko Central Clin. Hosp., USSR Ministry R.R.'s, 1934—; head 3d chair therapy Central Postgrad. Med. Inst., Moscow, 1936—. Decorated Order of Lenin. Mem. USSR Acad. Med. Sci. (hematology problems commn.), All-Union (bd. mem.), Moscow (bd. mem.), therapeutic socs., Znanie Soc. (med. and biol. sect.), also internat. assns. therapeutics and hematologists. Author: Basic Metabolism and Its Clinical Importance; Essays on Hot Climate Hygiene in the Conditions of Central Asia, 1935; Clinical Aspects and Treatment of Malaria, 1946-48; Diseases of the Blood and Hemopoietic System, 1948; Infectious Hepatitis, 1949; Problems and Scientists, 1949; Outline of Rational Chemotherapy, 1951; Lectures on Rheumatism, 1956; co-author: Tropical Diseases of Central Asia, 1935; Clinical Hematology, 1955; Diseases in Hot Countries, 1959; Audible Systems of Acquired Cardiac Defects, 1964. Co-editor Internal Diseases sect. Large Med. Ency., 2d edit.; mem. editorial council Antibiotics, Clin. Medicine. Research and numerous publs. on hematology, pathology in warm countries, chemotherapy, antibiotics, rheumatism, malaria, history of medicine; did intrathoracic blood transfusion and designed needle for sterile puncture with protective shield; introduced diagnostic marrow puncture in visceral leishmaniasis to replace splenic puncture; developer methods of diagnostic puncture of liver and lymph nodes in cases of neoplasms. Address: Central Postgrad. Med. Inst., pl. Vosstaniya 1-2, Moscow, USSR.*

KAST, Alfred, German physician; b. 1856; introduced phenacetine into therapeutics, 1887; credited with introducing sulfonal as medication, circa 1888. Died 1903.

KASTEN, Frederick H., Am. biologist; b. N.Y.C., Mar. 7, 1927; s. Isaac and Anna (Goldblum) K.; B.A., U. Houston, 1950; M.A., U. Tex., 1954, Ph.D., U. Tex., 1954; m. Agnes Marie Garrison, Feb. 2, 1949; children—Frederick L., Stephen D., Donald J., Glenn T. Cancer research scientist Roswell Park Meml. Inst., Buffalo, 1954-56; asst. prof. zoology Tex. A. and M. U., College Station, Tex., 1956-61; NSF sr. fellow Justus-Liebig U., Giessen, Germany, 1961-62; NSF sr. fellow Institut de Recherches Scientifique sur le Cancer, Ville Juif, France, 1962, NIH spl. research fellow, 1962-63; dir. dept. ultrastructural cytochemistry Pasadena (Cal.) Found. for Med. Research, 1963-67, research coordinator, since 1967—; adjunct associate prof. anatomy U. So. Cal. Sch. Medicine, Los Angeles, 1967—; asso. clin. prof. pathology Loma Linda (Cal.) U. Sch. Medicine and Grad. Sch., 1968—. First Feulgen Meml. lectr., 1962. Mem. A.A.A.S., Electron Microscopy Society of America, Histochem. Soc., Am. Soc. for Cell Biology, Tissue Culture Assn., Am. Assn. for Cancer Research, Am. Soc. Zoologists, Biol. Stain Com., German Histochem. Soc. Editor: (with W. Sandritter) One Hundred Years of Histochemistry in Germany, 1964; also articles. Research in histochemistry on nature Feulgen reaction for DNA; developed new fluorescent Schiff-type reagts.; detected dye contaminants in pyronin dyes, nucleolar lesions in virus-infected cells; determined nucleoprotein synthetic patterns in synchronized mammalian cells, electron microscopy of cultured cells. Home: 582 Prescott St., Pasadena 91104. Office: 99 N. El Molino Av., Pasadena, Cal. 91101.*

KASTLER, Alfred, French physicist; b. Guebwiller, Upper Rhine May 3, 1902; s. Frederick and Anna (Frey) K.; teaching degree, also D.Phys. Sci.; hon. degrees univs. Louvain and de Bise; m. Elise Cosset, Dec. 27, 1924; children—Daniel, Mireille, Claude-Yves. Prof. Lycees de Mulhouse, Colmar, Bordeaux, France, 1931-36; asst. to sci. faculty U. Bordeaux, 1931-36; lectr. sci. faculty Clermont-Ferrand, 1936-38; prof. faculty scis. U. Bordeaux, 1938-41; asso. prof., then prof. faculty scis. U. Paris (France), 1941—; Francqui prof. U. Louvain (France), 1953-54. Recipient grand prize de la Recherche sci. decerne par l'Academie des scis., 1956; grand prize sci. de la Ville de Paris 1963; Nobel prize in physics, 1966; decorated officer Legion of Honor (France). Mem. Inst. Acad. Sci., Flemish Royal Acad. Spl. research on magnetic resonance and optical pumping; developed optical resonance method using light beam and radio frequency field to study energy levels of atoms in excited states. Home: 1 rue du Val-de-Grace, Paris 5e, France.

KÄSTNER, Abraham Gotthelf, German mathematician, astronomer; b. Leipzig, Germany, Sept. 27, 1719; s. Abraham Kästner; studied under his father, and his uncle, G. Pommer; became dozent U. Leipzig, 1739, asst. prof. math., 1746; named prof. math. and physics U. Göttingen (Germany), 1756; dir. Göttingen Obs., 1756-1800. Fellow Royal Soc., 1789. Author: Anfangsgründe der Mathematik, 4 vols., 1758-69; Anfangsgründe der Arithmetik Geometric ebenen . . . , 1759; Analysis of Finite Space, 1760; Elements of Astronomy, 1772-74; Memoirs on Geometry, 1790-91; Geschichte der Mathematik, 4 vols., 1796-1800; also poems and other lit. works. 1st mathematician to write comphrehensive history of math.; 1st to define trigonometric and hyperbolic functions as pure numbers also to show reality of irreducible roots; popularized astronomy and calculus in Germany. Died Göttingen, June 20, 1800.

KASTNER, Georges-Eugène-Frédéric, physicist; b. Strasbourg, France, 1852. Author: Expériences nouvelle sur les flammes chantantes 1873; les Flammes chantantes, 1875; Application de gaz d'éclairage au pyrophone, 1875. Inventor pyrophone. Died Bonn, Germany, 1882.

KASTRUP, Hans, German surgeon; b. Dortmund, Sept. 18, 1909; s. Heinrich and Karoline (Trippe) K.; ed. univs. Bonn, Dusseldorf, Marburg; M.D., prof. medicine; m. Ruth Kuckelke, 1937; children—Christa, Hans-Joachim, Ursula. With Med. Acad. Dusseldorf, Inst. Pathology, U. Hamburg; prof., chief physician dept. surgery Municipal Hosp., Bad Oeynhausen. Mem. German Soc. Surgery, N.W. German Assn. Surgeons, Internat. Coll. Surgeons. Research and publs. on surgery of thorax, stomach, extremities. Home: Bad Oeynhausen-Lohe. Office: Stadt. Krankenhaus, Bad Oeynhausen, Germany.*

KASYANENKO, Vladimir Grigorevich, Russian anatomist, morphologist; b. Sept. 29, 1901; grad. Kiev Vet. and Zootech. Inst., 1926. Instr., Kiev Vet. and Zootech. Inst., 1926-38, prof., 1938-50; head dept. evolutionary morphology Inst. Zoology, Ukrainian Acad. Scis., 1947-50, dir., 1950-64. Decorated Order of Lenin. Mem. Ukrainian Acad. Scis., All-Union (dep. chmn. 1958—), Ukrainian (chmn. 1956—) socs. morphologists. Author: The Motorial and Support Apparatus of Horses: A Functional Analysis, 1947; Regularities of Adaptive Transformations in the Limb Joints of Mammals, 1956. Research in comparative morphology. Address: Ukrainian Acad. Scis., Vladimirskaya 55, Kiev, Ukrainian SSR, USSR.

KASYMOV, Abdul Guseinali oglu, Russian hydrobiologist; b. Lenkoran, Azerbaudzhan, USSR, Mar. 5, 1929; s. G. A. and N. A. Kasymov; ed. Pedagogical Inst. Azerbauzhan, 1959; postgrad. student Inst. Zoology, Acad. Sci. USSR, Leningrad; m. Valida Sadych, May 12, 1933; 1 child, Afag Abdul kyzy. mem. staff Sch. Lab. Hydrobiology, Inst. Zoology Acad. Scis., Azerbaudzhan, 1954-61, chief lab., 1961—; prof. chair zoology Invertebrata State U. Azerbaudzhan, 1957—. Mem. Hydrobiol. Soc. USSR (pres. Azerbaudzhan). Author: Hydrofauna of Lower Kura and Mingchaur Reservoir; also articles. Research on structure of fauna of Caucasian reserviors, history of origin and reconstrn. of forage reserve. Home: Aga-nejmatulla, 51 N8, Baku 52. Office: 6-chrebtoviija, 5 Inst. Zoology, Baku 73, USSR.*

KATAYAMA, Ryosuke, Japanese orthopedic surgeon; b. Mie Prefecture, Japan, Feb. 12, 1901; s. Sanshichi Jo and Iwao (Jo) K.; grad. Jikei U. Sch. Medicine, 1927; m. Kiyoe Yamakawa, Oct. 25, 1932; Kenichi (Mrs. Yahashi), Sachiko (Mrs. Yohashi), Ryuji (Mrs. Yamamoto), Kayoko (Mrs. Yamamoto), Kuniaki, Hiromi, Masahiro. Dir., Tokyu Hosp., Tokyo, 1945—; prof. orthopedics Jikei U. Sch. Medicine, Tokyo, 1945—. Recipient award Service for Crippled Children. Author: Orthopaedic I, Orthopaedic II; Operative Orthopaedics. Research on chemotherapy and operative treatment of bone and joint Tb, functional recovery of tuberculous joints. Home: 3-16-17 Zalmokuza Kamakura City, Japan. Office: 1-45-6, Kitasenzoku, Otaku, Tokyo, Japan.*

KATAYAMA, Yoshiwo, Japanese educator; b. Gumma-ken, Japan, Nov. 16, 1903; s. Kataro and Machi (Matsui) K.; M.S. in Agr., Kyoto U., 1931; Ph.D., Tokyo U., 1937; m. Toshi Shimamoto, Mar. 12, 1934; children—Setsuko (Mrs. Takashi Ban), Akiko. Vice asst. Kyoto U. 1931-33; vice-asst. Tokyo U., 1933-39; technician Dept. Agr. Exptl. Sta., Manchukuo, 1939-43; prof. Miyazaki Agrl. Coll., 1943-50; prof. Miyazaki U., 1950—, dir. library, 1965—. Mem. Genetics Soc., Breeding Soc., Bot. Soc. Author: Plant Breeding, 1956; also numerous articles. Research in cytogenetics and plant breeding, formation of aegilotricum or crossing problems and X-rayed derivatives of wheat or rice including mono- or polyhaploid, cytoplasmic inheritance in some variegated strains in cereals and change of chlorophyll a and b contents in various rice and other materials, allium cytology, cotton improvement, alkaloid reaction in yellow lupine, pharbitis genetics and strains tests of peanut and other crops. Home: 24-5 Funatsuka-cho Miyazaki-shi. Office: 100, Funatsukacho, Miyazaki-shi, Miyazaki-ken, Japan.*

KATCOFF, Seymour, Am. nuclear chemist; b. Chgo., Aug. 19, 1918; s. Nathan and Jennie (Jaffe) K.; B.S., U. Chgo., 1940, Ph.D., 1944; m. Edith Judith Laporte, May 23, 1951; children—Don J., Joel. Asso. chemist U. Chgo. Metall. Lab., 1943-45; staff mem. Los Alamos Sci. Lab., 1945-48; sr. chemist Brookhaven Nat. Lab., Upton, N.Y., 1948—; Weizmann fellow Weizmann Inst. Sci., Rehovot, Israel, 1958-59. Mem. Am. Chem. Soc., Am. Phys. Soc., A.A.A.S. Research, numerous publs. primarily in field of high energy nuclear reactions by radiochem. and nuclear emulsion techniques; determined yields, nuclear properties, recoil ranges of fission products, decay schemes various radioactive isotopes. Home: 200 N. Prospect Av., Patchogue, N.Y. 11772. Office: Brookhaven Nat. Lab., Upton, N.Y. 11973.*

KATER, Henry, English physicist, astronomer; b. Bristol, Eng., Apr. 16, 1777; s. Henry K.; student law sch.; student math. Royal Mil. Coll., Sandhurst, Eng. Staff for pendulum expts. at chief stas. Trigonometrical Survey Eng.; asso. with Arago, Mathieu, Colby in geog. observation studies; asst. to William Lambton on expdn. to survey land between Malabar and Coromandel. Mem. French Acad. Scis., 1819; Fellow Royal Soc., 1814. Recipient Copley medal, 1817. Author: An Account of the Construction and Verification of Certain Standards of Linear Measure for the Russian Government, 1832; Treatise on Mechanics, 1834. Determined length of second pendulum (Kater's pendulum) (led to increased accuracy in Brit. system of weights and measures); invented floating collimator for determining line of collimation of a telescope (Flotina collimator), 1825; research on type of steel best suited for compass needle; improved method of dividing astron. circles on prin. of beam compass; prepared a set of standard measures which was adopted by Russian govt. Died London, Eng., Apr. 26, 1835.

KATES, Morris, Canadian biochemist; b. Galati, Rumania, Sept. 30, 1923; s. Samuel and Toby (Cohen) K.; came to Can., 1924, naturalized, 1930; B.A., U. Toronto, 1945, M.A., 1946, Ph.D., 1948; m. Pirkko Helena Makinen, June 14, 1957; children—Anna-Lisa, Marja, Ilona. Research asst. Banting Inst., U. Toronto, 1948-49; with NRC, Ottawa, Ont., Can., 1949-68, sr. research officer div. bioscis., 1961-68; guest worker Nat. Inst. Med. Research, London, Eng., 1959-60; prof. chemistry dept. U. Ottawa, since 1968.—. Fellow Chem. Inst. Can.; mem. Canadian Biochem. Soc., Biochem. Soc. (London), Am. Chem. Soc., Am. Soc. Biol. Chemists, Am. Oil Chemists Soc. Editorial bd. Chemistry in Can., 1962-65, Jour. Lipid Research, 1965—, Chem. Phys. Lipids, 1966. Research, numerous publs. on chem. synthesis of lecithin, structure of alkaloid gelsemine, enzymatic hydrolysis of phospholipids and glycolipids in plants, isolation of plant phospholipids and glycolipids, biosynthesis of phospholipids in plants and microorganisms, chem. synthesis of glycerol diether-lipids, discoverer new class of alkyl glycerol diether-lipids in halophilic bacteria. Home: 1723 Rhodes Crescent, Ottawa 8. Office: Dept. Chemistry, Ottawa 2, Ont., Can.*

KATO, Eiichi, Japanese physician; b. Tokyo, Japan, May 3, 1924; s. Grenichi and Yasuko (Hisaoka) K.; M.D., Keio U., 1947, Dr. Med. Sci., 1955; Rockefeller fellow Yale, 1955-56; m. Emiko Kazama, Nov. 3, 1947; children—Akiko, Yoichi. Faculty, Keio U., Tokyo, 1948—, lectr., 1964—, chief physician, univ. health service, 1964—. Recipient Waksman Found. research awards, 1960, 64. Mem. Internat. Japanese socs. Nephrology, Gastroenterology, Angiology, Japanese socs. Medicine, Clin. Metabolism, Asian Soc. Cardiology. Author: Physiological Basis and Clinical Practice of Electrolyte and Body Water Balance, 1965; also numerous articles. Research on pathophysiol. changes of body water and electrolyte in various disease states. Home: 4 Daikyo-cho, Shinjuku-ku, Tokyo Japan.*

KATO, Ken, Japanese physician; b. Tokyo, Japan, Sept. 16, 1914; s. Ryoitsu and Sumiko (Santa) K.; M.D., Keio U., 1938, D.Med.Sci., 1951; m. Kazuko (Funaki), Oct. 26, 1941; children—Sadako, Hiroshi, Takayuki. With Keio U., Tokyo, 1938-63, asso. prof. ophthalmology, 1956-63; vis. fellow ophthalmology U. Pa., Phila., 1961-62; prof. Nihon U., Tokyo, 1963—, dir. ophthal. dept. Surugadai Hosp., 1963—. Mem. Japan Diabetic Soc., Japanese Coll. Angiology, Japanese Ophthal. Soc., Japanese Soc. Psychosomatic Medicine. Author: (with M. Matsui) Retinal Changes of Systemic Hypertension and Diabetes Mellitus, 1966; (with M. Uyemura) Atlas of Diseases of the Ocular Fundus, 1961; also numerous articles. Measurement of velocity of capillary blood stream in retina; statis. and histopath. observations on retinal changes of hypertensive and diabetic patients in Japan; founder psychosomatic ophthalmology in Japan. Home: 2-3-7 Yagisawa, Hoy-shi, Tokyo, Japan.*

KATO, Masao, Japanese metallurgist, nuclear scientist; b. Hyogoprefecture, Japan, June 23, 1916; s. Teisuke and Fujiko (Shimakawa) K.; grad. U. Tokyo (Japan), 1940, Ph.D., 1952; m. Ayako Ikeda, Dec. 4, 1942; children—Hiroko, Masayuki. Staff, Furukawa Electric Co. Ltd., 1940-42; asso. prof. metallurgy U.

Tokyo, 1942-62, prof. isotope engring. Inst. Indsl. Sci., Faculty Engring., 1960—, Inst. Space and Aero. Sci., 1964—; lectr. Isotope Sch., Japan Atomic Energy Research Inst., 1958—, Yokohama Nat. U., 1962—, Adviser, Isotope Center, Japan Atomic Energy Research Inst., 1961—. Mem. Japan Radioisotope Assn. (dir.), Japan Inst. Light Metals (Light Metal medal 1954, dir.), Japan Inst. Metals. Author: Aluminum, 1954; Industrial Applications of Radioisotopes, 1957; also numerous articles. Research on metallurgy of aluminum, lead, zinc alloys, mechanism of corrosion reactions of zinc and aluminum alloys, electrolytic refining of copper and nickel, isotope techniques and their applications to indsl. fields; developed first in situ measurement of sand transport on sea bed by radioisotope tracer technique. Home: 16-19 Minami-5 chome, Koenji, Suginami-ku, Tokyo, Japan.*

KATO, Tosio, mathematician; b. Tochigiken, Japan, Aug. 25, 1917; s. Shoji and Shin (Sakamoto) K.; B.Sc. U. Tokyo, 1941, D.Sc., 1951; m. Mizue Suzuki, Mar. 29, 1944. With U. Tokyo, 1943-62, prof., 1951-62; prof. math. U. Cal. at Berkeley, 1962—. Recipient Asahi award, 1960. Mem. Am. Math. Soc., Math. Soc. Japan, Phys. Soc. Japan. Author: Perturbation Theory for Linear Operators, 1966. Research, publs. in functional analysis and differential equations, with applications to math. physics and engring.*

KATO, Yogoro, Japanese chemist; b. Kariya, Japan, 1872; grad. Doshisha U., Kyoto, Japan, later Kyoto U., 1903; D.Sc., 1910; asst. prof., then prof. Tokyo Inst. Tech., now emeritus prof., became chief natural resources chem. inst. lab., 1939, named hon. prof., 1942; now dean tech. dept. Shinshu U.; past mem. research staff Mass. Inst. Tech. Recipient Ranju medal, Japanese Govt. Author books including: Principle of Electric Chemistry; Analytic Chemistry; Industrial Chemistry. Research in chem. solutions at high temperatures; invented oxidized metal magnet; developed method of processing alumina.

KATSH, Seymour, Am. pharmacologist, immunologist, educator; b. N.Y.C., Jan. 13, 1918; s. Julius and Sara (Levine) K.; B.A., N.Y. U., 1944, M.S., 1948, Ph.D., 1950; m. Grace Finkelstein, Apr. 17, 1946; children—Sara, Judith, Naomi. Instr., U. Mass., 1949-52; vis. asso. biologist Brookhaven Nat. Lab., 1950-51; research fellow Cal. Inst. Tech., 1952-55, Carnegie Instn. Washington, 1955-58; faculty dept. pharmacology U. Colo. Med. Center, Denver, 1958—, prof., 1965—, acting chmn. dept., 1964—. Fellow A.A.A.S., Soc. for Exptl. Biology and Medicine, Soc. Study Growth and Devel., Am. Assn. Immunologists; mem. Endocrine Soc., Am. Assn. Anatomists, Am. Soc. Zoologists, Am. Physiol. Soc., Sigma Xi, Beta Lambda Sigma. Research, numerous publs. on reproductive biology, control of problems in fertility and sterility. Home: 485 S. Krameria St., Denver 80222.*

KATSNELSON, Zakhar Saulovich, Russian histologist, biologist; b. Irbit (now Sverdlovsk Oblast), 1903; grad. Med. Faculty, Perm U., 1926; postgrad. Leningrad U., 1930-32; D.Med. Sci., 1936. Asst., lab. asst. dept. histology Perm U., 1924-26; asst. dept. zoology Urals Poly. Inst., Sverdlovsk, 1926-28, sr. asst., 1928-30, also instr. biology, head dept. Urals-Siberian Communist U.; asso. lab. exptl. zoology USSR Acad. Sci., 1930-32, head Lab. Exptl. Biology, Kazakhstan Sta., Alma-Ata, 1932; asst. dept. gen. biology Leningrad U., asst. dept. histology Lesgaft Inst. Phys. Culture, 1930-32; head chair gen. biology 3d Leningrad Med. Inst., 1932-40, Naval Med. Acad., Leningrad, 1940-48; asst., lectr., later dep. head chair histology and embryology 1st Leningrad Med. Inst., 1933-38; prof., 1936—; head chair histology and embryology Leningrad Vet. Inst., 1948—. Mem. All-Union Soc. Anatomists, Histologists and Embryologists (bd. mem. 1946—, dep. chmn. Leningrad dept. 1955—). Author: Types of Animal Organization, Vols. 1-2, 1939; 100 Years of Cell Study, 1939; Lectures on General Biology, 1940, 41, 42, 43, 44; Introduction to the System of Animals and Plants: Manual for a General Biology Course, 1945; Cell Theory, 1959; Protoplasm, 1962; Historical Development of the Cell Theory, 1963; co-author: Practical Studies in Histology and Embryology, 1959; A Practical Guide to Histology and Embryology, 1963; co-author, editor: Theodor Schwann: Microbiological Research, 1939. Research and numerous publs. on histogenesis, reactive lesions, regeneration of different types of epithelium, age changes in striated muscles of vertebrates, use of chem. tests on fetuses of amphibians to study histogenetic factors, amitosis, comparative histology of farm animals, fur-bearing animals and birds, methodology and history of cell theory. Address: Leningrad Vet. Inst., Chernigovskaya ulitsa 5, Leningrad, USSR.

KATSUKI, Shibanosuke, Japanese physician; b. Fukuoka, Japan, June 30, 1907; s. Sadataro and Sei Katsuki; M.D., Kyushu U., 1930, D.M.S., 1935; m. Akiko Katsuki, Nov. 16, 1937; 1 son, Motoya. Med. Faculty, Kyushu U., Fukuoka, Japan, 1935-50, prof. internal medicine, 1956—, dir. Neurol. Inst., 1964—. Mem. Am. (corr.), Japanese neurol. socs.; Japanese Diabetic Assn., Japanese Nephrology Assn., Japanese Soc. Endocrinology, Deutsche Gesellschaft für Neurologie, Am. Acad. Neurology (hon. corr.). Author: Neurological Examination, vols. I, II, 1962; also numerous articles. Research on control mechanism of diencephalic brain on endocrine function, cerebrovascular diseases in Japan especially on true frequency of cerebral hemorrhage and infarction; confirmed that incidence of cerebral infarction in Japan is increasing. Home: 6-36 Arato-3-chome, Fukuoka-City, Japan.*

KATSURA, Shigetoshi, Japanese physicist; b. Sendai, Japan, Apr. 29, 1922; s. Shigehiro and Atsuko (Ito) K.; B.Tech., Tohoku U., Sendai, 1944; Ph.D., 1958; m. Chiyo Teramoto, May 24, 1949; children—Shigeki, Noriko. Faculty, Tohoku U., 1949—, prof. dept. applied sci., 1961-64, prof. dept. applied physics, 1964—; vis. prof. U. Ore., Eugene, 1961-63. Mem. Phys. Soc. Japan, Inst. Elec. Communication Engring. Japan, Am. Phys. Soc. Author: Introduction to Statistical Mechanics, 1960; also articles. Research on theory of condensation by cluster expansion, ferro and antiferromagnetism of Heisenberg model, diffraction of electromagnetic waves by plate and hole, calculation of virial coefficients of imperfect gases. Home: 3-7 Kunimi 3-chome, Sendai, Japan.*

KATTSOFF, Louis Osgood, Am. mathematician; b. Phila., May 7, 1908; s. Samuel and Beka (Kutock) K.; A.B., U. Pa., 1929, M.A., 1930, Ph.D. (Harrison fellow), 1934; m. Hilda Polsky, Aug. 5, 1934; 1 dau., Anita (Mrs. Henry Kaplan). Faculty, U. N.C., 1935-57, Harpur Coll., Endicott, N.Y., 1957-59; prof. math. Boston Coll., Chestnut Hill, Mass., 1959—. Sage scholar Cornell U., 1932-33; DuPont research fellow U. Va., 1934-35. Mem. Am. Math. Assn., Phenomenological Soc., Am. Philos. Soc. Author: Philosophy of Mathematics, 1946; Elements of Philosophy, 1956; Logic and Nature of Reality, 1957; Physical Science and Physical Reality, 1958; Modern Trigonometry, 1960; Unified Algebra, 1962; Finite Mathematics, 1965; Making Moral Decisions, 1965. Contbr. numerous articles to math. jours. on abstract algebra, symbolic logic; foundations of mathematics. Home: 16 Chesley Rd., Newton Centre, Mass. 02159. Office: Carney Hall, Boston Coll., Chestnut Hill, Mass.

KATZ, Amrom Harry, Am. physicist; b. Chgo., Aug. 15, 1915; s. Max and Lena (Saffro) K.; B.A., U. Wis., 1939; postgrad. U. S. Dept. Agr. Grad. Sch., Mass. Inst. Tech., U. Cal. at Los Angeles; m. Louise Klibanow, Nov. 8, 1940; children—Barbara Susan, Deborah Ruth, Michael William. Statis. asst. U. S. Census Bur., 1939-40; jr. physicist to chief physicist Aerial Reconnaissance Lab., USAF, 1940-49, chief physicist, 1949-54; phys. scientist, mem. sr. staff RAND Corp., Santa Monica, Cal., 1954—. Cons., mem. govt. agys.; instr. math. and statistics U. Dayton, 1941-45; lectr. U. So. Cal., Cal. Inst. Tech., U. Cal. at Los Angeles, 1957—; prof. in residence, polit. sci. U. Cal. at Los Angeles, 1963—, sr. fellow Nat. Security Studies Program, 1963—; mem. participant various confs.; mem. bd. sponsors War-Peace Report Mag., 1961—; mem. bd. adv. Disarmament and Arms Control, 1963, Jour. Arms Control, 1963. Recipient Meritorious Civilian Service award USAF, 1947. Fellow Am. Geog. Soc., Soc. Photog. Instrumentation Engrs. (G. W. Goddard award 1963), Am. Soc. Aeros. and Astronautics; mem. Optical Soc. Am., Am. Soc. Photogrammetry, Soc. Photog. Sci. Engrs., Inst. Strategic Studies (London). Research, publs. on analysis and design of aerial reconnaisance systems, new methods of photo-interpretation, theories of strategy, tactics, disarmament and arms control, inspection problems, astronautics and observation satellites. Home: 1410 Jonesboro Dr., Los Angeles 90049. Office: 1700 Main St., Santa Monica, Cal. 90406.*

KATZ, Bernard, physiologist; b. Leipzig, Germany, Mar. 26, 1911; s. Max and Eugenie (Rabinowitz) K.; M.D., U. Leipzig, 1934; Ph.D., U. London, 1938, D.Sc., 1943; m. Marguerite Penly, Oct. 27, 1945; children—David, Jonathan. Beit Meml. Research fellow, 1938-39; Carnegie Research fellow, Sydney, Australia, 1939-42; asst. dir. biophys. research U. Coll., London, Eng., 1946-50, reader, 1950, prof., head biophysics dept., 1952—. Lectr. univs., socs. Mem. Agrl. Research Council, 1967—. Recipient Garten prize U. Leipzig, 1934; Feldberg award, 1965; Baly medal Royal Coll. Physicians, 1967; Copley medal Royal Soc., 1967. Fellow Royal Soc., 1952 (council 1964-65, v.p. 1965); fgn. mem. Royal Danish Acad. Scis. and Letters. Author: Electric Excitation of Nerve, 1939; Nerve, Muscle and Synapse, 1966; also articles. Research on nerve and muscle function especially transmission of impulses from nerve to muscle fibers. Office: Univ. Coll., Gower St., London, Eng.*

KATZ, Joseph Jacob, Am. chemist; b. Detroit, Apr. 19, 1912; s. Abraham and Stella (Assnin) K.; B.Sc., Wayne State U., 1932; Ph.D., U. Chgo., 1942; m. Celia S. Weiner, Oct. 1, 1944; children—Anna, Elizabeth, Mary, Abram. Research chemist Metall. Lab., Manhattan Dist., U. Chgo., 1943-46; sr. chemist Argonne (Ill.) Nat. Lab., 1946—; lectr. U. Chgo., 1963—. Guggenheim fellow, 1957-58. Mem. Am. Chem. Soc. (award for nuclear applications to chemistry 1961, sec.-treas. div. phys. chemistry 1966—), A.A.A.S., Phi Beta Kappa, Sigma Xi. Author: (with G. T. Seaborg) The Chemistry of the Actinide Elements, 1957; (with A. Crewe) Research, U.S.A., 1965. Am. editor Jour. Inorganic and Nuclear Chemistry, 1955—. Contbr. numerous articles to sci. jours. Contbns. to neptunium and plutonium chemistry, fluorine and liquid hydrogen fluoride studies; deuterium isotope studies in chemistry and biology. Home: 5658 Blackstone Av., Chgo. 60637. Office: 9700 S. Cass Av., Argonne, Ill. 60439.*

KATZ, Leo, Am. math. statistician; b. Detroit, Mich. Nov. 29, 1914; s. Max and Mollie (Gastman) K.; B.S. in Elec. Engring., Lawrence Inst., 1936; M.A., Wayne U., 1938; Ph.D., U. Mich., 1945; postdoctoral study U. Cal., 1948, 52-53, U. NC., 1950, Stanford, 1952-53; m. Jean Prepsky, Sept. 5, 1936; children—Michael, Daniel. Statistician Mich. Dept. Labor, 1941-42; mathematician Gen. Motors Corp., 1942-46; asst. prof. mathematics, asso. prof. statistics Mich. State U., 1946-56, prof., head dept. statistics, 1956-63, dir. statis. lab. 1959—; vis. prof. biostatistics Coll. Medicine, U. Cin., 1963-66, prin. investigator Nat. Sci. Found. Research, 1958—; vis. lectr. statistics NSF, 1963-65; cons. editor Wadsworth Pub. Company, 1963-66; statis. cons. European sci. liaison officer for mathematics and statistics, Office Naval Research, U. S. Navy, 1959-60; Ford Found. vis. distinguished prof. U. N.C., 1961-62. Fellow Inst. Math. Statistics, Am. Statis. Assn., Royal Statis. Soc., A.A.A.S.; mem. Am. Math. Soc., London Math. Soc., Psychometric Soc., Math. Assn. Am., Biometric Soc., Am. Assn. U. Profs., Inst. Math. Statistics (council), Sigma Xi. Contbr. articles profl. jours. Directed research on math. models for study of social organization theory, 1952-56; power of 2 x 2 test of independence, 1957-59; theory of math. statistics and probability, 1958-66. Home: 425 Durand St., East Lansing, Mich.*

KATZ, Leon, nuclear physicist, educator; b. Ukraine, Russia, Aug. 9, 1909; s. Jacob and Malka (Katz) K.; B.Sc., Queen's U., 1934, M.Sc., 1936; Ph.D., Cal. Inst. Tech., 1942; m. Georgina May Caverly, Jan. 4, 1941; children—Sylvan, Zender, David, Faye. Research engr. Westinghouse Electric Corp., Pitts., 1942-46; asso. prof. physics U. Sask. (Can.), Saskatoon, 1946-52, prof., 1952—, dir. Accelerator Lab., 1961—, head dept. physics, 1965—. Mem. Sci. Council Can. Fellow Royal Soc. Can., Am. Phys. Soc.; mem. Phys. Soc., Canadian Assn. Physicists (past pres.), Sigma Xi. Research, publs. on thermodynamics, especially specific heats of gases; nuclear physics, especially photo-nuclear reactions. Home: 12 Weir Crescent, Saskatoon, Sask., Can.*

KATZ, Louis Nelson, medical researcher; b. Pinsk, Poland, Aug. 25, 1897; s. Harry J. and Sarah (Rosenberg) Kates; came to U. S. 1900, naturalized, 1904; A.B., Western Reserve U., 1918, M.D., 1921, M.A., 1923, Doctor of Science honoris causa, 1965; m. Aline Grossner, June 15, 1928; one son, Arnold Martin. Intern and asst. res. medicine Cleveland City Hospital, 1921-23; director of Cardiovascular Institute, Michael Reese Hospital and Med. Center, Chgo., 1930—; assistant professor physiology U. of Chicago, 1930-41 professorial lecturer physiology since 1941. Attending physician Michael Reese Hosp. since 1947. Served as private, S.A.T.C., 1918. Pres. III Inter-American Cardiol. Congress 1948. Recipient Lasker award Am. Heart Assn., 1956; permanent hon. pres. Inter-Am. Cardiol. Soc. Fellow A.C.P., A.A.A.S.; mem. Am. (pres. 1951-52, chmn. sci. council, 1952-53), Chgo. (pres. 1954-57) heart assns., Am. Soc. Clinical Investigation. Central Soc. Clin. Research, Am. Society Study Arteriosclerosis (pres. 1954-55), Internat. Cardiol. Society, A.M.A., Am. Physiol. Soc. (pres. 1956-57), Phi Beta Kappa, Sigma Xi. Author: Elements of Electrocardiographic Interpretation, 1932, 3rd edit., 1944; Electrocardiography, 1941; co-author: Experimental Atherosclerosis, 1953; Clinical Electrocardiography, 1956; Nutrition and Atherosclerosis, 1958. Mem. editorial board Acta Cardiologica; editor sect. cardiovascular system Biol. Abstracts, Phila.; bd. editors Circulation, 1958—. Contbr. to nat. internat. med. jours. Research on interpretation of electrocardiogram; studies of arrhythmias, atherosclerosis. Home: 601 E. 32d St. Office: Michael Reese Hosp. and Med. Center, 29th and Ellis Av., Chgo.

KATZ, Sidney, chemist; b. Winnipeg, Man., Can., Aug. 17, 1909; s. David Benjamin and Mollie (Newburger) K.; B.S., U. Man., 1934, M.S., 1935; Ph.D., McGill U., 1937; m. Syma Frieden, July 24, 1937; children—Robert, Barry. Came to U. S., 1940, naturalized, 1948. Research chemist Manhattan Project, Chgo., 1944-45; supr. phys. chemistry Inst. Gas Tech., Chgo., 1945-52; instr. math. Ill. Inst. Tech., adviser IIT Research Inst., 1952—. Mem. Am. Chem. Soc., Sigma Xi, Phi Lambda Upsilon. Research on chem. reaction rates, thermodynamics of reactions, optical properties of small particles. Home: 8833 S. Euclid Av., Chgo. 60617. Office: 10 W. 35th St., Chgo. 60616.*

KATZ, Sol, Am. physician; b. N.Y.C., Mar. 29, 1913; s. Samuel and Bessie (Levy) K.; B.S., Coll. City N.Y., 1935; M.D., Georgetown U., 1939; m. Beatric Paul, Nov. 16, 1946; children—Paul, Rita, Judith. Practice medicine, specializing in internal medicine, Washington; chief pulmonary disease div. D.C. Gen. Hosp., 1945-59; chief med. service VA Hosp., Washington, 1959—; lectr. Med. Sch. George Washington U., 1959—; prof. medicine, Kober lectr. Med. Sch. Georgetown U., 1965—. Cons. NIH, Walter Reed Army Hosp. Diplomate Am. Bd. Internal Medicine. Mem. A.C.P., Am. Coll. Chest Physicians. Author: (with Stephen Sulavik) Pleural Effusions, 1963. Contbr. numerous articles to profl. jours.

Research, publs. primarily in field of chest diseases and antibiotics. Home: 3909 Ridge Rd., Annandale, Va. 22003. Office: VA Hosp., 50 Irving St. N.W., Washington 20422.*

KATZELL, Raymond Abraham, Am. psychologist; b. Bklyn., Mar. 16, 1919; s. Abraham and Fannie (Skoblow) K.; B.S., N.Y. U., 1939, M.A., 1941, Ph.D., 1943; m. Mildred Engberg, May 11, 1953. Faculty, N.Y. U, N.Y.C., 1942-43, 57—, prof., head, dept. psychology, 1963—; personnel psychologist Adj. Gen.'s Office, U. S. War Dept., N.Y.C., 1943-45; faculty U. Tenn., 1945-48, Syracuse U., 1948-51; v.p. Richardson, Bellows, Henry & Co., N.Y.C., 1951-57. Cons. bus., govt. agys., 1945—; chmn. adv. council psychology N.Y. State Dept. Edn., 1965-67. Mem. Am. (pres. indsl. div. 1960-61), Eastern, N.Y. State (pres. 1957-58) psychol. assns., A.A.A.S. Demonstrated that preparatory muscular set is an indication of intent to perform a psychomotor act; showed that morale may be described behaviorally as the math. factor common to divers behaviors of employees; explained basis of leader effectiveness. Home: 1 Barry Dr., Glen Cove, N.Y. 11542.*

KATZEN, Elliott Dexter, Am. aerodynamicist; b. Balt., Apr. 30, 1920; s. Louis and Sadie (Glass) K.; B.S., U. Md., 1943; M.S., U. Minn., 1947; m. June B. Davis, June 4, 1944; children—Phillip E. Sandra S., Sheila L. Staff, Ames Research Center, NASA Moffett Field, Cal., 1947—, sect. head, 1958-59, asst. br. chief, 1959-68, br. chief, 1968—. Asso. fellow Am. Inst. Aeros. and Astronautics. Research, publs. in aerodynamics and flight mechanics aiding in design and improvement of aircraft, missiles and spacecraft. Home: 455 Ferne Av., Palo Alto Cal. 94306. Office: NASA-Ames Research Center Moffett Field, Cal. 94035.*

KATZIN, Leonard Isaac, Am. chemist; b. Eau Claire, Wis., Jan. 18, 1915; s. Morris and Ida (Stein) K.; A.B., U. Cal. at Los Angeles, 1935; Ph.D., U. Cal. at Berkeley, 1938; m. Alice Ginsburg, Sept. 2, 1938; children—Ruth N., Martha R., Lisbeth E., Judith H. Research fellow U. Cal. at Berkeley, 1938-40; biologist USPHS, 1940-42; research fellow radiology U. Rochester (N.Y.) Med. Sch., 1942-43; chemist, asst. sect. chief metall. lab. U. Chgo., 1943-46, vis. prof. chemistry, 1956-57; sr. scientist, group leader Argonne (Ill.) Nat. Lab., 1946—. Lectr. chemistry U. Ill. Extension, 1961; exchange scientist Atomic Energy Research Establishment, Harwell, Eng., 1958-59. Mem. Am. Chem. Soc. (councilor 1962-67), Am. Phys. Soc., A.A.A.S. Author: (with G. T. Seaborg) Production and Separation of Uranium 233: Survey, 1951; Production and Separation of Uranium 233: Collected Papers, 1952; also numerous articles. Developed isolation procedures for gram amounts synthetic, fissionable uranium 233; elucidated radioactive decay series; demonstrated chemistry inorganic salts in organic solvents; research on bonding properties metal ions, polarized light spectroscopy; inventor atomic energy processes.*

KATZMAN, Philip Aaron, Am. biochemist; b. Omaha, May 18, 1906; s. Aaron and Anna (Gorelick) K.; A.B., Kalamazoo Coll., 1927; Ph.D., St. Louis U., 1932; m. Lillian Milder, Aug. 6, 1933; children—Marshall B., Aron S. Mem. faculty St. Louis U., 1932—, prof. biochemistry, 1952—, mem. bd. grad. studies, 1959—, univ. council, 1965-67. Lectr. U. Kan., 1954; vis. prof. U. S.D., 1963; mem. hormone adv. bd. U. S. Pharmacopia, 1940-50. Mem. Am. Soc. Biol. Chemists, Endocrine Soc., Am. Chem. Soc., A.A.A.S., Soc. Exptl. Biology and Medicine, Am. Assn. U. Profs., Royal Soc. Medicine, Sigma Xi. Contbr. numerous articles to profl. jours. Developed methods for preparing gonad-stimulating hormone which occurs in human pregnancy urine; determined biol. properties of estrogens and developed methods for their chem. determination; preparation and use of beta-glucuronidase from E. coli; research in antibiotics, antihormones. Home: 7612 Cornell Av., University City, Mo. 63130.*

KATZNELSON, Harry, microbiologist; b. Starodroga, Russia, Nov. 7, 1912; s. Keva and Necha (Levine) K.; B.S.A., U. B.C., 1934; M.S., State Coll. Wash, 1937; Ph.D., Rutgers U., 1939; postgrad. (fellow) Cornell U.; m. Lillian Rose Edelson, July 6 1941; children—Carla Jan (Mrs. Patrick Arthur Crawley), Peter Richard. Bacteriologist div. bacteriology and dairy research Sci. Service, Can. Dept. Agr., Ottawa, Ont., 1940-50, sr. bacteriologist, 1950-54, prin. bacteriologist, 1954-55, chief bacteriology div., 1955-59, dir. Microbiology Research Inst., 1959-65. Del., 2d UN Internat. Conf. on Peaceful Uses Atomic Energy, Geneva 1958; chmn. microbiology sect. IX Internat. Bot. Congress, Montreal, 1959. Fellow A.A.A.S., Royal Soc. Can.; mem. Canadian Soc. Microbiologists (2d v.p.), Am. Soc. for Microbiologists (past chmn. soil microbiology sect.), Agrl. Inst. Can., Profl. Inst. Can., also others. Research and numerous publs. on soil microbiology, metabolism and diagnosis of bacterial plant pathogens, diseases of honey-bee and their control. Died Feb. 10, 1965.*

KATZOFF, Samuel, aerodynamicist; b. Balt., Aug. 3, 1909; s. Meyer and Rachel (Snyder) K.; B.S., Johns Hopkins, 1929, Ph.D., 1934. With NASA Langley Research Center, Hampton, Va., 1936—, sr. staff scientist, 1964—; faculty U. Va., Va. Poly. Inst. Recipient Spl. Service award Langley Research Center, 1963. Mem. A.A.A.S., Am. Inst. Aeros. and Astronautics. Editor: Thermal Radiation of Solids, 1965. Research, numerous publs. in x-ray determination of molecular arrangement in liquids, theory of wall effects in wind-tunnel testing, theory of flow through axial compressors and turbines, thermal models of spacecraft, application of heat pipes and vapor chambers to thermal control of spacecraft. Home: 222 Regent St. 1, Hampton 23369. Office: Langley Research Center, Hampton, Va. 23365.*

KAUFFMAN, Robert Giller, Am. biologist; b. St. Joseph, Mo., Dec. 29, 1932; s. Elmer Eugene and Ruth (Giller) K.; B.S., Ia. State U., 1954; M.S., U. Wis., 1958, Ph.D., 1961; m. Phyllis Ann Smith, July 10, 1955; children—Rebecca Ruth, Ellen Marie. Instr., U. Wis., 1959-60; asst. prof. U. Ill., 1961-66; asso. prof. U. Wis., Madison, 1966—. Mem. Am. Meat Sci. Assn., Am. Soc. Animal Sci. Contbr. articles to sci. jours. Demonstrated that presence of intramuscular lipid and younger age enhanced palatability of pork; developed anatomical guides for studying muscle, lipid and bone in ovine and porcine carcasses; showed that eye lens of pig continues to grow during severe dietary restrictions and that total nitrogen content of the lens can be used to estimate chronol. age; studied effects of dietary stress, sex, maturation, protein intake on accumulation of lipids in porcine muscles.*

KAUFFMANN, Fritz Josua, bacteriologist; b. Preussisch Stargard, Jan. 15, 1899; s. Albert and Eva (Gottschalk) K.; ed. U. Griefswald, U. Hamburg, U. Munich; m. Hertha Ramann, Jan. 26, 1932. Staff, Robert Koch Inst., Berlin, 1923-33; faculty Statens Seruminstitut, Copenhagen, Denmark, 1933—, chief dept., prof. bacteriology. Chief, Internat. Salmonella Center, 1938-64. Recipient chevalier Danish Dannebrog; Belgian Leopold Ordre; Aronson prize, 1958; Madsen-Legat prize, 1962; Paul Ehrlich prize, 1964. Mem. Royal Danish Acad. Scis., numerous other profl. orgns. Author: Die Bakteriologie der Salmonella-Gruppe; The Bacteriology of Enterobacteriaceae; also numerous articles. Studies on classification of enterobacteriaceae. Home: 27 Amager Faelledvej, Copenhagen, Denmark.*

KAUFMAN, Allan Nathan, Am. physicist; b. Chgo., July 21, 1927; s. Justin and Millie (Low) K.; Ph.B., U. Chgo. 1947, B.S., 1949, M.S., 1951, Ph.D., 1953; m. Louise E. Lazarus, Apr. 14, 1957; children—Joel, Janet. Staff, Lawrence Radiation Lab., Livermore, Cal., 1953-64; sr. lectr. U. Cal. at Los Angeles, 1964-65; asso. prof. physics U. Cal. at Berkeley, 1965-67, prof., 1967—. Fellow Am. Phys. Soc. Research, publs. in plasma physics, electromagnetic properties of matter. Home: 7041 Devon Way, Berkeley, Cal. 94705.*

KAUFMAN, Frederick, chemist; b. Vienna, Austria, Sept. 13, 1919; s. Erwin and Else (Pollack) K.; came to U. S., 1940, naturalized, 1946; Ph.D., Johns Hopkins, 1948; m. Klara Simonyi, Nov. 2, 1951; 1 son, Michael S. Phys. chemist U. S. Army Ballistic Research Labs., Aberdeen Proving Ground, Md., 1948-51, chief phys. chem. sect., 1951-60, chief chem. physics br. Interior Ballistics Lab., 1960-64; lectr. Johns Hopkins, 1948-64; prof. chemistry U. Pitts., 1964—. Cons. indsl. and govt. labs., 1964—. Recipient Rockefeller Pub. Service award, 1955; R. H. Kent award, 1958; U. S. Army Research and Devel. Achievement award, 1962. Mem. Am. Chem. Soc., Am. Phys. Soc., Faraday Soc., Am. Geophys. Soc., Internat. Combustion Inst., Phi Beta Kappa, Sigma Xi. Research in chem. kinetics of combustion reactions, upper atmosphere reactions, atom recombinations, chemiluminescent processes, energy transfer processes. Home: 5854 Aylesboro Av., Pitts. 15217.*

KAUFMAN, Harold Frederick, Am. sociologist; b. nr. Greenville, O., May 6, 1911; s. Charles E. and Trecy (Valentine) K.; B.A., U. Mo., 1938, M.A., 1939; Ph.D., Cornell U., 1942; m. Lois Cook, June 8, 1939; 1 son, Harrell Lynn. Instr., U. Mo., 1942-45; sociologist U. S. Forest Service, Mont., 1945; asst. prof. U. Ky., 1945-48; prof. sociology Miss. State U., State College, 1948—, head dept. sociology and rural life, 1948-61; vis. lectr. Columbia, 1952-53, U. Wis., 1954. Fulbright Research scholar, India, 1961. Mem. Am., So. (past pres.), Rural (past pres.), Indian sociol. socs., Am. Assn. U. Profs., Adult Edn. Assn., European Soc. for Rural Sociology, Author numerous monographs, articles and book revs. to tech. jours., chpts. to books. Research on ways of studying community rank and prestige people, ways studying community stratification, structure and improvement programs. Home: 204 N. Nash St., Starkville, Miss. 39759. Office: Box 5161, State College, Miss. 39762.*

KAUFMAN, Herbert, Am. polit. scientist; b. N.Y.C., Sept. 21, 1922; s. Benjamin Harry and Gertrude (Meltzer) K.; B.S. in Social Sci., Coll. City N.Y., 1942; M.A., Columbia, 1946, Ph.D., 1950; m. Ruth L. Davis, Mar. 19, 1967. Research analyst Pres.'s Com. on Civil Rights, 1947; research asso. Inst. Pub. Adminstrn., N.Y.C., 1948-49; lectr. govt. Coll. City N.Y., 1951-53; faculty Yale, 1953—, prof. polit. sci., 1963—, chmn. dept., 1964-67. Cons. to fed., state, local govts. Co-recipient Fruin-Colnon award Nat. Municipal League, 1960. Fellow, Center for Advanced Study in Behavioral Scis., 1959-60. Mem. Internat., Am. (mem. council 1966-67) polit. sci. assns. Author: The Forest Ranger: A Study in Administrative Behavior, 1960; Politics and Policies in State and Local Governments, 1963; (with Wallace S. Sayre). Governing New York City, 1960. Office: Dept. Polit. Sci., Yale U., New Haven 06520.*

KAUFMAN, Hyman, Canadian mathematician; b. Lachine, Que., Can., Feb. 2, 1920; s. Solomon and Anna (Sabesinsky) K.; B.Sc., McGill U., 1941, M.Sc., 1945, Ph.D., 1948; postgrad. Brown U., Yale; m. Sylvia Van Straten, June 21, 1959. Research geophysicist Continental Oil Co., Ponca City, Okla., 1949-50; mem. theory group Lab. for Electronics, Boston, 1950-52; faculty McGill U., Montreal, Que., 1952—, prof. math., 1963—. Cons. to pvt. cos. Mem. Am. Math. Soc., Math. Assn. Am., I.E.E.E., Canadian Math. Congress, Soc. for Indsl. and Applied Math., Soc. for Exploration Geophysics, Canadian Operations Research Soc., Operations Research Soc. Am. Author: (with G. E. Roberts) Table of LaPlace Transforms, 1966; also articles. Research on properties correlation function in statis. communication theory, spl. properties differential equations. Home: 533 Querbes St., Montreal 8, Que., Can.*

KAUFMAN, M(oses) Ralph psychiatrist; b. nr. Beltz, Bessarabia, Russia, Oct. 5, 1900; s. Jakov and Sarah (Straker) K.; came to U. S., 1925, naturalized, 1937; M.D., McGill U., 1925; m. Ida Elizabeth Esack, Sept. 2, 1925; children—Paul Bettina Deborah. Intern Manhattan State Hosp., Ward's Island, N.Y., also asst. Vanderbilt Clinic, N.Y. City, 1925-26; resident neurol. staff Montefiore Hospital, 1926-27; assistant senior physician Boston Psychopathic Hospital, 1927-28; Commonwealth Research Fellow Harvard, Vienna, 1928-31; clinical director McLean Hosp., Waverly, Mass., 1931-33; chief psychiatrist Mt. Sinai Hosp., N.Y.C., 1945— dir. dept. psychiatry Inst. Psychiatry; prof. psychiatry Mount Sinai Medical Sch., 1964—; clin. prof. psychiatry Columbia Physicians and Surgeons College, 1948—; dean Page and William Black Post-Graduate School Medicine. Chairman medical advisory board of Hebrew University, Hadassah Med. Sch., Jerusalem, Israel. Diplomate Bd. of Psychiatry and Neurology, 1935. Fellow A.M.A. New York Academy Medicine (vice president), American Psychiatric Association (vice president 1963-64); mem. Am. Psychoanalytic Assn. (past pres.) Assn. Research Nervous Mental Disease, Am. Psychosomatic Soc., Harvey Soc., Mass. Psychiatric Soc., N.Y. Psychoanalytic Soc., N.Y. Soc. Clin. Psychiatry, A.A.A.S. Home: 1170 Fifth Av. Office: Mt. Sinai Hosp. N.Y.C. 10029.

KAUFMAN, Nathan, Canadian physician; b. Lachine, Que., Can., Aug. 3, 1915; s. Solomon and Anna (Sabesinsky) K.; B.Sc., McGill U., 1937, M.D., C.M., 1941; m. Rita Friendly, Sept. 10, 1946; children—Naomi, Michael, Miriam, Hannah, Judith. Faculty, Western Res. Med. Sch., 1948-60, asso. prof. pathology, 1954-60; prof. pathology Duke U. Sch. Medicine, 1960-67; prof., head dept. pathology Queen's U., Kingston, Ont., Can., 1967—. Sci. adv. com. and grants rev. subcom. United Health Found., 1964—; cons. pathologist VA Hosp., Durham, N.H., 1960-67, Canadian Forces Hosp., Kingston, 1968—. Diplomate Am. Bd. Pathology. Mem. Am. Soc. Clin. Pathologists, Am. Assn. Pathologists and Bacteriologists, Canadian Med. Assn., Internat. Acad. Pathology (council), A.A.A.S., Soc. Exptl. Biology and Medicine, Tissue Culture Assn., Am. Assn. for Study Liver Diseases, Am. Assn. Cancer Research, Am. Soc. Exptl. Pathology, N.Y. Acad. Scis., Am. Soc. Cell Biology, Am. Soc. Clin. Nutrition, Am. Inst. Nutrition, Am. Soc. Cytology, Am. Inst. Biol. Scis., Canadian Inst. Pathologists, Ont. Med. Assn., Ont. Assn. Pathologists, Sigma Xi. Asso. editor Lab. Investigation, 1952-66; asso. editor Am. Jour. Pathology, 1967, editorial bd., 1967—. Research iron absorption and metabolism, use of ethionine, an antimetabolite as a model for study of iron loading, altered iron metabolism, and altered tissue affinity for iron. Home: 84 Runnymede Rd., Kingston, Ont., Can.*

KAUFMAN, Seymour, Am. biochemist; b. Bklyn., Mar. 13, 1924; s. Charles and Anna (Blank) K.; B.S., U. Ill., 1945, M.S., 1946; Ph.D., Duke, 1949; m. Elaine Elkins, Feb. 6, 1948; children—Allan, Emily, Leslie. Fellow dept. pharmacology N.Y. U. Med. Sch., 1949-50, mem. faculty, 1950-54; chemist Lab. Cellular Pharmacology, Nat. Inst. Mental Health, NIH, Bethesda, Md., 1954-56, chief sect. on cellular regulatory mechanisms, 1956—. Mem. Am. Chem. Soc., Am. Soc. Biol. Chemists, Harvey Soc. Editor, Jour. Biol. Chemistry, 1964—; Archives of Biochemistry and Biophysics, 1964—. Research, publs. on mechanism of enzymatic Alpha-Ketoglutarate oxidation and synthesis of high-energy phosphate compounds accompanying this oxidation, mechanism of enzymatic hydroxylation of phenylalanine; discovered new hydroxylation coenzyme. Office: NIH, Bethesda, Md. 20014.*

KAUFMANN, Albert Rudolph, Am. metallurgist; b. Phila., Nov. 20, 1911; s. John Jacob and Anna Frances (Brieger) K.; B.S., Lafayette Coll., 1933; Sc.D., Mass. Inst. Tech., 1938; m. Gertrude Elizabeth Moakler, July 16, 1937; children—Albert Rudolph, Ann Marie, Nancy E. (Mrs. William Strathearn), Joan. Staff, Mass. Inst. Tech., Cambridge, 1937-55, asso. prof., 1948-55; v.p., tech. dir. nuclear metals div. Textron, Inc., West Concord, Mass.,

1955—. Mem. Am. Phys. Soc., Am. Soc. Metals, Am. Inst. Mining and Metall. Engrs., Am. Nuclear Soc. Contbr. chpts. to books, articles to profl. jours. Research on coextrusion method for prodn. of fuel elements; developed practical applications for new metals and materials. Home: 84 Maple St., Lexington, Mass. 02173. Office: Nuclear Metals div. Textron, Inc., West Concord, Mass. 01781.*

KAUFMANN, Berwind Petersen, Am. biologist; b. Phila., Apr. 23, 1897; s. Rudolph H. T. and Ida (Petersen) K.; B.S., U. Pa., 1918, M.A., 1920, Ph.D., 1925; m. Jessie Thomson McCulloch, Apr. 12, 1924; children—Berwind N., Carl B., Anders J. Prof. biology Southwestern U., Memphis, 1926-28; prof., head dept. botany U. Ala., 1928-36; resident investigator dept. genetics Carnegie Instn. Washington, 1937-60, acting dir., 1961-62, dir., 1962; prof. zoology and botany, sr. research scientist U. Mich., Ann Arbor, 1962—; guest investigator Brookhaven Nat. Lab., 1955-62. Cons. USPHS, 1944—; exec. com. NRC, 1957-62, mem. biology council, 1954-62. NRC fellow in biol. scis. Cal. Inst. Tech., 1932-33. Mem. Nat. Acad. Scis., Genetics Soc. Am. (past pres.). Author: (with M. Demerec) Drosophila Guide, 1940; also numerous articles. Research in cytology, cytogenetics plants and animals, radiation biology; proposed theory that chromosomes higher plants and animals are double stranded helical entities at all stages div. Home: 2650 Heather Way, Ann Arbor, Mich. 48104.*

KAUFMANN, Carl, German gynecologist; b. Malmedy, Germany, Aug. 21, 1900; s. Karl and Suzanne (Raufenstrauch) K.; received M.D.; became prof. U. Berlin, 1936, prof., clin. dir. U. Marburg, 1946; joined U. Cologne (Germany), 1954; dir. U. Women's Clinic, Cologne. Advanced use of estrogenic hormone, 1932. Office: Universitäts, Frauenklinik, Cologne, Germany.

KAUFMANN, Hans Paul, German chemist; b. Frankfort, Oct. 20, 1889; s. Paul and Susanne (Fenner) K.; ed. univs. Jena, Heidelberg, Berlin; Ph.D., D.Natural Sci. honoris causa; m. Marianne Sinzinger, Mar. 20, 1922; 1 son, Hans-Jürgen. Instr., U. Jena, 1919; asso. prof., full prof. at Münster and Berlin, 1931; prof. chemistry and pharmaceutics, dir. Central Research Inst. Fatty Bodies, Münster. Recipient Nomann Gold medal. Mem. German Assn. Fatty Bodies (pres.), Am. Chem. Soc., Am. Oil Chemists Soc. Author: Arzneimittel Synthese; Handbuch der Fett-Analyse; Fett-Seifen Anstrichmittel; Die Ernährungsindustrie. Home: Lortzingstrasse 10. Office: Pinsallee 76, Münster/Westf., Germany.

KAUFMANN, Henri Léon, French cardiologist; b. Maisons-Alfort, July 6, 1913; s. Bernard and Estelle (Reinhold) K.; M.D., Faculty Medicine, Sorbonne, Paris; m. Adèle-Anne-Marie Martin, Oct. 16, 1944; children—Martine, Jean, Claudine. Intern, Paris hosps., 1936, med. asst., 1946; head cardiology clinic Faculty Medicine, 1945, lab. chief, 1955; head physician cardiology service Preventive Inst., 1956. Lectr., Broussais Hosp., Columbia. Mem. Cardiology Soc., Allergy Soc., Renal Pathology Soc. Author: Maladie d'Osler; Notions pratiques de vectrographie, also numerous publs. on cardiovascular and gerontological diseases. Address: 25, rue de la Faisanderie, Paris 16, France.

KAUFMANN, Herbert Josef, pediatric radiologist; b. Frankfort a/Main, Germany, June 10, 1924; s. Eugene B. and Margarethe (Fuchs) K.; Bacc., Basel (Switzerland) Med. Sch., 1944, M.D., 1951; grad. Swiss Fed. Med. Bd., 1950; m. Anita K. Böhme, June 26, 1953; children—Michael, Daniel. Chief childrens O.P.D.-B.C.H., chief coordinator for polio, instr. pediatrics Boston U. Med. Sch., 1955-58; staff Children's U. Hosp., Basel, 1958—, head physician pediatric radiology, 1965—; privat docent for pediatric radiology U. Basel Med. Faculty, 1963—. Mem. Soc. Pediatric Radiology N.Am. (corr.), European Soc. Pediatric Radiology (pres. 1966-67), Royal Soc. Medicine (affiliate), Swiss Med. Soc. Author: Röntgenbefunde am kindlichen Becken b. angeborenen Skelettaffektionen u. chromosomalen Aberrationen, 1964; also numerous articles. Editor: Progress in Pediatric Radiology, 1967. Research on neonatal transumbilical angiography, neonatal toxicity problems of drugs; diagnostic importance of pelvic X-ray configuration in systemic skeletal disorders; X-ray appearance of Candida-Esophagitis. Home: 47 Feierabendstr. CH 4000, Basel. Office: Children's U. Hosp., CH-4000, Basel, Switzerland.*

KAUFMANN, Walther, German physicist; b. Lutzen, Dec. 7, 1887; s. Paul and Adelheid (von Merkel) K.; ed. U. Stuttgart, U. Hanover; D.Eng.; m. Gertrud Kleemann, Nov. 16, 1916; m. 2d, Carola Dalheimer, May 27, 1941; 1 dau., Ursula. With industry, 1913-19, instr., 1919; prof. mechanics at Hanover, 1920; prof. mechanics and aerodynamics U. Munich, 1932, prof. emeritus, 1955—. Mem. Bavarian Acad. Sci., Verein Deutscher Ingenieure, Soc. Applied Math. and Mech. Sci. Author: Mechanik starrer Körper, 1927; Statik der Tragwerke, 1957; Angewandte Hydromechanik, 1963. Address: Lindenstrasse 15, Lochhambei-Munich, Germany.

KAUFMANN, William Weed, Am. polit. economist; b. N.Y.C., Nov. 10, 1918; s. Charles Barnard and Nettie (Cramer) K.; B.A., Yale, 1939, M.A., 1947, Ph.D., 1948; m. Julia Thorns Alexander, Feb. 23, 1962. Instr., Yale, 1948-51, Ford Found. prof., 1959-60; faculty Princeton, 1951-55; mng. editor World Politics, Princeton, N.J., 1951-55; sr. staff mem. Rand Corp., Santa Monica, Cal., 1956-59, head social sci. dept., 1960-61; prof. Mass. Inst. Tech., Cambridge, 1961—. Cons. to govt., industry. Author (with G. Craig, R. Hilsman, K. Knorr), editor: Military Policy and National Security, 1956; author: The McNamara Strategy, 1964. Studies on problems of nuclear deterrence, limited war, and def. of NATO area, resulting in recommendations concerning the force structure and strategy of U. S. Home: 100 Memorial Dr., Cambridge, Mass. 02142.*

KAULBERSZ, Jerzy, Polish physiologist; b. Kalisz, Poland, July 19, 1891; s. Tadeusz and Emilia (Bogdanska) K.; student U. Munich (Germany), 1908-11; Dr.phil.nat., U. Freiburg (Germany), 1912; postgrad. U. Tübingen (Germany), 1914, U. Vienna (Austria), 1915-18; Dr.med., U. Krakow (Poland), 1920; m. Zofia Markiewicz, May 4, 1957. Asst., physiol. Inst. Krakow, 1922-30, docent, 1930-34, prof. physiology, head dept. physiology, 1934-60; prof., head dept. physiology Higher Sch. Phys. Edn., Cracow, 1960—; research asso., vis. Wayne U., Detroit, 1940-47, Vassar Coll., Poughkeepsie, N.Y. 1942. Rockefeller Found. fellow, 1958-59. Named Knight of Cross of Polonia Restituta, 1958. Mem. Am., Polish (pres. Krakow br. 1938—) physiol. socs. Author: Physiology of Vegetative Functions, 1953; Physiology of the Motion and of the Sensation, 1958; also numerous articles. Research on resistance of red blood cells at high altitudes, absence of urogastrone in urine after hypophysectomy and adrenalectomy, enterogastrone localization in different parts of small and large intestines. Address: 16 Grzegorzecka, Krakow, Poland.*

KAUNTZE, William Henry, Brit. surgeon; b. 1887; s. W. G. Kauntze; B.A., M.D., Ch.B., Victoria U., Manchester, Eng.; M.B., B.S., London U.; m. Edith Davidson Johnston, 1919; 1 dau. House surgeon Manchester Royal Infirmary, 1911-12; resident med. officer Salford Union Infirmary, 1912-13; med. officer W. African Med. Staff, 1914-16; served with W. and E. African Forces, 1914-19; sr. bacteriologist, dep. dir. lab. services Kenya Colony, 1919-32; dir. med. services, Uganda, 1932-41; asst. med. adviser Colonial Office, 1941-44, dep. med. adviser, 1944; med. adviser to Sec. State for Colonies, from 1944. Fellow Royal Coll. Physicians; mem. Royal Coll. Surgeons. Author: (with N. Jewell) Handbook of Tropical Fevers, 1931. Died Nov. 4, 1947.

KAUTSKY, John H., educator; b. Vienna, Austria, Mar. 5, 1922; s. Felix and Marta (Mueller) K.; came to U. S., 1939, naturalized, 1943; B.A., U. Chgo., 1946, M.A., 1947; M.A., Harvard, 1949, Ph.D., 1951; m. Lilli Unger, Oct. 17, 1947; children—Catherine C., Peter K. Research analyst Dept. State, Washington, 1951-53; research asso. Mass. Inst. Tech., 1953-54; vis. asst. prof. govt. U. Rochester, 1954-55; faculty Washington U., St. Louis, 1955—, prof. polit. sci., 1963—. Research fellow Russian Research Center, Harvard, 1958-59; guest prof. Inst. Advanced Studies, Vienna, 1964. Rockefeller Found. grantee, 1958-59. Mem. Am. Polit. Sci. Assn., (sec. 1965-66), Midwest Conf. Polit. Scientists, Am. Assn. U. Profs., Phi Beta Kappa. Author: Moscow and the Communist Party of India, 1956; Political Change in Underdeveloped Countries, 1962; Communism and the Politics of Development, 1968; also articles. Editorial bd. Midwest Jour. Polit. Sci., 1962-65.*

KAUZMANN, Walter, Am. chemist; b. Mount Vernon, N.Y., Aug. 18, 1916; s. Albert F. and Julia M. (Kahle) K.; B.A., Cornell U., 1937; Ph.D., Princeton, 1940; m. Elizabeth A. Flagler, Apr. 1, 1951; children—Charles Peter, Eric Flagler, Katherine Elizabeth Julia. Westinghouse Research fellow, 1940-42; jr. fellow Harvard, 1942; with Explosives Research Lab., Bruceton, Pa., 1942-44, Manhattan Project, Los Alamos, 1944-46; faculty chemistry Princeton, 1946—, David B. Jones prof., 1963—, chmn. dept., 1964—. Recipient Linderstrom-Lang medal, 1966. Guggenheim fellow, 1957. Mem. Nat. Acad. Scis., Am. Acad. Arts and Scis., Am. Chem. Soc., Am. Phys. Soc., Am. Soc. Biochemists, A.A.A.S., Fedn. Am. Scientists, Am. Assn. U. Profs., Sigma Xi. Author: Quantum Chemistry, 1957; Thermal Properties of Matter I: Kinetic Theory of Gases, 1966, II: Thermodynamics and Statistics, 1967; also articles. Research in phys. chemistry of proteins, theory of optical rotatory power, quantum chemistry. Home: 4 Newlin Rd., Princeton, N.J. 08540.*

KAVANAUGH, William Harrison, Am. mech. engr.; b. Williamsport, Pa., Aug. 19, 1873; s. Daniel and Emma (Ramsey) K.; M.E., Lehigh U., 1894; m. Julia Sara Vogt., Feb. 20, 1896; children—Emma Cosette (Mrs. William Harry Regelman), William Ramsey. Prin., Miners and Mechanics Inst., Freeland, Pa., 1894-95; mercantile business, Williamsport, 1895-97; instr. mech. engring. U. Ill., 1897-98; draftsman and chief draftsman Pa. R.R., 1898-1901; instr. mech. engring. and asst. prof. exptl. engring. U. Minn., 1901-07, prof. exptl. engring., 1907-16; prof. exptl. engring. U. Pa., Phila., from 1916; also cons. engr. Mem. Internat. Jury of Awards, Panama-Pacific Expn., San Francisco, 1915, Sesquicentennial Expn., Phila., 1926; mem. commn., Chgo., to report on "mushroom" system of concrete constrn., 1913; designed Exptl. Engring. Lab., U. of Minn. Died May 6, 1939.

KAVESH, Robert Allyn, Am. economist; b. N.Y.C., Sept. 12, 1927; s. Samuel and Pearl (Berlin) K.; B.S., N.Y. U., 1949; A.M., Harvard, 1950, Ph.D., 1954; m. Ruth Freidson, June 24, 1951; children—Richard, Laura, Andrew. Asst. prof. econs. Dartmouth, 1953-56; bus. economist Chase Manhattan Bank, N.Y.C., 1956-58; prof. econs. and finance Grad. Sch. Bus. Adminstrn., N.Y. U., N.Y.C., 1958—. Mem. Am. Finance Assn. (sec.-treas. 1961—), Regional Sci. Assn. (past sec.), Am. Econ. Assn. Author: (with W. F. Butler) How Business Economists Forecast, 1965; also articles. Developed methodology and procedure to link econ. and financial counter parts bus. activity.*

KAVETSKY, Rostislav Yevgenevich, Russian pathophysiologist; b. Samara, 1899; grad. Med. Faculty, Samara U., 1925; D.Med. Sci., 1938. With Moscow Postgrad. Med. Inst., 1925-31; dir. dept. exptl. oncology Bogomolets Inst. Exptl. Biology and Pathology, Kiev, 1931-53; sr. asso. Inst. Clin. Physiology, Ukrainian Acad. Sci., 1933-46, dir., 1946-52, Presidium mem., chmn. dept. biol. sci., 1952-60, dir. lab. compensatory and protective functions Inst. Physiology, 1953-60; head chair path. physiology Kiev Postgrad. Med. Inst., 1937-55; dir. lab. radioactive isotopes Kiev Research Inst. Radiology and Oncology, 1953-60; dir. Ukrainian Inst. Exptl. and Clin. Oncology, 1960—. Presidium mem. learned council Ukrainian Ministry Health; chmn. Commn. for Coordinating Cancer Research in Ukraine. Decorated Order of Lenin. Mem. Ukrainian Acad. Sci., Ukrainian Soc. Med. History (bd. mem.), Ukrainian Soc. Oncologists (dep. chmn.) Author: (monographs) Tumors; Alkali-Acid Balance; Respiration of the Tissue, 1947; The Tumor and the Body, 1962; co-author: The Sources of Soviet Medicine, 1954; The Tumor Process and the Nervous System, 1958; The Nature of Malignant Tumors, 1959. Mem. editorial bd. Large Med. Ency., 2d edit., Ukrainian Soviet Ency.; mem. exec. bd. Med. Jour. of Ukrainian Acad. Sci. Research and publs. on oncology, etiology and pathogenesis of malignant tumors, metabolism of tumor tissue and tumorous body, protective role of connective tissue, pathogensis of tumor process, role of nervous system and disturbances of endocrine regulation in devel. of malignant neoplasms, increased resistivity of body to tumors, chemotherapy and use of radioactive isotopes for diagnosis and treatment of tumor processes, tests of body response, methods of affecting mesenchyma, nervous and endocrine systems, molecular mechanism of carcinogenesis, history of medicine. Address: Research Inst. of Exptl. and Clin. Oncology, Vasilkovskaja 65, Kiev, Ukrainian SSR, USSR.

KAWAISHI, Kunio (Rinkei) Japanese physician; b. Hiroshima Prefecture, Japan, 1895; grad. Tokyo U., 1921; received M.D. Became asst. prof. Aichi Med. Coll., 1923, Prof., Taihoku (Taipei) U., 1938, dir. hosp. of Taihoku U., 1944, prof. Hiroshima U., 1949; became pres. Hiroshima Med. Coll., 1952—; dean med. dept. Hiroshima U., 1953—. Research on surgery internal organs, transfusion of animal blood to humans.

KAWAKAMI, Izumi, Japanese biologist; b. Kagoshima, Japan, June 25, 1912; s. Kasunosuke and Himo Kawakami; D.Sc., Kyoto (Japan) U., 1945; m. Itue Kawakami, Aug. 18, 1936; children—Chineko (Mrs. Hideo Maki), Kanji, Shoko, Yukito. Asso. prof. Kyoto (Japan) U., 1949; prof. Kyushu U. Fukuoka, Japan, 1949—. Mem. Soc. Cell Biology (committeeman), Soc. Electron Microscopy (committeeman). Author: Causality of Development, 1945; also articles. Research on developmental mechanism of sensory organs, active agts. of vertebrate tissues in primary induction. Office: Hakozaki, Fukuoka, Japan.*

KAWAKAMI, Tasae, Japanese fisheries scientist; b. Niigata, Japan, July 10, 1911; s. Kaichi and Toku (Takahashi) K.; student Imperial Fisheries Inst., 1929-35; doctorate Kyoto U., 1952; m. Sige Tihara, Mar. 23, 1941; children—Yoko, Kyoko, Mariko. Asst., Imperial Fisheries Inst., Tokyo, 1935—; faculty Kyoto Imperial U., 1947—, prof. applied physics 1953—. Author: Applied Mathematics for Agriculture and Fishery Science, 1958; also numerous articles. Research in applied physics, applied math., statistics for fishery sci.; mechanics of fishing gear, math. and statis. analysis of fishery resources, inshore oceanography; fishing techniques and tactics. Home: 944 Suyama, Maizuru, Kyoto-fu, Japan.*

KAWAMURA, Yojiro, Japanese neurophysiologist; b. Tokyo, Japan, Nov. 7, 1921; s. Kosaku and Kaneki Kawamura; M.D., Osaka U., 1940, D.M.S. in Physiology, 1952; m. Haru Ischizawa, Mar. 21, 1947; children—Yoko, Kiyoshi. Faculty, Osaka (Japan) U., 1946—, prof. oral physiology, 1959—, dir. Bio-Med. Library, 1966—. Vis. prof. physiology U. Cal. at Los Angeles Med. Center, 1960. Rockefeller Found. research fellow, 1960. Mem. Internat. Brain Research Orgn., Physiol. Soc. Japan, Internat. Assn. for Dental Research (councilor Japanese div.). Author: New Oralphysiology, 1957; Clinical Oralphysiology, 1959. Contbr. numerous articles to profl. jours. Introducer neurophysiology to dental sci., developer modern oralphysiology. Home: 18, 2-chome, Higashimachi, Hotarugaike, Toyonaka, Osaka, Japan.*

KAWASAKI, Chikataro, Japanese pharmacist; b. Kyoto, Japan; grad. Tokyo U., 1929; m. Misako Hattori. Successively instr. Tokushima Higher Tech. Sch., Tokyo Pharm. Coll.; later prof. Osaka U. Research and publs. on effect of acidity on sterene, colestelin, vitamins; simplified methods for detecting vitamins A and B1 in urine.

KAWASE, Osamu, Japanese pathologist; b. nr. Kyoto, Japan, July 12, 1909; s. Tai and Tuya (Kanematu) K.; M.D., Kyoto U., 1934, D.M.Sc., 1951; m. Sumi Zenno, June 1, 1946; children—Kuni, Atusi. Faculty, Kyoto U., 1934-46, prof. pathology Med. Coll., 1939-46; prof. pathology Osaka Women's Med. U., 1947-51, Inst. Constnl. Medicine, Kumamoto (Japan) U., 1951—. Mem. Japanese Path. Soc., Japanese Cancer Soc., Japanese Soc. for Electron Microscopy, Gerontological Soc. Am. Research, publs. on electron micros. structure of elastic fiber and arterial elastic membrane; visualization of fine structural disintegrating processes of the latter in atherosclerotic and hypertensive arterial lesions and age-related changes in infancy for establishment of initial stages of these lesions. Home: 1-12-9 Ooemati, Kumamoto. Office: Inst. Constl. Medicine, Kumamoto U., Kumamoto, Japan.*

KAWASSIADES, Constantine, Greek chemist; b. Velimachi, Nov. 14, 1904; s. Theodossius and Vassiliki N. Kawassiades; ed. U. Athens, U. London; Ph.D.; m. Anna Papadopoulos, Oct. 1, 1950; 1 son, Theodossius. Prof., U. Athens, 1935; prof. U. Salonika, 1940, prof. Faculty Indsl. Studies, 1957, dean Faculty Scis., 1944, 63, rector, 1955-56. Mem. Greek Atomic Energy Commn. Mem. Hellenic Union Chemists, Ramsay Soc., Chemistry Soc., N.Y. Acad. Scis. Author: Reduction by Hydrogen; Corrosion by Water Containing Aluminum Sulphate; The Rate of Absorption of Carbon Monoxide by Solutions of Cuprous Chloride; General and Inorganic Chemistry; General and Electronic Chemistry. Home: Odos Angheláki 1. Office: Panepistemion Thessaloniki, Thessaloniki, Greece.

KAY, G(eorge) F(rederick), geologist; b. Virginia, Ont., Can., Sept. 14, 1873; s. Joseph Sidney and Elizabeth Marshall (Rae) K.; B.A., U. Toronto, 1900, M.A., 1902, LL.D., 1936; fellow, U. Chgo., 1903-04, Ph.D., 1914; hon. D.Sc., Cornell Coll., Mt. Vernon, Ia., 1935; m. Bethea Hopper, Dec. 1902; children—Marjorie Kay McLaughlin, Marshall, Calvin. Prin. pub. schs., Ont., 1892-94; geologist, Lake Superior Power Co., 1900-02; asst. geologist, Ont. Bur. Mines, 1903; asst. prof. geology and mineralogy, U. Kan., 1904-07; prof. geology State U. Ia., Iowa City, from 1907, head dept. geology, 1911-34, dean Coll. Liberal Arts, 1917-1941; jr. geologist U. S. Geol. Survey, from 1907; state geologist of Ia., 1911-34. Fellow A.A.A.S. (v.p. 1929), Geol. Soc. Am., Ia. Acad. Sci. (pres. 1929). Research and publs. on mineral resources of southwestern Oregon and Bering Coal Field; Pleistocene geology; history of Pleistocene deposits; Gumbotil; pre-Illinoian Pleistocene geology of Iowa; ages of drift sheets; mapping the Aftonian and Yarmouth interglacial horizons in Iowa; significance of post-Illinoian, pre-Iowan loess; classification and duration of the Pleistocene period. Died Iowa City, July 19, 1943.

KAY, Jerome Harold, Am. thoracic and cardiovascular surgeon; b. St. Cloud, Minn., Mar. 17, 1921; s. Louis and Frances (Kerlan) K.; A.A., U. Cal. at Los Angeles, 1941; B.A., U. Cal. at San Francisco, 1943, M.D., 1945; m. Adrienne Levin, June 15, 1950; children—Gregory, Stephen, Karen, Cathy, Robert, Richard. Faculty, U. So. Cal., Los Angeles, 1956—, asso. prof. surgery, chief cardiac surgery, 1958—; head physician thoracic surgery Los Angeles County Gen. Hosp., 1958—; chief cardiac surgery St. Vincent's Hosp., Los Angeles, 1958—. Diplomate Am. Bd. Surgery, Am. Bd. Thoracic Surgery. Fellow A.C.S., Am. Coll. Cardiology, Am. Coll. Angiology; mem. Am., Cal., Pan Am. med. assns., N.Y. Acad. Scis., Am. Soc. Artificial Internal Organs, Johns Hopkins Med. and Surg. Assn., Soc. U. Surgeons, Am. Assn. Thoracic Surgery, Pan Pacific, Los Angeles surg. socs., Am. Therapeutic Soc., Am. Thoracic Soc., Am. Fedn. Clin. Research, Am. Coll. Cardiology, Assn. Am. Med. Colls., Am. Coll. Chest Physicians, Internat. Cardiovascular Soc. (N. Am. chpt.), Soc. Thoracic Surgeons (founder), and others. Research, publs on treatment of cardiac arrest, homotransplants, methods of open-heart surgery; inventor heart-lung machine—new disc valve. Office: 318 S. Alvarado St., Los Angeles 90057.*

KAY, John, inventor; b. Bury, Eng., July 16, 1704; m. Miss Moll; a son, Robert; remained in native town of Bury until mob broke into his home, 1753; fled to France. Improved reeds for looms; invented fly-shuttle which could be operated rapidly with one hand, 1733, comb for weavers loom, 1738; patentee batting machine for removing dust from wool; 1st great improvements in manufacture of cloth. Died France, 1764.

KAY-SHUTTLEWORTH, James Philips, English physician; b. Rochdale, Eng., July 20, 1804; s. Robert Kay; M.D., U. Edinburgh, Scotland, 1827; student anatomy, Dublin, Ireland; D.C.L. (hon.), Oxford; m. Janet Shuttleworth; 4 sons, 1 dau. Clk. med. wards Royal Infirmary, Scotland; practice medicine, Manchester, Eng. Mem. Royal Commn. for Advancement Sci., 1870. Recipient Fothergillian gold medal Royal Humane Soc. Mem. Royal Med. Soc. Scotland (sr. pres.). Author: De motu musculorum, 1827; The Defects in the Condition of Dispensaries, 1832; The Physiology, Pathology and Treatment of Asphyxia, (standard textbook for over half century), 1834. Pioneered in profl. medicine; founder clinics during cholera outbreak of 1832; began pupil-tchr. ednl. movement. Died London, Eng., May 26, 1877.

KAYA, Seiji, Japanese physicist; b. Kanagawa Prefecture, Japan, 1899; grad. Tohoku U., 1923; studied under Kotaro Honda; doctorate. Became prof. Tokyo U., 1943, dean Faculty Sci., 1949. Elected v.p. Sci. Conf. Japan, 1954. Research on metal physics, magnetism in iron.

KAYE, John, see Caius, John.

KAYSEN, Carl, Am. economist; b. Phila., Mar. 5, 1920; s. Samuel and Elizabeth (Resnick) K.; A.B. with highest honors in Econs., U. Pa., 1940; postgrad. Columbia, 1940-42; M.A., Harvard, 1947, Ph.D., 1954; m. Annette Neutra, Sept. 13, 1940; children—Susanne Neutra, Laura Neutra. Jr. fellow Soc. Fellows, Harvard, 1947-50, faculty, 1950-66, Littauer prof. polit. economy, 1964-66, asso. dean Grad. Sch. Pub. Adminstrn., 1961-66; dir. Inst. for Advanced Study, Princeton, N.J., 1966—. Dep. spl. asst. to Pres. Kennedy for nat. security affairs, 1961-63; cons. White House Bur. of Budget, 1961—; cons. pvt. cos. schs. Recipient Social Sci. Research Council Demoblzn. award, 1946-47. Guggenheim fellow, 1955-56. Mem. Council on Fgn. Relations (chmn. U. S. fgn. aid policy com.), Am. Econ. Assn., Econometric Soc., Am. Acad. Arts and Scis., Am. Philos. Soc., Phi Beta Kappa. Author: United States v. United Shoe Machinery Corporation, an Economic Analysis of an Anti-Trust Case, 1956; (with F. Sutton, S. E. Harris, J. Tobin), 1952; (with D. F. Turner) Anti-Trust Policy, 1959; (with Franklin M. Fisher) Demand for Electricity in the United States, 1962; also articles. Theoretical and applied work on nature and functioning of markets in a modern indsl. economy, relations of govt. and bus. and problems of maintaining a competitive economy social functions of business corp., ideological attitudes of Am. bus. community. Home: 97 Olden Lane, Princeton, N.J. 08540.*

KAYSER, Heinrich Gustav Johannes, German physicist; b. Bingen on Rhine, Germany, Mar. 16, 1853; s. Johann Jacob and Dorothea Amelia (von Metz) K.; ed. U. Strasbourg (France); Ph.D., Berlin, Germany, 1879; m. Mrs. Prümm, 1887. Asst., Phys. Inst., Berlin, 1878-85; prof. Technische Hochschule, Hanover, Germany, 1885-94; prof. physics, Bonn, Germany, 1894-1920. Fellow Royal Soc., 1911. Author: Introduction to Spectral Analysis, 1883; Handbook on Spectroscopy, 5 vols., 1899-1912; Tabelle der Schwrangungszahlen, 1925; Tabelle der Hauptlinien der Linienspektren aller Elemente, 7 vols., 1926; also treatise on electron theory, 1905. Measured velocities of sound in various media; research in spectroscopy; discovered (with C. D. T. Runge) existence of series of spectra of elements, and developed formula connecting wave length of lines with constants depending on elements (formed basis for later devels. in atomic spectroscopy). Died Bonn, Oct. 14, 1940.

KAZACHKOVOSKII, Oleg Dmitrievich, Russian nuclear physicist; b. 1915; grad. Dnepropetrovskii State U., 1938; became postgrad. Dnepropetrovskii Physico-Tech. Inst., 1941; Dr. Physico-Math. Scis., 1958. With Inst. Theoretical and Exptl. Physics, Acad. Scis. USSR, 1946-48; staff Physics Inst. for Utilization Atomic Energy, State Com. Council Ministers, USSR, 1948—; dir. Sci. Research Inst. Atomic Energy Reactors, Melekess. Recipient Lenin prize, 1960. Research on physics of nuclear reactors, fast-moving neutrons.

KAZAKEVICH, Iosif Yevseevich, Russian surgeon, traumatologist; grad. State Inst. Med. Knowledge, Leningrad, 1921. Asst., later lectr. dept. traumatology and orthopedics Leningrad Med. Inst., 1923-35, head chair, 1935-37; prof., 1935—; sr. asso. Leningrad Inst. Traumatology and Orthopedics, 1937-45; head chair traumatology and orthopedics Vilnius U., 1951—; chief traumatologist Lithuanian Ministry Health, 1951—, also dep. dir. for sci. work Inst. Tb and mem. learned council. Mem. Lithuanian Republic Soc. Surgeons (bd. mem.), Lithuanian Republic Soc. Phthisiologists (bd. mem.), Lithuanian Republic Soc. Traumatologists and Orthopedists (chmn.), Internat. Tb Union. Author numerous works including Arthroplasty of Major Joints; Manual of Orthopedics; Injuries of the Spinal Column. Address: Vilnius University, Universitetskaya ulitsa 3, Vilnius, Lithuanian SSR, USSR.

KAZAL, Louis Anthony, Am. biochemist; b. Newark, July 2, 1912; s. Leon A. and Anna (Wojnarowicz) K.; B.Sc., Seton Hall U., 1935; Ph.D., Rugers U., 1940; m. Marie A. Barry, Jan. 17, 1942; children—Marianna, Susan, Alicia, Louis Anthony. With Merck Sharp & Dohme, Inc., Glenolden and West Point, Pa., 1940-56, dir. biol. devel., 1950-54, mgr. tech. information, 1954-55, tech. asst. to med. dir., 1955-56; head sect. plasma fractionation Cardeza Found., Jefferson Med. Coll., Phila., 1956—, asso. dir., 1960—, asso. prof. medicine, 1960—, asso. prof. physiology, 1964—. Fellow A.A.A.S., N.Y. Acad. Scis.; mem. Am. Chem. Soc., Soc. for Exptl. Biology and Medicine, Am. Soc. Biol. Chemists, Phila. Hematology Soc., Physiol. Soc. Phila., Sigma Xi, Phi Lambda Epsilon. Author: L. M. Tocantins, Grune, Stratton) Blood Coagulation, Hemorrhage and Thrombosis, 1964; also numerous articles Research on mechanisms blood coagulation and factors influencing its rate, blood groups, methods separation blood components. Office: 1015 Sansom St., Phila. 19107.*

KAZANSKY, Boris Aleksandrovich, Russian organic chemist; b. Apr. 25, 1891; grad. Moscow U., 1918. Prof., Moscow U., 1935—; organizer, dir. catalytic synthesis lab. Inst. Organic Chemistry, USSR Acad. Sci., 1936-54, dir. Inst. Organic Chemistry, 1954—. Mem. Soviet delegation Internat. Congress on Theoretical and Applied Chemistry, Zurich, 1955. Decorated Order of Lenin; recipient Stalin prize, 1949. Mem. USSR Acad. Sci. Author: Catalytic Conversion of Hydrocarbons, 1957. Mem. editorial bd. Reports of USSR Acad. Sci. Research on catalytic conversions of hydrocarbons; discovered reaction of conversion of naphthenic hydrocarbons to aromatic hydrocarbons in presence of platinized carbon at 300-310° Centigrade, 1936; demonstrated that paraffins simultaneous capable of conversion to hydrocarbons of cyclopentane series in presence of platinum, 1954; devised (with G. S. Landsberg) method of detailed study of benzine. Address: Inst. Organic Chemistry, USSR Acad. Sci., Leninsky prospect 47, Moscow, USSR.

KAZANSKY, Valeriy Ivanovich, Russian surgeon, oncologist; b. 1894; grad. Petrograd Mil. Med. Acad., 1919; D.Med. Sci. Asst., Prof. S. I. Spasokukotsky's Surg. Clinics, 2d Moscow Med. Inst., Central Inst. Hematology and Blood Transfusion, 1935-38; sr. asso. Central Inst. Hematology and Blood Transfusion, 1938-40, head surg. clinic, 1941-52; chief oncologist USSR Ministry Health, 1947—; head 3d chair surgery Central Postgrad. Med. Inst., USSR Ministry Health, Moscow, 1953—. Mem. All-Union Soc. Oncologists (bd. mem.). Author: Material on the Study of the Hemopoietic Function of the Spleen, 1931; Splenectomy in Werlhoff's Disease, 1935; Trans-Pleural Resection of the Thoracic Portion of the Esophagus in Cancer, 1948, 2d edit., 1951. Co-editor Surgery sect. Large Med. Ency., 2d edit. Research and numerous publs. on surgery, hematology and oncology; 1st in USSR to perform radical operations for cancer of esophagus successfully; developer surg. methods for treating this cancer, also methods for anesthetizing anterior mediastinum in stenocarditis; 1st in USSR to perform and introduce omentocardiopexy for treating coronary insufficiency; developer original method to vascularize myocardium by suturing omentum to heart. Address: Central Postgrad. Med. Inst., pl. Vosstaniya 1-2, Moscow, USSR.

KAZARNOVSKII, Isaak Abramovich, Russian chemist; b. Sept. 29, 1890; grad. Zurich (Switzerland) U., 1914; mem. staff Karpov Physico-Chem. Inst. Moscow, from 1922. Recipient Stalin prize, 1941. Mem. USSR Acad. Scis. (corr.). Author: Structure of Inorganic Perioxides, 1940; Higher Oxides of Potassium; (with S. I. Raikhchtein) Inorganic Peroxides, 1947; (with G. P. Nikol'skii, T. A. Abletsova) New Oxide of Potassium, 1949; Kinetics of Spontaneous Decay of Ozonide of Potassium, 1956; Isotope Exchange of Oxygen between a Free Hydroxyl Radical and Water, 1956. Research on formation and structure of chlorides and metallic oxides; developed methods for prodn. sodium peroxide and anhydrous aluminum chloride from clays, for indsl. regeneration of air; discovered various sodium perioxides, potassium ozonides. Office: L. Karpov Physico-Chem. Inst., Moscow, USSR.

KEAN, Benjamin Harrison, Am. physician; b. Chgo., Dec. 2, 1912; s. Harrison and Tillie (Rhodes) K.; A.B., U. Cal. at Berkeley, 1933; M.D., Columbia, 1937. Practice medicine specializing in tropical medicine, N.Y.C., 1947—; faculty Cornell U. Med. Coll., N.Y.C., 1952—, clin. prof. tropical medicine, 1965—; staff N.Y., Drs. hosps.; dir. parasitology lab. N.Y. Hosp., 1952—. Cons. Ford Found. Travelers Health Inst., Devel. & Research Corp., Leonard Wood Meml. Diplomate Am. Bd. Pathology, Am. Bd. Microbiology. Fellow A.C.P. (life), Coll. Am. Pathologists; mem. A.M.A., Am. Soc. Clin. Pathologists, Am., Royal, N.Y. socs. tropical medicine, Am. Assn. Pathologists and Bacteriologists. Author: (with Breslau) Parasites of the Human Heart, 1964; (with Tucker) Traveler's Health Guide, 1965, Traveler's Medical Guide for Physicians, 1966; also numerous articles. Research on malaria, amebiasis, toxoplasmosis, schistosomiasis, nature and prevention diarrhea of travelers. Home: 435 E. 70th St., N.Y.C. 10021. Office: 728 Park Av., N.Y.C. 10021.*

KEANE, Austin, Australian mathematician; b. Sydney, Australia, Aug. 19, 1927; s. Phillip Walter and Annie (Driscoll) K.; B.Sc. with 1st class honors, 1949, M.Sc., 1951; Ph.D., U. New South Wales, 1955; m. Lorna Ada Hartman, May 20, 1952; children—Phillip Austin, Joanne. Faculty, U. New South Wales Sydney, 1954-57, asso. prof., 1959-61; research scientist Atomic Energy Research Establishment, Harwell, Eng., 1958; head theoretical physics A.A.E.C.R.E., Lucas Heights, New South Wales, 1961-64; prof. math. Wollongong (New South Wales) Coll., 1964—. Mem. Royal Soc. New South Wales (councillor 1963-66), Australian Math. Soc., Am. Nuclear Soc., Royal Astron. Soc. Editor: (with S. A. Senior) Complementary Maths, 1959, Mathematical Methods,

1961. Author: Integral Transforms, 1965; also articles. Research on finite elastic strain theory with applications to compression of planets; flow of air through coal mines allowing for porosity of walls; absorption of neutrons by resonances in nuclear reactor; devel. large computer codes for calculation of neutron spectrum in a reactor. Home: 743, Port Hacking Rd., Caringbah New South Wales. Office: Wollongong U. Coll., Wollongong, New South Wales, Australia.*

KEAST, Allen, zoologist; b. Sydney, Australia, Nov. 15, 1922; s. James Alfred and Mary Winning (Campbell) K.; B.Sc. with 1st Class Honors in Zoology, U. Sydney, 1951, M.Sc., 1953; M.A., Harvard, 1954, Ph.D. (Peter Brooks Saltonstall fellow), Harvard, 1955. Australian curator Australian Mus., 1952-53, curator, 1956-60; asst. prof. biology Queen's U., Kingston, Ont., Can., 1961-64, asso. prof., 1965-66, prof. zoology, 1967——. Research fellow Am. Mus. Natural History. Fellow Am. Ornithologists' Union (corr.); mem. Royal Australian Ornithol. Union (past pres.), Australian New Zealand Assn. for Advancement Sci., A.A.A.S., Linnean Soc. New South Wales, Am. Soc. Mammalogists, Royal Zool. Soc. New S. Wales. Author: Window to Bushland, 1959; Bush Birds, 1960; Australia and the Pacific Islands. A Natural History, 1960; also articles. Editor: Biogeography and Ecology in Australia, 1959. Evolutionary studies on continental vertebrates, Australian and Pacific biogeography, speciation in Australian birds, ecophysiology of desert adaptations, moult physiology of birds, evolution of ecol. specializations in African herbivores, Gt. Lakes fishes, factors governing evolution of faunas. Address: Biology Dept., Queen's U., Kingston, Ont., Can.

KEATE, Thomas, surgeon; b. 1745; asst. to John Gunning, St. George's Hosp., elected surgeon, 1792; became surgeon-gen. army, 1793; examiner Coll. Surgeons, 1800-02, served as master, 1902, 09, 18, resigned, 1813; surgeon to Prince of Wales (George IV), also Chelsea Hosp. Fellow Royal Soc., 1794. Author: Cases of Hydrocele and Hernia, 1788; also articles. First to tie subclavian artery for aneurysm. Died July 5, 1821.

KEATING, William Hypolitus, Am. mineral. chemist; b. Wilmington, Del., Aug. 11, 1799; s. John and Eulalia (Deschapelles) K.; grad. U. Pa., 1816, A.M., circa 1820; m. Elizabeth Bollman. Prof. mineralogy and chemistry U. Pa., 1822-27; geologist and historiographer of Maj. Stephen H. Lang's 2d expdn. of 1823; editor Conversations on Chemistry, 1824; a founder Phila. & Reading R.R.; founded Franklin Inst. of Pa., 1824, one of 1st mgrs., elected prof. chemistry. Mem. Am. Philos. Soc. (sec.). Largely responsible for discovery minerals including red zinc ore, franklinite, disulfite zinc carbonate. Died London, Eng., May 17, 1840.

KEATS, Theodore Eliot, Am. physician; b. New Brunswick, N.J., June 26, 1924; B.S., Rutgers U., 1945; M.D., U. Pa., 1947; m. Margaret E. McNamara, Aug. 17, 1949; children—Matthew M., Ian S. Faculty, U. Cal. Sch. Medicine, San Francisco, 1953-56, U. Mo., 1956-63; vis. prof. Karolinska Inst., Stockholm, Sweden, 1963-64; prof., chmn. dept. radiology U. Va. Sch. Medicine, Charlottesville, 1964——. Commonwealth Fund fellow. Mem. A.M.A., Am. Coll. Radiology, Assn. U. Radiologists, Am. Roentgen Ray Soc., Radiol. Soc. N.Am., Soc. Pediatric Radiology, Phi Beta Kappa, Sigma Xi, Alpha Omega Alpha. Author: (with Lee B. Lusted) Atlas of Roentgenologic Measurement, 1959, 67; also numerous articles. Research in Roentgen diagnosis, especially pediatric and cardiac radiology. Home: 1102 Rugby Rd., Charlottesville, Va. 22901.*

KECHT, Bruno Frank Michael, Austrian physician; b. Linz/Danube, Austria, Sept. 24, 1904; s. Hugo and Alice (Broedermann) K.; M.D., U. Innsbruck, 1928; m. Edith Girsig, July 23, 1948; children—Gert, Bernd, Irene. Practice medicine, specializing in ear, nose, throat, Linz, Austria, 1946——; cons. physician Linz children's and accident hosps.; prof. U. Vienna, since 1940——. Mem. Austrian, German ear, nose, throat, assns., Linz Chamber Physicians. Author: Oto-Laryngologie bei Schädelverlzungen, 1966; also numerous articles. Address: 26 Untere Donaulände, Linz, Austria.*

KECK, David Daniels, Am. plant taxonomist; b. Omaha, Oct. 24, 1903; s. Ezra David and Anne (Daniels) K.; A.B., Pomona Coll., 1925, M.A., 1926; Ph.D., U. Cal. at Berkeley, 1930; m. Marjorie Birdelle Stacy, June 1, 1928; children—Carol Joy (Mrs. Robert Allen Conrad) Katherine Eloise, (Mrs. James Norman Wilson), Elsbeth Marjorie (Mrs. W. A. Reynolds). Staff, Carnegie Instn. Washington, 1926-50, mem. div. plant biology, Stanford, Cal., 1929-50; with N.Y. Bot. Garden, N.Y.C., 1951-58, acting dir., 1958; program dir. systematic biology NSF, Washington, 1958-62, dep. div. dir. biol. and med. scis., 1962——. Panelist, systematic biology program NSF. 1954-58, Nat. Acad. Scis.-NRC, 1959——. Recipient (with J. Clausen, William M. Hiesey) Mary Soper Pope medal, 1949, Distinguished Service award, N.Y. Bot. Garden, 1963. Fellow A.A.A.S., Cal. Acad. Scis. (hon. life); mem. Bot. Soc. Am. (past v.p.), Am. Soc. Plant Taxonomists (past pres.), Cal. Bot. Soc., Soc. for Study Evolution, Am. Soc. Plant Taxonomists, Torrey Bot. Club, Am. Soc. Naturalists, Am. Inst. Biol. Scis. Author: (with J. Clausen, William M. Hiesey) Experimental Studies on the Nature of Species, vol. I, 1940, vol. II, 1945, vol. III, 1948; (with P. A. Munz) A California Flora, 1959; also numerous articles. Developed prins. exptl. systematics; revised numerous genera western Am. flowering plants. Home: 7227 Marbury Ct., Bethesda, Md. 20034. Office: 1800 G St., N.W., Washington 20550.*

KECK, James Collyer, Am. physicist; b. N.Y.C., June 11, 1924; s. Charles and Anne (Collyer) K.; B.A., Cornell, 1947, Ph.D., 1951; m. Margaret Ramsey, Sept. 6, 1947; children—Robert Lyon, Patricia Anne. Research asst. Cornell, 1951-52; sr. research fellow Cal. Inst. Tech., 1952-55; prin. scientist Avco-Everett Research Lab., Everett, Mass., 1955-65, dep. dir., 1960-64; Ford prof. engring. Mass. Inst. Tech., 1965——. Mem. Am. Phys. Soc., Phi Beta Kappa, Sigma Xi, Phi Kappa Phi. Research and publs. on high energy photonuclear reactions, develop. of diffusion, variational, and statis. theories for calculating the rates of elementary atomic and molecular reactions; measurement of the radiation from shock-heated air. Home: 52 Harold Parker Rd., North Andover, Mass. Office: 77 Massachusetts Av., Cambridge, Mass. 02139.*

KEDROV-ZIKHMAN, Russian agrochemist; b. Dec. 1, 1885; grad. Kiev U., 1913; later Dr.agr. and chem. sci. Agrl. chemist labs. and exptl. stas., Kiev Oblast, 1913-20; prof. Belorussian Agr. Acad., 1921-31; prof. Moscow Timiryazev Agr. Acad., 1931-41; acad. sec. dept. natural and agrl. sci. Belorussian Acad. Sci.; dir. lab. soil liming All-Union Inst. Fertilizers, Agrotech. and Agropedology, 1931——. Mem. learned council USSR Ministry Agr.; mem. Com. for Introducing Chemistry to Nat. Economy of USSR; chmn. agrochem. and pedological sect. House of Scientists USSR Acad. Sci. Mem. Belorussian Acad. Sci., All-Union Lenin Acad. Agrl. Sci. Research and numerous publs. on application of chalk to soil, role of magnium in chalk fertilizers. Office: A. USSR, Minsk, Belorussian SSR, Prosp. im. Stalina, 108, AN Bel. SSr.

KEEBLE, Sir Frederick William, Brit. botanist; b. Mar. 2, 1870; ed. Caius Coll., Cambridge; Sc.D., Cambridge U.; m. Matilde Marie Cecile Maréchal, 1898 (dec. 1915); 1 dau.; m. 2d, Lillah McCarthy, 1920. Controller horticulture food prodn. dept. Bd. Agr., 1917-19, asst. sec., 1919; dir. Royal Hort. Soc.'s Garden, Wisley, 1914-19; Sherardian prof. botany, also fellow Magdalen Coll., Oxford, 1920-27; Fullerian prof. Royal Instn., 1938; adviser in agr. Imperial Chem. Industries. Fellow Royal Soc., 1913. Author: Plant Animals; Practical Plant Physiology; Life of Plants; Fertilizers and Food Production; Polly and Freddie; Science Lends a Hand in the Garden, 1929; also articles. Co-author: Hardy Fruit Production. Died Oct. 19, 1952.

KEEFER, Chester Scott, Am. physician, med. educator; b. Altoona, Pa., May 3, 1897; s. John Henry and Gertrude (Scott) K.; B.S., Bucknell U., 1918, M.S., 1922, Sc.D., 1944; M.D., Johns Hopkins, 1922; Sc.D., Boston U., 1944, Bates Coll., 1962; m. Jean Balfour, Aug. 11, 1928; 1 dau., Ishbel McGill (Mrs. Carl B. Lyle, Jr.). Faculty, Johns Hopkins, 1925-26, U. Chgo., 1926-28, Peiping (China) Union Med. Coll., 1928-30, Harvard Med. Sch., Boston, 1930-40; Wade prof. medicine Boston U. Sch. Medicine, 1940-63, Wade prof. emeritus, cons. to dean and curriculum com. on med. edn., 1963——; physician-in-chief, also dir. Robert Dawson Evans dept. clin. research and preventive medicine Mass. Meml. Hosps., 1940-59, dir. med. center, 1956-61, sr. vis. physician, 1959-63, cons., mem. emeritus Robert Dawson Evans Meml., 1963——. Spl. asst. health and med. affairs to sec. Dept. Health, Edn. and Welfare, Washington, 1953-55, cons. div. pub. welfare, 1960-61. Recipient U. S. medal of Merit, 1946; His Majesty's medal for Freedom Gt. Britain, 1946; Isaac Ridgway Trimble medal Medico-Chirurg. Soc. Balt.; others. Fellow A.C.P. (pres. 1960); mem. Am. Philos. Soc., Am. Acad. Arts and Scis., N.Y. Acad. Scis., A.M.A., Am. Soc. Clin. Investigation, Am. Clin. and Climatol. Assn. (pres. 1962-63), Assn. Am. Physicians, Royal Coll. Physicians Gt. Britain (affiliate), Assn. Medica Argentina (fgn. corr.), Phi Beta Kappa, Sigma Xi, Alpha Omega Alpha. Author numerous books, pamphlets on medicine, infectious diseases and med. edn. Home: 71 Upland Rd., Brookline, Mass. 02146. Office: 80 E. Concord St., Boston 02118.*

KEELER, James Edward, Am. astronomer; b. La Salle, Ill., Sept. 10, 1857; s. William F. and Anna E. (Dutton) K.; B.A., Johns Hopkins, 1881; studied under Quincke, Bunsen, Helmholtz, Karper, Runge, at Heidelberg and Berlin, Germany, 1883-84; m. Cora S. Matthews, June 16, 1891. Mem. expdn. to observe eclipse Naval Obs., Fla., to Central City, Colo., July 29, 1878; became asst. to Langley, Allegheny Obs., 1881, succeeded Langley, 1891; became asst. astronomer Lick Obs., 1886, astronomer, 1888, dir., 1898. Recipient Rumford medal, Henry Draper medal. Fellow Royal Astron. Soc. Pacific; mem. Nat. Acad. Scis. Author: Spectroscopic Observations of Nebulae, 1894; Memoirs to Royal Astronomical Society of England; also articles. Co-editor, Astrophys. Jour. Determined wavelength of green line of nebular spectra; measured rotation of Saturn's rings, 1895; his observations that inner boundary of rings had slower period than outer boundary confirmed Maxwell's theory that rings of Saturn consisted of meteoric particles; observed radial velocities of Orion nebula and 13th planetary nebulae, showing they are in motion; observed 120,000 nebulae and concluded spiral nebulae is normal type, 1898. Died San Francisco, Aug. 12, 1900.

KEELEY, Leslie E., physician; b. Kings County, Ireland, May 4, 1834; s. Thomas H. and Maria (Enraught) K.; M.D., Rush Med. Coll., Chgo.; LL.D., Sancti Ludivici U., St. Louis, 1891; m. Mary Elizabeth Dow, June 1888. Practice medicine, Dwight, Ill.; founder Keeley Inst., Dwight, 1880. Author: (pamphlets) A Popular Treatment on Drunkeness and the Opium Habit and Their Successful Treatment with Double Chloride of Gold, the Only Cure, 1890; The Non-Heredity of Inebriety, 1896. Introduced method of treating alcoholism and addiction to morphine, opium, and others using strychnine and gold chloride (Keeley cure or gold cure). Died 1900.

KEEN, A(ngeline) Myra, Am. paleontologist; b. Colorado Springs, Colo., May 23, 1905; d. Ernest B. and Mary (Thurston) Keen; A.B., Colo. Coll., 1930; M.A., Stanford, 1931; Ph.D., U. Cal. at Berkeley, 1934. Research asso. geology Stanford, 1934-36, curator paleontology, 1936——, faculty, 1954——, prof. paleontology, 1965——. Cons. Paleontol. Research Instn. John Simon Guggenheim fellow, 1964-65. Fellow Paleontol. Soc., Geol. Soc. Am., Cal. Acad. Scis.; mem. Am. Malacological Union (past pres.). Author: (with Herdis Bentson) California Tertiary Marine Mollusca, 1944; Sea Shells of Tropical West America, 1958; Marine Molluscan Genera of Western North America, 1963. Research, numerous publs. on tertiary, recent mollusca West Coast of Ams. Home: 2241 Hanover St., Palo Alto, Cal. 94306. Office: Dept. Geology, Stanford, Cal. 94305.*

KEEN, William Williams, Am. surgeon; b. Phila., Jan. 19, 1837; s. William W. and Susan (Budd) K.; A.M., Brown U., 1859; M.D., Jefferson Med. Coll., 1862, Sc.D., 1912; LL.D., Brown, 1891, Northwestern, and Toronto, 1903, Edinburgh U., 1905, Yale, 1906, St. Andrews U., 1911, U. Pa., 1919; Ph.D., U. Upsala, 1907; Sc.D., Harvard, 1920; Dr. honoris causa U. Paris, 1923; studied in Europe, 1864-66; m. Emma Corinna Borden, 1867 (dec. 1886); children—Corrine (Mrs. Walter J. Freeman), Florence, Dora (Mrs. Geo. W. Handy), Margaret (Mrs. Howard Butcher, Jr.). Practiced in Phila., from 1866; conducted Phila. Sch. of Anatomy, 1866-75; lectr. pathol. anatomy Jefferson Med. Coll., 1866-75, prof. surgery, 1889-1907; prof. artistic anatomy Pa. Acad. Fine Arts, 1876-89; prof. surgery Woman's Med. Coll., 1884-89. Mem. NRC in World War I. Recipient Colver-Rosenberger medal of honor Brown U.; decorated officer Order Crown (Belgium), Legion of Honor (France). Hon. fellow A.C.S., Boston Surg. Soc. (Bigelow Gold medal), Royal Coll. Surgeons (Eng., Edinburgh, Ireland), Italian Surg. Soc.; asso. fellow Am. Acad. Arts and Scis.; mem. Am. Surg. Assn. (pres. 1899), A.M.A. (pres. 1900), Am. Philos. Soc. (pres. 1907-17), Internat. Congress Surgery, Paris (pres. 1920), Congress Am. Physicians and Surgeons (pres. 1903). Author: Surgical Complications and Sequels of Typhoid Fever, 1898; Animal Experimentation and Medical Progress, 1914; Treatment of War Wounds, 1917; Surgical Operations on President Cleveland, 1917. Editor: Health's Practical Anatomy, 1870; Diagrams of the Nerves of the Human Body, by W. H. Flower, 1872; American Health Primers, 1879-80; Holden's Medical and Surgical Landmarks, 1881; Gray's Anatomy, 1887; Americam Text-Book of Surgery, 1892, 1903; I Believe in God and in Evolution, 1922; Everlasting Life, 1924; Keen's System of Surgery, 8 vols., 1906-21. Successfully tapped ventricles of the brain, 1889; treated spastic torticollis by division of spinal accessory nerve and posterior roots of upper 3 spinal nerves, 1890; credited with successfully removing a meningioma, 1888. Died June 7, 1932.

KEENAN, Joseph Henry, Am. thermodynamicist; b. Wilkes-Barre, Pa., Aug. 24, 1900; s. Joseph Henry and Wilhelmina (Maurer) K.; B.S., Mass. Inst. Tech., 1922; LL.D., U. Glasgow, 1965; m. Isabel Morrison, Jan. 30, 1924; children—Esther Marie (Mrs. John W. Carr, III), Matthew Arnold. Steam turbine engr. Gen. Electric Co., Schenectady, 1922-28; asst. prof. mech. engring. Stevens Inst. Tech., 1928-34; asso. prof. Mass. Inst. Tech., 1934-39, prof., 1939——, head dept. mech. engring., 1958-61; Fulbright lectr. Cambridge U., Imperial Coll. Sci. and Tech., 1951. Del. Internat. Confs. on Properties of Steam, 1929——. Fellow Am. Acad. Arts and Scis.; hon. mem. Am. Soc. M.E. (Worchester Reed Warner medal); mem. Inst. Aero. Scis., Am. Soc. Engring. Edn., Am. Assn. U. Profs., Tau Beta Pi, Sigma Xi. Author: Steam Tables and Mollier Diagram, 1930, Thermodynamic Properties of Steam (with F. G. Keyes), 1936, Thermodynamics, 1941; Gas Tables (with J. Kaye), 1948; (with G. N. Hatsopoulos) Principles of General Thermodynamics, 1965. Research and publs. on improvements in the exposition of thermodynamics and statis. mechanics; also devel. of the availability concept. thermodynamics, fluid mechanics. Patented equipment for processing coffee, cocoa; devices for separating dust from gas streams. Home: 11 Howells Rd., Belmont Mass. 02178. Office: Mass. Inst. Tech., Cambridge, Mass. 02139.*

KEENAN, Philip Childs, Am. astronomer; b. Bellevue, Pa., Mar. 31, 1908; s. Charles G. and Evelyn (Childs) K.; B.S., U. Ariz., 1929, M.S., 1930; Ph.D., U. Chgo., 1932. Instr., Yerkes Obs., U. Chgo., 1936-42; physicist Bur. Ordnance, Navy Dept., 1942-46; faculty Ohio State U., Columbus, also Ohio Wesleyan U., Deleware, 1935-36, 46—, asso. prof. astronomy, 1948-56, prof., 1956—. Mem. Am., Royal astron. socs. Author: (with W. W. Morgan, E. Kellman) An Atlas of Stellar Spectra, 1942; also numerous articles. Research in stellar spectral classification. Home: 5303 N. High St., Columbus, O. 43214. Office: Perkins Obs., Delaware, O 43015.*

KEENEY, Arthur Hail, Am. ophthalmologist; b. Louisville, Jan. 20, 1920; s. Arthur Hale and Eugenia (Hail) K.; B.S., Coll. William and Mary, 1941; M.D., U. Louisville, 1944; M.Sc., U. Pa., 1952, D.Sc., 1955; m. Virginia Alice Tripp, Dec. 27, 1942; children—Steven Harris, Lee Douglas, Martha Blackledge. Resident surgeon Wille Eye Hosp., Phila., 1949-51; dir. ophthalmic research, instr. to asso. prof. ophthalmology U. Louisville Sch. Medicine, 1951-65; area cons. ophthalmology to VA, 1954—; ophthalmologist-in-chief Wille Eye Hosp. and Research Inst., Phila., 1965—. Recipient Bedell award, 1951; Zentmayer award, 1951. Fellow Am. Ophthalmology Soc.; mem. Soc. Oftal. de Columbia, Am. Acad. Ophthalmology and Otolaryngology, A.M.A., Am. Assn. Automotive Medicine (pres.). Author: Chronology of Ophthalmic Development, 1951; Lens Material in the Prevention of Eye Injuries, 1957. Contbr. numerous articles to med. jours. on capillary mechanisms in glaucoma; macular disease; strabismus; dyslexia; Newcastle virus disease; safety lens materials. Home: 8320 Seminole Av., Phila. 19118. Office: 1601 Spring Garden St., Phila. 19130.

KEEPEN, Vladimir Petrovich, see Köppen, Wladimir Peter.

KEEPIN, George Robert, Jr., Am. physicist; b. Oak Park, Ill., Dec. 5, 1923; s. George Robert and Erlene (Bennett) K.; Ph.B., U. Chgo., 1945; B.S., Mass. Inst. Tech., 1946, M.S., 1947; Ph.D., Northwestern U., 1949; m. Madge Mary Twomey, June 13, 1948; children—Robert, William, Ardis, Mavis, Denice. Postdoctoral fellow AEC, U. Cal. at Berkeley, 1949-50; physics staff U. Minn., 1950-52; research physicist Los Alamos Sci. Lab., 1952—, group leader pulsed neutron research, 1966—. Staff officer Internat. Atomic Energy Agy. confs.; cons., mem. coms. profls. orgns. Fellow Am. Phys. Soc., Am. Nuclear Soc.; mem. Fedn. Am. Scis., Sigma Xi. Author: Physics of Nuclear Kinetics, 1965; also chpts. other books, articles in profl. jours. Developed technique for measurement of neutron energy spectra; delayed neutron and reactor physics research; inventor techniques for internat. nuclear safegaurds, inspection and surveillance of fissile materials. Home: 155 Tunyo St. Office: Los Alamos Sci. Lab., Los Alamos 87544.*

KEESOM, Willem Hendrik, Dutch physicist; b. Texel Island, 1876; became prof. U. Leyden (Netherlands), 1923. Produced solid helium, 1926; measurements on expansion coefficients of solid bodies at very low temperatures; specific melting and evaporation heat of hydrogen and helium.

KEETON, William Tinsley, Am. zoologist; b. Roanoke, Va., Feb. 3, 1933; s. William Ivie and Doris (Tinsley) K.; A.B., U. Chgo., 1952, S.B., 1954; M.S., Va. Poly. Inst., 1956; Ph.D., Cornell U., 1958; m. Barbare Sue Orcutt, Aug. 9, 1958; children—Lynn Sue, Nancy Lee. Instr., Radford Coll., 1956; Schuler-Gage fellow Cornell U., 1956-58, faculty, 1958—, asso. prof. biology, 1964—. Cons., N.Y. State Edn. Dept., 1961—. Fellow A.A.A.S. (council 1963—); mem. Soc. for Study Evolution, Soc. for Systematic Zoology, Am. Entomol. Soc., Am. Soc. Zoologists, Animal Behavior Soc., Phi Kappa Phi, Sigma Xi, Gamma Alpha (nat. recorder 1960—). Author: A Taxonomic Study of the Milliped Family Spriobolidae, 1960; Biological Science, 1967; also articles. Research in evolutionary biology with particular emphasis on roles geographic and behavioral isolation in process speciation of animals, evolution of behavior, homing behavior. Home: 5 Snyder Heights, Ithaca, N.Y. 14850.*

KEFFER, Frederic, Am. physicist, educator; b. Anaconda, Mont., May 23, 1919; s. Robert and Placie (Munter) K.; B.S., Wash. State U., 1945; Ph.D., U. Cal. at Berkeley, 1952; m. Lore Maria Sanders, Apr. 24, 1949; children—Thomas, Leslie. Faculty physics U. Pitts., 1952—, prof., 1959—, dept. chmn., 1963—; vis. prof. physics U. Cal. at Berkeley, 1959-60. Cons. Westinghouse Research Labs., 1952-55, Bell Telephone Labs., summer 1961. Mem. A.A.A.S., Am. Phys. Soc. Contbr. articles on theory of ferromagnetism, antiferromagnetism, magnetic resonance, spin waves, double refraction, superexchange, dipolar interactions, magnetic anisotropy to profl. publs. Home: 235 Woodside Rd., Pitts. 15221.*

KEGEL, Otto H., mathematician; b. Bethlehem, Pa., July 20, 1934; s. Friedrich Otto and Margarete (Winkelmann) K.; Ph.D., U. Frankfurt/Main (Germany), 1961; m. Waltraut Wallner, Apr. 6, 1961; 1 dau., Hannah. Vis. asst. prof. U. Chgo., 1963-64; docent U. Frankfurt/Main, 1966—. Mem. Am., German, French, London math. socs. Contbr. articles in field to tech. jours. Address: 40 Alt-Roedelheim, Frankfurt/Main, West Germany.*

KEGELES, Gerson, Am. chemist; b. New Haven, Apr. 23, 1917; Alex and Jennie (Wilder) K.; B.S. in Chemistry, Yale, 1937, Ph.D., 1940, postgrad. 1940-41; postgrad. U. Wis., 1945-47; m. Bertha Webber, Apr. 16, 1944; children—Winifred, Lawrence, Stanley, Gloria, Joyce. Research phys. chemist Nat. Cancer Inst., 1947-51; mem. faculty Clark U., 1951—, prof. chemistry, 1956—, chmn. dept., 1960-63; cons. to industry, 1954—. Mem. Am. Acad. Arts and Scis., A.A.A.S., Am. Assn. U. Profs., Am. Chem. Soc., Am. Soc. Biol. Chemists, Biophys. Soc., Washington Acad. Scis. Research and publs. primarily on optical methods to facilitate the study of biol. materials, especially protein molecules. Home: 14 Tahanto Rd., Worcester, Mass. 01602.*

KEHL, William Louis, Am. physicist; b. Canton, O., Apr. 14, 1915; s. Herman E. and Lela (Shepherd) K.; A.B., Washington and Jefferson Coll., 1938; M.S., State U. Ia., 1941; postgrad. U. Pitts.; m. Marie E. Clark, Nov. 28, 1942; children—Dennis W., Lorraine E. With Gulf Research & Devel. Co., Pitts., 1941—, sr. research physicist, 1962—. Mem. Am. Crystallographic Assn. (sec. 1964—), Am. Phys. Soc., Pitts. Diffraction Soc., Am. Soc. for Testing Materials. Research, publs. on phase behavior inorganic mixed oxide systems, structural and catalytic properties mixed oxides and solid solutions metals; developed improved oil well casing perforator, improved catalysts for petroleum processing. Home: 1942 Fox Chapel Rd., Pitts. 15238. Office: P.O. Drawer 2038, Pitts. 15230.*

KEHOE, Robert Arthur, Am. physician; b. Georgetown, O., Nov. 18, 1893; s. Jeremiah W. and Jessie (Jones) K.; student Ohio State U., 1914-15; B.S., U. Cin., 1918, M.D., 1920; m. Lucile Marshall, June 11, 1920; 1 dau., Kathleen. Pathologist, Jewish Hosp., Cin., 1920-24; faculty U. Cin. Coll. Medicine, 1921-49, research prof., 1939-49, prof. occupational medicine, dir. Inst. Indsl. Health, 1948-65, dir. dept. preventive medicine and indsl. health, 1948-65, prof. emeritus occupational medicine, 1964—; dir. Kettering Lab., 1930-65. Med. dir. Ethyl Corp., 1925-58; nat. cons. USAF Surgeon Gen. U. Cin. Grad. Sch. Arts and Scis. fellow, 1937—. Recipient Knudsen award; 1948-49; Tech. and Sci. Socs. Council Cin. award, 1952; Pub. Health Fedn. Cin. award, 1963. Fellow A.M.A. (vice chmn. council on indsl. health 1949—), Am. Pub. Health Assn. (governing council 1955—), Am. Acad. Occupational Medicine (pres. 1949, award 1957); mem. Royal Soc. Health, Indsl. Med. Assn., A.A.A.S., Am. Physiol. Soc., Am. Indsl. Hygiene Assn., Am. Chem. Soc., Air Pollution Control Assn. Am. (Mellon award 1960), Sigma Xi, others. Research, numerous publs. field occupational health, toxicity of indsl. products, solvents, metals, organic compounds, behavior of certain mineral elements, trace metals, especially lead and fluorides. Home: 345 Resor Av., Cin. 45220. Office: Kettering Lab. U. Cin. Coll. Medicine, Eden Av., Cin. 45219.

KEHR, August Ernest, Am. geneticist; b. Frankfort, Ky., Mar. 2, 1914; s. Carl Adolf and Anna Esther (Heller) K.; B.S., Cornell U., 1936, M.S., 1947, Ph.D., 1950; m. Mary Louise Coon, Dec. 26, 1942; 1 dau., Janet Marie. Tchr. sci. and agr., Unadilla, N.Y., 1936-42, Hudson, N.Y., 1942-47; research asst. plant breeding Cornell U., 1947-50; asso. prof. horticulture La. State U., 1950-54; with Dept. Agr., also prof. horticulture Ia. State U., 1954-58; asst. br. chief, vegetables and ornamentals research br. Dept. Agr., Beltsville, Md., 1958-65, br. chief, 1965—. Mem. fellowship evaluation com. NSF, 1957-59. Mem. Am. Genetics Assn. (pres. 1964—), Soc. Study Growth and Devel., Am. Potato Assn., Am. Soc. Hort. Sci., Sigma Xi, Phi Kappa Phi, Gamma Sigma Delta. Contbr. articles profl. jours., chpts. in books. Home: 10202 Green Forest Dr., Silver Spring, Md. 20903. Office: Plant Industry Station, Beltsville, Md. 20705.*

KEHR, William Ralph, Am. agronomist; b. Blue Earth, Minn., July 1, 1921; s. John J. and Mary G. (Reintjes) K.; B.S., U. Minn., 1943, Ph.D., 1949; M.S., Kan. State U., 1947; m. Marie C. Sterner, June 22, 1946; children—Mary A., James W., Donald R. Plant breeder Cal. Packing Corp., Rochelle, Ill., 1949-53; research agronomist crops research div. Agr. Research Service, U. S. Dept. Agr., prof. agronomy U. Neb., Lincoln, 1953—. Fellow Am. Soc. Agronomy (dir.), A.A.A.S.; mem. Crop Sci. Soc. Am. (dir.), Am. Dehydrators Assn. (research council), Sigma Xi, Alpha Zeta, Gamma Sigma Delta. Research, numerous publs. in genetics and improvement of alfalfa and grain crops, breeding methods, insect and disease resistance, quantitative inheritance, chem. composition, genetics and cytogenetics of self- and cross-fertility, mgmt. systems of alfalfa. Home: 1410 N. 37th St., Lincoln, Neb. 68503.*

KEIDEL, Wolf-Dieter, German physiologist; b. Geimersheim, Dec. 14, 1917; s. Georg and Fanny (Dotterweich) K.; M.D.; m. Ursula Ludwig, Sept. 4, 1943; children—Dorothea, Ingrid, Matthias. Dir. Physiol. Inst., U. Erlangen/Nuremberg, 1961—, also full prof. physiology. Mem. German Physiol. Soc. (1st sec.), N.Y. Acad. Scis., Orlas-Ilbro-Paris Coll. Author: Kybernetik in Einzeldarstellungen; Vibrationsreception; Physiologie des N.S. von Standpunkt der Regelungslehre. Coeditor: Kybernetik (mag.), Zeitschrift für experimentelle Medizin. Research on electrophysiology of sense organs, cybernetics, physiology of central nervous system. Home: am Meilwald 23. Office: Physiological Institute, University of Erlangen/Nuremberg, Universitätstrasse 17, Erlangen, Germany.

KEIL, Karl Louis, German meteorologist; b. Wuppertal-Elberfeld, Germany, Nov. 7, 1898; s. Karl Guido and Wilhelmine (Schmits) K.; Ph.D, U. Giessen; m. Ilse Koppen, Dec. 3, 1926; children—Eberhard, Karl, Otto. Asst., collaborating scientist Prussian Obs., Lindenberg; with Prussian Meteorol. Inst., Berlin; govt. counselor Reichsamt für Wetterdienst, Berlin; cons. German Meteorol. Service, head library and information sects., Bad Kissingen; dir. climate atmospheric sect. German Central Office Meteorology, Offenbach. Mem. German Meteorology Soc., German Soc. Geophysics. Author: Handwörterbuch der Meteorologie, 1950, also articles. Editor: Meteorol. Rundschau, Springer-Verlag, Heidelberg. Address: Schopenhauerstrasse 24, 605 Offenbach/M., Germany.

KEIL, Werner, German zoologist, ornithologist; b. Giessen-Lahn, June 5, 1925; s. Karl and Wilhelmine (Link) K.; Ph.D., Justus-Liebig U., Giessen/Lahn; m. Erika Schmitt, Feb. 16, 1952; 1 child, Jutta. Asst., Obs. Applied Ornithology, Hessia, Rhenania, Pfalz, Sarre,, 1952-63 dir., 1964—. First pres. Ornithol. Observing Sta. Untermain, Frankfort. Recipient Silver medal Union Def. of Ornithology. Mem. German Soc. Ornithologists. Author: Ernährungsbiologie von Singvögeln; Steigerung der siedlungsdichte von Vöeln; Abwehr von Vogelschäden in der Forts-und Landwirtschaft; Untersuchungen zum Einsatz von Vögel im Rahmen der biologischen Schädlingsbekämpfung. Research on applied ornithology. Address: Steinauer Strasse 44, Frankfurt/Main, Germany.

KEILHOTZ, Gerald W., Am. chemist; b. West Chester, Pa., Aug. 7, 1912; s. Watson H. and Bessie (Woolens) K.; B.A., Lincoln U.; M.A., U. Utah; Ph.D., Ore. State U.; m. Barbara J. Jackson, June 1948; children—Candace, Carol, Michelle. Instr., Ore. State U., 1941-42; research chemist Gen. Chem. Co., 1942-44, Sylvania Electric Co., 1944-48; instr. U. Tenn., 1948-52; with Oak Ridge Nat. Lab., 1948—, reactor chemistry div., 1958—. Mem. Am. Chem. Soc., Am. Nuclear Soc., Sci. Research Soc. Am. Research, publs. in micro analysis, microwave spectroscopy, isotopic analysis, radiation damage, effects of radiation on corrosion of structural materials by molten fluorides, high temperature measurements, fission product release, transport and deposition. Home: 7121 Cheshire Dr., Knoxville, Tenn. 37919. Office: P.O. Box X, Ridge 37830.*

KEILIN, David, biologist; b. Moscow, USSR, Mar. 21, 1887; M.A., Magdalene Coll., Cambridge; prof. biology Cambridge U., 1931-52, also dir. Moltena Inst. Fellow Royal Soc. (Copley medal 1951), French Acad. Scis. (asso. fgn.). Research on biochemistry of higher Diptera; demonstrated existence of cytochrome (pigment found in some yeast and bacteria cells). Office: Molteno Inst., Cambridge, Eng.

KEILL, James, physician; b. Scotland, Mar. 27, 1673; M.D., Cambridge, Eng.; began practice medicine, Northhampton, Eng., 1703. Fellow Royal Soc., 1711. Author: The Anatomy of the Human Body, 1698; An Account of Animal Secretion, the Quantity of Blood in the Human Body, and Musuclar Motion, 1708. Supported mech. or iatro-math. sch. medicine; applied math. calculus to physiology; estimated quantity of blood in human body, blood flow, and force of heart. Died July 16, 1719.

KEILL, John, Brit. mathematician; b. Edinburgh, Scotland, Dec. 1, 1671; Incorporated M.A., Balliol Coll., Oxford, Eng., 1694, D.M., Oxford, 1713; M.A., Edinburgh U.; m. Mary Clements; 1 son. Became lectr. exptl. philosophy Hart Hall, Oxford, 1700, Savilian prof. geometry, 1713; treas. of Palatines, 1709-11; decypherer to Queen Anne, 1712-13. Mem. Commn. for Longitude. Fellow Royal Soc., 1700. Author: Introductio ad veram physicam, 1701; On the Laws of Attraction, 1708; On the Newtonian Solution of Kepler's Problem, 1708; Euclidis elementorum libri, 1717; Introduction ad veram astronomiam, 1718. Defended Newton's claim as inventor of differential calculus in answer to Leibnitz, 1705; application of Huyghen's theorems of centrifugal force to explain figure of earth; studied cohesion and elasticity; urged revival of study of Euclid, 1717. Died Oxford, Aug. 31, 1721.

KEIM, Christopher Peter, Am. physicist; b. Tecumseh, Neb., Apr. 6, 1906; s. Jacob Henry and Mary (Pohlenz) K.; A.B., Neb. Wesleyan U., 1927, Sc.D., 1959; M.S., U. Neb., 1932, Ph.D., 1940; m. Lucille Parli, June 25, 1929; children—Virginia Ann (Mrs. Clark Hayes), Robert Christopher. Engr., Lincoln Tel. & Tel. Co. (Neb.), 1928-31; prof. phys. scis. York (Neb.) Coll., 1933-37; research fellow U. Neb., 1937-40; mem. chemistry faculty Tulsa U., 1940-41; research engr. Sylvania Electric Products Corp., Salem, Mass., 1941-42; research fellow Mellon Inst., Pitts., 1942-44; sr. physicist Tenn. Eastman Corp., Oak Ridge, 1944-47, Union Carbide Corp., 1947; dir. stable isotope research and prodn. div. Oak Ridge Nat. Lab., 1947-57, dir. tech. information div., 1957—; dir. Mgmt. Services, Inc. Recipient Alumni Achievement award Neb. Wesleyan U., 1959. Fellow Am. Phys. Soc., Tenn. Acad. Sci., A.A.A.S.; mem. Am. Chem. Soc., Am. Nuclear Soc., Sigma Xi, Phi Lambda Upsilon, Sigma Pi Sigma. Research, publs. on physics and chemistry of surfaces, stable isotopes separation by electromagnetic process, preparation and handling of tech. information. Home: 102 Orchard Lane, Oak Ridge 37830. Office: Oak Ridge Nat. Lab., Oak Ridge 37831.*

KEIR, James, Brit. chemist; b. Stirlingshire, Scotland, Sept. 29, 1735; s. John and Magdalene (Lind) K.; ed. Edinburgh (Scotland) U.; m. Susanna Harvey, 1770; 1 dau., Amelia. Began chemistry, studies, Staffordshire, Eng., 1768; charge engring. works of Boulton and Watt, 1778; founder (with Alexander Blair) factory for prodn. alkali from sulphates of potash and soda, 1780; founder (with Blair) Tividale Colliery, 1794. Fellow Royal Soc., 1785, Royal Coll. Surgeons; mem. Lunar Soc. Birmingham. Author: treatise on different kinds of elastic fluids or gases, 1779; also articles; translated Macquer's Chemical Dictionary, 1776. Invented metal (almost indentical to muntzmetal) which could be forged or wrought when hot or cold; investigated differences between gases; contbd. to discovery of electro-plate process. Died West Bromwich, Eng., Oct. 11, 1820.

KEIRSTEAD, Burton Seely, Canadian educator; b. Woodstock, N.B., Can., Nov. 17, 1907; s. Wilfred Currier and Gertrude (Seely) K.; student Oxford; LL.D., U. N.B., 1948; m. Marjorie Stella Brewer, Aug. 2, 1933; children—Russell Clive, Pamela Jane. Prof. econs. U. N.B., 1931-42; prof. econs. McGill U., 1942-54, chmn. dept., 1948-52; prof. U. Toronto, 1954—; vis. research prof. Dalhousie U., 1940-41, Oxford U., 1950-51, U. W.I., 1960-61; guest lectr. Mass. Inst. Tech., Harvard, Johns Hopkins, Cambridge, Oxford, London, Liverpool univs. Fellow Royal Soc. Can.; mem. Canadian Assn. Econs. and Polit. Sci., Royal Econ. Soc. Author: Theory of Economic Change, 1948; Capital, Interest and Profits, 1959; numerous others; contbr. to Ency. Française. Devel. a theory of profit and its application. Home: 31 Alexander St., Toronto 5, Ont., Can.*

KEISLER, Howard Jerome, Am. mathematician; b. Seattle, Dec. 3, 1936; s. Harry Benton and Marion (Siegel) K.; B.S., Cal. Inst. Tech., 1959; Ph.D., U. Cal., Berkeley, 1961; m. Lois Joyce Hoffman, May 1, 1959; children—Randall Benjamin, Jeffrey Michael, Thomas David. Gen. Electric fellow U. Cal., Berkeley, 1959-60, NSF fellow, 1960-61, vis. prof., Los Angeles, 1967—; mathematician Inst. For Def. Analysis, Princeton, N.J., 1961-62; faculty U. Wis., Madison, 1962—, prof., 1967—; vis. research asso. Princeton, 1961-62. Alfred E. Sloan Found. fellow, 1966—. Mem. Am. Math. Soc., Assn. For Symbolic Logic. Author: (with C. C. Chang) Continuous Model Theory, 1966. Research in math. logic, particularly model theory and set theory. Home: 6318 Masthead Dr., Madison, Wis. 53705.*

KEITER, Friedrich Albin, anthropologist; b. Vienna, Austria, Nov. 26, 1906; s. Albin and Ella (Braun) K.; ed. univs. Vienna, Keil, Graz; Ph.D., M.D.; m. Margarete Heese, Mar. 30, 1937; 1 dau., Ute. Anthropologists, U. Kiel, 1929-32; agrégé, 1933; with U. Hamburg, 1934-39; dir. anthropology labs., prof. anthropology and human genetics U. Wurzburg, 1939-46. Vis. lectr. in U. S., 1960; lectr., Porto Alegre, Brazil, 1962. Mem. various med. assns. in Germany, U. S., Brazil. Author 10 vols. on anthropology, also numerous articles. Research on human genetics, behavioral anthropology. Home: Lenhartzstrasse 13, Hambourg. Office: Breitenfelderstrasse 62, Germany.

KEITH, Arthur, Am. geologist; b. St. Louis, Sept. 30, 1864; s. Harrison Alonzo and Mary Elizabeth (Richardson) K.; A.B., Harvard, 1885, A.M., 1886, postgrad. in sci., 1887; m. Elizabeth Marye Smith, June 29, 1916. Mem. Mass. State Topog. Survey, 1886-87; asst., U. S. Geol. Survey, in Tenn. work, 1887; in charge mapping party, Tenn., 1888-95; apptd. geologist, 1895; placed in charge areal and structural geology of U. S., and of Geologic Atlas of U. S., 1907, in charge East of 100th meridian, 1913-21; made spl. investigation in U. S. and Canada of Appalachian mountain structure, stratigraphy, and earthquakes. Prof. grad. dept., U. Tex., 1926. U. S. del. to Internat. Congress Geologists, 1913, to Internat. Geophys. Union. Mem. Geol. Soc. Am. (pres. 1927), NRC (chmn. div. geology and geography, 1928-31; chmn. coms. on earthquakes and fellowships), Nat. Acad. Scis. (treas.), A.A.A.S., Am. Assn. Petroleum Geologists, Am. Inst. Mining and Metall. Engrs., Seismol. Soc. Am., Am. Acad. Arts and Scis., Am. Assn. Geographers, Washington Acad. Scis. (v.p.), Am. Geophys. Union. Author: Geologic Atlas of U. S., Harpers Ferry Folio, 1894, and other folios, 1896-1907 (U. S. Geol. Survey); also many geol. articles. Died Feb. 7, 1944.

KEITH, Sir Arthur, Brit. anatomist, anthropologist; b. Old Machar, Aberdeen, Feb. 5, 1866; s. John and Jessie (Macpherson) K.; ed. Aberdeen, London, Leipzig univs.; M.D.; m. Cecilia Gray, 1899. Became prof. Royal Coll. Surgeons, 1908; Fullerian prof. physiology Royal Instn., 1917-23, sec., 1922-26, pres., 1926-29; rector U. Aberdeen, 1930-33. Fellow Royal Soc., 1913; pres. Royal Anthrop. Inst., 1912-14, Brit. Assn., 1927. Author: Introduction to the Study of Anthropoid Apes, 1896; Human Embryology and Morphology, 1901; Ancient Types of Man, 1911; The Human Body, 1912; Antiquity of Man, 1915; Menders of the Maimed, 1919; Engines of the Human Body, 1925; Nationality and Race, 1920; Religion of a Darwinist, 1925; Concerning Man's Origin, 1927; New Discoveries Relating to the Antiquity of Man, 1931; Darwinism and its Critics, 1935; (with T. D. McGown) Stone-age of Mt. Carmel Human Fossil Remains, 1939; Essays on Human Evolution, 1946; A New Theory of Evolution, 1948; An Autobiography, 1950; Darwin Revalued, 1955. Important work in the reconstruction of prehistoric man from remains; discoveries led to modification of Darwinian theory; extensive research on origin of the Sumerians, also research on origin of man, tuberculosis after 1933. Died Downe, Kent, Jan. 7, 1955.

KEITH, John Dow, Canadian physician; b. Can., Feb. 23, 1908; s. John Mitchell and Georgina (Haddow) K.; m. Mary MacLaren Carson, July 2, 1938; children—Nancy (Mrs. Stuart Smith), Christina, John. Physician in charge cardiac clinic and dept. Hosp. for Sick Children, Toronto, 1938—; prof. pediatrics. Pres., Canadian Heart Found., 1957-58, bd. dirs., 1958-64, sr. adv. council, 1964—; bd. dirs. Internat. Cardiology Found., 1962-64, med. adv. bd., 1964—. Fellow Am. Coll. Cardiology; mem. Canadian (sec. 1952-57), Am. heart assns., Canadian Pediatric Soc., Canadian, Ont. med. socs., Canadian Cardiovascular Soc., Acad. Medicine Toronto, Am. Acad. Pediatrics, Pediatric Research Soc. Author: (with Rowe, Vlad) Heart Diseases in Infancy and Childhood, 1958, 67; also numerous articles. Home: 555 University Av., Toronto, Ont., Can.*

KEITH, McKenzie Lawrance, geochemist; b. Edmonton, Alta., Can., Oct. 12, 1912; s. Homer P. and Ethel (Gracey) K.; B.Sc., U. Alta., 1934; M.Sc., Queens U., Ont., Can., 1936; Ph.D., Mass. Inst. Tech., 1939; m. Mary Emma Skavlem, Sept. 21, 1940; children—David, Andrew, Douglas, John, Laurie. Came to U. S., 1947, naturalized, 1951. Asst. prof. Queens U., 1940-47; petrologist Geophys. Lab., Washington, 1947-50; prof. geochemistry Pa. State U., University Park, 1950—; geol. field work for Geol. Survey Can., Ont. Dept. Mines, various mining cos. Fellow Geol. Soc. Am.; mem. Mineral. Soc. Am., Geochem. Soc. Research, publs. on geochemistry and petrology including origin of alkaline rocks, high-low inversion of quartz, trace element and isotopic indicators of marine and fresh-water sediments, methods of geochem. prospecting. Home: 525 Kemmerer Rd., State College, Pa. 16801. Office: Mineral Sci. Bldg., University Park, Pa. 16802.*

KEITH, Nathaniel Shepard, Am. electro-metallurgist; b. Boston, July 14, 1838; s. Bethuel and Elizabeth P. K.; ed. Dover, N.H., and New York; tech. edn. in mining and elec. engring.; ed. as physician but never practiced. Mining and metal. engr. in Colo., 1860-69; then elec. and electro-metall. engr.; inventor, with many patents in these lines; scientific editor Electrical World, 1884-85. Mfr. elec. appliances, as applied to mining and metallurgy, San Francisco, 1884-93; expert in same line in Eng. till 1897; sec.-treas. Am. Venture & Mines Corp. Organizer and 1st sec. Am. Inst. E.E., 1884. Author: Magnetic and Dynamo-Electric Machines, 1884. Died Jan. 28, 1925.

KEITH, Norman Macdonnell, physician; b. Toronto, Ont., Can., June 8, 1885; s. George and Agnes (Dow) K.; B.A., U. Toronto, 1908; M.D., Johns Hopkins, 1911; m. Edna I. Alexander, Aug. 31, 1920; children—Helen Mary (Mrs. Thomas Frederick Kling), Janet Isabel (Mrs. Wilbourn Coupery Shands), Alexander M. Faculty, Johns Hopkins Med. Sch., 1914-17, U. Toronto, 1919-20; asst. Mayo Clinic, Rochester, Minn., 1920-23, head, sect. medicine, 1923-36, sr. cons., 1936-50; prof. medicine Mayo Grad. Sch. Medicine, 1936-50. Mem. Assn. Am. Physicians (past chmn. sect. pharmacology and therapeutics), A.M.A., Am. Soc. Clin. Investigation, Am. Physiol. Soc. (pres. circulation sect. 1939), Am. Soc. Exptl. Pathology, Am. Soc. Pharmacology and Exptl. Therapeutics, Central Interurban Clin. Club, Central Soc. Clin. Research, Fedn. Am. Socs. Exptl. Biology, Sigma Xi. Research, publs. on wound shock, surg. shock, diseases of kidneys, heart and circulation, inorganic metabolism including potassium, diuresis, dehydration, blood vol. and hypertension. Home: 615 8th Av. S.W., Rochester 55901. Office: 200 1st St. S.W., Rochester, Minn. 55901.*

KEITH, Thomas Byron, Am. nutritionist; b. Bowling Green, Mo., Apr. 7, 1897; s. Andrew Nolan and Lottie (Hays) K.; B.S., U. Ida., 1924; M.S., U. Ill., 1926; Ph.D., Pa. State U., 1933; m. Helen Woods Bryson, June 7, 1931; children—Helen Woods (Mrs. Tren Arthur Williamson), Thomas Byron, Patricia Dean (Mrs. Wendell V. Tangborn). Faculty, Pa. State Coll., 1928-42; asso. prof. Mont. State Coll., 1946-47; faculty nutrition U. Ida., Moscow, 1947-66, ret., 1966. Mem. nutrition sub-com. beef cattle nutrition, NRC, 1950-65. Mem. Am. Soc. Animal Scis., Am. Inst. Nutrition, Sigma Xi, Gamma Sigma Delta. Research, numerous publs. on manganese requirements pigs, protein requirements ewes, pigs, and finishing beef cattle, antibiotics in pig and lamb rations, exptl. designs in animal nutrition, vitamin A deficiencies in beef cattle. Home: 933 E. 7th St., Moscow, Ida. 83843.*

KEKULÉ VON STRADONITZ, Friedrich August, German chemist; b. Darmstadt, Germany, Sept. 7, 1829; student architecture U. Giessen (Germany); studied chemistry under Dumas, Wurtz, Gerhardt. Became prof. Heidelberg, Germany, 1856, Ghent, Belgium, 1858, Bonn, Germany, 1867; became dir. Inst. Chemistry, Bonn, 1865. Fellow Royal Soc., 1875 (Copley medal, 1865). Mem. Internat. Chem. Congress, French Acad. Scis. Author: Chemie der benzolderivate, 1867; Lehrbuch der organischen Chemie, 1861; Anschütz, 1929; Buch der grossen Chemiker, 1930. Research on structure of organic compounds; proposed tetravalence of carbon and chain formation of alphatic series, 1858; redefined organic chemistry as chemistry of carbon compounds; originated ring or closed-chain theory of structure of benzene molecule (of fundamental importance in modern chemistry, especially devel. synthetic dyes); research on succinic, tartaric acid, unsaturated dicarboxylic acids. Died Bonn, July 13, 1896.

KELDYSH, Ludmíla Vsévolodovna, Russian mathematician; b. Orenburg, USSR, Mar. 11, 1904; d. Acad. V. Keldysh; grad. Moscow U., 1925; Dr.Physico-Math. Sci., 1941. With Moscow Aviation Inst., 1930-34; with Math. Inst., USSR Acad. Sci., 1934—, sr. asso., 1948—; dir. Physico-Math. Inst., 1941—. Author: The Structure of B-Set, 1945; Continuous Mapping of Compacta, 1947; Enlarging Zero-Dimensional Mapping, 1951; Monotonic Mapping of a Cube onto a Cube of Greater Dimensions, 1957; Conversion of a Monotonic Irreducible Mapping into a Monotonic Open Mapping and Enlarging Monotonic Open Maps of a Cube, 1957; Topological Imbeddings and Pseudoisotropy, 1966. Research on theory of real variable functions, theoretical geometrical topology. Office: Mathematichesky Institut AN SSR, Vavilova 42, Moscow, USSR.*

KELDYSH, Mstislav Vsevolodovich, Russian mathematician, mech. engr.; b. Jan. 28, 1911; grad. Moscow U., 1931; D.Physico-Math. Sci., 1938. Lectr., Moscow U., 1932—, now prof.; formerly with Central Inst. Aerohydrodynamics; head dept. applied math. Math. Inst., USSR Acad. Sci., 1960—, past acad. sec. dept. physico-math. sci. Decorated Order of Lenin (6); recipient State prize, 1942, 46, Lenin prize, 1957. Mem. USSR (Presidium mem. 1953-60, v.p. 1960-61, pres. 1961—), Czechoslovak, Armenian (hon.) acads. sci. Research and publs. on hydromechanics, aerodynamics, theory of vibration (widely used in mil. aviation for jet planes and computations for guided missiles), theory of waves on surface of heavy liquid. Office: USSR Acad. Sci., Leninsky prospect 14, Moscow, USSR.*

KELEMEN, George, physician; b. Budapest, Hungary, May 1, 1890; s. Adof K. and Hermina (Drucker) K.; student U. Goettingen (Germany); M.D., U. Budapest, 1913. Came to U. S., 1940, naturalized, 1946. Private docent U. Pecs, 1926-37, asso. prof., 1937-40; asso. prof. Tufts U. Med. Sch., 1940-46, research asso., 1946-47; instr. Harvard Med. Sch., 1946-53, research asso., 1953-56; emeritus clin. prof. surgery U. So. Cal. Sch. Medicine, 1964—; v.p. staff L.I. Hosp., Boston, 1961, pres., 1962; staff Mass. Eye and Ear Infirmary, 1943-56, Mass. Gen. Hosp., 1949-56; research asso. Mass. Eye and Ear Infirmary, 1956-60, research cons. otolaryngology, 1957-63; dir. temporal bone lab. Los Angeles Found. Otology, 1963—; mem. sci. adv. staff Inst. Laryngology and Voice Disorders, Los Angeles, 1966—. Recipient Hogyes prize U. Budapest, 1912; cavaliere ufficiale Order Italian Crown, 1938; diploma of merit Fedn. Otolaryn. Socs. Argentina, 1946. Numerous research grants including NIH, 1953-56, AEC, 1961, Central Bur. Research Am. Otol. Soc., 1961-65, Def. Research Found., 1964-67; Fulbright researcher A, Germany, 1946. Diplomate Am. Bd. Otolaryngology. Fellow A.M.A., A.C.S.; mem. Hungarian Soc. Natural Scis., Royal Soc. Medicine Budapest (past v.p. sect. otology), Hungarian (past 1st sec.), German otolaryn. socs., Am. Acad. Ophthalmology and Otolaryngology (Honor award 1963), PanAm. Assn. Oto-Rhino-Laryngology, Am. Acad. Neurology, N.Y. Acad. Scis., Internat. Soc. Biophonetics; corr. mem. French, Danish, Polish otolaryn. socs., French Phoniatric Soc., Hellenic Oto-Neuro-Ophthal. Soc.; hon. mem. Otolaryn. Soc. Madrid, Italian, Peruvian, Chilean, Hungarian, Colombian otolaryn. socs. Editorial staff Acta Oto-Laryngologica, Practica Oto-Rhino-Laryngologica, Folia Oto-Laryngologica Orientalia, Archivio Italiano per le Malattie della Trachea, Bronchi, Esofago, Parma, Ressegna Internazionale di Oto-Rhino-Laringologia, Zeitschrift für Laryngology, Rhinologie, Otologie, Stuttgart, Logos, Archives Otolaryngology, Research, publs. on biochemistry, gen. surgery, comparative anatomy and physiology, otolaryngology. Home: 2430 Ocean View Av., Los Angeles 90057. Office: 2122 W. 3d St., Los Angeles 90057.*

KELL, Nikolai Gierogievich, Russian geodesist; b. Jan. 20, 1883; grad. Petrograd Mining Inst., 1915. Topographer, Kamchatka Expdn., Russian Geog. Soc., 1908-10; instr. Ural Mining Inst., Sverdiovsk, 1917-22, rector, 1919-20; prof. Petrograd Mining Inst., 1923—; became dir. Leningrad Lab. Aerial Surveying Methods, USSR Acad. Scis., 1947. Mem. Russian Geog. Soc., USSR Acad. Scis. (corr.). Author: Maps of the Volcano of Kamchatka, 1928; The Graphic Method in Operations with Errors and Assumptions-Distributions, 1948; Advanced Geodesy and Geodetic

Work, 1932-33; Photography and Photogrammetry, 1937; Utilization of Results from Topographico-Geodesic Work for Engineering Purposes, 1950; Measurement Deciphering of Aerial Photos in Field Conditions, 1959; (with V. G. Zdanovich, K. A. Zvonarev, A. N. Belolikov, N. A. Gusev) Higher Geodesy, 1961. Research in geodesy, photogrammetry; developed methods for aerial photogrammetry which could be applied to geog. and geol. mapping. Office: Lab. Aeromethods, USSR Ministry Geology and Mineral Conservation, Birzhevoi, Proyezd 6, Leningrad, V-164, USSR.

KELLAND, Philip, English mathematician; b. 1808; s. Philip Kelland; B.A., Queens Coll., Cambridge, Eng., 1834, M.A., 1837; m. Miss Pilkington; m. 2d, Miss Boswell. Became prof. math. U. Edinburgh, Scotland, 1838, prof. natural philosophy, 1852-56. Fellow Royal Soc., 1838, Royal Soc. Edinburgh (pres. 1878-79), Soc. Arts Edinburgh (pres. 1853-54). Author: Theory of Heat, 1837; The Elements of Algebra, 1839; Lectures on the Principles of Demonstrative Mathematics, 1845; Elements of Geometry, 1859; Lesson on Physics, 1872; Introduction to Quaternian, 1873. Research and publs. on non-Euclidean geometry. Died 1879.

KELLAWAY, Charles Halliley, Australian physician; b. Jan. 16, 1889; s. Alfred Charles Kellaway; ed. Melbourne U.; M.B., B.S., 1911; M.D., 1913; M.S., 1914; Royal Soc. Foulerton student Nat. Inst. Med. Research, Physiology Sch., Oxford, also U. Coll. Hosp. Med. Sch., 1920-23; m. Eileen Ethel Scantlebury, 1919; 3 sons. Dir.-in-chief Wellcome Research Instn., 1944—; became registrar Melbourne (Australia) Hosp., 1913; named acting prof. anatomy Adelaide (Australia) U., 1915, acting prof. physiology, 1919; asst. med. unit U. Coll. Hosp., 1923; dir. Walter and Eliza Hall Inst. Research pathology, Melbourne, 1923-44; specialist physician Royal Melbourne Hosp., 1925-43. Chmn., State Cancer Adv. Com., 1929. Recipient Walter Burfitt prize Royal Soc. New S. Wales, 1932. Became fellow Royal Soc., 1940, Royal Coll. Physicians, Royal Australasian Coll. Physicians; mem. Royal Coll. Physicians. Research and publs. on anaphylaxis, immunity, adrenal glands hydatid antigens, pharmacology of staphylococcal toxin, Australian snake venoms, ascending renal infection, liberation of active substances in various cell injuries. Died Dec. 13, 1952.

KELLAWAY, Peter, neurophysiologist; b. S. Africa, Oct. 20, 1920; s. Cecil Lauriston and Doreen (Joubert) K.; came to U. S., 1937, naturalized, 1947; student Haileybury Coll., Australia, 1936-37; A.B., Occidental Coll., 1941, A.M., 1942; Ph.D., McGill U., 1947; m. Jo Ann Barbieri, Apr. 26, 1958; children—Judianne, David Peter, Kevin Sean. Lectr. physiology McGill U., 1946-47; faculty Baylor U., Houston, 1947—, prof., 1961—; dir. dept. clin. neurophysiology Meth. Hosp., Houston, 1949—. Mem. Am. Neurol. Assn., Am. Acad. Neurology, Am. Physiol. Soc., Am. Epilepsy Soc., Am. Electroencephographic Soc., Am. Soc. History Medicine, Am. League Against Epilepsy. Author: (with H. Conn) Convulsive Disorder, 1958, (with Chao, Druckman) Convulsive Disorders of Children, 1959; also numerous articles esp. on development of clinical electroencephalography for diagnosis in infants and young children. Home: 627 E. Friar Tuck St. Office: Baylor U. Coll. Medicine, Houston.*

KELLER, Edward Clarence, Jr., Am. geneticist; b. Freehold, N.J., Oct. 8, 1932; s. Edward Clarence and Pauline (Van Sickle) K.; B.Sc., Pa. State U., 1956, M.Sc., 1959, Ph.D., 1961; m. Helen Elizabeth Baylor, July 7, 1950; children—Edward Clarence III, Kim Lorie. Faculty, Pa. State U., 1956-61, NIH fellow, 1959-60; NIH fellow, research asso. U. N.C. Med. Sch., 1961-64; asso. prof. U. Md., College Park, 1964-67; mgr. data systems and biostatistics biol. systems div. NUS Corp., Hawthorne, Cal., 1967—; cons. N. Am. Aviation Corp., various govtl. agys. Mem. A.A.A.S., Am. Inst. Biol. Sci., Biometrics Soc., Ecol. Soc. Am., Genetics Soc. Am., Nat. Assn. Govt. Engrs., Pa. Acad. Sci., Soc. For Study Evolution, Gamma Sigma Delta, Phi Epsilon Phi, Sigma Xi, Phi Sigma. Research, publs. on genetic homeostasis in Drosophila; showed presence of interchromosomal interaction; also that levels of a specific enzyme are under polygenic control. Home: 317 Calle Mayor, Redondo Beach, Cal. 90277. Office: NUS Corp., 3309 W. El Segundo Blvd., Hawthorne, Cal. 90250.*

KELLER, Ferdinand, Swiss archeologist; prehistorian; b. Marthalen, Dec. 24, 1800; tchr. Tech. Inst., Zurich; founder Zurich Antiquarian Soc. Author: Pfahlbauberichte, 1854; Römische Ansiedlungen in der Ostschweiz, 1860; Archäologische Karte der Ostschweiz, 1873. Discovered and recognized importance of the pile dwellings of Obermeilem lake region in the history of man. Died Zurich, July 21, 1881.

KELLER, Hans Heinrich, Swiss mathematician; b. Wetzikon, Switzerland, Feb. 14, 1922; s. Heinrich J. J. and Emma (Schoch) K.; student U. Zurich (Switzerland), 1941-47, Ph.D., 1950; m. Silvia Maria Schnoz, Mar. 25, 1959. Faculty, U. Zurich, 1954—, prof. math., 1967—; vis. asso. prof. U. Mich., Ann Arbor, 1962-63. Mem. Schweizerische Mathematische Gesellschaft, Naturforschende Gesellschaft Zürich, Am. Math. Soc. Research, publs. on inequalities of Phragmén-Lindelöf type in potential theory, differential equations in Banach spaces, canonical transformations (of analytical mechanics), differential calculus in topological vector spaces, limit spaces. Home: 11 Stapferstrasse, 8006 Zürich, Switzerland.*

KELLER, Herbert Bishop, Am. mathematician; b. Paterson, N.J., June 19, 1925; s. Isaac and Sally (Bishop) K.; B.E.E., Ga. Inst. Tech., 1945; M.S., N.Y. U., 1948, Ph.D., 1954; m. Loretta Schertz, Feb. 1, 1953; children—Debra S., Steven S. Instr. physics Ga. Inst. Tech., 1946-47; faculty math. dept. Sarah Lawrence Coll., 1951-53; faculty N.Y. U., 1955-67, asso. prof. Courant Inst. Math. Sci., 1959-61, prof. applied math., 1961-67, asso. dir. AEC Computing and Applied Math. Center, 1964-67; vis. prof. Cal. Inst. Tech., 1965-66; prof. applied math. Cal. Inst. Tech., Pasadena, 1967—. Mem. Am. Math. Soc., Soc. Indsl. and Applied Math. (editor Jour. on Numerical Analysis, asso. editor Jour. on Applied Math.), Assn. for Computing Machinery (editor Monograph Series), Math. Assn. Am., A.A.A.S. Author: (with E. Isaacson) Analysis of Numerical Methods, 1966; Numerical Methods for Two-Point Boundary-Value Problems, 1968. Contbr. articles to sci. jours. on numerical analysis; fluid mechanics; electro-magnetic theory; nuclear reactors. Devel. numerical methods which enable computing machines to solve problems of interest to sci. and tech. Home: 1526 Gaywood Dr., Altadena, Cal. 91001. Office: Firestone Labs., Cal. Inst. Tech., Pasadena, Cal.*

KELLER, John Carlos, Am. entomologist; b. Arvada, Colo., July 16, 1918; s. Nicholas M. and Nell (Scott) K.; B.S., Colo. A. and M. Coll., 1941; M.S., U. Ill., 1947; m. Jean Carlstrom, Jan. 21, 1948; children—Jeffrey D., Julie L. Research entomologist Julius Nyman Co., Denver, 1947-48; research entomologist U. S. Dept. Agr., 1948-63, dir. Western Cotton Research Lab. Cotton Research Center, Phoenix, 1965—; head entomology sect. IAEA, Vienna, Austria, 1963-65. Mem. A.A.A.S., Entomol. Soc. Am. Contbr. numerous articles to profl. jours. Devel. granular insecticides, housefly baits; discovery of boll weevil feeding stimulants and attractants from cotton, female boll weevil attractant from male weevils. Home: 321 E. Manhattan Dr., Tempe, Ariz. 85281. Office: 4207 E. Broadway, Phoenix 85040.*

KELLER, Joseph Bishop, Am. mathematician; b. Paterson, N.J., July 31, 1923; s. Isaac and Sally (Bishop) K.; B.A., N.Y. U, 1943, M.S., 1946, Ph.D., 1948; m. Evelyn Fox, Aug. 29, 1963; children—Jeffrey M., Sarah N. Prof. math. Courant Inst. Math. Scis. N.Y. U., 1948—. Mem. Am. Math. Soc., Am. Phys. Soc. Studies, publs. on geometrical theory of diffraction; fluid dynamics; acoustics; kinetic theory; applied mathematics; geometrical optics. Home: 4 Washington Sq. Village, N.Y.C. 10012. Office: 251 Mercer St., N.Y.C. 10012.

KELLER, Robert, Swiss physician; b. Frauenfeld, Switzerland, Sept. 20, 1922; s. Robert Wilhelm and Fanny (Lerch) K.; M.D., U. Zurich, 1951; m. Maria Stocker, July 1953. Research fellow dept. pharmacology U. Zurich (Switzerland), 1945-55, head Lab. Exptl. Cytology, cons. exptl. allergy dept. dermatology, 1963—, lectr. immunology dept. med. microbiology, 1966—; research fellow dept. hygiene and applied physiology Fed. Sch. Tech., Zurich, 1955-62; gen. practice medicine, Zurich, 1951—. Mem. Royal Soc. Medicine (London, affiliate), N.Y. Acad. Scis., Histamine Club, Internat. Union Physiol. Scis., Collegium Internationale Allergologicum, Swiss Soc. Physiology, Physiol. Chemistry and Pharmacology. Author: Tissue Mast Cells in Immune Reactions, 1966. Research, numerous publs. in biochemistry of tissue mast cells, mechanisms involved in release of histamine and other biologically active substances from these cells by immunological reactions.*

KELLER, Walter, German pediatrician; b. Heidelberg, Feb. 7, 1894; s. Ludwig and Clara (Barth) K.; M.D., Ph.D.; m. Hanna Steidl, Apr. 14, 1924; 1 child, Lore. Asso. prof., head Mayence Clinic; dir. Univ. Clinic for Children, Giessen; full prof., dir. Children's Clinic, U. Fribourg, prof. emeritus, 1962—. Recipient Heubner medal in pediatrics. Mem. Leopoldina German Acad., N.Y. Acad. Scis., German Soc. Infantile Medicine, Am. Coll. Chest Physicians. Editor: Lehrbuches für Kinderheilkunde; co-editor various profl. mags. Research and publs. on Tb, allergy, virology, contagious diseases. Died Dec. 20, 1967.

KELLER, William Edward, Am. physicist; b. Cleve., Mar. 11, 1925; s. Bert and Sylvia (Rose) K.; B.A., Harvard, 1945, M.A., 1947, Ph.D., 1948; m. Esther Hellerman, June 27, 1947 (div. Dec. 1959); children—Jean Elizabeth, Carol Ann, Margaret Isabell, William Eric; m. 2d, Helen Strong, Apr. 29, 1961. Teaching fellow Harvard, Cambridge, Mass., 1945-48; research asso. Ohio State U., Columbus, 1948-50; staff mem. Los Alamos Sci. Lab., 1950—. Fellow Am. Phys. Soc.; mem. Am. Chem. Soc., Phi Beta Kappa, Phi Lambda Upsilon. Exptl. research on infrared spectroscopy of condensed molecules, thermodynamic properties of simple gases and liquids at low temperatures, hydrodynamics or superfluid liquid helium. Home: P.O. Box 7, Santa Cruz, N.M. 87567. Office: P.O. Box 1663, Los Alamos 87544.*

KELLERMAN, Karl Frederic, Am. bacteriologist; b. Göttingen, Germany, Dec. 9, 1879 (parents Am. citizens); s. William Ashbrook and Stella V. (Dennis) K.; student Ohio State U.; S.B., Cornell, 1900; D.Sc., Kans. Agrl. Coll., 1923; m. Gertrude Hast, Aug. 17, 1905; 1 son, Karl Frederic. Asst. in botany Cornell, 1900-01; asst. physiologist Bur. of Plant Industry, U. S. Dept. Agr., 1901-04; physiologist in charge lab. plant physiology, 1905-06; physiologist in charge soil bacteriology and water purification investigations, 1906-14; asst. chief of bur., 1914-17; asso. chief of bureau, 1917-33; chief Div. Plant Disease Eradication and Control, Bur. Entomology, from 1933. Mem. Fed. Hort. Bd., 1914-24; organized, 1915, and thereafter directed co-operative campaign of the Gulf States and Bur. of Plant Industry to eradicate citrus canker, one of the most contagious of all known diseases of citrus trees; mem. NRC, from 1917, sec. agrl. com., from 1918. Organized Jour. Agrl. Research, 1913, chmn. editorial com., 1913-24. Died Aug. 30, 1934.

KELLEY, Edward, alchemist, astrologer; b. Worcester, Eng., Aug. 1, 1555; brought up as apothecary, at early age acquired some skill in chemistry; said to have studied at Oxford under alias; m. Jane Kelley. Lawyer in London; pilloried for fraud or coining at Lancaster, 1580; skryer (seer) to John Dee, went with him to Prague, stayed with him at Emperor Rudolph II's court, imprisoned by Rudolph as imposter, won his favor later and knighted, but resumed old practices and again imprisoned; parted from Dee, 1588, remained in Germany; lost his life attempting to escape from prison, 1595; his Latin treatises on philosopher's stone issued 1676; mentioned in Hudibras. Included among the various works attributed to him are the Fragmenta aliquiot edita a Cambacis, 1647; Edwardi Kelluii Epistolae ad Edwardum Dyer; Tractatus duo egregii de Lapide Philosophorum, 1676. An alchemist and charlatan, he pretended to make gold by secreting his brother in a chest with a false bottom and then introduced gold into crucible when all had left room. Died Prague, Bohemia, 1595.

KELLEY, Harold Harding, Am. psychologist; b. Boise, Ida., Feb. 16, 1921; s. Harry H. and Maude M. (Little) K.; B.A., U. Cal. at Berkeley, 1942, M.A., 1943; Ph.D., Mass. Inst. Tech., 1948; m. Dorothy J. Drumm, Jan. 4, 1942; children—Ann R. (Mrs. Jonathon Rodnick), Sten, Laura Megan. Study dir., lectr. Research Center for Group Dynamics, U. Mich., Ann Arbor, 1948-50; asst. prof. dept. psychology Yale, 1950-55; prof. U. Minn., Mpls., 1955-61; fellow Center for Advanced Study in Behavioral Scis., Stanford, Cal., 1956-57; prof. psychology U. Cal. at Los Angeles, 1961—. Mem. Am. Psychol. Assn., Am. Sociol. Assn. Author: (with C. I. Hovland, I. L. Janis) Communication and Persuasion, 1953; (with J. W. Thibaut) The Social Psychology of Groups, 1959; also articles. Research on factors affecting conformity to group norms, processes by which interdependent persons work out satisfactory adjustments, negotiation and bargaining processes, social perception and communication. Home: 21634 Rambla Vista, Malibu, Cal. 90265. Office: Dept. Psychology, U. Cal. at Los Angeles 90024.*

KELLEY, Vincent Cooper, Am. geologist; b. Seattle, July 6, 1904; s. Seth Foster and Sara (Cooper) K.; B.A., U. Cal. at Los Angeles, 1931; M.S., Cal. Inst. Tech., 1932, Ph.D., 1937; m. Carol Harwood, 1929 (div. 1944); 1 son, David F.; m. 2d, Anne Robertson, Aug. 15, 1946; children—Anne Robertson, John R., Robert B., Paul. Faculty, U. N.M., 1937—, now prof., chmn. dept. geology; geologist U. S. Geol. Survey, 1938—. Cons. Humble Oil & Refining Co., 1959—; cons., panelist NSF, 1962—. Fellow A.A.A.S., Geol. Soc. Am.; mem. N.M. Mining Assn. (dir.), N.M. Geol. Soc., Soc. Econ. Geology (councilor), Am. Assn. Petroleum Geology, Am. Inst. Mech. Engrs. Author: Iron Ore Deposits of New Mexico, 1949; Geology of Caballo Mountains, New Mexico, 1952; Regional Tectonics of the Colorado Plateau, 1955; (with N. J. Clinton) Fracture Systems and Tectonic Elements of Colorado Plateau, 1960. Contbr. numerous articles in field to sci. jours. Research on geology of N.M., Wyo., Colo., tectonics of Colo. plateau, joints in sedimentary rocks, occurrences of iron ores, base metals, uranium, nature of monoclines. Home: 606 Vassar St. N.E., Albuquerque 87106.*

KELLEY, Walter Pearson, Am. chemist; b. Franklin, Ky., Feb. 19, 1878; s. John William and Mahala Eliza (Mayes) K.; B.S., Ky. State U., 1904; M.S., Purdue, 1907; Ph.D., U. Cal., 1912, LL.D., 1950; D.Sc., U. Ky., 1958; m. Sue Kathrine Eubank, Aug. 6, 1913. Asst. chemist Purdue Agrl. Expt. Sta., 1905-08; chemist Hawaiian Agrl. Expt. Sta., 1908-14; prof. agrl. chemistry U. Cal. at Berkeley, 1914-38, prof. soil chemistry, 1938-48, prof. emeritus, 1948—; cons. Gulf Research and Devel. Co.; U. Cal. and U. S. Govt. del. 3d Congress Internat. Soc. Soil Sci. Oxford, Eng., 1935. Fellow A.A.A.S., Am. Soc. Agronomists (pres. 1930), Am. Mineral. Soc.; mem. Am. Chem. Soc., Western Soc. Naturalists, Internat., Western socs. soil sci., Soil Sci. Soc. Am., Nat. Acad. Scis., Sigma Xi, Phi Beta Kappa, Phi Lambda Upsilon, Alpha Tau Omega. Author (monographs): Cation Exchange in Soils; Alkali Soils, their formation, properties and reclamation. Died May 19, 1965.

KELLICOTT, William Erskine, Am. biologist; b. Buffalo, N.Y., Apr. 5, 1878; s. David Simmons and Valeria Erskine (Stowell) K.; Ph.B., Ohio State U., 1898; Ph.D., Columbia, 1904; m. Mary Hicks,

Sept. 11, 1901. Asst., tutor, and instr. zoology, Barnard Coll. (Columbia), 1901-06; prof. biology, Goucher Coll., Balt., from 1906. In charge embryological instrn. Marine Biol. Lab., Woods Hole, Mass., from 1908; asst. statistician U. S. Food Adminstrn., Washington, from 1917. Fellow, Kahn Found. for Fgn. Travel of Am. Tchrs., 1912-13. Author: Social Direction of Human Evolution, 1911; Text-Book of General Embryology, 1913; Outlines of Chordate Development, 1913. Died Jan. 29, 1919.

KELLOGG, Albert, Am. physician, botanist; b. New Hartford, Conn., Dec. 6, 1813; s. Isaac and Aurilla (Barney) K.; M.D., Transylvania U. First botanist resident in Cal.; surgeon, botanist U. S. Govt.'s 1st expdn. to Bering Sea; a founder Cal. Acad. Scis., 1853, contbr. to its Proceedings, 1855. Author: Forest Trees of California, 1882. Died Alameda, Cal., Mar. 31, 1887.

KELLOGG, Arthur Remington, Am. biologist; b. Davenport, Ia., Oct. 5, 1892; s. Rolla Remington and Clara Louise (Martin) K.; A.B., U. of Kansas, 1915, M.A., 1916; student U. of Calif., 1916-17, Ph.D., 1928; m. Marguerite Evangel Henrich, Dec. 21, 1920. Taxonomic asst., Mus. of Birds and Mammals, U. of Kan., 1913-16; teaching fellow dept. zoölogy, U. of Calif., 1916-19; field asst. Bur. Biol. Survey, U. S. Dept. Agr., summers 1915, 16, 17, asst. biologist, 1920-24, asso. biologist, 1924-28; asst. curator, div. of mammals, U. S. Nat. Museum, 1928-40, curator, 1941-48; dir. U. S. Nat. Museum, 1948-62; asst. sec. Smithsonian Instn., Wash., 1958-62, research asso., 1962—; research asso. Carnegie Instn. of Wash., 1921-43. Member board govs. Crop Protection Inst., 1935-36, Nat. Research Council (div. biology and agr.), Washington, D.C., 1930-54 (vice-chmn. 1945-47). Advisory Committee Chemical-Biological Coordination Center, 1946-52; Pacific Science Board, 1946-52. Am. mem. com. of experts on whaling, League of Nations, Berlin, 1930; apptd. del. Internat. Conf. for Regulation of Whaling, London, 1937, Olso, Norway, 1938, London, 1938. London, 1944-45 (chmn. Am. delegation), Washington, D. C., 1946 (chmn. Am. delegation and chmn. conf.); London, 1949 (chmn. Am. delegation); U. S. commr. Internat. Whaling Commn., Oslo, Norway, 1950, Capetown, 1951, chmn. of commn. (1952-54), London, 1952, 53, 56-63, Tokyo, 1954, Moscow, 1955, The Hague, 1958, Sandefjord, 1964; del. internat. com. establish Internat. Hylean Amazon Inst., Brazil, 1947. Spl. detail to Brazil, 1943; dir. Canal Zone biol. Area, 1945-46. Arctic Research Lab. Advisory Bd., Navy Dept., 1948-51. Mem. Am. Soc. Mammalogists (past pres.), Paleaobiol. Soc. of Washington (pres. 1935-36), Biol. Soc. Wash., Am. Soc. Naturalists, Am. Assn. Anatomists, Soc. Systematic Zoology, Soc. Study Evolution, Acad. Natural Sciences of Phila. (corr.), Nat. Parks Assn. (trustee, 1931-49), Geol. Soc. Am. (fellow), Nat. Acad. Sci., Zool. Society of London, Am. Philos. Soc., American Academy of Arts and Sciences, Sigma Xi. Conglist. Club: Cosmos. Contbr. sci. jours. Authority on whales; taxonomic studies on amphibians and mammals, fossil and recent cetaceans, sirenians and pinnipeds. Home: 5305 28th St. N.W. Office: U. S. Nat. Museum, Smithsonian Instn., Washington.

KELLOGG, Herbert Humphrey, Am. metallurgist; b. N.Y.C., Feb. 24, 1920; s. Herbert H. and Gladys (Falding) K.; B.S., Columbia, 1941, M.S., 1943; m. Jeanette Halstead, July 20, 1940; children—Thomas Bartlett, Jane Falding, David Humphrey, Elizabeth Ann. Asst. prof. mineral preparation Pa. State U., State College, 1942-46; faculty Columbia, N.Y.C., 1946—, prof. extractive metallurgy, 1956—. Chmn. titanium adv. com. Office Def. Mblzn., 1954-58. Recipient Best Paper award extractive metals div. Am. Inst. Metall. Engrs. Mem. Am. Inst. Mining, Metall. and Petroleum Engrs. (chmn. extractive metallurgy div. 1958), Am. Chem. Soc., Electrochem. Soc., Am. Soc. for Metals, Sigma Xi, Tau Beta Pi. Research, numerous publs. on application thermodynamics and kinetics to extractive metall. chemistry at high temperature, systems metal-sulfur-oxygen, metal-metal chloride, fused-salt chemistry. Home: Closter Rd., Palisades, N.Y. 10964. Office: Columbia, N.Y.C. 10027.*

KELLOGG, James Lawrence, Am. biologist; b. Kewanee, Ill., Sept. 15, 1866; s. Hosmer L. and Emily (Platt) K.; B.S., Olivet (Mich.) Coll., 1888; Ph.D., Johns Hopkins, 1892; hon. A.M., Williams Coll., 1900; m. Ida M. Archambeault, June 16, 1892; children—Emilie (Mrs. E. A. Carrier), Louise, Helena (Mrs. W. E. Wright), Margaret (Mrs. J. R. Nelson). Prof. biology Olivet Coll., 1892-99; asst. prof. biology Williams Coll., Williamstown, Mass., 1899-1903, prof., 1903-34 (emeritus). Specialized in anatomy and habits of Lamellibranchiate mollusks. Author: (brochure) Oyster Culture Experiments for State of Louisiana, 1904; Shell-Fish Industries, 1910. Died July 8, 1938.

KELLOGG, John Harvey, Am. surgeon; b. Tyrone, Mich., Feb. 26, 1852; s. John Preston and Ann Jeanette K.; ed. State Normal Sch.; M.D., Bellevue Hosp. Med. Coll., N.Y. U., 1875; later studied in Europe; LL.D., Olivet (Mich.) Coll., Lincoln Meml. U.; Dr. Pub. Service, Oglethorpe U., 1937; m. Ella E. Eaton, Feb. 22, 1879. Practiced in Battle Creek, Mich., from 1875; supt. and surgeon Battle Creek (Mich.) Sanitarium, from 1876. Mem. Mich. State Bd. Health, 1878-90, 1912-16. Founder, pres. Race Betterment Found., 1906, Battle Creek (Mich.) Coll.; founder and med. dir. Miami-Battle Creek Sanitarium, Miami Springs, Fla. Fellow A.C.S., A.A.A.S., Royal Soc. Medicine (Eng.), A.M.A., Nat. Geog. Soc.; mem. Am. Pub. Health Assn.; corr. mem. Société d'Hygiène de France. Author: Plain Facts, 1877; Home Book of Modern Medicine, 1880; Man, the Masterpiece, 1885; Art of Massage, 1895; The Stomach, 1896; Rational Hydrotherapy, 1900; Light Therapeutics, 1910; Colon Hygiene, 1912; Neurasthenia, 1915; Health Series of Physiology and Hygiene (joint author), 1915; Health Question Box, 1917; New Method in Diabetes, 1917; Autointoxication, 1918; The Itinerary of a Breakfast, 1918; The New Dietetics, 1921; Tobaccoism, 1922; The Natural Diet of Man, 1923; How to Have Good Health, 1932; also many tech. papers and articles. Editor Good Health Mag., from 1873. Inventor med. and surg. apparatus and instruments, electric light bath (for therapeutic purposes); research and publs. in nutrition. Died Jan. 16, 1945.

KELLOGG, Paul Jesse, Am. physicist, educator; b. Tacoma, Nov. 6, 1927; s. Jesse I. and Irma (Kennedy) K.; B.S., Mass. Inst. Tech., 1950; Ph.D., Cornell U., 1955; m. Dorothy Del Bourgo, June 10, 1951 (div. Nov. 1964); children—Kenneth, David. Research fellow U. S. Naval Research Lab., 1955-56; research asso. U. Minn., Mpls., 1956-57, faculty, 1957—, prof. physics, 1964—. Fellow Am. Phys. Soc.; mem. Am. Geophys. Union, Am. Assn. U. Profs. Research, publs. on theory of radiation belt, shock wave around earth, shock waves in plasma. Home: 122 Arthur Av. S.E., Mpls. 55414.*

KELLOGG, Remington, Am. biologist, paleontologist; b. Davenport, Ia., Oct. 5, 1892; s. Rolla Remington and Clara Louise (Martin) K.; A.B., U. Kan., 1915, M.A., 1916; Ph.D., U. Cal. at Berkeley, 1928; m. Marguerite E. Henrich, Dec. 21, 1920. With Bur. Biol. Survey, Dept. Agr., 1920-28; curator U. S. Nat. Mus., 1928-48, dir., 1948-62; assistant secretary of the Smithsonian Instn., Washington, 1958-62, research asso., 1962—; research asso. Carnegie Instn., Washington, 1921-43. U. S. commr. Internat. Whaling Commn., 1950—, past chmn., 1952-55. Recipient U. Kan. Alumni award, 1947. Fellow Geol. Soc. Am. Zool. Soc. London; mem. Nat. Acad. Sci., Am. Philos. Soc., Am. Acad. Arts and Sci., Soc. Naturalists, Soc. Vertebrate Paleontology, Am. Soc. Mammalogists (past pres.), Soc. for Study Evolution, Assn. Anatomists, Internat. Hylean Amazon Inst. (Am. del. internat. commn. Belem, Brazil 1947). Author: Review of the Archaeoceti, 1936; also numerous articles on recent and fossil marine mammals. Research on geologic history and evolutionary devel. Cetacea, Pinnipedia and sirenia, their importance in correlating marine geologic formations widely separated at present with oceans; occurrence identical forms or fauanas as an indication of contemporaneity in geologic time. Home: 5305 28th St., N.W., Washington 20015. Office: Smithsonian Instn., Washington 20560.*

KELLOGG, Vernon Lyman, Am. zoologist; b. Emporia, Kan., Dec. 1, 1867; s. Lyman Beecher and Abigail (Homer) K.; A.B., U. Kan., 1889, M.S., 1892; Cornell, 1891; U. Leipzig, 1893, 97; U. Paris, 1904; LL.D., U. Cal., 1919, Brown U., 1920; Sc.D., Oberlin Coll., 1922; m. Charlotte Hoffman, Apr. 27, 1908; 1 dau., Charlotte Jean. Asst. and asso. prof. entomology U. Kan., 1890-94; prof. entomology, lectr. bionomics, Stanford U., 1894-1920. Dir., in Brussels, of Am. Com. for Relief in Belgium, 1915-16; asst. to U. S. food adminstr., 1917-19; chief of mission to Poland, spl. investigator in Russia, and other service in Europe with Am. Relief Adminstr., 1918-21; permanent sec. NRC, Washington, 1919-31, and chmn. div. on indsl. relations, 1919-29, sec. emeritus, 1932. Trustee Rockefeller Found., 1922-33, Brookings Inst., and other orgns. Mem. Nat. Acad. Scis., various other Am. and European sci. socs. Decorated by govts. of France, Belgium, Poland. Author: American Insects, 1904; Animal Studies (with D. S. Jordan and H. Heath), 1905; Darwinism Today, 1907; Evolution and Animal Life (with D. S. Jordan), 1907; Insect Stories, 1908; Scientific Aspects of Luther Burbank's Work (with D. S. Jordan), 1909; The Animals and Man, 1911; Economic Zoölogy and Entomology (with R. W. Doane), 1915; Losses of Life in Modern Wars and Race Deterioration (with G. Bodart), 1916; Headquarters Nights, 1917; The Food Problem (with A. R. Taylor), 1917; Fighting Starvation in Belgium, 1918; Germany in the War and After, 1919; Herbert Hoover—The Man and His Work, 1920; Nuova, the New Bee, 1921; Human Life as the Biologist Sees It, 1922; Mind and Heredity, 1923; Evolution, 1924; Reading with a Purpose—Biology, 1925. Contbd. to knowledge of evolutionary history of birds; was leading authority on Mallophaga order of birds. Died Aug. 8, 1937.

KELLOGG, William Welch, Am. meteorologist; b. New York Mills, N.Y., Feb. 14, 1917; s. Frederick S. and Elizabeth (Walcott) K.; B.A., Yale, 1939; postgrad. U. Cal. at Berkeley, 1940-41; M.A., U. Cal. at Los Angeles, 1942, Ph.D., 1949; m. Elizabeth Thorson, Feb. 14, 1942; children—Karl S., Judith L., Joseph W., Jane E., Thomas W. Pilot-weather officer USAAF, 1941-46; research asst., research asso., asst. prof. Inst. Geophysics, U. Cal. at Los Angeles, 1946-52; with Rand Corp., Santa Monica, Cal., 1947-64, head planetary scis. dept., 1959-64; asso. dir. Nat. Center for Atmospheric Research, also dir. Lab. Atmospheric Scis., Boulder, Colo., 1964—. Mem. earth satellite panel Internat. Geophys. Year, 1956-59; mem. space sci. bd. Nat. Acad. Scis., 1961—, com. on atmospheric scis., 1962—, com. for internat. years of quiet sun, 1962—; mem. adv. group on supporting tech. for operational meteorol. satellites NASA-U. S. Weather Bur., 1964—; rapporteur for meteorology of high atmosphere, commn. on aerology World Meteorol. Orgn., 1965—; chmn. internat. commn. on meteorology of upper atmosphere Internat. Union Geodesy and Geophysics, 1960—; mem. sci. adv. bd., office aerospace research USAF, 1965—; chmn. meteorol. satellite com. Advanced Research Projects Agy., 1958-59. Recipient (with S. M. Greenfield) Spl. award for pioneering work in planning of meteorol. satellite Am. Meteorol. Soc., 1961; Risseca award for contbn. to human relation in sci. Jewish War Vets. U. S. A., 1962-63. Mem. Am. Geophys. Union (chmn. com. on internat. participation for sect. on meteorology 1965—), Am. Meteorol. Soc. (council mem. 1960-63), Am. Inst. Aeros. and Astronautics (com. on physics upper atmosphere and space 1960—), Am. Astron. Soc., Sigma Xi. Research on meteorology, dynamics and turbulence of upper atmosphere, use of rockets and satellites for atmospheric research, prediction of radioactive fallout and dispersal, applications of infrared techniques, atmospheres of Mars and Venus. Home: 445 College Av., Boulder, 80302. Office: National Center for Atmospheric Research, Boulder, Colo. 80301.*

KELLOGG, Winthrop Niles, Am. psychologist; b. Mt. Vernon, N.Y., Apr. 13, 1898; s. Henry Niles and Florence Adele (Bowns) K.; student Cornell U., 1916-17; A.B., Indiana U., 1922; A.M., Columbia, 1927, Ph.D., 1929; Sc.D., Denison U., 1964; m. Luella Dorothy Agger, Dec. 1920; children—Jack Stanley (dec.), Donald Agger, Shirley Mae. Mem. faculty Columbia, 1927-29; faculty Ind. U., 1929-50; prof. exptl. psychology, Fla. State U., 1950-63, 1965—; research scientist, cons. Stanford Research Inst., Menlo Park, Cal., 1962-65; mem. NRC, 1951-54; research fellow Social Sci. Research Council, 1931-32. Decorated Croix de Guerre (France). Fellow A.A.A.S., Am. Psychol. Assn.; mem. Southeastern Psychol. Assn. (pres. 1958), Phi Beta Kappa, Sigma Xi, Kappa Sigma, Sigma Delta Chi. Author or co-author: An Experimental Comparison of Psycho-physical Methods, 1929; An Experimental Evaluation of Equality Judgments in Psychophysics, 1930; The Ape and the Child; 1933; Methods in Psychology, 1948; Readings in Learning, 1953; Porpoises and Sonar, 1961. Associate editor: The Psychol. Record. Pioneered in sonar in dolphins and echolocation in the blind; research and publs. on psychophys. methods, developmental psychology, animal behavior, the neural activity involved in the conditioning and learning process. Home: P.O. Box 1418, Sarasota, Fla. 33578.*

KELLY, Anthony, English physicist; b. Hillingdon, Middlesex, Eng., Jan. 25, 1929; s. Vincent Gerald and Violet (Vaughan) K.; B.S., U. Reading, 1949, B.S. (Spl.), 1950; Ph.D., U. Cambridge, 1953; m. Christina Margeret Dunleavie, July 14, 1956; children—Mary-Clare, Anthony Paul, Andrew Julian, Stephen Charles Vincent. Research asso. U. Ill., 1953-55; Imperial Chem. Industries research fellow U. Birmingham, 1955-56; faculty Northwestern U., 1956-59; lectr. metallurgy U. Cambridge (Eng.), fellow Churchill Coll., 1959—. Cons. Dow Chem. Co., U. S. Steel Co., U.K. Atomic Energy Authority, Atomic Weapons Research Establishment, Ministry Aviation. Fellow Inst. Physics; mem. Am. Inst. Mining and Metall. Engrs. Author: Strong Solids, 1966. Research, numerous publs. on mech. strength of solids; established principle of Fibre Reinforcement. Home: 20 Rutherford Rd., Cambridge, Eng.*

KELLY, E(verett) Lowell, Am. psychologist; b. Kokomo, Ind., Nov. 15, 1905; s. Alva E(lmont and Maude (Vickrey) K.; B.S., Purdue U., 1926; A.M., Colo. Coll. Edn, 1928; Ph.D., Stanford, 1930; Doctor Sci., Purdue U., 1955; m. Lillian Isaacs, Dec. 25, 1938; children—Patricia Ann (Mrs. Paul Klinger), Paul Alan, Pamela Jane. High sch. prin. and teacher Taiban (New Mexico) public schools, 1926-27; associate professor psychology, director of admissions University of Hawaii, 1930-32; Social Science Research Council fellow, study in Germany and Austria, 1932-33; chmn. dept. psychology U. of Conn., 1933-39; asso. prof. psychology, dir. psychol. clinic Purdue U., 1939-42; br. chief clin. psychologist V.A., 1946-47; prof. psychology U. of Mich., 1946—, chairman department of psychology, 1958-62; director Bureau Psychol. Services, Inst. Human Adjustment since 1950; cons. U. S. Pub. Health Service, U. S. Navy, V.A.; advisor to dir. Selective Service; dir. selection U. S. Peace Corps, Washington, 1961-62. Recipient Letter of Commendation (with ribbon), Sec. of the Navy, 1945. Mem. Am. (bd. dirs. 1949-52, pres. 1954-55), Mich. (pres. 1948-49) psychol. assns. Research in clin. psychology, personality measurement, personnel selection and tng. techniques; developed tests for prediction marital compatibility. Home: 2559 Blueberry Lane, Ann Arbor, Mich.

KELLY, George Alexander, Am. psychologist; b. Perth, Kan., Apr. 28, 1905; s. Theodore Vincent and Elfleda (Merriam) K.; B.A., Park Coll., 1926; M.A., U. Kan., 1928; B.Ed., U. Edinburgh (Scotland),

1930; Ph.D., U. Ia., 1931; m. Gladys Dorothy Thompson, June 3, 1931; children—Jacqueline (Mrs. George Edward Sharples), Joseph Vincent. Faculty, Ft. Hays Kan. State Coll., 1931-43; asso. prof. U. Md., College Park, 1945-46; prof. Ohio State U., Columbus, 1946-65; Riklis prof. behavioral sci. Brandeis U., Waltham, Mass., 1965-66. Diplomate Am. Bd. Examiners in Profl. Psychology (past pres.). Mem. Am. Psychol. Assn. (past pres. cons. div., past pres. clin. div., ann. award for distinguished contbn. clin. div. 1965). Author: Psychology of Personal Constructs, 2 vols., 1955; also articles. Devel. theory of psychology known as psychology of personal constructs. Home: 14 Hickory Lane, Framingham, Mass. 01701. Office: Dept. Psychology, Brandeis U., Waltham, Mass. Died Mar. 6, 1967.*

KELLY, Howard Atwood, Am. surgeon; b. Camden, N.J., Feb. 20, 1858; s. Henry Kuhl and Louisa Warner (Hard) K.; B.A., U. Pa., 1877, M.D., 1882, LL.D., 1907; LL.D., Aberdeen, 1906, Washington and Lee U., 1906, Washington Coll., 1933, Johns Hopkins, 1939; m. Laetitia Bredow, June 27, 1889; children—Olga Elizabeth Bredow, Henry Kuhl, Esther Warner (Mrs. Henry G. Seibels), Friederich Heyn, Howard Atwood, William Boulton, Margaret Kuhl (Mrs. Douglas Warner), Edmund Bredow, Laetitia Bredow (Mrs. Winthrop K. Coolidge). Founder, Kensington Hosp., Phila.; asso. prof. obstetrics U. Pa., 1888-89; prof. gynecology and obstetrics Johns Hopkins U., 1889-99, gynecology, 1899-1919, emeritus from 1919; gynecol. surgeon Johns Hopkins Hosp. 1899-1919; surgeon and radiologist Howard A. Kelly Hosp. from 1892; hon. curator Div. of Reptiles and Amphibians, U. of Mich. Hon. fellow learned socs. of Scotland, Ireland, Russia, Eng.; fellow So. Surg. and Gynecol. Soc., Am. Radium Soc., A.C.S., A.A.A.S., Brit. Gynecol. Soc., Md. Acad. Scis., Am. Geog. Soc.; mem. Am. Gynecol. Soc. (pres. 1912), Am. Urol. Assn., Roentgenol. Soc., Seaboard Med. Assn., Am. Mus. of Natural History, Am. Soc. Ichthyologists and Herpetologists, numerous other Am. and European med. socs. Author: Operative Gynecology (2 vols.), 1898, 1906; The Vermiform Appendix and its Diseases (with Elizabeth Hurdon), 1905; Walter Reed and Yellow Fever, 1906, 07, 23; Gynecology and Abdominal Surgery (edited with C. P. Noble), Vol. I, 1907, Vol. II, 1908; The Stereo Clynic, 84 sects., from 1908; Medical Gynecology, 1908, 1912; Appendicitis and Other Diseases of the Vermiform Appendix, 1909; Myomata of the Uterus (with T. S. Cullen), 1909; Cyclopedia of American Medical Biography (2 vols.), 1912; Some Am. Med. Botanists, 1913; Diseases of the Kidneys, Ureters, and Bladder (with C. F. Burnam—2 vols.), 1914, 1922; American Medical Biographies (with W. L. Burrage), 1920; A Scientific Man and the Bible, 1925; Gynecology, 1928; Dictionary of American Medical Biography (with W. L. Burrage), 1928; Electrosurgery (with Grant E. Ward), 1932; also many sci. articles. Introduced method of aeroscopic exam. of bladder, 1893, catheterization method of individual ureters in male, 1898, use of wax-tipped bougies for locating urinary-tract stones, 1900; invented rectoscope, 1903; one of 1st to use radium in cancer treatment. Died Jan. 12, 1943.

KELLY, John Forrest, elec. engr.; b. Carrick-on-Suir, Ireland, Mar. 28, 1859; s. Jeremiah and Kate (Forrest) K.; B.S. in physics and chemistry, Stevens Inst. Tech., 1878, Ph.D., 1881; m. Helen Tischer, 1892. Chemist in lab. of Thomas A. Edison, 1879; electrician N.Y. factory Western Elec. Co., 1879-82, with U. S. Electric Lighting Co., 1882, 1884-86; with Parker Electric Lighting Co., 1882-84; electrician Newark shops of Westinghouse Elec. Co. until 1892; with Stanley Lab. Co., 1892-95; cons. engr. for Stanley Electric Mfg. Co., Pittsfield, and Stanley Instrument Co., Great Barrington, Mass., until 1905; founded Telelectric Co., 1905, pres. until 1910. Recipient John Scott medal Franklin Inst., for electric player-piano 1909. Fellow Am. Inst. E.E. Introduced Cooke-Kelly food drying process; patentee in field of utilization of electricity, covering apparatus for generating, transmitting, distributing and measuring; first to produce stable iron for electromagnetic purposes; pioneer in long distance, high tension transmission work. Died Oct. 15, 1922.

KELLY, Mervin J., Am. physicist; b. Princeton, Mo., Feb. 14, 1894; s. Joseph Fenemore and Mary Etta (Evans) K.; B.S., Missouri Sch. of Mines and Metal. 1914; M.S., U. of Ky., 1915, D.Sc. 1946; Ph.D., U. Chgo., 1918; D.Eng., U. Mo., 1936, N.Y. U., 1955, Bklyn. Poly. Inst., 1955, Wayne State U., 1958, Princeton, 1959; LL.D., U. Pa., 1954; Dr. Honoris Causa, U. Lyons, 1957; D.Sc., U. Pitts., 1959, Case Inst. Tech., Stevens Institute Technology, 1959; m. Katharine Milsted, November 11, 1917; children—Mary (Mrs. Robert von Mehren), Robert Milsted. Physicist, research dept. of Western Electric Company, 1918-25; physicist, Bell Telephone Labs., New York, N.Y., 1925-28; dir. vacuum tube development, 1928-34, development dir. Transmission Inst. and electronics, 1934-36, dir. of research, 1936-44, exec. v.p., 1944-51, pres., 1951-58, chmn., 1959, dir., 1944-59, research mgmt. cons. IBM Corp., 1959-—; research mgmt. cons., dir. Bausch & Lomb, Inc., 1959-62; director Tungsol Electric, Incorporated, Prudential Insurance Co. of Am., Bausch & Lomb Optical Co. Chmn. com. Sec. Def. continental def.; spl. cons. Nat. Aeros. and Space Adminstrn.; mem. adv. com. Sci. Manpower; chmn. subcom. role industry coll. N.Y.C. bd. edn.; chmn. com. for Sec. Commerce on Nat. Bur. Standards, chairman of the visiting com., chairman com. evaluation of all research and engineering activities; chairman of the task force on research Hoover Commn.; chmn. Dept. of Air Force sci. adv. bd.; v. chmn. Dept. Navy naval research adv. com. Member bd. trustees Stevens Inst. Tech. Awarded Presdl. Certificate of Merit; Indsl. Research Inst. award, 1954; Christopher Columbus Internat. Communication prize, 1955; Air Force Assn. trophy award; James Forrestal medal Nat. Security Indsl. Assn., 1959; John Fritz medal for achievements in electronics, 1959; Mervin J. Kelly award in telecommunication, Am. Inst. E.E., 1960, Am. Inst. E.E. and NEMA Golden Omega Award, 1960, Joint Engring. Socs. Hoover Medal, 1961. Fellow Rochester Mus. Arts and Scis., 1960; mem. Am. Acad. Arts and Scis., Swedish Royal Acad. Sci., M.I.T. Corp.; chmn. adv. council, dept. elec. engring. Princeton; mem. advisory committee, department of electricity, University Rochester. Fellow Am. Phys. Soc., Am. Inst. E.E., Inst. Radio Engrs., Acoustical Soc. of Am., Am. Philos. Soc., Nat. Acad. Sci., Sigma Xi, Eta Kappa Nu, Tau Beta Pi. Research in applications of physics in electronics. Home: 2 Windemere Terrace, Short Hills, N.J. Office: 590 Madison Av., N.Y.C. 22.

KELLY, Patrick Joseph, Am. physician; b. Mpls., Feb. 12, 1926; s. Charles A. and Beulah (Ward) K.; B.S., St. Lawrence U., N.Y., 1945; M.D., St. Louis U., 1949; M.S., U. Minn., 1958; m. Sara Maynard, May 20, 1950; children—Patrick Joseph, Charles Michael, Robert Troy, Mary Margaret, Kathleen, Sara, Nancy, John. Cons. orthopedic surgery Mayo Clinic Mayo Grad. Sch. Medicine, Rochester, Minn., 1957-—, asso. prof. orthopedic surgery, 1963-—. Am. traveling fellow to Gt. Britain for Am. Orthopedic Assn., 1963. Mem. Am. Acad. Orthopedic Surgeons, Orthopedic Research Soc. Contbr. chpt. to Circulation of Bone, 1962. Asso. editor: Jour. Bone and Joint Surgery, 1961-—. Research, numerous publs. in anatomy and physiology circulation of bone, metabolic bone disease, infections of bones and joints. Home: 816 9th Av. Office: 200 First St., Rochester, Minn. 55901.*

KELLY, William, Am. inventor; b. Pitts., Aug. 21, 1811; s. John and Elizabeth (Fitzsimons) K.; m. Mildred A. Gracy, circa 1847. Jr. mem. firm McShane & Kelly, 1846; developer Suwanee Iron Works & Union Forge, mfg. sugar kettles; granted U. S. patent for original invention Bessemer process (a process of air-boiling of steel which made possible inexpensive soft steel), 1857; built 1st successful fitted converter at Cambria Iron Works, 1859; founder axe mfg. firm, Louisville, Ky., 1861; honored by Am. Soc. for Steel Treating with bronze tablet at site of Wyandotte Iron Works, 1925. Died Louisville, Feb. 11, 1888.

KELMAN, Arthur, Am. plant pathologist; b. Providence, Dec. 11, 1918; s. Philip and Minnie (Kollin) K.; B.S., U. R.I., 1941; M.S., N.C. State U., 1946, Ph.D., 1949; grad. student U. Wis., 1947-48; m. Helen Moore Parker, June 22, 1949; 1 son, Philip Joseph. Mem. faculty N.C. State U., Raleigh, 1948-65, prof., 1957-65, W. N. Reynolds distinguished prof. plant pathology, 1961-65; prof., chmn. dept. plant pathology U. Wis., Madison, 1965-—; vis. investigator Rockefeller Inst., 1953-54. Cons. United Fruit Co., 1958-64; vis. lectr. Am. Inst. Biol. Sci., 1961-62. Mem. Am. Phytopathol. Soc. (pres. 1966-67), Am. Soc. Foresters, Soc. Gen. Microbiology, Am. Soc. Microbiology, N.C. Acad. Sci., A.A.A.S., Am. Inst. Biol. Sci., Sigma Xi, Alpha Zeta, Gamma Sigma Delta, others. Author: The Bacterial Wilt Caused by Pseudomonas solanacearum, 1953; (with A. Husain) Tissue is Disintegrated in Plant Pathology, 1959; (with I. Buddenhagen) Biological and Physiological Aspects of Bacterial Wilt, 1964. Asso. editor Phytopathology, 1953-56. Research and publs. on devel. of standard methods of greenhouse screening for resistance in major hosts to bacterial wilt; studies on factors influencing virulence and resistance to bacterial plant pathogens. Address: Dept. Plant Pathology, U. Wis., Madison, Wis. 53706.*

KELMAN, Harvey, Am. psychiatrist; b. Mpls., June 2, 1934; s. Hyman and Lillian (Bank) K.; B.A. cum laude, U. Minn., 1956, M.D., 1961; postgrad. Washington Psychoanalytic Inst.; m. Rosalind Schribman, Mar. 23, 1958; children—Naomi Kelman, Ruth. Staff psychiatrist Patuxent Instn., Silver Spring, 1965-—. Cons. Alcoholism Clinic, Montgomery County Health Dept., Rockville, Md. Mem. Am., Minn. psychiat. assns., Assn. Washington Psychoanalytic Soc., A.M.A. Research, publs. on homotransplantation in inbred mice, electroshock therapy as treatment for narcotic withdrawal symptoms, physiologic correlates of emotional reactions, application of learning theory to psychotherapy. Address: 509 Lamberton Dr., Silver Spring, Md. 20902.*

KELMAN, Herbert Chanoch, psychologist, educator; b. Vienna, Austria, Mar. 18, 1927; s. Leo and Lea (Pomeranz) K.; came to U. S., 1940, naturalized, 1950; B.A., Bklyn. Coll., 1947; B.H.L., Sem. Coll. Jewish Studies, 1947; M.S., Yale, 1949, Ph.D., 1951; m. Rose Brousman, Aug. 23, 1953. Research asst. Yale, 1947-51; research fellow Johns Hopkins, 1951-54; research psychologist Nat. Inst. Mental Health, Bethesda, Md., 1955-57; lectr. social psychology Harvard, 1957-62; prof. psychology U. Mich., Ann Arbor, 1962-—, chmn. doctoral program in social psychology, 1966-67. Fellow, Center for Advanced Study in Behavioral Scis., Stanford, Cal., 1954-55, Inst. Social Research, Oslo, Norway, 1960-61, Western Behavioral Sci. Inst., La Jolla, Cal., 1964. Recipient A.A.A.S. Socio-Psychol. award, 1956. Mem. Soc. for Psychol. Study Social Issues (pres. 1964-65), Am. Psychol. Assn., Am. Sociol. Assn., Am. Assn. Pub. Opinion Research, Internat. Studies Assn. Editor, contbr. International Behavior: A Social-Psychological Analysis, 1965. Research, numerous publs. in devel. and exptl. testing theoretical framework for analysis of behavior change resulting from social influence with spl. emphasis on antecedents and consequents of changes differing in depth and stability. Home: 400 Barton Shore Dr., Ann Arbor, Mich. 48105.*

KELNER, Albert, Am. biologist; b. Phila., Sept. 7, 1912; s. Jacob Michael and Eva (Maimon) K.; A.B., U. Pa., 1940; M.Sc., N.C. State Coll., 1942; Ph.D., Pa., 1943; m. Adelyn Uswald, Dec. 14, 1946; children—Robert Michael, Margaret Andrea, Carole Lisa. Fellow, William Pepper Lab. clin. medicine U. Pa. Med. Sch., 1943-46; bacteriologist L.I. Biol. Lab., Cold Spring Harbor, N.Y., 1946-49; spl. research fellow Biol. Labs., Harvard, 1949-51; faculty Brandeis U., Waltham, Mass., 1951-—, prof. dept. biology, 1961-—, chmn. grad. com. on microbiology, 1955-56. Recipient Finsen Gold medal Internat. Congress Photobiology, 1964. Mem. Genetics Soc., Radiation Research Soc., A.A.A.S., Soc. Am. Microbiologists, Soc. for Gen. Microbiology. Research, publs. on induction mutations by X-rays and ultraviolet light in micro-organisms yielding variants producing new antibiotics, action ultraviolet and visible light on cells, mechanisms for repair of damage caused by ultraviolet light to DNA bacteria and other cells; discovered photoreactivation. Home: 303 Florence Rd., Waltham, Mass. 02154.*

KELSCH, Achille-Louis-Félix, French physician; b. Schiltighem, France, 1841; became physician Val-de-Grace, France, 1866, instr. epidemiology, 1870; prof. path. anatomy Faculty Medicine, Lille, France, until 1882; went to Paris; staff depts. epidemiology and diseases found in army Sch. Applied Medicine, Val-de-Grace; became dir. Inst. Higher Medicine, 1906; dir. Sch. Health Service, Lyons, France, Sch. of Val-de-Grace. Author: (with Kiener) Traité des maladies des pays chaude; Traité des maladies épidémiques, étiologie et pathogénie des maladies infectieuses, 1894-1906. Research on liver, kidneys, tropical diseases. Died Paris, 1911.

KELSER, Raymond Alexander, Am. bacteriologist, immunologist; b. Washington, D.C., Dec. 2, 1892; s. Charles and Josie Mary (Potter) K.; D.V.M., George Washington U., 1914; A.M., Am. U., 1922, Ph.D., 1923; diploma hon. asso. Royal Coll. Veterinary Surgeons, London; m. Eveline Harriet Davison, Sept. 5, 1914; 1 dau., Evelyn Rae (Mrs. John Andrew Allgair). Bacteriologist for H. K. Mulford Co., Glenolden, Pa. 1914, for U. S. Dept. Agr., 1915-18; served as lt. Veterinary Corps, U. S. Army, 1918, capt., 1918-20, commd. capt. regular army, 1920, advanced through grades to brig. gen., 1942, ret., 1946; chief of veterinary lab. Letterman hosp., San Francisco, 1918; comdg. officer Army Veterinary Lab., Phila., 1919-20; chief vet. lab. sect. Army Med. Sch., Washington, also instr. bacteriology, 1921-24, 1928-33; mem. U. S. Army Med. Dept. Research Bd., Manila, P.I., 1925-28; instr. in pathology and bacteriology, Grad. Sch. of Arts and Sciences, Am. U., 1930-33; research fellow in bacteriology Harvard U. Med. Sch., 1933-35; mem. Army Med. Research Bd., Ancon, C.Z., 1935-38; dir., veterinary div. Office of Surgeon Gen. U. S. Army, 1938-46; prof. bacteriology, dean of faculty, Sch. Veterinary Medicine, U. Pa., Phila., 1946-52. Mem. A.A.A.S., Am. Acad. Tropical Medicine, Am. Pub. Health Assn., Am. Vet. Med. Assn. (v.p.), Am. Assn. Pathologists and Bacteriologists, Nat. Acad. Sci., Soc. Am. Bacteriologists, Am. Soc. Exptl. Pathology, Am. Soc. Tropical Medicine, Washington Acad. Scis., Assn. Mil. Surgeons U. S., N.Y. Acad. Sci., Phi Zeta, Sigma Xi. Decorated D.S.M.; recipient Gorgas Medal; XII Internat. Vet. Congress prize; George Washington U. Alumni Achievement award. Author: Manual of Veterinary Bacteriology, 1927; also articles. Discovered transmission of encephalomyelitis by mosquitoes, developed vaccine for rinderpest, resulting in eradication of worst stock plague in Philippines; developed spore vaccine for anthrax that was improvement over Pasteur vaccine; perfected fixation test of great value for detecting botulius toxin in canned foodstuffs; studied degenerative arthritis in man and horse. Died Apr. 16, 1952.

KELVIN, Lord William Thomson, Scottish mathematician, physicist; b. Belfast, Ireland, June 26, 1824; s. James Thomson and Margaret (Gardener) K.; ed. Glasgow U., 1934-41, Peterhouse Coll., Cambridge, 1841-45, U. Paris, 1845; M.A., LL.D., D.C.L., D.Sc., M.D.; m. Margaret Crum, 1852; m. 2d, Frances Blandy, 1874. Fellow, Peterhouse Coll., 1846-52, 72; prof. natural philosophy Glasgow U., 1846-99, chancellor, 1894; electrician for Atlantic cables, 1857-58, 65-66, also participated in elec. work for other large cables, 1869, 73, 75, 79. Fellow Royal Soc. London (pres. 1890-95), 1851, Royal Soc. Edinburgh (pres. 4 times); mem. Brit. Assn. (pres. Edinburgh 1871),

French Acad. (fgn. asso.), Berlin Acad. Scis. (fgn.). Author: Electrostatics and Magnetism, 1 vol.; Mathematical and Physical Papers, 3 vols.; Popular Lectures and Addresses, 3 vols.; A Treatise on Natural Philosophy in conjunction with Professor P. G. Tait; Tables for facilitating the use of Sumner's Method at Sea; The Baltimore Lectures. Did important work in almost every br. of phys. scis.; one of 1st to teach physics in lab. as well as lecture hall; had part in making possible 1st transatlantic telegraph cable; estimated age of earth from solidification as between 20 and 400 million years; estimated age of sun (with Helmholtz); calculated energy radiated from sun's surface; modified Rayleigh's wave theory of light; developed Joule's doctrine of convertibility of heat and work, developed a dynamical theory of heat; propounded doctrine of dissipation of energy; collaborated with Joule in investigating Joule-Thomson effect; advocated absolute scale of measurements; proposed absolute (or Kelvin) scale of temperature; investigated electric oscillation by math. analysis, studied electric currents in cables; invented mirror galvanometer, electrometer, dynamometer, gyrostat, siphon recorder, Kelvin standard ballance, curb transmitter (with Jenkin); studied electrodynamic properties of metals and math. theory of magnetism; applied theory of elasticity to tides in solid earth; introduced Bell telephone into Gt. Britain; developed improved mariner's compass (still used today); invented tide predictor, tide gauge and harmonic analyzer. Died Largs, Ayr, Dec. 17, 1907.

KEMBALL, Charles, Brit. chemist; b. Edinburgh, Mar. 27, 1923; s. Charles Henry and Janet (White) K.; M.A., Sc.D., Cambridge U.; m. Kathleen P. Lynd, Dec. 27, 1956; children—Mary Lynd, Alan Charles, Heather Jane. Fellow, Trinity Coll., Cambridge, 1946-54; asst. in phys. chemistry Cambridge U., 1951-54; prof. phys. chemistry Queen's U., Belfast, Northern Ireland, 1954-66; prof. phys. chemistry U. Edinburgh, 1966—. Recipient Corday-Morgan medal and prize Chem. Soc., Ipatieff prize Am. Chem. Soc. Fellow Royal Soc.; mem. Royal Soc. Edinburgh, Royal Inst. Chemistry (Deldola medal). Research and publs. on adsorption and heterogeneous catalysis. Address: 10 Shrewsbury Dr., Belfast 9, Northern Ireland.

KEMBLE, Edwin Crawford, Am. physicist; b. Delaware, O., Jan. 28, 1889; s. Duston and Margaret (Day) K.; student Ohio Wesleyan U., 1906-07; B.S., Case Sch. Applied Sci., 1911, D.Sc., 1931; A.M., Harvard, 1914, Ph.D., 1917; Ed.D., R.I. Coll. Edn., 1957; m. Harriet Mary Tindle, Sept. 8, 1920; children—Robert Day, Jean Allen. Asst. instr. physics Carnegie Inst. Tech., 1911-13; engring. physicist Curtiss Motor Corp., 1917-18; instr. physics Williams Coll., 1919; faculty physics Harvard, 1919—, prof., 1930-57, prof. emeritus, 1957—, chmn. dept physics, 1940-45. Chmn. com. on molecular spectra in gases, NRC, 1923-27, mem. exec. com. div. phys. sci., 1938-40. Guggenheim fellow, 1927. Mem. Nat. Acad. Sci. (chmn. physics sect. 1945-48), Sigma Xi, Tau Beta Pi, Phi Kappa Psi. Author: (with others) Report on Molecular Spectra in Gases, 1926; Fundamental Principles Quantum Mechanics, 1937. Research on band spectra, fundamentals of quantum theory, philosophy of sci. Home: 8 Ash St. Place, Cambridge, Mass. 02138.*

KEMKES, Berthold Paul, German microbiologist; b. Frankfort, Germany, Mar. 5, 1901; s. Gerhard and Karoline Kemkes; ed. U. Frankfort, U. Leipzig; M.D.; m. Brigitte Fenner, Dec. 31, 1954; children—Bettina, Birgit. Agrégé, 1926; med. asst., 1926-29; town physician, 1929-35; dir. Inst. Hygiene, U. Frankfort, 1935-45; dir. Inst. Hygiene, U. Giessen, 1943—, titular prof., 1951—. Mem. German Soc. Hygiene and Microbiology. Author: Leitfaden der medizinischen Mikrobiologie und Parasitologie, 1948; Allgemeine Hygiene, 1950; Grundzüg der sozialen Gesundheitsfürsorge, 1952. Home: Freiligrathstrasse 3. Office: Friedrichstrasse 16, Giessen, Germany.

KEMP, James Furman, Am. geologist; b. N.Y.C., Aug. 14, 1859; s. James Alexander and Caroline (Furman) K.; A.B., Amherst, 1881, hon. Sc.D., 1906; E.M., Columbia, 1884; LL.D., McGill, 1913; m. Kate Taylor, Sept. 5, 1889; children—James Taylor, Philip Kittredge, Katherine Furman. Instr. and asst. prof. geology Cornell U., 1886-91; adj. prof. geology Columbia U., N.Y.C., 1891-92, prof., from 1892. Mgr. and scientific dir. New York Bot. Gardens, from 1898; conducted state and fed. surveys of geology Adirondack Mountains, 1890-92; cons. geologist Heckler Copper Co. Mem. Am. Inst. Mining and Metall. Engrs. (hon.), Mining and Metall. Soc. Am. (gold medalist), Nat. Acad. Scis. (pres.), A.A.A.S. (v.p.), Am. Philos. Soc., N.Y. Acad. Scis., Soc. Econ. Geologists, Geol. Soc. Am. (founder 1888). Author: Handbook of Rocks, 1896; Ore Deposits in the United States, 1893. Founder mag. Econ. Geology. Investigated deposits of lead and zinc ores in So. Mo.; research in pre-Cambrian geology of N. Am. continent; made detailed study of pegmatites. Died Nov. 17, 1926.

KEMP, Norman Everett, Am. biologist; b. Otisfield, Me., June 20, 1916; s. John and Marian (Foster) K.; B.S., Bates Coll., 1937; Ph.D., U. Cal. at Berkeley, 1941; postgrad. Yale; m. Ruth Estelle Robinson, Nov. 26, 1942. Instr., Wayne U., 1946-47; faculty U. Mich., Ann Arbor, 1947—, prof. zoology, 1961—; vis. investigator Rockefeller Inst., 1958; vis. colleague, Hawaii Marine Lab., 1965. Mem. Marine Biol. Lab., Woods Hole, Mass., 1957—. Mem. A.A.A.S., Am. Assn. Anatomists, Am. Soc. Zoologists, Am. Soc. Cell Biology, Soc. Developmental Biology, Electron Microscopy Soc. Am., Internat. Inst. Embryology, Mich. Acad. Sci. Arts and Scis. (past sec.), Mich. Electron Microscopy Forum. Research, publs. on differentiation digestive tract in frogs, respiratory metabolism heartless tadpoles, synthesis of yolk in developing oocytes, electron microscopic studies cortical changes following fertilization frogs' eggs, metamorphic changes induced in skin and skeleton frog tadpoles treated with thyroxine, regeneration in coelenterates and in aquatic vertebrates, somatic effects gamma irradiation. Home: 2576 Roseland Dr., Ann Arbor, Mich. 48103.*

KEMP, Stanley Wells, botanist; b. June 14, 1882; s. Stephen Kemp; ed. Trinity Coll., Dublin; Sc.D., Dublin; m. Agnes Green, 1913; 1 dau. Asst. naturalist Dept. Agr. and Tech. Instrn. for Ireland, Fisheries Br., 1903-09; supr. zool. sec. Indian Mus., Calcutta (later Zool. Survey India), 1910-24; zoologist Abor Punitive Expdn., 1911-12; dir. research to discovery com. Colonial Office, 1924-36; sec. Marine Biol. Assn. U.K., also dir. Plymouth Lab., from 1936. Fellow Royal Soc., 1931. Contbr. papers on Crustacea to sci. jours. Died May 16, 1945.

KEMPE, Sir Alfred Bray, English mathematician; b. Kensington, Eng., July 6, 1849; s. John Edward Kempe; M.A., Trinity Coll., Cambridge (Eng.) U.; D.C.L., Durham, Eng.; m. Mary Dowman (dec. 1893); m. 2d, Ida White; 2 sons, 1 dau. Barrister Inner Temple and Western Circuit, 1873; chancellor Diocese of London, from 1912; also chancellor dioceses Southwell, St. Albans, Peterborough, Chichester, Chelmsford. Fellow Royal Soc., 1881 (treas., v.p. 1899-1919). Author: How to Draw a Straight Line, 1877. Demonstrated that any algebraic curve can be described by link-motion; studied fundamental conceptions of symbolic logic and geometry, also relations between geometrical theory of points and logical theory of classes. Died Apr. 21, 1922.

KEMPER, William Doral, Am. soil physicist; b. Lethbridge, Alta., Can., Nov. 29, 1928; s. Fredrick George and Leona (Peterson) K.; came to U. S., 1946, naturalized, 1965; B.S., Brigham Young U., 1950; M.S., N.C. State Coll., 1952, Ph.D., 1954; m. Ida Low, Dec. 27, 1954; children—Greta, Lisa, Kristen, Denise, Fredrick William. Grad. research asst. N.C. State Coll., 1950-54; soil surveyor Canadian Govt., Alta., 1954; soil scientist U. S. Dept. Agr., prof. agronomy Colo. State U., Ft. Collins, 1954—. Cons. Petroleum Research Corp., 1957-63. Fellow Am. Soc. Agronomy; mem. Soil Sci. Soc. Am., Sigma Xi, Gamma Sigma Delta. Contbr. numerous articles to sci. jours. Developed theory for salt sieving effect in soils and compacted clays; demonstrated, developed theory for osmotic movement of water in soils; outlined and applied equations for diffusive and viscous movement of ions and water in soils; research in soil structure and soil water conservation. Home: 600 Juniper Lane, Ft. Collins, Colo. 80521.*

KEMPF, Paul, German astronomer; b. Berlin, Germany, 1856; ed. Heidelberg, Germany; Ph.D., Berlin, 1878; became dir. Obs., Potsdam, Germany, 1894. Author: Studies on Ptolemy's Theory and the Rotation of the Moon, 1878; Meteorological Observations made at Potsdam, 1881-93; Studies on the Mass of the Planet Jupiter, 1887; (with G. Müller), Research on the Absorption of Stellar Light by the Earth's Atmosphere, 1898; Photographic View of the Northern Skies, 1903. Determined mass of Jupiter, 1887; calculated period of sun's rotation, 1892; discovered wave lengths of 300 lines in solar spectrum, 1895; publs. on absorption of sunlight by earth's atmosphere, 1898; photog. survey of No. skies, 1903. Died Potsdam, 1920.

KEMPNER, Joseph, Am. aerospace engr.; b. Bklyn., Apr. 25, 1923; s. Arthur and Anna (Richman) K.; B.Aero. Engring. summa cum laude, Poly. Inst. Bklyn., 1943, M.Aero. Engring., 1947, Ph.D. in Applied Mechanics, 1950; m. Carol F. Brown, Jan. 12, 1947; children—Robert M., Marien A. Research fellow Poly. Inst. Bklyn., 1944; aero. engr. NACA, 1944-47; with Poly. Inst. Bklyn., 1947—, prof. applied mechanics, aerospace engring., 1957—, chmn. undergrad. aerospace studies, 1962-64, asst. dir. research, 1962-63, dir. applied mechanics, 1964-66, head dept., 1966—. Cons. to indsl. and govt. research labs. Fellow Am. Inst. Aeros. and Astronautics (asso.), N.Y. Acad. Scis.; mem. Tau Beta Pi, Sigma Xi, Sigma Gamma Tau. Research, numerous publs. in structural mechanics shear lag, concentrated loads on reinforced and unreinforced cylindrical shells, creep bending and buckling linear and nonlinear beams, plates and shells, ablation nose cones, stress, buckling and postbuckling analyses unreinforced and reinforced circular and noncircular cylindrical shells. Home: 1163 E. 13th St., Bklyn. 11230.*

KEMPNER, Walter, physician; b. Berlin, Germany, Jan. 25, 1903; s. Walter and Lydia (Rabinowitsch) K.; ed. U. Berlin, U. Heidelberg; M.D. Research asso., asst. to Prof. O. Warburg, Kaiser Wilhelm Inst., Berlin; asst. in medicine U. Berlin; prof. medicine Duke, Mem. A.M.A., A.C.P., Am. Soc. Physiology. Author: Stoffwechsel der Entzündung, 1930; Wirkung von Blausäure und Kohlenoxyd auf die Buttersäuregärung, 1933; The Role of Oxygen Tension in Biological Oxidations, 1939; Treatment of Kidney Disease and Hypertensive Vascular Disease with Rice Diet, 1944; Radical Dietary Treatment of Hypertensive and Arteriosclerotic Vascular Disease, Heart and Kidney Disease and Vascular Retinopathy, 1954. Creator rice treatment for hypertension and arteriosclerosis, diseases of heart, kidneys, retina, sugar diabetes and obesity. Address: Duke Medical Center, Durham, N.C.

KEMPSTER, Walter, physician; b. London, Eng., May 25, 1841; s. Christopher and Charlotte (Treble) K.; came to U. S. in infancy; M.D., L.I. Med. Coll., 1864; m. J. L. J. Poessel, June 28, 1913. Acting asst. surgeon U. S. Army, 1864-65; asst. supt. N.Y. State Asylum for Idiots, 1866-67; asst. physician N.Y. Hosp. for Insane, Utica, 1867-73; supt. Northern Hosp. for Insane, Wis., 1873-84; asst. editor Am. Jour. of Insanity, 1874-84; commr. of health, Milw., 1894-98; prof. mental diseases in Wis. Coll. Phys. and Surg. Author: Hospital Gangrene in Army Hospitals, 1866; Entero-Colitis in United States Army, 1866. First physician in U. S. to make systematic microscopic exam. of brains of insane, and to photograph same through microscope, 1867. Died Aug. 22, 1918.

KENCH, James Edward, biochemist; b. Bollington, Eng., Dec. 9, 1911; s. James William and Frances (Newmarch) K.; B.Sc. with honors, U. Bristol (Eng.), 1933, M.Sc., 1935, Ph.D., 1938, D.Sc., 1953; M.B., Ch.B., U. Manchester, (Eng.), 1950; m. Betty Noyce Bast, Oct. 2, 1939; children—John, Ann, Peter. With Long Ashton Research Sta., U. Bristol (Eng.), 1934-38; chem. pathologist, Sir Halley Stewart fellow biochemistry dept. hematology U. Manchester, 1941-52, sr. lectr. biochemistry, 1953-55, sr. lectr., 1955-57, reader chem. pathology, 1958; prof. chem. pathology U. Cape Town (S. Africa), 1958—. Cons. biochemist Groote Schuur Hosp., Cape Town, 1958—; mem. Internat. Commn. on Neurochemistry, 1962—. Fellow Royal Inst. Chemistry; mem. Biochem. Soc., Path. Soc. Gt. Britain, Path. Soc. S. Africa. Research, publs. on biochem. effects of heavy metals, degradation of haem compounds to bile pigments, porphyrin metabolism, proteins in serum and urine in human disease, their formation and structure, amino acids and peptides in human malnutrition, patterns of enzymes in disease. Home: 3 Annerley Rd., Cape Town, S. Africa.*

KENDALL, Arthur Isaac, Am. bacteriologist; b. Somerville, Mass., May 7, 1877; s. Isaac Brooks and Alice Rebecca (Fitz) K.; B.S., Mass. Inst. Tech., 1900; Ph.D., Johns Hopkins, 1904; Dr.P.H., Harvard Med. Sch., 1911; Sc.D., U. So. Cal., 1911; m. Gertrude Mary Woods, Dec. 21, 1904; 1 dau. Instr. bacteriology Johns Hopkins, 1902-04; instr. preventive medicine and hygiene Harvard Med. Sch., 1909-12; prof. bacteriology Northwestern U. Med. Sch., 1912-24, dean, 1916-24; prof. bacteriology and pub. health Washington U. Med. Sch., 1928-42, emeritus prof., 1942—. Chmn. yellow fever commn. Internat. Health Bd., 1917. Recipient Service medal Panama Commn.; fellow Rockefeller Inst., 1907-09. Mem. Soc. Exptl. Biology and Medicine. Fellow Am. Acad. Pediatrics. Author: Bacteriology: General Pathological and Intestinal, 3d edit., 1928; Civilization and the Microbe, 1923; Studies in Bacterial Metabolism, 1912-45. Research in intestinal bacteriology, chemistry of bacterial action, filterable bacteria, bacterial metabolism. Home: 7227 Olivetas Av., La Jolla, Cal.

KENDALL, Edward Calvin, Am. biochemist; b. South Norwalk, Conn., Mar. 8, 1886; s. George Stanley and Eva Frances (Abbott) K.; B.S., Columbia, 1908, M.S., 1909, Ph.D., 1910, D.Sc, 1951; D.Sc. (hon.), U. Cin., 1922, Yale, Western Res, Williams Coll., Nat. U. Ireland, Columbia, 1951, Gustavus Adolphus Coll., St. Peter, Minn., 1963; m. Rebecca Kennedy, Dec. 30, 1915; children—Hugh, Roy, Norman, Elizabeth (Mrs. Fred Cornwall). Research chemist Parke, Davis, and Co., Detroit, 1910-11; investigations on thyroid gland St. Luke's Hosp., N.Y. City, 1911-14; head sect. biochemistry Mayo Clinic, 1914, prof. physiologic chemistry Mayo Found., (U. Minn.), 1921-51; visiting prof. chemistry, James Forrestal Research Center, Princeton University, New Jersey, since 1952. Recipient: John Scott prize and Premium awarded by City of Phila., 1921 (researches in thyroxin); Chandler medal, Columbia U., 1925; Squibb award for outstanding research in endocrinology, 1945; Lasker award (jointly with Dr. Hench), Am. Public Health Assn., 1949; Page One award (jointly with Dr. Hench), Newspaper Guild New York, 1950; John Phillips Meml. award, Am. Coll. Phys. in Boston, 1950; Research Corp. award, by Research Corp. of N.Y., 1950; Remsen Meml. award, by Md. sect. Am. Chem. Soc., 1950; Research award, Am. Pharm. Mfrs. Assn., 1950; Passano award for 1950 (with Dr. Hench), Passano Found., San Francisco, 1950; Medal of Honour from Canadian Pharm. Mfrs. Assn., 1950; Nobel Prize in Physiology and Medicine (with Dr. Philip S. Hench and Dr. Tadeus Reichstein), 1950; Dr. C. C. Criss award (jointly with Dr. Hench), 1951, Award of Merit (with Dr. Hench) from Masonic Found. Med. Research and Humane Welfare, 1951; Cameron award (with Prof. Reichstein), U. Edinburgh, 1951; Heberden Soc. award, London, 1951; The Kober award, Association of American Physicians, 1952; Alexander Hamilton medal, Alumni of Columbia Coll., 1961. Mem. Am. Philos. Soc., Am. Acad. Arts and Scis., Am. Soc. Biol. Chemists (pres. 1925-26), Am. Physiol. Soc., Am. Soc. Exptl. Pathology, Am. Soc. Exptl. Biology and Medicine, Am. Chem. Soc., Harvey Soc., A.A.A.S., Assn. Am. Physicians, Assn. Study Internal Secretions, Nat. Acad. Scis., New York Academy of Sciences, Swedish Society. Author: Thyroxine, 1929. Contbns. include isolation of thyroxine, crystallization of glutathione and establishment of its chemical structure, separation and identification of a series of compounds from the adrenal cortex; prepared cortisone by partial synthesis (with Merck & Co, Inc.), investigated effects of cortisone and of ACTH on rheumatoid arthritis and in rheumatic fever (with Drs. Hench, Slocumb and Polley). Home: 3 Queenston Place, Princeton, N.J.

KENDALL, Ernest George, Am. metallurgist; b. N.Y.C., May 12, 1929; s. Ernest and Martha (Currie) K.; B.S. in Metall. Engring., Poly. Inst. Bklyn., 1949, M.S., 1950; Ph.D., U. Ky., 1957; m. Elva Matula, June 11, 1951; children—Walter Scott, Kenneth Richard. Metallurgist, Nat. Lead Co., 1950-51, Titanium Metals Corp. Am., Henderson, Nev., 1951-55, Atomics Internat. div. N.Am. Aviation, Canoga Park, Cal., 1957-61; head metallurgy and ceramics dept. Aerospace Corp., Los Angeles, 1961—. Instr., Bklyn. Poly. Inst., 1949-50, U. Ky., 1956-57, U. Cal. at Los Angeles, 1960—. Mem. Am. Soc. Metals, Am. Inst. Metall. Engrs., Am. Ceramic Soc., Am. Inst. Aeros. and Astronautics, Sigma Xi, Alpha Chi Sigma, Alpha Chi Rho. Research, publs., patents in field of metallurgy, refining of titanium metal, graphite coatings for melting zirconium, thermal shock resistant carbides. Home: 716 Indiana Ct., El Segundo, Cal. 90245. Office: P.O. Box 95085, Los Angeles 90045.*

KENDALL, James Madison, Am. physicist; b. Dallas, Sept. 27, 1899; s. William Bowdoin and Emma (Ganger) K.; B.S., Rice Inst., 1925; M.S., U. Md., 1953; m. Leona Kepley, Aug. 4, 1927; children—James Madison, Betty Ann (Mrs. Robert O. Teeg). Radio engr. Gen. Electric Co., 1925-30, RCA, 1930-34; instrumentation devel. with Geophys. Research Corp., 1934-41, U. S. Naval Ordnance Lab., Silver Spring, Md., 1941-62; engring. specialist Jet Propulsion Lab., Cal. Inst. Tech., Pasadena, 1962—. Mem. Inst. Aeros. and Astronautics, Instrument Soc. Am., Soc. Exploration Geophysicists. Research, publs. on high fidelity record reprodn.; developed bottom hole pressure gauge to determine performance and potential of oil wells; underwater microphone for absolute measurement of underwater sound intensities; instrumentation for measurement and recording of aerodynamic quantities in supersonic wind tunnels; absolute radiometer for measuring the intensity of sun light. Home: 629 Linda Vista Av., Pasadena, Cal. 91105.*

KENDALL, Peter Calvin, English mathematician; b. Yorkshire, Gt. Britain, June 1934; s. Roland and Jessie (Kershaw) K.; B.Sc., Queen Mary Coll., U. London (Eng.), 1955, Ph.D., 1959; m. Hazel Mary Fowler, Aug. 2, 1957. Faculty, Bedford Coll., U. London, 1958-63; lectr. applied math. U. Sheffield (Gt. Britain), 1963-65, prof., 1965—. Vis. scientist High Altitude Obs., Boulder, Colo., 1961-62, 64, Geophys. Inst., College, Alaska, 1965; vis. sr. research asso. Goddard Space Flight Center, Md., 1965. Mem. Royal Astron. Soc., Inst. Math. and its Applications, London Math. Soc., Computing Soc. Author: (with C. Plumpton) Magnetohydrodynamics I, 1963; (with D. E. Bourne) Vector Analysis, 1967; also articles. Research on motion electrically conducting fluids and charged particles in magnetic fields, applications math. and numerical methods to magnetoplasma diffusion in ionosphere. Home: Well House, The Green, Curbar, Sheffield. Office: Dept. Applied Math., The Univ., Sheffield 10, Gt. Britain.*

KENDEIGH, S(amuel) Charles, Am. ecologist, ornithologist; b. South Amherst, O., Dec. 18, 1904; s. Milo Cornelius and Clara (Gillman) K.; A.B., Oberlin Coll., 1926, A.M., 1927; Ph.D., U. Ill., 1930; m. Dorothy Elizabeth Sutton, Sept. 6, 1930; children—Katherine Jane (Mrs. Sidney P. Little), Donald Charles. Research asso. Baldwin Bird Research Lab., 1925-39; faculty Western Res. U., 1930-36; faculty U. Ill., Champaign, 1936—, prof., 1948—. Recipient Brewster award Am. Ornithol. Union, 1951. Mem. Ecol. Soc. Am. (past pres.), Wilson Ornithol. Soc. (past pres.), Nature Conservancy (founder), Am. Zool. Soc. Author: Physiology of the Temperature of Birds, 1932; The Role of Environment in the Life of Birds, 1934; Territorial and Mating Behavior of the House Wren, 1941; Bird Population Studies in the Coniferous Forest Biome, 1947; Parental Care and its Evolution in Birds, 1952; Animal Ecology, 1961; also numerous articles. Research, publs. on nesting behavior birds, ann. variations in animal populations, bird communities in relation to types of vegetation, regulation of bird temperatures, energy requirements animals for existence and control yearly cycle activities. Home: 1116 W. Healey St., Champaign, Ill. 61820.*

KENDREW, John Cowdery, English molecular biologist; b. Oxford, Eng., Mar. 24, 1917; s. Wilfrid George and Evelyn May Graham (Sandberg) K.; B.A., Trinity Coll., Cambridge U., 1939, M.A., 1943, Ph.D., 1949, Sc.D., 1962. With Ministry Aircraft Prodn., 1940-45; sci. adviser allied air comdr. in chief, South East Asia, 1944; dep. chmn. Med. Research Council Lab. for Molecular Biology, Cambridge U., 1947—; fellow of Peterhouse, Cambridge U., 1947—; reader Davy-Faraday Lab., Royal Instn., London, 1954-68; sci. adviser Ministry Def., 1960-64; editor-in-chief Jour. Molecular Biology, 1959—. Decorated comdr. Order Brit. Empire; recipient (with Max Perutz) Nobel prize in chemistry, 1962. Fellow Royal Soc., 1960; fgn. hon. mem. Am. Acad. Arts and Scis.; hon. mem. Am. Soc. Biol. Chemists; mem. Brit., Am. biophys. socs., Internat. Orgn. Pure and Applied Biophysics (v.p. 1964). In work with myoglobin determined structure of a protein in general outline (1957) and atomic detail (1959); observed alpha-helix arrangement of the polypeptide chain, thereby confirming Pauling's earlier description. Home: Guildhall, 4 Church Lane, Linton, Cambridgeshire, Eng. Office: Peterhouse, Cambridge, Eng.*

KENDRICK, James Blair, Jr., Am. plant pathologist; b. Lafayette, Ind., Oct. 21, 1920; s. James Blair and Violet (McDonald) K.; B.A., U. Cal. at Berkeley, 1942; Ph.D., U. Wis., 1947; m. Evelyn May Henle, May 17, 1942; children—Janet Blair, Douglas Henle. Mem. staff and faculty U. Cal. at Riverside, 1947—, prof. plant pathology and plant pathologist, 1961—, chmn. dept., 1963-68, asst. to chancellor, 1960-61; v.p. for agrl. scis. U. Cal., 1968—; NSF sr. postdoctoral fellow, U. Cambridge (Eng.) and Rothamsted (Eng.) Exptl. Sta., 1961-62; spl. research diseases vegetable crops. Participant 10th Internat. Bot. Congress, Edinburgh, Scotland, 1964. Fellow A.A.A.S.; mem. Am. Phytopathol. Soc. (bd. editors jour. 1965-68, councilor-at-large 1968—), Am. Inst. Biol. Scis., Phi Beta Kappa, Sigma Xi. Research and publs. on control of root diseases of beans; identification of air pollution as a major cause of injury to plants; soil-borne fungi responsible for seed and root decays of beans. Address: University Hall, 2200 University Av., Berkeley, Cal. 94720.*

KENEN, Peter Bain, Am. economist; b. Cleve., Nov. 30, 1932; s. Isaiah Leo and Beatrice (Bain) K.; A.B., Columbia U., 1954; M.A., Harvard, 1956, Ph.D., 1958; m. Regina Horowitz, Aug. 21, 1955; children—Joanne Lisa, Marc David, Stephanie Hope. Faculty, Columbia, 1957—, prof. econs., 1964—, chmn. dept. econs., 1967—. Cons., U. S. Treasury, 1962—, Bur. Budget, 1964—. Recipient David A. Wells prize Harvard, 1958-59. Mem. Am. Econ. Assn., Royal Econ. Soc., Econometric Soc., Acad. Polit. Sci., Council Fgn. Relations. Author: British Monetary Policy and the Balance of Payments, (1951-1957), 1960; Giant among Nations, 1960; Money, Debt and Economic Activity (with A. G. Hart), 1961; International Economics, 1964; also articles. Research on econ. basis for internat. trade (especially role of investment in human capital), internat. financial problems and policy. Home: 652 Cumberland Av., Teaneck, N.J. 07666. Office: 501 Fayerweather Hall, Columbia U., N.Y.C. 10027.*

KENKICHI, Hanai, Japanese mathematician; flourished 1850. Author: Seisan Sokuchi (A Short Course in Western Arithmetic) (1st Japanese math. work based on European methods; contained discussion of Napier's computing rods), 1856.

KENNAWAY, Sir Ernest Laurence, English chemist, physiologist; b. May 23, 1881; s. Laurence James Kennaway; ed. Univ. Coll., London, New Coll., Oxford, Middlesex Hosp.; M.A.; M.D., Oxford U.; D.Sc., London; m. Nina Marion Derry, 1920. Dir. Chester Beatty Research Inst., Royal Cancer Hosp. (Free), London; prof. exptl. pathology U. London. Recipient Walker prize Royal Coll. Surgeons; Baly medal, 1937; Anna Fuller Meml. prize (with others), 1939; Garton medal and prize Brit. Empire Cancer Campaign, 1946; Osler Meml. medal, 1950. Fellow Royal Soc., 1934 (Royal Medal 1941), Royal Coll. Physicians; hon. mem. several fgn. socs. Author: Some Religious Illusions in Art, Literature and Experience, 1953; also papers on cancer and biol. chemistry. Discovered prodn. of cancer by pure hydrocarbons, thus supporting theory chem. agts. cause much human cancer; showed hydrocarbon in heated coal tar is responsible for many cases of indsl. cancer. Died Jan. 1, 1958.

KENNEDY, Alex. Mills, Brit. physician; s. Alexander K.; M.D., U. Glasgow. Held positions including examiner in medicine univs. Wales, Glasgow, dir. med. unit and sr. hon. physician Cardiff Royal Infirmary, hon. cons. physician to other hosps. in S. Wales; sr. asst. to prof. medicine and pathology U. Glasgow; resident physician Glasgow Royal Infirmary; dean faculty medicine U. Wales; dir. research dept. Royal Maternity and Women's Hosp., Glasgow; finally emeritus prof. medicine U. Wales; hon. cons. physician United Cardiff Hosps. Fellow Royal Coll. Physicians. Author: Cerebro-spinal Fever; The Etiology, Symptomatology, Diagnosis and Treatment of Epidemic Cerebro-spinal Meningitis; Parasitology for Medical Students; Medical Case-taking; also many papers in profl. jours. Died Sept. 24, 1960.

KENNEDY, Alexander, Scottish physician; b. Jan. 16, 1909; s. John and Nora (Sykes) K.; ed. City of London Sch., St. Thomas's Hosp. Med. Sch., U. London, Johns Hopkins Med. Sch.; Rockefeller travelling fellow in neuropsychiatry, 1936; M.D., London, M.B., B.S.; m. Helen Emily Mary Walter, 1934, 3 sons; m. 2d, Joanna Birkett, 1944, 1 son. Held positions including physician Maudsley Hosp., physician for psychol. medicine Maida Vale Hosp. for Nervous Diseases; prof. psychol. medicine U. Edinburgh, 1955-60; physician to Royal Infirmary, Edinburgh; clin. dir. Royal Edinburgh Hosp. for Mental and Nervous Diseases; psychiat. adviser, clin. dir. for research S.E. Scotland Regional Hosp. Bd.; prof. psychol. medicine U. Durham, 1947-55. Recipient Gold medal Royal Medico-Psychol. Assn., 1938. Fellow Royal Soc. Medicine, Royal Coll. Physicians, Royal Coll. Physicians Edinburgh, Brit. Psychol. Soc.; mem. Alfred Adler Med. Soc. (pres.), Scottish Assn. Psychiat. Social Workers (pres.), Eugenics Soc. (v.p.), Royal Medico-Psychol. Assn. (v.p.), Australasian Assn. Psychiatrists (hon.), Royal Soc. Health, Internat. Med. Commn. on Boxing (pres.), Brit. Med. Students' Assn. (pres.), Electrophysiol. Technologists' Assn. (pres. 1959). Author: The Organic Reaction Types, 1948; The Modern Approach to Juvenile Delinquency, 1948; Report on the Mental Hygiene Services of the State of Victoria, 1949; Hysteria and its Treatment, 1950; also numerous tech. papers. Died June 11, 1960.

KENNEDY, Byrl James, Am. physician; b. Plainview, Minn., June 24, 1921; s. Arthur Sylvester and Anna (Fassbender) K.; B.S., U. Minn., 1943, B.A., 1943, B.M., 1945, M.D., 1946; M.S., McGill U., 1951; m. Margaret Bradford Hood, Oct. 21, 1950; children—Sharon Bradford, Scott, Grant. Practice medicine, specializing in internal medicine, Mpls., 1952—; asst. prof. medicine U. Minn. Med. Sch., 1952-57, asso. prof., 1957-67, prof. medicine, 1967—. Mem. A.M.A., Am. Assn. Cancer Research, A.A.A.S., Central Soc., Endocrine Soc., Am. Fedn. Clin. Research Am. Soc. Clin. Oncology. Research, numerous publs. in growth patterns cancer, treatment advanced cancer and leukemia with chem. agts. and hormones. Home: 1949 E. River Rd., Mpls. 55414.*

KENNEDY, Clarence Hamilton, Am. entomologist; b. Rockport, Ind., June 25, 1879; s. Albert Hamilton and Emma Dorinda (Tennant) K.; A.B., Ind. U., 1902, A.M., 1903, D.Sc., 1950; A.M., Stanford, 1915; Ph.D., Cornell U., 1919; m. Lydia June Findley, Mar. 16, 1927; children—Bruce Albert Hamilton, Mary Janet. Asst. in biology, Cornell U., 1915, instr. in limnology, 1916-17; instr. zoology N.C. State Coll., 1918-19; instr. entomology Ohio State U., 1919-21, asst. prof., 1921-30, asso. prof., 1930-33, prof., 1933-49. Fellow A.A.A.S., Entomol. Soc. Am. (pres. 1935), Royal Entomol. Soc. of London, Ohio Acad. Sci.; mem. Am. Assn. Econ. Entomologists, Ind. Acad. Sci., Ecol. Soc. Am. (mng. editor Annals 1929-45, mem. editorial bd. ecol. monographs 1934-40), Am. Soc. of Naturalists, Société Entomologique de France (Paris), Gamma Alpha, Sigma Xi (pres. Ohio Chpt. 1926-27), Gamma Sigma Delta. Author: Methods for Study of the Internal Anatomy of Insects, 1933; also articles in biol. jours., Ency. Brit. Died June 6, 1952.

KENNEDY, Donald, Am. biologist; b. N.Y.C., Aug. 18, 1931; s. William Dorsey and Barbara (Bean) K.; A.B., A.M., Ph.D., Harvard; m. Barbara Jeanette Dewey, June 11, 1953; children—Laura Page, Julia Hale. Teaching fellow, NSF fellow, Harvard, 1952-56; faculty Syracuse (N.Y.) U., 1956-60, faculty Stanford, 1960—, prof., exec. head, 1965—. Editor biology W. H. Freeman Co., San Francisco, 1962—. Mem. Soc. Gen. Physiologists, Am. Physiol. Soc., Am. Inst. Biol. Scis. (chmn. edn. com. 1963—). Author: (with W. H. Telfer) The Biology of Organisms, 1965; also articles. Editorial bd. Jour. Exptl. Zoology, 1965—, Zeitschrift vergl. Physiologie. 1966—, Jour. Neurophysiology, 1967—. Research on synaptic transmission in arthropod central nervous systems, reflex physiology in arthropods, invertebrate vision. Home: 680 Lowell Av., Palo Alto, Cal. 94301. Office: Dept. Biol. Sci., Stanford U., Stanford, Cal.*

KENNEDY, Eugene Patrick, Am. biochemist; b. Chgo., Sept. 4, 1919; s. Michael and Catherine (Frawley) K.; B.S., DePaul U., 1941; Ph.D., U. Chgo., 1949; M.S., Harvard, 1960; m. Adelaide Majewski, 1943; children—Lisa, Sheila, Kathy. Faculty, U. Chgo., 1951-60, prof. dept. biochemistry and Ben May Lab., 1956-60; Hamilton Kuhn prof. biol. chemistry Harvard Med. Sch., Boston, 1960—, head dept., 1960-65. Recipient Glycerine Research award, 1956. Mem. Am. Acad. Arts and Scis., Nat. Acad. Scis., Am. Soc. Biol. Chemists, Am. Chem. Soc. (Paul-Lewis award 1958). Research, numerous publs. on biochemical function of mitochondria; Coenzymes, pathways of biosynthesis; mechanisms of membrane function. Home: 63 Buckminster Rd., Brookline, Mass. 02146. Office: 25 Shattuck St., Boston 02115.*

KENNEDY, Eugene Richard, Am. bacteriologist; b. Scranton, Pa., July 3, 1919; s. Thomas A. and Margaret (Culkin) K.; B.S., U. Scranton, 1941; M.S., Cath. U., 1943; Ph.D., Brown U., 1949; m. Marjorie Anne Giblin, July 24, 1945; children—Anne, Michael, Christine. Peterson fellow Brown U., 1948, Arnold fellow, 1948-49; faculty Cath. U., Washington, 1949—, professor bacteriology, 1966—. Cons. bacteriologist Providence Hosp., Washington, 1954-58, staff microbiologist; 1958—. Diplomate Pub. Health and Med. Lab. Bacteriology Am. Bd. Microbiology. Mem. Am. Soc. For Microbiology, Washington Acad. Scis., A.A.A.S., Sigma Xi. Contbr. articles to tech. jours, Cath. Youth Ency. Research on Vi antigen typhoid, gram reaction, autogenous vaccines, staphylococci. Home: 11804 Saddlcrock Rd., Silver Spring, Md. 20902. Office: Biology Dept., Catholic U., Washington 20017.*

KENNEDY, Joseph Camp Griffith, Am. statistician; b. Meadville, Pa., Apr. 1, 1813; s. Thomas Ruston and Judith (Ellicott) K.; student Allegheny Coll., 1829, LL.D., 1862; m. Catharine Morrison, Oct. 21,

1834, 4 children. Owner, editor Crawford (Pa.) Messenger; sec. bd. engaged in preparations of plan for taking 7th census, 1849; apptd. superintending clk. U. S. census, 1850; sec. U. S. Commn. to World's Fair, London, Eng., 1851; commr. Internat. Exhbn., London, some years later; authorized in interests of census work to go abroad to examine statis. systems in Europe; chiefly responsible for orgn. 1st Internat. Statis. Congress, Brussels, Belgium, 1853, mem. 2d, 4th congresses, 1855, 60; superintending clk. 8th census, 1860-65. Died Washington, July 13, 1887.

KENNEDY, Robert, Scottish surgeon; b. Glasgow, Dec. 20, 1865; student univs. Edinburgh, Berlin; grad. U. Glasgow; M.A., M.D., D.Sc.; m. Janet Lyon Miller, 1895; 1 dau. Surgeon to out-door dept. Glasgow Victoria Infirmary, 1892-96; asst. surgeon Lock Hosp., 1893-04; staff Western Infirmary, 1896-1911, asst. surgeon, 1900-10, surgeon, lectr. clin. surgery, 1910-11; examiner in surgery St. Andrews U., 1901-04; lectr. applied anatomy U. Glasgow, 1906-1911, St. Mungo prof. surgery, ex-officio surgeon Royal Infirmary, from 1911; surgeon and cons. surgeon in charge orthopedic dept. Scottish Nat. Red Cross Hosp., Bellahouston, 1917-19; dir., cons. surgeon Shakespeare Hosp., Glasgow, 1918-19; surgeon-in-chief Bellahouston Hosp., Glasgow, 1919-22. Author: On the Regeneration of Nerves, 1897; On the Restoration of Co-ordinated Movements after Nerve-Crossing with Interchange of Function of the Cerebral Cortical Centres, 1901; Experiments on the Restoration of Paralysed Muscles by means of Nerve Anastmosis, 3 parts, 1911, 14, 15; also papers in med. publs. Died June 3, 1924.

KENNEDY, Robert Michael, Am. chemist; b. Scranton, Pa., July 6, 1915; s. Michael A. and Gertrude (McGee) K.; B.S., U. Scranton, 1937; M.S., Pa. State U., 1938, Ph.D., 1941; m. Grace Elizabeth Corcoran, Nov. 14, 1942; children—R. Michael, Brian John, Marie Therese. With Sun Oil Co., Marcus Hook, Pa., 1941—, dir. research, devel. div., 1959—. Mem. Am. Petroleum Inst., Ind. Research Inst., Am. Chem. Soc., Soc. Automotive Engrs., Am. Inst. Chem. E., Soc. Chem. Industries, Soc. Plastics Industries, Assn. Research Dirs., N.Y. Acad. Sci., Sigma Xi, Phi Lambda Upsilon. Research in areas of catalysts, distillation, alkylation, thermodynamic properties of organic compounds, organic chemistry, combustion phenomena. Home: Timber Lane, Newtown Square, Pa. 19073. Office: P.O. Box 426, Marcus Hook, Pa. 19061.*

KENNEDY, William Dempsey, Am. chemist; b. Chgo., Dec. 10, 1914; s. Clarence Clyde and Jessie (Waggy) K.; B.A., Stanford, 1936, Ph.D., 1940; m. Jean E. Young, Aug. 19, 1941. Research asst. Harvard, 1940-41, research asso., 1941-43; research supr. Woods Hole (Mass.) Oceanographic Inst., 1943-46; sr. research chemist research labs. Tenn. Eastman Co., Kingsport, 1946-54, research asso., 1954-61, div. head phys. div., 1961-66, asst. dir. research, 1966—. Recipient Pres.'s Certificate for Merit, 1946. Mem. Am. Chem. Soc., A.A.A.S., Tenn. Acad. Sci., Am. Ordnance Assn. Research, publs. on thermodynamic properties of organic compounds, explosives and shock-wave phenomena, properties of plastics. Home: 2116 Memorial Dr. Office: Research Labs., Tenn. Eastman Co., Kingsport, Tenn. 37662.*

KENNELLY, Arthur Edwin, elec. engr.; b. Bombay, East India, Dec. 17, 1861; s. David Joseph and Cathrine (Heycock) K.; ed. Univ. Coll. Sch., London; hon. Sc.D., U. Pitts., 1895, U. Toulouse (France), 1922; hon. A.M., Harvard, 1906; Dr.Engring., U. Darmstadt, 1936; m. Julia Grice, July 22, 1903; 1 son, Reginald Grice. Telegraph operator in Eng., 1876; asst. electrician in Malta, 1878; chief electrician of a cable repairing steamer, 1881; sr. ship's electrician Eastern Tel. Cable Co., 1886; prin. elec. asst. to Thomas A. Edison, 1887-94; asso. with Edwin J. Houston in firm Houston & Kennelly, cons. elec. engrs., Phila., 1894-1901; prof. elec. engring. Harvard, 1902-30; prof. elec. engring. Mass. Inst. Tech., 1913-24, dir. elec. engring. research, chmn. faculty, 1917-18. Exchange prof. to French univs. in applied science, 1921-22; research asso. Carnegie Instn., 1924-30; 1st vis. elec. engring. lectr. Iwadare Found., Japanese univs., 1931. Engr. in charge laying Vera Cruz-Frontera-Campeche cables for Mexican Govt., 1902. Recipient premiums Brit. Inst. E.E., Howard Potts Gold medal and Longstreth Silver medal Franklin Inst. Mem. NRC, Am. Inst. E.E. (pres. 1898-1900, Edison Gold medal 1933), Soc. Promotion Metric System of Weights and Measures (pres. 1904), Illuminating Engring, Soc. (pres. 1911), I.R.E. (pres. 1916, Gold medal 1932), Am. Acad. Arts and Scis. (v.p.) U. S. Nat. Com. of Internat. Elec. Commn. (hon. pres.), Tau Beta Pi. Author: The Application of Hyperbolic Functions to Electrical Engineering Problems, 1911; Artificial Electric Lines, 1917; Electrical Vibration Instruments, 1923; Vestiges of Premetric Weights and Measures Persisting in Metric-System Europe, 1926-27; Electric Lines and Nets, 1928. Conducted important research on electromagnetism; discovered (simultaneously with Oliver Heaviside) a layer of ionized air in upper atmosphere, now generally known as Heaviside layer, Kennelly-Heaviside layer, or the ionosphere, 1902; also studied alternating currents; applied advanced math. to understanding of elec. circuits contbd. to establishment of standards and units of elec. measurement. Died Boston, June 18, 1939.

KENNETT, Terence James, Canadian physicist; b. Toronto, Ont., Can., Aug. 8, 1927; s. Edward John and Winnifred (Druce) K.; B.Sc., McMaster U., 1953, M.Sc., 1954, Ph.D., 1956; m. Kathleen Margret Hay, Aug. 16, 1949; children—David Robert, Karen Dana. Asso. physicist Argonne Nat. Lab., 1956-59; faculty McMaster U., Hamilton, Ont., 1959—, prof. physics, 1966—. Mem. Am. Phys. Soc. Research, numerous publs. in nuclear spectroscopy, neutron reactions, devel. of nuclear instrumentation techniques of data acquisiton and reprodn. Home: 6 Dundana Av., Dundas, Ont., Can.*

KENNEY, Francis T., Am. biochemist; b. Springfield, Mass., Mar. 16, 1928; s. Edward M. and Mary (Byrnes) K.; B.S., St. Michael's Coll., 1950; M.S., U. Notre Dame, 1953; Ph.D., Johns Hopkins, 1957; m. Rose Ann Rescigno, Aug. 25, 1951; children—Jeffrey, Ellen. Instr., St. Michael's Coll., Winooski Park, Vt., 1950-51; research asso. pediatrics Cornell U. Med. Coll., 1957-59; biochemist Oak Ridge Nat. Lab., 1959—. Mem. Am. Soc. Biol. Chemists, Am. Chem. Soc., A.A.A.S., Sigma Xi. Research, publs. in mechanisms controlling nucleic acid and enzyme synthesis in mammalian systems, mechanism hormone action, mechanisms controlling enzymic differentiation during mammalian devel. Home: 919 W. Outer Dr., Oak Ridge 37830. Office: Biology Div., Oak Ridge Nat. Lab., Oak Ridge, 37831.*

KENNGOTT, Gustav Adolf, mineralogist; b. Breslau, Germany) (now Wroclaw, Poland), Jan. 6, 1813; research staff Hofmineralien Kabinett, Vienna, Austria; prof., Zurich, Switzerland. Author: Lehrbuch der Mineralogie, 1857; Lehrbuch der reinen Krystallographie, 1846; (with others) Handwörterbuch der Mineralogie, Geologie und Paläontologie, 1882-87. Research on mineralogy, crystallography, and petrology, mineral composition of Switzerland. Died Lugano, Switzerland, Mar. 14, 1897.

KENNICOTT, Robert, Am. naturalist; b. New Orleans, Nov. 13, 1835; s. Dr. John Albert and Mary (Ransom) K. Naturalist, made comprehensive natural history survey of So. Ill., 1855; a founder Chgo. Acad. Scis., 1856, became curator and trustee, circa 1863, later dir.; organized natural history mus. for Northwestern U., 1857; assembled and labeled collection from Cal. at Smithsonian Instn., Washington, 1858-59; explorer, made 1st expdn. to Canadian Arctic (now Manitoba, Can.), gathered collections of fauna, 1859; apptd. by Western Union Telegraph Co. to lead expdn. to Alaska to survey for overland telegraph line to Asia, 1865. Died Ft. Nulato, Alaska, May 13, 1866.

KENNY, Alexander Donovan, Am. pharmacologist; b. London, Eng., Mar. 4, 1925; s. Alexander and Alice (Barton) K.; B.Sc., Imperial Coll., U. London, 1945; M.S., Institutum Divi Thomae, Cin., 1949, Ph.D., 1950; m. Dorothy Marie LeTang, Aug. 19, 1950; children—Alexander Leo, Mary Alice, Virginia Ann, Peter Donovan. Came to U. S., 1947, naturalized, 1956. Chemist, Mactaggart & Evans, Ltd., London, 1945-47; sr. chemist Metabolism Labs., U. Coll. Hosp., London, 1950-51, temporary chief chemistry lab. Mass. Gen. Hosp., Boston, 1952; asst., instr. Harvard Sch. Dental Medicine, 1952-54, instr. pharmacology, 1954-55, asso. Harvard Med. Sch., 1955-59; faculty W.Va. U., Morgantown, 1959-67, prof. pharmacology, 1965-67; prof. pharmacology, sr. investigator Space Scis. Research Center, U. Mo., Columbia, 1967—. USPHS Spl. fellow Royal Postgrad. Med. Sch., London, 1967-68. Mem. Am. Chem. Soc., Endocrine Soc., Biochem. Soc. (Gt. Britain), Am. Assn. Pharmacology and Exptl. Therapeutics, Biophys. Soc., Soc. Endocrinology (Gt. Britain), Sigma Xi. Editorial bd. Endocrinology, 1965—. Research, numerous publs. in endocrinology, endocrine aspects bone and calcium metabolism, actions parathyroid hormone, thryocalcitonin, adrenal medullary hormones on bone and calcium metabolism. Home: 1006 Danforth Dr., Columbia, Mo. 65201.*

KENNY, Elizabeth (Sister Kenny), Australian nurse; b. Warrialda, New South Wales, Australia, Sept. 20, 1886; d. Michael and Mary (Moore) Kenny; privately educated; grad. St. Ursulas Coll., Australia, 1902. Nurse in Australian bush country, 1911-14; mil. nurse with Australian Army, 1914-18; research work in infantile paralysis from 1911; came to U. S., 1940, to present concept of symptoms of infantile paralysis to research inst. Author: Infantile Paralysis and Cerebral Diplegia, 1927; Treatment of Infantile Paralysis in Acute Stage, 1941; Kenny Concept of Infantile Paralysis and Its Treatment, 1942; And They Shall Walk (autobiography), 1943; Physical Medicine Concerning the Disease Infantile Paralysis, 1945; My Battle and Victory, 1951. Suggested treatment for anterior poliomyelitis, using hot, moist packs in acute stage, with passive exercise started early and followed by active exercise as soon as possible, 1937. Died Toowoomba, Queensland, Australia, Nov. 30, 1952.

KENT, Albert Frank Stanley, English physiologist; b. Mar. 26, 1863; s. George Davies and Anne (Rudgard) K.; ed. Magdalen Coll., Oxford; final honors Sch. Natural Sci., 1886; M.A.; D.Sc., Oxford; m. Theodora Hobson, 1904. Demonstrator physiology Owens Coll., Manchester and Victoria U., 1887-89; with U. Oxford, 1889-91; with St. Thomas' Hosp., London, 1891-95; prof. physiology U. Coll., Bristol, Eng., 1899-1909, U. Bristol, 1909-18; founder X-ray dept. St. Thomas' Hosp., Clin. Research Lab., U. Coll.; lectr., bacteriology, bacteriologist Bristol Royal Infirmary; organizer, dir. dept. indsl. adminstrn. Manchester Municipal Tech. Coll. Author: Blue Book on Industrial Fatigue, 1916. Editor: Physiology of Industrial Organization (Amar), 1918. Editor-in-chief for Gt. Britain, Jour. Indsl. Hygiene. Research and numerous publs. on physiology of heart; described muscular band with nerve fibers separating auricle from ventricles of heart. Died Mar. 30, 1958.

KENT, A(rthur) Atwater, Am. inventor; b. Burlington, Vt., Dec. 3, 1873; s. Prentiss J. and Mary Elizabeth (Atwater) K.; ed. Worcester Poly. Inst.; hon. E.E., U. Vt., 1924; Dr. Engring., Worcester Poly. Inst., 1926; D.Sc., Tufts Coll., 1927; m. Mabel Lucas, 1906; children—Arthur Atwater, Elizabeth Brinton (Mrs. William Laurens Van Alen), Virginia Tucker (Mrs. Kent Catherwood) and Jonathan Prentiss (adopted). Established Atwater Kent Mfg. Works, for manufacture of telephones and small volt meters, Phila., 1902; added manufacture of the unisparker (for which John Scott Medal was awarded 1914), panoramic sights, clinometers, fuse setters and angle of sights for the army during World War; started mfg. radio receiving sets, 1922. Mem. com. on general problems of radio broadcasting, 3d and 4th Nat. Radio Confs., 1924, 25; sponsored multi-station broadcasting of world's greatest musical artists, beginning 1925; sponsored Nat. Radio Audition to discover talented singer. Restored and presented the historic Franklin Inst. Bldg. to City of Phila. as a mus. (now Atwater Kent Mus.), 1938. Died Mar. 4, 1949.

KENT, Donald P., Am. gerontologist; b. Phila., June 4, 1916; s. Ralph and Ida (Peterson) K.; B.S., Pa. State Coll., 1940; M.A., Temple U., 1945; Ph.D., U. Pa., 1950; m. Marion H. Clime, Aug. 30, 1941; children—Marion H., Martha H. Instr. sociology U. Pa., 1945-50; asso. prof. sociology U. Conn., 1950-57; dir. Inst. Gerontology, 1957-61; spl. asst. to sec. health, edn. and welfare, Washington, 1961-63; dir. U. S. Office Aging, Washington, 1963-65; prof., head dept. sociology and anthropology Pa. State U., University Park, 1965—. Chmn., Conn. Commn. on Services for Elderly Persons, 1957-61; vice chmn. Pres.'s Council on Aging, 1961-63. Recipient Distinguished Service award Inst. for Ret. Profls. Fellow Gerontological Soc. (pres. psychol. and social sect. 1966-67), Am. Sociol. Soc. Author: The Refugee Intellectual, 1953; also articles. Editor-in-chief: Gerontologist, 1967—. Research on Negro aged, adjustment of aged, aged and automotive safety, profl. exec. retirement, integration of minority groups in larger society. Home: 926 Outer Dr., State College, Pa. 16801. Office: Dept. Sociology and Anthropology, Pa. State U., University Park, Pa. 16802.*

KENT, George Cantine, Am. zoologist; b. Kingston, N.Y., July 25, 1914; s. George C. and Charlotte (Delamater) K.; A.B., Maryville Coll., 1937; M.A., Vanderbilt U., 1938, Ph.D., 1942; m. Lila Mae Carringer, June 8, 1937; 1 dau., Susan Carolyn (Mrs. David Thomas Rovee). Faculty, La. State U., Baton Rouge, 1942—, prof. zoology, 1956—, chmn. zoology and physiology, 1959-63. Mem. Am. Inst. Biol. Scis., A.A.A.S., Am. Soc. Zoologists, Soc. for Exptl. Biology and Medicine, N.Y. Acad. Scis., Endocrine Soc., Assn. Southeastern Biologists (past pres.), Sigma Xi, Omicron Delta Kappa. Author: Comparative Anatomy of the Vertebrates, 1954; Comparative Anatomy of the Vertebrates, 1965; Practical Anatomy of the Dogfish, Necturus and Cat, 4th edit., 1962; also articles, chpt. in book. Research on cytology pituitary gland Triturus viridescens, physiology reprodn. golden hamster, estrous cycles, prolactin levels, circadian rhythms in prolactin release, decidual cell responses, duration and exptl. modification pseudopregnancy. Home: 482 Standford Av., Baton Rouge 70808.*

KENT, George Clarence, Am. plant pathologist; b. Keene, N.H., July 28, 1910; s. Harry Llewellyn and Ursula (Dickinson) K.; student N.M. A. and M. Coll., 1927-30; B.A., Oxford (Eng.) U., 1933; Ph.D., Ia. State Coll., 1936; m. Ruth Victoria Olson, July 30, 1938; children—Ann Elizabeth, George Alan, Thomas Richard. Faculty, Ia. State Coll., 1937-45; prof. plant pathology Cornell U., Ithaca, N.Y., 1945—, head plant pathology, 1950—, acting head botany, 1961-63, head botany, 1963-64; vis. prof. U. Philippines, Coll. Agr., 1952-54, 65. Cons. U. S. Army Biol. Center, 1949—, U. S. Dept. Def., 1960-63. Rhodes scholar, 1930-33. Fellow Am. Phytopath. Soc.; mem. Am. Assn. Rhodes Scholars, Bot. Soc. Am., Sigma Xi, Phi Kappa Phi. Author: (with I. E. Melhus) Elements of Plant Pathology, 1939; also articles. Research on cereal diseases, gen. plant pathology. Home: 408 Klinewoods Rd., Ithaca, N.Y. 14850.*

KENT, Jack Thurston, Am. mathematician, astronomer; b. Sardis, Tenn., Sept. 26, 1908; s. John Franklin and Daisy (Craven) K.; A.B. cum laude, Lambuth Coll., 1930; M.A. in Math., U. Ark., 1931; postgrad. Ohio State U.; m. Pauline Elizabeth Oates, May 27, 1936; children—Christopher Dee (Mrs. Marion H. Johnson), David Harbet. Faculty, Ft. Smith Jr. Coll., 1933-35, Ark. Poly. Coll., 1935-36; faculty Tex. A. and M. U., College Station, 1936—, asso. prof. math., 1952—. Participant, NSF Inst. Programs, 1958-65; dir. Moonwatch team, 1956-59. Ford Faculty fellow, 1951-52. Fellow Royal Astron. Soc.; mem. Am. Math. Soc., Am. Assn. U. Profs., A.A.A.S., Am. Soc. for Engring. Edn., Am. Astron. Soc., Astron. Soc.

Pacific, Tex. Ass. Coll. Tchrs., Nat. Geog. Soc., Internat. Oceanographic Found., Nat. Audubon Soc., Nat. Wildlife Soc., Omicron Phi Tau, Pi Mu Epsilon, Phi Kappa Phi. Author: Unified Mathematics for High School Seniors, 1934. Research on double stars (binaries) with spl. reference to orbit computation; work in celestial mechanics, with spl. reference to orbit computation, as applied to digital computers; spl. work in applied math. especially differential equations, with specialties of Bessel Functions. Home: 823 S. Rosemary Dr., Bryan, Tex. 77803.*

KENT, Paul Welberry, Brit. biochemist; b. Doncaster, Eng., Apr. 19, 1923; s. Thomas William and Marion (Cox) K.; B.Sc. with 1st class honors in Chemistry, Birmingham (Eng.) U., 1944, Ph.D., 1946; M.A., Oxford (Eng.) U., 1951, D.Phil., 1953, D.Sc., 1966; m. Rosemary Elizabeth Boutflower Shepherd, Aug. 23, 1952; children—Anthony M., Deborah J. B., Richard F., Peter J. Asst. lectr. Birmingham U., 1946-48; Imperial Chem. Industries fellow, 1948-50; vis. fellow Princeton, 1948-49; demonstrator biochemistry Oxford U., 1950-—, Lee's reader, tutor, fellow Christ Church Coll., 1955-—; research asso. Harvard, 1967. Recipient Rolleston prize, 1952. Fellow Royal Inst. Chemistry; mem. Biochem. Soc., Chem. Council, Chem. Soc. London, Soc. Gen. Microbiology, Sigma Xi. Author: (with M. W. Whitehouse) Biochemistry of the Amino-Sugars, 1955; also numerous articles. Research on chem. structures and enzymes in biol. formation of protein-polysaccharide complexes in mammalian epithelia, cartilage and bone; synthesized new types of sugars containing fluorine. Home: 37 St. Giles St., Oxford, Eng.*

KENT, William, Am. mech. engr.; b. Phila., Mar. 5, 1851; s. James and Janet (Scott) K.; A.B., Central High Sch., Phila., 1868, A.M., 1873; M.E., Stevens Inst. Tech., 1876; Sc.D., Syracuse U., 1905; m. Marion Weild Smith, Feb. 25, 1879. Editor of Am. Manufacturer and Iron World, Pitts., 1877-79; held positions as mech. engr. and supt. in iron and steel works, steam boiler business and torsion balance scale factory, 1879-90; cons. engr., from 1890; asso. editor Engineering News, 1895-1903; dean L. C. Smith Coll. Applied Sci., Syracuse U., 1903-08; editor Indsl. Engring., 1910-14. Lectr. steam engring., Newark Tech. Sch., 1888-95; also Lectr. Purdue, Bklyn. Inst., Franklin Inst. (Phila.), Worcester Poly. Inst., W.Va. U., U. Ill., Cornell, Stevens Inst. Mem. N.J. State Commn. on Pollution of Streams, 1898-99. Author: The Strength of Materials, 1879; The Mechanical Engineer's Pocket Book, 1895, 9th edition, 1915; Investigating an Industry, 1915; Bookkeeping and Cost Accounting for Factories, 1918. Invented weighing machinery, water-tube boilers, smokeless furnaces. Died Sept. 18, 1918.

KENTMANN, Johann, German geologist, physician; b. Dresden, Germany, Apr. 21, 1518; student medicine, U. Leipzig (Germany); M.D., Bologna, Italy, 1549; a son, Gottlieb. Promoted to magister, Leipzig, 1845. Author: De omni rerum fossilium genere, 1565-66; Nomenclature rerum fossilium, 1565; Calculorum qui in corpore ac membris hominum, 1565. Compiled 1st catalogue of fossils based on Agricola's system of classification; first detailed description of gallstones, 1565. Died June 15, 1574.

KENWORTHY, Walter, Am. zoologist; b. Phila, Jan. 19, 1922; s. James H. and Anna (Goldmann) K.; B.A., Temple U., 1943; Ph.D., U. Pa., 1953; m. Regina Dombrowski, Aug. 5, 1944; children—James, Paul. Faculty, Brown U., 1952-64, asso. prof. biology, 1958-64; dean coll., prof. biology Wheaton Coll., Norton, Mass., 1958-—. Mem. Genetics Soc. Am., Soc. for Study Evolution, Am. Soc. Mammalogists, Am. Soc. for Cell Biology. Fellow A.A.A.S. Research on natural control of pigment cell activity in those mammals which turn white in winter; on mechanism of x-ray action on genetic systems. Home: 30 E. Main St., Norton, Mass. 02766.*

KEPLER, Johannes (also: Keppler), astronomer; conceived at 4:37 a.m., May 16, 1571, b. 2:30 p.m., Dec. 27, 1571 (this according to horoscope he cast for himself at age 25), Weil-der-Stadt, a free imperial city in Duchy of Württemberg, Swabia; s. (and 1st-born child) Heinrich and Katharina (Guldenmann) K. who were married on May 15, 1571; ed. at Latin School in Leonberg, from 1579, seminary grammar school in Adelberg from 1584, sr. prepatory seminary in Maulbronn, 1586-89 (all in Duchy of Württemberg); B.A., U. Tübingen, 1588; returned to Maulbronn for 1 year; then (1589-94) studied philosophy, math., astronomy (under Michael Mästlin), theology in arts and theological faculties, U. Tübingen; M.A., 1591; m. Barbara Müller (already twice widowed at age 23, with one dau., Regina, by 1st marriage), whose father lived on estate, Mühleck in township of Gössendorf, Apr. 27, 1597 (d. 1611); 5 children: Heinrich, 1598, Susanna, 1599, Susanna, 1602, Friedrich, 1604, Ludwig, 1607; m. 2nd. Susanna Reuttinger, Oct. 30, 1613; 7 children: Margareta, 1615, Katharina, 1617, Sebald, 1619, Cordula, 1621, Fridmar, 1623, Hildebert, 1625; Anna Maria, 1630. District mathematician and tchr. of math. and astronomy at Protestant boy's seminary, Graz, capital of Austrain province of Styria, 1594-1600, when he was expelled by Catholics; asst. to Tycho Brahe at Benatky Castle, near Prague, 1600-01; succeeded Brahe as Imperial Mathematician (to Emperors Rudolph II and Ferdinand II), 1601-30 (lived in Prague 1601-12); district mathematician for states of Upper Austria, 1512-28 (resided at capital, Linz, 1612-26; at Ulm, Swabia, 1626-27; at Prague, 1627-28); private mathematician to Wallenstein, Duke of Friedland, in Duchy of Sagan, Silesia, 1628-30. Author: Calendarum und Prognosticum, 1591-99; Prodromus dissertationum cosmographicarum continens mysterium cosmographicum . . . , 1596; De fundamentis astrologiae certioribus, 1602; Judicium de trigono igneo, 1603; Ad vitellionem paralipomena quibus astronomiae pars optica traditur, 1604; De Stella nova in pede serpentarii, 1606; Astronomia nova . . . sea physica coelestis tradita commentariis de motibus stellae martis, 1609; Mercurius in sole, 1609; Narratio de observatis a se quatuor Jovis satellibus erronibus, 1610; Tertius interveniens, 1610; Dissertatio cum nuncio sidereo, 1610; Dioptrice, 1611; Nova stereometria doliorum, 1615; Epitomae astronomiae copernicanae, 1618-21; De cometis, 1619; Harmonice mundi, 1619; Chilias logarithmorum, 1624; Tabulae rudolphinae, 1627; Admonitio ad astronomos, 1629; Somnium seu astronomia lunari, 1634. Animated by a passion for exactitude, sought throughout his life to uncover the fundamental harmony of the world, and, as a result, laid the foundations of modern astronomy; found (1596) that between the spheres carrying six planets, one could fit the five regular solids, which seemed to be underlying reason for distances of planets from sun; using precise observations of Tycho Brahe and much patience and persistence in calculation, discovered Kepler's laws: 1. planets move in elliptical orbits with sun in one focus (thus ending tradition of more than 2000 years which held planetary motions must be explained by circles); 2. radius vector from sun to any planet sweeps over equal areas in equal times; 3. the squares of the periods of revolution of any two planets are as the cubes of their mean distance from the sun (Kepler's Laws were foundation on which Isaac Newton erected his system of the world); to explain why planets move as they do, devised a celestial physics, postulating rotation of sun (soon to be confirmed by Galileo's telescopic observations on sun spots) and a magnetic attraction between sun and planets; held gravity to be a mutual attraction between like bodies; related angular velocities of planets to musical intervals, thus presenting a polyphonic celestial harmony or music of the spheres; investigated nature of vision; introduced concept of ray of light; studied nature of light; explained reflection, studied refraction and contributed to theory of telescopes; investigated generation and reception of light by celestial bodies; investigated parallax and theory of eclipses; made use of logarithms; investigated volumes of solids of revolution, esp. volumes generated by revolution of conics; regularly prepared astrological forecasts. Died Regensburg (Ratisbon), Bavaria, Nov. 15, 1630.

KEPP, Richard Hans Kurt, German gynecologist; b. Hermannstadt, Feb. 7, 1912; s. Friedrich and Gabriele (Hochmeister) K.; M.D., Klausenberg; m. Annemarie Schneider, June 27, 1941; children—Konstanze, Christian. With Martin Luther Hosp., Hermannstadt, Univ. Clinic Gynecology, Gottingen; dir. Univ. Clinic Gynecology, Giessen. Mem. German Assn. Gynecologists, German Radiology Soc. Author: Grundlagen der Strahlentherapie; Gynaekologische Strahlen Therapie; Blutbildung und Blutumsatz beim Foeten und Neugeborenen; Aktuelle Probleme des Morbus haemolyticus neonatorum. Research on malignant tumors, radiation therapy, chemotherapy. Home: Klinikstrasse 28. Office: Klinikstrasse 32, Giessen, Germany.

KEPPELER, Gustav (Johannes), German chemist; b. Heilbronn, Germany, Feb. 27, 1876; Ph.D., Heidelberg, Germany, 1900. Asst., inorganic chem. inst., Tech. U. Darmstadt, 1900-07, became lectr., 1903; apptd. lectr. chem. tech. Tech. U., Hanover, Germany, 1908, prof., 1912; dir. war lab. of a mil. communications zone, 1915-16; named asso. prof. for tech. utilization of bog (later also of fuels, glass and ceramics); became prof., 1922; named prof. tech. chemistry, dir. Tech.-Chem. Inst., 1933, became emeritus, 1947, but continued teaching until 1952. Hon. fellow Brit. Soc. Glass Tech.; mem. German Ceramic Soc. (hon.; Seger Meml. medal 1939), German Glass Tech. Soc. (Otto Schott Meml. medal 1933). Research on glass tech., ceramic sci., utilization of bogs. Died Hanover, Aug. 21, 1952.

KERANEN, Jaakko, Finnish geophysician; b. Paltamo, Finland, June 1, 1883; s. Adolf and Anna (Lemonen) K.; B.A., U. Helsinki, 1910, M.A., 1912, Ph.D., 1927, Sc.D., 1958; m. Siiri Ingeborg Pajari, Dec. 16, 1914; children—Anna Liisa (Mrs. Lars Haagann), Elsa (Mrs. Jaakko Kenttimda), Lauri Veikko. Research, publs.on meteorology, magnetic activity, climatology; pioneer in establishment aviation weather services, synoptic weather services, agrl. meteorology. Home: 1A Tope Linsicatu, Helsinki 26, Finland.*

KERCKHOFF, Alan Chester, Am. sociologist; b. Lakewood, O., Mar. 14, 1924; s. August Edward and Helen (Heacock) K.; student Kent State U., 1942-43; A.B., Oberlin Coll., 1949; M.A., U. Wis., 1951, Ph.D., 1954; m. Sylvia Beth Stansbury, June 11, 1949; children—Steven Paul, Sharon Louise. Asst. prof. sociology Vanderbilt U., Nashville, 1953-55, 57-58; asst. to dir. Office for Social Sci. Programs, Air Force Personnel and Tng. Research Center, San Antonio, 1955-57; faculty, dir. grad. studies in sociology Duke, Durham, N.C., 1958-—, prof. sociology, 1964-—. NSF Sr. Postdoctoral fellow, 1965-66. Mem. Am. Assn. U. Profs., Am. Sociol. Assn., So. Sociol. Soc. Author: (with Kurt W. Back) The June Bug: A Study of Hysterical Contagion, 1968; also articles. Research on effects of pre-established or emergent definitions of situation on behavior of persons in interaction, especially family relations. Home: 1511 Pinecrest Rd., Durham, N.C. 27705.*

KERCKRING, Theodor, physician, anatomist; b. Hamburg, Germany, 1640; student Leyden, Netherlands; studied under Sylvius de las Boë and Spinoza; practiced medicine, Amsterdam, Netherlands; physician to Grand Duke of Tuscany in Hamburg, Germany; studied anatomy with Ruysch. Fellow Royal Soc., 1677. Author: Commentarius in Currum Triumphalen Antimonnii Basilii Valentini, 1665; Spicilegium anatomicium, 1670; Anthropogeniae ichnographia, 1671; Opera omnia, 1717, 29. Opposed iatrophysics and microscopy; described mucous folds of small intestine (Kerckring's folds), 1670; discovered a small part of occipital bone (Kerckring's ossicle); determined nature of postmorten residues, 1670; verified the fact that ovaries of female mammals produce eggs; prepared antimonic acid from potassium antimonate; developed method of preserving corpses by covering them with varnish. Died Nov. 2, 1693.

KERKER, Milton, Am. chemist; b. Utica, N.Y., Sept. 25, 1920; s. Samuel and Sarah (Cohen) K.; A.B., Columbia, 1941, M.A., 1947, Ph.D., 1949; m. Reva Stemerman, June 16, 1946; children—Ruth Ann, Martin Joseph, Susan Lee, Joel Leon. Faculty, Clarkson Coll. Tech., Potsdam, N.Y., 1949-—, prof. chemistry, 1956-—, chmn. chemistry dept., 1960-—, dean Sch. Sci., 1964-—; Unilevr vis. prof. U. Bristol (Eng.), 1967-68. Welch Found. lectr., 1965. Ford Found. fellow, 1952-53. Fellow Optical Soc. Am.; mem. Am. Chem. Soc. (chmn. div. colloid and surface chemistry, 1965-66), History Sci. Soc., Soc. for History Tech., A.A.A.S. Author: Electromagnetic Scattering, 1963; also numerous articles. Editor-in-chief Jour. Colloid and Interface Sci., 1965-—. Research on light scattering by colloidal particles, light scattering by heteropoly compounds. Home: 4 Hillcrest Dr., Potsdam, N.Y. 13676.*

KERLIN, Isaac Newton, Am. physician; b. Burlington, N.J., May 27, 1834; s. Joseph and Sarah (Ware) K.; M.D., U. Pa., 1856; m. Harriet C. Dix, 1865, 4 children. Asst. supt. Pa. Tng. Sch. for Feeble-minded Children at Elwyn, nr. Media, Pa., 1856-62, supt., 1864-93; a founder Nat. Assn. Supts. of Instn. for Feeble-minded 1876, sec., 1876-93; framed draft of bill to provide instns. in Western part of Pa. Author: The Manual of Elwyn, 1891. Pioneer in psychopathology, care and treatment of mentally deficient children and adults. Died Oct. 25, 1893.

KERMAN, Arthur Kent, physicist; b. Montreal, Que., Can., May 3, 1929; s. Samuel and Ida Kerman; B.Sc., McGill U., Montreal, 1950; Ph.D., Mass. Inst. Tech., 1953; m. Enid Ehrlich, Dec. 21, 1952; children—Ben, Daniel, Elizabeth, Melissa. NRC fellow Cal. Inst. Tech., Pasadena, 1954-55; Canadian NRC fellow Inst. for Theoretical Physics, Copenhagen, Denmark, 1954-55, research staff, 1955-56; faculty Mass. Inst. Tech., 1956-—, prof. physics, 1964-—; exchange prof. U. Paris (France), 1961-62. Cons. Ednl. Services, Inc., 1959-—, Los Alamos Sci. Labs., 1961-—, Brookhaven Nat. Lab., Upton, N.Y., 1964-—. Guggenheim fellow, 1961-62. Mem. Am. Phys. Soc., Canadian Assn. Physicists. Research, publs. on theory of nuclear structure and nuclear reactions. Home: 37 Rangeley Rd., Winchester, Mass. 01890. Office: Mass. Inst. Tech., Cambridge, Mass. 02139.*

KERMAN, Herbert David, Am. radiologist; b. Chgo., July 24, 1917; s. Abraham Bernard and Sarah (Horberg) K.; A.B., Duke, 1938, M.D., 1942; m. Ruth Harriet Rice, Jan. 14, 1942; children—David, Jeffry, Stephen, Michael. Faculty, Duke, 1949; faculty U. Louisville Sch. Medicine, 1949-56, dir. radioisotope lab., 1952-56; radiologist Halifax Dist. Hosp., Daytona Beach, Fla., 1956-—; also cons. to hosps. Radiologist staff med. div. Oak Ridge Inst. Nuclear Studies, 1950-52, cons., 1952-—. Diplomate Am. Bd. Radiology. Fellow Am. Coll. Radiology; mem. A.M.A., So. Med. Assn., A.A.A.S., Brit. Inst. Radiology, Radiol. Soc. N.Am., Am. Radium Soc., Fedn. Am. Scientists, Soc. Nuclear Medicine. Research, publs. on radioisotopes and radiation in diagnosis and treatment. Home: 2616 S. Pennisula St., Daytona 32018. Office: Halifax Dist. Hosp., Daytona, Fla. 32015.*

KERN, Bernard Donald, Am. physicist; b. New Castle, Ind., Oct. 31, 1919; s. William B. and Cecile (Hudson) K.; B.S., Ind. U., 1942, M.S., 1947, Ph.D. (AEC fellow), 1949; m. Nedda Wisler Burdsall, Aug. 20, 1946; children—Richard B., Jonathan K., Arthur R. Sr. physicist Oak Ridge Nat. Lab., 1949-50; faculty U. Ky., Lexington, 1950-—, prof. physics, 1958-—; vis. physicist U. S. Naval Radiol. Def. Lab., San Francisco, 1957-58; prof. under AID, Institut Teknologi Bandung (Indonesia), 1961-62. Recipient Alumni Research award U. Ky., 1959. Fellow Am. Phys. Soc.; mem. Am. Inst. Physics, Ky. Assn. Physics Tchrs. (pres. 1964-65), Sigma Xi. Research in measurements beta and gamma ray spectroscopy, proton and neutron induced nuclear reactions. Office: Dept. Physics and Astronomy, U. Ky., Lexington, Ky. 40506.*

KERN, Fred, Jr., Am. physician, educator; b. Montgomery, Ala., Sept. 9, 1918; s. Fred and Rose (Helburn) K.; B.A., U. Ala., 1939, postgrad. Sch. Medicine; M.D., Columbia, 1943; m. Bernie Cronheim, June 17, 1942; children—Katherinne, Patricia, David. Faculty, U. Colo. Sch. Medicine, Denver, 1952—, prof. medicine, 1965—, head div. gastroenterology, 1959—; attending physician Colo. Gen. Hosp.; cons. VA Hosp., Denver, Fitzsimmons Gen. Hosp. Diplomate Am. Bd. Internal Medicine. Fellow A.C.P.; mem. Am. Fedn. for Clin. Research, A.A.A.S., Harvey Soc., Am. Gastroent. Assn., Am. Assn. for Study Liver Disease, Am. Inst. Nutrition, Am. Soc. for Clin. Nutrition, Central Soc. for Clin. Research, Alpha Omega Alpha. Author: (with K. R. Hammond) Teaching Comprehensive Medical Care, 1959. Editorial bd. Gastroenterology, 1964—. Research, publs. on digestion and absorption of carbohydrates and lipids, role of bile salts in fat absorption, physiology of digestive tract, liver disease. Home: 740 Krameria St., Denver 80220. Office: 4200 E. 9th Av., Denver 80220.*

KERN, John William, Am. geophysicist; b. Mansfield, O., Dec. 16, 1930; s. William S. and Charlotte (Coleman) K.; student Pasadena City Coll., 1948-51; B.S., U. Cal. at Berkeley, 1956, M.A., 1958, Ph.D., 1960; m. Louise Catherine Olds, Mar. 15, 1962; 1 son, Wesley Edward. Phys. sci. RAND Corp., Santa Monica, Cal., 1960-64; asst. prof. physics U. Houston, 1964-66, asso. prof., 1966—. Mem. A.A.A.S., Am. Phys. Soc., Am. Geophys. Union, Internat. Sci. Radio Union. Research, publs. on theory of effects of mech. stress on remanent magnetisation of rocks; charge separation theory of auroras and electrojets; theory of polar magnetic storms; interactions between solar wind and geomagnetic field; effects of interstellar neutral hydrogen on solar wind.

KERN, Vincenz Sebastien von, see von Kern, Vincenz Sebastien.

KERN, Walther, German pharmacologist; b. Sprockhövel, Dec. 18, 1900; ed. U. Kiel, U. Iena; D.Eng., pharmacist. Asst., div. head Pharmacology Inst., Brunswick; prof., dir. Inst. Applied Pharmacology. Mem. German Soc. Pharmacology (pres. Westphalia-Lippe group), German Soc. Pharmacy, Soc. Cosmetic Chemists, German Soc. Chemistry, German Soc. for Lipid Research, Real Academia Madrid, Rhein-Westphalia Group for History Pharmacology (pres. subcom.). Author: Angewandte Pharmazie, 1953. Address: Hauptstrasse 76, Sprockhövel, Germany.

KERN, Werner Josef, German chemist; b. Tiengen-Baden, Feb. 9, 1906; s. Konrad and Anna (Haungs) K.; Ph.D. in Chemistry, U. Fribourg-en-Brisgau; m. Elfriede Baier, July 13, 1935; children—Rolf, Rosemarie, Herbert. Asst., instr., research chemist, prof., dir. Inst. for Organic Chemistry, U. Mainz. Mem. various sci. socs. Author: Anleitung zur organischen qualitativen Analyse, 1955; Die organische Chemie der Kunststoffe, 1963, also numerous works on polymerization, redoxpolymerization, macromolecular reactions, auto-oxydation. Home: Beuthenerstrasse 5, 65 Mainz-Rhein. Office: Inst. for Organic Chemistry, Johannes Gutenberg University, Mainz, Germany.

KERNAN, Roderick Patrick, Irish biophysicist; b. Dublin, Ireland, May 20, 1928; s. Dermod and Pauline (Hickey) K.; B.Sc. with honors, U. Coll., Dublin, 1951, M.Sc., 1952, Ph.D., 1956, D.Sc., 1964; m. Mary Cecily Kavanagh, Nov. 27, 1956; children—Susan Mary, Niall Justin. Fellow, Med. Research Council, 1953-57; research asso. Rockefeller Inst., N.Y.C., 1957-58; Med. Research Council fellow U. Coll., Dublin, 1958-66, asso. prof. physiology, 1966—; Rae prof. biochemistry Royal Coll. Surgeons, Dublin, 1965-66. Mem. adv. com. on metabolism and endocrinology to Med. Research Council; examiner in biochemistry to Royal Coll. Physicians of Ireland, 1966—. Fellow Royal Irish Acad. Medicine; mem. Royal Irish Acad., Physiol. Soc., Brit. Biophys. Soc. Author: Cell K, 1965; also articles. Research on origin of elec. charge across muscle cell membranes, influence of nerve on active transport of electrolytes in muscle; discovered that active pumping of sodium ions out of muscle fibres generated a P.D. across cell membrane causing potassium ions to move passively into cell, also that insulin stimulated sodium pump in muscle. Office: Physiology Dept., U. Coll., Merrion St., Dublin 2, Ireland.*

KERNER, Anton Ritter von Marilaun, Austrian botanist; b. Mautern, Lower Austria, Nov. 12, 1831. Prof. at Innsbruck and Vienna. Known for work in systematic botany, ecology and plant geography. Died Vienna, June 22, 1898.

KERNER, Christian Andreas Justinus, German physician; b. Ludwigsburg, Germany, Sept. 18, 1786; student natural history, Tübingen; M.D., 1809; Became physician, Wildbad, settled at Weinberg, 1818; dist. med. officer Weinberg. Author: Neue Beobachtungen uber die Würtenberg so häufig verfallen den todtlichen Vergiftungen durch den Genus geräucherter Würste, 1820; Das Fettgift, oder die Fettsäure, und ihre Wirkungen auf den thierischen Organismus. Ein Beytrag zur Untersuchung des verdorbenen Würsten giftig wirkenden Stoffes, 1822; De functi surgriderum partum auns; Eine Enschunung aus dem Kachtgebrete der Natur, 1836. First description of botulism during study of sausage poisonings, 1820; research on animal magnetism; publs. on clairvoyance and somnabulism, 1829. Died Feb. 21, 1862.

KERNKAMP, Milton Fredrick, Am. plant pathologist; b. St. Paul, Sept. 16, 1911; s. John Henry and Laura (Mahle) K.; B.S., U. Minn., 1934, M.S., 1938, Ph.D., 1941; m. Kathryn Griffith, June 27, 1936 (dec. May 1962); m. 2d, Marjorie Gerlich, Dec. 19, 1962. Mem. faculty U. Minn., 1936-41, 46—, prof. plant pathology, 1956—, head dept. plant pathology, 1961—; plant pathologist Dept. Agr., Meridian, Miss., 1941-42, collaborator Coop. Rust Lab., St. Paul, 1961—. Mem. Am. Phytopathol. Soc. (treas.-bus. mgr. 1967—, asso. editor jour. 1947-49, sec.-treas. 1948-51, v.p. 1952, pres. N. Central div. 1953), A.A.A.S., Am. Assn. U. Profs., Minn. Acad. Sci., Am. Soc. Agronomy, Am. Inst. Biol. Scis., Sigma Xi, Gamma Alpha, Gamma Sigma Delta. Mem. editorial bd. Minn. Sci., 1949-52, chmn., 1957-61. Research and publs. on forage crop diseases and corn smut fungus, Ustilago maydis, which helped to demonstrate that virulence of pathogens changed from time to time because of genetic changes. Home: 2466 N. Albert St., St. Paul 55113.*

KERNOHAN, James Watson, physician; b. Antrim, N. Ireland, Oct. 1, 1896; s. William James and Anne Elizabeth Kernohan; M.D., D.P.H., Queen's U., Belfast, N. Ireland, D.Sc., 1966; m. Eleanor M. Fletcher, Dec. 30, 1932; 1 dau., Eleanor Jean (Mrs. Robert O. Ryedell). Came to U. S., 1922, naturalized, 1938. Fellow pathology Mayo Found., 1922, staff, 1925-62; prof. pathology Mayo Grad. Sch. Medicine, U. Minn., Rochester, 1936-61; dir. Lab. Neuropathology, Barrow Neurologic Inst., Phoenix, 1962-64. Diplomate Am. Bd. Pathology (past pres.). Mem. A.M.A., Am. Assn. Pathologists and Bacteriologists, Am. Soc. Clin. Pathologists, Am. Assn. Neuropathologists, Minn. Med. Soc., Mich. Path. Soc. (hon.), Brit. Med. Assn. Author: (with A. Uihlein) Sarcomas of the Brain, 1962; (with Slooff, MacCarty), Intramedullary Tumors of Spinal Cord, 1964; also numerous articles. Research on diseases of central nervous system especially tumors of brain and spinal cord. Home: 918 9th Av. S.W., Rochester, Minn. 55901.*

KERNOT, W. C., engr.; b. Rochford, Essex, Eng., 1845; s. C. Kernot; M.A., M.C.E., Melbourne (Australia) U. Went to Australia, 1851; engr. Geelong, Coliban (both Australia) waterworks, 1865; lectr. Melbourne U., 1868, prof. engring., from 1883, founder metall. dept., 1893; with F. Ormand developed Working Men's Coll., Melbourne. Chmn. 2 juries on machinery Internat. Exhbn. of 1880; mem. Royal Commn. on Ry. Bridges, New S. Wales, Australia, 1882-84. Fellow Royal Geog. Soc.; mem. Instn. Civil Engrs., Royal Soc. Victoria (pres.). Author: Common Errors of Iron Bridge Design. Contbr. papers on balance beams, telescope tubes, aneroid barometers, strength of columns, others. Observed transit of Venus, 1874; (with Mr. Brennan) developed controllable fish torpedo, 1876-78. Died Mar. 14, 1909.

KERR, Frederick William Lawson, physician; b. Buenos Aires, Argentina, Mar. 25, 1923; s. Frederick Lawson and Madeleine (Wotton) K.; Baccalaureat, Nat. Coll. Buenos Aires, 1942; M.D. with honors, U. Buenos Aires, 1949; m. Patricia Jordan, Apr. 24, 1953; 1 dau., Kathryn. Came to U. S., 1952, naturalized, 1957. Instr., Washington U., St. Louis, 1956-58; faculty neurol. surgery State U. N.Y., Syracuse, 1958-59, clin. asst. prof., 1958-59; asst. prof. neurol. surgery U. Minn. Grad. Sch., Rochester, 1961-66, asso. prof., 1966—; cons. neurol. surgery Mayo Clinic, Rochester, 1960—. Acting chief neurosurgery VA Hosp., Syracuse, 1958-59. Diplomate Am. Bd. Neurol. Surgery. Mem. Am. Assn. Anatomists, A.C.S., A.M.A., Harvey Cushing Soc., Pan Am. Med. Assn., Sigma Xi. Research, publs. on pathways and mechanisms pain in face and head, central nervous system control viscera. Home: Mounted Route 72, Rochester, Minn. 55902.*

KERR, John, Scottish physicist; b. Glasgow, Scotland, Dec. 17, 1824; ed. Glasgow U.; M.A., LL.D.; m. Marion Balfour; 3 sons, 4 daus. Math. lectr. Free Ch. Tng. Coll., Glasgow. Fellow Royal Soc., 1890 (Royal medal 1898). Author: Elementary Treatise on Rational Mechanics, 1867; also papers on electro-optical subjects. Discovered Kerr effect (change produced in polarized light by reflection from polished pole of electromagnet. Died Glasgow, Aug. 18, 1907.

KERR, Sir John Graham, Brit. zoologist; b. Hertfordshire, Sept. 13, 1869; s. James Kerr; ed. Edinburgh, Cambridge univs.; m. 1st, Elizabeth Mary, 1903 (dec. 1934), 2 sons, 1 dau., 2d, Isobel Macindoe, 1936. Made zool. explorations in South America especially Pilcomayo River, 1889-91, expdn. to Gran Chaco to investigate the habits and life history of South Am. lungfish (Lepidosiren), 1896-97; demonstrator animal morphology U. Cambridge, 1897-1902; regius prof. zoology U. Glasgow, 1902-35. Chmn. advisory com. Fishery Research, 1941-49. Fellow Royal Soc., London 1909, Glasgow, 1909; pres. Royal Phys. Soc. Edinburgh, 1906-09, Royal Philos. Soc. Glasgow, 1925-28; Recipient Neill prize, 1904, Walsingham medal, 1898, Linnean Gold medal, 1955. Author: Primer of Zoology, 1921; Textbook of Embryology, 1919; Zoology for Medical Students; Evolution, 1926; An Introduction to Zoology; A Naturalist in the Gran Chaco; also papers. Pioneered in war camouflage according to biol. principles, originated protective coloration of ships by countershading and strongly contrasting patches (dazzle), 1914. Died Apr. 21, 1957.

KERR, Kathel Bedortha, Am. biologist; b. Chgo., Jan. 28, 1908; s. Norman and Lottie (Austin) K.; A.B., Oberlin Coll., 1929; M.S. (Jackson-Johnson fellow) Washington U., St. Louis, 1931; Sc.D., Johns Hopkins, 1936; m. Ruth Story Sidebotham, July 21, 1932; children—Norman Story, Kathel Austin. Instr., Washington U., St. Louis, 1932-34; NRC fellow Johns Hopkins, Balt., 1936-37; asst. bacteriologist U. Tenn., 1937-38; jr. zoologist NIH, Washington, 1938-40, asst. zoologist, N.Y.C., 1942-43; zoologist Cal. State Dept. Health, San Francisco, 1940-42; asso. parasitologist Inst. Inter-Am. Affairs, Vitoria, Brazil, 1943-44, parasitologist, 1944-45; helminthologist Salsbury Labs., Charles City, Ia., 1945-61, mgr. developmental research, 1961—. Chmn. sci. adv. com. Animal Health Inst., 1960-62. Fellow A.A.A.S.; mem. Ia. Acad. Sci., Am. Soc. Tropical Medicine and Hygiene, Am. Soc. Parasitology, Soc. Toxicology. Research, publs. on immunity to helminth parasites, poultry helminths, anthelminthics for poultry, toxicology vet. drugs. Patentee compounds for removal poultry helminths. Home: 709 2d Av. Office: 500 Gilbert St., Charles City, Ia. 50616.*

KERR, Paul Francis, Am. mineralogist; b. Hemet, Calif., Jan. 12, 1897; s. Joseph and Emma (Fowler) K.; B.S., Occidental Coll., 1919, D.Sc., 1960; Ph.D., Stanford U., 1923; m. Helen Richardson Squire, Sept. 2, 1924; children—Paul Squire, Ruth (Mrs. William Jakoby), Nancy (Mrs. Peter Del Grande). Mem. faculty Stanford, 1923-24; faculty Columbia, 1924—, prof. mineralogy, 1940—, Newberry prof., 1952—, emeritus, 1965—; research coordinator, 1950—; vis. Carnegie prof. S.A., 1941; guest prof. U. Oslo, Norway, 1960; cons. to brs. of govt. and instns. 1944; Chmn. com. Carnegie Endowment Internat. Peace, 1946; UN cons. Geneva Conf. on Peaceful Uses of Atomic Energy, 1955. Recipient K.C. Li award, 1957. Fellow Geol. Soc. Am. (v.p. 1947), Am. Geog. Soc., Mineral. Soc. Am. (sec. 1933-43, pres. 1946); mem. Soc. Econ. Geology, Am. Inst. Mining Engrs., N.Y. Acad. Science, A.A.A.S. Mineral Soc. Great Britain, Geol. Soc. Belgium (corr.), Bavarian Acad. Sci., Sigma Xi, Phi Beta Kappa. Author: Thin Section Mineralogy, 1933; Optical Mineralogy, 1941; Tungsten Mineralization in U. S.; Natural Occurrence of Uranium and Thorium, 1955, Uranium Alteration and Mineralization, Marysvale, Utah, 1957. Research and publs. on applications of geology and mineralogy; studies of Uranium-bearing veins, AEC, and quick clay USAF. Home: 501 W. 120th St., N.Y.C. 10027.*

KERR, Robert Bews, Canadian physician; b. Hamilton, Ont., Can., Aug. 30, 1908; s. George R. and Helen (Bews) K.; B.A., U. Toronto (Ont.), 1930, M.A., 1936, M.D., 1933. Successively became jr. demonstrator dept. medicine U. Toronto, 1939, sr. demonstrator, 1945, asst. prof., head dept. therapeutics, 1947, asso. prof., 1948; became prof., head dept. medicine U. B.C. (Can.), Vancouver, 1950. Fellow Royal Coll. Physicians and Surgeons Can., Royal Coll. Physicians. Research on diabetes mellitus; suggested (with C. H. Best, W. R. Campbell, A. A. Fletcher) combining insulin with zinc to slow down rate of absorption, 1936. Office: Dept. Medicine, U. B.C., Vancouver 8, B.C., Can.

KERR, Stanley Elphinstone, Am. chemist; b. Hopewell, N.J., Mar. 30, 1894; s. James Robert and Marion (Ross) K.; B.Sc. in Chemistry, U. Pa., 1917, M.S., 1918, Ph.D., 1925; m. Elsa Reckman, Jan. 1, 1922; children—Marion M. (Mrs. Roger S. Miller), Dorothy Ann (Mrs. Philip C. Jessup, Jr.), Malcolm H., Douglas S. Chmn. dept. biochemistry Am. U. Beirut, Lebanon, 1925-64, distinguished prof. emeritus, 1964—. Researcher, Harvard Med. Sch., 1932, U. Frankfurt, 1930, Sloane Kettering Inst., N.Y.C., 1950, Med. Coll. Va., 1957. Decorated Order of Merit, Chevalier de l'Ordre des Cedres, Republic of Lebanon. Mem. Am. Soc. Biol. Chemists, Alpha Omega. Research, publs. on permeability of red blood cells to sodium and potassium, in vivo fixation of brain by freezing, isolation from brain and quantitative determination of phosphocreatine, adenosine triphosphate, glycogen, triphosphoinositide, and changes resulting from autolysis, biosynthesis and metabolism of acid-soluble purine nucleotides. Home: 109 Cedar Lane, Princeton, N.J. 08540.*

KERR, Warwick Estevam, Brazilian geneticist; b. Santana de Parnaiba, S.P., Brazil, Sept. 9, 1922; s. Américo Caldas Kerr and Barbara Chaves de Oliveira; Agronomist, Escola Superior de Agricultura, Luiz de Queiroz, U. Sao Paulo, 1945, Ph.D., 1948; m. Lygia Sansígola, Mar. 7, 1947; children—Florence, Lucy, Américo, Jacira, Ligia Regina, Tania, Hélio. Biologist dept. genetics Escola Superior de Agr., Luiz de Queiroz, U. Sao Paulo, 1946-48, faculty, 1948-58, prof., head dept. genetics Faculdade de Medicina, Ribeirão Preto, Brazil, 1964—; prof., head dept. biology Faculdade de Filosofia, Ciencias e Letras de Rio Claro (Brazil), 1959-64; vis. prof. U. Cal. at Davis, 1951, Columbia, 1951-52. Sci. dir. State of Sao Paulo Sci. Found., 1962-64. Recipient André Dreyfus Nat. prize genetics, 1956; Catharina Prosdocimo Nat. prize genetics, 1965. Mem. Brazilian Soc. Genetics (pres. 1964-66), Brazilian Assn. for Advancement Scis., (sec. 1965—), Brazilian Acad. Scis., Author:

Scientific and Practical Beekeeping (with Erico Amaral); also numerous articles. Discovered genetic system of queen determination in bees of genus Melipona; research on biol. characteristics and genetics of bees. Home: Faculdade de Medicina de Ribeirão Preto, Casa 22, Ribeirao Preto, SP. Brazil.*

KERR, Washington Caruthers, Am. geologist; b. Guilford County, N.C., May 24, 1827; s. William M. and Euphence (Doak) K.; grad. U. N.C., 1850, Ph.D., 1879, LL.D., 1885; attended Lawrence Sci. Sch., Harvard; m. Emma Hall, 1853. Prof., Marshall U., Tex., 1851-52; with Nautical Almanac office, Cambridge, Mass., 1852; prof. chemistry and geology Davidson Coll., N.C., 1856-65; chemist, supt. Mecklenburg Salt Co., nr. Charleston, S.C., 1862; state geologist N.C., 1864-82; lectr. geology U. N.C., 1869-84; chief So. div. U. S. Geol. Survey, 1882-83. Author: Report of the Geological Survey of North Carolina, vol. 1, 1875, vol. 2, 1881; Minerals and Mineral Localities of North Carolina, 1881; Ores of North Carolina, 1888. Produced 1st basic geol. map of N.C., 1882, showed that state had been affected by glaciation; one of 1st to call to attention the phenomenon of soil creep. Died Asheville, N.C., Aug. 9, 1885.

KERR, Willard Augusta, Am. psychologist; b. Brookport, Ill., Feb. 25, 1918; s. Arthur Thomason and Augusta (Golightly) K.; B.E., So. Ill. U., 1939; M.S., Purdue U., 1940, Ph.D., 1942; m. Eleanor Moseley, Sept. 9, 1939; children—Karen (Mrs. Charles Miller Jensen), Douglas M., John Cameron. Indsl. music research dir. RCA, Camden, N.J., 1942-44; faculty Tulane U., 1946-47, Ill. Inst. Tech., 1947-67; lectr. Northwestern U., 1952-62; research asso. Indsl. Relations Center, U. Chgo., 1959-64; head dept. psychology Middle Tenn. State U., Murfreesboro, 1967—. Cons. industry and univs. Mem. Am., Midwestern psychol. assns., Chgo. Psychol. Club (past pres.), Sigma Xi. Author: (with F. Dunbar) Theory and Problems of Industrial Pyschology, 1966; also articles. Editor: Psychometric Affiliates, 1951—, William James Press, 1964—. Inventor tear (rip) method of opinion polling; developed standardized paper and pencil measurement of empathic ability; developed two accepted theories of accident causation. Home: 1620 E. Main St., Murfreesboro, Tenn. 37130.*

KERR, William, Am. nuclear engr., educator; b. Sawyer, Kans., Aug. 19, 1919; s. William and Maria Louise (Gill) K.; B.S. in Elec. Engring., U. Tenn., 1942, M.S., 1947, Ph.D., 1954; m. Ruth Duncan, Apr. 28, 1945; children—William Duncan, John Gill, Scott Winston. Faculty, U. Tenn., 1942-44, 46-48; faculty U. Mich., Ann Arbor, 1948—, prof., 1958—, chmn. dept. nuclear engring., 1961—, dir. Mich. Meml.-Phoenix project, 1965—. Supr. nuclear energy project AID, 1956-65; cons. to industry, 1954—. Mem. I.E.E.E., Am. Nuclear Soc., Am. Soc. for Engring. Edn., Sigma Xi, Eta Kappa Nu, Phi Kappa Phi, Tau Beta Pi. Contbr. articles to sci. jours. Developer beta-ray microscope for high resolution location radioisotopes, 1953. Research in kinetic behavior nuclear reactors, nuclear reactor shielding. Home: 2009 Hall St., Ann Arbor, Mich. 48104.*

KERSHBAUM, Alfred, Am. physician; b. Phila., Oct. 14, 1912; s. Samuel S. and Bertha (Silver) K.; B.S., Temple U., 1935, M.D., 1938; m. Judith Lowenthal Kurtz, June 25, 1939; children—Kenneth L., Susan K., Marjorie K. Research in cardiology Phila. Gen. Hosp., 1939-44, staff, 1942—, asst. chief div. cardiology, 1957—; practice medicine specializing in cardiology, Phila., 1945—. Fellow Council on Arteriosclerosis, Am. Heart Assn., 1962—; Nat. Heart Inst. grantee, 1961—; A.M.A. Edn. and Research Found. co-grantee, 1964—. Mem. Am. Fedn. for Clin. Research (charter), Soc. Nuclear Medicine, A.A.A.S., Heart Assn. Southeastern Pa., A.M.A. Contbg. author: Coronary Heart Disease, 1963; Atherosclerotic Vascular Disease, 1967. Research, publs. on atherosclerotic vascular disease, coronary heart disease, lipid metabolism, relationship of tobacco smoking to coronary heart disease, atherosclerosis, and lipid metabolism; discovered that smoking caused moblzn. of free fatty acids causing rise in blood levels; contbn. to knowledge of relationship of smoking and nicotine to lipid and catecholamine metabolism. Office: 1800 S. 55th St., Phila. 19143.*

KERST, Donald William, Am. physicist; b. Galena, Ill., Nov. 1, 1911; s. Herman Samuel and Lilian (Wetz) K.; B.A., U. Wis., 1934, Ph.D., 1937, D.Sc., 1961; D.Sc., Lawrence Coll., 1942; Dr. Honoris Causa, U. Sao Paulo, Brazil; m. Dorothy Birkett, Aug., 1940; children—Marilyn Elizabeth, Stephen Marshall. Mem. faculty U. Ill., 1938-57; war work, Los Alamos N.M., 1943-45; tech. dir., Midwestern Univs. Research Assn., 1953-57; with John Jay Hopkins Lab., Gen. Dynamics Corp., 1957-62; E. M. Terry prof. physics U. Wis., 1962—. Recipient John Scott award, 1946, John Price Wetherill medal Franklin Inst., 1950. Mem. Nat. Acad. Scis. (Comstock prize). Research and publs. primarily on devel. of betatron; nuclear physics; thermonuclear research; x-ray tubes and radiography; particle accelerators, plasmas. Home: 1506 Wood Lane, Madison, Wis. 53705.

KERSTEN, Martin, German physicist; b. Zittau, Apr. 28, 1906; s. Carl and Magdalene (Lamprecht) K.; ed. Technische Hochschule, Berlin; Ph.D. in Engring.; m. Cläre Roth, July 16, 1932; children—Brigitte, Susanne, Reinhard, physicist, Siemens & Halske AG, 1930-46; prof. exptl. physics Dresden (Germany) Technische Hochschule, also U. Jena (Germany), 1946-51; head labs. Vacuumschmelze AG, 1951-55; prof. Aachen Technische Hoschschule, 1955-61; became pres. Fed. Physico-tech. Establishment, 1961. Mem. German Soc. Physicists. Research and publs. on ferromagnetism, magnetic substances, tech. processes of magnetization. Home: Steinweg 19, 33 Braunschweig. Office: Bundesallee 100 (PTB), 33 Braunschweig, W. Germany.

KERTZ, Walter Gustav, German geophysicist; b. Remscheid, Feb. 29, 1924; s. Gustav I. and Elisabeth (Heinzelmann) K.; student U. Bonn, 1943-44; Dipl. Math., U. Göttingen, 1948, Dr.rer.nat., 1950; m. Ruth Friedrich, Jan. 3, 1950. Dozent, U. Göttingen (Germany), 1958-59; vis. asso. prof. geophysics N.Y. U., 1959-60; prof. Technische Hochschule, Braunschweig, Germany, 1961—. Mem. Seismol. Soc. Am., Deutsche Geophys. Gesellschaft, Meteorol. Gesellschaft Hamburg, Am. Geophys. Union, European Assn. Exploration Geophysicists, Soc. Exploration Geophysicists, Am. Meteorol. Soc., Wissenschaftlicher Beirat Deutscher Wetterdienst. Research, publs. on atmospheric tides, geomagnetic variations, electromagnetic induction in Earth, statistics of geophys. time series analysis. Home: Pestalozzist. 2, 33 Braunschweig, Western Germany.*

KERWIN, Larkin, Canadian physicist; b. Quebec, Que., Can., June 22, 1924; s. Timothy John and Catherine (Lonergan) K.; B.Sc., St. Francis Xavier U., 1944; M.S., Mass. Inst. Tech., 1946; D.Sc., Unibersité Laval., Quebec, 1949; m. Maria Guadalupe Turcot, June 10, 1950; children—Lupita, Alan, Larkin, Terence, Rosa Maria, Gregory, Timothy. Lectr., St. Francis Xavier U., 1944, U. Toronto (Ont., Can.), 1945; research physicist Geotech. Corp., Cambridge, Mass., 1945; faculty Université Laval, 1946—, prof., 1956—, chmn. dept. physics, 1961-67, dir. Atomic Physics Lab., 1960—, vice dean Faculty of Sci., 1967—. Chmn. adv. com. on electronic research Def. Research Bd. Can., 1959-66, chmn., 1961-64. Recipient Lt.-gov.'s medal, 1941; Gen. Gov.'s medal, 1944; Prix David, Province Que., 1951. Fellow Royal Soc. Can.; mem. Assn. canadienne-francaise pour l'avancement des Scis. (Médaille Pariseau 1964), Assn. canadienne des physiciens (past pres.) Assn. canadienne des radiologistes, Am. Phys. Soc., Canadian Assn. U. Tchrs., Internat. Union Pure and Applied Physics (asso. sec. gen. 1963—), Mexican Phys. Soc., Corp. des ingenieurs professionnels de Québec. Author: Atomic Physics—An Introduction, 1963; Introduction a la Physique Atomique, 1964; contbg. author: Mass Spectroscopy, 1963; also articles. Developed various instruments; research on energy factor simple atoms and molecules and their efficiency in producing collision-induced reactions. Home: 2166 Parc Bourbonniere, Sillery, Québec, 6. Office: Université Laval, Department de Physique, Qué. 10, Can.*

KERWIN, Richard Martin, Am. microbiologist; b. West Chester, Pa., Apr. 5, 1922; s. Charles Martin and Jane (Friday) K.; A.B., Dartmouth, 1947; M.S., U. N.H., 1949; Ph.D., Pa. State U., 1956; m. Mary Elizabeth Anderson, Apr. 6, 1957; children—Allison, Charles II, Sean. Sr. research scientist Wyeth Labs., Radnor, Pa., 1952—. Mem. Franklin Inst., A.A.A.S., N.Y. Acad. Scis., Soc. Indsl. Microbiologists, Am. Soc. Microbiologists, Am. Chem. Soc. Discovery, investigation new antibiotics, enzymes, drugs effective in cancer, enzymes used in modification steroids. Home: 116 E. Ashbridge St., West Chester, Pa. 19380. Office: Wyeth Labs., Inc., P.O. Box 8299, Phila. 19101.*

KERZE, Frank, Jr., Am. chem. engr.; b. Chgo., Nov. 17, 1908; s. Frank and Therese (Lovrencic) K.; B.S., Case Inst. Tech., 1929; M.A., Columbia, 1938; M.Chem.Engring., N.Y. U., 1941; m. Shirley Hymowitz, 1958. Sr. chem. engr. OSRD, 1941-42; instr. N.Y. U., 1942-46; with Oak Ridge Nat. Lab., 1946-50, Dept. Navy, 1950-55; with AEC, 1955—, Fuels and Materials Br., 1965—. Mem. Am. Chem. Soc., Am. Soc. Metals, Am. Nuclear Soc., Phi Lambda Upsilon. Editor: (with B. Lustman) Metallurgy of Zirconium, 1955. Research, publs. patents in field of chem.-metall.-materials engring. in devel. components and processes; participant in devel. Materials Test Reactor, reactors for submarines Nautilus, Seawolf, also Heavy Water Organic Cooled Reactor. Office: Fuels and Materials Br., AEC, Washington 20545.*

KESLER, Thomas Lingle, Am. geologist; b. Salisbury, N.C., Dec. 29, 1908; s. Thomas Monroe and Lillie (Lingle) K.; B.S. in Geology U. N.C., 1929, M.S., 1930; m. Margaret Alice Menges, July 2, 1932; children—Stephen Edward, Susan (Mrs. George Zell Porter). Asst. geologist Shell Petroleum Corp., Houston, 1930-31; engr. U. S. Dept. Agr., Western states, 1934-36; geologist U. S. Geol. Survey, Southeastern U. S., 1936-46, Thompson-Weinman & Co., Southeastern and N. Central States, 1946-52; sr. geologist U. S. Steel Corp., Tenn. and Va., 1952-53; chief geologist Foote Mineral Co., U. S., Can., Mexico, 1953-62; cons. geologist, N. and S. Am., 1962-65; chief geologist minerals and chems. Philipp Corp., Menlo Park, N.J., 1965-67; chief geologist Engelhard Minerals & Chems. Corp., Edison, N.J., 1967—. Fellow Geol. Soc. Am.; mem. Soc. Econ. Geologists, Am. Inst. Mining, Metall. and Petroleum Engrs., Am. Inst. Profl. Geologists, Geol. Soc. Washington. Research, publs. on geology indsl. minerals, stratigraphy and structure So. Appalachian region. Office: Engelhard Corp., Edison, N.J. 08817.*

KESSEL, Edward Luther, Am. entomologist; b. Osborne, Kan., Apr. 27, 1904; s. George Grant and Hattie (Flenniken) K.; student Greenville Coll., 1922-24; B.S., U. Cal. at Berkeley, 1925, M.S., 1928, Ph.D., 1936; m. Marguerite Berta Baldwin, Apr. 19, 1935; children—Margaret Mary (Mrs. Claude E. Hunter), Leonard Joseph. Instr. zoology, entomology Marquette U., 1928-30; faculty U. San Francisco, 1930—, prof., chmn. dept. biology, 1938—, editor Wasmann Jour. Biology, 1939—; asso. curator diptera Cal. Acad. Scis., 1950—, editor tech. papers, 1950—. Fellow Cal. Acad. Scis., A.A.A.S.; mem. Soc. Systematic Zoology (mem. council 1954-56), Pacific Coast (pres. 1947), Kan. entomol. socs., Sigma Xi, Phi Sigma. Author: The Embryology of Fleas, 1939; (with Alan Stone et al) A Catalog of the Diptera of America North of Mexico, 1965; also articles. Research field embryology of insects, especially of fleas, biosystematics of flies; contbns. to our knowledge of evolution of balloon-making habits of balloon flies, and giving of gifts to female by male, have resulted from research; authority on platypezidae published on systematics of N. and S. Am. and African forms, describing several new genera and numerous new species; discoveries pertaining to biology of these flies deal with immature stages, food and mating habits, activity-rest periodicity timed by angle of sun and attraction to smoke by smoke flies of genus Microsania. Home: P.O. Box 265, Novato, Cal. 94947. Office: Dept. Biology, U. San Francisco, San Francisco 94117.*

KESSEL, Richard Glen, Am. biologist; b. Fairfield, Ia., July 19, 1931; s. Oscar G. and Hazel (Humston) K.; B.S. summa cum laude, Parsons Coll., 1953; M.S., U. Ia., 1956, Ph.D. (George Lincoln Selley scholar, Bodine fellow), 1959. Asst. prof. Bowman Gray Sch. Medicine, 1959-61; faculty U. Ia., Iowa City, 1961—, asso. prof. zoology, 1964—. Postdoctoral fellow NIH, 1960-61. Recipient Career Devel. award Nat. Inst. Gen. Med. Sci., 1964—. Mem. Am. Assn. Anatomists, Am. Soc. for Cell Biology, Am. Soc. Zoologists, Electron Microscope Soc. Am., Soc. for Study Devel. and Growth, A.A.A.S. Research, publs. on structure and functon organelle systems in cells, mechanisms secretion, morphogenesis cell membrane systems, methods yolk formation oocytes, nuclear cytoplasmic interactions. Home: 2714 Ferndale Dr., Iowa City 52240.*

KESSEN, William, Am. psychologist; b. Key West, Fla., Jan. 18, 1925; s. Herman Lowery and Maria (Lord) K.; B.S., U. Fla., 1948; M.S., Brown U., 1950; Ph.D., Yale, 1952; m. Marion Lord, June 10, 1950; children—Judith, Deborah, Anne, Peter Christopher, Andrew Lord, John Michael. USPHS postdoctoral fellow Child Study Center, Yale, 1952-54, faculty, 1954—, now prof. psychology, research asso. pediatrics. Fellow Center Advanced Study in Behavioral Scis., 1959-60; mem. com. intellective processes research Social Sci. Research Council, 1959-64. Mem. A.A.A.S., Am. Psychol. Assn., Soc. Research Child Devel., Am. Assn. U. Profs., Sigma Xi. Author: (with G. Mandler) The Language of Psychology, 1959; The Child, 1965. Research on competence of newborn human infant, perception, attention and interest in infants, ability of sch. children and adults to process variability in environment. Home: 3 Halstead Lane, Branford, Conn. 06405. Office: 333 Cedar St., New Haven 06510.*

KESSLER, Edwin III, Am. meteorologist; b. N.Y.C., Dec. 2, 1928; s. Edwin and Marie (Weil) K.; A.B., Columbia, 1945; M.S., Mass. Inst. Tech., 1952, Sc.D., 1957; m. Lottie Catherine Menger, May 28, 1950; children—Austin Rainier, Thomas Russell. Chief synoptic meteorology sect. Weather Radar br. Air Force Cambridge Research Lab., Bedford, Mass., 1954-61; dir. atmospheric physics div. Travelers Research Corp., Hartford, Conn., 1961-64; dir. Nat. Severe Storms Lab., Research Labs., Environmental Science Services Administration, Norman, Okla., 1964—; adj. prof. meteorology U. Okla. Mem. A.A.A.S., Am., Royal meteorol. socs., Sigma Xi. Research, publs. on use of radar for precipitation forecasting and severe storm warning and severe storm research, radar data processing. Home: 2641 Butler Dr. Office: 1616 Halley Av., Norman, Okla. 73069.*

KESSLER, Henry H., Am. physician; b. New N.J., Apr. 10, 1896; s. Simon and Bertha (Portugese) K.; A.B., Cornell U., Ithaca, N.Y., 1916, M.D., Cornell U. Med. Sch., New York, N.Y., 1919; A.M., Columbia, 1932, Ph.D., 1934; m. Jessie Winnick, Dec. 25, 1918; children—Sanford, Jerome, Joan. In practice as orthopedic surgeon, Newark, N.J., since 1920; attending orthopedic physician, Newark City Hosp., Newark Beth Israel and Hosp. for Crippled Children, Hasbrouck (N.J.) Hosp. Cons. to UN on rehabilitation facilities in Yugoslavia, 1951, Indonesia, 1954, Philippines, 1955-56; cons. to Internat. Soc. for Welfare of Cripples on rehabilitation facilities in S. Africa, 1955; survey on rehabilitation needs and facilities in foreign countries, 1947-53; del. by Pres. of U. S. to Internat. Congress of Industrial Accidents in Budapest, 1928, Geneva, 1931, Brussels, 1935, Frankfort, 1938. Mem. N.J. Rehabilitation Commn. Diplomate Am. Bd. Orthopedic Surgery. Fellow Am. Coll. Surgeons, Am. Acad. Orthopedic Surgeons, Am. Pub. Health Assn., Am. Med. Assn. Commd. lt. comdr., U.S.N.R., 1936; on active duty since 1941; in South Pacific, 1943; formerly served as captain U.S.N.R. serving as chief, amputation center, U. S. Naval Hospital, Mare Island, Calif. Hunterian lecturer, London, Eng., 1935. Recipient gold medal Am. Acad. Orthopedic Surgeons, 1936; Am. Design award, 1944; Advertising Club award, Newark, 1945; recipient E. J. Ill award, 1951; Physicians award Pres. Com. on Employment Physically Handicapped, 1952; award Nat. Conf. Christians and Jews, 1953; Lasker Award, 1954; Gold Medal award by Holland Soc. of N.Y., 1956; Speidel award by Internat. Coll. of Surgeons, 1956; World Vets. Fed. award, 1956; Philippine Legion of Honor, 1956; William G. Anderson award, 1956; Order Phoenix Citation (Greece), 1962; Red Cross award (Madrid), 1964. Fellow Internat. Coll. Surgeons; past pres. Internat. Soc. for Welfare of Cripples; dir. Kessler Inst. for Rehabilitation. Mason. Author: Accidental Injuries, 1931; Occupational Disability Legislation, 1932; The Crippled and Disabled, 1934; Cineplasty, 1947; Rehabilitation of the Physically Handicapped, 1947; The Principles and Practices of Rehabilitation, 1950; Low Back Pain in Industry, 1955; Peter Stuyvesant and His New York, 1959. Noted for work in engring. of surg. and mech. techniques for attachment of artificial limbs. Home: 53 Lincoln Park, Newark 2.

KESSLER, Wayne Vincent, Am. bionucleonicist; b. Milo, Ia., Jan. 10, 1933; s. Joseph Edward and Genevieve (Frueh) K.; student Itasca Jr. Coll., 1951-52; B.S., N.D. State U., 1955, M.S., 1956; Ph.D., Purdue U., 1959; m. Olive Beatrice Buremaster, Sept. 10, 1953; children—Katherine Marie, Karl Matthew. Faculty, Purdue U., Lafayette, Ind., 1960—, asso. prof. health physics, 1964—, prof. bionucleonics, 1968—, mem. Purdue Trace Level Research Inst.; asst. prof. pharm. chemistry N.D. State U., 1959-60; vis. scientist Ind. Acad. Sci. Recipient Lederle Pharmacy Faculty award, 1962. Fellow A.A.A.S.; mem. Am. Pharm. Assn., Acad. Pharm. Scis., Am. Chem. Soc., Sigma Xi, Rho Chi, Phi Lambda Upsilon, Phi Kappa Phi. Research on analytical and biol. applications of radioactive tracers, body composition studies by potassium-40 measurements. Home: 2825 Forrest Lane, Lafayette, Ind. 47904.*

KESTIGIAN, Michael, Am. chemist; b. Charlton, Mass., Sept. 1, 1928; s. Vartan and Rose (Topelian) K.; B.S. with honors, U. Mass., 1952; M.S., U. Conn., 1954, Ph.D., 1956; m. Jean Anne French, Aug. 21, 1949; children—Michael Brian, Mark Craig. Teaching asst. U. Mass., 1951-52; with U. Conn., 1952-55, research asst., 1953-55; research chemist E. I. du Pont de Nemours & Co., Wilmington, Del., 1956-58; mem. tech. staff RCA Labs., Princeton, N.J., 1958-62; sr. staff scientist Sperry Rand Research Center, Sudbury, Mass., 1962—. Recipient RCA Labs. Achievement award, 1960. Mem. Am. Chem. Soc., Am. Ceramic Soc., Electrochem. Soc., Sigma Xi, Phi Lambda Upsilon (pres. 1954-55), Sigma Pi Sigma. Contbr. numerous articles to sci. jours. Preparation of inorganic materials containing elements in unusual valence states; preparation, purification and single crystal growth of electronically active materials; preparation of laser single crystals. Patentee in field. Home: P.O. Box 226, Stow, Mass. 01775. Office: 100 North Rd., Sudbury, Mass. 01776.*

KESTIN, Joseph, engr., educator; b. Warsaw, Poland, Sept. 18, 1913; s. Paul and Leah (Aizensztat) K.; Dipl. Ing. Engring., U. Warsaw, 1937; D.I.C., Imperial Coll., U. London, 1946, Ph.D., 1946; M.A., Brown U., 1955, D.Sc., U. London, 1966; m. Alicja Drabienko, Mar. 12, 1949; 1 dau., Anita Susan. Faculty, Polish U. Coll., London, Eng., 1944-52, prof., head dept. mech. engring., 1947-52; asso. prof. Brown U., Providence, 1952-55, prof. engring., 1955—; prof. associe U. Paris, Paris, France, 1966. Mem. Nat. Acad. Scis. adv. panel to Bur. Standards, 1960-66; mem. U. S. Del. to Internat. Conf. on Properties of Steam, 1954, 1956, 1963—; cons. Blaisdell Pub. Co., Electric Boat div. Gen. Dynamics Corp., David Taylor Model Basin. Recipient Instn. Mech. Engrs. London Water Arbitration prize, 1948. Mem. Instn. Mech. Engrs., Am. Soc. M.E., Sigma Xi, Tau Beta Pi. Author: A Course in Thermodynamics, 1966. Tech. editor Jour. Applied Mechanics, 1956—. Research, numerous publs. on most precise measurements of viscosity of gases and gaseous mixtures and their interpretation in terms of statis. mechanics, theory of oscillating disk viscometer, effects of free-stream turbulence and vibrations on heat-transfer rates, heat transfer across turbulent boundary layers, applications of thermodynamics to continuum mechanics. Home: 140 Woodbury St., Providence 02906.*

KESZTHELYI, Lajos, Hungarian physicist; b. Kaposvar, Hungary, Feb. 15, 1927; s. Jozsef and Terez (Virth) K.; Diploma Physicist, Roland Eotvos U., 1950, D.Phys. Scis., 1962; m. Sara Landori, Aug. 23, 1951. Staff mem. Central Research Inst. Physics, Budapest, Hungary, 1954-63, head Lab. Nuclear Physics, 1963—; demonstrator Roland Eotvos U., 1950-57. Mem. Roland Eotvos Phys. Soc. (award 1957). Author: Atoms and Atomic Particles, 1959; Scintillation Counters, 1963 (pub. in Hungarian); also numerous articles. Research in nuclear physics, photonuclear reaction, reaction on light nuclei, g-factors of excited states by means of internal magnetic fields, search for new Mössbauer-isotopes, phase transitions in solids by means of Mössbauer effect. Home: 5 Lorantffy Zs. Office: Konkoly Thege, Budapest, Hungary.*

KETCHAM, Alfred Schutt, Am. physician; b. Newark, N.Y., Oct. 7, 1924; s. Colston E. and Ellen (Schutt) K.; B.S., Hobart Coll., 1945; M.D., U. Rochester, 1949; m. Elsie Jane Chase, July 13, 1946; children—Sue Ellen, Wendy Jane, Sally Lyn, Jill Anne, Jeff Terry, Dana Kay. Surg. resident USPHS Hosp., San Francisco, Seattle, 1950-55; chief surgery USPHS Indian Hosp., Talihina, Okla., 1955-57; practice medicine, specializing in surgery, Bethesda, Md., 1957—; sr. investigator surgery br. Nat. Cancer Inst., 1957-62, chief surgery br., 1962—. Diplomate Am. Bd. Surgery. Mem. A.C.S., Am. Assn. Cancer Research, Soc. Head and Neck Surgeons, James Ewing Soc., Soc. Plastic and Reconstructive Surgery, Am. Radium Soc. Research, numerous publs. on factors influencing local control and distant dissemination of cancer. Home: 1006 Brice Rd., Rockville, Md. 20852. Office: Bldg. 10, NIH, Bethesda, Md. 20014.*

KETCHUM, Bostwick Hawley, Am. biol. oceanographer; b. Cleve., Jan. 21, 1912; s. Harold B. and Bertha (Mason) K.; A.B., St. Stephens Coll., Columbia, 1934; D.Sc., Bard Coll., 1964; Ph.D., Harvard, 1938; m. Emily King, Sept. 1, 1936; children—Carl, Paul, Sara Ann. Research asst. Biol. Labs., Harvard, 1938-39, lectr. biol. oceanography, 1960—; instr. L.I., U., 1939-40; with Woods Hole (Mass.) Oceanographic Inst., 1940—, sr. oceanographer, 1954-63, sr. scientist, 1963—, asso. dir., 1962—. Ofcl. U. S. participant 2d UN Internat. Conf. on Peaceful Uses Atomic Energy, 1958, Sci. Conf. on Disposal Radioactive Wastes, 1959. Mem. Am. Soc. Limnology and Oceanography, Ecol. Soc. Am., A.A.A.S., Am. Inst. Biol. Scis. (com. on hydrobiology), Phi Beta Kappa, Sigma Xi. Author: (with A. C. Redfield) Marine Fouling and Its Prevention, 1952. Research, numerous publs. on nutrition of marine plants, fertility of sea, circulation of coastal waters and estuaries. Home: 27 Buzzards Bay Av. Office: Woods Hole Oceanographic Instn., Woods Hole, Mass. 02543.*

KETCHUM, Milo Smith, Am. civil engr.; b. Burns, Ill., Jan. 26, 1872; s. Smith and Martha Ann (Clement) K.; B.S., U. Ill., 1895, C.E., 1900; Sc.D., Colo. Sch. Mines, 1926, U. Colo., 1927; m. Mary Esther Beatty, Sept. 17, 1903; children—Martha Esther (Mrs. N. C. Debevoise), Elizabeth Jane, Milo Smith, Jr. Instr. civil engring. U. Ill., Urbana, 1895-97, asst. prof., 1899-1903; bridge and structural engr. Gillette-Herzog Mfg. Co., 1897-99; contracting mgr. Am. Bridge Co., Kansas City, 1903-04; prof. civ. engring. U. Colo., 1904-19, dean Coll. Engring., 1905-19; prof. civil engring. and dir. dept. U. Pa., 1919-22; dean of Coll. of Engring. and dir. Engring. Expt. Sta., U. Ill., from 1922. Asst. dir. U. S. Govt. explosives plants, in charge of constrn. of smokeless powder plant at Nitro, W.Va., 1918-19. Mem. Crocker and Ketchum, cons. engrs., Denver, Colo., 1909-10. Author: Surveying Manual (with Prof. W. D. Pence), 1900, 5th edit., 1931; Design of Steel Mill Buildings, and The Calculation of Stresses in Framed Structures, 1903, 4th edit., 1921; Design of Walls, Bins and Grain Elevators, 1907, 3d edit., 1918; Design of Highway Bridges of Steel, Timber and Concrete, 1908, 2d edit., 1920; Design of Mine Structures, 1912. Structural Engineers' Handbook, 1914, 3d edit., 1924. Died Dec. 19, 1934.

KETELAAR, Jan Arnold Albert, Dutch chemist; b. Amsterdam, Netherlands, Apr. 21, 1908. s. Albert Jans and Lisette (Struycken) K.; Ph.D., U. Amsterdam, 1933; postgrad. U. Leyden (Netherlands), 1934-37, Cal. Inst. Tech., 1937; m.; children—Hubert Jan, Frederick Cornelis Johannes, Gysbert Jan. Lectr. chem. crystallography U. Leyden, 1936-40, reader phys. chemistry, 1940-41; prof. phys. chemistry U. Amsterdam, 1941-60, prof. electrochemistry, 1960—, dean Faculty Scis., 1953-57; coordinator research Kon. ZoutKetjen, 1960—. Mem. Nat. Bd. Edn., 1960—. Mem. Royal Netherlands Acad. Scis., Royal Netherlands Chem. Soc. Author: Chemische Binding, 1947; Chemical Constitution, 1953; Liaisons et Propriétés Chimiques, 1960; Chemische Konstitution, 1964; also numerous articles. Research on X-ray crystal determinations, infra-red absorption and reflection spectra, solid solutions, simultaneous transitions in compressed gases and liquids, electrochemistry of molten salts. Home: 91 Markeloseweg, Ryssen, Netherlands. Office: U. Amsterdam, Hengelo, Netherlands.*

KETHAM, Johannes von, see von Ketham, Johannes.

KETSKHOVELI, Nikolai Nikolaevich, Russian botanist; b. Tkviavi, USSR, Jan. 7, 1898; grad. Tiflis U., 1925. Asst., Tiflis U., 1923-29, lectr., 1929-30; dean Biol. Faculty, prof. botany Pedagogical Inst., 1930-33; head chair botany Georgian Agr. Inst., Subtropical Inst., and other higher edn. instns.; prof., head Faculty Botany, dean Faculty Natural Sci. (later Faculty Biology), Tbilisi U., 1933-45, rector, 1945—; sci. organizer, dir. dept. geobotany Tbilisi Bot. Inst., 1933-38; chmn. dept. agr. sci. Georgia Acad. Sci., 1942-56. Named Hon. Sci. Worker, Georgian SSR, 1945. Mem. Georgia Acad. Sci. (mem presidium 1941—). Author: The Basic Types of Vegetative Cover in Georgia, 1936; Cultivated Plant Zones of Georgia, 1957; The Flora of Georgia, 8 vols., 1941-52. Editor: Atlas of the Cultivated Flora of Georgia, 3 vols., 1939-51. Head numerous bot. expdns.; collected material demonstrating interrelationships of steppe and forest; research in flora of Georgia, weed erradication. Office: A. USSR, Tbilisi, Gruz. SSR, AN Gruz.SSR.

KETTEL, Karsten, Danish physician. b. Copenhagen, Denmark, Dec. 30, 1899; s. Rasmus Andersen and Johanne (Dahlerup) K.; grad. U. Copenhagen, Med. Sch., 1924, Thesis, 1930; m. Else Jespersen, Mar. 8, 1951. Asst., U. Copenhagen, 1924-39; chief surgeon otol. dept. Frederiksborg County Hosp., Hillerd, Denmark, 1943—. Otologic cons. Helsingor Hosp., 1939-62. Mem. Danish (past pres.), Brazilian (hon.) otol. socs. Author: Cold-hemagglutinins in Human Serum, 1930; Peripheral Facial Palsy, Surgery and Pathology, 1959; also numerous articles. Research on surgery of facial nerve. Home: Bukkeballevej 81, Rungsted, Denmark. Office: 34 Slotsgade, Hillerod, Denmark.*

KETTERING, Charles Franklin, Am. engr., inventor; b. nr. Loudonville, Ashland Co., O., Aug. 29, 1876; s. Jacob and Martha (Hunter) K.; E.E. in M.E., Ohio State U., 1904, E.D., 1929; hon. degrees from 29 other instns.; m. Olive Williams, Aug. 1, 1905 (dec. 1946); 1 son, Eugene Williams. Began with Star Telephone Co., Ashland, later with Nat. Cash Register Co., Dayton, O.; organized Dayton Engring. Labs. Co. (Delco) for mfg. own inventions; co-organizer The Dayton Metal Products Co. and Dayton-Wright Airplane Co., 1914; served for 27 yrs. as v.p. Gen. Motors Corp. and gen. mgr. research labs. div.; dir. Nat. Cash Register Co., Ethyl Corp., Mead Corp., Moraine Devel. Co.; chmn. bd. Winters Nat. Bank and Trust Co., Flexible Co. Founder, chmn. bd. Charles F. Kettering Found., for researches in natural sciences, including work on chlorophyll and photosynthesis, artificial fever therapy, and cancer; pres. Thomas A. Edison Found.; dir. Sloan-Kettering Inst. for Cancer Research; co-founder Moraine Park Sch., Dayton; trustee Ohio State U., Antioch Coll., Coll. Wooster (O.), U. Miami (Fla.), So. Research Inst.; served as chmn. Nat. Inventors Council, Nat. Patent Planning Com., engring. and indsl. research div. NRC; mem. sci. adv. bd. NRC; chmn. exec. com., Centennial of Engring., 1952. Fellow Nat. Acad. Scis.; mem. Soc. Automotive Engrs. (pres. 1918), A.A.A.S. (pres. 1945), several other scientific and engring. socs. Scientific work includes invention of automotive starting, lighting, and ignition systems, electrified cash register, credit systems, and accounting machines; invented and marketed small generating unit for lighting farmhouses, etc.; originated and guided researches resulting in higher octane gasolines, including tetraethyl lead, extraction of bromine from sea water, high compression automobile engines, improved automobile finishes, nontoxic and noninflammable refrigerant, elec. refrigeration, improved Diesel engines (applications of which include powering of railroads); contbr. to or responsible for other devels. of indsl. importance. Died Nov. 25, 1958.

KETTLE, Edgar Hartley, English pathologist; b. London, Apr. 20, 1882; s. Edgar and Mary (Hartley) K.; ed. St. Mary's Hosp. Med. Sch.; M.B., M.S., 1907; M.D., 1910; m. Marguerite Pam, 1918. Worked under Ludwig Aschoff, Freiburg, Germany, 1911; became dir. pathology 3d London Gen. Hosp., Wandsworth, 1916; became pathologist, lectr. St. Mary's Hosp., 1918, later dir. inst. pathology and med. research; pathologist Cancer Hosp.; prof. pathology Welsh Nat. Sch. Medicine, U. Wales, 1924-27, St. Bartholomew's Hosp. Med. Sch., U. London, 1927-34; dir. dept. pathology Brit. Postgrad. Med. Sch., from 1934; cons. pathologist St. Bartholomew's Hosp., Queen Alexandra Mil. Hosp., Milbank; examiner in pathology to several Brit. univs. Mem. Med. Research Council (com. on indsl. solvents, indsl. pulmonary disease, radiology). Fellow Royal Soc., 1936, Royal Coll. Physicians; mem. Path. Soc. Gt. Britain and Ireland, Brit. Med. Assn. Author: The Pathology of Tumours; also numerous papers. Research in histopathology, on exptl. rheumatism, cancer, splenomegaly, infection, immunity; gave 1st detailed description of actual lesions of gas gangrene, 1919; investigated silicosis and its relation to pulmonary Tb. Died London, Dec. 1, 1936.

KETTUNEN, Kalevi, Finnish physician; b. Iisalmi, Finland, Oct. 28, 1918; s. August and Justiina (Lyyra) K.; M.D., U. Helsinki, (Finland), 1946; radiologist U. Hosp. Helsinki, 1950; m. Inkeri Heikkinen, Aug. 2, 1947. Radiologist, Maria Hosp., Helsinki, 1951-59, chief radiologist, 1959—; docent radiology U. Helsinki, 1959—, acting prof. diagnostic radiology, 1962. Decorated Cross of Freedom with Red Cross. Mem. Comité Internat. de Radiologie, 1959-62. Mem. World orgn. U. Profs. Med. Radiology. Research, publs. on influence of hormones on composition of blood, radiotherapy of malignant tumors, roentgendiagnosis of tumors. Home: 23, Apollonkatu, Helsinki, Finland.*

KETY, Seymour Solomon, Am. physiologist; b. Phila., Aug. 25, 1915; s. Louis and Ethel (Snyderman) K.; A.B., U. Pa., 1936, M.D., 1940; m. Josephine Gross, June 18, 1940; children—Larence Philip, Roberta Frances. NRC fellow Harvard, Boston, 1942-43; from asso. to asst. prof. dept. pharmacology Sch. Medicine, U. Pa., 1943-48, prof. clin. physiology Grad. Sch. Medicine, 1948-61; sci. dir. Nat. Insts. Mental Health and Neurol. Diseases and Blindness, NIH, Bethesda, Md., 1951-56, chief Lab. Clin. Sci., Nat. Inst. Mental Health, 1956-61, 1962-67; prof.

psychiatry Harvard Med. Sch., 1967—; dir. psychiat. research labs. Mass. Gen. Hosp., 1967—. Henry Phipps prof. psychiatry, chmn. dept. psychiatry Johns Hopkins U. Sch. Medicine, Balt., 1961-62. Med. dir. USPHS, 1953-55. Recipient Theobold Smith award A.A.A.S., 1949; Max Weinstein award United Cerebral Palsy Research Found., 1954; Distinguished Service award U. S. Dept. Health, Edn. and Welfare, 1958; Stanley R. Dean Research award, U. Mich., 1962; lectr. various univs., instns. Mem. Nat. Acad. Scis., Am. Physiol. Soc., Am. Soc. For Pharmacology and Exptl. Therapeutics, Am. Soc. for Clin. Investigation, Soc. for Psychiat. Research, Assn. for Research in Nervous and Mental Disorders (pres. 1965), Am. Psychopath. Assn. (pres. 1965). Editor-in-chief; Jour. Psychiat. Research, 1959—. Research on chelation of lead with citrate in lead poisoning, cerebral circulation and energy metabolism, biochem. studies in schizophrenia. Address: Dept. Psychiatry, Mass. Gen. Hosp., Boston 02114.*

KEUP, Wolfram, biochemist, neurosurgeon; b. Berlin, Germany, Apr. 7, 1920; s. Erich and Erna (Szagunn) K.; ed. univs. Berlin, Innsbruck, Frankfort; M.D. Med. asst. Med. Clinic, Frankfort U., 1948-52; asst., biochem. inst. Berlin U., 1952-53, psychiat. clinic Berne U., 1953-54; with Neuro-surg. Clinic, Psychiat. Clinic of Zürich, 1954, N.Y. State Psychiat. Clinic, 1954-55, pharmacology dept. Columbia U.; med. asst. Psychiat. Hosp., Marens-Fribourg, Switzerland, then Bklyn. State Hosp., State U. N.Y., 1956-58; dir. research, 1958—. Author: Biochimie en Schizophrenie, 1954; Forme facultative de phenylketonurie, 1955; Insuline amorphe et hyaluronidase pour le traitement des psychoses, 1957; Le groupe de phenothiazines et la fonction hépatique, 1959; Un complèxe LSD-protéine et la psychose modèle, 1959. Address: 681 Clarkson Av., Bklyn.

KEVAN, Douglas Keith McEwan, entomologist; b. Helsinki, Finland, Oct. 31, 1920; s. Douglas K. and Gwynneth M. (Paine) K.; B.Sc. with 1st class honours in zoology, U. Edinburgh (Scotland), 1941; Associateship, Imperial Coll. Tropical Agr., U. W.I., Trinidad, 1943; Ph.D., U. Nottingham (Eng.), 1956; m. Kathleen Edith Luckin, Sept. 11, 1943; children—Peter G., Martin K., Simon M. Entomologist, H.M. Colonial Service, Trinidad, 1941-43, Kenya Dept. Agr., 1943-48, acting sr. entomologist Uganda Dept. Agr., 1945; Brit. mil. adminstr., civilian locust-liaison officer, Somalia, 1946-47; 1st head zoology sect. U. Nottingham (Eng.) Sch. Agr., 1948-58; prof. entomology Macdonald Coll., McGill U., 1958—, chmn. dept. entomology and plant pathology, 1958-64, chmn. entomology, 1964—, chmn. Lyman Entomol. Mus., 1960—. Fellow Royal Soc. Edinburgh, Royal Entomol. Soc. London; mem. Inst. Biology, Entomol. Soc. Can., Entomol. Soc. Am., Entomol. Soc. Que., Canadian Soc. Zoologists, Am. Entomol. Soc., Assn. Brit. Zoologists, Systematics Assn., Soc. Systematic Zoology, Assn. Applied Biologists, Brit. Assn. for Entomology, Sigma Xi, numerous others. Author: Soil Animals, 1962; also numerous articles. Editor, contbg. author: Soil Zoology, 1955. Research on Orthoptera, especially taxonomy of Pyrgomorphidae and Gryllidae, soil fauna especially smaller arthropods. Home: 20 Woodridge Crescent, Beaconsfield, Que. Office: Dept. Entomology, Macdonald Coll., Ste Anne de Bellevue, Que., Can.*

KEY, Ben Witt, Am. physician; b. Cuthbert, Ga., June 11, 1883; s. Benjamin Witt and Adelea (Allen) K.; B.A., Vanderbilt U., 1905; M.D., U. Pa., 1909; M.A., Columbia, 1925; m. Sarah Webb, Feb. 25, 1913; 1 dau., Sarah (Mrs. James O. Watts). With N.Y. Eye and Ear Infirmary, 1909-1936; practiced medicine, N.Y., from 1912. Fellow A.C.S. Contbr. numerous articles to med. jours. Noted eye surgeon, diagnostician, researcher; one of 1st surgeons to transplant human cornea. Died N.Y.C., June 5, 1940.

KEY, Charles Aston, English surgeon; b. Southwark, Eng., 1793; s. Thomas and Margaret (Barry) K.; apprenticed to father, 1810; student United Borough Hosps., also Guys; studied under Astley Cooper, 1815-18; m. Anne Cooper, 1818. Demonstrator anatomy St. Thomas Hosp., 1819-23; became asst. surgeon to Guys, 1821, surgeon, 1824. Fellow Royal Soc.; mem. Royal Coll. Surgeons. One of earliest London surgeons to use ether; publs. on hernia, lithotomy. Died 1849.

KEY, Joe Lynn, Am. plant physiologist; b. Troy, Tenn., Sept. 10, 1933; s. Joe Lee and Ethel (Culp) K.; B.S., U. Tenn., 1955; M.S., U. Ill., 1957, Ph.D., 1959; m. Connie Lou Clark, Sept. 8, 1956; children—Diana Lynn, Debra Lea. NSF fellow U. Cal. at Davis, 1959-60, asst. prof. botany, 1960-62; faculty Purdue U., Lafayette, Ind., 1962—, prof. plant physiology, 1966—. Mem. Am. Soc. Plant Physiologists, A.A.A.S., Sigma Xi. Research, publs. on hormonal regulation of plant growth and devel. especially cellular and molecular mechanisms of hormone action; pioneered advances on hormone action in relation to control of RNA and protein biosynthesis. Home: 11 Colony Rd., West Lafayette, Ind. 47906.*

KEY, John, see Caius, John.

KEYES, Charles Rollin, Am. geologist; b. Des Moines, Dec. 24, 1864; s. Calvin W. and Julia (Davis) K.; B.S., State U. Ia., 1887, A.M., 1890; Ph.D., Johns Hopkins, 1892. Asst. U. S. Geol. Survey, 1889-90; palaeontologist of Mo., 1890-92; asst. state geologist of Ia., 1892-94; dir. Mo. Geol. Survey, 1894-97; geol. travel in Europe, Asia and Africa, 1897-98; pres. N.M. State Sch. of Mines, 1902-06; foreign travel, 1906-07, 1926; pres., gen. mgr. mining cos. and other corps.; cons. mining engr. from 1890. Fellow Geol. Soc. Am., A.A.A.S., Am. Inst. Mining Engrs., Mining and Metall. Soc. Am., Ia. Acad. Scis. Author: Geological Formations, 1892; Coal Deposits, 1893; Organization of Geological Surveys, 1894; Paleontology of Missouri, 1894; Maryland Granites, 1895; Origin and Classification of Ore Deposits, 1900; Genesis of Lake Valley Silver Deposits, 1907; Ozark Lead and Zinc Deposits, 1909; Deflation, 1910; Mid-Continental Eolation, 1911; Bibliography of Geology, 1913; Mechanics of Laccolithic Intrusion, 1922; Orogenic Consequences of a Diminishing Rate of Earth's Rotation, 1922; Astronomical Theory of Glaciation, 1925; numerous memoirs and essays; also contbr. to scientific, tech. and ednl. jours. Editor Pan-Am. Geologists, from 1922. Died May 18, 1942.

KEYES, Edward Lawrence, Am. surgeon; b. Charleston, S.C., 1843; s. Erasmus O. and Caroline (Clarke) K.; ed. Yale, 1863; M.D., N.Y. U., 1866; student, Paris, France; became physician Bellevue Hosp., N.Y.C., 1868. Mem. Am. Assn. Genito-Urinary Surgeons (founder). Author treatise on surg. diseases of genito-urinary organs, syphilis. Pioneered in male genito-urinary surgery; discovered treatment for syphilis, 1877. Died 1924.

KEYES, Frederick George, physical chemist; b. Kingston, Can., June 24, 1885; s. John George and Margaret (Williams) K.; B.S., R.I. Coll., 1906; M.S., Brown Univ., 1907, Ph.D., 1909; D.Sc. from Yale University, 1934, D.Sc., R.I. Coll., 1942, Brown U., 1963; m. Gabrielle Alice Bowers, Dec. 27, 1923. Assistant in chemistry, Brown U., 1906-07, instr., 1908-09, G.A.R. research fellow, 1909; research asso. in physical chemistry, Mass. Inst. Tech., 1910-13; chief engr. Cooper-Hewitt Electric Co., 1913-16; asso. prof. phys. chemistry research, 1916-19, prof., 1920-1950, dir. research lab. phys. chemistry, 1920-1945. in charge dept. chemistry, 1922-1945, prof. emeritus and lecturer since 1950, all at Massachusetts Institute Tech. Chmn. U. S. commn. Internat. Steam Properties Com. Commd. capt. U. S. Army, Dec. 1917, major, Sept., 1918; dir. Sci. and Control Lab., A.E.F., at Puteaux, France, 1918-19; citation by comdr.-in-chief A.E.F., June 1919; lt. col. Chem. Warfare R.C. Recipient. Theodore William Richards Medal Award, 1942; American Society Mech. Engrs. Award, 1948. Fellow Am. Acad. Arts and Sciences, Am. Physical Soc., A.A.A.S.; mem. Am. Chem. Society, National Acad. Sciences, Am. Soc. Refrigerating Engrs. (hon.), Phi Sigma Kappa, Sigma Xi. Author: The Thermodynamic Properties of Ammonia, 1916; Thermodynamic Properties of Steam (with J. H. Keenan), 1936. Contbr. numerous papers on investigations of the thermodynamic properties of matter, kinetic theory of matter and applications of thermodynamics to chemical equilibria problems. Home: 15 Berkeley St., Cambridge, Mass. 02138.

KEYFITZ, Nathan, sociologist; b. Montreal, Can., June 29, 1913; s. Arthur and Anna (Gerstein) K.; B.Sc., McGill U., Montreal, 1934; Ph.D., U. Chgo., 1952; m. Beatrice Orkin, Oct. 8, 1939; children—Barbara Lee, Robert Norman. Census clk., statistician, sr. research statistician Dominion Bur. Statistics, Govt. of Can., 1936-59; dir. Colombo Plan Bur., Colombo, 1956-57; prof. sociology University Toronto, 1959-63, University Chgo., 1963—, chairman sociology department, 1965—. Tech. assistance assignments for UN, Burma, 1951, Indonesia, 1953-54, Argentina, 1960, Santiago, Chile, 1963. Board dirs. Social Science Research Council, 1959-64. Fellow Royal Soc. of Can., Royal Statis. Soc.; mem. Am. Statis. Assn. (chmn. social statistics sect. 1961), Canadian Polit. Sci. Assn. (chmn. sociology and anthropology sect. 1961), Inter-Am. Statis. Inst., Internat. Statis. Inst. Contbr. articles profl. jours. Principal work in mathematics of population; use of computer in population analysis. Home: 1200 E. Madison Park, Chgo.*

KEYNES, John Maynard, English economist; b. Cambridge, Eng., June 5, 1883; s. Neville and Florence Ada (Brown) K.; student Eton Sch.; M.A., King's Coll., Cambridge U.; Dr. (hon.) U. Oslo, Columbia; m. Lydia Lopokova, 1925. With India office Brit. govtl. service, 1906-08; mem. Royal Commn. on Indian Finance and Currency, 1913-14; with Brit. Treasury, 1915-19; prin. clk., 1917-19; prin. rep. at Paris Peace Conf., 1919; dep. of chancellor of exchequer on Supreme Allied Econ. Council; mem. Com. on Finance and Industry, 1929-31; editor Econ. Jour., 1912-41; lectr. King's Coll., Cambridge; visited U. S. leaders during Depression for consultation on recovery policy, and after World War II for negotiations on credit to Gt. Britain. Created first Baron of Tilton, 1942; fellow Brit. Acad.; fellow and bursar King's Coll. dir. Bank of Eng.; mem. Econ. Adv. Council; chmn. Arts Theatre, Cambridge; pres. Cambridge Union Soc., 1905; chmn. Arts Council Gt. Britain; pres. Royal Econ. Soc. Author: Indian Currency and Finance, 1913; The Economic Consequences of the Peace, 1920; Treatise on Probability, 1921; Revision of the Treaty, 1922; Monetary Reform, 1924; Laissez-Faire and Communism, 1926; Treatise on Money, 1930; (with Karl Pribram and E. J. Phelan) Unemployment as a World Problem, 1931; Essays in Persuasion, 1932; Essays in Biography, 1933; Means to Prosperity, 1933; General Theory of Employment, Interest and Money, 1936; How to Pay for the War, 1940. Revolutionized econ. theory with ideas for large-scale govt. planning within a free-enterprise framework; believed in flexible govt. expenditure to stabilize nat. income; favored various social programs and progressive taxes as means to more equal distbn. of wealth. Died Tilton, Eng., Apr. 21, 1946.

KEYNES, Richard Darwin, English physiologist; b. London. Eng., Aug. 14, 1919; s. Geoffrey Langdon and Margaret (Darwin) K.; M.A., Cambridge U., 1949, Ph.D., 1949, Sc.D., 1963; m. Anne Pinsent Adrian, Jan. 20, 1945; children—Adrian Maynard, Randal Hume, Roger John, Simon Douglas. Faculty, U. Cambridge (Eng.), 1949-60, lectr. physiology, 1953-60; head physiology dept., A.R.C. Inst. Animal Physiology, Babraham, 1960-64, dir., 1965—. Fellow Trinity Coll., 1948-52; fellow Peterhouse, 1952-60; fellow Churchill Coll., 1961—; fellow Eton Coll., 1963—. Fellow Royal Soc., 1959; mem. Physiol. Soc., Soc. for Exptl. Biology, Am. Biophys. Soc., Biophys. Soc. U.K. Contbr. numerous articles to tech. jours. Pioneered use radioactive isotopes Na24 and K42 in measuring movements sodium and potassium during nerve impulse; research on active transport ions in excitable tissues, mechanism additive discharge in electric organ, biochemistry electric organ, thermodynamics nerve and electric organ. Home: 3 Herschel Rd., Cambridge. Office: A.R.C. Inst. Animal Physiology, Babraham, Cambridge, Eng.*

KEYS, Ancel Benjamin, Am. physiologist; b. Colorado Springs, Colo., Jan. 26, 1904; s. Benjamin Pious and Caroline (Chaney) K.; B.A., U. Cal. at Berkeley, 1925, M.A., 1928, Ph.D., 1930; Ph.D., Cambridge (Eng.) U., 1936; m. Margaret Edith Haney, Sept. 20, 1939; children—Caroline Ann (Mrs. Julian D'Andrea), Henry Michael, Martha Jane. Nat Research fellow, Copenhagen, Denmark, Cambridge, 1930-32; lectr., demonstrator. Cambridge U., 1932-33; instr. Harvard, 1933-36; faculty Mayo Found., 1936-37, asso. prof., 1937-39; faculty U. Minn., Mpls., 1937—, prof. physiology, 1939-46, prof. pub. health, dir. physiol. hygiene, 1946—. Spl. asst. to sec. of war, 1941-43; head mission officer Office Lend-Lease, 1943-44; responsible investigator OSRD, 1943-46; expert cons. WHO, FAO, 1950—. Recipient U. S. Army-Navy Appreciation award; named Comdr. Order of Lion (Finland), 1965; Sr. Fulbright fellow Oxford (Eng.) U., 1951-52; Spl. USPHS fellow Naples (Italy), U., 1963-64; named Hon. citizen Gioia Tauro, Italy, 1957. Hon. mem. N.Y. Diabetes Assn., Acad. Medicine Rome, Acad. Medicine Chirurg. Piceno (Italy), Cardiological Soc. India, Endocrinol. Soc. India; mem. Am. Physiol. Soc., Am. Soc. Biol. Chemistry, Am. Epidemiological Soc., Am. Heart Assn., Am. Inst. Nutrition, Am. Soc. Clin. Nutrition, Am. Pub. Health Assn., Soc. for Exptl. Biology and Medicine, Brit. Nutrition Soc., Am. Coll. Cardiology. Author: (with others) The Biology of Human Starvation, 2 vols., 1950; (with Margaret H. Keys) Eat Well and Stay Well, 1959, rev., 1963; The Benevolent Bean, 1967; also numerous articles. Co-editor: Cardiovascular Epidemiology, 1956. Research on gill function and osmoregulation in fish, high altitude physiology, human starvation, epidemiology diet and heart disease in U. S., S. Africa, Italy, Spain, Finland, Japan, Yugoslavia, Greece, Holland, Eng., Belgium, diet and blood lipids in man. Home: 3270 Owasso Heights, St. Paul 55112. Office: Lab. Physiol. Hygiene, Stadium, U. Minn., Mpls. 55455.*

KEYS, David Arnold, Canadian physicist; b. Toronto, Ont, Can., Nov. 4, 1890; s. David Reid and Erskine (McLean) K.; student Upper Can. Coll., U. Munich, Germany, 1910-11; B.A., U. Toronto, 1915, M.A., 1916; Ph.D., Harvard, travelling fellow, 1920-21; Ph.D., Cambridge U., 1922; D.Sc. (hon.), McMaster U., Hamilton, Ont., 1947, McGill U., 1947, Toronto U., 1953, Ottawa U., 1957; honorary fellow, Trinity College, 1958; LL.D., Mount Allison University, 1958; L.H.D., Lawrence Inst. of Tech., 1959; married May Irene Freeze, 1921; one son, John David. Asst. physicist, later physicist antisubmarine div. Brit. Admiralty, 1918-19; teaching fellow Harvard, 1919; asst. prof. McGill U., 1922-26, asso. prof., 1926-29, professor, 1929-41, Macdonald prof. physics, 1941-47, dir. R.A.F. radio course, 1941-43, Army course, 1943-44; physicist U. S. Bur. Mines, 1927-28; cons. physicist Geol. Survey Can., 1929, geophysicist, 1929-30; mem. council Nat. Research Council Can., Ottawa, 1945, v.p. (sci.), 1947-55 in charge Atomic Energy project, Chalk River, Ont.; chmn. project coordinating com. Atomic Energy of Can. Ltd., 1952-53; sci. adviser to the pres., 1953-66, AECL Overseas rep., London, 1960-61. Canadian rep. on NATO Science Com., 1961-63. Recipient gold medal Canadian Assn. Physicists, 1964. Fellow Royal Soc. Can., Am. Phys. Soc., Chem. Inst. Can. (hon.), A.A.A.S.; mem. Engring. Inst. Can. (hon.). Author: Applied Geophysics (with A. S. Eve), 1929, rev. edit. 1954; Heat, Light and Sound, 1934; College Physics (with R. M. Sutton), 1944, rev. edit. 1955. Research in applied geophysics and piezoelectricity, esp., measurements of explosion pressures in sea water. Home: 817 River Rd., Deep River, Ont., Can.

KEYSER, Cassius Jackson, Am. mathematician; b. Rawson, O., May 15, 1862; s. Jacob B. and Margaret Jane (Ryan) K.; B.S., Ohio Normal U., 1883; studied law, Ann Arbor, Mich., and Kenton, O., 1883-85; B.S.,

U. Mo., 1892, LL.D., 1914; postgrad. U. Mich., 1894; A.M., Columbia, 1896; Ph.D., 1901, Sc.D., 1929; L.H.D., Yeshiva Coll., 1942; m. Ella Maud Crow, Aug. 19, 1885; m. 2d, Sarah Porter Youngman, Apr. 26, 1929. Prin. and supt. pub. schs., Ohio and Mo., 1885-90; instr. math U. Mo., 1891-92, State Summer Schools, Kirksville, Mo., 1892; apptd. Thayer scholar in math. Harvard, 1892; prof. math. State Normal Sch., New Paltz, N.Y., 1892-94; instr. math. Smith Acad. Washington U., St. Louis; 1894-95, Barnard Coll., 1897-1900; tutor, Columbia U., N.Y.C., 1897-1900, instr., 1900-03, adj. prof., 1903-04, Adrian prof. math. 1904-27, head dept., 1910-16, Adrain prof. emeritus from 1927; prof. U. Cal., 1911, 15, exchange prof., 1916. Fellow A.A.A.S.; mem. Am. Math Soc. Author: The New Infinity and the Old Theology, 1915; The Human Worth of Rigorous Thinking, 1916; Mathematical Philosophy, 1922; Thinking About Thinking, 1926; The Pastures of Wonder, 1929; Humanism and Science, 1931; Mathematics and the Dance of Life, 1937. Asso. editor of Scripta Mathematica. Contbd. particularly to geometry, founds. of math. and logic, philosophy of math. Died May 8, 1947.

KEYT, Alonzo Thrasher, Am. physician, physiologist; b. Higginsport, O., Jan. 10, 1827; s. Nathan and Mary (Thrasher) K.; M.D., Med. Coll. Ohio, 1848; m. Susannah Hamlin, 1848, 7 children. Inventor multigraph sphygmometer cardiograph; important contbr. to knowledge of circulation, perfected clin. methods of diagnosis of diseased conditions of circulation; papers published under title Sphygmography and Cardiography, Physiological and Clinical, 1887. Died Nov. 9, 1885.

KHABERMAN, Kharald Martovich, Russian entomologist; b. Dec. 19, 1904; grad. Tartu U., 1931. Asso., Tartu U., 1931-40, prorector, 1945-48; with Estonian Council Min., 1940-45; acad. sec. dept. biol. and med. sci. Estonian Acad. Sci., 1957-64, dir. Inst. Zoology and Botany, 1946—. Mem. Estonian Acad. Sci. Author: Postglacial Colonization of Estonian Territory by Immigrant Fauna and Problems of Zoogeographical Zoning, 1953; The Structure and Dynamics of Lowland Swamp Mesofauna in the Estonian SSR, 1956; The Application of Michurin's Theories in Ecological Geography, 1956; The Inexhaustible Spring, 1962; Observations on the Competition for Food Between Bream, Ruff and Perch, 1963. Research on hydrobiology and entomology, insect fauna of Estonia, fresh-water and salt-water waterways. Address: A USSR, Tallinn, Est. SSR, Akademiya nauk Est. SSR.

KHALEQUE, Kazi Abdul, Pakistani pathologist, physician; b. Jessore, Pakistan, Dec. 1, 1919; s. K. A. Samad and Zobeda Begum; M.B., Calcutta (India) U., 1944; D.B., Manchester (Eng.) U., 1951; postgrad. pathology Group Lab., Central Lab. Research Council, also Inst. Tropical Medicine; m. Begum A. Shamstalaat, Aug. 17, 1944; children—Saad Munira, K. Zulfiquer Mamun. Faculty, Dacca (Pakistan) Medical Coll., 1947—, prof. pathology, 1959—; prof. pathology Inst. Post-Grad. Medicine, Dacca, 1966—. Mem. Pakistan Med. Council, 1964—, Pakistan Med. Resarch Council, 1960-63. Fellow Coll. Pathology (London, Eng.). Author: Practical Pathology, 1954; also articles. Established biol. pregnancy test in Bufo. melanostictus; research on preparation of urease from melon seed and antigen for kala-azar, hepatic granuloma, Ramadhan fasting, hemoglobinopathy, rhinosporidiosis, tropical eosinophilia, anthrax. Home: 145 E, Dhanmondi R.A., Rd. 2, Dacca 2, Pakistan.*

KHALILOV, Zaid Ismailogly, Russian mathematician; b. Jan. 14, 1911; grad. Azerbaijan Pedagogical Inst., Baku, 1932; D.Physico-Math. Sci., 1946. Instr., Tbilisi Inst. Rail Transport Engring., 1932-41; instr. Azerbaijan U., 1940—; prof., 1946—; dir. Inst. Physics and Math., Azerbaijan Acad. Sci., 1950-57, acad. sec. dept. physico-math. and tech. sci., 1962—. Mem. Azerbaijan Acad. Sci. (v.p. 1957-60, pres. 1962—), Azerbaijan Znanie Soc. (chmn. 1962—). Author: The Principles of Functional Analysis, 1949; A Method of Solving Miscellaneous Problems, 1952. Research and publs. on differential and integral equations of math. physics. Address: Azerbaijan Acad. Sci., Baku, Azerbaijan SSR, USSR.

KHANIN, Mikhail Nikiforovich, Russian pathophysiologist; b. 1902; grad. Med. Faculty, Saratov U., 1927; D.Med. Sci. Asst. dept. path. physiology Saratov Med. Inst., 1928-32, lectr., 1932-41; head chair path. physiology Tashkent Med. Inst., 1941—. Mem. learned med. council Uzbek Ministry Health. Mem. All-Union (bd. mem.), Uzbek (bd. mem.), Tashkent City (chmn.) socs. pathophysiologists. Author: Pathogenesis of Circulatory Disorders in Blood Transfusion Shock, 1941; Development of Radiation Sickness in Animals During X-Ray Irradiation, 1962. Mem. editorial council Path. Physiology and Exptl. Therapy. Address: Tashkent Med. Inst., ulitsa Karla Marksa 88, Tashkent, Uzbekistan SSR, USSR.

KHARADZE, Yergeniy Kirillovich, Russian astrophysicist; b. Oct. 31, 1907; D.Physico-Math. Sci. Dir., Abastumani Astrophys. Obs., 1932—; prof. Tbilisi U., 1949—, also rector. Mem. Georgian Acad. Sci. Author: Catalogue of the Light Indices of 14,000 Stars and a Study of Light Absorption in the Galaxy on the Basis of the Color Surplus of Stars, 1952; Astronomy, 1957. Research and publs. on interstellar absorption of starlight, law of spatial distbn. of interstellar matter in galaxy. Address: Tbilisi University, prospect I. Chavchavadze 45, Tbilisi, Gruz. SSR, USSR.

KHARASCH, Morris Selig, chemist; b. Kremenetz, Ukrania, Aug. 24, 1895; s. Selig and Louise (Kneller) K.; S.B., U. Chgo., 1917, Ph.D., 1919; m. Ethel May Nelson, June 24, 1923; children—Robert Nelson, Elizabeth Janet. Chemist, Sprague Meml. Inst.. 1917-18, nat. research fellow in organic chemistry, 1919-22; asso. prof. organic chemistry, U. Md., 1922-24, prof., 1924-28; asso. prof. organic chemistry, U. Chgo., 1928-39; prof. since 1939; Carl William Eisendrath prof. chemistry, 1935—, Gustavus F. and Ann M. Swift distinguished service prof. in Gas Flame Div., U. S. Army, 1918; work on toxic gases at Johns Hopkins, later at Edgewood Arsenal; apptd. by sec. of war cons. to C.W.S., 1926; official investigator Nat. Defense Research Com. since 1941; cooperative expert on thermochemistry for Internat. Critical Tables. Presidential Merit Award, 1947; John Scott medalist for contributions to American agriculture, 1948; Theodore Richards Medalist, 1952. Mem. American Chemical Society, Nat. Acad. Scis., Phi Beta Kappa, Sigma Xi. Awarded patents along pharm. lines and treatment of fungus diseases of small grains, isolation of active principle of ergot (ergotocine), peroxide effect in organic reaction; atom mechanisms in organic reactions. Contbr. tech. articles in chem. jours. Died Oct. 7, 1957.

KHARITON, Yuliy Borisovich, Russian physicist, phys. chemist; b. Feb. 27, 1904; grad. Leningrad Poly. Inst., 1925. With lab. Leningrad Physicotech. Inst., 1921-27; asso. Inst. Chem. Physics, USSR Acad. Sci., 1931—. Decorated Order of Lenin. Mem. USSR Acad. Sci. Co-author: The Problem of the Chain Solution of the Basic Uranium Isotope, 1939; The Chain Decay of Uranium under the Influence of Slow Neutrons, 1940; author: The Problem of Impact Detonation, 1940. Research and publs. on alpha particle scintillation method, elaborated theory of initiation and propagation of explosive detonation, with Ya. B. Zeldovich completed 1st calculation of chain reaction of uranium fission. Address: Inst. of Chemical Physics, USSR Acad. Sci., Vorobevskoe sh. 2, Moscow, USSR.

KHARKEVICH, Aleksandr Aleksandrovich, Russian physicist; b. St. Petersburg, 1904; grad. Leningrad. Electrotech. Inst., 1930. Formerly with Central Radio Lab., Mil. Electrotech. Acad., Inst. Communications Engring. and Physicotech. Inst., USSR Acad. Sci., Leningrad; prof., 1938—; instr. Lvov Poly. Inst., 1944-48; with Inst. Physics, Ukrainian Acad. Sci., 1948-52; asso. Inst. Problems of Information Transmission, USSR Acad. Sci., 1952—, Moscow Inst. Electrotech. Communications, 1952—. Decorated Order of Lenin. Mem. USSR, Ukrainian (corr.) acads. sci. Research and numerous publs. on elec. engring. and communications. Address: Moscow Inst. of Electrotech. Communications, Aviamotornaya ulitsa 8, Moscow, USSR. Deceased.

KHARKEVICH, Dimitri Alexandrovich, Russian pharmacologist; b. Leningrad, USSR, Oct. 30, 1927; s. A. A. and T. D. (Ott) K.; candidat med. scis., First Med. Sch., Leningrad, 1953, Dr. Med. Scis., 1960; m. N. Krilova, Jan. 19, 1953; 1 son, Dimitri. Asst., dept. pharmacology First Med. Inst., Leningrad, 1953-55; research fellow Inst. Pharmacology and Chemotherapy, Moscow, 1955-57; docent dept. pharmacology First Med. Inst., Moscow, 1957-62, prof., 1962-64, chmn., 1964—. Mem. USSR Soc. Pharmacologists, USSR Acad. Med. Sci. (corr. mem., chief dept. scis. coordination 1967—). Author: Ganglionic Agents, 1962; Ganglion-blocking and Ganglion-Stimulating Agents, 1967; Pharmacology of -like Agents, 1967. Research on Pharmacology of autonomic ganglia and neuromuscular transmission, action of ganglion, blocking agts. and muscle relaxants on synaptic transmission, relationship between chem. structure and ganglion or neuromuscular blocking activity. Home: 11-80 Begovaja, Moscow D-284, USSR.*

KHASTGIR, Satis Ranjan, Indian physicist; b. Buswan, Bihar, India, Sept. 5, 1898; s. Satya Ranjan and Souranalini (Ghosh) K.; B.Sc. with honors in Physics, Calcutta (India) U., 1919, M.Sc., 1921; Ph.D. in Physics, Edinburgh (Scotland) U., 1924, D.Sc., 1926; m. Anila Bose, Apr. 14, 1934; 1 son, Shiva Ranjan. Prof. physics Presidency Coll., Calcutta, 1926-27; lectr. U. Coll., Colombo, Ceylon, 1928-30; reader physics Dacca U., 1931-48, physics head, 1945-48, dean Faculty Sci., 1947-48; faculty Banaras Hindu U., 1948-58, prof., head physics dept., 1956-58; Khaira prof., head dept. physics Calcutta U., 1958-63, dean Faculty Sci., 1963; prof., head dept. physics Bose Inst., Calcutta, 1964—. Fellow Nat. Inst. Scis.; mem. Indian Sci. Congress (past pres. physics sect.). Author: (monograph) Wireless, 1950; also numerous articles. Research (with Prof. C. G. Barkla) on J-phenomenon in X-rays, 1922-26; studied crystal rectification, artificial ionosphere, silent A/C discharges, polarization of radio waves in ionosphere, fading of radio signals, ionospheric absorption, waveforms of sferics, frequency spectrum of sferics, VLF-propagation. Address: Purva Palli, Santi ni Ketan, Bolpur, India.*

KHAZANOV, Anisim Timofeevich, Russian pathoanatomist; b. 1903; grad. Leningrad Inst. Med. Knowledge, 1928; D.Med. Sci., 1948. Prosector, Obukhov Hosp., Leningrad, asst. dept. path. anatomy 3d Leningrad Med. Inst., 1932-40; lectr. chair path. anatomy Leningrad Naval Med. Acad., 1940-48, prof., 1948-53; prof., 1948—; head chair path. anatomy Arkhangelsk Med. Inst., 1954-60, Ternopol Med. Inst., 1960—. Chief pathoanatomist Ternopol Oblast, 1960—. Author over 50 publs. including Pathological Anatomy and Pathogenesis of Chronic Unspecific Pneumonia, 1948. Address: Ternopol Med. Inst., Teatralnaya ulitsa 2, Ternapol, Ukrainian SSR, USSR.

KHELKVIST, German Avgustovich, Russian geologist; b. Oct. 5, 1894; grad. Tomsk Technol. Inst., 1923. Employed in oil industry, 1924-50; with various research establishments, 1950-56; prof. Moscow Petroleum Inst., 1956-58; dir. Sakhalin Comprehensive Research Inst., 1957—. Recipient Atlin Prize, 1958, Orders of Lenin, twice. Corr. mem. USSR Acad. Scis. (mem. presidium Siberian dept.). Author: Zonal Oilfields and Prospecting Methods, 1944; The Geological Structure of Zonal Oilfields, 1946; General Geology and the Geology of Oil, 1951; The Geology of Oil and Gas Deposits, 1955; The Principles of Oil and Gas Geology, 1957. Research on formation of oil and gas deposits; prospecting methods; introduced concept of zoned oil beds. Office: Sakhalin Complex Sci. Research Inst., USSR Acad. Scis., Yuzhno-Sakhalinsk, Sakhalin, USSR.

KHITRIN, Lev Nikolaevich, Russian thermophysicist; b. Feb. 20, 1907; grad. Moscow U., 1930. With All-Union Combustion Engring. Inst., 1931-41; instr. Moscow U., 1936—; asso. Power Engring. Inst., USSR Acad. Sci., 1945—; prof., 1953—. Recipient Stalin prize, 1950. Mem. USSR Acad. Sci. (corr.). Author: The Combustion of Carbon, 1949; Basic Characteristics of the Carbon Combustion Process, 1953; Comprehensive Fuel Utilization in Power Technology, 1956; The Physics of Combustion and Explosion, 1957; co-author: Use of Fuel in Power Technology: Ways of Effectively Utilizing Fuel, 1956. Editor: Gas Combustion at Electric Power Plants and in Industry, 1960; Combustion at Reduced Pressures and Some Problems of Flame Stabilization in Single-Stage and Two-Stage Systems, 1960. Research and publs. on combustion physics, theory of heterogeneous combustion, intensive methods of combustion and and comprehensive power technol. methods of fuel utilization. Home: Novopeschanaya utilza 17. Office: Power Engring. Inst., USSR Acad. Sci., Leninsky prospect 19, Moscow, USSR.

KHLOPIN, Nikolái Grigórievich, Russian histologist; b. July 28, 1897; s. G. V. Khlopin; grad. Mil. Medicine Acad., 1921, U. Petrograd, 1922. Mil. medicine, 1921-55; maj.-gen. Med. Forces, USSR Army; staff Oncological Inst., 1928-38, Inst. Exptl. Medicine, 1932-54; staff Oncology Inst., Acad. Med. Scis. USSR, Leningrad, 1955—. Author: General Biological and Experimental Foundations of Histology; also articles in histology and oncology. Recipient Stalin prize, 1947. Developed theory of divergent evolution of tissues, phylogenetic classification of tissue system.

KHODYAKOV, Nikolay Dmitrievich, Russian otorhinolaryngologist; b. Rostov-on-Don, 1898; grad. Med. Faculty, Don U., Rostov, 1924; Cand. Med. Sci., 1935; D.Med. Sci., 1936. Asst. dept. otorhinolaryngology 1st Leningrad Med. Inst., 1930-35; prof., 1935—; head chair otorhinolaryngology Ivanovo Med. Inst., 1935-47, Riga Med. Inst., 1947—. Chif otorhinolaryngologist Latvian SSR, 1948—; mem. learned council Latvian Ministry Health, 1948—. Mem. All-Union (bd. mem. 1958—), Latvian (chmn.) socs. otorhinolaryngologists. Author over 50 publs. including Incidence and Morphology of Pathological Lesions in Experimental Inflammation of the Middle Ear, 1936. Mem. editorial council Herald of Otorhinolaryngology. Address: Riga Med. Inst., b. Podonyu 12, Rita, Latvia SSR, USSR.

KHOLDIN, Semen Abramovich, Russian oncologist; b. Yanovka (now Odessa Oblast), 1896; grad. Med. Faculty, Novorossiysk U., Odessa, 1920, Odessa Clin. Inst., 1923; D.Med. Sci. Asso., Odessa Inst. Path. Anatomy, 1923-26; head 1st surgery dept. Leningrad Inst. Oncology, USSR Acad. Med. Sci., 1926—; asst., later lectr. Leningrad Postgrad. Med. Inst., 1937-44; head chair oncology Leningrad Postgrad. Med. Inst., 1944—. Decorated Order of Lenin. Mem. USSR Acad. Med. Sci. (corr.), All-Union (bd. mem.), Leningrad City and Oblast (learned sec.) socs. oncologists. Author: Hemangiomata and their Treatment, 1935; Electrosurgical Resections and Anastomoses in the Gastrointestinal Tract, 1941; Malignant Neoplasms of the Rectum, 1955. Co-editor: (symposia) Problems of Oncology, Problems of Modern Oncology; mem. editorial bd. Problems of Oncology. Research and numerous publs. on exptl. oncology, cancer of breast, gastrointestinal tract and rectum. Address: Leningrad Postgrad. Med. Inst., ulitsa Saltykova-Shchedrina 41, Leningrad, USSR.

KHOMENTOVSKII, Aleksandr Stephanovich, Russian geologist; ed. Siberian Technol. Inst., Tomsk. Dir. geol. research parties, 1930-37; faculty Krasnoyarsk State Pedagogical Inst., 1938-41; staff Yuzhuraluglerazvedka trust, Orensburg, 1943-54; chmn. geology and useful minerals Saratov State U., 1955-57, Perm Mining Inst., 1957-60; chmn. presidium Far Eastern br. USSR Acad. Scis., 1960—. Mem. USSR

Acad. Scis. Classification, tectonics, formation and distbn. of coal deposits in Siberia and Urals. Office: USSR Acad. Scis., Leninskii Prospekt, 14, Moscow, USSR.

KHORANA, Har Gobind, chemist; b. Raipur, India, Jan. 9, 1922; s. Shri Ganpat Rai and Shrimati Krishna (Devi) K.; B.S., Punjab U., 1943, M.S., 1945; Ph.D., Liverpool (Eng.) U., 1948; m. Esther Elizabeth Sibler, 1952; children—Julia, Emilie, Dave Roy. Head organic chemistry group B.C. Research Council, 1952-60; vis. prof. Rockefeller Inst., N.Y.C., 1958—; prof., co-dir. Inst. Enzyme Research, U. Wis., Madison, 1960—, prof. dept. biochemistry, 1962-64—, Conrad A. Elvehjem prof. life scis., 1964—; vis. prof. Stanford, 1964. Mem. adv. bd. Biopolymers. Recipient Merck award Chem. Inst. Can., 1958, Gold medal Profl. Inst. Pub. Service Can., 1960. Fellow Chem. Inst. Can.; mem. Nat. Acad. Sci. Author: Some Recent Developments in the Chemistry of Phosphate Esters of Biological Interest, 1961. Mem. editorial bd. Jour. Am. Chem. Soc., 1963—. Research and numerous publs. on chem. methods for synthesis of nuecleotides, coenzymes, and nucleic acids; elucidation of the genetic code. Home: Thorstrand Rd., Madison 53705. Office: 1702 University Av., Madison, Wis. 53706.*

KHOTSYANOV, Lev Kipriyanovich, Russian hygienist; b. Pskov, 1889; grad. Imperial Mil. Med. Acad., St. Petersburg, 1911; D.Med. Sci. Dist. pub. health officer, 1918-29; asso., later head lab. indsl. studies Inst. Labor Hygiene and Occupational Diseases, USSR Acad. Med. Sci., 1929-51, head hygiene dept., 1951-66; lectr. Central Postgrad. Med. Inst., 1931-48, head chair labor hygiene, 1948-56, prof., 1935—. Mem. hygiene commn. learned council RSFSR Ministry Health; mem. hygiene com. USSR Ministry Health. Mem. USSR Acad. Med. Sci. Author: Labor Hygiene in the Machine-Building Industry, 1941, 2d edit., 1947; Principles of Industrial Ventilation; Hygienic Principles of Industrial Ventilation and Its Operation; co-author: Labor Hygiene. Mem. editorial bd. Labor Hygiene and Occupational Diseases; mem. editorial council Hygiene and Sanitation; co-editor Hygiene sect. Large Med. Ency., 2d edit.; chief editor Labor Hygiene Manual, 1963. Home: Leningradsky prospect 75a, 34, Moscow A-57. Office: Inst. Labor Hygiene and Occupational Diseases, USSR Acad. Med. Sci., Meyerovsky pr. 31, Moscow, USSR.

KHRENOV, Konstantin Konstaninovich, Russian engr.; b. Feb. 25, 1894; grad. Petrograd Electrotech. Inst., 1918. Lectr., Petrograd Electrotechn. Inst., 1921-25; faculty Moscow Electromech. Inst. R.R. Engrs., 1928-47, prof., 1933-47; with Moscow Higher Tech. Sch., 1931—; staff Inst. Electrowelding, Ukraine Acad. Sci.; prof. Kiev Poly. Inst., 1947—. Recipient Stalin Prize, 1946; named honored scientist of Ukranian SSR. Mem. Ukranian SSR Acad. Scis. (mem. presidium 1953—, sec. tech. sci. dept. 1961—), USSR Acad. Scis. (corr.). Author: Underwater Electrowelding and Cutting of Metals, 1946; An Electric Welding Arc, 1949; The Welding, Cutting and Soldering of Metals, 1952; The Technology of Electric Arc Welding, 1940; Automatic Electric Arc Welding, 1949; New Developments in Informational Technology, 1949; Ceramic Fluxes for Automatic Arc Welding, 1954. Research on electric welding of metals; developed methods for underwater electric welding and metal cutting used in bridge and ship repair. Office: Acad. Scis. Ukranian SSR, Vladimitskaya, Ulitsa 54, Kiev, Ukranian SSR.

KHRISTIANOVICH, Sergei Alekseievich, Russian mech. engr.; b. Nov. 27, 1908; grad. Leningrad U., 1930. With Leningrad Hydrological Inst., 1930-37, Zhukovsky Central Inst. Aerodynamics and Hydrodynamics, 1937-53; with Inst. Chem. Physics, USSR Acad. Sci., 1956—, dir. Inst. Theoretical and Applied Mechanics, Siberian dept., 1963—. Recipient Stalin Prize, 1942, 46, 52, Order of Lenin, 4 times. Mem. USSR Acad. Scis. (mem. presidium 1956—, v.p. Siberian dept. 1961—). Research and publs. on mechanics of liquids and gases, plasticity, aerodynamics, gas flow at supersonic speeds, filtration; solved problem of spreading and reflecting of waves, 1938 and applied it to hydrotech. structures; solved problem of determining surface tension. Office: Siberian Br. of USSR Acad. Scis., Novosibirsk, Siberia, USSR.

KHRISTOV, Georgi Rankov, Bulgarian chemist; b. Gulabnik-Pernik, Bulgaria, Feb. 17, 1896; s. Ranko Khristov and Anastasia (Taseva) K.; grad. U. Sofia (Bulgaria), 1921; m. Gjulka Stoianova Radenkova, June 25, 1927; children—Boyan, Stoyan, Dimitar. Faculty dept. organic chem. tech. U. Sofia, 1922-42, asso. prof. chemistry, 1938-42; head dept. organic chemistry, prof. Inst. Tech., Sofia, 1942-58; dir. Inst. Chemistry, Bulgarian Acad. Scis., 1947-57, head dept. oil and fat chemistry and tech. Inst. Organic Chemistry, 1957-65. Recipient Dimitrov prize for scis., 1961; Orden Cyril and Methodius, 1956; Golden order of labour, 1961. Mem. Union Sci. Workers in Bulgaria, Bulgarian Acad. Scis. Author: Chemistry and Technology of Oil and Fats, 1952; also articles. Research on elaidinization of unsaturated fatty acids and oils using sulfur, phosphorus and alkyl-nitrates, method for refining of vegetable oils using dry washing soda, method for elimination of methoanol from alcoholic liquors, paper chromatographic separation of critical pairs of higher fatty acids. Home: 48, Pozitano St., Sofia 3. Office: Institut of Organic Chemistry, Sofia 13, Bulgaria.*

KHRUSCHOV, Grigorii Knonstantinovich, Russian cytologist, histologist; b. Mar. 3, 1897; grad. Moscow U., 1919. Staff, Moscow U., 1919-33; prof. Moscow Animal-Vet. Inst., 1933-45; dir. Inst. Cytology, Histology and Embryology, 1939-49; prof. 2d Moscow Med. Inst., 1945—; dir. Severtsov Inst. Morphology Animals, 1949—. Recipient Mechnikov Prize, 1949; named Honored Scientist, R.S.F.S.R. Corr. mem. USSR Acad. Scis. Author: Physical Properties of the Living Cell and Methods of Their Investigation, 1930; Role of Leucoyctes in Restoration Processes in Tissue, 1945; Leucocytic Systems of Mammals and Their Evolution, 1951. Research in comparative and exptl. histology and cytology. Died Dec. 22, 1962.

KHRUTSKY, Yevdokim Timofevich, Russian vet. surgeon, physiologist; b. 1903; grad. Vitebsk Vet. Inst., 1928; Cand. Biol. Sci., 1938; D.Biol. Sci., 1950. Vet. surgeon Omsk Vet. Inst., 1928-30, asst., 1930-38, lectr. dept. animal physiology, dir., 1938-40; head chair animal physiology Orenburg Agrl. Inst., 1940, 45—, former dean Vet. Faculty; dep. dir. for sci. work Stavropol Agrl. Inst., 1941-45. Mem. All-Union Soc. Physiologists (chmn. Orenburg dept., editor Proc.), All-Union Znanie Soc. (chmn. Orenburg dept.). Author over 50 publs. including Nervous Humoral Regulation of the Motor Activity of the Multi-Chamber Stomach of Calves and Lambs, 1950; Physiology of Farm Animals. Address: Orenburg Agrl. Inst., ulitsa Chelyuskintsev 18, Orenburg, RSFSR, USSR.

KHUMAL, Arnold Konstantinovich (real name Tudeberg), Russian mathematician; b. Mar. 10, 1908; grad. Tartu U., 1929; D.Physico-Math. Sci., 1946. Instr. Tartu U., 1929-40, prof., 1940—; prof. Tallinn Poly. Inst., 1944-47; dir. Inst. Physics, Math. and Mechanics, Estonian Acad. Sci., 1947-49. Mem. Estonian Acad. Sci. (v.p. 1953—). Author: Über die Theorie und die Anwendungsmethoden der Quadraturreihen, 1933; Ortoganalsystem von Polynomen und Extremumprobleme der Interpolationsrechnung, 1935; Die theoretische Reichweite der Faltungskonstruktionen, 1943; Nomographic Solution of Algebraic Equations, 1954; co-author: Course in Descriptive Geometry, 1947-50. Research and publs. on theory of interpolation, nomography and descriptive geometry. Address: Estonian Acad. Sci., Tallinn, Estonian SSR, USSR.

KHUNRATH, Heinrich (or Kunrath, Kunraht, Cunrath, Cunrad, Conrad), German alchemist; b. Leipzig, Saxony, 1560; student medicine, Basel, grad. 1588. Author: Von Hylealischen . . . chaos, 1598; Magnesia Catholica Philosophorum, 1599; Wahrhafter Bericht von philosophischen Athanor und dessez Gebrauch und Nutzen . . . , 1603; Hochnützliche, unumgärgliche, und gar nothwendige Drey Frazen die . . . Curation . . . Sandes, Grieses, Steins, . . . , 1607; De igne Magorum Philosophorumque secreto externo et visibili . . . , 1608; Amphitheatrum Sapientiae, 1609. Follower of Paracelus; believed in transmutation and prolongation of life by elixir; called alchemical sulphur, terra pinguis unctuosa. Died Leipzig, 1605.

KHUSU, Amaliya Pavlovna, Russian mathematician; b. Markovo, Leningrad Oblast, 1922; grad. Leningrad U., 1949. Asso. Leningrad dept. Math. Inst., USSR Acad. Sci., 1949—. Author: A Mathematical and Statistical Account of the Irregularities of a Surface Profile during Grinding, 1954; Certain Functionals Applied in Technological Processes, 1956; Certain Functionals on Random Fields, 1957; An Accelerated Statistical Method of Checking the Quality of Refractory Products, 1959. Address: Math. Inst., USSR Acad. Sci., Fontanka 25, Leningrad D-II, USSR.*

KIBEL, Ilya Afanasevich, Russian mathematician; b. Oct. 19, 1904; grad. Saratov U., 1925. Asso., Main Geophys. Obs., 1925-43; with Moscow Central Weather Forecasting Inst., 1943—, prof., 1949—. Stalin prize, 1941. Mem. USSR Acad. Scis. (corr.). Author: The Use of Equations of Baroclinic Liquid Mechanics in Meteorology, 1940; Temperature Distribution in the Earth's Sphere, 1943; An Introduction to the Hydro-Dynamic Methods of Short-Term Weather Forecasting, 1957; co-author: Theoretical Hydromechanics, 1963. Research and publs. on hydromech. and dynamic meteorology. 1st to develop and get spl. solutions for simplified closed system of equations of dynamic meteorology; worked out 1st approximation solution of problem precalculating pressure and temperature field for period of several days, 1940; his work has important role in modern weather forecasting methods and synthesis for climatic theory. Address: Moscow Central Weather Forecasting Inst., Bolshevistskaya ulitsa 13, Moscow, USSR.

KIBYAKOV, Aleksey Vasilevich, Russian physiologist; b. 1899; grad. Med. Faculty, Kazan U., 1927; D.Med. Sci., 1939. Asst., Kazan Med. Inst., 1930-36, head chair normal physiology, 1936-56, prof., 1936—; head chair normal physiology Pavlov 1st Leningrad Med. Inst., 1956—. Mem. USSR Acad. Med. Sci. (corr.). Author: The Mechanism of Vasodilation in Stimulation of Antidromic Nerves, 1931; Humoral Transfer of a Stimulus from One Neuron to Another, 1933; The Sympathetic Nervous System and Chromaffin Tissue, 1949; The Nature of the Regulator Effect of the Sympathetic Nervous System, 1950. Research and numerous publs. on function of mediators in nervous activity; discovered significance of mediators in regulating functional condition of central nervous system and peripheral organs; demonstrated role of mediators in transmitting stimuli in inter-nervous synapse by perfusion of superior cervical sympathetic ganglion in cats and spinal cord in frogs. Address: Pavlov 1st Leningrad Med. Inst., ulitsa L. Tolstogo 6-8, Leningrad, USSR.*

KIDD, Benjamin, Brit. social philosopher; b. Sept. 9, 1858; s. Benjamin Kidd; no formal univ. edn.; m. Maud Emma Isabel Perry, 1887; 3 sons. Mem. Staff Inland Revenue Office, resigned civil service after success of first book; travelled for economic studies through U. S., Can., 1898, South Africa, 1902; special corr. London Times. Author: Social Evolution, 1894; The Control of the Tropics, 1898; Principles of Western Civilisation, 1902; contbr. article Application of Doctrine of Evolution to Social Theory, Encyclopedia Brittanica, 1902, also article on sociology, 1911; The Significance of the Future in the Theory of Organic Evolution, 1906; Individualism and After, 1908; The Two Principal Laws of Sociology, 1909. Held that constant struggle exists between individualistic reasoning and group interest, also that religion is needed to curb individual selfishness and foster social progress: opposed socialism, discounted reason. Died Croydon, England, Oct. 2, 1916.

KIDD, John, English physician; b. London, Sept. 10, 1775; s. John and (Burslem) K.; M.D., Guy's Hosp., 1804; B.A., Christ Church, Oxford, Eng., 1797, M.A., 1800, M.D., 1801, M.D., 1804; m. Miss Savery; 4 daus. Chem. reader, Oxford, 1801, 1st Aldrichian prof., 1803-22, became Regius prof. medicine, 1822; physician Radcliffe Infirmary, 1808-26, librarian, 1834-51. Author: Outlines of Mineralogy, 1809; Essay on the Imperfect Evidence in Support of a Theory of the Earth, 1815; Introductory Lecture to a Course in Comparative Anatomy, 1824; Observations on Medical Reform, 1841; Anatomy of the Mole Cricket. Improved sci. teaching at Oxford; enacted Dr. Kidd's Examination Statute for his M.B. degree; discovered napthalene in coal tar (leading to use of coal as a source for chems.). Died Oxford, Sept. 17, 1851.

KIDD, John Graydon, Am. physician, educator; b. Edgewood, Tex., July 20, 1908; s. John Franklin and Rose (Steed) K.; A.B., Duke, 1928; M.D., Johns Hopkins, 1932; m. Maudine Adams, July 23, 1930; children—Maudanne (Mrs. Peter Barriskill), Graydon, Elizabeth, Robert. Intern, Henry Ford Hosp., Detroit, 1932-33, asst. resident, 1933-34; asst. Rockefeller Inst. for Med. Research, N.Y.C., 1934-38, asso., 1938-42, asso. mem., 1942-44; prof. pathology Cornell U. Med. Coll., N.Y.C., 1944—; pathologist in chief N.Y. Hosp., N.Y.C., 1944—. Recipient Eli Lilly award. Fellow A.A.A.S., N.Y. Acad. Medicine (past chmn. sect. microbiology); mem. A.M.A., Am. Soc. Exptl. Pathology (past pres.), Am. Assn. Pathologist and Bacteriologist (past pres.), Am. Assn. Immunologists, Soc. Am. Bacteriologists, Soc. for Exptl. Biology and Medicine, Soc. Gen. Microbiology (Gt. Britain), Am. Assn. Cancer Research, Harvey Soc., N.Y. Path. Soc. (past pres.), Royal Soc. Medicine (Gt. Britain), Path. Soc. Gt. Britain and Ireland, Phi Beta Kappa, Sigma Xi. Editor: The Pathogenesis and Pathology of Viral Diseases; The Dynamics of Virus and Rickettsial Infections. Experimental cancer research in viral etiology, serology, immunology, factors producing regression of lymphoma cells. Home: 229 Pondfield Rd., Bronxville, N.Y. 10708; also Center, Tex. Office: N.Y. Hosp., Cornell Med. Center, 1300 York Av., N.Y.C. 10021.*

KIDDER, George Wallace, Jr., Am. biochemist, microbiologist; b. Oregon City, Ore., Dec. 29, 1902; s. George and Jennie (Moulton) W.; A.B., U. Ore. 1926; M.A., U. Cal. at Berkeley, 1929; Ph.D., Columbia U., 1932; M.A., Amherst Coll., 1949; Sc.D., Wesleyan U., Middletown, Conn., 1950; m. Ruth Elizabeth Rushworth, Aug. 30, 1930; children—George Wallace, III, Beverly Ruth (Mrs. Earl F. Strout), Rushworth Moulton. Instr., Coll. City N.Y., 1929-37; asst. prof. biology Brown U., 1937-46; faculty Amherst (Mass.) Coll., 1946—, Stone prof. biology, 1949—. Mem. microbiology tng. grant com. NIH, USPHS, 1955—; asso. mem. Armed Forces Epidemiology Commn., 1957—. Fellow Am. Acad. Arts and Scis., A.A.A.S., Am. Acad. Microbiology; mem. Am. Soc. Biol. Chemists. Editor: (with G. H. Bourne) Biochemistry and Physiology of Nutrition, 2 vols., 1950; Chemical Zoology, Vol. 1, 1966. Contbr. numerous articles to tech. jours. Research on biochemistry of protozoa, growth inhibitors, chem. syntheses of organic molecules as growth inhibiting compounds, chemotherapy exptl. cancers, new coenzymes, unconjugated pteridines. Home: 567 Main St., Amherst, Mass. 01002.*

KIDDER, Wellington Parker, Am. engr., inventor; b. Norridgewock, Me., Feb. 19, 1853; s. Wellington and Annie (Winslow) K.; ed. Eaton Prep. Sch.; m. Emma Louise Hinckley, Sept. 4, 1878. Patented improvement in rotary steam engines, 1868; studied applied mechanics and drawing, Boston; invented web adjustable press, 1874, which received diploma Mass. Charitable Mechanics Inst., 1878; Kidder press for printing consecutive numbering of railroad and other tickets from continuous roll; invented many other improvements in presses, typewriters, automobile appliances, etc. Cons. engr. Rochester Industries, Inc. (N.Y.), mfrs. of his later improvements in writing machines, from 1923. Died Oct. 2, 1924.

KIDDY, Charles Augustus, Am. animal physiologist; b. Boston, Apr. 14, 1931; s. Charles Stanley and Mary Agnes (O'Neill) K.; B.S., U. Mass., 1951; M.S., U. Wis., 1955, Ph.D., 1958; m. Beverly Elaine Fournier, Dec. 22, 1951; children—Kenneth Charles, Kathleen Mary, Christopher Jay, Thomas Lee, Michael Paul. Dairy herd improvement supr. Worcester County, Mass., 1953-54; research asst. U. Wis., 1954-55, instr. genetics dept., 1955-58; research dairy husbandman dairy cattle research br. animal husbandry research div. Agrl. Research Service, U. S. Dept. Agr., Beltsville, Md., 1958—. Mem. Am. Dairy Sci. Assn., Am. Soc. Animal Sci., A.A.A.S. Research and publs. on immunological aspects of reproductive failure in animals, analysis of composition of cow's milk, genetically controlled variation in milk and blood proteins; co-discoverer genetically controlled variation in alpha si-casein in cow's milk. Office: Nat. Agrl. Research Center, Beltsville, Md. 20705.*

KIDINNU, see Cidenas.

KIDSON, Edward, meteorologist; b. Bilston, Eng., Mar. 12, 1882; s. Charles Kidson; ed. Canterbury Coll., U. New Zealand; M.Sc., D.Sc.; m. Isabel Maria Dann, 1919. Magnetic observer, dept. terrestrial magnetism Carnegie Instn. Washington, 1908-15, 19-21; asst. dir. Australian Meteorol. Bur., 1921-27, dir., from 1927. Mem. Internat. Meteorol. Com. Fellow Royal Soc. New Zealand, Inst. Physics. Contbr. vol. on meteorology to Report Brit. Antarctic Expdn., 1907-09, also papers to jours. Research, work on magnetic survey in S.Am., Australia, also at sea. Died June 12, 1939.

KIDSTON, Robert, Brit. geologist; b. Bishopton, Scotland; s. Robert Alexander Kidston; ed. U. Edinburgh; LL.D., D.Sc.; m. Agnes Marion Christian Oliphant; 2 daus. Fellow Geol. Soc. (Murchison fund 1887, Murchison medal 1916), Royal Soc. Edinburgh (Neill medal 1890), Royal Soc., 1902; mem. several fgn. socs. Author: Catalogue of Palaeozoic Plants in the Geological Collection of the British Museum; Végétaux houilliers dans le Hainaut-Belge dans les collections du Musée royal d'Histoire Naturelle à Bruxelles, 1911; also many papers on carboniferous flora. Died July 13, 1924.

KIELAN-JAWOROWSKA, Zofia, Polish paleontologist; b. Sokolow, Poland, Apr. 25, 1925; d. Franciszek and Maria (Osinska) Kielan; M.Scis., U. Warsaw (Poland), 1949, D.Sc., 1953; m. Zbigniew Jaworowski, May 8, 1958; 1 son, Mariusz. Asst., Zool. Mus., Warsaw, 1945-46; faculty paleontology Warsaw U., 1948-53; asso. prof. Inst. Paleozoology, Polish Acad. Scis., Warsaw, 1953-60, dir., prof., 1961—. Recipient medal Polonia Restituta, 1966; Accademician, Polish Acad. Scis., 1967—. Fellow Soc. Vertebrate Paleontology, Polish Geol. Soc., Paleont. Soc. Author books, monographs, articles. Described new faunas of Devonian and Ordovician trilobites from Poland and Paleozoic Polychaete jaw apparatuses previously almost unknown in fossil state; organizer, supr. Polish-Mongolian paleontol. expdns. to Gobi Desert and Western Mongolia. Home: 141 Bonifacego, Warszawa 34. Office: 93 Zwirki i Wigury, Warszawa 22, Poland.*

KIELHÖFER, Erwin, German chemist; b. Schwegenheim, Germany, Sept. 29, 1899; s. Heinrich and Maria (Silbernagel) K.; student U. Wurzburg, 1919, U. Heidelberg, 1920-21; Ph.D., U. Berlin, 1923; licensed food chemist degree, U. Dresden, 1933; m. Ilse Schondorff, Aug. 5, 1935; children—Bernd, Jutta (Mrs. Andreas Keiser), Hansjörg. Mem. editorial staff Beilstein Handbuch der organischen Chemie, Berlin, 1924; asst. Research Inst. for Viniculture, Neustadt (Palatinate), 1925-26; head Wine Research Inst., Trier, 1927-64; ret., 1965. Mem. Fed. Com. for Wine Research, Accademia Italiana Vite e Vino, Research Assn. German Viniculture (chmn. 1953—). Author: Moderne Kellertechnik, 1935. Research on wine making, including sterilization and stblzn., sulfur dioxide and ascorbid acid, reduction of acidity. Address: 16 Unterm Pulsberg, Trier, West Germany.*

KIELICH, Stanislaw, Polish physicist; b. Czempin, Poland, Nov. 10, 1925; s. Józef and Jadwiga (Kurowska) K.; M.Sc., Poznan (Poland) U., 1955; D.Sc., Polish Acad. Scis., 1960; Docent Habilit., A. Mickiewicz U., 1964; m. Klara Rozwalka, May 7, 1949. Staff, Lab. Dielectrics, Inst. Physics, Polish Acad. Scis., Poznan, 1955—, adj., 1959-62, head research lab. optics of dielectrics, 1962—; docent A. Mickiewicz U., Poznan, 1964—, head molecular physics dept., 1966—. Recipient prize Ministry Higher Edn. for eminent achievements in sci. research, 1965. Fellow Société des Amis des Sciences de Poznan; mem. Polish Phys. Soc. Research and publs. on role of intermolecular interactions in various phenomena of molecular deformation and orientation, theory of variations in electric, magnetic and optical properties of isotropic media due to intense laser light beams, theory of non-linear light scattering and other multiphoton processes. Office: 6 Grunwaldzka, Poznan, Poland.*

KIELLAND, Christian C. G., obstetrician; b. Zululand, 1871; inventor Kielland obstet. forceps with articulation allowing blades to move over each other, circa 1915. Died 1941.

KIELLEY, W. Wayne, Am. biochemist; b. Fessenden, N.D., Feb. 9, 1916; s. William Wesley and Henrietta (Espest) K.; student Duluth Jr. Coll., 1934-35, Duluth State Tchrs. Coll., 1936-37; B.S., U. Minn., 1940, M.S., 1942, Ph.D., 1946; m. Ruth Tomi Kojima, Mar. 8, 1947; children—Tomi Louise, Howard Wayne. UPSHS fellow U. Pa., 1946-49; instr. dept. physiology Tufts U. Med. Sch., 1949-51; biochemist Lab. Cellular Physiology, NIH, Bethesda, 1951-61, head sect. cellular physiology, 1961—. Mem. Am. Soc. Biol. Chemists, Am. Chem. Soc., A.A.A.S., N.Y. Acad. Sci., Sigma Xi. Research, publs. on aerobic energy transfer processes in cells, chem. basis of muscular contraction. Home: 10012 Woodhill Rd., Bethesda 20034. Office: NIH, Bethesda, Me. 20014.*

KIELMEYER, Karl F., German biologist; b. Tübingen, Germany, Oct. 22, 1765; prof. Stuttgart (Germany) Caroline Academy, 1790-96; became curator Stuttgart Mus. Nat. History, 1792; prof. Acad. Tübingen, 1796-1816; named dir. Royal Bot. Gardens, Stuttgart, 1816; asso. with Humbolt; tchr. Cuvier. Author of a book on organic forces in the scale of beings, 1793. Considered father of nature philosophy; 1st to express in physiol. form law of parallelism between stages of individual devel. and stages of scale of beings (later Haeckel formulated it as fundamental biogenetic law); demonstrated that frogs pass through fish stage of devel., that mammals have reptile circulatory system in early devel. Died Sept. 24, 1844.

KIENBOCK, Robert, Austrian radiologist; b. Vienna, Austria, Jan. 18, 1871; Author: Radiotherapie, 1907; Röntgendiagnostik der Knocken-und Gelenskskrankheiten, 1933-42. Pioneered in radiology; described acute atrophy of bone using radiophotography (Kienbock's atrophy), 1901; research on traumatic cavity formation in spinal cord (Kienbock's disease), 1902. Died 1954.

KIEPENHEUER, Karl Otto, German astronomer; b. Weimar, Germany, Nov. 10, 1910; s. Gustav and Irmgard (Funcke) K.; ed. univs. Berlin (Germany), Paris (France); Ph.D.; m. Sabine Thilo, May 15, 1936; children—Jakob, Kaspar, Juliane, Magdalene. Became dir. Fraunhofer Inst. Solar Physics, Freiburg, Germany, 1942. Mem. Internat. Acad. Astronomy, Paris. Author: Die Sonne; also articles. Editor daily internat. sun cards. Research on solar magnetic fields, effect of earth's atmosphere on astron. observations, radio astronomy. Home: Schöneckstrasse 6, Freiburg i. B. Office: Fraunhofer Institut, Freiburg i. B., West Germany.

KIER, Porter Martin, Am. paleontologist; b. Pitts., Oct. 22, 1927; s. Samuel Martin and Mary (Kebler) K.; B.S., U. Mich., 1950, M.S., 1951; Ph.D., U. Cambridge (Eng.), 1954; m. Mary Ellen Lavely, Sept. 15, 1950; children—William McKee, Elizabeth Lavely. Asst. prof. U. Houston, 1956-57; curator Smithsonian Instn., Washington, 1957—. Fulbright scholar, 1951-52. Fellow A.A.A.S., Geol. Soc. Am.; mem. Paleontol. Soc. Research and publs. in systematics of fossil echinoids and living habits of recent echninoids. Home: 5104 Bradford Dr., Annandale, Va. 22003. Office: Smithsonian Instn., Washington 22003.*

KIERLAND, Robert Richard, Am. physician, dermatologist; b. Mahnomen, Minn., June 13, 1910; s. Peter E. and Elsie (Oschner) K.; B.S., U. Minn., 1932, B.M., 1932, M.D., 1933, M.S., 1939; m. Margaret Lytle, Nov. 2, 1934; children—Maricia (Mrs. John McCally), Peter L. Fellow dermatology and syphilology Mayo Found., 1937-40, staff, 1940—, head sect. dermatology, 1962—; prof. dermatology Mayo Found. Faculty, Mayo Grad. Sch. Medicine, U. Minn., Rochester, 1957—. Pres., Dermatology Found., N.Y.C., 1965—. Mem. spl. med. adv. group VA 1964—. Mem. Am. Acad. Dermatology (past sec.-treas., past pres.), Am. Dermatol. Assn. (sec. 1963—), A.M.A. (chmn. sect. dermatology 1966—), Sigma Xi. Editorial bd. Archives of Dermatology, 1966—, Jour. Tropical Dermatology, 1964—, venereal disease br. USPHS, 1939—. Research and numerous publs. on dermatology and syphilology. Home: Mounted Route 72. Office: Mayo Clinic, 200 1st St., S.W., Rochester, Minn. 55901.*

KIERNAN, Francis, surgeon, anatomist; b. Ireland, 1800; ed. St. Bartholomew Hosp.; studied in France, Italy. Gave pvt. demonstrations in anatomy; a founder London (Eng.) U., became mem. senate, examiner anatomy and physiology, 1862. Fellow Royal Soc., 1834 (Copley medal), Royal Coll. Surgeons (became v.p. 1864). Author: The Anatomy and Physiology of the Liver, 1833. Pioneered use of microscope in study of liver physiology and anatomy; described interlobular spaces in liver (Kiernan's spaces), 1833. Died Dec. 31, 1874.

KIERNAN, James George, Am. neurologist; b. N.Y., June 18, 1852; s. Francis and Mary (Aiken) K.; student Coll. City N.Y., 1868-71; M.D., N.Y. U., 1874; m. Jane Ann Trumper, Feb. 10, 1881 (dec. 1903); m. 2d, Grace Cole, Dec. 11, 1917. Asst. physician Ward's Island (now N.Y. State Insane) Hosp., 1874-78, and as officer N.Y. Neurol. Soc. was active in reforms brought about by that soc. in Am. psychiatry and neurology; asst. prof. nervous and mental diseases Northwestern U., Ill., 1881-82; insp. Nat. Bd. Health, 1882; supt. Cook County Insane Hosp., Chgo., 1885-89, and forced investigation of county charities, 1885, which led to the "boodle" trials and convictions of 1887; prof. forensic psychiatry, Kent Coll. of Law, Chgo., 1890-1902; prof. mental and nervous diseases Milw. Med. Coll., 1894-97; prof. neurology Chgo. Post-Grad. Sch., 1903-04; prof. med. jurisprudence, Dearborn Med. Coll., Chgo., 1904; prof. nervous diseases Ill. Med. Coll., 1906-08; prof. mental and nervous diseases Chgo. Med. Sch. Expert for the defense in Guiteau trial, 1881, Mooney trial, 1884, and in many other criminal and civil cases involving medico-legal issues. Died July 1, 1923.

KIESS, Carl Clarence, Am. astrophysicist; b. Ft. Wayne, Ind., Oct. 18, 1887; s. John Frederick and Florence (Fordney) K.; A.B., Ind. U., 1910, Sc.D., 1963; Ph.D. (Lick Obs. fellow), U. Cal., Berkeley, 1913; m. Harriet K. Knudsen, June 21, 1919; children—Margaret Florence (Mrs. Wasley S. Krogdahl), John Anthony (dec.), Norman Halvor, Edward Marion. Physicist, prin. physicist Nat. Bur. Standards, Washington, 1917-57, lectr. advanced optics Postgrad. Sch., 1930-50; professorial lectr. astronomy Georgetown U., Washington, 1947—. Recipient Donohoe medal Astron. Soc. Pacific for discovery of Comet 1911 b, 1911, Meritorious, Exceptional Service medals U. S. Dept. Commerce, 1949, 55, Manhattan Project award, 1946. Fellow A.A.A.S., Am. Phys. Soc., Optical Soc. Am.; mem. Am., Pacific astron. socs., Internat. Astron. Union. Measurement, analysis of wavelengths of spectra of sun, planets, and various elements; studies of energy states of uranium that led to its correct location in periodic table. Home: 2928 Brandywine St., Washington 20008. Office: Georgetown Coll. Obs., Washington 20007.*

KIESSLING, Roland Richard, Swedish metallurgist; b. Boras, Sweden, Mar. 1, 1921; s. Fritz R. and Adéla (Schuback) K.; Fil.Dr. in Chemistry, Uppsala (Sweden) U., 1950; m. Elsa Gustafsson, July 2, 1944; children—Ann-Charlotte, Rolf, Leif. Asst. prof. Uppsala U., 1950-51; research scientist Söderfors Bruk, 1951-54; head metall. dept. Swedish Atomic Energy Comp., 1954-60; prof., dir. Swedish Inst. Metal Research, Stockholm, 1960—. Mem. Swedish Acad. Engring. Scis., Swedish Metallographers Assn. (chmn. 1960—), Am. Inst. Mech. Engrs., Brit. Iron and Steel Inst., Inst. Metals, Am. Soc. Metallurgists. Author: (with N. Lange) Nonmetallic inclusions in Steel, Part I, 1964, Part II, 1966; also numerous articles. Research on metallography, materials properties, crystallography. Home: 37 Höguddsvägen, Lidingö, Sweden. Office: 48 Dr. Kristinas Väg, Stockholm, Sweden.*

KIGOSHI, Kunihiko, Japanese chemist; b. Tokyo, Japan, July 7, 1919; s. Sempachi and Misao (Ito) K.; D.Sc., U. Tokyo, 1953; m. Norko Hayashi, Oct. 14, 1944; children—Masako, Ikuko. Research asst. Phys. and Chem. Research Inst., Tokyo, 1942-46; research asst. Meteorol. Research Inst., Tokyo, 1946-50; asst. prof. Gakushuin U., Tokyo, 1950-54, prof. chemistry, 1954—, dir. Radiocarbon Lab., 1959—. Mem. Chem. Soc. Japan, Japan Analytical Chem. Soc., Am. Geophys. Union, A.A.A.S. Author (in Japanese): (with Y. Miyake) Radiochemistry, 1956; Chronology of Radioactivities, 1965; also articles. Research on radiocarbon datings of geol. and archaeol. samples from Japan, Australia, Pacific islands; studies of relation between hist. variation in atmospheric radiocarbon concentration and variation in strength of geomagnetism. Home: 8-4, Higashi 3, Shibuyaku. Office: 5-1, Mejiro 1, Toshimaku, Tokyo, Japan.*

KIHARA, T., Japanese physicist; b. Tokyo, Japan, May 5, 1917; B.S., Tokyo U., 1941, D.Sc., 1948. Prof. physics U. Tokyo, 1958—. Research on intermolecular forces; Kihara model explains equation of states of gases; self-crystallizing molecular models explain structures of molecular crystals; in plasma physics obtained theoretical expression for transport coefficients of plasmas. Home: Oi 7-23-4 Shinagawa-ku, Tokyo, Japan.*

KIKOIN, Isaak Konstantinovich, Russian physicist; b. Mar. 28, 1908; grad. Kalinin Poly. Inst., Leningrad, 1931. Instr. higher tech. univs., Leningrad, Sverdlovsk, 1931-44; prof. Physics Engring. Inst., 1944—. Decorated Order of Lenin (3); Stalin prize, 1942. Mem. USSR Acad. Scis. Co-author: The Physics of Metals, 1934; Use of a Cluster of Tritium Ions for Determining the Density of Deuterium Plasma, 1963. Research and publs. on elec. and magnetic properties of metals and semi-condrs.; discovered photomagnetic effect in semi-condrs. (Kikoin's Effect), dependence of electric resistance in molten metals on temperature; developer apparatus for measuring strong direct currents, proved that positron charge is absolutely equal to electron charge. Address: Physics Engring. Inst., ulitsa Kirova 21, Moscow, USSR.

KIKUCHI, Chihiro, Am. physicist; b. Seattle, Sept. 26, 1914; s. Naoki and Mitsue (Ichinomiya) K.; B.S. in Physics, U. Wash., 1939, Ph.D., 1944; M.A. in Math., U. Cin., 1943; m. Grace Keiko Fujii, June 8, 1946; children—Naomi, Carl Hideki, Gary Eiji. Instr., Haverford Coll., 1943-44; faculty Mich. State U.,

1944-53; research physicist Naval Research Lab., 1953-55; head Solid State Physics Lab., U. Mich., Ann Arbor, 1955-59, prof. nuclear engring., 1959—; vis. prof. Brookhaven Nat. Lab., Upton, N.Y., 1951-52, Tsing Hua Nat. U., Taiwan, 1964; cons. Atomics Internat., Canoga Park, Cal., 1962, Lincoln Lab., Mass. Inst. Tech., 1963. Research on ruby maser, electron spin resonance, quantum electronics, radiation effects in solids. Home: 260 Sumac Lane, Ann Arbor, Mich. 48105.*

KIKUCHI, Dairoku, Japanese mathematician, seismologist; b. Tokyo, Japan, Mar. 17, 1855; student U. Coll. Sch., London, Eng., 1867-68, 71-73, St. Johns Coll., Cambridge, Eng., 1876-77. Prof. math. Imperial U. Tokyo, 1877-98, pres., 1898-1901; minister edn., 1901-03; pres. Imperial U. Kyoto (Japan), 1908; became pres. Imperial Acad. Tokyo, 1909. Created Baron, 1902. Founder, 1st sec. Earthquake Investigation Com. Mem. Seismol. Soc. Japan (Founder). Studied causes and course of earthquakes, after-shocks. Died 1917.

KIKUCHI, Masashi, Japanese physicist; b. Tokyo, Japan, 1902; s. Dairoku Kikuchi; grad. Tokyo U., 1926; doctorate. Became prof. Osaka U., 1931; also prof. Tokyo U. Recipient Acad. Prize, 1932, medal of culture, 1951. Author: Outline of Atomic Physics; Atomic Physics; Compositions of the Atom. Research in nuclear physics.

KIKUTH, Walther, physician; b. Riga, Latvia, Dec. 21, 1896; s. Martin K. and (Schultz) K.; student medicine Königsberg (now Kaliningrad, USSR), Freiburg; M.D.; m. Helga Timmermann, 1928. Asst., Hamburg Tropic Inst., 1924-29; field work, Brazil, 1927-38; became head Chemo-therapeutic Inst., Farbenfabriken Bayer, 1929; named acting dir. Hygiene Inst., prof. hygiene and microbiology, Düsseldorf, Germany, 1946. Lecture tours in Egypt, Eng., Italy, Spain, Balkans, S.Am. Recipient Gold Medal, 1934; Silver Medal Med. Sci., Jena, Germany; Bernhard Noeht medal. Mem. Royal Soc. Medicine, Tropical Medicine, Kaiserlich Leopoldinisch. Karolinische, Deutsche Acad. Halle, Royal Med. Acad. Barcelona (corr.), Royal Med. Acad. Madrid (corr.). Research and publs. in exptl. therapy, viruses, tropical medicine, gen. hygiene; introduced quinacrine hydrochloride for treatment of malaria, 1932. Home: 224 Himmelgeisterstr., Düsseldorf, Germany.

KILBOURNE, Edwin Dennis, Am. virologist; b. Buffalo, July 10, 1920; s. Edwin I. and Elizabeth (Alward) K.; A.B., Cornell U., 1942, M.D., 1944; m. Joy Schmid, Dec. 20, 1952; children—Edwin Michael, Richard Schmid, Christopher Norton, Paul Alward. Asst., Rockefeller Inst., 1948-51; faculty Tulane U., 1951-55; faculty Cornell U., Med. Coll., N.Y.C., 1955—, prof. pub. health, dir. div. virus research, 1961—; mem. staff N.Y. Hosp., 1955—. Mem. commn. on influenza Armed Forces Epidemiological Bd., 1965—; mem. Health Research Council N.Y.C., 1968—. Mem. Harvey Soc., So., Central (emeritus) socs. clin. research, A.A.A.S., Am. Assn. Immunologists, Soc. Exptl. Biology and Medicine, Am. Soc. Clin. Investigation (emeritus), N.Y. Acad. Sci., N.Y. Acad. Medicine, Am. Pub. Health Assn., Am. Thoracic Soc., Assn. Am. Physicians, Am. Soc. Microbiology, Infectious Diseases Soc. Am. Author: (with Wilson G. Smillie) Preventive Medicine and Public Health, 3d edit. 1963. Research and publs. on hormonal influences, genetic studies, and exptl. transmission of viruses, especially influenza. Home: 446 Hillcrest Rd., Ridgewood, N.J. 07450. Office: 1300 York Av., N.Y.C. 10021.*

KILBY, Clinton Maury, Am. physicist; b. Suffolk, Va., Nov. 1, 1874; s. Wallace and Margaret (Tynes) K.; A.M., Randolph-Macon Coll., 1896; postgrad. U. Chgo., 1903, Columbia, 1904; Ph.D., Johns Hopkins, 1909; m. Jean McDonald Graham, June 11, 1912. Instr. math. Randolph-Macon Coll., 1894-96; prin. pub. schs., 1896-98; master in math. and physics Woodberry Forest Sch., 1898-1905; lecture asst. in physics Johns Hopkins, 1908-09; instr. physics Lehigh U., 1909-10; prof. physics and astronomy Randolph-Macon Woman's Coll., from 1910. Fellow A.A.A.S.; mem. Am. Assn. Physics Teachers, Am. Phys. Soc., Optical Soc. Am., Va. Acad. Sci., Phi Beta Kappa, Gamma Alpha. Author: Laboratory Manual of Physics, 1912; The Constellations, 1918; Definitions and Fundamentals of Physics, 1921; Introduction to College Physics, 1929; Elements of Optics, 1943; Redetermination of Wave-lengths of the Arc and Spark Lines of Titanium, Manganese and Vanadium; The Effect of Capacity and Self-induction on the Wave-lengths of the Spark Lines. Died Mar. 13, 1948.

KILIAN, Charles-Constant-Wilfrid, French geologist; b. June 15, 1862, Schiltigheim, France; prof. geology and mining Faculty Scis., Grenoble, France; made mission to study earth tremors, Andalusia, 1885. Mem. French Acad. Scis. Research and publs. on geology of Alps of Eastern and So. France, geology of Jura, valley of Doubs, mountain of Lure, Upper Jurassic, lower chalk lands of South and of France; contbr. to geol. map of France; studied earthquakes of Andalusia. Died Grenoble, Sept. 30, 1925.

KILLE, Frank Ralph, Am. biologist, ednl. adminstr.; b. Salem, O., Aug. 2, 1904; s. Ellsworth I. and Retta (Vogel) K.; B.S., Coll. Wooster, 1926, LL.D., 1954; S.M., U. Chgo., 1929, Ph.D., 1934; m. Frances Kerby, Aug. 24, 1929; 1 son, J. Frank. Tchr. biology and chemistry Hubbard (O.) High Sch., 1926-28; faculty Birmingham-So. Coll., 1929-31; asst. in zoology, U. Chicago, 1931-33; with U. Rochester, 1933; faculty Swarthmore Coll., 1934-45, asso. prof. zoology, 1943-45; dean men, prof. zoology Carleton Coll., Northfield, Minn., 1945-46, dean coll., prof. zoology, 1946-58; asso. commr. for higher and profl. edn. State of N.Y., Albany, 1958-67; spl. asst. to commr. for sci. and tech. N.Y. State Edn. Dept., Albany, 1967—. Fellow A.A.A.S.; mem. Am. Soc. Zoologists, Am. Inst. Biol. Sci., Am. Assn. Anatomists, Marine Biol. Lab. Corp., Phi Beta Kappa, Sigma Xi. Research on regeneration in Holothurians (sea cucumbers), exptl. embryology. Home: 340 Albany-Shaker Rd., Loudonville, N.Y. 12211. Office: 800 N. Pearl St., Albany, N.Y. 12204.*

KILLEEN, John, Am. mathematician; b. Guam, Mariana Islands, July 28, 1925; s. John Patrick and Amy (Whitfeld) K.; A.B., U. Cal., Berkeley, 1949, M.A., 1951, Ph.D., 1955; m. Marjorie W. Lyman, Feb. 18, 1950; children—Michael, Joan, Sean, Katharine, Jack, Ann. Mathematician, Lawrence Radiation Lab., Berkeley, Cal., 1950-55, Livermore, Cal., 1957—; tech. staff Bell Telephone Labs., Murray Hill, N.J., 1955-56; research asso. Courant Inst. Math. Scis., N.Y. U., 1956-57. Cons., NASA Goddard Space Flight Center, 1961—, Advanced Kinetics, Costa Mesa, Cal., 1965—; lectr. U. Cal., Davis, 1963—. Mem. Am. Phys. Soc., Am. Geophys. Union, Am. Math. Soc., Math. Assn. Am., Sigma Xi. Contbr. articles to sci. jours. Application of computational methods to problems in plasma physics and controlled fusion research. Home: 1528 Campus Dr., Berkeley, Cal. 94708. Office: Box 808, Livermore, Cal. 94551.*

KILLIAN, Gustav, German laryngologist; b. Mainz, Germany, June 2, 1860; M.D., Freiburg, Germany; studied laryngology under Jurasz, Heidelberg. Lectr. laryngology, Freiburg; became prof. laryngology, Berlin, Germany, 1897. Contbr. numerous articles on nose, ethmoidal region, frontal sinus to tech. jours. Pioneered in bronchology and laryngology; introduced direct bronchoscopy; 1st to remove fgn. body from bronchi; originated numerous operations for diseases of throat and nose passages, including suspension laryngoscopy, 1912; modified esophagoscope for removal aspirated bone from trachea. Died Feb. 24, 1921.

KILLIAN, Hans Frans Edmund, German surgeon; b. Fribourg, Aug. 5, 1892; s. Gustav and Helen (Hein) K.; ed. univs. Fribourg, Munich; M.D.; m. Trude Bornhauser, Sept. 23, 1939; 2 sons. Asst., Munich, also Robert Koch Inst., Berlin; vol. Pharmacological Inst., Munich; asso. prof. surgery clinic U. Fribourg; dir., head surg. clinic U. Breslau; cons. surgeon 16th army, Lower Silesia; chief surgeon, dir, physician-in-chief Baden-Baden Hosp. Mem. Schlaraffia Soc. Author 12 books on medicine, other works on gen. surgery, anesthesiology. Home: Riedbergstrasse 24, Frieburg/Br. Office: Krankenhausstrasse 83, Donauerschingen/Baden, W. Germany.

KILMISTER, Clive William, Brit. mathematician; b. Epping, Essex, Eng., Jan. 3, 1924; s. William and Doris (Cookson) K.; B.Sc., Queen Mary Coll., U. London, 1944, M.Sc., 1948, Ph.D., 1950; m. Peggy Joyce Hutchins, Mar. 18, 1955; children—Andrew Clive, Sally Jane, Penelope Jill. Faculty, King's Coll., U. London, 1950—, prof. math., 1966—. Fellow Inst. Math.; mem. Brit. Soc. for Philosophy Sci. (treas.), London Math. Soc., Edinburgh Math. Soc., del Circolo Matematico di Palermo. Author: (with G. Stephenson) Special Relativity for Physicist, 1958; (with B. O. J. Tupper) Edington's Statistical Theory, 1963; Hamiltonian Dynamics, 1964; (with J. E. Reeve) Rational Mechanics, 1966; The Environment in Modern Physics, 1965; Men of Physics: Sir Arthur Eddington, 1966; Language, Logic and Mathematics, 1967; also articles. Editor: Generalized Electro-dynamics and the Structure of Matter (S. R. Milner), 1963. Studies on modern physics (quantum theory and relativity), founds. of each part of physics. Home: 11 Vanbrugh Hill, Blackheath, London, S.E.3., Eng.*

KIMATA, Masao, Japanese microbiologist; b. Tokyo, Japan, Nov. 18, 1907; s. Seiichi and Mitsu (Kobayashi) K.; grad. Imperial Fisheries Inst., Tokyo, 1930; D.Sc., Hokkaido Imperial U., 1942; m. Kimiko Oya, Jan. 3, 1941; Asst. bacteriology Imperial Fisheries Inst., 1930-35, asst. prof., 1935-46, prof., 1946—; prof. microbiology, dept. fisheries Kyoto U., Maizuru, Kyoto, Japan, 1947—, chief dept. fisheries, 1966—. Author: Science of Food Putrefaction, 1944; Science of Food Preservation, 1949; also numerous articles. Research on marine microbiology and food microbiology. Home: 964 Amarube-shimo, Maizuru, Kyoto. Office: Nagahama, Maizuru, Kyoto, Japan.*

KIMBALL, Allyn Winthrop, Am. statistician; b. Buffalo, Oct. 2, 1921; s. Allyn W. and Ethel (Manson) K.; B.S., U. Buffalo, 1943; postgrad. (fellow) Mass. Inst. Tech., 1943; Ph.D., N.C. State Coll., 1950; m. Evelyn Marie Lay, June 16, 1944; children—Keith Allan, Lynn Ellen. Instr., U. Buffalo, 1946; analytical statistician USAF Sch. Aviation Medicine, Randolph Field, Tex., 1948-50; chief statistics sect., math. panel Oak Ridge Nat. Lab., 1950-60; prof., chmn. dept. biostatistics, prof., chmn. dept. statistics, prof. biomath. Johns Hopkins, Balt., 1960-66, dean Faculty Arts and Scis., 1966—. Cons. USPHS, 1960—. Trustee Asso. Univs., Inc., 1962—. Fellow Am. Statis. Assn., A.A.A.S.; mem. Inst. Math. Statistics, Biometric Soc. Contbr. articles to tech. jours. Research on biomed. phenomena through math. and statis. analysis. Home: 1106 Hampton Garth St., Towson, Md. 21204. Office: Charles and 33d Sts., Balt. 21218.*

KIMBALL, George Elbert, Am. chemist; b. Chicago, July 12, 1906; s. Arthur Gooch and Effie Gertrude (Smallen) K.; B.S., Princeton, 1928, A.M., 1929, Ph.D., 1932; m. Alice Thurston Hunter, June 22, 1936; children—Prudence Bradstreet, Thomas Redington, Susanna Goodhue, Martha Smallen. Nat. research fellow, Mass. Inst. Tech., 1933-35; instr. physics, Hunter Coll., 1935-36; instr. and asst. prof. chemistry, Columbia, 1936-46; asso. prof., 1946-47, prof. chemistry, 1947-56; sci. adv. Arthur D. Little, Inc., 1956-61, v.p. 1961—. Examiner, reader coll. entrance exam. bd. NRC com. on contract catalysis in operations research since 1952. Dep. dir. Operations Research Group, Office Chief of Naval Operations; alternate mem. Joint Army-Navy Vision Com., 1945-46; mem. panel on underwater ord., 1945-51, Office Sec. of Def. since 1948; cons. Office Chief of Naval Operations since 1946, Office Sec. of Defense since 1949. Reilly lecturer, U. of Notre Dame, 1949. Mem. sci. adv. group U. S. Army Combat Devel. Command. Fellow American Phys. Society, New York Academy of Science. Member Operations Research Society America (president 1964-65), National Academy Scis.; Am. Chem. Society, U. S. Army Science Adv. Panel, Statis. Adv. Com. of U. S. Census, Phi Beta Kappa, Sigma Xi. Republican. Unitarian (trustee and president Hackensack Church 1948-51). Club: Cosmos (Washington). Author: Quantum Chemistry (with H. Eyring and J. Walter), 1944; Methods of Operations Research (with P. M. Morse), 1951. Asso. editor Jour. of Chemical Physics. Contbr. numerous articles to scientific jours. Research on electrochemistry; high polymers; reaction kinetics; molecular structure; quantum mechanics; operations research. Home: 20 Everett Av., Winchester, Mass. Office: Arthur D. Little, Inc., 35 Acorn Park, Cambridge 40, Mass.

KIMBALL, Gilman, Am. surgeon; b. New Chester, N.H., Dec. 8, 1804; s. Ebenezer and Polly (Aiken) K.; M.D., Dartmouth, 1827; studied surgery in Paris with Guillaume Dupuytren, 1829; m. Mary Dewar; m. 2d, Isabella Defries; 1 child. Surgeon, Lowell (Mass.) Corp. Hosp.; prof. surgery Vt. Med. Coll., 1844; prof. surgery Berkshire Med. Coll., Pittsfield, Mass., 1845. Vice pres. Mass. Med. Soc., 1878; pres. Am. Gynecol. Soc., 1882-83. Pioneer in gynecol. operations; 1st surgeon to remove tumor of uterus by abdominal incision; performed 1st successful removal of ovarian tumor. Died July 27, 1892.

KIMBALL, James Henry, Am. meteorologist; b. Detroit, Feb. 12, 1874; s. Charles Henry and Alice (Jordan) K.; student Mich. Agrl. Coll., 1891-94, B.S., 1912; student U. Mich., 1894-95; M.A., U. Richmond, 1914; Ph.D., N.Y. U., 1926; hon. Sc.D., Mich. State Coll., 1934. With U. S. Weather Bureau from 1895, serving in all sections of U. S. and in West Indies, finally in charge N.Y. office; meteorol. adviser, New York, for French High Commn., and instr. for flying units, 1917-18; faculty lectr. on aero. meteorology N.Y. U., 1941. Fellow Am. Meteorol. Soc., Inst. Aero. Scis.; mem. N.Y. Acad. Scis., Nat. Inst. Social Sciences (v.p.), D'Honneur Ligue Internationale des Aviateurs. Decorated Legion of Honor, France; Crown of Italy. Prepared first North Atlantic weather maps for transatlantic flying; cons. meteorologist for Lindbergh and subsequent Am. and European transatlantic fliers. Died Dec. 21, 1943.

KIMBALL, Richard Fuller, Am. biologist; b. Balt., Feb. 1, 1915; s. Edward N. and Marjorie (Pennington) K.; A.B., Johns Hopkins, 1935, Ph.D., 1938; m. Isabel Jepson Carter, Sept. 7, 1940; children—Isabel Jepson, Sherry Norris. Sterling fellow Yale, 1938-39; faculty Johns Hopkins, 1939-47; biologist Oak Ridge Nat. Lab., 1947-66, dir. biology div., 1967—. Lectr., U. Tenn., 1954—. Guggenheim fellow Karolinska Inst., Stockholm, Sweden, 1957-58. Fellow A.A.A.S.; mem. Genetics Soc. Am., Radiation Research Soc., Am. Soc. Cell Biology, Soc. Protozoologists, Am. Soc. Zoologists, Am. Soc. Naturalists, Research Soc. Am., Soc. for Study Evolution, Internat. Assn. for Radiation Research (sec.-treas. 1966—). Asso. editor Radiation Research, 1957-58; editorial bd. Jour. Protozoology, 1963—, Mutation Research, 1964—. Contbr. numerous articles to tech. jours. Research on genetics of mating types in ciliates, radiation and chem. mutagenesis in Paramecium, repair premutational damage, radiation biology of ciliates, autoradiographic and cytophotometric studies on cell cycle in ciliates. Home: 102 Norton Rd., Oak Ridge 37830. Office: Biology Div., Oak Ridge Nat. Lab., P.O. Box Y, Oak Ridge 37830.*

KIMBARA, Atsushi, Japanese elec. engr.; b. Hamamatsu, Japan, Mar. 2, 1902; grad. Tokyo U., 1925; postgrad., Eng., Germany, France, Am. Became engr., Communications Ministry, 1925; named prof. dept. tech. Nagoya U., 1941, head Research Inst. Atmospherics, 1941; participant joint research project on atmospherics, 1928—. Author: Devices for Removal of Atmospherics. Research on wireless transmission of pictures, 1936; removal of communication jamming, typhoon forecasting.

KIMBLE, George Herbert Tinley, geographer; b. London, Eng., Aug. 2, 1908; s. John Herbert and Minnie Jane (Dickerson) K.; B.A., King's College, London, 1929, M.A., 1931; Ph.D. summa cum laude, U. of Montreal, 1948; m. Dorothy Stevens Berry, July 20, 1935; children—Stephen, Gillian. Came to U. S., 1950. Lecturer in geography, Univ. Coll. of Hull, Eng., 1931-36, Univ. of Reading, Eng., 1936-39; first prog. of geography and chmn. dept., McGill U., 1945-50; dir. Am. Geog. Soc., N.Y. City, 1950-53; director survey Tropical Africa, 20th Century Fund, 1953-60; Rushton Foundation lecturer, 1952; Borah Foundation lecturer U. Ida., 1956; Haynes Found. lectr. U. Redlands, 1966; professor geography Ind. U., 1957—, chmn. dept., 1957-62; research dir. U. S. Geography Project, 20th Century Fund, 1962—. Fellow Royal Geog. Soc., Royal Meteorol. Soc., Am. Meteorol. Soc., Hakluyt Soc.; sec., treasurer Internat. Geographical Union, 1949-56; chairman of commission on Humid Tropics, 1956-61; member of Association American Geographers. Author: Geography in the Middle Ages, 1938; The World's Open Spaces, 1939; The Shepherd of Banbury, 1941; Canadian Military Geography, 1949; The Weather, 1951; The Way of the World, 1953; Our American Weather, 1955; Le Temps: D'hier, D'aujourd'hui, de Demain, pub. 1957; Tropical Africa, 2 volumes, published 1960. Editor: Esmeraldo de Situ Orbis, 1937; (with D. Good) Geography of the Northlands, 1955. Studied movement of Doldrum Belt in tropics for naval maneuvers; studied wave and wind conditions on European beaches; assisted in establishment Arctic Research Center. Home: R.F.D. 1, Solsberry, Ind. Office: Geography Dept. Ind. U., Bloomington, Ind.

KIMELDORF, Donald Jerome, Am. physiologist; b. Salt Lake City, Jan. 9, 1920; s. Samuel H. and Esther (Miller) K.; B.A., Reed Coll., 1942; M.A., U. Ore., 1944; Ph.D., U. Cal. at Los Angeles, 1947; m. Fay Irene Tamkin, Sept. 5, 1943; children—Martin Ralph, Howard Alex, Lloyd Jerome. Research fellow U. Cal. at Los Angeles, 1946-47; instr. biology U. Ore., Eugene, 1948; sr. investigator physiology U. S. Naval Radiol. Def. Lab., San Francisco, 1948-51, head physiology sect., 1951-54, head physiology-psychology br., 1955-67, acting head biomed. div., 1965-66; prof. radiation biology Ore. State U., 1967—. Mem. A.A.A.S., Am. Physiol. Soc., Radiation Research Soc., Sigma Xi. Author: (with E. L. Hunt) Ionizing Radiation: Neural Function and Behavior, 1965; also numerous articles. Asso. editor Radiation Research, 1964—. Discovered effects ionizing radiations on nervous system; research on use of radiation as conditional stimulus to motivate behavior, detection small doses radiation by living systems, long term and persistent effects radiation on growth, metabolic functions, tumors, cataracts, life span, radiation induced alterations in capacity to withstand environmental stresses. Address: Radiation Center, Ore. State U., Corvallis, Ore. 97331.*

KIMMEL, Donald Loraine, Am. neuroanatomist; b. Juniata, Pa., July 12, 1906; s. Jesse Melvin and Clara (Cutshall) K.; B.S., Gettysburg Coll., 1929; M.S., U. Mich., 1935, Ph.D., 1938. m. Jessie Elizabeth Brown, June 16, 1934; children—Donald Loraine II, Charles B., Richard O. High sch. sci. tchr., Swedesboro, N.J., 1929-36; food chemist Aurff Packing Corp., 1936; fellow in anatomy U. Mich., 1936-37, demonstrator, 1937-38; instr. La. State U. Sch. Med., 1938-42, asst. prof., 1942-43; asso. prof. anatomy Baylor U. Med. Sch. 1943; asso. prof. Temple U. Sch. Med., 1943-50, prof., 1950-57; prof., chmn. dept. anatomy Chgo. Med. Sch., 1957-66; prof., chmn. dept. anatomy W.Va. U. Med. Center, 1966—. Recipient A.M.A. award for demonstration of urinary bladder innervation, 1955. Mem. Am. Assn. Anatomists, A.A.A.S., Am. Neurol. Assn., Acad. Neurology, Phi Beta Kappa, Sigma Xi, Beta Beta Beta. Author: (with R. N. Baillif) The Structure and Function of the Human Body, 1945. Studies and publs. on devel. of nervous system with particular emphasis on embryological devel. of cranial and spinal nerves, and on devel. of sympathetic nervous system; studies on nerve regeneration and on recovery of bladder function following damage to and regeneration of its nerve supply; studies on sensory pathways in brain and spinal cord. Home: 609 Valley View St., Morgantown, W.Va. 26505.*

KIMOTO, Seiji, Japanese surgeon; b. Hiroshima, Japan, 1907; grad. Tokyo U., 1931; M.D., 1939. Named asst. prof. med. dept. Tokyo U., 1944, prof., 1953—. Research and publs. in heart, blood vessel and other internal organs; surg. treatment of heart diseases, 1943; originated hibernation anesthesia (using selective refrigeration of brain) for visible operation of heart; conducted operations to cut away aneurysm of aorta.

KIMURA, Katsumi, Japanese chemist; b. Okazaki, Aichi, Japan, Aug. 31, 1931; s. I. and K. (Kaneiwa) K.; B.S. Nagoya U., 1954, M.S., 1954, Dr.S., 1959; m. F. Goto, May 3, 1956; children—Keiko, Isao. Researcher, U. Tokyo (Japan), 1959-63; prof. dept. chemistry Faculty Engring. Sci., Osaka U., Toyonaka, Osaka, Japan, 1963—; research asso. Cornell U., Ithaca, N.Y., 1960-62. Mem. Chem. Soc. Japan (prize 1966). Research, publs. on determination of molecular structures of various organic and inorganic compounds in gaseous phase by an electron diffraction method; investigated experimentally and theoretically electronic absorption spectra of benzene derivatives and free radicals. Home: Godo-Shukusha-742, Nozue-1, Itami, Hyogo. Office: 1-1, Machikaneyama, Toyonaka, Osaka, Japan.*

KIMURA, Motoharu, Japanese physicist; b. Saitama Pref., Japan, June 23, 1908; s. Ryojiro and Rin (Tominaga) K.; grad. Tokyo (Japan) Imperial U., 1932, Ph.D., 1945; m. Masako Yamauti, June 20, 1939; children—Yuko, Matiko, Mihoko. Research asst. Earthquake Research Inst., Tokyo U., 1932-34; research mem. Inst. Phys. and Chem. Research, Tokyo, 1934-50; prof. physics Faculty Sci, Tohoku U., Sendai, Japan, 1950—, prof. nuclear physics Grad. Sch. Nuclear Sci., 1957—, dir. Lab. Nuclear Sci., 1966—. Mem. Phys. Soc. Japan. Author several books on neutron physics, nuclear physics, and radiation; also numerous articles. Research on method of gravity comparison by wireless, neutron physics, photonuclear reactions, resonance scattering of epithermal neutrons, extinction effect in diffractive scattering of thermal neutrons by crystals; constrn. of 300 MeV linear electron accelerator. Home: 13-20 Mukai Yama 3, Sendai, Japan.*

KIMURA, Motoo, Japanese geneticist; b. Okazaki, Japan, Nov. 13, 1924; s. Itsusaku and Kana (Kaneiwa) K.; M.Sc., Kyoto U., 1947; Ph.D., U. Wis., 1956; D.Sc., Osaka U., 1956; m. Hiroko Mino, Jan. 15, 1957; 1 son, Akio. Asst., Kyoto U., 1947-49; research mem. Nat. Inst. Genetics, Mishima, Japan, 1949-57, lab. head, 1957-64, head dept. population genetics, 1964—; vis. mem. U. Wis. Math. Research Center, 1961-62, vis. prof., 1966; vis. prof. U. Pavia, Italy, 1963, 65. Recipient Weldon Meml. prize, 1965. Mem. Genetics Soc. Japan (prize 1959), Biometrics Soc. Author: Outline of Population Genetics, 1960; Diffusion Models in Population Genetics, 1964; also numerous articles. Succeeded for 1st time in constructing mathematically entire process of change in gene frequency in finite populations under natural selection; contbns. toward modern devel. of math. theory of population genetics. Home: 7-24 Kiyozumi-cho, Mishima. Office: Nat. Inst. Genetics, Yata, Mishima, Shizuoka-ken, Japan.*

KINARD, Frank Efird, Am. physicist; b. Newberry, S.C., Jan. 15, 1924; s. James C. and Katherine (Efird) K.; A.B., Newberry Coll., 1946, A.B., 1947; M.S. in Physics, U. N.C., 1950, Ph.D., 1954; m. Mary Angelyn McNease, July 25, 1952; children—Sally Garner, Anne Dreher, James McNease. Instr. physics U. N.C., 1949-52; physicist E. I. DuPont de Nemours and Co., Savannah River Lab., Aiken, S.C., 1953-63, head univ. relations office, 1963-67; exec. dir. S.C. Commn. on Higher Edn., Columbia, 1967—. Mem. Am. Nuclear Soc. (mem. edn. com. 1965—), Am. Phys. Soc., Sigma Xi. Contbr. articles to tech. jours. Research on electron energy distbn. in sea-level cosmic radiation, neutron reactor physics. Home: 801 Albion Rd. Office: S.C. Commn. on Higher Edn., 2712 Millwood Av., Columbia, S.C. 29205.*

KINARD, Fredrick William, Am. physiologist; b. Leesville, S.C., Oct. 14, 1906; s. Jacob William and Annie (Schroder) K.; B.S., Clemson U., 1927; M.S., U. Va., 1932, Ph.D., 1933; M.D., U. Tenn., 1945; m. Lina Barr, Sept. 12, 1929; children—Anne DeZou (Mrs. T. E. Davenport), Fredrick William. Faculty, Med. Coll. S.C., Charleston, 1927-30, 33—, prof. physiology, 1953—, dean Sch. Grad. Studies, 1965—. Joint recipient Jefferson Research medal S.C. Acad. Sci., 1936. Mem. Am. Soc. Zoology, Am. Physiol. Soc., Soc. Exptl. Biology and Medicine, Sigma Xi, Alpha Omega Alpha, Rho Chi, Phi Kappa Phi. Research in blood borne factors in red and white blood cell formation; toxicology of tungsten compounds; ethyl alcohol metabolism including liver enzyme participation. Home: 472 Huger St., Charleston, S.C. 29403.*

KINCAID, John Franklin, Am. chemist; b. Blackwell, Mo., Feb. 27, 1912; s. John Randall and Rose (Rich) K.; A.B., Central Coll., Fayette, Mo., 1934; M.A., George Washington U., 1936; Ph.D., Princeton, 1938; m. Nancy Ange, June 28, 1938; children—James Randall, John Peter, Thomas Franklin. Tchr., U. Rochester, 1938-42; group leader Explosives Research Lab., Carnegie Inst. Tech., 1942-45; research scientist Gen. Electric Corp., 1945-46; supr. Rohm & Haas Co., Phila., 1946-58; div. head Inst. for Def. Analyses, Washington, 1958-62; cons. research and devel. Internat. Minerals and Chem. Corp., Skokie, Ill., 1962-63, v.p., 1963—; dir. Kabeyun, Inc. Chmn., U. S. Naval Propellant Plant Adv. Bd. Recipient Naval Ordnance Devel. award, 1945; Presdl. Certificate of Merit, 1948. Mem. Am. Chem. Soc., Indsl. Research Inst., Newcomen Soc., Research Dirs. Assn. Chgo. Research and publs. on nature of liquids, toxicology and mechanism of chem. reactions, including both theoretical and exptl. studies; discovered new processes for mfr. of nickel carbonyl and for preparation of solid propellant grains in large sizes. Home: 620 Chatham Rd., Glenview, Ill. 60025. Office: 5401 Old Orchard Rd., Skokie, Ill. 60076.*

KINCL, Fred A. Am. chemist; b. Everett, Mass., Dec. 13, 1922; s. George V. and Ludmila (Zidlicka) K.; student Poly. Inst., Prague, 1941-43; RN Dr., Charles U., Prague, 1948, D.Sc., 1967; m. Lada Stemberg, June 21, 1946; children—Peter A., Norman M. Jr. sci. officer Nat. Chem. Lab., Poona, India, 1950-53; research chemist, research coordinator Syntex, S.A., Mexico, 1953-62; asst. dir. vet. medicine Syntex Corp., Palo Alto, Cal., 1962-66; asst. dir. biomed. div. Population Council, Rockefeller U., N.Y.C., 1967—. Syntex Research fellow reprodn. physiology Hormonlaboratoriet, Karolinska Sjukhuset, Stockholm, Sweden. Mem. Am. Chem. Soc., Chem. Soc. (London), Mexican Endocrine Soc., Royal Soc. Medicine, N.Y. Acad. Scis. Co-author: Methods in Hormone Research, 1965; Internation Encyclopedia of Pharmacology and Therapeutics. Contbr. numerous articles to profl. jours. Research on preparation of synthetic steroid hormones, pharmacology of synthetic steroid hormones, mechanism of action of anti-fertility drugs. Home: 500 E. 77th St., N.Y.C.*

KINDLE, Edward Martin, Am. paleontologist, geologist; b. nr. Franklin, Ind., Mar. 10, 1869; s. Martin V. and Tabitha Ann K.; A.B., Ind. U., 1893, LL.D., 1939; M.S., Cornell, 1896; Ph.D., Yale, 1899; m. Margaret Ferris, Dec. 31, 1901; children—Winona Helen, Leroy Ferris, Cecil Haldane, Edward Darwin, Virginia Tomlinson, Margaret Crane, Madeleine Barton, Katharine, Charlotte. Instr. geology Ind. U., 1894-95; mem. Cornell expdn. to Greenland, 1896; asst. geologist Ind. Geol. Survey, 1898-1900; asst. geologist U. S. Geol. Survey, 1900-08, paleontologist, 1908-10, geologist, 1911; paleontologist in charge of paleontology, Geol. Survey of Can., Ottawa, Ont., 1912-38. Spl. lecturer in geology U. London (Eng.), 1928. Fellow Geol. Soc. Am., Am. Geog. Soc., Royal Soc. of Can. Author: (with A. K. Miller) Bibliographic Index of North American Devonian Cephalopoda. Research and publs. on Devonian formations; pioneer in research on characteristics and conditions of accumulation of sedimentary deposits, particularly recent sediments of lakes and seas. Died Aug. 29, 1940.

KINDLER, Hermann Julius Karl, German chemist; b. Lissa-Breslau, Sept. 7, 1891; s. Carl and Hedwig (Pauli) K.; Ph.D., U. Breslau; m. Margarethe Kübel, Nov. 8, 1924; children—Hildegard, Wolfgang. Became agrégé, U. Hamburg, 1923, dir. dept. chemistry and pharmacology, 1928, prof., dir. pharmacological-chem. dept., 1948—; prof., dir. pharmacological-chem. dept. U. Innsbruck (Austria), 1941-48. Mem. Gesellschaft Deutscher Naturforscher und Ärzte. Author: Studien über den Mechanismus chemischer Redaktionen unter besonderer Berücksichtigung von Oxydo-Reduktionen; Neue Wege zum Aufban von pharmakologisch wichtigen Verbindungen. Home: Bileweg 3, Wentorf bei Hamburg. Office: Schloss Reinbek bei Hamburg, W. Germany.

KINDRED, James Ernest, Am. histologist; b. Boston, Oct. 20, 1893; s. Ernest Winters and Etta (Woodburn) K.; A.B., Tufts Coll., 1914; A.M., U. Ill., 1915; Ph.D., 1918; m. Bessie Beatrice Gillis, July 1, 1924; 1 son, Robert Gillis. Faculty, Western Res. U., Cleve., 1918-23; faculty U. Va., Charlottesville, 1923—, prof. anatomy, 1938-64, prof. emeritus, 1964—. Civilian cons. Chem. Warfare Service, U. S. Army, 1944-45. Recipient Pres. and Visitors Research prize U. Va., (with H. Silvette) 1944, alone, 1950. Fellow A.A.A.S.; mem. Am. Assn. Anatomists, Am. Soc. Zoologists, Soc. for Exptl. Biology and Medicine, Marine Biol. Lab., Va. Acad. Sci., Phi Beta Kappa, Sigma Xi, Alpha Omega Alpha. Author: (with Harvey Ernest Jordan) Textbook of Embryology, 1926; also articles. Research on devel. fish skulls, invertebrate blood cells, hematology fetal and older rats, action nitrogen mustards on hemopoietic organs rats and dogs, abnormal twinning in man, cyclopia in man, eventration viscera in monsters of man, comparative hematology, chromosomes of human fetal ovaries, description of normal human embryo, quantitative methods for measurement blood cell prodn. Home: 2010 Hessian Rd., Charlottesville, Va. 22903.*

KING, Albert Freeman Africanus, physician; b. Oxfordshire, Eng., Jan. 18, 1841; s. Edward and Louisa (Freeman) K.; M.D., Columbian (now George Washington) U., 1861; M.D., U. Pa., 1865; A.M., U. Vt., 1883, LL.D., 1904; m. Ellen A. Dexter, Oct. 17, 1894. Practiced medicine, specializing in obstetrics, Washington; being present in theatre, attended Pres. Lincoln immediately after his assassination; prof. obstetrics Columbian U., Washington, 1871-1913, dean Medical Sch., 1879-94, prof. obstetrics, U. Vt., 1871-1913; obstetrician George Washington U. Hosp. Fellow A.M.A. Author: A Manual of Obstetrics, 1st edit., 1882, 11th edit., 1910; Mosquitoes and Malaria, 1883. Died Dec. 13, 1914.

KING, Charles Glen, Am. biochemist; b. Entiat, Wash., Oct. 22, 1896; s. Charles C. and Mary (Bookwalter) K.; B.S., Wash. State U., 1918, D.Sc., 1950; M.S., U. Pitts., 1920, Ph.D., 1923, D.Sc., 1950; D.P.S., Denison U., 1961; D.Sc., Drexel Inst. Tech., 1954; m. Hilda Bainton, Sept. 11, 1919; children—Dorothy (Mrs. Harold T. Hammel), Robert B., Kendall W. Mem. faculty U. Pitts., 1920-42, prof., 1930-42, dir. Buhl Found. Sci. Program, 1935-42, sci. dir. Nutrition Found., 1942-55, exec. dir. Nutrition Found., 1955-61, pres., 1961-63, trustee, 1963—; research asst. Columbia U., N.Y.C., 1926-27, vis. prof., 1942-46, prof., 1946-62, prof. emeritus, 1962, asso. dir. Inst. Nutrition Scis., 1963-66, lectr., 1966—; dir. grants adminstrn. St. Luke's Hosp.

Center, 1966—. Mem. food and nutrition bd. Nat. Acad. Scis.-NRC, 1941; pres. 5th Internat. Congress of Nutrition, 1960; sci. adv. bd. Robert A. Welch Found., 1954-64. Bd. dirs. Milbank Meml. Fund, Boyce Thompson Inst. for Plant Research; exec. bd. Am. Pub. Health Assn., 1950-62, pres., 1961-62. Recipient Grocery Mfrs. Assn. award, 1944; John Scott award, 1949; Nichols Appert award, 1955. Bd. Trustees Alumnus award Wash. State U., 1964; Conrad A. Elreherr award, 1966. Mem. Am. Chem. Soc. (Pitts. award 1943, Charles F. Spencer award 1963), Nat. Acad. Scis., Internat. Union Nutritional Scis. (pres. 1966—), Am. Soc. Biochemists (pres. 1954-55), Am. Inst. Nutrition (pres. 1949-50), Phi Beta Kappa, Sigma Xi, Omicron Delta Kappa, Phi Sigma, Lambda Chi Alpha. Contbr. numerous articles to profl. jours., sci. books. Isolated vitamin C, 1932, proof of its source, 1948; research on functions of vitamin C, on structure of fats. Home: 39 Devon Rd., Bronxville, N.Y. 10708. Office: St. Lukes Hosp. Center, Amsterdam Av. and W. 114, N.Y.C. 10025.*

KING, Clarence, Am. geologist; b. Newport, R.I., Jan. 6, 1842; s. James Rives and Florence (Little) K.; grad. Sheffield Scientific School, Yale, 1862. Crossed continent on horseback and joined Cal. Geol. Survey, 1863, remaining until 1866; made paleontol. discoveries which furnished the evidence upon which the accepted age of gold-bearing rocks was determined. Originated the plan and comd. the expdn. for geol. survey of 40th parallel, under auspices of army engr. dept., 1867-72; exposed Arizona "diamondfields" fraud, 1872; suggested and organized U. S. Geol. Survey, served as dir., 1878-81; practice as mining engr., from 1881. Mem. Nat. Acad. Scis., Am. Inst. Mining Engrs., Geol. Soc. London. Author: Systematic Geology (Vol. I, Professional Papers of Engr. Dept., U. S. A.). Mountaineering in Sierra Nevada; Report of Geological Exploration of the Fortieth Parallel, 6 vols., 1870-80; Catastrophism and the Evolution of Environment, 1877. Introduced system of denoting topography by contour lines into mapping; theorized on age of earth upon basis of rate of cooling of molten magmas. Died Phoenix, Dec. 24, 1901.

KING, David Thane, physicist; b. Wellington, New Zealand, Jan. 16, 1923; s. Gilbert and E. Lois (Thane) K.; B.Sc., Victoria Coll., U. New Zealand, 1944, M.Sc., 1947; Ph.D., U. Bristol (Eng.), 1951; m. Margaret Joyce Asboe, July 25, 1950; children—Christopher Larmor, Roger Allan, Judith Eleanor, Cynthia Margaret. Came to U. S., 1952, naturalized, 1960. Physicist, Nucleonics div. Naval Research Lab., Washington, 1952-55; with U. Tenn., 1955—, prof., 1961—; mem. users' group Argonne (Ill.) Nat. Lab., 1962—. Cons., Oak Ridge Nat. Lab., 1965—. Mem. Am., Italian phys. socs., Sigma Xi. Contbr. articles to sci. jours. Research on interactions of high-energy particles with atomic nuclei by photog. method. Home: 1113 Park Hill Circle, Knoxville, Tenn. 37921.

KING, Edward S(kinner), Am. astronomer; b. Liverpool, N.Y., May 31, 1861; s. Nathaniel and Cornelia C. (Skinner) K.; A.B., Hamilton Coll., 1887, A.M., 1890, Sc.D., 1927; m. Kate Irene Colson, July 23, 1890; children—Harold Skinner, Margaret Wight, Everett Tryon (dec.). With Harvard Coll. Obs., Cambridge, Mass., from 1887, observer in charge Harvard Sta. on Mt. Wilson, Cal., 1889; asst. prof. astronomy, 1913-26, Phillips prof., Harvard, 1926—. Fellow Am. Acad. Arts and Scis., A.A.A.S. Obtained first photog. observation of the occultation of a star, also first photograph of spectrum of the aurora; perfected method of obtaining circular photographic images of the stars without visual guiding of the telescope; devised method of transforming prismatic to normal spectra, both photographically and mechanically; determined photographic magnitudes of bright stars and planets, from images photographed out of focus; made photographic measures of the light of the moon, also of the sun; has maintained systematic tests of photographic plates, 1896—; also determined photovisual magnitudes of stars, and derived color indices of planets. Died Sept. 10, 1931.

KING, Elmer Richard, Am. physician; b. West Liberty, O., May 15, 1916; s. John Oliver and Clara (Yoder) K.; B.A., Ohio State U., 1939, M.D., 1941; postgrad. Duke, 1948-49; m. Marie Hutchings Pace, Mar. 29, 1941; children—Elmer Richard, Marsha Ann, John Michael, Suzanne. Staff, Operation Sandstone, Eniwetok, 1947, med. div. Oak Ridge Inst. Studies, 1949-50; chmn. dept. nuclear medicine USN Med. Sch., Bethesda, Md., 1953-61; chief radiology U. S. Naval Hosp., Bethesda, 1958-61; prof., chmn. div. radiation therapy and nuclear medicine Med. Coll. Va., Richmond, 1961-65, prof., chmn. dept. radiology, 1965—; clin. faculty George Washington U. Med. Sch., 1954-61, Georgetown U. Med. Sch., 1954-61. Cons., VA., 1961—; chmn. com. on radiology NRC, 1962—; mem. med. adv. com. U. S. AEC, 1962—; mem. sci. adv. com. Armed Forces Inst. Pathology, 1962—; Recipient Spl. Commendation, Surgeon Gen., U. S. Navy, 1961, Alumni Achievement award Ohio State U. Sch. Medicine, 1966. Diplomate Am. Bd. Radiology. Fellow Am. Coll. Radiology; mem. Am. Radium Soc., Am. Roentgen Ray Soc., Radiol. Soc. N.Am., Soc. Nuclear Medicine, Assn. U. Radiologists, Assn. Am. Med. Colls. Author: A Manual for Nuclear Medicine, 1961; (with Thomas G. Mitchell) Atomic Medicine, 1961; also numerous articles, chpts. in books. Devel. and techniques of care and mgmt. of nuclear casualties; devel. method of sampling chem. constituents of patient's blood to estimate radiation dosage; new approaches to mgmt. of concer including combinations of radiation, surgery and chemotherapeutic agts; research on relationships of immunology to cancer in reference to defs. by patient's body against invading cancer cells; evaluation and applications of new radioisotopes in diagnosis of disease; new applications of isotopes and isotopic compounds in treating of malignancies. Home: 10300 Cherokee Rd., Richmond, Va. 23235.*

KING, Franklin Hiram, Am. agriculturalist; b. Whitewater, Wis., June 8, 1848; s. Edmund and Deborah L. K.; grad. State Normal Sch., Whitewater, 1872; spl. course Cornell, 1876-78; m. Carrie H. Baker, 1880. On Wis. Geol. Survey, 1873-76; prof. natural scis. River Falls State Normal Sch., 1878-88; prof. agrl. physics U. Wis., Madison, 1888-1901; chief Div. of Soil Mgmt., U. S. Bur. Soils, 1901-04. Author: The Soil, 1895; Principles of Agricultural Irrigation and Farm Drainage, 1899; Physics of Agriculture, 1900; Ventilation for Dwellings, Rural Schools and Stables, 1908. Pioneer in making relief maps and models; studied water requirements of crops, irrigation, drainage and soil solutions; gave detailed account of soil conditions in Orient; responsible for various farm improvements, including round soil, new method of barn ventilation. Died Aug. 4, 1911.

KING, Gladys Florence Smith, Am. biologist. b. Buffalo, July 8, 1917; d. Philip N. and Rosa (Schulenberg) Smith; B.A., U. Buffalo, 1939; M.A., U. Cin., 1940, Ph.D. (Laws fellow in physics), 1942; postgrad. (Adam T. Bruce Postdoctorate fellow biology), Johns Hopkins, 1942-43; m. Joseph E. King, Sept. 17, 1941; children—Sally, Peter. Histologist, U. S. Dept. Agr., New Orleans, 1943-48; marine biochemist U. Miami, Sarasota, Fla., 1948-49; asso. chemist U. Hawaii, Honolulu, 1949-58; asst. prof. Chaminade Coll., Honolulu, 1958-59; research asso. George Washington U. Med. Sch., 1959-61; faculty Trinity Coll., Washington, 1961—, prof. biology, 1967—. Sigma Xi Research grantee, 1963-64, 67-68 Fellow A.A.A.S.; mem. Am. Inst. Biol. Scientists, Am. Soc. Zoologists, Am. Micros. Soc., Sigma Xi. Contbr. articles to tech. jours. Research on plant histochemistry and plant growth substances, cytochemistry of kidneys of Rana pipiens capable of spontaneous tumors. Home: 5114 Wehawken Rd., Washington 20016.*

KING, Harold, Welsh pharmacologist; b. N. Wales, Feb. 24, 1887; s. Herbert and Ellen Elizabeth King; ed. Univ., Bangor, Wales; M.Sc., Wales, 1911, D.Sc., 1920; m. Elsie M. Croft, 1923; 1 son. With indsl. bursary, 1911-12; chemist Tar and Ammonia Products, Beckton, 1911-12, Burroughts Wellcome & Co., 1912-19; head chemistry div. Nat. Inst. for Med. Research, 1910-50. Recipient Hanbury medal, 1941, Addingham Gold medal, 1952. Fellow Royal Soc., 1933, Royal Chem. Soc. Contbr. papers to chem. and biochem. jours. Research on synthesis of penicillin, cure for rickettsial infections, alkaloids of med. interest; isolated tubocurarine chloride from curare, 1935; resolved lyoscine and its components; made (with Otto Rosenheim) theoretical contbns. to chemistry of steroids. Died 1956.

KING, Harold, Am. surgeon; b. Bedford, Ind., Aug. 12, 1922; s. Harry D. and Sarah (Miller) K.; student Ind. U., 1940-43; M.D., Yale, 1946; m. Betty Jane Fink, 1953; children—Edward Dale, James Allen. Practice medicine, specializing in surgery, Indpls., 1955—; prof. surgery Ind. U. Med. Center, 1955—; teaching, cons. VA Hosp. Mem. A.M.A., A.C.S., Am. Assn. Thoracic Surgery, Soc. Vascular Surgery, Soc. U. Surgeons, Chest Club, Central Surg. Assn., Am. Heart Assn., Internat. Cardiovascular Soc., Internat. Soc. Surgery, Soc. Surgery Alimentary Tract. Contbr. numerous articles to profl. jours. Described increased susceptibility to infections that occur after splenectomy in infancy; developed methods for mech. by-pass of aorta during cross clamping. Home: 5430 Channing Rd., Indpls. 46226. Office: 1100 W. Michigan St., Indpls. 46208.*

KING, Henry Eugene, Am. psychologist; b. Wilmington, Va., Sept. 24, 1922; s. John Burwell and Lucy (McGehee) K.; B.A., U. Richmond, 1942; M.A., Columbia U., 1943, Ph.D., 1948; m. Kathleen Mary Young, Dec. 18, 1948; children—Anne Henry, Peter Overton. Lectr. psychology Columbia U., 1948-49; sr. scientist N.Y. State Brain Research Project, 1948-49; asso. prof. psychiatry Tulane U. Med. Sch., 1949-60; prof. psychology U. Pitts. Sch. Med., 1960—; chief psychologist Western Psychiat. Inst. and Clinic, Pitts., 1960—. Fellow Am. Psychol. Assn., A.A.A.S.; mem. Am. Physiol. Soc., Am. Psychosomatic Soc., Soc. Exptl. Biology and Medicine, Am. Psychopathol. Assn. Author: Psychomotor Aspects of Mental Disease, 1954; (with others) Studies in Schizophrenia, 1956; (with Columbia Greystone Assos.) Selective Partial Ablation of the Human Frontal Cortex, 1949; (with others) Studies in Topectomy, 1956. Studies, publs. on the relationship of brain to behavior; psychomotility as a correlate of central nervous system orgn.; an analysis of retention of sensory experience in adult human. Home: 900 College Av., Pitts. 15232. Office: 3811 O'Hara St., Pitts. 15213.*

KING, Ivan Robert, Am. astronomer; b. Far Rockaway, N.Y., June 25, 1927; s. Myram and Anne (Franzblau) K.; A.B., Hamilton Coll., 1946; A.M., Harvard, 1947, Ph.D., 1952; m. Alice Greene, Nov. 21, 1952; children—David, Lucy, Adam, Jane. Instr. astronomy Harvard, 1951-52; mathematician Perkin-Elmer Corp., Norwalk, Conn., 1951-52; methods analyst U. S. Dept. Def., Washington, 1954-56; with U. Ill., 1956-64; asso. prof. astronomy U. Cal., Berkeley, 1964-66, prof., 1966—, chmn. astronomy dept., 1967—. Mem. Am. Astron. Soc. (councillor, 1963-66), Internat. Astron. Union. Contbr. numerous articles in field to sci. jours. Study of structure of stellar systems.*

KING, John, Am. eclectic physician; b. N.Y.C., Jan. 1, 1813; s. Harman and Marguerite (La Porte) K.; grad. Reformed Med. Coll. of City N.Y., 1838; m. Charlotte M. Armington, 1833; m. 2d, Phebe Rodman, 1853. Sec. 1st nat. conv. Reform Med. Practitioners, 1848; became prof. materia medica and therapeutics, Memphis, 1849; prof. obstetrics Eclectic Med. Inst., Cin., 1851. Pres. Nat. Eclectic Med. Assn., 1878; 1st pres. Ohio Eclectic Med. Assn. Author: (most notable work) The American Dispensatory, 1852; American Obstetrics, 1853; Women: Their Diseases and Treatment, 1858; The Microscopist's Companion, 1859; The American Family Physician, 1860; Chronic Diseases, 1866; The Urological Dictionary, 1878. Leader of reform in Am. med. therapeutics; a founder eclectic sch. medicine; introduced to gen. use oleoresin of iris (1st of resin class of drugs); introduced podophyllin, hydrastis, sanguinaria. Died North Bend, O., June 19, 1893.

KING, John A., Am. zoologist; b. Detroit, June 22, 1921; s. Royal Ernest and Matie (Neupert) K.; B.A., U. Mich., 1943, M.S., 1948, Ph.D., 1952; m. Joan W. McGinty, June 25, 1949; children—Christopher Lee, Andrea Jean. NIH Postdoctoral fellow Jackson Lab., Bar Harbor, Me., 1951-53, staff scientist, 1953-61; prof. zoology Mich. State U., East Lansing, 1961—. Mem. Animal Behavior Soc. (past sec.), A.A.A.S., Am. Inst. Biol. Scis., Am. Psychol. Assn., Am. Soc. Zoologists, Am. Soc. Mammalogists, Ecol. Soc Am., Brit. Ecol. Soc. Research, publs. on devel. and evolution of behavior in mammals, mammalian population dynamics and social behavior. Home: 1460 Oscoda Rd., Okemos, Mich. 48864. Office: Dept. Zoology, Mich. State U., East Lansing, Mich. 48823.*

KING, John Albert, Am., chemist; b. Columbus, Ind., Sept. 6, 1916; s. Perry and Mary E. (Holland) K.; student Wabash Coll., 1933-35 ; A.B., Ind. U., 1938; M.S., U. Minn., 1940, Ph.D., 1942; m. Martha Lee Frank, June 10, 1941; children—Mary Lee (Mrs. William D. E. Coulson), Clarissa Anne, John Albert, II, Dennis P. DuPont fellow, U. Minn., 1941-42, Merck fellow, 1942-43; sr. chemist Winthrop Chem. Co. div. Sterling Drug., 1943-46; dir. chem. research Warren Lambert Pharm. Co., 1946-57; dir. research, gen. mgr. research div. Armour & Co., Chgo., 1957-60; mgr. research and devel. agrl. div. Am. Cyanamid Co., Princeton, N.J., 1960—. Mem. A.A.A.S., Am. Chem. Soc., Chem. Soc. London, Am. Inst. Chemistry, N.Y. Acad. Scis., Soc. Chem. Industry, Chemists Club, Sci. Research Soc. Am., Phi Lambda Upsilon, Sigma Xi. Contbr. numerous articles to tech. jours. Research on chemistry Vitamin E, syntheses of vitamin A, amino acids, chemistry rauwolfia alkaloids, mechanism of Willgerodt reaction, mechanism of acylative carboxylation, reaction of alpha-amino acids. Patentee pharm. drugs and compounds. Home: 90 Battle Rd. Office: P.O. Box 400, Princeton, N.J. 08540.*

KING, Joseph Edwin, Am. biologist; b. Washington, Ohio, Oct. 12, 1914; s. Harry James and Alma (Dobbins) K.; B.S., Wilmington Coll., 1939; M.S., U. Cin., 1941; student U. N.C., 1942; m. Gladys F. Smith, Sept. 17, 1941; children—Sally Ann, Peter Edward. With U. S. Fish and Wildlife Service, 1943—, fishery research biologist Gulf Coast, 1943-49, Hawaii, 1949-59, asst. chief br. marine fisheries, 1959-63, chief, Washington, 1963—; dist. dir. Central Atlantic States dist. Am. Inst. Fishery Research Biologists, 1961-64. Fellow Am. Inst. Fishery Research Biologists; mem. Ecol. Soc., Soc. Limnology and Oceanography, Am. Fisheries Soc., Am. Ornithologist Union, Sigma Xi. Studies, publs. on shrimp biology, zooplankton and forage fish distbn. and abundance in Central Pacific; tuna food habits; occurrence and distbn. of sea birds in central Pacific. Home: 5114 Wehawken Rd., Washington 20016. Office: U. S. Bur. Comml. Fisheries, Washington 20240.*

KING, L. D. Percival, Am. nuclear physicist; b. Williamstown, Mass., Dec. 29, 1906; s. J. Percival and Edith (Seyerlin) K.; B.S. in Mech. Engring. with honors in Physics, U. Rochester, 1930; postgrad. Mass. Inst. Tech.; Ph.D. in Physics, U. Wis., 1937; m. Edith M. Bork, Dec. 25, 1935; children—Nicholas S. P., Lydian B. L. Asst. in physics, fellow Mass. Inst. Tech., 1931-33, U. Wis., 1933-37; instr., Purdue U., 1937-42, OSRD fellow Manhattan Dist., 1942-43; staff Los Alamos Lab., 1943—, group leader in three divs. 1944-57, asst. div. leader reactor div., 1958-60, chmn. Rover Flight Safety Office, 1961—. Tech. dir. U. S., AEC, Geneva Atoms for Peace Conf., 1958, adviser U. S. delegation, 1955, 58; tech. adviser U. S. del. UN, 1958. Fellow Am. Phys. Soc.; mem. Am. Nuclear Soc. (dir. 1955-59, nat. program chmn. 1955-56), A.A.A.S., Sigma Xi, Gamma Alpha, Theta Delta Xi. Contbr. articles to tech. jours. on nuclear physics, reactors and nuclear rockets; patentee nuclear energy devices. Home: P.O. Box 98, Route 1, Sante Fe. Office: P.O. Box 1663, Los Alamos.

KING, Leonard William, English archeologist; b. London, Dec. 8, 1869; s. Robert and Mary (Scarborough) K.; ed. Rugby, Kings Coll., Cambridge; m. Anna Burke, 1906; 1 son, 1 dau. Conducted Brit. Museum's excavations at Nineveh, 1903-04; lectr. Assyrian and Babylonian archaeology, U. London, 1910-15, prof., 1915-19; asst. keeper Egyptian and Assyrian antiquities Brit. Mus., 1913-19. Author: Babylonian Magic and Sorcery, 1896; Cuneiform Texts in the British Museum, 1896-1914; Assyrian Chrestomathy, 1898; (with Wallis Budge) Guide to the Babylonian and Assyrian Antiquities in the British Museum, 1900; Letters of Hammurabi, 1898-1900; Babylonian Religion and Mythology, 1899; Assyrian Language, 1901; Annals of Kings of Assyria, 1902; Seven Tablets of Creation, 1902; Studies in Eastern History, 1904-07; (with R. C. Thompson) Sculptures and Inscriptions of Darius at Behisun, 1907; (with H. R. Hall) Egypt and Western Asia, 1907; History of Sumer and Akkad, 1910; Boundary-Stones in the British Museum, 1912. Collected valuable inscriptions shedding light on early Assyria, also contbd. to growth and classification of knowledge about these people. Died London, Aug. 20, 1919.

KING, Lester Snow, Am. physician, pathologist, historian of medicine; b. Cambridge, Mass., Apr. 18, 1908; s. Myron L. and Sophie (Snow) K.; ed. Harvard U., A.B. 1927, M.D. 1932; m. Marjorie C. Meehan, Dec. 23, 1931; children—Alfred, Frances (Mrs. Donald E. Widmann). Teaching fellow Harvard U., 1933-35; Moseley Travelling Fellow, 1935-36; asst. Rockefeller Inst. Med. Research, Princeton, 1937-40; pathologist Fairfield State Hosp., 1940-42; asst. in pathology, Yale U., 1940-42; pathologist Ill. Masonic Hosp., Chgo., 1946-63; clin. prof. pathology U. Ill.; senior editor Jour. Am. Med. Assn., 1963—; Professorial lectr. History of Medicine U. Chgo., 1964—; mem. editorial bd. Jour. Hist. of Medicine, 1963—. Recipient Boerhaave medal, 1964; mem. A.M.A., Am. Assn. Hist. Medicine. Author: Medical World of the 18th Century, 1958; Growth of Medical Thought, 1963. Research and numerous publs. on anatomy of nervous system, blood-brain barrier, virus diseases of the nervous system, demyelinating diseases; studies on med. history of 17th and 18th centuries, with special reference to history of ideas as applied to medicine. Home: 4204 N. Greenview Ave., Chgo. 60613. Office: 535 N. Dearborn St., Chgo. 60610.*

KING, Louis Vessot, Canadian physicist; b. Toronto, Ont., Can., Apr. 18, 1886; s. Alonso and Louisa (Vessot) K.; ed. McGill U., Montreal, Que., Can., Cambridge, Eng.; M.A., Cambridge; D.Sc., McGill U.; m. Lillian Dorothea Neville; Became asst. prof. physics McGill U., 1912, asso., 1915, prof. physics, 1920; ret.; research in submarine acoustics for elec. and submarine com. Brit. Inventions Bd., 1915-20. Recipient Howard N. Potts Gold medal Franklin Inst., 1919, Flavelle medal Royal Soc. Can., 1934. Fellow Royal Soc., 1924, Am. Phys. Soc.; mem. Am. Math. Soc., Am. Astron. Soc., Royal Soc. Can. (became sectional pres. 1919). Author: On the Numerical Calculations of Elliptic Functions and Integrals. Research and publs. on fog-alarms, intensity and distbn. of light from sky, molecular structure, electron theory, distbn. and penetration of radio waves into earth, radiation field, electromagnetic properties of wireless antennae; invented hot-wire anemometer; measurements of air velocity using hotwire method, sound at sea. Died June 6, 1956.

KING, Oscar A., Am. neurologist; b. nr. Peru, Ind., Feb. 22, 1851; s. Timothy Lewis and Mary M. (Wright) K.; M.D., Bellevue Hosp. Med. Coll. (N.Y. U.), 1878; m. Minerva Guernsey, 1887. Second and 1st asst. physician Wis. State Hosp. for Insane, 1879-82; attended lectures U. of Vienna and clinics in Allgemeinen Krankenhausen, 1880-81; spl. studies in neurology and psychiatry under Meynert, Leidersdorf, Weiss and Benedict. Prof. mental and nervous diseases U. Ill. Coll. Physicians and Surgeons, Chgo., 1882, later prof. neurology, psychiatry and clin. medicine, 1894, vice dean, from 1900; pathologist and cons. alienist Wis. state charitable and penal instns., 1895; prof. neurology Post-Grad. Med. Sch.; chief dept. of neurology West Side Free Dispensary; asso. mem. med. staff Cook County Hosp. Founder, pres., med. dir. Oakwood Retreat, Lake Geneva, Wis., pvt. sanitarium for care of the insane, 1883; founded, Lake Geneva Sanitarium, 1896, combined the two sanitaria, 1901. Applied the toxin of erysipelas effectively in the treatment of 23 nearly consecutive cases of mania and melancholia, 1896. Died Sept. 11, 1921.

KING, Richard Wayne, Am. physicist; b. Moline, Ill., June 21, 1924; s. Chauncey Morton and Rosa (Hamm) K.; A.B., Washington U., St. Louis, 1948, Ph.D., 1952; m. Patricia Ann Casady, Feb. 23, 1946; children—Jeffrey Thomas, Christopher Kenyon. Research physicist Nuclear Devel. Assos., White Plains, N.Y., 1952-53; physicist NRC, Washington, 1953-55; with Purdue U., 1955—, prof. 1961—, head physics dept., 1966—. Cons., Inst. for Def. Analysis, Washington, 1960-61. Alfred P. Sloan Research Found. fellow, 1959-65. Mem. Am. Phys. Soc., Am. Assn. U. Profs., Sigma Xi, Pi Mu Epsilon. Contbr. articles to sci. jours. Research in nuclear physics and elementary particles. Home: 305 Leslie Av., West Lafayette, Ind. 47906. Office: Physics Bldg., Purdue U., Lafayette, Ind. 47907.*

KING, Robert Bainton, Am. neurosurgeon; b. Pitts.; s. Charles Glen and Hilda (Bainton) G.; student U. Pitts., 1940-42, U. Mich., 1942-43; M.D., U. Rochester, 1946; m. Molly Gibbs, Aug. 25, 1951; children—Nancy, Susan, Kimberly. Faculty, Washington, U., St. Louis, 1951-57, asso. prof. neurosurgery, 1957; prof. neurol. surgery State U. N.Y., Upstate Med. Center, Syracuse, 1957—; attending surgeon State U. Hosp., Syracuse, Syracuse Meml. Hosp., 1957—; cons. surgeon Syracuse VA Hosp., Syracuse Psychiat. Hosp., 1957—, Crouse-Irving Hosp., 1964—, Chenango Meml. Hosp., Norwich, N.Y., Taylor Brown Hosp., Waterloo, N.Y., 1963—. Markle scholar, 1951-56. Mem. Harvey Cushing Soc. Congress Neurol. Surgeons, Neurol. Soc. Am., So. Neurosurg. Soc., N.Y. Acad. Scis., Research Soc. Neurol. Surgeons, A.C.S., Central N.Y. Surg. Soc., Acad. Neurol. Surgeons, Onondaga County Med. Soc., Soc. Neurol. Surgeons, Sigma Xi. Contbr. articles to tech. jours. Research on neurophysiology, somatosensory system concentrating on trigeminal nerve. Home: Maple Dr., Fayetteville, N.Y. 13066. Office: 750 E. Adams St., Syracuse, N.Y. 13210.*

KING, Robert Charles, Am. geneticist; b. N.Y.C., June 3, 1928; s. Charles J. and Amanda (McCutchen) K.; B.S., Yale, 1949, Ph.D., 1952; m. Violet Mejia DeLeon, Feb. 20, 1953; children—Donald Vaughan, Archie Tom, Amanda Tessa. With biology dept. Brookhaven Nat. Lab., 1951-55; faculty Northwestern U., Evanston, Ill., 1956—, prof., 1964—, acting chmn. biology dept., 1964-65; NSF sr. postdoctoral fellow U. Edinburgh, 1958; vis. investigator, fellow Rockefeller U., 1959; NSF sr. postdoctoral fellow div. entomology Commonwealth Sci. and Indsl. Research Orgn., Canberra, Australia, 1963. Fellow A.A.A.S.; mem. Genetics Soc. Am., Am. Soc. Zoologists, Am. Soc. Naturalists, Histochem. Soc., Soc. Devel. Biology, Am. Soc. Cell Biology, Electron Microscopy Soc. Am., Sigma Xi. Author: Genetics, 1962; Dictionary of Genetics, 1968. Contbr. numerous articles to sci. jours. Research on genetic control of oogenesis in Drosophila melanogaster. Home: 2104 Evert Ct., Northbrook, Ill. 60062. Office: Biology Dept., Northwestern U., Evanston, Ill. 60201.*

KING, Ronold Wyeth Percival, Am. applied physicist; b. Williamstown, Mass., Sept. 19, 1905; s. James Percival and Edith Marianne Beate (Seyerien) K.; A.B., U. Rochester, 1927, S.M., 1929; Ph.D., U. Wis., 1932; student U. Munich, Germany, 1928-29, Cornell U., 1929-30; m. Justine Merrell, June 22, 1937; 1 son, Christopher Merrell. Asst. in physics U. Rochester, 1927-28; Am.-German exchange student, 1929-30; White fellow in physics Cornell U., 1929-30; U. fellow in elec. engring. U. Wis., 1930-32, research asst., 1932-34; instr. physics Lafayette Coll., 1934-36, asst. prof., 1936-37; Guggenheim fellow, Berlin, 1937-38; with Harvard U., 1938—, prof. applied physics, 1946—. Guggenheim Fellowship, Europe, 1958. Fellow Am. Acad. Arts and Scis., Am. Phys. Soc., A.A.A.S., I.E.E.E.; mem. Modern Lang. Assn., Am. Assn. U. Profs., Bavarian Acad. Scis. (contbg. mem.), Phi Beta Kappa, Sigma Xi. Author: Electromagnetic Engineering, Vol. I, 1945; Transmission Lines, Antennas and Wave Guides (with A. H. Wing and H. R. Mimmo), 1945; Transmission-Line Theory, 1955; Theory of Linear Antennas, 1956; Scattering and Diffraction of Waves (with T. T. Wu), 1959. Pioneered in a coordinating program of theoretical and exptl. research in applied electromagnetism; developed maj. aspects of electromagnetic theory of linear radiating, receiving, and scattering antennas and arrays, and of the asso. transmission lines; introduced techniques for measuring important properties of antennas and transmission lines. Home: 92 Hillcrest Pkwy., Winchester, Mass. 01890. Office: Gordon McKay Lab., Harvard U., Cambridge, Mass. 02138.*

KING, Tsoo E., biochemist; b. Shanghai, China, Jan. 14, 1917; s. Y. Z. and Sein (Chen) K.; B.S., Nat. Central U., Nanking, China, 1936; M.S., Ore. State U., 1948, Ph.D., 1949; postdoctoral work U. Wis., 1949-50; m. Shu Lu, Feb. 10, 1946. Came to U. S., 1947, naturalized, 1956. Instr., Kweiyang Med. Coll., Kweiyang, China, 1937-38; demonstrator Peking (China) Union Med. Coll., 1938-42; chemist Peicomb Chem. Works, Shanghai, 1942-46; from asst. prof. to prof. Ore. State U., Corvallis, 1950-57, 58—, asst. dir. Sci. Research Inst., 1962—. Recipient Gov.'s N.W. Sci. award, 1962. Guggenheim fellow, NSF sr. fellow Cambridge (Eng.) U., 1956-58. Mem. Am. Biol. Chemist Soc., Am. Chem. Soc., Chem. Soc. (Eng.), Biochem. Soc. (Eng.), Faraday Soc. (Eng.) Sigma Xi. Author: (with H. Mason, M. Morrison) Oxidases and Related Redox Systems, 2 vols., 1965; also numerous articles. Research on chem. synthesis vitamin derivatives, microbial metabolism carbohydrates and energy producing systems, electron transport, coupled energy transfer in mammalian heart and liver tissues. Office: Ore. State U., Corvallis, Ore. 97331.*

KING, William Frederick, astronomer; b. Stowmarket, Eng., Feb. 19, 1854; s. William King and Ellen (Archer) K.; B.A., U. Toronto (Ont., Can.), 1875; LL.D.; m. Augusta Florence Snow, 1881; 2 sons, 1 dau. Named Dominion land and topographical surveyor, 1875; became insp. surveys Dept. Interior, 1881, chief insp., 1886; named chief astronomer Dept. Interior Can., 1890; His Majesty's commr. Internat. Boundary between Can. and U. S. under Treaties of 1892, 03, 06, 08; mem. Internat. Waterways Commn., 1904-07; named dir. Dominion Astron. Obs., 1905; supt., Geodetic Survey Can., 1909.Bd. examiners Dominion Land Surveyors, 1885—. Recipient Gold medal in Math., U. Toronto, Fellow Royal Astron. Soc. Can. (became hon. pres. 1906), Royal Soc. Can. (pres. 1911-12), A.A.A.S. Contbr. articles to sci. jours. Died Apr. 23, 1916.

KING-HELE, Desmond George, Brit. space scientist, mathematician; b. Seaford, Eng., Nov. 3, 1927; s. Sydney George and Bessie (Sayer) K.-H.; student Epsom Coll., 1941-45; B.A. with 1st class honors in math., Trinity Coll., Cambridge U., 1948, M.A., 1952; m. Marie Therese Newman, Aug. 31, 1954; children—Carole Margaret, Sonia Julie. Staff, Royal Aircraft Establishment, Farnborough, Eng., 1948—, sr. prin. sci. officer, space dept., 1960—. Fellow Royal Soc. London, Inst. Math. and its Applications, Royal Astron. Soc.; mem. Internat. Acad. Astronautics. Author: Shelley: His Thought and Work, 1960; Satellites and Scientific Research, 1960, 2d edit., 1962; Erasmus Darwin, 1963; Theory of Satellite Orbits in an Atmosphere, 1964; Observing Earth Satellites, 1966; The Essential Writings of Erasmus Darwin, 1968. Research, publs. on devel. of theory of earth satellite orbits; studying changes in satellite orbits to determine the properties of earth's upper atmosphere and gravitational field. Home: 3 Tor Rd., Farnham, Surrey, Eng. Office: Royal Aircraft Establishment, Farnborough, Hants, Eng.*

KINGDON, Kenneth Hay, physicist; b. Montego Bay, Jamaica, May 13, 1894; s. William Lionel and Mary Elizabeth (Randall) K.; B.A., McMaster U. (Can.), 1914; Ph.D., U. Toronto (Can.), 1920; D.Sc., Union Coll. (N.Y.), 1946; m. Leah Mae Zeh, Nov. 1, 1924; 1 dau., Mary Elizabeth. Research asso. Gen. Electric Research Lab., Schenectady, 1920-46, sect. mgr., 1954-59; tech. mgr. Knolls Atomic Power Lab., Schenectady, 1946-54; tech. cons. Schenectady, N.Y., 1959—. Discovered cesium ion emission, 1924; separated uranium isotopes, 1940; researched antisubmarine weapons, 1941-44. Address: 1174 Phoenix Av., Schenectady 12308.*

KINGMAN, John Frank Charles, English mathematician; b. London, Aug. 28, 1939; s. F. E. T. and M. E. (Harley) K.; B.A., U. Cambridge (Eng.), 1960, M.A., 1964; m. Valerie Cromwell, Dec. 16, 1964. Fellow Pembroke Coll., Cambridge, 1961-65; faculty U. Cambridge, 1962-65, lectr., 1964-65; reader U. Sussex (Eng.), 1965-66, prof. math. and statistics, 1966—; vis. lectr. U. Western Australia, 1963. Fellow Royal Statis. Soc.; mem. Inst. Math. Statistics, London Math. Soc., Internat. Assn. for Statistics in Phys. Scis., Internat. Statis. Inst. Author: (with S. J. Taylor) Introduction to Measure and Probability, 1966; also articles. Research on math. theory of probability and related subjects. Office: U. Sussex, Brighton, Sussex, Eng.*

KINGSFORD, Thomas, inventor; b. Wickham, Kent County, Eng., Sept. 29, 1799; s. George and Mary (Love) K.; m. Ann Thomson, 1818; m. 2d, Elizabeth Austen, 1839; 1 son, Thomson. Came to U. S., 1831; supt. starch factory William Colgate & Co., Harsimus, N.J., 1833-46, developed method of producing starch from corn (rather than wheat), 1842; manufactured starch in own plant, Bergen, N.J., 1846-48; founder, owner Oswego (N.Y.) Starch Factory, 1848-69; produced cornstarch suitable for food purposes, 1850. Died Oswego, Nov. 28, 1869.

KINGSLEY, J(ohn) Sterling, Am. biologist; b. Cincinnatus, N.Y., Apr. 7, 1853; s. Lewis and Julia A. (Kingman) K.; A.B., Williams Coll., 1875; Sc.D., Princeton, 1885; postgrad. U. Freiburg, 1891-92; m. Mary Emma Read, Jan. 31, 1882. Curator, Peabody Acad. Sci., 1876-78; asst. U. S. Entomol. Commn., 1877-80; curator Worcester Natural History Soc., 1881-82; prof. zoology Ind. U., 1887-89; prof. biology U. Neb., 1889-91, Tufts Coll., 1892-1913; prof. zoology U. Ill., 1913-21. Editor of Standard Natural History, 1882-86; Am. Naturalist, 1884-96; Journal of Morphology, 1910-20. Author: Elements of Comparative Zoology, 1896; Vertebrate Zoology, 1899; Guides for Vertebrate Dissection, 1907; Comparative Anatomy of Vertebrates, 1912, 3d edit., 1926; Vertebrate Skeleton, 1925. Translator: Hertwig's Manual of Zoölogy, 1902, revised 1912. Died Aug. 29, 1929.

KINGSTON, Robert Hildreth, Am. physicist; b. Somerville, Mass., Feb. 13, 1928; s. Alexander Haddon and Martha (Aitcheson) K.; B.S., Mass. Inst. Tech., 1948, M.S., 1948, Ph.D., 1951; m. Ruth Marilyn Ahara, Apr. 19, 1952; children—Robert Edward, Susan Elizabeth, Margaret Jane, Katherine Ahara. Staff physicist Bell Telephone Labs., Murray Hill, N.J., 1951-52; staff mem. Lincoln Lab., Mass. Inst. Tech., Lexington, 1952-61, leader optics and infrared group, 1961—; vis. asso. prof. elec. engring. Stanford, 1964-65. Chmn. spl. group on optical masers U. S. Dept. Def., 1962—. Fellow Am. Phys. Soc. I.E.E.E. Editor: Semiconductor Surface Physics, 1957. Research on X-ray emission bands of potassium and calcium, theory of transient response in semiconductor diodes, germanium surface properties and effects on diodes and transistors, low temperature semiconductor switching components, parametric amplifiers at microwaves, theory of parametric interactions at optical frequencies, 2d harmonics in semiconductor lasers; 1st use of solid-state maser in radar astronomy. Home: 4 Field Rd., Lexington, Mass. 02173.*

KINGZETT, Charles Thomas, Brit. chemist; b. Oxford, Eng., 1852; s. C. Kingzett; ed. Oxford; m. Adeline Froggatt, 1878 (dec. 1934); m. 2d, Mina Lillian Briggs; 2 sons, 1 dau. Founder Sanitas Co., Ltd. Fellow Chem. Soc., Inst. Chemistry (a founder 1877). Author: The History, Products, and Processes of the Alkali Trade, 1877; Animal Chemistry, or the Relations of Chemistry to Physiology and Pathology, 1878; Nature's Hygiene and Sanitary Chemistry, 1880, 5th edit., 1907; A Pocket Dictionary of Hygiene, 1898, 2d edit., 1904; Chemistry for Beginners and School Use, 1917, 4th edit., 1922; Chemical Encyclopaedia, a Digest of Chemistry and its Industrial Applications, 1920, 5th edit., 1932; also articles. Inventor various products based on research on ozone and oxidation of terpenes and essential oils. Died July 29, 1935.

KINMONTH, John Bernard, Brit. surgeon; b. Clare, U.K., May 9, 1916; s. G. H. and D. A. (Daly) K.; M.B., B.S., U. London, 1938, M.S., 1962; m. Kathleen Godfrey, June 26, 1946; children—Ralph John, Fergus William, Margaret Delia, Claudia Kathleen. House surgeon, registrar, resident surgeon St. Thomas's Hosp., 1938-43, dir. dept. surgery, surgeon Med. Sch., 1955—; research fellow Harvard Med. Sch., 1948-49; asst. dir. surg unit St. Bartholomews Hosp., 1950-54; prof. surgery U. London, 1955—; Hunterian prof. Royal Coll. Surgeons Eng., 1954, Sims prof., 1962-63. Mickle fellow U. London, 1949. Fellow Royal Coll. Surgeons Eng.; mem. Physiol. Soc., Surg. Research Soc., Internat. (v.p. since 1966), European (pres. 1966-67) cardiovascular socs., Med. Research Soc. Author: Vascular Surgery (with Rob, Simeone), 1962; also numerous articles. Devel. of lymphography (visualization of lymph system) in living humans; demonstrated movements of human lymph vessels; studied intralymphatic direct treatment of cancer. Home: 70 Ladbroke Rd., London W.11., Office: St. Thomas's Hosp., London, Eng.*

KINNE, Otto, German biologist; b. Bremerhaven, Germany, Aug. 30, 1923; s. Heinrich Otto and Kate (Koch) K.; student U. Tübingen (Germany), 1947-49; Dr.rer.nat., U. Kiel (Germany), 1952, Habilitation 1958; m. Eva-Maria Diettrich, Nov. 14, 1953; 1 son, Stephan. Research zoologist U. Cal. at Los Angeles, 1957-58; asst. prof. zoology U. Toronto (Ont., Can.), 1958-60, asso. prof., 1960-62; dir., prof. Biologische Anstalt Helgoland, Hamburg, Germany, 1962-65, leading dir., prof., 1965—. Guggenheim fellow, 1961. Editor: Crustaceana, Internat. Jour. Crustacean Research, 1960—, Helgoländer wissenschaftliche Meeresuntersuchungen, 1962—, Meteor-Werk, 1966—, Marine Biology, Internat. Jour. Life in Oceans and Coastal Waters, 1967—. Research, numerous publs. on exptl. ecology of marine and brackish water living animals, effects of temperature and salinity on function and structure of living systems. Home: 74 Erlenweg, 2083 Halstenbek, Pinneberg, Germany. Office: Biologische Anstal Helgoland, 2, Hamburg 50, Palmaille 9, Germany.*

KINNEAR, Wilson Sherman, Am. civil engr.; b. Circleville, O., May 25, 1864; s. Richard and Mary Hall (Crow) K.; A.B., U. Kan., 1884, C.E., 1907; m. Carrie M. Nichols, Nov. 13, 1887; children—Mrs. Carmen Johnston, Lawrence Wilson. In ry. constrn. in Middle West and Chile, S.Am., 1884-90; asst. engr., asst. chief engr., asst. supt., asst. gen. supt., chief engr. and asst. gen. mgr. Mich. Central R.R., 1890-1910; pres. Kansas City Terminal Ry. Co., 1910-12; pres. U. S. Realty & Improvement Co., N.Y., 1912-17; sr. partner W. S. Kinnear & Co., cons. engrs., N.Y., 1918. Recipient Norman medal Am. Soc. C.E., 1912. As chief engr. Detroit River Tunnel Co., built electrically operated tunnel under Detroit River, between Detroit and Windsor, Can., 1 3/4 miles in length, with largest cross-sectional area of any subaqueous tunnel in the world at the time. Died Aug. 8, 1941.

KINNEN, Edwin, Am. elec. engr.; b. Buffalo, Mar. 9, 1925; s. Albert and Anna (Kumpf) K.; B.S. in Engring., U. Buffalo, 1949; M.S. in Elec. Engring., Yale, 1950; Ph.D., Purdue U., 1958; m. Ellen Underwood, Nov. 29, 1952; children—Susan, Janet, Peter, Andrew. Research scientist Research Lab., Westinghouse Electric Corp., Pitts., 1950-55; faculty Purdue U., 1955-59, asst. prof., 1958-59; asst. prof. U. Minn., Mpls., 1959-63; asso. prof. U. Rochester (N.Y.), 1963—; lectr. U. Pitts., 1952-55; engring. cons., 1956-63. Mem. I.E.E.E., Am. Assn. U. Profs., A.A.A.S. Research and publs. on math. theories of system dynamics and automatic control devices, elec. properties of biol. tissues, devel. elec. impedance plethysmograph for monitoring thoracic blood flow; sampled data systems and nonlinear systems. Home: 20 Green Hill Lane, Pittsford, N.Y. 14534. Office: U. Rochester, Rochester, N.Y. 14627.*

KINNERSLEY, Ebenezer, elec. experimenter; b. Gloucester, Eng., Nov. 30, 1711; s. William Kinnersley; M.A. (hon.), Coll. of Phila., 1757; m. Sarah Duffield, 1739, 2 children. Came to Am., 1714; ordained to ministry Baptist Ch., 1743; asso. with Benjamin Franklin, Edward Duffield, Philip Synge, Thomas Hopkins in experiments with elec. fire; rediscovered Dr. DuFaye's 2 contrary electricities of glass and sulphur (led to verification of truth of positive-negative theory); delivered 1st recorded exptl. lectures on electricity in Fanneuil Hall, Boston, 1751; elected chief master Coll. of Phila., 1753, prof. English and oratory, 1755-75. Mem. Am. Philos. Soc. Demonstrated heat could be produced by electricity, invented elec. air thermometer, 1755. Died Phila., July 4, 1778.

KINNEY, Thomas DeArman, Am. physician; b. Franklin, Pa., Mar. 19, 1909; s. Thomas Andrew and Minnie (DeArman) K.; A.B., U. Pa., 1931; M.D., Duke, 1936; m. Eleanor Roberts, July 9, 1939; children—Thomas Roberts, Eleanor DeArman, Hannah Chase, Janet Shepard. Dir. pathology Cleve. Met. Hosp. (formerly Cleve. City Hosp.), 1950-60; sr. cons. Crile VA Hosp., Cleve., 1947-60; faculty Boston U., 1940-41, Tufts Coll. Med. Sch., 1943-44, Harvard, 1944-47, Western Res. U. Sch. Medicine, 1947-60; prof., chmn. dept. pathology Duke, Durham, N.C., 1960—. Cons., mem. adv. councils VA, NRC, Nat. Acad. Scis., USPHS. Mem. Am. Soc. Exptl. Pathology, Am. Assn. Pathologists and Bacteriologists, Internat. Acad. Pathology, Soc. Geog. Pathology, Am. Soc. Clin. Pathologists, Coll. Am. Pathologists, Tissue Culture Assn., A.M.A., Soc. Exptl. Medicine and Biology, Am. Soc. Cell Biology, Am. Soc. Clin. Nutrition, Am. Assn. Cancer Research, Intersoc. Cytology Council. Numerous publs. in sci. jours. on nutritional pathology, particularly in area of iron metabolism. Home: 3120 Devon St., Durham, N.C. 27707.*

KINO, Eusebio Francisco, explorer; b. Segno, Italy, circa 1645. Entered Jesuit order, Freiburg, Germany, 1665; assigned to fgn. mission being organized in Spain, 1678; arrived Mexico City, 1681; joined Atondo expdn. to lower Cal. as head Jesuit mission, 1682; hdqrs. at Mission Dolores, 1687-1711; founded missions in San Miguel, Magdalena, Altar, Sonoita, Santa Cruz, San Pedro river valleys; discovered, wrote 1st description of Casa Grande River; instrumental in return of Jesuits to Cal. peninsula, 1697; his maps of S.W. territory, jours. are among earliest records of the area. Author: (treatise) Exposicion Astronomica de el Cometa, 1681. Died Magdalena, Mexico, Mar. 15, 1711.

KINO, Gordon Stanley, elec. engr.; b. Melbourne, Australia, June 15, 1928; s. William Hector and Sybil (Cohen) K.; B.Sc. with 1st honors in Math., London (Eng.) U., 1948, M.Sc., 1950; Ph.D. in Elec. Engring., Stanford, 1955; m. Dorothy Beryl Lovelace, Oct. 30, 1955; 1 dau., Carol Ann. Jr. scientist Mullard Research Labs., Salfords, Eng., 1947-51; research asst. Stanford, 1951-55, research asso. Microwave Lab., 1957-61, prof. elec. engring. dept., 1961—; tech. staff Bell Telephone Labs., Murray Hill, N.J., 1955-57. Cons., Varian Assos., Palo Alto, Cal., 1957—, Standard Telephone Labs., Harlon, Essex, Eng., 1960—. Fellow I.E.E.E.; mem. Am. Phys. Soc. Research and publs. on microwave tubes, plasmas and solid state; designed electron gun used in crossed field tubes. Home: 867 Cedro Way, Stanford, Cal. 94305.*

KINOSHITA, Kameki, Japanese geologist; b. 1896; grad. geology Tokyo U., 1912; D.Sc. Successively asst. engr. Sendai Mining Sta., staff sci. dept. Tohoku U., supr. Fukuoka Mine, lectr. Kyushu U., engr. Geology Research Inst. of Commerce and Industry Dept.; prof. Kyushu U., 1933—, also at Kita-Kyushu (N. Kyushu) U. Fgn. Studies. Author books including: General Principles of Mineralogy; Minerals and Mineral Deposits in the South Sea Area. Research on aluminum deposits in Europe; studies in Thailand, Malay, Manchuria.

KINOSHITA, S., Japanese microbiologist; b. Japan, Oct. 31, 1915; grad. dept. agr. U. Tokyo, 1941. Microbiologist research lab. Kyowa Fermentation Industry Co. Ltd., Tokyo, Japan. Recipient award Japanese Acad. Sci. Research, numerous publs. on bacterial phage, fermentative prodn. of various amino acids and nucleic acids on an indsl. scale. Home: Tokyo Research Lab., Kyowa Fermentation Industry Co. Ltd., 712 Honinachida, machida Shi, Tokyo, Japan.*

KINOSHITA, Toichiro, physicist; b. Tokyo, Japan, Jan. 23, 1925; s. Tsutomu and Fumi (Ueda) K.; B.S., Tokyo U., 1947, Ph.D., 1952; m. Masako Matsuoka, Oct. 14, 1951; children—Kay, June, Ray. Came to U. S., 1952, naturalized, 1965. Mem. Inst. for Advanced Study, Princeton, N.J., 1952-54; Postdoctoral fellow Columbia U., N.Y.C., 1954-55; faculty Cornell U., Ithaca, 1955—, prof. physics, 1964—. Mem. Am. Phys. Soc. Contbr. articles to tech. jours. Theoretical study of structure of elementary particles and their mutual interactions. Home: 5 Winthrop Pl., Ithaca, N.Y. 14850.*

KINSEY, Alfred Charles, Am. zoologist, biologist; b. Hoboken, N.J., June 23, 1894; s. Alfred Seguine and Sarah Ann (Charles) K.; B.S., Bowdoin Coll., 1916; Sc.D., Harvard, 1920, Sheldon traveling fellow, 1919-20; m. Clara Bracken McMillen, June 3, 1921; children—Don (dec.), Anne, Joan, Bruce. Asst. in zoology Harvard, 1917-18, in botany, 1918-19; asst. prof. zoology, Ind. U., Bloomington, 1920-23, asso. prof., 1923-29, prof., from 1929, also Waterman research asso.; in charge biol. exploration in Mexico and Central Am., 1931-32, 35-36; in charge study on human sex behavior, supported jointly by Ind. U., Rockefeller Found. and NRC, from 1942. Mem. A.A.A.S., Am. Entomol. Soc., Assn. Econ. Entomologists, Ind. Acad. Sci., N.E. Bot. Club, Am. Soc. Geneticists, Am. Soc. Naturalists, Am. Sociol. Soc., Am. Assn. Marriage Counselors, Acad. Natural Sci., Am. Zoöl. Soc., Am. Iris Soc., Phi Beta Kappa, Sigma Xi. Author: An Introduction to Biology, 1926; Field and Laboratory Manual in Biology, 1927; The Gall Wasp Genus Cynips—A Study in the Origin of Species, 1930; New Introduction to Biology, 1933 (revised 1938); Workbook in Biology, 1934; The Origin of Higher Categories in Cynips, 1936; Methods in Biology, 1937; Edible Wild Plants of Eastern North America (with M. L. Fernald), 1943; Sexual Behavior in the Human Male (with W. B. Pomeroy and C. E. Martin), 1948; Sexual Behavior in the Human Female (with W. B. Pomeroy, C. E. Martin and P. H. Gebhard), 1953. Research in early career on gall wasps of Mexico and Central Am., with particular emphasis on life histories, gall formations, evolution, taxonomy and geog. distbn.; most widely known for his extensive studies investigating human sexual behavior, pub. 1948, 53; his program is being continued at Inst. Sex Research, Inc., Bloomington. Died Bloomington, Aug. 25, 1956.

KINSEY, Victor Everett, Am. ophthalmic physiologist; b. Pitts., Aug. 4, 1909; s. William L. and Grace (Everett) K.; B.S., U. Pitts., 1931, Ph.D., 1935; m. Irene Mackey, Aug. 22, 1935. Biophysicist, Cancer Research Labs., U. Pa., 1935-36; biochemist Sharp and Dohme Co., Phila., 1936-38; fellow ophthalmology U. Pitts. Sch. Medicine, 1938-40; faculty Howe Lab. Ophthalmology, Harvard Med. Sch., 1940-50; asst. dir. research Kresge Eye Inst., prof. ophthalmic chemistry Wayne State U. Sch. Medicine, Detroit, 1950-68; prof. Inst. Biol. Scis. Oakland U., Rochester, Mich., 1968—. Mem. numerous coms., adv. bds. on eye diseases and blindness. Recipient Warren Triennial award Mass. Gen. Hosp., 1944, Proctor Medal award Assn. Research in Ophthalmology, 1952, Albert Lasker award Lasker Found., 1956, award Modern Medicine, 1959. Asso. mem. Am. Acad. Ophthalmology and Otolaryngology; mem. Am. Chem. Soc., Am. Soc. Biol. Chemists, Assn. Research in Ophthalmology (chmn. 1961), mem. Sigma Xi; hon. mem. New Eng., Detroit ophthal. socs. Author: (with T. E. Hopkins) Index to Iron & Steel Patents, 1930, also supplement. Asso. editor several ocular jours. Contbr. articles to sci. jours. Research in ocular physiology and biochemistry; studies metabolism and transport mechanisms in the ocular lens; dir. controlled clin. investigation of role of oxygen as an etiologic agt. in prodn. blindness in premature infants. Home: 6916 Dublin Fair Rd., Troy, Mich. 48084. Office: Inst. Biol. Scis., Oakland U., Rochester, Mich. 48063.*

KINSLEY, Albert Thomas, Am. veterinary pathologist; b. Independence, Ia., Feb. 26, 1877; s. John and Jane Elizabeth (Footitt) K.; B.Sc., Kan. State Agrl. Coll., 1899, M.Sc., 1901; postgrad. U. Chgo., summer, 1901; D.V.S., Kan. City Vet. Coll., 1904; m. Anna Louisa Smith, Sept. 4, 1901; 1 son, Albert Smith. Instr. bacteriology Kan. State Agrl. Coll., 1899-1901; pathologist and dir. museum Kansas City Vet. Coll., 1904-18, later pres., until 1941; organizer, 1909, and pres. Am. Serum Co. Del. to Internat. Veterinary Congress, London, 1930. Mem. Am. Vet. Med. Assn. (pres. 1921-22), Mo. Valley Med. Assn. (pres. 1909-10). Author: Veterinary Pathology, 1910; Diseases of Swine, 1914; Swine Practice, 1920. Originator use of bacteria in vet. medicine. Died Dec. 8, 1941.

KINZEL, Augustus B(raun), Am. metallurgist; b. N.Y.C., July 26, 1900; s. Otto and Josephine (Braun) K.; A.B., Columbia, 1919; B.S., Mass. Inst. Tech., 1921, D.Met., Eng. 1922; D.Sc., University of Nancy (France); Dr. Engring., N.Y. Univ., 1955; D.Sc., Clarkson Coll. Tech., 1957; D. Hon. Causa, U. of Nancy, France, 1963; m. 1927 (divorced); children—Carol (Mrs. Charles Uht), Doris (Mrs. Richard Campbell), August F., Angela (Mrs. John W. Talbot), Helen (Mrs. William Murray Hawkins, Jr.); m. 2d, Marie MacClymont, May 3, 1945. Metallurgist Gen. Electric Co., Pittsfield, Mass., 1919-20, 22-23, Henry Disston & Sons, Inc., Phila., Pa. 1923-26; lectr., instructor extension courses in advanced metallurgy Temple University, 1925-26; research metallurgist Union Carbide & Carbon Research Labs., Incorporated, N.Y.C., 1926-28, group leader, 1928-31. chief metallurgist, 1931-45, vice pres., 1945-48, president, 1948—, vice president Electro Metall. Company a division 1944-54; dir. research Union Carbide Corporation 1954-55, v.p.-research, 1955—; trustee System Devel. Corp., Santa Monica, Cal., Mitre Corp., American Optical Co. (Southbridge, Mass.); dir. Gen. Am. Investors Co., (Inc., N.Y.C., Beckman Instruments, Inc., Fullerton, Cal. Chmn. governing council Courant Inst. Math. Scis. N.Y. U.; trustee Cal. Inst. Tech., Salk Inst. Biol. Studies, pres. 1965—. Miller Inst. Basic Research Sci. University California; chairman The Engring. Found. Bd. N.Y.C. 1945-48; v.p., dir. Berkshire Farm for Boys, Canaan, New York. Recipient Howe Memorial Lecture, Am. Inst. Mining and Metall. Engrs., 1952; Sauveur Lecture, Am. Soc. for Metals, 1952; Morehead medal, Internat. Acetylene

Assn., 1955; Stevens Inst. Tech. Power Metallurgy Medal award, 1959; Indsl. Research Inst. medal, 1960; James Douglas Gold Medal award, Am. Inst. Mining, Metall. and Petroleum Engrs., 1960. Mem. Mass. Inst. Tech. Corp. Fellow N.Y. Acad. Scis., Royal Soc. Arts, Metall. Society; mem. Eurospace (hon.), Engrs. Joint Council (hon., pres. 1960-61), Am. Philos. Soc., Soc. Chem. Industry (mem. exec. com. Am. sect.), Adv. Council Indsl. Research and Devel. N.Y., Nat. Acad. Scis., Am. Institute Mining Metallurgy and Petroleum Engineers (p.p.), Am. Welding Soc. (dir., Adams lecturer, 1944), Am. Soc. for Testing Materials, Am. Soc. for Metals (Campbell lecturer 1947, Burgess Meml. lectr. 1956, past chmn. N.Y. sect.), Soc. Automotive Engrs., Am. Iron and Steel Inst., Internat. Inst. Welding (v.p.). Sr. author: Alloys of Iron and Chromium, vols. 1 and 2, 1937 and 1940. Contbr. articles on metallurgy and engring. to tech. pubs. Holder numerous patents on metallurgy and engring. Extensive metallurgy research; with Union Carbide, developed techniques for comml. prodn. of titanium, columbium, chromium, calcium and manganese; developed first low-alloy engring. steels; contbd. to devel. stainless steel; devleoped vandium refining process. Home: 28 E. 63d St., N.Y.C. 10021. Office: 270 Park Av., N.Y.C.

KIORTSIS, Vassilios-Clitos, Greek zoologist; b. Thessaloniki, Greece, Oct. 6, 1925; s. George and Helen (Assimaki) K.; student U. Athens (Greece) Med. Sch., 1945; Licencié ès Sc., U. Geneva (Switzerland), 1949, Dr.ès Sc. Biol., 1953; m. Ismine-Melpomeni Vosniakou, Dec. 30, 1954; children—Georges, Paul, Demetrius. Instr. dept. zoology U. Geneva, 1949-52, chief instr., 1954-56, privatdozent, 1956-59, chargé de cours, 1959-60, vis. prof., 1960-61; instr. Mil. Med. Sch. Thessaloniki, 1953; asso. prof. zoology Nat. U. Athens, 1961-64, prof., dir. Zool. Lab. and Mus., 1964—. Mem. Council Fisheries, Greek Ministry Industry, 1962-65; dir. Advanced Study Inst. on Regeneration, 1964; v.p. Hellenic Inst. Oceanographic and Fisheries Research, 1965-67. Eisenhower Exchange fellow, 1967. Mem. Swiss Zool. Soc., Swiss Genetic Soc., Swiss Acad. Natural Scis., Hellenic Biol. Soc., Mediterranean Assn. for Marine Biology and Oceanology, Internat. Council Museums. Author: (with H. A. L. Trampusch) Regeneration in Animals and Related Problems, 1965; also articles. Research in expl. embryology, regeneration, teratology, mutagenesis induced by radiation, zooplankton. Home: 41, Ag.Georgiou St., Neon Psychikon, Athens, Greece.*

KIPPER, Aksel Yanovich, Russian astrophysicist; b. Nov. 5, 1907, grad. Tartu U., 1930. With Tartu Astron. Obs., 1930-44; faculty Tartu U., 1941—, prof., 1946—; dir. Inst. Physics and Astronomy, Estonian Acad. Sci., 1950—. Mem. Estonian Acad. Sci. (v.p. 1946-50). Author: Non-Stationary Magneto-Hydrodynamic Processes in Stars, 1954; The Luminescence of Gas Nebulae, 1955; Non-Stationary Stars, 1955; A Gravitational Paradox, 1962. Research on astronomy and quantum mechanics, dynamics of cepheids, biphoton processes, origin of magnetic fields of sun and stars, radiation processes of solar and stellar atmospheres. Address: Inst. Physics and Astronomy, Estonian Acad. Sci., Tallinn, Estonian SSR, USSR.

KIPPING, Frederic Stanley, Brit. chemist; b. Manchester, Eng., 1863; s. James Stanley and Julia (Duval) K.; student Owens Coll.; Ph.D., U. Munich (Germany); D.Sc., London; D.Sc. (hon.), Leeds, Eng.; m. Lily Holland; 2 sons, 1 dau. Chemist Manchester Corp. Gas Works, 1882-85; asst. to Prof. Perkin, Heriot-Watt Coll., Edinburgh, 3 years; lectr. chem. dept. Central Tech. Coll., 6 years; prof. chemistry Univ. Coll., Nottingham, Eng. Recipient Longstaff medal Chem. Soc., 1909. Fellow Royal Soc. (council 1911, Davy medal 1918), 1897. Author 1st textbook dealing solely with organic chemistry, 1894; also papers. Research on stereoisomerism in silicon atoms; synthesized silicon compounds; his work led to use of parallel carbon compounds in prodn. of new high temperature resistant plastics. Died Criccieth, Wales, May 1, 1949.

KIPRIANOV, Andrey Ivanovich, Russian organic chemist; b. 1896; grad. Kharkov U., 1919. With Kharkov U., until 1941, prof., 1940—, Kiev U., 1944—; dir. Inst. Organic Chemistry, Ukrainian Acad. Sci., 1945-59. Mem. Soviet delegation Internat. Congress on Theoretical and Applied Chemistry, Zurich, 1955; mem. Soviet delegation Internat. Congress on Indsl. Chemistry, Belgium, 1959. Recipient Stalin prize, 1942. Mem. Ukrainian Acad. Sci. Author: Color and Symmetry in the Structure of Organic Dyes, 1940; Cyanine Dyes from Phenazine Products, 1960; co-author: The Effects of the Solvent on the Coloring of Organic Dyestuffs, 1940; The Color Organic Dyes and the Systematic Arrangements of Their Molecules, 1950. Research and publs. on organic dyes; founded new rules governing range and suitability of dyes as sensitizers. Address: Kiev University, Vladimirskaya ulitsa 64, Kiev 17, Ukrainian SSR, USSR.

KIRBY, Daniel Bartholomew, Am. ophthalmologist; b. Cleve., Apr. 12, 1891; s. Daniel Bartholomew and Esther A. Robinson (Whitaker) K.; A.B., John Carroll U., 1912, A.M., 1914, LL.D., 1948; M.D., Western Res. U., 1916; postgrad. Harvard, Pa. U., Cornell U.; m. Cecilia Katherine Hahn, June 9, 1923; children—Mary Elizabeth (Mrs. J. Dukes Wooters, Jr.), Mother Joan Kirby, Cecilia (Mrs. Peter Mullen), Janet Whitaker. Resident surgeon Bellevue Hosp., N.Y.C., 1921-23, later surgeon-in-chief dept. ophthalmology; asso. of Dr. John M. Wheeler, 1923-28; practice medicine specializing in ophthalmology from 1923; prof. dept. ophthalmology, Coll. of Medicine, N.Y. U.; cons. surgeon in ophthalmology N.Y. Eye & Ear Infirmary and Manhattan Eye, Ear & Throat Hosp., New Rochelle Hosp. Diplomate Am. Bd. Ophthalmology (chmn.). Fellow N.Y. Acad. Med., A.C.S.; mem. A.M.A., Am. Ophthal. Soc., Am. Acad. Ophthalmology and Oto-Laryngology, Harvey Soc., ophthal. socs. of Brazil, Argentina, Chile, Peru, Cuba, France, Greece. Recipient Schneider prize in ophthalmology. Author: Surgery of Cataract, 1950; Advanced Surgery of Cataract, pub. 1955. Research and publs. on subjects including diseases of crystalline lens, devel. of system of intracapsular cataract extraction. Died Dec. 27, 1953.

KIRBY, William, English entomologist; b. Witnesham, Eng., Sept. 19, 1759; s. William and Lucy (Meadowe) K.; B.A., Caius Coll., Cambridge, Eng., 1781, M.A., 1815; m. Sarah Ripper, 1784; m. 2d, Charlotte Rodwell, 1816. Took holy orders, 1732; received sole charge of Barham, Eng., and remained until his death. Fellow Linnean Soc. (charter), Royal Soc., 1818; mem. Entomol. Soc. (became pres. 1837). Author: Introduction to Entomology, 1815-26; Monographia apium anglia, 1802; On the History, Habits, and Insects of Animals, 1835; also numerous articles. First study of English bees, 1802; founded new insect order, Strepsiptera, 1811. Died July 4, 1850.

KIRBY, William Murray Maurice, Am. physician; b. Springfield, S.D., Nov. 21, 1914; s. William McLeod and Era (Keeling) K.; B.S., Trinity Coll., Hartford, 1936; M.D., Cornell U., 1940; m. Georgiana H. Dole, Apr. 12, 1944; children—Barbara Dole, Philip Keeling, Richard Murray. Instr., Stanford, 1947-49; faculty U. Wash., Seattle, 1949—, prof. medicine, 1955—. Mem. numerous nat. profl. coms. Mem. Assn. Am. Physicians, Am. Soc. Clin. Investigation, others. Research and numerous publs. on infectious diseases, primarily concerning new antibiotics. Home: 5656 N.E. Keswick Dr., Seattle 98105.*

KIRCH, Christfried, German astronomer; b. Guben, Germany, Dec. 24, 1694; s. Gottfried and Marie-Marguerite (Winckelmann) K.; student Joachim Coll., Berlin, also Leipzig, Germany, Danzig, Poland; succeeded his father as dir. Royal Obs. Prussia. Mem. French Acad. Scis., St. Petersburg Acad. Scis. Author: Observationes astronomiae selectiores, 1730; Ephémérides. Detailed astron. observations in western Hemisphere, 1714-16; studied Mongols and Tartars. Died Berlin, Mar. 9, 1740.

KIRCH, Gottfried, German astronomer; b. Guben, Brandenburg, Dec. 18, 1639; ed. at Leipzig; m. 2d, Marie-Marguerite Winkelmann, 1692; 1 son, Christfried. Called to Berlin by Frederick I to head new obs., 1706, also royal astronomer. Mem. Berlin Acad. Sci. Author: Ephemeridum motuum caelestium annus primus, 1681-1702; Relation succincte de la nouvelle comete, 1683; Calendarium christianum, Judaicum et turcicum, 1685, pub. each year until his death. Studied comet of 1680; discovered comet of 1702. Died Berlin, July 25, 1710.

KIRCH, Maria Marguerite Winckelmann, German astronomer; b. Panitzch, Germany, Feb. 25, 1670; d. Reverend Winckelmann; student under Christopher Arnold; m. Gottfried Kirch, 1692; 1 son, Christfried, 3 daus., including Christine. Assisted her husband in astron. studies; after his death accepted protection of Baron of Krosick, 1710. Author: Von der Conjunction der Sonne des Saturni und der Venus, 1709; The Position of Jupiter and Saturn, 1712; also several almanacs, astron. tables, articles on conjunction of planets. Discovered comet of 1702, an aurora borealis, 1707. Died Berlin, Dec. 29, 1720.

KIRCHER, Athanasius, mathematician, physicist, chemist; b. Fulda, Germany, May 2, 1602; ed. Jesuit Würzburg (Germany) Sem., 1618; ordained priest, 1628; prof. math. philosophy and Oriental langs., Würzburg; studied antiquities, Avignon; prof. math. Collegium Romanum, Rome, Italy, 1635-43. Author: Ars magnesia, 1631; Magnes sive de arte magnetica, 1641; Ars magna lucis et umbrae, 1646, 71; Musurgia universalis, 1650; Lingua Aegyptiaca restituta, 1643; Mundus subterraneus (which describes volcanoes, origin of springs and other geol. phenomena), 1665; Phonurgia, 1673. Studied steam pressure, magnetism, sound and light; invented Aeolian harp, magic lantern, speaking tube; through work in microscope suspected the existence of microbes and suggested invisible bodies caused infectious diseases; pioneered in deciphering hieroglyphics; investigated subterranean forces of volcanoes. Died Rome, Nov. 28, 1680.

KIRCHHEIMER, Franz Waldemar, German geologist; b. Müllheim, Baden, Germany, July 1, 1911; s. Franz and Stephanie (Christ) K.; Dr.Phil., U. Giessen (Germany), 1933; m. Lina Suppan, Aug. 10, 1945; 1 son, Franz. Prof., U. Stüttgart (Germany) hon. prof. geology, 1954—; hon. prof. U. Heidelberg (Germany), 1960—, U. Freiburg (Germany), 1961—; dir. Geol. Survey Baden-Württemberg, 1947—, pres., 1954—; commissary of state, 1946-47; counsellor Ministry of Economy, Baden-Württemberg, 1952—. Recipient Liebig medal U. Giessen, 1959. Mem. acads. scis. in Halle, Heidelberg, Vienna. Hon. fellow Bot. Soc. Edinburgh. Author: Grundzüge einer Pflanzenkunde der deutschen Braunkohlen, 1937; Die Laubgewächse der Braunkohlenzeit, 1957; Das Uran und seine Geschichte, 1963; Die Bergbaugepräge aus Baden-Württemberg, 1967; also numerous articles. Research on paleontology of plants, geology and history of mining, numismatics. Home: 1 Gerbett. Office: 5 Albert, 78 Freiburg, Germany.*

KIRCHHOF, Johann K., German physician; b. Bremen Germany, Mar. 17, 1909; s. Josef Heinrich and Ludwina (Vollmer) K.; state exam. U. Berlin, 1936, Dr.med., 1937; m. Maria-Antonie Borucki, Dec. 15, 1947; children—Corinna, Christoph, Sixtus, Arabella, Silvia. Faculty, U. Bonn (Germany), 1948-59, became prof. neurology, psychiatry, 1958; lectr. dir. neurol. clinic U. Izmir (Turkey), 1958-65. Mem. German Soc. Neurology. Author: Das menschliche Antlitz im Spiegel organischnervöser Prozesse, 1960; also articles, numerous sci. films. Research on thalamus-syndromes in clin. features of delirium tremens, defective schizophrenias; Werdnig-Hoffmann disease; cinematographical research on neurol. disturbances; mimic research. Address: 178 Koblenzer Strasse, Bonn, West Germany.*

KIRCHHOFF, Alfred, German geographer; b. Erfurt, Germany, May 23, 1838. Tchr. secondary schs., Berlin, Germany; prof. U. Halle (Germany), 1873-1904. Editor: (jour). Unser Wissen von der Erde, 1885-93 (often called Kirchhoff's Collection). Research and publs. on methodology and content of geog. instrn., on geog. distbn. of plants, animals, gen. geography of Europe. Died Mockau/Leipzig, Germany, Feb. 8, 1907.

KIRCHHOFF, Gustave-Robert, German physicist; b. Königsberg, Germany, Mar. 12, 1824; prof. physics, Breslau, 1850-54; Heidelberg, 1854-74; Berlin, 1875-87. Fellow Royal Soc., 1875. Discovered (with Bunsen) spectrum analysis (leading to modern spectrography); also discovered cesium and rubidium; studied electric currents, propagation of light; formulated Kirchoff's laws of electricity. Died Berlin, Oct. 17, 1887.

KIRCHHOFF, Heinz, German radiologist, gynecologist; b. Wilhelmshaven, June 4, 1905; s. Bernhard and Adelheid (Lemcke) K.; M.D.; m. Ilse Dormann; children—Marlis, Renate, Elke, Bernd. Worked with Prof. Robert Schroder, Kiel, also Leipzig, Germany; maternity dir., prof. gynecology U. Göttingen (Germany). Author: Das lange Becken; other works on gynecology, childbirth, radiology. Address: Kirchweg 5, Göttingen, W. Germany.

KIRILLIN, Vladimir Alekseevich, Russian physicist; b. Jan. 20, 1913; grad. Moscow (USSR) Energetics Inst., 1936; D.Sc. Lectr., Moscow Power Engring. Inst., 1938-41, 43-52, prof., 1952-53; dep. minister higher edn. USSR, 1954-55; dep. chmn. State Com. New Tech., 1955; now head Lab. on High Temperatures, Moscow Energetics Inst. Head dept. sci. of univs., tech. schs., colls. Central Com., Communist Party, 1955—. Recipient Stalin prize, 1951, Lenin prize, 1959, Order of Lenin, 1963. Mem. USSR Acad. Scis. (v.p. 1963—). Author: Cycles of Internal Turbine Combustion, 1949; Basis of Experimental Thermodynamics, 1950; Thermodynamic Properties of Gases, 1953; Enthalpy and the Specific Heat of Tungsten in the Temperature Range of 0-2400 C., 1962. Research on thermal and phys. properties of heat carriers in power plants; instrumental in devel. of new standards on water and steam for super-pressure steam electric power stas. Home: ulitsa Kazakova, 29, Moscow. Office: Moscow Energetics Inst., Moscow, USSR.

KIRK, Charles Townsend, Am. geologist; b. Francisco, Ind., June 22, 1876; s. David Henry and Martha Jane (Townsend) K.; B.S., U. Okla., 1904, A.M., 1905; Ph.D., U. Wis., 1911; m. Bessie Keller, Aug. 22, 1906; children—Ora Jane, Betty Clare, David Keller, Florence Nell. Instr. in geology State Sch. of Mines, Butte, Mont., 1906-08; instr. in corr. sch. U. Wis., 1908-10; instr., asst. prof. geology Hunter Coll., N.Y.C., 1910-13; prof. geology U. N.M., and state geologist, 1913-17; field geologist U. S. Geol. Survey, 1910-11, 34-36, 38-39. Installed Okla. Mineral Exhibit, La. Purchase Expn., St. Louis, 1904; judge of exhibits Panama Pacific Internat. Expn., San Francisco, 1915. Mem. A.A.A.S., Geol. Soc. Am., Am. Inst. Mining and Metall. Engrs., Am. Assn. Petroleum Geologists, N.Y. Acad. Scis. (sec. geol. sect. 1913), Sigma Xi, Phi Kappa Phi. Author: Pennsylvanian-Permian Contact through Oklahoma, 1904; Mineralization in the Copper Veins at Butte, Montana, 1912; The Geography of New Mexico, 1917; Significant Features in Western Coal Deposits; Steep Subsurface Folds versus Faults, 1926; also revs. and articles. Conducted explorations mining, petroleum and natural gas lands, also dam sites in Rocky Mountains, Mid-continent and Spanish America, from 1909. Died June 1, 1945.

KIRK, Edwin, Am. geologist; b. Richland, S.D., Dec. 6, 1884; s. Nathan Allen and Caroline (Freeman) K.; B.S., Columbia U., 1907, Ph.D., 1911; m. Page Taylor, June 26, 1913; children—Mary Mann (Mrs. James C. Mayer), Edwin Roger. Named jr. geologist, U. S. Geol. Survey, 1909, asst. geologist, 1911, asso. geologist, 1913, paleontologist, 1914, geologist, 1920. Fellow Geol. Soc. Am.; mem. A.A.A.S., Paleontol. Soc. Am. Research and publs. on fossils, especially fossil crinoids, Paleozoic echinoderms. Died Washington, Nov. 16, 1955.

KIRK, John Esben, physician; b. Kallundborg, Denmark, Nov. 8, 1905; s. Ole Christian Mathias and Augusta Dorthea (Bang) K.; M.D., U. Copenhagen (Denmark), 1929, Ph.D., 1936; m. Irma Louise Muser, May 28, 1934; children—Ermalynn (Mrs. George Belmont Wuestling), Peter, Lillian (Mrs. John Charles Arlt), Thomas, Paul. Came to U. S., 1947, naturalized, 1949. Practice internal medicine, Roskilde, Copenhagen, Holstebro, Denmark, N.Y.C., St. Louis; dir. City Health Labs., Copenhagen, 1936-39; chief physician Holstebro County (Denmark) Hosp., 1939-47; faculty Washington U. Sch. Medicine, St. Louis, 1947—, prof. medicine, 1964—, dir. research div. gerontology, 1947—. Fellow A.C.P.; mem. Royal Soc. Medicine, Harvey Soc., Gerontological Soc. (past treas.), Am. Soc. for Study Arteriosclerosis (past dir.), Am. Soc. for Study Permortal Condition (past pres.), Westjutland Med. Soc. (Denmark; founder, hon. mem.). Author: Amino Acid and Ammonia Metabolism in Liver Diseases, 1936; Acidosis: Clinical Aspects and Treatment with Isotonic Sodium Bicarbonate Solution, 1942; (with S. A. Kvorning) Hypometabolism: A Clinical Study of 308 Consecutive Cases, 1946; Human Intestinal Gas and Meteorism: a Clinical Review with a Report of Quantitative Gas Analyses, 1947; Hand Washing: Quantitative Studies on Skin Lipid Removal by Soaps and Detergents Based on 1500 Experiments, 1966; also numerous articles. Research on treatment of acidosis with isotonic sodium bicarbonate solution, blood ammonia levels in cirrhosis, renal tubular secretion of uric acid, enzymes of arterial disease. Home: 7514 Oxford Dr., Clayton, Mo. 63105. Office: 5600 Arsenal St., St. Louis 63139.*

KIRK, Norman Thomas, Am. surgeon; b. Rising Sun, Md., Jan. 3, 1888; s. Thomas and Anna May (Brown) K.; M.D., U. Md. 1910, S.D.; S.D., Davidson Coll.; M. Anne Duryea, Sept. 1917; 2 daus. Entered U. S. Army, 1912, advanced through grades to brig. gen., became surgeon-gen. Recipient Distinguished Service medal So. Surg. Assn., 1945. Fellow Am. Coll. Surgeons (gov.), A.C.P. Author: Amputations, Operative Technique, 1924. Dealt with health problems of U. S. Army during World War I; worked chiefly with neuropsychiatric cases, procurement of surg. instruments, treatment and prevention of malaria, also expert on amputation (name given to certain type of tendoplastic amputation). Died 1960.

KIRK, Paul Leland, Am. criminalist, chemist. b. Colorado Springs, Colo., May 9, 1902; s. Elliott Hartman and Mary Ellen (Mitchell) K.; A.B., Ohio State U., 1924; M.S., U. Pitts., 1925; Ph.D. in Bio-chemistry, U. Cal., 1927; m. Reba Louise Charlton, July 7, 1928; children—Mary Elise (Mrs. Norman Charles Janke), Anita Louise (Mrs. Arnold Raymond Kaplan). Mem. faculty U. Cal. at Berkeley, 1929—, prof. biochemistry and criminalistics, 1951-54, prof. criminalistics, 1954—; owner, cons. chemist firm Paul L. Kirk, PH.D. and Assos., Berkeley, 1934—; mem. staff research Manhattan Project, 1942-45; in charge microchem. research and devel. Hanford Engr. Works, Richland, Wash., 1944-45. Hon. fellow Am. Inst. Chemists, N.Y. Acad. Sci., Koninklejke Vlaamse Academie Voor Wetenschappen, Lettern En Schone, Kunsten van Belgie; mem. A.A.A.S., Am. Chem. Soc., Am. Soc. Biol. Chemists, Am. Soc. Criminology, Internat. Assn. Forsensic Toxicologist (charter), Austrian Microchem. Soc. (hon.), others. Research and publs. on dissociation constants, microscopy of various amino acids; quantitative drop analysis; forensic chemistry and gas chromatography. Home: 1064 Creston Rd., Berkeley 94708. Office: 372 Colusa Av., Berkeley, Cal. 94707.*

KIRK, Raymond Eller, Am. chemist; b. Hamilton County, Neb., June 24, 1890; s. Joseph Alexander and Virginia Eads (Eller) K.; student Neb. State Normal, Kearney, 1910-13; B.S., U. Neb., 1915; M.S., Ia. State Coll., 1917; postgrad. U. Chgo., 1919; Ph.D., Cornell, 1927; m. Beth Sibley, June 30, 1920; children—Virginia, Josephine Alvira. Instr. in chemistry Ia. State Coll., 1917-20; asst. prof. chemistry U. Minn., 1920-27, asso. prof., 1927-29; prof., head dept. chemistry Mont. State Coll., and state chemist, 1929-31; prof., head dept. chemistry, Poly. Inst., Bklyn., 1931-55, also dean grad. sch., from 1944. Dir. Shellac Research Bur., 1936-42. Fellow Am. Inst. Chemists, A.A.A.S.; mem. Am. Chem. Soc., Am. Assn. Univ. Profs., Sigma Xi, Phi Lambda Upsilon, Phi Kappa Phi, Gamma Alpha, Delta Sigma Rho. Author: Laboratory Manual in Inorganic Chemistry (with M. C. Sneed), 1927. Co-editor Encyclopedia of Chemical Technology (15 volumes), 1947-56. Medical bd. editors Inorganic Synthesis, 1939. Contbr. to sci. jours. Died Feb. 6, 1957.

KIRK, William Gordon, animal scientist; b. Kirkton, Ont., Can., Jan. 26, 1898; s. Joseph and Annie (Switzer) K.; Asso. certificate, U. Sask., 1920, B.S., 1928; M.S., Ia. State U., 1931, Ph.D., 1934; m. Bessie Stewart McEown, Apr. 16, 1936. Vice dir. in charge Range Cattle Expt. Sta., Ona, Fla., 1941-65, animal scientist, 1965-68; vis. prof. animal sci. U. Mandalay, Burma, 1961-62. Mem. A.A.A.S., Am. Soc. Animal Sci., Range Mgmt. Soc. Am., Agr. Inst. Can., Am. Grassland Council, Soil and Crop Sci. Soc. Fla., Fla. Acad. Sci., Gamma Sigma Delta, Phi Kappa Phi. Research on utilization of citrus feeds, sugarcane bagasse, shrimp meal and urea for maintenance and fattening beef cattle, mineral supplements for swine and cattle, crossbreeding in comml. beef prodn. grazing value of Fla., pasture species. Home: 205 N. 8th Av., Wauchula, Fla. 33873. Office: Range Cattle Sta., Ona, Fla. 33865.*

KIRKALDY, John Samuel, Canadian metallurgist; b. Victoria, B.C., Can., May 13, 1926; s. Samuel N. G. and Onnalee I. (Greig) K.; B.A.Sc., U. B.C., 1949, M.A.Sc., 1951; Ph.D., McGill U., 1953; m. Annie Mary Berg, Sept. 3, 1952; children—Barbara, John, Lorne, Jennifer. Asst. prof. McGill U., 1954-57; faculty McMaster U., Hamilton, Ont., Can., 1957—, prof. metallurgy, 1963—, chmn. dept., 1962-66. Robert S. Williams lectr. Mass. Inst. Tech., 1964; NSF lectr. U. Denver, 1965; sr. fellow Royal Soc. Can., 1963-64. Mem. Canadian Inst. Mining and Metals, Am. Soc. Metals, Am. Inst. Mining, Metall. and Petroleum Engrs., A.A.A.S., Canadian Assn. Physicist, Inst. Metals (U.K.), Iron and Steel Inst. (U.K.). Author: (with R. G. Ward) Aspects of Modern Ferrous Metallurgy, 1963. Research and numerous publs. on application of irreversible thermodynamics to metall. and biol. processes, multicomponent diffusion, theory of stability of equilibrium and irreversible systems. Home: 31 Floresta Ct., Ancaster, Ont. Office: McMaster U., Hamilton, Ont., Can.*

KIRKBRIDE, Thomas Story, Am. physician; b. Morrisville, Pa., July 31, 1809; s. John and Elizabeth (Story) K.; M.D., U. P., 1832; m. Ann Jenks, 1839 (dec. 1862); children—Mrs. Thomas George Morton, Joseph John; m. 2d, Eliza Butler, 1866; children—Franklin Butler, Thomas Storey, Elizabeth Butler, Mary Butler. Resident physician Friends' Asylum for Insane, Phila., 1832, Pa. Hosp., 1833-35; physician-in-chief, supt. Pa. Hosp. for Insane, 1840-83; trustee 1st state hosp. in Pa., also Pa., Instn. for Blind. Mem. Coll. Physicians, Assn. Med. Supts. of Am. Instns. for Insane (an organizer, pres.). Author: Rules and Regulations for the Pennsylvania Hospital for the Insane, 1850; Construction, Organization and General Arrangements of Hospitals for the Insane, 1854. Pioneer in humane treatment of the mentally ill; insisted that insanity should be treated like other diseases, with equal hope of recovery; stressed value of hosp. treatment, occupations, amusements and surroundings, as well as recognition of personal dignity, as factors in the treatment of mental illness. Died Phila., Dec. 16, 1883.

KIRKENDALL, Walter Murray, Am. physician; b. Louisville, Mar. 31, 1917; s. Charles Allen and Margaret (Caplinger) K.; student Coll. Liberal Arts, U. Louisville, 1934-37, M.D., 1941; m. Margaret Jane Allen, Mar. 31, 1948; children—William Charles, James Allen, Matthew John, Thomas Murray, David Edwin, Nancy Jane, Mary Margaret, Kathryn Ann, Joseph Howard, Michael Bruce. With University Hosp., Ia. City, 1949-52; chief med. service VA Hosp., Iowa City, 1952-58; asso. prof. medicine U. Ia., 1958-59, prof. medicine, 1959—; dir. Cardiovascular Research Labs. U. Ia. Hosps., 1958—. Cons., VA Hosp., Iowa City, Nat. Heart Inst., NIH; chmn. council for high-blood pressure research Am. Heart Assn., 1966—; Louis Mark lectr. Am. Coll. Chest Physicians, 1963. Mem. Am. Assn. U. Profs., A.C.P., Central Clin. Research Club, Am. Fedn. Clin. Research, Central Soc. Clin. Research, Alpha Omega Alpha. Contbr. numerous articles to profl. jours. Clin. studies related to hypertension and renal disease. Home: 430 Brown St. Office: U. Ia. Hosps., Iowa City 52240.*

KIRKES, William Senhouse, English physician; b. Holker, Eng., 1823; M.D., Berlin, 1846; joined Bartholomew's Hosp., 1841, became med. registrar, demonstrator morbid anatomy, asst. physician, 1854, physician, 1864. Author: (with W. Baley) Recent Advances in the Physiology of Motion, 1848; (with James Paget) Hand-book of Physiology, 1848. Described effects of detached fibrinous deposits of heart mixing with circulating blood, 1852. Died 1864.

KIRKHAM, Don, Am. soil physicist, educator; b. Provo, Utah, Feb. 11, 1908; s. Francis Washington and Martha (Robison) K.; student U. Utah, 1925-27; A.B. with honors in physics, Columbia, 1933, A.M., 1934, Ph.D. in Physics, 1938; Dr. honoris causa in Agrl. Scis., Royal Agrl. U., Ghent, Belgium, 1963; m. Mary Elizabeth Erwin, Sept. 2, 1939; children—Victoria, Mary Beth, Don Collier. Asst. in physics Ernest Kempton Adams Precision Lab., Columbia, 1934-37; asst. prof., asso. Agrl. Expt. Sta., Utah State U., 1937-40; physicist Bur. Ordnance, USN, 1940-46; asso. prof. physics George Washington U., 1946. prof. soils and physics Ia. State U., Ames, 1946-59, Curtiss Distinguished prof. agr., prof. agronomy and physics, 1959—, dir. Water Resources Research Inst., 1964—. Land reclamation adviser Turkish Govt., 1959; adviser Internat. Atomic Energy Agy., Vienna, Austria, 1961; Ford Found. land reclamation cons., UAR, also lectr. Alexandria (Egypt) U., 1961; mem. council Sodic Soils Symposium, Budapest, Hungary, 1964. Fulbright prof., Netherlands, 1950-51; Fulbright prof., Guggenheim fellow, Belgium, 1957-58. Fellow Am. Phys. Soc., Am. Soc. Agronomy; mem. Am. Geophys. Union (award for paper on hydrology 1952), Soil. Sci. Soc. Am. (Stevenson award 1951), Am. Math. Assn., Netherlands Soc. Agrl. Research, Internat. Soil Sci. Soc. (del. Bucharest 1964), Ia. Acad. Scis., Sigma Xi, Phi Kappa Phi, Gamma Sigma Delta. Cons. editor Soil Sci., 1953—; asso. editor Water Resources Research, 1964—. Research, numerous articles on artificial drainage of land, measurement soil permeability and moisture, groundwater movement, soil aeration and structure. Home: 2109 Clark Av., Ames, Ia. 50810.*

KIRKLAND, Archie Howard, Am. entomologist; b. Huntington, Mass., June 4, 1873; s. Charles Henry and Jane Elizabeth (Parsons) K.; B.S., Mass. Agrl. Coll., 1894, M.S., 1896. Asst. entomologist Mass. Hatch Expt. Station, 1892-94; asst. entomologist Mass. Bd. Agr., Gypsy Moth Commn., 1894-1900; entomologist Bowker Insecticide Co., 1900-05; state supt. gypsy moth work, 1905-09; agt. and expert U. S. Bur. of Entomology, from 1911. Mem. Assn. Econ. Entomologists (pres. 1906). Perfected methods of controlling gypsy and brown tail moths, employing improved insecticides, spraying mechanisms and importation of parasites. Died 1931.

KIRKLIN, Byrl Raymond, Am. radiologist; b. Gaston, Ind., Sept. 22, 1888; s. John Walter and Sarah Lavina (McCreery) K.; B.S., Ind. U., 1926, M.D., 1914; m. Gladys Marie Webster, June 3, 1915; children—John Webster, Mary Webster. Radiologist Muncie (Ind.) Home Hosp. and private practice, 1916-25; radiologist Mayo Clinic, Rochester, Minn., 1926-54, chmn. sects. on Radiology, prof. radiology, Mayo Found., U. Minn., 1936-54. Sr. cons. to surgeon gen., U. S. Army and to USAF. Diplomate Am. Bd. Radiology (trustee, mem. adv. bd. med. specialties). Fellow A.C.P., Am. Coll. Radiology (pres. 1942-43), A.M.A., Internat. Coll. Surgeons (hon.); mem. Am. Roentgen Ray Soc. (pres. 1937-38), Radiol. Soc. N.Am., Am. Assn. Gastro-enterologists, Central Soc. Clin. Research, Am. Assn. Ry. Surgeons, Sigma Xi, Phi Rho Sigma; corr. mem. Academia Nacional de Medicina Republic de Colombia; hon. mem. Die Deutsche Röntgen-Gesellschaft (Germany), Royal Soc. Medicine (Eng.). Assn. Gastroenterologists of Paris (France), others. Contbr. to med. jours. Died Mar. 2, 1957; buried Rochester, Minn.

KIRKMAN, Hadley, Am. anatomist; b. Richmond, Ind., Mar. 14, 1901; s. Lee and Leila (Hadley) K.; A.B., State U. Ia., 1925; S.M., U. Chgo., 1929; Ph.D., Columbia Coll. Phys. and Surg., 1937; m. Gladys Tracy, Apr. 5, 1942; 1 dau., Tracy Leigh. Asst. prof. Ohio U., 1928-29; instr. N.Y. Med. Coll., 1929-32, Columbia Coll. Phys. and Surg., 1934-36; faculty Stanford Sch. Medicine, 1936—, prof. anatomy, 1949—. NSF fellow Chester Beatty Inst. for Cancer Research, Royal Cancer Hosp., London, Eng., 1957-58; USPHS Spl. fellow, 1958-59. Mem. A.A.-A.S., Histochem. Soc., Am. Assn. Anatomists, Am. Assn. for Cancer Research, N.Y. Acad. Scis., Am. Assn. U. Profs., Royal Soc. Medicine, Sigma Xi. Research, numerous publs. on functional histology pituitary gland, kidney, intestine, gall bladder, significance globule leucocytes in def. mechanisms, hormone-tumor relationships, epidemiology tumors in Syrian hamsters, cancer. Home: 623 Cabrillo Av., Stanford 94305.*

KIRKMAN, Thomas Penyngton, English mathematician; b. Bolton, Eng., 1806; B.A., U. Dublin, 1833, M.A., 1850; self-taught in math.; priest, 1836-45. Fellow Royal Soc., 1857. Research and publs. on pluquaternions, group theory, math. mnemonics, topology. Died 1895.

KIRKPATRICK, Clifford, Am. sociologist; b. Fitchburg, Mass., Oct. 22, 1898; s. Edwin Asbury and Florence (Clifford) K.; A.B., Clark Coll., 1920, A.M., 1922; Ph.D. (Harrison fellow), U. Pa., 1925; m. Mazelle Van Cleave, Apr. 24, 1959; children—Judith K., Meredith Kay, Laird Clifford. Prof. sociology U. Minn., 1935-49; chmn. dept. sociology Ind. U., Bloomington, 1949-53, prof. sociology, 1953—. Research dir. interviewing team Morale div. U. S. Strategic Bombing Survey, Germany, 1945; v.p. Nat. Council Family Relations, 1956; mem. adv. bd. Sociol. Abstracts, 1958—. Recipient Burgess award Nat. Council Family Relations, 1965. Guggenheim fellow, 1936-37. Mem. Am. Sociol. Assn. (past v.p.), Am. Assn. Marriage Counselors, Sociol. Research Assn. (past pres.), Ohio Valley Sociol. Assn. (past pres.). Author: Intelligence and Immigration, 1926; Religion in Human Affairs, 1929; Nazi Germany: Its Women and Family Life, 1938; What Science Says About Happiness in Marriage, 1947; The Family: As Process and Institution, 1955, 2d edit., 1963. Contbr. chpts. to Statistics in Social Studies, 1930; Man and His World 1932; War in the Twentieth Century, 1940; Premarital Dating Behavior, 1959; Ency. of Social Work, 1965; Internat. Ency. of the Social Sci., 1968. Research, publs. on effect of lang. handicaps; attitude measurement and familial roles; adjustment in marriage; religion and the social order; sexual aggression among coll. students. analysis of effects of social systems on family life. Home: Kirkcliff, Rural Route 2, Bloomington, Ind. 47401.*

KIRKPATRICK, Edwin Asbury, Am. psychologist; b. Peoria, Ia., Sept. 29, 1862; s. Francis Asbury and Catharine (Bradbury) K.; B.S., Ia. State Coll., 1887, M.Ph., 1889, scholar, 1889-90, fellow, 1890-91, Clark U.; m. Florence May Clifford, Aug. 20, 1895;

children—Marian Myrna, Clifford, Alice May, Ralph Leonard; m. 2d, Annis Louise Kinsman, May 2, 1927. Instr., Winona (Minn.) State Normal Sch., 1892-97; dir. child study dept., Fitchburg (Mass.) State Normal Sch., 1898-1928. Lectr. on edn., Smith Coll., 1905, Columbia, 1906, U. Chgo., 1907, U. Va., 1910, Cornell U., 1911, U. W.Va., 1913, U. Vt., 1914, Boston U., 1922. Recipient gold medal as collaborator of child-study exhibit made at St. Louis Expn., 1904. Author: Fundamentals of Child Study, 1903, revised 1929; Genetic Psychology, 1909; The Individual in the Making, 1911; Fundamentals of Sociology, 1916; Studies in Psychology, 1918; Imagination and Its Place in Education, 1919; Conduct Problems, 1930, Set B, 1931; The Sciences of Man in the Making, 1932; Mental Hygiene for Effective Living, 1934. Died Jan. 4, 1937.

KIRKPATRICK, Harry Allister, Am. physicist; b. Wessington, S.D., Nov. 8, 1891; s. Alexander and Helen Iowa (Harmon) K.; B.S., Occidental Coll., 1914; Ph.D., Cal. Inst. Tech., 1931; m. Anna Ethel Anderson, May 17, 1917 (dec. Jan. 1943); children—Helen Margaret (Mrs. C. Guy Warfel), Joann Louise; m. 2d, Olive Edith Hutchison, Dec. 31, 1950. Asso. in Physics U. Cal., Los Angeles, 1924-28; faculty Occidental Coll., Los Angeles, 1929—, chmn. dept. physics, 1937-41, prof., 1942-57, emeritus prof., 1957—; asst. prof. U. Hawaii, 1931-35; research scientist U. So. Cal. Research fellow Cal. Inst. Tech., 1935-41, research asso., 1958-63; staff mem., asso. head div. Radiation Lab., Mass. Inst. Tech., 1941-45; physicist Naval Ordnance Test Sta., Cal., 1946; hon. fellow U. Wis., Madison, 1958-59. Fellow Am. Phys. Soc.; mem. A.A.A.S., Am. Assn. Physics Tchrs. (pres. So. Cal. chpt. 1950-51), Sigma Xi, Sigma Pi Sigma, Alpha Tau Omega. Research, publs. on X-ray spectroscopy, Compton scattering of X-radiation by light elements, rules grating spectroscopy of X-rays and ultraviolet. Home: 5340 Kincheloe Dr., Los Angeles 90041.*

KIRKWOOD, Daniel, Am. astronomer; b. Harford County, Md., Sept. 27, 1814; s. John and Agnes (Hope) K.; A.M. (hon.), Washington Coll., 1849; m. Sarah McNair, 1845. Prin., Lancaster (Pa.) High Sch., 1843-49, Pottsville Acad., 1849-51; prof. math. Del. Coll., 1851-56, pres., 1854-56; prof. math. Ind. U., 1856-65, 1867-86; prof. math. and astronomy Jefferson Coll., Canonsburg, Pa.; apptd. lectr. Leland Stanford Jr. U., 1891. Mem. Am. Philos. Soc. Published his formula for rotation periods of planets in Proceedings of A.A.A.S., 1849, article on comets and meteors in Danville Quarterly Review, 1861. Author: Meteoric Astronomy, 1867; Comets and Meteors, 1873; The Asteroids, 1887. Explained unequal distbn. of asteroids in ring system of Saturn in terms of Kirkwood gaps, caused by perturbing effect of its satellites; his views subjected Laplace's theory to penetrating criticism. Died Riverside, Cal., June 11, 1895.

KIRKWOOD, John Gamble, Am. chemist; b. Gotebo, Okla., May 30, 1907; s. John Millard and Lillian (Gamble) K.; student Cal. Inst. Tech., 1923-25; S.B., U. Chgo., 1926, D.Sci., 1954; Ph.D., Mass. Inst. Tech., 1929; Sc.D., U. Brussels, 1959; m. Lillian Gladys Danielson, Sept. 5, 1930 (div. 1952); m. 2d, Platonia P. Kaldes, Mar. 11, 1958. 1 son, John Millard. Nat. research fellow Harvard, 1929-30; internat. research fellow, Leipzig and Munich, 1931-32; research asso., Mass. Inst. Tech., 1930-31, 1932-34; asst. prof. chemistry Cornell, 1934-37, asso. prof. U. Chgo., 1937-38; Todd prof. chemistry Cornell, 1938-47; Arthur Ames Noyes prof. chemistry Cal. Inst. Tech.; Sterling prof., chmn. dept. chemistry, dir. div. science, from 1956; Yale, Lorentz prof. theoretical physics, U. Leiden, 1959. Recipient Am. Chem. Soc. award in pure chemistry, 1936, Theodore William Richards medal Am. Chem. Soc., 1950, Gilbert Newton Lewis medal, 1953. Fellow Am. Acad. Arts and Scis., Am. Phys. Soc.; mem. Am. Chem. Soc., Nat. Acad. Scis. (fgn. sec.), Am. Philos. Soc., N.Y. Acad. Sci., Sigma Xi. Contbr. articles to scientific jours. Died Aug. 9, 1959; buried New Haven.

KIRRMANN, Albert, French organic chemist; b. Strasbourg, France, June 28, 1900; s. Emile and Melanie (Woelflin) K.; Licencié, Ecole normale Supérieure, 1921, Agrégé, 1923; Docteur ès Sciences, U. Paris, 1928; m. Suzanne Pinet, Oct. 8, 1927; 1 son, Jean-Michel. Prof., U. Bordeaux (France), 1930-34, U. Strasbourg (France), 1934-54, U. Paris, 1955; asst. dir. Ecole Normale Supérieure, Paris, 1955—. Decorated officier la Légion d'Honneur. Mem. Société Chimique de France, Société de Chimie Physique. Author: La Chimie d'hier et d'aujourd'hui, 1928; Chimie organique, 3 vols., 1927; also numerous articles. Research on organic reaction mechanisms, allylic rearrangements, amino aldehydes and their rearrangement, Grignard compounds, reaction kinetics. Home: 45 rue d'Ulm, Paris 5e, France.*

KIRSCHMER Otto, German physicist; b. Ingelfingen, Mar. 24, 1898; s. Karl and Lina (Braun) K.; ed. Munich Tech. Sch.; 1 child, Horst. Asst., Munich Tech. High sch.; dir. Kaiser-Wilhelm Inst.; asso. with constrn. of hydraulic power, Obernach, Walchensee; prof., Dresden; dir. Central Hydraulic Labs. of France, Maisons-Alfort; dir. M.A.N. Werk Gustarsburg; prof., Darmstadt, Germany. Author many works on hydraulic power. Home: Rosenhöhweg 15, 6100 Darmstadt. Office: Hochschulstrasse 1, 6100 Darmstadt, W. Germany.

KIRSHENBAUM, Abraham David, Am. chemist; b. N.Y.C., July 8, 1919; s. Samuel and Kate (Smotkin) K.; B.S., Coll. City N.Y., 1941; M.S., Polytech. Inst. Bklyn., 1945; m. Sarah Schrebrank, Sept. 19, 1943; children—Gerald Steven, Seymour Lester. Research scientist Manhattan Project, Columbia U., N.Y.C., 1941-46; research chemist Houdry Process Corp., Linwood, Pa., 1946-48, Standard Oil Devel. Co., Linden, N.J., 1948-49; project supr. Research Inst., Temple U., Phila., 1949-68, Picatinny Arsenal, Dover, N.J., 1968—. Mem. Am. Chem. Soc., A.A.A.S., Phi Lambda Upsilon. Publs. and patents in lubrication oil additives, tracer and exchange reactions, isotopes, inorganic fluorine chemistry, catalysis, decomposition studies, rubber, explosives and propellants, high vacuum techniques; gaseous burning velocity and calorimetric studies; high temperature phys. measurements and chem. reactions, free radicals, rare gas compounds.*

KIRSNER, Joseph Barnett, Am. physician; b. Boston, Sept. 21, 1909; s. Harris and Ida (Waizer) K.; M.D., Tufts U., 1933; Ph.D., U. Chgo., 1942; m. Minnie Shneider, Jan. 6, 1934; 1 son, Robert S. Mem. faculty U. Chgo., 1935—, prof. medicine, 1951—. Mem. numerous nat. profl. coms.; mem. adv. council Nat. Inst. Arthritis and Metabolic Diseases. Mem. Am. Gastroenterol. Assn. (governing bd.), A.M.A., others. Editorial bd. Annals Internal Medicine. Contbr. numerous articles to profl. jours. Research on various disorders of gastrointestinal system, especially peptic ulcers, nature, course, and mgmt. of inflammatory bowel disease; application of exfoliative cytology to cancer diagnosis. Home: 5805 S. Dorchester Av., Chgo. 60637.*

KIRTLAND, Jared Potter, Am. physician, naturalist; b. Wallingford, Conn., Nov. 10, 1793; s. Turhand and Mary (Potter) K.; M.D., Yale, 1815; attended Med. Dept., U. Pa., 1814; m. Caroline Atwater, May 22, 1815; m. 2d, Hannah Tousey, 1825. Probate judge, Wallingford, 1818; mem. Ohio Legislature, 1828-34; reformed penitentiary system, substituted indsl. work for inmates in place of previous idleness; trustee Western Res. Coll., 1833-35; prof. theory and practice of medicine Med. Coll. Ohio, Cin.; pres. Ohio Med. Conv., 1839, 49; a founder Cleve. Med. Coll., 1843, prof. theory and practice of medicine, 1843-64, prof. emeritus, 1864-77. Organized Cleve. Acad. Natural Sci. (reorganized as Kirtland Soc. Natural Scis. 1869), 1845, pres. until 1875; mem. A.A.A.S., Nat. Acad. Scis., Am. Philos. Soc. Editor Ohio Family Visitor, 1851-58. Research and publs. on climatology, insects, birds, fishes of Ohio; conducted intensive studies of mollusks, discovering that freshwater mollusks are bisexual; discovered parthogenesis in silkworm moth. Died Rockport, O., Dec. 10, 1877.

KIRWAN, Richard, Irish chemist, natural philosopher; b. Cloughballymore, Ireland, 1733; s. Martin and Mary (French) K.; ed. under Jesuits. France; LL.D., Dublin (Ireland) U., 1794; m. dau. of Sir Thomas Blake, Feb. 1757; children—Mrs. Trimleston, Mrs. Hill. Returned to Ireland, 1755; called to Irish bar, 1766; studied sci. in London, 1777-87; settled in Dublin, 1787. Recipient Copley medal 1782. Fellow Royal Soc., 1780; mem. Irish Acad. (became pres. 1799), French Acad. Scis. Author: Elements of Mineralogy, 1784; Essay on Phlogiston and the Constitution of Acids, 1787; An Estimate of the Temperature of Different Latitudes, 1787; Essay on the Analysis of Mineral Waters, 1799; Geological Essays, 1799. Proponent of phlogiston theory; Essay on Phlogiston trans. into French with a refutation of each chapter appended to it; Kirwan converted to antiphlogistic view, 1791; 1st to classify minerals using their chem. composition; suggested hydrogen is identical to phlogiston; one of 1st to study strontia; studied Irish geology and mineralogy; involved in controversy over Huttonian theory. Died Dublin, Ireland, June 1, 1812.

KIRWIN, Thomas Joseph, Am. urologist; b. Frederick, Md., 1891; s. James John and Margaret Mary (Surplus) K.; Ph.C., B.S., U. Mich., 1910; postgrad. U. Wis. 1912-13; M.D., Tulane, 1916; postgrad. Cornell, 1917; M.A. in anatomy, Columbia, 1923; M.S. in surgery, Yale, 1929; m. Margaret Hughes, Sept. 8, 1917; 1 dau., Ruth Ann (Mrs. William S. McLean). Chief of clinic, adjunct vis. urologist James Buchanan Brady Found., Dept. Urology, N.Y. Hosp.; attending genito-urinary surgeon, N.Y.C. Hosp., Welfare Island; cons. urologist Coney Island Hosp. (Bklyn.), Benedictine Hosp. (Kingston), Monmouth Meml. Hosp. (Long Branch, N.J.), South Nassau Communities Hosp.; prof. urology N.Y. Med. Coll.; dir. urology Flower and Fifth Avenue Hosp., Met. Hosp.; dir. Bird S. Coler Hosp. Diplomate Am. Bd. Urology. Fellow N.Y. Acad. Medicine, A.C.S., Internat. Coll. Surgeons; mem. A.M.A., Am., Italian, Internat. urol. socs., Rumanian Soc. Obstetrics, Gynecology and Urology, Societas Jacoponica Urologiae (hon.), Nu Sigma Nu, Sigma Xi. Author: (with Dr. O. S. Lowsley) Textbook of Urology, 1926, Urology for Nurses, 1936, Clinical Urology (2 vols..), 1940; chpts. in Oxford Loose Leaf Surgery, Ency. of Medicine; Cyclo. Medicine; also numerous articles in sci. jours. Devised instruments known as the Kirwin vesical neck resector, Kirwin prostatic resector, Kirwin lithotrite, Kirwin radon seed implanter, Kirwin measuring device for bladder tumor, Kirwin Automatic resectoscope, Kirwin cystoscope. Died Aug. 18, 1959.

KISER, Clyde Vernon, Am. demographer; b. Bessemer City, N.C., July 22, 1904; s. Augustus Burton and Minnie May (Carpenter) K.; A.B. (Mangum medal 1925), U. N.C., 1925, A.M., 1927; Ph.D. (Richard W. Gilder fellow 1930-31), Columbia, 1932; m. Louise Venable Kennedy, Feb. 24, 1934 (dec. Mar. 1954). Research fellow Milbank Meml. Found. N.Y.C., 1931-33, research asso., mem. tech. staff, 1933-62, sr. mem. tech. staff, 1962—; statis. cons. USPHS, 1936; vis. research asso., sr. research demographer Office Population Research, Princeton, 1942—; adj. prof. sociology N.Y.U., 1945-56. Chmn. coordinating com. Nat. Health Survey, 1936; chmn. com. N.Y.C., Internat. Population Conf., 1961; mem. census adv. com. Nat. Com. Vital Statistics. Recipient Grant Squires prize Columbia, 1940. Fellow Am. Statis. Assn.; mem. Am. Eugenics Soc. (pres. 1963—), Population Assn. Am (pres. 1952-53), Eastern Sociol. Soc. (v.p. 1959-60, Internat. Union Sci. Study Population (chmn. U. S. nat. com. 1958-61), Am. Pub. Health Assn., Am. Acad. Polit. and Social Sci., Am. Sociol. Soc. Author: Sea Island to City; A Study of St. Helena Islanders in Harlem and Other Urban Centers, 1932; Group Difference in Urban Fertility, 1942; (with Grabill and Whelpton) The Fertility of American Women, 1958. Editor: Research in Family Planning, 1962; (with Whelpton) Social and Psychological Factors Affecting Fertility, vols. I-V, 1946-58. Research and publs. on population trends, including a field study of Negro migration and investigations and analyses of ofcl. data on trends and differentials in family size. Home: 261 Hawthorne Av., Princeton, N.J. 08540. Office: Milbank Meml. Fund, 40 Wall St., N.Y.C. 10005.*

KISLIUK, Paul, Am. physicist; b. Phila., Feb. 22, 1922; s. Max and Sue (Pogust) K.; B.S., Queens Coll., 1943; M.A., Columbia U., 1947, Ph.D., 1952; m. Joy Rachin, Mar. 19, 1950; children—Amy, Margaret, Thomas, William. Mem. tech. staff Bell Tel. Labs., Murray Hill, N.J., 1952-62; with Aerospace Corp., Los Angeles, 1962—, sr. scientist, 1966—; vis. lectr. U. Cal., Los Angeles, 1965-66. Mem. Am. Phys. Soc., I.E.E.E. Publs. on microwave spectroscopy of gases, elec. discharges, interaction of gases with metal surfaces, lasers, optical spectroscopy of crystalline materials. Home: 532 Las Casas Av., Pacific Palisades, Cal. 90272. Office: P. O. Box 95085, Los Angeles 90045.*

KISSER, Josef George, Austrian botanist; b. Vienna, Austria, Sept. 19, 1899; s. Josef and Josefine (Feltl) K.; Dr.phil., U. Vienna, 1922; prof. h.c., U. Santa Maria, Rio Grande do Sul, Brazil, 1963; Dr. h.c., U. Göttingen, 1968; m. Mary Tuerlin, Feb. 28, 1925; children—Heinrich, Peter, Felix. Asst., Inst. for Plant Physiology, U. Vienna, 1920-30, Biol. Research Inst. Acad. Sci., Vienna, 1930-34, State Mus. for Natural History, 1934-36; prof., dir. Bot. Inst. and Garden, U. Agr., Vienna, 1936-38, 45—, rector, 1950-51, 62-65; sci. dir. Austrian Wood Research Inst., 1953-64. Recipient Österreichisches Ehrenkreuz für Wissenschaft und Kunst I. Klasse, 1960. Author: Leitfaden der botanischen Mikrotechnik, 1926; also numerous articles. Devel. new methods for micros. preparation of plant material; research on micros. wood structure especially in connection with wood tech., fume damage of plants and analysis of their physiol. causes. Home: 93, Baumgarten-Strasse, A-1140 Wien, Austria.*

KISSINGER, David George, Am. biologist; b. Reading, Pa., July 26, 1933; s. John Howard and Edith (Bixler) K.; B.A., Columbia Union Coll., 1954; M.S., U. Md., 1955, Ph.D., 1957; M.P.H., U. Cal. at Berkeley, 1958; m. Bessie Marchant Van Tassel, Aug. 21, 1955; children—Deborah Anne, David Ronn. Head biology dept. Oakwood Coll., Huntsville, Ala., 1958-60, Atlantic Union Coll., South Lancaster, Mass., 1960—. Am. Philos. Soc. grantee, 1959; Sigma Xi grantee, 1958, 61; NSF grantee, 1961, 63, 66. Mem. Entomol. Soc. Am., Entomol. Soc. Washington, Soc. Systematic Zoology, Sigma Xi. Author: Curculionidae of America North of Mexico, a Key to the Genera, 1964. Devised system of keys to permit identification of 300 species of one genus and the 400 genera from U. S.; classified weevils from N.Am. and the world. Home: 43 Bigelow Rd., South Lancaster, Mass. 01561.*

KISSINGER, Henry Alfred, polit. scientist; b. Furth, Germany, May 27, 1923; s. Louis and Paula (Stern) K.; A.B. summa cum laude, Harvard, 1950, M.A., 1952, Ph.D., 1954; m. Anne Fleischer, Feb. 6, 1949 (div. July 1964); children—Elisabeth, David. Faculty, Harvard, Cambridge, Mass., 1954—, dir. Def. Studies Program, 1958—, faculty mem. Center for Internat. Affairs, 1960—, prof. dept. govt., 1962—; study dir. nuclear weapons, fgn. policy Council Fgn. Relations, 1955-56; dir. Spl. Studies Project Rockefeller Bros. Fund, 1956-58; cons. state Dept. 1965—, U. S. Arms Control and Disarmament Agy., 1961-67, Rand Corp., 1961—, Nat. Security Council, 1961-62. Recipient citation for nuclear weapons, fgn. policy Overseas Press Club, 1958, Woodrow Wilson prize for best book in fields of govt., politics, internat. affairs, 1958. Mem. Am. Acad. Arts and Scis., Am. Polit. Sci. Assn. Author: A World Restored: Castlereagh, Metternich and the Restoration of Peace, 1812-1822, 1957; Nuclear Weapons and Foreign Policy, 1957; The Necessity for Choice: Prospects of American Foreign Policy, 1961; The Troubled Partnership: A Reappraisal of the Atlantic Alliance, 1965; Problems of National Strategy, 1965. Studies, publs. on U. S. fgn. policy and diplomacy; nuclear weapons and the arms race. Home: 419 Beacon St., Boston 02115. Office: 6 Divinity Av., Cambridge, Mass. 02138.*

KISSLINGER, Carl, Am. geophysicist; b. St. Louis, Aug. 30, 1926; s. Fred and Emma (Tobias) K.; B.S., St. Louis U., 1947, M.S., 1949, Ph.D., 1952; m. Millicent Ann Thorson, Mar. 27, 1948; children—Susan, Karen, Ellen, Pamela, Jerome. Faculty, St. Louis U., 1949-—, prof. geophysics, geophys. engring., 1961-—, chmn. dept., 1963-—. UNESCO expert in seismology, chief tech. adviser Internat. Inst. Seismology and Earthquake Engring., Tokyo, Japan, 1966-67. Mem. Am. Geophys. Union, Soc. Exploration Geophysicists, Seismol. Soc. Am., Am. Soc. Engring. Edn., Sigma Xi. Research on mechanism by which explosions generate seismic waves, seismic wave generation and propagation, seismic instrumentation. Home: 7434 Melrose Av., St. Louis 63130. Office: P.O. Box 8020, College Station, St. Louis 63156.*

KISTEMAKER, J. Dutch physicist; b. Kolhorn, Netherlands, Apr. 23, 1917; s. G. H. and E. (Blaauboer) K.; student Sch. Commerce, Alkmaar; student physics, math., U. Leyden; m. J. C. Keizer, 1943; 2 daus., 1 son. Asst. prof. N. Bohr's Inst., Copenhagen, Denmark, 1946-47; research physicist FOM-Found., 1947-53; dir. FOM-Inst. Atomic and Molecular Physics, Amsterdam, 1953-56; prof. Leyden U. 1956-—. Mem. sci. bd. Lab. Space Research, Utrecht, Netherlands. Mem. Nederlandse Natuurkundige Vereniging, Svenska Fysikersamcundet, Am. Inst. Physics. Author: (with J. Bigeleisen, A. O. Nier) Isotope Separation, 1957; also numerous articles. Editorial bd. Physica. Research in low temperature physics, including isotherms, temperature scale; isotope separation; ion source devel.; mass spectrometry; electronic and atomic collison physics. Home: Victorieplein 21 A/II, Amsterdam/zuid, Netherlands. Office: Kruislaan 407, Amsterdam/Watergraafsmeer, Netherlands.*

KISTIAKOWSKY, George Bogdan phys. chemist; b. Kiev, Russia, Nov. 18, 1900; son Bogdan and Mary (Berenstam) K.; came to U. S., 1926, naturalized citizen, 1933; student Kiev (Russia) pub. schs. and high sch. to 1918; Ph.D., U. of Berlin, 1925: D.Sc., Harvard, 1955, Williams Coll., 1958, Oxford 1959, U. Pa., 1960, U. Rochester, Carnegie Tech., 1961, Princeton U., Case Institute of Technology, 1962; m. Hildegard Moebius, 1926 (divorced 1942); 1 dau., Vera; m. 2d, Irma E. Shuler, 1945; m. 3d, Elaine Mahoney, 1962. Internat. research fellow, Princeton, 1926-28, research asso., 1928-30; asst. prof. chemistry, Harvard, 1930-33, asso. prof., 1933-37, prof. chemistry, 1937-59, 61-—; spl. asst. to The Pres. for sci. and tech., 1959-61; chmn. sci. bd., dir. Itek Corp. 1961-director The Cabot Corp. Mem. Pres.' science adv. com., 1957-63. Recipient Medal for Merit, 1946; Nichols medal, 1947; Brit. Medal for Service in Cause of Freedom, 1948; Priestly award, 1958; Willard Gibbs meda 1960; Medal Freedom, 1961; Ledlie prize, 1961, Parsons medal, 1963. Hon. fellow Chem. Soc. (London); mem. Royal Soc., Nat. Acad. Scis. (v.p. 1965-69), Am. Acad. Scis., Am. Chem. Soc., Am. Philos. Soc, Contbr. numerous articles to sci. jours. Research in chem. kinetics, thermodynamics of organic molecules, shock and detonation waves, molecule spectroscopy. Address: 12 Oxford St., Cambridge, Mass.

KISTNER, David Harold, Am. biologist; b. Cin., July 30, 1931; s. Harold Adolph and Hilda (Gick) K.; A.B., U. Chgo., 1952, S.B., 1956, Ph.D., 1957; m. Alzada Carlisle, Aug. 6, 1957; children—Alzada Hilda, Kymry Marie Carlisle. Research, teaching asst. U. Chgo., 1953-55, 56-57; research asso. Chgo. Acad. Scis., 1957; instr. U. Rochester, N.Y., 1957-59; prof. biology Chico (Cal.) State Coll., 1959-—. Research asso. Field Mus. Natural History, 1967-—. Fellow Chgo. Natural History Mus., 1955-56, Guggenheim Found., 1965-66; NSF grantee, 1959-—. Mem. Entomol. Soc. Am., Lepidopologists Soc., Soc. Study Evolution, Soc. Study Systematic Zoology, Am. Soc. Zoologists, Am. Soc. Naturalists, A.A.A.S., Sigma Xi, others. Research in Africa, Pacific, Asia; studies, publs. on mechanisms of integration of fgn. organisms into ant and termite socs.; behavior and phylogeny of Staphylinidae, including new genera and new species. Home: 3 Canterbury Circle, Chico, Cal. 95926.*

KISTNER, Robert William, Am. physician; b. Cin., Aug. 23, 1917; s. Alfred C. and Gertrude (Thienes) K.; B.A., U. Cin., 1938, M.D., 1942; m. Georgia Walker Golde, Aug. 26, 1943; children—Dana Ann, Robert William, Stephan P., Peter A. Instr., U. Cin., 1946-50, N.Y. State Coll. Medicine, 1951-52; faculty Harvard Med. Sch., Boston, 1952-—, asst. prof. obstetrics and gynecology, 1960-—. Sr. gynecologist and obstetrician Boston Hosp. for Women. Diplomate Am. Bd. Obstetrics and Gynecology. Fellow A.C.S., Am. Coll. Obstetricians and Gynecologists, Boston Obstet. Soc.; mem. A.M.A., Soc. Gynecol. Investigation, Am. Fedn. Clin. Research, Am. Fertility Soc., Endocrine Soc. Author: Principles and Practice of Gynecology, 1965; (with S. J. Behrman) Progress in Infertility, 1967; Practical Gynecologic Endocrinology, 1967; also numerous articles. Research on oral contraceptives; originator pseudopregnancy treatment of endometriosis, also induction of ovulation with clomiphene citrate; treatment of metastic endometrial cancer with progestin. Home: 454 Grove St., Needham, Mass. 02192.*

KIT, Saul, Am. biochem. virologist, educator; b. Passaic, N.J., Nov. 25, 1920; s. Isadore and Minnie (Dardick) K.; B.A. with highest honors in Biochemistry, U. Cal. at Berkeley, 1948, Ph.D., 1951; m. Dorothy Anken, Sept. 28, 1945; children—Sally, Malon, Gordon. Research biochemist U. Tex. M.D. Anderson Hosp. and Tumor Inst., 1953-55, asst. prof., 1956-57, asso. biochemist, 1957-60, biochemist, chief sect. nucleoprotein metabolism, 1961-62; faculty Baylor U. Coll. Medicine, Houston, 1956-—, prof. biochemistry, head div. biochem. virology, 1962-—; asso. prof. U. Tex. Postgrad. Sch. Medicine, 1959-61. Recipient Research Career award Nat. Insts. Allergy and Infectious Diseases, 1963. Mem. Am. Soc. Biol. Chemists, Am. Chem. Soc., Am. Soc. for Microbiology, Am. Assn. for Cancer Research, Am. Soc. for Cell Biology (treas. 1965-—), Phi Beta Kappa, Sigma Xi. Contbr. numerous articles to profl. jours. Asso. editor Cancer Research; editorial bd. Internat. Jour. Cancer. Research on induction of enzymes by animal viruses, mechanism of viral carcinogenesis, molecular biology, nucleic acids, biochemistry of cancer. Home: 11935 Wink Dr., Houston 77025.*

KITABATAKE, Takashi, Japanese physician; b. Aomori City, Japan, May 6, 1928; s. Ushitaro and Tsue (Kitabatake) Omori; grad. Hirosaki U. Sch. Medicine (Japan), 1952; M.D., Nagoya (Japan) U., 1958; m. Keiko Jone, Aug. 28, 1957; children—Sakae, Chiho, Masako. Asso. prof. Nagoya City U., 1963-64, lectr., 1964-—; chief dept. radiotherapy and exptl. radiology Aichi Cancer Center, Nagoya, 1964-67; prof. Niigata U., 1967-—. Mem. Japan Radiol. Assn., Japan Soc. Cancer Therapy, Japan Cancer Assn. Author: Clinical Radiology, 1964; Radiation Biology, 1964; also numerous articles Research on techniques and dose distbn. in moving field supervoltage therapy, sex ratio of offspring in X-ray workers, statis. studies on radiation carcinogenesis and shortening of life span. Home: 1-183 Sekiya-Honson-cho, Niigata. Office: Asahimachidori 1-757, Niigata, Japan.*

KITAIBEL, Paul, Hungarian chemist, botanist; b. Nagy-Marton, Hungary, 1757; student chemistry and botany Buda U.; became prof. botany and chemistry at Pest, Hungary, 1794. Author: Rare Plants of Hungary; General Botany of the Hungarian Empire; Hydrography of Hungary. Discovered tellurium (at the same time as Müller), 1789; systematic studies of Hungarian flora. Died Budapest, Hungary, 1817.

KITAMURA, Kanehiko, Japanese dermatologist; b. Kanagawa Prefecture, Japan, 1899; grad. Tokyo U., 1923; M.D. Became prof. Kumamoto Med. Coll., 1929; apptd. prof. Nagasaki Med. Coll., 1935; prof. Tokyo U., 1946. Author: Small Book of Dermatology. Research on an insect of Yangtze River, China; pigment and metabolism of skin; developed formula for diagnosis and treatment venereal diseases.

KITASATO, Shibasaburo, Japanese bacteriologist; b. Kumamoto, Japan, Dec. 20, 1856; ed. Med. Sch., U. Tokyo (Japan) 1883; studied under Robert Koch, Germany, 1885. Became dir. Research Inst. Infectious Diseases, 1891; named prof. Bacteriological Inst., U. Tokyo, 1892, placed in charge Bacteriological Inst., 1893; founder Kitasoto Lab., 1914; dean Med. Faculty Kei U. Mem. Imperial Acad., Japan Med. Assn. (pres.). Discovered (simultaneously with A. Yersin in Paris) infectious agt. of bubonic plague, 1894; isolated bacilli causing anthrax, dysentary, tetanus; developed (with E. A. von Behring) effective diptheria anti-toxin; worked in social hygiene. Died Wakanojo, Japan, June 13, 1931.

KITAZATO, Shibasaburo, see Kitasato, Shibasaburo.

KITCHEN, Sumner Wendell, Am. physicist; b. Somerville, Mass., Sept. 17, 1921; s. Albert Harold and Mabel (Kingston) K.; A.B., Oberlin Coll., 1943; Ph.D., N.Y. U., 1951. Physicist, Frankford Arsenal, Phila., 1943-46; instr. N.Y. U., 1946-50; group leader Earnest O. Lawrence Radiation Lab., Berkeley, Cal., 1950-54; mgr. nuclear analysis Knolls Atomic Power Lab., Schenectady, 1964-—. Mem. Am. Nuclear Soc. (chmn. N.E. N.Y. 1963-64), Am. Phys. Soc., N.Y. Acad. Sci., Sigma Xi. Research and publs. on design and devel. of nuclear reactors for naval applications; resonant cavity design for linear accelerator; electron attachment and collision phenomena for ionized particles; coinventor nondestructive test for stress corrosion cracking in non-ferrous metals; participant in devel. recoilless rifle. Home: 1311 Garner Av., Schenectady 12309. Office: P.O. Box 1072, Schenectady 12301.*

KITSON, Sir Albert Ernest, Brit. geologist; b. 1868; s. John K.; ed. in geology, mining and surveying Univ. and Sch. Mines, Melbourne; m. Margaret Legge Walker, 1910 (dec. 1920); m. 2d, Elinore Almond Ramage, 1927, 2 sons. In charge topog. and geol. surveys coalfields Geol. Survey Victoria, 1899; sr. geologist, 1903; prin., Mineral Survey, So. Nigeria, 1906-11; dir. Geol. Survey, Gold Coast, 1913-30. Govt. del. 12th-15th Internat. Geol. congresses, Toronto, 1913, Brussels, 1922, Madrid, 1926, S. Africa, 1929, World Power confs., London, 1924, Basle, 1926, London, 1928, Berlin, 1930, Empire Mining and Metall. Congress, Can., 1927. Recipient Wollaston Fund award, 1918, Lyell medal, 1927 (both Geol. Soc. London). Fellow Geol. Soc., Royal Geog. Soc., Royal Empire Soc.; mem. Geologists Assn. (pres.), Instn. Mining and Metallurgy (mem. council), Mineral. Soc., Australasian Inst. Mining and Metallurgy, Royal Soc. Victoria, Imperial Inst. (adv. council), Soc. Econ. Geologists U. S. A. (adv. council). Research and numerous publs. on geology, geography, mining, water-power, natural history and travel; traveled through Australasia, Can., U. S., Africa, Brit. Isles, Europe, examining geology and mineral fields; important discoveries include black and brown coalfields and oil-shale deposits in Nigeria, manganese, diamond and bauxite fields on Gold Coast, glacial and fossil beds in Victoria, Tasmania and Gold Coast. Died Mar. 8, 1937.

KITTEL, Charles, Am. physicist; b. N.Y.C., July 18, 1916; s. George Paul and Helen (Lemler) K.; student Mass. Inst. Tech., 1934-36; B.A., Cambridge U., 1938; Ph.D., U. Wis., 1941; m. Muriel A. Lister, June 23, 1938. Guggenheim fellow, 1946, 57, 64; research physicist Bell Telephone Labs., 1947-51; prof. physics U. Cal., Berkeley, 1951-—. Governing bd. Am. Inst. Physics, 1956-59. Recipient Oliver Buckley solid state physics prize, 1957. Fellow Am. Phys. Soc. (chmn. div. solid state physics 1957, council 1958-62), Am. Acad. Arts and Scis.; member Nat. Acad. Scis. Author: Introduction to Solid-state Physics, 1953; Elementary Statistical Physics, 1958; Quantum Theory of Solids, 1963; Mechanics (with M. A. Ruderman and W. D. Knight), 1965. Contbd. to application of microwave resonance methods to solid-state physics; research on theory of ferromagnetism, ultrasonics and math. physics. Home: 89 San Mateo Rd., Berkeley 7, Cal.

KITTEL, John (Howard), Am. metallurgist; b. Ritzville, Wash., Oct. 9 1919; s. John Charles and Mary (Wood) K.; B.S. in Metall. Engring., Wash. State U., 1943; m. Betty Anne Quackenbush, May 23, 1943; children—Loren Howard, Marian Elizabeth, Alan William, Jack Meredith. Metall. engr. NASA (formerly NACA), Cleve., 1943-47, aero. research scientist, 1950-51; asso. metallurgist Argonne Nat. Lab., Chgo., 1947-50, Argonne, Ill., 1951-61, sr. metall. engr., 1961-—. U. S. del. UN Internat. Conf. on Peaceful Uses Atomic Energy, Geneva, Switzerland, 1958, 64; cons., vis. lectr. Japan Atomic Energy Research Inst., Tokyo, Tokai-Mura, 1963. Mem. Am. Nuclear Soc. (dir. 1967-—; certificate merit for contbns. to tech. nuclear materials 1965), Am. Soc. for Metals, Am. Inst. Mining and Metall. Engrs., Research Soc. Am. Contbr. numerous articles to tech. jours. Research on high temperature materials for gas turbines and jet engines, effects nuclear irradiation on reactor materials; devel. improved nuclear fuels. Patentee fuel element designs for advanced fast breeder reactors. Office: 9700 S. Cass Av., Argonne, Ill. 60439.*

KITTLE, Charles Frederick, Am. surgeon; b. Athens, O., Oct. 24, 1921; s. Frederick F. and Ida (Falls) K.; A.B. with honors, Ohio University, Athens, O., 1944, Doctor of Laws, 1967; M.D. with honors University Chgo., 1945; M.S. in Surgery, U. Kan., 1950; m. Jeane Mignon Groenier, Mar. 23, 1945; children—Candace Mignon, Bradley Dean, Leslie Jeane, Brian David. Intern U. Chgo. Clinics, 1945-46; resident gen. and thoracic surgery U. Kan. Med. Center 1948-52; spl. tng. radio-isotopes for med. use Oak Ridge Inst. Nuclear Studies, 1950, cons. med. div., 1950-55; mem. faculty U. Kan. Sch. Medicine, 1950-66, asso. prof. surgery, lectr. history medicine, 1959-66; cons. thoracic surgery VA Hosp., Wadsworth, Kan., 1954-57, cons. gen. surgery, 1957-60; attending gen. surgery VA Hosp., Kansas City, Mo., 1954-66, VA Hosp., Wichita, Kan., 1955-62; prof. surgery, head sect. thoracic and cardiovascular surgery U. Chgo. Clinics, 1966-—; spl. research cardiovascular surgery, control of blood flow. Served as lt. (j.g.) USNR, 1946-48. Clin. fellow Am. Cancer Soc., 1950-52; Markle scholar med. scis., 1953-58. Diplomate Am. Bd. Surgery, Am. Bd. Thoracic Surgery. Mem. A.A.A.S., Am. Assn. History Medicine, Am. Assn. Thoracic Surgery, Am. Coll. Cardiology (bd. govs. Kan. 1963-66), A.C.S. (bd. govs. vascular surgery 1965-68), Am. Heart Assn. (chmn. program com. cardiovascular surgery, 1965-68, exec. com. cardiovascular surgery council 1962-—, asso. editor bull. 1964-66), A.M.A., Am. Physiol. Assn., Central Surg. Soc., Chgo. Med. Soc., Am. Surg. Assn., Internat. Cardiovascular Soc. (sec. 1965-—), Internat. Soc. Surgery, Soc. Clin. Surgery, Soc. Vascular Surgery, Soc. Univ. Surgeons (pres. 1966-67), Phi Beta Kappa, Sigma Xi. Home: 4950 Chicago Beach Dr., Chgo. 60615. Office: 950 E. 59th St., Chgo. 60637.

KITTOKU, Isomura (Iwamura) (Yoshinori), Japanese mathematician; author: Ketsugisho, (attempts measurement problems using rough approach to integral calculus); studied magic squares, magic circles, magic wheels; showed surface of a sphere is equal to pi times the square of diameter.

KITTREDGE, Joseph, Am. forest ecologist; b. Marblehead, Mass., Nov. 26, 1890; s. Joseph and Mabel Thorndyke (Hodges) K.; A.B. cum laude, Harvard, 1912, M.F., 1913; Ph.D., U. Minn., 1931; m. Dorothea C. Davis, Dec. 31, 1919. Forest asst. U. S. Forest Service, Ida., Mont.; 1913-17, chief Forest Investigations Office, Washington, 1919-23; silviculturist Lake States Forest Expt. Sta., St. Paul, 1923-31; prof. forestry U. Cal., Berkeley, 1932-54. Fellow A.A.A.S.; mem. Ecol. Soc. Am., Soc. Am. Foresters, Am. Soil Survey Assn., Internat. Soc. Soil Sci., Am. Geophys. Union, Sigma Xi. Author: Forest Influences, 1948; also articles. Research in forest planting, growth and silviculture of aspen, forest type, site and soil relations in lake states and effects of forests on losses of water by interception and evaporation; evolution of snow cover in sugar pine-fir zone; accumulation of litter and properties of forest floor in San Dimas Exptl. Forest, Cal. Home: 2663 Tallant Rd., Santa Barbara, Cal. 93105.*

KITTSLEY, Scott Loren, Am. chemist; b. Port Washington, Wis., Feb. 7, 1921; s. Harry Martin and Irene (Scott) K.; B.S., U. Wis., 1942; M.S., Western Res. U., 1944, Ph.D., 1945; m. Helen Ruth Jung, Aug. 23, 1946. Faculty, Marquette U., Milw., 1945-—, prof. chemistry, 1961-—, chmn. chemistry dept., 1957-62. Fellow A.A.A.S., Am. Inst. Chemists; mem. Am. Chem. Soc., Wis. Acad. Sci., Arts and Letters, Sigma Xi. Author: Physical Chemistry, 1955, 2d edit., 1963; also articles, book revs. Co-discoverer system consisting of 8 liquid phases in stable equilibrium; research chem. thermodynamics. Home: 839 N. Marshall St., Milw. 53202.*

KITZMILLER, James Blaine, Am. zoologist; b. Toledo, June 30, 1918; s. James B. and Ella (Halpin) K.; B.S., De Sales Coll., 1939; M.S., U. Mich., 1942, Ph.D., 1948; m. Mary Elizabeth Kiel, Oct. 25, 1943; children—Mary Kathleen, James Michael, Christine Ann, John Joseph. Faculty, U. Ill., Urbana, 1948-—, prof. zoology, 1959-—, chmn. dept., 1957-64. Cons., WHO, 1958-—, NIH, 1958-—. Mem. Am. Soc. Zoologists, Genetics Soc. Am., Soc. for Study Evolution, Am. Entomol. Soc., Am. Mosquito Control Assn. Author: Laboratory Exercises in General Zoology, 1959; also numerous articles. Research on genetics and cytogenetics of mosquitos especially Anopheles, using salivary gland chromosomes to reconstruct evolutionary relationships in closely related species groups. Home: 1406 Mayfair St., Champaign, Ill. Office: Morrill Hall, U. Ill., Urbana, Ill. 61801.*

KIVEL, Bennett, Am. physicist; b. Bklyn., Apr. 29, 1928; s. David L. and Grace (Schneider) K.; B.S., U. Fla., 1949, M.S., 1951; Ph.D., Yale, 1954; m. Deborah Chesler, June 21, 1956; children—Michael J., George T. Staff, Los Alamos Sci. Lab., 1951-55; prin. research scientist Avco-Everett Research Lab., Everett, Mass., 1955-—. Mem. ballistic missile def. adv. com. Advanced Research Project Agy. Research in nuclear physics, biol. effects of x-rays, spectra, physics. of air. Home, 146 Woodcrest Dr., Melrose, Mass. Office: 2385 Revere Beach Pkwy., Everett, Mass.*

KIVI, Erkki Ilmari, Finnish biologist; b. Helsinki, Finland, June 6, 1929; s. Lyydia Kivi; M.Sc. (Fil. maist.) U. Helsinki, 1951, Lic.Sc. (Fil.lis.), 1959, Ph.D., 1962; m. Hilkka Anneli Palonen, July 10, 1954; children—Paivi Anneli, Timo Ilmari. Plant breeder Plant Breeding Inst., Hankkija Wholesale Soc., Helsinki, 1952-—, dir., 1963-—; spl. lectr. U. Helsinki, 1964. NRC Can. fellow, 1961. Mem. Eucarpia. Research, publs. on use of artificially induced mutations in plant breeding, resistance to sprouting in head, breeding for quality of oil crops, agronomic and tech. value of spring wheat. Address: Tammisto, Helsingin pit., Finland.*

KIWISCH VON ROTTERAU, Franz, Bohemian gynecologist; b. Klattau/Bohemia, Apr. 30, 1814. Prof. in Prague and Würzburg. Author: Geburtskunde, 1851; Klinische Vorträge über spezielle Pathologie und Therapie der Krankheiten des weiblichen Geschlects, 2 vols., 1951-53; Die Krankheiten der Wöchnerinnen, 2 vols., 1940. Made gynecology and obstetrics into independent speciality. Died Prague, Bohemia, Oct. 24, 1852.

KIZER, Donald Earl, Am. biochemist; b. nr. Vinton, Ia., Oct. 12, 1921; s. Earl and Alma (Pardun) K.; B.S. in Chemistry, Upper Ia. U., 1947; M.S. in Bacteriology, Purdue U., 1950; Ph.D., U. N.C., 1954; m. Dora Lee Belding, Apr. 2, 1942; children—Paul, Karol, Nikola, Steven, Lisa, Donna. Research chemist biomed. div. The Samuel Roberts Noble Found., Ardmore, Okla., 1954-56, sr. research chemist, 1956-60, research asso., 1960-—. Mem. Am. Soc. Microbiology, Am. Chem. Soc., A.A.A.S., N.Y. Acad. Scis., Am. Assn. Cancer Research, Soc. for Exptl. Biology and Medicine, Am. Soc. Biol. Chemistry. Contbr. articles to tech. jours. Research on biochemistry of genesis of liver tumors which established existence of liver adenylic acid deaminase. Home: 825 H St. N.W. Office: P.O. Box 878, Ardmore, Okla. 73401.*

KJELDAHL, Johan Gustav Christoffer, Danish chemist; b. Denmark, 1849; dir. Carlsberg Labs., Copenhagen, Denmark; devised methods for determining different sugars in presence of each other; devised Kjeldahl method of determining protein by converting nitrogen to ammonia (important in modern agr. and bio-chemistry), 1893. Died 1900.

KLAATSCH, Hermann, German anthropologist; b. Berlin, Prussia, Mar. 10, 1863; ed. U. Heidelberg, U. Berlin; M.D. Prof. in Heidelberg; prof. anthropology; dir. Anthrop. Inst., ethnol. dept.; curator Anat. Inst., U. Breslau. Author: Der Werdegang der Menschheit und der Entstehung der Kulturen, 1920. Research on prehistoric skeletons; studied evolutionary history of man out of primitive stages. Died Eisenach, Germany, July 5, 1916.

KLAGER, Karl, chemist; b. Vienna, Austria, May 15, 1908; s. Leopold and Barbara (Strasser) K.; Diplome chemist, U. Vienna, 1931, Ph.D., 1934; m. Elisabeth Ramona Graven, Dec. 26, 1938; 1 son, Peter. Came to U. S., 1949, naturalized, 1955. Prodn. chemist Neuman, Arad, Roumania, 1934-36; pharm. research Chinoin, A. G. Budapest, Hungary, 1936-38; research chemist I. G. Farbenindustrie, Ludwigshafen, Germany, 1939-48; research and tech. operation mgr. Aerojet-Gen. Corp., Sacramento, 1949-—. Recipient U. S. Navy's Distinguished Pub. Service award, 1958. Mem. Am. Chem. Soc., Am. Rocket Soc., Research Soc. Am., Am. Inst. Chem. Engrs. Contbr. articles to tech. jours. Patentee in field. Research in high energy solid and liquid propellants, nitro compounds, nitropolymers, explosives; developed new high-energy chems. and synthesis procedures, new process for preparation unsymmetrical dimethylhydrazine for liquid rocket applications. Home: 4110 Riding Club Lane, Sacramento 95825. Office: P.O. Box 1947, Sacramento 95809.*

KLAGES, Friedrich August, German chemist; b. Bad Harzburg, Feb. 2, 1904; s. Heinrich and Frieda (Leonhardt) K.; ed. univs. Fribourg, Munich; Ph.D.; m. Rita Weinzierl, Aug. 16, 1929. Agrégé, 1935; became prof. chemistry, 1941. Mem. Deutscher Alpenverein. Author: Lehrbuch der organischen Chemie; Einführung in die organischen Chemie. Home: Schackstrasse 5/V, 8 München 22. Office: Karlstrasse 23, 8 München 2, W. Germany.

KLAGES, Ludwig, psychologist; b. Hanover, Germany Dec. 10, 1872. Tchr. at Kilchberg, Switzerland, from 1919; dir. pvt. seminars. Author: Prinzipien der Charakterologie, 1910; Handschrift und Charakter, 1917; Vom Wessen des Bewusstseins, 1921; Vom Kosmogonischen Eros, 1922; Einführung i. d. Psychologie der Handschrift, 1924; Zur Ausdruckslehre und Charakterkunde, 1927; Der Geist als Widersacher der Seele, 2 vols., 1929; Vom Sinn des Lebens, 1940; Mensch und Erde, 1956. Founder expressive psychology and sci. graphology; his vitalistic philosophy related to those of James and of Bergson. Died Kilchberg, July 29, 1956.

KLAHR, Carl Nathan, Am. physicist; b. Pitts. July 3, 1927; s. Samuel and Bertha Klahr; B.S., Carnegie Inst. Tech., M.S., D.Sc.; M.S. in Math. Econs., Mellon Grad. Sch. Indsl. Adminstrn., 1950; m. Dec. 28, 1953; children—Adina Beth, Jonathan, Phillip. Faculty, Carnegie Inst. Tech., 1947-49; asso. Cowles Column. for Research, 1949-50; research physicist Westinghouse Research Lab. 1950-52; project mgr. Nuclear Devel. Corp. Am., 1952-57; dept. head TRG, Inc., 1957-61; sr. asso. Fundamental Methods Assoc., Inc., N.Y.C., 1961-—. AEC fellow. Mem. Am. Phys. Soc., I.E.E.E., Operations Research Soc. Am., Am. Nuclear Soc., Am. Inst. Aeros. and Astronautics. Contbr. numerous articles to tech. jours. Patentee in field. Inventor, developer neutron transmutation doping for fabricating semiconductor devices, dustwall for hypervelocity impact shielding, nuclear reactor concepts for high economy, operations research methods for indsl. control and mgmt., reliability analyses, radar propagation. Office: 31 Union Square W., N.Y.C.*

KLAMANN, Dieter Paul Max, German chemist; b. Berlin, Germany, May 27, 1924; s. Kurt and Käthe (Mau) K.; 1. Staatspr., Technische Hochschule Wien (Austria), 1947, 2. Staatspr., 1950, Dr.techn., 1951; m. Elfriede Schönewald, Oct. 25, 1957; children—Jörg-Dieter, Uwe Matthias. Asst., Technische Hochschule Wien, 1949-54; faculty Tech. U. Berlin, 1956-—, prof. applied organic chemistry, 1965-—; prof. U. Hamburg (Germany), 1966-—; dir. research Esso A.G. Germany, 1960-—. Recipient C. G. Krafft medal Technische Hochschule Wien, 1951. Mem. Lubrication Fundamentals Com. Contbg. author: Ecyclopädie der Techn. Chemie, 1964. Research and publs. on chemistry of organic sulfur and nitrogen compounds, carbenes, chemistry and tech. detergents, lubeoils, polymers, petrochem. processes. Home: 24 Forsthöhe, 2104 Hamburg 92, West Germany.*

KLAPPENBACH, Miguel Angel, Uruguayan zoologist; b. Soriano, Uruguay, May 4, 1920; s. Leopoldo and Amada (Childe); m. Berta Ester Baptista, May 25, 1946; children—Sergio Gustavo, Celta Sonia. Staff, curator labs., Mus. Natural History Montevideo, Uruguay; curator zool. lab. Inst. Research Biol. Sci., Montevideo; tchr., 1958-—. Fellow, Conselho Nacional de Pesquisas de Brasil, Guggenheim Found., 1964. Mem. Soc. Uruguayan Zoologists, Soc. Argentina Ciencas Nat., Soc. Uruguayan Malcologists, Assn. Latino-Am. Herpetology and Ichhtiology. Research, publs. on zool. systematics (molluscs, amphibians, reptiles; from Uruguay and adjacent countries. Home: 1959 Constituyente Ap. 24. Office: 652 Buenos Aires, Montevideo, Uruguay.*

KLAPPER, Margaret Strange, Am. physician; b. New Orleans, May 23, 1914; d. William R. and Bessie (Cragin) Strange; B.S., Newcomb Coll., 1935; M.D., Tulane U., 1939; m. Clarence Edward Klapper, Apr. 17, 1946. Physician, instr. hygiene U. S.C., 1941-42; with Tulane U., New Orleans, 1942-46, instr., 1944-46; faculty Med. Coll. Ala., Birmingham, 1946-61, asst. prof. internal medicine, 1948-61; faculty U. Ala. Med. Center, Birmingham, 1951-—, asso. prof. clin. dentistry, 1959-—, asso. prof. medicine, 1961-66, prof. medicine, 1966-—, asst. dean for students, 1962-67, asso. dean for students, 1967-—, dir. med. outpatient div., 1957-68, dir. div. continuing med. edn., 1962-—. Diplomate Am. Bd. Internal Medicine. Fellow A.C.P.; mem. A.M.A., So. Med. Assn. (past chmn. sect. on medicine), Jefferson County Med. Soc., Med. Assn. State of Ala., Am. Fedn. for Clin. Research (sr.), Phi Beta Kappa, Sigma Xi, Alpha Omega Alpha, Alpha Epsilon Iota. Research and publs. on hypertension and relationship of hypertension and pyelonephritis. Home: 1619 Lakewood Dr., Birmingham, Ala. 35216.*

KLAPROTH, Martin Heinrich, German chemist; b. Wernigerode, Germany, Dec. 1, 1743; apothecary's apprentice; studied under Pott, Marggraff, Rose, in Berlin; at least 1 son, Heinrich Julius. Became apothecary, Hanover, Germany, 1766, Berlin, 1768-70; became prescriptionist, Danzig, Poland, 1770; set up own lab., 1780; apptd. tchr. chemistry Arty. Sch., 1792; named 1st prof. chemistry U. Berlin 1810. Fellow Royal Soc., 1795; mem. French Acad. Scis., Acad. Scis. Berlin (san. bd). Author: Beiträge sur chemischen Kenntnis der Mineralkörper, 1795; Chemisches Wörterbuch, 1807. Editor: (with Formey) Prussian Pharmacopeia, 1799. Father of analytical chemistry; improved methods of mineral analysis by bringing sample to constant weight through drying and ignition; discovered zirconia, 1789, uranium, 1789, chromium, 1798, cerium oxide, 1803; named tellurium (already discovered); pioneer in application of chemistry to archaeology. Died Berlin, Jan. 1, 1817.

KLASON, Johan Peter, Swedish chemist; b. Apr. 4, 1848; s. Kristoffer Adam and Elsa Helena (Billing) Claesson; Diploma, Lund U.; Ph.D., 1874; m. Marie Louise Hill; at least 1 child, Torsten. Lab prof. chem. instn., Lund, 1887; pres. bd. dirs. Lund Gas and Water Works; controller Klippans Brewery; prof. chem. tech. Chem. Inst. of Tech. U., Stockholm, 1890; prin. chem. tech. sch.; tchr. at Skogshogskolan. Hon. mem. Acad. Engrs. (medal 1943). Author over 100 articles. Research in lignin. Died Jan. 1, 1937.

KLATZO, Igor, physician; b. St. Petersburg, Russia, Oct. 9, 1916; s. George and Maria (Slussareff) K.; M.D., U. Freiburg, 1947; M.S., McGill U., 1951; m. Barbara Inwood, Feb. 10, 1952; children—Marie Luisa, Michael. Came to U. S., 1956, naturalized, 1962. Allan Blair Meml. fellow various centers U. S., Mexico, Europe, 1954-56; practice medicine, specializing in neuropathology, Bethesda, Md., 1956-—; head clin. neuropathology Nat. Inst. Neurol. Diseases and Blindness, 1956-67, acting chief lab. neuroanat. scis., 1967-—. Recipient Weil award Am. Assn. Neuropathology, 1965. Mem. Am. Assn. Neuropathologists (past pres.), World Fedn. Neurology Commn. Neuropathology (corr.), German Assn. Neuropathologists and Neuroanatomists (corr.). Editor (with Seitelberger) Brain Edema, 1967. Contbr. numerous articles to profl. jours. Studies on behaviour of blood-brain barrier, brain edema, exptl. prodn. neurofibrillary degeneration. Home: 10300 Rockville Pike, Rockville, Md. 20852. Office: NIH, Bethesda, Md. 20014.*

KLAUBER, Laurence Monroe, Am. elec. engr.; b. San Diego, Cal., Dec. 21, 1883; s. Abraham and Theresa (Epstein) K.; A.B., Stanford, 1908; LL.D., U. Cal. at Los Angeles, 1941; m. Grace Gould, Nov. 29, 1911; children—Alice K. (Mrs. David M. Miller), Philip Monroe. With San Diego Gas & Electric Co., 1911-—, pres., 1946-49, chmn. bd., 1949-54, cons., 1954-—; dir. Klauber-Wangenheim Co., San Diego. Lectr. biology Stanford, 1949-—; cons. curator reptiles San Diego Zool. Soc., 1922-—; hon. curator reptiles San Diego Soc. Natural History. Fellow Acad. Zoology (India, hon.), I.E.E.E., A.A.A.S. (past pres. Western div.); mem. Cal. Acad. Scis. (hon.), Herpetologists League, Am. Geog. Soc., Am. Mus. Natural History (corr.), Am. Soc. C.E., Am. Soc. M.E., Am. Chem. Soc., Pacific Coast Elec. Assn. (past pres.), Am., Pacific Coast (past pres.) gas assns., Am. Math. Soc., Math. Assn. Am., Soc. for Indsl. and Applied Math., Seismol. Soc. Am., Am. Meteorol. Soc., Am. Soc. Ichthyologists and Herpetologists (past pres.), San Diego Zool. Soc. (past pres.), So. Cal. Acad. Scis., Am. Ecol. Soc., Western Soc. Naturalists (past pres.), Soc. Systematic Zoology (past pres.), Internat. Soc. Toxinology, Am. Statist. Assn., Inst. Math. Statistics, Biometric Soc., Cal., San Diego hist. socs., Soc. for Bibliography Natural History, Sigma Xi, Tau Beta Pi. Author: Rattlesnakes: Their Habits, Life Histories and Influence on Mankind, 2 vols., 1956; also numerous articles on reptiles, elec. distbn. Inventor elec. distbn. apparatus. Home: 233 W. Juniper St., San Diego 92101. Office: Electric Bldg., San Diego 92112.*

KLAUS, Karl Karlovich (or Claus), chemist; b. Dorpat, Estonia, 1796; ed. Dorpat; pharmacist; asst. chemist, prof. chemistry U. Dorpat. Author (in Russian): Studies on Platinum Metals, 1854. Research on platinum residual metals; credited with discovering ruthenium, 1845. Died Dorpat, 1864.

KLAVINS, Janis Viliberts, pathologist; b. Rugaji, Latvia, May 6, 1921; s. Janis and Ida A. (Liepins) K.; student Latvian U., 1941-44; M.D., Christian Albrechts' U., 1948, Ph.D., 1959; m. Minjona Krumins, July 4, 1950; children—Ilze N., Lize K., Janis P., Filips K. Came to U. S., 1951, naturalized, 1957. Faculty, Western Res. U., 1955-60, asst. prof. pathology, 1960; faculty Duke U. Med. Center, 1960-65, prof., 1963-65; clin. prof. pathology State U. N.Y. Downstate Med. Center, Bklyn., 1965—; pathologist-in-chief Bklyn.-Cumberland Med. Center, 1965—. Mem. Royal Soc. Medicine, Am. Soc. Exptl. Pathology, Coll. Am. Pathologists, Internat. Coll. Pathologists, Am. Soc. Cytologists, Soc. for Exptl. Biology and Medicine, A.A.A.S., N.Y. Acad. Sci., Am. Assn. U. Profs., Am. Assn. Pathologists and Bacteriologists, Sigma Xi. Research, numerous publs. on tissue alterations due to amino acid imbalance, amino acid composition in serum, amino acid metabolism, iron absorption and deposition in tissues, hormonal effects on exptl. carcinogenesis. Home: 5 Broadmoor Rd., Scarsdale, N.Y. 10585. Office: 121 DeKalb Av., Bklyn. 11201.*

KLEBELSBERG-THUMBURG, Raimund, geologist; b. Brixen, Dec. 14, 1886; s. Konrad von Klebelsberg-Thumburg; ed. U. Munich, Ph.D., U. Vienna; Dr. honoris causa, U. Heidelberg, 1936; m. Martha Ferrari, 1917. Became lectr. U. Innsbruck (Austria), 1915, prof. geology and paleontology, 1921, rector, 1933-34, 42-45, also head Geol. Paleontol. Inst. Mem. Acad. Sci. Vienna, Acad. Scis. Munich, Russian Geog. Soc. Author: Beiträge zur Geologie Westturkestans, 1922; Geologischer Führer durch die Südtiroler Dolomiten, 1928; Geologie von Tirol, 1935; Handbuch der Gletscherkunde und Glazialgeologie, 2 vols., 1948-49. Editor, Zeitschrift für Gletscherkunde, 1927-57. Research on glacier geology; expdns. in Central Asia, So. Spain, Greece. Home: Schillerstrasse 13, Innsbruck, Austria.

KLEBER, Wilhelm Philipp Andreas, German mineralogist; b. Karlsruhe, Germany, Dec. 15, 1906; s. Josef and Philippine (Mohrschulz) K.; Dr.phil. nat., U. Heidelberg, 1931; m. Irmgard Knopf, Nov. 21, 1953; 1 dau., Gisela. Docent, U. Heidelberg, 1937-40; docent U. Bonn, 1940-43, unscheduled prof., 1943-53; prof. crystallography, mineralogy, Humboldt U., Berlin, 1953—, dir. Mineral. Inst. Recipient Nat. prize II; Alexander von Humboldt medal. Mem. German Assn. Crystallography, German Soc. Geol. Scis., German Mineral. Soc., others. Author. Angewandte Gitterphysik, 3d edit., 1960; Einführung in die Kristallographie, 9th edit., 1966; Kristallchemie, 1963; also numerous articles. Research on crystal chemistry, morphology, crystal growth, epitaxy. Home: 15 Ostendorfstrasse, 117 Berlin-Köpenik. Office: 43 Invalidenstrasse, 104 Berlin, East Germany.*

KLEBER, Willy, German chem. engr.; b. Louvain, Oct. 9, 1909; s. Anton and Louise (Keyser) K.; ed. univs. Bonn, Berlin; Ph.D.; m. Hedwig Ploberger, Aug. 21, 1940; 1 son, Michael. With Inst. Fermentation, 3 years; tech. dir. brewery and by-product food prodn., 3 years; tech. dir. meat and food prodn., 6 months; chief, testing sta. for a brewery, 4 years; sta. dir. Munich brewery. Author: Ueber die Umwandlung des Humulons beim Kochen mit Bierwürze; Zietgemässes Hopfenkochen; Die Bedeutung des rH in der Brauerei; Ueber die kolloide und biologische Stabilität der Biere; Zur Auflösung des Malzes. Home: Agnesstrasse 56a, München. Office: Romanstrasse 41, München 19, W. Germany.

KLEBS, Edwin Theodor Albrecht, German physiologist; b. Königsberg, Germany, 1834; studied under Kathke and Helmholtz in Königsberg, under Kölliker and Virchow in Würzburg, Germany; asso. prof., Bern, Switzerland; apptd. prof., Würzburg, 1873; in Zurich, 1882-93; came to U. S., 1895; prof. path. anatomy Rush Med. Coll., Chgo.; with pvt. lab., Hanover, Germany, 1900; with Orth's Inst., Berlin, 1905-13. Author (in German): Pathology of Gun-shot Wounds, 1873. Introduced microscope technique of paraffin impregnation, 1869; research on bacteriology of malaria, anthrax; 1st to produce Tb in cattle by feeding them infected milk, 1873; produced syphilis in apes, 1878; discovered causative bacillus of diphtheria (Klebs-Löffler bacillus, later isolated by F. Löffler), 1883. Died Bern, Switzerland, Oct. 23, 1913.

KLEBS, Georg, German botanist; b. Neidenburg, East Prussia, Oct. 23, 1857; ed. univs. Königsberg, Strasbourg, Würzburg Tübingen. Prof. botany univs. Basel, 1887, Halle, 1898, Heidelberg, 1907. Mem. acads. sci. Heidelberg, Turin. Author: Die Bedingungen der Fortpflanzung bei einigen Algen und Pilzen, 1896; Über die Rhythmik in der Entwicklung der Pflanzen, 1911. Research on lower plants and biology of plant propagation; recognized special characters of Dinophyceae Chrysophyceae and Xanthophyceae, 1883-84. Died Heidelberg, Germany, Oct. 15, 1918.

KLECHKOVSKY, Vsevolod Mavrikievich, Russian agrochemist; b. 1900; grad. Moscow Timiryazev Agrl. Acad., 1929. Instr., Moscow Timiryazev Agrl. Acad., 1930-55, prof., 1955—. Recipient Stalin prize, 1952. Mem. All-Union Lenin Acad. Agrl. Sci. Co-author: A Guide to Practical Work in Agrochemistry, 1937; The Use of Fertilizers in the Non-Black Earth Region, 1954; The Effect of Radiation on the DNA Composition of Plants, 1962. Research and publs. on crop nutrition using radioactive isotopes and nuclear irradiation; developer instruments for such research. Address: Moscow Timiryazev Agrl. Acad., Novoe sh. 51, Moscow, USSR.

KLECZKOWSKI, Alfred Alexander Peter, biochemist; b. Leningrad, USSR, Dec. 4, 1908; s. Alfred and Isabella (Rinkervitch) K.; M.D., Cracow (Poland) U., 1936; Ph.D., London (Eng.) U., 1943; m. Janina H. Mirzynska, July 20, 1932; 1 dau., Helen Ann. With Rothamsted Exptl. Sta., Harpenden, Eng., 1939—, sr. prin. sci. officer, 1961—. Fellow Royal Soc., 1962; mem. Biochem. Soc., Brit. Soc. Immunology, Sigma Xi. Research and numerous publs. on serological reactions, kinetics and energy requirements of effects of ultra violet radiations on viruses, enzymes, antibodies, antigens and proteins, biochemistry and biophysics of plant viruses, applications of statistics to problems in plant virology and photobiology. Home: 7 Grasmere Av. Office: Rothamsted Exptl. Sta., Harpenden, Herts, Eng.*

KLEEN, Werner Julius, German physicist; b. Hamburg, Germany, Oct. 29, 1907; s. Johann and Hedwig (Rosahl) K.; ed. univs. Göttingen, Heidelberg; Ph.D.; M. Johanna Elisabeth Roos, Aug. 3, 1933; children—Eva-Maria, Uwe, Jan. With Telefunken Soc. Berlin, 1931-46, wireless telegraph co., Paris, 1946-50, Nat. Inst. of Electronics of Madrid, 1950-52; dir. research Siemens and Halske, Munich, 1952—. Hon. prof. Munich Tech. High sch. Recipient plaque Swedish Tech. Inst., Stockholm, Gauss-Weber medal U. Göttingen. Mem. Assn. German Electro-technicians, Nachrictentechn. Ges., German Physics Soc., N.Y. Inst. Electronics Engrs. Author books on electronic tubes, electronic micro-waves, solid state physics. Home: Denningerstrasse 36, München. Office: Balanstrasse 78, München, W. Germany.

KLEEREKOPER, Herman, zoologist; b. Amsterdam, Netherlands, Dec. 19, 1910; s. J. Gerrit and Elizabeth (van Es) K.; M.A., U. Mich., 1948, Ph.D., 1950; D.Sc., Université de Paris, 1957; m. Coby van Neck, Apr. 3, 1933; children—Koos, Catie Mrs. J. R. Allan). Assistant prof. Escola Superior de Agricultura, Vicosa, Brazil, 1934; hydrobiologist State of Sao Paulo, Brazil, 1936-38, U. Sao Paulo, 1938-40; limnologist, dir. Biol. Sta. of Lagoa dos Quadros, Ministry Agr., Brazil, 1940-47; faculty McMaster U., Hamilton, Ont., Can., 1949-68, prof., 1953-68, chmn. dept. biology, 1953-59; prof. Inst. Life Sci., Tex. A. and M. U., College Station, 1968—. U. S. Dept. State fellow, 1947; NRC Can. fellow, 1956; Nuffield Found. fellow, 1960. Mem. Am. Soc. Zoologists, Physiology Soc. Can., Canadian Soc. Zoologists, Ecol. Soc. Am., A.A.A.S., Am. Soc. Limnology and Oceanography, Internat. Soc. Limnology, Animal Behavior Soc., Royal Soc. Medicine London (asso.). Author: Introduction to the Study of Limnology, 1944; Etude Limnologique de la Chimie des Sediments de Fond des Lacs de l'Ontario medidional, 1957; Olfaction in Fishes, 1968. Research, numerous publs. on hearing of fish, olfactory discrimination in fish, orientation through hearing in fish, orientation through olfaction in fish. Home: 2300 Morningstar Dr., Bryan, Tex. 77801. Office: Inst. Life Sci., Tex. A. and M. U., College Station, Tex. 77843.*

KLEIN, Abraham, Am. physicist; b. N.Y.C., Jan. 10, 1927; s. Phillip and Ida (Warshofsky) K.; B.A., Bklyn. Coll., 1947; M.A., Harvard, 1948, Ph.D., 1950; m. Murielle Leslie Pollack, June 18, 1950. children—Julia Meredith, Hilary Beth. Instr., Harvard, 1950-52, jr. fellow Soc. Fellows, 1952-55; faculty U. Pa., Phila., 1955—, prof. physics, 1958—; vis. prof. U. Paris (France), 1961-62. Cons. to pvt. cos. NSF fellow, 1961-62; Alfred Sloan fellow, 1961-63; recipient Distinguished Alumni award Bklyn. Coll., 1965. Fellow Am. Phys. Soc. (asso. editor Phys. Rev. 1965—), Italian, Japanese phys. socs., Am. Assn. U. Profs., A.A.A.S. Contbr. numerous articles to tech. jours. Research on theory of elementary particles, theory of many particle system in quantum mechanics, theory of nuclear structure. Home: 1313 Morris Rd., Wynnewood, Pa. 19096. Office: David Rittenhouse Lab., U. Pa., Phila. 19104.*

KLEIN, Claude Alois, physicist; b. Strasbourg, France, Nov. 4, 1925; s. Louis Claude and Elisabeth (Diebold) K.; B.S., U. Strasbourg, 1948, M.S., 1949; E.E., Ecole Supérieure d'Electricité, Sorbonne, Paris, France, 1951; Ph.D., Ecole Normale Supérieure, Sorbonne, 1955; m. Geneviève Marie-Louise Steinmetz, Jan. 2, 1950; 1 dau., Carine. Came to U. S., 1957, naturalized, 1964. Research engr. Ordnance Corps., French Army, Paris, 1951-55; asst. to mgr. div. mil. applications French AEC, Paris, 1955-57; prin. scientist research div. Raytheon Co., Waltham, Mass., 1957—; vis. lectr. U. Lyon, France, 1953-54, Ecole des Hautes Etudes Pratiques, Sorbonne, 1954-56, Lowell (Mass.) Technol. Inst., 1962-63. Mem. adv. com. Am. Carbon confs., 1965—. Mem. Am. Inst. Physics, I.E.E.E., Am. Inst. Aeros. and Astronautics, U. S. Naval Inst. Mem. editorial adv. bd. Nucleus, 1960—. Research, publs. on ionization and nonequilibrium conditions in rocket-engine exhausts, theory nuclear forces, hardcore and spin-orbit interactions from phase-shift analysis, radiation damage in semi-conductors; pioneered research in pyrolitic graphite; described and interpreted excitation mechanisms in electron-beam pumped lasers. Home: 9 Churchill Lane, Lexington, Mass. 02173. Office: 28 Seyon St., Waltham, Mass. 02154.*

KLEIN, Edward Emanuel, bacteriologist; b. Osijek, Yugoslavia, 1844; ed. Vienna; m. Sophia Amelia, 1877; went to Eng., 1872; with Brown Instn., 1872-97; lectr. advanced bacteriology St. Bartholomew's Hosp., London; with med. dept. Local Govt. Bd., 1871-1925. Fellow Royal Soc., 1875. Author: The Bacteriology and Etiology of Oriental Plague, 1906; also textbooks on histology and bacteriology. Opposed Koch's views on etiology of cholera; credited with being 1st to suggest a streptococcus causes scarlet fever, 1887. Died 1925.

KLEIN, (Christian) Felix, German mathematician; b. Düsseldorf, Germany, Apr. 25, 1849; student Bonn U., Leipzig, Göttingen (all Germany); M.D., Vienna, Austria, 1868; became asst., phys. inst., Bonn, 1866; lectr., Göttingen, 1871; named prof., Erlangen, Germany, 1872, Munich Tech. High Sch., 1875, Leipzig, 1886. Rep. Göttingen, World's Columbian Expn., Chg., 1893. Fellow Royal Soc., 1885 (Copley medal 1912); mem. French Acad. Scis., 1897. Author: Vorlesungen über das Ikosaeder und die Auflösung der Gleichungen vom fünften Grade, 1884; Vorlesungen über die Theorie der elliptischen Modalfunktionen, 1890-1912; Über lineare Differentialgleichungen der 2. Ordnung, 1894; Vorträge über ausgewählte Fragen der Elementargeometrie, 1895; Theorie des Kreisels, 2 vols., 1897-98; (with Fricke) Vorlesungen über die Theorie der automorphen Funktionen, 1897; Vorlesungen über die Entwicklung der Mathematik in 19 Jahrhundert, 1926-27. Became co-editor Mathematischen Annalen, 1875, Encylopädie der mathematischen Wissenschaften, 1898. Pioneer in work on Riemann geometry, hyperelliptic functions, differential functions; developed program in Erlangen for studying equivalence in transformation groups, thus unifying diverse forms of geometry, 1872; worked on application of math. to physics; influenced math. in U. S. through his numerous Am. pupils. Died Göttingen, June 22, 1925.

KLEIN, Fritz Shalom, Israeli chemist; b. Erfurt, Germany, Jan. 4, 1920; s. Ludwig E. and Elfriede (Braun) K.; M.Sc., Hebrew U., Jerusalem, 1942, Ph.D., 1947; m. Mussia Rotman, Sept. 19, 1949; children—Shelly, Ittai. Research asst. Hebrew U., Jerusalem, 1946-47; research scientist Israel Ministry Agr., Tel-Aviv, 1948-50; research asso. Weizmann Inst. Sci., Rehovoth, Israel, 1951-53, sr. scientist, 1954-63, asso. prof., 1964—; guest scientist Brookhaven Nat. Lab., Upton, N.Y., 1956-57, 1962-63; guest scientist Nat. Bur. Standards, Washington, 1963-64. Mem. Chem. Soc. London, Faraday Soc. London, Chem. Soc. Israel. Editor: High School Textbook in Chemistry, 1966. Research, numerous publs. on mechanism of isotopic exchange reactions in aqueous solutions, kinetic isotope effects in gas phase reactions, fast atom-molecule reactions, mechanism of ion-molecule reactions in the gas phase, method for enrichment of deuterium by isotopic exchange. Home: 1 Meonot Wolfson, Rehovoth, Israel.*

KLEIN, James Henry, Am. chem. engr.; b. Boston, Oct. 12, 1920; s. August C. and Maree (Keeling) K.; B.S. in Mech. Engring., Mass. Inst. Tech., 1943, M.S., 1943, Sc.D., in Chem. Engring., 1950; m. Mary Elizabeth Brown, Jan. 8, 1955; children—Carolyn, Cynthia, Frederic, Alison. Engr., Union Oil Co., 1943-46, Quinton Engrs., Ltd., 1946-47; research asso. Mass. Inst. Tech., 1947-50, engr., bus. mgr. Lexington project, 1948; project engr. Am. Research & Devel. Co., 1950-51; tech. dir. Inst. Inventive Research div. S.W. Research Inst., 1951-53; cons. engr., Boston, 1953-56, Denver, 1958—; tech. staff Ramo-Wooldridge, 1956-57; dir. engring. Stanley Aviation, 1958; dir. research Cook Batteries, 1959-63; v.p. research, co-founder, dir. Frost Engr. Devel. Co., 1961-64; v.p., treas., dir. Aerospace Engring. Sales Co., Denver, 1961—; v.p., dir. Umbroiler Co., Denver, 1958—; pres., treas. Klein Aerospace, Denver, 1966—; dir. Motion, Inc., Uptime Corp., Parrish Electronics. Cons. research founds., pvt. cos. Mem. Research Soc. Am., Am. Inst. Chem. Engrs., Am. Chem. Soc., Sigma Xi. Contbr. articles to tech. jours. Pioneered research in large-scale mass algae culture; developed lift slab bldg. constrn. techniques, high speed aircraft emergency escape systems; invented electro-chem. energy storage and conversion systems; invented and developed instrumentation and test equipment for med. research; devel. and comml. application computer equipment, electronic instrumentation, solar energy conversion systems. Home: 1852 Crestridge Dr., Littleton, Colo. 80120. Office: P.O. Box 1056, Englewood, Colo. 80110.*

KLEIN, James Raymond, Am. biochemist; b. Parkton, Md., May 8, 1908; s. Christian H. and Mildred (Lytle) K.; A.B., Duke, 1935, Ph.D., 1939; m. Lenetta Garrett, Sept. 2, 1932; children—Benjamin Garrett, Catherine Elizabeth. Instr., Duke Med. Sch., Durham, N.C., 1939-42; asst. prof. U. Coll. Medicine, Chgo., 1942-47; sr. biochemist Brookhaven Nat. Lab., Upton, N.Y., 1947—. Mem. Am. Soc. Biol. Chemists, Soc. for Exptl. Biology and Medicine, Radiation Research Soc. Contbr. numerous articles to tech. jours. Research in metabolism of various tissues, particularly central nervous tissue, liver, erythrocytes. Home: 100 Roseland Lane, East Pathogue, N.Y. 11772. Office: Biology Dept., Brookhaven Nat. Lab., Upton, N.Y., 11973.*

KLEIN, Johann Friedrich Carl, German mineralogist; b. Hanau-sur-Main, Germany, Aug. 15, 1842; prof. mineralogy, univs. Heidelberg, Göttingen, Berlin (all Germany); mem. French Acad. Scis. Research on crystallography. Died Berlin, June 23, 1907.

KLEIN, Karlheinz Odo Maria, Austrian physician; b. Vienna, Austria, Nov. 2, 1931; s. Karl and Maria (Rabe) K.; M.D., U. Vienna, 1955; m. Paula Nestl, July 27, 1959; children—Peter, Michael. Staff, Wilhelminen Hosp., Vienna, 1955—, resident in internal medicine, 1960—; practice medicine, Vienna, 1964—. Recipient Reward of City of Vienna, 1961; Cardinal Innitzer Fond grantee for sci. publs., 1962; 2d Eiselsberg prize (with Siedek), 1965. Mem. Viennese Soc. Physicians, Soc. Internal Medicin, Van Swieten Soc. Research, numerous publs. on dynamographical circulation analysis for research of pathphysiol. mechanisms, diagnosis and treatment of hypertension, arteriosclerosis and other diseases of arteries; oral hormonal cytology in male, male climacteric and otosclerosis; new verification of disturbances of vegetative system and how to influence it. Home: 78 Böcklin, Vienna 1020. Office: 15 Schönborn, Vienna 1080, Austria.*

KLEIN, Lawrence Robert, Am. economist; b. Omaha, Sept. 14, 1920; s. Leo Byron and Blanche (Monheit) K.; B.A., U. Cal. at Berkeley, 1942; Ph.D., Mass. Inst. Tech., 1944; M.A., Lincoln Coll., Oxford U., 1957; m. Sonia Adelson, Feb. 15, 1947; children—Hannah, Rebecca, Rachel, Jonathan. Faculty, U. Chgo., 1944-47; research asso. Nat. Bur. Econ. Research, 1948-50; faculty U. Mich., 1949-54; research asso. Survey Research Center, 1949-54, Oxford Inst. Statistics, 1954-58; mem. faculty U. Pa., 1958—, prof., 1958—, Univ. prof., 1964-68, Benjamin Franklin prof., 1968—; vis. prof. Osaka U., 1960, U. Colo., 1962, City U. N.Y., 1962-63, Hebrew U., 1964, Princeton U., 1966, U. Cal. at Berkeley, 1968. Prin. investigator econometric model project Brookings Instn., 1963—; cons. Canadian Govt., 1947 UN Com. Trade and Devel., E. I. du Pont de Nemours & Co., Inc. Fellow Econometric Soc. (past pres.), Am. Acad. Arts and Scis.; mem. Social Sci. Research Council (fellow 1945-46, 47-48, chmn. com.), Am. Econ. Assn. (John Bates Clark medalist 1959), Am. Assn. U. Profs. Author: The Keynesian Revolution, 1947; Textbook of Econometrics, 1953; An Econometric Model of the United States, 1929-1952, 1955. Editor: Internat. Econ. Rev., The Brookings Quar. Econometric Model of the U. S. Developed new methods of estimating econ. relations from observed data and techniques. Home: 1317 Medford Rd., Wynnewood, Pa. 19096. Office: University of Pa., Phila. 19104.*

KLEIN, Mark Joseph, Am. phys. metallurgist; b. Dubuque, Ia., May 27, 1921; s. August J. and Martha (Powers) K.; B.S., U. S. Mil. Acad., 1944; B.S., U. Nev., 1950; M.S., U. Wash., 1951; M.S., Stanford, 1960, Ph.D., 1961; m. Joyce E. Walta, June 11, 1952; children—David, Jeffry, Eric. Research metallurgist Pyremont Co., San Carlos, Cal., 1961; sr. scientist, research fellow Battelle Inst., Columbus, O., 1962-66; sr. staff research engr., Solar, San Diego, 1966—. Mem. Am. Soc. for Metals, Am. Inst. Mining, Metall. and Petroleum Engrs., Sigma Xi. Research and publs. on characterization of antiferromagnetic structure and behavior of interstitial atoms in metals using anelastic methods, effect of shock loading on imperfections in atomic structure of ceramics. Home: 3249 Loma Riviera Dr., San Diego 92110. Office: 2200 Pacific Hwy., San Diego 92112.*

KLEIN, Martin Jesse, Am. physicist; b. N.Y.C., June 25, 1924; s. Adolph and Mary (Neuman) K.; A.B., Columbia U., 1942, M.A., 1944; Ph.D., Mass. Inst. Tech., 1948; m. Miriam June Levin, Oct. 28, 1945; children—Rona Faith, Sarah Maeve, Nancy Ruth. Grad. asst. Columbia U., 1942-44; participant war research on sonar, Columbia U., 1944, Navy Operations Research Group, OSRD, 1945; research asso. physics Mass. Inst. Tech., 1946-49; with dept. physics Case Inst. Tech., Cleve., 1949-67, prof. physics 1960-67; prof. physics and history of science, Yale U., 1967—. NRC Postdoctoral fellow in physics Dublin Inst. for Advanced Studies, Ireland, 1952-53; Guggenheim fellow U. Leyden, Netherlands, 1958-59. Fellow Am. Phys. Soc., A.A.A.S., mem. Hist. Sci. Soc. Publs. in statis. mechanics, ferromagnetism, history of modern physics. Editor: Collected Scientific Papers of Paul Ehrenfest, 1959.*

KLEIN, Paul Günther, microbiologist; b. Sighisoara, Rumania, Oct. 23, 1919; s. Albert K. and Frida (Petrovits; K.; bachelor's degree, Brukenthalgymnasium, Sibiu, Rumania, 1938; student U. Cluj (Rumania), 1938-41, U. Tübingen (West Germany), 1941-42; M.D., U. Heidelberg (Germany), 1947; m. Hildegard Ludwig, Feb. 14, 1949; children—Joachim, Dorothea, Alexander, Stella. Asso., Inst. Pathology, Karlsruhe Hosp., 1948-52; asst. prof. Inst. Hygiene, Med. Acad., Düsseldorf, Germany, 1952-56, asso. prof., 1959-61; research asso. dept. pathology Cornell U. Med. Sch., N.Y.C., 1956-59; prof. med. microbiology, head dept. Johannes Gutenberg U. Med. Sch., Mainz, W. Germany, 1961—. Mem. Robert-Koch Institut (hon.), Royal Soc. Medicine London (affiliate), Collegium Internationale Allergologicum. Author: (with R. Böhmig) Pathologie und Bakteriologie der Endocarditis, 1953; Bakteriologische Grundlagen der Chemotherapeutischen Laboratoriumspraxis, 1957; also articles. Research on bactericidal and bacteriostatic effects of chemotherapeutical drugs, analysis of complement action, identification and description of new components, immunospecific staining of tissular antigen-antibody complexes in allergic disease. Home: 11 Luisenstrasse, Mainz-Gonsenheim, Germany. Office: Institut für Med. Mikrobiologie, 1 Langenbeckstrasse, 65 Mainz, Germany.*

KLEIN, Ralph, Am. phys. chemist; b. Pitts., Jan. 24, 1918; s. Ben and Rena (Heilbron) K.; B.S., Carnegie Inst. Tech., 1938; M.S., U. Minn., 1940; Ph.D., U. Pitts., 1950; widower; children—Kenneth, Douglas. Chemist, U. S. Bur. Mines, 1938-39, CWS, U. S. Army, 1940-41, U. S. Bur. Mines, 1945-55, Olin-Mathieson Chem. Corp., New Haven, 1955-60; chief surface chem. sect. Nat. Bur. Standards, Washington, 1960—. Recipient Meritorious Service award Dept. Commerce, 1964; Sci. and Tech. Fellow, 1964-65. Mem. Am. Chem. Soc., Am. Phys. Soc., Washington Acad. Sci., Sigma Xi. Publs. on research on establishment of bond energy in formaldehyde; synthesis of diborane; studies with field electron emission; reactions in cryogenic temperature region with atomic hydrogen. Home: 2215 Westview Dr., Silver Spring, Md. 20910. Office: Nat. Bur. Standards, Washington 20234.*

KLEIN, William H., Am. meteorologist; b. N.Y.C., Apr. 16, 1919; s. Abraham and Rae (Rosen) K.; B.S., Coll. City N.Y., 1938, M.S., 1940; M.S., Mass. Inst. Tech., 1944; postgrad. Am. U., 1951-61; Ph.D., N.Y. U., 1964; m. Jenny Witkowski, Dec. 26, 1942; children—Richard M., David B. Research forecaster U. S. Weather Bur., Suitland, Md., 1946-58, chief devel. and testing sect., 1959-64, dir. techniques devel. lab., Silver Spring, Md., 1964—. Recipient Meritorious Silver Service medal U. S. Dept. Commerce, 1964. Mem. Am. Meteorol. Soc., Am. Geophys. Union, Washington Acad. Scis., Royal Meteorol. Soc., A.A.A.S. Contbr. articles to profl. lit., monographs. Developed objective, computerized methods of predicting precipitation, surface temperature, and sea level pressure for 1 to 5 days in advance; contbns. to synoptic climatology. Home: 7921 Maryknoll Av., Bethesda, Md. 20034. Office: U. S. Dept. Commerce, ESSA, Weather Bur., Silver Spring, Md. 20910.*

KLEINER, Israel Simon, Am. biochemist; b. New Haven, Apr. 8, 1885; s. Isaac Lyon and Helen (Bretzfelder) K.; Ph.B., Yale, 1906, Ph.D., 1909; Sc.D., N.Y. Med. Coll., 1960; m. Alma Kempner, Mar. 27, 1912; children—Ruth Alma (Mrs. Arnold M. Glantz), Richard. Instr. physiol. chemistry Tulane U. Med. Coll., 1909-10; asst. Rockefeller Inst. for Med. Research, 1910-14, asso., 1914-19; prof. biochemistry N.Y. Med. Coll., 1919-60, dean, 1921-25, dir. dept., 1948-60; researcher Cold Spring Harbor, 1928-33. Cons. chemist Flower-Fifth Av. Hosp., N.Y.C., 1919-60. Recipient Alumni award N.Y. Med. Coll., 1959. Diplomate Am. Bd. Clin. Chemistry. Fellow Am. Assn. Clin. Chemists (Van Slyke award 1959); mem. Am. Physiol. Soc., Am. Soc. Biol. Chemists, Am. Soc. Pharmacology, Am. Inst. Nutrition, Soc. for Exptl. Biology and Medicine. Author: (with James M. Orten) Human Biochemistry, 1945; (with L. B. Dotti) Laboratory Instructions in Biochemistry, 1940; also numerous articles. Research in endocrine secretions, especially insulin. Home: 27 W. 96th St., N.Y.C. 10025. Office: 106th St. and Fifth Av., N.Y.C. 10029.*

KLEINERT, Theodore N., chemist; b. Vienna, Austria, Mar. 25, 1897; s. Theodore and Amalia (Herusch) K.; Dipl.Chem., Tech. U. Vienna, 1921, D.Sc., 1924; m. Rotrud Guggitz, Nov. 10, 1951 (dec. Feb. 1965); children—Eva L., Diane S. Chemist, Salpetreplant Erwa, 1921-22; asst. prof. Tech. U. Vienna, 1924-30; indsl. cons., 1930-40; research asso. Schlesische Cellulose & Papierfabriken E. Schoeller & Co., Germany, 1941-42; dir. research and devel. Lenzinger Zellwolle and Papierfabrik A.G., Austria, 1943-53; div. head Pulp Paper Research Inst. Can., Montreal, Pointe Claire, Que., 1954-62, prin. scientist, 1960-62; indsl. cons., 1962—. Mem. Austrian Assn. Pulp and Paper Chemists (K. Kellner award 1955), Canadian Pulp and Paper Assn. (I. H. Weldon award 1965), T.A.P.P.I. Research, numerous publs. and patents in cellulose, lignin, and wood chemistry, viscose process, prodn. artificial fibres, degradation of high polymers by free-radical mechanisms, devel. rapid pulping processes. Home and office: 120 Embleton Crescent, Pointe Claire, Que., Can.*

KLEINHOLZ, Lewis Hermann, Am. biologist; b. N.Y.C., May 18, 1910; s. Jacob Karl and Fannie (Geller) K.; B.Sc., Colby Coll., 1930, D.Sc. (hon.), 1953; M.A., Harvard, 1935, Ph.D., 1937; mem. faculty Colby Coll., 1930-33, Harvard, 1940-41; dir. invertebrate zoology staff Marine Biol. Lab., Woods Hole, Mass., 1949-53; faculty Reed Coll., Portland, Ore., 1946—, prof. biology, 1950—. Research investigator biology, Naples, Italy, 1937, 58, 60, 62, Plymouth, Eng., 1960; vis. research prof. U. Lund, Sweden, 1961-62. Guggenheim fellow, 1945-46; Fulbright fellow, 1951-52; NSF fellow, 1958-59. Fellow N.Y. Acad. Sci.; mem. Marine Biol. Assn. U.K., Marine Biol. Assn. India, Am. Soc. Zoologists, Phi Beta Kappa, Sigma Xi. Author: (with others) Selected Invertebrate Types, 1950, Physiology of Crustacea, 1960. Studies in physiology of pigmentary effectors; biochem. separation, purification of neurosecretory hormones.*

KLEINSCHMIDT, Hans Adolf, German pediatrician; b. Elberfeld, June 18, 1885; s. Eduard and Mathilde (Tillmans) K.; ed. univs. Fribourg, Kiel, Munich, Bonn; M.D.; m. Marie Nebelthau, Oct. 4, 1912; children—Gisela, Siegmar, Eberhard. Med. asst., instr., Marburg, also Berlin; clinic dir. U. univs. Hamburg, Cologne, Göttingen (all Germany). Hon. mem. Univs. Santiago, Conception. Recipient Bernard O'Higgins award; Emil von Behring prize U. Marburg; Albrecht von Haller medal U. Göttingen. Mem. various Am. and European pediatric and Tb orgns. Author: Milchanaphylaxie; Bildung von Diphterieantitoxin; Fettangereicherte Säuglingsnahrung; Rückbildungfahige Lungenverschattungen bei Tuberkulose; Lehrbuch der Kinderheilkunde. Address: Karlstrasse 41, Bad Honnef, W. Germany.

KLEINZELLER, Arnost, Czechoslovakian biochemist; b. Ostrava, Czechoslovakia, Dec. 6, 1914; s. Arnold and Josefa (Schongut) K.; M.D., U. Brno., Czechoslovakia, 1938; Ph.D., U. Sheffield, Eng., 1939-41; Rockefeller fellow U. Cambridge, 1942; D.Sc., U. Prague, 1959; m. Lotte Reuter, Apr. 2, 1943; children—Anna, Jana. Head lab. cell metabolism State Inst. Health, Prague, Czechoslovakia, 1945-48; faculty Tech. U., Prague, 1948-52, Charles U., Prague, 1952-55; head lab. cell metabolism Czechoslovak Acad. Sci., Prague, 1956—; vis. prof. dept. physiology U. Rochester (N.Y.) Sch. Medicine & Dentistry, 1966—. Mem. exec. com. Internat. Cell Research Orgn., 1962-67, chmn. panel IV 1962—. Med. Research Council fellow, 1943-44. Mem. Internat. Soc. Cell Biology (exec. com.), Czechoslovak Nat. Com. Biol. Scis. (hon. sec.), Biochem. Soc. Gt. Britain, Soc. Gen. Microbiology Gt. Britain, Academia Leopoldina, Gt. Britain Physiol. Soc. (asso.). Author: (with Vrba and Malek) Manometric Methods, 1954; Membrane Transport & Metabolism, 1961; (with Malek, Longmuir, Cerkasov, and Kovac) Manometric Methods, 1964. Research in metabolic processes in cells and tissues; mechanism of fatty acid formation in yeast cells; transport of electrolytes and sugars across cell membranes. Home: 8, Kratka, Prague 10, Czechoslovakia. Office: Czechoslovak Acad. Scis., Lab. for Cell Metabolism, Malá Plynarni, Prague 7, Czechoslovakia.*

KLEIST, E. Georg von, see von Kleist, E. Georg.

KLEITMAN, Nathaniel, physician; b. Kishinev, Russia, Apr. 26, 1895; s. Nathaniel and Pessia (Galanter) K.; came to U. S., 1915; B.C., City Coll. N.Y., 1919; A.M., Columbia, 1920; Ph.D., U. Chgo., 1923; m. Paulena Schweizer, June 14, 1927; children—Hortense, Ester. Instr. physiology and pharmacology U. Ga., 1920-22; NRC fellow U. Utrecht (Netherlands), U. Paris (France), U. Chgo., 1923-25; asst. prof. physiology U. Chgo., 1925-29, asso. prof., 1929-50, prof., 1950—. Cons., investigator Naval Med. Research Inst., Bethesda, Md., 1949. Recipient Harris medal City Coll. N.Y. Alumni Assn., 1947. Mem. Am. Physiol. Soc., Soc. Exptl. Biology, Inst. Medicine, Chgo. Author: Sleep Characteristics, 1937; Sleep and Wakefulness, 1939; also numerous articles. Pioneer in research on physiol. aspects of sleep, diurnal cycle of temperature, conditioned reflexes. Home: 222 Washington Av., Santa Monica, Cal. 90403.

KLEIWEG DE ZWAAN, Johannes Pieter, Dutch anthropologist; b. July 4, 1875; s. Hendrik and Adriana Jacoba (Kleiweg) de Zwaan; ed. univs. Leiden, Amsterdam, Berlin, Paris; M.D.; m. H. F. J. Vellema, Feb. 12, 1912; children—Pieter Johannes, Helene, Fredrika, Johanna. Prof. agrege, prof. U. Amsterdam; prof. Inst. Tropical Scis., Asmterdam; prof. anthropology and prehistory. Mem. Internat. Inst. Anthropology (v.p.), Ned. Nationaal Bur. voor Anthropologie Netherlandish Nat. Bur. Anthropologis, Kon. Ned. Aardrijkskundig Genootschap. Author: Durck Zentral Sumatra; Die Insel Vias bei Sumatra; Volkerkundliches und Geschichtliches uber die keilkunde der Chinesen und Japaner. Address: Noolseweg 30, Blaricum, Netherlands.

KLEMA, Ernest Donald, Am. nuclear physicist; b. Wilson, Kan., Oct. 4, 1920; s. William W. and Mary Bess (Vopat) K.; A.B. in Chemistry, U. Kan., 1941, M.A., in Physics, 1942; postgrad. Princeton, 1942, U. Ill., 1946-49; Ph.D. in Physics, Rice U., 1951; m. Virginia Clyde Carlock, May 23, 1953; children—Donald David, Catherine Marion. Staff scientist Los Alamos Sci. Lab., 1943-46; sr. physicist Oak Ridge Nat. Lab., 1950-56, prin. physicist, 1958; asso. prof. nuclear engring. U. Mich., 1956-58; prof. nuclear engring. Northwestern U., Evanston, Ill., 1959—, chmn. dept. engring. scis., 1960—. Chmn. subcom. neutron standards and measurement NRC, 1958-62; cons. Oak Ridge Nat. Lab., 1956—, Argonne (Ill.) Nat. Lab., 1959—. Fellow Am. Phys. Soc., Am. Nuclear Soc.; mem. Am. Soc. Engring. Edn., Phi Beta Kappa, Sigma Xi, Pi Mu Epsilon, Alpha Chi Sigma. Contbr. articles to tech. jours. Measurement fission cross sects.; research on gamma-gamma angular correlations, Q values, angular distbns. charged-particle reactions, thermal neutron parameters; developed empirical nuclear model, silicon semiconductor particle detectors and applications to study nuclear reactions. Home: 2435 Pomona Lane, Wilmette, Ill. 60091. Office: Tech. Inst., Northwestern U., Evanston, Ill. 60201.*

KLEMENS, Paul Gustav, physicist; b. Vienna, Austria, May 24, 1925; s. Walter and Ida (Klug) K.;

B.Sc., U. Sydney, 1946, M.Sc., 1948; D.Phil., Oxford U., 1950; m. Ruth H. Wiener, July 30, 1950; children—Michael W., Susan M. Prin. research officer div. physics Nat. Standards Lab., Sydney, Australia, 1950-59; with Westinghouse Research Labs., Pitts., 1959-67; chmn. dept. physics U. Conn., 1967—. mgr. transport properties of solids, 1964—. Vis. prof. U. Leiden, Netherlands, 1963-64. Fellow Inst. Physics, Am., Gt. Britain phys. socs. Researcher theory of transport properties and non-equilibrium processes in solids, 1948—. Home: 232 Ferguson Rd., Manchester, Conn. 06040. Office: Dept. Physics, U. Conn., Storrs, Conn. 06268.*

KLEMENT, Alfred William, Jr., Am. radiobiologist; b. Bryan, Tex., Nov. 8, 1923; s. Alfred William and Elizabeth (England) K.; student Tex. A. and M. Coll., 1941-42; B.S., U. S. Mil. Acad., 1945; postgrad. U. S. Naval Postgrad. Sch.; M.Bioradiology, U. Cal. at Berkeley, 1952; m. Frances Irene Bralley, Sept. 5, 1945; children—Carmen Elizabeth, Alfred William III. Commd. 2d lt. U. S. Army, 1945, advanced through grades to capt., 1951; nuclear effects engr. U. S. Army Chem. and Radiol. Labs., 1952-53; biophysicist U. S. Army Chem. Corps Med. Labs., 1953-54; res., 1957; radiation effects specialist div. Biology and Medicine U. S. AEC, Washington, 1957-59, sci. analyst, 1959-62, nuclear fallout specialist, 1962-65, asst. chief Fallout Studies br., 1965, sci. liaison dir. Office Canal Studies Nev. Operations Office, Las Vegas, 1966—. Mem. test adv. panel Nev. Test Orgn., 1958; commd. res. USPHS, 1959—. Recipient High Quality Performance award, 1964. Diplomate Am. Bd. Health Physics. Fellow Am. Pub. Health Assn.; mem. Health Physics Soc., A.A.A.S., Conf. Radiol. Health. Editor: Radioactive Fallout from Nuclear Weapons Tests, 1962; (with V. Schultz) Radioecology, 1963, Radioecology of Aquatic Organisms, 1966; Radioactive Fallout from Nuclear Weapons Tests, 1966. Research in radiol. protection, biophysics, effects of nuclear weapons, nuclear safety situations, radiobiology, medicine, environmental radiation, civil def. and radiol. physics, environmental scis., natural radioactivity, radioecology and effects of radiation. Home: 2877 Robar Dr., Las Vegas 89109. Office: U. S. AEC, P.O. Box 1676, Las Vegas, Nev. 89101.*

KLEMENT, Fedor Dmitrievich, Russian physicist; b. June 13, 1903; grad. Leningrad U., 1934; D.Physico-Math. Sci. Asso., learned sec. research inst., lectr., head lab. Leningrad U., from 1934; prof., rector Tartu U., 1951—; sect. head Inst. Physics and Astronomy, Estonian Acad. Sci., 1953—. Mem. Estonian Acad. Sci. Author: Formation Processes of Crystal Phosphoruses and Some Methods of Their Manufacture, 1956; Luminescent Methods of Physicochemical Analysis, 1958. Research and publs. on luminescence of solids. Address: Tartu University, ulitsa Yulikooli 18, Tartu, Estonian SSR, USSR.

KLEMIN, Alexander, aero. engr.; b. London, Eng., May 15, 1888; s. Albert and Dora (Clemens) K.; B.S., London U., 1909; S.M., Mass. Inst. Tech., 1915; LL.D., Kenyon Coll., 1934; D.Eng., N.Y. U., 1950; m. Ethel Murton, 1921; 1 dau., Diana. Came to U. S., 1913, naturalized, 1917. In charge aeronautics dept. Mass. Inst. Tech., 1916-17; officer in charge Research Dept. Army Air Service, McCook Field, Dayton, O., during World War I; cons. aero. engr., 1919-24; in charge Daniel Guggenheim Sch. of Aeronautics, N.Y. U., 1924-41; research prof. N.Y. U., 1942-44; lectr. aeros. Princeton, 1934-35; cons. engr. Aero. Br., Dept. of Commerce, 1927-29. Mem. aerodynamics and rotation wing aircraft coms. NACA. Winner of army and navy airplane design competitions; Dr. Alexander Klemin ann. award established by Am. Helicopter Soc., 1951. Fellow Inst. Aero. Scis.; mem. Am. Soc. M.E., Soc. Automotive Engrs., Royal Aero. Soc., Am. Helicopter Soc. (pres. 1949-50). Helicopter editor Aero Digest; editor Handbook of Aeronautical Engineering. Author: Text-book of Aeronautical Engineering, 1918; If You Want to Fly, 1926; Simplified Aerodynamics, 1927; Airplane Stress Analysis, 1927; The Helicopter Adventure, 1947; (with A. T. McPherson) Engineering Uses of Rubber, 1956; also articles. Designed 1st amphibian landing gear used in U. S., 1921; research on rotary-wing aircraft, helicopters. Died Mar. 13, 1950.

KLEMM, Alfred, German chemist; b. Leipzig, Feb. 15, 1913; s. Wilhelm and Erna (Kröner) K.; ed. univs. Leipzig, Dresden, Munich; Ph.D.; m. Hannelore Rothe, 1939; children—Silvia, Michael, Alfred, Tamina, Ebba, Aja, Imma, Otto. With Kaiser-Wilhelm Inst. Chemistry, 1939—; prof. U. Mainz, mem. Max-Planck Inst. Chemistry, Mainz, 1958—. Mem. German, Am. phys. socs., Am. Chem. Soc. Editor: Zeitschrift für Naturforschung. Research and publs. on phys. chemistry of isotopes, especially molten salts. Home: Beuthenerstrasse 25. Office: Saarstrasse 23, Mainz, Germany.*

KLEMM, LeRoy Henry, Am. chemist; b. Maple Park, Ill., July 31, 1919; s. Henry J. and Anna (Reines) K.; B.S., U. Ill., 1941; M.S., U. Mich., 1943, Ph.D., 1945; m. Christine Jones, Dec. 27, 1945; children—Richard A., Rebecca J., Ann C. Research chemist Am. Oil Co., Texas City, Tex., 1944-45; research asso. Ohio State U., 1946; instr. Harvard, 1946-47; faculty Ind. U., 1947-52, asst. prof., 1949-52; faculty U. Ore., Eugene, 1952—, prof. chemistry, 1963—; vis. prof. U. Cin., 1965-66; sci. asst. to dean liberal arts, dept. dir. Office Sci. and Scholarly Research, U. Ore., 1960-62. Guggenheim fellow Med. Research, Lab., London, Eng., also Eidg. Tech. Hochschule, Zurich, Switzerland, 1958-59. Mem. Am. Chem. Soc., Am. Assn. U. Profs., A.A.A.S., Sigma Xi. Research and publs. on syntheses of cancer-producing and inhibiting compounds, anti-malarial drugs, aromatic hydrocarbons and their nitrogen and sulfur analogs, lignans, correlation of molecular geometry with reactions on surfaces-heterogeneous catalysis, chromatography, electroreduction, complexation. Home: 1926 Moss St., Eugene, Ore. 97403.*

KLEMM, Wilhelm Karl, German inorganic chemist; b. Buhrau, Jan. 5, 1896; s. Wilhelm and Ottilie (John) K.; Ph.D., U. Breslau; Dr. honoris causa, Technische Hochschule, Darmstadt; m. Lina Arndt, Dec. 22, 1949. Agrégé, 1927; prof. at large, 1929; full prof., dir. Inst. Inorganic Chemistry, Danzig Tech. High Sch., 1933; with Inst. Inorganic Chemistry, U. Kiel, 1947, U. Munster, 1951; prof. emeritus, 1964. Recipient Liebig medal, Lavoisier medal French Acad. Chemistry. Mem. German Soc. Chemists (Carl Duisberg plaque), Austrian Assn. Chemistry, Chem. Soc. France, Indian Acad. Sci. Author: Anorganische Chemie; Magnetochemie; Experimentelle Einfuhrung in die anorganische Chemie, also numerous articles. Home: Redigerstrasse 86. Office: Hindenburgplatz, Munster/Westfalia, Germany.

KLEMOLA, Erkki Olavi, Finnish physician; b. Helsinki, Finland, June 13, 1910; s. Vaino and Frida (Kierinkivaara) K.; M.D., U. Helsinki, 1942; m. Aino Laakso, July 30, 1940; children—Arto, Liisa, Jukka, Pertti. Faculty, U. Helsinki, 1946—, prof. infectious diseases, 1960—; chief infectious diseases unit Aurorar Hosp., Helsinki, 1954—. Mem. Finnish Soc. Internal Medicine (past pres.). Research, numerous publs. on demonstration of causal role of cytomegalovirus in infectious-mononucleosis-like diseases without positive heterophil agglutination test, that subclin. cytomegalovirus infections transmitted via fresh blood are common in man. Home: 56 B. Runeberginkatu, Helsinki. Office: Aurora Hosp., Helsinki, Finland.*

KLEMPERER, Paul, pathologist; b. Vienna, Austria, Aug. 2, 1887; s. Alios and Jenny (Ippen) K.; M.D., Vienna, 1912; D.Sc. (hon.), U. Chgo., 1959; m. Margit Freund, 1926; 2 sons. Naturalized Am. citizen. Asst. prof. pathology Med. Sch., Loyola U., Cal., 1922-23; asso. prof. Postgrad. Sch., N.Y. U., 1923-26; prof. Coll. Phys. and Surg., Columbia U., 1932-55; vis. prof. Albert Einstein Coll. Medicine, Yeshiva U., 1955—; pathologist Mt. Sinai Hosp., N.Y.C., 1926-55, cons. pathologist, 1955—. Mem. Soc. Clin. Pathologists, Assn. Pathology and Bacteriology, Rheumatologic Soc. Denmark. Research and publs. on path. anatomy, collagen diseases; described (with A. D. Pollack, G. Baer) diffuse scleroderma (a collagen disease), 1942. Home: 1215 Fifth Av., N.Y.C. 29.

KLENCKE, (Philipp Friedrich) Hermann, German physician; b. Hanover, Germany, 1813; ed. Leipzig, Germany, Hanover; pioneer in prodn. exptl. Tb; demonstrated Tb may be transmitted by cow's milk to man, 1846; popularized med. developments. Died 1881.

KLENDSHOJ, Niels C(hristian), biochemist; b. Denmark, Feb. 1, 1902; s. Niels Christian and Ane Marie (Jensen) K.; B.A., Borgerdydskolen, Copenhagen, 1921; postgrad. Royal Polytech. Inst., U. Copenhagen, 1926; M.D., U. Buffalo, 1937; m. Annette Roberta Heinz, Feb. 24, 1933; children—Ole, John, Arne; m. 2d, June Esther Davis, Feb. 4, 1963. Came to U. S., 1927, naturalized, 1935. With Arner Co., Inc., Buffalo, 1927—, exec. v.p., 1927-52, pres., 1952-59; chmn. bd. Strong Cobb Arner, Inc. (merger Arner Co. and Strong Cobb and Co.), 1959-63, pres., chmn. bd., chief exec. officer, 1963—, also chmn. bd. Strong Cobb Arner of Can., Ltd.; biochemist Buffalo Gen. Hosp., 1937—, now dir. dept. biochemistry, asst. attending physician, mem. med. bd., clin. prof. toxicology Sch. Medicine, State U. N.Y., Buffalo. Past mem. joint legislative com. on narcotics State N.Y. Author: Fundamentals of Biochemistry in Clinical Medicine, 1953. Co-discoverer blood substances B and O.; practical application of specific blood factors in transfusions; contbns. in fields of analytical and clin. toxicology. Home: 80 Elmhurst Rd., Snyder, N.Y. 14226. Office: 100 High St., Buffalo. 14203.*

KLESTADT, Bernard, electronic engr.; b. Büren, Germany, Jan. 31, 1925; s. Rudolf and Ida (Alexander) K.; came to U. S., 1938, naturalized, 1944; student Lawrence Inst. Tech., 1942-43, N.Y. U., 1946-47; B.S., Columbia U., 1949, M.S., 1950; Ph.D., U. So. Cal., 1958; m. Bernice Florence Hersch, July 5, 1956; 1 son. Elec. engr. Aircraft Radiation Lab., Wright Air Devel. Center, Dayton, O., 1949; asst. project engr. Sperry Gyroscope Co., Great Neck, N.Y., 1950; mem. tech. staff Systems Devel. Labs., Hughes Aircraft Co., Culver City, Cal., 1950-60, sr. scientist space systems div., asst. mgr. flight control systems dept., El Segundo, Cal., 1960-65, mgr. control systems dept. missile systems div., Canoga Park, Cal., 1966—. Lectr. elec. engring. U. So. Cal., 1956-58. Hughes Staff Doctoral fellow, 1956-58. Mem. N.Y. Acad. Scis., Research Soc. Am., I.E.E.E. (sr.). Developed automatic control system for falcon air-to-air guided missiles, methods for analysis nonlinear servomechanisms and for synthesis linear systems; invented method space vehicle guidance and navigation, space vehicle control system, elec. device for generating nonlinear functions, level rejecting differential amplifier. Home: 16821 Ivyside Pl., Encino, Cal. 91316. Office: Hughes Aircraft Co., Canoga Park, Cal. 91304.*

KLIEMKE, Ewalt, German psychologist; b. Berlin, Sept. 27, 1898; s. Ernst and Martha (Tandecki) K.; ed. univs. Berlin, Heidelberg, Königsberg; Ph.D.; m. Ilse Rostosky-Breitkreutz. Psychologist to handicapped; instr.; specialist in pathology, corporal, moral and spiritual adjustment of handicapped persons; cofounder psychology for handicapped in Germany. Mem. German Assn. for Adjustment, Internat. League against Epilepsy. Contbr. articles to profl. publs. Address: Neurological Clinic of Dr. Schnieder, 7704 Gallingen Krs., Konstanz, Germany.

KLIEWE, Heinrich, German physician, biologist; b. Beckum, Westphalia, Sept. 7, 1892; s. Heinrich and Ida (Hunke) K.; M.D.; m. Anneliese Arnold, Mar. 21, 1925; children—Charlotte, Ilse, Ruth, Uta. Instr., U. Heidelberg; dir. investigation infectious diseases; prof. U. Giessen; prof., asst. dir. U. Mainz, prof. emeritus, 1961—. Mem. German Soc. for Struggle against Parasites, others. Author: Leitfaden Entseuchung und Entwesung, 1951; Wein und Gesundheit, 1962; Biol. Kampfstoffe, 1963, also over 200 articles on hygiene and bacteriology. Address: Langenbeckstrasse 1, Mainz, Germany.

KLIGERMAN, Morton M., Am. radiologist; b. Phila., Dec. 26, 1917; s. Samuel and Dorothy (Medvene) K.; B.S., Temple U., 1938, M.D., 1941, M.S., 1948; M.A., Yale, 1958; m. Barbara Bray Coleman, Mar. 14, 1956; children—Hilary, Roger B. C. Wilcox (foster child), Thomas A., Valli à C. Practice medicine, specializing in radiology, Phila., 1941-48, N.Y.C., 1948-58, New Haven, 1958—; faculty Temple U., 1947-48, Columbia, 1948-58; prof., chmn. dept. radiology Sch. Medicine Yale, 1958—; asst. to attending radiologist Columbia-Presbyn. Hosp., 1948-58; research collaborator Brookhaven Nat. Lab., Upton, L.I., N.Y., 1965—; cons. West Haven VA Hosp., 1958—. Recipient Distinguished Alumnus award Temple U., 1964. Diplomate Am. Bd. Radiology. Mem. Am. Assn. Cancer Research, Am. Soc. Therapeutic Radiologists (pres.-elect), Am. Thoracic Soc., Am. Trudeau Soc., Assn. U. Radiologists, Am. Coll. Radiology, Am. Radium Soc., Am. New Eng. roentgen ray socs., Assn. U. Radiologists (pres. elect. 1967-68), Am. Thoracic Soc., N.Y. Acad. Scis., Am. Cancer Soc. (trustee Conn.), A.M.A., Radiol. Soc. N. Am., Radiation Research Soc., Soc. Head and Neck Surgeons. Editor: (with H. W. Jacox) Yearbook of Radiology, 1953-64. Clin. studies, publs. in chemo-, radio-, and surg. therapy of cancer; sensitization, combination tumor therapy. Home: 274 Ogden St., New Haven 06511. Office: 333 Cedar St., New Haven 06510.*

KLIMAN, Bernard, Am. endocrinologist; b. Boston, Aug. 8, 1931; s. Jacob and Minnie (Burack) K.; A.B. magna cum laude, Harvard, 1951, M.D. magna cum laude, 1955; m. Phyllis Arlene Rice, June 16, 1955; children—Joyce, David. Research asso. NIH, Bethesda, Md., 1956-58, clin. investigator, 1960-62, now prin. investigator Extramural Grant program asso. in medicine Harvard Med. Sch., 1962—; mem. staff Mass. Gen. Hosp., Boston, 1962—; radioisotope cons. New Eng. Nuclear Corp. Diplomate Am. Bd. Internal Medicine. Mem. Endocrine Soc., Am. Fedn. Clin. Research. Contbr. articles to sci. jours. Devel. of radioisotope techniques for accurate measurement of function of steroid producing glands, especially adrenal cortex and testis; pioneer in devel. of double isotope method for measurement of aldosterone, prin. regulator of salt content of body in man. Home: 48 Glendale Rd., Newton Center, Mass. 02159. Office: Endocrine Unit, Mass. Gen. Hosp., Boston 02114.*

KLIMEK, Rudolf Marian, Polish physician, biochemist; b. Cracow, Poland, Dec. 12, 1932; s. Sylwester Józef and Józefa (Kolodziej) K.; Physician Diploma with distinction, Med. Acad., Cracow, 1955, M.D., 1961; m. Ewa Kownacka, Apr. 21, 1961; 1 son, Marek. Staff, Med. Acad. Cracow, 1952—, adj., asso. prof. I Clinic Obstetrics and Gynecology, 1957-65, chief Central Endocrinology Labs., 1965—. Mem. European Soc. Biochem. Pharmacology, Polish Soc. Biochemistry, Polish Endocrinological Soc., Polish Gynecol. Soc. Author: (with W. Król) Oxtocin and Its Analogues, 1964; also articles. Discovered and isolated cytochromes from sulfur bacteria; research on enzyme mechanisms in pregnancy; discovered Hypothalamosis post-graviditatem syndrome. Home: 6 Rzeznicza, Cracow, Poland.*

KLINE, Donald Edgar, Am. physicist; b. DuBois, Pa., Aug. 28, 1928; s. Ford Clark and Esther (Swanson) K.; B.S. in Physics, Pa. State U., 1951, M.S., 1953, Ph.D., 1955; m. Marilyn B. Leis, Nov. 23, 1949; children—Lorin Jon, Lyndon James, Lee Jerome. Faculty, Pa. State U., University Park, 1954-56, 61—, prof. nuclear engring., 1965—; vis. physicist Nuclear Reactor Facility, 1956-57; staff research physicist HRB-Singer, Inc., State College, Pa., 1957-61. Dir. Chemcut Corp., State College. Cons., NASA, 1963—. Mem. Am. Phys. Soc., Am. Nuclear Soc., Am. Soc. Engring. Edn., Am. Assn. U. Profs., Phi Beta Kappa, Phi Kappa Phi, Phi Eta Sigma, Sigma Pi Sigma, Pi Mu Epsilon, Sigma Xi. Author: (with A. Jacobs, F. J. Remick) Basic Principles of Nuclear Science and Reactors, 1960; also articles. Research on dynamic mech. spectroscopy polymers, radiation dosimetry, radioactive tracer techniques, radiation effects in materials, polymer physics. Home: Box 426-D Branch Rd., State College, Pa. 16801. Office: Hammond Bldg., University Park, Pa. 16802.*

KLINE, Milton Vance, Am. psychologist; b. Bklyn., Mar. 25, 1923; s. Joseph M. and Lillian (Zimmerman) K.; B.A., Pa. State U., 1944, Ed.D., 1947; M.A., Columbia, 1945; postgrad. N.Y. U.; m. Dorothy Weller, Feb. 25, 1951; 1 dau., Jill. Clin. psychologist Salvation Army, N.Y.C., 1945-46; clin. psychologist N.Y. Consultation Service, Vocational Adv. Service, N.Y.C., 1946-48; chief clin. psychologist Mental Hygiene div. Westchester County Health Dept., White Plains, N.Y., 1948-53; research project dir. psychology dept. L.I. U., 1947-56; cons. research psychologist N.Y. U. Prosthetic Research Lab., 1955-56; lectr. med. hypnosis Seton Hall Med. Coll., 1960-62; pvt. practice hypnoanalysis, psychotherapy, N.Y.C., 1953-—; pres. Inst. for Research in Hypnosis; dir. Morton Prince Clinic for Hypnotherapy. Recipient Bernard B. Raginsky award. Mem. Soc. Clin. and Exptl. Hypnosis (pres. 1961-63, Gold medal award), Am. Psychol. Assn., Am. Psychosomatic Soc., Acad. Psychosomatic Medicine, Am. Med. Writers Assn., A.A.A.S., N.Y. Acad. Sci. Author: Hypnodynamic Psychology, 1956; A Scientific Report on Search for Bridey Murphy, 1957; Freud and Hypnosis, 1958; The Nature of Hypnosis, 1963; Clinical Correlations of Experimental Hypnosis, 1964; Hypnosis and Psychodynamics, 1966. Publs. on investigations into nature of hypnosis and hypnotic behavior, hypnosis in psychotherapy, treatment of psychosomatic disturbances. Home: 121 Palmer Lane, Thornwood, N.Y. 10594. Office: 345 F. 58th St., N.Y.C. 10019.*

KLINE, Morris, Am. mathematician; b. Bklyn., May 1, 1908; s. Bernard and Sarah (Spatt) K.; Ph.D., N.Y. U., 1936; m. Helen Mann, Sept. 4, 1939; children—Elizabeth M., Judith M., Douglas M. Research asst. Inst. for Advanced Study, Princeton, N.J., 1936-38; instr., N.Y. U., 1938-42; physicist Signal Corps. Engring. Labs., Belmar, N.J., 1942-45; faculty N.Y. U., N.Y.C., 1945-—, prof. math., 1952-—, dir. div. electromagnetic research Courant Inst. Math. Scis., 1946-66, chmn. undergrad. math., Washington Sq., 1959-—. Fulbright lectr., Germany, 1958-59. John Simon Guggenheim fellow, 1958-59. Mem. Am. Math. Soc., Math. Assn. Am., I.E.E.E. Author: Mathematics in Western Culture, 1953. Mathematics and the Physical World, 1959; Mathematics, A Cultural Approach, 1962; Electromagnetic Theory and Geometrical Optics, 1965; also articles. Research in radio wave behavior; inventor in radio engring. Home: 1024 E. 26th St., Bklyn. 11210. Office: 251 Mercer St., N.Y.C. 10012.*

KLINE, Nathan Schellenberg, Am. psychiatrist; b. Phila., Mar. 22, 1916; s. Ignatz and Florence (Schellenberg) K.; B.A., Swarthmore Coll., 1938; M.D., N.Y. U., 1943; postgrad. Harvard, New Sch. Social Research, Washington Sch. Psychiatry, Princeton, Rutgers U.; m. Margot Hess, June 29, 1942; 1 dau., Marna. Res. officer USPHS, 1944-46; mem. staff Union County Mental Hygiene Soc. Clinic, Plainfield, N.J., 1946-47, VA Hosp., Lyons, N.J., 1946-50; dir. research Worcester (Mass.) State Hosp., 1950-52, Rockland State Hosp., Orangeburg, N.Y., 1952-—; faculty Columbia Coll. Phys. and Surg., 1948-—, asst. clin. prof. psychiatry dept., 1957-—; dir. psychiat. services Bergen Pines County Hosp., Paramus, N.J., 1963-—. Co-founder, dir. research Centre de Psychiatrie et Neurologie, Port-au-Prince, Haiti, W.I., 1959-—; mem. med. adv. council Medico, 1962-—, also trustee. Recipient Henry Wisner Miller award Manhattan Soc. Mental Health, 1963; Albert Lasker Clin. Research award Am. Pub. Health Assn., 1964, Albert Lasker award, 1957; certificates Japanese Soc. Psychiatry and Neurology, 1965, U. Indonesia, 1965; named grand officeur Legion d'Honneur et Merite (Republic of Haiti), knight Gt. Cross Serenissimi Mil. Order of St. Mary the Glorious, Internat. Cath. Order, 1961; knight grand comdr. Liberian Humane Order of African Redemption, 1963, others. Diplomate Am. Bd. Psychiatry and Neurology. Fellow Am. Coll. Neuropsychopharmacology (pres. 1967), Collegium Internationale Neuropsychopharmacologicum, Royal Soc. Medicine (Eng.), N.Y. Acad. Scis., N.Y. Acad. Medicine, A.C.P., A.A.A.S., Am. Psychiat. Assn.; mem. Am. Psychol. Assn., Am. Chem. Soc., Indian Psychiat. Assn., Nat. Health Edn. Com., Am. Therapeutic Soc., Société Moreau de Tours (France), World Fedn. Mental Health, Psychiat. Research Found., Pan Am. Med. Assn., A.M.A., Sigma Xi, others. Author: Psychopharmacology Frontiers, 1959; (with H. E. Lehmann) Handbook of Psychiatric Treatment in Medical Practice, 1962, Psychopharmacology, 1965. Mem. adv. editorial bd. Psychopharmacologia, 1958-—, Internat. Jour. Social Psychiatry, 1958-—; internat. bd. editors Excerpta Medica, 1955-—. Research, publs. on early use of psychopharm. treatment of mentally and emotionally ill; cultural factors in psychiat. disorders; use of computer systems in clin. research. Home: 1155 Sussex Rd., Teaneck, N.J. 07666. Office: 40 E. 69th St., N.Y.C. 10021.*

KLINGENBERG, Martin, German biochemist; b. Rostock, Dec. 5, 1928; s. Paul Friedrich and Henny (Duncker) K.; student U. Berlin; diploma U. Heidelberg, 1952, Ph.D., 1954; Nat. Acad. Sci. fellow U. Pa., Phila., 1954-56; m. Sylvaine Handrich, July 25, 1954; children—Elisabeth, Catherine, Anne. Asst., Physiol. Chem. Inst., U. Marburg (Germany), 1956-66, asst. prof., 1966-67; prof. phys. biochemistry, U. Munich, 1967-—, dir. physiol. and phys. chem. inst., 1967-—. Recipient Adolf Frick medal, 1964. Research, numerous publs. on energy-linked reactions in mitochondria, detection of reversibility of oxidative phosphorylation and electron transport in respiratory chain, pathways of hydrogen, phosphate transfer, mechanism of regulatory enzymes. Home: 33 Zeppelin Strasse, 355 Marburg, Germany. Office: 33 Goethe Strasse, 8 Munich, West Germany.

KLINGENSTIERNA, Samuel, mathematician, physicist; b. Tolefors, Sweden, Aug. 18, 1689; ed. Uppsala, Sweden; visited Germany, France, Eng., 1727; became prof. math. U. Uppsala, 1728, prof. physics, 1750. Mem. French Acad. Scis., Royal Soc., 1730. Uppsala, Acad. Scis. Stockholm. Research on integral calculations and manner of determining size of earth, before 1727; discovered achromatic glass. Died Stockholm, Oct. 26, 1785.

KLINKERFUES, Wilhelm, German astronomer, physicist; b. Hofgeismar, Mar. 29, 1827; asst. to Gauss; prof., dir. Göttingen (Germany) Obs. Author: Theoretische Astronomie, 1871; Die Prinzipien der Spektralanalyse, 1879. Research in theoretical astronomy, on determination of orbits of comets and planets; discovered 6 comets; inventor device by which gas lamps could be turned on and off from central gas works. Died Göttingen, Jan. 28, 1884.

KLIPPART, John Hancock, Am. agriculturalist; b. Canton, O., July 26, 1823; s. Henry and Eve (Henning) K.; m. Emeline Rahn, 1847, at least 1 child. Corr. sec. Ohio State Bd. Agr., 1856-78; mem. Am. Bd. Commrs. of London Internat. Exhbn., 1861-62; sent by Ohio Bd. Agr. to observe agrl. methods in Europe, 1865; asst. geologist Ohio Geol. Survey, 1869; del. from Ohio to Nat. Agrl. Conv., Washington, 1872; state fish commr. Ohio, 1873; edited reports of Ohio Bd. Agr., 1857-77. Author: The Wheat Plant, 1860; Principles and Practice of Land Drainage, 1861. Research and publs. on sheep and sheep culture, agrl. edn., dairy husbandry. Died Oct. 24, 1878.

KLITGAARD, Howard Maynard, Am. physiologist; b. Harlan, Ia., Oct. 16, 1924; s. Andrew and Gladys (Havick) K.; student U. Denver, 1942-43; B.A., U. Ia., 1949, M.S., 1950, Ph.D., 1953; m. Anna Plazak, Nov. 17, 1945; children—Andrew, Margaret, Michael, Patricia, Diana. With U. Ia., 1949-53, instr. physiology, 1951-53; with Marquette Sch. Medicine, 1953-—, asso. prof. physiology, 1960-66, prof., 1966-—, vice chmn. dept. physiology, 1967-—, chmn. research and grad. tng. com., 1961-—; lectr. Deaconess Hosp., Milw.; cons. VA Hosp., Wood, Wis. Mem. Am. Physiol. Soc., Soc. Exptl. Biology and Medicine, Am. Soc. Zoologists, Endocrine Soc., Sigma Chi. Contbr. articles to profl. jours. Research in endocrines and metabolism, radioisotope methodology; responsiveness of various animal tissues to thyroid hormones; correlation studies of changes in tissue oxygen consumption with tissue enzyme activities and electromicroscope changes in mitochondria, indicating presence of biochem.-structural relationship. Home: 9073 N. Silverbrook Lane, Milw. 53223.*

KLOFT, Werner Jakob, German zoologist; b. Erbach, June 17, 1925; s. Friedrich and Helene (Rauch) K.; ed. univs. Berlin, Prague, Wurzburg; Ph.D.; m. Erika Sturm, June 21, 1947; children—Iris, Arnfried. Asst., 1949-56; agrégé, 1st asst., 1956-63; prof. at large Inst. Applied Zoology, U. Wurzburg, 1963-—. Mem. German Zoology Soc., German Soc. Applied Entomology, Internat. Union for Study Social Insects, Agra Acad. Zoology. Research and numerous publs. on interrelationship of parasitic insects and host tissues, radioisotopes, zoology, entomology. Home: Hofstattstrasse 6, Veitschöchheim. Office: Röntgenring 10, Wurzburg, Germany.

KLOHS, Murle William, Am. medicinal chemist; b. Aberdeen, S.D., Dec. 24, 1920; s. William H. and Lowell (Lewis) K.; B.S., U. Notre Dame, 1947; Riker fellow Harvard, 1950; m. Dolores Catherine Borm, June 16, 1946; children—Wendy C., Linda L. Chemist, Harrower Labs., Glendale, Cal., 1947; with Rexall Drug Co., Los Angeles, 1947-49; sr. chemist Riker Labs., Northridge, Cal., 1949-57, dir. medicinal chemistry, 1957-—; tech. adviser 20th Century Fox Film Corp., 1952. Mem. Am. Chem. Soc., N.Y. Acad. Sci., Soc. Econ. Botany, Am. Pharmacognosy Soc. Research, publs. on alkaloids of various Veratrum and Rauwolfia species and constituents of Piper methysticum, synthetic medicinal drugs; isolation and structural elucidation of physiologically active constituents of folklore medicinal plants. Home: 4616 Conchita Way, Tarzana, Cal. 91356. Office: 19901 Nordhoff St., Northridge, Cal. 91326.*

KLOKE, Adolf, German agrl. chemist; b. Paderborn, Germany, May 29, 1921; s. Heinrich and Therese (Buschmeier) K.; diploma U. Göttingen, 1949, Dr.agr., 1951; m. Waltraud Schmidt, Sept. 18, 1951; children—Adolf, Hildegard, Monika. Sci. asst. Inst. for Agrl. Chemistry, Göttingen, 1951-59; lectr. agrl. chemistry U. Göttingen, 1960-62; lectr. plant nutrition Tech. U. Berlin, 1962-—; dir. Inst. for Non-parasitic Plant Diseases, Fed. Biol. Inst. for Agr. and Forestry, Berlin, 1959-—. Author: Die Humusstoffe des Bodens als Wachstumsfaktoren, 1963; also numerous articles. Research on non-parasitic plant diseases, including nutrition disturbances, deficiency diseases, damage from air pollution, from cultivation errors, from pesticides, from wrong fertilization. Home: 6 Marinesteig, 1 Berlin 38. Office: 19 Konigin-Luise-Strasse, 1 Berlin 33, Germany.*

KLOOSTERMAN, Hendrik Douwe, Dutch mathematician; b. Achtkarspelen, Netherlands, Apr. 9, 1900; s. Bindert and Maryke (van der Molen) K.; Ph.D., U. Leyden; m. M. H. Träger, Aug. 30, 1932. Lectr. mathematics, U. Leyden, 1932; prof. mathematics, U. Leyden, 1947. Mem., Academy of Amsterdam. Author numerous articles. Investigations of theta functions under modular group; binary modular congruence groups; finite differences. Home: 52 van Oldenbarneveltstraat, Leyden, Netherlands. Office: 46 Stationsweg, Leyden, Netherlands.*

KLOPPING, Hein Louis, chemist; b. Utrecht, Holland, Feb. 7, 1921; s. Louis J. H. and Anna A. (America) K.; B.A. in Chemistry, U. Utrecht, 1941, B.A. in Pre-Med. Sci., 1942, M.A. in Chemistry, 1946, Ph.D., 1951; m. Elizabeth P. van Silfhout, July 20, 1950; children—Martin L, Karin E. Came to U. S., 1951, naturalized, 1960. Instr., U. Utrecht, 1945-47; engr. Standard Oil Co., N.J., N.Y., Tex., 1947; research chemist, group leader Inst. Organic Chemistry, Toegepast Natuurwetenschappelijk Onderzoek, Utrecht, 1947-51; research chemist E. I. du Pont de Nemours and Co., Inc., Wilmington, Del., 1951-55, research scientist, 1955-62, research asso., 1962-—. Recipient Freedoms Found. award, 1964. Mem. Royal Netherlands Chem. Soc., Research Soc. Am., Am. Chem. Soc. Author: Organic Compounds with Fungicidal Activity, 1948; Chemical Constitution and Antifungal Action of Sulphur Compounds, 1951; also articles. Patentee in field. Research on organic synthesis, cancer, electro-physiol. studies on odor perception in insects; isolation, characterization biol. active materials from natural sources; inventor new drugs and agrl. chems. Home: 31 Shellburne Dr., Wilmington, Del. 19803. Office: Du Pont Exptl. Sta., Wilmington, Del. 19898.*

KLOPSTEG, Paul E(rnest), Am. physicist, engr., sci. adminstr.; b. Henderson, Minn., May 30, 1889; s. Julius and Magdalene (Kuesthardt) K.; B.S. in Elec. Engring., U. Minn., 1911, M.A., 1913, Ph.D., 1916; Sc.D. (hon.), Northwestern U., 1942, Conn. Wesleyan U., 1948; m. Amanda Toedt, June 11, 1914; children—Marie (Mrs. J. M. Graffis), Ruth (Mrs. J. C. Reed), Irma (dec.). Asst. prof. physics U. Minn., 1917; advt. mgr. Leeds & Northrup Co., Phila., 1918-21; with Central Sci. Co., Chgo., 1921-44, pres., 1930-44; prof. applied sci. and dir. research Northwestern U. Tech. Inst., Evanston, Ill., 1944-54, prof. emeritus, 1954-—. Cons. on research, edn., 1958-—; mem. numerous govt. agy. coms.; asst., asso. dir. NSF, 1951-58; chmn. bd. govs. Argonne Nat. Lab., 1949-50; mem. exec. com. bd. trustees U. Corp. for Atmospheric Rsearch, 1963-66. Recipient Modern Pioneers award Nat. Assn. Mfrs., 1939; Merit medal with Presdl. citation, 1948; Regent's Outstanding Achievement medal U. Minn., 1950. Fellow A.A.A.S. (pres. 1958), Am. Phys. Soc.; mem. Am. Inst. Physics (chmn. bd. govs. 1940-47), Am. Meteorol. Soc. (hon.), Nat. Acad. Scis. (chmn. com. on artificial limbs 1945-55, chmn. com. on atmosphere scis., citation com., 1964), Am. Optical Soc., Am. Assn. Physics Tchrs. (co-founder 1930, pres. 1953). Author: Turkish Archery and the Composite Bow, 1947; The Indispensable Tools of Science, 1960; also articles. Contbr. articles to Ency. Brit. Patentee in field. Research covers moving coil galvanometers; hydrogen ion concentrations; measurement of short time intervals; design of scientific apparatus. Address: 828 Apple Tree Lane, Glenview, Ill. 60025.

KLOSA, Josef, German chemist; b. Halbendorf-lez-Oppein, June 16, 1921; s. Franz and Anastasia (Wotzka) K.; Ph.D., U. Breslau; m. Evelyn Czayka, Dec. 27, 1950; children—Ellen, Clement, Daniela. Asst., Chem. Inst., U. Gottingen, 1945; dir. Undersleben Kyffh. Labs., 1946-49; dir. sci. div. Lessing Chem. Inst., Erfurt, 1949-50; chemist Schering-Ost, 1950-52; dir. Asal, Berlin, 1953-58, lab. dir., 1958-—. Mem. German Chem. Assn., Soc. for Study and Chemistry Fatty Bodies, Blood Research Soc. Author: Praktikum der Organische Chemie, 1950; Entwicklung und Chemie der Heilmittel, 3 vols., 1952; Organische Chemie, 1952; Antibiothika, 1952. Address: Jänickestrasse 13, Berlin 37, Germany.

KLOSKOWSKA, Antonina, Polish sociologist; b. Piotrków, Poland, Nov. 7, 1919; d. Vincent and Cecilia (Szretter) K.; M.A., U. Lodz (Poland), 1948, Ph.D., 1950; Docent Ministry Higher Edn., 1957. Faculty, U. Lodz, 1946-—, asso. prof., 1966-—, head chair history sociology, 1961-—; asst. prof. Center History and Sociology of Culture, Polish Acad. Sci., 1954-—. Mem. Polish (mem. bd.), Internat. (mem. com. on family research) sociol. assns. Author: Machiavelli as a Humanist, 1954; Mass Culture: Criticism and Defense, 1964; also articles. Asst. editor Przeglad Socjologiczny, 1959-—. Analysis of sociol. framework of mass culture and its different styles; content analysis of mass media and mass media publs. in Poland; models and patterns of family life. Home: Uniwersytecka 3, Lodz, Poland.*

KLOSOVSKY, Boris Nikodimovich, Russian neurosurgeon, morphologist, physiologist; b. 1898; grad. Med. Faculty, Azerbaijan U., 1925; D.Med. Sci., 1939. Head lab. for brain devel. study Brain Inst., 1930-49; head morphological (later tumor surgery dept.) Burdenko Inst. Neurosurgery, USSR Acad. Med. Sci., 1932-48, active mem., 1962—, chief dept. brain devel. in congenital and hereditary diseases, 1963—; head psychoneurol. clinic and lab. for study devel. of brain Inst. Pediatrics, 1950—; head lab. pathophysiology Brain Inst. Neurology, 1952—. Recipient Stalin prize, 1951. Author: The Development of the Cerebral Capillaries, 1949; Cerebral Blood Circulation, 1951; The Development of the Child Brain, 1954; Blood Circulation in the Brain, 1963; Excitatory and Inhibitory States of the Brain (with E. N. Kosmarskaya), 1963; co-author: Methods of Experimentally Removing All Vascular Plexi of the Cerebrum, 1956. Mem. editorial bd. Problems of Neurosurgery. Research and numerous publs. on relation between nerve cells and capillaries; 1st to describe some of brain's conduction paths; discovered and described vestibular nucleus; suggested 1st bulbotomy operation (with N. N. Burdenko); 1st to induce exptl. swelling and wrinkling of cerebrum, explained mechanism of this phenomenon; discovered spl. glandular organ developing at site of vascular plexi in lateral cerebral ventricles; discovered growth of cerebral capillaries using his method of impregnating cerebral vessels. Address: Inst. Pediatrics, USSR Acad. Med. Sci., Lomonosovsky prospekt 2/40, Moscow B-296, USSR.

KLOSTERMANN, Gerald Franz, dermatologist; b. Blankenburg am Harz, Germany, Jan. 22, 1920; s. Franz Julius and Elisabeth (Meurer) K.; M.D., Göttingen U., 1945; m. Christel Daalmann, Feb. 5, 1945; children—Manfred, Gerald, Eberhard. Staff, U. Göttingen, West Germany, 1945-53, 55—, asso. prof. dept. dermatology and venerology Faculty Medicine, 1958—; staff U. Erlangen, 1953-55. Mem. Deutsche Dermatologische Gesellschaft. Author: Pigmentflecken-polypose-Klinibiologische Studien am sogenannten Peutz-Syndrom, 1950; Der Diagnostische Blick-Atlas Zur Differential-Diagnose innerer Krankheiten (Südhof, Tischendorf), 1964, Spanish, 1965, French, English, Hungarian 1966. Research, publs. on clin., histological and therapeutical research of skin and allied diseases. Home: 12 Plauener Str., Göttingen, Niedersachsen, Deutsche Bundesrepublik. Office: 3 von Siebold-Str., Gottingen, Niedersachsen, W. Germany.*

KLOTZ, Arthur Paul, Am. physician; b. Milw., Sept. 28, 1913; s. Paul Oscar and Christine (Kratt) K.; B.S., M.D., U. Chgo., 1938; m. Margaret Ruth Pollard, Mar. 16, 1941; children—Stephen, Suzanne, John, Peter. Instr. medicine Billings Hosp., U. Chgo., 1951-54; chief sect. gastroenterology Med. Center, U. Kan., Kansas City, 1954—, faculty, 1954—, prof. medicine, 1962—. Cons. to hosps. Mem. A.M.A. (rep. to sci. exhibit 1963—). Contbg. author: Gastorenterologic Medicine, 1967; also numerous articles. Editorial bd. Am. Jour. Digestive Diseases, 1964—. Research in gastroenterology. Home: Rural Route 2, Box 246, Olathe, Kan. Office: U. Kan. Med. Center, Kansas City, Kan. 66103.*

KLOTZ, Irving Myron, Am. chemist; b. Chgo., Jan. 22, 1916; s. Frank and Mollie (Nasatir) K.; B.S., U. Chgo., 1937, Ph.D., 1940; Fellow Northwestern U., Evanston, Ill., 1940-41, mem. faculty, 1941—, prof. chemistry, 1950—, Morrison prof. chemistry, 1963—. Lalor fellow Marine Biol. Lab., Woods Hole, Mass., 1947-48, corp. mem. , 1947—, trustee, 1957-65; chmn. biophysics and biophys. chemistry study sect. NIH, 1963-66. Recipient Eli Lilly award Am. Chem. Soc., 1949. Mem. Am. Soc. Biol. Chemists, Am. Chem. Soc., A.A.A.S., Phi Beta Kappa, Sigma Xi. Author: Chemical Thermodynamics, 1950, 64; Energetics in Biochemical Reactions, 1957, 67; also numerous articles. Research in protein structure and behavior, biochem. energetics. Home: 2400 N. Lakeview Av., Chgo. Office: Dept. Chemistry, Northwestern U., Evanston, Ill.*

KLOTZ, Oskar, Canadian pathologist; b. Preston, Ont., Can., Jan. 21, 1878; s. Otto and Marie (Widenman) K.; M.B., U. Toronto, 1902; fellow in pathology, McGill U., 1903-05, M.D., C.M., 1906; postgrad. U. Bonn, 1904-05, U. Freiburg, 1908; m. Stella M. Scovil, Mar. 4, 1908. Demonstrator, lectr. McGill U., 1905-10; prof. pathology, and bacteriology U. Pitts., 1910-20; prof. pathology, Sao Paulo, Brazil (Rockefeller Found.), 1920-23; prof. pathology and bacteriology U. Toronto, from 1923. Dir. Magee Pathol. Lab., Pathology Labs., Toronto Gen. Hosp., from 1923; cons. pathologist, Hosp. for Sick Children, Toronto, from 1923; hon. cons. in pathology, Dept. of Health, Ont., from 1934; mem. Yellow Fever Commn., 1926, 28. Author: Arteriosclerosis, 1912. Died Nov. 3, 1936.

KLOTZ, Otto, Canadian astronomer, seismologist; b. Preston, Ont., Can., Mar. 31, 1852; ed. U. Toronto (Ont.), U. Mich.; D.Sc., LL.D.; m. Marie Widenmann, 1873; 3 sons. Surveyor, explorer in Canadian N.W.; made exploratory survey on 2000-mile canoe trip to Hudson's Bay, 1884; astronomer, from 1885; with Alaska Boundary Survey, 1893-94; dir. Dominion Obs. Canadian del. to internat. seismol. meetings at various fgn. cities, from 1907, to Internat. Astronomers' Union, Rome, 1922; chmn. nat. com. Can., Internat. Astron. Union, 1920. Fellow Royal Astron. Soc., Royal Soc. Can., New Zealand Inst. (hon.); mem. Am. seismol. Soc. (pres. 1920-21). Research and publs. on astronomy, gravity, terrestrial magnetism, seismology; completed 1st astronomic girdle of world, in Brit. Empire, 1903-04. Died Dec. 28, 1923.

KLOVE, Hallgrim, psychologist; b. Moss, Norway, June 22, 1927; s. Olav B. and Gudrun B. (Skram) C.; Candidatus Psychologiae, U. Oslo (Norway), 1952; m. Borghild M. Olsen, Aug. 1, 1953; children—Brynjulv, Kari, Erik. From instr. to asso. prof. Ind. U. Med. Sch., Indpls., 1954-62; dir. neuropsychol. lab., asso. prof. neurology U. Wis. Med. Sch., Madison, 1962—. Cons. VA hosps., 1963—. Mem. Am., Midwestern, Norwegian psychol. assns., Am. Acad. Neurology, Internat. Neuropsychology Soc., A.A.A.S., Am. Assn. U. Profs., Sigma Xi. Research and publs. on brain-behavior relationships; relationship between behavioral variables and neurol., bioelec. data; measurement methods for motor, sensory, intellectual changes secondary to cerebral disease or injury. Home: 2400 Mayflower Dr., Middleton, Wis. 53562. Office: 1954 E. Washington Av., Madison, Wis. 53704.*

KLUCKHOHN, Clyde Kay Maben, Am. anthropologist; b. Le Mars, Iowa, Jan. 11, 1905; s. George Wesley and Katherine (Swanzey) K.; student Culver Mil. Acad., 1918-21; Lawrenceville (N.J.) Sch., 1921-22; Princeton, 1922; A.B., U. Wis., 1928; student U. Vienna, 1931-32; M.A., Oxford (Rhodes Scholar), 1932; Ph.D., Harvard, 1936; L.H.D., U. New Mexico, 1949; m. Florence Rockwood, Oct. 15, 1932; 1 son, Richard Paul Rockwood. Asst. prof. anthropology U. New Mexico, also research asso. archaeology Sch. Am. Research, 1932-34; instr. anthropology Harvard, 1935-37, asst. prof., 1937-40, asso. prof., 1940-46, apptd. prof., 1946, dir. Russian Research Center, 1947-54. Fellow Center for Advance Study in Behavioral Science, 1954-55; cons. Gen. MacArthur's hdqrs., 1946-47; staff mem. Sch. Overseas Adminstrn., Harvard, 1943-44; dep. chief joint morale survey War Dept. and Office War Information, 1944-45; chmn. div. anthropology and psychology NRC, 1956-58; mem. adv. com. to Fgn. Service Inst., Dept. State, 1956-60. Hon. fellow Royal Anthrop. Inst. London; mem. Nat. Acad. Scis., Am. Philos. Soc., Am. Acad. Arts and Scis.; Am. Anthrop. Assn. (pres. 1947). Author: To the Foot of the Rainbow, 1927; Beyond the Rainbow, 1933; (with Dorothea Leighton) The Navaho, 1946, Children of the People, 1946; (with H. Murray) Personality in Nature Society and Culture, 1948; Mirror for Man, 1949; (with E. Z. Vogt and Leonard McCombe) Navaho Means People; Culture: A Critical Review of Concepts and Definitions; (with R. Bauer and A. Inkeles) How the Soviet System Works, 1956. Did extensive field work on Navajo Indians; contbd. to study of culture and personality. Died 1960.

KLUCKHOHN, Florence Rockwood (Mrs. Clyde K. M. Kluckhohn), Am. educator; b. Bluffs, Ill., Jan. 14, 1905; d. Homer G. and Florence (McLaughlin) Rockwood; A.B., U. Wis., 1927; Ph.D., Radcliffe Coll., 1941; m. Clyde K. M. Kluckhohn, Oct. 15, 1932; 1 son, Richard Paul Rockwood. Tchr., Wellesley Coll., 1940-48; research analyst OWI, Washington, 1944-45; lectr. dept. social relations, asso. Lab. Social Relations, Harvard, Cambridge, Mass., 1948—. Mem. staff Children's Med. Center, Boston, 1956-61. Mem. Am. Acad. Polit. and Social Scis., Am. Sociol. Assn., Am. Anthrop. Assn., Soc. Applied Anthropology, Phi Beta Kappa. Author: The American Family, Past and Present and America's Women, 1952; (with Fred L. Strodtbeck) Variations in Value Orientations, 1961, also articles, chpts. in textbooks. Developed theory and research methods for analysis of universals and variations in the basic values of human beings of differing socs.; utilization of these theories and testing methods for research on family systems, ednl. systems, emotional maladjustments, mental illness, and cultural integration and change. Home: 50 Fresh Pond Pkwy., Cambridge, Mass. 02138.*

KLUGE, Werner, German physicist; b. Leipzig, Apr. 24, 1902; s. Alfred and Gertrud (Schell) K.; ed. Stuttgart and Munich Technische Hochschule; engring. diploma; D.Eng.; m. Gerda Pupp, Sept. 30, 1933; children—Günter, Gerhard, Irene. Physicist, Berlin A.E.G., 1929; instr. Danzig Technische Hochschule, 1938; full prof. U. Prague Technische Hochschule, 1943; prof. Inst. für Gasentladungstechnik. Mem. Am. Phys. Soc., German Soc. Physics. Research and publs. on high temperatures, plasma, photoelectronics, conversion of energy. Home: Köllestrasse 36. Office: Breitscheidstrasse 2, Stuttgart, Germany.

KLÜKEN, Nobert, German physician; b. Krefeld, May 4, 1920; s. Wilhelm and Maria (Roether) K.; M.D., Martin Luther U., Halle/Wittenberg, 1945. Faculty, U. Saarland; 1955-64; faculty Essen Clinic of U. Munster, 1957—, prof., 1964—. Mem. Deutsche Gesellschaft für Angiologie (gen. sec.), Deutsche Gesellschaft für Phlebologie (mem. bd.), Essener Dermatologische Gesellschaft (sec.), Deutsche Gesellschaft für Dermatologie, College of Angiology, Tropical Dermatologia; hon. mem. Société françoise de phlébologie. Author: Fortschritte der Angiologie, 1957; Akrale Gefässreaktionen, 1959; Westafrika, 1964; also numerous articles. Research in angiology. Home: 43 Croussstrasse, 415 Krefeld/Nordrehin, Germany. Office: 55 Hufelsanstrasse, 43 Essen, Germany.*

KLUMPP, Theodore George, Am. physician; b. N.Y.C., May 15, 1903; s. Charles and Marie (Haye) K.; B.S., Princeton, 1924; M.D., Harvard, 1928; D.Sc., Phila. Coll. Pharmacy, 1943; m. Virginia Allan 1926 (div.); children—Maralys, Russell; m. 2d, Virginia Morgan, 1934; children—Virginia-Ann, Karla, Kathleen, Theodore George. Instr. internal medicine Yale Med. Sch., 1932-36, asst. clin. prof. medicine, 1936; asso. physician New Haven Hosp., 1932-36, chief hematological clinic, dir. med. lab., 1933-36; attending physician Gallinger Municipal Hosp., 1941; chief drug div. Food and Drug Adminstrn., Fed. Security Agy., Washington, 1936-41; dir. drugs, foods and phys. therapy, also sec. council pharmacy and chemistry A.M.A., Chgo., 1941-42; pres., dir. Winthrop-Stearns, Inc., 1942—, Winthrop Products N.J., 1947-51; bd. govs. Sterling-Winthrop Research Inst., 1946—; chmn. bd. Winthrop-Stearns Can., Ltd., 1951—. Mem. various industry-govt. adv. coms., 1942—; mem. study com. on fed. aid to pub. health Commn. on Inter-Govtl. Relations, Washington, 1954—; mem. Nat. Adv. Council on Vocational Rehab., Washington, 1955-59; mem. oral precipitation adv. com. Bur. Narcotics, U. S. Dept. Treasury, 1955—; mem. com. on aging, Nat. Social Welfare Council. Chmn. bd. govs. Nat. Vitamin Found. First recipient Spl. award for distinguished service Am. Pharm. Mfrs. Assn. Fellow A.M.A., Soc. for Clin. Investigation; mem. Nat. Pharma. Council, (pres. 1953—), A.C.P., Drug Mfrs. Assn. (v.p. 1952-55), Found. Tropical Medicine, N.Y. Acad. Medicine. Studies in therapeutics, materia medica, med. pharmacology, hematology, physiol. chemistry. Office: Winthrop Labs., 90 Park Av., N.Y.C.

KLÜPFEL, Walther Gustav Otto, German geologist; b. Heidelberg, May 28, 1888; s. Josef and Luise (Lösch) K.; Ph.D. in Geology and Paleontology, U. Strasbourg; m. Gabriele Hajek, Mar. 6, 1920; m. 2d, Luise Schmidt, Dec. 15, 1961. Asst. geologist Deutsch-Luxembourg Bergwerk Algemeine Gesellschaft, Metz, Seismos Gesellschaft Hanover; asso. prof. U. Giessen; head research geologists, prof. U. Marburg, U. Giessen. Recipient Gold medal Metz Acad., 1911. Research and publs. on stratigraphy, others. Address: Bergstrasse 35, 63-Giessen, Germany.

KLÜVER, Heinrich, psychologist; b. Holstein, Germany, May 25, 1897; s. Wilhelm and Dorothea (Wübbers) K.; student U. of Hamburg (Germany) and U. of Berlin, 1920-23; Ph.D., Stanford University, 1924; M.D. (honorary), Univ. of Basel, 1965; m. Cessa Feyerabend, Feb. 4, 1927. Came to U. S., 1923, naturalized, 1934. Instr. psychology U. of Minn., 1924-26; fellow Social Science Research Council, 1926-28; research psychologist Behavior Research Fund, Chicago, 1928-33; asso. mem. Otho S. A. Sprague Memorial Inst., U. of Chicago, 1933-38, mem. Otho A. Sprague Memorial Inst., 1938-46, prof. exptl. psychology, div. biol. sci., 1938-57, Sewell L. Avery distinguished service prof. biol. psychol., 1957-62, emeritus, 1963—; mem. program project com. Nat. Inst. Gen. Med. Scis., NIH, 1963-66; prin. investigator Nat. Inst. Mental Health, 1963—. Recipient Karl Spencer Lashley award in neurobiology Am. Philos. Soc., 1960; Samuel W. Hamilton award, Am. Psychopath. Assn., 1963; gold medal award Am. Psychological Found., 1965. Member Nat. Acad. of Sciences, Am. Academy Arts and Sciences, Am. Psychol. Assn., A.A.A.S., American Physiol. Soc., Am. Neurological Assn., Soc. Exptl. Psychologists, Midwestern Psychol. Assn., Neurosciences Research Program, Chicago Psychological Club, Optical Society of America, Society for Exptl. Biology and Medicine, Inst. Medicine of Chicago, Psychonomic Society, Soc. of Biol. Psychiatry, Am. Soc. Human Genetics, N.Y. Acad. Sci., Assn. Research Nervous and Mental Disease, Am. Philos. Soc., Am. Society for Cybernetics, Internat. Brain Research Orgn., Royal Society of Medicine, Pavlovian Soc. Exptl. Investigation Behavior, American Academy of Neurology. Author: An Experimental Study of the Eidetic Type, 1926; Mescal, 1928; Behavior Mechanisms in Monkeys, 1933. Asso. editor: Jour. of Psychol., 1935—, Jour. Comp. Physiol. Psychology, 1947-50, Exptl. Neurology, Psychopharmacologia, Jour. Genetic Psychology, Perspectives in Biology and Medicine, 1959—, Genetic Psychology Monographs, 1960—, Neuropsychologia. Contbr. articles to sci. jours. Discovered Klüver-Bucy Syndrome; determined role of brain in vision; discovered free porphyrins in central nervous system; new staining technique for nervous tissue (Klüver-Barrera method); devised method of equivalent and non-equivalent stimuli for analysis of behavior. Home: 5312 Blackstone Av., Chgo. 15. Office: U. Chgo., Chgo. 37.*

KLUYVER, Albert Jan, Dutch microbiologist; b. Breda, Netherlands, June 3, 1888; s. Jan Cornelis and Marie (Honigh) K.; Chem.E., Tech. U., Delft, Netherlands, 1910, Dr. Tech. Sci., 1914; postgrad. U. Vienna, 1911; D.Sc., Ia. State Coll., 1932; hon. doctorates U. Louvain (Belgium), 1953, Rutgers U., 1954; m. Helena Johanna van Lutsenburg Maas, July 29, 1916; children—Antoinette, Jan Cornelis, Marie, Coenraad Theodore, Clasine Albertine. Asst. prof. tech. botany, Tech. U., Delft, 1910-16; indsl. adviser Netherlands Indian Govt., Buitenzorg, Java, 1916; chem. adviser Vegetable Oil Mill Concern, Bandoeng, Java, 1919-21; prof. microbiology Tech. U., Delft from 1921. Mem. Koninklyke Akademie van Wetenschappen Amsterdam, 1926; Ridder Orde Nederlandsche Leeuw, 1930; mem. Koninklyke Vlaamsche Academie van Wetenschappen (Belgium), 1938. Recipient Emil Christian Hansen gold medal (Copenhagen), 1946; Copley medal Royal Soc. (London), 1954. Fgn. mem. Academia Scientiaram Fennica (Helsinki); pres. Koninklyke Nederlandse Akademie van Wetenschappen, 1947-54; hon. mem. Soc. Gen. Microbiology (Gt. Britain), Soc. Am. Bacteriologists, N.Y. Acad. Scis.; fgn. asso. Nat. Acad. Scis., Washington, 1950; fgn. mem. Royal Soc., London; fgn. hon. mem. Am. Acad. Arts and Scis.; corr. mem. Bot. Soc. Am.; mem. Netherlands Soc. Microbiology, Chem. Soc., Botanical Soc. Died May 14, 1956.

KNABE, Joachim, German pharm. chemist; b. Wilsdruff, Dresden, Germany, Jan. 18, 1921; s. Peter and Theodora (Priebs) K.; pharm. exam. Technische Hochschule Braunschweig, 1950, Dr.rer.nat., 1954; m. Elisabeth Weber, Aug. 28, 1952; children—Christine, Friedrich-Joachim. Faculty, Technische Hochschule Braunschweig, 1959-65; prof. U. Saarland, Saarbrücken, 1965—, dir. Inst. Pharm. Chemistry, 1965—. Mem. Deutsche Pharmazeutische Gesellschaft, Gesellschaft deutscher Chemiker, Chem. Soc. (London), Am. Chem. Soc. Author: (with H. Auterhoff) Lehrbuch der Pharmazeutischen Chemie, 1962; also articles. Research on alkaloids especially isoquinoline; analysis of pharms., resolution of racemic barbituric acids, rearrangement of dihydro-isoquinoline. Home: 95 Feldmannstrasse, 66 Saarbrücken, Germany.*

KNAPOWSKI, Stanislaw Karol, mathematician; b. Poznan, Poland, May 19, 1931; s. Roch Boleslaw and Zofia (Krysiewicz) K.; diploma, Poznan U., 1952, Ph.D. in Math., 1957; Magister, Wroclaw (Poland), 1954. Instr., Poznan U., 1954-57; Rockefeller Found. scholar, Cambridge, Eng., 1957-58; lectr. Poznan U., 1958-62; asst. vis. prof. Tulane U., 1962-63; Gast prof. Philipps U., Marburg/Lahn, Germany, 1964-65; faculty U. Fla., Gainesville, 1965-66; prof. U. Miami, Coral Gables, Fla., 1966—. Alfred P. Sloan Found. scholar, 1966—. Mem. Polish, Am. math. socs. Research, publs. on distbn. of primes, distbn. primes in arithmetical progressions and prime ideals in algebraic fields, distbn. of values of characters and other number-theoretic functions, linear processes of approximation. Home: 825 Coral Way, Coral Gables, Fla. 33134.*

KNAPP, Edward Alan, Am. physicist; b. Salem, Ore., Mar. 7, 1932; s. J. Gardner and Lucille (Moore) K.; A.B., Pomona Coll., 1954; Ph.D., U. Cal. at Berkeley, 1958; m. Jean Elaine Hartwell, June 27, 1954; children—Sandra, David, Robert, Mary Lucille. Staff, Los Alamos Sci. Lab., U. Cal., Los Alamos, N.M., 1958—, group leader, 1965—. Mem. Am. Phys. Soc. Contbr. articles to tech. jours. Research on high current proton linear accelerator structures and design, small angle photmeson prodn. to yield information on virtual pion states, plasma physics research in feasibility of controlled fusion power including plasma injection, magnetic trapping. Home: 227 El Conejo St., Office: Box 1663, Los Alamos 87544.*

KNAPP, Herman, physician; b. Dauborn, Prussia, Mar. 17, 1832; s. Johann K. (mem. German Reichsrath, etc.); M.D., U. of Giessen, 1854; studied in Germany, London and Paris; m. Adolfine Becker, 1864; m. 2nd. Hedwig Sachsowsky, 1878; 3 children. Prof. ophthalmology, Heidelberg, 1864-68; has practiced in New York, 1868—; prof. ophthalmology, New York U. Med. Coll., 1882-88; same chair, Coll. Phys. and Surg. (Columbia), 1888-1902 (emeritus). Founded New York Ophthalmic and Aural Inst., 1869; established Archives of Ophthalmology and Otology, 1869. Important role in development physiological optics; research on corneal curvature, 1859; optical constants, 1866. Died Mamaroneck, N.Y., Apr. 30, 1911.

KNAPP, Leslie William, Am. ichthyologist; b. Port Byron, N.Y., Nov. 17, 1929; s. Ronald Glen and Harriet (Chappell) K.; B.S. Cornell U., 1952, Ph.D., 1964; M.A., U. Mo., 1958; m. Betty Lou Hunt, Aug. 3, 1957. Supr. vertebrates Oceanographic Sorting Center, Smithsonian Instn., Washington, 1963—. Mem. Am. Soc. Ichthyologists and Herpetologists, Am. Fisheries Soc., Am. Soc. Systematic Zoologists, Am. Soc. Zoologists, Am. Inst. Biol. Scis., A.A.A.S. Research freshwater fishes of Eastern U. S., particularly family Percidae, Indo-Pacific family Platycephalidae. Home: Box 313A, Owings, Md. 20836. Office: Oceanographic Sorting Center, Smithsonian Instn., Washington 20560.*

KNAPP, Ludwig Friedrich, German chemist; b. 1814; became prof., Giessen, Germany, 1847, Brunswick, Germany, 1863. Author: Lehrbuch der chemischen Technologie, 2 vols. Research on prodn. of ultramarine, indsl. problems of tanning, glass, cement; developed tanning method using metallic salts which led to modern quick-process tanning with chromium salts. Died 1904.

KNAPP, Robert Hampden, Am. psychologist; b. Portland, Ore., Apr. 16, 1915; s. Joseph B. and Cornelia (Pinkham) K.; B.A., U. Ore., 1938; M.A., Harvard, 1940, Ph.D., 1948; M.A. (hon.), Wesleyan U., 1956; m. Johusia Nelson, Mar. 19, 1945; children—Robert H., Abigail, Hiatt, Sarah. Dir. div. propaganda research Mass. Com. on Pub. Safety, Boston, 1941-42; area operations officer Morale Operations Br. OSS, Washington, 1943-46; mem. faculty Wesleyan U., 1946-53, 54—, prof. psychology, 1955—, chmn. dept., 1956—; dep. dir. Behavioral Scis. div. Ford Found., 1953-54; staff psychologist Inst. Personality Assessment & Research, U. Cal., Berkeley, 1957-58. Dir. sci. faculty fellowship panel NSF, Assn. Am. Colls., 1956—; cons. founds., schs., hosps., 1957—. Mem. Am., Eastern, Conn. (pres. 1956-57) psychol. assns., Am. Orthopsychiat. Assn., Am. Soc. for Esthetics, Am. Assn. Humanistic Psychology, Phi Beta Kappa, Sigma Xi. Author: (with H. B. Goodrich) Origins of American Scientists, 1953; (with J. J. Greenbaum) The Younger American Scholar, 1954; Origins of the American Humanistic Scholar, 1964; also articles. Prin. research on characteristics of scholarly elites, psychology of temperament and aesthetic preference, and attitudes toward time. Home: 171 Mt. Vernon St., Middletown, Conn. 06457.*

KNAPP, Seaman Asahel, Am. agriculturist; b. Schroon Lake, N.Y., Dec. 16, 1833; s. Bradford and Rhoda (Seaman) K.; A.B., Union Coll., N.Y., 1856; LL.D., Upper Ia. U., 1882, Baylor U., 1908; D.Sc., Ia. State Coll. Agr. and Mech. Arts, 1909; m. Maria E. Hotchkiss, Aug. 6, 1856; children—Maria, Herman, Bradford, Seaman Arthur, Helen Knapp. Tchr., later asso. mgr. Ft. Edward Collegiate Inst., 1856-63; asso. mgr. Ripley Female Coll., Vt., 1863-65; pres. Ia. State Coll. for Blind, 1869-75; established Western Stock Jour. and Farmer, Cedar Rapids, Ia., 1872; prof. agr. Ia. State Agrl. Coll. 1879-86, pres., 1883-84; visited Japan, China and Philippines, for U. S. Dept. Agr., to report on resources of islands, 1898-99, Puerto Rico, 1900, Japan, China, P.I., Ceylon, India, Burma and H.I. 1901-02; in charge of Farmers Co-operative Demonstration Work in Southern states for U. S. Dept. Agr., 1902-10. Made important contbns. to sci. farming and stock breeding in U. S.; improved rice culture in Southwest; demonstrated methods for curbing boll weevil in Tex. Died 1911.

KNAUS, Hermann Hubert, Austrian physiologist; b. Sankt Veit-Glan, Oct. 19, 1892; s. Fritz and Amalia (Schebath) K.; M.D., U. Graz, 1920; m. Rizica Stankonvic, Oct. 15, 1932; 1 dau., Ingeborg Juliane. Asst., Univ. Clinic Gynecology, Graz, 1923; agrégé U. Graz, 1927; asso. prof. German U., Prague, 1930, full prof., 1934; dir. Gynecology Clinic, Prague, until 1945. Mem. Royal London Soc. Medicine, also others. Author: Die Physiologie der Zeugung des Menschen, also numerous other publs. Address: Stadiongasse 6, Vienna I, Austria.

KNAUTH, Christian, German botanist; b. Halle, Germany, 1654; s. Christoph Knauth; prof., Halle. Author: Methodus plantarum genuora, qua differentiae genericae, tam summae quam subalternae, ordine, digeruntur, 1706. Listed in alphabetical order 1,476 plant species, 1865; tried to modify Rivinus' system. Died 1716.

KNAYSI, Georges Abdallah, microbiologist; educator; b. Hasbaya, Lebanon, June 21, 1898; s. Abdallah Saad and Saleemy (Mattar) K.; student College de la Sagesse, 1911-13, Universite St. Joseph, 1913-20 (both Beirut, Lebanon), Ecole Duvigneau de l'Anneau, Paris, France, 1920-21; B.S., Cornell U., 1924, M.S., 1925, Ph.D., 1929; D.Sc., St. Bonaventure, 1952; m. Adele Catherine Maosha, Apr. 23, 1939; children—George Albert, Edmund Joseph, Fareed Anthony. Came to U. S., 1921, naturalized, 1929. Faculty, Cornell U., 1926—, prof. bacteriology, 1944-65, prof. microbiology, 1965—. Tech. adviser bacteriology Dept. Army, 1946-49; cons. bacteriology RCA Labs., 1947-48, U. S. Dept. Agr., 1960; Fulbright lectr. U. Paris Faculty Medicine, 1953. Mem. Am. Soc. Microbiology, A.A.-A.S., Am. Dairy Sci. Assn., Societe Francaise de Microbiologie, Societe de Pathologie Comparee (France), N.Y. Acad. Scis., Am. Acad. Microbiology, Sigma Xi, Phi Kappa Phi. Author: Elements of Bacterial Cytology, 1944, 2d edit., 1951. Research, publs. chiefly on orgn. and structure of bacterial cell and spore. Home: 119 Ithaca Rd., Ithaca, N.Y. 14850.*

KNEASS, Strickland Landis, Am. mech. engr.; b. Phila., Jan. 7, 1861; s. Strickland and Margaretta Sybilla (Bryan) K.; C.E., Rensselaer Poly. Inst., 1880; m. Mary Stewart Edwards, Oct. 22, 1888; children—Strickland, Edward, George Bryan. With William Sellers & Co., Phila., from 1883, became mgr. injector dept., 1895, v.p., from 1926. Recipient awards Chgo. Expn., 1893, St. Louis Expn., 1904, John Scott medal and premium, Franklin Inst., 1900. Author: Practice and Theory of the Injector. Died Nov. 24, 1928.

KNECHT, David Jordan, Am. physicist; b. Elgin, Ill., June 2, 1930; s. Walter Grote and Erma (Jordan) K.; B.S., U. Ill., 1951, M.S., 1952; Ph.D., U. Wis., 1958; m. Dzidra Jirgensons, June 1, 1957; 1 son, Frederick Jordan. Research asso. U. Wis., 1958-59; project officer USAF Spl. Weapons Center, 1959-61, research physicist, 1961-62; sci. dir. Space Physics br. Air Force Weapons Lab., 1962-64; research physicist Air Force Cambridge Research Lab., Bedford, Mass., 1964—. Mem. Am. Geophys. Union, Am. Phys. Soc. Contbr. numerous articles infield to profl. jours. Research in low energy nuclear physic, particles and fields in space. Home: 60 Sweetwater Av. Office: Air Force Cambridge Research Labs., L. G. Hanscom Field, Bedford, Mass. 01730.*

KNECHT, Robert William, Am. physicist; b. Ogdensburg, N.Y., May 4, 1928; s. Robert B. and Wilma (Rosenbaum) K.; B.S. in Physics, Union Coll., 1949; m. Edith Lillian Miller, Feb. 16, 1952; children—Susan Lillian, Sarah Jane, Scott Arthur. Physicist, Nat. Bur. Standards, Washington, Alaska, Boulder, Colo., 1949-65, chief, ionosphere research and propagation div., 1963-65; now dir. space disturbances lab. Inst. for Telecommunication Scis. and Aeronomy, Environmental Sci. Services Adminstrn., Boulder. Recipient Gold medal Dept. Commerce, 1965. Mem. Am. Geophys. Union, Research Soc. Am., A.A.A.S. Author: (with others) Space Physics. Research on ionospheric physics, including study of lunar tidal effects in ionosphere, application of radio pulse sounding technique to studies of topside of ionosphere using satellites, solar-terrestrial relations, including prediction of terrestrial effects. Home: 1702 Mariposa St., Boulder 80302. Office: Space Disturbances Lab., Inst. for Telecommunication Scis. and Aeronomy, Environmental Sci. Services Adminstrn., Boulder, Colo. 80301.*

KNEELAND, Samuel, Am. physician, naturalist; b. Boston, Aug. 1, 1821; s. Samuel and Nancy (Johnson) K.; A.B., Harvard, 1840, A.M., M.D., 1843; m. Eliza Curtis, 1849. A founder Boylston Med. Sch., 1847; physician to Boston Dispensary, 1845-47; demonstrator anatomy Harvard Med. Sch., 1851-53; surgeon U. S. Army, 1862-66, in charge Univ. Hosp., New Orleans, also gen. hosps. of Mobile, Ala.; mem. corp. Mass. Inst. Tech., prof. zoology and physiology, 1869-78, sec. of corp. and faculty, 1865-78. Recipient Boylston prize for thesis on contagiousness of puerperal fever, 1846. Mem. Am. Acad. Arts and Scis. (sec.). Author: Hydrotherapy, 1844; Science and Mechanism, 1854; The Wonders of the Yosemite Valley, 1871; Volcanoes and Earthquakes, 1888. An editor Annual of Scientific Discovery, 1866-69. Translator: Andry's Diseases of the Heart, 1847. Died Hamburg, Germany, Sept. 27, 1888.

KNEISSL, Max, German geodesist; b. Munich, Germany, Sept. 9, 1907; s. Sebastian and Walburga (Schmidbauer) K.; student Munich Inst. Tech.; Dr.-Ing. E.h., 1957; Dr.techn.E.h., 1966; m. Marianne Heigenmoser, 1936; children—Max, Ilse, Brigitte. Various ofcl. surveying positions, 1936-49; faculty Geod. Inst., Munich Inst. Tech., 1934-41, 48—, prof., dir. Geod. Inst., 1949—. Recipient Bavarian Service order. Mem. Internat. Assn. Geodesy (mem., pres. numerous internat. commns., study groups), Bavarian Acad. Sci. (head numerous commns.), Am. Geophys. Union, numerous others. Author: Handbuch der Vermessungskunde; also numerous articles. Research in geodesv: triangulation, adjustment, levelling, gravity, surveying, Instruments, etc. Address: 38 Maria-Ward-Strasse, 8 Munich 19, West Germany.*

KNESER, Hans Otto, German physicist; b. Berlin, July 1, 1901; s. Adolf and Laura (Booth) K.; ed. univ. and tech. schs. of Breslau and Munich; D.Eng.; m. Gertrud Gross, Oct. 12, 1933; children—Lorenz, Christof, Thomas, Stefan. Asst., U. Marbourg; asso. prof. U. Berlin, U. Tübingen; full prof. U. Stuttgart. Research and publs. on physics of relaxation processes. Home: Stirnbrandstrasse 6. Office: Widerholdstrasse 13, Stuttgart, Germany.

KNIGHT, Charles Mellen, Am. chemist; b. Dummerston, Vt., Feb. 1, 1848; s. Joel and Fannie Maria (Duncan) K.; A.B., Tufts Coll., 1873; A.M., 1878; post-grad. Harvard and Mass. Inst. Tech., 1874; Sc.D., Buchtel Coll., 1897; m. May Acomb, Aug. 31, 1882 (died Oct. 31, 1930); children—Maurice Acomb, Hal Greenwood, Helen Lillian. Prof. natural sci. Buchtel Coll. (now U. Akron, Ohio), 1875-83, phys. sci., 1883-1907, chemistry, 1907-13, prof. emeritus from 1913, dean, 1902-13, acting pres, 1896-97; Knight Chem. Lab. named for him, 1909. Fellow A.A.A.S.; mem. Am. Chem. Soc., Phi Beta Kappa, Phi Sigma Alpha. Established 1st academic course in rubber chemistry in U. S., 1909. Died July 3, 1941.

KNIGHT, Claude Arthur, Am. biochemist; b. Petoskey, Mich., Oct. 17, 1914; s. Claude and Wilhelmina (Hedt) K.; B.S., Alma Coll., 1936; Ph.D., Pa. State U., 1940; m. May Marion Ferry, June 9, 1943; children—Thomas, Susan, Robert. From asso. prof. to prof. biochemistry, virology, molecular biology U. Cal., Berkeley, 1948—; dir. NSF Summer Inst. Molecular Biology, 1964-67; cons. USPHS, 1965—. Mem. Am. Soc. Biol. Chemists, N.Y. Acad. Scis., A.A.A.S., Sigma Xi. Author: Chemistry of Viruses, 1963. Contbr. numerous articles to profl. jours. Research in change in composition virus proteins and nucleic acids with mutation, subunit concept of virus protein structure, amino acid sequence of viral protein, incorporation of host material into animal virus particles, location of genes in viral nucleic acid by biochem. methods.*

KNIGHT, Colin Sutherland, English chemist; b. Brighouse, Yorkshire, Eng., July 18, 1932; s. Sutherland and Edith (Broomhead) K.; B.Sc., U. London, 1953, M.Sc., 1955, Ph.D., 1960; m. Mavis Crossley, Mar. 24, 1956; children—Phillip Sutherland, Penelope Jane. Research fellow Bradford Inst. Tech., 1955-57; research chemist W. & R. Balston Ltd., Maidstone, Kent, Eng., 1958-60, chief research chemist, 1960-65, projects mgr., 1965—. Mem. Royal Inst. Chemistry (asso.). Research, publs. on chromatographic systems, discoverer new and improved techniques of traditional and ion-exchange paper chromatography, studies on effect on chromatographic behavior of chem. constn. and structural characteristics of media. Inventor, co-inventor, patentee chromatographic media. Home: 2 Arran Rd. Office: Springfield Mill, Sandling Rd., Maidstone, Kent, Eng.*

KNIGHT, Gowin, English natural philosopher; b. Corringham, Lincolnshire, baptised Sept. 10, 1713; s. Robert and Elizabeth Knight; held demyship of Magdalen Coll., Oxford, 1735-46, M.A., 1739, M.B., Oxford, 1742. Settled in London, practiced as physician; began magnetical research before 1744; experimented with new compass in voyage aboard Fortune, 1751; 1st prin. librarian of Brit. Mus., 1756. Fellow Royal Soc., 1747 (Copley medal, 1747). Author: Attempt To Demonstrate That All the Phenomena of Nature May Be Explained by Attraction and Repulsion, 1748; his papers on magnetism collected, pub. 1758; number of papers in Philos. Trans. His improved compass adapted in Royal Navy, 1752, rendered important if unrecognized services to navigation. Died June 8, 1772.

KNIGHT, Henry Granger, Am. chemist; b. Bennington, Kan., July 21, 1878; s. Edwin Richard and Elva Maud (Edwards) K.; A.B. in chemistry U. Wash., 1902, A.M., 1904; postgrad. U. Chgo., 1902-03, 06; Ph.D., U. Ill., 1917; m. Nelly Dryden, June 28, 1905; 1 son, Richard Dryden. Asst. prof. chemistry U. Wash., 1903-04; prof. chemistry and state chemist, U. Wyo., 1904-10; dir. Wyo. Expt. Sta., and Farmers' Insts., Wyo., 1910-18, dean Coll. of Agr., 1912-18; dean and dir. Okla. Agrl. Coll., Stillwater, 1918-21; hon. fellow Cornell U., 1921-22; dir. and research chemist Expt. Sta. W.Va. U., 1922-27, dean Coll. of Agr., 1926-27; chief Bur. of Chemistry and Soils, 1927-39, chief Bur. Agrl. Chemistry and Engring., U. S. Dept. of Agr., Washington, from 1939. Fellow Am. Inst. Chemists (pres. 1933-35); mem. Am. Chem. Soc., Sigma Xi, Phi Beta Kappa, Alpha Zeta, Phi Lambda Upsilon, Phi Kappa Phi. Writer of various monographs on qualitative analysis; research work on potable waters, effect of alkali upon seeds, food adulterations, Wyo. forage plants, soil nitrogen, wool, poisonous plants, effect of alkali upon cement drainage experiments, digestion experiments, soil acidity. Died July 13, 1942.

KNIGHT, James Allen, Am. psychiatrist; b. St. George, S.C., Oct. 20, 1918; s. Thomas Samuel and Carolyn (Carn) K.; A.B., Wofford Coll., 1941; B.D., Duke, 1944; M.D., Vanderbilt U., 1952; M.P.H., Tulane U., 1962; m. Sally Templeman, June 8, 1963. Asst. resident pediatrics Duke U. Hosp., 1953-54; resident psychiatry Tulane U. Service Charity Hosp., 1955-58; mem. faculty Baylor U. Coll. Medicine, 1958-61; faculty Tulane U. Sch. Medicine, 1961—, prof. psychiatry, asso. dean, 1964—; Harkness prof. psychiatry and religion, dir. program psychiatry and religion Union Theol. Sem., N.Y.C., 1963-64. Travelling fellow WHO at C. G. Jung Inst., 1961. Mem. Am. Psychiat. Assn., Am. Acad. Psychoanalysis, Am. Pub. Health Assn., Acad. Religion and Mental Health. Author: A Psychiatrist Looks at Religion and Health, 1964; Allergy and Human Emotions, 1967; Motivations in Play, Games and Sports, 1967; For the Love of Money, 1968. Research on psychosomatic medicine; interrelationships of psychiatry and religion; motivation; suicide. Home: 7450 Pearl St., New Orleans 70118.*

KNIGHT, Jonathan, Am. physician; b. Norwalk, Conn., Sept. 4, 1789; s. Jonathan and Ann (Fitch) K.; grad. Yale, 1808, M.A., 1811, M.D. (hon.), 1818; m. Elizabeth Lockwood, Oct. 1813. Licensed to practice medicine by Conn. Med. Soc., 1811; asst. prof. anatomy and physiology Yale, 1813, prof. surgery, 1838, founder Med. Sch.; sec. Conn. Med. Soc., 1817; pres. meetings of Nat. Med. Conv. which formed A.M.A., 1846, 47; pres. A.M.A., 1853. 1st surgeon to cure aneurisms by compression, 1848. Died New Haven, Conn., Aug. 25, 1864.

KNIGHT, Thomas Andrew, Brit. botanist; b. Ludlow, Eng., Aug. 12, 1759; s. Thomas Knight; ed. Balliol Coll., Oxford; m. Frances Felton, 1791; children—Frances (Mrs. Thomas Dendarves Stackhouse Acton), Lady William Rause Boughton, Mrs. Francis Walpole. Recipient Copley medal, 1806; Knightian medal founded in his honor. Fellow Royal Soc., 1805. Author: A Treatise on the Culture of the Apple and Pear; On the Manufacture of Cider and Perry, 1797; Pomona Herfordiensis, 1811; also papers. Produced many new seed varieties of fruits; studied vegetable physiology; inventor machine (named after him) which established principle of geotropism. Died May 11, 1838.

KNIGHT-JONES, Elis Wyn, Brit. zoologist; b. Hanford, Staffs, U.K., Mar. 7, 1916; s. William Ellis and Maudd (Cotterill) J.; B.Sc., U. Coll. N. Wales, Bangor, 1938, postgrad.; D.Phil., Jesus Coll., Oxford (Eng.), U., 1949; m. Mary Morgan-Jones, Dec. 9, 1939; children—Peter, Philip, Carolyn. Sr. sci. officer Fisheries Lab., Burnham-on-Crouch, U.K., 1947-49; zoologist, dep. dir. Marine Sci. Labs., Menai Bridge, U.K., 1949-56; prof. zoology U. Coll. Swansea (U.K.), 1956—. Fellow Inst. Biology Contbr. numerous articles to tech. jours. Discovered gregariousness in larval oysters, barnacles, tube worms; found and described new marine leeches and tube worms; codemonstrator that a marine animal can perceive tidal changes in hydrostatic pressure. Home: 15 Derwen Fawr Rd., Swansea, U.K.*

KNIKER, William Theodore, Am. immunologist, pediatrician; b. Seguin, Tex., Aug. 30, 1929; s. Theodore Siegfried and Alice (Steger) K.; A.A., Tex. Lutheran Coll., 1948; B.A., U. Tex., 1950, M.D., 1953; m. Maria Lydia Bobleter, May 1, 1965; children—Dorothy Ann, Judith Elise. Faculty U. Ark. Med. Center, Little Rock; 1958—, asso. prof. pediatrics and microbiology, 1967—, asst. dir. clin. research unit NIH, 1965—. Named Outstanding Hosp. Resident, U. Tex. Med. Br., 1957; Nat. Trudeau Soc. Research fellow, 1959-63; Research fellow immunology Scripps Clinic and Research Found., LaJolla, Cal., 1962-65; NIH Research Career Devel. award, 1968. Mem. So. Soc. for Pediatric Research, Nat. Trudeau Soc. (past mem. nat. com. on mycobacterial antigens), Am. Soc. Exptl. Pathology, Am. Soc. Nephrology, Sigma Xi. Contbr. articles to tech. jours. Separated and classified antigens of Mycobacterial groups of bacteria; demonstrated diagnosis of milk allergy by blood tests; elucidated mechanism of prodn. of vascular and renal disease by circulating antigen-antibody complexes. Home: 9701 Brooks Lane, Little Rock 72205.*

KNIPLING, Edward F(red), Am. entomologist; b. Port Lavaca, Tex., Mar. 20, 1909; s. Henry J. and Hulda (Rasch) K.; B.S., Tex. A. and M. Coll., 1930; M.S., Ia. State Coll., 1932, Ph.D., 1947; D.Sc., Catawba Coll., 1962, N.D. State U., 1967; m. Phoebe R. Hall, July 21, 1934; children—Edwina H., Anita A., Edward B., Gray D., Ronald R. With Dept. Agr., 1931—, beginning as jr. entomologist, successively asst. entomologist, asso. entomologist, sr. entomologist, prin., chief insects affecting man and animal research br., 1931-53, dir. entomology research div. Agrl. Research Service, 1953—; dir. entomology research Orlando Lab., Fla., 1943-46. Recipient U. S. Typhus Commn. medal, President's Medal for Merit, King's Medal for cause of freedom, alumni centennial award Ia. State Coll., 1958; Hoblitzelle nat. award in agrl. sciences, 1960; Rockefeller Pub. Service award, 1966; Nat. Medal of Sci., 1966; several distinguished service awards. Member A.A.A.S., Am. Soc. Tropical Medicine and Hygiene, Entomological Soc. Am., Washington Acad. Sci., Nat. Mosquito Assn., Nat. Pest Control Assn., Am. Assn. Econ. Entomologists (pres. 1952), also member of the National Academy of Scis., Sigma Xi. Club: Cosmos. Author articles on entomology research. Research on taxonomy, biology, and control of insects. Home: 2623 Military Rd., Arlington 7, Va. Office: Entomology Research Div., Plant Industry Station, Beltsville, Md.

KNIPP, Charles Tobias, Am. physicist; b. Napoleon, O., Aug. 13, 1869; s. Frederick F. and Pauline (Youche) K.; A.B., Ind. U., 1894, A.M., 1896; Ph.D., Cornell U., 1900; studied Cavendish Lab., Cambridge, Eng., 1910-11, 26-27; m. Frances Winona Knause, June 25, 1896; children—Pauline Louise, Frances Mary, Julian Knause, Barbara Matilda. Instr. physics Ind. U., 1893-1900, asst. prof. 1900-03; asst. prof. physics U. Ill., 1903-15, asso. prof., 1915-17, prof. exptl. electricity, 1917-37, emeritus, from 1937; vis. prof. physics, Rollins Coll., Winter Park, Fla., 1942-1945. Mem. adv. subcom. physics for Century of Progress Expn., Chgo., 1933. Fellow A.A.A.S.; mem. Am. Phys. Soc., Am. Inst. E.E., Optical Soc. Am., Soc. for Promotion Engring. Edn., Ill. Acad. Sci. (pres. 1921-22), Ind. State Acad., Sigma Xi (pres. 1930-31), Phi Beta Kappa, hon. mem. Tau Beta Pi, Eta Kappa Nu, Epsilon Chi, Synton. Designer of demonstration apparatus in physics; invented (with H. A. Brown) Alkali-vapor detector tube for use in radio; inventor of simple alpha-ray track apparatus; also of efficient mercury vapor vacuum pumps, electrodeless elec. discharge, cold-cathode rectifier. Died July 6, 1948.

KNIPP, Julian Knause, Am. physicist; b. Champaign, Ill., Jan. 18, 1910; s. Charles T. and Frances (Knause) K.; B.A., U. Ill., 1931; M.S., Harvard, 1932, Ph.D., 1935; m. Christine Simmons, Apr. 6, 1943; children—Julia Simmons, Barbara Lynell, Pauline Evelyn. Instr. physics Purdue U., 1936-41, asst. prof., 1941-46; asso. prof. Ia. State Coll., 1946-48, prof., 1948-55; prof., chmn. Tufts U., Medford, Mass., 1955-68, asso. dean liberal arts, 1967-68, dean Faculty Arts and Scis., 1968—; staff mem. Radiation Lab., Mass. Inst. Tech., 1942-45. Guggenheim fellow, Copenhagen, Denmark, 1948-49; NSF fellow, Rome, Italy, 1961-62. Fellow Am. Phys. Soc., Am. Acad. Arts and Scis. Epsicopalian. Research on theory of nucleus, theory of microwave tubes, high energy physics. Home: 46 Hancock St., Lexington, Mass. 02173.*

KNIPPER, Helmut Ludwig, German zoologist; b. Neu-Isenburg, July 20, 1914; s. Philipp and Katharina (Pons) K.; ed. U. Frankfort, U. Berlin; Ph.D.; m. Almut Janssen, Dec. 23, 1949; children—Hartmut, Jan-Herbert. Asst., U. Mainz; with Technische Hochschule, Darmstadt; aide, later curator Bremen Mus.; chief curator natural sci. collections at Karlsruhe. Mem. German Zoology Soc., Frankfort Senckenbergische Naturforschende Gesellschaft, Naturalists Assn. Research and publs. on molluscs, palaearctic and African orthoptera. Home: Spöcker Weg 42, 7501 Eggenstein. Office: Erbporinzenstrasse 13, 75 Karlsruhe, Germany.

KNISELEY, Richard Newman, Am. chemist; b. Wichita, Kan., Jan. 23, 1930; s. Clarence W. and Bertice (Newman) K.; student Kansas City (Mo.) Jr. Coll., 1947-49; A.B., U. Mo., 1951; M.S., Ia. State U., 1954; m. Martha L. Lane, June 22, 1951; children—Richard Newman, Mark Albert. Research asst. Inst. for Atomic Research, Ia. State U., Ames, 1951-55, research asso., 1955-59, asso. chemist, 1959-67, chemist, 1967—. Mem. Optical Soc. Am., Soc. Applied Spectroscopy, Coblentz Soc., Phi Lambda Upsilon, Sigma Xi. Contbg. author: Scott's Standard Methods of Analysis, 1965; Trace Analysis, 1966. Research and publs. on spectroscopic excitation, flames and flame mechanisms, molecular structure of nitrogen-containing compounds utilizing infrared absorption techniques, X-ray and electron microprobe studies, spectrographic techniques; invented new burners for premixed oxyacetylene flames. Home: 322 Hilltop Rd., Ames. Ia. 50010.*

KNISELY, Melvin Henry, pathologic physiologist; b. Hillman, Mich., June 17, 1904; s. Samuel Henry and Flora Belle (Hagerman) K.; A.B., Albion Coll., 1927; Ph.D., U. Chgo., 1935, Rockefeller Found. (Gen. Edn. Bd.) fellow, 1934-37; psotgrad. U. Chgo., 1934-36, U. Copenhagen, Denmark, 1936-37; m. Verona Butzer, June 29, 1928; children—Robert August, Susan Jane. Asst. zoology, botany, bacteriology, 1923-27; tchr. physics Bessemer (Mich.) High Sch., 1927-30; asst. prof. anatomy U. Chgo., 1937-45, asso. prof., 1945-48; chmn. dept. anatomy Med. Coll. U. S.C., 1948—; research cons. pathologic physiology maleria U. Tenn., 1940-42. Mem. (charter) Am., European Microcirculatory confs., Am., S.C. acads. sci., Am. Assn. Anatomists, Am. Soc. Zoologists, Am. Physiol. Soc., Assn. Am. Med. Colls., Am. Soc. Tropical Medicine and Hygiene, Fedn. Am. Socs. Exptl. Biology, Soc. Rheology, Am. Assn. Neuropathologists, Phi Beta Kappa, Sigma Xi. Research, publs. on microcirculation including study circulatory disorders in mammals, gen. pathologic circulatory physiology, blood cell pathology, diseases of liver, spleen. Home: 608 Harbor View Rd., Charleston, S.C. 29407.*

KNOBEIÉVSKII, Sergéi Tíkhonovich, Russian physicist; b. Apr. 26, 1890; grad. Moscow U., 1913. Mem. staff Inst. Popular Economy, Moscow, 1919-23, All-Union Electro-Tech. Inst., Moscow, 1923-29, State Inst. Non-Ferrous Metals, 1929-41; staff Moscow U., 1926—, prof., 1935—. Corr. mem. USSR Acad. Sci. Research on atomic structure of metals and alloys, including changes in their structure at plastic deformation.

KNOBIL, Ernst, physiologist; b. Berlin, Germany, Sept. 20, 1926; s. Jakob and Regina (Seidmann) K.; B.S., Cornell U., 1948, Ph.D. (Schering fellow endocrinology), 1951; m. Julane Hotchkiss, July 11, 1959; children—Erich Richard, Mark, Nicholas, Katharine. Came to U. S., 1940, naturalized, 1945. Asst. zoology Cornell U., 1948-49; Milton Research fellow Harvard, 1951-53, faculty Med. Sch., 1953-61; Richard Beatty Mellon prof. physiology, chmn. dept. U. Pitts. Sch. Medicine, 1961—; spl. research pituitary gland, endocrinology reprodn. John and Mary R. Markle Found. scholar med. scis., 1956-61; cons. USPHS, NRC. Fellow A.A.A.S.; mem. Am. Soc. Zoologists, Soc. Exptl. Biology and Medicine, Am. Assn. Anatomists, Am. Physiol. Soc. (Bowditch lectr. 1965, vis. scientist U. Notre Dame, Ohio U., Dennison U.), Endocrine Soc. (Ciba award 1961), Soc. Endocrinology (Great Britain), Royal Soc. Medicine (overseas affiliate), Sigma Xi. Mem. editorial bd. Am. Jour. Physiology and Endocrinology, Ann. Rev. Physiology. Research on mechanisms of hormone action; reproductive physiology; physiology of the pituitary gland. Home: W. Waldheim Rd., Pitts. 15215.*

KNOCK, Frances Engelmann, Am. surgeon, chemist; b. Chgo., July 8, 1921; d. William Fred and Frances (Tietze) Engelmann; B.S., U. Chgo., 1940, Ph.D., 1943; M.D., U. Ill., 1954; m. Theodore E. Knock, Oct. 23, 1943. Asso. organic chemist Armour Research Found., Chgo., 1943-46, organic chemist, 1946-49, chem. cons., 1950-51; dir. Nat. Registry Rare Chems., 1945-50; chem. cons. Merck Sharp & Dohme, 1954-56; adj. surgery Presbyn.-St. Luke's Hosp., 1959-61; dir. Knock Research Found., Chgo., 1955—; practice surgery, Chgo., 1959—; staff Presbyn.-St. Luke's Hosp.,

Chgo., VA Hosp., Hines, Ill.; faculty U. Ill., 1959—, asst. prof. surgery, 1961—. Cons. E. R. Squibb & Sons, 1964—. Swift fellow U. Chgo., 1943. Diplomate Am. Bd. Surgery, Nat. Bd. Med. Examiners. Fellow A.A.A.S., Inst. Chemists, Am. Geriatrics Soc., Sci. Council, Internat. Coll. Angiology; mem. A.M.A., Am. Chem. Soc., N.Y. Acad. Scis., Am. Inst. Chemists, Am. Coll. Angiology. Author: Anticancer Agents, 1966; articles. Devel. rapid setting resins for dentistry and neurosurgery; research on revascularization of heart for coronary artery disease by radical cardioomentopexy, synthesis of monomers, catalysts and promoters for new plastics, synthesis of improved sulfhydryl inhibitors for clin. cancer chemotherapy; proposed chromosomal residual protein-sulfhydryl groups in control of genes and in clin. cancer chemotherapy; improved drugs and shielding materials for neutron capture therapy of cancer; research on antibodies as peptizing agts. for cytotoxic colloids in cancer therapy. Home: 416 Country Lane, Glenview, Ill. 60025. Office: 30 N. Michigan Av., Chgo. 60602.*

KNOLL, Fritz Friedrich, Austrian botanist; b. Gleisdorf, Oct. 21, 1883; s. Friedrich and Maria (Carl) K.; Ph.D., Graz U.; m. Sophie Helsegg, Nov. 5, 1923; children—Friedl, Hans. Agrégé, U. Graz, 1912, U. Vienna, 1914; prof. botany German U. of Prague, 1923; prof. botany, dir. Bot. Inst. and Bot. Garden, U. Vienna, 1933-45, rector, 1938. Recipient Archduke Rainer medal Vienna Bot. Zoology Soc.; named laureat Imperial Acad. Sci., Vienna. Mem. Vienna Acad. Sci. (sec. gen.), also others. Author: Insekten und Blumen, 1921-26; Die Biologie der Blüte, 1956; Osterreichischer Naturforscher, Arzte und Techniker, 1957; Fortpflanzung und Vermehrung der Gewächse, 1963. Research on systematic and gen. botany, interrelationships between flowers and pollinating factors. Home: Stubenbastei 1, Vienna I. Office: Dr. Ignaz Seipel-Platz, Vienna I, Austria.

KNOLL, Josef Gabriel, German agronomist; b. Ulm, June 26, 1899; s. August and Maria (Eberle) K.; ed. Hohenheim Landwirtschaft Hochschule; diploma in agronomy; Ph.D. agrege; m. Liselotte Wenger, Feb. 19, 1938; children—Gabriele, Peter. Full prof. Inst. Agr., Sowing and Culture, U. Leipzig; founder, dir. Research Inst. for Culture Highlands, Baden; dir. prodn. and protective div. for plans FAO, Rome; full prof., dir. Fgn. Agronomy Inst., Hohenheim. Mem. German Agronomy Soc., German Land Devel. Inst. Author: Die Wiesentypen des württembergischen Unterlandes; Feldfutterbau; Die Gräser; World Aspects of Plant Protection. Home: Schlossergasse 22, Ulm/D. Office: Stuttgart-Hohenheim, Schloss, Germany.

KNOLLMAN, Gilbert Carl, Am. physicist; b. Cleve., Mar. 14, 1928; s. Paul C. and Louise (Heidebrink) K.; student Case Inst. Tech., 1945-46; B.S., Ga. Inst. Tech., 1949, M.S., 1950, Ph.D., 1961; postgrad. Stanford, 1965-66; m. Lorraine Gordon, Mar. 14, 1959; children—Kristi Jean, Katrina Paulet. Instr., Ga. Inst. Tech., 1949-50, research physicist, Research Inst., 1950-62, asst. prof. math. dept., 1952-60; research scientist Lockheed Research Labs., Palo Alto, Cal., 1962-63, staff scientist, 1963-64, sr. staff scientist, 64—, sr. mem., 1966—, cons. Lockheed Cal. Co., 1964—; asso. Palo Alto Research Observer Program, 1965—. Fellow Am. Phys. Soc., A.A.A.S., Acoustical Soc. Am., Am. Inst. Physics, Am. Assn. Physics Tchrs., Sigma Xi, Sigma Pi Sigma. Simulation studies; publs. on torpedo-countermeasure systems; analysis of acoustic wake characteristics; spl. underwater acoustic transmission problems; research on guided missile countermeasure techniques; research satellite protection systems; studies on coupled orbital bodies and objects ejected from orbiting rendezvous; vibration modes of piezoelectric crystals; relaxation in non-Newtonian liquids; propagation in random media; structure and properties of classical and quantum liquids. Home: 705 Charleston Ct., Palo Alto 94303. Office: 3251 Hanover St., Palo Alto, Cal. 94304.*

KNOPF, Eleanora Bliss (Mrs. Adolph Knopf), Am. geologist; b. Rosemont, Pa., July 15, 1883; d. Tasker Howard and Eleanora (Anderson) Bliss; grad. Baldwin Sch., 1900; A.B., A.M., Bryn Mawr Coll., 1904; Ph.D., U. Cal. at Berkeley, 1912; m. Adolph Knopf, June 23, 1920. Demonstrator, Geol. Lab., Bryn Mawr (Pa.) Coll., 1905-06; asst. curator Geol. Mus., Bryn Mawr, 1904-05, 09-10; aide U. S. Geol. Survey, Washington, 1912-17, asst. geologist, 1917-18, asso. geologist, 1918-28, geologist, 1928-53; geologist Md. Geol. Survey, Balt., 1917-20, 21-24; courtesy fellow Johns Hopkins, 1917-18; individual geol. research igneous and structural geology Rocky Mountains Mont., 1929—; research asso. dept. earth scis. Stanford, 1951—. Mem. or chmn. coms. NRC; vis. lectr. Yale, Harvard, 1934, Stanford, 1953, 55. Recipient citation Bryn Mawr Coll., 1960. Fellow Geol. Soc. Am.; mem. Washington Acad. Sci., Am. Geophys. Union, Cal. Acad. Sci., Sigma Xi. Author: Physiography of Baltimore County, Md., 1929; (with A. I. Jonas) Geology of Baltimore County, Md., 1929; (with E. Ingerson) Structural Petrology, 1938; Geology of Stissing Mountain Area, New York State, 1962. Introduced rock fabric analysis and structural petrology in U. S.; renaissance of exptl. rock deformation in U. S.; study of structure in Piedmont belt Eastern U. S., Rocky Mountains Mont., Spanish peaks Rocky Mountains Colo. Home: 130 Hardwick St., Woodside, Cal. 94062.*

KNOPF, Otto Heinrich, German astronomer; b. Hildburghausen, Saxe-Meiningen, Sept. 24, 1856; s. Hermann and Emmi (Braungart) K.; Ph.D., U. Jena, 1880; m. Hedwig Straubez, 1893; children—Walter, Herbert, Hilde. Asst. lectr. Friedrich Wilhelm Gymnasium, Berlin, 1880-81; computer Nat. Obs. of Cordoba (Argentina), 1881-83; asst. Berlin Astron. Calculating Inst., 1884-89; asst. Jena Obs., 1889, dir., 1900-27; privat dozent Jena U., 1893, extraordinary prof., 1897, prof. astronomy, 1923-27, emeritus, from 1927. Mem. Astron. Inst. Berlin. Author: Die Schmidtsche Sonnentheorie, 1893; Wahrscheinlichkeitsrechung, 1923; Mathematische Himmelskunde, 1925; Die geographischen Koordinaten von Jena, 1927; Christentum und Wissenschaft, 1936; Die Astronomie an der Universität Jena, 1937, also numerous articles. Research on determination of mean density of earth; applied Schmidt's theory of sun to spectroscopic determination of solar period of rotation; studied orbits of comets and planets; wrote on probability theory, math. astronomy and relation of Christianity and sci. Died 1941/42.

KNOPF, S. Adolphus, physician; b. Halle-on-the-Saale, Germany, Nov. 27, 1857; s. Adolphus and Nanina (Bock) K.; A.B., U. Paris (Sorbonne) 1890; M.D., Bellevue Hosp. Med. Coll., N.Y. U., 1888; Faculty of Medicine, U. Paris, 1895; m. Perle Nora Dyar, Oct. 19, 1889 (dec. 1931); m. 2d, Julia Marie Off, Oct. 6, 1935; children—Gertrude, Lucille, Adolphus. Prof. medicine, dept. phthisiotherapy, N.Y. Post-Grad. Med. Sch., Columbia U., 1908-20; sr. visiting phys. Health Dept.'s Riverside Tb Hosp., 1906-22; hon. dir. Gaylord Farm Sanitorium, Wallingford, Conn.; hon. pres. med. bd. Bruchesi Tb Inst., Montreal; attending Tb specialist USPHS, 1920-22; cons. physician to Riverside Hosp. (N.Y.), St. Gabriel's (N.Y.) Sanatorium for Consumptives, West Mountain Sanatorium, at Scranton, Pa., others. Fellow N.Y. Acad. Medicine, Assn. Mil. Surgeons U. S., Soc. Med. Jurisprudence, Am. Heart Assn., Am. Soc. for Psychical Research; hon. mem. Am. Assn. for Thoracic Surgery, Am. Tb Assn.; hon. v.p. Brit. Congress on Tb; govt. del. Internat. Prison Congress, Budapest, Internat. Tuberculosis Congress, Paris; v.p. sect. V of Tb Congress, Washington, 1908; ofcl. del. 4th Internat. Congress on Sch. Hygiene, Buffalo, 1913; apptd. U. S. rep. to Internat. Union Against Tb, The Hague, 1932; founder N.Y.C. and Nat. Tb assns. Author: Tuberculosis as a Disease of the Masses and How to Fight It (transl. into 27 langs.), 1901; Tuberculosis, a Preventable and Curable Disease, 1913; A History of the National Tuberculosis Association, 1922; Heart Disease and Tuberculosis, 1922. Leader in combating Tb; developed spl. diaphragm to help patients with apical Tb (known as Knopf's treatment). Died N.Y.C., July 15, 1940.

KNOPF, William Cleveland, Jr., Am. engr.; b. Louisville, Dec. 13, 1910; s. William Cleveland and Anne (Flood) K.; B.S., Washington and Lee U., 1932; M.S., Vanderbilt U., 1941; postgrad. Ill. Inst. Tech., 1941-43; Ph.D., Northwestern U., 1950; m. Mary Gene Herren, Jan. 18, 1941; children—Katherine (Mrs. James Stadelman), Gene Miller. Asst. dir. research Internat. Minerals and Chem. Corp., Skokie, Ill., 1954-59; tech. dir. U. S. Industries Tech. Center, Pompano Beach, Fla., 1959-60; grad. faculty U. Fla. Extension Div., 1960-61; chmn. oceanographic and hydrographic engring. task force Space Era Edn. Study, Fla. Bd. Control, 1962-63; prof., research prof. elec. engring. U. Fla., Gainesville, 1961-63; chmn. dept. elec. engring. and div. oceanographic engring. U. Miami Coral Gables, Fla., 1963-65, dean Sch. Engring., 1965—. Tech. and mgmt. cons. Nat. Acad. Sci.-NRC, 1947—, also mfg. cos. Mem. A.A.A.S., I.E.E.E., Am. Phys. Soc., Am. Soc. Engring. Edn., Instrument Soc. Am., Sigma Xi, Eta Kappa Nu. Contbr. numerous articles to profl. jours.; also classified papers. Home: 1350 Mendavia Av., Coral Gables, Fla. 33146.*

KNOPOFF, Leon, Am. geophysicist; b. Los Angeles, July 1, 1925; s. Max and Ray (Singer) K.; student Los Angeles City Coll., 1941-42; B.S., Cal. Inst. Tech., 1944, M.S., 1946, Ph.D., 1949; m. Joanne Van Cleef, Apr. 9, 1961; children—Katherine Alexandra, Rachel Anne, Michael Van Cleef. Faculty, Miami U., Oxford, O., 1948-50; faculty U. Cal., Los Angeles, 1950—, prof. physics, 1960—, prof. geophysics, 1957—, research musicologist, 1961—; prof. geophysics Cal. Inst. Tech., Pasadena, 1962-63; vis. prof. Technische Hochschule Karlsruhe (Germany), 1966. Chmn., U. S. Nat. Com. for Upper Mantle Project, 1963—; sec.-gen. Internat. Com. for Upper Mantle Project, 1963—. NSF fellow, Cambridge (Eng.) U., 1960-61; recipient Canadian Internat. Cooperation Year medal, 1965. Mem. A.A.A.S., Royal Astron. Soc., Am. Phys. Soc., Seismol. Soc. Am., Acoustical Soc. Am., Am. Geophys. Union, Nat. Acad. Scis., Am. Acad. Arts and Scis. Contbg. author: High Pressure Physics and Chemistry, 1962; Physical Acoustics, Vol. 3b, 1965; contbg. author The Earth Today, 1961; The Earth's Mantle, 1967; also numerous articles. Research on applied maths., elastic wave propagation, theoretical high pressure physics, geophysics of earth's deep interior, seismic observations of crustal and upper mantle structure, nature of earthquake focal mechanism, universal relations among phys. properties of solids, electronic music, computer music. Office: Physics Dept., U. Cal. at Los Angeles 90024.*

KNORR, Ludwig, German chemist; b. Munich, Germany, Dec. 2, 1859; ed. Munich, Heidelberg, Erlangen, Würzburg (all Germany); became instr. chemistry U. Erlangen, 1885; tchr., Würzburg; titular prof. U. Jena (Germany). Research in organic chemistry; discovered antipyrine, 1883; isolated and synthesized quinoline and pyrazole, by 1884; investigated keto and enol tautomers, morphine alkaloids. Died Jena, June 5, 1921.

KNOTT, Cargill Gilston, physicist; b. Penicuik, Scotland, June 30, 1856; s. Pelham Knott; ed. Edinburgh U.; D.Sc.; LL.D., St. Andrews; m. Mary Dixon, 1885; 1 son, 3 daus. Asst. to chair natural philosophy Edinburgh U., 1879-83, reader in applied math., from 1892; prof. physics Imperial U., Japan, 1883-91; conducted Magnetic Survey of Japan, 1887; Thomson lectr. United Free Ch. Coll., Aberdeen, 1905-06, 13-14. Fellow Royal Soc., 1920, Royal Soc. Edinburgh (Keith prize for work on magnetic strains 1897, gen. sec.). Author: Electricity and Magnetism; Physics; Physics of Earthquake Phenomena; Memoir of Professor P. G. Tait; Four-Figure Mathematical Tables. Contbg. author: Dictionary of Applied Physics, 1922; Ency. Brit. Revised Kelland and Tait's Quaternions, 3d edit., 1904. Contbr. papers on seismology, electricity, magnetism to jours. Editor meml. vol. Napier Tercentenary Celebration, Edinburgh, 1914. Died Oct. 26, 1922.

KNOWER, Henry McElderry, Am. anatomist; b. Balt., Aug. 5, 1868; s. Edward C. K. and Mary D. (McElderry) K.; A.B., Johns Hopkins, 1890, Adam T. Bruce fellow, 1895-96, Ph.D., 1896; m. Virginia DuBarry, Feb. 16, 1897; children—Henry DuB., Virginia (Mrs. William A. Moore). Instr. biology, Williams Coll., 1896-97; instr. anatomy Johns Hopkins, 1899-1908, asso., 1908-09; lectr. anatomy U. Toronto (Ont., Can.), 1909-10; prof. anatomy U. Cin., 1910-25; vis. prof. anatomy U. Ga., 1924-26; prof. anatomy U. Ala., 1926-29; guest in research, Wistar Inst. Anatomy, Phila., 1929-30; asso. prof. anatomy, Albany Med. Coll., from 1930; research fellow in biology Yale U., 1932-33, research asso. in biology, 1933-37. Asst. in zoology Marine Biol. Lab., Woods Hole, Mass., 1897, librarian, 1910-20; a founder, mng. editor Am. Jour. Anatomy, 1901-22; founder and first editor Anat. Record, 1906. Researches in embryology and anatomy of Termites; lymphatic and vascular systems of frog embryos, by experiment and injections; muscles of human heart. Died Jan. 10, 1940.

KNOWLES, Lucius James, Am. inventor; b. Hardwich, Mass., July 2, 1819; s. Simeon, Jr. and Lucetta (Newton) K.; A.M. (hon.), Williams Coll., 1865; m. Eliza Adams; m. 2d, Helen Strong. Perfected Knowles Safety Steam-boiler feed regulator, 1840, began mfg., 1859; began daguerreotype bus., Worcester, Mass., 1841; inventor machine for spooling thread, 1843; engaged in mfg., New Wooster, Mass., 1843-45; partner (with Harrison H. Sibley) to operate Old Draper Mill, mfg. cotton warp, Spencer, Mass., 1846; extended activities to include woolen mill, 1853-59; patented improvements in looms, 1856; erected bldg. near cotton factory to manufacture boiler-feeder water regulator, 1862; mem. Mass. Ho. of Reps. from Warren, New Braintree and West Brookfield, 1862, 65; made steam pumps, exptl. looms, 1863; propr. Knowles Steam Pump Co., L. J. Knowles & Bro. Loom Works (consol. as Crompton & Knowles Loom Works, 1897). Trustee Worcester Free Inst. Tech., 1871. Died Washington, Feb. 25, 1884.

KNOWLES, Robert, zoologist; b. 1883; s. J. Knowles; ed. Downing Coll., Cambridge, St. Mary's Hosp., London; m. Mary Evelyn Montgomery, 1907; 3 daus. Civil surgeon, Jhansi, 1908-12; asst. dir. Pasteur Inst. India, Kasauli, 1912-14; dir. Pasteur Inst., Shillong, Assam, 1916-20; organizing sec. Calcutta Sch. Tropical Medicine and Hygiene, 1920, prof. protozoology, from 1921. Recipient Silver Jubilee medal, 1935. Licentiate Royal Coll. Physicians. Fellow Asiatic Soc. Bengal; mem. Royal Coll. Surgeons. Asst. editor Indian Med. Gazette, 1922-28, editor, 1928-33. Contbr. papers on snake bite, malaria, kala-azar, surra, rat-bite fever, intestinal amoebiasis to med. jours. Died Aug. 3, 1936.

KNOWLTON, Charles, Am. physician; b. Templeton, Mass., May 10, 1800; s. Stephen and Comfort (White) K.; M.D., Dartmouth Med. Coll., 1824; m. Tabitha Stewart, Apr. 17, 1821. Wrote Elements of Modern Materialism (one of 1st Am. books philos. materialism), 1829; pioneer advocate of birth control; wrote Fruits of Philosophy; or, The Private Companion of Young Married People (1st edition anonymous 1832, 2d edit. signed 1833), imprisoned and prosecuted for its contents which are earliest Am. printed matter on birth control; prosecution made book bestseller with 9 Am. editions before 1839; book was popular in Eng. (1st edit. 1834), became subject of famous legal case involving sexual edn. Died Winchendon, Mass., Feb. 20, 1850.

KNOWLTON, Frank Hall, Am. botanist, paleontologist; b. Brandon, Vt., Sept. 2, 1860; s. Julius Augustus and Mary Ellen (Blackmer) K.; B.S., Middlebury Coll., 1884, M.S., 1887, hon. Sc.D., 1921; Ph.D., Columbian (now George Washington) U., 1896; m. Annie Stirling Moorhead, Sept. 27, 1887 (dec. 1890); 1 dau., Margaret; m. 2d, Rena Genevieve Ruff, Oct. 3, 1893; 1 son, Frank Lester. Aid, U. S. Nat. Mus., 1884-87, asst. curator botany, 1887-89, asst. palaeontologist, 1889-1900; palaeontologist U. S. Geol. Survey, 1900-07, geologist, from 1907; prof. botany Columbian U., 1887-96. Recipient Copley medal, 1917. Fellow Geol. Soc. Am. (v.p. 1917-18), Paleontol. Soc. Am. (pres. 1917-18). Editor The Plant World, 1897-1904 (founder); asst. on Century Dictionary, writing definitions in botany; had whole charge of botany for Standard Dictionary, for which he prepared about 25,000 definitions; assisted in preparing bot. definitions for new edit. (1900) of Webster's Dictionary; prepared bot. matter for The Jewish Ency. Author: Birds of the World, 1900; Catalogue Mesozoic and Cenozoic Plants of North America, 1919; Plants of the Past, 1926. Specialized in micros. study of internal structure of fossil plants, identified many new fossil species; pioneer in study of theory of geologic climates. Died Nov. 22, 1926.

KNOWLTON, George Franklin, Am. entomologist; b. Farmington, Utah, July 28, 1901; s. Benjamin Franklin and Minerva E. (Richards) K.; B.S., Utah State Agrl. Coll., 1923, M.S., 1925; postgrad. U. Minn., 1925-26; Ph.D., Ohio State U., 1932; m. Mary Watkins, July 29, 1925; children—Betty Jean (Mrs. Hubert E. Collmar), Kathryn Marie (Mrs. W. Doyle Cazier). Asst. entomologist Utah State U., Logan, 1925-30, asso. prof., 1930-45, prof., 1945-67, prof. entomology emeritus, 1967—; entomologist in charge insect and rodent control U. S. Army Engrs., 1944-45. Fellow A.A.A.S., Utah Acad. Scis., Arts and Letters, Herpetologists League, Entomol. Soc. Am.; mem. Entomol. Soc. Washington, Pacific Coast, Central States entomol. socs., Soc. Systematic Zoologists, Ecol. Soc. Am., Entomol. Soc. Kan., Sigma Xi, Epsilon Sigma Phi. Research and numerous publs. on insect control, insect food of lizards and birds, insect ecology, range entomology. Home: 304 East 4th North, N. Logan, Utah, 84321.*

KNOWLTON, Thomas, Brit. botanist; b. Eng., 1692; at least 1 son, Charles; supt. botanic garden of Dr. Sherard, Kent, Eng.; entered service of Richard Boyle, 1728. Bot. genus of order Ranunculaceae named in his honor. First in Eng. to discover moor-ball fresh water algae (called Aegagrophila by Linneus); discovered site of ancient city of Delgoricia, nr. Pocklington, Yorkshire, Eng. Died 1782.

KNOX, John Marshall, Am. physician; b. Dallas, Apr. 11, 1925; s. John Marshall and Katie (Dickie) K.; B.S., Tex. A. and M. U., 1944; M.D., Baylor U., 1949; m. Lullene Powell Knox, Dec. 18, 1948; children—Lynda Lee, Jane Ann, John Marshall, Byron. Jr. clin. instr. dermatology U. Mich., 1954-55; instr. dermatology and syphilology Baylor U. Coll. Medicine, 1955, asst. prof., 1955-59, asst. prof. dermatopathology, 1957-60, asso. prof. dermatology and syphilology, 1959-63, asso. prof. dermatopathology, 1960—, prof., chmn. dermatology and syphilology, 1963—; dir. Venereal Disease Control, City of Houston, 1957-65; spl. cons. USPHS Communicable Disease Center, 1958—; chief dermatology Ben Taub Gen. Hosp., VA Hosp. (both Houston); sr. attending physician Meth. Hosp., Houston. Diplomate Am. Bd. Dermatology. Mem. Soc. for Investigative Dermatology (dir. 1960-65), Am. Social Health Assn. (dir. 1962—), Am. Venereal Disease Assn. (pres.-elect 1965-68), So. Med. Assn. (chmn. dermatology sect. 1965-66), Am. Acad. Dermatology and Syphilology (dir. 1967—), Am. Dermatol. Assn., Am. Fedn. for Clin. Research, A.M.A., Am. Pub. Health Assn., Am. Soc. Dermatopathology, Assn. Profs. Dermatology, Internat. Soc. Tropical Dermatology, Pacific Dermatologic Assn., Soc. for Exptl. Biology and Medicine, Soc. for Investigative Dermatology, Alpha Omega Alpha. Bd. editors, cons. editor Dermatology Digest; editorial bd. Soc. for Academic Achievement, Cutis. Research and numerous publs. on cutaneous parasitic and infectious diseases, effects of sunlight on skin, treatment of syphilis. Home: 419 Blalock, Houston 77024. Office: 1200 Moursund Av., Houston 77025.*

KNOX, Robert, Scottish anatomist, ethnologist; b. Edinburgh, Sept. 4, 1791; s. Robert and Mary (Sherer) K.; M.D., Edinburgh, 1814; m., 1824; 6 children, including Edward. Asst. surgeon in army, 1815-32; sci. research at Cape, 1817-20; conservator mus. comparative anatomy and pathology Edinburgh Coll. Surgeons, 1825-31; anat. lectr., Edinburgh, 1826. Fellow London Ethnol. Soc. (named hon. curator mus. 1862). Author: The Edinburgh Dissector, 1837; Anatomy of the External form of Man, 1849; The Races of Men, 1850; A Manual of Human Anatomy, 1853; also papers. Added to knowledge of structure and physiology of eye. Died Dec. 20, 1862.

KNOX, Robert, Brit. radiologist; b. Leith, Scotland; M.B., C.M., M.D., Edinburgh U.; diploma in med. radiology and electrology Cambridge U.; student Guy's Hosp.; hon. radiologist King's Coll. Hosp., Royal Seabathing Hosp.; dir. elec. and radiotherapeutic dept. Cancer Hosp. (Free), London; cons. radiologist Gt. No. Central Hosp., Chelsea Hosp. for Women; hon. cons. radiologist Queen Alexandra Hosp., Milbank. Licentiate Royal Coll. Physicians. Fellow Royal Soc. Medicine, Med. Soc. London, Royal Photog. Soc.; mem. Royal Coll. Surgeons Eng., Royal Coll. Physicians London, Instn. Elec. Engrs., Brit. Med. Assn., Brit. Inst. Radiology (v.p.), Röntgen Soc. (pres.); hon. fellow, mem. fgn. sci. socs. Author: Radiography, X-Ray Therapeutica and Radium Therapy, 2d edit., 1917, 4th edit., 1924; also articles. Co-editor Brit. Jours. Radiology. Died Sept. 21, 1928.

KNOX, Robert Seiple, Am. physicist; b. Franklin, N.J., July 13, 1931; s. Harvey Stoll and Laura (Seipie) K.; B.S. in Engring. Physics, Lehigh U., 1953; Ph.D. in Physics and Optics, U. Rochester, 1958; m. Mirta I. Borges, Sept. 1, 1954; children—Bruce Robert, Wayne Harvey, Lee Benjamin. Research asso. U. Ill., 1958-59; research asst. prof., 1959-60; faculty U. Rochester (N.Y.), 1960—, asso. prof. dept. physics, 1963—. Cons., Argonne Nat. Lab., 1959—, Naval Research Lab., 1960—, Xerox Corp., 1963—. Fellow Am. Phys. Soc.; mem. A.A.A.S. Author: Theory of Excitons, 1963; (with A. Gold) Symmetry in the Solid State, 1964; (with D. L. Dexter) Excitons, 1965; also articles. Research on atomic spectra and structure, absorption and luminescence spectra ionic and molecular crystals, band structure of nonmetallic solids.*

KNOX, William Jordan, Am. physicist; b. Pomona, Cal., Mar. 21, 1921; s. Reginald Langenberger and Kate (Jordan) K.; B.S. in Chemistry, U. Cal., Berkeley, 1942, Ph.D. in Physics, 1951; m. Barbara Louise House, Sept. 27, 1948; children—William Jordan, Margaret Louise, Sarah Ann, Reginald Vernon. Chemist, technologist Atomic Energy Project, U. Chgo., Oak Ridge, Hanford, Wash., 1942-46; asst. prof. physics, research asso. Yale, New Haven, 1951-53, 55-60; physicist div. research AEC, Washington, 1953-55; asso. prof. physics U. Cal., Davis, 1960-66, prof., 1966—, chmn. dept., 1963-66; dir. Crocker Nuclear Lab., 1966-67. Cons., AEC, 1955-56. Fellow Am. Phys. Soc.; mem. A.A.A.S., Am. Assn. Physics Tchrs., Am. Assn. for UN, Sigma Xi, Delta Tau Delta. Contbr. articles to physics jours. Patentee in field. Exptl. researcher in nuclear physics, 1948—; plutonium and fission product chemistry; developer programs for ednl. and research assistance to univs. by AEC, 1953-55. Home: 731 Elmwood Dr., Davis, Cal. 95616.

KNUDSEN, Hans Lottrup, Danish physicist; b. Holstebro, Dec. 8, 1917; s. Oluf and Nicoline (Zornig) K.; ed. Tech. U., Denmark, Harvard; M.Sc., Ph.D. in civil engring.; m. Anna Lisa Frydendahl, May 22, 1943; children—Sten, Bo, Jes. Instr., Chalmers Tech. U., Gothenbourg, Sweden, 1942-43; engr. Sieverts Cable Factory, Sundbyberg, Sweden, 1943-46; instr. Tech. U. Denmark, 1948-54, lectr., 1950-54, prof. electromagnetic theory, 1954—, also dir. electromagnetic theory labs. Mem. Danish Acad. Tech. Scis., Danmarks Naturvidenskabelige Samfund, Dansk Ingeniorforening, Matematisk Forening, Fysisk Forening, Inst. Elec. and Electronic Engrs. Author: Antenna Systems with Complete or Partial Rotational Symmetry, 1952. Research on electromagnetic field, antenna theory, theory of wave guides and applied math. Home: Aamosebakken 28, Virum. Office: Lundtoftevej 100, Lyngby, Denmark.

KNUDSEN, Martin, Danish physicist; b. Feb. 15, 1871; s. Jorgen and Stine Knudsen; ed. Odense, Copenhagen; Dr. phil. h.c., Lund, Sweden; m. Ellen Ursin, 1903; 3 sons, 2 daus. Docent in physics U. Copenhagen, 1901, prof. physics, 1912-41, rector, 1927-28. Danish del. Internat. Council for Exploration of Sea, 1902-48, v.p., 1933-48. Recipient H. C. Orsted medal, 1916, Agassiz medal, 1936. Mem. Royal Danish Acad. (sec. 1917-46), Internal. Assn. Phys. Oceanography (pres. 1930-36), Internat. Union Pure and Applied Physics (v.p. 1923-34), various fgn. socs. Publs. on phys. and hydrographical topics. Died May 27, 1949.

KNÜLLE, Willi, zoologist; b. Wismar, Germany, Apr. 6, 1927; s. Anton and Martha (Gierhan) K.; student U. Rostock, 1946-48; Dr. rer. nat., U. Kiel, 1951; m. Dorothea Rother, Dec. 22, 1955. Scientist, Institut für Grunlandschädlinge der Biologischen Bundesanstalt, Oldenburg, Germany, 1951-55, Institut für Vorratsschutz der Biologischen Bundesanstalt, Berlin-Dahlem, Germany, 1955-62; prof. Ohio State U., Columbus, also Ohio Agrl. Research and Devel. Center, Wooster, 1963—. Mem. A.A.A.S., Deutsche Zoologische Gesellschaft, Am. Inst. Biol. Scis., Am. Soc. Zoologists, Entomol. Soc. Am. Contbr. articles to tech. jours. Research on environmental physiology, particularly water vapour sorption of arthropods, physiol. ecology of mites. Home: 569 Bloomington. Office: Ohio Agrl. Research and Devel. Center, Wooster, O. 44691.*

KNUNYANTS, Ivan Lyudvigovich, Russian organic chemist; b. June 4, 1906; grad. Moscow Higher Tech. Sch., 1928. With Moscow Higher Tech. Sch., 1928-31; asso. Inst. Organic Chemistry, USSR Acad. Sci., 1931—. Recipient Stalin prize, 1943, 48, 50. Mem. USSR Acad. Sci. Author: (with O. V. Kildsheva) Methods of Introducing Fluorine into Organic Compounds, 1946; Some Theoretical Problems of Modern Organic Chemistry, 1953; Modern Experimental Methods in Organic Chemistry, 1960; The Secret of the Life Force, 1962. Research and publs. on organic fluorine compounds; discovered aliphatic oxide conversion used in mfr. of vitamin B1 and anti-malaria preparations; many of his discoveries, such as photosensitizers, used in industry. Address: Inst. Organic Chemistry, USSR Acad. Sci., Leninsky prospect 31, Moscow, USSR.

KOANA, Jun, Japanese physicist; b. Tokyo, 1907; grad. Tokyo U., 1930. Apptd. instr. Tokyo U., 1933; D.Sc., 1946. Became asst. prof. 1st Higher Sch., 1936; became prof. Tokyo U., 1945. Recipient Asahi sci. prize, 1953-54. Author book on photography. Research in cell analysis, optics including devel. chem. analysis of cells using rays directed on them.

KOBAKHIDZE, David Nestorovich, Russian ecologist, zoologist; b. Kutaisi, Georgia, 1911; grad. All-Union Inst. Subtropical Crops, Tbilisi, 1932; postgrad. All-Union Inst. Plant Protection, Leningrad, until 1935; D.Biol. Sci. With Inst. Zoology, Georgian Acad. Sci., 1935—, prof., 1941—, dir., 1951; head chair zoology Pushkin Pedagogical Inst., Tbilisi. Author: Water as an Important Factor in the Biocenotic Dynamics of the Sunken Part of the Kolkhida Depression, 1940; The Principal Types of Biocenosis in the Central Part of the Kolkhida Depression, 1941; An Analysis of Terrestrial Biocenoses in the Central Part of the Kolkhida Depression, 1943; A New Understanding of the Concept of Biocenosis, 1946; The Basis of a Method of Laying Out Tea Plantations from the Point of View of Entomological Criteria, 1950; The Location of the Mottled Scorpion on the Slopes of the Main Caucasian Mountain Range Facing the Black Sea, 1960; A New Species of Pseudo-Scorpion from the Batumi Botanical Gardens, 1960; A New Subspecies of Pseudo-Scorpion from the Caucasus, 1963. Address: Pushkin Pedagogical Inst., prospect Chavchavadze 32, Tbilisi, Gruz. SSR, USSR.

KOBAYASHI, Hideshi, Japanese biologist, endocrinologist; b. Yonago, Japan, Mar. 16, 1919; s. Joichi and Tora (Baba) K.; B.D., grad. U. Tokyo, 1942, Ph.D., 1954; m. Tatsuko Narumi, May 5, 1950; children—Kozue, Shigeto, Makito. Faculty, Zool. Inst., U. Tokyo, 1950—, asso. prof., 1951—; research asso. dept. zoology Columbia U., 1957-59. Mem. internat. com. Symposium on Comparative Endocrinology, 1959—. NSF sr. fgn. scientist Wash. State U., 1964-65. Mem. Zool. Soc. Japan, Anat. Soc. Japan, Ornithol. Soc. Japan (award 1963), Endocrinological Soc. U.S.A. Editorial mem. Jour. Gen. and Comparative Endocrinology, 1960—, Endocrinologia japonica, 1962—; mng. editor Zool. Mag., 1962-64. Research, numerous publs. on ultrastructure of vertebrate hypothalmaic median eminence (important in regulation of function of anterior pituitary gland). Home: 3-16-17, Kamisaginomiya, Nakrano-ku, Tokyo, Japan.*

KOBAYASHI, Shoshichi, mathematician; b. Kofu, Japan, Jan. 4, 1932; s. Kyuzo and Yoshie (Obi) K.; B.S., U. Tokyo, 1953; student U. Paris, 1953-54, U. Strasbourg, 1953-54; Ph.D., U. Wash., 1956; m. Yukiko Ashizawa, May 11, 1957; children—Sumire, Mei. Mem. Inst. for Advanced Study, Princeton, N.J., 1956-58; research asso. Mass. Inst. Tech., 1958-60; asst. prof. U. B.C., 1960-62; faculty U. Cal., Berkeley, 1962—, prof., 1966—. Mem. Am., French math. socs., Math. Soc. Japan, Canadian Math. Congress. Author: Foundations of Differential Geometry, vol. 1, 1963. Global studies of manifolds with differential geometric method. Home: 421 Michigan Av., Berkeley, Cal. 94707.*

KOBAYASHI, Tatsuo, Japanese physicist; b. Okayama, Japan, 1886; grad. Tokyo U., 1918; Sc.D., 1924. Became asst. prof. Tokyo U., 1919, prof., 1927, prof. emeritus, 1947—; named dir. Nat. Fire Prevention Research Inst., 1948. Author: Studies in the Formation of Low and High Atmospheric Pressures. Research in meteorology, exptl. physics.

KÖBEL, Jakob (Kobelius, Jakob), German mathematician; b. Oppenheim, Germany, 1470; studied with Copernicus, Cracow, Poland; tchr. arithmetic; printer. Author: Rechenbiechlin, 1514; Geometrey von künstlichen Feldmessen, 1516; Eyn neiuw Visir Buchlin, 1515; Mit der Kryden, 1520. His arithmetic texts were important in 16th century. Died Oppenheim, Germany, Jan. 31, 1533.

KOBELIUS, Jakob, see Köbel, Jakob.

KOBELL, Franz von, see von Kobell, Franz.

KOBER, Carl L., radio and missile engr.; b. Vienna, Austria, Nov. 22, 1913; s. Leopold and Maria (Cremer) K.; Ph.D., U. Vienna, 1935; Dr. phil. habil., Tech. U. Vienna, 1938; m. Christiana Futschig, Mar. 26, 1942; children—Wolfgang, Peter C. Came to U. S., 1949, naturalized, 1958. Project engr. Elin A. G., Vienna, Austria, 1936-39; chief radar lab. GEMA, Berlin, Germany, 1940-45; mgr. automotive accessories Proschwitz Industrie Works Kufstein, Austria, 1945-48; asso. prof. applied physics Tech. U., Vienna, 1948-49; mgr. automotive accessories Secowerk, Vienna, 1948-49; cons. Aircraft Radiation Lab., Wright Air Development Center, 1949-53, cons. Armament Lab., 1953-55; dir. devel., mech. div. Gen. Mills, Inc., 1955-57; v.p. missiles and ordnance AVCO/Crosley, 1958-61; dir. Advanced Program div. Martin Co., Denver, 1961-62, dir. advanced tech., 1962—. Fellow Inst. Radio Engrs., Inst. Aerospace Scis. (asso.), mem. Am. Phys. Soc., Am. Ordnance Assn. Pioneered in radar in Germany; devel. work high resolution radars for reconnaissance and precision bombing, also radar inertial systems. Address: care Martin Co., Denver.*

KÖBERLE, Fritz, physician; b. Vienna, Austria, Oct. 1, 1910; s. Friedrich and Katharina (Steinberger) K.; Med.Doctor, U. Vienna, 1934; priv.dozent U. Vienna, 1941; m. Elisabeth Hafferl, Aug. 15, 1936; children—Gottfried, Roland, Krista, Erika. Asst., Pathologic Institut, U. Vienna, 1934-41, asso. prof.; chief pathology Inst. St. Pölten, 1946-53; prof., chmn. dept. pathology Med. Sch., Ribeirao Preto, U. Sao Paulo (Brazil), 1953—. Expert trypanosomiasis WHO, 1965—. Mem. Soc. Brasilian Pathologists, Deutsche Pathol. Gesellschaft, Latin-Am. Path. Soc., Brasilian Soc. Gastroenterology, Brasilian Soc. Cardiology. Research, numerous publs. on Chagas' disease proving its neurogenic nature, neurovegetative system in aged people; discovered etiology and pathogenesis of megaesophagus and megacolon in S. Am. Address: Faculdade Medicina, Ribeirao Preto, Est. Sao Paulo, Brazil.*

KOBZAREV, Yurii Borisovich, Russian radio engr.; b. Dec. 8, 1905; grad. Khar' Kov U., 1926. With Physico-Tech. Inst. of USSR Acad. Scis., Leningrad, 1926-43; prof. Moscow Energetics Inst., 1944-55; corr. mem. USSR Acad. Scis., 1953—, with Inst. Radio Engring. and Electronics, 1955—. Recipient Stalin prize, 1941. Publs. include: Parameters of piezoelectric crystal resonators, 1929; Peculiarities of crystal resonators; Representation of a tube characteristic by a power series, 1933; (with A. Ageev) Transient processes in resonance amplifiers, 1935; The theory of a tube generator with two degrees of freedom, 1950. Developed frequency stblzn. by means of quartz crystals in tube generators, 1926-31; worked on theory of oscillation of oscillator plates; studied phenomena in non-linear systems and indicated the high efficiency of quasi-linear method of treating these phenomena based on concept of complex amplitudes and resistance; active role in devel. radar. Office: Institute of Radio Engineering and Electronics of USSR Acad. of Sciences, Mokhovaya 11, K-9, Moscow, USSR.

KOCENT, Alexandr, phys. chemist; b. Charbin, China, June 30, 1922; s. Jan and Nina (Gudnina) K.: Dr. Rer. Nat., Charles U., Prague, Czechoslovakia, 1952; C.Sc., Polarographical Inst. of J. Heyrovsky, Prague, 1958; m. Miroslava Bobysutová, Nov. 7, 1951; 1 dau., Alexandra. With Inst. Med. Chemistry, Charles U., Prague, 1949-52; with biochem. dept. Cancer Research Inst., Brno, Czechoslovakia, 1952—, head physico-chem. lab., 1961—. Lectr. phys. biochemistry J. E. Purkyne U., Brno, 1961—. Author: (with M. Rybák, Z. Brada, I. M. Hais) Säulenchromatographie an Celluloseionaustauscher, 1966. Research, publs. on mechanism of Brdicka reaction (clin. polarographical test), theory of gradient elution chromotography. Home: 10 Kneslova. Office: 7 Zluty Kopec, Brno, Czechoslovakia.*

KOCH, Fred Conrad, Am. biochemist; b. Chgo., May 16, 1876; s. Frederick and Louise (Fischer) K.; B.S., U. Ill., 1899, M.S., 1900; Ph.D., U. Chgo., 1912; m. Elizabeth Miller, Sept. 7, 1922. Instr. chemistry, U. Ill., 1900-02; research chemist Armour & Co., 1902-09; asst. prof. physiol. chemistry U. Chgo., 1912-16, asso. prof., 1916-24, prof., 1924-40, chmn. dept. biochemistry, 1926-41, Frank P. Hixon distinguished service prof. biochemistry, 1941-42; research chemist, Armour and Co., from 1942. Mem. A.M.A., Soc. Biol. Chemists, Am. Chem. Soc., A.A.A.S., Soc. Exptl. Biology and Medicine, Assn. for Study of Internal Secretions, Sigma Xi, Gamma Alpha, Phi Chi, Phi Lambda Upsilon. Author: Practical Methods in Biochemistry, 1934; also many papers. Isolated (with C. R. Moore and T. F. Gallagher) 1st potent testicular extract containing male sex hormone, 1929. Died Jan. 26, 1948.

KOCH, Gerhard, German human geneticist; b. Neubrandenburg, Germany, Feb. 7, 1913; s. Hermann and Emma (Augustin) K.; student medicine U. Rostock (Germany), 1932-34, U. Königsberg (Germany), 1934-35, U. Breslau (Germany), 1935-38; m. Ana Maria Cudell, Apr. 27, 1950. Asst. neurol. clinics U. Rostock, 1939-40, U. Göttingen (Germany), 1947-48, U. Tübingen (Germany), 1949, U. Lisbon (Portugal), U. Oporto, 1949-52; asst. inst. for human genetics U. Münster/Westphalia, 1952-65, dir. dept. human genetic and psychoneurol. research, 1952-65, lectr. human genetics, 1954-65; dir. Inst. Human Genetics and Anthropology, U. Erlangen-Nuremberg (Germany), 1965—. Sci. colleague Kaiser-Wilhelm Inst. für menschliche Erblehre und Eugenik, 1943-45, Kaiser Wilhelm Institut für Hirnforschung, 1946-47. Mem. Am. Soc. Human Genetics. Author: Krampfbereitschaft, 1955; Humangenetik vol. V, 1 (with P. E. Becker), 1966; also numerous articles. Home: 105/III Hartmannstrasse, 852 Erlangen, Germany.*

KOCH, Guillaume-Daniel-Joseph, German botanist; b. Kusel, Mar. 5, 1771; studied medicine, Jena, Marburg (both Germany); became prof. botany and medicine, Erlangen, Germany, 1824. Author: Synopsis florae germanicae et helveticae (modeled after Grenier and Gordon's Flore de France), 1835-37. Died Erlangen, Nov. 14, 1849.

KOCH, Herman William, Am. physicist; b. N.Y.C., Sept. 28, 1920; s. John William and Elizabeth (Hirsch) K.; B.S., Queens Coll., 1941; M.S., U. Ill., 1942, Ph.D., 1944; m. Margaret Giles, Feb. 3, 1945; children—John, Kathleen, Donald, Robert, Russell. Asst. prof. physics U. Ill., 1944-49; research physicist, Oak Ridge, Tenn., 1945-46; chief high energy radiation sect. Nat. Bur. Standards, U. S. Dept. Commerce, Washington, 1949-62, chief div. radiation physics, 1962—. Named Alumnus of Year, Queens Coll., 1960; recipient Gold medal U. S. Dept. Commerce, 1962. Fellow Am. Phys. Soc.; mem. Nat. Com. Radiation Protection, Sigma Xi, Phi Kappa Phi, Gamma Alpha. Pioneered in the application of high energy x-rays to nuclear research; responsible for the devel., realization and calibration of standard instruments and methods for radiation protection, units and measurement. Home: 2922 Stanton Av., Silver Spring, Md. 20910. Office: Nat. Bureau of Standards, Washington 20234.*

KOCH, Jorgen, Danish physicist; b. Copenhagen, Apr. 6, 1909; s. Harald and Gerda Wienberg (Andersen) K.; ed. U. Copenhagen, Danzig, Berlin tech. colls.; Ph.D., D.Eng.; m. Greta Kjaer, May 3, 1945; children—Ida, Henning. Asst., U. Copenhagen, 1941, prof., 1954, prof. physics, dir. Physics Labs., 1957—. Mem. European Council on Nuclear Research, 1953-54, Health Service Council for Protection against Radiation. Mem. Denmark Soc. Natural Sci., Danish Acad. Tech. Sci., Am. Phys. Soc., Royal Acad. Sci. for Internat. Orgn. Pure and Applied Biophysics (nat. acad.). Author: Decay Curves of Uranium and Thorium Fission Products, 1939; Electromagnetic Separation of Noble Gas Isotopes and Their Use in Some Nuclear and Spectroscopic Experiments, 1953; Electromagnetic Isotope Separation and the Application of Electromagnetically Enriched Isotopes, 1957. Research on electromagnetic isotope separators and related ion accelerators and their applications to nuclear, atomic and solid state physics. Home: Juliane Mariesvej 20. Office: H. C. Osted Inst., Universitetsparken 5, Copenhagen, Denmark.

KOCH, Lauge, Danish geologist; b. Kjaerby, Denmark, July 5, 1892; s. Carl and Elisabeth (Knaver) K.; mag.sc., U. Copenhagen, 1920, Ph.D., 1929; m. Edith Nielsen, 1944; 3 sons, 1 dau. Prof. mineralogy Mineral. Mus., Copenhagen; dir. Geol. Services Eastern Greenland. Mem. French Acad. Scis. (corr.). Author: Nord on Gronland, 1925; Carboniferous and Triassic Stratigraphy of East Greenland, 1931; Geologie von Gronland, 1935; Survey of North Greenland, 1940. Discovered (with L. Myluis, Ericksen) farthest eastern point of Greenland; 1st geol. description of No. part of Greenland.

KOCH, Marie Louise, Am. med. microbiologist; b. Balt., June 16, 1899; d. Charles J. and Gertrude (Brossel) Koch; B.Sc., Johns Hopkins, 1921, M.Sc., 1935, Ph.D., 1949; m. C. E. McGuigan, June 1, 1922 (div. Nov. 1934); 1 dau., Dorothy (Mrs. Edward C. H. Schmidt); m. 2d, D. A. Sprosty, Feb. 20, 1937 (dec.). Asst. bacteriologist Md. Dept. Health, Balt., 1917-19; asst. serologist Phipps Clinic, Johns Hopkins Hosp., Balt., 1920-21; jr. toxicologist Chem. Warfare Service, Edgewood, Md., also bacteriologist Nat. Path. Lab., Detroit, 1922-23; bacteriologist Hynson, Wescott & Dunning, Balt., 1935-37; asst. prof. dept. microbiology and immunology Marquette U. Sch. Medicine, Milw., 1949—, now ret.; chief microbiology VA Center, Wood, Wis., 1950—, now ret. Cons. Milw. Blood Center, 1951—. Commended by adminstr. Vets. Affairs, 1962. Diplomate Am. Bd. Microbiology. Fellow Royal Soc. Health (Eng.), Am. Thoracic Soc., Am. Soc. for Microbiology, Am. Pub. Health Assn., N.Y. Acad. Scis. Research, publs. on selective media for C. diphtheriae, lab. diagnosis gonorrhea in female, drug-resistance staphylocci and streptococci, epidemiology hosp.-acquired staphylococcal infections, differentiations mycobacteria, epidemiology, therapy and pathogenesis hosp. infections caused by commonly occurring enteric gram-negative bacilli, their in vitro susceptibility to commonly used antibiotics and new antibiotics.*

KOCH, Richard, Am. pediatrician; b. N.D., Nov. 24, 1921; s. Valentine and Barbara (Fisher) K.; A.B., U. Cal. at Berkeley, 1958; M.D., U. Rochester, 1951; m. Kathryn Jean Holt, Oct. 2, 1943; children—Jill, Thomas, Christine, Martin, Leslie. Mem. staff Children's Hosp. of Los Angeles, 1952—, dir. mental retardation program, 1955—; faculty U. So. Cal., 1955—, asso. prof. pediatrics, 1962—. Mem. A.M.A., Am. Assn. on Mental Deficiency (pres. elect 1967), Am. Acad. Pediatrics, Nat. Acad. Retarded Children, Western Soc. Pediatric Research. Research, numerous publs. on mental retardation and relation to pediatrics. Home: 6334 W. 80th St., Los Angeles 90045. Office: 4614 Sunset Blvd., Los Angeles 90027.*

KOCH, Robert, physician, pioneer bacteriologist; b. Klausthal, Hanover, Dec. 11, 1843; M.D., U. Göttingen, 1866; m. Emmy Frantz, 1871 (div.). Asst. physician hosp. of Hamburg; practiced medicine, Langenhagen, 1866, Rackswitz, Wolstein, 1872-80; army physician Franco-Prussian War, 1870-71; mem. Berlin Dept. Health, 1880; studied cholera in Egypt and India 1883; prof. U. Berlin, dir. Inst. Hygiene, 1885; dir. Inst. for Infectious Diseases, Berlin, 1891-1904. Recipient Nobel Prize for physiology and medicine, 1905. Mem. French Acad. Scis. (fgn. asso.) 1903. Author: Gesamellte Werke, 2 vols., 1912. Introduced method of making bacterial smears and fixing them with heat, 1877; 1st to isolate and obtain pure culture of anthrax bacillus, 1882; identified comma bacillus as cause of Asiatic cholera, 1883; discovered organism responsible for Egyptian ophthalmia, 1883; produced tuberculin and erroneously thought it to be cure for Tb, later showed it to be of value in diagnosing the disease, 1890; studied rinderpest in South Africa, developed means of vaccinating against it, 1896; investigated bubonic plague in Bombay, 1897, malaria and sleeping sickness in Africa; discovered method of transmission of bubonic plague and sleeping sickness; elaborated gen. methods of bacteriological research and developed method of disinfection; developed means of cultivating bacteria in liquid and solid media, laying found. for rational system of bacterial culturing; established rules for properly identifying causative agts. of various diseases. Died Baden-Baden, Baden, May 27, 1910.

KOCH, Ruprecht, German radiation biologist; b. Berlin, Germany, Dec. 12, 1915; s. Carl and Kate (Goetz) K.; Dr.degree U. Berlin, 1945; m. Erna Schmiga, Aug. 19, 1947; children—Thomas Patrick, Rebecca Patricia. Staff, Pharmacol. Labs., Byk-Guldenwerke, Berlin, 1942-45, Path. Inst., Berlin U., 1945-46; faculty Freiburg U., 1946-47, Radiol. Inst., 1953— prof., 1962—, diätendozent, 1960—; staff U. Hosp. für Pediatrics, Marburg (Germany) U., 1947-53. Recipient Silver medal for sci. merits Med. Faculty U. Ferrara (Italy); 1960; Roentgen-Price, U. Giessen Natural Sci. Faculty, 1964. Mem. German Roentgen Soc., German Soc. for Biophysics, European Assn. for Radiation Research, Brit. Assn. for Radiation Research. Research, numerous publs. on radiation biology, tumorbiology, biophysics, energy transfer, toxicology, pharmacology. Home: 24 Händelst. D 78 Freiburg/Brsg.Fe.Rep. Germany.*

KOCHAN, Ivan, immunologist; b. Tudorko, Ukraine, Aug. 20, 1923; s. Wolodymyr and Alexandra (Dula) K.; student U. Lwiw (Ukraine), 1942-44; cand. med. U. Munich (Germany), 1948; M.Sc., U. Man. (Can.), 1955; Ph.D., Stanford, 1958; m. Tatiana Sawycka, Jan. 15, 1949; children—Andrew, Mark, Christine. Came to U. S., 1955, naturalized, 1960. Biologist, Stanford Research Inst., 1956-58, research asso. dept. med. microbiology, 1958-60; faculty Baylor U., Dallas, 1960-67, chmn. dept., prof. microbiology, 1962-67; prof. microbiology Miami U., Oxford, O., 1967—. NIH Research and Tng. grantee, 1962-67. Diplomate Am. Bd. Microbiology. Fellow Am. Trudeau Soc., Am. Soc. Microbiology, Am. Thoracic Soc., Sigma Xi. Research, publs. on role delayed hypersensitivity in Tb and homograft immunity, transfer delayed hypersensitivity by cell extract. Home: 11032 Creekmere Dr., Dallas 75218.*

KOCHANSKI, Adam, mathematician; b. Dobrzyn, Poland, early 17th century; tchr., Mainz, Germany, Florence, Italy, Olmutz (now Czechoslovakia); mathematician, librarian to John III Sobieski, king of Poland. Author: (dissertations in Annales de Leipzig) Analecta mathematica sive Theoreses mechanicae de natura machinarum fundamentalium, 1668, Considerationes et observationes physico-mathematicae circa diurnam Telluris vertiginem, 1682, Observationes cyclometricae ad facilitandam praxim accommodatae. Died 1695.

KOCHEL, Carl Walter, German geologist; b. Leipzig, Sept. 15, 1898; s. Richard and Helen (Hofmann) K.; ed. U. Leipzig, Clausthal Acad.; Ph.D.; m. Edith Kurlbaum, May 21, 1931; children—Andreas, Franz, Martin, Valentin. Agrégé at Leipzig; prof. Johns Hopkins, Balt.; asso. prof. at Leipzig; with Seismos Gesellschaft mit beschrankter Haftung, Hanover, Bochum Research Works; full prof. geology and paleontology at Marburg/Lahn. Mem. German, Vienna geol. socs., Geol. Assn., Soc. for Devel. Natural Sci. Marburg. Author: Zur Straat und Tektonik Bulgariens, 1927; Bayr. Berge zw. Lech und Loisach, 1931; Umbau der nördl. Kalk alpen, 1956. Schiefergeb. und hess. Senke um Marburg, 1958. Address: Deutschhaustrasse 10, Marburg, Germany.

KOCHENDÖRFER, Albert, German metallurgist; b. Stuttgart, Germany, Dec. 18, 1908; s. Albert and Karoline (Dörr) K.; Ph.D. in Tech. Sci., U. Cologne; m. Anna Kob, Apr. 2, 1938; 1 son, Eberhard. With Max-Planck Inst. for Metall. Research, Stuttgart, 1937; prof. agrégé Stuttgart Technische Hochschule, 1942, prof. at large, 1949; with Max Planck Inst. for Iron Research, 1951—; prof. U. Cologne, 1952—. Mem. German Phys. and Tech. Sci. Soc. Author: Plastizität von Kristallen und metallischen Werkstoffen, 1941; Physikalische Grundlagen der Formänderungfestigkeit der Metalle, 1963. Research on plasticity and strength of solids. Home: Baddenberg, 403-Ratingen. Office: Max-Planckstrasse 1, 4-Dusseldorf, Germany.

KOCHER, Louis, Moroccan entomologist; b. Monflanquin, Feb. 15, 1894; s. Louis and Elisabeth (Bastian) K.; baccalaureate Lycée Michelet, Paris; m. Emilya Djourachkovitch, Apr. 1, 1920; children—Yvonne, Pierre, Georgette, Cécile. With St. Cyr Mil. Sch., 1912-47, ret. in grade col.; entomologist, 1947—. Mem. French Entomol. Soc. Author: Catalogue commenté des coléoptères du Maroc; Localisations nouvelles ou intéressantes de coléoptères marocains; Contribution à l'étude des chrysomélidés du Maroc. Home: Bellile par Montpezat (lot-et-Garonne). Office: 8, place Alaouite, Rabat, Morocco.

KOCHER, (Emil) Theodor, Swiss surgeon; b. Bern, Switzerland, Aug. 25, 1841; M.D., U. Bern; studied surgery, Berlin, Paris, London; student of Lister; prof. surgery U. Bern, 1872-1911; dir. surg. clinic, Bern. Pres. 1st. Surg. Congress, Brussels, Belgium, 1904. Recipient Nobel prize in physiology and medicine, 1909. Mem. German Soc. Surgery. Author: Chirurgische Operationslehre, 1892; Die Therapie des Kropfes, 1904. Pioneer in work on physiology, pathology and surgery of thyroid gland; 1st to remove thyroid, 1876; early work in application of asepsis; devised method for setting dislocated shoulder, also for correcting hernia; skilled in surgery of brain, stomach, gall bladder. Died Bern, July 27, 1917.

KOCHERGIN, Ivan Grigorevich, Russian surgeon; b. 1903; grad. Med. Faculty, 2d Moscow U., 1903; D. Med. Sci. Asst., lectr. dept. surgery Gorky Med. Inst., 1934-39, dir., 1937-40, head chair surgery, 1940-—, prof., 1940-—; prof. chair clin. surgery 2d Moscow Med. Inst., 1940-51; dep. minister health, chmn. learned med. council USSR Ministry Health, 1952-56, dep. minister health, collegium mem., 1959-—, also chmn. council for coordination research and exploitation sci. discoveries. Dir., 5th All-Union Congress Dermatologists and Venerologists, Leningrad, All-Union Conf. Postgrad. Med. Inst. Dirs., Moscow, 1959, All-Union Conf. Postgrad. Med. Studies and Specialization, 1963. Decorated Order of Lenin. Mem. USSR Acad. Med. Sci. (corr.), Internat. Assn. Surgeons. Author: Preparation of the Surgeon's Hands and the Operational Area, 1941; Doctor's thesis on Medicine in 1941-45, Parts 1-2, 1947-49; The Doctor's Calendar for 1962, 1961. Dep. chief editor Large Med. Ency., 2d edit., 1955; dep. editor Herald of USSR Acad. Med. Scis.; mem. editorial bd. New Surg. Archives. Research and numerous publs. on surgery, pub. health orgn., orgn. med. tng.; developer (with S. I. Spasokukotsky) method of preoperative preparation of hands with 0.5 percent ammonium hydroxide solution; developer method to determine localization of metallic fgn. bodies in body (radioprobe), hexenal intraintestinal anesthesia, 1944. Address: USSN Ministry Health, Rakhmanovsky p. 3, Moscow, USSR.

KOCHESHKOV, Ksenofont Aleksandrovich, Russian chemist; b. Dec. 12, 1894; grad. Moscow U., 1922. Prof., Moscow U., 1935-—. Recipient Stalin prize, 1948. Mem. USSR Acad. Sci. (corr.). Author: Organic Gallium Compounds in the Arz Ga Class and Their Dioxanates, 1963; Co-editor series on Synthetic Methods in the Field of Organometallic Compounds, 1949. Research and publs. on chemistry of organometallic compounds; discovered new methods to synthesize organic compounds of lead, tin, silicon, alkaline metals, zinc, antimony and bismuth. Address: Moscow University, Leninskie gory, Moscow, USSR.

KOCHETKOV, Nikolay Konstantinovich, Russian organic chemist; b. 1915; grad. Moscow Inst. Fine Chem. Tech., 1939. Lectr., reader Moscow U., 1945-59, became prof., 1956; became dep. dir., chief lab. on hydrocarbons and nucleotides USSR Acad. Med. Scis., 1959, also head dept. organic synthesis Pharmacology and Chemotherapy Research Inst. Corr. mem. USSR Acad. Med. Sci., USSR Acad. Sci. Author: The Chemistry of beta-Chlorovinylketones, 1955; Indole Derivatives, 1957; Cycloserine and Related Compounds, 1958; Pyrrolizxdine Alkaloids, 1961; Thin-layer Chromatography of Cerebrosides, 1961. Research and numerous publs. on synthesis of pharmacological and chemotherapeutic compounds, relationship between chem. structure and biol. effect in various categories of organic compounds; synthesized antituberculosis antibiotic cycloserine. Office: Inst. Natural Compound Chemistry, USSR Acad. Med. Sci., 14, Solyanka, Moscow, USSR.

KOCHINA, Pelageia Iakovlevna (Polubarinova-Kochina), Russian hydrodynamicist; b. Astrakhan, USSR, May 13, 1899; grad. Petrograd. U., 1921; Dr.physicomath. sci. With Main Geophys. Obs., 1919-21; staff Trans. Inst., also Inst. Marine Engring., 1921-27; asst. lectr. Leningrad U., 1927-34, prof., 1934-—; asso. Math. Inst., then Mechanics Inst., USSR Acad. Sci., 1953-57; head dept. applied hydrodynamics Inst. Hydrodynamics, Siberian Acad. Sci., USSR; dep. Leningrad City Sov.; dept. Moscow City Sov., 1947-—. Recipient Stalin prize, 1958, Order Lenin, 1960. Author: An Integral Flow Equation for Tanks of Constant Depth, 1938; Problems of the Surface Movement of Ground Waters, 1942; Theory of Movement of Ground Waters, 1952. Research on theory of filtration, dynamic meteorology and theory of tides in basins, movement of ground waters. Home: Leninski Prospekt 7, Moscow. Office: Inst. Mechanics, USSR Acad. Scis., Leningradskii Prospekt 7, Moscow, USSR.

KOCZY, Fritz F., oceanographer; b. Vienna, Austria, June 16, 1914; s. Victor and Auguste (Swoboda) K.; Ph.D., U. Vienna, 1939; m. Brita Gunvor Bresky, Mar. 4, 1943. Came to U. S., 1957, naturalized, 1963. Faculty, Oceanographic Inst. Göteborg, 1939-47, 55-56; oceanographer Swedish Deep Sea Expdn., 1947-48; faculty oceanography Bd. Swedish Fisheries, Göteborg, 1948-55, 56; faculty Inst. Marine Sci., Miami, Fla., 1957-—, prof. marine sci., 1959-—. Mem. com. on oceanography Nat. Acad. Scis.-NRC, 1957-—; cons. under sea tech. NSF, Franklin GNO Corp., 1959-—. Mem. Am. Soc. Limnology and Oceanography, Am. Geophys. Union, A.A.A.S., Geochem. Soc., Soc. for Clay Research, Sweden. Contbg. author Nuclear Radiation in Geophysics, 1962; The Seas, vol. 3, 1963; Ency. Geophysics, 1965. Editor: Geochemica et Cosmochimca Acta 1965; editorial bd. Marine Geology 1964; adv. editorial bd. Earth and Planetary Science Letters, 1966; Ency. Sci. and Tech., 1960-62; International Series in Earth Science, 1965. Publs. in fields of geochemistry, phys. oceanography, chem. oceanography, geology, geomorphology, geochromology. Died Apr., 1967.

KODA, Shigeyasu, Japanese metallurgist; b. Tokyo, Japan, Jan. 27, 1907; s. Masakichi and Kuniko (Inaba) K.; B.Sc., U. Tokyo, 1932, D.Eng., 1951; m. Kiyoko Noguchi, Apr. 18, 1940; children—Yoko, Momoko, Shigeharu. Staff, Research Lab., Furukawa Electric Engring. Co., Tokyo, 1933-42; prof. dept. metallurgy Faculty Engring., Hokkaido U., Sapporo, Japan, 1943-58; prof. Research Inst. Iron, Steel and Other Metals, Tohoku U., Sendai, 1958-—; head dept. metal physics Nat. Research Lab. for Metals, Tokyo, 1966-—. Mem. Japan Inst. Metals (v.p. 1966-—, award 1952), Phys. Soc. Japan, Iron and Steel Inst. Japan, Inst. Metals (Eng.). Author: Metal Physics, 1964; also numerous articles. Research on non-ferrous phys. metallurgy; electron microscope observations of precipitation process of many non-ferrous alloys especially aluminum alloys; explanation of their hardening mechanism from direct observation of interaction between precipitates and moving dislocations. Home: 76 Eifuku-cho, Suginami-ku, Tokyo, Japan. Office: Tohoku U., Sendai, Japan.*

KODAIRA, Yoshio, Japanese physicist; b. Nagano Prefecture, Japan, 1903; grad. physics dept. Tokyo U., 1925; D.Sc., 1934. Joined Central Meteorol. Obs., 1925, later became engr.; studied in France, Norway, U.S.A., 1934; apptd. chief research dept. Central Meteorol. Obs., 1943, prof. Tokyo U., 1944; later dir. Osaka Dist. Meteorol. Obs.; dir. Tokyo Dist. Meteorol. Obs., Transp. Ministry, 1946-—; apptd. chief Meteorol. Research Inst., 1958. Author: Physical Mathematics; Application of Trigonometrical Series.

KODAMA, Kazunobu, Japanese chemist; b. Matsuyama, Japan, Oct. 6, 1914; s. Kinjiro and Misao (Suzuki) K.; M.S., Tokyo U., 1937, Dr.Sci., 1962; m. Yoneko Mimura, Feb. 3, 1944; children—Nobutake, Akiko, Tomofumi. Asst. researcher Shanghai (China) Sci. Inst., 1937-45; chief librarian Inst. Phys. Chem. Research, Tokyo, Japan, 1947-53; chief analytical sect. Nagoya (Japan) Municipal Indsl. Research Inst., 1953-63, head chemistry dept., 1963-—; prof. Tokyo Jikeikai Med. Coll., 1946-47. Mem. Am. Chem. Soc., Chem. Soc. Japan, Japan Soc. Analytical Chemistry. Author: Methods of Quantitative Inorganic Analysis, An Encyclopedia of Gravimetric, Titrimetric and Colorimetric Methods, 1963; also articles. Colorimetric determinations of platinum metals and adjacent metals with organic reagts., polarimetric determination of boric acid with mannitol or tartaric acid. Home: 3-23, Nagaike-cho, Showa-ku, Nagoya. Office: 3-24, Rokuban-cho, Atsuta-ku, Nagoya, Japan.*

KOEBERLE, Eugène, French surgeon, gynecologist; b. Sélestat, France, 1828; ed. Strasbourg, France; prof. surgery Faculty Medicine, Strasbourg. Author: Des cysticerques chez l'homme, 1861; Documents pour servir à l'histoire des fermeurs, 1865. Pioneer in ovariotomy (1st attempted 1862); performed 1st successful excision of uterus and ovaries for tumor, 1863; studied bladder worms; developed bladder operations. Died Strasbourg, 1915.

KOEBRICH, Gert, German chemist; b. Barcelona (Spain), Aug. 8, 1929; s. Konrad Albert and Nenny (von der Becke) K.; Dr.phil., U. Marburg/Lahn, 1956; m. Ursel Brötje, Dec. 29, 1959; children—Karen L., Sabine G. Docent, U. Heidelberg, 1962-—, sci. adviser, 1964-—. Mem. Soc. German Chemists, Am. Chem. Soc. Contbr. articles in field. Research on low-temperature chemistry of organometallic intermediates; synthetic work on small-ring compounds and polyenes. Address: 69 Kurfürstenanlage, 69 Heidelberg, West Germany.*

KOECHLIN-SCHLUMBERGER, Joseph, French geologist; b. 1796; mayor, Mulhouse, France; author studies on geology of Alsace, Vosges, Dauphine; made geol. maps of Upper and Lower Rhine. Died 1863.

KOEHLER, James Frederic, Am. physicist; b. Chgo., Oct. 21, 1904; B.S., Princeton, 1926, M.A., 1928, Ph.D., 1934; m. Mary Gray Peck, June 30, 1930; children—George F., Nancy (Mrs. Spahr Hull). Asso. prof. physics, Smith Coll., Northampton, Mass., 1931-40; research asso. Radiation Lab., Mass. Inst. Tech., 1940-42; head radar design Bur. Aeros., USN, Washington, 1942-45; dir. govt. and indsl. research Philco Corp., Phila., 1946-58. Mem. adv. council on research and devel. Chief Signal Officer U. S. Army, Washington, 1953-61; cons. to comdg. gen. U. S. Army Research and Devel. Lab., Ft. Monmouth, N.J., 1958-61. Fellow Am. Phys. Soc.; mem. I.E.E.E. Research in electronics, flourescence. Patentee radar display. Home: Bristol Rd., Damariscotta, Me.

KOEHLER, Otto David Waldemar, German zoologist; b. Insterburg (E. Prussia), Dec. 20, 1889; s. Eduard and Karoline (Heinrici) K.; student U. Freiburg; dr., U. Munich, 1911; m. Annemarie Deditius, Oct. 9, 1920; 1 dau., Barbara (Mrs. Rupprecht Lauck); m. 2d Gisela Hauchecorne, Sept. 18, 1955. Prof., U. Munich, 1923-25; prof., dir. Zoöl. Inst., U. Königsberg (now Kaliningrad, USSR), 1925-45, U. Freiburg im Breisgau, 1946-60, emeritus, 1958-—. Mem. Königsberger Gelehrte Gesellschaft Leopoldina, Societas pro fauna et flora fennica. Editor Zeitschrift für Tierpsychologie, 1935-—; co-editor Behaviour, 1946-—. Research and numerous publs. on developmental physiology and genetics of sea urchins; physiology of senses; animal orientation; contbns. to ethology. Address: Zoologisches Institut, Katharinenstrasse, Freiburg im Breisgau, West Germany.*

KOEHLER, René, French zoologist; b. St.-Die, France, 1860; physician; prof. zoology Lyons (France) Faculty Scis.; organizer deep-sea expdn. in Bay of Biscay. Mem. French Acad. Scis. Author: La faune de France. Research on echinoderms. Died 1931.

KOEHN, Carl James, Am. nutritionist, biochemist; b. Madison, Wis., Dec. 3, 1910; s. Carl John and Katherine (Meyer) K.; B.S., U. Wis., 1932, M.S., 1934, Ph.D., 1936; m. Irene Neoma Senty, May 21, 1937; 1 son, James Howard. Prof. nutrition Auburn (Ala.) U., 1936-42, 45-46; commd. 1st lt. U. S. Army, 1942, advanced through ranks to lt. col., 1964; chief nutrition br. Office Surg. Gen., Washington, 1948-52; comdg. officer U. S. Army Med. Nutrition Lab., Denver, 1954-59; ret., 1964; chief nutrition sect. S.W. Found. for Research and Edn., San Antonio, 1964-65; nutrition cons., 1965-—. Dir. nutrition surveys Nfld., Bermuda, 1943, India, 1944, Burma, 1945, Korea, 1953, N. Korean prisoners of war UN, Chinese prisoners of war, Chedje-do, 1953. Mem. 1st Internat. Congress Biochemistry, Cambridge, Eng., 1949. Diplomate Am. Bd. Nutrition. Mem. Am. Inst. Nutrition, Am. Chem. Soc., A.A.A.S., Sci. Research Soc. Am., Sigma Xi. Contbr. articles to sci. jours. Demonstrated that vitamin B2 was composed of riboflavin and niacin, enlargement parotid gland in human malnutrition, reduction in adenine-flavin system in liver as a result burn injury; developed standard for comml. canned dog foods, rations for baboons; established relationship between vitamin a and beta-carotene. Address: 1500 Riverside Dr., Austin, Tex. 78743.

KOELLE, George Brampton, Am. pharmacologist; b. Phila., Oct. 8, 1918; s. Frederick Christian and Emily Mary (Brampton) K.; B.Sc., Phila. Coll. Pharmacy and Sci., 1939, D.Dc. (hon.), 1965; Ph.D., U. Pa., 1946; M.D. (Chalfant fellow), Johns Hopkins, 1950; m. Winifred Jean Angenent, Feb. 6, 1954; children—Peter Brampton, William Angenent, Jonathan Stuart. Bio-assayist LaWall & Harrisson, Phila., 1939-42; asst. prof. pharmacology Columbia, Coll. Phys. & Surg., N.Y.C., 1950-52; prof. pharmacology U. Pa. Grad. Sch. Medicine, Phila., 1952-—, dean, 1957-59, chmn. dept. pharmacology Sch. Medicine, 1959-—, Elmer Holmes Bobst prof. pharmacology, 1963-—. Chmn. pharmacology and exptl. therapeutics study sect. NIH, 1965-68; cons. McNiel Labs., Ft. Washington, Pa., 1951-66. Fellow A.A.A.S., N.Y., Acad. Sci.; mem. Am. Soc. Pharmacology and Exptl. Therapeutics (pres. 1965-66), Histochem. Soc., Soc. Biol. Psychiatry, Internat. Union Pharmacology (sec.-gen. 1966-—), Sigma Xi, Alpha Omega Alpha. Editor: Cholinesterases and Anticholinesterase Agents (Springer-Verlag), 1963; Pharmacol. Revs., 1959-62. Contbr. articles, revs. to profl. lit. Developed histochem. methods for localization of cholinesterases and monoamine oxidase; investigated mechanisms of synaptic and neuroeffector transmission; pharmacology of anticholinesterase agts. and other drugs. Home: 132 Park Av., Swarthmore, Pa. 19081.*

KOELSCHE, Giles Alexander, Am. physician; b. Ashland, Ore., Sept. 3, 1908; s. Charles Lewis and Flora (Deiner) K.; B.S., M.D., Loma Linda U., 1931; M.S. in Medicine, Ph.D. in Medicine, Mayo Grad. Sch. Medicine, 1935; m. Mary Allen, June 26, 1930; children—Edward G., Marilyn Anne. With Mayo Grad. Sch. Medicine U. Minn., Rochester, 1937-—, asso. prof., 1962-—; cons. medicine Mayo Clinic, 1937-—. Sec. sci. and ednl. council Allergy Found. Am. Fellow Am. Acad. Allergy, Am. Coll. Allergists (pres. 1960), Internat. Assn. Allergologists (exec. com.), Am. Coll. Chest Physicians, A.M.A., Sigma Xi. Research, publs. on mgmt. of asthma, hay fever and chronic hives, immunologic problems that may be involved in transplantation of organs from one individual to another, factors which may lead to devel. of antibodies to organs in one's body with consequent damage to them, role that immunologic mechanisms may play in devel. of and resistance to cancer. Home: 41 Skyline Dr., Rochester, Minn. 55901.*

KOENIG, Alexander, ornithologist; b. St. Petersburg, Russia, Feb. 20, 1858; prof., Bonn, Germany. Author: Avifauna Spitzbergensis, 1911; Avifauna Aegyptiaca, from 1922. Research on birds of Africa and of North; donator research inst. which houses his collection, Bonn. Died Mecklenburg, Germany, July 16, 1940.

KOENIG, Frederick Otto, Am. chemist; s. Friedrich Otto and Margarette (Purington) K.; A.B., Harvard, 1922; Ph.D., Munich, 1929; m. Marie-Luise Gressmann, Dec. 30, 1929 (div. Oct. 1954); children—Frederick, George; m. 2d, Inge Rabes, Dec. 16, 1955; 1 son, Franklin. Faculty, U. Cal., Berkeley, 1931; faculty Stanford, 1931-—, prof. chemistry, 1943-—. NRC fellow, 1930. Mem. Am. Chem. Soc., A.A.A.S., History Sci. Soc. Research in chem. thermodynamics; effects of external fields on electromotive force; electrolytic double layer and the electrocapillary curve. Home: 1945 Cowper St., Palo Alto, Cal. 94301.

KOENIG, George Augustus, chemist; b. Willstätt, Baden, Germany, May 12, 1844; s. Johannes and Margaretha (Pfotzer) K.; mech. engr. degree Poly. Sch., Karlsruhe, 1859-63; student U. Heidelberg,

1863-65, U. Berlin, 1865-67, A.M., Ph.D., Heidelberg, 1867; Mining Acad., Freiburg, 1867-68; m. Wilhelmina Marquart, Oct. 7, 1869. Came to U. S., 1868; manufactured sodium stannate from tin scraps in Phila.; chemist Tacony Chem. Works, Phila., 1868-72; made exam. of old mines in Mexico, winter of 1870-71; asst. prof. chemistry and mineralogy U. Pa., 1872-79, mineralogy and geology, 1879-92; prof. chemistry, Mich. Coll. of Mines, Houghton, from 1892. Discoverer of minerals: Hydrotitanite, Randite, Leydyite, Alaskaite, Beegerite, Bementite, Footeite, Paramelaconite, Mezapilite, Mohawkite, Keeweenawite, Stibiodomykite, Melanochalcite; reexamined many other minerals and discovered diamond in meteoric iron. In 1885 published chromometric methods, a development of blow-pipe analysis, and in 1897 a new way of assaying without muffle; patented an assay furnace to carry out the methods; patented chlorination of low-grade silver and gold ores, 1881; worked on preparation of artificial crystals of arsenides. Author: Chemistry Simplified, 1905. Died Jan. 15, 1913.

KOENIG, Harold, Am. physician; b. N.Y.C., Mar. 16, 1921; s. David and Anna (Burg) K.; B.S., Rutgers U., 1942; M.S., Northwestern U., 1945; M.B., Chgo. Med. Sch., 1946, M.D., 1947; Ph.D., U. Pa., 1948; m. Ruth Steinman, Feb. 11, 1945; children—Susan Mae, Leslie Ellen. Instr. anatomy U. Pa. Med. Sch., 1947-48, asst. prof., 1948-49; asst. prof. neuroanatomy Chgo. Med. Sch., 1949-53, asso. prof., 1953-54; faculty neurology Northwestern U. Med. Sch., 1956—, prof., 1963—; chief neurology service VA Research Hosp., Chgo., 1956—. Fellow Acad. of Neurology; mem. Am. Neurol. Assn., Am. Assn. Anatomists, Am. Assn. Neuropathologists, Histochem. Soc., Internat. Neurochem. Soc., Am. Soc. for Cell Biology, Soc. for Exptl. Biology and Medicine. Contbr. numerous articles to sci. jours. Investigations in neurochemistry, neurohistochemistry, exptl. neuropathology, including structure and composition of lysososomes, neurobiol. action of nucleic acid antimetabolites, nucleic acid and protein metabolism in nervous system and biology of glia. Home: 910 N. Lake Shore Dr., Chgo. 60611. Office: 333 E. Huron St., Chgo. 60611.*

KOENIG, Karl Rudolph, German physicist; b. Königsberg, Germany, Nov. 26, 1832; ed. at Königsberg, pupil of J. B. Villaume. Settled in Paris, 1852, later became naturalized French citizen; founded workshop, Paris, 1858, where he built musical instruments and manufactured acoustical equipment; became an authority on acoustics. Author: Quelques expériences d' acoustique, 1882; Catalogue des appareils d'acoustique. Invented mamometric capsule for studying air vibrations; studied speed of sound, vibratory motion, normal tuning forks, muted sound, and acoustic percussions. Died Oct. 2, 1901.

KOENIG, Robert, German mathematician; b. Linz/Donau, Apr. 11, 1885; s. Josef and Marie (Bredow) K.; ed. U. Vienna, U. Gottingen; Ph.D.; m. Charlotte Schnell, Mar. 17, 1925; children—Helmut, Wolfgang. Instr. at Leipzig, 1911; asso. prof. U. Tubingen, 1914-22; full prof. U. Münster/Westphalia, 1922-27, U. Iena, 1927-45; full prof. U. Munich, 1947-55, full prof. emeritus, 1955—. Mem. Leipzig, Munich acads. scis. Author: Elliptische Funktionen, 1928; Mathemat. Grundalgen der Höheren Geodäsie, 1951. Research on functional theories. Address: Adelheidstrasse 21, Munich, Germany.

KOENIG, Samuel, mathematician, b. Buedingen, Germany, July 31, 1712; pupil of Jean Bernoulli and Wolf; instr., sec. to Marquis of Chatelet; named prof. math., philosophy, natural law, La Haye, 1749; mem. French Acad. Scis. Author: Über den Irreductibeln Fall, 1749; also numerous memoirs. Follower of Leibniz; friend of Voltaire and Réaumur; quarreled with Maupertuis on principle of least action; supported Leibnizian doctrine of conservation of "living forces." Died Zuilestein, Netherlands, July 22, 1757.

KOENIGS, Gabriel-Xavier-Paul, French mathematician; b. Toulouse, France, Jan. 17, 1858; ed. l'École Normale Supérieure, 1882; became mem. Faculty Besançon, (France), 1883, Faculty Toulouse, 1885; apptd. prof. mechanics Sorbonne, 1886, prof. mech. and exptl. physics, 1897; also head lectr. l'Ecole Normale Supérieure. Mem. French Acad. Scis. Author: Leçons de l'agrégation classique de mathématiques. Développements nouveau sur la géométrie, 1891; Sur les lignes géodésiques, 1893; La géométrie reglée et ses applications, 1895; Leçons de cinematique professeés à la Sorbonne, 1897; Introduction a une théorie nouvelle des mécanismes, 1905. Research in infinitesimal geometry, mechanics, gen. cinematics. Died Paris, Oct. 29, 1931.

KOESTER, Louis Julius, Jr., Am. physicist; b. Galveston, Tex., Jan. 25, 1925; s. Louis Julius and Cecilia (Lott) K.; student U. Tex., 1941-43; B.S., U. Wis., 1947, Ph.D., 1951; m. Marion Anna Mayer, May 19, 1946; children—Barbara Jane, Valerie Ann. Research asso. to asso prof. physics U. Ill., Urbana, 1951-63, prof. 1963—. Cons. Phys. Sci. Study Com., 1957-61, Argonne (Ill.) Nat. Lab., 1960—. Mem. Am. Phys. Soc., Am. Assn. U. Profs., Sigma Xi. Research on nuclear physics, electronics, and high energy physics. Home: 1013 W. Clark, Champaign, Ill. 61822. Office: Physics Dept., U. Ill., Urbana, Ill. 61803.*

KOFFKA, Kurt, psychologist; b. Berlin, Germany, Mar. 18, 1886; s. Emil and Luise (Levy) L.; ed. U. Edinburgh, 1904-05, U. Freiburg, 1908-09; U. Wurzburg, 1909-10; Ph.D., U. Berlin, 1908; m. Elisabeth Ahlgrim, July 21, 1923. Asst., U. Wurzburg, 1909-10, Acad. (now U.) Frankfort, 1910-11; privat-dozent U. Giessen, 1911-18, prof., 1918-27; vis. prof. Cornell U., 1924-25, U. Wis., 1926-27; William Allan Neilson research prof. Smith Coll., 1927-32, prof. psychology, 1932-41, also editor Studies in Psychology; co-founder, editor Psychologische Forschung, 1922-35. Author: Growth of the Mind, 1924; Principles of Gestalt Psychology, 1935. Founder (with Wertheimer and Köhler), also considered the systematizer of Gestalt sch. of psychology, which emphasized the whole rather than various parts, and suggested that the mind perceives situations as a whole rather than collection of parts; believed that mental processes must be studied in terms of integrated systems of phenomena, whose totality is not exactly equivalent to sum of parts; his basic interests were in developmental, genetic and ednl. psychology. Died Northampton, Mass., Nov. 22, 1941.

KOFFLER, Henry, microbiologist; b. Vienna, Austria, Sept. 17, 1922; B.S., U. Ariz., 1943; M.S. (Alumni Research Found. fellow), U. Wis., 1944, Ph.D. (NRC fellow), 1947; m. Phyllis Pierson, June 21, 1946. Faculty, Purdue U., Lafayette, Ind., 1947—, prof. bacteriology, 1952—, coordinator research dept. biol. sci., 1949-59, head dept., 1959—, asst. to dean Grad. Sch., 1957-59, asst. dean, 1959-60. Cons. to govt. agys., industry, univs., 1958—; mem. panel interdisciplinary cooperation Commn. on Undergrad. Edn. in Biol. Scis., 1965—, commr. Commn. on Undergrad. Edn. in Biol. Scis., 1696—, vice chmn., mem. exec. com. 1966—, chmn., 1967—. Guggenheim fellow Sch. Medicine, Western Res. U., 1953-54; recipient research award Purdue U., 1952, Eli Lilly & Co. award, 1957. Fellow Ind. Acad. Sci, A.A.-A.S., Am. Acad. Microbiologists. mem. Am. Soc. Biol. Chemists, Am. Biophys. Soc., Am. Soc. Exptl. Biologists, Am. Chem. Soc., Bot. Soc. Am., Am. Soc. Microbiologists, Brit. Soc. Gen. Microbiology, German Soc. Hygiene and Microbiology, Soc. for Cell Biology, Am. Inst. Biol. Scis., Sigma Xi. Research, numerous publs. on molecular biology of microorganisms, fungal cell walls, bacterial flagella, thermophilic bacteria; developed antibacterial circulin. Home: 131 Sunset Lane, West Lafayette, Ind. 47906. Office: Dept. Biol. Scis., Purdue U., Lafayette, Ind. 47907.*

KOFINK, Karl Walter, German physicist; b. Crispenhofen, Nov. 23, 1907; s. Heinrich and Lucie (Ehmann) K.; Ph.D., U. Berlin; m. Helen Bruhns, Oct. 3, 1938; 1 son, Alexander. Asst., Inst. Theoretical Physics, Berlin-Charlottenburg Technische Hochschule, 1934; with U. Frankfort, 1935, prof., 1940; with Stuttgart Technische Hochschule, 1948; prof. at large, 1952, full prof. chair structure of matter, 1961. Mem. German Soc. Physicists, Soc. for Math. and Applied Mechanics, Am. Phys. Soc. Research and publs. on quantum mechanics, Dirac's theory of electron, quantum statistics and infra-red radiation, supersonic aerodynamics and neutron transport theory. Home: Bruehlstrasse 21, Karlsruhe-Durlach. Office: Haputstrasse 54, Karlsruhe-Rintheim, Germany.

KOFOED-HANSEN, Otto Mogens, Danish physicist; b. Frederiksberg, Denmark, Apr. 25, 1921; s. Olaf and Anna (Hansen) K.-H.; Cand. Mag. et Mag. Scient., U. Copenhagen (Denmark), 1945, Dr.Phil., 1952; m. Anne Lise Skovgaard, Mar. 24, 1946; children—Anne Marie, Mogens. Goldsmith, Konrad Knudsen, Copenhagen, 1937-41; research asso. U. Copenhagen, 1946-54, asso. prof., 1955-56; research asso. Columbia, N.Y.C., 1954-55; head physics dept. Danish AEC 1956—; prof. reactor physics Copenhagen Tech. U., 1960—. Recipient Silver medal Danish Assn. Craftsmen, 1941; Gold medal U. Copenhagen, 1949. Mem. Danish Acad. Tech. Sci., Royal Danish Soc. Sci. and Letters, Inst. Strategic Studies, London. Author books including The Negotiators, 1964; also articles, chpts. in books. Devel. and constrn. instruments for investigations of radioactivity, determination of fundamental constants in beta decay; criteria for self-sustaining thermonuclear reactions; theoretical investigation in fluid mechanics; impact of sci. on soc.; energy resources. Home: 10 Langagervej, Himmelev, Roskilde, Denmark. Office: Physics Dept., Danish Atomic Energy Commn., Riso Roskilde, Denmark.*

KOFOID, Charles Atwood, Am. zoologist; b. Granville, Ill., Oct. 11, 1865; s. Nelson and Janette (Blake) K.; A.B., Oberlin, 1890, Sc.D., 1915; A.M., Harvard, 1892, Ph.D., 1894; Sc.D., U. Wales, 1920; LL.D., U. Cal., 1937; m. Carrie Prudence Winter, June 30, 1894. Tchr. Oberlin Acad., 1888-90; teaching fellow, Oberlin Coll., 1890-91; instr. vertebrate morphology, U. Mich., 1894-95; supt. Biol. Sta., U. Ill., Havana, Ill., 1895-1900; asst. prof. zoology U. Ill., 1897-1900; supt. Natural History Survey of Ill., 1898-1900; asst. prof. U. Cal., Berkeley, 1900-04, asso. prof. histology and embryology, 1904-10, prof. zoology, 1910-36, head dept., 1910-19, 23-36, vis. prof. biology, Tohoku Imperial U., Japan, 1930. Acting dir. San Diego Biol. Sta., 1905-06; asst. naturalist, U.S.S. Albatross, Coast of Cal., 1904; asso. naturalist, Agassiz expdn. to eastern tropical Pacific, U.S.S. Albatross, 1904-05; asst. dir. San Diego Research, 1912, Scripps Inst. Biol., 1900-24. Asso. editor Am. Naturalist, 1897-1908; zool. editor Jour. Applied Microscopy, 1900-04; Am. editor Internationale Revue des Hydrobiologie, 1910; editor U. Cal. Publs. in Zoology, 1909-36; asso. editor Jour. Morphology, 1920-24, Am. Jour. Hygiene, 1920, Zool. Acta (Stockholm), 1920, Isis, 1931. Fellow Nat. Acad. Scis., Am. Philos. Soc., Phila. Acad. Natural Science, A.A.A.S., Am. Acad. Arts and Scis., Royal Soc. Tropical Medicine and Hygiene, Cal. Acad. Scis., Am. Pub. Health Assn.; mem. Washington Acad. Scis., Western Soc. Naturalists, Am. Soc. Tropical Medicine, Am. Soc. Naturalists, Am. Micros. Soc., Assn. Am. Anatomists, Am. Soc. Zoologists, Am. Soc. Entomologists, History of Sci. Soc. Author: Biological Stations of Europe; also bulls., articles. Made spl. study of parasitology and amoebiasis; invented plankton net, water sampler for deep sea work, self-closing plankton net for horizontal towing. Died May 30, 1947.

KOFT, Bernard Waldemar, Am. microbiologist; b. Hammonton, N.J., Nov. 21, 1921; s. Emil N. and Berenice (Olensky) K.; B.S., Rutgers U., 1943; M.S., U. Pa., 1947, Ph.D., 1950; m. Betty J. Ward, June 11, 1944; children—Susan, Helen, David, Daniel, Paul. Faculty, Jefferson Med. Coll., Phila., 1950-57, asst. prof., 1953-57; faculty Rutgers U., New Brunswick, N.J., 1957—, prof. bacteriology, 1967—. Mem. A.A.A.S., N.Y. Acad. Sci., Am. Soc. for Microbiology (sec. N.J. br. 1964-66), Am. Inst. Biol. Sci., Sigma Xi. Research, publs. on biosynthesis folic acid, mode action sulfonamides, microbial degradation alkyl benzene sulfonate. Home: 196 Hardenberg Lane, East Brunswick, N.J., 08816. Office: Rutgers U., New Brunswick, N.J. 08903.*

KOGAN, Abram Aronovich, Russian obstetrician, gynecologist; b. 1899; grad. Med. Faculty, 2d Moscow Med. Inst., 1924; D.Med. Sci., 1939. Intern, asst., lectr. obstetrics and gynecology clinic 2d Moscow Med. Inst., also asso. other Moscow hosps., 1924-40; prof., head chair obstetrics and gynecology Tashkent Med. Inst., 1940—. Mem. learned med. council Uzbek Ministry Health. Mem. All-Union (bd. mem.), Uzbek (chmn.) socs. obstetricians and gynecologists. Mem. editorial council Obstetrics and Gynecology; coeditor Obstetrics and Gynecology sect. Large Med. Ency., 2d edit. Research and numerous publs. on use of blood transfusions and amniotic fluid in obstetrics and gynecology, endocrine diseases in women, role of nervous system in pathology of female genital region, pathogenesis of cancerous metastasis. Address: Tashkent Med. Inst., ulitsa Karla Marksa 103, Tashkent, Uzbekistan SSR, USSR.

KOGAN, Aleksandr Borisovich, Russian physiologist; b. 1912; grad. Rostov Med. Inst., 1932; D. Biol. Sci. Asso. dept. physiology Rostov Med. Inst., 1935-37; prof., head chair human and animal physiology Biology and Soil Faculty, Rostov U., 1947—. Recipient Pavlov prize USSR Acad. Sci. Mem. Pavlov All-Union (bd. mem.), Rostov City (bd. mem.) socs. physiologists. Author: Practical Guide to Comparative Physiology, 1959; Technique of Physiological Experiments, 1967; Electrophysiology, 1968; Principles of Physiology of Higher Nervous Activity. Mem. editorial council Sechenov Jour. Physiology. Research and numerous publs. on electro and neurophysiology, electroencephalography, neurocybernetics, biophysics, also other aspects of neurophysiology, mechanism of compound subcortical and conditioned reflexes, nature of nervous excitation and inhibition. Address: Rostov University, ulitsa Fridrikha Engelsa 115, Rostov-on-Don, RSFSR, USSR.

KOGER, Marvin, Am. geneticist; b. Colgate, Okla., May 18, 1915; s. James E. and Mary A. (Edwards) K.; B.S., N.M. State U., 1939; M.S., Kan. State U., 1940; Ph.D., U. Mo., 1943; m. Mildred Dorothy Dykes, Jan. 22, 1938; children—Thomas M., Dorothy Elizabeth (Mrs. G. H. Taft), Marva Jo, David Charles. From instr. to asso. prof. N.M. State U., 1943-51; prof., geneticist U. Fla., Gainesville, 1951—. Named Alpha Zeta Prof. of Year, 1966. Mem. Am. Soc. Sci., Am. Dairy Sci. Assn., Am. Inst. Biol. Sci., Genetics Soc. Am., Sigma Xi, Gamma Sigma Delta. Author: (with T. J. Cunha, A. C. Warnick) Crossbreeding Beef Cattle, 1963; also numerous articles. Research on genetic improvement of cattle in unfavorable environments. Home: Route 3, Box 278, Gainesville, Fla. 32601.*

KOHL, Hans, German physiologist; b. Bad Kreuzbach, July 13, 1902; s. Martin and Margarete (Linz) K.; M.D., U. Bonn; m. Margot Naporra, Oct. 28; children—Gabriele, Michaela. Prof. honoris causa, U. Bonn Polyclinic; chief physician intern sect. St. Peter Hosp., Bonn, 1945—. Mem. European Soc. for Hematology. Author: (monograph) Aminosäuren: Ihre theoretische und praktische Bedeutung für die klinische Therapie. Research on insulin, diabetes, coagulation of blood, physiol. implications of old age, amino acids. Home: Haus am Johannisberg, Oberkassel-Bonn. Office: St. Petrus Krankenhaus, Bonner Talweg 4-6, Bonn, Germany.

KÖHLER, Alban, German radiologist; b. Pesta, Germany, Mar. 1, 1874; radiologist, Wiesbaden, Germany. Author: Grenzen des Normalen und Anfänge des Pathologischen im Röntgenbild, 1910. Introduced method to teleroentgenography of heart in which more nearly parallel rays are utilized by placing plate about 7 feet from X-ray tube, 1905; described osteochondrosis of tarsal navicular bone (Köhler's disease, or Köhler's tarsal scaphoiditis), 1908. Died 1947.

KOHLER, Elmer Peter, Am. chemist; b. Egypt, Pa., Nov. 6, 1865; s. Lewis A. and Elizabeth (Newhardt) K.; A.B., Muhlenberg Coll., Allentown, Pa., 1886, A.M., 1889, LL.D., 1918; Ph.D., Johns Hopkins, 1892. Asso. in chemistry Bryn Mawr Coll., 1892-1897, asso. prof., 1897-1900, prof., 1900-12; prof. chemistry Harvard, from 1912. Fellow Am. Acad. Arts and Scis., Nat. Acad. Scis., Imperial Acad. Scis. of Halle. Died May 24, 1938.

KOHLER, G. A. Edward, Am. inventor; b. Phila., Pa., Feb. 17, 1864; s. Ignatius and Anna C. (Fischer) Kohler; ed. U. Pa., class 1886; m. Mary Ward Everett, Oct. 11, 1899; children—Franklin W., John Bowen. With U. S. Constrn. Co., Chgo., 1887-88, then with Peabody, Daniels & Co.; became sec. and gen. mgr. of a saw mfg. co.; entered electrical business, 1890, organized firm of Kohler Bros., 1891, of which became sole propr. 1910; dir. Kohler Aviation Corp. Invented and manufactured devices for operating newspaper presses, including magazine reels, known under the name Kohler System. Died Apr. 29, 1932.

KOHLER, George Oscar, Am. chemist; b. Milw., Apr. 9, 1913; s. Oscar Charles and Thora (Zachariasen) K.; B.S., U. Wis., 1934, M.S., 1936, Ph.D., 1938; m. Christine Gilchrist, Oct. 5, 1940; children—Cynthia Ann (Mrs. Robert E. Castner), Sylvia, William Mark. Postdoctorate fellow U. Wis., 1938-39; with Cerophyl Labs., Inc., Kansas City, Mo., 1939-56, v.p. research, 1952-56; with U. S. Dept. Agr., Albany, Cal., 1956—, lab. chief field crops lab., 1956—. Mem. Am. Chem. Soc., Am. Assn. Cereal Chemists, Inst. Nutrition, Am. Oil Chem. Soc., A.A.A.S., N.Y. Acad. Scis., Nat. Acad. Sci. (mem. cereals com. of food and nutrition bd.). Contbr. numerous articles to profl. jours. Developmental research on thyroactive proteins, protein quality, processing and dehydration of forage crops, nutritional value of forages, oilseed meals and cereal grains, carotenoid pigments; discovery of grass juice factor, an unidentified vitamin in leaves; improved methods of amino acid analysis; biol. availability of amino acids and carbohydrates. Home: 2259 Tamalpais Av., El Cerrito, Cal. 94530. Office: 800 Buchanan St., Albany, Cal. 94710.*

KÖHLER, Harro, Austrian veterinarian; b. Plauen, Dec. 12, 1918; s. Franz and Marie (Rossbach) K.; M.D. in Vet. Medicine; m. Ursula Strempel, July 24, 1943; children—Joachim, Sabina. Asst., Vet. Med. Sch., Vienna, 1946-53, instr., 1953-57, dir. Path. Inst., 1957—. Mem. German Pathology Soc., Soc. for Study Vet. Pathology, Austrian Soc. Vet. Medicine. Research and publs. on diseases of poultry and swine, virus and nutritional diseases. Home: Rudolf-Bärenhartgasse 10, Vienna XVII. Office: Linke Bahngasse 11, Vienna III, Austria.

KOHLER, Werner, German microbiologist; b. Dresden, Germany, Mar. 24, 1929; s. Alfred and Elsa (Richter) K.; D.Sc., Friedrich Schiller U., Jena, Germany, 1953; M.D., U. Rostock (Germany), 1954; m. Marie-Luise Harz, Feb. 18, 1956; children—Bernhard, Michael. Docent, Inst. Hygiene, Rostock, 1954-58; head dept. med. microbiology Inst. Microbiology and Exptl. Therapy Jena, German Acad. Scis., 1958—; asst. prof. med. microbiology U. Jena, 1958—. asst. Inst. Microbiology and Hygiene, Bad Elster, Germany, 1952-53, dir., 1966—. Author: (with H. J. Otte) Die Praxis der Resistenzj und Spiegelbestimmungen zur antibiotischen Therapie, 1958; Die Serologie des Rheumatismus und der Streptokokkeninfektionen, 1963; (with H. Mochmann) Grundriss der Medizinischen Mikrobiologie, 1964; also numerous articles. Research on serology, biochemistry and epidemiology of streptococci, pathogenesis of rheumatic fever, serology of Pseudomonas aeruginosa. Home: 28 Adolf Reichweinstrasse, Jena, 69, German Democratic Republic.*

KÖHLER, Wolfgang, psychologist, b. Reval, Esthonia, Jan. 21, 1887; s. Franz and Minni (Girgensohn) K.; student Gymnasium, Wolfenbüttel, 1896-1905, Tübingen, Bonn and Berlin univs. Germany, 1905-09; came to U. S., 1935; m. Lili Harleman, July 9, 1927; 1 dau., Karin. Privatdozent, Frankfurt on Main, Germany, 1912-21; dir. anthropoid sta., Tenerife, Canary Islands, 1913-20; prof. psychology, Göttingen U., Germany, 1921-22; prof. psychology and philosophy Berlin U., Germany, 1922-35; prof. psychology Swarthmore Coll., 1935-58; research prof. Dartmouth, 1958—. Mem. Am. Psychol. Assn., Soc. Exptl. Psychologists, Acad. Arts and Scis. (hon.), Am. Philos. Soc., Nat. Acad. Scis. Author: Die Physischen Gestalten in Ruhe and im Stationären Zustand, 1920; Mentality of Apes, 1925; Gestalt Physchology, 1929; Place of Value in a World of Facts, 1938; Dynamics in Psychology, 1940. Investigated psychology of perception and learning; brain physiology; rational determination in humans and animals; suggested electric currents in brain tissue, 1938. Died June 11, 1967.

KOHLRAUSCH, Friedrich Wilhelm Georg, German physicist; b. Rinteln, Germany, Oct. 14, 1840; s. Rudolf Kohlrausch; ed. Erlangen, Göttingen (both Germany). Became docent Frankfort (Germany) Physikalischer Verein, 1864; became prof. physics, Göttingen, 1866, Zurich (Switzerland) Polytechnikum, 1870, Würzburg, Germany, 1875, Strasbourg (now in France), 1880; named hon. prof. U. Berlin, 1900. Pres. Imperial Physico-tech. Inst., Berlin, 1895. Fellow Royal Soc., 1895. Author: Leitfaden der pratkischen Physik (widely used handbook on practical phys. method), 1870; über den Leitungswiderstand des Quecksilbers, 1888; Das Leitvermögen der Elektrolyte in besondere der Lösungen, 1898. Improved measuring methods in physics, especially in magnetism and electricity; constructed measuring instruments; research on elasticity of matter, electrolytic conductivity, reflection of light, definition of Ohm and electro-chem. equivalents. Died Marburg, Germany, Jan. 17, 1910.

KOHLRAUSCH, Rudolf-Hermann-Arndt, German physicist; b. Göttingen, Germany, Nov. 6, 1809; children include Fredrich Wilhelm Georg; tchr. math. and physics, Lüneberg, Rinteln, Cassel, Marburg (all Germany); became prof. physics U. Erlangen (Germany), 1857. Author: (with Weber) Electrodynamischen Massbestimmungen, 1857. Pioneer in work in electric currents which led to absolute system of elec. measurement; made (with Weber) 1st mech. measurement of electric current; measured electroscopic force with electrometer, 1848; measured velocity of light, 1855; formulated math. theory for currents named for him, 1857. Died Erlangen, Mar. 9, 1858.

KOHLS, Glen Milton, Am. entomologist; b. Vesta, Minn., Oct. 23, 1905; s. Martin Fredrick and Annie (Engeman) K.; B.S., Mont. State Coll., 1929; M.S., U. Minn., 1937; D.Sc. (hon.), Mont. State U., 1966; m. Jeannette Lettice Johnson, Oct. 7, 1931; children—Carol Virginia (Mrs. Richard W. Banghart), Margaret Ann (Mrs. Christopher Cottle), Glenda Sue. Lab. asst. Mont. Bd. Entomology, 1927-31; entomologist USPHS, Rocky Mountain Lab., Hamilton, Mont., 1931-49, sr. sanitarian, sanitarian dir., 1941—. Recipient medal U. S. A. Typhus Commn., 1945. Fellow A.A.A.S.; mem. Entomol. Soc. Am., Am. Soc. Parasitologists, Entomol. Soc. Washington, Soc. Systematic Zoology. Asst. editor Jour. Parasitology, 1954-64. Research, numerous publs. on classification of ticks and their role in causation of disease, tick borne disease of man and animals, scrub typhus in New Guinea, Burma, India, China. Home: 620 S. 3d St. Office: Rocky Mountain Lab., Hamilton, Mont. 59840.*

KOHLSCHÜTTER, Arnold, German astronomer; b. Halle, July 6, 1883; s. Ernst and Helen (Spielberg) K.; ed. univs. Heidelberg, Berlin, Strasbourg, Gottingen; Ph.D.; m. Charlotte Franz, 1913; 1 dau., Ingeborg. With Hamburg-Bergedorf Obs., Mt. Wilson Obs., Pasadena, Cal.; prof. astronomy, dir. Obs., U. Bonn. Mem. Bonn Astron. Soc. Author: Methode der spektroskopirchen Parallaxen zur Entferungsbestimmung weit entfernter Sterna; Stern-Katalog. Home: Königstrasse 16. Office: Stermwarte Bonn, Bonn, Germany.

KOHLSTAEDT, Kenneth George, Am. physician; b. Indpls., May 10, 1908; s. Albert W. and Flora E. (Kuerst) K.; B.S., Ind. U., 1930, M.D., 1932; m. LaVerne Freeman, July 14, 1935; children—Emilie (Mrs. William Rubright), Kathleen (Mrs. Anthony G. Laos). Asst. supt. Indpls. City Hosp. 1936-44; dir. Lilly Lab. for Clin. Research, Marion County Gen. Hosp., Indpls., 1945-60; dir. clin. research div. Eli Lilly & Co., Indpls., 1954-60, exec. dir. med. research, 1960-64, v.p. med. research, 1964—; prof. medicine Ind. U. Sch. Medicine, 1952—. Diplomate Am. Bd. Internal Medicine. Mem. Am. Heart Assn. (award of merit 1959), A.C.P. (regent 1964—), Am. Fedn. Clin. Research, A.M.A., Central Soc. Clin. Research, Am. Clin. and Climatol. Soc., Marion County Med. Soc., Ind. State Med. Assn., Soc. for Exptl. Biology and Medicine, N.Y. Acad. Scis., Alpha Omega Alpha. Contbr. articles to tech. jours. Co-discoverer angiotensin; research on ion exchange resin for treatment hypertension and congestive heart failure, kidney extracts and other antihypertensive agts., clin. pharmacology; developed technique for intra-arterial infusion blood for treatment shock. Home: 645 E. 80th St., Indpls. 46240. Office: 740 S. Alabama St., Indpls. 46206.*

KOHMAN, Truman Paul, Am. chemist; b. Champaign, Ill., Mar. 8, 1916; s. Edward F. and Marguerite (Villepigue) K.; A.B. cum laude, Harvard, 1938; Ph.D., U. Wis., 1943; m. Jane Sievers, Oct. 13, 1945; children—Leslie Jane, Paulette Lynn, Steven Truman. Teaching, research asst. U. Wis., 1938-42; research staff U. Chgo., 1942-44; chemist Hanford Engr. Works, DuPont Co., Richland, Wash., 1944-45; staff Argonne Nat. Lab., 1945-46; fellow Inst. Nuclear Studies, 1946-48; faculty Carnegie Inst. Tech., Pitts., 1948—, prof. chemistry, 1957—, program chmn. earth and astron. scis., 1963-67; vis. prof. Indian Inst. Tech., Kanpur, India, 1962-63. Cons. various corps. and instns. NSF fellow, 1957-58. Fellow Am. Phys. Soc., A.A.A.S.; mem. Am. Chem. Soc., Am. Geophys. Union, Geochem. Soc., Meteoritical Soc., Fedn. Am. Scis., Sigma Xi, Alpha Chi Sigma, Gamma Alpha, others. Research, publs. on theory and techniques of radioactivity, radioactive products of cosmic-ray bombardment of meteorites, meteorite crater investigations, nuclear phenomena in geochemistry. Home: 440 Greenhurst Dr., Pitts. 15243.*

KOHN, Alan Jacobs, Am. zoologist; b. New Haven, July 15, 1931; s. Curtis I. and Harriet (Jacobs) K.; A.B., Princeton, 1953; Ph.D., Yale, 1957; m. Marian S. Adachi, Aug 28, 1959; children—Lizabeth, Nancy, Diane, Stephen. NSF fellow, Yale, 1955-56; research asso. Hawaii Marine Lab., U. Hawaii, 1954-56; W. W. Anderson fellow Bingham Oceanographic Lab., Yale, 1958; asst. prof. zoology Fla. State U., 1958-61; faculty U. Wash., Seattle, 1961—, prof., 1967—; biologist Yale Seychelles Expdn., 1957-58; mem. Internat. Indian Ocean Expdn., 1963. Fellow A.A.A.S.; mem. Am. Malacological Union (v.p. 1966, past chmn. Pacific div.), Ecol. Soc. Am., Am. Soc. Zoologists, Soc. Systematic Zoology, Brit. Ecol. Soc., Marine Biol. Assn. U.K., Malacological Soc. London, Malacological Soc. Japan. Research and publs. on amount of organic matter produced by photosynthetic organisms on coral reefs; factors permitting large numbers of similar species of marine gastropod molluscs to maintain stable populations on coral reefs; factors determining abundance and species diversity of population of marine gastropod genus Conus, chem. nature and pharmacology of venom of Conus, taxonomy of genus Conus using new chronological approach. Home: 10203 241st Pl. S.W., Edmonds, Wash. 98020. Office: Dept. Zoology, U. Wash., Seattle 98105.*

KOHN, Moriz, see Kaposi, Moritz.

KOHN, Walter, physicist; b. Vienna, Austria, Mar. 9, 1923; B.A., U. Toronto (Ont., Can.), 1945, M.A., 1946, LL.D., 1967; Ph.D. in Physics, Harvard, 1948, Indsl. physicist Sutton Horsley Co., Can., 1941-43; geophysicist Koulomzine, Que., 1944-46; instr. physics, Harvard, 1948-50; asst. prof. Carnegie Inst. Tech., 1950-53, asso. prof., 1953-57, prof., 1957-60; prof. physics U. Cal. at San Diego, 1960—. NRC fellow, 1951; NSF fellow, 1958; Guggenheim fellow, 1963; recipient Oliver Buckley prize, 1961; NSF sr. postdoctoral fellow, 1967. Fellow Am. Phys. Soc. (counselor-at-large 1968—). Research on electron theory of solids, collision theory. Office: Dept. Physics, U. Cal. at San Diego, La Jolla, Cal. 92038.

KOHONEN, Teuvo Kalevi, Finnish physicist; b. Lauritsala, Finland, July 11, 1934; s. Väinö and Tyyne (Koivunen) K.; M.Sc., Tech. U. Helsinki, 1957, Dr. Eng., 1962; m. Elvi Anneli Trast, May 16, 1959; children—Virpi, Eevi, Jussi. Research asso. Finnish Atomic Energy Com., 1959-62; faculty Tech. U. Helsinki, 1963—, prof. tech. physics, 1965—; cons. engr. electronics, 1964—. Sec., Research Council for Tech. Scis., Helsinki, 1963-64. Mem. Finnish Acad. for Tech. Scis. Author: Organization of Scientific Research in Finland, 1964; also articles. Research on electrons and positrons, devel. time measuring methods for ultrashort intervals of time, design of digital and other electronic instruments for indls. processes. Home: Hakamäki 2 E 71, Tapiola, Finland. Office: Tech. U. Helsinki, Otaniemi, Finland.*

KOHUT, Heinz, physician; b. Vienna, Austria, May 3, 1913; s. Felix and Else (Lamol) K.; M.D., U. Vienna, 1938; m. Elizabeth Meyer, Oct. 9, 1948; 1 son, Thomas August. Came to U. S., 1940, naturalized, 1945. Practice medicine, specializing in psychiatry, Chgo., 1950—; mem. Inst. Psychoanalysis, Chgo., 1953—; professorial lectr. dept. psychiatry U. Chgo., 1957—. Diplomate Am. Bd. Psychiatry and Neurology. Fellow Am. Psychiat. Assn.; mem. A.M.A., Internat. (v.p. 1965—), Am. (A. A. Brill medal 1965, past pres.) psychoanalytic assns. Author: (with F. D. Seitz) Concepts and Theories of Psychoanalysis, 1963. Research, publs. on psychology of art, music, lit.; basic theory of psychoanalysis; theory and clin. importance of narcissism. Home: 5805 S. Dorchester Av., Chgo. 60637. Office: 180 N. Michigan Av., Chgo. 60601.*

KOIZUMI, Shiro, Japanese mathematician; b. Ueda, Japan, Nov. 15, 1902; s. Kisuke and Kiyo (Shomura) K.; grad. elec. engring. dept. Sch. Sci. and Engring., Waseda U., Tokyo, 1927, Dr.Engring., 1941. Faculty, Waseda U., 1929—, prof. math., 1943—, head prof. dept. applied physics, 1951-60, dir. sci. and engring. research lab., 1958-60. Mem. Am. Phys. Soc. Author: Theory of Function, Algebra of Matrices, 1953; also articles. Research on operational calculus, theory of circuits, mehrpoleitungstheorie. Home: 1691-27 Johsui-Shinmachi Kodairaschi, Tokyo, Japan.*

KOJIMA, Ken-Ichi, geneticist; b. Toyama, Japan, Sept. 17, 1930; s. Seiji and Masako (Hara) K.; B.S., Kyoto (Japan) U., 1953; Ph.D., N.C. State U., 1958; m. Chizuko Yoshimura, May 1, 1955; 1 son, Kenji. Faculty, N.C. State U., Raleigh, 1959-67, prof. genetics, 1964-67; prof. U. Tex., Austin, 1967—; vis. lectr. Kyoto U., 1963; vis. prof. Brown U., 1966. Recipient Sigma Xi Research award, 1960. Mem. Genetics Soc. Am., Soc. Am. Naturalists, Soc. for Study Evolution, Biometric Soc. Author: (with H. Kihara, H. Suemoto) Statistical Genetics, 1959; also articles. Research on genetics of populations as they relate to evolution. Home: 2802 Mountain Laurel Dr. Office: Dept. Zoology, U. Tex., Austin, Tex. 78712.*

KOJIMA, Tadanobu, Japanese physicist; b. Osaka, Japan, Dec. 1, 1923; s. Tomokichi and Toku (Segawa) K.; B.Sc., Osaka U., 1947, D.Sc., 1957; m. Kazuko Ishiwatari, Feb. 19, 1953; children—Junko, Hiroko. Faculty, Osaka City U., 1957-63, asst. prof., 1962-63, chief theoretical physics div. Research Inst. for Atomic Energy, 1957-63; asst. prof. Kobe (Japan) U., 1964-65; prof. Tohoku U., Sendai, Japan, 1965—.

Mem. Phys. Soc. Japan, Am. Phys. Soc. Author: Solid State Physics for Chemists, 1958; Interfacial Phenomena and Lattice Imperfections, 1959; also articles. Theory and expt. on lattice imperfections in ionic solids especially on color centers in alkali halides. Home: 3-9-9 Mukaiyama, Sendai, Japan.*

KOK, Bessel, biophysicist; b. Hardinxveld, Netherlands, Nov. 7, 1918; s. Johannes Evert and Cornelia (Grondys) Kok; Ph.D., U. Utrecht (Holland), 1948; m. Cornelia Hendrika Vogelesang, Feb. 12, 1943; children—Cornelia (Mrs. Bliss Forbush III), Johannes, Allart. With biophysics research group physics inst. Utrecht U., 1942-49, Plant Phys. Inst., U. Wageningen (Holland), 1949-58; biology div. Carnegie Inst. Washington, Stanford, Cal., 1951-52; scientist, head dept. Research Inst. for Advanced Studies, Balt., 1958—, asso. dir., 1964—. Recipient Charles F. Kettering award, 1963. Mem. Am. Soc. Biol. Chemists, Am. Soc. Plant Physiologists. Research, numerous publs. in photosynthesis and related areas, quantum yields, primary photochem. conversion, interactions with respiration, electron transport. Home: 6236 Bellona Av., Balt. 21212. Office: 1450 S. Rolling Rd., Balt. 21227.*

KOKAME, Jun, Japanese physicist; b. Kyoto, Japan, Jan. 18, 1925; s. Teinosuke and Toyo (Itoh) K.; B.Sc., Kyoto U., 1947, D.Sc., 1962; m. Ayako Kawashima, Jan. 10, 1953; children—Eri, Chiaki. Research asso. Inst. for Chem. Research, Kyoto U., 1947-65; asso. prof. Inst. for Nuclear Study, U. Tokyo (Japan), 1965—. Mem. Phys. Soc. Japan, Am. Inst. Physics. Research, publs. in nuclear reaction mechanisms and nuclear structure of light nuclei, scattering of proton and alpha particles; constrn. cyclotron at Kyoto U. Home: RA24, 32, Chuo-2-chome, Nakano-ku, Tokyo, Japan.*

KOKOBUO, Seiji, Japanese zoologist; b. Saitama Prefectre, Japan, 1889; grad. agrl. dept. Hokkaido U., 1911; Sc.D., 1930. Became asst. prof. fishery dept. Hokkaido U., 1930; named prof. Tohoku U., 1950; also dir. Asamushi Marine Lab.; ret., 1953; later became prof. Nihon U. Author: Planktology; Ocean Biology. Research on floating micro-organisms.

KOLAJA, Jiri Thomas, sociologist; b. Brno, Czechoslovakia, Oct. 21, 1919; s. Maxmillian and Valerie (Elias) K.; student Masaryk U., 1945-47, U. Chgo., 1949-51, Cornell U., 1953-54; m. Beata Osiecka, Apr. 8, 1963. Faculty, Talladega (Ala.) Coll., 1954-58, U. Ky., Lexington, 1958-66; prof. McMaster U., Hamilton, Ont., Can., 1966-68; prof. dept. sociology State U. Coll., Brockport, N.Y., 1968—. Mem. Am. Sociol. Assn., Internat. Inst. Sociology. Author: A Polish Factory, 1961; Worker's Council: The Yugoslav Experience, 1965; also numerous articles. Research on problem self-government in work orgns. Office: Dept. Sociology, State U. Coll., Brockport, N.Y. 14420.*

KOLB, Anton, German biologist; b. Erasbach, Nov. 5, 1915; s. Peter and Anna Kolb; Ph.D., U. Erlangen; m. Elisabeth Unterburger, 1948; children—Otto, Peter, Barbara. Asst., Zool. Inst., U. Erlangen, 1945-54; asso. prof. Bamburg Ecole Superieure, 1952-54, full prof., 1958—, rector, 1961-62; dir. Natural Sci. Mus., Bamberg. Mem. German Zoology Assn., German Assn. Mammalogy, Assn. Doctors and Physicians of Erlangen. Author: Die Ammoniten als Dibranchiata. Research and publs. on biology of mammals, especially bats, fossil tracks in Jura limestone. Home: Ebrardstrasse 44, Erlangen. Office: Jesuitenstrasse 2, Bamberg, Germany.

KOLBE, Adolph Wilhelm Hermann, German chemist; b. Elliehausen, Germany, Sept. 27, 1818; pupil of Wöhler, Göttingen, Germany; became asst. to Robert Bunson, Marburg, Germany, 1842, to Lyon Playfair, London, 1845; named prof., Marburg, 1851; prof., Leipzig, Germany, from 1865. Fellow Royal Soc., 1877. Author: Ausführliches Lehrbuch der organischen Chemie, 2 vols., 1855-64; Zur Entwicklungsgeschichte der theoretischen Chemie, 1881; Kurzes Lehrbuch der anorganischen Chemie, 1883. Editor: Handwörterbuch der Chemie; Jour. für praktische Chemie. Research on constitution of organic radicals (led to theory of valency); opponent of Kekulé's structural theory; introduced term synthesis; 1st to synthesize acetic acid, salicylic acid, 1848; research on organic acids, decomposition by electrolysis, synthesis of cacodyl; discovered Kolbe reaction, thus making possible preparation of salicylic acid in quantity, 1859, and eventual low-cost prodn. of aspirin; discovered synthesis of corallin from phenol, 1861; discovered nitromethane, also predicted existence of secondary and tertiary alcohols, 1872. Died Leipzig, Nov. 25, 1884.

KOLBEZEN, Martin J., Am. organic chemist; b. Pueblo, Colo., Apr. 16, 1914; s. Martin and Mary (Stepan) K.; B.S., Colo. State U., 1939; M.S., U. Utah, 1941, Ph.D., 1950; m. Fredrica Ruth Jones, Apr. 26, 1953. Research asst. U. Utah, 1939-41, research fellow, 1948-50; NRC fellow Ia. State Coll., 1941-42; chemist U. S. Bur. Mines, 1942-44; with U. Cal. Citrus Expt. Sta., Riverside, 1950—, chemist, lectr., 1964—. Mem. Am. Chem. Soc., Am. Inst. Biol. Scis., Am. Phytopath. Soc., Sigma Xi, Phi Kappa Phi. Research, publs. on chemistry and mode of action of pesticides, residue analyses, devel. analytical methods; climatic and biol. breakdown of pesticides; post-harvest preservation, decay prevention in fresh produce. Home: 4721 Monroe St., Riverside, Cal. 92504.*

KOLDERUP, Niels-Henrik, Norwegian geologist; b. Bergen, Jan. 31, 1898; s. Carl Fredrik and Kitty (Pedersen) K.; ed. univs. Oslo, Bergen, Göttingen, Vienna; Ph.D.; m. Else Krabbe, June 9, 1928; children—Lilly, Kitty, Else Grethe, Gedske. Lectr. geology U. Bergen, 1919, head seismol. sta., 1930, prof., 1939. Mem. Norwegian Acad. Sci., London Soc. Geologists. Research and publs. on applied geology, petrology, seismology. Home: Olav Rye 17. Office: Geological Institute, P.B. 2642, Bergen, Norway.

KOLESNIKOV, Boris Pavlovich, Russian geog. botanist, forestry engr.; b. St. Petersburg, 1909; grad. Agrl. Faculty, Far Eastern U. (now Far Eastern Forestry Inst.), Vladivostok, 1931, postgrad. dept. dendrology, until 1934. With dept. forestry Far Eastern Forestry Inst., 1931-34; asso. bot. sect. Far Eastern br. USSR Acad. Sci., 1934-35, sr. asso., 1935-39, sr. asso. forestry group No. Base, 1939-41, head bot. study group Far Eastern Mountain Taiga Sta., 1941—, head lab. forestry Far Eastern br, 1951, now asso. Inst. Biology, Urals br. Author: The Vegetation of the Tachinzhan Mountain Range, 1937; The Vegetation of the Khabarovsk Kray, 1948; The Flora and Fauna of the Southern Half of the Soviet Far East, 1949; Nature Conservation in Poland, 1962; co-author: The Forest Resources of the Maritime Kray, 1947. Research and publs. on dendrology, phytocenology of forests, origin, devel. and present dynamics of forests in Far East. Address: Inst. Biology, Urals Br., USSR Acad. Sci., Sverdlovsk, RSFSR, USSR.

KOLESNIKOV, Ivan Stepanovich, Russian surgeon; b. 1901; grad. Leningrad Mil. Med. Acad., 1931; postgrad. Kirov Mil. Med. Acad., from 1931; D.Med. Sci., 1946. Jr., sr. instr. Fedorov Hosp. Surgery Clinic, Kirov Mil. Med. Acad., Leningrad, until 1952; head chair hosp. surgery Kirov 1st Mil. Med. Acad., Leningrad, 1953—. Sci. dir. surg. dept. Leningrad Tb Research Inst., RSFSR Ministry Health. Recipient Lenin prize, 1961. Author: Surgical Treatment of Lung Cancer, 1957; Resection of the Lungs in Tuberculosis, 1958; Resection of the Lungs, 1960; Soviet Medical Experience in the Great Patriotic War 1941-45, Vols. 9-10; co-author: Operative Treatment of Deep Burns, 1960. Mem. editorial council Surgery; co-editor Surgery sect. Large Med. Ency., 2d edit.; co-editor An Atlas of Gunshot Wounds. Devised Kolesnikov Test (for detection hidden hemolysis). Address: Kirov 1st mil. Med. Acad., ulitsa Lebedeva 6, Leningrad, USSR.

KOLESOV, Semen Georgievich, Russian vet. surgeon; b. 1903; grad. Moscow Vet. Inst., 1933; D.Vet. Sci. Asso., State Sci. Control Inst. Vet. Preparations, USSR Ministry Agr., 1933—, head lab. anthrax and Listerellosis preparations, 1937—, dep. dir for sci. work, 1953-58, dir., 1958—, also mem. sci. tech. council. Chmn., Interdepartmental Anthrax Commn. Mem. All-Union Lenin Acad. Agrl. Sci. (coordinating council). Author: Drying of Microorganisms and Biopreparations; Anabiosis of Pathogenic Microorganisms. Mem. editorial bd. Vet. Ency., Veterinarian. Address: USSR Ministry of Agr., Sovkhoz Mikhaylovsky, Mikhaylovskoe 1, Moskovskaya oblast, USSR.

KOLHÖRSTER, Werner, physicist, astronomer; b. 1887. With Potsdam Obs., Germany. Ascended 9000 miles in balloon and observed radiation to be 12 times greater than at sea level. Died 1946.

KOLIN, Alexander, biophysicist; b. Odessa, Russia, Mar. 12, 1910; s. Rudolph Alexandrovich and Luba (Gershberg) K.; student U. and Inst. Tech., Berlin, Germany, 1929-33; Doctorate in Physics, German U. Prague (Czechoslovakia), 1934; m. Renée Marie Laure Bourcier, Aug. 8, 1951. Came to U. S., 1934, naturalized, 1940. Faculty, Coll. City N.Y., 1941-44, Columbia, 1944-46, N.Y. U., 1945-46, U. Chgo., 1946-56; faculty U. Cal. at Los Angeles, 1956—, prof. biophysics, 1961—. Recipient John Scott award for invention electromagnetic flow meter, 1965; Albert F. Sperry medal, 1967. Mem. Am. Phys. Soc., Am. Assn. Physics Tchrs., Marine Biol. Lab., A.A.A.S., Biophys. Soc., Am. Physiol. Soc., Sigma Xi. Author: Physics: Its Laws, Ideas and Methods, 1950; also numerous articles. Developed electromagnetic method blood flow determination, density gradient, pH gradient and conductivity gradient methods electrophoresis, continuous flow eclectrophoresis stabilized by electromagnetic rotation; discovered electromagnetophoresis. Home: 1298 Stradella Rd., Los Angeles 90024.*

KOLLAR, Edward Joseph, Jr., Am. psychiatrist; b. Streator, Ill., Oct. 22, 1918; s. Edward Joseph and Myrtle (Blakemore) K.; B.S. cum laude in Zoology, U. Ill., M.D. with high honors, 1945; M.S. in Zoology, Yale, 1942; m. Ann McKelvey, Aug. 4, 1950; children—Judith Ann, Peter Randall, Jeffrey Joseph, Alexandra. Chief psychosomatic service Wadsworth VA Gen. Med. and Surg. Hosp., Los Angeles, 1955-60, cons. psychiatrist, 1960—; faculty U. Cal. at Los Angeles, 1955—, vice-chmn. dept. psychiatry, 1960—, asso. prof., 1965—, asso. med. dir. Neuropsychiat. Inst., 1960—. Cons. to armed forces, USPHS, 1952—. Diplomate Am. Bd. Psychiatry and Neurology. Fellow Am. Coll. Psychiatry, Am. Psychiat. Assn. (chmn. com. disaster and civil def.); mem. So. Cal. Psychiat. Assn., Group for Advancement Psychiatry, Soc. Med. Consultants to Armed Forces, Alpha Omega Alpha, Phi Beta Kappa. Research, publs. on ulcerative colitis patients, super-obese patients, electro encephalographs of the aged, measurement of stress in starvation and sleep deprivation, group behavior of chimpanzees. Home: 1200 Tigertail Rd., Los Angeles 90049.*

KOLLATH, Rudolf Johannes Gustav, German physicist; b. Hohensalza, Dec. 21, 1900; s. Johannes and Margarete (Bion) K.; ed. in engring. Danzig Tech. Inst.; m. Bettina Lange, Aug. 2, 1952; children—Jurgen, Peter. With AEG Research Inst., Berlin, 1928-41; chem. work, Norway, 1941-44; with Betatron devel., London, 1944-46; cons. sci. U. Hamburg, 1948-53; prof., dir. Inst. Exptl. Physics, U. Mainz, 1953—. Mem. German Physics Soc. Author: Wirkungsquerschnitt gegenüber langsamen Elektronen und Ionen; Sekündärelektronen-Emission von 1937-41; Teilchenbeschleuniger, 1962. Research on electrons and ions of low energy and their interaction with matter, particle accelerators. Home: am Eselweg 51. Office: Jacob Welder Weg 11, Mainz, Germany.

KOLLATH, Werner Georg, German microbiologist; b. Gollnow, Germany, June 11, 1892; s. Georg and Marie (Riedel) K.; student univs. of Leipzig, Freiburg, Berlin, Kiel, Leipzig, 1911-20; M.D., 1920; m. Elisabeth Rossdeutscher. Asst. psychiatry U. Marburg (Germany), 1920-22; asst. Hygiene Inst., U. Breslau (now Wroclaw, Poland), 1923-25; prof. hygiene and bacteriology U. Rostock (Germany), 1934-45; dir. Mecklenburg Med. Exam. Office, until 1945; anti-epidemic officer, 1945-46; forced to leave univ., 1946; fled to W. Germany, 1947; with Karolinska Inst., Stockholm, 1948-49; Locarno, 1949-50; lectr., S.Am., 1950-51; ret., in Freiburg, 1951—. Mem. Med. Soc. Santiago (Chile), German Dental Soc. (corr.). Author: Lehrbuch der Hygiene, 1937; Die Ordnung unserer Nahrung, 1942; Der Vollwert der Nahrung, 1949-60; Zivilizationsbedingte Krankheiten und Todesuvachen, 1958. Developed nutritional foods, especially breakfast foods, various dietary aids.

KOLLE, Frederick Strange, plastic surgeon; b. Hanover, Germany, Nov. 22, 1871; s. John and Bertha (Schaare) K.; M.D., L.I. Coll. Hosp., 1893; hon. M.E., Nat. Coll. Electro-therapeutics, 1897; m. Loretto Elaine Duffy, 1899. Asst. physician Contagious Disease Hosp., Bklyn., 1894; practiced in Bklyn., from 1894. One of 1st X-ray investigators in U. S., 1896; chief instr. dept. electro-therapeutics, Elec. Engring. Inst., New York, 1896-1900; asso. editor Electrical Age, 1897-1902; radiographer to M.E. Hosp., Brooklyn. Inventor: Radiometer; Kolle X-ray coil switching devices; dentaskiascope; oesophameter; folding fluoroscope; X-ray printing process; Kolle focus tube; direct-reading X-ray meter and many instruments used in plastic surgery. Author: The Recent Roentgen Discovery, 1896; The X-Rays, Their Production and Application, 1896; Medico-Surgical Radiography; Subcutaneous Hydrocarbon Prostheses, 1908; Plastic and Cosmetic Surgery; Hakon Jare, 1911. Died May 10, 1929.

KOLLE, Wilhelm, German bacteriologist; b. Hannover, Germany, Nov. 2, 1868; worked at Koch's Inst., Berlin, several years; engaged in rinderpest work, S. Africa; became prof., Bern, Switzerland, 1902. Author: Lehrbuch der klinischen Untersuchungsmethoden, 1902. Editor: (with Wassermann) Handbuch der pathogenen Mikroorganismen, 3 edits. Research in bacteriology, immunology, chemotherapy; used (with Pfeiffer) killed cultures against typhoid, 1896; treated (with Ritz) exptl. syphilis in rabbits with bismuth, 1919; introduced Asiatic cholera vaccine. Died Wiesbaden, Germany, May 10, 1935.

KOLLER, Carl, ophthalmologist; b. Schüttenhofen, Bohemia (now Susice, Czechoslovakia), Dec. 3, 1857; grad. U. of Vienna, 1882; m. Laura J. Blum, Oct. 30, 1893; children—Louis Richard, Hortense (Mrs. James H. Becker). Came to U. S., 1888. Ophthalmic surgeon Mt. Sinai Hosp. and other instns., N.Y.C. Recipient Gold medals U. Heidelberg (Germany), Am. Acad. Ophthalmology, Otology and Rhinology. Mem. A.M.A., Am. Ophthal. Soc. (Gold medal), N.Y. Acad. Medicine (Gold medal); hon. mem. Soc. of Physicians of Vienna, Soc. of Physicians of Budapest, Royal Acad. of Medicine, Rome, Am. Socs. of Physiology and Pharmacology. Introduced cocain as a local anesthetic in eye operations, 1884; thus inaugurating era of local anaesthesia, for operations in the various branches of medicine and surgery; wrote many articles pertaining to biology and ophthalmology. Died N.Y.C., Mar. 21, 1944.

KOLLER, Theodor Hermann, gynecologist, obstetrician; b. Winterthur, Apr. 19, 1899; s. Hermann and Dora (Aeby) K.; ed. univs. Lausanne, Frankfort, Zurich; M.D.; m. Gertrud von Seutter, 1932; children—Hans, Theodor, Margrit. Titular prof., Zurich, 1941; full prof. gynecology and obstetrics, Basel, 1942, dean Med. Faculty, 1951. Mem. Swiss, Paris, Italian socs. obstetrics and gynecology, German Acad. Naturalists (Leopoldina), Swiss Soc. Natural Sci., German Gynecol. Soc., others. Author: Mutter und Kind, 1939-63; Lehre der Geburtschilfe, der Gynäkologie und der Geburtschilflichen Operationen, 1948-56; Thrombose und Embolie in Geburtshilfe und Frauenheilkunde, 1957. Research on endocrinology, operative gynecology, reanimation of cancer, thromboembolic diseases, family planning. Home: Am Ausserberg 8, Riehen b/Basel. Office: Schanzenstrasse 46, Basel, Switzerland.

KÖLLIKER, Rudolf Albert von, see VON KÖLLIKER, Rudolf Albert.

KOLLMANN, Julius, German anatomist, anthropologist; b. Dillingen, Feb. 24, 1834; grad. U. Munich, 1859; docent U. Munich, 1862; extraordinary prof. U. Basel, 1870, ordinary prof. anatomy, from 1878, emeritus, 1913. Author: Plastische Anatomie, 1886; Lebrbuch der Entwicklungsgeschichte, 1898; Handatlas der Entwicklungsgeschichte, 2 vols., 1907. Attempted to reconstruct features of earlier humanity; saw in the Pygmy the oldest human type. Died Basel, June 24, 1918.

KOLLROS, Jerry John, embryologist; b. Vienna, Austria, Dec. 29, 1917; s. Jakub J. and Terezie (Hruby) K.; came to U. S., 1920, naturalized, 1926; S.B., U. Chgo., 1938, Ph.D., 1942; m. Catharine Zenker Lutherman, Sept. 19, 1942; children—James Carl, Peter Richard. Research asst. U. Chgo., 1943-45, instr., 1945-46; faculty U. Ia., Iowa City, 1946—, prof. embryology, 1957—, chmn. dept. zoology, 1955—. Mem. cell biology study sect. NIH, 1960-64; cons. animal embryology Am. Coll. Dictionary, 1946, 64. Mem. Am. Soc. Zoologists (past treas.), Am. Assn. Anatomists (exec. com. 1962-66), Soc. for Study Growth and Devel., Soc. for Cell Biology, Am. Soc. Naturalists, Soc. for Exptl. Biology and Medicine, Am. Assn. U. Profs., Phi Beta Kappa, Sigma Xi. Research, publs. on maturation amphibian nervous system in presence thyroid, regenerative capacities developing brain, spinal cord, limbs in tadpoles. Home: 331 Melrose Ct., Iowa City, Ia. 52240.*

KOLM, Henry Herbert, physicist; b. Vienna, Austria, Sept. 10, 1924; s. Richard and Ella (Jellenik) K.; came to U. S., 1939, naturalized, 1941; S.B., Mass. Inst. Tech., 1950, Ph.D., 1954; m. Elizabeth Olmstead Cushing, June 6, 1953; children—Margaret, Juliet, Edna, Cornelia. Lab. research Mass. Inst. Tech., Cambridge, 1950—, physicist Nat. Magnet Lab., 1958—, mem. dirs. group, 1958—. Mem. Am. Inst. Physics, Am. Phys. Soc., Sigma Xi. Author: (with B. Lax, F. Bitter, R. Mills) High Magnetic Fields, 1962; also articles. First observed quantized vorticity in superfluid helium; generated highest pulsed magnetic field, also highest continuous magnetic field; developed method for using helium above its critical point as heat transfer medium at very low temperatures; inventor switch for very high electric currents. Home: Weir Meadow Rd., Wayland, Mass. 01778. Office: Mass. Inst. Tech., Cambridge, Mass. 02139.*

KOLMOGOROV, Andrey Nikolaevich, Russian mathematician; b. Apr. 25, 1903; grad. Moscow U., 1925. Former asst. prof. Natural Sci. Faculty, Paris U.; dean Mechanics and Math. Faculty, Moscow U., 1931—; sr. asso. Math. Inst., USSR Acad. Scis. Decorated Order of Lenin; recipient Stalin prize, Baltsan prize, 1962. Mem. USSR Acad. Scis., Nat. Acad. Scis. Author: Basic Concepts of Probability, 1936; Irrational Numbers, 1941; Newton and Modern Mathematical Thinking, 1946; co-author: Introduction of the Theory of Functions of a Real Variable, 3d edit., 1938; Algebra, part I, 1939. Dep. chief editor Reports of USSR Acad. Sci. Research and publs. on theory of real variable functions. Address: Math. Inst., USSR Acad. Scis., 1-y Akademichesky prospect 28, Moscow, USSR.

KOLODNER, Ignace I(zaak), mathematician; b. Warsaw, Poland, Apr. 12, 1920; s. Israel and Brucha (Gornostajski) K.; student U. Nancy (France), 1937-39; Diplome d'Ingenieur, U. Grenoble (France), 1940; Ph.D., N.Y.U., 1950; m. Ethel Zelnick, June 10, 1948; children—Richard David, Paul Robert; m. Dorothy C. Thomas, Apr. 15, 1968. Came to U. S., 1943, naturalized 1944. Mem. staff Courant Inst. Math. Scis., N.Y. U., 1948-56; prof. math. U. N.M., 1956-64; prof. math., head dept. Carnegie-Mellon U., 1964—. Mem. Am. Math. Soc., Am. Phys. Soc., Math. Assn. Am., Soc. Indsl. and Applied Math., Soc. Natural Philosophy, Am. Assn. U. Profs. Author numerous articles in field. Study of integral and differential equations; math. physics. Home: 5307 Northumberland St., Pitts. 15217.

KOLODNEY, Morris, chem. engr.; b. N.Y.C., Sept. 24, 1911; s. David L. and Eva (Gisses) K.; B.S., Coll. City N.Y., 1932, Chem.E., 1933; M.S., Columbia, 1936, Ph.D., 1939; m. Edith Sylvia Charney, Oct. 5, 1935; children—Steve E., Elizabeth E. Instr., Coll. City N.Y., 1934-43; resident scientist Manhattan Dist., Los Alamos Sci. Lab., 1943-45, asst. plant supt., 1945-46; faculty Coll. City N.Y., 1946—, prof. electrochem. and metallurgy, 1953—. Cons. to pvt. cos. Recipient Townsend Harris medal Coll. City N.Y. Alumni Assn., 1965. Mem. Am. Inst. Chem. Engrs., Am. Inst. Mech. Engrs., Am. Soc. Metals, Electrochem. Soc., Am. Assn. U. Profs., Inst. Mining, Metall. and Petroleum Engrs., Sigma Xi. Contbr. articles to tech. jours. Developed electrolytic method for making plutonium metal; research on high temperature materials for nuclear reactors. Home: 187 Wales Av., River Edge, N.J. 07661. Office: Coll. City N.Y., N.Y.C. 10031.*

KOLODZIEJCZAK, Jerzy Michal, Polish physicist; b. Warsaw, Poland, June 17, 1935; s. Stanislaw and Helena (Kobryn) K.; doctor physics Warsaw U., 1958, docent habil., 1963; m. Elzbieta Barbara Golebiewska, Sept. 9, 1958; 1 dau., Stella Malgorzata. Staff, Warsaw U., 1953-58, lectr., 1962—; staff Inst. Physics, Polish Acad. Scis., 1958—, head solid state spectroscopy div., 1964—, sci. dir. Inst. Physics, 1967—; staff magnet lab. Mass. Inst. Tech., 1961-62. Head solid state physics council physics com. Polish Acad. Scis., 1966—. Research and publs. on physics of semiconductor, transport phenomena, scattering processes, band structures, magneto-optical effects in semiconductors. Home: 8 Dlugosza. Office: 37 Zielna, Warsaw, Poland.*

KOLOMENSKY, Andrey Aleksandrovich, Russian physicist; b. 1920; grad. Gorky U., 1942; D.Physico-Math. Sci. Asso., later sr. asso. Physics Inst., USSR Acad. Sci., 1946—, sect. head, 1957—; prof. Moscow U., 1959—. Del., Internat. Conf. on High Energy Accelerators, Dubna, 1963. Recipient Lenin prize, 1959. Author: Research on the Theory of Particle Movement in Modern Circular-Orbit Accelerators, 1960. Research and publs. on accelerators, electrodynamics of anisotropic media and electron plasma; co-developer 10-Bev proton synchrotron. Address: Physics Inst., USSR Acad. Sic., Leninsky pr. 53, Moscow, USSR.

KOLOMIYCHENKO, Mikhail Isidorovich, Russian surgeon; b. Shpola (now Cherkassy Oblast), 1892; grad. Med. Faculty, Kiev U., 1919; D.Med. Sci. Intern, asst., lectr. dept. surgery Kiev Med. Inst., 1922-35, dep. dir. for studies, 1930-35, head chair gen. surgery, 1955—; dep. dir. for studies, head chair surgery Kiev Postgrad. Med. Inst., 1935-55. Chief surgeon Ukrainian Ministry Health, 1950-59. Decorated Order of Lenin (2). Mem. Internat. Assn. Surgeons, All-Union (hon. past bd. mem.), Ukrainian (dep. chmn.) socs. surgeons, Kiev United Med. Soc. (cofounder, past sec.). Mem. editorial council Surgery, Clin. Surgery. Research and publs. on surg. treatment of pericarditis, cardiospasm, pancreatitis, suppurative conditions of pleura, esophagoplasty, heart injuries, history of surgery, med. deontology; among 1st to do heart operations in Ukraine. Address: Kiev Med. Inst., b. Tarasa Shevchenko 18, Kiev, Ukrainian SSR, USSR.

KOLOS, Wlodzimierz, Polish physicist; b. Pinsk, Poland, Sept. 9, 1928; s. Pawel and Elzbieta (Saszko) K.; M.Sc., U. Poznan, 1950; Ph.D., U. Warsaw, 1953; m. Maria Wozna, Oct. 5, 1952; children—Anna, Robert. Research asst. U. Poznan, 1950-51, U. Warsaw, 1951-53; with Inst. Phys. Chemistry, Inst. Physics, Polish Acad. Sci., Warsaw, 1953-59; faculty Inst. Nuclear Research, Warsaw, 1959-66; faculty U. Warsaw, 1961—, prof. theoretical chemistry, 1965—. Vis. asso. prof. physics U. Chgo., 1965-67. Mem. Polish Phys. Soc., Polish Chem. Soc.; annual prize of International Academy of Quantum Molecular Science, 1967. Studies, publs. on quantum-mech. calculations of electronic structure of small molecules, especially hydrogen molecule. Home: 36 Pawia, Warsaw 1, Poland.*

KOLOSOV, Nikolai Grigorievich, Russian histologist; b. Apr. 29, 1897; grad. Kazan U., 1924. With dept. histology Kazan U., 1924-40; head chair histology Stalingrad, 1940-45; prof. Saratov Med. Inst., 1945-50; staff Physiology Inst., USSR Acad. Sci., 1950—; prof. histology and embryology Leningrad U., 1953—. Corr. mem. USSR Acad. Sci., USSR Acad. Med. Scis. Author: Some Chapters on the Autonomic Nervous System, 1948; The Innervation of the Viscera and the Cardiovascular System, 1954; Afferent Innervation of the Human Alimentary Canal, 1962. Editorial bd. Archives Anatomy, Histology, and Embryology. Research in neurohistology, double innervation of alimentary canal and pelvic organs, afferent innervation of vegetative ganglia and vegetative neurons, structure of autonomous nervous system and its interactions with central nervous system. Office: Leningrad U., Leningrad, USSR.

KOLOSVARY, Gabriel, Hungarian zoopaleontologist; b. Kolozsvár, Hungary, Aug. 18, 1901; s. Valentin and Alice (Stein) K.; Dr. philosophiae scientiarum summa cum laude, U. Szeged (Hungary), 1926; m. Gizella Vida, May 22, 1926; children—Valentin, Gabriel. Staff, Inst. Systematic Zoology, U. Szeged, 1923-29, dir., prof., 1954—; museologist Hungarian Nat. Mus., Budapest, Hungary, 1929-54. Decorated Order of Hungarian Hydrobiol. Assn., 1964; Order Eminent Pedagogist, 1956. Mem. Hungarian Acad. Sci., Société Linnéenne de Lyon, Société Biol. Hungary. Aurhor: Opilioniden of Hungary, 1929; Fisheries of the Tisza-Valley, 1938; also numerous monographs, articles. Editor: Tiscia, 1965—. Research on natural philosophy, arachnology, marine biology, paleontology, hydrobiology of fresh water. Home: 15 Boross József, Szeged, Hungary.*

KÖLREUTER, Joseph Gottlieb, German botanist; b. Sulz/Neckar, Germany, Apr. 27, 1733; ed. Berlin, Leipzig (both Germany); M.D., Tübingen, Germany, 1775; prof. natural history, Karlsruhe, Germany, 1768-1806, dir. bot. gardens, 1764-86. Author: Vorläufige Nachricht von einigen das Geschlecht der Pflanzen betreffender Versuchen und Beobachtungen, 1761-66. Pioneer in plant hybridization; recognized importance of insects and wind in plant fertilization; applied Linnaeus' sexual system to lower plants; early work on question of sexes in plants. Died Karlsruhe, Nov. 12, 1806.

KOLSTAD, George Andrew, physicist; b. Elmira, N.Y., Dec. 10, 1919; s. Charles Andrew and Rose (Haesloop) K.; B.S. in Physics magna cum laude, Bates Coll., 1943; Ph.D., Yale, 1948; m. Christine Joyce Stillman, July 22, 1944; childdren—Charles Durgin, Martha Rae, Peter Kenneth. Research asst. Eastman Kodak Co., Rochester, N.Y., 1937-39; faculty Yale, 1948-50, Sheffield-Loomis Fellow, 1947-48; physicist research div. U. S. AEC, Washington, 1950—. Mem. European-Am. Nuclear Data Com., 1960—; U. S. del. Internat. Nuclear Data Com., IAEA, 1962—. Recipient U. S. AEC Spl. Service award 1961. Fellow Am. Phys. Soc.; mem. Sigma Xi, Phi Beta Kappa. Research in pure and applied physics, mathematics; mass spectroscopy; radar countermeasures; a linear accelerator for prodn. neutrons and high intensity gamma rays. Home: Oak Hill, Laytonsville, Md. 20760. Office: U.S. AEC, Washington, 20545.

KOLSTAD, Per, Norwegian gynecologist; b. Porsgrunn, Norway, Aug. 21, 1925; s. Peter Lund and Ida (Larsen) K.; M.D., U. Oslo, 1950, Dr.med., 1964; m. Ottilia Gjärdman, Apr. 19, 1953; children—Agneta, Kim. With Norwegian Army Mobile Surg. Hosp., Korea, 1951-52; resident Drammen Hosp., 1952-54, Olso U. Hosp., 1945-55, Tönsberg Hosp., 1955-59, Norwegian Radium Hosp., 1959-63; research fellow Norwegian Cancer Soc., 1962-64, with Norsk Hydro's Inst. for Cancer Research, 1962-63, Karolinska Sjukhuset, Kvinnokliniken, Stockholm, 1963-65; asst. chief gynecol. dept. Norwegian Radium Hosp., Oslo, 1965-67, chief, 1967—. Nat. pres. World Assn. for Gynecol. Cancer Prevention. Mem. Oslo, Norwegian gynecol. socs., World Fedn. for Obstetrics and Gynecology, No. Assn. Obstetrics and Gynecology. Author: Therapeutic Abortion, 1956; Vascularization, Oxygen Tension and Radiocurability in Cancer of the Cervix, 1964; Cytologic, Vascular and Histologic Patterns of Dysplasia Carcinoma, 1966. Research on place of therapeutic abortion in family planning; diagnosis of early forms of cervical cancer especially place of colposcopy and cytology; treatment of gynecol. cancer. Home: 5 Montebellobk, Oslo. Office: Norwegian Radium Hosp., Oslo, Norway.*

KOLSTER, Frederick August, engr.; b. Geneva, Switzerland, Jan. 13, 1883; s. August and Henrietta (Valeton) K.; grad. Harvard, 1908; m. Lillian Ricker, 1908; children—Muriel, Fredericka. Worked in lab. of John Stone, 1902-08, radio-telephony dept. De Forest Telegraph and Telephone Co., 1908-11; radio specialist, organizer and chief radio sect. U. S. Bur. Standards, 1912-31; research engr. Internat. Telephone and Telegraph, 1931-38; cons., from 1938. Mem. Am. Inst. E.E., I.R.E., Am. Geol. Soc. Author: Distributed Capacity of Coils; Measurement of Logarithmic Decrement; The Radio Compass and Its Application to Navigation; Generation and Utilization of Ultra-Short Waves in Radio Communication. Inventor: Kolster decremeter (device to determine wavelength and logarithmic decrement of radio transmission, primarily from ships); Kolster radio compass (instrument to determine bearing of lost radio beacon or radio-equipped ship which was superior to other early compasses); installed 1st lighthouse radio beacons, Ambrose, Uire Island, Sea Girt, 1913. Died circa 1943.

KOLTHOFF, Izaak Mauritis, analytical chemist; b. Almelo, Holland, Feb. 11, 1894; s. Moses and Rosetta (Wysenbeek) K.; Ph.D., U. of Utrecht, Holland, 1918; D.Sc. (hon.), U. Chgo.; Hon. Dr. Degree U. Groningen, 1964. Asst. in chem., U. of Utrecht, 1915-18, conservator, 1918-24, privat docent, 1924-27; prof. analytical chem. and head dept. U. Minn., 1927-62, prof. emeritus, 1962—. Chmn. com. analytical chem. NRC. Comdr. in the order of Oranje-Nassau, 1947; recipient William H. Nichols medal, New York sect. Am. Chem. Soc.; Fisher award Am. Chem. Soc., also Minnesota award, Williard Gibbs medal, 1964; Anachem award 1961; Polarographic medal Polarographic Soc., England, 1964; Kolthoff medal Acad. Pharm. Sci., 1967. Fgn. mem. Royal Flemish Acad. Belgium; corr. mem. Royal Soc. Sci. and Lettres of Bohemia; hon. mem. Finnish Chem. Soc., Czechoslav Chem. Soc., corr. of the Royal Acad. of Scis., Amsterdam; fellow A.A.A.S.; mem. Am. Pharm. Assn. (hon.), Israeli (hon.), Am. chem. socs., Am. Electrochem. Soc., Dutch Chem. Soc., Dutch Pharm. Soc., Internat. Union Pure and Applied Chemistry (v.p.), others. Nat. Acad. Scis., Am. Acad. Arts and Scis., Author: Konduktometrische Titrationen, 1923; The Theory of Indicators, 1926; Potentiometric Titrations, 1932; Volumetric Analysis (2 vols.), 1929, (3 vols.), 1958; pH and Electrometric Titrations, 1941; Textbook on Quantitative Inorganic Analysis, 1935, 43, 52; Acid-Base Indicators, 1936; Polarographic Analysis and Amperometric Titrations, 1941-52; Emulsion Polymerization 1955. Assoc. ed. Jour. Am. Chem. Soc. Co-editor: Treatise on Analysis Chemistry. Contbr. to tech. publs. Research in phys. chemistry; volumetric analysis; kinetics and mechanism of emulsion polymerization; theory of acids and bases. Home: Campus Club, U. Minn., Mpls. 55455.*

KOMAR, Anton Panteleymonovich, Russian physicist; b. Jan. 30, 1904; grad. Kiev Poly. Inst., 1930. With Physico-Mech. Inst., USSR Acad. Sci., Leningrad, 1932-36, with Urals br., 1936-47, with Physics Inst., 1947-50, sr. asso. Physicotech. Inst., 1950—, dir., 1950-57; head chair exptl. nuclear physics Leningrad Poly. Inst., 1951—. Recipient Stalin prize, 1952. Mem. Ukrainian Acad. Sci. Author: The Geometry of Plastic Crystal Deformation Based on Laue Radiograms, 1936; The Ferro-Magnetic Properties of Alloys and the Remote Sequence of Atoms, 1947; Experiments with an Electron and Ion Projector, 1956. Research and publs. on phase transitions in metals, research and devel. accelerators, also various problems of nuclear physics. Address: Physicotech. Inst., Politekhnicheskaya ulitsa 3, Leningrad, USSR.

KOMATSU, Yusaku, Japanese mathematician, biologist; b. Kanazawa City, Japan, 1914; grad. Kanazawa Med. Coll., 1937, Tokyo U., 1940; D.Sc., 1945; M.D., 1948. Became instr. sci. dept. Tokyo U., 1944, also instr. agr. dept., 1947; named prof. Tokyo Tech. U., 1949. Author: Heredity of Blood Type and its Application; Outline of Mathematical Biology. Research in theory of functions, mass genetics.

KOMCZYNSKI, Ludwick, Polish pathologist; b. Jaskotki Village, Poznan, Poland, Aug. 16, 1909; s. Wladyslaw and Jadwiga (Gruszczynska) K.; M.B., U. Poznan, 1935, M.D., 1946; Dozent, Central Qualifying Commn., Warsaw, Poland, 1954; m. Krystyna Janowska, June 27, 1947; children—Hanna Matgorzata, Katarzyna. From asst. to sr. research asst., U. Poznan, 1935-52; chmn. dept. path. anatomy Med. Acad., Biatystok, Poland, 1952—, prof., 1960—, dean Faculty Medicine, 1953-60, rector, 1962—. Decorated Polonia Restituta Cross, Golden Cross Desert, Medal Millenium Poland, Medal for Distinction in Med. Labour. Mem. Polish Soc. Scientists, Anat.-Path. Soc., European Soc. Pathology, Soc. Scientists Biatystok. Editor: Autopsy-Technik, I, II, III; Histopathological Study for Student, I, II, III. Research and publs. on polymers in biology and medicine, morphological changes in organisms, pathology of blood vessels, exptl. oncology. Home: 12 Curie Skiodowska, Biatystok, Poland.*

KOMETIANI, Petr Antonovich, Russian biochemist; b. Oct. 16, 1901; grad. Tbilisi U., 1925. With various geoagrl. and zoovet. insts., 1927-34; prof. Tbilisi U., 1934—; head biochemistry dept. Inst. Physiology, Georgian Acad. Sci., 1934—. Mem. Georgian Acad. Sci. Author: Biology Today and Tomorrow, 1959. Research and publs. on biochemistry of muscle and nerve tissues; made diagram showing possibility of electron conductivity in living tissues, showed ways to reaminate adenylic system. Address: Tbilisi University, prospect Chavchavadze 45, Tbilisi, Gruz. SSR, USSR.

KOMISSARENKO, Vasiliy Pavlovich, Russian endocrinologist, path. physiologist; b. Chernyakhovo (now Kiev Oblast), 1907; grad. Kiev Med. Orderlies Sch., 1926, Kharkov Med. Inst., 1932; postgrad. in path. physiology Ukrainian Research Inst. Exptl. Endocrinology, 1932-35; D.Med. Sci., 1941. Asst. dept. path. physiology Kharkov Med. Inst., 1932-38; dir. Ukrainian Research Inst. Exptl. Endocrinology, 1935-37, Ukrainian Inst. Endocrinology and Organotherapy, 1937-39; dir. lab. endocrinology Inst. Exptl. Biology and Pathology, Ukrainian Ministry Health 1940-41, 44-52, former chief endocrinologist; prof., 1942—, head chair path. physiology Kiev Med. Inst., 1950-55; dir. lab. for endocrine functions Bogomolets Inst. Physiology, Ukrainian Acad. Sci., 1953—. Mem. Ukrainian Acad. Sci., All-Union Soc. Endocrinologists (chmn. 1962—). Author: The Pathogenesis of Insulin Shock, 1943; Material on the Active Mechanism of Insulin, 1943; Introduction to the Clinical Aspects of Disease of the Endocrine Glands, 1950; Adrenal Cortex Hormones and Their Role in the Body's Physiological and Pathological Processes, 1956; Splenin, 1960. Mem. editorial bd. Med. Affairs, Problems of Endocrinology, Sci. and Life. Research and numerous publs. on effect of hormones on body, effect of insulin on central nervous system and adrenalin on cardiovascular system; suggested new hormone preparations kortikotonin and splenin. Address: Inst. Physiology, Ukrainian Acad. Sci., ulitsa Bogomoltsa 4, Kiev, Ukrainian SSR, USSR.

KOMMANDEUR, Jan, Dutch chemist; b. Amsterdam, Netherlands, Nov. 29, 1929; s. Jan and Hendrika (Jorna) K.; B.Sc., U. Amsterdam, 1951, M.Sc., 1955, D.Sc., 1958; m. Elisabeth Eickholz, Mar. 2, 1951; children—Jan, Olivier, Jacob Christiaan. Fellow NRC Can., Ottawa, Ont., 1955-57; chemist Union Carbide Corp., Parma Research Lab., Parma, O., 1958-61; prof. phys. chemistry U. Groningen (Netherlands), 1916—. Mem. Dutch Chem. Soc. Research, publs. on photoconductivity of anthracene, organic semiconductors (perylene-iodine), magnetic and elec. properties of donor-acceptor complexes, electron spin resonance of small radicals and radical pairs, ionic conductivity in iodine. Home: 18 Lekstraat, Groningen, Netherlands.*

KOMMEDAHL, Thor, Am. plant pathologist; b. Mpls., Apr. 1, 1920; s. Thorbjorn and Martha (Blegen) K.; B.S., U. Minn., 1945, M.S., 1947, Ph.D., 1951; postgrad. Bethel Coll.; m. Faye Lillian Jensen, Aug. 5, 1951; children—Kris Alan, Siri Lynn, Lori Anne. Instr., U. Minn., 1946-51; asst. prof. Ohio State U., Columbus, also Ohio Agr. Research and Devel. Center, Wooster, 1951-53; faculty U. Minn., St. Paul, 1953—, prof. plant pathology, 1963—. Guggenheim Found. fellow Waite Agr. Research Inst., U. Adelaide (S. Australia), 1961-62; cons. Minn. Dept. Agr., 1954-60. Fellow A.A.A.S.; mem. Weed Sci. Soc. Am. (award of excellence 1964), Am. Phytopath. Soc., Bot. Soc. Am., Mycol. Soc. Am., Am. Assn. U. Profs., Minn. Acad. Sci., Sigma Xi, Gamma Alpha, Phi Epsilon Phi, Gamma Sigma Delta. Editor-in-chief Phytopathology, 1964-67. Research, publs. on factors affecting flax wilt, role damaged flax seed, role root exudates from peas on infection by pea wilt organism, role quackgrass residues in toxicity to field crops, soil microorganisms in crop sequences in relation to cereal root rots, role seed infection in establishing root and stalk rot of corn, biol. control of seedling blight of corn. Home: 1840 Roselawn Av. W., St. Paul 55113.*

KOMP, William H. Wood, entomologist; b. Yokohama, Japan, Mar. 16, 1893; s. Frederick and Carrie Joanna (Wood) K.; brought to U. S., 1895; derivative citizenship; student Mass. State Coll., 1911, N.Y. U., 1912; B.S., Rutgers U., 1916, M.S., 1917, D.Sc., 1955; fellow in agr. Cornell, 1917; m. Mildred Crowell Sept. 1, 1914; 1 dau., Anita (Mrs. Harry M. Williams). Ensign, USPHS, 1918, advanced through grades to capt., 1944, malaria control; vis. staff mem. Gorgas Meml. Lab,, Republic of Panama, 1931-47; traveling rep. Pan-Am. San. Bur., 1937; research in malaria, from 1921; loaned to Rockefeller Found. Internat. Health Div. for research on yellow fever, Colombia, 1936; cons. on malaria Inst. Inter-Am. Affairs, from 1942. Chmn. com. on entomology Pan-Am. San. Conf., Rio, 1942. Fellow A.A.A.S., Am. Soc. Tropical Medicine and Hygiene; mem. Am. Mosquito Control Assn., Nat. Malaria Soc., Sociedad Venezolano de Ciencias Naturales (corr. mem.), Am. Acad. Tropical Medicine, Isthian Med. Soc. (Panama), Chi Psi. Author: The Anopheline Mosquitoes of the Caribbean Region, bull., 1941; discovered Anopheles Darlingi (malaria mosquito) in Central Am., 1940. Died Dec. 7, 1955; buried New Brunswick, N.J.

KONDO, Sohei, biophysicist; b. Kurume, Japan, May 7, 1922; s. Masayuki and Yosie (Morohuzi) K.; B.Sc., Kyoto U., 1945, D.Sc., 1958; m. Kyoko Tukamoto, Apr. 30, 1954; children—Taisei, Yuko, Mari. Chief elec. engr. Western Japan Spinning Co., Hukuokaken, 1946-49; tchr. math. Kurume, Kumamoto, Kyoto High Sch., 1949-54; research fellow dept. physics Kumamoto and Kyoto U., 1951-54; researcher Nat. Inst. Genetics, Misima, Japan, 1955-59, head radioisotope Lab., 1959-63; prof. fundamental radiology Osaka U., Faculty Medicine, 1963—; vis. research biology div. Oak Ridge Nat. Lab., 1958-59, 1963-64, 1965, 66. Mem. Japanese Radiation Research Soc., Biophys. Soc. Japan, Genetics Soc. Japan, Phys. Soc. Japan, Radiation Research Soc., Health Physics Soc., Biophysics Soc. U. S. Research, numerous publs. on surface tension work, application of glass to accurate radiation dosimetry extended to alpha and beta rays and neutrons, radiation resistance mechanisms possessed by living things. Home: 6-2-13 Habikigaoka, Habikinosi, Osaka-hu, Japan. Office: 33 Joan-cho, Kita-ku, Osaka, Japan.*

KONDRATIEV, Viktor Nikolaevich, Russian phys. chemist; b. Feb. 1, 1902; grad. Leningrad Poly. Inst., 1924. Asso., Physicotech. Inst., USSR Acad. Sci., 1924-31, sr. asso. Inst. Chem. Physics, 1931—, bur. mem. dept. chem. sci., 1960—. Del., Geneva Internat. Conf. on Peaceful Uses Atomic Energy, 1955. Decorated Order of Lenin (2); recipient State prize, 1946; Bernard Lewis medal, 1965; August-Wilhelm von Hoffmann Denkmühe, 1967; 50-year Soviet Power Jubilee medal, 1967. Mem. USSR Acad. Sci. Author: Photochemistry, 1933; Elementary Chemical Processes, 1936; Free Hydroxyl, 1939; Spectroscopic Study of Gaseous Chemical Reactions, 1944; The Structure of Atoms and Molecules, 1946; Kinetics of Chemical Gas Reactions, 1958; Chain Reactions and Combustion and Explosion Processes, 1956; co-author: Electronic Chemistry, 1927; Elementary Energy Exchange Processes in Gases, 1933. Research and publs. on chem. kinetics, molecular spectroscopy, elementary processes in chem. conversions, structure of matter, photochemistry. Home: Leninskii prospekt 30 fl. 105. Office: Inst. Chem. Physics, USSR Acad. Sci., Vorobyekskoye chaussée 2-b, Moscow V-334, USSR.*

KONECNY, Gottfried Leopold, photogrammetrist; b. Troppau, Czechoslovakia, June 17, 1930; s. Franz Joseph and Olga (Sperner) K.; Dipl.Ing., Tech. U. Munich (Germany), 1957, Dr.Ing., 1960; M.S., Ohio State U., 1955, postgrad., 1955-56; m. Lieselotte Angstwurm, Aug. 27, 1958; children—Suzanne, Gottfried Ewald. Faculty, U. N.B., Fredericton, Can., 1959-65, prof., head dept. surveying engring., 1965—. Cons., N.B. Govt., 1959—; mem. various coms. for Canadian and provincial govts. Fulbright scholar, 1954-55; N.R.C. Sr. Research fellow, 1966-67. Registered profl. engr., N.B. Mem. Am. Geophys. Union, German Assn. Surveying, German, Am. socs. photogrammetry, Glaciological Soc., Can. Inst. Surveying, Can. Assn. U. Tchrs. Author: (with others) Handbuch für Vermessungkunde, Vol. Photogrammetry, 1967; also articles. Research on error theory convergent photography, methods combined triangulation-trialateration versus traversing studies. Home: 302 Fulton Av., Fredericton, N.B., Can.*

KONEN, Heinrich Mathias, German physicist; b. Cologne, Germany, Sept. 16, 1874; s. Heinrich and Anna (Dengler) K.; Ph.D., Bonn, 1897, M.D. honoris causa, 1947; ed. U. Cologne; m. Maria Landger, Aug. 11, 1908; children—Heinrich, Therese. Asst. physics inst., Bonn, 1899; under tchr. at gymnasium, 1900; asst. physics inst., privat docent U. Bonn, 1902, prof. physics, dir. phys. inst., 1920, rector emeritus, 1934; extraordinary prof., Munster, 1905, prof., 1919; minister edn. of North Rhine and Westphalia, 1946-48. Recipient Goethe medal. Fellow Accad. Pontificia dei Linci; mem. Notgemeinschaft der Deutschen Wissenschaft (v.p.). Author: (with Kayser) Spectroskopie, 1900-34; Geschichte der Gleich t2-Du2 = 1, 1902; Atlas der Emissionsspektren, 1905; Leuchten des Gase und Dämpfe, 1913; (with Geiger and Scheel) Handbuch der Physik, 1927-29; Physikalische Plaudereien, 1937. Contbd. to devel. of spectroscopy. Died Bonn, Germany, Dec. 31, 1948.

KONEV, Sergey Vasilievitsh, biophysicist; b. nr. Bryansr, Russia, Jan. 14, 1931; s. Wasiliy Nicolaevith and Yulia (Nicolaeni) K.; student soil and biology dept. Moscow State U., 1949-54, also postgrad. student chair biophysics soil and biology dept.; married; 2 children. Sr. scientist Biophys. Lab., All-Union Cattle Breeding Inst., Moscow, 1957-59, Lab. Biophysics and Isotopes, Byeloryssian Acad. Scis., Minsk, USSR, 1959—; lectr. gen. biophysics and bioenergetics Byeloryssian State U. Recipient Silver medal Exhbn. Achievements Nat. Economy USSR, 1965. Author: Electronically Excited States in Biopolymers, 1965; also articles. Research and publns. on ultraviolet luminescence, migration and conservation of energy in proteins; demonstrated possibility of stimulating cell mitosis by ultraviolet light. Address: Kavaleriyskaya 10 11, Minsk, USSR.*

KONIG, Arthur, physicist, physiologist; b. 1856; founder (with Ebbinghaus) Zeitschrift; author 8000 titles; contbd. Bibliography of Vision, to Physiologische Optik (Helmholtz), 2d edit., 1896; worked out curves of brightness as function of wave length, 1884, 1901; established color sensitivity curves for trichromatic and for dichromatic (color-blind) vision, 1886, 92; determined manner in which Weber fraction varies with visual intensity, 1888; established purpose of visual purple in rod vision, 1894. Died 1901.

KöNIG, Emanuel, Swiss physicist, physician; b. Basel, Switzerland, Nov. 1, 1658; M.D., Basel; prof. Greek, Basel, 1695-1703, prof. physics, 1703-11, prof. theoretical medicine, from 1711; mem. (under name Avicenna) Academia Naturae Curiosorum. Author: Regnum animale . . . , 1682; Regnum minerale . . . (on metals and preparations of acids, made free use of chem. symbols which were explained in a plate, theorized that nitrous particles from air are plant constituent), 1686; Regnum vegetabile . . . , 1688; . . . Seu thesaurus . . . , 1693. Died Jan. 30, 1731.

KöNIG, Johann Gerald, botanist; b. Livonia, Switzerland, 1728; pupil of Linnaeus; corresponded with Banks, Retzius; visited Iceland and wrote on its vegetable prodn., 1765; also visited E. Indies. Author: De indigenorum remediorum ad morbos curves regioni indemicos expeignandos efficacia, 1773; also several bot. treatises. Completed N. Burma's work on flora of E. Indies. Died India, 1785.

KöNIG, Lothar Alfons, German physicist; b. Ludwigshafen a.Rh., Germany, July 31, 1928; s. Alfons and Anna (Bullinger) K.; Diplom-Physiker, Johannes Gutenberg U., Mainz, Germany, 1953, Dr.rer.nat., 1955; m. Edith Hemel, June 7, 1957; children—Andreas, Stephan, Gisela. Sci. Asst. Max-Planck-Inst. für Chemie (Otto Hahn Inst.), Mainz, 1953-62; head radiation protection service Gesellschaft für Kernforschung mbH., Karlsruhe, Germany, 1962—. Mem. Deutsche Physikalische Gesellschaft, Gesellschaft Deutscher Naturforscher und Ärzte, Health Physics Soc., Fachverband für Strahlenschutz. Author: (with F. Everling, J. Mattauch, A. H. Wapstra) Nuclear Data Tables, Part I, Part II, 1960; also articles. Research on focusing properties of mass spectrometers and correction of their image defects; computation of relative nuclidic masses, consistent set of energies liberated in nuclear reactions; problems of nuclear safety, airborne effluents from nuclear installations, surface contaminations, reactor accident analysis. Home: 12 Otto-Hahn-Strasse, 7501 Leopoldshafen, Germany. Office: 5 Weberstr., 75 Karlsruhe, German Fed. Republic.*

KöNIGSBERGER, Leo, mathematician; b. Posen, Germany, Aug. 15, 1837; prof., Greifswald, Heidelberg, Dresden (all Germany), also Vienna, Austria. Author: Theorie der elliptischen Funktionen, 2 vols., 1874; Allgemeine Untersuchungen aus der Theorie der Diffenentialgleichungen, 1882. Research on theory of functions, differential equations, mechanics. Died Heidelberg, Dec. 15, 1921.

KONINGSBERGER, Victor Jacob, Dutch botanist; b. Buitenzorg, Feb. 10, 1895; s. Jacob and U. M. (Hellendoorn) K.; Ph.D. in Math. and Physics, U. Utrecht; m. M. J. Dijkstra, July 25, 1922; children—Mathilde, Victor, Tjoord, Johanna. Asst., 1924-26; dir. culture dept. exptl. sta. for Java sugar industry, 1926-34; prof. U. Utrecht, 1934—. Mem. Royal Inst. Tropics in Amsterdam (pres. 1951—), Royal Dutch Acad. Sci., Royal Dutch Union Scientists (hon.). Research and publs. on sugar cane, plant physiology. Home: H. Segherslaan 20, Bilthoven. Office: Lange Nieuwstraat 106, Utrecht, Netherlands.

KONNO, Kiyoshi, Japanese physician; b. Wakuyamachi, Miyagiken, Japan, May 19, 1924; s. Takuma and Yasue (Sakurai) K.; M.D., Tohoku U., 1943; m. Hideko Hashimoto, May 25, 1950; children—Mieko,

Keiko, Yasuo. Chest physician Research Inst. for Tb and Leprosy, and affiliated hosp., Sendai, Japan, 1957-—. Instr. medicine Tufts U. Med. Sch., Boston, 1956-57; faculty Tohoku U. Sch. Medicine, 1958-—, asso. prof., 1964-—; vis. prof. pub. health Cornell U. Med. Coll., N.Y.C., 1965-66. Recipient Gold medal Japanese Bacteriological Soc., 1965. Fellow Am. Coll. Chest Physicians; mem. Am., Japanese thoracic socs., Japanese Tb Soc. Contbr. numerous articles to profl. jours. Inventor biochem. method to differentiate tubercle bacilli from other kinds of acid-fast bacilli (Niacin Test), 1956; found that tubercle bacilli produce large amount of niacin (nicotinic acid) compared with other bacilli. Home: 40-31, Chomeizaka, Odawara, Sendai. Office: Research Institute for Tb and Leprosy, Kitayobancho, Sendai, Japan.*

KONO, R. Japanese virologist; b. Tokyo, Japan, Oct. 4, 1915; s. Kenzo and Aya (Mitsukuri) K.; M.D., Tokyo U., 1940, D.M.S., 1950; M.P.H., Johns Hopkins, 1953; m. Hanako Sassa, Apr. 28, 1942; children—Masako, Ritsuko, Momoko. Microbiologist, Inst. for Infectious Diseases, Tokyo, 1940; lectr. virology Inst. Pub. Health, Tokyo, 1958-63; prof., dir. Inst. for Virus Research, Kyoto U., 1958-63; dir. Central Virus Diagnostic Lab., NIH, Tokyo, 1963-—; cons. virus lab. service in Korea, WHO, 1966. Mem. Japan Virologist Soc. Author textbooks, numerous publs., research on epidemiology of virus in Japan especially on polio and influenza; discovered that some harmless viruses residing in human intestinal tract interfered growth of dangerous virus like polio agt., elucidated a mechanism of local intestinal resistance to polio-virus which appeared after adminstrn. of Sabin vaccine. Home: 87-18 Kunitachi, Kunitachi-shi, Tokyo. Office: NIH, 284 Kamiosaki Chojamaru, Shinagawa-ku, Tokyo, Japan.*

KONOBEEVSKY, Sergey Tikhonovich, Russian physicist; b. Apr. 26, 1890; grad. Moscow U., 1913; D.Physico-Math. Sci. Instr., Moscow Inst. Nat. Econs., 1919-23; with All-Union Electrotech. Inst., 1923-29; Inst. Non-Ferrous Metals, 1929-41; instr. Moscow U., 1926-35, prof., 1935-48; asso. various instns. USSR Acad. Sci., 1948-—. Decorated Order of Lenin. Mem. USSR Acad. Sci. (corr.). Author: Metal Crystallization during Solid State Conversions, 1937; Phase Conversion Theory, 1943. Research and publs. on x-ray structural investigation of metals and alloys, study of structural change during plastic deformation, phase conversions, alloy-aging, hard solution decay, irradiation effects on materials. Address: USSR Acad. Sci., Leninsky prospect 14, Moscow, USSR.

KONOPACKI, Mieczyslaw, Polish histologist; b. Wielum, Poland, Apr. 16, 1880; s. Bronislaw and Antoinette (Zdrojewska) K.; ed. Faculty Sci., Warsaw, Faculty Medicine, Cracow, Poland, also at Lwow, Poland; M.D.; m. Bronislawa Jakimaviecz, 1907. studied under Prof. Mitrofanoff, Warsaw, 1901-03, under Profs. Godlewski and Kostanecki, Cracow, 1903-07. Became maj. in command mil. hosp. of Infant Jesus, Warsaw, 1920; asst. to Prof. Szymonowicz, Lwów, 1907-14; with Prof. K. Heider, Innsbruck, Austria, 1912-13; became asso. prof. Lwów, 1914; named dir. Inst. Histology and Embryology, Warsaw, 1916, prof., 1921-—, also curator, in charge orgn. vet. studies, 1918-23; prof. Dental Inst., 1920-—. Mem. Soc. Polish Biology, Anat. Zool. Polish Soc. (v.p.), Soc. Polish Naturalists (v.p.), Soc. Sci. Warsaw, Polish Acad. Sci. Author: Über den Atmungsprozess bei den Regenwürmen, 1927; also articles. Research in histology, embryology, micromorphology. Office: 6 rue Lwowska, Warsaw, Poland.

KONOPICKY, Kamillo Theodor, ceramist; b. Vienna, Austria, Jan. 25, 1906; s. Theodor and Valerie (Loos) K.; Dr.Ing., Tech. U. Vienna, 1927; m. Martha Pelikan, May 17, 1937; children—Marlis, Gisela, Klaus, Erich. With Alterra-AG, Luxemburg, 1930-33, Martin-Pagenstecher AG, Germany, 1933-41, Boehler-AG, Austria, 1941-45, Oest. Am. Magnesit-AG, Austria, 1945-50, Metallwerk Plansee, Austria, 1950-52; dir. Research Inst. for Fire Protection Industry, Bonn, 1952-—; prof. Tech. U. Berlin, 1964. Mem. ceramic socs., Europe, Eng., U. S. Author: Refractories, 1957; also numerous articles. Research on corrosion, refractories, cermets. Home: 159 Lothar, Bonn. Office: 27 Helmholtz, Bonn, Germany.*

KONOPKA, Edward Alexander, Am. microbiologist; b. Newark, Oct. 13, 1919; s. Raymond and Josephine (Krolikowska) K.; B.S., U. Ky., 1942; m. Margaret Allen Buchanan, Mar. 1, 1945; children—Edward Allen, Nancy Gale, Connie, Susan. Microbiologist, Merck & Co., Rahway, N.J., 1945-47; dir. Montclair Biochem. Lab. (N.J.), 1947-48; sr. bacteriologists Ciba Pharm. Co., Summit, N.J., 1948-63, asso. dir. chemotherapy, 1963-—. Mem. Am. Soc. Microbiology, Soc. Indsl. Microbiology, Research Soc. Am., Theobald Smith Soc. Research and publs. on chemotherapy diseases caused by bacteria, fungi, protozoa. Home: 23 Sussex Rd., Murray Hill, N.J. 07971. Office: Ciba Pharm. Co., Summit, N.J. 07901.*

KONOVALOV, Nikolay Vasilevich, Russian neuropathologist; b. 1900; grad. Med. Faculty, 1st Moscow U., 1924, postgrad. in neuropathology; D.Med. Sci., 1937. Head chair nervous diseases 3d Moscow Med. Inst., 1934-36; sci. dir. neurol. dept. N.A. Semashko Hosp., Ministry R.R.'s, 1935-47; prof., 1939-—; dir. Inst. Neurology, USSR Acad. Med. Sci., 1948-—, also chmn. problem commn. on basic neurophysic diseases. Recipient Lenin prize, 1961. Mem. USSR Acad. Med. Sci. (v.p. 1950-53). Author: Hepato-Lenticular Degeneration, 1948. Co-editor Neurology sect. Large Med. Ency., 2d edit.; mem. editorial bd. S.S. Korsakov Jour. Neuropathology and Psychiatry. Research and numerous publs. on neurology; 1st to determine connection between liver diseases and pathology of nervous system; described torulous meningitis (caused by yeast fungus), described clin. aspects, pathomorphology and pathogenesis of hepatolenticular degeneration (called Westfal-Wilson-Konovalov's Disease in USSR). Home: Novoslobodskaya 57/5. Office: Inst. Neurology, USSR Acad. Med. Sci., ulitsa Shchipok 6-8, Moscow Zh-54, USSR.

KONSTANTINOV, Boris Pavlovich, Russian physicist; b. July 6, 1910; grad. Leningrad Poly. Inst., 1929; D.Physico-Math. Sci. With Leningrad Electrophys. Inst., also other instns., 1930-40; with Physicotech. Inst., USSR Acad. Sci., 1940-58, dir., 1958-—; prof. Leningrad Poly. Inst., 1947-—; rector U. Sci. Knowledge for Tchrs., Leningrad, 1961-—. Recipient Lenin prize, 1959. Mem. USSR Acad. Sci. Chief editor Jour. Tech. Physics, 1958-—. Research and publs. on theoretical and applied acoustics and phys. chemistry, equations for nonlinear acoustics; self-oscillating phenomena, processes of sound formation in musical instruments, acoustic signalling apparatus and developer theory for these phenomena, 1935-43; developer quantitative theory of air-screw sound, 1934; phenomena of viscosity and thermal conductivity on diffusion and absorption of sound in organic media, irregular waves in tubes, hollow absorption. Address: Physicotech. Inst., Politeckhnicheskaya ulitsa 2, Leningrad, USSR.

KOOI, Jacob, Dutch chemist; b. Velsen, Netherlands, May 23, 1922; s. Jan and B. E. (Hoogbruin) K.; Fin.Ex. cum laude, Municipal U., Amsterdam, Netherlands, 1952, Ph.D., 1956; m. Geertruida I. Nuijt, Apr. 1, 1952; children—Hilde B., Jan., Jacob G., Geerten S. Scientist, Found. Fundamental Research Matter, at Joint Est. Nuclear Energy, Kjeller, Norway, 1952-56; staff Reactor Centrum, Nederland, 1956-59, dept. head, 1958-59; dept. head EURATOM (European Atomic Energy Commn.), Brussels, Belgium, 1959-62, Mol, Belgium, 1962-65, Inst. Nuclear Phys. Research, Amsterdam, 1965-—. Mem. Royal Dutch Chem. Soc., Soc. Advancement Med. and Phys. Scis. Editor, Actinides Revs., 1967-—. Research and publs. on radiochemistry, chemistry of plutonium especially transplutonium elements. Address: Chem. Labs., U. Leiden, Leiden, Netherlands.*

KOOIJMAN, Eduard Cornelis, Dutch chemist; b. Amsterdam, Netherlands, Oct. 27, 1916; s. Adrian and Joanne (Goedraad) K.; Ph.D., Amsterdam U., 1946; m. Joanne Bernaerts, Dec. 9, 1958; children—Carel F., Joanne H., Paul E. Research chemist Royal Dutch Shell Labs., Amsterdam, 1944-58; research fellow Inst. Internat. Edn., Cal. Inst. Tech., 1947-48; prof. organic chemistry Leiden (Netherlands) U., 1958-—, dean Faculty Scis., 1967-—, also cons. Mem. Am., Royal Dutch (pres.) chem. socs. Research and publs. on reaction mechanisms, free radical processes and aromatic substitutions. Home: 31 Hugo de Grootstraat, Leiden, Netherlands.*

KOOP, Charles Everett, Am. surgeon; b. Bklyn., Oct. 14, 1916; s. John Everett and Helen (Apel) K.; A.B., Dartmouth Coll., 1937; M.D., Cornell U., 1941; Sc.D., U. Pa., 1947; m. Elizabeth Flanagan, Sept. 19, 1938; children—Allen vanBenschoten, Norman Apel, David Charles Everett, Elizabeth. Surgeon-in-chief Children's Hosp. of Phila., 1948-—; prof. pediatric surgery U. Pa. Sch. Medicine, 1959-—; editor-in-chief Jour. of Pediatric Surgery, 1965-—; cons. U. S. Navy, 1964-—. Mem. Am. Surg. Assn., Soc. U. Surgeons, Am. Acad. Pediatrics, Am. Coll. Surgeons, British Assn. Pediatric Surgeons, Internat. Soc. Surgery, Societe Francaise de Chirurgie Infantile, A.M.A., Deutschen Gesellschaft für Kinderchirurgie. Contbns. and publs. in surg. physiology, plasma substitutes, physiology of the surg. neonate, tech. advances in pediatric surgery. Home: 614 Righters Mill Rd., Narberth, Pa. 19072. Office: 1740 Bainbridge St., Phila. 19146.*

KOOPMANS, Tjalling Charles, economist; b. 'S Graveland, Netherlands, Aug. 28, 1910; s. Sjoerd and Wijtske (van der Zee) K.; M.A., U. Utrecht, 1933; Ph.D., U. Leiden, 1936; hon. doctorate econs. Netherlands Sch. Econs., 1963; m. Truus Wanningen, Oct. 1936; children—Anne W., Henry S., Helen J. Came to U. S., 1940, naturalized, 1946. Lectr. Netherlands Econ. U., Rotterdam, 1936-38; specialist financial sect. League of Nations, 1938-40; research asso. Sch. Pub. and Internat. Affairs, Princeton, also spl. lectr., sch. bus. N.Y.U., 1940-41; economist Penn Mutual Life Ins. Co., 1941-42; statistician Combined Shipping Adjustment Bd. and Brit. Mcht. Shipping Mission, 1942-44; faculty U. Chgo., 1944-54; prof. econs. Yale, 1955-—, dir. Cowles Found. for research econs., 1961-—; Frank W. Taussig prof. econs. Harvard, 1960-61. Fellow Econometric Soc. (pres. 1950), Inst. Math. Statistics (council 1952-54), Am. Statis. Assn.; mem. Am. Econ. Assn. (v.p. 1966), Internat. Statis. Inst. Author: Three Essays on the State of Economic Science, 1957. Editor: Statistical Inference in Dynamic Economic Models, 1950; Activity Analysis of Production and Allocation, 1951; co-editor: Studies in Econometric Method, 1953. Research and publs. on math. models of optimal econ. growth; math. formulation of criteria for optimality of growth; theory of efficient allocation of resources; methods for statis. estimation of econ. relationships. Address: Cowles Found. for Research in Economics, Yale U., 30 Hillhouse Av., New Haven.*

KOPACZEWSKI, Ladislaus (or Wladislas), chemist; b. Leczyca, Poland, June 27, 1886; began study in Bern, Switzerland, 1907; Dr. nat. phil., Fribourg, 1911; M.D., Paris, 1919; with Pasteur Inst., Paris, 1911-19; prof. Inst. Advanced Studies, Belgium, also Paris, from 1921. Author: Traité de biocolloidologie, 1930-38; Physico-chimie du lait, 1950. Research on colloids, catalysers, history of chemistry. Died 1953.

KOPFERMANN, Hans, German physicist; b. Wiesbaden, Germany, Apr. 26, 1895. Became privat docent U. Berlin, 1932, apl. prof. Technische Hochschule, Berlin-Charlottenburg, 1937; became prof. U. Kiel, 1932, U. Göttingen, 1942; prof., dir. Physics Inst., U. Heidelberg, 1953. Author: Kernmomente, 1940. Publs. on nuclear moments, improved spectroscopic methods; introduced molecular radiation method to nuclear physics.

KOPLIK, Henry, physician; b. N.Y., Oct. 28, 1858; s. Abraham S. and Rosalie K.; A.B., Coll. City N.Y., 1878; M.D., Coll. Phys. and Surg., Columbia, 1881; m. Stephanie Schiele, Nov. 1902. Specialist in disease of infancy and childhood; attending physician Good Samaritan Dispensary, St. John's Guild; cons. pediatrist to Mt. Sinai Hosp., Hebrew Orphan Asylum, Hosp. for Deformities, Jewish Maternity Hosp. (all N.Y.C.). Mem. permanent com. Congress Child Welfare; hon. mem. med. societies Vienna and Budapest. Author: Diseases of Infancy and Childhood, 1902 (4th edit.). Discovered early diagnostic sign for measles (Koplik's spots), also a bacillus of whooping cough; established 1st milk depot for infants in U. S. Died Apr. 30, 1927.

KOPP, Hermann Franz Moritz, German chemist, crystallographer; b. Hanau, Germany, Oct. 30, 1817; s. Johann Heinrich Kopp; student U. Heidelberg (Germany); grad. U. Marburg (Germany), 1838; pvt. docent Giessen (Germany) U., 1841-43, asso. prof., 1843-52, prof., 1852-63, head (with Heinrich Will) lab., 1852-53; prof. U. Heidelberg, 1863-92. Author: Geschichte der Chemie, 1843-47; Beiträge zur Geschichte der Chemie, 1869-75; Die Alchemie in älterer und neuer Zeit, 1886; Entwicklung der Chemie in der neueren Zeit. Pub. (with Liebig and Wohler) Annalen der Chemie und Pharmacie, 1851. A founder of phys. chemistry; studied onatomic and molecular volumes; 1st to set forth concept of specific volume, 1844; research on specific heat of liquids and solids; verified Neuman's law of additivity of molecular heat; 1st to measure carefully boiling points of organic substances; built differential barometer, 1837; noted as historian of chemistry. Died Heidelberg, Feb. 20, 1892.

KOPP, Johann Heinrich, German physician; b. Hanau, Germany, Sept. 17, 1777; student, Rinteln, Marburg, Jena (all Germany), 1797-1800; practiced medicine, Rödelheim, Frankfort/Main, Germany, 1801; became dist. physician, Amtes Schwarzenfels, 1802, prof. chemistry, physics and natural history Lyceum, 1807; named mem. staff Alstadt hosps., 1813, surgeon, 1814. Author: Versuch einer Darstellung des gelben Fiebers . . . , 1805; Medizinische Topographie der Stadt Hanau, 1807; Jahrbuch der Staatsarzneikunde, 1808-20; über körperliche Verletzungen . . . , 1812; Aerztliche Bemerkungen . . . , 1825; Denkwürdigkeiten in der ärzlichen Praxis, 4 vols., 1830-45. Described enlarged thymus (Kopp's asthma), 1808, an early case of thymus death in infants, 1830. Died Hanau, 1858.

KOPPANYI, Theodore, pharmacologist; b. Gyongyos, Hungary, Dec. 26, 1901; s. David and Giselle (Heller) K.; student U. Budapest (Hungary), 1918-20; Ph.D., U. Vienna (Austria), 1923. Came to Am., 1923, naturalized, 1929. Instr., U. Chgo. Med. Sch., 1923-27; asst. prof. pharmacology Syracuse U., 1927-29; research asso. pharmacology Cornell U., 1929-30; prof. pharmacology, chmn. dept. Georgetown U. Med. and Dental Sch., 1930-—. Recipient Raskob award, 1957, Vicennial medal Georgetown U., 1950, Fulbright award, 1959, Georgetown U. medal of honor, 1964. Fellow Washington Acad. Sci.; mem. Honor Med. Soc. (hon.), Assn. Medica Argentina, A.A.A.S., Am. Physiol. Soc., Am. Soc. Pharmacology and Exptl. Therapeutics, Soc. for Exptl. Biology and Medicine, Acad. Medicine. Author: Modern Problems of Biology, 1925; Conquest of Life and Toxicology or the Effects of Poisons, 1936; also numerous articles. Discovered regeneration in central nervous system; co-founder sci. drug turnover; pioneered research in effects barbiturates; analyzed effects analeptics in treatment exptl. barbiturate poisoning; introduced effective uses analeptics in clinic; research on basic pharmacologyautonomic nervous system. Home: 2712 Wisconsin Av. N.W., Washington 20007.*

KÖPPEN, Wladimir Peter (or Keepen, Vladimir Petrovich), meteorologist, climatologist; b. St. Petersburg, Russia, Sept. 25, 1846; ed. Heidelberg, Leipzig univs. With Russian Meteorol. Service, 1872-73, then Deutsche Seewarte, Hamberg, from 1875, research work, from 1879. Author: Die Klimate der geologischen Vorzeit, 1924; Grundriss der Klimakunde, 1931; also editor (with Rudolf Geiger) Handbuch der Dlimatologie. Research on climate of ocean areas and of upper-air; classified natural regions according to rainfall and temperatures. Died Graz, Austria, June 22, 1940.

KOPROWSKA, Irena Grasberg, Am. physician; b. Warsaw, Poland, May 12, 1917; d. Henryk and Eugenia (Cui) Grasberg; M.D., Warsaw U., 1939; m. Hilary Koprowski, July 14, 1938; children—Claude Eugene, Christopher Dorian. Asst. pathologist Rio de Janeiro (Brazil) city hosps., 1942-44; research asst. dept. pathology Cornell U. Med. Coll., N.Y.C., 1945-46; research fellow applied immunology Pub. Health Research Inst. City N.Y., 1946-47; asst. pathology N.Y. Infirmary, N.Y.C., 1947-49; research fellow Cornell U. Med. Coll., 1949-54; asst. prof. pathology State U. N.Y. Downstate Med. Center, also dir. cytology lab. dept. pathology Kings County Hosp., Bklyn., 1954-57; prof. pathology, head cytology sect. Hahnemann Med. Coll. and Hosp., Phila., 1957—. Research grantee USPHS, 1954—, Damon Runyon Fund, 1955-56, Cancer Control, 1958—, Am. Cancer Soc., 1958-61. Mem. A.M.A., Am. Soc. Cytology, A.A.A.S., Am. Soc. Exptl. Pathology, Am. Soc. Clin. Pathologists, Coll. Am. Pathologists, Internat. Acad. Pathology, Internat. Acad. Cytology. Research, publs. on devel. exptl. models for comparative studies of chem. induced cancer in mouse uterus and in tissue culture; devel. methodology for teaching clin. cytology. Home: 334 Fairhill Rd., Wynnewood, Pa. 19096. Office: 235 N. 15th St., Phila. 19102.*

KOPROWSKI, Hilary, physician; b. Warsaw, Poland, Dec. 5, 1916; s. Paul and Sarah (Berland) K.; B.A., Nicolas Rey Sch. Luth. Congregation, Warsaw, 1933; M.D., U. Warsaw, 1939; grad. Warsaw Conservatory Music, 1939; m. Irena Grasberg, July 14, 1938; children—Claude Eugene, Christopher Dorian. Came to U. S., 1944, naturalized, 1950. Research asst. dept. exptl. pathology U. Warsaw, 1936-39; staff Yellow Fever Research Service, Rio de Janeiro, Brazil, 1940-44; staff research div. Am. Cyanamid Co., Pearl River, N.Y., 1944-46; asst. dir. viral and rickettsial research Lederle Lab., Pearl River, N.Y., 1946-57; dir. Wistar Inst., Phila., 1957—; prof. microbiology Grad. Sch. Arts and Scis., U. Pa., 1957—, Wistar prof. research medicine U. Pa., 1957—. Cons. WHO, 1950—; mem. NRC, 1963-66. Decorated Commandeur Ordre du Merite pour la Recherche et I'Invention (France), Chevalier Order Royal de Lion (Belgium); recipient Alvarenga prize Coll. Physicians Phila., 1959. Fellow N.Y. Acad. Medicine, Am. Acad. Microbiology; mem. Am. Assn. Anatomists, Am. Soc. Microbiologists, Biochem. Soc. Eng., Am. Soc. Immunologists, Am. Soc. Tropical Medicine, Am. Assn. Cancer Research (dir.), N.Y. Acad. Scis. (pres. 1959, trustee 1960—). Asso. editor Ergebnisse der Mikrobiologie, 1965—. Research on virus diseases; discovered living virus vaccine against rabies, hog cholera, poliomyelitis. Home: 334 Fairhill Rd., Wynhewood, Pa. 19096. Office: 36th and Spruce Sts., Phila. 19104.*

KORANYI, Baron Alexander (or Sándor), physician; b. Budapest, Hungary, 1866; s. Baron Fredrich von Koran and Maluine (von Bónis) K.; ed. Budapest U., also in Germany; M.D. Author: (with Richter) Physics, Chemistry and Medicine, 1907. Research on osmotic pressure in pathology of heart disease, also in pathology and diagnosis of kidneys; introduced cryoscopy of urine as criterion of kidney function, 1894. Died 1944.

KÖRBLER, Juraj, Yugoslavian oncologist; b. Zagreb, Yugoslavia, Oct. 2, 1900; s. Gjuro and Hermina (Pavic) K.; M.D., Faculty Medicine, U. Freiburg, Breisgau, Germany, 1923; m. Mira Heissinger (dec.); 1 son, Jurica; m. 2d, Zora Puhl, June 25, 1966. Chief, dir. Oncologic Inst., Zagreb, 1931—; chief cancer research div. Inst. Med. Research, Yugoslav Acad., Zagreb, 1950—; prof. radiation therapy Faculty Medicine U. Zagreb, 1943—. Mem. Croatian Cancer Soc. (pres), Croatian Soc. History Medicine (pres.), Societé française d'histoire de la Medecine, Physikalischmedizinische Sozietat, Erlangen. Author: Klinische Krebsprobleme, 1940; Der Spinnerinnenkrebs, 1962; Iatrophsika, 1957. Discovered lip cancer in flax spinning women; research, publs. on various forms of cancer; introduced therapeutic innovations to Yugoslavia. Home: 35 Gajeva, Zagreb, Yugoslavia.*

KORDES, Ernst, chemist; b. St. Petersburg, Jan. 19, 1900; s. Richard and Bertha (Jacoby) K.; Ph.D., U. Göttingen; children—Erhard, Harald. Asst. at Konigsberg, Berlin, Leipzig, 1925-30; qualified at Leipzig, 1931; prof. at large, Leipzig, 1938; full prof. German U., Poznan, Poland, 1941-45, U. Iena, 1945-53; prof. structural chemistry U. Bonn, 1953—. Mem. German Chem. Soc., Bunsen Soc., German Mineralogy Soc., Geochemistry Soc., German Soc. Glass Tech. Translator: Geochemie in ausgewählten Kapiteln, 1930. Research and publs. on crystal chemistry, refraction, heterogeneous equilibrium, glass. Home: Kirchweg, Impekoren bei Bonn. Office: Meckenheimer Allee 168, Bonn, Germany.

KOREN, Kristian Johannes, Norwegian physicist; b. Halden, Norway, Nov. 8, 1911; s. Christian Johannes and Helga (Sollie) K.; B.Sc., Tech. U. Dresden, 1935, M.Sc., 1937; m. Lilli Emilie Dam, May 5, 1962; children—Peter Christian, William Jörgen. Physicist, State Inst. Radiation Hygiene, Oslo, Norway, 1938-41, Woolwich Armament Office, London, Eng., 1942-44; sr. physicist State Instn. Radiation Hygiene, Oslo, 1946-56, dir., 1956—; asst. prof. Oslo U., 1956—. Govt. del. OECD, 1958—. Mem. Nordic Soc. for Radiation Protection (pres. 1966—), Phys. Soc., Norwegian Radiologic Soc., Radiation Protection Bd. Author: Straler, liv og helse, 1961; also articles. Research on application of radiation physics in radiology, radiation protection of individuals. Home: 15 Michelets vei, Lysaker, Norway. Office: 70 Ullernchausseen, Oslo, Norway.*

KORF, Richard Paul, Am. mycologist; b. Bronxville, N.Y., May 28, 1925; s. Frederick and Emilie (Krug) K.; B.Sc., Cornell U., 1946, Ph.D., 1950; m. Kumiko Tachibana, June 27, 1959; children—Noni, Mia, Ian, Mario. With Cornell U., 1947—, prof. mycology, 1961—; lectr. botany dept. U. Glasgow, Scotland, 1950-51; Fulbright research prof. Yokohama (Japan) Nat. U., 1957-58. NSF fellowship, 1957-58. Mem. Botan. Soc. Am., Mycol. Soc. Am. (sec.-treas.), Internat. Assn. Plant Taxonomists. Research and publs. on classification and distbn. of cup-fungi particularly in Northeast Am., Japan, Southeast Asia, involving collecting fungi in frequently unexplored areas, describing new species and reinterpreting old species in modern terms. Address: Plant Pathology Herbarium Cornell U., Ithaca, N.Y. 14850.*

KORFF, Ernst Oskar Albin, German psychologist; b. Wuppertal, Oct. 30, 1902; s. Julius and Sophie (Fieseler) K.; ed. univs. Cologne, Frankfort, Fribourg; diploma in psychology; m. Fine Anna Leniger, July 4, 1942; children—Brigitte, Anne, Jutta. Asst., Institut für Wirtschaft-und Begabungsforschung, Wuppertal-Barmen; dir. Graphologisches und werbepsychologisches Institut, Siemsns Co.; personnel dir. for profs. in industry, 1937-44; founder (with wife) Institut für angewandte Psychologie. Author: Menschen beurteilen und Menschen führen; Arbeitstechnik für vielbeschaftigte; Zur Psychologie des leitenden Angestellten; Betrichspsycholisches Taschenbuch; Verkaufspsycholcgisches Taschenbuch; Reden, Diskutieren, Verhandeln. Editor: Praktische Psychologie. Research on psychology in human relations, group dynamics, psychodiagnostics in industry and vocational guidance. Address: 2073 Lütjensee Bez Hamburg, Haus Heidberg, Germany.

KORHONEN, Unto Kalervo, Finnish physicist; b. Viipuri, Finland, Jan. 20, 1915; s. Matti and Anna (Soininen) K.; M.Philosophy, U. Helsinki (Finland), 1944, Ph.D., 1951; m. Helvi Putkinen, Nov. 1, 1964; children—Matti, Juha, Marja-Leena, Leena, Olavi Isaac. Faculty, U. Helsinki, 1945—, docent physics, 1954—; prof. physics Tech. U. Helsinki, 1957—; dir. Tech. U. Tampere, 1965—. Decorated SL.K, VR 4. Mem. Phys. Soc. Finland (past chmn.), Sci. Workers Union Finland (chmn. 1961—), Tech. Soc. Finland. Contbg. author: Handbook of Technology, 1965. Research, publs. on devel. methods to obtain accurate exptl. knowledge about electron distbn. of atoms and ions belonging to crystal lattices. Home: 4 A 4 Suvikuja, Tapiola, Finland. Office: Tech. U., Otaniemi, Finland.*

KORKHAUS, Gustav, German dental surgeon; b. Cologne, Germany, Jan. 4, 1895; s. Emil and Wilhelmine (Stöbe) K.; ed. U. Berlin, U. Bonn; m. Margot Sprengholz. Head orthodontic dept. Municipal Clinic for Dental Treatment Sch. Children, 1927-34; became lectr. U. Bonn., 1929; chief orthodontic dept. U. Clinic and Polyclinic for Diseases of Teeth, Mouth, and Jaw, U. Bonn, 1934-46, became prof., 1935, asst. clinic dir., 1946, prof., dir. Dental Clinic, 1948—; also chief Postgrad. Edn. Center Rhenish Dental Surgeons. Mem. Leopoldina, Academia Odontologia, (Buenos Aires), German Acad. Natural Scientists, European Orthodontics Soc., German Soc. for Teeth, Mouth, and Jaw Medicine, German Soc. for Jaw Orthodontics. Author: Moderne orthodontische therapic, 3d edit., 1955; contbg. author: Handbuch der Zahnkielkunde, vol. 4, 1939. Developed orthodontia in Germany. Office: 16 Zitelmannstr., Bonn., Germany.

KORKIN, Alexander Nikolaevich, Russian mathematician; b. Russia, Feb. 19, 1837; s. Nikolai Korkin; prof. U. St. Petersburg; research on integral equations, theory of numbers; solved (with E. T. Zolotarev) problem of exact boundaries of minimum of positive sq. forms with 4 and 5 variables, 1877. Died Aug. 19, 1908.

KORKISCH, Johann, Austrian chemist; b. Vienna, Austria, Dec. 19, 1929; s. Willy and Anna (Schipek) K.; Ph.D., U. Vienna, 1957; m. Gertrude Erla, Sept. 23, 1957; 1 son, Thomas. Asst., Analytical Inst., U. Vienna, 1957—, lectr., 1964—, dozent, 1966—; staff U. Cal. Scripps Instn., La Jolla, 1961-62; resident research asso. U. Chgo.-Argonne (Ill.) Nat. Lab., 1967-68. Recipient Pregl-prize in microchemistry, 1966. Mem. Am. Chem. Soc., Austrian Chem. Soc. Author: Modern Methods for the Separation of Rarer Metal Ions, 1968; also numerous articles. Devel. analytical methods based on ion exchange and spectrophotometry for use in field of atomic energy and trace analysis; research on ion exchange behavior and separation of metal ions in mixed aqueous-organic media. Home: 5 Joachimstalerplatz, Vienna 16, Austria.*

KORMAN, Nathaniel Irving, Am. systems engr.; b. Providence, Feb. 23, 1916; s. William and Tillie (Jacobs) K.; B.S., Worcester Poly. Inst., 1937; M.S., Mass. Inst. Tech., 1938; Ph.D., U. Pa., 1958; m. Ruth Kaplan, Apr. 6, 1941; children—Michael A., Robert S. With RCA, 1938—, dir. med. plans, Princeton, N.J., 1966, mgmt. cons., 1967—. Charles A. Coffin fellow, 1937. Fellow I.E.E.E., Sigma Xi. Research pioneer, patents, articles in field of systems engring.; application systems engring. to missile systems, graphics systems, med. electronics. Address: 371 Riverside Dr., Princeton, N.J. 08540.*

KORMILEV, Nicholas Alexander, entomologist; b. Yalta, Russia, Jan. 29, 1901; s. Alexander N. and Catherine (Sakulin) K.; A.E., U. Zagreb (Yugoslavia), 1926; m. Magdalena Senkute, Feb. 20, 1952. Came to U. S., 1957, naturalized, 1962. Entomologist, Institute de Ciencies Naturales, Buenos Aires, Argentine, 1948-52; tchr. entomology Colegio Maximo, San Miguel, Buenos Aires, 1952-56; pvt. practice entomology, Bklyn., 1957—. Decorated Order St. Sava (Yugoslavia). Mem. Washington, N.Y. entomol. socs. Contbr. numerous articles to profl. jours. Research on hemiptera, particularly families Aradidae and Phymatidae. Home: 365 Lincoln Pl., Bklyn. 11238.*

KORN, Arthur, German physicist; b. Breslau, Germany (now Wroclaw, Poland), May 20, 1870; Ph.D., Leipzig, Germany, 1890; also studied at Paris, Berlin, Würzburg, Munich; prof. U. Munich, 1895-1908; U. Berlin, 1914-39. Author: Freie und erzwungene Schwingungen, 1910; Handbuch der Phototelegraphie und Telautographie, 1911. Studies in math. physics, theory of potentials; credited with 1st telegraphic transmission of photograph over circuit (to Nuremberg from Munich and back), 1904. Died Jersey City, U. S., Dec. 12, 1945.

KORN, Doris Elfriede, German mineralogist; b. Zwickau, May 30, 1904; d. Otto and Elsa (Junghanns) Korn; Ph.D. in Natural Sci., U. Heidelberg; m. Benno Schachner, Aug. 28, 1938; 1 dau., Melitta. Agrégé, U. Aix-la-Chapelle, 1933, full prof., 1949; dir. Mineralogy Inst. Mem. German Mineral. Assn., Geol. Assn., Geol. Soc. France, Soc. Geologists and Economists, Geochem. Soc., Am. Mineral. Soc. Research and publs. on structure of minerals and ores, mechanism of mineral formation, genesis of veins of minerals. Home: Muffeterweg 21. Office: Wüllnerstrasse 2, Aachen, Germany.

KORN, Edward David, Am. biochemist; b. Phila., Aug. 3, 1928; s. Joel and Carrie (Goldman) K.; A.B., U. Pa., 1949, Ph.D., 1954; m. Muriel E. Fisher, June 23, 1950; children—Elizabeth G., Sarah H. Sr. investigator Nat. Heart Inst. NIH, Bethesda, Md., 1953—. Mem. Am. Soc. Biol. Chemists, A.A.A.S. Research, publs. on enzymes and mechanisms for transport of lipids in mammals, biosynthesis and structure of mucopolysaccharides; biosynthesis of unsaturated fats; biochemistry of phagocytosis and chem. basis of electron microscopy. Home: 9412 Balfour Ct. Office: Nat. Heart Inst., Bethesda, Md. 20014.*

KORN, Granino Arthur, elec. engr.; b. Berlin, Germany, May 7, 1922; s. Arthur and Elizabeth (Friedlaender) K.; came to U. S., 1939, naturalized, 1944; B.A., Brown U., 1942, Ph.D. in Physics, 1948; M.A., Columbia, 1943; m. Theresa M. McLaughlin, Sept. 3, 1948; children—Anne Marie, John McLaughlin. Project engr. Sperry Gyroscope Co., Great Neck, L.I., N.Y., 194648; head, analysis group Curtiss-Wright Corp., Columbus, O., 1948-49; staff engr. Lockheed Aircraft Corp., Burbank, Cal., 1949-52; self-employed cons., Los Angeles, 1952-57; prof. elec. engring. U. Ariz., Tucson, 1957—; dir. Computer Research, Inc., Lafayette, Cal. Mem. Simulations Councils Inc., Internat. Analog Computer Assn., Sigma Xi. Author: Random-process Simulation and Measurements, 1966, Basic Tables in Electrical Engineering, 1966; (with Mrs. Korn) Electronic Analog Computers, 1952, Mathematical Handbook for Scientists and Engineers, 1961, 68, Electronic Analog and Hybrid Computers, 1964, Manual of Mathematics, 1967; (with H. Huskey) Computer Handbook, 1962; (with M. Klerer) Digital Computer User's Handbook, 1967, also chpts. in engring. handbooks. Devel. fast analog hybrid computing systems, digital simulation. Home: 6801 Opatas St., Tucson, Ariz. 85715.*

KORNBERG, Arthur, Am. biochemist; b. Bklyn., Mar. 3, 1918; s. Joseph and Lena (Katz) K.; B.S., Coll. City of N.Y. (N.Y. State scholarship), 1937; M.D., U. Rochester (Buswell scholarship), 1941; m. Sylvy R. Levy, Nov. 21, 1943; children—Roger, Thomas Bill, Kenneth Andrew. Intern medicine Strong Meml. Hosp., Rochester, N.Y., 1941-42; commd. officer, U.S.P.H.S., 1942, advanced through grades to med. dir., 1951; staff mem. Nat. Inst. Health, Bethesda, Md., 1942-52, nutrition sect., div. physiology, 1942-45, chief sect. enzymes and metabolism, 1947-52; guest research worker depts. chemistry and pharmacology coll. medicine, N.Y. U., 1946, dept. biol. chemistry med. sch., Washington U., 1947, dept. plant biochemistry U. Cal., 1951; formerly prof., head dept. microbiology, med. sch. Washington U.; now prof., head dept. biochemistry Stanford U. Served lt. (j.g.), med. officer, U. S. Coast Guard, 1942. Recipient Paul-Lewis award in enzyme chemistry, 1951; co-recipient of Nobel prize in physiology and medicine (with S. Ochoa), 1959. Mem. Am. Soc. Biol. Chemists, Am. Chem. Soc., Harvey Soc., Am. Soc. Clin. Investigation, Nat. Inst. Health (Biochemistry study sect.), Nat. Acad. Scis., Am. Philos. Soc. Author: Enzymatic Synthesis of DNA, 1961. Discovered deoxyribonucleic acid (DNA) polymerase, an enzyme which is able to reproduce new DNA template; this discovery demonstrated a rational enzymatic mechanism of replication of genetic material; earlier work elucidated the synthesis of DPN and FAD, two coenzymes. Home: 365 Golden Oak Dr., Alpine Hills, Portola Valley, Cal. Office: Stanford U., Sch. Medicine, Palo Alto, Cal.

KORNBERG, Hans Leo, biochemist; b. Herford, Germany, Jan. 14, 1928; s. Max and Margarete (Silberbach) K.; B.Sc. with honors in chemistry, U. Sheffield (Eng.), 1949, Ph.D. (John Stokes research fellow), 1953; M.A., U. Oxford (Eng.), 1958, D.Sc., 1961; m. Monica Mary King, Oct. 6, 1956; children—Julia Margaret, Rachel Elizabeth, Jonathan Paul and Simon Alexander (twins). Mem. staff M.R.C. unit cell metabolism research U. Oxford, 1955-60; research asso. Harvard Med. Sch., 1958; prof. biochemistry, head dept. U. Leicester (Gt. Britain), 1960—. Commonwealth Fund fellow, 1953-55. Fellow Royal Soc., 1950, Inst. Biology; mem. Biochem. Soc. (1st Colworth Medal 1963), Soc. Gen. Microbiology, Soc. Gen. Physiology. Author: (with Hans Krebs) Energy Transformations in Living Matter, 1957. Research, publs. on pathways whereby micro-organisms utilize C2 and C3 compounds as sole carbon and energy source, manner in which operation of these pathways is regulated; discoverer glyoxylate cycle, 1957, glycerate pathway, 1959, dicarboxylic acid cycle, 1960, fine control mechanisms regulating activity of isocitrate lyase, 1963, pep-carboxylase, 1965, pepsynthase, 1965. Home: 2 Woodland Av., Leicester, Gt. Britain.*

KORNBERG, Harry Alexander, Am. biologist; b. Chgo., Aug. 18, 1914; s. Louis and Anna (Benson) K.; B.S. in Chemistry, U. Ill., 1938; M.S., Wash. State U., 1940; Ph.D. in Biochemistry, U. Tex., 1942; m. Anne Wendling, June 5, 1937; children—Anne, Dianne, Gregory. Abbott-Glidden-Dilly fellow Northwestern U., 1942-43; asst. prof. chemistry Ore. State Coll., 1946-47; head biology div. Gen. Electric Co., Richland, Wash., 1947-50, mgr. biology operation, 1950-64; mgr. biology dept. Battelle-N.W., Richland, 1965—. Chmn. subcom. on inhalation hazards, PEAR, Nat. Acad. Scis., NRC, 1956-64. Mem. Am. Chem. Soc., Am. Inst. Biol. Scis., Health Physics Soc., Radiation Research Soc., Soc. for Exptl. Biology and Medicine (past chmn. N.W. sect.), Soc. for Toxicology, Am. Pub. Health Assn. Research and publs. on radiation biology, including cellular biology, entomology and field ecology. Home: 1734 Horn Av. Office: Biology Dept., Battelle-N.W., P.O. Box 999, Richland, Wash. 99352.*

KORNER, Antoine, botanist; b. Mautern, 1831; became prof. botany Polytechnic Sch., Budapest, Hungary, 1858; named prof. natural history U. Innsbruck (Austria), 1860; apptd. dir. Bot. Garden, also Mus. Vienna, 1878. Author: La flore des pays du Danube, 1863; Études sur les limites supérieures de la flore de montagne dans les alpes autrichiennes, 1863-67; Herbier des prairies autrichiennes, 1863-70; Culture des plantes alpestres, 1864; L'économie alpestre dans le Tyrol, 1868; Les jardins botaniques, 1874; La végétation de la Hongrie centrale et orientale et de la Transylvanie, 1875. Died Vienna, 1898.

KORNER, Asher, Brit. biochemist; b. London, Feb. 7, 1927; s. Solomon and Hetty (Copeland) K.; B.A. with 1st class honors, Trinity Coll., U. Cambridge, 1951, M.A., Ph.D., 1955; m. Shirley Austin, June 29, 1952; children—Deborah, Simon, Joseph, Jessica. Med. Research Council scholar dept. biochemistry U. Cambridge, 1951-54, John Jaffe's student Royal Soc., 1954-55, faculty, 1955-65, lectr. biochemistry, 1960-67; Rockefeller fellow dept. physiol. chemistry U. Cal. at Berkeley, 1955-56; prof. biochemistry U. Sussex (Eng.), 1967—; fellow, dir. studies in biochemistry, tutor for research students Clare Coll., 1960-66. Mem. Biochem. Soc., Biophys. Soc., Diabetic Soc., A.A.A.S. Research, publs. on mechanism of biosynthesis of proteins and ribonucleic acids in mammalian tissues, control of these processes by hormones, carcinogens and other agts. Home: King Henry's Rd., Lewes, Eng. Office: Sch. Biology, U. Sussex, Falmer, Sussex, Eng.*

KÖRNER, Wilhelm (or Guglieme), chemist; b. Kassel, Germany, 1839; prof., Milan, Italy; pioneer in isomers of polysubstituted benzene; gave formula for pyridine, 1869; developed basic absolute method for determining structure of benzene products; synthesized (with Menozzi) asparagine, 1874. Died 1925.

KORNETSKY, Conan, Am. psychologist; b. Portland, Me., Feb. 9, 1926; s. Alex and Ida (Rosenberg) K.; B.A., U. Me., 1948; M.S., U. Ky., 1951, Ph.D., 1952; m. Marcia Smargon, June 5, 1949; children—David, Lisa. Asst. psychologist Drug Addiction Research Center, Lexington, Ky., 1949-52; research psychologist Nat. Inst. Mental Health, Bethesda, Md., 1952-59; faculty Boston U. Sch. Medicine, 1959—, research prof. pharmacology and psychiatry, 1962—. Mem. psychopharmacology study sect. Nat. Inst. Mental Health, 1963—, ad hoc. com. psychol. factors in smoking Am. Cancer Soc., 1961-62. USPHS fellow, 1959-62; recipient NIH Career award, 1962—. Mem. Am. Soc. Pharmacology and Exptl. Therapeutics, Am. Psychol. Assn., Am. Coll. Neuropsychopharmacology, Collegium Internationale Neuro-Psychopharmacologicum, Eastern Psychol. Assn., Sigma Xi. Mem. editorial bd. Psychopharmacologia, 1961—; editorial adv. bd. Internat. Reviewer of Neurobiology, 1965—. Studies, publs. on effects of drugs on human, animal behavior; roles of anxiety in pain, immune response to morphine, reticular formation in schizophrenia. Home: 7 Rumford Rd., Lexington, Mass. 02173. Office: Boston U. Sch. Medicine, 80 E. Concord St., Boston 02118.*

KORNEV, Petr Georgievich, Russian surgeon; b. 1884; grad. Med. Faculty, Moscow U., 1908; D.Med. Sci., 1913. With bone Tb sanatorium and surg. hosp., 1919-36; lectr. 1st Leningrad Med. Inst., 1921-35, prof., 1935-40; head chair osteoarticular Tb, Leningrad Postgrad. Med. Inst., 1936—; dir., later sci. dir. Leningrad. Inst. Surg. Tb, RSFSR Ministry Health, 1930—. Decorated Order of Lenin; recipient Stalin prize, 1952. Mem. USSR Acad. Med. Sci., Internat. Assn. Surgeons, All-Union Soc. Surgeons (hon.), All-Union Soc. Phthisiologists (hon. chmn. 1957—), Leningrad Soc. Orthopedists and Traumatologists (hon. chmn. 1957—). Author: Transplantation and Growth of Bones, 1927; Problems of Wound Osteomyelitis, 1947; Osteoarticular Tuberculosis, 1951, 2d edit., 1952; The Clinical Aspects and Treatment of Osteoarticular Tuberculosis, 1959. Mem. editorial council Orthopedics, Traumatology and Prosthetics. Research and numerous publs. on osteoarticular Tb. Address: Leningrad. Inst. Surgical Tb, Leningrad-Lesnoe, USSR.

KORNHAUSER, William, Am. sociologist; b. Chgo., Feb. 5, 1925; s. Arthur and Beatrice Kornhauser; A.B., U. Chgo., 1948, M.A., 1950, Ph.D. (Livingston fellow, Social Sci. Research council fellow), 1953; m. Ruth Rosner, Sept. 18, 1949; 1 dau., Anne. Instr., Columbia U., 1952-53; faculty U. Cal. at Berkeley, 1953—, prof. sociology, 1964—; research asso. center Internat. Studies, Princeton, 1961-62. Fellow Center for Advanced Study in Behavioral Scis., Stanford, Cal., 1954-55; U. Cal. Faculty fellow, 1956. Mem. Am. Assn. U. Profs., Am. Sociol. Assn., Pacific Sociol. Assn., Author: The Politics of Mass Society, 1959; Scientists in Industry, 1962; also articles. Asso. editor Am. Sociol. Rev., 1966—. Research on social conditions democracy and totalitarianism, rebellion and revolution. Home: 1570 Hawthorne Terrace, Berkeley, Cal. 94708.*

KORNICKER, Louis Sampson, Am. zoologist; b. N.Y.C., May 23, 1919; s. Howard and Lena (Kohen) K.; B.S., U. Ala., 1941, B.S. in Chem. Engring., 1942; M.A., Columbia, 1954, Ph.D., 1957; m. Beatrice Nyman, Oct. 29, 1952; children—Lance, Steven, William. Asst. dir. Inst. Marine Sci., U. Tex., Port Aransas, 1957-60; geologist Office of Naval Research, Chgo., 1960-61; faculty Tex. A. & M. U., College Station, 1961-64; with div. crustacea, U. S. Nat. Mus., Smithsonian Instn., Washington, 1964, now curator. Fellow Geol. Soc. Am., A.A.A.S.; mem. Soc. Systematic Zoologists, Washington Paleontol. Soc., Am. Assn. Petroleum Geologists. Participant expdns. to Indian Ocean, Gulf Mexico, Caribbean Sea, Bahamas to study ecology of coral reefs and asso. biota; research, publs. in systematics and ecology of marine ostracoda. Home: 10400 Lake Ridge Dr., Oakton, Va. 22124. Office: Smithsonian Instn., Washington 20560.*

KORNMANN, Peter, German phycologist; b. Frankfurt/Main, Germany, Oct. 23, 1907; s. Georg and Elisabeth (Flühr) K.; Dr.phil.nat., U. Frankfurt/Main, 1931; m. Hilde Koch, Mar. 25, 1937; 1 son, Gerhard Reimer. Asst., Bot. Inst., U. Frankfurt/Main, 1931-36; chief bot. sect. Biol. Anstalt, Helgoland (Germany), 1936-39, 1950—. Editor: Ectocarpaceen-Studien (P. Kuckuck), 1964. Research and numerous publs. on marine algae including, lifecycles, taxonomy, development physiology. Address: Biologische Anstalt, 2192 Helgoland, West Germany.*

KORNMÜLLER, Alois Eduard, German physician; b. Brüx, Bohemia, Oct. 19, 1905; s. Alois and Bertha (Fritsch) K.; M.D., U. Prague; m. Elfriede Kahle, July 27, 1934; children—Ursula, Klaus, Joachim, Renata. With Kaiser Wilhelm Inst. for Encephaletic Research, 1930-45, Kaiser Wilhelm Inst. (now Max-Planck Inst. for Encephaletic Research), Güttingen, 1945—. Mem. German Electroencephalographic Soc., German Soc. Doctors and Physicians, German Physiology Soc., Anglo-German Soc. Physicians. Author: Die bioelektrischen Erscheinungen der Hinrindenfelder, 1937; Klinische Elektroenkephalographie, 1944; Die Elemente der nervösen Tätigkeit, 1947. Research on elements of nervous activity, electrobiology, physiology of senses, function of neuroglia, neurohormones, pathogenesis, therapy of psychoses. Home: Nonnensteig 26. Office: Bunsenstrasse 10, 34 Göttingen, Germany.

KORNS, Robert Fulton, Am. epidemiologist; b. Tai An Fu, China, Oct. 4, 1912 (parents Am. citizens); s. John H. and Bess (Pennywitt) K.; A.B., Dartmouth, 1934; M.D., Johns Hopkins, 1937, Dr.P.H., 1939; m. Esther L. Weinbach, June 30, 1938; children—Julianne, Nancy, Thomas, Robert Fulton, Stephen. Epidemiologist, N.Y. State Dept. Health, Albany, 1940-44, dir. div. of communicable disease control, 1945-53; dep. dir. Nat. Polio Vaccine Field Trial, Ann Arbor, Mich., 1953-55; asst. commr. for pub. health research devel. and evaluation N.Y. State Dept. Health, Albany, 1955—. Asso. prof. preventive medicine Albany Med. Coll., 19——. Mem. Am., Internat. epidemiologic socs. Studies, publs. on characteristics, control of communicable and non-communicable diseases. Home: Simmons Rd., Glenmont, N.Y. Office: 84 Holland Av., Albany, N.Y.*

KOROBEINIK, Juzi Fyodorovitsch, Russian mathematician; b. Rostov-on-Don, July 18, 1930; s. Fyodor N. and Anastasia (Veremenko) K.; candidate of sci., Rostov-on-Don State U., 1952, D.Sc., 1966; m. Nina B. Zavyalova, Apr. 21, 1962. Postgrad. of chair math. analysis Rostov-on-Don State U., 1952-55, asst. chair math. analysis, 1955-59, reader chair math. analysis, 1959—. Research and publns. on theory of complex variable and theory of differential equations (the properties of functional series) in the complex domain; infinite systems of differential and integral equations; differential equations of infinite order. Home: N 165/5 Kominterna. Office: 105 Engels Str., Rostov-on-Don, USSR.*

KOROPOV, Viktor Mikhaylovich, Russian veterinarian; b. 1906; D.Vet. Sci. Asst., dept. path. physiology Omsk Vet. Inst., 1928-31; head chair path. physiology Troitsky Vet. Inst., 1932-44, dir., 1941—; dir., head chair path. physiology Moscow Zool. and Vet. Inst., 1944—; rector Moscow Vet. Acad., 1948—. Mem. All-Union Acad. Agrl. Sci. (bur. mem. vet. sect.). Author numerous works including Veterinary Training in the USSR; Manual of Pathological Physiology; History of Veterinary Medicine in the USSR. Mem. editorial bd. Vet. Sci. Address: Moscow Vet. Acad., Moscow-Kuzminki, USSR.

KOROTKOV, Aleksey Andreevich, Russian organic chemist; b. Feb. 25, 1910; grad. Leningrad Chem. Tech. Inst., 1931. With synthetic rubber plants, 1931-45; asso. All-Union Research Inst. Synthetic Rubber, 1945—, Inst. High-Molecular Compounds, USSR Acad. Sci., 1953—. Mem. USSR Acad. Sci. (corr.). Research and publs. on polymerization of butadiene and isoprene by organometallic compounds and alkali metals. Died Feb. 5, 1967.

KOROVIN, Yevgeniy Petrovich, Russian geobotanist; b. Moscow, 1891; grad. Moscow U., 1917. Instr., lectr. Inst. Botany and Zoology, Uzbekistan Acad. Sci., 1920-32, dir., 1943-48, dir. Inst. Botany, 1950-52, Presidium mem., chmn. biol. dept. agrl. sci., 1955, 57; a founder, prof., head chair Central Asian U., Tashkent, 1932—. Decorated Order of Lenin. Mem. Uzbek Acad. Sci. Author: The Vegetation of Central Asia, 1934; Outline History of the Development of Vegetation in Central Asia, 1935; co-author: The Life of the Desert, 1936. Co-editor: The Flora of Uzbekistan, 1959. Research and numerous publs. on floristics and systematics, ecology; described over 100 new species and 8 new genera of Central Asian flora. Address: Central Asian University, ulitsa K. Marksa 32, Tashkent, Uzbekistan SSR, USSR.

KORR, Irvin Morris, Am. physiologist; b. Phila., Aug. 24, 1909; s. Samuel P. and Anna (Goldberg) K.; B.A., U. Pa., 1930, M.A., 1931; Ph.D., Princeton, 1935; m. Margot Lindsay, June 13, 1939; 1 son, David Lindsay. Procter fellow Princeton, 1935-36; faculty N.Y. U., 1936-42; research OSRD, 1942-45; prof. physiology, chmn. div. physiol. scis. Kirksville (Mo.) Coll. Osteopathy and Surgery, 1945—. Fellow A.A.A.S.; mem. Am. Physiol. Soc., Soc. Exptl. Biology and Medicine, Harvey Soc., Am. Inst. Biol. Scis., Assn. Am. Med. Writers, Sigma Xi. Research and numerous publs. in bioluminescence, cellular metabolism, comparative physiology of kidney, interchange through nervous system between musculoskeletal system and internal organs; demonstrated transfer of nerve-cell cytoplasm to muscle cells and possible trophic role; segmental disturbances in sympathetic nervous system. Home: 1105 S. Halliburton St., Kirksville, Mo. 63501.*

KORRINGA, Jan, physicist; b. Heemstede, Netherlands, Mar. 31, 1915; s. Harm and Antje (Hemelrijk) K.; M.A., Leiden U., 1939, Ph.D., Delft U., 1942; m. Johanna M. Regnault, Jan. 2, 1943; children—Maarten, Wouter, Derk-Jan. With Delft U., 1941-46, lectr., asst. prof., 1944-46; with Leiden U., 1946-53, asso. prof., 1951-53; with Ohio State U., 1953—, prof., 1956—; cons. Oak Ridge Nat. Lab., 1957—; cons. Chevron Research Co., La Habra, Cal., 1955—; vis. prof. Besancon, France, 1962-63. Mem. Am. Phys. Soc., Netherlands Phys. Soc. Research and publs. on theory of solids in particular metals; magnetic resonance; statist. physics. Home: 157 Rustic Pl., Columbus, Ohio 43214.*

KORRINGA, P(ieter), Dutch biologist; b. Heemstede N.H., Netherlands, Feb. 16, 1913; s. Harm and Antje (Hemelrijk) K.; Midegree cum laude, U. Amsterdam (Netherlands), 1937, D.degree, cum laude, 1940; m. Janna Kostermans, Sept. 30, 1941; children—Jacobine Annetta (Mrs. Juan Narbona), Anita Brita Suzanna. Biologist, Netherlands Inst. for Fishery Investigations, IJmuiden, 1937—, dir., 1957—; asso. prof. hydrobiology Amsterdam U., 1961—. Cons. oyster culture U. S., 1948, S.Africa, 1953. Named Officer Orange Nassau (Netherlands). Mem. Internat. Council Exploration of Sea (v.p. 1965—). Author: Experiments and Observations on Swarming, Pelagic Life, and Setting in European Flat Oyster, 1941; also numerous articles. Research on early stages of devel. of oyster; shell disease; analysis of lunar periodicity in breeding of marine organisms. Home: 8 Feithlaan, Driehuis-IJmuiden, N.H. Office: 1 Haringkade, P.O. Box 68, IJmuiden, N.H., Netherlands.*

KORSAKOFF, Sergei Sergeievich, Russian psychiatrist; b. Vladimir Province, 1854; grad. U. Moscow, 1875; M.D., 1887. Physician, Preobrazhenskii Hosp., 1875; mem. Kozhevnikov's staff Clinic for Nervous Diseases, 1876-79; private docent U. Moscow, 1888, appdt. supt. psychiat. clinic, also prof. extraordinarius, 1892; joint-founder Moscow Soc. Neuropathologists and Psychiatrists, Russian Assn. Psychiatrists and Neurologists. Author: Ob alkoholnom paralichije, 1887; Kurs psikhiatri, 1893. One of founders of Russian psychiatry; described psychosis (Korsakoff's syndrome) associated with chronic alcoholism; credited with establishment of concept of paranoia, invention of classification system for mental diseases, and with freeing mental patients from restraint in straight-jackets and similar devices. Died 1900.

KORSCHELT, Eugen, German zoologist; b. Zittau, Sept. 28, 1858. Prof., Marburg. Mem. German Zool. Soc. (rector 1914-15). Author: Lehrbuch der vergleichenden Entwicklungsgeschichte der wirbellosen Tiere, 2 vols., 1890-1910; Der Gelbrand, 2 vols., 1924-29. Editor: Zoolog. Anzeig; Handwörterbuch der Naturwissenschaften, 10 vols., 1912, also numerous articles. Died Marburg, Dec. 22, 1946.

KORSGAARD, Vagn, Danish physicist; b. Ribe, Sept. 8, 1921; s. Niels and Magda (Sorensen) K.; M.Sc., Tech. U., Copenhagen; m. Inger Rydgaard, July 21, 1945; children—Svend, Ellen. Engring. cons., 1945; instr. Tech. U. Denmark, 1953, prof., founder chief lab. thermal isolation, 1964. Recipient Esso prize, 1954. Mem. Soc. Danish Engrs., Am. Soc. Heating, Refrigerating and Air Conditioning Engrs. Research and numerous publs. on thermal insulation, materials and application, environmental engring., thermal design of bldgs. Home: Stengaardsminde, Horsholm. Office: Hjortekoervej 99, Lyngby, Denmark.

KORSHAK, Vasiliy Vladimirovich, Russian organic chemist; b. Jan 9, 1909; grad. Moscow Inst. Chem. Tech., 1931; D.Chem. Sci. Asso., Inst. Organic Chemistry, USSR Acad. Sci., 1932-54, asso. Inst. Elemental-Organic Compounds, 1954-—, dep. dir., 1958-—, also lab. head; prof. Moscow Inst. Chem. Tech., 1942-—. Recipient Stalin prize, 1949, 51. Mem. USSR Acad. Sci. (corr.). Co-author: Tetraethyl Lead, 1946; The Synthesis and Study of High-Molecular Compounds, 1953; author: The Chemistry of High-Molecular Compounds, 1950; Principal Methods of Synthesizing High-Molecular Compounds, 1953; Methods of High-Molecule Chemistry, 1955; Research on Phosphoro-Organic Polymers, 1960. Research and publs. on chemistry of high-molecular compounds, mechanism of Friedel and Crafts' reaction, methods of synthesizing organic substances. Address: Moscow Inst. Chem. Tech., Miusskaya pl. 5/II, Moscow, USSR.

KORSHUNOV, Boris Georgijevich, Russian chemist; b. Kineshma, Jvanovo Oblast, USSR, Mar. 8, 1930; s. Georgij Alexejevich and Alexandra (Krilova) K.; Candidate of Tech. Scis., Inst. Fine Chem. Tech., 1955; m. Marine Sergejevna Gasdanova, June 1, 1952; children—Michail, Dmitrij. Asst. prof. Inst. Fine Chem. Tech., Moscow, USSR, 1955-60, asso. prof., 1960-—. Author: Chemistry and Technology of Rare and Dispersed Metals. Research and publs. on physico-chem. founds. of chlorine metallurgy. Home: Zagoryanskaya Station, 21, Narechnaya Ulitsa, Moskow Oblast, USSR. Office: Institute of Fine Chemical Technology, Malaya Pirogovskaya Ulitsa, 1, Moscow, USSR.*

KORSHUNOVA, Olga Stepanovna, Russian microbiologist; b. Moscow, USSR, 1909; d. Stepan Aelksandrovitch and Vera (Sokolova) Korshunova; grad. Moscow State U., 1930, Candidate Med. Scis., 1946. Staff, Mechnikov Inst., Moscow, 1931-37, All-Union Inst. Exptl. Medicine, 1938-42; head exptl. prodn. dept. Sci. Research Control Inst., Moscow, 1943-45; with Gamaleya Inst. Epidemiology and Microbiology, USSR Acad. Med. Sci., Gamaleya, 1946-—, Recipient State medal for labor valor. Mem. Znanije. Author: (with Piontkovskaya, Zhmajeva) Ixodid Vectors of Rickettsial Diseases, 1952; Human Diseases with Natural Foci, 1960; (with Piontkovskaya) Experimental Findings on Tick-Borne Typhus in the Krasnoyarskaya Territory, 1966; also articles. Research on tick-borne typhus, isolated strains of R. sibirica, epidemic outbreaks of nephroso-nehritis in Primorje area in S. Ukraine, Karpat and Tula region, isolated rickettsiea from patients blood, ticks and fleas. Home: Moscow, Festivalnaya ul. 7, USSR. Office: Gamaleya Inst. Epid. Microb., USSR Acad. Med. Sci., Ul. Gamaleya 2, USSR.*

KORTING, G. W., German dermatologist; b. Hindenburg, Germany, Sept. 23, 1919; s. Hans and Gertrud (Brouder) K.; student univs. Berlin, Breslau; Dr. med.; m. Johanna Kühnert, Aug. 30, 1942; 1 son, Hans-Christian. Asst., chief physician U. Tübingen, 1950-61, asst. prof., 1953-61; prof. dermatology, head skin clinic U. Mainz, Germany, 1961-—. Hon. mem. several dermatol. assns. Author: (with Bode) Textbook of Dermatology, 1962; Therapy of Skin Diseases, 1967; also numerous articles. Home: 3 Eselsweg, Mainz-Bretzenheim. Office: 1 Langenbeckstrasse, 65 Mainz, West Germany.*

KORVIN-KROUKOVSKY, Boris Viacheslav, physicist; b. Koslov, Russia, Feb. 6, 1895; s. Viacheslav I. and Serafima (Strelnikov) K.; grad. Mil. Engring. Coll., Russia; M.S., Mass. Inst. Tech., 1921; M.S. (hon.), Stevens Inst. Tech.; m. Eugenie A. Novitsky, Oct. 1915. Aero. engr. Aeromarine Plane and Motor Co., Keyport, N.J., 1921-24; chief engr., v.p. in charge engring. Edo Corp., College Point, N.Y., 1925-47; research prof. Davidson Lab., Stevens Inst. Tech., Hoboken, 1947-59; ret., 1959. Fellow Am. Inst. Aeros. and Astronautics; mem. Soc. Naval Architects and Marine Engrs., Royal Inst. Naval Architects. Author: Theory of Seakeeping, 1961; also articles. Developed design stability calculations for airplanes, series of standardized floats for seaplanes, calculations for prediction of ship motions in waves; theory of devel. of sea waves under action of wind. Home: P.O. Box 247, East Randolph, Vt. 05041.*

KORYAKIN, Ivan Sergeevich, Russian hygienist; b. 1898; grad. Med. Faculty, Kazan U., 1927; D.Med. Sci., 1949. Head chair communal hygiene Health-Hygiene Faculty, Kazakhstan Med. Inst., 1946-51, dep. dir. for sci. work, 1953-55, head chair gen. hygiene, 1953-—, dir., 1955-—; prof., 1950-—. Chmn. hygiene com. Kazakhstan Ministry Health; dep. chmn. learned med. council, mem. hygiene com. USSR Ministry Health. Mem. All-Union (bd. mem.), Kazakhstan Sci. (bd. mem.) socs. hygienists. Mem. editorial bd. Pub. Health Service of Kazakhstan; mem. editorial council Hygiene and Sanitation. Research and numerous publs. on communal hygiene. Address: Kazakhstan Med. Inst., ulitsa Furmanova 54, Alma-Ata, Kazakhstan SSR, USSR.

KORYTNYK, Wsewolod (Walter), chemist; b. Caslav, Czechoslovakia, Apr. 21, 1929; s. Luka and Valentyna (Makarenko) K.; B.Sc. 1st class honors, U. Adelaide (Australia), 1954, Ph.D., 1957; m. Olena Kozak, Aug. 17, 1957; children—Natalie, Christine, Peter. Came to U. S., 1958, naturalized, 1965. Postdoctoral research fellow U. Adelaide, 1957-58, Purdue U., Lafayette, Ind., 1958-59; research chemist U. S. Dept. Agr., Pasadena, Cal., 1959-60; sr. cancer research scientist Roswell Park Meml. Inst., Buffalo, 1960-68, asso. cancer research scientist, 1968-—; research asso. U. Cal. at Berkeley, 1967-68. Commonwealth Sci. and Indsl. Research Orgn. Australian studentship, 1955-58; Am. Cyanamid fellow, 1958-59. Mem. Am. Chem. Soc., A.A.A.S., Chem. Soc. (London), Shevchenko Sci. Soc. Research and publs. on carbohydrate chemistry, flavan chemistry, synthetic chemistry, vitamin B6. Home: 49 Glendale Dr., Tonawanda, N.Y. 14150. Office: 666 Elm St., Buffalo 14203.*

KORZHINSKY, Dmitriy Sergeevich, Russian geologist, petrographer; b. Sept. 13, 1899; grad. Leningrad Mining Inst., 1926. With Central Geol. Survey Research Inst., 1926-37; asst., lectr., prof. Leningrad Mining Inst., 1929-40; with Inst. Geol. Sci., USSR Acad. Sci., 1937-56, asso. Inst. for Geology of Ore Deposits, Petrography, Mineralogy and Geochemistry, 1956-—, bur. mem. dept. geology and geog. sci., 1960-—. Decorated Order of Lenin; recipient State prize, 1946, Lenin prize, 1958. Mem. USSR Acad. Sci. (Karpinsky prize 1949). Author: Outline of Metasomatic Processes, 1955; Physico-Chemical Principles of the Analysis of Mineral Parageneses, 1957; Research and publs. on physicochem. analysis of mineralization processes, Pre-Cambrian crystalline rocks and their concomitant minerals in Yakutia and Eastern Siberia, skarn ore deposits of Urals and Central Asia. Address: Inst. for Geology of Ore Deposits, Petrography, Mineralogy and Geochemistry, USSR Acad. Sci., Staromonetny p. 35, Moscow Zh-17, USSR.

KORZYBSKI, Alfred Habdank (Skarbek), scientist, author; b. Warsaw, Poland, July 3, 1879; s. Ladislas Habdank K. and Countess Helena (Rzewuska) K.; ed. Warsaw Gymnasium and Warsaw Poly. Inst.; grad. Study in Germany, Italy, U. S.; m. Mira Edgerly, Jan. 1919. Came to U. S. 1919, naturalized, 1940. Managed family estates in Poland; teacher of mathematics, physics, French and German in Warsaw, Poland; served with cav. and bodyguard heavy arty., also attached to Intelligence Dept., Russian Gen. Staff; sent to U. S. and Canada as artillery expert; sec. Polish-French Mil. Commn. in U. S., 1918; recruiting officer Polish-French Army, U. S. and Can., 1918; war lecturer for U. S. Govt.; sec. Polish Commn. (Labor Sect.), League of Nations, 1920; writer and lecturer; became pres. and dir. Inst. of General Semantics, Chicago, 1938, now Lakeville, Conn. Fellow American Assn. Advancement Sci.; member Am. Math. Soc., Chicago Soc. for Personality Study, Assn. for Symbolic Logic, N.Y. Acad. Sciences, Soc. for Applied Anthropology. Author: Manhood of Humanity—The Science and Art of Human Engineering, 1921; Science and Sanity, An Introduction to Non-aristotelian Systems and General Semantics, 1933, 3d edit., 1947; also many scientific papers. Founded school of psychological-philosophical semantics as system of understanding human behavior. Address: Institute of General Semantics, Lakeville, Conn.

KOSER, Stewart Arment, Am. bacteriologist; b. Harrisburg, Pa., Mar. 30, 1894; s. Alexander Stewart and Ella L. (Arment) K.; Ph.B., Yale, 1915, M.A., 1917, Ph.D., 1918; m. Hilda Marion Croll, Aug. 23, 1927; 1 dau., Marion Aimee (Mrs. George M. Armstrong). Bacteriologist U. S. Bur. Chemistry, U. S. Dept. Agr., 1919-23; asst. prof. U. Ill., Urbana, 1923-28; faculty U. Chgo., 1928-—, prof. bacteriology, 1943-59, prof. emeritus, 1959-—. Mem. Am. Soc. Microbiologists (past councilor), Soc. Ill. Bacteriologists (past pres., Pasteur award 1949), Am. Soc. Microbiology, Am. Acad. Microbiology, A.A.A.S., Internat. Assn. Dental Research, Soc. Exptl. Biology and Medicine. Author: Laboratory Directions for General Bacteriology, 1931; Vitamin Requirements of Bacteria and Yeasts, 1968; also numerous articles. Research on differentiation species in colon group bacteria, studies bacillus botulism, characteristics paratyphoid and dysentery groups bacteria, nutrition and metabolism of bacteria with spl. reference to vitamin requirements, oral microorganisms. Home: 5710 Drexel Av., Chgo. 60637.*

KOSHLAND, Daniel Edward, Jr., Am. biochemist; b. N.Y.C., Mar. 30, 1920; s. Daniel Edward and Eleanor (Haas) K.; B.S., U. Cal., at Berkeley, 1941; Ph.D., U. Chgo., 1949; m. Marian Elliott, May 25, 1945; children—Ellen, Phyllis, James, Gail, Douglas. Chemist Shell Chem. Co., Martinez, 1941-42; research asso. Manhattan Dist. U. Chgo., 1942-44; group leader Oak Ridge Nat. Labs., 1944-46; predoctoral fellow U. Chgo., 1946-49; postdoctoral fellow Harvard, 1949-51; staff Brookhaven Nat. Lab., Upton, N.Y., 1951-65; affiliate Rockefeller Inst., N.Y.C., 1958-65; prof. biochemistry U. Cal., Berkeley, 1965-—; mem. editorial bds. Jour. Biol. Chemistry, Biochimica et Biophysica Acta, Accounts Chem. Research, Jour. Molecular Pharmacology; Walker Ames lectr. U. Wash., 1964; O. M. Smith lectr. U. Okla., 1963. Mem. Nat. Acad. Scis. (chmn. panel), Am. Chem. Soc., Am. Soc. Biol. Chemists, A.A.A.S., Am. Acad. Arts and Scis. Research in mechanism of enzyme action; devised reagents and theoret. tests to correlate the structure of enzymes with their physiol. function; showed enzyme specificity and control properties can be related to flexibility of the molecule. Home: 3991 Happy Valley Rd., Lafayette, Cal. 94549. Office: Biochemistry Dept. U. Cal., Berkeley, Cal. 94720.*

KOSKI, Walter Stankiewicz, Am. chemist; b. Phila., Dec. 1, 1913; s. Bruno Stankiewicz and Helen (Laskowski) K.; Ph.D., Johns Hopkins, 1942; m. Helen Ireton Tag, May 11, 1940; children—Carol Lee, Ann Louise, Nancy, Phyllis. Research chemist Hercules Powder Co., 1942-43; group leader Los Alamos Sci. Lab., 1944-47; faculty Johns Hopkins, 1947-—, prof. phys. chemistry, chmn. dept., 1955-—; physicist Brookhaven Nat. Lab., Upton, N.Y., 1947-48. Mem. Am. Chem. Soc., Am. Phys. Soc. Research, numerous publs. on chemistry boron hydrides, application stable and radioactive isotopes, effects radiation on matter, electron spin resonance, mass spectroscopy and nuclear chemistry. Home: 809 E. Seminary Av., Towson, Md. 21204. Office: Charles and 34th Sts., Balt. 21218.*

KOSMIDER, Stanislaw, Polish physician; b. Dobczyce, Poland, Mar. 7, 1927; s. Andrew and Josephine (Figlewicz) K.; Dipl. physician, Jagiellonica U., Cracow, Poland, 1950; dr. dissertation Sil. Med. Acad., 1951; m. Charlotte Boldyriw, May 23, 1953; children—George, John. Staff, Silesian Med. Acad., Zabrze, Poland, 1950-—, asst. prof., 1964-—. Sci. cons. Coal Mine Inst., Katowice, Poland, 1964-—. Recipient State Sci. prize on radiobiol. research, 1964, Mark of distinction for work in med. service, 1965. Mem. Polish Med. Soc., Polish Internat. Medicine Soc., Polish Soc. Cardiology. Research, numerous publs. on mechanism of ionizing rays in living organisms, early diagnosis of radiation lesions, new drugs in post-radiation diseases, pathogenesis of lead and mercury poisoning and role of metals in atheromatic process, metabolism of heart muscle; bio- and histochem. investigations of enzyme activity in blood and tissues. Home: 2 Marksa, Zabrze 8, Poland.*

KOSMODAMIANSKY, Vladimir Nikolaevich, Russian microbiologist; b. Uglich, 1889; grad. St. Petersburg Mil. Med. Inst., 1914; D.Med. Sci., 1935. Head health epidemiology dept. Perm Guberniya Dept. Health, intern infectious diseases dept. Perm Guberniya Hosp., head bacteriology lab. Perm Out-Patients Tb Clinic, 1919-23; asst. infectious diseases clinic Perm U., 1920-21; asso. epidemiology dept. Inst. Exptl. Medicine, 1922; asst. Leningrad Med. Inst., 1924-28, sr. instr. 1928-34, acting head chair microbiology, 1930-34, head chair microbiology, 1934-—; prof., 1932-—. Co-author: Guide to Practical Studies in Microbiology, 1930; Textbook of Medical Microbiology, 1932. Author numerous works, including Bacteriology and Pathogenesis of Tuberculosis, 1950. Address: Leningrad Med. Inst., ulitsa Lva Tolstogo 6-8, Leningrad, USSR.

KOSOLAPOFF, Gennady Michael, chemist; b. Viatka, Russia, Sept. 2, 1909; s. Michael Paul and Alexandra Vasily (Shikhova) K.; came to U. S., 1924, naturalized, 1942; B.S. in Chem. Engring., Cooper Union, 1932; M.S. in Biochemistry, U. Mich., 1933, Sc.D. in Chemistry, 1936; m. Dorothea W. Bouton, July 20, 1934; children—Alexandra, Michael, Patricia. Research chemist Libby-Owens-Ford Glass Co., 1936-38, Monsanto Chem. Co., 1938-48; faculty Auburn U., 1948-—, research prof., 1953-—. Research cons. indsl. and govtl. agys. 1943-—; dir. chem. research F. J. Seiler Research Lab., Office of

Aerospace Research, Colo., 1964; cons. AEC; lectr. chemistry of phosphorus, 1949-54. Mem. Am. Chem. Soc., Chem. Soc. London. Author: Organophosphorus Compounds, 1950; also sects. books and encys. Abstractor: Chem. Abstracts. Home: 422 N. College St., Auburn, Ala. 36830.*

KOSS, Leopold George, pathologist; b. Danzig, Poland, Oct. 2, 1920; M.D., U. Bern, Switzerland, 1946; m. Lydia Simone Palla; children—Michael S., Andrew C., Richard P. Came to U. S., 1947, naturalized, 1952. Attending pathologist, chief cytology service Meml. Hosp. Cancer and Allied Diseases, N.Y.C., 1961—; head sect. cytopathology Sloan Kettering Inst. for Cancer Research, 19——; asso. prof. pathology Sloan Kettering div. Cornell Med. Coll., 19——; cons. dept. health N.Y. State Hosp. for Spl. Surgery. Recipient Wien award, Goldblatt award, Alfred P. Sloan award, Papanicolaou award. Mem. Am. Soc. Cytology (pres., 1962), Am. Assn. Pathol. Bact., Am. Soc. Clin. Pathologists, James Ewing Soc., A.M.A., Internat. Acad. Pathology, Internat. Acad. Cytology, others. Author: Diagnostic Cytology, 1961. Publs. on pathology and cytology of cancer with spl. interest in early forms of human cancer and effects of anticancer drugs. Office: 444 E. 68th St., N.Y.C. 10021.*

KOSSEL, Albrecht, chemist; b. Rostock, Switzerland, Sept. 16, 1853; ed. Strasbourg, France, Rostock; at least 1 son, Walther; asst. to Felix Hoppe-Seyler, Strasbourg, 1877-81; became asst. prof. Berlin Physiol. Inst., 1883; named prof. physiology, Marburg, Germany, 1895, Heidelberg, Germany, 1901. Recipient Nobel prize for medicine and physiology, 1910. Author: Untersuchungen über die Nukleine und ihre Spaltungsprodukte, 1881; Die Gewebe des menschlichen Körpers und ihre mikroskopische Untersuchung, 1889-91; (with Behrens and Schiefferdecker) Leitfaden für medizinisch Chemische Kurse, 1917. Research on cells and their nuclei; discovered nucleic acids and their breakdown purine and pyrimidine products; isolated adenine, guanine; discovered histidine in spermatozoa, 1896, also action of nucleic acids on bacteria; discovered (with Steudal) thymine, 1900. Died Heidelberg, July 4, 1927.

KOSSEL, Walter, physicist; b. Berlin, Germany, Jan. 4, 1888; s. Albrecht Kossel; Ph.D., U. Heidelberg (Germany), 1911; hon. doctorate U. Halle (Germany), 1944. Became prof. physics U. Kiel (Germany), 1921, U. Danzig (Poland), 1932-45, Tübingen, Germany, from 1947. Recipient Max Planck medal, 1944; Univ. medal U. Kiel, 1948. Developed theory of electrovalence in which he distinguished between ionized compounds and non-ionized substances; pointed out stability of external layers of 8 electrons in atoms; investigated spectra of X-rays and gamma rays. Died 1956.

KOSTEN, Cornelis Willem, physicist; b. Rotterdam, Holland, Jan. 29, 1913; s. Bartel K. and Anna (Rijsdam) K.; student Tech. U. Delft (Holland), 1930-36, Ph.D., 1942; m. Veren Kosten, Sept. 8, 1938; children—Charlotte, Berthold. Research fellow Rubber Found., Delft, 1936-42; lectr. Tech. U. Delft, 1942-51, extraordinary prof., 1951-58, prof., 1958—; acoustical cons.; founder, 1st editor-in-chief internat. jour. acoustics Acustica, 1951-56. Mem. tech. physics dept. TNO-TH. Fellow Acoustical Soc. Am. Author: Elastic Properties of Rubber, 1942; (with C. Zwikker) Sound Absorbing Materials, 1959. Research and publn. on theory of sound absorbing materials, behavior and application; contbn. to internat. standardization in acoustics. Home: 4 Mijnbouwstratt, Delft. Office: 1 Lorentzweg, Delft, The Netherlands.*

KOSTENKO, Mikhail Polievkovich, Russian elec. engr.; b. Dec. 16, 1889; grad. Petrograd Poly. Inst., 1918; at Petrograd., later Leningrad Poly. Inst., 1918-30; prof. Central Asian Indsl., Tashkent, 1929-32, 42-44 cons., tchr., Rumania, Hungary, Bulgaria, Poland; dir. Inst. electromechanics Acad. Scis. Del., Paris Conf. on Large-Scale Elec. High-Tension Systems. Recipient Stalin prize, 1949, 51, Lenin prize, 1958; named honored sci. worked Uzbek USSR, 1944. Mem. USSR Acad. Scis. (presidium). Author: Alternating Current Commutator Machines, 1933; Turbo-generators, 1939; Electrical Machines, 1944-49; Electrical Machine Building, 1953. Contbd. to devel. of basic elec. machines produced in Russia; developed theory of transformers, polyphase asynchronous and commutating machines; worked out original scheme of commutative generators for alternating current. Address: Inst. of Electromechanics, USSR Acad. of Scis., Dvortsovaya Naberezhnaya, 18, Leningrad, USSR.

KÖSTER, Rolf, German geologist; b. Plön, Germany, May 31, 1929; s. Hellmuth and Elisabeth (Meyer) K.; Diplom-Geologe, U. Kiel (Germany), 1954, Dr. rer.nat., 1956; m. Dorothea Maier, May 14, 1959. Sci. asst. U. Kiel, 1954-61, prof. Universidad Austral de Chile, Valdivia, 1962-64; dozent U. Kiel (Germany), 1964-67, prof., 1967—. Mem. German Geol. Soc., Geologische Vereinigung, Geol. Soc. Sweden, Sociedad Geologica de Chile. Research, publs. on tectonics, seismology and volcanology, eustatic and isostatic sea-level changes, glacial morphology and glacial tectonics. Home: 19 Jägersberg, Kiel, Schleswig-Holstein 23, Germany.*

KÖSTER, Werner Otto, German physicist; b. Hamburg, Nov. 22, 1896; s. Paul and Elsa (Grallert) K.; ed. U. Friburg, U. Göttingen; Ph.D.; D.Eng. honoris cause, U. Berlin; m. Ilse Kerschbaum, May 12, 1923; children—Albrecht, Helma, Eberhard. Asst., Kaiser Wilhelm Inst. for Metall. Research, Dusseldorf, 1922-24; metallurgist Schweizerische Metallwerke Selve and Co., Thun, Switzerland, 1924-27; asst. Research Inst., Vereinigte Stahlwerke Algemeine Gesellschaft, Dortmund, 1927-33; dir. research labs. Deutsche Edelstahlwerke Algemeine Gesellschaft, Krefeld, 1933-34; dir. Kaiser Wilhelm Inst. and Max-Planck Inst. for Metall. Research, 1934—; rector Stuttgart Tech. Coll., 1957-58; with Max-Planck Soc. for Sci. Devel., 1960-63. Recipient Heyn medal Deutsch Gesellschaft für Metallkunde, medal Platinum Inst. Metals, French Soc. Metallurgy. Mem. German, Japan, London (hon.) insts. metals, German Acad. Natural Sci., Leopoldina Soc., German Soc. Metallurgy, Am. Soc. Metals, Verein Deutscher Eisenhüttenleute, Instituto del hierro y del acero (Madrid), Japanese Inst. Iron and Steel. Author 260 works in Reine und angewandte Metallkunde series. Editor: Zeitschrift für Metallkunde, 1936—. Research in thermic treatment and phys. properties of metals and alloys; devel. of permanently magnetic alloys and carbon-free steels. Home: Eduard-Pfeiffer-Strasse 79. Office: Seestrasse 75, Stuttgart, Germany.

KOSTERLITZ, Hans Walter, physiologist; b. Berlin, Germany, Apr. 27, 1903; s. Bernhard and Selma (Lepman) K.; M.D., U. Berlin, 1929; Ph.D., Aberdeen U., Scotland, 1936, D.Sc., 1944; m. Johanna Maria Gresshoener, Mar. 5, 1937; 1 son, John Michael. Asst., First Med. Clinic, U. Berlin, 1928-33; staff U. Aberdeen, 1936—, reader in physiology, 1955—; vis. lectr. Harvard Med. Sch., 1953-54. Fellow Royal Soc. Edinburgh; mem. Physiol. Soc., Brit. Pharmacological Soc., Soc. for Endocrinology. Contbr. numerous articles to profl. jours., chpts. to books. Research on mechanism of assimilation of galactose; discovered galactose-l-phosphate; effects of nutrition and hormones on composition and structure of live; physiology and pharmacology of transmission at autonomic neuroeffector junctions; motility of small intestine; mode of action of morphine and morphine-like drugs. Home: 16 Glendee Terrace, Cults, Aberdeen, Scotland.*

KOSTIC, Aleksander, Yugoslavian histologist; b. Belgrade, Yugoslavia, 1893; s. Djordje and Katarina (Putnik) K.; ed. Belgrade, also U. Strasbourg (France); m. Smilja Joksic. Lectr. med. faculty U. Belgrade, prof. histology and embryology, from 1924. Mem. Paris Assn. Anatomists, Paris Soc. Biology. Publs. on histology, exptl. Blastophtoria, expts. with gland hibernae, sexual questions.

KOSTOV, Ivan, Bulgarian mineralogist; b. Plovdiv, Bulgaria, Dec. 24, 1913; s. Kosta and Fanka (Apostolova) K.; student U. Sofia (Bulgaria), 1932-36; D.I.C., A.R.S.M., Royal Sch. Mines, U. London (Eng.), 1941; m. Maria Slavova, June 22, 1947; children—Fanny, Russian. Faculty, U. Sofia, 1938—, prof. mineralogy and crystallography, 1953—, head chair mineralogy and crystallography, 1953—; academician Bulgarian Acad. Scis., 1966—, head dept. mineralogy and crystallography, 1961—. Cons. Geol. Survey of Bulgaria, 1955-56. Recipient Order of Red Banner of Labour, 1964. Mem. Bulgarian Geol. Soc. (pres. 1960-62), Bulgarian Assn. Natural Scis. (v.p. 1958—), Union Sci. Workers in Bulgaria, Mineral. Soc. Gt. Britain, Geol. Soc. London (Fgn.), Commn. on New Minerals. Author: Crystallography, 1955; Mineralogy, 1957; also numerous articles. Research on origin of minerals, relationship between mineral deposits and igneous activity, special distbn. of zeolites and their relation to sulphide mineralization, crystal habit variation of minerals and its genetic significance; originated new classification of minerals based on geochem. and crystallochem. principles. Home: 53 Shishman, Sofia, Bulgaria.*

KOSTRZEWSKI, Jozef, Polish archeologist; b. Weglewo, Poland; Feb. 25, 1885; s. Stanislaw, and Elizbieta (Bronkanska) K.; ed. U. Wroclaw (Poland), U. Cracow (Poland), U. Berlin (Germany); Ph.D., U. Berlin; Dr.h.c., Jagiellonian U., Cracow, Mickiewicz, U., Poznan, Poland; m. Jadwiga Wroblewsk, 1914. Became prof., Lwow U., 1918, Poznan U., 1919; became prof., organizer Inst. Prehistory, Poznan U., 1919, dean Faculty Letters, 1926-27, prof., 1945-50, 56-60. Mem. Internat. Congress Prehistoric Sci. and Protohist. Sci. (mem. permanent council, Baltic Archeologists (mem. permanent council), Polish Acad. Learning, Polish Acad. Scis., Warsaw, Poznan sci. socs., Polish Archeol. Soc. (founding pres.). Author: Wielkopolska in Prehistory, 1914; From the Mesolithic Age to the Migrations of Peoples, 1939-48; Les Origines de la civilisation polonaise, 1947; Prehistory of Poland, 1949; Lusation Culture in Pomerania, 1958; Zur Frage der Siedlungsstetigkeit in der Vorgeschichte Polens, 1964. Opposed German theory of inferiority of Slavic culture and origin of Slavic peoples. Office: Biskupinska 1, Poznan, Poland.

KOSTYUK, Grigoriy Silovich, Russian psychologist; b. 1899. Prof., Kiev Pedagogical Inst.; dir. Ukrainian Psychol. Research Inst. Del., Internat. Congress Psychologists, Can., 1954, Belgium, 1957. Mem. RSFSR Acad. Pedagogical Sci. (corr.), RSFSR (v.p.), All-Union (chmn. Ukrainian dept.) socs. psychologists, Ukrainian Znanie Soc. (Presidium mem.). Mem. editorial bd. Problems of Psychology; editor: Outline History of Russian Psychology in the 19th Century, 1955; Psychology, 1955; Problems of General Psychology, 1955; The 1955 Ukrainian Republican Psychological Conference. Research and publs. on ednl. psychology, history of and theoretical psychology, pupil assimilation of sch. material. Address: Kiev Pedagogical Inst., b. Tarasa Shevchenko 22-24, Kiev, Ukrainian SSR, USSR.

KOSYGIN, Yuriy Aleksandrovich, Russian geologist; b. Jan. 22, 1911; grad. Moscow Petroleum Inst., 1931. With Inst. Mineral Fuels, USSR Acad. Sci., later Moscow Petroleum Inst., 1935-41; sr. asso. Geol. Inst., USSR Acad. Sci., 1945—. Mem. USSR Acad. Sci. (corr.). Author: The Petroleum Deposits of Turkmenia, 1933; The Tectonic Principles of Petroliferous Areas, 1952; The Salt Tectonics of Platform Areas, 1956; The Tectonics of Petroliferous Areas, 1958; The Principles of Ancient Platform Formation and the Location of Marginal Upheavals in the Structure of the Siberian Palteau, 1960. Research and publs. on tectonics of platforms and marginal sagging in petroliferous areas, helped determine laws of location of petroleum deposits in salt domes; co-compiler tectonic maps of USSR, 1952, 56. Address: Geol. Inst., USSR Acad. Sci., Pyzhevsky p. 7, Moscow, USSR.

KOTANI, Shozo, Japanese microbiologist; b. Hyogo, Japan, Nov. 29, 1922; s. Keizo and Jo (Tsuchiya) K.; M.D., Osaka U. Med. Sch., 1945, D.M.S., 1950; m. Yaeko Nakamura, May 23, 1949; children—Masako, Yoichi. Research asso. dept. pub. health Osaka U. Med. Sch., 1950-52, asso. prof., 1952-56; prof. dept. bacteriology Nara U. Med. Sch., 1956-64; prof. dept. microbiology Osaka U. Dental Sch., 1964—. Mem. Japan Bacteriol. Soc. (councillor 1957—). Contbr. numerous articles to sci. jours. Discovered two cell-wall lytic enzymes, one of which is active against lysine type cell walls and other is lytic on diaminopimelic acid type cell walls of lysozyme-resistant microorganisms; studied chem. and immuno-biol. properties of cell walls of various pathogenic and useful bacteria, including S. aureus, C. diphtheriae, L. plantarum, BCG and others by use of enzymes stated above. Home: 172-6 Sakura, Mino, Osaka, Japan.*

KOTEL'NIKOV, Vladimir Aleksandrovich, Russian radio engr.; b. Aug. 24, 1908; grad. Moscow Inst. Energetics, 1931. With Radio Engring. and Electronics Inst. of USSR Acad. Scis., dir., 1954—, academician, 1953—. Recipient Stalin prize, 1943, 46. Author: (with A. M. Nikolaev) Foundations of Radio-Engineering, I, 1950, II, 1954. Introduced concept of potential error stability as characteristic of given method of transmission; his method of analysis has had wide application and great significance for devel. new methods of radio communication; under his direction a multi-channel telephone-telegraphic line of radiocommunication on single frequency side band was worked out. Home: 1-Aya Cheremushkinskaya, 3, Moscow. Office: Institute of Radio Engineering and Electronics, USSR Acad. of Sciences, Mokhovaya ulitsa 11, K-9, Moscow, USSR.

KOTHARI, Laxman Singh, Indian physicist; b. Udaipur, Rajasthan, India, Apr. 13, 1926; s. Daulat Singh and Sujan (Surana) K.; B.Sc. with honors, St. Stephen's Coll., U. Delhi (India), 1946, M.Sc. (India govt. fellow), 1948; India govt. fellow U. Cambridge, Eng., 1951-54; Ph.D., U. Bombay (India), 1961; m. Sushila Vardya, Feb. 16, 1951; children—Vandana, Jyoti, Ranjan. Staff, Atomic Energy Establishment, Trombay, Bombay, 1955-61; faculty Panjab U., Chandigarh, India, 1961-62; reader physics U. Delhi, 1962-63, Sir Shri Ram prof. physics, 1963—. Mem. panel Internat. Atomic Energy Agy., Vienna, Austria. Fellow Nat. Inst. Scis. India; mem. Am. Nuclear Soc. Research, publs. on theoretical neutron physics; (with Singwi), explained scattering of thermal neutrons from graphite and developed a method for studying neutron thermalisation in crystalline moderators. Home: 5 University Rd., Delhi, India.*

KOTIN, Paul, Am. pathologist, physician; b. Chgo., Aug. 13, 1916; s. Elias and Rose (Spunt) K.; B.S., U. Ill., 1937, M.D., 1940; m. Helen Tepper, May 6, 1939; children—Joel Tepper, David Bernard. Practice medicine, specializing in pathology, Los Angeles, 1946-48, Bethesda, Md., 1962—; attending staff pathologist Los Angeles County Hosp., 1951-62; faculty U. So. Cal., 1961-62, prof. pathology, 1959-60, Paul Peirce prof. pathology, 1960-62; with Carcinogenesis Studies br. Nat. Cancer Inst., 1962—, sci. dir., 1966—, dir. div. Environmental Health Scis., NIH, 1967—; dir. Nat. Environmental Health Scis. Center in N.C., 1967—. Vis. prof. oncology U. Wis., 1959-60; vis. prof. pathology Duke, U. N.C. Chmn. Gordon Research Conf. on Cancer, 1965. NSF fellow, 1959-60. Mem. A.M.A., Am. Assn. Cancer Research, Am. Assn. Pathologists and Bacteriologists, N.Y. Acad. Scis., A.A.A.S., Am. Soc. Exptl. Pathology, Internat. Acad. Pathology, Am. Inst. Biol. Scis., Am. Soc. Cell Biology, Coll. Am. Pathologists, Sigma Xi. Identification, publs. on carcinogens in polluted air and cigarette smoke and their interaction with viruses in cancer prodn.; lung def. studies. Home: Iris Lane, Chapel Hill, N.C. 27514. Office: P.O. Box 12233, Research Triangle Park, N.C.

KOTO, Bunjiro (or Bundjiro), Japanese geologist, seismologist; b. Tsuwano (now in Shimane), Japan, Mar. 4, 1856; s. Samurai Jisei Koto; grad. geol. labs. Tokyo Imperial U., 1879; postgrad. Munich (Germany) U., 1882; Ph.D., Leipzig (Germany) U., 1884; m. Ume Arichi. Mem. faculty phys. sci. U. Tokyo, also with geol. survey, 1884-86, prof., coll. sci., from 1886, prof. geology and paleontology, grad. faculty, 1899, 1921, lectr., coll. sci., until 1923. Mem. Earthquake Investigation Com., 1892. Corr. mem. geol. socs. London, Am.; pres. Geol. Soc. Japan, 1909. Author Japanese textbooks on mineralogy, phys. geography, geology of China, Korea and Japan, earthquakes, volcanoes, petroleum, Pacific Ocean basis; also monograph The Great Eruption of Sakurajima in 1914, in Jour. Geol. Soc. Japan, 1916. Monograph of 1916 (with those of Omori) gives most complete picture ever pub. of phys. and geol. aspects of single volcanic outbreak, removed doubt about dislocation of crust in earthquake. Died Tokyo, Mar. 8, 1935.

KOTON, Mikhail Mikhailovich, Russian chemist; grad. Leningrad State U., 1935. With Leningrad Inst. High Pressures, 1934-36; with Leningrad Pediatric Med. Inst., 1936-39, prof., 1946—, also head dept. gen. and analytic chemistry; head dept. gen. and analytic chemistry Leningrad Physico-Tech. Inst., USSR Acad. Scis.; with Leningrad Inst. High-Molecular Weight Compounds, USSR Acad. Scis., 1952—, dep. dir., 1959-60, dir., 1960—; prof. Leningrad Poly., 1952-60. Recipient Stalin prize, 4 times. Corr. mem. USSR Acad. Scis. Research and publs. on chemistry of organic, metallo-organic and high molecular weight compounds. Office: Inst. High Molecular Weight Compounds, USSR Acad. Scis., Birzhevoy Prospekt 6, Leningrad, USSR.

KOTTER, Fred Ralph, Am. physicist; b. Salt Lake City, Dec. 8, 1915; s. Fred O. and Bertha (Midgley) K.; B.S., U. Utah, 1937; M.A., George Washington U., 1940; Sc.D., Mass. Inst. Tech., 1955; m. Lora Norman, Mar. 28, 1949; children—Loralee, Wade, Nola, Shauna, Virginia. Physicist, Nat. Bur. Standards, Washington, 1942-47, 55—. From instr. to asst. prof., Mass. Inst. Tech., 1948-54. Fellow Am. Phys. Soc.; mem. I.E.E.E., Sigma Xi, Phi Beta Kappa. Research in precise elec. measurements. Home: 12921 Crisfield Rd., Silver Spring, Md. 20906. Office: 401 HVL, Nat. Bur. Standards, Washington 20234.*

KÖTTGEN, Ulrich, German pediatrician; b. Cologne, Germany, June 30, 1906; s. Arnold and Hedwig (Wiskott) K.; M.D.; m. Beate von Grabe, 1934; children—Angelika, Eckart, Sabine, Rainer, Wolfram. Asst., chief physician Univ. Pediatric Clinic of Iena, 1934-36; head physician Munster U. Pediatric Clinic, 1937-48; dir. U. Mainz Pediatric Clinic, 1949—. Author: Statistische Untersuchungen zum kindlichen Rheumatismus; Rheumatisches Fieber im Kindesalter, also others. Address: Langenbeckstrasse 1, Mainz, Germany.

KOTTHAUS, Adolf, German ichthyologist; b. Meiersberg, Germany, May 4, 1909; s. Emil and Elisabeth (Bathe) K.; student U. Jena (Germany) 1928-1930, U. Vienna (Austria), 1930; Dr.phil., U. Greifswald (Germany), 1932; m. Emmy Budde, June 29, 1939; children—Udo, Rainer. Biologist Marine Biol. Sta., Büsum, 1933-34; sci. asst. Biologische Anstalt Helgoland, 1934-36; fisheries biologist Oceanografski Inst., Split, Yugoslavia, 1937-38; sci. asst. Biologische Anstalt Helgoland, 1939-53, chief dept. ichthyology, 1954-67, asst. dir., 1966-67, sci. dir. dept. ichthyology, 1967—. Mem. Deutsche Wissenschaftliche Kommission für Meeresforschung, 1954—; mem. various coms. Internat. Council Exploration of Sea, Copenhagen, Denmark, 1954—. Research and numerous publs. on fisheries biology and ichthyology especially on ocean perch, sole, plaice and fishes of Indian Ocean. Home: 350 Fibigerstrasse, 2 Hamburg 62. Office: 9 Palmaille, 2 Hamburg 50, Germany.*

KOTTKE, Frederick James, Am. physician; b. Hayfield, Minn., May 26, 1917; s. George G. and Harriet (Davidson) K.; B.S., U. Minn., 1939, M.S., 1941, Ph.D., 1944, M.D., 1945; m. Astrid M. Erling, May 27, 1939; children—Jane, Mary, Thomas. Faculty, U. Minn., 1941—, prof. phys. medicine, 1953—, head dept. phys. medicine, and rehab., 1952—, Baruch fellow phys. medicine, 1945-47. Chmn. program com. 3d Internat. Congress Phys. Medicine, 1960—; mem. Minn. Bd. Health, 1964-67. Diplomate Am. Bd. Phys. Medicine and Rehab. (chmn. 1963—). Mem. Am. Congress Phys. Medicine and Rehab. (past pres., Distinguished Service Key 1961), Am. Physiol. Soc., A.A.A.S., Am. Rehab. Found., Am. Assn. U. Profs., A.M.A., Soc. Exptl. Biology and Medicine, Nat. Rehab. Assn. Author: (with F. H. Krusen, P. M. Ellwood) Handbook of Phys. Medicine and Rehabilitation, 1965; also numerous articles. Research in explt. renal hypertension and its control with ergot derivatives, kidney function, temperature regulation, high altitude tolerance, circulation and quantitative assessment cardiac output and cardiac work, bone growth disturbance following acute anterior polio-myelitis, action sympathetic nervous system on bone growth in dogs and humans. Home: 2741 Drew S., Mpls. 55416.*

KOTYZA, Frantisek, Czechoslovak physician; b. Prostejov, Czechoslovakia, Oct. 10, 1906; s. Karel and Eleonora (Rozehnalová) K.; student Med. Faculty, Charles U., Prague, Czechoslovakia, 1924-30, Otolaryn. Clinic, 1924-30; D.Sc., 1967; m. Drahomíra Ocenasková, July 29, 1939; children—Jaromír, Vojtech, Vladivoj, Jirí. Asst. to prof. A. Precechtel, 1933-39; ordinarius, primarius otolaryng. dept. hosp. in Havlíckuv Brod, Czechoslovakia, 1939-48; prof. Med. Faculty, Plzen, Czechoslovakia, 1948—; chief Otolaryn. Clinic, Plzen, 1948—; dean Med. Faculty, Charles U., Plzen, 1950-52, vice dean, 1959-61. Mem. sci. council Ministry Common Health, 1953-66. Recipient medal of honor Charles U., 1965. Mem. Czechoslovak (past pres.), Polish (fgn.), French (fgn. corr.) otolaryn. socs. Author: (with others) Otolaryngology, 1953; (with A. Precechtel, R. Hladky, K. Sedlácek) Fundamentals of Otolaryngology, 1959; also numerous articles. Editorial bd. Ceskoslovenská Otolaryngologie, 1952—, Acta Universitatis Carolinae, 1961; pres. editorial bd. Acta Med. Pilsensia, 1960—. Research on arytaenoidectomy by external approach in cases of bilateral recurrent paralysis, permanent communication between frontal sinus and nasal cavity, tri-fenestral hearing, radiotherapy for chemodectoma ossis temporalis; clin. utilization of function of fenestra rotunda for improvement of hearing in tympanoplastic operations. Home: 18 Hálkova ul., Plzen, Czechoslovakia.*

KOURGANOFF, Vladimir, astronomer; b. Moscow, USSR, Mar. 3, 1912; s. Alexander and Sophie Kourganoff; D.Sc., Sorbonne; m. Ruth Moj, Oct. 15, 1935; children—Jean-Claude, Nadia. Instr., researcher Nat. Center Sci. Research, 1940-52; dir., prof. astronomy labs. Faculty Sci., Lille, 1952-61; prof. Faculty Sci., Paris, 1961—. Vis. prof. U. Cal. at Berkeley, 1958. Author: La part de la mécanique céleste dans la découverte de Pluton, 1941; Basic Methods in Transfer Problems, 1952; La recherche scientifique, 1958; L'astronomie fondamentale élémentaire, 1961. Research on theory of neutron diffusion, radiative transfer, basic phys. concepts, math. methods, psycho-sociology of sci. research. Address: Astronomy Service, Faculty of Sciences, Osray (Seine et Oise), France.

KOURIAS, Basile (Georges), Greek physician; b. Ambelakia, Greece, Mar. 5, 1903; s. Georges Basile and Aphrodite (Haliambalia) K.; Doctor degree, U. Athens, 1928; postgrad. in Austria, 1936, Germany, 1937, U. S. A., 1949; m. Catherine S. Lanaras, Sept. 25, 1949; children—Aphrodite, Georges, Stavros. With Red Cross Hosp., Athens, Greece, 1932—, asso. dir. surg. dept., 1937-52, dir., 1953—, permanent trustee, 1960—, faculty Nurses Sch.; faculty U. Athens, 1937—, asso. clin. prof., 1948—. Recipient several awards, including Cross of Saint Marc, 1955; Golden Cross Helenic, Red. Cross, 1959; Distinguished services Bulgarian Red. Cross, 1966. Hon. fellow A.C.S., German, Bulgarian socs. surgeons; mem. Hellenic Soc. Surgery (past pres.), Internat. Soc. Echinoccosis (past pres.), Soc. Internat. Chirurgie, Academie Chirurgie (France), Soc. Italian Chirurgia, Soc. Française Chirurgie. Author: (with K. Stucke) Atlas Per-Operative Cholangiography, 1966; also numerous articles. Research on peptic ulcer surgery, biliary surgery, pre- and post-operative cholangiography, surgery in echinococcus diseases, tentanus, pancreatic surgery, frostbite, geriatric surgery, hiatus-hernia surgery, hosp. infections. Home: II, Herodou Attikou, Athens 138. Office: Red Cross Hosp., Athens, 607, Greece.*

KOURILSKY, Raoul Georges, French physician; b. Bombon, July 19, 1899; s. Michel and Eugénie (Oudart) K.; M.D., U. Paris; D. honoris causa, U. Que., U. Montreal; m. Simone Develay, Mar. 26, 1931; children—Françoise, François, Marie-Thérèse, Elisabeth, Philippe, Olivier. Clinic chief, asst. to Prof. Sergent, 1927-34; chief service Hosps. of Paris, 1934; prof. agrégé internal pathology U. Paris, 1944, prof. chair clin. medicine, 1958; dir. Immunopathology Center, St. Anthony Hosp.; vis. prof. N.Y. U. Research and publs. on respiratory passages, Tb, acute infections, retarded hypersensibility. Home: 7, rue Bayard, Paris 8. Office: Hopital Saint-Antoine, 184, rue du Faubourg-Saint-Antoine, Paris 12, France.

KOUTRAS, Demetrios Anastasios, Greek physician; b. Athens, Greece, Jan. 25, 1930; s. Anastasios D. and Phaedra (Adamidou) K.; diploma Athens U. Sch. Medicine, 1955, doctorate, 1958. With dept. clin. therapeutics Athens U., 1955-59, 62-63, asst. prof. medicine, 1965—; Brit. Council scholar Glasgow (Scotland) U. at Western Infirmary, 1959-60, lectr. medicine, 1961; staff Alexandra Hosp., Athens, 1926-63, dir. thyroid clinic, 1964—; vis. scientist NIH, Bethesda, Md., 1963-64. Mem. Greek Endocrine Soc. (sec. gen.), other Greek and internat. med. socs. Author: (with E. J. Wayne, W. D. Alexander) Clinical Aspects of Iodine Metabolism, 1964; also articles. Research on thyroid gland especially iodine metabolism, which has facilitated diagnosis of thyroid disease and provided more rational interpretation of tests for thyroid function; studies on causes of non-toxic goiter in Glasgow, endemic non-toxic goiter in Ky., U. S., and Greece. Home: 14, Patriarchou Ioakim, Athens 139. Office: Alexandra Hosp., Vas. Sofias Str., Athens 611, Greece.*

KOUTS, Herbert John Cecil, Am. physicist; b. Bisbee, Ariz., Dec. 18, 1919; s. Oliver Allen and Lillian (Niemeyer) K.; B.S., La. State U., 1941, M.S., 1946; Ph.D., Princeton, 1952; m. Hertha Isla Pretorius, Feb. 2, 1942; children—Anne Elizabeth, Catherine Jennifer. Sr. physicist Brookhaven Nat. Lab., Upton, N.Y., 1950—. Mem. U. S. Adv. Com. on Reactor Physics, 1957-62, U. S. Adv. Com. on Reactor Safeguards, 1962-66, European-Am. Com. on Reactor Physics, 1961—. Recipient E. O. Lawrence award AEC, 1963. Fellow Am. Nuclear Soc. (dir. 1965—). Research and publs. on behavior neutrons in nuclear reactors; improved performance of research reactors. Home: S. Country Rd., Brookhaven, N.Y. 11719. Office: Brookhaven Nat. Lab., Upton, N.Y. 11973.*

KOUWENHOVEN, William Bennett, Am. elec. engr.; b. Brooklyn, N.Y., Jan. 13, 1886; s. Tunis Garet Bergen and Phebe Florence (Bennett) K.; E.E., Brooklyn Poly. Inst., 1906, M.E., 1907; Diplom Ingenieur, Karlsruhe Technische Hochschule, Baden, Germany, 1912, Doktor Ingenieur, 1913; Doctor of Laws, Johns Hopkins University, 1962; m. Abigail Baxter Remsen, June 22, 1910; 1 son, William Gerrit. Asst. in physics, Brooklyn Poly. Inst., 1906-07, instr. in physics and elec. engring., 1907-10; instr. in elec. engring., Washington U., 1913-14; same, Johns Hopkins, 1914-17, asso. in elec. engring., 1917-19; engring. supt. Winchester Repeating Arms Co. (leave of absence from Johns Hopkins), 1919-20; asso. prof. elec. engring., Johns Hopkins, 1919-30, prof. and asst. dean Sch. of Engring., 1930-38, prof. and dean school of engineering, 1938-54, professor emeritus, 1954—; lectr. in surgery Johns Hopkins Sch. Medicine, 1956—; cons. engineer U. S. Bureau of Mines, U. S. Bureau Standards. Instr., rank of capt., R.O.T.C., World War. Recipient Ludwig Hekton Gold medal, A.M.A., 1961; Edison Medalist, American Institute E.E., 1962. Fellow A.A.A.S., Am. Inst. E.E. (v.p. 1931-33; dir. 1935-39; chmn. Baltimore sect. 1922-31); mem. Am. Society for Testing Materials, American Welding Society, Executive Committee of Research Council, Nat. Acad. Sci., Am. Heart Assn., Nat. Assn. of Arbitration. Clubs: Johns Hopkins Baltimore Engineers, Gibson Island; Engineers (New York). Contributor of numerous articles and papers on electrical measurements, electric shock, magnetic analysis, electric welding, effect of electricity on heart, elec. defibrillation, closed chest cardiac massage, etc., to Trans. Am. Inst. Elec. Engring., Elec. World, Proc. Am. Soc. for Testing Materials, Welding Journal, med. and surg. jours., etc. Important work in the application of electrical science to medicine; developed countershock for the cure of heart fibrillation, also method of cardiopulmonary resuscitation. Home: 334 St. Dunstan's Rd., Balt.

KOUYOUMZELIS, Theodore George, Greek physicist; b. Kydoniai (Ayvalik) Asia Minor, Dec. 1, 1906; s. George and Polyxene (Petridou) K.; M.Chem., U. Athens (Greece), 1928, Dr.Sc., 1932; m. Stella Grigoriou, Nov. 21, 1940; 1 dau., Alexandra. Research fellow U. Munich (Germany), 1935, 37, U. Heidelberg (Germany), 1936, U. Manchester (Eng.), 1950, AEC schs. at Harwell, Eng., 1951, Oak Ridge, 1955; faculty U. Athens, 1926-58, extraordinary prof., 1949-58; prof., head physics lab. Tech. U. Athens, 1958—, dean faculty chem. engring., 1961—; prof. Royal Naval Acad., Piraeus, Greece, 1947—; sec.-gen. Greek AEC, 1954-60. Greek permanent rep. CERN, Geneva, Switzerland, 1955—. Decorated comdr. Order Phoenix, Order King George I. Mem. Greek Chem. Soc., Greek Phys. Soc., Greek Nuclear Soc., Am. Phys. Soc., Am. Nuclear Soc. Author: Nuclear Physics, 1947; Theoretical Electricity, 1948; Alternating Currents, 1948; Vibrations and Waves, 1948; Elements of Physics and Elementary Nuclear Physics, vol. I-IV (with S. Peristerakis), 1960; also articles. Research on Raman effect and structure of ions, glass and water molecular vibrations, gamma rays and their interaction with matter, counting and electronic devices, fallout measurement. Home: 23, Pindou str., Filothei-Athens, Greece.*

KOVACIC, Peter, Am. chemist; b. Wylandville, Pa., Aug. 1, 1921; s. Marko and Barbara (Radinich) K.; A.B., Hanover Coll., 1943; Ph.D., U. Ill., 1946; postgrad. Mass. Inst. Tech., 1946-47; D.Sc., Hanover Coll., 1964; m. Dorothy Fehrenbacher, June 29, 1946; children—Lynn, Jan, Paul, Don, Eric, Ken. Instr., Columbia, 1947-48; research chemist Du Pont Co., 1948-55; faculty Case Inst. Tech., Cleve., 1955—, prof. chemistry, 1965—. Mem. Am. Chem. Soc., Am. Assn. U. Profs., A.A.A.S., Sigma Xi, Phi Lambda Upsilon, Alpha Chi Sigma. Research and publs. on aromatic substitution, metal halides in organic chemistry, novel polymerization methods, chemistry N-halamines, chemistry cancer. Home: 6900 Wilson Mills Rd., Gates Mills, O. 44040. Office: University Circle, Cleve. 44106.*

KOVACS, Bela Adalbert, pharmacologist, educator; b. Nagykoros, Hungary, Aug. 28, 1921; s. Joseph and Amalia (Szabo) K.; M.D., Med. U., Szeged, Hungary, 1946; Ph.D., U. London, 1961; m. Eva M. Juhasz, Aug. 14, 1952. Faculty dept. pharmacology Med. U., Szeged, 1948-56; research fellow dept. organic chemistry U. Basle, 1956-57, Nat. Inst. for Med. Research, London, Eng., 1957-61; faculty McGill U., Montreal, Que., Can., 1961—, asso. prof. dept. pharmacology and exptl. medicine, 1964—. Mem. Am., Canadian, Brit. pharmacology socs. Research and publs. on naturally occurring antiallergic substances in plant and animal tissues. Home: 3644 Ontario Av., Montreal, Que., Can.*

KOVACS, Kalman, Hungarian endocrinologist, pathologist; b. Szeged, Hungary, July 11, 1926; s. Kalman and Aranka (Szekeres) K.; M.B.Ch.B., U. Med. Sch., Szeged, 1950; M.D., Hungarian Acad. Scis., Budapest, 1957, D.Med.Sc., 1963; Ph.D., U. Sch. Liverpool (Eng.) U., 1966; m. Eva Horvath, Sept. 18, 1962. Faculty, U. Med. Sch. Szeged, 1945——, sr. lectr. dept. internal medicine, 1963——; research asst. dept. pathology U. Liverpool, 1960, Crosby Research fellow dept. pathology, 1964-65. Mem. Hungarian Endocrine Soc. (sec. 1962——), Hungarian Physiol. Soc., Hungarian Oncological Soc., Hungarian Path. Soc., Brit. Endocrine Soc. Author: Die Rolle des Hypothalamus-Adenohypophsensystems in Wasserhaushalt, 1963; (with M. Julesz) Therapie und pathophysiologische Grundlagen der endokrinologischen Krankheiten, 1966; also numerous articles. Induction of pituitary necrosis in rats by means of drugs, drug and hormonal relationships in rats. Home: 22 Lenin. Office: Dept. Internal Medicine, U. Med. Sch., Szeged, Hungary.*

KOVALENKOV, Valentin Ivanovich, Russian engr.; b. USSR, Mar. 25, 1884; grad. Petersburg Electro-Tech. Inst., 1909, Petersburg U., 1911. Staff, Inst. Automation and Remote Control, USSR Acad. Scis., 1940-48, dir. Lab. in Solving Problems in Wire Communications, 1946-56. Named honored scientist Russian Soviet Fed. Socialist Republic, 1935; recipient Stalin prize, 1941. Corr. mem. USSR Acad. Scis. Author: Theory of Transmission in Electro-communication Lines, 2 vols., 1937-38; Basis of Theory on Magnetic Chains and its Use in Analysis of Relay Schemes, 1940; Electro-magnetic Processes Established Along Wire Lines, 1945. Chief work deals with theory of wire transmission of communication, also work on origination of telephone translation, from 1909; inventor in area of electrotechnics of sound movies. Office: Lab. in Solving Problems in Wire Communications, USSR Acad. Scis., Moscow, USSR.

KOVALEV, Nikolay Nikolaevich, Russian mech. engr.; b. Feb. 22, 1908; grad. Leningrad Tech. Inst., 1933. Engr., Leningrad Metal Plant, 1933——, chief designer, 1945-59; instr. Leningrad Poly. Inst., 1945-59, Leningrad Turbine Boiler Research Inst., 1959——. Recipient Stalin prize, 1946, 51, Lenin prize, 1959. Mem. USSR Acad. Sci. (corr.). Co-author: Hydroturbine Operation, 1941; author: Hydroturbines, 1961. Research and publs. on hydroturbine design; dir. work on reversible blade hydroturbines for Uglich and Rybinsk hydropower plants, 1937-39, later work on hydroturbines for Mingechaur, Tsimlyanskaya and Kuybyshev hydropower plants. Address: USSR Acad. Sci., Central Turbine Boiler Research Inst., Polytechnitsheskaya 24, Leningrad K-21, USSR.*

KOVALEVSKAYA, Sofya Vasiliyevna, see Kovalevski, Sonya.

KOVALEVSKI, Aleksandr Onufriyevich, embryologist; b. Dünaburg, Russia, Nov. 7, 1840; ed. Heidelberg, Tübingen (both Germany); pupil of Haeckel; prof. embryology St. Petersburg (Russia) U.; mem. St. Petersburg, French acads. scis. Author: Istorija razvitija Amphioxus lanceolatus, 1865; Anatomie des Balanoglossus delle Chiaje, 1866; Entwickelungsgeschichte der einfache Ascidien, 1866; Entwickelungsgeschichte des Amphioxus lanceolatus, 1867; Coeloplana metschnikowii, 1882; Étude sur l'anatomie de l'Acathobdella peledina, 1896. Introduced Darwinism into Russia; evolved 1st acceptable hypothesis of relation between vertebrates and invertebrates, also established phylum Chordata, during 1870s; research on embryology of worms, arthropods, brachiopods, coelenteratas. Died St. Petersburg, Nov. 22, 1901.

KOVALEVSKI (or KOVALEVSKAYA), Sonya (or Sophie), mathematician; b. Moscow, Russia, Jan. 15, 1850; d. Korvina-Krukovskaya; ed. Heidelberg, Germany, Göttingen, Germany; studied under Karl Weierstrass, also under Kirchhoff; m. Vladimir O. Kovalevski, 1868. Prof. higher math. dept. analysis Stockholm (Sweden), 1884-91. Recipient Prix Bordin, Paris Acad. Mem. St. Petersburg Acad. Scis. Author: Der Privatdozent, 1877; Der Kampf ums Glück, 1887; Kindheitserinnerungen, 1890. Research on partial differential equations, differential equations, Abelian integrals, theory of movement of solid body around an immovable point. Died Stockholm, Feb. 10, 1891.

KOVALSKII, Aleksandr Alekseievich, Russian physico-chemist; b. Sept. 10, 1906; grad. Leningrad Poly. Inst., 1930. Instr., Inst. Chem. Physics, USSR Acad. Sci., 1929-47, prof., 1947-57, dir. Inst. Chem. Kinetics and Combustions, Siberian dept., 1957——. Corr. mem. USSR Acad. Scis. Author: The Induction of Homogeneous Reactions in Gas by the Use of Solid Catalysts, 1946. Research on high energy particles, kinetics of chem. reactions, mechanism of heterogeneous catalytic reactions. Office: A. USSR, Novosibirsk, Sovetskaya ul. 20, Sibirskoe otdelenie Akademii nauk SSSR.

KOVALSKY, Aleksandr Alekseevich, Russian phys. chemist; b. Sept. 10, 1906; grad. Leningrad Poly. Inst., 1930. Instr., Inst. Chem. Physics, USSR Acad. Sci., 1929-47, prof., 1947-57, dir. Siberian dept. Inst. Chem. Kinetics and Combustion, 1957——; prof., 1947——. Mem. USSR Acad. Sci. (corr.). Research and publs. on kinetics of chem. reactions and nuclear physics, mechanism of heterogeneous catalytic reactions, high-energy particles. Address: Siberian Dept., USSR Acad. Sci., Sovetskaya ulitsa 20, Novosibirsk, USSR.

KOVANOV, Vladimir Vasilevich, Russian surgeon; b. 1909; grad. 1st Moscow Med. Inst., 1931, postgrad., 1931-34; M.D., 1946. Asst. dept. topographical anatomy 1st Moscow Med. Inst., 1934-38, asst. dept. faculty surgery, 1938-42, lectr. dept. operative surgery and topographical anatomy, 1946-47, head chair operative surgery and topographical anatomy, 1947——, dir., 1956——; prof., 1947——. Recipient Spasokukotsky prize, 1948. Mem. USSR Acad. Med. Sci., All-Russian (bd. mem.), All-Union (bd. mem.) socs. surgeons. Author numerous works including: Intracardiac, Intra-Arterial and Intravenous Administration of Drugs, 1946; New Ways of Injecting Drugs, 1948; Surgical Anatomy of the Lower Limbs, 1963; co-author: Surgical Anatomy of the Upper Limbs, 1964. Editor: Nursing Manual, 1964; dep. editor Exptl. Surgery. Address: 1st Moscow Med. Inst., B. Pirogovskaya ulitsa 2-6, Moscow, USSR.

KOVARIK, Alois Francis, Am. physicist; b. Spillville, Ia., Mar. 8, 1880; A.B., U. Minn., 1904, A.M., 1907, Ph.D. in Physics, 1909; Harling fellow Manchester U., 1910-11, D.Sc., 1916; A.M. (hon.), Yale, 1925, Charles U., Prague, Czechoslovakia, 1932. Instr. physics Decorah Inst., Ia., 1896-1900; faculty U. Minn., 1902-16, asso. prof., 1915-16; prof. physics Yale, 1925-48; mem. Manhattan Dist. project Yale div. Nat. Def. Research Com., 1942-43; mem. com. NRC, Radioactivity Congress, Brussels, 1910. Recipient medal Charles U., Prague, 1925. Mem. Am. Phys. Soc., Math. Soc. Am., Am. Meteorol. Soc., various other math. and phys. socs., U. S. and Czechoslovakia. Research on mobility of ions in gases, absorption and reflection of beta particles, periods of radioactive substances, recoil atoms, statis. method for alpha and beta particles, gamma rays, x-rays, electronics, radioactivity, Decorah ice cave, age of earth, automatic registration of rays. Died 1965.

KOVDA, Viktor Abramovich, Russian pedologist; b. Dec. 29, 1904; grad. Kuban Agrl. Inst., Krasnodar, USSR, 1927. Sci. worker Soil Inst., USSR Acad. Scis., from 1931, dir. Inst. Botany and Pedology Uzbek br., 1941-42; prof. Moscow (USSR) U., 1939-41, also currently; faculty Moscow Hydromelioratine Inst., 1943-48. Recipient Stalin prize, 1951. Dir., Dept. Natural Sci., UNESCO Secretariat; Russian del. Internat. Symposium of Problems of Desert Zones, Teheran, Iran, 1959. Corr. mem. USSR Acad. Scis. Author: Solonchak and Solonetz Soils, 1937; The Origin and Conditions of Salt Soils, 2 vols., 1945-47; Soils of the Northwestern Caspian Depression, 1950; Geochemistry of the USSR Deserts, 1954; The Mineral Composition of Plants and Soil Formation, 1956; Irrigation of the Deserts and Oases of Egypt, 1959; Space and Time Variations in Natural Atmospheric Radioactivity, 1960. Mem. editorial bd. jour. Pedology. Research on soils so. USSR; elucidated origin of solonetz and solonchak soils and suggested methods for their improvement; work has had important applications in reclamation of new land areas and constrn. of irrigation systems in USSR. Home: M. Yakimanka, 3, Moscow. Office: Moscow U.

KOWALESKI, (Hermann Waldemar) Gerhard, mathematician; b. Pommern, Germany, Mar. 27, 1876; Ph.D., Leipzig, Germany, 1898; prof., Greifswald; became asso. prof. math. U. Bonn (Germany), 1909; named prof. math. U. Prague (Czechoslovakia), 1912, U. Dresden (Germany), 1920; prof., Munich, Germany. Author: Determinanten-theorie, 1901; Grundzüge der Differential-und Integralrechnung, 1906; Einführung in der analytik Geometrie, 1910; Die Klassike Probleme der Analyse des Unendlichen, 1910. Research on transformation groups, interpolation, analytical geometry. Died Gräfelfing, Germany, Feb. 21, 1950.

KOWALEWSKI, Konstanty Piotr, physician; b. Petersburg, Russia, June 16, 1913; s. Ludwik and Wladislawa (Dabrowski) K.; M.Ph., U. Warsaw, Poland, 1935; Ph.D., M.D., U. Brussels, Belgium, 1947; m. Paule M. L. Bernier, Feb. 4, 1949; children—Barbara, Bernard. Came to Canada, 1952, naturalized, 1958. Research asst. endocrinology Univ. Hosp., Brussels, 1947-52; research fellow Cancer Soc. U. Alta., Edmonton, Can., 1953-55, asst. prof., 1955-59, asso. prof. exptl. surgery, 1959-67, prof., 1967——. adminstr. surg.-med. Research Inst. U. Alta., chief div. surg. research, 1962——. Mem. A.C.P., Am. Endocrine Soc., Soc. Exptl. Biology and Medicine. Research and publs. field vitamin A; gastric physiology and histamine; anabolic steroids. Home: 11650 74th Av., Edmonton, Office: Surg.-Med. Research Inst., U. Alta., Edmonton, Alta., Can.*

KOWALSKI, Edward, Polish hematologist; b. Sepolno, Poland, Mar. 14, 1914; s. Jerry and Helen (Katski) K.; M.S., Poznan (Poland) U., 1935; M.D., Warsaw (Poland) U., 1946; m. Maria Kopec, 1950; children—Barbara, Jan. With Warsaw U., 1946-57, lectr. biochemistry, 1947-52; sci. dir. Inst. Hematology, Warsaw, 1957-60; head dept. radiobiology and health protection Inst. Nuclear Research, Warsaw, 1955——. Mem. Internat. Soc. Hematology, Internat. Soc. Biochemistry, Czeckoslovakian Med. Soc. (hon.), others. Author: Biochemistry of Blood Platelets, 1967; also numerous articles. Research on blood coagulation and fibrinolysis; fibrinogen degradation products and their significance in hemostasis; platelet functions and their disturbances; erythropoesis and biosynthesis of heme, erythropoetin; bone metabolism and bone seeking radionuclides; influence of ionizing radiation on enzyme systems. Home: 64 Filtrova, Warsaw. Office: 16 Dorodna, Warsaw, Poland. Died Apr. 15, 1967.*

KOWARSKI, Lew, physicist; b. St. Petersburg, Russia, Feb. 10, 1907; s. Nicholas and Olga (Vlassenko) K.; degree in Chem. Engring., U. Lyons (France), 1928; B.Sc., U. Paris (France), 1931, Ph.D. in Phys. Sci., 1935; m. Kathe A. Freundlich, July 23, 1948; 1 dau. (from previous marriage), Irène Denise (Mrs. Gérard Hacques). Tech. sec., design engr. Société Le Tube D'Acier, Paris, 1929-37; research fellow Nat. Centre Sci. Research, France, 1937-40; personal asst. to prof. F. Joliot, Radium Inst. and Coll. de France, 1934-40; temporary civil servant Dept. Sci. and Indsl. Research, Cambridge, Eng., Chalk River, Ont., Can., 1940-46; dir. sci. services French Atomic Energy Commn., 1946-52; staff CERN (European Orgn. for Nuclear Research), 1953——, sr. physicist, 1965——; vis. prof. nuclear engring. Purdue U., Lafayette, Ind., 1963-65. Adviser, French delegation UN AEC, 1946-48; lectr. Conservatoire des Arts et Métiers, Paris, 1951-53; sci. adviser European Nuclear Energy Agy., 1956——; prof. Nat. Inst. Nuclear Sci. and Tech., Saclay, France, 1966——. Decorated Officier de la Légion d'Honneur. Fellow Am. Nuclear Soc.; mem. Société de Physique, Phys. Soc., Am. Phys. Soc., Research, numerous publs. on measurement surface migrations on growing crystals, new methods of gas network design, absorption of medium-energy neutrons; (with others) 1st evidence of emission of neutrons in fission; research on chain-reaching media; constrn. first Canadian, and first two French reactors; devel. data-processing devices. Home: 40 ave. William Favre, Geneva. Office: CERN, Geneva 23, Switzerland; also ENEA, 38 Blvd. Suchet, Paris 16, France.

KOYU (or Mitsuyoshi), Yoshida Schichibei, Japanese mathematician; b. 1598; studied under Mori Kambe Shigeyoshi; author: Jinko-ki (oldest existing Japanese math. work; explains operations on soroban, including square and cube root). Determined value of pi to be 3.16.

KOZHEVNIKOV, Aleksei Yakovlevich, Russian neurologist; b. Ryazin, Russia, 1836; M.D., Moscow U., 1860; student in Germany, Eng., Switzerland, Charcot's lab., France. Docent in nervous and mental diseases Novo-Ekaterininskii Hosp.; 1st neuropathologist, 1st alienist Moscow Faculty; instr. gen. pathology and therapeutics, 1871-74; prof. nervous and mental diseases Moscow U.; founder neurol. clinic, Moscow, Moscow Soc. Neurology and Psychiatry. Author: Shovnik statei po neuropathologii i psikhiatri, 1890; Kurs nervnikh bolieznei; lekstsii, 1892. Editor: Korsakov's Journ. Described form of epilepsy, 1893-94 (known as Kozhevnikov's epilepsy); did classic work on lathyrism, 1894; described familial spastic diplegia, 1895; wrote on neuropathology of nuclear ophthalmoplegia, myasthenia and bulbar paralysis; while at Charcot's lab. showed that in amyotrophic lateral sclerosis, nerve degeneration in form of corps granuleux could be followed up to motor cortex. Died 1902.

KOZHEVNIKOV, Petr Vasilevich, Russian dermatologist, venereologist; b. Rostov-on-Don, 1898; grad. Med. Faculty, Don U., Rostov-on-Don, 1922; D.Med. Sci., 1926. Intern, asst., lectr. chair skin and venereal diseases Don U., 1922-31; head chair skin and venereal diseases Tomsk Med. Inst., 1932-37, Turkmenia Med. Inst., 1937-47, Kirov Postgrad. Med. Inst., Leningrad, 1948——; prof., 1933——; sci. dir. State Research Inst. for Skin and Venereal Diseases, Leningrad, Leningrad Oblast Skin and Venereal Outpatients Dept. Del., Internat. Congress on Venereal Diseases and Treponemiasis, Washington, 1956. Mem. USSR Acad. Med. Sci. (corr.). Author: Atypical Pyodermitis, 1946; The Treatment of Skin Diseases, 1950; co-author: The Theory of Skin Leishmaniosis; Borovsky's Disease, 1953. Editor: Herald of Dermatology and venereology; co-editor Dermatology and Venereology sects. Large Med. Ency., 2d edit.; mem. editorial bd. Med. Excerpts. Research and numerous publs. on anatomy and physiology of skin in normal and pathol. conditions, lymphatic canaliculi in epithelium, fragility of skin vessels, dermo-galvanic reflex, body response and skin reaction in various types of dermatosis, exptl. leishmaniosis, atypical pyodermitis and eczematous reactions, classification and nomenclature of dermatosis; proved mercury in artery can cause deep buttock gangrene; described blue dermographism; distinguished two forms of dermal leishmaniosis, 1941 (developed method of prophylactic vaccination against this disease). Address: Kirov Postgrad. Med. Inst., ulitsa Saltykova-Shchedrina 41, Leningrad, USSR.

KOZINSKI, Andrzej Wladyslaw, biologist; b. Skierniewice, Poland, Oct. 1, 1925; s. Marian and Jozefina (Jasinski) K.; M.D., U. Warsaw, 1951, Ph.D. biochemistry, 1956; m. Patti Austin Barnes, Mar. 23, 1961; children—Catharine, Peter, Mark. Came to U. S., 1957, naturalized, 1963. Research asst. State Hygienic Inst., Warsaw, Poland, 1949-51; research asso. Inst. Biochemistry, Warsaw, Poland, 1951-53; asst. prof. dept. microbiology U. Warsaw, 1955-57; asst. prof. Inst. Microbiology Rutgers U., 1959; NIH fellow dept. biology Johns Hopkins, 1962; asso. mem. Wistar Inst. 1962-66; asso. prof. dept. med. genetics U. Pa., Phila. 1966——, Mem. Soc. Biol. Chemistry,

Biophys. Soc. Author (with others) Wirusologia Ogolna, 1958. Publs. on first described the molecular mechanism of mating as a breaking and reunion of DNA molecules; recombination as an enzymatic process. Home: 6817 Cresheim Rd., Phila. 19119.*

KOZLOV, Pëtr (Peter) Kuzmich, Russian explorer, archaeologist; b. Oct. 3, 1863. Took part in expdns. of N. M. Przevalsky, M. V. Pevcow, V. I. Roborovski; comdr. expdn. to Mongolia-Tibet, 1899-1901; leader Mongolian-Sychuan expdn., 1907-09; led Mongolian-Tibetan expdn., 1923-26. Mem. USSR Acad. Scis., Russian Geog. Soc. Author: Mongolia and Khan; Mongolia and the Dead City of Khara Khoto, 1923. Discovered important information relating to crography, geology, climatology, flora and fauna of Tibet, 1899-1901; gathered important information about eastern Tibetan tribes, 1899-1901; discovered ruins of ancient city of Khara Khoto in Gobi Desert, excavations here revealed important material about spiritual and cultural life of Tailageins, 1907-09; explored Karakorum, capital of empire of Genghis Khan, 1899-1901. Died Sept. 26, 1935.

KOZLOWSKI, Leon, Polish geologist; b. Rembieszyce, Poland, 1892; ed. U. Cracow (Poland), U. Tübingen (Germany); Ph.D. in Geology, Geography, Classical Archeology. Became asso. prof. U. Cracow, 1920; named asst. prof. U. Lwow (Poland); 1928; lectr. until 1930; polit. positions with Ministries of Finance and Agr. Author: Grand-Pol. at the Period of Stone; Ethnological Problem of Prehistoric Periods; Period of Stone in the Downs of the East of Little Poland; The Paleotho Period in Poland; The Neolithic Period in Poland; Die Steinzeit im Donaugebiet der Kleinpolnischen Höhe; Die ältere Steinzeit in Polen; Studies of Paleothique in the North of France, Belgian and England. Mem. numerous expdns.; research in Poland, Europe in gen., Baltic Province, Balkan, N. of France. Office: Warsaw Council Ministries, Warsaw, Poland.

KOZLOWSKI, Theodore Thomas, Am. plant physiologist; b. Buffalo, May 22, 1917; s. Theodore and Helen (Zamiara) K.; B.S., Syracuse U., 1939; postgrad. Mass. Inst. Tech., 1940-42; M.A., Duke, 1941, Ph.D. (research fellow) 1947; m. Maude Peters, June 29, 1954. Faculty, U. Mass., Amherst, 1947-58, prof., head dept. botany, 1950-58; prof. dept. forestry U. Wis., Madison, 1958-—, chmn. dept., 1961-64; NSF vis. scientist Soc. Am. Foresters, 1963-64, 65-66. Rapporteur U. S. physiology sect. World Consultation on Forest Genetics and Tree Improvement, Stockholm, Sweden, 1962; cons. Nat. Sci. Found., 1967-—. Sr. Fulbright Research scholar Oxford (Eng.) U., 1964-65. Mem. Bot. Soc. Am. (past chmn. Northeastern sect.), Am. Soc. Plant Physiology (past chmn. Northeastern sect.), Ecol. Soc. Am., Soc. Am. Foresters, Am. Inst. Biol. Scientists, Societas Forestalis Fenniae, Scandinavian Soc. Plant Physiology, Internat. Union Forest Research Orgnzns., Sigma Xi, Phi Kappa Phi, Phi Sigma. Author: (with P. J. Kramer) Physiology of Trees, 1960; Water Metabolism in Plants, 1964; also numerous articles. Editor: Tree Growth, 1962; Water Deficits and Plant Growth, 1968. Plant physiology editor Am. Midland Naturalist, 1965-—; editorial bd. Forest Sci., 1959-—, Ecology, 1968-—. Research on physiology and ecology of woody plants; effects of water deficits, biocides on plant growth. Home: 5134 Juneau Rd., Madison, Wis. 53705.*

KOZUSZNIK, Boguslaw, Polish physician; b. Sucha Dolna, Poland, Jan. 29, 1910; s. Jozef and Amalia (Nowokowna) K.; M.D., Charles U., Prague, 1935; m. Zdenka Urbanowna, Aug. 1, 1942. Staff, Miners Hosp., Petrkovice-Ostrawa, 1935-37, dir. hosp., 1937-39; chief med. officer Polish Mcht. Navy in U.K., 1941-45; under sec. state Polish Ministry Health, 1946-59; head of chair of orgn. pub. health and social medicine Postgrad. Sch., Med. Acad., Warsaw, 1959-—. Mem. Polish Parliament, 1947-52. Chief redactor Monthly Rev. Pub. Health, 1962-—; chmn. Polish nat. com., standing adv. com. UNICEF, mem. exec. bd.; exec. bd. WHO, 1948-50; mem. State Council for Peaceful Atomic Energy Utilization. Decorated Polonia Restituta; Order of White Lion of Czechoslovakia. Mem. Purkyne Med. Assn., Assn. of Hygiene USSR, Polish Med. Assn. (exec. bd.), Polish Assn. Social Medicine (chmn. 1966-—). Research, numerous publs. on pub. health, social medicine related to sociology, state san. legislation. Home: 16 I Armii W.P. 16, Warsaw. Office: 22 Chocimska, Warsaw, Poland.*

KOZYREV, Nikolay Aleksandrovich, Russian astrophysicist; b. St. Petersburg, 1908; grad. Physico-Math. Faculty, Leningrad U., 1928; D. Physico-Math. Sci., 1948. Instr. marine astronomy Frunze Naval Coll., 1931-—; asst. dept. math. Leningrad Inst. R.R. Engring., 1931-—; prof. Pokrovsky Pedagogical Inst., Leningrad, 1931-—; sr. asso. Main Astron. Obs., USSR Acad. Sci., Pulkovo, 1931-—. Author: The Radiant Equilibrium of Extended Stellar Photospheres, 1934; Sources of Stellar Energy and the Theory of the Internal Structure of Stars, 1948; The Secret of the Aristarchus Crater, 1963. Research and publs. on theory of structure of attenuated stellar atmospheres accounting for number of characteristics of hot stars; formulated new cosmogonic theory postulating that math. value representing passage of time is constant source of energy (repudiates current concept that energy emitted by stars originates from thermonuclear reaction occurring in stars themselves); discovered active lunar volcano and Aurora Borealis on Venus, 1958. Address: Main Astron. Observatory, USSR Acad. Sci., Leningrad-Pulkovo, RSFSR, USSR.

KRAATZ, Charles Parry, Am. pharmacologist; b. Rochester, N.Y., Aug. 6, 1906; s. Charles Henry and Elizabeth (Parry) K.; A.B., Berea Coll., 1926; A.M., U. Ky., 1932; Ph.D., U. Cin., 1936; m. Susan Elizabeth Haynes, Feb. 26, 1938; children—Margaret, Susan. Faculty, Chgo. Med. Sch., 1936-43, asso. prof., 1943; asst. prof. physiol. pharmacy Okla. A. and M. Coll., 1946-47; faculty Jefferson Med. Coll., Phila., 1947-—, prof. pharmacology, 1955-—. Mem. Am. Soc. Pharmacology and Exptl. Therapeutics, Soc. for Exptl. Biology and Medicine. Research and publs. on reproductive pharmacology, effects drugs on excitable tissues, botulinum toxin. Home: 1000 Lincoln Av., Springfield, Pa. 19064. Office: 1025 Walnut St., Phila. 19107.*

KRAATZ, Walter Charles, Am. zoologist; b. Milw., Jan. 20, 1893; s. Charles M. and Emma (Valentine) K.; B.A., U. Wis., 1918; M.A., Ohio State U., 1920, Ph.D., 1923; m. Julia W. Jaster, June 17, 1923; children—Charles Harold, Ernest Edward. Faculty, Ohio State U., 1922-23, Miami U., Oxford, O., 1923-24; faculty U. Akron (O.), 1924-—, prof. biology, head dept. biology, 1934-58, emeritus prof., 1959-—. Mem. Ohio Acad. Sci. (past pres.), A.A.A.S., Am. Inst. Biol. Scis., Am. Soc. Zoologists, Soc. for Study Evolution, Am. Soc. Limnology and Oceanography, Sigma Xi, Phi Sigma, Omicron Delta Kappa. Author: Laboratory Manual for General Zoology, 1934; Evolution Not Irreligious, 1925; also numerous articles. Research on water insects, plankton organisms, food of fishes, anatomy and devel. intestine minnow. Address: 1437 Bryden Dr., Akron, O. 44313.*

KRABBE, Knud H., Danish neurologist; b. Denmark, Mar. 3, 1885; s. Harald and Kristin (Hustru) K.; M.D. Specialized in neurology, Copenhagen, Denmark; lectr. Copenhagen U., 1915-27, 29-38; practice medicine, specializing in neurology, Copenhagen; neurologist St. Joseph's Hosp., Copenhagen; chief physician neurol. dept. Kommune Hosp., 1933-—. Corr. mem. Paris, Estonian, Phila. neurol. socs. Author: Les maladies des glandes endocrines, 1925; L'organe sous-commissural du cerveau, 1925; Morphogenesis of the Brain in Reptiles, 1939; Morphogenesis of the Brain in Lower Mammals, 1942; Morphogenesis of the Brain in Rodentia, Prosimiae and Edentates, 1944; Morphogenesis of the Brain in Hyracoidea, Ungulata, Carnivora and Pinnipedia, 1947; La lipomatose circonscrite multiple, 1944. Described familial infantile diffuse sclerosis (Krabbe's disease), 1934; demonstrated that many meningeal hemorrhages are caused by cerebral vascular malformation. Home: Dronn Tvaerg 6, Copenhagen, Denmark.

KRACIK, Jiri, Czechoslovakian physicist; b. Jicin Czechoslovakia, Apr. 14, 1929; s. Eduard and Emilie (Dolezalova) K.; E.E., Czechoslovak Tech. U., Prague, 1952, Doctor in Math. and Phys. Scis., 1964; m. Bozena Holzhauserova, Apr. 8, 1931; children—Marketa, Jiri. Chief asst. faculty elec. engring. Czechoslovak Tech. U., 1950-57, asso. prof., head Phys. Inst., vice dean Phys. Inst. Radiotech. Faculty, Podebrady, 1957-63, prof., head Phys. Inst., vice dean for sci., Prague, 1963-—, mem. sci. council, 1952-66. Mem. Phys. Council Czechoslovak Acad. Scis. Co-author: Electrical Discharges, 1964; Technical Physics, 1964, Plasma Physics, 1966. Research and publs. on discharge of gases, low-pressure, high-pressure, acceleration of shock waves, plasmatic clusters, resonance conditions of acceleration of plasma clusters and theory of spiraling discharge. Home: 17 Ceskomalinska. Office: 1902 Technicka, Prague, Czechoslovakia.*

KRADER, Lawrence, Am. anthropologist; b. N.Y.C., Dec. 8, 1919; s. Nathan and Mollie (Linder) K.; B.A., City Coll. N.Y., 1941; Ph.D., Harvard, 1954; m. Barbara Lattimer, May 14, 1952. Research asso., bur. social sci. research Am. U., 1953-56, prof. anthropology Am. U., 1959-63; head China program U. S. Bur. Census, Washington, 1956-58; head Asia div. Ohio State U., 1963-64; prof. anthropology Syracuse U., 1964-67; prof. anthropology City U. N.Y., 1967-—. Sec. gen. Internat. Union of Anthropol. and Ethnol. Scis., 1964-—; mem. com. on internat. relations in behavioral scis. NRC-U. S. Nat. Acad. Sci. 1966-—. Fellow A.A.A.S., Am. Anthropol. Assn.; mem. Internat. Social Sci. Council, Internat. Council for Philosophy and Humanistic Studies. Author: Social Organization of Turk-Mongol Nomads, 1963; Peoples of Central Asia, 1963, 2d edn., 1967; Anthropology and Early Law, 1967; Formation of the State, 1967. Research and publs. on devel. of an anthropol. concept of culture as it integrates the relevant concepts of anthropol. theory and those of related disciplines, law, etc. Home: 205 W. 89th St., N.Y.C. 10024.*

KRAEMER, Henry, Am. pharmacologist; b. Phila., July 22, 1868; s. John Henry and Caroline K.; student Girard Coll., 1877-83; Ph.B., Columbia, 1895; Ph.G., Phila. Coll. Pharmacy, 1889, hon. Pharm.M., 1912; Ph.D., U. Marburg, 1896; m. Minnie A. Behm. Instr. pharmacognosy, Coll. Pharmacy of City of N.Y., 1890; prof. botany and pharmacognosy Sch. of Pharmacy, Northwestern U., 1896-97; prof. botany and pharmacognosy and dir. micros. lab. Phila. Coll. Pharmacy, 1897-1917; prof. pharmacognosy U. Mich., 1917-20; dir. Kraemer Sci. Lab., from 1920; dir. Era courses in pharmacy from 1921. Mem. com. revision of U. S. Pharmacopoeia and chmn. sub-com. botany and pharmacognosy, from 1900. Mem. council on pharmacy and chemistry A.M.A., 1913-21; pres. Am. Conf. Pharm. Faculties, 1917, Phila. Author: A Text-book of Botany and Pharmacognosy, 1902, 07, 08, 10; Applied and Economic Botany, 1914, 16; Scientific and Applied Pharmacognosy, 1915, 20. Editor Am. Jour. Pharmacy, 1898-1917. Died Sept. 9, 1924.

KRAEPELIN, Emil, German psychiatrist; b. Neustrelitz, Mecklenberg, Germany, Feb. 15, 1856; M.D., 1878; Us. Würzburg, Munich; student of W. M. Wundt at Leipzig. Apptd. prof., U. Dorpat, 1886; U. Heidelberg, 1891; U. Munich, 1903-26; also founded research inst. at Munich, 1917. Author: Lehrbuch der Psychiatrie (4 vol.), 1883; Einführung in der physhiatrische Klinik (3 vol.), 1901; Über geistige Arbeit, 1903; Hundert Jahre Psychiatrie. Systematized psychiatry; classified mental diseases in terms of symptoms, development and outcome; divided these diseases into dementia-praecox and manic-depressive groups; studied effects of alcohol and fatigue on mental processes. Died Munich, Germany, Oct. 7, 1926.

KRAEVSKY, Nikolay Aleksandrovich, Russian pathoanatomist; b. 1905; grad. Med. Faculty, 1st Moscow U., 1928, postgrad. in path. anatomy, until 1930; D.Med. Sci., 1942. Asst., lectr., later prof. chair path. anatomy 2d Moscow Med. Inst., 1931-54; dir. path. anatomy lab. Central Inst. Hematology and Blood Transfusion, USSR Ministry Health, 1939-51, cons., 1951-54; dep. dir. for sci. work, head path. anatomy lab. Int. Biophysics, USSR Acad. Med. Sci., 1954-60, acad. sec. dept. med. and biol. sci., 1960-—; prof., head chair path. anatomy Central Postgrad. Med. Inst., 1954-—; sci. dir. path. anatomy dept. Botkin Gen. Hosp. Del., Internat. Conf. on Morphology of Precancerous States, Perugia, 1961. Mem. USSR Acad. Med. Sci., All-Union (bd. mem.), Moscow (chmn.) socs. pathoanatomists. Author: Rheumatic Granulomata of the Lung, 1946; Acute Leukosis Theory, 1950; Outline Pathological Anatomy of Radiation Sickness, 1957; co-author: Acute Leukosis, 1952. Editor: The Effects of Radioactive Strontium on the Animal Body, 1961; mem. editorial bd. Archives of Pathology, Bull. Exptl. Biology and Medicine, Pathology and Morphology sects. Large Med. Ency., 2d edit.; mem. editorial council Problems Hematology and Blood Trasnfusion. Research and numerous publs. on oncology, pathology of infectious diseases, rheumatism, hemopoietic system, pathology of mil. traumas. Address: Central Postgrad. Med. Inst., pl. Vosstaniya 1-2, Moscow, USSR.

KRAFFT, Georg Wolfgang, German physicist; b. Tuttlingen, Germany, July 15, 1701; ed. U. Tübingen (Germany); 1 son, Wolfgang Ludwig; tchr. math. and physics, gymnasium, St. Petersburg, Russia, 1725-44; prof. math. and physics U. Tübingen, 1744-54. Mem. acads. St. Petersburg, Berlin. Author: Experimentarum physicorum praecipuorum brevis descriptio, 1740; La Maison de glace, 1741; Institutiones geometriae sublimioris, 1753; Praelectiones academicae publicae in physicam theoreticum, 1753-54. Died Tübingen, June 12, 1754.

KRAFFT, Johann, see Crato von Crafftheim, Johannes.

KRAFFT, Johann Daniel (or Kraft), German chemist; from Dresden, Germany; publicized Hennig Brand's discovery of phosphorus on visits to Netherlands, Eng., Am.; performed expts. with phosphorus for Robert Boyle and other mems. Royal Soc., 1677; letters collected in library at Hanover, Germany.

KRAFFT-EBING, Baron Richard von, German neurologist, psychiatrist; b. Mannheim, Baden, Germany, Aug. 14, 1840; became prof. psychiatry at Strasbourg, 1872, Graz, 1873, Vienna, 1889. Author: Psychopathia Sexualis, 1886. Chiefly remembered today for his study of case histories of sexual abnormalities, of which he helped initiate sci. discussion and study; authority on psychol. aspects of mental disorders and their medicolegal relations. Died Mariagrun, nr. Graz, Austria, Dec. 22, 1902.

KRAFT, Johann Daniel, see Krafft, Johann Daniel.

KRAFT, Kurt, German chemist; b. Nuremberg, Jan. 22, 1907; s. Leo and Ida (Koch) K.; ed. Wurzburg Coll., univs. Heidelberg, Munich, Berlin; Ph.D., agrege; m. Suzanne Freudenberg, Feb. 15, 1944; children—Andrea, Cornelia. Asst., U. Munich, 1935-37; head research Knoll Algemeine Gesellschaft, 1938-47; prof. at large chemistry U. Heidelberg, 1947; asso. Carl Freudenberg, Weinheim, 1953. Mem. Am. Chem. Soc., German Soc. Chemists. Research and publs. on fluorine combinations, synthesis of resinous acids. Home: Im Gabelacker 8, Heidelberg. Office: Postfach 208 d. Fa. Carl Freudenberg, Weinheim-Bergstrasse, Germany.

KRAHL, M(aurice) E(dward), Am. biochemist; b. Cambridge City, Ind., Sept. 17, 1908; s. Michael and Bertha (Hastings) K.; A.B. (Rector scholar), DePauw U., 1929; Ph.D. (Eli Lilly & Co. fellow), Johns Hopkins, 1932; m. Barbara Dodson, June 4, 1932 (dec. 1962). Research chemist Eli Lilly & Co., Indpls. 1933-44; faculty Coll. Phys. and Surg., Columbia, 1944-

46, Washington U., 1946-53; prof. physiology U. Chgo., 1953—; Banting lectr. English Diabetic Assn., 1950. Trustee Marine Biol. Lab., Woods Hole, Mass. Mem. Am. Assn. Cancer Research. Am. Soc. Biol. Chemists, Am. Soc. Pharmacology and Exptl. Therapeutics, Harvey Soc., Soc. Exptl. Biology and Medicine, Soc. Gen. Physiologists, Am. Diabetes Assn., Biochem. Society (Brit.), Phi Beta Kappa, Sigma Xi. Author: The Action of Insulin on Cells, 1961. Editorial bd. Biol. Bull., 1954-57, Diabetes, 1959-64, Endocrinology, 1964—. Research and publs. on chem. basis of hormone and drug action; discoverer of direct action of insulin on adipose tissue and of dinitrophenol effects on cell div.; contbr. to structure of penicillin. Home: 5805 S. Dorchester Av. 60637.*

KRAINTZ, Leon, Am. physiologist; b. Johnstown, Pa., Oct. 3, 1924; s. Franz and Mary (Peterlin) K.; A.B., Harvard, 1950; M.A., Rice Inst., 1952, Ph.D. (NSF fellow), 1954; m. Frances Whitcomb, Aug. 27, 1949; children—Donna, Franz, Erika. Faculty, U. Tex., Houston, 1954-62; vis. instr physiology Baylor U., 1956-62; vis. prof. biology U. St. Thomas, Houston, 1957-64; prof. biology Rice U., 1962-63; faculty U. B.C., Vancouver, 1963—, prof. oral biology, 1966—, prof. physiology, 1966—. Fellow A.A.A.S.; mem. Am., Canadian physiol. socs., Endocrine Soc., Soc. Nuclear Medicine, Soc. Exptl. Biology and Medicine, Internat. Assn. Dental Research. Research and publs. on studies with radioactive isotopes, radiocalcium, radioiodine and radiochromium, physiology of thyroid and parathyroid glands and calcium metabolism, cerebrospinal fluid physiology, salivary gland physiology.*

KRAJINA, Zvonimir, Yugoslavian physician; b. Sibenik, Yugoslavia, Jan. 12, 1923; s. Marijan and Marija (Delic) K.; student Med. Faculty, U. Zagreb (Yugoslavia), 1941-46; m. Andelka Pusic, Oct. 7, 1944; children—Ljubomir, Andelko. With Otolaryngol. Clinic, Zagreb, 1947—, prof., 1964—. Research, publs. on allergy in otorhinolaryngology, influences of different factors (cold, dust) on mucosa of upper respiratory tract, surg. treatment of vasomotor rhinities by neurectomy of vidian nerve, cancer in otorhinolaryngology. Home: 30 Vocarsko naselje, Zagreb, Yugoslavia.*

KRAKAU, Carl Erik Torsten, Swedish ophthalmologist; b. Helsingborg, Sweden, July 27, 1921; s. Erik Viktor and Svanhild (Schlyter) K.; Med.Lic., U. Lund (Sweden), 1948; m. Marie-Louise Amelie von Leithner, Apr. 4, 1953. Docent, U. Lund, 1953-66; head dept. exptl. ophthalmology U. Eye Clinic, Lund, 1966—. Mem. Assn. for Eye Research (London, Eng.). Research and publs. on frequency analysis of EEG, biophys. and biomath. analyses of ERG and EMG problems, time series analysis of vision, ophthalmologic research instruments. Home: 1 Bengt Lidforss w., Lund. Office: Dept. Exptl. Ophthalmology, Univers. Eye Clinic, Lund, Sweden.*

KRAL, Vojtech Adalbert, physician; b. Prague, Czechoslovakia; Feb. 5, 1903; s. Samuel and Paula (Loewy) K.; M.D., German U. in Prague, 1927; m. Katherine Neumark, Aug. 15, 1935. Clin. asst. German U. in Prague, 1928-38, asst. prof. psychiatry and neurology, 1935-39; dir. Sanatorium for Nervous and Mental Diseases in Prague, 1945-48; fellow Lady Davis Found., Montreal, Que., Can., 1949-50; lectr. dept. psychiatry McGill U., Montreal, 1950-55, dir. gerontological unit Allen Meml. Inst., 1955—, asso. prof., 1960—; psychiatrist Royal Victoria Hosp., Montreal, 1960—. Fellow Am. Psychiat. Assn., Gerontological Soc., Am. Geriatrics Soc., A.A.A.S.; mem. Am. Psycopath. Assn., Assn. for Research in Nervous and Mental Disease, Canadian, Que. psychiat. assns., Canadian Neurol. Soc., Am. Acad. Neurology, Canadian Med. Assn., Montreal Medico-Chirurg. Soc., Montreal Clin. Soc., Soc. for Projective Techniques. Research, numerous publs. on amnestic syndrome, types of memory impairment in aged, adrenocortical function in psychoses of aged. Home: 4145 Blueridge Crescent, Montreal, Que., Can.*

KRAMER, Benjamin, Am. physician; b. Chatin, Russia, Aug. 15, 1888; s. Israel and Clara (Greenberg) K.; M.D., N.Y. U., 1909; M.S., State U. Ia., 1917; m. Sara Farber, Sept. 18, 1949; 1 son (by adoption), David I. Asst. prof. physiology Ia. State Med. Sch., 1913-17; asso. prof. pediatrics Johns Hopkins Med. Sch., 1917-25; chief pediatrics Bklyn. Jewish Hosp., 1925-55, now chief emeritus; prof. clin. pediatrics L.I. Coll. Medicine, 1926-55; dir. pediatrics Maimonides Hosp., Bklyn., 1955-64, now dir. emeritus; clin. prof. pediatrics State U. N.Y., Downstate Med. Center, Bklyn., 1950—, prof. emeritus, 1956—; cons. pediatrician to various hosps., 1955—. Mem. A.M.A., N.Y. State, Kings County med. socs., Am., Bklyn. acad. pediatrics, N.Y. Acad. Scis., N.Y. Acad. Medicine, Harvey Soc., Am. Soc. Clin. Investigation, Am. Pediatric Soc., Soc. for Pediatric Research (hon.), Am. Chem. Soc., Am. Soc. Biol. Chemists, Am. Physiol. Soc., A.A.A.S., Am. Heart Assn., Fedn. for Socs. for Exptl. Biology and Medicine, Alpha Omega Alpha. Author: Practical Pediatrics, 1958; also numerous articles. Research on methods for analyzing calcium in blood and its relation to convulsions, chemistry of bones. Home: 35 Sutton Pl., N.Y.C. 10022. Office: 943 Union St., Bklyn. 11215.*

KRAMER, David N., Am. chemist; b. Balt., Oct. 25, 1920; s. Morris and Bessie (Handelman) K.; B.A., Johns Hopkins, 1942; Ph.D., U. Md., 1949; m. Anne Kay, Oct. 17, 1943; children—Avi, Shira, Alisa, Kimi. With U. S. Indsl. Chem., Balt., 1942; with USPHS, 1944-46; faculty Ohio State U., Columbus, 1949-51; with Chem. Corps., 1951—, br. chief, 1953—. Commended by Dept. Army, 1953, 57, 65. Mem. Am. Chem. Soc., A.A.A.S., Sigma Xi, Rho Xi. Patentee in field. Research and publs. on organic synthesis, enzymology, microanalytic techniques, chemistry of dyestuffs, mechanisms of reactions, fluroescence, electrochem. procedures, bioinstrumentation, radiation chemistry, mechanisms of electron-transfer. Home: Wiltonwood Rd., Stevenson, Md. 21153. Office: Edgewood Arsenal, Edgewood, Md.*

KRAMER, G., Austrian botanist; flourished 1720; army physician; combined Tournefort's and Rivinus' systems of classification; proposed that orange or lime juice would cure scurvy, 1720; opposed the ideas of Camerarius.

KRAMER, Irvin Raymond, Am. metallurgist; b. Balt., Sept. 18, 1912; s. Robert and Rose (Samet) K.; B.S. (Md. state scholar), Johns Hopkins, 1935, M.S., 1945, D.Eng. (Univ. fellow), 1951; m. Gertrude Kaplan, Oct. 27, 1935; 1 dau., Marcia Sue. Chemist, Am. Radiator and Standard San. Corp., Balt., 1935-38; metallurgist, head, spl. alloys sect. Naval Research Lab., 1938-46, phys. scientist, head, materials br. Office Naval Research, 1946-51 (both Washington); asst. to pres. Horizons Titanium Corp., Princeton, N.J., 1951-53; v.p. Mercast Corp., N.Y.C. 1953-55; prin. scientist, chief, materials research Martin Marietta Corp., Balt., 1955—. Bd. advisers Materials Sci. & Engring., internat. jour., 1963—. Recipient Civilian Meritorious award USN, 1945. Johns Hopkins fellow, 1961-63. Mem. Am. Soc. for Metals, Am. Inst. Mining and Metall. Engrs., Inst. Metals, London, Sigma Xi. Research, publs. on plastic deformation and fracture of metals; influence of surface on deformation and fracture characteristics; effect of environment on mech. behavior of materials; work on alloy steel devel. Home: 2107 Northcliff Dr., Balt. 21209. Office: Mail No. 3033, Martin Co., Balt. 21203.*

KRAMER, Morton, biostatistician; b. Baltimore, Mar. 20, 1914; s. David and Sarah (Valenstein) K.; A.B., Johns Hopkins U., 1934, Grad. Sch. of Chemistry, 1934-35; Sc.D., Johns Hopkins Univ. Sch. of Hygiene and Pub. Health, 1939; m. Pauline Weinstein, Sept. 24, 1939; children—Barry Kenneth, James Lawrence, Nancy, Richard. Student asst. biostatistics, 1937-38, Johns Hopkins U. Sch. of Hygiene and Pub. Health; instr. in preventive medicine New York Univ. Med. Sch., 1938; statistician N.Y. State Dept. of Health, Albany, N.Y., 1939-40; asst. prof. of biostatistics Sch. of Tropical Medicine, and statistician, Insular Dept. of Health, San Juan, P.R., 1940-42; econ. analyst Treasury Dept., Washington, 1942-43; tuberculosis statistician, Cleveland and Cuyahoga County, 1943-46; asso. in biostatistics Western Reserve U. Sch. of Medicine, 1943-46; chief of information and research, Office of Internat. Health Relations, U.S.P.H.S., 1946-49, chief biometrics br. Nat. Inst. Mental Health since 1949; commd. sr. scientist (reserve), USPHS, 1948; consultant mental health World Health Orgn., 1959—, mem. expert panel health statistics, 1961—. Adviser, U. S. Delegation to Fifth Session Interim Commn. of World Health Orgn., Geneva, Switzerland, 1948; adviser, U. S. Delegation to Sixth Session Interim Commn. of World Health Orgn., June 1948; adviser, U. S. Delegation to First World Health Assembly, Geneva, Switzerland, June-July, 1948. Fellow of Am. Statis. Assn., Am. Pub. Health Assn., American Psychiatric Association (honorary), also American Orthopsychiatric Association; member of Phi Beta Kappa. Contbr. articles to sci. publs. Applied statistical method to clinical, laboratory and epidemiological research on mental disorders; designed and stimulated nat. and internat. research on factors related to mental disorders in various population groups and on evaluation of programs to prevent and control such disorders. Home: 9612 Sutherland Road, Silver Spring, Md. Office: Nat. Inst. of Mental Health, U. S. Pub. Health Service, Bethesda, Md.*

KRAMER, Paul Jackson, Am. plant physiologist; b. Brookville, Ind., May 8, 1904; s. LeRoy and Minnie (Jackson) K.; A.B., Miami U. (O.), 1926, D.Litt., 1966; M.S., Ohio State U., 1929, Ph.D., 1931; m. Edith Sara Vance, June 24, 1931; children—Jean Jackson (Mrs. A. F. Findeis), Richard Vance. Faculty, Duke, Durham, N.C., 1931—, prof., 1945-54, James B. Duke prof., 1954—; v.p. ecol.-physiol. sect. 8th Internat. Bot. Congress, 1954; dir. Sarah P. Duke Gardens, program dir. regulatory biology NSF, 1960-61. Recipient award Soc. Am. Foresters, 1961. Mem. Bot. Soc. Am. (past pres., Merit certificate 1956) Am. Inst. Biol. Scis. (past pres.), Am. Soc. Plant Physiologists (past pres.), A.A.A.S., Nat. Acad. Scis., Am. Acad. Arts and Scis. Author: Plant and Soil Water Relationships, 1949; (with T. T. Kozlowski) Physiology of Trees, 1960. Studies, publs. on water absorption by plants, inter-relationships between water and salt absorption; environmental influences on plant growth. Home: 2251 Cranford Rd., Durham, N.C. 27706.*

KRAMER, Philip, Am. internist; b. Chgo., Jan. 26, 1915; s. Maurice and Bessie (Fish) K.; student U. Chgo., 1932-35, Sch. Medicine, 1935-37, M.D., 1939; m. Marjorie Houghton Tate, Nov. 9, 1944. Practice medicine specializing in gastroenterology, Boston, 1946—; staff Mass. Meml. Hosp., 1948-58, 58—, U. Hosp., 1965—; chief gastroenterology VA Hosp., Syracuse, N.Y., 1958; chief gastrointestinal clinic Boston City Hosp., 1959—; faculty Upstate Med. Center, Syracuse, 1958; asso. prof. medicine Boston U. Sch. Medicine, 1965—. Cons., N.E. Hosp., Boston, 1954—, Framingham (Mass.) Union Hosp., 1954—. Diplomate Am. Bd. Internal Medicine. Mem. Am. Fedn. Clin. Research, Am. Gastroenterology Assn., A.M.A., A.C.P., Am. Physiol. Soc. Editorial bd. Am. Jour. Digestive Diseases, 1957-63. Research and publs. in esophageal, gastric and small intestinal motility and effect varying sodium intakes, hormones, drugs and foods on ileal excreta; described use mecholyl as physiologic and diagnostic procedure in cardiospasm and adverse effect anti-cholinergic drugs in obstructive peptic ulcer disease; codiscovered lower esophageal ring dysphagia. Home: 34 Strawberry Hill Rd., Natick, Mass. 01760. Office: 750 Harrison Av., Boston 02118.*

KRAMERS, Hendrik Anthony, Dutch physicist; b. Rotterdam, Holland, Dec. 17, 1894; prof. Leiden (Holland) U. Mem. French Acad. Scis. Research and numerous publs. on quantum physics, including intensity of atomic spectra, normalizing problems of charging and of mass; predicted (with Heisenberg) Raman effect. Died Oegstgeest, Holland, Apr. 24, 1952.

KRAMISH, Arnold, Am. nuclear physicist; b. Denver, June 6, 1923; s. John I. and Sarah Kramish; B.S., U. Denver, 1945; M.A., Harvard, 1947; postgrad. Stanford, 1947-48; m. Vivian Ruth Raker, Aug. 19, 1952; children—Pamela Lynn, Robert Samuel. With AEC, Washington, Los Alamos, Oak Ridge, 1946-51; with RAND Corp., Washington, Santa Monica, Cal., 1951—, now sr. staff mem. Cons., AEC, NSF, OECD, European Econ. Community; vis. prof. polit. sci. dept. U. Cal. at Los Angeles, 1965-66; Council Fgn. Relations research fellow, study dir. Study Group on Peaceful Uses Atomic Energy, 1958; Mem. Atomic Indsl. Forum, Inst. for Strategic Studies. Joint author, mem. bd. editors The Effects of Atomic Weapons, 1950; author: (with Eugene M. Zuckert) Atomic Energy for Your Business, 1956; Atomic Energy in the Soviet Union, 1960; The Peaceful Atom in Foreign Policy, 1963, also numerous articles. Research on tech. of nuclear energy, especially radiometers. Home: P.O. Box 35, Pacific Palisades, Cal. 90272. Office: 1700 Main St., Santa Monica, Cal. 90406.*

KRAMP, Christian, mathematician, med. writer; b. Strasburg, France, July 10, 1760. Prof. at Strasburg. Mem. French Acad. Sci. Author: Analyse de Réfractions Astronomiques et Terrestres, 1798. Died Strasburg, May 14, 1826.

KRASILNIKOV, Nikolay Aleksandrovich, Russian microbiologist; b. Medvedovka (now Kaluga Oblast), Dec. 18, 1896; grad. Med. Faculty, Leningrad U., 1926; postgrad. in microbiology USSR Acad. Med. Sci., 1926-29; D.Biol. Sci., 1937. Sr. asso., later prof. Inst. Microbiology, USSR Acad. Sci., 1929—, also head dept. interaction of microorganisms; prof., 1938—; head chair soil biology Moscow U., 1953—. Recipient Stalin prize, 1951, Mechnikov prize, 1951. Mem. USSR Acad. Sci. (corr.). Author: Classification Key to Actinomycetales, 1941; Microbiological Principles of Bacterial Fertilizers, 1945; Classification Key to Bacteria and Actinomyces, 1949; Actinomyces, Antagonists and Antibiotic Substances, 1950; Soil Bacteria and the Higher Plants, 1958. Mem. editorial bd. Antibiotics, Microbiology, Nature. Research and numerous publs. on geology, agr., soil, marine and med. microbiology and antibiotics; made one of 1st Soviet antibiotics Mycetin, 1938-39 (used to treat infectious diseases); advocated use of Actinomyces antibiotics for Tb and diphtheria, 1939; made classification key for bacteria and Actinomyces based on principles of evolution and phylogeny of organisms; devised classification key for Actinomyces, 1941, for bacteria, 1949; developer methods to select and identify antagonists capable of synthesizing antibiotics. Address: Inst. Microbiology, USSR Acad. Sci., Leninsky prospect 33, Moscow, USSR.

KRASNOGORSKII, Nikolái Ivánovich, Russian pediatrician, physiologist; b. July 25, 1882; grad. USSR Acad. Med. Scis., 1945; studied under I. P. Pavlov; doctorate, 1911. Prof. naval-mil. Med. Acad. Recipient Stalin prize, 1952. Research in pathophysiology of higher nervous system of children.

KRASNOVSKII, Alexander Abramovitch, Russian photobiochemist; b. Odessa, USSR, Aug. 26, 1913; D.Sc., Acad. Scis. USSR. Staff, A.N. Bakh Inst. Biochemistry, Moscow, now head lab. photobiochemistry; faculty Moscow U. Corr. mem. Acad. Scis. USSR. Author: The Contemporary Concept of Photosynthesis, 1946; The Photochemistry of Photosynthesis, 1947. Research, numerous publs. on reversible photochem. reactions of chlorophyll its analogs and derivatives, mechanism of electron transfer in photosynthesis, aggregated and monomeric forms of plant pigments in organisms, mechanism of photobiol. processes. Office: B 71 Leninsky prospect, Moscow, USSR.*

KRASNOW, Frances, Am. biochemist; b. N.Y.C., Oct. 16, 1894; d. Raphael and Sara Rifka (Lubarsky) Krasnow; B.S. magna cum laude, Barnard Coll., 1917; A.M., Columbia, 1917, Ph.D., 1922; m. Marcus Thau, Dec. 25, 1930; 1 dau., Hudelle. Mem. research and teaching staff dept. biochemistry Columbia Coll. Phys. and Surg., 1917-32; asst. dir. Sch. Dental Hygiene, Guggenheim Dental Clinic, N.Y.C., 1932-44, head dept. research, 1944-52; research dir. Universal Coatings, Inc., Newark, 1952—; cons. in biochemistry, 1952—. Mem. Am. Chem. Soc., A.A.A.S. (life), A.M.A. (life), Am. Soc. for Microbiology, Internat. Assn. for Dental Research (life, pres. N.H. sect. 1947-49, editor 1949-60), Am. Inst. Chemists, Am. Assn. Clin. Chemists, N.Y. Acad. Medicine (life), N.Y. Acad. Scis. (life), Phi Beta Kappa, Sigma Xi. Research and numerous publs. on growth requirements and harmful effects to human organs, tissue and blood systems from microorganisms; biochem.-physiol. studies correlating dentomed. findings for levels in saliva and blood of acid, alkali, calcium, magnesium, phosphorus, cholesterol, lechithin and protein to warn beginning of dental carries, permature aging, incipient generalized lowered functioning. Home: 405 E. 72d St., N.Y.C. 10021. Office: 10 Av. C, Newark 07114.*

KRASTANOV, Ljubomir, Bulgarian physicist; b. Pleven, Nov. 15, 1908; s. Krastan Evtimov and Raina (Andreeva) K.; student Sofia (Bulgaria) U., 1927-31, D.Sc., 1951; student Geophys. Inst., Leipzig, Germany, Geophys. Obs., Wansdorf, Germany, 1940; m. Victoria Marinova Markova, Feb. 4, 1945; 1 dau., Marina. Asst., Central Inst. Meteorology; faculty Sofia U., 1946—, prof. meteorology, 1951-59, head chair meteorology and geophysics, since 1960—; head dept. physics atmosphere Inst. Meteorology, Bulgarian Acad. Sci., 1948—, academician, 1960—; dir. Central Inst. for Meteorology, 1954-59; dir. Inst. Geophysics, 1959—. Mem. adv. com. WHO, 1963—. Hon. mem. Hungarian, Czechoslovakian, USSR acads. sci.; mem. Bulgarian Acad. Sci. (pres., minister 1962—), World Meteorol. Orgn., Hungarian Meteorol. Soc., Union Sci. Workers in Bulgaria. Research, numerous publs. on. microphysics of clouds and phase transition in atmosphere condensation and crystalization processes and possibilities of their stimulation. Home: 66 Asparuh, Sofia, Bulgaria.*

KRATKY, Otto, Austrian chemist; b. Vienna, Austria, Mar. 9, 1902; s. Rudolf and Leopoldine (Gartner) K.; Dipl.Ing., Tech. U. Vienna, 1927, Dr.techn., 1929; m. Gerda Korte, July 30, 1942; children—Lorle, Christoph. Staff, Kaiser-Wilhelm-Institut, Berlin-Dahlem, Germany, 1928-33, 40-46; head dept. phys. chemistry, I. Chem. Universitatslaboratorium, 1934-40, dozent, 1937; prof., head Inst. for Physikal. Chemie der Deutschen Technischen Hochschule, Prag; prof., head Inst. Physikal Chemie, U. Graz (Austria), dean philos. faculty, 1947-48, rector univ., 1956-57. Recipient Wolfgang Ostwald-preis Deutschen Kolloidges, 1961; Prechtl-medal Techn. Hochschule Vienna, 1965. Mem. Österreichische Akademie der Wissenschaften (Haitinger prize 1936, Erwin Schrodinger prize 1964), Deutsche Akademie der Naturforscher Leopoldina, Jugoslawische Akademie der Wissenschaften und Kunste. Research, publs. on contbns. to problem of supermolecular structure of high polymers, numerous papers on exptl. methods and theory of X-ray small angle scattering, applications in field of proteins, high polymers in solid state and in solution, inorganic colloids. Home: 32 Leonhardgurtel. Office: 5, Halbarthgasse, Graz, Styria, Austria.*

KRATZENSTEIN, Christian Gottfried, iatro-physicist; b. Wernigerode, Germany, 1723; prof. physics, Copenhagen, Denmark, 1754. Author: Medical Application of Electricity, 1745. Invented stethoscope, circa 1780; studied cloud formation, natural philosophy. Died 1795.

KRATZER, F(rank) H(oward), Am. nutritionist; b. Baldwinsville, N.Y., Jan. 24, 1918; s. Richard N. and Frances (Watts) K.; B.S., Cornell U., 1940; Ph.D., U. Cal. at Berkeley, 1944; m. Georgina Garbutt, July 7, 1946; children—James, Paul, Charles. Staff dept. poultry husbandry U. Cal. at Berkeley, 1940-44; asso. prof. Colo. A. and M. Coll., Ft. Collins, 1944-45; faculty U. Cal. at Davis, 1945—, prof. poultry husbandry, 1955—. Recipient Nat. Turkey Fedn. Research award, 1949, Am. Feed Mfrs. Research award, 1960. Mem. Poultry Sci. Assn., World's Poultry Sci. Assn., Am. Chem. Soc., Am. Inst. Nutrition, Soc. Exptl. Biology and Medicine, Biochem. Soc., A.A.A.S. Research and publs. on amino acid and vitamin requirements turkeys, growth inhibitors for chickens and turkeys, mineral requirements chickens and turkeys, influence chelates on mineral utilization. Home: 10 Parkside Dr., Davis, Cal. 95616.*

KRATZER, Nicholas, mathematician; b. Munich, Bavaria, 1487; ed. U. Cologne, U. Wittemberg; B.A., Corpus Christi Coll., Oxford, 1523; M.A., 1523. Brought to Eng. by Erasmus; tutor in household of Sir Thomas More; fellow Corpus Christi Coll., Oxford, 1517, lectr. astronomy, became math. reader; apptd. astronomer and horologer to Henry VIII, 1519. Left 2 books in manuscript, Canones Horopti and De Compositione Horologiorum. Designed sundials and other instruments. Died 1550.

KRAUPP, Otto, Austrian pharmacologist; b. Krems, Austria, Oct. 23, 1920; s. Josef and Josefine (Schwarz) K.; M.D., U. Vienna, 1944, Ph.D. in Chemistry, 1952; m. Adele Pachinger, Nov. 24, 1945; children—Ursula, Johanna, Theresia, Martin, Bettina. Faculty, U. Vienna, 1947—, asso. prof. pharmacology, 1962—. Mem. Austrian pharmakopoea Com., 1961—. Mem. Austrian Biochem. Soc. (past pres.). Research, numerous publs. on neuromuscular pharmacology; discovered potassium release from musculature by depolarizing muscle relaxants; devel. new muscle relaxant; study and devel. new long acting anticholinesterases; devel. drug acting on coronary vessels for treatment of coronary disease. Home: 17. Pötzleinsdorferstrasse, Vienna 18, Austria.*

KRAUS, Charles August, Am. chemist; b. Knightsville, Ind., Aug. 15, 1875; B.S., Kan., 1898, postgrad., 1898-99, 1900-01; postgrad. Johns Hospkins, 1899-1900; Ph.D., Mass. Inst. Tech., 1908; several hon. degrees. Instr. physics Cal., 1901-04; asst. Mass. Inst. Tech., 1904-08, research asso. in phys. chemistry, 1908-12, asst. prof. phys. chemistry research, 1912-04; prof. chemistry, dir. lab. Clark U., 1914-24; prof. chemistry, dir. chem. research Brown U., 1925-51, emeritus research prof., 1951—; cons. chemist fixed nitrogen lab. U. S. Dept. Agr., 1922-24, 35—; lectr. Harvard, 1929, Ohio State U., 1932. Cons. chemist U. S. Bur. Mines, 1918, CWS, 1918-19; vice chmn. div. chemistry and chem. tech. NRC, 1931-32, chmn., 1932-33, field sec., fellowship bd. in phys. sci., 1934-35. Recipient Gibbs medal, 1935, Richards medal, 1936, Franklin medal Franklin Inst., 1938. Mem. Nat. Acad., Chem. Soc. (Nichols medal 1923, pres. 1939), Philos. Soc., Phys. Soc., Am., Washington acads. sci. Research on elec. apparatus and conducting systems, liquid ammonia, crit. phenomenon, electrolytic and non-aqueous solutions, metals and their solutions, metallo-organic compounds, phase rule, free radicals, germanium and gallium, gaseous equilibria, vacuum seals, conduction in glass, gases and chem. processes. Died 1957.

KRAUS, Eric Bradshaw, meteorologist, oceanographer; b. Liberec, Bohemia, Mar. 22, 1913; s. Paul and Bertha (Frank) K.; student U. Vienna (Austria), 1930-32; Ph.D., Charles U., Prague, Czechoslovakia, 1946; postgrad. U. Bergen (Norway), 1939-40; m. Heather Bradshaw Johnson, Mar. 5, 1942; children—Nigel, Sibella, Deborah. Sr. research officer Commonwealth Sci. and Indsl. Research Orgn., Sydney, Australia, 1946-49; lectr. Sydney U., also operations research officer Quantas Airlines (formerly Brit. Commonwealth Pacific Airlines), 1949-52; with Snowy Mountains Hydro-Electric Authority, Cooma, Australia, 1952-61; Rossby fellow Woods Hole (Mass.) Oceanographic Instn., prof., dir. Inst. Atmospheric Sci., U. Miami, 1966—. chief UN Tech. Assistance Mission, Brit. E. Africa, 1955-56; vis. prof. meteorology Yale, 1961-63. Mem. Am., Royal Meteorol. socs. Pioneered successful cloud-seeding expts.; research and numerous publs. on recent climatic changes in tropics, air-sea interactions and thermoclines. Home: 3575 St. Gaudens Rd., Coconut Grove, Fla. 33133. Office: P.O. Box 9115, U. Miami, Coral Gables, Fla. 33124.*

KRAUS, Ernst Carl, German geologist; b. Freising/Obb., July 10, 1889; s. Carl and Natalie (Herwig) K.; Ph.D., U. Munich; m. Theodora Baer, Oct. 18, 1919; children—Friedrich, Helmut, Gertrud, Elisabeth, Theodora. Instr., prof., full prof. U. Riga; full prof. U. Munich, now prof. emeritus. Recipient Hanstille medal, 1959, S. von Bubnoff medal, 1963. Mem. Real Academia Cordoba, Leopoldina Soc. Author: Der alpine Bauplan, 1936; Baugeschichte der Alpen, 1951; Vergleichende Baugeschichte der Gebirge, 1951; Entwicklung der Kontinente und Ozeane, 1959, also over 180 works on gen. geology, hydrogeology. Editor: Geolog. Archiv, 1923-26. Home: Verdistrasse 40, Munich/Obermenzing. Office: Institut für Allgem. und Angewandte Geologie und Mineralogie, Luisenstrasse 37/1, Munich, Germany.

KRAUS, Friedrich, Austrian physician; b. Badenback, Bohemia, May 31, 1851; student U. Prague (Czechoslovakia); M.D., U. Vienna (Austria), 1882. Asst., Physiol.-Chem. Inst., Prague, 1882-85; privat docent internal pathology U. Prague, 1888-90; asst. to Kohler clinic Allgemeines Krankenhaus, Vienna, 1890; became asst. prof. U. Vienna, 1893; named prof. med. pathology and therapeutics clinic, Graz, Austria, 1894. Author: Über Ermüdung als Mass der Constitution und über Säure Autointoxication; Krankheiten der Mundhöhle und Speiseröhre; Krankheiten der Sogenannten Blutdrüsen. Research on alkalis and oxidation of sugar in blood. Died 1936.

KRAUS, Herbert W., Austrian neurosurgeon; b. Mahrisch Ostrau, Nov. 20, 1910; s. Karl and Eleonore (Himmelbauer) K.; M.D., U. Vienna (Austria) 1934; m. Gertrud Oesterreicher, June 14, 1941; children—Renate, Lore. Became asso. with Pharmacology Inst., Vienna Med. Clinic, 1935; with med. clinic U. Vienna, until 1958, became dean neurosurg. clinic, 1962; agrégé in surgery, 1949; named titular prof. surgery, 1958; chief of med. staff surgery Kaiser Franz-Josef Hosp., Vienna, 1958—. Fellow Chartered Inst. Secs., Internat. Coll. Angiology (sci. council); mem. German Surgeons Soc., German Soc. Neurosurgery, Austrian Surgery and Neurosurgery Soc. Author works on surgery and neurosurgery. Co-editor: Breitnerischen Operationslehre. Home: Weygrasse 5, Vienna III. Office: Allgemeines Krankenhaus, Alsenstrasse 4, Vienna IX, Austria.

KRAUS, Otto, German mineralogist; b. Nuremberg, Feb. 25, 1905; s. Conrad and Emilie Charlotte (Seelos) K.; ed. U. Munich, Tech. High sch. Munich; Ph.D.; m. Else Heck, Feb. 26, 1938; 1 son, Heimuth. Asst., U. Munich, 1931-44, prof. mineralogy, 1937-39, 46-49, prof., 1944—; dir. Bayerische Landesstelle für Naturschutz, Munich, 1949—; instr. protection of nature Munich Tech., 1956—. Mem. Nature Protection League (pres.), German Assn. for Protection of Natural Waters, Bavarian Bot. Soc. Author: Bis zum letzten Wildwasser, 1960; (filrm) Natur in Gefahr; also works on mineralogy, chemistry of crystals, conservation of nature, ecology. Home: Ungererstrasse 58, München. Office: Bayerische Landesstelle für Naturschutz, Maximilianstrasse, 39, München, W. Germany.

KRAUS, Rudolf, Austrian bacteriologist, immunologist; b. 1868; studied under R. Paltauf, Vienna; successively dir. Bacteriol. Inst., Buenos Aires, Argentina, State Serum Inst., Sao Paulo, Brazil, State Serotherapeutic Inst., Vienna. Discovered bacterial precipitins, 1897; research in bacteriology and serology. Editor handbooks on immunity. Died 1932.

KRAUSE, Alfons, Polish chemist; b. Strzelno, Poland, Nov. 6, 1895; s. Stanislaw and Rozalia (Meyer) K.; Dr.phil., Berlin U., 1918; D.Sc., Central Commn. Qualification, Warsaw (Poland), 1950; m. Marta Pommer, July 23, 1924; 1 son, Lucjan. Pub. sch. tchr., Berlin, 1916-19; indsl. chemist Duisburg, Germany, 1919-20; hygiene and water chemist, Poznan, Poland, 1920-21; faculty Poznan U., 1921—, prof., head Inst. Inorganic Chemistry, 1930-39, 45—; prof. Polish clandestine univ., Warsaw, 1940-44. Recipient State Sci. prize, 1955; Gold cross of Merit, 1958; 10th anniversary medal, 1959. Mem. Polish Acad. Sci., Polish Chem. Soc., Copernicus Soc. Author: Poznanskie Towarzystwo Przyjació Nauk, 1956; also numerous articles. Research on colloid dispersed gases, new ferromagnetic ferrites, silver ferrites and ferric hydroxides, their structure and catalytic behavior, respiratory ferment using mixed hydroxide of Iron (III), copper (II) and magnesium; developed theory of radical structure of solids, explaining reactions occurring on surface of solids; research on trace catalysis; discovered that nearly all organic compounds are specific peroxidase catalysts, decoloring indigocarmine in hydrogen perioxide solution. Home: 4 Rycerska, Poznan, Poland.*

KRAUSE, Ernst Henry, Am. physicist; b. Milw., May 2, 1913; s. Ernst and Martha (Strege) K.; B.S., U. Wis., 1934, M.S., 1935, Ph.D., 1938; m. Constance Fraser, June 29, 1939; children—Margaret Bird, Katharine Louise, Carol Marjorie, Susan Fraser. Asso. dir. research Naval Research Lab., Washington, 1938-54; dir. research labs., missile systems div. Lockheed Aircraft Corp., Van Nuys, Cal., 1954-55; pres., chmn. bd. Systems Research Corp., Van Nuys, 1955-56; v.p., mem. bd. dir., Aeronutronic Systems, Inc., Newport Beach, Cal., 1956-60; dir. tech. staff, aeronutronic div. Ford Motor Co., Newport Beach, 1960-62; v.p., gen. mgr. Aerospace Corp., San Bernardino, Cal., 1962—. Mem. sci. adv. bd. Redlands (Cal.) U., 1964—, Orange State Coll., Fullerton, Cal., 1964—; bd. dir., World Affairs Council-Inland Empire. Recipient, Distinguished Civilian Service award, USN, 1956. Fellow Am. Phys. Soc., Am. Inst. Aeros. and Astronautics (asso. fellow); Tau Beta Pi, Sigma Xi. Home: 1919 Glenwood Lane, Newport Beach, Cal. 92660. Office: 1111 Mill St., San Bernardino, Cal. 92402.*

KRAUSE, Ernst Ludwig, biologist; b. Zielenzig, Poland, Nov. 22, 1839; settled in Berlin, 1866. Author: Werden und Vergehen, 1876. Pub. (with Charles Darwin, Ernst Haeckel) jour. Kosmos, 1877-82. Propagated Darwin's theory of evolution. Died Eberswalde, Germany, Aug. 24, 1903.

KRAUSE, Fedor, surgeon; b. Friedland, Schlesien, (formerly Germany), Mar. 10, 1857; prof., Berlin, Germany, from 1901; devised Krause's operation for extradural excision of Gasserian ganglion for trigeminal neuralgia. Died Badgastein, Austria, Sept. 22, 1937.

KRAUSE, Wilhelm, German anatomist; b. Hannover, Germany, July 12, 1833; s. Karl Friedrich Krause; student medicine, Göttingen, Germany, Berlin, Germany, Vienna, Austria, Zurich, Switzerland. Became prof., Göttingen, 1860, Berlin, 1892. Author: Handbuch der Anatomie des Menschen, 1899-1905; Die motorischen Nervendplatten, 1869. First description of nerve end organs in skin; described intermediate disk of striated muscle (Krause's membrane). Died Berlin, 1910.

KRÄUSEL, Richard Oswald Karl, German botanist; b. Breslau (now Wroclaw, Germany), Aug. 29, 1890; s. Carl Friedrich-Wilhelm and Katharina (Barthl) K.; Dr.phil., U. Breslau, 1913; Dr. Sci. hon.caus., U. Durham, 1963; m. Johanna Mathilde Luise Wellenstein, June 3, 1915; children—Hildegard (Mrs. Wolfgang Müller-Stoll), Irmgard, Wolfgang. Prof.,

Goethe High Sch., Frankfurt/Main, 1920-21; faculty U. Frankfurt/Main, 1921—, prof. botany, paleonbotany, 1928—; curator dept. paliobotany Senckenberg Research Inst. and Natural History Mus., Frankfurt/Main, 1938-46, head bot. and paleobot. dept., 1946—. Recipient Askenasy prize; Iron Medal, 1950, Senckenberg Natural Sci. Soc.; medal 2d Panam. Congress Geology, 1957; O.A. Derby medal, Rio de Janeiro Div. geol. min. 1957; A. H. Dumont medal Belgian Geol. Soc.; Kirstenbosch medal, 1963. Mem. Internat. Assn. Plant Taxonomy, Internat. Assn. Wood Anatomy, Senckenberg Natural Sci. Soc., German Bot. Soc. (v.p.), Paleontol. Soc. (v.p.), Leopoldina, German Col. Soc., Geol. Assn., Am. Bot. Soc. (corr.); hon. mem. Internat. Orgn. Paleobotany, Paleobot. Soc. India. Author: Paläobotanische Untersuchungsmethoden, 1928; Versunkene Floren, 1950; (with H. Merxmüller, H. Notdurft) Mitteleuropäische Pflanzenwelt I, 1956, II, 1960; also numerous articles. Research on fossil plants of different ages, tertiary and mesozoic flora; discovery of new primitive plants in Devonian and Mesozoic flora of Godwana countries and their paleoclimatic importance. Address: 5 Danneckerstrasse, Frankfurt/Main, West Germany.*

KRAUSHAAR, William Lester, Am. physicist, educator; b. Newark, Apr. 1, 1920; s. Lester A. and Helen (Ousterhoudt) K.; B.S., Lafayette Coll., 1942; Ph.D., Cornell U., 1949; m. Margaret Freidinger, Feb. 27, 1943; children—Mark Jourdan, Susan, Andrew Woolman. Physicist, Nat. Bur. Standards, Washington, 1942-45; asso. prof. physics Mass. Inst. Tech., Cambridge, 1956-65, prof., 1962-65; prof. physics U. Wis., Madison, 1965—. Fellow Am. Phys. Soc., Am. Astron. Soc., Internat. Astron. Union, Am. Acad. Arts and Scis. Author: (with Uno Ingard) Introduction to Mechanics, Matter and Waves, 1960. Research and publs. on astrophysics; study of cosmic X and gamma radiation. Home: 1003 Oak Way, Madison, Wis. 53705.*

KRAUSKOPF, Konrad Bates, Am. geologist, b. Madison, Wis., Nov. 30, 1910; s. Francis Craig and Maude (Luvan) B.; student Whittier (Cal.) Coll., 1927-28; A.B., U. Wis., 1931; Ph.D. in Chemistry, U. Cal., 1934; Ph.D. in Geology, Stanford, 1939; m. Kathryn Isabel McCune, Jan. 1, 1936; children—Karen, Frances, Karl, Marion. Instr. chemistry, U. Cal., 1934-35; instr. phys. scis., Stanford, 1935-39, asst. prof. geology, 1939-42, asso. prof., 1942-50, prof. geochemistry, 1950—, asso. dean School of Earth Sciences, 1960—. Guggenheim Fellowship for Geochem. Research, Fulbright Grant for Geochem. Research in Norway, 1952-53; geologist, U. S. Geol. Survey, 1943—. Served as chief, geog. sect. of C-2, U. S. Army, GHQ Tokyo, 1947-48. Awarded Citation for meritorious civilian service, Tokyo, 1948; recipient Day medal, Geol. Society Am., 1960. Member A.A.A.S., Geol. Soc. Am. (pres. 1966-67), Nat. Acad. Scis., Am. Geophys. Union, Am. Geological Inst. (pres. 1963-64), Phi Beta Kappa, Sigma Xi, Sigma Nu, Alpha Chi Sigma, Phi Kappa Phi, Gamma Alpha. Author: Fundamentals of Physical Science, 1941; (with Arthur Beiser) The Physical Universe; Introduction to Geochemistry. Research on physical chemistry of ore solutions; petrology of metamorphic and igneous rocks; trace elements in sea water and in sedimentary rocks. Home: 806 La Mesa Dr., Menlo Park, Cal.

KRAUSPE, Carl August, German pathologist; b. Insterburg, Germany, July 19, 1895; s. Franz and Elisabeth (Heinrici) K.; ed. U. Königsberg (now Kaliningrad, USSR), U. Berlin, U. Leipzig (Germany); M.D. in Surgery honoris causa Perouse U., 1961. m. Ilse Ploch, Sept. 3, 1925; children—Eva-Maria, Sabine-Elisabeth. Asst., Leipzig, also Königsberg; instr., then prof. Leipzig, also Berlin; prof. Königsberg; dir. Kassel Bacterial Pathology Inst.; became prof. U. Hamburg, 1948. Author: Die Gasoedeme des Menschen, 1959. Mem. Deutschen Gesellschaft für Pathologie (editor Verhandlungen 1954-62), Natural Research Soc. Research and numerous publs. on pathology infectious diseases, pathology of bones and joints. Home: Brunsberg 39b, 2 Hamburg-Lockstedt, West Germany. Office: Pathologisch. Institut d. Universitat, Martinstrasse 52, Hamburg 20, West Germany.

KRAUSS, William Ernest, Am. biochemist; b. Newark, Jan. 11, 1899; s. Ernest and Amelia (Trimborn) K.; B.S., Cornell U., 1922, Ph.D., 1926; D.Sc., Coll. of Wooster, 1966; m. Mildred Stratton, Feb. 2, 1924; children—William Charles, Roscoe Ernest. Instr. animal husbandry Cornell U., 1922-26; asst., dept. dairy industry Ohio Agr. Exptl. Sta., 1926-31, asso., 1931-41, chief dept., 1941-47, asso. dir. sta., Wooster, O., 1948-65, asso. dir. Ohio Agr. Research and Devel. Center, Wooster, 1965—. Mem. food and nutrition bd. Nat. Acad. Sci., 1943-49, chmn. com. on milk, 1943-49, chmn. agrl. bd., 1950-63, mem. bd., 1949-64. Recipient Distinguished Nutrition award Distillers Feed Research Council, 1964. Mem. Am. Dairy Sci. Assn. (Borden award 1939), Am. Soc. Animal Sci., Am. Inst. Nutrition, Am. Soc. Biol. Chemists, Phi Kappa Phi, Sigma Xi, Gamma Sigma Delta. Numerous publs. on research on nutritional requirements of dairy animals, food value of milk, feeding value of forages, developed methods for fortifying milk with vitamin D and proved efficiency of such in treatment and prevention of rickets; minor effects of pasteurization on nutritive value of milk. Home: 2404 Cleveland Rd., Wooster, O. 44691.

KRAUSS, Wolfgang, German oceanographer; b. Kuehnhaide, Germany, Jan. 1, 1931; s. Max Arthur and Martha (Seltmann) K.; student U. Berlin; diploma in Meteorology, U. Hamburg, 1953, D.Natural Sci., 1955; m. Ameliese Kruass, June 27, 1957; 1 son, Matthias. Oceanographer, U. Hamburg, 1955-57; staff Kiel (Germany) U., 1957—, prof., 1962—, head sect. theoretical oceanography. Author: Einführung in die Theoretische Ocenographie, 2 vols., 1966; also articles. Research on theory of internal waves in ocean, dynamics of stratified media; devel. instruments for measurement of internal waves. Home: 82a Dorfstrasse Russee b. Kiel, Germany.*

KRAUT, Heinrich Albrecht, German chemist; b. Stuttgart, Germany, Sept. 2, 1883; s. Heinrich and Mariane (Leipheimer); ed. Tubingen (Germany) U., U. Munich; Ph.D.; M.D. honoris causa; m. Clare Holzinger, Nov. 28, 1921; children—Max, Elisabeth. Joined chemistry staff U. Munich, 1925; became head chem. div. Kaiser Wilhelm Institut für Arbeitsphysiologie, 1931, hon. prof., head food physiology dept., 1956; named prof. U. Munster (Germany), 1951; dir. Max Planck Institut für Ernährungsphysiologie. Mem. Gesellschaft Deutscher Chemiker, Deutsche Gesellschaft für Ernährung (hon. pres.), Gesellschaft Deutscher Chemiker. Research in microbiology of algae, nutrition, especially in developing nations. Home: Harnackstrasse 61b. Office: Rheinlanddam 201, Dortmund, West Germany.

KRAUZ, Cyril Casimir, Czechoslovakian engr.; b. Moravany, July 21, 1883; s. Josef and Mary (Copiach) K.; ed. Tech. U., Lwow, Poland; studies in Eng., France, Germany; Ph.D.; m. Gabriela Durdilova, 1905. Asst. engr. Inst. Agrochemistry, Dublany, Poland, 1905; became asst. prof. organic chemistry, Tech. U., Prague, Czechoslovakia, 1908; became prof. organic chemistry and tech. Tech. U., Brno, Czechoslovakia, 1913; chief engr. dept. for smokeless powder factory explosives, Blumau nr. Vienna, 1915-18; became prof. explosives and organic Tech., Tech. U., Prague, 1920. Expert, insp. explosives, Prague; mem. Examining Commn. for Civil Engrs., Prague. Mem. Masaryk Acad. Work, Soc. Czechoslovak Engrs., Czechoslovak Chem. Soc., Faculty Scis. (London), Am. Chem. Soc., Internat. Bur. Work, Inst. Mil. Scis. (mem. council), Internat. Assn. Indsl. Property, Bohemian Acad. Sci., Masaryk Acad. Work (pres.), Assn. Czechoslovak Chem. Engrs. (pres.), Czechoslovak Assn. for Testing Materials and Constructions (v.p.). Author: Textbook of Organic Elementary Analysis; Organic Technology; Testing of Black Powder; Technology of Organic Compounds; Testing of Glue and Gelatine; Testing of Turpentine Oil; also articles. Research in organic chemistry and tech., explosives. Office: C.S.R. Kaderakovska 1, Prague XIX, Czechoslovakia.

KRAVIS, Irving Bernard, Am. economist; b. Phila., Aug. 30, 1916; s. Nathan and Ethel (Gelgood) K.; B.S., U. Pa., 1938, M.A., 1939, Ph.D., 1947; m. Lillian Beatrice Panzer, June 22, 1941; children—Robert, Marcia, Ellen, Nathan. Instr. econs. Whitman Coll., 1941-42; economist Dept. Labor, 1946-48; asso. prof. U. Mass., 1948-49; mem. faculty U. Pa., 1949—, prof. econs., 1956—, chmn. dept., 1955-58, 62-67, asso. dean Wharton Sch. Finance and Commerce, 1958-60; cons. govtl. and internat. agys., 1949—. Mem. research staff Nat. Bur. Econ. Research, 1962—. Ford Found. fellow, 1960-61; Guggenheim fellow, 1967-68. Mem. Am. Econ. Assn., Royal Econ. Soc. Author: Domestic Interests and International Obligations, 1963. Research and publs. on studies comparing prices and income of different countries and studies dealing with trade policies and factors determining trade flows. Home: 438 Warick Rd., Wynnewood, Pa. 19096. Office: Dept. Economics, Univ. Pennsylvania, Phila. 19104.*

KRAVITZ, Edward, Am. lit. scientist; chemist; b. Phila., Sept. 1, 1920; s. Abraham and Zina (Lishinsky) K.; B.Sc., Phila. Coll. Pharm. and Sci., 1948, M.Sc., 1949, D.Sc., 1952. Supervisory research microbiologist-chemist U. S. Govt., 1952-58; asst. dir. hosp. clin. lab., asso. clin. pathology Woman's Med. Coll. Pa., Phila., 1958-59; asst. to dir. clin. lab. Eastern Pa. Psychiat. Inst., Phila., 1958-59; dir. Newark Med. Lab., 1959-64; dir. labs. Bissell Hosp., Wilmington, Del., 1961, lit. scientist, 1961—; mng. editor Cancer Research Phila., 1964—; sr. staff mem. Fels Research Inst., vis. lectr. biochemistry Dental Sch., Temple U., Phila., 1964—. Commended for sci. achievement Dept. Navy, 1953. Fellow Am. Pub. Health Assn., Am. Inst. Chemists, A.A.A.S.; mem. Conf. Biologic Editors. Co-developer method analysis uric acid in serum, realistic bioassay method for testing surface disinfection; co-author improved method micro-analysis bilirubin in newborn. Home: 5720 Wissahickon Av., Phila. 19144.*

KRAVTCHENKO, Julien, mathematician, mech. engr.; b. Vornneje, USSR, June 7, 1911; s. Alexis and Helene (Krzypkowska) K.; licence in scis., Faculty Scis., U. Paris (France), 1932, D.Sc., 1941; m. Suzanne Dobelmann, May 6, 1944; children—Jean Francois, Pierre-Yves. Lectr., Faculty Scis., U. Grenoble (France), 1944-47, titular prof., 1947—, dir. Labs. Fluid Mechanics, 1952—. Sci. cons. Center Nuclear Studies, Grenoble, 1958—. Decorated chevalier Legion of Honor, officer Palmes Acad. Mem. Univ. Assn. Fluid Mechanics, Math. Soc. France, Am. Math. Soc. Editor: (Henri Villat) Leçons sur les Fluides Visqueux, 1943. Research and numerous articles on theories of viscous fluids, waves and tides, drainage of underground pools of water, also theory of potential. Home: Bibl. Univers., 1 Blvd. Maréchal Lyuatey, Grenoble, France. Office: Labs. Fluid Mechanics, 4446 Av. Félix Viallet, Grenoble, France.*

KRAWCZYNSKI, Stefan, German physicist; b. Schwetzkau, Germany, Aug. 9, 1929; s. Henry and Tony (Loreck) K.; Diplom, U. Munich, Dr. rer. nat., 1955; m. Johanna Marinus, Sept. 23, 1964; 1 dau., Saskia. Asst., Max Planck Inst. Physics, Göttingen, 1955-56; head hot chemistry group decontamination sect. Gesellschaft für Kernforschung, Karlsruhe, Germany, 1956-61; became mem. sub.-com. German Atomic Energy Commn., 1961; named head reprocessing, decontamination EURATOM, 1961; Author: Radioactive Wastes, 1967; also articles. Research on catalysis, heavy water reactor construction, waste treatment, dry and wet reprocessing. Home: 76 Artillerie, Jülich 517. Office: 517. Jülich Kernforschung, Jülich, West Germany.*

KRAYER, Otto (Hermann), pharmacologist; b. Kondringen, Germany, Oct. 22, 1899; s. Hermann and Frieda (Wolfsperger) K.; student U. Freiburg (Germany), 1919-20, 22-23, 24-25, M.D., 1926, M.D. (hon.), 1957; student U. Munich (Germany), 1920-22, U. Berlin, 1923; M.A. (hon.), Harvard, 1942; M.D. (hon.), U. Gottingen (Germany), 1962; m. Erna Philipp, July 27, 1939. Came to U. S., 1937, naturalized, 1944. Asst. pharmacology U. Freiburg, 1926-27; faculty U. Berlin, 1927-33, prof. pharmacology, 1932-33, acting head dept., 1930-32; Rockefellow U. Coll., London, Eng., 1934; vis. prof., head dept. pharmacology Am. U., Beirut, Lebanon, 1934-37; faculty Harvard, 1937-66, head dept. pharmacology, 1939-66, prof. pharmacology, 1951-54, Charles Wilder prof., 1954-63, G.A. Pfeiffer prof., 1964-66, prof. emeritus, 1966. Mem. adv. com. USPHS pharmacology study sect., 1950-54; mem. sci. adv. com. Mass. Gen. Hosp., 1959-61; mem. adv. com. U. S. nat. com. Internat. Union Phys. Scis., 1959-65. Recipient Torald Sollman award in pharmacology Am. Soc. for Pharmacology and Exptl. Therapeutics, 1961, Schmiedeberg-Plakette, Deutsche Pharmakologische Gesellschaft, 1964. Mem. Nat. Acad. Sci., Am. Acad. Sci., Am. Soc. Pharmacology and Exptl. Therapeutics (past pres.); hon. mem. Argentine, Brit., Finnish, German pharm. socs. Editor: Arzneiverordnungslehre (P. Trendelenburg), 1931, 38, 52; Die Hormone II (P. Trendelenburg), 1934; Ergebnisse der Physiologie, biologischen Chemie und experimentellen Pharmakologie, 1933-35, 49—. Research and numerous publs. on physiology and pharmacology of circulation and autonomic system. Home: 27 Howland Rd., West Newton, Mass. 02165.*

KRAZER, Adolf, German mathematician; b. Zusmarshausen, Germany, Mar. 15, 1858; Ph.D., U. Würzburg (Germany), 1881; prof. math. Technische Hochschule, Karlsruhe, Germany, 1902-06. Author: Neue Grundlagen einer Theorie der allgemeinen Thetafunktion, 1892; Abelsche Funktion und allgemeinen Thetafunktion, 1921; Zur Geschrieben d. graphischen Darstellung von Funktionen, 1915. Research on gen. functions, especially theta functions. Died Karlsruhe, Aug. 7, 1926.

KREBS, Artur, engr., inventor; b. 1847; officer French infantry; detached to establish air sta., Chalais-Meudon, 1879; worked with Charles Renard; built and tested Le France (dirigible), 1st round flights, 1884; built multi-pole electric motor; built electric motors for submarines Gymnote, 1888, Gustave Zédé, 1893; inventor carburetor, other automotive equipment. Died 1935.

KREBS, Edwin Gerhard, Am. biochemist; b. Lansing, Ia., June 6, 1918; s. William Carl and Louise (Stegeman) K.; B.A., U. Ill., 1940; M.D., Washington U., St. Louis, 1943; m. Virginia Frech, Mar. 10, 1945; children—Sally, Robert, Martha. NIH research fellow biochemistry Sch. Medicine Washington U., 1946-48; faculty U. Wash., Seattle, 1948—, prof. biochemistry, 1957—; cons. biochemistry Madigan Gen. Hosp., Tacoma, 1960—; mem. physiol. chemistry study sect. NIH, 1963—; mem. research allocations com. Wash. Heart Assn., 1965—. John Simon Guggenheim fellows, 1959-66. Mem. Am. Soc. Biol. Chemists, A.A.A.S., Am. Chem. Soc., Sigma Xi. Research and publs. on carbohydrate metabolism in skeletal muscle, role of glycogen in muscle, specific enzymes of carbohydrate metabolism. Home: 2209 N.E. 177th St., Seattle 98155.*

KREBS, Sir Hans Adolf, biochemist; b. Hildesheim, Germany, Aug. 25, 1900; s. Dr. Georg and Alma (Davidson) K.; ed. univs. Gottingen, Freiburg, Munich, Berlin; M.D., Hamburg U., 1925; M.A., Cambridge, U. 1934; Hon. doctorates U. Chgo., 1954, Freiburg U., 1955, U. Paris, 1956, Glasgow U., 1958, Sheffield U., 1959, London U., 1959, Jerusalem U., 1960, Humboldt U., Berlin, Germany, 1960, Leicester U., 1963, Granada Univ., 1964, Leeds U., 1964, U. Pa., 1965; m. Margaret C. Fieldhouse, 1938; 3 children. Asst. to prof. O. H. Warburg, Kaiser Wilhelm Inst., Berlin, 1926-30; instr. medicine Freiburg U., 1932; Rockefeller research student Cambridge U., 1933-34, demonstrator in biochem., 1934-

35; lectr. pharmacology U. Sheffield, Eng., 1935-38, biochemistry, 1938-45, became prof., dir. research in cell metabolism, 1945; Whitley prof. biochemistry Oxford U., 1954—. Recipient (with F. A. Lipmann) Nobel Prize in medicine and physiology, 1953; Lasker Award, Am. Pub. Health Assn. Fellow Royal Soc., 1947 (Royal medal 1954, Copley medal 1961); fgn. asso. Nat. Acad. Scis. U. S. A. Contbr. articles to jours. Important contributions to the understanding of metabolic pathways; discovered ornithine cycle of urea synthesis in mammalian liver, 1932, tricarboxylic-acid cycle (now known as Kreb's cycle, an essential intermediate step in oxidation of foodstuffs), other discoveries concerning D-amino acid oxidase, glutamine synthesis, purine synthesis in birds, and mechanisms involved in metabolism regulation. Home: 27 Abberbury Rd., Iffley, Oxford, Eng.

KREBS, John Sparrow, Am. biophysicist; b. New Orleans, Dec. 3, 1924; s. Bernard L. and Flora (Sparrow) K.; B.S., La. State U., 1948; Ph.D., U. Cal. at Berkeley, 1958; m. Mary Jane Beneditz, Apr. 20, 1956. Research asst. La. State U. Sch. Medicine, 1948-49; NRC fellow, 1949-52; radiol. biologist U. S. Naval Radiol. Def. Lab., San Francisco, 1953-—. Mem. Radiation Research Soc., A.A.A.S., Am. Physiol. Soc. Research on biol. recovery from radiation injury, physiol. changes in ageing, effect of ionizing radiation on ageing, liver function and excretion of dyes used in study liver function. Home: 226 Juanita Way, San Francisco 94127. Office: U. S. Naval Radiol. Def. Lab., San Francisco 94135.*

KREBS, Norbert, geologist; b. Leoben, Austria, Aug. 29, 1876; became privat dozent U. Vienna, 1909; prof. U. Würzburg (Germany), 1917, U. Frankfort am Main (Germany), 1918, U. Freiburg, 1926; prof. U. Berlin, 1927-43, became emeritus, 1943. Author: Die Ostalpen und das heutige Österreich, 2 vols., 1927. Studied Austrian Alps. Died Berlin, Dec. 5, 1947.

KREBS, Richard Peake, Am. research engr.; b. Waterloo, Ia., Feb. 20, 1916; s. Rudolph R. and Pearne (Peake) K.; B.Sc. in Elec. Engring., U. Cin., 1938, M.S., 1941, Ph.D., 1942; m. Carol Mae Frank, Nov. 28, 1952. Physicist, NACA, Cleve., 1942-44; research scientist NACA-NASA, Cleve., 1944—. Mem. Am. Inst. Aeros. and Astronautics. Research on electronic instruments for biophysics, aircraft engine testing, turbojet engine components, dynamics rocket fuel systems, heat rejection in space. Home: 18850 E. Shoreland Dr., Rocky River, O. 44116. Office: 21000 Brookpark Rd., Cleve. 44135.*

KREBTZ, Johannes, Austrian physician; b. Vienna, Apr. 14, 1897; M.D., U. Vienna; m. Ilse Kaser, May 10, 1937; children—Elisabeth, Peter, Marie, Martha. Head physician Linz Hosp., 1930-42; head cancer research service of Vienna, 1950—; sec. Vienna Cancer Research Center, 1945—; founder Diagnostic Center of Soc. against Cancer. Author: Therapie und Praxis; also numerous articles. Editor, Krebsarzt, 1946—. Address: Universitätsstrasse 11, Vienna I, Austria.

KRECKER, Frederick Hartzler, Am. biologist; b. Tokyo, Japan, Dec. 21, 1881 (parents Am. citizens); A.B., Princeton, 1904, Ph.D., 1909; M.A., Cornell U., 1906; m. Margaret Ellen Brown, Aug. 22, 1916; children—Frederick M., Margaret E. (Mrs. Landis D. Baker). Prof. biology Marietta Coll., 1909-14; prof. zoology Ohio State U., 1914-29; prof., head dept. zoology Ohio U., Athens, 1929-52, prof. emeritus, 1952—, acting dir. Stone Biol. Lab., 1915-36. Fellow Am. Soc. Zoologists; mem. Ohio Acad. Sci. (past pres., exec. sec. jr. div., life mem., Distinguished Service award 1960), A.A.A.S., Am. Inst. Biol. Scis., Sigma Xi, Gamma Alpha. Author: General Zoology, 1934; also numerous articles. Editor, Ohio Jour. Sci., 1918-29. Research in animal regeneration, limnology.*

KRÉE, Paul, French mathematician; b. Belfort, France, Sept. 13, 1933; s. Louis Krée; student arts and crafts Chalons sur Marne, 1951-55; agrégé in math., Faculty Scis., U. Paris, 1957, D.Sc., 1965; m. Mirella Patenzan, June 25, 1966. Asst. lectr. Faculty of Scis., U. Paris, 1963-65; lectr. Faculty Scis., U. Nice (France), 1965—. Mem. Am. Math. Soc., Soc. Math. France, Soc. Former Students Sch. Arts and Crafts. Author: Introduction aux mathématiques et applications fondamentales, 1968; also numerous articles. Research on problems associated with Fourier transformation, also problems associated with integral equations and singular integrals. Home: 61 Mt. Boron, Nice (06), France. Office: Faculty Scis., Nice (06) France.*

KREEVOY, Maurice Mordecai, chemist; b. Boston, Aug. 28, 1928; s. Edward Phillip and Jennie (Gildisheim) K.; B.S., U. Cal., Los Angeles, 1950; Ph.D., Mass. Inst. Tech., 1954; m. Raye Gladys Schwartz, Mar. 29, 1953; children—Edith Pamela, William Seth. NSF fellow U. Utah, 1955-56; faculty U. Minn., Mpls., 1956—, prof. chemistry, 1964—, Sloan Found. fellow, 1960-64, NSF Sr. Postdoctoral fellow, 1962-63; cons. Gen. Mills, Inc., 1958—. Mem. Am. Chem. Soc., Chem. Soc. (London), Faraday Soc. Research, publs. on secondary, solvent, kinetic, hydrogen isotope effect; contbr. to theory structural effects on reactivity; pioneer process to remove detergent from waste water. Home: 2836 W. River Rd., Mpls. 55455.*

KREFFT, Gerhard Albert William, German ichthyologist; b. Hamburg, Germany, Mar. 30, 1912; s. Paul and Margarethe (Meusel) K.; student Techn. U., Braun-schweig, Germany, 1931, U. Würzburg (Germany), 1932; D.rerum naturalium, U. Hamburg, 1938; m. Ingeborg Lippert, Apr. 18, 1947; children—Sybille, Sabine, Susanne. Sci. asst. Zool. Mus., Hamburg, 1938-39, Zentralanstalt fur Fischerei, Hamburg, 1945-53; sci. officer Institut fur Seefischerei der Bundesforschungsanstalt für Fischerei, Hamburg, 1953-—, sr. sci. officer, 1964—. Fisheries and ichthyological expert Internat. Council for Exploration of Sea, Copenhagen, Denmark, 1955—, Internat. Commn. N. West Atlantic Fisheries, 1960-64, UNESCO, 1966—. Mem. Deutsche Zoologische Gesellschaft, Am. Soc. Ichthyologists and Herpetologists. Author: Die Schildkroten—Braunschweig, 1949; Gewinnung und Verwertung der Haut. Handbd. Seefisch. Nordeuropas 9, 1956; also numerous articles. Research on systematics and geog. distbn. of marine fishes especially sharks, skates and deap-sea fishes of Eastern Atlantic and Western Indian Ocean, fish communities on continental slopes and submarine ridges in N.E. Atlantic; discovery and first description of some fish species. Home: 10 b Kedenburgstrasse, Hamburg 70. Office: 9 Palmaille, Hamburg, 50 Germany.*

KREIDL, Alois, physiologist; b. Gratzen, Bohemia, Jan. 18, 1864; Research on physiology of nervous system, auditory organ, eye; 1st exptl. work (with J. P. Karplus) on hypothalamus, 1909. Died Vienna, Austria, Dec. 6, 1928.

KREIDL, Norbert Joachim, physicist; b. Atzgersdorf, Austria, July 3, 1904; s. Ignaz and Hilda (Krenn) K.; student U. Bern, Switzerland, 1925; Ph.D., U. Vienna, 1927; m. Melanie Schreiber, Sept. 22, 1934; children—John, Maria-Martina, Tobias. Came to U. S., 1938, naturalized, 1944. Postdoctoral fellow Kaiser Wilhelm Inst. Silikatforschung, Berlin, 1928-29; physicist to gen. mgr. Schreiber and Nephews Glass Works, Rapotin, Czechoslovakia, 1929-38; asso. State Glass Inst., Hradec Kralove, Czechoslovakia, 1929-38; faculty Pa. State U., 1938-43; sect. head to dir. material research and devel. Bausch & Lomb, Inc., Rochester, N.Y., 1943-66, cons., 1941-43, 65-—; prof. ceramics Rutgers U., New Brunswick, N.J., 1964-66; prof. ceramic engring. U. Mo., Rolla, 1966-—; Cons. various bus. firms, 1939—; dir. Linden Labs., Boalsberg, Pa., 1955—; adviser to Nat. Bur. Standards, 1957-63, Alfred U., 1963-66. Recipient Toledo Glass & Ceramics award, 1966. Mem. Internat. Commn. on Glass (v.p.), Am. Ceramic Soc. (trustee), Optical Soc. Am. (div. chmn.), Am. Chem. Soc., (div. chmn.), Brit. Soc. Glass Tech., Sigma Xi. Contbg. author Handbook of Glass Manufacturing (F. V. Tooley), 1960; (with J. R. Hensler) H. Hausner's Modern Materials, 1958; (with J. Rood) Vol. I Applied Optics (R. Kingslake), 1965; (with E. Lell and J. R. Hensler) Progress in Ceramic Science (J. Burke), 1966. Research, publs. on structure and composition of glass, its optical properties; color, radiation effects on glass; high pressure ceramics; gen. glass tech.*

KREIL, Karl, Austrian astronomer, meteorologist; b. Ried, Austria, Nov. 4, 1798; ed. Kremsmünster Abbey; successively became asst. dir. Vienna Obs., 1827, adj. La Brera Milan (Italy) Obs., 1831; joined Prague Obs., 1838, became dir., 1845; named prof. physics, Vienna, 1851; dir. Center Terrestrial Magnetism, Vienna, 1851. Author: Versuch, den Einfluss der Mondes auf der atmosphärischen Zustand unserer Erde zu erkennen, 1841; Über die Natur und Bewegung der Kometen, 1843; Über den Einfluss der Alpen auf der Äusserung der magnetischen Erdkraft, 1850; Einfluss der Mondes auf der magnetische Deklination, 1852. Founder sci. of terrestrial physics; meteorol. observations of Europe; studied earth's magnetic field, motion of comets. Died Vienna, Dec. 21, 1862.

KREIN, Mark Grigorievich, Russian mathematician; b. Apr. 3, 1907; postgrad. Odessa Inst. Pub. Edn.; Dr.physico-math. sci., 1939. Instr. Odessa U., 1933-34, prof., 1934-41; staff Kharkov U., Inst. Math. and Mechanics, 1934-40; staff Inst. Math., Ukraine Acad. Sci., 1940-41, 44-51; prof. Kubybyshev Ind. Inst. and Aviation Inst., 1941-44; prof. Odessa Inst. Naval Engring., 1944-45; prof. Odessa Civil Engring. Inst., 1954—. Author: The Theory of Self-conjugated Extensions of Semi-Limited Hermitian Operators and its Application, 1947; P. L. Chebyshev and A. A. Markov's Ideas on the Theory of Limiting Integral Values and their Further Development, 1951; The Theory of Linear Non-Self conjugated Operators, 1960; The Spectral Function of a Self-conjugated Operator in a Space with Indefinite Dimensions, 1963; co-author: Oscillatory Matrices and Nuclei and Minor Vibrations of Mechanical Systems, 1950. Corr. mem. Ukranian Acad. Sci. Office: USSR, Odessa, Ukr.SSR, ul Didrikhsona 4, Inzhenerno-stroitelny institut.

KREINDLER, Arthur; b. Rumanian neurologist; b. 1900; prof. Bucharest (Rumania) U.; mem. Acad. Socialist Republic Rumania, numerous fgn. sci. acads. and socs. Author: Epilepsy, 1957; Neurology, 1957; Asthenic Neurosis, 1961; (with M. Steriade) La physiologie de la physiopathologie du cervelet, 1958. Research on therapeutics of asthenic syndrome.

KREJCI, Karl Julius Emil, geologist; b. Gmünd, Austria, Apr. 15, 1898; s. Anton and Emilie (Graf) K.; ed. U. Vienna, Abo Acad., U. Berlin; Ph.D.; m. Hanna Hoppe, Dec. 28, 1922; children—Maja, Kaj, Uta, Erika, Volker. Prof., Canton U.; cons. to China mine expdn.; prof. Tech. Sch. Mining; dir. Continental Petroleum Soc.; tech. asst. dir. for oil co., Lisbon, Portugal; prof. Frankfort (Germany) U. Mem. Assn. German Geologists, Geologists Assn., Naturalists, Assn., Geochem. Soc., Geologists Union. Research and numerous publs. on oil, vulcanism, geochemistry, ores in Sweden, oil in Rumania. Home: Hügelstrasse 200, Frankfurt/Main, West Germany.

KREMENTZ, Edward Thomas, Am. physician; b. Newark, Apr. 30, 1917; s. Albert Martin and Agnes Templeton (Alguier) K.; B.A., Wesleyan U., Conn., 1939; M.D., U. Rochester, 1943; m. Carolyn Butler, Oct. 5, 1946; children—Edward Thomas, Anne Butler, Cynthia Aiguier, David George, Elizabeth Avery. Asso. surgeon Yale, 1948-50; faculty Tulane U., New Orleans, 1953—, prof. surgery, 1961—, cancer teaching coordinator, 1953—, dir. Cancer Clin. Research Center, Charity Hosp., 1961—. Research Career award Nat. Cancer Instn., NIH, 1962. Mem. A.M.A. (co-recipient Hektoen Gold Medal 1959), La. State Med. Soc., Am. Cancer Soc., A.A.A.S., Am. Assn. for Cancer Research, A.C.S., Am. Soc. Clin. Oncology, Am., So. (Shipley Gold medal 1964) surg. assns., Soc. for Exptl. Biology and Medicine, Soc. U. Surgeons, Societe Internationale de Chirurgie. Contbr. numerous articles to med. jours. Research on behavior and treatment cancer, transplantation studies homologous and heterologous endocrine and malignant tissues. Home: 500 Walnut St., New Orleans 70118.*

KREMER, Gerhard, see Mercator, Gerhardus.

KREPS, Yevgeniy Mikhaylovich, Russian physiologist; b. Apr. 30, 1899; grad. Mil. Med. Acad., 1923, postgrad., 1925-31. Head physiol. lab. Murmansk Biol. Sta., 1923-33; with Cambridge and Plymouth Biol. Sta., Eng., 1930-31; prof. Leningrad U., 1934-37; head lab. comparative physiology and biochemistry Pavlov Inst. Physiology, USSR Acad. Sci., 1935-60, dir. Inst. Evolutionary Physiology, 1960—. Mem. USSR Acad. Sci. (corr.). Author: Comparative Biochemistry of Muscular Activity, 1933; Nervous Activity and the Biochemical Evolution of the Brain in Ontogenesis 1952; Phospholipids in the Nervous System, 1956; New Data on the Evolution of the Phosphorus Metabolism of the Sea, 1959; Oxyhemometry, 1959. Mem. editorial staff Progress in Modern Biology. Research and publs. on comparative physiology and biochemistry of nervous system and respiratory function of blood, occupational physiology of divers and marine chemistry; evolved method of early diagnosis of sepsis on basis of carbonic anhydrase content of blood, also method of oxyhemometry. Address: Inst. Evolutionary Physiology, USSR Acad. Sci., pr. Maklina, Leningrad, USSR.

KRESS, Wilhelm, inventor, engr.; b. 1846; worked in Vienna, Austria; designed a helicopter, 1877; began making elastic-powered model airplanes after bird forms, 1877; built full-size machine mounted on 2 floats, with 3 wings in tandem, rear elevator and rudder, 2 airscrews driven by petrol motor, 1898-99, tested and wrecked, 1901. Died 1913.

KRESZE, Guenter Heinz Joachim, German chemist; b. Berlin, Germany, Sept. 7, 1921; s. Paul and Dora (Mill) K.; Dr.-Ing., Tech. U. Berlin, 1951; m. Elisabeth Ehrhardt, May 25, 1946; children—Georg-Burkhard, Jan-Michael. Docent, Tech. U. Berlin, 1955-61; asso. prof. Munich Inst. Tech., 1961-64, prof., 1964—. Mem. Soc. German Chemists, Am. Chem. Soc., Leopoldina. Author: Physikalische Methoden in der organischen Chemie, 1962; also numerous articles. Research on synthetic organic chemistry (Heterocyclics, sulphur compounds); mechanism of some chem. reactions. Home: 91c Mozartstrasse, 8012 Ottobrunn, Germany. Office: 21 Arcisstrasse, 2 Munich, West Germany.*

KRETCHMER, Norman, Am. pediatrician; b. N.Y.C., Jan. 20, 1923; s. Emanuel and Sue (Gross) K.; B.S., Cornell U., 1944; M.S., U. Minn., 1945, Ph.D., 1947; M.D., State U. N.Y. Coll. Medicine, N.Y.C., 1952; m. Muriel Reiter, Sept. 10, 1942; children—Pamela Sue, Paul Jay, Steven David. With U. Minn., 1944-47, L.I. Coll. Medicine, 1948-52; faculty U. Vt., 1947-48, Bklyn. Coll., 1950-55, Cornell U., 1953-59; prof., exec. head pediatrics Stanford (Cal.), 1959—; physician in chief pediatrics Stanford Hosp., 1959—. Mem. med. adv. council FDA, 1964—; spl. cons. WHO, 1967—, also numerous com. memberships. Diplomate Am. Bd. Pediatrics. Fellow A.A.A.S.; mem. Am., Chilean (hon.), Czechoslovakian (hon.) pediatric socs., Am. Soc. for Biol. Chemists, Am. Soc. for Human Genetics, Am. Soc Clin. Investigation, Assn. Am. Med. Colls., Harvey Soc., Histochem.Soc., Soc. Growth and Devel., Soc. for Pediatric Research (pres. 1967-68), Western Soc. for Pediatric Research (pres. 1966-67), Western Soc. for Clin. Research, Am. Acad. Pediatrics, Am. Chem. Soc., Sigma Xi, others. Mem. editorial bd. Pediatrics, 1958-64, Biologia Neonatorum, 1961—, Excerpta Medica, 1961— Advances in Pediatrics 1967—; mem. adv. bd. Stanford Med.

Bull. Research and publs. on biochem. studies of devel., biosynthesis of pyrimidine; intestinal digestion of carbohydrates. Home: 843 Sonoma Terrace, Stanford, Cal. 94304. Office: Dept. Pediatrics, Stanford U. Sch. Medicine, 300 Pasteur Dr., Palo Alto, Cal. 94304.*

KRETOVICH, Vatslav Leonovich, Russian biochemist; D. Biol. Scis. Staff, Technol. Inst. Food Industry, Moscow, USSR, A. N. Bakh Inst. Biochemistry, USSR Acad. Scis., from 1962. Corr. mem. Acad. Scis. Author: Principles of the Biochemistry of Plants (A. N. Bakh prize 1958), 1956; The biosynthesis of dicarboxylic amino acids and enzymatic transformations of amides in plants, 1958; Biochemistry of Grain and Bread, 1958; Enzymatic synthesis of glutamic acid and phenylalanine in plants, 1959; Interaction of lipases and lipoxidases in the oxidation process of plants, 1959; Serine synthesis from pyruvic acid in plants, 1961. Authority on plant biochemistry; investigated enzymatic properties and processes of various foods (such as wheat, soy beans, sunflower seeds).

KRETSCHMAR, Wolfram Georg Christian, German parasitologist; b. Breslau, Germany, Apr. 9, 1932; s. Paul Gustav and Minna (Hertel) K.; Dipl. Biologe, U. Hamburg, Germany, 1959, Dr.rer.nat. magna cum laude, 1959; student U. B.C., 1953-54; habilitation, U. Tubingen, 1966; m. Christa Karberg, Mar. 26, 1959; children—Martin, Peter, Michael. With Inst. Tropical Medicine, U. Tubingen, Germany, 1959—, chief dept. protozoology, 1959—. Author: Reallexikon der Medizin, 1966; also articles. Co-editor Zeitschrift fur Tropenmedizine und Parasitologie. Research on immunity to protozoan infections, especially malaria, physiology of malaria parasites; demonstrated that milk diet inhibits growth of malaria parasites in exptl. and human malaria due to its lack of p-aminobenzoic acid, and that resistance of infants to malaria in endemic regions may be attributable to dietary factors. Home: 13 Bohnenbergerstrasse, 74 Tubingen, Germany. Office: Tropenmedizinisches Institut der Universitat, 11 Wilhelmstrasse, 74 Tubingen, Germany.*

KRETSCHMER, Ernst, German psychiatrist; b. Wüestenrot, Germany, Oct. 8, 1888; ed. U. Tübingen, U. Munich; M.D.; Ph.D. honoris causa, Bonne, Germany, 1946; m. Luise Pregizer, 1915; 2 children. Became asst. Nervenklin, 1913; joined faculty U. Tübingen, Germany, 1918; became prof., dir. Nervenklin, Marburg, Germany, 1926; named prof., dir. Neurol. Clinic, Tübingen, 1946. Mem. Gen. Med. Soc. for Psychotherapy (hon. pres.), Soc. for Constl. Research (pres.), German Acad. Sci. Berlin, Bavarian Acad. Sci., Norwegian Acad. Scis. Oslo. Recipient Golden medal Med. Faculty, Würzburg U., 1943; Golden Kraeplin Medal, 1956. Author: Körperbau und Charakter; Geniale Menschen, 1929; Medizinische Psychologie, 1922; Hysterie, Reflex und Instinkt; Psychotherapeut. Studien; Die Sensitive Beziehungswahn, 1918. Developed classifications of cycloids (pyknics, extroverts) and schizoids (leptosomics, introverts), in accord with his theory (which has not been widely accepted) that body types and personality characteristics are related, 1921. Died Feb. 8, 1964.

KRETZULESCU, Nicolae, Rumanian physician; b. 1812; participant revolution of 1848; militant for Union of Rumanian Principalities; minister, prime minister under reign of Alexandru Ioan Cuzs, 1859-1866. Mem. Rumanian Acad. Initiator med. edn. in Rumania. Died 1900.

KREUZER, Ferdinand Josef Alexander, physiologist; b. Lucerne, Switzerland, July 11, 1919; s. Ferdinand and Katharina (Muff) K.; student U. Fribourg (Switzerland), 1940-42; M.D., U. Zurich (Switzerland), 1946, M.D. thesis, 1950, Ph.D., 1952; m. Christa L. A. Paustian, Jan. 22, 1955; 1 dau., Michaela Maria. Faculty, U. Fribourg, 1946-55; Swiss Acad. Med. Sci. research asso. in Kiel, Germany, Cambridge, Eng., U. Pa., Phila., 1952-54; faculty Dartmouth, Med. Sch., 1955-61, asso. prof., 1958-61; prof. physiology, head dept. U. Nijmegen, (Netherlands), 1961—; dir. research Cardiopulmonary Lab., Hitchcock Found., Mary Hitchcock Meml. Hosp., Hanover, N.H., 1957-60. Cons., VA Hosp., White River Junction, Vt., 1957-61. Mem. Am. Physiol. Soc., N.Y. Acad. Scis., Physiol. socs. Netherlands, Switzerland, France. Author: Modellversuche zum Problem der Sauerstoffdiffusion in den Lungen, 1953; also numerous articles. Research on diffusion of oxygen in blood and kinetics of chem. reaction of oxygen and carbon dioxide with hemoglobin, polarographic determination of oxygen in blood and gas, particularly with catheter electrodes in vivo, alveolararterial gas pressure differences and diffusing capacity of the lung, high altitude physiology, hemodynamics and viscosity of blood. Home: 64 Ubbergseveldweg, Nijmegen, Netherlands.*

KREWER, Semyon Efimovich, physicist; b. Moscow, Russia, Mar. 10, 1915; s. Efim A. and Anna (Turgel) K.; Dipl. Ing. Physics Technische Hochschule, Berlin, Germany, 1937; postgrad. Columbia, 1940; m. Elsa Silberstein, July 22, 1939; 1 dau., Julie Ann. Came to U. S., 1938, naturalized, 1943. Asst. to Enrico Fermi, Leo Szilard at Columbia, 1938-39; tech. dir. Krewer Research Lab., Point Lookout, N.Y., 1953—; with Photovolt Corp., N.Y.C., 1940-67, dir. research, v.p., 1959-67. Mem. A.A.A.S., Optical Soc. Am. Assisted in early atomic fission experiments; developed and designed a number of scientific instruments, pupillograph, dermofluorometer, radiation detection densitometer, gloss and reflection meters, phmeters, colorimeters, fluorometers, super sensitive lightmeters, photographic meters. Home: 514 Cathedral Pkwy., N.Y.C. 10025. Office: 75 Freeport Av., Point Lookout, N.Y. 11569.*

KREWSON, Charles Fleming, Am. biochemist; b. Hatboro, Pa., May 17, 1905; s. William Darragh and Emma (Lewis) K.; B.S., Pa. State U., 1927; M.S., U. Ky., 1932; Ph.D., U. Wis., 1940; m. Hazel Mae Hakes, Apr. 9, 1928; children—Marilyn Sheldon (Mrs. Harry Megargee Edenborn), Charles Fleming III, Stephen Karl. Instr. chemistry U. Ky., 1927-38, 1939-41; research asst. U. Wis., 1938-39; research chemist USDA Agrl. Research Service, Phila., 1941—. Recipient Wisconsin Alumni Research Found. award, 1938, Phila. sect. Service award, 1958. Mem. Am. Chem. Soc., Phila. Organic Chemists Club (chmn. 1947-48), Am. Oil Chemists Soc., A.A.A.S., Sigma Xi, others. Author (with John Q. Griffith, Jr., Joseph Naghski) Rutin and Related Flavonoids, 1955. Research and publs. on natural products, developed rutin, a flavonoid drug used for correction of blood-vascular disorders; invented process for preparation of the pellagra preventative vitamin factor, niacinamide; plant growth-regulator contbns. to prodn. of weed-killers, development of Ammo-1618, a growth-inhibitor antagonistic to gibberelic acid, discovered regulators of high specificity using optically active amino acids; discovered evidence that biol. oxidation of a fatty acid molecule may occur at the gamma as well as beta position; developed epoxy-bearing seed oils to the stage of comml. prodn.; discovered a lipase enzyme system in ironweed producing specific activity for the number two position of fatty acid glyceride molecules. Home: 1830 Woodland Rd., Abington, Pa. 19001. Office: 600 E. Mermaid Lane, Phila. 19118.*

KREYBERG, Leiv, Norwegian pathologist; b. Bergen, Norway, May 22, 1896; s. Peter Christian and Elisabeth (Konow) K.; M.B., 1921, M.D., 1929; became univ. scholar, 1932; m. Emmie Louise Klem, Jan. 12, 1924; 2 sons, 2 daus. Began practice medicine, 1921; house surgeon, Levanger, 1924-26; became prosector histology U. Oslo (Norway), 1926, prof. pathology, 1938—; chief lab. pathology Norwegian Radium Hosp., 1926-32. Chmn., Cancer Commn. Mem. Biol. Soc. Oslo (pres. 1928-30), Path. Soc., Sci. Acad. Oslo; Norwegian Med. Assn. (chmn.), Union Internationale Contre le Cancer (ofcl. rep.), Biologisk Selskap (pres. 1930-31); hon. mem. Indian Path. Soc., Société Francaise D'Angiologie et d'Histopathologie. Author: Kreft som Biologisk and og Socialmedisinsk Sporgsmal, 1937; Malgive Svulster, 1937; Atombomens Biologiske Virkninger, 1947; Legebok for Sjomenn, 1947, 52, 58. Research and publs. in exptl. pathology, lung cancer typing and role of smoking; problems of war medicine; correlation of path. and clin. aspects as early diagnosis. Home: 79 Munkedamsvei, Oslo, Norway.*

KRICK, Irving Parkhurst, Am. meteorologist; b. San Francisco, Calif., Dec. 20, 1906; s. H. I. and Mabel (Royal) K.; B.A., U. of Calif., 1928; M.S., Calif. Inst. Tech., 1933, Ph.D., 1934; m. Jane Clark, May 23, 1930; 1 dau., Marilynn; m. 2d, Marie Spiro; 1 son, I. P. Krick, II; Asst. mgr. radio station KTAB, 1928-29; pianist in concert and radio work, 1920-30; meteorologist since 1930; set up 1st modern airline weather forecasting service for Western Air Express, 1932; became mem. staff Calif. Inst. Tech., 1933, asst. prof. meteorology, 1935-38, asso. prof. and head dept. 1938-48; organized and pres. Am. Inst. of Aerological Research, 1948, and Water Resources Development Corp. in 1950, set up meteorology dept. for Am. Air Lines, Inc., 1935; cons., 1935-36; developed private weather forecasting serv., supplying information to various cos.; developed long-range weather forecasting method covering periods up to 6 mos. Served as lt. Coast Arty. Corps, U. S. Army, 1928-36; commd. ensign U.S.N.R., 1938; major U. S. Army Air Corps, Weather Directorate, Weather Central Division, unit comdr. of Long Range Forecast Unit A, 1942-43. Received Junior Chamber of Commerce Distinguished Service award; chosen one of 10 outstanding men under age 35 by U. S. Chamber Commerce. Mem. Am. Meteorol. Soc., A.A.A.S., Inst. Aero. Scis. Co-author: Sun, Sea and Sky, 1954. Writer articles on weather analysis and forecasting and its application to agrl. and bus. industries. Home: 328 Mt. View Pl. Office: 611 S. Palm Canyon Dr., Palm Springs, Cal. 92262.*

KRIEG, Aloysius, German microbiologist, insect pathologist; b. Frankfurt/Main, Germany, Apr. 5, 1927; s. Willy and Helena (Reiser) K.; Dr.rer.nat., U. Frankfurt/Main, 1951; m. Hilde Braungardt, May 3, 1952; children—Ingrid, Thomas, Lothar. Sci. asst. Carl Zeiss, Göttingen (Germany), 1952-54; sci. advisor Inst. Biol. Control, Fed. Biol. Adminstrn., Darmstadt (Germany), 1954—. Author: Grundlagen der Insektenpathologie, 1961; Bacillus thuringiensis, 1961; also numerous articles. Research on insect pathogenic bacteria and insect viruses; on rickettsiac, including reprodn. cycle and germinative transmission. Home: 108 Heinrich-Strasse. Office: 61 Kranichsteiner-Strasse, Darmstadt, West Germany.*

KRIEG, Wendell Jordan, Am. neuroanatomist; b. Lincoln, Neb., Apr. 13, 1906; s. Frank Wendell and Rosabelle (Jordan) K.; B.S. in Medicine, U. Neb., 1928; M.S., N.Y. U., 1931, Ph.D. in Anatomy, 1935; m. Roberta Elaine Neill, June 28, 1952; 1 son, Rupert. Instr., U. Neb., 1928; faculty N.Y. U., 1929-44, asst. prof., 1938-44; faculty Northwestern U. Med. Sch., Chgo., 1944—, prof. neurology, dir. Inst. Neurology, 1946-48, prof. anatomy, 1948—. Propr., Brain Books, Evanston, Ill., Author numerous books including: Functional Neuroanatomy, 1942, 66; Connections of the Frontal Cortex of the Monkey, 1954; Brain Mechanisms in Diachrome, 1955; Polychrome Atlas of the Brain Stem, 1960; Letters to My Son, 1960; Connections of the Cerebral Cortex, 1963; also articles. Research on rat and monkey cerebral cortex, extensive graphic reconstrn. internal brain anatomy in 3-D; inventor stereotaxic machines. Home: 1236 Hinman St., Evanston, Ill. 60202. Office: 303 E. Chicago Av., Chgo. 60611.*

KRIEGEL, Heinz Otto Werner, German radiobiologist; b. Brandenburg, Germany, May 19, 1922; s. Otto Friedrich and Anna (Wehe) K.; Dipl.Biologe, Humboldt U., Berlin, Germany, 1951, Dr.rer.nat., 1953; m. Elli Anna Maria Grabbert, Aug. 29, 1955; children—Matthias K., Stefan K., Ulrike. E. Research staff dept. for biol. cancer research Inst. for Medicine and Biology, Deutsche Academie der Wissenschaften, Berlin-Buch, 1951-57; team leader unit for research with radioisotopes Heiligenberg (Germany) Institut, 1957-67; leader dept. Nuclear biology, Soc. for Radiation Research, Munich (Germany), 1967—; dozent. Med. Faculty, U. Freiburg (Germany) 1966-67, U. Munich, 1967—. Recipient Richard Glocker prize, 1965, Berliner Röntgengesellschaft prize, 1966. Mem. European Soc. for Radiation Biology, Soc. for Nuclear Medicine. Research, publs. on distbn., incorporation, decorporation of radioisotopes in body; analysis of effects of radiation on embryonic devel.; cancer research. Home: 6 Riemer-schmidstrasse, Munich, Germany. Office: Soc. for Radiation Research, Munich, Germany.*

KRIEGER, Robert Harrison, Am. aero. engr.; b. Balt., Jan. 10, 1921; s. Raymond Buchheimer and Anna (Whitehurst) K.; student U. Va., 1943-44; B.S. in Aero. Engring., U. Mich., 1946; postgrad. Johns Hopkins, 1947; m. Vivian Burnette Johnson, Mar. 5, 1949; children—Scott Harrison, Kenneth Robert, Marlene Ann. With Glenn L. Martin Co., Balt., 1941-43, 1946-48; aero. research engr. Ballistic Research Labs., Aberdeen, Md., 1948-56, head supersonic wind testing, 1956-65, chief supersonic wind tunnels br., Aberdeen Proving Ground, Md., 1965—. Mem. aircraft and missiles projects allocation and priority group NASA, 1958—; mem. adv. group Engring. Design Handbook, 1956—. Mem. Sigma Xi, Tau Beta Pi. Research in aerodynamics, boundary layer and flow separation phenomena, heat transfer, guided missile stability and control. Home: 636 Sadler St., Aberdeen, Md. 21001. Office: Wind Tunnels br. Ballistic Research Labs., Aberdeen Proving Ground, Md. 21005.*

KRIES, Johannes von, see von Kries, Johannes.

KRIGBAUM, William Richard, Am. chemist; b. Beardstown, Ill., Sept. 29, 1922; s. Daniel Dwight and Ella (Sutton) K.; B.S., James Millikin U., 1944, Sc.D., 1966; M.S., U. Ill., 1948, Ph.D., 1949; m. Esther Jean Wolfe, July 14, 1945; children—Mary Kathryn, Janet Ann, Lynn Carol. Faculty, Duke, 1950—, prof., 1962—. Cons. E. I. duPont de Nemours & Co., 1955—. Alfred P. Sloan Research fellow, 1956-60; NSF Sr. Postdoctoral fellow Centre des Recherches sur les Macromolecules, Strasbourg, France, 1959-60, Institut fur physikalische Chemie, Graz, Austria, 1966-67. Fellow Am. Phys. Soc.; mem. Am. Chem. Soc., Am. Crystallographic Assn., Sigma Xi. Research, numerous publs. on characterization of high polymers by phys. chem. methods, such as x-ray diffraction, light scattering, osmotic pressure, solution thermodynamics; relation of molecular structure to phys. properties of elastomers and crystalline polymers. Home: 2504 Wilson St., Durham, N.C. 27706.*

KRIKORIAN, Oscar Harold, Am. chemist; b. Fresno, Cal., Nov. 22, 1930; s. Hagop Bedros and Aghavnie (Mardirosian) K.; B.S. in Chemistry, Fresno State Coll., 1952; Ph.D., U. Cal. At Berkeley, 1955; m. Marilyn Ann Kooyumjian, July 18, 1953; children—Deborah, Cheryl. Research chemist U. Cal. Lawrence Radiation Lab., Berkeley, 1952-55, chemist, Livermore, Cal. 1955—. Mem. Am. Chem. Soc. Research and publs. on optical and mass spectrometric identification high temperature gaseous molecules, measurement enthalpies vaporization molecules. Home: 22 Rio Del Ct., Danville, Cal. 94526. Office: P.O. Box 808, Livermore, Cal. 94551.*

KRILL, Alex Eugene, Am. ophthalmologist; b. Cleve., Oct. 20, 1928; s. Samuel William and Betty (Rosner) K.; B.S. in Math., Western Res. U., 1950; M.D., Ohio State U., 1954; m. Suzanne Altschul, Nov. 31, 1964. Ophthalmology spl. trainee USPHS, U. Ill., Chgo., 1960-61; ophthalmology postgrad. fellow U. Mich., Ann Arbor, 1961; faculty U. Chgo., 1961—, now prof. ophthalmology. Research, publs. on nature of various retinal diseases, color vision, studies on carriers of sex-linked defects to determine effect in female. Home: 5470 S. Everett Av., Chgo. 60615.*

KRIMBAS, Coastas, Greek agronomist, geneticist; b. Athens, Greece, Sept. 12, 1932; s. Basil and Aristea (Issyhaki) K.; ed. U. Lausanne, Switzerland, U. Paris, Columbia U.; Ph.D. in Phys. and Natural Scis.; m. Micheline Grobety, 1954; 1 dau., Alexandra. Became prof. genetics Athens Agronomic Sch., 1960. Rockefeller Found. scholar, 1962-63; scholarships French and Am. govts. Mem. Greek Biology Soc., Royal Inst. Agronomic Research. Research and publs. on genetics. Home: 56, rue Skoufa, Athens, Greece.

KRIMER, (Johann Franz) Wenzel, German physician; b. Datschitz, Germany, 1795; ed. Vienna, Austria; M.D., U. Halle (Germany), 1818; asst. to Scherer, Vienna; became pvt. docent, Bonn, Germany, 1820; practice gen. medicine, Aachen, Germany. Author: Untersuchungen über die nächste Ursache des Hustens, mit Beziehung auf die Lehre vom Athemholen und vom Croup Herausgegeben von F. Nasse, 1819; Physiologische Untersuchungen, 1820; Versuch einer Physiologie des Blutes, 1823; Anleitung zu zweckmässigen . . . , 1824; Ueber die radicale Heilung der Harnröhrenverengrungen und deren Folgen . . . , 1828; Mémoires, 1933. Introduced 1st aid pack for soldiers with 2 bandages, cloth and lint: studies on cure of pulmonary Tb. by surgery. Died Nov. 1834.

KRIMM, Samuel, Am. physicist; b. Morristown, N.J.; Oct. 19, 1925; s. Irving and Ethel (Stein) K.; B.S., Poly. Inst. Bkln., 1947; M.A., Princeton, 1949, Ph.D., 1950; m. Marilyn Neveloff; children—David, Daniel. With U. Mich., Ann Arbor, 1950—, prof. physics 1963—. Mem. exec. com. div. high polymer physics Am. Phys. Soc.; cons. pvt. industries. Research on crystallinity of polymers, infrared spectra of polyments, structure of fibrous proteins. Home: 410 Brookside St., Ann Arbor, Mich.*

KRISEMENT, Otto, German theoretical physicist; b. Leverkusen, Germany, Aug. 21, 1920; s. Otto and Margarete (Jacob) K.; Dr., U. Cologne (Germany), 1950; m. Esther Hoppe, Dec. 29, 1958; children—Vera-Marcelle, Esther-Nicole. Lectr., Technische Hochschule Aachen (Germany), 1960—; prof. U. Münster (Germany), also dir. Instituts für Metallforschung, Lehrstuhl für Theoretische Metallphysik, 1965—; sci. co-worker Max-Planck-Institut für Eisenforschung, Düsseldorf, Germany, 1951—; Research and publs. on theory of the liquid state. Home: 10 Schiller Strasse, 4401 Roxel, Germany. Office: 107 Steinfurter Strasse, 44 Münster, Germany.*

KRISHABER, Maurice, physician; b. Fuketchegy, Hungary, Apr. 13, 1836; student medicine, Vienna, Austria, Prague, Czechoslovakia, Paris, France. Became naturalized French citizen, 1872; began practice medicine, Paris, 1864. Author: De la neuropathie cérébro-cardiaque, 1873; (with Maruiac) Des laryngopathies pendant les premières phases de la syphilis, 1876; Sur le cancer du larynx, 1879. Founder (with Isambert, Ladreit of Lacharriere) Annales des maladies de l'oreille et du larynx, 1875. Research in laryngology, nervous diseases; one of founders of modern treatment of diseases of larynx; described neuropathy affecting nerves of sensation and heart (Krishaber's disease). Died Apr. 10, 1883.

KRISHNA, Narayana Pillai, Indian zoologist; b. Thiruvalla, Kerala, India, June 23, 1921; s. K. Pillai Narayana and K. Amma Kalyani; B.Sc., Sci. Coll., Trivandrum, 1942; M.Sc., Ph.D., Marine Biol. Lab., Trivandrum, 1965; m. Pappykutty Amma Sarada, Dec. 25, 1959; children—S. Radha, S. Jayalekshmi. Mem. faculty Marine Biol. Lab., Trivandrum, 1947—, officer in charge, 1962—. Recipient monograph award Forest Research Inst., Dehra Dum, India. Mem. Marine Biol. Assn. India (a founder), Bombay Natural History Soc. Research and publs. on taxonomy of marine crustacea of Kerala waters, marine fauna of Kerala waters. Home: Mathilbhagom, Thiruvalla, Kerala, India. Office: Marine Biol. Lab., Trivandrum-7, India.*

KRISHNA, Padmanabhan, Indian physicist; b. Madras, India, Jan. 13, 1938; s. N. Padmanabhan and N. Maitreyi; B.Sc., Holkar Coll., Indore, 1957; M.Sc., Delhi U., 1959; Ph.D., Banaras Hindu U., Varanasi, 1962; m. Minakshi Raval, May 11, 1960; children—Anuradha, Anjali. Commonwealth Sci. and Indsl. Research Orgn. research fellow Banaras Hindu U., 1949-61, U.G.C. research fellow, 1961-63, lectr. physics, 1963-66, now reader physics, dept. metallurgy. Brit. Council bursar, dept. metal physics Cavendish Lab., Cambridge (Eng.) U., 1964. Mem. Electron Microscope Soc. India. Author: (with Ajit Ram Verma) Polymorphism and Polytypism in Crystals, 1966. Discovered and determined by X-ray diffraction some long-period structures of silicon carbide that cannot be accounted for by dislocation theory of polytypism; found surface features of these crystals as determined by optical microscopy at variance with internal structure as determined by X-ray diffraction. Home: B.4.F.2. Ladies Colony, Banaras Hindu Univ., Varanasi-5, U.P., India.*

KRISHNA MURTI, Coimbatore Ramdorai, Indian biochemist; b. Coimbatore, India, Mar. 3, 1923; s. Coimbatore Krishnaswamy and Meenakshy Ramdorai; student St. Thomas Coll., Trichur, India; B.Sc. with honors, Univ. Coll. Sci. and Tech., Waltair, 1942; Ph.D., Indian Inst. Sci., Bangalore, 1948; m. Brinda Pattabhi, Aug. 21, 1946; children—Lakshmy Bharathi, Shridevi, Ramachandran. Chemist, Dunlop Rubber Tyre Co., Shagunge, 1942, Standard Pharm. Works, Calcutta, 1942-43; research asst., biochemistry dept. Indian Inst. Sci., Bangalore, 1945-50; with Central Drug Research Inst., Lucknow, India, 1950—, sr. sci. officer I, 1957-62, asst. dir., scientist-in-charge, 1962—. Hon. research asso. Univ. Coll., London, 1957-58; vis. scientist Nat. Inst. Med. Research, London; recognized tchr. for Ph.D. by Madras, Bombay, Poona, Agra, Aligarh (all India) univs. Mem. Assn. Sci. Workers India, Soc. Biol. Chemists, Assn. Food Technologists (both India), A.A.A.S. Research, publs. on milk-clothing and bacteriolytic enzymes of plant latices; developed technique for osmotic lysis of gram negative bacteria for study of enzyme localization; purified and studied nature of several enzymes of Vilerio cholerae; developed processes for preparation of protein hydrolysalis from oilcakes and digestive enzymes from indigenous sources. Home: C414 Havelock Rd., Lucknow. Office: Central Drug Research Inst., Chattan Manzil Palace, Lucknow, UP, India.*

KRISHNAN, Genta, Indian zoologist; b. Madras, India, Apr. 17, 1911; s. G. B. and G. (Rajamma) K.; B.A. with honors, Presidency Coll., Madras, 1931; M.Sc., U. Madras, 1947; Ph.D., U. Manchester (Eng.), 1950; m. Rajalakshmi, Jan. 22, 1951; children—G. S. Ramanan, G. S. Narayan, G. Ravi. Faculty, Presidency Coll., 1937-51, asst. prof., 1942-51; faculty U. Madras, 1951—, prof. zoology, 1959—. Mem. Zool. Soc. India, Indian Sci. Congress Assn. Research and publs. on chem. nature of arthropod cuticles, hardening of cuticle of some arthropods, hardening by sulphur bonds of arthropod cuticle. Home: 4 Gengureddy Rd., Madras-8, Madras, India.*

KRISHNAN, Maharajapuram Sitaram, Indian geologist; b. Tanjore, Madras, India, Aug. 24, 1898; s. Sitaram and Akhilandam (Srikrishna) K.; M.A., Tanjore and Presidency Coll., Madras, 1921; Ph.D., London (Eng.) U., 1924; m. Lakshmi Iyer, June 1, 1919; children—Akhilandeswari (Mrs. A. P. Subramaniam), Kalyani, Jayalakshmi. With Geol. Survey India, 1924-49, dir., 1951-55; dir. Indian Bur. Mines, 1949-51; mineral adviser Govt. of India, 1955-57; dir. Indian Sch. Mines, 1957-58, Nat. Geophys. Research Inst., 1961-63; prof. geology and geophysics Andhra (India) U., 1959-61; Hon. prof. geology Osmania U., Hyderabad, India, 1964—. Mem. Indian Sci. congress (past pres.), Indian Acad. Sci. (v.p. 1965—), Indian Geophys. Union (pres. 1965—). Author: Geology of India and Burma, 1943; also numerous articles, monographs. Research on econ. geology especially iron ores of India, tectonics of S.E. Asia, mineral resources of Madras, exploration and devel. of lignite deposit in Madras. Home: 12-5-30 Tarnaka-Moulali Rd., Hyderabad, India.*

KRISS, Joseph P., Am. internist; b. Phila., May 15, 1919; s. Max and Sima (Charny) K.; B.S. cum laude, Pa. State Coll., 1939; M.D., Yale, 1943; m. Regina Tarlow, June 15, 1948; children—Eric, Paul, Mark. Instr., Yale, 1944-45; research fellow Washington U., 1946-48; faculty Sch. Medicine, Stanford, Palo Alto, Cal., 1948—, prof. medicine and radiology, 1962—, dir. div. nuclear medicine 1958—; also cons. hosps. Commonwealth Fund fellow, vis. sci., Eng., 1957-58. Diplomate Am. Bd. Internal Medicine, Nat. Bd. Med. Examiners. Mem. Soc. Nuclear Medicine, Endocrine Soc., Western Soc. Clin. Research, Am. Fedn. Clin. Research, Western Assn. Physicians, A.A.A.S., Phi Beta Kappa, Phi Kappa Phi, Alpha Epsilon Delta, Alpha Omega Alpha. Research and publs. on salt and water metabolism, renal and adrenal physiology, thyroid and pituitary disorders, sickle cell anemia, multiple myeloma, bioassay of pituitary thyrotropic hormone, med. applications of radioactive isotopes, mechanism of erythrocyte mfr. and regulation, metabolism of halogenated pyrimidines. Home: 735 Mayfield St., Stanford, Cal. 94305.*

KRISTOFFERSEN, Thorvald, bacteriologist; b. Nykoebing F., Denmark, May 6, 1919; s. Rasmus and Anna (Rasmussen) K.; B.S., Royal Vet. and Agrl. Coll., Copenhagen, 1944; M.S., Ia. State U., 1949, Ph.D., 1954; m. Ruth M. Sprecher, Jan. 2, 1948; 1 son, Erik Paul. Came to U. S., 1946, naturalized, 1950. Research asst. Ia. State U., 1946-54; faculty Ohio State U., Columbus, 1955—, prof., 1963—. Recipient Pfizer-Paul Lewis award, 1965. Mem. Am. Dairy Sci. Assn., Am. Soc. Microbiology, A.A.A.S., Am. Assn. U. Profs., Sigma Xi, Gamma Sigma Delta. Research and publs. on biochemistry of cheddar cheese ripening, factors affecting flavor formation in cheese and compounds contbg. to flavor, fast-cure process for cheese, spreadability of butter, effectiveness of chem. sanitizers. Home: 1470 Osborn Dr., Columbus 43221. Office: 2121 Fyffe Rd., Columbus, O. 43210.*

KRITCHEVSKY, David, biochemist; b. Kharkov, Russia, Jan. 25, 1920; s. Jacob and Leah (Kricewski) K.; came to U. S., 1923, naturalized, 1929; B.S., U. Chgo., 1939, M.S., 1942; Ph.D., Northwestern U., 1948; m. Evelyn Sholtes, Dec. 21, 1947; children—Barbara Ann, Janice Eileen, Stephen Bennett. Chemist, Ninol Labs., Chgo., 1939-46; Postdoctoral fellow Fed. Inst. Tech., Zurich, Switzerland, 1948-49; biochemist Radiation Lab., U. Cal. at Berkeley, 1950-52, Lederle Lab., Pearl River, N.Y., 1952-57; biochemist Wistar Inst., Phila., 1957—; prof. biochemistry Sch. Vet. Medicine, U. Pa., Phila., 1965—. Mem. USPHS study sect. Nat. Heart Inst., 1964—; chmn. research com. Spl. Dairy Industry Bd., 1963—. Recipient Research Career award Nat. Heart Inst., 1962. Mem. Am. Soc. Biol. Chemists, A.A.A.S., Am., Swiss, Brit. chem. socs., Soc. Exptl. Biology and Medicine, Arteriosclerosis Council, Am. Heart Assn., Am. Soc. Oil Chemists (chmn. methods com. 1963-64), Internat. Soc. Fat Research. Author: Cholesterol, 1958; also numerous articles. Editor: (with G. Litwack) Actions of Hormones on Molecular Processes, 1964. Co-editor, Advances in Lipid Research, 1963—. Research on role vehicle when cholesterol and fat produces atherosclerosis in rabbits, effects saturated and unsaturated fat, deposition orally administered cholesterol in aorta man and rabbit; pioneered use radioactive cholesterol for metabolic expts. Home: 136 Lee Circle, Bryn Mawr, Pa. 19010. Office: Wistar Inst., 36th and Spruce Sts., Phila. 19104.*

KRIVANEK, Jerome Oldrich, Am. biologist; b. Chgo., Dec. 27, 1924; s. Joseph Anton and Bessie (Urse) K.; B.S., U. Ill., 1948, M.S., 1951; Ph.D., U. Fla., 1955; m. Robin Cooley, Sept. 15, 1953; children—Jennifer Beth, Joseph George. Instr. Newcomb Coll., Tulane U., 1955-57, asst. prof., 1957-60; faculty U. South Fla., 1960—, asso. prof. biology, 1961—, chmn. zoology, 1961, 63, 64. Recipient Research prize Assn. Southeastern Biologists, 1955. Fellow A.A.A.S.; mem. Am. Inst. Biol. Scis., Growth Soc., Am. Soc. Zoologists, Am. Assn. U. Profs., Sigma Xi, Phi Sigma. Research and publs. on analysis of behavior in annelids and embryonic forms, biochemistry of devel. in slime mold, taxonomy and natural history of fresh-water and marine gastrotrichs. Home: 2802 Gaines St., Tampa, Fla. 33618.*

KRNJEVIC, Kresimir, physiologist; b. Zagreb, Croatia, Sept. 7, 1927; s. Juraj and Nada (Hirsl) K.; M.B., Ch.B., Edinburgh (Scotland) U., 1949, B.Sc., 1951, Ph.D., 1953; m. Jeanne W. Bowyer, Sept. 27, 1954; children—Peter Juraj, Nicholas John. Beit Mml. fellow Edinburgh U., 1952-54; research asso., asst. prof. U. Wash., 1954-56; vis. fellow Australian Nat. U., 1956-58; prin. sci. officer A.R.C. Inst. Animal Physiology, 1959-64; vis. prof. McGill U., Montreal, Que., Can., 1964-65, dir. Wellcome dept. anesthesia research, 1965—. Recipient Ellis prize Edinburgh U., 1952. Mem. Physiol. Soc., Brit. Med. Assn., Montreal Neurol. Soc. Contbg. author: Neurochemistry, 1962; also numerous articles. Asso. editor Exptl. Brain Research, 1965—. Research in neurophysiology, nerve trunks, nerve-muscle transmission and spinal cord, chem. transmitters which transmit information between individual nerve cells in brain. Home: 653 Belmont Av., Montreal 6, Que., Can.*

KROEBEL, Werner Adolf Johannes, German physicist; b. Berlin, Apr. 7, 1904; s. Adolf and Katharina (Hilbig) K.; Sc.D., U. Göttingen, 1929; m. Alma Kroebel Boker, Feb. 19, 1930; 1 son, Reinhart. Head dept. for Tekade, Nürnberg, Germany, 1935-38; head dept. communication scis. U. Kiel (Germany), 1938-46, dir. Electronics Inst., 1946-47, prof. applied physics, 1946—, dir. Inst. Applied Physics, 1946—. Mem. Deutsche Physik Gesellschaft, Gesellschaft Deutsche Naturforscher und Arzte, others. Research, numerous publs. on impulse technic, electronics in phys. research, oceanography, radio-astronomy, nuclear physics. Home: 2308 Wehrbergallee 43, Preetz, Germany. Office: 23 Ohlshausenstr., Kiel, Germany.*

KROEBER, Alfred L., Am. anthropologist; b. Hoboken, N.J., June 11, 1876; s. Florence and Johanna (Mueller) K.; A.B., Columbia U., 1896, A.M., 1897, Ph.D., 1901, Sc.D. Yale, 1946; LL.D., California, 1951; D.H.L. (honorary), Harvard University, 1952, Columbia, 1953; married Henriette Rothschild, May 24, 1906 (dec.); m. 2d Theodora Kracaw Brown, Mar. 26, 1926; children—Clifton, Theodore (stepsons), Karl, Ursula. Asst. in English, 1897-99, fellow in anthropology, 1899-1900, Columbia U., anthropological expedns., 1899-1901, New Mexico, 1915-20, Mexico, 1924, 30, Peru, 1925, 26, 42; instr. anthropology, 1901, asst. prof., 1906, asso. prof., 1911, prof. 1919-46, curator, 1908-25, dir., 1925-46, Anthropol. Museum, U. of Calif. Ethnol. exploration in Calif. since 1900; curator anthropology, 1900, and 1903-11, Calif. Acad. Sciences; research asso. Chgo. Natural History Mus., 1925—; vis. prof. Harvard, 1947-48, Columbia, 1948-52, Brandeis, 1954, Yale, 1958; fellow Center for Advanced Study Behavioral Sciences, 1955-56. Founder and president, 1917, Am. Anthrop. Assn.; pres. Am. Folk Lore Society, 1906. Mem. Linguistic Soc. of America (pres. 1940), Nat. Acad. Sciences, Am. Philos. Soc., Am. Academy Arts and Sciences, American Ethnological Society. Recipient of the Huxley Medal, Royal Anthropological Institute, 1945; Viking Medal, 1946. Author: Zuni Kin and Clan; Peoples of the Philippines; Anthropology 1923, 1948; Handbook of Indians of Cal., 1925; Culture Element Distribution, 1936; Cultural and Natural Areas, 1939; Peruvian Archaeology; Configurations of Culture Growth, 1944; The Nature of Culture, 1952; Anthropology Today, 1953; Style and Civilization, 1957. Leading American anthropologist; authority on Indians of North and South America, especially of western U. S., Mexico and Peru; stimulated use of statistical method in anthropology; studied nature of culture; defined society as interaction of groups and individuals. Died Paris, Oct. 5, 1960.

KROG, John Olav, Norwegian physiologist; b. Meldal, Aug. 29, 1918; s. Olav and Anna (Resell) K.; Cand. real., U. Oslo (Norway), 1949, Dr. philos., 1964; m. Hildur Nygard, June 24, 1949; children—Jo, Tor, Siri. Tchr. in basic sci. Oslo Tchrs.' Coll., 1950; research physiologist, exchange visitors program USPHS, Anchorage, Alaska, 1950-54; research fellow circulatory physiology Ulleval Hosp., U. Oslo, asst. prof. (prosektor), inst. exptl. med. research, 1962-65, univ. prof., chem. zoophysiology, 1965—; vis. research prof. U. Ky. Med. Sch., Lexington, 1964-65. Internat. Union Physiol. Scis. rep. to Sci. Commn. on Oceanographic Research. Mem. Internat. Biophys. Union Scandinavian Physiol. Soc., A.A.A.S., Arctic Inst. N. Am., European Microcirculatory Soc., Aerospace Med. Assn. Author comparative study of effect of cervical sympathetic stimulation on cerebral blood flow. Research and publs. on thermoregulation and peripheral circulation in Arctic animals, regulation of cerebral circulation, design of polarographic oxygen electrodes, methods for extracorporal circulation and regional hypothermia of brain in surgery. Home: Box 1051, Blindern, Oslo, Norway.*

KROGH, (Schack) August (Steenberg), Danish physiologist; b. Grenna, Denmark, Nov. 15, 1874; student Lund, Sweden, Göttingen, Germany, Budapest, Hungary; Masters Degree, 1899; m. Marie Jörgensen, 1905; 1 son, 3 daus. Became asst. to Bohr, Copenhagen, Denmark, 1908, prof. zoology, 1909-45. Recipient Nobel prize in medicine and physiology, 1920. Fellow Royal Soc., 1937. Author: Anatomy and Physiology of Capillaries, 1923; Osmotic Regulations in Aquatic Animals, 1939; Comparative Physiology of Respiratory Mechanisms, 1941. Research on body's respiration, metabolism and circulation, effect of all-meat diet on Eskimo; discovered regulation of motor mechanism of capillaries; showed diffusion and not secretion, accounts for all oxygen taken into blood. Died Copenhagen, Sept. 13, 1949.

KROGMAN, W(ilton) M(arion), Am. physical anthropologist; b. Oak Park, Ill., June 28, 1903; s. Wilhelm Claus and Lydia Magdalena (Wriedt) K.; Ph.B., Univ. Chicago, 1926, A.M., 1927, Ph.D., 1929; Cleveland Foundation fellow in anatomy, Western Reserve, 1928-29; Fellow Nat. Research Council, Royal Coll. of Surgeons, London, Eng., 1930-31; LL.D. (honorary), Baylor University, 1955; married Virginia Lane, Dec. 22, 1931 (divorced 1944); children—Marian Knox, William Lane; m. 2d, Mary H. Winkley, Apr. 18, 1945; children—John Winkley, Mark Austin. Instr. phys. anthropology, U. Chgo., 1929-30; asso. prof. anat. and phys. anthropology, Western Res., 1931-38, U. of Chgo., 1938-47; prof. physical anthropology, div. grad. medicine Univ. of Pa., 1947—, also chmn. department of phys. anthropology; dir. Phila. Center for Research in Child Growth since 1947; Walker Ames lecturer in anthropology, Univ. of Wash., summer 1944. Fellow A.A.A.S. (sec., sec. H., 1933-44, chmn., 1948-49), mem. Nat. Acad. Scis., Am. Assn. Physical Anthropologists (past pres.), Soc. Research Child Devel. (past pres.), Phi Beta Kappa, Sigma Xi. Specialist in skeletal identification in medico-legal problems. Researcher in the physical growth and devel. sch. child. Home: 1040 Cornell Av., Drexel Hil, Pa. 19026. Office: Div. Grad. Medicine, U. Pa., Phila. 19104.

KROGMANN, David W., Am. biochemist; b. Washington, Oct. 21, 1932; s. Rudolph Francis and Cecilia (O'Dea) K.; A.B., Cath. U. Am., 1953; Ph.D., Johns Hopkins, 1957; m. Loretta I. Kurek, June 20, 1958; children—Michelle, Patricia, Paul. Postdoctoral fellow McCollum-Pratt Inst., Johns Hopkins, Balt., 1957-58; research asso. in biochemistry U. Chgo., 1958-60; asst. prof. chemistry Wayne State U., Detroit, 1960-63, asso. prof., 1963—. Recipient Career Devel. award USPHS, 1963—. Mem. Fedn. Am. Socs. for Exptl. Biology, Am. Soc. Plant Physiologists. Author: (with W. E. Powers) Biochemical Dimensions of Photosynthesis, 1965; also articles. Research on elucidation of energy conversion mechanism of photosynthesis through isolation and identification of individual reactants and catalysts from green plant material studied in cell-free partial reactions of photosynthesis. Home: 25547 Briar Dr., Oak Park, Mich. 48237.*

KROGSTAD, Blanchard Orlando, Am. zoologist; b. Sletten, Minn., Oct. 6, 1921; s. Jens and Julia (Sonstelie) K.; B.A., Bemidji (Minn.) State Coll., 1946; M.A., U. Minn., 1948, Ph.D., 1951; m. Doris Jane Van Winkle, Dec. 23, 1946; children—Jineen Elyse, Rolf Gregory, Bruce Elliott. Faculty, St. Olaf Coll., Northfield, Minn., 1951-54; faculty dept. biology U. Minn. at Duluth, 1954—, prof., 1963—, asst. prof., asso. prof. U. Minn. Biol. and Forestry Sta., Itasca Park, 1951-59. Staff mem. Rockefeller Found., Chapingo, Mexico, 1963-64. Mem. Ecol. Soc. Am., Entomol. Soc. Am., Am. Inst. Biol. Scis., Am. Assn. U. Profs., Minn. Acad. Sci., Sigma Xi. Research on ecology of insects, population dynamics, description of immature insects, life history studies of parasitic insects. Home: 5705 Juniata St., Duluth, Minn. 55804.*

KROH, Oswald, German psychologist; b. Dec. 15, 1887; s. Hermann and Luise (Grauel) K.; ed. U. Munich, U. Marburg; m. Lucie Seiffert, 1923. Tchr. elementary and secondary schs.; became lectr. U. Göttingen, 1921; named prof. U. Braunschweig, 1922, U. Tübingen, 1923, U. Munich, 1938, U. Berlin, 1942; prof., dir. Psychol. Inst. for Psychol. Research, Free U. Berlin, 1949—. Mem. Bavarian Acad. Scis., German Acad. Natural Scientists. Author: Subjektive Anschauungsbilder bei Jugendlichen, 1922; Phasen der Jugendentwicklung, 1926; Die Psychologie des Grundschulkindes, 22 edit., 1944; Psychologie der Oberstufe, 10th edit., 1944; Experimentalle Beiträge zur Typenkunde, 3 vols., 1929, 32, 34; Revision der Erziehung, 2d edit., 1954. Office: 24 Wichernstr., Berlin-Dahlem, Germany.

KROHN, Robert Dean, Am. physicist; b. Milw., Dec. 26, 1919; s. Hugh K. and Pauline (Matthiesen)) K.; B.S. in Elec. Engring., U. Wis., 1942, M.S. in Physics, 1946; m. Barbara Rockwell, May 31, 1943; children—Douglas, Bruce, Jacqueline, Diane. Asst., NDRC, Madison, Wis., 1942-43; staff Los Alamos Sci. Lab., 1943—, task unit comdr. Eniwetok Proving Grounds, 1951-52, responsible reviewer to AEC, 1958—, alternate group leader tech. information group, 1956—. Designed electrostatic generator, radiation detection instruments; research on nuclear physics measurements, atomic weapon surveillance techniques. Home: 5 Kiowa Lane, Los Alamos 87544. Office: Box 1663, Los Alamos 87544.*

KROL, Wladyslaw, Polish physician; b. Zagorzyce, Distr. Miechów, Poland, Apr. 20, 1915; s. Stanislaw and Maria (Mucha) K.; Physician, Cracow (Poland) Clinic Internal Diseases, 1939, M.D., 1948, dozent, 1954; m. Marie, Dec. 8, 1942; 1 son, Richard. Founder dir. out-patient ward thyroid diseases Cracow Dist., 1951-59, outpatient ward endocrinological diseases, 1959-62, out-patient ward cardiological diseases, 1961, dir. First Clinic Internal Diseases, 1965—; faculty Cracow Med. Acad., 1954—, vice dean Med. Faculty, 1965—, editorial com. Med. Academie Jubilee Publs. for 600th Anniversary, 1961-64. Recipient govt. decorations Golden Order for Merit, Silver Order for Merit, medal for 10th Anniversary of People Republic of Poland, medal for 100th Anniversary of the Poland Millenium. Mem. Polish Soc. Endocrinology (chmn. Cracow 1961-65), Polish Soc. Cardiology (co-founder Cracow, pres. 1965—), Polish Med. Soc. (past vice chmn. Cracow br.), Polish Soc. Internists (past vice chmn. Cracow br.), French Cardiologists Soc. (fgn.), Internat. Soc. Internal Diseases. Co-author: Differential Diagnostics of Internal Diseases, 1968; also articles, chpts. in textbooks. Research on etiopatogenesis of atheromatosis; showed correlation between hypophyseo-suprarenal hyperfunction and the decrease in the lipemic indicators of the serum, essential role of hypertension and feeding in atheromatosis; use of self-constructed model for artificial causing of myocardial infarctus; established norms of venous tension; showed (with Tochowicz) arterial tension basic and casual. Home: 10 Tarlowska, Cracow. Office: 17 Kopernika, Cracow, Poland.*

KROLIKOWSKI, Wojciech, Polish physicist; b. Warsaw, Poland, July 16, 1926; s. Stafan and Kazimiera (Tabaczynska) K.; M.Sc., Warsaw U., 1950, Ph.D., 1952, D.Sc., 1957; m. Zofia Zienkiewicz, June 28, 1950; children—Jan, Malgorzata. Staff, Warsaw U., 1950—, prof. theory elementary particles, 1965—; mem. Inst. for Advanced Study, Princeton, N.J., 1961; vis. prof. Middle E. Tech. U., Ankara, Turkey, 1965; prof. Internat. Centre for Theoretical Physics, IAEA, Trieste, 1966. Grantee, Swiss Fed. Inst. Tech., Zurich, 1956-57; recipient Polonia Restituta. Author: (with W. Rubinowicz) Theoretical Mechanics, 1955; also articles. Formulated (with J. Rzewuski) relativistic one-time many body problem in field theory; developed hypothesis of intrinsic exclusion principle and partial conservation of tensor current. Home: Pulawska 1 A, Warszawa 12, Poland.*

KROLL, Norman M., Am. physicist; b. Tulsa, Apr. 6, 1922; s. Cornelius and Grace (Aaronson) K.; A.B., Columbia, 1942, A.M., 1943, Ph.D., 1948; m. Sally Sharlot, Mar. 15, 1945; children—Linda Ruth, Cynthia Anne, Heather Roma, Ira Joseph. Faculty, Columbia, 1949-62; prof. U. Cal., San Diego, 1962—, chmn. dept. physics, 1963-65. NRC fellow, 1948-50; Guggenheim fellow, 1955-56; Fulbright scholar, 1955-56; NSF Sr. fellow, 1965-66. Fellow Am. Phys. Soc., Phi Beta Kappa, Sigma Xi. Home: 2457 Calle del Oro, La Jolla, Cal. 92038.*

KRÖLLER, Ernst Heinrich Wilhelm, German chemist; b. Wiesbaden, Germany, May 16, 1920; s. Wilhelm and Juliane (Roos) K.; chem. diploma Technische Hochschule Berlin (Germany), 1941, Dr.rer.nat., 1945; food chem. exam U. Frankfurt/Main (Germany), 1948; m. Helene Maria Gattinger, Jan. 28, 1950. Acad. asst. Technische Hochschule Berlin, 1941-45; acad. colleague State Chem. Testing Office of Hessia, Frankfurt/M., 1947-52, acad. asst., 1952-54; dir. chem. adviser, 1954-58; sci. councillor Fed. Health Office, Berlin, 1958-63, chief sci. councillor, 1963-67, sci. dir., 1967—, lab. dir., 1958—. Mem. Gesellschaft Deutscher Chemiker (chmn. Berlin sect. food chemistry group 1965—), Deutsche Gesellschaft für Fettwissenschaft e.V. Contbg. author: Handbuch der Lebensmittelchemie, Vol. II, 1967, Residue Revs., vol. 12, 1966. Research and publs. on new analysis methods for plant protective means in foods and for food additives; metabolites from fumigation substances in foods; cancerogenic pyrolysis products in smoking and cancerogenic alkylizing compounds in food. Home: 23b Haderslebener Strasse, 1 Berlin 41, Office: 82/84 Unter den Eichen, 1 Berlin 45, Germany.*

KROMPHARDT, Wilhelm Martin Justin, German economist; b. Schönebeck/Elbe, Germany, May 30, 1897; s. Hermann Georg Justin and Anna (Eiselen) K.; student univs. Greifswald, Marburg, Halle, Bonn; Dr.sc.pol., U. Kiel, 1924; m. Ilse Helene Tetzner, Dec. 30, 1924; children—Ellen Ilse Anna, Jürgen Alfred Justin, Inge Johanne Friederike (Mrs. Jürgen Rode). Faculty U. Münster/Westphalia, 1926-37, asso. prof., 1931-37; asso. prof. U. Rostock, 1937-41, prof., 1941-46; prof. econs. Hanover Inst. Tech., 1946-49, U. Göttingen, 1949-56, U. Heidelberg, 1956—. Mem. econ. council German Fed. Ministry Econs., Fed. Ministry Agr., 1948—. Rockefeller fellow, 1932-33. Mem. Heidelberg Acad. Scis., Assn. for Social Politics 1872, Econometric Soc., Royal Econ. Soc., German Soc. Sociology, German Statis. Soc., Soc. Social Progress, numerous others. Author: Die Systemidee im Aufbau der Casselschen Theorie, 1927; Marktspaltung und Kernplanung in der Volkswirtschaft, 1947; (with Henn, Förstner) Lineare Entscheldungsmodelle, 1962; also articles in field. Research on econ. theory, bus. cycle and growth theory, econ. policy and European econ. Integration. Address: 40 Uferstrasse, 69 Heidelberg, West Germany.*

KRON, Gerald Edward, Am. astronomer; b. Milw., Apr. 6, 1913; s. Edmund and Letty (Dietrich) K.; B.S., U. Wis., 1933, M.S., 1934; Ph.D., U. Cal. at Berkeley, 1938; m. Katherine Carson Gordon, Apr. 22, 1946; children—Donald Gordon, Richard Gordon, Jenny Caroline, Virginia Carson, Charles Edmund. With Lick Obs., U. Cal. at Mt. Hamilton, 1938-65, astronomer, 1952-65; dir. U. S. Naval Obs., Flagstaff (Ariz.) Sta., 1965—; hon. research asso. Australian Nat. U., Canberra, 1961. Mem. Am., Royal astron. socs., Astron. Soc. Pacific (past pres.), Internat. Astron. Union. Research in fields eclipsing binary stars, globular clusters, applications multi-color photometry, direct electrographic rec. type image tubes; developed improved methods photoelectric detection and measurement light. Home: 416 Bertrand St., Office: U. S. Naval Obs., P.O. Box 1149, Flagstaff, Ariz. 86001.*

KRONBERGER, Hans, Brit. physicist; b. Linz, Austria, July 28, 1920; s. Norbert and Olga (Kellner) K.; B.Sc., U. Durham (Eng.), 1944; Ph.D., U. Birmingham (Eng.), 1948; m. Joan Iliffe, 1951 (dec. 1962); children—Paul D. Hanson (stepson), Zoe E., Sarah J. Mem. staff, Atomic Energy Team, 1944-46, Atomic Energy Research Establishment, Harwell, Eng., 1946-51, indsl. group atomic energy dept., 1951; head U.K. Atomic Energy Authority Diffusion Plant Labs., Capenhurst, 1953-56; chief physicist indsl. group U.K. Atomic Energy Authority, Risley, 1956-58, dir. research and devel. devel. and engring. group, 1958-60, scientist in chief reactor group, 1961—. Decorated Comdr. Order Brit. Empire. Fellow Royal Soc., Inst. Physics. Publs. on devel. methods of separation of stable isotopes including uranium-235; research on tech. of diffusion plants and nuclear reactors; assessment of utilization of plutonium; organized U.K. devel. program on desalination. Home: No. 3, Smith's Lawn, Holly Rd. S., Wilmslow, Cheshire, Eng. Office: U.K. Atomic Energy Authority, nr. Warrington, Lancaster, Eng.*

KRONECKER, (Karl) Hugo, physiologist; b. Germany, 1839; student medicine Berlin, Germany, Heidelberg, Germany, Pisa, Italy; worked under Kühne; studied under Carl Ludwig, Leipzig (Germany) Lab.; became tchr. physiology, Bern, Switzerland, 1884; pres. Marcy Inst. Founder, Internat. Congress Physiology, 1898. Fellow Royal Soc., 1909. Research on cardiac and muscular physiology, vasomotor nerves. Died 1914.

KRONECKER, Leopold, German mathematician; b. Leignitz, Germany, Dec. 7, 1823; Ph.D., U. Berlin, 1845; m. Fanny Prausnitzer, 1848. Became prof., Berlin, 1883. Fellow Royal Soc., 1884. French Acad. Scis. Author: Grundzüge einer rein arithmetischen Theorie der algebraischen Grossen, 1882; Über den Zahlbegriff, 1887; Gesammelte Werke (editor Kurt Hensel) 1895-1931. Editor, Journal für die reine und angewandte Mathematik. Died Berlin, Dec. 29, 1891.

KRÖNIG, August Karl, German physicist; b. Schildesche, Germany, Sept. 3, 1822; Lectr. Kölnischen Gymanasium, Berlin. Author: Dissertation de acidi chromici salibus crystallinis, 1845; Neue Methode zur Vermeid von Rechenfehlern, 1855; also articles. Founder (with Clausius) kinetic theory of gases; showed contradiction between computed velocity of molecules and length of time odor of pungent gas can be detected, 1856. Died 1879.

KRÖNLEIN, Rudolf Ulrich, Swiss surgeon; b. Stein, Switzerland, Feb. 19, 1847; studied under Langenbeck; prof., Zurich, Switzerland. Studied surg. technique, especially brain surgery; introduced craniometer; advanced cerebral topography; 1st to operate successfully for ruptured appendix. Died Zurich, Oct. 26, 1910.

KRONSBEIN, John, Am. physicist, metallurgist; b. Ohiowa, Neb., Nov. 27, 1904; s. William and Anna (Thyssen) K.; student U. Bonn (Germany), 1924-25; Ph.D. in Pure Math., U. Leipzig (Germany), 1930;

Dr.rer.nat. in Physics, U. Jena (Germany), 1931; m. Peggy Ethel Gilding, June 22, 1934 (dec.). Research and devel. engr. Incandescent Heat Co. Ltd., Birmingham, Eng., 1931-32; tech. dir. Brightside Plating & Mfg. Co., Birmingham, 1932-38; mng. dir. Electrochem. Engring. Co., London, Eng., 1938-45; research and devel. chief Hoosier Cardinal Corp., Evansville, Ind., 1945-48; prof., dir. Sch. Engring., Evansville Coll., 1948-58; prof. mech. engring., theoretical physics, metallurgy and materials engring., U. Fla., Gainesville, 1958—; staff physicist Deaconess Hosp., Evansville, 1950-58; chief engr. Chgo. Midway Lab., Chgo. U., 1950-51. Mem. London Math. Soc., Am. Phys. Soc., Math. Assn. Am. Contbr. numerous articles to jours., chpt. on copper to monograph, Ency. for Electrochemistry. Research in electrochemistry, pure math., geometry, conformal mapping, relativity, unified field theory, applied topology. Home: 2530 N.W. 12th Av., Gainesville, Fla. 32601.*

KROPOTKIN, Peter, Russian polit. theorist, geographer; b. Moscow, Nov. 27, 1842; ed. for army. Mem. wealthy family; entered corps of pages to tsar, 1857; mil. ofcl. in Siberia, 5 years, headed several geog. expdns. at time, resigned, 1867; entered U. St. Petersburg; sec. Russian Geog. Soc.; explored glacial deposits in Finland and Sweden, 1871; became asso. with internat. Workingman's Assn., 1872, turned from socialism to anarchism; arrested and imprisoned for subversive propaganda, 1874; escaped, fled to Eng., Switzerland, Paris, Switzerland, 1876, expelled from Switzerland, went to France, arrested in Lyons, 1883, imprisoned but pardoned by Georges Clemenceau; settled in Eng., 1886-1917, returned to Russia at outbreak of revolution; supported moderate govt. of Kerensky; ret. from polit. activity after triumph of Bolsheviks. Author: The Place of Anarchy in Socialist Evolution, 1886; The Conquest of Bread, 1888; Anarchism: Its Philosophy and Ideal, 1896; The State: Its Part in History, 1898; Modern Science and Anarchism, 1903. Recognized authority of geography, Russian life and lit., sociology, outstanding theorist of anarchism; his concepts based on evolutionist beliefs; held that man and society do not necessarily develop from lower to higher stages; felt that man could impede natural or normal devel. of body or society by asserting his will in certain ways, when this resistance to natural evolution becomes too great a crisis occurs in body (illness), in society (revolution); concluded that revolutions were natural and necessary aids to evolution; advocated decentralization of industry; believed that pvt. property and state allied with it destroy freedom and must be destroyed themselves to reestablish freedom which was present prior to existence of either of these instns. Died Russia, Nov. 8, 1921.

KROPP, Benjamin Nathan, anatomist; b. Boston, May 24, 1899; s. Nathan and Bessie (Wise) K.; B.S., Harvard U., 1923, A.M., 1924, Ph.D., 1927; m. Laurel Germaine Choquette, July 24, 1934; children—Mary Esther, Jonatha Elisabeth (Mrs. Robert P. Ceely), Laura Ann, Josefa Hannah-Bayla. Instr. zoology Harvard U., 1927-29, Sheldon travelling fellow, 1929-31; investigator Kaiser Wilhelm Inst. Biologie, Berlin-Dahlem, Germany, 1929-31; research fellow gen. physiology Harvard U., 1932-34; instr. anatomy Boston U. Med. Sch., 1934-35; supr. syphilis investigation Boston City Hosp., 1935; instr. anatomy, research fellow obstetrics Harvard Med. Sch., 1936-38; with Queen's U., Canada, 1938—, emeritus prof., 1964—; prof. anatomy La. State U. Sch. Medicine, 1964-65; prof. anatomy, head dept. Ghana Med. Sch., Accra, Ghana, also advisor in med. edn. Ghana Med. Sch., 1965-67; prof. anatomy U. W.I., Kingston, Jamaica, 1967—. Mem. Am. Assn. Anatomy, Am. Soc. Zoology, Canadian Assn. Anatomy, Canadian Soc. Physiology, Am. Assn. Phys. Anthropology, N.Y. Acad. Sci., A.A.A.S., Sigma Xi. Author: Laboratory Guide for Histology and Embryology, 1959. Research, publs. on in utero devel. fetal reactions; developmental abnormalities of devel. in newborn; instrumentation for microscopy. Home: 385 Bath Rd., Kingston, Ont., Can. Office: Dept. Anatomy, U. W.I., Kingston 7, Jamaica, W.I.*

KROSCH, (Karl) Heinz, German physician; b. Rosslau, Germany, July 5, 1917; s. Franz and Frida (Wietschke) K.; state med. exam., 1945; m. Christine Schulze, Aug. 18, 1954; children—Angela, Claudia. With II. med. clinic U. Halle, 1949-62, dep. dir. I. med. clinic, 1962-64, prof., dir., 1964—. Recipient Gold Hufeland medal; Purkyne medal Czechoslovakia. Mem. German Soc. Internal Medicine, German Soc. Circulation Research, Soc. Internal Medicine, Work Group Osteology East Germany, U. Halle Med. Sci. Soc. Internal Medicine (chmn.). Editor: (with W. Kaiser) Aktuelle Probleme der inneren Medizin I, 1964, II, 1965. Research, numerous publs. on blood viscosity, aortic arc syndrome, pericarditis, orthostasis electrocardiogram, mineral metabolism, enzyme diagnosis of heart infarct. Home: 17 Ernestusstrasse. Office: 22 Leninallee, Halle, East Germany.*

KROTKOV, Fedor Grigorevich, Russian hygienist; b. Maslovo (now Ryazan Oblast), 1896; grad. Leningrad Mil. Med. Acad., 1926, postgrad., until 1929; D. Med. Sci. Jr., sr. instr. chair hygiene Leningrad Mil. Med. Acad., 1929-31, prof., 1932-37; prof., head chair radiation hygiene Moscow Central Postgrad. Med. Inst., 1937—, head chair mil. hygiene, 1944-57; dir. Inst. Aviation Medicine, 1935-47; USSR dep. minister health, 1946-47; acad. sec. dept. hygiene, microbiology and epidemiology USSR Acad. Med. Sci., 1944-50. USSR rep. WHO, 1946. Decorated Order of Lenin (2). Mem. USSR Acad. Med. Sci., All-Union Hygiene Soc. (bd. mem. 1947-56, dep. chmn. 1956—). Author: Handbook on Military Hygiene, 1933, 2d edit., 1939; Textbook on General Hygiene, 1947; co-author: Military sanitation Manual, 1932; Military Hygiene, 1936. Editor Hygiene sect., chmn. com. on med. radiology Large Med. Ency., 2d edit.; editor Med. Radiology; mem. editorial council Hygiene and Sanitation. Research and numerous publs. on field hygiene, food and water supply to troops. Address: Moscow Central Postgrad. Med. Inst., pl. Vosstaniya 1-2, Moscow, USSR.

KRUCK, Thomas, German chemist; b. Aichach, Bavaria, Germany, Jan. 28, 1934; s. Thomas and Zäzilia (Eberl) K.; Abitur Oberrealschule Schrobenhausen, Bavaria, 1953; Diplomchemiker, Technische Hochschule München (Germany), 1959, Doktor der Naturwissenschaften, 1967; m. Juliane Finsterer, Dec. 12, 1932; children—Ruperta, Thomas. Dozent inorganic chemistry Technische Hochschule München, 1964-66; prof., sci. counselor U. Köln (Germany), 1966—. Mem. Deutsche Chemische Gesellschaft. Research and publs. on cationic carbonyl complexes and related compounds; discovered system of PF3-complexes. Home: 15/I Germaniastrasse, München, Bavaria, West Germany. Office: 47 Zülpicher Strasse, Köln, West Germany.*

KRUEGER, Albert Paul, Am. biometerorologist; b. Butte, Mont., Mar. 17, 1902; s. James and Lillian (Ornsten) K.; A.B. with great distinction, Stanford, 1925, M.D., 1929; m. Rose Alberta Margolis, Sept. 30, 1922 (dec. 1952); children—James Samuel, Elsie Louise (Mrs. Thoni Cherian); m. 2d, Mildred Freeman, Feb. 20, 1953. Instr., asst. prof. dept. bacteriology and exptl. pathology Stanford, 1927-29; asso. gen. physiology Rockefeller Inst. Med. Research, 1929-31; faculty U. Cal., Berkeley, 1931—, prof., 1938-57, emeritus prof., 1957—, lectr. medicine U. Cal. Sch. Medicine, 1931-57, emeritus lectr., 1957—, chmn. dept. bacteriology, 1946-52; sci. dir. Naval Biol. Lab. U. Cal., 1950-54, chmn. bd. advisers, 1947-57, chmn. bd. med. consultants, 1962—. Cons. Oakland Naval Hosp., 1952-60; cons. infectious diseases Cowell Meml. Hosp. U. Cal., 1931-57; vis. staff U. Cal. Hosp., 1931-57; cons. council on pharmacy, chemistry A.M.A., 1931-39. Recipient Internat. Soc. Biometerology William S. Petersen award, 1966. Diplomate Am. Bd. Microbiology. Fellow Cal. Acad. Sci., Royal Soc. Health, Am. Acad. Microbiology, Am. Pub. Health Assn., A.M.A.; mem. Soc. Exptl. Biology and Medicine, Cal. Acad. Medicine, Am. Assn. Immunologists, Am. Soc. Microbiology, Soc. Gen. Physiologists, Assn. Mil. Surgeons U. S., Internat. Soc. Biometerology, Air Pollution Control Assn., Am. Inst. Med. Climatology (v.p.), Phi Beta Kappa, Sigma Xi, Alpha Omega Alpha. Research field basic properties of bacterial viruses; developed undenatured bacterial antigens; studies, numerous publs. field biol. changes induced by air ions. Home: 1770 Arch St., Berkeley, Cal. 94709.*

KRUEGER, Felix, German psychologist; b. Posen, Aug. 19, 1874; prof., Buenos Aires, then Leipzig, 1919-38, and Halle. Author: Das Wesen der Gefühle, 1928; Lehre von dem Ganzen, 1948; Zur Philosophie und Psychologie der Ganzheit, 1953. Founder Leipzig sch. of Ganzhiet-psychology, which stressed entirety of mind and the many assorted factors that assert influences upon it. Died Basel, Feb. 25, 1948.

KRUEGER, Peter J., Canadian chemist; b. Altona, Man., Can., Nov. 11, 1934; s. Jacob J. and Elizabeth (Friesen) K.; B.Sc. with honors in Chemistry, U. Man., 1955, M.Sc., 1956; D.Phil., (Shell postgrad. scholar, U. Man. Travelling fellow) Oxford (Eng.) U., 1958; m. Dorothy Isabel Lashley, July 18, 1959; children—Kathryn Elisabeth, Vivian Louise. Asst. prof. chemistry U. Alta. (Can.), Calgary (now ind. U. Calgary 1966), 1959—, prof., head chemistry dept., 1966—; vis. scientist NRC of Can., 1966-67. Recipient Coblentz award in spectroscopy, 1967. NRC Can., Postdoctoral fellow, 1958. Fellow Chem. Inst. Can.; mem. Spectroscopy Soc. Can. (nat. dir. 1964—), Am. Chem. Soc., Chem. Inst. Can., Coblentz Soc., Royal Inst. Chemistry. Research molecular structure and conformations with infra-red spectroscopy, molecular interactions in solution. Home: 88 Brown Crescent, N.W., Calgary, Alta., Can.*

KRUESI, John, inventor; b. Speicher, Switzerland, May 15, 1843; m. Emily Zwinger, 1871, 8 children. Came to U. S., 1870; became mechanic and engr. to Thomas Edison, 1871; built many of inventions designed by Edison; foreman machine shop, Menlo Park, N.J., 1877; built 1st Edison phonograph, 1877; helped perfect incandescent lamp; designed machinery for manufacture of electric lighting equipment; supt. Edison Machine Works, N.Y.C., 1881-86; patented waterproof and insulated method for laying underground cables, 1882; gen. mgr., chief mech. engr. Edison Machine Works, Schenectady, N.Y., 1886-95, chief engr., 1895-99. Died Schenectady, Feb. 22, 1899.

KRUG, Hans Joachim, German geographer; b. Berlin, Apr. 6, 1893; s. Ludwig and Marie (Bartsch) K.; ed. in Prussian Cadet Corps; m. Margarete Blumhardt, Aug. 18, 1942. Officer, 1911-37; consul to Peru, China. Mem. Berlin Geog. Soc., Berlin Anthropology Soc., Geography Soc. Berlin (pres.). Author: Wanderungen und Wandlungen in China; Götterthrone im Urwald; Autralien und Ozeanien; Bertelsmann Länderkunde; Westermann geogr. Lexikon; contbg. author: Europa ohen die Mitte. Home: Fasanenstrasse 42, 1 Berlin 15. Office: Gesellschaft für Erdkunde, Grunewaldstrasse 35, 1 Berlin, West Germany.

KRÜGER, Friedrich Rudolf Ludwig, German biologist, chemist; b. Siegen, Germany, Aug. 18, 1902; s. Rudolf and Eva (Schaper) K.; ed. univs. of Heidelberg, Kiel, Fribourg; Ph.D., chemistry diploma; m. Minka Ferger, Apr. 2, 1929; children—Rolf, Alfred, Ingrid, Margot. Successsively asst., agrégé, cons. scientist Helgoland Biol. Inst. Author: Ultrastruktur der Ciliaten-Trichochysten; Stoffwechsel des Schweinspulwurmes; Funktion des Hämaglobins; Experimentelle Biologie von Arenicola; Mathematische Behandlung biologischer Temperaturfunktionen. Research in math. of growth; comparative physiology of animals, especially metabolism of invertebrate animals. Home: Elbterrasse 20, 2000 Hamburg-Blankenese, West Germany. Office: Palmaille 9, 2 Hamburg-Altona, West Germany.

KRÜGER, Otto Wilhelm Karl, German mineralogist; b. Berlin, Aug. 28, 1897; s. Karl and Elise (Schreiber) K.; ed. U. Berlin, Tech. Coll. Berlin; Ph.D. Became asst., mineralogy inst. U. Berlin, 1923, prof., 1930—. Mem. German Geog. Soc., Berlin, Am. Geog. Soc., Technogeog. Soc. (pres.), Regional Inst. Tech. Berlin (pres.). Address: Stendelweg 22, 1 Berlin 19, W. Germany.

KRUGER, P(eter) Gerald, Am. physicist; b. Deer Creek, Minn., May 7, 1902; s. Fred W. and Minnie (Knaup) K.; A.B. in Physics, Carleton Coll., 1925; postgrad. (Am. Exchange fellow), U. Berlin, Germany, 1926-27; Ph.D., Cornell U., 1929; m. Erika Wackenroder, Aug. 26, 1931. Heckscher research fellow Cornell U., Ithaca, N.Y., 1930-31; faculty U. Ill., Urbana, 1931—, prof. physics, 1939—, acting head dept. physics, 1938-46; NRC fellow, Berlin, 1929-30. Fellow Am. Phys. Soc., Am. Physics Tchrs. Assn.; mem. Midwestern Univs. Research Assn. (founder 1953, dir. labs., Madison, Wis. 1956-57), Sigma Xi, Rho Kappa Phi. Researcher in field atomic and nuclear physics, 1926—. Home: 914 W. Union St., Champaign, Ill. 61820.*

KRUIZINGA, Pieter, Dutch geologist; b. Eenrum, Nov. 13, 1885; s. Edze and Hilegiena (Venhuizen) K.; Ph.D., Groningen, Netherlands; m. Alida Katharina Smit, Dec. 19, 1914; children—Edze, Ernst. Curator Delft Geol.-Mineral. Mus.; prof. geology and paleontology Delft Tech. U.; prof. Cath. U. Nimegue. Mem. Royal Soc. Geology and Mineral Exploitation Netherlands (hon.). Research and publs. on fossils, animals, plants, ores, geol. phenomena. Home: Parkflat de Valkenburcht falt 138, Oosterbeek. Office: Univ. Laborat. der Universiteit, Driehuizenverweg 200, Nijmegen, Netherlands.

KRUKOWSKI, Wlodzimierz, Polish engr.; flourished circa 1920-30; s. Antoni and (Chmielewska) Krukowski; studied electrotechnics Tech. U., Damstadt; D.Eng.; m. Miss Wasilkowska. Prof. electrotech. measurement, chmn. electrotech. lab. Tech. U., Lwow; sci. collaborator Office for Measures, Warsaw, in charge works at Polish standard for elec. units; active in working out specifications Union Polish Electrotech. Engrs.; permanent del. to Polish electrotech. com. in adv. com. 13 on measuring instrument Internat. Electrotech. Commn. Mem. Union Polish Electrotechnicians, Internat. Conf. on Electric High Tension. Author: Vorgänge in der Scheibe eines Induktionzählers und der Wechelstrumkompensator als Hilfsmittel zu deren Erforschung, 1920; Grundzüge der Zählertechnik, 1930, also numerous articles. Research on elec. measurement; invented number of improved elec. measuring devices.

KRUMBOLTZ, John Dwight, Am. psychologist; b. Cedar Rapids, Ia., Oct. 21, 1928; s. Dwight John and Margaret E. (Jones) K.; B.A., Coe Coll., 1950; M.A., Tchrs. Coll., Columbia, 1951; Ph.D., U. Minn., 1955; m. Helen Brandhorst, Aug. 22, 1954; children—Ann, Jennifer. Counselor, tchr. W. Waterloo (Ia.) High Sch., 1951-53; instr. U. Minn., 1954-55; faculty Mich. State U., 1959-61, asso. prof. ednl. psychology, 1959-61; faculty Stanford, 1961—, prof. edn. and psychology, 1966—. Recipient Outstanding Research award Am. Personnel and Guidance Assn., 1959, 66. Mem. Am. Ednl. Research Assn. (v.p. 1966—), Am. Psychol. Assn., Am. Personnel and Guidance Assn., A.A.A.S. Author: (with W. W. Farquhar, C. G. Wrenn) Learning to Study, 1960; also articles. Editor: Revolution in Counseling: Implications of Behavioral Science, 1966; Learning and the Educational Process, 1965. Research on behavior modification procedures in counseling. Home: 933 Valdez Pl., Stanford, Cal. 94305.*

KRÜMMEL, Otto, German oceanographer; b. Exin, Germany, July 8, 1854; prof. geography Kiel, also Marburg, Germany; mem. (with Hensen) Atlantic expdn. Author: Der Ozean, 1886; Handbuch der Ozeanographie, 2 vols., 1907-11. Died Cologne, Oct. 12, 1912.

KRÜNITZ, Johann Georg, German physician; lectr., Frankfort/Oder, Germany; later writer, Berlin; editor Ökonomisch-technologische Encyclopädie, from 1773, also other encyclopedic works. Died 1796.

KRUPNOV, Yevgeniy Ignatevich, Russian archeologist; b. Mozdok (now North Ossetian ASSR), 1904; student ethnol. dept. North Caucasian Pedagogical Inst., 1924-27; grad. History Faculty, Moscow U., 1930; D.Hist. Sci., 1960. Jr., sr. asso. State Hist. Mus., Moscow, 1930-51; sr. asso. Inst. Archeology, USSR Acad. Sci., 1937—, dep. dir., 1951-60, head Neolithic and Bronze Age sect., 1960—, dir. archeol. research in No. Caucasus, 1930's—. Recipient Lenin prize, 1963. Author: The Caucasus in the First Millenium B.C., 1936; The Kayakent Tomb-Monument of Ancient Albania, 1940; History of Ingushetia from Ancient Times to the 18th Century, 1941; The Northeastern Caucasus in the First and Second Millenia B.C., 1946; The Ancient History and Culture of Kabardinia, 1957; New Findings on the Archeology of the Northern Caucasus, 1958; Ancient History of the Northern Caucasus, 1960. Address: Inst. of Archeology, USSR Acad. Sci., 1-ya Cheremushkinskaya ulitsa 19, Moscow, USSR.

KRUPP, Alfred, German metallurgist; b. Essen, Germany, Apr. 26, 1812; s. Friedrich Krupp; founder steel and armaments factory, Essen, 1846. Invented crucible steel process; produced 1st war materials of cast and forged steel, superior gun barrels, 1844, howitzers, 1855, cannons with rifled barrels, round bolt breech, 1864; developed new Bessemer steel. Died July 14, 1887.

KRUSE, Heinrich, German physician; b. Uetersen, June 24, 1906; s. Michael and Helena Feuerschutz) K.; ed. univs. Kiel, Fribourg, Munich, Konigsberg; M.D.; m. Margarete Adloff, June 14, 1946. Health officer; chief physician, country hosp.; mil. hygienist; asst. Göttingen Inst. Hygiene; dir., prof. Pub. Health Inst.; cons. for local hygiene WHO. Mem. Berlin Soc. Medicine and Natural Sci. Author: Wasser; Desinfektion un Schädlingsbekämpfung; other works on gen. hygiene. Home: Marinesteig 37, Berlin 38. Office: Corrensplatz 1, Berlin 33, W. Germany.

KRUSE, Walther, German bacteriologist; b. Berlin, Germany, Sept. 8, 1864; prof., Bonn, Germany, Königsberg (now Kaliningrad, USSR), Leipzig, Germany. Author: Allgemeine Mikrobiologie, 1910; Einführung in die Bakteriologie, 1920. Discovered (with Shiga) dysentery bacillus. Died Sept. 1, 1943.

KRUSENSTERN, Baron Adam Ivan von, see von Krusenstern, Baron Adam Ivan.

KRUTA, Vladislav Otto, Czechoslovakian physiologist; b. Belá p. Bezdezem, Czechoslovakia, June 27, 1908; s. Afred and Marie (Konetopska) K.; M.D., Charles U., Prague, 1931, D.Sc., 1962; postgrad. Faculté des Sciences, Sorbonne, Paris; m. Emmy Bahuault, Nov. 27, 1937; children—Venceslas, Aline, Vladislav. Faculty, Prague (Czechoslovakia) U., 1930-38, 46—, prof. applied physiology Med. Sch., 1946—; chmn. dept. physiology Med. Faculty Hradec Králové, 1948-51; prof. physiology Med. Faculty, U. J.E. Purkyne, Brno, Czechoslovakia, 1951—. Adviser in postwar relief matters Czechoslovakian Govt., 1943-45; chief Med. Post-War Relief Activities, Ministry Health, Prague, 1945-48. Mem. Czechoslovak Physiol. Soc. (past pres.), Czechoslovak Biol. Soc. (past pres.), Internat. Acad. History of Medicine, Société de Biologie Paris (corr.). Recipient Czechoslovak War medal, 1945; Ministry Health Sci. Council award, 1961. Author: (with Z. Hornof, V. Seliger) Uvod do fysiologie telesnych cviceni, 1953; Med.Dr.Jiri Prochaska (1749-1820). Zivot, dilo, doba, 1956; (with others) Ucebnice fysiologie, 3 vols., 1955-59, rev. 2 vols. 1960, 62; Nekolik pohledu na srovnávaci fysiologii srdce a krevniho obehu, 1958; M.-J.-P. Flourens, J.-E. Purkyne et les débuts de la physiologie de la posture et de l'équilibre, 1964. Editor: Jan Evangelista Purkyne, 1951-65; Kpocatkum vedecké zivotni dráhy J. E. Purkyne. Korespondence s práteli z prazskych let 1815-1823, 1964; also numerous articles. Exposition of regular variations of strength of heart muscle contractions due to changes of interval between contractions and of frequency, and mechanism of gradation of heart muscle contractions, comparative physiology of the heart and circulation, history of physiology. Home: Tvrdeho 13, Brno, Czechoslovakia.*

KRUZHILIN, Georgii Nikitich, Russian combustion engr.; b. June 6, 1911; grad. Leningrad Physico-Mech. Inst., 1934. With Leningrad Central Boiler and Turbine Inst., 1933-46, with Urals dept., 1936-38; with USSR Acad. Scis., 1946—, with Power Engring. Inst., 1955—, dir. Krzhizhanovsky Power Engring. Inst., 1960—. Mem. USSR Acad. Scis. (corr.). Research and publs. on determination of ratio of heat emission along a surface; calculation of thermal boundary layers; theory of heat emission in steam condensation and boiling liquid. Office: A. USSR, Moskva, Leninsky prosp. 19, Energetichesky institut AN SSSR.

KRWAWICZ, Tadeusz Jan, Polish ophthalmologist; b. Lwow, Jan. 15, 1910; s. Jan and Michalina (Jasinski) K.; grad. medicine Jan. Kazimierz U., Lwow, 1938, D.Medicine, 1939; m. Zdzislawa Hebal, May 11, 1944; 1 son, Leslaw. Asst., Eye Clinic, Jan Kazimierz U., 1939; dir. Ophthalmol. Clinic, Med. Acad., Lublin, Poland, 1948—. Mem. Polish Acad. Scis., 1967—. Recipient Polish State prize 1st degree, 1966; also numerous high Polish state distinctions. Mem. Polish Ophthalmol. Soc., Société Francaise d'Ophthalmologie, Oxford Ophthalmol. Congress, Soc. Cryo-Ophthalmology, Polish Med. Alliance (hon.). Author: (with another) Modern Advances of Cataract Surgery, 1963; Management of Complications in Eye Surgery, 1965; also numerous articles. Research on histochemistry of cornea; developed new techniques of keratoplasty for optical reasons, new methods of storing corneal tissue for transplantation; originated cryosurgery of cataract and cryotherapy, especially treatment of herpes simplex keratitis by application of low temperature. Home: 49 Krakowskie Przedmiescie, Lublin, Poland.*

KRYLOV, Aleksandr Petrovich, Russian petroleum engr.; b. Aug. 14, 1904; grad. Leningrad Mining Inst., 1926; D.Sc. Joined faculty, Moscow Petroleum Inst., 1933, became prof., 1949; named dept. dir. All-Union Sci. Research Petroleum Inst., also chief lab. Inst. Petroleum, USSR Acad. Scis., 1953. Recipient Stalin Prize, 1949, Lenin Prize, 1962. Mem. Acad. Sci. (corr.). Coauthor: Course on the Working of Petroleum Deposits, 1940; The Scientific Principles of Working Petroleum Deposits, 1948; The Exploitation of Petroleum Deposits, 1949; Elasto-Plastic Conditions in Petroleum Layers, 1955. Research on method of exploiting oil deposits. Office: Dorogomilovskaya nab. 1/2, Moscow USSR; also Moscow Petroleum Inst., USSR Acad. Scis., Moscow, USSR.

KRYLOV, Vladimir Ivanovich, Russian mathematician; b. Dec. 15, 1902; grad. Leningrad U., 1928; D.Physico-Math. Sci., 1951. Instr., Leningrad U., 1930-52, prof., 1952-57; prof., 1952—; head lab. approximation calculations Inst. Physics and Math., Belorussian Acad. Sci., 1956—. Mem. Belorussian Acad. Sci. Author: The Functions of Regular Values in a Semi-Plane, 1939; co-author: Methods of Approximation in Advanced Analysis, 1952. Research and publs. on theory of function of complex variables, computer math. Address: Inst. Physics and Math., Belorussian Acad. Sci., Minsk, Belorussian SSR, USSR.

KRYNSKI, Stefan, Polish microbiologist; b. Wilno, Poland, Aug. 1, 1914; s. Kazimierz and Felicja (Barwinska) K.; physician U. Poznan (Poland), 1938; dr. med., Jagellon U., Kraków, Poland, 1946; docent habil, Med. Sch. Gdansk (Poland), 1951. Staff, Weigl Inst., Lwów, Poland, 1940-44, dir., Lublin, 1944-46; faculty Med. Sch. Gdansk, 1946—, prof. microbiology, 1954—. Cons., Marine and Tropical Medicine Inst., Gdansk, 1946-55. Decorated Golden Cross of Merit. Mem. Polish Microbiol. Soc., Soc. Francaise de Microbiologie, Polish Med. Soc., Polish Soc. Parasitology. Research and numerous publs. on biology of infection and toxic action of R.prowazeki in lice, biology of S. aureus and hosp. cross-infection; discovered new antibiotic, tetaine. Home: 1-d Hibnera, Gdansk, Poland.*

KRYSPIN-EXNER, Kornelius Wilhelm Johann, Austrian physician; b. Vienna, Austria, June 14, 1926; s. Botho and Helene (Laimer) K-E.; M.D., Vienna U., 1954. Asst., Inst. for Psychiatry, Vienna U., 1954—; head Rehab. Center for Alcoholics, Vienna, 1961—. Mem. Austrian Fedn. for Mental Health (mem. working com.), Gesellschaft D.Aerzte in Vienna. Author: Psychosen u.Prozessverlaufe des Alkoholismus, 1966. Publs. on description of specific types of alcoholism; exptly. induced alcoholic psychoses; manic-depressive illnesses; mechanism of tranquilizer addiction. Home: 29 Billrothstrasse, Vienna 1190. Office: 23 Spitalgasse, Vienna 1090, Austria.*

KRZYWOBLOCKI, Maria Zbigniew von, applied mathematician; b. Lemberg, Poland, July 29, 1904; s. Benedikt and Stanislawa (Zimmer) K.; B.S., Lemberg Inst. Tech., 1926, diplom-ingenieur, 1936; postgrad. U. Warsaw (Poland), 1937-39; M.S. in Aeros., Bklyn. Poly. Inst., 1943, Ph.D. in Aeros., 1944; M.S. in Math., Brown U., 1945; M.A. in Math., Stanford, 1946; Sc.D., U. Lille (France), 1955; m. Cecylia Bizowski, Mar. 26, 1946. Came to U. S., 1942, naturalized, 1950. Instr. math. Lemberg Inst. Tech., 1928-33, instr. mech. lab., 1933-36, instr. aerodynamic lab., 1933-34, vis. asst. prof., 1936-39; designer Lemberg Wooden Airplane and Glider Co., 1928-33; lectr. Lemberg Aviation Sch., 1929-36; research engr., Instr. Lemberg Aero. Inst., 1934-36; chief design group Lublin Aircraft Co. (Poland), 1936-37; designer, chief stress analyst Polish Nat. Aircraft Co., Warsaw, 1937-39; chief group Hydravions Amiot Co., Cherbourg, France, 1939-40; aircraft engr. Canadian Aircraft Co., Ft. William, Can., 1940-41; aircraft engr., stress analyst Noorduyn Aircraft Co., Montreal, Can., 1941-42; teaching asso. Bklyn. Poly. Inst., 1942-44; asso. prof. U. Ill., 1946-49, prof., 1949-60; prof. space scis., Mich. State U., East Lansing, 1960—. Cons. to numerous govtl., bus. and ednl. instns. Recipient Best Tchr.'s award U. Ill., 1949. Fellow Am. Phys. Soc.; mem. Am. Math. Soc., A.A.A.S., Soc. Engring. Sci., Indian Soc. Theoretical and Applied Mechanics, Sigma Xi, Alpha Gamma Rho, Pi Tau Sigma. Author of a series of 4 books on construction of wooden planes, 1935-38; Bergman's Integral Operator Technique in the Theory of Compressible Fluid Flow, 1960. Research and publs. on applied math., fluid dynamics, celestial mechanics and relativity. Home: 165 Milford St., East Lansing, Mich. 48823.*

KUBASCHEWSKI, Oswald, phys. chemist; b. Berlin, Germany, July 13, 1912; s. Dr. iur Kurt and Ruth (Barnewitz) K.; student Rostock U., 1930-32; Dr. phil.nat., Freiburg (Germany) U., 1935; Dr.phil.habil. Technische Hochschule, Stuttgart, 1942; m. Ortrud von Goldbeck, Sept. 13, 1949; children—Peter, Karin. Sci. asst. Kaiser Wilhelm Inst. für Metallforschung, Stuttgart, 1936-45; lectr. Technische Hochschule, Stuttgart, 1942-47; research chemist Nat. Phys. Lab., Teddington, Middlesex, Eng., 1947-50, staff, 1950—, sr. prin. sci. officer, 1955—; vis. faculty Pitts. U., 1964, Central Inst. for Indsl. Research, Oslo, Norway, 1966, Technische Hochschule, Aachen, Germany, 1966-67. Decorated Jurat de St. Émilion. Mem. Deutsche Bunsengesellschaft, Faraday Soc., Inst. Metals. Author: (with F. Weibke) Thermochemie der Legierungen, 1943; (with E. L. Evans) Metallurgical Thermochemistry, 1951; (with B. E. Hopkins) Oxidation of Metals and Alloys, 1953; (with J. A. Catterall) Thermochemical Data of Alloys, 1956; also numerous articles. Research on thermochem. properties, oxidation of, diffusion in metals and alloys and application to assessment of practical problems. Home: 44 Manor Dr., Surbiton, Surrey, Eng. Office: Nat. Phys. Lab., Teddington, Middlesex, Eng.*

KUBICZ, Józef, Polish physician; b. Swilcza, Rzeszów, Poland, Oct. 26, 1906; s. Walenty and Aniela (Czach) K.; D. Degree in Philosophy, Jan. Kazimierz U., Lwów, Poland, 1930, postgrad. Med. Faculty; Dypl. Phys., Med. Inst. in Lwów, 1941; m. Aleksandra Mrozek, Oct. 15, 1955; 1 son, Ziemowit. Bacteriol. controller Typhus Inst., Lwów, 1941-42; asst. Children Clinic, Lwów, 1943-45; sr. asst. Dermatol. Clinic, Wroclaw, 1946; asst. prof. U. Wroclaw, 1954-65, asso. prof., 1965—, dir. Clinic Skin and Venereal Diseases, 1961—; dean Med. Faculty, Med. Sch. Wroclaw, 1956-60. Mem. Wroclaw Sci. Soc., Polish Dermatol. Soc., Polish Physiol. Soc., Polish Med. Soc. in Wroclaw. Research, publs. on use of young lice fed on skin used as gen. nutrient medium for isolating rickettsia and viruses, pharmacol. activation of vegetable oils by elec. diffusing of metals into oils and heating them with magnesium filings or metallic titanium; new pathologic skin conditions, neurodermatitis profunda, adenosis sebacea eruptiva, syndromatology, exoelectrons in water, body fluids, expiration air, breakdown of hydrogen peroxide and oxydation processes in body fluids as source of exoelectrons; electric catalytic elements. Address: Chalubinskstr. 1, Wroclaw, Poland.

KUBIKOWSKI, Piotr, Polish pharmacologist; b. Krasilov, Poland, Nov. 19, 1903; s. Jan and Maria (Sokolowski) K.; Faculty of Medicine, U. Lwow, 1928, M.D., 1930, docent od pharmacology, 1939; pharmacology studies in Vienna, Austria, Berlin, Germany, Paris, France, 1936-37; WHO fellow U. S. A., 1947-48; m. Zofia Stefanski, Apr. 17, 1948. Prof., head pharmacology dept. Med. Acad. Warsaw (Poland), 1948—, vice-rector, 1960—; dir. Inst. Drugs, Warsaw, 1951—. Mem. sci. com. Ministry of Health and Welfare. Decorated Comdr. Cross, Polonia Restituta, others. Mem. Polish Physiol. Soc. (v.p.), Polish Pharm. Soc. (pres. 1965—); hon. mem. numerous fgn. sci. socs. Author 4 books on pharmacology and toxicology. Research, publs. on pharmacology of vegetative nervous system, toxicology of drugs and other chem. compounds, pharmacology of hypotensive drugs, neuro-and psychopharmacology. Home: 8 Kredytowa, Warsaw. Office: 26/28 Krakowskie, Przedmiescie, Warsaw 64, Poland.*

KUBITSCHEK, Wilhelm, Austrian archeologist; b. Pressburg (now Bratislava, Czechoslovakia), June 28, 1858; prof., Graz, Vienna (both Austria); dir. coin collection. Author: Rundschau über ein Zuinquennium der antiken Numismatik, 1896; Ausgewählte romische Medaillons der kaiserlichen Müzensammlung in Wien, 1909. Died Vienna, Oct. 2, 1936.

KUBO, Masaji, Japanese physical chemist; b. Tokyo, Japan, Feb. 21, 1911; grad. Tokyo U., 1933. Instr., Keijo U., Seoul, Korea, Tokyo U.; became prof. Nagoya U., 1948; guest research worker Harvard, 1951; Recipient prize Japan Sci. Assn. Author: Outline of Quantum Chemistry. Research in electric osmosis, quantum chemistry.

KUBOTA, Bennosuke, Japanese chemist; b. Chiba Prefecture, Japan, 1885; grad. Tokyo U., 1910; D.Sc. Prof., 8th Higher Sch., Nagoya, Japan; staff mem. Sci. Research Inst., mem. emeritus, 1947—; prof. Tokyo U., 1934-47, emeritus prof., 1947—; now prof. Chiba Comml. Coll., also instr. Waseda U. Recipient Akurai prize Japan Chem. Acad., 1925. Author: Study of Contact Reaction of Inorganic Catalyzer; Standard Organic Chemistry; Theory of Contact Reaction. Research in organic chemistry, especially contact reaction.

KUCERA, Clair Leonard, Am. ecologist; b. Belle Plain, Ia., Apr. 30, 1922; s. Charles J. and Emma (Krafka) K.; B.S., Ia. State U., 1947, M.A., 1948, Ph.D., 1950; m. Elizabeth A. Tremmel, July 18, 1946; children—Ron, Kim, Carol, Gary. Faculty, U. Mo., Columbia, 1950—, prof., 1960—, chmn. dept. botany, 1960-64, dir. Tucker Prairie Research Sta., 1958—. Mem. A.A.A.S., Am. Jr. Botanists, Ecol.

Soc. Am., Sigma Xi. Author: Grasses of Missouri, 1961. Studies, publs. on grassland productivity in undisturbed state, assessing the role of root systems in prairie on devel. soil humus; how roots break down in the cyclic process of growth and decay. Home: 500 Rockhill Rd., Columbia, Mo. 65201.*

KÜCHEMANN, Dietrich, aerodynamicist; b. Göttingen, Germany, Sept. 11, 1911; s. Rudolf and Martha (Egener) K.; Dr.rer.nat., Göttingen U., 1936; m. Helga Janet Praefcke, Sept. 19, 1936; children—Christine Friederike, Dietmar Erich, Eva Beate. Aero. research Aerodynamische Versuchsanstalt, Göttingen, 1936-46; aero. research Royal Aircraft Establishment, Farnborough, Hants, Eng., 1946—, head aerodynamics dept., 1966—. Decorated Comdr. of Order of Brit. Empire, 1964. Fellow Royal Soc., 1963, Royal Aero. Soc. (Silver medal 1962); asso. fellow Am. Inst. Aeros. and Astronautics. Author: (with J. Weber) Aerodynamics of Propulsion, 1953; also numerous articles. Research on aerofoil theory, swept wing design, aerodynamics of propulsion, waverider aircraft for hypersonic flight; devel. slender wings for supersonic flight leading to Concord aircraft project. Home: 32 Echo Barn Lane, Farnham, Surrey, Eng. Office: Royal Aircraft Establishment, Farnborough, Hampshire, Eng.*

KÜCHENMEISTER, Gottlob Friedrich Heinrich, German parasitologist; b. Buchheim, Germany, Jan. 22, 1821; Author: Über Cestoden im allgemeinen und die des Menschen in besondere, 1853; Die Parasiten des Menschen, 1855; Die in und an dem Körper des Menschen vorkommenden Parasiten, 2 vols., 1855; Luthers Krankengeschichte, 1881. Research in parasitology, cestoids; discovered effect of Peruvian balsam on scabies. Died Dresden, Germany, Apr. 13, 1890.

KUDO, Tatsuyuki, Japanese neurosurgeon; b. Tokyo, Japan, Sept. 2, 1911; s. Itsuki and Sawa (Nakano) K.; M.D., Keio U. Med. Sch., 1935; m. Kazue Koyama, Oct. 18, 1942; children—Akiko, Asahiko, Fumio. Prof. surgery, chief neurol. surgery Keio U. Med. Sch., Tokyo, 1961—. Mem. Japan Neurosurg. Soc. (pres. 1966), Japan Surg. Soc. (councilor 1946—), Japanese Soc. Neurology. Editor: Occlusion of the Circle of Willis, 1966. Contbr. numerous articles to sci. jours. Developed hypothermic device to obtain near 0°C cerebral temperature by perfusion of artificial solution through cerebral vascular beds, over ninety minutes of absolute cerebral circulatory arrest was achieved and highly vasculartumors were removed safely in dry surg. field; described occlusion of circle of Willis, a new disease found in Japanese. Home: 7 1-chome Tokiwadai Itabashi, Tokyo. Office: 35 Shinanomachi Shinjuku, Tokyo, Japan.*

KUDREV, Todor, Bulgarian plant physiologist; b. Village Bebrovo, Bulgaria, Dec. 12, 1922; s. Georgi Stanev and Maria Todorova (Haidutova) K.; Agriculturist, Sofia U., 1947, C.Sc., Faculty Agr., 1958; D.Sc., Agrl. Acad. Sci., 1968; m. Nadiia Dimitrova Khristova, Nov. 8, 1954; children—Georgi Todorov (dec.), Maria Todorova, Todor Todorov. With Central Agrl. Research Inst., 1948-55; with Plant Prodn. Inst., 1955-66, sr. sci. officer, head plant physiology dept., 1959-66; head plant physiology dept. Inst. Genetics and Plant Breeding, Sofia, 1967—; tchr. mgmt. U. Sofia, 1963—. Mem. Assn. Scientists of Bulgaria. Contbr. numerous articles to profl. jours. Research on recovering drought-damaged plants by means of growth substances. Home: 32 Buzemska St., Opitno pole, Sofia 18. Office: Inst. Genetics and Plant Breeding, Sofia 13, Bulgaria.*

KUEBLER, William Frank, Jr., Am. chemist; b. Kansas City, Mo., Feb. 21, 1916; s. William Frank and Leonne (Scott) K.; A.A., Jr. Coll. Kansas City, 1935; B.A., U. Kan., 1937; m. Dorcas Naomi Rittenhouse, July 25, 1945; children—Carole Lynn, Diann Celeste. Pharm. chemist Peerless Serum Co., Kansas City, Kan., 1938-40; pharm. tablet maker George A. Breon & Co., Kansas City, Mo., 1940-42; chief control chemist Thompson-Hayward Chem. Co., Kansas City, Mo., 1941-42, research chemist, 1946-51; research chemist Jensen-Salsbery Labs. div. Richardson-Merrell, 1951-54, group leader, 1954-57, dir. pharm. research and devel., 1957-59, research adminstr., 1959-61, asst. to research dir. project planning and analysis, 1961-65; asst. to tech. dir. Marion Labs., Inc., Kansas City, Mo., 1965—. Fellow Am. Inst. Chemists; mem. Am. Chem. Soc. (membership chmn. Kan.-Western Mo. chpt. 1964-65), Am. Statis. Assn., Biometrical Soc., Kansas City Applied Math. Soc. Contbr. articles to tech. jours. Application statis. methodological techniques; research in pharm. chemistry. Home: 8506 Woodson Dr., Overland Park, Kan. 66207. Office: 10236 Bunker Ridge Rd., Kansas City, Mo. 64137.*

KUEHL, Frederick Albert, Am. biochemist; b. Malden, Mass., Nov. 9, 1916; s. Fred A. and Helen (Wohlleben) K.; A.B., Wesleyan U., Malden, Mass., 1939; M.A., Ph.D., Harvard, 1942; m. Jeah Vail, June 7, 1941; children—Kathie, Carol, Fred, Clinton, Donald, Steven. With NDRC, U. Pa., 1942; dir. bioorganic chemistry, Merck & Co., Rahway, N.J., 1943—. Mem. Am. Chem. Soc., Fedn. Am. Socs. for Exptl. Biology. Contbr. articles to tech. jours. Research on isolation and structure of streptomycin and other antibiotics, vitamin B12, hormones, catacholamines. Home: 47 Wigwam Rd., Locust, N.J. 07760. Office: Merck, Sharp & Dohme Research Labs., Rahway, N.J.*

KUEHNE, Martin Erick, Am. chemist; b. Floral Park, L.I., N.Y., May 29, 1931; s. Martin Ludwig and Ruth (Protze) K.; B.A., Columbia, 1951, Ph.D., 1955; M.A., Harvard, 1952; m. Hannelore E. Naumann, Aug. 15, 1953; 1 son, Stephen Eric. Sr. chemist Ciba Pharm. Co., Summit, N.J., 1955-61; faculty U. Vt., Burlington, 1961—, asso. prof. chemistry, 1965—. Boese fellow, 1954; Alfred P. Sloan Found. fellow, 1965-67. Mem. Am. Chem. Soc., Sigma Xi, Phi Lambda Upsilon. Research in new synthetic organic reactions; total syntheses of natural products; structure determinations of natural products; medicinal chems. Home: 169 South Cove Rd., Burlington, Vt. 05401.*

KUENEN, Donald Johan, Dutch zoologist; b. Leyden, Netherlands, Mar. 19, 1912; s. Johannes Petrus and Dora (Wicksteed) K.; Dr.'s degree, U. Leyden, 1939; m. Louisa Janssens, Sept. 29, 1939; children—Johannes, Andrina (Mrs. Otto Strack), Dora. Asst. Zool. Lab., 1933-39; entomologist Shell Research Lab., Amsterdam, 1939-41, Lab. Fruit Growing Research, Wilhelminadorp, Netherlands, 1941-49; sci. officer Zool. Lab., Leyden, 1949-50; prof. zoology U. Leyden, 1950—, dean Faculty Scis., 1961-63, rector, 1964-65, prorector. 1966—. Decorated knight Order Netherlands Lion, 1966. Hon. fellow Royal Entomol. Soc.; mem. Internat. Union for Conservation Nature (exec. bd. 1963—). Author: (with others) Zoology for Medical Students, 1967; also articles. Research on control of orchard pests, relations of plant to pests as factor in control, ecol. problems of insect control, nature conservation on ecol. basis. Home: 43 Cobetstraat, Leyden, Netherlands.*

KUENEN, Philip Henry, geologist; b. Dundee, Scotland, July 22, 1902; s. J. P. and D. (Wicksteed) K.; Dr., U. Leyden (Netherlands), 1925; Dr.h.c., U. Dublin (Ireland) 1956, U. Krakow (Poland), 1964; m. C. Pyzel, Apr. 2, 1932 (dec. 1967); children—Charlotte (Mrs. S. Manger), Madeline (Mrs. E. Dommering), Patricia, Diederik. Asst., Geol. Inst., Leyden, 1926-34; mem. Snellius Deep-Sea Expdn., Moluccas, 1929-30; faculty U. Groningen (Netherlands), 1934—, prof., 1943—, dean sci. faculty, 1951-52, rector, 1960. Decorated Ridder Nederlandse Leeuw, 1962; recipient medal Société géologique de Belgique, 1952, Geologische Genootschap Netherland, 1954, Penrose Geol. Soc. Am., 1961, Geologische Vereinigung, 1966. Mem. Royal Netherl. Acad. Sci., Am. Acad. Arts and Scis. (hon.), N.Y. Acad. Scis. (hon.), also hon. mem. several geol. socs. Author: Marine Geology, 1950; also other books, numerous articles. Research on marine geology, including coral reefs; expts. on tectonics, volcanoes, sediment structures; particle abrasion in rivers and by wind; turbidity currents that flow down underwater slopes due to high density of sedimentary load; application of results to ancient deposits and recent deep-sea deposits. Home: 214 Houteniaan, Groningen, Holland.*

KUENTZEL, Lester Everette, Am. phys. chemist; b. Ft. Worth, Feb. 12, 1911; s. Albert Paul and Arva (Ambler) K.; A.B. in Chemistry, U. Colo., 1936, Ph.D. in Phys. Chemistry, 1941; m. LaVerne Violet Carpenter, Aug. 22, 1934; children—Sandra Lee, Garth Lester. Head phys. measurements sect. Wyandotte Chems. Corp. (Mich.), 1941-45, head phys. analytical dept., 1945-56, sr. research chemist, 1956-60, research asso., 1960—. Mem. free radicals project Nat. Bur. Standards, 1956-58, cryogenic researcher NASA, 1958-63; dir. data storage and retrieval programs for Am. Soc. Testing and Materials and Nat. Reference Data Systems, Nat. Bur. Standards, 1963-67. Mem. Am. Soc. for Testing and Materials (vice chmn. com. E-13, 1958-67, award merit 1960), Am. Optical Soc., Am. Chem. Soc., Coblentz Soc. (bd. govs. 1964-67), Assn. Analytical Chemists. Author: Molecular Formula Index of Published Infrared Spectra, 1962; Coden for Periodical Titles, 1963, 66; also articles. Patentee in field. Invented variety paint remover formulations; developed method indexing spectroscopic and X-ray data in IBM cards. Home: 1773 15th St., Wyandotte 48192. Office: Wyandotte Chems. Corp., Wyandotte, Mich. 48193.*

KUENTZEL, Ward Edward, Am. chemist; b. Taylor Twp., Minn., Oct. 17, 1893; s. Julius A. and Augusta (Steinke) K.; B.S., U. Minn., 1917; M.S., George Washington U., 1924, Ph.D., 1928; m. Ethel Roberts, May 11, 1918; children—Dorothy Jean, (Mrs. James Leon), Elizabeth Jane (Mrs. Robert Courtney Maulbetsch). Chemist, Bur. Chemistry, Washington, 1917-18, nitrate div. Enlisted Ordnance Dept., 1918-19, Fixed Nitrogen Research Lab., Washington, 1919-30, Whiting Labs., Standard Oil Co. (Ind.), 1930-58; mem. High Pressure Assos., Autoclave Engrs., Erie, Pa., 1958—. Mem. Am. Chem. Soc., Alpha Chi Sigma, Phi Lambda Upsilon. Contbr. articles to tech. jours. Patentee in field. Co-inventor magne-dash autoclave. Research on petroleum refining. Address: 1246 Edgewood Dr., Lakeland, Fla. 33803.*

KUENZLER, Edward Julian, Am. ecologist; b. West Palm Beach, Fla., Nov. 11, 1929; s. Edward and Flora (Jeske) K.; B.S., U. Fla., 1951; M.S., U. Ga., 1953, Ph.D., 1959; m. Jutta Gertraud Koslowski, Sept. 4, 1965. Research asst. marine biology Woods Hole (Mass.) Oceanographic Instn., 1959-63, asso. scientist, 1963-65; asso. prof. dept. environmental scis. and engring. U. N.C., Chapel Hill, 1965—. Mem. Am. Soc. Limnology and Oceanography, Ecol. Soc., A.A.A.S. Contbr. articles to tech. jours. Research on natural populations birds, spiders, mussels, nutrient cycles in marine organisms, demonstrated phosphate-repressible phosphatases in marine algae. Home: Route 1, Box 254, Chapel Hill, N.C. 27514.*

KUFFLER, Stephen William, Am. neurophysiologist; b. Tap, Hungary, Aug. 24, 1913; s. William and Elizabeth K.; M.D., U. Vienna, 1937; A.M. (hon.), Harvard, 1959; M.D. (hon.), U. Bern, 1964; m. Phyllis M. Shewcroft, Apr. 25, 1943; children—Suzanne, Damien, Julian, Genie. Came to U. S., 1945, naturalized, 1954. Mem. Kanematsu Inst., Sydney, Australia, 1938-45; Coman fellow U. Chgo., 1946-47; asso. prof. Johns Hopkins Sch. Medicine, Balt., 1947-56, prof. ophthomic physiology and biophysics, 1956-59; prof. neurophysiology Harvard Med. Sch., Boston, 1959-64, Robert Winthrop prof. neurophysiology, 1964-66; Robert Winthrop prof. neurobiology, chmn. dept. neurobiology, 1966—. Non-resident fellow, Salk Inst., La Jolla, Cal., 1966; Harvey lectr., 1959; Ferrier lectr. Royal Soc. of London, 1965. Bd. dirs. Marine Biol. Lab., Woods Hole, Mass. Mem. Nat. Acad. Scis., Am. Acad. Arts and Scis., Physiol. Soc. London, Am. Physiol. Soc. Editor, Jour. Neurophysiology. Research and numerous publs. on orgn. of nervous system, synaptic transmission, sense organ physiology, neuroglia. Home: 9 Surrey Rd., Newton, Mass. 02158. Office: 25 Shattuck St., Boston 02115.*

KUGELMASS, Isaac Newton, Am. pediatrician; b. N.Y.C., June 28, 1896; s. Maurice M. and Sarah (Spirer) K.; B.S., Coll. City N.Y., 1916; M.A., Columbia U., 1917; Ph.D., Johns Hopkins, 1921; M.D., Yale, 1925; Docteur Spécial en Sciences Physiologiques, U. Brussels, 1922; m. Ella H. Fishberg, Aug. 18, 1934. Nutritionist, N.Y. State U. S. Food Adminstrn., 1916-17; sci. lectr. N.Y.C. Bd. Edn., 1916-17; instr. chemistry Columbia, 1917-18; prof. chemistry Howard Coll., 1918-20; cons. chem. warfare service, 1918-20; research asso. pediatrics Yale Med. Sch., 1922-25; dir. pediatric research Fifth Av. Hosp., 1927-35; attending pediatrician French, Riverside, Ruptured and Crippled hosps., 1930-40; chief pediatrician N.Y.C. Children's Hosp., 1940-45; chief Growth Clinic Ruptured and Crippled Hosp., 1940-45; chief immunology clinic French Hosp., 1945-50; dir. pediatrics Broad St. Hosp. and Pan-Am. Clinics, N.Y.C., 1945-50; cons. pediatrician Manhattan Gen., Monmouth Meml., Lyon Meml. hosps., Northwoods Sanatorium, Trondeau Sanatorium, Grand Central Hosp.; cons. N.Y.C. Dept. Health and Hosps., dir. Heckschen Inst. for Child Health. Fellow A.M.A.; mem. Harvey Soc., Royal Soc. Pub. Health, Am. Acad. Mental Deficiency, Soc. Exptl. Medicine and Biology, Soc. Research Child Devel., Am. Pub. Health Assn., Am. Therapeutic Soc., Assn. Am. Tchrs. Diseases Children, Nat. Tb Assn. Author numerous books including Biochemistry of Blood, 1960; Biochemical Diseases, 1962; Clinical Pediatrics, 1964; also articles. Research on buffer mechanisms and ion compositions of blood, diseases of blood, discovery of vitamin P, diseases of newborn and discovery of Kugelmass' sign of latent allergy in infants; nutrition in infancy. Home: 910 Park Av., N.Y.C. 10021. Office: 1060 Park Av., N.Y.C. 10028.*

KÜHL, Heinrich Max Arthur, German biologist; b. Weimar, May 12, 1909; s. Max and Margarete (Friedrich) K.; ed. univs. Kiel, Hamburg; Ph.D.; m. Christa (Schütte), Dec. 14, 1943. Dir. Laboratorium für Bewuchsforschung, Cuxhaven, then Fed. Research Labs. Mem. German Zoology Soc., Internat. Soc. Limnology, Tech. Soc. for Naval Constrn., Senckenbergische naturforschende Gesellschaft, Hamburg Assn. for Natural Sci. Research and publs. on biology of polluted water, water chemistry, marine ecology, aquarium ecosystems, biology and hydrography of estuaries, barnacles. Home: Siedelhof 16, Cuxhave. Office: Bei der Alten Liebe 1, Cuxhaven, W. Germany.

KUHLBRODT, Erich, German meteorologist; b. Berlin, Aug. 22, 1891; s. Gotthard and Pauline (Nitze) K.; Ph.D., U. Gerlin; m. Lucie Raehder, Nov. 2, 1929; children—Dietrich, Adelheid, Eckhard, Sigrun, Rüdiger. Metoerologist maritime obs., Hamburg, Germany, 1920-48, maritime meteorol. office of Hamburg, 1948-56; ret.; part time prof. at large U. Hamburg. Recipient Silver medal Maritime Obs.; meteorol. plaque. Mem. Meteorol. Soc. Hamburg (hon.), German Geophysics Soc., Hamburg Geog. Soc. Address: Haynstrasse 19, Hamburg 20, W. Germany.

KUHLENBECK, Hartwig, neurobiologist; b. Jena, Germany, May 2, 1897; s. Ludwig and Helene (Ayrer) K.; B.A., Domgymnasium, Naumburg, Germany, 1914; Ph.D., U. Jena, Germany, 1920, M.D., 1922; Sc.M.D. (hon.), Woman's Med. Coll. Pa., 1965; m. Ozelia Marguerite Proteau, Jan. 30, 1924. Practice medicine, Mexico City, Mexico, 1923-24; lectr. anatomy, neurology Tokyo (Japan) Imperial U., Japan, 1924-27; asst., privatdozent anatomy U. Breslau, Germany, 1927-33; vis. fellow anatomy U. Pa., 1934-35; prof. anatomy, chmn. dept Woman's Med. Coll. Pa., 1935-63, research prof. anatomy, 1963-66, research prof. neurology, 1966—. Cons. Armed Forces Inst. Pathology, 1947-55; fgn. sci. mem. Max-Planck Inst. for Brain Research, Frankfurt, Germany, 1963—. Fellow A.A.A.S., Coll. Physicians Phila.;

mem. Am. Assn. Anatomists, Am. Assn. Neuropathologist, Assn. Mil. Surgeons U. S., Japan Assn. Anatomists (hon.), Sigma Xi. Author: Lectures on Central Nervous System, 1927; The Human Diencephalon, 1954; Brain and Consciousness, 1957; Mind and Matter, 1961; Central Nervous System of Vertebrates, Vols. 1, 2, 1967. Contbr. numerous articles to sci. jours. Data on origin of cerebral cortex; morphologic evolution of brain; homologies of cell masses in forebrain of vertebrates on the basis of a new topologic notation applicable to all vertebrates from fish to man; topologic significance of morphologic homology concept; founds. of neurol. epistemology; proof that brain-mind problem is logically insolvable; definition of consciousness in terms of a perceptual space-time system; meaning of consciousness in neurol. epistemology. Office: Woman's Med. Coll. Pa., Phila. 19129.*

KUHLMANN, Charles-Frédéric, French chemist; b. Colmar, France, May 22, 1803; studied chemistry under Louis Nicholas Vauquelin; a son, Jules Frédéric. Applied chemist in industry, 1832-54; dir. mint, Lille, France; owner several chem. factories; owner Société anonyme de manufacture des produits chimiques, Lille. Mem. French Acad. Scis., Soc. Agr. Author: Recherches scientifiques et publications diverses, 1827; Cours de géologie; Experiences chimiques et agronomiques, 1847; Application des silicates alcalins et solubles au durcissement des pierres calcaires poreuses, 1855; also articles. Research on composition of dyes, process for preparation of baryta; studies on cristallogenic force, 1864-65; confirmed catalytic oxidation of cyanogen to oxides of nitrogen; discovered several esters. Died Lille, Jan. 27, 1881.

KUHLMANN, Jules Frédéric, French chemist; b. Lille, France, 1841; s. Charles-Frédéric Kuhlmann; dir. factories in Lille; collaborated in discovery of metal thallium. Died 1881.

KUHN, Adam, physician, botanist; b. Germantown, Pa., Nov. 17, 1741; s. Adam Simon and Anna (Schrack) K.; M.D., U. Edinburgh, 1767; m. Elizabeth Hartman, 1780, 2 children. Prof. materia medica and botany Coll. Phila. (now U. Pa.), 1768-89; physician Pa. Hosp., 1775-98; cons. physician Phila. Dispensary, 1786; a founder Coll. Physicians of Phila., 1787, pres., 1808; prof. theory and practice medicine U. of State Pa., 1789, prof. practice of physics, 1792-97; mem. Am. Philos. Soc.; 1st prof. botany in Am. colonies; Kuhnia eupatorioides named after him. Died Phila., July 5, 1817.

KÜHN, Alfred, German biologist; b. Baden-Baden, Germany, Apr. 22, 1885; s. Richard and Josehine (Behaghel) K.; Ph.D., 1938; Dr.Rer.Nat., h.c., Freiburg; Dr.h.c., Oslo, Norway; Dr.Med. h.c., Göttingen.; m. Margaret Geiges, 1914. Asst., Zool. Inst., Freiburg U.; lectr., Freiburg, 1908-18; asst. Zool. Inst., Berlin U., 1918-20; prof. zoology, dir. Zool. Inst., Göttingen (Germany) U., 1920-37; dir. Kaiser-Wilhelm Inst. Biology, Berlin-Dahlem, Germany, 1937-45; dir. Max-Planck Inst. Biology, Tübingen, Germany, 1945-64; prof. zoology, Tübingen, 1945-64. Author: Anleitungen zu tierphysiologischen Grundversuchen, 1917; Die Orientierung der Tiere im Raum, 1919; Morphologie der Tiere in Bildern I, 1921, II, 1926; Grundriss der Vererbungslehre, 3d edit., 1961; (with K. Henke) Genetische und entwicklungsphysiologische Untersuchungen, I-VII, 1929, VIII-XII, 1932, XIII-XIV, 1936; (with K. Grobben) Lehrbuch der Zoologie, 1932; Goethe und die Naturforschung, 1933; (with E. Caspari, E. Plagge) Über hormonale Genwirkungen, 1935; Vorlesungen über Entwicklungsphysiologie, 1955; Versuche zur Entwicklung eines Modells der Genwirkungen, 1956; Über die Wirkungsweise von Erbfaktoren, 1958; Genetisch bedingte Mosaikbildungen, 1959; (with B. Berg), Parabiosen als Mittel zur Aufklärung von Genwirkungen, 1962. Research on animal psychology, genetics, exptl. morphogenesis. Home: Spenmannstr. 34, Tubingen, German Fed. Republic.

KUHN, Eduard Jaroslav, Czechoslovakian physician; b. Bílsko, Czechoslovakia, Sept. 21, 1918; s. Eduard and Anna (Zatloukalová) K.; student Charles U., Prague, Czechoslovakia, 1937-39, M.D., 1946, candidate Scis., 1961, D.Sc., 1963; m. Jarmila Pavlicová, Jan. 23, 1923; children—Anna Jarmila, Eduard Vladimír. Reader internal medicine Charles U., 1948-52; staff Inst. Human Nutrition, Prague, 1953—, sr. research worker, 1963—, chief lab. for energetic metabolism, 1955—; lectr. Postgrad. Med. Inst., Prague, 1955—. Recipient various award for work in nutrition and metabolism J. E. Purkyne Med. Assn., Ministry of Health. Mem. J. E. Purkyne Med. Assn., Internat. Soc. for Biochem. Pharmacology. Author: Priority Follow-up Care of Peptic Ulcer Patients, 1956; Sleep and Metabolism, 1966; also numerous articles. Research on care of peptic ulcer patients, response of body metabolism to various emotional states, physical and mental efficiency. Home: K Novému dvoru 41, Prague 4-Lhotka. Office: Inst. Human Nutrition Prague 4-Krc., Czechoslovakia.*

KÜHN, Heinrich, German mathematician; b. Nov. 10, 1690; prof. math. Gymnasium, Danzig, Poland. Author: Meditationes de origine fontium et aguae putealis . . . , 1741; Tertamen de aequationibus cubicus quibuscunque perfecte resolvendis, 1771. Approached graphical representation of complex numbers. Died Danzig, Oct. 8, 1769.

KUHN, Heinrich Gerhard, physicist; b. Breslau, Germany, Mar. 10, 1904; s. Felix Wilhelm and Martha (Hoppe) K.; Dr.Phil., U. Göttingen, 1926; M.A., Oxford U., 1941; Dr.Sci. (hon.), U. Aux-Marseilles, 1958; m. Marie Bertha Nohl, Sept. 10, 1931; children—Anselm Thomas, Nicholas John. Faculty, U. Göttingen, 1926-33; faculty Oxford U., Eng., 1933—, fellow, tutor Balliol Coll., 1950—, reader in physics, 1955—. Cons. Atomic Energy Research Establishment, Harwell, 1958-59. Recipient Holweck prize French and English phys. socs., 1967. Fellow Royal Soc., 1954, Phys. Soc. London; mem. Am. Phys. Soc., Optics Soc. Am. Author: Atomspektren, 1934; Atomic Spectra, 1962; also numerous articles. Research on atomic and molecular structure; devel. atomic beam absorption technique. Home: 25 Victoria Rd., Oxford, Eng.*

KÜHN, Julius, German agronomist; b. Pulsnitz, Germany, Oct. 23, 1825; prof., dir. Agrl. Inst., U. Halle (Germany). Mem. French Acad. Scis. (corr.). Author: Traité de l'alimentation rationelle des betes bovines. Studies on bovine alimentation, vegetable pathology. Died Halle, Apr. 14, 1910.

KÜHN, Klaus, German biochemist; b. Breslau, Germany (now Wroclaw, Poland), May 1, 1927; s. Georg and Anneliese (Zoche) K.; Dipl. Chem., U. Munich, 1953; Dr.rer.nat., U. Heidelberg, 1955; m. Barbara Bleimund, Mar. 1, 1956; children—Thomas, Sabine, Gabriele. Lectr., U. Heidelberg, Germany, 1960-63, prof., 1963-66; sci. mem. dept. head Max Planck Inst. Protein and Leather Research, Munich, Germany, 1966—. Mem. Soc. Physiol. Chemistry, Soc. German Chemists. Research, numerous publs. on biochemistry of connective tissue, structure and metabolism of collagen, non-collagenous proteins and mucopolysaccharides of connective tissue. Home: 17 Lavendelweg, Munich 8. Office: Max Planck Institute, 46 Schillerstrasse, Munich 15, West Germany.*

KÜHN, Othmar, Austrian paleozoologist; b. Vienna, Nov. 5, 1892; s. Charles and Marie (Herdin) K.; Ph.D., U. Vienna; hon. doctorates univs. Athens, Bucharest; m. Vilma Hickl, Sept. 19, 1919; Dir., Mus. Natural History, Vienna; prof. U. Vienna, became dean, 1957, rector, 1960. Recipient Silver medal U. Jasi Roum, Gold medal U. Addis Ababa. Mem. Austrian, Greek, Serbian acads. sci., Paleontol. Soc. India, Austrian Geol. Soc. Author: Lexique international de stratigraphie, 1962; Fossilium catalogus intern.; Hydrozoa et Rudistae; Handbuch der Palaozologie; many other publs. Research on bryozoa, hydrozoa (corals), triasis, cretacious and tertiary faunas, rudists, ammonites, inocerames. Home: Dornbacherstrasse 65, Vienna 17, Austria.

KUHN, Richard, chemist; b. Vienna, Austria, Dec. 3, 1900; Ph.D., Munich, 1922; Dr.Phil. h.c. Prof., Swiss Tech. High Sch., Zurich, 1926-29; dir. chem. div. Kaiser Wilhelm Inst. for Med. Research, 1929—; faculty Heidelberg, 1926, prof. biochemistry, 1950—; v.p. Max Planck Inst., 1955—. Recipient Nobel Prize, 1938; Schule medal Stockholm Chem. Soc., 1950; Exner medal, 1952. Hon. fellow Indian Chem. Soc.; mem. German Chem. Soc. (pres.), Union Internat. de Chimie (v.p. 1938), Heidelberg Acad. Scis.; hon. mem. N.Y. Acad. Scis., Indian Soc. Biol. Chemistry, Soc. de Chimie Industrielle, London Chem. Soc., Polish Chem. Soc., Indian Nat. Inst. Sci., Verein Österreichische Akad. der Wissenschaften, Japanese Pharm. and Biochem. Socs. Author textbook on enzymes, 1927; also numerous articles. Research on carotinoids, vitamins; synthesized and discovered complete chem. formula for vitamin B6, 1939; synthesized vitamin A, 1937; isolated vitamin B2; discovered method for dissolving symplexes from plants using invert-soaps. Died Heidelberg, Ger., July 31, 1967.

KÜHN, Rudolf, German astronomer; b. Stuttgart, Germany, Jan. 27, 1926; s. Rudolf and Rose (Faber) K.; Ph.D.; m. Ursula Becker, Sept. 1, 1948; children—Christiane, Barbar, Rainer. With U. Munich Obs., 1948-56, Wendelstein Obs., 1950-54. Mem. Astronomische Gesellschaft. Author various books, papers, TV broadcasts. Address: Lamont 9, Munich, West Germany.

KÜHNE, Wilhelm Friedrich (Willy), German physiologist; b. Hamburg, Germany, Mar. 28, 1837; ed. Lüneburg, Germany, Göttingen, Germany, Paris, France; studied under Bernard. Staff, Berlin Path. Lab.; became prof., Heidelberg, Germany, 1871. Fellow Royal Soc., 1892. Author: Lehrbuch der physiologische Chemie, 1866; Über die peripherischen Endorgane der motorischen Nerven, 1862; Untersuchungen über das Protoplasma, 1864. Demonstrated living protoplasm is a liquid, 1864; introduced term enzyme for organic substances activating chem. changes; research on chemistry of digestion, physiology of muscles; isolated trypsin, 1874; proved muscle protein can be coagulated. Died Heidelberg, June 10, 1900.

KUHNHOLTZ-LORDAT, Georges Marie Ernest Frederic, French botanist; b. Motpellier, France, Jan. 8, 1888; s. Gerard and Marthe (Vernazobres) K.-L.; Ph.D., U.Montpelier; children—Herve, Brigitte. Head of works U. Montpellier; prof. botany Montpellier Sch. Agr; prof. Natural History Mus., until 1958; Mem. Agrl. Acad. France. Research and publs. on vegetal pathology, bot. geography, genetics. Home: 5, rue St. Vincent de Paul, Montpellier (Herault), France.

KUHN-SCHNYDER, Emil Eduard, Swiss paleontologist; b. Zurich, Switzerland, Apr. 29, 1905; s. Emil and Berta (Morf) K.-S.; Dipl. Tchr. Biology Swiss Fed. Inst. Tech., Zurich, 1927, Dr.sc. nat., 1932; m. Hanna Schnyder, Oct. 1, 1951. Became tchr. nat. sci. Bremgarten, AG, 1931; named chief asst., zool. mus. U. Zurich, 1940, lectr. comparative anatomy and paleontology, 1947, prof. paleontology, 1955, dean philos. faculty II, 1968, also dir. paleontol. inst. and mus.; prof. paleontology Swiss Fed. Inst. Tech., 1956—. Mem. natural sci. socs. Switzerland, Zurich, Swiss Geol. Soc., Swiss Paeleontol. Socs., Soc. Teilhard de Chardin, Paleontol. Soc. Am., Soc. Natural History Basel (hon.), Soc. Senckenberg Frankfort/Main (corr.), others. Author: (with K Hescheler) Die Tierwelt d. prähist. Siedl. d. Schweiz, 1949; Geschichte d. Wirbeltiere, 1953. Contbr. author: Evolut. d. Org. (G. Heberer), 1967. Research on Pleistocene mammals, Triassic reptiles, excavations in Trias of Monte San Giorgio, Switzerland. Home: 6, Ilgenstrasse, 8032 Zurich, Switzerland.*

KUILMAN, Luppo Willem, Dutch botanist; b. Lobith, Nov. 2, 1902; s. Marten and Maria (van Cuylenburgh) K.; Ph.D., U. Amsterdam (Netherlands); m. J.E.V.D. Berg, Nov. 19, 1930; children—Marten, Geertrui, Maria, Theodorus. With exptl. agronomy sta., Bogor, Java, 1930-49; prof. plant physiology, Bogor, 1948-54; with office of documentation Royal Tropical Inst. Amsterdam, 1954-59; prof. Free U. Amsterdam, 1955—. Mem. Royal Dutch Bot. Assn. Research and publs. on plant physiology of healthy and diseased rice, 1934-42, also on world wide food problems. Home: Koninginneweg 56Hs, Amsterdam Z, Netherlands.

KUIPER, Gerard Peter, astronomer; b. Harencarspel, The Netherlands, Dec. 7, 1905; s. Gerard and Anna (de Vries) K.; B.Sc., U. of Leyden, The Netherlands, 1927, Ph.D., 1933; m. Sarah Parker Fuller, June 20, 1936; children, Paul, Lucy. Came to U. S., 1933, naturalized, 1937. Research asst. astronomy, U. of Leyden, 1928-33; research fellow Lick Obs., U. Cal., 1933-35, research asso., 1935, lectr. Harvard, 1935-36; faculty U. Chgo., 1936-60, dir. of Yerkes and McDonald Obs., 1947-49, 57-60; head Lunar and Planetary Lab., U. Ariz., 1960—, prin. investigator NASA's ranger program. Named comdr. Order Orange Nassau (Netherlands); recipient Janssen Medal, French Astron. Soc.; Rittehouse medal. Mem. Nat. Acad. Scis., Am. Acad. Arts and Scis., Internat. Astron. Union, Am. Astron. Soc., Astron. Soc. Pacific, Royal Astron. Soc. London (asso.), Netherlands Soc. Scis. (fgn. mem.), Royal Netherlands Acad. Scis. (fgn. mem.). Author (with others) 3 atlases of the moon. Editor: The Atmospheres of the Earth and Planets, 1949, 52; The Solar System, 4 vols., 1953-61; Stars and Stellar Systems, 9 vols., 1960—. Home: 721 N. Sawtelle Av. Office: Lunar and Planetary Lab., U. Ariz., Tucson.*

KUIPER, Nicolaas Hendrik, Dutch mathematician; b. Rotterdam, Netherlands, June 28, 1920; s. Nicolaas and Martha (Kalle) K.; ed. Leyden U.; doctorate, 1946; m. Maria Agnete Kramers, July 4, 1947; children—Pieter Nicolaas, Anna Suzanne, Niels Jacob. Tchr. secondary sch., Dordrecht, 1942-47; asst. Inst. for Advanced Study, Princeton, N.J., 1947-49; instr. U. Tech., Delft, 1949-50; prof. U. Agrl. Science, Wageningen, 1950-62; apptd. prof. U. Amsterdam, 1962. Mem. Royal Netherlands Acad. Sci., Amsterdam, 1965. Author: Linear Algebra, 1962; also articles. Research on contractibility of unitary group of Hilbert space, also differential map preserving length of curves of surfaces in Euclidean 3-space. Home: 376 Hollandseweg, Wageningen, Netherlands. Office: 121 Nieuwe Achtergracht, Amsterdam C., Netherlands.*

KUIPERS, Lauwerens, mathematician; b. Dordrecht, Netherlands, Jan. 3, 1909; s. Hendrikus J. and Teuna (Swets) K.; Ph.D., Vrije Universiteit, Amsterdam, 1947; m. Francina Okker, July 11, 1934; children—Thea (Mrs. Marinus Slooter), Francina Cornelia (Mrs. Kees Ouwehand), Hendrik, Laurens. Prof. math. U. Indonesia, 1949-55; prof. math. U. Redlands, 1955-56, Technische Hogeschool, Delft, Netherlands, 1956-66; prof. math. So. Ill. U., Carbondale, 1966—. Mem. Malayan, Austrian, Am. math. socs., Société Math. de Belgique, Wisk. Genootsch. A'dam. Author: (with R. Timman) Analysis I, II Handboek der Wiskunde. Contbr. numerous articles to sci. jours. Research in theory of uniform distbn. modulol. Address: So. Ill. U., Carbondale, Ill. 62901.*

KUIVILA, Henry Gabriel, Am. chemist; b. Fairport Harbor, O., Sept. 17, 1917; s. Matt and Saima (Kujala) K.; B.S., Ohio State U., 1942, M.A., 1944; Ph.D., Harvard, 1948; m. Nancy M. Corn, June 13, 1943; children—Henry Gabriel, Nancy J., Ronald J. Faculty, U. N.H., 1948-64; prof., chmn. dept. chemistry State U. N.Y., Albany, 1964—. NSF Sr. fellow, Guggenheim fellow, 1959. Mem. Am. Chem. Soc., A.A.A.S., Am. Assn. U. Profs., N.Y. Acad. Scis. Contbr. numerous articles to profl. jours. Research in mechanisms of reactions, particularly of organoboron and organotin compounds, with emphasis on electrophilic and free radical reactions. Home: 36 E. Bayberry Rd., Glenmont, N.Y. Office: 1223 Western Av., Albany, N.Y. 12203.*

KUJALOVA, Vera, Czechoslovakian biologist; b. Prague, Czechoslovakia, June 16, 1925; d. Francis Kujal and Jarmila (Jiránková) Kujalová; Mag.pharm., Charles U., Prague, 1948, Dr.rer.nat., 1951, Ph.D., 1962. Staff dept. biol. control medicaments, Nat. Inst. Health, Prague, 1952-56; research worker physiol. dept. Inst. Nutrition, Prague, 1956—. Recipient prize sci. council Ministry of Health, 1963. Mem. J. E. Purkyne's Med. Soc., Physiol. Sect. Author: (with P. Fábry, R. Petrásek, E. Holecková), Adaptation to Changed Pattern of Food Intake, 1962; also articles. Research on adaptation phenomena, intestinal resorption, metabolism of adipose tissue. Home: 3/244 Vojtesská, Prague. Office: 801 Budejovická, Prague, Czechoslovakia.*

KUKARKIN, Boris Vasilevich, Russian astronomer; b. Oct. 30, 1909; D. Physico-Math. Sci. Asso., Sternberg Astron. Inst., 1932—, dir., 1952-56; prof. Moscow U., 1951—. Dep. chmn. astron. council USSR Acad. Sci., 1957—. Mem. Internat. Astron. Union (v.p. 1955-61). Author: Studies of the Structure and Development of Stellar Systems Based on the Study of Variable Stars, 1949; General Catalogues of Variable Stars, 1948, 58; co-author: A General Catalogue of Variable Stars: Supplements, 1949-68. Editor (bull.) Variable Stars, 1928—. Research and publs. on variable stars, structure and devel. of stellar systems. Address: Moscow University, Sternberg State Astron. Inst., Moscow V-234, USSR.*

KUKKAMÄKI, Tauno Johannes, Finnish geodesist; b. Heinola, Finland, Oct. 11, 1909; s. Juho and Aina (Roni) K.; M.S., Turku (Finland) U., 1930, Ph.D., 1933; m. Anna-Liisa Järvinen, Sept. 13, 1936; children—Tuuli, Pilvi. State geodesist Finnish Geodetic Inst., Helsinki, 1935-55, acting dir., 1955-63, dir., 1963—; research asso. Ohio State U., 1952-54; leader sci. expdns. in W. Africa, Argentina, Greenland, Holland. Decorated Finnish Cross of Independence 4 class and 3 class with oak leaves, Comdr. Lion of Finland. Mem. Geog. Soc. Finland (past pres.), Internat. Assn. Geodesy (past v.p.). Author: Untersuchungen über die Meterendmasse aus geschmolzenem Quarz, 1933; also numerous articles. Research on atmospheric refraction, measurement geodetic standard base lines with Väisälä interference comparator. Home: Runebergink. 40 A 13. Office: Hämeentie 31, Helsinki 50, Finland.*

KULEBAKIN, Viktor Sergeevich, Russian elec. engr.; b. Oct. 18, 1891; grad. Moscow Higher Tech. Sch., 1914; D.Sc.; became mem. faculty Moscow Higher Tech. Sch., 1917, prof., 1921; named prof. Air Force Engring. Acad., 1923; organizer All-Union Electrotech. Inst., Moscow Energy Inst., Inst. Automatics and Telemechanics of Acad. Scis. Recipient Stalin prize, 1950. Corr. mem. Acad. Scis. Author: Testing of Electric Machines and Transformers, 2d edit., 1935; Electric Apparatus, 1932; (with A. M. Senkevich) Electrical Equipment of Airplanes, Part I, 1945; (with L. M. Snideev, V. D. Nagorskii) Electrification of Airplanes, 1952. Research on electronic computers, automatic regulation and design of regulators, elec. ignition of air craft engines; laid found. for elec. engring. in Russian aviation; contbd. to electric locomotion in mines. Home: B. Khariton'evskii p. 12/1, Moscow. Office: Air Force Engring. Acad., Moscow, USSR.

KULKA, Johannes Peter, physician; b. Vienna, Austria, Feb. 7, 1921; s. Ernest Walter and Anna (Jolles) K.; came to U. S., 1933, naturalized, 1938; A.B., Cornell U., 1941; M.D., Johns Hopkins, 1944. Faculty, Harvard Med. Sch., 1947—, asst. clin. prof. pathology, 1961-67, asso. clin. prof., 1967—; pathologist Robert B. Brigham Hosp., Boston, 1955—; asso. pathol. staff Peter Bent Brigham Hosp., Boston, 1961—. Diplomate Am. Bd. Pathology. Mem. Am. Soc. Pathology and Bacteriology, Am. Soc. Exptl. Pathology, Internat. Acad. Pathology, Am. Coll. Pathology, Soc. Investigative Dermatology, Am. Rheumatism Assn., Soc. for Exptl. Biology and Medicine, Microcirculatory Soc. Author: (with Leon Sokoloff, Lent C. Johnson, Roger Terry) Syllabus to set of Lantern Slides Illustrating Pathology of Rheumatic Diseases, 1956. Editorial bd. Arthritis and Rheumatism, 1960—. Systematic morphogenetic study of earliest obtainable lesions in rheumatoid arthritis; devel. quick-freezing technique for demonstrating terminal vascular bed in a life-like state; exptl. reproduction of rheumatoid granulomas in rats; research in cold injury and rheumatoid diseases. Home: 358 Harvard St., Cambridge, Mass. 02138. Office: 125 Parker Hill Av., Boston 02120.*

KULP, John Laurence, Am. chemist; b. Trenton, N.J., Feb. 11, 1921; s. John and Helen (Gill) K.; B.S., Wheaton Coll., 1942; M.S., Ohio State U., 1943; M.A., Princeton, 1944, Ph.D., 1945; m. Helen Masterson, June 17, 1944; children—Ruth, Ellen, John, James. Chemist, Nat. Aniline div. Allied Chem. & Dye Corp., 1942; phys. chemist Manhattan Project, Princeton, 1944-45; Kemp fellow Columbia, 1945-46, fellow, 1946-47, faculty, 1947-65, prof. geochemistry, 1958-65, adj. prof., 1965-67; pres. Isotopes, (a Teledyne Co.), Westwood, N.J., 1964—. NSF Sr. Postdoctoral fellow, 1958. Fellow Geol. Soc. Am., Mineral. Soc. Am., A.A.A.S. (Cleveland award 1951); mem. Am. Chem. Soc., Phys. Soc., Am. Assn. Petroleum Geologists, Am. Geophys. Union. Research, numerous publs. on isotope geology, radiochemistry, geochemistry of trace elements, nuclear geophysics, age determination of geologic materials, chemistry and structure of clay minerals, adsorption phenomena, geochemistry of ore deposits, world-wide fallout of atomic debris, effects of nuclear weapons, differential thermal analysis. Home: 40 Valley View Av., Ridgewood, N.J. 07656. Office: 50 Van Buren Av., Westwood, N.J. 07675*

KULPE, Oswald, German psychologist; b. Candau in Courland, Latvia, 1862; ed. gymnasium at Libau; studied under Wundt, Leipzig, 1881, history at Berlin, later Dorpat, studied under C. E. Muller, Göttingen, 1883. Returned to Wundt at Leipzig, 1886, became dozent, 1888, then Wundt's 2d asst., apptd. extraordinary prof., Leipzig, also prof. Wurzburg, 1894, Bonn, 1909. Author: Grundriss der Psychology (among first text-books on exptl. psychology, states basic concepts of the new psychology and defines psychology as sci. of facts of experience), 1893; Einleitung in die Philosophie, 1895; die Philosophie der Gegenwart in Deutschland, 1902. Among chief representative of the new psychology; responsible for movement (now known as doctrine of Wurzburg sch.) which encouraged shift in emphasis in psychology from consciousness to conduct; made exptl. studies on attention, abstraction, and psychophysics. Died 1915.

KUMAHARA, Yuichi, Japanese physician; b. Miyazaki, Japan, Jan. 2, 1925; s. Yuzo and Masae (Goto) K.; M.D., Osaka (Japan) U., 1950, Ph.D., 1957; m. Sayuri Kumahara, Nov. 25, 1954; children—Chizuko, Hideo. Clin. asso. Osaka U. Hosp., 1957-59, asso. dir. Central Lab. for Clin. Investigation, 1965—; research asso. Harvard Med. Sch., 1959-60; faculty Osaka U. Med. Sch., 1960—, asso. prof. medicine, 1965—. Mem. Japanese Soc. Internal Medicine, Japanese Endoclinological Soc., Japanese Soc. Clin. Pathology. Author: Current Topics in Thyroid Research, 1965; also reports, other books. Research on pathogenesis of Graves disease; discovered abnormal thyroid stimurator in pituitaries of Graves' disease cases. Home: 1-6, Yamanoue-takara, Toyonaka, Osaka-fu, Japan. Office: 3, Dojima-hama, Fukushima-ku, Osaka, Japan.*

KUMAKAWA, Muneo, Japanese physician; b. 1858; adopted by Soetsu Kamakawa; grad. Tokyo U., 1883; studied biochemistry Berlin U.; became prof. Tokyo U., 1891, dean Med. Faculty, 1917. Japanese del. World Med. Conf., Madrid, Spain, 1903. Mem. Imperial Acad. Research on albumen requirements in food, origin of chemically pure vitamin B, determination of fat generation by albumen. Died 1918.

KUMATE, Jesus, Mexican physician; b. Mazatlán, Sin., Mexico, Nov. 12, 1924; s. Efrén and Josefina (Rodriguez) K.; B.Sc., Mil. Med. Sch., 1946; M.D., Polytechnic Inst., Mexico, 1948-50, Ph.D. in Biochemistry, 1963; m. Bertha Guerra, Mar. 2, 1957. Acad. staff Children's Hosp., Mexico, 1953-66, research coordinator, 1961—; faculty Mil. Med. Sch., Mexico, 1949-54, asso. prof. biochemistry, 1952-54; faculty Mexico Nat. U., Mexico City, 1955—, asso. prof. infectious diseases, 1960—. Recipient F. Gomez award Child Hosp., 1965. Mem. Nat. Acad. Medicine (Carnot prize 1962), Nat. Acad. Pediatrics. Research, numerous publs. on mechanisms of pathogenicity in gram-negative bacteria, especially in diahrrhea of newborns and infants, typhoid fever of children, mechanisms of protein transfer across human placenta. Home: 64 Macuiltepec, Mexico D.F. 21. Office: 162 Márquez, Mexico D.F. 7, Mexico.*

KUMAZAWA, Zen-an, Japanese chemist; b. Edo, Japan, 1845; s. Ryotai Kumazawa; studied under Bankei Otsuki; studied medicine, Germany, beginning in 1870; became asst. prof. Tokyo (Japan) U., 1877; became chief engr. Osaka Cement Co., 1886; became instr. Osaka Kyoritsu Med. Sch., 1890; chief engr. Osaka Ryuso Co., Kansai dist.; instr. Osaka Higher Tech. Sch. Author: Methods of Chemical Analysis; Common Mineralogy; Survey on Physiography. Studies in devel. biochromate kalium; improved manufacture of sulphate. Died 1906.

KUMERLOEVE, Richard Arthur Hans, German zoologist; b. Liepzig, Germany, Sept. 5, 1903; s. Arthur and Palma (Schenk) K.; Ph.D., U. Leipzig; m. Gertraude Doller, Mar. 30, 1935; 1 son, Arnd. Dir. nat. zoology and ethnography museums, Dresden, Germany; dir. Zoology Inst., Dresden; 1st dir. Vienna (Austria) Nat. Sci. Museums; 8 expdns. to Middle and Nr. East; researcher Am. U. of Beirut (Lebanon); African expdns. for protection of nature. Mem. sci. assns. in Germany, Turkey, Hungary. Author: Zur Kenntnis der Avifauna Kleinasiens, 1962; Notes on the Birds of the Lebanese Republic, 1962; Vergleichende Untersuchunger über das Gonadensystem weiblicher Vögel, 1930-31; various other works. Home: Hubert Teissner-strasse 7, 8032, München-Gräfelfing, West Germany.

KUMINS, Charles Arthur, Am. chemist; b. N.Y.C., Jan. 13, 1915; s. Louis and Francis (Negrem) K.; B.S., Coll. City N.Y., 1936; M.S., Poly. Inst. Bklyn., 1941; m. Annabelle Ringgold, June 15, 1946; children—Lawrence Charles, Noel, Douglas Paul. Chemist titanium div. research lab. Nat. Lead Co., 1937-41; group leader Wyandotte Chem. Co., 1941-42; asst. research dir. Inter-chm. Corp. Central Research Lab., Clifton, N.J.; now dir. research Bruning div. Addressograph-Multigraph Corp. Chmn., Am. Chem. Soc. Symposia, 1963-65. Recipient 1st Prize, Roon Found. award, 1965. Mem. A.A.A.S., Am. Chem. Soc. Contbr. articles to profl. jours., chpts. to books. Patentee in field. Discovered mandelic acid as spl. reagent for zirconium analysis; research on long range interactions between crystallites and polymers, diffusion and transport through polymers, free volume theories, new polymers of iron molecular sieves and zeolites. Home: Butternut Rd., Barrington Hills, Ill. 60010. Office: 1800 W. Central Rd., Mt. Prospect, Ill. 60056.*

KUMLER, Warren Donald, Am. chemist; b. Seven Mile, O., Feb. 18, 1905; s. Charles Augustus and Maude (Yingling) K.; B.S., Antioch Coll., 1929, M.S., 1930; Ph.D., U. Cal. at Berkeley, 1934; m. Alice McClintock, July 28, 1927; children—Archie Ann (Mrs. Herbert W. Suelzle), Joan Mary. Instr., Deep Springs (Cal.) Coll., 1929, dean, 1930-32; faculty Sch. Pharmacy, U. Cal. at San Francisco, 1935—, prof., 1950—, chmn. dept. pharm. chemistry, 1959-65, associate dean, 1965—; vis. prof. Cal. Inst. of Tech., 1952, Imperial Coll., U. London, 1959; Knapp vis. prof., research lectr. U. Wis., 1958; research Oxford U., 1958; cons. OSRD, 1943-46; mem. panel NIH. Fellow N.Y. Acad. Scis., A.A.A.S.; mem. Am. Chem. Soc., Am. Pharm. Assn., Sigma Xi, Rho Chi, Phi Delta Chi, Contbg. Author: Textbook of Organic Medicinal and Pharmaceutical Chemistry, 1949-61. Research and publs. on dipole moments, spectra, dissociation constants, molecular structural sunscreens, pharm. chemistry. Home: 33 Teresita Blvd., San Francisco 94217.*

KUMM, H. Karl William, geographer; b. Hanover, Germany, Oct. 19, 1874; s. H. William and M. W. (Kistenbrugge) K.; grad. Gymnasium Osterode, 1894; studied Harley Coll., London, 1896-98, Heidelberg, 1900, Jena, 1902-03; Ph.D., U. Freiburg, 1903; also studied Cambridge and Paris, and Arabic in Egypt, Hausa in Tripoli; m. Lucy Guinness, 1900 (dec. 1906); children—Henry W., Karl Grattan; m. 2d, Frances Gertrude Cato; children—Lucy Gertrude, Frederick John. Began as explorer in Africa, 1899; first white man to traverse the north-central African divide between Congo Shari and the Nile; founded missions, schs., hosps.; founder and mng. dir. Bd. for Medical Edn. and Research in Africa, also of Sudan United Mission (22 denominations); made maps for Peace Conf., Paris; later resided in Pacific Beach, Cal. Author: The Lands of Ethiopia, 1910; African Missionary Heroes and Heroines, 1917. Engaged in hort. experiments with Passifloraceae in Southern Cal. Died Aug. 22, 1930.

KUMM, Henry William, physician; b. Wiesbaden, Germany, Feb. 26, 1901; s. H. Karl Wilhelm and Lucy (Guinness) K.; S.B., Haverford Coll., 1921; M.D., Johns Hopkins, 1925, Dr.P.H., 1928; diploma in tropical medicine and hygiene, Conjoint Bd. Eng., 1927; m. Annie Joyce Beale, June 15, 1928; children—William Howard, Joceline (Mrs. Charles E. Alexander), Frederick Guinness. Came to U. S., 1915, naturalized, 1945. Staff mem. Internat. health div. Rockefeller Found., 1928-50, rep. Cost Rica, C.Am., 1939-40, El Salvador, 1941-42, Panama, 1942-43, Colombia, 1943-44, Italy, 1944-45, dir. Yellow Fever Lab., Rio de Janeiro, Brazil, 1945-50; asst. dir. research Nat. Found. for Infantile Paralysis, 1951-53, dir., 1953-58; asso. dir. for med. and natural scis. Rockefeller Found., N.Y.C., 1959-64. Cons. malaria control Surgeon Gen., Allied Control Commn., Italy, 1944-45. Decorated Order So. Cross, (Brazil); recipient Duncan medal, Lalcaca medal London Sch. Hygiene and Tropical Medicine, 1927. Mem. Royal, Am. socs. tropical medicine and hygiene, A.A.A.S., Am. Pub. Health Assn., Am. Acad. Preventive Medicine. Contbr. numerous articles on investigation and control yellow fever, malaria, yaws, poliomyelitis, insect vectors yellow fever and malaria to tech. jours. Home: The Rafters, Chocorua, N.H. 03817.*

KUMMELL, Hermann, German surgeon; b. Korbach, Germany, 1852; discovered Kummel-Verneuil disease caused by fracture of vertebra, 1891; one of 1st to practice appendectomies; 1st to apply X-rays to treatment of lupis vulgaris. Died Hamburg, Germany, 1937.

KUMMER, Ernst Eduard, German mathematician; b. Sorau, Germany, Jan. 29, 1810; Ph.D., U. Halle, 1828-31; prof. math. U. Breslau (now Wroclaw, Poland), 1842-55, U. Berlin, 1856-84; prof. Berlin War Coll. Prof., Liegnitz Gymnasium, 1832-42; Fellow Royal Soc., 1863; mem. French Acad. Scis. (Grand prize 1857), Royal Berlin Acad., Acad. Scis. Berlin. Author: De numeris complexis qui unitatis radicibus et numeris integris realibus constant, 1844. Research on hypergeometric series and transformations; invented Kummer's surfaces in geometry; originated system of ideal numbers leading to devel. field theory numbers. Died Berlin, May 14, 1893.

KUMMEROW, Fred August, food chemist; b. Berlin, Germany, Oct. 4, 1914; s. August E. and Helene (Rieck) K.; came to U. S., 1923, naturalized 1932; B.S. in Chemistry, U. Wis., 1939, M.S. in Biochemistry, 1941, Ph.D., 1943; m. Amy L. Hildebrand, June

24, 1942; children—Max F., Jean M., Kay M. Mem. faculty Clemson (S.C.) Coll., 1943-45, Kans. State U., Manhattan, 1945-50; prof. food chemistry U. Ill., Urbana, 1950—. Mem. Am. Heart Assn., Soc. Exptl. Biology and Medicine, Am. Inst. Nutrition, Am. Soc. Biol. Chemists, Am. Chem. Soc., A.A.A.S., Gerontological Soc., Am. Oil Chemists Soc., Am. Dairy Sci. Assn., Am. Poultry Sci. Assn., Inst. Food Technology, Phi Kappa Phi. Author: The Metabolism of Lipids as Related to Atherosclerosis, 1965, also articles. Mem. editorial bd. Jour. Dairy Sci., 1958—. Research on nutritive values of heated fats; relationship between fats and serum cholesterol, linoleic acid metabolism, fatty acid derivatives of vitamin B6. Home: 28 Montclair Rd., Urbana, Ill. 61801.*

KUN, Ernest, chemist; b. Sopron, Hungary, Oct. 22, 1919; M.D., U. Budapest, 1943. Came to U. S., 1946, naturalized, 1952. Spl. research fellow USPHS, U. Wis., 1953-55, asst. prof. enzyme chemistry, 1955-56; faculty U. Cal., San Francisco, 1956—, prof. biochemistry, pharmacology, sr. mem. Cardiovascular Research Inst., 1965—. Established investigator Am. Heart Assn., 1956; cons. NSF. Recipient Research Career award USPHS, 1964—. Fellow Brit. Biochem. Soc.; mem. A.A.A.S., Am. Soc. Biol. Chemists (Travel award 1966), Am. Soc. Pharmacology and Exptl. Therapeutics, Am. Chem. Soc., Soc. Biol. Medicine, Sigma Xi. Editorial bd. Molecular Pharmacology. Research, numerous publs. on enzyme chemistry, study of structure, function and control of enzymes, synthesis and mechanism of action of fluorine containing specific enzyme inhibitors; discovered glycolic oxidase, thiol pyruvate transsulfurase, mitochondrial malate dehydrogenase. Home: 95 Sunview Terrace, San Francisco.*

KUNA, Martin, chemist; b. Gr. Schutzen, Czechoslovakia, May 4, 1909; s. Samuel and Anna (Jurka) K.; came to U. S., 1913, naturalized, 1921; B.S., N.Y. U., 1933, M.S., 1937, Ph.D., 1949; m. Esther Ruth Ewart, Oct. 28, 1939; children—Ann Elizabeth, Richard Martin. With Rockefeller Inst. for Med. Research, N.Y.C., 1926-40; research chemist Merck & Co., Inc., Rahway, N.J., 1941-45; sr. research investigator Wyeth Inst. in Phila., 1945-47; sr. biochemist Oak Ridge Nat. Lab., 1947-49; research asso. Sharp & Dohme, Inc., Glenolden, Pa., 1949-51; dir. biol. research Nepera Chem. Co., Yonkers, N.Y., 1951; prin. research investigator Bristol-Myers Products, Bristol-Myers Co., Hillside, N.J., 1952—. Faculty, U. Tenn., Knoxville, 1948; lectr. biology N.Y. U., 1962-63. Fellow N.Y. Acad. Scis.; mem. Am. Chem. Soc., A.A.A.S., Electrochem. Soc., Faraday Soc., Soc. Cosmetic Chemists, Internat. Assn. Dental Research, Sigma Xi. Research and publs. on sugars, amino acids, antibiotics, phys. measurements, intermediary metabolism, radiochemistry, nucleid acids, amidines, analgesics. Home: 442 Otisco Dr., Westfield, N.J. 07090. Office: 225 Long Av., Hillside, N.J. 07207.*

KUNCKEL, Johann (Kunkel, Kunkel von Löwenstern), German chemist; b. Hutten, Germany, 1630; chemist, pharmacist for Dukes Charles and Henry of Lauenbourg, 1654; became dir. lab. and manufacture of glass for Elector of Brandebourg, Berlin, Germany, 1670; named adviser to Royal Mines, Stockholm, Sweden, 1693; prof. U. Wittenberg (Germany). Author: Nützliche Observationes oder Ammerckungen, 1676; Chymische Ammerckungen, 1677; Collegium Physico-Chymicum Experimentale, 1716; Über Curiose Chymische Tractätlein, 1721; Ars vitravia, 1679; Laboratorium Chemicum, 1716. Rediscovered process for preparation of phosphorus (previously used by Brandit); studied effect of light on plants, artificial rubies, compositions of salts, volatile oils, fermentation and putrification; improved process of glassmaking; recognized that air is necessary for fire. Died Stockholm, 1703.

KUNCKEL D'HERCULAIS, Philippe-Alexandre-Jules, French naturalist; b. Paris, 1843; asst. Mus. Paris; founder lab. for study, Buenos Aires, Argentina. Author: Recherches sur le développement et l'organisation des volucelles, 1875 (prize by Acad. Scis.); Études sur les acridiens et leurs migrations en Algérie, 1895. Studied devel. of two-winged insects, anatomy of insects, especially evolution of crickets and their methods of combat. Died Conflans-Sainte-Honorine, France, 1918.

KUNDRAT, Hans, Austrian physician; b. Vienna, Austria, Oct. 6, 1845; M.D., U. Vienna, 1868; became asst. to Rokitansky, 1868; named prof. path. anatomy U. Graz (Austria), 1877, U. Vienna, 1882. Author: Die Selbstverdauungsprocesse der Magenschleimhaut, 1877; Arhinencephalie als typische Art von Missbildung, 1882; Die Porencephalia, 1882. Studies on malignant diseases of lymphoid tissue; described lymphosarcoma (Kundrat's disease), 1893. Died Apr. 25, 1893.

KUNDT, August (Adolph Eduard Eberhard), German physicist; b. Schwerin, Germany, Nov. 18, 1839; ed. U. Berlin; prof. physics Strasbourg, France, 1872, Berlin, Germany, 1888, Zurich (Switzerland) Poly., 1867. Author: Vorlesungen über Experimentalphysik, 1903. Research on gases, including heat conductivity and friction, rotation of plane of polarization; studies in dispersion of light, optical properties of metals; invented Kundt's tube for determination of velocity of sound in gases. Died Israelsdorf, Germany, May 21, 1894.

KUNDU, Mukul Ranjan, Indian physicist; b. Calcutta, India, Feb. 10, 1930; s. M. L. and Monoroma Kundu; B.Sc. with honors, Calcutta U., 1945, M.Sc., 1951; Sc.D., U. Paris, 1957; m. Ranu Paul, Sept. 9, 1958; children—Krishna, Rina. Research asst. Calcutta U., 1951-54; French govt. scholar Paris Obs., 1954-56; research asso. Paris and Meudon Observatories, 1956-58; sr. research fellow Nat. Phys. Lab., New Delhi, 1958-59; asso. research physicist U. Mich., Ann Arbor, 1959-63; asso. prof. astronomy Cornell U., Ithaca, N.Y., 1961-65; asso. prof. physics Tata Inst. Fundamental Research, Bombay, 1965—. Fellow Royal Astron. Soc.; mem. Inst. Elec. Engrs., Internat. Astron. Union, Am. Astron. Soc., Sigma Xi. Author: Solar Radio Astronomy, 1965; also articles. Research in radio astronomy with particular reference to sun and sun-earth relationships. Home: 41 Kenilworth, Peddar Rd., Bombay 26. Office: Tata Inst. Fundamental Research, Bombay 5, India.*

KUNIEDA, Hiroshi, Japanese botanist; b. Shizuoka Prefecture, Japan, 1888; grad. Tokyo U., 1914; D.Agr. Asst. prof., prof. Tokyo U., ret., 1948; prof. Aoyama Gakuin U., Tokyo. Mem. Bot. Assn., Japan Marine Assn. Research and publs. on aquatic plants, seaweed. Died Tokyo, Feb. 13, 1954.

KUNKEL, Hans Adam, German biophysicist; b. Berlin, Germany, Nov. 16, 1910; s. Georg and Irmgard (Oehlrich) K.; student U. Königsberg (Germany), 1931-33; Dipl.Phys., U. Berlin, 1946; Dr.rer.nat., U. Heidelberg (Germany), 1958; m. Hilde Sachter, Feb. 8, 1940; children—Hans Peter, Angela. Asst. physicist Gynecol. Hosp., U. Göttingen, 1950; head radiobiol. dept. Gynecol. Hosp., U. Hamburg (Germany), 1951-62, dir. Radiobiol. Inst., prof. biophysics, 1962—. Mem. German Soc. Biophysics (pres. 1967—), Deutsche Physikalische Gesellschaft, European Soc. for Radiation Biology. Research and numerous publs. on application of radioisotopes in medicine and biology, dosimetry and RBE of high energy electrons, betatron-therapy, genetics effects of ionizing radiations, chem. radioprotection. Home: 20 Lenhartzstrasse. Hamburg 20, Germany.*

KUNKEL, Johann, see Kunckel, Johann.

KUNKEL, Robert, Am. horticulturist; b. Holladay, Utah, Dec. 7, 1912; s. George Frank and Margarita (Pardonner) K.; B.S., Utah State Agrl. Coll., 1937; Ph.D., Cornell U., 1945; m. Norma Mackay, June 28, 1941; children—Douglas Frank, Maeva, Ralph Mackay, George Robert, Charlotte. Prof., Colo. State U., Ft. Collins, 1945-56; acting prof. Cornell U., Ithaca, N.Y., 1956-57; horticulturist Wash. State U., Pullman, 1957—. Fellow A.A.A.S.; mem. Am. Soc. Hort. Sci., Am. Inst. Biol. Sci., Am. Potato Assn., Western Soc. Soil Sci., Sigma Xi, Alpha Zeta. Research and publs. on nutritional factors affecting processing quality and physiol. diseases of potatoes. Home: 311 Fountain St., Pullman, Wash. 99163.*

KUNKEL VON LÖWENSTERN, see Kunckel, Johann.

KUNN, Richard, chemist; b. Vienna, Austria, Mar. 12, 1900; s. Richard and Angelika (Rodler) K.; student U. Vienna; Promotion, U. Munich, 1922; m. Daisy Hartmann, Dec. 18, 1928; children—Daisy, Peter, Hans-Jürg, Heide (Mrs. Christoph Kaempf), Elfriede (Mrs. Schindler). Prof., U. Zürich (Switzerland), 1926-29, ETH, 1929-50; prof. U. Heidelberg (Germany) U., 1950—; dir. Max-Planck-Inst. for Med. Research, Heidelberg, 1928—. Recipient Nobel prize, 1939. Research, numerous publs. on natural compounds, theoretical problems of organic chemistry, including polyene, kumulene, carotine, lactoflavin (vitamin B2), adermin (vitamin B6). Home: 41 Wilckensstr. 69, Heidelberg, Baden, Germany.*

KUNO, Hisashi, volcanologist; b. Tokyo, Japan, Jan. 7, 1910; s. Kamenosuke Kuno and Tome Yoshikawa; student U. Tokyo Geol. Inst., 1929-32, D.Sc., 1948; m. Kimiko Kuno, Nov. 26, 1940; children—Takashi, Shizuko. Asst. U. Tokyo, 1933-39, asst. prof., 1939-55, prof., 1955—; internat. vis. sci. Am. Geol. Inst., 1961; vis. prof. U. Minn., 1964. Recipient Japan Acad. prize, 1954. Fellow (hon.) Geol. Soc. Am.; fellow Am. Geophys. Union, Mineral. Soc. Am.; mem. Internat. Assn. Volcanology (pres.), Volcanological Soc. Japan (pres.), Geol. Soc. Japan, Geol. Soc. London, Nat. Acad. Sci. (fgn. asso.). Spl. study volcanoes of Japan and Hawaii. Home: 1-129, Seki-machi, Nerima-ku, Tokyo. Office: Geological Inst., U. of Tokyo, Tokyo, Japan.

KUNRATH, Conrad, see Khunrath, Heinrich.

KUNTH, Karl Sigismund, German botanist; b. Leipzig, Germany, June 18, 1788; went to Paris, 1813; became prof. U. Berlin, 1819; vice dir. Berlin Bot. Gardens. Mem. French Acad. Scis. Author: (with Humbolt) Nova genera et species plantarum, 7 vols., 1815-28; Enumeratio plantarum omnium hucusque cognitarum secundum familias naturales disposita, 5 vols., 1833-50. Classification and description of plants collected by Humbolt and Bonpland; studied monocotyledons. Died Berlin, Mar. 22, 1850.

KUNTZ, Albert, Am. anatomist; b. Batesville, Ind., Mar. 19, 1879; s. Andrew and Barbara (Butz) K.; A.B., Morningside Coll., Ia., 1904; Ph.D., State U. Ia., 1910; M.D., St. Louis U., 1918; m. Emma S. Magdsick, Aug. 28, 1912; 1 dau., Elizabeth Louise (Mrs. R. Hollis Hamstra). Prof. sci. Charles City (Ia.) Coll., 1905-08; fellow State Univ. of Ia., 1909-11, instr. in animal biology, 1911-13; asst. prof. anatomy St. Louis U. Sch. of Medicine, 1913-16, asso. prof., 1916-19, prof. anatomy, 1919-30, prof. microanatomy, dir. dept., 1930-46, prof. anatomy, dir. dept., from 1946. Mem. A.A.A.S., Am. Soc. Zoölogists, Am. Assn. Anatomists, Soc. Exptl. Biology and Medicine, Am. Neurol. Soc. (asso.), Sigma Xi, Alpha Omega Alpha, Pi Gamma Mu. Author: The Autonomic Nervous System, 1929, 4th edit., 1953; Neuro-Anatomy, 1931, 5th edit., 1950; The Neuroanatomic Basis of Surgery of the Autonomic Nervous System, 1949; Visceral Innervation and Its Relation to Personality, 1951; also articles. Died Jan. 19, 1957.

KUNTZ, James Eugene, Am. plant pathologist; b. Leipsic, Ohio, Aug. 14, 1919; s. Edward C. and Bessie (Sherrard) K.; B.A., Ohio Wesleyan U., 1941; M.S., U. Wis., 1942, Ph.D., 1945; m. Helen H. Hartley, Dec. 28, 1942; children—James Edward, David William, Patricia Susan. Plant pathologist, plant breeder Wis. Seed Co., Racine, 1945-46; plant pathologist U. Wis., Madison, 1946—. Mem. Am. Phytopath. Soc., Soc. Am. Foresters, A.A.A.S., Phi Beta Kappa, Sigma Xi. Research and publs. on diseases of forest trees and their control, including selection and breeding for disease resistance; vascular wilt diseases, stem cankers, decay, use of herbicides in forest practice. Home: 905 Harrison St., Madison, Wis. 53706.*

KUNTZ, Robert Elroy, Am. parasitologist; b. Lawton, Okla., Feb. 23, 1916; s. Justin S. and Lulu (Graves) K.; Asso.Sci., Cameron State Sch. Agr., 1936; B.S., U. Okla., 1938, M.S., 1940; Ph.D., U. Mich., 1947; m. Nedra L. Campbell, Sept. 14, 1938; children—Lisa A., Karl J., Eric C. Commd. ensign USN, 1943, advanced through grades to capt., 1962; parasitologist Naval Med. Research Unit, Cairo, Egypt, 1948-53, Taipei, Taiwan, 1957-62; parasitologist, instr. Naval Med. Sch., Bethesda, Md., 1953-57; parasitologist Naval Med. Research Inst., Bethesda, 1963-64; ret., 1964; chmn. dept. parasitology S.W. Found. Research and Edn., San Antonio, 1964—; leader several expdns., field studies on Taiwan, Philippines, Offshore Islands Taiwan; leader U. S. Naval Expdn. to N. Borneo, 1961. Mem. Am. Inst. Biol. Scis., Am. Soc. Parasitologists, Am. Micros. Soc. Author: Snakes of Taiwan, 1962. Contbr. numerous articles to profl. jours. Research on biology of parasites in field with emphasis on those that cause disease in man and domestic animals; discovered that lower mammals play role as reservoirs of infection to man by parasite diseases. Home: 144 Cadillac Dr., San Antonio 78213. Office: S.W. Found. for Research and Edn., San Antonio 78206.*

KUNTZE, Otto, German botanist; b. Leipzig, Germany, June 23, 1843; ed. U. Berlin (Germany), U. Leipzig; toured world, 1874-76; founded herbariums, Kew, Eng., Berlin, 1884-90; traveled in Europe, N. Africa, S.Am., Transcaucasia, 1891-92, S. Africa, 1894; ret. 1895. Author: Revisio generum plantarum, 3 vols., 1891-98; Lexicon generum phanerogamarum, 1903. Studied flora and fauna of Sargasses and Singapore, also India; studied Latin nomenclature, cinchona; revised bot. terminology according to Paris congresses. Died San Remo, Piedmont, Italy, Jan. 27, 1907.

KUNTZMANN, Jean, French mathematician; b. Pont a Mousson, France, June 1, 1912; s. Louis and Blanche (Vuillemot) K.; student Ecole Normale Superieure, Paris, 1931-34; D.Sc., 1940; m. Marguerite Coudert, 1938; 3 children. Lectr., then prof. Faculté des Sciences de Grenoble (France), 1945—, also dir. service applied math.; dir. calculus lab. of univ. Mem. French Assn. Operational Research, Math. Soc. France, I.E.E.E. Author 7 books (2 in English transl.); also articles. Research in math. for the engr., numerical analysis, Boolean algebra. Home: 30, Boulevard Foch, Grenoble 38, France. Office: B.P. 7, Saint Martin d'Heres, 38, France.*

KUNZ, Bruno Friedrich Johann, Austrian geophysicist; b. Vienna, Jan. 25, 1906; s. Johann Baptiste and Hedwig (Stanzl) K.; Ph.D., U. Vienna; m. Leopoldine Belza, Nov. 13, 1927; children—Bruno, Reinhold. Career as geophysicist; physicist-in-chief, several mineral oil firms; contractor for geophys. research in drilling and boring. Mem. Geophys. Exploration Soc., German Geophys. Soc., European Assn. Geophys. Exploration, Vienna Geol. Soc. Contbr. papers to jours. Made study-voyages to N. and S.Am.; specialist in exploration of oil, ore, salt mining, water supply; conducted explorations in Africa. Home: Schottenfeldgasse 61, Vienna VII. Office: Schwarzenbergplatz 16, Vienna I, Austria.

KUNZ, George Frederick, Am. gem expert; b. N.Y., Sept. 29, 1856; s. J. G. and Marie Ida (Widmer) K.; ed. Cooper Union; hon. A.M., Columbia, 1898; Ph.D., U. Marburg, 1903; Sc.D., Knox, 1907; m. Sophia Hanforth, Oct. 29, 1879 (dec. 1912). Vice pres., gem expert Tiffany & Co., jewelers, N.Y.C., from 1879. Spl. agt. U. S. Geol. Survey, 1883-1900; in charge dept. mines, Paris Expn., 1889, Kimberley Expn.,

1892, Chicago Expn., 1893; hon. spl. agt. of mines, Atlanta Expn., 1895, Omaha Expn., 1898; on spl. investigation of Am. pearls U. S. Fish Commn. 1892-98; del. from U. S. to Internat. Congress, Paris, 1900; radium commr. St. Louis Expn., 1904; spl. agt. in charge of precious stones, 12th U. S. Census. Decorated by govts. of France, Norway and Japan. Hon. mem. Chambre Syndicale Pierres Précieuses, Paris; research curator, precious stones, Am. Mus. Natural History; founder and hon. pres. Museums of the Peaceful Arts (now Mus. of Science and Industry), N.Y.C. Pres. Am. Metric Assn. Author: Gems and Precious Stones of North America; The Book of the Pearl (with Charles H. Stevenson), 1908; The Curious Lore of Precious Stones, 1913; E. Roty and His Work, 1914; Magic of Jewels, 1915; Ivory and the Elephant, 1915; Shakespeare and Precious Stones, 1916; The Ring, 1917. Died June 29, 1932.

KUNZE, Richard Ernest, physician, naturalist; b. Altenburg, Saxony, Apr. 7, 1838; s. John Jacob and Adelaide K.; M.D., Eclectic Med. Coll. of N.Y., 1868. Practiced in N.Y. until removing to Ariz., 1896; made studies and investigations in med. botany and insect fauna of Ariz. Originator of Ariz. cactus farm, and exporter of cacti to the bot. gardens of the world. Corr. sec. Eclectic Med. Coll. of New York. Died Feb. 10, 1919.

KUNZLER, J. E., Am. physicist; b. Willard, Utah, Apr. 25, 1923; s. John J. and Freida (Meier) K.; B.S., U. Utah, 1945; student Purdue U., 1945-46; Ph.D., U. Cal., Berkeley, 1950; m. Lois McDonald, Dec. 29, 1950; children—Carol Ann, Marilyn, Bonnie, Kim. Teaching asst. Purdue U., 1945-46; teaching asst. U. Cal., Berkeley, 1946-47, research asst., 1947-50, research asso., 1950-52; mem. tech. staff Bell Telephone Labs., Murray Hill, N.J., 1952-61, head metal physics research dept., 1961—. Recipient John Price Wetherill award Franklin Inst., 1964. Fellow Am. Phys. Soc., Am. Chem. Soc., A.A.A.S. Contbr. numerous articles to sci. jours. Low temperature solid state research, including investigations of elec., thermal, magnetic properties of high purity metals and other solids; discovery of high-current, high magnetic field capabilities of Nb_3Sn and other superconductors leading to practical high field superconducting magnets. Home: R.D. 2, Washington, N.J. 07882. Office: Mountain Av., Murray Hill, N.J. 07971.*

KUO SHOU CHING, Chinese astronomer, mathematician; b. 1231; flourished Mongol Ct., 1262. Author: Shou-shi-li, pub. 2d. half 14th century in Yüan Annals, Yüan Shih; also many works on astronomy. Computed new calendar (Shou-shih calendar), 1280; probably introduced spherical trigonometry to China; determined winter solstice of 1280; built instruments, including an armillary sphere, compendium instrument; many astron. observations. Died 1316.

KUPALOV, Petr Stepanorich, Russian physiologist; b. Oct. 13, 1888; M.D., Petrograd Mil. Acad. Medicine, 1918; student of Pavlov. Mem. sci. mission Prof. Hill's Physiol. Lab., London, Eng., 1928-30; dep. to Pavlov in charge sci. work Pavlov Physiol Dept., staff All-Union Inst. Exptl. Medicine, USSR Acad. Med. Sci., 1925—; staff mem., head, chair normal physiology 1st Pavlov Med. Inst., Leningrad, also dir. physiol lab. State Roentgenological Inst., 1931-52. Organizer, sec. All-Union Congress Physiologists; mem. 13th, 15th, 19th, 20th internat. congresses Physiologists, Neurol. Congress, Brussels, 1957. Recipient Pavlov Gold medal. Mem. USSR Acad. Med. Scis., Leningrad Soc. Physiologists, Biochemistry, and Pharmacologists (chmn. 1950—), Royal Soc. Physiologists, Eng., All-Union Pavlov Physiol. Soc. (chmn. central council 1959—). Author: The Effects of a System of Rhythmic Conditioned Reflexes on the Formation and Existence of a New Conditioned Reflex, 1933; Physiological Study of the Higher Manifestations of Life Activity in Animals, 1946; Experimental Neuroses in Animals, 1952; Some Problems of the Physiology of Higher Nervous Activity, 1956. Publisher: Lectures on Pavlovian Physiology. Coeditor physiology sect. Lage Med. Ency., 2d edit.; chief editor Pavlov Jour. Higher Nervous Activity; editorial bd. Bull. Exptl. Biology and Medicine, Sechenov Physiol. Jour. Founder Physiol. sch. which has continued teaching of Pavlov. Address: Inst. of Exptl. Medicine, Solyanka 14, Moscow, USSR.

KUPCHAN, S. Morris, Am. chemist; b. N.Y.C., Nov. 10, 1922; s. Hyman and Rebecca (Fishgold) K.; B.S., Coll. City N.Y., 1941; M.A., Columbia U., 1942, Ph.D., 1945; m. Nancy R. Slater, Mar. 28, 1954; children—Clifford, Charles. Research asso., lectr. Columbia U., 1945-48; research fellow, instr. chemistry Harvard, 1948-55; faculty U. Wis., Madison, 1955—, prof., 1961—; mem. pharmacology, exptl. therapeutics study sect. NIH, 1963-67, medicinal chemistry study sect., 1967—. Recipient Ebert prize, 1960, Research Achievement award in natural products Am. Pharm. Assn., 1965. Mem. Am. Chem. Soc., Am. Pharm. Assn., Chem. Soc. London, Soc. Econ. Botany (past pres.), N.Y. Acad. Scis. Contbr. numerous articles to profl. jours. Research, publs. on chemistry biologically-active natural products, derived synthetic drugs, alkaloids, tumor inhibitors, anti-malarials, hypotensives, synthetic organic chemistry, intramolecular catalysis in enzyme model systems. Home: 3404 Circle Close, Madison, Wis. 53705.*

KUPFERBERG, Alfred Ballen, Am. microbiologist; b. Bklyn., Nov. 25, 1913; s. Isaac and Sarah (Ballenberg) K.; B.S. in Edn., U. No. Ill., 1939; M.S., U. Ia., 1941; Ph.D. in Microbiology, Rutgers U., 1951; m. Beatrice Halpern, Nov. 27, 1942; children—Paul, Susan. Research staff dept. obstetrics and gynecology State U. Ia., 1939-41; asso. microbiologist Ortho Research Found., Raritan, N.J., 1946-51, dir. div. microbiology, 1951—. Fellow Am. Acad. Microbiology; mem. Am. Soc. for Microbiology (councilor N.J. br. 1964—), Soc. for Tropical Medicine and Hygiene, Am. Chem. Soc., Soc. for Exptl. Biology and Medicine, Mycol. Soc. Am., Internat. Study Group for Control Human Trichomoniasis, N.Y. Acad. Scis., A.A.A.S., Internat. Soc. for Tropical Dermatology. Contbr. articles to tech. jours. Research on chemotherapy infectious diseases, antibiotics, nutrition protozoa, prodn. enzymes by microorganisms, isolation and charaterization proteins. Home: 2 Clair Dr., Someville, N.J. 08876. Office: div. Microbiology, Ortho Research Found., Raritan, N.J. 08869.*

KUPFFER, Adolph Theodore, see de Kupffer, Adolph Theodore.

KUPPERMANN, Aron, chem. physicist; b. Sao Paulo, Brazil, May 6, 1926; s. Jacob and Mary (Feldman) K.; Came to U. S., 1953, naturalized, 1965; B.Sc. in Chem. Engring., U. Sao Paulo, 1948; Ph.D. in Phys. Chemistry, U. Notre Dame, 1956; m. Roza Davidson, Jan. 21, 1951; children—Baruch, Miriam, Nathan, Sharon. Asst. prof. Instituto Technologico de Aeronautica, Sao José dos Campos, Brazil, 1950-51; faculty U. Ill., Urbana, 1955-63, asso. prof., 1961-63; prof. chem. physics Cal. Inst. Tech., Pasadena, 1963—; research asso. Argonne (Ill.) Nat. Lab., 1957, Inst. Atomic Energy, Sao Paulo, 1960. Cons., Jet Propulsion Lab., Pasadena, 1965—. Recipient Centennial Sci. award U. Notre Dame, 1965. Mem. Internat. Union Radiation Research (councilor for chemistry 1965—), Am. Chem. Soc., Am. Phys. Soc., Faraday Soc., Radiation Research Soc. (councilor for chemistry 1966-68), A.A.A.S., Am. Assn. U. Profs., Sigma Xi, Phi Lambda Upsilon, Alpha Chi Sigma. Editorial bd. Jour. Phys. Chemistry, 1965—. Research and publs. on chem. dynamics, radiation chemistry, photochemistry, electron scattering, molecular beams, theory of chem. reactions. Home: 2487 Morslay Rd., Altadena, Cal. 91001. Office: Chemistry Dept., Cal. Inst. Tech., Pasadena, Cal. 91109.*

KUPRADZE, Viktor Dmitrievich, Russian mathematician; b. Kutaisi, Nov. 2, 1903; grad. Tbilisi U., 1927; D.Physico-Math. Sci., 1935. Instr., Leningrad U., 1933-34; with Math. Inst., USSR Acad. Sci., 1934-35; dir. Math. Inst., Georgian Acad. Sci., 1936-41, acad. sec. dept. math. and physics, 1964—, also mem. Presidium; prof. Tbilisi U., 1936—, rector, 1954-58; Georgian minister edn., 1944-53. Mem. Georgian Acad. Sci. Author: Basic Problems of the Mathematical Theory of Diffraction, 1950; Boundary Problems of Oscillation Theory and Integral Equations, 1950; co-author: A General Mixed Boundary Problem of the Theories of Elasticity and the Theory of Potential, 1963. Research and publs. on differential and integral equation theory, math. physics and theory of elasticity; developer theory of boundary problems of stable oscillations in elastic bodies, solved problems of electromagnetic wave diffraction. Address: Tbilisi University, Chavchavadze 1, Tbilisi, Gruz. SSR, USSR.

KUPREVICH, Vasiliy Feofilovich, Russian botanist; b. Klanniki (now Minsk Oblast), Jan. 24, 1897; grad. Moscow Pedagogical Inst., Inst. for Advanced Pedagogical Tng., 1931; D.Biol. Sci., 1941. Sr. Asso. Inst. Biology, USSR Acad. Sci., 1934-38, head lab., 1938-49, dir. Bot. Inst., 1949-52. Mem. USSR Acad. Sci. (corr.), Belorussian (pres. 1952—) acads. sci. Author: Cereal Parasites; Problems and Methods of Studying Diseases in Agricultural Crops; Virus Diseases of Potatoes. Former editor News of Belorussian Acad. Sci. Discovered extracellular enzymes in obligate parasites; proved possibility of heterotrophenous feeding of higher plants in natural environments. Address: Belorussian Acad. Sci., Minsk, Belorussian SSR, USSR.

KUPRIANOFF, Johann Michael, food engr.; b. St. Petersburg, Russia, Dec. 7, 1904; s. Michael Andrei and Maria (Rogozine) K.; student Poly. Inst. Leningrad (Russia), 1922-24; Dipl.Engr., Karlsruhe (Germany) Inst. Tech., 1928, Dr. Engring., 1931; m. Maria Mueller, July 7, 1949; children—Michael, Tanja (Mrs. Dietrich Edel), Peter. Chief engr., dir. refrigeration dept. Robert Bosch Co., Stuttgart, Germany, 1934-46; dep. dir. Fed. Research Inst. Food Refrigeration, Karlsruhe, 1946-48, dir., 1948—; prof., dir. Inst. Food Engring., U. Karlsruhe, 1961—. Mem. German AEC. Recipient Order Merit for Research and Invention (comdr.); Carl von Linde gold medal; medal French Nat. Fed. Food Industries. Mem. Internat. Inst. Refrigeration (pres. tech. bd. 1963—), Assn. German Engrs. (chmn. food engring. group), German Research Council, Heidelberg Acad. Scis., Phi Tau Sigma; hon. mem. Spanish Soc. Bromatology. Author: Dry Ice, 1953; (with Plank) Small Refrigerating Machine, 1960; (with Steinle, Plank) Refrigerants, 1956; (with Lang) Irradiation and Contamination of Foodstuffs, 1960; also numerous articles. First thermodynamic three-phase diagrams for carbon dioxide in gas, liquid and solid state; thermodynamic analysis and diagrams of refrigerants; refrigeration machinery; preservation and processing of foodstuffs. Home: 12 Südliche Hildapromenade. Office: 12 Kaiserstrasse, 75 Karlsruhe, West Germany.*

KURCHATOV, Igor Vasilievich, Russian nuclear physicist; b. Russia, 1903; grad. Crimean U., Simferopol; student Baku Polytechnical Inst. Joined staff Physico-Technical Inst., Acad. Scis. of USSR, 1925, dir. nuclear physics lab. Physico-Technical Inst., supervised building of cyclotrons; research elec. relations in materials which possess quality of spontaneous polarization; observed nuclear fission provoked by neutron bombardment, 1933; research spontaneous fission of uranium. Elected to Supreme Soviet, Communist Party, 1950. Dir. Atomic Energy Inst., Acad. Scis. USSR. Decorated Order of Lenin; recipient Stalin prize. Mem. Acad. Scis. of USSR (presidium of acad.). Author: Splitting the Atomic Nucleus; 1935. Died Feb. 1960.

KURDIMOV, Georgii Viacheslavovich, Russian physicist, metallurgist; b. Feb. 14, 1902; grad. Leningrad Poly. Inst. 1926; with physicotech. inst. USSR Acad. Scis., 1925-32; asso. Dnepropetrovsk Physicotech. Inst., also prof., head chair physics of metals Dnepropetrovsk U., 1932-41; dir. lab. of physics of metals Ukrainian Acad. Scis., 1946-51; dir. inst. metallography and physics of metals Central Research Inst. of Ferrous Metallurgy, Moscow, 1944—. Mem. USSR, Ukrainian acads. scis. Recipient Stalin prize, 1949. Author: The Heat Treatment of Steel in the Light of X-Ray Studies, 1932; General Laws of Phase Transformations in Eutectoid Alloys, 1936; On the Kinetics of the Conversion of Austenite into Martensite at Low Temperatures; The Use of X-Ray Structure Analysis Methods for Study of the Tempering of Chilled Steel, 1950. Authority on annealing and tempering of steel, also on phenomena of phase conversions; specialist on properties and structure of martinite; discovered new class of phase conversions in solids. Address: Acad. of Scis., Vladimirskaya Ulitsa 54, Kiev, USSR.

KURE, Shuzo, Japanese physician; b. 1865; s. Koseki Kure; grad. Tokyo U., 1890; asst. prof. Tokyo U., after grad.; sent for study mental diseases to Germany, Austria, 1897; prof. Tokyo U., also chief physician Sugamo Mental Hosp., 1901-25. Recipient decoration from Dutch Govt. Author: Life and Achievements of Dr. Siebold; also med. treatises, biographies of physicians. Introduced new methods for treatment of mentally disturbed patients. Died 1932.

KUREPA, (Georges) Duro R., Yugoslavian mathematician; b. Glina, Yugoslavia; Aug. 16, 1907; s. Rade Vasilj and Andelija (Mrakovic) K.; dipl., Philos. Faculty, U. Zagreb (Yugoslavia), 1931; m. Nada Jagic; children—Snezana, Radan, Nada, Jurij-Arhimed. With Philos. Faculty, U. Zagreb, 1931-32, 38-65, prof., 1946-65; prof. Faculty Sci Beograd (Yugoslavia), 1965—. Dr. ès sci. math., Faculty Sci., Paris, 1935; postgrad., Warsaw, Poland, 1937; with Inst. for Advanced Studies, Princeton, N.J., 1959-60; dir. Math. Seminar and Inst., Zagreb, 1944-65. Mem. Soc. Math., Physics and Astronomy Croatia (pres.), Union Socs. Math., Physics and Astronomy Yugoslavia (pres.), Internat. Com. Math. Instrn. (v.p. 1952-62), Math. Inst. (pres.), Instrn. Council Serbia (pres.), Yugoslavian Acad. Sci. and Arts Zagreb (corr.), Soc. Math. de France, Soc. Math., Physics and Astronomy Serbia, Am. Math. Soc. Author: Theory of Sets, 1951; What are Sets, 1960; Higher Algebra, 1965; also numerous other books, articles. Studies in systematic study of ordered sets called trees and ramifications (scope, formulation of hypothesis or postulate in connection with Suslin problem, introduction principles, duality and reciprocity considerations); introduced non-numerical ecart and various functions (cellularity, stellarity) to topology; studies in matrix theory. Home: 7 Zagrebacka, Belgrade, Yugoslavia.*

KURIAN, Chaligathu Varghese, Indian oceanographer; b. Mavelikara, India, Mar. 11, 1919; s. K. and T. C. (Annamma) K.; B.Sc., Kerala U., Trivandrum, 1942, M.Sc., 1947; D.Sc., Belgrade (Yugoslavia), 1955; m. J. Amminy, Sept. 12, 1948; children—Soosy, George. Research scholar, Trivandrum, 1943-47, demonstrator, 1947-48; faculty U. Kerala, 1948-—, prof., 1965—, head oceanography dept., 1962-—; scholar Split Instn. Oceanography, 1954-56, Plymouth Lab., Surrey, Eng., 1955-56, Oceanographic Instn., Paris, France, 1955. Research and publs. on bottom fauna and bottom deposits of S.W. coast of India, Cumacea, plankton of Adriatic sea. Home: 1541/XXIII, Foreshore Rd., Ernakulam, India.*

KURIHARA, Kenneth Kenkichi, economist; b. Kutchan, Japan, Jan. 8, 1910; s. Kichizo and Natsu (Koshida) K.; came to U. S., 1930, naturalized, 1963; B.A., Ohio Wesleyan U., 1935; M.A., Oberlin Coll., 1936; Ph.D., State U. Ia., 1942; Dr. Econs., Hitotsubashi U., 1958; m. Kyo Kurosawa, Mar. 19, 1948. Mem. faculty Princeton, 1946-47; faculty Rutgers U., New Brunswick, N.J., 1947—, prof. econs., 1960-68; distinguished prof. econ. theory State U. N.Y., Binghamton, 1968—; guest lectr. numerous univs. U. S. and abroad, 1953-66. Fulbright lectr., Japan, 1965. Recipient Christian R. and Mary Lindbark Found. award for Distinguished Teaching,

Rutgers U., 1963. Fellow Royal Econ. Soc.; mem. Am. Econ. Assn., Artus. Arthor: Monetary Theory and Public Policy, 1950; Introduction to Keynesian Dynamics, 1956; The Keynesian Theory of Economic Development, 1959; National Income and Economic Growth, 1961; Applied Dynamic Economics, 1963; Macroeconomics and Programming, 1964. Editorial mem. Indian Jour. Econs., 1960——. Research in monetary-fiscal analysis and policy; nat. income analysis; econ. fluctuations and growth analysis; internat. econ. relations; Keynesian and post-Keynesian econs. Home: 3205 Knapp Rd., Vestal, N.Y. 13901. Office: Econs. Dept., State U. N.Y., Binghamton, N.Y. 13901.*

KURIR, Anton, forestry scientist; b. Ragusa, Dalmatia, Oct. 8, 1909; s. Markus and Klara (Bohm) K.; engring. diploma U. Zagreb, 1938; Agr. Dr., Agr. U. Vienna, 1943; Dr.rer.nat., U. Vienna, 1945; m. Inge Hintermayer, Apr. 29, 1945; children—Franka, Markus. Asst., Inst. Entomology, U. Zagreb, 1938-42, Inst. Forestentomology Agr., U. Vienna, 1942-43, Inst. Zoology, 1944-45, Inst. Forestentomology Agr., 1945-61; prof., chmn. forest entomology and forest protection Agr. U. Vienna, 1961——, prof. forest hygiene, 1966——. Mem. Austrian Ct. of Patents. Mem. Internat. Union Forest Research Orgns., Internat. Assn. U. Profs. and Lectrs., German Soc. for Applied Entomology. Author: Wichtige forstschadliche Insekten, 1947; also numerous articles. Research on zoology, forest entomology, forest protection, hygiene of forests, wood-destroying insects, insect control and pub. health, radiology. Home: 20 Ley-Str., 1200 Vienna, Austria.*

KURIYAMA, Sutezo, Japanese chemist; b. Wakayama Prefecture, 1891; grad. Kyushu U., 1916; D.Eng., 1928. Staff engr. Osaka Indsl. Lab., 1916; asst. prof. Kyushu U., later prof., emeritus prof., 1955——. Adviser, Kokoku Jinken (Rayon) Pulp Co. Research and publs. on organic synthetic and high molecular chemistry, especially fiber and plastics.

KURKJIAN, Charles Robert, Am. ceramist; b. Wanamassa, N.J., Dec. 7, 1929; s. John G. and Esther (Grunke) K.; B.S., Rutgers U., 1952; Sc.D., Mass. Inst. Tech., 1955; m. Dorothy A. Mazzarella, June 5, 1955; children—Karen, Lynne, Charles R. Postdoctoral fellow Mass. Inst. Tech., 1955-57, Sheffield (Eng.) U., 1957-59; mem. tech. staff Bell Telephone Labs., Murray Hill, N.J., 1959——. Mem. Am. Ceramic Soc., Soc. Glass Tech. Research in structure and properties of inorganic glass, mech. relaxation and flow behavior and spectroscopy. Home: 82 Harrison Brook Dr., Basking Ridge, N.J. 07920. Office: Mountain Av., Murray Hill, N.J. 07971.*

KURLAND, Leonard Terry, Am. epidemiologist; b. Balt., Dec. 24, 1921; s. Ellis M. and Sara (Shain) K.; B.A., Johns Hopkins U., 1942, Dr.P.H., 1952; M.D. with Honors, U. Md., 1945; M.P.H. cum laude, Harvard, 1948. Served with USPHS, 1946-64; chief epidemiology br. Nat. Inst. Nervous Diseases and Blindness, NIH, Bethesda, Md., 1955-64; head sect. med. statistics, epidemiology and population genetics Mayo Clinic, Rochester, Minn., 1964——. Cons. collaborative and field program NIH, 1965——; prof. epidemiology Mayo Grad. Sch. Medicine, 1965——. Diplomate Am. Bd. Pub. Health and Preventive Medicine. Fellow A.M.A., Am. Pub. Health Assn.; mem. Am. Soc. Human Genetics, Am. Assn. Neuropathology (co-recipient ann. award for outstanding paper 1961), Multiple Sclerosis Socs. U. S. (adv. council), Multiple Sclerosis Soc. Can. (adv. council), Am. Epidemiological Soc., Internat. Epidemiological Assn., Japanese Clin. Soc. Neurology (exec. council), Soc. Neuroepidemiology. Research and publs. on epidemiology and genetics of neurol. diseases, ophthalmologic disorders, malignancies, stroke. Home: 1165 Plummer Circle, Rochester 55901. Office: 200 1st St. S.W., Rochester, Minn. 55901.*

KURODA, Rokuro, Japanese chemist; b. Tokyo, Japan, Oct. 7, 1926; s. Washizo and Sato (Yamada) K.; B.S., Tokyo U. Edn., 1950, Ph.D., 1957; m. Satiko Furubayashi, Nov. 30, 1958; children—Youzo, Keiko. Faculty chemistry dept. Tokyo U. Edn., 1953-65, reader, 1960-65; prof. analytical chemistry U. Chiba (Japan), 1965——; radiochemist U. Mich., 1962-63. Recipient Tokyo U. Edn. award for geochem. work, 1956. Mem. Japan Soc. for Analytical Chemistry (editorial bd. 1964——), Spectroscopic Soc. Japan (counselor 1965——), Am. Chem. Soc., Chem. Soc. Japan. Research and numerous publs. on radioactivation analysis and ion-exchange chromatography, trace elements in rocks and meteorites, analytical chemistry of rare earths and scandium. Home: 3-115-3 Ohmiyamae, Suginami-ku, Tokyo, Japan.*

KUROSH, Aleksandr Gennadievich, Russian mathematician; b. Jan. 19, 1908; grad. Smolensk U., 1928; Ph.D. in Physico-Math. Sci. Instr., Moscow U., 1930-37, prof., 1937——. Mem. Moscow Math. Soc. (pres.). Author: The Theory of Groups, 1944; An Advanced Algebra Course, 5th edit., 1956; Free Sums of Multioperator Algebra, 1960; Direct Decompositions in Algebraic Categories, 1960. Address: Moscow University, Leninskie gory, Moscow, USSR.

KURSANOV, Andrey Lvovich, Russian plant physiologist, biochemist; b. Nov. 9, 1902; s. L. N. Kursanov; grad. Physico-Math. Faculty, Moscow U., 1926. With Research Inst. of Sugar Industry, 1929-44; instr. Moscow Timiryazev Agrl. Acad., 1929-38; with Inst. Biochemistry, USSR Acad. Sci., 1935——, dir. Inst. Plant Physiology, 1952——, bur. mem. dept. biol. sci., until 1963; prof. Moscow U., 1944——. Decorated Order of Lenin. Mem. USSR Acad. Sci. (Presidium mem. 1960——). Author: The Circulation of Organic Matter in Plants, 1952; The Biological Synthesis of Disaccharides, 1954. Research and publs. on plant metabolism, action of enzymes in live plant and carbon dioxide assimilation of root system, tea leaf tannins, photochemistry, enzymology and physiology of diseased plants, metabolism in biol. media devoid of cell structure, plant reaction to environment proving primacy of environment over generic principle; established relation of carbon dioxide assimilation to outflow of plastic matter from leaves. Address: Inst. Plant Physiology, USSR Acad. Sci., Leninsky prospect 33, Moscow, USSR.

KURSANOV, Dmitriy Nikolaevich, Russian organic chemist; b. Apr. 13, 1899; grad. Moscow U., 1924. With Moscow Textile Inst., 1930-47; prof., 1936——; asso. Inst. Organic Chemistry, USSR Acad. Scis., 1943——, Inst. Sci. Information, 1953——. Recipient Lenin prize, 1963. Mem. USSR Acad. Sci. (corr.). Research and publs. on organic reaction mechanisms, reciprocal intramolecular action of atoms in organic compounds. Address: Inst. Organic Chemistry, USSR Acad. Scis., Vavilov St. 28, Moscow, USSR.*

KÜRSCHAK, J., Hungarian mathematician; b. 1864. Extended properties of absolute values for field of ordinary complex numbers to elements of abstract field, which had far reaching consequences in arithmetic and algebra, 1913. Died 1933.

KURSE, Paul Frederick, Jr., Am. chemist; b. Buffalo, Aug. 25, 1921; s. Paul Frederick and Miriam (Wise) K.; student Bowdoin Coll., 1939-40; B.S., U. Me., 1944; Ph.D., U. Tex., 1951; m. Elizabeth Jean Kronzer, June 3, 1949; 1 son, Paul Frederick, III. Devel. chemist Gen. Electric Co., Pittsfield, Mass., 1944-46; staff Sandia Corp., Albuquerque, 1950-52; sr. research chemist Samuel Roberts Noble Found., Inc., Ardmore, Okla., 1952-54, asst. dir. biomed. div., 1954-65, dir. biomed. div., 1965——. Bd. dirs. Carter County Guidance Clinic, So. Okla. Ambulance Service. Mem. Am. Chem. Soc. (past chmn. Okla. sect.), Am. Assn. for Cancer Research (past pres. S.W. sect.), A.A.A.S., Soc. for Exptl. Biology and Medicine, Tissue Culture Assn., Sigma Xi, Phi Lambda Upsilon. Contbr. articles to tech. jours. Research on synthesis of organic compounds; developed differentiation tests for primary, secondary, tertiary alcohols and amines, perfusion system for replicate animal cell cultures; research on multiple-layered populations of animal cells. Home: P.O. Box 1774. Office: Box 878, Ardmore, Okla. 73401.*

KURSHAKOV, Nikolay Aleksandrovich, Russian physician; b. Kronshtadt, 1886; grad. St. Petersburg Mil. Med. Acad., 1910; D.Med. Sci., 1912. With dept. diagnostics St. Petersburg (later Petrograd) Mil. Med. Acad., 1910-25; prof., 1925——; prof. chair clin. propedeutics and therapeutics Voronezh U. (now Voronezh Med. Inst.,), 1925-35; prof., head clinic hosp. therapeutics Moscow Oblast Clin. Inst., 1935-42, 45-55; head chair and clinic of hosp. therapy 1st Moscow Med. Inst., 1942-43, 50-51; chief clinician USSR Ministry Health, 1945——; sci. dir. Burdenko Main Mil. Hosp. of Soviet Army; head clin. dept. Inst. Biophysics, USSR Acad. Med. Scis., also bur. mem. clin. dept. Decorated Order of Lenin. Mem. USSR Acad. Med. Scis. (corr.), All-Union Soc. Therapeutists. Author: Normal and Pathological Blood Circulation, 1933, 2d edit., 1947; The Peripheral Heart, 1930; Acute Radiation Sickness, 1956. Mem. editorial bd. Med. Abstracts Jour.; mem. editorial council Therapeutic Archive, Clin. Medicine; co-editor Internal Diseases sect. Large Med. Ency., 2d edit. Research and numerous publs. on physiology and pathology of blood circulation, alimentary dystrophy, hematology, metabolism, rheumatism, endocarditis, radiation sickness. Home: 3-ya Meshchanskaya 61. Office: Inst. Biol. Physics, USSR Acad. Med. Scis., Leninsky prospect 14, Moscow, USSR.

KURSUNOGLU, Behram, physicist; b. Bayburt, Turkey, Mar. 14, 1922; s. Ismail and Hanife (Esenulku) K.; B.Sc., Edinburgh U., 1949; Ph.D., Cambridge U., 1952; m. Sevda Arif, Sept. 25, 1952; children—Sevil, Ayda, Ismet. With Cornell U., 1952-54, Yale, 1955, Turkish Gen. Staff on Atomic Matters, 1956-58, Turkish Atomic Energy Commn., 1956-58; dean faculty nuclear scis. and tech. Middle East Tech. U., Ankara, Turkey, 1956-58; vis. prof. physics U. Miami, Coral Gables, Fla., 1954-55, prof. physics, 1958——, dir. Center for Theoretical Studies, 1965——. Cons., Oak Ridge Nat. Lab., 1962-64. Fellow Am. Phys. Soc.; mem. Am. Assn. Physics Tchrs., Am. Assn. U. Profs., Sigma Xi. Author: Modern Quantum Theory, 1962. Contbr. numerous articles to profl. jours. Research on theory of gen. relativity, internal and external symmetries in particle physics, theory of Brownian movement, diffusion and turbulence in plasma physics. Home: 6200 Leonardo St., Coral Gables, Fla. 33134.*

KURTI, Nicholas, physicist; b. Budapest, Hungary, May 14, 1908; s. Charles and Margaret (Pinter) K.; ed. Paris, Berlin; M.A., Ph.D.; m. Georgiana Shipley, Sept. 24, 1946; children—Susannah, Camilla. Sci. asst., Breslau, Oxford; instr. lectr. Oxford (Eng.) U., researcher Brasenose Coll. Recipient Holweck prize and medal, Fritz London prize. Fellow Royal Soc., 1956; mem. Inst. Physics. Research and publs. on low temperature physics, magnetism, technological and biol. applications of low temperature. Home: 38 Blanford Av., Oxford. Office: Clarendon Lab., Oxford, Great Britain.

KURU, Masaru, Japanese surgeon; b. Mie Prefecture, 1902; grad. Tokyo U., 1926; M.D., 1933. Prof., Kanazawa Med. U., 1941, Osaka U., 1945——. Recipient prize for research Japan Imperial Acad., 1949. Author: Itami (Pain); Muroku (Record of Dream). Research on cancer, substance on which stomach, breast and various cancers live and develop, preliminary conditions of cancer, determination of cell groups in connection with transmission of sensation of pain and temperature in posterior part of spinal cord.

KURZEPA, Stanislaw Jósef, Polish pharmacologist; b. Warsaw, Poland, July 4, 1922; s. Jan and Helena (Starenga) K.; Doctor's degree, Warsaw U., 1959, Docent's degree, 1965; m. Zofia Wroblewska, Dec. 16, 1950; 1 son, Pawel Stanislaw. Staff, Pharmacological Lab., Soc. Apothecaries, 1948-51, dept. pharmacology Med. Acad., Warsaw, 1950-52; staff dept. pharmacology Inst. Mother and Child, Warsaw, 1954——; staff dept. pharmacology Inst. Pharmacy, Warsaw, 1957——; staff dept. pharmacology Oxford (Eng.) U., 1962. Mem. Polish Physiol. Soc., Polish Pharmacological Soc., Polish Pharm. Soc., European Soc. Biochem. Pharmacology. Author: Drugs Treatment in Pediatry, 1965; also articles chpt. in book. Research on biochem. mechanisms of changes of reactivity to drugs during postnatal ontogenesis. Home: 19 Krucza, Warsaw. Office: 25 Madalinskiego, Warsaw, Poland. Died Warsaw, Jan. 21, 1968.*

KURZWEG, Hermann Herbert, physicist; b. Germany, Jan. 1, 1908; s. Paul O. and Bertha (Friede) K.; student Tchrs. Coll., 1928; Ph.D., Univ. Leipzig (Germany), 1933; m. Erna Michaelis, Apr. 21, 1934; children—Ulrich, Diemut (Mrs. Robert W. Balentine), Lutz. Came to U. S., 1946, naturalized, 1954. Asst. instr. U. Leipzig, 1934; research sci. Carl Zeiss Optical Works, 1935-36; chief research dir. German Supersonic Wind Tunnel Labs., Peenemuende, 1937-43, chief research, dep. dir. aeroballistics lab., Kochel, 1943-45; cons., div. chief U. S. Naval Ordnance Lab., White Oak, Md., 1946-52, chief aeroballistic research dept., 1952-56, asso. tech. dir. aeroballistic research, 1956-60; asst. dir. research NASA, 1960, dir. research, 1961——, also liaison mem. div. math. of Nat. Acad. Sci., 1963-66; part-time lectr. dept. aero. engring. U. Md. Grad. Sch., 1947-60; chmn. indsl. and profl. adv. council, dept. aero engring. Pa. State U., 1962-67; mem. com. aeroballistics Navy Bur. Ordnance, 1947-58, chmn., 1958-60; mem. conf. com. Nat. Conf. Adminstrn. Research, 1958-63; mem. adv. panel aero. Office Asst. Sec. Def., 1958-61; mem. fluid dynamics panel Adv. Group Aero. Research and Devel., NATO, 1953——, chmn. panel, 1963——. Recipient Meritorious Civilian Service award U. S. Navy, 1951. Fellow Am. Inst. Aero. and Astronautics, Washington Acad. Scis., Philos. Soc. Washington. Research and publs. on original work in hydrodynamics; in supersonic aerodynamics, fin stabilization of supersonic missiles; devel. supersonic and hypersonic wind tunnels and firing ranges, contbns. to high-speed ballistics publs. Home: 731 Quaint Acres Dr., Silver Spring, Md. 20904. Office: 600 Independence Av. S.W., Washington.*

KUSAKA, Yuzuru, Japanese chemist; b. Kobe, Japan, Oct. 19, 1925; s. Senji and Tsuyuko (Kunimura) K.; B.S., Faculty Sci., Kyoto (Japan) U., 1947, D.Sci., 1959; m. Teru Naoki, Apr. 17, 1952; children—Miki, Hiroyuki. Faculty, Konan U., Kobe, Japan, 1954——, prof. chemistry, 1962——; vis. prof. Tsing Hua U., Hsinchu, China, 1966-67. Postdoctral fellow U. Mich., Ann Arbor, 1960-61. Mem. Japan Chem. Soc., Japan Soc. for Analytical Chemistry, Japan Radioisotope Assn., Geochem. Soc. Japan. Author: (with others) Rapid Radiochemical Separations, 1961; also articles. Radioactivation analysis using low-level neutron source, 14 MeV neutron generator, with atomic reactor; rapid radiochem. separations. Home: 74, Chimori-1, Suma-ku, Kobe, Japan.*

KUSCER, Ivan, physicist; b. Vienna, Austria, June 17, 1918; s. Ljudevit and Marjeta (Kautsky) K.; diploma U. Ljubljana (Yugoslavia), 1941, doctor degree, 1951; m. Martina Kramberger, July 26, 1952; children—Samo, Andrej. Faculty, U. Ljubljana, 1945——, prof. physics, 1963——; vis. staff physics dept. U. Birmingham (Eng.), 1952-53, nuclear engring. dept. U. Mich., Ann Arbor, 1963-64, Brookhaven Nat. Lab., Upton L.I., N.Y., 1964-65. Author: (with A. Moljk) Fizika, 1958; also articles. Research on theory of neutron transport and of radiative transfer, kinetic theory of gases, hydrology. Home: 15 Koblarjeva, Ljubljana, Yugoslavia.*

KUSCH, P(olykarp), physicist; born Blankenburg, Germany, Jan. 26, 1911; s. John Matthias and Henrietta (van der Haas) K.; B.S., Case Inst. Tech., 1931, D.Sc., 1956; M.S., U. Ill., 1933, Ph.D., 1936; D.Sc., Ohio State University, 1959; D.Sc., Gustavus Adolphus Coll., St. Peter, Minn., 1963; came

U. S., 1912, naturalized, 1922; m. Edith Starr McRoberts, Aug. 12, 1935 (dec. 1959); children—Kathryn, Judith, Sara; m. 2d, Betty Jane Pezzoni, 1960; dau., Diana. Asst. U. Ill., 1931-36; assistant University of Minnesota, 1936-37; instructor Columbia University, 1937-41; engineer Westinghouse, 1941-42; research associate Columbia U., 1942-44; mem. tech. staff Bell Telephone Labs., 1944-46; asso. prof. physics Columbia U., 1946-49, prof., 1949—, chairman of department of physics, 1949-52, 60-63. Recipient Nobel prize in physics (with W. E. Lamb), 1955; fellow of Center for Advanced Study in the Behavioral Sciences, 1963-64. Fellow Am. Phys. Soc.; mem. Nat. Acad. Scis. Research in atomic, molecular and nuclear physics. Important work with mass spectroscopy; by passing atomic beams through a magnetized vacuum, gave precise determination of magnetic moment of the electron, 1949. Home: 450 Riverside Dr., N.Y.C. 10027.

KUSHNER, Samuel, Am. chemist; b. Auburn, N.Y., Apr. 25, 1915; s. Joseph and Minnie (London) K.; B.S. in Chemistry, U. Mich., 1939, M.S., 1940, Ph.D., 1942; m. Ruth Ellen Paul, May 7, 1937; children—Joseph Werner, Carolyn Ann, Ellen Diane, Michael James. Staff, Lederle Labs., Pearl River, N.Y., 1942—, unit leader, 1955-56, dept. head chemotherapy, 1956—. Mem. Am. Chem. Soc., N.Y. Acad. Scis., Sigma Xi, Phi Lambda Upsilon. Contbg. author: Medicinal Chemistry, 1960. Patentee in field. Research and publs. on tropical diseases; synthesized medicine for treatment filariasis; demonstrated importance of pyrazinamide for treatment of Tb.; synthesized fragments of tetracycline. Home: 138 Highview Av., Nanuet, N.Y. 10954. Office: Lederle Labs., Pearl River, N.Y. 10965.*

KUSS, Georges, French physician; b. 1867; did fundamental research in Tb of children, described a complex lesion (known as primary complex, Kuss-Ghon focus or Ghon complex); established that children of tubercular parents are not born with Tb, but prone to be affected later, 1898. Died 1936.

KUSSMAUL, Adolf, German physician; b. Graben, Germany, 1822; student Heidelberg, Germany; M.D., Wurzburg, Germany, 1855. Served in armed services, 1848-49; Kandern, 1850-53; asso. prof., Heidelberg, 1855-59; prof. Clinic, Erlangen, Germany, 1859-63, Fribourg-en-Brisgau, Germany, 1863-76, Strasbourg, France, 1876-88. Author: Untersuchungen über den constitutischen Mercurialismus, 1861; Über die Behandlung der Magenerweiterung, 1869; Die Störungen der Sprache, 1877. Described inflammatory disease of arteries (Kussmaul's disease), 1866; 1st to use esophagoscope clinically, 1867; 1st to mention use of stomach pump, 1869; described labored breathing in diabetic coma (Kussmaul's sign), 1874; research leading to devel. ophthalmoscope. Died 1902.

KUSTER, Eberhard, microbiologist; b. Bonn, Germany, Oct. 22, 1918; s. Ernst and Gertrud (Winkelmann) K.; Diplom-Landwirt, U. Göttingen, 1946, Dr.-rer.nat., 1949; m. Waltraut Dierker, Sept. 26, 1945; children—Sylvia, Eckart. Head microbiology unit Inst. Soil Biochemistry, Agrl. Research Centre, Braunschweig-Völkenrode, Germany, 1949-59; head dept. indsl. microbiology U. Coll., Dublin, Ireland, 1959—. Sec., Internat. Subcom. on Taxonomy of Actinomycetes, 1958—. Mem. Royal Irish Acad. Dublin, German Bot. Soc., German Soc. for Hygiene and Microbiology, Soc. for Gen. Microbiology, Am. Soc. for Microbiology, Soc. Chem. Industry. Research, publs. on transformation of soil organic matter to humic acids by micro-organisms, microbiology of peat and its industrial uses, thermophilic micro-organisms, taxonomy of actinomycetes. Home: 5 Roebuck Crescent, Clonskeagh, Dublin 14, Ireland.*

KÜSTER, Ernst, German botanist; b. Breslau, Germany (now Wroclaw, Poland), June 28, 1874; ed. U. Munich, U. Leipzig, U. Berlin; Ph.D., Halle, Germany, 1900. Mem. staff univs. of Munich, Halle Kiel, Bonn (all Germany); prof. botany, dir. botany Inst. and Gardens, Giesen (Germany) U., 1920—. Mem. Royal Micros. Soc. London, Dutch Acad. Natural Sci.; corr. mem. acads. sci. of Switzerland, Vienna, Budapest, Inst. Bologna. Author: Pathologische Pflanzenantomie, 1903; Die Gallen der Pflanzen, 1911; Botanische Betrachtungen über Alter und Tod, 1921; Anleitung zur Kultur der Mikroorganismen, 1922; Pathologie der Pflanzenzelle, 2 vols., 1929, 37; 100 Jahre Tradescantia 1933; Ergebnisse und Aufgaben der Zellmorphologie, 1942; Experimentelle Zelleforschung, 1948.

KÜSTNER, Friedrich, German astronomer; b. Mehlem, Germany, Aug. 22, 1856; prof., dir. Bonn (Germany) Obs.; Author: Untersuchung uber die Veränderlichkeit der Polhöhe, 1888; Spektrographische Bestimmung der Sonnenparallaxe, 1905; Katalog von 10663 Sternen, 1908; Photographische Vermessungen von Sternhaufen, 1920. Detected variations in elevation of the pole; photog. measurements of nebulae. Died Bonn, Oct. 15, 1936.

KUSUNOKI, Kou, Japanese geophysicist; b. Hokkaido, Japan, Nov. 18, 1921; s. Masaichi and Tsune (Mori) K.; B.Sc., Hokkaido U., Sapporo, Japan, 1944, postgrad., 1946, D.Sc., 1962; m. Sachiko Arakawa, Apr. 20, 1949; children—Jun, Shoji. Research scientist oceanographic sect. Inst. Low Temperature Sci., Hokkaido U., 1944-46, asso. prof. geophysics, 1946-65; chief 1st research lab. polar div. Nat. Sci. Mus., Ueno Park, Tokyo, 1966—. Mem. Japanese Antarctic Research Expdn., 1956-57. Mem. Nat. Antarctic Com., 1956-58, 64—. Mem. Glaciological Soc., Am. Geophys. Union, Japan Oceanographical Soc., Arctic Inst. N.Am. Research and publs. on phys. and chem. properties of sea ice in Hokkaido, phys. and chem. oceanography of Arctic Ocean and Antarctic Ocean, atmospheric turbulence, measurement of water content of fog, photointerpretation of Antarctic sea ice and land ice, measurements of ice drift in Okhotsk sea and Greenland Sea. Home: Kamisaginomiya 5-27-28 Nakano-ku, Tokyo. Office: Nat. Sci. Mus., Ueno Park, Tokyo, Japan.*

KUTHE, Karlheinz, German phytopathologist; b. Celle, Mar. 8, 1908; s. Heinrich and Paula (Weinsheimer) K.; Ph.D., U. Giessen (Germany); m. Friedel Held, Jan. 7, 1933; children—Hiltrud, Gernot, Ehrhard, Ulrich, Michel, Dietrich. Asst., Phytopath. Inst., Landesberg-Warthe, 1933-37; dir. Heppenheim sect. Office for Safeguard of Plants, Giessen, 1937-38; dir. Office for Safeguard of Plants, Salzburg, 1939-45; engaged in 1st actual work for protection of plants, 1945-50; with research inst. for work on potato bug (Colo. beetle), Darmstadt, 1950-53; dir. Giessen Service for Safeguard of Plants in Hessia, 1953—. Mem. Assn. German Phytopathologists, Religionsgemeinschaft Deutsche Unitarier. Research and publs. on Carpocapsa pomonella, Venturia inaequalis, also on simplicication of processes for plant protection, preventive methods against potato diseases. Home: Rabenweg 36, 63 Giessen. Office: Eichgartenallee 1, 63 Giessen, West Germany.

KUTKA, Mikulás, Czechoslovakian physician; b. Jovro, Czechoslovakia, Dec. 17, 1926; s. Vladimír and Agatha (Flenko) K.; M.U.Dr., Comenius U., Bratislava, Czechoslovakia, 1951, M.D., 1954; C.Sc., Slovak Acad. Scis., 1962; m. Anna Cizmár, Aug. 14, 1965. Asst. Inst. Physiology, Med. Faculty, Bratislava, 1950-51; physician dept. medicine mil. hosp., 1952-55; sci. worker Inst. Endocrinology, Slovak Acad. Scis. Lab. for Use Radioisotopes, Bratislava, 1956—, in charge dept. thyroid pathophysiology, 1962—; courses physiology at Comenius U., 1951-62; chief sci. investigator research contract with IAEA, 1965—. Mem. Commn. State Research Problem for Use of Radioisotopes in Medicine, 1964—. IAEA fellow, Rome, Italy, 1962-63. Editorial bd. Endocrinologia Experimentalis, 1964—. Research and publs. on some compounds in organisms using radioisotopes, math. models of kinetics, effect of surg. trauma on thyroid and kidney function. Home: 22 Vidlicova, Bratislava, Office: 1/a Obrancov mieru, Bratislava, Czechoslovakia.*

KÜTZING, Friedrich-Traugott, German botanist; b. Rittebourg, Germany, 1807; student natural scis., Halle U. (Germany); became pharmacist; later traveled in So. Europe, studying flora of Adriatic Coast; tchr. natural sci., Nordhausen, Germany, 1838-83. Author: Sinopsis diotomearum, 1833; Phycologia generalis, 1843; Species algarum, 1849; Éléments d'une philosophie de la botanique, 1851-52; also paper on yeast plant, 1837. Studied marine plants of Mediterranean and Adriatic seas, 1835; independently discovered yeast cell and studied its chemistry; described numerous microscopic living organisms in infusions of various plants and solutions of organic compounds; studied philosophy of botany; gave theories on origin of species (similar to those later given by Darwin). Died Nordhausen, 1893.

KUTZSCHER, Edgar Walter, physicist; b. Leipzig, Germany, Mar. 21, 1906; s. Arno Fritz and Maria (Scherbe) K.; Ph.D. in Physics, U. Berlin, 1933; Dr.phil.habil. in Applied Physics, Inst. Tech. Berlin, 1936; Dr.Engring. (h.c.), Tech. U., Hannover, 1963; m. Edith Hildegard Wagner, Dec. 1, 1945; children—Detlef Kurt, Bernd Michael. Came to U. S., 1947, naturalized, 1954. Physicist, German Dept. Def., 1934-37; Asst. prof. Inst. Tech., Berlin, Germany, 1937-45; dir. research Electroacustic Co., Kiel, Germany, 1937-45; dir. U. Extension, Flensburg, Germany, 1946-47; physicist U. S. Navy, 1947-51, Santa Barbara (Cal.) Research Center, 1951-53; dept. head Lockheed Aircraft Corp., Burbank, Cal., 1954—. Recipient Todt prize, Republic of Germany, 1944. Mem. Optical Soc. Am. Contbr. articles to tech. jours. Patentee in field. Research on photoconductive cells, infrared, optical and acoustical systems; discovered photoconductivity of lead sulphide. Home: 15450 Briarwood Dr., Sherman Oaks, Cal. 91403. Office: Lockheed Aircraft Corp., Burbank, Cal.*

KUWADA, Hikaru, Japanese geneticist; b. Osaka City, Japan, Mar. 15, 1918; s. Keiji and Kito USUI K.; B.S. in agriculture Kyoto Imperial U., 1941, Ph.D., 1961; m. Sumie Yamada, Dec. 16, 1945; children—Noriko, Hiroshi. Asst., Kyoto Imperial U., 1942-49; prof. Kagawa Agrl. Coll., 1949-54, prof., 1954—, librarian, Br. Library Faculty of Agriculture, 1958—, councilor, 1964—. Mem. Japanese Soc. Breeding, Genetic Soc. Japan, Crop Sci. Soc. Japan, Japanese Soc. Hort. Sci. Studies, numerous publs. on interspecific and intergeneric hybridization and newly artificially raised amphidiploid plants in Malvaceae. Home: 1737 Kita-cho, Takamatsu City, Japan. Office: 2393 Ikenobe, Miki-cho, Shikoku, Japan.*

KUWATA, Tsutomu, Japanese chemist; b. Fukuyama City, Hiroshima Prefecture, Oct. 24, 1902; grad. Tokyo U., 1925. Instr. chemistry Tokyo U., 1925-32, asst. prof., 1932-41, prof., 1941—. Recipient Blue Ribbon Medal of Honor for work on aromatic essence, 1950. Author: Oil and Fats Chemistry; Solvents, others. Research on oil, fats, lubricants, perfume.

KUWATA, Yoshinari, Japanese botanist; b. Osaka, Japan, 1882; grad. Tokyo U., 1908; D.Sc.; postgrad. in U. S. and Eng. Prof., Kyoto U., 1923-43, prof. emeritus, 1943—. Mem. Japan Acad. (prize for research in spiral structure of chromosomes 1953). Author: Structure of Chromosomes; Dissolution of Prasenchyma, also others.

KUZIN, Aleksandr Mikhaylovich, Russian biochemist; b. Moscow, 1906; grad. Moscow U., 1929; D.Biol. Sci. Head lab. isotopes and radiation biol. dept. USSR Acad. Sci., 1951, now head lab. chemistry of plant substances Inst. Biochemistry, also prof. Mem. USSR Acad. Scis. (corr.). Author: Modern Concepts in the Chemistry of Simple Carbons, 1939; Organic Catalysts in Sugar Synthesis, 1940; Specific Polysaccharides, 1940; The Chemistry and Biochemistry of Pathogenic Microbes, 1946; The Chemistry of Antigens, 1947. Address: Inst. of Biochemistry, USSR Acad. Scis., Leninsky prospect 33, Moscow USSR.

KUZMA, Yurij Bohdanovich, Russian chemist; b. Lviv, Ukraine Ukrainian SSR, Oct. 26, 1934; s. Bohdan and Olga Kuzma; chemist, U. Lviv, 1955, candidate chem. scis., 1963; m. Olena Topinko, July 1956; 1 child, Zvenyslava. Lab. asst. U. Lviv, 1955-59, research student, 1959-62, asst., 1962-66, asst. prof., 1966—, dean chem. dept., 1967—. Mem. Mendeleev Chem. Soc. USSR. Contbr. papers to sci. publs. Research in crystal chemistry of intermetallic compounds, phase equilibria. Home: 100 Margarynov, Lviv, Ukrainian SSR.*

KUZNETSOV, Iurii Alekseievich, Russian geologist; b. Apr. 19, 1903; grad. Tomsk U., 1924; faculty Tomsk Poly. Inst., 1930—, prof., 1938—; with Geol. Survey Inst., until 1938, then with Ind. Inst.; asso. with Siberian geol. survey. Corr. mem. USSR Acad. Scis. Author: The Petrology of the Pre-Cambrian of the South Yenisey Mountain Ridge, 1941; The Origin of Magmatic Rocks, 1955; The Facies of Magmatic Rocks, 1955; The Special Role of Granitoid Intrusions in the History of Magmatism in the Altay-Sayan Plicated Area, 1960. Research on stratigraphy, techtonics, petrology, metallogeny of Altay, Kuznets, Eastern Sayan and Yenisey mountain ridges; systematized characteristics of tectonic structures. Address: Tomsk Polytechnic Inst., USSR Acad. Scis., Tomsk, USSR.

KUZNETSOV, Sergei Ivanovich, Russian microbiologist; b. Poretskoe (now Vladimir Oblast), 1900; grad. Moscow State U., 1923, postgrad., 1925-31. Chemist-bacteriologist at hydrobiol. sta., 1920-25; asst. Moscow State U., 1925-31; chief microbiol. lab. of limnological sta., 1931-41; sr. sci. worker Lublin plant for decontamination of sewer water, 1941-46; with USSR Acad. Sci. Inst. Microbiology, 1942, head dept. microorganisms, 1946—, corr. mem. acad., 1960—. Mem. Biology of Reservoirs, Verkhnye-Nikolskoye Acad. Scis. Author publs. including: The question of the possibility of radiosynthesis, 1956; (with others) An experiment on suppressing the development of sulfate-reducing bacteria in the oil strata of the Kabriva field, 1957; Principal results in the investigation of the microflora of oil deposits, 1957; The geological activity of microorganisms, 1959; (with Zhadin, Timofeevresovsky) The role of radioactive isotopes in solving the problems of hydrobiology, 1958; (with Kabanova, Pishchurina) Fluorescent antibodies and their use in cytology and microbiology, 1957; (with Sokolova) Contributions to the physiology of thiobacillus thioparus, 1960; (with Pantskhava) Effect of methane-forming bacteria on increasing the electrochemical erosion of metals, 1961. Research on geol. activity, physiology of microbes. Home: M. Kolkhoznaya pl. 1/3, Moscow. Office: Institute of Microbiology of USSR Academy of Sciences, Leninskii Prospekt, 33, Moscow, USSR.

KUZNETSOV, Valeriy Alekseevich, Russian geologist; b. Apr. 12, 1906; grad. Tomsk Geol. Survey Inst., 1932. With West Siberian geol. instns., 1932-44; asso. Mining and Geol. Inst., West Siberian Br., USSR Acad. Sci., 1945—; mem. USSR State Com. for Coordinating Sci. Research. Mem. USSR Acad. Sci. (corr., Presidium's prize 1946, 53). Author: Basic Problems Relating to the Stratigraphy and Tectonics of Central and Western Tuva, 1953; The Geotectonic Zoning of the Altay-Sayan Plicated Area, 1954; Symmetrical Forms of Bodies of Natural Abrasion, 1963; co-author: The Hyperbasic Rocks of Tuva, 1955. Research and publs. on mercury deposits, metallogeny of mercury, tectonic structure and magnetism in Altay, Tuva and Altay-Sayan mountain area, laws governing distbn. of mercury deposits in Western Siberia. Address: USSR State Com. for Coordinating Sci. Research, ulitsa Gorkogo 11, Moscow, USSR.

KUZNETSOV, Vladimir Dmitrievich, Russian physicist; b. May 13, 1887; grad. St. Petersburg U., 1910. Mem. various instns. of higher learning, Tomsk, 1911—; mem. staff Tomsk U., 1917—, prof., 1920—;

organizer Tomsk Physico-Tech. Inst., 1929, now asso. Decorated Order of Lenin; recipient Stalin prize, 1942. Mem. USSR Acad. Scis. (corr.). Author: Solid-State Physics, 5 vols., 1937-49; Crystals and Crystallization, 1953; The Surface Energy of Solid Bodies, 1954; Accretions during Cutting and Friction, 1956. Editor: Trans. of Siberian Physico-Tech. Inst. at Tomsk, 1959. Research on properties of solid bodies and phenomena arising in solid bodies in process of their manufacture; worked out basic phys. theory for rapid cutting of metals. Address: Presidium of Siberian Dept., USSR Acad. Scis., Novosibirsk, USSR.

KUZNETSOV, Yefrem Aleksandrovich, Russian petrographer; b. 1892; grad. Moscow U., 1917. Instr., Moscow Mining Acad., Moscow Geol. Survey Inst., Moscow U., 1917-39; head chair petrography Moscow U., 1939——. Co-author: Petrographical Provinces of the USSR, 1936; author: The Tectonics of the Central Urals, 1941; The Petrography of Magmatic and Metamorphic Rocks, 1956. Research and publs. on geology and petrography in Central Urals, stratigraphy of northern Pay Khoy and Yugorsky Peninsula, Urals greenstone belt, alkaline rocks of Kaslya and Kyshtym. Address: Moscow University, Leninskie gory, Moscow, USSR.

KVASNITSKII, Aleksei Vladimirovich, Ukrainian physiologist; b. Feb. 24, 1900; grad. Kamenets-Podolsky Agrl. Inst., 1925; tchr. agr., Vinnitsa; sci. staff Hog-breeding Inst., Poltava, 1931-41; prof. Poltava Agrl. Inst., 1942-52; dir. lab. for physiology of farm animals, physiol. inst. Ukrainian SSR Acad. Scis.; became dir. lab. Poltava State Exptl. Stat., 1956. Mem. Ukrainian Acad. Sci., Ukrainian Acad. Agrl. Sci. Author: Problems of the Physiology of Digestion in Pigs, 1951; New Developments in the Physiology of Animal Reproduction, 1950; (with V. A. Koniukova) Applications of I. P. Pavlov's Principles to Livestock Breeding, 1954; The State and Prospects of the Introduction of Artificial Insemination of Pigs, 1960; How To Avoid Mistakes in Artificial Insemination, 1960. Introduced the original method in inter-breed transplantation of fertilized egg cells in farm animals. Address: Shvedskaya mogila, Selskokhozyaystvennaya opytnaya stantsiya, A. USSR, Poltava, Ukr. SSR.

KVATER, Yefim Ilich, Russian gynecologist; b. 1894; student Med. Faculty, Zurich U.; grad. Med. Faculty, Kazan U., 1920; D.Med. Sci., 1935. Intern, asst. clinic obstetrics and gynecology 1st Moscow U., 1921-30, lectr., 1931-34; head chair obstetrics and gynecology Health Hygiene Faculty, Sechenov 1st Moscow Med. Inst., 1934-52; prof., 1934——; head Clinic Physiology and Pathology of Pregnancy, 1956——; head endocrinological clinic Inst. Obstetrics and Gynecology, RSFSR Ministry Health, Moscow; sci. dir. Moscow Maternity Home Number 12. Mem. All-Russian Soc. Obstetricians and Gynecologists (bd. mem.). Author: Hormone Diagnosis and Treatment in Obstetrics and Gynecology, 1956, 2d edit., 1961. Mem. editorial bd. Med. Abstracts Jour.; dep. editor Obstetrics and Gynecology sect. Large Med. Ency., 2d edit. Research and publs. on gonorrhoea in women, role of sex hormones in normal and path. menstrual cycle; described Trichomonas diseases of urethra and paraurethral ducts in woman; introduced use of mammofizin preparation; suggested methods of hormone therapy of metrorrhagia, labor stimulation, conservative treatment of uterine myoma. Home: Mozhayskoe sh. 52-70. Office: Inst. Obstetrics and Gynecology, B. Pirogovskaya ulitsa 2-6, Moscow, USSR.

KWAPINSKI, Jerzy, microbiologist, serologist; b. Piotrkow, Poland, July 15, 1920; s. Boleslaw and Helena (Chvedcak) K.; M.D., Warsaw U., 1948, C.Sc. in Microbiology, 1956. Asso. prof. microbiology Med. Acad., Warsaw, 1950-56; WHO Research fellow State Serum Inst., Copenhagen, London Hosp. Med. Coll. U. London, 1957-58; chief sect. microbiology pneumocon. research unit S. African Inst. Med. Research, Johannesburg, 1958-60; head sect. microbiology U. New Eng., Australia, 1961-66; prof. microbiology and immunology Miss. State U., Starkville, 1966——. Fellow U. Vienna, U. Bonn; WHO research fellow, 1966. Author numerous books the most recent being: Immunology of Rheumatism, 1962; Compendium of Microbiology, 1963; Methods of Serological Research, 1965; Analytical Serology of Microorganisms, 1967. Contbr. numerous articles to profl. jours. Devel. anti-tuberculous vaccines obtained by transformation; elucidation of serological relationships between higher bacteria and their serological classification; devel. new serological methods. Address: P.O. Box 2862, State College, Miss. 39762.*

KWIETNIEWSKI, Casimir, Polish zoologist; b. Warsaw, Poland, Apr. 3, 1873; s. Wladyslaw and Maria (Grabowska) K.; ed. Poly. Sch. Zurich, U. Munich, U. Jena; Dr. Phil.; m. Maria Grosser, 1909; Asst., zool. inst., Messina, Italy, 1900-05; asst. lectr., prof. incaricato U. Padova (Italy), 1900-05; docent U. Pisa (Italy), 1907; prof. zoology and comparative anatomy U. Lvov (Poland), from 1907, dean philos. faculty, 1922-23. Mem. Sci. Soc. Lvov, Anat. and Zool. Soc., Biol. Soc. Research and publs. on anatomy and classification of Actinaria corals, anatomy, embryology and classification of Mollusca Pteropods, micros. anatomy of skin of Elasmobranche fishes, embryological devel. of urogenital system of mammals. Address: Dwernickiego str. 11, Lvov, Poland.

KYAN, John Howard, inventor; b. Dublin, Ireland, Nov. 27, 1774; s. John Howard Kyan; trained to take charge of copper mines, owned by his father, Wicklow; joined Greave's Vinegar Brewery, London. Author: The Elements of Light and their Identity with those of Matter radiant or fixed, 1838. Began studies on preservation of wood, 1812; discovered and patented use of bichloride of mercury or corrosive sublimate for preservation of wood (Kyanising process), 1832. Died Jan. 5, 1850.

KYAS, Otto, Czechoslovakian agrl. chemist; b. Brno, Czechoslovakia, Dec. 9, 1876; s. Jan and Anna (Dvoràk) K.; ed. Coll. Brno, 1884-94; grad. Tech. U.; m. Helen Kalda, 1910. Chemist, distillery at Hodolany, nr. Olomouc, 1899; chief lab., sugar plant at Bedihost, nr. Prostejov, 1899-1900; with Agrl. Research Inst., Brno, 1900——, dir. inst., 1918——, chief of state inst. for protection against adulteration of food-stuffs, 1925——, also asst. prof. agrl. chemistry Brno Tech. Hochschule. Mem. Masaryk Acad. Sci., Prague (pres. agr. and forestry div. 1923-25), Internat. Soc. Soil Sci. (del. congress, Munich 1941), Assn. Agrl. Expt. Insts. and Stas. in C.S.R. (past v.p.), Deutsche Chemische Ges., Société Chimique de France, Am. Chem. Soc., Chem. Soc. Eng. Research and publs. on biochemistry, plant nourishment, fertilization of soils, control of fertilizers, feeding stuffs. Address: Kvetná ul. 19, Brno, Czechoslovakia.

L

LAASONEN, Pentti Johannes, Finnish mathematician; b. Helsinki, Finland, Mar. 14, 1916; s. G. J. and Mildred (Stadius) L.; M.A., 1937, Ph.D., 1944, U. Helsinki; m. Saara Saastamoinen, Apr. 27, 1945; children: Helena, Annikki, Inkeri, Marjatta. Chief mathematician, State Airplane Factory, Tampere, 1943-48; prof. of Strength of Materials, Technical U. Helsinki, 1948-61; prof. Applied Math., U. Helsinki, 1961-62; Prof. Math., Tech. U. Helsinki, 1962——; chief mathematician, State Institute for Technical Research, 1949-67. Decorated Commander of Order of Finnish Lion; IV class Cross of Freedom. Mem., Acad. of Scis. (treas. 1961——); rep. of math., State Committee of Sci., 1964——; mem. several Finnish and fgn. socs. Investigations in differential equations, theory of elasticity. Home: Mäkipellontie 6B, Helsinki 32, Finland. Office: Technical University, Helsinki-Otaniemi, Finland.

LA BACH, James Oscar, Am. chemist; b. Newcastle, Ind. Apr. 22, 1871; s. James Mayer and Cornelia Esther (Ryker) L.; student Carleton Coll.; B.S., U. Tenn., 1895, M.S., 1897; m. Mary Shepherd Parker, Oct. 10, 1917. Asst. chemist, food nutrition work, U. Tenn., 1895-97; chemist Procter & Gamble Co., Cin., 1897-1901; chief chemist Ky. Food and Drug Control Labs., at Ky. Agrl. Expt. Sta., U. Ky., Lexington, from 1901, also head of food and drug control dept., 1916-18; dir. labs., Ky. State Bd. Health, from 1918; collaborating chemist, U. S. Bur. Chemistry. Died Aug. 31, 1922.

LABARRAQUE, Antoine-Germain, French chemist; b. Oloron, France, 1777; studied under Chaptal. Requisitioned into the grenadiers of La Tour d'Auvergne; named pharmacist in chief, Spanish Hosp., Berra, Italy; went to Montpellier, France, and from there to Paris. Author: Sur la dissolution du phosphore, 1805; Sur les electuaires, 1805; De l'emploie des chlorures de sodium et de chaux, 1825; Manière de se servir du chlorure de sodium, 1825; Notes sur les asphyxies produites par les émanations des fosses d'aissences, 1825. Prepared a solution of chlorinated soda for disinfectant (formerly known as Labarraque's solution, or sodium hypochlorite solution, now as Javelle water), 1820. Died Paris, 1850.

LA BARRE, Weston, Am. anthropologist; b. Uniontown, Pa., Dec. 13, 1911; s. Isaac Weston and Artemisia Van Meter (Hannah) LaB.; A.B. summa cum laude, Princeton, 1933; Ph.D. with honors, Yale, 1937; m. Maurine Boie, July 9, 1939; children—John Boie Keasbey, David Quinton Lefebvre, Louise Anne Stephens. Fellow Santa Fe Lab. Anthropology, 1935, Am. Mus. Natural History, also Yale Inst. Human Relations, 1935-36; Sterling fellow, Yale, 1937-38; Social Sci. Research Council post-doctoral research tng. fellow Menninger Clinic, 1938-39; community analyst War Relocation Authority, Topaz, Utah, 1943; mem. faculty Rutgers U., 1939-43; faculty U. N.C. Med. Sch., Durham, 1955——; prof. anthropology, Duke, Durham, 1958——. Permanent cons. com. adolescence Group for Advancement Psychiatry, 1959——; mem. various coms. NRC, 1948-50; mem com. racial minorites Am. Orthopsychiat. Assn., 1961. Recipient Roheim Meml. award Post-doctoral Inst. Psychoanalytic Research, 1958. Guggenheim fellow, 1946; NSF fellow, 1962, 66-67; Viking Fund fellow, 1962-63. Fellow Am. Anthrop. Assn., Current Anthropology; mem. Phi Beta Kappa, Sigma Xi. Author: The Peyote Cult, 1938; Aymara Indians, 1948; The Human Animal, 1954; Materia Medica of the Aymara, 1960; They Shall Take Up Serpents, 1962; also numerous articles, revs. specializing in ethnography, holistic synthesis of cultural and phys. anthropology with psychoanalysis; specialist in native narcotics, primitive and folk religions, psychoanalytic culture and personality studies. Home: Mt. Sinai Rd., Route 1, Durham, N.C. 27705.*

LA BAUME PLUVINEL, Eugène-Aymar, French astronomer; b. Paris, Nov. 6, 1860; mem. French Acad. Scis., 1932. Author: La photographie au gélatinobromure d'argent, 1891. Research in astron. physics; made detailed observations of 9 eclipses, 1889-1932; pioneer in photography of planetary surfaces in monochromatic light, 1909. Died Vic-sur-Cère, France, July 18, 1938.

LABBÉ, Léon, French surgeon; b. Merlerault, France, 1832; ed. U. Caen (France), also Paris; M.D., 1861; qualified as surgeon, 1863. Became surgeon of hosps., 1864; named prof. Faculty of Medicine, Paris, 1864; elected senator of Orne, France, 1892, 1900 (caused passage of Léon Labbé law which required antityphus vaccination in the French army). Mem. Acad. Medicine, French Acad. Scis. 1903. Author: Quelques Réflexions au sujet du traitiment des fistules génito-urinaires chez femme, 1869; Traité des tumeurs benignes du sein, 1876; Leçons de clinique chirurgicale, 1876; Opération de la fourchette, 1876; (with C. Remy) Traité des fibromes de la paroi abdominale, 1888. Showed gastric operations could be performed by all surgeons following well formulated rules; (with E. Guyon) used morphine as a preanesthetic medication, 1872. Died Paris, Mar. 21, 1916.

LABBE, Robert Ferdinand, Am. biochemist; b. Portland, Ore. Nov. 12, 1922; s. Frerdinand D. and Leona Ruth (Buchholz) L.; B.S., U. Portland, 1947; M.S., Ore. State U., 1949, Ph.D., 1951; m. Norma Lee Wiley, Aug. 21, 1955; children—Sharlene Renee, Yvonne Denise, Valerie Jean. AEC fellow Columbia U., 1951-53; faculty U. Ore. Med. Sch., 1953-57, asst. prof., 1955-57; research faculty U. Wash. Med. Sch., Seattle, 1957——, asso. prof., 1960——; vis. asst. prof. Enzyme Inst., U. Wis., 1957; vis. scientist div. plant industry Commonwealth Sci., and Indsl. Research Orgn., Canberra, Australia, 1965. Mem. Am. Soc. Biol. Chemists, Am. Chem. Soc., A.A.A.S., Am. Soc. for Cell Biology, Western Soc. for Clin. Research, Western Soc. Pediatric Research. Research and publs. on intermediary metabolism, biol. oxidation, heme biochemistry, inborn errors of metabolism, mental retardation. Home: 3705 N.E. 187th St., Seattle 98155.*

LABES, Mortimer Milton, Am. chemist; b. Newton, Mass., Sept. 9, 1929; s. Phillip and Anna (Seifer) L.; A.B., Harvard, 1950; Ph.D., Mass. Inst. Tech., 1954; m. Mary Lawson Wright, June 23, 1953; children—Karen, Daniel, David, Jonathan, Peter, Miriam. Research chemist Sprague Electric Co., 1953-57; sr. research chemist Franklin Inst. Research Labs, Phila., 1957-59, sr. staff chemist, 1960-61, lab. mgr., 1961-62, tech. dir., 1962-67. Prof. chemistry Drexel Inst. Tech., Phila., 1966——. Mem. Am. Chem. Soc., Am. Phys. Soc., Chem. Soc. London, Sci. Research Soc. Am. Editor: (with D. Fox, A. Weissberger) Physics and Chemistry of Organic Solid State, Vol. I, 1963, Vol. II, 1965, Vol. III, 1967. Co-editor, Molecular Crystals, 1966——. Research and publs. on electronic properties of simple and polymeric organic crystals, kinetics of organic reactions, novel detection systems, molecular complexes, novel polymerizations. Home: Box 534 S. Roberts Rd., Rosemont, Pa. 19010. Office: Drexel Inst. Tech., Phila. 19104.*

LABHART, Heinrich, Swiss phys. chemist; b. Switzerland, June 27, 1919; s. Heinrich and Hedwig (Nussbaumer) L.; student Swiss Fed. Inst. Tech., 1938-46; m. Yvonne Vuillien, 1948; children—Thomas, Martin, Christoph. Staff, Abteilung für industrielle Forschung, ETH, Zurich, Switzerland, 1947-51; with CIBA, Ltd., Basel, Switzerland, 1951-64; lectr., 1951-58; asso. prof., Basel, 1958-64; prof., head phys. chemistry Inst., U. Zurich, 1964——. Mem. bd. Swiss Nat. Sci. Found. Publs. on theoretical and exptl. work on light absorption of organic molecules, exptl. method for determination of orientation of optical transition moments in molecules and for evaluation of charge distbn. in electronically excited molecules; contbd. to phys. understanding of photochem. reactions. Home: 430a Steinradstrasse, 8704 Herrliberg, Switzerland. Office: 76 Ramistrasse, 8001, Zurich, Switzerland.*

LA BILLARDIÈRE, Jacques-Julien Houtou de, French botanist; b. Alencon, France, Oct. 28, 1755; explored Cyprus, Syria, Mt. Lebanon; naturalist in expdn. sent in search of La Pérouse, 1791. Author: Icones plantarum Syriae, 1791-1812; Novae Hollandiae plantarum specimen, 1804-06; Relation du voyage á la recherche de La Pérouse pendant les années 1791-1792, 1800. First bot. study of plants of So. hemisphere; bot. research in Mediterranean, around Australia and Java. Died Jan. 8, 1834.

LA BLANCHÉRE, Pierre-René-Marie-Henri Moulin du Coudray, French naturalist; ed. Sch. Forestry; b. La Fleche, France, 1821; a son, René de. Author: Plantes et animaux, 1867; Nouveau dictionnaire général des peches, 1867; Voyage au fond de la mer, 1868; les Oiseaux utiles et les oiseaux nuisibles, 1870; la Peche en eau douce, 1881; Récits de peche; et de voyages, 1885; les Amis des plantes et leurs ennemis, 1885; Choses et autres, causeries de l'oncle Tobie, 1885; les Animaux racontés par eux-memes, 1885. Prepared (with Faye) 1st photographs of sun, Paris, 1855; introduced use of dry collodion. Died 1880.

LA BOË, see Sylvius de la Boë, Franciscus.

LABORDE, Jean-Baptiste-Vincent, French physician; b. Buzet, France, 1831; chief physiol. work on staff Faculty of Medicine. Research in physiology, including heart, respiratory, cerebral localization and bulbaires, reflexes, temperature; studied supplicies; originated method of rhythmical tractions of the tongue for eliciting the respiratory reflex in cases of apparent death. Died Paris, 1903.

LABORIT, Henri-Marie, physician; b. Hanoi, Indo-China, Nov. 21, 1914; s. Henri-Ferdinand and Denise (de Sauniere) L.; B.A., Carnot U., Paris, France, 1931; med. studies Naval Health Service, Bordeaux, 1934, M.D., 1938; m. Genevieve de Saint-Mart, Dec. 19, 1936; children—Marie-Christine, Marie-Noëlle, Henri-Philippe, Jacques, Jean. Hosp. extern, 1935, intern, 1936; commd. lt. French Navy, 1937, advanced through grades to colonel, 1963; physician torpedo boat, Sirocco, 1939; physician small warships, also surg. dept. at Dakar, Oran, 1941-45; surgeon 4th Cruiser Div., Anzio and So. France; asst. surgeon, Toulon, 1945-47; chief physician and surgeon French Navy, 1948-60, director med. physiol. research, 1951-60; maitre de recherches des Services de Sante des Armées, 1960—. Attached to the technical section on research and study Health Service of French Army, Val-de-Grace Hospital, Paris, 1951; staff Mission in Indo-China for application artificial hibernation war casualties, 1953. Decorated Knight Legion of Honor, Croix de Guerre (France); recipient Albert Lasker award American Pub. Health Assn., 1957. Mem. Acad. Surgery (nat. asso.). Author books, numerous sci. articles. Introduced use of artificial hibernation in surgery and use of chlorpromazine as a tranquilizer for pregnant women. Home: 93 rue de la Santé, Paris. Office: Hôpital Boucicaut, 78 rue de la Convention, Paris 15, France.

LABOULAYE-MARILLAC, Pierre-Charles-Madeleine de, French chemist; b. Billom, 1771; lt. col. during Revolution; offered himself as a hostage for Louis XVI; emigrated later and served in army of princes; returned to France under Consulat; became dir. Manufacture des Gobelins, 1817; later comptroller of expenses of house of king. Author: Mémoires sur les couleurs inalterables pour la teinture, 1814. Chem. research on fabric dyeing. Died 1824.

LABOULBÈNE, Jean-Joseph-Alexandre, French physician; b. Agen, France, 1825; prof. history of medicine Faculty of Paris (France); mem. Acad. Medicine. Author: Faune entomologique française, 1854; Recherches chimiques et anatomiques sur les affections pseudo-membrative et histologique; Histoire de la médecine, 1878; l'Hôpital de la Charité, à Paris, 1879. Died Paris, 1898.

LA BRECQUE, Germain Claude, entomologist; b. Drummondville, Que., Can., July 9, 1922; s. Josaphat L. and Rezeda (Theroux) LaB.; came to U. S., 1923, naturalized, 1943; B.S., Tufts U., 1949; M.S., U. Tenn., 1950; Ph.D., Ohio State U., 1957; m. Virginia A. Behnke, Dec. 31, 1952; children—Donald M., Denise K. Med. entomologist entomology research div. insects affecting man and animals, research br. U. S. Dept. Agr., Orlando, Gainesville, Fla., 1951—. Cons. Naval Med. Research Unit 3, Cairo, UAR, 1952-53, 65—, Sudan, Africa, 1964; asso. prof. U. Fla., Gainesville, 1964—. Mem. Entomol. Soc. Am., Fla. Entomol. Soc. (award for contbns. to insect control by chemasterilants 1963) Sigma Xi. Editor: (with Carroll N. Smith) Insect Chemosterilants, 1967; also numerous articles. Pioneered demonstration that satisfactory control sand flies could be obtained by chem. larvicides; discovered and proved physiol. resistance to organophophorus insecticides in house flies in Fla.; demonstrated potentialities of chemosterilant technique for practical use in control of house flies. Home: 213 N.W. 29th St., Gainesville 32601. Office: Entomology Research Div., U. S. Dept. Agr., Box 1268, Gainesville, Fla. 32601.*

LABY, Thomas Howell, Australian physicist; b. Victoria, Australia, 1880; student U. Sydney (Australia), Emmanuel Coll., Cambridge, Eng.; M.A., Sc.D., M.A., U. Melbourne (Australia); m. B. Littlejohn, 1914; 2 daus. Demonstrator chemistry U. Sydney, 1900-05; prof. physics, Wellington, New Zealand, 1909-15; prof. natural philosophy U. Melbourne, 1915-44; adviser Australian Govt. on X-rays and radium, 1928-37; mem. Radio Research Bd., 1928-40; chmn. Optical Munitions Panel of Australia. Fellow Royal Soc., 1931; mem. Australasian Assn. for Advancement Sci. (pres. sect. A), Inst. Physics (pres. Australia 1939-41). Author: (with Kaye) Physical and Chemical Constants, 9th edit.; also papers on X-rays, heat, reflection atmospherics by ionosphere, precision measurements of fundamental phys. constants. Editor: (with A. B. Edge) Principles and Practice Geophysical Surveying. Died June 21, 1946.

LABZOFFSKY, Nicholas Andrew, virologist; b. Verchneudinsk, Russia, Jan. 6, 1910; s. Andrew and Alexandra (Grudinin) L.; B.S.A., U. B.C., Vancouver, 1934; D.V.M., Ont. Vet. Coll., Guelph, Can., 1939; D.V.Sc., U. Toronto (Ont.), 1943; m. Janet Shelsy, June 30, 1945. Chief virologist, head virus lab. Pub. Health Lab. Service, Ont. Dept. Health, Toronto, 1943—. Mem. Am. Soc. for Microbiology, Canadian Soc. Microbiologists. Fellow Royal Soc. Health. Research, numerous publs. on antigenic structure of tibacillus and some fungi, tularemia among beavers and muskrats, sero-diagnosis of gonorrhea, leptospirosis, trichinosis, toxoplasmosis, psittacosis, environmental factors affecting survival of equine encephalomyelitis virus, immunofluorescence in diagnosis of viral infections, efficacy of poliomyelitis vaccine, prevalence of viral infections in Ont. Home: 106 Exbury Rd., Downsview, Ont. Office: Pub. Health Labs., Box 9000, Terminal A, Toronto, Ont., Can.*

LACAILLE, Nicolas Louis de, French astronomer; b. Rumigny, France, Mar. 15, 1713. Educated for church, became abbé; protege of Duke of Bourbon; worked with Cassini in French survey; employed (with Maraldi) to plot geographically French coasts between Nantes and Bayonne, 1738; became asst. on commn. to verify meridian, 1739; became prof. math. Collège Mazarin, Paris, 1739; became dir. sci. expdn. to Cape of Good Hope, 1750, and remained for 4 years. Mem. French Acad. Scis. Author: Leçons élémentaires de mathematiques, 1741; Leçons élémentaires de mécaniques, 1743; Leçons élémentaires d'astronomie geometrique et physique, 1746; Leçons élémentaires d'optique, 1750; Astronomiae fundamenta, 1757; Tables de logarithmes, 1760; Tabulae solares, 1758; Coelum australe stelliferum, 1760; Australe stelliferum, 1763; Journal historique du voyage fai au cap de Bonne-Esperance, pub., 1763; also numerous articles. Determined positions of nearly 10,000 stars at Cape Town, S. Africa, and formed 14 new constellations; 1st to measure a S. African arc of meridian; determined lunar and solar parallax; observations on inclination and declination of magnetized alloys; made more accurate estimate of distance of moon, taking into account that earth is not a perfect sphere, 1761; his so. observations with similar observation in no. hemisphere led to more accurate values for distances of moon, Venus, Mars. Died Paris, Mar. 21, 1762.

LACASSAGNE, Antoine Marcellin Bernard, French physician; b. Villerest, France, Aug. 29, 1884; s. Alexandre and Magdeleine (Rollet) L.; ed. Lycée Ampere, degree in natural scis.; M.D., Faculty Medicine and Scis., Lyons. Chief of lab., later head of services Pasteur Inst.; dir. biol. sect. Radium Inst., 1937-55, hon. dir., 1955—; prof. Coll. of France, 1941, later hon. prof. Recipient UN prize for work on cancer, 1962. Mem. French Acad. Scis., 1949, Nat. Acad. Medicine, Acad. Surgery. Notable work in radiology, cancerology, hormonology; discovered that breast cancer can be induced in animals by estrone benzoate, 1932. Office: Institut du Radium, 26 rue d'Ulm, Paris 5, France.

LACASSAGNE, Jean Alexandre Eugène, French physician; b. Cahors, France, 1843; held chair in dept. legal medicine Faculty of Lyons (France); author of Précis de médecine judiciaire, 1878; founder Archives d'anthropologie criminelle. Died Lyons, 1924.

LACAZE-DUTHIERS, (Felix) Joseph Henri de, French anatomist, biologist, zoologist; b. Montpezat, May 15, 1821; M.D., Paris; physician, founder seaside stas. in maritime zoology, Roscoff, 1873, Banyuls; named prof. zoology, Lille, France, 1854; became prof. malacology at Muséum, 1865, Sorbonne, Paris, 1868. Fellow Royal Soc., 1897; mem. French Acad. Scis., Acad. Medicine. Author: Histoire naturelle du corail, 1863; Histoire de l'organisation et du développement . . . du dentale, 1858; le Monde de la mer et ses labratoires, 1889; also numerous articles. Research on anatomy of mollusks, reprodn. among various types of insects; founder exptl. zoology in France. Died La Fons, France, July 21, 1901.

LACEY, William Noble, Am. chem. engr.; b. San Diego, July 25, 1890; s. David Sherman and Charlotte (Noble) L.; B.A., Stanford, 1911, Chem.E., 1912; M.S., U. Cal., Berkeley, 1913, Ph.D., 1915; m. Ruth Elder, Aug. 11, 1917 (dec. Nov. 1960); children—Margaret (Mrs. P. V. Laylander), David Sherman, II; m. Madeline Hawley McClellan, July 15, 1963. Research chemist Giant Powder Co., San Francisco, 1915; research asso. in phys. chemistry Mass. Inst. Tech., 1916; faculty Cal. Inst. Tech., 1916—, prof. chem. engring., 1931-62, prof. emeritus, 1962—, dean grad. studies, 1946-56, dean faculty, 1961-62. Cons. in patent litigation, 1930—. Recipient Hanlon award Nat. Gas Assn. Am., 1964, Lucas medal Am. Inst. Mining and Metall. Engrs., 1947, Presdl. certificate of merit Pres. Truman, 1948. Mem. Am. Chem. Soc., Am. Inst. Chem. Engrs. Author: Instrumental Methods Chemical Analysis, 1924; (with B. H. Sage) Volumetric and Phase Behavior of Hydrocarbons, 1939, Thermodynamic Properties of Hydrocarbons, 1950, Some Properties of Hydrocarbons, 1955, Thermodynamics of One-Component Systems, 1957; (with W. H. Corcoran) Introduction to Chemical Engineering Problems, 1960; also numerous articles. Pioneered research on exptl. determination of thermodynamic properties of hydrocarbons under conditions corresponding to those found in underground reservoirs of petroleum and natural gas. Home: 2928 2d Av., San Diego 92103.*

LA CHAMBRE, Marin Cureau de, see de la Chambre, Marin Cureau.

LA CHAPELLE, Abbé de, French mathematician; b. 1710; became royal censor. Fellow Royal Soc., 1747; mem. Acad. Lyons, Acad. Rouen. Studied geometry, especially conic sections; invented cork diving suit. Died Paris, 1792.

LA CHAPELLE, Edward Randle, Am. glaciologist; b. Tacoma, May 31, 1926; s. Edward K. and Mirtie (Randles) La C.; B.S. in Physics and Math., U. Puget Sound, 1949, Sc.D. (hon.), 1967; m. Dolores Greenwell, June 4, 1950; 1 son, Randall R. Snow and avalanche researcher Swiss Fed. Inst., 1950-51; with Am. Geog. Soc., 1952-56, project field leader, 1954-56; avalanche hazard forecaster U. S. Forest Service, 1952—; sr. scientist Blue Glacier project U. Wash., Seattle, 1957-68, vis. asso. prof. geophysics, since 1968—. Mem. Am. Geophys. Union (chmn. research com. on glaciers), Glaciological Soc. (gov.). Research and publs. on mass and heat budgets of temperate glacier, basal sliding mechanism and flow of glaciers, developer techniques for snow avalanche forecasting and control. Home: 12004 84th St. N.E., Kirkland, Wash. 98033. Office: Dept. Atmospheric Scis., U. Wash., Seattle 98105.*

LA CHAPELLE, Marie Louise Dugès, French physician; d. Louis Dugès; m. M. Lachapelle, 1792; assisted her mother who was sage-femme, head maternity services l'Hôtel-Dieu, and succeeded her after 1795. Author: Pratique des accouchements (based on 50,000 cases), 3 vols. Made numerous improvement in lying-in hosps. Died Paris, Oct. 4, 1821.

LACHER, John Robert, Am. chemist; b. Montrose, Colo., May 26, 1911; s. Walter and Nellie (Freese) L.; A.B., U. Colo., 1933; Ph.D., Harvard, 1936; m. Dorothy E. Dick, Apr. 11, 1931; children—Carol (Mrs. Lorrence Greenlee), John W. Research asst. Harvard U., 1937-44; instr. Brown U., 1938-41; research chemist Mallinchrodt, St. Louis, 1941-44; research chemist DuPont, Willington, 1944-45; prof. U. Colo., 1945—; chief scientific advisor Research and Engring. Command U. S. Army Chem. Corps., 1952-53. Mem. Am. Chem. Soc., Assn. Harvard Chemists, Phi Beta Kappa, Sigma Xi. Research, publs. in thermochemistry and photochemistry. Home: Geneva Park, Boulder, Colo. 80302.*

LACHMAN, Ernest, anatomist; radiologist; b. Glogau, Germany, Feb. 8, 1901; s. Nathan and Olga (Efrem) L.; M.D., U. Breslau, 1925; m. Anya Havkin, May 10, 1930. Came to U. S., 1934, naturalized, 1940. Asst. physician Rudolf Virchow Hosp., Berlin, Germany, 1928-30, asst. chief in radiology, 1930-33; asst. prof. anatomy U. Okla. Sch. Medicine, 1934-39, asso. prof., 1939-45, prof. anatomy, chmn. dept., 1945-68, prof. radiology, 1964—, Regents' prof. anatomy, radiology, 1967—. Diplomate Am. Bd. Radiology, Fellow A.M.A.; mem. Am. Coll. Radiology, Okla. State, Okla. County med. assns., Am. Assn. Anatomists, Am. Assn. Phys. Anthropologists, Soc. Exptl. Biology and Medicine, Sigma Xi, Phi Beta Pi, Alpha Omega Alpha. Author: Case Studies in Anatomy, 1965. Studies and numerous publs. on method of vitalizing teaching of gross anatomy by introduction of clin. case studies utilizing various audiovisual ednl. tools; research on x-ray anatomy of skeleton and viscera; life history of cranial sutures; topographical anatomy; X-ray biology and genetics; problems in diagnostic and therapeutic clin. roentgenology; anat. pathways in spread of tumors. Home: 920 N.E. 50th St., Oklahoma City 73105.*

LACHMANN, Frank Michael, psychologist; b. Breslau, Germany, Dec. 9, 1929; s. Hans and Kate (Landsberg) L.; came to U. S., 1938, naturalized, 1944; B.A., N.Y. U., 1951; M.A., Northwestern U., 1953, Ph.D., 1955; certificate in psychoanalysis Postgrad. Center for Mental Health, 1964; m. Annette Schamroth, July 15, 1962. Staff psychologist Montrose VA Hosp., 1955-56; sr. psychologist N.Y. Regional Office, VA, N.Y.C., 1956-60; asso. psychologist Postgrad. Center for Mental Health, N.Y.C., 1960—; research psychologist Jewish Bd. Guardians, N.Y.C., 1961—; lectr. psychology Pace Coll., 1957-58, Hunter Coll., 1958-60, Postgrad. Center for Mental Health, 1964—. Cons., Psychiat. Treatment Center, N.Y.C., 1963—. Diplomate Am. Bd. Examiners in Profl. Psychology. Mem. Am., N.Y. State, Eastern psychol. assns., Am. Orthopsychiat. Assn., N.Y. Soc. Clin. Psychology, A.A.A S. N.Y. State scholar, 1947-57; USPHS fellow, 1952-53; 54-55. Research on anxiety symptoms in psychiat. patients, dream recall and its relationship to other psychol. variables. Home: 221 W. 82 St., Office: 333 W. End Av., N.Y.C. 10023.*

LACK, David Lambert, English zoologist; b. London, July 16, 1910; s. Harry Lambert and Kathleen (Rind) L.; B.A., Cambridge U., 1932, M.A., 1935, Sc.D., 1948; m. Elizabeth Twemlow Silva, July 9, 1949; children—Peter Christopher, Andrew John, Martin Paul, Catherine Mary. Biology master Dartington Hall Sch., Devon, 1933-40; mem. radar staff Army Operational Research Group, 1940-45; dir. Edward Grey Inst. Field Ornithology, Oxford U., 1945—, fellow Trinity Coll., 1963—, Recipient Brewster medal Am. Ornithol. Union, 1948; Godman-Salvin medal Brit. Ornithol. Union, 1959. Fellow Royal Soc., 1951; mem. Brit. Ecol. Soc. (pres. 1964-65), Internat. Ornithol. Congress (pres. 1962-66). Author: Life of Robin, 1943; Darwin's Finches, 1947; Natural Regulation of Animal Numbers, 1954; Swifts in a Tower, 1956; Evolutionary Theory and Christian Belief, 1957; Enjoying Ornithology, 1965; Population Studies on Birds, 1966; also numerous articles. Studies of life history of robins, evolution of finches, reasons for ecol. differences between bird species, migration of Brit. birds as analyzed by radar. Office: Edward Grey Inst. Field Ornithology, Oxford, Eng.*

LACKMAN, David Buell, Am. immunologist; b. Plymouth, Conn., Dec. 29, 1911; s. Edward R. and Rena D. (Buell) L.; B.S., U. Conn., 1933; Ph.D., U. Pa., 1937; m. Bernita E. Dowd, June 10, 1947; children—Gloria, Laurena, Arloa, Lisa. Instr. bacteriology, asso. pub. health and preventive medicine U. Pa., 1937-41; with USPHS, Rocky Mountain Lab., Hamilton, Mont., 1941—, head serological tech. sect., 1946—. Vis. lectr. U. Mont., 1951—; cons. in virology Mont. Bd. Health, 1966—. Diplomate Am. Bd. Microbiology. Fellow Am. Pub. Health Assn., Am. Acad. Microbiology; mem. Soc. Am. Microbiologists, Mont. Acad. Scis., Assn. Am. Immunologists, Am. Soc. Profl. Biologists. Demonstrated serologic activity nucleic acids; purified Q fever vaccine tested in man and developed for human use; adapted radioactive precipitation test to study Q fever. Home: 429 Daly Av. Office: Rocky Mountain Lab., Hamilton, Mont. 59840.*

LA CONDAMINE, Charles Marie de, French physicist, math. geographer; b. Paris, Jan. 28, 1701; joined army at age of 17; left to study sci.; became asst. chemist French Acad. Scis., 1730; made various voyages along coasts of Africa and Asia, equatorial S.Am.; went (with Godin and Bouguer) to measure arc of Meridian, Peru, 1735; made 1st sci. exploration of Amazon; visited Italy to determine length of Roman foot, 1757. Fellow Royal Soc., 1748; mem. French Acad. Scis. Author: Relation abrégée d'un voyage fait dans l'interieur de l'Amérique méridionale, 1747; la Figure de la terre, déterminée par les observations de MM. de la Condamine et Bouguer, 1749; Mésure des trois premiers degrés du méridien dans l'hémisphère austral, 1751; Histoire des pyramides de Quito, 1751; Journal du voyage fait par ordre du roi à l'Équateur, 1751; Trois mémoires sur l'inoculation, 1754, 58, 65; Histoire de l'inoculation de la petite vérole, 1773; Lettres et Mémoires; also verse. Brought back 1st information on india-rubber; improved determination of shape of earth by measuring an arc of meridian in Peru; discovered earth to be an oblate spheroid which supported Newton's ideas; discovered effect of mountains in deflecting pendulum; observed topography of Eduador; advocated inoculation against smallpox; advocated use of universal unit of length (after his death metric system was introduced to France). Died Paris, Jan. 28, 1774.

LACORDAIRE, Jean Théodore, French entomologist; b. Recey-sur-ource, France, Feb. 1, 1801; began career in business; went to S.Am., 1825, later 3 trips to study insects; prof. zoology Liège Belgium, 1836-70. Author: Introduction à l'entomologie, 1834-37; Faune entomologique des environs de Paris, 1835; Histoire naturelle des insectes, 14 vols., 1854-68; Genera des coléoptères, 12 vols. (last three written by Chapuis), 1854-76; also other publs. on natural history. Research on beetles. Died July 18, 1870.

LACROIX, François Antoine Alfred, French mineralogist; b. Mâcon, France, Feb. 4, 1863; student Mâcon, Coll. de France; Docteur ès Scis., Sorbonne, 1889; hon. degrees from universities at Oxford, Montreal, Oslo, Liège, Brussels, Athens. m. dau. of Ferdinand Fouqué. Asst. to prof. Ferdinand Fouqué; prof. mineralogy Musée d'histoire naturelle, Paris, 1893-1936; dir. mineralogy lab. École des Hautes Études, from 1896; led expdns. to French Indo-China, W. Africa, Antilles. Mem. French Acad. Scis., 1904 (perpetual sec. 1914); Legion of Honor. Fellow Royal Soc., 1919. Author: (with A. Michel-Lévy) Les minéraux des roches, 1890; Les enclaves des roches volcaniques, 1893; Mineraloge de France et de ses colonies, 5 vols., 1893-1913; Le granite des Pyrénés, 1898; La montagne Pelée et ses eruptions, 1904; Mineraloge de Madagascar, 1922-25; numerous publs. Studied Volcano of Mt. Pelée after eruption of 1902, Vesuvius after eruption of 1906; studied meteorites; discovered 47 minerals. Died Paris, Mar. 16, 1948.

LACROIX, Sylvestre-François, French mathematician; b. Paris, Apr. 28, 1765; studied under Monge; became prof. dept. math. Sch. Marine Guards, Rochefort, France, 1782; named prof. l'École militaire, l'École artillerie Besancon, l'École normale, l'École des quatre Faculty Scis., prof. Collége de France. Mem. French Acad. Scis., 1799. Author: Eléments de géométrie déscriptive, 1796; Traité du calcul différential et du calcul intégral, 3 vols., 1799; Traité des différences et des séries, 1800; Introduction à la géométrie mathématique et critique et à la géométrie physique, 1801; Traité élémentaire du calcul des probabilities, 1816; Traité du calcul différentiel et intégral, 1820. Introduced term, Naperian logarithms, differential coefficient; 1st definition of definite and indefinite integral; discovered independently some processes in descriptive geometry; formulated criteria for distinguishing between singular solutions and particular solutions of differential equations of the 1st order; research on theory of functions of a real variable. Died Paris, May 25, 1843.

LACY, Paul Eston, Am. pathologist; b. Trinway, O., Feb. 7, 1924; s. Benjamin Lemert and Amy Cass (Cox) L.; B.A. cum laude, Ohio State U., 1945, M.D. cum laude, 1948, M.Sc., 1948; Ph.D., U. Minn., 1955; m. Emelyn Ellen Talbot, June 7, 1945; children—Paul Eston, Steven Talbot. Fellow in pathology Mayo Clinic, Rochester, Minn., 1951-55; postdoctorate fellow Nat. Cancer Inst., Washington U. Med. Sch., St. Louis, 1955-56, faculty, 1956—, Mallinckrodt prof., chmn. dept. pathology, 1961—; pathologist-in-chief Barnes and Allied Hosps., also St. Louis Children's Hosp., 1961—. Banting Meml. lectr. Brit. Diabetes Assn., 1963. Mem. Am. Assn. Exptl. Pathologists, Am. Diabetes Assn., St. Louis Path. Soc., Internat. Acad. Pathology, N.Y. Acad. Scis., Am. Assn. Pathologists and Bacteriologists, Am. Assn. Anatomists, Royal Soc. Medicine (asso. mem.), Sigma Xi, Phi Eta Sigma, Alpha Epsilon Delta, Phi Beta Kappa, Alpha Omega Alpha. Contbr. articles to profl. jours. Identification of islet cell types by electron microscopy; determination of mechanism of secretion of islets of Langerhans. Home: 63 Marshall Pl., Webster Groves, Mo. 63119. Office: 660 S. Euclid, St. Louis 63110.*

LACY, Raymond Elmer, Am. electronic engr.; b. Camden, N.J., Mar. 17, 1916; s. Elmer and Irene (Stetser) L.; B.S. in Elec. Engring., Drexel Inst. Tech., 1938; M.S., N.Y. U., 1940, postgrad. 1950-51; postgrad. Bklyn. Poly. Inst., 1941-42; Johns Hopkins, 1950-54, Columbia U., 1956, La Salle U., 1957, Fairleigh Dickenson Coll., 1964-65. Cadra engr. Potomac Elec. Power Co., 1933-35; Prodn. engr. Philco Corp. Corp., 1935; cons. engr. Chubbuck & Patrick, Phila., 1937; instr. elec. engring. N.Y. U., 1938-40; radio design engr. U. S. Signal Corps Lab., 1940-47; dep. chief communication research and devel., 1947-57; dir. communication research U. S. Army Electronic Command, 1957—. Mem. mgmt. panel Lincoln Lab., 1955—. Harvard fellow, 1941. Mem. I.E.E.E., Armed Forces Communication Electronic Assn., Internat. Sci. Radio Union (mem. commn. 2 U. S. nat. commn. 1960-65), Nat. Acad. Sci., Sigma Xi, Tau Beta Pi, Eta Kappa Nu. Author: Applications of Conformal Mapping, 1939; Table of Operational Transforms, 1940; also articles. Research in electronic equipment used for communications and surveillance for U. S. Army. Home: 360 Nutswamp Rd., Red Bank, N.J. 07701. Office: U. S. Army Electronics Command, Inst. for Exploratory Research, Ft. Monmouth, N.J.*

LAD, Robert Augustin, Am. chemist; b. Chgo., May 8, 1919; s. Frank James and Rose (Augustin) L.; student Wright Jr. Coll., Chgo., 1935-37; B.S., U. Chgo., 1939, M.S., 1941, Ph.D., 1946; m. Dolores R. Terwoord, Nov. 7, 1944; children—John, Thomas, Frank, Kathryn, Mary Clare, Margaret, Matthew, Robert J., Cecilia. Research scientist nat. def. research com. Office Naval Research, Manhattan Projects U. Chgo., 1941-46; aero. research scientist NACA, NASA, Cleve., 1946—, head materials chemistry sect., 1952-59, chief Chem. Phys. br. NASA Lewis Research Center, 1959—; cons., lectr. High Sch.-Sci. Seminar Program, Cleve., 1958-61; mem. panel on solid state scis. NRC. Recipient Jr. award Nat. Assn. Corrosion Engrs., 1954. Mem. Am. Phys. Soc., Sigma Xi. Research, publs. on chemistry hydrides of boron, aluminum, mass spectrometry hydrocarbons, corrosion molten salts, phys.-chem. factors influencing mech. deformation ionic solids, characterization of surface defects on single crystal surfaces. Home: 3555 Rocky River Dr., Cleve. 44111. Office: 21000 Brookpark Rd., Cleve. 44135.*

LADD, George Edgar, Am. econ. geologist; b. Haverhill, Mass., July 23, 1864; s. George W. and Eliza A. (Priest) L.; A.B., Harvard, 1887, A.M., 1888, Ph.D., 1894; studied German univs. 2 yrs.; m. Mary O. Hammond, May 24, 1889. Asst., and asst. geologist, U. S., Mo. and Tex. Geol. surveys, 1887-92; asst. in geology Harvard, 1892-94; asst. geologist and chemist Ga. Geol. Survey, 1895-96; dir. and prof. geology and mining Sch. of Mines and Metallurgy, U. Mo., 1897-1908; pres. Okla. Sch. Mines and Metallurgy, 1908-13; pres. N.M. Coll. Agr. and Mech. Arts, 1913-17; econ. geologist, Bur. of Pub. Roads, U. S. Dept. Agr., from 1917; lectr. on engring. geology U. of Md., 1921. Died Dec. 23, 1940.

LADD, George Trumbull, Am. psychologist; b. Painesville, O., Jan. 19, 1842; s. Silas T. and Elizabeth (Williams) L.; A.B., Western Reserve U., 1864, A.M., 1867; grad. Andover Theol. Sem., 1869; (D.D., 1879, LL.D., 1895, Western Reserve; A.M., Yale, 1881; LL.D., Princeton, 1896); m. Cornelia A. Tallman, Dec. 8, 1869; m. 2d, Frances Stevens, Dec. 9, 1895. Preached, Edinburg, O., 1869-71; pastor Spring St. Congl. Ch., Milwaukee, Wis., 1871-79; prof. mental and moral philosophy, Bowdoin Coll., 1879-81, Yale, 1881-1905, emeritus prof., 1905. Lecturer Andover Theol. Sem., 1879-81; lectured at Doshisha and in Summer School in Japan, 1892 and 1899, on invitations from Imperial Ednl. Soc. and Imperial U. of Tokyo; several times lecturer, and conducted Grad. Sem. in ethics, Harvard, 1895-96; lecturer U. of Bombay, and at Calcutta, Madras, Benares, India, 1899-1900; del. World's Congress Psychologists, Paris, 1900; lecturer Western Reserve U. and State U. of Ia., 1906; at Imperial univs., colls., and pvt. univs., Japan, 1907; guest and unofficial adviser of Prince Ito, in Korea, 1907; lecturer in Honolulu, 1907, Coll. for Women, Western Reserve, 1908. Decorated Order Rising Sun, 3d class, 1899, 2d class, 1907; gold medal, Imperial Ednl. Soc. of Japan, 1907 (hon. mem.). Author: Elements of Physiological Psychology, 1887; Introduction to Philosophy, 1890; Psychology, Descriptive and Explanatory, 1894; Primer of Psychology, 1894; Philosophy of Mind, 1895; Outlines of Descriptive Psychology, 1898; A Theory of Reality, 1899; Philosophy of Conduct, 1902; Philosophy of Religion (2 vols.), 1905; Knowledge, Life and Reality, 1909; What Can I Know?, 1914; The Secret of Personality, 1918. Known for textbooks; systematizer of Am. psychol. concepts; pioneered new psychology in U. S. Died Aug. 8, 1921.

LADD, Harry Stephen Am. geologist; b. St. Louis, Jan. 1, 1899; s. Charles Pomeroy and Alice (Bemis) L.; A.B., Washington U., St. Louis, 1922; M.S., U. Ia., 1924; Ph.D., 1925; m. Jane Dobyne Mahler, Sept. 4, 1934; children—Nicholas B., David A. Research fellow Bishop Mus., Honolulu, 1925-26, 28; asst. prof. geology U. Va., Charlottesville, 1926-29; geologist Venezuela Gulf Oil Co., Maracaibo, 1929-31; research asso. U. Rochester (N.Y.), 1934-36; engring. geologist Nat. Park Service, Atlanta, Richmond, Va., 1936-39; geologist U. S. Geol. Survey, Washington, 1940—. Mem. sci. staff Operation Crossroads, 1946-47; research affiliate U. Hawaii, 1965. Mem. Geol. Soc. Am. (v.p. 1955), Paleontol. Soc. (pres. 1954), Am. Assn. Petroleum Geologists, A.A.A.S. (v.p. 1965), Soc. Econ. Paleontologists and Mineralogists. Research and publs. on Tertiary mollusca, geology of Pacific Islands, origin of coral reefs. Home: 3905 Leland St., Chevy Chase, Md. 20015. Office: U. S. Nat. Mus., Washington 20560.*

LADD, Robert Boyd, Am. operations research scientist, statistician; b. Bishop, Tex., Mar. 16, 1915; s. Arlington Ringgold and Pauline Amanda Ladd; B.A., Tex. Coll. Arts and Industries, 1935, M.A., 1939; postgrad. U. Tex., 1938-41, George Washington U., 1942-46; m. Cornelia Calver, May 27, 1967; children—(by previous marriage) Margaret Ellen, Robert Jameson. Statistician, U. S. Army Transp. Corps., Washington, 1943-49, U. S. Bur. Budget, Washington, 1949-53; operations analyst Operations Research Office, Johns Hopkins, Chevy Chase, Md., 1953-55; research mgr. for Indsl. econs. S.W. Research Inst., San Antonio, 1955-58; research project chmn., cons. ORD, Bethesda, Md., 1958-61; v.p. Surveys and Research Corp., Washington, 1961—; sr. staff mem. Nat. Council Marine Resources and Engring. Devel., Exec. Office of Pres., Washington, 1967—. Cons. to numerous fed. agys.; statis. adviser to Republic of China, Taiwan. Fellow Am. Statis. Assn. (mem. nat. council 1957-59), Royal Statis. Soc., A.A.A.S.; mem. Washington Operations Research Council (past pres.), Soc. for Internat. Devel., Operations Research Soc. Am. Author: Southwest Resources Handbook, 1957-58; Maritime Resources for Security and Trade, 1963; also articles. Operations research applications in transp. studies abroad, econ. feasibility and cost benefit methods and applications especially in marine scis., scil manpower, research and devel. in Mainland China including econ. effects. Home: 717 College Pkwy., Rockville, Md. 20850. Office: 1030 15th St. N.W., Washington 20005.*

LADD, William Sargeant, Am. physician; b. Portland, Ore., Aug. 16, 1887; s. William Mead and Mary Lyman (Andrews) L.; B.S., Amherst Coll., 1910; M.D., Columbia, 1915; m. Mary Richardson Babbott, June 5, 1913; children—Frances Wood, William Sargent, Anthony Thornton, John. Asst. physician, Presbyn. Hosp., N.Y.C., 1917-19, 1924-31; instr. Columbia, 1917-19, 1921-24, asso. 1924-31; instr. Johns Hopkins U. and asst. in medicine Johns Hopkins Hosp., 1919-21; asst. prof. medicine, Cornell, 1931-32, prof. medicine, 1932-42, asso. dean Med. Coll., 1931-35, dean 1935-42 professor of clinical medicine since 1942, asst. vis. physician Bellevue Hosp., 1932-35; attending physician N.Y. Hosp., 1935. Mem. A.A.A.S., A.M.A., Soc. for Exptl. Biology, Soc. of Mammalogists, Harvey Soc., Alpha Omega Alpha, Alpha Delta Phi. Contbr. articles to med. jours. and mountaineering mags. Died Sept. 16, 1949.

LADD-FRANKLIN, Christine, Am. psychologist, logician; b. Windsor, Conn., Dec. 1, 1847; d. Eliphalet and Augusta (Niles) Ladd; grad. Vassar Coll., 1869, LL.D., 1887; studied at Johns Hopkins, 1878-82, under Prof. Sylvester and others, also at Göttingen and Berlin; fellow in math. Johns Hopkins, 1879-82; m. Fabian Franklin, Aug. 24, 1882; 1 dau., Margaret. One of asso. editors Baldwin's Dictionary of Philosophy and Psychology, 1901-02; lectr. in logic and psychology, Johns Hopkins 1904-09, Columbia U., N.Y.C., from 1910. Author: Colour and Colour Theories (collected papers), 1928. Pub. Ladd-Franklin theory of color vision (accounting for devel. of man's color sense), 1892; contbd. doctrine of antilogism to logic. Died Mar. 5, 1930.

LADENBURG, Albert, German chemist; b. Mannheim, Germany, July 2, 1842; studied under Bunsen dan Kekulé; m.; became prof., Kiel, Germany, 1873, Breslau (now Wroclaw, Poland), 1889. Recipient Davy medal. Mem. French Acad. Scis., 1909. Author: Handbook of Chemistry, 13 vols.; History of Chemistry, 1869; also numerous articles. Synthesized piperidine, pyridine, coniine; discovered formula for tropine, composition of ozone, asymmetric nature of nitrogen in some organic compounds; studied organic silicon compounds; demonstrated equivalence of hydrogen atoms in benzene and the structure of ortho, meta and para compounds in benzene (proposed prismatic structure), 1st to isolate hyosine or scopolamine. Died Breslau, Aug. 15, 1911.

LADENBURG, Rudolf Walter, physicist; b. Kiel, Germany, June 6, 1882; s. Albert and Margarete (Pringsheim) L.; Ph.D., U. Munich, 1906; postgrad., Cambridge, Eng., 1906-07; m. Else Uhthoff, Aug. 15, 1911; children—Margarete (Mrs. Fritz Eichenberg), Kurt, Eva Marie (Mrs. Ewald Mayer). Came to U. S., 1931. Univ. instr. and prof. at Breslau, 1908-25; scientific hon. mem. Acad. Göttingen, at U. of Berlin, 1925-31; Brackett research prof. physics, Princeton U., 1931-50. Contbr. scientific books and articles. Died Apr. 3, 1952.

LADERMAN, Jack, Am. mathematician; b. N.Y.C., Jan. 6, 1914; s. Joseph and Fannie (Freeman) L.; B.S., Coll. City N.Y., 1934; M.A., Columbia U., 1935, Ph.D., 1953; m. Marion F. Shupnik, June 8, 1947; 1 son, Julian D. Statistician, FCC, N.Y.C., 1937; analyst N.Y. State Dept. Labor, 1938; mathematician Computation Lab., Nat. Bur. Standards, 1939-40, 46-49; statistician Chem. Corps., 1940-46; research scientist Columbia U., 1949-51; mathematician Office of Naval Research, Washington, 1951-54, Office of Naval Research, N.Y.C., 1954-—; tech. dir. Service Bur. Corp., N.Y.C., 1962-63; adj. asso. prof. math. N.Y. U., 1957-64. Cons. to pvt. industry, 1956-— Recipient many awards from City of N.Y., U. S. Govt. Mem. Am. Math. Soc., Inst. Math. Statistics, Am. Statis. Assn., A.A.A.S., Biometric Soc., Operations Research Soc. Am., Soc. for Indsl. and Applied Math., Inst. Mgmt. Scis. Author: (with others) Series National Bureau of Standards Mathematical Tables, 20 vols., 1939-50; also articles. Research in math. statistics, numerical analysis, operations research. Home: 2630 Kingsbridge Terrace, Bronx, N.Y. 10463. Office: 207 W. 24th St., N.Y.C. 10011.*

LADIK, János József, Hungarian quantum chemist; b. Budapest, Hungary, June 2, 1929; s. Ferenc Lajos Lichtmann and Livia Kertész; Dipl. Chem.Ing., Tech. U. Budapest, 1948, Dr., 1967, Candidate Chem. Sci., 1967; m. Éva Kovács, Dec. 31, 1955; children—Annamária, Judit. Asst. prof. Phys. Chemistry Inst., Tech. U. Budapest, 1952-57; jr. sci. officer State Inst. for Hygiène, Budapest, 1957-58; sci. officer Central Research Inst. for Chemistry, Hungarian Acad. Scis., Budapest, 1959-64, head theoretical group, 1965-—; vis. sr. research scientist quantum chemistry group Uppsala (Sweden) U., 1963, 65; mem. biophysics com. Hungarian Acad. Scis., 1965-—. Mem. Hungarian Phys. Soc., Hungarian Chem. Soc., Hungarian Biophys. Soc., Hungarian Biochem. Soc. Numerous publs. on 1st calculation of relativistic and Lamb shift corrections terms of hydrogen molecule; calculation of electronic structure of DNA/band structure, ultraviolet spectrum, conductivity, double-well potential of H-bonds of base pairs; interpretation of mutagenic effect of ultraviolet radiation; research on probable mechanism for starting of DNA duplication, extension of quantum chemistry. Home: XII Sólyom-u 10, Budapest. Office: II. Pusztaszeri ut 57-69, Budapest, Hungary.*

LADMAN, Aaron Julius, Am. anatomist; b. Jamaica, N.Y., July 3, 1925; s. Thomas and Ida (Sobin) L.; student Miami U., Oxford, O., 1942-43; A.B., N.Y. U., 1947; postgrad. U. Cin., 1948-49; Ph.D., Ind. U., 1952; m. Barbara Powers, Dec. 26, 1948; children—Susan Elizabeth, Thomas Frederick. Teaching fellow anatomy U. Cin., 1948-49, Ind. U., 1949-52; with Harvard Med. Sch., 1952-61, asso., 1955-61; asso. prof. U. Tenn. Med. Units, 1961-64; vis. asso. prof. Yale, 1964; prof., chmn. dept. anatomy U. N.M. Sch. Medicine, Albuquerque, 1964-—. Bd. dirs. Rehab. Center Albuquerque. Research fellow Am. Cancer Soc., 1952-55; USPHS Spl. Research fellow, 1955-57; recipient Research Career Devel. award USPHS, 1962-64. Mem. Soc. for Exptl. Biology and Medicine, Am. Assn. Anatomists, Am. Soc. for Cell Biology, Electron Microscope Soc. Am., Internat. Soc. for Cell Biology, Endocrine Soc., Histochem. Soc., Am. Soc. Zoologists, A.A.A.S., Sigma Xi. Asso. editor Anatomical Record, 1967-—. Research and publs. on structure and cellular chemistry of conducting and transporting tissues. Home: 2617 Cutler Av. N.E., Albuquerque 87106.*

LA DUE, John Samuel, Am. physician; b. Minot, N.D., Sept. 6, 1911; s. Samuel John and Edith (Mann) La D.; A.B., U. Minn., 1932; S.M. in Medicine, 1940, Ph.D. in medicine, 1941; M.D., Harvard, 1936; m. Margaret Ruth Stokes, Apr. 24, 1937. 1937-40, Adj. staff physician Mpls. Gen. Hosp., 1940-41; asst. vis. physician Charity Hosp., New Orleans, 1941-43, vis. physician, 1943-45, dir. Lung Sta., 1943-45; clin. asst. Meml. Hosp., N.Y.C., 1945-46, asst. attending physician, 1946-48, asso. attending physician, 1948-—; physician to outpatient dept. N.Y. Hosp., 1946-47, asst. attending physician, 1948-—; teaching fellow Med. Sch. U. Minn., 1937-40, instr. medicine, 1940-41, lectr. La. State U. Sch. Medicine, 1941-43, asst. prof. medicine, 1943-45; clin. asst. Cornell U., Med. Coll., 1945-46, instr. medicine, 1946-47, asst. prof. clin. medicine, 1948-58, asso. prof. clin. medicine, 1958-—; dir. Heart Sta., Meml. Center, N.Y.C., 1955-57, dir. cardiovascular sect., 1957-—; dir. cardiovascular sect. Sloan Kettering Inst., 1957-—; practice medicine, specializing in internal medicine, N.Y.C., 1945-—. Recipient, Hektoen medal A.M.A., 1955, Alfred P. Sloan award, Alfred P. Sloan Found., 1955, Cummings Humanitarian award, 1963, 64; Molina lectr., Manila, 1963. Diplomate. Am. Bd. Internal Medicine. Fellow N.Y. Acad. Sci.; mem. Am. Coll. Cardiology (pres. 1962-63), Am. Fedn. Clin. Research (pres. 1947), A.M.A., A.C.P., Am. Coll. Chest Physicians, Acad. Nacional de Medicinas do Brazil, Harvey Soc., Soc. Exptl. Biology and Medicine, Am. Heart Assn., N.Y. Heart Assn., N.Y. Acad. Medicine, James Ewing Soc., Harvard Med. Soc. N.Y., Acad. Nacional de Medicina de Peru (hon.), Philippine Heart Assn., Sigma Chi, Sigma Xi. Episcopalian. Office: 34 E. 67th St., N.Y.C. 10021.*

LAENNEC, René Théophile Hyacinthe, French physician; b. Quimper, France, Feb. 17, 1781; ed. under his uncle, a physician in Nantes, France, later in Paris; M.D., École de Santé, Paris, 1804; Became physician Hôpital Necker, 1806, chief physician, 1816; joined staff Hôpital Beaujon, 1812, Salpetrière, 1814; became prof. medicine Collège de France, 1822; named prof. clin. medicine Charité, 1823. Mem. Royal Acad. Medicine. Author: Traité de la auscultation médiate et les maladies des poumons et du coeur, 1819. Editor, Jour. de Médecine, 1814. Founder chest medicine; invented stethoscope; used auscultation in diagnosis and correlated sounds heard in living patients with diseases discovered after their death; related symptoms to lesions causing them. Died Kerlouanec, France, Aug. 13, 1826.

LAEVASTU, Taivo, oceanographer; b. Vihula, Estonia, Feb. 26, 1923; s. Wilhelm Granfeldt and Martha (Vogt) Lengi; student U. Kiel (Germany), 1945-47; B.S., U. Gothenburg (Sweden) also U. Lund (Sweden), 1951; M.S., U. Wash., 1954; Ph.D., U. Helsinki (Finland), 1960; m. Irene Merikanto, Sept. 1, 1949; children—Eva, Pia, Steeve. Asst. oceanographer Inst. Gothenburg, 1949-51; fisheries officer Swedish Salmon Com., 1952-53; research asso. U. Wash., 1953-55; oceanographer FAO UN, Rome, Italy, 1955-62; asso. prof. U. Hawaii, Honolulu, 1962-64; chief oceanographer U. S. Fleet Numerical Weather Facility, Monterey, Cal., 1964-—. Mem. Am. Geophys. Union, Am. Soc. Limnology and Oceanography, Am. Meteorol. Soc., Sigma Xi. Author: Factors Affecting the Thermal Structure of the Sea, 1960; (with I. Hela) Fisheries Hydrography, 1962; Manual of Methods in Fisheries Research, 1965; also numerous articles. Pioneered research and application of synoptic heat exchange computations and establishment 1st numerical synoptic oceanographic analysis and forecasting; research on size and mass distbn. cosmic dust, fisheries oceanography and bio-and geo-chem. circulation trace elements in sea. Home: 7 Ralston Dr. Office: U. S. Naval Postgrad. Sch., Monterey, Cal. 93940.*

LAFAY, Bernard, French physician; b. Malakof, Sept. 8, 1905; s. Auguste and N. (Chambon) L.; M.D., Ph.D., U. Paris; m. Paulette Manonviller, Nov. 25, 1931; children—Francoise, Philippe. Began practice medicine, Paris, 1931; elected 2d sector Paris municipal council, 1945, 53, 5th sector, 1959, v.p. council, 1947-48; sec., radical party, 1946-48; senator from Seine dist., 1946-51; dep. of Seine, 1951-58; sec. of state in charge pub. works, 1952; sec. of state for econ. affairs, 1953-54; with ministry pub. health and population, 1955-56; senator from Seine, mem. group of senators of democratic left, 1959. Decorated Order of Pub. Health. Mem. Acad. of Medicine. Research and publs. on cancerous tumors. Address: 123, rue de Longchamp, Paris 16, France; also La Bernadière, Bazoches (Seine et Oise), France.

LAFFERTY, James Martin, Am. physicist; b. Battle Creek, Mich., Apr. 27, 1916; s. James V. and Ida M. (Martin) L.; student Western Mich. U., 1934-37; B.S. in Engring. Physics, U. Mich., 1939, M.S. in Physics, 1940, Ph.D. in Elec. Engring., 1946; m. Eleanor J. Currie, June 27, 1942; children—Martin C., Ronald J., Douglas J., Lawrence E. Physicist, Eastman Kodak Research Lab., Rochester, N.Y., 1939, Gen. Elec. Research Lab., Schenectady, 1940, Carnegie Instn. Washington, 1941-42; research asso. Gen. Elec. Research Lab., Schenectady, 1942-54, mgr. phys. studies sect., 1955-—. Recipient Devel. award Bur. Naval Ordnance, 1946, U. Mich. Distinguished Alumnus citation, 1953. Fellow Am. Phys. Soc., I.E.E.E.; mem. Am. Vacuum Soc. (dir., sec.), Sigma Xi, Phi Kappa Phi, Iota Alpha. Editor, contbr. to Dushman's Scientific Foundations of Vacuum Technique, 1962. Asso. editor Jour. Vacuum Sci. and Tech., 1966-—. Contbr. articles to sci. jours. Patentee in field. Inventor lanthanum boride cathode, 1950, hot cathode magnetron ionization gauge, 1961, triggered vacuum gap, 1966. Home: 1202 Hedgewood Lane, Schenectady 12309. Office: P.O. Box 8, Schenectady 12301.*

LAFFITTE, Paul Frederic, French chemist; b. Marseille, France, Jan. 1, 1898; s. Léon André and Marie (Begou) L.; student Faculté des Science, Paris, 1916; Licencié ès sci., U. Paris, 1921; D.Sc., Faculté des Sciences, Paris, 1925; m. Françoise C. Horn, Feb. 19, 1929; children—Marc, Anne-Marie (Mrs. Larnaudie), Monique (Mrs. Angenleux). Asst., Faculté des Sciences, U. Paris, 1921-29, prof. Sorbonne, 1941-—; prof. Faculté des Sciences, U. Nancy (France), 1929-41. Decorated comdr. des palmes académiques; recipient Gold medal Soc. for Encouragement Nat. Industry; Sir A. Egerton Gold medal Combustion Inst.; laureat French Acad. Scis. Mem. French Petroleum Inst. (pres. 1954-—), Chem. Soc. France (past pres.), Soc. Phys. Chemistry (past pres.), Nat. Council Sci. Research (mem. cons. com.). Author: La détonation des mélanges gazeux; 1938; (with Henry Brusset) L'hydrogène, les gaz inertes, les halogènes, 1954; also numerous articles. Research in phys. and inorganic chemistry, including chem. kinetics, mechanism of reactions, oxidation of gaseous mixtures, combustion, detonation, and explosions, oxidation of metals, phase diagrams in inorganic chemistry. Home: 118 Rue d'Assas, Paris VIe F, France.*

LAFON, Jean Claude, French physician; b. Rouen, France, Dec. 17, 1922; s. Jacques and Eva (de Visme) L.; Doctor, Med. Sch., U. Strasbourg (France), 1952; m. Yvonne Koenig, Aug. 10, 1946; children—Jean-Marie, Christine, Jerome, Anne-Celine, Florence, Olivier, Beatrice, Nathalie, Laure. Research fellow sect. physiology. Centre National de la Recherche Scientifique, 1955-—; tchr. phonetics U. Humanities, Grenoble, France, 1961-—. Gen. sec. Institut Audiophonology, Lyon, France, 1958-—. Author: Message and Phonetics, 1961; The Phonetic Test, 1966; also numerous articles. Research on speech perception and transmission, methodology and apparatus, pathology of speech, voice and hearing. Home: 18 Grande rue de St. Rambert, Lyon. Office: Institut D'Audiophonology, Lyon, France.*

LAFON, Jean Pierre, French mathematician; b. Ville Franche, France, Oct. 12, 1929; s. Albert and Anne (Comes) L.; ed. École Normale Supérieure, 1949-52, agrégé; m. Monique Augé, July 23, 1955; children: François, Dominique, Martin. With National Center for Scientific Research, 1954-59; instr., Clermont Ferrand and Montpellier, 1960-67; prof. Fac. of Sci., Toulouse, 1967-—. Awarded Palms of Academy. Mem., French Math. Soc.; Am. Math. Soc. Author several articles. Studied endoplismic rings of finite type; Heuselian rings; relationships between Heusel's theory and preparation theorem. Home: 7 ruel 'Infirmerie, Montpellier, Hérault, France. Office: Faculté des Sciences, Toulouse, Haute Garonne, France.*

LAFOND, Eugene C., Am. oceanographer; b. Bridgeport, Wash., Dec. 4, 1909; s. William N. and Bessie (Imes) LaF.; A.B. in Chemistry, San Diego State Coll., 1932; D.Sc. in Oceanography, Andhra U., 1956; m. Katherine Wagner Gehring, Sept. 4, 1935; children—William Gehring, Robert Eugene. Research asst. Scripps Instn. Oceanography, U. Cal. at La Jolla, 1933-41; supervisory oceanography Navy Electronic Lab., San Diego, 1941-—, head marine environment div., 1960-—; prof. oceanography, Andhra U., Waltair, India, 1952-53, 55-56; specialist oceanography U. S. State Dept., 1956-57; sr. scientist U.S.S. Skate N. Pole Expdn., 1958, U. S. Dept. State-Scripps Instn. NAGA Expdn., 1960; chief scientist Woods Hole Oceanographic Instn., U. S. program in biology Internat. Indian Ocean Expdn., 1962-63; dep. dir. Office Oceanography, UNESCO, Paris, France, 1963-64. Mem. A.A.A.S., Am. Soc. Limnology and Oceanography, Am. Geophys. Union, Internat. Assn. Phys. Oceanographers, Internat. Union Geodesy and Geophysics, Maritime Research Soc. San Diego, Sigma Xi. Author: Processing Oceanographic Data, 1951; also numerous articles. Research on nature of internal waves in sea, sea level, thermal structure, deep submersibiles, arctic Ocean, Bay of Bengal, South China Sea, upwelling motion in ocean, fixed platforms, slicks, water turbidity, underwater sound, sea water chemistry. Home: 4505 Santa Cruz Av., San Diego 92107. Office: U. S. Navy Electronics Lab., San Diego 92152.*

LA FORGE, Laurence, Am. geologist; b. N.Y.C., Sept. 17, 1871; s. Abiel Teeple and Margaret Swain (Getchell) L.; A.B., Harvard, 1899, A.M., 1900, Ph.D., 1903; m. Fannie Agnes Carryer, June 28, 1893 (dec. July 1924); 1 dau., Helen Grace (Mrs. Henry Gilmore Brousseari); m. 2d, Kate Louise Harbaugh, Sept. 8, 1930. Instr. astronomy, Alfred U., 1896-97; Austin teaching fellow Harvard, 1902-03; geologist U. S. Geol. Survey, 1901-05, 14-27; aid in geology U. S. Nat. Museum, 1905-08; research asso. Harvard, 1932-38; prof. geology Suffolk U., 1939-40; prof. geology Teachers Sch. of Sci., 1934-48; also prof. geology, Tufts Coll., 1942-45; cons. geologist, 1908-14, and from 1927. Fellow Am. Geog. Soc., N.Y. Acad. Scis.; mem. Am. Geophys. Union, Am. Forestry Assn., Nat. Geog. Soc., A.A.A.S., U. S. Infantry Assn., Soc. Am. Mil. Engrs., Am. Museum Natural History. Died May 29, 1954.

LAG, Jul, Norwegian agronomist; b. Flesberg, Nov. 13, 1915; s. Torstein and Joran (Brattas); doctorate Agrl. Coll. Norway; m. Ingrid Brenner, Aug. 8, 1956; children—Marit, Torleiv. Prof., dir. Soil Sci. Inst. Norway, 1949-—, rector coll., 1958-62. Mem. Norwegian Acad. Sci., Indian Soc. Soil Sci., Finnish Forestry Soc. Research and publs. on influence of geol. conditions on soil formation, soil genesis and productivity in relation to soil-forming factors. Address: Vollebekk, As, Norway.

LAGACHE, Daniel, French psychologist; b. Paris, Dec. 3, 1903; s. Louis and Marthe (Brien) L.; m. Marianne Hossenlopp, July 17, 1954; children—Mariel, Catherine, Elisabeth, Agnes. Ph.D., M.D., U. Paris; agrégé in philosophy, 1928; hon. doctorate U. Montreal; Became head clinic Psychiat. Hosp., 1935; apptd. prof. U. Strasbourg (France), 1937, Sorbonne, 1947; dir. psychiat. inst. U. Paris; founder, dir. psycho-sociol. labs. Mem. Medico-psychol. Soc., French Psychol. Soc., Psychiat. Evolution Group, Psychoanalysis Soc. France, Nat. Council Sci. Research, Internat. Psychoanalysis Assn. Author: Les hallucinations verbales et la parole, 1934; La jalousie amoureuse, 1947; La psychanalyse, 1955. Home: 240 bis, bd. Saint-Germain, Paris 7, France.

LA GALLA, Julius Caesar, Italian physician, philosopher; b. Padula, Kingdom of Naples, 1576; studied at Naples; prof. philosophy Roman Coll., 1597-1624. Author: De phoenominis in orbe lunae . . . , 1612; Treatise on Comets, 1613; De Immortalitate animorum (attempted to prove Aristotle admitted the immortality of the soul), 1621. 1st description of Bologna phosophorus (phosphorescent barium sulphide); explained phosophorescence is due to absorption and re-emission of light; summarized Democritos's ideas. Died Rome, Mar. 15, 1624.

LA GARAYE, Count Claude-Toussaint Marot, French chemist; b. Rennes, France, 1675; m., 1701; founder hosp. in chateau at Brittany, France. Author: Chimie hydraulique pour extraire les gels essentiels des végétaux, 1746. Contbd. to devel. of hosps. in France; organized hosp. system in Brittany; discovered practical method of preparing cinchona, also process for obtaining extracts from medicinal plants. Died 1755.

LA GARDE, Louis Anatole, Am. surgeon; b. Thibodaux, La., Apr. 15, 1849; s. Jules Adolph and Aurelia (Daspit) L.; student La. Mil. Acad., 1866-68; M.D., Bellevue Hosp. Med. Coll., 1872; m. Frances Neely, Mar 4, 1879. Apptd. asst. surgeon U. S. Army, 1874, advanced through grades to col. Med. Corps, 1910; retired 1913, recalled to active duty and served during World War I. Participated in Sioux Indian War, 1876; comd. Divisional Reserve Hosp., 5th Army Corps, Siboney, Cuba, 1898; in charge evacuation of sick and wounded to Northern hosps.; prof. mil. surgery N.Y. U., from 1900; comdt. U. S. Army Med. Sch., 1910-13; mem. Nat. Bd. Med. Examiners. Mutter lectr. Coll. Physicians, Phila., 1902. Author: Gunshot Injuries, 2d edit., 1916. Research on septic bullets and septic powders; demonstrated ineffective material not destroyed by firearms. Died Mar. 7, 1920.

LAGASCA, Mariano, Spanish botanist; b. Encinacorba, Spain, 1776; studied medicine, Valencia, Spain; dir. bot. gardens, Madrid; prof. botany U. Madrid; made expdn. to classify Spanish flora, 1803. Author: Elenchus plantarum quae in horto regio botanico Martritensi. Discovered a species of cudbear lichen native to Spain used to prepare litmus paper. Died Madrid, 1839.

LAGEMANN, Robert Theodore, Am. physicist; b. Marion, O., Aug. 31, 1912; s. Theodore Frederick and Martha (Niemann) L.; A.B., Baldwin-Wallace Coll., 1934, D.Sc., 1962; M.S., Vanderbilt U., 1935; Ph.D. (teaching fellow), Ohio State U., 1940; m. Margie Plummer, July 30, 1938; 1 son, Robert Conrad. grad. asst. Ohio State U., 1936-39, research fellow, 1939-40; instr. Marshall Coll., 1940-41; research physicist OSRD also Manhattan Project, Columbia, 1942-43; faculty Emory, 1941-51; Landon C. Garland prof. physics, Vanderbilt U., 1951-65, dean Grad. Sch., 1965-—. Mem. council Oak Ridge Inst. Nuclear Studies, 1949-51, 53-67. Mem. Am. Phys. Soc. (treas. S.E. sect. 1950-57, chmn. 1957-58), Ga. (pres. 1951-52), Tenn. (pres. 1959) academics of science, Am. Assn. Physics Tchrs., A.A.A.S., Nat. Sci. Tchrs. Assn. (treas. 1957-59). Author: (with R. W. Schulz) Physics for the Space Age, 1961; Physical Science: Origins and Principles, 1963. Research and publs. on separation of the uranium isotopes by gaseous diffusion, on the elucidation of the structure of molecules using the techniques of absorption of infrared light, and on understanding the mechanisms (relaxation phenomena) whereby sound is absorbed in various compounds. Home: 2206 Hampton Av., Nashville 37215.*

LAGERSPETZ, Kari Yrjö Henrik, Finnish animal physiologist; b. Helsinki, Finland, Sept. 16, 1931; s. Yrjö and Lyyli (Levanon) L.; Cand.Phil., U. Helsinki, 1954, Mag.Phil., 1955; Lic.Phil., U. Turku, 1958, D.Phil., 1960; m. Kirsti Ahlman, June 19, 1954; children—Yrjö Eerik, Juhani Henrik, Olli Markus, Mikko Kari. Instr. zoology U. Helsinki, 1952-53, 56; faculty U. Turku (Finland), 1954-—, research fellow NRC for Scis., 1962-64, prof. physiol. zoology, 1964—, dir. zoophysiol. lab., dept. zoology, 1964-—. Author: Eläin ja kone (The animal and the machine), 1966; also articles. Research on postnatal devel. of thermoregulation in mice, temperature acclimation in poikilotherms, physiol. basis of aggressiveness (with Kirsti Lagerspetz, Rauno Tirri), theoretical biology, especially teleological mechanisms in regulation. Home: 10 C Kerttulinkatu, Turku 2, Finland.*

LAGERSTROM, Paco Axel, applied mathematician; b. Oskarshamn, Sweden, Feb. 24, 1914; s. Paco Harald and Karin (Wiedemann) L.; Filosofie Kandidat, U. Stockholm, 1935, Filosofie Licenciat, 1939; postgrad. U. Muenster (Germany), 1937-38; Ph.D. in Math., Princeton, 1942. Came to U. S., 1939, naturalized, 1949. Instr., Princeton, 1941-44; flight test engr., aerodynamicist Bell Aircraft, 1944-45; aerodynamicist Douglas Aircraft, 1945-46, cons., 1946-66; research asso., staff engr. Jet Propulation Lab., Cal. Inst. Tech., Pasadena, 1946-47, faculty, 1947-—, prof. aeros., 1952-—; vis. prof. mechanics U. Paris (France), 1960-61. Guggenheim fellow, 1960-61; decorated Palmes Académiques (France). Mem. Am. Math. Soc. Contbg. author: Handbook of Engineering Mechanics, 1962; Princeton Series on High Speed Aerodynamics and Jet Propulsion, vol. IV, 1964; also articles. Research in theories supersonic flight, flow viscous fluids, astrodynamics, math. perturbation methods. Home: 57 San Miguel Rd., Pasadena, Cal. 91105.*

LAGNY, Thomas Fantet de, see de Lagny, Thomas Fantet.

LAGOW, Herman Edward, Am. physicist, space engr.; b. Whitney, Tex., Oct. 18, 1923; s. Arthur C. and Nettiemae (DeGraefeinreid) LaG.; B.A., Baylor U., 1953; postgrad. in physics and math. U. Md., George Washington U.; m. Rita Esther Brial, Dec. 21, 1945; children—Peter, Joanna, Karen, Roger, Lynn, Andrew. Joined Naval Research Lab. as jr. physicist, 1943; trans. with group upper air physicists to NASA, 1958, now asst. dir. for systems reliability, head environmental com. Goddard Space Flight Center, Greenbelt, Md. Chmn. environmental sub-com. Sci. Program for Project Vanguard; chief scientist in charge 3 atmospheric structure IGY rockets successfully fired at Ft. Churchill, Can.; project mgr. Heavy IGY satellite (now known as Explorer VII); U. S. del. IGY confs., Toronto, 1957, Moscow, 1958, 1st Internat. Space Sci. Symposium, Nice, France, 1960, 2d Symposium, Florence, Italy, 1961. Authority in study of properties of atmospheric gases at high altitudes; designed resistance thermometers for 1st flight measurements of rocket skin temperatures; other research and publs. on upper atmosphere, measuring devices, rocket and satellite tech. Home: 9336 Harvey Rd., Silver Spring, Md. 20910. Office: Goddard Space Flight Center, Greenbelt, Md.*

LAGOWSKI, Joseph John, Am. chemist; b. Chgo., June 8, 1930; s. Joseph Thomas and Helen (Kasprzycki) L.; B.S., U. Ill., 1952; M.S., U. Mich., 1954; Ph.D., Mich. State U., 1957, Cambridge (Eng.) U., 1959; m. Jeanne Wecker Mund, Feb. 13, 1954. Teaching fellow U. Mich., 1952-54; asst. demonstrator U. Cambridge, 1958-59, supr. inorganic chemistry, 1958-59; asst. prof. U. Tex., Austin, 1959-63, asso. prof. chemistry, 1963-67, prof., 1967-—. Mem. Am. Chem. Soc., Chem. Soc., Sigma Xi, Phi Lambda Upsilon. Author: (with G. W. Watt, L. F. Hatch) Chemistry, 1964, Chemistry in the Laboratory, 1964; The Structure of Atoms, 1964; The Chemical Bond, 1966; also articles. Research on non-aqueous solvent systems, chemistry liquid ammonia solutions, spectroscopy, organometallic pi systems, non-carbonoid aromatic systems, fluorine chemistry, fluoroorganometallic systems, chemistry of mercury. Home: 1114 W. 22d St., Austin, Tex. 78701.*

LAGRANGE, Joseph Louis, Comte de, French mathematician, physicist, astronomer; b. Jan. 25, 1736, Turin, Piedmont, Italy; s. Marie Thérèse Gros L.; ed. Turin College; m. Mlle. Lemonnier; prof. math., Artillery School, Turin, 1755; apptd. (by Frederick the Great) Dir. math. division, Berlin Acad. Scis. (succeeding Euler) 1766-86; ret. to Paris, 1787; pres. comm. for reform of weights and measures, 1793; prof. math., École Normale, 1795; 1st prof. geometry, École Polytechnique, 1797; founder and mem. Turin Acad. Scis., 1788; mem. French Acad. Scis., 1772 (prizes 1764, 1766, 1772, 1774, 1778); Fellow Royal Soc., 1791; Grand Officer, Legion of Honor; apptd., by Napoleon, senator, then Count of the Empire. Author: Théorie des variations séculaires des éléments des planètes, 1781; Mécanique analytique, 1788; Théorie des fonctions analytiques, 1797; Traité de la résolution des équations numériques de tous dégrés, 1798 (contains proof of theorem that every algebraic equation has a root); Leçons sur le calcul des fonctions, 1801; many scientific papers. Found general solution of isoperimetrical problems, in course of which he further developed calculus of variations; made many important contributions to solution of partial differential equations, to theory of numbers, to theory of equations, and to applications of calculus to physical and astronomical problems; solved problems originally proposed by Fermat; proved irrationality of pi; resolved to employ only algebraic methods in endeavor to systematize mechanics; showed mechanics could be founded on principle of least action; also used principle of virtual velocities; made mechanics a branch of analysis; gave Lagrangian form to equations of motion; studied perturbations, three-body problem, libration of moon, motion of Jupiter's satellites, cometary problems, Kepler's problem, nature and propagation of sound, vibration of strings, echoes, beats, hydrodynamics; all investigations characterized by rigor and generality; applied differential calculus to theory of probability and theory of errors; influential in causing adoption of decimal system. Died Paris, France, April 10, 1813.

LAGRANGE, René François, French mathematician; b. Charbonnières, France, Aug. 25, 1895; s. Alfred and Anne (Communal) L.; Ph.D., agrégé; m. Antoinette Martineau, June 4, 1930; children—Jacques, Jean Louis, Michel. Lectr., Rennes, Lille (both France); prof. Faculty Scis., Dijon, France; prof. physics and indsl. chemistry Ecole Supérieure. Research and publs. on differential absolute functions of Legendre, differential geometry, solid geometry, combined analysis. Home: 7, rue du Chateau, Dijon. Office: Faculté des Sciences, Dijon, France.

LA GRECA, Marcello, Italian zoologist; b. Cairo, Dec. 8, 1914; s. Stanislao and Clotilde (Paggi) La G.; Ph.D., U. Naples (Italy); m. Christina Ferrara, Nov. 18, 1939; 1 son, Luciano. Prof. zoology Faculty Scis., prof. gen. zoology Faculty Medicine, U. Catania (Italy). Mem. Italian Acad. Entomology, Italian Zool. Union, Italian Entomol. Soc., Italian Soc. Biogeography, Naples Naturalists Soc., Accademia Gioenia, Catania. Research and publs. on systematics of orthoptera and mantodea, gen. zoology, zoogeography, soil ecology, morphology of arthropods, acridology. Home: via Basile 6, Catania. Office: Istituto di Zoologia, via Androne 81, Catania, Italy.

LAGRONE, Cyrus Wilson, Jr., Am. psychologist, educator; b. Paint Rock, Tex., Jan. 8, 1911; s. Cyrus Wilson and Truda (Gough) L.; B.A., E. Tex. State Coll., 1930; M.A., U. Tex., 1932; Ph.D., 1942; Ford fellow, vis. scholar Columbia, 1954-55; m. Mary Katherine Dodson, June 2, 1951. Instr., Lee Coll., Baytown, Tex., 1934-40, U. Tex., Austin, 1940-42, summers 1934-42; asst. prof. U. So. Cal., Los Angeles, 1946-47; prof., chmn. dept. psychology Tex. Christian U., Fort Worth, 1947-—. Diplomate in clin. psychology Am. Bd. Examiners Profl. Psychology. Fellow Am. Psychol. Assn.; mem. Am. Assn. U. Profs., Southwestern, Tex. (pres. 1950-51) psychol. Assns., Alpha Chi, Phi Delta Kappa. Research, publs. on visual accuracy; prediction of delinquency; sex, personality differences. Home: 3200 Spanish Oak Dr., Fort Worth 76109.*

LAGUCHEV, Sergey Sergeyevich, Russian histologist-cytologist; b. Moscow, USSR, Oct. 15, 1918; s. Sergey Dametievich and Evgenia (Schephina) L.; grad. First Moscow Med. Inst., 1941, Candidat Med. Sci., 1953; D.Med. Sci., Acad. Med. Scis., 1966; m. Anna Ivanovna Jukova, June 13, 1958; 1 son, Aleksey Sergeyevich. Med. officer Inst. Morphology, Acad. Med. Scis., Moscow, 1941-47, jr. research asst., 1948-51; asst. chair histology and embryology First Moscow Med. Inst., 1951-57; sr. research asst., lab. histology Inst. Exptl. Biology, Acad. Med. Scis., Moscow, 1958-60, head group exptl. morphology of cell, 1960-63, mgr. lab. histophysiology, 1964-—. Decorated Order Patriotic War 2d degree, 1941-45. Mem. Moscow Soc. Investigators of Nature, Soc. Anatomists, Histologists and Embryologists USSR. Author: (with M. Subbotiu) Histological Technique, 1954, The Principles of General Histology, 1950; also articles, chpts. in books. Demonstration of separate systems of hormonal regulation in female reproductive organs; disturbances of sexual rhythms of mitosis as the first symptom of breast cancer. Home: 7 Festivalnaya, Moscow A-195. Office: 8 Baltiyskaya, Moscow A-135, USSR.*

LAGUERRE, Edmond, French mathematician; b. Bar-le-Duc, France, Apr 9, 1834; prof. École polytechnique, Coll. de France; mem. French Acad. Scis., 1884. Author: Note sur la résolution des équations, 1880; Théorie des équations numeriques, 1884; Recherches sur la géométrie de direction, 1885; Oeuvres, edited by C. Hermite and others, 2 vols., 1898-1905. A founder of modern geometry; worked on analytical and infinitesimal geometry, theory of algebraic equations; contbd. to devel. of linear systems; developed theory of curves based on circle composed of 2 cycles; considered a straight line to be composed of 2 semi-lines; distinguished indivisible curves from those composed of 2 trajectories running in opposite directions. Died Bar-le-Duc, Aug. 14, 1886.

LAGUNA, Andres de, see de Laguna, Andrés.

LAHEE, Frederic Henry, Am. geologist; b. Hingham, Mass., July 27, 1884; s. Henry Charles and Selina Ida Mary (Long) L.; A.B., Harvard, 1907, A.M., 1908, Ph.D. in Geology, 1911; m. Louie Karr Hodge, Dec. 23, 1912; children—Genevieve, Henry, Ruth Holden, John Aspinwall. Asst. instr. in geology Harvard and Radcliffe Coll., 1906-09, instr. geology, 1909-12; instr. geology Wellesley, part time, 1910-18; instr. in geology Mass. Inst. Tech., 1912-15, asst. prof., 1915-18; geologic aid U. S. Geol. Survey, 1915; with Sun Oil Co. 1918-55, chief geologist, in charge geol., paleontol., chem. and geophysical work, 1920-47, geological and research counselor 1947-55, ret. 1955; cons. geologist, 1955-—; cons. geologist White River Natural Gas Co., 1959-—; adviser to Petroleum Information, Inc. of Denver; expert witness on gas-rate cases, involving testimony on exploratory drilling. Chmn. com. on exploration for Dist. III, Petroleum Adminstrn. for War, 1942-45. Recipient Sidney Powers Meml. award Am. Assn. Petroleum Geologists, 1953. Mem. Geol. Soc. Am., Geog. Soc. Am., Am. Assn. Petroleum Geologists (editor 1929-32; pres. 1932-33; rep. on NRC 1936-40, chmn. com. on statistics of exploratory drilling 1935-55, chmn.

adv. com. on radioactive mineral exploration 1952), Soc. Econ. Geologists, Am. Inst. Mining Metall. and Petroleum Engrs., Am. Petroleum Inst. (chmn. com. on oil reserves 1946-55), Mid Continent Oil and Gas Assn., Phi Beta Kappa. Author: Field Geology, 1916, 23, 31, 41, 51, 61; tech. articles on metamorphism, structural and petroleum geology. Proposed "Lahee Method" for statistical recording of data on holes drilled for oil and/or gas. This plan is used throughout N. Am. and in many foreign countries. Address: 7219 Kenny Lane, Dallas 30.

LAHEY, Frank Howard, Am. surgeon; b. Haverhill, Mass., June 1, 1880; s. Thomas and Honora Frances (Powers) L.; M.D., Harvard, 1904; Sc.D., Tufts, 1927, Boston U., 1943, Northwestern U., 1947; LL.D., U. Cin., 1951; m. Alice Wilcox, Apr. 15, 1909. Surgeon Long Island Hosp., 1904-05, Boston City Hosp., 1905-07; resident surgeon Haymarket Sq. Relief Sta., 1908; instr. in surgery Harvard Med. Sch., 1908-09, 1912-15; asst. prof., later prof. surgery Tufts Med. Sch., 1913-17; prof. clin. surgery Harvard Med. Sch., 1923-24; surgeon in chief N.E. Baptist hosps.; dir. surgery The Lahey Clinic, Boston. Fellow A.C.S. (bd. govs.) Royal Coll. Surgeons (Eng., hon.); mem. Am. and Internat. surg. assns., Am. Assn. for Study of Goitre, A.M.A. (pres 1942), Société des Chirurgiens de Paris. Author of Lahey Clinic Number (Surg. Clinics of N. America), pub. yearly. Contbr. numerous articles on surg. subjects. Died June 27, 1953.

LA HIRE, Philippe de, French mathematician; b. Paris, France, Mar. 18, 1640; s. Laurent de La H.; studied painting in Rome, Italy; later studied sci. and math.; student of Gaspard Desarques; married twice; 8 children, including Gabriel Philip. Cartographer for govt.; sent to prepare gen. map of kingdom and to continue measurement of meridian begun by Picard; apptd. prof. Collège de France, 1682; taught at Acad. Architecture; employed to take levels for aqueducts of Louis XIV. Mem. French Acad. Scis., 1678. Author: New Elements of Conic Sections, 1679; A Large Treatise on Conic Sections in folio, 1685; Astronomical Tables, 1702; School of Land Surveyors, 1692; Treatise on Mechanics, 1695; Treatise on Gnomonics or Dealling. Research and publs. on exptl. astronomy, physics, natural history, geometry, roulettes, graphic methods, epicycloids, conchoids, magic squares; established elements of plane analytic geometry; tried to show that all theorems of Apollonios' conic sections can be derived from circle with Desargues' method of central projection. Died Paris, Apr. 21, 1718.

LAHIRY, Nripendra lal, Indian food technologist; b. Mymensingh, Bengal, India, Mar. 1, 1914; s. Kumud Chandra and Basant Kumari (Devi) L.; B.Sc., Presidence Coll., Calcutta, 1935; M.Sc., Dacca U., 1937; research fellow biochemistry Indian Inst. Sci., Bangalore, 1938-43; postgrad. London, Manchester univs. (Eng.) 1946-47; Sc.D., Dept. Food Tech., Mass. Inst. Tech., Cambridge, 1952; m. Smt. Rani, July 5, 1942; children—Rana, Benu. Chief chemist, prodn. mgr. Biochemical Products Ltd., Bombay, India, 1943-46; co-lectr. biochemistry, organic chemistry Am. U., Washington, 1953-55; vis. asst. prof. Coll. Medicine, Howard U., Washington, 1956-57; asst. dir. Central Food Tech. Research Inst., Mysore, India, from 1957, apptd. chmn. Discipline of Meat, Fish, Poultry Tech., 1963. Fellow A.A.A.S.; mem. Assn. Food Technologists India, (asso.) Royal Inst. Chemistry of Gt. Brit. and Ireland, Sigma Xi. Contbr. articles to jours. Developed several high protein foods to combat malnutrition, based on soya bean, peanut, fish, meat, poultry, also process for mfg. vegetable milk powder from 2 or more vegetable oilseeds; research on tech. aspects of preparing protein hydrolysate, fractionation of fish muscle proteins, cold storage. Home: P. 23a/41C/1 Diamond Harbour Rd., New Alipore, Calcutta, West Bengal, India. Office: Central Technological Research Institute, V.V. Mohalla P.O., Mysore-2, India.*

LAHMANN, Heinrich, German physician; b. Bremen, Germany, Mar. 30, 1860; head of a sanatorium nr. Dresden, Germany; studied med. climatology; founder of sci. dietetics. Died Friedrichstal/Radeberg, Germany, June 1, 1905.

LAI, Gianpaolo, Italian psychiatrist; b. Forli, Italy, May 27, 1931; s. Benedetto and Cia (Poni) L.; D. medicine, U. Bologna (Italy), 1957; postgrad. psychoanalysis U. Lausanne (Switzerland), 1960-66; m. Pierrette Lavanchy, Nov. 21, 1966. Asst., Psychiat. Clinic, U. Bologna (Italy), 1957-58; staff psychiat. clinic U. Lausanne, 1960-66, chef de clinique adj., 1964-66, chef de clinique, 1966; practice medicine specializing in psychoanalysis, Milan, Italy, 1966—. Leading work groups, Milan, Turin; guest Swiss Soc. Psychoanalysis. Research and publs. on psychotropic drugs and their mechanism of action, group psychotherapy and individual psychotherapy of psychotic patients, role of psychoanalysis in formation of psychiatrist. Address: 4 via Carroccio, Milan, Italy.*

LAIDLAW, Alexander Hamilton, physician, b. nr. Lanark, Scotland, July 11, 1828; s. Alexander and Margaret (Hamilton) L.; grad. Phila. Central High Sch., 1845; grad. Hahnemann Homeopath. Med. Coll., Phila., 1861; also studied in colleges allopathy, hydropathy, electropathy, eclecticism, and hypnotism; m. Anna Turner Sites, Oct. 28, 1865. In practice of medicine, 1856-1905; prin. Mauch Chunk High Sch., 1851-52, Monroe Grammar Sch., Phila., 1853-60; supt. pub. schs., Hudson County, N.J., 1868-69; prof. anatomy N.Y. Homeopath. Med. Coll., 1868; retired from practice, 1905. Author: American Pronouncing Dictionary of the English Language, 1859; Curability of Consumption, 1861. Died 1908.

LAIDLAW, Harry Hyde, Jr., Am. educator; b. Houston, Apr. 12, 1907; s. Harry Hyde and Louisa (Quinn) L.; B.S., La. State U., 1933, M.S., 1934; Ph.D., U. Wis., 1939; m. Ruth Grand Collins, Oct. 26, 1946; 1 dau., Barbara Scott. With U. S. Dept. Agr., 1929-34, 35-39; research asst. La. State U., 1934-35; research asst. Wis. Alumni Research Found., 1937-39; prof. biol. sci. Oakland City Coll., 1939-41; state apiarist Ala. Dept. Agr. and Industries, 1941-42; faculty U. Cal., Davis, 1947—, prof. entomology, apiculturist, 1959—, asso. dean Coll. Agr., 1960-64. Fellow A.A.A.S.; mem. Genetics Soc. Am., Entomol. Soc. Am., Am. Soc. Zoologists, Am. Soc. Naturalists, Sigma Xi. Author: (with J. E. Eckert) Queen Rearing, 1950. Devel. publs. on techniques for artificial insemination, mutation, breeding of honey bees. Home: 761 Sycamore Lane, Davis, Cal. 95616.*

LAIDLAW, Sir Patrick Playfair, Brit. physician; b. Glasgow, Scotland, Sept. 26, 1881; s. Robert and Elizabeth (Playfair) L.; ed. St. John's Coll., Cambridge, 1900-04, Guy's Hosp., London, from 1904; studied at Vienna, Freiburg, 1913-14. Demonstrator in physiology Guy's Hosp., Sir William Dunn lectr. pathology, 1913; researcher Wellcome Physiol. Research Labs., Herne Hill, 1909-13; staff mem. Med. Research Council, Nat. Inst. for Med. Research, 1922, dep. dir., head dept. pathology and bacteriology of Inst., from 1936. Linacre lectr., 1935; Rede lectr., 1938; hon. fellow St. John's, Cambridge, 1940. Fellow Royal Soc. (Royal medal 1933), 1927, Royal Coll. Physicians. Research on pharm. actions of histamine (with Henry Dale); showed that tyramine both deaminized and oxidized by acid of liver (with A. J. Erwins), found method to produce indolethylamine by synthesis, studied its metabolic breakdown (with Erwins); 1st described use of hydrogen in spongy platinum to remove last traces of oxygen from culture tube and so greatly improve conditions for growing anaerobic bacteria, 1915; worked on histamine shock and physiol. relations (with A. N. Richards of Phila.), leading to improved treatment of wounded in World War I (their analysis since has been shown to be partly in error); bacteriological and virus research after 1922; found evidence supporting Carre's claim that filterable virus causes distemper; research on means of inducing immunities; laid found. for Laidlaw's research on epidemic influenza, showed (with C. H. Andrewes and Wilson Smith) that infective agt. in human epidemic influenza is virus; produced new method (with Clifford Dobell) to grow parasitic amoeba profusely, utilizing rice starch as carbohydrate for growing organisms. Died London, Mar. 19/20, 1940.

LAIDLER, Keith James, chemist; b. Liverpool, Eng. Jan. 3, 1916; s. George J. and Hilda (Findon) L.; B.A., Trinity Coll., Oxford U., 1937, M.A., 1956, D.Sc., 1956; Ph.D., Princeton, 1940; m. Mary Cabell Auchincloss, June 22, 1943; children—Margaret Cabell, Audrey Auchincloss, James Reid. With Catholic U. Am., 1946-55, asso. prof., 1948-55; prof. chemistry U. Ottawa (Ont., Can.), 1955—, chmn. dept. chemistry, 1960-66, vice-dean faculty sci., 1962-66; Commonwealth vis. prof. U. Sussex, Falmer, Eng., 1966-67. Fellow Chem. Inst. Can., Royal Soc. Can.; mem. Faraday Soc., Chem. Soc. Author: (with S. Glasstone and H. Eyring) The Theory of Rate Processes, 1941; Chemical Kinetics, 1950; Introduction to the Chemistry of Enzymes, 1954; The Chemical Kinetics of Excited States, 1955; The Chemical Kinetics of Enzyme Action, 1958; Reaction Kinetics, Vol. 1, Homogeneous Gas Reactions, Vol. 2 Reactions in Solution, 1963; Chemical Kinetics, 2d edit. 1965; Principles of Chemistry, 1966. Numerous publs. on kinetics, mechanisms of gasphase, surface, solution, enzyme reactions, with studies on environmental influences. Home: 551 Mariposa Av., Rockcliffe Park, Ottawa 2, Ont., Can.*

LAINE, Veikko Aatos Ilmari, Finnish physician; b. Turku, Finland, June 20, 1911; s. Frans A. and Fannie (Aalto) L.; M.D., U. Helsinki, 1937; m. Onerva Aurora Brusila, Aug. 20, 1936 (dec. 1954); children—Pekka Ilmari, Pirkko Onerva (Mrs. Henrik Lilius), Marja Onerva, Matti Ilmari; m. 2d, Aino Kaarina Lahti-Nuuttila, Feb. 5, 1956. Gen. practice Pub. Health Service, 1937-39; house physician in internal medicine and psychiat. clinics, 1939-45, in rheumatic and internal medicine hosps., 1945-50; med. dir. Rheumatism Found. Hosp., Heinola, Finland, 1950—; lectr. rheumatology U. Helsinki, 1963—. Mem. Finnish Assn. Rheumatologists (chmn. 1959—). Author: Reumatologi, 1963; Reumataudie, 1963; also numerous articles. Research in clin. rheumatology and epidemiology of rheumatic diseases, preventive surgery in rheumatoid arthritis. Home: Heinola, Finland. Office: Rheumatism Found. Hosp., Heinola, Finland.*

LAINSON, Ralph, parasitologist; b. Beeding, Eng., Feb. 21, 1927; s. Charles Harry and Annie May (Denyer) L.; B.Sc. with honors in Zoology, Brighton (Eng.) Tech. Coll., 1952; Ph.D., London Sch. Hygiene and Tropical Medicine, 1955; D.Sc., U. London, 1965; m. Ann Patricia Russell, Sept. 28, 1957; children—Karen Susan Lainson, Amanda Jane, Stephen Paul. Lectr. parasitology London Sch. Hygiene and Tropical Medicine, 1955-59, Nat. Insts. Health research staff, 1962-65; sr. sci. officer Leishmaniasis research unit Brit. Honduras, C. Am., 1959-62; sr. sci. officer Wellcome Parasitology Unit, Belem, Brazil, 1965—. Fellow Royal Soc. Tropical Medicine and Hygiene; mem. Soc. Protozoology, Soc. Parasitologists. Research and publs. on epidemiology and life cycles of parasitic protozoa especially human infections; demonstrated role of forest rodents as reservoirs of human cutaneous leishmaniasis in Central and South Am.; co-worker in 1st exptl. transmission of Leishmania to man by Phlebotomus in New World. Address: The Wellcome Parasitology Unit, Instituto Evandro Chagas, Av. Almirante Barroso, Belem, Brazil.*

LAIRD, Anna Kane, Am. biologist; b. N.Y.C., June 7, 1922; d. Pierre and Lillie (Moinehan) Laird; B.A., U. Pa., 1943, M.D., 1946; Ph.D., U. Wis., 1952; m. A. D. Barton, Aug. 12, 1950; 1 son, David Laird. Postdoctoral fellow in cancer research USPHS, 1949-51, Am. Cancer Soc., 1952-54; asso. biologist Argonne (Ill.) Nat. Lab., 1954-58, asso. pathologist, 1958—. Fellow A.A.A.S.; mem. Am. Soc. Cell Biology, Am. Soc. Exptl. Pathology. Research and publs. in math. analysis of growth of animals and their parts; introduced algebraic transformations of growth equation that permit growth curves of widely different animal forms to be superimposed, yielding information on nature of biol. time, heritable factors in normal growth, and properties of malignant growth. Home: 1320 Turvey Rd., Downers Grove, Ill. 60515. Office: Argonne Nat. Lab., Argonne, Ill. 60439.*

LAIRD, Donald Anderson, Am. applied psychologist; b. Angola, Ind., May 14, 1897; s. Allen Max and Grace (Anderson) L.; B.A., U. Dubuque, 1919, Sc.D. (hon.), 1927; Ph.D., U. Ia., 1923; m. Hilda Drexel, Nov. 18, 1916 (dec. 1938), 1 son, David Drexel (dec. 1944); m. Eleanor Childs Leonard, Apr. 18, 1940. NRC fellow in biol. sci. Yale, 1923-24; asst. prof. to head dept. psychology Colgate U., Hamilton, N.Y., 1924-39; dir. Ayer Found. for Consumer Analysis, Phila., 1939-40; cons., author, 1940—. Diplomate in Industrial Psychology, Am. Bd. Examiners in Profl. Psychology. Fellow A.A.A.S.; mem. Am. Psychol. Assn., Authors Guild. Originated part of Colgate Mental Hygiene Tests, 1925; 1st measurements of city noise, 1929; pioneer in indsl. psychology. Author: numerous books including Psychology of Selecting Employees, 1925; Technique of Handling People, 1943; Technique of Getting Things Done, 1947; Practical Business Psychology, 1951, 66; Techniques for Efficient Remembering, 1960; How to Get Along with Automation, 1964. Address: Route 2, N. River Rd., Lafayette, Ind. 47901.*

LAIRD, Elizabeth Rebecca, Canadian physicist; b. Owen Sound, Ont., Can. Dec. 6, 1874; d. John G. and Rebecca (Lapierre) Laird; B.A., U. Toronto, 1896; Ph.D., U. Bryn Mawr, 1901; student U. Berlin, Germany, 1898-99, Cambridge U., 1909, U. Würzburg, 1913-14, U. Chgo., 1919; D.Sc., U. Toronto, 1927; LL.D., U. Western Ont., 1954. Instr. math Ont. (Can.) Ladies Coll., Whitby, 1896-97; asst. in physics Mt. Holyoke Coll., 1901-02, instr., 1902-04, prof., head dept., 1904-40, emeritus prof., 1940—; physicist on radar devel. U. Western Ont., 1940-45, hon. prof. physics, 1945-53. Sarah Berliner Research fellow, 1913-14. Fellow A.A.A.S., Am. Phys. Soc.; mem. Optical Soc., History of Sci. Soc., Am. Assn. U. Profs., Am. Assn. Physics Tchrs., Canadian Assn. Physicists. Research and publs. on properties of soft X-rays in spectral region between the extreme ultra violet and known X-rays. Home: 9 Grosvenor St., London, Ont., Can.*

LAIRD, Wilson Morrow, Am. geologist; b. Erie, Pa., Mar. 4, 1915; s. Charles William and Elizabeth (Morrow) L.; B.A. cum laude, Muskingum Coll., 1936, D.Sc., 1964; M.A., U. N.C., 1938; Ph.D., U. Cin., 1942; m. Reba Allene Laird, Aug. 8, 1938; children—Douglas, David, Donald, Dorothy. Grad. asst. U. N.C., 1936-38, U. Cin., 1938-40; geologist Pa. Geol. Survey, 1936, 37, 40; asst. prof. geology U. N.D., Grand Forks, 1940-42, asso. prof., 1942-46, prof., 1946—. Geologist, State of N.D., 1941—; geologist U. S. Geol. Survey, 1944-45; cons. geologist, 1944-53. Fellow Geol. Soc. Am.; mem. Am. Assn. Petroleum Geologists (President's award 1947), Paleontol. Soc., A.A.-A.S., Am. Soc. Limnology and Oceanography, Assn. Am. State Geologists (pres. 1951), N.D. Acad. Sci. (pres. 1952), Sigma Xi, Sigma Gamma Epsilon. Research and numerous publs. on geology of N.D., Mont., Alaska, oil and gas conservation especially of Devonian system, glacial geology in N.D. and adjacent area, Alaska. Home: 2800 University Av., Grand Forks, N.D. 58201.*

LAITINEN, Herbert August, Am. chemist; b. nr. Wadena, Minn., Jan. 17, 1915; s. Nestor Alfred and Minnie (Nikkari) L.; B.Chemistry, U. Minn., 1936, Ph.D. (Shelvin fellow), 1940; m. Marjorie Violet Gorans, June 15, 1940; children—Kenneth, Richard, Roger. Faculty, U. Ill., Urbana, 1940—, prof. chemistry, 1947—, head div. analytical chemistry, 1953—. Chmn. com. on analytical chemistry NRC, 1963-66. Recipient Fisher award in analytical chemistry Fischer Sci. Com., 1961. Guggenheim fellow, 1953, 62. Mem. Am. Chem. Soc. (nat. council 1961—), Internat. Union Pure and Applied Chemistry, A.A.A.S., Electrochem. Soc., Sigma Xi, Phi Lambda Upsilon, Alpha Chi Sigma. Author: (with I. M. Kolthoff) Ph and Electrotitrations, 1941; Chemical Analysis, 1960. Researcher in analytical chemistry and electrochemistry, 1936—. Home: 609 Hessel Blvd., Champaign, Ill. 61820.*

LAITINEN, Lauri Veli, Finnish neurosurgeon; b. Utajärvi, Finland, Dec. 14, 1928; s. Eino A. and Sanni (Moilanen) L.; M.D., U. Helsinki (Finland), 1953, Ph.D., 1956; m. Kerstin Holmberg, Dec. 26, 1955; children—Lena Alice, Jens. Hon. clin. asst. Maida Vale Hosp., London, Eng., 1960-61; cons. neurosurgeon U. Central Hosp., Helsinki, 1962—; lectr. neurosurgery U. Helsinki, 1965—. Author: Craniosynostosis, 1956; also articles. Research on growth of skull bones, intracranial arterial aneurysms, stereotaxic treatment of Parkinson's disease, torticollis and familial tremor, devel. monopolar impedance technique for investigation cerebral blood flow. Home: Apollonk. 23 B. Helsinki 10. Office: Topeliuksenk 5, Helsinki 26, Finland.*

LAJTHA, Abel, biochemist; b. Budapest, Hungary, Sept. 22, 1922; s. Laszlo and Rose (Hollos) L.; Ph.D., U. Budapest, 1945; m. Marie Snyder, Nov. 25, 1953; children—Terry, Kathryn. Came to U. S., 1948, naturalized, 1954. Asst. prof. U. Budapest, 1945-47; fellow Royal Instn., London, Eng., 1947-48; asst. prof. Inst. Muscle Research, Woods Hole, Mass., 1948-49; research asso., asst. prof. Columbia U., 1949—; asso. research scientist N.Y. State Psychiat. Inst., 1949-61, asso. research scientist, 1956-61; prin. research scientist N.Y. State Research Inst. for Neurochemistry and Drug Addiction, Ward's Island, N.Y.C., 1961, dir., 1966—. Cons., N.Y. Sch. Psychiatry, 1958—, VA Hosp., East Orange, N.J., 1964—, Willowbrook State Neuroendocrine Research Unit, 1965—. Mem. Am. Acad. Neurology, Am. Soc. Biol. Chemists, A.A.A.S., Biochem. Soc., Collegium Internationale Neuropsychopharmacologium, Internat. Soc. Neurochemistry, Internat. Brain Research Orgn., Sigma Xi. Editor, Jour. Neurochemistry, 1962—, Internat. Rev. Neurobiology, 1964—, Brain Research, 1966—. Contbr. numerous articles to tech. jours. Research on brain barrier system, role of cerebral membranes in effect of drugs and path. changes, mechanism of cerebral protein synthesis and breakdown, alterations in brain proteins during function; measured rates of protein turnover; established instability brain proteins. Home: 2395 Palisade Av., N.Y.C. 10463. Office: N.Y. State Research Inst. for Neurochemistry and Drug Addiction, Ward's Island, N.Y.C. 10035.*

LAJTHA, Laszlo George, hematologist, radiobiologist; b. Budapest, Hungary, May 25, 1920; s. Lazlo John and Rose (Hollos) L.; M.D., U. Budapest, 1944; D.Phil., Oxford (Eng.) U., 1950; m. Gillian Macpherson Henderson, Aug. 28, 1954; children—Christopher Alastair, Adrian James. Asst. prof. dept. physiology U. Budapest, 1943-46; Brit. Council scholar Radcliffe Infirmary, Oxford, 1947-49; research asso. Churchill Hosp., Oxford, 1950-62; dir. Paterson Labs., Christie Hosp. and Holt Radium Inst., Manchester, Eng., 1962—; reader in pathology U. Manchester Med. Sch., 1962—. Mem. sci. adv. com. Brit. Empire Cancer Campaign. Mem. Brit. Soc. Hematology, Brit. Inst. Radiology, Med. Research Soc., Assn. for Radiation Research, Radiation Research Soc., Brit. Soc. Cell Biologists. Author: The Use of Isotopes in Haematology, 1961; also numerous articles. Research on devel. methods of bone marrow culture and high resolution autoradiography, mechanisms of differentiation in bone marrow cells, effects of radiation on nucleic acid metabolism and on bone marrow autoradiographic analysis of human chromosome duplication, cell population kinetics. Home: 5 Carrwood Rd., Wilmslow, Cheshire, Eng. Office: Wilmslow Rd., Manchester, Eng.*

LAKE, Richard, Brit. physician, surgeon; b. Dec. 14, 1861; s. Thomas Lake; m. Mildred Pelly; 1 dau.; m. 2d, Ellen Sapsworth, 1918. Surgeon, Nat. Aid Soc., Brit. Red Cross, Sudan Campaign, 1885; served in Servo-Bulgarian War, 1885-86; surgeon, laryngologist Mt. Vernon Hosp. for Consumption, 1897-1904; cons. surgeon diseases ear, nose and throat Univ. Coll. Hosp., Royal Ear Hosp., Seamen's Hosp.; surgeon Royal Ear Hosp., 1898-1926; Geoffrey E. Duveen lectr. otology U. London, 1924-28. Fellow Royal Coll. Surgeons; mem. W. London Medico-Chirurg. Soc. (pres. 1907-08), Royal Soc. Medicine (pres. otol. sec. 1913-14). Author: Laryngeal Phthisis; Handbook of Diseases of the Ear, 5th edit. with E. A. Peters; International Directory, Laryngologists and Otologists, 2d edit.; Contributions to Art and Science of Otology; Fishing Memories; The Grayling. Translator: Cerebellar Abscess (Neuman). Died Nov. 1, 1949.

LAKOVLEV, Nikolai Nikolaevich, Soviet geologist, paleontologist; b. Apr. 27, 1870; mem. staff Geologic Com. (now All-Union Geology Inst.), Leningrad, USSR, 1895—, dir. 1923-26; prof. Leningrad Mining Inst., 1900-30. Recipient Karpinskii Prize, USSR Acad. Sci., 1948. Mem. USSR Acad. Sci. (corr.). First paleoecologic studies in Russian invertebrates; participated in study of Donetz Coal Basin, 1892; geol. surveys in Urals, Caucasus, Transcaucasia.

LALAND, Soren, Norwegian biochemist; b. Bergen, Norway, May 23, 1922; s. Per and Mary (Moe) L.; Ph.D., U. Birmingham (Eng.), 1952; m. Elizabeth Eva Parry, Aug. 19, 1951; children—Per Anthony, Pal Henry, Christian Soren. Mag. sci. in chemistry U. Oslo (Norway), 1947-51, prof. biochemistry, 1956—. Cons., Nyegaard & Co. A/S, Oslo, 1952. Research and publs. on carbohydrate chemistry including identification of carbohydrate components in nucleic acids, bluish fluorescent substance present in liver and blood, biosynthesis of bacterial polypeptides in particular gramicidins; demonstrated that vitamin A is attached to prealbumin; with others isolated and identified ADP as substance which aggregates blood platelets. Home: 40, Bjerkasen, Blommenholm, Norway. Office: U. Oslo, Blindern, Norway.*

LALANDE, Joseph Jérôme Le François de, French astronomer; b. Bourg, France, July 11, 1732; studied law, also became pupil of Delisle and P. C. Lemonnier, Paris. Became prof. astronomy Coll. de France, 1762; named dir. Paris Obs., 1795. Fellow Royal Soc., 1763; mem. French Acad. Scis., 1753, Berlin Acad. Berlin. Author: Traité d'astronomie, 2 vols., 1764, enlarged edit., 4 vols., 1771-81; Voyage d'un français en Italie, 1769; Astronomie des dames, 1785; Abrégé de navigation, 1793; Histoire céleste française (50,000 stars placed), 1801; Bibliographie astronomique (with history of astronomy 1781-1802), 1803; concluding 2 vols. Histoire des mathématiques (J. E. Montucla), 2d edit., 1802; also various papers. Editor: Connoissance des temps, 1759-74, 1794-1807. Made observations on lunar parallax from Berlin, in concert with N. L. de Lacaille from Cape of Good Hope, 1751; calculated (with A. C. Clairaut) return of Halley's comet, 1759; pub. corrected edit. Halley's tables with history of comet, 1759; provided best planetary tables of 18th century; observed Neptune 50 years before Leverrier without realizing it was a planet; noted for publs. on transit of Venus in 1769; popularized astronomy; instituted Lalande prize for main astron. performance of each year, 1802. Died Paris, France, Apr. 4, 1807.

LALANNE, Léon-Louis-Chrétien, French engr., inventor; b. Paris, July 3, 1811; student École Polytechnique, Paris, 1829-31; gen. insp. civil constrn.; dir. Nat. Sch. Civil Engring.; mem. French Acad. Scis. Author: Mémoire sur l'arithmo-planimètre, 1840. Inventor arithmoplanimetry on basis of laws of similarity, 1839, also algebraic balance for resolving 7th degree numerical equations. Died Paris, Mar. 12, 1892.

LALICH, Joseph John, Am. pathologist; b. Slunj, Jugoslavia, Nov. 23, 1909; s. Joseph and Agnes (Pozega) L.; B.S., U. Wis., 1932, M.D., 1937; m. Margaret Hanstein, Feb. 8, 1941. Research fellow medicine, 1938-42; instr. pathology U. Wis., 1946-48, asst. prof., 1948-49, asso. prof. pathology, 1949-52, prof., 1952—. Cons. VA Hosp. Mem. Am. Assn. Pathologists and Bacteriologists, Am. Soc. Exptl. Pathology. Research, numerous publs. on role of aciduria in prodn. of hemoglobin-uric nephrosis, mechanism of aortic rupture in exptl. lathyrism, exptl. prodn., of aortic aneurysms in rats, exptl. prodn. of infarcts in rats. Home: 6306 Mound Dr., Middleton, Wis. Office: 470 N. Charter St., Madison, Wis. 53706.*

LALLEMAND, André, French astronomer; b. Cirey, Sept. 9, 1904; Ph.D., U. Strasbourg (France); agrégé; m. Ancel, Aug. 2, 1928; children—Francois, Denis. Astronomer, Strasbourg Obs.; apptd. to Paris Obs. Research on photoelectric phenomena, mainly as applied to astronomy. Home: 61, av. Niel, Paris, 17. Office: 61, av. de l'Observatoire, Paris 14, France.

LALLEMAND, Claude-François, French surgeon; b. Metz, France, Jan. 26, 1790; ed. Paris; prof. external clin. medicine, Montepellier, France; physician to Mehemet Ali, viceroy of Egypt; mem. French Acad. Scis. Author: Recherches anatomico-parthologiques sur l' encéphale, 1824. Studied meningitis and related brain diseases; developed cure for vesico-vaginal fistulas, Died July 23, 1854.

LALLEMAND, Jean-Pierre (called Charles), French engr., inventor; b. St. Aubin-sur-Aire, France, Mar. 7, 1857; gen. insp. mines; dir. Nat. Bur. Surveying; mem. French Acad. Scis., Bur. Longitudes. Inventor medimaremeter for measuring level of ocean despite tides. Died Bussy, France, Feb. 1, 1938.

LALLY, Vincent Edward, Am. meteorologist; b. Brookline, Mass., Oct. 13, 1922; s. Michael James and Ellen (Dolan) L.; student Boston Coll., 1940-42; B.S., U. Chgo., 1944; B.E.E., Mass. Inst. Tech., 1948, M.S., 1949; m. Marguerite Mary Tibert, June 4, 1949; children—Dennis Vincent, Marianne Theresa, Stephen James. Engr., Bendix-Friez, Towson, Md., 1949-51; sect. chief Air Force Cambridge Research Center, Bedford, Mass., 1951-58; research mgr. Tele-Dynamics, Phila., 1958-61; program head Nat. Center for Atmospheric Research, Boulder, Colo., 1961—; asso. editor Jour. Applied Meteorology. Mem. Am. Inst. Aero. and Astronautics, A.A.A.S., I.E.E.E., Research Soc. Am., Am. Meteorol. Soc., Am. Geophys. Union, Sigma Xi. Contbr. articles to sci. jours. Developed and tested concepts of superpressure ghost balloons which hold all duration records for sustained flight at all altitudes. Patentee in fields of telemetry, space communications, meteorol. instruments. Home: 4330 Comanche Dr., Office: Nat. Center for Atmospheric Research, Boulder, Colo. 80302.*

LA LOUBÈRE, Antoine de (Lalouère, Lalouvère), French mathematician; b. Rieux, France, 1600; mem. Soc. of Jesus; lectr. on math., rhetoric, theology, humanistic subjects coll. at Toulouse, France. Author: Geometria veterum promota de cycloide, 1660; Elementa tetragonismica, 1651; Propositio 36a excerpta, 1659; Veterum geometria, 1660. Research on quadrature of circle, hyperbola, cycloid; developed method in which tangent is defined as direction of a moving point (later applied in kinematics). Died 1664.

LAM, Conrad Ramsey, Am. surgeon; b. Oglesby, Tex., Aug. 19, 1905; s. Elie and Inez (Hitt) L.; A.B., Hardin-Simmons U., 1927; M.D., Yale, 1932; M.S. in Surgery, U. Mich., 1938; Doctor Honoris Causa, U. Puebla (Mex.), 1956; m. Marian Melbourne Smith, Aug. 2, 1941; children—Marjorie, Richard, Janet, Douglas. Asso. surgeon Henry Ford Hosp., Detroit, 1938-48, surgeon-in-charge div. thoracic surgery, 1948—. Recipient Keeter Alumni award, Hardin-Simmons U., 1952; Order of J. Fernandez Madrid, med. dept. Colombian Army, 1959. Mem. Detroit Acad. Surgery (pres. 1953), Mich. Acad. Sci., Arts, Letters (pres. 1960), Detroit Physiol. Soc. (pres. 1947), Bolivian Soc. Surgeons (hon.). Editor: Cardiovascular Surgery, 1955. Research on treatment of burns, blood clotting, thoracic and cardiovascular surgery, 1st operation to replace thoracic aorta for aneurysm. Home: 28130 Westbrook St., Farmington, Mich. Office: Henry Ford Hosp., Detroit 48202.*

LAM, Herman Johannes, Dutch botanist; b. Veendam, Jan. 3, 1892; s. Anske and Margien (Winter) L.; Ph.D., U. Utrecht (Netherlands); m. B. Moorrees, Apr. 22, 1922; children—Ellen, Hermine. Botanist, Java Bot. Gardens, Buitenzorg, 1919-33; dir. Nat. Bot. Mus. (Rijksherbarium), 1933-62; prof. botany U. Leyde (Netherlands), 1933-62, rector, 1958-59. Research and publs. on taxonomy of higher plants, phytogeography, phylogeny. Home: Rijnburgerweg 127 Leyden, Netherlands.

LAMANNA, Carl, Am. bacteriologist; b. Bklyn., Dec. 1, 1916; s. Frank and Margaret (Ottavi) L.; B.S., Cornell U., 1936, M.S., 1937, Ph.D. 1939; m. Ruth Weed, June 6, 1942; children—Carla Susan, Roger Weed. Instr., Wash. State U., 1940-41, Ore. State U., 1941-42, La. State U. Sch. Medicine, 1942-44; br. chief CWS, Frederick Md., 1944-48; asso. prof. Johns Hopkins Sch. Pub. Health, 1948-57; sci. dir. Naval Biol. Lab., U. Cal., Berkeley, 1957-61; dep. sci. adviser life scis. div. Office of Chief Research and Devel. Dept. Army, Washington, 1961—; WHO vis. prof. microbiology Inst. Hygiene U. Philippines, Manila, 1954-55. Fellow N.Y. Acad. Scis., Washington Acad. Scis., Am. Acad. Microbiology, A.A.A.S. Author (with M. F. Mallette) Basic Bacteriology, 1953, 2d edn., 1959, 3d edn., 1965. Contbr. numerous articles to sci. jours. Pioneer in field of crystallization of bacterial toxin; chem. differences between bacterial spores and vegetative cells; identification of aerobic spore forming bacteria; nature of gram stain; oral toxicity of bacterial poisons. Home: 3812-37th St. N., Arlington, Va. 22207. Office: Office Chief Research and Devel., Dept. Army, Washington 20310.*

LAMANON, Robert de Paul de, see de Lamanon, Robert de Paul.

LAMARCK, Jean Baptiste Pierre Antoine de Monet, Chevalier de, French naturalist; b. Bazantin, France, Aug. 1, 1744; studied for priesthood Jesuit Coll., Amiens; entered army at age 17; received officer's commn. for bravery; mil. career ended due to poor health; studied medicine, meteorology, botany, Paris, soon devoted full time to botany; mem. Royal Acad., keeper Royal Garden, 1774; wrote book of French flora establishing his popularity, 1778; admitted to Acad. Scis., 1779; apptd. botanist to king, 1781; travelled through Europe studying natural history, 1781-82; held bot. appointment at Jardin du Roi, 1788-93; prof. invertebrate zoology Mus. Natural History, Paris, from 1793. Author: Flore française, 1778; Dictionnaire de botanique, 1783-96; Système des animaux sans vertèbres, 1801; Recherches sur l'organisation des corps vivants, 1802; Philosophie zoologique, 1809; Histoire naturelle des animaux sans vertèbres, 1815-22; others. Did extensive work with living and fossil invertebrata, considered founder of modern invertebrate zoology; 1st to distinguish vertebrate from invertebrate on basis of vertebral column; established classification of great group of animals; devised one of earliest hypothesis explaining evolutionary devel. of life; maintained that species not fixed but that they changed and developed, also that certain portions of body developed or atrophied according to amount of use or disuse they got; believed that evolution was result of attempts made by animals to change (believed that needs, habits and wants of animals changed leading to changes or modifications of various organs); thought such acquired characteristics were passed on to descendants; pioneer in bringing concept of evolution into field of biology; popularized word biology. Died Paris, Dec. 18, 1829.

LAMARE, Pierre Joseph Henri, French geologist; b. Paris, Feb. 6, 1894; s. Paul and N. (Delacourtie) L.; Ph.D., LL.M., U. Paris; m. France Parant, June 9, 1953. Prof., Nat. Inst. Agronomy; became mem. Faculty Scis., Bordeaux, France, 1950. Mem. Acad. Sci. Madrid, Spanish Research Council, French Geol. Soc. (past pres.), Geography Soc., Assn. French Geographers. Research and publs. on geol. structure, phys. geology, petrology wells in explored regions; explored geology of Africa, Arabia, Venezuela, French and Spanish Pyrenees, 1922-23, 28-29. Office: Faculté des scis., 20, cours Pasteur, Bordeaux (Gironde), France.

LAMARQUE, Paul-Jean-François, French physician, radiologist; b. Bazas, France, July 23, 1894; s. Eusebe and Anne (Laulan) L.; M.D., Ph.D., U. Bordeaux; agrégé Faculty Medicine, 1923; m. Suzanne Brunet, Jan. 19, 1921; children—Francoise, Monique, Jean-Louis. Became titular prof., chair electro-radiology, 1947; apptd. to chair cancerology, faculty medicine Montpellier (France) U.; dir. Regional Anti-Cancer Center; specialist in electro-radiology in hosps. Mem. Superior Council on Hygiene; mem. cancer sect. Ministry Health; mem. cons. com. biology AEC. Research and publs. in radiology. Home: Mas de Rouel, Montpellier. Office: 5, rue Boussairolles, Montpellier (Herault), France.

LAMARTINIÉRE, Germain-Pichaut de, see de Lamartinière, Germain-Pichaut.

LAMB, Arthur Becket, Am. chemist; b. Attleboro, Mass., Feb. 25, 1880; s. Louis Jacob and Elizabeth Camerden Townsend (Becket) L.; A.B., A.M., Tufts, 1900, Ph.D., 1904, D.Sc., 1922; A.M., Harvard, 1903, Ph.D., 1904; postgrad. univs. Leipzig, 1904, Heidelberg, 1905; m. Blanche Anne Driscoll. Dec. 27, 1923. Instr. electrochemistry Harvard, 1905-06; asst. prof. chemistry N.Y. U., 1906, asso. prof., 1907, prof., 1909-12, also dir. Havemeyer Chem. Lab.; asst. prof. chemisty Harvard, 1912-20, prof., 1920-48, prof. emeritus, 1949-52, dir. chem. lab., 1912-47, dean Grad. Sch. Arts and Scis., 1940-43. mem. U. S. Fixed Nitrogen Mission, 1919; dir. Fixed Nitrogen Research Lab., Washington, 1919-21. Recipient Am. Chem. Soc., Nichols medal, 1943, Priestly medal, 1949, Austin M. Patterson award, Dayton Sect., 1951; Ballou medal, Tufts Coll., 1944. Fellow A.A.A.S. (v.p. 1933) Chem. Soc. London (hon.); mem. Am. Chem. Soc. (pres. 1933 editor Jour. 1917-49), Am. Acad. Arts and Scis., Am. Philos. Soc., Am. Electrochem. Soc., Washington Acad. Scis., Nat. Acad. Sci., Deutsche Chemische Gesellschaft, Deutsche Bunsen-Gesellschaft, Phi Beta Kappa, Alpha Chi Sigma, Phi Lambda Upsilon. Research in electrochemistry, isomerism in cumaric acid derivatives, Thomson effect in electrolytes, heat of absorption, crystallogenetic adsorbents, liquid junction potentials. Died May 15, 1952.

LAMB, Daniel Smith, Am. physician; b. Phila., May 20, 1843; s. Jacob Matlack and Delilah Mick (Rose) L.; A.B., Central High Sch., Phila., 1859, A.M., 1864; M.D., Georgetown U., 1867; LL.D., Howard U., 1913, D.Sc., 1923; m. Elizabeth Scott, May 20, 1868 (dec.); children—Mrs. Lillie Fraley Carney, Robert Scott; m. 2d, Isabel Haslup, July 3, 1899; children—Ella (dec.), Delilah (dec.). Acting asst. surgeon U. S. Army, on duty at Army Med. Museum, 1868-92, pathologist to Army Med. Mus., 1892-1920; prof. materia medica and then anatomy, Howard U., 1873-1923; prof. gen. pathology U. S. Coll. of Vet Surgeons, 1894-1900. Conducted postmortem examinations of Pres. Garfield, Henry Wilson, Senator Brooks, and the assassin, Guiteau. Mem. Assn. Am. Anatomists (v.p., sec.), Assn. of Acting Asst. Surgeons U. S. Army (pres.), Med. Soc. D.C. (pres. 1901), Washington Acad. Scis. (pres. 1904-05). Author: History of Medical Department Howard University, Washington, 1900. Editor Washington Medical Annals, History Medical Society of D.C. Died Apr. 21, 1929.

LAMB, Sir Horace, English mathematician, physicist; b. Manchester, Eng., Nov. 27, 1849; s. John and Elizabeth (Rangeley) L.; B.Sc., Trinity Coll., Cambridge, Eng., 1872; hon. degrees, Glasgow, Scotland, Oxford, Eng., Cambridge, Dublin, Ireland, Manchester, Eng., St. Andrews, Scotland, Sheffield, Eng.; m. Elizabeth Foot, 1875; 3 sons, 4 daus. Fellow, lectr. Trinity Coll., Cambridge, 1872-75; prof. math. U. Adelaide (Australia), 1875-85; prof. pure, later applied math. Owens Coll., Victoria U., Manchester, 1885-1920; Rayleigh lectr. Cambridge, 1920-34. Cons. to Admiralty, 1914-18; mem. Aero. Research Com., 1921-27. Hon. fellow Trinity Coll., Cambridge. Recipient Copley medal, De Morgan medal, 1911. Fellow Royal Soc., 1884 (Royal medal 1902); mem. Brit. Assn. (became pres. 1925), London Math. Soc. (pres. 1902-04), Manchester Lit. and Philos. Soc. Author: Mathematical Theory of the Motion of Fluids, 1878; A Treatise on the . . . Motion of Fluids, 1879; Hydrodynamics, 1895; Infinitesimal Calculus, 1897; Dynamical Theory of Sound, 1910; Statics: Including Hydrostatics and Elements of the Theory of Elasticity, 1912; Dynamics, 1914; Higher Mechanics, 1920. Studied deflection of gravity by tidal loading of earth's surface, propagation of earthquake tremors on surface of an elastic solid, theory of tides and waves, motion of perforated solids in perfect liquid, oscillations of a viscous spheroid, elastic deformation of plates and shells, wave propagation, elec. induction. Died Cambridge, Dec. 4, 1934.

LAMB, Lawrence Edward, Am. physician; b. Fredonia, Kan., Oct. 13, 1926; s. John Robert and Dell M. (Ross) L.; M.D., U. Kan., 1949. Fellow, Emory U., 1953-54; Am. Heart Assn. Research fellow, Dr. Pierre Duchosal, Geneva, Switzerland, 1954-55; dir. cardiology dept. internal medicine Sch. Aviation Medicine, Randolph AFB, 1955-57, chief, 1957-62, dir. consulation services, 1959-62; chief clin. sci. div. Sch. Aerospace Medicine, Brooks AFB, Tex., 1962-64, chief aerospace med. scis. div., 1964-66, prof. internal medicine, 1958-66; prof. medicine Baylor U. Coll. Medicine, Houston, 1966—. Cons., Project Mercury, NASA, 1960-64, cons., dir. life scis., 1965—. Recipient Dept. Def. Distinguished Civilian Service medal, 1962. Diplomate Am. Bd. Internal Medicine. Fellow Am. Coll. Cardiology (past trustee), A.C.P., Am. Coll. Chest Physicians, Aerospace Med. Assn. (Arnold D. Tuttle award 1959), Am. Coll. Clin. Pharmacology and Chemotherapy, Heart Assn. (mem. com. on criteria and methods epidemiological studies cardiovascular diseases 1963) Author: Fundamentals of Electrocardiography and Vectorcardiography, 1957; Electrocardiography and Vectorcardiography. Instrumentation, Fundamentals, and Clinical Applications, 1965; also numerous articles. Research on normal values in electrocardiography, abnormal cardiovascular findings in asymptomatic Am. males, med. evaluation for space crews, med. information handling, bed rest and simulated weightlessness. Office: Dept. Medicine, Baylor U. Coll. Medicine, Houston 70025.*

LAMB, Willis E(ugene), Jr., Am. physicist; b. Los Angeles, July 12, 1913; s. Willis Eugene and Marie Helen (Metcalf) L.; B.S., U. Cal., 1934, Ph.D., 1938; D.Sc., U. Pa., 1953; M.A., Oxford U., 1956; M.A., Yale University, 1961; L.H.D., Yeshiva University, 1965; m. to Ursula Schaefer, June 5th, 1939. Mem. faculty Columbia, 1938-52, prof. physics, 1948-52; prof. physics Stanford, 1951-56; Wykeham prof. physics and fellow New Coll., Oxford U., 1956-62; Ford prof. physics Yale, 1962—; Morris Loeb lectr. Harvard. 1953-54. Cons. Philips Labs., Bell Telephone Labs., Perkin-Elmer, NASA. Visiting com. Brookhaven Nat. Lab. Recipient (with Dr. Polycarp Kusch) Nobel prize in physics, 1955; Rumford Premium, Am. Acad. Arts and Scis., 1953; Research Corp. award, 1955; Guggenheim fellowship, 1960-61, Yeshiva Award, 1962. Fellow Am. Phys. Soc., N.Y. Acad. Scis.; hon. fellow Inst. Physics and Phys. Soc. (Guthrie lectr. 1958); mem. Nat. Acad. Scis. Research in theoretical physics, on atomic and nuclear structure, microwave spectroscopy, fine structure of hydrogen and helium, magnetron oscillators, masers and lasers; in precise experiments with hydrogen measured a splitting of energy levels (Lamb shift), 1947; developed theories of beta decay and of interactions of neutrons. Office: Sloane Physics Lab., Yale U., New Haven.*

LAMBERG, Bror-Axel, Finnish physician; b. Helsinki, Finland, Mar. 1, 1923; s. Axel Evert and Anna (Perkowsky) L.; cand.med., Helsinki U., 1946, lic.med. (leg. Physician), 1949, Dr.med. and chir., 1953; m. Carin Anita Emilia Olin, Dec. 21, 1947; 1 dau., Christel Johanna Emilia. Staff, U. Helsinki, 1952—, asst. chief physician I dept. medicine, 1962-65, asso. clin. prof. medicine III dept. medicine, 1965—; asst. physician med. dept. Maria Hosp., Helsinki, 1955-59; dir. Minerva Found. Inst. for Med. Research, 1959—. Mem. adv. bd. Gesellschaft für Nuclearmedizin, 1963—, Ciba Found., 1967—. Decorated Liberty Cross with swords, 1944. Mem. Am. Thyroid Assn. (corr.), European Thyroid Assn. (dir.), European Soc. Clin. Investigation, Royal Soc. Medicine (affiliate), Finnish Soc. f. Endocrinol. (chmn. 1968—), Finnish Soc. for Nuclear Medicine (chmn. 1966—), Finnish Soc. for Internal Medicine (past sec.). Author: Radioactive Phosphorus as Indicator in a Chick Assay of Thyrotropic Hormone, 1953; also numerous articles. Developed method for assay of thyrotropic hormone using thyroid uptake of radioactive phosphorus as indicator; demonstrated increasing effect of thyrotropic preparations on uptake of radioactive sulphur by lachrymal gland in guinea-pig, decreasing effect of thiazides on renal excretion of calcium in normal, hypo- and hypercalcemic subjects and lact of effect in adrenal insufficiency; studies on diagnostic and therapeutic problems of thyroid disease in Finland; clin. and exptl. studies on metabolic effects of thyroid hormones. Home: 11 Norra Mossavägen 17, Grankulla, Finland. Office: III Dept. Medicine, 4 Haartmaninkatu, Helsinki 29, Finland.*

LAMBERT, Aylmer Bourke, English botanist; b. Bath, Eng., Feb. 2, 1761; s. Edmund and Bridget (Bourke) L.; student St. Mary Hall, Oxford (Eng.) U.; m. Catherine Bowater. Founded herbarium of 30,000 specimens, Boyton, Eng., 1802. Fellow Linnean Soc. (charter, v.p. 1796-1842), Royal Soc., 1791. Author: A Description of the Genus Cinchona, 1797; Monograph of the Genus Pincus; An Illustration of the Genus Cinchona, 1821; also articles. 1st description of Carduis tuberosus, Centaurea nigrescens. Died Jan. 10, 1842.

LAMBERT, Charles Irwin, Am. psychiatrist; b. Argyle, Wis., Dec. 6, 1877; s. Furniss and Mary Wasley (Reynolds) L.; grad. Ia. State Tchrs. Coll., 1897; B.S., State U. Ia., 1901, M.A. and M.D., 1903; postgrad. Harvard and Munich; m. Bess Ann Coomer, Oct. 9, 1907; children—John Pierce, Robert Reynolds, Elizabeth Ann; m. 2d, Florence B. Gilpin, April 29, 1940. Instr. pathology Coll. of Medicine, State U. Ia., 1904; asst. in neuropathology N.Y. State Psychiat. Inst., 1905-13; pathologist Manhattan State Hosp., 1910-13; asst. dir. Bloomingdale Hosp., 1913-22; chief of Vanderbilt Clinic (psychiatric dept. Columbia U.), 1922-28; asso. prof. psychiatry Coll. Phys. and Surg., Columbia, from 1922; prof. psychiat. edn. Tchrs. Coll., Columbia; med. dir. Four Winds Sanitarium, Katonah, N.Y. Mem. A.M.A., Am. Psychiat. Assn., Sigma Xi. Died Apr., 1954.

LAMBERT, Edward Howard, Am. physiologist; b. Mpls., Aug. 30, 1915; s. Harry Edward and Ida (Peterson) L.; B.S., U. Ill., 1936, M.S., 1938, M.D., 1939, Ph.D. in Physiology (fellow), 1944; m. Louise Augusta Rueckheim, Oct. 12, 1940. Instr. med. tech. and biol. scis. Herzl Jr. Coll., Chgo., 1941-42; asso. in medicine, research physiologist OSRD project U. Ill. Coll. Medicine, 1942-43; research asst. acceleration lab. Mayo Aeromed. Unit, 1943-45; cons. physiology Mayo Clinic, Rochester, Minn., 1945—; faculty Mayo Grad. Sch. Medicine, U. Minn., Rochester, 1945—, prof. physiology, 1958—. Cons., U. S. Navy, 1948, 52-53, USPHS, 1955; mem. adv. bd. Myasthenia Gravis Found., 1959—; mem. adv. group NIH, 1963—. Recipient Presdl. certificate merit, 1947. Mem. Am. Physiol. Soc., Am. Neurol. Assn., Am. Assn. Electromyography and Electrodiagnosis (pres. 1957-58, council 1954—), Am. Electroencephalographic Soc., Aerospace Med. Assn. (Arnold D. Tuttle award 1952), Minn. Acad. Scis., Minn. Soc. Neurol. Scis., Soc. for Exptl. Biology and Medicine, A.A.A.S., Sigma Xi, Phi Chi, Alpha Omega Alpha, Phi Kappa Phi. Editorial bd. Am. Jour. Physiology, also Jour. Applied Physiology, 1965—. Research and numerous publs. on effects of acceleration on man, 1943-46, neuromuscular disorders, 1947—. Home: 202 14th St., N.E. Office: Mayo Clinic, Rochester, Minn. 55901.*

LAMBERT, Fred Dayton, Am. botanist; b. Muscatine, Ia., Oct. 28, 1871; s. Daniel Meader and Ellen (Scudder) L.; Ph.B., Tufts Coll., 1894, A.M., Ph.D., 1897; postgrad. U. Freiburg (Germany), 1910-11, Naples Zoöl. Sta., 1911; m. Mary Anna Ingalls, June 6, 1903; 1 dau., Elizabeth Allen. Asst. in biology Tufts Coll., Mass., 1896-97, instr., 1897-98, instr. natural history, 1899, asst. prof. biology, 1904, prof. botany, from 1913; submaster Edward Little High Sch., Auburn, Me., 1898-99. Died Feb. 21, 1931.

LAMBERT, Johann Heinrich, German mathematician, physicist, astronomer, philosopher; b. Mülhausen, Alsace, Apr. 26, 1728; s. of a tailor; began as clerk; became tutor to Swiss family, Chur, 1748; also tutor of Iseln, Basel, Switzerland; Bavarian building official; founder, mem. Bavarian Acad. Scis., Munich; elected to Berlin Acad. Scis., 1761; confirmed by Frederick II as ordinary member, 1765; with Berlin Acad., 1763-67; apptd. head of Prussian Architectural Council by Frederick II, 1764; editor, Ephemeris (Prussian astronomical almanac or calendar), Author: Les propriétés remarquable de la route de la lumière par les airs, 1758; Photometria, 1760; Neues Organon, 1764; Beiträge zur Gebrauch der Mathematik und deren Anwendung, (2 vols.), 1765-72; Pyrometrie, 1779; Theories der Parallellinien, 1786; Deutscher Gelehrter Briefwechsel (4vol.), 1781-84; Zur Theorie der Parallellinien (questioned validity of parallel axiom), 1786. Research in photometry, esp. 1st principles of direct light; light passing through transparent media; light reflected from polished surfaces; scattering of light; comparative luminosities of heavenly bodies; relative intensities of colored light and shadows; a founder of physiological optics; discovered method of measuring light intensity and absorption (lambert, unit of light, named after him, also theorem which applies inverse squares to illumination of surfaces); found Newton's corpuscular theory of light more easily understood but Euler's (or Huygen's) wave theory corresponded better to nature of things; proved heat of fire exists, not as light, but as obscure heat (thus supported Scheele's conclusions); deduced law according to which force of magnetic pole varies with distance (after Coulomb); proved pi is irrational; introduced hyperbolic functions sinh and cosh; studied fixed compass constructions, spherical trigonometry; constructed color pyramid; formulated theorem on motion of planets; one of 1st to realize there are other stellar systems besides Milky Way; recommended use of new symbols in meteorology; measured coefficient of expansion of air; probably named and did early experiments in hygrometry; worked on map projections. Died Berlin, Prussia, Sept. 25, 1777.

LAMBERT, Robert Frank, Am. elec. engr.; b. Warroad, Minn., Mar. 14, 1924; s. Fred Joseph and Nutah (Gibson) L.; student Case Inst. Tech., 1944-45, U. Wis., 1945-46: B.E.E., U. Minn., 1948, M.S., 1949, Ph.D., 1953; m. June Darlene Flatten, June 30, 1951; children—Cynthia Marie, Susan Ann, Katherine Cheryl. Faculty, U. Minn., Mpls., 1949-54, 55—, research dir. elec. engring. dept., 1957—, prof. elec. engring., 1958—, asso. dean Institute of Technology, 1967—; visiting asst. prof. Mass. Inst. Tech., 1954-55. Cons. editor elec. engring. Internat. Text Co., 1960—; acoustical cons. Mem. I.E.E.E., Acoustical Soc. Am., Am. Soc. Engring. Sci., Am. Soc. Engring. Edn., Sigma Xi, Tau Beta Pi, Eta Kappa Nu, Gamma Alpha. Author: (with others) Noise Reduction, 1960, Acoustical Fatigue, 1964. Contbr. numerous articles to profl. jours. Research, publs. on design, devel. acoustic measurement devices, random signal processing equipment, spl. computers for test scoring. Home: 2503 Snelling Curve, St. Paul 55113. Office: Elec. Engring. Dept., U. Minn., Mpls.*

LAMBERT, Walter Davis, Am. geodesist; born at West New Brighton, N.Y., Jan. 12, 1879; s. Walter and Elizabeth Bigelow (Davis) L.; A.B., Harvard,

1900, A.M., 1901; studied U. of Pa.; Sc.D. (honorary), Ohio State University, 1957; married Bertha Brown, June 18, 1917 (deceased October 1959). Instructor of mathematics, Purdue University, 1901-02; instructor mathematics and astronomy, University of Me., 1902-04; computer Coast and Geodetic Survey, 1904-07; Harrison fellow in mathematics, U. of Pa., 1907-08; instr. mathematics, same univ., 1908-11; computer, later senior and principal mathematician, and chief, section of gravity and astronomy, U. S. Coast and Geodetic Survey, retired 1949; cons. Inst. of Geodesy, Photogrammetry and Cartography, at Ohio State University. Served as first lieutenant engrs., U. S. Army, Sept. 1917-May 1919; with 101st Engr. Regt., A.E.F., and with Engr. Purchasing Office, in France, 15 mos. Decorated D.S.M., U. S. Dept. of Commerce. Fellow A.A.A.S.; member of Mathematical Assn. of America, Am. Geophys. Union (William Bowie medal, 1949) Nat. Acad. Scis. Institut de France (corr.), Am. Astron. Soc., Seismological Soc. of America, Philos. Soc. Washington (ex-pres.), Washington Acad. Sciences (ex-v.p.), Internat. Astron. Union, Internat. Assn. Geodesy (hon. pres.), Phi Beta Kappa, Sigma Xi, Unitarian. Club: Cosmos. Author of various ofcl. publs. U. S. Coast and Geodetic Survey. Contbr. to Ency. Britannica, handbooks on geodesy and geophysics and sci. jours. Home: P.O. Box 1025, Canaan, Conn. 06018.*

LAMBERT-BEY, engr.; b. Valenciennes, France, 1804; made expdn. to Egypt, 1840; became dir. Cairo (Egypt) Poly. Sch., 1845; worked on Egyptian project for reinforcing Nile banks; gave impetus to constrn. of Suez Canal by promoting Lepère's ideas; returned to Paris, 1851. Died 1864.

LAMBERTYE, Léonce de, see de Lambertye, Léonce.

LAMBIN, Suzanne, French microbiologist; b. Nantes, France, Aug. 1, 1902; d. René and Valentine (Perthuy) L.; ed. univs. Nantes, Paris; Ph.D. Asst. in physics Nantes Sch. Medicine and Pharmacy; became pharmacist U. Paris, 1925; apptd. asst. in microbiology, 1928; named head of works, 1945; lectr. agrégé natural scis., 1949; prof. microbiology Faculty Pharmacy, Paris, 1951—. Decorated Order Pub. Health. Mem. Assn. Microbiologists, Acad. Pharmacy, Soc. Biology. Research and publs. on evolution of bacterial cultures and trial methods of antiseptic substances. Home: 15, rue Saussier-Leroy, Paris 17. Office: 4, ave. de l'Observatoire, Paris 6, France.

LAMBTON, William, Brit. geodesist; b. Grosby Grange, Eng., 1756; served as lt. Brit. army, India, from 1798; founder Trigonometric Bur. India. Fellow Royal Soc., 1817; mem. French Acad. Scis., Asiatic Soc. Author: The Effect of Machines in Motion, 1798. In his survey of India, determined latitudes, longitudes, altitudes of many locations, 1815; made important observations on terrestrial refraction. Died Hingah-Ghaut, India, Jan. 20, 1823.

LAMÉ, Gabriel, French mathematician; b. Tours, France, July 22, 1795; ed. École Polytechnique, Paris; engr. in chief of mines, Russia; named prof. physics École Polytechnique, 1832; prof. prabability Sorbonne; mem. French Acad. Scis., Bur. Longitudes. Author: Leçons sur la théorie de l'élasticité, 1852; Leçons sur les functions inverses des transcendantes et la surfaces isothermes, 1857. Leçons sur les coordonnées curvilignes, 1859; La théorie analytique de la chaleur, 1861; also papers. Research on gear wheels, 1874, vibration of light in transparent bodies, 1834, isotherm surfaces in homogenous bodies by temperature equilibrium, 1837, isostatic surfaces in homogeneous solid bodies by elasticity equilibrium, 1841; introduced new math. functions named after him to solve problems of temperature equilibrium in ellipsoids. Died Paris, May 1, 1870.

LAMEERE, Auguste Alfred Lucien Gaston, Belgian zoologist; b. Lxelles-les-Bruxelles, Belgium, June 12, 1864; student Free U. Brussels; prof. zoology Free U. Brussels until 1934. Mem. French Acad. Scis. Author: Faune de Belgique, 3 vols., 1895-1907; Précis de zoologie, 6 vols., 1927; also notes and memoirs. Research on secondary sexual characteristics of the Prionides, insects of the Sahara desert, especially their social life. Died Ixelles-les-Bruxelles, May 6, 1942.

LA MER, Victor Kuhn, Am. chemist; b. Leavenworth, Kan., June 15, 1895; s. Joseph Secondule and Anna Pauline (Kuhn) La M.; A.B., U. of Kan., 1915; Ph.D., Columbia, 1921; student U. of Chicago, 1916, Cambridge University, England, 1922-23, Copenhagen, Denmark, 1923; Sc.D. (hon.), Clarkson College, 1962; m. Ethel Agatha McGreevy, July 31, 1918; children—Luella Belle (Mrs. A. P. Slaner), Anna Pauline (Mrs. Alex Burgo), Eugenia Angelique (dec.). Chemist and high sch. teacher chemistry, 1915-16; research chemist Carnegie Inst. of Washington, 1916-17; asst. chemistry Columbia U., 1919-20, successively instr., assistant and asso. prof., prof., 1935-61, prof. emeritus, 1961—, sr. researcher mineral engring., Columbia, 1963—; Fulbright professor U. Copenhagen, 1953; vis. prof. Stanford U., 1931, Northwestern U., 1928. Priestley lecturer, Pa. State Coll., 1932; Fulbright lecturer, Australia, 1959; distinguished lectr. Shell Devel. Co., 1963; lectr. NSF Colloid Sch., U.S.C., 1962-63, NSF Colloid School of Lehigh University, 1965. Recipient Presidential Certificate of Merit, 1948; Kendall award, Am. Chem. Soc., 1956. Mem. Royal Danish Acad. Sci., Nat. Acad. of Scis., Div. mem. Nat. Defense Research Com. Mem. Jury of Award, Nichols medal, 1934-38 (chairman 1934, 37); president Leonia Civic Conference, 1941. First lt. Sanitary Corps, U. S. Army, 1917-19. Hon. professor San Marcos U., Lima, Peru; mem. Royal Belgian Academy Arts Letters and Science. Fellow New York Academy of Science (v.p. 1939-41; treasurer 1943, president 1949); member Am. Chem. Soc., American Physical Society, Faraday Soc. (Eng.), Natl. Acad. of Sciences, Sigma Xi, Phi Lambda Upsilon, Phi Chi, Epsilon Chi. Translator and editor of Fundamentals of Physical Chemistry by Arnold Eucken (with Eric Jette), 1925. Asso. editor of Jour. Chem. Physics, 1933-36. Editor-in-chief Jour. Colloid Sci. Author: Retardation of Evaporation by Monolayers; also numerous articles. Gave first experimental verification of Huckel theory of electrolytes (with J. N. Bronsted, 1924); discovery of higher order Tyndall Spectra, 1941. Home: 353 Moore Av., Leonia, N.J. Office: Columbia U., N.Y.C. 27.*

LAMERTON, Leonard Frederick, English biophysicist; b. Southampton, Eng., July 1, 1915; s. Alfred and Florence (Mason) L.; B.Sc., U. Coll. Southampton, 1936, M.Sc., 1938; Ph.D., U. London, 1947, D.Sc., 1964; m. Mora Macleod, July 24, 1965. Mem. staff Inst. Cancer Research, U. London (formerly Royal Cancer Hosp.), 1939—, prof. biophysics, 1960—. Mem. U.K. delegation UN Sci. Com. on Effects of Atomic Radiation, 1960-64. Mem. Brit. Inst. Radiology (pres. 1957-58), Hosp. Physicists Assn. (pres. 1961). Studies, publs. on radiation protection, cell population kinetics. Home: 10 Burgh Mount, Banstead, Surrey. Office: Inst. Cancer Research, Surrey, Sutton, Surrey, Eng.*

LAMÉTHERIE, Jean-Claude de, see de Lamétherie, Jean-Claude.

LA METTRIE, Julien Offroy de, French physician, botanist, philosopher; b. St. Malo, France, Dec. 25, 1709; studied with Jesuits, Caen, France. doctorate in medicine, Reims, France. Became French army surgeon, 1742; forced to flee to Leiden, Netherlands, because of his writings, 1746; then went to Prussia where Frederick II gave him asylum, Berlin. Author: Observationes de médecine pratique, 1743; Histoire naturelle de l'ame, 1745; La Faculté vengée, 1747; l'Homme machine, 1748; L'Homme plante, 1748; Réflexions philosophiques sur l'origine des animaux, 1750; Les Animaux plus que machines, 1750; Vues physiologiques sur l'organisation végètale et animale; L'Art de jouir, 1751. Leader French materialistic thought in most extreme form. Died Berlin, Nov. 11, 1751.

LAMLA, Ernst Anton Heinrich, German physicist; b. Berlin, Nov. 21, 1888; s. Anton and Anna (Krüger) L.; Dr.phil., U. Berlin, 1912; m. Henriette Scharr, Oct. 22, 1917; children—Gudrun, Ingrid. Faculty, U. Berlin, 1918-27, headmaster, 1924-27; oberschulrat, 1927-32; v.p. province Bd. Schs. and Colls., Magdeburg, Germany, 1932-33; physicist in pvt. practice, 1933-47; headmaster Classical Coll., hon. prof., U. Göttingen (Germany), 1947-54. Mem. Sci. Examination Bd., Berlin, 1928-32; mem., head Examination Bds., 1947-64. Mem. German Phys. Soc., Soc. German Scientists and Physicians. Author: Hydrodynamics of Relativity Theory, 1912; Compendium of Physics for Scientists and Medical Students, 1925; Schulreform, 1930; Max Planck, 1955; Statistical Methods in Physics, 1960. Editor, Naturwissenschaften, 1950-66. Research and publs. on relativity theory including light clocks, 1967, quantum theory especially interference of X-rays and electrons, aerodynamics, math., philos. and edn. Home: 21, Friedrich-Jenner-Str., 34 Göttingen/Germany.*

LAMOND, Donald Ross, Australian physiologist; b. Nowra, New South Wales, Australia, Apr. 7, 1928; s. Ross Bernard and Ellen (Madden) L.; B.V.Sc., Sydney U., 1949; B.Agr.Sci., U. New Zealand, 1954, M.Agr. Sci. with honors, 1955; Ph.D., U. Sydney, 1958; D.V. Sc., 1966; m. Jennifer Alvia Pursell, June 8, 1957; children—Jacqueline, Michael, Stephen, Alexander. Lectr. animal husbandry U. New Eng., Armidale, New South Wales, 1959-60; research scientist div. animal physiology Commonwealth Sci. and Indsl. Research Orgn., St. Lucia, Australia, 1961—, officer in charge Beef Cattle Research Unit, Brisbane, 1964—. Author: Dairy Cattle Husbandry, 1961, also numerous articles. Research on endocrine control of female mammalian reprodn.; use of hormones in animal prodn.; factors influencing fertility in farm animals. Address: care John A. Wyatt & Co., 74 Eagle St., Brisbane, Australia.*

LAMONT, Johann von, astronomer; b. Braemar, Scotland, Dec. 13, 1805; student Scottish Benedictine monastery of St. James, Ratisbon. Joined Royal Obs., Bogenhausen, 1827, and remained until his death, became asst. to dir., 1828, dir., 1835; founder magnetic obs., Bogenhause, 1840; prof. astronomy U. Munich (Germany), 1852-79; made magnetic surveys of France and Spain, 1856-57, N. Germany, Denmark, 1858, also Bavaria. Mem. Royal Astron. Soc., Royal Soc. Edinburgh, Fellow Royal Soc., 1852. Author: Handbuch des Erdmagnetismus, 1849; Astronomie und. Erdmagnetismus, 1851; Handbuch des Magnestismus, 1849; also numerous articles. Published Observationes Astronomicae (from obs.), 10 vols.; Annalen der Sternwarte, 34 vols., Jahbüchlicher, 4 vols., 1838-41. Made zone-observations of 34,674 small stars between latitudes +27° and −33°; described ghost micrometer, 1839; discovered decennial magnetic period, 1850, earth currents, 1862; determined orbital elements of satellites Enceladus and Tethys of Saturn, periods of 2d and 4th satellites of Uranus. Died Munich, Aug. 6, 1879.

LAMONTE, Francesca R(aymond), ichthyologist; B.A. and Certificate of Music, Wellesley Coll. Asso. curator, dept. fishes and aquatic biology, Am. Museum of Natural History, now curator emeritus, mem. numerous expdns. Del. XI Internat. Congress Zoologists, Padua, Italy; mem. fisheries com. World's Fair, N.Y.C., 1939-40. Fel. A.A.A.S., N.Y. Zool. Soc.; mem. N.Y. Acad. Scis., Soc. Systematic Zool., Outdoor Writers Assn. of Am., Internat. Game Fish Assn. (exec. com.), Am. Soc. Ichthyologists and Herpetologists. Editor (with J. T. Nichols), Fieldbook Fresh Water Fishes of N.A., by R. Schrenkeisen, 1938; (with B. Fitzgerald) Game Fish of the World, 1949; (with I. Gabrielson) The Fisherman's Ency., 1950. Author: (with M. R. Welch) Vanishing Wilderness, 1934; North American Game Fishes, 1945; Marine Game Fishes of the World, 1952; Giant Fishes of the Ocean, 1966; translator The Pale Mountains, by K. F. Wolff, 1926. Home: 11 E. 87th St., N.Y.C. 28.*

LAMOTTE, Martial, French botanist; b. 1821; dir. Clermont-Ferrand Mus. Author: Prodrome de la flore du plateau central de la France, 1877-81. Died 1883.

LAMOTTE, Maxime Georges, French zoologist; b. Paris, June 26, 1920; s. Georges and Denise (Huguet) L.; Ph.D., École normale supérieure; agrégé in natural scis.; m. Francoise Polonovski, Aug. 28, 1944; children—Alain, Bertrand, Christine, Nicole, Martine, Sylvie. Prof. zoology Faculty Scis., Lille, France; prof. U. Paris Faculty Scis. Author: Introduction à la biologie quantitative; Initiation aux méthodes statistiques en biologie. Home: 116, bd. Raspail, Paris 6. Office: 24, rue Lhomond, Paris 5, France.

LAMOUROUX, Jean-Vincent-Félix, French botanist, zoologist; b. Agen, France, May 3, 1779; prof. natural history Caen (France) Faculty Scis.; mem. French Acad. Scis. Author: Observations sur plusieurs espèces de fucus, 1805; Historie des polypiers coralligènes flexibles, 1816; Historie naturelle des zoophytes, 1824. Research on sea plants and animals, especially fucus and polyps; discovered 140 new zoophyte species. Died Caen, Mar. 26, 1825.

LAMPADIUS, Wilhelm August, German chemist; b. Hehlen, Germany, Aug. 8, 1772; prof., Freidburg, Germany; founder 1st European gas bus. Author: Handbuch der allgemeinen Hütterkunde, 2 vols., 1801-10; Handwörterbuch der Hütterkunde, 1817; Erläuternde Experimente über dem Grundlehren der allgemeinen und Mineralchemie, 2 vols., 1809-10. Discovered carbon disulphide, 1796. Died Freiburg, Apr. 13, 1842.

LAMPARIELLO, Giovanni, Italian mathematician; b. Capua, Jan. 29, 1903; s. Pasquale and Amalia (Russo) L.; Ph.D., U. Bologna (Italy); m. Irma Braidotti, May 15, 1939; children—Silvana, Elena, Francesco, Paolo. Asst., instr. infinitesimal calculus U. Rome; prof. mechanics; instr. math. U. Messina. Collaborator, Italian Ency. Research and publs. on history of science, math., theory of relativity. Home: via Felice Cavallotti 119, Rome. Office: via Cesare Battisti 251, Messina, Italy.

LAMPÉ, Arno Eduard, German physician; b. Kleinheubach, Feb. 8, 1886; s. Eduard and Julie (Lettenmeyer) L.; M.D., U. Munich (Germany); m. Melek Ziya, Nov. 30, 1956. Asst., Abderhalden council; head physician von Romberg council Munich clinic, also Neuwittelsbach Clinic, Munich; head commr. clinics I and II, U. Munich, named hon. prof. internal medicine, 1961. Mem. German Soc. Medicine, German Med. and Nat. Research Soc., Internat. Soc. Medicine and Cybernetics, Friends of Bavarian Acad. Scis., Geothe Wiemar Soc., Frankfort Free Fund, Holderlin Soc. Research and publs. on physiology, internal medicine, especially diseases of digestive system, dietetics. Address: Widenmayerstrasse 23/2, Munchen 22, W. Germany.

LAMPE, Isadore, physician; b. London, Eng. Nov. 16, 1906; s. Joseph and Anna (Tamarkin) L.; A.B., Western Res. U., 1927, M.D., 1931; Ph.D., U. Mich., 1938; m. Rae Ethel White, Oct. 23, 1943; children—William H., Matthew M. Instr. radiology U. Mich., 1934-38, asst. prof. radiology, 1938-43, asso. prof., 1943-49, prof., 1949—, chief radiation therapy div., 1939—. Mem. Am. Club Therapeutic Radiologists (pres. 1962), A.M.A., Am. Roentgen Ray Soc., Am. Radium Soc., Radiol. Soc. N.Am., Phi Beta Kappa, Sigma Xi, Alpha Omega Alpha. Author: (with F. J. Hodges, J. E. Holt) Radiology for Medical Students, 1947, 4th edn., 1965. Contbr. numerous articles to sci. jours. Studies in the field of clin. radiation treatment, chiefly of cancerous diseases. Home: 1600 Newport Rd., Ann Arbor, Mich. 48103.*

LAMPEN, Heinrich, German physician; b. Freckenhorst, June 25, 1914; s. Leopold and Anna (Horstrup) L.; ed. U. Frankfort (Germany); m. Maria Nover, May 16, 1950; children—Thomas, Gertrud, Beata-Maria. Substitute prof. U. Frankfort, asst., chief physician, clinic; chief physician St. Francis Hosp. Mem. German, Internat. socs. internal medicine. Research and publs. on internal medicine, arterial hypertension, cardiology, nephrology. Home: Lina Oetker Steasse 2, Beilefeld. Office: Kiskerstrasse 26, Bielefeld, W. Germany.

LAMPERT, Seymour, Am. aero. engr.; b. Bklyn., Mar. 5, 1920; s. Max and Esther (Bakst) L.; B.S., Ga. Inst. Tech., 1943; M.S., Cal. Inst. Tech., 1947, Aero. Engr., 1948, Ph.D., 1954; m. Shirley Ruth Axelrod, Mar. 21, 1948; children—Rachel Beth, David Aaron, Martin Daniel. Research scientist NACA, Moffett Field, Cal., 1944-51; research engr. Cal. Inst. Tech., Pasadena, 1951-54; lectr., research asso. U. So. Cal., Los Angeles, 1954-58; chief engr. Odin Assos., Pasadena, Cal., 1956-57; mgr. aerospace mechanics Ford Aeronutronic, Newport Beach, Cal., 1957-62; dir. advance entry and ballistic systems N.Am. Aviation, Downey, Cal., 1962—. Faculty, Ga. Inst. Tech., Atlanta, 1943-44; cons. underwater ordnance Naval Ordnance Test Sta., Pasadena, 1951-52. Mem. Am. Geophys. Union, Am. Inst. Aeros. and Astronautics, N.Y. Acad. Sci. Research, publs. on supersonic wing theory; pioneer in solution of camber and twist of swept wings at subsonic speeds; devel. analytical techniques and methods used for spacecraft design. Home: 13032 Eton Pl., Santa Ana, Cal. 92705. Office: 12214 Lakewood Blvd., Downey, Cal. 90241.*

LAMPI, Eugene, Am. physicist; b. Chisholm, Minn., Apr. 19, 1917; s. Elias and Marya (Luoma) L.; B.E.E., U. Minn., 1940, Ph.D., 1949; m. Virginia Ruth Leatherman, Aug. 10, 1945; children—Mary Catherine, Ruth Virginia. Physicist, Naval Ordnance Lab., Washington, 1941-46; research fellow U. Minn., Mpls., 1949-57; mem. tech. staff Bell Tel. Labs., Allentown, Pa., 1957—. Fellow Am. Phys. Soc.; mem. Am. Inst. Elect. Electronic Engrs., Sigma Xi, Eta Kappa Nu. Research on neutron scattering, accelerator design, semiconductor device devel. Home: 709 N. 7th St., Emmaus, Pa. 18049. Office: 555 Union Blvd., Allentown, Pa. 18103.*

LAMPITT, Leslie Herbert, Brit. chemist; b. Sept. 30, 1887; s. Daniel Lampitt and Eliza (Haywood) L.; student U. Birmingham (Eng.), 1906-11; D.Sc. Dir., chief chemist J. Lyons and Co. Ltd.; chief chemist La Meunerie Bruxelloise, Brussels, Belgium, 1911-14. Hon. sec. Brit. Nat. Com. for Chemistry. Fellow Royal Inst. Chemistry; mem. Soc. Chem. Industry (past pres.; medal 1943, internat. medal 1949) Internat. Union Pure and Applied Chemistry (hon. treas.), Société de Chimie Industrielle (Lavoisier medal 1950), Chemists Club N.Y. (hon. fgn. mem.), Inst. Chem. Engrs. Contbr. articles to tech. jours. Died June 3, 1957.

LAMPLAND, Carl Otto, Am. astronomer; b. Dodge County, Minn., Dec. 29, 1873; s. Ole Helleckson and Beret (Skartum) L.; B.S., Valparaiso U., 1899; A.B., Ind. U., 1902, A.M., 1905, LL.D., 1930; m. Verna Basil Darby, Feb. 8, 1911. Prin. Bloomfield (Ind.) High Sch., 1902; astronomer Lowell Obs., Flagstaff, Ariz., from 1903; mem. Lowell Obs. Eclipse Expdn. to Kan. 1918, also solar eclipse expdn. to Ensenada, Mexico, 1923; exchange prof. astronomy Princeton, 1929. Recipient medal for photographs of planet Mars, Royal Photog. Soc. Gt. Brit., 1907. Was asst. to Percival Lowell in visual observations of the planets Venus, Mars, Jupiter and Saturn; gave much attention to devel. of photography of delicate detail on planetary surfaces; prin. work of later years was photog. observations of planets, satellites, comets, nebulae, novae, and star fields; discovered many variable stars, and changes in nebulae; measurements of radiation from planets and determination of planetary temperatures (early work with W. W. Coblentz, continued and extended later with assistance of V. D. Lampland); transmission of the earth's atmosphere (with A. Adel); investigations in connection with trans-Neptunian planet Pluto. Fellow Am. Acad. Arts and Scis., A.A.A.S.; mem. Am. Astron. Soc. (council), Internat. Astron. Union (com. on planets and nebulae), Astron. Soc. Pacific, Société Astronomique de France, Astronomische Gesellschaft, Am. Physical Soc., Math Assn. America, Am. Philos. Soc., Am. Math. Soc., Soc. for Research on Meteorites, Sigma Xi, Phi Beta Kappa; hon. mem. Sociedad Astronomico de Mexico. Contbr. to astron. jours. Died Dec. 14, 1951.

LAMPORT, Harold, Am. physiologist; b. N.Y.C., Feb. 16, 1908; s. Arthur Matthew and Sadie (Payson) L.; student U. Gottingen, Germany, 1928-29; B.S. cum laude, Harvard, 1929; M.D., Columbia U., 1934; m. Golden R. Siwek, Mar. 27, 1933; children—Anthony Matthew, Stephanie Payson (Mrs. James Carson Nohrnberg). Dir., Koster Research Lab., N.Y., 1937-39; research asst. neurology Columbia U., N.Y.C., 1939-40, research asso., 1940-42; faculty Yale Sch. Medicine, New Haven, 1940-67, asso. prof. physiology 1944-65, lectr., 1965-67; research prof. Mt. Sinai Sch. Medicine, 1966—; Research asso. Pierce Found., New Haven, 1942-43, med. dir. mil. research projects, 1943-45; asst. sec. com. on missile casualties NRC, 1943-47; responsible investigator OSRD, 1942-45. Recipient Ultrasonic Med. Research prize Birtcher Found., 1955. Fellow A.A.A.S., N.Y. Acad. Medicine; mem. Am. Physiol. Soc., Biophysics Soc., Soc. Rheology, N.Y. Acad. Scis., Phi Beta Kappa, Sigma Xi. Contbr. articles to sci. lit. Inventor pneumatic lever suit for altitude and black-out protection of aviators, 1942. Home: Conn. 06880. Office: Mt. Sinai Sch. Medicine, Fifth Av. and 100th St., N.Y.C. 10029.*

LAMSON, Charles Henry, Am. inventor; b. Augusta, Me., Sept. 17, 1847; s. Joseph S. and Eunice E. (Winslow) L.; ed. Exeter, N.H.; m. Elizabeth H. Cox, July 27, 1874. Engaged in business as watchmaker and optician. Inventor of luggage carriers for bicycles and of novel types of kites and aeroplanes for use in meteorol. observations; also of flying machine; was the first to obtain an Am. patent for a method of tilting or warping the wings of kites and aeroplanes for the purpose of balancing them in flight. Died May 1930.

LAMSON, George Herbert, Jr., Am. zoologist; b. Malden, Mass., Apr. 8, 1882; s. George Herbert and Sarah (Liscombe) L.; B.Agr., Conn. Agrl. Coll., 1902; B.S., Mass. Agrl. Coll., 1903; M.S., Yale, 1905; postgrad. Wesleyan U., 1903-04; m. Kate Arroll, July 27, 1909; 1 son, Arroll L. With Conn. Agrl. Coll., Storrs, from 1906, prof. zoölogy, from 1910, dean div. agrl. sci., 1928; econ. zoölogist Storrs Expt. Sta. Investigator in applied agrl. science. Died 1931.

LAMSON-SCRIBNER, Frank, Am. botanist; b. Cambridgeport, Mass., Apr. 19, 1851; s. Joseph S. and Eunice E. (Winslow) Lamson; adopted at age of 3 by family named Scribner; B.S., Me. State Coll. Agr., 1873; LL.D., U. Me., 1920; m. Ella Augusta Newmarch, Dec. 25, 1877; children—Allen, Frank, Louise; m. 2d, Marjorie Fleming Anderson, Aug. 12, 1913. Teacher, public schools, Maine; clerk to secretary Maine State Board of Agriculture 2 yrs.; officer Girard Coll. Phila., 1876-85; taught botany in summer schs. of sciences; spl. agt. in charge mycol. sect., bot. div., U. S. Dept. of Agr., 1885-86, chief sect. of vegetable pathology, 1887-88; prof. botany, U. of Tenn., 1888-94; dir. Tenn. Agrl. Expt. Sta., 1890-94; chief div. of agrostology U. S. Dept. Agr., 1894-1901; chief Insular Bur. of Agr., P.I., 1901-04. As spl. agt. and expert on exhibits had charge of the preparation and display of exhibits at a large number of fairs and expns. made by the Agrl. Dept., 1904-22, including the St. Louis Expn., 1904; Portland, Ore., 1905; Seattle, Wash., 1909; Buenos Aires, Argentina, 1910; Turin, Italy, 1911; Lethbridge, Can., 1912; Inter. Refrigeration Expn., Chicago, 1913; San Francisco Expn., 1915, at which served as mem. Govt. Exhibit Bd. by presdl. apptmt. Conducted series of war-time exhibits for food conservation in coöperation with the State Fairs throughout the country, 1917-20; apptd. by sec. of state, director of exhibits for U. S. Com. to Brazilian Centennial Expn., Rio de Janeiro, 1922-23; personal asst. to dir. Commercial Museum, Phila., 1924; spl. asst. to U. S. Com. and supervisor of U. S. Govt. Exhibits, Sesquicentennial Expn., Phila., 1926-27; dir. Hist. Museum of Cumberland County Hist. Soc., Carlisle, Pa., 1927-28; advisory mem. science planning com. of "Chicago Century of Progress Exposition 1933." Decorated Chevalier du Mérite Agricole (France), 1889. Author: Weeds of Maine, 1869; Fungus Diseases of the Grape and Other Plants, 1890; Grasses of Tennessee, 1894; American Grasses (3 vols.), 1897-1900. Translator: (with E. A. Southworth) The True Grasses (from the German), 1890. Genus of grasses, Scribneria, named in his honor. Died Washington, Feb. 22, 1938.

LAMY, Bernard, French philosopher, mathematician; b. Le Mans, France, 1645; prof. philosophy, Angers, France; Oratorian father. Author: Traité de mecanique, 1679; Traité de la grandeur en général, 1680; Entretiens sur les sciences, 1684; widely read treatises on geometry, arithmetic, analysis. Taught Descartes' philosophy in spite of persecution from Aristotelians in Angers; advocated Pascal's principles of composition of matter. Died 1716.

LAMY, Claude August, French chemist; b. Neris, France, 1820; ed., Paris.; prof. physics, Lille, France; prof. chemistry, Paris. Isolated thallium (independently of Crookes), 1862. Died Paris, 1878.

LAMY DE LA CHAPELLE, Edouard, French botanist; b. Limoges, France, 1803; Author: Exposition systematique des lichens de Cauterets, 1884; Invasion dans la Haute-Veinne de la maladie de la vigne dite le mildiou. Studied flora of Haute-Veinne region of France, phanerogramic plants, lichens, mosses, other cryptogams. Died 1886.

LANA-TERZI, Francesco, see de Lana, Francesco.

LANBE-EICABAUM, Wilhelm, German psychiatrist; b. Hamburg, Germany, Apr. 28, 1875; instnl. psychiatrist, Berlin, Tübingen, Hamburg, Germany. Author: Genie, Irrsinn und Ruhm, 1928; Das Genie-Problem, 1931. Studied problem of genius which he saw as a result of sociol. rather than biol. factors; studied relation between genius and psychosis. Died Hamburg, Sept. 4, 1950.

LANCASTER, Peter, mathematician; b. Appleby, Westmorland, Eng. Nov. 14, 1929; s. John Thomas and Emily (Kellett) L.; B.Sc., U. Liverpool, Eng., 1952, M.Sc., 1956; Ph.D., U. Singapore, 1964; m. Edna Lavinia Hutchinson, Sept. 3, 1951; children—Jane, Jill, Joy. Applied mathematician English Elec. Co. Ltd. div. Warton Aerodrome, Lancashire, Eng. 1952-57; sr. lectr. applied math. U. Singapore, 1957-62; prof. math. U. Calgary, Alta., Can., 1962—; vis. prof. math. Cal. Inst. Tech., 1965-66. Mem. Canadian Math. Congress, Am. Math. Soc., Soc. for Indsl. and Applied Math. Author: Lambda Matrices and Vibrating Systems, 1966; Theory of Matrices, 1968. Contbr. articles to sci. jours. Contbns. to advances in theory of matrices and applications of operator theory to problems in theory of vibrations and in numerical analysis. Office: Dept. Math., U. Calgary, Alta., Can.*

LANCEREAUX, Étienne, French physician; b. Brécy-Brières, France, Nov. 27, 1829; certified physician, 1869; aggregation, 1877. Author: Des affections syphilitiques, 1861; Atlas d'anatomie pathologique, 2 vols., 1872; Traité des maladies du foie et du pancréas, 1899. Research on disorders of nervous system caused by syphilis, 1861; discovered relationship between diabetes mellitus and a pancreas disease (pancreatic diabetes), 1877; other research on alcoholism, malaria, arteriosclerosis. Died Paris, Oct. 1910.

LANCIANI, Rudolfo Amadeo, Italian archaeologist; b. Rome, Italy, Jan. 1, 1846; ed. Collegio Romano, U. Rome; LL.D., Aberdeen, Scotland, Harvard, Glasgow, Scotland; D.C.L., Oxford; Ph.D., Wurzburg, Germany, Rome; m. Mary Ellen Rhodes, 1875; m. 2d, Theresa, Duchess of San Teodoro, 1920. Became sec. Archeol. Commn. Rome, 1872; named dir. excavation, prof. ancient topography U. Rome; commd. to design and supervise constrn. Archaeol. Park, Rome; rebuilt Forma Urbis of Septimius Severus; royal commr. for reorgn. Center of Rome. Recipient Gold. medal R.I.B.A. Mem. French Acad. Scis. (corr.); mem. Lincei Acad., Luca Acad., Inst. Berlin, Archeol. Soc. Brussels, Royal Belgian Acad. Author: Ancient Rome in the Light of Modern Discoveries, 1888; Pagan and Christian Rome, 1892; Forma urbis Romae, 1892; The Ruins and Excavations of Ancient Rome, 1897; Destruction of Ancient Rome, 1899; New Tales of Old Rome, 1901; Wanderings in the Roman Campagna, 1909; Ancient and Modern Rome, 1925; also numerous articles including plan of ancient Rome. Discoveries include House of Vestals, Basilica Julia, Imperial Palace of Palatine, baths of Caracalla, temple of Jupiter Capitolinus, Hadrian's villa below Tivoli, Trajan's harbor at Ostia. Died Rome, May 21, 1929.

LANCISI, Giovanni Maria, Italian physician; b. Rome, Italy, Oct. 26, 1654; ed. Rome; with S. Spirito Hosp. from student days until his death; physician to Popes Innocent XI, Innocent XII, Clement XI; apptd. by pope to investigate increase in sudden deaths in Rome, 1706. Fellow Royal Soc., 1706. Wrote monographs on influenza, cattle plague (Rinderpest), malaria, sudden death, aneurysms. First description of valvular vegetation, 1707, of syphilitic heart disease, 1728; drained some of worst parts of Pontine Marshes; pointed out malaria was prevalent in districts where mosquitoes were numerous and advocated quinine for its cure; pointed out syphilitic nature of aneurysms; described syphilis of heart. Died Jan. 20, 1720.

LANCZOS, Cornelius, physicist, educator; b. Szekesfehervar, Hungary, Feb. 2, 1893; s. Carolus Loewy and Adele (Hahn) L.; B.Sc., U. and Polytechnicum Budapest (Hungary), 1916; Ph.D., U. Szeged (Hungary), 1921; D.Sc., Trinity Coll., Dublin, Ireland, 1962; m. Ilse Hildebrand, Feb. 19, 1955; 1 son, Elmar (by previous marriage). Instr., U. Freiburg (Germany), 1921-24, U. Frankfurt (Germany), 1924 31; fellow U. Berlin (Germany), 1928-29; prof. Purdue U., 1931-46; sr. research engr. Boeing Airplane Co., Seattle, 1946-49; staff Nat. Bur. Standards, Inst. for Numerical Analysis, U. Cal. at Los Angeles, 1949-53; with N. Am. Aviation, 1953-54; sr. prof. physics Dublin Inst. for Advanced Studies, 1954—. Lectr. various univs. Fellow A.A.A.S., Am. Phys. Soc.; mem. Am. Math. Soc., Math. Assn. Am. (Chauvenet prize 1960), Soc. for Indsl. and Applied Math., Royal Irish Acad., Sigma Xi, Sigma Pi Sigma (hon.). Author: The Variational Principles of Mechanics, 1949, 2d edit., 1962; Applied Analysis, 1957; Linear Differential Operations, 1961; Albert Einstein and the Cosmic World Order, 1965; The Fourier Series and its Applications, 1965. Discovered application of Chebyshev polynomials, 1938, quadratic action principle gen. relativity, 1931—, unified field theory, 1957—, boundary value problems, 1961—. Home: 129 Lower Baggot St., Dublin 2, Ireland.*

LAND, Charles Henry, dentist; b. Simcoe, Ont., Can., Jan. 11, 1847; s. John Scott and Sarah (Hayden) L.; studied dentistry under Dr. J. B. Meacham, of Brantford, Can., 1864-66, and in offices of Drs. M. B. Sherwood, L. P. Haskell and W. W. Allport, Chgo., 1868-71; m. Evangeline Lodge, Apr. 28, 1875 (dec. 1919). In dental practice, Detroit, from 1871. Author: Scientific Adaptation of Artificial Dentures, 1885; A Study in Aesthetic Dentistry, 1911. Invented a gold and a porcelain inlay system, and process of artificially replacing enamel on defective human teeth; also gas and oil burners and furnaces, incandescent grates and furnaces adapted for same. Died Aug. 3, 1922.

LAND, Edwin Herbert, Am. inventor; b. Bridgeport, Conn., May 7, 1909; s. Harry M. and Martha G. Land; prep. edn., Norwich Acad.; student Harvard; Sc.D.,

Tufts, 1947, Poly. Inst. Bklyn., 1952, Colby Coll., 1955, Harvard, 1957, Northeastern U., 1959, Carnegie Inst. Tech.. 1964, Yale, 1966, Columbia, 1967; LL.D., Bates Coll., 1953, Washington U., 1966, U. Mass., 1967; m. Helen Maislen, 1929; children—Jennifer, Valerie. During coll. yrs. began development of means for polarization of light as an applied science; invented polarizer used as camera filter; established business with George W. Wheelwright, 3d, Boston, Mass., 1935; organized Polaroid Corp., Cambridge, Mass., 1937, becoming pres., chmn. bd. and dir. of research; invented camera that delivers finished photograph immediately after exposure is made, 1947; also now provides black and white transparencies; present research includes automobile headlight system and three-dimensional pictures. During World War II conducted research leading to development of new weapons and war materials, including plastic optical lenses for devices for seeing at night, filters for preadapting eyes of personnel for night duty, new types of lightweight stereoscopic rangefinders and an infinity optical ring sight used on anti-aircraft guns and bazookas; also served as consultant on missiles, Div. 5, Nat. Defense Research Com., and consultant, Div. 2, and adviser on guided missiles to U. S. Navy; directed Navy plant for development and manufacture of new kind of missile. Cons.-at-large Pres.'s Sci. Adv. Com.; mem. Nat. Commn. on Tech., Automation and Econ. Progress, 1965-66. Fellow of the Sch. for Advanced Study, vis. inst. prof. Mass. Inst. Technology, 1956—; William James lectr. psychology Harvard, 1966-67. Recipient numerous awards, including Hood medal of Royal Photographic Soc., Cresson medal, also Howard N. Potts medal of the Franklin Institute, Phila., Scott medal of Philadelphia City Trusts; Rumford Medal, A.A.A.S., 1945; Holley Medal. Am. Soc. Mech. Engrs., 1948; Duddell medal, Brit. Phys. Soc., 1949; Progress medal Soc. Photog. Engrs., 1955; Progress medal Royal Photog. Soc. Gt. Britain, 1957; F. W. Brehm Meml. Lecture medal, 1957; Presdl. Medal of Freedom, 1963; Nat. Medal of Sci., 1967. Fellow Photog. Soc. Am., Am. Acad. of Arts and Sciences (pres. 1951-53), Royal Photog. Soc. Gt. Britain, Royal Micros. Soc. (hon.), Soc. Photog. Scientists and Engrs. (hon.); mem. Am. Philos. Soc., National Academy of Sciences, Optical Society of Am. (hon. mem., past dir.), Nat. Acad. Engring., N.Y. Acad. Scis., German Photog. Soc. (hon.), Sigma Xi. Home: 163 Brattle St. Office: 730 Main St., Cambridge, Mass.*

LAND, James Edward, Am. chemist; b. Filbert, S.C., Jan. 5, 1915; s. William E. and Edna (Pursley) L.; B.S., Clemson Coll., 1935; M.S., Tulane U., 1938; Ph.D., U. N.C., 1949; m. Jeanetta Thomas, May 29, 1943. Faculty, Sch. Chemistry, Auburn (Ala.) U., 1938—, prof., 1955—. Cons., U. S. Army Propulsion Lab., Redstone Arsenal, 1962—. Mem. Am. Chem. Soc. (exec. com. Southeastern Regional 1956—), Chem. Soc. (London), Ala. Acad. Sci., Sigma Xi, Phi Kappa Phi, Phi Lambda Upsilon. Contbr. articles to tech. jours. Research on preparation and determination of formation constants of niobium coordination complex compounds; determined mean ionic activity coefficients electroylytes in solution, differential thermal analysis. Home: 261 Cary Dr., Auburn, Ala. 36830.*

LANDACRE, Francis Leroy, Am. anatomist; b. Hilliards, O., Feb. 13, 1867; s. Joseph Perry and Sarah Jane (Dobyns) L.; student Ohio Wesleyan U., 1887-91; A.B., Ohio State U., 1895; Ph.D., U. Chgo., 1914; m. Frances Ward Yeazell, Dec. 17, 1901; children—Katharine Anita, Elizabeth Wade. Prof. embryology, Ohio Med. U., 1896; prof. histology and embryology, Ohio Med. U. and Starling-Ohio Med. Coll., 1902-14; asst. in zoölogy and entomology Ohio State U., 1895-1900, asso. prof., 1902, prof., 1908, prof. anatomy, 1914—; lectr. neurology U. Cal., 1924-27. Author: A Laboratory Guide in Zoölogy, 1904; A Laboratory Guide for Vertebrate Dissections, 1918. Died Aug. 23, 1933.

LANDAHL, Herbert Daniel, math. biologist; b. Fancheng, China, Apr. 23, 1913; s. C. W. and Alice (Holmberg) L.; A.B. magna cum laude, St. Olaf Coll., 1934; S.M., U. Chgo., 1936, Ph.D., in Math. Biophysics, 1941; m. Evelyn C. Blomberg, Aug. 23, 1940; children—Carl D., Carol A., Linda C. Research asst. U. Chgo., 1937-41, faculty, 1942—, prof. math. biology, 1949-56, prof. math. biology 1965—, acting chmn. com. on math. biology, 1965—, Cons. to govt. agys. Recipient certificate OSRD, 1945, USPHS Research Career award, 1962—. Fellow A.A.A.S.; mem. N.Y. Acad. Scis., Biophys. Soc., Ill. Acad. Sci., Biometric Soc. (mem. council 1960-64, adv. bd. 1965—). Author: (with A. S. Householder) Mathematical Biophysics of the Central Nervous System, 1945; also numerous articles. Math. studies of excitation, respiration, growth; diffusion and transport; studies on aerosol inhalation, compartment theory, neural net theories of psychol. phenomena. Home: 2256 W. 112th St., Chgo. 60643.*

LANDAU, Edmund, German mathematician; b. Berlin, Germany, Feb. 14, 1877; s. Leopold Landau; studied in Berlin and Munich, 1893-99; Dr.Phil., Berlin, 1899; hon. Dr.Phil., U. Oslo (Norway), 1929. Prof. math. U. Berlin, 1901-09, U. Göttingen (Germany), 1909-34. Author: Handbuch der Lehre von der Verteilung der Primzahlen, 2 vols, 1909; Vorlesungen über Zahlentheorie, 3 vols., 1927; Grundlagen der Analysis, 1930; also numerous articles. Research in analytic number theory, especially distbn. of prime numbers and prime ideals. Died Berlin, Feb. 19, 1938.

LANDAU, Lev Davidovich, Russian physicist; b. Baku, USSR, Jan. 22, 1908; s. David and Lubov Landau; grad. Leningrad State U., 1927; D.Physico-Math. Sci., 1934; m. 1939; 1 son. Faculty, Leningrad Physico-Tech. Inst.; 1928-31; head theoretical dept. Kharkov Physico-Tech. Inst., 1931-37, prof. physics, 1935—; head theoretical dept. Inst. Phys. Problems, USSR Acad. Scis., Moscow, 1937—; prof. Moscow State U., 1948—. Recipient Stalin prize, 1946, F. London prize, 1960, Max Planck medal, 1960, Lenin prize for sci., 1962; Nobel prize for physics, 1962. Fellow Brit. Royal Soc.; mem. USSR, Danish, Netherlands, Nat. acads. sci., Am. Acad. Arts and Scis., Phys. Soc. London. Author: Continuum Mechanics, Hydrodynamics and the Theory of Elasticity, 1944; Selected Scientific Papers, 1964; co-author: Field Theory, 2d ed., 1948; Quantum Mechanics, Part I, 1948; Statistical Physics (Classic and Quantum), 1951; Course of Theoretical Physics, 7 vols., 1938-62. Research and numerous publs. on solid state, low temperature and nuclear physics, quantum field theory; developer thermodynamic theory of 2d order phase transitions in solid bodies, macroscopic theory of superfluidity of liquid helium and predicted possibility of sound wave propagation in liquid helium at 2 different speeds; theory of combined inversion and double-component neutrino. Died Apr. 1, 1968.

LANDAUER, Carl, economist; b. Munich, Germany, Oct. 15, 1891; s. Abraham and Elsbeth (Feuchtwanger) L.; student U. Munich also U. Berlin (Germany), 1909-14; Ph.D., U. Heidelberg (Germany), 1915; LL.D., U. Cal. at Berkeley, 1961; Dr. rer. pol. h.c., U. Hamburg, 1966; m. Hilde Stein, Feb. 16, 1916; children—Ilse (dec.), Gerti (Mrs. Edmund de Schweintz Brunner), Walter, Ernest. Came to U. S., 1933, naturalized, 1940. Research asst. Inst. Econs., Kiel, 1915-16; with German War Food Adminstrn., 1916-19; journalist, 1920-26; faculty Sch. Bus. Adminstrn., Berlin, 1926-33, asso. prof., 1932-33; editor German Economist, 1926-33; lectr. econs. U. Cal. at Berkeley, 1934-36, prof. econs., 1936-59, emeritus, 1959—; with U. Hamburg, (Germany), 1962-63. Cons., U. S. Dept. State, U. Berlin, 1949-50. Recipient Fulbright Teaching award, 1959. Mem. Am. Econ. Assn., Am. Hist. Assn., Assn. for Comparative Econs., Nat. Planning Assn. Author: Grundprobleme der funktionellen Verteilung des wirtschaftlichen Wertes, 1923; Planwirtschaft und Verkehrswirtschaft, 1931; Theory of National Economic Planning, 1944; (with Elizabeth Valkenier, Hilde Landauer) European Socialism, 2 vols, 1959; Comparative Economic Systems, 1964; also articles. Research on theory econ. planning in market economies; clarification determining factors in struggles between evolutionary and revolutionary socialism; analysis of methodological problems in comparing different econ. systems. Home: 1317 Arch St., Berkeley, Cal. 94708.*

LANDAUER, Edmond, Belgian management scientist; b. Brussels, Belgium, Nov. 19, 1883; D.Sc., Brussels; traveled extensively. Mgr., textile firm, Roumania; textile consultant in Belgium, France, England, Denmark; principal organizer, 2nd Internat. Management Congress, Brussels, 1925; sec.-general, Internat. Committee Scientific Management, 1927; vice-chmn., 1932. Commander of Order of Crown of Roumania; chevalier of Order of Leopold. Author: L'Organisation Scientifique (collected articles), 1935. Developed theory of rational purchasing policy; assimilated and disseminated Am. management ideas and practices. Died July 1934.

LANDAUER, Rolf William, physicist; b. Stuttgart, Germany, Feb. 4, 1927; s. Karl and Anna (Dannhauser) L.; came to U. S. 1938, naturalized 1944; S.B., Harvard, 1945, A.M., 1947, Ph.D., 1950; m. Muriel Jussim, Feb. 26, 1950; children—Karen Lyalli, Carl Hollis, Thomas Andrew. Theoretical solid state physicist, Lewis Lab., NACA, Cleve., 1950-52; research physicist IBM, Poughkeepsie, N.Y., 1952-60; with IBM Research Center, Yorktown Heights, N.Y., 1960—, dir. phys. sci. 1962-66, asst. dir. research, 1966—. Served with USNR 1945-46. Fellow Am. Phys. Soc.; mem. I.E.E.E. (sr.), Phi Beta Kappa, Sigma Xi. Diversified contributions to theoretical solid state physics and computer device theory. Home: 67 Round Hill Dr., Briarcliff Manor, N.Y. Office: P.O. Box 218, Yorktown Heights, N.Y. 10598.*

LANDAUER, Walter, geneticist; b. Manheim, Germany, July 15, 1896; S. Friedrich and Charlotte (Ziegler) L.; student U. Frankfurt (Germany), 1919; Dr.Phil. Nat., U. Heidelberg (Germany), 1922; m. Elly T. Bernstein, Sept. 4, 1964. Came to U. S., 1924, naturalized, 1940. Asst., Zool. Inst., U. Heidelberg, 1922-24; faculty U. Conn., Storrs, 1924—, prof., head dept. animal genetics, 1928-64, prof. emeritus, 1964—; hon. research asso. dept. zoology and comparative anatomy Univ. Coll., London, Eng. 1964-67, dept. animal genetics, 1967—. Recipient Borden award Poultry Sci. Assn., 1954. Mem. A.A.A.S., Genetics Soc. Am., Am. Soc. Zoologists, Am. Soc. Naturalists, Soc. for Study Devel. and Growth, Internat. Inst. Embryology, Teratology Soc., Am. Acad. Arts and Scis., Am. Assn. U. Profs., Am. Com. for Protection Fgn. Born, Phi Beta Kappa (hon.), Sigma Xi. Publs. on genetic studies of inherited defects, suppression of harmful mutations; research in body heat regulation, origin of congenital malformations caused by genes and specific chems. Home: 16 Highgate Av., London N. 6, Eng.*

LANDE, Alfred, physicist; b. Elberfeld, Germany, Dec. 13, 1888; ed. U. Marburg, U. Göttingen; Ph.D., U. Munich, 1914. Privat docent, Frankfurt, 1920-22; prof. Tübingen U., 1922-31; prof. physics Ohio State U., Columbus, 1931—. Fellow Phys. Soc.; mem. A.A.-A.S. Author: Quantum mechanics, 1951; From Dualism to Unity in Quantum Physics, 1960; New Foundations of Quantum Mechanics, 1967. Authority on atomic structure and quantum theory, spectral lines, Zeeman effect, multiplet theory. Office: Dept. of Physics, Ohio State University, Columbus, O. 43210.

LANDECKER, Werner Siegmund, sociologist; b. Berlin, Germany, Apr. 30, 1911; s. Adolf and Hertha (Kahn) L.; J. U.D., U. Berlin, 1936; Ph.D., U. Mich., 1947; m. Marjory Victoria Records, July 15, 1942; children—John Records, Thomas Franklin. Came to U. S., 1937, naturalized, 1944. Faculty, Ind. U., 1941-42; faculty U. Mich., Ann Arbor, 1942—, prof. sociology, 1961—. Mem. Am. Sociol. Assn., Ohio Valley Sociol. Soc., Mich. Acad. Sci., Arts and Letters. Author: (with Freedman, Hawley, Lenski and Miner) Principles of Sociology, 1952. Contbns. to analysis of social integration and social stratification. Home: 1113 Ferdon Rd., Ann Arbor, Mich. 48104.*

LANDEN, John, English mathematician; b. Peakirk, Eng., Jan. 23, 1719; began as surveyor; land agt. to William Wentworth, 1762-88. Fellow Royal Soc., 1766; mem. Spalding Soc. Author: Mathematical lucubrations, 1755; A Discourse Concerning the Residual Analysis, 1758; Animad versions on Dr. Stewart's Computation of the Sun's Distance from the Earth, 1771; Mathematical Memoirs, 1780; also articles. Studied points connected with fluxionary calculus; showed roots of cubic equation can be derived using infinitesimal calculus; discovered theorem on hyperbolic arc in terms of two elliptic arcs (named after him), 1775; solved problem of spinning of top; explained Newton's error in calculating effects of precession; studied elliptic functions, residual analysis, Landen's point, Landen's transformations named after him. Died 1790.

LANDER, Harry, physician; b. Edinburgh, Scotland, Sept. 8, 1928; s. Harry and Jean (Crichton) L.; student U. Western Australia, 1946; M.B.B.S., U. Adelaide (So. Australia), 1951; m. Constance Margaret Price, Feb. 3, 1953; children—Andrew Harry, Elizabeth Margaret, David Owen. Lectr., U. Adelaide, 1954, research fellow medicine, 1955-57; sr. lectr., 1960-64, reader medicine, 1965—; asst. to Nuffield prof. medicine Radcliffe Infirmary, Oxford, Eng., 1958-59; vis. fellow div. hematology dept. internal medicine Washington U. Sch. Medicine, St. Louis, 1965; hon. physician Royal Adelaide Hosp., 1960—. Royal Australasian Coll. Physicians Travelling fellow in medicine and allied scis., 1958; Commonwealth fellow Harkness Found. N.Y., 1965. Fellow Australasian Coll. Physicians, Australasian Coll. Physicans (past mem. state com.); mem. Australian Soc. for Med. Research (past fed. dir.), Internat., Asian-Pacific, Australian socs. hematology, Australian Soc. for Med. Research. Research and publs. on arsenic poisoning, neurol. disorders, techniques of gastrointestinal biopsy, blood platelet structure, function and kinetics. Home: 2A Highfield Av., St. Georges, South Australia 5064. Office: U. Adelaide, Adelaide, S. Australia, Australia.*

LANDES, Herbert Ellis, Am. urologist; b. Greencastle, Ind., Oct. 5, 1894; s. Albert and Mary Louise (Ellis) L.; A.B., De Pauw U., 1917; M.S., U. Chgo., 1919; M.D., Rush Med. Coll., 1922; grad. study U. Vienna, 1930, Johns Hopkins Hosp., 1930-32; m. Wyota Ann Ewing, Sept. 4, 1918; children—Mary Louise, John Ewing. Intern Presbyn. Hosp., 1921-22; clin. prof. urology, sch. medicine Loyola U., 1932, prof. 1934—, chmn. dept. urology, 1939—; sr. attending urologist Mercy Hosp., 1932, chmn. dept. urology, 1939—; cons. urologist Chgo. Municipal Tb Sanitarium, Elgin (Ill.) State Hosp., Little Co. of Mary Hosp., Chgo., Burlington R.R., Lewis Meml. Hosp., Chgo. Diplomate Am. Bd. Urology. Fellow A.C.S.; mem. A.M.A., Ill. State Med. Soc., Inst. Medicine, Am., Chgo. urology assns., Sigma Xi, Phi Beta Kappa. Contbr. to textbooks on urology; also articles. Died Sept. 24, 1959.

LANDES, Kenneth Knight, Am. geologist, educator; b. Seattle, May 10, 1899; s. Henry and Bertha (Knight) L.; B.S., U. Wash., 1921; A.M., Harvard, 1923, Ph.D., 1925; m. Susan Elizabeth Beach, Sept. 20, 1924; children—Walter Henry, Robert Kenneth, Katherine Flora (Mrs. Charles A. Reinke, Jr.). Prof. geology, dept. chmn. U. Kan., 1926-41; geologist State of Kan., 1926-41; prof. geology, dept. chmn. U. Mich., Ann Arbor, 1941—. Cons. U. S. Geol. Survey, 1921-51; cons. for various corps., UN, 1951—. Fellow Inst. Petroleum (London); mem. Geol. Soc. Am. (v.p. 1946), Mineral. Soc. Am. (pres. 1945), Am. Assn. Petroleum Geologists (hon., editor 1951-53), Am. Inst. Mining, Metall. and Petroleum Engrs., A.A.A.S. (v.p. 1951), Geol. Soc. Mich. (pres. 1945), Soc. Econ. Geologists, Soc. Econ. Paleontologists and Mineralogists, Sigma Xi, others. Author: Petroleum Geology, 1951; (with R. C. Hussey) Geology and

Man, 1948; also numerous articles. Home: 1005 Berkshire Av., Ann Arbor, Mich. 48104.*

LANDING, Benjamin Harrison, Am. physician; b. Buffalo, Sept. 11, 1920; s. Benjamin Harrison and Margaret (Crohen) L.; A.B., Harvard, 1942, M.D., 1945; m. Selma Ruth Baernstein, June 4, 1949; children—Benjamin H., Susan L., William M., David A. Dir. pathologist Cin. Childrens Hosp., 1953-61; pathologist, dir. labs., Los Angeles Childrens Hosp., 1961—; instr. pathology Harvard Med. Sch., Boston, 1950-53; faculty U. Cin. Coll. Medicine, 1953-61; prof. pathology and pediatrics U. So. Cal. Sch. Medicine, Los Angeles, 1961—. Mem. Am. Assn. Pathologists and Bacteriologists, Internat. Acad. Pathology, Histochem. Soc., Endocrine Soc., Am. Pediatric Soc., Soc. for Pediatric Research. Author: (with Sidney Farber) Tumors of Cardiovascular System, 1956; also numerous articles. Research on pediatric pathology emphasizing histochem. study endocrine and metabolic diseases. Home: 1063 Bethany Rd., Burbank, Cal. 91504. Office: 4650 Sunset Blvd., Los Angeles 90027.*

LANDIS, Eugene Markley, Am. physiologist; b. New Hope, Pa., Apr. 4, 1901; s. Henry Garges and Sadie (Markley) L.; B.S., U. Pa., 1922, M.S., 1924, M.D., 1926, Ph.D., 1927; M.S. (hon.), Yale, 1938; m. Elizabeth Barrett Gerhard, May 29, 1934; 1 dau., Barbara Gerhard (Mrs. Jerry Amos). Faculty, U. Pa., Phila., 1920-39, asso. in medicine, 1931-35, research asso. in pharmacology, 1932-39, asst. prof. medicine, 1935-39; prof. internal medicine, head dept. U. Va., 1939-43; George Higginson prof. physiology, head dept. Harvard Med. Sch., Boston, 1943—, chmn. div. med. scis., 1949-52. Sec. com. on aviation medicine NRC, 1940-46, chmn. subcom. on acceleration, 1941-46. Recipient Phillips medal A.C.P., 1936. Mem. Nat. Acad. Scis., Am. Acad. Arts and Scis., Am. Soc. Clin. Investigation (pres. 1943-43), Am. Physiol. Soc. (pres. 1952-53), Phi Beta Kappa, Sigma Xi, Alpha Omega Alpha, Alpha Mu Pi Omega. Editor: Circulation Research for Am. Heart Assn., 1963-66. Research, numerous publs. on capillary physiology, circulation, renal diseases, hypertension, pressures and fluid movements in single blood capillaries by micromethods. Home: 519 Washington St., Brookline, Mass. 02146.*

LANDIS, Phillip Sherwood, Am. chemist; b. York, Pa., July 29, 1922; s. Roy Hampton and Catherine (Trimmer) L.; B.S., Franklin & Marshall Coll., 1943; M.S., U. Ky., 1947; Ph.D. (Socony Mobil fellow), Northwestern U., 1958; m. Vivian Center, Aug. 25, 1944; children—Bryan Hayden, Michael Eugene. Chemist, Cities Service Refining Corp., Lake Charles, La., 1944-46; research chemist Mobil Oil Corp., Paulsboro, N.J., 1947-55; with Mobil Research and Devel. Corp., Princeton, N.J., 1958—, sr. research asso., 1966—. Mem. Am. Chem. Soc. (chmn. South Jersey sect. 1960-61), N.J. Acad. Scis. (treas. 1967—), Sigma Xi. Patentee, publs. on catalysis of organic reactions by crystalline aluminosilicates; synthesis, devel. rust inhibitors for gasolines, fuel oils, turbine oils, and lubricating oils. Home: 458 Queen St., Woodbury, N.J. 08096. Office: Mobil Research and Devel. Corp., Box 1025, Princeton, N.J. 08540.*

LANDIS, Walter Savage, Am. chemist; b. Pottstown, Pa., July 5, 1881; Met.E., Lehigh U., 1902, M.S., 1906, Sc.D., 1922; postgrad. U. Heidelberg, 1905-06, Aachen, 1909. Asst. in metallurgy Lehigh U., 1902-04, instr., 1904-07, asso. prof., 1910-12; chief technologist Am. Cyanamid Co., 1912-22, v.p., 1922—. Recipient Chem. Indsl. medal Am. sect. Soc. Chem. Industry, 1936, Perkin medal, 1939, medal Am. Inst. Chemists, 1943. Mem. Mineral and Metal Engring. Soc., Soc. Chem. Engrs., Am. Chem. Soc., Electrochem. Soc. Research on nitrogen fixation, electric smelting, heavy chems.

LANDIS, William Weidman, Am. mathematician; b. Coatesville, Pa., Feb. 15, 1869; s. Isaac Daniel and Anna Mary (Davis) L.; Ph.B., Dickinson Coll., 1891, A.M., 1894; student Johns Hopkins, 1891-94, Sc.D., 1907. Prof. math. Thiel Coll., Pa., 1894-95; prof. math. and astronomy. Dickinson Coll., since 1895. Mem. Am. Math. Soc., Am. Math. Assn., Phi Beta Kappa. Contbr. to jours. Died Apr. 8, 1942.

LANDOIS, Leonard, German physiologist; b. Münster, Germany, Dec. 1, 1837; grad., Greifswald, Germany, 1863; named prof. physiology, Greifswald, 1872. Author: Lehrbuch der Physiologie, 1879; Lehrbuch der Physiologie des Menschen, 12th edit., 1909. also articles on blood, pulse and transfusion, optical phenomena, electromagnetism. Discovered animal serum will hemolyze human blood, 1875: originated test for carbon monoxide in blood. Died Greifswald, Nov. 17, 1902.

LANDOLT, Hans, phys. chemist; b. Zurich, Switzerland, Dec. 5, 1831; studied with Mitscherlich, Rose and Bunsen; prof. at Bonn, Aachen, and Berlin (all Germany). Author: (with Börnstein) Tables, 1883. Research on optical refractivity of organic compounds, constancy of weight during chem. reactions; improved polarization apparatus. Died Berlin, Mar. 15, 1910.

LANDOR, A. Henry Savage, anthropologist; b. Florence; s. Charles Savage L.; studied and traveled in East and U. S. First white man to reach both sources of Brahmaputra River, Tibet, and establish their exact position, 1897; 1st white man to settle geog. problem that no range higher than Himalayas existed north of Brahmaputra River in Tibet; 1st white man to explore Central Mindanao Island and discover the white tribe (Mansakas); held world's record in mountaineering (until Duke of Abruzzi's expdn.), having reached altitude of 23,490 feet on Mt. Lumpa, Nepal, 1899; went overland from Russia to Calcutta, across Persia, Afghanistan and Beluchistan, 1901-02; spent a year cruising in Philippine and Sulu Archipelagos, visiting some 400 islands, 1903; crossed Africa at its widest part, 1906; crossed S. Am. from Rio de Janeiro to Lima, over unexplored Central Brazil and the Andes, 1910-12. Fellow Royal Geog. Soc.; mem. Royal Instn., École d'Anthropologie of Paris (hon. corr.), Geog. Soc. Marseilles (hon. mem., gold medallist), Italian Anthrop. Soc. (hon.), Internat. Anthrop. Inst., Paris (founder mem.). Author: Alone with the Hairy Ainu, or 3800 Miles on a Pack-saddle; Corea, or the Land of the Morning Calm; A Journey to the Sacred Mountain of Siao-ou-tai-shan; In the Forbidden Land, 1898; China and the Allies, 1901; Across Coveted Lands, 1902; The Gems of the East, 1904; Tibet and Nepal, 1905; Across Widest Africa, 1908; The Americans in Panama, 1910; An Explorer's Adventures in Tibet, 1910; Across Unknown South America, 1913; Mysterious South America, 1914. Inventor 2 types of improved armoured cars, new type of rigid airship, device for destroying barbed-wire entanglements, armoured motorcycle with mitrailleuse. Died Dec. 26, 1924.

LANDOUZY, Louis Joseph, French physician; b. Reims, France, Mar. 27, 1845; s. Marc-Hector Landouzy; ed. Med. Sch. Reims, later Faculty Medicine Paris; M.D., 1876; m. Louise Richet. Prof., med. clinic, dean Faculty Medicine Paris; hosp. physician Laennec Hosp., Paris. Mem. French Acad. Scis., Acad. Medicine. Author: La Typhobacillose; Asthme et tuberculose; Tuberculose chez les enfants du premier age; Tuberculoses professionnelles; Hérédité de la tuberculose; Tuberculose: maladie sociale; Cent ans de physiologie, 1808-1908; Empoisonnements non professionnels par l'aniline; Localisations angiocardiaques typhoidiques; Atrophies myopathiques: myopathie type Landouzy Déjerine; Paralyses dans les maladies aigues; (with D. H. Labbé); Enquete sur l'alimentation d'une centaine d'ouviers et d'employés parisiens, 1905; Education alimentaire rationnelle, 1911; Les Sérothérapies, 1898; (with Jayle); Glossaire médical, 1902; (with Gautier, Moureu, de Lannay), Crénothérapie, cures hydro-minérales, 1910; (with L. Bernard) Éléments d'Anatomie et de Physiologie médi claes, 1913; (with Jean Heitz) Le Substratum scientifique de la Balnéothérapie. Research and numerous publs. on serotherapy, syphilis, typhoid fever, nervous diseases, Tb; showed relationships between pleuresy and Tb; described spirochetal jaundice, 1883, various muscular atrophies (Landouzy's sciatica, Landouzy-Dejerine atrophy). Died Paris, May 9, 1917.

LANDOVITZ, Leon Fred, Am. physicist; b. Bklyn., May 24, 1932: s. Philip and Ethel (Leibman) L.; A.B., Columbia U., 1953, Ph.D., 1958; m. Renee M. Altman, July 19, 1959. NSF fellow Inst. Advanced Study, 1957-58; research asso. Brookhaven Nat. Lab., Upton, L.I., N.Y., 1958-60; asst. prof. physics Belfer Grad. Sch. Sci., Yeshiva U., N.Y.C., 1960-62, prof., since 1967—. Cons. editor physics books Reinhold Pub. Co. Mem. Am., Italian phys. socs., Sigma Xi. Research and publs. in field of elementary particles, astrophysics. Home: 16 Locust Dr., Great Neck, L.I., N.Y. 11021. Office: Belfer Grad. Sch. Sci., Yeshiva U., 186th St. and Amsterdam Av., N.Y.C. 10033.*

LANDRETH, David, Am. agriculturist; b. Phila., Sept. 15, 1802; s. David and Sarah (Arnell) L.; ed. common schs., Phila.; m. Elizabeth Rodney, 1825; m. 2d, Martha Burnet, 1842. Became propr. of his father's nursery and seed bus., Phila., 1828; a founder Pa. Hort. Soc., v.p., 1829-36; became publisher Illustrated Floral Mag., 1832; constructed nursery and arboretum, Bristol, Pa., 1847; a founder Farmers' Club of Pa., 1847; active in agrl. experimentation, cattle breeding, 1850's, 60's; pres. Phila. Soc. for Promotion of Agr., 1856; experimented with steam plowing, digging and chopping, 1870's; pub. Am. edit. Dictionary of Modern Gardening (George W. Johnson), 1847. Died Bristol, Feb. 22, 1880.

LANDRUM, Robert D(allas), Am. chem. engr.; b. Terre Haute, Ind., Feb. 8, 1882; s. James Wesley and Kate (Tolbert) L.; B.S., Rose Poly. Inst., 1904, M.S., 1909, Ch.E., 1914; m. Ethel Price Sherwood, Sept. 1, 1908 (dec. 1935); children—Sherwood, Robert James, Kate Tolbert; m. 2d, Margaret Elizabeth Carr, 1937; 1 dau., Peggy Ann. Chemist and enameler. Columbian Enameling & Stamping Co., 1904-07; asst. prof. chemistry, U. Kan., 1907-10; chem. engr. Lisk Mfg. Co., Canandaigua, N.Y., 1910-13; cons. engr. Mich. Enameling Works, Kalamazoo, Gen. Stamping Co., Canton, O., 1913-14; chem. engr. and mgr. service dept. Harshaw Fuller & Goodwin Co., mfrs. indsl. chemicals, Cleve. 1914-22; v.p. Vitreous Enameling Co., also Vitreous Steel Products Co., Cleve., 1922-25; gen. mgr. ceramic materials div. Titanium Alloy Mfg. Co., Cleve. 1925-32; sales mgr. Harshaw Chem. Co., Chgo., 1932-35, manager spl. div., 1935-42, mgr. tech. sales, 1942—. Fellow A.A.A.S., Am. Ceramic Soc. (pres. 1924-24); mem. Am. Chem. Soc., Am. Inst. Chem. Engrs., Soc. Chem. Industry, Keramos, Société de Chimie Industrielle, France, Chgo. Drug and Chem. Assn. Author: Enamel, 1918; Bibliography and Abstracts of Literature on Enamels, 1929; also numerous articles and tech. papers. Deceased.

LANDRY, Jean Baptiste Octave, French physician; b. Limoges, France, Oct. 10, 1826; M.D., 1854; physician Hydrotherapeutic Establishment, Auteuil, France. Author: Recherches sur les causes et les indications curatives des maladies nerveuses, 1855; Traité complet des paralysies, 1859. Described acute ascending spinal paralysis which usually begins in the muscles of the feed (Landry's paralysis), 1859. Died Auteuil, 1865.

LANDS, Alonzo Mitchell, Am. pharmacologist; b. Lesterville, Mo., Aug. 19, 1904; s. Edward Mitchell and Bertha (Airsman) L.; B.S., State Tchrs. Coll., Memphis, 1926; M.S., U. Kan., 1930, Ph.D., 1934; m. Bernice Rubenstein, May 17, 1926; children—Beatrice (Mrs. Robert L. Easterly), William E. M., John S. Faculty, U. Ky., 1929-31, U. Kan., 1931-34; asst. prof. Georgetown U., 1934-41; pharmacologist Frederick Stearns & Co., Detroit, 1941-47; asst. dir. biology div., dir. pharmacology group Sterling-Winthrop Research Inst., Rensselaer, N.J., 1947—. Mem. Am. Physiol. Soc., Am. Soc. for Pharmacology and Exptl. Therapeutics, Soc. Exptl. Biology and Medicine, N.Y. Acad. Sci., A.A.A.S. Research, publs. on investigations of the autonomic nervous system and actions of drugs on it; use of drugs in treatment of cardiovascular disorders, asthma, other diseases. Home: Box 225, New Baltimore, N.Y. 12124. Office: Sterling-Winthrop Research Inst., Rensselaer, N.J. 12144.

LANDSBERG, Grigorii Samuilovich, Russian physicist; b. 1890; ed. Moscow U. With Moscow U., 1923-45, 47-51; joined staff Inst. Physics, USSR Acad. Scis., 1934, also chmn. com. for spectroscopy. Recipient Stalin prize, 1941, Order of Lenin, twice. Author: Intermolecular Forces and Combined Diffusion of Light, 1938. Research in optics. Office: Acad. Scis. USSR, Moscow, USSR.

LANDSBERG, Helmut Erich, meteorologist; b. Frankfurt am Main, Germany, Feb. 9, 1906; s. Georg Julius, and Klara (Zedner) L.; Ph.D., U. Frankfurt, 1930; m. A. Frances Simpson; 1 son, Bruce S. Came to U. S. 1934, naturalized 1938. Supr., Taunus Obs., Germany, 1931-34; asst. prof. geophysics Pa. State U., 1934-41; asso. prof. meteorology U. Chgo., 1941-43; operations analyst, cons. USAAF, 1943-45; exec. dir. Com. on Geophysics and Geography, Research and Devel. Bd., 1946-51; dir. geophys. research directorate USAF Cambridge Research Center, Mass., 1951-54; dir. climatology U. S. Weather Bur., Washington, 1954—. Vis. prof. U. Md., 1964—; dir. Environmental Data Service, U. S. Environmental Sci. Service Administrn., 1965—. Recipient Gold medal Dept. Commerce, 1960. Fellow A.A.A.S., Am. Acad. Arts and Sci., Meteorit. Soc., Am. Geophys. Union (pres. meteorol. sect. 1956-59); mem. Am. Meteorol. Soc. (counselor 1952-60, v.p. 1962-64, Spl. award in bioclimatology 1964). Author: Physical Climatology, 1958. Editor: Advances in Geophysics, 1952—; editor-in-chief World Survey of Climatology, 1962—. Home: 5107 53d Ave., Washington 20031. Office: U. S. Weather Bur., Washington 20235.*

LANDSBERGER, Henry Adolf, sociologist; b. Dresden, Germany, Aug. 5, 1926; s. Ernst and Annie (Winter) L.; B.Sc., London Sch. Economics, 1948; Ph.D., Cornell U., 1954; m. Betty H. Hatch, June 9, 1951; children—Margaret Ann, Samuel Ernest, Ruth Elizabeth. Came to U. S., 1949, naturalized, 1960. Research asso. Sch. Indsl. and Labor Relations, Cornell U., Ithaca, N.Y., 1954-55, asst. prof., 1956-60, asso. prof., 1960-64, prof., 1964—; asst. research officer Oxford (Eng.) Inst. Statistics, U. Oxford, 1955-56; asst. dir. Cornell Soc. Sci. Research Center, 1956-58; hon. mem. Faculty Econ. Sci., U. Chile, Santiago; prof. Latin-Am. Faculty of Social Sci., UNESCO, 1962-64. Mem. Am. Sociol. Assn., Am. Psychol. Assn., Indsl. Relations Research Assn., Am. Rural Sociol. Assn., Latin Am. Studies Assn. Author: Hawthorne Revisited, 1958; Iglesia, Intelectuales y Campesinos, 1967. Contbr. articles to sci. jours. Research on peasant movements and rural orgn., labor leaders, personnel managers, young workers, mediation process, small groups, orgn. theory, methodology. Home: 225 Berkshire Rd., Ithaca, N.Y. 14850.*

LANDSBOROUGH, David, Scottish naturalist; b. Dabry, Scotland, Aug. 11, 1779; ed. Dumfries Acad., U. Edinburgh; m. Margaret M'Leish, 1817; 4 sons, including William, David; 3 daus. Ordained to ministry Ch. of Scotland, 1811; joined Free Kirk and became minister of Saltcoats, 1843. Mem. Linnean Soc. (asso.). Author: Excursions to Arran Ailsa Craig, and the two Cumbraes, 1847; Popular History of British Sea-Weeds, 1849; Popular History of British Zoophytes or Corallines; also articles. Discovered Ectocarpus landsburgii (an algae), also nearly 70 species of plants and animals in Scotland. Died Sept. 12, 1854.

LANDSTEINER, Karl, pathologist; b. Vienna, Austria, June 14, 1868; s. Leopold and Fanny (Hess) L.; M.D., U. of Vienna, 1891; D.Sc., U. of Chicago, 1927; hon. D.Sc., Cambridge U., England; hon. M.D., Université Libre de Bruxelles, Belgium; D.Sc., Harvard; m. Helene Wlasto, 1880; 1 son, Earnest. Pathologist,

U. of Vienna, 1909-19; mem. Rockefeller Inst. for Med. Research, 1922-39, now emeritus. Recipient Nobel prize in physiology and medicine, for discovery of human blood groups, 1930; Paul Ehrlich medal, 1930; Chevalier Legion of Honor (France); Dutch Red Cross medal, 1933. Member Nat. Acad. Sciences, Royal Swedish Academy Science, Danish Academy Science, Deutsche Akademie Naturforscher (Halle, Germany), Am. Assn. Immunologists (pres. 1929), Swedish Med. Soc., Harvard Soc., Am. Philos. Soc., Société Belge de Biologie (Bruxelles), Am. Soc. of Naturalists; hon. mem. Pathol. Soc. of Gt. Britain and Ireland, Vienna Med. Soc.; fellow New York Acad. Medicine; hon. fellow Royal Soc. of Medicine (London); hon. mem. Pathol. Soc. of Philadelphia, Reale Academia delle Scienze (Modena, Italy); corr. mem. Med. Chirurgical Soc., Edinburgh. Author: The Specificity of Serological Reactions (helped establish immunochemistry as br. of sci.), 1936. Contbr. papers on immunology, bacteriology and pathology; chemistry of antigens; human blood groups; etiology of poliomyelitis; etiology of paroxysmol hemoglobinuria, syphilis; devised method of preparation of alcoholic extract of antigen for Wasserman test; (with others) discovered blood groups M, N, and MN; his studies led to the creation of the ABO system of classifying blood and made transfusions medically safe by making it possible to give recipients blood matching their own blood. Died N.Y.C., June 26, 1943.

LANDTMAN, Bernhard, Finnish physician; b. Helsinki, Nov. 17, 1916; s. Gunnar and Alice Landtman; M.D., U. Helsinki, 1944; m. Brita Thune, Nov. 11, 1950. Apptds. at hosps. in stockholm, 1945, 49-50, London, 1948, Baltimore 1952-62, also in India, WHO, 1950-51. Mem. Finnish (chmn. 1964-65), Brit. (hon.) pediatric assns., Finnish Med. Soc. (chmn. 1966). Research, numerous publs. on pediatrics and pediatric cardiology. Home: 15 Tunturikatu. Office: 11 Stenback St., Helsinki, Finland.*

LANDY, Maurice, Am. immunologist; b. Cleve., Mar. 8, 1913; s. Joseph and Rose (Eisenstein) L.; A.B., Ohio State U., 1934, M.A., 1934, Ph.D., 1940; m. Reba D. Altman, June 1, 1947. Chief dept. bacteriology Gen. Biochem. Corp., Cleve., 1936-42; chief dept. bacteriology Wyeth Inst., Phila., 1946-47; chief typhoid research unit, chief dept. bacteriology and immunology Army Med. Service Grad. Sch., Washington, 1947-56; chief immunology sect. Nat. Cancer Inst., NIH, Bethesda, Md., 1956-62, chief lab. immunology, 1962-67, chief allergy and immunological br. Nat. Inst. Allergy and Infectious Diseases, 1967—. Sci. adviser A.M.A. Inst. for Biomed. Research. Recipient Dept. Health, Edn. and Welfare Superior Service award, 1966. Mem. Reticuloendothelial Soc. (asso. editor Jour.). Author: Bacterial Endotoxins, 1963; also numerous publs. Asso. editor Jour. Immunology, Jour. Bacteriology. Isolation and characterization of typhoid antigens, immunology of enteric infection, biologic properties of bacterial endotoxins, mechanisms of immune response, biologic agts. effecting immunosuppression. Home: 9010 Bradgrove Dr., Bethesda 20034. Office: Nat. Inst. Allergy and Infectious Diseases, NIH, Westwood Bldg., Bethesda, Md. 20014.*

LANE, Alfred Church, Am. geologist; b. Boston, Jan. 29, 1863; s. Jonathan A. and Sarah D. (Clarke) L.; A.B., Harvard, 1883, A.M., Ph.D., 1888; student U. Heidelberg, 1885-87; Sc.D., Tufts Coll., 1913; m. Susanne Foster Lauriat, Apr. 15, 1896; children—Lauriat, Frederic Chapin, Harriet Page (Mrs. C. D. Rouillard). Instr. math. Harvard, 1883-85; petrographer Mich. State Geol. Survey and instr. Mich. Coll. Mines, 1889-92; asst. state geologist of Mich., 1892-99, state geologist, 1899-1909; Pearson prof. geology and mineralogy Tufts Coll., 1909-36, ret. on account of tchrs. oath, elected prof. emeritus, 1936. Spl. lectr. on econ. geology, U. of Mich., 1904; 1st cons. in science, Library of Congress, 1929. Fellow A.A.A.S., Am. Acad. Arts. and Scis. (v.p.), Geol. Soc. Am. (pres. 1931); mem. NRC, Am. and Boston mineral socs., Am. Inst. Mining and Metall. Engrs., Harvard Engring. Soc. Boston Natural History Soc., Am. Forestry Assn., Bond Astron. Club (pres. 1933-35), Lake Superior Mining Inst. (treas. 1893) Canadian Mining Inst. (corr.), Geol. Soc. of Belgium (hon.), other socs. Editor and part author books, also reports of Geol. Survey of Mich., Canada and U. S. Author: Die Korngrosse der Auvergnosen. Died Apr. 16, 1948.

LANE, Sir (William) Arbuthnot, Brit. surgeon; b. Ft. George, Scotland, July 4, 1856; s. B. Lane; ed. Stanley House, Bridge of Allan, Guy's Hosp.; M.B., M.S., London; m. Charlotte Briscoe, 1884; (dec. 1935); children—William, Rhonda, 2 other daus.; m. 2d, Jane Mutch, 1935. Became demonstrator anatomy Guy's Hosp., 1882, mem. staff, 1888; surgeon Hosp. for Sick Children, Great Ormond St., 1883-1916; cons. surgeon Aldershot Command, 1914-18; founder Queen Mary's Hosp. for plastic surgery, Sidcup, Eng., also New Health Soc. (1st organized body to deal with social medicine), 1925. Fellow Royal Coll. Surgeons. Author: Operative Treatment of Chronic Constipation, 1909; New Health for Everyman, 1932; books on operative treatment for cleft palate, fractures; also numerous papers. Eminent abdominal surgeon noted for operation for cleft palate at 1 day old, for treatment of simple fractures, for removal of large gut (focus of sepsis in his opinion); developed Lane technique of aseptic surg. excellence; devised 1st system of plates and screws to replace wires for joining fractured limbs, 1892-1905; inventor Lane's kink to treat intestinal paralysis. Died London, Jan. 16, 1943.

LANE, Cecil Taverner, physicist; b. Luton, Eng., Sept. 10, 1904; s. Cecil W. and Lillian (Spivey) L.; B.Sc., McGill U., Montreal, Que., Can. 1926, M.Sc., 1927, Ph.D., 1929; M.A. (hon.), Yale, 1950; m. Josephine Almyra Ball, Mar, 16, 1935; children—Richard Taverner, Barbara Ainsworth. Came to U. S. 1932, naturalized 1940. Research fellow, U. Munich, Germany, 1930-32, Yale, 1933-34; from biophysicist to asst. prof. physics, Yale, 1934-42; research, devel. Walter Kidde & Co., Belleville, N.J., 1942-45; asso. prof. Yale, 1945-50, prof., 1950—. Fellow Am. Phys. Soc. Author: Superfluid Physics, 1962. First to liquefy helium in quantity in U. S., 1941; many discoveries in field of low temperature physics and cryogenic tech., 1946—. Home: 1281 Ridge Rd., North Haven, Conn. Office: Sloane Physics Lab., Yale U., New Haven.*

LANE, Charles Edward, Am. marine biologist; b. Riverton, Wyo., Dec. 17, 1909; s. Charles E. and Clare (Bark) L.: A.B., U. Wis., 1931, M.A., 1933, Ph.D., 1935; m. Betsy C. Ohnewald, Oct. 11, 1931; 1 son, Frank A. Faculty U. Wichita, 1936-42, asso. prof., 1940-42; research coordinator Borden Co., N.Y.C., 1946-49; prof. marine sci. U. Miami (Fla.), 1949—. Cons., Office Naval Research, 1955-60. Mem. N.Y. Acad. Scis., Am. Physiol. Soc., Am. Soc. Zoologists, Internat. Soc. for Toxinology, Soc. for Exptl. Biology and Medicine. Author: (with M. F. Guyer) Animal Biology, 1964; also numerous articles. Defined toxicity of certain marine invertebrates and characterized toxins; described life history and mode of action of marine wood-boring organisms; research in mechanism of salt and water balance in marine invertebrates. Home: 6295 Chapman Field Dr., Miami 33156. Office: U. Miami, Rickenbacker Causeway, Miami, Fla. 33149.*

LANE, Clarence Guy, Am. physician; b. Billerica, Mass., Oct. 21, 1882; s. Albert Clarence and Estella Josephine (Davis) L.; A.B., Harvard, 1905, M.D., 1908; m. Mary Rivers McHarry, May 31, 1919; 1 son, Robert. Intern Worcester City Hosp., 1908-10; in gen. practice, Woburn, Mass., 1910-14; specialist in dermatology, Boston, 1914—; mem. dept. dermatology Mass. Gen. Hosp., 1920-47, chief of dept., 1936-47, tchr. in dept. dermatology Harvard Med. Sch., 1922-47, head of dept., 1936-47; clin. prof. dermatology, 1939-47, emeritus, 1947—. Recipient Cutter medal, Phi Rho Sigma, 1948. Mem. A.M.A., Am. Bd. Dermatology and Syphilology (dir. and sec., 1932-43; pres. 1944-45), Nat. Com. on Indsl. Dermatoses, Am. Dermatol. Assn. (dir. 1927-35, pres. 1935, sec. 1925-30), N.E. Dermatol. Soc., N.Y. Acad. Medicine. Editor: Vol. X of Practitioners Medical Library, 1935; contbr. about 70 articles to med. jours. Editorial bd. N.E. Jour. Medicine, Archives of Dermatology and Syphilology. Died Mar. 12, 1954.

LANE, Harlan Lawson, Am. psychologist; b. Bklyn., Aug. 19, 1936; s. Benjamin and Dorothy Edith L.; A.B., Columbia U., 1958, A.M., 1958; Ph.D. (Nat. Inst. Mental Health fellow), Harvard, 1960. Mem. faculty U. Mich., Ann Arbor, 1960—, prof. psychology, 1967—, dir. Center for Research on Lang. and Lang. Behavior, 1965—. Recipient Distinguished Service award U. Mich., 1963. Fellow Am. Psychol. Assn., A.A.A.S.; mem. Psychonomic Soc., Am. Speech and Hearing Assn., Assn. on Mental Deficiency. Author: (with D. J. Bem) A Laboratory Manual for the Control and Analysis of Behavior, 1964; also articles. Extensive research on the perception and prodn. of speech in speaker's native lang. and acquisition of these skills in course of learning a fgn. lang.; lectr., developer techniques and devices for facilitating second-lang. learning. Home: 2180 St. Francis Dr., Ann Arbor, Mich. 48104.*

LANE, Jonathan Homer, Am. physicist; b. Geneseo, N.Y., Aug. 9, 1819; grad. Yale, 1846. Tchr. sem., Castleboro, Vt., 1846-47; with U. S. Coast Survey, 1847-48, supt., 1869-80; asst. examiner Patent Office, 1848-51, prin. examiner, 1851-57; expert and counsellor in patent cases, 1857-60 or 61; became involved in oil devel., Venango County, Pa.; pvt. bus., Washington, 1866-69; verifier standards Office Weights and Measures; mem. U. S. expdn. to observe total solar eclipse, Des Moines, 1869. Mem. A.A.A.S., Nat. Acad. Scis. Research and publs. on electric currents, compression and expansion of gases, phys. constitution and temperature of sun, density of solar gas, mech. prodn. of low temperature; invented optical telegraph, mercury horizon, electric governor, machine for finding real roots of higher equations using electric currents in a set of small induction coils; formulated Lane's law which states gaseous bodies may grow hotter from heat generation by contraction as they cool. Died Washington, May 3, 1880.

LANE, Malcolm Daniel, Am. biochemist; b. Chgo., Aug. 10, 1930; s. Malcolm Daniel and Helga (Nielsen) L.; B.S., Ia. State U., 1951, M.S., 1953; Ph.D., U. Ill., 1956; m. Patricia Sonquist, Jan. 18, 1931; children—Claudia J., Malcolm Daniel. Asso. prof. Va. Poly. Inst., Blacksburg, 1956-61, prof. 1962-64; sr. postdoctoral fellow Max Planck Institut für Zellchemie, Munich, Germany, 1963; asso. prof. biochemistry N.Y. U. Sch. Medicine, N.Y.C., 1964—. Mem. Am. Inst. Nutrition (Mead Johnson award 1966), Am. Soc. Biol. Chemists, Harvey Soc., Sigma Xi. Research and publs. on enzymatic carboxylation; carbohydrate metabolism; lipid metabolism; enxymatic mechanisms of propionyl CoA carboxylase, acetyl CoA carboxylase, ribulose diphosphate carboxylase, phosphoenol pyruvate carboxylase, phospoenolpyruvate carboxykinase, biotin. Home: 2 Washington Sq. Village, N.Y.C. 10012.*

LANE, William Carr, Am. physician; b. Fayette County, Pa., Dec. 1, 1789; s. Presley Carr and Sarah (Stephenson) L.; student Jefferson Coll., Chambersburg, Pa.; grad. Dickinson Coll., Carlisle, Pa.; postgrad. Med. Dept., U. Pa.; m. Mary Ewing, Feb. 26, 1818; 2 children. Served as surgeon's mate at Ft. Harrison during Creek War, 1813; post surgeon, 1816; resigned from army, 1819; became q.m. gen. of Mo., 1822; 1st mayor St. Louis, 1823-29, 38-40; served as surgeon in Black Hawk War, 1832; a founder Mo. Med. Coll., 1840, also prof. obstetrics; apptd. gov. N.M. Territory, 1852; returned to practice medicine, St. Louis, 1853. Author: Water for the City (advocated municipal waterworks for St. Louis), 1860. Died St. Louis, Jan. 6, 1863; buried St. Louis.

LANEY, Francis Baker, Am. geologist; b. nr. Springfield, Mo., Apr. 9, 1875; s. John Baker and Jane (Alexander) L.; B.S., Drury Coll., Springfield, 1902; M.A., U. Wis., 1905; Ph.D. in Geology, Yale, 1908; m. Minnie D. Towner, Sept. 1910; 1 son, Francis T. Began as asst. geologist, N.C., 1903; asst. curator applied geology, Nat. Museum, Washington, 1908-10; geologist U. S. Geol. Survey, from 1914; microscopist and mineralogist U. S. Bur. Mines, 1914-19; metallographist Central Checking Lab. of Ordnance Dept. U. S. Army, Pitts., 1917-18; head dept. of geology, U. Ida., Moscow, from 1920. Died Apr. 23, 1938.

LANFRANCHI (Lanfranco, Lanfranc, Lanfrancus Mediolanensis), Italian surgeon; probably b. in Milan, Italy; disciple of William of Saliceto, practiced medicine, Milan until circa 1290; then practiced in Lyons, France and other provincial cities of France; went to Paris, circa 1295. Mem. Paris Surgeon's Guild, Confrérie de Saint Côme. Author: Chirurgia parva; Chirurgia magna, 1296. Renewed practice and teaching of surgery in medieval Italy and France; stressed importance of studying anatomy of wounded organ; insisted on clinical teaching; 1st to describe cerebral concussion, 1296; 1st to differentiate between hypertrophy and cancer of the breast, 1296; distinguished between hemorrhoids of arteries and veins; indicated pressure points; discouraged use of probes in diagnosis. Died circa 1306.

LANG, Andrew, sci. writer, philosopher; b. Selkirk, Scotland, Mar. 31, 1884; s. John and Jane Plenderleath (Sellar) L.; ed. St. Andrews U., hon. Ph.D., 1885; Balliol Coll., Oxford (Eng.) U., hon. D.L.; m. Leonora Blanche Alleyne. Became journalist, London, Eng., during late 1870's; 1st Gifford lectr. St. Andrew's U., 1888. Fellow Royal Brit. Acad.; mem. Soc. Psychical Research London (pres. 1911). Author: Literature and Religion, 2 vols., 1887; Book of Dreams and Ghosts, 1897; Making of Religion, 1898; Magic and Religion, 1901; also numerous others. Made 1st full statement of anthrop. method applied to comparative study of mythology (Fortnightly Review, May, 1873); proved folklore to be foundation on which lit. mythology rests; made numerous experiments in crystal gazing (scrying), also took special interest in dowsing. Died July 20, 1912.

LANG, Anton, plant biologist; b. Petersburg, Russia, Jan. 18, 1913; s. George F. and Vera (Davidov) L.; Dr.Nat.Sci., U. Berlin (Germany), 1939; m. Lydia Kamendrowsky, Apr. 24, 1946; children—Peter Lang, Michael, Irene. Came to U. S., 1950, naturalized, 1956. Sci. asst. Kaiser-Wilhelm Inst. Biology, Berlin, Tübingen, Germany, 1939-49; research asso. McGill U., Montreal, Que., Can., 1949; vis. prof. Tex. A. and M. Coll., 1950; research and sr. research fellow Cal. Inst. Tech., Pasadena, 1950-52, prof. biology, head Earhart-Campbell Plant Research Labs., 1959-65; faculty U. Cal. at Los Angeles, 1952-59, asso. prof., 1955-59; prof. botany Mich. State U., East Lansing, 1965—; dir. Mich. State U./AEC Plant Research Lab., 1965—. Lady Davis Found. fellow, 1949; NSF Sr. Postdoctoral fellow, 1958-59. Fellow Am. Assn. Advancement Sci.; member Nat. Acad. Sciences, Leopoldina Acad. Naturalists and Physicians, Am. Soc. Plant Physiologists (past v.p.), Bot. Soc. Am., Soc. Developmental Biology, Sigma Xi. Editor: Ency. of Plant Physiology, Vol XV Development and Differentiation, 1965. Contbr. numerous articles to tech. jours. Research on factors controlling flowing in plants, effects of environment on plant growth, action of plant hormones. Home: 1538 Cahill Dr., East Lansing, Mich. 48823.*

LANG, Arnold, Swiss zoologist, anatomist; b. Oftringen, Switzerland, June 18, 1855; ed. U. Geneva, U. Jena (Germany); degree, 1876. Privatdocent zoology, Bern, Switzerland, 1876-78; asst. zool. sta., Naples, Italy, 1878-85; named Ritter prof. phylogeny, Jena, 1886; became prof. zoology and comparative anatomy, Zurich, 1889; rector Zurich U., 1898-99. Author: Die Polycladen (Seeplanarien) des Golfes von

Neapel, 1884; Lehrbuch des vergleichen Anatomie der werbellosen Tiere, 4 vols., 1888-94; Handbuch der Morphologie des wirbellosen Tiere, 1912; Ueber den Einfluss der Festsitzenden. Research on morphology and genetics. Died Zürich, Nov. 30, 1914.

LANG, Erich Karl, physician; b. Vienna, Austria, Dec. 7, 1929; s. Johann H. and Cecilia (Felkel) L.; came to U. S., 1949, naturalized, 1960; M.S., Columbia, 1951; M.D., U. Vienna, 1953; m. Nicoli J. Miller, Apr. 21, 1956; 1 son, Erich Christopher. Resident in radiology Johns Hopkins Hosp., Balt., 1956-59, instr., 1959-61; radiologist Meth. Hosp., Indpls., 1961——. Asst. prof. radiology Vanderbilt U., Nashville, 1963. Fellow Billroth Med. Soc.; mem. Am. Coll. Radiology, Radiol. Soc. N.Am., A.M.A., Ind. Med. Soc., Ind. Roentgen Ray Soc., Johns Hopkins Med. and Chirurg. Soc., Soc. Nuclear Medicine, Am. Coll. Chest Physicians. Research in arteiographic demonstration of tumors of genitourinary tract. Home: 2045 W. 56th St., Indpls. 46208. Office: care Meth. Hosp., N. Capitol Av., Indpls. 46207.*

LANG, Karl Niklaus, Swiss physician, geologist; b. Lucern, Switzerland, Feb. 18, 1670; M.D., Rome, 1692; M.D., Lucern; practiced medicine, Lucern, from 1709. Mem. French Acad. Scis., Royal Prussion Soc. Scis. Author: Historia naturalis lapidum figuratorum Helvetiae (one of earliest Swiss mineral. and paleon. studies), 1708; Beschreibung des sehr schälichen Genuss der Korn-Zapffen in dem Brot, 1717. Founder Lucern Mus. Natural History. Died May 2, 1741.

LANG, Stewart M., Am. ceramist; b. Altoona, Pa., Mar. 27, 1921; s. Louis and Celia (Goldfisher-Zimet) L.; B.S. in Ceramics, Pa. State U., 1942; m. Edith Parish, May 9, 1948; children—Frederick D., Stephen A. Quality control and estimating engr. Asphalt Protected Steel Co., Washington, Pa., 1942; ceramic engr. Nat. Bur. Standards, Washington, 1946-57, asst. sect. chief, 1955-57; sr. ceramist Nuclear Materials and Equipment Corp., Apollo, Pa., 1957-59; sr. research scientist Owens-Ill. Tech. Center, Toledo, 1959——. Cons. to pvt. cos. Recipient U. S. Dept. Commerce Meritorious Service Silver medal, 1957. Fellow Am. Ceramic Soc. Research and publs. on reactions and phys. properties of pure and mixed oxide materials; design and devel. high temperature research instrumentation. Home: 2341 Cheltenham Rd., Toledo 43606. Office: 1700 N. Westwood Av., Toledo 43607.*

LANG, Tzu-Wang, Chinese cardiologist; b. Chekiang, China, Apr. 15, 1929; s. Wang-Chieh and Chun-Haiang (Chang) L.; M.D., Nat. Def. Med. Center, Taiwan, 1955; m. Winnie Weiken Chi, Apr. 15, 1959; 1 son, Teh-Yuan-Lang. Physician, Army 53d Hosp., Kinmen, Taiwan, 1956-67; asst. resident Army Gen. Hosp., Taipei, Taiwan, 1957-60, chief resident in cardiology, 1960-61, attending physician, chief cardiovascular diagnostic lab., 1961-63; instr. medicine Nat. Med. Center, 1958-63; research fellow Am. Coll. Cardiology Scholar's Program, Cedars-Sinai Med. Research Inst., Los Angeles, 1963-65; research fellow Los Angeles County Heart Assn., 1964-65; attending physician in cardiology, research fellow in charge cardiovascular research Kohlberg's Med. Research Lab., Vets. Gen. Hosp., Shih-Pai, Taipei, 1966——; chief cardiology Tri-Service Gen. Hosp., Taipei, 1967——. Recipient Gold Medal of Hero for Overcoming Difficulties in Med. Sci., 1962. Mem. Republic of China Soc. Cardiology (sec.-gen. 1963), Chinese Med. Assn., Internat. Soc. Cardiology, Am. Coll. Cardiology. directly upon myocardium. Address: Sect. Cardiology, Author: Birth Control, 1959; (with E. Corday) Cardiac Arrhythmias, 1966. Publs. on research in preparation of autotransplanted heart-lung and cerebral venous shunt for pharmacological assay of cardiovascular drugs; first group (with E. Corday) to define antiarrhythmic action of diphenylhydantoin (dilantin) directly upon mycardium. Address: Sect. Cardiology, Tri-Service Gen. Hosp., P.O. Box 7440, Taipei, Taiwan, Republic of China.*

LANGDELL, Robert Dana, Am. physician; b. Pomona, Cal., Mar. 14, 1924; s. Walter Irving and Florence (Reichenbach) L.; student Pomona Coll., 1941-43; M.D., George Washington U., 1948; m. Alice Evelyn Pritt, June 3, 1948; children—Robert Dana, Sara Ellen. Fellow dept. pathology U. N.C. Sch. Medicine, 1949-51, faculty, 1951-54, 56——, prof., 1964——. Sr. Research fellow USPHS, 1957-62. Recipient Research Career Devel. award USPHS, 1962-67. Mem. A.A.A.S., A.M.A., Am. Acad. Pathologists and Bacteriologists, Soc. for Exptl. Biology and Medicine, Am. Soc. Clin. Pathologists, Am. Soc. Exptl. Pathology, Am. Assn. Blood Banks, Coll. Am. Pathologists. Research, publs. on blood coagulation, especially hemophilia and hemophilia-like disorders; developed widely used test for partial thromboplastin time and assay for antihemophilic factor; determined plasma clearance of procoagulants and transfusion therapy of bleeding disorders. Home: 11 William Circle, Chapel Hill, N.C. 27514.*

LANGDON-BROWN, Sir Walter, Brit. physician; b. Bedford, Eng., Aug. 13, 1870; s. Rev. Dr. Brown (biographer of Bunyan); ed. St. John's Coll., Cambridge, Eng.; M.A., M.D., U. Cambridge; D.Sc. (hon.), Oxford (Eng.) U.; LL.D., Dalhousie U., Nat. U. Ireland; m. Eileen Presland; m. 2d, Freda Hurry. Became sr. physician Yeomanry Hosp., Pretoria, Transvaal, 1900; sr. med. officer Provident Mut. Life Assurance Assn.; cons. physician St. Bartholomew's Hosp., Met. Hosp.; prof. physic U. Cambridge; fellow Corpus Christi Coll. Fellow Royal Coll. Physicians (Croonian lectr., sr. censor, Harveian orator, Linacre lectr.), Royal Coll. Physicians in Ireland (hon.); pres. sect. medicine Brit. Med. Assn. Author: Physiological Principles in Treatment, 8th edit., 1943. Sympathetic Nervous System in Disease, 2d edit., 1921; The Endocrines in General Medicine, 1927; other med. works. Died Oct. 3, 1946.

LANGE, Arthur Leslie, Am. geophysicist; b. Plainfield, N.J., Mar. 25, 1926; s. Moritz and Margaret (Schutte) L.; B.S., Stanford, 1951; m. Judy M. Hayford, Mar. 19, 1966. Physicist, U. S. Office Naval Research, Stanford, Cal., 1950-52; speleologist Western Speleological Inst., Santa Barbara, 1952-56; geophysicist Newmont Exploration Ltd., Danbury, Conn., 1956-58; geophysicist Stanford Research Inst., Menlo Park, Cal., 1959——; speleologist, trustee Cave Research Assos., Castro Valley, Cal., 1958——. Mem. Soc. Exploration Geophysicists. Editor Caves and Karst, also Cave Studies, 1958——. Contbr. articles to sci. jours. Studies on cave origin; devel. of theory of changing geometry of caves; detection of caves by magnetics; earthquake aftershock investigations; storage and retrieval of geologic information. Home: Box 416. Office: Stanford Research Inst., Menlo Park, Cal. 94025.*

LANGE, Bernhard Karl Maria, German phytopathologist; b. Oldenburg, Germany, Dec. 21, 1910; s. Bernhard Johannes and Elisabeth (Hagen) L.; student U. Innsbruck (Austria), 1929-30, U. Köln (Germany), 1930-31, U. Halle/Saale (Germany), 1932-34; Dr.-philos. U. Graz (Austria), 1934; m. Maria Elisabeth Mousset, 1937; children—Monika (Mrs. Herbert Ippensen), Gudrun (Mrs. Eberhard Patrzek), Roswith (Mrs. Martin Kleiss). Asst., Plant Protection Service, Oldenburg, 1934-37; sec., dir. Plant Protection Service, WeserEms, Oldenburg, 1937-64, chief lab. approved pesticides, 1964——. Mem. research approving formulations on rodent control and rodent biology Biologische Bundesanstalt Braunschweig, 1954——; mem. Oberlandwirtschaftsrat, 1965——. Mem. Deutsche Phytomedizinische Gesellschaft, 1966. Author: (with Wilhelm Holz) Fortschritte in der chemischen Sihädlingsbekämpfung, 1962; (with Friedrich Zacher) Vorratsschutz gegen tierische Schädlinge, 1964; also numerous articles. Research on nematodes biology, insect control, resistance and biology, rodent control and biology, control of water voles and rats, pests in grain and milling products. Home: 98 Rostocker-Str. Office: 8 Ratsherr-Schulze-Str., 2900, Oldenburg, Germany.*

LANGE, Carl Georg, Danish psychologist, physician; b. Vordingbord, Denmark, 1834; student U. Copenhagen; studied under Moritz Schiff, U. Florence (Italy). Sent by Danish govt. to investigate gen. health and hygiene in Greenland, 1863; named lectr. path. anatomy U. Hosp., Copenhagen, 1877. Author: Contributions to the Knowledge of the Chronic Inflammation of the Spinal Cord, 1873; Ueber Germütsbewegungen, 1887; Contributions to the Physiology of Sensual Pleasure, 1899. Developed (with William James) James-Lange theory of emotion; earliest description of acute bulbar paralysis, 1868; 1st to interpret and explain posterior spinal sclerosis as a secondary degeneration; described pathogenesis of tabes dorsalis (locomotor ataxia). Died 1900.

LANGE, Friedrich Albert, German philosopher, sociologist, economist; b. Sept. 28, 1828; ed. Duisburg, Zurich, Switzerland, Bonn, Germany; instr., Bonn, 1855-58; pub. anti-Bismark newspapers; named instr., Zurich, 1869, prof., 1870; prof., Marburg, Germany, 1872-75. Author: Die Arbeitefrage, 1865; Die Grundlagen der mathematisch Psychologie, 1865; Logische Studien, 1877; History of Materialism, 1877. A Neo-Kantian; refuted materialism; supported working-class movement. Introduced Darwinistic sociology and philosophy of history to Germany. Died 1875.

LANGE, Johannes, German physician; b. Lowenberg, Germany, 1485; M.A., U. Leipzig (Germany), 1514; M.S., U. Pisa (Italy), 1522; student U. Ferrara (Italy), U. Bologna (Italy); physician to 4 electors in Heidelberg, Germany for 40 years after graduation. Author: Epistolae medicinales, 1544. 1st description of chlorosis, 1554. Died 1565.

LANGE, Kurt, physician; b. Berlin, Germany; Oct. 31, 1906; s. Georg and Pauline (Neumann) L.; student U. Heidelberg (Germany), 1924-25; M.D., Berlin Med. Sch., 1930; m. Helen Marcus, June 16, 1936; children—Peter, Monica. Came to U. S., 1939, naturalized, 1944. Instr., Berlin Med. Sch., 1931-33; faculty N.Y. Med. Coll., N.Y.C., 1940——, asso. prof. pediatrics, 1951——, prof. medicine, 1964——; staff Flower and Fifth Av., Met., Bird S. Coler, Chenango hosps., Recipient prize State of Prussia, 1929; Hectoen Gold medal for research, 1966. Fellow A.C.P., Am. Coll. Cardiology, A.M.A. (Bronze medal arigial research 1946); mem. Soc. for Exptl. Pathology, Soc. for Exptl. Biology and Medicine, Harvey Soc., Am. Fedn. for Clin. Research (Sr.), Am. Heart Assn., Alpha Omega Alpha. Research, numerous publs. on fluorescein method to trace circulation, pathology and therapy lesions due to cold climate, immunologic basis kidney disease, steroid therapy nephrosis, immunohistology kidney diseases, steroid therapy chronic glomerulonephritis. Home: 519 E. 86th St., N.Y.C. 10028. Office: 11 E. 68th St., N.Y.C. 10021.*

LANGE, Morten, Danish botanist; b. Odense, Denmark, Nov. 24, 1919; s. Jakob and Leila (Larsen) L.; M.Sc. in Botany, U. Copenhagen, 1945; postgrad., U. Mich., 1947; Dr.Phil., U. Copenhagen, 1952; m. Bodil Simonsen, Nov. 24, 1943; children—Jakob, Lene. With U. Copenhagen, 1949——, prof. botany, 1958——, vice curator Bot. Mus., 1955——, dir. Inst. Thallophyta, 1958——; mem. Parliament, 1960——, v.p., 1964——. Editor: Botanisk Tidsskrift, also Dansk Botanisk Arkiv, 1949——. Research on mycology, including distbn. of fungi in arctic areas, ecology of macromycetes, sexual behavior in Coprinus, Danish hypegean fungi. Home: 66 Ronnebaervej, Holte, Denmark. Office: 140 Gothersgade, Copenhagen, Denmark.*

LANGECKER, Hedwig, German pharmacologist; b. Schluckenau, Jan. 29, 1894; s. Leo and Rosa (Pickhart) L.; M.D., German U. Prague. Asso. prof. pharmacology, toxicology, pharma-codynamy, Prague, also Free U. Berlin; now prof. emeritus. Mem. German Soc. Pharmacology and Endocrinology, Med. Soc. Physiol. Chemistry. Research and publs. on pharmacology and pharmacodynamy. Address: Schöningstrasse 1/11, Berlin 65, W. Germany.

LANGEN, Eugen, German engr.; b. Cologne, Germany, 1833; ed. Polytechnic, Karlsruhe, Germany; founder (with Guilleaume) works for electric lighting, (with N. A. Otto) works to manufacture gas engines, Deutz, nr. Cologne, 1869. Aided N. A. Otto in prodn. of his gas engine; originated suspended ry. Died 1895.

LANGENBECK, Bernhard Rudolph Konrad von, German surgeon; b. Parding-büttel, Hanover, Germany, Nov. 8, 1810; s. Neffe von Konrad Johann Martin Langenbeck; M.D., Göttingen, Germany, 1834; postgrad., London, Paris. Became prof. surgery, dir. Friedrich Hosp., Kiel, Germany, 1842; served as army surgeon German-Danish War, 1848; dir. Clin. Inst. for Surgery and Ophthalmology, Berlin, 1848-82; tchr. of Billroth, Gurit. Mem. German Surg. Soc. (founder). Author: Chirurgische Beobachtungen aus dem Kriege, 1874. Founder, Archiv fur klinische Chirrugie. Discovered Candida albicans, 1839; devised 21 new operations, including reconstruction of nose (rhinoplasty), closure of cleft palate, methods of excising ankle, knee, hip, wrist, elbow, shoulder, lower jaw, plastic surgery of lip; reformed mil. surgery by use of bone resections rather than amputations when possible. Died Weisbaden, Germany, Sept. 29, 1887.

LANGENBECK, Conrad Johann Martin, German ophthalmologist; b. Horneburg, Germany, 1776; became private docent, Göttingen, Germany, 1802, asso. prof., 1804-14, prof. anatomy and surgery, 1814; named surgeon to Acad. Hosp., 1802; founder surgery and ophthalmology clinic, 1807; named surgeon-gen. Hanoverian Army, 1814; founder, anat. theater, 1828-29. Author: Treatise on Lithotomy, 1802; Bibliotek für die Chirurgie, 8 vols., 1802-28; Treatise on Diagnosis and Therapy of Surgical Diseases, 4 vols., 1822-50. Originated operation of iridencleisis for constrn. of artificial pupil by implanting a slip of iris in a corneal incision, 1811. Died 1851.

LANGENBERG, Donald Newton, Am. physicist; b. Devils Lake, N.D., Mar 17, 1932; s. Ernest George and Fern (Newton) L.; B.S., Ia. State U., 1953; M.S., U. Cal. at Los Angeles, 1955; Ph.D. (NSF fellow), U. Cal. at Berkeley, 1959; m. Patricia Ann Warrington, June 20, 1953; children—Karen Kaye, Julia Ann, John Newton, Amy Paris. Electronics engr. Hughes Research Labs., Culver City, Cal., 1953-55; acting instr. U. Cal. at Berkeley, 1958-59; faculty U. Pa., Phila., 1960——, prof., 1967——; maître de conférence associé École Normale Supérieure, Paris, France, 1966-67. NSF Postdoctoral fellow, 1959-60; Alfred P. Sloan Found. fellow, 1962-64; Guggenheim fellow, 1966-67. Fellow Am. Phys. Soc., Sigma Xi. Research and publs. on solid state physics including electronic band structure in metals and semiconductors, quantum phase coherence effects in superconductors. Home: 118 Penarth Rd., Bala-Cynwyd, Pa. 19004. Office: Dept. Physics, U. Pa., Phila. 19104.*

LANGENSTEIN, Heinrich von, see von Langenstein, Heinrich.

LANGER, James Stephen, Am. physicist; b. Pitts., Sept. 21, 1934; s. Bernard F. and Liviette (Roth) L.; B.S., Carnegie Inst. Tech., 1955; Ph.D., U. Birmingham (Eng.), 1958; m. Elinor G. Aaron, Dec. 21, 1958; children—Ruth, Stephen, David. Faculty, Carnegie Inst. Tech., Pitts., 1958——, prof., 1967——. NSF Postdoctoral fellow U. Cal., San Diego, 1961-62; vis. asso. prof. Cornell U., 1966——. Mem. Am. Phys. Soc., A.A.-A.S. Research in theory of phase transitions, condensation, superconductivity, properties of disordered systems. Home: 2488 Mt. Royal Rd., Pitts. 15217.*

LANGER, Karl Ritter, Austrian anatomist; b. Vienna, Apr. 15, 1819; prof. zoology, Budapest; prof. anatomy, Vienna. Discovered and mapped lines of normal elasticity in skin (Langer's lines, Langer's fissure lines), 1861, lymphatics in periosteum. Died Vienna, Dec. 7, 1887.

LANGER, Lawrence M(arvin), Am. physicist; b. N.Y.C., Dec. 22, 1913; s. Alexander H. and Anna (Brown) L.; B.S., N.Y. U., 1934, M.S., 1935, Ph.D., 1938; m. Beatrice Fisher, July 5, 1936. Instr., Ind. U., Bloomington, 1938-41, asst. prof., 1941-47, asso. prof. 1947-52, prof., 1952—, acting chmn. physics dept., 1961-62, 65, chairman of the physics department, 1966—. Sci. cons. on delivery of bomb to Hiroshima, War Dept., 1945; expert cons. AEC, 1948-61; cons., observer Greenhouse atom bomb tests, Marshall Islands, 1951; adv. cons. nuclear data project Nat. Acad. Sci.-NRC, 1957-60; dir. project on nuclear spectroscopy Office Naval Research, 1963—. Fellow Am. Phys. Soc.; mem. Sigma Xi, Sigma Pi Sigma. Research and numerous publs. on nuclear physics, counting equipment, microwave radar, underwater sound, ballistics, nuclear weapons, discovered 1st unique, spectrum shape; contbr. to design, devel., delivery 1st A-bomb. Home: 1342 Southdowns St., Bloomington, Ind. 47403.*

LANGER, Rudolph Ernest, Am. mathematician; b. Boston, Mar. 8, 1894; s. Charles R. and Johanna (Rockenbach) L.; B.S., Harvard, 1917, M.A., 1920, Ph.D., 1922; m. Louise A. Stemler, Oct. 12, 1918; Faculty, Dartmouth, 1923-26; asst. prof. Brown U., 1926-27; prof. U. Wis., Madison, 1927—; vis. prof. Harvard, 1936, Stanford, 1936, Ohio State U., 1931, U. Tex., 1946. Dir., Math. Research Center, U. S. Army, Madison, Wis., 1956-64. Recipient medal and citation for outstanding civilian service U. S. Army, 1964. Contbr. articles to math. research jours. Study of asymptotic solutions of ordinary differential equations; boundary problems in ordinary differential equations; Stoke's theorem. Home: 822 Miami Pass, Madison, Wis. 53711.*

LANGERHANS, Paul, German pathologist; b. Berlin, July 25, 1847; prof. Freiburg, Germany. Author: Beitrag zur mikroskopischen Anatomie der Bauchspeicheldrüse, 1869. Discovered irregular islands of cells in the pancreas which produce insulin (islets of pancreas, also islets or islands of Langerhans), 1869. Research on path. anatomy. Died Funchal/Madeira, July 20, 1888.

LANGERMAN, Johann, German physician; b. 1768; head med. affairs, Prussia; a founder German sch. psychology; responsible for reforms in care of psychotics. Died 1832.

LANGEVIN, Paul, French physicist; b. Paris, Jan. 23, 1872; studied under J. J. Thompson, Cambridge, Eng., in 1890's; Ph.D. under Pierre Curie, Sorbonne, Paris, 1902; tchr. physics Collège de France, 1904-09; became prof. gen. and exptl. physics Sorbonne, 1909; dir. École de physique et de chimie industrielles. Fellow Royal Soc., 1928; mem. French Acad. Scis., French Acad. Marine. Popularized Einstein's theories for France; his research on ultrasonic sound forms basis of modern sonar sci.; studies on molecular structure of gases, especially Brownian movement, secondary X-rays, relativity, magnetic theory; developed theory of magnetism which explained some of Curie's work and accounted for paramagnetism and diamagnetism of gases, 1905. Died Paris, Dec. 19, 1946.

LANGFELD, Herbert Sidney, Am. psychologist; b. Phila., July 24, 1879; s. Charles and Flora R. L.; A.B., Central High Sch., Phila., 1897; student Haverford Coll., 1897-98; Ph.D., Berlin, 1909; Dr. de l'Univ., U. Montreal, 1954; m. Florence Hoffman Purdy, Oct. 6, 1904; m. 2d, Mary Brita Bergland, June 11, 1932, Sec. naval attaché Am. Embassy, Berlin, 1902-03; research fellow. Harvard, 1909-10, instr., 1910-15, asst. prof., 1915-22, asso. prof., 1922-24, acting dir. Psychol. Lab., 1917-19, dir., 1919-22; prof. and dir. psychol. lab., Princeton, 1924-47; Stuart prof. of psychology, 1937-47, emeritus, 1947—. Fellow A.A.A.S. (v.p. 1931); member Am. Psychol. Assn. (sec. 1917-19, pres. 1930. hon. pres. New York branch, 1935-36. Research sec. Y.M.C.A., in France, 1918. Sec. gen. Internat. Congress Psychology, 1945-50, Internat. Union Scientific Psychology, 1950-54; v.p. sect. exptl. psychology Internat. Union Biology, 1955—. Fellow N.Y. Acad. Scis.; chairman committee on international relations division anthropology and psychology, Nat. Research Council. Mem. Pontifical Acad. Sciences, Phi Beta Kappa, Author: On the Psychophysiology of a Prolonged Fast, 1914; The Aesthetic Attitude, 1920. Joint author: An Elementary Laboratory Course in Psychology, 1916. Joint author and editor: Problems of Personalities (author and editor); Psychology, A Factural Textbook, 1935; A Manual of Psychological Experiments, 1937; Introduction to Psychology, 1939. Joint editor: Psychology for the Fighting Man, 1943. Collaborator: Psychology for the Armed Services, 1945. Co-editor: Foundations of Psychology, 1948; History of Psychology in Autobiography, 1952. Editor Psychol. Monographs, 1931-34; editor Psychol. Rev., 1934-47. Died Feb. 25, 1958.

LANGFELDT, Gabriel, Norwegian psychologist; b. Kristiansand, Dec. 23, 1859; s. Gerhard and Amalie N. Langfeldt; M.D., U. Oslo (Norway); m. Else Nilssen, Aug. 2, 1932; children—Else Marie, Jan Gabriel. First asst. Bergen mental hosp.; police physician, Bergen; 1st asst. then chief physician U. Vinderen Oslo Psychiat. Clinic; prof. U. Oslo, 1940—. Mem. Royal Med. Assn. London, Norwegian Acad. Scis., Am., Los Angeles assns. psychology. Author: The Endocrine Glands and Autonomic Systems in Dementia Praecox, 1926; The Schizophreniform States, 1937; The Jealousy Syndrome, 1956; A Study of the Philosophy and Life of Albert Schweitzer, 1961. Research in clin. and forensic psychiatry. Home: Björnveien 34, Oslo. Office: Univ. Psychiat. Clinic, Vinderen Oslo, Norway.

LANGFORD, Herbert Gaines, Am. physician; b. Columbia, S.C., Apr. 22, 1922; s. Herbert and Bessie Martin (Gaines) L.; B.S., U. S.C., 1942; M.D., Med. Coll. Va., 1945; m. Martha Allen Johns, Feb. 18, 1952; children—William Herbert Gaines, Francis Page Johns, Robert Nelson, Ellen Rosewell. Faculty, Med. Coll. Va., 1952-55, also practice medicine specializing in cardiology, Richmond, Va., 1952-55; faculty U. Miss. Med. Center, Jackson, 1955—, prof. medicine, 1964—, asso. prof. physiology, 1963—, chief, endocrinology and hypertension div., 1955—. Guest investigator dept. Regius prof. medicine U. Oxford (Eng.), 1962. Mem. Am. Fedn. Clin. Research, Endocrine Soc., So. Soc. Clin. Research (counselor 1962-65), Am. Diabetes Assn., Am. Physiol. Soc., Am. Heart Assn., Soc. Exptl. Biology and Medicine, Council High Blood Pressure Research (med. adv. bd.), Am. Soc. Clin. Investigation, Alpha Omega Alpha. Contbr. to Pheochromocytoma, 1953, also articles. Research in blood pressure control, hypertension, and endocrinology. Home: 1746 Brecon Dr., Jackson, Miss. 39211.*

LANGFORD, Raymond Robert, Canadian limnologist; b. Warkworth, Ont., Can., Sept. 22, 1908; s. Robert Albert and Minnie (Curtis) L.; B.Sc. with honours in Biology, U. Sask., 1932; Ph.D. in Zoology, U. Toronto, 1936; m. Greta May Strangeways, July 3, 1937; children—Elizabeth May (Mrs. Arthur McClellan), Robert Wesley. Faculty dept. zoology U. Toronto (Ont., Can.), 1932—, prof., 1964—, asso. chmn. dept. zoology, 1964, acting chmn., 1965, dir. Fisheries Lab., 1945-55. Cons., Dept. Lands and Forest Research Div., 1945-66. Mem. Internat. Assn. Limnology, Am. Soc. Limnology and Oceanography, Canadian Soc. Zoologists. Contbr. articles to sci. jours. Research on productivity of fresh waters, nutrient chems., prodn. of algae as plants fed upon by microscopic animals, in turn fed upon by fish, zooplankton populations and their vertical, horizontal and seasonal variations.

LANGHAAR, Henry Louis, Am. mech. engr.; b. Bristol., Conn., Oct. 14, 1909; s. Louis and Elizabeth (Williamson) L.; B.S. in Mech. Engring., Lehigh U., 1931, M.S., 1933, Ph.D., 1940; m. Isabelle Babcock, May 25, 1937; children—Bonnie, Sally L. (Mrs. Charles Stewart Havens), Kayri. Test engr. Ingersoll-Rand Corp., Phillipsburg, N.J., 1933-36; seismographer Carter Oil Co., Tulsa, 1936-37; math. instr. Purdue U., Lafayette, Ind., 1940-41; structures engr. Consol-Vultee Aircraft Corp., San Diego, 1941-47; prof. theoretical and applied mechanics U. Ill., Urbana, 1947—; prof. Indian Inst. Tech., Kharagpur, 1960-62. Fellow Am. Soc. M.E.; mem. A.A.A.S., Sigma Xi. Author: Dimensional Analysis and Theory of Models, 1951; (with A. P. Boresi) Engineering Mechanics, 1959; Energy Methods in Applied Mechanics, 1962; also articles. Research on theory dimensional analysis, viscous flow in region following entrance to a tube, stress analysis of aircraft structures, theory of elasticity, theories of plates and shells, analysis buckling. Home: 803 S. Anderson St., Urbana, Ill. 61801.*

LANGHANS, Theodor, German anatomist, pathologist; b. Usingen, Naussau, Germany, 1839; ed. at Göttingen, Berlin, Würzburg (all Germany); worked in Switzerland; described the cytotrophoblast, the inner cellular layer of trophoblast (Langhans' layer; the cells of the layer are known as Langhans' cells), 1870; described the giant cells found in tubercles of Tb, 1867. Died 1915.

LANGLEY, John Newport, English physiologist; b. Newbury, Eng., Nov. 2, 1852; s. John and Mary (Groom) L.; Sc.D., St. John's Coll., Cambridge; LL.D., St. Andrews; ScD. (hon.), Dublin; M.D. (hon.), Groningen; hon. Dr., Strasbourg; fellow Trinity Coll., Cambridge; m. Vera K. Forsythe-Grant, 1902; 1 dau. Lectr., Trinity Coll., Cambridge U., univ. lectr., 1884-1903; became prof. physiology, 1903. Became fellow Royal Soc., 1883, v.p., 1904-05, recipient Royal medal, 1892. Recipient Baly medal Royal Coll. Physicians, 1903, Andreas Retzins medal Swedish Soc. Physic, 1912. Mem. Internat. Congress Physiology, Neurol. Soc. Gt. Britain (pres. 1893), Brit. Assn. (pres. physiol. sect. 1899), Soc. Belge de Biol., Soc. Royal de Sci. Med. et Nat. de Bruxelles, Portuguese Soc. Natural Sci.; corr. mem. Société de Biologie Paris, Phys. Med. Soc. Florence, Imperial Mil. Acad. Medicine Petrograd; hon. mem. Soc. Neurol. Kasan, Swedish Soc. Physicians, Soc. de Biol. Buenos Aires, Am. Physiol. Soc.; fgn. mem. Acad. Roy. de Méd. de Belgique, Kong. Danske Vidensk. Selskab. Copenhagen, R. Accad. d. Lincei Rome, Svenska Vetensk. Akad., Stockholm, Vetensk. Soc. Upsala, Bavarian Acad. Sci. Munich. Author: Autonomic Nervous System, 1921; On Reflex Action from Sympathetic Ganglia, 1894. Editor, Jour. Physiology, 1890-1906, Research on sympathetic nervous system; demonstrated secretory granules accumulate when gland is not secreting and are discharged when secretion begins; observations on action of philocarpine on heart, 1874; introduced term autonomic nervous system, 1898, parasympathetic nervous system, sympathetic nervous system, 1905; disproved various conceptions concerning involuntary nervous system.

LANGLEY, Samuel Pierpont, Am. astronomer, physicist; b. Roxbury, Boston, Aug. 22, 1834; grad. Boston High School; D.C.L., Oxford; D.Sc., Cambridge, Eng.; LL.D. Harvard, Princeton, Yale, U. Wis.; U. Mich.; Ph.D., Stevens Inst. Tech. Practice architecture and civ. engring.; asst. Harvard Observatory, 1865; later asst. prof. mathematics U. S. Naval Acad.; dir. Allegheny Observatory, 1867, where, 1869, he founded the system of railway time service from observatories, which has since become general, and where he devised the bolometer (for measuring distbn. of heat in spectrum of sun, determined transparency of atmosphere to different solar rays and measure of increase of their intensity at high altitudes), and other apparatus. Organized expdn. to Mt. Whitney, 1881, where he reestablished the solar constant and discovered an entirely unsuspected extension of the invisible solar spectrum. Has carried out extended expts. on the problem of mech. flight. Established Astrophysical Observatory and the Nat. Zool. Park, Washington; sec. Smithsonian Instn., Washington, from 1887. Mem. Nat. Acad. of Sciences, Royal Soc. London (fgn.), Royal Soc. Edinburgh, A.A.A.S. (pres. 1887). Has been awarded Janssen medal, Inst. of France; Rumford medal of the Royal Soc. of London, and of Am. Acad. Arts and Sciences; Henry Draper medal, Nat. Acad. Science, etc., and numerous others. Author: The New Astronomy, 1888; Researches on Solar Heat; Experiments in Aerodynamics, 1891; Internal Work of the Wind, 1893; On the Possible Variation of the Solar Radiation, 1905. Trustee Carnegie Instn. First to give clear explanation, substantiated by expt., of way in which birds can soar without appreciable motion of their wings; carefully wrought aerodynamic principles showing how air would support thin wings of particular shapes (work theoretically correct, but used structural materials of insufficient strength, so planes were not able to make successful flights); important drawing of gt. sunspot of Dec. 1873; made 1st flights of mechanically propelled heavier-than-air machines in world; designed a motor-driven model flying-machine which made a flight of 3000 ft. Died Aiken, S.C., Feb. 27, 1906.

LANGLO, Kaare, meteorologist; b. Bergen, Norway, Oct. 7, 1913; s. Einar and Borhild (Langlo) Olsen; Honours degree in maths. and natural scis., Oslo (Norway) U., 1941, Dr.Philos., 1953; m. Else Lökke, June 4, 1943; children—Erik, Anne Lise. Sci. asst. No. Light Obs., Tromsö, Norway, 1939-43; meteorologist later chief div., Norwegian Meteorol. Inst., Oslo, 1943-52; chief administrv. div. World Meteorol. Orgn., Geneva, Switzerland, 1952-53, chief tech. div., 1953-67, dir., 1968—. World Meteorol. Orgn. rep. on Sci. Com. on Antarctic Research, 1962—, Sci. Com. on Oceanic Research, 1962—, Internat. Ozone Commn., 1963—, Internat. Commn. on Polar Meteorology, 1964—. Mem. Norwegian Geophys. Soc. Research and publs. on ann. and day-to-day variations in atmospheric ozone; 1st radiowave reflection study of ionosphere at high latitude from 2 stas. simultaneously synoptic studies of relations between atmospheric ozone and frontal cyclones. Home: 10 Château Banquet, Geneva. Office: World Meteorol. Orgn., Geneva, Switzerland.*

LANGLOIS, Jean-Paul, French physiologist; b. Paris, 1862; aggregation, 1898; prof. Conservatory of Arts and Crafts; dir. Revue générale des Sciences, 1910-23. Research on physiol. prodn. of heat by animal bodies, pulmonary circulation, physiology of suprarenal glands, hygiene. Died 1923.

LANGLOIS, Thomas Huxley, Am. ecologist; b. Detroit, Feb. 19, 1898; s. Arsene Ferdine and Tiney (Mosher) L.; B.S., U. Mich., 1924, M.S., 1925; Ph.D., Ohio State U., 1935; m. Marina Louise Holmes, Apr. 19, 1924 (dec. 1963); children—Elmire Marina (Mrs. Howard Conklin, Jr.), Esther Ferdina (Mrs. George Daniel Stewart), Florence Dhilve, Sarah Zoe (Mrs. Edward Bradford Titchener), Caroline Lizette (Mrs. Allan Wilson Smith); m. 2d, Bertha Witmer Elliott, June 20, 1964. Asst. fisheries biologist Mich. Dept. Conservation, 1924-29; chief fish mgmt. and propagation Ohio Div. Conservation, 1930-46; asst. dir. Franz Theodore Stone Lab., Ohio State U., 1936-37, dir., 1938-55, prof. zoology, 1938-55, research prof., 1956-64, emeritus research prof., 1964—. Rep., Natural Resources Council Am., 1940—. Fellow Ohio Acad. Sci.; mem. Am. Fisheries Soc. (past pres.), Am. Soc. Limnology and Oceanography, Toledo Zool. Soc. (hon. trustee), Am. Assn. U. Profs., Am. Soc. Zoologists, Sigma Xi. Author: The Western End of Lake Erie and its Ecology, 1954; (with Marina H. Langlois) South Bass Islands and Islander, 1948; also numerous articles. Developed methods controlling cannibalism of bass in rearing ponds; research on relationships of people on South Bass Island on Lake Erie to their natural resources, structures of icesheets of Lake Erie and their relation to methods of formation; discovered ecol. changes taking place in Lake Erie, predicted present conditions and proposed method of limiting pollution; planned fish mgmt. of Portage River, O.; studied bird migration via islands of Lake Erie. Home: 63 W. Beaumont Rd., Columbus, O. 43214.*

LANGMUIR, Alexander Duncan, Am. epidemologist; b. Santa Monica, Cal., Sept. 12, 1910; s. Charles Herbert and Edith (Ruggles) L.; A.B., Harvard, 1931;

M.D., Cornell U., 1935; M.P.H., Johns Hopkins, 1940; m. Sarah Ann Harper, June 1, 1940; children—Ann Ruggles (Mrs. William Randall Fowler, Jr.), Paul Harper, Susan Davis, Lynn Adams, Jane Adams (dec.). Med. cons. in pneumonia control N.Y. State Dept. Health, 1937-39, asst. dist. state health officer, 1940-41; dep. commr. Westchester County (N.Y.) Health Dept., 1941-42; epidemiologist Commn. on Acute Respiratory Disease, U. S. Army, 1942-46; asso. prof. epidemiology Johns Hopkins, 1946-49; chief epidemiology br. Communicable Disease Center, USPHS, Atlanta, 1949—. Recipient Distinguished Service medal Dept. Health, Edn., Welfare, 1958; Bronfman prize for pub. health achievement, 1965. Research and publs. in pneumonia and respiratory diseases, influenza, airborne infection, poliomyelitis, biol. warfare def.; organized Epidemic Intelligence Service Program for surveillance of communicable diseases of nat. importance. Home: 783 Houston Mill Rd., Atlanta 30329. Office: 1600 Clifton Rd., Atlanta 30333.*

LANGMUIR, Arthur Comings, Am. chemist; b. Evanston, Ill., Feb. 7, 1872; s. Charles and Sadie (Comings) L.; Ph.B., Columbia, 1893; Ph.D., U. Heidelberg (Germany) 1895; m. Alice Dean, 1896; children—John, Ruth (Mrs. Van de Water). Asst. to Prof. C. F. Chandler, N.Y., 1895-96; head chemist Ricketts and Banks, N.Y., 1896-1900; instr. Pratt Inst., 1898-1900; became chemist, supt. Marx and Rawolle, Bklyn., 1900-20; research and cons. chemist, 1920-41; chemist U. S. Shellac Importers Assn., London, 1924, Calcutta, 1925; named chmn. Am. Chem. Soc. Commn., London, 1910. Fellow A.A.A.S.; mem. Am. Chem. Soc. (became chmn. N.Y. sect. 1912), Soc. Chem. Industries (v.p. 1917-19), Am. Inst. Chem. Engrs. Research and numerous articles on shellac and glycerine; founded internat. methods for shellac analysis and glycerine analysis. Died May 14, 1941.

LANGMUIR, Irving, Am. physicist, chemist; b. Brooklyn, N.Y., Jan. 31, 1881; s. Charles and Sadie (Comings) L.; Met. E., Columbia Sch. of Mines, 1903; Ph.D., U. of Göttingen, 1906; D.Sc., Northwestern, 1921, Union U., 1923, Columbia, 1925, Kenyon Coll., 1927, Princeton, 1929, Lehigh U., 1934, Harvard, 1938, Oxford, 1938, Rutgers, 1941, Queen's Coll. (Canada), 1941; D.Ing., Tech. Hochschule, Berlin, 1929; LL.D., Edinburgh, 1921, Johns Hopkins, 1936, U. of Calif., 1946; m. Marion Mersereau, Apr. 27, 1912; children—Kenneth, Barbara. Instr. chemistry, Stevens Inst., Hoboken, N.J., 1906-09; physical chem. research, Research Lab. of Gen. Electric Co., Schenectady, N.Y., since 1909, now consultant of Research Lab.; engaged in development of gas filled tungsten lamps, electron discharge apparatus, condensation high vacuum pump, atomic hydrogen welding, work on monomolecular films and surface chemistry, cloud physics, including weather modification; (with Lewis) evolved an atomic theory; developed vacuum tube for fluorescent images of sufficient intensity to be photographed, 1940; applied monomolecular layer method to measuring of molecular sizes of viruses and toxins, 1941; numerous patents radio engring.; study of atomic structure shed light on meaning of isotopes, etc.; also in 1917-18, on devices for submarine detection at Naval Exptl. Sta., Nahant, Mass. Lecturer, London, 1938, Hitchcock Foundation lecturer, U. of Calif., 1946; mem. bd. trustees, State U. of N.Y., Sept. 1948-50. Fellow Royal Soc., 1935, A.A.A.S. (president 1941), Am. Physical Soc., Indian Acad. Sci. (hon.); mem. Am. Philosophical Soc., Am. Chemists Society (pres. 1929), Nat. Acad. Sciences, Am. Acad. Arts and Sciences, Royal Soc. Upsala, corresponding mem. Académie des Sciences, Paris, 1951, Tau Beta Pi, Phi Lambda Upsilon, Sigma Xi; also hon. mem. Royal Instn., Chem. Soc. of London, Royal Physiog. Soc. (Lund), Academia Brasileria de Sciencias (Brazil), Société de Chimie Industrielle. Was awarded Nichols medal by N.Y. sect. of American Chem. Soc. for researches on chem. reactions at low pressures, 1915; Hughes medal, Royal Soc. London, for researches in molecular physics, 1918; Nichols medal for researches on atomic structure, 1920; Rumford medal, Am. Acad. Arts and Sciences, for researches on thermionic phenomena, 1920; Cannizzaro prize, Royal Acad. of Lincei (Rome), 1925; Perkin medal, 1928; Sch. of Mines medal, Columbia U. in recognition of achievements in science and invention, 1929. Chandler medal, 1930; Willard Gibbs medal, 1930; Popular Science Monthly prize, 1932; Nobel prize in Chemistry, 1932, for work in surface-chemistry; Franklin medal by the Franklin Institute, 1934; Holly medal, by American Society of Mech. Engrs., 1934; award under John Scott Medal Fund, 1937, by Board of City Trusts, Phila., Egleston medal Columbia Engring. Schs. Alumni Assn., 1939; Nat. Pioneer award Nat. Assn. Mfrs., 1940; Faraday Medal award Elec. Engrs. of Gt. Britain, 1943; Medal for Merit, U. S. Army and Navy, 1948; John Carty Medal Nat. Acad. Scis., 1950. Contbr. to jours. Died Aug. 17, 1957.

LANGMUIR, Robert Vose, Am. elec. engr.; b. White Plains, N.Y., Dec. 20, 1912; s. Dean and Ethel (Irimey) L.; A.B., Harvard, 1935; Ph.D. in Physics, Cal. Inst. Tech., 1943; m. Bertha Lee Lord, May 19, 1939; children—Alan, Ann (Mrs. John C. Urey). Physicist, Consol. Engring. Corp., Pasadena, Cal., 1939-42, Gen. Electric Co., Schenectady, 1942-48; with Cal. Inst. Tech., Pasadena, 1948—, prof. elec. engring, 1952—. Cons. Space Tech. labs., Redondo Beach, Cal. 1958—. Fellow Am. Phys. Soc., mem. I.E.E.E. Participant in construction of electron synchrotrons, 1946-60, including constrn. 1st synchrotron in U. S.; devel. mass spectrometer. Author: Electromagnetic Fields and Waves, 1960. Home: 1855 Homewood Dr., Altadena, Cal. 91001. Office: 1201 E. Cal. St., Pasadena, Cal. 91109.*

LANGNER, Siedfried Wolfgang, German sylviculturist; b. Chemnitz, Sept. 26, 1906; s. Friedrich and Anna (Sacher) L.; ed. Dresden, Germany; doctorate in forestry engring.; m. Margarete Bieler, July 23, 1939; children—Wolf, Kurt, Arthur. A founder, dir. Inst. Forest Genetics; also prof. Author: Kreuzungsversuche mit Larix euoopea und L. Leptolepsis; Beiträge zur Lösung des Problems der Befruchtungsverhältnisse im Wald mittels einer Mendelpaltung; Improvement Through Individual Tree Selection and Testing Seed Stands and Clonal Seed Orchards; Einige Lärche, Larix Mill. Handbuch der Pflanzenzuchtung. Founder, coprop. Silvae Genetica, other internat. mags. Research on inbreeding, inter- and intraspecific hybridization. Home: 207 Grosshansdorf (Holstein), Lurup 2. Office: Instutut für Forstgenetik, Sieker Landstrasse 2, 207 Schmalenbeck und Ahrensburg (Holstein), W. Germany.

LANGSDORF, Alexander, Jr., Am. physicist; b. St. Louis, May 30, 1912; s. Alexander S. and Elsie (Hirsch) L.; B.A., Washington U., St. Louis, 1932; Ph.D., Mass. Inst. Tech., 1937; m. Martyl S. Schweig, Dec. 31, 1941; children—Suzanne M., Alexandra. NRC fellow U. Cal., at Berkeley, 1938; instr. Washington U., 1939-43; physicist Argonne (Ill.) Nat. Lab., 1943—. Bd. dirs. Ednl. Found. Nuclear Sci. Recipient Alumni award Washington U., 1958. Fellow Am. Phys. Soc., Am. Nuclear Soc. Inventor diffusion cloud chamber; isolated several fission products, made cyclotron arc improvements; helped build first heavy water reator (pile), measured neutron cross sects., built first pile oscillator for neutron studies; electrostatic generator constn. and use in fast neutron physics, fast neutron cross-sects., scattering, and polarization measurements, theory of design of such expts, fast neutron collimation and shielding. Home: Route 1, Box 228, Roselle, Ill. 60172. Office: 9700 S. Cass Av., Argonne, Ill. 60440.*

LANGSDORFF, Georg Heinrich von, German physician; b. Wöllstein, Germany; med. doctorate, Göttingen, Germany, 1797; physician to Prince Christian of Waldeck and accompanied to him to Portugal where he served with the prince; later served with English troops in Portugal; joined as personal physician Russian Count Rezánov, 1801, accompanied him on trips from Copenhagen, Denmark, 1803, to Sitka, Alaska, 1805-06, to Cal., 1806; later went to Brazil; then returned to Germany. Mem. Imperial Acad. Scis. St. Petersburg (corr.). Observed and collected plant and animal life on voyage to Cal. some of which are in St. Petersburg Mus. Died 1852.

LANGSLEY, Donald Gene, Am. psychiatrist; b. Topeka, Oct. 5, 1925; s. Morris J. and Ruth (Pressman) L.; B.A., State U. N.Y., 1949; M.D., U. Rochester, 1953; m. Pauline Doris Royal, Sept. 9, 1955; children—Karen Jean, Dorothy Ruth, Susan Louise. USPHS Career tchr. psychiatry U. Cal. Sch. Medicine, San Francisco, 1959-61; with U. Colo. Sch. Medicine, Denver, 1961-68, asso. prof. psychiatry, 1965-68; prof., chmn. psychiatry U. Cal., Davis, 1968; dir. inpatient service Colo. Psychopathic Hosp., Denver, 1961-68, dir. Family Treatment Unit, 1963-68; attending psychiatry VA Hosp., Denver, 1962-68; cons. psychiatry Colo. State Hosp., Pueblo, 1962-68; mem. Colo. State Mental Health Planning Com., 1963-68; cons. Dept. Def. Indsl. Security Program, 1966—: cons. Gov.'s Commns. on Mental Health and Criminally Insane Facilities, 1962-66. Diplomate Am. Bd. Psychiatry and Neurology; fellow Am. Psychiat. Assn.; mem. A.M.A., Colo. Psychiat. Soc., Assn. Advancement of Psychotherapy. Contbr. articles, abstracts, revs. to profl. jours. Research, publs. in area of family psychology, family therapy, treatment of family crises, treatment methods in psychiatry, techniques for teaching; measurement of effects of teaching psychiatry to med. students. Office: U. Cal. Sch. Medicine, Davis, Cal. 95616.*

LANGSTON, Hiram Thomas, surgeon; born Rio de Janeiro, Brazil, Jan. 12, 1912 (derivative citizenship); s. Alva B. and Louise F. (Diuguid) L.; student Collegio Batista, Rio de Janeiro, 1926-28, Georgetown (Ky.) Coll., 1929; A.B., U. Louisville, 1930, M.D., 1934; M.S. in Surgery, U. Mich., 1941; m. Helen M. Orth, June 22, 1941; children—Paula F., Thomas O., Carol E. Asst. resident surgery U. Hosp., Ann Arbor, Mich., 1937-38, resident surgery, 1938-39, resident thoracic surgery, 1939-40, instr. thoracic surgery, 1940-41; faculty Northwestern U., 1941-42, 46-48, Wayne U., 1948-52; asso. prof. surgery U. Ill. Coll. Medicine, Chgo., 1952-62, prof. surgery, 1963—; chief surgeon Chgo. Tb Sanitarium, 1952—; cons. thoracic surgery VA hosp., Hines, Ill., 1952—; staff St. Joseph's, Grand Henrotin, Augustana, hosps., Chgo.; cons. 5th Army, VA, Washington; adv. council Thoracic Surgery Am. Coll. Surgeons. Diplomate Am. Bd. Surgery, Bd. Thoracic Surgery (a founder). Fellow A.C.S.; mem. A.M.A., Chgo. Tb Soc., Am. Assn. Thoracic Surgery (sec. 1956-61), Am. Thoracic Soc., Nat. Tb Assn., Am. Coll. Chest Physicians, Am., Pan-Pacific, Central, Western surg. assns., Theta Kappa Psi, Alpha Omega Alpha. Mem. editorial bd. John Alexander Series, 1956—, Jour. Cardiovascular Surgery, 1960—. Research and publs. on surgery on pulmonary cancer and Tb.; the pathology of hemothorax; wound healing-sensitivity to catgut; autologous blood transfusion. Home: 952 Pine Tree Lane, Winnetka, Ill. 60093. Office: 55 E. Washington St., Chgo. 60602.*

LANGSTON, James Horace, Am. chemist; b. Garrison, Tex., Oct. 8, 1917; s. James Horace and Ruth (Green) L.; B.A., Stephen F. Austin State Coll., 1937; M.A., U. N.C., 1939, Ph.D., 1941. Research chemist Columbia Chem. div. Pitts. Plate Glass Co., Barberton, O., 1941-46; faculty Clemson (S.C.) Coll., 1946-58, prof. textile chemistry, 1951-58; prof. chemistry, head dept. Samford U., (formerly Howard Coll.), Birmingham, Ala., 1958—; Fulbright prof. chemistry Central U. and Nat. Poly. Sch., Quito, Ecuador, 1959-60. Chem. cons. to pvt. cos., 1950-58. Fellow Am. Inst. Chemists; mem. Am. Chem. Soc., Sigma Xi. Contbr. articles to tech. jours. Discovered numerous new organic chem. compounds; research on polymerization, catalysis, pharmacological effects, textile finishing. Home: 1008 Vista Circle, Birmingham, Ala. 35216.*

LANKESTER, Sir Edwin Ray, Brit. zoologist; b. London, Eng., May 15, 1847; s. Edwin and Phebe (Pope) L.; student Downing Coll., Cambridge, 1864-66, Christ Ch., Oxford, 1866-68; M.A., LL.D., D.Sc., Oxford; hon. D.Sc., Leeds, Eng.; hon. fellow Exeter Coll. Radcliffe Travelling fellow, 1870; became fellow, lectr. Exeter Coll., 1872; prof. zoology and comparative anatomy U. Coll., London, 1874-90; Regius prof. natural history Edinburgh, Scotland, 1882; Linacre prof. comparative anatomy Oxford, 1891-98; dir. natural history depts. Brit. Mus., 1898-1907; emeritus prof. zoology and comparative anatomy U. London. Recipient Royal medal Royal Soc., 1885; Copley medal, 1913; Darwin-Wallace medal Linnean Soc., 1908; Linnean medal, 1920. Became fellow Royal Soc., 1875, v.p., 1882-96; fgn. asso. French Acad., 1910. Mem. Marine Biol. Assn. (founder 1884, pres. 1892); corr. mem. Imperial Acad. Scis., Petrograd, Royal Bohemian Soc., N.Y. Acad. Scis., Lincei Acad., Royal Belgian Acad., Am. Nat. Acad. Sci., Am. Philos. Soc., Biol. Soc. Paris, Acad. Scis. Phila. Author: A Monograph of the Cephalaspidian Fishes, 1870; Comparative Longevity, 1871; Degeneration, 1880; Limulus, an Arachnid, 1881; Spolia Maris, 1889; The Advancement of Sciences (collected essays), 1889; Monograph of the Okapi; Atlas of 48 Plates, 1910; Extinct Animals, 1905; The Kingdom of Man, 1907; Science from an Easy Chair, 1910, 2d series, 1912; Diversions of a Naturalist, 1915; Science and Education, 1919; Secrets of Earth and Sea, 1920; Great and Small Things, 1923. Editor: Quar. Jour. Micros. Sci., 1860—; joint editor; Sci. Memoirs Thomas Henry Huxley; editor: A Treatise on Zoology, 1900-09. Research on protozoan parasites and embryology of Mollusca, comparative anatomy of animals; introduced term, pronephros, circa 1885; discovered flint implements in Suffolk showing presence of skilled workers in Pliocene. Died London, Aug. 15, 1929.

LANNELONGUE, Odilon Marc, French surgeon; b. Castera-Verduzan, France, Dec. 4, 1840; Dr.medicine, Paris, 1867; became physician hosps., 1869; joined Faculty Scis., Paris, 1869, became prof. pathology, 1888; became dep. of Condom, France; 1893. Mem. Académie de médecine, French Acad. Scis., French Surg. Soc. Author: De l'ostéomyélite aiguë pendant la croissance, 1879; Abces froids et tuberculose osseuse, 1881; (with Ménard) Tuberculose vertébrale, 1888; Affections congénitale, 1881; (with Achard) Traité des Kystes congénitaux, 1888, also numerous articles. Treated cretinism by transplanting thyroid gland from an animal to a human, 1890. Died Paris, Dec. 22, 1911.

LA NOË, Gaston-Ovide de, see de la Noë.

LANPHIER, Edward Howell, Am. physician, physiologist; b. Madison, Wis., May 29, 1922; s. Ira Burton and Beatrice (Howell) L.; student Carleton Coll., 1940-42, Dartmouth, 1943-44; B.S., U. Wis., 1946; M.S. in Pharmacology, M.D., U. Ill., 1949. Commd., lt. comdr. M.C., USN, 1951, advanced through grades to lt. comdr., 1957; staff Exptl. Diving Unit, Washington, 1952-58; asso. prof. physiology Sch. Medicine, State U. N.Y., Buffalo, 1959—. Mem., Theta Delta Chi, Phi Beta Pi. Editor: United States Navy Diving Manual, Part I., 1959; co-editor: The Science of Skin and Scuba Diving, 1957. Research and publs. on physiology of a hyperbaric environment. Home: 50 Mapleview Dr., Buffalo 14226.*

LANSBERGEN, Philip van, see van Lansbergen, Philip.

LANSING, Ambrose, Egyptologist; b. Cairo, Egypt, Sept. 20, 1891; s. Joseph McCarrell and Isabella (Strang) L.; came to U. S., 1904, citizen by birth; A.B., Washington Jefferson Coll., 1911; student Leipzig (Germany) U., summers 1912-14; L.H.D., Bowdoin, 1948; m. Caroline Cox, Feb. 27, 1923; 1 son, Cornelius. With Met. Mus. Art, N.Y.C., 1911—, field archeologist in Egypt, 1911-22, asst. curator Dept. Egyptian Art, 1922-26, asso. curator, 1926-39, curator Dept. Egyptian Art, 1939—. Mem. vis. com. Semitic and Egyptian Civilization, Harvard. Fellow Am. Acad. Arts and Scis.; mem. Am. Oriental Soc., Oriental Inst. Chgo., Archeol. Inst. Am. Egypt Exploration Soc., Am. Museums Assn. Author: Egyptian Expdn. Reports (supplements to Bull. Met. Mus. Art), 1917-36, also contbr. articles to Bull. Died May 28, 1959.

LANSING, John Ernest, Am. chemist; b. Brookline, Mass., June 3, 1878; s. John Arnold and Florence (Stetson) L.; A.B., Harvard, 1898, A.M., 1900; traveled in Europe, 1898-99; m. Lucy Caroline Wells, June 27, 1907 (dec. Oct. 1916); m. 2d, Josephine Camp Belcher, July 3, 1918; children—John Belcher, Edward Stickney, Marion Frances. Instr. natural scis. Phillips Acad., Andover, Mass., 1901-05; asst. prof. chemistry Hobart Coll., 1905-06, prof., 1906—, also registrar, 1914-21, acting pres., 1941-42. Mem. Am. Chem. Soc. Author: Laboratory Experiments in Chemistry, 1908, rev., 1944; A Short Course in Qualitative Analysis, rev., 1948, retitled Lansing's Qualitative Analysis, rev., 1958. Died Sept. 28, 1958.

LANSTON, Tolbert, Am. inventor; b. Troy, O., 1844; s. Nicholas Randall and Mary Jane (Wright) L.; LL.B., Columbian Coll., (now George Washington U.), 1867; m. Beatty G. Herdel, 1866; 1 son, Aubrey. Admitted to D.C., bar, 1867; apptd. clk Pension Office, Washington, 1865-87, became chief clk (3d officer in rank). Recipient Cresson Gold medal Franklin Inst. Phila., 1896. Patentee Lanston/monotype (type forming and composing machine which revolutionized printing process), 1887; inventor machine which added numbers across or down a column, mail lock, hydraulic dumb-waiter, adjustable or removable calk horse-shoe. Died 1913.

LANYON, Wesley Edwin, Am. ornithologist; b. Norwalk, Conn., June 10, 1926; s. William J. and Frances A. (Merrill) L.; A.B. in Zoology, Cornell U., 1950; Ph.D., U. Wis., 1955 m. Vernia E. Hall, Jan. 29, 1951; children—Cynthia Hall, Scott Merrill. Mem. faculty U. Ariz., 1955-56; asst. prof. Miami U., Oxford, O., 1956-57; asst. curator birds Am. Museum Natural History, N.Y.C., 1957-63, asso. curator, 1963-68, curator, 1968—, resident dir. Kalbfleisch Field Research Sta. of museum, 1958—; adj. prof. biology City U., N.Y., 1968—, mus. expdns. to numerous fgn. countries. Fellow Am. Ornithologists Union; mem. Cooper, Wilson ornithol. socs., Eastern Bird Banding Assn., Ecol. Soc. Am., Soc. Study Evolution, Soc. Systemic Zoology, Linnaean Soc. N.Y., Sigma Xi. Contbr. profl. jours. Home: R.D. 1, Deer Park Av., Huntington, L.I., N.Y. 11743. Office: American Museum Natural History, N.Y.C. 10024.*

LANZA, Gaetano, Am. engr.; b. Boston, Sept. 26, 1848; s. Gaetano and Mary Ann (Paddock) L.; C.E., U. Va., 1869, B.Sc., 1870, C.E., M.E., 1870; m. Jennie D. Miller, Jan. 27, 1891. Asst. instr. math. U. Va., 1870-72; instr. and asst. prof. Mass. Inst. Tech., 1872-75, prof. theoretical and applied mechanics, 1875-1911, in charge dept. mech. engring., 1883-1911, prof. emeritus, 1911; spl. cons. Baldwin Locomotive Works. Fellow Am. Acad. Arts and Scis. Author: Applied Mechanics, 1885-1905; Dynamics of Machinery, 1911. Died 1928.

LANZETTA, John Thomas, Am. psychologist; b. N.Y.C., Apr. 24, 1926; s. Michael and Anna (Del Judice) L.; B.S., Lafayette Coll., 1948; Ph.D., U. Rochester, 1952; m. Jane Louise Warner, Jan. 14, 1946; children—Joan L., Susan A., Thomas M., Patricia A., Donna L., Michael J. Research asso. U. Rochester, 1951-53; supr. social psychology USAF, 1953-56; faculty U. Del., Newark, 1956-65, prof., 1959-61, dir. Center for Research Social Behavior, 1959-65, chmn. dept. psychology, 1960-63, H. Rodney Sharp Prof. Psychology, 1961-65; prof. Dartmouth, Hanover, N.H., 1965—. Liaison scientist Office Naval Research, London, Eng., 1962-63; mem. adv. panel psychology and social scis. Dept. Def., 1960; mem. surgeon gen. adv. panel neuropsychiatry USN, 1962—; cons. govt. agys., industry. Research fellow Social Sci. Research Council, 1957. Mem. Eastern Psychol. Assn., Social Sci. Research Council (com. on transnat. soc. psychology 1963—), N.Y. Acad. Scis., A.A.A.S., Sigma Xi, Psi Chi. Research on group performance, cognitive processes, decision making, communication systems. Home: Meadow Lane, Hanover, N.H.*

LANZL, Lawrence Herman, Am. physicist; b. Chgo., Apr. 8, 1921; s. Hans and Elsa (Seitz) L.; B.S., Northwestern U., 1943; M.S., U. Ill., 1947, Ph.D., 1951; m. Elizabeth E. Farber, Sept. 18, 1947; children—Eric L., Barbara J. Asst., dept. astronomy, interim instr. dept. physics Northwestern U., 1941-43; jr. physicist Manhattan Project, U. Chgo., also Los Alamos Sci. Lab., 1944-45; research asst., dept. physics U. Ill., 1946-50; asso. physicist naval reactor Argonne Nat. Reactor, 1951; asso. prof., dept. radiology and Argonne Cancer Research Hosp., U. Chgo., 1951—, on leave to Internat. Atomic Energy Agy., Vienna, Austria, 1967-68. Mem. electron dosimetry Internat. Commn. Radiol. Units and Measurements, 1964—; mem. dosimetry sub-com., radiation study sect., div. research grants NIH, 1957—. Recipient certificate of appreciation for Manhattan Project work War Dept., 1945. Mem. Am. Assn. Physicists in Medicine (pres. elect, dir.) A.A.A.S., Am. Phys. Soc., Hosp. Physicists Assn. (Gt. Britain), Radiation Research Soc., Health Physics Soc., others. Author: (with others) Moving Field Radiation Therapy, 1962. Editor: (with others) Radiation Accidents and Emergencies in Medicine, Research and Industry, 1965; also numerous reports and articles. Performed physics expts. on mock-up of Nautilus submarine reactor; extraction electron beam from betatron, devel. Cobalt-60 for radiation cancer therapy; devel. heterogeneous phantoms for dosimetry; studies on multiple scattering of high energy electrons; angular distbn. of 17 Mev X-rays; 1st use of betatron for cancer therapy; devel. and use of 40 Mev linear accelerator for electron therapy. Home: 411 Douglas St., Park Forest, Ill. Office: 950 E. 59th St., Chgo. 60637.*

LA PALOMBARA, Joseph, Am. polit. scientist; b. Chgo., May 18, 1925; s. Louis and Helen (Teutonico) La P.; A.B., U. Ill., 1947, M.A., 1950; M.A., Princeton, 1952, Ph.D., 1954; M.A. (hon.), Yale, 1965; m. Lyda Ecke, June 22, 1947; children—Richard Dean, David D., Susan Dee. Faculty, Ore. State U., 1947-50, asst. prof., 1947-50; instr. Princeton, 1952; faculty Mich. State U., 1953-64, prof. polit. sci., 1958-64; chmn. dept., 1958-63; prof. Yale, 1964—. Cons., Twentieth Century Fund, 1965—, Ford Found., 1965—; staff Social Sci. Research Council, 1966—. Decorated Order of Merit, Italian Republic, 1964; Rockefeller fellow, 1963-64; fellow Center for Advanced Study in Behavior Scis., 1961-62. Mem. Internat., Am. (exec. com. 1967—) polit sci. assns. Author numerous books including: Guide to Michigan Politics, 1956, rev., 1960; The Italian Labor Movement, 1957; Le Elites politiche, 1961; Interest Groups in Italian Politics, 1964; Italy: The Politics of Planning, 1966; also articles. Co-editor, contbg. author: Elezioni e comportamento politico in Italia, 1962; Bureaucracy and Political Development, 1963; Political Parties and Political Development, 1966. Research on European polit. systems, comparative bureaucracy, polit. devel. in emerging nations. Home: 19 Windsor Rd. E., North Haven, Conn. 06473. Office: Hall Grad. Studies, Yale, New Haven, 06511.*

LAPAZ, Lincoln, Am. mathematician, meteoriticist; b. Wichita, Kan., Feb. 12, 1897; s. Charles Melchior and Emma (Strode) L.; A.B., Fairmont Coll., 1920; A.M., Harvard, 1922; Ph.D. (NRC fellow), U. Chgo., 1928; m. Leota Rae Butler, June 18, 1922; children—Leota Jean, Mary Strode (Mrs. Harry L. Baldwin, Jr.). Instr. math. Fairmont Coll., 1917-20, Harvard, 1921-22, Dartmouth, 1922-25, U. Chgo., 1929-30; faculty Ohio State U., 1930-45, prof., 1942-45; prof., head dept. math. and astronomy, dir. Inst. Meteoritics, U. N.M., Albuquerque, 1945-53, dir. div. astronomy, dir. Inst., 1953-62, dir. Inst., 1962—; research mathematician OSRD, 1943-44; tech. dir. Operations Analysis Sect., Hdqrs. 2d Air Force, 1944-45. Adv. council World Who's Who in Sci. Fellow A.A.A.S., Meteoritical Soc. (pres. 1941-46); mem. Am. Math. Soc., Math. Assn. Am., Internat. Astron. Union, Am. Astron. Soc., Royal Astron. Soc. Can., Astron. Soc. Pacific, Am. Meteor Soc. (dir. S.W. sect. 1948—), N.Y., Wis. acads. scis., Sigma Xi. Author: (with Carmichael, Weaver) The Calculus, 1938, Physics and Medicine of the Upper Atmosphere, 1952, Advances in Geophysics, vol. 4, 1958, Space-Nomads, 1961. Catalog of the Collections of the Institute of Meteorites, 1965; also numerous articles. Asso. editor, Meteoritics, 1953-54. Discovery Norton County meteorite, largest aerolite so far recovered in world. Office: Inst. Meteorites, U. N.M., Albuquerque.*

LAPENNA, Marino, surgeon, radiologist; b. Hainfeld, Austria, Mar. 16, 1900; s. Lino and Dolores (Marani) L.; M.D., U. Rome; m. Simonetta Zanni, Dec. 12, 1925; children—Maria Lodovica, Florenza. Prof. radiology; dir. Istituto Radiologico Ospedali Ruiniti di Trieste (Italy). Mem. various sci. assns. Research and publs. in field. Home: piazza Oberdan 4, Trieste. Office: piazza Ospedale, Trieste, Italy.

LA PÉROUSE, Jean François de Galaup, Comte de, French naturalist, explorer; b. Guo, Aug. 22, 1741; s. Victor-Joseph and Marguerite (de Ressequier) La P.; ed. for naval service. Became guard, marines, 1756; served in campaigns against Hudson Bay Co., New Eng., 1778-83; named comdr. expdn. of discovery for French govt., 1785; visited N.W. coasts Am.; explored N.E. coasts Asia; sailed from Botany Bay, 1788; English capt. Dillon discovered his ships had been wrecked off Vanikoro, an island N. of New Hebrides. Author: Voyage autour du monde, 4 vols., 1797, English transl., 3 vols., 1798, new French edit. 1 vol., 1888. Discovered Sakhalin and Yezo in N.E. coasts of Asia were separate islands. Died circa 1788.

LAPERSONNE, Félix de, see de Lapersonne, Félix.

LAPEYRE, Noël, French surgeon; b. Montpellier, France, Dec. 25, 1887; s. Constant and Claudine (Devic) L.; M.D., U. Montpellier; m. Magdeleine Rochefort, June 26, 1920; children—Pierre, Robert, Christian. Became asst. in physiology U. Montpellier, 1908, head surg. clinic, 1913, agrégé in surgery, 1920, prof., gynecol. clinic, 1932, prof., surg. clinic, 1943-59, hon. prof., adminstr., 1949; surgeon in charge St.-Eloi clinics. Address: 2, rue Delpech, Montpellier (Hérault), France.

LA PEYRONIE, François Gigot de, French surgeon; b. Montpellier, France, Jan. 15, 1678; surgeon l'Hôtel-Dieu de Montpellier; became army surgeon, 1704, surgeon of duke of Chaulnes, Paris, 1714; named chief surgeon hôpital de la Charité; became 2d royal surgeon, 1717; apptd. surgeon to Louis XV, 1736; demonstrator anatomy Jardin du Roi, Paris; obtained rights equivalent to univ. doctors for surgeons; converted his castle to a hosp. Mem. French Acad. Scis., French Acad. Surgery (a founder). Contbr. articles to jours. Introduced improved methods of treatment and better adminstrv. regulation to army; improved anal fistula operations; asso. various movements with various areas of brain; studied resection of strangulated hernia. Died Versailles, France, Apr. 24, 1747.

LAPEYSSONNIE, Léon, French epidemiologist; b. Montpellier, France, Oct. 16, 1915; s. L. F. and Maria (Bruchet) L.; B.S. with honors, Montpellier Med. Faculty, 1933, M.D. with honors, 1941; M.S., Montpellier Sci. Faculty, 1958; postgrad. Pasteur Inst.; m. Juliette Euzière, Dec. 24, 1939; children—Jeanne-Marie, Francoise, Jacques, Marie, Cecile, Bernard. Chief, Mobile Field Unit for Sleeping Sickness, Africa, 1942-47; chief research unit for Trypanosomiasis, West Africa, 1949-52; chief Army Lab. Unit, Haiphong, Viet Nam, 1953-54; prof. preventive medicine, dean Pondicherry Med. Coll., India, 1954-57; prof. preventive medicine Sch. Tropical Medicine, Marseille, 1958-63; chief of mission WHO, Iran, 1964-66, WHO prof. preventive medicine, Tunis, Tunisia, 1966—. Mem. neisseria subcom. Internat. Com. on Bacteriol. Nomenclature, 1963. Fellow Royal Soc. Tropical Medicine and Hygiene London, French Soc. Microbiology, Paris Soc. Pathologie Exotique, French Soc. Hygiene (sci. council 1964). Author: Elements D'Hygiene et de Sante Publique sous les Tropiques, 1962; also numerous articles. Research on human trypanosomiasis and its chemotherapy and prophylaxis; discovery of Japanese B encephalitis in India; epidemiology and control of cerebrospinal meningitis in Africa and discovery of a single shot method for its treatment. Home: Villa Marie, Chemin Pioch Boutonnet, Montpellier, France. Office: WHO, Tunis, Tunisia.*

LAPICQUE, Louis, French physiologist; b. Epinal, France, Aug. 1, 1866; s. Auguste and Marie (Richardot) L.; ed. Coll. Epinal; Sorbonne, Paris, 1884; Lic.Sc., 1886; M.D., 1895; Sc.D., 1897; m. Marcelle de Heredia, 1902; 1 son. Named chief lab. Faculty of Medicine, 1894; became lectr. Sorbonne, 1899, prof. gen. physiology, 1919-36; became prof. gen. physiology Nat. Mus. Natural History, 1911; pres., dir. Institut Marey, 1936-41. Named hon. mem. Faculty of Medicine, Santiago, Chile. Hon. mem. Acad. Scis. Brazil, Am. Acad. Arts and Scis., Physiographical Soc. Lund, Polish Acad. Scis., German Acad. Natural Scis., Romanian Acad., English Physiol. Soc., Italian, Argentine physiol. socs; mem. Nat. Acad. Medicine, Assn. Biol. Chemistry, Soc. Phys. Chemistry, Psychol. Soc., Société de biologie de Paris (past pres.), French Acad. Scis., French Acad. Medicine. Introduced term chronaxie (now usually chronoxy) which designates duration of time a current twice as strong as the galvanic threshold must flow to excite the test tissue, 1909. Died Paris, Dec. 6, 1952.

LAPIN, Peter Ivanovich, Russian botanist, dendrologist; b. Penza, USSR, Jan. 29, 1909; s. I.S. and E.P. (Suzjumova) L.; grad. Leningrad (USSR) Sylviculture Acad., 1931; cand.biol.sci., All-Union Research Inst. Plant Industry, 1936; m. E.P. Krivosheina, May 5, 1931; children—Edgar, Lidia. Staff, Central Research Inst. Sylviculture, 1929-31; head specialist sylviculture Irkutsk region, 1931-33; sr. scientist dendrology dept. Nikita State Bot. Garden, Yalta, 1933-39; sr. scientist Komaron Bot. Inst., U.S.S.R. Acad. Sci., Leningrad, 1939, staff Main Bot. Garden, 1945—, chief dendrology dept., 1948—, dep. dir., 1951-66; dir. Leningrad Region Seed Testing Sta., 1939-41. Recipient Red Star, 1945, medals for pub. service, 1942, 45, 65; medal for sci. work, 1952. Mem. Internat. Hort. Soc., Moscow Soc. Plant Protection, Moscow Greening Soc. Author: (with others) Trees and Shrubs, 1959; also articles. Research on prins. of evaluation of perspective for introduction of trees according to their rhythm of growth. Home: Moscow, Botanicheskava 33, fl. 25 USSR. Office: Main Bot. Garden, USSR Acad. Sci., Botanicheskaya 4, Moscow, USSR.*

LAPLACE, Ernest, Am. surgeon; b. New Orleans, July 9, 1861; s. Basil and Eugenie (Sauvage) L.; A.B., Georgetown Coll., 1880, A.M., 1887, LL.D., 1895; M.D., U. La., 1884, Faculté de Médécine of Paris, 1886; pupil of Pasteur, Lister, Koch, and Billroth; m. Catherine Borsch, 1902. Prof. surgery Medico-Chirurg. Coll., Phila., 1892—; prof. surgery Grad. Sch. Medicine U. Pa.; surgeon Phila., Medico-Chirurg., Polyclinic, and Misericordia hosps.; cons. surgeon State Hosp. for Criminal Insane. Inventor 1st forceps for Intestinal Anastomosis. Died May 15, 1924.

LAPLACE, Pierre Simon, Count, then Marquis de, French mathematician, astronomer, physicist; b. Mar. 28, 1749, Beaumont-en-Auge, Normandy (Calvados), France; ed. at military school in Beaumont, and in Caen, 1765-67. Prof. math. at École militaire, Paris, 1768; examiner to Royal Artillery, 1784; prof. math., École normale, Paris; minister of interior under Napoleon, 1799; then made senator, 1799 (chancellor of senate, 1803); mem., Bureau des longitudes; became count, 1806; marquis, 1817. Mem., French Acad. Scis., 1773; mem., Académie française, 1816 (pres. 1817); Fellow, Royal Soc., 1789; grand officer, Legion of Honor. Author: Théorie du mouvement et de la figure elliptique des planètes, 1784; Théorie des attractions des sphéroides et de la figure des planètes, 1785; Exposition du système du monde, 1796; Traité de méchanique céleste, 1799-1825; Théorie analytique des probabilités, 1812; Essai philosophique sur les probabilités, 1814; many scientific articles. Because of his work in mamt. astronomy, has often been called the Newton of France; studied perturbations of planets and demonstrated stability of solar system; showed mean motions invariable or subject only to small periodic perturbations; discovered causes of inequality of motions of Jupiter and Saturn; investigated theory of Jupiter's satellites, acceleration of moon's mean motion, aberrations in movements of comets, theory of tides; developed famous nebular hypothesis of origin of solar system; studied capillary action, electricity, equilibrium of rotating fluid mass; developed theory of surface tension based on molecular attraction in liquid; corrected Newton's theory of velocity of sound in gases; with Lavoisier, conducted experiments of specific heat and heat of combustion, showed respiration a form of combustion, and laid foundations for thermochemistry; made great contributions to theory of probability; with Legendre, introduced use of partial differential equations into study of probability; made much use of potential functions which he showed were solutions of famous partial differential equation since named after him. Died Paris, France, Mar. 5, 1827.

LAPOINTE, Henri, Canadian physician; b. Quebec, Que., Can., Feb. 2, 1901; s. Georges and Loretta (Levasseur) LaP.; B.A., Que. Sem., 1920; B.M., Laval U., 1922, M.D., 1925; m. Marcelle Cote, June 17, 1930 (dec.); children—Andre, Pascal; m. 2d, Michelle Rufiange, May 4, 1961. Dir. radiology dept. Infant-Jesus Hosp., Quebec, 1935-66; dir. dept. radiology faculty medicine Laval U., Quebec, 1961—; dir. dept. radiology St. Sauveur Hosp., Val D'Or, Que., 1966—. Diplomate Royal Coll. Physicians and Surgeons, Coll. des Medecins et chirurgiens. Fellow Am. Coll. Radiology; mem. Canadian Assn. Radiologists (past pres.), Radiol. Soc. N.Am., Canadian Med. Assn., Association des Medecins de Langue Francaise du Can., Association Canadienne Francaise de Radiologie, Association des Radiologiste de la Province de Que., Société medicale de Que., Société Medicale des Hopitaux Universitaires de Quebec, Am. Radium Soc., Association Canadienne Francaise pour L'Avancement des Sciences, Soc. Nuclear Medicine. Research and publs. in diagnostic and therapeutic radiology. Home: 169 Cadillac Bourlamaque, Val D'Or, Abitibi East Que., Can.*

LAPORTE, Otto, physicist; b. Mainz, Germany, July 23, 1902; s. Wilhelm and Anna (Geyl) L.; Ph.D., U. Munich (Germany), 1924; m. Adele Pond, Oct. 16, 1959; children—Claire, Irene. Came to U. S., 1924, naturalized, 1936. Internat. Edn. Bd. fellow Nat. Bur. Standards, 1924-26; faculty U. Mich., Ann Arbor, 1926—, prof. physics, 1940—; sci. attache Am. Embassy, Tokyo, Japan, 1954-56, 61-63. Fellow Am. Phys. Soc. Discovered parity law for atomic spectra; pioneered observation spectra from shock heated gases. Home: 959 Forest Rd., Ann Arbor, Mich. 48105.*

LAPP, Ralph Eugene, Am. physicist; b. Buffalo, Aug. 24, 1917; s. Henry R. and Lilley (Grammel) L.; B.S., U. Chgo., 1940, Ph.D., 1945; m. Jeannette F. DeRome, Sept. 6, 1956; children—Christopher Warren, Nicholas DeRome. Asst. to dir., Metall. Lab., Chgo., 1943-45; asst. dir. Argonne (Ill.) Nat. Lab., 1945-46; sci. advisor War Dept., Washington, 1946-47; exec. dir. Atomic Energy, Research and Devel. Bd., Dept. Def. Washington, 1948; head nuclear physics div. Office Naval Research, Washington, 1949; dir. Nuclear Sci. Service, Washington, 1950-61; sec., bd. mem. Quadri-Sci. Inc., Washington, 1961—. Member of the American Institute of Physics, American Geophysical Union, A.A.A.S., Washington Philos. Soc., Washington Acad. Scis., Phi Beta Kappa, Sigma Xi. Author: Nuclear Radiation Physics, 1948; The New Force, 1953; Atoms and People, 1955; Radiation, 1957; The Voyage of the Lucky Dragon, 1958; Roads to Discovery, 1960; Kill and Overkill, 1962; Man and Space, 1961; Matter, 1963; The New Priesthood, 1965; The Weapons Culture, 1968. Research in cosmic radiation, nuclear physics, mass spectroscopy, radioactive fallout; study of mil. aspects of atomic energy and social impact of sci. revolution. Home: 7215 Park Terrace Dr., Alexandria, Va. 22307. Office: 1028 Conn. Av., Washington.*

LAPPARENT, Albert-Auguste Cochon de, see de Lapparent, Albert-Auguste Cochon.

LAPPARENT, Marie-Jacques Cochon de, see de Lapparent, Marie-Jacques Cochon.

LAPPERT, Michael Franz, chemist; b. Brno, Czechoslovakia, Dec. 31, 1928; s. Julius and Cornelie (Beran) L.; B.Sc., U. London, 1949, Ph.D., 1951, D.Sc., 1960. Mem. faculty No. Poly., London, 1952-59, sr. lectr., 1955-59; tchr. U. London, 1957-59; lectr. Manchester Coll. Sci. and Tech., Manchester U., 1959-61, sr. lectr., 1961-64; reader in chemistry U. Sussex, Brighton, Eng., 1964—. Cons. to Midland Silicones Ltd. Fellow Royal Inst. Chemistry, Chem. Soc. London; mem. Am. Chem. Soc. Author: (with G. J. Leigh) Developments in Inorganic Polymer Chemistry, 1962. Editorial bd. Jour. Organometallic Chemistry; sr. editor Phys. Inorganic Chemistry. Research and numerous publs. on boron, organometallic and coordination chemistry. Home: 4 Varndean Gardens, Brighton 6, Eng.*

LAPPO, Arkadii Ivanovich, Russian geneticist; b. Feb. 8, 1904; grad. Byelorussian Agrl. Acad., 1927. Staff, Byelorussian Agrl. Acad., until 1941; dir. selection sect. Inst. for Socialist Agr., Byelorussian Acad. Sci., 1944-49, acad. sec. sect. biol. and agrl. scis., 1953-56; acad. sec., Byelorussian Acad. Agrl. Scis., 1957—. Mem. Byelorussian Acad. Agr., Byelorussian Acad. Sci.

LAPUSAN, Ioan, Rumanian physician; b. Cublesul Somesan, Raionul Gherla, Regiunea Cluj, July 28, 1919; s. Nicolae and Marie (Turdeanu) L.; M.D., Faculty Medicine, Bucharest, Rumania, 1946; specialist physician, infectious diseases, 1959; sr. physician, pub. health, 1961; m. Ana Alexandrescu, Feb. 14, 1948. Sect. head Ministry Pub. Health and Social Assistance, 1948-49; head dept. health statistics and demography Inst. Hygiene, 1949—. Asst. lectr., lectr., dept. biostatistics I.M.F., Bucharest, 1949-52; hosp. specialist Hosp. Infectious Diseases, Bucharest, 1955-60. Named excellent worker in health-med. activity, 1954. Mem. Union of Socs. of Med. Sci. Author: (with P. Muresan) Elementary Notions of Biostatistics, 1951; also papers. Studies of health conditions of inhabitants of Rumania concerning statistics of hosps., morbidity in pub. instns., infectious, occupational and mental diseases; research on standards and norms for health assistance, med. control in dispensaries of chronic outpatients, evidence system and statis. reporting, hygiene-health indicators, efficiency indicators, activity and econ. efficiency of health units, demographic studies. Home: 2 Intrarea Legendei, raionul 1 Mai, Bucharest. Office: 1-3 Dr. Leonte, raionul 16 Februarie, Bucharest, Rumania.*

LAPWORTH, Charles, geologist-paleontologist; b. Farringdon, 1842; s. James Lapworth; ed. Madras Coll., St. Andrews, Scotland, 1875-81; LL.D., M.Sc.; M.Sc., Birmingham U.; LL.D., U. Aberdeen, Scotland, 1883, U. Glasgow (Scotland), 1912; m. Janet Sanderson; 2 sons, 1 dau. Sch. master, Galashiels, Scotland; prof. geology and physiography Birmingham U., 1831-13, emeritus prof., 1914—. Cons. geologist mining and civil engring. Water Bills of Birmingham, Gloucester, Harrogate, Leicester; mem. Royal Coal Commn., 1902-04. Recipient Wilde medal Manchester Philos. Soc., 1905. Became fellow Royal Soc., 1888, recipient Royal medal, 1891. Fellow Geol. Soc. (Bigsby medal 1887, Wollaston medal 1889, pres. 1902-04); mem. Brit. Assn. (became pres. sect. C 1892). Editor: Page's textbooks on geology and phys. geography; (monograph) British Graptolites, 1900-08; Intermediate Textbook of Geology, 1899. Research and publs. on So. uplands of Scotland leading to solution of many problems in lower Palaeozoic rocks of Britain; studies in Graptolites and lower Palaeozoic rocks; proposed term Ordovician for strata between Cambrian and Silurian. Died Mar. 13, 1920.

LAQUEUR, Ernst, physiologist, biochemist; b. Obernigh, nr. Trebnitz, Germany, 1901. Discovered estrogenic activity in urine of males (with E. Dingemanse, P. C. Hart and S. E. de Jongh), 1927; isolated corticosterone (with de Fremery, Reichstein, Spanhoff, and Uyldert), 1937; also investigated physical chemistry of casein, autolysis, luna pathology, vision physiology, female sex hormones; isolated male sex hormone (testosterone), 1935. Died Switzerland, Aug. 19, 1947.

LAQUEUR, Ludwig, ophthalmologist; b. Fürstenberg, Germany, July 25, 1839; student Paris, Berlin; M.D., 1860. Became privat-docent, Berlin, 1860; asst. at Liebreich's Ophthalmol. Hosp., Paris, 1863-69; named asst. prof. U. Strasbourg (France), 1872, prof. ophthalmology, 1877. Author: Études sur les affections sympathiques de l'oeil, 1869; also articles. Introduced physostigmine as a miotic in the treatment of glaucoma, 1876. Died 1909.

LA QUINTINE, Jean de, see de la Quintine, Jean.

LA RAMÉE, Pierre de, see Ramus, Petrus.

LARCAN, Alain, French physician; b. Nancy, France, Feb. 25, 1931; s. Jean and Colette (Fruhinsholz) L.; Interne des Hôpitaux, 1952; D. en méd., 1957; m. Thérèse Giraud, May 29, 1954; children—Dominique, Pascale. Work in internal medicine, especially metabolic diseases and resuscitation, Faculté de Médecine, Nancy, also Centre hospitalier, Nancy; dir. univ. preventive medicine. Med. cons. for civil defense. Author: Le coeur des hémopathies; (with Coirault, Davidou), Les polyradiculonévrites; (with C. Huriet) L'ECG dysmétabolique; (with P. Vert) L'acidosétose diabétique; (with P. Michon) Agression et réanimation en médecine interne; (with G. Ablard) Le RAA de l'adulte; also numerous articles. Research on diabetic acidosis and coma, insulin therapy, nephropathies of Waldenström's diseases, leukoses, cardiac insufficiency, parenteral nourishment. Home: 7 Rue Victor Hugo, Nancy 54, France.*

LARDNER, Dionysius, physicist; b. Dublin, Ireland, Apr. 3, 1793; m. Cecilia Flood, 1815 (div. 1849); 3 children; m. 2d, Mary Heaviside; 2 children. Prof. astronomy and physics U. London, 1827-40; gave popular sci. lectures in all prin. cities of U. S., 1840-45. Fellow Royal Soc., 1828. Author: Treatise on Algebraical Geometry, 1823; On the Differential and Integral Calculas, 1825; Museum of Science and Art, 12 vols.; also publs. on hydrostatics, pneumatics, mathematics, algebraic geometry, differential and integral calculus, steam engine. Died Naples, Italy, Apr. 29, 1859.

LARDY, Henry Arnold, Am. biochemist; b. Roslyn, S.D., Aug. 19, 1917; s. Nicholas and Elizabeth (Gebetsreiter) L.; B.S., S.D. State U., 1939; M.S., U. Wis., 1941, Ph.D., 1943; m. Annrita Dresselhyrs, Jan. 21, 1943; children—Nick, Diana, Jeffrey, Michael. NRC fellow Banting Inst., Toronto, Ont., Can., 1944-45; faculty U. Wis., Madison, 1945—, prof. biochemistry, 1950—, chmn. research dept. Enzyme Inst., 1950—. Recipient Neuberg medal Am. Soc. European Chemists, 1956. Mem. Am. Chem. Soc. (chmn. biol. div. 1958, Lewis award 1949), Am. Soc. Biol. Chemistry (pres. 1964), Soc. Cell Biology, Nat. Acad. Sci., Am. Acad. Arts and Scis. Editor: Respiratory Enzymes, 1948; (with Boyer and Myrback) The Enzymes, 8 vols., 1959-63. Research and numerous publs. on metabolism of mammalian spermatozoa, oxidative phosphorylation, energy transfer systems, mode action antibiotics, function hormones, path of carbon in gluconeogenesis. Home: Thorstrand Rd., Madison, Wis. 53705.*

LARGETEAU, Charles-Louis, French astronomer; b. Mouilleron-en-Pareds, France, July 22, 1791; ed. École Polytechnique; sec. Paris Obs.; various positions as engr., geographer, geodetist; mem. Bur. Longitudes, French Acad. Scis., 1847. Author: Table de précession, d'observation et de mutation pour les étoiles principales, 1839; Table pour le calcul des syzgies écliptiques, 1843. Developed method for calculating eliptic conjunction of 2 celestial bodies; research on mutation of some stars. Died Pouzauges, France, Sept. 11, 1857.

LARGUIER, Everett Henry, Am. mathematician; b. New Orleans, Jan. 26, 1910; s. Henry Joseph and Ella (Foley) L.; student Loyola U., New Orleans, 1930-32; A.B. (President's scholar), St. Louis U., 1934, M.S. (President's scholar), 1935, Ph.L., 1937, S.T.L., 1942; Ph.D., U. Mich., 1947. Prof., Spring Hill Coll., Mobile, Ala., 1937-38, chmn. dept., 1947—, trustee, 1952-58, asst. St. Louis U., 1938-42, vis. prof., 1949; spl. lectr. Loyola U., New Orleans, 1942. Mem. Math. Assn. Am., Am. Math. Soc., Societe Mathematique de France, Sigma Xi. Author: Fundamental Concepts of Mathematics, 1950, rev., 1960; Elements of Calculus, 1959. Research and publs. primarily in probability and topology, founds. of math. and philosophy. Home: 4307 Old Shell Rd., Mobile, Ala. 36608.*

LARGUS, Scribonius, physician; flourished 1st century; studied in Rome in the time of Tiberius, A.D., 1-50; ct. physician to Emperor Claudius; drew up list of 271 prescriptions, circa A.D. 47; empiricist in method.

LARIONOV, Andrei Nikolaevich, Russian elec. engr.; b. July 16, 1889; grad. Moscow Tech. Coll., 1919. Tchr., Moscow Tech. Coll., 1919-30; also with All Union Electro-Tech. Inst., 1921-41; assisted in orgn. Moscow Inst. Energetics, 1930, became prof., 1933; with USSR Acad. Scis. Inst. Automation and Telemechanics, 1953—, corr. mem. acad., 1953—. Author: Utilization of Electricity in Aviation and Motor Transport, 1954; (with others) Basis of Electrical Equipment in Aircraft and Automobiles, 1955, Hysteresis of Electric Motors, 1956; Selection of single optimum frequencies for autonomous systems of alternating current with special elements and electric machines of automation, 1957. Research on theory, calculation and constrn. of spl. electric machines and electric drives; proposed a 3-phase bridge scheme of current rectification, 1924; took part in planning elec. equipment in airplane Maxim Gorki in solution of problems of starting turbo and hydro generators in power plants, and electrification of oil fields. Home: Krasnokazarmennaya, 12. Office: Institute of Automation and Remote Control, USSR Academy of Sciences, Kalanchevskaya Ulitsa 15-a, Moscow, USSR.

LARIONOV, Leonid Fedorovich, Russian physician; b. Tobolsk, Siberia, USSR, Aug. 2, 1902; s. Fedor and Maria (Protopopova) L.; grad. Tomsk State U., 1925; Cand. Med. Scis., Leningrad U., 1930, D.Med. Scis., 1937; m. Anna Fedorovna Artemieva, June 1, 1929; children—Galina (Mrs. Konstantinov), Vyacheslav. Asst. dept. path. physiology U. Tomsk, 1925-27, 1st Med. Inst., Leningrad, 1935-40; research worker lab. cancer research Inst. Roentgenology, Leningrad, 1927-40, head lab. cancer research, 1945-51; prof. dept. path. physiology Med. Inst., Minsk, 1940-41; head lab. exptl. therapy of cancer Inst. Oncology, Leningrad, 1945-51; head lab. exptl. cancer chemotherapy Inst. Exptl. Clin. Oncology, USSR Acad. Med. Sci., Moscow, 1951—. Cons. clin. cancer chemotherapy various med. instns., 1952—. Recipient Stalin prize for devel. of chemotherapy of Hodgkin's disease, 1951. Mem. USSR Acad. Med. Sci. (corr.), Royal Soc. Medicine (London), Am. Assn. Cancer Research (corr.). Author: Endocrine Glands and Cancer, 1938; Chemotherapy of Hodgkin's Disease and Leukemia, 1951; Cancer Chemotherapy, 1962 (English edit. 1965); also articles. Research on exptl. and clin. cancer chemotherapy; co-inventor anti-cancer drugs Novembichin, 1952, Sarcolysin, 1954,

Dopan, 1955, Asaline, 1958; elaborated chemotherapeutic methods and established curability of Hodgkin's disease, multiple myeloma and oesophagus carcinoma. Home: 36, kv. 104, Frunzennskaya Naberejnaya, Moscow G-146. Office: 6, Kashirskoye Chaussee, Moscow M-478, USSR.*

LA RIVE, Auguste Arthur de, physicist; b. Geneva, Switzerland, Oct. 9, 1801; s. Charles Gaspard L. R.; children include: William. Prof. Natural Philosophy, Geneva Academy, 1823-30; spent 1830-36 in London and Paris; returned to Geneva to edit the Bibliothèque universelle de Genève, 1836; edited Archives de l'Électricté (5 vols.), 1841-5; special envoy to Eng. to secure declaration to prevent French aggression, 1860. Fellow Royal Soc., 1846; French Acad. Soc. (prize), 1830; mem. other orders and societies. Author: Mémoire sur les Caustiques, 1824; Théorie de la Pile Voltaique, 1836; Traité de l'électricite théorique et appliquée (3 vols.), 1854-8; numerous articles. Studied specific heat of gases and temperature of earth's crust, 1823-30; presented new theories on Aurora Borealis; gilded brass and silver electrolytically, 1840; designed electro-chemical condenser, 1843; studied magnetism and electricity; proposed double current theory; opposed contact theory, maintaining that chemical action was sole source of electrification; studied passage of electricity through gases; proved that ozone is formed by electrical sparks in pure oxygen; measured conductivity of solutions; studied heat conductability of wood, 1823; discovered principle of sine galvanometer; studied relation of temp. to electrical conductivity in a wire. Died Marseilles, France, Nov. 27, 1873.

LA RIVE, Charles Gaspard de, Swiss surgeon, chemist, physicist; b. Geneva, Switzerland, 1770; student medicine, natural scis., Edinburgh, Scotland; a son, Auguste Arthur; one of leaders in Geneva's struggle for independence against Napoleon; became mem. provisional council of Switzerland, 1813; became pres. of both councils of republic, 1813; a founder Mus. Natural History, also Botanic Garden, Geneva. Author: Institutions de géométrie enrichies de notes critiques et philosophiques sur la nature et les développements de l'esprit humain, 1746; Traité des sections coniques appliquées a la pratique de differents arts, 1750; le Ventriloque ou l'Engastrimythe, 2 vols. (gave explanation of ventriloquism), 1772; Observations sur la conversion de l'amidon en sucre; Essai sur la théorie des proportions chimiques et sur l'influence chimique de l'electricité. One of 1st to demonstrate effect of electric current on magnetic needle; studied Voltaic electricity and built Voltaic pile, 1820; invented galvanometer. Died 1834.

LA RIVERS, Ira, Am. biologist; b. San Francisco, May 1, 1915; s. Ira John and Yvonne (Groulx) La R.; B.S. in Zoology, U. Nev., 1937; postgrad. N.C. State Coll.; Ph.D. in Entomology, U. Cal. at Berkeley, 1948; m. Marian Byrd Ballinger, Dec. 21, 1951; 1 son, Ira John III. With U. S. Bur. Biol. Survey, 1936, U. S. Bur. Plant Quarantine, 1937, 42, Nev. State Dept. Agr., 1939; entomologist U. Nev., Reno, 1939-42, faculty, 1948—, prof. biology, 1961—, chmn. dept. biology, 1953-63; with Cal. Dept. Agr., Yermo, Cal., 1942. Fellow Oahu Micros. Soc., Acad. Zoology India; mem. Biol. Soc. Nev., Entomol. Soc. Washington, Colepteroust Soc., Western Soc. Naturalists, Cal., So. Cal. acads. sci., Soc. Systematic Zoology, Soc. Vertebrate Paleontology, Am. Malacological Union, Ecol. Soc. Am., Entomol. Soc. Author: General Zoology Laboratory Manual, 1955; Fishes and Fisheries of Nevada, 1962; Algae of Nevada; also numerous articles. Research, descriptions aquatic water bug family Naucoridae, Western U. S. terrestrial beetles Tenebrionidae, Nev. algae, Tertiary cold-blooded vertebrates from Nev.; research on Nev. arthropods, endemism in Gt. Basin fishes; chromatographic and electrophoretic research sund dune biota. Home: New Dog Valley Rd., Verdi, Nev. 89439. Office: Box 8096, Reno 89507.*

LA RIVIÉRE, Roch Le Bailiff, Sieur de, see de la Rivière, Roch Le Baillif, sieur.

LARK-HOROVITZ, Karl, physicist; b. Vienna, Austria, July 20, 1892; s. Moritz and Adelle (Hofmann) Horovitz; Ph.D., U. Vienna, 1919; m. Betty Friedlaender, July 26, 1916; children—Caroline Betty, Karl Gordon. Came to U. S., 1925, naturalized, 1936. Asst. tchr. physics U. Vienna, 1919-25; internat. research fellow U. Toronto, 1925-26, U. Chgo., 1926, Rockefeller Inst., 1926-27, Stanford, 1927-28; prof. physics, Purdue U., since 1928, dir. Phys. Lab. since 1929, head dept. physics since 1931. Fellow Am. Phys. Soc., A.A.A.S. (gen. sec. 1948-51); mem. Am. Assn. Physics Tchrs., Sigma Xi. Contbr. articles to sci. jours. Died Apr. 14, 1958.

LARKIN, Edgar Lucien, Am. astronomer; b. La Salle County, Ill., Apr. 5, 1847; s. Herman I. and Jane L.; ed. La Salle Co., Ill.; m. Alice A. Everman, Apr. 29, 1869. Opened New Windsor (Ill.) Obs., dir., 1880-88, Knox Coll. Obs., 1888-95; dir. Lowe Obs., Echo Mountain, Calif., 1900—. Author: Radiant Energy, 1903; Within the Mind Maze, 1911; The Matchless Altar of the Soul, 1916; Popular Studies in Recent Astronomy. Died Oct. 11, 1924.

LARKS, Saul David, physiologist, biophysicist, bioengr.; b. London, Eng., June 16, 1910; s. David and Sarah (Edelman) L.; came to U. S., 1915, naturalized, 1930; B.S. in Elec. Engring., U. Ill., 1943; M.S., Northwestern U., 1951; Ph.D. in Biophysics, U. Cal. at Los Angeles, 1956; m. Golda Gezuk, June 13, 1931; children—Leonard, Anita. With U. Cal. at Los Angeles, 1955-61, asst. prof., 1958-61; prof. elec. engring. Marquette U., Milw., 1961-66; prof. vet. physiology U. Mo., 1966—. Cons. to pvt. cos. Mem. I.E.E.E. (adminstrv. com. bioengring. group 1965—, editorial adv. bd. 1958—), Am. Inst. E.E. (chmn. comm. on elec. tech. in medicine and biology 1962), Internat. Fedn. for Med. Electronics and Biol. Engring. (councillor for U. S. 1963-65), Biophys. Soc., Engrs. Joint Council (coms. 1964—), A.A.A.S., N.Y. Acad. Sci., Sigma Xi, Eta Kappa Nu, Tau Beta Pi. Author: Electrohysterography, 1960; Fetal Electrocardiography, 1960; also numerous articles. Discovered repetitive pattern in human electrohysterogram, elec. onset of labor; pioneered research in fetal physiology, fetal electrocardiography; discovered method for calculation of elec. axis of fetal heart; research in perinatal physiology, reproductive biology. Home: Villa Capri, 1725 E. Broadway, Columbia, Mo. 65201.*

LARMI, Teuvo Kaarlo Ilmari, Finnish surgeon; b. Turku, Finland, July 4, 1924; s. Hugo Karl Emil and Ragnhild Hella (Eino) L.; M.D., U. Helsinki, Finland, 1949, D. Medicine and Surgery, 1954, Docent in Surgery, 1964; m. Margit Petri, Apr. 22, 1950; children—Lea Margareta, Eva Katrina, Elsi Helena. Mem. staff U. Central Hosp., 1955-60, Maria Hosp., 1960-63, Aurora Hosp., 1963-64 (all Helsinki); expert surgeon States Accident Agy., 1959-64; head surgeon Tampere Central Hosp., 1964-65; prof. surgery, head dept. surgery U. Oulu, Finland, 1965—, mem. bldg. com. Oulu U. Hosp. Mem. spl. consultation com. Central Med. Bd. Research, publs. on physiology of lungs, thoracic and abdominal surgery, immunology of tissues. Home: Kirkkokatu 11 A 25, Oulu, Finland. Office: Oulun Yliopisto, Kirurgian Klinikka, Kontinkangas.*

LARMOR, Sir Joseph, Brit. mathematician, physicist; b. Magheragall, County Antrim, Ireland, July 11, 1857; s. Hugh and Anna (Wright) L.; ed. Royal Belfast Academical Instn., Queen's Coll., Belfast, St. John's Coll., Cambridge; M.A., D.Sc., London; Sc.D. (hon.), Cambridge, Oxford, Queen's, Dublin; LL.D, Glasgow, Aberdeen, Birmingham, St. Andrews, London Centenary; D.C.L., Durham. Prof. natural philosophy Queen's Coll., Galway, also Queen's U. in Ireland, 1880-85; lectr. math. Cambridge U., 1885-1903; Lucasian prof. math., 1903-32; examiner in math. and natural philosophy U. London. Fellow Royal Soc., 1892, sec., 1901-12, v.p., 1912-14, recipient Royal medal, 1915, Copley medal, 1921; fellow Royal Soc. Edinburgh, St. John's Coll., Royal Asiatic Soc., French Acad. Scis., 1920; mem. Cambridge Philos. Soc. (pres., Hopkins prize 1897), London Math. Soc. (De Morgan medal 1914, pres. 1914-15), Am. Acad. Arts and Scis. (hon.), Lit. and Philos. Soc. Manchester (Wilde medal), Asiatic Soc. Bengal, Calcutta Math. Soc., R. Cornwall Poly. Soc., Am. Philos. Soc. (fgn.), Société Hollandaise des Sciences, Washington Acad. Scis., U. S. Nat. Acad. Scis. (fgn. asso.), R. Accad. dei Lincei (Rome), Inst. di Bologna (corr.). Author: Aether and Matter, 1900; Memoirs on mathematics and physics, repub., 1927-29; Collected Edits. of memoirs and corr. of Sir G. G. Stokes, Lord Kelvin, James Thomson, J. Clerk Maxwell, G. F. FitzGerlald, H. Cavendish. Worked on math. problems in electrodynamics and thermodynamics; studied atomic structure. Died May 19, 1942.

LARNED, Joseph Gay Eaton, Am. inventor; b. Thompson, Conn., Apr. 29, 1819; s. George and Anna (Spalding) L.; grad. Yale, 1839; m. Helen Lee, May 9, 1859. Tchr. classics Chatham Acad., Savannah, Ga., 1839; pvt. tchr., Charleston, S.C., 1840; took charge of acad. in Waterloo, N.Y., 1841; tutor Yale, 1842; wrote articles for New Englander, Conn., 1845-46; studied law at home, admitted to Conn. bar, 1847; moved to N.Y.C., 1854; in partnership with Wellington Lee to manufacture steam fire engines at Novelty Iron Works, N.Y., 1855-63; invented steam fire engine and demonstrated it in N.Y.C. and other cities; became asst. insp. of ironclads for Navy Dept. in charge of work at Green Point, Bklyn., 1863; returned to law practice, N.Y.C., after Civil War. Author: A Quarter-Century Record of the Class of 1839, Yale College, 1865. Died N.Y.C., June 3, 1870.

LAROCHE, Gilles Joseph Régent, biochemist; b. Lauzon, Que., Can., July 11, 1922; s. Charles E. and Jeannette (Octeau) LaR.; L.Sc., U. Montreal, 1947; M.Sc., McGill U., 1951; Ph.D., U. Wash., 1957; m. D. Catherine Macrae, Aug. 26, 1961; Came to U. S., 1952, naturalized, 1959. Lab instr., U. Montreal (Que.), 1947-48, 51-52, lectr., 1948-52; lab instr. McGill U., 1949-52; research asst. U. Wash., Seattle, 1953-57; asst. physiologist U. Cal. at Berkeley, 1957-58, research biochemist, 1958-66, research adviser, 1959-66; vis. scientist Am. Physiol. Soc., 1965-66; dir. Narragansett (R.I.) Marine Sport Fish Lab., U. R.I., 1966-67; acting chief fish and invertebrate sects. Nat. Marine Quality Lab., U. S. Dept. Interior, 1967—. Mem. A.A.A.S., Am. Soc. Zoologists, N.Y. Acad. Sci., Endocrine Soc., Am. Physiol. Soc., Royal Soc. Medicine, Soc. Nuclear Medicine, Sigma Xi. Author: Fish Nutrition, 1967; also numerous articles. Established correlation between thyroid activity and collagen synthesis in fish. Home: 111 Roger Williams Dr., North Kingstown, R.I. 02852. Office: Nat. Marine Water Quality Lab., P.O. Box 277, West Kingston, R.I. 02892.

LA ROCHE, René, Am. physician; b. Phila., Sept. 23, 1795; s. René and Marie Jeanne (de la Condemine) La R.; M.D., U. Pa., 1820; m. Mary Jane Ellis, 1824. Tchr. medicine in summer sch. U. Pa.; edited N.Am. Med. and Surg. Jour., Phila., 1826-31; active Coll. of Physicians; pres. Path. Soc. of Phila. Author: Yellow Fever, Considered in Its Historical, Pathological, Etiological and Therapeutical Relations, 2 vols., 1855. Died Phila., Dec. 9, 1872.

LARON, Zvi, physician; b. Cernauti, Rumania, Feb. 6, 1927; s. Moshe Langberg and Rosa (Feller) L.; M.D., Hebrew U., Jerusalem, Israel, 1953; m. Tova Shuisha, Sept. 26, 1954; children—Avidan, Daphna. Research fellow pediatrics Harvard Sch. Medicine Mass. Gen. Hosp., Boston, 1956-57; head pediatric metabolic and endocrine service Beilinson Med. Center, Petah Tikva, Israel, 1957—; lectr. Faculty Sci., Tel Aviv (Israel) U., 1963—; asso. prof. pediatrics, 1966—; faculty Tel Aviv Postgrad. Med. Sch., 1965—. Recipient Ioffe prize in endocrinology, 1962. Mem. Israel Soc. Pediatrics and Endocrinology, European Pediatric Endocrine Soc., Royal Soc. Medicine. Contbg. author: Somatic Growth, 1966; Handbuch der Kinderheilkunde, Band I, 1966. Research, numerous publs. on skin turgor as index of dehydration, interaction of cortico-steroids with other hormones, similarity of bovine and human growth hormones, discovery of new inborn error of metabolism of growth hormone. Home: 265 Modein, Ramat Gan, Israel, Office: Beilinson Hosp., Petah Tikva, Israel.*

LAROSE, Paul, Canadian chemist; b. Montreal, Que., Can., Apr. 26, 1898; s. Ludger and Lydie (Webb) L.; B.Sc., McGill U., 1920, M.Sc., 1923, Ph.D., 1925; m. Marie Antoinette Picard, July 2, 1925. Chemist, No. Elec. Co., Montreal, 1920-22; prof. Laval U., Quebec, Que., 1925-30; research chemist Nat. Research Council, Ottawa, Ont., Can., 1930-63; ret., 1963. Mem. Inst. Textile Sci., Fiber Soc. Research and numerous publs. on properties of textile fibers, their structure and reactivity with dyes and acids; devel. of testing methods for textiles.

LARREY, Dominique Jean, French surgeon; b. Baudean, France, July 8, 1766; student Saint-Joseph Hosp., Toulouse, France. Served at Hôtel-Dieu, Paris; became surgeon in navy, 1788, in army, 1792; became dir. ambulances Italian Army, 1797; founder Med. Sch., Milan, Italy; organized Health Service, Egyptian Army, 1798; founder Center Med. Study, Cairo, Egypt; became surgeon Imperial Guard, 1801; surgeon-in-chief to Army of Napoleon. Mem. French Acad. Scis., 1829, French Acad. Medicine. Author: Dissertation sur les amputations des membres à la suite des coups de feu, 1797; Mémoires de médecine et de chirurgie militaire, 1836; Clinique chirurgicale, 5 vols., 1830-36; Mémoires de chirurgie militaire et compagne, 1812-22. Developed method of disarticulation of humerus at shoulder joint by incision extending from acromion to center of the axilla (Larrey's amputation); 1st description of trench foot, 1813; 1st to recognize therapeutic value of maggots for treating infected wounds, 1812, the contagiousness of trachoma, 1802; created flying ambulance. Died Lyon, July 25, 1842.

LARREY, Baron Félix-Hippolyte, French surgeon; b. Paris, Sept. 18, 1808; s. Dominique Larrey; M.D., Belgium; aggregation, Paris; army med. supv., pres. consultative com. on army sanitation; became chief Army Med. Corps., 1859; prof. clin. surgery Val-de-Grace Hosp., Paris; cons. physician to Napoleon III. Mem. French Acad. Scis., 1867, Acad. Medicine (pres. 1863), Soc. Surgery, Council Health. Author: Du meilleur traitement des fractures du col du fémur, 1835; Sur la methode analytique en chirurgie, 1841; Diagnostique et curabilité du cancer, 1954; De étherisation sous le rapport de la responsabilité médicale, 1857. Pioneer in use of ether; specialist in fractures of neck and femur; research on cancer. Died Paris, Oct. 8, 1895.

LARRIEU, Marie Josette Boubee, French physician; b. Pau, France, Sept. 30, 1926; d. Jean L. and Simone (Bastoul) Boubee; M.D., Sch. Medicine, Paris, France, 1952; m. Hubert Larrieu, July 3, 1948; children—Catherine, Nathalie. Asst. dir. Centre Nat. de Transfusion, Paris, 1955-57; research asst. Centre de Recherches sur les Maladies du Sang, I.N.S.E.R.M., Paris, 1957—. Mem. Internat. Soc. Hematology, Internat. Soc. Transfusion. Author: (with J. P. Soulier), 1954; also numerous articles. Research on physiopathology of blood coagulation and hemostasis; differentiation of hemophilia into two groups, Willebrand's disease, fibrinolytic system of platelets and leucocytes; effect of fibrinogen degradation products on platelets and blood coagulation; modifications of platelets and clotting factors after irradiation and during bone marrow grafts in leukemic patients. Home: 1 Université. Office: 2, Place du Dr. Fournier (Hôpital Saint-Louis), Paris, France.*

LARSEN, Borge, Danish biochemist; b. Copenhagen, Denmark, July 23, 1925; s. Jens Peter and Inger (Juhl) L.; Magister Scientiarum, U. Copenhagen, 1953; m. Johanne Hojlund Kondrup, Dec. 28, 1949; children—Johannes, Inger, Annemarie, Sune. Research asst. Inst. Pathology, U. Copenhagen, 1954-58; research fellow

Harvard Med. Sch., also Mass. Gen. Hosp., Boston, 1958-60; research asso. Biochemistry Inst., U. Arhus (Denmark), 1960-63, research asso., Cancer Research Inst., 1964—. Research and publs. on properties of serum proteins, especially interactions of various compounds with serum albumin. Home: 7 Mariedalsvej, Braband, Denmark. Office: Cancer Research Inst., Kommunehospitalet, Arhus, Denmark.*

LARSEN, Edwin Merritt, Am. chemist, educator; b. Milw., July 12, 1915; s. Howard Reynolds and Ella (Tees) L.; B.S., U. Wis., 1937; Ph.D., Ohio State U., 1942; m. Kathryn Marie Behm, Aug. 17, 1946; children—Robert, Lynn, Richard. Chemist, Rohm and Haas, Phila., 1937-38; teaching asst. Ohio State U., 1938-42; faculty dept. chemistry U. Wis., Madison, 1942-43, 46—, prof., 1958—. Vis. prof. U. Fla., 1958, Fulbright lectr. Technische Hochschule, Vienna, Anorganische Institut, 1966-67. Mem. A.A.A.S., Am. Chem. Soc. (chmn. Wis. sect.), Wis. Acad. Scis., Arts and Letters, Sigma Xi, Phi Lambda Upsilon. Author: Transitional Elements, 1965. Contbr. articles to profl. jours. Research on transitional elements, titanium, zirconium, thorium, hafnium. Home: 109 Standish Ct., Madison, Wis. 53705.*

LARSEN, Esper Signius, Jr., Am. petrologist; b. Astoria, Ore., Mar. 14, 1879; s. Esper Signius and Louisa (Pauly) L.; B.S., U. Cal. at Berkeley, 1906, Ph.D., 1908; m. Eva Audrey Smith, 1910; children—Clark Smith, Esper Signius III. Asst. petrographer Geophys. Lab., Carnegie Instn., Washington, 1907-09; staff U. S., Geol. Survey, 1909, in charge petrology, 1918-23, cons. geochemistry and petrology br., 1949—; prof. petrography Harvard, 1923-49, prof. emeritus, 1949—. Mem. Nat. Acad. Scis., Am. Acad. Arts and Sci., Am. Inst. Mining, Metall. and Petroleum Engrs., Geol. Soc. Am. (Penrose medal), Mineral. Soc. Am. (pres. 1928, Roebling medal), Soc. Econ. Geologists, Mineral. Soc. Gt. Britain, Geol. Soc. London. Research and numerous publs. on volcanics of San Juan region of southwestern Colo., igneous rocks of Highwood Mountains of Mont., batholithic rocks of So. Cal.; detailed optical properties of over 600 minerals (with H. Berman), still foremost reference work on subject; developed Larsen method for age determination of rocks. Office: 3930 Connecticut Av., Washington.

LARSEN, Helge, Norwegian biochemist; b. Aalesund, Apr. 25, 1922; s. Oscar and Olga (Grytnes) L.; Ph.D. in Chem. Engring., Tech. U. Norway; m. Olga Helene Frisvoll, Mar. 25, 1948; children—Jan Helge, Anne Kathrine. Head research Norwegian Council. Sci. and Indsl. Research, 1947-55; instr. microbiology Tech. U. Norway, 1951-55; prof. biochemistry U. Norway, 1955—. Prof. bacteriology U. Cal. at Berkeley, 1961-62. Hon. mem. Royal Norwegian Acad. Scis., Norwegian Acad. Tech. Contbr. articles on biochemistry, microbiology. Research in microbial and marine biochemistry. Home: Biskop Skaars gate 8, Trondheim. Office: Narges Tekniske Hogskole, Trondheim, Norway.

LARSEN, Joseph Reuben, Jr., Am. insect physiologist; b. Ogden, Utah, May 21, 1927; s. Joseph Reuben and Charolette (Anderson) L.; B.S., U. Utah, 1950, M.S., 1952; Sc.D., Johns Hopkins, 1958; m. Shauna Lucy Stewart, Dec. 15, 1948; children—Pamela B., Deborah L., Jennifer A. Entomologist biol. warfare U. S. Army Chem. Corps, 1952-54; faculty U. Pa., 1958-62, U. Wyo., 1962-63; asso. prof. entomology U. Ill., Urbana, 1963—. Mem. A.A.A.S., Entomol. Soc. Am., Am. Soc. Zoology, Electron Microscope Soc. Am., Am. Micros. Soc. Author: A Laboratory Manual in Biology, 1965. Research in endocrine relationship in insects, insect histology, microanatomy, electron microscope studies of insect chemoreceptors and insect nervous tissue; physiology of insect sensory receptors and insect nervous systems. Home: 2506 William St., Champaign, Ill. 61820. Office: Dept. Entomology, U. Ill., Urbana, Ill. 61801.*

LARSEN, Kai, Danish botanist; b. Hillerod, Denmark, Nov. 15, 1926; s. Axel and Elisabeth (Hansen) L.; B.Sc., Ferderiksborg St. Coll., U. Copenhagen (Denmark), 1946, M.Sc., 1952; m. Anna G. Steneker, 1949; children—Ann-Margrit, Hanne; m. 2d, Anna D. Frandsen, Apr. 23, 1957; children—Lisbeth, Christian. Sci. asst. bot. lab. Copenhagen U., 1952-55, censor in botany, 1962—; sci. asst. Royal Danish Soc. Pharmacology, 1955-62, asst. prof., 1962-63; prof. botany Aarhus (Denmark) U., 1963—, head bot. Inst., dir. Herbarium, 1963—; mem. expdns. to Greenland, 1947, Canary Island, 1958, Thailand, 1958, 61-62, 63, 66. Decorated Officer, Crown of Thailand, 1967. Mem. Nordic Soc. Taxonomic Botany (chmn. 1967). Plant genus Larsenia (acanthaceae) named in his honor. Author numerous books and articles on taxonomic botany and cytology. Editor, contbg. author series: studies in the Flora of Thailand, 1959—. Bot. exploration of Thailand; relic status of Canarian plants, relationships in genus of Lotus. Home: 6 Store Torv, Aarhus, Denmark.*

LARSEN, Poul Lauritz, plant physiologist; b. Höjby, Denmark, Oct. 30, 1909; s. Lauritz and Anna (Jörgensen) L.; Ph.D., U. Copenhagen; m. Ellen Mikkelsen, Jan. 30, 1932; children—Oluf, Bodil, Agnete. Instr. microscopy and microbiology labs. Danish Inst. Tech., 1936-49; prof. plant physiology U. Chgo., 1949-52; prof. botany U. Bergen (Norway), 1952—, dir. Botany Inst., 1955—. Mem. Norwegian Acad., Leopoldina, Royal Acad. Denmark, Scandinavian Assn. Plant Physiologists (became pres. 1964), Soc. for Advancement Sci. Bergen, Sigma Xi. Author: Plantelivet, 1948; Planternes väkststoffer, 1962. Research and publs. on geotropism, growth regulators, heart of beechwood, formation of solid material in plants. Home: Professorveien 14, Minde, Bergen, Norway.

LARSH, Howard William, Am. mycologist; b. East St. Louis, Ill., May 20, 1914; s. John E. and Maggie Larsh; B.A., McKendree Coll., 1936; student Wash. U., 1936-37; M.S., U. Ill., 1938, Ph.D., 1941; m. Georgia Lee Thomson, Sept. 4, 1938; 1 son, Jonathan. With U. Ill., 1937-41, U. Okla., 1941-43; plant pathologist U. S. Dept. Agr., Fayetteville, Ark., 1943-46; chmn. dept. U. Okla., 1945-52; with USPHS, Kansas City, 1950-51, cons., 1952—; research prof. U. Okla., 1962—, chmn. dept., 1967—. Cons. Mo. State Sanatorium, Mt. Vernon, McKnight State Sanatorium, Carlsbad, Tex., VA, Oklahoma City. Diplomate Am. Acad. Microbiology. Fellow A.A.A.S., Okla. Acad. Sci., Am. Pub. Health Assn.; mem. Bot. Soc. Am., Am. Soc. Microbiology, Reticuloendothelial Soc., Soc. Exptl. Biology and Medicine, Am. Soc. for Microbiology, Sigma Xi, Lambda Tau, Phi Sigma. Contbr. numerous articles to sci. jours. Home: 611 Broad Lane, Norman, Okla. 73069.*

LARSON, Bruce Linder, Am. biochemist; b. Mpls., June 24, 1927; s. Leif Raeder and Olive Hazel (Linder) L.; B.S., U. Minn., 1948, Ph.D., 1951; m. Marjorie Helen Hersleth, Sept. 24, 1954; children—Eric Martin, David Bruce, Brian Linder. Research asst. U. Minn., 1948-51; instr. U. Ill., 1951-53, asst. prof., 1953-59, asso. prof., 1959-66, prof. biol. chemistry dept. dairy sci., 1966—. Contbr. Internat. Biochem. Congresses, Russia, 1961, Japan, 1967; Fulbright lectr. Argentina, 1965. Recipient Am. Chem. Soc. Borden award, 1966. Mem. Am. Soc. Biol. Chemistry, Am. Chem. Soc., Am. Soc. Cell Biology, Am. Dairy Sci. Assn., A.A.A.S., Sigma Xi. Research and numerous articles on milk proteins and biochem. mechanisms concerned with metabolism of milk; devel. of methods for in virto cultivation of secretory cells of mammary gland and study of biochem. differentiation and dedifferentiation in these cells.*

LARSON, Carroll Bernard, Am. orthopedic surgeon; b. Council Bluffs, Ia., Sept. 10, 1909; s. Charles Bernard and Ida Caroline (Skarin) L.; B.S., U. Ia., 1931, M.D., 1933; m. Nadine West Townsend, Nov. 17, 1934; children—John Weston, Carroll Bernard, Charles Jeffrey, Lori. Mem. orthopedic staff Harvard Med. Sch., Mass. Gen. Hosp., 1940-50; prof., head dept. orthopedic surgery U. Ia. Coll. Medicine, 1950—, dir. div. rehab., 1953—. Chmn. med. adv. bd. Shriner's Hosps. Crippled Children. Decorated Royal Order St. Olaf (Norway). Diplomate Am. Bd. Orthopedic Surgery. Fellow A.C.S.; mem. A.M.A., Am. Acad. Orthopedic Surgeons (pres. 1966-67). Am. Orthopedic Assn. (hon.), Am. Rheumatism Association. Mem. editorial bd. Jour. Bone and Joint Surgery. Home: Route 2, The Meadows. Office: University Hosp., Iowa City*

LARSON, Charles Philip, Am. physician; b. Eleva, Wis., Aug. 15, 1910; s. Clarence Philip and Louise (Steig) L.; B.A., Gonzaga U., 1931; M.D., C.M., McGill U., 1936; certificate of proficiency U. Mich., 1939; m. Margaret Ida Kobervig, Dec. 6, 1944; children—Charles Philip, Lillian Louise (Mrs. Philip Randolph), Charles Palmer, Christine, Elizabeth, Paul M., Laurence A. Dir. labs. Tacoma Gen. Hosp., 1939—; sr. partner Larson, Wicks, Whitaker, Reberger, Tacoma, 1947—; faculty pathology U. Ore. Med. Sch., 1939-41; asst. clin. prof. pathology U. Wash., Seattle, 1948—, asst. prof. biology U. Puget Sound, 1960—. Cons. in pathology to hosps., local govt., armed forces, U. Tex., 1939—. Mem. Coll. Am. Pathologists (pres. 1958-59), Am. Cancer Soc. (pres. Wash. div. 1958), Internat. Assn. Forensic Pathologists (pres. 1957), Pacific N.W. Soc. Pathologists (pres. 1947), N. Pacific Soc. Neurology and Psychiatry (pres. 1947), Internat. Congress Clin. Pathology (v.p.), Am. Soc. Clin. Pathologists, Nat. Safety Council. Author: Manual of Neuropathology, 1940. Research and publs. on cancer of liver in trout, death of Olympia oysters, air embolism acute mental hosp. deaths. Home: 2018 N. 30th St. Office: 315 S. Kay St., Tacoma 98405.*

LARSON, Daniel Lewis, Am. physician; b. Sioux City, Ia., Nov. 28, 1921; s. Charles G. and Selma (Daniels) L.; M.D., Columbia U., 1946; m. Josephine Storey, Feb. 23, 1962. John and Mary Markle scholar Columbia U., 1950-57; faculty, 1950—, asst. prof. medicine, 1959—; asst. attending physician Presbyn. Hosp., N.Y.C., 1965—; dir. medicine St. Barnabas Hosp., N.Y.C., 1965—; practice internal medicine, N.Y.C.; chief Rheumatic Fever Clinic, Vanderbilt Clinic, N.Y.C. 1955—. Med. cons. UN, 1960—. Mem. A.C.P., Harvey Soc., Soc. for Clin. Investigations, Am. Rheumatic Soc., Am. Assn. Immunology. Author: Systemic Lupus Erythematos, 1961; Arthritis and Rheumatism, 1962; also articles. Immunochem. and clin. research on patients with rheumatic diseases. Home: 2441 Webb Av., Bronx 10468. Office: St. Barnabas Hosp., 183d and 3d Av., Bronx, N.Y. 10457.*

LARSON, John Augustus, psychiatrist; b. Shelbourne, N.S., Can., Dec. 11, 1892; s. Lars and Lucina Antonina (Mack) L.; A.B., Boston U., 1914, A.M., 1915; Ph.D., U. Cal., 1920, grad. Sch. Criminology, 1923; M.D., Rush Med. Coll., 1928; m. Margaret Steele Taylor, Aug. 9, 1922 (dec. 1960); 1 son, John William Earle. Faculty, Boston U., 1914-15, U. So. Cal., 1915-16, U. Cal., 1916-20, Johns Hopkins, 1927-28, U. Ia. Med. Sch., 1929-30, U. Chgo., 1922-36; research psychologist State of Ill., 1923-27, asst. criminologist, 1930-36; asst. dir. Psychopathic Clinic, City of Detroit; dir. Seattle Mental Hygiene Clinic; chief psychiatrist N.M. State Hosp., 1946-47; supt. Ariz. State Hosp., Phoenix, 1947-49, Logansport (Ind.) State Hosp.; clin. dir., dir. research Wabash Valley Sanitarium Hosp., Lafayette, Ind., 1955-57; supt. Maximum Security Hosp., also prison psychiatrist Tenn. State Prison, Nashville, 1957-61; psychiat. dir. Huron (S.D.) Mental Health Center, 1961-63; now researcher and med. writer, Nashville. Life fellow Am. Psychiat. Assn., Am. Orthopsychiat. Assn.; life mem. A.M.A., Am. Med. Writers Assn.; mem. World Psychiat. Assn., Acad. Criminal Interrogation, Internat. Assn. Police Psychiatry, Johns Hopkins Med. and Surg. Assn., Internat. Soc. Criminal Identification, Norse Civic Assn., Sigma Psi, Alpha Kappa Kappa, others. Author: Larson Single Fingerprint System, 1923; (with George W. Harry, Leonardo Keeler) Lying and its Detection, 1932. Research, publs. on invention of cardio-pneumo-psychograph, commonly known as lie detector (1921); on psychol. aspects of deception, crime, alcoholism, cerebral disorders. Address: 652 Ezell Rd., Nashville.*

LARSON, Paul, Danish industrialist; b. Denmark, 1859; Danish industrialist from Aalborg; invented modern cement prodn. machines especially the rotation furnace. Died 1935.

LARSON, Paul Stanley, Am. pharmacologist; b. Cannon Falls, Minn., June 9, 1907; s. John Frank and Anna (Bergquist) L.; student Modesto Jr. Coll., 1926-28; A.B., U. Cal., Berkeley, 1930, Ph.D., 1934; m. Dorothy Gardiner Smith, Mar. 12, 1936; 1 son, John Gardiner. Faculty, Georgetown Med. Sch., Washington, 1934-39; pharmacologist Frederick Stearns, Detroit, 1940-41, lectr. pharmacology Wayne U. Med. Sch., Detroit, 1940-41; faculty Med. Coll. Va., Richmond, 1939-40, 41—, Haag prof., chmn. dept. pharmacology, 1963—. Mem. various coms. AEC, NRC, A.M.A., NIH. Mem. Am. Soc. Pharmacology & Exptl. Therapeutics, Am. Physiol. Soc., Soc. Exptl. Biology and Medicine, A.A.A.S., N.Y., Va. acads. medicine, Am. Chem. Soc., Royal Soc. Medicine, Am. Coll. Clin. Pharm. and Therapeutics, Soc. Toxicology (pres. 1963-64). Author (with H. B. Haag, H. Silvette) Tobacco: Experimental & Clinical Studies, 1961. Research and publs. in protein, purine and mineral metabolism, toxicology of pesticides and food additives, biol. actions of nicotine and tobacco. Home: 3044 Stratford Rd., Richmond, Va. 23225.*

LARSON, William Earl, Am. soil scientist; b. Creston, Neb., Aug. 7, 1921; s. John William and Lela (Craig) L.; B.S., U. Neb., 1944, M.S., 1946; Ph.D., Ia. State U., 1949; m. Ruthelaine Thomsen, June 15, 1947; children—Larry W., Stephen E., Suzanne L., Kathy M. Asst. prof. agronomy Ia. State U., Ames, 1949, prof. soils, 1954-67; prof. soils U. Minn., Mpls., 1967—; soil scientist U. S. Dept. Agr., Mont. State Coll., 1950-54, Ia. State U., 1954-67, U. Minn., 1967—. sr. research fellow Australian-Am. Edn. Found., Adelaide, 1965-66. Recipient certificate of Merit U. S. Dept. Agr., 1958, Best Paper awards Soil Conservation Soc. Am., 1961, 63. Fellow Am. Soc. Agronomy, Soil Conservation Soc. Am., A.A.A.S.; mem. Soil Sci. Soc. Am., Soil Conservation Soc. Am. Research and publs. on mgmt. of soils. Home: 3334 Richmond Av., St. Paul 55112.*

LARSON, Winford Porter, Am. bacteriologist; b. Poy Sippi, Wis., Mar. 7, 1880; s. Charles J. and Mary (Peterson) L.; student Milton Coll., 1896-97, Union Coll., Lincoln, Neb., 1897-99; M.D., U. Ill. Coll. Medicine, 1904; grad. study U. Berlin, 1906-08, Sorbonne, Paris, 1909-10; m. Alma E. Meldal, Apr. 4, 1908; children—Lorna G., Douglas M. Instr. bacteriology U. Minn., 1911-12, asst. prof. bacteriology, 1912-15, asso. prof., 1915-18, prof. and head of dept. since 1918. Mem. Soc. Am. Bacteriologists, Am. Soc. Immunologists, Assn. Pathologists and Bacteriologists, Soc. Exptl. Biology and Medicine, N.Y. Acad. Scis. (asso.). Contbr. to bacteriology. Died Jan. 1, 1947.

LARSSON, Karl Erik, Swedish physicist; b. Varnum, Sweden, Feb. 24, 1923; s. Axel Hjalmar and Agnes (Eriksson) L.; M.S., Uppsala (Sweden) U., 1948; Ph.D., Stockholm (Sweden) U., 1955; m. Kerstin Lindquist, Oct. 4, 1947. Staff, Research Inst. for Nat. Def., Stockholm, Sweden, 1948-50, AB Atomenergi, Stockholm, 1950-55, 56-61; Brookhaven Nat. Lab., Upton, L.I., N.Y., 1955-56; with Royal Inst. Tech., Stockholm, 1961—, now chmn. dept. tech. physics. Contbg. author: Thermal Neutron Scattering; Simple Dense Fluids. Research, publs. in neutron physics including neutron techniques, neutron flux studies, neutron spectrometry, atomic and molecular dynamics in solids and liquids. Home: 13 Kattilmunds Väg, Rotebro, Sweden. Office: Royal Inst. Tech., Stockholm, Sweden.*

LARSSON, Yngve Axel Andreas, Swedish physician; b. Jan. 9, 1917; s. Yngve Gustav R. and Elin (Bonnier) L.; M.D., Karolinska Inst., Stockholm, Sweden, 1943; m. Gerd Franzen, July 23, 1940; children—Yngve Mikael, Soren Yngve, Betty Cecilia, Simon Yngve; m. 2d, Ulla Barnholdt, May 15, 1965; children—David Joakim, Yngve Jonas. Research fellow Karolinska Inst.,

1952-56, asso. prof. pediatrics, 1956; head diabetes services dept. Crown Princess Lovisa's Children's Hosp., Stockholm, 1957-64; dir. Ethio-Swedish Pediatric Clinic, Addis Ababa, Ethiopia, 1965—; prof. pediatrics Haile Selassie I Univ., 1965—. Mem. Swedish, Ethiopian med. assns. Author: Morphology of the Pancreas and Glucose Tolerance Under Different Experimental Conditions, 1956; (with G. Sterky, G. Christiansson) Long-Term Prognosis in Juvenile Diabetes, 1962; The Management of Diabetes, 1954, 58, 61; also numerous articles. Research in peripheral circulation in man, its disorders and their treatment, childhood diabetes, its clin. manifestations, treatment, complications and prognosis, exptl. diabetes, morphology of pancreas after ligation of bile duct and pancreatic duct, clin. studies of obesity in children, effect of exercise in normal and disabled teenagers. Address: Ethio-Swedish Pediatric Clinic, P.O. Box 1768, Addis Ababa, Ethiopia.*

LARTET, Édouard Armand Isidore Hippolyte, French paleontologist, archaeologist; b. St. Guirauld, Gers, France, Apr. 15, 1801; studied law, Auch and Toulouse. Began excavating fossils nr. Auch, 1834; continued exploration of French caves, Dordogne and Vézère valleys and Aurignac; prof. paleontology Mus. of Jardin des Plantes, 1869-71. Fgn. mem. Geol. Soc. London, 1857; pres. Société Géologigue and Société d'Anthropologie, 1866, Internat. Congress Archaeology and Prehistoric Anthropology, 1867. Author: The Antiquity of Man in Western Europe, 1860; New Researches on the Coexistence of Man and of the Great Fossil Mannifers Characteristic of the Last Geological Period, 1861; (with W. H. Christie) Relique aquitanicae, 1865. A founder of modern paleontology; described Pliopithecus antiquus, 1836, Dryopithicus, 1850; discovered Cro-Magnon Man, 1868; responsible for archaeol. collection in Mus. St. Germaine; demonstrated that man lived in times of Pleistocene mammals. Died Seissan, France, Jan. 28, 1871.

LA RUE, Carl Downey, Am. botanist; b. Williamsville, Ill., Apr. 22, 1888; s. Abraham Chronister and Charlotte Parthena (Bates) La R.; B.S., Valparaiso U., 1910, A.B., 1911; A.B., U. Mich., 1914, A.M., 1916, Ph.D., 1921; research, Harvard, 1936-37; m. Evelina Brown Forman, June 1, 1914; children—Adrian Jan Pieters, Anna Virginia, Charlotte Evelina, Carl Forman. Instr. botany, Syracuse U., 1916-17, botanist Hollandsch-Amerikaansche Plantage Maatschappij, Sumatra, 1917-20; instr. botany, asst., asso. prof. U. Mich., 1920-44, prof., 1944—; mem. Mich. Biol. Sta. staff, 1925-50; research Fed. Expt. Sta., P.R., 1951; specialist, rubber investigation, in charge S.Am. rubber expdn. U. S. Dept. Agr., 1923-24; specialist, co-dir. Ford Morot Amazon-Tapajos Expdn., 1926-27; agt., rubber investigations, in charge expdns., Bolivia, Nicaragua, Mexico U. S. Dept. Agr., 1940-41, prin. specialist U. S. Dept. Agr., 1943-44. Fellow A.A.A.S.; mem. Am. Soc. Naturalists, Bot. Soc. Am., Sullivant Moss Soc., Torrey Bot. Club, Mich. Acad. Sci., Arts, Letters (sec., 1923), Sigma Xi. Author Agrl. Dept. bulls.; contbr. tech. articles to bot. jours. Died Aug. 19, 1955.

LASAGNA, Louis Cesare, Am. physician; b. N.Y.C., Feb. 22, 1923; s. Joseph and Carmen (Boccignone) L.; B.S., Rutgers U., 1943; M.D., Columbia U., 1947; m. Helen Chester Gersten, July 20, 1946; children—Nina, David, Maria, Kristin, Lisa, Peter, Christopher. Instr., John Hopkins, 1950-52, faculty, 1954—, asso. prof. medicine, 1957—, asso. prof. pharmacology, 1959—; research asso. Harvard, 1953-54. Mem. Am. Soc. for Pharmacology and Exptl. Therapeutics, Am. Soc. for Clin. Investigation, Assn. Am. Physicians. Author: The Doctors' Dilemmas, 1962; also numerous articles. Research on analgesics, placebo phenomenon, methodology of clin. trials. Home: 1903 South Rd., Balt. 21209.*

LASALLE, Joseph Pierre, Am. mathematician; b. State College, Pa., May 28, 1916; s. Leo Joseph and Aline (Mistric) LaS.; B.S., La State U., 1937; Ph.D. (Henry Laws fellow), Cal. Inst. Tech., 1941; m. Eleanor Seip, June 12, 1942; children—Nannette Maria, Marc Joseph. Instr., U. Tex., 1941-42, Radar Sch., Mass. Inst. Tech., 1941-42; research asso. Princeton, 1943-44, Cornell U., 1944-46; faculty Notre Dame, 1946-47; 48-58, prof. math., 1951-58; asso. dir. Math. Center, Research Inst. for Advanced Studies, Balt., 1958-64; prof., dir. Center for Dynamical Systems, Brown U., Providence, 1964—, also chmn. div. applied mathematics, 1968—. Fellow A.A.A.S.; mem. Soc. for Indsl. and Applied Math. (past pres.), Am. Math. Soc., Math. Assn. Am. (Chauvenet prize 1956). Author: (with N. Haaser, J. Sullivan) Introduction to Analysis, Vol. 1, 1959; (with S. Lefschetz) Stability by Liapunov's Second Method with Applications, 1961, Recent Soviet Contributions to Mathematics, 1962. International Symposium on Nonlinear Differential Equations and Nonlinear Mechanics, 1963; (with N. Haaser, J. Sullivan) Intermediate Analysis, Vol. 2, 1964; (with Jack K. Hale) Differential Equations and Dynamical Systems, 1967; also articles. Pioneered devel. of modern theory of optimal control; contbns. to theory of differential equations and dynamical systems and to theory of stability. Home: 78 Gov. Bradford Dr., Barrington, R.I. 02806.*

LASAULX, Arnold von, see von Lasaulx, Arnold.

LASÉGUE, Ernest Charles, French physician; b. Paris, Sept. 5, 1816; M.D., 1846; Studied under Trousseau. Became head clinic, 1852; joined Faculty Paris, 1853; became physician hosps., 1854; in charge complementary courses on mental diseases and neurology dept., 1862, 65, 66; named prof. pathology, 1867; became prof. hosp. clinics l'hôpital de la Pitie, Paris, 1870. Mem. French Acad. Medicine. Author: Traité des angines, 1848; Traité de l'auscultation de Laenac; Traité de la goutte d Lydenham (translation); lessons and memoires compiled by H. Blum, 2 vols., 1884; also articles on dyspepsia, mental and nervous disorders, diabetes. Described persecution mania (Lasègue's disease), 1852; proved elevation of extended lower extremity in sciatica causes pain along sciatic nerve, 1881; sign of Laseque (nervous disorder), syndrome of Laseque (syndrome in hysteria) named after him. Died Paris, Mar. 20, 1883.

LASFARGUES, Etienne Yves, biologist; b. Milhars, France, May 5, 1916; s. Emmanuel and Rosa (Durand) L.; B.S.-Philosophy, Lycee Voltaire, Paris, France, 1936; postgrad. Nat. Vet. Sch. of Alfort, France; D.V.M., Paris U., 1942; m. Jennie Carmella DiFine, July 15, 1950; children—John Emmanuel, Michele Rose-Marie. Came to U. S., 1955, naturalized, 1962. Roux Research fellow Pasteur Inst., Paris, 1942-44, asst., 1944-50, head lab., 1950-55; research fellow Am. Cancer Soc. Inst. for Cancer Research, Phila., 1947-50; asso. microbiology Columbia, N.Y.C., 1955-59; asst. prof. Coll. Phys. and Surg., 1959—. Recipient Jensen prize Nat. Acad. Medicine, Paris, 1945; named hon. head lab. Pasteur Inst., 1956. Mem. European Tissue Culture Club, Internat. Soc. for Cell Biology, Tissue Culture Assn., Am. Assn. for Cancer Research, Am. Soc. for Cell Biology. Research, numerous publs. on tissue cultures applied to study microbial toxins and allergic phenomena, cell growth factors, replication mammary carcinoma virus in tissue culture, transformation normal mammary cells, mechanism mouse mammary carcinogensis; pioneered cell line human mammary carcinoma, method cultivation normal mammary tissues.

LASHAS, Vladas Laurinovich, Russian physiologist; b. 1892; grad. Med. Faculty, Tartu U., 1918. Asst., later lectr. dept. physiology, a founder Med. Faculty, Kaunas U., 1922-25, prof., 1926-50; prof. chair physiology Kaunas Med. Inst., 1950—. Mem. USSR Acad. Med. Sci. (corr.), Lithuanian Acad. Sci. (acad. sec. dept. natural sci. 1946-62). Author: Anaphylaxis, 1926; Principles of the Science of Nutrition, 1945; Vitamins, 1946; The Endocrine Glands, 1946; Physiology of the Muscles and the Nervous System, 1950; Practical Studies in Physiology, 1954; Physiology of Analysers, 1954; Blood and Blood Circulation, 1958. Co-editor Physiology sect. Large Med. Ency., 2d edit. Research and publs. on anaphylaxis and anaphylactic shock, dietary problems; deviser method to determine resorption of crude proteins from intestines. Address: Kaunas Med. Inst., ulitsa Michkevichausa 7, Vilnius, Lithuanian SSR, USSR.

LASHKAREV, Vadim Yergenevich, Russian physicist; b. Oct. 7, 1903; grad. Kiev Inst. Pub. Edn. (now Kiev U.), 1924. Instr., Kiev Poly. Inst., 1925-30; with Leningrad Physicotech. Inst., USSR Acad. Sci., Leningrad. Poly. Inst., 1930-35; head dept. semicondrs. Inst. Physics, Ukrainian Acad. Sci., 1939-60, dir. Inst. Semicondrs., 1961—; dir. Lab. Photoelec. Phenomena; prof. Kiev U., 1939-57. Mem. Ukrainian Acad. Sci. Author: Electron Diffraction, 1933; Photoelectromotive Forces in Semiconductors, 1952; Solar Batteries, 1958; Solid-State Physics, 1959. Research and publs. on exptl. and theoretical electron diffraction and semicondr. physics. Address: Inst. Semiconductors, Ukrainian Acad. Sci., Vladimirskaya ulitsa 54, Kiev, Ukrainian SSR, USSR.

LASHLEY, K(arl) S(pencer), Am. psychologist; b. Davis, W.Va., June 7, 1890; s. Charles Gilpen and Margaret Blanche (Spencer) L.; A.B in Zoology, W.Va. U., 1910; M.S. in Bacteriology, U. Pitts., 1911, D.Sc., 1936; Ph.D. in Genetics, Johns Hopkins, 1914, Johnston scholar, 1915-16, LL.D., 1953; M.A. (hon.) Harvard, 1942; D.Sc., U. Chgo., 1941, Western Res. U., 1951, U. Pa., 1955; m. Edith Ann Baker, 1918 (dec. 1948); m. 2d, Dr. Claire Imredy Schiller, 1957. Instr. psychology U. Minn., 1917-18, asst. prof. psychology, 1920-21, asso. prof., 1921-24, prof., 1924-26; investigator U. S. Interdepartmental Social Hygiene Bd., 1919-20; acting prof. psychology U. Chgo., summer 1925, prof. psychology, 1929-35; acting prof. psychology Columbia, summer 1926; research psychologist Behavior Research Fund of Inst. for Juvenile Research, Chgo., 1926-29; lectr. univs. London, Berlin, Moscow, 1932; prof. psychology Harvard, 1935-37, research prof. neuropsychology, 1937-55, emeritus, 1955-58; dir. Yerkes Labs. of Primate Biology, 1942-55, emeritus, 1955-58. Recipient Howard Crosby Warren medal in psychology Soc. Exptl. Psychologists, 1937; Daniel Giraud Elliot Medal in zoology Nat. Acad. Scis., 1943; William Baly Medal in physiology Royal Coll. Physicians, 1953. Mem. Nat. Acad. Scis., Am. Philos. Soc., Am. Acad. Arts and Scis., Am. Psychol. Assn. (pres. 1929), Soc. Am. Naturalists (pres. 1947), Am. Soc. Zoologists, Am. Physiol. Soc., Soc. Exptl. Psychologists, Am. Soc. Human Genetics, NRC (div. anthropology and psychology 1927-30, 32-35); hon. mem. Am. Neurol. Assn., Harvey Soc., N.Y. Acad. Scis., Brit. Assn. Study Animal Behavior; fgn. mem. Royal Soc. London, Brit. Psychol. Assn. Contbr. more than 100 articles and monographs (including Brain Mechanisms and Intelligence, 1929) on structure and functions of brain, comparative psychology, animal behavior, instincts of birds and primates, learning, and genetics; specialist in genetic psychology; discovered that localization of brain functions is characterized by little precision or persistence; inventor jumping techniques for rats. Died Aug. 7, 1958.

LASHOF, Richard Kenneth, Am. mathematician; b. Phila., Nov. 9, 1922; s. Samuel and Anna (Sandler) L.; B.S., U. Pa., 1943; Ph.D., Columbia, 1954; m. Joyce Ruth Cohen, June 11, 1950; children—Judith, Carol, Daniel. Faculty, U. Chgo., 1954—, prof. math., 1964—, chmn. dept., 1967—. NSF Sr. Postdoctoral fellow, 1960-61. Mem. Am. Math. Soc., Math. Assn. Am. Research, publs. on devel. of theory of smoothing of high dimensional surfaces and curvature of surfaces; studied algebraic machinery for distinguishing topological spaces, especially those with multiplicative properties. Home: 4812 S. Greenwood Av., Chgo. 60615.*

LASKER, Gabriel Ward, anthropologist; b. Huntington, Eng., Apr. 29, 1912; s. Bruno and Margaret (Ward) L.; student, U. Wis., 1928-30; A.B., U. Mich., 1934; M.A., Harvard, 1940, Ph.D., 1945; m. Bernice Antoville Kaplan, July 31, 1949; children—Robert Alexander, Edward Meyer, Ann Titania. Faculty, Wayne State U. Sch. Medicine, Detroit, 1946—, prof., 1964—; asst. dir. ednl. resources in anthropology project U. Cal., Berkeley, 1960-61. Fulbright research scholar, Peru, 1957-58. Fellow, Am. Anthropol. Assn., A.A.A.S. (vice president and chmn. anthropology sect.); mem. Conf. Biol. Editors (chmn. 1963—), Am. Assn. Phys. Anthropologists (pres. 1963—), Am. Assn. Anatomists, Am. Soc. Human Genetics, Soc. for the Study of Human Biology. Author: The Evolution of Man, 1961; Human Evolution, 1963. Editor: The Processes of Ongoing Human Evolution, 1960; Yearbook of Physical Anthropology, 1945—; (with David G. Mundlebaum, Ethel M. Albert) The Teaching of Anthropology, 1963. Editor: Human Biology, 1953—. Research on phys. anthropology, especially relation between community size, breeding practices, and frequency of genetic characteristics. Office: 1400 Chrysler Expressway, Detroit 48207.*

LASKI, Harold J., Brit. polit. scientist; b. Manchester, Eng., June 30, 1893; s. Nathan and Sarah L.; ed. New Coll., Oxford; B.A., 1914; LL.D., U. Athens; m. Frida Kerry, 1 dau. Lectr. in history McGill U., Can., 1914-16; lectr. Harvard, 1916-20; Henry Ward Beecher lectr. Amherst Coll., 1917; Harvard lectr. Yale, 1919-20; with London Sch. Econs., 1920-50; lectr. Magdalene Coll., Cambridge, 1922-25; prof. polit. sci. U. London, 1926-50; vis. prof. Yale, 1931, Storrs lectr., 1933; Donnellun lectr. Trinity Coll., Dublin, 1936. Vice chmn. Brit. Inst. Adult Edn., 1921-30; exec. Fabian Soc., 1922-36; mem. Indsl. Ct., 1926-50, Lord Chancellor's Com. on Delegated Legislation, 1929, Deptl. Com. on Local Govt., 1931, Com. on Legal Edn., 1932, council Inst. Pub. Adminstrn.; chmn. exec. com. Labour party, 1936-49, chmn., 1945-46. Author: The Problem on Sovereignty, 1917; Authority in the Modern State, 1919; Political Thought from Locke to Bentham, 1920; Foundations of Sovereignty, 1921; Letters of Burke (editor), 1922; The Defense of Liberty against Tyrants (editor), 1924; Autobiography of J. S. Mill (editor), 1924; A Grammar of Politics, 1925; Communism, 1927; Liberty in the Modern State, 1930; The Dangers of Obedience, 1930; An Introduction to Politics, 1931; Studies in Law and Politics, 1932; The Crisis and the Constitution, 1932; Democracy in Crisis, 1933; The State in Theory and Practice, 1935; The Rise of European Liberalism, 1936; Parliamentary Government in England, 1938; The American Presidency, 1940; Reflections on the Revolution of Our Time, 1943; Faith, Reason, Civilisation, 1944; The American Democracy, 1948. A leading proponent of left-wing polit. outlook in modern English-speaking world; urged reform based on econ. realities rather than abstract theories of state, 1920's-30's; offered arguments in favor of pluralistic state as opposed to monistic. Died Mar. 24, 1950.

LASKIN, Daniel M., Am. oral surgeon; b. Ellenville, N.Y., Sept. 3, 1924; s. Nathan and Flora (Kaplan) L.; student N.Y. U., 1941-42; B.S., Ind. U., 1947; M.S., U. Ill., 1951; m. Eve Pauline Mohel, Aug. 25, 1945; children—Jeffrey, Gary, Marla. Faculty, U. Ill., Chgo., 1949—, prof. dept. oral and maxillofacial surgery, 1960—, asso. head dept., 1962—, clin. prof. surgery, 1961—, dir. temporomandibular joint center, 1963—; attending oral surgeon Edgewater Hosp., Bethesda Hosp., Mt. Sinai Hosp. (all Chgo.). Diplomate Am. Bd. Oral Surgery. Fellow Am. Coll. Dentists, A.A.A.S.; mem. Ill. Splt. Bd. Oral Surgery, Am. Dental Assn., Am. Soc. Oral Surgeons (editor Newsletter 1965—), Internat. Assn. Dental Research, Am. Dental Soc. Anesthesiology, Am. Soc. Exptl. Pathology, Am. Assn. Dental Editors, Omicron Kappa Upsilon, Sigma Xi, others. Editor sect. Jour. Oral Surgery, 1966—. Research and publs. on connective tissue physiology and pathology, particularly cartilage and bone metabolism, craniofacial growth, and pathology of temporomandibular joint. Home: 3844 Enfield St., Skokie, Ill. 60077. Office: 2440 W. Peterson St., Chgo. 60645.*

LASLEY, John Wayne, Jr., Am. mathematician; b. Burlington, N.C., Sept. 22, 1891; s. John Wayne and Elizabeth (Abernethy) L.; A.B., U. N.C., 1910, M.A., 1911; Ph.D., U. Chgo., 1920; m. Edna Rain Millikan, Sept. 9, 1924; 1 son, John Wayne III. Faculty, U. N.C., Chapel Hill, 1911-55, prof. math., acting head, dept. math., 1954-55. Recipient Tanner award for excellent teaching U. N.C., 1962. Mem. Phi Beta Kappa, Sigma Xi, Sigma Chi. Author (with others): Theory of Relativity, 1924; Introductory Mathematics, 1933, Differential Calculus for Students of Economics, 1940, also articles. Contbns. to differential geometry. Home: 523 E. Rosemary St., Chapel Hill, N.C. 27514.*

LASORSA, Giovanni, Italian statistician; b. Giovinazzo, Nov. 22, 1900; s. Domenico and Maria (Taldone) L.; m. Giulia Maraviglia, Feb. 12, 1936; children—Maria Paola, Claudia, Caterina, Daniela, Chiara, Andrea. Prof. statistics U. Naples (Italy). Recipient Gold medal for culture and teaching. Mem. Internat. Inst. Statistics. Home: piazza Quinto Cecilio 3, Roma, Italy. Office: via Partenope 36, Naples, Italy.

LASSAIGNE, Jean-Louis, French chemist; b. Paris, 1800; studied under Vauquelin; named prof. Alfort Sch., 1828; recipient award for perfecting and contributing to pottery enamel, 1831. Author: Abrégé élémentaire de chimie organique et inorganique, 2 vols., 4th edit., 1846. Discovered the mineral delphinite, phosphoric ether, pyrogenic acids, pyrocitric acids, phosphoric acid; research on carbonization of organic material, arsenic salts, animal chemistry. Died 1859.

LASSAR, Oskar, German dermatologist; b. Hamburg, Germany, Jan. 11, 1849; ed. U. Heidelberg, (Germany), U. Göttingen (Germany), U. Strasburg (France), U. Würzburg (Germany); M.D., 1873, Served as lt. Franco-Prussian War; asst. physiol. Inst., Göttingen; asst. Path. Inst., Breslau (now Wroclaw, Poland), 1875-78; established practice of dermatology, Berlin, 1878; named privat-docent, 1880; founded pvt. hosp. and dispensary for dermatology and syphilis; founder Lassar's Shower-bath in which poor could get bath for 2 cents, 1883; asso. Robert Koch, Prussian Bd. Health. Contbr. essays to med. jours. Mem. Berlin Dermatol. Soc. (a founder). Editor, Dermatologische Zeitschrift. Introduced zinc oxide paste for treating eczema, 1889. Died 1908.

LASSEK, Arthur Marvel, Am. anatomist, educator; b. Eau Claire, Wis., Sept. 17, 1902; s. Joseph Edward and Caroline (Wold) L.; B.S., U. Wash., 1924-28; M.S., Northwestern U., 1929, Ph.D., 1931; M.D., U. Tenn., 1936; m. Dorothy Adelaide Jones, Aug. 26, 1928 (dec. Jan. 1965); children—Arthur Marvel, William Day. Asst. in anatomy U. Wash., 1926-27; teaching fellow dept. anatomy Northwestern U., 1928-31; asst. prof. anatomy U. S.D., 1931-32; prof. anatomy, head dept. Med. Coll. S.C., 1933-48; prof., head dept. anatomy Boston U. Med. Sch., 1948-65, now professor emeritus, since 1965—. Member of Sigma Xi, Alpha Omega Alpha, and Phi Rho Sigma. Author: The Pyramidal Tract: Its Status in Medicine, 1954; The Human Brain: Primitive to Modern, 1957; Human Dissection: Its Drama and Struggle, 1958. Research, numerous articles, neurol. tracts on motor system and med. edn. Home: 30 Hilltop Av., Lexington, Mass. 02173.

LASSELL, William, English astronomer; b. Dolton, Eng., June 18, 1799; ed. Rockdale Acad.; LL.D., Cambridge, 1874; 1 dau., Caroline. Apprenticed in mcht's office, Liverpool, Eng., 1814-21; became brewer, Liverpool, 1825; built obs., Starfield, nr. Liverpool, and moved it to Bradstones, 1854; built reflecting telescope with 4 feet aperture, 1859-60, which he mounted and worked with at Valetta, 1861-64; set up obs. nr. Maidenhead, Eng. Fellow Royal Soc., 1849 (Royal medal 1858); mem. Royal Astron. Soc. (Gold medal 1849, pres. 1870-72). Built 1st adaptation to reflectors of equatorial plan of mounting and observed the paths of comets further than was possible at any pub. obs.; verified discovery of Neptune, 1847; discovered Triton, satellite of Neptune, 1846, Hyperion, 8th satellite of Saturn, 1848, Ariel and Umbriel, 2 nearest satellites of Uranus, 1851; catalogued many new nebulae. Died Maidenhead, Oct. 5, 1880.

LASSEN, Niels Alexander, Danish clin. physiologist; b. Copenhagen, Denmark, Dec. 7, 1926; s. H. C. A. and J. A. (Kahler) L.; grad. medicine U. Copenhagen, 1951; specialist in internal medicine, 1962; specialist in clin. physiology, 1967; m. Edda Sveinsdottir, Aug. 27, 1955; children—Henrik, Anders, Jens. Clin. work in internal medicine and neurology in hosps. in Copenhagen, 1951-62; chief lab. clin. physiology Bispebjerg Hosp., Copenhagen, 1962—; prof. cardiopulmonary lab. Columbia U. at Bellevue Hosp., N.Y.C., 1962; vis. prof. dept. clin. physiology Lund (Sweden) U., 1965. Mem. Danish. Soc. Internal Medicine, Danish Soc. Clin. Biochemistry and Physiology. Research and numerous publs. on cerebral and muscle circulation including methods of measurements using radioactive inert gases, krypton and xenon. Home: 14 Rosavej. Office: Bispebjerg Hosp., Copenhagen, Denmark.*

LASSETTRE, Edwin Nichols, Am. phys. chemist; b. nr. Forsyth, Ga., Oct. 26, 1911; s. Carlos Edwin and Jennie J. (Nichols) L.; B.S., Mont. State Coll., 1933; Ph.D., Cal. Inst. Tech., 1938; m. Ilse R. Sturies, Dec. 22, 1951. Faculty, Ohio State U., 1937-62. prof. chemistry, 1950-62; staff fellow Mellon Inst., Pitts., 1962—, also mem. adv. com.; group leader, research scientist Manhattan Project Columbia, 1944, Carbide & Carbon Chems. Corp., 1945; cons. AEC. Mem. Am. Chem. Soc., Am. Phys. Soc. Study of selection rules in excitation of atoms and molecules by electron impact; determination of generalized oscillator strengths and study of forbidden transitions. Home: 224 E. Waldheim Rd., Pitts. 15215. Office: 4400 Fifth Av., Pitts. 15213.*

LASSONE, Joseph Marie François, French chemist, physician; b. Carpentras, France, July 3, 1717; studied medicine L'hospice de la Charité; studied surgery under Morand; pvt. physician to Marie Leczinska, later to Louis XVI, and Marie-Antoinette; physician Faculty of Paris; mem. Faculty of Medicine, Paris. Recipient (with Lecat) prize for work on cancer of womb, 1738. Mem. French Acad. Scis., 1742, Royal Soc. Medicine (founder). Author: Méthode éprouvée pour le traitement de la rage, 1776; also articles. Research on surgery, anatomy, natural sci., chemistry; discovered carbonic oxide (carbon monoxide) which he produced by reducing calx of zinc with charcoal, 1776. Died Paris, Dec. 8, 1788.

LASSUS, Pierre, French surgeon; b. Paris, Apr. 11, 1741; prof. history of legal medicine, then prof. external pathology Faculty of Paris; surgeon to daus. of Louis XV; mem. French Acad. Scis., 1795. Author: Dissertation sur la lymphe, 1773; Pathologie chirurgicale, 1795. Died Paris, Mar. 16, 1807.

LASSWELL, Thomas Ely, Am. sociologist; b. St. Louis, Oct. 29, 1919; s. Gus and Miriam (Ely) L.; student Westminster Coll., 1936-38; A.B., Ark. Coll., 1940; postgrad. Columbia; M.S., U. So. Cal., 1947, Ph.D., 1953; m. Marcia Lee Eck, May 29, 1950; children—Marcia Jane, Thomas Ely, Julia Lee. Faculty, Pepperdine Coll., 1950-54, Grinnell Coll., 1954-59; prof. sociology U. So. Cal., Los Angeles, 1959—, chmn. dept. sociology, anthropology, 1965-66; vis. prof. Stephen F. Austin State Coll., 1950, 56, 58, Whittier Coll., 1957, Northwestern U., 1960, Pomona Coll., 1967—. Asso. dir. Nat. Tchr. Corps, 1966—; cons. social psychology of architecture to Deasy & Bolling, Caudill, Rowlett & Scott, U. Cal. (Irvine). Fellow Am. Sociol. Assn.; mem. Pacific Sociol. Assn., Nat. Council Family Relations, Phi Kappa Phi, Alpha Kappa Delta. Author: Class and Stratum, 1965; (with John H. Burma, Sidney H. Aronson) Life in Society 1965. Asso. editor: Sociology and Social Research, 1960—. Contbr. numerous articles, book revs. to profl. jours. Research in identification, isolation and definition of variables related to social class as conceptualized from several different perspectives; pioneered in application of social psychol. analysis of archtl. problems. Home: 875 Hillcrest Dr., Pomona, Cal. 91766. Office: U. So. Cal., Los Angeles 90007.*

LASSWITZ, Kurd, German philosopher; b. Breslau, Germany (now Wroclaw, Poland), 1848. Author: Geschichte der Atomistik vom Mittelafter bis Newton, 1890; Wirklichkeiten, Beiträge zum Weltverstandnis, 1900; Auf zwei Planeten, 1897; also poetic and utopian novels including: Nie und Immer, 1902. Died Gotha, Germany, Oct. 17, 1910.

LAST, Raymond Jack, surgeon; b. So. Australia, May 26, 1903; s. John and Mildred Louisa (Rundle) L.; M.D., M.B., U. Adelaide (Australia); B.S., Fellow Royal Coll. Surgeons, Gt. Britain; m. M. S. Milne, 1939. Surgeon to Haile Selassie, Ethiopia; prof. applied anatomy, dir. Royal Coll. Surgeons Eng. Mem. Brit. Med. Assn., Anat. Soc. Gt. Britain. Author: Wolff's Anatomy of Eye and Orbit, 1961; Aids to Anatomy, 1962; Anatomy: Regional and Applied, 1963; also articles on applied anatomy. Address: 38 Lincoln's Inn Fields, London W.C. 2, England.

LATHAM, Charles, Brit. mining engr.; b. Birkdale, Eng., 1868; s. James Beven Latham; ed. Sch. Mines, Wigan, Eng.; articled to Moss Hall Coal Co., Wigan, asst. gen. mgr., 1891-93; dir. mining Univ. Coll., Nottingham, Eng., 1893-1902; Dixon lectr. mining U. Glasgow, 1902-07, prof. mining, from 1907. Examiner mining U. Manchester, 1908. Mem. Inst. Mining Engrs. Author: Colliery Winding Machinery; Coal Cutting by Machinery; Detection and Estimation of Inflammable Gases in Mine Air by means of Flame Caps. Died Sept. 27, 1917.

LATHAM, Earl, Am. polit. scientist; b. New Bedford, Mass., Oct. 28, 1907; s. Artemas Leroy and Irene (Ganson) L.; B.A., Harvard, 1931, postgrad. Law Sch., 1932-33, Ph.D., 1939; M.A. (hon.), Amherst Coll., 1949; m. Margaret Victoria Perrier, Aug. 21, 1935; children—Peter Samuel, Susan Gail. Instr., tutor dept. govt. Harvard, 1936-40; faculty U. Minn., 1940-48, asso. prof. polit. sci., 1944-48; Joseph B. Eastman prof. Amherst (Mass.) Coll., 1948—; staff U. S. Bur. Budget, 1942-46. Mem. exec. council Adminstrv. Conf. U. S., 1961-62. Cited by Woodrow Wilson Found., 1956; recipient Sr. Research award in Am. Govtl. Affairs, Social Sci. Research Council, 1961-62, David DiLloyd prize Harry S. Truman Library Inst., 1967. Fellow Am. Acad. Arts and Scis.; mem. Am. (past v.p.), New Eng. (past pres.) polit. sci. assns., Am. Assn. U. Profs. Author: The Group Basis of Politics, 1952; The Politics of Railroad Coordination, 1959; The Communist Controversy in Washington, 1966; also articles. Editor: The Philosophy and Policies of Woodrow Wilson, 1958. Empirical analyses of group basis of decision-making in a pluralistic democracy and social roots of polit. action using as research situations an attempt in Congress to reverse a Supreme Ct. decision, an effort to diminish the power autonomy of r.r. corps. by govtl. persuasion and phenomenon of McCarthyism. Office: Dept. Polit. Sci., Amherst Coll., Amherst, Mass. 01002.*

LATHAM, John, Brit. ornithologist; b. Eltham, Eng., June 27, 1740; s. John Latham; studied anatomy under Hunter, London; M.D., Erlangen, Germany, 1795; m. 1763; m. 2d, Miss Delamott, 1798; 1 son. Practiced medicine at Dartfort, until 1796, then moved to Romsey, Eng. and studied nature. Fellow Soc. Antiquaries, Royal Soc., 1775; mem. Linnean Soc. (a founder 1788). Author: A General Synopsis of Birds, 3 vols., 1781-85; Index Ornithologicus sive systema ornithologiae, 2 vols., 1790; General History of Birds, 10 vols., 1821-24, Index, 1828. Authority on assigned names of species; listed several new genera and species of birds. Died Romsey, Feb. 4, 1837.

LATHAM, Peter Wallwork, Brit. physician; b. Wigan, Eng., Oct. 21, 1832; s. John Latham; student Glasgow U., (scholar 1855) Caius Coll., Cambridge, Eng., St. Bartholomew's Hosp., London; M.A., M.D.; m. Jamina Burns M'Diarmid, 1863 (dec. 1877); 1 dau.; m. 2d, Marianne Frances Bernard, 1884. Became fellow Downing Coll., 1860; councillor Royal Coll. Physicians, 1886-87, censor, 1887-89, sr. censor, 1894-95; named Harveian orator, 1888; Downing prof. medicine Cambridge, 1874-94; asst. physician Westminster Hosp.; physician Addenbrooke's Hosp., Cambridge, 1863-99; cons. physician. Research and publs. on phthisis, headache, formation of uric acid in animals, pathology of rheumatism, gout, diabetes. Died Oct. 29, 1923.

LATHAM, Robert Gordon, English philologist, ethnologist; b. Billingborough, Eng., Mar. 24, 1812; s. Thomas Latham; B.A., King's Coll., Cambridge (Eng.) U.; 1832, M.A.; M.D., U. London, 1842; Prof. English lang. and lit. U. Coll., London, 1839-41; held appointments, London hosps., 1842-49; named lectr. forensic medicine Middlesex Hosp., 1842, asst. physician, 1844; resigned med. posts, 1849; named dir. ethnol. dept. Crystal Palace, 1852; received pension on civil list, 1863. Author: Norway and the Norwegians, 2 vols., 1840; Treatise on the English Language, 1841; History and Etymology of the English Language, 1851; Natural History of the Varieties of Man, 1850; Logic and its Application to Language, 1856; The Ethnology of Europe, 1852; Elements of Comparative Philology, 1862; Dictionary of the English Language, 2 vols., 1867-70; Descriptive Ethnology, 2 vols., 1859; Nationalities of Europe, 2 vols., 1863. Revised Johnson's Dictionary, 1870. Developed theory that Aryan race originated in Europe rather than Central Asia. Died Putney, Eng., Mar. 9, 1888.

LATHE, Grant Henry, biochemist; b. Grand Forks, B.C., Can., July 27, 1913; s. Frank Eugene and Annie (Smith) L.; B.Sc., McGill U., 1934, M.Sc., 1936, M.D. C.M., 1938, Ph.D., 1947; m. Margaret Eleanor Brown, 1939; 1 son, Robert; m. 2d, Joan Frances Hamlin, 1950; children—Susan, Richard, Caroline. Imperial Chem. Industries Research fellow Oxford U., London, 1946-48; lectr. dept. chem. path. Guy's Hosp. Med. Sch., London, 1948-49; biochemist Bernhard Baron Meml. Research Lab., Queen Charlotte's Maternity Hosp., London, 1949-57; prof. chem. path. U. Leeds (Eng.), 1957—; cons. pathologist Leeds Gen. Infirmary, 1957—. Research and numerous publs. on biochem. mechanisms for excreting bile pigments and poor devel. of these in newborn infants; molecular sieve techniques for chem. separations. Home: 14 Lidgett Park Rd., Leeds 8, Yorkshire, Eng.*

LATHROP, Katherine Austin, Am. radiobiologist; b. Lawton, Okla., June 16, 1915; d. William Colson and Gladys (Gray) Austin; student Cameron Coll., 1932-34; B.S., Okla. State U., 1936, B.S., M.S., 1939; m. Clarence A. Lathrop, Aug. 2, 1938; children—Jane Ellen, Suzanne (Mrs. Malcolm Crenshaw Moore, Jr.), Laura Eugene, David Austin, Julia Louise. Research asst. U. Wyo., 1942-44; jr. chemist bio-med. div. Manhattan Project, Metall. Lab., U. Chgo., 1945-47; asso. biochemist Argonne (Ill.) Nat. Lab., 1947-54; research asso., instr. Argonne Cancer Research Hosp., dept. surgery U. Chgo., 1954-59, research asso., asst. prof., 1959-67, associate prof. dept. radiology, since 1967—. Mem. Radiation Research Soc., Soc. Nuclear Medicine, Tissue Culture Assn., A.A.A.S., Ill. Acad. Sci., Sigma Xi. Research and numerous publs. on devel. of radioisotopes for med. uses including therapy and diagnosis. Home: 5514 Woodlawn Av., Chgo. 60637.*

LATIES, George G., plant physiologist; b. Russia, Jan. 17, 1920; s. Simon Gregory and Marusia (Glushanok) L.; B.S., Cornell U., 1941; M.S., U. Minn., 1942; Ph.D., U. Cal. at Berkeley, 1947; m. Betsy Henderson, June 22, 1947. Research fellow Cal. Inst. Tech., 1947-49, 50-52, sr. research fellow, 1955-58; asst. prof. botany, research botanist U. Mich., Ann Arbor, 1952-55; faculty U. Cal. at Los Angeles, 1959—, prof. physiology, 1963—. Guggenheim fellow biology, 1966-67. Mem. Am. Soc. Plant Physiologists (v.p.), Bot. Soc. Am., Scandanavian Soc. Plant Physiology, A.A.A.S., Soc. Gen. Physiologists, Soc. for Growth and Devel. Asso. editor Plant Physiology. Research, publs. on biochem. changes during differentiation and growth in plant tissues and their relation to salt transport phenomenon; elucidation dependence of plant membrane permeability of metabolic state of tissues. Home: 1207 Tigertail Rd., Los Angeles 90049.*

LATIMER, Homer Barker, Am. anatomist; b. Rock Creek, Ohio, Mar. 2, 1882; s. Homer Dykman and Luella (Berker) L.; A.B., U. Minn., 1907, A.M., 1908, Ph.D., 1921; postgrad. U. Berlin, U. Chgo.; m. Emily Myra Longfellow, Dec. 21, 1929; 1 dau., Margaret Elizabeth (Mrs. Elmer Merle Straight). Prof. biology Charles City (Ia.) City Coll., 1908-10; with U. S. Bur. Fisheries, 1910-11; prof. zoology Neb. Wesleyan Coll., 1911-16; asso. prof. anatomy U. Neb., 1918, prof., 1918-26; prof. anatomy U. Kan., Lawrence, 1926-52, prof. emeritus, 1952; vis. prof. U. Tex., 1942, U. Mo., 1955-56, U. Ark., 1961-62. Fellow A.A.A.S.; mem. Am. Assn. Anatomists, Am. Soc. Zoologists, Sigma Xi. Numerous publs. on research in comparative neurology, pre- and post-natal growth, quantitative anatomy. Home: 820 Missouri St., Lawrence, Kan. 66044. Died Mar. 22, 1966.

LATIMER, Wendell Mitchell, Am. chemist; b. Garnett, Kan., Apr. 22, 1893; s. Walter and Emma (Mitchell) L.; A.B., U. Kan., 1915; Ph.D., U. Cal., 1918; m. Bertha Eichenauer, Aug. 1, 1917; 1 son, Walter R.; m. 2d, Glatha Hatfield, June 16, 1926; children—Eleanor Ann, Robert Milton. Lectr. and demonstrator in chemistry U. Cal., 1918-21, asst. prof., 1921-23, asso. prof., 1924-31, prof., 1931—, asst. dean Coll. Letters and Science, Calif., 1923-24, dean Coll. Chemistry, 1941-50, asso. dir. radiation lab., 1949—. Dir. Manhattan Eng. Dist. Contract on Chemistry of plutonium, 1943-47. Guggenheim fellow, Munich, 1930. Mem. Chem. Corps Research Council, 1947-51, Commn. of Electrochemistry of Internat. Union of Pure and Applied Chemistry, Acad. Polit. Sci., Nat. Acad. Scis., Am. Chem. Soc. (Nichols award N.Y. sect. 1955), A.A.A.S., Electrochem. Soc., Sigma Xi. Author: A Course in General Chemistry (with W. C. Bray), 1923; Reference Book of Inorganic Chemistry (with J. H. Hildebrand), 1929; Oxidation Potentials, 1938. Contbr. to chem. publs. Editor Prentice-Hall chemistry series. Asso. editor Jour. Chem. Phys., 1933, Chem. Rev., 1940. Research in organic chemistry, entropy of aqueous ions, thermodynamics. Died July 6, 1955.

LATNER, Albert Louis, English clin. biochemist; b. London, Dec. 5, 1912; s. Harry and Miriam (Gordon) L.; B.Sc. with 1st class honors, U. London, 1932, M.Sc., 1933; D.Sc., U. Liverpool (Eng.), 1958, M.D., 1948, Ch.B., 1939; m. Gertrude Franklin, Sept. 3, 1936. Asst. lectr. dept. physiology U. Liverpool, 1933-36, 39-41; sr. reg. chem. pathology Brit. Postgrad. Med. Sch., 1946-47; lectr. chem. pathology U. Durham (Eng.), 1947-55, reader med. biochemistry, 1955-61; prof. clin. biochemistry U. Newcastle upon Tyne, 1961—. Mem. Nat. Commn. for Clin. Chemistry, Joint Examinations Bd. Mastership in Clin. Biochemistry. Fellow Royal Coll. Physicians London, Coll. Pathologists; mem. Internat. Fedn. for Clin. Chemistry (mem. com. on standards), Assn. Clin. Biochemists (mem. council), Biochem. Soc., Physiol. Soc., Path. Soc., Assn. Clin. Pathologists, Brit. and Irish Group for Human Tumour Investigation, Brit. Assn. for Cancer Research. Contbg. author: Advances in Clinical Biochemistry, vol. 9, 1966. Research, numerous publs. on intrinic factor and absorption of vitamin B12, effect of oncogenic viruses on isoenzymes in tissue culture, hemoglobin biosynthesis, wax metabolism in cochineal insect, disturbances of liver metabolism; isoenzyme studies including clin., diagnostic and biol. significance, various cancers, especially cancer of cervix. Home: Ravenstones, Rectory Rd., Newcastle upon Tyne 3, U.K.*

LATOUR, Marius; engr. in Belgium; pioneered series connection accumulator motor for alternating current of normal frequency, 1901.

LA TOURRETTE, Marc-Antoine-Louis Claret de, see de la Tourrette, Marc-Antoine-Louis Claret.

LATREILLE, Pierre André, French zoologist; b. Brives, France, Nov. 29, 1762; adopted by Abbé Haüy, 1778; studied at Lemoine Coll., Paris; studied entomology, Brives, for 2 years; ordained. 1786; became head entomol. dept. Muséum d'Histoire Naturelle, Paris, 1799, succeeded Lamarck as prof. entomology, 1829. Mem. French Inst. Author: Course d'entomologie, 2 vols., 1831-33; Histoire naturelle des singes, 2 vols., 1801; Histoire naturelle des reptiles, 4 vols., 1802; Histoire naturelle générale et particulière des crustacés et des insectes, 14 vols., 1802-05; Genera crustaceorum et insectorum, 4 vols., 1806-09; Considérations sur l'ordre naturel des animaux; Familles naturelles du règne animal. Father modern entomology; established a number of insect orders; assisted Lamarck and Cuvier in some of their works. Died Paris, Feb. 6, 1833.

LATROBE, Benjamin Henry, engr., architect; b. Fulneck, Eng., May 1, 1764; s. Benjamin and Anna Margaret (Antes) L.; studied architecture under Samuel Pepys Cockerell, Eng., 1788-89; m. Lydia Sellon, 1790, 4 children including Henry, Lydia; m. 2d, Mary Hazlehurst, May 2, 1800; children—John, Benjamin Henry. Executed his 1st independent archtl. work Hammerwood Lodge, East Grinstead, Sussex, Eng., circa 1787; later became surveyor of police force of London (Eng.); came to Am., 1796; cons. on improvement of nav. on James River, 1796; designed prison on principle of solitary confinement, Richmond, Va., 1797; completed exterior of Va. State Capitol (designed by Jefferson); designed Bank of Pa., Phila., 1798; designed and engineered project for pumping Phila. water supply from Schuylkill River by using pumps operated by steam engines, 1799; undertook improvement of nav. on Susquehanna River; designed several houses in Phila.; apptd. surveyor of public bldgs. by Pres. Jefferson, 1803; commd. to design South Wing of Capitol Bldg. which would contain U. S. Ho. of Reps.; did much work on Washington and N.Y.C. naval yards, 1804; apptd. engr. Chesapeake and Del. Canal, 1804; became partner of Robert Fulton, Robert R. Livingston and Nicholas J. Roosevelt to build steamboat to navigate Ohio River, 1812 (project collapsed after Fulton's death, 1815); worked on reconstrn. of Capitol Bldg. and White House, Washington (after destruction by British), 1815-17; adviser to Thomas Jefferson on design of Pavillions V and III for U. Va., 1817; after death of son Henry, completed building of water works, New Orleans, 1817. Author: View of the Practicability and Means of Supplying the City of Philadelphia with Wholesome Water, 1799. Died of yellow fever, New Orleans, Sept. 3, 1820.

LATTA, Harrison, Am. physician; b. Los Angeles, Apr. 5, 1918; s. Smith and Elisa (Keating) L.; A.B in Chemistry, U. Cal. at Los Angeles, 1940; M.D., Johns Hopkins, 1943; m. Mae Nye, June 17, 1941; children—William, John, Gilbert, Ann. Research fellow Harvard Med. Sch. also Children's Med. Center, 1948-49; research asso. dept. biology Mass. Inst. Tech., 1949-51; asst. prof. pathology Western Res. U., 1951-54; faculty U. Cal. at Los Angeles, 1954—, prof. pathology, 1960—. Cons. pathology USPHS, 1965—, VA Center, Los Angeles, 1954—. Mem. Am. Assn. Pathologists and Bacteriologists, Am. Soc. for Exptl. Pathology, Am. Soc. for Cell Biology, Internat. Acad. Pathology, Phi Beta Kappa, Alpha Omega Alpha. Contbr. numerous articles to tech. jours. Research on hypersensitivity and diseases of kidney, biophysics, electron microscopy. Home: 325 Arno Way, Pacific Palisades, Cal. 90272. Office: Dept. Pathology, U. Cal., Los Angeles 90024.*

LATTA, Randall Kirk, Am. entomologist; b. Morning Sun, Ia., Aug. 7, 1905; s. James Kirk and Eva (Upson) L.; A.B., Ia. Wesleyan Coll., Mt. Pleasant, 1927, D.Sc., 1953; M.S., Ia. State U., 1928; m. Helen Emaline Kincaid, Jan 14, 1928. Research entomologist Bur. Entomology and Plant Quarantine, U. S. Dept. Agr., Santa Cruz, Cal., 1928-29, Whittier, Cal., 1929-30, Sumner, Wash., 1930-36, Babylon, N.Y., 1937, Washington, 1938-46, El Paso, Tex., 1947-48, Beltsville, Md., 1949-51, chief div. stored-product insects investigation, 1951-54; head stored-products insects sect. Agrl. Marketing Service, Washington, 1954-58; entomology adv. AID/Jamaica, Kingston, 1958-63; agrl. officer entomology FAO, Montego Bay, Jamaica, 1963-65; cons., Washington, 1965—. Mem. Entomol. Soc. Am., Washington Entomol. Soc., Washington Acad. Scis., Theta Kappa Nu. Contbr. articles on commodity treatments for products regulated by plant quarantines, fumigation of stored products for insect control, heat-generated insecticidal aerosols, nematodes of Jamaica, relation of plant nematodes to lethal yellowing disease of coconut palms. Address: 2122 California St. N.W., Washington 20008.*

LATTIMER, John Kingsley, Am. physician; b. Mt. Clemens, Mich., Oct. 14, 1914; s. Irrie Eugene and Gladys (Lenfestey) L.; B.A., Columbia Coll., 1935; M.D., Columbia U., 1938, Sc.D., 1943; m. Jamie Elizabeth Hill, Jan. 23, 1948; children—Evan, Jon, Gary. Chief urology U. S. Army Gen. Hosps., Germany, 1943-46; with Columbia U., 1946—, prof., chmn. dept. urology, 1955—; dir. urol. service Presbyn. Hosp., N.Y.C., 1955—; dir. urol. service Babies Hosp., N.Y.C., 1955—; dir. urol. service Delafield Hosp., N.Y.C., 1955—; dir. Squier Urol. Clinic, N.Y.C., 1955—; dir. research unit for renal Tb, U. S. VA Hosp., Bronx, N.Y., 1955—. Sr. cons. urology Harlem Hosp., also others. Recipient Smith prize, 1943; also numerous prizes for sci. exhibits A.M.A., Am. Urol. Assn. Mem. Soc. Pediatric Urology (pres.), A.C.S. (chmn. adv. council on urology, gov.) Am. Acad. Pediatrics (chmn. com. pediatric urology), N.Y. Acad. Medicine (chmn. sect. on urology), Am. Urol. Assn. (pres. N.Y. sect.), Harvey Soc., A.M.A., numerous others. Research and publs. in treatment of kidney Tb., cancer of prostate, urol. conditions in children; med. edn. innovations in urology. Office: Columbia U. Med. Sch., N.Y.C. 10032.*

LATYSHEV, Georgiy Dimitrievich, Russian physicist; b. Bezhitsa, 1907; grad. Petrograd. Poly. Inst., 1929; D.Physico-Math. Sci., 1940. Asso., Physicotech. Inst., Ukrainian Acad. Sci., Kharkov, 1930-41, with Inst. Physics and Math., 1941-43; head dept. physics, Kharkov, Ufa, Leningrad, 1930-58; with Physico-tech. Inst., USSR Acad. Sci., 1943-52; dir. Inst. Nuclear Physics, Kazakhstan Acad. Sci., 1958—. Recipient Stalin prize, 1949. Mem. Ukrainian (corr.), Kazakhstan acads. sci. Research and publs. on physics of atomic nucleus; 1st in USSR to disintegrate lithium atom nucleus. Address: Inst. Nuclear Physics, Kazakhstan Acad. Sci., Alma-Ata, Kazakhstan SSR, USSR.

LAUBACH, Charles, Am. geologist, archaeologist; b. Durham, Pa., Aug. 29, 1836; s. Anthony L.; grad. Collegiate Inst., Easton, Pa., 1860; studied phrenology; studied medicine with Dr. H. A. Benton, Saratoga, obstetrics with Dr. Jacob Ludlow, Easton; m. Jane Raub, Mar. 29, 1860. Practiced med. electricity and homeopathy while giving phrenological lectures and delineations of character, also engaged in farming; began sci. investigations, 1865, devoted prin. attention to them, 1870—. Corr. mem. U. Pa. archaeol. and palaeontol. dept.; mem. Anthropol. Club, Phila.; charter mem. Bucks County Hist. Soc. Author: History of Durham Township, 1887; Geology of Bucks County, Pa., in Warner's History of Bucks County; Prehistoric Man in the Delaware Valley, 1880. Died 1904.

LAUBE, Gustave-Charles, geologist; b. Teplitz, Bohemia, 1839; prof. geology U. Prague (Czechoslovakia); dir. efforts that recovered dried-up wells and springs of Teplitz. Died Prague, 1923.

LAUBENDER, Walther, German pharmacologist; b. Freising-Obb, Dec. 26, 1898; s. Franz and Eugenie (Fuchs) L.; ed. univs. Munich, Erlangen, Frankfort, Heidelberg (all Germany); M.D.; m. Margarethe Wuth, Feb. 24, 1923; m. 2d, Erna Albert, Aug. 3, 1961. Asst., Swiss Inst. Mountain Physiology; pharmacology, Geissen, 1925-26; asst., Frankfort, 1926-43; instr., Frankfort, also Marburg, Germany; called to Gratz, Austria; asst. prof., U. Frankfort, prof., head exptl. medicine, 1956—. Mem. German Soc. Pharmacology, Senckenberg Research Soc., Frankfort. Home: Louise-Seherstrasse 17, Diez-Lahn. Office: Abt. fur Experimentelle Medizin, Universitat, Senckenberg-Anlage, Frankfurt am Main, W. Germany.

LAUBENGAYER, Albert Washington, Am. chemist; b. nr. Salina, Kan., Feb. 22, 1899; s. Charles August and Tabitha (Schwilk) L.; B.Chemistry, Cornell U., 1921, Ph.D., 1926; m. Grace Louise Ware, Aug. 30, 1930; children—Susan Jane (Mrs. Thomas G. Cowing), Nancy Carol (Mrs. Daniel Smothergill). Instr., Ore. State Coll., Corvallis, 1921-23; faculty Cornell U., Ithaca, N.Y, 1923—, prof. chemistry, 1936-66, emeritus prof., 1966—, acting head dept. chemistry, 1959, 64. Cons. to pvt. cos. Mem. Am. Chem. Soc. (One of Ten Most Outstanding Inorganic Chemists in U. S. A., Chgo. sect. 1947), Am. Figure Skating Assn., Sigma Xi, Alpha Chi Sigma, Theta Xi. Author: Laboratory Experiments and Problems in Gen. Chemistry, 1933, rev., 1936; General Chemistry, 1949, rev., 1957; also numerous articles. Research on chemistry germanium, boron, aluminum, gallium, indium, chrominium, fluorine; synthesis and characterization properties, structure and bonding of compounds of these elements; research on function and properties donor-acceptor molecular adducts boron and aluminum compounds with amines and pyrolysis these adducts to build inorganic polymers, hydrous oxide systems. Home: 235 Berkshire Rd., Ithaca, N.Y. 14850.*

LAUBENTHAL, Florin, German neurologist; b. Mayen/Rhine, Oct. 22, 1903; s. Anton and Therese Klöppel; ed. univs. Bonn, Munich; M.D.; m. Francis Bruck, Nov. 16, 1931; children—Florin, Werner, Kurt. Prof. neurology and psychiatry U. Münster; asst., physician-in-chief Univ. Clinic Neurology; head physician, municipal clinic neurology and psychiatry, Essen, Germany. Author: Zwischenhirnsyndrome, 1948; Hirn und Seele, 1953; Leitfaden der Neurologie, 1963; Missbrauch und Sucht, 1964; also articles. Home: Holsterhausenstrasse 189, Essen. Office: Hufenlandstrasse 55, Essen, W. Germany.

LAUBEUF, Maxime, French marine engr.; b. Poissy, France, Nov. 23, 1863; chief engr. naval constrn.; mem. French Acad. Scis., 1920, French Marine Acad. Author: Sous-marins et submersibles, 1915; Sous-marins, torpilles et mines, 1923. Built 1st modern submarine (Narval), weighing 120 tons, with steam engine, accumulators with electric motor; carried out plans for 1st French navy ship with turbine (Voltigeur, a torpedo boat chaser). Died Cannes, France, Dec. 23, 1939.

LAUBIER, Lucien Claude, French biologist; b. Lille, France, Sept. 22, 1936; s. Jean Victor and Suzanne (Benhamou) L.; Licencié ès scis. naturelles, Paris Faculty Scis., 1956, ès scis. naturelles, 1965; Asst., Arago Lab., Paris Faculty Scis., 1956-60, head research, 1960-64, asst. dir., 1964—. Mem. Zool. Soc. France. Author: Le coralligène des Albères; Monographie biocénotique; also articles. Systematic studies of polychaetous amelids, described 10 new species; research on systematics of parasitic copepods of invertebrates, hard-bottom variety of calcareous red algae. Address: Laboratoire Arago, 66, Banylus-sur-Mer, France.*

LAUDON, Lowell Robert, Am. geologist; b. Redwood Falls, Minn., Feb. 4, 1905; s. A. R. and Florence (Baker) L.; student Ia. State Tchrs. Coll., 1925-27; B.A., U. Ia., 1928, M.A., 1929, Ph.D., 1930; m. Florence Mildred Stanzel, Mar. 21, 1930; children—Thomas S., Richard B., Robert C., John Lowell. Asst. prof. U. Tulsa, 1930-41, asst. registrar, dean men, 1939-41; prof. geology, chmn. geology dept. U. Kan., Lawrence, 1942-48; prof. geology U. Wis., Madison, 1949—. Geol. cons. Humble Oil Co. Research Center, Houston, 1954—. Mem. Am. Assn. Petroleum Geologists (Distinguished lectr. 1961), Geol. Soc. Am., Paleontol. Soc. Am., Sigma Xi. Research, publs. on evolution and history fossil crinoidea, Mississippian stratigraphy and paleotectonic history western N.Am. Home: Route 1, Waunakee, Wis. 53597. Office: Sci. Hall, U. Wis., Madison, Wis.

LAUE, Max Theodor Felix von, German physicist; b. Pfaffendorf bei Koblenz, Germany, Oct. 9, 1879; s. Julius and Minna (Zerrenner) L.; Univs. Strasburg, Göttingen, Munich, 1898-1903; Dr. deg., U. Berlin, 1903; hon. D.Sc Manchester, 1936, Chicago, 1948; m. Magdalene Degen, Oct. 6, 1910; children—Theodor Hermann, Hildegard Minna (Mrs. Kurt Lemcke). Privat dozent U. Berlin, 1906-09, U. Munich, 1909-12; extraordinary prof. U. Zurich, 1912-14; prof. U. Frankfurt am Main, 1914, U. Berlin, 1919-43; acting dir. Kaiser- Wilhelm-Inst. Physics, 1921—. Max-Planck-Inst. Physics, 1943-51; dir. Fritz-liaber-Inst. of the Max-Planck-Gessellschaft, Berlin-Dahlem, 1951-59, emeritierter director (director emeritus), 1959—. Recipient Nobel prize in physics for discovery X-ray interferences, 1914; Planck-Medailie, 1932. Orden Pour le mérite für Künste and Wissenscaft, 1952; Grosses Verdienstkreuz mit stern der Bundesrepublik, 1953; Offizierskreuz der Ehrenlegion der französischen Republik, 1957. Mem. academies of science; Berlin, New York, Vienna, Washington, Rome and numerous others; Fellow Royal Soc., 1949. Author: Relativitàtstheorie. I, 1911, II 1921; Rögenstrahlinterferentzen, 1941; Materiewellen und ihre Interferezen, 1943; Geschichte der Physik, 1947; Theorie der Supraleitung, 1947. First used crystals as diffraction grating for x-rays proving electromagnetic nature of x-rays, allowing x-ray wave length to be measured, and providing technique for study of atomic structure of crystals and polymers; research on theory of relativity; asserted theory of superconductivity and of relativistic thermodynamics; worked on quantum theory, Compton effect, Einstein-Bohr equation, disintegration of atoms. Died Berlin, West Germany, April 23, 1960.

LAUFER, Arthur Russell, Am. physicist; b. N.Y.C., June 21, 1920; s. Frederick and Bertha (Hara) L.; A.B. magna cum laude, Bklyn. Coll., 1940; M.S., Yale, 1947; Ph.D., N.Y. U., 1949; m. Ethel Marie Halvorsen, Sept. 4, 1947; children—Arthur Russell, Christopher Alan, John Stewart. Instr. physics Yale, 1941-44, Mich. State Coll., 1944-45, N.Y. U., 1947-49; asst. prof. physics U. Mo., Columbia, 1949-52, asso. prof., 1952-53; phys. sci. adminstr. Office Naval Research, Pasadena, Cal., 1953-58, chief scientist, 1958-67, dep. dir. and chief scientist, since 1967—. Head transducer design sect. USN Underwater Sound Lab., summer 1951; research asso. Cal. Inst. Tech., 1954-55. NSF grantee, 1952; Outstanding Achievement award USN, 1957, 61. Mem. Am. Phys. Soc., Am. Assn. U. Profs., Sigma Xi. Author: Yale Physics Laboratory Manual, 1944; Lansing Unlimited, 1947; also articles. Editor-in-chief Procs., founder, chmn. nat. exec. com. Infrared Information Symposia. Determined causes of time lags in nuclear counters; proved Faraday's definitions of "paramagnetism" and "diamagnetism" covered only certain spl. cases; devised new basic method for absolute calibration of acoustic hydrophones; determined frequency dependence of acoustic cavitation, causes of ultrasonic luminescence; established technique for launching artificial meteors. Home: 1260 Sierra Madre Villa Av., Pasadena 91107. Office: 1030 E. Green St., Pasadena, Cal. 91101.*

LAUFER, Berthold, anthropologist, Orientalist; s. Max and Eugenie (Schlesinger) L.; b. Cologne, Germany, Oct. 11, 1874; ed. U. Berlin, 1893-95, Sem. for Oriental Languages, Berlin, 1894-95; Ph.D., U. Leipzig, 1897; LL.D., U. Chgo., 1931; m. Bertha Hampton. Came to U. S., 1898. Leader Jesup N. Pacific expdn. to Saghalin Island and Amur region of Eastern Siberia for exploration of ethnology of native tribes, 1898-99, of Jacob H. Schiff expdn. to China for culture-hist. investigations and ethnol. collections, 1901-04; asst. in ethnology Am. Mus. Natural History, 1904-06; lectr. anthropology Columbia, 1905, anthropology and East Asiatic langs., 1906-07; asst. curator East-Asiatic div. Field Mus. Natural History, 1908—, asso. curator Asiatic ethnology, 1911—, curator of anthropology, 1915. Leader Blackstone expdn. to Tibet and China, 1908-10, Marshall Field expdn. to China, 1923. Pres. Am. Oriental Soc., 1930-31; mem. com. on promotion of Chinese studies, com. on promotion of Japanese studies Am. Council Learned Societies. Author: Skizzen der monogolischen Literatur, 1907; Skizzen der Amnaschurischen Literatur, 1908; Chinese Pottery of the Han Dynasty, 1909; Jade, 1912; Descriptive Account of the Collection of Chinese, Tibetan, Mongol, and Japanese Books in the Newberry Library, 1913; Chinese Clay Figures, 1914; Beginning of Porcelain in China, 1917; also many monographs. Asso. editor Am. Jour. Archaeology; spl. corr. Nat. Library, Peiping, China. Research on ethnology, archaeology, art, and philology of Asia, history of cultivated plants and domesticated animals; made extensive collections of oriental books and mss. for Newberry and John Crerar libraries, Chgo. Died Chgo., Sept. 13, 1934.

LAUFER, Hans, developmental biologist; b. Gruenberg, Germany, Oct. 18, 1929; s. Sol and Margarete (Freundlich) L.; came to U. S., 1939, naturalized, 1944; B.S., Coll. City N.Y., 1952; M.A., Bklyn. Coll. N.Y., 1953; Ph.D. (W. D. Janes fellow), 1957; m. Evelyn Green, Oct. 31, 1953; children—Jessica Kari, Marc Reed, Leonard Justin. E. G. Conklin Meml. fellow Marine Biol. Lab., Woods Hole, Mass., 1956, Lalor fellow, 1962, 63; NRC fellow Carnegie Instn. Washington, 1957-59; asst. prof. biology Johns Hopkins, 1959-65; asso. prof. zoology and entomology U. Conn., Storrs, 1965—; vis. scholar developmental biology Western Res. U., Cleve., 1962. Mem. corp. Marine Biol. Lab. Fellow N.Y. Acad. Scis.; mem. Am. Soc. for Cell Biology, Am. Soc. Zoologists, Soc. for Developmental Biology, Am. Inst. Biol. Scis., A.A.A.S., Am. Assn. U. Prof. Research, numerous publs. on analysis macromolecular changes during salamander limb regeneration and insect metamorphosis, nucleocytoplasmic interactions during insect devel. including studies of relationship of hormones and chromosomal puffs to diptera salivary gland function and devel. Home: 49 Constance Dr., Manchester, Conn. 06002. Office: Dept. Zoology and Entomology, Life Sci. Bldg., U. Conn., Storrs, Conn. 06268.*

LAUFFER, Max Augustus, Am. biophysicist, educator; b. Middletown, Pa., Sept. 2, 1914; s. Max Augustus and Elsie M. (Keiper) L.; B.S., Pa. State U., 1933, M.S., 1934; Ph.D., U. Minn., 1937; m. Dorothy L. Easton, Dec. 20, 1936 (dec.); 1 son, Edward William; m. 2d, Erika R. Erskine, Mar. 21, 1964; children—Susan Keiper, Max Erskine, John Erskine. Began career as grad. asst., instr. biochemistry U. Minn., 1935-37; fellow, asst., asso. Rockefeller Inst., Princeton, N.J., 1937-44; faculty U. Pitts., 1944—, prof., 1947—, head dept. biophysics, 1949-56, dean div. natural scis., 1956-63, Andrew Mellon prof. biophysics, chmn. dept., 1963-67. Vis. prof. U. Bern (Switzerland), 1952; sci. adv. com. Boyce Thompson Inst. for Plant Research, Inc., 1953—; mem. program project com. Nat. Inst. Gen. Med. Scis., 1961-63, chmn., 1962-63; mem. Gen. Med. Scis. Council, NIH, 1963-67; sci. adv. bd. Delta Regional Primate Research Center, Tulane U., 1964-67. Recipient Outstanding Achievement award U. Minn., 1964. Fellow N.Y. Acad. Sci.; mem. Am. Chem. Soc. (Eli Lilly research award in biochemistry, 1945, Pitts. award 1958), Fedn. Am. Scientists (chmn. Pitts. br. 1958-59), Biophys. Soc. (pres. 1961), Am. Soc. Biol. Chemists, Soc. Exptl. Biology and Medicine, Phi Beta Kappa, Sigma Xi, Alpha Zeta, Alpha Chi Sigma, Gamma Alpha, Phi Kappa Phi, Phi Eta Sigma, Gamma Sigma Delta, Phi Lambda Upsilon, Sigma Pi Sigma, Phi Sigma. Bd. editors Archives Biochemistry and Biophysics, 1944-54; Biophys. Jour., 1960-64; co-editor Advances in Virus Research, 1953—. First to determine size, shape of tobacco mosaic virus, extensive study of phys. properties of viruses; publs. on research. Home: 1273 Folkstone Dr., Pitts. 15243.*

LAUGHLIN, Harry Hamilton, Am. biologist; b. Oskaloosa, Ia., Mar. 11, 1880; s. George Hamilton and Deborah Jane (Ross) L.; B.S., N. Mo. State Normal Sch., 1900; M.S., Princeton, 1916, Sc.D., 1917; M.D. (hon.) U. Heidelberg (Germany), 1936; m. Pansy Bowen, Sept. 13, 1902. Tchr. agr. N. Mo. State Normal Sch., 1907-10; supt. Eugenics Record Office (div. Dept. Genetics of Carnegie Instn. Washington) from its orgn., 1910, until 1921, in charge 1921-40; eugenics expert Com. on Immigration and Naturalization, House of Rep., 1921-31; eugenics asso. Psychopathic Lab., Municipal Ct., Chgo., 1921-30; U. S. immigration agt. to Europe, Dept. Labor, 1923-24; in charge of researches on genetics of thoroughbred horse since 1923; dir. Survey of Human Resources of Conn., 1936-38. Mem. Galton Soc., Eugenics Research Assn. (sec.-treas. 1917-39); Am. Soc. Internat. Law, Am. Statis. Assn., Am. Eugenics Soc. pres., 1927-28). Asso. editor Eugenical News, 1916-39. Editor A Decade of Progress in Eugenics, 1934. Author: Mitotic Stage Duration (Carnegie Instn.), 1918; State Institutions for the Defective, Dependent and Delinquent Classes (Bur. of Census), 1919; Eugenical Sterilization in the United States (Municipal Court, Chicago), 1922; Analysis of America's Modern Melting Pot (U. S. Govt. publ.), 1923; Europe as the Emigration-Exporting Continent and the United States as the Immigration-Receiving Nation (U. S. Govt. publ.), 1924; The General Formula of Heredity, 1933; Immigration Control, 1934; Racing Capacity in the Thoroughbred Horse, 1934; Probability Resultant, 1934; Conquest by Immigration, 1939; Current Studies on Race Conditions in the United States. Died Jan. 26, 1943.

LAUGIER, André, French chemist; b. Lisieux, France, 1770; chief, bur. powders and saltpeters Pub. Health Commn.; tchr. chemistry, mil. hosp. sch., Toulon, France, l'École centrale, Var, France, chem. sch., Lille, France; named prof. natural history Sch. Pharmacy, 1803, later dir. Mem. Acad. Medicine. Author: Cours de chimie générale professé au Jardin du Roi, 1828. Research on processes for separating cobalt from nickel, iron from titanium, cerium from iron, also for converting gum juice into milk sugar. Died Paris, 1832.

LAUGIER, Henri, French physiologist; b. Mane, Aug. 5, 1888; s. Albert and Marie (Coulomb) L.; M.D., Ph.D., U. Paris. Prof. physiol. study Conservatory Arts and Trades, 1929-38; dir. Nat. Center for Sci. Research, 1938-40; prof. physiology U. Montreal, 1940-43; rector Alger Acad., 1943-44; dir. cultural relations Ministry Fgn. Affairs, 1944-46; acting sec.-gen. UN, 1946-51; prof. gen. physiology Sorbonne, Paris, 1937—. Research and publs. on physiology, hygiene. Address: 55, rue de Babylone, Paris 7, France.

LAUGIER, Paul-Auguste Ernest, French astronomer; b. Paris, Dec. 22, 1812; s. André Laugier; student École Polytechnique, 1832-34, Paris Obs., 1934; astronomer Paris Obs.; mem. Bur. Longitudes, French Acad. Scis., 1843. Author: Calcul des éléments paraboliques de la comète de 1840; Recherches sur la rotation du soleil, 1841; Découverte d'une nouvelle comète, 1842; Sur les tachic du soleil, 1942; Recherches sur le pendule, 1845. Died Paris, Apr. 5, 1872.

LAUGIER, Stanislas, French surgeon; b. Paris, Jan. 28, 1799; s. André Laugier; M.D., 1828; aggregation, 1829; surgeon Paris hosps.; prof. clin. surgery U. Paris; mem. French Acad. Scis., 1868, Acad. Medicine. Author: Amputation de la cuisse, 1841; Des lésions de la moelle epinière, 1848. Research on path. physiology of Asian cholera; noted for operations for cataracts and lesions of spinal cord, also work on amputations. Died Paris, Feb. 15, 1872.

LAUNAY, Louis Alphonse Auguste de, see de Launay, Louis Alphonse Auguste.

LAUNOIS, Pierre-Émile, French physician; b. Miremont, France, 1856; certified physician, Paris, 1895; became aggregate prof., 1898; research in histology, on hormones, diseases of spleen; described (with M. Cleret) gigantism caused by excessive pituitary secretion (Launois', or Laumois-Cleret, syndrome), 1910. Died Paris, 1914.

LAUNOY, Léon, French biologist; b. St.-Maixent, Aug. 27, 1876; s. Aimé and Léonie (Babineau) L.; ed. U. Paris, Mus. Natural History; Ph.D.; m. Jeanne Serée, Sept. 7, 1904; children—Jean-Louis, André, Margeurite. Asst. Pasteur Inst., Paris, 1903-20; became mem. Faculty Pharmacy, Paris, 1903, prof. zoology and physiology, 1937-43; dir. Med. Biology Found., 1903-56; founder, dir. Vet. Medicine Works, 1928-54. Recipient 3 Inst. awards for works in cytology, physiology, exotic pathology. Mem. Acad. Medicine, Biology Soc., Acad. Pharmacology (pres.), French Soc. Therapeutics and Pharmacodynamy. Research and publs. on chem. vaccination against sleeping sickness. Address: 17, rue de Lorraine, Saint Germain-en-Laye (Seine et Oise), France.

LAURENBERG, Peter, philosopher, physician, mathematician, physicist; b. Rostock, Germany, Aug. 26, 1585; s. Wilhelm Laurenberg; prof. philosophy, medicine physics, math., poetry, Rostock. Author: Apparatus plantarius (on nomenclature, classification of species, planting and uses of bulbous and tuberous plants), 1632; Synopsin aphorismorum chymiatricorum . . . , 1642. Translated Drebbel's works into Latin. Died May 13, 1639.

LAURENBERG, Wilhelm the Younger, physician, botanist; b. 1547; from Rostock, Germany; s. Wilhelm Laurenberg; practiced medicine, Copenhagen, Denmark. Author: Botanotheca, sive modus conficiendi herbarium vivum (plants divided into natural sects.), 1626; Historia descriptionis aetitidis, sive lapidis aquilae (on eagle stone), 1627. Died 1612.

LAURENCE, George Craig, Canadian physicist; b. Charlottetown, Can., Jan. 21, 1905; s. Harold Forbes and Florence (Dexter) L.; B.Sc., Dalhousie U., 1925, M.Sc., 1927, LL.D.; Ph.D., Cambridge U., 1931; D.Sc., U. Sask., Queens U.; m. Elfreda Elizabeth Blois, Jan. 2, 1931; children—Janet Patricia (Mrs. Stewart Buchanan), Judith Elizabeth (Mrs. Peter Jost). Dir. chemistry and engring. div. Nat. Research Council Can., 1949-52; dir. chemistry and engring. div. Atomic Energy of Can. Ltd., 1952-55, dir. reactor research and devel. div., 1955-61; pres. Atomic Energy Control Bd. Can., 1961—. Chmn., Adv. Com. on Reactor Safety, 1956—. Fellow Royal Soc. Can.; mem. Canadian Assn. Physicists (pres. 1952-53), Radiol. Soc. N.Am., Am. Coll. Radiology, Am. Nuclear Soc. Research and publs. on reactor safety practices, pollution control. Home: 1 Beach Av., Deep River, Ont. Office: 107 Sparks St., Ottawa, Ont., Can.*

LAURENCE, John Zacharias, English physician; b. Eng., 1830; (with Robert Charles Moon) described syndrome characterized by hypogenitalism, obesity, retinitis pigmentosa, mental deficiency (Laurence-Moon-Biedle syndrome), 1866. Died 1874.

LAURENCE, Kurt Michael, pediatric neuropathologist; b. Berlin, Germany, Aug. 7, 1924; s. Gustav and Grete (Heymann) Loebenstein) L.; B.A. with honors in Natural Scis., Cambridge U., 1947; M.B.Ch.B., U. Liverpool (Eng.), 1950; m. E. R. Settle, July 9, 1949; children—Stephen, Amanda, Elizabeth. Registrar pathology Portsmouth and Isle of Wright Path. Service, 1953-55; research fellow in hydrocephalus and spina bifida Hosp. for Sick Children, London, 1955-59; sr. lectr. pediatric pathology Welsh Nat. Sch. Medicine, also hon. cons. pathologist United Cardiff (Wales) Hosps., 1959—; pathologist in charge research labs. dept. child health Welsh Nat. Sch. Medicine, 1959—. Dir. Cytogenetics Unit for Wales, 1959—. E. G.

Furside scholar Cambridge U., 1963; Erasmus Wilson demonstrator Royal Coll. Surgeons London, 1959. Fellow Royal Soc. Medicine; mem. Assn. Clin. Pathologists, Coll. Pathologists, Brit. Soc. Neuropathology. Contbg. author: Congenital Abnormalities of Infancy, 1963. Research, publs. on neurol. diseases of children, especially pathogenesis, clin. problems, genetics, epidemiology and sociology of hydrocephalus, spina bifida and pulmonary disease. Home: Springside, Pen-Y-Turnpike, Dinas Powis, Glamorgan, U.K. Office: Llandough Hosp., Penarth, Glamorgan, U.K.*

LAURENT, Auguste, French chemist; b. La Folie, Haute Marne, France, Nov. 14, 1807; studied under Dumas; ed. Sch. Mines; Dr.Sci., 1837; at least 1 son, Mathieu-Paul. Asst., École central; early in life was mining engr.; prof. chemistry Bordeaux, 1838-46; worked with Gerhardt in Paris; became assayer Paris Mint, 1848; founder (with Gerhardt) monthly reports fgn. chem. works. Mem. French Acad. Sci., 1845. Author: Methode de chimie, 1854; Theory of Derived Radicals, 1843. Made (with Gerhardt) a systematic classification of organic compounds known as nucleus theory (which replaced dualistic theory of Berzelius), 1837; distinguished between equivalent, atomic and molecular weights, 1843; discovered anthracene, 1832, anthraquinon, 1835, phthalic acid, 1836, phthalic anhydride and phthalimide; identified phenol as carbolic acid, 1841; demonstrated relation of ethers to oxides and alcohol to water, 1846; formulated geometrical configuration for carbon radicals; built saccharimeter to determine strength of sugar solution by rotation of plane of polarization of transmitted light (thus becoming one of the earliest chemists to work with stereochemistry); Laurent's acid named after him. Died Paris, Apr. 15, 1853.

LAURENT, (Paul Mathieu) Hermann, French mathematician; b. Bordeaux, France, Sept. 2, 1841; s. Auguste Laurent; ed. l'École polytechnique, l'École d'application de Metz. Répétiteur, Polytechnic Sch., Paris. Author: Traité d'algebre, 3 vols., 1867; Traité de mécanique rationelle, 2 vols., 1871; Traité d'analyse, 7 vols., 1885-91; Théorie des séries, 1864; Théorie des residus, 1865; Théorie des équations differentielles ordinaires simultanées, 1873; Théorie élémentaire des fonctions elliptiques, 1882; Théorie des jeux de hasard, 1893; Théorie et pratique des assurances sur la vie, 1895; also articles on math., especially calculus of variations, differential equations, probability elimination, complex quantities, heat conduction, elliptic functions, theory of residues. Extended Cauchy's theorem on series by introducing Laurent expansion of functions. Died Paris, Feb. 19, 1908.

LAURENT, Raymond Ferdinand, herpetologist; b. Wasmes, Belgium, May 16, 1917; s. Armand Charles and Blanche (Carpentier) L.; candidate en sciences biologiques U. Libre de Bruxelles, 1935, licencié en sciences zoologiques, 1937, docteur en science, zoologiques, 1940; m. Louis Elisa Fenaux, June 16, 1960; children—(by previous marriage) Alain-Jean-Armand, Philippe Claude-Armand-Luigi, Veronique-Blanche-Giulietta. Staff Fonds Nat. de la Recherche Scientifique, Bruxelles, Belgium, 1941-47, chercheur qualifié, 1945-47; attaché au Musée du Congo, Tervuren, 1947-49, conservateur-adjoint, 1949—; chargé de mission Institut pour la Recherche Scientifique en Afrique Centrale, Uvira, Congo, 1949-51, 55-56, Astrida, Rwanda, 1951-52; prof. U. Officielle du Congo Belge, Elisabethville, 1957-61; Investigator herpetology NSF grantee Mus. Comparative Zoology, Harvard, 1961-64; investigator in herpetology Fundacion Instituto Miguel Lillo, Tucumàn, Argentina, 1964—. Recipient Prix Lamarck de la'Académie Royale Belgique, 1943. Mem. Soc. Am. Ichthyologists and Herpetologists. Research, numerous publs. on osteology and systematics of African Amphibia, systematics of African amphibians and reptiles especially from Congo and Angola, tree-frog genus Hyperolius; developed new biometrical method for discovery of best ratio characters. Home: Lola Mora, Villa Marcos Paz, Tucumán. Office: Fundacion Miguel Lillo, 205 Miguel Lillo, Tucumán, Argentina.*

LAURENT, Torvard Claude, Swedish biochemist; b. Stockholm, Sweden, Dec. 5, 1930; s. Torbern and Bertha (Svensson) L.; M.B., Karolinska Institutet, Stockholm, 1950, License in Medicine, 1958, M.D., 1958; m. Uilla B. G. Hellsing, Oct. 9, 1953; children—Birgitta, Claes, Agneta. Faculty, Karolinska Institutet, 1949-58; research fellow Retina Found., Boston, 1953-54, asst. dir., 1959-60, asso., 1960-61; established investigator Swedish Med. Research Council, 1961-63; asst. research prof. Uppsala (Sweden) U., 1964-66, prof. med. chemistry, 1966—; asst. research prof. Boston U., 1960-62. Research, publs. on physiology of connective tissue, ophthalmic biochemistry, methods to separate biol. macromolecules, physicochem. properties of polysaccharides and their physiol. role. Home: Hävelvägen 9, 75247 Uppsala. Office: Medicinsk-Kemiska institutionen, Uppsala U., Uppsala, Sweden.*

LAURENZI, Luciano, Italian archeologist; b. Trieste, Italy, June 30, 1902; s. Giuseppe and Leopolda (Vorell) L.; ed. U. Bologna (Italy), Italian Sch. Archeology, Athens; m. Antonietta Barattini, Apr. 20, 1928; children—Alessandro, Donatella. Prof. archeology U. Bologna; asso. prof., titular prof. U. Pisa (Italy); prof. archeology, insp., dir. roles on antiquity State of Italy; govt. commr. Archeol. Inst. Rome; dir. Rodi Archeol. Inst.; dir. Italian Archeol. Sch., Athens; dir. Civic Mus., Bologna. Recipient Gold medal of merit of arts and culture. Mem. Lincei Nat. Acad., Bologna Acad. Scis. Home: via Lame 69, Bologna. Office: via Musei 8, Bologna, Italy.

LAURIE, Victor William, Am. chemist; b. Columbia, S.C., June 1, 1935; s. Victor Hugo and Kathleen (Rice) L.; B.S., U. S.C., 1954; Ph.D., Harvard, 1957; m. Ans Gompel, Dec. 10, 1965. Research asso. Nat. Bur. Standards, Washington, 1957-59; NSF postdoctoral fellow U. Cal. at Berkeley, 1959-60; asst. prof. chemistry Stanford, 1960-66; asso. prof. chemistry Princeton, 1966—. Alfred P. Sloan fellow, 1963—. Mem. Am. Chem. Soc., Am. Phys. Soc., A.A.A.S., Phi Beta Kappa. Mem. editorial com. Ann. Revs. of Phys. Chemistry, 1964—. Research in molecular structure and spectroscopy. Home: 18 N. Stanworth Dr., Princeton, N.J.

LAURIKAINEN, Kalervo Vihtori, Finnish physicist; b. Pielisjärvi, Finland, Jan. 6, 1916; s. Viktor Gregorius and Hilja (Karhunen) L.; Cand.Phil., U. Helsinki, 1940, Ph.D., 1950; m. Aila Onerva Annikki Siikanen, Sept. 13, 1942; children—Petri Kalervo, Erkki Johannes. Lectr. math. scis. Inst. Tech., Turku, Finland, 1946-56; asso. prof. physics U. Turku, 1956-60; prof. nuclear physics U. Helsinki, 1960—, head dept. nuclear physics, 1961—, head dept. computing center, 1961—; rector North-Carelian Summer U., Joensuu, Finland, 1966—. Mem. adminstrv. bd. NORDITA, 1960—. Mem. Finnish (past pres.), Am. phys. socs. Author: Introduction to Modern Physics, 1961; Problems in Modern Physics, 1967; also articles. Research on properties of deuteron and two-nucleon scattering with emphasis on hypotheses concerning nuclear forces. Office: 20 Siltavuorenpenger, Helsinki 17, Finland.*

LAURILA, Simo Heikko, geodesist; b. Kangasala, Finland, Jan. 15, 1920; s. Juho H. and Fanny (Heikkila) L.; B.Sc., Finland's Inst. Tech., 1946, M.Sc., 1948, Ph.D., 1953; m. Mary Isabel Kouki, Feb. 10, 1946; children—Helena (Mrs. Douglas Schmidt), Marketta, Kristina, Taina, Paula. Came to U. S., 1953, naturalized, 1961. Research asso. photogrammetric bur. Govt. of Finland, Helsinki, 1948-53; geodesist Finnish Geodetic Inst., Helsinki, 1953; research asso. Ohio State U., Columbus, 1953-59, asso. research supr., 1960-66, faculty, 1956-66, prof. geodesy, 1962-66; research specialist N.Am. Aviation Corp., Columbus, 1966; head geodesy program U. Hawaii, Honolulu, 1966—, prof. geodesy, 1966—. Lectr., Finland's Inst. Tech., 1950-52, Finland's Mil. Acad., 1950-52. Recipient award Tech. Found. Advanced Sci., Finland, 1960; Kaarina and W. A. Heiskanen award Ohio State U., 1965. Mem. Am. Soc. Photogrammetry, Am. Geophys. Union. Author: Electronic Surveying and Mapping, 1960; also articles. Research on monocular and stereophotogrammetry, geodesy and electronic surveying, propagation electromagnetic waves. Home: 707 N. Kainalu Dr., Kailau, Oahu, Hawaii 96734. Office: Inst. Geophysics, U. Hawaii, Honolulu 96822.*

LAURITSEN, Charles Christian, physicist, educator; b. Holstebro, Denmark, Apr. 4, 1892; s. Thomas and Maria (Nielsen) L.; came to U. S., 1916, naturalized, 1928; grad. Odense (Denmark) Tekniske Skole, 1911; Ph.D., Cal. Inst. Tech., 1929; m. Sigrid Henriksen, May 21, 1915; 1 son, Thomas. Practiced elec., radio engring., 1918-26; faculty physics Cal. Inst. Tech., Pasadena, 1930—, prof. physics, 1935-62, prof. emeritus, 1962—. Cons. various U. S. govt. and mil. offices. Recipient Gold medal Am. Coll. Radiology, 1931; Naval Ordnance Devel. award, 1945, Medal for Merit, 1948; USAF certificate of appreciation, 1953; Distinguished Civil Service award U. S. Army, 1957; U. S. Army certificate of appreciation, 1955; Capt. R. D. Conrad award, 1958; certificate of commendation Fleet Ballistic Missile Program, USN, 1960; Kommander of Dannebrog (Denmark), 1953. Fellow Am. Phys. Soc. (past pres.), Coll. Radiol.; mem. Nat. Acad. Sci., Am. Philos. Soc., A.A.-A.S., Royal Soc. Copenhagen (fgn.), Sigma Xi. Contbr. numerous articles to profl. jours. Died Apr. 13, 1968.

LAURITSEN, Thomas, physicist; b. Copenhagen, Denmark, Nov. 16, 1915; s. Charles C. and Sigrid (Henriksen) L.; came to U. S., 1916, naturalized, 1942; B.S., Cal. Inst. Tech., 1936, Ph.D., 1939; m. Margaret Laura Solum, May 28, 1946; children—Eric, Margaret Ann, Charles. Faculty, Cal. Inst. Tech., Pasadena, 1945—, prof. physics, 1955—; vis. prof. U. Mexico, 1957. Recipient Presdl. certificate of merit for outstanding contbn. to war effort, 1948; Fulbright award Inst. Theoretical Physics U. Copenhagen, 1952-53, NSF Sr. Postdoctoral fellow, 1963-64. Fellow Am. Phys. Soc., Royal Danish Acad. Sci. and Letters; mem. Sigma Pi Sigma. Author: (with Richtmeyer and Kennard) Introduction to Modern Physics, 1955. Research, numerous publs. in energy levels of light nuclei, exptl. studies of nuclear reactions. Home: 1559 Rose Villa St., Pasadena, Cal. 91106.*

LAURITZEN, C(hristian), German gynecologist, endocrinologist; b. Rendsburg, Germany, Dec. 6, 1923; s. Christian and Ella (Fredeland) L.; student U. Berlin (Germany), 1942-45; M.D., U. Kiel (Germany), 1949; m. Brigitte Schoreit, Mar. 18, 1954; children—Christine, Constanze. Asst. gynecologic clinic U. Kiel, 1950-63, faculty, 1961—, prof. obstetrics and gynecology, sci. counselor, 1965—, head physician, 1963—; research fellow Clinic and Hormone Lab., Karolinska Hosp., Stockholm, Sweden, 1958. Mem. German Soc. Gynecology and Obstetrics, German Soc. Endocrinology, German Soc. Fertility and Sterility. Author: (with E. Diczfalusy) oestrogene beim Menschen, 1961; (with A. Klopper) Gynecologic Endocrinology, 1968; also articles. Research on biologic effects of hormones and hormone metabolites in therapy, pathogenesis of icterus neonatorum, endocrinology of pregnant, fetal and neonatal states. Home: 36 Niemannsweg, Kiel (23), Western Germany.*

LAUSCHER, Friedrich, Austrian meteorologist; b. Vienna, Aug. 4, 1905; s. Killian and Agnes (Gradinger) L.; Ph.D., U. Vienna; m. Adele Wittmann, Sept. 8, 1936. Asst. U. Vienna, 1928; became asst. Central Inst. Meteorology and Geodynamics Vienna, 1929, dir., 1935—; instr. meteorology Vienna Poly. Sch., 1957—. Mem. Austrian Soc. Meteorology, Sonnblick Assn., Soc. Geography, Biophysics Soc. Author many publs.; contbr. to Wetter und Leben. Home: Zehanthofgasse 25, Vienna 9. Office: Hohe Warta 38, Vienna 9, Austria.

LAUSON, Henry Dumke, Am. physiologist; b. New Holstein, Wis., Aug. 20, 1912; s. Henry Detlef and Lydia (Dumke) L.; student Mission House Coll., 1930-31; B.S., U. Wis., 1936, Ph.D., 1939, M.D., 1940; m. Eleanor Catchis, Sept. 4, 1936. OSRD fellow in physiology and medicine N.Y. U. Coll. Medicine, 1942-43, instr., 1943-46; asso. Rockefeller Inst. for Med. Research, 1946-50; asso. prof. Cornell U. Med. Coll., 1950-55; prof., chmn. dept. physiology Albert Einstein Coll. Medicine, Yeshiva U., N.Y.C., 1955—. Cons., Nat. Inst. Arthritis and Metabolic Diseases, NIH, Bethesda, Md., 1961-65. Mem. Harvey Soc. (past sec.), Am. Physiol. Soc., Soc. for Exptl. Biology and Medicine, Am. Soc. for Clin. Investigation, Sigma Xi, Phi Beta Kappa, Alpha Omega Alpha. Research, publs. on pituitary-ovarian relationships, blood circulation in shock, pressure rec. in human heart, mechanisms edema formation, protein excretion in nephrosis, control concentration antidiuretic hormone in blood plasma. Home: 40 Barrow St., N.Y.C. 10014.*

LAUSSEDAT, Aime, French inventor; b. Moulins, France, Apr. 19, 1817; student Polytech. Sch., also at Metz, France. Asst. to Faye in astronomy, geodesy Polytech. Sch., 1853-55, titular prof., 1856, dir. studies, 1880; became dir. Conservatoire des Arts et Métiers, 1881. Mem. French Acad. Scis., 1894. Author: Iconometrie et Metrophotographie, 1891; Histoire de la cartographie, 1892; Les applications de la perspective au lever des plans, 1893. 1st to use photography in map production; made map of Mont-Valerien using projected outlines (led Henri Joseph Vallot to photogrammetrically survey over 72,000 hectars, including Mont-Blanc range); his son (Col. Laussedat) built photo-rangefinder, 1903 and later built photo theolite; inventor numerous astron. instruments. Died Mar. 19, 1907.

LAUSTER, Franz, German physicist; b. Frankfort, Germany, June 16, 1897; s. Heinrich and Kathinka (Geist) L.; Ph.D.; m. Margarete Sehlbrede, June 2, 1928; children—Jurgen, Wolfgang, Thomas. Lab. asst.; dir. fabrication, electronics industry; sec. gen. Verband Deutscher Elektrotechniker. Mem. Lichttechnische Gesellschaft. Author: Leitfaden der Elektrowärmetechnik, 1963; Manuel d'Electrothermie Industrielle, 1967; various publs. on ultraviolet rays, electric ovens, transmission of heat. Home: Gluckstrasse 12, Hanau/Main, W. Germany.

LAUTERJUNG, Karl Heinz, German physicist; b. Leichlingen, Germany, May 10, 1914; s. Karl Friedrich and Hedwig (Feeth) L.; Dr.Phil., U. Köln, 1941; m. Anneliese Hassbach, Dec. 19, 1942; children—Karl Lutz, Friedrich Gerd. Faculty, U. Köln, 1948-56, U. Heidelberg, 1956-60; staff Max Planck Institut, Heidelberg, 1954-57; dir. Institut für Kernphysik, U. Köln, 1960—. Mem. Deutsche Physikalische Gesellschaft. Studies, publs. on beta-decay, nuclear spectroscopy, nuclear reactions, atomic physics, cosmic ray physics, weak interactions. Home: 6 Schallstrasse, 5 Köln-Lindenthal, Germany.*

LAUTH, Ernest Alexandre, Alsatian anatomist, physiologist; b. Strasbourg, France, May 14, 1803; student medicine; made sci. trips to Germany, Eng., Switzerland; prof. Faculty of Strasbourg. Recipient Experimental Physiology prize Inst. France. Author: Structure et les usages des Vaisseaux lymphatiques, 1824; Manuel de l'anatomiste, 1829; Observationes anatomicae de parte cephalica nerve sympathici, 1831; Memoire sur le testionle humaine, 1832; Variété de la distribution des muscles chez l'homme, 1833; Exposition et application des sources des connaissances physiologiques, 1836. Described venous sinus of sclera (circular canal nr. the junction of sclera with cornea giving rise to anterior ciliary veins) known as Lauth's canal, also Schlemm's canal), 1829; also described by Friedrich S. Schlemm, 1830. Died 1837.

LAVAL, Carl Gustaf Patrik de, see de Laval, Carl Gustaf Patrik.

LAVAL, P., French physician; b. Espalion, France, Feb. 6, 1913; m. Hélène Cottin, Dec. 22, 1938. Prof. physiopathology of respiratory tract Marseille (France) Faculty Medicine, 1959—, also head clinic, dir. inst. pneumonophthistic research and studies; head pneumology dept. Regional Center for Cancer Research, 1953—. Laureate, Acad. Medicine. Mem. French Soc. Tb, Soc. Respiratory Pathology, Internat. Union Against Tb, French Soc. Fight Against Cancer. Author: Le traitement de la tuberculose pulmonaire, 1963; many other publs. Contbg. author: Feuillets cliniques, 3 vols., 1958. Research on bronchial cancer, pulmonary immunology, pulmonary fibrosis, respiratory physiopathology, tumors of pleura. Home: 22, rue Ed. Rostand, Marseille. Office: Hopital Sainte-Marguerite, Marseille, France.*

LA VALLÉE-POUSSIN, Charles-Jean-Gustave-Nicolas de, see de la Vallée-Poussin, Charles-Jean-Gustave-Nicolas.

LAVATER, Henry, Swiss physician; b. Zurich, Switzerland, 1560; studied medicine in various acads. in Germany and Italy; a son, Jean-Henri. Prof. medicine and math., Zürich; dir. Caroline Coll.; named mem. Swiss delegation sent to Henry IV, 1595. Author of publs. including: Defensio medicorum . . . , 1610; Epitome philosophiae naturalis, ex Aristotelis, 1621. Defended the medicine of Galen against Sala; compiled work of natural philosophy from Aristotle; studied motion of earth. Died Zurich, 1623.

LAVEN, Hannes, German geneticist; b. Dremmen, Germany, Feb. 10, 1913; s. Gerhard and Clara (Nobis) L.; student U. Cologne, U. Bonn, U. Berlin; Dr.-rer.nat., U. Konigsberg, 1939; m. Brunhilde Tolkmitt, Jan. 5, 1940; children—Gerhard, Burkhard, Reinhard. Asst. in zoology Berlin U., 1939-40; asst. Tropeninstitut, Hamburg, 1946-53; asst. Max Planck Inst., Tübingen, 1953-59; prof., dir. Inst. for Genetics, Mainz U., 1959—; vis. prof. U. Ill., 1958-59, 61-62. Cons. to WHO. Recipient Genetics prize Deutsche Forschungsgemeinschaft, 1957. Publs. on devel., genetics of mosquito Culex pipiens, detection and explanation of cytoplasmic incompatibility. Home: am Eselsweg 26, 65 Mainz, Germany.*

LAVENDA, Nathan, Am. biologist; b. N.Y.C., Dec. 10, 1918; s. Zukin and Etie (Weinstein) L.; B.S., City Coll. N.Y., 1942; M.Sc., N.Y. U., 1947, Ph.D., 1952; m. Harriet Rebecca Zukowsky, June 24, 1961; children—Elaine, David, (by previous marriage) Bernard, Ronald, Iris, Stuart, Steven. With U. S. Fish and Wildlife Service, N.Y.C., 1944-47, USPHS, N.Y.C., 1947-50; faculty N.Y. U., 1952-56, Howard U., 1952-56; research asso. Albert Einstein Coll. Medicine, also asst. to dir. research lab. Jewish Meml. Hosp., 1956-60; asso. prof. physiology State Coll. at North Adams, Mass., 1961-67; faculty Wis. State U., Oshkosh, 1967—. Mem. Am. Soc. Zoologists, A.A.-A.S., Am. Assn. U. Profs., N.Y. Acad. Scis. Research, publs. on life history of sea bass, new anat. site for medullary adrenal, changes in blood picture of mammals with female sex hormones, demonstration of viruses in human cancer using an original technique for detection of tissues; traced fiber tracts from retina to hypothalmus of brain. Home: 1001 Cherry St., Oshkosh, Wis.*

LAVER, Myron Bertrand, physician; b. Bucharest, Rumania, Aug. 17, 1926; s. Bertrand and Raissa (Sklar) L.; came to U. S., 1941, naturalized, 1943; student Earlham Coll., 1948; M.D., U. Basle, Switzerland, 1956; m. Liselotte M. Aeschlimann, May 6, 1954; children—Kim B., Karin A., Lesley M. Faculty, Harvard Med. Sch., 1960—, asst. clin. prof. anesthesia, 1966—; staff Mass. Gen. Hosp., Boston, 1960—. Author: (with others) Respiratory Care, 1965. Contbr. articles to med. jours. Developer methodology for clin. assessment blood gases, 1960; research in lung complications. Home: 39 Walnut Rd., Weston, Mass. 02193. Office: Mass. Gen. Hosp., Fruit St., Boston 02114.*

LAVERAN, Charles Louis Alphonse, French physician, parasitologist; b. Paris, June 18, 1845; M.D., U. Strasbourg (France), 1867; prof. mil. hygiene and parasitology Val-de-Grace (France) Med. Sch., 1883-96; staff Pasteur Inst., Paris, 1897-1922, founder lab. tropical diseases, 1907. Recipient Nobel prize in physiology and medicine, 1907. Fellow Royal Soc., 1916, mem. Acad. Medicine, French Acad. Scis., 1895, Soc. Exotic Pathology (founder 1908). Author: Traité des maladies et épidémies des armées, 1875; Traité des fièbres palustres, 1884; Traité d'hygiene militaire, 1896; Traité du paludisme, 1898; (with F. Mesnil) Trypanosomes et trypanosomiases, 1904; also numerous textbooks and sci. papers. Research in leishmaniasis, trypanosomiasis, other protozoal diseases; discovered parasite (hematozoaire) of human malaria, 1880. Died Paris, May 18, 1922.

LAVES, Fritz Henning, crystallographer; b. Hannover, Germany, Feb. 27, 1906; s. Georg and Margret (Hoppe) L.; student U. Innsbruck, 1924, U. Göttingen, 1924-26; Dr. phil., U. Zurich, 1929; m. Melitta Druckenmüller, Apr. 7, 1938; children—Grazia (Mrs. Hans Schicht), Charlotte, Katarina. Asst., dozent U. Göttingen, 1930-44; prof. U. Halle, 1944-45, U. Marburg, 1945-49, U. Chgo., 1949-54, U. Zurich (Switzerland), 1954—, Fed. Inst. Tech., Zurich, 1954—; dir. Inst. Crystallography and Petrography, Zurich, 1954—. Mem. German, Swiss, Am. mineral. socs. Research, numerous publs. on crystal structure of metallic compounds and feldspars. Home: 8 Faehnlibrunnstrasse, Küsnacht, Zurich. Office: 5 Soneggstrasse, Zurich, Switzerland.

LAVES, Kurt, astronomer; b. Lyck, Germany, Aug. 24, 1866; s. Hermann Karl and Julie (Krahnefeld) L.; student U. Koenigsberg, 1886-87, U. Berlin, 1887-91, A.M., Ph.D., 1891; m. Luise Moshagen, Aug. 25, 1896. Asst., Royal Obs., Berlin, 1892-93; docent reader, asst. and asso. in astronomy U. Chgo., 1893-97, instr., 1897-1901, asst. prof., 1901-08, asso. prof., since 1908. Fellow A.A.A.S.; mem. Astronomische Gesellschaft (Leipzig), Astron. and Astrophys. Soc. America, Am. Math. Soc. Died Mar. 25, 1944.

LAVIER, Georges, French physician; b. Dijon, France, June 2, 1892; ed. univs. Dijon, Paris; M.D.; m. Lucienne George, Apr. 8, 1933. Became asst. Faculty Medicine, 1920; agrégé in medicine, 1926; prof. Faculty Medicine, Lille, France; sec. gen., inst. tropical medicine Faculty Medicine, Paris. Mem. State Commn. Soc. Nations for Human Trypanosomiasis. Mem. Hygiene Council, Soc. Hygiene Commn., Acad. Medicine, Savarin Acad. Research and publs. on parasitic tropical diseases. Home: 12, ave. de l'Observatoire, Paris 6. Office: 15, rue de l'École-de-Medicine, Paris 6, France.

LAVINE, Leroy Stanley, Am. orthopaedic surgeon; b. Jersey City, Oct. 28, 1918; s. Max and Katherine (Miner) L.; A.B., N.Y. U., 1940, M.D., 1943; m. Dorothy Kopp, Feb. 14, 1946; children—Michael, Nancy. Resident instr. in orthopaedic surgery Ind. U. Coll. Medicine, Indpls., 1951-52; faculty State U. N.Y. Downstate Med. Center, Coll. Medicine, Bklyn., 1952—, prof., co-head div. orthopaedic surgery, 1965—; adj. prof. grad. sch. N.Y. U., 1966—. Cons. Am. Mus. Natural History; cons. in orthopaedic surgery Bklyn. VA Hosp., 1965—. Fellow A.A.A.S., N.Y. Acad. Scis., A.C.S., Am. Acad. Orthopaedic Surgeons, N.Y. Acad. Medicine; mem. Am. Assn. Physiol. Anthropologists, Orthopedic Research Society, American Orthopedic Association International Soc. orthopedic Surgery and Traumatology, Sigma Xi. Research, publs. on basic molecular structure, growth and metabolism of bone; demonstrated that there is a stress induced elec. effect in hard tissues whose origin appears to be in long chain fibers; use of comparative biol. approach to show biochem. mechanisms of mineralization have been elucidated, especially using a marine protozoan. Home: 375 East Shore Rd., Great Neck, N.Y. 11023. Office: 2035 Lakeville Rd., New Hyde Park, N.Y. 11234.*

LAVOISIER, Antoine Laurent, French chemist; b. Paris, Aug. 26, 1743; s. Jean Antoine and Emilie (Punctis) L.; student Mazarin Coll., 1754-61; B.Law, 1763; licentiate, 1764; studied astronomy under N. L. de Lacaille, chemistry under G. F. Rouelle, botany under Bernard de Jussieu; m. Marie Anne Pierrette Paulze, Dec. 16, 1771. Asst. in preparation geol. map of France; became dir. govt. gunpowder works, 1776; named farmer-gen., 1779; mem. commn. to establish uniform weights and measures, 1790; became commissary of treasury, 1791. Fellow Royal Soc., 1788. Asso., French Acad. Scis., 1768 (prize for plan to improve street lighting in Paris 1766). Author: Opuscules physiques et chimiques; Reflexions sur le phlogiston, 1783; Traité élémentaire de chimie, 1789; Mémoires de chimie; (with Simon de Laplace) Sur la chaleur (helped lay founds. of thermochemistry), 1783; (with de Morveau, Berthollet, Fourcroy) Méthode de nomenclature chimique (presents chem. nomenclature on which present system is based), 1787; Mémoires de chimie, pub. 1805; also numerous articles. Founder modern chemistry; introduced effective quantitative methods into chemistry; disproved phlogiston theory by discovering relation between combustion and oxygen, 1775; named oxygen and claimed to have discovered it with Priestley and Scheele; on basis of work by Priestley and Cavendish, discovered role of oxygen in plant and animal respiration, 1790; divided substances into elements and compounds; gave theory of formation of chem. compounds; made expts. to determine composition of water and other compounds; explained formation of salts and acids; determined composition of nitric and sulphuric acid; 1st to make water-gas; inventor gasometer. Guillotined in Paris, May 8, 1794.

LAVOLLAY, Jean Albert, French chemist; b. Argenteuil, May 12, 1907; s. Jean and Leonie (Babin de Grandmaison) L.; Ph.D., U. Paris; m. Elisabeth Priewisch, Oct. 17, 1940; children—Jean-Max, Isabelle. Became research dir. Nat. Sci. Research Center, 1942; prof. agr. and biol. chemistry Nat. Conservatory Arts and Scis., 1945—; instr. Sorbonne. Mem. Acad. Agr. France. Research and publs. on chemistry and physiology. Home: 46, rue de Dunkerque, Paris 9. Office: 292, rue Saint-Martin, Paris, France.

LAVORSKII, Vasilii Ivanovich, Soviet geologist, paleontologist; b. Jan. 10, 1875; grad. Mining Sch., Dombrov, St. Petersburg Mining Inst., 1913. Began as miner, 1893; worked in Urals; past mem. staff Geologic Com. (later All-Union Geology Inst.). Author: (with P. I. Butov) Kuznetsk Coal Basin (Przhevalskii prize USSR Geographic Soc.), 1927; Stromatoporoidea of USSR, Part I, 1955. Recipient Stalin prize, 1946. Discovered deposits of coking coal in So. part of basin; a surveyor of Kuzbass.

LAVRENKO, Eugienii Mikhailovich, Russian geobotanist; b. Feb. 24, 1900. With Bot. Gardens, Kharkov, 1921-28; asst. prof. Kharkov Agrl. Inst., 1929-31, prof., 1931-34; with Bot. Inst. of USSR Acad. Scis., 1934—, dir., 1946—, corr. mem. acad., 1946—. Author: A History of Flora and Vegetation in the USSR Based on Current Data on Plant Distribution, 1938; The Steppes of the USSR, 1940; The Phytogeosphere, 1949; The Age of the Botanical Regions of Non-Tropical Eurasia, 1951; The Steppes and Agricultural Lands in Steppe Areas, 1956; The Long-Term Plan for a Geographical Network of Reservations in the USSR, 1958. Developed new classification for steppe vegetation of USSR; investigated zoning and compiled vegetation maps; introduced concept of phytogeosphere as part of biosphere. Address: V. L. Komarov Institute of Botany, USSR Academy of Sciences, Ulitsa Popova, 2, Leningrad, P-22, USSR.

LAVRÉNTIEV, Mikhail Alekseievich, Russian mathematician; b. Kazan, USSR, Nov. 19, 1900; grad. Moscow U., 1922; D.Phys.-Math. Scis., 1933; D.Tech. Scis., 1932. Instr. Moscow higher ednl. instns., 1921-31; prof., Moscow U., 1931-41; head dept. theory of functions, Math. Inst., USSR Acad. Scis., 1934—; dir. Inst. Math. and Mechanics, Ukrainian Acad. Sci., 1939-48, acad. v.p., 1945-48; dir. Inst. for Precision Mechanics and Computer Engring., USSR Acad. Sci., 1950-53, acad. sec. phys.-math. sci. sect., 1951-53, 54-57, acad. pres., 1957—, chmn. presidium Siberian br., 1957—; prof. Novosibirisk U., 1958—; dir. Inst. Hydrodynamics, Siberian Acad., 1958—. Mem. sci. expdns. to Dikson Island, Kamchatka, to study possible use of subterranean heat. Recipient Stalin Prize, 1946, 49, Order of Lenin, Order of Red Banner of Labor. Mem. Ukrainian Acad. Scis., USSR Acad. Scis. Author: The Theory of Conformal Mappings, 1934; Some Properties of Schlicht Functions with Application to the Theory of Streams, 1938; A General Problem in the Theory of Quasi-Conformal Mappings of Plane Domains, 1947; The Basic Theorem of the Theory of Quasi-Conformal Mappings of Plane Domains, 1948. Studies on theory of functions of complex variables, applied hydrodynamics, nonlinear waves in hydrodynamics, mechanics of continuous mediums; developed theory of quasiconformal mapping, theory based on observation of cumulative change in explosions. Home: Siberian Dept., USSR Acad. Scis., 20 Sovietskaya St., Novosibirsk, USSR.

LAVROVSKII, Konstantin Petrovich, Russian organic chemist; b. Dec. 31, 1898; grad. Moscow U., 1926. With State Sci. Research Oil Inst., 1930-34, became prof. there, 1933; worked in oil industry; with USSR Acad. Scis., 1942—, formerly at Inst. Mineral Fields, later Inst. of Oil, corr. mem. acad., 1953—. Recipient M. V. Frunze and N. D. Zelinskii prizes. Coauthor: Catalytic Pressure Cracking of Cyclic Hydrocarbons, 1952; Gaseous Paraffin Refining in High-Speed Cracking, 1954; Physiocochemical Studies of High-Speed Cracking, 1955; The Interaction of Iron Ores and Methane in a Fluidized Bed, 1957; A Method of Studying Fast Reactions in Turbulence Reactors by Means of Labeled Atoms, 1960. Research in field of chemistry and technology of oil refining and organic catalysis; developed comml. prodn. of aviation gasoline; studies of catalytic hydrocarbon transformations were theoretical basis for prodn. unsaturated gases and high octane fuels. Home: 1st Donskoi PR. 15. Office: USSR Academy of Sciences Institute for Petrochemical Synthesis, Leninskii Prospekt, 29, Moscow, USSR.

LAW, Lloyd William, Am. exptl. pathologist; b. Ford City, Pa., Oct. 28, 1910; s. Craig Smith and Cora (Whiteley) L.; B.S., U. Ill., 1931; A.M., Harvard, 1935, Ph.D., 1937; m. Bernette Bohen, May 5, 1942; children—Lloyd William, David Bradford. Instr., Charleston (Ill.) High Sch., 1931-33; research asso. Harvard, 1936-37, Stanford, 1937-38; with Jackson Meml. Lab., Bar Harbor, Me., 1938—, sci. dir., 1946-47, trustee, 1947—; mem. staff Nat. Cancer Inst., Bethesda, Md., 1947—, head carcinogenesis sect., 1950—. Bd. sci. advisers Roswell Park Meml. Inst. for Cancer, Buffalo, Hektoen Inst., Chgo., WHO, Am. Cancer Soc. Recipient Anne Frankel Rosenthal A.A.A.S. Cancer Research award, 1957; G.H.A. Clowes Meml. award, 1965. Mem. A.A.A.S., Am. Assn. Cancer Research (dir., past pres.), Soc. Exptl. Biology and Medicine, Am. Soc. Exptl. Pathology, N.Y. Acad. Scis., Genetics Soc. Am., Royal Soc. Medicine, Italian Cancer Soc. Research, publs. on basic mechanisms involved in change from a normal to a cancerous cell in animals. Home: 9810 Fernwood Rd., Bethesda 20034. Office: Nat. Cancer Inst., Bethesda, Md. 20014.*

LAW, Russell Robin, Am. physicist; b. Hampton, Ia., Jan. 11, 1907; s. Chalmer Andrew and Barbara (McDowell) L.; B.S., Ia. State U., 1929, M.S., 1931; D.Sc., Harvard, 1933; m. Myra Harms, Oct. 19, 1929; children—Lucy Chappell (Mrs. D. J. Webster), John Townsend. Faculty, Ia. State U., 1929-31; research asso. geophysics Harvard, 1933-34; research physicist radiotron div. RCA Mfg. Co., Harrison, N.J., 1934-41, mem. tech. staff RCA Labs., Princeton, N.J., 1941-53; asst. to v.p. CBS-Hytron, Danvers, Mass., 1953-54, dir. research and devel., 1954-57; dir. new product devel. Hughes Products Group, Culver City, Cal., 1957-58; asso. dir. Hughes Research Labs., Malibu, Cal., 1958-61; spl. cons. Hughes Aircraft Co. aerospace group, Culver City, 1961—. Spl. cons. U. S. Dept. Def., 1956—. Fellow I.E.E.E.; mem. Am. Phys. Soc., Sigma Xi, Tau Beta Pi, Eta Kappa Nu, Phi Sigma Kappa. Asso. editor Transactions on Electron Devices, I.E.E.E., 1960—. Contbr. numerous tech. articles; patentee electronics field. Home: 17352 Sunset Blvd., Pacific Palisades, Cal. 90272. Office: Florence and Teale Sts., Culver City, Cal. 90232.*

LA WALL, Charles Herbert, Am. pharmacist, chemist; b. Allentown, Pa., May 7, 1871; s. John Jacob and Emma Jane (Boas) La Wall; grad. coll. prep. course, State Normal Sch., Bloomsburg, Pa., 1888; Ph.G., Phila. Coll. Pharmacy, 1893, Ph.M., 1905; Phar.D. (hon.), U. Pitts., 1919; Sc.D., Susquehanna U., 1920; m. Millicent Saxon Renshaw, June 5, 1907. Comml. chemist, 1894-1903; instr. pharmacy Phila. Coll. Pharmacy and Science, 1900-06, asso. prof., 1906-18, prof. theory and practice of pharmacy, dean, 1918—. Chemist Pa. Dept. Agr., Bur. Food, 1904—, Pa. State Pharm. Examining Bd., 1905-12, 1914—, Pa. Health Dept., 1906-18; food inspection chemist U. S. Dept. Agr., 1907-12; mem., officer U. S. Pharmacopoeia Revision Com., 1910—; mem. revision com. Nat. Formulary, 1906-29; pres. Am. Assn. Coll. Pharmacy, 1923. Recipient Remington medal, 1928. Mem. Am. (pres. 1919), Pa. (pres. 1911) pharm. assns. Author: Four Thousand Years of Pharmacy, 1927. Joint Author: Leffmann and La Wall's Organic Chemistry, 1905. Collaborating editor U. S. Dispensatory, 1917—, Remington Practice Pharmacy, 1918—. Died Dec. 7, 1937.

LAWALREE, André Gilles Celestin, Belgian botanist; b. Terwagne, Feb. 2, 1921; s. Narcisse and Maria (Istas) L.; Ph.D., U. Louvain (Belgium); m. Anne Collaris, Apr. 28, 1951; children—Marie-Rose, Dominique, Sabine. Asst. curator, asst. dir. labs., later dir. lab. State Bot. Garden, Brussels, Belgium. Recipient Crepin prize Royal Soc. Botany Belgium, prize Belgian Royal Acad. Scis. Mem. Royal Bot. Soc., Graphia, Bot. Soc. Brit. Isles (hon.), Soc. Luxembourg's Naturalists, Luxembourg Inst. Sci. Author: Flore des ptéridophytes de Belgique; Flore des spermatophytes de Belgique. Home: 3, rue Van Elderen, Brussels 16. Office: 236, rue Royale, Brussels 3, Belgium.

LAWES, Sir John Bennet, English agriculturist; b. Rothamsted, Eng., Dec. 28, 1814; s. John Bennet Lawes; ed. Oxford, D.C.L., 1892; LL.D., Edinburgh, 1877; D.Sc., Cambridge, 1894; engaged in farming, from 1834; founder 1st factory for manufacture of mineral superphosphate, Deptford, 1843, built larger factory, Barking Creek, 1857, sold bus., 1872; manufactured citric and tartaric acid, London, circa 1870-1899; founder (with J. H. Gilbert) Rothamsted agrl. expt. sta., 1843; introduced allotment gardens for Harpenden villages, 1852. Recipient Albert Gold medal Soc. Arts, 1894. Fellow Royal Soc., 1854 (Royal medal 1867); mem. Royal Agrl. Soc. (council, v.p., trustee). Author: Compensation for Unexhausted Manures, 1883; (with H. J. H. Gilbert) The Rothamsted Memoirs on Agricultural Chemistry and Physiology, 7 vols., 1893-99, also articles on Rothamsted expts. Patentee 1st superphosphate fertilizer, 1842, thus founding artificial manure industry. Died Rothamsted, Aug. 31, 1900.

LAWFORD, John Bowring, surgeon; b. Montreal, Que., Can., 1858; s. Frederick Lawford; M.D., M.Ch., LL.D., McGill U.; postgrad. St. Thomas's Hosp., London, Eng. Cons. surgeon Royal London Ophthalmic Hosp., Moorfields; cons. ophthalmic surgeon St. Thomas's Hosp; mem. med. appeal bd. Royal Navy. Fellow Royal Coll. Surgeons Eng., Royal Soc. Medicine (v.p. ophthalmology sect. London), Am. Acad. Ophthalmology and Otolaryngology (hon.); mem. Ophthal. Soc. U.K. (pres.), Council Brit. Ophthalmologists (pres.), French Soc. Ophthalmology (hon.). Contbr. articles to Ency. Medica, also to ophthal. jours. Died Jan. 3, 1934.

LAWLAH, John Wesley, Am. radiologist; b. Bessemer, Ala., Aug. 12, 1904; s. John Wesley and Mattie (Lindsey) L.; B.S., Morehouse Coll., 1925, D.Sc. (hon.), 1941; M.S., U. Wis., 1929; M.D., U. Chgo., 1932; m. Leora Frances McCarrell, Feb. 9, 1933; children—Evelyn Frances (Mrs. Ernest D. Fears), John W. 3d. Practice medicine specializing in radiology Chgo., 1935-41, Washington, 1941—; radiologist Provident Hosp., Chgo., 1935-36, med. dir., supt., 1936-41; dean Med. Sch., prof. radiology Howard U., Washington, 1941-46, clin. prof. radiology, 1946—; supt. Freedmen's Hosp., Washington, 1942-44, radiologist, 1946—, mem. commn., 1955—. Mem. panel med. experts WHO, 1946; cons. radiologist Mt. Alto Vet.'s Hosp., Washington, 1957—, D.C. Gen. Hosp., 1957—; mem. Pub. Health Av. Council, D.C. Commrs., 1961-64. Recipient certificate for meritorious service Selective Service, 1942-45. Mem. Nat. Conf. Hosp. Adminstrs. (sec.-treas. 1936-41), Alpha Omega Alpha, Sigma Sigma, Alpha Phi Alpha, Sigma Pi Phi. Research, publs. in embryology, cardiovascular roentgenology. Home: 32 Bryant St., N.W. Washington 20001. Office: 2208 Georgia Av., N.W. Washington 20001.*

LAWRENCE, Barbara (Mrs. William Edward Schevill), Am. zoologist; b. Boston, July 30, 1909; d. Harris Hooper and Theodora (Eldredge) Lawrence; B.A., Vassar Coll., 1931; m. William Edward Scheville, Dec. 23, 1938; children—Lee, Edward. Staff, Mus. Comparative Zoology, Harvard, 1931—, asso. and acting curator mammals, 1942-52, curator mammals, 1952—. Mem. Am. Soc. Mammalogists, A.A.-A.S., Phi Beta Kappa, Sigma Xi. Research, numerous publs. on systematics mammals, functional anatomy head and throat in cetaceans, skeletal anatomy certain artiodactyls, identification dogs and related mems. genus Canis. Home: Garfield Rd., Concord, Mass. 01742. Office. Mus. Comparative Zoology, Harvard, Cambridge, Mass. 02138.*

LAWRENCE, Ernest Orlando, Am. physicist; b. Canton, S.D., Aug. 8, 1901; s. Carl Gustavus and Gunda (Jacobson) L.; student St. Olaf Coll., Northfield, Minn., 1918-19, A.B., U. of S.D., 1922; A.M., U. of Minn., 1923; student U. of Chicago, 1923-24; Ph.D., Yale, 1925; Sc.D., U. of S.D., 1936; Princeton, Yale, Stevens Inst. Tech., 1937, Harvard, U. of Chicago, Rutgers U., 1941. McGill U., Montreal, Can., 1946; LL.D., U. of Mich., 1938, U. of Pa. 1942 Sc.D., Univ. B.C., 1947, U. So. Cal., 1949, University San Francisco, 1949; LL.D., U. Glasgow, 1951; m. Mary Kimberly Blumer, 1932; children—John Eric, Margaret Bradley, Mary Kimberly, Robert Don, Barbara Hundale, Susan. Nat. Research fellow, Yale University, 1925-27, assistant professor physics, 1927-28; associate professor physics, U. of Calif., 1928-30, prof., 1930—, dir. Radiation Lab., 1936—. Awarded Elliott Cresson medal, Franklin Inst., 1937; Research Corp. prize and plaque, 1937; Comstock prize, Nat. Acad. Sciences, 1937; Hughes medal, Royal Soc. (Eng.), 1937; Nobel prize in physics, 1939; Duddell medal, The Phys. Soc., 1940; William K. Dunn award, American Legion, 1940; Holley Medal, American Society of Mechanical Engineers, 1942; Medal for Merit, 1946; Medal of Trasenster, 1947; Officier de la Legion d'Honneur, 1948; Faraday Medal, 1952; Annual award Am. Cancer Society, 1954. Board Foreign Scholarships, 1947. Mem. Solvay Conf., Brussels, 1933; mem. (hon.) U.S.S.R. Acad. Scis., 1943, Royal Swedish Acad. Scis., 1952, Royal Irish Acad., 1948. Mem. bd. of trustees, Carnegie Institution of Washington, 1944. Fellow American Physical Society, A.A.A.S., American Acad. of Arts and Sciences; Hon. Fellow Royal Soc. of Edinburgh, The Phys. Soc., Indian Acad. Sci.; mem. Nat. Acad. Scis., Am. Philos. Soc., Phi Beta Kappa, Sigma Xi, Gamma Alpha. Contbr. to Proc. Nat. Acad. Scis., Physical Review. Research in nuclear physics; invented cyclotron (using this, made researches into structure of atom, made transmutation of certain elements, produced artificial radioactivity), 1931; used radiation in study of problems in biology and medicine; made possible 1st extensive clin. use of neutron for cancer-therapy, prodn. of radio-phosphorus and other radio-isotopes for med. use, prodn. of radioiodine for 1st therapeutic use in successful treatment of hyperthyroidism; during World War II, instrumental in isolating uranium 235 (used in atomic bomb). Died Palo Alto, Calif., Aug. 27, 1958.

LAWRENCE, George Newbold, Am. ornithologist; b. N.Y.C., Oct. 20, 1806; s. John Burling and Hannah (Newbold) L; m. Mary Ann Newbold, 1834. Became interested in study of birds, circa 1820, collected some 8,000 stuffed birds over the years (collection later became property of Am. Mus. Natural History); entered father's wholesale drug firm, N.Y.C., 1822, later became partner, then head of firm, 1835; devoted later years to complete study of ornithology; became interested in neotropical birds, circa 1858, became expert on birds of W.I. and Central Am.; a founder, hon. mem. Am. Ornithologists' Union; hon. mem. Zool. Soc. of London, Brit. Ornithologists' Union, others; a founder Coll. of Pharmacy of City of N.Y. Author: (with Spencer F. Baird, John Cassin) report of N.Am. birds published in Vol. IX of Reports of Explorations to Ascertain the . . . Route for a Railroad from the Mississippi River to the Pacific Ocean, 1858. Died N.Y.C., Jan. 17, 1895.

LAWRENCE, H(enry) Sherwood, Am. physician, educator; b. N.Y.C., Sept. 22, 1916; s. Victor John and Agnes (Whalen) L.; A.B., N.Y. U., 1938, M.D., 1943; m. Dorothea Wetherbee, Nov. 13, 1943; children—Dorothea, Victor, Geoffrey. Faculty, N.Y. U. Sch. Medicine, N.Y.C., 1947—, John Wyckoff fellow in medicine, 1948-49, dir. student health, 1950-57, head infectious disease and immunology div., 1959—, prof. medicine, 1961—, co-dir. med. services, 1964—, vis. physician U. Hosp., 1964—; cons. medicine Manhattan VA Hosp., 1964—, infectious disease program com. VA Research Service, 1960—; cons. Allergy and Immunology Study sect. USPHS, 1960-63, chmn., 1963—. Asso. mem. commn. on streptococcal and staphylococcal diseases Armed Forces Epidemiological bd. Dept. Def., 1956—; mem. coms. Nat. Acad. Scis.-NRC, 1957—; mem. allergy and infectious disease panel Health Research Council, City N.Y., 1962—. Commonwealth Found. fellow Univ. Coll., London, Eng., 1959; Research Career Devel. awardee USPHS, 1960-65. Diplomate Am. Bd. Internal Medicine. Fellow A.C.P.; mem. Assn. Am. Physicians, Am. Soc. for Clin. Investigation, Am. Assn. Immunologists, Soc. for Exptl. Biology and Medicine (editorial bd. Proc.), Harvey Soc. (sec. 1957-60), Peripatetic Clin. Soc., Infectious Diseases Soc. (charter), Royal Soc. Medicine (affiliate, Eng.), Internat. Transplantation Soc. (chmn. constn. com.), Societe Francaise D'Allergie (corr.), Alpha Omega Alpha. Editor: Medical Clinics of North America, 1957; Cellular and Humoral Aspects of Hypersensitive States, 1959; editorial bd. Transplantation, Ann. of Internal Medicine. Research, publs. on characterization transfer factor in delayed allergy, mechanisms tissue damage and homograft rejection in man. Home: 343 E. 30th St., N.Y.C. 10016.*

LAWRENCE, John Hundale, Am. physician; b. Canton, S.D., Jan. 7, 1904; s. Carl Gustavus and Gunda (Jacobson) L.; A.B., U. S.D., 1926, D.Sc., 1942; M.D., Harvard, 1930; Dr. honoris causa, U. Bordeaux (France), 1958; D.Sc., Cath. U. Am., 1959; m. Amy McNear Bowles, June 20, 1942; children—John Hundale, Amy Sheldon, James Bowles, Steven Ernest. Asso. physician New Haven Hosp., 1934-37; instr. internal medicine Yale, 1934-37; faculty U. Cal. at Berkeley, 1937—, prof. med. physics, 1948—, dir. Donner Lab., 1948—, physician-in-chief Donner Pavilion, 1954—, asso. dir. Lawrence Radiation Lab., 1959—. Lectr., vis. prof. numerous instns., socs.; cons. govt. agys. Recipient Caldwell medal Am. Roentgen Ray Soc., 1941; Sir James Mackenzie Davidson medal Brit. Inst. Radiology (London), 1947; certificate appreciation U. S. War and Navy Dept., 1948; medal U. Bordeaux, 1958; Silver Cross Royal Order Phoenix, Greece, 1962; medal Pasteur Inst., Paris, France, 1963. Guggenheim fellow, 1958. Fellow A.C.P.; mem. Royal Soc. Medicine, Am. Soc. Clin. Investigation, Am. Clin. Climatol. Assn., Soc. Nuclear Medicine, Western Assn. Physicians, Am. Physiol. Soc., Harvard Med. Alumni Assn. (pres. 1945-46, 62-63), Phi Beta Kappa, Sigma Xi, Alpha Omega Alpha, others. Author: Polycythemia: Physiology, Diagnosis and Treatment, 1955; (with B. Manowitz, B. Loeb) Radioisotopes and Radiation: Recent Advances in Medicine, Agriculture and Industry, 1964. Editor: (with John Gofman) Advances in Biological and Medical Physics, 1948—, Advances in Atomic Medicine, 1965; editorial bd. Acta Haematologica, 1953—, Acta Isotopica, 1960—, Am. Jour. Roentgenology, Radium Therapy, and Nuclear Medicine, 1950—, Cancer, 1948—, Jour. Nuclear Medicine, 1959—. Research on isotopes in medicine, diseases of blood, high altitude physiology, metabolism normal and cancer tissues, biol. effects of radiation, heavy particles; introduced isotopes into med. therapy, 1936; demonstrated 1st substance (estrogen) to give some protection against radiation injury; discovered (with assos.) narcotic properties xenon, 1946; pioneered use neutrons, protons and alpha particles in med. therapy. Home: 220 Glorietta Rd., Orinda, Cal. 94563. Office: Donner Lab., U. Cal., Berkeley, Cal. 94720.*

LAWRENCE, John Seward, Am. physician; b. Smithfield, Va., Aug. 15, 1896; s. John Walter and Anna (Pulley) L.; B.A., U. Va., 1917, M.D., 1921; m. Nell Fain, Sept. 10, 1929; children—Anne, Jean, John. Faculty, Vanderbilt U. Sch. Medicine, 1927-28, Harvard Med. Sch., 1928-29, U. Rochester Sch. Medicine and Dentistry, 1929-48; faculty U. Cal. Med. Center, Los Angeles, 1947-63, prof. medicine emeritus, 1963—. Cons. univs., hosps., VA. Diplomate Am. Bd. Internal Medicine. Mem. Am. Clin. and Climatol. Assn., A.C.P., Am. Fedn. Clin. Research, Am. Heart Assn., A.M.A., Am. Soc. Clin. Investigation, Assn. Am. Physicians, Am., Internat., Interurban socs. hematology, Los Angeles County Med. Assn., Los Angeles Acad. Medicine, Pacific Interurban Clin. Assn., Western Assn. Physicians. Publs. in gen. field hematology especially physiology and dynamics of white blood cells. Home: 1331 W. Highland Av., Redlands, Cal. Office: Dept. Medicine, U. Cal. Los Angeles Sch. Medicine, Los Angeles 90024.*

LAWRENCE, Merle, Am. physiol. psychologist; b. Remsen, N.Y., Dec. 26, 1915; s. George William and Alice (Bowne) L.; A.B., Princeton, 1938, M.A., 1940, Ph.D., 1941; m. Roberta Harper, Aug. 8, 1942; children—Linda Alice (Mrs. Thomas Plichta), Roberta Harper (Mrs. James Henderson), James Bowne. Asst. prof. Princeton, 1946-50, asso. prof., 1950-52; asso. in research Lempert Inst. Otology, N.Y.C., 1946-52; asso. prof. dept. otorhinolaryngology U. Mich. Med. Sch., 1952-57, research asso. Inst. Indsl. Health, 1952—, prof. otorhinolaryngology, physiology, psychology, 1957—, dir. Kresge Hearing Research Inst., 1961—. Mem. Acoustical Soc. Am., Am. Physiol. Soc., A.A.A.S., Am. Otol. Soc. (award of merit 1967), Am. Rhinol., Laryngol., Otol. Soc., Am. Acad. Ophthalmology and Otolaryngology (award of merit 1965). Author: Experiments in Human Behavior, 1949; (with E. G. Wever) Physiological Acoustics, 1954; (with M. Alpern, D. Wolsk) Sensory Processes, 1967; also numerous articles. Editorial bd. Archives of Otolaryngology, 1961—; rev. editor Jour. Auditory Research, 1960—. Contbns. to understanding of middle and inner ear physiology, acoustic distortion properties of ear, circulation of inner ear fluids and innervation of tympanic muscles. Home: 2611 Hawthorne Rd., Ann Arbor, Mich. 48104.*

LAWRENCE, Robert Standish, Am. physicist; b. Worcester, Mass., Oct. 28, 1925; s. Ralph S. and Maude (Hayward) L.; B.S. in Physics, Worcester Poly. Inst., 1949; M.S., Yale, 1950; m. Pamela A. Hanford, July 28, 1951; children—Vernon R., Viki A. Physicist, Nat. Bur. Standards, Washington, 1948-52, Boulder, Colo., 1952-65, Environmental Sci. Services Adminstrn., Boulder, Colo., 1965—. Mem. Am. Astron. Soc., Am. Geophys. Union, Internat. Sci. Radio Union. Observation and interpretation of effects of upper atmosphere upon radio waves, using artificial earth satellites and radio-astron. sources, atmospheric effects on optical propagation. Home: Salina Star Route. Office: ESSA/ITSA, Boulder, Colo. 80302.*

LAWRENCE, Walter, Jr., Am. surgeon; b. Chgo., May 31, 1925; s. Walter and Violette (Matthews) L.; student Dartmouth, 1943-44; Ph.B., U. Chgo., 1944, S.B., 1945, M.D., 1948; m. Susan Shryock, June 20, 1947; children—Walter Thomas, Elizabeth, William Amos, Edward Gene. House officer in surgery Johns Hopkins Hosp., 1948-50; with Meml. Center N.Y., 1951-66, asso. attending surgeon, 1962-66; with Cornell Med. Coll., 1957-66, clin. asso. prof. surgery, 1963-66; with Sloan-Kettering Inst., 1956-66, asso. mem., 1960-66; prof. surgery, chmn. div. surg. oncology Med. Coll. Va., Richmond, 1966—. Recipient Sloan award, 1964. Mem. N.Y. Acad. Medicine, N.Y. Acad. Sci., Royal Soc. Medicine, A.M.A., A.A.A.S., A.C.S., Am. Assn. Cancer Research, Soc. U. Surgeons, Halsted Soc., Harvey Soc., James Ewing Soc., Soc. Surgery Alimentary Tract, Johns Hopkins Med. and Surg. Assn., N.Y. Surg. Soc., N.Y. Cancer Soc., N.Y. Clin. Soc., N.Y. State Med. Soc., others. Contbr. numerous articles to med. jours. Clin. studies surgery of neoplasms and surgery of gastrointestinal tract; exptl. surgery and physiology with particular emphasis on gastric surgery, liver dysfunction, regional chemotherapy for cancer and organ transplantation. Home: 6501 Three Chopt Rd. Richmond 23226. Office: 1200 E. Broad St., Richmond, Va. 23219.*

LAWRENCE, Sir William, English surgeon; b. Cirencester, Eng., July 16, 1783; ed. sch., Elmore, Eng.; apprenticed to John Abernath, St. Bartholomews Hosp., 1799; m. Louisa Trevor Senior; 1 son, 2 daus. Apptd. lectr. demonstrator St. Bartholomews Hosp., 1813; apptd. surgeon London Infirmary for Diseases of Eye, 1814, Royal Hosps. of Bridewell and Bethlehem, 1815; surgeon St. Bartholomew's Hosp., 1824-65, became lectr., 1829; apptd. prof. anatomy and surgery Coll. Surgeons, 1815. Recipient Jacksonian prize Coll. Surgeons. sgt. surgeon to Queen Victoria, beginning 1857. Fellow Royal Soc., 1813; mem. French Acad. Scis., Royal Coll. Surgeons (became pres. 1846, 55), Med. Chirurg. Soc. (became pres. 1831), Inst. France (corr.). Author: The Treatment of Hernia, 1806; Anatomico-Chirurgical Values of the Nose, Mouth, Larynx and Fauces, 1809; Lectures on Comparative Anatomy, Physiology, Zoology and Natural History of Man, 1816-18; An Introduction to Comparative Anatomy and Physiology, 1816; Treatise on Diseases of the Eye, 1833; also articles. Anticipated evolutionary doctrines of Weismann and Darwin; pioneered in comparative anatomy; studied ophthalmic surgery. Died London, July 5, 1867.

LAWROSKI, Stephen, Am. chem. engr.; b. Scranton, Pa., Jan. 17, 1914; s. Alex and Nancy (Lutzka) L.; B.S., Pa. State U., 1934, M.S., 1939, Ph.D., 1943; m. Helen Wilson, Sept. 14, 1947; children—Nancy Ann, Stephen Wilson. Research asst. Pa. State U., 1934-43; with Standard Oil Devel. Co. (now Esso Research & Engring. Co.), Elizabeth, N.J., 1943-44; with Manhattan Project, Chgo., 1944-46, Oak Ridge Nat. Lab., Tenn., 1946-47; with Argonne Nat. Lab., Ill., 1947—, asso. lab. dir., 1963—. Mem. gen. adv. com. AEC, Washington, 1964—; cons. U. S. Army Chem. Corps., Edgewood Arsenal, Md., 1961—. Evan Pugh scholar, 1934; Louise Carnegie scholar, 1932. Mem. Am. Nuclear Soc. (bd. dirs. 1955-58). Editorial adv. bd. AEC Reactor Handbook, 1950—; editor Reactor Fuel Processing Quar. Nuclear Research and devel. on separations processes for nuclear reactor fuels; patentee in petroleum tech., purification and recovery of nuclear reactor fuels, ore purification, fluidized bed metal denitration. Home: 144 S. Sleight St., Naperville, Ill. 60540. Office: Argonne Nat. Lab., Argonne, Ill. 60440.*

LAWS, Bernard Courtney, English naval architect; ed. Royal Coll. Sci., London; D.Sc., London; Dr.-ès-Sc., Paris.; m. Eugenie Miller Watson, 1900; 1 son. Apprentice, H. M. Dockyard, Portsmouth, Eng.; in charge Designing Office, Fairfield Co.; naval architect to Le Chantier Naval de Nicolaieff, Russia; spl. surveyor for research Lloyd Registers, later cons. for research Lloyds Register Shipping; cons. Tungum Alloy Cos.; examiner shipbldg. City and Guilds of London Inst.; examiner Instn. Civil Engrs. Nat. scholar, recipient George Stephenson Gold medal Instn. Civil Engr. Fellow Inst. Physics; mem. Inst. Naval Architects, London Math. Soc. Author: Questions in Naval Architecture; Stability and Equilibrium of Floating Bodies. Publs. on inventions of instruments for measuring and recording strains in structures. Died Sept. 17, 1947.

LAWS, John William, English diagnostic radiologist; b. Luton, Eng., Oct. 25, 1921; s. Robert Montgomery and Lucy (Ibbottson) L.; M.B., Ch.B., Sheffield (Eng.) U., 1944; m. Pamela King, Apr. 7, 1945; children—James, Susan. Sr. lectr. diagnostic radiology Royal Postgrad. Med. Sch. London, also radiologist Hammersmith Hosp., London, 1955-67; dir. radiology King's Coll. Hosp., London, 1967—, examiner radiology fellowship examination; mem. fellowship bd. Faculty Radiologists. Fellow Faculty Radiologists, Royal Coll. Physicians; mem. Brit. Inst. Radiology (Barclay prize 1964), Brit. Soc. Gastroenterology, Royal Soc. Medicine. Dep. editor Brit. Jour. Radiology. Research, publs. on pulmonary emphysema, correlation of radiol., clin. and path. data in diseases of small intestine, peripheral vascular system of lungs; radiol. studies of patients with malabsorption due to small intestinal diseases; devel. test for disaccharidase deficiency. Home: 5 Frank Dixon Way, Dulwich Village, London S. E. 21. Office: Dept. Radiology, King's Coll., Hosp., Denmark Hill, London S.E. 5, Eng.*

LAWSON, Alexander, engraver; b. Ravenstruthers, Scotland, Dec. 19, 1773; m. Elizabeth Scaife, June 6, 1805; 3 children, including Oscar A., Mary Lockhart. Went to Liverpool to work for his brother, 1788; moved to Manchester, Eng., 1789, experimented in engraving; came to U. S., 1794; worked for Thackara and Vallance, Phila., 1794-96; set up own engraving business, 1796; did series of 4 plates illustrating Thomson's Four Seasons, 1797; engraved plates for supplemental volumes of Thomas Dobson's Ency., 1803; partner with J. J. Barralet for short time, engraved plates for The Powers of Genius, by Rev. John Blair Linn, 2d edit., 1802; friend of naturalist Alexander Wilson, agreed to do plates for less than $1 a day for Wilson's American Ornithology, 9 vols., 1808-14; engraved various portraits; made plates for American Ornithology; or the Natural History of Birds Inhabiting the United States Not Given by Wilson, by Charles Lucien Bonaparte, 4 vols., 1825-33; his work presented in exhbn. of 100 notable Am. engravers at N.Y. Public Library, 1928. Died Phila., Aug. 22, 1846.

LAWSON, Andrew Cowper, geologist; b. Anstruther, Scotland, July 25, 1861; s. William and Jessie (Kerr) L.; A.B., U. Toronto, 1883, A.M., 1885, D.Sc., 1923; Ph.D., Johns Hopkins, 1888; LL.D., U. Cal., 1935; D.Sc., Harvard, 1936; m. Ludovika von Jantsch, Nov. 30, 1889 (dec. Dec. 1929); children—Andrew Werner, William Eric, Ludovico, James Albert; m. 2d, Isabel R. Collins, Jan. 5, 1931. Geologist Geol. Survey Can., 1882-90; prof. mineralogy and geology U. Cal., 1890-1928, dean Coll. of Mining, 1914-18, ret., 1928. Del. Geol. Congress, London, 1888, St. Petersburg, 1897, Toronto, 1913, Madrid, 1926; chmn. Calif. Earthquake Investigation Commn., 1906; chmn. geol. and geog. NRC, 1923. Hayden medalist Acad. of Natural Science, 1935. Fellow Geol. Soc. Am. (pres. 1926; Penrose medalist 1938), Soc. Econ. Geologists, A.A.A.S., Am. Acad. Arts and Scis., Seismol. Soc. Am. (pres. 1909-10), Nat. Acad. Scis., Am. Philos. Soc., Am. Assn. Petroleum Geologists. Contbr. numerous geol. papers and monographs. Research in Archean geology, geomorphology, petrography, econ. geology, seismology; gave new interpretation of Pre-Cambrian, 1887; classified Pre-Cambrian levels of rocks; studied unique geol. conditions in Cal.; investigated causes of earthquakes; gave 1st field course in geology in Am. West. Died June 16, 1952.

LAWSON, Andrew Werner, Jr., Am. physicist; b. San Francisco, Mar. 3, 1917; s. Andrew Werner and Teresa (Harrison) L.; A.B., Columbia, 1936, Ph.D., 1940; m. Mary Ann Wyman, Apr. 7, 1945; children—Andrew, Ann, Katharine. Asst. physics Columbia, 1936-39, U. fellow, 1939-40; instr. U. Pa., 1940-41, asst. prof., 1941-44; staff mem. Radiation Lab., Mass. Inst. Tech., 1944-46; asst. prof. Inst. for Study of Metals, physics dept. U. Chgo., 1946-47, asso. prof., 1947-50, acting chmn. physics dept., 1948-50, prof. 1950-61, chmn. dept., 1950-56, asso. dir. Inst. for Study of Metals, 1952-56; prof., chmn. physics dept. U. Cal., Riverside, 1961-64, prof., 1961—. Cons. Du Pont Co., Wilmington, Del., 1947—, Gen. Atomic, Bourns, Riverside, 1961—. Recipient Presdl. Certificate of Merit, 1947; One of Ten Outstanding Young Men award, U. S. Jr. C. of C., 1952. Mem. Am. Phys. Soc., Am. Geophys. Union, Soc. Automotive Engrs., Am. Assn. Physics Tchrs., Phi Beta Kappa, Sigma Xi, Delta Upsilon. Research in solid state physics, high pressure physics, high polymer physics. Home: 6101 Hawarden Dr., Riverside, Cal.*

LAWSON, George (McLean), Am. physician; b. Middle Haddam, Conn., May 26, 1898; s. George Newton and Ida Louise (McLean) L.; student Bates Coll., 1915-19; M.D., Yale, 1924; D.P.H., Harvard, 1933; m. Gladys Holmes, May 6, 1922; 1 son, David Herbert Otis. Intern and asst. resident pediatrics Yale, 1923-24; mem. commn. for Study of Whooping Cough, Boston, 1924-28; bacteriologist Mass. Gen. Hosp., 1927-29; instr., bacteriology Harvard Med. Sch., 1927-29; prof. bacteriology U. Louisville Med. Sch., 1929-32, prof. pub. health and bacteriology, 1932-37; epidemiologist and vital statistician Louisville Health Dept., 1932-37; prof. of preventive medicine and bacteriology U. Va., since 1937. Fellow Am. Pub. Health Assn.; mem. Soc. Am. Bacteriologists, Soc. Pediatric Research, A.A.A.S., Va. Acad. Sci., Albemarle County Med. Soc., Albermarle Tb Assn. Vis. Nurses Assn., Sigma Xi. Author: Sect. on whooping cough in Diagnostic Procedures and Reagents (with P. Kendrick, J. Miller), 2d. ed., 1945. Contbr. articles in field of bacteriology and communicable disease control. Died Sept. 20, 1951.

LAWSON, Isaac, physician; b. Scotland; M.D., Leiden (Netherlands), U., 1737; studied medicine and botany under Herman Boenhaave and Van Royen; became physician in Brit. Army. Linnaeus named genus, Lawsonia (henna of East), in his honor. Printed (with Gronovius): Systema Naturae (Linnaeus), 1735. Died Dosterhout, Netherlands, 1747.

LAWSON, James Glen, Brit. physician; b. Dundee, Scotland, Sept. 30, 1920; s. James and Mary (Small) L.; M.B., Ch.B. with distinction, U. St. Andrews (Scotland), 1943, M.D., 1955; m. Jean Grassick Scorgie, Mar. 28, 1953; children—James Glen, Michael Clark. Asst., U. St. Andrews, 1947-51; lectr. obstetrics and gynecology U. Aberdeen (Scotland), 1951-56; sr. lectr. Welsh Nat. Sch. Medicine, U. Wales, 1956-65; cons. obstetrician and gynecologist United Cardiff (Wales) Hosps., 1965—, dir. cervical cytology, 1964—. Fellow Royal Coll. Obstetricians and Gynecologists. Contbg. author: Combined Textbook of Obstetrics and Gynecology, 1957; Cancer of Cervix, 1959; Dyplasia, 1964; Obstetrics in General Practice, 1966. Research, publs. on cancer of cervix uteri, epidemiological analysis, cytological surveys, enzyme studies on vaginal fluid, orgn. scheme to screen cytologically entire female population of Cardiff and Area. Home: 31 Llandennis Av. Office: 73a Cathedral Rd., Cardiff, Wales.*

LAWSON, James Llewellyn, physicist; b. Pasumalai, S.India, Dec. 17, 1915; s. James Hoy and Frances (Jones) L.; B.A., U. Kan., 1935, M.A., 1936; Ph.D., U. Mich., 1939; m. Jane Kraft, June 16, 1940; children—Carol Jean, Nancy Louise (Mrs. Marvin), James Robert. Research asso. Gen. Electric Research Lab., Schenectady, 1945-56, mgr. electron physics dept., 1956-65, mgr. information scis. lab. Research and Devel. Center, 1965—. Mem. Am. Inst. Physics, A.A.A.S., I.E.E.E. Author: (with G. E. Uhlenbeck) Threshold Signals, 1950. Research in nuclear physics, beta ray spectroscopy, electron accelerators, photon scattering, electronics, microwaves, radar systems, electronic measurements, signal processing and measurement. Home: 2532 Troy Rd., Schenectady 12309. Office: P.O. Box 8, Schenectady 12301.*

LAWSON, Paul Bowen, entomologist; b. Sitapur, India, Aug. 18, 1888; s. James Chapell and Ellen (Hoy) L.; came to U. S., 1903; student Oberlin Coll., 1905-06; B.S., John Fletcher Coll., University Park, Ia., 1909; M.S., U. Kan., 1917, Ph.D., 1919; m. Sarah Alice Cooper, July 20, 1910 (dec.); children—Lois Marguerite (Mrs. Purdy F. Meigs), Lila Alice (Mrs. Charles E. Smith); m. 2d, Elizabeth C. Rupp, June 16, 1941. Instr. biology John Fletcher Coll., 1910-15; instr. entomology, U. Kan., Lawrence, 1916-20, asst. prof., 1920-21, asso. prof., 1921-22, prof. and asst. dean, 1922-29, prof. and asso. dean, 1929-33, prof., dean, from 1934. Fellow Entomol. Soc. Am.; mem. Kan. Acad. Sci., Phi Beta Kappa, Sigma Xi. Editor of Jour. of Kan. Entomol. Soc. Died Mar. 30, 1954.

LAWTON, Alfred H(enry), Am. physician; b. Carson, Ia., July 26, 1916; s. George A. and Cora Anna L.; A.B., Simpson Coll., 1937, Doctor of Science (honorary), 1958; Master of Science, Northwestern, 1939, B.M., 1940, M.D., 1941, Ph.D., 1943; m. Mary Ellen Swick, Aug. 20, 1940; children—George W., Dianna M., Lola J. Rotating interne, Passavant Hosp., Chicago, 1940-41, surgical residency, Henry Ford Hosp., Detroit, 1941-42; surg. (lt. comdr.) U.S.P.H.S., Apr. 1942-Nov. 1946; asst. prof. physiol. and pharmcol. and of med., The Med. Sch., Univ. Ark., 1946-47; dean, sch. of med., and prof. physiology and pharmacology, Univ. N.D., 1947-48; chief research div., research and edn. service, dept. med. and surg., Vets. Adminstrn., Washington, 1948-51; director of medical research, DCS-D, Hq. USAF, 1951-55; cardiologist and dir. research lab., VA Center, Bay Pines, Fla., 1955-60, asst. dir. profl. services for research and edn., 1960-62; dir. research accidents and aging study group div. accident prevention USPHS, St. Petersburg, 1962-64; dir. human devel. study center Nat. Inst. Child Health and Human Devel., U. S. Dept. Pub. Health and Welfare, 1964—; asst. clin. prof. medicine George Washington U., 1948-55. Fellow A.A.A.S., Aerospace Med. Assn., Am. Geriatrics Soc., Gerontol. Soc., Inc., Am. Pub. Health Assn.; mem. Am. Chem. Soc., A.M.A., Am. Heart Assn., Acad. Religion and Mental Health, Fla. Council on Aging, Am. Soc. Pharmacology and Exptl. Therapeutics, Am. Physiol. Soc., S.A.R., Sigma Xi, Phi Beta Pi, Lambda Chi Alpha, Pi Kappa Delta, Sigma Tau Delta. Contbr. profl. jours. Home: 627 Riverhills Drive, Temple Terrace, Fla. 33617. Office: U. South Fla., Tampa, Fla. 33620.*

LAX, Benjamin, physicist; b. Miskolz, Hungary, Dec. 29, 1915; B.M.E., Cooper Union, 1941; Ph.D. in physics Mass. Inst. Tech., 1949; m., 1942; 2 children. Came to U. S., 1926; mech. engr. U. S. Engring. Office, 1941-42; cons. Sylvania Elec. Products, Inc., 1946; with Cambridge Research Center, U. S. Air Force, 1946-51; mem. staff, solid state group Lincoln Lab. Mass. Inst. Tech., 1951-53, head ferrites group, 1955-57, asso. head communications div., 1957-58; head solid state div., 1958-60, dir. Francis Bitter Nat. Magnet Lab., from 1960, prof. physics, from 1965, asso. dir. Lincoln Lab., 1964-65. Asso. editor Jour. Applied Physics of Am. Inst. Physics, 1957-59, Microwave Jour. U. S., Physics Review of Am. Physics Soc., 1960-63; mem. solid state science panel Nat. Acad. Scis.; mem. advisory com. Nat. Bur. Standards. Recipient Buckley prize, 1960. Fellow Physics Soc., Am.

Acad. Radar. Pioneer in studies of cyclotron resonance in semiconductors; developed theoretical techniques for gyrotropic media; related research in microwave discharges, ferrites, plasma physics, magnetospectroscopy, high magnetic fields, excitons. Address: Francis Bitter Nat. Magnet Lab., Mass. Inst. Technology, Cambridge, Mass. 02139.

LAX, Melvin, Am. physicist; b. N.Y.C., Mar. 8, 1922; s. Morris and Rose (Hutterer) L.; B.A. (N.Y. State Regents scholar, Charles Hayden scholar), N.Y. U., 1942; S.M. (Applied Math. fellow), Mass. Inst. Tech., 1943; Ph.D., 1947; m. Judith Heckelman, June 26, 1949; children—Ruth Laurie, David Alan, Jonathan Robert, Naomi Ilyssa. Research physicist Underwater Sound Lab., Mass. Inst. Tech., 1942-45, research asso. physics dept., 1947; faculty physics Syracuse (N.Y.) U., 1947-55, prof., 1955; mem. tech. staff Bell Telephone Labs., Murray Hill, N.J., 1955—, head theoretical physics dept., 1962-64. Cons. Naval Research Lab., 1951-55; lectr. Princeton, 1961, Oxford (Eng.) U., 1961-62. Fellow Am. Phys. Soc. Bd. editors Phys. Rev., 1958-60. Research in acoustics, meson theory, fluctuation theory, elec., optical, magnetic, vibrational and transport properties of solids. Home: 12 High St., Summit, N.J. 07901. Office: Bell Telephone Labs., Murray Hill, N.Y. 07971.*

LAX, Peter D., Am. mathematician; b. Budapest, Hungary, May 1, 1926; s. Henry and Clara, L.; B.A., N.Y. U., 1944, Ph.D., 1949; postgrad. Stanford, Tex. A. and M. U.; m. Anneli Cahn; children—John, James David. Prof. math. N.Y. U.; prof. math. Courant Inst. Math. Scis., 1958—, dir. AEC Computing and Applied Math. Center, 1963—. Recipient Ford award, 1966; Mem. Am. Math. Soc. Author: (with R. Philipps) Scattering Theory, 1967; also articles. Home: 300 Central Park W., N.Y.C. 10024.*

LAYARD, Sir Austen Henry, archeologist; b. Paris, Mar. 5, 1817; s. Henry P. L.; studied Eng., France, Switzerland; D.C.L., Oxford U., 1849. Traveled through Orient, 1839; began exploring ruins in Mesopotamia, 1845; later excavated at Nimrud and Kuyunjik, which he identified as site of Ninevah; returned to Eng. to attend Oxford; made 2d expdn. to investigate ruins in Babylonia and S. Mesopotamia, 1849; turned to politics, elected Liberal mem. for Aylesbury, 1852; lord rector Aberdeen U., 1855; undersec. fgn. affairs, 1861-66; apptd. trustee Brit. Mus., 1866, chief commr. works in W. E. Gladstone's govt., 1868; ambassador at Constantinople, 1877-80; retired to Venice, wrote on Italian art. Decorated Grand Cross of Bath, 1878. Publs. include: Ninevah and its Remains, 1848; Inscriptions in the Cuneiform Character from Assyrian Monuments, 1851; Discoveries in the Ruins of Ninevah and Babylon, 1853; The Ninevah Court in the Crystal Palace, 1854. Recognized in field of archaeology primarily for discovery and excavation of mound of Nimrud. Died London, July 5, 1894.

LAYCOCK, Thomas, physiologist; b. Wetherby, Eng., 1812; ed. U. Coll., London; studied anatomy and physiology under Lisfranc and Velpeau, Paris, 1834; M.D., Göttingen, Germany, 1839. Became lectr. clin. medicine Gork Sch. Medicine, 1846; named prof. practice physic Edinburgh U., 1855. Fellow Royal Soc. Edinburgh; mem. Brit. Assn. (became sec. 1844), Royal Coll. Surgeons. Author: A Treatise on the Nervous Diseases of Women, 1840; Lectures on the Principles and Methods of Medical Observation and Research, 1856; The Social and Political Relations of Drunkenness, 1857; Mind and Brain, 1859; also numerous articles. Translator, editor: Principles of Physiology (J. A. Unger), 1851; A Dissertation on the Functions of the Nervous System (G. Prochaska). Developed plan of polit. medicine (now known as state medicine); 1st to propose the theory of the reflex action of the brain, 1844; 1st to apply theory of evolution to devel. nervous centers in animal kingdom to man. Died Sept. 21, 1876.

LAYNE, Donald Sainteval, biologist; b. Lime Ridge, Que., Can., Apr. 5, 1931; s. John Graham and Kathleen (Alkinson) L.; B.S., McGill U., 1953, Ph.D., 1957; m. Edith Common, Apr. 25, 1959; children—Donald Graham, Kathleen Renate, Geoffrey Haddon. With Worcester Found. for Exptl. Biology, Shrewsbury, Mass., 1959—, dir. steroid tng. program, 1964—; asso. prof. biology Clark U., Worcester, Mass., 1964—. Mem. Canadian Biochem. Soc., A.A.A.S., Am. Physiol. Soc. Research, publs. in isolation metabolites natural estrogens, metabolism norsteroids. Home: 20 Spring Circle, Shrewsbury, Mass. 01545. Office: Worcester Found. for Exptl. Biology, Shrewsbury, Mass. 01545.*

LAYNE, James Nathaniel, Am. zoologist; b. Chgo., May 16, 1926; s. Leslie Joy and Harriet (Hausman) L.; student Chgo. City Jr. Coll., 1946-47; B.A. with distinction in Zoology, Cornell U., 1950, Ph.D., 1954; m. Lois Virginia Linderoth, Aug. 26, 1950; children—Linda Carrie, Kimberly, Jamie Linderoth, Susan Nell. Asst. prof. zoology So. Ill. U., 1954-55; faculty U. Fla., Gainesville, 1955-63, asso. prof. biology, 1959-63; asst. curator biol. sci. Fla. State Mus., Gainesville, 1955-59, asso. curator biol. sci., 1959-63; asso. prof. zoology Cornell U., Ithaca, N.Y., 1963-67; curator Archbold Expdns., Am. Mus. Natural History, Lake Placid, Fla., 1967—. Vis. scientist primate ecol. sect. lab. perinatal physiology Nat. Inst. Neurol. Diseases and Blindness, 1961; research asso. Fla. State Mus., 1963—; dir. Archbold Biol. Sta., Lake Placid, 1967—. Mem. Am. Inst. Biol. Scis., A.A.A.S., Am. Soc. Mammalogists, Ecol. Soc. Am., Soc. for Study Evolution, Soc. Systematic Zoologists, Am. Soc. Zoologists, Wildlife Disease Assn., Wildlife Soc., Animal Behavior Soc. Research, publs. on ecology, behavior, physiology and distbn. of mammals, including several species of N.Am. small animals and phylogeny of genus Peromyscus; 1st observations on behavior of Amazon dolphins. Home: Route 1, Box 307. Office: Archbold Biol. Sta., Lake Placid, Fla. 33852.*

LAYTON, Laurence Laird, Am. allergologist, biochemist; b. Boomer, W.Va., Mar. 8, 1914; s. John Wister and Eva (Nutter) L.; B.Sc., W. Va. Inst. Tech., 1937; M.Sc., W.Va. U., 1939; Ph.D., Pa. State U., 1942; m. Lisa Philips, Oct. 18, 1941; children—Thomas Nutter, Annalisa Laird, Lawrence John, Deborah Huddleston. Research phys. chemist Eastman Kodak Co., 1942-43; asst. prof. inorganic chemistry U. Md., 1943-46; asst. prof. biochemistry Johns Hopkins, 1946-51; chief chem. warfare div. Chem. Corps., Dugway Proving Ground, Utah, 1951-54; asso. dir. research and devel. USN Propellant Facility, Indian Head, Md., 1954-57; head allergens investigation Western Regional Research Lab., Agr. Research Service, Albany, Cal., 1959-63, head physiologically active compounds investigation, 1963—. Recipient Superior Service award USN, 1956; Superior Service award U. S. Dept. Agr., 1963. Mem. Am. Chem. Soc., Am. Acad. Allergy, Soc. Expt. Biol. Medicine, Internat. Primatological Soc., Sigma Xi. Contbg. author: Traité de Biologie Appliquée, 1964; also numerous articles. Research on discovery role sulfate in wound healing and inhibition by cortisone, diagnostic and research test for allergies by transfer of blood serum to monkeys. Home: 670 San Luis Rd., Berkeley, Cal. 94707.*

LAYTON, Wilbur L., Am. psychologist; b. Atlantic, Ia., Mar. 26, 1922; s. Charles Emory and Helen Marie (Dorsey) L.; B.S., Ia. State U., 1943; M.A., Ohio State U., 1947, Ph.D., 1950; m. Gloria M. Madsen, Aug. 26, 1944; children—Gregory Jon, Patricia Lynn, Charles Frederick. Mem. faculty U. Minn., 1948-59; psychol. cons. to bus. and industry, 1950—; prof. psychology, head dept. Ia. State U., 1960-67, vice president for student affairs, 1967—; vis. prof. colls. Field selection officer Peace Corps., 1963. Mem. Gov. Minn. Adv. Com. Vocational Rehab., 1955-59, Gov. Ia. Com. Mental Health, 1961-62. Diplomate Am. Bd. Examiners Profl. Psychology. Fellow A.A.A.S., Am. Psychol. Assn.; mem. Am. Civil Liberties Union, Am. Coll. Personnel Assn., Nat., Ia. rehab. assns., Nat. Council Measurement in Edn. (pres. 1966), Am. Personnel and Guidance Assn., Am. Ednl. Research Assn., Psychometric Soc., Psychonomic Soc., Psi Chi (pres. Ia. State U. chpt. 1942-43). Author: Counseling Use of the Strong Vocational Interest Blank, 1958; (with others) Testing in Guidance and Counseling, 1963. Editor: The Strong Vocational Interest Blank: Research and Uses, 1960; cons. editor Jour. Counseling Psychology, 1966. Research and publs. on devel. techniques of psychol. measurement and research in scholastic aptitude, interest and personality measurement; developer Minn. Scholastic Aptitude Test and co-developer Minn. Counseling Inventory. Home: 3604 Roos Rd., Ames, Ia. 50010.*

LAYZER, David, astronomer; b. Cleve., Dec. 31, 1925; s. Hilary and Rhea (Volk) L.; A.B., Harvard, 1947, Ph.D., 1950; m. Jean Isobel Walker, Aug. 18, 1959. Faculty Harvard, 1953—, prof. astronomy, 1960—. Mem. Internat. Astron. Union, Am., Royal astron. socs., Am. Acad. Arts and Scis. Bok prize 1960. Research, publns. on atomic physics, ionosphere physics, cosmogony. Home: 42 Fairmont St., Belmont 78, Mass. Office: Harvard Coll. Observatory, 60 Garden St., Cambridge, Mass. 02138.

LAZAN, Benjamin Joseph, Am. engr.; b. N.Y.C., Apr. 20, 1917; s. Samuel and Pauline (Breson) L.; B.S., Rutgers U., 1938; M.S., Harvard, 1939; Ph.D., Pa. State U., 1942; m. Jeannette Waxler, Dec. 25, 1939; children—Gilbert, Douglas. Asst. prof. Pa. State U., 1939-42; with Sonntac Sci. Corp., 1942-46; prof., head materials engring. dept. Syracuse (N.Y.) U., 1946-51; prof., head dept. mech. and materials engring. dept. U. Minn., Mpls., 1951—, asso. dean Inst. Tech., prof., head dept. aeros. and mechanics. Cons. engr. indsl. firms, USAF. Fellow Am. Soc. M.E. (Noble award 1943), Am. Acoustical Soc.; mem. Am. Soc. Metals (Henry Howe medal 1951), Am. Soc. Testing and Materials (C. Dudley medal 1949), Soc. Exptl. Stress Analysis (past pres.), Phi Beta Kappa, Sigma Xi, Tau Beta Pi. Research, numerous publs. in rheological and strength properties of materials, damping and fatigue of metals, high temperature properties. Died June 29, 1966.

LAZAR, Norman Henry, Am. physicist; b. Bklyn., June 21, 1929; s. Robert and Beatrice (Krugman) L.; B.S., Coll. City N.Y., 1949; M.S., Ind. U., 1951, Ph.D. in Physics, 1953; m. Katie Williams, Nov. 18, 1961. Group leader physicist Oak Ridge Nat. Lab., 1953—. Fellow Am. Phys. Soc. Research in physics of plasmas asso. with controlled nuclear reactions; research in nuclear structure; devel. use of scintillation spectrometers. Home: 410 Virginia Rd., Oak Ridge 37830. Office: P.O. Box Y, Oak Ridge 37831.*

LAZARENKO, Boris Romanovich, Russian inventor; b. 1910; grad. Moscow U., 1936; D.Tech. Sci. With All-Union Electrotech. Inst., 1935-42, Research Inst., USSR Ministry Elec. Industry, 1942-48; dir. Central Research Lab. Elec. Processing of Metals, USSR Acad. Sci., 1948-61; dir. Inst. Energetics and Automation, Moldavian Acad. Sci., 1961-64, dir. Inst. Applied Physics, 1964—. Recipient Stalin prize, 1946. Mem. Moldavian Acad. Sci. Co-author: Electric Metal Erosion, 1944-46; Physics of the Spark Method Metalworking, 1946; The Electric Spark Working of Metals, 1950. Deviser new tech. methods for metal plating, method to pulverize conducting materials; co-developer (with N. L. Lazarenko) electric spark method of metal working. Address: Inst. Applied Physics, Moldavian Acad. Sci., Kishinev, Moldavian SSR, USSR.

LAZAREV, Boris Georgievich, Russian physicist; b. Aug. 6, 1906; grad. Leningrad Poly., 1930. Joined staff Leningrad Phys.-Tech. Inst. after grad.; staff Ural Phys.-Tech. Inst., 1932-37; staff Phys.-Tech. Inst., Ukrainian Acad. Sci., 1937—. Recipient Stalin prize, 1951. Mem. Ukrainian Acad. Scis. Research and publs. in low-temperature physics; 1st to measure nuclear paramagnetism in condensed state.

LAZAREVIC, Djordje, Yugoslavian phys. metallurgist; b. Thessalonica, Yugoslavia, Oct. 23, 1920; s. Peter L. and Fanija (Markovic) L.; Diploma, Faculty Tech., 1951; M.S., U. Birmingham (Eng.), 1960; m. Milunovic Ljubica, June 28, 1958; children—Peter, Vera. Staff, Inst. Nuclear Scis., Boris Kidric, Vinca, Yugoslavia, 1951—, asst. prof. reactor materials, 1963, 65. Decorated Order of Labor. Mem. Serbian Chem. Soc. Research, publs. on phys. metallurgy and radiation damage of uranium and its alloys; cooperative influence of burn-up rate and secondary phase precipitates on radiation stability of metallic uranium fuels; modified Jominy test for oxydizable metals; radiation rig designs. Home: 34 Alekse Nenadovica, Belgrade, Yugoslavia. Office: Inst. Nuclear Scis., P.O. Box 522, Vinca, Yugoslavia.*

LAZAROW, Arnold, Am. anatomist; b. Detroit, Aug. 3, 1916; s. George and Rose (Brown) L.; B.S., U. Chgo., 1937, M.D., 1941, Ph.D., 1941; m. Jane S. Klein, Dec. 15, 1940; children—Paul, Normand. Research asso. U. So. Cal., 1943; faculty Western Res. U., Cleve., 1943-54; prof., head dept. anatomy U. Minn., 1954—. Mem. nat. adv. council USPHS, 1961-65; trustee Marine Biol. Lab., 1961—. Mem. Am. Assn. Anatomists (exec. com. 1963-67), Am. Chem. Soc., Am. Diabetes Assn. (council 1956-62), Soc. Exptl. Biology and Medicine, Phi Beta Kappa, Sigma Xi, Alpha Omega Alpha (Joseph A. Capp prize 1942), Omicron Kappa Upsilon (hon.). Research, publs. on insulin synthesis, storage and release by pancreatic islet tissue; factors controlling prodn. and course of exptl. diabetes, cytochem. and histochem. studies on cell constituents. Home: 221 Woodlawn Av., St. Paul 55105. Office: 262 JacH, U. Minn., Mpls. 55455.*

LAZARUS, David, Am. physicist; b. Buffalo, Sept. 8, 1921; s. Barney B. and Lillian (Markel) L.; B.S., U. Chgo., 1942, M.S., 1947, Ph.D., 1949; m. Betty Jane Ross, Aug. 15, 1943; children—Barbara, William, Mary Ann, Richard. Instr. electronics U. Chgo., 1942-43; research asso. radio research lab. Harvard, 1943-45; electronics engr. U. Chgo., 1946-49, instr. physics, 1949; faculty physics U. Ill., Champaign, 1949—, prof., 1960—; vis. scientist Am. Inst. Physics, N.Y.C., 1962. Cons. phys. sci. study com. Halicrafters Co., Chgo., 1957. Gen. Electric Co., Cin., 1960, Gen. Atomic, La Jolla, Cal., 1962-63, Addison-Wesley Pub. Co., Reading, Mass., 1964. Author: (with H. de Waard) Modern Electronics, 1966. Research on physics of solids; and on antennas and electronic control system. Home: 502 W. Vermont Av., Urbana, Ill. 61801.*

LAZARUS, Richard Stanley, Am. psychologist; b. N.Y.C., Mar. 3, 1922; s. Abe and Mathilda (Marks) L.; A.B., Coll. City N.Y., 1942; M.S., U. Pitts., 1947, Ph.D., 1948; m. Bernice Newman, Sept. 2, 1945; children—David, Nancy. Asst. prof. Johns Hopkins, 1948-53; asso. prof. Clark U., Worcester, Mass., 1953-57; faculty U. Cal. at Berkeley, 1957—, prof. psychology, 1959—. Psychol. cons. VA, 1951—. USPHS Spl. fellow Waseda U., Tokyo, Japan, 1963-64; USPHS grantee, 1953—. Diplomate Am. Bd. Examiners. Mem. Am. (rep. from div. 8 1963-66), Western, Cal. State psychol. assns., A.A.A.S., Sigma Xi. Author: (with G. W. Shaffer) Fundamental Concepts in Clinical Psychology, 1952; Adjustment and Personality, 1961; Psychological Stress and the Coping Process, 1966; also numerous articles. Editor: Foundations of Modern Psychology, 14 vols., 1963-66; adv. editor Jour Cons. Psychology, 1963-66, Psychophysiology, 1966—. Research on psychol. defs. against threat, personality characteristics determining stress reactions, stress reduction, gen. stress theory. Home: 3255 Hamlin Dr., Lafayette, Cal. 94549. Office: U. Cal., Berkeley, Cal. 94720.*

LAZARUS, Sydney Simon, physician; b. Glasgow, Scotland, 1919; s. David and Fanny (Levy) L.; B.S. cum laude Coll. City N.Y., 1938; M.D., Chgo. Med. Sch., 1943; M.Sc., Queen's U., Ont., Can., 1956; m. Esther Seinfeld, 1945; 1 dau., Ann. Research fellow dept. labs. Jewish Chronic Disease Hosp., Bklyn., 1949-

51, asst. to dir. labs., 1953-54, fellow clin. pathology, 1956-57; asst. resident pathology Coney Island Hosp., Bklyn., 1951-52; sr. fellow in pathology Queen's U. and Kingston (Ont., Can.) Gen. Hosp., 1954-56; vis. asst. prof. pathology Albert Einstein Coll. Medicine, 1957-60; chief pathology, asst. dir. Isaac Albert Research Inst. and dept. labs. Jewish Chronic Disease Hosp., Bklyn., 1957—. Fellow Am. Bd. Pathology; mem. A.M.A., Kings County Med. Soc., Am. Diabetes Assn., Am. Assn. Pathologists and Bacteriologists, Am. Soc. Exptl. Pathology, Histochem. Soc., Am. Soc. Cell Biology, Coll. Am. Pathologists, N.Y. Soc. Electron Microscopy. Publs. in exptl. morphology, histochemistry, electron microscopy; mechanism of insulin secretion from pancreatic islet cells. Home: 1401 Ocean Av., Bklyn. Office: 86 E. 49th St., Bklyn. 11203.*

LAZEAR, Jesse William, Am. physician; b. Balt., May 2, 1866; s. William and Charlotte (Pettigrew) L.; A.B., Johns Hopkins, 1889; M.D., Columbia, 1892; postgrad. Pasteur Inst., Paris, France; m. Mabel Houston, 1896; 2 children. Physician, Johns Hopkins Hosp., Balt., 1895; asst. surgeon in U. S. Army at Columbia Barracks, Quemados, Cuba, 1900; mem. Yellow Fever Commn. with Maj. Walter Reed, bitten while in charge of mosquitos; died helping to show that mosquitos were carriers of yellow fever, Quemados, Sept. 25, 1900.

LAZENBY, William Rane, Am. horticulturist; b. Bellona, N.Y., Dec. 5, 1850; s. of Charles and Isabella L.; B. Agr., Cornell, 1874; M.Agr., Ia. Agrl. Coll., 1887; m. Harriet E. Akin, Dec. 15, 1896. Instr. and asst. prof. botany and horticulture Cornell, 1874-81; prof. botany and horticulture Ohio State U., Columbus, 1881-92, horticulture and forestry, 1892-1910, prof. forestry, from 1910; forest engr. Biltmore Forest Sch., 1912. Dir. Ohio Agrl. Expt. Sta., 1882-87, sec. Ohio Med. U., 1894-1914; lectr. before farmers' institutes, 1881-1906. Mem. Soc. Promotion Agrl. Sci. (pres., 1895-97), Am. Pomol. Soc. (v.p., 1905—); Ohio Acad. Sci. (pres. 1902-03), A.A.A.S. (v.p. 1896). Deceased.

LAZURUS, Moritz, German psychologist, philosopher; b. Filehne, Posen, Sept. 15, 1824; s. rabbinical scholar; ed. U. Berlin, 1850. Prof., U. Bern, 1860-66; prof. philosophy Kriegsakademie, Berlin, 1868-73; prof. U. Berlin, 1873-96. Named Geheimrath, 1894. Author: Das Leben der Seele, 2 vols., 1856-57; Über den Ursprung der sitten, 1860; Über die Ideen in der Geschichte, 1865; Zur Lehre von den Sinnestauschungen, 1867; Ideale Fragen, 1875; Erziehung und Geschichte, 1881; Unser Standpunkt, 1881; Über die Reize des Spiels, 1883; Der Prophet Jeremias, 1894; Die Ethik des Judentums, 1898. Founder (with H. Steinthal) jour. Zeitschrift fur Völkerpsychologie und Sprachwissenschaft, 1859. Developed concept of group mind; stressed need for psychology to study society as well as individual; stressed identification of individual with his group. Died Meran, Tyrol, Apr. 13, 1903.

LEA, Isaac, Am. malacologist; b. Wilmington, Del., Mar. 4, 1792; s. James and Elizabeth (Gibson) L.; LL.D. (hon.), Harvard, 1852; m. Frances Carey, 1821; children—Mathew Carey, Henry Charles, 1 dau. Became mem. Acad. Natural Sciences of Phila., 1815, pres., 1858-63; partner in father-in-law's publishing firm, 1821-51. Mem. A.A.A.S., Am. Philos. Soc. (v.p.). Author: Observations on the Genus Unio, 13 vols., 1827-74; Synopsis of the Family of Naiades, 1836. Concentrated on studies of fresh-water mollusks, became recognized authority in field, described more than 1800 species mollusks, recent and fossil. Died Phila., Dec. 8, 1886.

LEA, Mathew Carey, Am. chemist; b. Phila., Aug. 18, 1823; s. Isaac and Frances (Carey) L.; studied chemistry under James C. Booth; m. Elizabeth Jaudon, July 14, 1852; m. 2d, Eva Lovering. Pioneered use of photography in study of chemistry in U. S.; mem. Franklin Inst., Nat. Acad. Scis. Author: A Manual of Photography, 1868. Died Phila., Mar. 15, 1897.

LEACH, Albert Ernest, Am. chemist; b. Boston, Apr. 7, 1864; s. John Brooks and Mary Pamelia (Bellows) L.; S.B., Mass. Inst. Tech., 1886; m. Martha Hughes Thompson, Sept. 2, 1890. Expert in patent causes, 1887-92; asst. analyst Mass. State Bd. Health, 1892-99, chief analyst, 1899-1907, having charge of analysis of food and drugs for adulteration; chief U. S. Food and Drug Inspection Lab., Denver, from 1907. Author: Food Inspection and Analysis, 1904, 2d edit., 1909. Died 1910.

LEACH, Roland Melville, Jr., Am. nutritionist; b. Framingham, Mass., Aug. 27, 1932; s. Roland Melville and Mabel (Golden) L.; B.S., U. Me., 1954; M.S., Purdue U., 1956; Ph.D., Cornell U., 1960; m. Ethel Richards, Dec. 28, 1954; children—Raymond, Gary. Grad. asst. Purdue U., 1954-56; grad. asst. Cornell U., Ithaca, N.Y., 1956-59, asst. prof. animal nutrition, 1960—; chemist, research chemist U. S. Plant, Soil and Nutrition Lab., Ithaca, 1959—. Mem. Am. Inst. Nutrition, Poultry Sci. Assn. Research, publs. on mineral nutrition of animals, metabolic role of trace elements in bone formation. Home: R.D. 2. Office: U. S. Plant Soil and Nutrition Lab., Ithaca, N.Y. 14850.*

LEACH, Samuel Ashworth, English chemist; b. Blackburn, Eng., Oct. 26, 1932; s. Samuel and Elizabeth (Ashworth) L.; B.Sc., U. Manchester, 1953, Ph.D., 1956; m. Jean Marie Howson, July 30, 1958; children—Bridget, Dominic, Harriet, Matilda, Benedict. Postdoctoral fellow U. Rochester (N.Y.), 1956-58; mem. Med. Research Council external staff Dental Sch., U. Manchester, 1958-63; chemist Unilever Research Lab., Isleworth, Eng., 1963-66; sr. lectr. in dental sci. Dental Sch., U. Liverpool, 1966—; hon. adviser in dental research U. Coll. Dental Sch., London, 1965—. Mem. Biochem. Soc., Internat. Assn. Dental Research, European Orgn. for Research on Fluorine and Caries Prevention. Publs. on role of fluorides in reducing dental decay, formation of dental plaque on human teeth, biochemistry and bacteriology of saliva, inorganic chemistry of calcium phosphates. Home: 12 Bushbys Lane, Formby, Liverpool. Office: Dental Sch., U. Liverpool, Liverpool, Eng.*

LEACH, William Elford, Brit. naturalist; b. Plymouth, Eng., 1790; student medicine under Abernathy, St. Bartholomew's Hosp., London; M.D., Edinburgh, 1812. Became asst. librarian, natural history dept. Brit. Mus., 1813, asst. keeper, 1821; ret. to live in Italy, 1821. Fellow Royal Soc., 1816; mem. Linnean Soc., also numerous other socs. in Eng., France, Am. Author: The Zoological Miscellany, 1814-17; Malacostraca podophthalma Brittaniae, or a Monograph on the British Crabs, 1815-16; Systematic Catalogue of the Specimens of the Indigenous Mammalia and Birds that are preserved in the British Museum, 1816; A Synopsis of the Mollusca of Great Britain (edited by J. E. Gray), 1852; also articles in jours., encys. Research on entomology and malacology, especially crustacea; developed system of arrangement in conchology and entomology, 1813. Died Palazzo San Sebastiano, Italy, Aug. 25, 1836.

LEACHMAN, Robert Briggs, Am. physicist; b. Lakewood, O., June 11, 1921; s. Milton George and Wilma (Rothenbecker) L.; B.S., Case Inst. Tech., 1942; Ph.D., Ia. State U., 1950; m. Lenore Jean Collins, June 16, 1946; children—Elaine Alice, Mark Collins, Gregg Milton. Staff mem. radiation lab. Mass. Inst. Tech., Cambridge, 1943-45; instr. Ia. State U., Ames, 1947-48; staff mem., group leader Los Alamos Sci. Lab., 1950-67, head cyclotron research, 1957-67; prof., head dept. physics Kan. State U., 1967—. Prof. U. N.M., Los Alamos, part-time 1964-65. Guggenheim fellow Nobel Inst., Stockholm, Sweden, 1955-56; Fulbright fellow Inst. Theoretical Physics, Copenhagen, Denmark, 1962-63. Fellow Am. Phys. Soc. Research, numerous publs. in physics of nuclear fission, atomic stopping properties of very heavy ions, nuclear reactions particularly fluctuations in reaction cross sections. Home: 1609 Sunny Slope Lane, Manhattan Kan. 66502.*

LEACOCK, Eleanor Burke (Mrs. Richard Leacock), Am. anthropologist; b. Weehawken, N.J., July 2, 1922; d. Kenneth and Lily (Batterham) Burke; B.A. cum laude, Barnard Coll., 1944; M.A., Columbia, 1946, Ph.D., 1952; m. Richard Leacock, Dec. 27, 1941 (div. May 1962); children—Elspeth, Robert, David, Claudia; m. 2d, James Haughton, Aug. 1966. Research asst. dept. psychiatry Cornell U. Med. Coll., 1952-55; lectr. Queen's Coll., 1955-56, Coll. City N.Y., 1956-60; spl. cons. behavioral scis. sect. div. gen. health services U. S. Dept. Health, Edn. and Welfare, 1957-58; sr. research asso. Bank St. Coll. Edn., 1958-65; asso. prof. anthropology Poly. Inst. Bklyn., 1963-67, prof., 1967—. Lectr., Washington Sq. Coll., N.Y. U., 1960-61. Fellow Am. Anthrop. Assn.; mem. A.A.A.S. (sec. sect. H 1962—), Am. Indian Ethnohist. Soc. (past sec.-treas.). Author: (with Martin Deutsch, Joshua Fishman) Bridgeview Study, Toward Integration in Suburban Housing, 1964; also articles, book revs. Editor: Ancient Society (Lewis Henry Morgan), 1963. Research on nature property relations among Eastern Canadian Indians, analysis nature status considerations regarding staying or moving in an interracial neighborhood, epidemiology mental illness, exposition limitation traditional statis. studies illness as related to class, nationality and urban residence theories cultural deprivation in relation to edn. Home: 65 Horatio St., N.Y.C. 10014.*

LEADBETTER, Charles, inventor, astronomer; flourished 1728; gauger Royal Excise; later tchr. math., navigation, astronomy at Hand and Pen, London. Author: A Treatise of Eclipses, 1727; Astronomy of the True System of the Planets Demonstrated, 1728; Astronomy of the Satellites of the Earth, Jupiter, and Saturn . . . , 1729; The Royal Gauger, 1739; The Young Mathematicians' Companion, 1739. Invented slide-rule, 1750; one of 1st to comment on Newton; made new tables for planetary motion; showed some of the properties of stereographic projection.

LEADBETTER, Wyland F., Am. surgeon; b. Livermore Falls, Me., Jan. 9, 1907; s. Charles K. and Maude E. (Randall) L.; B.S., Bates Coll., 1928; M.D., Johns Hopkins, 1932; m. Lois A. Billings, June 13, 1934; children—Emily, Charles, Wyland. asst. pathology Johns Hopkins Hosp., 1933-34, asst. resident urology, 1934-36, resident urology, 1937-38; resident urology Ancker Hosp., St. Paul, 1936-37; urologist Lahey Clinic, Boston, 1938-39; practice medicine specializing in urology, Boston, 1939-42, 45—; consulting urologist VA Hosp., Boston; chief urol. service Mass. Gen. Hosp., Boston, 1954; cons. to various hosps. former prof. urology Tufts U. Med. Sch.; now clin. prof. surgery Harvard Med. Sch. Trustee Bates Coll. Former mem. Am. Bd. Urology. Fellow A.C.S., Am. Acad. Arts and Scis.; mem. Am. Bd. Urology, A.M.A., Am. Surgical Soc., Am. Assn., Genito-Urinary Surgeons, Internat. Soc. Urology, Am. Urol. Assn., N.E. Surg. Soc., Clin. Soc. Genito-Urinary Surgeons, Phi Beta Kappa, Alpha Omega Alpha. Home: 166 Marlborough St. Office: Mass. Gen. Hosp., Boston, Mass.*

LEADER, Robert Wardell, Am. comparative pathologist; b. Tacoma, Jan. 16, 1919; s. Robert J. and Edith (Wardell) L.; B.S., Wash. State U., 1952, D.V.M., 1952, M.S., 1955; m. Florence Helen Ranger, Dec. 23, 1940; children—Cheri (Mrs. Carl Muir), Mary, Robert. Instr. pathology Wash. State U., Pullman, 1952-55; postdoctoral fellow virus lab. U. Cal. at Berkeley, 1955-56; faculty Wash. State U., 1956-65, prof., 1964-65; asso. prof., dir. comparative pathology lab., Rockefeller U., N.Y.C., 1965—. Mem. N.Y. Acad. Scis., N.Y. Pathology Soc., Am. Soc. for Exptl. Pathology, Am. Coll. Vet. Pathology, Internat. Acad. Pathology. Research, publs. in human-animal comparative pathology, including study of analogous patterns of diseases of chronic degenerative and connective tissue group.*

LEADER, Solomon, Am. mathematician; b. Spring Lake, N.J., Nov. 14, 1925; s. William and Mary (Weinstein) L.; B.S., Rutgers U., 1949; A.M., Princeton, 1951, Ph.D., 1952; m. Elvera Grutter, Dec. 17, 1960; children—Jeremy, Shana, Rachel. Faculty, Rutgers U., New Brunswick, N.J., 1952—, prof. math., 1961—. Mem. Am. Math. Soc., Math. Assn. Am., Sigma Xi. Research, publs. on functional analysis, general topology. Home: 15 Louise Dr., Milltown, N.J. 08850. Office: Math. Dept., Rutgers U., New Brunswick, N.J. 08903.*

LEAF, Albert Lazarus, Am. forest soil scientist; b. Seattle, May 16, 1928; s. Aron and Dora Marie L.; B.S.F., U. Wash., 1950, M.F., 1952; Ph.D., U. Wis., 1957; m. Wilma Lorraine Parker, June 15, 1952; children—Melody Anne, Dawn Marie, Douglas Parker. Staff, Pacific N.W. Forest and Range Expt. Sta., U. S. Forest Service, Portland, Ore., 1951; research forester U. Wash., Seattle 1951-52; research asst. U. Wis., Madison, 1952-57; faculty State U. Coll. Forestry, Syracuse (N.Y.) U., 1957—, prof. silviculture, 1965—. Fulbright grantee U. Helsinki (Finland) Sch. Forestry, 1954-55. Fellow A.A.A.S.; mem. Council on Fertilizer Application (past chmn. forestry com.), Soc. Am. Foresters, Soil Sci. Soc. Am., Internat. Soc. Soil Sci., Am. Soc. Agronomy, Wisconsin Acad. of Science, Arts and Letters, New York Acad. Scis., Sigma Xi, Phi Sigma, Xi Sigma Pi. Author: (with D. P. White) Forest Fertilization, 1956; also articles. Research on forest soils and tree nutrition; fertilization, nursery soils, chemistry of soil of forests. Home: 124 Haddonfield Dr., Dewitt, N.Y. 13214. Office: State U. Coll. Forestry, Syracuse U., Syracuse, N.Y. 13210.*

LEAF, Boris, physicist; b. Yokohama, Japan, Mar. 4, 1919; s. Aron L. and Dora (Guralsky) L.; came to U. S., 1922, naturalized, 1936; B.S. summa cum laude, U. Wash., 1939; Ph.D., U. Ill., 1942; m. Genevieve Lukman, Aug. 28, 1947; children—Evelyn Marie, David Alexander, Michael Leon. Research asst. NDRC, U. Ill., 1942-43; faculty U. Ill., 1943-44; asso. chemist metall. lab. U. Chgo., 1944-45; Frank B. Jewett fellow Yale, 1945-46; mem. faculty Kan. State U., 1946-65; prof., chmn. dept. physics State U. N.Y., Cortland, 1965—. Mem. faculty scis. Free U. Brussels (Belgium), 1958-60; research physicist U. S. Naval Radiol. Def. Lab., San Francisco, summers 1963, 64, 66. Fellow Am. Phys. Soc.; mem. A.A.A.S., Am. Assn. U. Profs. (nat. council 1963-66), Sigma Xi, Phi Beta Kappa, Pi Mu Epsilon, Phi Lambda Upsilon. Studies in statis. mechanics and thermodynamics. Home: 28 Melvin Av., Cortland, N.Y. 13045.*

LEAF, Cecil Huntington, English surgeon; b. Feb. 19, 1864; s. Frederick Henry L.; M.A., M.B., B.C., Trinity Coll., Cambridge, Eng.; m. Fanny Grierson; 1 dau. Mem. staff London Hosp., from 1887, filled offices of house surgeon, house physician, clin. asst., clin. ophthalmic asst., demonstrator anatomy, Med. Sch., 1894-98; surgeon Cancer Hosp. and Gordon Hosp., London; research worker Brown Inst. Fellow Royal Coll. Surgeons, Royal Soc. Medicine; mem. Anat. Soc., West London Medico-Chirurg. Soc. Research and publs. on causes and prevention of breast cancer; diseases of rectum and anus; lancet expts. with chloroform; surg. anatomy of lymphatic glands. Died Oct. 5, 1910.

LEAHY, Arthur Herbert, Brit. mathematician; b. Corfu, May 25, 1857; s. Arthur and Harriet (Tabuteau) L.; ed. Trinity Coll., Dublin, Ireland, Pembroke Coll., Cambridge; B.A. as 9th Wrangler, 3d class Class Tripos, 1881; M.A., 1884; m. Margaret Nourse, 1913; 1 son, 1 dau. Instr., Royal Mil. Acad., Woolwich, Eng., 1882-83; math. master Bradfield Coll., 1883-85; became fellow Pembroke Coll., 1887, bursar, 1888-92, math. lectr., 1887-92; prof. math. U. Sheffield (Eng.), also Firth Coll., Sheffield, 1892-1922, dean Faculty Pure Sci., 1905-11, dean Faculty Arts, 1919-22, pub.

orator, 1912-22. Mem. Sheffield Lit. and Philos. Soc. (became pres. 1909), Brit. Assn. (became v.p. sect. A 1910). Author: The Courtship of Ferb, 1902; Heroic Romances of Ireland, 1905. Research and publs. on oscillatory actions in ether, functions on spherical harmonics. Died May 16, 1923.

LEAKE, Chauncey D., Am. pharmacologist, physiologist; born Elizabeth, N.J., September 5, 1896; s. Frank Walker and Helen Caroline (Luttgen) L.; Litt.B., Princeton, 1917; M.S., University of Wisconsin, 1917, Ph.D., 1923; L.H.D. (hon.), Kenyon; D.Sc. (honorary), Women's Medical College; married to Elizabeth Nancy Wilson, October 1, 1921; children—Chauncey, Wilson Walker. Instr. physiology, 1919-23, asst. prof. pharmacology, U. of Wis., 1923-28; prof. pharmacology (organized dept.), U. of Calif., 1928-42; became exec. vice pres. in charge of medical branch U. Tex., 1942-55; former prof. of pharmacology Ohio State U.; coordinator student research tng., lectr. pharmacology, history and philosophy of medicine, U. Cal. Med. Center, San Francisco, 1962-—; prof. med. jurisprudence Hastings Coll. of Law, 1963-—; vis. mem. Inst. Advanced Study, Princeton, 1950, 52, 56; former lectr. on human relations and librarian, U. of Cal. Medical Sch.; Clendening lectr. U. Kansas, 1951. In Chem. Warfare Service, 1918-19. Cons. Nat. Res. Council, U.S.P.H.S. Received spl. award Internat. Anesthesia Research Soc., 1928. Hon. fel. Am. Coll. Dentists; mem. A.A.A.S. (v.p., chmn sect. Sci., 1940; pres., 1960), A.M.A. (chmn. sect. Pharmacology, 1937), Am. Physiol. Soc., Soc. Exptl. Biol. Medicine (chmn. Pacific sect., 1936-38; pres. 1962-63), Am. Assn. History of Medicine (pres. 1961-62), History of Science Soc. (pres. 1936-39), Cal., Texas acads. of Sci., Philos. Soc. Texas, Am. Acad. Arts and Scis., Am. Soc. Pharmacology (pres. 1958-59). Hon. cons. Army Med. Library, 1946-50 (pres. 1948), Sci. Adv. Earuch Com., 1946-49. Editor: Percival's Med. Ethics, 1927; Tex. Rep. Biol. Med.; asso. editor, Arch. Internat. Pharmaco. Author: Letheon 1947; Can We Agree?, 1950; Old Egyptian Medicine, 1952; Ashbel Smith and Yellow Fever, 1951. Some Founders of Physiology, 1956; The Amphetamines, 1958. Translator: Harvey's De Motu Cordis, 1928. Contbr. over 300 articles med. and sci. periodicals. Research on central nervous system activity, narcotics, chemotherapy, and general principles of functional activity; introduced divinyl ether as an anesthetic for brief operations, 1930; introduced carbarsone as antiamebic; studies of history and philosophy of medicine. Office: U. Cal. Med. Center, San Francisco 22.

LEAKE, John, English mathematician; flourished 1650-86; established as math. practitioner, London, by 1650; one of 6 men called to resurvey London after Gt. Fire, 1666; named 1st master math. sch. Christ's Hosp., 1673-77. Editor, translator, several math. textbooks. Invented rule for solving spherical triangles.

LEAKE, Norman Hansford, Am. chemist; b. Proffit, Va., Dec. 29, 1917; s. Perry Hansford and Lydia (Cox) L.; B.S., U. Va., 1940, M.S., 1943, Ph.D., 1946; m. Gwen Bennett, Aug. 17, 1946; children—Hansford Bennett, Ann Randolph, John Churchill. With Rohm & Hass Co., Phila., 1946-49, S. E. Massengill Co., Bristol, Tenn., 1949-59; faculty Bowman Gray Sch. Medicine, Winston-Salem, N.C., 1959-—, asso. pharmacology, 1961-—, research asso. reproductive biology, 1965-—. Mem. Am. Chem. Soc., Am. Inst. Chemists, A.A.A.S., Sigma Xi, Alpha Chi Sigma. Research, publs. and patentee in field of organic compounds synthesized for potential use; most recent research concerned with carbohydrate and lipid metabolism during pregnancy. Home: 435 Westover Av., Winston-Salem, N.C. 27104.*

LEAKE, William Martin, English archaeologist; b. London, Jan. 14, 1777; s. John Martin and Mary (Calvert) L.; ed. Royal Mil. Acad., Woolwich, Eng.; hon. D.C.L., Oxford, 1816; m. Elizabeth Wray, 1838. Served to col. Brit. army, 1794-1815; traveled through Asia Minor and Cyprus, 1800; went to Turkish army in Egypt via Athens, Cyprus and Syria, 1801; employed in survey of Egypt until 1802; visited Greece, 1805 and remained there engaged in surveys, exploration and diplomatic negotiations, until 1807; returned to Greece for Brit. Govt., 1808; returned to Eng., 1809; studied Swiss mil. instns., 1815; returned to Eng. and ret. as col., 1815. Fellow Royal Soc., 1815, Royal Geog. Soc.; mem. Berlin Acad. Scis. (hon.), Inst. France, (corr.). Author publications including: The Topography of Athens, 1821; Journal of a Tour in Asia Minor, 1824; Travels in the Morea, 1830; Travels in Northern Greece, 1835; Numismata Helennica, 1854-59; Researches in Greece, 1814; Topography of Athens and the Demic, 1821, 1829; Numismata Hellenica, 3 vols., 1859. Collected coins, gems, vases and inscriptions; presented his collection of marbles to Brit. Mus., 1839; research on topography of Athens, Morea, No. Greece. Died Brighton, Eng., Jan. 6, 1860.

LEALE, Charles Augustus, Am. physician; b. N.Y.C., Mar. 26, 1842; s. William Pickett and Anna Maria (Burr) L.; student of Dr. Austin Flint, Sr., Dr. Frank H. Hamilton; attended med. and surg. clinics, N.Y.C.; M.D., Bellevue Hosp. Med. Coll., 1865; m. Rebecca Medwin Copcutt, Sept. 3, 1867; children—Annie (dec.), Lilian, Medwin, Marion, Loyal, Helen (Mrs. James Harper). Apptd. and served full term as med. cadet U. S. Army; took charge of ward for wounded officers, exec. officer U. S. Army Gen. Hosp., Armory Sq., Washington, 1865; 1st surgeon to reach President Lincoln after he was shot, Apr. 14, 1865, placed in charge of the President by Mrs. Lincoln; prolonged his life and remained continuously with him until he died, Apr. 15; studied Asiatic cholera in Europe, 1866; in practice, N.Y.C. 1866-—; in charge of children's class Northwestern Dispensary, 1866-71; physician Central Dispensary 2 yrs.; pres. St. John's Guild, 1891, 92; chmn. Floating Hosp. Com.; reorganized Children's and Seaside hosps. Pres. Northwestern Med. and Surg. Soc., 1872, N.Y. County Med. Assn., 1885, 86. Died June 13, 1932.

LEAMAN, William Gilmore, Am. cardiologist; b. Phila., Sept. 3, 1898; s. William Gilmore and Eleanor (Pelly) L.; M.D., U. Pa., 1922; postgrad. med. sch. Harvard, 1929; m. Ann Hankins, Jan. 7, 1925; 1 dau., Nancy Ann; m. 2d, Eleanor Lowe, Nov. 6, 1948. Intern U. Pa. Hosp., 1922-24, resident in pathology, 1924-25; faculty Woman's Med. Coll., 1930-63, 1930-35, asst. prof. med., 1935-42, prof. med. and chmn. dept. med., 1942-59, prof. medicine in charge of cardiology, 1959-— prof. emeritus, 1963-—; staff Phila. Gen. Hosp., 1931-—; Hosp. No. Div., 1940-58, pres. med. bd., 1952-53; cons. cardiologist Misericordia Hosp., Meml. and Northeastern hosps., VA Hosp. (all Phila.); vis. lectr. U. Pa. Grad. Sch. Medicine, 1954-58; dir. post-grad. course in cardiology A.C.P., 1946, 49, 51, 53; mem. state Pa. Grad. Edn. Inst., 1950-53. Recipient distinguished service award, Am. Heart Assn., 1967; Pa. Heart Assn., 1957. Diplomate Am. Bd. Internal Medicine. Fellow A.C.P., Am. Assn. Hist. Medicine (treas. 1941-46), A.M.A., Am. Geriatrics Soc., A.A.A.S., Phila. Coll. Physicians; mem. Heart Assn. Am. Heart Assn. (dir. 1954-55, fellow council clin. cardiology), Assn. Am. Med. Colls., Phila. Physiol. Soc., Pa. Trudeau Soc., Med. Soc. State Pa. (editor Cardiovascular Briefs 1956-58). Author: Management of the Cardiac Patient, 1940; Editor: Management of Common Cardiac Conditions, 1946 Tratamento dos Cardiacos, 1945. Home: P.O. Box 463, Unionville, 19375.*

LEAMING, Jacob Spicer, Am. agriculturist; b. Ohio, Apr. 2, 1815; s. Christopher and Margaret S.; m. Lydia Ann Van Middlesworth, Mar. 1, 1839; 7 sons, 2 daus.; developed Leaming corn, a variety of corn which matures early with a good yield; it won many prizes at the Paris Expn., 1878. Died 1885.

LEATHEM, James Hain, Am. endocrinologist; b. Lebanon, Pa., Apr. 10, 1911; s. James and Sarah (Hain) L.; B.S., Lebanon Valley Coll., 1932, D.Sc., 1965; postgrad. U. Pitts.; M.A., Princeton, 1936, Ph.D., 1937; postgrad. U. Mich. Med. Sch.; m. Anne Bradshaw Leathem, June 12, 1946. Proctor fellow in biology Princeton, 1937-38; research asso. anatomy Columbia Coll. Phys. and Surg., 1938-41, 43-45; faculty Rutgers U., New Brunswick, 1941-46, 46-—, prof. zoology, 1948-—, asst. dir. Bur. Biol. Research, 1959-62, asso. dir., 1962-64, dir., 1965-—; instr. Middlesex Gen. Hosp., New Brunswick, 1942-45; vis. prof. U. Tenn., 1951, Princeton, 1957-58, U. Colo., 1963. Mem. research career awards NIH, 1962-—; mem. biology and agr. adv. bd. NRC, 1962-—. Sr. Faculty fellow NSF, 1963-65; recipient Distinguished Research award adv. bd. Rutgers Research Council, 1964. Mem. Am. Phys. Soc., Endocrine Soc., Soc. Endocrinology, Soc. for Exptl. Biology and Medicine, Am. Assn. Anatomists, Am. Assn. Cancer Research, Am. Soc. for Study Sterility, Am. Soc. Animal Sci., Am. Soc. Zoology. Editor: Reproductive Physiology and Protein Nutrition, 1959. Research, numerous publs. on reproductive physiology, hormonal control protein metabolism and protein nutrition and hormone action. Home: 610 S. 1st Av., Highland Park, N.J. 08904. Office: Bur. Biol. Research, New Brunswick, N.J. 08903.*

LEATHERS, Wailer S(mith), Am. physician; b. nr. Charlottesville, Va., Dec. 4, 1874; s. James Addison and Bettie Elizabeth (Pace) L.; diploma U. Va., 1892, M.D., 1895; LL.D., U. Miss., 1924; grad. work, Johns Hopkins, 1896, U. Chgo., 5 summers, also Biol. Lab., L.I., summer, 1897, Marine Biol. Lab., Woods Hole, Mass., summer 1900, Harvard Med. Sch., summer 1906; LL.D., Tulane U., 1938; m. Ola Price, Nov. 14, 1906; 1 dau., Lucy Dell. Head dept. chemistry Miller Sch. of Va., 1896-97; prof. biology U. S.C., 1897-99; prof. biology U. Miss., 1899, prof. physiology, 1903, prof. physiology and hygiene, 110-24, also organized and served as dean of Med. Sch., 1390-24; prof. preventive medicine and public health, Vanderbilt U., since 1294, asso. dean Med. Sch., 1927, dean since 129. 8Dir. pub. health of Miss., 119; exec. officer State Bd. Health, mem., 1917-24; Fellow A.M.A., Am. Pub. Health Assn. (pres. 1940-41), A.A.-A.S. (v.p. 1928), Assn. Am. Med. Colls. (pres. 1942-43), Soc. Med. Officers of Health (Eng.), Royal San. Inst. (Eng.); mem. Am. Soc. Tropical Medicine; So. Med. Assn. (pres., 1922-28). Phi Beta Kappa, Sigma Xi (charter mem.). Contbr. numerous sci. articles to med. jours. Died Jan. 26, 1946.

LEATHES, John Heresford, English biochemist; b. London, 1864; s. Stanley and Matilda (Butt) L.; ed. Winchester Coll., New Coll., Oxford, M.A., B.M., B.Ch., 1893; ed. Guys Hosp., London U., Berne, Switzerland, 1895-97, Strasbourg, 1897-99; D.Sc., Sheffield U., 1933, Manchester U., 1935. Demonstrator in physiology Guys Hosp., London, 1894; prof. U. London, 1899; lectr. physiology St. Thomas's Hosp., 1899-1909; asst. Lister Inst., 1901-09; prof. path. chemistry Toronto, Ont., Can., 1909-14; prof. physiology Sheffield U., 1915-33, emeritus, 1933-56. Fellow Royal Soc., 1911, Royal Coll. Physicians, Royal Coll. Surgeons; mem. Physiol. Soc., Biochem. Soc., Am. Soc. Biochemistry. Author: Problems in Animal Metabolism, 1905; (with H. S. Raper) Monograph on Fats, 1910, 20, also articles. First to describe alkaline tide of urine, 1919. Died Sept. 14, 1956.

LÉAUTÉ, Henri, physicist; b. Honduras, 1847; prof. for mechanics École polytechnique, Paris, France; research on mech. problems including friction of rotating waves, cinematography, the theory of the lamellar brake, the control of the water turbine, vibrations of mech. systems; determined the kinetic characteristic of machines. Died 1916.

LÉAUTÉ, Henry-Charles-Victor-Jacob, mathematician, physicist; b. Badelize, Brit. Honduras, Apr. 26, 1847 (of French parents); ed. Poly. Sch.; D. ès scis., 1876; became tutor mechanics Poly. Sch., 1877; named prof., 1895; dir. Indsl. Soc. Telephones, from 1893; mfg. engr. French govt.; mem. French Acad. Scis., 1890. Author: Note sur le degré et la classe d'une courbe parallèle à une courbe donnée, 1876; Transmission par cables métaliques, 1895. Research in pure analysis, later in electricity. Died Paris, Nov. 5, 1916.

LEAVELL, Hugh Rodman, Am. physician; b. Louisville, Nov. 17, 1902; s. Hugh Nelson and Harriet (Rodman) L.; B.S., U. Va., 1925; M.D., Harvard, 1926; Dr.P.H., Yale, 1940; m. Esta Carter Holt, Nov. 22, 1958; children—Hugh Nelson, Barbara Hazard, Rodman. Practice medicine specializing in internal medicine, 1930-37; dir. health Louisville and Jefferson County, Ky., 1934-44; dep. dir. European Regional Office, UNRRA, London, 1944-45; asst. dir. for med. sci. Rockefeller Found., 1945-46; prof. pub. health practice Harvard Sch. Pub. Health, 1946-63; Cons. on health Ford Found., New Delhi, India, 1956-57, 63-—; mem. Bd. Health, Cambridge, 1952-55, Boston, 1958-63; mem. Mass. Pub. Health Council, 1960-63. Recipient Prentiss award in health edn. Cleve. Health Mus., 1963. Mem. Am. (pres. 1954-55), Mass. (Lemuel Shattuck award 1961, pres. 1962-63), pub. health assns., Assn. Schs. Pub. Health (pres. 1961-63), Nat. Health Council (pres. 1955-56), Phi Beta Kappa, Alpha Omega Alpha, Delta Omega, Phi Kappa Phi. Author: Textbook on Preventive Medicine for the Doctor in His Community, 1953. Contbr. articles to social scis. in health, health edn., health adminstrn. to tech. jours. Home: 51 Lodi Estate, New Delhi. Office: 32 Ferozshah Rd., New Delhi 1, India.*

LEAVENWORTH, Francis Preserved, Am. astronomer; b. Mt. Vernon, Ind., Sept. 3, 1858; s. Seth M. and Sarah (Nettleton) L.; A.B., Ind. U., 1880, A.M., 1888; m. Jennie Campbell, Oct. 11, 1883; children—Mary Louise, Francis Maury, Richard Ormond. On staff Cin. Obs., 1880-82; asst. McCormick Obs., U. Va., 1882-87; dir. Haverford Coll. Obs., 1887-92; asst. prof. astronomy U. Minn., Mpls., 1892-97, prof., 1897-1927. Author: Double Star Observations, 1888; proc. Haverford Coll. Observatory, 1891; Parallax, LI, 1196, 1892; Photographs of Eros for Solar Parallax, 1902. Died Nov. 12, 1929.

LEAVITT, Frank McDowell, Am. inventor; b. Athens, O., Mar. 3, 1856; s. John McDowell and Bithia (Brooks) L.; M.E., Stevens Inst., 1875, E.D., 1921; m. Gertrude Goodsell, Nov. 8, 1893. Designer steam steering-gear, 1876; head draftsman Bliss & Williams, Bklyn., 1877-81; master mechanic Tex.-Mexican Ry., 1881-82; mgr. Graydon & Denton Mfg. Co., Jersey City, 1882-84; became asst. supt. E. W. Bliss Co., Bklyn., 1884, later chief engr.; undertook introduction of the Whitehead torpedo into U.S.N., 1890, also installed plant of U. S. Projectile Co. for manufacture shells and other projectiles used in war; invented Bliss-Leavitt torpedo, used U. S. Navy, having a range of 12,500 yards; inventor of an automatic can-making machine, a press for producing all kinds of hollow pressed ware. Died Aug. 6, 1928.

LEAVITT, Henrietta Swan, Am. astronomer; b. Lancaster, Mass., July 5, 1868; d. George R. and Henrietta (Kendrick) L.; grad. Soc. for Collegiate Instrn. of Women (now Radcliffe Coll.), 1892. Staff, Harvard Coll. Obs., from 1902, became head dept. photog. stellar photometry. Developed methods for determination of photog. magnitudes of variable stars; determined magnitudes of stars in sequences; discovered 4 novae, 2,400 variable stars; discovered that brightest of cluster variables in Magellanic Clouds has longest periods of variation, thereby introducing period-luminosity relationship for estimating distances of certain stars. Died Cambridge, Mass., Dec. 12, 1921.

LEAVITT, Sheldon, Am. physician; b. Grand Rapids, Mich., Apr. 9, 1848; s. David Sheldon and Martha Ann L.; M.D., Hahnemann Med. Coll., Chgo., 1877; postgrad. in Germany, Eng. and France; m. Marcella E. Smith, 1871; children—C. Franklin, Florence Belle. Prof. obstetrics Hahnemann Med. Coll., 1880-98; prof. gynecology Chgo. Homeo. Med. Coll., 1898-1902. Author: The Science and Art of Obstetrics, 1882; Psycho-Therapy, 1903; Paths to the Heights, 1908; The Psychic Solution of the Problem of Cure, 1909; Volo-therapy, 1917; Living a Century, 1928. Died Feb. 1, 1933.

LEAVITT, William Grenfell, Am. mathematician; b. Omaha, Mar. 19, 1916; s. Frederick W. and Mattie (Knapp) L.; A.B. (with high distinction), U. Neb., 1937, M.A., 1938; postgrad. Princeton; Ph.D., U. Wis., 1947; m. Genevieve Ruth Tollefsen, Dec. 27, 1941; children—Carolyn (Mrs. Henry B. McEwen), Elizabeth, Robert. Instr., U. Wis., 1946; faculty U. Neb., Lincoln, 1947-—, chmn. dept. math., 1954-64, prof., 1956-—. Faculty research fellow, 1952; NSF fellow, 1959-60. Mem. Am. Math. Soc., Math. Assn. Am. (bd. govs. 1957-60), Sigma Xi, Phi Beta Kappa. Research, publs. in abstract algebra, primarily in areas of Theory of Rings and Theory of Modules; originator ring invariant; module type, and its generalization, maxit of a ring; most recent contbns. in theory of radicals, particularly on properties of the Kurosh lower radical. Home: 2101 Park Av., Lincoln, Neb. 68502.*

LE BAILLIF, Alexander-Claude-Martin, French physician; b. St.-Fargeau, France, 1764. Author: Notice sur la construction du sidéroscope, 1827. Perfected Charles' microscope; developed glass micrometer, new galvanmeter, dry cells; inventor device for detection of iron or steel splinters in eye. Died 1831.

LEBEDEV, Aleksandr Alekseevich, Russian physicist; b. 1893; grad. Petrograd U., 1916. Prof., Petrograd U., 1916-24, Leningrad U., 1924-—; dept. head State Optical Inst., 1957. Decorated Order of Lenin (4); recipient Stalin prize, 1947, 49. Mem. USSR Acad. Sci. Author: Polimorphism and Fritting of Glass, 1921; The Polarization Interferometer and Its Applications, 1931; Structural Changes in Glass, 1953. Chief editor News of USSR Acad. Sci. Physics Series. Research and publs. on electronic optics, process of optical glass fritting; built polarization interferometer, model of electron microscope, 1931. Address: Leningrad University, Universitetskaya n. 7-9, Leningrad, USSR.

LEBEDEV, Pëtr Nikolajevich, Russian physicist; b. Feb. 24, 1866, Moscow; studied under Kundt at Strasbourg, France; became prof. physics, Moscow, 1892. Author: Experimental Research on Light Pressure, 1901. Proved the existence of extremely small pressure exerted by light and measured it (confirmed Maxwell's theories); research on earth's magnetism. Died Moscow, Mar. 1, 1912.

LEBEDEV, Sergei Alekseevich, Russian elec., radio engr.; b. Nov. 2, 1902; grad. Moscow Tech. Coll., 1928. With All-Union Electrotech. Inst., 1928-45; dir. Inst. Electroengring., Ukrainian SSR Acad. Scis., 1946-51, academician, 1945; dir. Inst. Exact Mechanics and Computation Techniques of USSR Acad. Scis., 1953-—, academician, 1953-—; prof. Moscow Physico-Tech. Inst., 1952-—. Participant Joint Computer confs., Phila., 1958, 59. Decorated Hero Socialist Labor, 1956; Order Lenin, 1962. Author: Stability of Electrical Systems in Parallel Operation, 2d edit., 1934; Electronic Computers, 1955; The Artificial Stability of Synchronous Machines, 1960. Directed constrn. high speed computers; developed theory of artificial stability of synchronous machines. Home: Novopeschanaya, 17. Office: Institute of Precision Mechanics and Computation Techniques, USSR Academy of Sciences, Leninskii Prospekt, 51, Moscow, USSR.

LE BÉGUE DE PRESLE, Achille-Guillaume, French physician; b. Pithiviers, France, 1735; physician, friend to J. J. Rousseau. Author: Pronostics utiles au labourer et au voyageur, 1770; Notice sur les derniers jours de J. J. Rousseau, 1778. Collaborator Bibliotèque physico-économique, 1786-92. Studied hypochondria, hysteria, med. needs of travelers and workers; disseminated med. sci. through works written for educated laymen. Died Paris, 1807.

LE BEL, Joseph Achille, French chemist; b. Péchelbronn, France, Jan. 21, 1847; ed. Ecole Polytechnique, Paris; student of Balard, 1871, Wurtz, 1872. Dir. oil operations, Péchelbronn, until 1889; pvt. research, Paris. Fellow Royal Soc., 1911; mem. French Acad. Scis., 1929. Author: Cosmologie rationelle, 1929. Investigated optical activity and fermentation; announced theory of asymmetric carbon atom, 1874; credited as a founder of stereochemistry; showed that replacement of radical in optically active compound by radical already present destroyed optical activity; sought to determine existence of optically active compounds of nitrogen, 1891. Died Paris, Aug. 6, 1930.

LEBEL, Norman Albert, Am. chemist; b. Augusta, Me., Mar. 22, 1931; s. Joel and Rose (Croteau) LeB.; A.B., Bowdoin Coll., 1952; Ph.D., Mass. Inst. Tech., 1957; m. Marie Constance Ouellette, Sept. 1, 1952; children—Mark Andrew, Norma Jean, Carl Philip. Chemist, Merck & Co., Rahway, N.J., 1952-54; faculty Wayne State U., Detroit, 1957-—, prof. chemistry, 1964-—. Cons. Ethyl Corp., 1963-—. Mem. Am. Chem. Soc. (sec.-treas. organic div. 1964-—), Chem. Soc. (London), Sigma Xi, Phi Lambda Upsilon. Research, publs. on mechanisms and stereochemistry of organic reactions, new synthetic methods. Home: 17517 Pennington St., Detroit 48221.*

LEBER, Ferdinand Joseph von, see von Leber, Ferdinand Joseph.

LEBERT, Hermann, physician, pathologist; b. Breslau (now Wroclaw, Poland), June 9, 1813; M.D., Zurich, Switzerland, 1834; became prof. medicine, Zurich, 1853; prin. prof. medicine, Breslau, 1859-74; mem. French Acad. Scis. Author: Traité d'anatomie pathologique, 1855-60. First to describe brain abscess accurately, 1856; research on bone diseases; made microscopic studies of tumors and inflammation. Died Bex, Switzerland, Aug. 1, 1878.

LEBESGUE, Henri Leon, French mathematician; b. Beauvais, France, June 28, 1875; ed. École normale supérieure; Ph.D., U. Paris, 1897. Became agregé prof. math. scis. Lycée Nancy (France), 1897; lecture master Faculty Sci., Rennes, France, 1902-06; prof. Faculty Sci., Poitiers, France, 1906-10; lectr. master Faculty Sci., Paris, 1910-19, prof., 1920-21; became prof. Coll. France, 1921. Recipient le Prix Saintour, 1917. Fellow Royal Soc., 1934; mem. French Acad. Scis., London Math. Soc. (hon.). Author: Lecons sur l'intégration et la recherche des fonction primitives, 1904, 28; Lecons sur les séries trigonométriques, 1906; also articles. Research on theory of functions of real variables; formulated new definition of definite integral; originated pavement theorem. Died Paris, July 26, 1941.

LE BLANC, Félix, chemist; b. Florence, Italy, 1813 (of French parents); ed. Sch. Mines, Paris; worker in Dumas' lab., 1839-45; became asst. in chemistry École Polytechnique, 1846; named prof. École Centrale des Arts et Manufactures, 1854. Author: Cours de chimie analytique, 1875; (with Limousin and Schmitz) Rapport sur le matériel des arts chimiques de la pharmacie et de la tannerie, 1883. Research on carbon dioxide, compressed air, lead oxide. Died 1886.

LEBLANC, Hugh Linus, Am. polit. scientist; b. Alexandria, La., Oct. 30, 1927; s. Moreland Paul and Carmen (Haydell) LeB.; B.A. in Govt., La. State U., 1948; M.A., U. Tenn., 1950; Ph.D. in Polit. Sci., U. Chgo., 1958; m. Shirley Jean Smith, Feb. 28, 1953; children—Leslie Ann, Alexander Hugh. Research asso. Council of State Govts., Chgo., 1950-52; faculty George Washington U., Washington, 1955-—, prof. polit. sci., 1963-—. Lectr., U. S. Civil Service Commn., 1961-—; cons. Dept. Interior, 1960-61. Mem. Am., So., D.C. (pres. 1966-67) polit. sci. assns., Am. Acad. Polit. and Social Sci., Am. Soc. Pub. Adminstrn. Author: (with R. D. Campbell) An Information System for Urban Planning, 1965; (with Campbell, M. Mason) Shoreline Recreation Resources of the United States, 1962. Developed method for collecting data useful in planning process; documented constituency influences on voting decisions of state senators. Home: 3403 Barger Dr., Falls Church, Va. 22044. Office: Dept. Polit. Sci., George Washington U., Washington 20006.*

LEBLANC, Maurice (Charles Leonard Armand Maurice), French inventor, engr.; b. Paris Mar. 2, 1857; engr. Eastern R.R. Co. Mem. French Acad. Scis. Author: Sur la stabilité de la marche des commutatrices, 1901; La lampe en quartz, 1910; L'arc éléctrique, 1922. Built Panchahuteur (rotary converter), 1894; developed theory of commutator motor; discovered method of starting 3-phase motor (using cage type armature for resistance regulation of motor's vibrations), 1900; 1st to demonstrate currents of adjustable frequency using accumulator motors from asynchronous motors; built condenser with pumps for drawing out air and injecting water for turboelectric generators (used by Westinghouse-Leblanc Co.). Died Paris, Oct. 27, 1923.

LEBLANC, Nicolas, French chemist; b. Issoudun, France, 1742; ed. medicine and surgery; pupil of Darcet and Rouelle. Surgeon to Duke of Orleans, from 1780; began pvt. chem. expts., 1789; set up factory for mfg. sodium carbonate, nr. St. Denis, 1792; dir. powders and saltpeters Legislative Assembly. Awarded prize for method of mfg. sodium hydroxide and sodium carbonate French Acad. Scis., 1783. Author: Memoires sur la fabrication du sel ammoniac et la soude, 1798; Crystal Technology, 1786. Research on crystallization of neutral salts; developed LeBlance Process for making soda from salt. Died St. Denis, Jan. 10, 1806.

LEBLOND, Charles P., anatomist; b. Lille, France, Feb. 5, 1910; s. Oscar and Jeanne (Des Marchelier) L.; L.Sc., U. Lille, 1932; M.D., U. Paris, 1934, D.Sc., 1945; Ph.D., U. Montreal, 1942; m. Gertrude Sternschuss, Oct. 22, 1936; children—Philippe L., Paul N., Pierre F., Marie Pascale. Asst. in histology U. Lille and U. Paris, 1934-35; Rockefeller fellow dept. anatomy Yale, 1935-37; in charge biology div. Lab. Synthese Atom, Paris, France, 1937-40; research fellow U. Rochester, 1940-41; with McGill U., Montreal, Que., Can., 1941-43, asst. prof. histology, 1942-43, 46-—, prof. anatomy, 1948-—. Fellow Royal Soc. Can., Am. Assn. Anatomy, Canadian Assn. Anatomists, Royal Soc. London. Author: L'Acide Ascorbique dans les Tissues et sa Detection, 1936; Radioautography as a Tool in the Study of Protein Synthesis, 1965. Devel. publs. on modern method of radioautography and its use in study of hormone, protein, and sugar formation in cell. Home: 68 Chesterfield St., Montreal, Que., Can.*

LEBLOND, Guillaume, French mathematician; b. Paris, 1704; named prof. math. to pages of grand stable of King, 1736, to King's children, 1751; sec. to Victoire. Author: Essai sur la castramétation, 1748; L'arithmétique et la géometrie de l'officier; also articles; contbr. to l'Encyclopedie. Died Versailles, France, 1781.

LE BON, Gustave, French social psychologist, physician; b. Nogent-le-Rotrou, France, May 7, 1841; ed. in medicine. Practiced medicine, Paris, from 1870; chief doctor with volunteer ambulance corps, 1870; traveled to India to study Buddhist monuments, 1884. Author: L'homme et les sociétés, 2 vols., 1881; La Civilisation des Arabes, 1884; Les monuments de l'Inde, 1891; Les lois psychologiques de l'évolution des peuples, 1894; La psychologie des foules, 1895; L'évolution des forces, 1899; Psychologie de l'education, 1904; L'évolution de la matière, 1905; Les opinions et les croyances; genèse-évolution, 1911; La révolution française et la psychologie des révolutions, 1912; Psychologie des temps nouveaux, 1920; L'évolution actuelle du monde, 1927. Noted for analysis of crowd behavior and unconscious motivation; formulated a collective and racist social psychology; stressed influence of emotion on history. Died Marnes-la-Coquette, nr. Paris, Dec. 13, 1931.

LEBON, Philippe, French chemist; b. Brachay, France, 1767; causeways bridge engr., Angouleme; prof. Sch. Bridges and Causeways, Paris; invited to help prepare ceremony for Napolean's coronation. Author: Thermolampes ou poêles qui chauffent, 1801. Began expts. with illuminating gas, 1797; received patents for use of gas in lighting, 1799; developed thermolamp (gas burning lamp); studied vapor machines; described steam-boilers, condensation by injection. Died Paris, 1804.

LEBOUR, George Alexander Louis, Brit. geologist; b. 1847; s. Alexander Lebour; ed. Royal Sch. Mines; M.A., D.Sc.; m. Emily Hodding. Lectr. geol. surveying, Durham Coll. Sci. (now Armstrong Coll.), Newcastle-upon-Tyne, Eng., 1873-79, became prof. geology, 1879, vice prin., 1902. Fellow Geol. Soc. (Murchison medal 1904); mem. Brit. Assn. (mem. several research coms.) Author: Geology of Northumberland and Durham, 1878-89; Geological Map of Northumberland, 1877. Sub-editor fgn. geology Geol. Record, 1874-80. Research and numerous publs. on geology of Durham, heat conductivity of rocks, carboniferous geology, underground temperature, relation between pub. health and geology, fgn. and Brit. coal fields. Died Feb. 7, 1918.

LEBOWITZ, Joel Louis, physicist; b. Taceva, May 10, 1930; B.S., Bklyn. Coll., 1952; M.S., Syracuse U., 1955, Ph.D., 1956; m. Estelle Mandelbaum, June 21, 1953. Came to U. S., naturalized. NSF postdoctoral fellow, Yale, 1956-57; faculty Stevens Inst. Tech., 1957-59; faculty Yeshiva U., N.Y.C., 1959-—, prof. physics, 1965-—, also acting chmn. dept. physics Belfer Grad. Sch. Sci., 1964-—. Mem. A.A.A.S., Am. Assn. Physics Tchrs., Am. Phys. Soc., Am. Assn. U. Profs., Sigma Xi. Research and publs. on statis. mechanics of equilibrium and nonequilibrium processes; low temperature physics; solid state physics and quantum theory of measurement. Home: 600 W. 218th St., N.Y.C. 10034.*

LEBRIE, Stephen Joseph, Am. physiologist; b. Long Island City, N.Y., Jan. 23, 1931; s. Stephen J. and Ann (Fascinati) LeB.; B.S., L.I. U., 1953; M.A., Princeton, 1955, Ph.D., 1956; m. Elizabeth Kuhta, Mar. 31, 1956; children—Stephanie Lizette, Beth Ann. Research asso. Harrison dept. surg. research U. Pa., 1956-57; faculty Tulane U., New Orleans, 1957-66, asso. prof. physiology, 1963-66; asso. prof. physiology Ohio State U., Columbus, 1966-—. Recipient Lederle Med. Faculty award, 1959-62. Mem. Am. Physiol. Soc., Am. Assn. U. Profs., Sigma Xi. Contbr. articles to tech. jours. Research, publs. on hormone action, function lymphatic system, basic renal physiology water snakes, site hormone action renal tubules.*

LE BRIS, Jean Marie; b. 1817; capt. in coast guard; founded glider flying in France; tested glider with movable wings nr. Dovarnenez by pulling it with a wagon. Died 1872.

LE CAT, Claude Nicolas, French physician, physiologist; b. Blerancourt, France, Sept. 6, 1700; med. degrees, Reims, France, 1732. Apptd. surgeon, physician to M. de Tresson, Archbishop of Rouen, France, 1729; settled at Rouen, 1733; surgeon in chief Hosp. at Rouen; established pub. sch. surgery and anatomy, Rouen, 1736; began course exptl. philosophy, Rouen, 1746; prof. anatomy surgery, Rouen. Fellow Royal Soc., 1739; mem. Royal Acad. Rouen (sec.), Royal Acad. Surgery (won all prizes 1732-38), French Acad. Scis., Acad. Surgery Paris, Acad. Madrid (corr.). Author: Traite des sens, 1740; Lettres concernait l'operation de la taille, 1749; Recueil de pièces sur l'opérat de la taille, 1749-53; Traite des sensations et des passions en generale, 1766; Physiologie, 1767; Cours abrege d'osteologie, 1768; Traité de la nature du fluide des nerfs, 1765. Invented new instruments for lithotomy; perfected operations for lachrymal fistula; introduced to France, Cheselden's procedure for operation on bladder. Died Rouen, Aug. 20, 1768.

LECÈNE, Paul, French surgeon; b. Paris, France, 1878; named hosp. surgeon, 1907; became agrégé prof. surg. pathology Faculty of Medicine, Paris, 1921. Author: Les tumeurs du rein; Les tumeurs de l'intestin grele; Manuel de pathologie chirurgicale; Évolutions de la chirurgie. Studied surg. pathology including tumors of kidney and small intestine. Died Paris, 1929.

LECH, Christer, Swedish mathematician; b. Stockholm, Sweden, Mar. 30, 1926; s. Halvar and Ragna (Siwerson) L.; Ph.D., U. Stockholm, 1960; m. Britta Källman, May 18, 1966; 1 dau., Eva. Faculty, U. Stockholm, 1960—, prof., 1965—. Work in algebraic geometry. Home: 8 Höjdstigen, Lidingö, Sweden. Office: Dept. of Math., Univ. Stockholm, Box 23144, Stockholm 23, Sweden.*

LE CHATELIER, Henry-Louis, French management scientist metallurgist, chemist; b. Oct. 8, 1850; ed. Collège Rollin, École Polytechnique, École des Mines, Paris. Apptd. prof. general chemistry, École des Mines, 1878; prof. mineral chemistry, Collège de France, 1888; prof. general chemistry, Collège de France, 1897-1907; inspector-general of mines, 1907; prof. general chemistry, Sorbonne, 1907-25; commissioned by Minister of Armaments to work in armaments-producing factories, 1914-18; presided over 2nd French Management Congress, 1924; led French delegation, 1st Internat. Management Congress, Prague, 1924. Recipient Prix Jerome Ponti, 1892; Prix La Caze, 1895; gold medal, Internat. Comm. for Sci. Management, 1929. Mem., Comité National de l'Organisation Française, Taylor Soc., Am. Soc. of Mech. Engr.; French Acad. Scis.; pres., Mineralogical Soc., Soc. for Encouragement of National Industry, Soc. of Physics; fgn. mem., Netherlands Soc. of Scis., Berlin Acad. Scis.; Commander Legion of Honour; hon. pres. Internat. Congress of Mining, 1935. Author: Science and Industry, 1925; Taylorism, 1934; Method in the Experimental Sciences, 1836; translated works of J. W. Gibbs into French; numerous pub. in physical sciences, esp. in Revue de Métallurgie (which he founded). Introduced scientific management into France; taught application of Taylor's principles to industry; research on specific heats of gases at high temperatures, mass action in explosion reactions, freezing point curves, chemistry of silicates, chemistry of cement and glass, chemistry and physics of flames, thermodynamics; discovered law of reaction for the effect of pressure and temperature on equilibrium; devised optical pyrometer, ry. water brake. Died Miribel-les-Echelles, Sept. 17, 1936.

LE CHATELIER, Louis, mine engr.; b. 1815; research on counterweight including disturbances in the motion of locomotive, 1849; tested products with aluminum content; introduced the use of the Siemens furnace for the manufacture of gas; patentee steel prodn. identical to that of Martin, 1863. Died 1873.

LECHER, Ernst, Austrian physicist; b. Vienna, Austria, 1856; prof. physics U. Vienna. Measured dielectricity constants; studied electromagnetic waves in circuit with a condenser and parallel wires using a glass tube with rarified gas; determined stationary electric waves along the wires and proved position of maxima changed when a dielectric was put into the system. Died 1926.

LECHNER, Helmut Eduard Viktor, Austrian physician; b. Graz, Austria, May 19, 1927; s. Eduard and Regina (Weingraber) L.; Doctor Medizin, U. Graz, 1951; Specialist for Neurology and Psychiatry, Neurol. Inst. Queen Sq., London, 1954; m. Maria Bauer, Jan. 3, 1954; children—Beringaria, Janette. Head physician U. Nervenklinik, U. Graz, 1958—, dozent, 1959—; dir. dept. for applied neurophysiology Landeskrankenhaus, Graz, 1964—. Mem. Austrian (chmn.), Brit. EEG socs., Royal Soc. Medicine London, Internat. Soc. Microcirculation. Author: Scientific and Clinical Research of the Cerebral Circulation (with Bertha, Eichorn, Auell), 1962; (with Martin) Rhoencephalography, 1963; also articles. Research on diagnosis, therapy of head injuries, memory disturbances, photogenic epilepsy, disturbances due to circulatory neurochem. defects in brain, rheography, clin. application of electroencephalogram. Home: Geidorfgürtel 46, Graz, Austria.*

LECLAINCHE, Auguste-Louis-Emmanuel, French veterinarian; b. Piney, France, Aug. 29, 1861; dir. Ministry of Agr.; named prof. Toulouse (France) Sch. Vet. Medicine, 1891. Founder, Internat. Office for Study Epizootic Diseases. Mem. French Acad. Scis. (became pres. 1937), French Acad. Surgery, French Acad. Agr. Author: (with Nocard) Histoire de la médicine vétérinaire; Traité des maladies microbiennes des animaux domestiques. Research on infectious and parasitic diseases, epidemiology; discovered new treatments for the diseases of house pets; established doctorate and aggregation for vet. sci. in France. Died Paris, Nov. 26, 1953.

LECLANCHÉ, Georges, French chemist; b. Paris 1839; invented electric cells with porous containers, 1867, cells with pressed plates, 1873, movable, 1876-78, elec. mercury contact, galvanic cell which is named after him. Died 1882.

LE CLERC, Daniel, Swiss physician; b. Geneva, Switzerland, Feb. 4, 1652; student Montpellier, France, Paris; practiced in Geneva; counsellor of republic; gave up practice medicine for pub. life, 1704; mem. French Acad. Scis., 1699. Author publs. including: complete Surgery, 1695; History of Medicine, 1696; Historia naturalis et medica latorum lumbricorum intra homines . . . nascentium, 1715; (with Manget) Bibliothèque anatomique. Died Geneva, June 8, 1728.

LECLERC DU SABLON, Albert-Mathieu, French botanist; b. Bagnols-sur-Cèze, France, Mar. 25, 1859; prof. botany Faculty of Sci., Toulouse; mem. French Acad. Scis., 1920. Author: Recherches sur la déhiscence des fruits à pericarpe sec, 1884; Cours de botanique, 1900. Research on fruit and grain of phanerogams, reproductive organs of cryptogams, digestion of reserve food by roots. Died Vénéjan, France, Mar. 18, 1944.

LECLERG, Erwin Louis, Am. biometrician; b. St. Louis, Feb. 16, 1901; s. Charles F. and Lillian (Dunn) LeC.; B.S., Colo. State U., 1924; M.S., Ia. State U., 1925; Ph.D., U. Minn., 1932; m. Phylis M. Robertson, Aug. 10, 1926; 1 son, Robert E. Staff, U. S. Dept. Agr., 1924-64, asst. plant pathologist Colo. Agr. Expt. Sta., 1925-30, research coordinator, 1948-54, dir. biometrical services Agr. Research Service, Beltsville, Md., 1954-64; chmn. dept. biol. scis. Grad. Sch. Dept. Agr., 1955-64, mem. council, 1949-64; prin. budget examiner Bur. Budget, 1946-48. Fellow A.A.A.S., Royal Statis. Soc. London, Washington Acad. Sci.; mem. Am. Soc. Agronomy, Bot. Soc. Washington, Am. Phytopath. Soc., Potato Assn. Am., Crop Sci. Soc., Biometric Soc., Am. Inst. Biol. Scis., Sigma Xi, Gamma Alpha, Sigma Chi. Research, publs. on rootrot diseases sugar beets, breeding for disease resistance in potatoes, statis. methodology. Home: 6804 40th Av., University Park, Hyattsville, Md. 20782.*

L'ECLUSE, Charles, see Clusius, Carolus.

LECOINTRE, Georges, French geologist; b. La Chapelle Blanche, Oct. 19, 1888; s. Pierre and Henriette (Delamarre de Monchaux) L.; ed. univs. Nancy, Paris (both France); Ph.D.; m. Solange Chenu de Mangou, Dec. 15, 1920; 1 child, Colegge. Asso. with Bur. Geol. and Mineral Research, Paris. Research and publs. on geology and Morocco and middle west region of France. Home: 17, av. de Saxe 7. Office: 74, rue de la Federation, Paris, France.

LECOMTE, (Paul) Henri, French botanist; b. Saint-Nabord, France, Jan. 8, 1856; became prof. phanerogamy Paris Mus. Natural History, 1906; mem. French Acad. Scis., 1917. Author: Les bois d' Indochine; Les bois de Madagascar, 1907; participated founding catalogue of colonial flora, especially of Indochina, 1907.

LECOMTE, Jean, French physicist; b. Paris, Aug. 5, 1898; s. Réné and Marguerite (Laniel) L.; hon. doctorates from various fgn. univs.; m. Marguerite Turquet Bravard de la Boissière, June 2, 1958; children—Gilles, Martine, Béatrice, Philippe. Dir. research. Nat. Center for Sci. Research. Mem. French Acad. Scis. Research and publs. on infrared radiation. Address: 6, rue de l'Alboni, Paris 16, France.

LECOMTE, Jean, Belgian physiologist; b. Mons (Liege), Belgium, May 15, 1921; s. Georges and Hortense (Bernard) L.; ed. U. Liege; med. doctor, 1946; agrégé, 1956; prof. physiology U. Liege, 1959. Research in anaphylaxis. Office: Institut Leon Fredericq-Physiologie, 17 Place Delcour, Liege, Belgium.*

LECONTE, John, Am. physicist; b. Liberty County, Ga., Dec. 4, 1818; s. Louis and Ann (Quarterman) LeC.; grad. Franklin Coll. (now U. Ga.), 1838; M.D., Coll. Physicians and Surgeons, N.Y.C., 1841; m. Eleanor Graham, June 20, 1841. Prof. physics and chemistry Franklin Coll., 1846-55; prof. physics U. So. Cal., 1856-69; became prof. physics U. Cal., 1869, also pres., 1869, 75-81. Mem. Nat. Acad. Scis. Author: (papers) Experiments Illustrating the Seat of Volition in the Alligator, 1845; On the Influence of Musical Sounds on the Flame of a Jet of Coal-gas, 1858; On Sound Shadows in Water, 1882; Physical Studies of Lake Tahoe, 1883, 84. Discovered sensitivity of flame to mus. vibrations, 1857. Died probably Berkeley, Cal., Apr. 29, 1891.

LECONTE, John Lawrence, Am. entomologist; b. N.Y.C., May 13, 1825; s. John and Mary Anne (Lawrence) LeC.; grad. Mt. St. Mary's Coll., 1842; M.D., Coll. Physicians and Surgeons, N.Y.C., 1846; m. Helen Grier, Jan. 10, 1861. Investigated and published papers on entomology and zoogeography, 1846-61; surgeon of volunteers, M.C., U. S. Army, 1861-65, later lt. col., med. insp.; studied minerology, geology and entomology, 1865-83; chief clk. U. S. Mint, Phila., 1878-83; an incorporator Nat. Acad. Scis.; pres. A.A.-A.S., 1874. Author: On the Classification of the Carabidae of the United States, 1862-73, later edit. with G. H. Horn; List of Coleoptera of North America, 1866; The Rhynchophora of America North of Mexico, 1876; also papers Coleoptera of Europe and North America, 1848, The Rhynchophora of America North of Mexico, 1876, Classification of the Coleoptera of North America, 1883. Research and publs. on entomology, mineralogy, geology, ethnology, fossil mammals, radiates; 1st biologist to map faunal areas of western U. S. Died Phila., Nov. 15, 1883; buried Phila.

LE CONTE, Joseph, Am. geologist, natural scientist; b. Liberty County, Ga., Feb. 26, 1823; grad. Franklin Coll., U. Ga., 1841; A.M., 1845; Coll. of Phys. and Surg., N.Y., 1845; Harvard, B.S., 1851; LL.D., U. of Ga., 1879; Princeton, 1896; m. Caroline Elizabeth Nesbit, Jan. 14, 1847. In Confederate service as chemist of med. lab. nitre and mining operations, during Civil War; prof. geology U. Cal., 1869-96. Author: Religion and Science (a course of Sunday lectures); Elements of Geology (text-book for univs.); Sight, or Principles of Monocular and Binocular Vision; A Compend of Geology for High Schools; Evolution and Its Relation to Religious Thought. Died 1901.

LE CONTE, Joseph Nisbet, Am. mech. engr.; b. Oakland, Calif., Feb. 7, 1870; s. Joseph and Caroline Elizabeth (Nisbet) L.; B.S., U. Cal., 1891, LL.D., 1945; M.M.E., Cornell U., 1892; m. Helen Marion Gompertz, June 10, 1901 (dec. Aug. 1924); children—Helen Malcolm, Joseph; m. 2d, Adelaide Elizabeth Graham, Feb. 16, 1929. Asst. in mechanics U. Cal., 1892-95, instr. in mech. engring., 1895-1903, asst. prof., 1903-12. prof. engring. mechanics, 1912-30, prof. mech. engring., 1930-37, emeritus since 1937. Collaborator with Yosemite Nat. Park Adv. Bd., 1943. Mem. Am. Soc. M.E., Sigma Xi, Phi Beta Kappa. Author: Elementary Treatise on Mechanics of Machinery, 1902; Hydraulics, 1926. Contbr. tech. articles to mags. Made pioneer explorations in the high Sierra of Calif. between 1890 and 1910, including 1st ascents of 6 peaks between 13,000 and 14,500 feet elevation. Died Feb. 1, 1950.

LECOQ, Henri, French botanist; b. Avesnes, France, Apr. 11, 1802; dean Faculty Scis., Clermont. France; cons. mem. French Inst.; dir. Clermont-Ferrand Bot. Gardens; mem. French Acad. Scis., 1859. Author: Principes élémentaires de botanique, 1828; De la toilette et de la coquetterie des végétaux, 1847; Botanique populaire, 1862; Considérations sur les phénomenes glaciaires de l'Auvergne, 1871; Étude de la géographie botanique de l'Europe, 1854. Popular publs. on classification of plants, description and reproductive organs of common flowers, plants used for fodder; studied geology of Auvergne, especially glacier formations. Died Clermont-Ferrand, Aug. 4, 1871.

LECOQ DE BOISBAUDRAN, Paul Émile, see de Boisbaudran, Paul Émile Lecoq.

LÉCORCHÉ, French physician; b. St.-Mardsen-Othe, France, 1830; became aggregate physician Faculty Medicine, 1869; physician Paris hosps., 1872. Author: Du diabète sucré chez la femme, 1855; Les altérations athéromateuses des artères, 1869; Traité du diabète, 1877; Traité de l'albuminurie et du mal de Bright, 1888; Traité théorique et pratique de la goutte. Research on achrestic diabetes, arteriosclerosis, Bright's disease, gout.

LECORNU, Léon-François-Alfred, French physicist; b. Caen, France, Jan. 13, 1854; gen. insp. mines; became prof. mechanics Ecole polytechnique, 1904, also École Supérieure d'Aeronautique, École des Mines; mem. French Acad. Scis., 1910 (pres. 1930). Author: Dynamique appliqué, 1908; Étude géometrique sur l'equilibre et la descente rectiligne de l'aéroplane, 1909. Research in geometry and analysis resulted in resolution of problems of surfaces and geometrical properties on continuous space; investigated variable length pendulum, movement of projectile in resistant space, gyratory motion of fluids; worked to improve flywheels, springs, gear wheels, aircraft stabilizers. Died St.-Aubin-sur-Mer, France, Nov. 13, 1940.

LECOUNT, Edwin Raymond, Am. pathologist; b. Fond du Lac, Wis., Apr. 1, 1868; student Carroll Coll., 1887-88; M.D., Rush Med. Coll., 1892; grad. study Johns Hopkins Hosp., 1893-94, Pasteur Inst., Paris, 1896, Berlin, 1905. Prof. pathology Rush Med. Coll., 1892—; attending pathologist Cook County, Presbyn., St. Luke's, St. Elizabeth's and St. Anthony's hosps. Pres. Am. Assn. Pathologists and Bacteriologists, Assn. Cancer Research. Died 1935.

LECOUTEUR, Kenneth James, physicist; b. Jersey, Channel Islands, Sept. 16, 1920; s. Philip and Eva (Gartrell) LeC.; B.A., St. John's Coll., Cambridge (Eng.) U., 1941, Ph.D., 1948; m. Enid Domville, July 14, 1950; children—Caroline, Penelope, Mary. Sci. officer T.R.E., Eng., 1941-45; fellow St. John's Coll., Cambridge, 1945-48; Turner and Newall fellow Manchester (Eng.) U., 1947-49; sr. lectr., reader Liverpool (Eng.) U., 1949-56; prof. theoretical physics Australian Nat. U., Inst. Advanced Studies, Canberra, 1956—. Fellow Australian Acad. Sci.; mem. Cambridge Philos. Soc., Australian Math. Soc., Australian Computer Soc. Research, publs. on field theory of elementary particles, efficient extraction of particle beams from large synchro-cyclotrons, theory of nuclei especially statis. model of nuclear disintegrations. Home: 12 Hutt, Canberra A.C.T., Australia.*

LECUIRE, Jean François Michel, French neurosurgeon; b. Toulon, Oct. 7, 1912; s. Emile and Gabrielle (Julien) L.; M.D., U. Lyons (France); m. Marie-Madeleine Marche, Aug. 27, 1945; children—Roger, Michel, Francois, Laurence. Intern Lyons Hosp., 1933; accepted to Faculty, 1939; became head surg. clinic, 1942; fellow Montreal Neurol. Inst., 1945-46; head works on anatomy, 1946-49; became prof. agrégé neurosurgery Faculty Medicine, Lyons, 1949, prof. neurosurgery, 1963; apptd. neurosurgeon Lyons Hosp., 1958; named head neurosurg. services Neurol. Hosp., 1962. Recipient Silver medal for mil. health service. Mem. Lyons Surg. Soc., Internat. Soc. Surgery, Soc. French Neurosurgery. Address: 19, rue Malesherbes, Lyon (Rhône), France.

LE DANTEC, Félix-Alexandre, French biologist; b. Plougastel-Daoulas, France, 1869; ed. École normale supérieur, 1885-88, Inst. Pasteur. Mem. Pavie's mission to Laos, 1889-90; Pasteur sent him to Brazil to establish lab. for study yellow fever, 1893; lecture master Faculty Lyons, (France); placed in charge gen. embryology course Sorbonne, Paris, 1899. Author: La digestion intracellulaire; La matière vivante, 1895; Théorie nouvelle de la vie, 1896; Evolution individuelle et hérédité, 1898; la Sexicalité, 1899; Lamarchiens et Darwiniens, 1900; Le confit, 1901; Traité de biologie, 1903; Introduction à la pathologie general, 1906; L'atheïsme, 1906; La crise du transformisme, 1910; L'egoisme, base de toute société, 1911; La Science de la vie, 1912. Research on cancer; developed idea of functional assimilation from Lamarck's notions of use and disuse. Died Paris, 1917.

LEDBETTER, Myron Calvert, Am. biologist; b. Ardmore, Okla., June 25, 1923; s. Robert H. and Adaline (Moore) L.; B.S., Okla. State U., 1948; M.S., U. Cal. at Berkeley, 1951; Ph.D., Columbia, 1958. Tchr., Blackwell (Okla.) Pub. Schs., 1951-52; research asst., asst. plant anatomist Boyce Thompson Inst. for Plant Research, Inc., Yonkers, N.Y., 1952-60; guest asso. Brookhaven Nat. Lab., Upton, L.I., N.Y., 1958-60, cell biologist, 1965—; guest investigator, fellow, research asso. Rockefeller U., N.Y.C., 1960-61; research asso. Harvard, 1961-65. Mem. A.A.A.S., Am. Inst. Biol. Scis., Am. Soc. for Cell Biology, Bot. Soc. Am., Electron Microscope Soc. Am., N.Y. Acad. Scis., Sigma Xi. Contbr. articles to tech. jours. Research, publs. on plant anatomy, physiologically dwarfed internodes seedings, pathology ozone on plant foliage, feeding damage plant tissues by an insect, autoradiography, distbn. tritiated glycine and thymidine in rusted bean leaves, distbn. radioactive fluorine-18 and stable fluorine-19 in tomato plants, morphology microtubules and their relation to plant cell walls; co-discoverer microtubules in plant cells; co-inventor thin metal aperturer for electron microscopy. Home: P.O. Box 145, Port Jefferson, L.I., N.Y. 11777. Office: Brookhaven Nat. Lab., Upton, L.I., N.Y. 11973.

LEDEBOUR, Karl Friedrich, see von Ledebour, Karl Friedrich.

LEDEN, Ido Evert, Swedish chemist; b. Kattarp, Sweden, Nov. 27, 1912; s. Josef Lorens and Edith (Nilsson) L.; Ph.D., U. Lund (Sweden), 1943; m. Majken Louise Tornberg, Apr. 10, 1939; children—Ido Lorenz, Bo, Lars Gunnar. Asst. prof. Lund U., 1943-45, prof. phys. chemistry, 1960—, dean Faculty Chemistry, 1964—; asst. prof. Chalmers Int. Tech., Gothenburg, Sweden, 1946-60; lectr. Hvitfeldtska Gymnasium, Gothenburg, 1946-60; Mem. com. Council Europe, 1963-65. Mem. Chem. Soc. Lund (past chmn.), Royal Physiographic Soc. Lund, Swedish Chem. Soc. (RNO 1960, GM Sci., 1960, Teaching, 1968). Author: (with J. Lundberg) Kemi für Gymnasiet, 1953; (with S. Andersson) Kemi für Gymnasiet, 1966; Kemi, 1967; also articles. Research on complex species in solutions, equilibrium constants, calorimetric determinations of enthalpy, kinetics. Home: 3 Stationsv., Akarp, Sweden. Office: Kemicentrun, Lund, Sweden.*

LE DENTU, Jean-François-Auguste, surgeon; b. La Basse-Terre, Guadeloupe, 1841; certified surgeon, 1872; asst. to chmn. clin. surgery l'Hôtel Dieu, Paris, France, 1876-77; prof. clin. surgery l'Hopital Necker; mem. Acad. Medicine. Author: Des anomalies du testicule, 1869; Traité de chirurgie clinique et opératoire, 10 vols., 1901. Died Paris, 1926.

LEDERBERG, Joshua, Am. geneticist; b. Montclair, N.J., May 23, 1925; s. Zwi and Esther (Gold) L.; B.A. with honours Columbia, 1944, postgrad. med. sch., 1944-46; Ph.D., Yale, 1948, Sc.D., 1960; m. Marguerite S. Kirsch, Apr. 5, 1968. Asst. prof. genetics U. Wis., 1947-50, asso. prof., 1950-54, prof. 1954-58, reorganized dept. med. genetics, 1957, chmn., 1958-59; prof. genetics and biology, exec. head dept. genetics Med. Sch., Stanford, 1959—; dir. Kennedy Labs. for Molecular Medicine, 1962—. Vis. prof. bacteriology U. Cal. at Berkeley, 1950; Fulbright vis. prof. bacteriology Melbourne (Australia) U., 1957. Recipient (with G. W. Beadle, E. L. Tatum) Nobel prize in medicine and physiology, 1958. Mem. Nat. Acad. Scis. Discoveries concerning mechanism of genetic recombination and orgn. of genetic material of bacteria; pioneer in field of bacterial genetics. Office: Dept. of Genetics, Stanford Univ., School of Medicine, Palo Alto, Cal. 94304.*

LEDERER, Edgar, biochemist; b. Vienna, Austria, June 5, 1908; s. Alfred and Friederike (Przibram) L.; Ph.D., U. Vienna, 1930; D.es Sc., Faculty of Sci., Paris, 1938; m. Hélène Fréchet, June 16, 1932; children—Marianne, Sylvie (Mrs. Jean-Marie Lehn), Florence, Pascal, Denis, Aline, Pierre. Research asst. Kaiser Wilhelm Inst., Heidelberg, 1930-33, Paris, 1933-35; dir. research Vitamin Inst., Leningrad, 1935-37; research fellow Centre National de la Recherche Scientifique, Paris, 1938, Lyon, 1940-47, Paris, 1947-54, dir. Inst. Chemistry Natural Compounds, Gif sur Yvette, France, 1960—; prof. biochemistry Faculty of Sci., Paris, 1954—. Decorated chevalier Legion d'Honneur; recipient Fritzsche award Am. Chem. Soc., A.W. von Hofmann award Chemische Gesellschaft, 2 awards Acad. Scis., Paris; Paul Karrer lectr., Gold medal, Zurich. Hon. mem. Société Chimique de Belgique, Chem. Soc. London, Am. Soc. Biol. Chemists, Biochem. Soc. Finland; mem. Royal Irish Acad., Leopoldina (Germany). Author: Les caroténoides des Plantes, 1935; Les Caroténoids des Animaux, 1935; Chromatographie, 1949; (with M. Lederer) Elsevier, 1953, 2d edit., 1957; (with others) Chromatographie, 1960. Research and numerous publs. on introduction of chromatography as an analytical and preparative method in organic chemistry, isolation and structural determination of numerous natural compounds, chem. and biochem. studies of biologically active complex compounds produced by tubercle bacillus and other microorganisms. Home: 9, Boulevard Colbert, Sceaux, Seine 92. Office: Institut Chimie, Centre National de la Recherche Scientifique, Gif sur Yvette, Essonne, France.*

LEDERER, Emil, German economist, sociologist; b. Pilsen, Bohemia, July 22, 1882; studied at several Austrian and German univs; doctor's degree in jurisprudence U. Vienna; m. Emy Seidler. Sec. trade unions, Lower Austria, 1907-10; named lectr. U. Heidelberg (Germany), 1912, prof., 1920; named dir. econ. div. Austrian Fed. Com. for Socialization, 1919; vis. prof. U. Tokyo, 1925-27; dismissed by Nazis and went to Japan, 1933; from there came to N.Y.C. where he taught at New Sch. for Social Research. Author numerous publs. including: Soziologie der Revolution, 1919; State of the Masses, 1940; Grundzüge der ökonomischen Theorie. Editor: Das Kartellproblem, 3 vols., 1930-32. Studied problems of social groups from a Marxist point of view; sociol. interpretation of econ. facts; believed classless soc. was a threat, co-existence of various groups makes soc.; rejected purely econ. interpretation of history. Died N.Y., May 29, 1939.

LEDERER, Francis Loeffler, Am. surgeon; b. Chgo., Sept. 18, 1898; s. Jacob and Frances (Loeffler) L.; S.B., U. Chgo., 1918; M.D., Rush Med. Coll., 1921; postgrad. U. Berlin and U. Vienna, 1925; m. Anne Pollock, Mar. 4, 1925; 1 son, Francis II. m. 2d, Adrienne Burman, June 17, 1959. Practiced in Chicago, 1921—; prof., head otolaryngology, emeritus U. Ill. Coll. Medicine; sr. staff Michael Reese, Grant, also Columbus hosps.; chief otolaryngol. service, Research and Ednl. Hosps. dir. otolaryngol. service, Ill. Eye and Ear Infirmary, Hines Hosp. (sr. cons.); cons. Presbyn.-St. Luke's Hosp.; nat. cons. surgeon gen. USAF. Cons. otolaryngol. U. S. Naval Hosp. Great Lakes, Ill., area cons. in audiology, V. A. Fellow A.C.S.; Internat. Coll. of Surgeons (pres.), mem. A.M.A., Inst. Med. laryngol. and otol. soc., Am. Otol. Soc., Am. Acad. Ophthalmology and Otolaryngology (pres.), Am. Otol., Rhinol. and Laryngol. Soc., Am. Laryngological Assn., Am. Bronchoesophagol Assn., Am. Coll. Chest Phys., Am. Assn. Mil. Surgeons, Sigma Xi, Alpha Omega Alpha, Phi Delta Epsilon, others. Chief asso. editor (otolaryngology) of Cyclopedia of Medicine, Surgery, Specialties. Author: Diseases of the Ear, Nose and Throat, 6th edit., 1953. Contbr. to jours. and books on subjects pertaining to ear, nose and throat specialty. Home: 199 E. Lake Shore Dr., Chgo. 60611. Office: 307 N. Michigan Av., Chgo. 60601.*

LEDERMAN, Leon Max, Am. physicist; b. N.Y.C., July 15, 1922; s. Morris and Minna (Rosenberg) L.; B.S., Coll. City N.Y., 1943; M.A., Columbia, 1948, Ph.D., 1951; m. Florence Gordon, Sept. 19, 1943; children—Rena S., Jesse A., Heidi R. Faculty, Columbia, N.Y.C., 1951—, prof. physics, 1958—, dir. Nevis Labs.; guest scientist Brookhaven Nat. Labs., Upton, L.I., N.Y., 1955—. Cons. Inst. for Def. Analysis, Washington, 1959—. Recipient National Medal of Science, 1965. Mem. Am. Italian phys. socs., Nat. Acad. Scis. Research, numerous publs. on elementary particle physics in gen.; observed parity violation in meson decays; discovered two neutrinos, antideuterons, long lived neutral k-meson. Home: 34 Overlook Rd., Dobbs Ferry, N.Y. 10522. Office: Dept. Physics, Columbia U., N.Y.C. 10027.*

LEDERMÜLLER, Martin Frobenius, German naturalist; b. Nuremberg, Germany, 1719. Author: Mikroskopische Gemüths und Augenergätzen, 3 vols., 1760-64. Known for researches with the microscope. Died 1769.

LEDINGHAM, Sir John C. G., bacteriologist; b. Boyndie, Banffshire, 1875; s. James and Isabella (Gardiner) L.; M.A., U. Aberdeen, 1895, B.Sc., 1900, M.B., 1902, D.Sc., LL.D., 1935; Sc.D., Dublin, Leeds; Fife Jamieson and Struthers Gold medallist in anatomy; Anderson scholar, 1902-04; postgrad. U. Leipzig, 1902-03, London Hosp., 1904-05; m. Barbara Fowler, 1913; 1 son, 1 dau. Asst. bacteriologist Lister Inst., London, 1905-08, chief bacteriologist, 1908-30, dir., 1931-43; prof. bacteriology U. London, emeritus from 1943; in charge bacteriol. dept. King George Hosp., Waterloo, 1915. Mem. med. adv. com. in Mediterranean; cons. bacteriologist, Mesopotamia, 1917; Harben lectr., London, 1924; Herter lectr., Balt., 1934; mem. Med. Research Council, 1934-38; pres. 2d Internat. Congress for Microbiology, London, 1936, hon. pres. 3d Congress, N.Y., 1939. Fellow Royal Soc., 1921, Royal Coll. Physicians. Author: (with J. A. Arkwright) Carrier Problem in Infectious Diseases, 1912; also numerous articles. Asso. editor System of Bacteriology, 1929-31. Studied role of carrier in spread of typhoid fever; proved existence of disease Kala-azar in China. Died Oct. 4, 1944.

LEDOUBLE, Anatole, French anatomist; b. Rocroy, France, 1848; interned in Paris hosps.; prof. anatomy l'École de medecine, Tours, France; mem. French Acad. Medicine. Author: Traité des variations du système musculaire de l'homme et de leur signification au point de vue de l'anthropologie zoologique, 1897; Rakilais anatomiste et physiologiste, 1899; Traité du variations des os du crane, 1903; Traité des variations des os de la face; Bossuel anatomiste et physiologiste. Research on morphological variation of human body; publs. on variation in muscular system in bones of head and face; studied laws of embryology. Died Tours, 1913.

LEDOUX, Lucien Gaspard Henri, Belgian biochemist; b. Brussels, Belgium, Oct. 22, 1928; s. Henri and Suzanne (Vlamynck) L.; Licence en Sciences chimiques with grande distinction, U. Libre Bruxelles (Belgium), 1950, Ph.D. with 1a plus grande distinction, 1954, D.Sc., 1966; m. Gilberte H. S. Vanderperre, Aug. 19, 1950; children—Yves, Michel, Claudine, Denis. Staff, Fonds national de la Recherche scientifique, U. Brussels, 1952-58; biochemist Centre d'Etudes de l'Energie nucléaire, Mol, Belgium, 1958—; expert Internat. Atomic Energy Agy., 1964. Recipient Prix Staes-Apring, Acad. des Sci. Mem. Sociétés de Biochimie et de Radiobiologie. Contbg. author: Nucleic Acid Researcher, 1965; also numerous articles. Research on permeability of living cells to proteins and nucleic acids. Home: 2 Donk, Oud-Turnhout, Belgium. Office: Steenweg Retie, Mol, Belgium.*

LEDOUX, Paul, Belgian astrophysicist; b. Forrières, Belgium, Aug. 8, 1914; s. Justin and Ida (Delperdange) L.; Licencié en Sciences physiques, U. Liege (Belgium), 1937, Doctorate en Sciences, 1946, Agrégé de l'Enseignement Supérieur, 1949; m. Aline Michaux, Feb. 14, 1939; 1 dau., Jacqueline (Mrs. Daniel Grossman). Belgian Govt. fellow Inst. for Theoretical Astrophysics, Oslo, Norway, also Stockholm (Sweden) Obs., 1939-40; Belgian Am. Found. fellow Yerkes Obs., Williams Bay, Wis., 1940-41, 46-47; with U. Liege, 1947—, prof., 1959—; vis. prof. U. Cal. at Berkeley, 1963. Recipient Prix Francoui, 1964. Mem. Royal Belgian Acad. Contbr. articles to jours., chpts. to books. Research in stellar structure, stellar stability, variable stars, gravitational stability and formation of stars; introduced semi-convective zones; discovered upper limit to mass stable stars depending on nuclear reactions present; tentative interpretation of beta Canis Majoris variable stars in terms of splitting of periods due to rotation. Home: 55 rue de la Faille, Liege, Belgium. Office: Institut d'Astrophysique, Cointe-Sclessin, Belgium.*

LEDRAN, Henri François, French surgeon; b. Paris, 1685; s. Henri L. Fellow Royal Soc., 1744. Author: Treatise on Lithotomy, 1730; Observations on Surgery, 1731; Practical Reflections on Gunshot Wounds, 1737. Expert on lateral method of lithotomy. Died 1770.

LEDUC, Gaston Gabriel, French economist; b. Hérisson, France, July 27, 1904; s. Gustave and Marie (Copéré) L.; Agrégé des sciences economiques, U. Paris, 1930; student Faculté Aix-en-Province, France; m. Lise Croquin, Dec. 24, 1930; children—Tony, Francois, Xavier, Edith, Sabine. Prof. econs. U. Caen (France), 1930-36, U. Rio de Janeiro (Brazil), 1936-38, U. Cairo (Egypt), 1938-41, U. Beirut (Lebanon), 1942-48; prof. econs. U. Paris (France), 1946—. Econ. and financial adviser Free French Delegation, Middle E., 1941-45. Decorated Legion de Honneur. Mem. Société economie politique (pres., gov.). Author: Theorie des prix de monopoli economie ou development; also numerous articles. Research on econ. theory and econs. of devel. especially applications to African economics. Home: 19 Bel-Air Presles 95, France. Office: Faculty Law and Econs., Paris, France.*

LEE, Benjamin, Am. surgeon; b. Norwich, Conn., Sept. 26, 1833; s. Alfred and Julia (White) L.; A.B., U. Pa., 1852; M.D., N.Y. Med. Coll., 1856; Ph.D., U. Pa., 1876; studied in Europe; m. Emma Hale White, Apr. 5, 1859. Prof. orthopedics Phila. Polyclinic, 1895-96; health officer, Phila., 1898-99; sanitarian Pa. State Bd. of Agr.; sec. Pa. Bd. Health, 1885-1905; sec. Pa. Quarantine Bd., 1893-1905; asst. to commr. of health of Pa., 1905——. Pres. Am. Acad. Medicine, 1884, Am. Orthopedic Assn., 1891-92, Conf. State and Provincial Bds. Health of N. America, 1898, Am. Pub. Health Assn., 1898; treas. Pa. Med. Soc. Author: Correct Principles of Treatment for Angular Curvature of the Spine, 1872; Tracts on Massage (translated from the German), 1885. Editor Am. Med. Monthly. Died July 11, 1913.

LEE, Charles Alfred, Am. physician; b. Salisbury, Conn., Mar. 3, 1801; s. Samuel and Elizabeth (Brown) L.; grad. Williams Coll., 1822; M.D., Berkshire Med. Inst., 1826; m. Hester Mildeberger, 1828; 9 children. Practiced medicine, N.Y.C., 1827; a founder No. Dispensary, N.Y.C., operator, 1827-32; mem. staff Greenwich Cholera Hosp., 1832-circa 1849; attending physician N.Y. Orphan Asylum, 1832-49; pathology Geneva Med. Coll., 1844-47; lectr. medicine Starling Med. Coll., O., 1847; prof. medicine U. Buffalo (N.Y.), 1848-60; lived, practiced medicine, Peekskill, N.Y., from 1850; made study European hosps. for U. S. Govt., 1862-63; sanitation adviser to U. S. Army, 1864-65. Author: (papers) Catalogue of Medicinal Plants in N.Y., 1848, Remarks on Wines and Alcohol, 1871. Editor: Conspectus of Pharmacopeoias . . . (A. T. Thomson), 1843; Principles of Forensic Medicine (W. A. Guy), 1845. A founder N.Y. Jour. Medicine, 1843, editor, 1846-53. Died Peekskill, Feb. 14, 1872.

LEE, Cheng-Chun, pharmacologist; b. Chiangtu, China, May 24, 1922; s. Jui-Sheng and Szu (Cheng) L.; B.S., Nat. Central U., Nanking, China, 1945, M.S., 1948; M.S., Mich. State U., 1950, Ph.D., 1952; m. Janice Yu-Che Wong, Feb. 9, 1958; children—James P., Raymond W. Came to U. S., 1948, naturalized, 1961. Pharmacologist, Lilly Research Lab., Indpls., 1952-63; sr. pharmacologist Midwest Research Inst., Kansas City, Mo., 1963-66, prin. pharmacologist, 1966-67, head pharmacology and toxicology, 1967——; lectr. U. Mo., Kansas City, 1964-66; asst. prof. U. Kan. Med. Center, Kansas City, 1966——. Mem. Am. Soc. Pharmacology and Exptl. Therapeutics, Soc. Toxicology, N.Y. Acad. Scis., Am., Chinese physiol. socs., Chinese Soc. Vet. Medicine. Research, numerous publs. on gen. pharmacology and toxicology of therapeutic agts., drug absorption, distbn., excretion and metabolism, antimicrobial agts. and chemotherapy, cholic acid and cholesterol metabolism, renal pharmacology, microsomal drug metabolizing enzymes. Home: 1005 W. 85th Terrace, Kansas City 64114. Office: 425 Volker Blvd., Kansas City, Mo. 64110.*

LEE, Choi Chuck, Canadian chemist; b. Vancouver, B.C., Can., Apr. 27, 1924; s. Joseph and Ho (See) L.; B.R. with great distinction in Chem.Engring., U. Sask. (Can.), 1947, M.Sc. in Chemistry, 1949; Sc.D., Mass. Inst. Tech., 1952; m. Amy Fong-Sam, Aug. 20, 1953; children—Kenneth, Karen, Kristina, Kathryn. Research asso. U. Sask., Saskatoon, 1952-54, faculty, 1955——, prof. chemistry, 1964——; asst. prof. U. B.C., Vancouver, 1954-55. Fellow Chem. Inst. Can., Chem. Soc. (London, Eng.); mem. Am. Chem. Soc., Am. Assn. Cereal Chemists, A.A.A.S. Research, numerous publs. on phys. organic chemistry, applications carbon-14 and tritium as tracers in studies rearrangements carbonium ions, deuterium kinetic isotope effects in solvolytic reactions, rearrangements induced by gamma-irradiation, cereal chemistry using radioactive isotopes as tracers. Home: 504 Leslie Av., Saskatoon, Sask., Can.*

LEE, David Bryon, Am. san. engr.; b. Douglasville, Ga., Sept. 23, 1907; s. W. A. and Mollie (Smith) L.; B.S. in Mech. Engring., U. Fla., 1932; M.S. in San. Engring., Harvard, 1937; m. Billie Rawls, July 28, 1939; children—David B., Susan Rawls. With Fla. State Bd. Health, Jacksonville, 1932-68, dist. engr., 1935-37, malaria control engr., 1937-41, chief san. engr., dir. Bur. San. Engring., 1942-68; with David B. Smith Engineers, Inc., since 1968——; with Inst. Inter-Am. Affairs, 1942-45; spl. cons. 1948——. Lectr. san. engring. U. Fla., Gainesville, 1959——; adviser U. S. WHO, 1949; mem. panel Pres.' Commn. Health Needs of Nation, 1952; mem. san. engring. com. NRC, 1952-56. Recipient Centennial award U. Fla., 1953; named one of Top Pub. Works Men of Year, Am. Pub. Works Assn., 1961. Fellow Fla. Engring. Soc. (Profl. award for exceptional service to engring., 1949, pres. 1952-53), Am. Pub. Health Assn.; mem. Am. Water Works Assn. (Fuller award 1954, chmn. Fla. sect. 1958-59), Am. Soc. C.E., Fla. Pub. Health Assn., Fla. Pollution Control Assn., Water Pollution Control Fedn. (Kenneth Allen award 1948, nat. dir.-at-large 1952-53, pres. 1954-55), Am. Acad. San. Engrs., Nat. Assn. Sanitation (hon.), Sigma Tau, numerous others. Home: 1321 Pinetree Rd., Jacksonville 32207. Office: 1321 Pinetree Rd., Jacksonville, Fla. 32207; also 2512 S.W. 34th St., Gainesville, Fla. 32607.*

LEE, Douglas Harry Kedgwin, physiologist; b. Bristol, Eng., Feb. 22, 1905; s. Harry Alfred and Ada (Hicks) L.; B.Sc., U. Queensland, 1925, M.Sc., 1927; M.B., B.S., U. Sydney, 1929, M.D., 1940, D.T.M., 1933; m. Dorothy Louise Yingling, Sept. 11, 1962; children—Shirley Dugmore, Roderick Kedgwin. Came to U. S., 1947, naturalized, 1953. Prof. physiology King Edward VII Coll. Medicine, Singapore, 1935-36, U. Queensland, Brisbane, Australia, 1936-48; prof. physiol. climatology Johns Hopkins, Balt., 1948-55; asso. sci. dir. for research Q.M. Corps., 1955-60; chief Occupational Health Labs., USPHS, Cin., 1960-66; asso. sci. dir. for planning Nat. Environmental Health Service Center, Research Triangle Park, N.C., 1966——. Cons. FAO, 1947-55. Fellow Royal Australian Coll. Physicians; mem. Am. Physiol. Soc. (editorial bd.), Internat. Soc. Biometeorology (pres.), A.A.A.S., N.Y. Acad. Scis., Sigma Xi. Author: Physiology of Tissues and Organs, 1947; Physiological Objectives in Hot Weather Housing, 1953; Manual of Field Studies in Domestic Animals, 1953; Climate and Economic Development in Tropics, 1957. Research, numerous publs. on effect of hot climates on man and animals and application to housing, clothing, industry and econ. devel. Home: 3204 Arrowwood Dr., Raleigh, N.C. 27604. P.O. Box 12233, Research Triangle Park, N.C. 27709.*

LEE, Everett S., Am. demographer-sociologist; b. Raines, S.C., Dec. 31, 1919; s. Alexander S. and Mattie (Scott) L.; A.B., U. Pa., 1949, M.A., 1950, Ph.D., 1952; m. Anne Schacht, Aug. 4, 1950; children—Dorothy, Deborah, John, Sarah. Faculty, U. Pa., Phila., 1952-66, asso. prof., 1956-66; prof., chmn. dept. sociology and anthropology U. Mass., Amherst, 1966. Mem. Nat. Com. on Vital and Health Statistics, 1965——. Harrison fellow U. Pa., 1948-50; Social Sci. Research Council fellow, 1950-51. Fellow A.A.-A.S. (mem. council 1963-65); mem. Am. Statis. Assn. (pres. Phila. chpt. 1964-66), Fellows in Am. Studies (sec. 1964-66), Population Assn. Am. (1st v.p. 1967-68), Internat. Union for Sci. Study Population, Am. Sociol. Assn., Am. Eugenics Assn. Author: (with Richard Easlerlin, Ann Miller) Population Redistribution and Economic Growth in the United States, 1870-1950, 1957; (with Benjamin Malzberg) Migration and Mental Disease, 1956; also articles. Research on migration, mental illness, Negro intelligence.*

LEE, James, Brit. botanist; b. Selkirk, Eng., 1715; gardener Sion House, nr. Brentford, Eng.; became nurseryman Vineyard, Hammersmith, Eng., 1760. Translated part of Linnaeus' works into English under the title of: Introduction to the Science of Botany, 1760. Introduced cultivation of fuchisia in Eng., also many exotic plants. Died July 1795.

LEE, James Paris, inventor; b. Hawick, Scotland, Aug. 9, 1831; s. George and Margaret (Paris) L.; ed. Galt, Ont., Can., Dumfries; m. Caroline Chrysler, 1852 (died 1888). Inventor Lee-Metford, Lee-Enfield, Lee-Straight Pull, other mag. rifles; manufactured his rifles in Milw., then with Remingtons in Ilion, N.Y., later with cos. in Hartford, Conn., London, Eng. Died Feb. 24, 1904.

LEE, Joseph C(hing-yuen), neuroanatomist, exptl. neuropathologist; b. Anchi, Fukien, China, Feb. 25, 1922; s. Fat Siew and Chi (Hsu) Lee; B.Sc., Lingnan U., China, 1947; M.Sc., U. Sask., Can., 1958, Ph.D., 1961, M.D., 1962; m. Katherine Kam-yuen So, July 30, 1954; 1 son, Paul Ying. Faculty, Lingnan U., Canton, China, 1947-54, U. Hong Kong, 1955-57, U. Sask., Saskatoon, 1958-63; faculty State U. N.Y., Buffalo, 1963——, asso. prof. anatomy, 1965——. Recipient Grand award Lingnan U., 1947; Lederle Med. Faculty award U. Sask., 1961-63. Nat. Found. for Neuromuscular Diseases N.Y. research grantee, 1962-65; Am. Cancer Soc. grantee, 1965——. Mem. Am., Canadian assns. anatomists, Canadian Fedn. Biol. Socs., Am. Soc. Cell Biology, Electron Microscopy Soc. Am., Am. Assn. Neuropathologists, A.A.-A.S., Royal Soc. Medicine, London, Cajal Club Neuroanatomists U. S. A. Author: Human Anatomy, 1951; (with Louis Bakay) Cerebral Edema, 1965. Studies, publs. on vascular permeability of brain, ultrastructural changes in nerves and muscles, cerebral edema, brain tumors. Home: 26 Sandhurst Lane, Buffalo 14221.*

LEE, Leroy William, Am. urol. surgeon; b. Omaha, July 29, 1914; s. Herman C. and Charlotte (Du Bois) L.; student U. Chgo., 1932-34; B.Sc., U. Neb., 1936, M.Sc. in Pathology, 1938, M.D., 1939; m. Dorothy M. Reed, Dec. 21, 1937; children—Marcia E. (Mrs. Clifford Hollestelle), Marie A. (Mrs. William R. Johnson), Harry R., Roger E. Practice medicine, specializing in urology, Omaha, 1945——; faculty U. Neb., Omaha, 1945——, prof., chmn. dept. urology, 1952——; sr. urologist U. Hosp., Omaha, 1952——; sr. urologist Clarkson Hosp., Omaha, 1952——, pres. staff, 1960. Cons. in urology SAC, Omaha, 1948——, VA Hosp., Omaha, 1952——, Douglas County Hosp., Omaha, 1960——. Mem. A.M.A., Neb. State Med. Assn. (councillor 1964——), Alumni Assn. U. Neb. (past pres.), Am. Urol. Assn. (exec. com. S. Central sect., 1964——), Sigma Xi, Alpha Omega Alpha, Phi Chi. Research on toxicology of nitrogen, rare earth oxides, and electric arch; devel. many new techniques in prostatic surgery, instruments, therapeutic agents in urology. Home: 2517 N. 55th St., Omaha 68104. Office: Doctors Bldg., Omaha 68131.*

LEE, Lieng-Huang, chemist; b. Fukien, China, Nov. 6, 1924; s. Bin and Da-Mei (Chang) L.; B.S., Nat. U. Amoy, China, 1947; M.S., Case Inst. Tech., 1954, Ph.D., 1955; m. Chiu-bin Wu, Feb. 8, 1949; children—Muriel Ei-hui, Daniel Chia-sen, Robert Minsen, Grace Mei-hui. Came to U. S., 1952, naturalized, 1964. Mem. staff Nantou Sugar Factory, Nantou, Taiwan, Chia-Yee Solvent Works, Chia-Yee, Taiwan, 1948-51, Rain Stimulation Research Inst., Taipei, Taiwan, 1951-52; research asso. Case Inst. Tech., Cleve., 1955-56; mem. staff Dow Chem. Co., Midland, Mich., 1958——, sr. research chemist, 1963——. Cons., lectr., vis. prof. univs. and industry, Taiwan, Formosa, 1956-58. Mem. Sci. Research Soc. Am. (past pres. Midland br.), Am. Chem. Soc., Sigma Xi. Research, publs. on discovery of the wettability and glass temperature relationship for polymers, and degradation theories for polycarbonate and epoxy resins; invented new method for rain stimulation, new polymers and monomers. Home: 4407 Huron Dr. Office: Plastics Lab., 1702 Bldg., Dow Chem. Co., Midland, Mich. 48640.*

LEE, Milton Oliver, Am. biologist; b. Conneaut, O., Sept. 8, 1901; s. Malcolm Anthony and Jennie Viola (Hazeltine) L.; A.B. Ohio State U., 1922. A.M., 1923, Ph.D., 1926; grad. work Marine Biol. Lab., 1923, 27; m. Helen Isabel Mayhew, June 21, 1923. Instr. Ohio State U., 1923, Ia. State Coll., 1925; research fellow in physiol., asso. and physiol. Meml. Found. Neuro-Endocrine Research, Harvard, 1927-42; chief labs. Worcester State Hosp., 1946-47; exec. sec., div. biology and agr., NRC, 1948-52, mem.-at-large, 1952——, exec. sec. Am. Inst. Biol. Scis., 1948-50; exec. dir. Fedn. Am. Socs. Explt. Biology, 1947-49, fedn. sec., 1949-60, exec. officer, 1963——; v.p. Am. Documentation Inst., 1952, pres., 1954; Treas., Inter. Fedn. of Documentation, 1952-62, mem. bur., 1951——; mem. nat., exec. coms.; co-chmn. orgn. com. Internat. Conf. Sci. Information, 1958; gen. sec. 5th Internat. Congress Nutrition, 1960; mem. panel sci. information, President's Sci. Adv. Com., 1961——. Sec.-treas. Sci. Manpower Commn., 1953. Mem. Am. Physiol. Soc. (exec. sec. 1947-56), Am. Inst. of Nutrition, Soc. for Exptl. Biology and Med., Assn. for Study of Internal Secretions (v.p., 1929), Gamma Alpha, Phi Beta Kappa, Sigma Xi. Mng. editor Endocrinology, 1935-42, Jour. Clin. Endocrinology, 1938-42, Am. Jour. Physiol., Physiol. Reviews, Jour. Applied Physiology, Jour. Neurophysiology Fedn. Proceedings. Home: Sanbornton, N.H.; also 4700 Locust Hill Circle, Bethesda, Md. 20014, Office: 9650 Wisconsin Av., Washington 14.*

LEE, Min Jai, Korean botanist; b. Hoeryung, Korea, Feb. 22, 1917; s. Yong-Suck and Sun-Chung (Yoon) L.; B.Pharm., Seoul (Korea) Pharm. Coll., 1939; B.S., Hokkaido Imperial U., Japan, 1942, D.Sc. 1961; m. He-Kyung Lee, Dec. 8, 1948; children—In-Sook, Chong-Bum, Chong-Ho, In-Kyung. Research mem. Inst. Sci. Research, Manchukuo, Manchuria, 1943-45; prof. Seoul Nat. U., 1947——, dir. Inst. Marine Biology, 1964——. Vice minister edn. Republic of Korea, 1961-62; vice chmn. com. of cultural property conservation Republic of Korea, 1961——, commr. Atomic Energy Com., 1964——; mem. Korean Nat. Com. for UNESCO, 1963-64. Recipient prize Korean Pharm. Soc., 1960. Mem. Republic of Korea Nat. Acad. Sci. (prize 1965), Am., Japanese, Korean (past pres.) bot. socs., Am., Japanese socs. plant physiologists, Korean Microbiol. Soc. Author: Pharmaceutical Botany, 1958; Plant Physiology, 1959; (with Soon-Woo Hong) General Botany, 1962; also articles. Research on effect of 2, 4-D against plant, cold resistance of plant, antibiotic substance, physiol. effects of radiation, physiol. study of Euglena, studies of chem. components and relationship to phylogeny of Algae, microflora of Han River, agarophyte from Korean Coast, studies of plant tumor. Home: 24 4-Ka, Myungryun-dong, Chongro-Ku, Seoul, Korea.*

LEE, Robert, Brit. gynecologist, obstetrician; b. Melrose, Scotland, 1793; s. John Lee; M.D., U. Edinburgh, (Scotland), 1814; student medicine, Paris, 1821-22; began practice medicine, London, Eng., 1817; physician to Prince Woronzow, gov.-gen. Crimea, 1824-26; became physician Brit. Lying-in-Hosp., 1827; named lectr. midwifery in Webb, 1829; became Regius prof. midwifery, 1834; lectr. midwifery and diseases of women St. Georges Hosp., London, 1835-66; Harveian lectr. Royal Coll. Physicians, 1864, ret., 1875. Fellow Royal Soc., 1830; mem. Edinburgh Coll. Surgeons, Royal Coll. Physicians. Publs. include: Clinical Midwifery, 1842; History of the Discoveries of the Circulation of the Blood, of the Gangalia, and Nerves and of the Action of the Heart, 1865; A Treatise on Hysteria, 1871. Described cervical nerve ganglion of uterus (Lee's ganglion), 1841; research on pathology of women, puerperal fever; prolonger dissections of ganglia and nerves of uterus. Died Surbiton Hill, Eng., Feb. 6, 1877.

LEE, Thomas George, Am. histologist; b. Jacksonville, N.Y., Nov. 27, 1860; s. Horace Cooper and Sarah Lavinia (Shaw) L.; B.S., M.D., U. Pa., 1886; postgrad. U. Würzburg, 1887; B.S., Harvard, 1892; U. Munich, 1892; m. Emma Louise Shaw, Dec. 21, 1887. Asst. in histology and embryology U. Pa., 1884-86; lectr., and dir. of lab. Yale, 1886-91; asst. in histology Radcliffe, 1891-92; prof. histology and embryology, dir. lab., U. Minn., 1892-1909, prof. anatomy, dir. Inst. Anatomy, 1909-13, prof. comparative anatomy, 1913-29, prof. emeritus, 1929. Asso. editor Anat. Record. Died Sept. 1, 1932.

LEE, Tsung-Dao, physicist; b. Shanghai, China, Nov. 25, 1926; s. Tsing-Kong Lee and Ming-Chang Chang; student Nat. Chekiang U., Kweichow, China, 1943-44, Nat. Southwest Associated U., Kunming, China, 1945-46; Ph.D., U. Chgo., 1950; m. Jeanette Chin, June 3, 1950; children—James, Stephen. Research asso. in astronomy U. Chgo., 1950; research asso. and lectr. in physics U. Cal., 1950-51; mem. Inst. Advanced Study, Princeton, N.J., 1951-53; asst. prof. physics Columbia, 1953-55, asso. prof., 1955-56, prof., 1956-60, 63—, Enrico Fermi prof. physics, 1964, adj. prof., 1960-62, prof. physics Inst. Advanced Study, Princeton, N.J., 1960-63; Loeb lectr. Harvard, 1957, 64. Recipient Albert Einstein award in sci. Yeshiva U., 1957, (with Chen Ning Yang) Nobel prize in physics, 1957. Mem. Nat. Acad. Sci. Important work with Yang on parity nonconservation (refuted law of parity, which stated that for each sub-atomic particle there exists somewhere in the universe another particle that is identical but of mirror-image configuration); research on statistical mechanics, astrophysics, hydrodynamics, nuclear and subnuclear physics, and field theory.*

LEE, William, inventor; b. circa 1589; student Christ's Coll., B.A., St. John's Coll., Cambridge (Eng.) U., 1583; M.A., 1586; Became curate, Calverton, Eng., 1589; settled in Rouen, France, at the request of Henry IV. Invented knitting machine (or stocking frame) (basis of all other machines in knitting and lace making), 1589. Died Paris, 1610.

LEE, Willis Thomas, Am. geologist; b. Brooklyn, Pa., Dec. 24, 1864; s. John C. and Louesa J. (Garland) L.; Ph.B., Wesleyan U., Conn., 1894, M.S., 1898; fellow in geology, U. Chgo., 1898-1900, Johns Hopkins, 1902-03; Ph.D., Johns Hopkins, 1912; m. Mary Ingham, 1900; children—Elizabeth Louesa, Dana Willis. Instr. science R.I. Coll. Agr. and Mechanic Arts, Kingston, R.I., 1894-95; prof. geology and biology Denver U., 1895-98; prin. high sch., Trinidad, Colo., 1900-02; with U. S. Geol. Survey, 1903—; head dept. geology U. Okla., 1919; lectr. geology Yale, 1919. Died June 16, 1926.

LEECH, Paul Nicholas, Am. chemist; b. Oxford, O., Aug. 12, 1889; s. William David and Ann Cora (Druley) L.; A.B., Miami U., 1910; Sc.M., U. Chgo., 1911, Ph.D., 1913; Pharm.M. (hon.), Phila. Coll. Pharmacy and Science, 1937; m. Esther O. Birch, Mar. 10, 1916; children—Esther Doris, Paul Nicholas. Research asst., U. Chgo., 1911-13; chemist A.M.A., 1913-24; dir. Chem. Lab. same, 1924—, also dir. sci. exhibit, 1922-30, editor Ann. Repts.; sec. Council on Pharmacy and Chemistry, 1932—; dir. div. of foods, drugs and phys. therapy, 1936—. Mem. Am. Chem. Soc. Died Jan. 14, 1941.

LEEDS, John, Am. mathematician, astronomer; b. Bay Hundred, Talbot County, Md., May 18, 1705; s. Edward and Ruth (Ball) L.; m. Rachel Harrison, Feb. 14, 1726; 3 children. Commr. justice of peace Talbot County, 1734; clk. Talbot County Ct., 1738-77; regular mem. commn. from Md. to mark off long-disputed Md.-Pa. boundary line, 1762; observed transit of Venus, obtained results published in Royal Soc. London's Philos. Trans., 1770; treas. Eastern Shore dist., Md., 1766; justice Provincial Ct., 1766; naval officer Port of Pocomoke, 1766; surveyor gen. Md., 1766-circa 1775, after 1783-90. Died Wade's Point, Md., Mar. 1790.

LEEDS, Sanford Edgar, Am. physician; b. San Francisco, Nov. 14, 1909; s. Louis and Amelia (Snoek) L.; A.B., U. Cal. at Berkeley, 1931, M.D., 1936; m. Syra Florence Nahman, Apr. 9, 1941; 1 son, Andrew Loren. Practice surgery, San Francisco, 1946—; asso. chief surgery Mt. Zion. Hosp., San Francisco, 1948—; asst. clin. prof. surgery U. Cal. Med. Sch., San Francisco, 1952—. USPHS, NIH grantee, 1948—. Mem. San Francisco County Med. Soc., A.M.A., Am. Thoracic Surg. Soc., A.C.S., Am. Heart Soc., Howard C. Naffziger, San Francisco, Pacific Coast surg. socs., Western Soc. for Clin. Research. Research, publs. on exptl. renal hypertension, transplantation organs, blood flow organs, exptl. patent ductus arteriosus, cardiac resuscitation, blood flow pump, lymph flow from heart and lungs, lymph flow in heart failure. Home: 3440 Washington St., San Francisco 94118. Office: 2211 Post St., San Francisco 94115.*

LEEDY, Daniel Loney, Am. wildlife biologist; b. Butler, O., Feb. 17, 1912; s. Charles Monroe and Bernice (Loney) L.; A.B. with honors in Geology, Miami U., Oxford, O., 1934, B.S. in Edn., 1935; M.Sc. in Zoology, Ohio State U., 1938, Ph.D., 1940; m. Barbara Emma Sturges, Nov. 25, 1945; children—Robert Raymond, Kathleen Eleanor. Instr. wildlife mgmt. Ohio State U., Columbus, 1940-42, leader Ohio Coop. Wildlife Research Unit, 1945-48; biologist in charge coop. wildlife research unit program Fish and Wildlife Service, U. S. Dept. Interior, Washington, 1949-57, chief br. wildlife research, 1957-63, chief div. research and edn. Bur. Outdoor Recreation, 1963-64, water resources research scientist Office Water Resources Research, 1965—. Mem. numerous coms. Dept. of Interior; mem. several coms. Nat. Acad. Sci.-NRC. Recipient Conservation award Am. Motors, 1958. Fellow A.A.A.S.; mem. Wildlife Soc. (pres.), Washington Biologists Field Club, numerous other profl. socs. Contbg. author: The Ring-necked Pheasant, 1945; Manual of Photographic Interpretation, 1960; Waterfowl Tomorrow, 1964. Contbr. articles to profl. publs. Home: 10707 Lockridge Dr., Silver Spring, Md. 20901. Office: Office Water Resources Research, U. S. Dept. of Interior, Washington 20240.*

LEEPER, Robert Ward, Am. psychologist; b. Braddock, Pa., Sept. 25, 1904; s. Charles Wilson and Carrie (Stevenson) L.; A.B., Allegheny Coll., 1925; M.A., Clark U., 1928, Ph.D., 1930; m. Dorothy Mildred Olson, Mar. 17, 1931; children—Alice (Mrs. Fred Attneave), Edward, David, Arthur. Instr. psychology Paine Coll., Augusta, Ga., 1926-27, U. Ark., Fayetteville, 1930-33, Cornell Coll., Mt. Vernon, Ia., 1934-37; faculty U. Ore., Eugene, 1937—, prof. psychology, 1949—, chmn. dept., 1954-63. NRC fellow, 1933-34; Guggenheim Found. fellow, 1948-49; Fulbright lectr. U. Aberdeen, 1955-56. Mem. Am., Western (pres. 1952) psychol. assns., Phi Beta Kappa, Sigma Xi. Author: (with Peter Madison) Toward Understanding Human Personalities, 1959; also articles, monographs, chpts. in books. Home: 2760 Agate St., Eugene, Ore. 97403.*

LEES, Charles Herbert, Brit. physicist; b. Glodwich, Oldham, July 28, 1864; s. John and Jane (Ogden) L.; ed. Owens Coll., Manchester, U. Strasbourg, City and Guilds Coll., London; D.Sc.; m. Evelyn May Savidge, 1902; 3 sons, 2 daus. Became sr. asst. lectr., demonstrator physics Owens Coll., 1891; named lectr. physics, asst. dir. phys. labs. U. Manchester, 1900; prof. physics U. London, also mem. delegacy for managing Goldsmiths' Coll.; mem. food investigation bd. Road Research Bd.; chmn. engring. com. Food Bd.; mem. safety, also air inventions com. Mines Research Bd. Fellow Royal Soc., 1906; hon. sec. Literary and Philos. Soc. Manchester, 1901-06; pres. Phys. Soc. London, 1918-20; v.p. Inst. Physics, 1921-23. Coauthor textbooks on practical physics; contbr. to Dictionary of Nat. Biography, other publs. Died Sept. 25, 1952.

LEES, David Bridge, Brit. physician; ed. Trinity Coll., Cambridge, Eng.; M.D.; cons. physician Hosp. for Sick Children, St. Mary's Hosp.; fellow Royal Coll. Physicians; mem. Royal Coll. Surgeons. Author: The Treatment of some Acute Visceral Inflammations, and other papers, 1905; The Physical Signs of Incipient Pulmonary Tuberculosis, and its Treatment by Continuous Antiseptic Inhalations, 1909. Died Aug. 16, 1915.

LEES, Edwin, English botanist; b. Worcester, Eng., 1800; ed., Birmingham, Eng.; began career as printer, stationer, Worcester; pub. Worcestershire Miscellany, 1829; founder Worcester Lit. and Sci. Inst., 1829. Fellow Linnean Soc., Geol. Soc. Author: Botany of the Malvern Hills, 1868; The Botanical Looker-out, 1842; Pictures of Nature, 1856; Botany of Worcestershire, 1867. One of 1st in Eng. to study the forms of brambles; Rubus leesii named in his honor. Died Oct. 21, 1887.

LEES, Frederick, English neurologist; b. Hemsworth, Yorkshire, Eng., Dec. 5, 1920; s. Frank and Mabel (Chappell) L.; M.P.S., U. Leeds, 1942; M.B.Ch.B., U. Sheffield, 1952, D.C.H., 1954; M.R.-C.P., London, 1954; m. Marie Constance Richardson, Sept. 18, 1948; children—Paul Tudor, Josephine Mary. House physician Royal Infirmary, Sheffield, 1952; house physician Children's Hosp., Sheffield, 1952-53, registrar, 1953-57; scholar French Inst. Hygiene, Paris Hosps., 1955; Churchill scholar, Copenhagen, 1956; sr. registrat neurology Manchester Royal Infirmary, 1957-60; sr. registrar neurology St. Bartholomew's Hosp., London, 1960-63; cons. neurologist Colchester, Chelmsford, South Essex hosps., 1963. Recipient Pleasance prize U. Sheffield, 1951, Walter S. Kay prize, 1950; West Riding Pactitioners prize, 1952. Fellow Royal Soc. Medicine (mem. council, neurology sect. 1964); mem. Assn. Brit. Neurologists, Brit. Med. Assn., Colchester, Chelmsford, Basildon med. socs. Author: Fundamentals of Current Medical Treatment, 1965. Publs. on devel. of soda water test for pyloric obstruction; gastric mucosal changes in anemia and after partial gastrectomy; studies of natural history of cervical spondylosis. Home: Canonium Cottage, Feering Hill, Kelvedon, Essex, Eng. Office: Creffield House, 2A Oxford Rd., Colchester, Essex, Eng.; also 138 Harley St., London, Eng.

LEES, George Martin, Brit. geologist; b. Ireland, Apr. 16, 1898; s. George Murray Lees; ed. St. Andrews Coll., Dublin, Royal Mil. Acad., Royal Sch. Mines, Vienna U.; Ph.D.; m. Hilda Frances Andrews, 1931; 1 son. Pol. officer Civil Adminstrn., Kurdistan, Iraq, 1919-20; mem. geol. staff Anglo-Iranian Oil Co. Ltd., 1921, chief geologist, 1931-53. Mem. Geol. Survey Bd., Colonial Geol. Survey Bd.; geol. adviser Burmah Oil Co. Ltd., Iraq Petroleum Co. Ltd. Recipient Sidney Powers gold medal Am. Assn. Petroleum Geologists, 1954. Fellow Royal Soc., 1948, Royal Geog. Soc., Geol. Soc. (pres. 1951-53, Bigsby medal 1943). Research and numerous publs. on petroleum geology. Died Jan. 25, 1955.

LEES, Samuel, Brit. mech. engr.; b. Aug. 26, 1885; s. Samuel Henry Lees; student Coll. Tech., Manchester, Eng.; B.A., St. John's Coll., Cambridge, 1909; M.A., M.Sc.; m. Elsie Elizabeth Mann, 1919; 2 sons. Fellow St. John's Coll., Cambridge; reader applied thermodynamics, faculty tech. U. Manchester, 1913-19; Hopkinson lectr. thermodynamics U. Cambridge, 1919-29; Chance prof. mech. engring. U. Birmingham, from 1931. Cons. mech. engr. Silica Gel Ltd., 1929-31. Mem. Instn. Mech. Engrs. Contbr. papers on heat and mechanics to sci. publs. Died Jan. 27, 1940.

LEES, Sidney, Am. bioengr.; b. Phila., Apr. 17, 1917; s. Charles K. and Rose (Segal) L.; B.S., Coll. City N.Y., 1938; S.M., Mass. Inst. Tech., 1948, Sc.D., 1950; m. Marjorie Berman, Sept. 17, 1946; children—David Eric, Paul Andrew, Eliot Jay. With U. S. Weather Bur., 1938-40, U. S. Signal Corps., 1940-43; faculty Mass. Inst. Tech., 1950-57, cons., 1957-62; prof. engring. Thayer Sch., Dartmouth, 1962-66; head bioengring. dept. Forsyth Dental Center, Boston, 1966—. Mem. Am. Soc. M.E., I.E.E.E., Am. Inst. Physics, Sigma Xi. Author: (with Draper, McKay) Instrument Engineering, 3 vols., 1952-55. Editor: Air, Space and Instruments, 1964. Research, publs. on understanding of theory of measurement; application to new devices including control of chem. processes and guidance of vehicles; devel. new instruments for meterology, also ultrasonic instruments for diagnostic applications in dentistry. Home: 50 Eliot Memorial Rd., Newton, Mass. 02158. Office: 140 Fenway St., Boston 02115.*

LEESON, Charles Roland, anatomist; b. Halifax, Eng., Jan. 26, 1926; s. Charles Ernest and Gladys (Stott) L.; B.A., M.A., Cambridge U., 1947, M.B., B.Chir., M.D., 1950; m. Marjorie Martindale, Apr. 24, 1954; children—Roland Mark, Nicola Jane, Christine Ann, David, Paula. Practice medicine, specializing in anatomy, Cardiff, Wales, 1955-58, Halifax, N.S., Can., 1958-62, Iowa City, 1963-66, Columbia, Mo., 1966—; faculty U. Wales, 1955-58, Dalhousie U., 1958-62; prof. anatomy U. Ia., 1963-66; prof., chmn. dept. anatomy U. Mo., Columbia, 1966—. Mem. Anat. Soc. Gt. Britain, Am., Canadian assns. anatomists, Electron Micros. Soc. Am. Author: (with T. S. Leeson), Histology, 1966; also numerous publs. on studies of histochemistry and electron microscopy of salivary and testicular glands. Home: 1019 Vegas Dr., Columbia, Mo. 65201.*

LEET, Lewis Don, Am. geophysicist, educator; b. Alliance, O., July 1, 1901; s. Kline F. and Lela Grace (Caskey) L.; student Columbia, 1919-20; B.S., Denison U., 1923, D.Sc., 1947; Ph.D., Harvard, 1930; m. Frances A. Brokaw, Feb. 7, 1925 (div.); children—Nancy Anne (Mrs. Richard Ryan), Robert Kline; m. 2d, Florence Anderson Blanchard, July 2, 1956; stepchildren—Brenda Elizabeth, Darrell Russell. Seismologist, Dominion Obs., Ottawa, Ont., Can., 1929-30, Geophys. Research Corp., 1930-31; in charge seismograph sta. Harvard, Cambridge, Mass., 1931—, faculty geology, 1931—, prof., 1946—, chmn. div. geol. scis., 1951-55. Cons. Isthmian Canal studies, Panama Canal, 1946-47; cons. Tecon Corp., Panama Canal, 1954, Office Chief Engrs., U. S. Army, 1947-48, Tex. Co., 1947, TVA, 1949. Mem. Geol. Soc. Am., Seismol. Soc. Am., Sigma Xi, Delta Upsilon, Phi Mu Alpha. Author: Practical Seismology and Seismic Prospecting, 1938; Causes of Catastrophe, 1947; Earth Waves, 1950; (with Sheldon Judson) Physical Geology, 1954; Vibrations from Blasting Rock, 1960; (with Florence Leet) The World of Geology, 1961, Earthquake—Discoveries in Seismology, 1965. Developed 3-component portable seismograph for rec. vibrations from explosions and indsl. sources, 1945.*

LEETE, Edward, chemist; b. Leeds, Eng., Apr. 18, 1929; s. Edward Cecil and Elsie (Griffith) L.; B.Sc., U. Leeds, 1948, Ph.D., 1950, D.Sc., 1965; m. Marie Rideough, Dec. 21, 1954; children—Peter Andrew, Allison Jane. Came to U. S., 1954, naturalized, 1961. Postdoctoral research fellow NRC Can., 1950-54; faculty U. Cal. at Los Angeles, 1954-58; faculty U. Minn., Mpls., 1958—, prof. chemistry, 1963—. Cons. medicinal chemistry study sect. NIH, USPHS, 1962-65. Alfred P. Sloan fellow, 1961, 64; Guggenheim fellow, 1965. Mem. Am. Chem. Soc., Chem. Soc. London, Soc. Chem. Industry, Am. Soc. Pharmacognosy, Sigma Xi, Phi Lambda Upsilon. Author: (with P. Bernfeld) Biosynthesis of Natural Substances, 1963; also numerous articles. Research on ways plants produce alkaloids and steroids, morphine, reserpine, strychnine, cocaine, cancer curing compounds, examination plants for medicinal agts. Home: 110 Bedford S.E. St., Mpls. 55414.*

LEEUWENHOEK (or Leuwenhoek), Antony van, Dutch microscopist; b. Delft, Oct. 24, 1632; received little formal edn.; owned drapery shop where he studied fibers using hand-made lenses; apptd. janitor Delft City Hall (a sinecure which he held for remainder of life); fellow Royal Soc., 1680; corr. mem. Paris Acad. Scis.; contbr. numerous papers to Philosophical Transactions of Royal Soc., Memoirs of Paris Acad. Scis. Ground lenses of great precision; built microscopes, described and illustrated observations made through them; studied the fine structure of teeth, muscles, skin, hair, and ivory; confirmed and extended Malpighi's earlier demonstrations of blood capillaries, 1668; gave first accurate description of red blood corpuscles, 1674; described, illustrated spermatozoa in dogs; described structure of yeast, also stems of monocotyledonous and dicotyledonous plants, 1680; contbd. to disproving theory of spontaneous generation, through his descriptions of the reproduction of weevils, fleas, eels, shellfish; gave first complete description of bacteria and protozoa, circa 1683; first to discover rotifers. Died Delft, Aug. 26, 1723.

LEEVY, Carroll Moton, Am. physician; b. Columbia, S.C., Oct. 13, 1920; s. Isaac Samuel and Mary (Kirland) L.; A.B., Fisk U., 1941; M.D., U. Mich., 1944; m. Ruth Secora Barboza, Feb. 4, 1956; children—Carroll Barboza, Maria Secora. Practice medicine specializing in internal medicine, Jersey City, 1948-54, 56-58; dir. clin. investigation and outpatient dept. Jersey City Med. Center, 1948-58; staff Jersey City Med. Center, 1948—; research asso. Thorndike Meml. Lab., Harvard, 1958-59; asso. prof. medicine, dir. div. hepatic metabolism and nutrition Seton Hall Coll., Medicine, 1958-62; prof., dir. div. hepatic metabolism and nutrition N.J. Coll. Medicine, Jersey City, 1962—. Cons. to hosps. Recipient Distinguished Service award N.J. Jr. C. of C., 1957; Alumni award Fisk U., 1966. Mem. Soc. for Exptl. Biology and Medicine, Am. Gastroent. Assn., Am. Assn. Study Liver Disease, Am. Inst. Nutrition, Am. Soc. Clin. Nutrition, A.C.P., A.A.A.S., A.M.A., Nat. Med. Assn. Author: Practical Diagnosis and Treatment of Liver Disease, 1957; Evaluation of Liver Function in Clinical Practice, 1965; also numerous articles. Editorial bd. Jour. Nat. Med. Assn., 1958—; Jour. Clin. Nutrition, 1964—. Research on pathogenesis cirrhosis alcoholic, hepatic DNA synthesis and regeneration, mechanism portal hypertension and mal-utilization vitamin and proteins in liver disease; developed an in vitro techniques to study hepatic nucleic acid synthesis in man; introduced ear densitometry for evaluating liver function. Home: 907 Av. C, Bayonne, N.J. 07002. Office: N.J. Coll. Medicine, 24, Baldwin Av., Jersey City, N.J. 07304.*

LEFÉBURE DE FOURCY, Louis-Etienne, French mathematician; b. St.-Dominique, France, Aug. 26, 1785; student Ecole polytechnique, 1903-05; served with ground arty. and engr. corps; became prof. differential and integral calculus Faculty Scis. Paris, 1838. Author: Leçons d'algèbre, 1826; Leçons de géométrie analytic, 1827; Traité de géométrie descriptive, 1829; Éléments de trigonométrie, 1830; Traité de plus grand commun diviseur algebrique et de l'elimination entre deux équations et deux inconnues, 1857. Specialized in spheric and rectilinear geometry, theory of straight line, algebraic common denominator. Died Paris, Mar. 12, 1869.

LE FEBVRE, Jean, French astronomer; b. Lisieux, France, 1650; self-educated; went to Paris upon Picard's request to continue La connaissance des temps, 1682; mem. French Acad. Scis., 1679 (excluded in 1701 because of LaHire's accusation of plagiarism of Tables Astronomiques). Author: La connaissance des temps, 1682-1701) Ephémérides calculées sur le meridien de Paris pour les années 1684, 1685. Known for accurate astron. instruments. Died Paris, 1706.

LEFEBVRE, Nicaise (or Nicolas Lefèvre), French chemist; b. Ardennes, France, circa 1610; ed. Acad. Sedan; at least 1 son, Philibert; succeeded William Dawson as prof. botany and chemistry at Jardin du Roi, Paris; apothecary for Louis XIV; became dir. lab., prof. chemistry to Charles II of Eng., also apothecary in ordinary to royal household, 1660; mgr. chem. lab. St. James, 1664-66. Fellow Royal Soc. Author: Chimie théorique et practique, 1660; (article) Disputatio de myrrhata potione, 1660; Discours sur le grand cordial de Sir Walter Raleigh, 1665. Theoretical views influenced by Paracelsian concepts; properties of chem. substances; 1st to formulate law of saturated solutions; discovered acetate of mercury in white crystals; considered nitre the universal chem. agt., engendered by sun. Died London, between 1669 and 1674.

LEFEBVRE, René François Benoit, French bacteriologist; b. La Bassée, Feb. 5, 1913; s. Maurice and Aline (Pillart) L.; ed. univs. Lille, Paris (both France); Ph.D.; m. Suzanne Ammeux, Aug. 23, 1945; children—Christian, Brigitte, Catherine. Head bacteriology lab., later dir. lab. med. biology Faculty Medicine Lille. Mem. Club Issera Nieuport, Belgium. Author: Etude du metabolisme microbien par l'indice chromique, 1944; other works on biology. Address: 189, rue de Solferino, Lille, France.

LEFEVRE, Paul Green, Am. cell physiologist; b. Balt., Dec. 27, 1919; s. Ralph Blaine and Josephine (Green) LeF.; A.B., Johns Hopkins, 1940, postgrad., 1940-41; Ph.D., U. Pa., 1945; m. Marian Ellison Willis, Jan. 1, 1948; children—Bradley Paul, Louise Victoria, Vanessa Fay, Ralph Sebastian. Faculty, U. Vt. Coll. Medicine, Burlington, 1945-52, asso. prof., 1950-52; asst. to chief med. br. U. S. AEC, Washington, 1952-55; scientist Brookhaven Nat. Lab. Med. Research Center, Upton, N.Y., 1955-60; prof. pharmacology U. Louisville, 1960-68; prof. physiology State U. N.Y., Stony Brook, 1968—. Mem. U. S. secretariat 1st Internat. Conf. on Peaceful Uses Atomic Energy, Geneva, Switzerland, 1955. Fellow A.A.A.S.; mem. Am. Physiol. Soc., Soc. Gen. Physiologists, Am. Soc. for Cell Biology, Marine Biol. Lab. Corp., Red Cell Club, Phi Beta Kappa, Sigma Xi. Author: Active Transport Through Animal Cell Membranes, 1955; also articles. Research on mechanisms by which cell membranes transport simple sugars back and forth between cell interior and exterior. Home: Tagliabue Rd., Shoreham, N.Y. 11786. Office: Health Scis. Center, State U. N.Y., Stony Brook, N.Y. 11790.*

LEFEVRE, Raymond James Wood, phys.-organic chemist; b. London, Eng., Apr. 1, 1905; s. Raymond James and Ethel May (Wood) Le F.; Ph.D., Queen Mary Coll., London, 1928, D.Sc., 1933; m. Catherine Gunn Tideman, Aug. 1, 1931; children—Ian, Nicolette (Mrs. Maxwell Thorpe). Faculty, U. Coll., London, 1928-46, reader organic chemistry, 1939-46; with U.K. Air Ministry, 1940-46, asst. dir. research and devel. Ministry Aircraft Prodn., London, 1944; head chem. dept. Royal Aircraft Establishment, Farnborough, Eng., 1944-46; prof. chemistry, head Chem. Sch., U. Sydney (Australia), 1946—. Found. fellow Australian Acad. Sci., 1954. Decorated Smith medal, 1952; Coronation medal, 1953. Fellow Royal Inst. Chemistry, Royal Soc., Royal Australian Chem. Inst. Author: Dipole Moments, 1938. Research, publs. on molecular stereo-chemistry especially electronic distbn., application of Kerr effect in chemistry. Home: 6 Aubrey Rd., Northbridge, Sydney, N.S.W., Australia.*

LEFÉVRE-GINEAU, Louis (chevalier d'Ainelle), French mathematician, physicist; b. Authe, France, Mar. 7, 1751; became prof. exptl. physics Coll. de France, 1788, relieved of duties when elected liberal mem. Chamber of Deputies, 1820-23. Mem. French Acad. Scis., 1795. Helped institute metric system in France as mem. commn. to establish decimal system. Died Paris, Feb. 3, 1829.

LEFFLER, John Edward, Am. chemist; b. Brookline, Mass., Dec. 27, 1920; s. Edward Gustav and Anna (Rose) L.; B.S., Harvard, 1942, Ph.D., 1948; duPont fellow Cornell U., 1948-49; m. Nell Christine Foust, Nov. 26, 1952. Instr., Brown U., 1949-50; faculty Fla. State U., Tallahassee, 1950—, prof. chemistry, 1957—. Mem. Chem. Soc., London, Eng., Am. Chem. Soc. Author: The Reactive Intermediates of Organic Chemistry, 1956; (with E. Grunwald) Rates and Equilibria of Organic Reactions, 1963. Discovered carboxy-inversion reaction of diacyl peroxides; developed phenomenological theory of substituent and solvent effects on reactivity; elucidated mechanisms of various organic, organo-iodine and organo-metallic reactions. Home: 2413 Miranda Av., Tallahassee 32304.*

LE FORT, Léon-Clémont, French surgeon; b. Lille, France, 1829; certified hosp. surgeon; became aggregate prof. Faculty Medicine, 1863, prof. operative medicine, 1873, prof. clin. surgery, 1882. Mem. Acad. Medicine. Author: De la résection du genou, 1864; Mémoire sur l'hygiène hospitalière en France et en Angleterre; Des maternités, 1866; La chirurgerie militarie et les societés de secours en France et à l' étranger, 1872; Manuel de médecine opératoire. Noted for Le Fort's operation for repairing prolapse of uterus. Died Loiret, France, 1893.

LEFORT, Marc, French chemist; b. Souvigny, France, Nov. 13, 1922; s. Henri and Marie (Dufal) L.; Licence, U. Paris, 1945, Docteur es Scis., 1950; m. Jacqueline Danton, Feb. 20, 1954; children—Anne, Vincent. Lectr. radiochemistry Inst. du Radium, Paris, 1956-60; prof. chemistry Faculte Scis., Paris, 1960-62; prof. nuclear chemistry, dir. nuclear chemistry lab. New Scis. Faculty, Orsay, France, 1962—. Mem. French Nat. Com. for Sci. Research. Mem. Am. Phys. Soc. Author: Nuclear Radiations, 1958; Chimie Nucleaire, 1965; Nuclear Chemistry, 1967; also numerous articles. Research on transient states in water and aqueous solutions under gamma and alpha rays, nuclear reactions induced by high energy particles, high energy fission, alpha clusters in nuclei, nucleae reactions induced by heavy multi-charged ions. Home: 12 Rue Villa Naude, Orsay 91, France.*

LEFSCHETZ, Solomon, mathematician; b. Moscow, Russia, Sept. 3, 1884; M.E., École Centrale, Paris, France, 1905; Ph.D., Clark U., 1911; m. Alice Berg Hayes, July 3, 1913. With Westinghouse Electric & Mfg. Co., Pittsburgh, Pa., 1907-10; instructor of mathematics University of Nebraska, 1911-13; instructor mathematics, U. of Kan., 1913-16, asst. prof., 1916-19, asso. prof., 1919-23, prof., 1923-25; visiting prof., Princeton U., 1924-25, asso. prof., 1925-28, prof., 1928-32, H. B. Fine Research prof., 1933-53, now emeritus; professor National University Mexico, 1954—; dir. RIAS Math. Center, 1958—; vis. prof. Brown U., 1964—. Member American Math. Soc. (pres. 1935-36), Math. Assn. America, A.A.A.S., Nat. Acad. Sciences, Am. Philos. Soc., Société Math. de France, Sociedad Mat. of Mexicana, Académie des Sciences de Paris (corr.), Royal Soc. (fgn.), Madrid Acad. de Sciencia (fgn.). Decorated Order Aztec Eagle; recipient Bordin prize French Academy, for work in algebraic geometry, 1919; Bocher prize Am. Math. Society, 1924; Feltrinelli prize Academia dei Lincei, 1956; Nat. Medal of Sci. for Math., 1965. Author of L'Analysis Situs et la Géométrie Algébrique, 1924; Topology, 1930; Algebraic Topology, 1942; Introduction to Topology, 1949; Algebraic Geometry, 1952; Differential Equations; Geometric Theory, 1958; (with J. P. Lasalle) Stability Theory by Liepunor's Direct Method. Work in algebraic geometry, topology, and differential equations. Home: 11 Lake Lane, Princeton, N.J.*

LEGAIT, Etienne Jules, French histologist; b. Chartres, France, Apr. 1, 1911; s. Pierre and Louise (Vernet) L.; Dr. en Médècine, U. Nancy (France), 1934; m. Hermance de Ramaix, Aug. 12, 1950; children—Jean Francois, Benoit, Amaury, Vincent. Licencie es sciences naturelles Agrege d'histologie et d'embryologie, 1946-50; prof. histology U. Fribourg (Switzerland), 1950-52; prof. histology U. Nancy, 1952—. Named Officer L'Ordre des Palmes académiques, 1956. Mem. Soc. de Biologie de Nancy (pres. 1959—), Assn. des Anatomists (sec.-gren. 1965—), Soc. Sciences de Nancy (pres. 1966—). Author: (with Florentin) Démonstrations d'Histologie, 1949; also numerous articles. Research on blood vessels of ependymal organs of 3d ventricle, hypothalamo-hypophyseal system, intermediary lobe of hypophysis. Home: 34 Rue N.D. de Lourdes, Nancy 54, France.*

LE GALLEY, Donald Paul, Am. scientist; b. Norwalk, O., June 30, 1901; s. Marion E. and Mabel (Nunamaker) LeG.; B.S., Heidelberg Coll., 1925; M.S., Pa. State U., 1930, Ph.D., 1935; m. Evangeline L. Cook, Sept. 1, 1927; children—David Marion, Paul Robert. Scientist, Office Naval Operations USN, 1950-56; sci. staff cons. electronic div. Gen. Motors Co., Milw., 1956-59; mem. tech. staff Space Tech. Labs., Redondo Beach, Cal., 1959—; cons. USAAF, 1944. Mem. Am. Inst. Aeros. and Astronautics, Am. Soc. Engring. Edn., Am. Astronautical Soc., Am. Geophys. Union, Sigma Xi, Sigma Pi Sigma. Editor, contbr. chpts. to Space Science, 1963; Space Exploration 1964; Space Physics, 1964. Research on x-rays, missile and space tech. Home: 32679 Seagate Dr., Palos Verdes Peninsula, Cal. 90274.*

LEGALLOIS, Julien Jean César, French physiologist; b. Dol, France, 1770; M.D., Sch. Medicine, Paris, 1801; student U. Dol; student medicine, Caen, France. Children include Eugéne. Author: Is the Blood Identical in all the Vessels through Which it Passes?, 1801; Experiments on the Principle of Life . . . , 1812. Research on relation of vagus nerve to respiration, proved bilateral sect. may produce bronchopneumonia, 1812; located center of respiration in medulla oblongata, 1812. Died 1814.

LEGENDRE, Adrien Marie, French mathematician; b. Toulouse, France, Sept. 18, 1752; ed. Collège Mazarin, Paris. Became prof. math., École Militaire, Paris, 1775-80; later at École Normale, councilor, U. Paris, 1808; replaced Lagrange at Bureau of Longitudes, 1812. Mem. French Acad. Scis., 1783; Fellow, Royal Soc., 1789. Author: L'attraction des ellipsoïdes, 1783; Éléments de géometrie, 1794; Essai sur la théorie des nombres, 1798; Nouvelle théorie des parallèles, 1803; Nouvelles méthodes pour la détermination des orbites des comètes, 1805; Traité des fonctions elliptiques et intégrales Eulériennes, 1826-29. Best known for his important work on elliptic integrals; also studied paths of projectiles in resisting media, attraction of ellipsoids, theory of numbers (announced law of quadratic reciprocity); first to publish (although not 1st to originate) method of least squares; proved pi is irrational, predicted it is transcendental; (with Laplace) created spherical harmonics; collaborated in preparation of centesimal trigonometric tables. Died Paris, France, Jan. 10, 1833.

LE GENTIL DE LA GALAISÈR, Guillaume-Joseph-Hyacinth-Jean-Baptiste, French astronomer; b. Coutances, France, Sept. 12, 1725; student of Delisle, Paris; mem. French Acad. Scis., 1753. Author: Voyage dans les mers de l'Inde, 1779-81. Sent by French Acad. to observe transit of Venus, Pondicherry, India, but failed to see it in 1761 or 69; studied astron. systems of Brahmins; measured obliquity of ecliptic at 23 degrees; research on nebulae, influence of moon of tides; drew up tables of refraction widely used in late 18th century. Died Paris, Oct. 22, 1792.

LEGER, Louis (-Urbain-Eugène), French biologist; b. Coches, France, Sept. 7, 1866; prof. zoology Grenoble (France) Faculty Scis., Grenoble Sch. Medicine. Mem. French Acad. Scis., 1928. Studied sexual phases of parasitic protozoa, their penetration into host, alternate cycles of living from 2 hosts. Died July 7, 1948.

LEGG, Thomas Percy, English surgeon; b. Leeds, Eng., 1872; ed. Yorkshire Coll., Victoria U., St. Bartholomew's Hosp., King's Coll. Hosp.; M.S., London. Surgeon, lectr. surgery, sr. surg. registrar, surg. tutor King's Coll. Hosp.; sr. resident med. officer, surgeon Royal Free Hosp.; cons. surgeon Italian Hosp., Pub. Dispensary, Drury Lane, Farnham Cottage Hosp., Frimley Cottage Hosp.; house surgeon St. Bartholomew's Hosp.; examiner in surgery Cambridge U. Fellow Royal Coll. Surgeons Eng. (ct. examiners), Royal Soc. Medicine, Med. Soc. London, Zool. Soc. Co-author: Cleft Palate and Hare lip. Contbr. articles on goitre treatment, plastic surgery, other surg. topics, to jours. Died Oct. 8, 1930.

LEGGO, Christopher, physician; b. Ottawa, Ont., Can., Jan. 23, 1897; s. William A. and C. Mary (Waugh) L.; M.D., McGill U., 1919; m. Helen Elizabeth Thym, Mar. 4, 1926; children—Helen, Christopher. Practice medicine specializing in preventive medicine, Palo Alto, Cal., 1954-58, Sacramento,

1958—; med. dir. Eastman Kodak Processing Lab., 1954-58; med. officer Cal. Personnel Bd., 1958—; asso. clin. prof. preventive medicine Stanford, 1954-58. Recipient citation for meritorious service Pres.'s Com. on Employment Physically Handicapped, 1962. Diplomate Am. Bd. Preventive Medicine and Occupational Health. Fellow Indsl. Med. Assn., Am. Coll. Preventive Medicine; mem. Western Indsl. Med. Assn. (past pres.), Delta Kappa Epsilon. Asso. editor: Industrial Medicine and Surgery, 1939-64; Archives of Industrial Health, 1953-58. Studies, publs. on psychiat. observations in industry; need for incorporation of sociology, anthropology, psychiatry into bus., professions, govt. Home: 1154 13th Av., Sacramento 95822. Office: 801 Capitol Mall, Sacramento 95814.*

LEGLER, John Marshall, Am. zoologist; b. Mpls., Sept. 9, 1930; s. Frederick William and Helen (Hertig) L.; student U. Keil (Germany), U. Heidelberg (Germany); B.A., Gustavus Adolphus Coll., 1953; Ph.D. in Zoology, U. Kan., 1959; m. Avis M. Johnson, Dec. 21, 1952; children—Austin F., Edward P., Gretchen T., Allison K. Asst. curator in charge herpetology U. Kan. Mus. Natural History, 1954-59; faculty U. Utah, Salt Lake City, 1959—, asso. prof. zoology, 1964—, curator herpetology, 1959—. Mem. Am. Soc. Icthyologists and Herpetologists, Herpetologists League (sec. treas. 1960-66, v.p. 1966—), Brit. Herpetological Soc., Soc. for Study Evolution, The Herpetologists' League (president 1968-69). Author: Natural History of the Ornate Box Turtle, Terrapene ornata ornata Agassiz, 1960; also articles. Research on classification and natural history reptiles and amphibians, chelonian morphology and histology reptilian integumentary organs. Office: Dept Zoology, U. Utah, Salt Lake City 84112.*

LEGRAIN, Marcel Charles, French physician; b. Paris, Oct. 13, 1923; s. Pierre and Germaine (Mermet) L.; M.D., Paris U., 1951; m. Colette Bonamy, Dec. 16, 1952; children—Sylvie, Anne, Pierre, Michel. Research fellow Harvard Med. Sch., 1951-52; physician Hôpitaux, Paris, 1960; named dir. research I.N.S.E.R.M., also chief dept. nephrology Foch Hosp., Suresnes, France, 1965. Mem. Société de Nephrologie (gen. sec.), Am. Soc. Artificial Internal Organs, Royal Coll. Medicine. Research and numerous publs. on treatment of renal insufficiency, peritoneal irrigation and renal homotransplantation. Home: 15 boulevard Lannes, Paris 16, France. Office: C.M.C. Foch Suresnes 92, France.*

LEGUERINAIS, Jacques Paul, French physician; b. Paris, France, July 31, 1919; s. Emile Léon and Lucienne (Crette) L.; Degree Physics, Chemistry and Biology, U. Paris, 1939, Degree Indsl. Hygiene, 1947, Medicine D., 1951; m. Cécile Louise Schackemy, July 29, 1953; 1 son, Patrick Dominique. Asst. collaborator cancer sect. Nat. Inst. Hygiene, Paris, 1951-57, chief cancer sect., 1958-63; surgeon dispensaries Seine dept., adminstr., surgeon Clinic Victor Hugo, Paris, 1963—. Named chevalier Order Pub. Health, 1956; officer Nat. Order of Merit, 1962. Author: Peut-on prévenir le Cancer?, 1963; also numerous articles, monographs. Research on epidemiology of cancer in France and French Africa, comparative study methods of treatment of cancer, possible prevention of cancer through studies of human carcino-genetic factors. Home: 15 rue Anatole France, 92, Sevres, France. Office 5bis rue du Dome, Paris XVI°, France.*

LEGUEU, Félix, French surgeon; b. Angers, France, 1863; became hosp. surgeon, Paris, 1895; passed agrégé exam, 1898; became agrégé prof. disorders urinary tract, 1912; mem. Acad. Medicine, Société d'urologie (founder). Author: Traité chirurgical d'urologie, 1910; la Chirurgerie de Necker, 1917; also other publs. Died Poissy, France, 1939.

LE GUYON, Robert, French bacteriologist; b. Oran, June 3, 1899; M.D., Ph.D.; m. Jeanne Ducreux. Agrégé, Faculty Medicine Strasbourg (France); senator from Loir-et-Cher, 1948-55; chair bacteriology, sch. medicine and pharmacy, 1955—; chair exptl. biology Rennes Faculty Medicine, 1960—. Mem. adminstrv. council Found. for Devel. of Inst. of Cancer. Research in virology and cancer. Address: 38, rue Nicolo, Paris 16, France.

LEHMAN, Edwin Partridge, Am. surgeon; b. Germantown, Pa., June 9, 1888; s. Benjamin N. and Emily (Partridge) L.; B.A., Williams, 1910; M.D., Harvard, 1914, John Harvard fellow, 1913-14; m. Margaret Maxwell, Oct. 1, 1921; children—Richard, Lois Ann. Surg. house officer Peter Bent Brigham Hosp., Boston, 1914-15; asst. resident surgeon Barnes Hosp., St. Louis, 1915-16, resident surgeon, 1919-20, asst. surgeon, 1922-28; asst. in surgery Washington U. Sch. of Medicine, St. Louis, 1916-20, instr. surgery, 1920-21, instr. clin. surgery, 1921-26, asst. prof., 1926-27, asso. prof., 1927-28, in charge lab. surg. pathology, 1916-17; asst. surgeon St. Louis Children's Hosp., 1924-28, St. Louis Jewish Hosp., 1927-28; cons. surgeon St. Louis Maternity Hosp., 1927-28; surgeon St. Louis City Hosp., 1920-27 (chief surgeon Unit No. 1, 1926-27); surgeon to out-patients Washington U. Dispensary, 1920-28; prof. surgery and gynecology, dir. dept. U. Va. 1928-50, prof. surgery, dir. dept. since 1950; chief surgeon and gynecologist U. Va. Hosp., 1928-50, chief surgeon since 1950. Certified mem. founders' group Am. Board Surgery. Fellow Am. (v.p. 1946), So. (pres. 1948) surg. assns., Internat. Surg. Soc., A.C.S. (chmn. cancer com.), Soc. Univ. Surgeons (hon.), Am. Assn. for Surgery of Trauma; mem. A.M.A., So. Med. Assn., St. Louis Assn. of Surgeons, St. Louis Med. Soc. (hon.), Phi Beta Kappa, Sigma Xi. Contbr. to med. publs., author various monographs, articles covering lab. investigation, clin. observation. Died May 27, 1954.

LEHMAN, Russell Sherman, Jr., Am. mathematician; b. Ames, Ia., Jan. 25, 1930; s. Russell Sherman and Ruth (Fehleisen) L.; B.S., Stanford, 1951, M.S., 1952, Ph.D., 1954; m. Lillian Kreling, June 21, 1952; children—Clifford, Anne, John, Helen, Andrew. Problem analyst Computing Lab., Ballistic Research Labs., Aberdeen Proving Ground, Md., 1955-56; Fulbright research grantee, Göttingen, Germany, 1956-57; faculty U. Cal. at Berkeley, 1958—, prof. math., 1966—. Cons. RAND Corp., 1954-65. Mem. Am. Math. Soc., Math. Assn. Am., Assn. for Symbolic Logic, Assn. for Computing Machinery. Research, publs. in partial differential equations, number theory, computing, dynamic programming, math. logic. Home: 79 Ardmore Rd., Berkeley, Cal. 94707.*

LEHMANN, Adolf Ludwig Ferdinand, Canadian chemist; b. Sparrow Lake, Muskoka, Dec. 22, 1863; s. Adelbert Louis Lehmann; student Ont. (Can.) Agrl. Coll., Guelph; B.S.A., Toronto, Ont.; Ph.D., U. Leipzig (Germany); LL.D., Alta., Can.; m. Mary Georgena Lovick, 1898 (dec.); m. 2d, Caroline Melissa Lovick, 1911; 2 sons, 1 dau. Asst., chem. dept. Ont. Agrl. Coll.; asst. chemist Dominion Exptl. Farms, Ottawa, Ont.; chemist Agrl. Exptl. Sta., New Orleans; lectr. organic chemistry Queen's U., Kingston, Ont.; founder, dir. state dept. agr. Govt. of Mysore, Bangalore, India, 1898-1908; prof. chemistry U. Alta., 1909-30. Contbr. to chem. and agrl. publs. Died Sept. 27, 1937.

LEHMANN, Donald Lewis, Am. biologist; b. Glendale, Cal., Dec. 25, 1924; s. Harry Frederick and Carolyn (Lemmon) L.; A.B., San Francisco State Coll., 1949; M.A., U. Pacific, 1950; Ph.D., Ore. State U., 1954; m. Pauline Helen Zembal, Feb. 12, 1952; children—Vivian, Robert, James, Michael, Carolyn. Faculty, U. Pacific, 1955-62, asso. prof. biology, 1958-62; faculty Whitman Coll., 1962—, prof. biology, 1966—, chmn. div. basic scis., 1963-66, dean faculty, 1966—. Fellow tropical medicine La. State U. Sch. Medicine, Central Am., E. Africa, 1959-61; Fulbright scholar, Peru, 1965. Mem. Royal Soc. Tropical Medicine, Soc. Protozoologists, Soc. Parasitologists. Research, publs. on physiol. biochem. and cultural differentiation of those parasites producing African sleeping sickness in man and Nagan in other mammals; detection of drug-fast strains of sleeping sickness organisms; cytochem., cultural and biochem. studies on blood parasites of cold-blooded vertebrates, life cycles and devel. blood parasites. Home: 250 Merrill Rd., Walla Walla, Wash. 99362.*

LEHMANN, Heinz Edgar, psychiatrist; b. Berlin, Germany, July 17, 1911; s. Richard and Emmi (Grönke) L.; student U. Freiburg, 1929-31, U. Marburg, 1930; M.D., U. Berlin, 1935; student U. Vienna, 1932; m. Annette Joyal, July 28, 1940; 1 son, Francois. Emigrated to Can., 1937, naturalized, 1948. Intern Martin Luther and Jewish hosps., Berlin, 1935, Childrens Meml. Hosp., Montreal, 1937; postgrad. work psychiatry and neurology under profs. K. Birnbaum and Simon, 1936; mem. psychiatric staff Douglas Hosp., (formerly Verdon Protestant Hosp.), Montreal, Can., 1937—, clin. dir., 1947—; mem. dept. psychiatry McGill University, 1948—, prof., 1965—; vis. prof. 1965—; U. Cin., 1961—. Cons., mem. various councils nat. mental health orgns. U. S. and Can. Recipient Page One award Newspaper Guild N.Y., 1956, Ann. Merit award Canadian Mental Health Assn., 1957, Albert Lasker award Am. Pub. Health Assn., 1957. Fellow Am. Psychiatric Assn.; mem. Canadian Psychiatric Assn., Montreal Medico-Chirurgical Soc. Author: (with N. Kline) Handbook of Psychiatric Treatment in Medical Practice, 1962. Contributor of articles to profl. jours. Home: 6603 Lasalle Blvd. Office: 6875 Lasalle Blvd., Verdun, Que., Can.*

LEHMANN, Hermann, biochemist; b. Halle, Germany, July 8, 1910; s. Paul and Bella (Apelt) L.; student U. Freiburg (Germany), U. Frankfurt (Germany), U. Berlin, U. Heidelberg (Germany); M.B., U. Basle (Switzerland), 1934; Ph.D., U. Cambridge, 1938, Sc.D., 1957; m. Benigna Norman Norman-Butler, Mar. 10, 1942; children—Susannah, Ruth, Paul, David. Research biochemist, Heidelburg, 1934-36; research on anemia, Uganda, Africa, 1947-49; pathologist Pulmonary Hosp., Kent, U.K., 1949-51; chem. pathologist St. Bartholomew's Hosp., London, 1951-63; univ. biochemist Adelen Crooke Hosp., Cambridge, 1963—; hon. dir. abnormal haemoglobin research unit Med. Research Council univ. dept. biochemistry Cambridge U., 1963—, fellow Christ's Coll., 1964—. Named hon. prof. U. Freiburg, 1964; recipient Rivers medal for anthrop. work Royal Anthropol. Inst., 1963. Fellow Royal Inst. Chemistry, Royal Coll. Physicians, Coll. Pathology. Author: (with R. G. Huntsman) Man's Haemoglobins, 1966; also numerous articles. Research on intermediary carbohydrate metabolism, nutritional anemias, hookworm anemia in Uganda; discovered (with Raper) differential distbn. of sidding in E. African hills, (with Cutbush) presence of sidding in India; discovered new variants of abnormal hemoglobins; 1st discovery of genetic variation of a serum protein, 1955 (pseudocholinesterase). Home: 22 Neuton Rd. Office: Addenbrooke's Hosp., Trumpington Str., Cambridge, Eng.*

LEHMANN, Karl-Otto, German physicist; b. Karlsruch, Germany, Mar. 10, 1903; s. Otto and Olga Lehmann; ed. Elektromaschinenbau, also Tech. Highsch. Karlsruhe; diploma engring., 1925; Ph.D., U. Frieburg, 1934; Asst., electrotech. inst. Karlsruhe Tech. Highsch., 1925-29, asst. physics, 1935-37; asst. in exptl. physics U. Fribourg, 1930-34; physicist Braunkohle und Benzingesellschaft, 1937-45; became prof. electrotech. Karlsruhe Coll., 1946; named dir. works Apparatebau Hundsbach, 1960. Pres. Community of Work for Electronic Optics, Karlsruhe; mem. com. for econ. research Ministry Economy, Bade-Wurtemburg. Research and publs. on magnetic amplification of measures and furnishing of current for measuring instruments. Home: villa Stroh, Leisbergstrasse 21, Baden-Baden. Office: Apparatebau Hundsbach, Hundsbach-Forbach, W. Germany.

LEHMANN, Otto, German physicist; b. Baden, Germany, Jan. 13, 1855; prof. physics U. Strasbourg (France); named prof. Technische Hochschule, Karlsruhe, Germany, 1889; mem. French Acad. Scis., 1912. Author: Über physikalische Isomerie, 1877; Molekularphysik, 1889; Les cristaux liquides, 1909. Discovered optical properties of liquid crystals such as ammonium oleate, molecular physics especially microscopic and microchemic research. Died Karlsruhe, June 17, 1922.

LEHNARTZ, Emil Friedrich Robert, German physician; b. Remscheid, June 29, 1898; s. Emil and Marie (Weisenfeld) L.; ed. univs. Frankfort, Fribourg; M.D.; m. Margarete Zimmermann, Dec. 31, 1927; m. 2d, Eva Königslow, Apr. 11, 1955. Head asst., Göttingen, Germany; prof. chemistry, Munster, Germany; dir. Inst. Physiol. Chemistry; rector U. Munster. Mem. Physiol. Chem. Socs., German Soc. Physiology, Am. Chem. Soc., Biochem. Soc., N.Y. Acad. Sci. Author: Einführung in die chemische Physiologie, 1937; Physiologische Chemie, 1951; Physiologische und pathologisch-chemische Analyse, 1953. Research on intermediary metabolism. Home: Malmedyweg 11, 44 Munster. Office: Waldeyerstrasse 15, 44 Munster, W. Germany.

LEHNER, Joseph, Am. mathematician; b. N.Y.C., Oct. 29, 1912; s. Louis and Rachel (Rosenbloom) L.; B.S., N.Y. U., 1938; M.A., U. Pa., 1939, Ph.D., 1941; m. Mary Beluch, Aug. 17, 1938; 1 dau., Janet Louise. Instructor, Cornell University, 1941-43; mathematician Kellex Corp., N.Y.C., 1943-46, Hydrocarbon Research Inc., N.Y.C., 1946-49; asso. prof. U. Pa., 1949-52; staff Los Alamos Lab., 1952-57; prof. math. Mich. State U., 1957-63; prof. U. Md., College Park, 1963—. Cons. Oak Ridge Nat. Lab., 1949-51, Sandia Corp., 1958-63; mathematician Nat. Bur. Standards, part-time 1963—. Fellow A.A.A.S.; mem. Am., London math. socs., Math. Assn. Am. Author: Discontinuous Groups and Automorphic Functions, 1964; A Short Course in Automorphic Functions, 1966; also articles. Research on analytic theory of numers, theory automorphic functions, neutron transport theory, theory of diffusion cascades. Office: Math. Dept., U. Md., College Park, Md. 20742.*

LEHNINGER, Albert L(ester), Am. biochemist; b. Bridgeport, Conn., Feb. 17, 1917; s. Albert O. and Wally Selma (Heymer) L.; B.A., Wesleyan U., 1939, D.Sc. (hon.), 1954; M.S., U. Wis., 1940, Ph.D., 1942; M.D. (hon.), U. Padva, Italy; m. Janet Wilson, Mar. 12, 1942; children—James Wilson, Erika Jan. Mem. faculty U. Wis., 1942-45, U. Chgo., 1945-52; DeLamar prof. physiol. chemistry, dir. dept. sch. medicine Johns Hopkins, 1952—; vis. prof. U. Frankfurt, Germany, 1951, U. Cambridge, Eng., 1951-52. Recipient Paul Lewis award Am. Chem. Soc., 1948, Distinguished Service award U. Chgo.; Guggenheim fellow, 1951-52, 64. Mem. Am. Chem. Soc., A.A.A.S., Am. Inst. Nutrition, Nat. Academy of Sciences, American Society of Biological Chemists, Am. Acad. Arts and Scis., Biochem. Soc. (London), Phi Beta Kappa, Sigma Xi, Phi Nu Theta. Author: The Mitochondrion, 1964; Bioenergetics, 1965. Research and publs. on the finding that mitochondria are the site of cellular and phosphorletion, that during electron transport in mitochondria there are structural changes and ion transport. Editorial bd. Jour. Biol. Chemistry. Home: Garrison Forest Rd., Owings Mills, Md. Office: Sch. Medicine, Johns Hopkins U., Balt.*

LEHR, David, physician; b. Austria, Mar. 22, 1910; s. Salomon and Esther (Buck) L.; M.D., U. Vienna (Austria), 1935; m. Helen Salzburg, Mar. 21, 1948; children—Karin Elizabeth, Jonathan Mathias. Came to U. S., 1939, naturalized, 1945. Asst., dept. pharmacology Med. Sch., U. Vienna, 1934-38; instr. pharmacology Royal U. Lund (Sweden), 1938-39; pharmacologist, research asso. dept. pathology Newark Beth Israel Hosp., 1939-42; faculty N.Y. Med. Coll., N.Y.C., 1941—, asso. prof. medicine, 1949—, asso. prof. pharmacology, 1949-54, chmn. research com. Met. Med. Center, 1954—, prof., dir. dept. pharmacology, 1954-56, prof., chmn. dept. physiology and pharmacology, 1956-64, prof., chmn. dept. pharmacology, 1964—; asso. attending physician Flower and Fifth Av. Hosps., 1949—, poison control officer, 1955—; vis. physician Met., Bird S. Coler hosps., 1954—. Mem. poison control adv. bd. City N.Y. Dept. Health,

1955—; vice chmn. panel on neurology and psychiat. disease Health Research Council City N.Y., 1961—; Claude Bernard prof. Inst. Exptl. Medicine and Surgery, U. Montreal, 1961. Recipient honors including 1st award for clin. research Med. Soc. State N.Y., 1949, certificate for sci. exhibit Assn. Am. Mil. Surgeons, 1949, Alumni medal N.Y. Med. Coll., 1965. Fellow A.C.P., A.A.A.S., N.Y. Acad. Medicine, N.Y. Acad. Scis., Am. Coll. Cardiology, A.M.A., Am. Coll. Clin. Pharmacology and Chemotherapy; mem. Am. Soc. for Pharmacology and Exptl. Therapeutics, Soc. for Exptl. Biology and Medicine, Sci. Research Soc. Am., Harvey Soc., Internat. Soc. Internal Medicine, Am. Soc. Arteriosclerosis, Assn. Am. Med. Colls., Am. Soc. for Exptl. Pathology, Am. Heart Assn. (council on arteriosclerosis), Pirquet Soc. (pres. 1958-60), Contin Soc. (hon.). Research, numerous publs. on myocardial, renal necrosis; methods for estimation of sulfonamides and serum calcium; discoverer of sulfonamide mixture principle. Home: 79 Lloyd Rd., Montclair, N.J. Office: Fifth Av. at 106th St., N.Y.C. 10029.*

LEHR, Marguerite (Anna Marie), Am. mathematician; b. Balt., Oct. 22, 1898; d. George and Margaret (Kreuder) Lehr; A.B., Goucher Coll., 1919; Am. Assn. U. Women fellow U. Rome, 1923-24; Ph.D. Bryn Mawr Coll., 1925; fellow by courtesy Johns Hopkins, 1931-32. Faculty dept. math Bryn Mawr Coll., 1924—, prof., 1955—; lectr. Swarthmore Coll., 1944, Institut Poincare, Paris, France, 1950. Vis. fellow Princeton, 1956-57; mem. Woodrow Wilson Fellowship Award Com., 1956-65. Recipient Distinguished Work Citation, Goucher Coll., 1954. Fellow A.A.A.S.; mem. Math. Assn. Am. (vis. lectr. 1958-59), Am. Math. Soc., Société de France, Phi Beta Kappa, Epsilon Chi. Publs. on algebraic geometry, humanist aspects of math; also lectr., cons. on curriculum and teaching of math. for state and pvt. orgns. Home: 110 E. William St., Salisbury, Md. 21801.*

LEHRMAN, Daniel S., Am. psychologist, behavioral scientist; b. New York, N.Y., June 1, 1919; B.S. City College, 1947, Ph.D. N.Y. U., 1919; m. 1943; 2 children. Asst. psychologist, Haskins Labs., 1945-47; lectr. psychology, N.Y. U., 1950; from asst. prof. to assoc. prof., Rutgers U., 1950-58; prof. 1958—; dir., Inst. Animal Behavior, 1959—; visiting prof. Yale U., 1957-58. Mem. Soc. Zoologists; Ecology Soc.; Psychological Assn.; Ornithological Union. Research on animal instinct; comparative and evolutionary study of animal behavior; hormones and behavior; investigated species of ring-neck dove for explanation of interaction of hormones, experience, and external stimuli in its behavior. Office: Institute of Animal Behavior, Rutgers University, Newark, N.J.

LEHRMAN, Leo, Am. chemist; b. N.Y.C., July 3, 1900; s. Morris and Elizabeth (Rumsch) L.; B.S., Coll. City N.Y., 1921; M.A., Columbia, 1923, Ph.D., 1926; m. Etta Cohn, Aug. 29, 1930; 1 dau., Barbara. Faculty, Coll. City N.Y., 1921—, prof. chemistry, 1954—; faculty Columbia, 1923-25, N.Y. U., 1925-29. Mem. Am. Chem. Soc., Sigma Xi, Phi Lambda Upsilon. Research, publs. on starch, coprecipitation, methods of analysis and corrosion. Home: 430 E. 86th St., N.Y.C. 10028.*

LEHTI, Raimo Armas, Finnish mathematician; b. Viipuri, Finland, Oct. 4, 1931; s. Armas Aarne and Elsa (Ahtiainen) L.; Cand.Phil., U. Helsinki, 1957, Dr.Phil., 1958. With Astron. Obs., Helsinki (Finland) U., 1955-64, observator, 1959-64, docent math., 1961—; faculty Tech. U. Finland, 1964—, prof., 1967—. Research, publs. on linear algebra, geometry, tensor calculus and relativity. Home: Kivimäentie 39, Helsinki 67, Finland.*

LEHTO, Olli Erkki, Finnish mathematician; b. Helsinki, Finland, May 30, 1925; s. Lauri and Hilma (Autio) L.; M.Sc., U. Helsinki, 1947, Ph.D., 1949; m. Eva Ekholm, Jan. 16, 1954; children—Riitta Leena, Erkki. With U. Helsinki, 1951-52, 56—, prof., 1961—; asst. Acad. Finland, 1953-56; vis. prof. U. Uppsala, 1958, U. Aarhus, 1961, U. Minn., 1962, Stanford, 1962, ETH, Zurich, Switzerland, 1964-65, U. Mich., 1967, U. Minn., 1968. Mem. Finnish Acad. Sci., Finnish Math. Soc. (pres. 1962—). Author monograh: (with K. I. Virtanen) Quasikonforme Abbildungen, 1965. Research, publs. on complex analysis. Home: Puistokaari 7A, Helsinki, Finland.*

LEIBFRIED, Guenther, German physicist; b. Fraulautern, Germany, June 10, 1915; s. Arthur and Ilse (Doerries) L.; Ph.D., U. Goettingen, 1939. Faculty, U. Goettingen, 1939-56; prof. U. Aachen (Germany), 1956—. Research, numerous publs. on solid state physics, crystal theory, radiation damage. Office: 55 Templergraben, Aachen, Germany.*

LEIBNIZ, Gottfried Wilhelm, Baron von (or Leibnitz), German mathematician, physicist; philosopher; b. Leipzig, July 1, 1646; s. Registrar and Prof. Frederick L. and his 3rd wife; taught himself Latin at 8 years of age, then Greek; studied law, math., and philosophy at Us. Leipzig, 1661-63, Jena, 1663-66, Altdorf (where he received doctorate in laws, 1666), Nuremberg; never married. Became sec., Rosicrucian Soc., Nuremberg, 1666; entered service of Arch-bishop Elector of Mainz, Johann Philipp von Schönborn, 1667; served as diplomat in Paris, 1672-76, where he met Huygens, Arnauld, Malebranche; also visited London, 1673 (met Oldenburg) and 1676; privy councilor, court advisor, librarian, historian to Dukes of Braunschweig-Lüneburg (Johann Friedrich, Georg Ludwig, Ernst August) at Hanover, 1676-1716; visited Vienna, 1712-14; became imperial privy councilor. Given title Freiherr; Fellow, Royal Soc., 1673; mem. French Acad. Scis., 1700; mem. and instrumental in founding Berlin Acad. Scis. (1st pres.-for life, 1700; ceased going to Berlin by 1711). Author: De principio individui, 1663; Dissertatio de arte combinatoria, 1666; Hypothesis physica nova, 1671; Discours de métaphysique (written 1686; pub. 1846); Système nouveau de la nature, 1695; Nouveaux essais sur l'entendement humain, 1704; Essais de théodicée sur la bonté de Dieu, la liberté de l'homme et l'origine du mal, 2 vols., 1710; La monadologie (written 1714, pub. 1st in Latin 1721); Principes de la nature et de la grace fondées en raison, 1714; hundreds of other articles and thousands of letters. Endeavored to develop a universal symbolism and attendant algebra so that one might calculate truth of any proposition (for this he is often regarded as forerunner of modern symbolic logic); invented the calculus (roughly simultaneously and independently, it appears, of Newton); introduced and developed much modern mathematical notation; proposed dot to indicate multiplication, decimal point, equal sign, colon for division and ratio; introduced numerical superscripts as exponents; devised elongated s as symbol for integration and the d for the differential (Leibniz's notation quickly adopted on continent of Europe); adopted rationalistic approach to science; as a mechanical philosopher, opposed action at a distance; held space (order of coexistences) and time (order of successions) are systems of relations among things, not absolute entities; postulated theory of monads (not atoms, as basic building block of universe); formulated appropriate measure of motion as product of mass and velocity squared; proposed conservation of vis viva (Kinetic energy); claimed to have found independently of Newton an inverse square law of planetary motion; his work in mechanics and dynamics closely connected with his basic metaphysical ideas (monads, pre-established harmony, ours the best of all possible worlds, law of continuity, i.e. nature admits no gaps, etc.); invented stepped cylinder calculating machine, which could multiply and divide; designed wagon wheels, improved ships' hulls and smokestacks; devised new type of nail; attempted to design windmill-operated pump for Harz mountain mines (project abandoned); studied geology, paleontology, chronology (held earth to be immensely old); said to have proposed existence of an unconscious; an irenicist, sought to bring about unity between Protestant and Catholic Churches; hoped to advance cultural contact between China and Europe with Russia serving as intermediary; urged Peter the Great to found the St. Petersburg Academy of Sciences; engaged in heated controversies with Newton and Newtonians on priority in invention of calculus, conservation of vis viva (with Clarke), nature of space and time, attributes of God, miracles, the mechanism of nature and its relation to God. Died Hanover, Germany, Nov 14, 1716.

LEICESTER, Henry Marshall, Am. chemist; b. San Francisco, Dec. 22, 1906; s. John Ferard and Elsie (Allen) L.; A.B., Stanford, 1927, M.A., 1928, Ph.D., 1930; m. Leonore Azevedo, Feb. 22, 1941; children—Henry Marshall, Martha Katherine, Margaret Anne. Instr. chemistry Oberlin Coll., 1930-31; chemist Carnegie Instn., Washington, 1932; research asso. Stanford, 1933-34; chemist Midgeley Found., Ohio State U., 1935-38; faculty Coll. Physicians and Surgeons, San Francisco, 1938—, prof. chemistry, 1948—. Mem. Am. Chem. Soc. (Dexter award in history of chemistry 1962), Internat. Assn. Dental Research, History Sci. Soc., Soc. for History Tech., A.A.A.S. Author: Biochemistry of the Teeth, 1949; (with H. S. Klickstein) Source Book in Chemistry, 1952; Historical Background of Chemistry, 1956; also numerous articles. Editor-in-chief, Chymia, 1951—. Research on biochemistry teeth in relation to mechanism fluoride tooth decay. Home: 560 Menlo Oaks Dr., Menlo Park, Cal. 94025. Office: 2155 Webster St., San Francisco 94115.*

LEIDHEISER, Henry, Jr., Am. chemist; b. Union City, N.J., Apr. 18, 1920; s. Henry and Margaret M. (Steinel) L.; B.S., U. Va., 1941, M.S., 1943, Ph.D., 1946; m. Virginia M. Townsend, Feb. 21, 1944; children—Margaret F., Henry III. Research asso. U. Va., 1946-49; with Va. Inst. Sci. Research, Richmond, 1949-68, dir., chief exec. officer, 1960-68; dir. Center for Surface and Castings Research, prof. chemistry Lehigh U., Bethlehem, Pa., 1968—. Mem. metallurgy adv. com. Oak Ridge Nat. Lab., 1957-61; chmn. Gordon Conf. on Corrosion, 1964; pres. Research Crystals, Inc.; dir. Diversified Investors, Inc. Recipient Oak Ridge Inst. Nuclear Studies Research award; Westinghouse Signal and Brake award Inst. Metal Finishing. Mem. Am. Chem. Soc., Electrochem. Soc. (Young Authors award), Am. Soc. Metals, Inst. Metals, Faraday Soc., Am. Soc. Testing Materials, Va. Acad. Sci. (J. Shelton Horsely Research award), Sigma Xi. Editor: (with Roscoe Hughes) Exploring Virginia's Human Resources. Research, numerous publs. on interaction of gases, liquids with metal surfaces. Home: 811 Delaware Av., Bethlehem, Pa. 18015.*

LEIDL, Werner, German veterinarian; b. Moss, Germany, June 3, 1925; s. Martin and Therese (Obermeier) L.; Dr.med.vet., U. Munich, 1951; m. Annemarie Friedrich, June 16, 1955; 1 son, Reiner. With U. Munich, 1952—, a.o. prof., 1963—; guest prof. Utah State U., 1958-59, U. Cairo, 1961-62. Mem. Deutsche Veterinarmedizinische Gesellschaft, Deutsche Gesellschaft zum Studium der Fertilität, Soc. for Study Fertility, Sigma Xi. Author: Klima und Sexualfunktion der Haustiere, 1959; also articles. Research on physiology and pathology of reprodn. in domestic animals, preservation of mammalian sperm cells, diagnosis and treatment of venereal diseases in domestic animals, andrology, artificial insemination. Home: 12 Königinstrasse, Munich, Germany.*

LEIDY, Joseph II, Am. physician; b. Phila., Pa., Apr. 11, 1866; s. Philip (M.D.) and Penelope Fontaine Maury (Polk) L.; A.B., Phila. Central High Sch., 1884, A.M., 1885; M.D., U. of Pa., 1887; m. Helen Redington Carter; children—Cornelia Carter (Mrs. J. Hamilton Cheston), Philip Ludwell, Carter Randolph, Joseph (dec.). Physician University Hosp., 1887-89, Pa. Gen. Hosp., 1889-91, Pa. Hosp. for Insane; surgeon Howard Hosp., 1891; Hamilton and Phila. Dispensary; phys. to Med. Clinic, Pa. Hosp., 1893-1903; asst. demonstrator pathology, anatomy and morbid histology, U. of Pa., 1892, and of anatomy, 1894. Fellow Coll. of Physicians, A.A.A.S. Acad. Natural Sciences. Official del. from U. S. Govt. and juror on hygiene, Paris, 1900; decorated Officier de l'Instruction Publique, France, 1900; official del. from U. S. Govt. to the Internat. Congress of Hygiene and Demographie, 1900; represented A.A.A.S. at transfer of home of Charles Darwin to British Nation, 1929. Recognized as foremost Am. anatomist of his time; showed conclusively that before the discovery of Am. by Columbus, the horse had lived and become extinct on Am. continent; discovered Trinchina spiralis (minute parasitic worm in pork) sometimes occurring in muscles of humans; 1st to suggest probability that certain parasitic forms communicated from animals to man might be one of previously unrecognized causes of pernicious anemia. Died July 7, 1932.

LEIFSON, Einar, biologist; b. Kristiandsand, Norway, Nov. 8, 1902; s. Alfred Theodor and Johanne (Midtgarden) L.; came to U. S., 1914, naturalized, 1919; B.S., N.D. State U., 1925; Ph.D., U. Chgo., 1929; M.D., McGill U., 1948; m. Margaret F. Fletcher, Aug. 2, 1930; 1 dau., Margot (Mrs. Roger L. Harvey). Faculty, U. S.D., 1937-46, Johns Hopkins, 1929-37; prof. microbiology Loyola U., Chgo., 1948—. Author: Bacteriology, 1942; Atlas of Bacterial Flagellation, 1960. Research, publs. on devel. of methods for isolation of bacteria in pure culture, methods for characterization and identification of bacteria; description of new species of bacteria and detailed study of several genera of bacteria; also detailed study of about 1000 marine bacteria and their taxonomic position. Home: 611 N. West St., Wheaton, Ill. 60187. Office: P.O. Box 1336, Hines, Ill. 60141.*

LEIGH, Thomas Francis, Am. entomologist; b. Beaumont, Cal., Mar. 7, 1923; s. Herbert Clement and Sarah Shirley (Robinson) L.; B.S., U. Cal., 1949, Ph.D., 1956; m. Nina Anatol Eremin, Nov. 26, 1954; children—Michael Andrew, Nicholas Jonathan. Entomologist, Westley Pest Control Assn., 1949-50; research asst. U. Cal. at Berkeley, 1950-53, asst., asso. entomologist, Davis, 1958—; asst., asso. entomologist U. Ark., 1954-58. Mem. Entomol. Soc. Am., Ecol. Soc. Am., A.A.A.S., Am. Inst. Biol. Scis., Sociedad Mexicana de Entomologia, Sigma Xi. Research, numerous publs. on biology and ecology of insects, mites and spiders of field crops, particularly cotton, control of destructive species by cultural, biol. and chem. methods, resistance in cotton to mites and insects. Home: 242 Pine St. Office: 17053 Shafter Av., Shafter, Cal. 93263.*

LEIGH, Townes Randolph, Am. chemist, b. Panola County, Miss., Oct. 26, 1880; s. Elbridge Gerry and Susie (Gattis) L.; B.S., Iuka Inst., 1901; A.B., Lebanon U., 1902; Ph.D., cum laude, U. Chgo., 1915; D.Sc. (hon.), Stetson U., 1941; m. Blanche Baird Winfield, Mar. 24, 1907. Vice pres., later pres. Mary Connor Coll., Paris, Tex., 1903-08; pres. Tex. Mil. Acad., 1904-06; head dept. science Ouachita (Ark.) Coll., 1907-09; head dept. chemistry Woman's Coll. of Ala., Montgomery, 1910-14; asst. prof. chemistry Carleton Coll., Northfield, Minn., 1915-17; head dept. of chemistry Georgetown (Ky.) Coll., 1917-20; head dept. of chemistry U. Fla., since 1920, dean Coll. of Pharmacy, 1923-33, dean Coll. Arts and Scis., including Sch. Pharmacy, 1933-48, acting v.p. 1934-46, v.p., 1946-48. State chemist, Fla., 1931. Mem. Revision Com. U. S. Pharmacopeia XI. Fellow Royal Soc. Arts, Am. Inst. Chemists; mem. Am. Pharm. Assn., Am. Chem. Soc. (Hertz medal 1932), Am. Assn. Colls. of Pharmacy (pres. 1931-32), many state profl. socs., Sigma Xi. Author of chem. and hist. pamphlets. Inventor Leigh fog screen for protection of vessels against submarines. Died Feb. 15, 1949.

LEIGHLY, John Barger, Am. geographer; b. Locust Grove, O., Nov. 6, 1895; s. Philip T. and Margaret (Reed) L.; A.B., U. Mich., 1922; postgrad. U. Stockholm, Ph.D., U. Cal. at Berkeley, 1927; LL.D., Central Mich. U., 1956; m. Katherine Edmonds, Jan. 4, 1929 (dec. July 1956); m. 2d, Barbara Hawn, June 11, 1959; children—Joseph, Margaret. Faculty U. Cal. at Berkeley, 1927-60, prof. emeritus, 1960—; vis. prof. univs. Stockholm, Uppsala, 1950, Hawaii, 1953; lectr. San Francisco State Coll., 1963—. With Soil Conservation Service, Ariz., N.M., Washington, part time 1935, 40, 41; with Weather Bur., Alaska, 1940;

sr. instr. USAAF Tech. Tng. Commn., 1943-44, meteorologist Weather Central, 1944-46. Hon. fellow Am. Geog. Soc. (Cullum medal 1964); mem. Am. Geophys. Union, Am. Name Soc., Assn. Am. Geographers (hon. pres. 1958), A.A.A.S., Phi Beta Kappa. Editor: The Physical Geography of the Sea (M. F. Maury), 1963; Land and Life (Carl Ortwin Sauer), 1963. Studies, publs. on action streams on surface of earth, phys. climatology, history earth scis. Home: 430 Vassar Av., Berkeley, Cal. 94708.*

LEIGHTON, Alexander Hamilton, Am. psychiatrist; b. Phila., July 17, 1908; s. Archibald Ogilvy and Gertrude Ann (Hamilton) L.; A.B., Princeton, 1932; M.A., Cambridge (Eng.) U., 1934; M.D., Johns Hopkins, 1936; m. Dorothea Cross, Aug. 17, 1937 (div. 1965); children—Dorothea Gertrude, Frederick Archibald. Social Sci. Research Council fellow Columbia, 1939-40; prof. sociology and anthropology Cornell U., Ithaca, 1947-66, prof. social psychiatry Med. Coll., 1956-66; prof. social psychiatry, chmn. dept. behavioral sci. Harvard Sch. Pub. Health, Boston, 1966—. Cons. WHO. John Simon Guggenheim Meml. fellow, 1946-47; fellow Center for Advanced Study Behavioral Scis., 1957-58; Reflective fellow Carnegie Corp., N.Y., Europe, 1962-63. Recipient Human Relations award Soc. for Advancement Mgmt., 1946. Fellow A.A.A.S., Am. Psychiat. Assn., Am. Anthrop. Assn.; mem. Am. Philos. Soc., Am. Psychopath. Assn., Sigma Xi, Phi Beta Kappa, Alpha Omega Alpha. Author: (with Dorothea C. Leighton) The Navaho Door, 1944; The Governing of Men, 1945; Human Relations in a Changing World, 1949; My Name is Legion, 1959; An Introduction to Social Psychiatry, 1960; (with C. Hughes, M. A. Tremblay, R. Rapport) People of Cove and Woodlot, 1960; (with D. Leighton, J. Harding, A. Macmillan) The Character of Danger, 1963; (with J. Clausen, R. Wilson) Psychiatric Disorder Among the Yoruba, 1963. Editor: (with others) Exploration in Social Psychiatry, 1957; (with J. M. Murphy) Approaches to Cross Cultural Psychiatry, 1965.*

LEIGHTON, Gerald, Brit. physician; b. Bispham, Eng., Dec. 12, 1868; s. J. Leighton; ed. Edinburgh (Scotland) U.; M.D., D.Sc., C.M.; m. Clara Gordon. Practised in Herfrodshire, Eng. until 1902; apptd. interim prof. pathology Royal Vet. Coll., Edinburgh, 1902; med. officer foods Dept. Health for Scotland. Fellow Royal Phys. Soc. Edinburgh; hon. mem. Woolhope Naturalists Club. Author: Botulism and Food Preservation, 1923; British Serpents; British Lizards; Reptile Studies: The Greatest Life; Scientific Christianity; (with Douglas) Meat Inspection and the Meat Industry; Huxley, His Life and Work, 1912; Embryology, 1912; Life of James Leighton; Handbook of Meat Inspection, 1924; The Principles and Practice of Meat Inspection, 1927; also articles. Editor: The Modern Veterinary Adviser, 1909. Editor, founder Field Naturalists Quar. Died Sept. 8, 1953.

LEIGHTON, Morris Morgan, Am. geologist; b. nr. Wellman, Ia., Aug. 4, 1887; s. Stephen Tibbetts and Jane (Wellman) L.; B.A., U. Ia., 1912, M.S., 1913; Ph.D., U. Chgo., 1916; D.Sc., U. So. Ill., 1954; m. Ada Hariette Beach, Aug. 12, 1913; children—Freeman Beach, Morris W., Richard T. Faculty geology U. Wash., Ia. State Tchrs. Coll., Ohio State U., U. Ill., 1915-23; geologist Ill. Geol. Survey, Urbana, 1919-23, chief, 1923-54, chief emeritus, 1954—. Mem. adv. com. U. S. Geol. Survey, 1943-58, cons. glacial geology, 1956—; mem. Ill. Mus. Bd., 1937-61, chmn., 1959-61; del. Internat. Geol. Congress, Mexico City, 1956. Named Distinguished Alumnus, U. Ia., 1947. Fellow Chgo. Geog. Soc., Geol. Soc. Am. (councillor 1937-40), A.A.A.S. (v.p. geology 1941); mem. Chgo. Acad. Sci. (hon.), Am. Assn. Petroleum Geologists (hon.), Soc. Econ. Geologists (hon., past pres.), Am. Assn. State Geologists (hon., past pres.), Ill. Mining Inst. (past pres.), Ill. Acad. Sci. (past pres.), Am. Geol. Inst. (past pres.), Sigma Xi. Author: Atlas of Illinois Resources, 1944; (with others) Loess Formations of the Mississippi Valley, 1950; also articles. Editor: Ill. Geologists Jour., 1949-54; bus. editor Jour. Econ. Geology, 1943—. Discovered Farmdale Glacial Substage, (with Louis L. Ray) Neb. and Kan. glaciations in No. Ky.; revised Wis. Glacial Stage classification; specialist glacial geology Upper Miss. River Valley, stratigraphy loess deposits along Miss. River. Home: 307 E. Florida Av., Urbana 61801. Office: Natural Resources Bldg., Urbana, Ill. 61803.*

LEIGHTON, Philip Albert, Am. chemist; b. Los Angeles, Aug. 9, 1897; s. Charles A. and Marie (Plattenburg) L.; A.B., Pomona Coll., 1920, A.M., 1923, Sc.D., 1962; A.M., Harvard, 1925, Ph.D., 1927; postgrad. U. Munich, London U.; m. Susan Case, July 6, 1922 (dec. Nov. 1938); 1 son, Philip Doddridge; m. 2d, Maria Blaisdell, Aug. 11, 1940. Instr. chemistry Harvard, 1927; Sheldon fellow U. Munich, 1927-28; instr. chemistry Stanford, 1928-29, prof., 1937-62, emeritus, 1962—, head chemistry dept., 1939-52, chmn. sch. phys. scis., 1941-42, dean phys. scis., 1946-50; chmn. bd. Metronics Assos., Inc., Palo Alto, Cal. Cons. Air Pollution Research Center, U. Cal., Riverside, 1962—. Recipient Air Pollution Control Assn. Chambers award, 1961; Conservationist of Year award Cal. Wildlife Fedn., 1966. Mem. Phi Beta Kappa, Sigma Xi. Author: Determination of the Mechanism of Photochemical Reactions, 1938; (with W. A. Noyes, Jr.) Photochemistry of Gases, 1941; Photochemistry of Air Pollution, 1961; also numerous articles. Research field mechanisms of photochem. reactions, free radical reactions, photochemistry of lower atmosphere, photochem. air pollution, transport and diffusion of airborne materials. Home: 2389 Alamo Pintado Rd., Solvarg, Cal. 93463. Office: 3201 Porter Dr., Palo Alto, Cal. 94304.*

LEIGHTON, Walter (Woods), Am. mathematician; b. Toledo, Sept. 6, 1907; s. Walter and Alice (Woods) L.; B.A., Northwestern U., 1931, M.A., 1931; A.M., Harvard, 1933, Ph.D., 1935; m. Mary Virginia McKee, Dec. 28, 1937; children—Walter McKee, Nancy Elizabeth. Faculty, Harvard, 1933-36, U. Rochester, 1936-37, Rice Inst., 1937-43; sr. research mathematician, Applied Math. Group, Columbia, 1943-44; dir. Applied Math. Group, Northwestern U., 1944-45; prof. math., chmn. dept. Washington U., St. Louis, 1946-54, Carnegie Inst. Tech., 1954-59; Elias Loomis prof. math. Western Res. U., 1959—; chief div. math. office Sci. Research USAF, 1953-54, cons., 1954-62, Mem. Am. Math. Soc., Math. Assn. Am. Circolo Matematicodi Palermo, Phi Beta Kappa, Sigma Xi, Sigma Nu. Presbyn. Author: Ordinary Differential Equations, 1963. Research and publs. on stability for ordinary differential equations, calculus of variations, continued fractions. Home: 2498 Newbury Dr., Cleveland Heights, O. 44118. Office: Mills Sci. Center, Western Res. U., Cleve. 44106.*

LEIKOLA, Erkki Ensio, Finnish pharmacist, biochemist; b. Artjärvi, Finland, Aug. 14, 1900; s. Albert Ferdinand and Mathilda (Landen) L.; grad. Helsingin Normaalilyseo, 1918; licentiate of med., 1928; M.D., 1929, U. Helsinki; m. Elli Suolahti, Apr. 10, 1934; children: Matti, Anto, Juhani, Marjatta. Asst., Dept. Med. Chem., U. Helsinki, 1927-34; docent, 1931-41, acting prof., 1931-33, U. Helsinki; prof. Pharmaceutical Chem. and principal Dept. of Pharmacy, U. Helsinki, 1938-67; managing dir., 1933-51, now chmn., Board of Pharmaceutical and Chemical Laboratories Lääketehdas Orion Oy, Helsinki. Mem.; 1943-48, chmn., 1948-50, State Board for Pharmaceutical Products; mem., Sci. Advis. Committee to State Med. Board, 1944-49; chmn., Finnish Permanent Pharmacopoeia Committee, 1945-63; mem., 1948-63, chmn. 1952, 1954, 1960, Scandinavian Pharmacopoeia Committee; chmn., Helsinki City Hosp. Board, 1954-64; chmn. Board of Helsinki U. Central Hosp., 1957-66; mem., Parliament, 1945-51, 1954-62. Mem., Finnish Pharmaceutical Soc.; Finnish Med. Soc. Duodecim (sec. 1935-37, mem. of board, 1938-42); Soc. of Finnish Chemists (mem. of board, 1939-48, chmn. 1941-42); Soc. of Finnish Pharmacists; Fédération Internationale Pharmaceutique. Author about 100 published articles. Specialist in pharmaceutical and medical chemistry. Home: 47 A Runeberginkatu, Helsinki 26, Finland.

LEIMANIS, Eugene, applied mathematician; b. Koceni, Latvia, Apr. 10, 1905; s. Ernest and Emma (Tomsons) L.; Mag.Math., U. Latvia, 1929; postgrad. (Latvian Culture Fund fellow), U. Leipzig, Obs. Copenhagen, (Morbergs Meml. Fund fellow), Paris; Dr. Rer. Nat. in Math., Hamburg, 1947; m. Zigrida Gipslis, Nov. 14, 1942; children—Ilze, Antra, Emils, Eva, Ruta, Emanuels. Faculty, U. Latvia, 1929-44, Greifswald U., 1944-45, Baltic U., 1946-48; faculty U. B.C., Vancouver, Can., 1949—, prof. math., 1955—. Adv. scientist Lockheed Missiles & Space Co., Palo Alto, Cal., summer 1962. Canadian Math. Congress Summer Research Inst. fellow, 1953, 55. Fellow A.A.A.S.; mem. Am., Indian, London math. socs., Math. Assn. Am., Canadian Math. Congress, Soc. Math. de France, Unione Mat. Italiana, Deutsche Math. Ver, Soc. for Indsl. and Applied Math., Am. Astron. Soc., Astron. Gesellschaft. Author: Theoretical Mechanics, 1940; (with Minorsky) Dynamics and Nonlinear Mechanics, 1958; General Problem of the Motion of Coupled Rigid Bodies About a Fixed Point, 1965. Mem. editorial bd. Archive for Rational Mechanics and Analysis, 1959—; reviewer Math. Revs., 1949—, Applied Mechanics Revs., 1950—. Research on theory of algebraic curves, ordinary differential equations, exterior ballistics, periodic solutions in celestial mechanics. Address: Dept. Math., U. B.C., Vancouver, B.C., Can.*

LEIMU, Reino Sulo, Finnish chemist; b. Turku, Finland, June 19, 1904; s. Anton Ferdinand and Ida (Vilenius) L.; Phil.Mag., U. Turku, 1931, Ph.D., 1935; m. Airi Annikki Valtonen, Feb. 18, 1940; children—Pekka Tapani, Sirkku Annikki. With U. Turku, 1929—, prof. chemistry, 1947—, dean faculty math. and Natural scis., 1951-56; rector Turku Summer U., 1966—. Mem. Finnish Acad. Scis., Assn. Finnish Chemists', Turku Chemists' Club. Author: Kinetics of Aliphatic Acid Halides, 1935; also articles. Research in organic synthesis and reaction kinetics, analysis of fats, use of fatty acids separated from Finnish tall oil in nutrition and as animal fodder ingredients. Home: 5 b A 18, Rauhankatu, Turku, Finland.*

LEIN, Allen, Am. physiologist; b. N.Y.C., Apr. 15, 1913; s. Ben and Nina (Elinson) L.; B.A., U. Cal. at Los Angeles, 1935, M.A., 1938, Ph.D., 1940; postgrad. U. Chgo.; m. Teresa R. LaFratta, Nov. 29, 1941; children—Laura, David. Research asso., instr. Ohio State U. Med. Sch., 1940-43, Muellhaupt scholar, 1940-41; asst. prof. Vanderbilt U. Med. Sch., 1946-47; faculty Northwestern U. Med. Sch., Chgo., 1947—, prof. physiology, 1961—, dir. student affairs, 1960-64, asst. dean, dir. grad. affairs, 1964—. Vis. prof. chemistry Cal. Inst. Tech., 1954-55. Guggenheim fellow Collège de France, 1958-59. Fellow A.A.A.S.; mem. Soc. Exptl. Biology and Medicine, Am. Physiol. Soc., Am. Chem. Soc., Endocrine Soc. Research, publs. on mode of action thyroid hormone, insulin system on carbohydrate and fat metabolism. Home: 2414 Brown Av., Evanston, Ill. 60201. Office: 303 E. Chicago Av., Chgo. 60611.*

LEINFELDER, Placidus Joseph, Am. physician; b. La Crosse, Wis., Aug. 26, 1905; s. Joseph James and Elizabeth (Stellfplug) L.; B.A., U. Wis., 1926, M.D., 1929; m. Jane Tenney, Jan. 24, 1929 (dec. 1959); children—Joseph Tenney, Mary Elizabeth (Mrs. Edmund E. Byrnes), Carl Jerome; m. 2d, Eloise Zeller, Mar. 19, 1960. Faculty, U. Ia. Coll. Medicine, Iowa City, 1934—, prof. ophthalmology, 1946—. Cons. Ia. Braille and Sight Sav. Sch., 1945—; chief cons. in ophthalmology Atomic Bomb Casualty Com., 1961—; mem. adv. com. on atomic bomb casualty com. NRC, 1963—; chmn. com. for classification of impairment visual function Rehab. Codes; asso. examiner Am. Bd. Ophthalmology, 1938—. Mem. Am. Assn. for Research in Ophthalmology, (prize paper 1940), Johnson County Med. Soc. (pres. 1957), Am. Ophthal. Soc., Sigma Xi, Pi Kappa Alpha. Research on physio-pathology of radiation cataract, cellular metabolism of retina and lens. Home: 440 Lexington Av., Iowa City 52241.*

LEININGER, Robert Irvin, Am. polymer chemist; b. Cleve., May 11, 1919; s. Irvin G. and Reta (Gensheimer) L.; B.S., Fenn Coll., 1940; M.S., Western Res. U., 1941, Ph.D., 1943; m. Ruby Amelia Osterman, May 22, 1942; children—Karen, Christine, Nels, Eric. Instr., Fenn Coll., 1940-43; research chemist Monsanto Chem., 1943-48; research chemist Battelle Meml. Inst., Columbus, O., 1948-51, asso. chief, 1951-60, chief, 1960-66, asso. mgr. dept. chemistry and chem. engring., 1966—. Mem. Am. Chem. Soc. (chmn. Columbus sect. 1967), Royal Soc. Medicine, Am. Assn. Textile Chemists and Colorists, Sigma Xi. Research, publs. on polymerization, copolymerization, radiation chemistry, and long-term properties of plastics, med. applications of plastics. Home: 1973 Milden Rd., Columbus 43221. Office: 505 King Av., Columbus, O. 43201.*

LEIPNIK, Roy Bergh, Am. mathematician; b. Los Angeles, May 6, 1924; s. Robert and Mary (Bergh) L.; A.A., U. Cal. at Berkeley, 1943, Ph.D., 1948; S.B., U. Chgo., 1944, S.M., 1946; fellow Inst. Advance Study, Princeton, 1948-50; m. Joan Hagist, Feb. 18, 1944; children—Karl, Erik, Mark. Faculty, U. Wash., Seattle, 1950-57; cons. math. U. S. Naval Ordnance Test Sta., China Lake, Cal., 1957—. Fulbright research prof. U. Adelaide, Australia, 1955, 63; vis. prof. U. Fla., 1961, 64; cons. to bus. firms, instns. Mem. Am. Math. Soc., Math. Assn. Am., Internat. Math. Soc., I.E.E.E., Inst. Math. Statistics. Author: (with T. Koopmans, H. Rubin, others) Dynamic Economic Models, 1950; (with H. S. Green) Foundations of Magnetohydrodynamics and Plasma Physics, 1967. Asso. editor Jour. Transp. Research, 1966—. Research, publs. on inversion of non-linear operators, statis. spectral analysis, information theory in quantum mechanics, traffic flow, vectorial pair-matching, selection and allocation, transport of plasma in magnetic fields. Home: 522B Nimitz St. Office: Code 607, Naval Weapons Center, China Lake, Cal. 93555.*

LEIPPER, Dale Frederick, Am. phys. oceanographer; b. Salem, O., Sept. 8, 1914; s. Robert and Myrtle (Cost) L.; B.S. in Edn., Wittenberg U., 1937; M.A., Ohio State U., 1939; Ph.D., U. Cal. at Los Angeles, 1950; m. Virginia Alma Harrison, May 14, 1942; children—Diane Louise, Janet Elizabeth, Bryan Robert, Anita Dale. Oceanographer, Scripps Inst. Oceanography, La Jolla, Cal., 1945-49; prof. dept. oceanography and meteorology Tex. A. and M. U., 1949—, head dept., 1949-64. Steering com. Earth Sci. Curriculum Project, Boulder, Colo., 1963—; mem. working group on oceanography sci. com. on Antarctic research Nat. Acad. Scis.-NRC Com. on Polar Research. Recipient Battalion award for contbn. to Tex. A and M. U., 1953. Mem. Tex. Acad. Sci. (pres. 1955), Am. Soc. Limnology and Oceanography (pres. 1957-58). Research on sea temperature as related to synoptic oceanography and sea-air interaction. Home: Route 3, Box 266 B, Bryan, Tex.*

LEIPUNSKII, Aleksandr Ilyich, Russian physicist; b. Dec. 7, 1903; grad. Leningrad Poly., 1926. Staff, Leningrad Physico-Tech. Inst., 1924-30; staff Physico-Tech. Inst., Ukrainian Acad. Sci., 1930-41, dir. Physics Inst., 1941-46; prof. Moscow Engring. and Physics Inst., 1946—. Recipient Lenin prize, 1960. Author: The Dispersion of Photo-Neutrons of Different Energies by Atomic Nuclei, 1940. Research and publs. on nuclear reactors, fast-moving neutrons, energy transmission, splitting of lithium and boron nuclei, slow neutrons, photoneutrons. Home: Zhitnaya 10. Office: Inzhenerno-tekhnichesky institut, Moscow, USSR.

LEIRIS, Michel, French ethnographer; b. Paris, France, 1901; s. Eugène and Marie Madeline (Caubet) L.; diploma in science Sorbonne, Paris; m. Louise Godon. Author: Hautmal; Aurosa; l'Afrique fantome, 1934; l'Age d'homme, 1939; la Règle de jeu; Biffures, 1948; Fourbis, 1955; Contasts de civilisation en Martinique et en Guadeloupe. Specialist in

African studies. Address: 55 bis, quai des Grands-Augustine, Paris 6, France.

LEISERSON, Avery, Am. polit. scientist; b. Madison, Wis., June 27, 1913; s. William Morris and Emily Nash (Bodman) L.; A.B., U. Ill., 1934; Ph.D. (Social Sci. Research Council fellow), U. Chgo., 1941; m. Winifred Smith, Sept. 18, 1936; children—Michael, Nancy, (Mrs. Edward H. Martin), John, Alan. With NLRB, 1938-40, Princeton, 1940-43, U. S. Bur. Budget, 1942-45; faculty U. Chgo., 1946-52; prof. polit. sci. Vanderbilt U., Nashville, 1952—, chmn. dept. polit. sci., 1952-65. Sr. Research fellow Social Sci. Research Council, 1962-63. Mem. Am. (v.p. 1967), So. (pres. 1966—) polit. sci. assns., A.A.A.S., Phi Beta Kappa. Author: Administrative Regulation, 1942; Parties and Politics, 1958; also articles. Editor: The American South in the 1960's, 1964; Jour. Politics, 1961—. Developed functional theory interest groups and organized labor in pub. adminstrn., behavioral framework for comparative analysis party systems; research on govt. process of policy making for sci. Home: 3605 Saratoga Dr., Nashville 37205.*

LEISHMAN, Sir William Boog, Brit. physician; b. Nov. 6, 1863; s. W. Leishman; ed. Glasgow (Scotland) U.; C.B., 1915; M.B., C.M.; LL.D., Glasgow U., McGill U., Montreal, Que., Can.; m. Maud Elizabeth Gunter, 1902; 1 son, 3 daus. Became surgeon Army Med. Service, 1887, advanced through grades to lt.-gen., 1923; served with Waziristan Expdn., 1894-95; asst. prof. pathology Army Med. Sch., Netely, 1900-03; prof. Pathology, Royal Army Med. Coll., 1903-13; European War, 1914-18; dir.-gen. Army Med. Service, 1923—; examiner pathology U. Oxford; examiner tropical medicine U. Cambridge. Mem. Yellow Fever Commn., W. Africa, 1913-15; mem. Med. and San. Adv. Com. for Tropical Africa, Colonial Office, 1913, Med. Research Council, 1913; chmn. foot and mouth diseases research com. Ministry Agr.; mem. Sci. Adv. Com. Brit. Empire Cancer Campaign, 1924. Fellow Royal Soc., 1910, Royal Coll. Physicians London, Royal Faculty Physicians and Surgeons Glasgow; mem. Royal Coll. Physicians Edinburgh. Contbr. articles to sci. jours. First successful cultivation of gonococcus, 1882; described intracellular forms of parasite causing Kala-azar (Leishman-Donovan bodies), 1900; developed vaccine for typhoid. Died June 2, 1926.

LEITCH, Archibald, Brit. pathologist; b. June 8, 1878; s. Neil Leitch Rothesay; ed. Glasgow (Scotland) U.; M.D.; grad. in medicine, 1902; m. Ethel MacLeod Lochhead, 1908; 1 son. Dir. Research Inst., Cancer Hosp., London; prof. exptl. pathology U. London; asst. to dir. Middlesex Cancer Research Labs., dir. Caird Research Lab., Dundee, Scotland; pathologist Cancer Hosp., London; Became research fellow St. Andrews U., Scotland, 1910. Mem. Home Office Com. Enquiry on Mule Spinners' Cancer; gen. sec. Internat. Cancer Com. Mem. Leeuwenhoek Vereeniging, Path. Soc. Gt. Britain and Ireland (charter), Assn. franc pour l'etude du Cancer (corr.), German Central Commn. for Cancer Research (corr.). Editor: Report of International Cancer Conference, 1928. Died Jan. 27, 1931.

LEITE LOPES, Jose, Brazilian physicist; b. Recife, Brazil, Oct. 28, 1918; Chem. Engr., U. Pernambuco, 1939; B.Sc., Fed. U. Rio de Janeiro, 1943; Ph.D., Princeton, 1946; 3 children. Prof. theoretical physics Fed. U. Rio de Janeiro, 1946—; prof. Brazilian Center Research Physics, Rio de Janeiro, 1949—, sci. dir., 1960-64. Mem. Inst. Advanced Study, Princeton, N.J., 1949-50; sr. research fellow Cal. Inst. Tech., Pasadena, 1956-57; prof. d'Orsay Faculty Scis., U. Paris, 1964-67; dir. div. phys. scis. Brazilian NRC, 1955-61, bd. mem., 1961-64. Mem. Brazilian Nat. Acad. Scis., Brazilian Assn. Advancement Sci., Brazilian Phys. Soc. Council. Author: Atomic Theory of Matter, 1958; Fondements de la Physique Atomique, 1967; also numerous articles. Research on shell nuclear effects of photonuclear reactions; relativistic study of Fock space; first estimation of induced pseudoscalar coupling in muon capture by protons. Home: 202 rua Luiz Cantanhede, Rio de Janeiro. Office: Brazilian Center for Research in Physics, 71 Av. Wenceslau Braz, Rio de Janeiro, Brazil.*

LEITER, Joseph, Austrian physician; b. Austria, 1824; invented electric esophagoscope, 1880; asst. to Max Nitze in constrn. of first cystoscope, 1877. Died 1892.

LEITER, Joseph, Am. biochemist; b. N.Y.C., May 14, 1915; s. Abraham and Eva (Mandelberg) L.; B.S. in Chemistry, Bklyn. Coll., 1934; postgrad. Columbia, George Washington U.; Ph.D., Georgetown U., 1949; m. Sarah Freedman, Oct. 8, 1939; children—Andrew B., Everett R. Chemist, Nat. Bur. Standards, 1935-38; chemist-biochemist, head bio-chem. sect. lab. Chem. Pharmacology, Nat. Cancer Inst., NIH, 1938-54; asst. chief lab. activities cancer chemotherapy Nat. Service Center, Nat. Cancer Inst., Bethesda, Md., 1955-61, chief drug evaluation br., 1961-63, chief cancer chemotherapy National Service Center, 1963-65; asso. dir. intramural programs Nat. Library Medicine, USPHS, Bethesda, Md., 1965—. Cons. com. on cell biology UNESCO, 1959-62. Research fellow Applied Biology and Medicine Harvard, 1938-39. Mem. Am. Chem. Soc., Am. Assn. for Cancer Research, A.A.A.S., Soc. for Pharmacology and Exptl. Therapeutics. Asso. editor Cancer Research, 1956—; cons. editor for U. S., European Jour. Cancer, 1964—. Research, publs. on prodn. cancer with atmospheric dusts, devel. chem. agts. for treatment cancer, mechanism action chemotherapeutic agts. Home: 4814 Essex Av., Chevy Chase, Md. 20015. Office: 8600 Rockville Pike, Bethesda, Md. 20014.*

LEITES, Samuil Moiseevich, Russian pathophysiologist; b. 1899; grad. Kharkov Med. Inst.; D.Med. Sci., 1936. Asst., lectr. dept. path. physiology Kharkov Med. Inst., 1923-30; prof., 1930—; head chair pathophysiology Smolensk Med. Inst., 1931-33, Alma-Ata Med. Inst., 1941-45; head chair pathophysiology Ukrainian Postgrad. Med. Inst., Kharkov, head path. chemistry dept. and clin. physiol. lab. Ukrainian Inst. Endocrinology, 1933-41; head clin. physiol. labs. Moscow Diet Clinic, 1945-51; head dept. path. physiology All-Union Inst. Exptl. Endocrinology, USSR Ministry Health, 1945-61, dep. dir. studies, 1961—; head chair pathophysiology Central Postgrad. Med. Inst. Decorated Order of Lenin. Mem. All-Union (v.p.), Moscow (bd. mem.) socs. pathophysiologists, All-Union (bd. mem.), Moscow (bd. mem.) socs. endocrinologists. Author: (monograph) The Physiology and Pathophysiology of Fat Metabolism; The Regulation of Fat-Carbohydrate Metabolism; Pathophysiology of Metabolism and Internal Secretion; also monographs of metabolism and endocrinology. Mem. editorial bd. Problems of Endocrinology and Hormone Therapy; mem. editorial council Archives of Pathology, Path. Physiology and Exptl. Therapy. Research and numerous publs. on pathophysiology and endocrinology, role of lipocaic in regulation of fat metabolism and its disturbances in exptl. and clin. pathology of endocrine glands, pathogenesis of alipotropic fat infiltration of liver, exptl. therapy of liver steatosis with lipotropic nutritive facors; discovered phenomena of autoregulation in fat-lipoid and nitrogen metabolism, analyzed their mechanism and nature of disturbance of metabolism autoregulation processes in exptl. pathology and in diseases of liver, kidneys and endocrine glands; distinguished two pathogenetic forms of diabetes (insular form and total pancreatic form); developer new method to obtain lipocaine (to treat diabetes and liver diseases). Address: Central Postgrad. Med. Inst., 2 Botkinsky pr. 7, Moscow A-284, USSR.*

LEITH, Charles Kenneth, Am. geologist; b. Trempealeau, Wis., Jan. 20, 1875; s. Charles A. and Martha E. (Gale) L.; B.S., U. Wis., 1897, Ph.D., 1901, LL.D., 1956; LL.D., Kenyon Coll., 1926; D.Sc., Lawrence Coll., 1930, Columbia, 1940, Stevens Inst. Tech., 1943; m. Mary E. Mayers, Jan. 6, 1898; children—Kenneth, Andrew. Asst. geologist U. S. Geol. Survey, 1900-05; asst. prof. geology U. Wis., 1902-03, prof., 1903-45. Professorial lectr. pre-Cambrian geology, U. of Chicago, 1905-17; adviser to numerous govt. agys. and pvt. orgns. Fellow Am. Acad. Arts and Scis., Geol. Soc. Am. (v.p. 1927; pres. 1933; Penrose medalist 1942); mem. Am. Inst. Mining and Metall. Engrs., A.A.A.S. (v.p. 1920), Wis. Acad. Scis., Arts and Letters, Nat. Acad. Scis., Am. Philos. Soc., Soc. Econ. Geologists (pres. 1925), Mining and Metall. Soc. Am. Canadian Mining Inst., British Inst. of Mining and Metall. (hon.) Geol. Soc. London (hon.). Author of books and articles on pre-Cambrian metamorphic, structural, economic geology, world minerals in their internat. relations, iron dists. of U. S., origin of iron ores. Latest book World Minerals and World Peace, 1943. Asso. editor Jour. Geology and Econ. Geology. Died Sept. 13, 1956; buried Madison, Wis.

LEITMANN, George, engring. scientist; b. Vienna, Austria, May 24, 1925; s. Josef and Stella (Fischer) L.; came to U. S., 1940, naturalized, 1944; B.S., Columbia, 1949, M.A., 1950; Ph.D., U. Cal. at Berkeley, 1956; m. Nancy Lloyd, Jan. 28, 1955; children—Josef L., Elaine M. Physicist, U. S. Naval Ordnance Test Sta., 1950-57; prof. engring. sci. U. Cal. at Berkeley, 1957—. Cons. on dynamics, dynamical systems, optimal control to industry and govt. Fellow Am. Inst. Aeros. and Astronautics (asso.). Author: Optimization Techniques, 1962; Introduction to Optimal Control, 1966; Topics in Optimization, 1966; also articles. Research in theory and application of optimization and optimal control of dynamical systems, variable mass systems, stability of solutions.*

LEITNER, Jack, Am. physicist; b. Bklyn., Oct. 18, 1931; s. Albert and Gertrude (Rubin) L.; B.A., Columbia, 1952, Ph.D., 1957; m. Diane Djivre, June 7, 1959; children—Lynn, Vicki. Research asso. Duke, 1956-58; faculty Syracuse (N.Y.) U., 1958—, prof. physics, 1965—; vis. physicist U. Cal. at Berkeley, 1959. Dir. Syracuse High Energy Physics Contract, Office Naval Research, 1963—; cons. IBM, 1961-64. Mem. Am. Phys. Soc., A.A.A.S., Italian Phys. Soc., N.Y. Acad. Scis., Am. Assn. U. Profs., Sigma Xi. Research, numerous publs. in elementary particles; co-discoverer 4 new particles; measured many strong and weak interaction properties mesons and hyperons. Home: 248 Scottholm Terrace, Syracuse, N.Y.*

LEIVO, William John, Am. physicist; b. New Castle, Pa., Sept. 11, 1915; s. Matthew and Emma (Poikonen) L.; B.S., Carnegie Inst. Tech., 1939, M.S., 1945, D.Sc., 1948; m. Ruth E. Bell, July 22, 1939; children—Charles Clinton, Frances Eleanor. Supt. bldg. constrn. M. Leivo & Sons, Inc., 1939-42; instr. physics Carnegie Inst. Tech., Pitts., 1942-49, asst. prof., 1949-55; prof. physics Okla. State U., Stillwater, 1955—. Fulbright lectr. U. Alexandria (Egypt), 1961. Fellow Am. Phys. Soc.; mem. Am. Assn. Physics Tchrs., Am. Assn. U. Profs., Sigma Xi, Sigma Pi Sigma. Research, publs. in solid state physics, semiconducting diamond and radiation effects in solids; determination of properties of color centers in alkali halide crystals produced by X-rays and high energy protons from synchrocyclotron. Home: 1021 W. Connell St., Stillwater, Okla. 74075.*

LEJARS, Félix, French surgeon; b. Unverre (Eure-et-Loire), France, 1863; certified surgeon, Paris, 1891; became aggregate prof. Faculté de médecine, Paris, 1892; prof. surg. clinic. Mem. French Acad. Medicine. Author: Chirurgie d'urgance, 1899; Diagnostic chirurgical. Died Paris, 1932.

LE JEUNE, Francis Ernest, Am. physician; b. Thibodaux, La., Aug. 26, 1894; s. Henry E. and Alice Rosa (Doucet) LeJ.; B.S., Jefferson Coll., 1914; M.D., Tulane U., 1920; M.S., Jefferson Coll., 1924; m. Anna Lynne Dodds, Dec. 27, 1922; children—Ann (Mrs. George T. Schneider), Francis Ernest. Prof., head dept. otolaryngology Tulane U. Sch. Medicine, New Orleans, 1936-53, prof. emeritus, 1958—; mem. staff Eye, Ear, Nose and Throat Hosp., New Orleans, 1922—; co-founder Ochsner Clinic and Found. Hosp., New Orleans, 1942, head, dept. otolaryngology, 1942-64; cons. Charity Hosp., 1953—. Dir. Am. Bd. Otolaryngology, trustee Alton Ochsner Med. Found. Mem. A.M.A., A.C.S. (bd. govs.), Am. Laryngol. Assn. (Caselberry award 1936, Newcomb award 1963, pres. 1963), Am. Bronchoesophogological Soc. (pres. 1967), Am. Acad. Ophthalmology and Otolaryngology (Honor key 1948, pres. 1966), Pan Am. Congress Otolaryngology, Am. Triological Assn. (pres. 1953), Phi Chi, Alpha Omega Alpha. Lectr., publs. on cancer larynx. Editorial bd. Laryngoscope 1945—. Contbr. articles to text books on otolaryngology. Home: 49 Audubon Blvd., New Orleans 70118. Office: 1514 Jefferson Hwy., New Orleans 70121.*

LEJEUNE, Jérôme Jean Louis Marie, French human geneticist; b. Montrouge, Seine, France, June 13, 1926; s. Pierre Ulysse and Marcelle (Lermat) L.; M.D., Faculty Medicine, Paris, France, 1951, Ph.D., 1960; m. Birthe Bringsted, May 1, 1952; children—Anouk, Damien, Karin, Clara, Thomas. Research staff Centre Nat. de la Recherche Scientifique, 1952-64, maitre, 1959-62, dir., 1962-64; prof. fundamental genetics Faculty Medicine Paris, 1964—. Genetic expert to WHO, 1962—, Sci. Com. on Atomic Radiation, 1957—. Recipient Kennedy award, 1962; Cognac-Jay award, 1963; named officer Order Nat. du Merite, 1964. Mem. Societé Francaise de Génétique, Royal Soc. Medicine. Author: (with R. Turpin, H. Jérôme) Les Chromosomes Humains, 1965; also numerous articles. Discovered first chromosomal aberration in man, trisomy 21 of mongolism, 1st translocation in man, 1st disease related to loss of a part of an autosome, cri du chat disease. Home: 31 Rue Galande Paris 5. Office: Institut de Progénèse, 15 Rue Ecole de Medecine, Paris 6, France.*

LEJSEK, Karel, Czechoslovakian physician; b. Vysoké Myto, Czechoslovakia, Oct. 20, 1932; s. Karel and Marie (Zásterová) L.; MUDr., Mil. Med. Faculty, Hradec Králové, Czechoslovakia, 1957; m. Jindriska Zápalová, Nov. 7, 1957; 1 dau., Regina. Surgeon, Mil. Hosp., Plzen, Czechoslovakia, 1957-60; sr. lectr. dept. chemistry and biochemistry Charles U. Med. Faculty, Hradec Králové, 1960—. Mem. Czechoslovak Biol. Soc. Research, publs. on lipid peroxidation after irradiation and partial hepatectomy, function and morphology of rat liver mitochondria after irradiation and partial hepatectomy. Address: 870 Simkova, Hradec Králové, Czechoslovakia.*

LELEIR, Luis F., biochemist; b. Paris, France, Sept. 6, 1906; s. Federico and Hortensia (Aguirre) L.; M.D., U. Buenos Aires (Argentina), 1932; D. honoris causa U. Paris (France), U. Granada (Spain); m. Amelia Zuberbühler, 1 dau., Amelia. Staff, Inst. Physiology, U. Buenos Aires, 1934-35, 37-43, asso. prof., head dept. biochemistry; research worker Biochem. Lab., Cambridge, Eng., 1936; research asso. U. Washington, St. Louis, 1944; research worker Columbia Coll. Phys. and Surg., 1944-45, Instituto Biología y Medicina Exptl., Buenos Aires, 1946; dir. Instituto de Investigaciones Bioquímicas, Fundación Campomar. Recipient awards Argentine Sci. Soc., Heley Hay Whitney Found., Severo Vaccaro Found., Argentina, Bunge and Born Found., Argentina. Mem. Nat. Acad. Scis. (fgn.), Am. Acad. Arts and Scis., Am. Philos. Soc., Argentine Soc. for Biochem. Research (chmn.). Author: Fisiología Humana; Renal Hypertension; also numerous articles. Research on oxidation of fatty acids and biochem. mechanisms of renal hypertension, isolation various diphosphate sugars, enzymes of metabolism, biosynthesis of saccharides. Home: 2574 Newton, Buenos Aires 25, Argentina.*

LELEVIER, R(obert) E(rnest), Am. physicist; b. Los Angeles, Nov. 7, 1923; s. Ernest and Catherine (Grusling) LeL.; Ph.D. (AEC fellow), U. Cal. at Los Angeles, 1951; m. Anna Marie Smith, Aug. 5, 1945; children—Robert Ernest, Jon Anton, Suzanne Marie. Staff, Lawrence Radiation Lab., Livermore, Cal., 1951-59; with RAND Corp., Santa Monica, Cal., 1957—, now sr. staff mem. Exploratory studies on gravitational collapse of stars and their approach to main sequence; studies of transient responses of ionosphere to pulses of radiation. Home: 961 Jacon Way, Pacific Palisades, Cal. 90272. Office: 1700 Main St., Santa Monica, Cal. 90406.*

LELIÈVRE, Claude H., French mineralogist; b. Paris, June 28, 1752; gen. insp. French mines; mem. French Acad. Scis., 1795. Author: Description de divers procédés pour extraire la soude du sel marin. Developed method for extracting sodium carbonate from salt water, 1795. Died Oct. 19, 1835.

LE LIÈVRE, Jean (Leporis, Joannes), French physician, anatomist; b. Semur, Côte d'Or, France; flourished 1400. Cleric, Diocese of Autun; master regent Med. Faculty, U. Paris, by 1392, dean, 1394-95; physician to theologian Nicolas de Clamanges, circa 1360-1434, also to Duke Louis D'Orléans and his son, Charles. Presided over earliest recorded anat. dissection in Paris, 1407. Died 1418.

LELOIR, Henri Camille C., French dermatologist; b. France, 1855. Author: Traité pratique et théorique de la lèpre (classic text on the pathology of leprosy), 1886; described lupus vulgaris erythematoides (erythematodes) which is called Leloir's disease, 1890; Died 1896.

LELONG, Jacqueline Ferrand, French mathematician; b. Alès, Feb. 17, 1918; d. Auguste and Jeanne (Chambon) Ferrand; Ph.D. in Math., agrégée in math. sci.; m. Pierre Lelong, Nov. 22, 1947; children—Jean, Henri, Francoise, Martine. Lectr., girls' sch., 1939-43; instr. gen. math. and rational mechanics Faculty Scis., Bordeaux, France, 1943-45; instr. differential and integral calculus Faculty Caen (France), 1945-57, titular prof., 1948; prof. Faculty Lille (France), 1948-56; prof. math. Sorbonne, 1957—. Recipient prize Girbal-Baral Inst. Research and publs. on theory of potential, Riemann spaces. Home: 95, bd. Jourdan, Paris 14. Office: 11, rue Pierre-Curie, Paris 5, France.

LELONG, Pierre Jacques, French mathematician; b. Paris, Mar. 14, 1912; s. Charles and Marguerite (Bronner) L.; Ph.D., agrégé; m. Jacqueline Ferrand, Nov. 22, 1947; children—Jean, Henri, Francoise, Martine. Prof., univs. Grenoble, Lille, Paris; cons. to pres. of Republic, 1959-61. Mem. Math. Soc. France. Research and publs. on functional analysis, information theory, analytic functions of several complex variables, orgn. of sci. research, econ. incidences. Home: 95, bd. Jourdan, Paris 14. Office: Institut Poincaré, 11, rue Pierre-Curie, Paris 5, France.

LEMAIRE, François Jules, French scientist; b. France, 1814; discovered antiseptic value of carbolic acid, 1860. Died 1886.

LEMAISTRE, Charles Aubrey, Am. physician; b. Lockhart, Ala., Feb. 10, 1924; s. John Wesley and Edith (McLeod) LeM.; A.B., U. Ala., 1944; M.D., Cornell U., 1947; m. Joyce Trapp, June 3, 1952; children—Charles F., William Sidney, Joyce Anne, Jean Helen. Intern, resident medicine N.Y. Hosp., 1947-49; research fellow Cornell U. Med. Coll., 1949-51, mem. faculty, 1951-54, asst. prof. medicine, 1953-54; mem. faculty Emory U. Sch. Medicine, 1954-59, prof. preventive medicine, chmn. dept., 1957-59; prof. medicine U. Tex. Southwestern Med. Sch., 1959—, asso. dean, 1965—; Cons. epidemiology Communicable Disease Center, USPHS, 1953—; cons. medicine VA, 1954-59, Baylor U. area med. cons. Atlanta area, 1958-59; vis. staff physician Grady Meml. Hosp., Atlanta, 1954-59, Emory U. Hosp., 1954-59, Parkland Meml. Hosp., Dallas, 1959—; med. dir. Woodlawn Hosp., Dallas, 1959-65; mem. Nat. Citizens Commn. Internat. Cooperation, 1965. Mem. A.M.A.-Edn. Research Found. com. research tobacco and health, 1964—; chmn. Gov. Tex. Com. Tb Eradication, 1963—. Chmn. steering com. Presbyn. Physicians for Fgn. Missions, 1960—; mem. Ministers Cons. Clinic, Dallas, 1961-62. Mem. Am. (v.p. 1964-65), So. (pres. 1963-64) thoracic socs., Nat. Tb Assn., A.M.A., Central Soc. Clin. Research, Alpha Omega Alpha. Contbg. author: A Textbook of Medicine, 10, and 11th edits., 1963, Ency Americana: Smoking and Health, 1966. Pharmacology in Medicine, 1958; translating author: The Tubercle Bacillus, 1955. Mem. editorial bd. Am. Rev. Respiratory Diseases, 1955-58. Home: 3600 Lindenwood Av., Dallas 75205. Office: 3819 Maple St., Dallas 75219.*

LEMARCHAND-BERAUD, Therèse Marie, Swiss chemist; b. Geneva, Switzerland, Sept. 25, 1928; d. Pierre and Annette (Léveque) Beraud; Licence-ès-Sciences chimiques avec mention biologique U. Geneva, 1950; Doctorat ès-Sciences Biochimiques U. Geneve et Lausanne, 1960; m. Lemarchand, Mar. 4, 1961; children—Pierre-Etienne, Isabelle. Asst. exptl. endocrinology lab. U. Geneva, 1951-52; research asso. Med. Clinic and Policlinic, Bavière Hosp., Liège, Belgium, 1951-52; research asso. Biochemistry Lab., U. Med. Clin., Lausanne, Switzerland, 1952—; staff Sorbonne Inst. Exptl. Physiology, also Collège de France, Paris, 1955. Mem. Société de Biologie médicale, Société suisse de Chimie clinique, Société Suisse d'Endocrinologie, Association Suisse des Jeunes Chercheurs, Association Européene de la Thyroide. Author: (with B. R. Scazziga) Physiopathologie thyroidienne, 1966; also articles. Research in endocrinology, especially biochemistry of thyroid hormones, their method of transport and their relation with pituitary gland and other glands. Home: 20 avenue de Valmont, 1010 Lausanne. Office: Clinique Médicale Universitaire, Hôpital Nestle, 1000, Lausanne, Switzerland.*

LEMBERG, (Max) Rudolf, biochemist; b. Breslau, Germany, Oct. 19, 1896; s. Arthur and Margarete (Wendriner) L.; Natura Tenaz, Breslau U., 1917, Ph.D. summa cum laude, 1921; Venia legendi in Chemistry, Heidelberg (Germany) U., 1930; m. Hanna Adelheid Claussen, Dec. 23, 1924. Asst., Chem. Inst., Breslau, U., 1921-24; chemist C. F. Boehringer & Sons, 1924-26; staff Heidelberg U., 1926—, prof., 1933-56, prof. emeritus, 1956—; head biochemistry dept. Inst. Med. Research, Royal N. Shore Hosp., Sydney, Australia, 1935—, asst. dir., 1953—; vis. prof. U. Pa., 1966. Rockefeller Found. fellow Inst. Biochemistry, Cambridge, Eng., 1930-32; Acad. Assistance Council fellow, 1933-35. Recipient Coronation medal, 1952; James Cook medal Royal Soc. New S. Wales, 1965; Britannica-Australia award in sci., 1965. Fellow Royal Soc. London, Australian Acad. Sci. (past mem. council, past v.p.), Royal Australian Chem. Inst. (H. G. Smith medal 1949); corr. mem. Heidelberg Akademie d'Wissenschaften, Academia Chirurgico, Anatomica Pergia; mem. Am. Chem. Soc., Biochem. Soc. London, Australian Biochem. Soc. (hon.), Am. Soc. Biol. Chemists (hon.). Author: (with J. W. Legge) Hematin Compounds and Bile Pigments, 1949; also numerous articles. Editor: (with J. E. Falk, R. K. Morton), Haematin Enzymes, 2 vols., 1961. Research on bile pigments, porphyrin derivatives, respiratory enzymes; discovery of phycobilins of photosynthetic algae. Home: 57 Boundary Rd., P. O. Wahroonga (Sydney) N.S.W. Office: Inst. Med. Research, Royal N. Shore Hosp., Sydney (P. O. Crows Nest), N.S.W., Australia.*

LEMBERT, Antoine, French surgeon; b. 1802; studied under the Parisian surgeon Baron Guilaume Dupuytren; invented an approximating suture which ensured the union of serous surface with serous surface in suturing intestine (Lembert's suture); 1826, this invention laid the founds. for modern intestinal and gastric surgery. Died 1851.

LEMCHE, Henning Mourier, Danish zoologist; b. Copenhagen, Denmark, Aug. 11, 1904; s. Soren Jacobsen and Inge (Mourier) L.; cand.mag. U. Copenhagen, 1927, dr.phil., 1937; m. Inger Sodemann, Apr. 17, 1932; children—Ib Sodemann, Kirsten Sodemann, Karen Sodemann (Mrs. Jens Scheel Vandel). Staff, Zool. Lab., Royal Vet. and Agrl. Coll., 1928-48, amanuensis I, 1946-48; staff Zool. Mus., U. Copenhagen, 1949—, curator, 1955—, lectr., 1962—. Mem. Internat. Commn. on Zool. Nomenclature, 1948—. Mem. Dansk Naturhistorisk Forening i Kjobenhavn (past mem. council, past pres.), Internat. Union Biol. Scis. (mem. Danish nat. com. 1950—), Unitas Malcologica Europaea (past pres.). Recipient Galathea medal, 1955. Author: Fra Molekyle til Menneske, 1945; also numerous articles. Research on color patterns in insects, sea slugs; discovered ancient-type mollusc Neopilina. Home: 2 Olesvej, Virum, Denmark. Office: 15 Universitetsparken, Copenhagen, Denmark.*

LEMÉHAUTÉ, Bernard Jean François, hydrodynamicist; b. Saint Brieuc, France, Mar. 29, 1927; s. Sylvestre F. and Louise E. (Redon) LeM.; Baccalaureat Math., U. Rennes, 1947; Licencié ès sciences Eng. Ecole Nationale Supérieure Electrotechnique et Hydraulique, Toulouse (France) U., 1951; Diplome études Supérieure Hydrodynamics cum laude U. Paris (France) 1953; Docteur ès sciences with highest distinction U. Grenoble (France), 1957; m. Pierrette A. Richalet, Oct. 27, 1930; children—Anne Louise, Patrick Paul. Came to U. S., 1961, naturalized, 1966. Hydraulic engr. Neyrpic-Sogreah, Grenoble, 1953-57; asso. prof. Ecole Polytechnique, Montreal, Que., Can., 1957-59; research prof. Queen's U., Kingston, Ont., Can., 1959-61; dir. geomarine div. Nat. Engring. Sci. Co., Pasadena, Cal., 1961-66; v.p. Tetra Tech., Inc., Pasadena, 1966—. Mem. Cal. Gov.'s Adv. Commn. Ocean Resources, 1966—; mem. com. on control engring. and inland waters Nat. Acad. Engring., 1965—. Recipient prize Société Hydrotechnique de France, 1953. Mem. Am. Geophys. Union, Am. Soc. C.E., Internat. Assn. for Hydraulic Research, Marine Tech. Soc., Permanent Internat. Assn. Navigation Congresses, Ingenieurs Civils de France. Author: Ocean Sciences, 1964; also articles. Research on hydrodynamics water waves, gravity waves generated by underwater nuclear explosions, resonance in harbor basins, unsteady flow motion through porous medium, hydraulic similitude and scale model study, sand transport by waves and currents. Home: 2159 Highland Vista Dr., Arcadia, Cal. 91006. Office: 630 N. Rosemead Blvd., Pasadena, Cal. 91107.*

LEMENRY, Louis, French chemist; b. Paris, 1675; s. Nicolas Lemenry; M.D., Paris, 1698; physician at Hôtel Dieu for 33 years; prof. chemistry Jardin du roi. Author publs. including: Reflexions physiques sur le défaut et le peu d'utilité des analyses ordinaires des plantes et des animaux, 1719; Mémoires sur les monstres, 1738-40; Mémoires sur le trou ovale, 1739. Died 1743.

LÉMERY, Jacques, French chemist; b. Paris, France, Jan. 1678; s. Nicolas Lemery; mem. French Acad. Scis., 1715. Author: Sur un nouveau phosphore, 1715; De l'action des sels sur différente matière inflammables, 1713. Research on phosphorus, effect salts and chems. on inflammable matter. Died 1721.

LÉMERY, Louis, French physician; b. Paris, France, Jan. 25, 1677; s. Nicolas Lemery; studied medicine under Tournetourt; 1st physician to queen of Spain; doctor Paris Faculty Medicine; prof. chemistry Royal Bot. Gardens. Mem. French Acad. Scis., 1700. Author: Dissertation sur la nature des os, 1704; Traite des aliments suivant les principes chimiques et mécaniques, 1702. Studied bone marrow and related diseases. Died Paris, June 9, 1747.

LEMERY, Nicolas, French chemist, physician; b. Rouen, France, Nov. 17, 1645; studied pharmacy under Rouen apothecary; student chemistry under Glazer, Jardin du Roi, under Verchant, Montpellier, France; M.D., U. Caen (France), 1683. Lectr. on chemistry, Montpellier; founder pharmacy, Paris; fled to Eng., 1681; returned to France, 1683; reopened pharmacy in Paris; began practice medicine, Paris, 1685; joined Roman Catholic Church, 1686, because revocation of Edict of Nantes, 1685, deprived him of right to practice; named apothecary to Louis XIV, Paris, 1686. Mem. French Acad. Scis., 1699. Author: Cours de chymie (popular text, went through many editions in French, Latin, German, English and Spanish, 1675-1757); Pharmacopée universelle, 1697; Dictionaire universel des drogues simples, 1698; Traité de l'antimoine, 1707. Studied theory of volcanoes; discovered iron in blood; numerous applications of chemistry to medicine. Died Paris, June 18, 1715.

LEMIERRE, André, French physician, bacteriologist; b. Paris, France, July 30, 1875; s. Alfred Lemierre; ed. École Monge, Faculty Medicine, Paris; studied under Widal; m. Jeanne Madeleine Richelot, Apr. 1, 1909; children—Jacqueline, Jacques, Alfred. Prof., Paris. Research and publs. on diseases of kidneys and infectious diseases; isolated (with J. Reilly, A. Laporte, M. Morin) streptobacillus moniliformis from patient with rat-bite fever, 1937. Died La Bernerie, France, Aug. 11, 1956.

LEMKAU, Paul Victor, Am. physician; b. Springfield, Ill., July 1, 1909; s. John Henry and Caroline (Stork) L.; A.B., Baldwin-Wallace Coll., 1931, Sc.D. (hon.), 1954; M.D., Johns Hopkins, 1935; Dr.P.H. (hon.), Dickinson Coll., 1958; m. Ruth Claire Roehm, Sept. 22, 1934; children—Ann (Mrs. William C. Houpt), Mary (Mrs. Charles L. Horn), Elizabeth Roehm, Philip John, Carolyn Claire. Faculty, Sch. Hygiene and Pub. Health Johns Hopkins, Balt., 1941—, prof., chmn. dept. mental hygiene, 1951—; chief, div. mental health Md. Dept. Health, Balt., 1949-53; dir. N.Y.C. Community Mental Health Bd., 1955-57. Mem. expert com. on mental health WHO, 1963—, cons. 1949—. Diplomate Am. Bd. Psychiatry and Neurology, Am. Bd. Preventive Medicine and Pub. Health. Fellow Am. Psychiat. Assn., Am. Pub. Health Assn.; mem. Nat. Mental Health Adv. Council, Nat. Assn. for Mental Health (dir. 1948-63), Philippine Soc. Psychiatry and Neurology (hon.). Author: Basic Issues in Psychiatry, 1958; Mental Hygiene in Public Health, 1949, 55. Home: 1510 Berwick Rd., Ruxton, Md. 21204. Office: 615 N. Wolfe St., Balt. 21205.*

LEMMON, Richard Millington, Am. chemist; b. Sacramento, Nov. 24, 1919; s. Dal Millington, and May Alice (Dunn) L.; A.B., Stanford, 1941; M.S., Cal. Inst. Tech., 1943; Ph.D., U. Cal. at Berkeley, 1949; m. Marguerite Hayward, July 9, 1949; children—Janet, Marilyn, Brian. Research chemist Lawrence Radiation Lab., U. Cal. at Berkeley, 1951—, associate director laboratory of chem. biodynamics, 1957—. Postdoctoral fellow USPHS, 1950-51; Guggenheim fellow, 1964-65. Mem. Am. Chem. Soc. (chmn. Cal. sect. 1961), A.A.A.S. Research, publs. on effects radiation on organic compounds, detailed mechanisms organic reactions. Home: 298 Los Altos Dr., Berkeley, Cal. 94708.*

LEMNIUS, Levinus (or Lemmens, Lieven, Lemse), Dutch chemist, physician; b. Zierikzee, Zeeland, Netherlands, May 20, 1505; studied theology, medicine under Vesalius, Gesner, Dodoens; grad. from Padua, 1525. Practiced medicine in Zirickzee. Author: De Miraculis . . . Naturae, 2 books, 1559, enlarged to 4 books, 1564; De Habitu et Consituttione Corporis, quam Graeci Krasin, Triuiales Complexionem vocant, Libri II, 1604; Similitudinum ac Parabolarum quae in Bibliis ex Herbis atque Arboribus desumuntur, di lucida explicatio . . . , 1594. Described distillation of brandy; preparation of pulvis praecipitatus; mentioned compass, pistols, potable gold, manufacture of salt, soap, potash, use of saltpetre in cooling wine. Died Zierikzee, July 1, 1568.

LE MOAL, Guy Léon Yves, French ethnologist; b. Paris, Oct. 6, 1924; s. Léon and Yvonne (Marterer) Le M.; ed. U. Paris, Mus. of Man; M.A.; diploma Inst. Ethnol. Research. Became asst. French Inst. of Black Africa, Dakar, 1949; founder Ifan Center of Upper Volta, dir., 1950—. Research and publs. on ethnology, sociology, religions. Home: 65, bd. Pasteur, Paris 15, France. Office: IFAN Center, Ouagadougou, Republic of Upper Volta.

LEMOIGNE, Maurice Auguste, French agronomist; b. Paris, Dec. 16, 1883; chief of services of fermentations, hon. prof. Pasteur Inst.; hon. prof. pres. com. Nat. Inst. Agronomy Research. Decorated order of merit in agr. Mem. French Acad. Scis., Acad. Agr. Address: 56, bd. Barbès, Paris 18, France.

LEMOINE, Émile-Michel-Hyacinthe, French mathematician, engr.; b. Quimper, France, 1840; ed. École polytechnique, Paris; chief Paris Gas Inspection Service, 1886-96; participated in founding math. and phys. socs. of France, French Assn. for Advancement Sci., Jour. de physique; co-founder, became editor l'Intermédiaire des mathématiciens, 1894; founder la Trompette. Research and publs. on properties of triangle and asso. points, lines, and circles, 1873, local probability, geometry, including Lemoine circle; developed system of geometrographics for numerically comparing geometric constrns. Died 1912.

LEMOINE, (Clément)-Georges, French chemist; b. Tonnerre, France, Jan. 16, 1841; named gen. insp. pub. works, 1881; prof. chemistry École polytechnique, Paris; mem. French Acad. Scis., 1899. Author: Études sur les equilibres chémiques, 1882; Études sur l'action chimique de la lumière, 1883; Vitesse de décomposition de l'eau oxygenée, 1914. Research on chem. equilibriums, esterification, iodhydric gas, hydrology; described catalytic effect of porous substances, 1874; originated adsorption hypothesis. Died Paris, Nov. 13, 1922.

LE MOINE, Sir James MacPherson, Canadian naturalist; b. Quebec, Can., Jan. 24, 1825; ed. Quebec; m. Harriet Mary Atkinson, 1856. Author: L'ornithologie du Canada, 1861; Les Pecheries du Canada, 1862; Maple Leaves, 6 vols., 1863-94; The Tourist's Note Book, 1870; Quebec Past and Present, 1876; The Scot in New France, 1879; The Chronicles of the St. Lawrence, 1879; Picturesque Quebec, 1882; Canadian Heroines, 1887; The Birds of Quebec, 1891; Monographies et Esquisses; Legends of the St. Lawrence, 1898; The Annals of the Port of Quebec, 1901. Died Feb. 5, 1912.

LEMON, Harvey B(race), Am. physicist; b. Chicago, Apr. 23, 1885; s. Henry Martyn and Harriet Ella (Brace) L.; B.A., U. of Chicago, 1906, M.S., 1910, Ph.D., 1912; m. Louise M. Birkhoff, Dec. 25, 1907; children—Harriet Birkhoff (Mrs. D. S. Moir), Doctor Henry Martyn. Began as assistant in physics, Univ. of Chicago, 1911, now prof. emeritus. Dir. sci. and edn., Mus. of Science and Industry, since 1950. Trustee, Lewis Institute and Illinois Institute Tech., 1930-42; advisor in physical science to editors Encyclopedia Britannica, 1944—. Served as captain Ordnance Dept., U. S. Army, 1918. Chief physicist Ballistic Lab., Aberdeen Proving Grounds, 1942-43. Fellow Am. Physical Society; mem. A.A.A.S., Am. Assn. Physics Teachers (pres. 1939-40), Delta Upsilon, Phi Beta Kappa, Sigma Xi, Sigma Pi Sigma. Unitarian. Clubs: Quadrangle (past pres.), Chicago Literary (past pres.). Author: From Galileo to Cosmic Rays, 1934; Cosmic Rays Thus Far, 1936; Analytical Experimental Physics (with M. Ference, Jr.), 1943; What We Know and Don't Know about Magnetism, 1945; From Galileo to the Nuclear Age, 1946. Research in spectroscopy (designed a spectrophotometer), on thermionic excitation-high resolution, and on basic principles of high degree activation of charcoal; adapted coconut-shell charcoal for gas masks. Home: 5801 Dorchester Av., Chgo. 60637. Office: Mus. of Sci. and Industry, Jackson Park, Chgo. 60615.

LEMON, Henry Martyn, Am. physician; b. Chgo., Dec. 23, 1915; s. Harvey Brace and Louise (Birkhoff) L.; B.S., U. Chgo., 1938, Sch. Medicine, 1936-38; M.D. cum laude, Harvard, 1940; m. Harriet Tuxbury Qua, May 3, 1941; children—Elizabeth Anne (Mrs. Wendell E. Carr), Harvey Brace, Stanley Moncrief, David Tuxbury, Jennifer Jane. With Boston U. Sch. Medicine, 1946-61, asso. prof. medicine, 1948-61; prof. internal medicine, dir. Eppley Cancer Inst., U. Neb. Coll. Medicine, 1961-68; cons. VA, Boston, Omaha, 1950—. Diplomate Am. Bd. Internal Medicine. Mem. James Ewing Soc., Am. Fedn. Clin. Research, Central Soc. Clin. Research, Neb., Mass. med. socs., A.M.A., A.C.P., Am. Assn. Cancer Research, Am. Soc. Clin. Oncology, Endocrine Soc., Phi Beta Kappa, Sigma Xi, Alpha Omega Alpha. Publs. on quantitative aspects of air-borne infection hemolytic streptococci, transplantations human cancers to hamster cheek pouch, cytologic diagnosis pancreas cancer, metabolism of steroids by cancer tissue; steroid, chemotherapy of human cancer, electrometric timing of human and animal ovulations; abnormal parathyroid function in cancer, abnormal estrogen balance in cancer of female breast lymphoma, leukemia epidemiology. Home: 10805 Poppleton Av., Omaha 68144.*

LEMON, Willis Storrs, physician; b. Villa Nova, Ont., Can., Feb. 8, 1878; s. George and Jane (Honey) L.; M.B., U. Toronto Faculty of Medicine, 1905; m. Ethel M. Haines, June 29, 1909; children—Katherine Ethel (Mrs. George A. Lord) Janette Louise, Willis Edward. Came to U. S., 1909, naturalized, 1917. Intern Toronto Gen. Hosp., 1905-06, Parry Sound Hosp., Ont., 1906-07; demonstrator in pathology and therapy U. Toronto, 1906-07; practice of medicine, Toronto, 1907-08; asso. physician Canadian Nat. Sanatorium for Tb, Gravenhurst, Ont., 1908-09; practiced in La Grange, Ill., 1909-17; asst. in sect. in div. of medicine Mayo Clinic, Rochester, Minn., 1917-18, head of sect., 1918-46; prof. medicine Mayo Found. of U. Minn., 1934-46. Recipient Gold medal (with Dr. S. W. Harrington) for exhibit at meeting of A.M.A., 1935. Mem. A.M.A., Minn., So. Minn. med. socs., Minn. Trudeau Med. Soc. (pres.), Central Interurban Clin. Club (v.p. 1935), Minn. Soc. Internal Medicine, Assn. Am. Physicians, Am. Soc. for Clin. Investigation, Am. Assn. for Thoracic Surgery, Sigma Xi. Contbr. chpt. The Nature of Postoperative Pulmonary Diseases, Prophylactic Measures and Treatment to (book) The Stomach and Duodenum by Eusterman and Balfour; also numerous articles to med. jours. Deceased.

LE MONNIER, Louis-Guillaume, French naturalist; b. Paris, June 27, 1717; s. Pierre LeMonnier; doctorate, U. Paris; prof. botany Jardin du Roi; physician to Louis XV, Louis XVI; ruined by Revolution; opened herb store in Montreuil. Mem. French Acad. Scis., 1743, Royal Soc. Author: Lecons de physique expérimentale sur l'equilibre des liquides, 1742; Lettre sur la culture du café, 1773. First to recognize that air is almost constantly electrically charged; research and publs. on Leiden jar, speed of electricity, magnetic needle. Died Montreuil, France, Sept. 7, 1799.

LEMONNIER, Pierre-Charles, French astronomer; b. Saint-Sever, France, June 28, 1675; children—Pierre-Charles, Louis-Guillaume, Adélaide (Mrs. Lagrange). Prof. philosophy Harcourt Coll., Paris. Mem. French Acad. Scis., 1736. Author: Cursus philosophiae, 1750; Observations faites par ordre du roi pour reconnaitre la distance de Paris à Amiens, 1757; Primiers traités élementaires de mathématiques dictés en l'Université de Paris, 1758. Prepared map of moon; perfected instruments and methods of observation for astronomy. Died Saint-Germain-en-Laye, France, 1757.

LEMONNIER, Pierre Charles, French astronomer; b. Paris, France, Nov. 23, 1715; m. Mademoiselle De Cussy; 3 daus. Made lunar map, which allowed him membership of Acad. Scis., at age 20; asst. to Maupertuis and Cliraut in measuring degree of meridian within polar circle, Lapland, at age 21; became prof. philosophy Coll. Harcourt, Paris; astronomer to Louis XV; Mem. Royal Soc., French Acad. Marine. Author: Histoire céleste, 1741; Theorie des comètes, 1743; Nouveau zodiaque, 1755; Observations de la lune de soleil, et des étoiles fixes, 1751-75; Louis de magnetismé, 1776-78. Made lunar observations for 50 yrs.; improved astron. measurements in France; introduced transit instrument at Paris Obs., also other modern Brit. instruments and methods; recorded observation of Uranus (before its discovery as planet); research on terrestrial magnetism and atmospheric electricity; gave positions of many fixed stars; recorded Saturn's influence on Jupiter's orbit. Died Héril, nr. Bayeux, France, May 31, 1799.

LEMPFERT, Rudolph Gustave Karl, Brit. meteorologist; b. Manchester, Eng., 1875; ed. Emmanuel Coll., Cambridge, Eng.; M.A., Cambridge; m. Marjorie Olive Hayward, 1916 (dec. 1953); 1 dau. Asst. master Rugby Sch., 1900-02; joined staff Meteorol. Office, 1902, asst. dir., 1919-38. Mem. Royal Meteorol. Soc. (became pres. 1930). Co-author: Life History of Surface Air Currents, 1906; author: Meteorology, 1920. Died June 24, 1957.

LENARD, Philipp Edward Anton, physicist; b. Pressburg, Austria-Hungary, June 7, 1862; s. Philipp and Antonia (Baumann) von L.; ed. U. Budapest, U. Vienna; studied under Helmholtz, Berlin, Bunsen and Quincke, Heidelberg; Ph.D., M.D., D.Eng. (hon.); m. Kath. Schlehner. Asst. to Prof. Quincke, 1887, to Prof. Heinrich Hertz, Bonn; privat docent U. Bonn, 1892; extraordinary prof. U. Breslau, 1894; docent Technische Hochschule, Aachen, 1895; prof. theoretical physics U. Heidelberg, 1896; prof. physics, dir. Phys. Inst., U. Kiel, 1898; dir. Phys. Inst., dir. Radiol. Inst., U. Heidelberg, 1907, 09-30. Recipient Nobel prize in physics for work on cathode rays, 1905, Franklin medal, Phila. Editor: Grosse Naturforscher, 1929; Deutsche Physik (Hertz), 4 vols., 1936-37; Wissenschaftliche Abhandlungen aus den Jahren 1886-1932, 2 vols., 1942-43. Work on capillarity, phosphorescence effects of ultraviolet light and electric capacity of bismuth; discovered that cathode rays observed in Crookes' tubes propagated as well in air at atmospheric pressure as in tube, keeping their properties, this work paved way for Roentgen and x-rays (Lenard rays named after him); 1st to prove that in photoelectric effect negative electricity released consists of electrons; 1st to show that electron required minimum amount of energy before it could produce ionization, realized atom must have structure and must consist largely of empty space. Died Messelhausen, Germany, 1947.

LENCHEK, Allen Martin, Am. physicist; b. Bklyn., July 8, 1931; s. Herman and Mildred (Goldstein) L.; B.S., U. Chgo., 1957; Ph.D., U. Md., 1963; m. Jacqueline Patentreger, Mar. 18, 1956; children—Armand, Rose Anne. Research asso. U. Md., 1962-63; fellow Observatoire de Paris (France), 1963-64; vis. scientist Royal Inst. Tech., Stockholm, Sweden, 1964; asst. prof. Grad. Research Center, Dallas, 1964-65; asst. prof. physics U. Md., College Park, 1965—. R. H. Goddard scholar, 1961-62; NATO fellow, 1963. Mem. Am. Phys. Soc., Am. Geophys. Union, Am. Assn. U. Profs., Fedn. Am. Scientists, Sigma Xi. Research, publs. on space physics, cosmic rays, geomagnetically trapped radiation and interplanetary medium. Home: 2C Eastway, Greenbelt, Md. 20770. Office: U. Md. Dept. Physics and Astronomy, College Park, Md. 20742.*

LENDE, Richard Allan, Am. physician; b. Canby, Minn., Aug. 14, 1924; s. Hector C. and Lela (Ennis) L.; B.S., Ore. State Coll., 1948; M.D., U. Ore., 1951; M.Sc., McGill U., 1956; m. Danielle Holmgren, Feb. 26, 1954; children—Leith, Lorinda, Kristin, Robert, Elizabeth, Patience. Nat. Found. for Infantile Paralysis fellow Montreal (Que., Can.) Neurol. Inst., 1955-56; Wis. Alumni Research Found. fellow U. Wis., 1959; faculty U. Colo. Med. Sch., 1959-64, asso. prof., 1964; prof., head subdept. neurosurgery Albany (N.Y.) Med. Coll., 1965—. Mem. Harvey Cushing Soc., Am. Physiol. Soc., Research Soc. Neurol. Surgeons, Am. Acad. Neurology. Research, publs. on function cerebral cortex mammals with emphasis on mode sensory and motor localization. Home: 110 Green Av., Castleton-on-Hudson, N.Y. 12033. Office: Subdept. Neurosurgery, Albany Med. Coll., Albany, N.Y. 12208.*

LENDI, Egon, geographer; b. Trente, Italy, Sept. 1, 1906; s. Franz and Barbara (Neudolt) L.; Ph.D., U. Vienna (Austria); m. Maria Leth, Sept. 3, 1935; children—Gerhard, Wolfgang, Herwig. Asst. curator mus.; instr., prof. University. Author: Salzburg-Atlas; various other works on geography. Home: Fürstenbrunstrasse 4, Salzburg. Office: Geographisches Institut der Universität, Wolf Dietrichstrasse 5, Salzburg, Austria.

LENG, Charles W(illiam), Am. entomologist; b. S.I., N.Y., 1859; B.S., Bklyn. Poly., 1877; pres. John S. Leng's Son and Co., 1888-1919 research asso. Mus. Natural History, 1911-41; made expdns. to Labrador, Nfld., Fla., P.R., Cuba; dir. Pub. Mus., S.I., 1919-41. Mem. Entomol. Soc., N.Y. Entomol. Soc. (pres., sec. 1911-30), S.I. Inst. (sec. 1881-85, 1917-41), Bklyn. Entomol. Soc. Author: Catalogue of the Coleoptera of North America, North of Mexico; (with W. S. Blatchley) Rhynocophora of Northeastern America; Lampyridae of West Indies. Research on geog. distbn. of insects. Died Apr. 6, 1941.

LENGEMANN, Frederick William, Am. phys. biologist; b. N.Y.C., Apr. 8, 1925; s. Peter and Dorathea (Wolter) L.; B.S., Cornell U., 1950, M.Nutritional Sci., 1951; Ph.D., U. Wis., 1954; m. J. Joan Doremus, Dec. 23, 1950; children—Frederick William, David Munson. Research asso. AEC agrl. research program Oak Ridge, 1954-55; asst. prof. chemistry U. Tenn., Memphis, 1955-59; asso. prof. radiation biology Cornell U., Ithaca, N.Y., 1959-62, prof. phys. biology, 1963—; biochemist div. biology and medicine AEC, Washington, 1962-63. Cons. div. radiol. health USPHS, 1960-64, Fed. Radiation Council, 1964. Mem. A.A.A.S., Am. Dairy Sci. Assn., Am. Inst. Nutrition, Sigma Xi, Phi Kappa Phi. Research, numerous publs. on calcium and strontium, absorption, metabolism and deposition in bone, mechanisms of secretion of iodine into milk; deviser system for predicting total intake of radioiodine, cesium and strontium by humans after single deposition of radioactive fallout. Home: R.D. 1, Freeville, N.Y. Office: Dept. Phys. Biology, N.Y. State Vet. Coll., Cornell U., Ithaca, N.Y.*

LENGFELD, Felix, Am. chemist; b. San Francisco, Feb. 18, 1863; s. Louis and Henrietta (Honisberg) L.; Ph.G., Coll. Pharmacy, U. Cal., 1880; spl. student U. Cal., 1884-86; Ph.D., Johns Hopkins, 1888; studied Zürich Polytechnicum, U. Liège, U. Munich. Prof. chemistry and assaying, S.D. Coll. Mines, 1890-91; instr. chemistry, U. Cal., 1891-92; instr. chemistry and asst. prof., U. Chgo., 1892-1901; sec., later pres. Lengfeld's Pharmacy, San Francisco. Author: Inorganic Preparations. Editor dept. pharmacy and chemistry Cal. State Med. Jour. Died 1938.

LENHER, Victor, Am. chemist; b. Belmond, Ia., July 13, 1873; s. Levi H. and Susan (Keller) L.; student Dickinson Coll., 1889-90; U. Pa., 1893, Ph.D., 1898; m. May Blood, Aug. 29, 1900; children—George, Sam. Asst. in chemistry U. Cal., 1893-96, Columbia, 1898-1900; instr. chemistry, Evening High Sch., N.Y., 1899-1900; asst. prof. gen. and theoretical chemistry U. Wis., Madison, 1900-04, asso. prof. chemistry, 1904-07, prof., from 1907. Mem. com. on uses of selenium and tellurium NRC, 1919-24, chmn., 1921-24. Author: Laboratory Experiments, 3 edits., 1902; The Electric Furnace (transl. of Moissan), 2 edits., 1904, 20. Discovered the unusual solvent properties of selenium oxychloride, 1919. Died June 12, 1927.

LENHOSSEK, Mihaly Michael von, Hungarian anatomist; b. 1863; ed. Tübingen (Germany) U., Budapest (Hungary) U.; prof. anatomy Budapest U.; mem. Hungarian Acad. Scis. (v.p.). Author: Die Entwicklung des Glaskorpers, 1903; Utmutatas anatomia gyakorlatokhoz, 1908; Azember helyeaterieszetben, 1914; Azember anatomiaja, 1922. Described ascending roots of vagus and glossopharyngeal nerves (bundle of Lenhossek), 1894. Died 1937.

LENIHAN, John Mark Anthony, Brit. physicist; b. Carlisle, Eng., June 23, 1918; s. Thomas Arthur and Frances (O'Donoghue) L.; B.Sc., U. Durham (Eng.), 1938, M.Sc., 1941; Ph.D., U. Glasgow (Scotland),

1949; m. Nancy Melrose, June 5, 1953; children—Francis John, Christine Jane. Staff, physics dept. U. Durham, 1940-45; lectr. natural philosophy U. Glasgow, 1945-48; physicist Western Infirmary Glasgow, 1948-52; regional physicist Western Regional Hosp. Bd., Glasgow, 1952——. Mem. Sec. of State's Adv. Com. on Physics, 1963——. Fellow Royal Soc. Edinburgh, Inst. Physics (past v.p.); mem. Royal Philos. Soc. Glasgow (v.p. 1963-66), Instn. Elec. Engrs. Author: Textbook of Electronics, 1948; Atomic Energy and its Applications, 1954; also articles. Editor: (with S. J. Thomson) Activation Analysis: Principles and Applications, 1965; (with J. F. Loutit, J. H. Martin) Strontium Metabolism, 1967. Made precise determination of speed of sound by electronic method; developed clin. uses of activation analysis, large scale applications of physics to medicine. Home: 1 Kingsborough Gardens, Glasgow W.2. Office: 9-13 W. Graham St., Glasgow C.4., Scotland.*

LENIN, Vladimir Ilich (orig. surname Ulyanov) (Nikolai), Russian political thinker, revolutionist; s. Ilya and Maria (Blank) U.; ed. U. Kazan, 1887 (expelled for participating in student demonstrations); U. St. Petersburg (passed law exam), 1891; m. Nadezhda Konstantinovna Krupskaya, 1898. Started law practice, Samara, Russia, 1892; moved to St. Petersburg, 1893; traveled to Switzerland, 1895; returned to Russia; arrested for Menshevik affiliations, 1895; exiled to eastern Siberia for 3 years; released, 1900; organized (with Plekhanov) underground Social Democratic Party for purpose of training leaders for working class revolution against Tsar, 1900; went to Switzerland to join with Martov in publishing newspaper Iskra (the Spark); at party congresses at Brussels and London opposed views of Martov and Menshiviks, 1903; became leader of Bolshevik wing of Russian Marxist movement, 1903-13; returned to Russia at outbreak of revolution, 1905; left Russia in 1907 for party meeting in London; declared Bolsheviks a separate party, Prague, 1912; enabled by German govt. to return to Russia after overthrow of Tsar; led overthrow of provisional govt. of Kerensky in Oct. Revolution, 1917; became chmn. of Council of People's Commissars, 1917; proclaimed formation of Third (or Communist) International, 1919; became head of newly formed Union of Soviet Socialist Republics, 1922; launched New Economic Policy, 1923. Author: The Development of Capitalism in Russia (under pseudonym Vladimir Ilin), 1899; Workers of the World Unite, 1897; What is to be done, 1902; Materialism and Empiriocriticism, 1909; Imperialism, the Highest Stage of Capitalism, 1916; The State and Revolution, 1917; Sochineniya (4th ed., 35 vols.), 1941-59. Founder and leader of Bolshevism; founder of Soviet Union; believed imperialism (epitomized in World War I) was final stage of capitalism; felt war would provide opportunity for proletariat revolt; urged proletariats to oppose war by internat. civil war against capitalist class; contributed to Marxist thought by concept of revolutionary party as highly disciplined unit, by his analysis of imperialism and colonial areas as targets for anti-capitalist action. Died Gorki, near Moscow, Russia, Jan. 21, 1924.

LENNARD-JONES, Sir John Edward, English chemist, physicist; b. Oct. 27, 1894; D.Sc., Manchester U. Sc.D., Cambridge U.; D.Sc. (hon.), Oxford U., 1954; m. Kathleen Mary Lennard, 1925; 1 son, 1 dau. Lectr. math. Manchester U.; reader in math. physics, prof. theoretical physics Bristol U., dean faculty sci., 1930-32; Plummer prof. theoretical chemistry U. Cambridge, 1932-53, also life fellow Corpus Christi Coll.; prin. Univ. Coll., N. Staffordshire, from 1953-54. Served as key scientist, chief supt. armament research, chief sci. officer, dir.-gen. sci. research Ministry Supply, 1939-45; Recipient Hopkins prize Cambridge Philos. Soc., 1953. Fellow Royal Soc., 1933, Davy medal, 1953; pres. Faraday Soc., 1948-50. Research and publs. on electronic structure of molecules, surface chemistry, chemistry of carbod, liquid structure, interatomic forces. Died Nov. 1, 1954.

LENOIR, Jean Joseph Étienne, inventor; b. Mussy-la-Ville, Belgium, Jan. 12, 1822; self educated; worked for enameler; recipient Montyon prize Inst. de France, prize Société d'Encouragement, Gold medal French Automobile Club, prize of Argenteuil in Hohe for new leather tanning methods. Invented electric brake for railroads, 1855, electric motor and complete signal system for railroads, 1856, mech. kneading trough and gov. for electric motor, 1857, 1st workable internal combustion engine, using illuminating gas as fuel, 1859, 1st automobile with internal combustion engine, 1862, method for tanning leather with ozone, 1880, motorboat with internal combustion engine, 1886; improved Parkes method for galvanoplastic reprodn. of high relief statues, 1851. Died Varenne-St. Hilaire, France, Aug. 4, 1900.

LENORMANT, Charles, French archeologist, historian; b. Paris, June 1, 1802; originally studied law; named insp. fine arts, 1825; accompanied Champollion to Egypt, 1828; named prof. history Sorbonne, U. Paris, 1835; mem. commn. to explore the Moreas, Greece; named curator Library of Arsenal, 1830, also Royal Library; became asst. curator Cabinet of Antiques at Louvre, 1832, curator printed books, 1837, dir. Cabinet medals, 1841; named prof. Egyptian archeology College de France, 1848. Author: des Artistes contemporains, 1833; Trésor de numismatique et de glyptique, 20 vols., 1834-50; Musée des antiquites egyptiennes, 1835-42; (with de Witte) Elite des monuments céramographiques, 4 vols., 1837-61; Introduction to Oriental History, 1838; also numerous articles on art and archeology. Died Nov. 24, 1859.

LENORMANT, François, French archeologist, historian; b. Paris, 1837; s. Charles Lenormant; sub-librarian at Inst., 1862-72; prof. archeology Bibliothèque nationale, 1874-83; numerous trips to Italy, Greece, Middle East; excavations at Eleusis and along Sacred Ways. Recipient prize in numis. Académie des inscriptions, 1856. Author: A Manual of Ancient History of the East, 1869-70; Les origines de l'histoire d'apres la Bible, 1880-84; Les sciences occultes en Asie; La monnaie dans l'antiquite, 1878; also articles on Greek archaeology. First to recognize the Semitic lang., Akkadian, from cuneiform inscriptions; studied ancient nations of the East especially origins and early forms of their civilizations. Died Paris, Dec. 10, 1883.

LE NORTE, André, French landscape architect; b. Paris, 1613; studied under Simon Vouet, a painter; studied in Italy, beginning in 1778; gardener to Louis XIV, Versailles, succeeded his father as supt. Tuileries, and planned Av. of Tuileries; named dir. Royal Gardens by Louis XIV; planned parks at Versailles, Dijon, Chartilly, Saint-Cloud, and others; landscaped Versailles Palace. Created jardin français as garden form. Died 1700.

LENOX-CONYNGHAM, Gerald Ponsonby, Brit. geophysicist, geodicist; b. Moneymore, County Derry, Ireland, Aug. 21, 1866; s. Sir William Fitzwilliam and Laura (Arbuthnot) L.-C.; ed. Edinburgh Acad., 1875-82, oyal Mil. Acad., 1883-85, Sch. Mil. Engring., Chatham, 1885-87; M.A.; m. Elsie Margaret Bradshaw, Nov. 15, 1890; 1 dau. Mem. Survey of India, 1889-21; supt. Trigonometrical Survey of India, 1912-21; reader in geodesy Cambridge U., 1922-47. Fellow Royal Astron. Soc., Royal Geog. Soc.; mem. Royal Soc. London. Research on gravity measurements in India; constructed pendulum apparatus (still in use for most precise long distance gravity connections); founded dept. geodesy and geophysics at Cambridge; developed teaching of surveying; his geodetic work in India combined practical importance for survey with wider importance for understanding structure of country. Died Oct. 27, 1956.

LENSE, Josef, mathematician; b. Vienna, Austria, Oct. 28, 1890; s. Ignaz and Josefine (Chocholauschek) L.; Ph.D., U. Vienna; m. Eugenie Krenbauer, Apr. 20, 1927. Asst., instr. U. Vienna; asso. prof., later titular prof. math. Munich (Germany) Tech. Coll., dir. math. inst., 1946-61. Mem. Bavarian Acad. Scis., German Union Mathematicians, Soc. for Applied and Mech. Math., Austrian Soc. Math., Circoló Mathematico di Palermo. Author: Reihenetwicklungen in der math. Physik; Kugelfunktionen. Research and publs. on higher math., theoretical astronomy and physics, math. axioms. Address: Herzogstrasse 62, Munich, W. Germany.

LENSKI, Gerhard Emmanuel, Jr., Am. sociologist; b. Washington, Aug. 13, 1924; s. Gerhard Emmanuel and Christine (Umhau) L.; B.A., Yale, 1947, Ph.D., 1950; LL.D., Wittenberg U., 1964; m. Jean Virginia Cappelmann, Aug. 14, 1948; children—Jean Christine, Robert Gerhard, Katherine Lee, Richard Eimer. Faculty, U. Mich., 1950-63; prof. sociology U. N.C., Chapel Hill, research prof. Inst. Research in Social Sci., 1963——, acting chmn. dept., 1965-66. Mem. com. sci. personnel Social Sci. Research Council, 1965-66. Mem. Am. Sociol. Assn. (chmn. MacIver award com. 1966), So. Sociol. Soc. Author: (with Freedman, Hawley, Landecker, Miner) Principles of Sociology, 1956; The Religious Factor, 1961; Power and Privilege, 1966. Contbns. to devel. synthetic theory of stratification; research on status inconsistency, behavioral effects of religion. Home: 404 Westwood Dr., Chapel Hill, N.C. 27514.*

LENTI, Camillo, Italian biochemist; b. Alexandria, Italy, Mar. 28, 1910; became univ. prof., 1954; chair biochemistry, dir. inst. biochemistry, univs. Sassari, Turin (Both Italy). Author: Manuale di biochimica. Address: Università degli Studi, Turin, Italy.

LENZ, Friedrich Alfred, German physicist; b. Herrsching, Mar. 21, 1922; s. Fritz and Emilie (Weltz) L.; ed. Berlin, Keil, Göttingen, Aachen; Ph.D.; m. Fredeke von Alvensleben, May 26, 1950; children—Gerlinde, Udo, Reimar. Advanced from asst. to instr., then prof. Mem. German Soc. for Electronic Microscopy, German Physics Soc. Research and publs. on electron scattering, electron optics, electron interferences. Home: Bohnenbergerstrasse 21, Tübingen. Office: Zeppelinstrasse 6, Tübingen, W. Germany.

LENZ, Heinrich Friedrich Emil, physicist; b. Dorpat (now Tartu, Estonia), Feb. 2, 1804; accompanied Otto of Kotzebue on his 2d trip around world, 1823-26; prof. U. St. Petersburg, Russia; mem. Acad. Scis. St. Petersburg (rector). Author: Manual of Physics, 1864; also articles. Research on conductivity of materials and its relation to temperature; discovered law relating heat and current (Joule's law), Lenz's law on direction of induced current in electromagnetic induction, dependence of elec. resistance on temperature. Died Rome, Feb. 10, 1865.

LENZEN, Victor F., Am. physicist; b. San Jose, Cal., Dec. 14, 1890; s. Theodore W. and Kate (Schnoor) L.; B.S., U. Cal. at Berkeley, 1913; Ph.D., Harvard, 1916; m. Esther Valee Hayden, July 13, 1935. Asst. in philosophy Harvard, 1917-18; with U. Cal. at Berkeley, 1918——, prof. physics, 1939-58, prof. emeritus, 1958——. Author: The Nature of Physical Theory, 1931; Procedures of Empirical Science, 1938; Causality in Natural Science, 1952; also articles. Research in philosophy of sci., history of physics, sci. work of C. S. Peirce. Office: Physics Dept., U. Cal., Berkeley, 94720.*

LEO AFRICANUS (or al-Hasan ibn-Muhammad al-Wazzan, also Johannes Leo), geographer; b. Granada, Spain, circa 1494; ed. Fez; traveled extensively on coml. and diplomatic missions in western N. Africa as early as 1512; in Sudan and Sahara regions, 1513-15; later visited Constantinople, Egypt, Arabia, Armenia Tabriz; ascended Nile from Cairo to Assuan; captured by pirates and sent to Rome, 1517, converted to Christianity, learned Italian and Latin. Author: Descrittione dell 'Africa (written in Arabic, Italian transl. 1550, only source of knowledge of geography of Sudan for many years, main source for knowledge of Islam in Europe until 19th century). Died Tunis, N. Africa, circa 1552.

LEODAMAS OF THASOS, mathematician; flourished circa 380 B.C.; studied under Plato; Plato invented method of analysis for him. Helped increased number math. theorems and arranged them in more sci. system.

LEON, mathematician; flourished circa 365 B.C.; studied under Neoclides. Author: Elements of Geometry (number of propositions and use of proved propositions superior to Hippocrates). Introduced or improved diorismi (definition, specification).

LÉON, Gustave, mining engr.; b. 1863; invented portable fire damp gauge consisting of a Wheatstone bridge, half of which was in pure air, the other half in the air to be tested. Died 1916.

LEON, Nelson, geneticist; b. Izmir, Turkey, Oct. 4, 1923; s. Rafael and Donna (Cohen) L.; student Istanbul, Turkey, Oct. 4, 1923; s. Rafael and Donna (Cohen) L.; grad. Istanbul (Turkey) U., 1948; postgrad. U. Rome (Italy). Fellow, Sao Paulo (Brazil) U., research fellow 1st Med. Clinic, Hosp. das Clinicas, Faculty of Medicine, 1953-63, asst. Lab. for Med. Genetics, 1960——. Recipient Vicente-Amato Neto Usofarma Lab. award. Mem. Sociedade Brasileira de Endocrinologia and Metabologia (founding). Research, publs. on etiopath. mechanisms of rare diseases, gonadal dysgenesis, effects of Klinefelter's syndrome, phenotypes of abnormal gentoypes. Home: 87 Nestor Pestana, Sao Paulo, S.P., Brazil.*

LEON OF THESSALONICA, Byzantine inventor, encyclopedist; flourished 829-67; archbishop of Thessalonica; prof. Univ. Magnause palace, Constantinople; author med. ency.; wrote on math., medicine, astrology, with spl. attention to surgery; inventor mech. devices of Magnaura; helped initiate Byzantine renaissance of 9th and 10th centuries.

LEONARD, Arthur Gray, Am. geologist; b. Clinton, N.Y., Mar. 15, 1865; s. Delavan Levant and Mary Louise (Raymond) L.; A.B., Oberlin, 1889, A.M., 1895; Ph.D., Johns Hopkins, 1898; m. Katherine Gue, Oct. 8, 1901. Spl. asst. Ia. Geol. Survey, 1893-97; prof. geology and related science, Western Coll., Toledo, Ia., 1894-96; asst. state geologist Ia. Geol. Survey, 1896-97, 1900-03; asst. prof. geology Mo. State U., 1899-1900; state geologist of N.D., and prof. geology U. N.D., Grand Forks, from 1903. Asst., U. S. Geol. Survey, 1905-07. Died Dec. 17, 1932.

LÉONARD, Camille, French mech. engr.; b. l'Aude, France, 1880; asst. to head engr. tech. dept. Société générale de constructions électriques et mécaniques; invented transmission system which transformed accumulator current to direct current with variable tension; this made it possible to control the motor of a machine under considerable stress. Died 1939.

LEONARD, Nelson Jordan, Am. chemist; b. Newark, Sept. 1, 1916; s. Harvey Nelson and Olga (Jordan) L.; B.S., Lehigh U., 1937, D.Sc., 1963; B.Sc., Oxford (Eng.) U., 1940; Ph.D., Columbia, 1942; m. Louise Cornelie Vermey, May 10, 1947; children—Kenneth Jan, Marcia Louise, James Nelson, David Anthony. Fellow, asst. chemistry U. Ill., Urbana, 1942-43, faculty, 1943——, prof., 1952——, head div. organic chemistry, 1954-63. Investigator antimalarial program, com. med. research OSRD, 1944-46; sci. cons., spl. investigator Field Intelligence Agy. Tech., U. S. Army, Dept. Commerce, 1945-46; mem. adv. panel chemistry NSF, 1948-61; mem. program com. in basic phys. scis. Alfred P. Sloan Found., N.Y.C., 1961-67. Rockefeller Found. fellow, 1950; Swiss-Am. Found. fellow 1953; Guggenheim Found. fellow, 1959, 67. Fellow Am. Acad. Arts and Scis.; mem. Nat. Acad. Scis., Am. Chem. Soc. (award for creative work in synthetic organic chemistry 1963), A.A.A.S., Chem. Soc. London, Swiss Chem. Soc., Gesellschaft Deutsch Chemiker, Phi Beta Kappa, Sigma Xi, Alpha Chi Sigma. Discovery new functionality in organic compounds by means of transannular reactions; synthesis of analogs of coenzymes, cofactors and nucleosides possessing biol. activity, organic synthesis involving small charged rings, synthesis of compounds possessing cytokinin activity; co-discoverer electrolytic method synthesis of medium size ring compounds; correlation of conformation and ultraviolet spectra of 1, 2-diketones. Home: 606 W. Indiana St., Urbana, Ill. 61801.*

LEONARD, Raymond Jackson, Am. geologist; b. Rosalia, Wash., 1887; s. Ralph Leonard; grad. mining Poly. Coll. Engring., Oakland, Cal., 1908; Ph.D., U. Minn.; m. Madge MacClung, 1912; children—Jackson, Robert. Engaged in mining, Republic, Wash., 1908-17; student, instr. Ore. State Coll., 1921-23, U. Minn., 1923-26; successively asst. prof., asso. prof., prof. geology U. Ariz., 1926-37, dean grad. coll., 1934-37. Research and publs. on hydrothermal alteration of certain silicate minerals, also other problems in chem. geology. Died Tucson, Nov. 20, 1937.

LEONARD, Robert Walton, Am. physicist; b. San Diego, Apr. 6, 1910; s. James Walton and Jessie Almira (Norton) L.; B.A., U. Cal. at Berkeley, 1935, M.A., 1936, Ph.D., 1940; m. Viola Groch, Dec. 18, 1936 (div.); children—James Walton, William Curtis, Katherine Marie. Mem. faculty U. Cal. at Los Angeles, 1941—, prof. physics, 1955—. Fellow Am. Phys. Soc., Acoustical Soc. Am. (pres. 1962-63). Research and publs. on propagation of sound in gases, liquids and solids; generation and measurement of high intensity sound; modeling techniques in arch. acoustics. Contbr. sections handbook. Home: 3320 Pacific Av., Venice, Cal. Office: Physics dept. U. Cal., Los Angeles 90024.*

LEONARD, Samuel Leeson, Am. endocrinologist; b. Elizabeth, N.J., Nov. 26, 1905; s. Harry L. and Elsie (Mould) L.; B.S., Rutgers U., 1927; M.S., U. Wis., 1929, Ph.D., 1931; m. Olive Lucile Rees, Aug. 26, 1934; children—David, Patricia (Mrs. Lawrence Hoard). NRC fellow Columbia Coll. Phys. and Surg., 1931-33; asst. prof. Union U., 1933-37, Rutgers U., 1937-41; faculty Cornell U., Ithaca, 1941—, prof. zoology, 1950—. Fellow A.A.A.S.; mem. Am. Soc. Zoologists, Am. Soc. Anatomists, Am. Soc. Physiologists, Endocrine Soc., Soc. for Exptl. Biology and Medicine. Research, numerous publs. on hormone inter-relationships, sperm transport, hormone control mammary glands, mechanisms hormone action by study enzyme changes in tissues. Home: 107 Eastwood Av., Ithaca, N.Y. 14850.*

LEONARD, Sherman John, Am. food technologist; b. Oakland, Cal., Apr. 10, 1916; s. Sherman and Mabel (White) L.; B.S., U. Cal., Berkeley, 1940; m. Anesiades Salati, Nov. 24, 1957; children (by previous marriage)—Susannah Grace (Mrs. William Thomas Waller), Sherman Charles, Joseph Shinn, Sarah Millicent, John Salati. Sales engr. Food Machine & Chem. Corp., 1940-43; asst. market specialist processed food div. food adminstrn. U. S. Dept. Agr., 1943-44; chief chemist Shuckle & Co., Inc., 1944-50; partner St. Clair Packing Co., 1950-51; with U. Cal., Davis 1951—, lectr., food technologist, 1967—. Adviser food tech. to various Latin Am. countries; internat. dir. FAO, UN, Tropical Center Food Research and Tech., Campinas, Brazil, 1964-65. Mem. Inst. Food Technologists, A.A.A.S., Sigma Xi. Research, publs. on food handling, processing and packaging of tropical and sub-tropical foods; new process and new product devel., including high-temperature, short-time heat treatment, aseptic processing, bulk handling, bulk processing, and bulk storage and fruit concentrates. Home: 1311 Cornell Dr., Davis, Cal. 95616.*

LEONARD, Warren Henry, Am. agronomist; b. New Sharon, Ia., July 5, 1900; s. Edward James and Zilla (Miller) L.; B.Sc., Colo. State U., 1926; M.Sc., U. Neb., 1930; Ph.D., U. Minn., 1940; m. Editha Todd, June 4, 1931; 1 dau., Kay. Asst. extension agronomist Colo. State U., Ft. Collins, 1926-27, asst. editor publs., 1928, faculty agronomy, 1929—, prof., 1946—; asst. in agronomy U. Neb., 1928-29. Asst. agronomist Colo. Agrl. Expt. Sta., 1929-34, asso. agronomist, 1934-46, agronomist, 1946—. Chief Agr. div. Natural Resources sect. SCAP, Tokyo, Japan, 1948-49; survey U. Peshawar (West Pakistan), FOA and Colo. State U., 1954; cons. in agrl. prodn. Mission to Libya, Internat. Bank for Reconstrn. and Devel., 1959. Recipient Profl. Achievement award Colo. State U. Alumni Assn., 1957. Fellow Am. Soc. Agronomy, A.A.A.S., Population Reference Bur.; mem. Genetics Soc. Am., Am. Genetics Assn., Genetics Soc. Japan, Genetics Soc. Can., Am. Statis. Assn., Biometrics Soc., Bot. Soc. Am., Am. Soc. Sugar Beet Technologists, Am. Inst. Biol. Scis., Japanese Soc. Breeding (hon.), Sigma Xi, Phi Kappa Phi, Gamma Sigma Delta. Author: (with John H. Martin) Principles of Field Crop Production, 1949; (with R. S. Whitney) Field Crops in Colorado, 1950; (with Donald R. Wood) General Field Crops Laboratory Manual, 5th edit., 1959; (with E. L. LeClerg and Andrew Clark) Field Plot Technique, 2d edit., 1962; (with John H. Martin) Cereal Crops, 1963; also numerous other tech. works on corn improvement, barley genetics, sorghum breeding, applied statistics, Japanese agr. Home: 525 Whedbee St., Ft. Collins, Colo. 80521.*

LEONARDI, Piero, Italian geologist, prehistorian; b. Valdobbiadene, Treviso, Italy, Jan. 29, 1908; s. Giuseppe and Caterina (Zuanon) L.; Dr.Rerum Naturalium, U. Padua (Italy), 1931; m. Elisa Giada, Dec. 8, 1935; children—Maria, Giuseppe, Antonia, Giovanni. Fellow geology and paleontology U. Padua, U. Ferrara, 1935, faculty, 1935, prof. geology, dir. Geology Inst., 1949—; dir. Istituto Ferrarese di Paleontologia Umana, 1950—. Recipient Nat. prize Accademia dei Lincei, 1958; Gold medal of culture and edn., 1963; named knight comdr. S. Silvestro Papa, 1964. Mem. Italian Geol. Soc. (past pres.), Italian Inst. Prehistory and Protohistory (past pres.), Acad. Lincei, Istituto Veneto di Scienze Lettere e Arti. Author: L'evoluzione dei viventi, 1950; Carlo Darwin, 1966; Le Dolomiti, 1967; Treatise of Geology, 1968; also numerous memoirs, articles. Research on stratigraphy and tectonics of Dolomites Region, paleontology of triassic invertebrates and permian vertebrates; developed new theory on evolution; excavated prehist. layers with discovery of new Mousterian culture, the Bernardinian; research on geology of moon. Home: 2521 S. Polo, Venice, Italy. Office: 32 Corso Ercole I°d 'Este, Ferrara, Italy.*

LEONARDO DA PISA, see Fibonacci, Leonardo.

LEONARDO DA VINCI (Leonardo di Ser Piero da Vinci), Italian artist, scientist; b. Vinci, Apr. 15, 1452; illegitimate s. Ser Piero da Vinci and Caterina. Raised by his father; apprentice under painter Andrea del Verrocchio, circa 1466-77; under patronage of Lorenzo de Medici, studied anatomy, astronomy, botany, math., engring., music, 1477-82; as protégé of Ludovico Sforza, Duke of Milan, served as court painter, chief engr., dir. pub. works, pageant-master (may also have been dir. of acad. arts and scis. founded by duke), 1482-99; went to Venice, when Ludovico was expelled from Milan by King Louis XII of France, 1499; became architect, engr. gen. for Cesare Borgia, Duke of Valentinois, travelled through most of central Italy as Borgia's chief mil. engr. during Romagna campaigns, 1502-03; returned to Florence, 1503; then to Milan (by invitation of Charles d'Amboise), 1506; apptd. painter, engr. in ordinary by Louis XII, 1507-13; entered the service of Medici, Rome, 1513, encountered difficulties with charges of impiety and body-snatching in connection with his anat. studies; subsequently accepted offer of King Francis I of France to come to Castle Cloux, nr. Amboise, France, 1516-19. Author: many notebooks in manuscript, generally written in mirror script; some of them have been published by later editors, e.g., Codex sul volo degli uccelli; Del motu e misura dell' acqua; De ludo geometrico; Tratto di anatomici; and Trattato della pittura. One of greatest, most versatile geniuses of all times; his art marked a new era in Italian painting; famous for his engring. accomplishments; designed gliders (based on observations of birds in flight) which have proven entirely practical, also designed a parachute, Martesana canal, irrigation systems and other engring. projects, models for canalization and control of Arno River; constructed first elevator; invented waterbellows, a mercury siphon clock, a steam gun; made archtl. studies on stresses of arches, walls, and columns; helped design Milan cathedral; made important studies of human anatomy based on his dissection of human cadavers; studied heart and valves; probably first to make wax castings of cerebral ventricles; considered founder of iconographic and physiologic anatomy; gave accurate representation of uterus and statuatory position of fetus in utero; knew of camera obscura; understood that falling bodies accelerate as they fall; grasped impossibility of perpetual motion; showed effective weight of body on inclined plane is proportional to angle of plane; knew force parallelogram, but failed to generalize its use; considered possibility of long-continued changes in structure of earth, and studied fossils; first to explain phenomenon of earth-light on moon; observed compression and resistance of air; described coalescence of liquid drops; compared densities of liquids by balancing columns of liquid in U-tube. Died Castle Cloux, nr. Amboise, May 2, 1519.

LEONE, Charles Abner, Am. zoologist; b. Camden, N.J., July 13, 1918; s. Charles and Hilda (Barnett) L.; B.S., Rutgers U., 1940, M.S., 1942, Ph.D., 1949; m. Madelyn Ann Danner, Nov. 20, 1941; children—Charles Timothy, Patricia Ann, Dennis Alan. Faculty zoology Rutgers U., 1940-49; faculty U. Kans., Lawrence, 1949—, prof. zoology, 1959—. Research asso. Argonne Nat. Lab., 1955-56, cons. bio-med. div., 1955-60; cons. Schering Corp., 1964—, Sun Oil Co., 1965—. Fellow A.A.A.S., N.Y. Acad. Sci.; mem. Am. Assn. Immunologists, Radiation Research Soc., Kans. Acad. Sci. Author: Ionizing Radiations and Immune Processes, 1962; Taxonomic Biochemistry and Serology, 1964. Contbns. to field of teaching grad. students with guidance to advanced degrees; research on effects of ionizing radiations of immune processes and classification of animals using serological means. Home: 450 Nebraska St., Lawrence, Kans. 66044.*

LEONE, Giovanni Battista Carcano, Italian anatomist; b. Italy, 1536; First clear description of lacrimal duct with its exact anat. position, 1574. Died 1606.

LEONHARD, Gustav, German mineralogist; b. Munich, Germany, Nov. 22, 1816; s. Karl Cäsar von Leonhard; prof., Heidelberg, Germany. Author: Handwörterbuch der topographischen Mineralogie, 1842; Grundzüge der Mineralogie, 1851; Die Mineralogie Badens, 1852; Grundzüge der Geognosie und Geologie, 1851. Died Heidelberg, Nov. 27, 1878.

LEONHARD, Karl, German physician; b. Edelsfeld, Germany, Mar. 21, 1904; s. Oskar and Julie (Maier) L.; m. June 8, 1931; children—Volkmar, Hildegard, Waltraud. Head physician Gabersee sanitarium, 1931-35, Erlangen sanitarium, 1935-36; asso. prof., head physician neurol. clinic, U. Frankfurt/Main, 1936-55; prof. neurol. clinic Med. Acad., Erfurt, 1955-57; prof. Charité neurol. clinic Humboldt U., Berlin, 1957—. Author: Die defektschizophrenen Krankheitsbilder, 1936; Involutive und idiopathische Angstdepressionen in Klinik und Erblichkeit, 1937; Gesetze und Sinn des Träumens, 1939; Ausdruckssprache der Seele, 1949; Grundlagen der Psychiatrie, 1948; Grundlagen der Neurologie, 1951; Manual de psiquiatria, 1953; Individualtherapie und Prophylaxe der hysterischen, anankastischen und sensohypochondrischen Neurosen, 1959; Biologische Psychologie, 4th edit., 1966; Individualtherapie der Neurosen, 2d edit., 1963; Kinderneurosen und Kinderpersönlichkeiten, 2d edit., 1963; Instinkte und Urinstinkte in der menschlichen Sexualität, 1964; Differenzierte Diagnostik der endogenen Psychosen, abnormen Persönlichkeitsstrukturen und neurotischen Entwicklungen, 1964; Prognostische Diagnose der endogenen Psychosen, 1964; Normale und abnorme Persönlichkeiten, 1964; Die Klinische Lokalisation der Hirntumoren in der Kritik der technischen, bioptischen und autoptischen Nachprüfung, 1965; also numerous articles. Research in neurology, psychiatry, psychology. Home: 23 Tiniusstrasse, Berlin-Heinersdorf. Office: 20-21 Schumannstrasse, 104 Berlin, East Germany.*

LEONICENO, Nicolo, Italian physician; b. Lonigo, Italy, 1428; doctorate physics, Padua, Italy. Tchr. math., moral philosophy, Ferrara, Italy, from 1464. Author: Plinii et aliorum plurum auctorum, que de simplicibus medicaminibus scripserunt, erores notati, 1491; De Plinii et plurium aliorum in medicina erroribus, 1492; Epidemia, quam Itali morbum Gallicum, Galli vero Neapolitanum vodant, 1497. Translator: Aphorisms (Hippocrates), 1509; The History of Dio Cassium, The Dialogues of Lucian; also works of Galen including, De motu musculorum. Research on herpetology; described syphilitic hemiplegia, 1497; recognized the fact that syphilis may produce visceral lesions; criticized Pliny's med. errors. Died 1524.

LEONTEV, Aleksey Nikolaevich, Russian psychologist; b. 1903; D.Pedag. Sci. With RSFSR Acad. Pedagogical Sci., v.p., 1960—; prof. Recipient Ushinsky medal. Author: The Development of Memory, 1931; An Outline of Mental Development, 1947; The Psychological Bases of Consciousness in Learning, 1947; The Nature and Formation of the Mental Characteristics and Processess in Man, 1955; The Development of the Mind, 1959; The Formation of Aptitudes, 1960; The Social Nature of Man's Mind, 1961; co-author: The Restoration of Movement, 1945. Sci. editor Man (children's ency.), 6th vol., 1960; editor: The Psychology of Creative Thought, 1960. Research and publs. on child and pedagogical psychology, theoretical and exptl. study of aptitudes. Address: RSFSR Acad. Pedagogical Sci., B. Polyanka 58, Moscow, USSR.

LEONTIEF, Wassily Wassily, economist; b. Leningrad, Russia, Aug. 5, 1906; s. Wassily W. and Eugenia (Becker) L.; M.S., U. Leningrad, 1925; Ph.D., U. Berlin (Germany), 1928; Dr. Honoris Causa, U. Brussels (Belgium), 1962; Dr. of Univ., York University; m. to Estelle Marks, December 23, 1932; 1 dau., Svetlana (Mrs. Paul Alpers). Came to U. S., 1931, naturalized, 1939. Research asso. U. Kiel, 1927-28; econ. adviser Chinese Govt., Nanking, 1928-29; research asso. Nat. Bur. Econ. Research, 1931; faculty Harvard, 1931—, Henry Lee prof. econs., 1946—. Cons. U. S. Dept. Labor, 1941-47, 61—, UN Sec.-Gen's Consultative Group Econ. and Social Consequences Disarmament, 1961-62. Decorated officer of the Legion d'Honneur (France); also Order Cherubim, U. Pisa (Italy). Hon. fellow Royal Statis. Soc. Gt. Britain; mem. Am. Philos. Soc., Am. Acad. Arts and Scis., Royal Econ. Soc., Internat. Statis. Inst. Author: The Structure of American Economy, 1951; (with others) Studies in the Structure of American Economy, 1953; Input-Output Economics, 1966; Essays in also numerous articles. Office: Harvard U., Littauer Center, Cambridge, Mass. 02138.*

LEONTOVICH, Mikhail Aleksandrovich, Russian theoretical physicist; b. Mar. 9, 1903; grad. Moscow U., 1923. With Biophys. Inst., later Kursk Magnetic Anomaly Research Commn., from 1923; asso. Inst. Physics, Moscow U., 1929—, prof., 1934-45, 55—; asso. Physics Inst., USSR Acad. Sci., 1934-41, 46-52, asso. Inst. Atomic Energy, 1951—, bur. mem. dept. physico-math. sci., 1960-63. Decorated Order of Lenin (2); recipient Lenin prize, 1958. Mem. USSR Acad. Sci. Author: Statistical Physics, 1941; (with M. L. Levin) The Theory of Oscillation Excitation in Antenna Vibrators, 1944; An Introduction to Thermodynamics, 1952. Research and publs.

on electrodynamics, optics, statis., radio and thermonuclear physics. Address: Moscow University, Leninskie gory, Moscow, USSR.

LEOPOLD, Aldo Starker, Am. wildlife biologist; b. Burlington, Ia., Oct. 22, 1913; s. Aldo and Estella (Bergere) L.; B.S., U. Wis., 1936; Ph.D., U. Cal. at Berkeley, 1944; m. Elizabeth Weiskotten, Aug. 7, 1938; children—Frederic Starker, Sarah Pendleton. Field biologist Mo. Conservation Commn., 1939-44, Pan Am. Union, 1944-46; faculty U. Cal. at Berkeley, 1946—, prof. zoology, 1946—, acting dir. Mus. Vertebrate Zoology, 1965—, dir. Wildlife-Fisheries Research Sta., Truckee, Cal., 1965—, asst. to chancellor, Berkeley, 1960-63. Chmn., Wildlife Mgmt. Adv. Com., 1961-64. Mem. Cal. Acad. Scis. (pres. 1959—), A.A.A.S., Am. Soc. Mammalogists, Wildlife Soc. (award for best wildlife publ. 1959, Aldo Leopold medal 1965), Am. Ornithologists' Union, Wilderness Soc., Am. Inst. Biol. Scis., Ecol. Soc., Am., Nat. Audubon Soc., Nature Conservancy, Wilson Ornithol. Soc. Author: Wildlife of Mexico: The Game Birds and Mammals, 1959; The Desert, 1961; Fauna Silvestre de Mexico, 1965; also numerous articles. Research on zoology, game animals and birds, wildlife surveys U. S., Alaska, Mexico, Australia, New Zealand, Africa. Home: 712 The Alameda, Berkeley, Cal. 94707.*

LEOPOLD, Luna Bergere, Am. hydraulic engr.; b. Albuquerque, N.M., Oct. 8, 1915; s. Aldo and Estella (Bergere) L.; B.S., U. Wis., 1936; M.S., U. Cal. at Los Angeles, 1944; Ph.D., Harvard, 1950; m. Carolyn Clugston, Sept. 6, 1940; children—Bruce Carl, Madelyn Dennette. With Soil Conservation Service, 1938-41, U. S. Engrs. Office, 1941-42, U. S. Bur. Reclamation, 1946; head meteorologist Pineapple Research Inst. of Hawaii, 1946-49; hydraulic engr. U. S. Geol. Survey, 1950—, chief hydrologist, 1957-66, sr. research hydrologist, 1966—. Served as capt., air weather service, USAAF, 1942-46. Recipient Distinguished Service award, Dept. of Interior, 1958; Kirk Bryan award Geol. Society Am., 1958; Veth medal, Royal Netherlands Geog. Soc., 1963. Mem. Am. Soc. C.E., Am. Meteorol. Society, National Academy of Sciences, American Geological Society, Am. Geophys. Union, Sigma Xi. Author: (with Thomas Maddock, Jr.) The Flood Control Controversy, 1954; Fluvial Processes in Geomorphology, 1964; Water, 1966; also tech. papers. Research on river morphology; erosion and sedimentation; rainfall characteristics; hydrology of arid regions. Home: 5705 Springfield Dr., Washington 16. Office: U. S. Geol. Survey, Washington 20242.

LEOPOLD OF AUSTRIA (Leopoldus ducatus Austriae filius), Austrian astronomer, meteorologist; flourished 2d half 13th century. Author: Compilatio de astrorum scientia. Used old concentric spheres theory to account for distances of planets; postulated 10 celestial spheres, the last 3 being spheres of fixed stars, signs, firmaments; publs. on meteorology including weather signs and rules for the meaning of thunder for each month.

LEOPORI, Nullo Glauco, Italian zoologist; b. Collesalvetti, Oct. 1, 1913; dir. inst. biology and gen. zoology, prof. zoology, faculty math., physics and natural sci., prof. gen. zoology, faculty vet. medicine U. Sassari (Italy). Address: Università degli Studi, Sassari, Italy.

LEOTAUD, Vincent, French mathematician; b. Diocese of La Val Louis Embran, France, 1595; prof. Coll. Dole; geometrical publs. include: Gometricae practicae elemta, 1631; Cyclomathia, 1663. Died 1672.

LE PAGE, Gerald Alvin, biochemist; b. Medicine Hat, Can., Oct. 9, 1917; s. Aubrey Johnson and Florence (Armstrong) Le P.; B.Sc. with honors in Chemistry, U. Alta. (Can.), 1940, M.Sc. in Biochemistry, 1941; Ph.D. in Biochemistry and Bacteriology, U. Wis., 1943; m. Marjorie H. Smith, Mar. 6, 1944; children—Gary S., Gaylis J. Came to U. S., 1945, naturalized, 1951. Profl. asso. U. Wis., Madison, 1943-44; bacteriologist-biochemist Lab. Hygiene FDA, Ottawa, Ont., Can., 1944-45; faculty U. Wis., 1945-58, prof. oncology, 1954-58; program dir. dept. biol. sci. Stanford Research Inst., Menlo Park, Cal., 1958-61, chmn. dept. biochem. oncology, 1961—. Mem. Am. Chem. Soc., Am. Assn. for Cancer Research, Am. Soc. Biol. Chemists, N.Y. Acad. Sci., Am. Coll. Clin. Pharm. and Chemotherapy, A.A.A.S. Research, numerous publs. on metabolism tissues normal and neoplastic, cancer chemotherapy, mechanism, resistance. Home: 88 Moulton Dr., Atherton, Cal. 94025. Office: 333 Ravenswood Av., Menlo Park, Cal. 94025.*

LEPAGE, Wilbur Reed, Am. elec. engr.; b. Kearny, N.J., Nov. 16, 1911; s. Wilbur Nicholas and Gertrude Elizabeth (Reedt) LeP.; E.E., Cornell U., 1933, Ph.D., 1941; M.S. in Physics, U. Rochester, 1939; m. Eveline Marie Jacobsen, June 9, 1936; 1 dau., Margaret Ann. Instr. engring. U. Rochester, 1933-38; grad. student, teaching asst. Cornell U., 1939-41; staff advanced development lab. RCA, 1941; physicist radiation lab., Johns Hopkins, 1942-45; sr. research engr. Stromberg Carlson Co., 1946; prof. elec. engring. Syracuse U., 1947—, chmn. dept., 1956—; initiated ann. Sagamore Conf. Elec. Engring. Edn., 1957. Cons. Signal Corps, U. S. Army, 1953. Trustee Electronic Engring. Found., 1959—. Fellow I.E.E.E.; mem. Am. Soc. Engring. Edn., Am. Assn. U. Profs., Sigma Xi. Author: Analysis Alternating Current Circuits, 1952; (with S. Seely) General Network Analysis, 1952; Complex Variables and the Laplace Transform for Engineers, 1961. Contbr. sect. electric filters encys. Sci. and Tech., sect. electric circuits Ency. Americana. Home: 217 DeWitt Rd., Syracuse, N.Y. 13214.*

LEPAIGE, Constantin-Jérôme, Belgian mathematician; b. Liège, Belgium, 1852; prof. U. Liège, became emeritus prof., 1923. Author: Essais de géometrie supérieure du toisième ordre, 1882-83; Sur les surfaces du troisième ordre, 1883. Research in analytic geometry, history math., theory of surfaces including constrn. 3d order surface passing through 19 given points. Died 1929.

LEPAWSKY, Albert, Am. social-polit. scientist; b. Chgo., Feb. 16, 1908; s. Morris and Rose (Devin) L.; Ph.B., U. Chgo., 1927, Ph.D., 1931; postgrad. (Social Sci. Research Council fellow) U. Hamburg, London (Eng.) Sch. Econs. and Polit. Sci., U. Berlin; m. Rosalind Almond, Apr. 17, 1935; children—Martha (Mrs. John Barker), Michael, Susan (Mrs. Saul Rosenstreich), Lucy. Faculty, U. Chgo., 1930-42, U. Ala., 1945-52; research dir. City Chgo. Law Dept., 1935-36; asst. dir. pub. adminstrn. Clearing House, Chgo., 1936-38; prof. polit. sci. U. Cal. at Berkeley, 1953—. Cons. UN, AID, TVA; dir. tng. Chgo. Met. Def. Com., 1941-42; mem. UN Mission Tech. Assistance to Bolivia, 1950-51. Mem. Am. (v.p. 1956-57), Western (pres. 1963-64) polit. sci. assns., Am. Soc. Pub. Adminstrn. (exec. council 1950). Author: Judicial System of Metropolitan Chicago, 1932; Home Rule for Metropolitan Chicago, 1935; (with Merriam, Parratt) Government of the Metropolitan Region of Chicago, 1933; State Planning and Economic Development, 1949; Administration: The Art and Science of Organization and Management, 1949; Agenda for International Development, 1961; also articles. Research in social scis. on polit. instns. and govtl. processes. Home: 2570 Cedar St., Berkeley, Cal. 94708.*

LEPESCHKIN, Eugene, cardiologist; b. Kazan, Russia, Apr. 15, 1914; s. Wladimir W. and Eugenia (Ryskin) L.; student Pasadena Jr. Coll., 1930-31, U. Cal. at Berkeley, 1931-32; M.D., U. Vienna, 1939; m. Julie Wilson, May 30, 1949; children—Tamara, Ludmilla, Nina. Came to U. S., 1927, naturalized, 1951. Research in cardiology, Burlington, Vt., 1947—; faculty U. Vt., 1947—, prof. medicine, 1966—. Recipient Research Career award Nat. Heart Inst., 1962—. Mem. Am. Physiol. Soc., New Eng. Cardiovascular Soc., Am. Coll. Cardiology (bd. govs. Vt.), Am. Heart Assn., Laennec Cardiovascular Sound Soc., Sigma Xi. Author: Das Elektrokardiogramm, 1941; Systematisches Hilfsbuch Der Elektrokardiographischen Diagnose, 1947; Modern Electrocardiography, 1950; (with F. D. Johnston) Selected Papers of F. N. Wilson, 1955; also numerous articles. Editor: (with Z. Z. Zal) Jour. of Electrocardiology, 1966. Research on U wave of electrocardiogram, changes of electrocardiogram in electrolyte imbalance, electrocardiographic response to exercise, T wave changes in terms of electrolyte and temperature gradients within heart muscle. Home: 75 Bilodeau Ct. Office: Cardiovascular Research Unit, De Goesbriand Hosp., Burlington, Vt. 05401.*

LEPESHINSKAIA, Olga Borisovna, Russian biologist; b. Aug. 18, 1871; grad. Rozhdestvensky Courses, 1897; grad. Lausanne, Switzerland, 1902; M.D., Moscow U., 1915. Staff labs. Tashkent, USSR, 1919-20, Moscow, 1920-26; staff Histological Lab., Timiriazev Biol. Inst., 1926—; staff Cytology lab. All-Union Inst. for Exptl. Medicine, USSR Acad. Med. Scis., 1936—, staff Inst. Exptl. Biology, 1949—, head lab. cytology, 1957—; prof. therapy Moscow U., 1915-19. Recipient Stalin prize, 1950, Order of Lenin, Order of Red Banner of Labor. Mem. USSR Acad. Med. Scis. Author: Membranes of Living Cells and their Biological Significance, 1952; A New Cell Theory and its Factual Basis, 1955. Research and numerous publs. on noncellular living matter, including its structure and function. Editorial bd. Progress in Modern Biology; editorial council Archives of Anatomy, Histology and Epidemiology. Office: Institut eksperimentalnoy biologii ANSSR, Leninsky prosp. 14, Moscow, USSR.

LÉPIN (or LIEPINIA), Lydia Karlovna, Russian physical chemist; b. Apr. 4, 1891; grad. Higher Women's Courses, Moscow, 1917. Staff, Inst. Popular Economy Plekhanov, 1917-30, Moscow Higher Tech. Sch., 1920-32, Voroshilov Acad. Chem. Def., 1932-41; prof. 1934—; staff Moscow U., 1920-30, 42-46; staff Latvian U., also Chemistry Inst., Latvian Acad. Sci., 1946—. Co-author: Inorganic Chemistry. An Introduction to Preparative Inorganic Chemistry, 1932; Surface Compounds and Surface Chemical Reactions, 1940; others. Research and publs. on action of chem. colloids in retarding corrosion, hydride formation, adsorption in solids, surface reactions, dispersion of solutes. Office: Gosudarstvenny universitet, Bulvar Raynisa 9, Riga, Lat. SSR, USSR.

LÉPINE, Jacques-Raphaël, French physician; b. Lyons, France, July 6, 1840; studied under Charcot; became hosp. physician, Paris, 1874; agrége Faculty of Paris, 1875; prof. med. clinic Faculté de médecine, Lyons; chief, sch., Lyons. Mem. French Acad. Scis. (corr.), French Acad. Medicine (asso.). Editor-in-chief, founder Revue de médecine; (with Landouzy) founder, Archives de médecine experimental. Research on disorders of nervous system, kidneys, also diabetes. Died Menton, France, Nov. 17, 1919.

LEPINE, Pierre Raphaël, French microbiologist; b. Lyons, France, Aug. 15, 1901; s. Jean and Elisabeth (Thyss) L.; M.D., Ph.D.; m. Marie-Madeleine Dollfus, Dec. 3, 1926; Became prof. Am. U. Beirut, Lebanon, 1925; named lab. chief Pasteur Inst., Paris, 1928, head virus service, 1940—; dir. Pasteur Inst., Athens, 1930-35; prof. U. Montreal, 1946-53. Fellow N.Y. Acad. Sci., mem. Nat. Acad. Medicine, Acad. Sci., Assn. Microbiologists, Epidemiologists and Hygienists of USSR (hon.), Royal Acad. Medicine Belgium (hon.). Research and publs. on viruses, electronic microscopy, tissue cultures, ecology and epidemiology of viruses, viral diseases of man and animals, viral vaccines; discovered and developed antipolio vaccine, 1957. Address: 25, rue du Docteur Roux, Paris 15, France.

LE PLAY, Pierre Guillaume Frédéric, French engr., economist; b. La Riviere-Saint-Sauveur (Calvados), Apr. 11, 1806; ed. Ecole Polytechnique. Apptd. head Permanent Com. Mining Statistics, 1834; engr.-in-chief, prof. metallurgy Sch. of Mines, 1840, Insp., 1848; apptd. to take charge orgn. of Exhbn. of 1855 by Napoleon III, later apptd. counsellor of state, commr. gen. of Exhbn. of 1867, senator of empire and grand officer Legion of Honor, 1867. Founder Société internationale des études pratiques d'economie sociale, 1856. Author: Les ouvriers européens, 1855; Le réforme sociale, 2 vols., 1864; L'organisation de la famille, 1871. Traveled through Europe studying social condition of working classes; accounted for social phenomena in terms of human nature (including culture), occupation, and phys. environment; one of 1st to recognize importance of sociology and its effect on econs.; founder social econs., which seeks human happiness in social and moral devel. as well as wealth and politics. Died Paris, Apr. 5, 1882.

LE POIVRE, Jacques F., Flemish mathematician; b. Flanders; worked at Le Mons, Flanders. Author: Traité des sections du cylindre et du cône, 1704. Introduced (independently of La Hire) new method of describing ellipses, parabolas, and hyperbolas in which they were no longer considered sects. of a cone or cylinder. Died 1710.

LE POOLE, Jan Bart, Dutch physicist; b. Amersfoort, The Netherlands, Apr. 1, 1917; s. Rudolf and Emily G. (Faure) Le P.; degree in engring. physics Tech. U. Delft, 1941, dr. tech. sci., 1954; m. Maria A. Croiset van Uchelen, Aug. 1, 1941; children—Rudolf S., Geertruida, Henry L., Caroline. With Royal Yeast and Spirit Factory, 1941-43, Philips Research Lab., 1942-48; faculty Tech. U. Delft, 1948—, prof. pure and applied physics, 1957—. Cons. Philips Eindhoven, Old Delft Optical Industries. Mem. Internat. Fedn. Socs. for Electron Microscopy (sec. gen. 1961—). Research, publs. on properties of electron lenses; devel. electron microscope with widely variable magnification and selected area electron diffraction; high voltage electron microscope; X-ray projection microscopy; inventor beam wobbler focusing aid, electron microscope with low chromatic aberration, time-of-flight mass spectrometer, achromatic electron diffraction, miniature electron lens. Home: 5 Nassaulaan, Delft, Holland.*

LEPORE, Joseph Vernon, Am. physicist; b. Detroit, Oct. 9, 1922; s. Joseph and Agda (Monstrom) L.; B.S. with honors in Physics, Allegheny Coll., 1943; Ph.D. in Nuclear Physics, Harvard, 1948. Faculty, Princeton, 1943-44, Ind. U., 1950-51; physicist, chemist Tenn. Eastman Corp., Oak Ridge, 1944-46; mem., AEC fellow Inst. for Advanced Study, Princeton, N.J., 1948-50; AEC fellow U. Cal., Berkeley, 1950, physicist Lawrence Radiation Lab., 1951—, lectr., 1954-65. Adviser to test dir. AEC, Nev., 1957. Mem. Am. Phys. Soc., Sigma Xi. Research in nuclear physics, quantum field theory, particle physics. Home: 712 Moraga Rd., Moraga, Cal. 94556. Office: Lawrence Radiation Lab., U. Cal., Berkeley, Cal. 94704.*

LEPORIS, Joannes, see le Lièvre, Jean.

LEPPER, Mark Hummer, Am. internist; b. Washington, June 12, 1917; s. Henry Albert and Georgianna (Hummer) L.; A.B., George Washington U., 1938, M.D., 1941; m. Joyce Mae Sullivan, June 7, 1941; children—Mark Roger, Joyce Hummer. Fellow medicine George Washington U., 1942-43, clin. instr. medicine, 1946-48, asso. medicine, 1949-50; asst. resident medicine Gallinger Municipal Hosp., Washington, 1943-44; pvt. practice, Washington, 1946-48; cons. internal medicine Walter Reed Gen. Hosp., Washington, 1948-50; med. supt. Municipal

Contagious Diseases Hosp., Chgo., 1950-52; asso. prof. medicine charge preventive medicine U. Ill., Chgo., 1952-55, prof. preventive medicine, head dept., 1955-66, prof. preventive medicine and community health, 1966—; exec. v.p. profl. and academic affairs Presbyn.-St. Luke's Hosp., Chgo., since 1966—. Diplomate Am. Bd. Internal Medicine. Fellow A.C.P.; mem. A.M.A., Am. Fedn. Clin. Research, Central Soc. Clin. Research, Am. Soc. Clin. Investigation, Central Research Club, Am. Pub. Health Assn., Assn. Am. Med. Colls., N.Y. Acad. Sci., Phi Beta Kappa, Sigma Xi. Home: 327 Hampton St., Hinsdale, Ill. Office: 1853 W. Polk St., Chgo. 60612.*

LEPPERT, George, Am. mech. engr.; b. Kansas City, Mo., July 17, 1924; s. George Alva and Leah (Oleve) L.; B.S., U. Wis., 1947; M.S., Ill. Inst. Tech., 1952, Ph.D., 1954; m. Helen Robinson, Jan. 16, 1961; children—Susan Adele, Donald George, Clinton Russell. Maintenance and constrn. engr. Monsanto Chem. Co., St. Louis, 1947-49; research engr. reactor engring. div. Argonne (Ill.) Nat. Lab., 1950-53; mgr. engring. research sect. Monsanto Chem. Co., Dayton, O., 1953-54; faculty Stanford, 1954—, prof. mech. engring., 1960—, dir. nuclear engring. div., 1956-60. Cons. to pvt. cos. NSF fellow, 1952. Mem. Am. Soc. M.E. (postgrad. student award 1953), Am. Nuclear Soc., Am. Inst. Aeros. and Astronautics, Am. Soc. Engring. Edn. Research, publs. on heat transfer especially application to nuclear reactors, boiling heat transfer, two-phase flow, plasma radiation and high temperature gas convection. Home: 931 Casanueva Pl., Stanford, Cal. 94305.*

LEPPIK, Elmar Emil, botanist, plant pathologist; s. Tenno and July (Barbits) L.; M.Botany, U. Tartu (Estonia), 1926; postgrad. U. Berne (Switzerland); D.Sc. (Rockefeller Found. fellow) Fed. Tech. U. Zurich (Switzerland), 1929; m. Lilly Hanson, Aug. 22, 1930; children—Virve (Mrs. Voldemar Vaher), Marys (Mrs. Andreas Lambert), Avo, Ülo. Came to U. S., 1950, naturalized, 1955. Prof. botany, plant pathology U. Tartu, 1930-44, State Hort. Sch., Freising, Germany, 1945-46; lectr. U. S. Army Agr. and Tech. Sch., Weihenstephen, Germany, 1946, Tech. U. Munich, 1946-50; prof. Augustana Coll., Sioux Falls, S.D., 1950-55; research scientist U. Minn., 1955-57; prof. Ia. State U., 1957-64; research botanist, plant pathologist U. S. Dept. Agr., Beltsville, Md., 1965—; guest investigator Tropical Research Inst., U. El Salvador, C.Am., 1953-54; external mem. biology faculty U. Poona (India), 1964—. Danforth Found. fellow, 1943. Fellow A.A.A.S.; mem. Soc. Naturalists U. Tartu (editor Annales and Archivum 1936-44), Zürich Bot. Soc. (corr.), Societas Zool. Bot. Vanamo Fennicae (corr.), N.Y. Acad. Sci. Author several textbooks and manuals on botany and plant pathology; also numerous articles, monographs. Developed theory evolutionary interrelationship between insects and plants, phylogenetic system for evolution of rust fungi in connection with their host plants. Home: 10400 46th Av., Office: New Crops Research Br., Plant Industry Sta., Beltsville, Md. 20705.*

LE PRINCE, Everett Franklin, inventor; b. France; 1 son, Adolphe, 1 dau. Worked in Leeds, Eng., N.Y.; inventor motion picture camera and projector including 16 lens camera patented in Am., single lens camera patented in Eng. Disappeared France, Sept. 1890.

LEPRINCE, Jean Baptiste, inventor, painter, engraver; b. Metz, France, 1734; painter and engraver of scenes in Russia and Siberia; decorated Imperial Palace, St. Petersburg; invented a copper plate engraving style, 1780. Died Saint-Denis-du Port, France, 1781.

LEPRINCE-RINGUET, Louis, French physicist; b. Alès (Gard), France, Mar. 27, 1901; s. Félix and Renée (Stourm) L-R.; ed. Polytech. Sch., Superior Sch. Electricity; m. Jeanne Motte, May 6, 1929; children—Bénédite (Mme. Ch. de la Roncière), Dominique, Renée-Noel (Mme. Jacques de Vathaire), Marie-Pascale (Mme. Bernard Dangy), Francois, Louis-Pentecote, Néril. Telegraph engr.; asst. to Duke of Broglie at Laboratoire de physique des rayons X (lab. X-ray physics); lab. dir. Sch. for Advanced Practical Studies; prof. physics Poly. Sch.; head dept. nuclear physics College of France, 1959—. Commissaire of atomic energy; v.p. sci. council European Center Nuclear Research. Recipient Eve Delacroix prize, 1958. Mem. French Acad. Scis., French Soc. Physics (pres. 1956). Author: Rayons cosmiques; Les inventeurs celebres, 1953; Le grandes decouvertes du XXe siecle, 1956; Des atomes et des hommes, 1958. Specialist in X-ray research. Home: 86, rue Grenelle, Paris 7. Office: École Polytechnique, 17 rue Descartes, Paris 5, France.

LEPSIUS, Karl Richard, German archaeologist, Egyptologist; b. Naumburg, Germany, Dec. 23, 1810; s. Karl Peter L.; ed. U. Leipzig, 1829, U. Göttingen, 1830; Dr., U. Berlin, 1833; postgrad., Paris, 1833, Rome, 1836; m. Elisabeth Klein, 1846; children include Johannes. Headed Friedrich Wilhelm IV's expdn. to Egypt, 1943-45; prof. U. Berlin, 1846; keeper Royal Library, Berlin, 1873; named Geheimer Oberregierungsrath, circa 1883. Author: Paläographie als Mittel fur der Sprachforschung, 1834; Lettre à M. Rosellini sur l'alphabet hiéroglyphique, 1837; Totenbuch der Ägypter, 1842; Inscriptiones unbricae et oscae, 1841; Denkmäler aus Ägypten und Äthiopien, 12 vols., 1849-60; Chronologie der Ägypter, 1849; Königsbuch der alten Ägypter, 1858; Nubische Grammatik mit einer Einleitung uber die Völker und Sprachen Afrikas, 1880. Discovered several monuments of Egyptian Old Kingdom, unknown pyramids, mastabas; 1st to measure Valley of Kings; found trilingual tablet, Decree of Canopus, 1866. Died Berlin, July 10, 1884.

LERAY, Jean, French mathematician; b. Chanteney, France, Nov. 7, 1906; ed. Normal Superior Sch.; D.Sc.; m. Marguerite Trumier, Oct. 20, 1932; children—Jean-Claude, Francoise, Denis. Prof., Faculty of Scis., Nancy, 1936-41, then Paris, 1941-47, Coll. France, from 1947. Mem. French Acad. Scis., French Math. Soc., Am. Math. Soc. Specialist on partial differential equations, functions, and topography. Home: 12, rue Pierre-Curie, Sceaux, Seine, France.

LERICHE, René, French surgeon; b. Roanne, France, Oct. 12, 1879; s. Ernest L. and Marie (Chanussy) L.; M.D., U. Lyons (France); m. Miss Calenborn, May 10, 1912; 1 dau., Jacqueline. Prof. clin. surgery, U. Strasbourg (France); prof. College de Frances; surgeon Am. Hosp., Neuilly, France. Mem. French Acad. Medicine, French Acad. Surgery. Author: Thérapeutique chirurgicale, 1925; la Chirurgie de la douleur, 1937; la Chirurgie á l'ordre de la vie, 1944; les Artères, 1945; les Maladies de la vasomotricité, 1945; la Thromboseartérielle, 1946; Philosophie de la chirurgie, 1951; les Bases de la chirurgie physiologique, 1955. Founder modern vascular surgery, sympathectomy; introduced periarterial sympathectomy for paresthesia and vasomotor disturbances (Leriche's operation), 1916-17. Died Cassis, France, Dec. 28, 1955.

LE RICHE, William Harding, physician; b. Dewetsdorp, S. Africa, Mar. 21, 1916; s. Josef Daniel and Georgina (Harding) le R.; B.Sc., U. Witwatersrand, Johannesburg, S. Africa, 1936, M.B., B.Ch., 1943, M.D., 1949; M.P.H., Harvard, 1950; m. Margaret Cardross, Dec. 11, 1943; children—Jenny, Robert, Nicole, Giles, Claire. Carnegie Research grantee Nutrition Survey, Pretoria, S. Africa, 1939-40; nutrition research S. African Inst. for Med. Research, Johannesburg, 1949-50; Rockefeller fellow. Harvard, 1952-54; cons. epidemiology Dept. Nat. Health and Welfare, Ottawa, Can., 1954-57; research med. officer Physicians Services, Inc., Toronto, Ont., Can., 1957; with U. Toronto, 1957—, prof. pub. health, 1959-62, prof., head dept. epidemiology and biometrics, 1962—. Cons. statistics Hosp. for Sick Children, Toronto; asso. staff U. Guelph, Ont., Can.; mem. med. research Def. Research Bd., Dept. Nat. Health and Welfare, Ont. Council of Health. Fellow Royal Statis. Soc., Am. Epidemiological Soc.; mem. Canadian, Brit., S. African med. assns., Am. Pub. Health Assn. Author: (with C. E. Balcom, G. van Belle) The Control of Infections in Hospitals, 1966; also numerous articles. Devel. growth and nutrition studies in S. Africa; clin. research on obesity; clin. drug trials; studies on epidemiology of hosp. infections and communicable diseases; epidemiologic studies on med. care and hosps.; current disease picture in Can. Home: 30 Golfdale Rd. Office: 150 College St., Toronto, Ont., Can.*

LERMAN, Leonard, Am. molecular biologist; b. Pitts., June 27, 1925; s. Meyer L. and Fanny (Hoffman) L.; B.S. in Chemistry, Carnegie Inst. Tech., 1945; Ph.D., Cal. Inst. Tech., 1950; m. Claire Lindegren, July 18, 1952; children—Averil, Lisa, Alexander. Research asst. explosive research lab. Carnegie Inst. Tech., 1945; Schenley fellow Inst. Radiobiology and Biophysics, U. Chgo., 1949-51; faculty U. Colo. Sch. Medicine, 1951-65; prof. molecular biology Vanderbilt U., Nashville, 1965—. Am. Cancer Soc. scholar, 1956-59; USPHS fellow, 1963-65. Mem. Am. Chem. Soc., Biophys. Soc., Genetics Soc. Am., Sigma Xi. Research on nature of antibodies, molecular basis of heredity, including specific alteration of DNA structure, origin of mutations, DNA replication, transfer of heredity by DNA. Home: 810 Curtiswood Lane, Nashville 37204.*

LERMAN, Sidney, physician, biochemist; b. Montreal, Que., Can., Oct. 6, 1927; s. Aaron and Rachel (Sivack) L.; B.Sc. magna cum laude, McGill U., 1948, M.D.C.M., 1952; M.S. in Biochemistry, U. Rochester, 1961; m. Marilyn Lois Frank, Apr. 14, 1957; children—Lora Rachel, Mark Jonas. J. B. Collip Found. fellow, McGill U., 1955-57; faculty U. Rochester (N.Y.) Sch. Medicine and Dentistry, 1957—, asst. prof. biochemistry, asso. prof. ophthalmology, 1962—; staff Strong Meml. Hosp., Rochester, 1957—, sr. asso. ophthalmologist, 1962—. Recipient award in Ophthalmology Rochester Acad. Medicine, 1958, J. B. Cramer Meml. award, 1958; certificate of merit Rochester Eye-Bank and Research Soc., 1961; Lucien Howe award N.Y. State Soc. Medicine, 1961, 62. Fellow A.C.S.; mem. Am. Acad. Ophthalmology, Am. Assn. Research in Ophthalmology, Am. Chem. Soc., A.M.A., Am. Soc. Biol. Chemists, Am. Soc. Human Genetics, Gerontological Soc., Am. Assn. U. Profs., N.Y. Acad. Sci. Author: Glaucoma: Chemistry, Mechanisms and Therapy, 1961; Cataracts: Chemistry, Mechanisms and Therapy, 1964; Ophthalmic Biochemistry, 1965; Basic Ophthalmology, 1966; also articles. Research in biochemistry, with spl. emphasis on metabolism of the normal and cataractous lens. Home: 135 Fairhill Dr., Rochester, N.Y. 14618.*

LERNER, Aaron Bunsen, Am. physician; b. Mpls., Sept. 21, 1920; s. Morris and Lena (Schneider) L.; B.A. cum laude, U. Minn., 1941, M.S., 1942, Ph.D., 1945, M.D., 1945; m. Marguerite Rush, June 21, 1945; children—Peter, Michael, Ethan, Seth. Am. Cancer Soc. fellow dept. biochemistry Western Res. U., 1948-49; asst. prof. dermatology U. Mich., 1949-52; asso. prof. U. Ore. Med. Sch., 1952-55; asso. prof. medicine Yale Sch. Medicine, 1955-59, prof. dermatology, 1959—. Recipient Thomas F. Andrews undergrad. award for research U. Minn., 1941. Mem. Am. Acad. Dermatology, Am. Dermatologic Assn., New Eng. Dermatology Assn., Sigma Xi, Alpha Omega Alpha, Phi Lambda Upsilon. Author: (with M. Lerner) Dermatologic Medications, 1954; also numerous articles. Research on skin pigmentation, pituitary hormones, pineal gland hormones, blood proteins. Home: Old Mill Rd., Woodbridge, Conn. 06525. Office: Yale Sch. Medicine, 333 Cedar St., New Haven 06510.*

LERNER, Daniel, Am. social scientist; b. N.Y.C., Oct. 30, 1917; s. Louis and Yetta (Swiger) L.; student Johns Hopkins, 1934-35; A.B., N.Y. U., 1938, M.A., 1939, Ph.D., 1948; student Harvard, Columbia, 1939-42, Stanford U., 1946; m. Jean Weinstein, May 16, 1947; children—Louise Alberta, Thomas, Amy. Free-lance feature writer, 1937-39; instr. Modern European history and lit., 1939-42; European rep. Library of Congress Mission, 1946-47; research asso. The Hoover Inst. and Library, Stanford U., also exec. sec. and research dir. Internat. Studies Project, The Hoover Inst., 1947—, prof. Stanford, also vis. prof. Columbia, 1951, prof. sociology Mass. Inst. Tech., 1953—, Ford prof. sociology and internat. communications, 1958, chmn. polit. and social sci., 1963-64; dir. Institut de Recherches Sociales, Paris, France, 1955, research asso., Inst. of War and Peace Studies; guest lectr. U. Paris (Sorbonne); Awarded Bronze Star Medal, Purple Heart (U. S.), Palmes Académiques, Officier d'Académie (France). Mem. Am. Sociol. Soc., Am. Polit. Sci. Assn., World, Am. assns. pub. opinion research, Am. Psychol. Assn. Author: Sykewar: Psychological Warfare Against Germany, 1949; Propaganda in War and Crisis, 1951; The Policy Sciences (with H. D. Lasswell), 1951; The Nazi Elite, 1951; LaQuerelle de La CED (with Raymond Aron), 1956; France Defeats E.D.C., 1957; The Passing of Traditional Society, 1958; The Human Meaning of the Social Sciences, 1959; Evidence and Inference, 1960; Quantity and Quality, 1961; Parts and Wholes, 1963; Cause and Effect, 1965. Research and publs. on formulated theoretical basis for strategy of truth in psychol. warfare; diagnosis of psychol. functions of empathy in the modernization of traditional socs. Home: 233 Grant Av., Newton Centre, Mass. Office: Mass. Inst. Tech. (53-369), Cambridge, Mass. 02139.*

LERNER, I. Michael, geneticist; b. Harbin, China, May 15, 1910; s. Michael and Cecelia (Sudja) L.; B.S.A., U. B.C., 1931, M.S.A., 1932, D.Sc. (hon.), 1962; Ph.D., U. Cal. at Berkeley, 1936; m. Ruth A. K. Stuart, June 9, 1937. Came to U. S., 1933, naturalized, 1942. Faculty dept. poultry husbandry U. Cal. at Berkeley, 1936-58, prof., 1951-58, prof. genetics, 1958—, chmn. dept., 1958-63; fellow Center for Advanced Study in Behavioral Scis. U. S. rep. Permanent Com. on Internat. Genetics Congress, 1953-62. Recipient Poultry Sci. Research prize, 1937; Belling prize, 1940; Borden award and medal, 1951; Mendel Centennial medal Czechoslovakian Acad. Scis. Fellow Eugenics Soc., Poultry Sci. Assn.; mem. Soc. Study Evolution (pres. 1964), Am. Soc. Naturalists (v.p. 1957), Nat. Acad. Sci., Am. Acad. Arts and Sci., Genetics Soc. Am., Am. Soc. Zoologists, Am. Genetics Assn., Acad. Georgofili (fgn. mem.), Italian Soc. Genetics. Author: Population Genetics and Animal Improvement, 1950; Genetic Homeostasis, 1954; Genetic Basis of Selection, 1958; (with H. P. Donald) Modern Developments in Animal Breeding: A Study of Their Genetic and Social Aspects, 1966; Heredity, Evolution and Society, 1968. Editor: Evolution, 1959-61. Home: 2507 Rose Walk, Berkeley, Cal. 94708.*

LEROUX, P. E., mining engr.; engr. Arts et Métiers, 1889; in 1910 built a compression compound mine locomotive, an air compressor for up to 30 atmospheres in 4 stages, scooping machines, bucket chain conveyors (for handling coal), also gobbing machine. Died 1943.

LE ROUX, Pierre Claude, French physicist; b. Douvres, Calvados, Apr. 26, 1891; s. Joseph and N. (Guillemenet) Le R.; ed. Lisieux Coll., U. Caen (France); M.A., Ph.D.; m. Marie-Louise Legueux, Sept. 26, 1926; 1 dau., Françoise-Jeanne. Asst., later head of works, prof. physics Faculty Scis. Caen; dir. Tech. Inst. Normandy. Mem. French Physics Soc., Union of Physicists. Author: Mesure de la viscosité de l'eau en valeur absolue et variation avec la temperature; Pléochroïsme dans certains cristaux et phenomènes d'absorption dans le spectre infrarouge. Home: 15 bis rue Isidore-Pierre, Caen. Office: Faculté des Scis. de Caen, Caen (Calvados), France.

LE ROY, André, French surgeon; b. Paris, Nov. 25, 1907; s. Edouard and Thérèse (Gernez) Le R.; ed. Faculty Medicine, Paris; m. Christiane Blanc, Dec. 15, 1943; children—Edouard, Patrick, Bernard, Isabelle, Martin. Chief surgeon Hôpital Notre Dame de Bon Secours. Mem. Acad. Surgery. Address: 4, rue de Commaille, Paris, France.

LE ROY, Charles, French physician, chemist; b. Paris, Jan. 12, 1726; s. Julien LeRoy; M.D., 1722; doctor, then prof. Montpellier (France) Faculty Medicine. Fellow Royal Soc., 1770; mem. French Acad. Scis., 1752. Author: De aquarum mineralium, 1758; De purgantibus, 1759. Research on scurvy, splenic fever, anatomy, especially respiration of turtle. Died Dec. 10, 1779.

LE ROY, Donald James, phys. chemist; b. Detroit, Mar. 5, 1913; s. Charles Lafayette and Emma Frances (McCoubrey) Le R.; B.A., U. Toronto (Ont., Can.), 1935, M.A., 1936, Ph.D., 1939; m. Lillice Marie Eyer Read, June 12, 1940; children—Rodney Lash, Robert James, Alexander Charles, John Donald. Research chemist Ont. Research Found., Toronto, 1939-40, NRC, Ottawa, Ont., 1940-44; faculty U. Toronto, 1944—, prof. phys. chemistry, 1950—, chmn. dept. chemistry, 1960—. Mem. NRC of Can., 1964—. Fellow Royal Soc. Can., Chem. Inst. Can.; mem. Am. Chem. Soc., Am. Phys. Soc., Faraday Soc. Author: (with F. E. W. Wetmore) Principles of Phase Equilibria, 1951; also numerous articles on phys. chemistry. Asso. editor Canadian Jour. Chemistry, 1961-66. Research on chem. kinetics atomic and free radical reactions. Home: 605 Oriole Pkwy., Toronto 12, Ont., Can.*

LE ROY, Edouard, French philosopher, mathematician; b. Paris, 1870; student l'École normale superieure, 1892; docteur ès sciences; agrégé math.; tchr. math. various lycées, Paris; replaced Henri Bergson, Collège de France, 1914, named nominal prof. philosophy, 1921. Mem. l'Académie des sciences morales, French Acad. Scis., 1945. Author: Dogme et critique, 1906; Henri Bugson, 1912; L'exigence idéaliste et le fait de l'evolution, 1927; Les origines humaines et l'evolution de l'intelligence, 1928; le Problème de Dieu, 1929. Tried to unite philosophy, sci. and religion; believed sci. gives only schematic representation of reality; the role of a philosopher is to rediscover primitive existence or being; only intuition can attain to concrete being. Died Paris, 1954.

LE ROY, George Veach, Am. physician; b. Wilkinsburg, Pa., Oct. 28, 1910; s. Frank Odell and Bess (Veach) LeR.; B.S., U. Pitts., 1932; M.D., U. Chgo., 1934; m. Maria A. Ignelzi, July 31, 1930; 1 dau., Georgia (Mrs. Haynes). Mem. med. faculty U. Chgo., 1935-38, 51-66, prof., 1954-66; research fellow Northwestern U. Med. Sch., Chgo., 1938-42, asso. prof., 1945-51; pvt. practice medicine, Chgo., 1946-51; med. dir. Met. Hosp., Detroit, 1966—; clin. prof. medicine Wayne State U., Detroit. Mem. life sci. com., space sci. bd. Nat. Acad. Sci.; mem. adv. com. on med. uses of isotopes AEC. Mem. A.M.A., Central Soc. for Clin. Research, Chgo. Soc. Internal Medicine, Radiation Research Soc., Assn. Am. Physicians, Alpha Omega Alpha. Studies, publs. on effects of ionizing radiation on man and animals; application of radioisotopes to clin. research; intermediate metabolism. Home: 9000 E. Jefferson Av., Detroit 48214. Office: 1800 Tuxedo Av., Detroit 48206.*

LE ROY, Jean-Baptiste, French physicist; b. Paris, Aug. 15, 1720; s. Julien LeRoy; dir. Royal Acad. Machines. Fellow Royal Soc., 1773; mem. French Acad. Scis., 1751. Research and publs. on electricity, especially electro-static devices; perfected lightning rods and aerometers; studied ways of circulating air in prisons and hosps. Died Jan. 21, 1800.

LEROY, Julien, French watchmaker; b. Tours, France, 1686; children—Pierre, Charles, Jean Baptiste, Julien David. Named horloger to the king, Louis XV, Paris, 1739. Recipient prize for building accurate chronometer. Invented horizontal clocks, repeater for watches; developed clock mechanism compensating for changes of temperature. Died 1759.

LEROY, Louis, Am. physician; b. Chelsea, Mass., Sept. 15, 1874; s. Charles L. A. and Lizzie F. (Somerby) L.; A.B., U. Nashville, 1900; M.D., Medico-Chirurgical Coll., Phila., 1896; m. Joe Carr, Jan. 11, 1922; children—Charles Louis, Joe Carr. Asst. in pathology Medico-Chirurg. Coll., 1896; prof. pathology and bacteriology Harvey Med. Coll. and Ill. Med. Coll., Chgo., 1899-1900; prof. pathology and bacteriology Vanderbilt U., 1899-1905; state bacteriologist and smallpox expert to State of Tenn., from 1897; pathologist Nashville City and St. Thomas hosps., Nashville, 1896-1904; prof. practice of medicine Coll. Physicians and Surgeons, Memphis, from 1905; prof. theory and practice of medicine, U. Tenn., from 1911; staff physician City and Bapt. Meml. hosps.; chief of staff Memphis Gen. Hosp., 1937; v.p. Tenn. Bd. of Health. Dir. rescue div. A.R.C., from Cairo to Rosedale, in Mississippi River flood, 1937; med. mem. Memphis and Shelby County Tb Commn., 1918-28. Diplomate Am. Bd. Internal Medicine, 1937. Fellow A.C.P.; mem. A.M.A., Am. Congress Tuberculosis. Author: Essentials of Histology, 1900; Smallpox Diagnosis, Treatment, etc., 1901; Pulmonary Tuberculosis. Asso. editor Examiner and Practitioner, N.Y. Died May 9, 1944.

LEROY, Pierre, French watchmaker; b. Paris, 1717; s. Julien Leroy; recipient prize for best method of measuring time at sea French Acad. Scis. Author: Etrennes chronométriques pour l'année, 1760. Discovered isochromism of spiral springs; developed detached escapement for watches, 1748. Died 1785.

LEROY D'ETIOLLES (or Etolles), Jean Jacques Joseph, French surgeon; b. France, 1798; recipient award for inventions French Acad. Scis., 1822; invented instruments used in lithotrity. Died Aug. 25, 1860.

LESAGE, Georges-Louis, Swiss physicist, mathematician; b. Geneva, Switzerland, June 13, 1724; ed. Paris, Geneva; med. degree; became prof. math., Geneva, 1750, Fellow Royal Soc., 1775; mem. French Acad. Scis. Author: Essais de chimie mécanique, 1758; Newtonian Lucretius, 1782; Fragments on Final Causes; Traité de physique mécanique, 1818. Invented and built elec. telegraph apparatus featuring 24 keys, each with a wire (protype of Holundermark electrometer), Geneva, 1774. Died Geneva, Nov. 9, 1803.

LESBRE, Michel, French chemist; b. Lyon, France, June 3, 1908; s. François Xavier and Sophie (Brunier); Dr. ès scis. in Physics, U. Lyons; m. Suzanne Guinand, Apr. 7, 1931; children—François, Jacques, Geneviève (Mrs. Puif), Jean, Bernard, Didier. In charge research Centre de la Recherche Scientifique, 1932; instr. Faculté de Besançon, France, 1937-40, Faculté des Sciences, Lyons, 1940-48; lectr. Faculté des Sciences, Toulouse, France, 1949-59; titular prof. organic chemistry, Toulouse, 1959—. Mem. Am. Chem. Soc., Chem. Soc. France. Author: Traité de chimie organique de V. Grigard, 1948; also numerous articles. Research in organotin and organogermanium chemistry. Home: 90, rue des 36 Ponts, Toulouse, France.*

LESCHOT, Georges, inventor; watchmaker in Geneva, Switzerland; improved manufacture of clocks in factories; invented rock drill with diamond point, 1862.

LESCOP, René Hervé Georges Marie, French physicist; b. Landerneau, Apr. 16, 1907; s. Hervé-René and Berthe (Torillec) L.; ed. U. Rennes, Ecole Polytechnique; m. N. Dubourg, 1934; children—Chantal, Dominique. Dir. armament studies, 1946-58; treas. gen. of Orne, 1948; dir. indsl. mechanics and electronics Central Adminstrn. for Industry and Commerce; sec. gen. AEC; dir. Maurice Bourges-Maunoury's cabinet; pres. council, 1957. Research on shooting torpedoes by air, light weapons; innovator indsl. prodn. of plutonium, also of 2d quinquennial plan for atomic energy. Home: 23, av. de Breteville, Neuilly-sur-Seine. Office: 23 av. Franklin Roosevelt, Paris 8, France.

LESCOT, Simon, surgeon; b. Paris, early 17th century; head physician, Genoa, Italy. Author: Dissertation sur la mycologie, 1684; Research and publs. on fungi, distbn. of veins and arteries; introduced Swammerdam's method of injecting colored wax in the vessels of cadavers into France. Died Genoa, 1690.

LE SEUR, Father Thomas, mathematician; b. Rethel, France, Oct. 1, 1703; Minim friar; lived in Italy; prof. math. Sapience Coll., Rome; mem. French Acad. Scis., 1745, Marine Acad. Editor: Principia (by I. Newton), Eléments de calcul intégral. Research and publs. on integral calculus, decomposition of equations into fractions. Died Rome, Sept. 26, 1770.

LESHER, Samuel Walter, Am. radiation biologist; b. Spokane, Wash., June 9, 1916; s. E. and Octavia (Dilg) L.; A.B., U. Ill., 1942; Ph.D., Wash. U., 1950; div.; 1 dau., Betty Jean (Mrs. Robert Mangers). Grad. asst. Ore. State U., 1946-48, Wash. U., 1948-50; asst. prof. U. Kan. Med. Sch., 1950-52; embryologist Mich. State U. and U. S. Dept. Agr. Regional Poultry Lab., 1952-54; radiation biologist div. biol. and med. research dept. Argonne (Ill.) Nat. Lab., 1954-68; dir. radiation biology Allegheny Gen. Hosp., Pitts., 1968—. Nat. Cancer Inst. Sr. Postdoctoral fellow Royal Cancer Hosp., Sutton, Eng., 1962. Mem. Am. Soc. Zoologists, Am. Soc. Cell Biology, Am. Soc. for Cancer Research, Gertontol. Soc., Radiation Research Soc., A.A.A.S., Sigma Xi. Research, numerous publs. on effects radiation and aging on cell proliferation, DNA synthesis, enzyme induction. Home: 2 Bayard Rd., Pitts. 15213. Office: Allegheny Gen. Hosp., 320 E. North Av., Pitts. 15212.*

LESIAK, Tadeusz, Polish chemist; b. Niewiadow, Poland, Sept. 7, 1925; s. Jan and Feliksa (Truszkiewicz) L.; grad. N. Copernicus U., Torun, Poland, 1954, D.Sc., 1960, Docent habilit., 1965; m. Maria Zofia Lipinski, July 5, 1958; 1 dau., Maria Jolanta. Asst., N. Copernicus U., 1958-60, adj., 1961-65, asst. prof. organic chemistry, 1966—. Research fellow Polish Chem. Soc., 1953. Research, publs. on new catalytic synthesis of heterocyclic compounds, indoles, coumarones and thianaphthenes, new methods for detection and determination of heterocycles mentioned. Home: 22a/8 Kraszewskiego, Torun, Poland.*

LESINS, Karlis Adolfs, geneticist; b. County Galgauska, Latvia, July 30, 1906; s. Janis and Anna (Lacis) L.; grad. U. Latvia, 1928; licentiate of agronomy Royal Agrl. Coll. Sweden, 1950; D.Sc., U. Alta., Can., 1957; m. Irma Kaminskis, Nov. 2, 1939. Headmaster, Sch. Agr., Bebrene, Latvia, 1931-41; supt. Agrl. Research Sta., Osupe, Latvia, 1941-44; agronomist Swedish Seed Assn., Uppsala, 1944-51; faculty U. Alta., Edmonton, Can., 1951—, prof. genetics, 1962—. Mem. Genetic Soc. Can., Am. Genetics Assn. Research, publs. on genus Medicago, cytology, inheritance and taxonomy. Home: 9727 65 Av., Edmonton, Alta., Can.*

LESLEY, J. Peter, Am. geologist; b. Phila., Sept. 17, 1819; grad. U. Pa., 1838, LL.D.; asst. geol. survey of Pa., 1838-41; student Princeton Theol. Sem., 1841-44, U. Halle, 1945; m. Susan Inches Lyman, Feb. 13, 1849. With Am. Tract Soc., 1845-47; pastor Congl. Ch., Milton, Mass., 1848-51, left ministry, became profl. geologist, Phila.; sec. Am. Iron Assn., 1855-59; prof. geology and mining and dean sci. faculty U. of Pa., 1872-78, became emeritus prof., 1886; chief geologist in charge of complete resurvey of Pa., begun, 1874; U. S. commr. to Paris Expn., 1867. Original mem. Nat. Acad. Scis.; mem. Am. Philos. Soc. (sec., librarian 1858-85), A.A.A.S. (pres. 1883-85). Author: Man's Origin and Destiny from the Platform of Sciences; Coal and Its Topography, 1856; The Iron Manufacturer's Guide, 1859; Paul Dreifuss, His Holiday Abroad. Made extensive researches in coal, iron and oil fields in U. S. and Can. Died 1903.

LESLIE, Sir John, Scottish physicist, mathematician; b. Largo, Scotland, Apr. 10, 1766; ed. St. Andrews (Scotland) U., Edinburgh (Scotland) U. Became tutor to 2 Americans in Va., 1789; superintended studies of Wedgwoods, 1790-92; named prof. natural philosophy, Edinburgh, 1819. Fellow Royal Soc., 1807 (Rumford medal, 1805). Author: An Experimental Inquiry into the Nature and Properties of Heat, 1804; Analytical Geometry and Geometry of Curved Lines, 1809-21; Elements of Geometry, Geometrical Analysis and Plane Trigonometry, 1809; A Short Account of Experiments and Instruments depending on the Relation of Air to Heat and Moisture, 1813; Philosophy of Arithmetic, 1817; Elements of Natural Philosophy, 1823; also articles; contbr. to Ency. Britannica. Built hygrometer, 1800, differential air thermometer used for radiation research, 1st artificial congelation, 1810; discoveries in radiation of heat, 1804; developed method of obtaining artificial ice, 1817; invented photometer; verified Laplacean equilibrium formula for phenomena of surface tension. Died Coates, Scotland, Nov. 3, 1832.

LESNY, Ivan, Czechoslovakian physician; b. Prague, Czechoslovakia, Nov. 8, 1914; s. Vincent and Milada (Krausova) L.; M.B., Charles U. Prague, 1948, M.D., C.Sc., 1956, D.Sc., 1965; m. Zdenka Zahradníková, Sept. 27, 1947 (div. Sept. 1954); children—Peter, Ivan. With Neurol. U. Hosp. Prague, 1938—, asst. physician, 1945—, chief physician children's div., 1950—; asst. prof. neurology Faculty Pediatrics, Charles U., 1955—; founder Czechoslovak Commn. for Child Neurology, also Czechoslovak EEG Commn., 1957. Cons. neurologist Motor Reedn. Centre for Cerebral Palsied Children, Zeleznice, Eastern Bohemia, 1953—. Mem. Am. Acad. for Cerebral Palsy (fgn. corr.). Author: Cerebral Palsy, 1959; Developmental Diagnosis in Child Neurology, 1965; also numerous articles. Research on treatment of myotonias with myastenic blood, descriptions of some modifications of infantile reflexes, eserin activation in EEG readings in children developmental signs in child neurology. Home: 585 Noskova, Prague 8, Czechoslovakia.*

LE SOUEF, W. H. Dudley, Australian zoologist; b. Victoria, Australia; s. A. A. C. Le Souëf; ed. Melbourne, Australia, Crediton Grammar Sch., Devon, Eng.; m. Edith E. Wadeson, 1888; 3 sons, 3 daus. Became dir. Zool. Gardens, 1901; also mgr. Royal Park, Melbourne. Fellow Am. Ornithol. Union (corr.); mem. Brit. Ornithol. Union, Royal Australian Ornithol. Union (hon.; pres.); corr. mem. Zool. Soc., Washington Park Zool. Soc. Author: Wild Life in Australia; Visit to Albatross Island, 1895; Description of Birds' Eggs from North Australia, 1896-98; Mound-building Birds of Australia, 1899; Visit to Queensland, 1891; co-author: Animals of Australia; Birds of Australia; also articles on ornithology and geography, photographs. Died Sept. 6, 1924.

L'ESPERANCE, Elise Strang, Am. physician; b. Yorktown, N.Y.; daughter Albert Strang, M.D., and Kate (Depew) Strang; M.D., Woman's Medical Coll. of N.Y. Infirmary for Women and Children, 1901; D.Sc., Woman's Med. Coll., Pa.; LL.D., Lindenwood Coll.; asst., dept. of pathology, Cornell U. Med. Sch., 1910-12, instr., 1912-20, asst. prof., 1920-32, asst. prof., dept. preventive medicine, 1944-50, professor emeritus since 1950. Mary Putnam Jacobi Fellow for research in tumor pathology, Munich, Germany, 1914. Resident, Babies Hospital, New York City, 1901-02; pathologist, also dir. lab., New York Infirmary for Women and Children, 1910-44; dir. Strang Tumor Clinic since 1937; pathol., New York, Harlem, and Manhattan Maternity Hosp.; bacteriologist and asst. pathol. Memorial Hosp., also dir. Strang Cancer Prevention Clinics, 1940-50; staff, Memorial Hospital and New York Infirmary, 1937-50; attending physician in preventive medicine Memorial Hospital 1948-50. Received Lasker Award, Am. Pub. Health Assn., 1951. Fellow N.Y. Academy Medicine; member Westchester Co. N.Y. County and New York State med. socs., New York Pathology Society, American Assn. Pathol. and Bacteriol., Am. Assn. Immunologists, Am. Radium Soc., Harvey Soc., Am. Cancer Soc., Woman's, Am. Med. Woman's and Am. med. assns., (hon.) Am. Radiologists Soc. Cons. editor: Jour. Am. Med. Woman's Assn. Established cancer prevention clinic which provided model for many similar centers, (with sister, Kate Depew Strang) tumor clinic N.Y. Infirmary for Women and Children, 1932. Died 1959.

LESPIEAU, (Pierre-Léon) Robert, French chemist; b. Paris, July 15, 1864; ed. École normale supérieure; prof. chemistry Paris Faculty of Scis., also Paris Central Sch. Arts and Manufactures. Mem. French Acad. Scis., 1934. Research in cryometry, ebulliometry, stereochemistry, change in wavelength of light, Raman effect; synthesized many organic compounds, especially polyalcohols. Died Cannet, France, Apr. 21, 1947.

LESPINASSE, Victor Darwin, Am. surgeon; b. Aurora, Ill., Dec. 2, 1878; s. Raymond and Clara Belle (Bradley) L.; M.D., Northwestern U. Med. Sch., 1901; m. Anna L. King, June 30, 1909; children—Victoire D., Victor King. Asso. prof. genito-urinary surgery, Northwestern U. Med. Sch.; urologist Wesley Meml. Hosp., Chgo. Fellow A.C.S.; mem. Am. Urol. Assn., A.M.A., Alpha Kappa Kappa. Awarded certificate of honor for method of blood vessel anastomosis, A.M.A., 1910; silver medal, New York meeting A.M.A., 1917; diploma for original exptl. work on spermatogenesis and sterility, A.M.A., 1920. Died Dec. 16, 1946.

LESQUEREUX, Leo, paleobotanist; b. Fleurier, Switzerland, Nov. 18, 1806; s. V. Aimé and Marie Anne Lesquereux; m. Sophia von Reichenberg. Became specialist on peat bogs; dir. peat bogs for Swiss Govt., 1844-48; came to Boston, 1848, worked with Louis Agassiz; expert on coal formation, surveyed Ky., Ill., Ind. and Miss. for coal deposits; mem. Nat. Acad. Sci. Author: Description of the Coal Flora of the Carboniferous Formation in Pennsylvania and Throughout the United States, 2 vols.. 1880-84. Died Columbus, O., Oct. 25, 1889.

LESSEN, Martin, Am. engr.; b. N.Y.C., Sept. 6, 1920; s. Philip and Lena (Sukornik) L.; B.S., Coll. City N.Y., 1940; M.S., N.Y. U., 1942; Sc.D., Mass. Inst. Tech., 1948; m. Elizabeth Scher, Aug. 27, 1948; children—Margot, Deborah, David. Mech. engr. U. S. Navy Dept., Naval Yard, N.Y.C., 1940-45; aero. research sci. NACA, 1948-49; faculty Pa. State U., 1949-53, U. Pa., 1953-60; prof. chmn. dept. mech. and aerospace scis. U. Rochester, 1960—, Yates Meml. prof. engring., 1967—. NSF sr. fellow, 1966-67; IBM grantee, 1948. Mem. Am. Soc. M.E., Am. Phys. Soc., Gesellschaft fur Angewandte Mathematik und Mechanik, Sigma Xi, Tau Beta Pi. Research and publs. on jet and boundary layer flows, fully-coupled thermal shock; wave propagation in plasmas; plasma acceleration by travelling wave fields; inventor statorless jet engine; researcher relativistic gas-dynamics. Home: 9 Idlewood Rd., Rochester, N.Y. 14618.*

LESSER, Gerald S., Am. psychologist; b. Jamaica, N.Y., Aug. 22, 1926; s. Nathan and Esther (Lacks) L.; B.A., Columbia, 1947, M.A., 1948; Ph.D., Yale, 1952; M.A. (hon.), Harvard, 1963; m. Stella Scharf, Aug. 15, 1953; children—Julie Anne, Nina Ruth, Theodore. Research psychologist N.Y. State Brain Research Assn., 1948-49; postdoctoral fellow Yale Child Study Center, 1952-54; asst. prof. Adelphi Coll., 1954-56; asso. prof. Hunter Coll. City U. N.Y., 1956-63; prof. edn. and devel. psychology, dir. lab. human devel. Harvard, 1963—, Bigelow prof. edn. Fellow Am. Psychol. Assn., Sigma Xi. Home: 12 Tower Rd., Lexington, Mass. 02173. Office: Larsen Hall, Harvard Univ., Cambridge, Mass. 02138.*

LESSLER, Milton A., Am. cell physiologist; b. N.Y.C., May 18, 1915; s. Lewis and May (Levitt) L.; B.S., Cornell U., 1937, M.S., 1938; Ph.D., N.Y. U., 1950; postgrad. Oak Ridge Inst. Nuclear Studies; m. Katherine A. Banks, Feb. 20, 1943; children—Barbara Ellen, Mark Alan, Roy W. Harris. Sci. tchr. Flatbush Inst., 1939-40; technician N.Y. State Health Dept., 1940-42; Pre and postdoctoral fellow NIH Cancer Inst., N.Y. U., 1946-51; faculty Ohio State U. Coll. Medicine, Columbus, 1951—, prof. physiology, 1963—. Cons. Instrument Co., Yellow Springs, O., 1963-66, Carl Zeiss, N.Y.C., 1966. Mem. Am. Assn. Cancer Research, Am. Physiol. Soc. (vis. scientist 1962-66, lectr. workshop 1958), Internat., Am. socs. for cell biology, Ohio Conf. Coll. Biology Tchrs., Radiation Research Soc., Sigma Xi. Author: (with E. G. Bourne, J. F. Danielli) International Review of Cytology, vol. 2, 1953, vol. 16, 1964; (with others) Radiobiology at the Intracellular Level, 1959. Research on oxidative metabolism cells by automated techniques, biol. effects low-level X-radiation, comparative studies erythrocytes and precursors. Office: 410 W. 10th Av., Columbus, O. 43210.*

LESSON, René Primevère, French biologist; b. Rochefort, France, Mar. 20, 1794; zoologist in chief Duperry Expdn. to Pacific, 1822-25; head pharmacist French Navy; named prof. med. physics and chemistry Rochefort Sch. Medicine, 1826. Mem. French Acad. Scis., 1833. Author: Voyage médical aufour du monde, 1829; Manuel d'ornithology domestique, 1834. Studied all animals except birds on Duperry expdn., ethnography of human races, domestic birds. Died Rochefort, Apr. 28, 1849.

LESTER, Charles Turner, Am. chemist; b. Covington, Ga., Nov. 10, 1911; s. Richard P. and Estelle (Rush) L.; A.B., Emory U., 1932, M.A., 1934; Ph.D., Pa. State U.; m. Marlyn Elizabeth Tate, June 23, 1936; children—Charles Turner, James G., III. High sch. sci. tchr., Thomson, Ga., 1934-35; instr. Oxford (Ga.) Coll., 1934-35; research chemist Am. Cyanamid, Bound Brook, N.J., 1941-42; faculty Emory U., Atlanta, 1942—, prof. chemistry, 1950—, chmn. dept. chemistry, 1954-57, dean Grad. Sch., 1957—. Mem. Am. Chem. Soc. (past chmn. Ga. sect., Herty medal 1965), Am. Inst. Chemists (past chmn. Piedmont chpt.), Phi Beta Kappa, Sigma Xi, Phi Kappa Phi. Research and publs. on effect of structure of certain kinds of organic molecules on their chem. reactions, especially with regard to ketones and esters; patentee use of certain vat dyes in nonalkaline solutions. Home: 281 Chelsea Circle, Decatur, Ga. 30030. Office: Emory U., Atlanta 30322.*

LESTER, Charles Willard, Am. physician; b. Saratoga Springs, N.Y., Sept. 14, 1892; s. James Westcott and Bertha (Dowd) L.; B.A., Williams Coll., 1914; M.D., Columbia, 1918; m. Marianne Groves Stebbins, June 5, 1920. Surg. pathologist Roosevelt Hosp., N.Y.C., 1921-29, attending thoracic surgeon, 1943-57, cons., 1957—; asst. attending surgeon Fifth Av. Hosp., N.Y.C., 1928-35; asst. attending to attending surgeon Bellevue Hosp., 1935-57; attending thoracic surgeon Hosp. for Spl. Surgery, 1937-54, cons., 1954—; asso. attending surgeon Triboro Hosp., N.Y.C., 1941-47; attending surgeon Municipal Sanitarium, Otisville, N.Y., 1937-50; cons. thoracic surgeon VA Hosp., Castle Point, N.Y., 1944—, Rockland State Hosp., Orangeburg, N.Y., Horton Meml. Hosp., Middletown, N.Y., St. Francis Hosp., Port Jervis, N.Y. Asso. prof. clin. surgery N.Y. U., 1933-49, N.Y. Med. Coll., 1948-57; asst. clin. prof. surgery Columbia, 1950-57. Diplomate Am. Bd. Surgery, Am. Bd. Thoracic Surgery. Fellow A.C.S., Am. Coll. Chest Physicians, N.Y. Acad. Medicine; mem. Am. Assn. Thoracic Surgeons, Am. Thoracic Soc., N.Y. Surg. Soc. Author: Thoracic Surgery, 1941; also articles. Research in thoracic surgery, especially of children including lung function studies after removal of a lung; non-deforming type thoracoplasty; devel. operations to correct chest deformities, particularly funnel chest and pigeon breast; research in paleopathology. Home: 320 E. 72d St., N.Y.C. 10021.*

LESTER, David, Am. biochemist; b. New Haven, Jan. 22, 1916; s. Asher and Esther (Rubin) L.; B.S., Yale, 1936; Ph.D., 1940; m. Ruth Weiss, Sept. 17, 1938; children—Anne D., James M. With Yale, 1940-62, research asso., 1946-62; prof. biochemistry Rutgers U., New Brunswick, N.J., 1962—. Fellow A.A.A.S.; mem. N.Y. Acad. Sci., Am. Soc. Pharmacology and Exptl. Therapy, Am. Indsl. Hygiene Assn., Japan Med. Soc. Alcoholic Studies (hon.). Author: (with L. A. Greenberg) Handbook of Cosmetic Materials; Interscience, 1954. Asso. editor Quar. Jour. Studies on Alcohol. Research, publs. on analytical methods for determination of ketone bodies and acetic acid, intermediary metabolism of aniline compounds, etiology of alcoholism. Home: 29 Forester Dr., Princeton, N.J. 08540. Office: Smithers Hall, Rutgers The State U., New Brunswick, N.J. 08903.*

LESTER, Gabriel, Am. microbiologist; b. New Haven, Aug. 21, 1920; s. Asher and Esther (Rubin) L.; B.S., U. Conn., 1948; M.S., Yale, 1950, Ph.D., 1951; m. Thelma Lillian Singer, July 3, 1943; children—Daniel F., Alan P. Mem. research staff Charles Pfizer & Co., Bklyn., 1952-54; faculty Yale, 1954-56; staff scientist Worcester Found. Exptl. Biology (Mass.), 1956-61; prof., chmn. dept. biology Reed Coll., Portland, Ore., 1961—. Fellow A.A.A.S.; mem. Am. Soc. Microbiology, Soc. Developmental Biology, Soc. Gen. Physiologists, Sigma Xi. Research, publs. in fields of enzymes, amino acids, hormones.*

LESTER, Oliver Clarence, Am. physicist; b. Morris County, Kan., Nov. 3, 1873; s. John Augustus and Mary Virginia (Watts) L.; A.B., Central Coll., Fayette, Mo., 1897, A.M., 1898; A.M., Yale, 1902, Ph.D., 1904; LL.D., Colo. Coll., 1941; m. Pynk Johnson, Sept. 8, 1897; children—Katherine Wheeler (Mrs. H. Laurence Humbley), Oliver Clarence, John Augustus. Prof. Latin and Greek, Hendrix Coll., Conway, Ark., 1897-98; asst. prof. Latin and Greek, Central Coll., 1898-1901; asst. in physics Yale, 1901-04, Loomis fellow in physics, 1903-04, instr. physics, 1904-07; prof. physics U. Colo., from 1907, dean Grad. Sch., from 1919, v.p., from 1931, also acting pres., 1922-23, 32-33; prof. physics U. Ind., 1942-43, U. Colo., 1943-49; physicist Colo. State Geol. Survey, 1914-18; dir. research Carnatite Products Co. (Vanadium Products Co.), Boulder, Colo., 1919-22. Fellow A.A.A.S. (pres. S.W. div. 1933-34), Am. Phys. Soc., Am. Assn., Am. Geog. Soc.; mem. Am. Soc. Engring. Edn., Am. Geophys. Union, Colo.-Wyo. Acad. Sci. (pres. 1929-30), Am. Assn. Physics Tchrs., Sigma Xi, Tau Beta Pi, Alpha Chi Sigma, Sigma Pi Sigma. Author: The Integrals of Mechanics, 1909; also various scientific papers. Died Sept. 28, 1951.

LESTER, Richard Garrison, Am. radiologist, physician; b. N.Y.C., Oct. 24, 1925; s. L. I. and Pauline (Smolan) L.; A.B., Princeton, 1948; M.D., Columbia, 1948; m. Marion Louise Kurtz, Jan. 17, 1949; children—Elizabeth P., Andrew W. Faculty, U. Minn. Med. Sch., Mpls., 1954-61; prof., chmn. dept. radiology Med. Coll. Va., Richmond, 1961-65, Duke Med. Sch., Durham, N.C., 1965—. Fellow Am. Coll. Radiology, Am. Coll. Chest Physicians (chmn. com. on cardiovascular roentgenology 1963—); mem. Soc. for Pediatric Radiology (past sec.-treas.), Assn. U. Radiologists (dir. 1965—), Am. Roentgen Ray Soc., Radiol. Soc. N.Am. Author: (with Jesse E. Edwards, Lewis S. Carey, Henry N. Newfeld) Congenital Heart Disease: Correlation of Pathologic Anatomy and Angiocardiography, 1965; also articles. Research on diagnosis congenital and acquired heart disease by X-ray methods, pediatric radiology, new methods rec. radiographic images, uses image intensification. Home: 2703 Montgomery St., Durham, N.C. 27705.*

LESTIBOUDOIS, (Thémistocle) Gaspard, French botanist, physician; b. Lille, France, Aug. 12, 1797; s. Jean-François Lestiboudois; M.D., Paris, 1818; prof. botany and zoology U. Lille; made voyage to study plague in Algeria, 1835. Mem. French Acad. Scis., 1845. Author: Études sur l'anatomie et la physiologie végétaux, 1840. Research on anatomy and physiology of plants, bubonic plague in Algeria. Died Paris, Nov. 22, 1876.

LESTRADET, Henri, French physician; b. Esternay, France, Feb. 17, 1921; s. Georges and Henriette (Tanneur) L.; Licencie es sciences Faculté des Sciences de Paris, 1942, Docteur en Medecine, 1945; m. M. A. Woimant, May 14, 1948; children—Francois, Anne, Luc, Odile, Claire, Marie. Fellow, Children's Hosp. Research Found., Cin., 1949-50; staff Hopitaux de Paris, 1960—, Faculté de Medecine, Paris, 1950—, prof. agrégé, 1962—; dir. Centre de Recherche sur le Diabete chez l'Enfant, 1964—; dir., physician for vacation camps for diabetic children in France. Decorated officier Ordre de la Santé Publique, 1961. Mem. Am. Diabetes Assn., European Club for Pediatric Research, Societé Francaise de Pediatrie, Diabetologues de Langue Française. Author: (with I. Royer) Traitement du Diabetes Infantile, 1958; (with A. Schaetz) Der Diabetes Mellitus, 1966; also numerous publs. Research on metabolic diseases in children, diabetes mellitus in children, exptl. studies on utilization of ketones bodies in acedosis, utilization of glucose and spontaneous feeding habits in diabetes, microcirculation in prediabetic children. Home: 7 Place du Tertre, Paris 18. Office: Hopital Herold 7 Place Rhin et Danube, Paris 19, France F. 75.*

LESUEUR, Charles Alexandre, naturalist; b. Le Havre, France, Jan. 1, 1778; s. Jean-Baptiste Denis and Charlotte Geneviéve (Thieullent) L.; attended Royal Mil. Sch., Beaumont-En-Auge, France, 1787-96. Mem. French scientific expdn. which explored coasts of Australia, 1800-04, took many zool. specimens back to France, including 2,500 new species; in West Indies, 1815-16; came to Am., 1816; made tour (with Am. geologist William Maclure) of much of interior Am., 1816-17, painted and collected specimens on trip; engraver and tchr. of drawing, Phila., 1817-26; curator Acad. Natural Scis. of Phila., 1817-25; tchr. drawing, New Harmony, Ind. (community founded by Robert Owen), 1826-37, also continued scientific work; lived in Paris, France, 1837-45; wrote 29 monographs on Am. fishes; engraved plates for scientific publs. Author: (with others) Voyage de Découvertes aux Terres Australes, 2 vols., 1807-16. Died Dec. 12, 1846.

LESUK, Alex, Am. biochemist; b. N.Y.C., July 23, 1917; s. Paul G. and Olga (Sereda) L.; B.S magna cum laude, Coll. City N.Y., 1936; Ph.D., Yale, 1939, Internat. Cancer Research Found. fellow, 1939-40; m. Mary Regina Franke, Nov. 6, 1943; children—Diana Mary, Elaine Mary. With Winthrop Labs., 1940-49, head biochemistry sect., sr. research biologist Sterling-Winthrop Research Inst., 1949— (both Rensselaer, N.Y.). Asso. prof. chemistry St. Bernardine of Siena Coll., Loudonville, N.Y., 1950—; lectr. Rensselaer Poly. Inst., Troy, N.Y., 1963. Fellow A.A.A.S., N.Y. Acad. Scis.; mem. Am. Chem. Soc., Phi Beta Kappa, Sigma Xi. Research, publs. on crystallization of human urokinase (blood-clot-dissolving medicinal agt.); preparation of therapeutic medicinal agts. (thyroid hormone, papain, silver proteinates, others); isolation and chem. structure determination of natural products; also patentee in field. Home: 37 Groesbeck Pl., Elsmere, N.Y. 12054. Office: Sterling-Winthrop Research Inst., Rensselaer, N.Y. 12144.*

LESZCZYNSKI, Zbigniew Kazimierz, Polish chemist; b. Poznan, Poland, Apr. 21, 1927; s. Tadeusz Tomasz and Kazimiera (Czypicka) L.; M.A., U. A. Mickiewicz, Poznan, 1950; M.Engring., Politechnic Sch., Warsaw, Poland, 1954; D.Sc., Inst. Gen. Chemistry, Warsaw, 1960; m. Halina Franke, Aug. 12, 1950; children—Jolanta, Malgorzata, Jerzy. With Chem. Indsl. Inst., Warsaw, 1950—, head asst. technol. dept., 1956-65, mgr. asst. chem. engring. and devel. div. Inst. Gen. Chemistry, Warsaw, 1966—; sci. adviser research lab. Inst. Car Transport, 1952-53; head research dept. Refinery and Petrochem. Factories, Plock, Poland, 1960-63. Decorated Gold Cross Merit; UN fellow U. Hull (Eng.), 1963-64. Mem. Chemists and Engrs. Assn. Chem. Industry, Chem. Industry, Chem. Assn. Germany. Author: Practice in Chemistry, 1955; also articles. Research on kinetics of oxidation and decomposition of its hydroperoxide of isopropylbenzen, reaction of butadien with deuterium, tech. of catalytic hydrogenation of nitrobenzen, one step synthesis of N-ethylaniline, maleic anhydride from benzene, synthesis of ethyllamines, conversion of hydrocarbons with sulpher. Home: 4a Bytomska, Warszawa 32. Office: 8 Rydygiera, Warszawa 86, Poland.*

LETAVET, Avgust Andreevich, Russian hygienist; b. Senuli, Latvia, 1893; grad. Med. Faculty, Moscow U., 1917. Sr. asst. dept. labor hygiene 2d Moscow Med. Inst., 1924-31, Central Inst. Labor Protection, 1925-27; head hygiene sect. Central Inst. Labor Hygiene and Indsl. Sanitation, RSFSR Peoples Commissariat Health, 1927-35; prof., 1931—; head chair indsl. hygiene Central Postgrad. Med. Inst., Moscow, 1931-55; head lab. for indsl. microclimate Central Inst. Labor Hygiene and Occupational Diseases, USSR Acad. Med. Sci., 1935-48, dir. Inst., 1948—, acad. sec. dept. hygiene, microbiology and epidemiology, 1957-60. Dep. chmn. Silicosis Commn., 1948—; chmn. Labor Hygiene Problem Commn. Decorated Order of Lenin (2); recipient State prize, 1949; Lenin prize, 1963. Mem. USSR Acad. Med. Sci., All-Union Soc. Hygienists (dep. chmn.), Internat. Assn. Occupational Medicine (v.p. permanent com. 1960—). Author: Hygiene Problems of Radiology, 1957; co-author: Labor Hygiene in Agriculture: A Manual for Physicians, 1960. Editor: The Toxicology of New Chemical Substances, Nos. 1, 2, 3, 1961; Labor Hygiene and Occupational Diseases; mem. editorial council Hygiene and Sanitation; coeditor Hygiene sect. Large Med. Ency., 2d edit. Research and numerous publs. on labor hygiene, effects of indsl. meteorol. conditions on body, radiation heat exchange between man and his environment, prophylactic measures against silicosis and radiation sickness, hygiene standards, labor hygiene in work with radioactive substances and radiation. Address: Central Inst. Labor Hygiene and Occupational Diseases, USSR Acad. Med. Sci., Meyerovsky pr. 31, Moscow, USSR.

LETICHE, John Marion, Am. economist; b. Uman, Russia, Nov. 11, 1918; s. Leon and Mary Letiche; came to U. S., 1941, naturalized, 1949; B.A., McGill U., 1940, M.A. summa laude, 1941; Ph.D., U. Chgo., 1951; m. Emily Kuyper, Nov. 17, 1945; 1 son, Hugo K. Faculty, U. Cal., Berkeley, 1946—, prof. econs., 1960—; vis. prof. (Smith-Mundt fellow) U. Aarhus, U. Copenhagen, Denmark, 1951-52. Adviser, cons. to various govt. agys. Guggenheim fellow, 1956-57. Mem. Am. Econ. Assn., Econometric Soc., Royal Econ. Assn. Author: Reciprocal Trade Agreements in the World Economy, 1948; Balance of Payments and Economic Growth, 1967. Editor: The System or Theory of the Trade of the World (Gervaise), 1954; A History of Russian Economic Thought, 1964. Researcher in econs. Home: 968 Grizzly Peak Blvd., Berkeley, Cal.*

LETORT, Maurice Joseph Amand, French physical chemist; b. Corps-Nuds, France, Dec. 15, 1907; s. Joseph Marie Emile and Berthe (Thebaud) L.; ed. Ecole Nationale Supéreure de Chimie, Paris, 1928; Dr. ès Scis., Sorbonne, Paris, 1937; m. Assia Lichterman, Aug. 20, 1936. Asst. phys. chemistry U. Liège (Belgium), 1932-36; prof. Institut Français de Prague, Czechoslovakia, also dir. sect. sci. and tech., 1937-39; lectr. Faculté des Sciences, Caen, France, 1941-43; prof. Faculté des Sciences, Nancy, France, 1943-58; dir. Ecole Nationale Supérieure des Industries Chimiques, U. Nancy, 1946-56; sci. dir. gen. Centre d'-Etudes et Recherches des Charbonnages de France, Paris, 1956—. Cons. on com. to Prime Minister, 1958-62. Decorated Officier de la Légion d'Honneur, Commandeur de l'Ordre national des Palmes Académiques, Commandeur de l'Ordre du Mérite pour la Recherche et l'Invention. Mem. Internat. Union Pure and Applied Chemistry (pres. div. phys. chemistry, v.p. union 1953-57, mem. exec. com. 1959-63), French Chem. Soc., French Acad. Scis., Phys. Chemistry Soc. (pres. 1963-64), Indsl. Chemistry Soc., Faraday Soc., Internat. Com. for Sci. Research and Tech. Author: Conceptions actuelles sur le mécanisme des réactions chimiques (cinétique), 1937; also numerous articles. Research on by-products of carbonization and macromoleular chemistry, chem. kinetics—pyrolysis and autoxidation of organic bodies, combustion of graphite and macropolymerization; 1st to discover polyacetaldehyde (macromolecule) could be prepared from solid state, 1948. Office: 35 rue Saint-Dominique, Paris 7, France.*

LETOURNEAU, Charles-Jean-Marie, French anthropologist; b. Auray, France, 1831; docteur en medecine, 1858; prof. history civilizations l'École d'anthropologie, 1885-1902; mem. la Société d'anthropologie de Paris (pres. 1886, sec. gen. 1887-1902). Author: L'evolution de la morale du mariage et de la famille, de la propriété, du commerce, de l'esclavage, d l'education; L'evolution religieuse; La guerre dans les diverses races humaines; La psychologie ethnique. Developed thesis that all sociology is based on ethnography. Died Paris, 1902.

LETSINGER, Robert Lewis, Am. chemist; b. Bloomfield, Ind., July 31, 1921; s. Reed A. and Etna (Phillips) L.; student Ind. U., 1939-41; B.S., Mass. Inst. Tech., 1943, Ph.D., 1945; m. Dorothy C. Thompson, Feb. 6, 1943; children—Louise, Reed, Sue. Asso., Mass. Inst. Tech., 1945-46; research chemist Tenn. Eastman Corp., 1946; faculty Northwestern U., Evanston, Ill., 1946—, prof. chemistry, 1959—. Guggenheim fellow, 1956. Mem. Am. Chem. Soc., A.A.A.S., Sigma Xi. Research, numerous publs. on stereochemistry, synthesis metallic organic compounds, catalysis, polymer supports organic reactions. Home: 316 3d St., Wilmette, Ill. 60091. Office: Northwestern U., Evanston, Ill. 60201.*

LETTS, Edmund Albert, Brit. chemist; b. Sydenham, Eng., Aug. 27, 1852; s. Thomas Letts and Emma (Horwood) L.; ed. King's Coll., London, U. Vienna, U. Berlin; D.Sc. (hon.). Became chief asst. Chem. Lab., U. Edinburgh (Scotland), 1872; named 1st prof. chemistry U. Coll., Bristol, Eng., 1876; prof. chemistry Queen's U., Belfast, Ireland. Fellow Royal Soc. Edinburgh (Keith prize 1890), Chem. Soc., Royal Inst. Chemistry. Author: The Pollution of Estuaries and Tidal Waters; Some Fundamental Problems in Chemistry—Old and New; also articles. Died Feb. 19, 1918.

LETTSOM, John Coakley, physician; b. W.I. nr. Tortola, V.I.; ed. Edinburgh, (Scotland), Paris, Netherlands; studied under John Fothergill, W. Cullen, Edinburgh; M.D., Leiden, Germany, 1796; practice medicine, London, from 1769; founder Gen. Dispensary, Aldersgate St., Royal Sea Bathing Hosp., Margate, Eng. Lettsomian lectures named in his honor. Fellow Royal Soc., 1773; mem. Royal Home Soc. (a founder), Med. Soc. London (founder). Author: The Natural History of the Tea-Tree, 1772; Naturalist's and Traveller's Companion, 1774; Life of Dr. Fothergill, 1783; also wrote a jour. of the 2d voyage of Capt. Cook. Supported Jenner on vaccination; pioneered study drug addiction; 1st descriptions of multiple neuritis, 1779-87. Died 1815.

LETULLE, Maurice, French physician; b. Mortagne, France, 1853; became hosp. physician, 1883; apptd. agrégé prof. Faculté de Paris, 1889; named prof. anatomy path., 1916; founder l'hôpital Boucieaut, a library-mus.; mem. French Acad. Medicine. Author: l'Anatomie pathologique des poumons; L'anatomie des lésions tuberculeuses. Research on path. anatomy. Died Paris, 1929.

LEUBA, James Henry, psychologist; b. Neuchatel, Switzerland, Apr. 9, 1868; s. Henri and Cecile (Sandoz) L.; B.Sc., Neuchatel, 1887; came to U. S., 1887; Ph.D., Clark U., 1895; studied Leipzig, Halle, Heidelberg, Paris, 1897-98; m. Berthe A. Schopfer, Jan. 6, 1806; children—Clarence James, Gladys Aline. Prof. psychology, Bryn Mawr Coll., 1889-1933, emeritus since 1933. Fellow A.A.A.S. Author: The Psychological Origin and the Nature of Religion, 1909; A Psychological Study of Religion, 1912; The Beliefs in God and Immortality, 1916; The Psychology of Religious Mysticism, 1925; God or Man?—a Study of the Value of God to Man, 1933. Ready for publication, The Reformation of the Churches. Died Dec. 8, 1946.

LEUBE, Wilhelm Oliver, German physician; b. Ulm, Germany, Sept. 14, 1842; prof. Würzburg, Germany. Improved stomach probe techniques; described form of septicemia without local infection, 1878. Died Langenargen, Germany, May 16, 1922.

LEUCIPPOS, Greek philosopher; b. Miletos, Greece; flourished circa 440 B.C. Disciple of Zeno, tchr. of Democritos. Author: The Great World System; The Mind (contained in Democritean corpus). Credited (with Democritos) as founder of atomistic theory; said by some to be 1st to state rule of causality (every event has natural cause); stated that bodies in circular motion move from center, motion is made possible by positing empty space, atoms are infinite in number and space infinite in extent.

LEUCKART, Karl Georg Friedrich Rudolf, German zoologist; b. Helmstedt, Germany, Oct. 7, 1822 or 23; ed. U. Göttingen (Germany). Joined faculty U. Giessen (Germany), 1850, U. Leipzig (Germany), 1870. Fellow Royal Soc., 1877. Author: Die Parasiten des Menschen, 2 vols., 1862-76; also numerous articles. Described morphology and life history of Taenia echinococcus, also relationship between hydatid cysts and tapeworms in dog, 1863-76; pioneered parasitology and animal ecology; described structure of sponges and classified them as coelenterates; distinguished between Coelenterata (jellyfish) and Echinodermata (starfish); demonstrated some human diseases were caused by multicellular creatures; 1st to study life histories of many parasitic worms. Died Leipzig, Germany, Feb. 6, 1898.

LEUPOLD, Friedrich, German physician; b. Würzburg, June 9, 1918; s. Ernst and Elisabeth, N. L.; M.D.; m. Anne Schrauff, Jan. 31, 1946; children—Eva, Martin. Physician-in-chief, dir. Johanniter Hosp., Reinhausen. Mem. German Soc. Internal Medicine, Soc. Physiol. Chemistry, German Soc. Medicine. Research and publs. on exchanges of lipids, radioisotope tech., gen. medicine, hormonal regulation, arteriosclerosis. Home: Berg. Gladbach, Siefen 13. Office: Johanniter Krankenhaus, Reinhausen, W. Germany.

LEUPOLD, Jacob, mechanic; b. Planitz Zwickau, Germany, 1674; ed. Jena, Germany, Wittenberg, Germany; operated mech. workshop, Leipzig, Germany; apptd. mem. Council Mines; councillor for commerce, Prussia. Mem. Royal Soc. Berlin. Author: Theatrum Machinarum (on machines, statics, hydrostatics, mech. scis.), 1724-39. Improved air pump; built multiplication machine; designed machine to raise water through expansion, 1725. Died 1727.

LEURECHON, Jean, French inventor, mathematician; b. Bar-le-Duc, France, 1591; Jesuit scholar. Author: Récréations mathématiques, 1626; also wrote on astronomy. Described steam sphere whose jets propelled a small mill. Died Pont-à-Mousson, France, Jan. 17, 1670.

LEURET, François, French psychiatrist, physician; b. Nancy, France, 1797; became chief physician of Bicetre, Paris, France; author: Psychological Fragments on Insanity, 1834; On the Moral Treatment of Insanity, 1840. Introduced name, fissure of Rolando, for Luigi Rolando who first described it, 1839. Died 1851.

LEUSCHNER, Armin Otto, Am. astronomer; b. Detroit, Jan. 16, 1868; s. Otto Richard and Caroline (Humburg) L.; grad. Royal Wilhelms-Gymnasium, Cassel, Germany, 1886; A.B., U. Mich., 1888, Sc.D. (hon.), 1913; grad. student Lick Obs., U. Cal., 1888-90; Ph.D., Berlin, 1897; Sc.D. (hon.), U. Pitts., 1900; LL.D., U. Cal., 1938; m. Ida Louise Denicke, May 20, 1896 (dec. Nov. 1941); children—Erida Louise, Richard Denicke, Frederick Denicke. Instr. math. U. Cal., 1890-92, asst. prof., 1892-94, asst. prof. astronomy and geodesy, 1894-98, asso. prof., 1898-1907, dir. Student Obs., 1898-1938, prof. astronomy and chmn. dept., 1907-38, dean Grad. Sch., 1913-18, 20-23, emeritus, 1938—. Recipient Watson Gold medal, Nat. Acad. Scis., 1916; Bruce Gold medal, Astron. Soc. Pacific, 1936; Rittenhouse medal, 1937. Fellow Cal. Acad. Scis., A.A.A.S., Seismol. Soc. Am., Internat. Geophys. Union, Astron. Soc. Pacific (pres. 1908, 36, 43); mem. Nat. Acad. Scis., NRC (exec. sec.), Am. Philos. Soc., Am. Math. Soc., Astronomische Gesellschaft, Am. Astron. Soc., Washington Acad. Scis., Sigma Xi, Phi Beta Kappa; fgn. asso. Royal Astron. Soc. of London; foreign mem. Royal Physiographical Soc., Lund, Sweden. Special field of investigation, theoretical astronomy; also perturbations of Watson asteroids; improvement in methods of determining preliminary orbits of comets and planets; perturbations of the Hecuba group of minor planets. Author: Beitrage zür Kometenbahnestimmung, Berlin, 1897; Short Methods of Determining Orbits from Three Observations; Tables of Minor Planets Discovered by James C. Watson; Research Surveys of 1091 Minor Planets; also papers on astron. subjects. Died Apr. 22, 1953.

LEUTWEIN, Friedrich, German geochemist; b. Berlin, Aug. 9, 1911; s. Friedrich and Dorothea (Sachau) L.; ed. univs. Giessen and Fribourg; Ph.D., agrégé, 1945; m. Ingrid Arning, June 14, 1939; children—Klaus, Marianne, Burkhard. Asst., Fribourg, 1936-39; engr., Mines Service, 1939-43; head central labs. Erzgebirge Mines, 1943-47; titular prof. mineralogy Frieberg Sch. Mines, 1947-58, rector, 1949-53; hon. prof. U. Hamburg, 1958-60; asso. prof. U. Nancy (France) Faculty Scis., 1960-62; dir. research, center sci. research Nancy Center for Geochem, and Petrographic Research, 1960—; Mem. Leipzig, German acads. sci., German Engring Soc., Bonn Geol. Soc., Engrs. and Miners, French Soc. Mineralogy and Crystallography, Geochem. Soc. (charter mem.). Research and publs. on geochemistry and metallogenesis. Address: Nancy B.P. 682, C.R.P.G. Vandoeuvre-lès-Nancy, France.

LEV, Maurice, Am. pathologist; b. St. Joseph, Mo., Nov. 13, 1908; s. Benjamin and Rose (Lev) L.; B.S., N.Y. U., 1930; M.D., Creighton U., 1934; M.A., Northwestern U., 1966; m. Lesley Beswick, Sept. 7, 1947; children—Benita, Peter. Faculty, U. Ill. Coll. Medicine, Chgo., 1939-42, 47-51, asso. prof., 1948-51, lectr., 1963—; asst. prof. Creighton U. Med. Sch., Omaha, 1946-47; faculty U. Miami Sch. Medicine, Coral Gables, Fla., 1951-57, prof. pathology, 1956-57; pathologist, dir. research labs. Mt. Sinai Hosp. Greater Miami, Miami Beach, Fla., 1951-57; dir. Congenital Heart Disease Research and Tng. Center, Hektoen Inst. for Med. Research, Chgo., 1957—; prof. pathology Northwestern U. Med. Sch., Chgo., 1957—; professorial lectr. U. Chgo. Sch. Medicine, 1959—. Cons. cardiovascular pathology Children's Meml. Hosp., Chgo., 1957—; career investigator and educator Chgo. Heart Assn., 1966—. Fellow A.M.A., Am. Coll. Cardiology; mem. Coll. Am. Pathology, Am. Soc. Clin. Pathology, N.Y. Acad. Scis., Am. Coll. Chest Physicians, Inst. Medicine Chgo., Am. Heart Assn., Council on Epidemiology. Author: (with A. Vass) Spitzer's Architectures of the Normal and Malformed Hearts, 1951; Autopsy Diagnosis of Congenitally Malformed Hearts, 1953; (with B. M. Gasul, R. A. Arcilla) Heart Disease in Children: Diagnosis and Treatment, 1966; also numerous articles. Research in pathology of congenital heart disease and conduction system. Home: 1014 Elmwood St., Evanston, Ill. 60602. Office: 637 S. Wood St., Chgo., 60612.*

LEVADITI, Constantin, bacteriologist, physician; b. Galatz, Rumania, Aug. 1, 1874; student Bucharest (Rumania) U.; M.D., Paris U.; Dr. h.c., Amsterdam U., Ghent (Belgium) U.; m. Miss Istrati; 2 children. Named prof. Pasteur Inst., 1926, also head lab.; sci. dir. Inst. Alfred Fournier. Recipient Cameron prize, Edinburgh, 1928, Phila. prize, 1929, Paul Erlich prize, 1931. Mem. Nat. Acad. Medicine. Author: La syphilis, 1909; Les ectodermoses neurotropes, 1922;

L'herpes et le zona, 1926; Le bismuth dans le traitement de la syphilis, 1924; Prophylaxie de la syphilis, 1935; Les ultravirus des maladies, 1938; Precis de virologie humaine, 1945; La penicilline et ses applications therapeutiques, 1945; La streptomycine et ses applications therapeutiques, 1948. Research on ultraviruses, Tb., antibiotics, certain spirochetal diseases, serological diagnosis, neurotropic viruses, treatment of syphilis with bismuth. Died 1953.

LEVAILLANT, François, ornithologist; b. Paramaribo, Surinam, 1753. Author: Voyages de Levaillant dans l'intérieur de l'Afrique, 1790-96; Histoire naturelle des oiseaux d'Afrique, 6 vols., 1796-1812. Traveled in S. Africa, studying natives and collecting birds, 1780-85. Died Sezánne, France, Nov. 22, 1824.

LEVAN, Johan Albert, Swedish chromosome cytologist; b. Göteborg, Sweden, Mar. 8, 1905; s. Emil Teodor and Amy (Gabrielsson) L.; Fil.Kand., U. Lund (Sweden), 1927, Fil.Lic., 1933, Fil.Dr., 1935, M.D. (hon.), 1958; m. Karin Linnéa Malmberg, Jan. 2, 1933; children—Hans Göran Fredrik, Gerda Maria Cecilia Torudd. Asst. cytogeneticist Swedish Sugar Co., 1933-39; head dept. chromosome research Svalöf Plant Breeding Inst., 1939-47; faculty U. Lund, 1947—, prof. cytology, 1961—; research asso. U. Cal., Berkeley, also U. S. Dept. Agr., Beltsville, Md., 1937; sr. research fellow Inst. for Cancer Research, Fox Chase, Pa., 1952, 54; vis. mem. Sloan-Kettering Inst., N.Y.C., 1955, 57; research scientist M.D. Anderson Hosp., Houston, 1959-60; vis. prof. U. Pa., Phila., 1961. Cons. cytogenetic dept. Inst. for Med. Research, Camden, N.J., 1961—. Recipient W. Westrup award, 1953; G. Dahlberg medal, 1963. Mem. Mendelian Soc., Royal Physiographic Soc., Soc. Zool. Bot. Fenn. Vanamo, N.Y. Acad. Scis., Swedish Soc. Surgeons and Physicians, Am. Assn. for Cancer Research, Royal Danish Acad. Sci., Letters, Soc. Sci. Fenn., Swedish Royal Acad. Research, numerous publs. on various aspects of chromosomes, 1929-38, cytotaxonomy of genus Allium, 1938-50, chromosome responses to changes in environment, analysis of c-mitosis (effect of colchicine), chromosomes in cancer, influence of viruses on chromosomes, 1950—. Home: 18 Pedellgatan, Lund, Sweden.*

LEVAVASSEUR, Léon, engr., painter; b. Cherbourg, France, 1863; invented 1st airplane motor (Antoinette) in V-form with condensation of the cooling water (mounted on a single decker with thickened wings and triangular struts and equipped with cross rudders). Died 1922.

LEVEN, Charles L., Am. economist; b. Chgo., May 2, 1928; s. Elie H. and Ruth (Reinach) L.; student Ill. Inst. Tech., 1945, 46-47, U. Ill., 1947-48; B.S. with honors, Northwestern U., 1950, M.A. (Ford Found. fellow), 1957, Ph.D. (Social Sci. Research Council fellow), 1958; m. Judith R. Danoff, Sept. 10, 1950; children—Ronald L., Robert M., Carol E. Research asst. Fed. Res. Bank, Chgo., 1949-51, economist, 1951-56; asst. prof. econs. Ia. State U., 1957-59; asst. prof. econs. and regional sci. U. Pa., 1959-62; asso. prof. econs. U. Pitts., 1962-65, asso. dir. Center for Regional Econ. Studies, 1963-65; prof. econs., also dir. Inst. for Urban and Regional Studies, Washington U., St. Louis, 1965—. Cons. to numerous univs., govt., industry; mem. com. regional accounts Resources for Future, Inc., 1959—, regional econs. adv. com. Dept. Commerce, 1963-66, adv. com. small area data Bur. Census, 1965—. Recipient Alumni Achievement award Ia. State Coll., 1958; Ford Found. award, 1959. Social Sci. Research Council grantee, 1960; Com. on Urban Econs. grantee, 1965; NSF grantee, 1967. Mem. Am. Econ. Assn., Econometric Soc., Regional Sci. Assn. (pres. 1963-64). Author: Theory and Method of Income and Product Accounts for Metropolitan Areas, 1957; (with P. Davidson, E. Smolensky) Aggregate Demand and Supply Problems, 1965; also articles. Research in application of techniques of social accounting to econ. analysis of met. areas. Home: 7042 Delmar St., St. Louis 63130.*

LEVENE, Phoebus Aaron (Theodore), chemist; b. Sagor, Russia, Feb. 25, 1869; s. Solom and Etta (Brick) L.; M.D., Imperial Mil. Med. Acad., St. Petersburg, Russia, 1891; spl. student univs. of Berne, Marburg, Berlin, Munich and Columbia; m. Anna M. Erickson, June 9, 1920. Came to U. S., 1892. Asso. in chemistry State Pathol. Inst., N.Y., 1896-1905; chemist Saranac Lab. for Study of Tb, Saranac Lake, N.Y., 1900-02; Herter lectr. in pathol. chemistry N.Y. U., 1905-06; asst. in chemistry Rockefeller Inst., 1905, mem., 1907-39, emeritus, 1939—. Recipient Willard Gibbs medal Chgo. sect. Am. Chem. Soc., 1931, William H. Nichols medal N.Y. sect., 1938. Contbr. papers on proteins, nucleins, carbohydrates, lipoids, problems of stereo-chemistry, also monographs, Hexosamines and Mucoproteins, 1925, Nucleic Acids, 1931. Research on proteins, hexosamines, stereochemistry; influenced by Kossel in interest in nucleic acids; isolated and identified carbohydrate portion of nucleic acid molecule; showed that ribose is contained in some nucleic acids, 1909, deoxyribose (hitherto unknown) in others, 1929; discovered how components of nucleic acids combine into nucleotides and how nucleotides combine into chains; laid basis for work by Todd. Died N.Y.C., Sept. 6, 1940.

LÉVEQUE, Pierre, French hydrographer; b. Nantes, France, Sept. 3, 1746; prof. hydrography Nantes Naval Sch.; also hydrographic engr.; mem. French Marine Acad., French Acad. Scis., 1801. Author: Table pour la determination des longitudes, 1776. Provided points of globe and basic calculations for determining longitude using the method of Lalande. Died Le Havre, France, Oct. 16, 1814.

LEVEQUE, William Judson, Am. mathematician; b. Boulder, Colo., Aug. 9, 1923; s. Earl Mehlum and Doris (Roberts) LeV.; B.A. summa cum laude, U. Colo., 1944; M.A., Cornell U., 1945, Ph.D., 1947; m. Viola Evelyn Johnson, Sept. 11, 1949; 1 son, Randall John. Benjamin Peirce instr. Harvard, 1947-49; faculty U. Mich., Ann Arbor, 1949—, prof. math., 1960—; exec. editor Math. Revs., Ann Arbor, 1965-66. Fulbright research fellow, Eng., 1951-52; Alfred P. Sloan research Fellow, Eng., Germany, 1957-60. Mem. Am., London math. socs., Math. Assn. Am. Author: Topics in Number Theory, vols. I, II, 1956; Elementary Theory of Numbers, 1962; also articles. Research in number theory primarily in diophantine approximation and probabilistic number theory. Home: 3094 Newcastle Rd., Ann Arbor, Mich. 48104.*

LEVER, John, Brit. physician; b. Plumstead, Eng., 1811; ed. Guy's Hosp.; M.D., Giessen, Germany, 1842; m. Miss Pettigrew, 1836 (dec. 1849); m. 2d, Miss Farebrother, 1849; 6 children. Surgeon, Bridgehouse-place, Newington Causeway, Eng., 1834-42; practiced medicine, London, 1842-58; asst. dept. midwifery Guy's Hosp., became lectr. on midwifery and physician-accoucher, 1845. Recipient Fothergillian medal for essay on diseases of uterus, 1843. Licenciate Soc. Apothecaries. Mem. Royal Coll. Surgeons, Hunterian Soc. (pres.). Author: Case of Hidrosis or Hidrotic Fever, 1837; also papers in Guys' Hosp. Reports. First to discover relationship between eclampsia (convulsions occurring in late stages of pregnancy and childbirth) and albuminuria. Died London, Dec. 29, 1858.

LEVER, Walter Frederick, physician; b. Erfurt, Germany, Dec. 13, 1909; s. Alexander and Edith (Hirschberg) L.; student U. Heidelberg (Germany), 1928-30; M.D., U. Leipzig (Germany), 1934; m. Frances Broughton, Sept. 28, 1940; children—Joan (Mrs. Harold H. Woollard III), Susan Howe. Came to U. S., 1935, naturalized, 1941. Practice medicine, specializing in dermatology, Boston, 1940—; faculty Med. Sch. Harvard, 1944-59, lectr., 1959—; prof. dermatology Tufts U. Sch. Medicine, 1959—, chmn. dept., 1961—; chief dermatology Boston Dispensary, 1959—; dir. dermatology Boston City Hosp., 1961—; bd. consultation Mass. Gen. Hosp.; cons. Peter Bent Brigham, Robert Breck Brigham hosps. Mem. gen. medicine study sect. NIH, 1959-63. Mem. A.M.A., Am. Dermatol. Assn., Am. Acad. Dermatology (past dir.), Soc. Investigative Dermatology (past dir.), New Eng., Austrian, Brit., Danish, Finnish, German, Greek, Indian, Japanese, Polish, Venezuelan, Yugoslav dermatol. socs., New Eng. Path. Soc. Author: Pemphigus and Pemphigoid, 1965; Histopathology of the Skin, 4th edit., 1967. Editorial bd. Archives of Dermatology, 1963—. Research on treatment of pemphigus with corticosteroids, studies of blister formation, description of bullous pemphigoid as a disease different from pemphigus; studies of origin and formation of skin tumors by histochemistry and electron microscopy; hyperlipemia and role of defective lipoprotein lipase activity as cause. Home: 632 Wellesley St., Weston, Mass. 02193. Office: Boston Dispensary, Boston 02111.*

LEVERENZ, Humboldt Walter, Am. chem. physicist; b. Chgo., July 11, 1909; s. Paul Frederick and Lydia (Humboldt) L.; A.B., Stanford, 1930; postgrad. U. Münster (Germany), 1930-31, Harvard Grad. Sch. Bus. Adminstrn., 1958; m. Edith Ruggles Langmuir, Nov. 30, 1940; children—Langmuir David, Edith Humboldt, Julia Bulkeley, Ellen Langmuir. Research scientist RCA, Camden, N.J., 1931-38, Harrison, N.J., 1938-42; with RCA Labs., Princeton, N.J., 1942—, asso. dir., 1961—; bd. dirs. Labs. RCA Ltd., Zurich, Switzerland. Mem. operating com. Indsl. Reactor Labs., Plainsboro, N.J., 1956-60, chmn., 1958; mem. materials adv. bd. Nat. Acad. Scis., 1964—. Recipient Frank P. Brown award Franklin Inst., 1954. Fellow Am. Phys. Soc., I.E.E.E., Optical Soc. Am.; mem. Am. Chem. Soc., Swiss Phys. Soc., Sigma Xi. Inventor luminescent materials and screens for fluorescent lamps, TV and radar kinescopes and metascopes; processes for phosphors (luminescent materials) and ferrites for electronics; co-developer of some of earliest practical secondary-electron emitters and magnetic ferrites. Home: 35 Westcott Rd. Office: RCA Labs., Princeton, N.J. 08540.*

LEVERETT, Frank, Am. geologist; b. Denmark, Ia., Mar. 10, 1859; s. Ebenezer and Rowena (Houston) L.; B.Sc., Ia. Agrl. Coll., 1885; Sc.D. (hon.), U. Mich., 1930; m. Frances E. Gibson, Dec. 22, 1887 (dec. July 1892); m. 2d, Dorothy C. Park, Dec. 18, 1895. Tchr. in pub. schs., 1878-79; taught natural scis., Denmark Acad., 1880-83; entered U. S. Geol. Survey, 1886, asst. geologist, 1890-1900, geologist, 1901-29. Lectr. glacial geology U. Mich. Fellow Geol. Soc. Am., A.A.-A.S., Geol. Soc. Washington; mem. Nat. Acad. Sci., Am. Philos. Soc., Nat. Geog. Soc., Am. Forestry Assn., Wash., Ia., Mich. and Wis. acads sci., Sigma Xi. Author: Water Resources of Illinois, 1896; Water Resources of Indiana and Ohio, 1897; The Illinois Glacial Lobe, 1899; Glacial Deposits of the Erie and Ohio Basins, 1901; Soils of Illinois, Report Ill. Bd. World's Fair Commrs.; Pleistocene Features and Deposits of the Chicago Area (paper); Flowing Wells and Municipal Water Supplies of the Southern Peninsula of Michigan, 1906; Water Supplies of the Eastern Portion of the Northern Peninsula of Michigan, 1906; Comparison of North American and European Glacial Formations, 1910; Surface Geology of Northern Peninsula of Michigan, 1911; Surface Geology of Southern Peninsula of Michigan, 1912; The Pleistocene of Indiana and Michigan and the History of the Great Lakes, 1915; Surface Formations and Agricultural Conditions of Northwestern Minnesota, 1915, Northeastern Minnesota, 1916, Southern Minnesota, 1918; Moraines and Shore Lines of the Lake Superior Region, 1929; Surface Geology of Northern Kentucky, 1929; Quaternary Geology of Minnesota and Parts of Adjacent States, 1932; Glacial Deposits in Pennsylvania, 1934; Pleistocene Beaches in the Huron, Erie and Western Ontario Basins, 1939; Stream Capture and Drainage Shiftings in the Upper Ohio Region, 1939. Leading Am. glacial geologist of his time; noted for thorough and accurate field observations; thoroughly explored Upper Mississippi Valley; discovered fossil specimen Sigillavia Leveretti; studied glaciers of Europe, 1908. Died Nov. 15, 1943.

LE VERRIER, Urbain Jean Joseph, French astronomer; b. Saint-Lô, Normandy, France, Mar. 11, 1811; ed. schs. Saint-Lô; student math. Collège Caen (France), 1828-30, Collège St. Louis, Paris, France (Prix de mathématiques speciales 1831), from 1831, École Polytechnique, from 1831; m. Mlle. Choquet, 1837. Worked (under Gay-Lussac) at Adminstrn. Tobaccos, to 1836; tchr. Collège Stanislas, 1836-37; asst. in astronomy École Polytechnique, from 1837; 1st holder of chair celestial mechanics (established for him by royal ordinance), Faculty Scis., U. Paris, from 1846; dir. Paris Obs., 1854-70, 72-77. Elected mem. legislative assembly, 1849; senator, from 1852. Founder Internat. Meteorol. Found. Fellow Royal Soc. (Copley medal 1846), 1847; mem. Association Scientifique de France (founder, pres.), French Inst. (astronomy sect.), Royal Soc. Göttingen (fgn. asso.). Decorated Legion Honor, comdr. Royal Order Dannebroga. Contbr. sci. articles to publs. Early sci. career in chem. research; turned to astronomy, calculated variations of planetary orbits for 200,000 year period (demonstrated stability of solar system); constructed tables of movements of sun, moon, planets (especially Mercury); postulated existence of planet between Mercury and sun (Vulcan), 1845; studied irregularities in motion of Uranus, and on basis of calculations predicted existence of unknown planet (seen by Galle, in Berlin, Sept. 23, 1846, within 1 degree of place Le Verrier indicated; later named Neptune); (unknown to Le Verrier) J. C. Adams in Eng. also obtained same result, now considered co-discoverers of Neptune. Died Paris, Sept. 23, 1877.

LEVEY, Stanley, Am. biochemist; b. Detroit, Sept. 15, 1915; s. Harry and Martha (Lieberstein) L.; B.S., Wayne State U., 1938; M.S., U. Mich., 1940, Ph.D., 1943; m. Jennie Schneider, Mar. 7, 1943; children—Norman Howard, Deborah Ann. Faculty, Wayne State U., 1943-51, asst. prof. biochemistry, 1951; faculty Western Res. U. Med. Sch., 1951-61, asso. prof. biochemistry, dept. surgery, 1963—; adj. asso. prof. biochemistry U. So. Cal. Sch. Medicine, 1961-63. Mem. Am. Soc. Biol. Chemistry, Am. Chem. Soc., Am. Inst. Nutrition, Soc. for Exptl. Biology and Medicine. Contbg. author: Complications in Surgery, 1960; also numerous articles. Asso. editor Nutrition Revs., 1959-63. Research on nutrition and chem. changes in man and animals as influenced by trauma, infection, surg. procedures. Died Cleve., Nov. 19, 1967.

LEVI, Doro, archeologist; b. Trieste, June 1, 1898; ed. univs. Florence, Rome; M.A. Became insp., then dir. antiquities of Florence, 1926; founder 1st Italian Archeol. mission into Mesopotamia and Kakzu, Assyria, 1931-32; prof. archeology and history of ancient art U. Cagliari, also supr. antiquities and art, Sardinia, 1935-38; dir. Italian Archeol. Sch., Athens; dir. Italian archeol. missions to Nr. East, 1947; organizer excavations in Crete, new Italian missions to Anatolia, Palestine, Libya; initiated excavations in Iasos. Mem. Academia dei Lincei, Greek Archeol. Soc., Am. Acad. Scis., German Archeol. Inst., Guggenheim Soc., Inst. for Advanced Studies, Princeton, N.J. Editor Ann. Athens Archeol. Sch., Art Bull. Italian Ministry Edn., Editor: La Parola del Passato. Address: Scuola Archeologica Italiana, 56 Leophoros Amalias St., Athens, Greece.

LEVI, Herbert Walter, Am. biologist; b. Frankfurt, Germany, Jan. 3, 1921; s. Ludwig and Irma (Hochschild) L.; came to U. S., 1938, naturalized, 1945; student Art Student League, 1938-39; B.S., U. Conn., 1945; M.S., U. Wis., 1947, Ph.D., 1949; m. Lorna K. Rose, June 13, 1949; 1 dau., Frances. Faculty extension div. U. Wis., Madison, 1949-56, asso. prof., 1955-56; asst. curator arachnology Mus. Comparative Zoology, Harvard, 1956-57, asso. curator, 1957-66, curator, 1966—. Sec., Rocky Mountain Biol. Lab., Crested Butte, Colo., 1959-66; v.p. Centre Internat. de Documentation Arachnologique, Paris, France, 1965—. Mem. Am. Soc. Zoologists, Soc. for Study Evolution, Soc. Systematic Zoology, Wildlife Soc., Am. Ecol. Soc., Am. Ornithol. Union, Am. Micros. Soc., Internat. Soc. Toxinology, Nature Conservancy (bd. govs. 1956-62). Research, publs. on Am. theridiid spiders including black widows. Home:

Wheeler Rd., Pepperell, Mass. 01463. Office: Mus. Comparative Zoology, Harvard, Cambridge, Mass. 02138.*

LEVI, Werner, polit. scientist; b. Halberstadt, Germany, Mar. 23, 1912; s. Gustav and Zipora (Petuchowski) L.; student univs. Berlin, Heidelberg, Geneva, Paris, Frankfort, 1930-34; J.D., U. Fribourg, 1934; M.A. (Shevlin fellow), U. Minn., 1943, Ph.D., 1944; m. Ilse Steuermann, July 24, 1936; children—Antonia J., Matthew D. Came to U. S., 1934, naturalized, 1944. Mem. faculty U. Minn., 1943-63; prof. polit. sci. U. Hawaii, Honolulu, 1963—; vis. prof. Melbourne U., 1947, Marburg U., 1948, New Delhi U., 1950, Geneva U., 1958-59, Hawaii U., 1963; sr. specialist East-West-Center, Honolulu, 1967. Recipient prize Com. Econ. Devel., 1958; Fulbright grantee, 1955, 66. Mem. Am. Polit. Sci. Assn., Am. Assn. U. Profs., Assn. Asian Studies, Assn. Am. Indians. Author: American-Australian Relations, 1947; Fundamentals of World Organization, 1950; Free India in Asia, 1952; Modern China's Foreign Policy, 1953; Australia's Outlook on Asia, 1958; also articles. Pioneered interdisciplinary study of internat. orgn.; among first to analyze India's fgn. policy; treated important aspects of Australian relations with U. S. A. and Asia; pioneered study of politics of Nepal. Home: 2400 Sonoma, Honolulu, Hawaii 96822.*

LEVI-CIVITA, Tullio, Italian mathematician, physicist; b. Padua, Italy, Mar. 29, 1873; prof. mechanics Sch. Engrs., Padua, 1898-1918; prof. mechanics U. Rome, 1918-38. Fellow Royal Soc., 1930; mem. French Acad. Scis., 1911. Author: Lezioni di meccanica razionale, 3 vols., 1923-27; Lezioni di calcolo differenziale assoluto, 1925; A Simplified Presentation of Einstein's Unified Field Equations, 1929; also articles. Research on pure geometry, hydrodynamics, absolute differential calculus (Einstein's studies in part depended on this work), tensor calculus and its applications to math. physics; introduced idea of parallel displacement, 1917; developed idea of curvature of spaces, covariant differentiation in Riemennian geometry. Died Rome, Dec. 29, 1941.

LEVI-STRAUSS, Claude, social anthropologist; b. Brussels, Belgium, Nov. 28, 1908; s. Raymond and Emma (Levy) Levi-S.; Agrégé de Philosophie, U. Paris (France), 1931, Docteur ès Lettres, 1948; Doctor Honoris Causa, U. Brussels, 1962, Oxford (England) University, 1964, Yale University, 1965, University Chicago, 1967; 1 son by previous marriage, Laurent; m. Monique Roman, Apr. 5, 1954; 1 son, Matthieu. Prof. sociology U. Sao Paulo (Brazil), 1935-38, New Sch. Social Research, 1941-45; cultural counsellor French embassy, Washington, 1946-47; asso. curator Musée de l'Homme, Paris, 1948-49; prof. Ecole des Hautes Etudes, Paris, 1950—, Coll. de France, Paris, 1959—. Served with French Army, 1939-40. Decorated Legion of Honor (France); recipient Viking Fund medal and award, 1965. Fgn. hon. mem. Royal Anthropl. Inst. Great Britain, Am. Philos. Soc., Am. Acad. Arts and Scis., Acad. Netherlands, Acad. Norway, British Acad., Nat. Acad. Sci. U. S. A. Author: Les Structures élémentaires de la parente, 1949; Tristes Tropiques, 1955; Anthropologie structurale, 1958; La Pensée sauvage, Le Totemisme aujourd'hui, 1962; Le Cru et le Cuit, 1964; Du Miel aux Cendres, 1967. Known for his intuitive, philos. approach to anthropology; believes that there is a logic to all human thought and that primitive peoples are logical rather than prelogical; extensive studies of myths as logical systems created by man to solve problems which face him. Home: Rue des Marronniers, Paris 16. Office: 11 Pl. Marcelin-Berthelot, Paris 5, France.

LEVICH, Veniamin Grigorevich, Russian theoretical physicist; b. Mar. 30, 1917; grad. Kharkov U., 1937. With Inst. Electro-Colloid Chemistry (now Inst. Phys. Chemistry), USSR Acad. Sci., 1937-58, dir. theoretical dept. Inst. Electrochemistry, 1958—; instr. Moscow Pedagogical Inst., 1940-49; prof., head chair Moscow Physics Engring. Inst., 1950—. Mem. USSR Acad. Sci. (corr.). Author: Statistical Physics, 1950; Physico-Chemical Hydrodynamics, 1952; Theory of a Double Electrical Layer in Concentrated Solutions, 1963. Research and publs. on applying methods of theoretical physics to physico-chem. processes in physico-chem. hydrodynamics. Address: Moscow Physics Engring. Inst., ulitsa Kirova 21, Moscow, USSR.

LEVIN, Abraham Louis, physician; b. Poland, Dec. 16, 1878; s. Jacob and Esther (Postawelsky) L.; M.D., Tulane Med. Coll., 1907; m. Bessie Goldman, 1911; 1 son, Irving Aaron. Served to capt. as gastroenterologist, U. S. Army, World War I.; prof. gastroenterology Tulane U.; clin. prof. medicine La. State U. Med. Center; chief gastroenterology Charity Hosp., New Orleans; sr. asso. gastroenterology Touro Infirmary; gasterologist to several hosps. Mem. A.M.A., So. Med. Assn., Am. Soc. Tropical Medicine, Am. Med. Editors and Authors Assn., Am. Assn. U. Profs. Contbr. numerous med. articles on gastroenterology to tech. jours. Invented nasal gastroduodenal tube for use in gastrointestinal operations (Levin tube), 1921; 1st to call attention to function of liver in reducing blood sugar (before the discovery of insulin). Died New Orleans, Sept. 16, 1940.

LEVIN, Ernest Maurice, Am. phys. chemist; b. Detroit, Dec. 25, 1914; s. Nathan P. and Anna (Rosenthal) L.; B.A., U. Cal. at Los Angeles, 1935, M.A., 1937; m. Doris Esther Frankel, Aug. 8, 1945; children—Ellen Rhoda, Fred Jay, Robert Alan. Phys. sci. aide Nat. Bur. Standards, Riverside, Cal., 1937-39, analystical chemist, San Francisco, 1939-44, phys. chemist, Washington, 1944—. Recipient Meritorious Service award (silver medal) U. S. Dept. Commerce, 1960. Fellow Am. Ceramic Soc. (S. B. Meyer Jr. award glass div. 1959), Am. Inst. Chemists, Washington Acad. Scis.; mem. Am. Chem. Soc., Brit. Ceramic Soc. Author: (with Howard F. McMurdie) Phase Diagrams for Ceramists, 1956, part II, 1959; (with Carl R. Robbins, H. F. McMurdie), 1964; also articles. Research on liquid immiscibility in oxide systems. Home: 7716 Sebago Rd., Bethesda, Md. 20034. Office: Nat. Bur. Standards, Washington 20234.*

LEVIN, Gilbert Victor, Am. san. engr.; b. Balt., Apr. 23, 1924; s. Henry I. and Lillian (Richman) L.; B.E., Johns Hopkins, 1947, M.S., 1948, Ph.D., 1963; m. Marian Bloomquist, Oct. 25, 1953; children—Ron, Henry, Carol. Jr. asst. san. engr. Md. Dept. Health, 1948-50; asst. san. engr. Cal. Dept. Pub. Health, 1950-51; pub. health engr. D.C. Dept. Pub. Health, 1951-56; v.p. Resources Research, Inc., Washington, 1956-63; dir. spl. research Hazleton Labs., Inc., Falls Church, 1963-65, dir. life systems div., 1965-67, dir.; pres. Biospherics Research, Inc., Washington, 1967—; biochemist D.C. Dept. San. Engring., 1962-63; research asst. dept. biochemistry Schs. Medicine and Dentistry, Georgetown U., 1952-61, clin. asst. prof. preventive medicine, 1952-60. Cons. U. S. Dept. Interior, 1963—. Mem. A.A.A.S., Marine Tech. Soc., Am. Soc. C.E., Am. Pub. Health Assn., Am. Water Works Assn., Water Pollution Control Fedn., Sigma Xi. Research, publs. on devel. method for searching for extraterrestrial life, san. engring., life scis., applied biology; invented rapid method for detection bacterial contamination in water, food, air, sewage treatment process to produce effluent low in phosphate. Patentee in field. Home: 9219 LeVelle Dr., Chevy Chase, Md. 20015. Office: 1246 Taylor St. N.W., Washington 20011.*

LEVIN, Isaac, physician; b. Sagor, Russia, Nov. 1, 1866; s. Salon and Etta (Brick) L.; M.D., Mil. Med. Acad., Petrograd, 1890; m. Sophie Bloch, Feb. 25, 1890; children—Ben Fenton, Charles Emmerson, Ralph Theodore. Came to U. S., 1891, naturalized, 1901. Began practice, Petrograd. 1890; asso. in pathology and cancer research Columbia, 1909-15; clin. prof. cancer research N.Y. U. since 1915; chief of cancer div. Monteflore Hosp., 1912-25; chief in radiology St. Bartholomew's Hosp., 1917-22; dir. N.Y.C. Cancer Inst., 1923-30; cons. in radiology Lebanon Hosp. since 1915. Fellow A.A.A.S.; mem. A.M.A., Am. Physiol. Soc., Am. Assn. Pathologists and Bacteriologists, Soc. Exptl. Biology and Medicine, Am. Assn. Cancer Research, Am. Radium Soc., Radiological Soc. N. Am., Am. Genetic Assn., Am. Med. Editors' and Authors' Assn., Harvey Soc., N.Y. Acad. Medicine. Editor Archives of Clinical Cancer Research, 1925-30. Contbr. many papers and articles in exptl. and clin. medicine. Died June 19, 1945.

LEVIN, Norman Lewis, Am. zoologist, parasitologist; b. Hartford, Conn., Mar. 31, 1924; s. Joseph and Fannie (Sosin) L.; B.S., U. Conn., 1948, M.S., 1949; Ph.D., U. Ill., 1956; m. Shirley Ginsberg, Sept. 3, 1950; children—Faye Deborah, Alan Jeffery. Faculty, U. Ill., 1953-57, Westminster Coll., Fulton, Mo., 1957-60; faculty Bklyn. Coll., 1960—, prof. biology, 1964—. La. State U. InterAm. fellow in tropical medicine, 1959. Fellow A.A.A.S.; mem. Am. Assn. U. Profs., Am. Soc. Parasitologists, Am. Soc. Tropical Medicine and Hygiene, Am. Micros. Soc., Am. Inst. Biol. Sci., Sigma Xi, Phi Sigma. Author sci. biographies Ency. Sci. and Tech., 1966—. Research, publs. on life cycle, morphology, distbn. various helminth parasites. Office: Bedford and Av. H, Bklyn. 11210.*

LEVIN, Nyman, English physicist; b. London, Eng., Feb. 17, 1906; s. Lewis and Anne (Lishonsky) L.; ed. Imperial Coll. Sci. and Tech., London; B.S., Ph.D. Dir. Admiralty Gunnery Establishment; dir. research and devel. for high precision industries; asst. dir. Atomic Weapons Research Establishment; dir. Atomic Weapons Research Establishment. Mem. Inst. Physics, Phys. Soc., Royal Soc. Arts, Royal Coll. Sci. (asso.). Home: The Dyke House, Aldermaston, Berkshire. Office: A.W.R.E., Aldermaston, Berkshire, England.

LEVIN, William C., Am. physician; b. Waco, Tex., Mar. 2, 1917; s. Samuel P. and Jeanette (Cohn) L.; B.A., U. Tex., 1938, M.D., 1941; m. Edna Lee Seinsheimer, June 23, 1941; children—Gerry Lee, Carol Lynn. Faculty, U. Tex. Med. Br., Galveston, 1944—, prof. medicine, 1965—, dir. Hematology Research Lab., Clin. Research Center and Blood Bank, 1946—. Cons. to surg. gen. U. S. Army, USAF, 1963—. Mem. A.M.A., Internat., Am. socs. hematology, Am. Fedn. for Clin. Research, Central, So. socs. for clin. research, Soc. for Exptl. Biology and Medicine, A.A.A.S., Am. Assn. for Cancer Research. Research, numerous publs. on genetic aspects of abnormal hemoglobins, characteristics of immunoglobulins in diseases of hypersensitivity and in malignant disorders, dynamics and characteristics of lymphocytes in disorders of immunity. Home: 1301 Harbor View Dr., Galveston, Tex. 77550.*

LEVINE, Albert K., Am. chemist; b. Bayonne, N.J., Sept. 13, 1917; s. Israel and Dora (Feldman) L.; B.S., Rutgers U., 1938, M.S., 1939, Ph.D., 1941; m. Marilyn Wurtzel, Dec. 13, 1953; children—Irving, Sally, David. Research chemist Schweitzer Paper Co., Spotswood, N.J., 1942-43; spectroscopist Amco Magnesium Corp., Wingdale, N.Y., 1943-44; chief chemist Mut. Chem. Co., Jersey City, 1944-47; faculty Bklyn. Coll., 1947-61; head luminescent materials, inorganic chemistry research Gen. Telephone & Electronics Labs., Bayside, N.Y., 1961-66; prof. chemistry, chmn. div. natural scis. Richmond Coll., N.Y.C., 1966—. Vis. lectr. solid state physics Imperial Coll. Sci., London, Eng., 1951-52. Mem. Am. Chem. Soc., Phi Beta Kappa, Sigma Xi. Editor: Lasers, Vol. I, 1966, Vol. II, 1968. Research, publs. primarily in field of synthesis and crystal growth of electronically active solids and on elucidation of mechanisms underlying phenomena they exhibit. Home: 10-46 Utopia Pkay., Beechhurst, N.Y. 11357. Office: Richmond Coll., S.I., N.Y.*

LEVINE, Arthur Sidney, Am. food scientist; b. Boston, Mar. 15, 1913; s. Hyman S. and Sarah (Barkin) L.; B.S., U. Mass., 1935, M.S., 1936, Ph.D., 1939; m. Sarah I. Toabe, Dec. 24, 1939; children—Peter D., Janet M. Food research Mass. Agrl. Expt. Sta., U. Mass., Amherst, 1936-46, prof. dept. food sci., tech., 1946-66; food cons., 1966—. Mem. Inst. Food Technologists, Packaging Inst., Am. Pub. Health Assn., Am. Soc. Microbiology, Research and Devel. Assos., Sigma Xi, Phi Kappa Phi, Phi Tau Sigma. Research, numerous publs. on action of acetic acid on microorganisms, food preservation, prodn., spoilage, packaging, composition, food poisoning. Address: 500 S. Pleasant St., Amherst, Mass. 01002.*

LEVINE, Lawrence, Am. biochemist; b. Hartford, Conn., July 18, 1924; s. Isadore and Mamie (Jacobs) L.; B.A., U. Conn., 1948; M.S., U. Mich., 1950; Sc.D., Johns Hopkins, 1953; m. Helen Van Vunakis, Jan. 31, 1958; children—Stephanie, Joanne. Instr., Johns Hopkins, 1954; sr. research asso. N.Y. State Dept. Health, 1955-57; faculty Brandeis U., Waltham, Mass., 1957—, prof. biochemistry, 1963—. Mem. research com. Boston Med. Found., 1966—; mem. research study sect. USPHS, 1964—. Mem. Am. Cancer Soc. (lifetime professorship), Soc. Biol. Chemists, Soc. Am. Immunologists, N.Y. Acad. Sci., Soc. Microbiologists. Research on chem. and phys. nature of complementarity between antigen and antibody.*

LEVINE, Milton Isra, Am. physician; b. Syracuse, N.Y., Aug. 15, 1902; s. David Levine and Daisy (Baum) L.; B.S., Coll. City N.Y., 1923; M.D., Cornell U., 1927; m. Jean H. Seligmann, June 14, 1936; children—Carol (Mrs. J. Gottfried Paasche), Ann. Practice medicine specializing in pediatrics, N.Y.C., 1929—; asst. clin. pediatrician N.Y. Nursery and Child's Hosp., 1929-32; asst. attending pediatrician, N.Y. Hosp., N.Y.C., 1932-52, asso. attending pediatrician, 1952-56, attending pediatrician, 1956—; dir. research child Tb, N.Y.C. Dept. Health, 1931-42, dir. com. for study poliomyelitis Bur. Labs., 1931, cons. pediatrician, 1942—; chmn. adv. com. on BCG vaccine against Tb, N.Y. State Dept. Health, 1947-55; pediatrician Children's Tb Clinic, Harlem Hosp., N.Y.C., 1933-42; asst. prof. clin. pediatrics Cornell U. Med. Coll., 1944-54, asso. prof. clin. pediatrics, 1954—. Mem. adv. com. Midcentury Whitehouse Conf. on Children and Youth, Washington, 1960. Fellow Am. Acad. Pediatrics, Am. Pub. Health Assn., Am. Coll. Chest Physicians; mem. A.M.A., Am. Trudeau So., Nat. Tb Assn., A.A.A.S., N.Y. Acad. Sci., Soc. Sci. Study of Sex, Phi Delta Epsilon. Author: (with Jean H. Seligmann) The Wonder of Life, 1940, A Baby is Born, 1949; Pulmonary Pathology in Infants and Children, 1965; (with Armand Mascia) Pulmonary Diseases and Anomalies of Infancy and Childhood, 1966; also numerous articles on disease of childhood, child psychology and sex edn. Research on childhood Tb with spl. work on prophylactic immunization against Tb. Home: 302 W. 12th St., N.Y.C. 10014. Office: 1111 Park Av., N.Y.C. 10028.*

LEVINE, Morton Ashur, Am. physicist; b. Boston, May 5, 1922; s. Bernard Isaac and Esther (Barsky) L.; B.S., U. Mass., 1947; M.S., Tufts U., 1951; m. Marilyn Landau, June 26, 1951; 1 dau., Lisa Gail. Staff, Los Alamos Sci. Lab., 1944-46; research asso. Tufts U., Medford, Mass., 1951-54; br. chief, plasma astrophysics br. Space Physics Lab., USAF Cambridge Research Labs., Hanscom Field, Mass., 1954—. Mem. Sigma Xi. Devel. several devices in field of controlled thermonuclear reactors; original theory and expts. on stabilized or H centered pinch and devel. of cross-magnetic field geometries currently known as minimum B magnetic fields. Home: 94 Hoitt Rd., Belmont, Mass. 02178. Office: USAF Cambridge Research Labs., L. G. Hanscom Field, Bedford, Mass. 01731.*

LEVINE, Norman Dion, Am. biologist; b. Boston, Nov. 30, 1912; s. Max and Adele (Daen) L.; B.S., Ia. State Coll., 1933; Ph.D., U. Cal. at Berkeley, 1937; m. Helen Marie Saxon, Mar. 2, 1935. Asst. dept. zoology U. Cal. at Berkeley, 1933-37; with U. Ill., Urbana, 1937—, prof. Coll. Vet. Medicine, 1953—, sr. mem. Center for Zoonoses Research, director of Center for Human Ecology; vis. prof. dept. microbiology U. Hawaii, Honolulu, 1962. Mem. tropical medicine and parasitology study sect. NIH, 1965—, chmn., 1966—. Mem. Am. Acad. Microbiology (gov. 1963-65), A.A.A.S. (committeeman-at-large), Am. Bd. Microbiology (gov. bd. 1959-64), Nat. Acad. Scis.-NRC, Soc. Protozoologists (pres. 1959-60), World Fedn. Parasitologists (council 1960-65), Soc. Nematologists, Soc. Exptl. Biology and Medicine, Am. Soc. Tropical Medicine and Hygiene, Ill. Acad. Sci., Am. Vet. Med. Assn., Am. Pub. Health Assn., Sigma Xi. Author: Protozoan Parasites of Domestic Animals and of Man, 1961; (with Virginia Ivens) The Coccidian Parasites (Protozoa, Sporozoa) of Rodents, 1965; Pavlovsky's Natural Nidality of Transmissible Diseases, 1966. Editor: Malaria in the Interior Valley of North America, 1964; Jour. Protozoology, 1965—. Research, publs. on taxonomy, structure and biology of protozoan and nematode parasites and animal diseases, relation of chem. structure to biol. activity of potential anthelmintics. Home: 702 LaSell Dr., Champaign, Ill. 61820.*

LEVINE, Philip, physician, immunologist; b. Kletzk, Russia, Aug. 10, 1900; s. Morris and Fay (Zirulick) L.; came to U. S., 1908, naturalized, 1917; B.S., Coll. City N.Y., 1919; M.D., Cornell U., 1923, M.A., 1925; m. Hilda Lillian Perlmutter, May 1, 1938; children—Phyllis Ann (Mrs. Harvey Klein), Mark Armin, Paul Karl (dec.) Victor Raphael. Asst. to asso. Rockefeller Inst. for Med. Research, 1925-32; faculty U. Wis. Med. Sch., 1932-35; bacteriologist, serologist Newark Beth Israel Hosp., 1935-44; dir. div. immunohematology Ortho Research Found., Raritan, N.J., 1944-65, emeritus professor, since 1965—; consultant to St. Michael's Hosp., Newark, Nassau Hosp., Mineola, L.I., N.Y., Muhlenberg Hosp., Plainfield, N.J. Recipient Mead Johnson award, 1942; Ward Burdick award, 1946; Lasker award 1946; Phi Lambda Kappa Grand award 1947; Passano Found. award, 1951; award of merit Netherlands Red Cross, 1959; Johnson medal for research and devel., 1960; 1st Oehlecker award German Soc. for Blood Transfusion, 1964. Fellow Am. Acad. Allergy, A.C.P.; mem. A.A.A.S., N.Y. Acad. Medicine, Am. Genetic Assn., Am. Assn. Immunologists, Soc. for Exptl. Biology and Medicine, Internat. Soc. Hematology, Internat. Soc. Blood Transfusion, Harvey Soc., Sigma Xi. Author numerous publs. on human blood groups, transfusions; author laws in N.J. and Wis. on blood tests in paternity disputes. Discovered Rh factor, 1939, Cellano factor and its genetic relationship to Kell factor, 3 new blood factors, 1951, (with Landsteiner) blood factors M, N, P; described cause of disease erythroblastosis fetalis in newborn, also isoimmunization through pregnancy. Home: 1068 Kenyon Av., Plainfield, N.J. 07060. Office: Ortho Research Found., Raritan, N.J. 08869.*

LEVINE, Pincus Philip, Am. veterinarian; b. N.Y.C., Aug. 25, 1907; s. Joseph and Emma (Abel) L.; B.S., Coll. City N.Y., 1927; M.S., Cornell U., 1932, D.V.M., 1932, Ph.D., 1937; m. Selma Hyman, Oct. 28, 1933; children—Joseph Jonathan, Seth Stanley. Tchr. pub. schs., N.Y.C., 1930-31; pathologist N.Y. State Conservation Dept., 1932-34; faculty N.Y. State Vet. Coll., Ithaca, 1934—, prof. poultry diseases, 1944—, head dept. avian diseases, 1961-66. Cons. govt. agys., Rockefeller Found. Guggenheim fellow 1947-48. Mem. Am. Assn. Avian Pathologists (president 1962-63), World Veterinary Poultry Association (vice president since 1959—), A.A.-A.S., Am. Soc. Zoologists, Am. Soc. Parasitologists, Soc. Exptl. Biology and Medicine, Am. Vet. Med. Assn., Poultry Sci. Assn., Am. Assn. Avian Pathologists, Internat. Acad. Pathology, Sigma Xi, Phi Kappa Phi, Phi Zeta, others. Research, publs. on parasitic, bacterial and viral diseases of poultry; discovered effectiveness of sulfonamides in control of coccidiosis in poultry, also of dipping fertile eggs in antibiotic solutions for control of mycoplasmosis in poultry. Home: 1872 Slaterville Rd., Ithaca, N.Y. 14850.*

LEVINE, Robert, Am. chemist; b. Boston, July 30, 1919; s. Louis Harry and Sarah (Heller) L.; B.A., Dartmouth, 1940, M.A., 1942; Ph.D., Duke, 1945; m. Dorothy Clair Meltzer, Aug. 6, 1950; children—Ruth Marcia, Barbara Susan, David Joel. Chemist, Mathieson Chem. Corp., Niagara Falls, N.Y., 1945-46; faculty U. Pitts., 1946—, prof. chemistry, 1959—; Cons. Monsanto Chem. Co., Schering Corp., Wyandotte Chem. Corp., Reilly Tar and Chem. Corp., others. Mem. Am. Chem. Soc., Am. Assn. U. Profs., N.Y. Acad. Scis. Sigma Xi, Phi Lambda Upsilon. Editorial bd. Jour. Heterocyclic Chemistry. Research and publs. on new methods for the synthesis of organic compounds; the mechanism of organic reactions; the synthesis of potential cancer chemotherapeutic agts. Home: 5734 Woodmont St., Pitts. 15217.*

LEVINE, Samuel Albert, physician; b. Lomza, Poland, Jan. 1, 1891; s. Abram and Anna (Scheinkopf) L.; came to U. S., 1894, naturalized, 1896; A.B., Harvard, 1911, M.D., 1914; Sc.D., Adelphi Coll., 1959; L.H.D., Yeshiva U., 1959; m. Rosalind Weinberg, June 20, 1926; children—Carol F. (Mrs. William B. Schwartz), Herbert J., Joan B. (Mrs. Simon Scheff). Asso. in medicine Peter Bent Brigham Hosp., Boston, 1914-15, asso. physician, 1919-40, physician, 1940-57, emeritus, cons. in cardiology, 1957—; asst. Rockefeller Hosp., N.Y.C., 1916-17; instr. medicine Harvard Med. Sch., Boston, 1920-30, asst. prof., 1930-48, clin. prof., 1948-57, emeritus clin. prof. medicine, 1957—; practice medicine specializing in cardiology, Brookline, Mass., 1919—. Samuel A. Levine professorship of medicine established at Harvard Med. Sch. by Charles E. Merrill, 1954; recipient 6th ann. salute to med. research City of Hope, 1963; Samuel Levine Cardiac Center dedicated at Peter Bent Brigham Hosp., 1965. Mem. Am. (Gold Heart award 1959), New Eng. (past pres.) heart assns., Am. Coll. Cardiology, Assn. Am. Physicians, Soc. Clin. Investigation, A.A.A.S., Internat. Cardiology Found. (lay council), Am. Acad. Arts and Scis., Alpha Omega Alpha; hon. mem. Brit., Australian, French, Italian, Brazil, Argentina heart socs. Author: Coronary Thrombosis, Its Various Clinical Features, 1929; Clinical Heart Disease, 1936; (with W. Proctor Harvey) Clinical Auscultation of the Heart, 1949; (with Bernard Lown) Current Concepts in Digitalis Therapy, 1954; also articles. Died Mar. 31, 1966.

LEVINE, Seymour, Am. psychologist; b. Bklyn., Jan. 23, 1925; s. Joseph and Rose (Reines) L.; B.A., U. Denver, 1948; M.A., N.Y. U., 1951, Ph.D., 1952; m. Barbara McWilliams, Feb. 14, 1949; children—Robert Thomsen, Leslie Ingrid, Alicia Margaret. Faculty, Boston U., 1952-53, Ohio State U., 1956-60; USPHS fellow clin. psychology Inst. Psychosomatic and Psychiat. Research and Tng., Michael Reese Hosp., Chgo., 1953-55, research asso., 1955-56; fellow dept. neuroendocrinology Inst. Psychiatry, Maudsley Hosp., London, 1961-62; asso. prof. psychiatry Stanford Sch. Medicine, Palo Alto, Cal., 1962—. Recipient Hoffheimer Research award 1961. Mem. Psychonomic Soc., Am. Psychol. Assn., Endocrine Soc., Soc. Exptl. Biology and Medicine, A.A.A.S. Author book, articles. Contbns. include behavioral effects of infantile stimulation in animals, effects of hormones and stress on behavior, elaborated process of sexual differentiation as a function of presence or absence of gonadal hormones in the newborn animal. Home: 927 Valdez Pl., Stanford, Cal.*

LEVINE, Seymour, Am. physician; b. N.Y.C., Mar. 13, 1925; s. Herman and Pearl (Kramer) L.; B.A. cum laude, N.Y. U., 1946; M.B., Chgo. Med. Sch., 1947, M.D., 1948; m. Lillian Konigsberg, June 20, 1945; children—Linda Grace, Sandra Beth. With USPHS, 1951-53; resident pathology N.Y. U., Bellevue Med. Center, 1953-54, Montefiore Hosp., N.Y.C., 1954-56; practice medicine, specializing in pathology, Jersey City, 1956-64, N.Y.C., 1964—; pathologist St. Francis Hosp., 1956-64; prof. pathology N.Y. Med. Coll., 1964—. Mem. Am. Assn. Path. Bacteriologists, Am. Assn. Neuropathologists, Soc. Exptl. Biology and Medicine, Am. Soc. Exptl. Pathology. Research, numerous publs. on infrared spectrophotometry in bacteriology, effects of various types of anoxia on nervous system, allergic encephalomyelitis, brain edema. Office: Bird S. Coler Hosp., Welfare Island, N.Y. 10017.*

LEVINE, Victor Emanuel, biochemist, nutritionist, explorer; b. Minsk, Russia, Aug. 4, 1892; s. Israel and Eva Leah (Meisels) L.; brought to U. S., 1898; B.A., Coll. City New York, 1909; M.A., Columbia, 1911, Ph.D., 1914; grad. study, Johns Hopkins, U. of Toronto; M.D., Creighton U., 1928. Mem. faculty Coll. Phys. and Surg., Columbia, 1913-16; asst. prof. organic chemistry, Fordham, 1915-16; dir. chem. lab., Beth Israel Hosp., N.Y. City, 1916-17; dir. Chem. and Path. Labs., N.Y. City, 1917-18; asst. prof. biol. chemistry, Sch. of Medicine, Creighton U., 1918-20, prof. biol. chemistry and nutrition and head of dept., 1920—, dir. Grad. Sch. of Chemistry 1928—; cons. USPHS, 1927-38, 39; vis. Fulbright prof. 1960-62; dir. health dept. Dwarfies Corp., Council Bluffs. Leader sci. expdn. to Arctic for Office of Naval Research, U. S. Navy, hdgrs., Arctic Research Lab., Point Barrow, Alaska, 1948. Fellow A.A.A.S., Am. Geog. Soc., Am. Inst. Chemistry, N.Y. Acad. Scis., Am. Pub. Health Assn., Royal Soc. Arts and Scis. (Gt. Britain), Brit. Inst. Philosophic Studies, Royal Anthrop. Inst., Gr. Brit. and Ireland, Internat. Dental Research Assn., Am. Med. Writers Assn., mem. nat., state and local profl. and sci. assns. and orgns. in med. and nutrition fields; officer of several. Arctic explorer primarily for biol. studies 1921—. Awarded honorary scroll by Columbia Graduate School Alumni Assn., 1937. Author and co-author several books 1929-35; (with C. P. Stewart and A. Stolman) Toxicology, 2 vols., 1961; Introducción a Toxicologia (Spanish), 1962. Contbr. articles on biology aspects of the Eskimo to Ency. Arctica. Died Sept. 29, 1963.

LEVINGER, Joseph Solomon, Am. physicist; b. N.Y.C., Nov. 14, 1921; s. Lee J. and Elma (Ehrlich) L.; B.S., U. Chgo., 1941, M.S., 1944; Ph.D., Cornell U., 1948; m. Gloria Edwards, Aug. 14, 1943; children—Samuel, Laurie, Louis, Joseph Lee. Jr. physicist, metall. lab. U. Chgo., 1942-44; physicist Franklin Inst., Phila., 1944-45; teaching asst. Cornell U., Ithaca, N.Y., 1946-48, instr. physics, 1948-51, AVCO vis. prof., 1961-64; faculty physics La. State U., Baton Rouge, 1951-60, prof., 1957-60; prof. physics Rensselaer Poly. Inst., Troy, N.Y., 1964—. Guggenheim fellow, 1957-58. Author: Nuclear Photodisintegration, 1960; Secrets of the Nucleus, 1967. Research in atomic nuclear and high energy physics. Home: Red Mill Rd., Rensselaer, N.Y. 12144.*

LEVINSON, Abraham, Am. physician; b. Aug. 25, 1888; s. Yehudah and Rebecca (Kreuger) L.; M.D., U. Ill., 1911; postgrad. U. Vienna, 1914; B.S., U. Chgo., 1917; studied Vienna and Berlin, 1923, 28, 30, 33; m. Ida Perlstein, 1912; children—Myrtle, Judith, Julian. Began practice, Chgo., 1911; sr. attending pediatrist Michael Reese and Mt. Sinai hosps.; attending pediatrist, chief staff, children's div. Cook County Hosp.; v.p. staff Michael Reese Hosp., 1932-34; prof. pediatrics Northwestern U. Med. Sch.; Recipient Gold medal for contbns. to med. sci. Phi Lambda Kappa, 1940. Am. Bd. Pediatrics. Mem. A.M.A. (certificate honor, class 1, exhibit on cerebrospinal fluid, 1932), Ill. (certificate of merit for exhibit of original sci. research 1936, 1940), Chgo. med. socs., Chgo. Pediatric Soc. (pres. 1935-36), Am. Acad. Pediatrics, Inst. Medicine, Am. Assn. Med. History, Sigma Xi. Author: Cerebro-spinal Fluid in Health and Disease, 1919, 3d edit., 1929; Tobias and His Work, 1924; Examination of Children, 1924, 2d edit., 1927; Textbook on Pediatric Nursing, 1925, 3d edit., 1944 (Hebrew translation, 1933); Pioneers of Pediatrics, 1936, 1943; The Mentally Retarded Child, 1952; also numerous articles for med. jours. pertaining to children's diseases, chpts. for various systems of pediatrics, also translated German chpts. on pediatrics. Condr. research in biochemistry and pediatrics. Died Sept. 17, 1955.

LEVINSON, Boris Mayer, psychologist; b. Kalvarijah, Lithuania, July 1, 1907; s. Moses and Rose (Lev) L.; came to U. S., 1923, naturalized, 1930; B.S. cum laude Coll. City N.Y., 1937, M.S., 1938; Ph.D., N.Y. U., 1947; m. Ruth Berkowitz, June 16, 1934; children—Martin, David. Psychologist, Adult Guidance Service, N.Y.C. Bd. Edn., 1937-38, chief psychologist, 1938-40; clin. psychologist Vets. Rehab. and Mental Hygiene Clinics, Bklyn. Jewish Hosp., 1944-48; supervising clin. psychologist Yeshiva U., 1951-53, chief psychologist, 1953-56, dir. Psychol. Center, 1956-63, faculty, 1951—, prof. psychology, 1956—; chief psychologist Jewish Meml. Hosp., N.Y.C., 1957-59. Lectr., Hunter Coll., 1949-60, Coll. City N.Y., 1947-53; cons. psychologist, 1938—; psychol. cons. homelessness N.Y. Community Council. Diplomate in clin. psychology. Fellow Am. Psychol. Assn., Soc. Projective Techniques; mem. Inter-Am. Soc. Psychology, N.Y. Soc. Clin. Psychologists, Eastern, N.Y. State psychol. assns., Inter-Am. Assn. Psychology, N.Y. Acad. Scis. Contbr. numerous articles in field to sci. jours. Pioneer in use of domestic pets in child psychotherapy, in psychol. studies of homeless men on Skid Row—theory of homeless personality, in study of psychol. traits of Jewish children and youth of traditional background. Home: 39-25 47th St., Sunnyside, L.I., N.Y. 11104. Office: Yeshiva U., 55 Fifth Av., N.Y.C. 10003.*

LEVINSON, Horace Clifford, Am. mathematician; b. Chgo., June 30, 1895; s. Salmon Oliver and Helen Haire; B.A., Yale, 1917; Ph.D., U. Chgo., 1922; Dr., U. Paris, 1923; m. Alma Prescott Wells, June 23, 1921; children—Alma Prescott (Mrs. Edward J. Curran Jr.), Ruth Bartlett. Instr. dept. math. Ohio State U., 1923; dir. research, treas., dir. Bernard Hewitt & Co., 1924-31; v.p., treas., dir. Rural Progress, Inc., Chgo., 1934-38; dir. research, treas., dir. L. Bamberger & Co., Newark, 1939-46; chmn. operations research com. Nat. Acad. Sci.-NRC, 1949-55; cons. operations research Arthur D. Little Inc., 1952-54. Mem. Am. Math. Soc., Operations Research Soc. Am., Sigma Xi. Club: Tavern (Chgo.). Author: (with E. B. Zeisler) The Law of Gravitation in Relativity, 1931; Your Chance to Win, 1939; The Science of Chance, 1950; Chance, Luck and Statistics, 1963. Research in gen. theory of relativity; application of sci. method to work (operations research). Home: Alewive Farm, Kennebunk, Me. 04043.*

LEVINSTEIN, Henry, physicist; b. Themar, Germany, Dec. 4, 1919; s. Moritz and Nanette (Mayer) L.; came to U. S., 1935, naturalized, 1943; B.S., U. Mich., 1942, M.S., 1943, Ph.D., 1947; m. Betty Glixon, June 24, 1962; children—Harvey, Richard, Robert. With U. Mich., 1942-47; faculty Syracuse (N.Y.) U., 1947—, prof. physics, 1955—. Univ. adviser Tex. Instruments, 1960—, Gen. Telephone & Electronics, 1967—; tech. adv. bd. Aerojet Gen. Corp., 1961—; cons. Westinghouse Elec., 1955-60, Office Army Research, 1959-62, Jet Propulsion Lab. Cal. Inst. Tech., 1963-66. IBM, 1965—, Mem. Am. Phys. Soc. (chmn. N.Y. sect. 1967—), Optical Soc. Am., Am. Assn. Physics Tchrs. Editor: Photoconductivity, Proc. of an Internat. Conf., 1962. Research, publs. on infrared radiation detectors and detection techniques. Home: 2654 E. Genesee St., Syracuse, N.Y. 13224.*

LEVITON, Alan Edward, Am. zoologist; b. Bklyn., Jan. 11, 1930; A.B., Stanford, 1949, A.M., 1953, Ph.D., 1960; m. Gladys Ann Robertson, June 30, 1952; children—David A., Charlotte A. Profl. lectr. Golden Gate Coll., San Francisco, 1953—; lectr. biology Stanford, 1962—, asso. curator zool. collections, div. systematic biology, 1962; asst. curator dept. herpetology Cal. Acad. Scis., San Francisco, 1957-60, asso. curator, 1960-61, curator, 1962—, chmn. dept., 1964—. Mem. Herpetologists League (pres. 1960-61), Am. Soc. Ichthyologists and Herpetologists, Am. Inst. Biol. Sci., Soc. Study Evolution, Am. Soc. Zoology, Herpetologists League, Sigma Xi. Research, numerous publs. in systematics and zoogeography of reptiles and amphibians of Asia, Nr. East; studies on classification of reptiles above species level. Home:

571 Kingsley St., Palo Alto, Cal. 94301. Office: Cal. Acad. Scis., Golden Gate Park, San Francisco 94118.*

LEVITSKII, Oleg Dmitrievich, Russian geologist; b. Mar. 19, 1909; grad. Leningrad Mining Inst., 1930; Geol. surveys Eastern areas of USSR for various orgns.; staff Inst. for Geology Ore Deposits, Petrography, Mineralogy, Geochemistry, USSR Acad. Scis., 1956—. Recipient Stalin prize, 1946. Corr. mem. USSR Acad. Sci. Research in wolfram deposits of Eastern Trans-Baikal region; tin deposits; geol. studies in Far E. and N.E. USSR.

LEVITZ, Mortimer, Am. chemist; b. N.Y.C., May 11, 1921; s. Hyman and Ida (Goldberg) L.; B.S., Coll. City N.Y., 1941; M.A., Columbia, 1947, Ph.D., 1951; m. Catherine Blum, June 1, 1947; children—Ellen Maud, Stuart Michael. Staff, Manhattan Project, S.A.M., Columbia, 1942-44, research asso., 1951-52; research asso. N.Y. U., N.Y.C., 1952-56, faculty, 1956—, professor obstetrics and gynecology, 1967—. Mem. endocrine study sect. NIH, 1966. Recipient Career Devel. award USPHS, 1962. Mem. Am. Chem. Soc., Am. Soc. Biol. Chemists, Endocrine Soc., Soc. for Gynecol. Investigation, Sigma Xi, Phi Lambda Upsilon. Editorial bd. Jour. Clin. Endocrinology and Metabolism, 1962—. Research, publs. on syntheses for carbon 14, hydrogen 3 labelled estrogens and estrogen conjugates, elucidation of estrogen metabolism in man; developed useful method for assaying plasma estriol in pregnancy, elucidation of genetic defect in maple sugar urine disease. Home: 64-53 215th St., Bayside, N.Y. 11364. Office: 550 1st Av., N.Y.C. 10016.*

LEVRET, André, French surgeon, obstetrician; b. Paris, 1703; held ofcl. appointments to all female brs. of royal family. Author: Observations sur les causes et les accidents de plusiers accouchements laborieux, 1747; Observations sur la cure radicale de plusieurs polypes de la matrice, de la gorge et du nez operée par de nouveaux moyens, 1749; The Accoucheur's Art demonstrated by Physical and Mechanical Principles, 1753. Studied extra-uterine pregnancy, placenta previa, delivery procedures for different presentations; improved obstetric forceps with cephalic and pelvic curves (Levret's forceps), 1747. Died Jan. 22, 1780.

LEVSHIN, Vadim Leonidovich, Russian physicist, b. Jan. 27, 1896; grad. Moscow U., 1918. Staff, Physics and Biophysics Inst., 1919-32; staff Physics Inst., Moscow U., 1930-35, prof., 1944—; staff Physics Inst., USSR Acad. Sci., 1934—; instr. Moscow Geol. Survey Inst., 1930-33, prof., 1933-35. Recipient Stalin prize, 1951, 52. Research and publs. in photoluminescence. Home: B.Kaluzhskaya 13, Moscow. Office: Fizichesky institut, AN SSSR, Miusskaya 3, Moscow 3-a, USSR.

LEVY, Arthur, Am. chemist; b. N.Y.C., Sept. 29, 1921; s. Morris and Mary (Knapp) L.; B.S. in Chemistry with honors, Queens Coll., 1943; M.S. in Phys. Chemistry, U. Minn., 1948; m. Rita V. Huch, Nov. 15, 1949; children—Mark, Patricia, Paul, Richard. Chemist, Los Alamos Nat. Lab., 1944-46, Brookhaven Nat. Lab., Upton, L.I., N.Y., 1950-51; aero. research engr. NACA, Lewis Labs., Cleve., 1948-50; asso. chief Battelle Meml. Inst., Columbus, O., 1951—. Mem. Am. Chem. Soc., Combustion Inst., A.A.A.S. Research, publs. on high explosive chemistry, radiation chemistry of aqueous systems, organo-metallic syntheses and decompositions, decomposition kinetics, flame and plasma chemistry, propellant chemistry, air pollution chemistry. Home: 614 Farrington Dr., Worthington, O. 43085. Office: 505 King Av., Columbus, O. 43201.*

LEVY, Barnet M., Am. dental pathologist; b. Scranton, Pa., Jan. 13, 1917; s. J. Julius and Sophia (Straus) L.; A.B., U. Pa., 1938, D.D.S., 1942; M.S. in Bacteriology and Pathology, Med. Coll. Va., 1944; m. Sylvia Salwen, June 30, 1940. Faculty, Med. Coll. Va., 1942-44, Washington U., St. Louis, 1944-49, Columbia Sch. Dentistry, 1949-57; prof. pathology U. Tex. Dental Br., Houston, 1957—, dir. Inst. for Dental Sci., 1964—. Mem. numerous coms. NIH, NRC. Diplomate Am. Bd. Oral Pathology (pres. 1965-66). Fellow Am. Acad. Oral Pathology, A.A.A.S.; mem. Am. Assn. Cancer Research, Am. Pub. Health Assn., Internat. Assn. for Dental Research (pres. 1965-66), Histochem. Soc., numerous others, Sigma Xi, Omicron Kappa Upsilon. Author: (with W. Shafer, M. Hine) Textbook of Oral Pathology, 1958. Research, numerous publs. on growth, malignant growth, causes and course of peridontal diseases in marmosets and other mammals. Home: 3736 Underwood St., Houston 77025.*

LEVY, Charles Kingsley, Am. radiation biologist; b. Boston, Dec. 25, 1924; s. Maurice Ely and Bertha (Becker) L.; B.Sc., George Washington U., 1948, M.Sc., 1951; Ph.D., U. N.C., 1956; m. Andrea Joyce Kniznick, Dec. 21, 1958; children—Brett Matthew, Adam Jonathan, Alison Victoria. Faculty, Vassar Coll., 1956-58; research collaborator Brookhaven Nat. Lab., Upton, N.Y., 1957-61; staff scientist Worcester Found., Shrewsbury, Mass., 1958-62; asso. prof. biology, also research asso. prof. radiobiology Boston U. Grad. Sch. and Med. Sch., 1962—; research asso. Mass. Gen. Hosp., 1963—. Cons. AEC, NASA; pres. Biotek, Inc., Boston, 1961—. Mem. Am. Physiol. Soc., Radiation Research Soc., Soc. Gen. Physiologists, N.Y. Acad. Scis., Am. Phys. Soc. Author: An Introduction to Electrophysiol. Techniques, 1966, also articles. Studies in biol. effects of radiation in mammalian cell systems, particularly physiol. alterations in sensory and neural coordinating mechanisms; demonstrated resistance of non-dividing cells to radiation; studies in extreme sensitivity of several types of sensory mechanisms to ionizing radiation. Home: 336 Singletary Lane, Framingham, Mass. 01701. Office: 2 Cummington, Boston 02215.*

LEVY, David Mordecal, Am. psychiatrist; b. Scranton, Pa., Apr. 27, 1892; s. Benno C. and Sarah (Breakstone) L.; A.B., Harvard, 1914; M.D., U. Chgo., 1918; m. Adele Rosenwald, June 2, 1927. Practice medicine specializing in psychiatry, Chgo., 1920-27, N.Y.C., 1927—; chief staff N.Y. Inst. for Child Guidance, 1927-33; organizer, chief cons. Attitude Study Project, N.Y.C. Health Dept., 1948—; attending psychiatrist Psychoanalytic Clinic for Tng. and Research, also N.Y. State Psychiatric Inst., N.Y.C., 1949—; mem. Inst. for Advanced Study, Princeton, N.J., 1951-53; tng. and supervising analyst Psychoanalytic Clinic for Tng. and Research, Columbia, 1944-62, prof. psychiatry, 1944-57, lectr., 1957-64; vis. lectr. Tulane U., New Orleans, 1955—; research prof. Yale, 1947-51; Recipient Thomas W. Salmon award, Salmon Com. on Psychiatry and Mental Hygiene, 1946; Samuel W. Hamilton award Am. Psychopath. Assn., 1954; Emil A. Gutheil Meml. award Assn. for Advancement Psychotherapy, 1963. Fellow N.Y. Acad. Medicine, N.Y. Acad. Scis., Am. Psychiat. Assn. (life; Agnes P. McGavin award 1965); mem. Assn. for Psychoanalytic and Psychosomatic Medicine (founding mem.), Am. Orthopsychiat. Assn. (founding mem. pres. Eastern sect. 1938-39), Am. Acad. Child Psychiatry (life mem.), N.Y. Psychiat. Assn. (past pres.), Am. Psychoanalytic Assn. (life mem., past pres.), Royal Soc. Medicine London (affiliate). Author: Sibling Rivalry, 1937; Maternal Overprotection, 1943; New Fields of Psychiatry, 1947; Behavioral Analysis, 1958; The Demonstration Clinic, 1959; also numerous articles. Home: 993 Fifth Av., N.Y.C. 10028. Office: 47 E. 77th St., N.Y.C. 10021.*

LEVY, Harry, Am. mathematician; b. Boston, Jan. 9, 1902; s. Philip and Jetta (Levine) L.; A.B., Harvard, 1920, A.M., 1923; Ph.D., Princeton, 1924; m. Lucretia Mae Switser, Aug. 1, 1928. Instr., Harvard, 1920-22, NRC fellow, 1924-27; faculty, U. Ill., Urbana, 1927—, prof. math., 1951—. Mem. Am. Math. Soc., Math. Assn. Am., Phi Beta Kappa, Sigma Xi. Author: Projective and Related Geometries, 1964; also articles. Asso. editor Am. Jour. Math., 1929-37; Transactions of Am. Math. Soc., 1948-53. Research on differential geometry. Home: 712 W. Nevada St., Urbana, Ill. 61801.*

LEVY, Hilton Bertram, Am. biochemist; b. N.Y.C., Sept. 21, 1916; s. Harry S. and Dorothy G. (Edelman) L.; B.S., Coll. City N.Y., 1935; M.A., Columbia, 1936; Ph.D., Poly. Inst. Bklyn., 1946; m. Nettie Zack, Jan. 18, 1942; children—Charles Eric, Harriet S. Chief chemist Gen. Sci. Labs., N.Y.C., 1937-41; research biochemist Meml. Hosp. Cancer and Allied Diseases, 1941-46; staff Overly Biochem. Research Found., N.Y.C., 1946-52; chief viral biochem. unit NIH, 1952—; staff Cancer Inst., Paris, France, 1963-64. Fellow A.A.A.S.; mem. Am. Assn. Immunologists, American Association of Biological Chemists, Soc. for Exptl. Biology and Medicine, Soc. Gen. Physiology, Soc. Microbiologists, de Langue Francaise, Am. Soc. Microbiologists. Contbr. numerous articles to tech. jours., revs. and chpts. to books. Demonstrated similarity plant and animal respiration; research on cell metabolism, effects virus infection on cell metabolism, virus replication, biochem. mechanisms in animal defs. against virus infection. Home: 9400 Linden Av. Office: NIH, Bethesda, Md. 20014.*

LEVY, Jacques Raphael, French astronomer; b. Jarny, France, Mar. 13, 1914; s. Paul and Laure Levy; ed. Normal Superior Sch., 1933-36; agrégé in math., 1936; Dr. Math. Scis., 1943. m. Marie Francoise Saint-Aubin, Sept. 5, 1946. Became prof. math., lycees of Orleans, Lorient, Nice, 1938-45; astronomer Paris Obs., 1945, charge of Meridian Service, 1948, titular astronomer, 1964. Editor-in-chief Bulletin Astronomique, 1966; charge of instruction astronomy and celestial mechanics Faculty of Scis., Paris, from 1947. Mem. Math. Soc. France, Internat. Astron. Union. Research, publs. on fundamental astronomy, and trajectories of 3-body problems. Address: 29, rue Boulard, Paris 14, France.

LÉVY, Jean Marc, French pediatrician; b. Bischwiller, France, Aug. 26, 1927; s. Arthur and Irène (Lévy) L.; M.D., Faculte de Medecine de Strasbourg (France), 1955; m. Nicole Rosenstiel, July 4, 1954; children—Michel, Fabienne. Staff, Hospices Civils, Strasbourg, 1950—, chef de clinique, 1955-59, asst., 1959-61, professeur agrégé pediatrics, 1961-63, maitre de conférences, staff physician, 1963—. Mem. Société française de Pédiatrie. Research, publs. on infectious pathology of newborn and child, malignant hemopathies in infancy, repercussion of maternal dysgravidity on newborn infant. Home: 39, Allée de Robertsau, Strasbourg 67, France.*

LEVY, Matthew Nathan, Am. physiologist; b. N.Y.C., Dec. 2, 1922; s. David Leonard and Rose (Flomenhaft) L.; B.S., Western Res. U., 1943, M.D., 1945; m. Ruth Selma Joseph, Mar. 30, 1946; children—Donald Jay, Garry Edward, James Robert. Faculty, Western Res. U., Cleve., 1948-52, Albany Med. Coll., 1953-57; dir. research St. Vincent Charity Hosp., Cleve., 1957—; asso. prof. physiology Western Res. U., 1967-67; asso. prof. Case Inst. Tech., 1957—; chief dept. investigative medicine Mt. Sinai Hospital, Cleveland, 1967—. Recipient Lederle Med. Faculty award, 1955-57. Mem. Am. Physiol. Soc., Am. Heart Assn., Soc. Exptl. Biology and Medicine, Ohio State Med. Assn., Sigma Xi. Research, numerous publs. on phys. determinants of blood flow through tissues, mech. behavior of heart reflex regulation of cardiac activity, extent of collateral circulation to cardiac muscle. Home: 2439 Elmdale Rd., Cleve. 44118. Office: 2351 E. 22d St., Cleve. 44115.*

LÉVY, Maurice, French mathematician, engr.; b. Ribeauville, France, Feb. 28, 1838. Insp. gen. bridges and rivers; prof. analytical and celestial mechanics College de France; prof. applied mechanics École centrale des arts et manufactures. Mem. French Acad. Scis., 1883. Wrote treatise on graphic statics; also publs. on hydrodynamics, elasticity, statics. Solved problem of calculation of pressure of earth masses; used principle of overloading elastic stress in calculating supports. Died Paris, Apr. 26, 1915.

LEVY, Milton, Am. biochemist; b. St. Louis, July 13, 1903; s. Herman Israel and Jennie (Leventhal) L.; student Mo. Sch. Mines, 1921-24; B.S. in Chem. Engring., Washington U., St. Louis, 1925; Ph.D., St. Louis U., 1929; m. Helen N. Class, Oct. 26, 1924; children—Rosalyn (Mrs. Sam H. Wood, dec.), Robert M. Fellow in med. scis. NRC, Harvard, 1929-30; faculty N.Y. U., Coll. Medicine, 1930—, prof., 1954-56, prof., chmn. dept. biochemistry Coll. Dentistry, 1956—. Cons. radioisotopes Manhattan Hosp., 1956—, Bronx Vet. Hosp., 1952—. Bd. sci. counselors Nat. Inst. Dental Research, Bethesda, Md., 1962-67; chmn. com. on selection NSF postdoctoral fellows in med. scis. NRC, Washington, 1962-66. Guggenheim fellow, 1956. Gt. Tchrs. award, N.Y. U. Alumni Assn., 1964. Mem. Research Soc. Am., Am. Chem. Soc., Am. Soc. Biol. Chemists, Soc. for Exptl. Biology and Medicine, N.Y. Acad. Sci., Sigma Xi, Omicron Kappa Upsilon. Research, publs. on protein structure, mustard gas, chem. embryology; inventor constriction pipette. Home: 39-95 48th St., Long Island City, N.Y. 11104.*

LÉVY, Paul Pierre, French mathematician; b. Paris, France, Sept. 15, 1886; s. Lucien and Alice (Wolff) L.; ed. Ecole Polytechnique, 1904-06; Dr.ès sci., Ecole des Mines de Paris, 1911; m. Suzanne, Jan. 6, 1913; children—Marie-Helene (Mrs. Laurent Schwartz), Denise (Mrs. Robert Piron), Jean-Claude. Prof., Ecole des Mines de Saint-Étienne, 1910-13; répétiteur Ecole Polytechnique, Paris, 1913-20, prof., 1920-59; prof. Ecole Nationale Supérieure des Mines, Paris, 1914-51; vis. prof. U. Cal. at Berkeley, 1950, 55, Columbia, N.Y., 1955, Catholic U. Am., Washington, 1962, also others. Mem. Acad. Scis., Paris, math. socs. France, Palermo, London, Royal Statistical Soc. Author: Leçons d'analyse fonctionelle, 1922, 51; Calcul des Probabilités, 1925; Théorie de l'Addition des Variables alèatoires, 1937-54; Processus stochastiques et mouvement brownien, 1948-65; also articles. Research on theory of probabilities, functional analysis (in the meaning of Volterra) and other problems of analysis (analytic functions, series, partial differential equations), and geometry. Address: 38 Av. Théophile Gautier, Paris, France.*

LEVY, Paul Warren, Am. physicist; b. Chgo., Mar. 17, 1921; s. Leon Osmers and Anna M. (Kyhl) L.; B.S., U. Chgo., 1943; Ph.D., Carnegie Inst. Tech., 1954; m. Phyllis Edythe Osterholm, Sept. 2, 1944; children—Stephen David, Franklin Parker, Claudia Loren, Lorraine Beverly. Jr. physicist Metall. Lab., U. Chgo., 1943-44; physicist Clinton Lab., Oak Ridge Nat. Lab., 1944-48; physicist Brookhaven Nat. Lab., Upton, N.Y., 1952—. Cons. for comml., govt. labs. Fellow Am. Phys. Soc.; mem. Optical Soc. Am. Research in solid state physics, effects of radiation on properties of non-metals. Home: 180 Candee Av., Sayville, L.I., N.Y. Office: Physics Dept., Brookhaven Nat. Lab., Upton, N.Y. 11973.*

LEVY, Robert Louis, Am. physician; b. N.Y.C., Oct. 14, 1888; s. Louis and Harriet (Strouse) L.; A.B., Yale, 1909; M.D., Johns Hopkins, 1913; m. Beatrice N. Straus, June 29, 1920; children—Barbara, Gerald Dun, Jessica. Practice medicine, specializing in cardiology, N.Y.C., 1922—; prof. clin. medicine Columbia, 1922-54, prof. emeritus, 1954—; cons. physician Presbyn. Hosp., N.Y.C., 1954—; cons. cardiologist Roosevelt, French, White Plains, Englewood hosps., to Sec. War, World War II. Recipient plaque N.Y. Acad. Medicine, 1962. Editor 2 books on cardiovascular diseases; editorial bd. Am. Heart Jour., Jour. Chronic Diseases 1958-65. Contbr. articles to med. jours. Home: 720 Park Av., N.Y.C. 10021.*

LÉVY-BRUHL, Lucien, French philosopher, psychologist; b. Paris, Apr. 10, 1857; ed. École normale supérieure, 1876; agrégé de philosophie, 1879; docteur és lettres, 1884. Prof. philosophy Lycée Louis-le-Grand, 1885-95; joined faculty Sorbonne, Paris, 1895, became maitre de conférences, 1895, placed in charge history modern philosophy, 1902, named adjoint prof., 1905, prof., 1908-39. Mem. Académie des sciences morales. Author: Les fonctions mentales dans les sociétés inférieures, 1910; La mentalité primitive, 1922; Le surnaturel et la nature dans la mentalité primitive, 1922; Jean Juarès, esquisse biographique, 1923. Editor: Revue philosophique de la France et de l'étranger. Noted for pioneering studies on psychology of primitive peoples; his work helped advance study of ethnography. Died Paris, 1939.

LEWANDOWSKY, Felix, dermatologist; b. Hamburg, Germany, Oct. 1, 1879; prof., Basel Switzerland; research on Tb, especially of epidermis. Died Hamburg, 1921.

LEWES, George Henry, Brit. psychologist; b. London, Apr. 18, 1817; ed. Eng.; studied philosophy, Germany. Wrote drama articles for lit. revs., 1840-49; actor Charles Dickens theatre co., other, companies, 1848-50; lectr. physiology Fox chapel, Finsbury, 1848; produced literary work, Germany, 1855; studied marine zoology, Ilfracombe, 1856; founder, editor Fortnightly Rev., 1865. Author: Biographical History of Philosophy, 1846; Life of Goethe, 1855; Sea Side Studies, 1858; Physiology of Common Life, 1859; Studies in Animal Life, 1862; Problems of Life and Mind, 1879; Study of Psychology, pub. 1879. Contbr. to devel. of empirical metaphysics; stressed introspection in psychology, using both subjective and objective methods, balancing physiol. data by sociol. data; regarded mind as unit similar to body with aspects that can be logically separated yet are not wholly distinct from each other. Died Apr. 18, 1878.

LEWIN, Georg Richard, German surgeon; b. Sondershausen, Schwarzburg-Sondershausen, 1820; studied in Berlin, Halle, Leipzig, Heidelberg, Vienna, Paris. Dir. sect. for venereal diseases Hosp. Charité, Berlin, 1863; extraordinary prof. dermatology, Berlin, 1868. Author: Die Inhalationstherapie in Krankheiten der Respirationsorgane, 2d edit., 1865; Die Behandlung der Syphilis durch subkutane Sublimatinjektion, 1869; Die Sklerodermie, 1895. First to remove laryngeal growth with air of laryngoscope, 1861. Died Berlin, 1896.

LEWIN, Kurt, psychologist; b. Mogilno, Germany, Sept. 9, 1890; s. Leopold and Recha (Engel) L.; ed. Kaiserin Augusta Gymnasium, 1903-08, U. of Freiburg, 1908, U. of Munich, 1909, U. of Berlin, 1909-14 (Ph.D.); m. Gertrud Weiss, Oct. 1928; children—Esther Agnes, Reuven Fritz, Miriam Anna, Daniel Meier. Came to U. S., 1932. Asst., Psychol. Inst., Berlin, 1921; privatdocent of philosophy, U. of Berlin, 1921-26, prof. of philosophy and psychology, 1926-33; visiting prof. of psychology, Stanford U., 1932-33; acting prof. of psychology, Cornell U., 1933-35; prof. of child psychology, Child Welfare Research Station, U. of Ia., 1935-44; dir. Research Center for Group Dynamics, Mass. Inst. Tech., since 1944; visiting prof., Harvard, 2d semester, 1938-39, 1939-40, Univ. of California, Berkeley, summer session, 1939. Counsellor, U. S. Department of Agriculture, Washington, D.C., since 1942; Office of Strategic Services, 1944-45; chief cons., Commission on Community Interrelations, N.Y., from 1944. Member American Psychol. Assn., Midwest. Psychol. Assn., Soc. for Psychol. Study of Social Issues (chmn.), Psychometric Soc., A.A.A.S., Sigma Xi. Author: A Dynamic Theory of Personality, 1935; Principles of Topological Psychology, 1936; The Conceptual Representation and Measurement of Psychological Forces, 1938; Studies in Topological and Vector Psychology I and II; Studies in Child Welfare, 16, No. 3, 1940; Resolving Social Conflicts, 1947. Influenced by Gestalt psychology; worked on problems of individual and group motivation in situational context; used models to demonstrate dynamics of group behavior; expanded field of psychol. investigation. Died Newtonville, Mass., Feb. 12, 1947.

LEWIN, Menachem, chemist; b. Sokoly, Poland, Mar. 26, 1918; s. Icchak and Fryda (Bialodvorski) L.; M.S. in Chemistry, Hebrew U., 1944, Ph.D. in Phys. Chemistry, 1947; m. Rachel Joachimovicz, Sept. 30, 1944; children—Dorith, Icchak, Judith. Prin. research officer Research Council Isreal, Tel Aviv, 1949-53; dir. research Inst. for Fibers and Forest Products Research, Jerusalem, Israel, 1954——. Guest lectr. fiber chemistry Hebrew U., 1963——; chmn., mem. numerous govt. profl. and planning coms., 1950——. Recipient Habif prize for outstanding research in cellulose chemistry Geneva U., 1959. Fellow Textile Inst.; mem. Israel (chmn. Jerusalem br.), Am. chem. socs., Am. Soc. Textile Chemists and Colorists, T.A.P.P.I., Am. Soc. for Testing and Materials. Editor, contbr. Problems in Wood Chemistry, 1957; mem. editorial bd. Israel Jour. Tech., 1959-bromine-chlorine systems, mechanisms of oxidation ——. Research, numerous publs. on phys. chemistry of and yellowing of cellulose and nitration of wood, crimp of cotton and wool fibers; inventor permanent flameproofing and rot resisting of wood and wood products by bromination of lignin in situ, shrink proofing of wool with bromates, bleaching of cellulose fibers with hypobromites, utilization of sisal wastes and extraction of hecogenin. Home: 16 Hillel St. Office: 3 Emek Refaim, Jerusalem, Israel.*

LEWIN, Ralph Arnold, microbiologist; b. London, Eng., Apr. 30, 1921; s. Maurice and Ethel (Michaelis) L.; B.A., Cambridge U., 1942, M.A., 1946; M.S., Yale, 1949, Ph.D., 1950; m. Joyce Mary Chismore, June 11, 1950 (div. 1966). Asst. research officer NRC, Halifax, N.S., Can., 1952-55; investigator Woods Hole (Mass.) Marine Biol. Lab., 1956-59; asso. prof. Scripps Inst. Oceanography, U. Cal., La Jolla, 1959-67, prof., 1967——. Recipient Darbaker prize Bot. Soc. Am., 1958. Mem. Internat. Esperanto Assn. (N.Am. del.), numerous sci. socs. Editor: Physiology and Biochemistry of Algae, 1962. Research, numerous publs. on algae and related organisms in pure culture, specifically problems dealing with sexuality, genetics, nutrition, metabolic products and motility. Home: 8481 Paseo del Ocaso, La Jolla, Cal. 92038.*

LEWIN, Seymour Z., Am. chemist; b. N.Y.C., Aug. 16, 1921; s. Charles and Ida (Lazaroff) L.; B.S., Coll. City N.Y., 1941; M.S., U. Mich., 1942, Ph.D., 1950; Professor Honoris Causa, Instituto Quimico de Sarria, Barcelona, Spain, 1962; m. Pearl Goldman, Oct. 17, 1943; children—David, Jonathan. Lectr., U. Mich., 1946-47, Research fellow, 1947-50; faculty N.Y. U., N.Y.C., 1950——, prof. chemistry, 1960——, chief scientist Conservation Center, 1961——. Cons. to govt. and industry, 1952——. Recipient K. Fajans prize U. Mich., 1957; Belgian-Am. Ednl. Found. fellow, 1963. Fellow N.Y. Acad. Scis. (A. Cressy Morrison prize 1954), Am. Inst. Chemists; mem. Am. Chem. Soc., Faraday Soc., Phi Beta Kappa, Sigma Xi. Author: Chemists' Dictionary, 1958; also numerous articles. Instrumentation editor Jour. Chem. Edn., 1959——. Patentee in field. Research in phys. properties, molecular structure; devel. instrumentation, instrumental techniques investigation and analysis. Home: 61-33 215 St., Bayside, N.Y. 11364. Office: N.Y. U., Washington Sq., N.Y.C. 10003.*

LEWIS, Agnes Smith, Brit. archaeologist; d. John Smith; LL.D., St. Andrews; D.D., Heidelberg; Litt.D., Dublin; hon. Phil.Doc., Halle; m. Samuel Savage Lewis. Author or editor publs. including: Eastern Pilgrims, Effie Maxwell, Glenmavis, Brides of Ardmore, Monuments of Athens (trans.), Glimpses of Greek Life and Scenery, Through Cyprus, Introduction to the Four Gospels from the Sinaitic Palimpsest, Some Pages of the Sinaitic Palimpsest retranscribed, A Translation of the Syriac Gospels, In the Shadow of Sinai, The Story of Ahikar, The Palestinian Syriac Lectionary of the Gospels, Palimpsest Fragments of Palestinian Syriac; Select Narratives of Holy Women from the Sinai Palimpsest, with a Translation, Codex Climaci Rescriptus, The Old Syriac Gosepels, or Evangelion da Mepharreshre, Leaves from Three Ancient Qurans, 1914, Apocrypha Syriaca Sinaitica, Acta Mythologica Aposlotorum in Arabic. Discovered number of manuscripts in Arabic, also portion of original Hebrew manuscript of Ecclesiastes (written circa 200 B.C.); discovered Syro-Antiochene, or Sinaitic Palimpsest (most ancient known manuscript of Four Gospels in Syriac), 1892; also other valuable Oriental manuscripts. Died Mar. 1926.

LEWIS, Albert Buell, Am. anthropologist; b. Clifton, O., June 21, 1867; s. Charles Boughton and Anna (McKeehan) L.; student U. Wooster, 1890-93; student U. Chgo., 1894-97, A.B., 1894; Ph.D., Columbia, 1906; m. Gertrude Louise Clayton, Dec. 23, 1915; 1 son, Edgar Bennett. Asst. in biology U. Chgo., summer 1894, histology, 1895, bacteriology, 1896; fellow U. Neb., 1897-99, instr., 1899-1902; asst. Field Mus., Chgo., 1907-08, asst. curator Melanesian ethnology, 1908-36, curator, 1937——. Head of Joseph N. Field S. Pacific Expdn., 1909-13, which visited in interest of anthrop. dept. of Field Mus., Fiji, New Caledonia, New Hebrides, Solomon Islands, Bismarck Archipelago, Admiralty Islands, New Guinea and Dutch Indies. Died Oct. 10, 1940.

LEWIS, (William) Arthur, economist; b. St. Lucia, B.W.I., Jan. 23, 1915; s. George and Ida (Barton) L.; student St. Mary's Coll., St. Lucia, 1924-29; B.Com., London Sch. Econs., 1937, Ph.D., 1940; M.A. (hon.), Manchester U., 1951; L.H.D., Columbia, 1954; LL.D., U. Toronto, 1959, Williams, 1959, Wales, 1960, Bristol, 1961, Dakar, 1962, Leicester, 1964 others; hon. fellow London Sch. Econs., 1959, Weizmann Inst., 1962; m. Gladys Isabel Jacobs, May 5, 1947; children—Elizabeth Anne, Barbara Jean. Asst. lectr., then lectr., reader London Sch. Econs., 1938-48; Stanley Jevons prof. polit. economy U. Manchester, 1948-59; prin., vice chancelor U. West Indies, 1959-63; prof. econs., internat. affairs Princeton, 1963——. Prin. Bd. Trade, then Colonial Office U.K., 1943-44; mem. Colonial Econ. Adv. Council U.K., 1945-49; dir. Colonial Devel. Corp. U.K. 1950-52; mem. Deptl. Com. on Fuel and Power U.K., 1951-52; econ. adviser Prime Minister of Ghana, 1957-58; dept. mng. dir. UN Spl. Fund, 1959-60; spl. adviser Prime Minister West Indies, 1961-62; dir. Indsl. Devel. Corp., Jamaica, 1962-63; dir. Central Bank of Jamaica, 1961-62. faculty Princeton, 1963——. Decorated Knight Bachelor, 1963. Mem. Royal Econ. Soc. (mem. council 1949-58), Manchester Statis. Soc. (pres. 1956), Econ. Soc. Ghana (pres. 1958), Am. Econ. Assn. (v.p. 1965); hon. fgn. mem. Am. Acad. Arts and Scis., Am. Philos. Soc. Author: Economic Problems of Today, 1940; The Principles of Economic Planning, 1949; The Economics of Overhead Costs, 1949; Economic Survey, 1919-39, 1950; The Theory of Economic Growth, 1955; Politics in West Africa, 1965; Development Planning, 1966. Developed the economic theory of dual societies, with modern and traditional sectors; researches into the expansion of a world economy in the last quarter of 19th century; techniques of planning econ. growth in poor countries. Home: 121 Broadmead, Princeton, N.J. 08540.*

LEWIS, Benjamin Marzluff, Am. physician; b. Scranton, Pa., Oct. 7, 1925; s. David Morgan and Margaret (Marzluff) L.; M.D., U. Pa., 1949; m. Marie McNamara, Dec. 29, 1956; children—David, Paul, Anne. Med. house officer Peter Bent Brigham Hosp., Boston, 1949-50; resident in medicine, 1952-53; research fellow in medicine Harvard Med. Sch., 1950-52; sr. asst. research fellow U. Pa., 1953-55; faculty Wayne State U., 1956-—, prof., 1962-—; pres. staff Receiving Hosp., Detroit, 1964-65; cons. Dearborn VA, Detroit Meml. hosps. Recipient Lederle Med. Faculty award, 1959; NIH Career Research award, 1962. Mem. Am. Soc. Clin. Investigation, Am. Physiol. Soc., Central Soc. Clin. Research, Sigma Xi. Research, numerous publs. abnormal function in heart disease, effects of low oxygen on circulation, methods for measuring pulmonary function and clin. course of various lung diseases.*

LEWIS, Dan, Brit. geneticist; b. Stoke-on-Trent, Eng., Dec. 30, 1910; s. Ernest Albert and Edith (Clarke) L.; B.Sc. with honors, U. Reading, 1935; Ph.D., London (Eng.) U., 1940, D.Sc., 1953; m. Mary Phoebe Eleanor Burry, Dec. 27, 1933; 1 dau., D. Hazel Cordelia (Mrs. David Jones). Research scholar Reading U., 1935; sci. officer John Innes Inst., 1935-49, head genetics dept., 1949-57; Quain prof. botany U. Coll., London, 1957-—, dean faculty sci., 1964-67. Rockefeller fellow Cal. Inst. Tech., 1955-56; vis. prof. genetics U. Cal. at Berkeley, 1961-62; Royal Soc. vis. prof. U. New Delhi (India), 1965-66; vice chmn. governing body Cambridge (Eng.) Plant Breeding Inst., 1963-—; Royal Soc. vis. lectr., USSR, 1967. Fellow Royal Soc., 1955. Editor: From Mendels Factors to the Genetics Code, 1965; Sci. Progress, 1962-—. Research, numerous publs. on genetic control of plant self-incompatibility, plant mutation induced by radiation, resistance of amino acid analogues in fungi. Home: 50 Canonbury Park N., London, Eng.*

LEWIS, Daniel Clark, Am. mathematician; b. Millville, N.J., Aug. 14, 1904; s. Daniel Clark and Florence (Davis) L.; A.B., Haverford Coll., 1926; A.M., Harvard, 1928, Ph.D., 1932; m. Florence Russell Harvey, Sept. 14, 1935; children—Ellen Russell, Sara Ross, Caroline Worrell. Instr. math. Lehigh U., 1929-31; NRC fellow, 1933-35; instr. Cornell U., 1935-39; asst. prof. U. N.H., 1939-41, asso. prof., 1941-46; prof. U. Md., 1946-48; prof. applied math. Johns Hopkins, Balt., 1948-—; research mathematician div. war research Columbia, 1943-45, Research Inst. for Advanced Study div. Martin Co., Balt., 1958-59. Cons. to industry, govt. lab. Mem. Am. Math. Soc., Math. Assn. Am., Soc. for Applied and Indsl. Math., Circolo Matematico di Palermo, Freethinkers Am., Phi Beta Kappa, Sigma Xi. Contbr. articles and revs. to math. jours. Editor: Am. Jour. Math., 1949-52; mem. staff reviewers Math. Revs., 1939-61. Research on ordinary differential equations in real domain, dynamical systems, four color map problem, partial differential equations; discovered and developed theory of autosynartetic solutions generalizing a theory of Poincare on periodic solutions of ordinary differential equations. Home: 5205 Putney Way, Balt. 21218.*

LEWIS, Dean (De Witt), Am. surgeon; b. Kewanee, Ill., Aug. 11, 1874; s. L. W. and V. Winifred L.; A.B., Lake Forest U., 1895; M.D., Rush Med. Coll., 1899; D.Sc., U. of Ireland; m. 2d, Norene Kinney, Dec. 26, 1927; children—Julianne, Dean, Mary Elizabeth. Asst. in anatomy Rush Med. Coll., 1900-01, asso., 1901-03, instr. in surgery, 1903-19, asso. prof., 1919-20, prof., 1920-24; prof. surgery U. Ill., 1925; prof. surgery Johns Hopkins U., also surgeon in chief Johns Hopkins Hosp., 1925-39. Editor Archives of Surgery, Internat. Surg. Digest. Editor: Practice of Surgery. Contbr. papers on hypophysis and on surgery of nerves to med. lit. Died Oct. 9, 1941.

LEWIS, Donald John, Am. mathematician; b. Adrian, Minn., Jan. 25, 1926; s. William J. and Ellanora (Masgai) L.; B.S., Coll. St. Thomas, 1946; Ph.D., U. Mich., 1950; m. Carolyn Dana Hauf, Dec. 28, 1953. Instr., Ohio State U., Columbus, 1950-52; NSF fellow, mem. Inst. for Advanced Study, Princeton, N.J., 1952-53; faculty U. Notre Dame (Ind.), 1953-61, asso. prof., 1957-61; faculty U. Mich., Ann Arbor, 1961-—, prof. math., 1963-—. Research fellow U. Manchester (Eng.), 1959-60, Cambridge (Eng.) U., 1960-61; NSF Sr. fellow, 1959-61; vis. sr. fellow U. Cambridge, 1966. Mem. Am., London math. socs. Author: Introduction to Algebra, 1966; also articles. Research on diophantine equations (primarily in many variables), arithmetic properties of polynomials, integer solutions of equations. Home: 1050 Wall St., Ann Arbor, Mich. 48105.*

LEWIS, Edward B., Am. biologist; b. Wilkes-Barre, Pa., May 20, 1918; s. Edward B. and Laura (Histed) L.; B.A., U. Minn., 1939; Ph.D., Cal. Inst. Tech., 1942; m. Pamela Harrah, Sept. 26, 1946;

children—Hugh, Glenn, Keith. Instr. Cal. Inst. Tech., 1946-48, asst. prof., 1948-49, asso. prof., 1949-56, prof. biology, 1956—. Mem. Nat. Adv. Com. on Radiation, 1958-61. Served to capt. USAAF, 1942-46. Rockefeller fellow, 1948-49. Mem. Genetics Soc. Am. (sec. 1962-64), Nat. Acad. Scis., A.A.A.S., Am. Soc. Human Genetics, Am. Soc. Naturalists. Editor: Genetics and Evolution, 1961. Research on genetics; somatic effects of radiation. Home: 805 Winthrop Rd., San Marino, Cal. Office: California Inst. Technology, Pasadena, Cal.

LEWIS, Edward Sheldon, Am. chemist; b. Berkeley, Cal., May 7, 1920; s. Gilbert Newton and Mary (Sheldon) L.; B.S., U. Cal. at Berkeley, 1940; M.A., Harvard, 1947, Ph.D., 1947; m. Fofo Catsinas, Dec. 21, 1955; children—Richard Peter, Gregory Gilbert. NRC fellow U. Cal. at Los Angeles, 1948; faculty Rice U., Houston, 1948—, prof. chemistry, 1962—, chmn. dept. chemistry, 1963-67; vis. prof. U. Southampton (Eng.), 1957. Mem. Am. Chem. Soc., Chem. Soc. (London), A.A.A.S., Am. Assn. U. Profs., Houston Philos. Soc., Phi Beta Kappa, Sigma Xi. Editor: (with A. Weissberger, S. Friess) Technique of Organic Chemistry, vol. 8, part 1, 1961, part 2, 1963. Research, publs. on mechanism of organic reactions especially involving expulsion of stable molecules nitrogen, sulfer dioxide and carbon dioxide, effect heavy hydrogen on rates of organic reactions. Home: 5651 Chevy Chase St., Houston 77027.*

LEWIS, Floyd John, Am. surgeon; b. Waseca, Minn., Nov. 26, 1916; s. Floyd John and Mable (Griesy) L.; B.S., U. Minn., 1938, M.S., 1941, M.D., 1942, Ph.D., 1950; m. Ruth Chesna, Dec. 31, 1945; children—Mary, Jean, Scott. Faculty surgery U. Minn., Mpls., 1950-56, asso. prof. surgery Northwestern U., Chgo., 1956-57, prof., 1957—; attending surgeon Passavant, Vets. Research hosps., Chgo. Mem. Am. Surg. Assn., Soc. Clin. Surgery, U. Surgeons, Am. Thoracic Assn., Halsted Soc. Research, publs. in cardiovascular surgery and hypothemia. Home: 1345 Ashland Av., Wilmette, Ill.*

LEWIS, F(rancis) Park, physician, oculist; b. Hamilton, Ont., Can., May 19, 1855; s. John W. and Hannah M. (Gavin) L.; M.D., Pulte Med. Coll., 1876; grad. N.Y. Ophthalmic Hosp., 1877; studied in London, Berlin and Vienna; m. Grace K. Moseley, 1889; children—Katharine Park, Dorothea Park, Frances Park. Practiced in Buffalo, 1876—. Recipient Leslie Dana medal Am. Acad. Ophthalmology and Oto-Laryngology, 1928; Chancellor medal U. Buffalo, 1933. Vice pres. Nat. Soc. for Prevention of Blindness, Internat. Assn. for Prevention of Blindness. Mem. editorial staff Am. Jour. Ophthalmology. Promoted internat. use of silver salt solution to save eyes of babies. Died Sept. 10, 1940.

LEWIS, George William, Am. aero. engr.; b. Ithaca, N.Y., Mar. 10, 1882; s. William H. and Edith (Sweetland) L.; M.E., 1908, Cornell U., M.M.E., 1910; Sc.D., Norwich U., D.Eng., Ill. Inst. Tech., 1944; m. D. Myrtle Harvey, Sept. 9, 1908; children—Alfred William, Harvey Sweetland, Myrtle Norlaine, George William, Leigh Kneeland, Armin Kessler. Instr. in engring. Cornell U., 1908-10; prof. engring. Swarthmore Coll., 1910-17; engr. in charge Clarke Thomson Research, 1917-19; sales mgr. Phila. Surface Combustion Co., 1919; exec. officer Nat. Adv. Com. Aeronautics, 1919-24, dir. aeronaut. research, Washington, 1924-47, research cons., 1947-48. Recipient Daniel Guggenheim medal, 1936; Spirit St. Louis medal, 1944. Hon. fellow Inst. Aero. Scis. (pres. 1939), Soc. Automotive Engrs. (v.p. 1931); mem. Am. Philos. Soc., Nat. Acad. Scis. Died July 12, 1948.

LEWIS, Gething Morgan, physicist; b. Swansea, U.K., June 1, 1912; s. Richard Howell and Annie (Rees) L.; B.Sc., in Physics, London (Eng.) U., 1935, B.Sc, in Math., 1937, M.Sc., 1939; Ph.D., Glasgow (Scotland), 1953; m. Elvira Evans, June 22, 1940. Staff demonstrator U. Coll. London, 1937-39; staff Royal Aircraft Establishment, Farnborough, also Royal Radar Establishment, Malvern, 1940-45; faculty Hull U., 1945-47; faculty Glasgow U., 1947—, sr. lectr., 1958—; staff CERN, Geneva, Switzerland, 1962-63. Fellow Inst. Physics, Royal Soc. Edinburgh. Research, publs. on electromagnetic waves, nuclear physics and fundamental particles. Home: 7 Banavie Rd., Glasgow, W.I., Scotland, U.K.*

LEWIS, Gilbert Newton, Am. chemist; b. Weymouth, Mass., Oct. 23, 1875; s. Frank W. and Mary B. (White) L.; student U. Neb.; A.B., Harvard, 1896, Ph.D., 1899; postgrad. univs. Leipzig, Göttingen (both Germany); D.Sc., U. Liverpool (Eng.), 1923, U. Wis., 1928, U. Chgo., 1929. m. Mary Hinckley Sheldon, June 20, 1912; children—Richard Newton, Margery, Edward Sheldon. Instr. chemistry Harvard, 1899-1900, 01-06; in charge weights and measures P.I., 1904-05; asst. prof. chemistry Mass. Inst. Tech., 1907-08, asso. prof., 1908-11, prof., 1911-12; prof. chemistry U. Cal. at Berkeley, from 1912; Silliman lectr. Author: (with M. Randall) Thermodynamics and the Free Energy of Chemical Substances, 1923; Valence and the Structure of Atoms and Molecules, 1923; The Anatomy of Science, 1926. Died Mar. 24, 1946.

LEWIS, (Frank) Harlan, Am. botanist; b. Redlands, Cal., Jan. 8, 1919; s. Frank Hooker and Mary (Smith) L.; A.A., San Bernardino Valley Coll., 1939; A.B., U. Cal. at Los Angeles, 1941, M.A., 1942, Ph.D., 1946; postgrad. Cal. Inst. Tech.; m. Margaret Ruth Ensign, Aug. 2, 1945; children—Donald Austin, Frank Murray. Faculty, U. Cal., Los Angeles, 1946—, prof. botany, 1956—, systematist in expt. sta., 1956-62, chmn. dept. botany, 1959-63, dean div. life scis., 1962—. Cons. NSF. NRC fellow, 1947-48; Guggenheim fellow, 1954-55. Fellow A.A.A.S.; mem. Soc. Study Evolution (pres. 1961), Internat. Orgn. Plant Biosystematists (v.p.), Bot. Soc. Am. (pres. Pacific div. 1959), Am. Inst. Biol. Sci., Am. Soc. Naturalists, Am. Soc. Plant Taxonomists, Genetics Soc. Am., Ecol. Soc. Am., Internat. Soc. Plant Taxonomists, Phi Beta Kappa, Sigma Xi. Author: (with Margaret Ensign Lewis) The Genus Clarkia, 1955. Editorial com. Am. Jour. Botany; editorial bd. Am. Naturalist. Research, publs. on evolution and systematics of flowering plants. Home: 609 S. Bundy Dr., Los Angeles 90049.*

LEWIS, Harold Walker, Am. physicist; b. Keene, N.H., May 7, 1917; s. Hiram Edwin and Lena (Ashton) L.; B.S., Middlebury Coll., 1938; M.A., U. Buffalo, 1940; Ph.D., Duke, 1950; m. Mary Anne O'Rourke, June 1, 1946; children—Barbara, Richard. Faculty, Duke, 1949—, prof. physics, 1959—, dean arts and scis., vice provost, 1963—. Fellow Am. Phys. Soc.; mem. Am. Assn. Physics Tchrs. Research in nuclear structure. Home: 1708 Woodburn Rd., Durham, N.C. 27705.*

LEWIS, Herman William, Am. geneticist; b. Chgo., July 10, 1923; s. Jacob and Mary (Weiss) L.; B.S., U. Ill., 1947, M.S., 1949; Ph.D., U. Cal. at Berkeley, 1953; m. Helen Simon, Dec. 29, 1942; children—David, Miriam. USPHS research fellow U. Cal., Berkeley, 1952-54; asst. prof. biology Mass. Inst. Tech., 1954-61; chmn. dept. life scis. Mich. State U., 1961-62; program dir. genetic biology program NSF, Washington, 1962-66, head cellular biology sect., 1966—. Mem. A.A.A.S., Genetic Soc. Am., Biophys. Soc., Am. Soc. Cell Biology. Research, publs. on specificity of genetic control on structure, rate of synthesis of proteins. Office: NSF, Washington 20550.*

LEWIS, Howard Phelps, Am. physician; b. San Francisco, Feb. 18, 1902; s. Edmund Phelps and Mary Edith (Howard) L.; B.Sc., Ore. State U., 1924; M.D., U. Ore., 1930; m. Wava Irene Brown, July 2, 1927; children—Richard Phelps, Thomas Howard. Faculty, U. Ore. Med. Sch., Portland, 1932—, prof. medicine, chmn. dept., 1947—. Mem. Residency Rev. Com. in Internal Medicine, 1960-65, chmn., 1963-65; mem. adv. council Nat. Heart Inst., USPHS, 1956-61. Recipient Meritorious Achievement award U. Ore. Med. Sch. Alumni Assn., 1963. Diplomate Am. Bd. Internal Medicine. Mem. A.A.A.S., Am. Clin. and Climatol. Assn., A.C.P. (pres. 1960), Am. Fedn. Clin. Research, Am. Heart Assn. (Merit award 1960), A.M.A. (past chmn. sect. internal medicine 1955), Assn. Am. Physicians, Assn. Profs. Medicine, N.Pacific Soc. Internal Medicine (past pres.), Pacific Interurban Clin. Club, Portland Acad. Medicine (past pres.), Soc. Med. Cons. Armed Forces, Western Assn. Physicians, Western Soc. Clin. Research, Sigma Xi, Delta Chi, Nu Sigma Nu, Alpha Omega Alpha. Author: The History and Physical Examination, 1958, 61, 66; also articles. Editorial bd. Archives Internal Medicine, 1961—, Circulation, 1961-65, 66—. Home: 2151 S.W. Laurel St., Portland, Ore. 97201.*

LEWIS, Isaac Newton, Am. inventor; b. New Salem, Pa., Oct. 12, 1858; s. James H. and Anne (Kendall) L.; grad. U. S. Mil. Acad., 1884; m. Mary Wheatley, Oct. 21, 1886; children—Richard Wheatley, Laura Anne, George Fenn, Margaret Kendall. Commd. 2d lt. 2d Arty., 1884; promoted through grades to col., 1913; ret., 1913. Mem. bd. on regulation of coast arty. fire, N.Y. Harbor, 1894-98; recorder Bd. of Ordnance and Fortification, Washington, 1898-1902; instr. and dir. Coast Arty. Sch., Ft. Monroe, Va., 1904-11. Made study of methods of mfr. and supply of ordnance materials, in Europe, 1900, resulting in complete re-armament of field arty. of U. S.; originator of plan for modern corps orgn. for artillery which was adopted by Congress, 1902; inventor Lewis machine gun which was in gen. use by allies throughout the World War; also inventor of numerous mil. instruments and devices in gen. use, including successful arty. range and position finder, a replotting and relocating system for coast batteries, time interval clock and bell system of signals, quick firing field gun and mount, quick reading mech. verniers, electric car lighting and windmill electric lighting systems. Died Nov. 9, 1931.

LEWIS, Jessica Helen (Mrs. Jack D. Myers), Am. physician; b. Harpswell, Me., Oct. 26, 1917; d. Warren Harmon and Margaret (Reed) Lewis; A.B., Goucher Coll., 1938; M.D., Johns Hopkins, 1942; m. Jack D. Myers, Aug. 31, 1946; children—Judith Duane, John Lewis, Jessica Reed, Elizabeth Read, Margaret Anne. USPHS Research fellow U. N.C., 1947-48, research asso. dept. physiology, 1948-55; asso. dept. medicine Duke Med. Sch., 1951-55; research asso. dept. medicine U. Pitts., 1955-58, faculty, 1958—, research asso. prof., 1965—. Mem. Am. Physiol. Soc., Am. Soc. Clin. Investigation, Am. Fedn. Clin. Research, Soc. for Exptl. Biol. Medicine, Internat. Soc. Hemotology, Am. Soc. Hemotology, Sigma Xi. Research, numerous publs. on mechanism blood coagulation, fibrinolysis, hemorrhagic and thrombotic diseases. Home: 2689 Oak Hill Dr., Allison Park, Pa. 15101. Office: Dept. Medicine, U. Pitts., Pitts. 15101.*

LEWIS, John Bryn, Brit. chemist; b. Swansea, Wales, U.K., Nov. 12, 1919; s. Aneurin and Gwladys (Thomas) L.; B.Sc., U. Coll. Swansea, U. Wales, 1941, D.Sc., 1966; m. Elizabeth McGowan Moffat, Sept. 27, 1948; children—Gwyneth Helen, David Moffat. Tech. officer Imperial Chems. Industry explosives div. Ardeer & Dalbeattle, Scotland, 1941-45; sr. prin. scientist chem. engring. div. Atomic Energy Research Establishment, Harwell, Berkshire, Eng., 1948—. Fellow Royal Inst. Chemistry; mem. Instn. Chem. Engrs. Research, publs. on mass transfer control in liquid-liquid extraction and solid dissolution, kinetics of carbon gasification. Home: 28 Fitzharry's Rd., Abingdon, Berkshire. Office: Bldg. 353 Atomic Energy Research Establishment, Harwell, Berkshire, Eng.*

LEWIS, Lena Armstrong, Am. physiologist; b. Lancaster, Pa., July 12, 1910; d. John Wythe and Lena (Armstrong) Lewis; A.B., Lindenwood Coll., 1931, LL.D., 1952; M.A., Ohio State U., 1938, Ph.D., 1940. Spl. post-doctorate fellow Cleve. Clinic Found., 1941-43, mem. staff research div., 1943—. Mem. Am. Physiol. Soc., Endocrine Soc., Soc. Exptl. Biology and Medicine, Atherosclerosis Council, A.A.A.S., Sigma Xi. Author: Electrophoresis in Physiology, 1950, 2d edit., 1960; also numerous articles. Research in physiol. factors regulating level of lipoproteins, proteins and lipids in blood serum and tissues and their relation to devel. of atherosclerosis. Home: 386 S. Belvoir Blvd., Cleve. 44121. Office: Cleve. Clinic Found., Cleve. 44106.*

LEWIS, Oscar, Am. anthropologist; b. N.Y.C., Dec. 25, 1914; B.S.S., Coll. City N.Y., 1936; Ph.D., Columbia, 1940; married 1937. Research asso. Yale, 1942-43; propaganda analyst U. S. Dept. Justice, 1943; social sci. Dept. Agr., 1944-45; Dept. State vis. prof., Havana, 1945-46; asso. prof. anthropology Washington U., St. Louis, 1946-48; prof. anthropology U. Ill., 1948—; cons. anthropologist Ford Found., India, 1952-54. Social Sci. Research Council fellow, 1952; Ford Found. grantee, 1952. Recipient Mass Media award Nat. Conf. Christian and Jews, 1962; Family Life Book award Child Study Assn. Am., 1961; Gallimard prize (France), 1964. Mem. Am. Anthrop. Assn. Author: Tepoztlan, Village in Mexico; Life in a Mexican Village; Tepoztlan Restudied; Five Families; Mexican Case Studies in the Culture of Poverty; Children of Sanchez; Pedro Martinez: A Mexican Peasant and His Family; Village Life in Northern India; La Vida: A Puerto Rican Slum Family, 1966. Research and publs. on spl. anthrop. approach to family studies; originated concept of a culture of poverty. Office: Dept. Anthropology, U. Ill., Urbana, Ill.*

LEWIS, Paul A., Am. pathologist; b. Chgo., Apr. 14, 1879; s. Clinton H. and Caroline (Hobart) L.; student U. Wis., 1897-98, Wis. Coll. Physicians and Surgeons, Milw., 1899-1901; M.D., U. Pa., 1904; m. Louise Durbin, Aug. 6, 1906; children—Janet, Hobart. Resident in pathology Boston City Hosp., 1904-05; asst. in Antitoxin Lab., Mass. Bd. Health, Boston, 1905-08; fellow in comparative pathology Harvard, 1906-08; asso. in pathology Rockefeller Inst., N.Y.C., 1908-10; dir. lab. Henry Phipps Inst., Phila., 1910-23; prof. exptl. pathology U. Pa., 1921-23; asso. mem. Rockefeller Inst. for Med. Research, N.Y., 1923—. Used (with Simon Flexner) cultures of a filtrable virus to transmit poliomyelitis to monkeys, 1909. Died June 30, 1929.

LEWIS, Philip Meriwether, Am. ophthalmologist; b. Cismont, Va., Aug. 10, 1898; s. Thomas Walker and Jane (Page) L.; M.D., U. Va., 1920; m. Ruth Marie Kitchins, Dec. 29, 1925; children—Jean Lewis (Mrs. Thomas Russell Price), Meriwether Lewis (Mrs. John Fargason III). Practiced medicine specializing in ophthalmology, Memphis, 1924—; faculty U. Tenn. Coll. Medicine, Memphis, 1925—. prof., chmn. div. ophthalmology, 1944—. Mem. Memphis and Shelby County Med. Soc. (past pres.), Am., Tenn. acads. ophthalmology and otolaryngology (past pres.), Am. Ophthal. Soc. (pres. 1966—), A.M.A., So. Tenn. State med. assns. Research, publs. on ocular findings in meningicoccic meningitis, drugs for treatment of gonorrhaeal conjunctivitis and keratits, cataract extraction with use of alphachymotrypsin. Home: 1960 N. Parkway, Memphis 38112. Office: 130 Madison Av., Memphis 38103.*

LEWIS, Ralph Oscar, Am. soil conservationist; b. Parsons, Kan., Feb. 7, 1905; s. Oscar B. and Nellie May (Meador) L.; B.S., Kan. State U., 1929; m. Margaret Lorraine Rutter, Feb. 14, 1930; children—Richard D., James A., Mary Margaret, Barbara Karen. Soil surveyor Kan. State Coll., Manhattan, 1929-32; county agrl. agt., Ellsworth, Kan., 1932-35; staff Soil Conservation Service, 1935-56, dep. dir. soil survey operations, 1954-56; soil sci. adviser ICA, Philippines, 1956-58; dep. agrl. officer, USOM, Cambodia, 1958-60, Thailand, 1960-61; agrl. officer, AID, Laos, 1962-64, soils adviser, Turkey from 1964. Mem. Soil Conservation Soc. Am., Soil Sci. Soc. Am., Internat. Soil Sci. Soc., Philippine Soil Sci. Soc., Am. Soc. Agronomy, Range Mgmt. Soc. Am. Home: 2200 Casey Rd., Campbell, Cal. 95008.*

LEWIS, Stephen Robert, Am. surgeon; b. Mt. Horeb, Wis., Aug. 26, 1920; s. Russell F., and Gertrude (Leffler) L.; B.A., Carroll Coll., 1941; M.D., Marquette U., 1944; m. Audrey Morton, June 12, 1948; children—Stephen Robert, Virginia Ann. Faculty, U. Tex. Med. Br., Galveston, 1950—, prof. surgery, 1961—, asst. dean medicine, 1958-62, dir. postgrad. edn., 1956—, chief plastic surgery, 1961—; cons. USAF, St. Mary's Infirmary, Galveston County Meml. Hosp. Fellow A.C.S.; mem. A.M.A., Tex. Med. Assn., Galveston County Med. Soc., Am. Soc. Plastic and Reconstructive Surgeons, Plastic Surgery Research Council, Am. Assn. Plastic Surgeons, Am. Assn. for Cleft Palate Rehab., Soc. Head and Neck Surgeons, Law Sci. Acad. Am. (past pres.), Assn. Am. Med. Colls., Edn. and Research Inst. Gen. Medicine, Am. Assn. for Surgery Trauma, Pan-Pacific Surg. Assn., Singleton, Tex. surg. socs., Southeastern Soc. Plastic Surgeons, Tex. Soc. Plastic Surgeons, Postgrad. Med. Assembly South Tex., U. Plastic Surgeons Travel Club, Sigma Xi, Mu Delta. Research, numerous publs. on lymphatics, burns; developer method of growing human epidermal cells in tissue culture. Home: 2902 Dominique St. Office: 915 Strand St., Galveston, Tex. 77550.*

LEWIS, Sir Thomas, Brit. pathologist, cardiologist; b. Wales, 1881; s. Henry Lewis; ed. U. Coll., Cardiff, Wales, U. Coll. Hosp., London; M.D., D.Sc., LL.D.; m. Lorna James, 1916; 1 son, 2 daus. Physician, U. Coll. Hosp., London; hon. cons. physician Ministry Pensions. Fellow U. Coll. London. Became fellow Royal Soc., 1918. Recipient Royal medal, 1927, Copley medal, 1941. Author: Mechanism of the Heart Beat, 1911; Clinical Disorders of the Heart Beat, 191; Clinical Electrocardiography, 1913; Lectures on the Heart, 1915; The Soldier's Heart and the Effort Syndrome, 1940; Blood vessels of the Human Skin, 1927; Diseases of the Heart, 1932; Clinical Science, 1934; Vascular Disorders of the Limbs, 1936; Pain, 1942; also numerous articles. Editor, Heart, also Clin. Sci. Research in electrocardiography; 1st description of auricular fibrillations, 1909; designated sinoauricular node as pacemaker of heart, 1910; introduced term, effort syndrome; 1917; described carotid sinus syndrome, 1932. Died Mar. 17, 1945.

LEWIS, Timothy Richards, Welsh bacteriologist, physician; b. Hafod, Wales, Aug. 31, 1841; s. William and Britania (Richards) L.; studied at U. Coll., 1863-66; M.B., C.M., 1867; m. Emily Francis Brown, Oct. 8, 1879. Apprenticed to Norbarth pharmacist at the age of 15; 4 years later went to London; joined dispensary German Hosp., Dalston; went with David Douglas to India to study cholera, for 5 years, 1869; apptd. spl. asst. to San. Commr., 1874, began gen. work San. Dept., 1879; became asst. prof., Neltey, 1833. Recipient Fellowes Silver medal U. Coll. Discovered nematoid worm (named Flori bancrofti, 1877, now Wucheria bancrofti), 1870, flagellate in rat blood (Trypanosoma lewisi), 1877; pioneered in tropical diseases, especially cholera, leprosy, pathology enteric fever. Died May 7, 1886.

LEWIS, Urban James, Am. biochemist; b. Flagstaff, Ariz., Apr. 28, 1923; s. Urban J. and Mary (Rozen) L.; B.A., San Diego State Coll., 1948; Ph.D., U. Wis., 1952; m. Loraine Joyce Chambers, July 16, 1950; children—Geoffrey Paul, Wayne Kent. Fellow, Med. Nobel Inst., Stockholm, Sweden, 1952-53; instr. U. Chgo., 1953-54; research asso. Merck & Co., Rahway, N.J., 1954-61; asso. mem. Scripps Clinic and Research Found., La Jolla, Cal., 1961—. Mem. Am. Soc. Biol. Chemists, Am. Chem. Soc., Sigma Xi. Research, publs. on isolation of biol. new form vitamin B12, isolation and characterization of elastase, helenine, crystallization of human growth hormone, conversion products of pituitary growth hormone and prolactin, pituitary gland of rat and mouse. Home: 5733 Skylark Pl., La Jolla 92037. Office: 476 Prospect St., La Jolla, Cal. 92037.*

LEWIS, Warren Burton, Am. chemist; b. Pomona, Cal., Feb. 24, 1918; s. Philo Guy and Helen (Thomas) L.; student Pomona Jr. Coll., 1935-37; B.A., Pomona Coll., 1940; M.A., U. Cal. at Los Angeles, 1942; Ph.D., Mass. Inst. Tech., 1948; m. Jane Dyer Sanford, June 7, 1942 (div. Apr. 1965); children—Catherine Louise, Warren Sanford, Carolyn Marie, Loretta Jane; m. 2d, Helen Maxine Stewart, Oct. 29, 1965. Staff mem. Los Alamos Sci. Lab., 1948—. Mem. Am Chem. Soc., Am. Phys. Soc. Research, publs. on radioactive tracer studies of electron transfer reactions, magnetic properties of plutonium compounds, nuclear magnetic resonance studies of rare earth salts in solution using 017, electron spin resonance of color centers in LiH, 017 nuclear magnetic resonance, paramagnetic relaxation in transition metal complexes in solution. Home: 1023 Opal St. Office: Box 1663, Los Alamos 87544.*

LEWIS, Warren Harmon, Am. cytologist, embryologist; b. Suffield, Conn., June 17, 1870; s. John and Adelaide E. (Harmon) L.; student Chgo. Manual Tng. Sch., 1886-89; B.S., U. Mich., 1894; M.D., Johns Hopkins, 1900; m. Margaret Reed, May 23, 1910; children—Margaret Nast, Warren Reed, Jessica Helen. Asst. in zoölogy, U. Mich., 1894-96; asst. in anatomy, Johns Hopkins U., 1900, instr., 1901-03, asso., 1903-04, asso. prof., 1904-13, prof. physiol. anatomy, 1913-40. Research asso., dept. embryology, Carnegie Instn. of Washington, 1919-40; mem. Wistar Inst. Anatomy and Biology, 1940-58, research. Member Mount Desert Island Biological Laboratory (pres. 1933-37). Recipient Harrison Prize Internat. Soc. Cell Biology, 1960. Fellow A.A.A.S.; mem. Internat. Soc. Cell Biology (hon. pres. 1962—), Nat. Acad. Scis., Am. Assn. Anatomists (pres. 1934-36), Am. Physiol. Soc., Am. Soc. Naturalists, Am. Philos. Soc., Internat. Soc. for Experimental Cytology (pres. 1939-47), Am. Assn. for Cancer Research; hon. mem. Tissue Culture Assn., La Société de Médecine de Gand; fgn. mem. Accademia Nazionale dei Lincei; hon. fellow Royal Micros. Soc. Author numerous important profl. papers relating to cytology and embryology. Co-author: General Cytology; The Structure of Protopalms. Editor of 20th to 24th (inclusive) edits. of Gray's Anatomy. Motion pictures of cells and eggs; with Margaret R. Lewis gave description of mitochondria, 1914. Home: Hamilton Ct. Hotel, Phila. 4. Ret.

LEWIS, Warren Kendall, Am. chem. engr.; b. Laurel, Del., Aug. 21, 1882; s. Henry Clay and Martha Ellen (Kinder) L.; S.B., Mass. Inst. Tech., 1905; Ph.D. in Chemistry, U. Breslau, Germany, 1908; D.Sc. (hon.), U. Del., 1937, Harvard, 1951, Bowdoin Coll., 1952; D.Eng. (hon.), Princeton, 1947; m. Rosalind Denny Kenway, Oct. 1909. With Mass. Inst. Tech., 1910—, prof. chem. engring., 1915-48; ret. Mem. Am. Chem. Soc., Am. Acad. Arts and Scis., Am. Inst. Chem. Engrs., Nat. Acad. Scis., Instn. Chem. Engrs. Great Britain (hon.), Am. Inst. Mining, Metall. and Petroleum Engrs. Conglist. Club: Chemists' (N.Y.C.). Contbr. tech. jours. on chem. engring. Author: Industrial Chemistry of Colloidal and Amorphous Materials, 1942. Research on distillation, thermal properties of liquids, and on colloids. Address: Mass. Inst. Tech., Cambridge, Mass.

LEWIS, Wilfrid Bennett, physicist; b. Castle Carrock, Cumberland, Eng., June 24, 1908; s. Arthur Wilfrid and Isoline (Steavenson) L.; B.A. with honors, Gonville and Caius Coll., Cambridge (Eng.) U., 1930, M.A., Ph.D., 1934; D.Sc., Queen's U., Kingston, Ont., Can., 1960, U. Sask. (Saskatoon, Can.), 1964, McMaster U., 1966; LL.D., Dalhousie U., Halifax, N.S., Can., 1960, Carleton U., Ottawa, Ont., 1962. Research fellow Cambridge U., 1934-40, demonstrator in physics, 1935-37, lectr., 1937-39; radar work Air Ministry and Ministry Aircraft Prodn., 1939-46, chief supt. Telecommunications Research Establishment, 1945-46; dir. div. atomic energy research NRC Can., Chalk River, Ont., 1946-52; v.p. research and devel. Atomic Energy Can. Ltd., Chalk River, 1952-63, sr. v.p. sci., 1963—. Canadian rep. sci. adv. com. UN, 1955—; mem. sci. adv. com. Internat. Atomic Energy Agy., 1958—, adv. com. on physics research Def. Research Bd. Decorated comdr. Order Brit. Empire, 1946; recipient Am. Medal Freedom, 1947. Fellow Royal Soc., 1945, Royal Soc. Can., Am. Nuclear Soc. (charter, dir., pres. 1961-62), Phys. Soc. London, Cambridge Philos. Soc., Am. Phys. Soc. Author: Electrical Counting, 1942; also numerous articles. Research in energies and fine structure of alpha rays, nuclear transmutations by ions accelerated by high voltage and cyclotron; discovered Lithium-8 alphas "scaling" counters, portable 2-way radio-telephone, airborne and ground radar and nav. systems; fluctuations in thermal radiation; nuclear power by neutron econ. fuel cycles especially in heavy water moderated reactors. Home: 13 Beach Av., Deep River, Ont. Office: Atomic Energy Can. Ltd., Chalk River, Ont., Can.*

LEWIS, Willard Deming, Am. applied physicist; b. Augusta, Ga., Jan. 6, 1915; s. Willard and Constance (Deming) L.; A.B., Harvard, 1935, A.M., 1939, Ph.D., 1941; B.A., Oxford (Eng.) U., 1938, M.A., 1945; L.H.D., Moravian Coll.; LL.D., Lafayette Coll., Rutgers U., Hahnemann Medical Coll.; m. Marian Carter Chapman, Nov. 1, 1941 (dec. July 1965); children—Caroine Carter, Constance Carter, Linda Deming, Catherine Doten, Marian Chapman. With Bell Telephone Labs., Inc., 1941-62, exec. dir. research communications systems, 1958-62; a founder, mng. dir. systems study center Bellcomm, Inc., 1962-64; pres. Lehigh U., 1964—. Dir. Riverside Mills, Augusta. Mem. Polaris Communications Com., 1958-63, Def. Industry Adv. Com., 1962-63; cons. Office Sci. and Tech., also Pres's Sci. Adv. Com., 1962—. Mem. Naval Research Adv. Com., 1964—, vice chmn., 1965. Mem. school bd., Little Silver, New Jersey, 1949-50. Mem. overseers committee to visit division engineering and applied sci., Harvard, 1954—; mem. council Harvard Found. Advanced Study and Research, 1959-64, chmn., 1961-62. Fellow I.E.E.E.; mem. Am. Phys. Soc., Assn. Computing Machinery, Inst. Mgmt. Scis., Am. Inst. Aero. and Astronautics, Am. Assn. Rhodes Scholars, Harvard Engring. Soc. Patentee in field. Research on magnetism; radar antennas; telephone switching; phonograph distortion; radio repeaters and microwave filters. Home: President's House, Lehigh Univ. Campus, Bethlehem, Pa. 18015.

LEWIS, William, English physician; b. Richmond, Eng., 1714; B.Physik, Christ Church, Oxford (Eng.) U., B.A., 1734, M.A., 1737, M.B., 1741, M.D., 1745. Practiced medicine at Kingston-on-Thames; lectr. to Prince of Wales, Kew, Eng. Recipient Gold medal Soc. for Improvement Arts, Mfg. Fellow Royal Soc., 1745. Author: A Course of Practical Chemistry . . . , 1746; The Pharmacopoeia of the Royal College of Physicians at Edinburgh . . . , 1748; Oratio in Theatro Sheldoniano . . . , 1749; An Experimental History of the Materia Medica . . . , 1761; Commercium philosophico-technicum . . . , 1763, 66; Experiments and Observations on American Potashes . . . , 1767; The New Dispensatory, 1753; other books, articles. Described fusible metal bath; discovered ether extracts gold from its solution in aqua regia; also that borax increases the solubility of tartar; studied nut galls as a source for ink; thought waters distilled from bitter almonds and bitter fruit kernels contain poisonous hydrocyanic acid; described preparation of glucose from honey. Died Richmond, Jan. 19 or 21, 1781.

LEWIS, William Bevan, Brit. pathologist; b. Cardigan, Eng., May 21, 1847; s. William Thomas and Jane Mansel (Bevan) L.; studied at Cardigan and Guy's hosps. Practiced medicine, Burry Port, for 4 years; mem. staff W. Riding Asylum, Wakefield, Eng., for 35 years, became med. dir.; prof. mental disease U. Leeds (Eng.) for 25 years. Described giant cells in precentral convolution of brain, 1878; illustrated cells of Betz (giant pyramidal cells of cerebral cortex) discovered and described by Vladimir A. Betz, 1878. Died Oct. 14, 1929.

LEWIS, William Cudmore McCullagh, Brit. chemist; ed. Queen's Coll. (now Queen's U.), Belfast, Ireland; grad. with 1st class honors, B.A., Royal U. Ireland, 1905; M.A. with 1st Honors, U. student in Exptl. Scis., 1906. Research staff chemistry Liverpool (Eng.) U., Heidelberg, Germany, U. Coll., London; became lectr., demonstrator chemistry U. Coll., London, 1910; Brunner prof. phys. chemistry U. Liverpool (Eng.), 1913-47. Fellow Royal Soc., 1926. Author: System of Physical Chemistry, 3 vols.; also articles on phys. chemistry. Died Feb. 11, 1956.

LEWIS, William M., Am. zoologist; b. Faison, N.C., Nov. 26, 1921; s. Mack Colin and Blanche (McGowan) L.; B.A., N.C. State Coll., 1942; M.S., Ia. State Coll., 1948, Ph.D., 1949; m. Sue D. Sparks, Feb. 14, 1942; children—William M., Robert Jeffery, Susan Lane, Steven McGowan, Catherine Lynn. Research aid U. S. Dept. Agr., Raleigh, N.C., 1939-42; research fellow Ia. State Coll., 1945-49; faculty So. Ill. U., Carbondale, 1949—, prof. zoology, 1959—, also dir. fishery research lab. Mem. Am. Fisheries Soc., Am. Inst. Fisheries, Sigma Xi. Author: Maintaining Fishes for Experimental and Instructional Purposes, 1962. Research, numerous publs. on diseases of fishes, fish farming techniques, biology of fishes. Home: 1211 Hill St., Carbondale, Ill. 62901.*

LEWIS, W(inford) Lee, Am. chemist; b. Gridley, Cal., May 29, 1878; s. George Madison and Sarah Adeline (Hopper) L.; A.B., Stanford, 1902; A.M., U. Wash., 1904; Ph.D., U. Chgo., 1909; m. Myrtilla Mae Lewis, Sept, 1907; children—Mrs. Mirlam Lee Reiss, Mrs. Winifred Lee Harwood. Asst. and instr. chemistry U. Washington, 1902-04; prof. chemistry Morningside Coll., Sioux City, Ia., 1904-06; instr. chemistry Northwestern U., 1909, asst. prof., 1914, asso. prof., 1917, prof. and head of dept., 1919-24. Asst. chemist U. S. Dept. Agr., 1908-10; city chemist, Evanston, 1912-18; cons. practice; dir. dept. of sci. research Inst. Am. Meat Packers since 1924. Fellow A.A.A.S., mem. Am. Chem. Soc., Sigma Xi. Developed poison gas Lewisite during World War I (never used in combat). Died Jan. 20, 1943.

LEWISON, Edward Frederick, Am. surgeon; b. Chgo., Feb. 11, 1913; s. Maurice and Julia (Trockey) L.; B.S., U. Chgo., 1932; M.D., Johns Hopkins, 1936; m. Elisabeth Oppenheim, 1939 (dec. Jan. 1947); 1 son, John Edward; m. 2d Betty S. Fleischmann, Mar. 21, 1948; children—Edward M., Robert S., Richard J. Pvt. practice medicine, specializing in surgery, Balt., 1945—; surgeon, chief Breast Clinic Johns Hopkins Hosp., also asst. prof. surgery Sch. Medicine. Diplomate Am. Bd. Surgery. Fellow A.C.S., Royal Soc. Medicine; mem. Am. Cancer Soc. (certificate merit 1953, pres. Md. div.). Author: Breast Cancer and Its Diagnosis and Treatment, 1955; also numerous articles. Research benign, malignant breast disease. Home: 7501 Park Heights Av., Balt. 21208. Office: 550 N. Broadway, Balt. 21205.*

LEWONTIN, Richard Charles, Am. geneticist; b. N.Y.C., Mar. 29, 1929; s. Max and Lillian (Wilson) L.; A.B. magna cum laude, Harvard, 1951; M.A., Columbia, 1952, Ph.D., 1954; m. Mary Jane Christianson, Apr. 1, 1947; children—David, Stephen, James, Timothy. Faculty, N.C. State Coll., 1954-58; faculty U. Rochester, 1958-64, prof. biology, 1962-64; prof. zoology U. Chgo., 1964—. Mem. U. Seminar Assn. Human Evolution Columbia, 1959—. NSF Research grantee, 1958-59, 63—, Office Naval Research, 1958-59, USPHS, 1958-63, AEC, 1960—. NSF predoctoral fellow, 1952-54, postdoctoral, 1955; Fulbright fellow, 1961-62, NSF Sr. postdoctoral fellow, 1961-62. Fellow Am. Acad. Arts and Scis.; mem. Soc. Study Evolution, Biometrics Soc.; mem., Nat. Acad. Scis. Author: (with others) Quantitative Zoology, 1961. Asso. editor: Evolution, 1959-63; Der Zuchter, 1965—; editor (with W. K. Baker) The American Naturalist, 1965—. Publs. on devel. of theoretical aspects of natural selection, genetic variation in natural populations.*

LEWTON-BRAIN, Lawrence, agriculturist; b. 1879; s. J. Lewton-Brain; ed. Firth Coll., Sheffield, Eng.; Found. scholar, Hutchinson student St. John's Coll., Cambridge, Eng.; B.A.; m. Annie O'Meara, 1904; 3 sons. Became demonstrator botany Cambridge U., 1900; named mycologist, agrl. lectr. Imperial Dept. Agr. for W.I., 1902; became asst. dir. div. pathology and physiology Hawaiian Sugar Planters' Assn., Expt. Sta., 1905, named dir. div., 1907; dir. agr. Federated Malay States, 1910—, Fed. Malay States and Straits Settlements, 1919—. Fellow Linnean Soc. Research and publs. on botany and pathology of tropical cultivated plants, including diseases of sugar-cane, cacao, cotton, breeding and hybridization of sugar cane. Died June 24, 1922.

LEWY, Robert Barnard, Am. physician; b. Chgo., Oct. 4, 1909; s. Alfred and Minnie (Barnard) L.; B.S., M.D., U. Chgo.; m. Evelyn Bluestone, Jan. 25, 1941; children—Alfred, Margery. Practice medicine specializing in otolaryngology, Chgo.; faculty U. Ill. Coll. Medicine, Chgo., 1937—, clin. asso. prof. otolaryngology. Cons. otolaryngologist Chgo. Municipal Contagious Disease Hosp., 1945—; lectr. numerous univs., U.S.A., Can., Europe. Mem. Am. Laryngol. Assn. Research, publs. on reestablishment of glottic function, comparative studies in pain control. Home: 7547 S. Merrill Av., Chgo. 60649.*

LEXELL, Anders Jean, mathematician, astronomer; b. Abo, Finland, Dec. 24, 1740; prof. math., St. Petersburg, Russia; mem. French Acad. Scis., 1776, acads. of St. Petersburg, Stockholm. Research in spherical geometry and trigonometry; Lexells' theorem states geometric locus of vertices of spherical triangles of same base and height is an arc of a small circle passing through points diametrically opposed to the extremities of the common base; observed 1st comet of short revolution period (which was named after him), 1770. Died St. Petersburg, Nov. 30, 1784.

LEY, Willy, paleontologist, engineer, space expert; b. Berlin, Germany, Oct. 2, 1906; s. Julius Otto and Frida (May) L.; student U. Berlin, 1927; Doctor of Humane Letters, Adelphi Coll., 1959; m. Olga Feldman, Dec. 24, 1941; children—Sandra, Xenia. Came to U. S., 1935, naturalized, 1944. Free lance writer since 1927; sci. editor newspaper PM, N.Y. City, 1940-44; research engr. Washington Inst. Tech., College Park, Md., 1944-47; information specialist office tech. services, Dept. Commerce, Washington, since 1947. Fellow Brit. Interplanetary Soc.; mem. German Rocket Soc. (founding mem. 1927, v.p. 1928-33), Inst. Aeronautical Sci., Am., Pacific rocket socs., Royal Astron. Soc. Can., Soc. Am. Mil. Engrs., A.A.A.S. Author: Die Moglichkei der Weltraumfahrt, 1928; Konrad Gesner, Leban und Werk, 1930; Grundriss einer Geschichte der Rakete, 1931; The Lungfish and the Unicorn, 1940; Bombs and Bombing, 1941; Shells and Shooting, 1942; The Days of Creation, 1942; Rockets, 1944; Rockets and Space Travel, 1947, rev. 1957, 61; Conquest of Space, 1948; Dragons in Amber, 1950; Lands Beyond (with L. S. de Camp), 1952; Engineer's Dreams, 1954; Salamanders and Other Wonders, 1955; The Exploration of Mars (with Wernher von Braun), 1956; Willy Ley's Exotic Zoology, 1959; Watchers of the Skies, 1963; Beyond the Solar System, 1964; articles mil. publs. and mags. Theorist on conditions in space and on other planets. Address: 37-26 77th St., Jackson Heights 72, N.Y.*

LEYBOURN, William, Brit. mathematician; b. 1626; said to have begun career as printer; tchr. math., profl. land surveyor, London; Author: (with Vincent Wing) Urania Practica (1st English book on astronomy), 1648; Planometria, or the Whole Art of Surveying the Land, 1650; The Compleat Surveyor, 1653; Arithmetic, Vulgar, Decimal, and Instrumental, 1659-78; The Line of Proportions or Numbers, 1667; Platform Guide Mate for Purchasers, Builders, Measurers, 17 1667; Cursus Mathematicus, 1690; Panarithmologia (1st English ready reckoner), 1793; Mathematical Institutions, 1704. Editor works of Gunter. Died circa 1700.

LEYBURN, James Graham, Am. sociologist, educator; b. Hedgesville, W.Va., Jan. 17, 1902; s. Edward Ridley and Nancy Granville (Harlan) L.; A.B., Duke, 1920, A.M., 1921, LL.D., 1962; A.M., Princeton, 1922, Ph.D., 1927. Instr. econs. and sociology Hollins Coll., 1922-24; instr. econs. and social instns. Princeton, 1924-25; faculty sociology Yale, 1927-47; prof. sociology Washington and Lee U., Lexington, Va., 1947—, dean U., 1947-56. Prin. mission officer Lend Lease Adminstrn., South Africa, 1943-44. Recipient Honneur et merite Republique d'Haiti, 1949; Anisfield-Wolf award, 1942. Mem. Am. Sociol. Assn., Am. Anthrop. Assn., Phi Beta Kappa, Sigma Xi. Author: Handbook of Ethnography, 1931; Frontier Folkways, 1935; The Haitian People, 1941; World Minority Problems, 1947; The Scotch-Irish: A Social History, 1962. Research, publs. on social history of Haiti, and Scotch-Irish in Scotland, Ulster, and Am. colonies. Home: 30 University Pl., Lexington, Va. 24450.*

LEYDEN, Ernst Viktor von, German physician; b. Danzig, Germany, Apr. 20, 1832; studied under Schönlein, also Traube; became successor to Frerich, 1885; prof. medicine U. Berlin; active in founding of sanitarium for TB; mem. French Acad. Scis. Author monographs on tabes dorsalis, poliomyelitis, respiration in fever. Described colorless, needle-like crystals of phosphate in sputum of bronchial asthma patients (Charcot-Leyden-Zenker or Charcot-Leyden crystals 1869); early description of myotomia congenita, 1876; 1st description of fatty infiltration of heart, 1882. Died Charlottenburg, Germany, Oct. 5, 1910.

LEYDIG, Franz, see von Leydig, Franz.

LEYONMARK, Gustave Adolf, Swedish mathematician; b. Saetesgaarden, Sweden, 1734; mem. Stockholm Acad. Scis. Author: Treatise on Positive, Negative, and Imaginary Roots of 3d and 4th Degree Equations; Method for Finding Factors Squared and Cubed (Square and Cubic) in 5th Degree Equations; On the Reduction of Exponential Equations to Algebraic Ones; On the Exponential Xx=A. Died Stockholm, 1815.

LEYPUNSKY, Aleksandr Ilich, Russian physicist; b. 1903; grad. Leningrad Poly. Inst., 1926; D. Physico-Math. Sci. With Leningrad Physicotech. Inst., 1924-30; with Physicotech. Inst., Ukrainian Acad. Sci., 1930-41, dir. Physics Inst., 1941-46; prof. Moscow Engrng. and Physics Inst., 1946—. Decorated Order of Lenin; recipient Lenin prize, 1960. Mem. Ukrainian Acad. Sci. Research and publs. on phenomena of energy transmission by excited atoms and molecules with free electron, 1926-27, artificial splitting of lithium and boron nuclei by protons, 1932, interactions of slow neutrons and photoneutrons with atomic nuclei, physics of fast-neutron nuclear reactors. Address: Moscow Engrng. and Physics Inst. . . . , ulitsa Kirova 21, Moscow, USSR.

LEYTON, Albert Sidney Frankau, Brit. pathologist; b. London, Jan. 5, 1869; ed. Gonville and Caius Coll., Cambridge; M.A., M.D., Sc.D., D.P.H.; m. Helen Gertrude Steward, 1909; 2 sons. Lectr. exptl. medicine U. Liverpool (Eng.); dir. Liverpool Cancer Research; prof. pathology U. Leeds (Eng.); dir. clin. lab. Addenbrookes Hosp., Cambridge; asst. physician Hosp. for Consumption. Fellow Royal Coll. Physicians (Goulstonian lectr.); mem. Physiol. Soc., Path. Soc. Research and publs. on nervous system, cancer, enteric fever, morbid histology. Died Sept. 21, 1921.

LEYTON, Albert Sidney Frankau, see Grünbaum, Albert Sidney Frankau.

L'HÉRITIER DE BRUTELLE, Charles-Louis, French botanist; b. Paris, June 15, 1746. Procurer for king; asst. to master of forest and rivers in generality of Paris; admitted to Ct. of Aids, 1775; twice apptd. judge civil tribunal of Paris after Revolution; mem. French Acad. Scis., 1790. Author: Stirpes novae aut minu cognitae (detailed classification of exotic plants); Sertum anglicum (description of plants in Royal Garden, Kew, Eng.), 1788; Flora of Peru (a compilation left in manuscript from notes and herbae of Dombey). Died Paris, Aug. 15 or 16, 1800.

LHOMME, Jean, French economist, sociologist; b. Bordeaux, France, Apr. 7, 1901; s. Jules and Jeanne (Boyer) L.; Lic. ès lettres, U. Toulouse (France), 1919, Lic. en droit, 1920; Doct. en Droit, U. Paris, 1925, Agrég. des Fac. de Droit, 1930; m. Simone Rougier, Dec. 17, 1935; children—Bernard, Francoise, Jacqueline. Prof. secondary schs., 1920-25; prof. econs. U. Algiers (Algeria), 1925-27, U. Lille (France), 1928-40; prof. U. Paris, 1940—; dir. studies l'Ecole pratique des Hautes Etudes, Sorbonne, Paris, 1948—. Decorated officier Legion d'Honneur. Author: Anciens et nouveaux impôts directs, 1925; Letaux de l'intéret et l'évolution économique, 1937; Le Problème des Classes, 1938; L'Impôt sur le revenu en Angleterre, 1939; Utilisation, gaspillage, prodigalité, 1946; La Politique sociale de l'Angleterre contemporaine, 1953; La grande Bourgeoisie au pouvoir, 1960; Pouvoir et Société économique, 1966. Dir. directorial com. Revue Economique. Research, publs. in financial matters, econ. and sociol. matters. Home: 14 Rue de Logelbach, Paris 17°, France.*

LHOTKA, John Francis, Am. histochemist; b. Butte, Mont., May 13, 1921; s. John Francis and Mary (Backowske) L.; B.A., U. Mont., 1942; M.S. (Stain fellow) Northwestern U., 1948, M.B., 1949, M.D., 1951, Ph.D., in Anatomy, 1953. Asst. prof. anatomy U. Okla. Sch. Medicine, Oklahoma City, 1951-55, asso. prof., 1955—. Recipient Zerbe award Am. Numis. Assn., 1960, medal merit, 1958. Fellow Histochem. Soc., Chem. Soc., Soc. for Exptl. Biology, Geriatric Soc., Am. Soc. Zoologists, Internat. Acad. Pathology, Am., Royal Numis. socs., Archeol. Inst. Am.; mem. Biol. Stain Commn., Am. Assn. Anatomists, Sigma Xi (sr.). Author: Introduction to East Roman Coinage, 1959; Empirical Guide to Medieval German Bracteates, 1958; Survey of Medieval Iberian Coinage (with P. K. Anderson), 1963; Medieval Feudal French Coinage, 1966; also numerous sci. and numis. articles. Research on localization tissue polysaccharides in degenerative processes, argyrophilic reactions, tissue heavy metal localizations, histochem. polysaccharide methods, histochem. research on human embryo. Office: 800 N.E. 13th St., Oklahoma City 73104.*

LHULLIER, Simon-Antoine-Jean, Swiss mathematician; b. Geneva, Switzerland, Apr. 24, 1750; Lived in Warsaw, Poland, Tübingen, Germany; became prof. math., Geneva, 1795. Recipient prize on nature of infinity Acad. Berlin, 1786. Fellow Royal Soc., 1791. Author several works including: Analogie entre les triangles, rectangles, rectilignes et spheriques; Gergonne's Annales Math., I, 1810-11; Exposition élément des principes des calculs supérieures, 1796. Studied measure and constrn. of polygons, polyhedrons, analogy between rectilinear and spherical right-sided triangles; developed theorem analogous to Pythagorean. Died Geneva, Mar. 28, 1840.

LHUYD, Edward, botanist; b. Carmarthenshire, Wales, 1660; s. Edward and Bridget (Pryse) L.; entered Jesus Coll., Oxford, 1682, M.A. honoris causas, Oxford, 1701. Keeper, Ashmolean Mus., Oxford, 1690-1709. Fellow Royal Soc., 1708. Author: Lythophylacii Britannici iconographia, 8 vols., 1699; Archaeologia Britannica, vol. I, 1707; Adversaria de fluviorum, montium, urbium in Britanniae nominbus, 1719; also numerous articles. First publs. on fossil plants in Eng. Pioneered in paleontology, including description of 1,600 Brit. fossil plants and animals; collected natural history specimens on 5-year trip to Wales; research in Celtic. Died Oxford, June 30, 1709.

LI, C. C., geneticist; b. Tientsin, China, Oct. 27, 1912; s. Jui and Liu L.; B.S., U. Nanking, 1936; Ph.D., Cornell U., 1940; m. Clara Lem, Sept. 20, 1941; children—Carol S., Steven M. Came to U. S., 1937, naturalized, 1956. Prof., U. Nanking, 1943-46; prof., chmn. dept. agronomy Nat. Peking U., 1946-50; prof. biometry U. Pitts., 1951—. Mem. Am. Statis. Assn., Am. Genetics Assn., Am. Soc. Human Genetics, Am. Pub. Health Assn., Biometrics Soc. Author: Population Genetics, 1955; Human Genetics, 1961; Introduction to Experimental Statistics, 1964; also numerous articles. Asso. editor Am. Naturalist, Am. Jour. Human Genetics, Biometrische Zeitschrift. Research in methodology in math. and statis. genetics with respect to human populations. Home: 24 Wilkins Rd., Pitts. 15221.*

LI, Chen Pien, physician; b. Hunan, China, Oct. 16, 1898; s. Koa F. and Chen (Tze) L.; M.D., Hunan-Yale Coll. Medicine, 1925; m. Han Chih Tang, May 1, 1929; children—Lotta (Mrs. Michael Chi), Raymond. Came to U. S., 1949, naturalized, 1955. Research worker Rockefeller Inst. for Med. Research, N.Y.C., 1929-31; faculty Peking (China) Union Med. Coll., 1931-32, Shanghai (China) Med. Coll., 1932-34, Chinese Army Med. Coll., 1937-47; med. bacteriologist in charge polio research Communicable Disease Center, Ala., 1951-55; chief, sect. virus biology div. biologics standards NIH, Bethesda, Md., 1955—. Fellow N.Y. Acad. Scis. Contbr. articles to profl. lit. Devised method for cultivation of vaccina virus; isolated an avirulent strain of type 1 poliovirus now used in Sabin poliovaccine; discovered antiviral and antibacterial paolins from abalone, oyster, calf thymus, clam. Home: 5023 Alta Vista Rd. Office: NIH, Bethesda, Md. 20014.*

LI, Choh Hao, chemist, endocrinologist; b. Canton, China, Apr. 21, 1913; s. Kan Chi and Miu-ching (Tsui) L.; B.S., U. Nanking, 1933; Ph.D., U. Cal. at Berkeley, 1938; M.D. (hon.), Cath. U. Chile, 1962; m. Annie Lu, Oct. 1, 1938; children—Wei-i, Anni-si, Eva. Came to U. S., 1935, naturalized, 1955. Instr. chemistry U. Nanking, 1933-35; faculty U. Cal. at Berkeley, 1938—, also prof. biochemistry, prof. exptl. endocrinology, dir. Hormone Research lab., U. Cal. at San Francisco, 1950—; vis. prof. various univs.; vis. scientist Children Cancer Research Found., 1955, 63—. Recipient Lasker award in basic med. sci., 1962. Guggenheim fellow, 1948. Mem. Am. Chem. Soc. (Cal. sect. award 1951), Am. Acad. Arts and Sci. (Emory Septennial prize 1955), Harvey Soc., Academia Sinica Rep. of China, Biol. Soc. Chile, Argentina Soc. Endocrinology and Metabolism, Am. Soc. Biol. Chemistry, Endocrine Soc. (Ciba award 1947), Biochem. Soc., London. Research, publs. on chemistry and biology of pituitary hormones, protein and polypeptide chemistry; accomplished isolation and identification of five hormones of anterior pituitary gland; synthesis of part of adrenocorticotropin (ACTH) molecule; description of complete structure of human growth hormone. Home: 901 Arlington Av., Berkeley, Cal. 94707. Office: Hormone Research Lab., U. Cal., San Francisco 94122.*

LI, Choh-lu, neurosurgeon, neurophysiologist; b. Canton, China, Sept. 19, 1919; s. Kiang-Chi and Mei-Ching (Isu) L.; M.D., Nat. Med. Coll., Shanghai, China, 1942; M.Sc., McGill U., Montreal, Que., Can., 1949, Ph.D., 1954; m. Isabel Dzung, Aug. 12, 1949; children—Claire-Ming, David Yuan, Ann-Ling. Sr. investigator neurophysiology McGill U., 1951-54; asso. neurosurgeon NIH, Bethesda, Md., 1954—. Mem. Internat. Brain Research Orgn., Am. Acad. Neurology, Am. Soc. Physiology, Research Soc. Am. Neurol. Surgeons, Am. Soc. Electromyography and Electrodiagnosis, A.A.A.S., Am. EEG Soc., Am. Soc. for Exptl. Biology, Chinese Med. Soc. U. S. A., Washington Acad. Neurosurgery. Editorial bd. Jour EEG and Neurophysiology, Life Scis.; sci. counselor Internat. Jour. Neuropharmacology, 1961—. Research, publs. on 1st use microelectrodes in study elec. activity single nerve cells in cerebrum, brain wave activity and discharge relationships, synaptic potentials, nerve-muscle activity relation in Parkinsonism. Home: 8421 Magruder Mill Ct., Bethesda, 20034. Office: Bldg. 10, NIH, Bethesda, Md. 20014.*

LI, Hsien-Wen, Chinese geneticist, plant breeder; b. Szechuan, China, Oct. 10, 1902; s. Chung Kuan and Nang Li; B.S.A., Purdue U., 1926; Ph.D., Cornell U., 1930; m. Elaine Cheng, Aug. 1, 1931; children—En-Che, Che-Yu, Che-chu, Hui-Tse. Asst. prof. Nat. Central U., Nanking, China, 1929-30; prof. Northeastern U., Mukden, Manchuria, 1930-31; prof. Honan U., 1931-35; prof., head dept. agronomy Nat. Wuhan U., Hupei, 1935-37; head dept. agr. Provincial Agr. Improvement Sta., Szechuan, Chengtu, China, 1937-41; dir. Sta. for Improvement Rice and Wheat, Szechuan, 1941-45; research fellow Inst. Botany, Academia Sinica, 1945—, Shanghai, 1945-48, Tainan, Taiwan, 1948-61, Nankang, Taipei, 1961—, dir., 1955—. Named Academician, Acade mia Sinica, 1948. Research, numerous publs. on cytogenetics of Italian millet, wheat, sugar cane and rice; improvement crop varieties in wheat, rice, Irish potatoes, sugar cane; prodn. erectoides in rice by use of atomic energy and mutagenic chems. Home: Inst. Botany, Academia Sinica, Nankang, Taipei, Republic of China.*

LI, Norman Chung, chemist; b. Foochow, China, Jan. 13, 1913; s. Pei Ting and Chiu (Lin) L.; B.S., Kenyon Coll., 1933; M.S., U. Mich., 1934; Ph.D., U. Wis., 1936; m. Hazel Su-chen Chou, Mar. 31, 1937; children—Peter, Paul, John, Mary, Catherine. Came to U. S., 1945, naturalized, 1955. Lectr., asso. prof., prof. Chinese univs., 1936-46; faculty St. Louis U., 1946-52, asso. prof., 1948-52; prof. chemistry Duquesne U., Pitts., 1952—. Tech. assistance expert Internat. Atomic Energy Agy.; cons. Argonne Nat. Lab., 1956-58; vis. prof. Tsing Hua U., Taiwan, 1964. Mem. Am. Chem. Soc., Soc. Applied Spectroscopy, A.A.A.S., Sigma Xi, Sigma Pi Sigma. Publs., research on nature of metal binding and hydrogen bonding in biol. systems, use of radioisotopes and nuclear magnetic resonance in coordination chemistry. Home: 5563 Beacon St., Pitts. 15217.*

LI, Shu-Hua, physicist; b. Chan-li, Hopei, China, Sept. 23, 1890; s. Wan-Kuei and Chang Li; Licencié ès Sciences Physiques, U. Paris (France), 1920, D.Sc., 1922; m. Wen-Tien Wang, Dec. 20, 1943; children —Chi-Chen (Mrs. Yin Long), Hsiao-Jun, Yu-Chen (Mrs. Tien-Hui Lin). Prof. physics Nat. U. Peking (China), 1922-28; acting pres. Nat. U. Peiping (China), 1928-29; v.p. Nat. Acad. Peiping, 1928; mem. Academia Sinica, 1948—; minister edn. China, 1931. Chinese del. Gen. Conf. UNESCO, 1945-47, 49, chief del., 1952; pres. Commn. Mixte des Oeuvres Franco-Chinoises, 1935-52; mem. council Palace Mus. Peiping, 1933-49. Named officier de la Légion d'Honneur (France), 1926, commandeur, 1949. Mem. Am. Phys. Soc., Am. Chem. Soc., N.Y. Acad. Scis. Author: Origin of Paper-Making, 1955; The South-Pointing Carriage and the Mariner's Compass, 1959; Origin of Printing in China, 1962; also numerous articles. Research on electric osmosis, phys. properties large molecules, history sci. Home: 501 W. 123d St., N.Y.C. 10027.*

LIACOPOULOS, Panayotis, biologist; b. Carditsa, Greece, Nov. 30, 1919; s. Constantin and Cleopatre (Contidis) L.; grad. Med. Sch., U. Athens (Greece), 1942; M.D., U. Paris, 1961; m. Monique Briot, Mar. 9, 1957; children—Christophe, Hélène. Fgn. asst. Broussais Hosp., Paris, 1951-53; research scientist Centre National de la Recherche Scientifique, 1954-61, maitre de recherche, 1961—. Mem. French Soc. Allergy, Brit. Soc. for Immunology, French Soc. Microbiology, French Soc. Immunology. Author: (with Pasteur Vallery-Radot, R. Wolfromm, J. Charpin, B. N. Halpern), Traité des Maladies Allergiques, 1963; also numerous articles. Research on pharmacological problems in allergy and mechanism of anaphylaxis and its inhibition; discovery of new method of inducing transplantation tolerance through non-specific inhibition of various immunological reactions. Home: 57 rue du Cherche-Midi, Paris (7°). Office: 96 rue Didot, Paris (14°) France.*

LIAPUNOV, Aleksandr Mikhailovich, Russian mathematician, mech. engr.; b. Yaroslavl, Russia, June 6, 1857; grad. U. St. Petersburg, 1880. Lectr., U. Kharkov, from 1885; worked in St. Petersburg, from 1902. Mem. Chamber of Sci. Studied, developed theory and created method of solving problems in areas of stability, equilibrium and motion in mech. systems, also research on hydrodynamics. Died Odessa, Mar. 11, 1918.

LIBAU, Andreas, see Libavius.

LIBAVIUS, Andreas (or Libau), German physician, chemist; b. Halle, Germany, circa 1540; M.D., Jena, Germany; became tchr., Ilmenau, Germany, 1581, Coburg, 1586; prof. history and poetry, Jena, 1586-91; town physician, Rothenburg, Germany, 1591-1607; became insp. schs. and gymnasium, 1592; 1st dir. Gymnasium Casimirianum, Coburg. Author of many works on chemistry and medicine the best known of which is Alchymia, 1597 and 1606, which described chemical methods and preparation of chemical substances; this work used as basis of Jean Beguin's Tyrocinium Chymicum, 1610, and thus may be ranked as font of 17th century chemical French text book tradition. Discussed analysis of mineral waters, the most detailed 16th century description of aqueous analysis; no less convinced of significance of chemistry for medicine than Paracelsus, Libavius nevertheless attacked the mysticism of the latter and emphasized practical nature of subject; similarly attacked Rosicrucians for mystical works, 1615-16; believed in possibility of transmutation; described dry reactions in assaying metallic ores, reactions and analytical tests for mineral waters and other substances; discovered stannic chloride (spiritus fumans Libavii), ammonium sulphate, glass of antimony; derived sulphuric acid by burning sulphur in moist air under glass bell, claiming to have obtained purer product than Baptista Porta; observed blue produced by copper in ammonia. Died Coburg, Germany, July 25, 1616.

LIBBY, Raymond Loring, Am. biophysicist; b. Portland, Me., Aug. 26, 1903; s. Samuel Butterfield and Ann (Frausing) L.; B.S., Rutgers U., 1925, M.S., 1937, Ph.D., 1938; m. Bessye Drury, Oct. 14, 1933; 1 dau., Barbara Jeanne (Mrs. Thomas C. Raup). Asst. dir. biol. research Lederle Labs., 1937-48; chief radiation biology div. atomic energy project U. Cal. at Los Angeles, 1948-50, prof. radiology, chief div. radiation biology Med. Center, U. Cal. at Los Angeles, 1955—. Cons. to pvt. cos., govt. agys.; adviser Govt. of Greece, 1961-62. Cited by Greek AEC, 1962. Fellow N.Y. Acad. Sci., A.A.A.S.; mem. Am. Assn. Immunologists, Soc. Nuclear Medicine, Soc. for Exptl. Biology and Medicine, Sigma Xi. Contbg. author: Nuclear Medicine, 1965. Research, publs. in immunochemistry, instrument for measuring antigen-antibody reactions; developed scintillation counting and scanning instrumentation for nuclear medicine; designed and developed extracorporeal irradiator for exptl. use in organ transplants and leukemia. Home: 20358 Big Rock Dr., Malibu, Cal. 90265. Office: 10833 LeConte Av., Los Angeles 90024.*

LIBBY, W(illard) F(rank), Am. chemist; b. Grand Valley, Colo., Dec. 17, 1908; s. Ora Edward and Eva May (Rivers) L.; B.S., U. Cal., 1931, Ph.D., 1933; Sc.D., Wesleyan U., 1955, Syracuse U., Trinity Coll. U. Dublin, Carnegie Inst. Tech., Georgetown U., Newcastle Upon Tyne, others; m. Leonor Hickey, Aug. 9, 1940 (div. 1966); m. to Leona Woods Marshall, 1966. children—Janet Eva, Susan Charlotte. Mem. faculty, U. Cal. at Berkeley, 1933-45; chemist Columbia U. War Research Div., 1941-45; prof. chemistry Inst. Nuclear Studies and dept. chemistry U. Chgo., 1945-54; prof. chemistry, dir. Inst. Geophysics and Planetary Physics U. Cal. at Los Angeles, 1959—; spl. vis. prof. University of Colorado, 1967. Mem. Gen. Adv. Com., 1950-54, AEC, 1960-62; mem. AEC, 1954-59, 60-62. Recipient Chandler medal, Columbia, 1954; Am. Chem. Soc. award, 1956; Remsen Meml. Lectr. award, 1955; N.Y.C. Coll. lectr. award, 1956; Elliott Cresson medal, Franklin Inst., 1957; Willard Gibbs medal, 1958; Albert Einstein award, 1959, Nobel Prize for Chemistry, 1960; Day medal Geol. Soc. Am., 1961; others. Guggenheim fellow, 1941, 51, 59-62. Fellow Am. Nuclear Soc.; mem. Am. Philos. Soc., Am. Acad. Arts and Scis., Nat. Acad. Scis., Heidelberg Acad. Sci., A.A.A.S., Am. Phys. Soc., Am. Chem. Soc., Royal Swedish Acad. Sci., Bolivian Soc. Anthropology, Am. Inst. Aeros. and Astronautics, Am. Geophys. Union, Geochem. Soc., So. Cal. Acad. Sci., Phi Beta Kappa, Pi Mu Epsilon, Alpha Chi Sigma, Sigma Xi, others. Author: Radiocarbon Dating, 1955. Mem. editorial bd. Science, 1962—. Research and publs. in chemistry of space program, radiochemistry particularly hot atom chemistry, tracer techniques and isotope tracer work; known for devel. of "atomic time-clock," a method for determining geol. age by measuring amount of radioactive carbon-14 in organic or carbon-containing objects; work on natural tritium and for its use in hydrology and geophysics. Home: 11901 Sunset Blvd. Los Angeles 90049.*

LIBERSON, Wladimir Theodore, physician; b. Kiev, Russia, Aug. 2, 1904; s. Jacob and Clara (Pessis) L.; student U. Moscow (Russia), U. Leningrad (Russia), 1921-24; M.D., Faculty Medicine, U. Paris (France), 1936; Ph.D. in Physiology U. Montreal (Que. Can.), 1950; m. Cathryn Raftis, Aug. 31, 1964; children—Annette, Helen (Mrs. Robert Parker). Research fellow Centre Nat. de la Recherche Scientifique, Henry Rousselle Hosp., Paris, 1932-36; instr., chief exptl. research Lab. Physiology Work, École des Hautes Études, U. Paris, 1934-40; research asst. Mt. Sinai Hosp., N.Y.C., 1941; staff Neuropsychiat. Inst., Hartford (Conn.) Retreat, 1941-48, dir. physiol. research lab., 1948-54; dir. lab. clin. neurophysiology Hartford Hosp., 1948-54; dir. research on electroencephalography and shock therapy VA Hosp., Northampton, Mass., 1953-55, asst. dir. profl. services for research, 1956-59; chief phys. medicine and rehab. State Vets. Hosp., Rocky Hill, Conn., 1956-59; chief phys. medicine and rehab. VA Hosp., Hines, Ill., 1959—; prof., chief sect. on phys. medicine and rehab. Loyola U. Stritch Sch. Medicine, Hines, 1959—. Cons., Little Co. of Mary Hosp., Evergreen Park, Ill., MacNeal Meml. Hosp., Berwyn, Ill., 1966—; research asso. Columbia, 1942-48; prof. École Libredes Hautes Études, New Sch. for Social Research, N.Y.C., 1944. Recipient Prix Bouchard, Société de Biologie de Paris, 1938; Gold medal award 3d Internat. Congress Phys. Medicine and Rehab., 1960. Mem. Am. EEG Assn., Am. Acad. Phys. Medicine and Rehab., Am. Congress Phys. Medicine and Rehab., Am. Soc. Electromyograohy and Electrodiagnosis, Am. Coll. Neuropsychopharmacology, A.M.A., Internat. Brain Research Orgn., Soc. Biol. Psychology. Author: Metabolisme et Obesite, 1936; Recherches Biometriques les EEG's Individuels and Les Frequence Cardiaque et l'Indice Systolique dans les Etats Affectifs, 1941; also numerous articles. Translator, editor: Neurological Mechanisms of Higher Vertebrate Behavior (J. S. Beritoff), 1965. Research on brief stimulus therapy, electrophysiol. brace for hemiplegics and motorized splinting brace. Address: P.O. Box 28, Hines, Ill. 60141.*

LIBMAN, Emanuel, Am. physician; b. N.Y.C., 1872; s. Fajbush and Hulda (Spivak) L.; A.B., Coll. City of N.Y., 1891; M.D., Coll. Phys. and Surg. (Columbia), 1894; post-grad. work, Berlin, Vienna, Munich, 1896-97, Berlin, 1903, 09, Johns Hopkins, 1906. Practiced Medicine, N.Y., since 1894; House physician Mt. Sinai Hosp., 1894-96, asso. pathologist, 1898-1923, attending phys., 1912-25, then cons. physician; cons. physician, Montefiore Hosp., Harlem Hosp., Beth Israel Hosp., Hosp. for Joint Diseases, French Hosp. (all N.Y.), Methodist-Episcopal Hosp., United Israel-Zion Hosp., Beth-El Hosp. (all Bklyn.). Fellow N.Y. Acad. Medicine; mem. Am. Assn. Pathologists and Bacterologists, A.A.A.S., Assn. Am. Physicians, Am. Soc. Advancement Clin. Investigation, A.C.P., Internat. Assn. Geog. Pathology, Internat. Med. Museums Assn., Pathol. Soc. N.Y., Harvey Soc., Am. Assn. Immunologists, Phi Beta Kappa; corr. mem. Société Française de Cardiologie. Contbr. papers appertaining to clin. medicine, pathology and bacteriology. Presented 1st proven case of aneurysm by impact, 1888; gave 1st description of growth bacteria on media containing sugar and serum, 1901; pointed out a peculiar action of streptococcus on blood on blood plates, 1905; demonstrated (with Herbert Louis Celler) that Streptococcus endocarditis is most common organism in subacute bacterial endocarditis, 1906; made (with R. Ottenberg) 1st clin. study of blood transfusions, 1915; studied coronary artery thrombosis, 1917-28; described (with B. Sacks) a form of valvular and mural endocarditis (Libman-Sacks disease), 1924; demonstrated analogies between sprue and pernicious anemia; made observations on sensitiveness of pain. Died June 28, 1946.

LIBRI-CARRUCCI DELLA SOMMAIA (Count Guillaume Brutus Icilius Timoléon), mathematician, bibliophile; b. Florence, Italy, Jan. 2, 1803; prof. U. Pisa (Italy); went to Paris, 1830, naturalized, 1833; became prof. analysis Faculty Scis. Paris, 1833; insp.-gen. pub. instrn., insp. libraries of France; after being accused of stealing books from French libraries, fled to London; Mem. French Acad. Scis., 1832. Author: Sur la théorie des nombres, 1820; Mémoires de mathématiques et de physique, 1827; L'historie des sciences mathématiques en Italie depuis la Renaissance jusqu'à la fin du XVII siècle, 1838-41; Souvenirs de la jeunesse de Napoléon, 1842; Lettres sur le clergé et la liberté de l'enseignement, 1844. Died Fiesole, Italy, Sept. 28, 1869.

LICETO, Fortunio (Liceti, Licetus of Rapallo, Fortunius), Italian physician; b. Rapallo, nr. Genoa, Italy, 1577; became prof. philosophy, Padua, Italy, 1609, prof. medicine, 1645, Author numerous publs. including: Treatise on the Nature of Monsters, 1616; De novis astris et cometis libri sex. Developed Aristotelian doctrine of activity of senses dominated by power of spirit. Died 1657.

LICHARDUS, Branislav, Czechoslovak physician; b. Palúdzka, Czechoslovakia, Dec. 1, 1930; s. Branislav and Zuzanna (Dankova) L.; MUDr. cum laude, Commenius U., Bratislava, Czechoslovakia, 1956; specialist internal medicine Ministry of Health, 1960; C.Sc. in Physiology, Czechoslovak Acad. Scis., 1963; m. Eva Kellerová, Sept. 5, 1953; children—Jana, Iva. Asst. dept. gynecology and obstetrics, dept. medicine State Hosp., Bratislava, Czechoslovakia, 1956-57; sci. worker, head lab. electrolytes Inst. Endocrinology, Czechoslovak Acad. Scis., 1958—; predoctoral student Inst. for Cardiovascular Research, Prague, Czechoslovakia, 1961-62; postdoctoral student dept. physiology U. Alta., Edmonton, Can., 1964-65. Recipient prize Czechoslovak Nephrological Soc., 1966. Mem. Czechoslovak Med. Soc. J. E. Purkyne, Internat. Nephrological Soc. Research, publs. on regulation of body fluids metabolism, hormonal regulation of sodium excretion, body fluids vol., devel. of CNS regulation of body fluids. Home: 3/A Suvorovova ul. Office: 1/A Ul.obrancov mieru, Bratislava, Czechoslovakia.*

LICHNEROWICZ, Andrew Léon, French mathematician; b. Bourbon, France, Jan. 21, 1915; s. Jean Joseph and Antoinette (Gressin) L.; student Ecole Normale Supérieure, 1933-36; agrégé in math., 1936; Docteur és Sciences, Paris U., 1939; m. Suzanne Magdelain, Nov. 26, 1942; children—Marc, Jacques, Jérome. Prof., U. Strasbourg (France), 1941-49, Sorbonne, Paris, 1949-52; prof. Collège de France, Paris, 1952—; vis. prof. Princeton, U. Ill.; U. Cal. at Berkeley. Decorated officer Légion d'Honneur, commandeur Ordre du Mérite; recipient Fubini prize, 1955. Mem. French Acad. Scis. (past pres.), Accademia dei Lincei (fgn.), Am. math. socs. Author: Problèmes globaux en mecanique relativiste, 1939; Algèbre linéaire, 1947; Théorie relativiste del a gravitation, 1954; Théme des Carnexias, 1955; Théorie globale des groupes d'holonomie, 1955; Géometrie des groupes de transformation, 1958; also numerous articles. Research on math. foundations of theory of gen. relativity, shock waves in relativistic hydrodynamics and magneto-hydrodynamics, differential geometry, notion of holonomy, curvature, and topol. invariants, correlations between geometry and topology. Home: 6 Avenue Paul Appell, Paris, France.*

LICHSTEIN, Jacob, Am. med. scientist, physician; b. Phila., July 4, 1908; s. Solomon and Clara (Barmach) L.; student U. Pa., 1925-28; M.D., Jefferson Med. Coll., 1932; m. Lilian Cohen, Dec. 24, 1939; children—Henry Alan, Gabrielle. Faculty, U. So. Cal. Sch. Medicine, 1946-63, asst. prof. clin. medicine, 1952-64; med. dir. Gay-Henry Gastrointestinal Research Found., Los Angeles; asst. prof. clin. medicine U. Cal. Coll. Medicine, Los Angeles, 1964-67, University of California at Los Angeles, 1968—. practice medicine, Phila., 1936-41, specializing in gastroenterology, Los Angeles, 1946—. Supr., Gastrointestinal Clinic, Cedars of Lebanon Hosp., 1955-57. Diplomate Am. Bd. Internal Medicine. Fellow A.C.P.; mem. Am. Fedn. Clin. Research (sr.), Am. Soc. Gastrointestinal Endoscopy, Am. Coll. Gastroenterology (1st prize sci. exhibit, 1959), So. Cal. Soc. Gastroenterology (past pres.), N.Y. Acad. Scis., American Gastroenterological Association. Research, publs. on effect anticholinergic and tranquilizing drugs on peptic ulcer and gastric secretion, psychosomatic interrelationships functional gastrointestinal disorders, clin. studies on unstable bowel, pharmacological studies anabolic steroids and pepsin inhibitors, methods gastrointestinal motility recs. Home: 3870 Latrobe St., Los Angeles 90031. Office: 6423 Wilshire Blvd., Los Angeles 90048.*

LICHT, Sidney, Am. physician; b. N.Y.C., Apr. 18, 1907; s. Herman and Julia (Frank) L.; B.S., Coll. City N.Y., 1927; M.D., N.Y. U., 1931; m. Elizabeth Schweitzer, Mar. 16, 1937; children—Vera Hermine, Phyllis Roxane, Jeffrey Louis. Instr. phys. therapy N.Y. U., 1937-42; faculty Columbia, 1938-42; lectr. phys. medicine Boston U., also Tufts U., 1947-53; asst. prof. phys. medicine Yale, 1955-64, curator phys. medicine collections Yale Med. Library, 1960—. Cons. VA, 1957-64. Recipient Medal of City of Paris, 1964. Hon. mem. phys. medicine socs. Argentina, Britain, Can., Denmark, France, Italy. Author: Music in Medicine, 1946. Editor: Electrodiagnosis and Electromyography, 1956; Therapeutic Heat and Cold, 1957; Therapeutic Exercise, 1958; Therapeutic Electricity, 1959; Massage, Manipulation and Traction, 1960; Medical Hydrology, 1963; Medical Climatology, 1964. Co-editor: Occupational Therapy, 1949; Physical Medicine in General Practice, 1951. Editor, Occupational Therapy and Rehab., 1946-51. Promotion of knowledge in phys. medicine and rehab. Address: 360 Fountain St., New Haven 06515.*

LICHTENBERG, Don Bernett, Am. physicist; b. Passaic, N.J., July 2, 1928; s. Milton and Ida (Krulewitz) L.; B.A., N.Y. U., 1950; M.S., U. Ill., 1951, Ph.D., 1955; m. Rita Kalter, Jan. 10, 1954; children—Naomi, Rebecca. Research asso. Ind. U., Bloomington, 1955-57, asso. prof., 1963-66, prof., 1966—; vis. prof. U. Hamburg, Germany, 1957-58; asst. prof. Mich. State U., 1956-61, asso. prof., 1961-63; physicist Stanford (Cal.) Linear Accelerator Center, 1962-63. Mem. Am. Phys. Soc., Phi Beta Kappa, Sigma Xi. Author: Meson and Baryon Spectroscopy, 1965; (with K. Johnson, J. Schwinger, S. Weinberg) Particles and Field Theory, 1965; also numerous articles. Home: 715 S. Fess, Bloomington, Ind. 47401.*

LICHTENBERG, Georg Christoph, German physicist; b. Ober Ramstaedt, Germany, July 1, 1742; student Göttingen, Germany; prof. U. Göttingen. Wrote criticism of Lavater's sci. of physiognomy, Vass's system of Greek pronunciation, Sturm und Drang writers. Author: Ample Connmantary on the Engravings of Hogarth, 1794-99. Discovered elec. dust figures, named for him, which are produced on a nonconducting plate by a point source of electricity, 1777; research in exptl. physics, especially electricity. Died Göttingen, Feb. 24, 1799.

LICHTENBERG, see von Lichtenberg.

LICHTENSTEIN, Parker Earl, Am. psychologist; b. Somerville, Mass., June 25, 1915; s. Robert Lee and Estella (Hyde) L.; B.S., U. Mass., 1939, M.S., 1941; Ph.D., Ind. U., 1948; m. Marion Ruth Locke, June 17, 1943; children—Parker Ross, Karen Ruth, Barbara Lee. Research asst. OSRD, Harvard, 1942, 45; faculty Ind. U., 1946-47; asst. prof. Antioch Coll., 1948-49; asso. prof. Denison U., Granville, O., 1949-51, prof. psychology, 1951—, chmn. dept., 1951-54, dean Coll., 1954—, trustee, sec. Research Found., 1954—. Cons. Knox Seminars in Ednl. Mgmt., 1964-65. Mem. Am. Conf. Acad. Deans (sec., editor 1962-65, chmn. 1966-67), Am., Midwestern psychol. assns., Am. Assn. U. Profs., A.A.A.S., Phi Beta Kappa, Sigma Xi, Omicron Delta Kappa, Pi Sigma Alpha, Phi Sigma Kappa. Asso. editor Psychol. Record, 1959—. Research, publs. on conditioned emotional responses in dogs, effects of pre-frontal lobotomy on acquisition, retention and relearning. Home: Orchard Dr., Granville, O. 43023.*

LICHTENSTEIN, Pearl Rubenstein, Am. astronomer; b. Boston, June 13, 1917; d. Louis and Lisa (Cheskis) Rubenstein; S.B., Mass. Inst. Tech., 1938; M.A., Radcliffe, 1940, Ph.D., 1942; m. Roland M. Lichtenstein, June 21, 1946; children—Ann, Walter. Research asso. Mass. Inst. Tech. Radiation Lab., 1942-46, Franklin Inst. Lab., Phila., 1946-47, Gen. Engrng. Lab., Gen. Electric Co., Schenectady, N.Y., 1948; research asso. in astronomy Rensselaer Poly. Inst., Troy, N.Y., 1957—. Mem. Am. Astron. Soc., Am. Geophys. Union, Phi Beta Kappa, Sigma Xi. Contbg. author: Microwave Propagation, 1946. Research, publs. on effects weather on tropospheric propagation microwaves, effects particles from sun on ionosphere and propagation radio waves through disturbed ionosphere. Home: 1200 Van Antwerp Rd., Schenectady 12309. Office: Rensselaer Poly. Inst., Troy, N.Y. 12181.*

LICHTHEIM, Ludwig, physician; b. Breslau, Germany (now Wroclaw, Poland), Dec. 7, 1845; studied medicine at univs. Berlin, Zurich, Breslau; grad. Berlin, 1867. Prof., Bern, Switzerland, Königsberg, Germany; Publs. include: The Treatment of Pleuritic Exudation, 1872; Disturbances in Circulation of Lungs and their Influences on Blood Pressure, 1876; Periodic Hemoglobinuria, 1878; also articles. Pioneered in modern neurology; studied path. anatomy, disturbed circulation in lungs, hydremia, atelactasis, leukemia, periodic hemoglobinuria, typhus fever, atrophy of pancreas, cardiac diseases, spinal cord changes in pernicious anemia, brain tumors, brain cysts, paralysis of abnormal muscles, progressive muscular atrophy, diagnosis of meningitis; one of 1st to perform puncture in brain for diagnostic purposes; 1st description of subcortical sensory aphasia (Lichtheim's disease), subacute combined degeneration of spinal cord (Lichtheim's syndrome); studied path. mold, Aspergillus, 1882. Died Bern, Jan. 13, 1928.

LICHTIN, Norman Nahum, Am. chemist; b. Newark, Aug. 10, 1922; s. James Jechiel and Clara (Greenspan) L.; B.S., Antioch Coll., 1944; M.S., Purdue U., 1945; Ph.D., Harvard, 1948; m. Phyllis Selma Wasserman, May 30, 1947; children—Harold Hirsh, Sara Marjorie, Daniel Albert. Faculty, Boston U., 1947—, prof. chemistry, 1961—; vis. chemist Brookhaven Nat. Lab., Upton, L.I., N.Y., 1957-58, research collaborator, 1958—. Guest scientist Weizmann Inst. Sci., Rehovoth Israel, 1962-63; vis. prof. Hebrew U., Jerusalem, 1962-63. NSF fellow, 1962-63. Fellow A.A.A.S.; mem. Am. Chem. Soc., Radiation Research Soc., Am. Assn. U. Profs., Sigma Xi. Research, publs. on mechanisms of chem. reactions, including reaction of atomic nitrogen with organic compounds; influence of high energy (ionizing) radiation on organic compounds; ionization processes and ionic reactions in solutions in liquid sulfur dioxide. Home: 195 Morton St., Newton Centre, Mass. 02159. Office: 685 Commonwealth Av., Boston 02215.*

LICHTWITZ, Leopold, physician; b. Olawa, Germany, Dec. 9, 1876; med. degree, Leipzig, Germany, 1901; named prof. U. Göttingen, 1913; dir. City Hosp., Altona, Germany, 1917-31; chief med. dir. Rudolf Virchow Hosp., Berlin, 1931-33; chief med. div. Montefiore Hosp., N.Y., 1933-42. Mem. German Congress on Internal Medicine (pres. 1933). Author: Formation of Gall and Kidney Stones, 1914; Clinical Chemistry, 1918; Diseases of the Kidney, 1921; Diseases of Metabolism, 1926-27; Functional Pathology, 1941; Nephritis, 1942; Pathology and Therapy of Rheumatic Fever, 1943; also chpts. in handbooks on pathology, therapy, metabolic diseases, numerous articles. Research on physiology and pathology of kidney, glands of internal secretion. Died New Rochelle, N.Y., Mar. 18, 1943.

LIDDEL, Urner, Am. physicist; b. Butler, Mo., Sept. 3, 1905; s. Robert Lee and Lutitia (Urner) L.; A.B., Central Coll., Mo., 1926; postgrad. U. Chgo.; Ph.D., George Washington U., 1941; D.Sc., Central Meth. Coll., Mo., 1963. Physicist, Fixed Nitrogen Research Lab., Dept. Agr., 1929-36, Am. Cyanamid Co., Stamford (Conn.) Research Labs., 1936-41; head physics and nuclear physics brs. Office Naval Research, 1946-52; dir. product devel. Bendix Aviation Corp., 1952-55; physicist Lab. Phys. Biology, Nat. Inst. Arthritis and Metabolic Diseases, NIH, Bethesda, Md., 1955-56; dir. acad. year and in-service insts. NSF, 1956-57; physicist Lab. Phys. Biology, Nat. Inst. Arthritis and Metabolic Diseases, 1957-60; asst. dir. Hughes Research Labs., Malibu, Cal., 1961-62; asst. dir. for gen. research Advanced Research Projects Agy., Office Sec. Def., Washington, 1960-61; asst. dir., chief scis. lunar and planetary div. Office Space Sci. and Applications NASA, Washington, 1962—. Fellow I.R.E. (past chmn. profl. group on med. electronics), Am. Phys. Soc., Phys. Soc. (London), A.A.-A.S., Am. Nuclear Soc. Washington sect., past dir., Philos. Soc. Washington (v.p. pres. 1965), Washington Acad. Scis. (past mem. com. on admissions), Internat. Fedn. for Med. Electronics, Am. Inst. Aeros. and Astronautics (asso.); mem. Biophys. Soc., Optical Soc., Sigma Xi. Recipient Distinguished Alumnus award Central Coll., 1950. Research on measurement forces hydrogen bonding in organic molecules, devel. and use infrared spectrophotometry, devel. and adminstrn. research. Home: 301 G St. S.W., Washington 20024. Office: 400 Maryland Av. S.W., Washington 20546.*

LIDDELL, Duncan, Scottish mathematician, chemist, physician; b. 1561; studied lang. and philosophy, Aberdeen, Scotland, math. medicine, Frankfurt/Oder, Germany, math., Silesia, Germany; M.Ph., Rostock, Germany, 1589; M.D., Helmstadt, Germany, 1596. Prof. math., Helmstadt, 1591-1603, named dean faculty philosophy, 1599, prorector, 1604; 1st physician Ct. of Brunswick; returned to Scotland, 1607; endowed professorship math. in Marischal Coll., Aberdeen, 1613. Author: Disputationum medicinalium liber, 1605, also posthumous edit. under title Universae medicinae compendium, 1720; Ars medica, 5 vols., 1608; De febribus libri tres, 1610; Tractatus de dente Aureo, 1628; Artes conservandi sanitatem libri duo (edited by D. P. Dun), 1651. 1st in Germany to teach astronomy of Copernicus and Tycho Brahe with Ptolemaic hypothesis. Died Dec. 17, 1613.

LIDDIARD, Edwin Andrew Guthrie, Brit. metallurgist; b. Eng., Nov. 30, 1903; s. Albert Guthrie and Martha (Mott) L.; B.Met., Sheffield (Eng.) U., 1925; B.A., Gonville and Caius Coll., Cambridge (Eng.) U., 1928, M.A., 1932; m. Christine Banham, Apr. 4, 1932; 1 dau., Edwina Christine (Mrs. Albert Wallace Grace). Lab. asst. Cammell Laird & Co., Sheffield, 1921-25; research metallurgist synthetic ammonia and nitrates Imperial Chem. Industries, 1928-32; asst. devel. officer Brit. Non Ferrous Metals Research Assn., Euston, 1932-37; research mgr. BNFMRA, 1934-46; dir. research, chief exec. dir. Fulmer Research Inst., Stoke Poges, Bucks, Eng., 1946—. Fellow Inst. Metallurgists, Inst. Physics; mem. Inst. Metals (past chmn. London sect.). Contbg. author: The Metallurgy of the Rarer Metals, 1952; The effect of Environment on Embrittlement in Metals, 1961. Research, publs. on devel. new alloys including aluminium-30% tin alloy, aluminium-copper-cadmium; corrosion especially aluminium alloys. Home: Hockley Rise, Stoke Poges. Office: Fulmer Research Inst., Stoke Poges, Bucks, Eng.*

LIDDLE, Grant Winder, Am. physician; b. American Fork, Utah, June 27, 1921; s. Parley H. and Elizabeth (Winder) L.; B.S., U. Utah, 1943; M.D., U. Cal., 1948; m. Clara Lucille Everett, Nov. 20, 1942; children—Kathryn L. (Mrs. James Caleb Wallwork), Annette, Rodger Alan, Robert, Patricia. USPHS research fellow metabolic unit, also instr. medicine U. Cal. Sch. Medicine, 1951-53; sr. asst. surgeon, surgeon, sect. on clin. endocrinology Nat. Heart Inst., 1953-56; asso. prof. medicine, chief endocrine service Vanderbilt U. Sch. Medicine, Nashville, 1956-61, prof. medicine, chief endocrine service, 1961—. Mem. endocrinology study sect. USPHS, 1958-62, mem. diabetes and metabolism tng. grants com., 1962-66, chmn., 1963-66. Fellow A.C.P.; mem. Am. Soc. for Clin. Investigation (sec.-treas. 1963-66), Endocrine Soc. (postgrad. com. 1960-63, mem. council 1962-65, Upjohn award 1962), So. Soc. for Clin. Research (pres. 1965-66), A.M.A., Am. Clin. and Climatological Assn., Assn. Am. Physicians, Am. Fedn. Clin. Research, Phi Beta Kappa, Sigma Xi, Phi Eta Sigma, Alpha Omega Alpha, Phi Kappa Phi. Mem. editorial bd. Jour. Clin. Endocrinology and Metabolism, 1960—. Research and numerous publs. in fields metabolism and endocrinology, particularly regarding ACTH, steroids, aldosterone secretion. Home: 6524 Jocelyn Hollow Rd., Office: Vanderbilt Univ. Sch. of Medicine, Nashville 37203.

LIDE, David Reynolds, Jr., Am. physicist; b. Gainesville, Ga., May 25, 1928; s. David Reynolds and Kate (Simmons) L.; B.S., Carnegie Inst. Tech., 1949; A.M., Harvard, 1951, Ph.D., 1952; m. Mary Ruth Lomer, Nov. 5, 1955; children—David Alston, Vanessa Grace, James Hugh, Quentin Robert. Fulbright scholar, Ramsay Meml. fellow Oxford (Eng.) U., 1952-53; research fellow Harvard, 1953-54; physicist Nat. Bur. Standards, Washington, 1954—, chief infrared and microwave spectroscopy sect., 1963—. Lectr. physics U. Md., College Park, 1956—; mem. physicist panel U. S. Bd. Civil Service Examiners, 1958—. NSF Sr. Postdoctoral fellow U. Coll., London, Eng., 1959-60. Recipient U. S. Dept. Commerce Silver medal 1965. Fellow Am. Phys. Soc., Washington Acad. Scis.; mem. A.A.A.S. Research, numerous publs. on microwave and infrared spectroscopy, molecular structure, internal rotation in molecules, free radicals, high-temperature chemistry. Home: 4604 Tournay Rd., Washington 20016. Office: Nat. Bur. Standards, Washington 20234.*

LIDICKER, William Zander, Jr., Am. biologist; b. Chgo., Aug. 19, 1932; s. William Z. and Frida (Schroeter) L.; B.S., Cornell U., 1953; M.S., U. Ill., 1954, Ph.D., 1957; m. Naomi Ishino, Aug. 18, 1956. Faculty, U. Cal. at Berkeley, 1957—, asso. prof. zoology, 1965—, asst. curator mammals, 1957-65, asso. curator, 1965—. Mem. Ecol. Soc. Am., Brit. Ecol. Soc., Japanese Soc. Population Ecology, Am. Soc. Mammalogists, Australian Mammal Soc., Soc. for Study Evolution, Soc. Systematic Zoology, Am. Soc. Zoologists, Am. Ornithologists Union, Am. Inst. Biol. Scis., A.A.A.S., Sigma Xi, Phi Sigma. Research, publs. on population ecology mammals, taxonomy Mexican and Australian mammals, hibernation physiology bats, mammalian genetics. Home: 108 Willow Lane, Berkeley, Cal. 94707.*

LIDZ, Theodore, Am. psychiatrist; b. N.Y.C., Apr. 1, 1910; s. Israel and Esther (Shedlin) L.; A.B., Columbia, 1931, M.D., 1936; M.A. (hon.), Yale, 1951; m. Ruth Wilmanns, Nov. 23, 1939; children—Victor Meyer, Charles Wilmanns, Jerome Shedlin. Faculty, Johns Hopkins, 1941-51, asso. prof. psychiatry, asst. prof. medicine, 1947-51; prof. psychiatry Yale, 1951—, chmn. dept., 1967—. psychiatrist-in-chief Grace-New Haven Community Hosp., 1951-61; Nat. Inst. Mental Health career investigator, 1961—. Recipient 1st Frieda Fromm-Reichmann award Acad. Psychoanalysis, 1961. Fellow Center for Advanced Studies in Behavioral Scis., 1965-66. Mem. Am. Psychiat. Assn. (past chmn. com. on med. edn. and Hofheimer award com.), Md. Psychiat. Soc. (past pres.), Am. Psychosomatic Soc. (past pres.), Am., Western New Eng. psychoanalytic socs. Author: The Family and Human Adaptation, 1963; Schizophrenia and the Family (with S. Fleck, A. Cornelion), 1966; through the Life Cycle, 1968; also articles. Initiated and investigated theory that etiology of schizophrenia is related to disturbed family environments; studies of brain damaged patients, toxic psychoses, psychosomatic disorders, combat neuroses. Home: Orchard Rd., Woodbridge, Conn. 06525. Office: 34 Park St., New Haven.

LIE, Kian Joe, parasitologist; b. Sukabumi, Indonesia, Nov. 25, 1916; s. Siong Pin and How (Thio) L.; Ph.D. cum laude, U. Indonesia, 1941, M.D., 1942; Dr. Tropical Medicine and Hygiene, U. London, 1950; m. Loan Eng Injo, May 20, 1948; children—Kok Tiong, Kok An. Lectr., Inst. Pathology, Sch. Medicine, Djakarta, Indonesia, 1942-47; sr. research asst. Inst. Tropical Hygiene and Parasitology, Leyden, Holland, 1947-49; prof. parasitology, gen. pathology Sch. Medicine, Djakarta, 1950-60; vis. prof. Hooper Found. U. Cal. Med. Center based Kuala Lumpur, Malaysia, 1960-64, research parasitologist Hooper Found. U. Cal. Med. Center, San Francisco, 1964—. Mem. expert adv. panel on parasitic diseases WHO. Recipient Jang Seng Ie medal, 1940; Eykman medal, 1949. Mem. Am. Soc. Tropical Medicine, Am. Soc. Parasitologists. Research, numerous publs. on various aspects of soil transmitted helminths of man, etiology and epidemiology of diarrhea among infants, human mycotic infections, trematode interaction within snail host.*

LIE, Marius Sophus, Norwegian mathematician; b. Nordfjordeif nr. Bergen, Norway, Dec. 17, 1842; Ph.D., Christiania (now Oslo) U., 1868; Began as asst. tutor Christiania U., named prof. math., 1872; became prof. geometry, Leipzig, Germany, 1886; returned to Christiania, 1898. Recipient Lobatchewsky prize. Fellow Royal Soc., 1895; mem. French Acad. Scis., 1892, London Math. Soc., Cambridge Philos. Soc. Publs. include: On a Geometric Transformation, 1870; Classification and Integration of Ordinary Differential Equations in X and Y with the Aid of a Group of P Transformations, 1884; On the Theory of Contact Groups, 1888; (with Engel) Theories of Groups of Transformation, 2 vols., 1888-90; Theorie der Transformationsgruppen, 1893. Research on minimal surfaces, non-Euclidean geometry, differential equations; developed theory of transformation groups; discovered (with Klein) Klein-Lie curve. Died Oslo, Feb. 18, 1899.

LIEB, Charles Christian, Am. pharmacologist; b. N.Y.C., Apr. 19, 1880; s. Charles Adam and Magdalena (Stephan) L.; student Columbia Inst., 1893-98; A.B., Columbia Coll., 1902, M.D., Coll. Phys. and Surg., 1906; post-grad. London U., 1908-09, Utrecht U., 1929; m. Henrietta Haaker, June 25, 1908. Instr., Coll. Phys. and Surg., Columbia, 1909-10, asst. prof. pharmacology, 1910-21, asso. prof., 1921-23, prof., 1923-29, Hosack prof. 1929-44, emeritus, 1944—. Mem. Physiol. Soc. (London), Am. Physiol. Soc., Soc. of Pharmacology and Exptl. Therapeutics, Soc. Exptl. Biology and Medicine, A.M.A., Phi Beta Kappa, Sigma Xi. Died Apr. 1956.

LIEB, Hans, Austrian biochemist; b. Weiz, July 20, 1887; s. Johann and Josefa (Schwarzl) L.; Ph.D.; m. Sophie Baltl, Aug. 12, 1918; children—Irmgard, Johannes, Herbert. Became asso. prof. biochemistry U. Graz (Austria), 1924, prof., dir. biochem. inst., 1931-62. Recipient Austrian award of honor for sci. and art; Wilhelm-Exner medal Austrian Profl. Assn. Mem. Austrian Acad. Sci., Vienna. Author: Mikrochemie (organisch); Physiologische Chemie. Address: Bergmanngasse 28, Graz, Austria.

LIEB, John William, Am. mech. engr., b. Newark, Feb. 12, 1860; s. John William and Christina (Zens) L.; E.D., M.E., Stevens Inst. Tech., 1880; m. Minnie F. Engler, July 29, 1886. Employed as draftsman, 1880-81; put in charge (by Edison) of installation of elec. equipment of old Pearl St., Edison Sta., and assisted in subsequent tests and expts. on this 1st elec. sta. in U. S., supplying current for incandescent lighting and power from an underground system, and on inauguration of regular service, Sept. 4, 1882, was apptd. 1st electrician Edison Electric Illuminating Co. of N.Y.; next installed mech. equipment, dynamos, and Edison Underground System at Milan, Italy, 1883, for Italian Edison Co.; later installed trolley system in Milan; returned to Edison Elec. Illuminating Co., 1894; v.p. and gen. mgr. N.Y. Edison Co.; also pres. Elec. Testing Labs. Died Nov. 1, 1929.

LIEBALDT, Gerhard Paul, German physician; b. Nuremberg, Germany, Feb. 16, 1917; s. Paul and Clara (Müller) L.; student Stuttgart (Germany) Inst. Tech., U. Erlangen (Germany); Dr.med., U. Tübingen (Germany), 1948; m. Gerda Sinnecker, Mar. 11, 1958; children—Claudia Irene Gerda, Ursula Marianne. Chief asst. Inst. Brain Research, U. Tübingen, 1954-64; dir. neuro-path. and clin.-chem. lab., neurol. clinic U. Wurzburg (Germany), 1965—. Mem. German socs. Pathology, Human Genetics and Anthropology, Assn. German Neuropathologists and Anatomists, Leopoldina, Med. Natural Sci. Assn. Tübingen, Physiko-Medica Wurzburg. Research, numerous publs. on influence of psychopharmaceutics on blood-brain-barrier, neuropath. effects of these drugs, cybernetics of schizoform psychoses, cerebral convulutions. Address: 8 Friedrich Spee-Strasse, 87 Wurzburg, West Germany.*

LIÉBEAULT, Ambroise Auguste, French physician; b. France, 1823; began study hypnotism, 1860; started practice of medicine Nancy (France) Hosp., 1864. Author: Du sommeil et des états analogues, 1886. Began study and practice of hypnotism; introduced psychotherapy which he used instead of hypnotic suggestion, 1889. Died 1904.

LIÉBECQ, Claude, Belgian biochemist; b. Liege, Belgium, Sept. 28, 1921; s. Georges Léon and Marguerite (Bastin) L.; M.D., U. Liège, 1947; m. Suzanne Hutter, Apr. 30, 1919; 1 son, Georges André. Brit. Council scholar, 1947-49; Belgian Am. Ednl. Found. Grad. fellow, 1949-50, Hon. Advanced fellow, 1964; prof. biochemistry U. Liège, 1950—, prof. gen. physiology, 1959—; vis. prof. physiology Vanderbilt U. Med. Sch., Nashville, 1964. Mem. Belgian Nat. Com. Biochemistry, 1953—. Decorated Officier de l'Ordre de Léopold. Mem. Fedn. European Biochem. Socs. (editor-in-chief European Jour. Biochemistry 1966—), Société chimique de Belgique (pres. 1965-67), Société Belge de Biochimie Belgische Vereniging voor Biochemie (v.p. 1966—), biochem. socs. of France, Gt. Britain, Germany, Physiol. Soc. Belgium and France, A.A.A.S. Research and numerous publs. on structure of green breakdown products of hemoglobin, biochemistry of fluoroacetate poisoning (utilized in radioprotection), carbohydrate metabolism of muscle tissue, properties of magnesium salts of adenine nucleotides, isolation and structure of adenosine tetraphosphate, influence of sulfhydryl radioprotectors on cellular respiration, redox potential of blood and hydroxylations in liver. Home: 87 Avenue de Péville, Grivegnée (Liège), Belgium.*

LIEBEN, Fritz, Austrian biochemist; b. Vienna, Feb. 25, 1890; s. Adolf and Mathilde Baronne (Schey) L.; Ph.D., U. Vienna (Austria); m. Gerda Seutter von Loetzen, Sept. 1921; children—Verena, Joachim, Wolfgang. Asst., later instr. U. Vienna, became prof., 1935, suspended for polit. reasons, 1938; lectr. U. Wash., St. Louis, Johns Hopkins, Mt. Sinai Hosp., 1941-53. Mem. Biochem. Soc., Soc. Austrian Chemists. Author: Geschichte der physiologischen Chemie; 1935; Vorstellungen vom Aufbau der Materie, 1953; many other works. Home: Pyrkerg 4 B/8, Vienna 19. Office: Währingerstrasse 10, Vienna 9, Austria.

LIEBER, Charles Saul, physician; b. Antwerp, Belgium, Feb. 13, 1931; s. Isaac and Lea (May) L.; M.D., Brussels U., 1955; m. Adele Tornhajm, Dec. 21, 1954; children—Colette Dianne, Daniel Charles. Came to U. S., 1958, naturalized, 1966. Research fellow Belgian Council for Sci. Research, 1956-58; research fellow in medicine Harvard and Thorndike Meml. Lab. of Boston City Hosp., 1958-60, instr., research asso., 1961-62, asso. in medicine, 1962-63; asso. prof. medicine, dir. Liver Disease and Nutrition unit Cornell U. and Cornell Med. Div., Bellevue Hosp., N.Y.C., 1963—; asso. attending physician N.Y. Hosp., N.Y.C.; attending physician Manhattan Vets. Hosp.; asst. vis. physician Meml. and James Ewing hosps., N.Y.C. Laureate of Belgian Govt. for Research, 1956. Mem. Am. Soc. Clin. Investigation, Am. Gastroent. Assn., Am. Inst. Nutrition, Am. Assn. Studies of Liver Diseases, Am. Soc. Clin. Nutrition, Am. Fedn. Clin. Research, A.A.A.S., Am. Oil Chem. Soc. Research, numerous publs. in role of ammonia produced from urea in prodn. of hepatic coma and gastric hypoacidity; demonstrated that alcohol, ind. of nutritional deficiencies, is cause of fatty liver; discovered alcoholic hyperuricemia and its mechanism, 1st known link between alcohol and gouty attacks. Home: 6 Johnson Av., Englewood Cliffs, N.J. 07632.*

LIEBER, Oscar Montgomery, Am. geologist; b. Boston, Sept. 8, 1830; s. Francis and Matilda (Oppenheimer) L.; ed. Goettingen, Germany. Participated in street fighting in Berlin, 1848, Dresden, Germany, 1849; returned to U. S., and settled in S.C.; Joined Confederate Army in Civil War; state geologist, Miss., 1850-51; geol. survey, Ala., 1854-55; dir. geol., mineral., and agr. survey State of S.C., 1856-60; geologist astron. expdn., Labrador, 1860. Author: The Assayer's Guide, 1862; The Analytical Chemist's Assistant; also articles, reports. Translator: Beispiele zur Vebung in der Analytischen Chemie (Wöhler), 1852. Grouped formations according to their lithological nature; studied itacolumite which he considered a true sandstone; opposed idea that minerals in veins are crevices filled by leeching action of water sedimentary rocks were the source of gold into which permeating rocks on either side; believed certain it was infused. Died Richmond, Va., June 27, 1862.

LIEBER, Paul, Am. mechanician, geophysicist; b. Balt., Oct. 11, 1918; s. David and Rose (Rodblatt) L.; student Johns Hopkins, 1937-39; M.S., Cal. Inst. Tech., 1941, Ph.D., 1951; m. Angela Bettina Vegara, Mar. 8, 1946; children—Michael, Leonardo, Joseph, Victoria, Jonathan. Supr. flutter and dynamics group Douglas Aircraft Co., Santa Monica, Cal., 1940-46; asst. prof. Poly. Inst. Bklyn., 1946-51; faculty Rensselaer Poly Inst., Troy, N.Y., 1951-57, prof. geophysics, 1953-57; Fulbright prof. Israel Inst. Tech., Haifa, 1957-58; prof. engring. sci. U. Cal. at Berkeley, 1957—. Cons. to govt. agys, pvt. cos. Recipient medallion Prime Minister Israel for services rendered as UNESCO expert, 1959-60. Mem. Am. Geophys. Union, A.A.A.S., Sigma Xi, Tau Beta Pi. Research, numerous publs. in aeroelasticity, inelastic constn. earth; invented acceleration vibration and impact damper; formulated and applied variational prins. for viscous flows; research in classical mechanics, universal constants physics. Home: 43 Dos Osos Rd., Orinda, Cal. 94563. Office: 6125 Etchevvery Hall, U. Cal., Berkeley, Cal. 94720.*

LIEBERKÜHN, Johann Nathaniel, German anatomist; b. Berlin, Sept. 5, 1711; ed. Halle, Jena, Leiden, Paris, London; practiced medicine, Berlin; fellow Royal Soc., 1741. Author: Dissertatio de fabrica et actione villorum intestinorum tenuium, 1745. Perfected Leeuwenhoek's solar or reflector, microscope, 1738; described tubular depressions in mucosa of small intestine (Lieberkühn's, or Galeati's glands), 1745; described Lieberkühn's ampulla (now believed non-existent); demonstrated lung cavity. Died Berlin, Oct. 7, 1756.

LIEBERMAN, Irving, Am. cell biologist; b. Bklyn., Oct. 25, 1921; s. Joseph and Florence (Levinson) L.; B.A., Bklyn. Coll., 1944; D.V.M., Middlesex Vet. Coll., 1945; M.S., U. Ky., 1948; Ph.D., U. Cal. at Berkeley, 1952; m. Evelyn May Becker, Oct. 9, 1947; children—James Irving, Mary Jill. Faculty, Miami U., Oxford, O., 1948-49, Washington U. Sch. Medicine, St. Louis, 1953-56; postdoctoral fellow Nat. Found. Infantile Paralysis, Bethesda, Md., 1952-53; faculty U. Pitts., 1956—, prof. microbiology, 1962-66, prof. anatomy, cell biology, 1966—. Mem. Fedn. Am. Socs. Exptl. Biology, Am. Soc. Cell Biology. Contbr. articles to profl. lit. Research on control mechanisms for DNA synthesis. Home: 6117 Reynolds St., Pitts. 15206.*

LIEBERMAN, Seymour, Am. biochemist; b. N.Y.C., Dec. 1, 1916; s. Samuel D. and Sadie (Levin) L.; B.A., Bklyn. Coll., 1936; M.S., U. Ill., 1937; Ph.D., Stanford, 1941; m. Sandra Spar, June 5, 1944; 1 son, Paul. Chemist, Schering Corp., 1938-39; Rockefeller scholar, 1939-41; spl. research asst. Harvard, 1941-45; asso. Sloan Kettering Inst., N.Y.C., 1945-50; faculty Columbia, N.Y.C., 1950—, prof. biochemistry, 1962—. Mem. endocrinology study sect. NIH, 1959-65. Mem. Am. Cancer Soc. (mem. panel on etiology 1956-60), Am., Swiss chem. socs., A.A.A.S., Am. Soc. Biol. Chemists, Am. Assn. for Cancer Research, Harvey Soc., Soc. for Exptl. Biology and Medicine, Endocrine Soc., Sigma Xi. Contbr. numerous articles to tech. jours. Asso. editor Jour. Clin. Endocrinology and Metabolism, 1963—. Research on biochemistry and metabolism steroid hormones. Home: 32-22 163d St., Flushing, N.Y. 11358. Office: 630 W. 168th St., N.Y.C. 10032.*

LIEBERMANN, Karl Theodore (or Carl), German chemist; b. Berlin, Feb. 23, 1842; pupil of Bunsen and Baeyer; became prof. Technische Hochschule, Charlottenburg, Germany, 1873, Kaiser Wilhelm Inst., Berlin, 1913. Recipient Perkin medal, 1906. Synthesized (with Graebe) alizarin, 1869, thus making possible devel. of German dye industry; discovered reaction between nitrous acid and phenols, also secondary amines named after him, 1874; showed (with P. Seidler) that chrysarobin is methylalhydroxyanthranol, 1878; research (with F. Giesel) on alkaloids of coca leaves led to improved tech. prodn. of cocaine, also to discovery of isomeric cinnamic acids, 1890-92; studied quercitin, rhamnetin, derivatives of naphthalene, anthracene. Died Berlin, Dec. 28, 1914.

LIEBERSON, Stanley, Am. sociologist; b. Montreal, Que., Can., Apr. 20, 1933 (parents Am. citizens); s. Jack and Ida (Cohen) L.; M.A., U. Chgo., 1958, Ph.D., 1960; m. Patricia Ellen Beard, 1960; children—Rebecca, David, Miriam. Faculty, U. Ia., 1959-61; asso. dir. Ia. Urban Community Research Center, 1959-61; faculty sociology U. Wis., Madison, 1961-67, prof., 1966-67; prof. sociology U. of Wash., Seattle, 1967—. Mem. com. sociolinguistics Social Sci. Research Council, 1965—; chmn. race, ethnic and related topics subcommittee Com. on Population Statistics, 1965—. Recipient Colver-Rosenberger Ednl. prize, 1960. Mem. Am., Midwest sociol. socs., Am. Statis. Assn., Population Assn. Am., Soc. For Study Social Problems. Author: (with others) Metropolis and Region, 1960; Ethnic Patterns in American Cities, 1963. Research, publs. on ecol., demographic analysis of various facets of race and ethnic relations such as segregation and assimilation, riots, Australia, linguistic pluralism, occupational composition; measures of linguistic diversity; banking aspects of met. dominance. Home: 3247 N.E. 104th St., Seattle 98125.*

LIEBIG, Justus von, German chemist; b. Darmstadt, Germany, May 12, 1803; s. Georg and Maria Karoline (Moserin) L.; Ed. U. Bonn (Germany), 1819; Ph.D., U. Erlangen (Germany), 1822; student chemistry under Gay-Lussac, Sorbonne, Paris; Became prof. chemistry U. Giessen (Germany), 1824, U. Munich, 1852. Mem. French Acad. Scis., 1842. Author: Handbuch der organischen Chemie, 1839; Die organischen Chemie in ihrer Anwendung auf Physiologie und Pathologie, 1842; Chemische Briefe,

1844; Die Grundsätze der Agrikulturchemie, 1855. Founder Annalen der Chemie. Introduced 1st student's chem. lab. in continental Europe, Giessen; pioneered research in physiol. chemistry; studies in prodn. of fat, fulminates, agrl. chemistry, chem. fertilizers, chloral, cyan compounds, acetaldehyde; perfected methods for quantitative organic analysis, especially by combustion; produced chloroform using chlorine and caustic alkali with alcohol; deduced existence of benzoyl radical; established (with Wöhler) theory of organic radicals; determined structure of hipparic acid; developed polybasic acids; divided foods into fats, carbohydrates, and proteins; stated fats and carbohydrates acted as fuel for body; studied (with Wöhler) uric acid; discovered method for silvering mirrors. Died Munich, Germany, Apr. 18, 1873.

LIEBLEIN, Seymour, Am. aerospace engr.; b. N.Y.C., June 17, 1923; s. David and Rose (Brandstein) L.; B.Mech. Engring., Coll. City N.Y., 1944; M.Aero. Engring., Case Inst. Tech., 1952. Research engr. NACA, Cleve., 1944-52, div. cons. 1952-58; with NASA, Cleve., 1958—, chief flow analysis br., 1960-66, chief V/STOL engine br., 1966—. Recipient NACA Exceptional Service medal, 1957; Aero-Space award Air Force Assn., 1967. Fellow Am. Inst. Aeros. and Astronautics associate fellow, recipient Goddard award 1967); mem. Am. Soc. M.E. (Gas Turbine Div. Ann. award 1961), Soc. Automotive Engrs. Research, publs. on axial-flow compressors for advanced jet engines; developed improved design methods and criteria, basic concept and design techniques for transonic compresser; research, publs. on basic theory, design concepts and meteroid protection for large radiators for use in dissipating waste-heat from power generating systems in space. Home: 20123 Lorain Rd., Fairview Park, O. 44126. Office: 21000 Brookpark Rd., Cleve. 44135.*

LIEBOWITZ, Benjamin, Am. physicist; b. N.Y.C., Feb. 20, 1890; s. Simon and Fannie (Unterberg) L.; E.E., Columbia, 1911, Ph.D., 1915; m. Virginia Liebowitz, Feb. 12, 1914; children—Naomi (Mrs. Ramsay Wood), Elsbeth. With Thomas A. Edison, 1917, Emil J. Simons, 1918, Publix Shirt Corp., 1922-32; pres. Trubenizing Process Corp., N.Y.C., 1953—. Research, publs. on much closer connection between classical and quantum mechanics than was formerly suspected, a new force in classical electro-magnetic theory; inventor fused collar and non-removable collar stay (for shirt industry). Home: Box 266, R.D. 1, New Canaan, Conn. 06840. Office: 171 Madison Av., N.Y.C. 10016.*

LIEBREICH, Matthias Eugène Oscar, German physician; b. Königsberg, Germany, 1839; became prof. pharmacology, Berlin, 1872. Introduced chloral as hypnotic, 1869; studied anesthetic properties of chloral hydrate, chloral butylic, ethylene chloride, mercury preparations. Died 1908.

LIEBREICH, Richard, ophthalmologist; b. Königsberg, Germany, June 30, 1830; ed. U. Königsberg, U. Berlin, U. Halle (Germany); M.D., 1853; postgrad. under Donders, Utrecht, Netherlands, under Brücke, Berlin. Asst. ophthalmol. inst. Berlin U., 1854-62; pvt. practice, Paris; became lectr., clinician St. Thomas Hosp., London, 1870; reduced med. activities to concentrate on art research, 1895. Author: Ophthalmoscopische Atlas, 1863; (with Laqueur) Recueil des travaux del la Société médicale allemande de Paris, 1865; Eine neue Methode der Cataractextraction, 1072; On the Use and Abuse of Atropin, 1873; School Life in its Influence on Sight and Figure, 1877; Clinical Lecture on Convergent Squint, 1874. Built large and portable ophthalmoscopes; studied path. changes in disc and vessels, diseases of optic nerve; 1st to use lateral illumination to examine eye, 1855. Died 1917.

LIEF, Harold Isaiah, Am. physician; b. N.Y.C., Dec. 29, 1917; s. Jacob F. and Mollie (Filler) L.; A.B., U. Mich., 1938; M.D., N.Y. U., 1942; certificate psychoanalysis Columbia, 1950; m. Myrtis Aline Brumfield, Mar. 3, 1961; children—Polly, Jonathan Felix, Caleb Brumfield, Frederick Victor. Asst. physician Columbia Coll. Physicians and Surgeons 1949-51, asso. psychoanalyst Columbia Psychoanalytic Clinic for Tng. and Research, 1950-53; faculty Tulane U., New Orleans, 1951—, prof., 1960—, dir. Hutchinson Meml. Psychiat. Clinic, 1951—; vis. physician Charity Hosp., New Orleans, 1951-60, sr. vis. physician, 1960—; vis. prof. U. Va., 1958. Commonwealth Fund fellow, 1963-64. Mem. Am. Acad. Psychoanalysis (trustee, chmn. exec. council, pres. elect.), Am. Psychosomatic Soc. (council, 1959-62), A.M.A., Am. Psychiat. Assn., A.A.A.S., Am. Sociol. Assn., Sigma Xi, Alpha Omega Alpha, others. Author: (with others) The Eighth Generation, 1960; (with others) The Psychological Basis of Medical Practice, 1963. Research, publs. on personality and culture of Am. Negroes, emotional devel. of med. students, psychosomatic illness, psychotherapy. Home: 1565 Exposition Blvd., New Orleans 70118. Office: 1430 Tulane Av., New Orleans 70112.*

LIENERT, Gustav, German psychologist; b. Michelsdorf, Dec. 13, 1920; s. Emil L.; M.D., Vienna. Asst., Inst. psychology U. Marburg; asso. prof. inst. psychology U. Hamburg; prof. inst. psychology Düsseldorf (Germany) Med. Acad. Mem. German Assn. Psychology, Internat. Coll. Neuropsychopharmacology, Psychol. Soc. Author: Testaufbau und Testanalyse; Verteilungsfreie Methoden in der Biostatistik. Research in psychopharmacology, psychol. statistics, differential and clin. psychology. Address: Himmelgeisterstrasse 127, Düsseldorf, W. Germany.

LIEPMANN, Hugo Carl, German neurologist; b. Berlin, Germany, 1863; Ph.D. in Philosophy, M.D., 1894; studied anatomy under Weigert, Wernicke; Asst. to psychiatrist Jolly; became psychiatrist Municipal Mental Instn. of Dalldorf, Berlin, 1899; became privatdozent U. Berlin, 1901; later dir. mental instn., Herzberge, Germany; collected 26 brains with case histories. Author: Ideenfluch (analysis of thinking process), 1904; also articles. Discovered there is a dominance of left cortical hemisphere in right-handed persons, 1904; studied word deafness, visual agnosia, echolalia; discovered isolated apraxia of left side of body as a symptom of involvement of corpus callosum; distinguished learned motor processes from ideation (processes bringing them into action); tried to relate phenomena to physiol. processes; while studying alcoholic delirium, discovered hallucinations can be induced artificially. Died 1925.

LIER, Frank George, Am. plant morphologist; b. N.Y.C., Feb. 19, 1913; s. Frank and Anna (Leitl) L.; A.B., Columbia, 1935, M.A., 1937, Ph.D., 1950; m. Ethel Margaret Moritz, Dec. 23, 1937. With Columbia, 1946—, prof., 1967—; reviewer bot. manuscripts Columbia U. Press. Mem. Bot. Soc. Am., Torrey Bot. Club (pres., 1964), Am. Bryological Soc., Am. Inst. Biol. Scis., N.Y. Acad. Scis., A.A.A.S., Sigma Xi. Contbr. numerous articles in field to profl. jours. Basic studies on cell wall devel., also cell division in cork and cork cambium tissues of plant stems. Home: 5282 Post Rd., Bronx, N.Y. 10471.*

LIERLE, Dean McAllister, Am. surgeon; b. Marshalltown, Ia., May 21, 1895; s. Fredrick Perkins and Laura (Ralston) L.; student Stanford, 1914-16; B.S., U. Ia., 1919, M.D., 1921, M.S., 1923; student Boston U., 1921-22, U. Pa., 1928, U. Vienna (Austria), 1929; D.Sc. (hon.), Wayne U.; m. Pauline Thompson, Feb. 24, 1923; children—Dean McAllister, Richard Burton, William Bayard. Practiced medicine, 1921; faculty U. Ia., 1921—, prof. otolaryngology and maxillofacial surgery, 1928—; now emeritus; cons. surgeon Gen. Diplomate Am. Bd. Otolaryngology (sec.-treas.). Fellow A.C.S. (v.p., regent); mem. Am. Hearing Soc., A.M.A., Am. Laryngol. Assn. (Newcomb award 1956, De Roaldes award 1960), Am. Acad. Ophthalmology and Otolaryngology (sec. home study course), Am. Coll. Allergists, Am. Assn. Univ. Profs., Audiology Found., Ia. Acad. Sci., Am. Laryngological, Rhinol. and Otolaryngol. Soc., Am. Otol. Soc., Soc. Research Child Devel., Am. Broncho-Esophagological Assn., Am. Bd. Plastic Surgery (founders group), Am. Coll. Chest Phys., Am. Speech Corr. Assn., A.A.A.S., Sigma Xi, Alpha Omega Alpha. Home: 5 Knollwood Lane, Iowa City.*

LIESE, Walter, German silviculturist; b. Berlin, Jan. 31, 1926; s. Johannes and Erika (Süvern) L.; ed. univs. Fribourg, Göttingen/Hanover, Munden; Ph.D.; m. Elsa Pabst, Mar. 14, 1952; children—Andreas, Stephan. Prof. wool biology; asst. expt. silviculture sta., Nordhein-Westphalia; sci. collaborator Usines Rütgers; asst., later instr. U. Fribourg; instr., prof. at large U. Munich; prof. U. Hamburg (Germany). Research and publs. on wood anatomy, wood pathology and preservation, forest botany, forestry and forest products in developing countries; expert on Indonesia and the Indies for FAO; specialist in electron microscopy. Home: Otternweg 11, 2055 Aumühle/Hamburg. Office: Schloss, 2057 Reinbek, W. Germany.

LIEUTAUD, Joseph, French physician, anatomist; b. Aix, France, June 21, 1703; doctorate, Aix; postgrad. Montpellier, France; prof. anatomy, Aix; joined Versailles (France) Royal Infirmary, 1750; named chief physician to King Louis XV, circa 1770, 1st physician, Louis XVI, 1774. Fellow Royal Soc., 1739; mem. French Acad. Scis., 1752. Author: Essais anatomiques contenant l'histoire exacte des parties qui composent l'homme, avec la manière de disséquer, 1742; Elementa physiologiae, 1745; Examen critique du traité de la structure du coeur, a Sénac, 1750; Synopsis universa praxeas medicae, 2 vols., 1765, also pub. as Précis de la médecine pratique, 1759, and Précis de la matière médicale, avec un traité des aliments et des boissons, 1766, Historia anatomica-medica sistens, numerosissima cadaverum humanorium exstispicia edente portal, 1767; Anatomie historique, 1776-77; also numerous articles, observations. Pioneered surg. anatomy; 1st description of trigone of urinary bladder (trigone of Lieutaud), 1742. Died Versailles, Dec. 6, 1780.

LIFE, Andrew Creamor, Am. botanist; b. Grant County, Ind., July 30, 1869; s. Christian and Ruth Ann (Elliott) L.; A.B., Ind. U., 1896, A.M., 1897; post grad. Washington U., St. Louis, 1904-06, U. Chgo., 1906-07, U. Cal., 1923-24, also several summers; m. Cora Mae Smith, June 20, 1917. Began as instr. botany Earlham Coll., 1904; instr. botany Washington U., 1904-06; asst. prof. botany U. So. Cal., 1907-10, asso. prof., 1910-15, prof., 1915—. Died Sept. 8, 1933.

LIFSHITS, Ilya Mikhailovich, Russian physicist; b. 1917; grad. Kharkov State U., 1936, Kharkov Mechanico-Mathine-Bldg. Inst., 1938, D.Sc., 1941; became sci. worker Ukrainian SSR Acad. Scis. Physico-Tech. Inst., Kharkov, 1937, sect. chief, 1941; named chmn. theoretical physics dept. Kharkov State U., 1944. Recipient Mandelshtam Meml. prize, 1952; Sincon Meml. prize Phys. Soc. London, 1961. Corr. mem. Ukrainian SSR Acad. Scis., USSR Acad. Scis. Author: On the Theory of X-rays Scattering by Crystals with Variable Structure, 1937; Optical Behavior on of Non-ideal Crystal Lattices in Infra-read, 1942; Short Elastic Waves Scattering in a Crystal Lattice, 1948; On a Problem of Perturbation Theory, Connected with Quantum Statistics, 1952; (with A. V. Pogorelov) On Determination of Dermi-surface and Velocities in Metals from Magnetic Susceptibility Oscillations, 1954; On the Theory of Magnetic Susceptibility in Metals at Low Temperatures, 1955; High-pressure anomalies of electron Properties of Metals, 1960; The Kinetics of Ordering at Phase Transition of Second Order, 1962; (with V. V. Slesov) On the Kinetics of Fidd Diffusional Decay of Superstaturated Solid Solutions, 1958, Dynamic Equilibrium of a Fog Cloud over a Liquid Surface, 1962. Research on dynamic theory of non-ideal crystals, electronic structure of metals, theory of solid state physics and low temperature physics. Address: Ukrainian SSR Acad. of Scis. Physico-Tech. Inst. of USSR Acad. of Scis., Yumovskii Tupik, 2, Kharkov 24, Ukrainian SSR.

LIGDA, Myron George Herbert, Am. research meteorologist; b. Oakland, Cal., Jan. 10, 1920; s. Paul and Edith (Griswold) L.; A.B., U. Cal. at Berkeley, 1942; S.M., Mass. Inst. Tech., 1948, Sc.D., 1953; m. Evelyn Elnora Dalke, Aug. 22, 1942; children—Richard Worthington, Carol Louise, Valorie Jean. With Mass. Inst. Tech., 1946-54; asso. prof. Tex. A. and M. Coll., College Station, 1954-58; sr. research meteorologist, lab. mgr. Stanford Research Inst., Menlo Park, Cal., 1958—. Cons. to govt. agys. Mem. Soc. Photog. Scientists and Engrs., Am. Meteorol. Soc., Am. Geophys. Union. Author: Astronautics for Science Teachers, 1965; also numerous articles, chpts. in books. Asso. editor Jour. Applied Meteorology, 1963—. Pioneered work on radar meteorology, used laser radar for atmospheric observations and probing; research on use artificial satellites for weather observation, study use learning machines for meteorol. analysis. Home: 23744 Arbor Av., Los Altos, Cal. 94022. Office: Stanford Research Inst., Menlo Park, Cal. 94025.*

LIGHTHILL, Michael James, physicist; b. Paris, France; Jan. 23, 1924; s. E. B. Lighthill; grad. Trinity Coll., Cambridge; D.Sc. (hon.), Liverpool, 1961, Leicester, 1965; m. Nancy Alice Dumaresq, 1945; 1 son, 4 daus. With aerodynamics div. Nat. Phys. Lab., 1943-45; fellow Trinity Coll., Cambridge, 1945-49; sr. lectr. math. U. Manchester (Eng.), 1946-50, Beyer prof. applied math., 1950-59; dir. Royal Aircraft Establishment, Farnborough, 1959-64; 1st pres. Inst. Math. and Its Applications, 1964—. Mem. Adv. Council on Tech., 1964, Natural Environment Research Council, 1965, Shipbldg. Inquiry Com., 1965. Fellow Royal Soc., 1953; mem. Am. Acad. Arts and Scis. (fgn.). Author: Higher Approximations in Aerodynamic Theory, 1954; Introduction to Fourier Analysis and Generalised Functions, 1958. Contbg. author: Surveys in Mechanics, 1956; Laminar Boundary Layers, 1964. Research in aerodynamics of aircraft theory of shock waves and theory of jet noise, on hydraulics and hydrodynamics, oceanography, applie math.; enlarged range of nonlinear problems for application to aerodynamics and other fields. Address: Imperial Coll. of Science and Tech., London S.W. 1, Eng.

LIGHTNER, Max William, Am. engr.; b. Loysville, Pa., Jan. 23, 1908; s. Clarence W. and Fannie (Heim) L.; B.S., Pa. State U., 1929; M.S., Carnegie Inst. Tech., 1930; m. Estelle Logan, May 23, 1931; children—Max William, Susanne (Mrs. Raymond Smith). Research engr. metall. adv. bd. Carnegie Inst. Tech., 1931-33, faculty, 1935-40; with U. S. Steel Corp., Pitts., 1933-42, 1944—, v.p. applied research, 1956-68, adminstrv. v.p. research and tech., 1968—; v.p. operations, dir. research Heppenstall Co., Pitts., 1942-44. Mem. Am. Soc. Metals (David Ford McFarland award 1950), Am. Inst. Mining, Metall. and Petroleum Engrs., Am. Iron and Steel Inst., Engrs. Soc. Western Pa., Air Pollution Control Assn., Am. Ordnance Assn., Indsl. Research Inst. Author: Modern Steels, 1939. Research, publs. on phys. chemistry of steelmaking; devel. of carbon and alloy steels, and heat-treatment of steel. Home: 80 Lebanon Hills Dr., Pitts. 15228. Office: 525 William Penn Pl., Pitts. 15230.*

LIHL, Franz, Austrian physicist; b. Pottenstein, Aug. 28, 1906; s. Franz and Ida (Woerle) L.; Ph.D., U. Vienna (Austria); m. Herta N., June 29, 1937; children—Gunter, Waltruat, Erika, Reinhard. Asst., Vienna, Frieberg; various positions in industry; prof., dir. inst. applied physics Vienna Tech. Coll., 1958—. Mem. German Metallurgy Soc., Austrian Soc. Physics, Chemico-Physics Soc. Research and publs. in radiol. physics, metall. physics, malleability, crystallization, recrystallization. Home: Belghofergasse 29, Vienna. Office: Karlspl. 13, Vienna, Austria.

LIKERT, Rensis, Am. social scientist; b. Cheyenne, Wyo., Aug. 5, 1903; s. George Herbert and Cornelia (Zonne) L.; A.B., U. Mich., 1926; Ph.D., Columbia, 1932; m. Charlotte Jane Gibson, Aug. 31, 1928; children—Elizabeth Jane (Mrs. James E. Pohlman), Patricia Anne (Mrs. Martin David). Faculty, N.Y. U., 1930-36; head, research dept. Life Ins. Agy. Mgmt. Assn., Hartford, Conn., 1935-39; head, div. program surveys Bur. Agrl. Econs., Washington, 1939-46; head, morale div. U. S. Strategic Bombing Survey, Washington, 1944-46; prof. psychology and sociology U. Mich., Ann Arbor, 1946—, dir. Survey Research Center, 1946-49, Inst. for Social Research, 1949—. Psychologist, N.Y., 1959—; cons. psychologist, Mich., 1962—; chmn. bd. Found. for Research on Human Behavior, 1962—. Recipient Paul D. Converse award U. Ill., 1955; James A. Hamilton Hosp. Adminstrs.' award, 1961; Pubs. award Orgn. Devel. Council, 1962; U. Mich. Distinguished Faculty Achievement award, 1962; Warren W. Stockberger award Soc. Personnel Adminstrn., 1963; McKinsey Found. Book award Am. Acad. Mgmt., 1962. Fellow Am. Psychol. Assn., Am. Statis. Assn. (pres. 1958), Inst. Math. Statistics, A.A.A.S.; mem. Am. Sociol. Assn., Sigma Xi, Phi Kappa Phi, Tau Beta Pi. Author: (with Gardner Murphy) Public Opinion and the Individual, 1938; New Patterns of Management, 1961; The Human Organization: Its Management and Value, 1967; Editor: (with Samuel Hayes) Some Applications of Behavioural Research, 1957. Contbns. to devel. improved methods for attitude measurement; improvement of sample survey techniques in social scis.; devel. theory of organizational behavior linking sociol. and psychol. concepts. Home: 1720 Morton Av., Ann Arbor, Mich. 48104.*

LIKHACHEV, Andrey Gavrilovich, Russian otorhinolaryngologist; b. Kovel (now Volhynia Oblast), 1899; grad. Med. Faculty, Voronezh U., 1922; D.Med. Sci., 1935. Intern, asst., sr. asst. clinic ear, nose and throat diseases Voronezh U., 1922-29; asst. clinic ear, nose and throat diseases 1st Moscow Med. Inst., 1929-30, dep. dir., 1930-34, acting prof., acting dir., 1934-35, dean and dep. dir., 1929-41, dir., 1942-48; prof., 1936—, now dir. ear, nose and throat diseases clinic; dep. dir. Moscow Central Research Inst. Otorhinolaryngology, 1936-38, dir., 1938-41; head chair ear, nose and throat diseases Sechenov 1st Moscow Med. Inst., 1936—; founder Central Research Inst. Otorhinolaryngology, 1936; chief otorhinolaryngologist RSFSR Ministry Health, 1946—. Mem. learned med. council USSR Ministry Health. Decorated Order of Lenin. Mem. All-Union (chmn. 1941—), All-Russian (bd. mem.), Moscow (dep. chmn.) socs. otorhinolaryngologists. Author: Ear, Nose and Throat Diseases, 1955, 58, 61. Dep. editor Herald of Otorhinolaryngology, Otorhinolaryngology sect. Large Med. Ency., 2d edit.; mem. editorial bd. Med. Abstracts Jour.; chief editor multi-vol. handbook on otorhinolaryngology. Research on principles of tissue therapy and penicillin therapy in otorhinolaryngology, otogenic intracranial complications, nasopharyngal cancer, traumatic injuries of nose and accessory sinuses; developer modification of plastic treatment after radical operation of temporal bone, also method of local anesthesia in surgery of nasopharyngal fibromata. Address: 1st Moscow Med. Inst., B. Pirogovskaya ulitsa 2-6, Moscow, USSR.

LIKHACHEV, Nikolai Victorovich, Russian virologist; b. Dec. 8, 1901; Staff, All-Union (Lenin) Acad. Agrl. Scis., 1956—; joined staff Moscow Zoo-Veterinarian Inst., 1929; with Ministry Agr., 1931—. Research in viral diseases of animals.

LILIENFELD, Abraham Morris, Am. epidemiologist; b. N.Y.C., Nov. 13, 1920; s. Joel and Eugenia (Kugler) L.; A.B., Johns Hopkins, 1941, M.P.H., 1949; M.D., U. Md., 1944; m. Lorraine Zemil, July 18, 1943; children—Julia, Saul, David. Intern, resident obstetrics Lutheran Hosp., Balt., 1944-46; asso. pub. health physician N.Y. State Dept. Health, 1949-50; dir. So. health dist. Balt. Health Dept., 1950-52; asst. prof. epidemiology Johns Hopkins, 1952-54, prof. chronic diseases, chmn. dept., 1958—; dir. gamma globulin evaluation center USPHS, 1952; chief dept. statistics and epidemiology Roswell Park Meml. Inst., 1954-58. Mem. Nat. Adv. Heart Council, 1962-66; research adv. council Am. Cancer Soc., 1964—; staff dir. President's Commn. Heart Disease, Cancer and Stroke, 1964-65; chmn. adv. council preventive medicine Md. Dept. Health, 1964—. Trustee Md. Heart Assn.; exec. bd. Md. div. Am. Cancer Soc. Research Career awardee NIH, 1962—. Fellow Am. Pub. Health Assn., Am. Coll. Preventive Medicine; mem. Am. Epidemiol. Soc., A.A.A.S., Am. Statis. Assn., Am. Sociol. Assn., Am. Soc. Human Genetics. Home: 6200 Gist Av., Balt. 21215.*

LILIENFIELD, Lawrence Spencer, Am. physiologist; b. Bklyn., May 5, 1927; s. Henry J. and Lee (Markman) L.; B.S., Villanova U., 1945; M.D., Georgetown U., 1949, M.S., 1954, Ph.D., 1956; m. Eleanor Russ, Oct. 22, 1950; children—Jan, Adele, Lisa. Faculty, Georgetown U., Washington, 1955—, chmn. dept. physiology and biophysics, 1963—, prof., 1964—. Established investigator Am. Heart Assn., 1958—. Recipient Research Career award USPHS, 1963. Helen Millenson Meml. fellow, 1955, 56. Fellow A.C.P., A.A.A.S.; mem. A.M.A., N.Y. Acad. Medicine, Am. Heart Assn., Am. Soc. for Clin. Investigation, Assn. Am. Med. Colls., Am. Physiology Soc., Biophys. Soc., Sigma Xi, Alpha Omega Alpha. Research, publs. on circulation in health and disease, treatment for high blood pressure, kidney function and its relationship to intra renal circulation. Home: 6304 Maiden Lane, Bethesda, Md. 20034. Office: 3900 Reservoir Rd. N.W., Washington 20007.*

LILIENTHAL, Howard, Am. surgeon; b. Albany, N.Y., Jan. 9, 1861; s. Meyer and Jennie (Marcus) L.; A.B. cum laude, Harvard, 1883, M.D., 1887; with McLean Asylum, br. Mass. Gen. Hosp., 1886; grad. Mt. Sinai Hosp., N.Y., 1888; m. Mary Harriss d'Antignac, Oct. 19, 1891 (dec. Mar. 1910); children—Mary d'Antignac (Mrs. Thompson Lawrence), Howard; m. 2d, Edith Strode, Nov. 7, 1911. Lectr. surgery, N.Y. Polyclinic, 1888, later cons. surgeon; surgeon, Mt. Sinai Hosp., 1892-1940, Bellevue Hosp., 1909-40; prof. clin. surgery Cornell U. Med. Coll. from 1917. Fellow A.C.S. (Founders Group), Am. Bd. Surgery, A.A.A.S.; mem. Am. Soc. Control of Cancer, A.M.A., Am. Surg. Assn., Am. Soc. Thoracic Surgery (pres.), N.Y. Soc. for Thoracic Surgery (pres.), N.Y. Surg. Soc. (pres.), Med. Soc. County of N.Y. (pres.), N.Y. Acad. Medicine, Société Internat. de Chirurgie, Académie de Chirurgie (corr.). Author: Imperative Surgery, 1900; Thoracic Surgery, 1925. Contbr. to Binnie's Treatise on Regional Surgery, 1917, Ochsner's Surgical Diagnosis and Treatment, 1920; more than 300 contbns. to surg. lit. Adv. editor Jour. Thoracic Surgery. Described Lilienthal's probe for detecting bullets and metal objects in body by means of elementary electric circuit, 1910; recorded a pneumonectomy for sarcoma of lung in a tuberculous patient, 1933. Died Apr. 30, 1946.

LILIENTHAL, Joseph Leo, Jr., Am. physician; b. N.Y.C., Nov. 1, 1911; s. Joseph Leo and Edna (Arnstein) L.; B.S., Yale, 1933; M.D., Johns Hopkins, 1937; m. Katherine Arnstein, June 25, 1937; children—Julia, Nina. Clin. clk. Nat. Hosp., Queen's Sq., London, Eng., 1937; house officer Presbyn. Hosp., N.Y.C., 1938-40; resident physician Johns Hopkins Hosp., 1940-42, physician since 1946; asso. prof. medicine Johns Hopkins, since 1946, prof. environmental medicine since 1950. Mem. Nat. Bd. Med. Examiners, chmn. medicine test com. Physiol. Study Sect., Nat. Inst. Health USPHS, 1953—. Fellow N.Y. Academy Sci.; mem. Am. Inst. Biol. Scis., Am. Clin. Climatol. Assn., Soc. Med. Consultants to Armed Forces, Assn. Am. Physicians, Am. Soc. Clin. Investigation, Am. Physiol. Society, Soc. Exptl. Biology and Medicine (mem. nat. council), Interurban Clin. Club, Am. Fedn. Clin. Research, Sigma Xi, Phi Beta Kappa. Research in neuromuscular and respiratory physiology. Died Nov. 19, 1955.

LILIENTHAL, Otto, German aero. engr.; b. Auklam, Germany, May 23, 1848; employed in indsl. works, Berlin. Author: Der Vogelflug als Grundlage der Fliegerkunst, 1889. Evaluated uses of ship siren and steam boiler; developed apparatus with 2 wings and vertical tail which he attempted to fly, 1877; developed means of achieving greatest horizontal stability while attempting to fly; 1st flight with gliders; 1st expts. with mech. air travel; built double decker glider; studied birds' wings as models for achieving flight in man. Died nr. Rhinow, Germany, Aug. 10, 1896.

LILIO, Luigi (or Aloysius Lilius), Italian physician, astronomer; b. Calabria, 1510. Cons. to Pope Gregory XIII on calendar reform; Gregorian calendar based on Lilio's system; determined approximate equation of lunar and solar years. Died 1576.

LILJEDAHL, Sten-Otto, Swedish surgeon; b. Motala, Sweden, Mar. 31, 1923; s. Nils and Edit (Svensson) L.; M.D., Karolinska Inst., 1955; m. Ingrid Anne Marie Fröjd, Aug. 31, 1948; children—Eva Christina, Nils Peter, Karin Margaretha. Docent in surgery Karolinska Inst., Stockholm, 1956-60, asso. prof. surgery, 1960—. Author: Treatment of Burns; Medicinsk Arshyh, 1966; Munsgards forlag; also numerous articles. Research on severe burns, elec. burns, pulmonary burns. Home: 7 Virebergsvägen, Solna, Sweden. Office: Karolinska Inst., Stockholm 60, Sweden.*

LILJEQUIST, Gösta Hjalmar, Swedish meteorlogist; b. Norrköping, Sweden, Apr. 20, 1914; s. Herman Julius and Thina (Carlsson) L.; B.Sc., Lund (Sweden) U., 1937; Fil.lic., Uppsala (Sweden) U., 1926, Fil.Dr., 1957; m. Ann-Margret Wallin, Oct. 20, 1945; 1 son, Gösta Lennart. Asst., Bornö Oceanographic Sta., Swedish Hydrog-Biol. Commn., 1937-38; hydrologist Swedish Meterol. Hydrological Inst., 1938-39, meteorologist, 1939-58; prof. meteorology Uppsala U., 1958—, dean math. phys. sect., 1964-66. meteorologist, Norwegian-Brit.-Swedish Antarctic Expdn., 1957-58; organizer, leader Swedish-Finnish-Swiss Expdn. to N.E. Land, 1960. Recipient medal Norwegian King, 1952; Riddare au Wordstjarne orden, 1963. Mem. Sallskapet for Amtropolzi Georap., Kungl. Svenska Vetenskapsakaderin, Kungl. Svenska Vetenskaps Societén. Author: Arktisk Utpost, 1960; Meteorology, 1962; Popular Meterology, 1966; also articles. Research on polar meteorology, including radiation, turbulent heat transfer, energy exchange of Antarctic ice-shelf; atmospheric optics, climatic variation; meteorol. edn. and tng.; penetration of light into sea, sea-ice conditions. Home: Observatorie parken, Uppsala, Sweden. Office: Meteorologiska Institutionen, Uppsala, Sweden.*

LILJESTRAND, Goeran, Swedish physiologist, pharmacologist, educator; b. Gothenburg, Sweden, Apr. 16, 1886; s. Erik Petter and Tekla (Carlberg) L.; candidate medicine U. Stockholm (Sweden), 1909, licentiate medicine, 1915, M.D., 1917, D.Sc. (hon.), 1952; M.D. (hon.) U. Dorpat (Estonia), 1932, U. Ghent (Belgium), 1955, U. Paris (France), 1958; m. Elsa Margareta Wretlind, Dec. 10, 1910 (dec. 1948); children—Brita (Mrs. Sven Grape), Ake, Margit (Mrs. Kjell Halvarson); m. 2d, Maud von Koch, Oct. 24, 1949; 1 son, Nils Goeran. Asst. prof. physiology Caroline Inst., Stockholm, 1917-23, asst. prof. physiology and pharmacology, 1923-27, prof. pharmacology, 1927-51, emeritus prof., 1951—. Sec., 12th Internat. Physiol. Congress, Stockholm, 1926; mem. bd. Inst. Pub. Health, 1938-42, Inst. Phys. Edn., 1938-42; sci. adv. bd. medicine Swedish Govt., 1931-48; mem. Nobel Com., 1938-51, sec. physiology and medicine, 1918-60, chmn. trustees Nobel Found., 1958—. Recipient Björken prize U. Uppsala, 1930; gold medal Swedish Govt., 19—, gold medal Caroline Inst., 1960. Hon. mem. Am., Brit., Scandinavian (sec. 1926-31, 35-51) physiol. socs., Brit. Pharmacol. Soc., Deutsche Pharmakol. Ges. (Schmiedeberg plaquette 1962), Swedish Med. Soc. (Alvarenga prize 1918, 23, 25, Regnell prize 1946), A.M.A., Acad. de Médecine Brussels, Soc. Philomatique (Paris), Alpha Omega Alpha; mem. Swedish (Gold medal 1965), Danish, Finnish acads. sci., Akad. Naturforscher, N.Y. Acad. Medicine. Author: Lehrbuch der Pharmakologie, 1936-59; (with others) Nobel, the Man and His Prizes, 1950, 62; (with others) Karolinska Institutets Historia 1910-60, 1960; (with B. Holmstedt) Readings in Pharmacology, 1963. Editor: Les Prix Nobel, 1953-65; Acta Physiologica Scandinavica, 1940-57. Contbr. numerous articles on respiration and circulation to profl. publs. Home: 15 Svedjevägen, Bromma, Stockholm. Office: Caroline Inst., Stockholm, Sweden.*

LILLEHEI, Clarence Walton, Am. surgeon; b. Mpls., Oct. 23, 1918; s. Clarence I. and Elizabeth (Walton) L.; B.S. with distinction, U. Minn., 1939, M.B., 1941, M.D., 1942, M.S. in Physiology (Ebin fellow, Rockefeller Found. fellow), also Ph.D. in Surgery, 1951; m. Katherine Ruth Lindberg, Dec. 31, 1946; children—Kimberle Rae, Craig Walton, Kevin Owen, Clark William. Faculty, U. Minn. Med. Sch., Mpls., 1951-68, prof. dept. surgery, 1956-68; surgeon-in-chief N.Y. Hosp., L. A. Stimson prof. surgery, chairman department of surgery Cornell University Medical College, since 1968—. Decorated Bronze Star medal; Orden del Merito Sanitario, Republic de Colombia; recipient Albert and Mary Lasker Found. award, 1955; Oscar B. Hunter Meml. award Am. Therapeutic Soc., 1958; officer Order Leopold, Belgium, 1960; Gairdner Found. Internat. award, 1963; others. Diplomate Am. Bd. Surgery, Am. Bd. Thoracic Surgery. Fellow Am. Coll. Cardiology (Susan and Theodore Cummings Humanitarian award 1963, pres. 1966-67), A.C.S., Am. Coll. Chest Physicians, Am. Heart Assn.; mem. A.A.A.S. (Ida B. Gould Meml. award 1956), Internat. Cardiovascular Soc., Soc. Exptl. Biology and Medicine, U. Surgeons, A.M.A. (Hektoen Gold medal 1957), Am. Assn. Thoracic Surgery, Am. Surg. Assn., Soc. Vascular Surgery, Soc. Thoracic Surgeons, Club Pan Americano De Doctores (Mexico), Royal Soc. Medicine (Great Britain), Royal Soc. Scis. (Sweden), Universidad Catolica de Chile, Sociedad De Cirujanos de Chile, Sociedad Chilena De Cardiologia, others, Sigma Xi. Research, publs. on defects of heart, gen. surg. procedures, mgmt. of cardiac patient. Home: 435 E. 70th St., N.Y.C. 10021. Office: Cornell U. Med. Center, N.Y. Hosp., 525 E. 68th St., N.Y.C. 10021.*

LILLEHEI, Richard Carlton, Am. surgeon; b. Mpls., Dec. 10, 1927; s. Clarence I. and Elizabeth (Walton) L.; B.A., U. Minn., 1948, B.S., 1949, B.M., 1951, M.D., 1952, Ph.D., 1960; m. Betty Jeanne Larsen, Dec. 20, 1952; children—Richard Carlton, Theodore J., John C., James L. Asst. prof. surgery U. Minn. Med. Sch., 1960, asso. prof., 1962—; practice medicine limited to cardiovascular, pulmonary and gen. abdominal surgery. Markle scholar in med. scis., 1960-65. Named Outstanding Young Man of Minn., Minn. Jr. C. of C., 1962; recipient Hektoen Gold medal A.M.A., 1964. Diplomate Am. Bd. Surgery, Bd. Thoracic Surgery. Fellow Am. Coll. Cardiology; mem. Am. Assn. for History Medicine, Am. Physiol. Soc., Am. Soc. for Artificial Internal Organs, Minn. Acad. Medicine, A.C.S., Soc. for Cryobiology, Internat. Cardiovascular Soc., Soc. for Exptl. Biology and Medicine, Soc. for Vascular Surgery, Central Surg. Assn., Soc. U. Surgeons, A.M.A., Soc. Alimentary Tract Surgeons, Phi Beta Kappa, Phi Rho Sigma, Alpha Omega Alpha. Contbr. articles profl. jours. Exptl. surgery in field transplantation and etiology and treatment of shock due to hemorrhage, infection and cardiac failure, with particular emphasis on cardiac failure due to myocardial infarction; also studies in field of transplantation and preservation of organs in exptl. lab. and in man. Home: 1814 Oliver Av. S., Mpls. 55405. Office: Box 388, Univ. of Minn. Hospitals, Mpls. 55455.

LILLER, William, Am. astronomer; b. Phila., Apr. 1, 1927; s. Carroll Kalbaugh and Catherine (Dellinger) L.; A.B., Harvard, 1949; A.M., U. Mich., 1951, Ph.D., 1953; m. Martha Locke Hazen, June 20, 1959; children—John Avery, Tamara Kay, Hilary Webb. Supt. meteor expdn. Harvard, Las Cruces, N.M., 1952-53, prof., Cambridge, Mass., 1960—, chmn. dept.

astronomy, 1960-66; faculty U. Mich., Ann Arbor, 1953-60, asso. prof. astronomy, 1959-60. Chmn., Com. on Astron. Motion Pictures, 1959-64. Guggenheim fellow, 1964-65. Fellow Royal Astron. Soc.; mem. Am. Astron. Soc., Internat. Astron. Union, Sigma Xi. Editor: Space Astrophysics, 1961. Research, publs. on spectroscopy and photometry gaseous nebulae, comets and stars. Home: 77 Snake Hill Rd., Belmont, Mass. 02178. Office: Harvard Coll. Obs., Cambridge, Mass. 02178.*

LILLEY, Arthur Edward, Am. astronomer; b. Mobile, Ala., May 29, 1928; s. Arthur L. and Susie (Jones) L.; B.S., U. Ala., 1950, M.S., 1951; Ph.D., Harvard, 1954; m. Marion Hewett Talbot, July 23, 1961; children—Kathryn, Pauline Saxon. With Naval Research Lab., Washington, 1954-57; faculty Yale, 1957-59; faculty Harvard, Cambridge, Mass., 1959—, prof. astronomy, 1963—; astronomer charge radio astronomy Smithsonian Astrophys. Obs., 1966—; vis. prof. physics Mass. Inst. Tech., 1966-67. Recipient Bausch and Lomb Sci. medal, 1946; Bart J. Bok prize Harvard, 1958; named One of Ten Outstanding Young Men, U. S. Jr. C. of C., 1961. Alfred P. Sloan fellow, 1958-64. Mem. Am. Acad. Arts and Scis., Am. Phys. Soc., Am. Astron. Soc., Internat. Astron. Union, Union Radio Scientifique Internationale, Sigma Xi, Pi Mu Epsilon. Originator devel. of radio telescopes to be flown in space, microwave absorption lines in spectra radio stars, detection of several spectral lines in radio astronomy; participated Mariner 2 Venus program. Home: 89 Fletcher Rd., Belmont, Mass. 02178. Office: Harvard Obs., 60 Garden St., Cambridge, Mass. 02138.*

LILLEY, George, Am. mathematician; b. Kewanee, Ill., Feb. 9, 1854; s. William and Harriet (Huntley) L.; ed. Knox Coll., Ill., 1869-73 (M.A.); student U. Mich., 1873-75; grad. Ill. Wesleyan U., Ph.D., 1882; M.A., 1886; M.A., Washington and Jefferson, 1878; LL.D., Chaddock Coll., 1886; m. Sophia Adelaide Munn, June 11, 1879. Engaged in business, Corning, Ia., 1878-80; pres. S.D. Agrl. Coll., 1884-87, prof. math. and engring., 1887-90; pres. Washington Agrl. Coll., and Sch. Science, 1890-93; prin. Park Sch., Portland, Ore., 1894-96; prof. math. U. Ore., 1897—. Died 1904.

LILLIE, Frank Rattray, zoologist; b. Toronto, Ont., Can., June 27, 1870; s. George W. and Emily (Rattray) L.; A.B., U. Toronto, 1891, D.Sc., 1919; fellow Clark U., 1891-92, U. Chgo., 1892-93, Ph.D., 1894; D.Sc., Yale, 1932, and Harvard, 1938; LL.D., Johns Hopkins, 1942; m. Frances Crane, June 29, 1895. Instr. zoölogy, U. Mich., 1894-99; prof. biology Vassor Coll., 1899-1900; asst. prof. zoology and embryology U. Chgo., 1900-02, asso. prof., 1902-07, prof since 1907, chmn. dept. zoology, 1911-35, dean dir. biol. scis., 1931-35. Head dept. zoology Marine Biol. Lab., Woods Hole, Mass., 1893-1907, assistant director, 1900-08, dir., 1908-26, pres., 1925-1942; pres. Woods Hole Oceanographic Inst., 1930-39. Fellow A.A.A.S. (v.p. 1914); mem. Nat. Acad. Scis. (pres. 1935-39), NRC (chmn. 1935-36), Am. Philos. Soc., Acad. Nat. Sci., Phila.; Société Belge de Biologie, Société de Biologie, Paris, Am. Soc. Naturalists (v.p. 1914, pres. 1915), Am. Soc. Zoologists (pres. Central Br. 1905-08), Assn. Am. Anatomists, Boston Soc. Natural History, Royal Soc. of Edinburgh, Zool. Soc. of London. Author: The Development of the Chick; The Woods Hole Marine Biological Laboratory, 1944. Contbr. to sci. jours. Mng. editor Biol. Bull., 1902-26; asso. editor Jour. Exptl. Zoology Physiol. Zoology. Died Nov. 5, 1947.

LILLIE, Ralph Dougall, Am. pathologist, histochemist; b. Cucamonga, Cal., Aug. 1, 1896; s. William A. and Ida F. (Howell) L.; A.B., Stanford, 1917, M.D., 1920; m. Ethel Astrup Christensen, May 14, 1920; children—Margaret A. (Mrs. Thomas R. Lusk), Susan Jane (Mrs. Robert Hendershott), Ida A. (Mrs. Herbert Herman), Mary W. (Mrs. Thomas E. Wallace). Officer, USPHS, 1920-60, chief div. pathology (now called lab. exptl. pathology), NIH, 1937-60, ret., 1960; research prof. pathology La. State U. Sch. Medicine, New Orleans, 1960—. Rep. nat. orgns. to Biol. Stain Commn., 1959-63, pres., 1959-63; Am. v.p. 2d Internat. Congress for Histochemistry and Cytochemistry, Frankfurt, Germany, 1964. Mem. Am. Assn. Pathologists and Bacteriologists, Am. Soc. Exptl. Pathology, A.M.A., Assn. Mil. Surgeons U. S., Internat. Acad. Pathology (pres. 1948), Histochem. Soc. (pres. 1958), Argentina Path. Soc. (hon.), others. Editor-in-chief Jour. Histochemistry and Cytochemistry, 1952-64. Research, publs. on pathology of infectious disease, intoxications, dietary deficiencies, pathogenesis of liver cirrhosis; histochem. reactions for various proteins and other biochems. Home: 338 Walnut St., New Orleans 70118. Office: 1542 Tulane Av., New Orleans 70112.*

LILLIE, Ralph Stayner, biologist; b. Toronto, Can., Aug. 8, 1875; s. George Waddell and Emily Ann (Rattray) L.; B.A., U. Toronto, 1896, Sc.D., 1936; grad. student U. Mich., 1896; Ph.D., U. Chgo., 1901; m. Helen Eva Makepeace, June 2, 1906; children—Frank Rattray, Walter Makepeace. Asst. in physiology Harvard, 1901-02, instr. physiology, 1905-06; instr. and adj. prof. physiology U. Neb., 1902-05; research asst. Carnegie Inst. Zoöl. Sta., Naples, Italy, 1904-05; Johnston scholar Johns Hopkins, 1906-07; instr., asst. prof. physiology and exptl. zoölogy U. Pa., 1907-13; prof. biology Clark U., 1913-20; biologist Nela Research Lab., Cleve., 1920-24; prof. gen. physiology U. Chgo., 1924-40, emeritus, 1940—; instr. and investigator general physiology Marine Biol. Lab., Woods Hole, Mass., 1902—. Fellow Am. Acad. Arts and Scis., A.A.A.S.; mem. Am. Physiol. Soc., Am. Soc. Biol. Chemists, Am. Soc. Naturalists, Soc. Exptl. Biology and Medicine, Am. Soc. Zoölogists, Am. Philos. Soc., Phi Beta Kappa, Sigma Xi. Spl. researches in fundamental properties of living substance and physiology of stimulation, growth, cell-division, radiation effects, philos. aspects of biology. Died Mar. 19, 1952.

LILLIE, Robert Jones, Am. poultry husbandman; b. Rochester, Minn., Apr. 15, 1921; s. Walter Ivan and Opal (Jones) L.; student Swarthmore Coll., 1940-42; B.S., Pa. State U., 1944; M.S., U. Md., 1946, Ph.D., 1949; m. Mary Ellen Guers, July 24, 1946; children—Elizabeth Jane, Kathryn Joann. Poultry husbandman U. S. Dept. Agr., Beltsville, Md., 1947—. Mem. standard diet subcom. NRC, 1954. Mem. Poultry Sci. Assn. (Research award 1950), Am. Inst. Nutrition, World's Poultry Congress, Sigma Xi, Gamma Sigma Delta. Contbr. numerous articles to tech. jours. Established folic acid requirements of chickens, nutrition-genetic relationship in vitamin D requirement for normal feather pigmentation; demonstrated antibiotics were more effective in old than new quarters, males require linoleic acid for normal fertility. Home: 4022 Foreston Rd. Office: Agrl. Research Center, Beltsville, Md. 20705.*

LILLY, Daniel McQuillan, Am. biologist; b. Central Falls, R.I., July 20, 1910; s. Edward P. and Ethel G. (Blount) L.; A.B., Providence Coll., 1931, M.S., 1936; Ph.D., Brown U., 1940; m. Catherine M. Kelly, June 19, 1948; 1 son, John K. From instr. to asst. prof. biology Providence Coll., 1932-42; instr. biology Shrivenham Am. U., Eng., 1945; from asst. prof. to prof. biology St. John's U., Jamaica, N.Y., 1946—, chmn. dept., 1957-61, dept. rep. in charge of grad. studies, 1961-65. Mem. A.A.A.S., Am. Soc. Zoologists, Soc. Protozoologists, N.Y. Acad. Sci., Assn. Mil. Surgeons, Sigma Xi. Research, publs. on vitamins, other growth factors of microorganisms; nucleic acid metabolism of protozoa. Home: 56-10 187th St., Flushing, N.Y. 11365. Office: Grand Central and Utopia Pkwy., Jamaica, N.Y. 11432.*

LIM, Robert Kho-Seng, physiologist; b. Singapore, Oct. 15, 1897; s. Boon-Keng and Margaret (Huang) L.; M.B., Ch.B., U. Edinburgh, 1919, Ph.D., 1920, D.Sc., 1924; D.Sc., Hong Kong U., 1951; m. Margaret Torrance, July 10, 1920 (dec.); children—Effie (Mrs. O. Philip Edwards), James T.; m. 2d, Tsing-Ying Tsang, July 2, 1946. Came to U. S., 1949, naturalized, 1955. Vis. research prof. U. Ill., 1949-50; prof., head dept. physiology Creighton U. Med. Sch., 1950-51; dir. med. scis. research Miles Lab., Elkhart, Ind., 1952—. Mem. Am., Brit., Chinese physiol. socs., Nat. Acad. Scis. Author: (with others) A Stereotoxic Atlas of the Dog's Brain, 1960. Research on gastroenterology, central nervous system, pain and analgesia. Home: 738 Marine Av., Elkhart, Ind. 46514.*

LIM, Thomas Pyung-Kee, physiologist; b. Seoul, Korea, June 1, 1924; s. Byung-OOk and Kee-Nam (Ohr) L.; M.D., Severance Union Med. Coll., Seoul, 1948; M.S., Northwestern U. Med. Sch., 1951, Ph.D., 1953; m. Kimberly Sung-Hai Kim, June 20, 1955; children—Diane Yung-Jin, Margaret Hyo-Jin, Elizabeth Sung-Jin. Came to U. S., 1950, naturalized, 1959. Research asso. Stanford Med. Sch., Palo Alto, Cal., 1953-54, Northwestern U., Med. Sch., Chgo., 1954-56; staff Lovelace Found. and Clinic, Albuquerque, 1956-61; dir. cardiopulmonary lab. Tucson Med. Center, 1961—; lectr. bioengring. U. Ariz., 1964—. Cons. Pima County Gen. Hosp., Tucson, 1961—. Mem. A.A.A.S., Am. Physiol. Soc., Am. Thoracic Soc., Am. Heart Assn., A.M.A., Am. Coll. Chest Physicians. Author: Cardiopulmonary Function Tests in Clinical Medicine, 1966; also articles. Research on body temperature regulation, respiration, circulation, cardiopulmonary function tests, spatial vectorcardiography. Home: 6025 E. Hampton St. Office: 5301 E. Grant S., Tucson 85716.*

LIMPER, Karl Esslinger, Am. geologist, educator; b. Wichita, Kan., Mar. 1, 1914; s. Henry William and Vera S. (Esslinger) L.; B.S., Beloit Coll., 1935, M.A., 1937; Ph.D., U. Chgo., 1953; m. Betty A. Buntin, Apr. 16, 1944; 1 dau., Layne A. Instr., Miami U., Oxford, O., 1939-41, faculty, 1947—, prof., chmn. dept. geology, 1956-60, dean Coll. Arts and Sci., 1960—; instr. Hamilton Coll., 1941-42. Fellow, A.A.A.S., Ohio Acad. Sci.; mem. Geol. Soc. Am., Paleontol. Soc. Am., Nat. Assn. Geol. Tchrs., Sigma Gamma Epsilon, Sigma Pi, Phi Sigma. Home: 134 Hilltop Rd., M.R. 50, Oxford, O. 45056.*

LIN, Chia-Chiao, mathematician; b. Fukien, China, July 7, 1916; s. Kai and Y. T. (Liang) L.; B.S., Nat. Tsing Ilua U., 1937; M.A., U. Toronto, 1941; Ph.D., Cal. Inst. Tech., 1944; m. Shou-Ying Liang, Dec. 21, 1946; children—Edward S., Lillian S. Came to the United States 1941. Assistant professor later associate prof. mathematics, Brown University, 1945-47; asso. prof. mathematics, Mass. Inst. Tech. 1947-53, prof., 1953-66, Inst. prof., 1966—. Guggenheim Fellow, 1954-55, 1960. Fellow Am. Acad. Arts and Scis., Inst. Aeros. and Astronautics, Am. Phys. Soc.; mem. Am. Astron. Society, Nat. Academy of Sciences, Am. Math. Soc. Author: The Theory of Hydrodynamic Stability, 1955. Research on stellar dynamics; hydrodynamics; astrophysical problems. Home: 8 French Rd., Weston, Mass. Office: Mass. Inst. of Technology, Cambridge, Mass.

LIN, Shao-Chi, aero. engr.; b. Canton, China, Jan. 5, 1925; s. Wei-Cheng and Ruth (Lai) L.; came to U. S., 1948, naturalized, 1960; B.Sc., Nat. Central U., Chunking, China, 1946; Ph.D., Cornell U., 1952; m. Lily Yuli Shao, July 30, 1955. Research engr. Bur. Aircraft Industry, China, 1947-48; research asso. Cornell U., Ithaca, N.Y., 1952-54, acting asst. prof., 1954; prin. research scientist AVCO-Everett (Mass.) Research Lab., 1955-64; vis. lectr. Mass. Inst. Tech., 1963; prof. engring. physics, U. Cal. at San Diego, 1964—. Mem. Am. Phys. Soc., Am. Geophys. Union, Am. Inst. Aeros. and Astronautics (Research award 1965). Contbg. author: Ency. Sci. and Tech., 1960. Research, publs. on re-entry physics, especially hypersonic flow theory, electrical and electromagnetic properties of ionized gases, shock wave phenomena. Home: 2614 Costebelle Dr., LaJolla, Cal. 92037.*

LIN, Tung Yen, civil engr.; b. Foochow, China, Nov. 14, 1911; s. Ting Chang and Feng Yi (Kuo) L.; B.S., in Civil Engring., Tangshan Coll., Chiaotung U., 1931; M.S., U. Cal. at Berkeley, 1933; m. Margaret Kao, July 20, 1941; children—Paul, Verna. Came to U. S., 1946, naturalized, 1951. Chief bridge engr., chief design engr. Chinese Govt. Rys., 1933-46; asst., asso. prof. U. Cal., 1946-55, prof., 1955—, chmn. div. structural engring., 1960-63; dir. structural lab., 1960-63; dir. T. Y. Lin & Assos., Van Nuys, Cal., T. Y. Lin Internat., also T. Y. Lin, Kulka, Yang and Asso., San Francisco, cons. structural engrs., 1952—; cons. to State of Cal., Def. Dept., also to industry. Chmn. World Conf. on Prestressed Concrete, 1957, Western Conf. on Prestressed Concrete Bldgs., 1960. Fellow Am. Soc. C.E. (Wellington award, Howard award); mem. Am. Soc. Engring. Edn., Internat. Fedn. for Prestressing, Am. Concrete Inst., Prestressed Concrete Inst. Author: Design of Prestressed Concrete Structures, 1955 (with B. Bresler) Design of Steel Structures, 1960. Contbr. articles profl. jours. Research on design of steel and prestressed concrete structures; bridge and structural engineering. Home: 8701 Don Carol Dr., El Cerrito, Berkeley, Cal. 94530. Office: 14656 Oxnard St., Van Nuys, Cal.; also 15 Vandewater St., San Francisco.

LINACRE, Thomas, Brit. physician, scholar; b. Canterbury, Eng., circa 1460; ed. Cathedral Sch., Canterbury; studied at Oxford, 1480-84; M.D., Padua, Italy. Ordained priest, 1520; practised medicine, London; physician to Henry VII and Henry VIII of Eng.; apptd. tutor Prince Arthur, 1501; lectr. Cambridge; founder readership in medicine Oxford; taught Erasmus, Sir Thomas More. Linacre professorship in anatomy, Oxford, named in his honor. Founder, 1st pres., Royal Coll. Physicians. Author: De emendata structura Latini sermonis, 1524. Translator: De sphaera (Proclus), 1499; De sanitate tuenda, 1517, Methodus medendi, 1519, De naturalibus facultatibus, 1523 (all Galen). One of 1st from Eng. to study Greek in Italy; introduced (with Colet) New Learning to Eng. Died Oct. 20, 1524.

LINARES, Enrique, Argentinian geologist; b. Buenos Aires, Argentina, Oct. 1, 1928; s. Enrique Pedro G. and Maria del Pilar (Cataluña) L.; D.Geology, U. Buenos Aires, 1956; postgrad. Yale, 1964-65; m. Raquel Eloisa Vazquez, May 8, 1954; children—Estela Beatriz, Liliana Raquel, Cecilia Sonia, Sandra Isbel. Geologist, AEC Argentina, Buenos Aires, 1954—, chief raw materials dept. Servicio Labs., 1960—; asst. econ. geology U. La Plata (Argentina), 1963-65; asso. prof. mineralogy U. Buenos Aires, 1966—. Recipient certificate of merit, USCG, 1958. Mem. Argentina Geol. Soc., Mineral. Soc. Am. Research, publs. on mineralogy, isotope geology and geochronology, identification and studies of uranium minerals, uranium deposits and origin of them. Home: Lugones 2862, Buenos Aires. Office: Ada. del Liberatdor 8250, Buenos Aires, Argentina.*

LINBERG, Boris Edmundovich, Russian surgeon; b. Shatsk (now Ryazan Oblast), 1885; grad. Med. Faculty, Moscow U.; Dr.Med.Sci., 1922. Dist. physician Krasny Kut, 1911-12, Kozlov, 1912-13; 2d physician Sratov R.R. Hosp., 1912-14, head surgeon, 1918-22; faculty Saratov U., 1912-14, asst. prof., 1918-22, lectr., 1922-23; physician 12th Inf., 1914-18; prof., head surgery dept. Smolensk U., 1923-33; head surgery dept. Moscow Med. Inst., 1933-42; dir. treatment thoracic wounds in evacuation hosps., 1941-45; head gen. surgery Moscow Stomatological Inst., 1951—. Cons., Moscow Oblast Clin. Research Inst., 1951—. Chmn., Expert Commn. on Surgery, 1937-58. Recipient Lenin prize, 1961, Order of Lenin, Order Red Banner of Labor. Mem. Moscow and All-Union Soc. Surgeons (mem. bd.) Author: Twenty Years of Experience in the Surgical Treatment of Chronic Pulmonary Emphysema, 1956; co-author: Cancer. Coauthor, editor: Problems of Thoracic Surgery, 6 vols., 1945. Co-editor surg. sect. Large Med. Ency., 2d edit. Founder 1st lung surgery dept. in USSR, 1933; introduced methods of osteosynthesis, herniotomy, lower jaw resection, costoversion thoracoplasty; devel. treatment for pulmonary abscesses using underwater drainage, treatment for accumulations of blood and air in pleural cavity from wounds; performed 1st lobectomy in bronchiectasis in USSR. Address: Meditsinsky stomatologichesky institut, Kalyaevskaya 18, Moscow, USSR.

LINCECUM, Gideon, Am. physician, naturalist; b. Hancock County, Ga., Apr. 22, 1793; s. Hezekiah and Sally (Hickman) L.; studied medicine privately, Ga., 1815-17; m. Sarah Bryan, Oct. 25, 1814. Commr. apptd. by Miss. Legislature to organize County of Monroe, 1821-22; Indian trader in Miss., 1823-circa 1827; practiced medicine, Cotton Gin Port, Columbus, Miss., 1830-48; owned plantation, Long Point, Tex., 1848-74, studied insects and made lengthy investigation of life of mound-building ants; corresponded with many fgn. naturalists, including Charles Darwin; sent specimens to Smithsonian Instn., Acad. Natural Scis., Phila., Jardin des Plantes, Paris, France; papers on insects published in Jour. of Proc. of Linnaean Soc., London, 1852, Proc. of Acad. Nat. Scis. of Phila., 1866. Died Long Point, Nov. 28, 1874; buried Long Point.

LINCICOME, David Richard, Am. biologist; b. Champaign, Ill., Jan. 17, 1914; s. David Rosebery and Olive Iola (Casper) L.; B.S., M.S., U. Ill., 1937; Ph.D., Tulane U., 1941; m. Dorothy L. Van Cleave, Sept. 1, 1941 (dec. 1953); children—Judith Ann, David Van Cleave; m. 2d, Margaret A. Stirewalt, Dec. 29, 1954. Guest scientist Naval Med. Research Inst., 1954-61; faculty Howard U., Washington, 1955—, now prof. zoology. Vis. scientist Nat. Inst. Arthritis and Metabolic Diseases, 1964-65; sec. Midwestern Conf. Parasitologists, 1949. Fellow N.Y. Acad. Scis., A.A.A.S.; mem. Am. Soc. Zoologists, Soc. Systematic Zoologists, Japanese, Brit. socs. parasitology, Am. Soc. Cell Biology, Am., German socs. parasitologists, Am. Inst. Biol. Scis., Am. Micros. Soc., Helminthological Soc. (sec. 1959, v.p. 1962), Royal Soc. Tropical Medicine and Hygiene (local sec. N.Am. 1964—), Fedn. Am. Socs. Exptl. Biology, Am. Physiol. Soc., Am. Soc. Tropical Medicine, Wildlife Disease Assn., Soc. Protozoologists, Am. Soc. Naturalists, Soc. Exptl. Biology and Medicine (sec. D.C. 1965—), Phi Beta Kappa, Sigma Xi. Author: (with others) Clinical Parasitology, 1957. Editor: Experimental Parasitology, 1951—; Internat. Rev. Tropical Medicine, 1961—; editorial bd. Parasitological Revs., 1963-65. Studies, numerous publs. on morphology, classification of worms; co-devel. clin. techniques for diagnosis of parasites; physiol. and molecular implications of host-parasite relationship. Home: 7118 Cedar Av., Takoma Park, Md. 20001; also "The Fort," Route 771, Seven Fountains, Va. 22653. Office: Dept. Zoology, Howard U., Washington 20001.*

LINCK, Albert John, Am. plant physiologist; b. Portsmouth, O., Aug. 18, 1926; s. Arthur John and Estella C. (Schreiner) L.; B.Sc., Ohio State U., 1950, M.S., 1951, Ph.D., 1955; m. Vandora G. Pierson, July 20, 1957; children—Erik John, Troy Gregory. Faculty, U. Minn., St. Paul, 1955—, prof. plant physiology, 1961—, asst. dir. Expt. Sta., 1966—. Dir. acad. year inst. in biology NSF, 1964—. Mem. A.A.A.S., Am. Inst. Biol. Scis., Am. Soc. for Plant Physiology, Scandinavian Soc. Plant Physiologists, Bot. Soc. Am., Minn. Acad. Sci., Internat. Soc. Plant Morphologists, Gamma Alpha (nat. pres. 1960-64, nat. editor 1964—). Research, publs. on movement organic and inorganic compounds in plants, effect plant growth regulators on growth and devel. in plants. Home: 4136 Reiland Lane, St. Paul 55112.*

LINCK, Johann Heinrich, German pharmacist; b. Leipzig, Germany, 1675; studied medicine at Copenhagen (Denmark); at least one son; established pharmacy, Leipzig, which was the best in Saxony. Fellow Royal Soc., 1718. Author: Dissertation sur le cobalt; De Stelles marines liber singularis, 1733. Died 1735.

LINCOLN, Azariah Thomas, Am. chemist; b. Montfort, Wis., June 25, 1868; s. Joseph Hollis and Margaret (Laird) L.; B.S., U. Wis., 1894, M.S., 1898, Ph.D., 1899; m. Jennette Emeline Carpenter, June 30, 1904. Instr. chemistry U. Cin., 1900-01; instr. chemistry U. Ill., 1901-03, asst. prof., 1903-08; asst. prof. chemistry Rensselaer Poly. Inst., 1908-12, prof. phys. chemistry, 1912-21; prof. chemistry Carleton Coll., 1921-23, chmn. dept. chemistry, 1923-39, emeritus, 1939—. Mem. Am. Chem. Soc., Am. Electro-Chem. soc., Soc. Chem. Industry, A.A.A.S., Sigma Xi. Author: Elementary Quantitative Analysis (Lincoln and Walton), 1907; Textbook of Physical Chemistry, 1918. Translator: (with David H. Carnahan) Theoretical Principles of the Methods of Analytical Chemistry, Based upon Chemical Reactions (from the French), 1910; General Chemistry, 1927. Died Mar. 31, 1958; buried Lancaster, Minn.

LINCOLN, Bert Hartzell, Am. chemist, chem. engr.; b. Van Buren, Ark., Sept. 6, 1900; s. Charles A. and Colia (Epps) L.; B.S., U. Ark.; grad. study U. Colo.; Ch.D., Acad. Sci. Arts, Brazil; m. Sara Alice Blackburn; children—Gilbert H., Sara Ann, John Charles. Asst. instr. U. Ark. and U. Colo.; state bacteriologist Dept. San. Engring., State of Colo., 1924; with Continental Oil Co., 1926-33; admitted to Patent Office bar, 1933; with The Lubrizol Dev. Corp., 1935-51; Continental patent adviser 1948-57; prof., head chemistry dept. So. State Coll., 1958—; Res. officer CWS 1926-41. Mem. (Nat.) Toluene Tech. Com., 1941-45; mem. Nat. Roster Sci. and Specialized Personnel; mem. Tech. Adv. Com., 1941-45; mem. Armed Service Forces Roster of Ammunition, 1941-45. Recipient Distinguished War Service award Petroleum Adminstr. for Wars, 1946. Fellow of Royal Soc. Arts (Eng.), Am. Inst. Chemists; mem. Am. Chem. Soc., A.A.A.S., Sigma Xi, Alpha Chi Sigma, Kappa Alpha, Alpha Phi Epsilon. Mem. Am. Petroleum Inst. com. on book, Phys. Contents of Hydrocarbons, 1939. Contbg. author: The Science of Petroleum, 1938. Author: Fundamental Chemical and Physical Forces in Lubrication, 1935; Practical Selection of Improved Lubricants, 1935; X-ray Diffraction Studies of Lubricants, 1936; Uses of Fats, Fatty Oils and Derivatives in the Petroleum Industry, 1940; Oxidation of Petroleum Lubricants, 1941; Changes Occurring in Oils and Engines from Uses, 1942; Determination of Salts in Addition Agents, 1943; Radio-Sulfur Traces of Sulfurized Lubrication Addition Agents, 1943. Has 108 U. S. patents, also many fgn. country patents, as inventor or co-inventor. Address: Southern State College, Magnolia, Ark. 71753.*

LINCOLN, Rufus Pratt, Am. physician; b. Belchertown, Mass., Apr. 27, 1840; s. Rufus S. and Lydia (Baggs) L.; A.B., Amherst Coll., 1862; M.D., Harvard, 1868; m. Caroline C. Tyler, 1869; 1 son, Rufus Tyler. Served to col. Mass. Volunteers, 1862-65; engaged in pvt. practice medicine, specializing in laryngology and intranasal surgery; developed technique for almost painless removal of semimalignant retronasal growths; a founder N.Y. Laryngol. Soc.; pres. Am. Laryngol. Assn. Died N.Y.C., Nov. 27, 1900.

LIND, Andrew William, Am. sociologist; b. Seattle, Oct. 27, 1901; s. John Andrew and Sophia (Johnson) L.; A.B., U. Wash., 1924; M.A., U. Chgo., 1925, Ph.D., 1931; m. Katherine Evelyn Niles, Sept. 12, 1935; children—Karen (Mrs. Michael Taylor), Alvin, Loren. Teaching fellow, U. Wash., 1923-24; research fellow U. Chgo., 1925-27; research asst. U. Hawaii, Honolulu, 1927-30, faculty, 1931-65, sr. prof. sociology, 1951-67, emeritus, 1968—, dean grad. sch., 1947-51, dir. Social Research Lab., 1942-61; vis. prof. Stanford, summer 1939, Fisk U., 1946-47, U. Wash., summer 1958, U. Colo., 1965; Fulbright research prof. U. Coll. W.I., Kingston, Jamaica, 1954; Fulbright lectr. Chulalongkorn U., Thailand, 1960-61; Fulbright research prof., New Guinea, 1968. Dir. Conf. on Race Relations in World Perspective, Honolulu, 1954. Mem. Hawaiian Acad. Sci. (pres. 1957-58), Phi Beta Kappa, Phi Kappa Phi, Sigma Chi. Author: An Island Community: Ecological Succession in Hawaii, 1938; Hawaii's Japanese: Experiment in Democracy, 1946; Hawaii's People, 1955. Editor, contbg. author: Race Relations in World Perspective, 1956. Home: 2609 Doris Place, Honolulu 96822.*

LIND, James, Scottish physician; b. Edinburgh, 1716; M.D., U. Edinburgh, 1748; apprenticed to George Langlands, 1731; served as surgeon in the navy; stationed at Minorca, 1739; served on coast of Guinea, W.I., Mediterranean, English Channel; apptd. to ship, Salisbury, 1746; lived in Edinburgh, 1748-58; physician to Naval Hosp., Haslar, 1758-94. Fellow Coll. Physicians Edinburgh (treas. 1757-58), Royal Soc., 1777. Author: A Treatise on Scurvy, 1754; An Essay on the Most Effectual Means of Preserving the Health of Seamen in the Royal Navy, 1757; An Essay on Diseases incidental to Europeans in Hot Climates, 1768. Father naval hygiene in Eng.; advocated ventilation of ships, better food, establishment hosp. ships in tropical waters, cleanliness and good ventilation in sick bays; described scurvy, 1753; discovered steam from salt water was fresh and recommended ships be supplied with water by distillation of salt water, 1761; reintroduced lemon juice as a scurvy preventative; studied typhus. Died Gosport, Eng., July 13, 1794.

LIND, Niels Christian, engr.; b. Copenhagen, Denmark, Mar. 10, 1930; s. Axel Holger and Karen (Larsen) L.; M.Sc., Royal Tech. U. Denmark, 1953; Ph.D., U. Ill., 1959; m. Veronica Claire Hummel, Nov. 29, 1957; children—Julie Wilhelmina, Peter Christian. Design engr. Dominia Ltd., Copenhagen, 1953-54; engr. Bell Telephone Co., Montreal, Que., Can., 1954-55; field engr. Drake-Merritt Co., Cut Throat Island, Labrador, 1955; design engr. Fenco Ltd., Montreal, 1956; faculty U. Ill., Urbana, 1956-60, asst. prof., 1959-60; faculty U. Waterloo (Ont., Can.), 1960—, prof. civil engring., 1963—. Mem. Canadian Standards Assn. (chmn. com. on light gage steel design 1964—), Am. Soc. C.E. Research, publs. in analysis stresses in pressure vessels, stability analysis for arches domes, and tall inelastic frameworks and reticulated shells; developed method to analyse deflections in structural frameworks. Home: 78 Roosevelt Av., Waterloo, Ont., Can.*

LIND, Samuel Colville, Am. chemist; b. McMinnville, Tenn., June 15, 1879; s. Thomas Christian and Ida (Colville) L.; A.B., Washington and Lee U., 1899, D.Sc. (hon.), 1939; S.B., Mass. Inst. Tech., 1902; Ph.D. U. Leipzig, 1905; postgrad. U. Paris, Inst. Radium Research, Vienna, Austria; D.Sc. (hon.), U. Colo., 1927, U. Mich., 1940, U. Notre Dame, 1963; m. Marie Holladay, Jan. 24, 1915; 1 son, Thomas Colville. Instr., asst. prof. phys. and inorganic chemistry U. Mich., 1905-13; research and adminstrn. U. S. Bur. Mines, Denver, 1913-16, Golden, Colo., 1916-20, Reno, 1920-23, chief chemist, Washington, 1923-25; asso. dir. Fixed Nitrogen Research Lab., U. S. Dept. Agr., 1925-26; dir. sch. chemistry U. Minn., 1926-35, dean inst. tech., 1935-47. Cons. Union Carbide Corp. (formerly Carbide & Carbon Chems. Co.), Oak Ridge, 1948-65; cons. in radioactivity Oak Ridge Nat. Lab., 1958-65. Recipient Nichols medal award, 1926; Priestley medal award, 1952. Fellow A.A.A.S.; mem. Nat., Minn., Tenn. acads. scis., Am. Philos. Soc., Am. Chem. Soc. (pres. 1940, Remsen Meml. lectr. 1947, East Tenn. sect. Lind lectureship 1954), Electrochem. Soc. (pres. 1927), Am. Phys. Soc., Am. Inst. Chem. Engrs., Nat. Adv. Health Council, Am. Soc. for Engring. Edn., Radiation Research Soc., Sigma Xi (Distinguished Service award 1955), Phi Beta Kappa (hon. mem.), Phi Lambda Upsilon (hon. mem.), Alpha Chi Sigma. Author: The Chemical Effect of Alpha Particles and Electrons, 1921, rev., 1928; (with G. Glockler) The Electrochemistry of Gases and Other Dielectrics, 1939; Radiation Chemistry of Gases, 1961. Editor: Jour. Phys. Chemistry, 1932-51. Editorial bd. Am. Chem. Soc. monographs, Internat. Critical Tables. Research in radioactivity and photochemistry; inventor of electroscope for radio measurements; originator of ionization theory of chem. effects of radium rays. Died Feb. 12, 1965.

LINDAHL, Eric Jean, Am. mech. engr.; b. Cleve., b. Mar. 8, 1905; s. Malcolm and Bertha (Stenstrom) L.; B.S., U. Wyo., 1932, M.S., 1933, M.E., 1946; m. Anna Louise Monfort, Dec. 27, 1939; children—Ralph Malcolm, Peter Magnus, Alfred Carl. Mem. U. S. Engr. Dept., Fort Peck, Mont., 1935-36; instr. Catholic U. Am., 1936-37, U. Md., 1937-39; faculty Ohio State U., 1939-47; head dept. mech. engring. U. Wyo., 1947-61, prof. mech. engring., 1961—; cons., 1940—. Mem. Am. Soc. M.E., Am. Soc. Heating, Refrigerating and Air Conditioning Engrs., Am. Soc. Engring. Edn., Am. Assn. U. Profs., Sigma Xi, Tau Beta Pi. Author: Hydraulic Machinery (with S. R. Beitler), 1947. Contbr. tech. articles engring. publs. Home: 1324 Steele St., Laramie, Wyo.*

LINDAHL, (John Harald) Josua, zoölogist; b. Kongsbacka, Sweden, Jan. 1, 1844; s. Johan and Mathilda (Rjoerklander) L.; A.B., Royal U. Lund (Sweden), 1863, A.M. and Ph.D., 1874; m. Sophie Pohlman, 1877. Accompanied Gwyn-Jeffreys and Carpenter as asst. zoölogist in Brit. deep-sea exploring expdn. in H.M.S. Porcupine, 1870; zoologist in charge of expdn. to Greenland, 1871, in Swedish warships Ingegerd and Gladan; docent in zoology U. Lund, 1874; sec. Royal Swedish delegation to Internat. Geog. Congress, Paris, 1875; sec. Royal Swedish commn. to Centennial Expn., Phila., 1875-77; prof. natural scis. Augustana Coll., Rock Island, Ill., 1878-88; curator Ill. State Mus. Natural History, Springfield, 1888-93; dir. Mus. Natural History, Cin., 1895-1906; mgr. Salubrin Lab., 1906—. Died Apr. 18, 1912.

LINDAUER, Martin, German zoologist; b. Wäldle/Garmisch, Germany, Dec. 19, 1918; s. Matthias and Katharina (Erhard) L.; Dr.nat.; Rockefeller fellow, 1952; m. Franziska Fleck, Dec. 21, 1943; children—Georg, Franziska, Martin. Asst. DFG, 1950; faculty U. Munich, 1955-63, asso. prof., 1961-63; prof., dir. zoöl. inst. U. Frankfort/Main, Germany, 1963—. Mem. Leopoldina, Am. Acad. Arts and Scis., Phi Beta Kappa. Author: Communication Among Social Bees, 1961; also numerous articles. Research on sensory physiology, orientation, mutual communication in honey bees. Address: 8 Flughafenstrasse, Frankfort /Main, West Germany.*

LINDBERG, Bengt Gustaf, Swedish chemist; b. Stockholm, Sweden, July 17, 1919; s. Henry Ragnar and Elsa (Tengvall) L.; Fil.Dr., Stockholm U., 1950; m. Ethel Jonsson, Dec. 28, 1947; children—Elsa, Gustaf. Faculty, Royal Inst. Tech., Stockholm, 1950-65, lectr. wood chemistry, 1955-65; prof. organic chemistry Stockholm U., 1965—; head wood chemistry dept. Swedish Wood Research Inst., 1955-65. Recipient Norblad-Ekstrand medal, RNO. Mem. Royal Swedish Acad. Scis. Research, numerous publs. in carbohydrate chemistry including isolation, characterization and synthesis of natural, low-molecular weight carbohydrates, chem. reactions during pulping processes, structure of polysaccharides from wood, fungi and bacteria. Home: 180 Asögatan, Stockholm Sö, Sweden.*

LINDBERG, Howard Avery, Am. physician; b. Chgo., May 30, 1910; s. Fritz Albin and Ella C. (Carlson) L.; B.S., Northwestern U., 1931, M.S., 1934, M.D., 1935; m. Joan Streeter, Nov. 24, 1938; children—Suzanne, Nancy, Jean. Practice medicine specializing in internal medicine-cardiovascular, Chgo., 1942—; med. dir. Peoples Gas Light & Coke Co., Chgo., 1943—; asso. prof. Northwestern Med. Sch., Chgo., 1948—; chief staff Northwestern Passavant Meml. Hosp., 1956-57. Cons. to pvt. businesses, U. S. Dept. Labor. Mem. Chgo. Heart Assn. (epidemiological com. 1957—), Sigma Xi, Sigma Alpha Epsilon, Phi Rho Sigma. Contbr. articles on coronary prevention to tech. jours. Home: 1112 Greenwood Av., Wilmette, Ill. 60091. Office: 670 N. Michigan Av., Chgo. 60611.*

LINDBERG, Olov Nils Hugo, Swedish biologist; b. Karslkrone, Sweden, July 26, 1914; s. Lars Johan and Elsa (Palmqvist) L.; Ph.D., U. Stockholm (Sweden), 1946; m. Karin Britta Gudmundsson, June 12, 1943; children—Pia, Gudmund. With Wenner-Gren Inst., Stockholm, 1952—, prof. exptl. zoology and zoophysiology, head inst., 1955—; research staff Stat. Biol., Roscoff, France, 1939, Washington U., St. Louis, 1946-47, Stat. Zool., Naples, Italy, 1949, Woods Hole, Mass., 1956, U. Cal. at Los Angeles, 1963-64. Mem. Internat. Cell Research Orgn. (chmn. panel II), Royal Acad. Sci. Editor: (with Ernster) Protoplasmatologia, Chemistry and Physiology of Mitochondria and Microsomes, 1954; (with T. W. Goodwin) Biological Structure and Function, vol. I, vol. II. Research, publs. on biochem. basis of subcellular orgn. and biol. generation of energy. Home: 18 Virebergsvägen, Solna, Sweden. Office: 16 Norrtullsgatan, Stockholm. Sweden.*

LINDBERG, Robert Benjamin, Am. microbiologist; b. Grand Rapids, Mich., Dec. 26, 1914; s. William S. and Gertrude (Tholen) L.; B.S., U. Mich., 1936, M.S., 1937, Ph.D., 1950; m. Louise Larrabee, Nov. 14, 1942; children—Ann Elaine, Orrin Henry. Commd. 2d lt. U. S. Army, 1942, advanced through grades to col., 1960; chief bacteriology br. 18th Med. Gen. Lab., 1944-47, 406th Med. Gen. Lab., Tokyo, 1950-53, chief dept. bacteriology Walter Reed Army Inst. Research, 1953-57, chief U. S. Army Lab., Europe, 1957-61, chief bacteriology br. U. S. Army Surg. Research Unit, Ft. Sam Houston, Tex., 1957—. Decorated Bronze Star medal. Mem. Soc. Am. Bacteriologists (pres. Washington br. 1956-57), Research and Engring. Soc. Am. (pres. S. Tex. br. 1966—), Am. Assn. Immunologists, Am. Soc. Microbiologists, N.Y. Acad. Sci., A.A.A.S., Am. Pub. Health Assn., Sigma Xi. Author: (with L. S. McClung) Manual of Methods in Bacteriology, Anaerobic Bacteria, 1957. Publs. on devel. Sulfamylon burn cream to prevent infection; discovery of histoplasmin and yeast antigen of Histoplasma used in diagnosis of mycotic infections; mgmt. war injuries. Home: 540 Wheaton Rd., San Antonio 78234. Office: U. S. Army Surg. Research Unit, Ft. Sam Houston, Tex. 78234.*

LINDBLAD, Bertil, astronomer; b. Orebro, Sweden, Nov. 26, 1895; s. Birger and Sara Gabriella (Waldenstrom) L.; Ph.D., Upsala U., 1920, U. Copenhagen, 1946; D.Sc., U. Mich., 1950; Dr. Astronomy, U. Torun, 1959; m. Dagmar Bolin, May 24, 1924; children—Bengt Jacob, Per Olof, Birgitta, Bo Sigurd. Study observatories Mt. Wilson, Harvard, Lick Obs., 1920-21; prof. astronomy Royal Swedish Acad. Scis., dir. Stockholm Obs., 1927; head tchr. astronomy U. Stockholm; Morrison research asso. Lick Obs., Cal., 1939. Recipient Medaille Jansen, Acad. of Scis., Paris, 1938, gold medal Royal Astron. Soc., 1948, Bruce medal Astron. Soc. of Pacific, 1953; decorated Comdr. Swedish Order of North Star, Order Leopold of Belgium. Mem. Royal Swedish Acad. Sci. (pres. 1938-39, 60-61), Swedish Natural Sci. Research Council (pres. 1950), Internat. Astron. Union (pres. 1948-52), Internat. Council Sci. Unions (pres. 1952-55), Physiolographic Soc. of Lund, Upsala Soc. Scis., Danish Soc. Scis., Finnish Soc. Sci., Acad. des Sciences of Paris, Royal Astron. Soc. London, Soc. des Sciences of Liege, Accademia dei Lincei of Rome, Am. Acad. Arts and Scis., Nat. Acad. Sci., Am. Astron. Soc., Polish Academy of Scis., Inst. de Coimbra, Portugal. Died June 26, 1965.

LINDBLOM, Charles Edward, Am. social scientist; b. Turlock, Cal., Mar. 21, 1917; s. Charles August and Emma (Norman) L.; B.A., Stanford, 1937; Ph.D., U. Chgo., 1945; m. Rose K. Winther, June 6, 1942; children—Susan, Steven, Eric. Instr. econs. U. Minn., 1939-46; faculty Yale, 1946—, prof. econs. and polit. sci., 1965—. Econ. adviser U. S. aid mission, India, 1963-65. Guggenheim fellow, 1961; Center for Advanced Studies in Behavioral Scis. fellow, 1954-55. Mem. Am. Polit. Sci. Assn., Am. Econ. Assn. Author: Unions and Capitalism, 1949; (with R. A. Dahl) Politics, Economics and Welfare, 1953; (with D. Braybrooke) A Strategy of Decision, 1963; The Intelligence of Democracy, 1965; also articles. Research on theory of complex decision making. Home: 9 Cooper Rd., North Haven, Conn. 06473. Office: Yale, New Haven 06520.*

LINDE, Jonas Otto, Swedish physicist; b. Mo, Angermanland, Sweden, May 1, 1898; s. Olof and Anna (Salomonsson) L.; student U. Lund (Sweden), 1919-24; Ph.D., Royal Inst. Tech., also U. Stockholm, 1939, D.Ph., 1940; m. Karin Mattsson, Oct. 14, 1933; children—Birgitta (Mrs. Sven Jörgen Vessman), Henrik, Magnus, Monica. With Royal Inst. Tech., Stockholm, Sweden, 1925—, prof. solid state physics, 1961-64. Mem. Royal Swedish Acad. Engring. Scis., Royal Acad. Scis. Stockholm. Research, publs. on electric and thermal properties of metallic conductors resulting in establishment of quantitative law for impurity of part of electric resistivity for certain class of alloys; established (with others) existence of order-disorder transformations of fundamental nature in alloys, 1925; discovered ordered alloy structures with regular phase-shifts in atomic arrangement type of which structure was evaluated for copper-Gold II, 1936. Home: 1 Fagelsträcket, Lidingö, Sweden. Office: Royal Inst. Tech., Stockholm 70, Sweden.*

LINDE, Karl Paul Gottfried von, German chemist; b. Berndorft, June 11, 1842; student Polytechnikum, Zurich, Switzerland, 1861-64; prof. applied thermodynamics Technische Hochschule, Munich. Research, patents, and publs. on refrigeration, heat; liquified (independently of Hampson) gases using Joule-Kelvin effect; founder 1st plant for liquefying air by this method, 1895; built machine for prodn. oxygen, 1896; used fractional distillation of liquid air to obtain oxygen on indsl. scale, 1901 (led to widespread use of oxygen in industries at present time). Died Munich, Germany, Nov. 16, 1934.

LINDE, Leonard M., physician; b. N.Y.C., June 1, 1928; s. Ben and Marsha (Weinberg) L.; B.S., U. Cal. at Davis, 1947; M.D., U. Cal. Sch. Medicine, San Francisco, 1951; m. Shirley Dann, Apr. 22, 1951; children—Bruce, Lauren, Brian, Peter. Fellow in pediatric cardiology U. Cal. at Los Angeles Med. Center, 1956-57, asst. prof. cardiology, dept. pediatrics, 1957-61, lectr. in physiology, dept. physiology, 1960, asst. prof., 1960-61, asst. prof. pediatrics and physiology, 1961-63, asst. prof. pediatrics (cardiology) and physiology, 1963—; vis. prof. pediatrics U. Tokyo (Japan), 1965. Pediatric cons. Crippled Children's Services, State of Cal., 1957—; cons. Crippled Children's Services, Child Cardiac Clinic, Imperial County, Cal., 1959—; nat. med. cons. to Surgeon Gen. USAF, 1965—. Recipient Ross award for pediatric research Western Soc. for Pediatric Research, 1962. Diplomate Am. Bd. Pediatrics (council on pediatrics), Am. Bd. Pediatrics, Am. Bd. Pediatric Cardiology; mem. Am. Physiol. Soc., Western Soc. for Pediatric Research (exec. council), Soc. for Pediatric Research, Alpha Zeta. Book reviewer pulmonary circulation, congenital heart disease Annals of Internal Medicine, 1965—. Research and numerous publs. in clin. aspects pediatric cardiology, studies of pulmonary circulation in health and disease, evaluation of phys. and mental devel. in children. Home: 2733 Manning, Los Angeles 90064. Office: Univ. of Cal. Sch. of Medicine, Los Angeles 90024.*

LINDEMAN, Verlus Frank, Am. zoologist; b. Ashton, Ill., Sept. 2, 1902; s. Henry and Tena (Kirsten) L.; B.S., N. Central Coll., 1926; M.S., State U. Ia., 1929, Ph.D., 1930; m. Dorothy Cawelti, Mar. 29, 1929; children—Robert Lindeman, Joan Sheriden. With State U. Ia., 1926-30, instr., 1929-30; faculty Syracuse (N.Y.) U., 1930—, prof., 1945—, acting chmn., 1943-45. Mem. Am. Soc. Zoologists, Soc. Exptl. Biology and Medicine, Am. Physiol. Soc., Sigma Xi (pres. Syracuse chpt. 1939). Research, publs. on neural enzymes, respiratory metabolism, amphibian metamorphosis and DDT and secondary sex characteristics, 5' Nucleotidase in connective tissue. Home: 112 Dewittshire Rd., DeWitt, N.Y. 13214.*

LINDEMANN, Erich, psychiatrist; b. Witten, Germany, May 2, 1900; s. Erich and Anna (Raeker) L.; Ph.D., U. Marburg and Giessen, Germany, 1922, M.D., 1926; m. Elizabeth Brainerd, Sept. 30, 1939; children—Jeffrey, Brenda. Rotating intern U. Cologne Med. Coll., 1925; resident physician dept. neurology U. Heidelberg, Germany, 1926-27; faculty U. Ia., 1927-35; Rockefeller fellow psychiatry, physiology Harvard, 1935-36, instr. psychiatry Harvard Med. Sch., 1937-41, asso. psychiatry, 1941-48, prof. psychiatry, 1954—, also instr. psychiatry, asso. prof. mental health Harvard Sch. Pub. Health, prof. psychiatry emeritus, 1965; vis. prof. psychiatry, Stanford U. Sch. Medicine, 1965—. Mem. expert panel mental health WHO, 1957—. Mem. Am. Psychiat. Assn. (chmn. com. 1952), Group for Advancement Psychiatry (chmn. com. 1950-53), NRC, Social Sci. Research Council, Am. Psychosomatic Soc. (exec. council 1954), Am. Assn. U. Profs., A.M.A., Am. Neurol. Assn., Am. Orthopsychiat. Assn., Am. Psychoanalytic Assn., Am. Psychol. Assn., Nat. Assn. Mental Health, Am. Anthrop. Assn., Am. Assn. Applied Anthropology, Am. Psychosomatic Assn., Assn. Research Nervous and Mental Diseases. Discovered the use of intravenous injections of sodium amytal for the purpose of establishing therapeutic contact with patients; described and demonstrated the significance the forms of path. grief; devel. model for the intro. of preventive psychiat. programs; also inc. social sci. concepts and methods into psychiat. tng. and research. Home: 170 Lake Shore Rd., Boston 02135. Office: Stanford Med. Center, Stanford, Cal. 94305.*

LINDEMANN, Ferdinand, German mathematician; b. Hannover, Germany, Apr. 12, 1852; named prof., Königsberg, Germany, 1883, Munich, 1893-23; also prof., Freiburg, Germany. Author: Untersuchungen über Geometrie, 2 vols., 1875-91. Research on projective geometry and Abelian functions; proved ratio, pi, is a transcendental number, 1882, and therefore it is impossible to square the circle using ruler-and-compass constrn.; developed method of solving equations of any degree using transcendental functions, 1892. Died 1939.

LINDEMANN, Kurt, German surgeon; b. Berlin, July 21, 1901; s. Paul and Marie (Dilgers) L.; ed. univs. Berlin, Kiel; m. Ursula Christa Strauch, Aug. 1, 1946; children—Jürgen, Joachim, Ursula. Prof. surgery and orthopediatrics, Göttingen, Germany; asso. prof., later prof. orthopediatrics, Heidelberg, Germany; dir. univ. clinic orthopediatrics, rector U. Heidelberg. Mem. Leopoldina Acad. Research; Italian Soc. Orthopedics and Traumatology, German Union for Retng. of Handicapped (pres.), German Union for Aid to Handicapped (pres.). Author: Handbuch der Orthopedie; Kurzgefasstes Lehrbuch der orthopädischen Krankheiten; Die Erkrankunger e der Wirbelsäule; Lehrbuch der Krankengymnastik; Ergebniss der Chirurgie und Orthopaedie; Zeitschrift für Orthopedie; also articles. Home: Im Hofert 1, Heidelbert-Schlierbach. Office: Orthopädische Universitatsklinik, Heidelberg-Schlierbach, W. Germany.

LINDEN, Henry Robert, chem. engr.; b. Vienna, Austria, Feb. 21, 1922; s. Dr. Fred and Edith (Lermer) L.; B.S., Georgia Sch. Tech., 1944; Master Chemical Engring., Poly. Inst. of Bklyn., 1947; Ph.D., Ill. Inst. Tech., 1952; div.; children—Robert, Debra. Came to U. S., 1939, naturalized, 1945. Chem. engr. Socony Vacuum Labs., 1944-47; with Inst. of Gas Tech., 1947—, successively supr. oil gasification, asst. research dir., asso. research dir., acting dir., 1947-56. research dir., 1956-61, director, 1961—; research asso. prof. Ill. Inst. Tech., 1954-61, adj. prof., 1961—; v.p., dir. Gas Devels. Corp. Recipient award of merit, operating sect. Am. Gas Assn. Mem. Am. Chem. Soc. (recipient H. H. Storch award; chmn. div. fuel chemistry 1967), American Institute Chem. Engrs., Institute of Fuel, American Gas Assn., Am. Petroleum Inst., Sigma Xi, Phi Kappa Phi, Tau Beta Pi, Phi Lambda Upsilon. Author tech. articles. Holder U. S. and fgn. patents in fuel tech. Devel. new system of correlation of petroleum and pure hydrocarbon properties based on carbon-hydrogen ratio; new theory of severe pyrolysis of petroleum and pure hydrocarbons; new processes for conversion of coal, oil shale and petroleum into low molecular weight paraffin hydrocarbons. Office: 3424 S. State St., Chgo. 60616.*

LINDEN, see van der Linden.

LINDENAU, see von Lindenau.

LINDENBAUM, Seymour Joseph, Am. nuclear physicist; b. N.Y.C., Feb. 3, 1925; s. Morris and Anne (Chanover) L.; A.B., Princeton, 1945; M.A., Columbia, 1949, Ph.D., 1951; m. Leda Isaacs, June 29, 1958. Asso. physicist Brookhaven Nat. Lab., Upton, N.Y., 1951-54, physicist, 1954-63, sr. physicist, 1963—, group leader high energy counter research group physics dept., 1954—. Vis. prof. physics U. Rochester, 1958-59; cons. various univs. and govt. labs., 1956—, Centre d'Etudes Nucleaires de Saclay, France, 1957, CERN, Geneva, 1962. Fellow Am. Phys. Soc.; mem. A.A.A.S. Discoverer nucleon isobars dominated high energy elementary particle interactions; originator Isobar model; inventor, with research group, of automatic counter hodoscope system and on-line computer technique. Home: Old Bridge Rd., Setauket, N.Y. 11785. Office: Brookhaven Nat. Lab., Upton, N.Y. 11973.*

LINDET, (Gaston-Aimé-) Léon, French chemist, agronomist; b. Paris, Apr. 10, 1857; docteur ès sciences, 1886; became substitute prof. Paris Agronomic Inst., 1888, prof. tech. agr., 1891; mem. French Acad. Scis., 1920, Acad. Agr. Author: La Bière, 1892; Le Froment et sa mouture, 1903. Research on food canning industry, French grape vines, composition of barley and malt used for brewing, fermentation and storing of bread, starch of seeds. Died Gaillon, France, June 15, 1929.

LINDGREN, Frank Tycko, Am. biophysicist; b. San Francisco, Apr. 14, 1924; s. Tycko and Grace (Lund) L.; B.A., U. Cal. at Berkeley, 1947, Ph.D., 1955; m. Helen Montgomery Darrow, Aug. 8, 1953. With Donner Lab. U. Cal., Berkeley, 1955—, research asso. in biophysics, 1956—. Mem. Phi Beta Kappa, Sigma Xi. Asso. editor Lipids, 1966—. Studies, publs. on phys., chem. properties of blood lipids; analytical methods in lipid biochemistry, ultracentrifugation, gas chromatography. Home: 2707 Rose St., Berkeley 94708. Office: Donner Lab., U. Cal., Berkeley, Cal. 94720.*

LINDGREN, Waldemar, geologist; b. Kalmar, Sweden, Feb. 14, 1860; s. Johan Magnus and Emma (Bergman) L.; ed. in Sweden; M.E., Sch. of Mines, Freiberg, Germany, 1883; D.Sc., Princeton, 1916, Harvard, 1935; m. Ottolina Allstrin, Mar. 8, 1886. Asst. geologist U. S. Geol. Survey, 1884-95, geologist, 1895-1915, chief geologist, 1911-12; William Barton Rogers prof. econ. geology Mass. Inst. Tech., in charge dept. geology, 1912-33. Author: Mineral Deposits. Promoted application of petrography and microscopy to study of ores and mineral deposits. Died Nov. 3, 1939.

LINDHEIMER, Ferdinand Jacob, b. Frankfurt-am-Main, Germany, May 21, 1801; s. Johan H. Lindheimer; attended U. Weisbaden, U. Bonn (Germany); m. Elenore Reinarz, 1846. Came to U. S., 1834; served in Tex. Army in war for independence; travelled throughout Tex. collecting bot. specimens, 1841-52; participated in exptl. communistic colony, Bettina, Tex., 1847; editor Neu Braunfelser (Tex.) Zeitung, 1852-70; his bot. work described in Plantae Lindheimerianae pub. in Boston Jour. of Natural History, 1845, 50. Died New Braunfels, Comal County, Dec. 2, 1879.

LINDHOLM, Einar, Swedish physicist; b. Göteborg, Sweden, July 20, 1913; s. Josef and Anna (Bostedt)

L.; Fil.mag., U. Lund (Sweden), 1935; Fil.lic., U. Stockholm (Sweden), 1939, Fil.doktor, 1942; m. Kerstin Sandvall, Oct. 25, 1942; children—Margareta, Bengt, Anders. Asso. prof. physics U. Stockholm, 1942-47; tchr. high sch., 1949-50; asso. prof. Chalmers Inst. Tech., Göteborg, 1950-56; prof. Royal Inst. Tech., Stockholm, 1956—. Contbr. articles to tech. jours. Developed theory of broadening of spectral lines which explained displacement, 1942; built tandem mass spectrometer for study of charge exchange between positive ions and molecules; worked out theory of recombination energies of positive ions; studied mass spectra and ion-molecule reactions of many small molecules and results interpreted using molecular orbital theory. Home: 5 Sigynvagen, Djursholm, Sweden. Office: Royal Inst. Tech., S-100 44 Stockholm, Sweden.*

LINDLEY, John, English botanist; b. Catton, Eng., Feb. 5, 1799; s. George Lindley; m. Miss Freestone, 1823; 2 daus., 1 son, Nathaniel. Became garden asst. Hort. Soc., 1822, asst. sec., 1828, vice sec., 1841, mem. council, hon. sec., 1858-62; prof. botany Univ. Coll., London, 1829-60, emeritus, 1860-65. Lectr. botany Apothecaries' Co., Chelsea, 1836-53. Honored in name of genus Lindleya of order Rosaaeae. Fellow Royal Soc., 1828 (Royal medal 1857); mem. French Acad. Scis., 1853, Agr. Soc. Author: A Synopsis of British Flora, arranged according to the Natural Order, 1829; An Introduction to the Natural System of Botany, 1830; The Genera and Species of Orchidaceous Plants, 1830-40; (with William Hutton) The Fossil Flora of Great Britain; Key to Structural and Systematic Botany, 1835, enlarged as Elements of Botany, 1841; The Theory of Horticulture, 1840, enlarged as The Theory and Practice of Horticulture, 1842; The Vegetable Kingdom, 1846; (with Paxton) The Flower Garden, 1851-53. Cofounder, editor Gardner's Chronicle, 1841-65; editor Jour. Hort. Soc., 1846-55. Research on roses and pomology of Britain; preferred classification system of A. L. de Jussieu to that of Linnaeus; outstanding taxonomist of his time; amassed orchid collection, later acquired by Kew Gardens. Died Chiswick, Eng., Nov. 1, 1865.

LINDNER, Eduard, Czechoslovakian physician; b. Prostejov, Czechoslovakia, Feb. 5, 1917; s. Eduard and Marie (Losová) L.; medicinae universae doctor Charles U., Prague, Czechoslovakia, 1945; Candidate Med. Scis., J. E. Purkyne U., Brno, Czechoslovakia, 1960; Habilitation, Palacky U., Olomouc, Czechoslovakia, 1965; m. Zofie Smolková, July 30, 1947; children—Zdenka, Jitka, Blanka. Staff, dept. surgery Clinic Gynecology, Olomouc, 1945-49, dept. gynecology, 1949—, clinic rep., 1953—, temporary clinic mgr., 1959-61. Mem. Czechoslovak spolecnosti J. E. Purkyne. Research, publs. on problems of fertility and sterility especially relationship of tocopherols. Home: 12, Nám. cs. armády, Olomouc. Office: Clinic of Gynaec., Olomouc, Czechoslovakia.

LINDNER, Erwin, German entomologist; b. Böglins, June 7, 1888; s. Otto and Anna (Kessler) L.; Ph.D., U. Munich; m. Freya Uhlenhuth, Oct. 22, 1918; 1 child, Elmar. Became asst. zoologist Württ. Naturaliensammlung, 1918; curator, curator-in-chief Staatl. Mus. fur Naturkunde, Stuttgart, Germany; with German expdn. to Gran Chaco, 1925-26; head 1st German Expdn. in E. Africa, 1951-52; visited S. and W. Africa, 1958-59. Recipient Fabricius medal. Mem. German Entomol. Soc., German Zool. Soc. Editor: Die Fliegen der paläarktischen Region. Home: Kräherwald 191, Stuttgart. Office: Arsenalplatz 3, Ludwigsburg, W. Germany.

LINDNER, Károly, Hungarian chemist; b. Budapest, Hungary, Jan. 29, 1921; s. Elek and Irma (Schopp) L.; Academic Candidate degree József Nádor Tech. U., Budapest, 1956; m. Edit Gottschall, Feb. 12, 1945; children—Mária, Károly, Ferene; m. 2d, Katalin Szotyori, Oct. 15, 1966. Demonstrator, Tech. U., Budapest, 1941-42; chemist Inst. Food Control Budapest, 1942-49; became head dept. food chemistry Inst. Nutrition, Budapest, 1949, now dep. dir. Vice pres. Council Food Hygiene, Ministry Health; mem. Council Food Analysis, Hungarian Acad. Sci. Author: Hungarian Food Composition Tables. Quantitative determination of free and protein amino acids in foodstuffs; demonstrated similarities in amino acid compositions of plant varieties, dominant role of its chief protein fraction. Home: 32/b Németvölgyi, Budapest. Office: 3/a Gyáli út, Budapest, Hungary.*

LINDNER, Katalin Szotyori, Hungarian chemist; b. Békéscsaba, Hungary, Dec. 30, 1920; d. István and Judit (Kesjár) Szotyori; ed. U. Scis., Szeged, Hungary; m. Sándor Szöke, Jan. 20, 1944; 1 dau., Katalin; m. 2d, Károly Lindner, Oct. 15, 1966. Chief lab. sugar factory Petöháza, 1944-49; research worker Research Inst. Sugar Industry, 1949-50; sr. clk. tech. dept. Directorate Sugar Industry, Ministry Food Industry, 1951-54; sci. research worker Inst. Nutrition, Budapest, 1954—. Research and publs. on beet sugar manufacture, influence of diet on ossification, composition of different organs, tissues and human milk; devel. of methylesterification method, method for determination of vitamin C in heat-treated and dried matters. Home: 32/b Németvölgyi, Budapest. Office: 3/a Gyáli út, Budapest, Hungary.*

LINDQUIST, Armin Hellmut, oceanographer, biologist; b. Riga, Latvia, Feb. 4, 1928; s. Alfred Heinrich and Wera (Fleissner) L.; Licenciat Philosophy, U. Stockholm (Sweden), 1957, D.Philosophy, 1961; m. Marianne Barrling, June 6, 1954; children—Susanne, Madeleine, Annika. Staff, Royal Bd. Fisheries, Gothenberg, 1955—, researcher Inst. Marine Research, Lysekil, Sweden, 1958—, dir., 1966—. Cons. UNESCO, S.Am., 1964. Research, publs. on water exchange and zooplankton distbn. in Baltic, long term variations in marine environment in Scandinavia, biology of clupeidae. Home: Kungstorget 7. Office: Inst. Marine Research, S-45300 Lysekil, Sweden.*

LINDQUIST, Juan C., Argentine agrl. engr.; b. Buenos Aires, Argentina, Feb. 9, 1899; s. Arvid and Juana (Dominique) L.; student U. La Plata (Argentina). With Faculty Agronomy, La Plata, prof. phytopathology; dir. Instituto de Botanica Spegazzini, La Plata. Mem. Fitosanitarias (dir.), Agronomy Argentine Soc., Agronomy Soc., Bot. Argentine Soc., Sociedad Latino Americana de Fitopatologia. Research, numerous publs. on Argentine rusts. Home: Calle 1 Number 430, La Plata, Argentina.*

LINDQVIST, Gunnar August, Finnish psychiatrist; b. Nurmijarvi, Jan. 16, 1904; s. August Konstantin and Hanna (Schmidt) L.; M.D., U. Helsinki (Finland); m. Sunna Hagman, June 21, 1934; children—Elisabeth, Peter. Physician-in-chief Niuvanniemi Psychiat. Hosp.; specialist in neurology and psychiatry. Mem. Finnish Acad. Medicine, Finnish Assn. Neurology and Psychiatry. Address: Niuvanniemi, Kuopio, Finland.

LINDROTH, Carl Hildebrand, Sweidsh entomologist, biogeographer; b. Lund, Sweden, Sept. 8, 1905; s. Hjalmar Axel and Stina (Hildebrand) L.; mag.phil., U. Stockholm, 1926; lic.phil., U. Uppsala (Sweden), 1929, dr.phil., 1932; m. Gun Bodman, May 21, 1931; children—Britta (Mrs. Dick Hansson), Claes, Robert, Magnus, Rolf. High sch. tchr., 1932-51; prof. entomology U. Lund (Sweden), 1951—, dir. Zool. Inst., 1956-61. Hon. mem. Zool. Bot. Gesellschaft Vienna, Royal Entomol. Soc. (London), Entomol. Soc. Stockholm; mem. Soc. Biology Tchrs. (past pres.), Swedish Soc. Oikos (past pres.), Entomol. Soc. Lund (pres. 1952—). Author: The Faunal Connections between Europe and North America, 1957; also numerous articles. Attempted to reconstruct Pleistocene history of terrestrial fauna of No. circumpolar region; studied influence of man on dispersal of organisms, location of refuges where animals and plants survived during glacial periods. Home: 2B Vintergatan, Lund, Sweden.*

LINDSAY, George Edmund, Am. biologist; b. Pomona, Cal., Aug. 17, 1916; s. Charles W. and Alice (Foster) L.; B.A., Stanford, 1951, Ph.D., 1956. Dir., San Diego Mus. Natural History, 1956-63, Cal. Acad. Scis., San Francisco, 1963—. Fellow Cal. Acad. Scis., San Diego Zool. Soc.; Cactus and Succulent Soc. Am.; mem. A.A.A.S. (exec. com. Pacific div. 1956-60), Bot. Soc. Am., Cal. Bot. Soc., San Diego Soc. Natural History (past dir., life mem.). Research, numerous publs. on classification and physiol. and morphological adaptation of desert plants, desert flora of Southwestern U. S. and Mexico especially Baja Cal. Home: 55 Chamasero St., San Francisco 94132. Office: Cal. Acad. Scis., San Francisco 94118.*

LINDSAY, James Bowman, elec. engr., philologist; b. Carmyllie, Scotland, Sept. 8, 1799; ed. St. Andrews U., 1822. Apprenticed as hand-loom weaver, Carmyllie; became lectr. math. and phys. sci. Watt Instn., Dundee, Scotland, 1829. Author: Chrono-Astrolabe, 1858; A Full Set of Astronomical Tables; incomplete compilation of pentecontaglossal dictionary. Patented wireless system of telegraphy in which water was used as an elec. conductor. Died June 29, 1862.

LINDSAY, John Ralston, physician, educator; b. Renfrew, Ont., Can., Dec. 23, 1898; s. John M. and Christena (Wright) L.; M.D., McGill U., 1925; postgrad. clinic, Zurich, Switzerland, Vienna, Austria; M.D. honoris causa, U. Uppsala (Sweden), 1963; m. Elizabeth Wood, Feb. 6, 1937 (div. 1955); children—Christena W. (Mrs. Robert Gritschke), Anne S. (Mrs. Don M. Greer, Jr.), Elisabeth W.; m. 2d, Dorothy Doyle Morrison, Mar. 1, 1962 (dec. 1964). Came to U. S., 1928, naturalized, 1936. Faculty, U. Chgo., 1928—, prof., head otolaryngology sect., 1940-64, Thomas D. Jones prof. surgery, head otolaryngology sect., 1964—, dir. Midwestern Temporal Bone Banks Center, 1961—. Diplomate Am. Bd. Otolaryngology (dir.). Fellow A.C.S.; mem. A.M.A. (chmn. ENT sect. 1955-56), Ill., Chgo. med. socs., Am. Acad. Ophthalmology and Otolaryngology (Gold Key award 1940, council 1962—, pres. 1964), Chgo. Laryngol. and Otol. Soc. (pres. 1947), Inst. Medicine Chgo., Am. Laryngol., Rhinol. and Otol. Soc. (pres. 1961-62), Otosclerosis Study Group (pres. 1953), Am. Otol. Soc. (pres. 1956-57, award of merit 1961), Am. Broncho-Esophagological Assn., Am. Laryngol. Assn., Triological Soc. (pres. 1962), Collegium Oto-Rhino-Laryngologicum Amicitiae Sacrum (pres. 1967; George E. Shambaugh prize 1959), Sigma Xi. Research, numerous publs. in pathology of ear and physiology of vestibular system. Home: 5801 Dorchester Av., Chgo. 60637.*

LINDSAY, Robert Bruce, Am. physicist; b. New Bedford, Mass., Jan. 1, 1900; s. Robert and Eleonora (Leuchsenring) L.; A.B., M.S., Brown U., 1920; Ph.D., Mass. Inst. Tech., 1924; Ed.D. (hon.), R.I. Coll., 1959; m. Rachel Tupper Easterbrooks, July 29, 1922; children—Robert, Evelyn Tupper (Mrs. Richard C. Roberts). Instr. physics Mass. Inst. Tech., 1920-22; fellow Am.-Scandinavian Found., Copenhagen, Denmark, 1922-23; instr. physics Yale, 1923-27, asst. prof., 1927-30; asso. prof. physics Brown U., Providence, 1930-36, Hazard prof., 1936—, chmn. dept., 1934-54, dir. ultrasonics lab., 1948-62, dean grad. sch., 1954—. Vis. prof. physics Poly. Inst. Bklyn., 1931-46; mem. adv. panel Nat. Bur. Standards, 1957—; U. S. mem. Internat. Commn. on Acoustics, 1963—; mem. physics survey com. Nat. Acad. Scis.-NRC, 1964-66. Fellow A.A.A.S. (v.p. 1958-59); mem. Am. Phys. Soc., Am. Acad. Arts and Scis. (v.p. 1957-59), Acoustical Soc. Am. (pres. 1956-57, editor-in-chief 1957—, gold medal 1963), Am. Assn. Physics Tchrs. (Distinguished Service citation 1963), Am. Inst. Physics (gov. bd. 1957—, exec. com. 1959—), Soc. Rheology, Am. Assn. U. Profs., Am. Math. Soc., Assn. Am. Univs. (pres. Assn. Grad. Schs. 1965-66), Philosophy Sci. Assn., History Sci. Soc., Phi Beta Kappa, Sigma Xi. Author: Acoustics, 1930; (with H. Margenau) Foundations of Physics, 1936; Physical Statistics, 1941; Physical Mechanics, 1950; Concepts and Methods of Theoretical Physics, 1951; Mechanical Radiation, 1960; also articles on atomic physics, acoustics, history and philosophy of physics. Home: 71 Vassar Av., Providence 02906.*

LINDSEY, Alton Anthony, Am. ecologist; b. Monaca, Pa., May 7, 1907; s. Earl C. and Lois (Whitmarsh) L.; B.S., Allegheny Coll., 1929; Ph.D., Cornell U., 1939; m. Elizabeth Smith, June 2, 1939; children—David Earl, Louise W. Biologist, Byrd Antarctic Expdn. II, 1933-35; instr. Am. U., Washington, 1937-40; asst. prof. U. Redlands (Cal.), 1940-42; asst. prof. U. N.M., Albuquerque, 1942-47; faculty Purdue U., Lafayette, Ind., 1947—, prof. biology, 1958—, ecologist Canadian Arctic Permafrost Expdn., 1951. Recipient Spl. Congl. medal, 1935. Fellow A.A.A.S., Ind. Acad. Sci.; mem. Ecol. Soc. Am., Izaak Walton League, Nature Conservancy, Wilderness Soc., Save-the Dunes Council, Phi Beta Kappa, Sigma Xi. Editor: Natural Features of Indiana, 1966. Publs. on systematization vegetational analysis methods in plant sociology, vegetation State of Ind. Home: 191 Drury Lane, West Lafayette, Ind. 47906. Office: Purdue U., Lafayette, Ind. 47907.*

LINDSLEY, Donald Benjamin, Am. physiol. psychologist; b. Brownhelm, O., Dec. 23, 1907; s. Benjamin Kent and Martha E. (Jenne) L.; A.B., Wittenberg Coll., 1929; M.A., U. Ia., 1930, Ph.D., 1932; D.Sc. (hon.), Brown U., 1958, Wittenberg U., 1959, Trinity Coll., Hartford, Conn., 1965; m. Ellen Ford, Aug. 16, 1933; children—David Ford, Margaret Roberts, Robert Kent, Sara Ellen (Mrs. Harold A. Lyons). Instr., U. Ill., 1932-33; NRC fellow Harvard Med. Sch., 1933-35, William James lectr., 1958; research asso. Western Res. U. Med. Sch., 1935-38; asst. prof. psychology Brown U., 1938-46; dir. psychology and neurophysiology labs. Bradley Hosp., East Providence, R.I., 1938-46; dir. war research project on radar operation OSRD, Yale, 1943-45; prof. psychology Northwestern U., 1946-51; prof. psychology and physiology U. Cal. at Los Angeles, 1951—, mem. Brain Research Inst., 1961—. Mem. sci. adv. bd. USAF, 1947-50; mem. undersea warfare com. NRC, 1951-64; U. Research lectr. U. Cal., Los Angeles, 1960. Recipient U. S. Presdl. certificate of merit, 1948. Guggenheim fellow, 1959. Mem. Am. (past pres. div. physiol. and comparative psychology, Distinguished Sci. Contbn. award 1959), Midwestern (past pres.), Western (past pres.) psychol. assns., Central Electroencephalographic Assn. (past pres.), A.A.A.S. (past v.p.), Western Electroencephalographic Soc. (past pres.), Nat. Acad. Scis. (past sect. chmn.), Am. Acad. Arts and Scis., Soc. Exptl. Psychologists, Am. Physiol. Soc., Am. Acad. Neurology, Internat. Brain Research Orgn., Am. Electroencephalographic Soc., Soc. for Exptl. Biology and Medicine. Research, numerous publs. on muscular contraction and reflex control movement, behavioral disorders in childhood, mechanisms of sleep, wakefulness and consciousness, brain mechanisms of perception, attention, learning and vision, intra-brain relationships; pioneered use of electroencephalogram in study brain devel. in infancy and childhood. Home: 471 23d St., Santa Monica, Cal. 90402. Office: Dept. Psychology, U. Cal. at Los Angeles, Cal. 90024.*

LINDSLEY, John Berrien, Am. physician; b. Princeton, N.J., Oct. 24, 1822; s. Philip and Margaret (Lawrence) L.; A.B., U. Nashville, 1839; M.D., U. Pa., 1843; D.D. (hon.) Princeton, 1856; m. Sarah McGavock, Nov. 9, 1857; 6 children. Ordained by Presbytery of Nashville (Tenn.), 1846; faculty med. dept. U. Nashville (1st school of kind south of Ohio River), 1850, induced trustees to buy Peabody Normal Coll., dean, 6 years, prof. chemistry and pharmacy, until 1873, chancellor, 1855, dean med. sch., 4 years; in charge of Confederate hosps. in Nashville during Civil War; organized Montgomery Bell Acad. as prep. sch., 1867; an organizer Tenn. Coll. Pharmacy, 1873, later prof. materia medica; health officer, Nashville, 1876-80; contbr. to Theol. Medium; published Mil. Annals of Tenn. Confederate. Died Nashville, Dec. 7, 1897.

LINDSTROM, David Edgar, Am. sociologist; b. Charleston, Neb., Apr. 14, 1899; s. Gustaf and Hulda (Lundgren) L.; B.Sc. in Agr., U. Neb., 1924; M.Sc., U. Wis., 1928, Ph.D., 1932; postgrad. U. Ill.; m. Mary Lucile Halverson, June 27, 1930; children—Judith (Mrs. Ralph A. Keller), David G., Daniel A. Field man U. S. Dept. Agr., 1919-20, U. S. Reclamation Service, 1921-22; instr. Neb. Sch. Agr., 1924-25; tchr. sci., Curtis, Neb., 1925; faculty U. Ill., Urbana, 1929-—, prof. sociology, 1943-—. Dir. rural research Internat. Christian U., Tokyo, 1953-55; faculty rural sociology and extension wing Coll. Agr., Jabalpur, India, 1960-62; vis. prof. Va. Poly. Inst., Blacksburg. Recipient Ill. Welfare Assn. Distinguished Service award, 1939; Swedish Social Sci. Council grantee, 1949-—; Distinguished Service award Nat. Congress Christians and Jews, 1955. Mem. Am. Country Life Assn. (past pres.), Midwest (past pres.), Rural sociol. socs., Am. Sociol. Assn., Am. Assn. U. Profs. Author: Rural Life and the Church, 1946; American Farmers' and Rural Organizations, 1948; American Rural Life, 1948; American Foundations of Religious Liberty, 1950; Rural Social Change, 1960; (with Emil Brunner) The Meeting of Sociology and Religion, 1962; also numerous articles. Research on farmers' organizational participation, devel. rural community schs., diffusion of farming and homemaking practices in Japanese rural villages, farmers' attitudes toward new social policies in Sweden, changes in rural communities in Sweden, tech. change in farming in India, ednl. and vocational needs of rural youth.*

LINDUSKA, Joseph Paul, Am. biologist; b. Butte, Mont., July 25, 1913; s. Joseph and Helen (Nettick) L.; B.A., Mont. State Sch. Mines, 1934; M.A., U. Mont., 1938; Ph.D., Mich. State U., 1949; m. Lilian Ruth Hopkins, Aug. 7, 1936; children—Joanne Ruth (Mrs. Kent S. Price III), James Joseph. Biologist, Mich. Game Dept., Lansing 1940-42; entomologist Bur. Entomology; Orlando, Fla., 1942-46; from protection leader insecticides, wildlife, to chief br. Game Mgt. U. S. Fish and Wildlife Service, Washington, 1947-56; dir. pub. relations, wildlife mgr. Remington Arms Co., Bridgeport, Conn., 1956-66; asso. dir. U. S. Bur. Sport Fish and Wildlife, 1966-—. Reient Conservation Edn. award Wildlife Soc.; Conservation Service award U. S. Dept. Interior. Mem. Outdoor Writers Assn. Am. (dir., chmn. conservation council, Jade of Chiefs award), Mason-Dixon Outdoor Writers (past pres.), Wildlife Soc. (past pres.), Sigma Xi, Phi Sigma. Editor: Waterfowl Tomorrow, 1964. Author numerous research publs. on entomology and ecology. Address: Bur. Sport Fish and Wildlife, Washington 25.*

LINDVALL, Frederick Charles, Am. elec. and mech. engr.; b. Moline, Ill., May 29, 1903; s. Gustav and Alma (Freeburg) L.; student U. Cal. at Los Angeles, 1920-22; B.S., U. Ill., 1924; Ph.D., Cal. Inst. Tech., 1928; D.Sc., U. Ireland, 1963; Doctor of Engineering, Purdue U., 1966; m. Janet Smith, Aug. 27, 1928; children—Charles Eric, Martha Joan (Mrs. Henry Parsons Erwin, Jr.), John Robert. Teaching fellow Cal. Inst. Tech., Pasadena, 1925-28, faculty, 1930-—, prof., 1942, supr. OSRD, 1942-45, chmn. div. engring. and applied sci., 1945-—; engr. Gen. Electric Co., 1928-30. Member board of dirs. Bell & Howell Corp., Preco Corp. Mem. pres.' medal com. NSF, 1962-65; cons. Pres.' Office Sci. and Tech.; mem. regional adv. bd. Bank of Am. Recipient Naval Ordnance Devel. award, 1944; Pres.' citation for merit, 1946. Fellow Am. Inst. E.E., Am. Soc. M.E.; member of the National Academy of Engineering, Am. Soc. for Engring. Edn. (pres. 1957-58), Engrs. Council for Profl. Devel. (chmn. edn. and accreditation com.), Nat. Acad. Scis. (adv. com. on Africa 1963-—), Sigma Xi, Tau Beta Pi. Contbr. articles on vacuum switching glow discharge phenomena, weld testing and radiography, high voltage phenomena, ry. equipment, vibration and dynamics to profl. jours. Home: 1224 Arden Rd., Pasadena, Cal. 91106.*

LINDWALL, Harry Gustave, Am. chemist; b. New Haven, Nov. 22, 1902; s. Gustav Jacob and Christine (Swenson) L.; B.S. Yale, 1923, Ph.D., 1926. Research chemist E. I. du Pont de Nemours & Co., 1926-28; faculty chemistry N.Y. U., 1928-59, prof., 1938-59, chmn. dept., 1937-55; tech. information scientist Olin Mathieson Chem. Corp., New Haven, 1959-67; chem. cons., 1967-—. Cons. for anti-malarial drugs Squibb Inst. for Med. Research, 1943, 44. Fellow Am. Inst. Chemists (past chmn. N.Y. chpt.); mem., Am. Chem. Soc., A.A.A.S., Sci. Research Soc. Am., Phi Beta Kappa, Sigma Xi, Tau Beta Pi, Alpha Chi Sigma, Phi Lambda Upsilon. Publs., research on chemistry and synthesis of indole, pyridine, quinoline compounds. Address: 111 Dessa Dr., Hamden, Conn. 06517.*

LINEHAN, Daniel, Am. geophysicist, dir. obs.; b. Beverly, Mass., Aug. 6, 1904; s. Daniel M. and Louise (Fogg) L.; A.B., Boston Coll., 1930, M.S., 1931; postgrad. theology Weston Coll., 1933-37; M.A. in Geology, Harvary, 1939, postgrad., 1940-45; D.H.L., Le-Moyne Coll., 1957; D.Sc, Holy Cross Coll, 1959, Lowell Technol. Inst., 1960. Joined Soc. of Jesus, 1924, ordained priest Roman Cath. Ch., 1936; instr. physics Coll. of Holy Cross, 1931-33; prof. physics Weston Coll., 1941-43; seismologist charge Weston (Mass.) Obs., 1934-50, dir., 1950-—; prof. geophysics Boston Coll., 1946-—, chmn. dept., 1948-63. Collaborator U. S. Coast and Geodetic Survey, 1940-—; cons. geophysicist Mass. Dept. Pub. Works, 1941-51, Mass. Dept. Pub. Health, 1940-—, U. S. Navy Dept. Antarctica, 1954-58; mem. UNESCO Seismol. Missions, 1961-64; distinguished lectr. Am. Assn. Petroleum Geologists, 1959; geophysicist Dow Expdn. to Arctic, summer, 1954, U. S. Navy Expdn. to Antarctica, 1954-55, U. S. Navy Operation Deepfreeze I, 1955-56, Deepfreeze III, 1957-58. Recipient Insignis medal Fordham U., 1958, Distinguished Pub. Service award U. S. Navy, 1958, Golden Plate award, 1962, Antarctic medal, 1965. Registered profl. engr., R.I. Mem. Seismol. Soc. Am., Internat. Union Geophysicists, Arctic Inst. N.Am., Brit. Glaciol. Soc., Scott Polar Soc., Am. Meteorol. Soc., Am. Inst. Mining, Metall. and Petroleum Engrs., Soc. Exploration Geophysicists, A.A.A.S., Am. Geog. Soc., Internat. Inst. Elec. and Electronic Engrs., Oceanographic Found., Am. Assn. Petroleum Geologists, Am. Mus. Natural History, Am. Polar Soc., European Assn. Exploration Geophysicists, Am. Radio Relay League, Cath. Commn. Intellectual and Cultural Affairs, Boston Mus. Sci., Seabees (hon.), Boston Soc. Civil Engrs., Boston Geol. Soc. (pres. 1947-49), New Eng. Water Works, Explorers Club. Pioneer in measurement thickness of ice at South Pole; introduced the applications of geophys. methods to other fields of science; established network of seismic stas. in Northeastern U. S. Address: Weston Observatory, 319 Concord Rd., Weston, Mass. 02193.*

LINELL, Martin Larsson, entomologist; b. Sweden, 1849; ed. in Sweden. Came to U. S., 1879; commd. to work on Coleoptera collection Nat. Mus., Washington, 1888. Credited with organizing Coleoptera specimens in nat. collection. Died 1897.

LING, Arthur Robert, Brit. biochemist; b. Esher, Eng., May 6, 1861; ed. Finsbury Tech. Coll.; M.Sc., Birmingham, Eng.; m. Harriet Emmeline Ella Eyles, 1886; 3 daus. Became cons. chemist B.E.R. Newlands, 1898-1903; ind. cons. chemist, 1903-20; prof. malting and brewing, also biochemistry fermentation U. Birmingham, 1920-31, emeritus prof., 1931-—; asst. chemist, later chief chemist London Beetroot Sugar Assn.; lectr. fermentation industries Sir. John Cass Inst., London. Fellow Inst. Chemistry. Editor, Jour. Inst. Brewing, 1895-1920. Research and publs. on organic chemistry, analytical chemistry, biochemistry, including chemistry of starch, glycogen with products of their hydrolysis with enzymes of barley and barley malt. Died May 14, 1937.

LING, Gilbert Ning, molecular biologist; b. Nanking, China, Dec. 26, 1919; s. Yen-tse and Chi-lan (Phi-lan Ho) L.; B.Sc. (Ku Meng-yü scholar) Nat. Central U., Chungking, China, 1943; postgrad. United S.W. U., Kumming, China; Ph.D. (Boxer Indemnity fellow), U. Chgo., 1948, postgrad. (Seymour Coman fellow); m. Shirley Wong, July 18, 1951; children—Mark, Timothy, Eva. Came to U. S., 1945, naturalized 1962. Instr., Johns Hopkins Sch. Medicine, 1950-53; faculty U. Ill. Sch. Medicine, Chgo., 1953-57, asso. prof., 1956-57; sr. research scientist dept. basic research Eastern Pa. Psychiat. Inst., Phila., 1957-62; dir. dept. molecular biology Pa. Hosp., Phila., 1962-—. Mem. Fedn. Am. Socs. for Exptl. Biology, Biophys. Soc. Author: A Physical Theory of Living State: The Association-Induction Hypothesis, 1962; also articles. Developed microelectrode, new phys. theory living cell; research on electropotential living cells, molecular mechanism in control cell functions by hormones and drugs. Home: 307 Berkeley Rd., Merion, Pa. 19066. Office: 8th and Spruce Sts., Phila. 19107.*

LINGAFELTER, Edward Clay, Jr., Am. chemist; b. Toledo, Mar. 28, 1914; s. Edward Clay and Winifred (Jordan) L.; B.S., U. Cal. at Berkeley, 1935, Ph.D., 1939; m. Roberta Crowe Kneedler, Apr. 30, 1938; children—Robert Edward, Thomas Edward, James Edward, Richard Edward, Daniel Edward. With U. Wash., Seattle, 1939-—, prof. chemistry, 1952-—, asso. dean Grad. Sch., 1960-—. Mem. Am. Chem. Soc., Am. Crystallography Assn., A.A.A.S. Research, publs. on solutions, molecular structures of paraffin-chain and coordination compounds. Home: 5323 27th Av. N.E., Seattle 98105.*

LINGANE, James Joseph, Am. chemist; b. St. Paul, Sept. 13, 1909; s. Vincent A. and Catherine (Hennessy) L.; B.Chemistry, U. Minn., 1935, Ph.D., 1938; A.M. (hon.), Harvard, 1946; m. Beatrice M. Kinderwater, May 28, 1938; children—Peter J., Mary B., Catherine B., Paul J. Instr., U. Minn., 1938-39, U. Cal. at Berkeley, 1939-41; faculty Harvard, 1941-—, prof. chemistry, 1952-—, chmn. dept. chemistry, 1956-59. Recipient Distinguished Achievement award U. Minn., 1959. Mem. Am. Chem. Soc. (Fisher award in analytical chemistry 1958), Soc. for Analytical chemistry (hon., London, Eng.), Am. Acad. Arts and Scis. Author: (with I. M. Kolthoff) Polarography, 1941, 52; Electroanalytical Chemistry, 1953, 58; Analytical Chemistry of Selected Metallic Elements, 1965; also numerous articles. Developed new methods chem. analysis particularly by electrochem. techniques. Home: 27 Locust Av., Lexington, Mass. 02173. Office: Harvard, Cambridge, Mass. 02138.*

LINGE, Kurt, German physicist; b. Heidelberg, Germany, Sept. 24, 1900; s. Karl and Margarete (Kühne) L.; engring. diploma Tech. Coll. Danzig; m. Vera Gatjens, July 6, 1939; children—Volker, Ute. Career as engr. and prof. Mem. Union German Engrs., Union Refrigeration Techniques Am. Research and publs. on refrigeration techniques, thermodynamics, climatization, techniques of measuring, air conditioning, two-phase flow. Home: Märchenring 6, Karlsruhe-Rüppurr. Office: Kaiserstrasse 12, Karlsruhe, W. Germany.

LINGENS, Franz Martin, German chemist; b. Elberfeld, Germany, Aug. 9, 1925; s. Johann Wilhelm and Adelheid (Eicker) L.; Dipl.-Chem., Dr.rer.nat.; m. Sibylle Kern, Aug. 4, 1961; 1 dau., Brigitte. Faculty, U. Tübingen, Germany, 1959-—, sci. adviser microbiol. chemistry, 1963-65, unscheduled prof., 1965-—. Mem. Soc. German Chemists, Soc. Biol. Chemistry, German Soc. Hygiene and Microbiology. Research, numerous publs. on chem. mutagenesis, biosynthesis of amino acids and vitamins, regulation phenomena in microorganisms. Address: 9 Brucknerweg, 74 Tübingen, West Germany.*

LINHART, Jiri George, physicist; b. Prague, Czechoslovakia, Apr. 13, 1924; s. Gustav and Anna (Cilek) L.; Dipl.Ing, Technische Hochschule, Prague, 1948; student physics and math. Karl's U., 1948; Ph.D., King's Coll., London U., 1952; m. Susanna Terrell Jan. 4, 1950; children—Anna, Illona, Jan, Natasha. Research physicist microwaves Brit. Thomson Houston Research Lab., Rugby, Eng., 1951-56; research physicist plasma application to accelerators, 1956-60; research dir. controlled fusion EURATOM-LGI, Frascati, Italy, 1960-—; prof. plasma physics for postgrad. students Rome U., 1964-—. Author: Plasma Physics, 1960; also articles. First to calculate Cherenkov radiation from charges moving near a di-electric boundary; 1st to study ultra-high density plasma; mem. of team 1st to generate reproducible megagauss magnetic fields; research on theory of plasma, exploding foils and their use in high voltage techniques. Home: 3 via di Salè. Office: Laboratorio gas ionizzati, Frascati, Italy.*

LINING, John, physician; b. Scotland, 1708; m. Sarah Hill, 1739; came to Am., 1730; practiced medicine, Charlestown (now Charleston), S.C.; made extensive studies of epidemic diseases; sent to Europe 1st sci. account of yellow fever in Am., 1748; studied effects of climate on metabolism; kept 1st published weather records in Am.; contbr. to publs. Royal Soc., other jours. Died Sept. 21, 1760.

LINK, Heinrich Friedrich, German biologist; b. Hildesheim, Germany, Feb. 2, 1767; prof. botany and med. botany, Berlin; dir. Royal Bot. Garden, Berlin. Fellow Royal Soc., 1842; mem. French Acad. Scis., 1828. Author: Histoire naturelle considérée comme commentaire du monde primitif et de l'Antiquité (with Hoffmannseeg), Flore du Portugal, 1809-40. Research on plant anatomy and physiology, plants of ancient Greece and Rome, flora of Portugal; showed influence of nature of soil on plant growth. Died Berlin. Jan. 1, 1851.

LINK, Henry Charles, Am. psychologist; b. Buffalo, Aug. 27, 1889; s. George and Martha (Kraus) L.; student Northwestern (now N. Central) Coll., 1908-10; A.B., Yale, 1913, A.M., 1915, Ph.D., 1916; m. Carolyn Crosby Wilson, May 2, 1917; children—James Wilson, Robert Frederick, Anne Luise. Psychologist since 1917; with U. S. Rubber Co., 1919-23, Lord & Taylor, N.Y., 1923-28, Gimbel's, Pitts., 1928-30; sec.-treas. Psychol. Corp., 1931-41, v.p., dir., since 1941; dir. Psychol. Service Center, N.Y., 1931-41; originator P.Q. or Personality Quotient, 1936; founder, dir. Psychol. Barometer (poll of public opinion and buying habits by 10,000 quarterly interviews), 1932-—; dir. Vocational Adjustment Bur., N.Y. Mem. Am. Psychol. Assn., Assn. Cons. Psychologists, Phi Beta Kappa. Author: Employment Psychology, 1919; Education and Industry, 1923; The New Psychology of Selling and Advertising, 1932; The Reutrn to Religion, 1936; The Rediscovery of Man, 1938; The Rediscovery of Morals, 1947; The Way to Security, 1951. Died Jan. 9, 1952.

LINK, Karl Paul, Am. biochemist; b. LaPorte, Ind., Jan. 31, 1901; s. George and Fredericka (Mohr) L.; B.S., U. Wis., 1922, M.S., 1923, Ph.D., 1925; postgrad. U. St. Andrews, Scotland, U. Graz, Austria, U. Zurich, Switzerland; m. Elizabeth Feldman, Sept. 20, 1930; children—John Kailin, Thomas Paul, Paul K. K. Faculty U. Wis., Madison, 1927-—, prof. biochemistry, 1930-—. Cons. chemistry and tech. plant products to govt. depts., industry including Wis. Alumni Research Found., 1940-—, Pabst Brewing Co., 1944-—. Recipient Cameron award U. Edinburgh, 1952; Albert Lasker awards Am. Pub. Health Assn., 1955, Am. Heart Assn., 1960; John Scott award City of Phila., 1959. Mem. Harvey Soc., Nat. Acad. Scis. (Kovalenka medal 1967), A.A.A.S., Am. Chem. Soc., Fedn. Am. Soc. for Exptl. Biology, Sigma Xi. Research, publs. on chemistry of sugars, disease resistance of plants; chemistry and biochemistry of coumarins; headed discovery and devel. of anticoagulant drug Dicumarol used to combat intravascular clotting, thrombosis and pulmonary embolism; discovered and developed warfarin for rodent control and sodium warfarin for clin. anticoagulant purposes. Home: 1111 Willow Lane, Madison, Wis. 53705.*

LINNAEUS, Carolus (Carl von Linné), Swedish botanist, taxonomist; b. Rashult, Province of Smaland, May 23, 1707; s. Nils and Christina (Brodersonia) L.; ed. U. Lund, 1727, U. Uppsala, 1728, studied medicine, Harderwijk, M.D., 1735; m. Sara Moraea, June, 1739; 1 son, Carl. Apptd. adjunctus to prof. Olaf Rudbeck, U. Uppsala, 1730; on expdn. sponsored by Acad. Sci. Uppsala, traveled 4,600 miles in Lapland and Upper Scandinavia, discovered 100 species of plants, 1732; taught methods of assaying ores, Uppsala, 1733; traveled through Province of Dalecarlia studying flora (at request of gov.), 1733; left Sweden to study medicine, 1735; visited Eng. and gathered plant specimens, 1736; practiced medicine, then became naval physician, Stockholm, 1738; apptd. prof. medicine, then botany U. Uppsala, 1741; expdn. through Öland and Gotland, by command of state (pub. results of journey containing index using for first time specific names in nomenclature), 1741-42; created Knight of Polar Star (first scientist to receive this honor), 1752. Fellow Royal Soc., 1753. Author over 180 works including: Systema Naturae, 1735; Genera Planetarum, 1737; Critica Botania, 1 Classes Plantarum, 1738; Species Plantarum, 1753; Genera morborum, 1763. Known as father of modern systematic botany; established method of plant classification based on sexual characteristics (stamens and pistils), which led to devel. of binary system of nomenclature (still in use today, system classes plants and animals by genus and species); major defender of anti-evolutionary doctrine of fixity of species (but probably changed his view during later life). Died Uppsala, Jan. 10, 1778.

LINNELL, Robert Hartley, Am. phys. chemist; b. Kalkaska, Mich., Aug. 15, 1922; s. Earl Dean and Constance Ruth (Hartley) L.; B.S., in Chemistry, U. N.H., 1944, M.S. in Inorganic Chemistry, 1948; Ph.D. in Phys. Chemistry (USN fellow), U. Rochester, 1950; m. Myrle Elizabeth Talbot, June 17, 1950; children—Charlene, Lloyd, Randa, Dean. Faculty, U. N.H., 1947, U. Beirut, 1950-54, U. Vt., 1958-61; v.p., tech. dir. Tizon Chem. Co., Flemington, N.J., 1955-58; lab. dir. Scott Research Labs., Perkasie, Pa., 1961-62; program dir. phys. chemistry NSF, Washington, 1962-65, with Office of Program Devel. and Analysis, 1965-67, department of scientific development, 1967—; cons. Reheis Chem. Co., Berkeley Heights, N.J., 1958-62. Research Corp. grantee, 1950-54, 58-61; NSF grantee, 1959-61; USPHS grantee, 1961-62; Am. Petroleum Inst. grantee, 1961-62. Mem. Am. Chem. Soc., A.A.A.S., Air Pollution Control Assn. Patentee in zirconium tech. field. Sci. contbns. in photochemistry, including air pollution photochemistry, chemistry of exhausts, hydrogen bonding, autoxidation. Home: Route 3, Gaithersburg, Md. 20760. Office: NSF, Washington 20550.*

LINNER, Edward Robert, Am. chemist; b. Buffalo, Oct. 12, 1899; s. Robert Mathias and Wilhelmina (Ode) L.; B.S., U. Buffalo, 1925; postgrad. U. Wis.; Ph.D., U. Minn., 1934; m. Celestia Davidson, Dec. 24, 1931; 1 dau., Janet (Mrs. Charles E. Townsend). Chemist, Acheson Graphite Co., Niagara Falls, N.Y., 1920-24; grad. asst. in chemistry U. Wis., Madison, 1924-28; instr. Lafayette Coll., Easton, Pa., 1928-31; instr. agrl. biochemistry U. Minn., Mpls., 1931-34; faculty dept. chemistry Vassar Coll., Poughkeepsie, N.Y., 1934—, prof., 1950-65, emeritus, 1965—, chmn. chemistry dept., 1959-61, chmn. admissions, 1950-55, dir. admissions, 1955-57; adj. prof. Faculty, Summer Inst. for High Ability High Sch. Students, NSF, 1963-65. Am. Scandinavian Found. J. G. Bergquist fellow Biochem. Inst., Uppsala, Sweden, 1949-50. Mem. Am. Chem. Soc., Am. Assn. U. Profs., New Eng. Assn. Chemistry Tchrs., N.Y. Acad. Scis., Sigma Xi. Research in adsorption from solution, activation processes for adsorption, history of sci. Home: 89 Raymond Av., Poughkeepsie, N.Y.*

LINNETT, John Wilfrid, Brit. chemist; b. Coventry, Eng., Aug. 3, 1913; s. Alfred Thirlby and Ethel (Ward) L.; D.Phil., Oxford U., 1937, M.A., 1938; m. Rae Ellen Libgott, Dec. 19, 1947; children—Simon, Sophia. Jr. research fellow Balliol Coll., Oxford, 1939-45; lectr. Brasenose Coll., Oxford, 1943-45, tutorial fellow Queen's Coll., 1945-65, demonstrator in inorganic chemistry Oxford U., 1938-62, reader, 1962-65; prof. phys. chemistry Cambridge U., 1965—; vis. prof. U. Cal. at Berkeley, 1964. Mem. com. Brit. sect. Combustion Inst. Henry fellow Harvard, 1937-38. Fellow Royal Soc., 1955; mem. Faraday Soc. (mem. council 1956-58, v.p. 1959-61, 65-67), Chem. Soc. (council 1960-62, hon. sec. 1962-65), N.Y. Acad. Scis. Author: Wave Mechanics and Valency, 1960; The Electronic Structure of Molecules, 1964; also numerous articles. Research on vibration of molecules and force constants, electronic structure of molecules, valency and chem. binding, quantum mech. calculations relating to valency, flame propagation and its mechanism, reactions of atoms at surface and in Gas Phase, gen. reaction kinetics. Home: 16 Brookside, Cambridge, Eng.*

LINNIK, Iurii Vladimirovich, Russian mathematician; b. Jan. 21, 1915; s. Vladimir P. Linnik; grad. Leningrad U., 1938, D.Phys.-Math. Scis., 1940. With Leningrad br. Math. Inst., USSR Acad. Scis., 1940—; prof. Leningrad U., 1944—. Recipient Stalin prize, 1947. Mem. USSR Acad. Scis. (corr.). Research on theory of numbers; gave estimation of smallest prime number in arithmetical progression with a large difference; work in calculus of probability, heterogeneous Markov chains, math. statistics. Address: Leningrad Section, Mathematical Institute, USSR Acad. Scis., Fontanki 25, Leningrad D-11, USSR.

LINNIK, Vladimir Pavlovich, Russian physicist; b. July 6, 1889; grad. Kiev U., 1914; with State Optical Inst., from 1926; later prof. Leningrad U.; academician USSR Acad. Sci., 1939. Recipient Stalin prize, 1946, 50. Publs. on device for interference investigation of reflecting objects under microscope (microinterferometer), 1933, device for interference investigations of microprofile of surface, (microprofilometer), 1945, interferometer for controlling large machine details, 1942, interference passage instrument, 1946, fundamental possibility of lessening influence of atmosphere on image of a star, 1957, statistically similar zones of a linear type, 1962, theory of statis. similar zones, 1962, asymptate of whole number third order matrices, 1962. Research on optics and its application in instrument-making industry; constructed double microscope, 1929, microinterferometer for controlling exactness of processing of surfaces, 1933, microscope for studying surface of red-hot bodies, interferometers for measuring double stars and angular diameter of sun; developed methods of lab. investigation and testing of optical devices; designed control devices for optical-mech. industry. Address: Dept. of Physics, Leningrad Univ., Leningrad, USSR.

LINSER, Hans, biochemist; b. Linz/Danube, Austria, Apr. 7, 1907; s. Hermann and Johanna (Schachermeyr) L.; Dr. phil., U. Vienna, 1930; m. Carola Koehler, Apr. 14, 1932; children—Gerhard, Christa (dec. 1960). Biol. researcher I.G. Farbenindustrie AG, Ludwigshafen, Germany, 1930-46; chief biol. research dept. and agrl. research sta. Österreichische Stickstoffwerke AG, Linz, 1947-60; prof., dir. inst. plant nutrition U. Giessen, Germany, 1960—. Mem. Soc. Plant Nutrition (pres. 1966—), Liebig Mus. Soc. (pres. 1966—), Research Bd. for Nutrition, Agr., Forestry. Author: Chemismus des Lebens, 1948; Handbuch der Pflanzenernährung und Düngung, 1963-68; others; also numerous articles. Research in plant biochemistry, fluorescence microscopy, fluorometry, growth substances, morphoregulators; introduced hormonal herbicides in Austria and use of ccc against lodging of cereals; improved baking quality of wheat. Home: 115 Landstrasse, Linz/Donau, Austria 4020. Office: 7 Braugasse, Giessen, West Germany 63.*

LINSER, Karl, German dermatologist; b. Pforzheim-Bade, Sept. 10, 1895; s. Karl and Emilie (Kälber) L.; ed. univs. Fribourg, Heidelberg, Würzburg; M.D.; hon. doctorate U. Liepzig; m. Erika May, July 14, 1928; children—Renate, Christian. Dermatologist, Dresden, Germany, 1926-47; with dermatol. clinic Friedrichstadt Hosp.; dir. univ. clinic dermatology, Leipzig, 1947-50; dir. univ. clinic dermatology U. Humboldt, Berlin, 1950-62, pres. central adminstrn. for pub. health, 1946-49, prof. emeritus, 1962—. Recipient Goethe prize, Berlin. Mem. Soc. Dermatology (hon.), German Med. Orgn., German Acad. Sci., German Soc. Clin. Medicine. Contbg. author textbooks on dermatology and venereology. Address: Wodan-Strasse 19, Berlin-Hessenwinkel, W. Germany.

LINSKENS, Hans-Ferdinand, Dutch botanist; b. Lahr, May 22, 1921; s. Albert and Maria (Bayer) L.; ed. univs. Berlin, Cologne, Bonn, Zurich; Ph.D.; m. Ingrid M. Rast, Sept. 23, 1955; children—Perpetua-Maria, Justinus-Albrecht, Benedictus-Leonhard, Peregrina-Gemma. Career as prof. and researcher. Mem. Royal Dutch Bot. Soc., Am. Soc. Plant Physiologists, Schweizerische Botan. Gesellschaft, Soc. German Botanists, Dutch Fedn. Electronic Microscopy. Author: Papierchromatographie in der Botanik; Praktikum der Papierchromatographie, 1961; Modern Methods of Plant Analysis, 1962; also papers. Research on biochemistry of fertilization and incompatibility, molecular developmental cytology. Home: Oosterbergweg 5, Beek-bij-Nijmegen. Office: Botan. Laborat. Universiteit, Driehuizerweg 200, Nijmegen, Netherlands.

LINSLEY, Earle Gorton, Am. entomologist; b. Oakland, Cal., May 1, 1910; s. Earle Garfield and Marguerite G. (Vesper) L.; B.S., U. Cal. at Berkeley, 1932, M.S., 1933, Ph.D., 1938; m. Juanita Marion Murdoch, Aug. 22, 1935; children—James Murdoch, Joan Linsley (Mrs. Michael Hill MacFarlane). Research asst. entomology U. Cal. at Berkeley, 1933-35, teaching asst. agr., Los Angeles, 1935-37, instr. entomology, jr. entomologist Cal. Agrl. Expt. Sta., Berkeley, 1939-43, asst. prof., asst. entomologist, 1943-49, asso. prof., asso. entomologist, 1949-53, prof., entomologist, 1953—, chmn. dept. entomology and parasitology, 1951-59, dean Coll. Agr., 1960—, asst. dir. Cal. Agrl. Expt. Sta., 1960-63, asso. dir., 1963—. Instr. Sch. Field Natural History, Nat. Park Service, 1938-41; sec. Am. Com. on Entomol. Nomenclature, 1943-48; collaborator U. S. Dept. Interior, 1953-55; research prof. Miller Inst. for Basic Research in Sci., 1960; participant Galapagos Internat. Sci. Project, 1964. Guggenheim Meml. Found. fellow Am. Mus. Natural History, 1947-48. Fellow A.A.A.S., Cal. (research asso. 1942—), So. Cal. acads. scis., Entomol. Soc. Am. (pres. 1952); mem. Entomol. Soc. Can., Ecol. Soc. Am., Soc. for Study Evolution, Soc. Systematic Zoology, Coleopterists Soc., Am. Forestry Assn., Am., Western socs. naturalists, Soc. Mexicana de Entomologia, Pacific Coast (pres. 1938-40), Kan. entomol. socs., Agrl. Forum, Sigma Xi, Alpha Zeta, Phi Sigma, Gamma Alpha, Phi Mu Delta. Author: (with Mayr, Usinger) Methods and Principles of Systematic Zoology, 1953. Editor: Pan-Pacific Entomologist, 1943-48. Research, publs. on taxonomic, ecol., behavioral, distribitional, evolutionary aspects of entomology. Home: 290 Alvarado Rd., Berkeley, Cal. 94705.*

LINTNER, Joseph Albert, Am. entomologist; b. Schoharie, N.Y., Feb. 8, 1822; s. George Ames and Maria (Wagner) L.; Ph.D. (hon.), U. State N.Y.; m. Frances C. Hutchinson, 1856. In business, N.Y.C., 1837-48, Schoharie, 1848-60; mfr. woolens, Utica, N.Y., 1860-68; began collecting insects, 1853; asst. in zoology N.Y. State Mus., 1868-74, head entomol. dept., 1874-80; N.Y. State entomologist, 1880-98; entomol. editor Country Gentleman for 25 years; pres. Assn. Econ. Entomologists, 1892. Author: (pamphlet) Entomological Contributions, 4 issues, 1872-79. Died Rome, Italy, May 5, 1898.

LINTON, Ralph, Am. anthropologist; b. Phila., Feb. 27, 1893; s. Isaiah Waterman and Mary Elisabeth (Gillingham) L.; B.A., Swarthmore Coll., 1915; M.A., U. Pa., 1916; Ph.D., Harvard, 1925; m. Adelin M. Hohlfeld, Aug. 31, 1934; 1 son, David Hector. Field work in archaeology, N.M., 1912, 17; Guatemala, 1913, N.J., 1915, Ill., 1916, Colo., 1919, Marquesas Islands, 1920-21, Ohio, 1924, Wis., 1932-33; in ethnology, Polynesia, 1920-22, Madagascar, 1925-27, S. Africa, 1928, Okla., 1934. Asst. curator of ethnology Field Mus. Natural History, 1922-28; prof. anthropology U. Wis., 1928-37; prof. anthropology Columbia, 1937-39; chmn. dept. of anthropology, 1939-43; Sterling prof. anthropology Yale since 1946. Recipient Viking medal, 1952; Huxley Meml. medal, 1954. Hon. fellow Royal Anthrop. Inst. Gt. Britain; mem. Am. Anthrop. Assn. (pres. 1946), A.A.A.S. (v.p. 1937), Nat. Acad. Scis. NRC, Social Sci. Research Council, Am. Council Learned Socs., Acad. Malgache (hon.), Phi Beta Kappa, Sigma Xi. Author: The Material Culture of the Marquesas Islands, 1924; Use of Tobacco Among North American Indians, 1924; The Archaeology of the Marquesas Islands, 1925. Guide to the Polynesian and Micronesian Collections, Field Museum, 1925; The Tanala, A Hill-Tribe of Madagascar, 1932; The Study of Man, an Introduction, 1936; Acculturation in Seven American Indian Tribes, 1940; Cultural Background of Personality, 1945; The Science of Man in the World Crisis, 1945; The Tree of Culture, 1955; Most of the World, 1949. Synthesized various anthrop. directions. Died Dec. 24, 1953.

LINVILLE, Thomas Merriam, Am. elec. engr.; b. Washington, Mar. 3, 1904; s. Thomas and Clara (Merriam) L.; E.E., U. Va., 1926; grad. Harvard. Advanced Mgmt. Program, 1950; m. Eleanor Priest, Nov. 25, 1939; children—Eleanor, Thomas Priest, Edward Dwight. With Gen. Electric Co., Schenectady, 1926—, mgr. exec. devel., 1951-53, mgr. sci. research, Research Lab., 1953-64, mgr. sci. research applications, 1964—. Chmn. devel. council for sci. Rensselaer Poly. Inst., 1958-66; mem. vis. com. Clarkson Inst. Tech., 1959-67. Recipient Charles A. Coffin award Gen. Electric Co. 1946; Engr. of Year award Schenectady Profl. Engrs. Soc., 1960. Fellow I.E.E.E. (past dir., rep. to Nat. Acad. Scis.-NRC 1959—), Am. Soc. M.E., A.A.A.S.; mem. Nat. (v.p. 1962-64, dir. 1952-64), N.Y. State (pres. 1954-55, trustee) socs. profl. engrs., I.E.E.E., Am. Soc. for Engring. Edn., Am. Acad. Polit. and Social Sci., Raven Soc., Theta Tau, Delta Upsilon. Developed theory and math. treatment resulting in new procedures for calculating operational impedances and transient performance of synchronous and d-c machines and networks; developed servo-mechanisms, control and propulsion systems for submarines, elec. systems for aircraft, computerized design methods, metal rolling equipment; inaugurated exec. devel. and edn. programs, sci. research operations and applications. Home: 1147 Wendell Av., Schenectady. Office: P.O. Box 1088, Schenectady 12301.

LINZ, Henry-Oscar, German geologist; b. Leipzig, Germany, 1848; participated in sci. expdn. to W. Africa, 1874; went to Tangiers, Senegal by way of Sahara and Timbuktu, 1879; went from Congo to Lake Tanganyika, beginning in 1885; prof. geography German U. Prague; dir. rev. Aus allen Weltteilen; mem. Vienna Geog. Soc. Vienna (became sec. gen. 1883). Author: Sketches of West Africa, 1878; Timbuktu, Voyage Across Morocco, the Sahara, the Soudan, 1884. First exploration of the Congo to Lake Tanganyika, 1885. Died Vienna, 1925.

LION, Kurt S., physicist; b. Kassel, Germany, May 17, 1904; s. Mority and Ida (Meyer) L.; Dipl.Ing., Techn. Inst. Darmstadt (Germany), 1928, Dr.-Ing., 1933; m. Elsa Strauss, Feb. 3, 1935; 1 son, John Rene. Came to U. S., 1941, naturalized, 1947. Asst. to dir. Turkish State U., Istanbul, 1935-37; chief lab. physics U. Fribourg (Switzerland), 1937-41; asso. prof. applied biophysics Mass. Inst. Tech., Cambridge, 1941—. Author: Instrumentation in Scientific Research, 1959. Research, numerous publs. and patents on instrumentation, particle acceleration, med. instrumentation, transducers, image intensification, gas analysis. Home: 9 Herbert Rd., Belmont, Mass. 02178. Office: 77 Massachusetts Av., Cambridge, Mass. 02139.*

LIONETTI, Fabian Joseph, Am. biochemist; b. Jersey City, Mar. 3, 1918; s. Anthony and Theresa (Petrelle) L.; A.B., N.Y. U., 1943, M.S., 1945; Ph.D., Rensselaer Poly Inst., 1948; m. Kathryn Grysko, Dec. 26, 1943; children—Karen, Fabian Joseph, Donald. Faculty, Boston U. Sch. Medicine, 1949-65, asso. prof. biochemistry, 1955-65; asso. mem. Inst. Health Scis., Brown U., Providence, 1965—, asso. dir. div. hematologic research, 1965—. Recipient Mathewson medal Am. Inst. Metals, 1952. Fulbright scholar, Italy, 1961. Mem. Am. Soc. Biol. Chemists, Am. Chem. Soc., N.Y. Acad. Scis., Am. Hematology Soc. Research, publs. on carbohydrate metabolism in human red blood cells, chem. pathways and enzymes of pentose sugars, physiology of red cell membranes. Home: 349 Central Av., Milton, Mass. 02187. Office: 12 George St., Providence 02912.*

LIOSNER, Lew, Russian biologist; b. Witebsk, Russia, Feb. 20, 1909; s. Dawid and Roza (Silberman) L.; student Moscow Med. Inst., 1927-30; D.Biol. Sci., Moscow U., 1940; m. Mary Woronzowa, July 15, 1941; 1 son, Alexander; m. 2d, Inna Markelowa, Jan. 10, 1958; 1 son, Wladimir. Sr. sci. collaborator Inst. Exptl. Morphogenesis, Moscow U., 1931-41, prof., 1959—; dozent biology Moscow Med. Inst., 1941-46; sr. sci. collaborator Inst. Exptl. Biology, Acad. Med. Sci., 1946-56, chief dept. growth and devel., 1956—. Author: (with M. Woronzowa) Physiological Regeneration, 1955; A Sexual Propagation, a Regeneration (trans. into English), 1957; Regeneration in Mammals, 1960; Compensatory Hypertrophy, 1963; Cell Renewal, 1966; also numerous articles. Research on analysis of conditions of regeneration of organs in Amphibia; revealing of capacity and of modes of regeneration in mammals. Home: 2 log. 35 Proezd Mchat, Moscow. Office: 8 Baltiyskaya, Moscow, USSR.*

LIOTTA, Domingo, physician; b. Diamante, Entre Rios, Argentina, Nov. 29, 1924; s. Domingo and Maria (Motta) L.; B.S., Nat. Coll. Uruguay, 1942; M.D., Nat. U. Cordoba, 1949, D.M.S., 1952; m. Olga Troncoso, Nov. 4, 1958; children—Mary Grace, Dominic, Mary Stella, Mary Olga, Carlos Augusto. Resident gen. surgery Sch. Medicine Nat. U. Cordoba, 1952-55, prof. anatomy, 1955-65, adscript prof. surgery, 1963—; resident gen., thoracic, cardiovascular surgery U. Lyon, 1956, 58-59; practice medicine, specializing in surgery, Cordoba, 1950—, Houston, 1961—; observer dept. artificial organs Cleve. Clin., 1961; Cardiovascular Surgery fellow Coll. Medicine Baylor U., 1961-64, asst. prof. surgery, project asso. dir. Artificial Heart Program, 1964—. Recipient Ann. award Argentina Assn. Surgery Accesit, 1952; Premio Quinquenio de la Revista Brasileira de Gastroenterologia, 1956; Schleussner Radiol. award Frankfurt (Germany), 1957; Young Investigator award Am. Soc. Cardiology, 1962, 63; others. Mem. Soc. Surgery Cordoba, Soc. Anatomists (France), Am. Soc. Artificial Internal Organs, Soc. Cryobiology, Soc. Surgery Rosario (Argentina), Assn. Advancement Med. Instrumentation. Author: (with others) La Duodenographie Hypnotique, 1963, Early Radiological Diagnosis of the Tumor of Pancreas, 1965, Mechanical Devices to Assist the Failing Heart, 1966, Heart Substitutes, 1966; also numerous articles. Research on hypotonic duodenography, radiol. technique to study head of pancreas, duodenum and region of ampulla of vater; pioneer in devel. and implantation 1st artificial heart in a human being and 1st paracorporeal circulatory pump. Home: 4058 Falkirk St., Houston 77025. Office: 1200 Moursund St., Houston 77025.*

LIOUVILLE, Joseph, French mathematician; b. St. Omer, France, Mar. 24, 1809; ed. L'école polytechnique. Became prof. L'école polytechnique, 1833; prof. Sorbonne, also Collège de France; dir. Bureau des longitudes. Fellow Royal Soc., 1850; mem. French Acad. Scis., 1839. Founder, Journal des mathématiques dures et appliquées, 1836, editor for 40 years. Research and numerous publs. on differential equations, boundary values, theories of applicability of surfaces and conformal transformations; 1st proof of existence of transcendental functions; introduced concept of geodesic curvature; originated theory of doubly periodic functions from gen. theorems in theory of analytic functions of a complex variable. Died Paris, Sept. 8, 1882.

LIPATOV, Sergey Mikhaylovich, Russian phys. chemist; b. Oct. 11, 1899; grad. Moscow Higher Tech. Sch., 1923; D.Chem. Sci., 1936. With chem. lab. 1st Moscow Cotton-Printing Plant, 1924-27, central lab. Ivanovo-Voznesensk Textile Trust, lectr. Ivanovo-Voznesensk Poly. Inst., 1927-29; head lab. L. D. Karpov Physico-Chem. Inst., Moscow, 1929-32; instr. Moscow Inst. Light Industry, 1931-34, prof., 1934—; asso. Colloidal and Electrochem. Inst., USSR Acad. Sci., 1938-41; head chair phys. and colloidal chemistry Moscow Textile Inst. Mem. sci. and tech. council USSR Ministry Food Industry. Recipient D. I. Mendeleev prize, 1927. Mem. Belorussian Acad. Sci. Author: The Colloid-Chemical Principles of Dyeing. 1929; The Theory of Colloids, 1934; High-Molecular Compounds: Lyophilic Colloids, 1934; High-Molecular Compounds, 1936; The Physical Chemistry of Colloids, 1948. Mem. editorial bd. Colloid Jour. Research and numerous publs. on adsorption of electrolytes on cellulose and insoluble crystals, high-molecular compounds. Address: Moscow Textile Inst., Donskaya ulitsa 62, Moscow, USSR.

LIPKIN, David, Am. chemist; b. Phila., Jan. 30, 1913; s. William and Ida (Zipin) L.; B.S. U. Pa., 1934; Ph.D., U. Cal. at Berkeley, 1939; m. Virginia M. Cohen, Apr. 25, 1955; children—John J., Robert C., Jeffrey A., Edward W. Research fellow U. Cal. at Berkeley, 1939-42; research chemist Atlantic Refining Co., Phila., 1934-36; research chemist Manhattan project Radiation Lab., U. Cal. at Berkeley, 1942-43; staff Manhattan Project, Los Alamos, 1943-46, group leader, 1945-46; faculty Washington U., St. Louis, 1946—, prof. chemistry, 1948—, dept. chmn., 1964—. Cons. Office Naval Research, 1951-53. Guggenheim Meml. fellow, 1955-56; sr. vis. fellow OEEC Labs. Virus research unit Agrl. Research Council, Cambridge, Eng., 1960. Mem. Am. Chem. Soc., N.Y. Acad. Scis., Sigma Xi, Tau Beta Pi. Research, publs. in fluorescence and phosophorescence of organic molecules, photochem. reactions of organic compounds at low temperatures in rigid media, chemistry and spectra of free radicals, chemistry of nucleosides, nucleotides, and nucleic acids. Home: 64 Arundel Pl., St. Louis 63105.*

LIPKIN, H(arry J(eannot), physicist; b. N.Y.C., June 16, 1921; s. Louis Isreal and Bertha (Bernhardt) L.; B.E.E., Cornell U., 1942; A.M., Princeton, 1948, Ph.D., 1950; m. Geraldine Weinberg, June 14, 1949; children—Naomi, David. Staff, Radiation Lab., Mass. Inst. Tech., 1942-46; prof. Weizmann Inst. Sci., Rehovoth, Israel, 1952—; vis. prof. U. Ill., Urbana, 1958-59, 62-63, Hebrew U., Jerusalem, Israel, 1956-58, Tel Aviv (Israel) U., 1965-66. Mem. Am., Israel, Italian phys. socs. Author: Beta Decay for Pedestrians, 1962; Lie Groups for Pedestrians, 1965; also numerous articles. Research in theoretical and exptl. nuclear physics, Mössbauer effect, elementary particle theory. Home: 3 Neveh Weizmann, Rehovoth, Israel.*

LIPMAN, Charles Bernard, plant physiologist; b. Moscow, Russia, Aug. 17, 1883; s. Michael Gregory and Ida (Birkhahn) L.; brought to U. S., 1889; B.Sc., Rutgers, 1904, M.S., 1909, Sc.D. (hon.), 1934; M.S., U. Wis., 1909; Ph.D., U. Cal., 1910; m. Marion Amesbury Evans, May 25, 1925; 1 dau., Georgia Evans. Instr. soil bacteriology U. Cal., 1909, asst. prof. soils, 1910; asso. prof., 1912, prof. soil chemistry and bacteriology, 1913-21, prof. plant nutrition, 1921-25, prof. plant physiology since 1925, dean Grad. Div. since 1923. Fellow A.A.A.S., Soc. for Research on Meteorites; mem. Am. Chem. Soc., Am. Soc. Plant Physiologists, Soc. Am. Bacteriologists, Cal. Bot. Soc., Cal. Acad. Sci., Soc. Exptl. Biology and Medicine, Sigma Xi, Phi Beta Kappa. Contbr. numerous papers on plant physiology, plant chemistry, plants and soils, soil and marine bacteria, bacteria in ancient rocks, meteorites. Asso. editor Plant Physiology, Jour. Bacteriology, Soil Science. Died Oct. 22, 1944.

LIPMANN, Fritz Albert, biochemist; b. Koenigsberg, Germany, June 12, 1899; s. Leopold and Gertrud (Lachmanski) L.; M.D., U. Berlin, 1924, Ph.D., 1927; m. Freda M. Hall, June 21, 1931; 1 son, Stephen Hall. Came to U. S., 1939, naturalized, 1944. Research asst. Kaiser Wilhelm Inst., Berlin, 1927-31; fellow Rockefeller Inst. for Med. Research, N.Y.C., 1931-32; research asso. Biol. Inst. of Carlsberg Found., Copenhagen, Denmark, 1932-39, Cornell U. Med. Sch., N.Y.C., 1939-41; research fellow Harvard Med. Sch., Boston, 1941-43, asso. in biochemistry, 1943-49; prof. biol. chemistry Mass. Gen. Hosp., Boston, 1949-57; prof. Rockefeller U., N.Y.C., 1957—. Recipient Mead-Johnson & Co. award, 1948; Carl Neuberg medal, 1948; Nobel prize in physiology and medicine (with H. A. Krebs), 1953; Nat. Medal of Sci., 1966. Fellow Danish Royal Acad. Scis.; mem. Nat. Acad. Sci.; Fellow Royal Soc., 1962. Contbr. numerous articles to profl. jours. Discovered coenzyme A; identified carbamyl phosphate as metabolic donor of carbamyl groups during urea and pyrimidine synthesis; characterization of metabolic sulfate activation for synthesis of chondroitin sulfate in cartilage; research in biosynthesis of protein. Home: 150 E. 18th St., N.Y.C. 10003.*

LIPNER, Harry, Am. physiologist; b. N.Y.C., Aug. 26, 1922; s. Samuel and Sarah (Linkoff) L.; B.S., L.I. U., 1942; M.S., U. Chgo., 1947; Ph.D., U. Ia., 1952; m. Ethel Lapis, Nov. 11, 1949; children—Laura Jean, Sandra Lea, William F., Michael A. Analyst, Internat. Vitamin Corp., 1942-43; research chemist Nat. Oil Prodn. Co., 1943-45; research asso. U. Ia., 1952; research fellow Nat. Cancer Inst., Bethesda, Md., 1952-54; instr. Chgo. Med. Sch., 1955; faculty Fla. State U., Tallahassee, 1955—, prof. physiology, 1965—. Mem. Am. Physiol. Soc., Soc. Gen. Physiologists, Am. Zool. Soc., Endocrine Soc., A.A.A.S., Sigma Xi. Research, publs. on mechanism by which ovum escapes from Graafian follicle in mamposed hypothesis concerning role of hormones in regmals, thyroid effect on carbohydrate metabolism; proulating escape of ovum. Home: 3214 Brookforest Dr., Tallahassee, Fla. 32303.*

LIPPARD, Vernon William, Am. physician; b. Marlboro, Mass., Oct. 4, 1905; s. William Charles and Lucy (Balcom) L.; B.S., Yale, 1926, M.D., 1929; D.Sc., U. Md., 1955; m. Margaret Isham Cross, Aug. 29, 1931; 1 dau., Lucy (Mrs. Robert Tracy Ryman). Instr. pediatrics Cornell U. Med. Coll., Ithaca, N.Y., 1932-35, asso. in pediatrics, 1935-39; asso. dean Columbia Coll. Physicians and surgeons, N.Y.C., 1939-46; dean, prof. pediatrics La. State U., New Orleans, 1946-49; dean, prof. pediatrics U. Va., Charlottesville, 1949-52, Yale, New Haven, 1952-67; asst. to pres. and fellows Yale Corp. for Med. Devel., New Haven, Conn., since 1967—. Director, N.Y. Study Commn. for Study Crippled Children, 1938-39; bd. med. cons. Oak Ridge Inst. Nuclear Studies, 1947-52, Brookhaven Nat. Lab., 1954-59; bd. counsellors Smith Coll., 1959—. Mem. Assn. Am. Med. Colls. (pres. 1954-55), Soc. for Pediatric Research, Assn. Am. Physicians, Sigma Xi, Alpha Omega Alpha. Research in immunology in infancy, med. sch. adminstrn. Home: 42 Lincoln St., New Haven 06511. Office: 333 Cedar St., New Haven 06510.*

LIPPERSHEY, Hans, Dutch optician; b. Wesel, Netherlands, 1587. Lens grinder, Middelburg, Holland. Devised one of first telescopes, 1608. Died 1619.

LIPPINCOTT, James Starr, Am. horticulturist, meterologist; b. Phila., Apr. 12, 1819; s. John and Sarah (Starr) L.; attended Haverford Coll., 1834-35; m. Susan Haworth Ecroyd, 1857; m. 2d, Anne E. Shephard, 1861. Farmer at Cole's Landing, later at Haddonfield, N.J.; invented vapor index for measuring humidity of air; meteorol. observer for Smithsonian Instn. at Cole's Landing, 1864-66, Haddonfield, 1869-70. Author: Universal Pronouncing Dictionary of Biography and Mythology, 1870. Contbr. articles to Reports of Commrs. of Agr., including: Climatology of American Grape Vines, 1862, Geography of Plants, 1863, Market Products of West New Jersey, 1865, Observations on Atmospheric Humidity, 1865, The Fruit Regions of the Northern United States and Their Climates, 1866. Died Greenwich, Cumberland County, N.J., Mar. 17, 1885.

LIPPINCOTT, Sarah Lee, astronomer; b. Phila., Oct. 26, 1920; s. George Eyre and Sarah Lee (Evans) L.; B.A., U. Pa., 1942; M.A., Swarthmore Coll., 1950. Research asst. Sproul Obs., Swarthmore (Pa.) Coll., 1942-50, research asso., 1951—, lectr., 1961—; mem. French Solar Eclipse Expdn., Öland, Sweden, 1954. Fulbright fellow Observatoire de Paris (France), 1953-54. Mem. Am. Astron. Soc. (mem. commn. on double stars, parallaxes and proper motions 1958—), Internat. Astron. Union. Author: (with Joseph M. Joseph) Point to the Stars, 1963; (with L. Lafore) Philadelphia the Unexpected City, 1965; also numerous articles. Research on stellar distances, nearby stars, mass determination from study double stars, accuracy trigonometric parallax, solar surface. Home: 510 Elm Av., Swarthmore Pa. 19081.*

LIPPINCOTT, William Thomas, Am. chemist; b. Balt., Apr. 4, 1924; s. William Thomas and Marie (Petry) L.; B.Sc., Capital U., 1948; Ph.D. (DuPont Teaching fellow), 1954; m. Shirley Sue Epperson, Jan. 1, 1953; children—William Thomas III, David B., Loralee L. Faculty, Capital U., 1948-53, asst. prof., 1951-53; asst. prof. Mich. State U., 1954-57; asso. prof. U. Fla., 1957-61; prof. Ohio State U., Columbus, 1961—, vice chmn. chemistry, 1964-66. Mem. Adv. Council on Coll. Chemistry, 1964—. Mem. Am. Chem. Soc. A.A.A.S. (mem. coop. com. on teaching sci. and math. 1962-65), Sigma Xi, Phi Lambda Upsilon, Alpha Chi Sigma. Author: (with A. B. Garrett, F. H. Verhoek) Textbook of Chemistry, 1968; also articles. Research on rates and mechanisms autoxidations, organometallic compounds, ednl. methods. Home: 4172 Rudy Rd., Columbus, O. 43214.*

LIPPMAN, Edmund Oskar, chemist; b. Vienna, Jan. 9, 1857; indsl. chemist; prof., Halle, Germany. Author: Der Zucker, 1878; Die Zuckerarten und ihre Derivate, 1882; Die Chemie der Zuckerarten, 1890; Geschichte der Zuckers, 1890; Abhandlungen und Vorträge zur Geschichte der Naturwissenschaften, 2 vols., 1906-13; Entstehung und Ausbreitung der Alchemie, 2 vols., 1919-31; Beiträge zur Geschichte der Naturwissenschaft und Technik, 1923. Discovered sugar (glucose plus water) already exists in beetroot and is not produced by refining; studied sugar chemistry, history of sci. Died Halle, Sept. 24, 1940.

LIPPMANN, Gabriel, physicist; b. Hollerich, Luxembourg, Aug. 16, 1845; ed. Paris, also Germany. Mem. staff various sci. labs., Germany; prof. math. physics Paris Faculty Sci., 1883-86, prof. exptl. physics, dir. research lab., 1883-1921. Fellow Royal Soc., 1896; mem. French Acad. Scis. Recipient Nobel prize for physics, 1908. Author: Relation Between Electric Phenomena and Capillaries, 1875; Cours de thermodynamique, 1886; Cours d'acoustique et d'optique, 1888. Inventor Lippmann capillary electrometer, 1873, galvanometer, coelostat, uranagraph, mercury electrodynamometer; developed 1st practical method of permanent color photography; used laminated film to produce 1st color photograph of spectrum; predicted piezo-electricity; studied polarization of electrolytes, batteries, electric dilatation of glass. Died at sea, July 31, 1921.

LIPPS, Theodor, German psychologist; b. Wallhalben, July 28, 1851; prof. successively at Bonn, Breslau, Munich. Author: Grundtatsachen des Seelenlebens, 1883; Raumästhetik, 1897; Ästhetik, 2 vols., 1903-06; Die ethischen Grundfragen, zehn Vorträge, 2d edit., 1905; Leitfaden der Psychologie, 3d edit., 1926. Maintained that mind is the sum of retained experience; most important concept was that of Einfühlung (projection of self into what is seen); worked on aesthetics, ethics, optical illusion. Died Munich, Oct. 17, 1914.

LIPSCHITZ, Rudolf Otto Sigismund, German mathematician; b. Königsberg (now Kaliningrad, Russia), May 14, 1832; ed. U. Königsberg; Doctorate, U. Berlin (Germany); m. Ida Pascha; prof. Breslau (now Wroclaw, Poland); became prof., Bonn, Germany, 1864, rector, 1874-75. Author: Bedeutung der Theoretischen mechanik, 1866; Wissenschaft und Staat, 1874; Lehrbuch der Analysis, 2 vols., 1877-80; also articles. Formulated Cauchy-Lipschitz theorem in connection with existence theorem for solutions of 1st order differential equation; studied trigonometric sines, spaces of n-dimensions, algebra, hydrodynamics; contributed to theory of numbers, Bessel's functions, Fourier series, analytical mechanics, potential theory. Died Bonn, Oct. 7, 1903.

LIPSCHUTZ, Alexander, physician; b. Riga, Latvia, Russia, Aug. 28, 1883; med. edn. Berlin, Göttingen, Zurich, 1901-07; M.D., Göttingen, 1907; Sci.D. h.c., La Habana, Cuba, 1963; m. Margaret Vogel-Leech, May 14, 1914; 2 daus. Lectr. physiology U. Bern (Switzerland), 1914-19; prof. physiology, dir. physiol. inst. U. Dorpat (Tartu, Estonia), 1919-26, U. Concepción (Chile), 1926-36; dir. Inst. Exptl. Medicine of Nat. Health Service, Santiago, Chile, 1937-60, ret., 1960, hon. mem. Nat. Health Service; academic mem. faculty medicine U. Chile, 1952—. Lectr. univs. and acads., Gt. Britain, France, Switzerland, Spain, Portugal, U. S., Can., S. Am., USSR, China; hon. prof. various S. Am. univs.; chief Chilean mission for study of Indians of Tierra del Fuego, 1946. Fellow Royal Anthrop. Inst.; hon. mem. Royal endocrinological, med. and biol. socs. of Chile, Argentina, Uruguay, Mexico, Spain, Italy; corr. mem. Acad. Sci. (Turin, Italy), acads. medicine Madrid, Turin and Mexico, Soc. Biology, Paris. Recipient Charles L. Mayer Cancer award Nat. Sci. Fund of Nat. Acad. Sci., 1944. Author: Allgemeine Physiologie des Todes, 1915; Die Pubertätsdrüse u. ihre Wirkungen, 1919; Internal Secretions of the Sex Glands, 1924; Transplantación Ovárica, 1930; (with Jaime Pi-Sunyer) Curso Prático de Fisiología, 1934-35; Steroid Hormones and Tumors, 1950; Steroid Homeostasis, Hypophysics and Tumorigenesis, 1957; El Problema Racial en la Conquista de America y el Mestizaje, 1963; Warum wir sterben, 1914. Research on sex hormones, 1915—, antagonism of sex hormones, 1923-25, exptl. tumorigenesis due to ovarian-hypophysical imbalance, 1925-35, uterine and abdominal tumours induced by oestrogens, 1938-50, antitumorigenic action of progesterone and other steroids, 1939—, exptl. ovarian tumours, 1946-66, contraceptives, 1962-66. Address: 366 Av. Hamburgo, Santiago, Chile.*

LIPSCHÜTZ, Benjamin, Austrian dermatologist; b. 1878; described a rapidly spreading non-venereal ulcer of the vulva believed to be caused by Bacillus crassi (Lipschutz ulcer), 1918. Died 1931.

LIPSCOMB, Alys Harris, Am. physician; b. Sassafras Ridge, Ky., July 13, 1915; s. Sydney Harris and Myrtle (Pledger) L.; B.A., U. Tenn., 1936, M.S., 1944, M.D., 1945. Faculty, U. Tenn. Coll. Medicine, Memphis, 1940—, asso. prof. clin. medicine, 1958—; cons. nuclear medicine Kennedy VA Hosp., Memphis, 1951—, Meth. Hosp., Memphis, 1955—. Diplomate Am. Bd. Internal Medicine. Fellow A.C.P.; mem. Shelby County, Memphis med. socs., A.M.A., Tenn. Diabetes Assn., Acad. Internal Medicine (pres. 1965—), Am. Soc. Nuclear Medicine, Mortar Bd., Sigma Xi, Chi Omega, Phi Kappa Phi, Alpha Epsilon Iota. Research, publs. on diabetes, thyroid disease and med. use radio isotopes. Home: 35 N. Cox St., Memphis 38104.*

LIPSCOMB, Paul Rogers, Am. orthopaedic surgeon; b. Clio, S.C., Mar. 23, 1914; s. Paul Holmes and Mary Emma (Rogers) L.; B.S., U. S.C., 1935; M.D., Med. Coll. S.C., 1938; M.S. in Orthopaedic Surgery, Mayo Found., U. Minn., 1942; m. Phyllis M. Oesterreich, July 20, 1940; children—Susan Lovering, Paul Rogers. Intern Cooper Hosp., Camden, N.J., 1938-39; asst. to staff orthopaedic surgery Mayo Clinic, 1941-43, staff asso., 1943—, v.p. staff, 1963; cons. Methodist, St. Mary's hosps., 1943—; mem. faculty Mayo Grad. Sch., U. Minn., 1944—, prof. orthopaedic surgery, 1961—. Diplomate Am. Bd. Orthopaedic Surgery. Mem. Am. Acad. Orthopaedic Surgery, Orthopaedic Research Soc., A.M.A., A.C.S., Am. Orthopaedic Assn., Am. Soc. Surgery Hand, Clin. Orthopaedic Soc. Internat. Soc. Orthopaedic Surgery and Traumatology, Am. Rheumatism. Assn., Central, Internat. orthopaedic clubs, Sigma Xi. Research and publs. on surg. treatment of rheumatoid arthritis of the hand and paralytic conditions of the upper extremity; surg. treatment of diseases of both hip joints and to surg. treatment of Volkmanns' Ischemic contracture. Home: Oakledge, M.R. 72, Rochester. Office: Mayo Clinic, Rochester, Minn.*

LIPSCOMB, William N(unn), Jr., Am. phys. chemist; b. Cleve., Dec. 9, 1919; s. William Nunn and Edna Patterson (Porter) L.; B.S., University Kentucky, 1941, D.Sc. (honorary), 1963; Ph.D., California Inst. of Tech., 1946; m. Mary Adele Sargent, May 20, 1944; children—Dorothy Jean, James Sargent, Phys. chemist OSRD, 1942-46; with U. Minn., 1946-59, successively asst. prof., asso. prof. and acting chief phys. chemistry div., prof. and chief phys. chemistry division, 1954-59; professor chemistry, Harvard University, 1959—, chairman department of chemistry, 1962-65. Member U.S.A. Nat. Com. for Crystallography, 1954-59, 60-63, 65-67; chmn. program com. Fourth International Congress of Crystallography, Montreal, 1957. Guggenheim fellow, Oxford U., Eng., 1954-55; NSF sr. postdoctoral fellow, 1965-66; Robert Welch Found. lectr., 1966, Howard U. distinguished lecture series, 1966. Recipient Harrison Howe Award in Chemistry, 1958. Member Mineral Society America, Am. Phys. Soc. Am. Chem. Soc. (chairman Minn. sect. 1949-50), Am. Crystallographic Assn. (pres. 1955), Am. Acad. Arts and Sciences, National Acad. Science, Sigma Xi, Phi Beta Kappa. Author: The Boron Hydrides, 1963; also articles chem. and phys. jours. Asso. editor: Jour. of Chemical Physics, 1955-57. Clarinetist, mem. Amateur Chamber Music Players. Research on valence theory; diffraction studies of molecules and crystals. Home: 26 Woodfall St., Belmont, Mass. 02178. Office: Dept. of Chemistry, Harvard U., Cambridge, Mass. 02138.

LIPSET, Seymour Martin, Am. sociologist; b. N.Y.C., Mar. 18, 1922; s. Max and Lena (Lippman) L.; B.S., Coll. City N.Y., 1943; Ph.D., Columbia U., 1949; m. Elsie Braun, Dec. 26, 1944; children—David, Daniel, Carola. Lectr., U. Toronto, 1946-48; asst. prof. U. Cal., Berkeley, 1948-50; asst. prof., asso. prof. Columbia, 1950-56; prof. U. Cal., Berkeley, 1956-66, Harvard, 1966—. Fellow Social Sci. Research Council; mem. Am. Acad. Arts and Scis., Am. Sociol. Assn. (council 1959-62), Am. Polit. Sci. Assn. Author: Agrarian Socialism, 1950; (with Martin Trow, J. S. Coleman) Union Democracy, 1956; (with Reinhard Bendix) Social Mobility in Industrial Society, 1959; Political Man, 1960; The First New Nation, 1963; also numerous articles. Research on comparative analyses of social stratification and polit. behavior which has sought to locate generalizations that hold across nat. boundaries, including factors related to social mobility, social requisites of polit. democracy; determinants of voting, factors affecting govt. of pvt. assns. Home: 162 Washington St., Belmont, Mass. 02178. Office: 580 Wm. James Hall, Harvard U., Cambridge, Mass. 02138.*

LIPSETT, Mortimer Broadwin, Am. physician; b. N.Y.C., Feb. 20, 1921; s. Theodore and Gertrude (Broadwin) L.; A.B., U. Cal. at Berkeley, 1943; M.S., U. So. Cal., 1947, M.D., 1951; m. Marie Louise Nieft, Dec. 18, 1948; children—Roger, Edward. USPHS fellow Sloan-Kettering Inst., N.Y.C., 1954-56, asst. mem., 1956-57; sr. investigator Nat. Cancer Inst., NIH, Bethesda, Md. 1947-60, asst. chief endocrinology br., 1960-65, chief endocrinology br., 1965—; asso. clin. prof. medicine Howard U. Sch. Medicine, 1961—. Recipient Sloan award for Cancer Research, Sloan-Kettering Inst., 1955. Mem. Am. Soc. Clin. Investigation, Am. Fedn. Clin. Research, Harvey Soc., Am. Assn. Cancer Research, A.C.P., Endocrine Soc., Alpha Omega Alpha. Editor: Gas Chromatography of Steroids, 1965. Research, numerous publs. on endocrinology of breast cancer, therapy of choriocarcinoma, devel. methods for measuring male sex hormones, testosterone, description hormone excretion in adrenal cancer, devel. of therapy. Home: 5102 Acacia Av., Office: NIH, Bethesda, Md. 20014.*

LIPTON, Morris Abraham, Am. physician; b. N.Y.C., Dec. 27, 1915; s. Theodore and Rose (Latt) L.; B.S., Coll. City N.Y., 1935; Ph.D., U. Wis., 1939; M.D., U. Chgo., 1948; m. Barbara Steiner Lipton, Dec. 28, 1940; children—Judith Eve, Susan Victoria, David Mark. Asso. dir. U. Chgo. Toxicity Lab., 1941-46, asst. prof., 1952-55; asst. prof. medicine Northwestern U. Med. Sch., 1955-59; faculty U. N.C. Med. Sch., Chapel Hill, 1959—, prof. psychiatry, 1963—. Mem. mental health and psychopharmacology study sects. Nat. Inst. Mental Health, 1960-65, chmn. psychopharmacology study sect. 1964—. Mem. Am. Coll. Neuropsychopharmacology (chmn. constn. com. 1965—), Am. Soc. Biol. Chemists, Am. Psychosomatic Soc., A.C.P. Research numerous publs. on biochem. enzymology, neuroendocrinology and psychiatry, mechanisms by which brain hormones are synthesized and regulated, effects alteration levels these hormones on behavior, effect drugs on neurohormones and behavior. Home: 2004 N. Lake Shore Dr., Chapel Hill, N.C. 27515.*

LIQUORI, Alfonso Maria, Italian chemist; b. Naples, Italy, Apr. 8, 1926; s. Costantino and Maria (Soave) L.; Ph.D., U. Rome; m. Laura Barcellini, Apr. 25, 1948; children—Maria Cristina, Costantino, Aldo. Joined faculty, 1957; prof. organic chemistry Faculty Sci. and Math., U. Bari (Italy); prof. phys. chemistry U. Naples (Italy), 1960-66; prof. dir. Inst. Phys. Chemistry, U. Rome, 1966—; dir. Nat. Center for Chemistry Macromolecules, Nat. Research Council. Mem. Accademia Pontaniana. Contbr. articles to tech. jours. Home: via Aniello Falcove 249, Naples, Italy. Office: Instituto di chemica fisica, Universita di Roma, Rome, Italy.

LISFRANC, Jacques, French surgeon; b. St. Paul, France, Apr. 2, 1790; surgeon at La Pitié. Author: Nouvelle méthode opératoire pour l'amputation partielle du pied, 1815. Developed many new operations, including method of disarticulating shoulder joint, 1815, excision of rectum, lithotomy in women, amputation of cervix uteri, partial amputation of foot at tarsometatarsal asticulation. Died Paris, May 13, 1947.

LISMAN, Henry, Am. physicist; b. Boston, July 3, 1913; s. David and Celia (Gersberg) L.; B.S. in Edn., Boston U., 1934, M.A., 1935, Ph.D., 1939; postgrad. Harvard; m. Rachel Lewit, Apr. 3, 1938; children—Meira (Mrs. Solomon Max), Elliot Cyril. Faculty Northeastern U., 1939-42; physicist Signal Corps Radar Lab., 1942-47; prof. math. Yeshiva U., N.Y.C., 1942—; physicist U. S. Army Electronic Labs., Fort Monmouth, N.J., 1948—. Mem. Am. Math. Soc., Am. Phys. Soc. Research on electromagnetic and acoutic wave propagation, photogrammetry microwaves, magnetic phenomena. Home: 1693 Selwyn Av., Bronx, N.Y. 10457. Office: Yeshiva Coll., 500 W. 185th St., N.Y.C. 10033.*

LISONEK, Peter Anthony, Czechoslovakian physician, anatomist; b. Olomouc, Czechoslovakia, Dec. 16, 1928; s. Anthony and Anna (Friml) L.; MUDR, Med. Sch., Palacky U., Olomouc, 1952, postgrad. examination for surg. specialization, 1958, C.Sc., 1960; m. Drahomira Smetana, Aug. 6, 1959; 1 son, Peter. Research worker, head dept. normal and exptl. neuromorphology Inst. Normal Anatomy, Palacky U. Med. Sch., 1964—. Mem. Czechoslovak Anthropologic Soc. (head Olomouc br.), Czechoslovak Soc. Anatomists, Czechoslovak Soc. Anthropologists, Czechoslovak Biol. Soc. Author: Distal Metaphysics of Juvenile Humerus from Surgical Standpoint, 1958; also articles. Ramification of phrenic nerve for non-injuring thoraco-diaphragmotomy, 1952; studies on prins. of morphologic remodellation after supracondylar fractures of humerus with new method of controlling perfection of reposition, 1957-60, osteometry of humerus and humeral septum during ontogenesis, including original theory of origin of supratrochlear foramen, 1958-59; exact description of vascularization of renal calices, pelvis and pelviureteral junction for surg. purposes, 1961-62; description 2 transitional forms of musculature of renal pelvis, 1963; hypopharyngeal diverticula including description of new form and role of vessels perforating musculature, 1964-65. Home: 5 Svermova, Olomouc, Czechoslovakia.*

LISSAJOUS, Jules Antoine, French physicist; b. Versailles, France, Mar. 4, 1822; student Ecole normale, 1841-44; docteur ès sciences, 1850. Prof. physics, Collège Saint-Louis, Paris, France; rector Académie de Chambery, France, 1874-75, Académie de Besançon, France, 1875-79. Mem. French Acad. Scis., 1879. Author: Étude optique des mouvements vibratoires, 1873. Important studies in acoustics, optics; inventor optical comparateur, 1857; invented device that shows visually Lissajous' figures (two simple harmonic motions acting at right angles to one another); studied transverse vibrations of flexible thin plates; system optical telegraphy (used during siege of Paris, 1870); collaborator with Foucault; completed works of Fresnel. Died Plombieres-les-Dijon, France, June 24, 1880.

LISSAK, Kalman, Hungarian physiologist; b. Budapest, Hungary, Jan. 13, 1908; s. Kalman and Aranka (Kozak) L.; M.D., U. Budapest, 1933; Privat Docent in Physiology, U. Debrecen, 1937; m. Janina Malachowska, Mar. 24, 1937; children—Stefania (Mrs. Tamas Fekete), Kalman. Faculty, U. Debrecen, 1933-43; Rockefeller Found. fellow dept. physiology Harvard Med. Sch., 1937-39; prof. physiology U. Med. Sch., Pecs, Hungary, 1943—. Mem. Hungarian Physiol. Soc. (pres.), Hungarian EEG Soc. (pres.), Hungarian Sci. Council of Phys. Edn. (pres.), Hungarian Acad. Scis. (pres. physiol. council), Internat. Union Physiol. Scis. (council). Author: (with E. Endroczi) Die Neuroendokrine Steuerung der Adaptationstätigkeit, 1960; The Neuroendocrine Control of Adaptation, 1965; also numerous articles. Research on adrenaline content of adrenergic neurons, humoral mediation in central nervous system, inhibitory substance of central nervous system, behavioral, neurophysiol. and electrophysiol. investigation of conditioning process, humoral control of animal behavior, neuroendocrine control of adaptation, central nervous control of pituitary function. Address: 80 Rakoczi St. Inst. Physiology, Univ. Med. Sch. of Pécs, Pécs, Hungary.*

LISSAUER, Heinrich, German neurologist; b. Neidenburg, Germany, Sept. 12, 1861; described spinal cord tract composed of fibers from dorsal roots and fasciculus proprius, (Lissauer's tract), 1885, atypical gen. paralysis with aphasia, monoplegia, convulsions (Lissauer's paralysis), published posthumously, 1901, Died Hallstatt, Germany, Sept. 21, 1891.

LISSNER, Josef Anton, German radiologist; b. Tütz, Pommern, Germany, Apr. 22, 1923; s. Frank A. and Elisabeth M. (Schulz) L.; Doktordiplom, U. Erlangen (Germany), 1951, Habilitation, 1960; m. Marianne H. Fonk, Feb. 2, 1951; children—Adelheid, Mechthild, Almuth, Radegund, Elisabeth Katharina. Asst. physician U. Klinik Erlangen, 1951-54; asst. radiologist, Med. U. Klinik Frankfurt, Germany, 1954-64; chief univ. physician Klinik fur Strahlentherapie und Nuklearmedizin, Frankfurt, 1964—; vis. staff Radiumhemmet, also Karolinska Sjukhuset, Stockholm, Sweden, 1960, Royal Infirmary, Edinburgh, Scotland, 1961, Gesundheitsministerium of Bundesrepublik Deutschland, 1962. Author: Flächen-und Elektrokymographie, 1962; (with A. Gebauer, O. Schott) Das Röntgenfernsehen, 1965; also articles. English editor X-ray Television, 1967. Research on heart movements with electrokymography, X-ray TV and its application to radiol. work. Home: 1 Sigmund-Freud-Str. Office: 14 Ludwig-Rehn-Str., 6000 Frankfurt/Main, West Germany.*

LIST, Georg Friedrich, economist; b. Wurttemberg, Germany, Aug. 6, 1789; s. Johannes and Mrs. (Schafer) L.; ed. U. Tubingen; m. Catherine (Seybold) Neidhard, circa 1818; 4 children. Entered public service of Wurttemberg, 1806; rose to ministerial undersec.; prof. administrn. and politics U. Tubingen, 1817; lost post, 1819, because of polit., econ. ideas; elected to Diet of Wurttemberg from Reutlingen, 1819; charged with sedition because of this advocacy of reforms, exiled; came to Am., 1825, naturalized; toured Atlantic states with Lafayette; became editor of Readinger Adler, Reading, Pa.; loomed as a leading advocate of protective tariff and "Am. System" with his writings, address to Pa. legislature, 1828, and dispute with Gov. W. B. Giles of Va.; developed anthracite deposits near Tamaqua, Pa.; organized Little Schuylkill Navigation, R. R. & Coal Co. (progenitor of modern Reading System), 1828, opened 1831; exec. agt. Dept. of State, went to Europe, planned to introduce Pa. anthracite coal there; U. S. consul, Baden, Germany, 1831-34, Leipzig, 1834-37, Stuttgart, 1843-45; founder Zollvereinsblatt, polit. economy jour., 1843; visited England in vain attempt to prepare comml. alliance between that nation and Germany, 1846; retired to Augsburg. Author: Outlines of American Political Economy, 1827; National System of Political Economy, 1841. An architect of econ. nationalism; believed nation's wealth depends on integrated devel. of productive forces, not individual accumulation; urged protective tariff for infant industries in emerging countries; advocated German customs union. Died a suicide, Tyrol, Nov. 30, 1846.

LIST, Hans, Austrian physicist; b. Graz, Austria, Apr. 30, 1896; s. Hugo and Anna Raab von Rabenau L.; Ph.D., Gratz Tech. Coll.; m. Elfriede Lenore Wachter, Oct. 30, 1937; children—Harald, Dagmar, Gerda, Helmut. Prof., U. Tunghi, China, 1926-32, Graz Tech. Coll., 1932-41; with Dresden Tech. Coll., 1941-45; owner, gen. mgr. Prof. Dr. Hans List Motor Studies Co., 1945—. Recipient Medal of Merit in bronze and silver with ribbon of Order of Merit; Austrian Order of Merit; award Town of Gratz; Akroyd Stuart prize, 1955. Mem. Austrian Acad. Sci., Instn. Mech. Engrs., Verein Deutscher Ingenieure, Austrian Acad. Sci., Automotive Engrs.' Soc. Author: Thermodynamik der Verbrennungskraftmaschine, 1938; Der Ladungswechsel der Verbrennungskraftmaschine, 1949-52. Home: Heinrichstrasse 126. Office: Elisabethstrasse 5, Graz, Austria.

LIST, Roland, cloud physicist; b. Frauenfeld, Switzerland, Feb. 21, 1929; s. August J. and Anna (Kaufmann) L.; dipl. Phys. ETH, Zurich, 1952, Dr. sc. nat. ETH, 1960; m. Gertrud K. Egli, Apr. 14, 1956; children—Beat R., Claudia G. Head, hail sect. Swiss Fed. Snow and Avalanche Research Inst., Weissfluhjoch-Davos, 1952-63; prof. physics (meteorology) U. Toronto (Ont., Can.), 1963—. Mem. Schweizerische Physikalische Gesellschaft, Canadian, Am. meteorol. socs., Canadian Assn. Physicists, Am. Geophys. Union, A.A.A.S. Research on formation and modification of precipitation, simulation expts., cloud dynamics, numerical modelling of clouds and precipitation. Home: 58 Olsen Dr., Don Mills, Ont., Can. Office: Dept. Physics, U. Toronto, Toronto 5, Ont., Can.*

LISTER, Joseph, (1st Baron Lister of Lyme Regis), Brit. surgeon, biologist; b. Upton, Essex, England, Apr. 5, 1827; s. Joseph Jackson and Isabella (Harris) L.; ed. U. Coll., London, B.A., 1847, M.B., 1852; hon. LL.D., D.Sc.; m. Agnes Syme, 1856. House surgeon to James Syme, Edinburgh, 1853-56; asst. surgeon Edinburgh Royal Infirmary, 1856; prof. surgery Glasgow U., 1860-69, surgeon to infirmary, 1861; prof. clin. surgery Edinburgh U., 1869-77; prof. clin. surgery King's Coll., London, 1877-93; sgt.-surgeon to Queen Victoria, 1878. Fellow Royal Soc., 1860 (pres. 1895-1900); pres. Brit. Assn. for Advancement of Science, 1896; 1st chmn. Brit. Assn. for Advancement of Science, 1896; 1st chmn. Brit. Inst. Preventive Medicine, 1891 (became Lister Inst. Preventive Medicine, 1903). Created baronet, 1893; raised to peerage, 1897; one of first to receive Order of Merit. Author: Early Stages of Inflammation, 1859; Ligature of Arteries and the Antiseptic System, 1869; The Germ Theory of Fermantative Changes, 1875; Collected Writings, 1909; numerous papers. Founder of aseptic surgery; introduced antiseptic procedure in surgery; advanced the use of carbolic acid as a germicide on wound dressings, 1865; initiated use of absorbable ligatures and of drainage tubes in treatment of wounds; first to isolate pure culture bacteria, 1874; conducted first original work on structure of iris; gave first clear description of dilating mechanism of pupil, 1853; described dorsal tubercle of radius (Lister's tubercule), 1865. Died Walmer, Kent, Feb. 10, 1912.

LISTER, Joseph Jackson, English optician, biologist; b. London, Jan. 11, 1786; s. John and Mary (Jackson) L.; self-taught; m. Isabella Harris, July 14, 1818; 4 sons, including Joseph, 3 daus.; wine mcht. Contbr. articles to tech. jours. Fellow Royal Soc., 1832. Tried to improve object-glass, 1824; discovered constrn. principles of modern microscope, 1830; developed law of aplanatic foci, 1830; discovered true form of mammalian red blood cell, 1834. Died Oct. 24, 1869.

LISTER, Martin, Brit. zoologist; b. Radclive, Eng., 1638; s. Sir Martin Lister; B.A., St. Johns Coll., Cambridge, 1659, M.A., 1662; med. studies, France until 1670; children—Susannah, Mary, Alexander. Practiced medicine, York, until 1683; apptd. 2nd physician in ordinary to Queen Anne, 1709. Mem. Royal Soc., 1671, Royal Coll. Physicians. Author: Historiae animalium Angliae, 1678; Historia conchyliorum, 1685-92; Novae ac curiosae exercitationes et descriptiones thermarum ac fontium medicatorum Angliae, 1682; also articles on spiders, mollusks, minerals. First Brit. conchologist. Died 1712.

LISTON, Robert, surgeon; b. Ecclesmachan, Scotland, Oct. 28, 1794; ed. U. Edinburgh, 1808-15. House surgeon Royal Infirmary, Edinburgh, 1814-16, surgeon, 1827-34; mem. staff London Hosp., 1816-18; surgeon, prof. clin. surgery U. Coll. Hosp., London, from 1835. Mem. Royal Coll. Surgeons Eng. (council 1840, bd. examiners 1846). Fellow Royal Soc., 1841. Author: The Elements of Surgery, 3 parts, 1831, 32, 40; Practical Surgery, 1837. Renowned for surg. skill; 1st surgeon in Eng. to remove scapula, circa 1834, also to use ether in operation, 1846; devised Liston's splint for thigh dislocations; proposed use of spl. mirror to inspect larynx, 1837. Died London, Dec. 7, 1847.

LIT, Alfred, Am. psychologist; b. N.Y.C., Nov. 24, 1914; s. Zachary Oscar and Elsie (Jaro) L.; B.S., Columbia, 1938, A.M., 1943, Ph.D., 1948; m. Imogene Speegle, Jan. 27, 1947. Individual practice optometry, N.Y.C., 1938-43; mem. faculty Columbia, 1946-56; research psychologist U. Mich., 1956-59; head, engring. psychology staff Bendix Systems Div., Ann Arbor, Mich., 1959-61; prof. psychology So. Ill. U., Carbondale, 1961—. Mem. com. on vision Armed Forces-NRC, 1961—. Am. Acad. Optometry grantee, 1949-51, NIH-USPHS grantee, 1962—, NSF grantee, 1962—. Fellow Am. Psychol. Assn., Am. Acad. Optometry, A.A.A.S.; mem. Assn. Schs. and Colls. Optometry, Optical Soc. Am., Psychonomic Soc., N.Y. Acad. Scis., Human Factors Soc., Sigma Xi. Research in human vision on problems in binocular depth discrimination, light and dark adaptation, color vision, visual reaction time, physiol. optics and visual electrophysiology. Home: 603 Emerald Lane, Carbondale, Ill. 62901.*

LITCHFIELD, John Hyland, Am. biochemist, microbiologist; b. Scituate, Mass., Feb. 13, 1929; s. Frank Albert and Alma (Hyland) L.; S.B., Mass. Inst. Tech., 1950; M.S., U. Ill., 1954, Ph.D. (Hackett fellow, univ. fellow), 1956; m. Dianne Chappell, Apr. 15, 1966. Chief chemist Searle Food Corp., Hollywood, Fla., 1950-51; research food technologist Swift & Co., Chgo., 1956-57; faculty Ill. Inst. Tech., Chgo., 1957-60; project leader bioscis. Battelle Meml. Inst., Columbus, O., 1960-62, asst. chief bioscis., 1962-64, chief, biochemistry and microbiology research, 1964—. Cons. to food industry, 1957-60. Fellow Am. Pub. Health Assn.; mem. Am. Chem. Soc., A.A.A.S., Am. Soc. for Microbiology, Inst. Food Technologists, Soc. for Indsl. Microbiology, Assn. Vitamin Chemists, Mycol. Soc. Am., Am. Inst. Biol. Sci., N.Y. Acad. Sci., Soc. Chem. Industry Gt. Britain, Inst. Biology Gt. Britain, Royal Soc. Health, Soc. for Gen. Microbiology Gt. Britain, Soc. for Applied Bacteriology Gt. Britain, Sigma Xi, Gamma Sigma Delta. Author: (with M. E. Parker) Food Plant Sanitation, 1962. Research in chemistry and tech. of food processing and preservation; devel. mass cultivation methods for producing microorganisms as a source of food; recovery of proteins from waste food materials; food and san. microbiology, including new methods for isolating enteric disease microorganisms; nutritional evaluation of food processing methods. Home: 1270 Fenceway Dr., Columbus 43224. Office: 505 King Av., Columbus, O. 43201.*

LITCHFIELD, Paul Weeks, Am. engr.; b. Boston, Mass., July 26, 1875; s. Charles M. and Julia (Weeks) L.; B.S., Mass. Inst. Tech., 1896; m. Florence Brinton, June 23, 1904; children—Mrs. Howard Hyde, Mrs. A. Wallace Denny. Supt., Goodyear Tire and Rubber Co., 1900-15, v.p., 1915-26, pres., 1926-40, chmn. bd. since 1930; pres. Goodyear Aircraft Corp. Mem. Soc. Automotive Engrs., Nat. Assn. Mfrs., Nat. Air Council, Rubber Mfrs. Assn. Am. (pres.). Writer of Autumn Leaves, Industrial Republic; Industrial Voyage (autobiography of P. W. Litchfield); also mag. articles. Leader in devel. of lighter-than-air craft; contbr. to automotive and aero. tech., also to prodn. of synthetic rubber during World War II. Died Mar. 8, 1958.

LITEANU, Candin, Rumanian chemist; b. Ciugudul de Sus, Rumania, July 6, 1914; s. Vasile and Ighifta (Boldor) L.; engr. Faculty Agr. Cluj (Rumania), 1937; bachelor's degree U. Cluj, 1941, Ph.D., 1945; m. Maria Mihalca, Oct. 14, 1944; children—Margareta, Victor, Ileana. Faculty, U. Cluj, 1941—, prof., 1951—, chmn. dept. analytical chemistry, 1964—; tech. adviser Chem. Factory, Terapia, Cluj, 1948-51; research fellow Inst. Chemistry, Acad. Socialist Republic Rumania, Cluj, 1949-53, sr. research fellow, 1953-54, head research sect., 1954—. Author: Quantitative Analytical Chemistry, Volumetry, 1956; (with R. Ripan, E. Popper) Qualitative Analytical Chemistry, 1954; also numerous articles. Theoretical problems of analytical chemistry, paperthermochromatography; various methods of analysis; electrolytical method for obtaining sulphur dioxide from gypsum dissolved in molten sodium chloride. Home: 12 Republicii, Cluj, Rumania.*

LITINOV, Nikolay Nikolaevich, Russian hygienist; b. 1894; grad. Med. Faculty, Kiev U., 1916; D.Med. Sci. Head, Stalingrad City and Oblast Health Epidemiology Sta., 1921-45; asst. chair hygiene Stalingrad Med. Inst., 1921-45, lectr., acting head, 1932-45; former asso. Stalingrad Oblast Inst. Epidemiology and Microbiology; asso. USSR Ministry Health, 1946-51; sr. asso. Inst. Gen. and Communal Hygiene, USSR Acad. Med. Sci., 1946-56, dir. Inst. Gen. and Communal Hygiene, 1956—; head course communal hygiene Central Postgrad. Med. Inst., 1946-52; head chair gen. hygiene 2d Moscow Med. Inst., 1952-59. Decorated Order of Lenin. Mem. USSR Acad. Med. Sci. (corr.), Russian (bd. mem.), All-Union (bd. mem.) socs. hygienists and med. officers of health, Purkinje Czech Med. Soc. (hon.). Co-author, editor: Reservoir Hygiene, 1961; author: Radiation Injuries to the Bone System, 1964. Mem. editorial council Hygiene and Sanitation. Address: Inst. Gen. and Communal Hygiene, USSR Acad. Med. Sci., Pogodinskaya ulitsa 10, Moscow, USSR.

LITMAN, Arnold Powell, Am. metallurgist; b. Kansas City, Mo., July 23, 1926; s. Phillip G. and Sarah (Powell) L.; B.S. in Metall. Engring., U. Ill., 1946; M.S. in Indsl. Mgmt., Ga. Inst. Tech., 1949; M.S. in Metall. Engring., U. Tenn., 1964; m. Beatrice Lazaroff, June 27, 1954; children—Ellen Rae, Ruth Ann, Judith Kay, Andrew Hays, Margaret Jo. Metallographer, Union Wire Rope Corp., Kansas City, Mo., 1946-48, chief metallurgist, 1949-51; metallurgist metals div. Olin-Mathieson Corp., East Alton, Ill., 1953-56, nuclear div. Black, Sivalls & Bryson, Inc., Tulsa, 1956-58; metallurgist metals and ceramics div. Oak Ridge Nat. Lab., 1958-66, group leader, 1966—. Chmn., 15th AEC Corrosion Symposium, 1966. Mem. Am. Soc. Metals (chmn. Oak Ridge chpt. 1966-67), Am. Inst. Mining, Metall. and Petroleum Engrs., Metall. Soc. Research, publs. on corrosion by liquid metals, salts and gases, nuclear reactor engring. materials, solid state bonding, manufacture and use of high carbon steel wire and wire rope; devel. mfg. methods for nuclear reactor and mil. components, mfg. methods for heat transfer components; inventor selective bonding agt. for solid state welding. Home: 105 Berwick Dr., Oak Ridge 37830. Office: P.O. Box Y, Oak Ridge 37831.*

LITOVITZ, Theodore Aaron, Am. physicist; b. N.Y.C., Oct. 14, 1923; s. Nathen and Mary (Freifeld) L.; B.A., Cath. U. Am., 1946, Ph.D., 1950; m. Charlotte Goldberg, Aug. 11, 1946; children—Toby Lynn, Gary Lane. Prof. physics Cath. U. Am., Washington, 1950—; cons. Georgetown Med. Sch., Washington, 1950-56, Naval Ordnance Lab., Bethesda, Md., 1952—. Indsl. cons., 1958—. Fellow Am. Phys. Soc., Am. Acoustical Soc. Author: (K. F. Herzfeld) Absorption and Dispersion of Ultrasonic Waves, 1959. Home: 904 Devere Dr., Silver Spring, Md.*

LITRICIN, Olga, Yugoslavian ophthalmologist; b. Kikinda, Yugoslavia, Mar. 12, 1918; d. Millovoje and Mara (Lang) Radin; M.D., U. Belgrade (Yugoslavia), 1942, D.Med.Scis., 1958; m. Triva Litricin, Nov. 4, 1940; children—Vera (Mrs. Arsen Semerad), Mirjana. Practice gen. medicine, 1942-46; resident in ophthalmology Ophthalmic U. Hosp., Belgrade, 1946-49; faculty U. Belgrade Sch. Medicine, 1949—, asso. prof. ophthalmology, 1960—. Mem. Yugoslav, French ophthalmic socs. Author: (with I. Stankovic, M. Blagojevic, M. Danic) Ophthalmology, 1965. Research, publs. on cancer in ophthalmology, infectious diseases to the eye, radiation damage to the eye, ophthalmic pathology, traumatology and surgery. Home: 7 Branka Cvetkovica. Office: 22 Visegradska, Belgrade, Yugoslavia.*

LITSKY, Warren, Am. microbiologist; b. Worcester, Mass., June 10, 1924; s. Joseph and Esther (Kreplick) L.; A.B., Clark U., 1945; M.S. (W. K. Kellogg Found. fellow), U. Mass., 1948; Ph.D., Mich. State U., 1951; m. Bertha Shulman Yanis, Aug. 27, 1965. Instr., Mich. State U., East Lansing, 1951; asst. research prof. U. Mass., Amherst, 1951-53, asso. research prof., 1953-56, research prof., 1956-61, commonwealth prof., 1961—, dir. Inst. Agrl. and Indsl. Microbiology, 1962—, mem. senate, 1964-67. Cons. USPHS, 1961—, numerous corps. Recipient Sayer prize for excellence in bacteriology Mich. State U., 1951. Fellow Am. Pub. Health Assn., Royal Soc. Health (Eng.), A.A.A.S.; mem. Am. Soc. Microbiology (dir. Conn. Valley br. 1958-63, editorial bd. Applied Microbiology 1960—), Soc. Indsl. Microbiology (editor newsletter), N.Y. Acad. Sci., Internat. Assn. Milk and Food Sanitarians, Soc. Applied Bacteriology (Eng.), Phi Tau Sigma. Research, publs. in field of pub. health bacteriology; developed methods for isolating indicator organisms for fecal pollution of water and foods; studies on gas sterilization for med. equipment; (with others) perfected new ultra high temperature method of pasteurization with modification for rapid vaccine prodn. using high temperature:

short time ratios; made 1st clin. study amphoteric terramycin. Home: 74 Taylor St., Amherst, Mass. 01002.*

LITTAUER, Uriel Z., Israeli biochemist; b. Tel-Aviv, Israel, Feb. 24, 1924; s. Franz and Regina (Lehrfruend) L.; M.Sc., Hebrew U., Jerusalem, Israel, 1950, Ph.D., 1954; m. Atida Eizenberg, July 12, 1946; children—Tahel, Michal. Faculty, Weizmann Inst. Sci., Rehovoth, Israel, 1949—, head biochemistry sec., 1964-66, prof. biochemistry, 1966—. vis. research fellow Washington U., St. Louis, 1955-56, Stanford, 1962, Harvard, 1963. Recipient Somach Sachs award for chemistry Weizmann Inst. Sci., 1958. Mem. Israel Biochem. Soc. (internat. sec.), Israel Chem. Soc., Biochem. Soc. Eng., European Molecular Biology Orgn. Editor in biology Holden-Day Pub. Co., 1963; editor European Jour. Biochemistry, 1966—. Research, publs. on nucleic acids and polynucleotides with spl. emphasis on the chem., physico-chem. and biol. properties of ribonucleic acids. Address: Weizmann Inst. Sci., Rehovoth, Israel.*

LITTELL, Frank Bowers, Am. astronomer; b. Scranton, Pa., Feb. 21, 1869; s. Henry Woolsey and Marie Antoinette (Bowers) L.; Ph.B., Wesleyan, 1891, Sc.D., 1919; A.M., Columbian (now George Washington) U., 1894; m. Josephine La Monte Mercereau, Apr. 9, 1902; children—Marion Mercereau, Charles Henry. Computer Naval Obs., 1891-96, 97-98, asst. astronomer, 1898-1901; tchr. math. Scranton High Sch., 1896-97; prof. math. USN, 1901-33. Mem. U. S. eclipse expdns. to Barnesville, Ga., 1900, Solok, Sumatra, 1901, Porta Coeli, Spain, 1905, Los Angeles (airship), 1925, expdn. to Sumatra, 1926; mem. U. S. party to determine Washington-Paris longitude, using radio signals, 1913-14; variation of latitude by photog. zenith tube, 1915-33; determination of world longitudes by radio, San Diego (Cal.) sta., 1926; made catalogue of 23,521 stars (with W. S. Eichelberger); vertical circle observations made with 5-inch altazimuth, 1898-1907, 16-33 (pub. 1939); Washington-Paris longitude by radio signals (with G. A. Hill). Fellow A.A.A.S., Royal Astron. Soc. (Eng.); mem. Philos. Soc. Washington, Am. Astron. Soc. (councillor 1928-31), Washington Acad. Scis., Sociedad Astronomica de Mex., Soc. Astron. de France, Internat. Astron. Union, Am. Geophys. Union Astron., Soc. of Pacific, Phi Beta Kappa. Spl. editor on astronomy Webster's New Internat. Dictionary and World Book Ency. Annual. Died Mar. 28, 1951; buried Arlington Nat. Cemetery.

LITTELL, Squier, Am. physician; b. Burlington, N.J., Dec. 9, 1803; s. Stephen and Susan (Gardner) L.; M.D., U. Pa., 1824; m. Mary Graff Emlin, 1834; 2 children. Practiced medicine, Phila.; specialist in ophthalmic surgery; surgeon Wills Hosp. for lame, halt and blind, 1834-64; mem. Coll. Physicians of Phila., 1836-86; editor Monthly Jour. of Fgn. Medicine, 1828-29, Banner of the Cross, 1839-41. Author: A Manual of Diseases of the Eye, 1837. Died Bay Head, N.J., July 4, 1886.

LITTLE, Arthur D(ehon), Am. chem. engr.; b. Boston, Dec. 15, 1863; s. Thomas J. and Amelia (Hixon) L.; ed. Mass. Inst. Tech.; Ch.D. (hon.), U. Pitts., 1918; hon. asso., Coll. Tech., Manchester, Eng., 1929; Sc.D., Manchester U., 1929, Tufts, 1930, Columbia, 1931; m. Henrietta Rogers Anthony, Jan. 22, 1901. Chemist and supt. of 1st mill in U. S. making sulphite wood pulp, 1884-85; chem. engring. practice, Boston, 1886—; pres. Arthur D. Little, Inc., 1909—; dir. Arthur D. Little Indsl. Corp. Founder of Chem. Engring. Practice, Mass. Inst. Tech.; organized Natural Resources Survey of Can. for C.P. Ry., 1916-17. Recipient Perkin medal, 1931. Fellow Am. Acad. Arts and Scis., A.A.A.S., Chem. Soc. (Eng.), Inst. of Fuel (London). Author: Chemistry of Paper Making (with R. B. Griffin), 1894; The Handwriting on the Wall, 1928; The Fifth Estate, 1925. Inventor processes of chrome tanning, electrolytic manufacture of chlorates, artificial silk, gas and petroleum; research on airplane dopes, acetone prodn., smoke filters, during World War I. Died Aug. 1, 1935.

LITTLE, Clarence C(ook), Am. biologist; b. Brookline, Mass., Oct. 6, 1888; s. James Lovell and Mary Robbins (Revere) L.; A.B., Harvard, 1910; S.M., Grad. Sch. of Applied Science (Harvard), 1912, Sc.D. 1914; LL.D., U. N.H., 1924, Albion (Mich.) Coll., 1925, U. N.M., 1929, Colby College, Waterville, Me., 1935; Litt.D., U. Me., 1932; Sc.D., U. Chgo., 1950, Boston U., 1951; L.H.D., Dickinson Coll., 1951; Ed.D., Marietta Coll., 1952; m. Katharine Day Andrews, May 27, 1911 (div. 1929); children—Edward Revere, Louise, Robert Andrews; m. 2d, Beatrice W. Johnson, 1930; children—Richard Warren, Laura Revere. Sec. to Corp. Harvard Univ., 1910-12; research asst. in genetics, 1911-13, research fel. in genetics (cancer commn.), 1913-17, asst. dean, 1916-17; overseer, 1942-48, 1955-61, all Harvard U.; faculty Harvard Med. Sch., 1917-25; asst. dir., Sta. for Exptl. Evolution, Carnegie Instn., 1921-22; pres. U. Me., 1922-25, U. Mich., 1925-29; dir. Jackson Meml. Lab., 1929-56, dir. emeritus, 1956—. Dir. Am. Cancer Soc., 1929-45. Sci. dir. Council for Tobacco Research, 1954—; mem. sci. adv. bd., Gesell Inst. Child Devel. 1952—. Officer, del., or mem. several confs. eugenics, euthanasia, birth control, genetics, 1921—. Recipient medal Am. Cancer Soc., 1950, Pioneers Hall of Fame, 1952. Fellow Nat. Acad. Scis.,

Am. Acad. Arts and Scis., N.Y. Acad. Medicine, A.A.-A.S., Nat. Inst. Social Scis.; mem. nat., state, tech., scientific and profl. socs. and orgns., officer or dir. several. Author: Civilization Against Cancer, 1939; Genetics, Biological Individuality, and Cancer, 1954; Inheritance of Coat Color in Dogs, 1956; Research and publs. on genetics, cancer research, edn. and social problems. Address: Ellsworth, Me.*

LITTLE, Edward Milton, Am. physicist; b. Cin., Oct. 14, 1897; s. George Henry and Jessie (Gearing) L.; B.S. in Physics, U. Wash., 1918, M.S., 1922; Ph.D., U. Ill., 1926; m. Geneva Tracy Millett, June 26, 1948; 1 son, Ronald Vernon. Lab. asst. Bur. Standards, Washington, 1918-19; tchr. high sch., Renton, Wash. 1919-20, Puyallup, Wash., 1921-22; faculty U. Mont., 1926-29, 31-40; research physicist in acoustics Bell Telephone Labs., N.Y.C., 1929-31; research physicist U. S. Forest Service, Missoula, Mont., 1934; war research in fire control, shock waves, Phila., Washington, Aberdeen, Md., Silver Spring, Md., 1940-49; prof. U. Md., 1948-50; head physics dept., asst. dir. geophys. Inst., U. Alaska, Fairbanks, 1950-52; research physicist sea ice physics Navy Electronics Lab., San Diego, 1952—. Mem. Am. Phys. Soc., Am. Assn. Physics Tchrs., A.A.A.S., Acoustical Soc., Am. Arctic Inst. N.Am., Glaciological Soc., Am. Geophys. Union, N.Y. Acad. Sci., Sigma Xi, Phi Delta Kappa, Sigma Pi Sigma. Research, publs. on ionization cesium vapor, shock wave oblique reflections, sea ice physics; invented visibility meter. Home: 1091 Sunset Cliffs Blvd., San Diego 92107. Office: Navy Electronics Lab., San Diego 92152.*

LITTLE, Elbert Luther, Jr., Am. botanist; b. Ft. Smith, Ark., Oct. 15, 1907; s. Elbert Luther and Josephine (Conner) L.; B.A., U. Okla., 1927, B.S., 1932; M.S., U. Chgo., 1929, Ph.D., 1929; m. Ruby Rema Rice, Aug. 14, 1943; children—Gordon Rice, Melvin Weaver, Alice Conner. Asst. prof. biology Southwestern State Coll., 1930-33; with Forest Service, U. S. Dept. Agr., 1934—, dendrologist, Washington, 1942—. Collaborator, U. S. Nat. Mus., 1965—; vis. prof. biology Va. Poly. Tech., 1966-67. Recipient Superior Service award U. S. Dept. Agr. 1960. Fellow Washington, Okla. Acads. scis.; mem. Am. Inst. Biol. Scis., A.A.A.S., Soc. Am. Foresters, Bot. Soc. Am., Am. Soc. Plant Taxonomists, Ecol. Soc. Am., Phi Beta Kappa, Sigma Xi, numerous others. Author: Check List of Native and Naturalized Trees of the United States, 1953; (with Frank H. Wadsworth) Common Trees of Puerto Rico and the Virgin Islands, 1964; also numerous articles. Research field trees and vegetation of U. S., tropical Am. Home: 924 20th St. S., Arlington, Va. 22202. Office: U. S. Forest Service, Washington 20250.*

LITTLE, Henry N., Am. biochemist; b. Portland, Me., Oct. 17, 1920; s. George Tappan and Bertha (Nelson) L.; B.S., Cornell U., 1942; M.S., U. Wis., 1946, Ph.D., 1948; m. Isabelle M. Billings, June 20, 1948; children—Nathan H., Joyce H., Dana W., Karen M. Research asso. dept. biochemistry U. Chgo., 1948-49; asst. prof. biology McCallum Pratt Inst., Johns Hopkins, 1949-51; asso. prof. chemistry U. Mass., Amherst, 1951-56, prof. chemistry, 1956-66, prof. biochemistry, 1966—. Mem. Am. Chem. Soc., A.A.A.S., Am. Soc. Biol. Chemists, Sigma Xi. Research, publs. on chemistry and biosynthesis of porphyrias and hemoproteins such as hemoglobin and chlorophyll. Home: 147 Red Gate Lane, Amherst, Mass. 01002.*

LITTLE, Homer Payson, geologist; b. Columbia, Conn., Aug. 3, 1884; s. Payson Elliot and Emma (Bascom) L.; A.B., Williams Coll., 1906; Ph.D., Johns Hopkins, 1910; LL.D., Clark U., 1954; m. Elizabeth Louise Thomson, June 24, 1911; children—Elbert Payson, John Bascom, Emma Elizabeth (Mrs. Robert F. Campbell), Ruth Mallory (Mrs. LeBaron R. Briggs III). Instr., prof. geology Colby Coll., 1910-20; exec. sec. div. geology and geography NRC, Washington, 1920-22; dean undergrads., prof. geology Clark U. Worcester, Mass., 1922-54, dean emeritus, prof. emeritus, 1954—. Lectr. in geology Bangor Theol. Sem., 1913, 16, 19, Worcester Poly. Inst., 1932-37; instr. geography John Hopkins, summer 1921. Fellow A.A.-A.S., Geol. Soc. Am.; mem. Archeol. Inst. Am. (past pres. Worcester soc.), Worcester Natural History Soc. (past pres.), Phi Beta Kappa, Phi Sigma Kappa, Gamma Alpha. Research and publs. on geology of Eastern shore Md., Anne Arende Count of Md.; gravels, sands and clay of Waterville, Me., regions, Rocky Mountain area Peneplains, glacial geology. Address: 36 Whitman Rd., Worcester, Mass. 01609.*

LITTLE, James, Irish physician; b. Newry, N. Ireland, s. Archibald and Mary (Coulter) L.; student Med. Sch., Trinity Coll., Dublin; M.D., LL.D., U. Edinburgh; M.D. (hon.), U. Dublin; m. Anah Murdoch; 2 sons, 1 dau. Became physician Adelaide Hosp., 1866; prof. practice medicine Royal Coll. Surgeons, Ireland, also fellow; Regius prof. physic Dublin U., from 1898; cons. physician, 4 Dublin hosps.; named hon. physician to His Majesty in Ireland, 1908; crown rep. for Ireland, Gen. Med. Council. Pres. Royal Coll. Physicians, Ireland, Royal Acad. Medicine, Ireland; mem. Royal Irish Acad. Author: First Steps in Clinical Study; Chronic Diseases of the Heart. Died Dec. 23, 1916.

LITTLE, James Maxwell, Am. educator; b. Commerce, Ga., Dec. 16, 1910; s. Claude and Cora (Quillian) L.; A.B., Emory U., 1932, M.S., 1933; Ph.D., Vanderbilt U., 1941; m. Louise Toepel, Aug. 20, 1933; children—Paul G., Nancy C., John M. Asst. in physiology Emory U. Med. Sch., 1933-35; instr. biochemistry Vanderbilt U., 1936-41; faculty Bowman Gray Sch. Medicine, 1941—, prof. pharmacology, chmn. dept., asso. physiology, 1963—. Mem. Am. Physiol. Soc., A.A.A.S., Soc. Exptl. Biology and Medicine, Am. Soc. Pharmacology and Exptl. Therapeutics, Am. Heart Assn., N.C. Forsyth County heart assns., Phi Beta Kappa, Sigma Xi. Author: An Introduction to the Experimental Method, 1961. Publs. on study of substances in urine increasing kidney salt excretion; co-discovery of methocarbamol, a muscle relaxant, effects of low oxygen tension on circulation, action of hormones on kidney function. Home: 3100 Buena Vista Rd., Winston-Salem, N.C. 27106.*

LITTLE, Noel Charleton, Am. physicist; b. Brunswick, Me., Dec. 25, 1895; s. George Thomas and Lilly (Lane) L.; A.B., Bowdoin Coll., 1917; A.M., Harvard, 1920, Ph.D., 1923; m. Marguerite Dorothy Tschaler, Sept. 28, 1923; children—Mary Thayer (Mrs. Richard T. Huggins), Clifford Charlton, Dana Anton. Faculty, Bowdoin Coll., Brunswick, Me., 1919—, prof., 1926—, chmn. physics dept. 1926-64. Vis. lectr. Harvard, summer 1947; research study Tubingen, Germany, 1927-28, Geneva, Switzerland, 1937, Stockholm, Sweden, 1956. Mem. Delta Kappa Epsilon. Author: Physics, 1953. Home: 60 Federal St., Brunswick, Me.*

LITTLE, Silas, Am. forester; b. Newbury, Mass., Jan. 17, 1914; s. Silas and Margaret (Brown) L.; B.S., Mass. State Coll., 1935; M.F., Yale, 1936, Ph.D., 1947; m. Marian Louise Norris, May 29, 1941; children—Silas III, Maynard Norris, Justin Frank, Margaret June, Dorothy Brown. With Northeastern Forest Expt. Sta., U. S. Forest Service, 1936—, prin. silviculturist, Upper Darby, Pa., 1965—. Mem. Soc. Am. Foresters, Ecol. Soc. Am. Research, publs. on ecology and silviculture loblolly pine and asso. species in Eastern Md., shortleaf and pitch pines, Atlantic white cedar and asso. species in so. N.J., yellow-poplar and asso. species in inner coastal plain and Piedmont. Home: Creek Rd., Moorestown, N.J. 08057. Office: 6816 Market St., Upper Darby, Pa. 19082.*

LITTLE, Silas, Jr., Am. forester; b. Newbury, Mass., Jan. 17, 1914; s. Silas and Margaret (Brown) L.; B.S., Mass. State Coll., 1935; M.F., Yale U., 1936, Ph.D., 1947; m. Marian Louise Norris, May 29, 1941; children—Silas III, Maynard Norris, Justin Frank, Margaret June, Dorothy Brown. With U. S. Forest Service, Northeastern Forest Expt. Sta., Upper Darby, Pa., 1936—, prin. silviculturist, 1961—. Named N.J. Forest Conservationist, 1966. Mem. Soc. Am. Foresters, Ecol. Soc. Am. Research, numerous publs. on ecology, silviculture, fire protection of several forest types in Coastal Plain and Piedmont of N.E., especially N.J., Md., Pa. Home: R.D. 2, Creek Rd., Moorestown, N.J. 08057. Office: 6816 Market St., Upper Darby, Pa. 19082.*

LITTLE, William John, Brit. surgeon; b. nr. London, 1810; ed. St. Omer Coll., France; M.D., Berlin, 1837; chief founder Orthopaedic Inst., (became Royal Orthopaedic Hosp.), London, 1838. Author: On Ankylosis or Stiff Joint, 1843; On the Nature and Treatment of the Deformities of the Human Frame, 1853; On Spinal Weakness and Spinal Curvatures, 1868. Described a congenital spastic paralysis of infants caused by insufficient devel. of pyramidal tracts (Little's disease), 1843, progressive muscular atrophy, 1853. Died 1894.

LITTLEHALES, George Washington, Am. hydrographic engr.; b. Schuylkill County, Pa., Oct. 14, 1860; s. William Henry and Margaret (Reber) L.; B.S., U. S. Naval Acad., 1883; C.E., Columbian (now George Washington) U., 1888; m. Helen Powers Hill, Jan. 26, 1896; children—Margaret Powers (Mrs. Philip G. Vondersmith), James Hill, George Reber. Hydrographic engr. U. S. Hydrographic Office, 1900-1932. Cons. hydrographer, dept. terrestrial magnetism Carnegie Inst., 1904-06; prof. nautical science George Washington U., 1913-27. Del. to various confs. and congresses. Fellow A.A.A.S.; mem. Washington Acad. Scis. (v.p. 1905, 12, 13), Philos. Soc. Washington (pres. 1905), Am. Soc. Naval Engrs., Am. Geophys. Union (v.p. 1926-29). Author: Development of Great Circle Sailing; The Methods and Results of the Survey of Lower California; Submarine Cables; The Magnetic Dip or Inclination; The Meridional Parts of the Terrestrial Spheriod, 1889; The Azimuths of Celestial Bodies, 1902; The Forms of Isolated Submarine Peaks, 1891; The Azimuth and the Hour Angle from the Latitude of the Observer and the Declination and Altitude of the Observed Celestial Body, 1903; A New and Abridged Method of Finding the Locus of Geographical Position and the Compass Error; Altitude, Azimuth, and Geographical Position, 1906; Geographical Position-Line Tables, 1915; Chart for Finding Geographical Position in Aerial Navigation, 1918; Tables of the Simultaneous Altitudes and Azimuths of Celestial Bodies, 1920; The Sumner Line of Position, furnished ready to lay down upon the chart by means of tables of simultaneous hour-angle and azimuth of celestial bodies, 1923; Finding Geographical

Position in the Region of the North Pole, 1925; Mechanical Means for Finding Geographical Position in Navigation, 1929; Both the Latitude and the Longitude of the Ship Found in a Single Operation from the Observation of the Altitude and the Azimuth of a Celestial Body, 1937. A founder Internat. Jour. Terrestrial Magnetism, 1896, asso. editor. Research and publs. in hydrography, oceanography and terrestrial magnetism; published about 3,000 charts used in nav. of vessels of world; editor math. tables for use of Am. seamen. Died Aug. 12, 1943.

LITTLEWOOD, John Edensor, Brit. mathematician; b. Rochester, Eng., June 9, 1885; s. Edward Thornton and Sylvia Maud (Ackland) L.; student Trinity Coll., Cambridge, Eng., 1903-08; M.A., Cambridge U., 1913, Sc.D. (hon.), 1965; LL.D., St. Andrew U., 1931; D.Sc. (hon.), U. Liverpool (Eng.), 1927. Richardson lectr. U. Manchester (Eng.), 1907-10; fellow Trinity Coll., 1908——, lectr., 1910-28; Cayley lectr. Cambridge U., 1920-28, Rouse Ball prof. math. 1928-50. Fellow Royal Soc., 1916. Recipient Royal medal Royal Soc., 1929, Sylvester medal, 1944, Copley medal, 1958; de Morgan medal London Math. Soc., 1939, Sr. Berwick prize, 1960. Fgn. or corr. mem. French, Royal Danish, Royal Dutch, Göttingen, Royal Swedish acads. Research, publs. on theory of real functions. Home: Trinity Coll., Cambridge, Eng.*

LITTRÉ, Alexis, French surgeon, anatomist; b. Cordes, France, July 21, 1658; studied at Ville Franche, Roverchand, Montpellier, Paris, all France. Physician med. faculty U. Reims (France); prof. anatomy U. Paris; physician in Paris; mem. French Acad. Scis., 1701. Contbr. articles to tech. jours. First to suggest colostomy for relief of intestinal obstruction (Littre's operation), 1710; 1st description of mucous glands of male urethra (Littre's glands), 1700, herniation of adiverticulum (Meckel's diverticulu or Littré's hernia), 1719. Died Paris, Feb. 3, 1725.

LITTRÉ, Émile, French physician, philologist; b. Paris, Feb. 1, 1801; ed. Lycée Louis-le-Grand. Author: Oeuvres d'Hippocrate, 1839-61; Histoire de la Langue française, 2 vols., 1862; Dictionnaire de la langue française, 4 vols., 1863-72, supplement, 1878. Translator numerous works, including: Le choléra oriental, 1832; La science au point de vue philosophique, 1873. Studied Arabic, Sanskrit, English, German; advocated positivism and applied it to study of lang.; claimed to have found organic basis of morality; studied oriental cholera. Died Paris, June 2, 1881.

LITTROW, Joseph Johann von, see von Littrow, Joseph Johann.

LITVAK, Jorge, Chilean physician; b. Antofagasta, Chile, May 2, 1929; s. Mauricio and Juanita (Lijavetsky) L.; B.Sc., Instituto Nacional, 1946; M.D., U. Chile, 1953; m. Paule Gasman Battier, Mar. 24, 1962; children—Joanna, Alejandra. Faculty U. Chile Med. Sch., Santiago, 1953——, asso. prof. medicine, 1962——, chief dept. endocrinology also Lab. Radioisotopes, Hosp. J. J. Aguirre, 1958——. Med. adviser Chilean Nuclear Energy Commn., 1965——. Research fellow medicine Harvard Med. Sch., 1956-57. Mem. Endocrine Soc., Am. Fedn. for Clin. Research, A.C.P., Sociedad Médica de Chile. Co-author: Endocrinología Clínica, 1967; also numerous articles. Research on bone metabolism using radioisotopes and collagen metabolism, parathyroid diseases and metabolic bone diseases; developed model based on digital and analogic computing. Home: 1805 Las Petunias, Santiago, Chile.*

LITYNSKI, Tadeusz, Polish agrl. chemist; b. Dzikow, Poland, June 2, 1901; s. Joseph and Maria (Wiehler) L.; Ph.D., Jagellonian U., 1926; widower; 1 dau., Barbara Krol. Faculty, Jagellonian U., Krakow, 1945——, prof. agrl. chemistry, 1956——, rector Coll. Agrl., 1959-62. Mem. Polish Acad. Sci., Internat. Soc. Soil Sci. Author: Soil Fertility and Fertilization, 1967. Research on fertilizing value of cement kiln dusts, Polish rock phosphates, ammonia, ureaformaldehyde preparations, contamination of Polish soils by strontium 90. Home: 5 Sobieski, Krakow, Poland.*

LITZENBERG, Jennings Crawford, Am. obstetrician; b. Waubeek, Ia., Apr. 6, 1870; s. William Denny and Lydia (Crawford) L.; B.Sc., U. Minn., 1894, M.D., 1899; postgrad. U. Vienna, 1909-10, later in Berlin, London, Glasgow and Dublin; m. Elizabeth Anna Fisher, June 3, 1902; children—Avis, Karl; m. 2d, Olga Hansen, Jan. 27, 1934. Gen. practice, 1900-10, specialized in obstetrics and gynecology since 1910; mem. Nicollet Clinic; instr. obstetrics U. Minn., 1901-07, asst. prof., 1907-10, asso. prof. obstetrics and gynecology, 1910-14, prof. and head of dept., 1914-38, emeritus since 1938; chief dept. obstetrics and gynecology Univ. Hosp.; obstetrician Eitel and Northwestern hosps.; cons. obstetrician and gynecologist Fairview Hosp. Mem. Am. Bd. Obstetrics and Gynecology. Fellow Am. Gynecol. Soc. (pres. 1940-41), Am. Assn. Obstetricians and Gynecologists and Abdominal Surgeons (pres. 1938), A.C.S., A.M.A., Hennepin County Med. Soc. (pres. 1919), Minn. Acad. Medicine (pres. 1932), Central Assn. Obstetricians and Gynecologists, Sigma Xi. Frequent contbr. on profl. topics. Died Aug. 15, 1948.

LITZINGER, Marie, Am. mathematician; b. Bedford, Pa., May 14, 1899; d. Rush and Katherine (O'Connell) L.; A.B., Bryn Mawr Coll., 1920, M.A., 1922; student U. Rome (Italy), 1923-24; Ph.D., U. Chgo., 1934. Tchr. Devon (Pa.) Manor Sch., 1920-22, Greenwich (Conn.) Acad., 1924-25; instr. Mt. Holyoke Coll., South Hadley, Mass., 1925-28, asst. prof., 1928-37, chmn. dept. math. since 1937, asso. prof., 1937-42, prof. since 1942, prof. math. John Stewart Kennedy Found., since 1948. Mem. Am. Math. Soc., Math. Assn. Am., Conn. Valley Sect. Assn. Tchrs. of Math. in New Eng. (pres. 1940-41), Sigma Xi. Author: (article) A Basis for Residual Polynomials in N Variables, 1935. Died Apr. 7, 1952.

LITZMAN, Carl C. T., German gynecologist, obstetrician; b. 1815; described kyphoscoliotic type of female pelvis which had been irregularly contracted by rickets, 1853, coxalgic female pelvis (deformity caused by hip joint disease), 1853; developed clin. classification of female pelvis, 1861. Died 1890.

LIU, Chi Kong, physician, cardiologist; b. Canton, China, Aug. 16, 1920; s. Ruck Fan and Hai Tao (Chao) L.; student Chung Ching U., Kumning, China, 1939-40; M.D., Nat. Central U., Chung Kiang, China, 1947; m. Milly Leong, Nov. 18, 1952; children—Patricia, Richard. Fellow hematology Stanford, 1950-51, fellow cardiology, 1952-53; practice medicine specializing in cardiology; staff Chgo. Med. Sch., 1953-58, chief cardiac catherization lab., 1954-58; dir. cardiopulmonary lab. Los Angeles County Gen. Hosp., Torrance, Cal., 1958——; asso. prof. medicine U. Cal. at Los Angeles, 1962——; chief cardiac catherization lab. Mt. Sinai Hosp., 1954-58. Fellow Am. Coll. Cardiology; mem. Am., Chgo., Los Angeles County, Long Beach (dir.) heart assns., Los Angeles County Med. Assn., A.M.A. Author: Gases in the Blood; Formulas used in Cardiac Catherization in Rheumatic Heart Disease; (with A. A. Luisada) Left Heart Catherterization in Rheumatic Heart Disease; also articles, monographs, chpts. in books. Discovered method to record heart sounds and murmurs in 4 chambers of heart, measurements blood vols. in arterial and venous segment lungs, research in kidney disease and patho-physiology of heart disease. Home: 26807 Grayslake Rd., Palos Verdes Peninsula, Cal. Office: 1000 W. Carson St., Torrance, Cal. 90509.*

LIU, Chien, physician; b. Canton, China, Mar. 6, 1921; s. Chu-tong and Ken-ju (Wong) L.; B.S., Yenching U., 1942; M.D., West China Union U., 1947; m. Beatrice Kwong, Sept. 20, 1947; children—G. Kim, J. Norman, O. Christine, Anthony G. Research fellow in medicine Johns Hopkins U., Balt., 1949-52; research asso. in bacteriology and immunology Harvard Med. Sch., Boston, 1952-55, asso., 1955-58, asst. prof., 1958; asso. prof. pediatrics U. Kan. Sch. Medicine, Kansas City, 1958-63, prof. medicine and pediatrics, head sect. infectious diseases U. Kan. Med. Center, 1963——; attending pediatrician Children's Mercy Hosp., Kansas City, Mo., 1958-62, chmn. dept. pediatrics, 1962-63. Research career award USPHS, 1963——. Diplomate Am. Bd. Pediatrics. Mem. Am. Assn. Immunologists, A.M.A., Soc. Pediatric Research, Am. Acad. Microbiologists, Infectious Disease Soc. Am. Research, numerous publs. in pathogenesis and diagnosis of viral infections; elucidation of etiology of primary atypical pneumonia devel. of rapid diagnostic methods for influenza, measles, respiratory viruses. Home: 8100 Tomahawk Rd., Prairie Village, Kan. 66208.*

LIU, Pinghui Victor, microbiologist; b. Feng Hsiang, Formosa, Feb. 9, 1924; s. Chineng and Chulan (Wang) L.; M.D., Tokyo Jikei-Kai Sch. Medicine, 1947; Ph.D., Tokyo (Japan) Med. Sch., 1957; m. Chiameng Judy Yen, July 10, 1959; 1 dau., Nancy J. Came to U. S., 1954, naturalized, 1964. Tech. adviser dept. bacteriology 406th Med. Gen. Lab., U. S. Army, 1947-54; research fellow dept. microbiology U. Louisville Sch. Medicine, 1956-57, faculty, 1957——, asso. prof. microbiology, 1962——. USPHS Sr. Research fellow, 1957-61; recipient Research Career Devel. award NIH, 1962——. Diplomate Am. Bd. Microbiology. Mem. Am. Soc. for Microbiology, New York Acad. Scis., A.A.A.S., Sigma Xi. Research, publs. on pathogenesis bacterial diseases, taxonomy pseudomonas and related organisms. Home: 47 Sycamore Rd., Jeffersonville, Ind. 47130. Office: 101 W. Chestnut St., Louisville 40202.*

LIU HSIN, Chinese scholar, astronomer; s. Liu Hsiang; imperial prince who flourished under Western Han; succeeded his father as imperial librarian, 7 B.C. Completed catalogue begun by his father, Han I-wen-chih, 6 B.C. Author: San-t'ungli (treatise on calendar). Introduced period of 23,639,040 years to reconcile tropical and sidereal years since the precession of equinoxes was unknown to him. Committed suicide, 22 A.D.

LIU HUI, Chinese mathematician; flourished in Kingdom of Wei toward end of period of 3 Kingdoms, 221-65. Wrote commentary, Arithmetic in Nine Sections, 263; Sea-island Arithmetical Classic. Studied arithmetic and geometry, including problem of measuring an island from a distance; used red computing rods for positive numbers and black for negative numbers.

LIU HUNG, Chinese astronomer; flourished under Later Han, circa 196; proved equator and ecliptic do not coincide, that solstitial points are not fixed, length of tropical year is not exactly 365 1/4 days.

LIU SHAO, Chinese anthropologist; flourished as high ofcl., 224. Author: Jen-wu-chih. Editor: Hsiaoching. Studied div. of mankind into classes according to their dispositions which were revealed by outward characteristics.

LIVE, Israel, microbiologist, educator; b. Zakrzewce, Austria, Apr. 26, 1907; s. Herman and Rose (Nagelberg) L.; came to U. S., 1928, naturalized, 1934; V.M.D., U. Pa., 1934, A.M., 1936, Ph.D., 1940; m. Anna Harris, Nov. 25, 1936; children—Theodore Ross, David Harris. Faculty, U. Pa. Sch. Vet. Medicine, Phila., 1934——, prof. microbiology, 1953-—. mem. expert panel on brucellosis WHO, 1950——. Diplomate Am. Bd. Microbiology. Fellow A.A.A.S., Am. Acad. Microbiology, Am. Coll. Vet. Microbiologists (charter); mem. Sigma Xi. Research on antigenic, serological tests for certain bacteria; diagnostic skin test, vaccine for Brucellosis; study of staph infections in man, animals. Home: 2414 Bryn Mawr Av., Phila. 19131.*

LIVEING, George Downing, Brit. chemist; b. Naylard, Eng., Dec. 21, 1827; s. Edward and Catherine (Downing) L.; student math. St. John's Coll., Cambridge (Eng.) U., 1850, natural scis., 1851; studied under Karl Rammelsberg, beginning in 1851; m. Catherine Rowland Ingram, 1860. Taught chemistry to med. students, 1852; named fellow, lectr. chemistry St. Johns Coll., 1853, professorial fellow, 1889, prof. chemistry, Cambridge, 1861; apptd. pres., 1911; named prof. chemistry Staff Coll., also Royal Mil. Sch., Sandhurst, 1860; ret., 1908. Recipient Davy medal, 1901. Fellow Royal Soc., 1869 (mem. council 1881-82, 1903-04). Contbr. articles to tech. jours. Author: Chemical Equilibriums, the Result of the Dissipation of Energy, 1885. Spectroscopic research with Sr James Dewar; studied thermodynamics in chemistry. Died Cambridge, Dec. 26, 1924.

LIVERSIDGE, Archibald, Brit. mineralogist; b. Turnham Green, Nov. 17, 1847; s. John Liversidge; ed. Royal Sch. Mines, Royal Coll. Chemistry; Royal Exhibitioner, 1867; Christ's Coll., Cambridge; M.A., Cambridge; LL.D., Glasgow, Scotland. Became instr. chemistry Royal Sch. Naval Architecture, 1867; named univ. demonstrator chemistry, Cambridge, 1870; prof. chemistry U. Sydney (Australia), 1872-1908, emeritus prof., 1908——, founder Faculty Sci., 1879, 1st dean, 1879-1894; founder Sch. Mines, 1890; asso. Royal Sch. Mines. Fellow Royal Soc., 1882, Royal Soc. Edinburgh (hon.), Inst. Chemists, Chem. Soc., Geol. Soc., Royal Geol. Soc.; mem. Royal Soc. New South Wales (hon. sec. 1875-1889, pres. 1886, 90, 1901), Philos. Soc. Cambridge, Phys. Soc. London, Mineral. Soc. Gt. Britain, Mineral. Soc. France, Royal Soc. Victoria, New Zealand Inst., Royal Hist. Soc. London, K. Leop.-Car. Akad. Halle, N.Y. Acad. Scis. (corr.), Royal Soc. Tasmania, Royal Soc. Queensland, Soc. d'Acclim. Mauritius, Edinburgh Geol. Soc., Australian Assn. for Advancement Sci., Soc. Chem. Industry (chmn. Sydney sect. 1903-05), Australasian Assn. for Advancement Sci. (hon. sec. 1888-1909, pres. 1888-90, v.p., 1890-——), Brit. Assn. (became a v.p. 1896). Author: The Minerals of New South Wales; also numerous articles. Research on chemistry and Mineralogy. Died Sept. 26, 1927.

LIVI, Livio, Italian mathematician; b. Rome, Italy, Jan. 2, 1891; s. Ridolfo and Luisa (Bacci) L.; Dr.Jur.; m. Rita Olga Paladini, 1924. Prof. statistics, Modena, Italy, 1914-21, Trieste, Italy, 1921-26; prof. demography Rome (Italy) U., 1926-38, prof. statistics, 1949——; prof. statistics Florence (Italy) U., 1928-49; dean Florence Superior Inst. Social Sci. Mem. Italian Inst. Anthropology (past pres.), Unione Internaz. Studi sulla popolazione (v.p.), Internat. Inst. Statistics. Author: Gli ebrei alla luce della statistica, 2 vols., 1919-20; I fattori biologici dell'ordinamento sociale, 1936; Le leggi naturali della popolazione, 1939; Storia demografica dei Rodi, 1944; Elementi di statistica, 11th edit., 1957; La vecchia e la nuova sociologia generale positiva, 1957. Address: 18 via A. Baldesi, Florence 18, Italy.

LIVINGOOD, Clarence S., Am. physician; educator; b. Elverson, Pa., Aug. 7, 1911; s. Clarence Adam and Eliza (Zerr) L.; B.S., Ursinus Coll., 1932; M.D., U. Pa., 1936; m. 2d Louise Sinclair Woelpper, Oct. 24, 1947; children—Wilson, Louise, Clarence Adam II, Susan, Elizabeth. Asst. prof. dermatology U. Pa., 1946-48; prof. dir. dept. dermatology U. Tex. Med. Sch., 1949-53; chmn. dept. dermatology Henry Ford Hosp., Detroit, 1953——. Mem. commn. on cutaneous diseases Armed Forces Epidemiology Bd., 1956——. Sec.-gen., XII Internat. Cong. Dermatology, 1962. Diplomate Am. Bd. Dermatology (mem. 1961——, sec. 1963——). Mem. Am. Dermatol. Assn. (dir. 1965——), Am. Acad. Dermatology (dir.), Assn. U. Profs. Dermatology, A.M.A. (past chmn. dermatology sect.), Detroit Dermatol. Soc. (past pres.), Soc. Investigative Dermatology (past pres.). Author: (with others) Manual of Dermatology, 1942; also chpts. in various textbooks. Editorial bd. Archives Dermatology, 1960——, Current Therapy; internat. editorial bd. sect. dermatology, venereology Excerpta Medica, 1962——. Research and numerous publs. in contact dermatitis, bacterial infections of skin, drug reactions, fungus infections. Home: 345 University Pl., Grosse Pointe, Mich. 48230. Office: 2799 W. Grand Blvd., Detroit 48202.

LIVINGOOD, John Jacob, Am. physicist; b. Cin., Mar. 7, 1903; s. Charles J. and Lily (Foster) L.; A.B., Princeton, 1925, M.A., 1927, Ph.D., 1929; m. Carolyn Zipf, June 25, 1934; children—Charles Albert, John Michael. Instr. physics, Princeton, 1929-32; research asso. Radiation Lab., U. Cal. at Berkeley, 1932-38; faculty instrn. physics dept. Harvard, Cambridge, 1938-42, research asso. Radio Research Lab., OSRD, 1942-45; asso. dir. research div. Collins Radio Co., Cedar Rapids, 1945-52; asso. dir physics div. Argonne (Ill.) Nat. Lab., 1952-56, dir. particle accelerator div, 1956-58, sr. physicist, 1958—. Fellow Am. Phys. Soc., Am. Nuclear Soc. Author: (with G. P. Harnwell) Experimental Atomic Physics, 1933; Principals of Cyclic Particle Accelerators, 1961. Research in line spectra, high power radio tubes, artificial radioactivity, particle acceleration. Accelerators, 1961. Home: 836 S. County Line Rd., Hinsdale, Ill. Office: Argonne Nat. Lab., Argonne, Ill.*

LIVINGSTON, Burton Edward, Am. plant physiologist; b. Grand Rapids, Mich., Feb. 9, 1875; s. Benjamin and Keziah (Lincoln) L.; B.S., U. Mich., 1898; Ph.D., U. Chgo., 1902; m. Grace Johnson, 1905 (div. 1918); m. 2d, Marguerite A. Brennan Macphilips, 1921. Asst. in plant physiology U. Mich., 1895-98; fellow and asst. in plant physiology U. Chgo., 1899-1905; soil expert U. S. Bur. Soils in charge of fertility investigations, 1905-06; staff mem. dept. bot. research Carnegie Instn., Washington, 1906-09; prof. plant physiology, Johns Hopkins U., 1909-32, prof. plant physiology and forest ecology, 1932-40, dir. lab. of plant physiology, 1913-40, emeritus prof. 1940. Mem. NRC. Fellow Am. Acad. Arts and Scis.; mem. Am. Philos. Soc., Bot. Soc. Am., Am. Soc. Naturalists (pres. 1933), Ecol. Soc.; Am., Am. Soc. Plant Physiologists (Hales prize award 1946, pres. 1934); permanent sec. A.A.A.S., 1920-31, gen. sec., 1931-34. Author: Role of Diffusion and Osmotic Pressure in Plants, 1903; Distribution of Vegetation in U. S., as Related to Climatic Conditions (with F. Shreve), 1921; also papers. Editor: Physiological Researches; English edit. Palladin's Plant Physiology, 1918; Botanical Abstracts, 1918-20. Inventor: porous cup atmometer (for measuring evaporation as climatic factor); radio-atmometer (for measuring sunshine); auto-irrigator (for automatic control of soil moisture in potted plants); instruments for measuring water-supplying and water-absorbing power of soils. Died Feb. 8, 1948.

LIVINGSTON, Milton Stanley, Am. physicist; b. Broadhead, Wis., May 25, 1905; s. Milton McWhorter and Sara (Ten Eyck) L.; A.B., Pomona Coll., 1926; M.A., Dartmouth, 1928, D.Sc., 1963; Ph.D., U. Cal. at Berkeley, 1931; m. Lois Emily Robinson, Aug. 8, 1930; children—Diane, Stephen T. Instr. physics Dartmouth, 1928-29; research asso. U. Cal. at Berkeley, 1932-34; asst. prof. physics Cornell U., Ithaca, N.Y., 1934-38; asso. prof. physics Mass. Inst. Tech., Cambridge, 1938-54, prof., 1954—. Accelerator project chmn. Brookhaven Nat. Lab., 1946-58; dir. Cambridge Electron Accelerator, Harvard-Mass. Inst. Tech., 1956-67, Nat. Accelerator Lab., 1967—. Mem. Am. Acad. Sci., Sigma Xi. Author: High Energy Accelerators, 1954; (with J. P. Blewett) Particle Accelerators, 1962; Development of High Energy Accelerators, 1966. Research and devel. of cyclotron and other accelerators; co-discoverer alternating gradient focusing; designer multibillionvolt accelerators. Home: 5505 Lakeside Dr., Lisle, Ill. 60532. Office: 1301 W. 22d St., Oak Brook, Ill. 60521.*

LIVINGSTON, Ralph, Am. chemist; b. Keene, N.H., May 16, 1919; s. David Israel and Annie (Finkelstein) L.; B.S., U. N.H., 1940, M.S., 1941; D.Sc., U. Cin., 1943; m. Zelda Becker, Oct. 24, 1943; children—Beverly Jean, Sally Maura, Donna Ruth, Stuart Aaron. Chemist, Metall. Lab., Manhattan Project, U. Chgo., 1943-45; group leader chemistry Oak Ridge Nat. Lab., 1945—, asso. dir. chemistry div., 1965—; prof. chemistry U. Tenn., Knoxville, 1964—; vis. prof. chemistry Cornell U., Ithaca, N.Y., 1961-62. John Simon Guggenheim fellow, also Fulbright Research scholar, France, 1960-61. Mem. Am. Chem. Soc., Am. Phys. Soc., A.A.A.S. Contbr. articles to tech. jours. Research on measurement moments radioactive nuclei by radiofrequency and microwave spectroscopy, fundamental measurement chlorine, bromine, iodine compounds by nuclear quadrupole spectroscopy, chem. free radicals by electron spin resonance. Home: 109 Valparaiso Rd., Oak Ridge 37830. Office: P.O. Box X, Oak Ridge 37830.*

LIVINGSTON, Robert (Stanley), Am. phys. chemist; b. San Francisco, Aug. 11, 1898; s. Andrew William and Laura (Hoag) L.; B.S., U. Cal., Berkeley, 1922, M.S., 1923, Ph.D., 1925; m. Savetta Leslie Chucovich, Aug. 22, 1931; 1 dau., Mary Elizabeth (Mrs. Michael Kaye). Faculty, U. Cal., Berkeley, 1926-27; faculty U. Minn., Mpls., 1927-67, prof., chief, div. phys. chemistry, 1959-67; prof. chemistry San Diego State Coll., 1967—. NDRC asst., 1942-45. Recipient Army-Navy certificate of appreciation, 1947. Howard Meml. fellow, 1925-26; Lalor Found. fellow, 1938-39; Guggenheim fellow, 1953-54. Mem. A.A.A.S., Am. Chem. Soc., Faraday Soc., Minn. Acad. Author: Physico Chemical Experiments, 1939. Exptl. studies of kinetics of inorganic reactions in aqueous solutions, radiation-chem. reactions in gas phase, photo-chem. reactions in solution, and chemiluminescence in solution; measurements of photochem. and spectroscopic properties of chlorophyll in vitro. Home: 5621 Ashland Av., San Diego 92120.*

LIVINGSTON, Robert Burr, Am. neurophysiologist; b. Boston, Oct. 9, 1918; s. William Kenneth and Ruth Forbes (Brown) L.; A.B., Stanford, 1940, M.D., 1944; m. Mandana Beckner, Dec. 21, 1954; children—Louise Shaw, Dana Ruth, Justyn Ann. Instr. physiology Yale Sch. Medicine, 1946-48, asst. prof., 1948-51; asst. research prof. Harvard, 1946-47; sr. fellow in neurology NRC, 1948-50; asso. prof. neurophysiology U. Cal. at Los Angeles Sch. Medicine, 1951-56, prof., 1956; sci. dir. Nat. Inst. Neurol. Diseases and Blindness, Nat. Inst. Mental Health, Bethesda, Md., 1956-61, chief lab. neurobiology, 1960-63, asso. chief div. research facilities and resources NIH, 1963-65; prof., chmn. dept. neuroscis. U. Cal. San Diego, LaJolla, 1965—. Asst. to pres. Nat. Acad. Scis., chmn. NRC, 1950-51; cons. NRC, 1951-55, NASA, 1958-62, Ency. Brit., 1958-60, World Book Ency., 1964-65; tchr. NIH Grad. Sch., 1963-65; lectr. Mid-Career course U. S. State Dept., 1957-64; founder Inst. for Policy Studies, 1961, Am. Soc. for Cybernetics, 1964. Decorated Bronze Star. Sr. fellow in neurology NRC, 1948-49; Holiday lectr. A.A.A.S., 1964. Editor: Narcotic Drug Addiction Problems, 1963; (with A. A. Imshenetsky and G. A. Derbyshire) Life Sciences and Space Research, 1963. Research on neurophysiol. mechanisms underlying behavior. Home: 8758 La Jolla Scenic Dr., La Jolla, Cal. 92037.*

LIVINGSTON, Robert Louis, Am. chemist; b. Ada, O., Nov. 15, 1918; s. Ralph and Ruth (Sink) L.; B Sc. Ed., Ohio State U., 1939; M.S., U. Mich., 1941, Ph.D., 1943; m. Virginia Capron, May 30, 1943; children—Douglas, Roy, Margaret, Bruce. Asso. chemist Naval Research Lab., Ann Arbor, Mich., 1944-46; faculty Purdue U., Lafayette, Ind., 1946—, prof. chemistry, 1954—, asst. dept. head, 1960—. Mem. Am. Chem. Soc. (sec. div. chem. edn. 1963—), Am. Crystallographic Assn., Sigma Xi, Phi Lambda Upsilon, Phi Delta Kappa, Phi Eta Sigma, Gamma Alpha. Author: (with F. D. Martin) Lecture Notebook and Recitation Questions, General Chemistry, pub. annually 1952-64. Research on molecular structure of gaseous molecules by electron diffraction. Home: 1001 Hillcrest St., West Lafayette, Ind. 47906.*

LIVINGSTON, Samuel, Am. pediatrician; b. Phila., July 14, 1908; s. Benjamin and Bertha (Mogul) L.; B.S. magna cum laude, Georgetown U., 1928, M.S. in Chemistry, 1930; M.D., Vanderbilt U., 1934; m. Anne Pruce, Sept. 5, 1936; children—Elaine (Mrs. Jay Soloman), Herbert. Pediatrician, Johns Hopkins Hosp, Balt., 1936—, electroencephalographer, 1948-51, asst. epilepsy clinic, 1936-46, dir., physician in charge epilepsy clinic, 1946—; faculty Johns Hopkins, 1939—, asso. prof. pediatrics, 1966—; staff Sinai Hosp., Balt., 1944—, asst. physician internal medicine, 1954—. Med. adv. bd. Partridge Sch., Springfield, Va., 1958—, Medic-Alert Assn. U. S., Chesapeake Epilepsy Soc., 1961—, Balt. chpt. Md. Assn. Retarded Children, 1965—; cons. on epilepsy. Recipient numerous awards for epilepsy research. Mem. Soc. Pediatric Research, Eastern Assn. Electroencephalographers, Am. Pub. Health Assn., Nat. Rehab. Soc., A.M.A., Am. Epilepsy Soc. Author: The Diagnosis and Treatment of Convulsive Disorders in Children, 1954; Living with Epileptic Seizures, 1963; Drug Therapy for Epilepsy, Antiepileptic Drugs: Usage, Metabolism and Untoward Reactions, 1966. Research, publs. on causes, treatment of epilepsy and convulsive disorders in children. Home: 6813 Park Heights Av., Balt. 21215.* Office: Johns Hopkins U., Balt. 21218.*

LIVINGSTONE, Daniel Archibald, Am. biologist; b. Detroit, Aug. 3, 1927; s. Harrison Lincoln and Elizabeth (Matheson) L.; student McGill U., 1944-45; B.Sc., Dalhousie U., 1950, M.Sc., 1950; Ph.D., Yale, 1953; m. Bertha Griffin Ross, June 21, 1952; children—Laura Ross, Mary Lisa, John Malcolm, Christina Ann, Elizabeth. NRC Postdoctoral fellow Cambridge (Eng.) U., 1953-54; NRC Postdoctoral fellow, spl. lectr. Dalhousie U., 1954-55; faculty U. Md., 1955-56; faculty Duke, 1956—, professor zoology, 1967—; limnologist U. S. Geol. Survey, intermittently 1956—. Guggenheim fellow, 1960-61. Mem. N.C. Inst. Sci., Am. Soc. Icthyologists and Herpetologists, Ecol. Soc. Am., Am. Soc. Limnologists and Oceanographers, Geochem. Soc., N.C. Acad. Sci., Internat. Assn. Fundamental and Applied Limnology, Assn. pour l'Etude Taxonomy de la Flore d'Afrique Tropicale, Sigma Xi. Zoology editor Ecol. Monographs, 1961-66. Research, publs. on distbn. freshwater fishes N.S., effects temperature and activity on oxygen consumption salmon embryos, reactions between sedimentary and dissolved phosphorus in lakes, origin and devel. lakes as ecol. systems, dissolved solids in lake and river water, vegetational changes Pleistocene in Alaska, N.S., E. Africa. Home: 626 Starmont Dr., Durham, N.C. 27705.*

LIVINGSTONE, David, explorer, physician; b. Blantyre, Scotland, Mar. 19, 1813; s. Neil and Agnes (Hunter) L.; student medicine Anderson Coll., Glasgow U., also in London; licentiate in medicine Faculty Physicians and Surgeons of Glasgow; m. Mary Moffat, 1844; 1 son, Oswell. Worked in cotton mill, from 1823; med. missionary London Missionary Soc., Africa, 1840-57; embarked for Cape of Good Hope, 1840; explored No. Kalahin Desert, 1849, Linyanti, Upper Zambesi, Lake Dilolo, Zambesi River to Indian Ocean, 1853-56; apptd. Brit. consul at Quelimane, 1858; explored (with Sir John Kirk) Zambesi River, 1858-64; also sources of Nile, 1866, lake regions of S. Africa, 1867-73; object of search by Henry M. Stanley (leader of expdn. to find Livingstone), found him 1871. Mem. French Acad. Scis., 1869. Author: Missionary Travels and Research in South Africa, 1857; The Zambesi and its Tributaries, 1865. Discovered Lake Ngami, 1849, Zambesi River, 1851, lakes Shir Wa and Nyasa, 1859, lakes Bangweulu and Mweru, 1867-68; described relapsing fever often resulting from tick bite, also tsetse fly and cattle disease caused by its bite, 1857; 1st to use arsenic in treatment of nagana (horse and cattle disease in Africa), 1858. Died No. Rhodesia, Africa, May 1, 1873.

LIVINGSTONE, Frank Brown, Am. anthropologist; b. Winchester, Mass., Dec. 8, 1928; s. Guy P. and Margery (Brown) L.; A.B. in Math., Harvard, 1950; M.A., U. Mich., 1955, Ph.D. in Anthropology, 1957; m. Carol Southworth Ludington, Aug. 13, 1960; 1 dau., Amy Fenner. NSF Postdoctoral fellow Liberian Inst. Tropical Medicine, Harbel, West Africa, 1957-58; faculty U. Mich., Ann Arbor, 1959—, asso. prof. anthropology, 1963—. Mem. Am. Anthrop. Assn., Am. Assn. Phys. Anthropologists, Am. Soc. Human Genetics, Royal Anthrop. Inst. Research, publs. on sickle cell gene and other abnormal hemoglobins in Liberia and West Africa; analysis hereditary blood defects and significance in human racial variation. Home: 471 Rock Creek Dr., Ann Arbor, Mich. 48104.*

LIZARS, John, Scotch surgeon; b. Edinburgh, Scotland, circa 1785; s. Daniel L.; M.D., Edinburgh U., 1810. Surgeon on board man-of-war; naval service on Portuguese coast during Peninsular War; taught anatomy and surgery, Edinburgh; named prof. surgery Royal Coll. Surgeons, Edinburgh, 1831; sr. operating surgeon (with Robert Liston) Royal Infirmary; pvt. practice surgery, Edinburgh, until his death. Fellow Royal Coll. Surgeons, Edinburgh, 1815. Author: A System of Anatomical Plates of the Human Body, with Description, 1822; Observations on Extraction of diseased Ovaria, illustrated by Plates coloured after Nature, 1835; System of Practical Surgery . . . 1835. Introduced operation for removal of upper jaw; first attempt to perform ovariotomy in Gt. Britain, 1825; Lizar lines named after him. Died Edinburgh, May 21, 1860.

LJIMA, Takeshi, civil engr.; b. Taiwan, Jan. 20, 1922; s. Ekichi and Makiyo (Kamada) I.; grad. civil engring. dept. Kyushu U., Fukuoka, Japan, 1946; D.Engring., Tokyo (Japan) U., 1957; m. Haruko Nagai, May 3, 1950; children—Aiko, Kaoruko, Kiyoo. Govt. officer, engr. Ministry Transp., research engr. Port and Harbor Technics Research Inst., 1946-57; chief research engr., 1957-62, head design standards div., 1962-64; prof. coastal engring. dept. hydraulix civil engring. Faculty Engring., Kyushu U., 1964—. Mem. Japan Soc. Civil Engring. Author: Method of Survey of Coast and Harbors, 1959; also articles. Research in coastal oceanography and engring.; devel. method measurement ocean waves and wave forecasting. Home: 1 chome, Higashimatubara, Imajuku, Fukuoka City, Japan.*

LLAURADO, Josep G(arcia), biomed. researcher; b. Barcelona, Spain, Feb. 6, 1927; S. José Garcia and Rosa Llaurado; B.A., B.S. summa cum laude, J. Balmes Inst., Barcelona, 1944; M.D. cum laude, U. Barcelona, 1950; M.S. in Biomed. Engring., Drexel Inst., 1963; m. Catherine Dorothy Entwistle, June 28, 1958 (dec. Nov. 1965); children—Thadd, Oleg, Montserrat; m. 2d, Deirdre Mooney, Nov. 9, 1966. Came to U. S., 1957, naturalized, 1966. Instr. medicine Med. Sch. Barcelona, 1950-52; Brit. Council scholar, asst. med. research Postgrad. Med. Sch., U. London (Eng.), 1952-54; asst. prof. Med. Sch., Otago, New Zealand, 1954-57; Hite Found. fellow U. Tex. M.D. Anderson Hosp., 1957-58; USPHS steroid biochemist U. Utah, Coll. Medicine, Salt Lake City, 1958-59; sr. endocrinologist pharmacology dept. Pfizer, Groton, Conn., 1959-61; asso. prof. physiology U. Pa., Phila., 1963-67; asso. prof. biomed. engring. Marquette U., Milw., 1967—; Rockefeller vis. prof. U. Valle, Colombia, S.Am., 1958. Mem. A.A.A.S., Endocrine Soc., Am. Physiol. Soc., I.E.E.E., Am. Soc. Pharmacological and Exptl. Therapeutics, Am. Assn. U. Profs., Société de Biométrie Humaine (Paris, France), Sigma Xi. Research, numerous publs. on constrn. 1st flame photometer in Spain; research on aldosterone and its role in metabolic alterations after surg. operation, prolongation of skin transplants by use synthetic adrenocortical hormones; demonstrated anti-inflammatory effects new synthetic steroids, effects of amphetamine on sodium and potassium excretion, connective tissue in arteries, computer compartmental analysis of sodium distbn. in tissues. Office: Neurosci. Research Labs., VA Hosp., Wood, Wis. 53193.*

LLEWELLYN, Donald Rees, chemist; b. Dursley, Eng., Nov. 20, 1919; s. Reginald George and Mable (Derrett) L.; ed. U. Birmingham, 1938-41, Oxford (Eng.) U., 1941-45, Cambridge (Eng.) U., 1945-46, U. Bangor, 1946-49, U. London (Eng.), 1949-56, U. Auckland, 1956-64, U. Waikato, 1964; m. Ruth Marian Llewellyn, Sept. 14, 1943; children—Joan Day, Dennis Day, Robert David Llewellyn. Vice-chancellor U. Waikato, Hamilton, New Zealand, 1964—. Mem. Atomic Energy Com., 1958—. Mem. New Zealand Inst. Chemistry (v.p. 1966—). Research, publs. on application of radioactive and stable isotopes to study reaction mechanism. Home: Cambridge Rd., Hamilton, New Zealand.*

LLOYD, David P(ierce) C(aradoc), Am. physiologist; b. Auburn, Ala., Sept. 22, 1911; s. Francis Ernest and Mary Elizabeth (Hart) L.; B.Sc., McGill U., 1932; B.A. (Rhodes Scholar), Oxford U., Eng., 1936, Dr. Philosophy, 1938, D.Sc., 1961, M.A., 1964; M.A. m. Kathleen Mansfield Elliott, Mar. 20, 1937; children—Marion Gwenellen, Owen H. T., Evan E. M.; m. 2d, Cynthia Meynell, Nov. 17, 1957. Departmental demonstrator physiology Oxford U., 1935-36; asst., asso. dept. med. research U. Toronto, 1936-39; with Rockefeller Inst. Med. Research, 1939-43, 46, mem., 1949—, prof., 1957—; asst. prof. physiology Yale, 1943-45; Johnston lectr. U. Minn., 1948; James Arthur lectr. Am. Mus. Natural History, 1958. Adv. com. Nat. Found. Infantile Paralysis, 1947-59; trustee Internat. Poliomyelitis Congress, 1948—. Awarded Reeve prize medicine U. Toronto, 1939. Mem. Acad. Sci., Am. Physiol. Soc., Physiol. Soc. Gt. Britain, A.A.A.S., Assn. Research Nervous and Mental Diseases, Harvey Soc. Contbr. chpt. Howell's Textbook of Physiology, rev. edit. 1946, 49, 55. Research and publs. on discoveries on relationship to synaptic transmission in the spinal cord to other systems of the body. Home: 425 East 63d St. Office: Rockefeller Inst. Med. Research, 66th St. and York Av., N.Y.C. 10021.*

LLOYD, E(dwin) Russell, Am. geologist; b. Lloydsville, W.Va., Nov. 3, 1882; s. Nimrod Wesley and Mary Magdalene (Bender) L.; grad. W.Va. Conf. Sem., 1901; A.B., Ohio Wesleyan U., 1905; elected Rhodes scholar from W.Va., at Oxford U., 1905, B.A., 1908, Burdett Coutts scholar, 1908-09; postgrad. U. Chgo., 1909-11; m. Helen Burnett Gardner, Jan. 7, 1920; children—Anne Gardner, Edwin Russell. Asst. and asso. geologist U. S. Geol. Survey, Washington, 1911-17, 18-19, in charge coal land classification, 1915-16; geologist Sinclair-Central Am. Oil Co., 1917-18; chief geologist Sinclair-Wyo. Oil Co., Casper, Wyo., 1919-21, Mid-Kansas Oil & Gas Co., Mineral Wells, Tex., 1921-23, Argo Oil Co., Denver, 1923-24; cons. practice, also chief geologist N.Y. Oil Co., Denver, 1924-26; dist. geologist Roxana Petroleum Corp., Roswell, N.M., 1927; cons. practice, Denver and Midland, Tex., 1928-32; dist. geologist, Superior Oil Co., Midland, 1932-36; cons. practice since 1936. Fellow Geol. Soc. Am.; mem. Am. Assn. Petroleum Geologists (hon. mem. asso. editor), Am. Geophys. Union, Sigma Xi. Deceased.

LLOYD, Francis Ernest, botanist; b. Manchester, Eng., Oct. 4, 1868; s. Edward and Leah (Pierce) L.; A.B., Princeton, 1891, A.M., 1895; student at Munich, 1898, Bonn, 1901; D.Sc. (hon.), U. of Wales, 1933, Masaryk U., 1938; m. Mary Elizabeth Hart, May 18, 1903; children—Mary Elizabeth, Francis Ernest Llewellyn, David Pierce Caradoc. Instr. William Coll., 1891-92; prof. biology and geology Pacific U., Ore., 1892-95, biology, 1895-97; adj. prof. biology Tchrs. Coll. (Columbia), 1897-1906; investigator Desert Bot. Lab., Carnegie Instn., Washington, 1906; instr. Harvard Summer Sch., 1907; cytologist Ariz. Expt. Sta., 1907; dir. dept. of investigation Continental-Mexican Rubber Co., 1907-08; prof. botany Ala. Poly. Inst., 1908-12; Macdonald prof. botany McGill U., Montreal, 1912-34, emeritus prof. botany since 1934; cons. U. S. Rubber Co., 1919-33. Bot. explorations Lumholtz expdn. to Mexico, 1890; Columbia expdn. to Puget Sound and Alaska, 1896; N.Y. Bot. Garden to Dominica, B.W.I., 1903, Java, Sumatra, F.M.S., 1919, S. Africa, 1929, 1935; Australia, New Zealand, 1935-36. Pres. bot. sect. Brit. Assn. Advancement of Science, 1933; fellow A.A.A.S. (sec.; v.p. 1923, sect. G), Royal Soc. Can. (pres. 1933), Am. Soc. Plant Physiology (pres. 1927; Barnes life mem.), Linnean Soc. of London, Edinburgh Bot. Soc. (hon.); mem. Phila. Coll. Pharmacy (hon.) Bot. Soc. Am., Torrey Bot. Club (asso. editor of Bull., 1899-1902; treas., 1902-06), Centro de Sciencios Letras e Artees, Campinas (Brazil), Czechoslovakian Bot. Soc., Sigma Xi. Author: The Teaching of Biology (botany) in the Secondary School (American Teachers Series), 1904; The Comparative Embryology of the Rubiaceae, 1902; The Physiology of Stomata, 1908; Gnayule, (a rubber plant of the Chihuahuan Desert), 1911; The Carnivorous Plants, 1942; also various other bot. papers, including studies on transpiration, stomata, tannin, rubber, cotton, growth, colloids, fluorescence, reproduction and carnivorous plants. Editor Plant World, 1905-08. Died Oct. 18, 1947.

LLOYD, Humphrey, Irish scientist; b. Dublin, Apr. 16, 1800; s. Bartholomew and Eleanor (McLaughlin) L.; B.A., Trinity Coll., Dublin, 1819; M.A., 1827; D.D., 1840; D.C.L., Oxford, 1855; m. Dorothea Bulwer, July 1840. Erasmus Smith prof. natural and exptl. philosophy Trinity Coll., Dublin, 1831-43, named sr. fellow, 1843, vice provost, 1862, provost, 1867, also in charge magnetic obs. Recipient Cunningham Gold medal Royal Irish Acad., 1862. Fellow Royal Soc., 1836, Royal Soc. Edinburgh; mem. Royal Irish Acad. (pres. 1846-51), Brit. Assn. (named pres. 1857). Author: A Treatise on Light and Vision, 1831; Two Introductory Lectures on Physical and Mechanical Science, 1834; The Elements of Optics, 1849; Elementary Treatise on Wave Theory of Light, 1857; The Doctrine of Absolutism, 1871; Treatise on Magnetism, General and Terrestrial, 1874. Research on optics and magnetism; originated method of producing interference fringes with a single mirror; studied internal conical refraction, including verification of Sir W. Hamilton's predictions on prodn. hollow cylinder of rays by internal conical refraction in biaxial crystals. Died Dublin, Jan. 17, 1881.

LLOYD, James, Am. physician; b. Oyster Bay, L.I., N.Y., Mar. 24, 1728; s. Henry and Rebecca (Nelson) L.; studied medicine under Dr. William Clark, Boston, 1745-50, obstetrics and surgery under William Smellie and William Cheselden, London, Eng., 1750-52; m. Sarah Corwin; at least 1 child, James. Began practice surgery, Boston, 1752; 1st physician to practice midwifery in Am.; early advocate of vaccination for smallpox; only noted physician to remain in Boston during Am. Revolution. Died Boston, Mar. 14, 1810.

LLOYD, John Uri, Am. pharmacist; b. Bloomfield, N.Y., Apr. 19, 1849; s. Nelson Marvin and Sophia (Webster) L.; ed. pvt. schs., Florence, Burlington and Petersburg, Ky.; Ph.M. (hon.), Phila. Coll. of Pharmacy, 1890; Ph.D., U. Ohio, 1897; LL.D., Wilberforce U., D.Sc. U. Cin., 1916; M.D., Eclectic Med. Coll. Cin., 1921; m. Adeline Meader, Dec. 27, 1876; m. 2d, Emma Rouse, June 10, 1880. Prof. pharmacy Cin. Coll. Pharmacy, 1883-87; prof. chemistry, 1878-—, pres., Eclectic Med. Inst., 1896-1904; pres. Lloyd Bros., Pharmacists, Inc. Studied dialect, superstition and folk-lore of No. Ky. Recipient Remington medal Am. Pharm. Assn., 1920. Revised Cleaveland's Pronouncing Med. Dictionary, 1881. Author: Etidorhpa, The End of Earth, 1895; Stringtown on the Pike, 1900; Warwick of the Knobs, 1901; Red Head, 1903; Scroggins, 1904. Has especially investigated plant chemistry and phytochemistry as applied to medicines, alkaloids, glucosids and proximate principles, precipitates in fluid extracts, and phenomena of capillarity. Editor: History of the Vegetable Drugs of the Pharmacopeia of the United States, 1911; A Study of Digitalis, 1912; and of Coca (with J. T. Lloyd), 1913; History of the Discovery of the Alkaloidal Affinities of Hydrous Aluminum Silicate, 1915; Continued Study in Adsorption—Kryptonine, 1916; Echinacca Angustifolia, 1917; A Study in Solvents, 1917; History of the Vegetable Drugs of the U. S. Pharmacopeia, 1920; Felix Moses, the Beloved Jew, 1930; Physics in Pharmacy (with Dr. Wolfgang Ostwald, Leipzig, Germany, 5 parts issued). Asso. editor Pharm. Rev., to 1909, Eclectic Med. Jour., Eclectic Med. Gleaner. Research work in colloidal chemistry. Died Apr. 9, 1936.

LLOYD, Lewis Ewan, Canadian nutritionist; b. Montreal, Que., Can., Feb. 3, 1924; s. Normal Lewis and Annie (Ewan) L.; B.Sc., McGill U., 1948, M.Sc., 1950, Ph.D., 1952; m. Pauline Marguerite Sharpe, June 3, 1950; children—David, Nancy, Robert, James. Research asso. Cornell U., Ithaca, N.Y., 1952-53; faculty McGill U., Montreal, 1953-67, prof., chmn. dept. animal sci., 1960-67; dir. U. Man. Sch. Home Econs., Winnipeg, Can., 1967—; research scholar Rowett Research Inst., Aberdeen, Scotland, 1958-59. Mem. Agrl. Inst. Can., Canadian Soc. Animal Prodn., Nutrition Soc. Can., Am. Inst. Nutrition, Am. Soc. Animal Sci., Nutrition Soc., U.K. Author: (with E. W. Crampton) Fundamentals of Nutrition, 1959. Research, publs. on value of foods as related to humans and to domestic animals. Home: 43 D'Arcy Dr., Winnipeg 19. Office: Sch. Home Econs., U. Man., Winnipeg, Man., Can.*

LLOYD, Marshall Burns, Am. inventor; b. St. Paul, Minn., Mar. 10, 1858; s. John and Margaret (Conmee) L.; ed. pub. schs., Meaford, Can.; m. First invention was combination bagholder and scale, for farmers; founded Lloyd Mfg. Co., mfrs. bedspring weaving machine, Mpls., 1900, later at Menominee, Mich.; invented new method and machinery for making thin seamless steel tubing; also new method for producing wicker articles and loom for weaving wickers; v.p. Automatic, Seamless Tubing Co. Died Aug. 10, 1927.

LLOYD, Ray Dix, Am. physicist; b. Salt Lake City, Mar. 10, 1930; s. Ray E. and Dixie (Penrose) L.; B.S., U. Utah, 1954, M.S., 1956; postgrad. U. Southwestern La., La. State U.; m. Louise Mortensen, July 10, 1954; children—Thomas R., Janna, Alan T., Chris R. Research asso. div. radiobiology anatomy dept. U. Utah, Salt Lake City, 1961-64, research instr., asst. physics group leader, 1964—. Mem. Radiation Research Soc., Health Physics Soc. Research, publs. on effects of ionizing radiation on living systems, prodn. and therapy neoplasia by internally deposited radioactivity, application radioactive tracers in medicine; design and application radiation counting systems. Office: Radiobiology Lab., U. Utah, Salt Lake City 84112.*

LLOYD, Samuel, Am. surgeon; b. Jersey City, Aug. 4, 1860; s. Gardner Potts and Emma (Disbrow) L.; B.Sc., Princeton, 1882; M.D., U. Vt., 1884, Coll. Phys. and Surg. (Columbia), 1886; m. Adele Ferrier Peck, June 11, 1888; children—Elisabeth (Mrs. Edward H. Wardwell), Augustine A. (Mrs. John Prince Hazen Perry), Samuel. Prof. surgery N.Y. Post Grad. Med. Sch.; attending surgeon Lutheran Hosp. of Manhattan. Fellow A.C.S. First to demonstrate that lung expansion can be maintained without adhesions and mech. apparatus. Died Dec. 19, 1926.

LLOYD, Stewart Joseph, chemist; b. Hamilton, Ont., Can., Sept. 12, 1881; s. Joseph and Sage (Peregrine) L.; B.A., U. Toronto, 1904; M.Sc., McGill U., 1906; Ph.D., U. Chgo., 1910; m. Edith Marian Dawson, Dec. 27, 1911; children—Frances Valentine, Virginia Edith, Edith Vane. Prof. chemistry and metallurgy U. Ala., 1909—, also dir. lab. and dean Sch. Chemistry, Metallurgy and Ceramics; cons. chem. engr. Ala. Power Co.; asst. state geologist, 1930—; acting state geologist of Ala., 1939-45. Mem. Am. Inst. Chem. Engrs., Am. Chem. Soc., Soc. Chem. Industry, Am. Inst. Chemists, Electrochem. Soc., Canadian Inst. Chemistry. Died Aug. 1959.

LLOYD, Trevor, geographer; b. London, Eng., May 4, 1906; s. Jonathan and Mary (Gordge) L.; B.Sc., Bristol U., 1929, D.Sc., 1949; Ph.D., Clark U., 1940; M.A. (hon.), Dartmouth, 1944; m. Joan Glassco, 1936 (div. Mar. 1965); children—Mona (Mrs. Viggo Mollerup), and Hugh Glassco. Specialist in school geography, Winnipeg, Man., Can., 1931-38; faculty Carlton Coll. (Minn.), 1942, Dartmouth, 1942-59; faculty McGill U., Montreal, 1959—, chmn. dept. geography, 1962-66. Chief, Geog. Bur. Govt. Can., 1947-48; consul for Can., Greenland, 1944-45. Fellow Arctic Inst. N.Am. (founder, chmn. bd. govs.); mem. Greenland Soc., Canadian Inst. Internat. Affairs, Assn. Am. Geographers, Canadian Assn. Geographers (pres. 1958-59), A.A.A.S. (sect. v.p. 1965), Geog. Soc. Finland (corr.). Author: Sky Highways, 1944; Canada and Her Neighbours, 1947; Southern Lands, 1954; (with D. J. Seiveright) Lands of Europe and Asia, 1957. Studies of econ. and social devel. of the Subarctic in all countries; stimulation of effective govt. policies in No. Can., encouragement of internat. cooperation in No. research devel. geography as a significant feature of Canadian edn. Home: 4875 Fulton St., Montreal, Que., Can.*

LLULL, Ramón, see Lullius, Raimundus.

LLWYD, Edward, see Lhuyd, Edward.

LOBACHEVSKI, Nikolai Ivanovich, Russian mathematician; b. Makariev, Russia, Oct. 11, 1793; ed. U. Kazan. Faculty U. Kazan, from 1812, asst. prof. math., 1814-16, extraordinary prof., 1816-23, prof., 1823-46, rector, 1827-46. Author (originally in Russian): Geometrical Investigations on the Theory of Parallel Lines, 1829-30; Principles of Geometry, 1829-30; Imaginary Geometry, 1835; New Principles of Geometry, 1835-38; Polnoe sobranie sociennij (5 vols.), 1846-51; Pangéométrie, 1855. Independently of J. Bolyai, introduced 1st comprehensive system of non-Euclidean geometry (1st conceived by Gauss); including concept that indefinite number of lines can be drawn in a plane parallel to given line through given point, 1826 (later Einstein showed the universe was non-Euclidean in structure); showed necessity of distinguishing between continuity and differentiability in math. Died Kazan, Russia, Feb. 24, 1856.

LOBECK, Armin Kohl, Am. geologist, geographer; b. N.Y.C., Aug. 16, 1886; s. Adolph Christian and Elmire Celeste (Voullaire) L.; A.B., Columbia, 1911, A.M., 1913, Ph.D., 1917; m. Bertha Merrill, Dec. 25, 1917; children—Elmire, Merrill. Instructor Phila. Coll. Pharmacy, 1911-14; asst. prof. U. of Wis., 1919-24, asso. prof., 1924-29; prof. geology, Columbia, 1929-54, emeritus prof. Bd. govs. Nature Conservancy, 1954—. Founder, 1922, pres., dir. Geographical Press. Mem. geog. sect., Am. Commn. to Negotiate Peace, Paris, World War I. Consultant War and State Depts., World War II. Neil Miner Medal, Assn. Geology Tchrs., 1956. Fellow A.A.A.S., N.Y. Acad. Science; mem. Geol. Soc. of America, Assn. of Am. Geographers, Sigma Xi. Author: Block Diagrams, 1924; Guide to Geology of Allegany State Park, 1927; Geology of Mammoth Cave National Park, 1928; Airways of America, 1933; Geomorphology, 1939; Military Maps and Photographs (with Maj. W. Tellington, U. S. Army), 1944; Geological Diorama of U. S.; Things Maps Don't Tell Us, 1956. Contbr. maps, block diagrams, guides and articles to geol. astronomy and geog. publs. Noted for physiographic descriptions of Europe, Asia, Africa and N. Am. Died Apr. 26, 1958.

LOBRY, Cornelius Adriann de Bruyen, Dutch chemist; b. 1857; prof. at Amsterdam. Research on reaction di-nitro compounds, action ammonia on carbohydrates, alcohols as solvents (leading to isolation of free hydroxylamine and hydrazine). Died 1904.

LOBSTEIN, Johann Georg C.F.M., German pathologist; b. Alsace, Germany, 1777; described a ganglion of great splanchnic nerve above the diaphragm (Lobstein's ganglion), 1823, osteogenesis imperfecta (osteopsathyrosis), a condition in which the bones are abnormally brittle and fracture easily (Lobstein's disease). Died 1835.

LOCHHEAD, Allan Grant, Canadian bacteriologist; b. Galt, Ont., Can., June 21, 1890; B.A., McGill U., 1911, M.Sc., 1912, Ph.D., 1919; Dr. Agr. (hon.), Giessen, Germany, 1963; m., 1919; 2 children. Lectr. bacteriology Macdonald Coll., McGill U., 1918-19; bacteriol. chemist Canadian Milk Prod., Toronto,

Ont., 1919-21, Malt Prod. Co., 1921-33; lectr. biochemistry Alberta, 1922-23; agrl. bacteriologist, central exptl. farm Canadian Dept. Agr., 1923-38, chief div. bacteriology, exptl. farms, 1923-38; with sci. service, 1953-55; research officer microbiol. inst., 1944-60; ret. Fellow Royal Soc. Can. (Flavelle medal 1958), A.A.A.S.; mem. Am. Soc. Microbiology, Am., Canadian (pres. 1953, hon. mem.), socs. microbiology, Brit. Soc. Applied Bacteriology (hon.). Authority on agrl. and food bacteriology, soil microbiology and growth factors; research on microbiol. aspects of dairying, food preservation, soil fertility, apiculture; introduced methods for classifying soil bacteria according to their nutritional requirements. Home: 389 3d Av., Ottawa 1, Ont., Can.*

LOCHHEAD, William, biologist; b. Apr. 2, 1864; s. W. and Helen (Campbell) L.; ed. McGill, Cornell univs; B.A., M.Sc.; m. Lilias Grant, 1889; 1 son. Tchr. sci, Perth, Galt, London collegiate insts., 1887-94, 95-98; prof. biology Ont. (Can.) Agrl. Coll., 1898-1906; prof. biology Macdonald Coll., McGill U., from 1906. Fellow A.A.A.S.; mem. Que. (pres. 1916-17), Ont. (pres. 1902) entomol. socs. Author: The San Jose and Allied Scales; The Hessian Fly; The Pea Weevil; Nature Studies; Spray-Calendar; Heredity and Genetics; Modern Biological Laws and Theories relating to Animal and Plant Breeding, 1911; Class Book of Economic Entomology, 1919. Editor Jour. Agr. for Que., 1908-20. Died Mar. 26, 1927.

LOCHTE-HOLTGREVEN, Walter, German physicist; b. Hamburg, Germany, Oct. 15, 1903; s. Theodor and Louise (Holtgreven) L.; ed. univs. of Marburg, Zurich, Gottingen; Ph.D., agrégé; m. Irene Kossel, Nov. 27, 1943; children—Herman, Albrecht, Gudula, Martin. Asst., U. Groningen, 1927-30; John Harling fellow, Manchester, Eng., 1930-34; asst. U. Kiel, 1934-39; with Meteorol. Service, 1939-41; staff Gatow Ballistics Inst., 1941-43; dir. exptl. physics Inst., U. Kiel (Germany), 1943—. Mem. Internat. Astron. Union, Deutscher Ausschuss für Spektroscopie, Deutsche Phsikalische Gesellschaft, Optical Soc. Am. Author: Production and Measurements of High Temperatures. Research and publs. on plasma physics, spectroscopy, optics. Home: Roonstrasse 10, Kiel, W. Germany.

LOCK, Frank Ray, Am. physician; b. Lake Charles, La., Oct. 30, 1910; s George T. and Delia (Moss) L.; A.B., Cornell U., 1931; M.D., Tulane U., 1935; m. Mary Frances Bonney, June, 26, 1936; children—Frank Ray, David McBrier, James Bonney. Fellow, obstetrics and gynecology Tulane U., 1938-40; fellow in gynecology and pathology Johns Hopkins, 1940; prof., dir. dept. obstetrics and gynecology Bowman Gray Sch. Medicine, Winston-Salem, N.C., 1941—; chief service N.C. Bapt. Hosp., Winston-Salem, 1941—; practice medicine specializing in obstetrics and gynecology, Winston-Salem, 1941—. Mem. Am. Coll. Obstetricians and Gynecologist (pres. 1964-65), Am. Gynecol. Soc. (pres. 1968-69), Am. Assn. Obstetrics and Gynecology, president 1965-66), Am. Gynecol. Club, Am. Bd. Obstetrics and Gynecology (asso. examiner 1950—), A.C.S. (adv. council on obstetrics and gynecology 1956-59), Am., N.C., Forsyth County med. socs., A.A.A.S., S. Atlantic Assn. Obstetricians and Gynecologists, So. Med. Assn. Home: 2841 Galsworthy Dr., Winston-Salem, N.C. 27106.*

LOCK, Robert Heath, botanist; b. Eton Coll., Windsor, Eng., 1879; s. J. B. Lock; ed. Gonville and Caius Coll., Cambridge (Eng.) U., M.A., Sc.D.; m. Bella Sidney Woolf, 1910. Apptd. sci. asst. to dir. Royal Botanic Gardens, Peradeniya, Ceylon, 1902, asst. dir., 1908, acting dir., 1909, 12; became curator Cambridge U. Herbarium, 1904; insp. His Majesty's Bd. Agr. and Fisheries, from 1913. Fellow Royal Hort. Soc. Author: Recent Progress in the Study of Variation, Heredity, and Evolution, 1906, 3d edit., 1911; Rubber and Rubber-Planting, 1913; also papers on, genetics, agr., botany. Research on Hevea rubber cultivation. Died June 25, 1915.

LOCK, William Owen, physicist; b. Cirencester, U.K., Aug. 8, 1927; s. William George and Dorothy (Cosslett) L.; B.Sc. with honors, Bristol (Eng.) U., 1948, Ph.D., 1952; m. Eleanor Regina James, Aug. 14, 1952; children—Nicholas William Hugh, Adrian Christopher Glyn, Evan Maxwell David. Imperial Chems. Industries research fellow U. Manchester (Eng.), 1952-53; lectr. physics U. Birmingham (Eng.), 1953-59; with CERN (European Orgn. for Nuclear Research), Geneva, Switzerland, 1959—, sr. physicist, leader nuclear emulsion group, 1960-65, coordinator sci. and engring. staff, head fellows and visitors service personnel div., 1965—, acting leader personnel div., 1968. Fellow Inst. Physics, Phys. Soc. London. Author: High Energy Nuclear Physics, 1960; also articles. Research on cosmic rays and high energy nuclear physics using particle beams from accelerators also nuclear emulsion technique. Home: 2 Chemin de Tavernay, Geneva. Office: Personnel Div, CERN, 1211 Geneva 23, Switzerland.*

LOCKE, John, Brit. philosopher; b. Somersetshire, Eng., Aug. 29, 1632; s. John and Agnes (Keene) L.; B.A., Oxford, 1656, M.A., 1658. Lectr. rhetoric, Greek, philosophy Oxford, 1660-64; studied medicine Oxford, 1666, also in France, 1675-79; tutor, adviser to Earl of Shaftesbury, from 1667; became sec. of presentations under Shaftesbury, 1672; fled to Holland, 1684, returned to Eng. during Revolution, became sec. of appeal, also adviser to govt. on coinage, 1689-1704; lived with Sir Francis and Lady Masham, Essex, from 1690. Fellow Royal Soc., 1668. Author: Letters on Toleration, 1689; Two Treatises on Government, 1690; Essay Concerning Human Understanding, 1690; Some Thoughts Concerning Education, 1693; The Reasonableness of Christianity, 1695; Known as father of Eng. empiricism, also as philos. forerunner of exptl. psychology; leading philosopher of freedom in his day; in political theory contradicted views of Thomas Hobbes and Robert Filmer; envisioned the state guided by natural law; forecast the labor theory of value; supported religious toleration; stressed necessity of tolerating opposite opinion; stated that matter harbored primary (material of science) and secondary (source of man's perceptions) qualities; repudiated doctrine of innate ideas, i.e. regarded mind at birth as blank slate (tabula rasa), on which experiences inscribed themselves. Died Essex, Eng., Oct. 28, 1704.

LOCKE, John, Am. inventor; b. Lempster, N.H., Feb. 19, 1792; s. Samuel Barron and Hannah (Russell) L.; M.D., Yale, 1819; m. Mary Morris, Oct. 25, 1825; several children. Asst. surgeon USN, 1818; curator botany Harvard, 1819-21; founded, conducted Cin. Female Acad., 1822-35; prof. chemistry and pharmacy Med. Coll. of Ohio, Cin., 1835-53; studied geology and paleontology of Ohio, Ill., Ia. and Wis., 1835-40, later studied terrestrial magnetism and electricity; invented a surveyors' compass, level, orrery; invented electro-magnetic chronograph which revolutionized method of longitude determination, for U. S. Coast Survey, 1844-48. Author: Outlines of Botany, 1819; also papers. Died Cin., July 10, 1856.

LOCKE, Michael, biologist; b. U.K., Feb. 14, 1929; s. R. H. and K. N. (Waite) L.; M.A., Cambridge (Eng.) U., 1955, Ph.D., 1956. Lectr. zoology U. W.I., Jamaica, 1956-61; faculty Western Res. U., Cleve., 1961—, prof. biology, 1967—. Mem. Soc. Developmental Biology (editor Symposia 1962—). Research, publs. on ultrastructure of insect-cells during devel., role of gradients in growth and devel. of insect-epidermis. Address: biology dept., Case Western Res. U., Cleve. 44106.*

LOCKETT, Mary Fauriel, pharmacologist; b. Lancashire, Eng., Mar. 17, 1911; d. Harry Duncan and Mary (Hicks) Lockett; M.B.B.S., Royal Free Hosp. Med. Sch., London, 1936, M.D., 1936; Ph.D., Newnham Coll., Cambridge (Eng.) U., 1940. Lectr. pharmacology U. Coll. London, 1941-47; sr. lectr. pharmacology U. Glasgow, 1947; head dept. physiology and pharmacology Chelsea Coll. Sci. and Tech., 1953-62; Wellcome research prof. in pharmacology U. Western Australia, 1963—. Jr. Beit Meml. fellow, 1936-40. Mem. Physiol. and Pharmacological Soc. Brit. and Australia, Endocrinological Soc. Britain and Australia, Renal Soc. Britain and Australia. Research, publs. on mechanism of action of sympathetic nervous systems of mammals and mechanisms which control water and salt content of animal body. Home: 9 Hardy St., Nedlands, Australia.*

LOCKHART, Luther Bynum, Jr., Am. chemist; b. Atlanta, Sept. 13, 1917; s. Luther Bynum and Louisa (Hamilton) L.; A.B. in Chemistry, Emory U., 1938; Ph.D. in Organic Chemistry, U. N.C., 1942; m. Elizabeth Jane Brodnan, July 7, 1951; children—Katherine, Thomas, Jean, Kent. Research asso. U. N.C., 1942-43; with chemistry div. U. S. Naval Research Lab., 1943—, head, phys. chemistry br., Washington, 1954—. Mem. Am. Chem. Soc., Am. Geophys. Union, Research Soc. Am., Washington Acad. Scis. Research, publs. in organic syntheses, study of multiple-layered reflection modifying coatings, measurement of natural and fission product radioactivity in the atmosphere, use of airborne radioactivity as tracers for meteorol. processes, study of radioactive aerosols, radiotracer studies of adsorption processes. Home: 6820 Wheatley Ct., Falls Church, Va. 22042. Office: 4555 Overlook Av. S.W., Washington 20390.*

LOCKHART, William Raymond, Am. bacteriologist; b. Carlisle, Ind., Nov. 25, 1925; s. William Elmer and Jewel (Johnson) L.; A.B., Ind. State Coll., 1949; M.S., Purdue U., 1951, Ph.D., 1954; m. Barbara Powers, Aug. 2, 1947; children—Laurence L.; Thomas R., Alan K., Sally Ann. Asst. bacteriologist Ohio River Valley Water Sanitation Commn., 1949; fellow U. Chgo., 1953-54; faculty Ia. State U., Ames, 1954—, prof. chmn. dept. bacteriology, 1960—. Fellow Am. Acad. Microbiology; mem. Am. Soc. for Microbiology (pres. N. Central br. 1964-65), Soc. for Gen. Microbiology, Sigma Xi. Editorial bd. Applied Microbiology, 1963-64; Jour. Bacteriology, 1965—; sect. editor Biol. Abstracts, 1956—. Research, publs. on bacterial growth particularly biochem. control. growth in bacteria, applications math. methods and computers to biol. classification.*

LOCKHEAD, John Hutchison, zoologist; b. Montreal, Que., Can., Aug. 7, 1909; s. John and Alida (Van Slyke) L.; M.A., St. Andrews (Scotland) U., 1930; B.A., Cambridge (Eng.) U., 1932, Ph.D., 1937; m. Margaret Szabo, Mar. 26, 1938; children—John Van Slyke, William Matthew. Came to U. S., 1938, naturalized, 1946. Sr. curator Cambridge U. Mus. Zoology, 1935-38, instr., 1936-38; asst. biologist Va. Fisheries Lab., lectr. biology Coll. William and Mary, 1941-42; faculty U. Vt., Burlington, 1942—, prof. zoology, 1951—. Mem. Marine Biol. Lab. Corp., Woods Hole, Mass., 1944—. Mem. Am. Soc. Zoologists, Sigma Xi. Contbg. author: Selected Invertebrate Types, 1950; The Physiology of Crustacea, 1961. Research, publs. on physiology and functional anatomy of crustacea including feeding mechanisms, functions of 2 types of eye, mechanisms for controlling swimming position, endocrine systems, molting reprodn.*

LOCKIE, L. Maxwell, Am. physician; b. Buffalo, June 30, 1903; s. John W. and Emma (Fink) L.; M.D., U. Buffalo, 1929, Ph.G., 1923, B.S. in Medicine, 1925; m. Margaret V. Van Volkenburg, July 14, 1934; children—L. Maxwell, George N. Faculty, State U. N.Y. Coll. Medicine, Buffalo, 1929—, prof., head dept. therapeutics, 1939-63, clin. prof. medicine, 1963—; staff Buffalo Gen., Childrens hosps. Cons. to hosps. Fellow A.C.P.; mem. Ligue Internat. contre le Rhumatisme (past U. S. councillor), Société Internationale de Medicine, Sociedad Medica de Estados Unidos de Norteamerica y Mexico (charter, past pres.), Socidad Argentina de Rheumatologica (hon.), Sociedad Mexicana de Rheumatologica (hon.), Archives Internamericanos de Rheumatologia (U. S. editor 1961—), Am. Rheumatism Assn. (past pres.), A.M.A., Am. Pharm. Assn., Am. Fedn. Clin. Research, Nat. Rehab. Assn., Am. Acad. Compensation Medicine, U. S. Pharmacopeia, Johns Hopkins Med. Soc., Mid-W. Conf. Rheumatic Diseases (past pres.), Am. Soc. Hosp. Pharmacists (hon.), Assn. Am. Med. Colls., Am. Therapeutic Soc., Nat. Soc. Clin. Rheumatologists, Alpha Omega Alpha, Nu Sigma Nu. Editorial bd. Am. Jour. Orthopedics, 1959—. Numerous publs. on diagnosis, treatment of gout and rheumatoid arthritis; devised high-fat diet to produce attacks; described use of B.A.L. in treatment of reactions to gold therapy. Home: 130 Morris Av., Buffalo 14214.*

LOCKWOOD, Charles Barrett, Brit. surgeon; prof. Royal Coll. Surgeons, 1887-89, 95, also fellow, mem. council; surgeon Gt. No. Central Hosp.; cons. surgeon St. Bartholomew's Hosp.; pres. Anat. Soc., Harveian Soc. London, Med. Soc. London. Author: Aseptic Surgery, 2d edit., 1899; A Radical Cure of Hernia, Hydrocele, and Varicocele, 1899; The Surgery and Pathology of Appendicitis, 2d edit., 1906; Cancer of the Breast, 1913. Described thickened region of orbital fascia between sheaths of inferior oblique muscles, Tenon's capsule, inferior rectus (Lockwood's ligament), 1885. Died Nov. 8, 1914.

LOCKWOOD, John Alexander, Am. physicist; b. Easton, Pa., July 12, 1919; s. Harold John and Elizabeth (Van Campen) L.; A.B., Dartmouth, 1941; M.S., Lafayette Coll., 1943; Ph.D., Yale, 1948; m. Jean Elizabeth Manville, Mar. 28, 1942; children—Heidi, Nancy, Jane. Mem. faculty U. N.H., 1948—, prof. physics, 1959—, chmn. dept., 1962-65; research and publs. on spl. research cosmic ray time variations, neutron measurements. Mem. Am. Phys. Soc., Am. Assn. Physics Tchrs., A.A.A.S., Am. Geophys. Union, Sigma Xi. Author articles in field. Home: 6 Valentine Hill Rd., Durham, N.H. 03824.*

LOCKWOOD, John Salem, surgeon; b. Shanghai, China, Oct. 2, 1907; s. William Wirt and Mary Rebecca (Town) L.; A.B., De Pauw U., 1928; M.D., Harvard, 1931; Med. Sc.D., Coll. Phys. and Surg. Columbia, 1937; m. Dorothy E. Tufts, Oct. 1, 1932; children—Elinor Towne, Marcia Robinson, Dorothy Tufts. Intern Presbyn. Hosp., N.Y.C., 1932-34, jr. and sr. fellow in surgery, 1934-37; instr. in surgery and fellow in surg. research U. Pa. Sch. Medicine, 1937-39, asst. prof. surg. research, acting dir. dept., 1942-44; asso. in surgery and dir. tumor clinic Hosp. of U. Pa., 1940-42; prof. surgery Yale, also asso. surgeon New Haven Hosp., 1944-46; prof. surgery Columbia U. and attending surgeon Presbyn. Hosp., 1946—. Civilian mem. com. on med. scis. Nat. Research and Devel. Bd., Nat. Mil. Establishment, 1948—. Recipient Presdl. Certificate of Merit, 1948. Mem. Am., N.Y. County med. assns., Soc. U. Surgeons (pres. 1948), Am. Soc. for Clin. Investigation, Am. Surg. Assn., Soc. Clin. Surgery, Phila. Coll. Physicians, N.Y. Surg. Soc., A.C.S., N.Y. Acad. Medicine, Harvey Soc., Am. Assn. for Cancer Research, Sigma Xi. Contbr. articles to med. jours. Mem. editorial bd. Christopher's Textbook of Surgery, Annals of Surgery. Died June 16, 1950.

LOCKYER, Sir Joseph Norman, English astrophysicist, astronomer; b. Rugby, England, May 17, 1836; s. Joseph Hooley and Anne (Norman) L.; ed. pvt. schs., and on Continent; hon. LL.D., Glasgow, Aberdeen, Edinburgh; D.Sc., Cambridge, Sheffield, Oxford; m. 1st, Winifred Trebinshon, 1858; 4 sons, 2 daus.; 2d, Thomazine Mary Browne Brodhurst, 1903. Became clerk in War Office, 1857, editor Army Regulations, 1865; apptd. sec. Duke of Devonshire's Com. on Science, 1870; chief of 8 govt. eclipse expdns., 1870-1905; Rede lectr. Cambridge, 1871, then in science and art dept., 1875; prof. Royal Coll. Science, 1881; dir. Solar Physics Obs., South Kensington, 1885-1913; dir. Hill Obs., Salcombe Regis, 1913-20. Fellow Royal Soc., 1869; mem. French Acad. Sci., 1873; pres. Brit. Assn. for Advancement of Science, 1903-04. Author: Elementary Lessons in Astronomy, 1870; Contributions to Solar Physics, 1873; Spectroscope and its Applications, 1873; Primer of Astronomy, 1874; Studies in Spectrum Analysis, 1878; Star-gazing, Past and Present, 1878; Chemistry of Sun, 1887; Movements of the Earth, 1887; Meteoritic Hypothesis,

1890; Dawn of Astronomy, 1894; The Sun's Place in Nature, 1897; Recent and Coming Eclipses, 1897; Inorganic Evolution, 1900; Stonehenge, and other British Stone Monuments Astronomically Considered; Education and National Progress, 1906-07; Surveying for Archeologists, 1909; Tennyson as a Student and Poet of Nature, 1910. Pioneer in spectroscopic study of sun; discovered (with Frankland) helium in sun's atmosphere, 1868; first to show that spectrum is also indicator of a substance's temperature; anticipated modern views on course of stellar evolution; founder-editor Nature jour., 1869. Died Salcombe Regis, Devon, England, Aug. 16, 1920.

LOCQUIN, Emile René, French chemist; b. Nevers, Oct. 22, 1876; s. Victor and Andrea (Bernot) L.; ed. faculties sci. of Lyons, Lille, Nancy, Paris (all France); Ph.D.; m. Germaine Freylon, Aug. 2, 1910; 1 son, Jacques. Asst. in chemistry, Sorbonne, Paris; instr., then prof. Faculty Sci., Lyons. Mem. French Acad. Scis. Author: Traité de chimie organique; also articles. Address: 283, rue de Créqui, Lyons (Rhône), France.

LOCY, William Albert, Am. zoologist; b. Troy, Mich., Sept. 14, 1857; s. Lorenzo D. and Sarah (Kingsbury) L.; B.Sc., U. Mich., 1881, M.Sc., 1884, post-grad., 1881-82, D.Sc. (hon.), 1906; fellow in biology, Harvard, 1884-85, U. Berlin, 1891; Ph.D., U. Chgo., 1895; m. Ellen Eastman, June 26, 1883. Prof. biology Lake Forest (Ill.) U., 1887-89, animal morphology, 1889-96; prof. physiology Rush Med. Coll., Chgo., 1891; prof. zoology Northwestern U., 1896—. Zool. Sta., Naples, 1902-03. Editor in charge of zoöl. articles, and author of several, in new Am. supplement to Ency. Brit. Author: Biology and Its Makers, 1908; The Main Currents of Zoölogy, 1918. Died Oct. 9, 1924.

LODENKÄMPER, Johannes Heinrich Hans, German bacteriologist; b. Bottrop, Germany, Dec. 23, 1907; s. Johann and Berhardine Schulte (Nienhaus) L.; ed. U. Marburg, Fribourg, Munich, Vienna, Hamburg; M.D.; m. Kaethe Gerwien, Sept. 23, 1907; 1 dau., Ingrid. Became asst. Chem.-Physics Inst. Kaiser Wilhelm, 1934; with Hygiene Inst., U. Frankfort, 1934-35; 1st asst. U. Königsberg (now Kaliningrad, USSR) Inst. Hygiene, 1935-45; chief physician U. Inst. Hygiene and Bacteriology, Wuppertal-Ronsdorft, staff Allgemeines Krankenhaus Sankt Georg' Inst. Bacteriology and Serology, Hamburg. Mem. German Hygiene and Microbiology Assn. Author: Entwicklung und heutiger Stand der Lehre von der Pleomorphie und Zyklogenie der Bakterien, 1939. Research and publs. on hepatitis, cholangitis, anaerobic infections, hygiene, thyroid autoantibodies, microbiology of teeth, serology, oral immunization. Address: Duvenwischen 16, Hamburg-Volksdorf, W. Germany.

LODGE, James Piatt, Jr., Am. chemist; b. Decatur, Ill., Feb. 4, 1926; s. James P. and Grace (Carr) L.; B.S., U. Ill., 1947; Ph.D., U. Rochester, 1951; m. Nancy Pickering Myers, Sept. 4, 1948; children—Martha P., Judith T., Susan P., Elizabeth H., Eric P. Asst. prof. chemistry Keuka (N.Y.) Coll., 1950-52; research chemist Cloud Physics Lab., U. Chgo., 1952-55; chief chem. research and devel. sect. Lab. Engring. and Phys. Scis., Div. Air Pollution USPHS, Robert A. Taft San. Engring. Center, Cin., 1955-61; program scientist Nat. Center for Atmospheric Research, Boulder, Colo., 1961—; affiliate prof. chemistry La. State Coll. Cons. air and indsl. hygiene lab. Cal. State Bd. Health, 1963—; mem. environmental sci. and engring. study sect., div. research grants NIH, 1962-66; tech. adv. com. Regional Air Pollution Control Agy., Denver, 1964-66; adv. com. on weather control Colo. Gov.'s Adv. Com., 1963-66. Fellow A.A.A.S. (commn. on air conservation 1962-64); mem. Am. Chem. Soc. (com. on air pollution 1958—, chmn. div. water, air and waste chemistry 1967), Am. Geophys. Union (chmn. com. on atmospheric chemistry 1957—), Am. Meteorol. Soc., Air Pollution Control Assn., Electron Microscope Soc. Am., Sigma Xi, Phi Kappa Phi, Phi Eta Sigma. Editor for western hemisphere Internat. Jour. Air and Water Pollution, 1959—. Publs. on research on chemistry of very fine particles; analysis of trace gases; air pollution chemistry. Home: 801 Circle Dr. Office: Nat. Center for Atmospheric Research, Boulder, Colo. 80302.*

LODGE, Sir Oliver Joseph, Brit. physicist; b. Penkhull, Eng., June 12, 1851; s. Oliver and Grace (Heath) L.; D.Sc., U. Coll., London; D.Sc. (hon.), Oxford, Cambridge, Manchester, Liverpool, Sheffield, Leeds, Adelaide, Toronto; LL.D.. St. Andrews, Glasgow, Aberdeen, Edinburgh; m. Mary Marshall, 1877; 6 sons, 6 daus. Prof. physics U. Coll., Liverpool, Eng., 1881-1900; prin. U. Birmingham (Eng.), 1900-19; ret., 1919. Recipient Rumford medal, Royal Soc., 1898; Alfred medal, 1919. Became fellow Royal Soc., 1887. Author: Elementary Mechanics, 1877; Modern Views of Electricity, 1889; Signalling without Wires; Lightning Conductors and Lightning Guards; Life and Matter, 1905; Electrons; Modern Views of Matter; The Substance of Faith; Man and the Universe, 1908; The Ether of Space, 1909; The Survival of Man; Raymond or Life and Death; Christopher, a Study in Human Personality; Making of Man, 1924; Atoms and Rays, 1924; Ether and Reality, 1925; Relativity, 1925; Electrical Precipitation, 1925; Talks about Wireless, 1925; Evolution and Creation, 1927; Modern Scientific Ideas, 1927; Science and Human Progress, 1927; Beyond Physics, 1930; Advancing Science, 1931; My Philosophy, Containing Final Views on the Ether of Space, 1933. Research in ultra-short waves, lightning, electromagnetic radiation, Hertzian waves, thermal conductivity, induction balance of D. E. Hughes; showed moving matter does not exert a drag on ether; attempted to reconcile sci. and religion. Died Lake, Eng., Aug. 22, 1940.

LODWICK, Gwilym Savage, radiologist; b. Mystic, Ia., Aug. 30, 1917; s. Gwilym Savage and Lucy A. (Fuller) L.; student Drake U., 1934-35; B.A., U. Ia., 1942, M.D., 1943; m. Patricia Ann Galligan, Malcolm Kerr, Terry Ann. Fellow in radiol. pathology Armed Forces Inst. Pathology, 1951; clin. asst. prof. radiology U. Ia., 1952-55; chief radiol. service VA Hosp., Iowa City, 1952-55; asso. prof. U. Ia. Coll. Medicine, 1955-56; prof. radiology Sch. Medicine, chmn. dept. Univ. Med. Center, Columbia, Mo., 1956—, acting dean, 1959, asso. dean Sch. Medicine, 1959-64. Cons. Ellis Fischel State Cancer Hosp., 1959—, Mo. Div. Mental Diseases, 1965—. Recipient Bronze medal Am. Roentgen Ray Soc. Fellow Am. Coll. Radiology; mem. Am. Registry Radiologic Technologists (trustee 1961—, pres. 1964-65), Radiologic Soc. N.Am. (Magna Cum Laude award for basic research 1964), Assn. U. Radiologists, A.A.A.S., A.M.A., Sigma Xi, Alpha Omega Alpha. Guest editor: Radiology and Skeletal Systems, Radiologic Clinics of North America, 1964. Contbr. articles to tech. jours. Developed methods to convert radiographic images into digital data for computer aided med. diagnosis; research on computer-mediated method for reporting and storing radiol. consultations and radiol. systems analysis, diagnosis bone diseases, prognostic significance radiographs bone tumors. Home: 208 W. Ridgeley Rd., Columbia, Mo. 65202.*

LOEB, Carl M., Jr., Am. metallurgist; b. St. Louis, Aug. 10, 1904; s. Carl M. and Adeline (Moses) L.; B.S., Princeton, 1926; M.S., Mass. Inst. Tech., 1928; m. Lucille H.S. Schamberg, Jan. 30, 1929; children—Constance (Mrs. George L. Cohn), Carl M. III, Peter K. Metallurgist, Central Alloy Steel Corp., Canton, O., 1928-29, Anaconda Copper Mining Co. (Mont.), 1929-32; metallurgist Am. Metal Climax, Inc., Detroit, 1932, dist. metallurgist, N.Y.C., 1933-37, v.p. in charge devel., 1937-54, dir., 1948—, mem. exec. com., 1960—; chmn. bd. Am. Thermocatalytic Corp., 1962—; dir. Kawecki Chem. Co., Alloys Research & Mfg. Co.; ltd. partner Carl M. Loeb, Rhoades & Co., 1955—. Author: (with others) Molybdenum, Steels, Irons, Alloys. Research, publs. on devel. cast forgeable molybdenum and use of molybdenum sulphide as a lubricant. Home: Whippoorwill Rd., Armonk, N.Y. 10504. Office: 510 E. 86th St., N.Y.C. 10028.*

LOEB, Jacques, biologist; b. Mayen, Germany, Apr. 7, 1859; s. Benedict and Barbara (Isay) L.; grad. Ascanisches Gymnasium, Berlin; studied medicine Berlin, Munich, Strassburg; M.D., Strassburg, 1884; state exam., Strassburg, 1885; asst. in physiology, U. of Würzburg, 1886-88, U. of Strassburg, 1888-90; biol. sta., Naples, 1889-91; (hon. D.Sc., Cambridge, Eng., 1909; M.D., Geneva, 1909; Ph.D., Leipzig, 1909); m. Anne L. Leonard, 1891. Asso. in biology, Bryn Mawr, 1891-92; asst. prof. physiology and exptl. biology, 1892-95, asso. prof., 1895-1900, prof., 1900-02, U. of Chicago; prof. physiology, U. of Calif., 1902-10; head div. gen. physiology, Rockefeller Inst. for Med. Research, 1910—. Fellow Am. Acad. Arts and Sciences; corr. mem. French Acad. Sci., 1914. Author: The Heliotropism of Animals and Its Identity with the Heliotropism of Plants, Würzburg, 1890; Physiological Morphology, I, 1891, II, 1892; Comparative Physiology of the Brain and Comparative Psychology, 1900; Studies in General Physiology, 1905; Dynamics of Living Matter, 1906; The Mechanistic Conception of Life, 1912; Artificial Parthenogenesis and Fertilization, 1913; The Organism as a Whole, 1916; Forced Movements, Tropisms and Animal Conduct, 1918; Proteins and the Theory of Colloidal Behavior, 1922; Regeneration, 1924. Leading mechanist in U. S.; tried to demonstrate that life phenomena can be reduced to physical chemical laws; applied theory of tropism in plants to simple animals; suggested human values might be produced by tropisms; noted for 1899 discovery of artificial fertilization of eggs and raising frogs so fertilized to maturity, 1908. Died Hamilton, Bermuda, Feb. 11, 1924.

LOEB, Leo, pathologist; b. Germany, Sept. 21, 1869; s. Benedict and Barbara (Isay) L.; studied natural sci. and medicine, univs. of Heidelberg, Berlin, Freiburg, Zurich, 1889-96; Sc.D. (hon.), Washington U., 1948; m. Georgiana Sands, Jan. 3, 1922. Research fellow McGill U., 1903; asst. prof. exptl. pathology U. Pa., 1904-10; dir. dept. pathology Bernard Skin and Cancer Hosp., St. Louis, 1910-15; prof. comparative pathology Washington U., 1915-24, prof. pathology, 1924-37, emeritus since 1937. Recipient John Phillips Meml. prize A.C.P., 1935. Fellow A.A.A.S.; mem. Am. Assn. Pathologists and Bacteriologists (pres. 1914-15), Am. Physiol. Soc., Soc. Cancer Research (pres. 1911), Internat. Assn. for Cancer Research (v.p.), Soc. Exptl. Medicine and Biology, Assn. Am. Physicians, Am. Philos. Soc., A.M.A., Washington Acad. Sci., Am. Assn. for Study of Goitre, Nat. Acad. Scis., French Soc. of Endocrinology (hon.). Author: The Venom of Heloderma (with collaborators), 1913; Edema, 1923; The Biological Basis of Individuality, 1945. Contbr. chiefly on tissue and tumor growth, tissue culture, psychology of generative organs, pathology of circulation, venom of Heloderma, analysis of exptl. ameobocyte tissue, internal secretions, biol. basis of individuality; credited with being 1st to culture cells artificially, 1898; produced by mech. stimulation decidual tissue in uterus of mice, 1907; studied relation of sex hormones to cancer; suggested intravenous use of copper colloidal solution in treatment of cancer, 1912; showed that oophorectocy reduces incidence of hereditary mammary cancer in mice, 1919. Died Dec. 28, 1959.

LOEB, Leonard Benedict, physicist; b. Zurich, Switzerland, Sept. 16, 1891; s. Jacques and Anne (Leonard) L.; came to U. S., 1891, naturalized, 1905; student U. Cal. at Berkeley, 1908-10, Columbia, 1910-11; B.S. in Chemistry, U. Chgo., 1912, Ph.D. in Physics, 1916; m. Lora Lane, June 9, 1928 (div. Nov. 1940); children—Anne Leonard (Mrs. Glen Bredon), Jacqueline Lora (Mrs. Paul Conner); m. 2d, Charlotte Pearson, May 9, 1941; children—Diana Benedict, Valerie Jean. Asst. physicist U. Chgo., 1914-16, NRC fellow in physics, 1919-23; lab. asst., asst. physicist U. S. Bur. Standards, 1916-17; asst. U. Manchester (Eng.), 1919; faculty physics U. Cal. at Berkeley, 1923—, prof., 1929-59, prof. emeritus, 1959—. Tech. cons. USN, Office Naval Research, U. S. Bur. Standards, USAF, Gen. Electric Co., Schenectady, Carpco Research and Engring., Inc., Jacksonville, Fla., Callery Chem. Co. (Pa.), Sandia Corp., Albuquerque; chmn. No. Cal. Selective Service Adv. Com. on Sci., Engring. and Specialized Personnel, 1954—; Kelvin lectr. Inst. Elec. Engring., London, Eng., 1947; mem. honor praesidium 5th Internat. Conf. on Ionization Phenomena in Gases, Munich, Germany, 1961. Named Useful Citizen, U. Chgo., 1944. Fellow Am. Phys. Soc. (sec. West Coast 1929-35), Phys. Soc. London; mem. German, French phys. socs., Phys. Soc. Japan, Geophys. Union, U. S. Naval Inst., Phi Beta Kappa, Sigma Xi. Author: Kinetic Theory of Gases, 1927, 34; Nature of a Gas, 1931; Fundamentals of Electricity and Magnetism, 1931, 38, 47; (with A. S. Adams) Development of Physical Thought, 1933; Atomic Structure, 1938; Fundamental Processes of Electrical Discharge in Gases, 1939; (with J. M. Meek) Mechanism of the Electric Spark, 1941; Basic Processes of Gaseous Electronics, 1955, 60; Static Electrification, 1958; Electrical Coronas; Their Basic Physical Mechanisms, 1965; also numerous articles. Asso. editor Jour. Applied Physics, 1950-53, Jour. Geophys. Research, 1964—. Home: 2615 Etna St., Berkeley, Cal. 94704.*

LOEB, Michel, Am. psychologist; b. Montgomery, Ala., Oct. 21, 1926; s. Lester Michel and Elsie (Mayer) L.; B.S., U. Ala., 1948; A.M., Columbia, 1949; Ph.D., Vanderbilt U., 1953; m. Margaret Gordon Cameron, Apr. 11, 1953; children—William Arthur, Stephen Alexander, Richard Lester. Research psychologist U. S. Army Med. Research Lab., Fr. Knox, Ky., 1952-56, USAF Personnel and Tng. Research Center, Tyndall AFB, Fla., 1956-57, semi-automatic ground environment operator Research Unit, USAF Personnel and Tng. Research Unit, Hanscom Field, Bedford, Mass., 1957-58; chief audition br. U. S. Army Med. Research Lab., Ft. Knox, Ky., 1959-68; faculty U. Louisville, 1959—, professor of psychology, 1968—. Mem. Am., Midwestern, psychol. assns., Acoustical Soc. Am., Psychonomic Soc., So. Soc. for Philosophy and Psychology, A.A.A.S. Research, publs. in psychoacoustics and vigilance. Home: 3601 Whitehall Ct., Valley Station, Ky. 40272.*

LOEB, Morris, Am. chemist; b. Cin., May 23, 1863; s. Solomon and Betty (Gallenberg) L.; A.B., Harvard, 1883; Ph.D., U. Berlin, 1887; student U. Heidelberg, 1887-88, U. Leipzig, 1888; Sc.D., Union U., 1911; m. Eda Kuhn, Apr. 3, 1895. Asst. to Prof. Wolcott Gibbs, Newport, R.I., 1888-89; docent Clark U., 1889-91; prof. chemistry N.Y. U., 1891—, dir. chem. lab., 1895-1906. Fellow N.Y. Acad. Scis., A.A.A.S.; mem. Am. Chem. Soc., N.Y. Elec. Soc., Am. Electrochem. Soc., Soc. Chem. Industry, German Chem. Soc. Contbr. articles on organic and phys. chemistry. Died Oct. 8, 1912.

LOEB, Robert Frederick, Am. physician; b. Chgo., Mar. 14, 1895; s. Jacques and Anne (Leonard) L.; student U. Chgo., 1913-15, Sc.D., 1951; M.D., Harvard, 1919; Docteur Honoris Causa, U. Strasbourg (France), 1951, U. Paris (France), 1952; LL.D., U. Wales, 1953; Sc.D., N.Y. U., 1955, Kenyon Coll., 1957, Oxford (Eng.) U., 1961; Columbia, 1961, Dartmouth, 1962, Trinity Coll., U. Dublin, 1962; LL.D., Amherst Coll., 1961; Bard prof. medicine Columbia, 1947-59, Bard prof. emeritus, 1959—; dir. med. service Presbyn. Hosp., N.Y.C., 1947-59, cons., 1959—; vis. prof. pro-tem Peter Bent Brigham Hosp., 1941; James Howard Means vis. physician Mass. Gen. Hosp., 1959; vis. Regius prof. Oxford (Eng.) U., 1961; vis. prof. medicine Cornell U., 1967. Mem. Pres.'s Sci. Adv. Com., 1950-53, 1959-62; mem. Nat. Sci. Bd., 1950-64; mem. adv. com. for biology and medicine U. S. AEC, 1958-64. Decorated Order of Brilliant Star with Grand Cordon (Republic of China); recipient Stevens Triennial prize Columbia, 1926; John and Samuel Bard award Bard Coll., 1952; 150th Anniversary award Mass. Gen. Hosp., 1960. Hon. fellow Royal Coll. Physicians (London), Royal Acad. Medicine Belgium, Royal Australian Coll. Physicians; mem. Nat.

Acad. Scis., Am. Acad. Arts and Scis., Am. Philos. Soc., Assn. Am. Physicians (past pres., Kober medal 1959), Am. Soc. for Clin. Investigations (past pres.), A.C.P. (master), Harvey Soc. (past pres.), Assn. Physicians Gt. Britain and Ireland (overseas hon. mem.), Brit. Med. Assn. (corr.), Société Médicale des Hopitaux de Paris (hon.), Norwegian Med. Soc., Danish Soc. Internal Medicine. Co-editor: Cecil-Loeb Textbook of Medicine, 1947; Martini's Principles and Practice of Physical Diagnosis. Research, numerous publs. on electrolyte physiology in health and disease states, control of renal sodium and potassium excretion by adrenal glands. Home: 950 Park Av., N.Y.C. 10028.*

LOEBLICH, Helen Nina Tappan, Am. paleontologist; b. Norman, Okla., Oct. 12, 1917; d. Frank Girard and Mary (Jenks) Tappan; B.S., U. Okla., 1937, M.S., 1939; Ph.D., U. Chgo., 1942; m. Alfred Richard Loeblich Jr., June 18, 1939; children—Alfred Richard III, Karen (Mrs. Jere Lipps), Judith Anne (Mrs. David Covey), Daryl. Instr. geology Tulane U., 1942-43; geologist U. S. Geol. Survey, Washington, 1943-45, 1947-57, Fullerton, Cal., 1957—; lectr. geology U. Cal. at Los Angeles, 1958-65, asso. research geologist, 1961-64, sr. lectr. geology, 1965-66, prof. geology, 1966—; research asso. Smithsonian Instn., Washington, 1954-57. John Simon Guggenheim Found. fellow, 1953-54. Mem. Geol. Soc. Am., Paleontol. Soc., Soc. Protozoologists, Phycological Soc. Am., Nat. Assn. Geology Tchrs., Phi Beta Kappa, Sigma Xi. Author: (with A. R. Loeblich Jr.) Treatise on Invertebrate Paleontology, part C, Protista 2, Foraminiferida, 2 vols., 1964. Research, publs. on morphology, distbn., systematics, nomenclature of foraminifers, thecamoebians, silicoflagellates, ebridians, tintinnids, coccolithophorids, radiolarians, chitinozoa, Cretaceous-Tertiary boundary, Cretaceous of Gulf Coast, Mesozoic of N. Alaska; Jurassic of S.D.; reclassification of foraminifera, fossil phytoplankton and effect on atmospheric changes of phytoplankton abundance. Home: 658 Lemon Hill Terrace, Fullerton, Cal. 92632. Office: Dept. Geology, U. Cal., Los Angeles 90024.*

LOEFER, John B., Am. biologist; b. Forest Junction, Wis., June 14, 1908; s. John and Amelia (Schubring) L.; A.B., Lawrence Coll., 1929; M.S., N.Y. U., 1931, Ph.D. in Protozoology, 1933; m. Ruth Orth, Sept. 3, 1934; children—Naomi Ruth (Mrs. Raymond L. Owens), Jeffrey J. Faculty, Berea Coll., 1935-43, Coll. City N.Y., 1935, 46, Bklyn. Coll., 1946, Trinity U., 1950-53; sr. research biologist, head, dept. exptl. biology S.W. Found. for Research and Edn., San Antonio, 1946-53; coordinator biol. scis. Office Naval Research, Pasadena, Cal., 1953—. Research fellow Cal. Inst. Tech., 1954-63; research zoologist U. Cal. at Los Angeles, 1953-55, research asso., 1955—. Fellow A.A.A.S., mem. Am. Physiol. Soc., Am. Soc. Zoologists, Soc. Exptl. Biology and Medicine, Soc. Protozoologists, Sci. Research Soc. Am., Sigma Xi. Publs., basic research on morphology, physiology, nutrition, and serology of free-living protozoa. Home: 133 W. Terrace St., Altadena, Cal. 91001. Office: 1030 E. Green St., Pasadena, Cal. 91101.*

LOEFFLER, Frank Joseph, Am. physicist, educator; b. Ballston Spa, N.Y., Sept. 5, 1928; s. Frank Joseph and Florence (Farrell) L.; B.S. in Engring. Physics Cornell U., 1951, Ph.D., 1957; m. Eleanor Jane Chisholm, Sept. 8, 1951; children—Peter, James, Margaret, Anne Marie. Research asso. Princeton, 1957-58; faculty Purdue U. Lafayette, Ind., 1958—, professor, 1962-66; vis. prof. Hamburg U. (Germany), 1963-64. Mem. Am. Phys. Soc., Tau Beta Pi, Sigma Xi. Publs., exptl. research in high energy particle interactions using both counter-spark chamber and bubble chamber techniques, prodn. resonant state from strong interactions and on-line data acquisition-processing systems. Home: 437 Maple St., West Lafayette, Ind. 47906.*

LOEPER, Maurice René, French physician; b. Paris, France, Dec. 27, 1875; s. Paul and Marthe (Considère) L.; M.D., Faculty Medicine; m. Marie Courtes-Lapeyrat. Became prof. therapeutics, Paris, 1927; mem. staff Paris hosps. Mem. Acad. Medicine (pres. 1953), Internat. Union Therapeutics (pres.). Research and publs. on pathology of digestion, nutrition, therapeutic medicine; introduced notion of chem. specificity in biology. Address: 39, rue Saint-Dominique, Paris 7, France.

LOESER, Lewis Henry, Am. physician; b. Montclair, N.J., Aug. 25, 1903; s. David and Estelle (Cohn) L.; student Tufts Coll.; M.D., 1927; m. Rhoda Sophie Levy, Aug. 3, 1930; children—John Davis, Individual practice neuro-psychiatry, Irvington, N.J., 1930—; asso. clin. prof. psychiatry N.Y. Med. Coll. 1950—; attending neuro-psychiatrist various hosps., Essex County, N.J. Chmn. med. bd. Nat. Multiple Sclerosis Soc. N.J. br., 1948—; med. dir. N.J. Assn. for Retarded Children, 1952—; sr. cons. neuro-psychiatry VA, 1946—; bd. mgrs. N.J. Diagnostic Center, Menlo Park, 1949-62, pres., 1958-62. Fellow Acad. Neurology, Am. Psychiat. Assn.; mem. Am. Group Psychotherapy Assn. (pres. 1954), N.J. Neuropsychiat. Assn. (pres. 1951), Tau Epsilon Phi. Researcher, publs. on group psychotherapy to various publs. Home: Shady Glen, West Orange, N.J. 07052. Office: 22 Ball St., Irvington, N.J. 07111.*

LOEVE, Michel, mathematician; b. Jan. 22, 1907; Docteur ès-sciences mathématiques, Sorbonne, Paris, France, 1941; Actuaire, I.S.F.A., Lyon, France, 1936. Prof. math. U. Cal. at Berkeley, 1948—. Author: Probability Theory, 1955; also articles. Research in probability theory.*

LOEW, Earl Randall, Am. physiologist; b. nr. Allegan, Mich., Nov. 17, 1907; s. Amos W. and Nora (Bond) L.; B.S., Mich. State U., 1929; M.S., Wayne State U., 1936; Ph.D. (Porter fellow), Northwestern U. Med. Sch., 1939; m. Reva M. Moored, June 12, 1934; children—Donald E., James B., Jane C., Betty J. Faculty, Wayne State Med. Coll., Detroit, 1929-41; sr. pharmacologist Parke Davis Co., Detroit, 1941-44; faculty U. Ill. Coll. Medicine, Chgo., 1944-48; prof., head, dept. physiology Boston U. Sch. Medicine, 1948—. Mem. research career award com. NIH, 1958-63. Mem. A.A.A.S., Am. Acad. Arts and Sci., Physiol. Soc., Pharmacological Soc., Soc. Exptl. Biology and Medicine (editorial bd. Proc. 1948-58). Research in histamine, antihistamine drugs; autonomic and adrenergic blocking drugs; gastric secretion and motility. Home: 60 Hull St., Newtonville, Mass. 02160. Office: 80 E. Concord St., Boston 02118.*

LOEWE, Lotte Luise Friederike, German chemist; b. Breslau, Germany, Nov. 7, 1900; d. Eugen and Helene (Druey) Loewe; Dr. Phil., U. Breslau, 1927. Asst., Chem. Inst., U. Breslau, 1927-33; sci. collaborator Chem. Inst., U. Zurich (Switzerland), 1934, U. Istanbul (Turkey), 1934-55, U. Basel (Switzerland), 1955-61, J. R. Geigy AG, Basel, from 1961; pvt. dozent chemistry U. Frieburg i. Br. (Germany), from 1956. Recipient Bundesverdienstkreuz, 1955. Mem. German, Swiss chem. socs., German, Swiss Acad. unions, Swiss Microanalytic Soc. Co-research, articles in field of organic chemistry, including work on uric acid, carotinoids, keto-enol-tautomery, diazomethane reactions, kinetics of ascorbic acid. Home: 10 Sommergasse, Basel. Office: J. R. Geigy AG, Basel, Switzerland.*

LOEWE, Walter Siegfried, pharmacologist; b. Fuerth, Bavaria, Aug. 19, 1884; ed. univs. Freiburg, Berlin, Munich, Strasbourg; M.D., 1908; m., 1919; 2 children. Head chem. lab., psychiat. clinic Leipzig, Germany, 1910-12; faculty Göttingen, 1912-21, asso. prof. pharmacology, 1918-21; prof. pharmacology, head dept. pharmacology and med. chemistry Dorpat, 1921-28; dir. lab. Municipal Hosps., Mannheim, 1928-34; mem. staff, lab. med. div. Mt. Sinai Hosp., 1934-36; research guest, dept. pharmacology Cornell Med. Coll., 1936-46; research prof. pharmacology U. Utah Coll. Medicine, Salt Lake City, 1946—. Prof. Heidelberg, 1928-33, hon. prof., 1929; staff Montefiore Hosp., 1936-43. Mem. Am. Acad., Pharmacological Soc., Soc. Exptl. Biology, German Pharmacological Soc. (hon.), German Soc. Endocrinology (hon). Research on lipids, narcosis, bioassay, quantitative pharmacology, sex hormones, drug synergism, laxatives, marihuana, anticonvulsants, antihistaminics, sympatholytics, sympathoganglionic stimulants; developed Anthallan for use in cases of hay fever, other noninfectious nasal disorders, skin diseases. Address: U. Utah Coll. Medicine, Salt Lake City.

LOEWENBERG, Benjamin Benno, German surgeon; b. Sonnenburg, Germany, 1836; studied in Berlin, Bonn (both Germany); discovered bacillus in nasal secretion of patient with ozena, 1884; later wrote on its pathology.

LOEWENSTEIN, Rudolph M(aurice), psychoanalyst, author; b. Lodz, Poland, Jan. 17, 1898; s. Maurice and Charlotte (Taube) L.; M.D., U. Berlin, 1923, U. Paris, 1935; m. Elisabeth R. Geleerd; children—Dominique-Thérèse, Elisabeth-Charlotte, Marie-Françoise, Richard Joseph. Came to U. S., 1942, naturalized, 1948. Asst. Berlin (Germany) Psychoanalytic Inst., 1923-25; asst. to Prof. Henri Claude, outpatient dept., clinic of psychiatry, med. faculty, U. Paris, 1925-39; tng. analyst, lectr. Paris (France) Psychoanalytic Inst., 1925-40; tng. analyst, lectr. N.Y. Psychoanalytic Inst., 1944—, v.p., 1948-50, pres., 1950-52; asso. clin. prof. psychiatry, med. sch., Yale, 1948-52; pvt. practice psychoanalysis, N.Y.C., 1943—. Decorated: Croix de Guerre (France). Mem. A.M.A., Am. (pres. 1957-58), Internat. (v.p. 1965-67), Paris (charter) psychoanalytical assns., Am. Psychiat. Assn., Am. Psychiat. Assn., N.Y. Soc. Clin. Psychiatry, A.A.A.S., Conf. on Jewish Social Studies. Author: Christians and Jews: A Psychoanalytic Study, 1951; Editor: Drives, Affects, Behavior, 1953. Research and publs. since 1923 on topics of psychoanalytic theory and clinic, theory of psychoanalytic technique, psychoanalytic ego psychology, application of psychoanalytic principles to social problems. Address: 1100 Madison Av., N.Y.C. 10028.*

LOEWI, Otto, pharmacologist; b. Frankfort-on-Main, Germany, June 3, 1873; s. Jacob and Anna (Willstatter) L.; student Us. Strassbourg and Munich, 1891-96, M.D., Strassbourg, 1896; Sc.D. (hon.), N.Y. U., 1944, Yale, 1951; Ph.D. (hon.), Graz, 1950; M.D. (honorary), univs. Graz, Frankfurt, 1953; m. Gulda Goldschmiedt, Apr. 5, 1908; children—Harold, Anna (Mrs. Ulrich Weiss), Victor, Geoffrey W. Asst. pharmacology U. of Marburg, 1897, privat-dozent, 1900-05; asso. prof. pharmacology U. Vienna, 1905-09; prof. pharmacology, Univ. at Graz, Austria, 1909-38; came to U. S., 1940, and since research prof. pharmacology, New York U. Coll. of Medicine; Dunham lecturer, Harvard, 1933; Harvey lecturer, New York, 1933; Ferrier lecturer, Royal Soc. London, 1935; Franqui Professor, Brussels, 1938-39; Dohme lecturer, Johns Hopkins, 1941; Walker-Ames visiting prof., U. Washington, 1942. Janeway lecturer, N.Y. 1945; Hughlings Jackson Memorial lecturer, McGill U., 1946; vis. scholar, U. of Va., 1948. Recipient (with Sir Henry Dale) Nobel prize in physiology and medicine, 1936; Cameron prize, Edinburgh, 1944; Austrian distinguished order for art and science, 1936. Hon. fellow Royal Soc. Edinburgh, N.Y. Acad. Medicine, fellow Academia dei Lincei Rome; fgn. mem. Royal Soc., 1954; corr. mem. Bavarian Acad. Sci.; hon. mem. Am. Pharmacological Soc., Physiol. Soc. London, Societa di Biologia (Rome), Royal Soc. of Physicians (Budapest); asso. mem. Société de Biologie (Paris and Brussels). Research on physiology of metabolism of kidney, heart, hormones, nervous system, and on chem. transmission of nervous impulse. Died N.Y.C., Dec. 25, 1961.

LOEWINSON-LESSING, Franz Julievitch, Russian geologist; b. St. Petersburg (now Leningrad), Russia, Mar. 10, 1861. Founder, Petrographical Inst., Acad. Arts and Scis. USSR. Wrote petrographic textbooks; also chem. classification eruptive rocks. Studied petrology and solcanism. Died Oct. 25, 1939.

LOEWUS, Frank Abel, Am. biochemist; b. Duluth, Minn., Oct. 22, 1919; s. David G. and Alyce (Abel) L.; B.S., U. Minn., 1942, M.S., 1950, Ph.D., 1952; m. Mary E. Walz, Dec. 26, 1947; children—Rebecca, David, Daniel. Research asst. U. Minn., St. Paul, 1947-51; research asso. biochemistry U. Chgo., 1952-55; chemist Western Regional Research Lab. U. S. Dept. Agr., Albany, Cal., 1955-64; prof. biology State U. N.Y. at Buffalo, 1964—. Studies on mechanisms of enzyme action, biosynthesis of ascorbic acid, plant cell wall carbohydrates, and related compounds found in plant tissues. Home: 35 Woodhaven Rd., Amherst, N.Y. 14226.*

LOEWY, Ariel Gideon, biochemist; b. Bucharest, Rumania, Mar. 12, 1925; s. Frederick and Eva Andree (Ellman) L.; B.S., McGill U., 1945, M.S., 1946; Ph.D., U. Pa., 1951; m. Karin Elisabeth Rademacher, Mar. 3, 1951; children—Michael, Andreas Frederick, Eva Susanna, Daniel Rademacher. Faculty dept. biology Haverford (Pa.) Coll., 1953—, chmn. dept., 1954—, prof., 1963—. Mem. exec. com. Commn. on Undergrad. Edn. in Biol. Scis.; mem. biology editorial bd. Holt, Rinehart & Winston, Inc. Mem. Am. Soc. Biol. Chemists, Am. Soc. Cell Biology, Sigma Xi. Author: (with Philip Siekevitz) Cell Structure and Function, 1963. Research in molecular mechanism of protoplasmic streaming, insoluble fibrin formation, muscle contraction. Home: 2 College Lane, Haverford, Pa. 19041.*

LÖF, George Oscar Gage, Am. chem. engr.; b. Aspen, Colo., Dec. 13, 1913; s. Anders J. O. and Zella (Cole) L.; B.S., U. Denver, 1935; Sc.D., Mass. Inst. Tech., 1940; m. Laura Davadell Scobey, Sept. 15, 1940; children—Larry O. D., Laura M., Lance G. A., Linnea D. Mem. faculty U. Colo., 1940-47; cons., tech. dir. Denver chem. plant, Colo. Fuel & Iron Corp., 1947-48; prof., chmn. chem. engring. dept., dir. Inst. Indsl. Research U. Denver, 1948-52; cons. chem. engr., Denver, 1952—; research asso. U. Wis., 1955-67; acting dir. Solar Lab., 1964-65; research asso. Resources for Future Inc., Washington, 1955—; prof. civil engring. dept. Colo. State U., 1967—; guest scientist Acad. Sci. USSR, 1964; mem. Com. Natural Resource Conservation and Devel., 1965. Pres. Umbroiler Co., 1957—. Registered profl. engr., Colo. Mem. Am. Chem. Soc., Am. Inst. Chem. Engrs. (chmn. nat. meeting 1953), Solar Energy Soc. (dir.), Am. Soc. Heating, Refrigeration and Air-Conditioning Engrs., Sigma Xi, Tau Beta Pi, Phi Lambda Upsilon. Co-author: Technology in American Water Development, 1959; (with another) Water Demand for Steam Electric Generation, 1966. Research and publs. on devel. of tech. and apparatus for the practical utilization of solar energy; developed solar house-heating system; tech. and econ. aspects of water use and its supply. Patentee solar cooker. Home: 6 Parkway Dr., Englewood, Colo. 80110. Office: Farmers Union Bldg., Denver 80203.*

LO FENG-LUH, Sir Chih Chen, Chinese mathematician; b. 1850; s. Lo Shao Tsung; ed. Imperial Naval Coll., Pagoda Anchorage, River Min; grad. Foochow Coll., 1872; m. Ouei, (dec. 1899); m. 2d, Kiping Ouei. Became attaché to 1st permanent Chinese Legation, London, Eng., 1877; transferred to Berlin, Germany, 1879-81; apptd. sec. to Viceroy Earl Li Hung Chang, 1882; made indsl. tour of Eng. and Scotland, 1899-1900; minister in Eng., to 1901; E.E. and M.P. of Emperor of China in Russia, 1901-02. Author: Problems in Nautical Astronomy and Navigation; Solutions of Problems by Indeterminate Equations. Died June 10, 1903.

LÖFFLER, Friedrich August Johannes, German bacteriologist; b. Frankfort/Oder, Prussia, June 24, 1852; student Wurzburg (Germany) U.; M.D., Friedrich Wilhelm U., Berlin. Mil. physician; prof. Greifswald; mem. staff Friedrich Wilhelm Inst., Berlin, 1884; worked with Robert Koch, 1879-84; dir. Robert Koch Inst., Berlin, from 1913. Author: Vorlesungen über der geschichte entwichlung der Lehre von den Bacterien, Vol. 1, 1887; Untersuchungen über die Bedeutung der Mikroorganismen für die enstehung der Diphtherie, 1884. Discovered bacillus of glanders and of swine erysipelas, 1882; isolated and made culture of Klebs-Loffler bacillus (organism causing diphtheria), 1884; introduced methylene blue as staining agt.; showed (with Paul Frosch) filterable virus caused foot-and-mouth disease, 1898, also developed protective serum against the disease. Died Berlin, Apr. 9, 1915.

LOFGREN, Clifford Swanson, Am. entomologist; b. St. James, Minn., July 29, 1925; s. Olaf P. and Lillie (Swanson) L.; B.A., Gustavus Adolphus Coll., 1950; M.S., U. Minn., 1954; Ph.D., U. Fla., 1967; m. Renee K. Liljenberg, Apr. 17, 1954; children—Richard O., Judith R., Cynthia J. Agriculturist, Plant Pest Control div. Agrl. Research Service, U. S. Dept. Agr., Hicksville, N.Y., 1954-55, entomologist Entomology Research div., Orlando, Fla., 1955-57, Plant Pest Control div., Gulfport, Miss., 1957-63, Entomology Research div., Gainesville, Fla., 1963—. Mem. Entomology Soc. Am., Am. Mosquito Control Assn., Fla. Entomol. Assn., Fla. Anti-Mosquito Assn. Contbr. numerous articles to sci. jours. Devel. methods for control of anthropods of med. importance with emphasis on insecticides, resistance studies, biocontrol; discovered effective bait for control of imported fire ant. Home: 1321 N.W. 31st Dr. Office: 1600 S.W. 23d Dr., Gainesville, Fla. 32601.*

LOFGREN, Edward Joseph, Am. physicist; b. Chgo., Jan. 18, 1914; s. Joseph E. and Charlotte (Bladholm) L.; A.B., U. Cal. at Berkeley, 1938, Ph.D., 1946; m. Lenore Nadaner, Oct. 1, 1938; children—Helen, Laurel, Claire. Physicist, Lawrence Radiation Lab., U. Cal. at Berkeley, 1940-44, 45-46, 48—, group leader in charge bevatron 1954—; group leader Los Alamos Sci. Lab., 1944-45; asst. prof. U. Minn., 1946-48. Fellow Am. Phys. Soc. Discovered (with co-workers) heavy particles in cosmic rays, 1948; high intensity ion source, 1954; devel. bevatron, 1954—; design studies of very high energy accelerators; research in electromagnetic isotope separation. Home: 990 Miller Av., Berkeley, Cal. 94708.*

LOFLAND, Hugh B., Jr., Am. biochemist; b. Rockwall, Tex., July 16, 1921; s. Hugh B. and Lydia (Benton) L.; B.S., Tex. A. and M. U., 1947, M.S., 1948; Ph.D., Purdue U., 1952; m. Norma Maie Handley, Aug. 21, 1942; children—Dee Anne (Mrs. Wilson Clark Lamb Jr.), Marian Lisa, Hugh Maxwell. Faculty, Bowman Gray Sch. Medicine, Winston-Salem, N.C., 1952—, professor pathology, dir. clin. chemistry labs., 1966—. Fellow Am. Soc. Study Arteriosclerosis; mem. Soc. Exptl. Biology and Medicine (sec.-treas. Southeastern sect. 1965-68), Am. Soc. Exptl. Pathology, Am. Inst. Nutrition, Am. Assn. Clin. Chemists, American Chem. Soc., Sigma Xi. Research and numerous publs. on fundamental mechanisms of diseases of arteries, especially metabolism of fatty substances in arterial wall; nutritional studies in man and exptl. animals relating to devel. aortic and coronary artery disease; basic biochemistry of fats; devel. analytical methods for use in diagnostic procedures. Home: 2820 Birchwood Dr., Winston-Salem, N.C. 27103.*

LOFTFIELD, Robert Berner, Am. biochemist; b. Detroit, Dec. 15, 1919; s. Sigurd and Katherine (Roller) L.; B.S., Harvard, 1941, M.A., 1942, Ph.D., 1946; m. Ella Bradford, Aug. 24, 1946; children—Lore, Eric, Linda, Norman, Bjorn, Curtis, Katherine, Earl, Allison, Ella-Kari. Faculty, Mass. Inst. Tech., 1946-48, Harvard Med. Sch. and Mass. Gen. Hosp., 1948-64; prof., chmn. dept. biochemistry U. N.M. Sch. Medicine, Albuquerque, 1964—. Mem. biochemistry study sect. USPHS, 1964—; mem. adv. com. on pathogenesis Am. Cancer Soc., 1964—. Damon Runyon fellow Karolinska Medicinska Inst., Stockholm, Sweden, 1952-53; Guggenheim fellow Med. Research Council, Cambridge, Eng., 1961-62; Mem. A.A.A.S., Am. Chem. Soc., Am. Assn. Cancer Research, Soc. Biol. Chemists, Marine Biol. Lab., Biophys. Soc. Research, publs. on organic reaction mechanisms, mechanism of protein biosynthesis. Home: 707 Fairway N.W., Albuquerque 81707.*

LOFTUS, Thomas Anthony, Am. psychiatrist; b. Mass., Aug. 4, 1913; s. Thomas Anthony and Helen Therese (Colligan) L.; A.B., LaSalle Coll., 1936; M.D., U. Pa., 1940; grad. Psychoanalytic Clinic for Research and Tng., Columbia, 1949; m. Mary M. Jackson, Nov. 20, 1943 (div. 1952); children—Thomas A., Robert J., m. 2d, Eleanor Ward Miranda, July 26, 1958; 1 son, Christopher M. Psychiatric studies N.Y. Hosp., Cornell Med. Sch., 1941; resident physician Payne Whitney Psychiat. Clinic, N.Y.C., 1941-44; asst. psychoanalyst Psychoanalytic Clinic Research and Tng., 1949-51; pvt. practice psychoanalytic psychiatry, 1946-57; staff St. Luke's Hosp., N.Y.C., 1955-57; faculty Jefferson Med. Coll., 1957-61, W.Va. Sch. Medicine, 1961-63; prof. psychiatry N.Y. Sch. Psychiatry, N.Y.C., 1963—, asst. dean, 1965—; cons. VA hospitals, Kings Bridge, New York, 1946-51, Montrose, New York, 1950-57, Clarksburg, W.Va., 1961-63. Diplomate Am. Bd. Psychiatry and Neurology, 1946. Mem. Am. Psychiatric Assn., Assn. Psychoanalytic Medicine, Acad. Psychoanalysis, A.M.A., Alpha Omega Alpha. Author: Diagnosis in Clinical Psychiatry, 1960. Research and publs. on psychoendocrinology, especially the relationship between emotion and reproductive behavior in human female. articles med. publs. Home: 55 E. 87th St., N.Y.C. 10028.*

LOGAN, Frank Anderson, Am. psychologist; b. Palatka, Fla., July 22, 1924; s. Frank Anderson and Bernice (Hilty) L.; B.A., State U. Ia., 1948, M.A., 1950, Ph.D., 1951; m. Julia Bingham Allen, July 1, 1948; children—Frank Anderson, III, Nancy Allen. Faculty, Yale, 1951-64; prof., chmn. dept. psychology U. N.M., Albuquerque, 1964—. Mem. study sect. NIH. Mem. Am., N.M., S.W. psychol. assns., A.A.A.S., Sigma Xi. Author: (with others) Behavior Theory and Social Science, 1955; Incentive, 1960; (with Wagner) Reward and Punishment, 1965; also articles. Research on behavior theory particularly relationship between learning and response speed. Home: 1300 Stagecoach Rd. S.E., Albuquerque.*

LOGAN, George Bryan, Am. physician; b. Pitts., Aug. 1, 1909; s. Arch Hodge and Amy (Dunlap) L.; B.S., Washington and Jefferson Coll., 1930; M.D., Harvard, 1934; M.S. in Pediatrics, U. Minn., 1940; m. Rhoda Kathryn Palmer, June 24, 1939; children—Kathryn P., Jane P. Instr. pediatrics U. Minn. Med. Sch., 1940; faculty Mayo Grad. Sch. Medicine, U. Minn., Rochester, 1940—, prof. pediatrics, 1962—; cons. sect. on pediatrics Mayo Clinic, 1940—. Fellow Am. Acad. Allergy (exec. com. 1968—), Am. Coll. Allergists; mem. A.A.A.S., A.M.A., Am. Acad. Pediatrics (Minn. chmn. 1958-61, dist. chmn., exec. bd. 1963-67, pres. 1967-68) Northwest Pediatric Soc. (pres. 1946; mem. sub-bd. pediatric allergy 1955-66, chmn. 1963-66), Am. Pediatric Soc., Am. Assn. Med. History, N.Y. Acad. Scis., Sigma Xi, Phi Gamma Delta, Nu Sigma Nu, others. Editorial bd. Am. Jour. Diseases Children, 1961—. Contbr. numerous articles to profl. jours. Home: 1115 Plummer Circle, Rochester, Minn. 55901.*

LOGAN, James, naturalist; b. Lurgan, County Armaugh, Ireland, Oct. 20, 1674; s. Patrick and Isabel (Hume) L.; m. Sarah Read, Dec. 9, 1714, 5 children. Came to Am., 1699; sec. to William Penn, 1699; sec. Province of Pa.; clk. Pa. Provincial Council, 1701-17, voting mem., 1702, pres. and sr. mem., chief exec. Province of Pa., 1736-38, commr. property and receiver-gen.; alderman Phila., 1717; mayor, 1722; justice Phila. County, 1726; judge Ct. Common Pleas, 1727; chief justice Pa. Supreme Ct., 1731-39; contbr. pvt. library of over 3,000 books to Phila. (forming basis of Phila. Public Library). Author: Experimenta et meletemata de plantarum generatione, 1738. Research in optics and on maize plants; astronomy, math. Died Stenton, Pa., Oct. 31, 1751.

LOGAN, John Henry, Am. physician; b. Abbeville, S.C., Nov. 5, 1822; s. John and Susan (Wilson) L.; grad. S.C. Coll., 1844, Charleston Med. Coll., circa 1849; m. Eliza Calhoun. Editor Banner (newspaper), Abbeville, circa 1845; practiced medicine, Greenwood, S.C., 1850's; served as surgeon Confederate Army during Civil War; became prin. Synodical Inst., Talladega, Ala., 1865; prof. chemistry Atlanta (Ga.) Med. Coll., from circa 1870; editor Atlanta Med. Jour. Author: Students' Manual of Chemico-Physics. Died Atlanta, Mar. 23, 1885.

LOGAN, Joseph Granville, Jr., Am. physicist; b. Washington, June 8, 1920; s. Joseph Granville and Lula (Briggs) L.; B.S., D.C. Tchrs. Coll., 1941; Ph.D. in Physics, U. Buffalo, 1955; m. Esther Taylor, June 30, 1944; children—Joseph Michael, Eileen Cecile. Physicist, Nat. Bur. Standards, 1944-47, Cornell Aero. Lab., 1947-58; head propulsion research dept. Space Tech. Labs., 1958-61; dir. aerodynamics and propulsion research lab. Aerospace Corp., 1961—. Mem. Am. Phys. Soc., Am. Inst. Aero. and Astronautics. Patentee in field of non-steady engines. Research and publs. on applications of shock tubers to hypersonic areodynamic; chem. kinetics; high temperature gas dynamics; re-entry aerodynamics. Home: 3652 Olympiad Dr., Los Angeles 90043. Office: Aerospace Corp., Los Angeles 90045.*

LOGAN, Richard Fink, Am. geographer; b. Great Barrington, Mass., June 1, 1914; s. James O. and Edith (Fink) L.; B.A., Clark U., 1936, M.A., 1937; M.A., Harvard, 1948, Ph.D., 1949; m. Estelle Field, Sept. 3, 1939; children—Sandra Margaret, Janet Catherine (Mrs. Harald Hanssen). Faculty, U. Cal. at Los Angeles, 1948—, prof., 1961—. Mem. Assn. Am. Geographers, Am. Geog. Soc., Assn. Pacific Coast Geographers, Cal. Council Geography Tchrs. Contbr. numerous articles to profl. jours. Specialist in desert areas, both phys. and cultural, field research in U. S., Mexico, So. and S.W. Africa, Sudan. Home: 943 6th St., Santa Monica, Cal. 90403. Office: Dept. Geography, U. Cal., Los Angeles 90024.*

LOGAN, Thomas Moldrup, Am. physician; b. Charleston, S.C., July 31, 1808; s. George and Margaret (Polk) L.; grad. Med. Coll. S.C., 1828; m. Susan Richardson; m. 2d, Mary Greely, 1864. Practiced medicine, Charleston, 1828-43, New Orleans, 1843-50, Sacramento, Cal., 1850-76; taught medicine Med. Coll. of S.C., Charleston; mem. staff Charity Hosp., New Orleans; active in fostering measures for public health protection; instrumental in founding Cal. Bd. Health, 1870, sec. until 1876; made studies of epidemiology and sch. hygiene; pres. Cal. Med. Soc. (1870), Agassiz Inst. of Sacramento, A.M.A. (1872). Author: (articles) History of Medicine in California, 1858; Report on the Medical Topography and Epidemics of California published in Trans. A.M.A., 1859. Died Sacramento, Feb. 13, 1876; buried Sacramento.

LOGAN, Sir William Edmond, Canadian geologist; b. Montreal, Que., Can., Apr. 20, 1798; s. William and Janet (Edmond) L.; grad. Edinburgh (Scotland) U., 1817; LL.D., U. Montreal. Entered counting house of uncle (Hart Logan), London, 1818; mgr. coppersmelting, coal mining works, Swansea, Wales, 1831-38; geol. expdn., Isle of Sheppey, France, Spain, circa 1832-35; studied phenomena of annual freezing of St. Lawrence River, 1841; expdn. coal fields, Pa. and N.S.; dir. Canadian Geol. Survey, 1842-70. Knighted, 1856. Decorated cross Legion of Honor, 1855. Fellow Royal Soc., 1851, Geol. Soc. (Wollaston medal). Author: (with Thomas Sterry Hunt) Geology of Canada, 1863; also numerous articles. First to show rocks in lower Can. were altered, crystallized palaeozoic strata (previously classed as primitive azoic); his geol. maps and sections were used for H. T. De La Beche's geol. maps of Britain. Died June 22, 1875.

LOGAN, William Newton, Am. geologist; b. Barboursville, Ky., Nov. 5, 1869; s. Henry Elderberry and Jane Elizabeth (Points) L.; B.A., M.A., U. Kan., 1896; Ph.D., U. Chicago, 1900; m. Janette Cecil DeBaun, Aug. 31, 1898; children—Lois Lucene (Mrs. R. E. Esarey), Harlan DeBaun. With Kan. Geol. Survey, summers, 1893-97; supt. pub. schs., Pleasanton, Kan., 1896-98; with Wyo. Scientific expdn., 1899; prof. geology and mineralogy, St. Lawrence U., Canton, N.Y., 1900-03; with N.Y. Geol. Survey, summer 1902; prof. geology and mining engring., Miss. Agrl. and Mech. Coll., and geologist Miss. Expt. Sta., 1903-16, also geologist State Geol. Survey, 1903-16, and dean Sch. of Science, A. and M. Coll., 1913-16; asso. prof. econ. geology, Ind. U., 1916-19, prof. and state geologist from 1919. Determined boundaries of Am. Upper Jurassic Sea ("Logan Sea"); first to suggest the biochem. theory of the origin of halloysite and other kaolin-forming minerals, Life mem. and fellow A.A.A.S.; fellow Geol. Soc. Am., Ind. Acad. Science, Royal Soc. Arts, London; mem. Am. Assn. State Geologists, Kan. Acad. Science, Miss. Acad. of Science, Sigma Xi, Phi Beta Kappa, Pi Gamma Mu; corr. mem. Soc. Geol. de France. Republican. Methodist. Kiwanian. Author: The Upper Cretaceous of Kansas, 1896; Invertebrates of Kansas Cretaceous, 1898; Brick Clays of Mississippi, 1907; Pottery Clays of Mississippi, 1909; Kaolin of Indians, 1919; Petroleum and Natural Gas of Indiana, 1920; The Elements of Practical Conservation, 1924; Geology of the Deep Wells of Indiana, 1926; Sub-Surface Strata of Indiana; also wrote Building Stone in Indiana, Mineral Wool in Indiana, Clay Products of Indiana, Geological Map of Indiana, Rock Products of Indiana, and numerous brochures, articles, etc. General editor Handbook of Indiana Geology. Deceased.

LOGAN, William Philip, Brit. statistician; b. Glasgow, Scotland, Nov. 2, 1914; s. Frederick and Elizabeth (Dowie) L.; ed. U. Glasgow, U. London; M.D., Ph.D.; m. Pearl Piper, July 21, 1940; children—Alan, Donald, Ian, Graham, Pamela, Mary. Chief med. statistics Gen. Register Office of Eng. and Wales, 1951-60; dir. statistics div. WHO, 1961—. Mem. Royal Soc. Medicine, Royal Coll. Physicians, Royal Soc. Statistics. Research and publs. on health statistics, epidemiology. Home: 11, rue Butini, Geneva. Office: OMS, Palais des Nations, 1211 Geneva 27, Switzerland.

LOHMANN, Heinrich Werner, German astronomer; b. July 18, 1911; s. Alfred and Gertrud (Engler) L.; Ph.D., U. Leipzig (Germany); m. Margarete Simon, Oct. 9, 1948. Began as asst., 1938; mil. service, 1939-45; resumed work, 1949; successively observer, 1954, astronomer, 1960, prof. at large, 1962, obs. astronomer in chief, 1962. Mem. Astronomische Gesellschaft. Research and publs. on galaxies, star clusters. Home: Römerstrasse 58b. Office: Mönchhofstrasse 12-14, Heidelberg, W. Germany.

LÖHNEYSS, Georg Engelhard von, see von Löhneyss, Georg Engelhard.

LOINGER, Angelo Giuseppe, Italian physicist; b. Verona, Italy, Apr. 1, 1923; s. August and Wanda (Remondini) L.; Dottore in Fisica, U. Bologna (Italy), 1947; m. Carla Enrica Colombo, Mar. 6, 1950; children—Guido, Eugenio. Faculty of scis. U. Pavia (Italy), 1950-60, prof. theoretical physics, 1961—; prof. U. Messina (Italy), 1960-61; researcher Istituto Nazionale di Fisica Nucleare, 1955—. Mem. Italian Phys. Soc. Research, publs. on founds. quantum mechanics, founds. statis. mechanics (ergodic theory). Home: 99 Viale Libertà, Pavia, Italy.*

LOISELEUR-DESLONGCHAMPS, Jean Louis Auguste, French botanist; b. Dreux, France, Mar. 24, 1774; Author: Flora gallica, 2 vols., 1806/07; Flore générale de la France, 1820; La rose, son histoire, sa culture, sa poésie, 1844. Classified indigenous plants and studied indigenous medicinal plants; genus Loiseleuria named in his honor. Died Paris, May 13, 1849.

LOIZOS, Antonios, Greek constrn. engr.; b. Pireus, Greece, 1916; s. Andreas and Maria (Gardelli) L.; ed. Athens Tech. U., Sch. Bridges and Hwys., Paris; Dr.Engring., Civil Engr.; m. Victoria Kouzounidou, 1949; 1 son, Andreas. With Ministry Transports, 1941-55; asst., 1941-55; prof. bridges Tech. U. Athens, 1955—; named sec. gen. Ministry Transports, 1958. Cons., Sch. Mcht. Marines. Recipient Medal of Exceptional Merit, Silver Cross. Mem. Internat. Assn. Bridges and Hwys., Internat. Assn. Mechanics of Rocks. Author: Wooden Constructions, 1948-58; Foundations, 1957-62; Applied Mechanics, 1958; also other publs. Home: 51, Odos Posseidonos, Pal. Phaleron, Athens, Greece.

LOKKI, Olli Kristian, Finnish mathematician; b. Helsinki, Finland, Apr. 28, 1916; s. Karl Olof and Anna (Lampen) L.; Ph.D., U. Helsinki (Finland), 1947; m. Kirsti Alopaeus, Aug. 20, 1941; children—Juhani, Kaarina, Heikki. Asst., U. Helsinki, 1942-50; faculty Tech. U., Helsinki, 1945—, prof. applied math., 1960—; docent U. Helsinki, 1952—. Cons. on math. statistics, operations research in medicine, industry and mil. Research, publs. on math., math. statistics, operations research. Home: 15 Temppelikatu, Helsinki 10, Finland.*

LOLLI, Giorgio, biochemist, physiologist; b. Ancona, Italy, Oct. 4, 1905; s. Fausto and Ada (Leoni) L.; M.D., U. Rome; m. Valeria N., Apr. 19, 1936. Research staff Lab. Applied Physiology, Yale, 1940-57, became med. dir. pioneer program Yale Clinic for Alcoholics, 1955; founder International Center for Psychodietetics, N.Y.C., Rome. Mem. A.M.A. Author: Alcohol in Italian Culture; Social Drinking. Research and publs. on phys., psychol., religious, ednl., legal, sociol. aspects of alcohol; also its effect on nutritrition; obesity, manutrition. Address: via Paisiello 26, Rome, Italy; also 1161 York Av., Suite 1-A, N.Y.C.

LOM, Jiří, Czechoslovakian protozoologist; b. Prague, Czechoslovakia, Oct. 24, 1931; s. Frantisek and Emilie (Pavlatová) L.; B.Sc., Charles U., Prague, 1954; Ph.D., Czechoslovak Acad. Scis., 1957; m. Jarmila Koterová, July 7, 1956; children—Ivana, Lucie. Staff, Protozool. Lab., Czechoslovakian Acad. Sci., 1958-61, staff Inst. Parasitology, Prague, 1962—; vis. research asso. prof. U. Ill., Chgo., 1965-66. Recipient several awards for sci. activity Czechoslovakian Acad. Sci. Mem. Soc. Protozoologists, Groupment de Protistologues de la Langue Francaise, Soc. Czechoslovak Parasitologists, Czechoslovak Zool. Soc. Editor: (with J. Ludvik, J. Vavra) Progress in Protozoology; editorial bd. Acta Soc. Zool. Bohemoslov, 1962—, Protistologica, 1965—. Research, publs. on distbn., morphology and significance of parasitic protoza of fish; electron micros. studies on devel. parasitic protozoa and comparative structure of ciliates; discovered many new species of parasitic ciliated protozoa. Home: 62 Stresovická, Prague 6. Office: Inst. Parasitology, Vinicná 7, Prague 2, Czechoslovakia.*

LOMBARD, Claude-Antoine, French physician; b. Dôle, France, Aug. 17, 1741; head surgeon Dôle Hosp; apptd. head surgeon Strasbourg (France) Hosp., 1792; mem. French Acad. Scis., 1796. Author: Complication du vice vénériennes, 1790; Remarques sur les lésions, 1791. Research on healing of wounds, sutures, prevention of infection, venereal diseases, lesions of head. Died Montmagny. France, Apr. 15, 1811.

LOMBARD, George Francis Fabyan, Am. behavioral scientist; b. Newton, Mass., Jan. 31, 1911; s. Percival H. and Isabel (Fabyan) L.; A.B., Harvard, 1933, M.B.A., 1935, D.C.S., 1941; m. Mary Esther Jackson, Sept. 9, 1937; children—Joshua, Esther (Mrs. Eric W. Danielson, Jr.), Marshal, Emily, Rachel Annabel. Asst. dean Harvard Grad. Sch., Bus. Adminstrn., Cambridge, Mass., 1936—, instr. indsl. research, 1940-42, asst. prof., 1942-46, asso. prof. human relations, 1946-52, prof., 1952—, asso. dean ednl. programs, 1962—. Mem. subcom. on rehab. com. on work in industry NRC, 1944-45; mem. adv. panel for research in human relations Office Naval Research, 1946-52. Mem. Am. Sociol. Assn., Soc. for Psychol. Study Social Issues, Soc. for Applied Anthropology. Author: Behavior in a Selling Group: A Case Study of Interpersonal Relations in a Department Store, 1955; also chpts. in books. Research on behavior in work groups, application of psychotherapy in bus. relationships. Home: 441 Glen Rd., Weston, Mass. 02193. Office: Harvard Business School, Boston 02163.*

LOMBARD, René, French chemist; b. Die (Drôme), France, July 7, 1912; s. Joseph and Amelie (Hocat) L.; student Ecole Normale Supieure, 1931-35; m. Yvonne Dupont, July 18, 1939; children—Claudine (Mrs. Alain Juillard), Jacques, Josette, Yves, Michel. Prof., Faculté des Sciences, Strasbourg, France, 1946—. Mem. Chem. Soc. France. Author: Produits resineux, 1946; also articles. Research in acids, autooxidation. Home: 15, rue Oberlin, Strasbourg, France.*

LOMBARDINI, Siro, Italian economist; b. Milan, Italy, July 3, 1924; s. Angelo and Teresa (Masarati) L.; Laurea in Econs., Catholic U. Milan, 1946; postgrad. London Sch. Econs., 1949, U. Chgo., 1950-51; Libera Docenza in Econs., 1954; m. Rosa Giusti, Sept. 7, 1958; children—Paolo, Giorgio, Chiara. Lectr., Cath. U., Milan, 1951-56, U. Modena, 1954-58; prof. Cath. U., 1957-63, U. Turin, 1963—; vis. prof. Cambridge (Eng.) U., 1964; condr. seminars, Paris, Harvard, Leningrad. Dir. sci. com. for econ. planning of Umbria region, 1959-63; dir. Istituto Ricerche Economiche e Sociali, Turin, 1958—; mem. comitato scientifico per la programmazione economica Ministry for the Budget; cons. Ministry for State Enterprises. Mem. Econometric Soc., Associazione Italiana degli Economisti. Author: Il monopolio nella teoria economica, 1953; Fondamenti e problemi dell'economia del benessere, 1954; L'analisi della domanda nella teoria economica, 1957; Appunti alle lezioni di Economia politica, 2 vols., 1961; Appunti alle lezioni di politica economica, 1964; La programmazione economica: idee, esperienze e risultati; also articles. Research on monopoly theory, including analysis of implications of the firms' devel. on the market structures and of origin of obstacles to entry of new firms into markets; studies on welfare econs. and planning, including proposal of a disaggregated model for analysis of efficiency of investment, 1954, discussion of limitations of current theory of welfare econs. due to their static character, methods of interdisciplinary studies for regional planning, a new tech. and hist. philosophy of econ. planning; research on theory of demand, including analysis of interdependences between individual demands, the time context of the analysis, some implications of indivisibilities of durable goods. Home: 128 Strada Roaschia, Chieri, Turin. Office: University of Turin, 10 Via Carlo Alberto, Turin, Italy.*

LOMBROSO, Cesare, Italian criminologist, anthropologist; b. Verona, Nov. 18, 1836; ed. univs. Pavia, Padua, Vienna; m., 1870, 1 son, 2 daus. Served as mil. physician; apptd. prof. psychiatry U. Pavia, 1862; became dir. insane asylum, Pesaro, 1871; apptd. prof. forensic medicine U. Turin, 1876, prof. psychiatry, 1896, prof. criminal anthropology, 1905. Author numerous works including: L'Uomo delinquente, 2 vols., 1876; La Pellagra in Italia, 1885; La Donna delinquente, 1893; Le Crime, causes et remedes, 1899; Delitti vecchi e delitti nuovi, 1902; After Death—What?, 1908. Regarded as founder of sci. of criminology; held that there is a criminal type, characterized by phys., mental and nervous deviations due to atavism and degeneration; strove for more constructive treatment of criminals; described nature of genius as degenerative condition somewhat analogous to insanity; traced origin of pellagra to diseased maize, 1872. Died Turin, Oct. 19, 1909.

LOMONOSOV, Mikhail Vasilievich, Russian chemist, astronomer, grammarian, poet, historian; b. Deniskova (now Lomonosov), nr. Archangel, Russia, Nov. 8, 1711; s. Vasilii Dorofeev L.; studied science in Moscow, 1731-35; at U. of Imperial Academy of Sci., St. Petersburg, 1736; then in Germany at Marburg (with C. Wolff), 1736, and at Freiberg, 1739; m. Elizabeth Zilch, 1740. Prof. at U. of Imperial Acad. Scis., 1745 (later rector); built laboratory for teaching and research, 1749; established factory for making colored glass and mosaics; Secretary of State, 1764; mem. St. Petersburg Acad. Scis., 1741; wrote several memoirs for Comentarii of St. Petersburg Acad. Urged importance of study of chemistry in Russia; stressed especially it be developed mathematically; noted increase in weight when lead converted into minum; strongly criticized pholgiston theory; explicitly stated idea of conservation of matter; may have influenced Lavoisier; held weight not proportional to mass (might change in same place if particles of an aggregate separated so as to expose more surface to action of gravity); regarded heat as form of motion of particles (hence often considered anticipator of kinetic theory of gases); developed theory of gases which assumes particles rotating and motion transmitted from particle to particle by friction between their rough surfaces; experimented on expansion of air by heat; 1st to record freezing of mercury; 1st to observe atmosphere of Venus (during solar transit, 1761); adopted Wolffian rationalistic approach to science; researches in metallurgy, geology, meteorology, cartography; translated textbooks into Russian; wrote grammar that reformed Russian literary language; pub. 1st history of Russia, 1760; (with Euler) helped found U. Moscow; often called founder Russian science; wrote poems, drama. Died St. Petersburg, Russia, Apr. 4, 1765.

LOMONT, John S., Am. math. physicist; b. Ft. Wayne, Ind., Aug. 26, 1924; s. C. F. and R. S. Lomont; B.S.Ch.E., Purdue U., 1946, M.S., 1947, Ph.D., 1951. Prof. math. Poly. Inst. Bklyn., 1960-65; prof. U. Ariz., Tucson, 1965—. Mem. Am. Math. Soc., Am. Phys. Soc., N.Y. Acad. Sci., Sigma Xi. Author: Applications of Finite Groups, 1959; also articles. Research on math. theorems relevant to quantum theory. Home: 8 Paseo Redondo, Tucson 85705.*

LONDON, Heinz, physicist; b. Bonn, Germany, Nov. 7, 1907; s. Franz and Louise (Hamburger) L.; student Bonn U., 1926-27, Berlin (Germany) Technische Hochschule, 1927-29, München (Germany) U., 1929-31; Dr.phil., Breslau (Germany) U., 1934; m. Lucie Meissner, Aug. 23, 1946; children—Louise Ann, Norah Frances, Martin Thomas, Frederic Charles. Research asso. Oxford (Eng.) U., also Bristol (Eng.) U., 1934-41; war researcher Dept. Sci. and Indsl. Research, 1941-46; physicist Atomic Energy Research Establishment, Harwell, Berks, Eng., 1946—, sr. prin. sci. officer, 1949-58, dep. chief scientist, 1958—. Recipient Simon Meml. prize, 1959. Fellow Royal Soc., 1961, Phys. Soc. Editor: Separation of Isotopes, 1961; Research, publs. on superconductivity, superfluid helium, isotope separation; invention method of continuously producing temperatures below 0.1°K by mixing isotopes of helium. Home: 44 Cumnor Hill, Oxford, Eng. Office: Atomic Energy Research Establishment, Harwell, Berks, Eng.*

LONDON, Irving Myer, Am. physician; b. Malden, Mass., July 24, 1918; s. Jacob A. and Rose (Goldstein) L.; A.B., Harvard, 1939, M.D., 1943; Sc.D. (hon.), U. Chgo., 1966; m. Huguette Piedzicki, Feb. 27, 1955; children—Robert L. J., David T. Faculty Columbia, 1947-55, asso. prof. medicine, 1954-55; staff Presbyn. Hosp., N.Y.C., 1946-55, asso. attending physician, 1954-55; prof., chmn. dept. medicine Albert Einstein Med. Coll., N.Y.C., 1955—; dir. med. service Bronx Municipal Hosp. Center, 1955—; vis. scientist Pasteur Inst., Paris, 1962-63. Mem. med. fellowship bd. Nat. Acad. Sci.-NRC, 1955-63, mem. subcom. on blood and related problems, 1955-64; mem. hematology study sect. USPHS, 1955-59, chmn. metabolism study sect., 1961-63; mem. bd. sci. counselors Nat. Heart Inst., 1964—, subcom. on intravenous alimentation Dept. Army, 1955-60; research council Pub. Health Research Inst. N.Y.C., 1958-63; bd. sci. cons. Sloan-Kettering Inst. Cancer Research, 1960—, chmn., 1964-65; vis. mem. panel on biol. sci. and advancement medicine Nat. Acad. Scis., 1966—, mem. bd. medicine, 1967—; adv. com. to dir. NIH, 1966—; hon. lectr. State U. N.Y., 1957, U. Cin., 1958, Med. Coll. Va., 1960, U. Colo., 1962, Georgetown U., 1966, U. Stockholm, 1964. Recipient Theobald Smith award A.A.A.S., 1953. Fellow Am. Acad. Arts and Scis; mem. Am. Soc. Clin. Investigation (pres. 1963-64), Am. Soc. Biol. Chemists, Am. Soc. Exptl. Biology and Medicine, Ass. Am. Physicians, Harvey Soc., Interurban Clin. Club, Internat. Soc. Hematology, Practitioners Soc., Phi Beta Kappa, Alpha Omega Alpha. Editorial bd. Physiology for Physicians, 1962—. Research on metabolism of erythrocyte; biosynthesis and metabolism of hemoglobin and bile pigment. Home: 4740 Iselin Av., Bronx, N.Y. 10471.*

LONDON, Julius, Am. meteorologist; b. Newark, Mar. 26, 1917; s. Philip and Sadie (Weinberger) L.; A.B. in Math., Bklyn. Coll., 1941; M.S. in Meteorology, N.Y. U., 1948, Ph.D., 1951; m. Dorothy Sibulsky, July 6, 1946; children—David, Richard. Meteorologist, U. S. Weather Bur., N.Y., 1942; faculty N.Y. U., 1947-61; prof. atmospheric physics U. Colo., Boulder, 1961—, chmn. dept., 1966—. Lectr., Columbia, 1954-55; vis. prof. Pa. State U., 1955; chmn. panel on ozone com. atmospheric scis. Nat. Acad. Scis., 1964-65. Mem. Am. (traveling lectr. 1958-61, 63), Royal meteorol. socs., Am. Assn. U. Profs., Am. Geophys. Union, Sigma Xi. Author: Weather and Climate, 1960. Research, publs. on radiative transfer in earth's atmosphere, including a study of atmospheric heat balance; studies of physics of upper atmosphere, photochemistry of atmospheric ozone; analysis of variations and geog. distbn. of total ozone amounts. Home: 245 Abbey Pl., Boulder, Colo. 80302.*

LONDON, Morris, Am. biochemist; b. N.Y.C., July 2, 1922; s. Harry and Sarah (Weizman) L.; B.S., Coll. City N.Y., 1946; M.S., Fordham U., 1948; Ph.D., Ohio State U., 1950; m. Beatrice Berwald, Mar. 27, 1948; children—Eric, Carole. Research fellow N.Y. U. Med. Sch., 1950; research asso. Columbia Coll. Phys. and Surg., 1951-56; chief chemist N. Shore Hosp. div. labs., Manhasset, N.Y., 1956—. Mem. Am. Chem. Soc., Harvey Soc., Am. Assn. Clin. Chemists, Am. Soc. Biol. Chemists. Research, numerous publs. on physico-chem. properties acid phosphatase and described its active center; demonstrated that uricase may be used to break down uric acid causing gout in vivo; isolated and crystallized prostatic acid phosphotase. Home: 226 W. Shore Dr., Kauneonga Lake, N.Y. 12749. Office: N. Shore Hosp., Community Dr., Manhasset, N.Y. 11030.*

LONDON, Perry, Am. psychologist; b. Omaha, June 18, 1931; s. Max and Rose (Novoselsky) L.; B.A., Yeshiva Coll., 1952; M.A., Columbia Tchrs. Coll., 1953; Ph.D., Columbia, 1956; m. Vivian Jacobson, June 3, 1954; children—Miriam D., Judith G., Susan E., Debra S. Mem. faculty U. Ill., 1959-63; vis. asso. prof. Stanford, 1962-63; asso. prof., dir. psychol. research and service center, U. So. Cal., 1963-66, dir. clin. psychology tng., 1964-66; tng. cons. VA, 1960—; civilian expert cons. Surgeon Gen. U. S. Army, 1965—. Recipient Career Scientist Devel. award, NIH, USPHS, 1966—. Fellow Am. Psychol. Assn.; mem. Psychonomic Soc., Internat. Soc. Clin. and Exptl. Hypnosis, A.A.A.S. Author: The Modes and Morals of Psychotherapy, 1964. Research and publs. on exptl. research in human personality, especially on hypnosis; developed measuring instrument for studying susceptibility of children. Home: 839 S. Holt Av., Los Angeles 90035. Office: 734 W. Adams Blvd., Los Angeles 90007.*

LONG, Carleton Curtis, Am. chem. engr.; b. Boulder, Colo., June 3, 1909; s. Jesse Dilman and Cora Gertrude (Curtis) L.; B.S. in Chem. Engring., U. Colo.,

1931, Ph.D. in Phys. Chemistry, 1935; M.A. in Chemistry, Stanford, 1932; m. Dorothy Belle Peters, Dec. 19, 1935; children—Dwight Everett, Douglas Alan, Carolyn Elizabeth. Research engr. zinc smelting div. St. Joseph Lead Co., Monaca, Pa., 1935-37, dir. plant research dept., 1937-55, dir. research, 1955—. Mem. Metall. Soc. (pres. 1960), Am. Inst. Mining, Metall. and Petroleum Engrs. (past dir., v.p. 1961-63), Am. Inst. Chem. Engrs., Am. Chem. Soc., Nat., Pa. socs. profl. engrs., Electrochem. Soc., Inst. Metals. Research in surface chemistry, photog. latent image, electrothermic winning of zinc and magnesium, fluidized bed purification and roasting of sulfide ores, dedusting of indsl. gases. Home: 138 College Av., Beaver, Pa. 15009. Office: Box A, Monaca, Pa. 15061.*

LONG, Charles Alan, Am. biologist; b. Pittsburg, Kan., Jan. 19, 1936; s. Dorsey Arnold and Mary Belle (Selig) L.; B.S., Kan. State Coll., 1957, M.S., 1958; Ph.D., U. Kan., 1963; m. Claudine Fern Lowder, Aug. 28, 1960; 1 son, Charles Alan. Research asst. Mus. Natural History, U. Kan., 1962-63; faculty U. Ill., 1963-66, asst. prof. life sci., 1965-66; asso. prof. dept. biology Wis. State U., Stevens Point, 1968—. Faculty fellow U. Ill., 1964. Mem. Am. Soc. Mammalogists (mem. com. on nomenclature 1966—). Author: The Mammals of Wyoming, 1965; also numerous articles. Editor: Album of North American Animals (with Clark Bronson, Vera Dugdale) 1966. Research on taxonomy of animals of Wyo., their Quarternary evolution and geography, morphology and systematics of badger, morphometric variation in mammals. Home: 3256 Welsby Av., Stevens Point, Wis. 54481.*

LONG, Crawford Williamson, Am. anesthetist, surgeon; b. Danielsville, Ga., Nov. 1, 1815; s. James and Elizabeth (Ware) L.; grad. Franklin Coll. (now U. Ga.), 1835; M.D., U. Pa., 1839; m. Caroline Swain, Aug. 11, 1842. Moved to N.Y., 1839, practiced as surgeon; moved to Ga., 1841, practiced surgery; accidently discovered that sulphuric ether could be used as anesthetic in early 1840's; performed operation on neck of James Venable for removal of cystic tumor, 1842, performed 5 other operations using this procedure before 1846; published results of these expts. in So. Med. and Surg. Jour., 1849; practiced medicine and surgery, Athens, Ga., 1850-58; probably 1st to use ether as an anesthetic, Jefferson, Ga., 1842. Died Athens, June 16, 1878.

LONG, Cyril Norman, Am. physiologist, biochemist; b. Nettleton, Eng., June 19, 1901; s. John Edward and Rose (Langdell) L.; B.Sc., U. Manchester (Eng.), 1921, M.Sc., 1923, D.Sc. in Physiology, 1932; M.D. C.M., McGill U. (Can.), 1928, Sc.D. (hon.), 1961; M.A. (hon.), Yale, 1936; Sc.D. (hon.), Princeton, 1946; M.D. (hon.), U. Venezuela, 1962; m. Hilda Gertrude Jarman, July 28, 1928; children—Barbara Rosemary (Mrs. R. P. Simons), Diana Elizabeth (Mrs. D. D. Hall). Came to U. S., 1932, naturalized, 1942. Faculty, U. Coll., London, Eng., 1923-25, McGill U., Montreal, Que., Can., 1925-32; dir. George S. Cox Med. Research Inst., also faculty U. Pa., 1932-36; faculty Yale, 1936—, Sterling prof. physiol. chemistry, 1938-52, chmn. dept., 1936-52, dean Sch. Medicine, 1947-52, Sterling prof. physiology, 1952—, chmn. dept. physiology, 1952-64. Cons. NIH, NSF, Army Med. Center; also mem. sci. advisers bd. Yerkes Regional Primate Research Center, Jane Coffin Childs Fund. Med. Research; Recipient certificate of appreciation Army and Navy Depts., 1948; Modern Medicine award for distinguished achievement Modern Medicine, 1959; Sci. award Pharm. Mfrs. Assn., 1959. Guggenheim fellow, 1955-56. Mem. Nat. Acad. Scis., Am. Philos. Soc., Am. Acad. Arts and Scis., Am. Assn. Physicians, Am. Soc. Clin. Investigation, Soc. Exptl. Biology and Medicine, Endocrine Soc. (Squibb award 1950, Schering award 1956), Am. (Banting medal 1951), Brit. (hon. life) diabetes assns., Argentine Soc. Biology and Medicine (hon.), Am. Physiol. Soc., Soc. Biol. Chemists, Physiol. Soc. Great Britain, Biochem. Soc. Great Britain, Sigma Xi, Alpha Omega Alpha, Nu Sigma Nu. Research, publs. on muscular exercise in man, function of hypothalmus, role of adrenal cortex in carbohydrate metabolism; mechanism of action of adrenal cortical hormones, factors influencing secretion of adrenal cortex. Home: 100 Old Farm Rd., Hamden, Conn. 06517. Office: 333 Cedar St., New Haven 06510.*

LONG, Earl Albert, Am. physicist-chemist; b. Altoona, Pa., July 2, 1909; s. Arthur Russell and Grace Emma (Beattie) L.; A.B., Catawba Coll., 1930, D. Sc. (hon.), 1951; M.S., Ohio State U., 1932, Ph.D., 1934; m. Marietta Susan Moss, Feb. 8, 1947; step children—George E. Shambaugh, Susan S. (dec.). Nat. Research fellow chemistry U. Calif., 1934-36, instr., 1936-37; fellow Lalor Found., 1937-38, asst. prof. U. Mo., 1938-43, asso. prof., 1943-45, prof. chemistry, 1945; with Los Alamos Lab., 1943-46, asst. dir., 1945; prof. inst. metals and dept. chemistry U. Chgo., 1946-61, dir. Inst. Study of Metals, 1957-61; asst. lab. dir. Gen. Atomic div. Gen. Dynamics Corp., 1960—. Fellow Am. Phys. Soc.; mem. Sigma Xi, Phi Lambda Upsilon. Research and publs. on thermodynamics, nuclear physics, low temperature and solid state physics, phys. adsorption, devel. of low temperature liquefaction systems, properties of liquid helium. Home: 1406 La Jolla Knoll, La Jolla, Cal.*

LONG, Esmond Ray, Am. pathologist; b. Chgo., June 16, 1890; s. John Harper and Catherine Belle (Stoneman) L.; A.B., U. of Chicago, 1911, Ph.D., 1919; M.D., Rush Medical College (U. of Chicago), 1926; post-graduate studies German U. of Prague; Sc.D. (hon.), Univ. Pa., 1948; m. Marian Boak Adams, June 17, 1922; children—Judith Baird (Mrs. J. L. Pincus), Esmond Ray. Asst. in pathology, U. of Chicago, 1911-13; asst. desert laboratory of Carnegie Instn. of Washington, at Tucson, Ariz., 1914-15; Trudeau fellow, Saranac Lab., Saranac Lake, N.Y., 1918, 20; instr. pathology, 1919-21, asst. prof., 1921-23, asso. prof., 1923-28, prof., 1928-32, U. of Chicago; prof. pathology, U. of Pa., 1932-55; also director lab., Henry Phipps Institute for Study, Treatment and Prevention of Tb., 1932-35, dir. Institute, 1935-55, emeritus professor of pathology, 1955—; consultant Leonard Wood Meml.; traveling representative Pan-Am. Sanitary Bur., 1941. Pres. Wistar Inst. of Anatomy and Biology, 1939-42. Chief cons. on tuberculosis, Office of the Surgeon Gen., U. S. Army, Wash., D.C., 1942-45; chief, Tuberculosis Sect., Pub. Health Br., Office of Mil. Govt., U. S. Zone, Germany, Aug. Nov. 1945; acting chief, Tuberculosis Service, Vets. Adminstrn., Nov. 1945-Jan. 1946; col. Med. Res., 1946; hon. retirement as colonel, 1950. Mem. Nat. Research Council 1932-55 (chmn. div. med. sci. 1936-39, chmn. subcommittee Tb, 1940-42, 46-47, 55-60, chmn. com. on vets medical problems 1956-60), Nat. Tb Assn. (pres. 1936-37, dir. med. research 1947-55), A.A.A.S. (v.p., chmn. sect. on med. scis. 1936-37), Am. Thoracic Soc., Assn. Am. Physicians, Am. Assn. Pathologists and Bacteriologists (pres. 1936-37), Am. Philosophical Society, American Association of History of Medicine (pres. 1939-41), A.C.P., Nat. Acad. Scis., Coll. of Physicians (Phila.); hon. mem. European and S. Am. med. and tuberculosis socs. Awarded Trudeau Medal, Nat. Tuberculosis Assn., 1932; Legion of Merit, 1946; Strittmatter award, Phila. Co. Medical Society, 1950; Phila. award, 1954. Mem. U. S. nat. commn. on UNESCO, 1951-54. Author: The Chem. of Tuberculosis, 1923; A History of Pathology, 1928; Selected Readings in Pathology, 1929; Antonio Benivieni: The Hidden Causes of Disease (with Charles Singer), 1954; Tuberculosis in the Army of the United States in World War II (with Seymour Jablon), 1955; History of the Therapy of Tuberculosis and the Case of Frédéric Chopin, 1955; Tuberculosis Medical Research NTA, 1904-55 (with Virginia Cameron), 1959; A History of American Pathology, 1962. Editor Internat. Jour. Leprosy, 1964—. Research on chemistry, nature, use of tuberculin in the diagnosis of tuberculosis. Contbr. med. jours. Home: 220 Locust St., Phila. 19106.*

LONG, Franklin Asbury, Am. chemist; b. Great Falls, Mont., July 27, 1910; s. F.A. and Ethel (Beck) L.; B.A., M.A., Mont. State U., 1932; Ph.D. in Chemistry, U. Cal., Berkeley, 1935; m. Marion Thomas, Aug. 12, 1937; children—Elizabeth, Franklin A. Fellow dept. chemistry U. Cal. Berkeley, 1933-35; instr. U. Chgo., 1936-37; faculty Cornell U., Ithaca, N.Y., 1937—, prof., chmn. dept. chemistry, 1950-60, faculty trustee, 1956-59, v.p. for research and advanced studies, 1963—; research supr. Explosives Research Lab., NDRC, Pitts., 1943-45; vis. chemist, cons. Brookhaven Nat. Labs., Upton, L.I., N.Y., 1946—. Chmn. div. chemistry and chem. tech. NRC, 1964—; mem. Pres.'s Sci. Adv. Com., Washington, 1961—; asst. dir. sci. and tech. U. S. Arms Control and Disarmament Agy., Washington, 1962-63; cons. to govt. agys., industry. Mem. Nat. Acad. Scis. Editorial bd. Jour. Phys. Chemistry, 1962—. Contbr. numerous articles on chemistry to sci. jours., encys., reference works. Home: 429 Warren Rd., Ithaca, N.Y. 14850.*

LONG, James Scott, Am. chemist; b. York, Pa., Aug. 11, 1892; s. William Lawrence and Anna (Wellensiek) L.; Ch.E., Lehigh U., 1914, M.S., 1915, D.Sc., 1957; Ph.D., Johns Hopkins, 1922; D.Sc., N.D. State Coll., 1963; m. Harriet Wilkinson, Sept. 5, 1916 (dec. Apr. 1947); children—James Scott II, Harriet June (Mrs. James Calvin McFerran IV); m. 2d, Grace Marie Foster, July 23, 1949; 1 son, George Lewis. Faculty, Lehigh U., 1915-34, prof., 1927-34; chem. dir., bd. dirs. Devoe & Raynolds, Louisville, 1934-57; faculty U. Louisville, 1934—, distinguished prof., 1956—; prof. emeritus U. So. Miss., 1964—; v.p. charge research Pan Am. Tung Research and Devel. League. Recipient Heckel award, 1953; Mattiello award, 1954. Fellow Am. Inst. Chemists, Royal Soc. Arts (Eng.); mem. Phi Beta Kappa, Sigma Xi, Tau Beta Pi, Omicron Delta Kappa. Contbr. numerous articles to Am. and European sci. jours. Author: (with H. V. Anderson) Chemical Calculations. Patentee in field. Headed group that created epoxy resin, solvent process for cooking resins. Home: 206 Southampton Rd., Hattiesburg, Miss. 39401.*

LONG, John Harper, Am. chemist; b. nr. Steubenville, O., Dec. 1856; s. John and Elizabeth (Harper) L.; B.S., U. of Kan., 1877; studied at Tübingen, Würzburg, and Breslau, 1877-80; Sc.D., Tübingen, 1879; m. Catherine Stoneman, Aug. 24, 1885. Prof. chemistry, Northwestern U. Med. Sch., from 1881; dean Northwestern U. Sch. Pharmacy, 1913-17. Mem. referee bd. of consulting scientific experts U. S. Dept. Agr.; mem. revision com. U. S. Pharmacopoeia. Author: Elements of General Chemistry, 1898, 4th edit. 1906; Text-Book of Analytical Chemistry, 1898, 3d edit. 1906; Laboratory Manual of Physiological Chemistry, 1894; Text-Book of Physiological Chemistry, 1905, 2d edit. 1909. Investigated influence of Sodium benzoate, copper and alum on health; pioneer chem. phase of med. edn. Died June 14, 1918.

LONG, John Paul, Am. pharmacologist; b. Albia, Ia., Oct. 4, 1926; s. John Edward and Bessie (Oswandel) L.; B.S., U. Ia., 1950, M.S., 1952, Ph.D., 1954; m. Marilyn Stookesberry, June 11, 1950; children—Jeffrey, John, Jane. Research pharmocologist Sterling Winthrop Research Inst., Rensellaer, N.Y., 1954-56; faculty U. Ia., Iowa City, 1956—, prof. pharmacology, 1963—. Mem. Am. Soc. Pharm. Exptl. Therapy (Abel award 1962), others. Research, publs. on synthesis and pharmacology of hermicholinum analogs, neuromuscular blocking actions of antibiotics, structure activity relationships of numerous series of autonomic agts. Home: 1105 Tower Ct., Iowa City 52240.*

LONG, Le Roy, Am. surgeon; b. Lincoln County, N.C., Jan. 1, 1869; s. William T. and Mary E. L.; M.D. with 1st honors, Louisville Med. Coll., 1893; m. Martha Downing, Apr. 29, 1896; children—Le Roy Downing, Wendell McLean. Demonstrator genitourinary diseases Louisville Med. Coll., 1894-95; dean and prof. surgery U. Okla. Sch. Medicine, 1915-31; chief Le Roy Long Clinic. Mem. Okla. Bd. Med. Examiners, 1911-15. Fellow Western Surg. Assn., A.C.S. (gov.). Died Oct. 27, 1940.

LONG, Perrin Hamilton, Am. physician; b. Bryan, O., Apr. 7, 1899; s. James Wilkinson and Wilhelmina Lillian (Kautsky) L.; B.S., U. Mich., 1924, M.D., 1924; M.D. (hon.), U. Algers, 1944; F.R.C.P., 1946; D.Sc., Trinity Coll., 1955; m. Elizabeth D. Griswold, Sept. 6, 1922; children—Perrin Hamilton, Jr., Priscilla Griswold. Resident physician, Thorndike Meml. Lab., Boston City Hosp., 1924-25, interne fourth med. service, 1925-27; vol. asst., Hygienic Inst., Freiburg, Germany, 1927; asst. and asso., Rockefeller Inst. for Med. Research, 1927-29; faculty Johns Hopkins U. Med. Sch., 1929-51, prof. preventive med., 1940-51; physician Johns Hopkins Hosp., 1940-51; chmn., prof., dept. medicine, coll. medicine State U. N.Y., N.Y.C., 1951-61, now emeritus; dir. U. div. med. service Kings Co. Hosp. Center, Bklyn., 1951-61, cons., 1961—, also chief dept. medicine; cons. VA, FDA, USPHS, also Dept. of Army. Med. advisor, O.C.D. Planning, 1948. Trustee Martha's Vineyard Hosp.; dir. Am. Field Service, 1959-64, Physician's Club, 1964—. Awarded Croix de Guerre 1918, Chevalier Legion of Honor, 1951 (both France); Legion of Merit, 1945, Hon. Officer (Mil. Div.) Order Brit. Empire, 1945. Fellow Royal Coll. Physicians, 1946; mem. Am. Soc. Clin. Investigation, Harvey Soc., Soc. Am. Bacteriologists, A.M.A., Assn. Am. Physicians, Zeta Psi, Alpha Omega Alpha, Sigma Xi. Author: Clinical and Experimental Use of Sulfanilamide, Sulfapyridine and Allied Compounds (with Eleanor A. Bliss), 1939; ABC's of Sulfonamide and Antibiotic Therapy, 1948. Contbr. many med. articles to jours. Editor-in-chief Resident Physician, Medical Times. Introduced med. use of sulfanilamide drugs into U. S. Died Dec. 17, 1965.

LONG, Robert Radcliffe, Am. meteorologist; b. Glen Ridge, N.J., Oct. 24, 1919; s. Clarence D. and Gertrude (Cooper) L.; A.B. in Econs., Princeton, 1941; M.S. in Meteorology, U. Chgo., 1949; Ph.D., U. Chgo., 1950; m. Cristina Nersing, Oct. 8, 1964; children—John R., Robert W. Faculty, Johns Hopkins, 1951—, prof. fluid mechanics, 1959—. Author: Fluid Models in Geophysics, 1956; Mechanics of Solids and Fluids, 1961; Engineering Science Mechanics, 1963; also articles. Research on motion of liquids and gases as influenced by density variations and by rotation of earth. Home: 802 Beaverbank Circle, Towson, Md. 21204. Office: Charles and 33d Sts., Balt. 21218.*

LONG, Ruby Pauline King, Am. physiologist; b. Chattanooga, June 10, 1914; d. James Lonnie and Metta Belle (Wrinkle) K.; B.S., Birmingham-So. Coll., 1935; M.S., U. Chgo., 1940; postgrad. (Ford fellow) U. Cal. at Berkeley, (Westinghouse fellow) Mass. Inst. Tech., (NSF fellow) U. N.D., (NSF fellow) U. Wyo.; m. Roger Winston Long, June 1, 1940; children—Eleanor Pauline (Mrs. Stanford Harmon Downey), Winston Huey, Barbara Louise, Kathy Jean. Sci. tchr., Moulton, Ala., 1935-37, Bessemer, Ala., 1937-38, Enterprise, Ala., 1938-40, Phillips High Sch., Birmingham, Ala., 1944-45; sci. tchr. Woodlawn High Sch., Birmingham, 1940-42, 48-58, 60—; tchr. Brooke Hill Sch. for Girls also U. Ala. Extension Center, Birmingham, 1947-48, Shades Valley, Birmingham, 1958-59. Named Most Outstanding Sci. Tchr. in State of Ala., Ala. Acad. Sci., 1960; Most Outstanding Biology Tchr. in State of Ala., Nat. Biology Tchrs. Assn., 1962. Mem. N.E.A., Ala. Edn. Assn., Birmingham Tchrs. Assn., Classroom Tchrs. Assn., Am. Zool. Soc., Kappa Delta Epsilon. Research on effect hypophysectomy on male mice, morphology and physiology. Home: 5323 10th Av. S. Office: 5620 1st Av. N., Birmingham, Ala. 35222.*

LONGACRE, Andrew, Am. physicist, Am. electronic engr.; b. Yonkers, N.Y., June 10, 1904; s. Frederick V.D. and Harriette (Blake) L.; B.S., Wesleyan U., Conn., 1926; M.S. in Physics, Princeton, 1929, Ph.D., 1933; m. Marian C. Sykes, June 30, 1934; children—Harriette (Mrs. John B. Phelps), Marian (Mrs. Bruce R. McCart), Sarah (Mrs. Robert W. Hazelton), Andrew. Instr., Phillips Exeter Acad., 1933-46; staff, Radiation Lab., Mass. Inst. Tech., 1941-46; physicist Oak Ridge Nat. Lab., 1946-48; asso. prof. physics U. Ill., Urbana, 1948-51, prof. physics, 1951-57, staff Control Systems Lab., 1951-57; staff Radiation Lab., U. Cal., Berkeley, 1949-51; prof. engring. Syracuse (N.Y.) U., 1957-—, dir. Def. Systems Lab., 1957-61. Mem. sci. and tech. adv. sect. G.H.Q., U. S. Army, Pacific, 1945, mem. research and devel. council U. S. Signal Corps, 1954-61; mem. sci. adv. panel, U. S. Army, 1957-—, chmn. electronics command adv. group, 1963-64, 1964-—. Recipient President's Certificate of Merit, 1948. Fellow Am. Phys. Soc., A.A.A.S., I.E.E.E.; mem. N.Y. Acad. Scis., Am. Soc. Physics Tchrs. Developed, patented basic radar devices and systems, especially noncoherent doppler techniques and systems. Home: 204 Clinton St., Fayetteville, N.Y. 13066.*

LONGACRE, Jacob J. James, Am. plastic surgeon; b. Northampton, Pa., July 26, 1907; s. Jacob E. and Anne Koehler (Beitel) L.; B.A. magna cum laude, Lehigh U., 1927; M.D. cum laude, Harvard, 1932; Ph.D. in Surgery, Cin. Gen. Hosp., 1939; m. Margaret Suzanne Gruen, Oct. 21, 1939; children—Frederick Gruen, James Gruen. Practice medicine specializing in plastic and reconstructive surgery, Cin., 1946-—; asso. attending staff Good Samaritan Hosp. Cin., 1940-—; sr. plastic cons. Vets. Hosp., Cin., 1946-—; dir. plastic surg. dept. Christ and Bethesda hosps., Cin., 1948-—; cons. St. Luke Hosp., Ft. Thomas, Ky., William Booth Meml., St. Elizabeth hosps., Covington, Ky. Chmn. profl. adv. com. for crippled children Welfare Dept. Ohio, 1955-65. Diplomate Am. Bd. Surgery, Am. Bd. Plastic Surgery (chmn. exam. com.). Mem. Am. Soc. Plastic and Reconstructive Surgery, Am. Assn. Plastic Surgeons, Am. Assn. for Cleft Palate Rehab. (pres. 1960-61), Ohio Valley Plastic Surg. Soc. (pres. 1960-61), Cin. Acad. Medicine, Am., Ohio med. assns., Am. Assn. Thoracic Surgery. Contbg. author Reconstructive Plastic Surgery, 5 vols., 1964. Contbr. numerous articles to profl. publs. Home: 3460 Oxford Terrace, Cin. 45220. Office: 2350 Auburn Av., Cin. 45219.

LONGCOPE, Warfield Theobald, Am. physician; b. Balt., Mar. 29, 1877; s. George von S. and Ruth (Theobald) L.; A.B., Johns Hopkins, 1897, M.D., 1901, LL.D., 1951; LL.D., St. John's Coll., 1934; D.Sc., U. Rochester, 1941; hon. doctorate U. Paris, 1945; m. Janet Percy Dana, Dec. 2, 1915; children—Barbara, Duncan, Mary Lee, Christopher. Resident pathologist Pa. Hosp. Phila., 1901-04, dir. Ayer Clin. Lab., 1904-11; asst. prof. applied medicine U. Pa., 1909-11; asso. prof. practice of medicine, Columbia, 1911-14, Bard prof., 1914-21; asso. vis. physician, Presbyn. Hosp., N.Y.C., 1911-14, dir. med. service, 1914-21; prof. medicine Johns Hopkins Med. Sch. and physician in chief Johns Hopkins Hosp., 1922-46. Fellow A.C.P., Am. Acad. Arts and Scis., Coll. Physicians Phila. (hon.); mem. Assn. Am. Physicians, A.M.A., Soc. Exptl. Biology and Medicine, Am. Soc. Clin. Investigation, Am. Soc. Exptl. Pathology, A.A.A.S., N.Y. Acad. Medicine, N.Y. Clin. Soc., Am. Clinic and Climatological Soc., Harvey Soc., Medico Chirurg. Faculty of Med., Balt. City Med. Soc., Nat. Acad. Scis.; hon. mem. Royal Soc. of Medicine, Société des Hôpital, Paris Hon. Died Apr. 25, 1953.

LONGDEN, Aladine Cummings, Am. physicist; b. Leesville, O., Feb. 19, 1857; s. Samuel and Adaline (Cummings) L.; A.B., DePauw U., 1881, A.M., 1884; postgrad. U. Chgo., 1897-99; Ph.D., Columbia, 1900; m. Jean I. Humble, Dec. 24, 1884. Instr. math. and English, St. John's Mil. Acad., Haddonfield, N.J., 1881-82; prof. math. Riverview Mil. Acad., Poughkeepsie, N.Y., 1882-83; prof. math. St. James Mil. Acad., Macon, Mo., 1883-84, prof. natural scis., 1884-88; prof. physics and chemistry State Normal Sch., Westfield, Mass., 1888-97; asst. in physics U. Chgo., 1898-99; instr. physics U. Wis., 1900-01; prof. physics and astronomy Knox Coll., 1901-26, emeritus. Died July 12, 1941.

LONGENECKER, Herbert Eugene, Am. biochemist; b. Lititz, Pa., May 6, 1912; s. Abraham S. and Mary Ellen (Herr) L.; B.S., Pa. State U., 1933, M.S., 1934, Ph.D., 1936; D.Sc., Duquesne U., 1951; LL.D., Loyola U., Chgo., 1963; m. Marjorie Jane Segar, June 18, 1936; children—Herbert Eugene, Marjorie S., Geoffrey H., Stanton L. NRC fellow, 1936-38; faculty, dean U. Pitts., 1938-55; v.p. U. Ill., 1955-60; pres. Tulane U., New Orleans, 1960-—. Mem. Commn. on Fed. Relations, Am. Council Edn., 1961-66, chmn., 1963-65; chmn. adv. com. Coll. and U. Presidents to Inst. Internat. Edn., 1963-66; mem. selection com. for high energy accelerator Nat. Acad. Scis., 1965-67; chmn. acad. adv. bd. U. S. Naval Acad., 1966-—. Trustee Nat. Commn. on Accrediting, 1961-68, pres., 1964-66; bd. dirs. Council So. Univs., 1960-—, pres., 1964-66; bd. dirs. Am. Univs. Field Staff, 1960-—, pres., 1962-64; bd. dirs. Inst. Def. Analyses, Council Financial Aid to Edn., S.W. Research Inst., Baking Research Found., Grad. Research Center of S.W., chmn. trustees Nutrition Found., 1965-—. Mem. Sigma Xi (exec. com. 1965-—). Research in chemistry and physiology of fats and oils. Home: 2 Audubon Pl., New Orleans 70118.*

LONGET, François-Achille, French physician; b. St. Germain-en-Laye, France, May 25, 1811; named prof. physiology Paris Faculty of Medicine, 1853; mem. French Acad. Scis., 1860, French Acad. Medicine. Author: Traité d'anatomie et de physiologie du système nerveux, 1842; Sur les fonctions de l'épiglotte, 1841. Introduced ideas on physiology of bulbs and cerebellar middle peduncies; studied functions of epiglottis; clarified difference between mixed, sensory and motor nerves. Died Bordeaux, France, Apr. 20, 1871.

LONGLEY, Albert Edward, Am. cytogeneticist; b. Paradise, N.S., Can., Mar. 12, 1893; s. Joseph Spurgeon and Tryphena (Kinley) L.; came to U. S., 1920, naturalized, 1927; B.Sc., Acadia U., 1920; M.A., Harvard, 1922, Ph.D., 1923; m. Pheobe Mae Palmer, Apr. 17, 1922; children—Mary Olga (Mrs. Edward Harold Coe). With U. S. Dept. Agr., 1923-60, geneticist, Washington, 1943-47, Pasadena, Cal., 1947-60; staff mem. Rockefeller Found., intermittantly 1960-64. Mem. A.A.A.S., Genetics Soc., Bot. Soc. Am., Sigma Xi, Gamma Alpha. Contbr. articles to tech. jours. Determined chromosome numbers in crop plants; research on morphology chromosomes maize and its relatives. Home: 2304 Mission Ct., Columbia, Mo. 65201.*

LONGLEY, (William) Warren, geologist; b. Paradise, N.S., Can., Apr. 8, 1909; s. Harold Graham and Mabel (Longley) L.; B.S., Acadia U., 1931; M.S., U. Minn., 1937, Ph.D., 1937; m. Anita Sallans, Sept. 15, 1935 (div. July 1956); children—William Warren, Margaret (Mrs. Lewis Beisner), Barbara (Mrs. Robert Hardifer); m. 2d, Betty Clark Johns, Sept. 10, 1957. Came to U. S., 1931, naturalized, 1953. Asst., geologist Canadian Geol. Survey, 1930-35; instr. Dartmouth, 1935-40; faculty U. Colo., Boulder, 1940-—, prof. geology, 1952-—; collaborator seismology U. S. Coast and Geodetic Survey, 1954-—. Geologist, Que. (Can.) Dept. Mines, 1936-43, 46, 50; cons. geologist Kennecott Copper Corp., 1944-45. Fellow Geol. Soc. Am., Geol. Assn. Can.; mem. N.Y. Acads Scis., Soc. Econ. Geologists, Soc. Exploration Geophysics, Seismol. Soc. Am., Am. Assn. Petroleum Geologists, Am. Geophys. Union, Am. Soc. Photogrammetry, Sigma Xi. Contbr. articles to tech. jours. Exploration and discovery mineral deposits in pre-Cambrian Shield Can. Home: 821 Spring Dr., Boulder, Colo. 80302.*

LONGLEY, W(illiam) H(arding), biologist; b. Paradise, N.S., Can., Oct. 27, 1881; s. Israel Manning and Ermina Judson (Morse) L.; A.B., Acadia U., Wolfville, N.S., 1901, D.Sc., 1931; grad. study, Provincial Normal Sch., Truro, N.S., 1902; A.B., Yale, 1907, A.M., 1908, Ph.D., 1910; m. Hazel Fowler Baird, Sept. 12, 1908; children—William Harding, Elizabeth Fowler Baird, John Prescott, James Baird. Came to U. S., 1906, naturalized, 1925. Instr. biology Yale, 1910; asso. prof. botany Goucher Coll., 1911-14, prof., 1914-17, prof. biology, from 1917; exec. officer Tortugas (Fla.) Marine Biol. Lab., Carnegie Instn. Washington, from 1923; collaborator marine invertebrates U. S. Nat. Mus., Died Mar. 10, 1937.

LONGMAN, Ivor Martin, mathematician; b. London, Eng., Mar. 8, 1923; s. Harry and Leah (Glassman) L.; B.A., Emmanuel Coll., Cambridge, 1943, M.A., 1948; D.Sc., Israel Inst. Tech., 1957; m. Rony Zur, Mar. 25, 1953; children—David, Shulamith, Benjamin. Exptl. asst. Radar Research and Devel. Establishment, Malvern, Eng., 1943-46; research scientist Brit. Aluminium Co., Falkirk, Scotland, 1948-49; sci. officer Brit. Iron and Steel Research Assn., Sheffield, 1949-51; research officer, Guest Keen & Nettlefolds Group Research Lab., Wolverhampton, 1952-53; research asst. Israel Inst. Tech., Haifa, 1953-55; research asst. Weizmann Inst., Rehovot, Israel, 1955-58, sr. scientist in applied math. dept., 1962-—; asst. research geophysicist Inst. Geophysics, U. Cal. at Los Angeles, 1958-62; Mem. Royal Astron. Soc., Am. Geophys. Union, Israel Math. Union, Seismol. Soc. Am. Mem. rev. staff Math. Revs., 1962-—. Research, publs. on solution of theoretical problems in geophysics, particularly seismology; devel. spl. math. computing techniques for solving problems with electronic computer. Home: 3 Harimon St., Rehovot, Israel.*

LONGMIRE, Conrad Lee, Am. physicist; b. Loyston, Tenn., Aug. 23, 1921; s. Henry Grant and Flossie (Hill) L.; B.S. in Engring. Physics, U. Ill., 1942; Ph.D. in Theoretical Physics, U. Rochester, 1948; Sc.D. (hon.), New Eng. Coll. Pharmacy, 1961; m. Theresa Maria Izzo, Nov. 28, 1943; children—Judith, Henry, Jonathan, Patrick, Jennifer, Maria, Matthew. Staff mem. radiation lab. Mass. Inst. Tech., 1942-46, Los Alamos Sci. Lab., 1949-62; vis. prof. physics U. Rochester, 1951, Cornell U., 1953-54; alternate div. leader theoretical div. Los Alamos Sci. Lab., 1956-—; cons. govt. agys., bus orgns.; cons. Presidents Sci. Adv. Com., 1960-—. NRC predoctoral fellow, 1946-48; recipient Ernest Orlando Lawrence award AEC, 1961; Meritorious Civilian Service commendation Dept. Air Force, 1965. Fellow Am. Acad. Arts and Scis.; mem. Am. Phys. Soc. Author: Elementary Plasma Physics, 1963. Bd. editors Physics of Fluids, 1960-—. Research and devel. of nuclear weapons; studies of effects of nuclear explosions, research in plasma physics, devel. of controlled thermonuclear reators. Home: 4756 Trinity Dr. Office: Los Alamos Sci. Lab., Los Alamos. 87544.*

LONGMIRE, William Polk, Am. surgeon; b. Sapulpa, Okla., Sept. 14, 1913; s. William Polk and Grace Mae (Weeks) L.; A.B., U. Okla., 1934; M.D., Johns Hopkins, 1938; m. Jane Cornelius, Oct. 28, 1939; children—William Polk III, Gill, Sarah Jane. Faculty, Johns Hopkins Med. Sch., 1943-48; prof. surgery, chmn. dept. U. Cal., Los Angeles, 1948-—; guest prof. Free U. Berlin (Germany) 1952-54, hon. prof., 1956; nat. cons. gen. surgery Office Surgeon Gen., U. S. Army, USAF, 1956-—. Cons. to hosps., state and govt. agys.; chmn. Western sect. Nat. Cancer Chemotherapy Study, 1962-65; chmn. Conf. Com. on Grad. Tng. in Surgery, 1963-65; chmn. surgery study sect. B NIH, 1961-—. Mem. A.C.S. (chmn. forum com. 1959-62), Cal. Med. Assn. (mem. sci. bd. 1962-—), Phi Beta Kappa, Alpha Omega Alpha, Sigma Xi (asso.). Editorial bd. Am. Surgeon, 1955-—, Jour. Surg. Research, 1960-—, Advances in Surgery, 1963, Cal. Medicine, 1963-—. Publs. in devels. in surgery of liver and biliary tract, homo-transplantation of tissues, coronary artery surgery. Home: 1525 Sorrento Dr., Pacific Palisades, Cal. 90272.*

LONGO, Carmelo, Italian mathematician; b. Rome, Italy, June 26, 1912. Joined faculty, 1956; became prof. analytic geometry with elements of projection and descriptive geometry and drawing Faculty Math., Phys. and Natural Scis., U. Parma (Italy), 1959; became prof. geometry Turin (Italy) Poly. Sch., 1961; instr. differential geometry and algebra Faculty Sci., U. Turin. Address: Politecnico, Turin, Italy.

LONGOBURGO, Bruno da, Italian surgeon; b. Cozenza, Italy; flourished 1252; ed. Salerno, Italy; practice medicine, Padua, Italy, Verona, Italy. Author: Chirurgia magna, 1252; writings on wounds, ulcers, eye diseases, polyp in nose, fractures, luxation, diseases of lips and mouth, cancer, edema, dropsy, hernia, stones and lithotomy, castration, erysipelas, dentistry.

LONGOMONTANUS, Christian Severin, Danish astronomer; b. Longberg, Jutland, Oct. 4, 1562; s. Sören (or Severin) ed. at Lemvig, Longberg and Viborg; M.A., Rostock, circa 1601; asst. to Tycho Brahe at Uraniborg, 1589-97; and in Prague, 1600; became rector Viborg, 1603; named prof. U. Copenhagen, 1605; founder Round Tower Obs., Copenhagen (built by King Christian IV), 1632. Author: Systematis mathematici, 1611; Disputatio de eclipsibus, 1616; Cyclometria e lunulis reciproce demonstrata, 1612; Astronomia danica, 1622; Disputationes quatuor astrologicae, 1622; Pentas problematum philosophiae, 1623; De chronolabio historico, 1627; Inventio quadraturae circuli, 1634; Coronis problematica ex mysteriis trium numerorum, 1637; Introductio in theatrum astronomicum, 1639; also others. Opposed Kepler's theory of elliptical orbits with a planetary theory based in part on that of Tycho Brahe. Died Oct. 8, 1647.

LONGSHORE, Hannah E., Am. physician; b. Montgomery County, Md., May 30, 1819; d. Samuel and Paulina Myers; grad. (1st class) Female Med. Coll. (now Woman's Med. Coll.), Phila., 1851; m. Thomas E. Longshore, Mar. 26, 1841; lectr.; to women on physiology and hygiene, apptd. demonstrator anatomy, Woman's Med. Coll., after graduation. Died 1901.

LONGSTAFF, George Blundell, Brit. entomologist, statistician; b. Feb. 12, 1849; s. George Dixon and Maria (Blundell) L.; ed. New Coll., Oxford, Eng., St. Thomas's Hosp.; M.A., M.D., Oxford U.; m. Sara Leam Dixon (dec. 1903); 2 sons, 2 daus.; m. 2d, Mary Jane, 1906. Mem. London County Council for Wandsworth, 1889-1903, chmn. building act. com. Fellow Royal Coll. Physicians, Soc. Antriquaries, Linnean Soc., Chem. Soc., Geol. Soc.; v.p. Royal Statis. Soc., Entomol. Soc. Author: Studies in Statistics, 1891; The Longstaffs of Teesdale and Weardale, 1907; Marriage and the Church, 1910; Butterfly Hunting in Many Lands, 1912. Died May 7, 1921.

LONGSTREET, William, Am. inventor; b. Allentown, N.J., Oct. 6, 1759; s. Stoffel and Abigail (Wooley) L.; m. Hannah Randolph, 1783; at least 6 children. Interested in mech. instruments at early age; moved to Augusta, Ga., 1783, began to work seriously on steam engines; given patent on steam engine he had constructed by Ga. Legislature, 1788; invented and patented breast-roller of cotton gins, before 1801, also designed portable saw mill; built small steam boat which ran on Savannah River, Ga., 1806. Died Augusta, Sept. 11, 1814.

LONGSTRETCH, Miers Fisher, Am. astronomer; b. Phila., 1819; M.D., U. Pa., 1856; began practice medicine, 1856; mgr. Friends Obs., Phila.; mem. Nat. Acad. Sci. (charter), Am. Philos. Soc. Died 1891.

LONGSWORTH, Lewis Gibson, Am. chemist; b. Somerset, Ky., Nov. 16, 1904; s. Lawrence Roscoe and Sarah Elizabeth (Nichols) L.; A.B., Southwestern Coll., Winfield, Kan., 1925; A.M., Kan. U., 1927, Ph.D., 1928; m. Helen Frances Cady, June 24, 1929; children—Anne Louise, Ralph Cady, Stella Caroline. Nat. Research Council fellow in chemistry, Rockefeller Inst., 1928-30; asst. Rockefeller Inst. for Med. Research, New York, 1930-39, asso., 1939-45, member since 1949. Member Nat. Academy Sciences, Am. Chemical Soc., N.Y. Acad. Sciences, Harvey Soc., Phi Beta Kappa, Sigma Xi. Contbr. numerous articles on chemistry to scientific jours. Research on optical methods in electrophoresis; transference phenomena in solutions of electrolytes and electrophoresis of proteins by the moving boundary method; thermal and isothermal diffusion. Home: 144-60 29th Av., Flushing, N.Y. Office: Rockefeller U., 66th St., and York Av., N.Y.C. 10021.

LONGUET-HIGGINS, Hugh Christopher, Brit. chemist; b. Lenham, Kent, Eng., Apr. 11, 1923; s. Henry Hugh Longuet and Albinia Cecil (Bazeley) L.-H.; student (scholar) Winchester Coll., 1935-41, (scholar) Balliol Coll., Oxford, Eng., 1941-45; M.A., Oxford U., D.Phil., 1949; research asso. U. Chgo., 1948-49; lectr., reader theoretical chemistry, U. Manchester (Eng.), 1949-52; prof. theoretical physics King's Coll., London, Eng., 1952-54; John Humphrey prof. theoretical chemistry U. Cambridge (Eng.), 1954—; fellow Corpus Christi Coll., Cambridge, 1954—; warden Leckhampton House, 1961—. Recipient Harrison Meml. prize Chem. Soc., 1950. Fellow Royal Soc., 1958, Chem. Soc., Cambridge Philos. Soc.; mem. Am. Acad. Arts and Scis. (hon. fgn.), Nat. Acad. Scis. (fgn. asso.). Research, numerous publs. on theoretical physics and chemistry, including molecular orbital theory, molecular spectroscopy and simple and polymeric liquid theories; discovered theory of conformal solutions and defined symmetry groups of non-rigid molecules. Home: 26 Dundas St., Edinburgh 3. Office: Dept. Machine Intelligence and Perception, Hope Park Sq., Meadow Lane, Edinburgh 8, Scotland.*

LONGUET-HIGGINS, Michael Selwyn, Brit. mathematician; b. Lenham, Kent, Eng., Dec. 8, 1925; s. Henry Hugh and A. Cecil (Bazeley) L.-H.; B.A., Trinity Coll., Cambridge (Eng.) U., 1946, Ph.D., 1951; m. Joan Redmayne Tattersall, Dec. 23, 1958; children—Ruth, Mark, John, Anne. Staff Admiralty Research Lab., Teddington, Eng., 1945-48; research student, sr. scholar Trinity Coll., Cambridge U., 1948-51, research fellow, 1951-55; Commonwealth Fund fellow Scripps Instn. Oceanography, 1951-55; staff Nat. Inst. Oceanography, Wormley, Eng., 1954—; vis. prof. math. Mass. Inst. Tech., 1958, U. Cal. at San Diego, 1961-62, U. Adelaide (Australia), 1964. Sr. lectr. U. Cal. at San Diego, 1965—. Recipient Rayleigh prize U. Cambridge, 1957. Fellow Royal Soc., 1963. London, Royal Astron. Soc.; mem. Challenger Soc. Research and publs. on a theory of generation of microseismo by 2d-order pressures in sea waves; enumeration of concave uniform polyhedra, theory of water waves, statis. properties of random functions and random surfaces; theoretical studies of ocean currents, planetary waves, elec. induction by ocean currents and waves. Home: Woodpeckers, The Av., Godalming, Eng. Office: Nat. Inst. Oceanography, Wormley, Godalming, Eng.*

LONGWELL, Chester Ray, Am. geologist, educator; b. Spalding, Mo., Oct. 15, 1887; s. John Kilgore and Julia (Megown) L.; B.A., U. Mo., 1915, M.A., 1916; Ph.D., Yale, 1920; LL.D., U. Mo., 1940; m.; 1 dau., Mari Louise (Mrs. Richard N. Walker); m. 2d, Irene Moffat, children—Flora May (Mrs. John R. Davis), Ray M. Tchr. pub. sch., Marion County, Mo., 1908-10; asst. geologist Okla. Geol. Survey, 1916; asst. geologist, geologist U. S. Geol. Survey, 1920—; faculty Yale, 1920-56, prof., 1929-56; research asso. Stanford, Cal., 1956—. Chmn. sect. geology and geography NRC, 1937-40. Mem. Nat. Acad. Sci., Am. Philos. Soc., Am. Acad. Arts and Scis., Am. Geol. Soc. (pres. 1949-50), A.A.A.S., Phi Beta Kappa, Sigma Xi. Author: (with A. Knopf, R. F. Flint) Physical Geology, 1932, 40, 48; (with R. F. Flint) Introduction to Physical Geology, 1955, 62. Research, publs. on field research and geologic mapping in Conn., So. Nev. Home: 1820 Mark Twain St., Palo Alto, Cal.*

LONGWORTH, Nicholas, Am. horticulturist; b. Newark, Jan. 16, 1782; s. Thomas and Apphia (Vanderpoel) L.; studied law in Judge Jacob Burnet's office, Cin.; m. Susan Connor, 1807; 4 children. Clk. brother's store, S.C.; moved to Cin., 1803; practiced law; defended horse thief in his 1st case for which he received 2 copper stills, traded these for 33 acres which were later valued at $2,000,000; entered real estate bus., became millionaire; became interested in horticulture, 1828, producted marketable wine from grapes he had raised; also interested in cultivating strawberries, discovered that stamintate and pistillate plants had to be interplanted if crop was to be successful; waged strawberry war with those who doubted his findings, 1842, wrote numerous articles on subject. Author: A Letter from N. Longworth . . . On the Cultivation of the Grape and Manufacture of the Wine, Also, On the Character and Habits of the Strawberry Plant, 1846. Died Feb. 10, 1863.

LONICERUS, Adam, German naturalist, zoologist; b. Marburg, Germany, Oct. 10, 1528; s. Johann Lonicer; studied medicine at Frankfort, Germany, Freiburg, Germany, math., Marburg; m. dau. of printer Egenolphe. Received position of pensioned naturalist, Frankfort, 1544, and held it for 32 years. Author: Botanicon sive historia plantarum, animalium, metallorum . . . , 1540; Methodus rei herbariae, 1550; Naturalis historlaeo pus novum, 1551-55; Künstliche Conterfeytunge der Bäume . . . , 1557; Venatus et avcupium, 1582. Classified fowls, magpies, wasps, bees, flies and bats as birds of prey; described cetaceans. Died Frankfort, May 19, 1586.

LONNQUIST, John Hall, Am. agronomist; b. Ashland, Wis., May 22, 1916; s. James Oscar and Ethel (Hall) L.; B.S., U. Neb., 1940, Ph.D., 1949; M.S., Kans. State Coll., 1942; postgrad., Ohio State U., Ia. State Coll.; m. Betty Claire Hanson, July 27, 1942; children—John Hall, Ladd, George, Tom, Kathleen, Kristine, Margaret Ann, Ken. Faculty, Pa. State Coll., State College, 1943; faculty U. Neb., Lincoln, 1943-67, Howard S. Wilson prof. agronomy, 1961-67, instr. advanced plant breeding Grad. Sch., 1955-67; dir. Internat. Maize Improvement Center, Mexico, 1967—. Cons., vis. lectr. in plant breeding, Latin Am., 1956, 61; adviser to Rockefeller Found. agrl. programs, Central, S.Am., 1955—. Recipient Distinguished Service to Agr. award Sta. KMMJ, 1962. Fellow Am. Soc. Agronomy (Crop Sci. award 1961); mem. Sociedade Brasileira de Genetica, Sigma Xi, Gamma Sigma Delta, Alpha Zeta. Research, publs. in corn breeding and genetics. Home: 2355 Calumet Ct., Lincoln, Neb. 68503. Office: CIMMYT, Londres 40, Mexico 6, D.F.*

LONSDALE, Dame Kathleen, crystallographer; b. Newbridge, S. Ireland, Jan. 28, 1903; d. Harry Frederic and Jessie (Cameron) Yardley; B.Sc. in Physics, Bedford Coll. for Women, London (Eng.) U., 1922; M.Sc., U. Coll., 1923; D.Sc., U. London, 1929, U. Wales, 1960, U. Leicester (Eng.), 1962, U. Manchester (Eng.), 1962, U. Lancaster (Eng.), 1967; LL.D., U. Leeds (Eng.), 1967, U. Dundee (Scotland), 1968; m. Thomas J. Lonsdale, Aug. 27, 1917; children—Jane (Mrs. Eric Goodwin), Nancy (Mrs. John Dawson), Stephen. Research asst. Royal Instn., London, 1922-24, 32-34, 37-42, Deverhulme fellow, 1935-37, Dewar fellow, 1944-46; faculty U. Coll., London, 1946—, prof. chemistry, 1949-68, fellow, 1952—; mgr., v.p. Royal Instn., 1961-64; mem. Court U., Essex, Eng., 1965—; fellow Bedford Coll., London U., 1966—. Decorated Dame Comdr. Order Brit. Empire. Fellow Phys. Soc. Inst. Physics, Chem. Soc., Am. Mineral. Soc., Royal Soc., 1945 (mem. council, v.p. 1960-61); mem. Internat. Union Crystallography (pres. 1966), Brit. Assn. Advancement Sci. (hon. sec. 1960-65, pres. 1967-68). Author: Structure Factor Tables, 1936; Crystals and X-Rays, 1949; International Tables for X-ray Crystallography, vol. I, 1952, (with N. F. M. Henry) vol. II, 1959, (with J. S. Kasper) vol. III, 1962; also numerous articles. Research on application of X-ray diffraction and other techniques to study organic molecules especially benzene, thermal vibrations, solid state chem. reactions, diamonds and urinary calculi. Home: 125 A Dorset Rd., Bexhill-on-Sea, Sussex, Eng.*

LONSDALE, William, Brit. geologist; b. Bath, Eng., Sept. 9, 1794; s. William and Mary (Wagstaffe) L.; Began geol. studies, 1815; curator Lit. and Sci. Inst. Bath; curator Geol. Soc. London, 1829-42; geol. researcher, from 1842. Recipient Wollaston medal for coral research, 1846. Fellow Geol. Soc. Surveyed, mapped geol. formations of Gloucestershire, 1832; studied fossils in no. and so. Devon, suggested so. Devon limestone fossils date from an age between Silurian and Carboniferous periods (led to Devonian System of classification by Murchison and Sedgwick 1839), 1837. Died Bristol, Eng., Nov. 11, 1871.

LOOF, Pieter Aart Albertus, Dutch zoologist; b. Haarlem, Netherlands, July 15, 1925; s. Pieter Cornelis and Anna (Zuurendonk) L.; B.Sc., U. Amsterdam (Netherlands), 1948, M.Sc., 1955; m. Alida Nellie Mastebroek, May 15, 1956; children—Nellie E., Pieter C., Remmelt, Cornelius A., Philippus L. Guest worker nematology Plantenziektenkundige Dienst, Wageningen, Netherlands, 1955-58; taxonomist nematology dept. Phytopathology Lab., Agrl. U., Wageningen, 1958—; tchr. biology Reformed Tchrs. Sem., Amersfoort, Netherlands. Mem. Soc. European Nematologists, A.A.A.S. Research, publs. on taxonomy of free-living and plant parasitic nematodes including descriptions of new genera and species, morphology and geography of nematodes. Home: 22 Burg. Prinslaan Ede, Netherlands. Office: 15 Geertjesweg, Wageningen, Netherlands.*

LOOFBOUROW, John Robert, Am. biophysicist; b. Cin., Nov. 1, 1902; s. John Wilson and Henrietta (Botts) L.; A.B., U. Cin., 1923; Sc.D., U. Dayton, 1936; m. Dorothea M. Gano, July 6, 1926; 1 son, John Wiltshire. Instr. physics U. Cin., 1925-29, research asso., 1929-35; prof. biophysics U. Dayton, 1935-36; research prof. Institutum Divi Thomae, 1935-40; asso. prof. Mass. Inst. Tech., 1940-45, prof. biophysics and exec. officer biol. dept., 1945—. Exec. sec. radar div., NDRC, 1942-46; cons. Crosley Radio Corp., 1924-34; chmn. spl. adv. bd. AEC, 1947-48. Fellow A.A.A.S., Am. Phys. Soc., Am. Acad. Scis.; mem. Chem. Soc. (London, Eng.), Phys. Soc. (London), Biochem. Soc. (London), Faraday Soc., Optical Soc. Am., Beta Theta Pi. Contbr. to sci. jours. Died Jan. 22, 1951.

LOOMIS, Alfred Lee, Am. physicist; b. N.Y.C., Nov. 4, 1887; s. Henry Patterson and Julia (Stimson) L.; grad. Phillips Acad., Andover, Mass., 1905; A.B., Yale, 1909; LL.B., Harvard, 1912; D.Sc., Wesleyan U., 1932; M.Sc., Yale U., 1933; LL.D., U. of Calif., 1941; children—Alfred Lee, William Farnsworth, Henry. Admitted to New York bar; member Winthrop & Stimson, 1916-20; vice president Bonbright & Co., 1919-33; director (physicist) Loomis Laboratories, Tuxedo Park, New York 1928-41; pres. Loomis Inst. for Scientific Research, Inc., 1930-65, chief div. 14 NDRC, 1940-47. Trustee Carnegie Instn. Wash. Gov., New York Hosp. Decorated Medal for Merit, 1948; King's Medal for Service, 1948. Mem. Am. Physical Soc., Am. Chem. Soc., Am. Astron. Soc., Royal Astron. Soc., Am. Philos. Soc. of Phila., Nat. Acad. Science, Inst. Radio Engrs. Studies of phys. and biol. effects of high frequency sound waves, Aberdeen chronography, precision measurements of time, and elec. potentials of the brain; pioneer in design of Loran Navigation System.*

LOOMIS, Charles Price, Am. sociologist; b. Broomfield, Colo., Oct. 26, 1905; s. George Foote and Sarah (Price) L.; B.S., N.M. Coll. Agr. and Mech. Arts, 1928; M.S., N.C. State Coll., 1929; Ph.D., Harvard, 1933; postgrad. U. Heidelberg, U. Koenigberg (Germany); m. Zona Kemp, May 27, 1950; children—Martha Jane, Laura Sarah, Elizabeth Ann, Vera Virginia. Research prof. Mich. State U., East Lansing, 1944—, head dept. sociology, anthropology, 1944-57. Mem. study com. Hosp. Facilities, Community Orgn. br. Insts. Health USPHS, 1956-58; mem. U. S. Commn. UNESCO, 1963—; program specialist Ford Found. in India, 1964-66. Mem. Am. (pres.), Rural sociol. assns., Soc. Applied Anthropology, Soc. For Study Psychol. Issues, Soc. Study Social Problems. Author: Turrialba, 1953; (with J. Allan Beegle) Rural Sociology, 1953, Rural Social Systems, 1950; (with Olen Leonard) Latin American Social Institutions, 1953; Social Systems, 1960; (with Zona K. Loomis) Modern Social Theories, 1962; also numerous articles. Developed theory of social cultural change explaining techniques used in leap frog change of some communist socs., conceptual scheme called the Processualy Articulated Structural Model for determining and specifying processes and elements involved in social action, some of interpersonal analyses procedures later used in study of diffusion of improved agrl. and med. practices. Home: 1155 Sabron Dr., East Lansing, Mich. 48823.*

LOOMIS, Elias, Am. mathematician, astronomer; b. Willington, Conn., Aug. 7, 1811; s. Heskel and Jerusha (Burt) E.; student Yale, 1830; studied in Paris; LL.D, U. N.Y., 1854. Prof. math. Western Res. Coll., 1837-44; prof. natural history U. N.Y., 1844-47, 49-60; prof. natural history Yale, 1860-89. Author: Analytical Geometry, 1850; Recent Progress in Astronomy, 1850; Introduction to Practical Astronomy, 1856; also numerous articles. Calculated difference in longitude between N.Y. and other cities; determined speed of electric current in wire; made observations for determining altitude of shooting stars; 1st in U. S. to observe Halley's comet on its return to perihelion, 1835; computed elements of its orbit. Died 1889.

LOOMIS, Elmer Howard, Am. physicist; b. Vermillion, N.Y., May 24, 1861; s. Hiram Warren and Adaline Sabra (Sayles) L.; A.B., Madison (now Colgate) U., 1883; student Göttingen and Strassburg univs., 1890-93, Ph.D., Strassburg, 1893; Sc.D. (hon.), Colgate, 1910; m. Mary E. Bennett, July 23, 1885 (dec. 1904); 1 son, Robert B.; m. 2d, Grace Eaton Woods, Oct. 12, 1911. Tchr. physics and chemistry Colgate Acad., Hamilton, N.Y., 1883-90; instr., asst. prof. and prof. physics. Princeton, 1894-1929; dir. N.J. Sanitarium for Th, McKinley Hosp., Trenton; dir. N.J. Anti-Tb Assn. Devised important improvement in methods of determining freezing points of dilute solutions. Died Jan. 22, 1931.

LOOMIS, Francis Wheeler, Am. physicist; b. Parkersburg, W.Va., Aug. 4, 1889; s. Charles Wheeler and Miriam (Nye) L.; A.B., Harvard, 1910, A.M., 1913, Ph.D., 1917; s. Edith Livingston Smith, July 24, 1922; children—Margaret (Mrs. Charles Ragan Weaver), Ann Livingston (Mrs. Robert Herman Selsbee), Miriam Nye (Mrs. Wallace Russell Baker). Asst. prof., asso. prof. physics N.Y. U., 1920-29; prof. physics, head dept. U. Ill., Urbana, 1929-57, dir. Control Systems Lab., 1952-59, prof. emeritus, 1959—; asso. dir. Radiation Lab., Mass. Inst. Tech., Cambridge, 1941-45, dir. project Charles and Lincoln Lab., 1951-52. Mem. Am. Phys. Soc. (pres. 1949), A.A.A.S. (v.p. sect. B 1948), Nat. Acad. Scis., Phi Beta Kappa, Sigma Xi. Contbns. include discovery of isotope effect in molecular spectra, analysis of vibrational and rotational structure of diatomic molecules of iodine and of alkalis by correlation of their spectra in fluorescene, in magnetic rotation and in absorption. Home: 804 W. Illinois St., Urbana, Ill. 61801.*

LOOMIS, Frederic Brewster, Am. biologist; b. Brooklyn, Nov. 22, 1873; s. Nathaniel H. and Julia R. (Brewster) L.; B.A., Amherst Coll., 1896; student U. Munich, 1897-99, Ph.D., 1899; m. Florence C. Calhoun, Sept. 7, 1904; children—Newell Calhoun, Frederic Brewster. Asst., biol. lab. Amherst Coll., 1896-97, instr. biology, 1891-1903, asso. prof., 1903-08, prof. comparative anatomy, 1908-17, prof. geology, 1917—. Dir. Amherst exploring expeditions to Big Bad Lands, S.D., Wasatch Basin, Wyo., Converse County, Wyo., Sioux County, Neb., Maine shell heaps, Patagonia, N.E. Colo., New Mexico and Fla. Fellow Am. Acad. Arts and Scis., Geol. Soc. Am. Author: Hunting Extinct Animals in the Patagonian Pampas, 1913; The Deseado Formation of Patagonia,

1915; Common Rocks and Minerals, 1923; Evolution of the Horse, 1926. Died July 28, 1937.

LOOMIS, Mahlon, Am. dentist, inventor; b. Oppenheim, N.Y., July 21, 1826; s. Nathan and Waitie (Jenks) L.; studied dentistry, Cleve., 1848; m. Achsah Ashley, May 28, 1856. Practiced dentistry, Earlville, N.Y., Cambridgeport, Mass., Phila., 1848-60; began to expt. in electricity, 1860, an early expt. was to force growth of plants by buried metal plates attached to batteries; carried on 2-way wireless communication over distance of 18 miles, 1868; 1st to use aerial in wireless telegraphy; erected towers for aerials; founded Loomis Aerial Telegraph Co. (inc. by Act of Congress, 1870), did not receive financial backing necessary to carry on his expts.; practiced dentistry from mid-1870's. Died Terre Alta, W.Va., Oct. 13, 1886.

LOOMIS, Ted Albert, Am. physician; b. Spokane, Wash., Apr. 24, 1917; s. George Albert and Sadie (Turner) L.; B.S., U. Wash., 1939; M.S., U. Buffalo, 1941, Ph.D., 1943; M.D., Yale, 1946; m. Marion Ruth Adams, Mar. 19, 1944; children—Bonnie Jeanne, Becky Anne. Practice medicine, specializing in pharmacology, Seattle, 1946—; faculty U. Wash. Sch. Medicine, 1947—, now prof. pharmacology; state toxicologist, 1955—. Mem. Soc. Toxicology, Soc. Pharmacology and Exptl. Therapeutics, Soc. Exptl. Biology and Medicine. Research, numerous publs. on pharmacology of anticoagulant drugs, pharmacology and toxicology of insecticides and pesticides.*

LOOMIS, Walter Earl, Am. plant physiologist; b. Makanda, Ill., June 12, 1898; s. Walter Scott and Mary Jane (Anderson) L.; student So. Ill. Normal U., 1917; B.S., U. Ill., 1921; M.S., Cornell U., 1922, Ph.D., 1924; m. Helen Mary Parke, Oct. 9, 1924; children—Walter David, Robert Simpson. Asst., U. Ill., 1921; grad. asst. Cornell U., 1921-24, NRC fellow in botany, 1924-25; asst. prof. U. Ark., 1925-27; asso. prof. Ia. State U., Ames, 1927-43, prof., 1943—, research prof. Ia. Agrl. Exptl. Sta., 1943—; vis. prof. Cornell U., 1955, Duke, 1962, U. N.C., Chapel Hill., 1964-65. Agt., U. S. Dept. Agr., Bur. Plant Industry, 1938, Bur. Econs., 1941; cons. Crop Protection Inst., 1945-48, Assn. Am. Railroads, 1951-60, Standard Oil Co. Ind., 1948-60; spl. sci. aide Rockefeller Found., Mexico, 1954; agrl. adviser Egypt and Italy, 1956, Iran, 1957. Fellow A.A.A.S., Ia. Acad. Sci. (chmn. bot. sect. 1942-43); mem. Am. Soc. Plant Physiologists (pres. 1942-43, Charles Reid Barnes award), Am. Inst. Biol. Sci., Am. Soc. Naturalists, Am. Soc. Hort. Sci., Am. Soc. Agronomy, Bot. Soc. Am., Scandinavian Soc. Plant Physiologists, Sigma Xi, Phi Kappa Phi. Author: (with C. A. Shull) Methods in Plant Physiologist, 1937; (with C. L. Wilson) Botany, 1957. Editor: (with J. Franck) Photosynthesis in Plants, 1949; Growth and Differentiation in Plants, 1953; editorial bd. Plant Physiology, 1953-59; editorial staff, Biol. Abstracts, 1945—. Research, publs. and teaching emphasizing plant translocation and growth. Home: 1003 N. Hyland Av., Ames, Ia. 50010.*

LOOMIS, William Farnsworth, Am. biochemist; b. Tuxedo Park, N.Y., Aug. 11, 1914; s. Alfred L. and Ellen (Farnsworth) L.; S.B. cum laude, Harvard, 1936, M.D. cum laude, 1941; m. Frances Whitman, Oct. 1, 1965; children—Joan, William, Jacqueline, Barton (by previous marriages); Rhonda, Deanna, Cynthia, Jefferson. Asst. prof. biology Mass. Inst. Tech., 1949; asst. dir. Rockefeller Found., 1949-52; dir. Loomis Lab., Greenwich, Conn., 1952-64; prof. biochemistry Brandeis U., Waltham, Mass., 1964—. Mem. Dept. Def. Com. on Biol. Warfare, 1949. Hon. fellow A.A.A.S., N.Y. Acad. Scis.; mem. Am. Soc. Biol. Chemistry, Am. Chem. Soc., Harvey Soc., Am. Zool. Soc., Soc. for Study Growth and Devel., Alpha Omega Alpha. Author: The God Within, 1967; also articles. Editor: (with H. M. Lenhoff) The Biology of Hydra, 1961. Research on aerobic phosphorylation in mitochondria, uncoupling action of dinitrophenol, hydra as tool for study of cellular differentiation, role of carbon dioxide in cytostasis and differentiation, origin of races as adaptations to ultraviolet-induced vitamin D biosynthesis. Home: 148 Plain Rd., Wayland, Mass. 01778. Office: Grad. Dept. Biochemistry, Brandeis U., Waltham, Mass.*

LOONEY, William Boyd, Am. radiobiologist, biophysicist; b. South Clinchfield, Va., Mar. 18, 1922; s. Isaac H. and Lucille (Boyd) L.; student Emory and Henry Coll., 1939-41; B.S., U. S. Naval Acad., 1944; M.D., Med. Coll. Va., 1948; Ph.D. (Spl. USPHS fellow), U. Cambridge (Eng.), 1960; m. Elizabeth Faris, May 21, 1955; 1 son, William Boyd, II. AEC Research fellow NRC, Argonne Nat. Lab., 1950-52; asst. U. Chgo., 1950-52; vis. research fellow Mass. Inst. Tech., 1955-57; clin. and research fellow Mass. Gen. Hosp., also Harvard Med. Sch., Boston, 1955-57; asst. prof. radiobiology Johns Hopkins, 1959-60; faculty U. Va. Sch. Medicine, Charlottesville, 1961—, asso. prof. radiobiology and biophysics, 1964—; asst. dir. in charge research Radioisotope Lab., U. S. Naval Hosp., Nat. Med. Center Bethesda, Md., 1952-55. Cons. to pvt. cos., govt. agys., Egyptian AEC. Mem. A.A.A.S., A.M.A., Assn. for Radiation Research (Gt. Britain), Va. Med. Soc., Med. Soc. D.C., Radiation Research Soc., Biophys. Soc., Fedn. Am. Sci., Am. Soc. for Cell Biology, N.Y. Acad. Scis., Royal Micros. Soc. (Gt. Britain), Genetics Soc. Am., Am. Assn. for Cancer Research, Sigma Xi. Research, publs. on continuous low level irradiation in man, strontium and calcium metabolism, in vivo studies DNA replication in higher replication in higher organisms, effects radiation on nucleic acid synthesis, biol. effects radiation on mammalian cells. Home: Suffolk Rd., West Leigh, Charlottesville, Va. 22901.*

LOOPER, Edward Anderson, Am. otolaryngologist; b. Silver City, Ga., Dec. 16, 1888; s. John Anderson and Jennie (Stewart) L.; M.D., U. Md., 1912; Emory U., 1908-10; Ophthal.D., U. Colo., 1913; m. Lola Patenall, Jan. 15, 1920; children—Edward A., Lola Elise, Sybil Ann. Began practice as specialist in eye, ear, nose and throat, Balt., 1913; prof. diseases of nose and throat U. Md., since 1921; laryngologist Univ. Hosp. since 1921; surgeon, Balt. Eye, Ear and Throat Hosp., since 1930; mem. exec. com. and mem. staff Woman's Hosp. since 1931; oto-laryngologist Md. State Sanatorium for Tb, since 1922; otolaryngologist, Eudowood Sanatorium for Tb, since 1920; laryngologist Md. Gen., St. Agnes, Franklin Sq., W. Balt. Gen., Nurses and Child's hosps.; mem. staff Union Meml., Mercy hosps.; cons. laryngologist Kernan Hosp. for Crippled Children; bronchoscopist and esophagoscopist Univ. Hosp. and U. S. Marine Hosp.; cons. otolaryngologist Provident Hosp.; otolaryngologist Balt. City Hosps. for Tb, Edward McCready Meml. Hosp., Crisfield, Md., Havre de Grace Hosp., Md. Fellow A.C.S.; mem. Am. Bronchoscopic Soc. (v.p.), Am. Rhinol., Laryngol. and Otol. Soc., Am. Acad. Ophthalmology and Oto-Laryngology, Am. Laryngol. Assn., Internat. Coll. Surgeons, Med. and Chirurg. Faculty of Md., So. Med. Soc., Am. Coll. Chest Physicians (asso.). Author: The Diagnosis and Treatment of Laryngeal Tuberculosis, 1937; also articles related to subjects in specialty. Died Jan. 14, 1953.

LOOSANOFF, Victor Leo, marine biologist; b. Russia, Oct. 3, 1899; s. Leo Ignatius and Lydia (Mihailovna) L.; came to U. S., 1922, naturalized, 1928; B.S., U. Wash., 1927; Ph.D., Yale, 1936; m. Tamara Alexis Dimitrieff, Jan. 22, 1928. Biologist, State Wash. Dept. Fisheries, 1927-31; chief marine biologist State Va., 1931; aquatic biologist U. S. Fish and Wildlife Service, 1931—; lab. dir. U. S. Bur. Comml. Fisheries Biol. Lab., Milford, Conn., 1931-62; sr. scientist U. S. Bur. Comml. Fisheries, Tiburon, Cal., 1962—; adj. prof. U. Pacific; lectr. U. Wash.; spl. sci. adviser Yale, Bingham Oceanographic Lab., 1947-62; prof. dept. zoology Rutgers U., 1953-59. Fellow A.A.A.S.; mem. Conn. (v.p. 1947-51), N.Y. acads. arts and scis., Nat. Shellfisheries Assn. (pres. 1948-49), Am. Soc. Zoologists, Am. Ecol. Soc., Am. Soc. Limnology and Oceanography, Atlantic, Pacific fisheries biologists, Gulf and Caribbean Fisheries Inst., Cal. Acad. Scis., Sigma Xi. Research, publs. on physiology, propagation and protection of comml. mollusks such as oysters and clams; developed methods of operations of shellfish hatcheries methods for protection of mollusks. Home: 17 Los Cerros Dr., Greenbrae, Cal. 94904. Office: U. Pacific Lab., Dillon Beach, Cal. 94929.*

LOPEZ IBOR, Juan José, Spanish psychoneurologist; b. Sollana, Apr. 22, 1908; s. Miguel and Vicenta (Ibor) L.; ed. univs. of Valencia, Madrid, Berlin, Munich, Zurich, Paris; M.D.; Dr. honoris causa U. San Marcos de Luina; m. Feb. 11, 1942; children—Juan, Miguel, Francisco, Javier, Alfonso, Maria José, Marta, Blanca, Luis, Alicia, Carlos, Sofia. Dir., U. Hosp.; chief services Neuropsychiat. Hosp.; chief dept. psychiat. research; prof. psychiatry U. Madrid. Recipient Order of Health; Order of Liberator (Venezuela). Mem. Royal Acad. Medicine, Royal Medico-Psychol. Assn. (hon.), Am. Neurol. Assn., Lisbon Acad. Sci., Am. Assn. Psychiatry, Austrian, French, German, Swedish, Swiss, Venezuelan socs. neurology. Home: calle Peguerinos 27, Puera de Hierro, Madrid. Office: Catedra de psiquiatria, Hospital clinico, Madrid 3, Spain.

LOPEZ-OCHOTERENA, Eucario, Mexican biologist; b. México, Mexico, Aug. 20, 1927; s. Eucario and Celia (Ochoterena) López; Dr. en Biologiá, Nat. Autonomous U. Mexico, 1949; m. Esperanza Barajas-Casso, Oct. 4, 1953. Tchr. biology prep. schs., 1953-64; prof. Faculty Scis., Nat. Autonomous U. Mexico, 1962—; prof. zoology Escuela Nacional de Ciencias Biologicas, I.P.N., 1965—; prof. Escuela Nacional de Antropología e Historia, 1965—. Mem. Colegio de Biologos de Mexico (past treas.), Colegio de Profesores de la Facultad de Ciencias (past pres.). Research, publs. on protozoology especially free-living and ciliates, photomicrography. Home: Sinaloa No. 75-11 México 7, D.F., Mexico.*

LOPUSZANSKI, Jan Tadeusz, Polish physicist; b. Lwow, Poland, Oct. 21, 1923; s. Wladyslaw Ignacy and Janina (Kuzmicz) L.; M.A., U. Wroclaw (Poland), 1950; Ph.D., Jagellonan U., Cracow, Poland, 1956; m. Halina Emilia Pidek, July 14, 1953; 1 son, Maciej. Faculty, U. Wroclaw, 1950—, asso. prof., 1959—, vice dean faculty math., physics and chemistry, 1957-58, dean, 1962-64; research scholar Instituut voor Theor. Physica, Utrecht, Holland, 1958, N.Y. U., 1960-61, Inst. for Advanced Study, Princeton, N.J., 1964-65. Decorated Silver Cross of Merit, Chivalrous Cross of Order of Polonia Restituta. Mem. Polish, Am. phys. socs., Polish Math. Soc. Author: (with A. Pawlikowski) Statistical Physics, 1965; also articles. Research on cosmic rays, statis. physics, quantum field theory. Home: 79 Braci Glerymski, Wroclaw 12, Poland.*

LORAIN, Paul Joseph, French physician; b. Paris, 1827; ed. Lyons, France; joined Faculty Medicine, Paris, 1860; named prof. history medicine of Paris, 1873. Recipient gold medal for work during cholera epidemic, 1866. Described infantile traits which persisted into adolescence and adulthood including phys., sexual and mental underdevel. (Lorain syndrome), 1871. Died 1875.

LORAINE, John Alexander, Brit. endocrinologist; b. Edinburgh, Scotland, May 14, 1924; s. L. D. and Ruth (Jack) L.; M.B.Ch.B., U. Edinburgh, 1946, Ph.D., 1949, D.Sc., 1959; m. Diana G. M. Macpherson, Oct. 19, 1960. Dir. clin. endocrinology research unit Med. Research Council, Edinburgh, 1961—; asso. endocrinology Western Hosp., Edinburgh, 1957—; hon. sr. lectr. dept. pharmacology U. Edinburgh, 1965—; hon. cons. endocrinologist Royal Infirmary, Edinburgh, 1962—; vis. prof. Donner Lab. and Donner Pavilion, U. Cal. at Berkeley, 1964. Fellow Royal Coll. Physicians Edinburgh; mem. Soc. Endocrinology, Soc. Study Fertility. Author: Hormone Assays and their Clinical Application, 1958, 2d edit. (with E. T. Bell), 1966. Editor (with E. T. Bell) Recent Research on Gonadotrophic Hormones, 1967. European editor, Vitamins and Hormones. Research and publs. on bioassay methods for hormones, especially gonadotrophins of pituitary and placental origin, clin. application of hormone assays, endocrinology of fertility, contraception and abnormal gynecolocial conditions, relationship of endocrinology to gastroenterology and nutrition. Home: 37 Buckingham Terrace. Office: Clin. Endocrinology Research Unit, 2 Forrest Rd., Edinburgh 1, Scotland.*

LORAND, Laszlo, biochemist; b. Gyor, Hungary, Mar. 23, 1923; s. Hugo and Margaret (Klein) L.; Absolutorium, U. Budapest (Hungary) Med. Sch., 1948; Ph.D. in Biomolecular Structure, U. Leeds (Eng.), 1951; m. Joyce A. Bruner, Nov. 28, 1953; 1 dau., Michele Alexandra. Came to U. S., 1952, naturalized, 1957. Demonstrator dept. biochemistry U. Budapest, 1946-48; research asst. U. Leeds, 1948-52; research asso., asst. prof. dept. physiology and pharmacology Wayne U. Med. Sch., Detroit, 1952-55; faculty Northwestern U., Evanston, Ill., 1955—, prof. biochemistry div. dept. chemistry, 1961—. Recipient USPHS Research Career award Nat. Heart Inst., 1962. Mem. Soc. Biol. Chemists, Am. Chem. Soc., Biochem. Soc. (London), Internat. Soc. Hematology, Soc. Gen. Physiology, Soc. for Exptl. Biology and Medicine, A.A.A.S., Am. Physiol. Soc. Research, numerous publs. on molecular basis blood clotting and muscle relaxation; elucidated chem. events underlying clotting of fibrinogen in blood.*

LORBER, John, pediatrician; b. Budapest, Hungary, Oct. 12, 1915; s. Akos and Elizabeth (Lichtmann) L.; B.A., U. Cambridge (Eng.), 1943, M.B., B.Chirurg., 1944, M.D., 1951; m. Joan R. I. Young, Apr. 14, 1945; children—Steven John, Diana. Research asst. U. Sheffield (Eng.), 1948-50, lectr. 1950-53, sr. lectr., 1953-57, reader, 1957—; WHO vis. prof. pediatrics U. Ceylon, Kandy and Colombo, 1965; vis. prof., lectr. univs. U. S. A., Can., Australia, New Zealand, India, Europe, S.E. Asia. Fellow Royal Coll. Physicians London, Royal Soc. Medicine London; mem. Brit. Pediatric Assn., Brit. Tb Assn., Assn. for Research into Congenital Hydrocephalus and Spina Bifida. Research, numerous publs. on causation, elimination and treatment in childhood Tb, genetics and therapy of hydrocephalus and spina bifida, pediatric neurology. Home: 305 Ecclesall Rd. S., Sheffield, Yorkshire, Eng.*

LORBER, Stanley, Am. physician; b. N.Y.C., Nov. 23, 1917; s. Samuel and Martha (Oberlander) L.; A.B., U. Pa., 1939, M.D., 1943; m. Selma Rosen, Aug. 16, 1945; children—Susan Alice, Betty Joyce, Jeffrey H. Asst. instr. medicine U. Pa., 1946-48; asso. Temple U. Sch. Medicine, 1948-54, faculty, 1954—, clin. prof. medicine, head dept. gastroenterology, 1963—; lectr. medicine U. Pa., 1960—. Cons. gastroenterology VA, 1955—. Diplomate Am. Bd. Internal Medicine. Mem. Pa., Philadelphia County med. socs., A.M.A., Am. Gastroent. Assn., Am. Phys. Soc., Sigma Xi. Contbr. articles to profl. jours. Motility studies of esophagus and stomach; research on pancreatic secretion and effects pancreas on gastric secretion, inhibition of gastric secretion, duodenal mechanism. Home: 908 Church Rd., Wyncote, Pa. 19095. Office: Temple U. Hosp., 3401 N. Broad St., Phila. 19140.*

LORBER, Victor, Am. physiologist; b. Cleve., Apr. 22, 1912; s. Samuel Zachary and Anna (Elconin) L.; B.S., U. Chgo., 1933; M.D., U. Ill., 1937; Ph.D., U. Minn., 1943; m. Friedel Melanie Mundstock, Mar. 12, 1937; children—Ruth, Peter, Margaret. Faculty, U. Minn., 1943-46, asst. prof., 1944-46; asso. prof. biochemistry Western Res. U., 1946-51, career investigator Am. Heart Assn., prof. biochemistry, 1951-52; career investigator, prof. physiology U. Minn., Mpls., 1952—. Mem. Am. Physiol. Soc., Soc. Exptl. Biology and Medicine, Minn., Am. heart assns. Research, publs. on over-all and intermediate energy metabolism in heart muscle; role of inorganic ions in heart muscle function.*

LORCH, Edgar Raymond, Am. mathematician; b. Canton de Vaud, Switzerland, July 22, 1907; s. Henry and Marthe (Racine) L.; came to U. S., 1917, naturalized, 1931; m. Else Petersen, July 31, 1937; children—Duncan, Madeleine, Ingrid; m. 2d, Maristella de Panizza, Mar. 25, 1956; children—Lavinia Edgarda, Douatella Livia. Faculty, Columbia, 1935—, prof. math., 1948—, chmn. Barnard Coll. sect., 1948-63; vis. prof. Carnegie Inst. Tech., 1950, Stanford, 1963, Middle E. Tech. U., Ankara, Turkey, 1965; vis. lectr. U. Rome, 1966. Fulbright lectr. Coll. de France, 1958. NRC fellow Harvard, 1933-34; Cutting Traveling fellow, Europe, 1934-35; Fulbright lectr. U. Rome, 1953-54. Mem. Am., French, Austrian math. socs., Math. Assn. Am., Phi Beta Kappa, Sigma Xi. Author: Spectral Theory, 1962; also articles. Research in theory Banch spaces and algebras, theory convex bodies, theory abstract integration, rings continuous functions. Home: 838 Riverside Dr., N.Y.C. 10032.*

LORD, Henry Curwen, Am. astronomer; b. Cin., Apr. 17, 1866; s. Henry Clark and Eliza Burnet (Wright) L.; Ohio State U., 1884-87; B.S., U. Wis., 1889; m. Edith L. Hudson, June 22, 1898. Asst. in math. and astronomy Ohio State U., 1891-94, asso. prof. astronomy, 1894-1900, prof., 1900—, dir. Emerson McMillin Obs., 1894—. Fellow Royal Astron. Soc. Deceased.

LORD, John Prentiss, Am. surgeon; b. Dixon, Ill., Apr. 17, 1860; s. John L. and Mary Louise (Warner) L.; M.D., Rush Med. Coll., 1882; post-grad. N.Y. Post-Grad. Coll., 1886; m. Minnie L. Swingley, Oct. 20, 1886; children—Frances Louise, Upton Prentiss. Practiced at Creston, Ill., 1882-86; settled at Omaha, 1886; surgery, exclusively, 1893—; prof. anatomy Creighton Med. Coll., 1892, prof. surgery, 1893-1913; prof. orthopedic surgery U. of Neb., 1913—; orthopedic surgeon Clarkson U., Methodist and St. Catherine's hosps.; attending orthopedic surgeon Lord Lister Hosp.; chief surgeon Neb. Orthopedic Hosp.; cons. orthopedic surgeon Convalescing Home for Crippled Children; dist. surgeon I.C. R.R.; cons. surgeon C.,R.I.&P. Ry. Pres. Omaha Midwest Clin. Assn., 1935. Fellow A.C.S.; mem. numerous profl. socs. Died Mar. 3, 1940.

LORD, Nathaniel Wright, Am. metallurgist; b. Cin., Dec. 26, 1854; s. Henry C. and Eliza (Wright) L.; E.M., Columbia Coll. Sch. Mines, 1876; became chemist and engr. Monte Grande Gold Mining Co., 1879; chemist Ohio Geol. Survey 1883-88; prof. metallurgy and mineralogy Ohio State U.; chemist in charge analysis of fertilizers for Ohio Bd. Agr. Author: Notes on Metallurgical Analysis. Contbg. author: Iron Manufacture of Ohio, vol. 5, Natural and Artificial Cements, vol. 6 (Ohio Geol. Survey). Died 1911.

LORD, Rexford D., Am. ecologist; b. West Reading, Pa., July 31, 1927; s. Rexford D. and Ella (Matheson) L.; B.A., Pa. State U., 1950; M.S., Tex. A. and M. U., 1953; Sc.D., Johns Hopkins, 1956; m. Julia S. Wood, Sept. 1, 1951; children—Rex H., Esther A.C., Raymond. Ecologist, Ill. Nat. History Survey, 1956-62, Pan-Am. Zoonoses Center, Pan Am. Health Org., Azul, Argentina, 1962-63, Tech. Analysis br. U. S. AEC, 1963-64; ecologist virology sect. Communicable Disease Center USPHS, Atlanta, 1964—. Mem. Ecol. Soc. Am., Am. Soc. Mammalogists, Wildlife Soc., Am. Ornithologists Union. Author: The Cottontail Rabbit in Illinois, 1963; also articles. Pioneer radio-telemetry of biol. information from wild animals; discovery technique for estimating age of mammals through growth rate of lens of eye; discovery increase litter size with increase in latitude in N.Am. mammals. Home: 1235 S. Indiancreek Dr., Stone Mountain, Ga. Office: 1600 Clifton Rd., Atlanta 30333.*

LORD, Richard Collins, Am. chemist; b. Louisville, Oct. 10, 1910; s. Richard C. and Katherine (Trimble) L.; B.Sc., Kenyon Coll., 1931, D.Sc., 1957; Ph.D., Johns Hopkins, 1936; m. Wilhelmina Van Dyke, June 5, 1943; children Diana Van Dyke (Mrs. Scott Adam), Susan Trimble, Margaret Collins, Catherine Lewis. NRC fellow U. Mich., 1937, U. Copenhagen, 1938; asst. prof. Johns Hopkins, 1939-42; tech. aide optics div. NDRC, 1942-46; faculty Mass. Inst. Tech., Cambridge, 1946—, prof. chemistry, 1954—, dir. Spectroscopy Lab., 1946—. Vis. lectr. Inst. Chemistry, London, Eng., 1964; Reilly lectr. Notre Dame U., 1958. Recipient Presdl. certificate merit, 1948; Pitts. Spectroscopy award, 1966. Guggenheim fellow, 1959-60. Fellow Optical Soc. Am. (past pres.), A.A.A.S., Am. Acad. Arts and Scis.; mem. Internat. Union Pure and Applied Chemistry (chmn. com. molecular structure spectroscopy 1961-67), Am. Chem. Soc., Soc. Applied Spectroscopy, Phi Beta Kappa, Sigma Xi. Author: (with G. R. Harrison, J. R. Loofbourow) Practical Spectroscopy, 1948; also numerous articles. Developed techniques infrared and Raman spectroscopy for research in molecular spectroscopy; elucidated intramolecular forces particularly in cyclic molecules; research on hydrogen bonding in molecules of biol. significance. Home: 16 Spafford Rd., Milton, Mass. 02186. Office: Spectroscopy Lab., Mass. Inst. Tech., Cambridge, Mass. 02139.*

LORENTE DE NO, R(afael), physiologist; b. Zaragoza, Spain, Apr. 8, 1902; s. Francisco Lorente and Maria de Nó (de Lorente); M.D.; U. Madrid, 1923; M.D., honoris causa, Upsala University, 1953; m. Hede Birfeld, Mar. 21, 1931; 1 dau., Edith. Came to U. S., 1931, naturalized, 1944. Asst., Inst. Cajal, Madrid, Spain, 1921-29; head dept. otolaryngology Valdecilla Hosp., Santander, 1929-31; neuro-anatomist Central Inst. Deaf, St. Louis, 1931-36; asso. physiologist Rockefeller Inst., N.Y. City, 1936-38, asso. mem., 1938-41, mem. since 1941; physiologist since 1924. Mem. Am. Physiol. Soc., Am. Assn. Anatomists, Am. Neurol. Soc., Nat. Academy Sciences, American Academy of Arts and Sciences. Author: A Study of Nerve Physiology: Studies from The Rockefeller Institute, 1947; also articles in sci. jours. Research on neurophysiology; anatomy of the central nervous system. Home: 325 E. 72d St. Office: Rockefeller Institute, 66th St. and York Av., N.Y. City.

LORENTZ, George, mathematician; b. Leningrad, USSR, Feb. 25, 1910; s. Rudolph F. and Milena (Chegodaev) L.; student U. Leningrad, 1928-31; Dr. rer. nat., U. Tübingen, 1945, Dr. rer. nat. habil., 1946; m. Tanny Belikov, Feb. 22, 1942; children—Rudolph Alexander H., Erica Mary, Irene, Olga, Katherine. Came to U. S., 1953, naturalized, 1959. Faculty, U. Leningrad, 1932-41, U. Frankfurt am Main, 1946-48, U. Tübingen, 1948-49, U. Toronto, 1949-53; prof. Wayne State U., 1953-58; prof. Syracuse (N.Y.) U., 1958—. Mem. Am. Math. Soc., Math. Assn. Am., Deutsche Mathem Vereinigung, Canadian Math. Congress. Author: Bernstein Polynomials, 1953; Approximation of Functions, 1966. Research, publs. primarily in fields of approximation, summation of divergent series by means of linear matrices, functional analysis. Home: 319 Southfield Dr., Fayetteville, N.Y. 13066. Office: Dept. Math., Syracuse U., Syracuse, N.Y. 13210.*

LORENTZ, Hendrik Antoon, Dutch physicist; b. Arnheim, Holland, July 18, 1853; Doctorate, Leiden U., 1875. Prof. theoretical physics Leiden U., 1878-1923; dir. Teyler Lab., Haarlem, from 1912; with Columbia, 1906, Collège de France, 1912-13; expeditions to New Guinea, 1907, 1909. Mem. Royal Soc. London. Recipient (with P. Zeeman) Nobel prize for physics, 1902; Rumford medal, 1908, Copley medal, Royal Soc. London, 1918. Author: Over de theorie der terugkaasting en der breking van het licht, 1875; hd Théorie électromagnetique de Maxwell et son application aux corps mouvants, 1892; Versuch einer Theorie der elektrischen und optischen Erscheinungen in bewegten Körpern, 1895; The Theory of Electrons, 1905. Authority on quantum physics; discoverer (independent of Fitzgerald) of Lorentz-Fitzgerald contraction, 1904; research on electromagnetism (led to Einstein's theory of relativity), gravitation, thermodynamics, radiation, kinetic theory; provided explanation of Zeeman effect; deduced mathematically the behavior of light, as indicated in J. C. Maxwell's electromagnetic theory, 1873; formulated electron theory of matter; by solving hydrostatic and hydrodynamic engineering problems contributed significant data for the drainage of the Zuider Zee. Died Haarlem, Holland, Feb. 4, 1928.

LORENZ, Adolf, orthopedic surgeon; b. Weidenau, Silesia, Apr. 21, 1854; s. Johann and Agnes L.; m. Emma Lecher, 1885; grad. U. Vienna, 1880. Prof. orthopedic surgery U. Vienna; worked under Theodor Billroth; surg. demonstrator at med. congresses, Berlin, 1895, Eng., U. S.; councillor Austrian Govt. Author: Orthopädie der Huftgelenks-Kontrakturen und Ankylosen, 1899; Das instrumentelle kombimierte Redressement der Huftgelenks Kontrakturen, 1898; Über die Heilung der angeboremen Hutgelenks-Verrenkung durch unblutige Einrenkung und functionelle Belastung, 1900; (with Saxl) Orthopaedics in Medical Practice, 1913. Father of German orthopedic medicine; developed bloodless method of reducing congenital dislocation of hip joint. Died Weidenau, Feb. 12, 1946.

LORENZ, Frederick Wharton, Am. physiologist; b. Berkeley, Cal., Dec. 30, 1908; s. William Frederick and Alice (Wharton) L.; B.S., U. Cal. at Berkeley, 1931, Ph.D., 1938; m. Ruth Elizabeth Ahnstrom, Apr. 2, 1932; 1 son, Erick Wharton. Faculty, U. Cal. at Davis, 1938—, prof. physiology, chmn. dept. animal physiology, 1964—. NSF Sr. fellow, 1957-58. Mem. Am. Chem. Soc., Am. Physiol. Soc., Am. Soc. Zoologists; Endocrine Soc., Soc. for Study Fertility (Eng.), Soc. for Exptl. Biology and Medicine (chmn. No. Cal. sect. 1966) Poultry Sci. Assn. (Research award 1945, Borden award 1960), Sigma Xi, Alpha Chi Sigma. Research, numerous publs. on hormones and lipogenesis, fertility factors, and spermatology, egg conditions in birds. Home: 810 Oeste Dr., Davis, Cal. 95616.*

LORENZ, Konrad Zacharias, zoologist; b. Vienna, Austria, Nov. 7, 1903; s. Adolf and Emma (Lecher) L.; student medicine N.Y.C., Vienna, M.D., 1928; Ph.D. in zoology, Vienna, 1933; Ph.D. (hon.), U. Leeds, Eng., 1962; M.D. (hon.), U. Basel, 1966, Yale, 1967; m. Margarethe Gebhardt, June 24, 1927; children—Thomas, Dagmar, Agnes (Mrs. Mario von Cranach). Asst. 2d, Anat. Inst., Vienna, 1928-35; lectr. comparative anatomy, psychology U. Vienna, 1937-38, asst. prof., 1938-40; prof. psychology, head dept. U. Königsberg, Germany, 1940-42; head Inst. Comparative Ethology, Altenberg, Austria, 1949-51; head research dept. comparative ethology Max-Planck Found., Buldern, Germany, 1951-54, vice dir. inst., 1954-61, Seewiesen, dir., 1961—. Lectr., U. Vienna; hon. prof. U. Münster, U. Munich. Recipient Golden medal Zool. Soc. N.Y., Preis der Stadt Wien, Goldene Wilhelm Bölsche Medaille, Osterreichisches Ehrenzeichen für Wissenschaft and Kunst. Mem. Austrian, Bavarian, Nat. (asso.) acads. sci., N.Y. Zoöl. Soc., Royal Soc., Am. Acad. Arts and Scis., numerous others. Author: On Aggression, 1966; Evolution and Modification of Behavior, 1966; King Solomon's Ring, 1952; Man Meets Dog, 1954. Pioneer in modern ethology; formulated school of study based on concept that an animal's behavior is a product of adaptive evolution. Address: Max-Planck-Institut, 8131 Seewiesen, Germany.*

LORENZ, Ludvig Valentin, physicist; b. 1829; prof. U. Copenhagen; became prof. Mil. Acad. Copenhagen, 1866. Mem. French Acad. Scis. Research on electromagnetism and optics; developed (with H. A. Corentz) formula relating refraction and specific density of medium. Died 1891.

LORENZ, Werner Carl Albert, German radiotherapist; b. Frankfort, Germany, Apr. 27, 1920; s. Carl and Martha (Kettner) L.; ed. univs. Göttingen, Königsberg (now Kaliningrad, USSR), Fribourg, Marburg; M.D.; m. Gertrud, Apr. 25, 1943; children—Uwe, Antje. Began practice medicine, specializing in radiology and radiotherapy, 1950; named lectr. med. radiography U. Mainz (Germany), 1954; became chief physician, later prof. at large, 1960, cons. scientist, 1962; asso. prof. radiology U. Frankfort (Germany), 1964—. Mem. Radiography Soc., German Med. Soc., German Atom Forum, German Biophysics Soc. Author: Strahlenschutz in Klinik und Arztlicher Prazis, 1961. Research and publs. on radiobiology, treatment of cancer. Home: Kaiserstrasse 341, Mainz, Germany. Office: Universitätsklinik für Strahlentherapie und Nuklearmedizin, Ludwig-Rehn-Strasse 14, Frankfort, W. Germany.

LORENZ, William Frederick, Am. psychiatrist; b. N.Y.C.; Feb. 15, 1882; s. Herman and Elise (Kuenzlen) L.; student N.Y. U., 1898-99; M.D., Bellevue Hosp. Coll., 1903; m. Ada Holt, May 21, 1915; children—Adrian Holt Vanderveer, William Frederick, Thomas Holt, Paul Kuenzlen, Joseph Dean. Med. intern Gen. Hosp., N.Y.C., 1903-05; med. staff Manhattan State Hosp., 1906-10; clin. dir. Wis. State Hosp., 1910-14; spl. expert, research investigation of Pellagra, USPHS, 1914-15; dir. Wis. Psychiat. Inst., 1915—; prof. psychiatry, U. Wis., 1915—. Pres. Wis. Service Recognition Bd., 1921-24; mem. Med. Council U. S. Vets.' Bur., 1923—; pres. Wis. Rehab. Bd., 1925; chmn. State Bd. Mental Hygiene, 1938—. Mem. A.M.A., Am. Psychiat. Assn., Assn. Research in Nervous and Mental Diseases, Central Psychiat. Soc., Milw. Neuro-psychiat. Soc., Sigma Xi. Contbr. new remedies for treatment of syphilis of central nervous system (with Dr. A. S. Loevenhart), 1920-25; work in promoting rehab. of disabled ex-service men; investigations in use of carbon dioxide gas in treatment of psychoses. Died Feb. 18, 1958; buried Madison, Wis.

LORENZINI, Stefano, physician; b. Italy in the 1600's. Author: Osservazioni inlorno alle torpedine, 1678. Showed the electric ray (fish) numbs its victims by direct touch only, 1678; observed cerlam lateral sense organs in Selace.

LORETA, Pietro, Italian surgeon; b. Ravenna, Italy, 1831; ed. U. Bologna (Italy); named anat. prosector U. Bologna, 1861, dir. surg. clinic, 1865, prof. surgery, 1868. Performed 1st successful pyloroplasty (operation of Loreta); emphasized cleanliness in surgery. Died 1889.

LORETI, Francesco, Italian anatomist; b. Bellano, Italy, Dec. 25, 1901; s. Lodovico and Antonietta (Venini) L.; ed. Pavia Faculty Medicine, Borromeo Coll., Pavia; m. Andreina Ricci, Apr. 8, 1943; children—Jacopo, Lorenzo, Franca. Staff, U. Ferrara (Italy), 1947-50; became prof. human anatomy U. Turin (Italy), 1950; dir. Inst. for Phys. Studies in Turin. Mem. Turin, Ferrara acads. medicne and sci., Assn. Doctors Como. Author textbooks. Research and publs. on embryology, histology, human and comparative morphology. Home: via Richelmy 19, Turin. Office: corso Massimio D'Azeglio 52, Turin, Italy.

LORGNA, Antonia Maria, Italian mathematician; b. Cerea, Italy, Oct. 18, 1735; rose to rank of col. of engrs.; then became prof. math. Mil. Sch., Verona, Italy. Fellow Royal Soc., 1788; mem. French Acad. Scis., Societa Italiana (founder 1782). Author: Della graduazione de' termometri a mercurio, 1765; Opuscula mathematica e physica, 1770. Research on math. hydraulics. Died Verona, June 28, 1796.

LORHAN, Paul Herman, Am. physician; b. Montclare, Pa., Apr. 7, 1908; s. George Joseph and Juliana (Straka) L.; A.B., Ohio State U., 1931; M.D., Creighton U., 1935; m. Grace Evelyn Hemphill, Mar. 15, 1941; children—Evelyn Paula (Mrs. Edward C. Cook), Paul Herman, Joseph, Lawrence, James. Faculty, U. Kan., 1938-58, prof. anesthesia, chmn. dept., 1948-58; prof. anesthesia, surgery U. Cal. Med. Center, Los Angeles, 1958—; chief dept. anesthesiology Harbor Gen. Hosp., Torrance, Cal., 1958—. Cons. U. S. Army Hosp., Ft. McArthur, San Pedro, Cal., 1960—, U. S. Naval Hosp., Long Beach, Cal., 1960—. Recipient Silver Plaque, U. Mendes Pelazo, Santander, Spain, 1953. Fellow Internat. Coll. Anesthesiologists; mem. Acad. Anesthesiologists, Am. Soc. Anesthesiologists, A.M.A., South western Surg. Assn., Cal., Los Angeles County med. assns., Cal. Soc. Anesthesiologists. Author: Geriatric Anesthesia, 1955; Growing Old Gracefully, 1965; also numerous articles. Editor: International Anesthesiology Clinics, 1964. Research on effects premedication on arterial oxygen tension, vasopressor pharmacology. Home: 913 Avenida Mirola, Palos Verdes Estates, Cal. Office: 1000 W. Carson St., Torrance, Cal. 90509.*

LORIA, Achille, Italian economist; b. Mantua, Italy, Mar. 2, 1857; studied law, Bologna, Italy, then econs., Rome, Berlin, London; prof. econs., Siena, Italy, 1881-91, Padua, 1891-1903, Turin, Italy, beginning in 1903. Author: Le basi economiche della costituzione soziale, 1886; La costituzione economica odierna, 1889; e la scienza, 1901; Il valore della moneta, 1891; La sociologia, 1900; La sintesi economica, 1900; La terra ed il sistema sociale, 1892; Il capitalismo Le basi economiche della costituzione societa, 1912; I fondamenti scientifica della riforma economica, 1922; Karl Marx, 1924; Corso di economia politica, 1927. Developed theory l'unité foncière (quantity of land necessary for each individual to sustain himself); believed in econ. determinism; studied problems of wealth distbn. and land tenure. Died San Giovanni, Italy, Nov. 6, 1943.

LORIA, Gina, Italian mathematician; b. Mantua, Italy, May 19, 1862; doctor math. scis. U. Turin (Italy), 1883; became prof. higher geometry U. Genoa (Italy), 1886, also in charge descriptive geometry. Author: Algebraic and Transcendental Special Curves, 1902; Past and Present Theories of Geometry, 1909; Storia delle matematiche, 3 vols., 1929-33; History of Mathematics in Hellenic Antiquity, 4 vols., 1929; also 4 vol. work from courses in geometry; collaborated in publ. of Complete Works (Fermat); Scientific Memoires of Tannery. Died Genoa, 1939.

LORIMER, Frank, Am. demographer; b. Bradley, Me., July 1, 1894; s. Addison Benjamin and Florence Olive (Livermore) L.; A.B., Yale, 1916; A.M., U. Chgo., 1921; B.D., Union Theol. Sem., 1923; Ph.D. Columbia, 1929; m. Faith Moors Williams, Aug. 21, 1922 (dec. September 1958); children (adopted)—Luba Joyce (Mrs. William C. Hill), David (dec.), Thomas; m. 2d, E. Peart Barris, Aug. 19, 1962; children—Francine, Deborah Andrea; Asso. dir. Abraham Lincoln Center, Chgo., 1920-21; minister N.Y. City Bapt. Mission Soc., 1922-25; asst. prof. philosophy, Wells Coll., Aurora, N.Y., 1927-28; lectr. Wellesley (Mass.) Coll., 1928-29; research dir. Bklyn. Conf. in Adult Edn., 1929-30; research fellow Eugenics Research Assn., 1930-34; sec. Population Assn. of America, 1934-39, pres., 1946-47; prof. sociology, Grad. Sch. Am. U. 1938-64, now emeritus; cons. to various brs. of govt.; adminstrv. dir. Internat. Population Union, 1948-56; research asso. Princeton U. Office Population Research, 1961-64; vis. prof. demography U. Philippines, 1964-66, Mem. Internat. Union for Sci. Study of Population (past pres.), Am. Statis. Assn., Am. Sociol. Soc. Author: The Growth of Reason: A Study of the Role of Verbal Activity in the Growth of the Structure of the Human Mind, 1929; The Making of Adult Minds in a Metropolitan Area, 1931; Dynamics of Population; Social and Biological Significance of Changing Birth Rates in the United States (with F. Osborn), 1934; Foundations of American Population Policy (with E. Winston and L. Kiser), 1940; Population of the Soviet Union: History and Prospects, 1946; Culture and Human Fertility, 1954; Demographic Information on Tropical Africa, 1961 Research and publs. on the analysis of population trends in U. S. and fgn. countries. Home: 395 Nut Plains Rd., Guilford, Conn. 06437.*

LÖRINC, Andor, Hungarian chemist; b. Sighet, Maramarossziget, Hungary, May 16, 1908; s. Sándor and Paula (Sichermann) Lieberman; Dip.Chem.Eng., German Tech. U., Brno, Czechoslovakia, 1931; Cand. Chem. Sci. with acad. distinction, Hungarian Acad. Sci., 1966; Dr.Techn., Tech. U., 1966; m. Eva Sporer, Mar. 4, 1939; children—Anita Katherine, Andrea Eva. Chemist colorist, sect. chief Hungarian Dyeing Works, Budapest, 1932-39; tech. mgr., chemist colorist factories Porin Puuvilla O.Y., Pori, Finland, 1939-41, Finska Sidenväveri AB, Helsinki, Finland, 1942-44; research chemist Royal Vet. High Sch., Stockholm, also tech. mgr., chem. colorist, Helsingborg, Sweden, 1944-47; chief colorist, sci. sect. chief Research Inst. for Organic Chem. Industry, Budapest, 1947-58, mgr., Central Research Lab. for Coloristics, 1959—. Vis. lectr. U. for Tech. Sci., also Inst. for Postgrad. Course for Engrs., Budapest, 1952-55, 63, 66—; ofcl. expert textile chems. and auxs. Hungarian Ind., Budapest, 1954—. Recipient Medal of Labour, 1960, Commemorative Medal of Carl Than, 1963. Mem. Hungarian Chem. Soc. (mem. presidium, also dep. sec.-gen. 1952—, pres. coloristic sect. 1958—). Author: Colouristical Problem in the Dye Industry, 1952; Theoretical and Practical Problems of Vat- and Naphthol-Dyeing, 1954; Investigations of the Character of AQ-Vat-Dyeing, 1955; Textile Colour Investigations, 1964 (with Péter) Textile Dyeing, 1962; (with Tettamanti) Investigations of Substant. of Naphthol-Comp., 1962; (with Mrs. Erdélyi), Textile Auxiliaries, 1963; also articles. Editor-in-chief Colouristical Rev., 1959—. Developed new method for evaluation of dispersity distbn. of vat dyestuffs and pigments, research method for study dyes; studied azo dyes, including structure and fastness values and components of some insoluble azo dyes; (with F. Péter) indirect polarographic determination of several tensides. Home: 2 II.Fény, Budapest. Office: 71 IV. Váciut Budapest-Ujpest, Hungary.*

LORINCZ, Albert Bela, physician; b. Budapest, Hungary, Sept. 6, 1922; s. Frank Coleman and Theresa (Csore) L.; B.S., U. Chgo., 1944, M.D., 1946; m. Ann Marie Callaghan, Mar. 23, 1946; children—Margaret Alice, Albert Gregory, Paul Francis, Ann Elizabeth, Thomas Andrew, Catherine Bernadette, Peter Henry. Commd. 1st lt. M.C., U. S. Army, 1946, advanced through ranks to lt. col., 1956; ret., 1958; asst. prof. U. Chgo. Sch. Medicine, 1958-61, prof., 1966—; prof. obstetrics and gynecology, chmn. dept. Creighton U. Sch. Medicine, Omaha, 1961-66; dir. depts. obstetrics and gynecology Booth Meml. Hosp., Archbishop Bergan Mercy Hosp., Omaha, 1961-66. Cons. to various hosps. Fellow A.C.S.; mem. Am. Coll. Obstetrics and Gynecologists, Central Assn. Obstetricians and Gynecologists, Am. Chem. Soc., Am. Fertility Soc., A.M.A., Sigma Xi. Contbr. articles to tech. jours. Beta-Glucuronidase activity in genital cancer, patterns protein excretion in pregnancy, free amino acids in breast cancer, renal function in pregnant women with acute hypertension, puerperal cerebral venous thrombosis, free amino acids female reproductive tissues, response malignancy to limiting amino acids. Home: 5634 S. Woodlawn Av., Chgo. 60637.*

LORINCZ, Allan Levente, Am. physician; b. Chgo., Oct. 31, 1924; s. Frank C. and Theresa (Csore) L.; B.S., U. Chgo., 1945, M.D., 1947; m. Lillian Irene Tatter, Feb. 2, 1952; children—Donald, Linda, Alice. Faculty, U. Chgo., 1951—, professor dermatology, 1967—, head, dermatology sect., 1960—. Chief dept. dermatology Walter Reed Army Inst. Research, Washington, 1954-56; nat. cons. in dermatology Surgeon Gen. USAF, 1962—, spl. cons. USPHS, 1961-64; mem. com. on cutaneous system NRC, 1962—. Mem. Chgo. Dermatol. Soc. (pres. 1963), Phi Beta Kappa, Sigma Xi, Alpha Omega Alpha. Researcher in physiology and biochemistry of skin, cutaneous fungous diseases, acne, psoriasis. Home: 9905 S. Kilbourn Av., Oak Lawn, Ill. 60453. Office: 950 E. 59th St., Chgo. 60637.*

LORING, Edward Greely, Am. ophthalmologist; b. Boston, Sept. 28, 1837; s. Edward Greely V. and Harriet (Boott) L.; attended Harvard, 1857-58, M.D., 1864; studied medicine, Florence and Pisa, Italy, 1858-61; m. Chevalita Jarves, Jan. 3, 1866; m. 2d, Helen Swift. Intern in ophthalmology Boston City Hosp., Mass. Charitable Eye and Ear Hosp., 1865-66; practiced medicine, Balt., 1866-67; in partnership with Dr. Cornelius Rea Agnew, 1867-73; practiced in N.Y.C., 1873-88; surgeon Bklyn. Eye and Ear Hosp., Manhattan Eye and Ear Hosp., N.Y. Eye and Ear Family; greatest achievement was improvement of ophthalmoscope. Author: A Text Book on Ophthalmoscopy, 1st vol., 1886, 2d vol., 1891. Died N.Y.C., Apr. 23, 1888.

LORING, Hubert Scott, Am. biochemist; b. Belize, Brit. Honduras, Nov. 19, 1908 (parents Am. citizens); s. Hubert Whittier and Jessie (Scott) L.; A.B., Pomona Coll., 1929; M.S., U. Ill., Ph.D., 1934; m. Dorothy Laxson, Aug. 1933; children—Margaret Elizabeth (Mrs. David Goodale), William. Faculty, Stanford, 1936—, prof. biochemistry, 1947—; Walker-Ames prof. U. Wash., 1952; vis. prof. U. Cal., Berkeley, 1953. Mem. Am. Soc. Biol. Chemists, Am. Chem. Soc. Contbr. chpt. to The Nucleic Acids, 1955. Research, numerous publs. in biochemistry and molecular biology of plant viruses, plant nucleases, aggregation and dissociation of tobacco mosaic virus components. Home: 716 Salvatierra St., Stanford, Cal. 94305.*

LORING, John Alden, Am. naturalist; b. Cleve., Mar. 6, 1871; s. Benjamin William and Nellie (Cohoon) L.; ed. Owego Free Acad., zoöl. gardens, Europe, 1 yr. Field naturalist U. S. Biol. Survey, 1892-97; curator animals N.Y. Zool. Park, 1897-1901; field naturalist in Europe for U. S. Nat. Mus. (broke previous records for field work by collecting and preserving skins of 913 mammals and birds in 63 days); field naturalist Smithsonian-Roosevelt sci. expdn. to Africa, 1909-10. Mem. Am. Ornithologists' Union, Biol. Soc. Washington. Author: Young Folks' Nature Field Book; African Adventure Stories; also articles relating chiefly to birds and mammals. Died May 8, 1947.

LORING, Ralph Alden, Am. physicist; b. Natick, Mass., June 16, 1897; s. Henry Everett and Florence Lindsay (Keith) L.; B.S., Dartmouth, 1920; A.M., Harvard, 1927; Ph.D., Ohio State U., 1932; m. Louise Adele Russell, Aug. 21, 1925; children—Elizabeth Louise, David Henry, Judith Adele. Instr. math. Dartmouth, 1920-22; asst. in physics Harvard, 1922-26, instr., 1926-27; instr. in physics Northwestern U., 1927-31; hon. research fellow in physics Ohio State U., 1932-34; asst. prof. U. Louisville, 1934-37, asso. prof. physics, 1937-43, prof. since 1943, head dept. physics since 1934; research with AEC, Oak Ridge, 1952. Mem. Am. Phys. Soc., A.A.A.S., Am. Assn. Physics Tchrs., Phi Beta Kappa, Sigma Xi. Contbr. articles on spectroscopy to phys. jours. Died Dec. 31, 1952.

LORRAIN, Paul, Canadian physicist; b. Montreal, Que., Can., Sept. 8, 1916; s. Joseph Alphonse and Marie-Ange (Le Bel) L.; B.A., U. Ottawa, 1937; B.Sc., McGill U., 1940, M.Sc., 1941, Ph.D., 1947; m. Dorothee Sainte-Marie, May 22, 1944; children—Francois, Denis, Claire, Louis. Research asso. Cornell, 1947-49; faculty U. Montreal, 1949—, head dept. physics, 1957-66, prof. physics; vis. prof. U. Grenoble (France), 1961-62. Mem. NRC Can., 1960-66. Mem. Royal Soc. Can., Am. Phys. Soc., Canadian Assn. Physicists (pres. 1964-65). Author: (with Dale R. Corson) Introduction to Electromagnetic Fields and Waves, 1962. Research, publs. on reactions producing protons in Philips Ionization Gauge type of ion source; devel. of an improved version of cascade high-voltage generator; determination of nuclear parameters by neutron scattering. Home: 5566 Queen Mary Rd., Montreal 29, Can. Office: Dept. de Physique, Université de Montreal, Case postale 6128, Montreal 3, Can.*

LORRY, Anne-Charles, French physician; b. Crosne, France, 1726; practiced in Paris; docteur régenet of faculty; physician to Louis XV in his last illness. Author: Tractus de morbis cutaneis, 1777; one of first works on dermatology; also brought out edit. Aphorisms (Hippocrates). Died 1783.

LORTIE, Marcel, Canadian forest pathologist; b. Quebec, Que., Can., Apr. 15, 1931; s. Joseph and Gerardine (Dubeau) L.; B.S., Laval U., 1956; Ph.D., U. Wis., 1962; m. Marguerite Ouellet, Sept. 1, 1956; children—Marie-Josee, Louis. Research officer forest pathology Can. Dept. Forestry, 1956-63; prof. forest pathology Laval U., Quebec, 1963—, head forest mgmt. dept., 1966—, mem. exec. com., 1965—; chmn. Corp. Forest Engrs. Que., 1966—. Recipient Gold medal for pub. service in sch. bds. Cath. Sch. Bds. Que., 1965. Mem. Canadian Phytopath. Soc., Canadian Inst. Forestry. Research on causes of disease on birch trees known as Birch Dieback, disease cycle of cankers on trees, heartwood discolorations of hardwood trees. Home: 141 du Temple, Quebec 5, Que., Can.*

LOSANA, Luigi, Italian chemist; b. Turin, Italy, Nov. 12, 1896; s. Ottavio and Eleonora (Buffa) L.; Ph.D. Past prof. chemistry and mineralogy Turin Poly. Sch. Author: Chimica applicata, 1938; Metallurgia, 1939; Chimica generale, 1941; Research and numerous publs. on applied chemistry, mineralogy, beryl., spl. steels, light metals. Address: Politecnico, Turin, Italy.

LöSCH, Friedrich, physician; b. Germany, in the 1840's; discovered the causative agt. in amebic dysentery, Entamoeba histolytica, and thus distinguished it from all other forms of dysentery, 1875.

LOSCHMIDT, Joseph, Austrian physicist; b. Putschirn, Bohemia, Mar. 15, 1821. Prof., Vienna. Author: Chemische Studien, 1861. First to calculate number of molecules in cubic centimeter of gas at 0°C. and atmospheric pressure, 1865 (called Loschmidt number, it is 2.687 x 1019); devised Loschmidt's paradox on entropy of a gas. Died Vienna, Austria, July 8, 1895.

LOSSE, Guenter Hermann, German chemist; b. Quedlinburg, Germany, June 5, 1928; s. Hermann Karl and Auguste (Böttcher) L.; B.S., U. Rostock, 1951; Ph.D., U. Halle, 1953; m. Annerose Ternow, May 14, 1959. Head asst. U. Halle, 1953-58, lectr., 1959-63, prof., 1964; prof. U. Dresden (Germany), 1964—. Research, numerous publs. synthesis of peptides, proteins. Home: 83 Zeunerstrasse, Dresden, Germany.*

LOSSEN, Heinz Josef Maria Alexander, German radiologist; b. Wiesbaden, Mar. 4, 1893; s. Hermann and Maria (Rosenhaus) L.; M.D.; m. Emmi Schnell, Jan. 29, 1944; children—Anneliese, Heinz-Georg, Eleonore-Alma. Chief radiology services Holy Spirit Hosp., Frankfort, Germany, 1918-45; head X-ray research William Georg Kerckhoff Inst., Bad Nauheim, Germany, 1931-36, 45-61; dir. Röntgen Inst., Johannes Gutenberg

Inst., Mainz, Germany, 1948-61; became dean Faculty Medicine, 1958. Recipient Röntgen prize Remschedi-Lennep, H. E. Albers-Schonberg medal German Röntgen Soc. Mem. German Röntgen Soc. (pres.; sec. gen. 1957—), Internat. Union against Tuberculosis. Author: Materialiensammlung der Unfälle und Schäden in Röntgenbetrieben, 1925-27; Konstrastmittel, 1938; Allgemeine Röntgendiagnostik, 1948; Ausübung der Heilkunde mit radioaktiven Stoffen, 1962. Address: Fischtorplatz 20/111, 65 Mainz, W. Germany.

LOSSEN, Karl August, German geologist; b. Kreuznach, Germany, 1841; studied mining engring. Berlin; grad., Halle, Germany, 1866; after graduation became asst. geologist Prussian Geol. Survey; instr. petrology Berlin Mining Acad.; named lectr., Berlin, 1870; became prof., 1882; apptd. prof. U. Berlin, 1886. Mineral, lossenite, named in his honor. Research and publs. on petrolography of Harz Mountains. Died Berlin, Feb. 24, 1893.

LOSSNITZER, Arno Georg Heinz, German biometeorologist; b. Görlitz, May 23, 1904; s. Willi and Hedwig (Heinss) L.; Ph.D., U. Leipzig; m. Lotte Brock, Mar. 5, 1927; children—Eva, Hans, Wolf-Dieter. Meteorologist for State and Reich; head meteorology services State of Baden; dir. Meteorol. Inst., U. Freiburg-Baden, West Germany. Mem. Soc. Agr. and Forests (pres. 1946-52). Editor, Archiv d. Wiss. Gesellschaft für Land und Forstwirtsch. Research and numerous publs. in rays, biometeorology. Home: Sonnhalde 71, Freiburg-Baden, W. Germany.

LOSTROH, Ardis June, Am. biologist; b. Malcolm, Neb., Dec. 21, 1925; d. Louis Henry and Huldah (Larson) Lostroh; A.B., U. Neb., 1950, M.A., 1952; Ph.D., U. Cal. at Berkeley, 1956; m. Maurice E. Krahl, Feb. 4, 1967. Jr. research endocrinologist Hormone Research Lab., U. Cal. at Berkeley, 1956-57; postdoctoral fellow Nat. Found., Coll. France, Paris, 1957-58; USPHS fellow dept. physiology U. Chgo., 1959-60; faculty U. Cal. Med. Center, San Francisco, 1961—, asso. prof. exptl. endocrinology, 1965—. Mem. Endocrine Soc., Tissue Culture Assn., Am. Physiol. Soc., Phi Beta Kappa, Sigma Xi, Alpha Lambda Delta. Research, publs. on growth and reprodn. using mammal pituitary hormones. Home: 2636 Warring St., Berkeley, Cal. 94704. Office: 1088 Health Scis., W., U. Cal. Med. Center, San Francisco 94122.*

LOTHEISSEN, Georg, surgeon; b. Geneva, Switzerland, 1868; s. Ferdinand Lotheissen; ed. U. Vienna (Austria); prof., Vienna. Author: Chirurgie der Speiseröhre, vol. 34 of Neue Deutsche Chirurgie; contbg. author: Ency. Surgery. Introduced ethyl chloride as an anesthetic; research on surgery of esophagus. Died 1941.

LOTKA, Alfred James, mathematician; b. Lemberg, Austria, Mar. 2, 1880; s. Jacques and Marie (Doebely) L. (parents Am. citizens); B.Sc., Birmingham (Eng.) U., 1901, D.Sc., 1912; M.A., Cornell U., 1909; grad. study, U. Leipzig (Germany), 1901-02, Johns Hopkins, 1922-24; m. Romola Beattie, Jan. 5, 1935. Asst. chemist Gen. Chem. Co., 1902-08, chemist, 1914-19; asst. in physics, Cornell U., 1908-09; examiner U. S. Patent Office 1909; asst. physicist U. S. Bur. Standards, 1909-11; editor Scientific Am. Supplement, 1911-14; supr. math. research, statis. bur. Met. Life Ins. Co., 1924-33, gen. supr., 1933-34, asst. statistician, since 1934. Fellow Am. Statis. Assn. (pres. 1942), A.A.A.S., Royal Econ. Soc., Population Assn. Am. (pres. 1938-39), Inst. Math. Statistics; mem. Internat. Union for Sci. Investigation of Population Problems, Internat., Inter-Am. statis. insts., Econometric Soc., Am. Math. Soc., Am. Pub. Health Assn., Swiss Actuarial Soc., Washington Acad. Sci., Sigma Xi. Author: Elements of Physical Biology, 1925; The Money Value of a Man (with L. I. Dublin), 1930, rev. edit. 1946; Length of Life (with L. I. Dublin), 1936; Théorie Analytique des Associations Biologiques, 1934-39; Twenty Five Years of Health Progress (with L. I. Dublin), 1937; also numerous publs. in sci. and tech. jours. on math. analysis of population, math. theory of evolution, acturial math. applied to problems of population and of indsl. replacement. Died Dec. 5, 1949.

LOTTIN, Victor-Charles, French navigator; b. Paris, Oct. 26, 1765; mem. Dumont-d'Urville's expdn. around the world; capt. frigate; mem. French Acad. Scis. Author: Notice sur les aurores boreales, 1839. Research on phys. scis. in Dumont-d'Urville expdn., especially on northern lights. Died Versailles, France, Nov. 22, 1852.

LOTZE, Franc Wilhelm, German geologist; b. Amelunxen, Germany, Mar. 27, 1903; s. Franz and Antonie (Buthe) L.; Ph.D., U. Göttingen; m. Maria Gutberlet, May 18, 1929. Successively asst., agrégé, asso. prof. at large; dir. Viennese Bur. for Agrarian Research; prof., dir. Geol. and Paleontol. Inst., U. Münster. Recipient Hans Stille medal German Geol. Soc. Mem. Research Group Westphalia, German Acad. Naturalists, Acad. Sci. and Lit., Mainz. Author: Steinsalz und Kalisalze. Home: 44 Münster/Westphalia, Staufenstrasse 44, W. Germany.

LOTZE, Rudolf Hermann, German philosopher, psychologist; b. Bautzen, May 21, 1817; M.D., U. Leipzig, 1838, Ph.D., 1838; Practiced medicine, Zittau, 1 year; became dozent faculties medicine and philosophy, Leipzig, 1839; chair of philosophy Gottingen, 1844-81; apptd. prof. U. Berlin, 1881. Author: Allgemeine Pathologie und Therapie als Mechanische Naturwissenschalten, 1842; Allgemeine Physiologie des körperiichen Lebens, 1851; Medizinische Psychologie oder Physiologie der seele, 1852; Physiologische Untersuchungen, 1853; Mikrokosmus, 1856-64, Systeme der Philosophie, 2 vols., 1874-79; Grundzüge der Psychologie, 1881. Elaborated philos. system of teleological idealism; aided in founding sci. of physiol. psychology; his theory of space-perception influenced found. of later empiristic theories; pioneer of the new psychology; many of his students became eminent psychologists. Died July 1, 1881.

LOUBATIERES, Auguste, French physiologist, pharmacologist; b. Agde, France, Dec. 28, 1912; s. Félix and Jeanne L.; D. en méd., Faculty Medicine Montpellier (France), 1938, D. ès scis., Faculty Scis., 1946; m. Suzanne Vivares, Sept. 11, 1940; children—Marie-France, Michèle, Geneviève. Became prof. applied physiology and pharmacodynamics Faculty Medicine Montpellier, 1952, prof. pharmacology and pharmacodynamice, 1967—, also dir. lab. applied physiology and pharmacodynamics, lab. pharmacology and pharmacodynamics. Recipient Ricaux prize Acad. Medicine, 1940, Specia prize, 1960; prix superior council sci. and tech. research French Acad. Scis., 1957, Osiris prize, 1960. Mem. French Acad. Medicine, Soc. Biology, Nat. Acad. Medicine Buenos Aires, Royal Soc. Medicine, London, N.Y. acads. scis., Royal Acad. Belgium. Contbr. papers to sci. publs. Discovered hypoglycemic sulfamides and antidiabetic sulfamides, cardiotonic properties of heptaminol and other cardiotonic substances; research on mechanism of hyperglycemic sulfamides. Home: Résidence des Sophoras, Allée des Sophoras, Montpellier, Hérault, France.*

LOUD, Warren Simms, Am. mathematician; b. Boston, Sept. 13, 1921; s. Roger Perkins and Esther (Nickerson) L.; S.B., Mass. Inst. Tech., 1942, Ph.D., 1946; m. Mary Louise Strasburg, Dec. 27, 1947; children—Margaret, Elizabeth, John. Instr., Mass. Inst. Tech., 1943-47, vis. fellow, 1955-56; faculty U. Minn., Mpls., 1947—, prof. math., 1959—; vis. prof. Math. Research Center, U. S. Army, 1959-60, Technische Hochschule, Darmstadt, Germany, 1964-65. Mem. Am. Math. Soc., Math. Assn. Am. (past mem. bd. govs.), Soc. for Indsl. and Applied Math. (editor Rev. 1961—), A.A.A.S. Research, publs. on existence and behavior periodic solutions of certain nonlinear differential equations. Home: 4253 Sheridan Av. S., Mpls. 55410. Office: Main Engring. Bldg., U. Minn., Mpls. 55455.*

LOUDERBACK, George Davis, Am. geologist; b. San Francisco, Apr. 6, 1874; s. Davis and Frances Caroline (Smith) L.; A.B., U. Cal., 1896, Ph.D., 1899, LL.D., 1946; m. Clara Augusta Henry, Oct. 3, 1899. Asst. in mineralogy U. Cal., 1897-1900, faculty, 1906—, prof., 1917-44, emeritus, 1944—, dean Coll. Letters and Scis., 1920-22, 30-39; prof. geology and mineralogy U. Nev., 1900-06; research asst. Carnegie Inst., 1903-05; In charge geol. expdn. interior of China for Standard Oil Company of N.Y., later for Chinese Govt., 1914-16; in charge coöp. war mineral investigation in Cal., for U. S. Geol. Survey, U. S. Bur. Mines and State Council Def., 1918-19; mem. Pacific Coast sub-com. geology of NRC, 1917-19. Fellow A.A.A.S., Geol. Soc. Am. (v.p. 1936); mem. Seismol. Soc. Am. (editor Bull. 1935—, pres. 1914, 29-35), Am. Inst. Mining and Metall. Engrs., Cal., Washington acads. scis., Am. Geog. Soc., Mineral. Soc. Am., Soc. Econ. Geologists (v.p. 1939), Am. Petroleum Geologists, Am. Geophys. Union, Am. Soc. Oceanography and Limnology, Phi Beta Kappa, Sigma Xi. Contbr. on geol. and mineral topics, especially on basin range struc. and west coast stratigraphy, faultlines and earthquakes; discoverer of Benitoite and other minerals; research on phys. history of Pacific coast, China, P.I. Died Jan. 27, 1957.

LOUDON, James, Canadian mathematician, physicist; b. Toronto, Ont., Can., 1841; B.A., M.A., LL.D., U. Toronto; LL.D., Queen's U., Princeton, Glasgow U.; D.C.L., Trinity U., Toronto; m. Julia M'Dougall, 1872. Became math. tutor Univ. Coll., Toronto, 1863, classic and math. tutor, 1864, math. tutor, dean residence, 1865, prof. math. and natural philosophy, 1875, univ. prof. physics, from 1887, univ. pres., from 1892. Fellow Royal Soc. Can. (pres. 1901-02). Contbr. math. and phys. papers to jours. Died Feb. 26, 1902.

LOUGH, James Edwin, Am. psychologist; b. Eaton, O., June 24, 1871; s. William Henry and Ester Green (Stubbs) L.; A.B., Miami U., 1891, A.M., 1894, Pd.D. (hon.), 1913; A.B., Harvard, 1894, A.M., 1895, Ph.D., 1898; m. Dora Albonetta Bailey, June 27, 1900; children—Edwin Bailey, Barbara Esther, Richard Colburn, Dorothea. Tchr. pub. schs., Ohio, 1891-93; instr. psychology Radcliffe Coll., 1894-98; Wellesley, 1897-98, Harvard, 1896-98; prof. psychology State Normal Sch., Oshkosh, Wis., 1898-1901; prof. exptl. psychology and method Sch. Edn., N.Y. U., 1901-27, sec. Sch. Pedagogy, 1902-17, Washington Sq. Collegiate Div., 1903-13 (originator collegiate div. 1903), dir. Summer Sch., 1902-22, dean Div. for Extramural Teaching, 1908-26; founder Univ. World Cruise, acting pres. on 1st cruise, 1926-27, ednl. dir. since 1926; pres. Am. Floating U., since 1932. Fellow A.A.A.S., Am. Geog. Soc.; mem. Am. Psychol. Assn., Am. Philos. Soc., Phi Beta Kappa. Author: Outline of Psychology for Teachers, 1902; Analyzing Yourself, 1916; Psychology for Teachers (with Benson, Skinner and West), 1926. Died June 3, 1952.

LOUGHLIN, Gerald Francis, Am. geologist; b. Hyde Park, Mass., Dec. 11, 1880; s. John Francis and Adelia (Lane) L.; S.B., Mass. Inst. Tech., 1903; Ph.D., Yale, 1906; m. Grace E. French, Aug. 22, 1906; 1 dau., Beryl Frances (Mrs. W. S. Burbank). Asst. in geology Mass. Inst. Tech., 1903-04, instr. geology, 1906-12; asst. Yale, 1904-06; with U. S. Geol. Survey, 1912, as chief nonmetals sect., div. mineral resources, 1917-19, chief metals sect., 1919, chief div. mineral resources, 1919-23, chief sect. geology metalliferous deposits, 1923-35, chief geologist, 1935-44, Spl. scientist since 1944. Mem. Soc. Econ. Geologists, Am. Mineral Soc., Am. Inst. Mining and Metallurgy Engrs., Geology Soc. Am., Washington Acad. Sci., A.A.A.S. Author: Clays of Conn. (Conn. Survey), 1906; Lithology of Connecticut (with J. Barrell), 1912; Gabbro and Associated Rocks Near Preston, Conn.; Geology and Ore Deposits of Tintic Utah (with W. Lindgren), 1917; Oxidized Zinc Ores of Leadville, Colo., 1919; Geology and Ore Deposits of Utah (with B. S. Butler), 1920. In charge preparation Mineral Resources of the United States (6 vols.), 1919-24; Geology and Ore Deposits of Leadville, Colo., 1926; Indiana Oolitic Limestone, Relation of Its Natural Features to Its Commercial Grading, 1929; Gold Reserves of the U. S. (with H. G. Ferguson and others), 1930; Geology and Ore Deposits of the Cripple Creek District (with A. H. Koschmann), 1935; geology and ore deposits of the Magdalena District, N.M. (with A. H. Koschmann), 1943; also papers on ore deposits, building stone and durability of concrete aggregates. Died Oct. 22, 1946.

LOUGHNAN, Frederick Charles, Australian geologist; b. Sydney, Australia, June 4, 1923; s. Aubert Sherman and Clare E. (Martyn) L.; B.Sc. with 1st class honours, Sydney U., 1950; Ph.D., U. New S. Wales, 1957; m. Margaret Anderson Armour; children—Kerry Ann, Karen Jane. Geologist, Geol. Survey New S. Wales, 1950-53; asso. prof. geology U. New South Wales, Kensington, Australia, 1953—. Cons. Quality Earths Pty. Ltd., Sydney, 1958—. Mem. Geol. Soc. Australia, Australian Inst. Mining and Metallurgy (asso.), Mineral. Soc. Am., Soc. Econ. Palaeontologists and Mineralogists, Australian Clay Minerals Soc., Australian Ceramic Soc. Research, publs. on chemistry and mineralogy of sedimentary rocks. Home: 4 Pindari Av., Castlecove, N.S.W. Office: U. New South Wales, I Barker St., Kensington, N.S.W., Australia.*

LOUGHNANE, Farquhar McGillivray, Brit. surgeon; b. 1885; ed. King's Coll., St. Thomas's Hosp.; m. Blanca Keatinge, 1927; house surgeon St. Thomas's Hosp., Royal Sea-bathing Hosp., Margate; resident med. officer Camberwell Infirmary; sr. house surgeon Royal Infirmary, Leicester; sr. hon. surgeon St. Mary's Hosp., Plaistow; hon. surgeon All Saints' Hosp.; asst. urol. surgeon Prince of Wales' Hosp.; urol. surgeon London County Council; cons. surgeon Hampton Cottage Hosp.; surgeon Emergency Med. Service. Fellow Royal Coll. Surgeons. Research and publs. on spinal Tb, genital Tb, Frost bite, renal sarcoma of infants, cystitis, haematuria, prostate, prostatic obstruction, renal aneurysm, transplantation of gracilis muscle, trans-urethral prostatectomy, stricture of urethra. Died July 14, 1948.

LOUGHRIDGE, Donald Holt, Am. physicist; b. Lincoln, Neb., Sept. 4, 1899; s. Edward Bryson and Carrie (Holt) L.; B.S., Cal. Inst. Tech., 1923, Ph.D., 1927; m. Ada V. Jacobs, 1918; children—Ernestine (Mrs. Erman Pearson), Donald Holt, Glenn Howard. Asst. prof. physics U. So. Cal., 1930-31; faculty U. Wash., 1931-48; chief scientist U. S. Army, 1948-51; asst. dir. reactor div. U. S. AEC, 1951-53; dean Northwestern Tech. Inst., 1953-56; head physics and math. dept. Gen Motors Research Labs., Warren, Mich., 1956-61; dir. applied research Aerospace Corp., 1961-64; prof. physics and math. Cal. State Coll., Palos Verdes, part-time 1964—. Mem. Sigma Xi, Tau Beta Pi. Research, publs. on x-rays. Home: 948 Granvia Altamira St., Palos Verdes Estates, Cal. 90274.*

LOUIS, Antoine, French surgeon; b. Metz, France, 1723; settled in Paris at early age; oracle, counsel in med. jurisprudence to tribunes; mil. surgeon; mem. French Acad. Surgery (sec.). Author: Dictionnarie de chirurgie, 2 vols.; 1772; also articles on surgery. Differentiated between death and suicide by hanging; studied limits of pregnancy; invented a surg. technique, Louis' Angle. Died 1792.

LOUIS, Henry, Brit. metallurgist; b. Dec. 7, 1855; ed. Royal Sch. Mines; Queen's scholar, Duke of Cornwall scholar; M.A., D.Sc.; m. Rosalie James, 1895. With iron and steel works, Nova Scotia, Swansea, also nr. Edinburgh; gold mining in S. Am., W. Africa; gold and diamond mining in Transvaal and S. Africa; tin and gold mining in Malay Peninsula, Siam, other locations in Far East; coal and iron ore mining in various locations; cons. mining and metall. engr. to numerous cos.; prof. mining Armstrong Coll., U. Durham, 1896-1923, William Cochrane lectr. metallurgy. Commr., Trinidad Asphalt Enquiry, 1902; mem. internat. jury Brussels Exhbn., 1910; mem. Royal Commn. on Mine Subsidence; del. Empire Congress

Mining and Metallurgy; del. Internat. Congress Mining, Metallurgy and Applied Geology at Liege, at Paris, 1935, also to Centenary of Geol. Survey, and 75th Anniversary Celebration Swedish Inst. Engrs., Stockholm, 1936. Fellow Imperial Coll. Sci. and Tech., Royal Inst. Chemistry, Geol. Soc.; mem. North of Eng. Inst. Mining and Mech. Engrs., Iron and Steel Inst. (pres., Bessemer medallist), Instn. Mining Engrs. (pres., hon. fgn. sec., medal of Inst.), Inst. Metallurgy, Coke Oven Mgrs. Assn. (hon.), Soc. Chem. Industry (pres.), Am. Iron and Steel Inst. (hon.), Am. Inst. Mining and Metall. Engrs. (sr.). Author: A Handbook of Gold Milling; The Dressing of Minerals; The Metallurgy of Tin; Ore Deposits (Philips and Louis); Electricity in Mining; The Production of Tin; Monographs on Metallurgy of Tin and of Platinum in the Mineral Industry; Traverse Tables and Coordinate Surveying (Louis and Caunt); Tacheometer Tables (Louis and Caunt); Mineral Valuation, 1923; The Preparation of Coal for the Market, 1928; Mineral Deposits, 1934; Unsolved Problems in Metal Mining, 1908; also numerous articles. Died Feb. 22, 1939.

LOUIS, Pierre Charles Alexandre, French physician; b. Champagne, France, Apr. 14, 1787; student medicine Reims; M.D., Paris, 1813. Practiced medicine in Odessa, Russia for a time, returned to Paris, 1820. With La Pitié, then Hôtel-Dieu; mem. acad. med. Author: Recherches anatomiques, pathologiques et thérapeutiques sur la fièvre typhoide et sur la phtisie pulmonaire, 1828; Recherches sur la saignée, 1828; Recherches sur la phtisie pulmonaire, 1829; Examen de l'examen de M. Broussais, 1834. Founder of med. statistics; kept records of many cases of different diseases and of different treatments; showed efficacy or uselessness of each kind of treatment (e.g. blood letting useless in pneumonia); provided basis for path. anatomy of lung tuberculosis and typhus; described angle formed by manubrium and body of sternum (known as Louis' angle), 1825; discovered pulmonary tuberculosis generally begins in left lung, also tuberculosis of any part of body is attended by localization in lungs; introduced term typhoid fever and described characteristic rose spots of typhoid fever, 1829. Died France, 1872.

LOUNASMAA, Olli Viktor, Finnish physicist; b. Turku, Finland, Aug. 20, 1930; s. Aarno Viktor and Maj Ingrid (von Hellens) L.; M.Sc., U. Helsinki (Finland), 1953; Ph.D., U. Oxford (Eng.), 1958; m. Vilma Astrid Inkeri Kupiainen, 1951; children—Kiti Meerit Irmeli, Marja Riitta Anneli. Asst. physics U. Helsinki, 1953-54, U. Turku, 1954-59; resident research asso. Argonne (Ill.) Nat. Lab., 1960-61, 62-63, 64-65; faculty tech. physics Tech. U., Helsinki, 1963—, prof., 1965—; docent physics U. Turku, 1959-64, U. Helsinki, 1962—. Mem. Finnish, Am., phys. socs., Acad. Tech. Scis. Finland. Research and publs. on thermodynamic properties of helium-4 near lambda curve; specific heat measurements of rare earth metals at low temperatures for study nuclear contribution; Mössbauer research using helium-3 helium-4 dilution refrigerator. Home: Ritokalliontie 21 B, Helsinki 33, Finland.*

LOUNSBURY, Mackenzie, Canadian chemist; b. Burks Falls, Ont., Can., Jan. 4, 1924; s. Morris and Nita (Patton) L.; B.Sc., McMaster U., 1945, M.Sc., 1947; D.Sc., U. Paris (France), 1949; m. Marion Margaret Graham, Sept. 16, 1949; children—Susan Elizabeth, Nancy Margaret. Staff, Atomic Energy Can., Ltd., Chalk River, Ont., 1949—, head mass spectrometry sect., 1953—, asso. research officer, 1954-61, sr. research officer, 1961—. French Govt. scholar, 1947-48; NRC scholar, 1948-49. Fellow Chem. Inst. Can. (past sect. chmn., nat. councillor 1965—); mem. Assn. Chem. Profession Ont., Canadian Assn. for Applied Spectroscopy. Research, publs. on mass spectrometry, natural isotopic abundances of chlorine, krypton, xenon and uranium, neutron capture and fission cross-sects. of actinide elements, nuclear changes in neutron-irradiated uranium. Home: 21 Laurier Av., Deep River, Ont. Office: Atomic Energy Can., Ltd., Chalk River, Ont., Can.*

LOUREIRO, Juan de, Portuguese botanist; b. circa 1715; mem. Soc. of Jesus; practice medicine, Cochin, China, also China; returned to Portugal after 36 years. Author: Flora of Cochin China (contained description of many new genera), 1790. Studied flora of Cambodia. Died 1796.

LOURIA, Donald Bruce, Am. physician; b. N.Y.C., July 11, 1928; s. Milton Roland and Lucy (Littauer) L.; B.S., Harvard, 1949, M.D., 1953; m. Barbara Watson, May 21, 1955; children—Dana, Charles. Asst. surgeon Nat. Inst. Allergy and Infectious Disease, NIH, USPHS, Bethesda, Md., 1955-57; faculty Cornell U. Med. Coll., Ithaca, N.Y., 1958—, asso. prof. medicine, 1964—, dir. infectious disease lab., 1958—; staff med. div. Bellevue Hosp., N.Y.C., 1958—; co-dir. infectious disease service Meml. Hosp., N.Y.C., 1964—. Pres., N.Y. State Council in Drug Addiction, 1965—. Mem. Am. Soc. Clin. Investigation, Am. Fedn. for Clin. Research, N.Y. County Med. Soc. Author: Nightmare Drugs, 1966; also numerous articles. Research on pathogenesis and immunology of deep-seated fungus infections, carcinogenic effects of fungus toxins, influence of alcohol and narcotics on susceptibility to infection, superinfections. Home: 501 E. 79th St., N.Y.C. 10021. Office: Bellevue Hosp., 1st Av. and 26th St., N.Y.C. 10016.*

LOUROS, Nicholas, Greek gynecologist, obstetrician; b. Athens, Greece, Mar. 6, 1898; s. Constantin and Efrosyni (Veropoulos) L.; ed. univs. Athens, Bern, Vienna, Munich, Berlin, Dresden; M.D.; m. Ioanna Mitsotakis, 1937. Became asst. prof. Friedrich Wilhelm U., Berlin, 1925; head research dept. Dresden Maternity Center, 1925-29; prof. U. Athens, 1930-35; prof. obstetrics and gynecology Marica Iliadi Maternity, Athens, 1935—; became dean Faculty Medicine, 1948; founder, chief Alexandra Maternity. Mem. hygiene council Ministry Pub. Health, 1935; expert female diseases and infant illnesses WHO, Geneva. Mem. Obstetrics Soc. Athens (past pres.), Greek Eugenics Soc., Internat. Coll. Surgeons, Leopoldina Acad. Contbr. numerous articles to sci. jours. Home: 5, rue Semitelou. Office: Maternité universitaire, rue K. Lourou, Athens, Greece.

LOUTIT, John Freeman, med. biologist; b. Perth, Australia, Feb. 19, 1910; s. John Freeman and Margaret (Broadfoot) L.; student U. W. Australia, 1928, U. Melbourne, 1929-30, U. London, 1934-35, student Oxford U., 1931-34, M.A., 1938, D.M., 1946; V.M.D. (hon.), Stockholm U., 1965; m. Thelma Salsbury, June 14, 1941; children—Andrew, Margaret, Sarah. First asst., registrar London (Eng.) Hosp., 1937-39; physician, emergency Med. Service, dir. S. London Blood Supply Depot, 1940-47; dir. Med. Research Council Radiobiol. Research Unit, Didcot, Berkshire, Eng., 1947—. Fellow Royal Coll. Physicians, Royal Soc., 1963. Decorated Comdr., Order Brit. Empire. Author: Irradiation of Mice and Men, 1962; (with H. S. Micklem) Tissue Grafting and Radiation, 1966; also numerous articles. Developed use of acid citrate dextrose medium to extend shelf life of stored blood, separated 2 causes of fragile red cells in haemolytic anaemia, congenital defect and acquired autoimmunity; demonstrated successful transplantation of haemopoietic cells between individual mice after massive irradiation. Home: Green Farm, Steventon, Berkshire. Office: Radiobiol. Research Unit, Harwell, Didcot, Berkshire, Eng.*

LOUTTIT, Chauncey McKinley, Am. psychologist; b. Buffalo, Oct. 9, 1901; s. William Henry and Susan (Bruman) L.; B.S., Hobart Coll., 1925; Ph.D., Yale, 1928; m. Laura Talcott, Aug. 23, 1926; children—Robert Irving, Richard Talcott. Research fellow Tng. Sch., Vineland, N.J., 1925; instr. Yale, 1925-28; research asso., psychol. clinic U. Hawaii, 1928-30; asst. prof. Ohio U., 1930-31; asst. prof. psychology Ind. U., 1931-38, asso. prof., 1938-40; dir. psychol. clinic, 1931-40; with U. S. Naval Med. Sch., 1940-41; asst. chief psychology div. Office of Coordinator of Information, 1941-42; asst. officer in charge quality control sect., tng. div. Bur. Naval Personnel, 1942-44; comdg. officer Naval Tng. Sch. (indoctrination), Camp Macdonough, Plattsburgh, N.Y., 1944; comdg. officer, service sch. command Naval Tng. Center, Bainbridge, Md., 1944-45; prof. psychology Ohio State U., 1945-46; dean faculty Sampson Coll., 1946-47; exec. dean, Galesburg Undergrad. Div., U. Ill., 1947-49, asst. to provost 1949-54; vis. prof. U. Ore., summer 1937, Ohio State U., 1939; prof., chmn. dept. psychology Wayne U., 1954—. Mem. Ind. Soc. Mental Hygiene (pres. 1941), Ind. State Conf. Social Work (bd. dirs., also exec. com., 1937-40). Fellow A.A.A.S., Am. Psychol. Assn.; mem. Am. Assn. for Applied Psychology (exec. sec., 1940-42; pres., 1943), Phi Beta Kappa, Sigma Xi. Author: Bibliography of Bibliographies in Psychology, 1900-1927, 1928; Handbook of Psychological Literature, 1933; Clinical Psychology, 1936, rev. edit., 1947. Editor: Directory of Applied Psychology, 1st edit., 1941, 2d edit., 1943; Psychological Abstracts, 1947; Professional Problems (with R. S. Daniel), 1953. Died May 24, 1956.

LOUVEAUX, Jean Louis, French biologist; b. Paris, Sept. 25, 1920; s. Jules Léon and Hermance (Bouchet) L.; student U. Rennes (France), 1942-45; Thesis in Biology, Paris, 1958; m. Josette Jeanne Godmé, July 27, 1942; children—Alain, Francoise, Catherine. Civil servant French Agrl. Ministry, Rennes, 1942-46; staff French Nat. Inst. for Agrl. Research, Bures sur Yvette, 1946—, research master, 1958-63, research dir., 1963, head Inst. for Research on Honey-Bees and Social Insects, 1963—; faculty Versailles French Nat. Hort. Coll., part-time, 1952—. Recipient Gold medal French Acad. Agr., 1960; named laureate SIMCA Found., 1960, officier du mérite agricole, 1961. Mem. French Bot. Soc., French Chem. Experts Soc., Internat. Union for Study Social Insects, English Bee Research Assn. Author: Carnivorous Plants and Hostile Plants, 1965; (with Anna Maurizio) Some European Bee Plants Pollens, 1965; also articles. Research on honey bee biology, bee botany and melisso-palinology; discovered laws followed by honey-bees when choosing various pollens they collect. Home: 1, rue de la Guyonnerie, Bures sur Yvette, 91. Office: Station de Recherches sur l'abeille et les Insectes sociaux. Bures sur Yvette, 91 France.*

LOUW, Jan Hendrik, South African surgeon; b. S. Africa, May 26, 1915; s. Gideon Johannes and Ellen (Hurter) L.; M.B., Ch.B., U. Cape Town (S. Africa), 1938, Ch.M. with honors, 1946; m. Catherina van Breda, Oct. 2, 1943; children—Robert Alexander, Katherine Mary, Eleanor Jean. Staff, U. Cape Town and Asso. Hosps., 1945-50, 52—, prof. surgery, chmn., dept. surgery, 1955—; Nuffield Dominion Travelling fellow Hosp. for Sick Children, London, 1951. Fellow Royal Coll. Surgeons Eng., A.C.S.; mem. S. African Coll. Physicians, Surgeons and Gynaecologists (council), S. African Med. Assn. (pres. Cape Western br. Silver medal), S. African Assn. Surgeons (chmn. Cape Western br.), Royal Soc. Medicine, Brit. Assn. Paediatric Surgeons. Author: (with Charles F. M. Saint) Surgical Note-Taking; also numerous articles. Research on morphology of biliary system, origin of congenital intestinal atresia, malformations of anus and rectum, Hirschsprung's disease, surgery of peptic ulceration pancreatic disease, new techniques in vascular surgery. Home: 15, Berram Rd., Rondebosch, Cape Town, S. Africa.*

LÖVE, Askell, biosystematist; b. Reykjavik, Iceland, Oct. 20, 1916; s. Sophus Carl and Thora (Jonsdottir) L.; B.A., Reykjavik Coll., 1937; M.Sc., U. Lund (Sweden), 1941, Ph.D., 1942, D.Sc., 1943; m. Doris Wahlén, Apr. 30, 1940; children—Gunnlaug (Mrs. Theodore D. Swanson), Lóa (Mrs. Gunner Kaersvang). Dir. Inst. Botany and Plant Breeding, U. Iceland, Reykjavik, 1945-51; asso. prof. botany U. Man., Winnipeg, Can., 1951-56; research prof. biosystematics U. de Montreal (Que., Can.), 1956-63; prof. botany U. Colo., Boulder, 1964—, chmn. dept. biology, 1966—. John Simon Guggenheim Meml. Found. fellow, 1963-64. Fellow Iceland Acad. Sci. and Letters, Mendelian Soc. Lund; mem. Internat. Orgn. Plant Biosystematists (pres. 1960-64, v.p. internat. com. on chemotaxonomy 1964—), Swedish Phytogeog. Soc. (corr.), A.A.A.S. Author several books including: (with Doris Löve) Cytotaxonomical Conspectus of the Icelandic Flora, 1956, Chromosome Numbers of Central and Northwest European Plant Species, 1961; also numerous articles. Editor: (with Doris Löve) North Atlantic Biota and Their History, 1963. Research on history and devel. arctic and Atlantic floras, geobot. significance polyploidy, sex determination mechanisms in higher plants. Home: 473 Harvard Lane, Boulder, Colo. 80302.*

LOVE, Augustus Edward Hough, Brit. mathematician, geophysicist; b. Weston-super-mare, Eng., Apr. 17, 1863; s. John Henry and Emily (Serle) L.; B.Sc., St. Johns, Cambridge, 1886; M.A., D.Sc. Fellow St. John's Coll., Cambridge, 1886-99; prof. theoretical mechanics and physics, Oxford, Eng., Sedleian prof. natural philosophy, 1899-1940. Recipient De Morgan medal, 1926, Royal medal 1909, Adams prize, 1911. Fellow Royal Soc., 1894 (Sylvester medal 1937); mem. London Math. Soc. (pres. 1912-19, sec. 1895-1910), Italian Accademia dei Lincei (asso.), Institut de France (corr.). Author: Some Problems of Geodynamics, 1911; A Treatise on the Mathematical Theory of Elasticity (gives gen. theory of stress and strain, equilibrium and stability conditions of elastic plates, shells and solids, bending and vibrations of beams, torsion of rods, transmission of force), 1892-93; Theoretical Mechanics; Elements of Calculus; also articles. Research on elasticity of solids and its application to problems of earth's crust, hydrodynamics and electromagnetism with related differential equations; discovered Love waves (distortional surface waves); developed theory of biharmonic analysis. Died Oxford, June 5, 1940.

LOVE, J. Grafton, Am. neurosurgeon; b. Elizabeth City, N.C., Oct. 7, 1903; s. William Thomas and Mary (Ball) L.; B.A., Wake Forest Coll., 1925; M.D., U. Pa., 1927; M.S. in Surgery, U. Minn., 1932; Sc.D., Make Forest Coll., 1952; m. Mary Elizabeth Terry, June 29, 1933; children—Benjamin Terry, J. Grafton, David Milton, II, Rebecca Marta. Fellow Mayo Grad. Sch. Medicine, 1929-33, staff, 1934—, head sect. neurol. surgery, 1955-63, also prof. neurol. surgery; sr. cons. sect. neurol. surgery Mayo Clinic, 1963—; prof. U. Minn., Rochester, 1951—; vis. prof. neurol. surgery Ohio State U., 1960, U. Cal. at San Francisco, 1963, U. Ala., 1964. Diplomate Am. Bd. Psychiatry and Neurology, Am. Bd. Neurol. Surgery. Mem. Harvey Cushing Soc. (past pres.), Minn. Soc. Neurol. Scis. (past pres.), Soc. Neurol. Surgeons (past pres.), A.M.A., Central Surg. Assn., Central Neuropsychiat. Assn., Soc. Minn. Med. Assn. Research and numerous publs. on neurol. surgery, especially diagnosis and treatment intractable backache and sciatic pain, detection and removal of brain tumors producing blindness, and those producing deafness and unsteadiness of gait; pioneered in surgery of protruded intervertebral disks, psychosurgery, surgery of tumors of pituitary glands. Home: 604 9th Av. S.W. Office: Mayo Clinic, 200 1st St. S.W., Rochester, Minn. 55901.*

LOVE, James Kerr, Scottish surgeon; b. Beith, Scotland, 1858; s. James Love; ed. Glasgow U.; M.D., LL.D.; m. Jane Corbett, 1887; 3 daus. Lectr. aural surgery U. Glasgow; aurist Glasgow Instn. for Deaf and Dumb; hon. aural surgeon Glasgow Royal Infirmary; examiner physiology Nat. Assn. Tchrs. of Deaf; Nat. Bur. lectr. on prevention of deafness, 1912; hon. fellow Royal Faculty Physicians and Surgeons, Glasgow, Am. Laryngol., Rhonol., and Otol. Soc.; pres. sect. otology Royal Soc. Medicine, 1924-25. Author: Deaf-Mutism, a Clinical and Pathological Study, 1896; Diseases of the Ear for Practitioners and Student, 1904; The Deaf Child: A Manual for Teachers and School Doctors, 1911; Diseases of the Ear in School Children: An Essay on the Prevention of Deafness, 1919. Editor: Helen Keller in Scotland (with intro.), 1933. Died May 30, 1942.

LOVE, R(obert) Merton, agronomist; b. Tantallon, Sask., Can., Jan. 29, 1909; s. John Norman and Ethel (Dafoe) L.; B.Sc., U. Sask., 1932, M.Sc., 1933; Ph.D. magna cum laude, McGill U., 1935; m. Eunice Olive Huskins, June 20, 1936; children—John Merton, David Huskins, Bruce William. Came to U. S., 1940, naturalized, 1947. Faculty, U. Cal., Davis, 1940—, prof., agronomist, 1951—, chmn. dept. agronomy, dir. grasses research, 1959—. Dir., Ganadera Internacional, S.A., Spain. Mem. organizing com. 6th Internat. Grassland Congress, 1952; Fulbright research scholar New Zealand-Australia, 1956-57, Greece, Portugal, Spain, Italy, Cyprus, 1967. Recipient Rockefeller Found. research travel grant to Greece, Europe, 1964; medallion award Am. Forage and Grassland Assn., 1966; Calouste Gulbenkian award (Portugal), 1967. Fellow Am. Soc. Agronomy (Stevenson award 1952, Service award 1966), A.A.A.S.; mem. Am. Inst. Biol. Scis., Am. Soc. Range Mgmt., Crop Sci. Soc. Am., Bot. Soc. Am., Genetics Soc. Am., Canadian Soc. Agronomy, Brit. Grassland Assn., Ecol. Soc. Am., Am. Forage and Grassland Council (dir.), Tropical Grassland Soc. (U.K.), Sociedad de Biologia do Rio Grande do Sul (hon., Brazil). Research, numerous publs. on cytogenetics of cereals; improvement of grazing and wild lands by modification of environment. Home: 740 Miller Dr., Davis, Cal. 95616.*

LOVE, Robert, pathologist; b. Glasgow, Scotland, Apr. 27, 1921; s. Robert and Mary (Mackervail) L.; M.B., Ch.B., U. Glasgow, 1944, M.D., 1950; m. Sheila M. C. Watt, Jan. 3, 1945; children—Carol A., Angus R., Duncan R. Came to U. S., 1950, naturalized, 1956. Practice medicine, specializing in pathology, Pearl River, N.Y., 1950-55, Bethesda, Md., 1955-60, Phila., 1960—; pathologist viral, rickettsial research Lederle Labs., 1950-55; pathologist Nat. Cancer Inst., 1955-60; prof. pathology Jefferson Med. Coll., dir. Grad. Tng. Program in exptl. Pathology, 1960—. Mem. health research facilities sci. rev. com. NIH. Mem. Am. Assn. Cancer Research, Am. Assn. Exptl. Pathology, Internat. Soc. Cell Biology. Asso. editor: Jour. Nat. Cancer Inst., 1959-60; Cancer Research, 1964—. Numerous publs. on pathology of cell in viral infections, viral therapy of cancer, structure and function of nucleolus, cytochemistry of nucleoproteins and effects of metabolism on cell differentiation. Home: 1309 Morris Rd., Wynnewood, Pa. 19096. Office: Jefferson Med. Coll., 1025 Walnut St., Phila. 19107.*

LOVEJOY, Arthur Oncken, philosopher; b. Berlin, Germany, Oct. 10, 1873; s. Wallace W. and Sarah (Oncken) L.; A.B., U. Cal., 1895, LL.D., 1924; A.M., Harvard, 1897; postgrad. U. Paris; LL.D., hon. doctorates U. Mo., 1939, Princeton, 1940, Kenyon Coll., 1941. Asso. prof. philosophy Stanford, 1899; prof. Washington U., 1901-08, U. Mo., 1908-10; prof. philosophy Johns Hopkins, 1910-38, emeritus, 1938-62; lectr. univs. Cal., Chgo, Harvard, Columbia, Swarthmore Coll. Mem. Am. Philos. Assn. (pres. 1915), Am. Assn. Univ. Profs. (pres. 1918, founder), Modern. Lang. Assn. Author: Bergson and Romantic Evolutionism, 1909; Revolt Against Dualism, 1930; (with G. Boas) Primitivism and related ideas in Antiquity, 1935; The Great Chain of Being, 1936; Essays in the History of Ideas; Reason, the Understanding, and Time; Reflections on Human Nature; Thirteen Pragmatisms and Other Essays. Chmn. editorial bd. Jour. History of Ideas. Noted for advocacy of study of history of ideas. Died Balt., Dec. 30, 1962.

LOVELL, Alfred Charles Bernard, Brit. radio astronomer; b. Oldland Common, Gloucestershire, Eng., Aug. 31, 1913; s. Gilbert and Emily Laura (Adams) L.; B.Sc., U. Bristol, 1934, Ph.D., 1936; M.Sc. (hon.), U. Manchester; D.Sc. (hon.), U. Leicester, 1961, U. Leeds, 1966; LL.D. (hon.), Edinburgh U., 1961, U. Calgary, 1966; m. Mary Joyce Chesterman, Sept. 14, 1937; children—Jennifer Susan (Mrs. John Driver), Julian Patrick Bryan, Judith Ann (Mrs. Peter Spence). Roger Paul, Elizabeth Philippa Bridget. Faculty, U. Manchester, 1936—, prof. radio astronomy, 1951—. With airborne radar group Air Ministry Research Establishment, 1939-45; mem. sci. research council, chmn. Astron. Space and Radio Bd. Decorated Order Brit. Empire, 1946; Duddell medal Phys. Soc., 1954; Knighted, 1961; Daniel and Florence Guggenheim Internat. Astronautics award, 1961; French Order of Merit, 1962; Churchill Gold medal Soc. Engrs., 1964. Fellow Royal Soc., 1955 (Royal medal 1960, mem. council 1963-65), Inst. Physics, Instn. Elec. Engrs. (hon.); mem. Am. Acad. Arts and Scis. (hon. fgn.), N.Y. Acad. Scis. (hon. life), Royal Swedish Acad. (hon.). Author: Science and Civilization, 1939; World Power Resources and Social Development, 1945; Radio Astronomy, 1951; Meteor Astronomy, 1954; (with R. Hanbury Brown) The Exploration of Space by Radio, 1957; The Individual and the Universe, 1958; The Exploration of Outer Space, 1961; (with Joyce Lovell) Discovering the Universe, 1963; The Story of Jodrell Bank, 1968; also articles. Founder, dir. Jodrell Bank Labs. for Radio Astronomy with world's largest radio telescope; research in exploration of radio emissions from space; tracking of lunar and space probes. Home: The Quinta, Swettenham, Congleton, Cheshire, Eng. Office: Nuffied Radio Astronomy Labs., Jodrell Bank, Macclesfield, Cheshire, Eng.*

LOVELL, John Harvey, Am. biologist; b. Waldoboro, Me., Oct. 21, 1860; s. Harvey Hinckley and Sophronia Caroline (Bulfinch) L.; A.B., Amherst, 1882, A.M., 1899; m. Lottie Evangeline Magune, Oct. 24, 1899; children—Harvey Bulfinch, Ralph Marston. Devoted many yrs. to original observations on floroecology of northern plants, their manner of pollination, insect visitors, distribution of flower colors; specialist in photography of flowers, natural size. Author: The Flower and the Bee; Plant Life and Pollination, 1918; The Honey Plants of North America, 1926. Biol. editor ABC of Bee Culture, issued triennially; contbr. daily illustrated articles on New Eng. plants to Boston Globe, other N.E. newspapers, 1926—. Died Aug. 2, 1939.

LOVELL, Robert (or Lovel), Brit. naturalist; b. Lapworth, Eng., circa 1630; s. Benjamin L.; B.A., Christ Church, Oxford, 1659, M.A., 1663; practiced medicine, Coventry, Eng. Author: Pabbotanolgia, 1659; Panzoologicomineralogia, 1661; also compilations on herbs, animals, and minerals. Died Nov. 1690.

LOVELL, Stanley Platt, Am. chemist; b. Brockton, Mass., Aug. 29, 1890; s. Gustavus C. and Ella M. (Platt) L.; Dartmouth, 1908; B.S. in Chemistry, Cornell U., 1912, M.S., 1914; m. Mabel Bigney, June 25, 1915; 1 son, Richard H. Chemist, Spencer Kellogg & Sons, Buffalo, 1915-20, E. I. DuPont de Nemours & Co., Inc., Wilmington, Del., 1920-29; v.p. Beckwith Co., Boston, 1927-47; pres. Lovell Chem. Co., Watertown, Mass., 1947—; dir. Raytheon Co., Lexington, Mass., Baird-Atomic Co., Cambridge, Mass., Millipore Filter Co., Bedford, Mass., Moore Drop Forge Co., Springfield, Mass., McCord Corp., Detroit. Dir. research and devel. OSS, 1942-47. Recipient Presdl. medal for merit, 1947. Mem. Am. Chem. Soc., Sigma Phi Epsilon. Research on precipitated colloids, synthetic waxes, filtration membranes. Home: 65 Prospect Park, Newtonville, Mass. Office: 118 Main St., Watertown, Mass. 02160.*

LOVEN, Sven Ludwig, zoologist; b. Stockholm, Sweden, Jan. 6, 1809; prof. Mus. Natural History, Stockholm. Fellow Royal Soc., 1885; mem. French Academy Scis. Author: Echinoconidae, 1888; also other publs. Studied marine fauna of N. coasts, Baltic Seas, especially polyps, worms, crustaceans, distbn. of birds. Died Karlsborg, Rumania, Sept. 4, 1895.

LOVERA, Giuseppe Antonino Carlo, Italian physicist; b. Turin, Italy, Nov. 13, 1912; s. Carlo and Giuseppina (Suppia) L.; Ph.D. in Physics and Math. and Phys. Sci. Asst. exptl. physics U. Turin (Italy), 1933-51; prof. exptl. physics; prof. terrestrial physics; titular prof. physics U. Modena (Italy), 1951-60; prof. physics Turin Poly. Sch., 1960—. Mem. Italian Soc. Geophysics and Meteorology, Nat. Acad. Scis., Letters and Arts at Modena. Research and publs. on cosmic radiation, especially geophys. aspects in nuclear physics. Address: Istituto di Fisica Sperimentale, Politecnico, corso Duca degli Abruzzi 24, Turin, Italy.

LOVERING, John Francis, Australian geochemist; b. Sydney, Australia, Mar. 27, 1930; s. George Francis and Dorothy (Mildwater) L.; B.Sc. with honors, U. Sydney, 1951, M.Sc., 1953; Ph.D., Cal. Inst. Tech., 1956; m. Jennifer Kerry FitzGerald, Dec. 18, 1954; children—Kathleen Erin, Matthew John, Adam Barns. Research scholar Cal. Inst. Tech., 1953-55; research fellow Australian Nat. U., Canberra, 1956-60, fellow, 1960-64, sr. fellow, 1964—; vis. prof. U. Rochester, N.Y., 1963-64. Cons. Consol. African Selection Trust. Mem. Internat. Assn. Geochemistry and Cosmochemistry (mem. council 1965—), Geochem. Soc. (regional v.p. 1960—), Geol. Soc. Australia (v.p. 1962-63, 66—). Numerous publs. on structure and chem. evolution of earth's crust and interior, meteorites, solar system and universe. Home: 38 Beauchamp St., Deakin, A.C.T., Australia.*

LOVERING, Joseph, Am. astronomer, mathematician; b. Charlestown, Mass., Dec. 25, 1813; s. Robert and Elizabeth (Simonds) L.; A.B., Harvard, 1833, M.A., 1836, LL.D., 1879, attended Harvard Divinity Sch., 1835-36; m. Sarah Hawes, 1844; 4 children. Tchr., Charlestown, 1833-35; lectr. natural philosophy Harvard, 1837, Hollis prof. math. and natural philosophy Harvard, 1838-88, dir. Jefferson phys. lab., 1884-88; mem. A.A.A.S. (sec. 1854-73, pres. 1873), Am. Philos. Soc., Nat. Acad. Sci., Am. Acad. Arts and Scis. (pres. 1880-87). Author: Aurora Borealis, 1873; also many papers. Editor: Electricity and Magnetism (Farrar). Editor Proc. A.A.A.S. In charge of computations for determining transatlantic longitude from telegraphic observations on cable lines (for U. S. Coast Survey), 1867-76. Died Cambridge, Mass., Jan. 18, 1892.

LOVERING, Thomas Seward, Am. geologist; b. St. Paul, May 12, 1896; s. Thomas D. and Estelle (Wilcox) L.; E.M., Minn. Sch. Mines, 1922; M.S., U. Minn., 1923, Ph.D., 1924; m. Corinne Gray, Oct. 11, 1919; 1 son, Tom G. Instr., U. Ariz., 1924-25; prof. U. Mich., 1934-41; with U. S. Geol. Survey, 1925-34, 41-46, 47-65, chief geochem. exploration sect., 1958-62, 66—. Vis. prof. U. Minn., Ariz., Tex.; mem. Pres. Materials Policy Commn., 1952; del., internat. confs. Recipient Distinguished Service gold medal U. S. Dep. Interior, 1959; Achievement award U. Minn., 1960; Jackling award Am. Inst. M.E., 1965; Penrose gold medal Soc. Econ. Geologists, 1965. Mem. Am. Inst. Mining Engrs., Nat. Acad. Sci., Geol. Soc. Am., Am. Geophys. Union, Geochem. Soc., Soc. Ecol. Geology, A.A.A.S., Clay Minerals Soc., Ariz. Geol. Soc., Sigma Xi, Tau Beta Phi. Author: Minerals in World Affairs, 1943; also numerous articles. Established criteria for recognizing presence of blind ore bodies deep below surface, geochemistry of rock alteration, mechanisms and geol. importance of sliica translocation into plants. Home: 10185 W. 25th Av., Denver 80215.*

LOVEWELL, Joseph Taplin, Am. chemist; b. Corinth, Conn., May 1, 1833; s. Nehemiah and Martha (Willis) L.; A.B., Yale, 1857, Ph.D., 1874; m. Margaret L. Bissell, Sept. 3, 1863; m. 2d, Caroline F. Barnes, June 30, 1885. Prin. Prairie du Chien (Wis.) Coll., 1859-63; prin. high sch. and supt. of city schs., Madison, Wis., 1863-64; prof. math. Wis. State Normal Sch., Whitewater, 1870-73; prof. chemistry and physics State Coll., Pa., 1875-77, Washburn Coll., Kan., 1878-99; sec. and librarian Kan. Acad. Sci. 1902-16, emeritus sec., 1916, also pres. 2 yrs., editor 5 vols. Trans. Died Sept. 11, 1918.

LOVTRUP, Soren, Danish zoophysiologist; b. Copenhagen, Denmark, Feb. 11, 1922; mag. scient., U. Copenhagen, 1945, dr.phil., 1953. Research asst. Carlsberg Lab., Copenhagen, 1946-54; asst. prof. U. Göteborg (Sweden), 1954-57, research asso., 1957-65; prof. U. Umea (Sweden), 1965—. Research, numerous publs. on embryology, cell physiology, various micromethods—devel. electromagnetic diver balance. Address: Dept. Biology, U. Umea, Umea 6, Sweden.*

LOW, Abraham Adolph, physician, psychiatrist; b. Baranow, Poland, Feb. 28, 1891; s. Lazar and Blossom (Wahl) L.; student U. Strasbourg, 1910-13; M.D., U. Vienna, 1919; m. Mae Willett, June 18, 1935; children—Phyllis Kay, Marilyn Carroll. Came to U. S., 1921, naturalized, 1927. Intern Allgemeines Krankenhaus, Vienna, Austria, 1919-20; practice medicine, N.Y.C., 1921-23, Chgo., 1923-25, neurology, 1925-30, psychiatry, 1931—; instr. neurology, med. sch. U. Ill., 1925-31, asst. prof. psychiatry, 1931-40 asst. dir. psychiat. inst., 1931-40, asso. prof. psychiatry, 1940, acting dir. psychiat. inst., 1940-41; asst. state alienist Ill., 1933-41; founder, med. dir. Recovery, Inc., 1937—. Fellow A.M.A., Am. Psychiat. Assn.; mem. Ill. Psychiat. Soc., Chgo. Neurol. Soc., Central Neuropsychiat. Assn., Am. Group Psychotherapy Assn. Author: Studies in Infant Speech and Thought, 1936; Techniques of Self-Help in Psychiatric After-Care, 3 vols., 1943; Mental Health Through Will-Training, 1950. Contbr. articles to profl. publs. Died Nov. 17, 1954.

LOW, Albert Peter, Canadian geologist; b. Montreal, Que., Can., May 24, 1861; s. John W. L.; grad. with 1st class honors McGill U., Montreal, 1882; LL.D., Queens U., 1907. Joined staff Geol. Survey, 1882; began exploration of No. Que., 1892; placed in charge Can.'s mineral exhbn., Paris, 1900; iron ore exploration Dominion Devel. Co., 1901-03; commanded expdn. to No. waters Geol. Survey, 1903; named dir. Geol. Survey Can., 1906; ret., 1908; dep. minister Dept. Mines, 1903-13. Fellow Geol. Soc. Am. (became v.p. 1916). Author: The Cruise of the Neptune, 1905; also articles on geology. Studied previously unexplored Labrador Peninsula including geography and geology (leading to discovery iron bearing sediments); discovered Labrador Peninsula had been center of Pleistocene continental ice sheet. Died 1942.

LOW, Barbara Wharton, biochemist; b. Lancaster, Eng., Mar. 23, 1920; d. Matthew and Mary Jane (Wharton) Low; B.A., Oxford U., 1942, M.A., 1946, D.Phil., 1948; m. Metchie J. E. Budka, July 13, 1950. Came to U. S., 1946, naturalized, 1957. Faculty, Harvard, 1948-56; asso. prof. biochemistry Columbia, N.Y.C., 1956-66, prof., 1966—. Rose Sidgwick Meml. fellow Am. Assn. U. Women, 1946-47; Spl. Rockefeller Found. fellow, 1947; professeur associé Université de Strasbourg, 1965; cons. USPHS, 1966—. Fellow Am. Acad. Arts and Scis.; mem. A.A.A.S., Am. Assn. U. Profs., Am. Chem. Soc., Am. Crystallographic Assn., Am. Soc. Biol. Chemists, Chem. Soc. (London), Sigma Xi. Contbr. chpts. to The Chemistry of Penicillin, 1949; The Proteins, 1953; Currents in Biochemical Research, 1956; The Chemical Basis of Heredity, 1957; Molecular Biology, 1960. Research, numerous publs. on structure of penicillin, albumin, insulin, other small molecules; devel. of density gradient column. Home: 551 W. 239th St., Bronx, N.Y. 10463. Office: 630 W. 168th St., N.Y.C. 10032.*

LOW, George Carmichael, Brit. physician; b. Oct. 14, 1872; s. S. M. Low; ed. Madras Coll., St. Andrews, Scotland; St. Andrews U.; Edinburgh (Scotland) U.; M.A., M.D., C.M.; m. Edith Nash, 1906. Sr. physician Hosp. Tropical Diseases, Endsleigh Gardens, Eng.; dir. div. clin. tropical medicine London Sch. Hygiene and Tropical Medicine. Mem. Sleeping Sickness Commn., Royal Soc., Uganda; supt. London Sch. Tropical Medicine. Mem. Colonial Adv. Med. Com. Craggs Research scholar for study filariasis, W.I.; recipient Mary Kingsley medal Liverpool (Eng.) Sch. Tropical Medicine, 1929. Fellow Royal Coll. Physicians, Zool. Soc., Royal Geol. Soc.; mem. Royal Soc.

Tropical Medicine and Hygiene (pres.), Brit. Ornithol. Union. Author: The Literature of the Charadriiformes, 2d edit., 1894-1928, 31. Contbg. author: Textbook of the Practice Medicine, 1941. Research and publs. on filariasis and other tropical diseases; discovered release of Filaria bancrofti through proboscis of mosquito. Died July 31, 1952.

LOW, George Michael, engr.; b. Vienna, Austria, June 10, 1926; s. Arthur and Gertrude (Burger) L.; B.Aero. Engring., Rensselaer Polytech. Inst., 1948, M.S. in Aero. Engring., 1950; m. Mary R. McNamara, Sept. 3, 1949; children—Mark S., Diane E., G. David, John M., Nancy A. With NASA, and predecessor, 1949—, dep. asso. adminstr. manned space flight, 1963-64, dep. dir. Manned Spacecraft Center, Houston, 1964-67, mgr. Apollo Spacecraft Program, 1967—. Recipient Outstanding Leadership medal NASA, 1962. Arthur S. Flemming award U. S. Jr. C. of C., 1963. Fellow Am. Astron. Society; associate fellow of the Am. Inst. Aero. and Astronautics. Research in the fields of aero-dynamic heating, boundary layer theory and transition, and internal flow in supersonic and hypersonic aircraft; responsible for over-all mgmt. and direction of manned space flights and the field centers directly asso. with these programs. Home: 504 Clearview Av., Friendswood, Tex. 77546. Office: NASA Manned Spacecraft Center, Houston 77058.*

LOW, Robert Cranston, Scottish dermatologist; b. Edinburgh, 1879; s. Thomas Low; M.B., Ch.B., Edinburgh U., 1902, M.A., 1924; postgrad., Breslau, Hamburg, Vienna, Paris, 1905-06; m. Evelyn Frances Henderson, 1914; m. 2d, Alice Armstrong Grant, 1922; 1 son. Resident physician Royal Maternity Hosp., Edinburgh, 1902-03; staff Royal Infirmary, Edinburgh, 1903, Royal Edinburgh Hosp. for Sick Children, 1903-04; clin. asst., skin dept. Royal Infirmary, 1904-05, named asst. physician, 1906, later cons. physician; lectr. skin diseases Edinburgh U. Fellow Royal Soc. Edinburgh, Royal Coll. Physicians Eng. Author: Carbonic Acid Snow as an aid to Diagnosis and Treatment of Diseases of the Skin, 1911; Anaphylaxis and Sensitisation, with special reference to the Skin and its Diseases, 1924; The Common Diseases of the Skin, 3d edit., 1938; also articles. Died Feb. 3, 1949.

LOW, William, physicist; b. Vienna, Austria, Apr. 25, 1922; s. Nachum and Erna (Rihalt) L.; B.A. with honors, Queens U., 1946, A.M., 1947; Ph.D., Columbia, 1950; m. Dvora Lederer, Dec. 19, 1948; children—Esther, Nachum, Avraham, Chava, Shimon. Instr. physics Hebrew U., Jerusalem, 1950-55, lectr., 1955-59, asso. prof., 1959-61, prof. exptl. physics, 1963—; vis. prof. univs., U. S., Can. Cons. Israel AEC, industries. Recipient Israel prize for Exact Sci., 1961; Rothschild prize for Physics, 1964. Guggenheim fellow, 1963-64. Mem. Am. Phys. Soc., Israel Phys. Soc., N.Y. Acad. Scis. (Cressy, Morrison awards for natural scis. 1956), Sigma Xi. Author: Paramagnetic Resonance in Solids, 1950; also numerous articles. Research on microwave spectrum of molecules, paramagnetic resonance, optical spectra of paramagnetic solids, optical spectra of transition elements, quantum electronics, shock wave phenomena. Home: 4 Hat Ibonim, Jerusalem, Israel.*

LOWBURY, Edward Joseph Lister, Brit. bacteriologist, pathologist; b. London, Dec. 6, 1913; s. Benjamin William and Alice (Hallé) L.; B.A., U. Coll., Oxford U., 1936, M.A., 1940, B.M., B.C.L., 1940, D.M., 1957; postgrad. London Hosp., 1937-39; m. Ruth Alison Young, June 14, 1954; children—Ruth, Pauline, Miriam. House physician, surgeon, London Hosp., 1940; jr. bacteriologist Pub. Health Lab., Cambridge U., 1940-42; specialist in pathology Royal Army Med. Corps, 1943-47; mem. sci. staff common cold research unit Med. Research Council, 1947-49, burns and indsl. injuries research unit, Birmingham, Eng., 1949—; hon. dir. Hosp. Infection Research Lab., Summerfield Hosp., also Birmingham Accident Hosp., 1963—. Hon. cons. bacteriologist, adviser hosp. infection Birmingham Regional Hosp. Bd.; hon. research fellow Birmingham U.; cons. hosp. infection WHO, U. S. A., 1965. Fellow Coll. Pathologists; mem. Inst. Accident Surgery, Path. Soc. Gt. Britain and Ireland, Soc. for Gen. Microbiology, N.Y. Acad. Scis. Research, numerous publs. on application of statistically designed clin. trials and lab. exptls., value of some new antibiotics and alternate method for prevention and treatment of burn infection, skin disinfection; improved method of culture of Pseudomonas, Clostridia. Home: 79 Vernon Rd. Office: Med. Research Council Unit, Accident Hosp., Birmingham, Eng.*

LOWDERMILK, Walter Clay, Am. forester-hydrologist; b. Liberty, N.C., July 1, 1888; s. Henry Clay and Helen (Lawrence) L.; A.B. (Rhodes scholar), Oxford, 1915; Ph.D., U. Cal., 1929; D.Tech. Sci. (hon.), Israel Inst. Tech., 1952; m. Inez May Marks, Aug. 15, 1922; children—William F., Winifred Esther (Mrs. Wilmot N. Hess). With U. S. Forest Service, 1915-17, 19-22, project leader erosion stream flow research, 1928-33; research prof. famine prevention project U. Nanking, China, 1922-27; asst. chief, chief research U. S. Dept. Agr. Soil Conservation Service, 1933-47; cons. to govts., UN Morroco, Algeria, Tunisia, 1947-57, Carnegie Corp. for Brit. Colonial Office, 1948-50, Japan, 1951, Save Redwoods

League. Recipient Medallion of Valor, Israel, 1961; Maccabee award, 1964. Fellow Soc. Am. Foresters, Soc. Soil Conservation (pres. 1941-43); mem. Am. Geophys. Union. Author: Conquest of Land Through 7000 Years, 1939; Tracing Land Use Across Ancient Boundaries, 1940; Palestine, Land of Promise, 1940; also numerous articles. Demonstrated that forest litter as a sponge in absorbing rain waters is insignificant in comparison with function of keeping rain waters clear, readily to infiltrate to full depth of soil mantles and into ground waters; founder San Dimas Exptl. Forest in Forest Hydrology; demonstrated geol. norm of erosion in excavation of river basins and controls of accelerated erosion in land uses. Home: 1620 LeRoy Av., Berkeley, Cal. 94709.*

LÖWDIN, Per-Olov, physicist; b. Uppsala, Sweden, Oct. 28, 1916; s. Erik Wilhelm and Eva (Östgren) L.; Fil.Lic., Uppsala, 1942, Fil.Doktor, 1948; m. Karin Vilhelmina Höök, Feb. 7, 1960; children—Per E. A., Per E. J. Elmsäter-Löwdin, Eva K. B. Elmsäter-Löwdin, Anna K. C. Lectr. mechanics and math. physics, Uppsala, 1942-48, asst. prof., 1948-52; asso. prof. Swedish Natural Sci. Council, Uppsala, 1955-60; prof. quantum chemistry, head dept. Uppsala U., 1960—. Vis. prof. Mass. Inst. Tech., U. Chgo., Duke; grad. research prof. chemistry and physics, dir. quantum theory project U. Fla., Gainesville, 1960—. Fellow Am. Phys. So Brit. phys. socs., Royal Acad. Arts and Scis., Uppsala, Norwegian Acad. Scis., Oslo, Royal Soc. Scis., Uppsala; mem. Internat. Acad. Molecular Quantum Science (v.p.). Editor: Hylleraas vol. Revs. Modern Physics, 1963; (with J. C. Slater, R. Pauncz) Mulliken vol. Jour. Chem. Physics, 1965; Advances in Quantum Chemistry series, 1964—; (with B. Pullman) Molecular Orbitals in Chemistry, Physics, and Biology, 1964; Quantum Theory of Matter, 1965—; Quantum Theory of Atoms, Molecules, and the Solid State, A Tribute to John C. Slater, 1966. Editor Internat. Jour. Quantum Chemistry, 1966—. Contbr. papers to sci. jours. Home: 12 St. Olofsgatan, Uppsala, Sweden.*

LOWE, Charles Upton, Am. physician; b. Pelham, N.Y. Aug. 24, 1921; s. Joseph and Edith (Rosenfeld) L.; B.S. cum laude, Harvard, 1942; M.S. cum laude Yale, 1945; m. Eileen Selma Josten, Nov. 2, 1955; children—Sarah Margaret, Elizabeth Edith, Josten Stephen, Susannah Cambria. NRC fellow U. Minn., 1948-50, asst. prof., 1950-51; faculty U. Buffalo, 1951-65, Buswell fellow, research prof. pediatrics, 1956-65; staff, dir. research Children's Hosp., Buffalo, 1951-65, prof. pediatrics U. Fla., Gainesville, 1965—, dir. Human Devel. Center, 1966. Mem. com. on infant nutrition NRC, 1963—. Mem. Am. Soc. for Clin. Nutrition, Am. Pediatric Soc., Am. Inst. Nutrition, N.Y. Acad. Scis., Am. Acad. Pediatrics (chmn. com. on nutrition 1957-60, 63-68), Royal Soc. Medicine (London, Eng.), Am. Soc. Clin. Research, Soc. for Pediatric Research, A.A.A.S., Soc. for Exptl. Biology and Medicine, Sigma Xi. Research, numerous publs. on nature and mgmt. cystic fibrosis pancreas in children, purine and pyrimidine metabolism, nutritional problems in children; described Lowe's Syndrome; developed therapeutic techniques for certain forms glycogen storage diseases and parenteral fluid therapy in children. Home: 312 N.W. 23d St., Gainesville, Fla. 32601.*

LOWE, Ephraim Noble, Am. geologist; b. Utica, Miss., May 5, 1864; s. Edmund F. and Emily M. (Peyton) L.; Ph.B., U. Miss., 1884, postgrad. 1890; M.D., Tulane U., 1892; postgrad. U. Chgo., summers 1905, 06, 07; m. Sarah Yeager, Nov. 28, 1895 (dec. 1898); children—Marguerite Emily, Edmund Peyton; m. 2d, Laura Edna Haley, May 14, 1904; 1 dau., Edna May. Engaged in pvt. geol. and biol. work in Colo., 1887-89; practiced medicine in Miss., 1892-93, in Colo., 1893-1902, also geol. work; asst. in geology and biology U. Miss., 1905-07, acting prof., 1907-08, prof., 1908-09, head dept. geology, 1909—; dir. Miss. Geol. Survey, 1909—. Fellow A.A.A.S., Am. Geog. Soc. Author: Plants of Mississippi; Economic Geography of Mississippi. Died Sept. 12, 1933.

LOWE, James, English inventor; apprenticed to Edward Shorter, master mechanic, 1813; m. eldest dau. of Mr. Barnes, May 30, 1825; a dau., Henrietta (Mrs. Frederick Vansittart). Joined crew of whaling ship, 1816. Patented improvements in propelling ships, 1838, which consisted of 3 parts, a segment of screw, segment of screw below watermark and totally immersed, segment of screw applied on axis below water, 1838; used 1st in Wizard, 1838, then in her majesty's steamships, Rattler and Phoenix. Died Oct. 12, 1866.

LOWE, Percy Roycroft, Brit. ornithologist; b. Jan. 2, 1870; s. John Rooe L.; ed. Jesus Coll., Cambridge, Guy's Hosp., London; B.A., M.B., B.C.; m. Harriette Dorothy Meade-Waldo, 1 dau. Worked with Leicester Gen. Infirmary; sr. house surgeon Derby County Infirmary; civil surgeon in charge H.R.H. Princess Christian's Hosp. Train, S. African War, 1899-1901; in med. charge Earl of Dunraven's Hosp. Yacht for Officers, operations in Mediterranean, World War I; made 6 voyages with Sir Frederic Johnston, making extensive collections of island birds (now in Brit. Mus.), 1903-09; keeper in charge ornithology Brit. Mus. Natural History, 1918; Mem. adv. com. for ornithology Home Office; mem. council Royal Soc. for

Protection Birds; Brit. rep. Internat. Ornithol. Congress, 1938; pres. Brit. Ornithologist Union; chmn. European and Brit. sects. Internat. Com. for Preservation Birds. Recipient Verner von Heidenstan Gold medal Royal Swedish Acad. Sci.; Salvin-Godman medal Brit. Ornithologists Union, 1946. Author: A Naturalist on Desert Islands; also many publs. on anatomy and classification of birds in sci. jours. Died Aug. 18, 1948.

LOWE, Peter, Scottish surgeon; b. Errol, Scotland, circa 1550; ed. Paris, circa 1565; m. Helena Weymis, 1603; children—John, James. Practiced medicine in France and Flanders, 22 years; master Paris Faculty Surgery; maj. surgeon Spanish Regiments, Paris, 1589-90; surgeon to King Henry IV; settled at Glasgow, Scotland, 1598, fought against med. charlatans; founded Glasgow Faculty Physicians and Surgeons, 1599. Author: The Whole Course of Surgery, 1597; An Easy, Certain and Perfect Method to Cure and Prevent the Spanish Sickness, 1596; The Poor Man's Guide, Treatise on Parturition. Died circa 1612.

LOWE, Thaddeus S. C., Am. inventor; b. Jefferson, N.H., Aug. 20, 1832; s. Clovis and Alpha Green L.; spl. studies in chemistry; m. Leontine A. Gachon, Feb. 14, 1855; 3 sons. Constructed balloons as an aid to study of atmospheric phenomena, 1856, 58-59, securing instruments from govt., invented and made other instruments for investigating upper air currents including an Altimeter for quickly measuring latitude and longitude without a horizon; built, largest aerostat yet made, 1859-60, made trip from Cin. to a point 900 miles distant, nr. S.C. coast, in 9 hours; 1861; sent 1st telegraphic message from balloon to Pres. Lincoln, 1861; chief aero. sect., army, during Civil War; invented system of signaling to comdr. of field batteries from high altitudes (used in Peninsula campaign, later by Brazilian Govt.), 1862; inventor compression ice machine, 1865, making first artificial ice in U. S.; built regenerative metall. furnaces for gas and petroleum fuel, 1869-72; invented and built water-gas apparatus, 1873-75, and later other machines; from 1897 inventing and putting into operation New Lowe Coke Oven system, for simultaneously producing gas and metall. coke, known as Lowe Anthracite; built Mt. Lowe Ry., 1891-94; established Lowe Obs. in Sierra Madre Mountains. Author: The Air Ship of New York, 1859; Navigating the Air, 1907. Died Jan. 1913.

LOWELL, Francis Cabot, Am. inventor; b. Newburyport, Mass., Apr. 7, 1775; s. John and Susanna (Cabot) L.; grad. Harvard, 1793; m. Hannah Jackson, Oct. 31, 1798, 1 dau., 3 sons, including John. Mcht. with William Cabot, Boston, 1793-1810; went to Eng. for health, 1810, studied textile mills; formed Boston Mfg. Co. which bought land at Waltham, Mass., designed from memory or invented machines for cotton factory (1st factory in world to unite all processes of cloth under one roof); lobbied for tariff of 1816 which protected his cloth mill; Lowell (Mass.) named after him. Died Boston, Aug. 10, 1817.

LOWELL, Francis Cabot, Am. physician; b. Boston, Aug. 6, 1909; s. Frederick E. and Isabel (Shaw) L.; B.S., Harvard, 1932, M.D., 1936; m. Elizabeth Shurcliff, Oct. 16, 1938; children—Francis Cabot, Charles Russell, Thomas Homer. Practice medicine, specializing in allergy, Boston, 1946—; asso. prof. medicine Boston U., 1945-58; asst. prof. medicine Med. Sch. Harvard, 1958—; physician Med. Service Mass. Gen. Hosp., 1958—, chief allergy unit, 1958—; area cons. VA Hosp., Boston, 1963—; mem. adv. council Nat. Inst. Allergy and Infectious Diseases, 1965—; sr. allergy cons. Soldiers Home, Chelsea, 1966—. Fellow A.C.P.; mem. Am. Acad. Allergy (past pres.). Contbr. numerous articles to profl. jours. Research in immunologic mechanisms in allergy and resistance to insulin, cigarette smoking as a cause of obstructive pulmonary disease emphysema, pulmonary function in bronchial asthma, nature and cause of heaves in horses, assessment of injections of allergenic extracts in treatment of hay fever and asthma, role of eosinophil in allergic disease, steroids in mgmt. of bronchial asthma. Home: Garfield Rd., Concord, Mass. 01742. Office: Mass. Gen. Hosp., Boston 02114.*

LOWELL, Percival, Am. astronomer; b. Boston, Mar. 13, 1855; s. Augustus and Katharine Bigelow (Lawrence) L.; A.B., Harvard, 1876; (LL.D.), Amherst, 1907, Clark U., 1909); m. Constance Savage Keith, June 10, 1908. Went to Japan 1883, and lived there from time to time till 1893; counsellor and foreign sec. to Korean Spl. Mission to U. S. Established Lowell Obs., 1894; undertook eclipse expdn. to Tripoli, 1900; sent expdn. to the Andes to photograph planet Mars, 1907. Received Janssen medal of French Astron. Soc., 1904, for researches on Mars; gold medal for work on Mars from Sociedad Astronomica de Mexico, 1908; has made discoveries on the planets Mercury, Venus, Saturn, and especially Mars; apptd. non-resident prof. astronomy, Mass. Inst. Tech., 1902. Fellow Am. Acad. Arts and Sciences. Author: Chöson—The Land of the Morning Calm, 1885; The Soul of the Far East, 1886; Noto, 1891; Occult Japan, 1894; Mars, 1895; The Solar System, 1903; Mars and Its Canals, 1906; Mars as the Abode of Life, 1908; The Evolution of Worlds,

1909; The Genesis of the Planets, 1916. Built astron. bos., nr. Flagstaff, Ariz., 1893-94; known for math. work predicting discovery of Planet X (discovered by C. W. Tombaugh, 1930 which was named Pluto), Proponent of theory that intelligent life exists on Mars. Died Nov. 13, 1916.

LOWENSTEIN, Milton Dewey, Am. architect; b. N.Y.C., June 16, 1898; s. Albert L. and Celia (Abraham) L.; B.A., Columbia, 1922, M.A., 1934; postgrad. College de France, Courtauld Inst., London; m. Francoise Dufour, Mar. 1952; children—Celia, David, Glarance, Guermantes. Architect, Internat. YMCA, Shanghai, China, 1923-24; archeologist Commission Des Monuments Historique, France, 1925-27; archtl. cons. N.Y. Dept. Social Welfare, 1929; housing adviser Camac St. Co., N.Y.C., 1936; sr. instr. State U. N.Y., Delhi, N.Y., 1953-56; architect, housing adviser, Indonesia, 1956-57; asst. prof., dir. Tech. Research Service Center, Ariz. State U., Tempe, 1958-63; co-dir. Indian Self-Help Housing Evaluation Project, HHFA, 1964-66. Mem. A.I.A. Research, numerous publs. in air navigation instruments, archtl. design for humid tropical climate, archtl. design for arid climates. Home: 8119 E. Indian School Rd., Scottsdale, Ariz. 85251.*

LOWENSTEIN, Otto, German psychiatrist, neurologist; b. Osnabrück, Germany, May 7, 1889; s. Julius and Henrietta (Grünewald) L.; ed. univs. Bonn, Göttingen (both Germany); M.D.; m. Martha Grünewald, May 22, 1920; children—Anne Elisabeth, Marie Dorothea. Agrégé in psychiatry and neurology, 1920-23; asso. prof. psychiatry and neurology, Bonn, 1923-31, prof., 1931-48, emeritus, 1948—; prof. clin. neurology N.Y. U., 1939-47; prof. clin. neurology Columbia, 1947-48, dir. lab. pupillography, 1949-62; ret.; cons., N.Y.C. Mem. Am. Acad. Neurologists, N.Y. Acad. Sci., Am. Soc. for Devel. of Sci. Author: Experimentalle Hysterielehre, 1923; Studien zut Physiologie und Pathologie der Pupillenstörungen mit bezonderer Berücksichtigung der Schizophrenie, 1933; Pupillenstörungen bei syphilitischen Erkrankungen des Zentrainervensystems, 1935; Der psychische Restitutioneffekt, 1937; Die Bedeutung psychiascher Vorgänge für das biologische Gesamtgeschehen, 1953; also articles. Home: 229 E. 79th St., N.Y.C. Office: 1199 Park Av., N.Y.C. 10028.

LOWENTHAL, David, Am. geographer; b. N.Y.C., Apr. 26, 1923; s. Max and Eleanor (Mack) L.; B.S., Harvard, 1943; M.A., U. Cal., Berkeley, 1950; Ph.D., U. Wis., 1953. Research analyst U. S. State Dept., 1945-46; faculty Vassar Coll., 1952-56; Fulbright research fellow U. Coll. W.I., Jamaica, 1956-57; research asso. Am. Geog. Soc., 1956—; U. seminars asso. Columbia, N.Y.C., 1957—. Hon. research fellow U. Coll., London, 1962-63; vis. prof. Harvard, Mass. Inst. Tech., Clark U., 1967. Recipient Herfurth award U. Wis., 1959; Geog. Soc. Chgo. award, 1960. Guggenheim fellow, 1965-66. Mem. A.A.A.S. (council), Assn. Am. Geographers (Meritorious Contbn. award 1960, chmn. program com. 1965), Am. Hist. Assn., Am. Anthrop. Assn., Am. Studies Assn., Internat. Geog. Union, Soc. History Discoveries. Author: George Perkins Marsh, Versatile Vermonter, 1958. Editor, contbr. The West Indies Federation: Perspectives on a New Nation, 1961; editor: Man and Nature, 1965; Environmental Perception and Behavior, 1967. Mem. editorial adv. bd. Internat. Ency. Social Scis., 1964—. Traced history of relations between man and environment and background of conservation movement; explored attitudes toward landscape and habits of perception in Eng. and Am. Home: 75 Riverside Dr., N.Y.C. 10024. Office: Broadway at 156th St., N.Y.C. 10032.*

LOWENTHAL, Leo, sociologist; b. Frankfurt/Main, Germany, Nov. 3, 1900; s. Victor and Rosy (Bing) L.; student U. Giessen (Germany), 1919-20, U. Heidelberg (Germany), 1920-21; Ph.D., U. Frankfurt, 1923; m. Marjorie Fiske, June 26, 1953; children—Daniel, Carol Fiske Lissance. Came to U. S., 1934, naturalized, 1940. Sr. research asso. Inst. Social Research, Frankfurt U., 1926-33; sr. research asso. Columbia, 1934-49, lectr. sociology, 1940-55; dir. research Voice Am., Dept. State-USIA, 1949-55; fellow Center for Advanced Study in Behavioral Sci., Stanford, 1955-56, 66-67; prof. sociology U. Cal. at Berkeley, 1956—. Mem. Am. Sociol. Assn., Am. Psychol. Assn., Am. Assn. for Pub. Opinion Research, Am. Studies Assn. Author: Prophets of Deceit, 1949; Literature and the Image of Man, 1957; Literature, Popular Culture and Society, 1961; (with S. M. Lipset) Culture and Social Character, 1961; also articles. Integration of internat. communications research into sociol. and anthropol. context; bringing hist. perspective to popular culture theory; pioneered sociology lit. in U. S. Home: 1967 Clay St., San Francisco 94109. Office: Dept. Sociology, U. Cal., Berkeley, Cal. 94720.*

LOWER, Richard, Brit. physician; b. Cornwall, Eng., 1631; s. Humphrey and Margery (Billing) L.; B.A., Oxford, 1653, M.A., 1655, M.B., M.D., 1665; studied chemistry with Peter Sthael; m. Elizabeth Billing; 2 daus. Practiced medicine, London; assisted Dr. Willis in anat. research. Fellow Royal Soc., 1667 Royal Coll. Physicians. Author: Diatribae T. Willisii de febribus vindicatio, 1665; Tractatus de corde, 1669; Dissertatio de origino catarrhi, 1672. First to transfuse blood directly from one animal to another 1665; credited with discovering scroll-like structure of heart muscle, 1669; showed heartbeat is due to contraction of muscular walls; instrumental in disproving theory that nasal secretions originate in pituitary gland, 1672; described tubercle on inner surface of right atrium of heart; proved that brightness of arterial blood is due to absorption of air; noted for anatomical research on brain with Willis. Died Jan. 17, 1691.

LOWER, William Edgar, Am. surgeon; b. Canton, O., May 6, 1867; s. Henry and Mary (Deeds) L.; M.D., Western Res. U., 1891; m. Mabel Freeman, Sept. 6, 1909; 1 dau., Mary. Practiced in Cleve., since 1892; asso. surgeon Lakeside Hosp., 1910-31; attending surgeon Lutheran Hosp. since 1896; dir. surgery Mt. Sinai Hosp., 1916-24; asso. prof. genito-urinary surgery Western Res. U., 1910-1931; a founder, dir. Cleve. Clinic Found., since 1921; surgeon Cleve. Clinic Hosp. since 1924. Mem. Am. Urol. Assn. (pres. 1914-15), Am. Assn. Genito-Urinary Surgeons (pres. 1922), Ohio State Med. Soc. (pres. 1915), Acad. Medicine, Cleve. (pres. 1909-10), Clin. Soc. Genito-Urinary Surgeons (pres. 1922), Interurban Surg. Soc. (pres. 1920-27), Soc. Clin. Surgery, Société Internationale de Chirurgie Urologie; fellow A.M.A., A.C.S., Am., So. surg. assns. Author: Anoci-Association (Crile and Lower), 1914; Surgical Shock and the Shockless Operation through Anoci-Association, 2d edit., 1920; Roentgenographic Studies of the Urinary System (Lower and Nichols), 1933. Died June 17, 1948.

LOWIE, Robert Harry, ethnologist; b. Vienna, Austria, June 12, 1883; s. Samuel and Ernestine (Kuhn) L.; brought to U. S. at age of 10; A.B., Coll. City N.Y., 1901; Ph.D., Columbia U., 1908; Sc.D., U. Chgo., 1941; m. Luella W. Cole, Aug. 23, 1933. Asst. dept. anthropology Am. Mus. Natural History, 1908-09, asst. curator, 1909-13, asso. curator, 1913-21; lectr. Columbia U., 1920-21; asso. prof. anthropology U. Cal., 1917-18, 21-25, prof., 1925-50; vis. prof. Yale, 1937; faculty research lectr. U. Cal., 1949; anthrop. expdns. to No. Plains Indians, 1906-14, 16, 31, Lake Athabaska, 1908, plateau tribes, 1914-15, Hopi, 1915, 16. Viking Fund medalist, 1947; Thomas H. Huxley Meml. lectr. and medalist, 1948. Fellow Am. Ethnol. Soc. (sec. 1910-20; pres. 1920-21), Royal Anthrop. Inst. (hon.); mem. NRC (chmn. div. anthropology and psychology 1931-32), Nat. Acad. Scis., Am. Philos. Soc., Am. Anthrop. Assn. (pres. 1935), Am. Folk-Lore Soc. (pres. 1916-17), German Ethnol. Soc. (hon.), Bavarian Acad. Scis. (corr. mem.), Société des Americanistes de Paris (corr. mem.), Phi Beta Kappa, Sigma Xi. Author: The Assiniboine, 1909; Social Life of the Crow Indians, 1912; Societies of the Crow, Hidatsa and Mandan Indians, 1913; The Sun Dance of the Crow Indians, 1915; The Age-Societies of the Plains Indians, 1916; Culture and Ethnology, 1917; Myths and traditions of the Crow Indians, 1918; Primitive Society, 1920; Primitive Religion, 1924; The Origin of the State, 1927; Are We Civilized?, 1929; Intro. to Cultural Anthropology, 1934, 40; The Crow Indians, 1935; The History of Ethnological Theory, 1937; The German People, 1945; Social Organization, 1948; Toward Understanding Germany, 1954; Indians of the Plains, 1954; A Practical Handbook for Planning a Trip to Europe (with Luella Cole), 1957; Robert H. Lowie, Ethnologist (Autobiography); Crow Texts; Dictionary of the Crow Language; Lowie's Selected Papers. Known for work on N. Am. Indians; authority on tribes of Plains; contbd. to devel. of Am. ethnol. theory. Died Sept. 21, 1957.

LOWIG, Carl, German chemist; b. Kreuznach, Germany, 1803, apothecary, Kreuznach, 1825; studied under Gmelin, Heidelberg, Germany; prof. chemistry, Zürich, later Breslau (now Wroclaw, Poland). Author monograph on properties of bromine and its compounds. Prepared bromine, 1825, but Balard published its discovery before he could produce more for complete examination. Died Breslau, 1890.

LOWIG, Henry F(rancis J(oseph), mathematician; b. Prague, Czechoslovakia, Oct. 29, 1904; s. Henry and Catherine (Chwojka) L.; dr. rerum naturalium, German U. Prague, 1928; D.Sc., U. Tasmania, 1951; m. Libby Barbara Otta, Sept. 7, 1949; children—Ingrid Henriette, Evan Henry Francis. Faculty, German U. Prague, 1935-38, U. Tasmania, 1948-57; asso. prof. math. U. Alta. (Can.), Edmonton, 1957-67, professor of mathematics, since 1967—. Member Canadian Math. Congress, Am. Math. Soc., Czechoslovak Soc. Arts and Scis. in Am., Canadian Assn. U. Tchrs. Contbns. to theory of Hilbert spaces, foundations of theory of algebras, difference equations, lattice theory. Home: 15212 81st Av., Edmonton, Alta., Can.*

LOWITZ, Tobias, chemist; b. Göttingen, Germany, 1757; prof. chemistry Imperial Acad. St. Petersburg (Russia). Discovered color changing property of carbon, 1785; crystallized acetic acid, 1789, grape sugar in honey, 1792. Died 1804.

LOWMAN, Edward Wynne, Am. physician; b. Orangeburg, S.C., July 16, 1915; s. Oscar Cyril and Geraldine (Cave) L.; B.S., The Citadel, 1936; M.D., Med. Coll. of S.C., 1940; M.S., U. Minn., 1950. Fellow, Mayo Found., 1948-51; clin. dir. Inst. Rehab. Medicine, N.Y. U. Med. Center, N.Y.C., 1952—, prof. rehab. medicine, 1962—; staff N.Y. U., Manhattan VA, Bellevue hosps. Diplomate Am. Bd. Phys. Medicine and Rehab. Mem. A.M.A., A.C.P., Am. Acad. Phys. Medicine and Rehab., Am. Congress Rehab. Medicine (pres. 1966-67), Am., N.Y. (pres. 1958-59) rheumatism assns. Contbr. chpts. to books, articles to profl. jours. Research in rehab. of rheumatoid cripple, market and devel. of mech. aids, orthoses, self-help devices for severely physically disabled. Home: UN Plaza, N.Y.C. 10017. Office: 400 E. 34th St., N.Y.C. 10016.*

LOWMAN, Frank George, Am. biogeochemist; b. Douglass, Kan., Dec. 22, 1921; s. Frank Benjamin and Lois (Simmons) L.; B.S., U. Wash., 1942, Ph.D., 1956; m. Cleora Agnes Adams, Sept. 19, 1942; children—Merryl Lee (Mrs. Ivan Nazario), Karen Kay, Leon Vince, Michael Eugene, Frans Barton. Research asst. Lab. Radiation Biology, U. Wash., Seattle, 1948-52, research asso., 1952-61; chief scientist, head, marine biology div. P.R. Nuclear Center, Mayaquez, 1961—. Mem. Sigma Xi, Phi Sigma. Research on marine biogeochemistry of trace elements and radionuclides; investigations in distbns. of radionuclides from nuclear devices in sea. Home: 40 Millonarios. Office: P.R. Nuclear Center, College Station, Mayaquez, P.R. 0708.*

LOWRANCE, Edward Walton, Am. anatomist; b. Ogden, Utah, June 17, 1908; s. Samuel Franklin and Edith M. (Sandusky) L.; student Utah Westminster Coll., 1926-28; A.B., U. Utah, 1930, A.M., 1932; Ph.D. (Rosenberg Research fellow), Stanford, 1937; postgrad. U. Kan.; m. Rhoda Elizabeth Patton, June 21, 1935; children—Margaret Ann (Mrs. Jack Hamilton), Janet Elizabeth. Faculty, Sch. Medicine U. Mo., Columbia, 1950—, prof. anatomy, 1955—, acting chmn. dept. anatomy, 1961-62. Fellow A.A.A.S.; mem. Am. Assn. Anatomists, Am. Micros. Soc., N.Y. Acad. Scis., Western Soc. Naturalists, Sigma Xi, Phi Beta Pi. Reviewer, Handbook of Circulation, 1958-60; mem. reviewing panel skeletal weights sect. Handbook of Biological Data, 1955—. Research in effect of temperature gradients, light, hydrogen ion concentration and centrifugal force on direction of differentiation of eggs of marine brown algae belonging to fucaceae, linear growth of rabbit tendon and long bones, quantitative characteristics of bones of juvenile and adult Virginia opossums, quantitative characteristics of human skeleton. Home: 103 Thistledown Dr., Columbia, Mo. 65201.*

LOWREY, Lawson Gentry, Am. psychiatrist; b. Centralia, Mo., Dec. 23, 1890; s. Ernest and Eupha Orme (Sappington) L.; student Bethany Coll., 1905-07; A.B., U. Mo., 1909, A.M., 1910; M.D. cum laude, Harvard, 1915. Pathologist Danvers State Hosp., 1914-17; 1st asst. physician and chief med. officer, Boston Psychopathic Hosp., 1917-20; asst. dir. Psychopathic Hosp., U. Ia., 1920-23; dir. demonstration child guidance clinic (in Dallas, Mpls., St. Paul, Cleve.), Nat. Com. for Mental Hygiene, 1923-27; dir. Inst. for Child Guidance, N.Y., 1927-33; attending physician N.Y. Neurol. Inst., 1932-37; psychiatrist, pediatric dept. New Rochelle (N.Y.) Gen. Hosp., 1933—; cons. in psychiatry Grasslands Hosp., 1930-44; psychiatrist Clinic for Gifted Childrens, N.Y. U., 1933-35, Bklyn. Hebrew Orphan Asylum, 1937-44; dir. Mental Hygiene Research Unit, Vocational Adjustment Bur., 1937-39; psychiatrist Traveler's Aid Soc., 1938-42; dir. Bklyn. Child Guidance Centre, 1940-45; asso. attending psychiatrist Vanderbilt Clinic, 1945—; asst. in anatomy U. Mo., 1909-10, asst. prof. anatomy, 1911-12; 1909-10; prof. anatomy and histology U. Utah, 1910-11; teaching fellow in histology and embryology Harvard, 1912-14, James Jackson Cabot research fellow neuropathology, 1915-16, fellow neuropathology, 1916-18, instr. in neuropathology, 1918-20, in psychiatry, 1918-20, in psychology, 1919-20; asst. and asso. prof. U. Ia., 1920-23; lectr. So. Meth. U., 1923, U. Minn. Med. Sch., 1923-24, Smith Coll. Sch. of Social Work, 1926-36, N.Y. Sch. of Social Work, 1930-36, N.Y. U. Sch. Edn., 1933-35, Hunter Coll., 1937; asst. clin. prof. psychiatry Columbia Coll. Phys. and Surg., 1945—. Mem. Assn. Anatomists, Assn. Pathologists and Bacteriologists, A.M.A., Am. Psychiat. Assn., A.A.A.S., Boston Soc. Neurology and Psychiatry, N.E. Psychiat. Soc., N.Y. Neurol. Soc. Acad. Polit. and Social Sci., Am. Orthopsychiatric Assn. (pres. 1928-30, editor, 1930-48), Am. Psychopathol. Assn., Sigma Xi, other socs. Author: Report of Kindergarten Project, 1937-39; Psychiatry for Social Workers, 1946. Editor: Monograph Series of American Orthopsychiatric Assn.; Institute for Child Guidance Studies. Editor Am. Jour. Orthopsychiatry, 1930-48. Died Aug. 1957.

LOWRY, James Victor, Am. physician; b. Milw., Nov. 15, 1913; s. Theodore and Mildred (Victor) L.; B.S., U. Wis., 1934, M.D., 1937; m. Ethelyn Hoyt, Oct. 13, 1936; children—Michael, Denis, Jane, Timothy. Mem. research staff NIH, Washington, 1940-43; chief psychiat. service USPHS Hosp., Ft. Worth, 1943, clin. dir. USPHS Hosp., Lexington, Ky., 1944-47, med. officer in charge, 1954-57, chief community services Nat. Inst. Mental Health, 1947-54, dep. chief Bur. Med. Service, USPHS, Washington, 1957-58, chief, 1958-64, asst. surgeon gen., 1957-64; dir. dept. mental hygiene State of Cal., Sacramento, 1964—. Diplomate Am. Bd. Psychiatry and Neurology. Mem. Am. (council), Central Cal. psychiat. assns., Alpha Omega Alpha, Sigma Sigma. Contbr. articles to profl. jours. Home: Box 131, El Macero, Cal. 95618. Office: 1500 5th St., Sacramento 95814.*

LOWRY, Oliver Howe, Am. pharmacologist; b. Chgo., July 18, 1910; s. Charles D. and Lydia P. (Hess) L.; student U. Innsbruck, Austria, also Freiburg, Germany, 1929-30; B.S., Northwestern, 1932; M.D., U. Chgo., 1937, Ph.D., 1937; m. Norma G. Van Ness, Apr. 13, 1935; children—Susan Lowry, Stephen, Charles, Emily; m. Adrienne C. Clark; 1 son, John. Engaged as instr. of biochemistry Harvard, 1937-42; asso. chief div. biochemistry and nutrition Pub. Health Research Inst. of N.Y., 1942-47; prof., head dept. pharmacology Washington U. Sch. Medicine since 1947. Mem. Am. Soc. Biol. Chemists, Am. Soc. Pharmacology and Exptl. Therapeutics, Nat. Acad. Sci. Research on chemistry of aging; tissue electrolytes; neurochemistry; histochemistry; nutrition and detection of nutritional deficiency. Home: 9735 Litzinger, Ladue, Mo. 63124. Office: Washington U. Med. Sch., St. Louis 10.

LOWRY, Thomas Martin, Brit. chemist; b. Yorkshire, Eng., Oct. 26, 1874; s. Edward Pearce and Jemima (Hofland) L.; D.Sc., Central Tech. Coll., London, 1899; hon. degrees, univs. Cambridge, Dublin, Brussels; m. Eliza Wood, 1904; 2 sons, 1 dau. Asst. to H. E. Armstrong, 1896-1913; lectr. chemistry Westminster Tng. Coll., 1904-13; became prof., head chemistry dept. Guy's Hosp. Med. Sch., London, 1913; prof. phys. chemistry Cambridge (Eng.) U., from 1920. Fellow Royal Soc., 1914; mem. Faraday Soc. (pres. 1928-30). Author: Optical Rotary Power, 1935; also numerous articles. Found that freshly prepared nitro-d-camphor solutions showed change of rotary power with time (mutarotation), 1898; studied polarimetry, reversible isomeric change, also prototropic isomeric change (led to an extended definition of acids and bases). Research (with Percy Corlett) on variation of rotatory power with wave-lengths. Died Cambridge, Nov. 2, 1936.

LOWSLEY, Oswald Swinney, Am. urol. surgeon; b. Santa Barbara, Cal., Sept. 4, 1884; s. Vincent and Willie Ann (Swinney) L.; A.B., Stanford, 1905; M.D., Johns Hopkins, 1912; grad. Bellevue Hosp., N.Y., 1915; children—Lydia Ann, David William, Martha Winifred, Oswald Swinney; m. 3d, Celeste Nocito, Aug. 29, 1949; m. 4th, Winifred Atherton, Jan. 17, 1953. Practiced in N.Y.C., since 1915; cons. urol. surgeon Ruptured and Crippled Peekskill, Monmouth Meml., Nassau County and Bloomingdale hosps., also to Stamford Hosp., St. Luke's Hosp. (Newburgh, N.Y.), Jamaica Hosp., Fitkin Meml. Hosp. (Spring Lake, N.J.), Norwalk (Conn.) Hosp., Flushing Hosp., St. Agnes Hosp. (White Plains, N.Y.) Englewood (N.J.) Hosp., Nat. Jewish Hosp. (Denver), Kings Hosp. (Bay Shore, L.I.), St. Clare's Hosp., N.Y. Hosp. (N.Y.C.), Jersey City Med. Center; pres., dir. Oswald Swinney Lowsley Found. urology; 1st dir. dept. urology, James Buchanan Brady Found. of N.Y. Hosp. and St. Clare's Hosp. Diplomate Am. Bd. Urology. Fellow A.C.S., Internat. Coll. Surgeon, N.Y. Acad. Medicine; mem. A.M.A., Am. (pres. 1941-42), Internat. urol. assns., Barcelona Urol. Soc., N.Y. State, N.Y. County, Costa Rican, Mexican, Venezuelan, Uruguayan med. socs., Osler Soc. N.Y., Royal Soc. Medicine Gt. Britain (corr.), numerous other European, S.Am. urol. and med. socs. Author: Embryology of the Prostate, 1912; A Textbook of Urology (with T. J. Kirwin), 1926; Urology for Nurses (with T. J. Kirwin), 1936; Clinical Urology (with T. J. Kirwin), 1940, 2d edit., 1943; also many articles on urol. subjects; operative treatment of the kidney and the embryology, anatomy, morphology, pathology and surgery of the prostate gland, tuberculosis of kidney, diverticulitis of bladder and urethra, and lesions of the ureter. Editor of Oxford Urological Surgery and Yearbook of Urology. Originator of Lowsley ribbon gut method of kidney operations and of operation for relief of impotence, also a new operation for hypospadias; inventor many operative instruments and tables; pioneer in surgery of genito-urinary organs under local and regional anesthesia. Died June 4, 1955.

LOWY, Alexander, Am. chemist; b. N.Y., Mar. 31, 1889; s. David and Fanny (Weiss) L.; B.S., Columbia, 1911, A.M., 1912, Ph.D., 1915; m. Dora Landberg, Dec. 23, 1915; children—Evelyn F., Muriel A., Alexander D. Asst. in electrochemistry, Columbia U., and teaching in N.Y.C. until 1918; prof. organic chemistry U. Pitts., from 1918. Mem. Am. Electrochem. Soc. (v.p. 1930-33, 38-41). Author: Introduction to Organic Chemistry, 1922; Study Questions in Organic Chemistry, 1923; Laboratory Methods in Organic Chemistry, 1926; Industrial Organic Chemicals and Dye Intermediates, 1935. Patentee numerous chem. research discoveries. Died Dec. 25, 1941.

LOWY, Peter Herman, chemist; b. Vienna, Austria, Jan. 3, 1914; s. Robert and Marianne (Rosenberg) L.; Doctorandum Chem., U. Vienna, 1938; m. Ruth Schlosburg, Aug. 30, 1940; children—Judith Ann, Robert William, Richard Frank. Came to U. S., 1938, naturalized, 1943. Food chemist, Rochester, N.Y., 1940-45; research asst. Cal. Inst. Tech., Pasadena, 1946-49, research fellow, 1949-65, sr. research fellow, 1965—. Mem. Am. Chem. Soc., Sigma Xi. Research, publs. on synthesis radioactive amino acids, amino acid and protein biosynthesis, isolation erythropoietin, Home: 188 S. Meridith Av., Pasadena, Cal. 91106.*

LOZIER, Clemence Sophia Harned, Am. physician; b. Plainfield, N.J., Dec. 11, 1813; d. David and Hannah (Walker) Harned; grad. Syracuse Med. Coll., 1853; m. Abraham Lozier, 1830; 1 son, Abraham. Moved to N.Y.C., lectured on hygiene and physiology in her home; founder N.Y. Med. Coll. and Hosp. for Women, 1863; reorganized coll., became dean and prof. gynecology and obstetrics, 1867-88; supporter Hosp. for Women; mem. Nat. Working Women's League, W.C.T.U., N.Y.C. Suffrage League (pres. 13 years), Nat. Woman's Suffrage Assn. (pres. 5 years). Author: (pamphlet) Childbirth Made Easy, 1870. Died N.Y.C., Apr. 26, 1888.

LOZIER, W(illiam) Wallace, Am. physicist; b. Humboldt., Ill., Aug. 3, 1906; s. Jesse Martin and Florence (Williams) L.; B.A., DePauw U., 1928; Ph.D., U. Minn., 1931; m. Mabel Elizabeth Schroer, Nov. 2, 1929; 1 son, John William. NRC fellow Princeton, 1931-33, research asst., 1933-35; faculty Columbia, 1935-36; with carbon products div. Union Carbide Corp., 1936—, research physicist, research lab., Parma, O., 1955—. Fellow Am. Phys. Soc.; mem. Illuminating Engring. Soc., Soc. Motion Picture and TV Engrs., Lambda Chi Alpha. Research, publs. on gaseous electronics, carbon arc tech., high temperature properties of graphite. Home: 18523 E. Shoreland Av., Rocky River, O. 44116. Office: 12900 Snow Rd., Parma, O. 44130.*

LOZNER, Eugene Leonard, Am. educator; b. Stamford, N.Y., Apr. 29, 1915; s. Samuel and Rebecca (Barnhard) L.; B.A., Columbia, 1933; M.D., Cornell U., 1937; m. Jean MacPherson Culver, July 3, 1942; 1 son, Eugene Culver. With Harvard, 1938-47, Syracuse U., 1947-50, prof. medicine, State U. N.Y., 1956—; cons. VA, Psychiat. Community, Chenango, St. Joseph's Hosps. Mem. Am. Soc. Clin. Investigation, A.C.P., Internat., Am. socs. hematology, N.Y. Acad. Medicine, A.A.A.S., Am. Fedn. for Clin. Research (pres. 1946-47), Phi Beta Kappa, Sigma Xi, Alpha Omega Alpha. Contbr. chpts. to Oxford Medicine, 1941; Blood Derivatives, 1947; Progress in Gynecology, 1946; Internal Medicine, 1951. Research, publs. on coagulation defects in hemophilia, use of Vitamin K and treatment of hemorrhagic diseases; discoverer hemorrhagic diseases due to anticoagulants; inventor refrigeration unit for distbn. of blood. Home: 7 Bradford Dr., Syracuse, N.Y. 13224.*

LUBARSCH, Otto, German pathologist; b. Germany, 1860; trained in pathology and medicine; became prof. pathology in Rostock (Germany), 1894; dir. Path. Anat. Inst. in Berlin, Germany; author: (with Ostertag) Ergebnisse der allgemeinen Pathologie und Pathologischen Anatonie, 1896; also published work on immunity and phagocytosis. Introduced term hypernephroma, 1894; described small crystals in the epithelial cells of the testis which resemble sperm crystals (Lubarsch's crystals), 1894. Died 1933.

LUBBOCK, Sir John, Brit. biologist; b. London, Apr. 30, 1834; s. 3d baronet of Avebury and Harriet Hotham; D.C.L., LL.D.; m. Ellen Hordern, 1856 (dec. 1879); m. 2d, Alice Augusta Laurentia Fox-Pitt-Rivers, 1884; 5 sons, including John Byron Fraser, 4 daus. Banker, head Robarts, Lubbock, and Co.; mem. Parliament from Maidstone, 1870-80; vice-chancellor U. London, 1872-80; mem. Royal Commn. on Advancement of Sci. Fellow Royal Soc., 1858 (v.p.); corr. mem. French Acad.; pres. Soc. Antiquaries, Sociol. Soc., Royal Micros. Soc., Entomol. Soc., Ethnol. Soc., Linnean Soc., Anthrop. Inst., Ray Soc., Statis. Soc., African Soc.; Internat. Inst. Sociology, Internat. Assn. Prehistoric Archaeology, Internat. Assn. Zoology; fgn. sec. Royal Acad. hon. mem. many fgn. sci. socs. Author: The Use of Life; The Beauties of Nature; The Pleasures of Life, 2 parts; Scientific Lectures; Fifty Years of Science; 1881; British Wild Flowers, Considered in Relation to Insects; Flowers, Fruits, and Leaves; The Origin and Metamorphoses of Insects; On Seedlings, 2 vols.; La vie des plantes; Ants, Bees, and Wasps; On the Senses, Instincts, and Intelligence of Animals; Chapters in Popular Natural History; Monograph on the Collembola and Thysanura; Prehistoric Times; The Origin of Civilisation and the Primitive Condition of Man; Buds and Stipules, 1898. Died May 28, 1913.

LUBBOCK, Sir John William, Brit. astronomer, mathematician; b. Westminster, Eng., Mar. 26, 1803; s. John W. and Mary (Entwisle) L.; grad. Trinity Coll., Cambridge, 1825; M.A., 1833; m. Harriet Hotham, June 29, 1833; 11 children, including John. Became partner in father's bank, 1815; named Bakerian lectr., 1836; 1st vice chancellor London U., 1837-42; assisted in establishment Brit. Almanac, 1827. Recipient Royal medal. Fellow Royal Soc., 1829 (treas. 1830-35, v.p. 1838-47). Author: Six Maps of the Stars, 1830; On the Theory of the Moon and the Perturbations of the Planets, 1833-61; An Elementary Treatise on Computation of Eclipses and Occultations, 1835; Elementary Treatise on the Tides, 1839; also articles. Compared tidal observations with theory; simplified methods in phys. astronomy; developed method for calculating cometary and planetary orbits, 1829; demonstrated stability of solar system; claimed to have reduced tabular errors of moon below those of observations. Died June 20, 1865.

LUBCKE, Ernst Carl Wilhelm, German physicist; b. Wolfenbüttel, Germany, Dec. 16, 1890; s. Georg and Helen (Eichler) L.; m. Lina Kloppenburg, July 25, 1918; children—Enno, Renate. ed. univs. of Heidelberg, Berlin, Gottingen; Ph.D. Collaborator Atlas Werke Bremen, Germany; staff research labs dynamo plant Siemens & Co.; became agrégé in tech. physics Brunswick Tech. Coll., 1929, asso. prof., 1936; named prof., dir. Physics Inst., U. Rostock (Germany), 1946; became hon. prof. Tech. U. Berlin, Germany. Mem. Assn. German Physics Socs., Assn. German Engrs., Union German Electrotechnicians, Am. Acoustics Soc. Author: Schallabwehr im Bau-und Maschinenwesen, 1940. Home: Westendallee 92D, 1 Berlin 19. Office: Jobensstrasse 1/111, 1 Berlin 12, W. Germany.

LUBIENIECKY, Stanislaus, astronomer; b. Racon, 1625; ed. Thorn; at least 2 daus. Gov. to Count of Niemirycz; travelled with him to Holland and France; minister of ch., Lublin, Poland. Author: Theatrum cometicum (account of every comet seen or recorded from the deluge to 1665), 1666-68; also numerous other publs. Died Hamburg, Germany, 1675.

LUBITZ, Joseph Morton, Am. physician; b. N.Y.C., July 1, 1910; s. Jacob and Lena (Pachenak) L.; B.A., N.Y. U., 1932; M.D., U. Munich (Germany), 1936; postgrad. U. Ill., 1943-45, Tulane U., 1945-46; m. Ina Maria Van Treeck, June 14, 1940; children—Joan (Mrs. John Grosz), Peter, Thomas. Chief lab. USPHS Marine Hosp., 1940-46, VA Hosp., Wood, Wis., 1946-57; practice medicine specializing in pathology, Milw., 1957—; asso. prof. pathology Marquette U. Med. Sch., Milw., 1954—. Mem. A.M.A., Am. Assn. Path. Bacteriology, Chgo. Path. Soc., Wis. Soc. Pathologists (pres. 1961), Med. Soc. Milwaukee County, A.A.A.S., A.C.P., State Med. Soc. Wis., Am. Assn., Blood Banks, Am. Soc. Clin. Pathology, Am. Coll. Pathologists, Sigma Xi. Research, publs. in serology, hematology, tropical medicine, liver disease. Home: 1950 Alverno Dr., Brookfield, Wis. 53005. Office: 7635 W. Oklahoma Av., Milw. 53215.*

LUC, Henri, French laryngologist; b. St. Omer, France, 1855; 2 daus., 1 son. Practiced in Paris; repaired war injuries; corr. fellow Am. Laryngol., Soc.; hon. mem. Am. Otol. Soc., Am. Laryngol., Rhinol. and Otol. Soc. Author: Lecons sur les supparations de l'oreille moyenne, 1900. Various surg. procedures in antrum, front and ethnoid sinuses named for him; originated treatment of severe disorders of maxillary sinus by providing an additional opening in anterior wall through a supradental fossa (Caldwell-Luc operation), 1889; developed various instruments. Died Sept. 25, 1925.

LUCAE, August, German otologist; b. 1835; 1st to evaluate sound transmission through cranial bones as a diagnostic aid in ear diseases, 1870. Died 1911.

LUCAS, Bernard George Budden, Brit. physician; b. Poole, Eng., Feb. 16, 1916; s. Edmund Hall and Lena (Budden) L.; student Middlesex Hosp. Med. Sch., London, Eng., 1933-38, Diploma in Anesthesiology, 1939; m. Alice Blackburn, Aug. 24, 1940; children—Meryl (Mrs. Alan Hobbs), Peter Bernard, Simon George, Andrew Edmund; m. Betty Druscilla Thompson, July 8, 1963. Research asst. anesthesia U. Coll. Hosp. Med. Sch., London, 1945-46, cons., 1948—, research asso. dept. mech. engring., 1966—. Cons. anesthetist Hosp. for Sick Children, London Brompton Hosp., London, 1946—, Nat. Heart Hosp., London, 1948—, King Edward VII Hosp., Midhunt, Eng., 1949—. Sci. adviser Fire Brigade Union, London and Vickers Ltd., London, 1963—; cons. Chem. Def. Establishment, Ministry Def., Porton, Eng., 1953-66. Licentiate Royal Coll. Physicians. Fellow Faculty Anaesthetists, Companion Instn. Mech. Engrs.; mem. Royal Coll. Surgeons, Thoracic Soc., Physiol. Soc., Royal Soc. Medicine, Hunterian Soc., Brit. Acad. Forensic Scis., Am. Soc. Anesthesiologists, Assn. Anesthetists, Biol. Engring. Soc. Research and publs. on oxygen lack, especially brain damage, deliberate hypothermia for cardiac surgery, high pressure oxygen for clin. purposes; invented med. apparatus for resuscitation and other fields of biomed. engring. Home: 8 Cathcart Hill, London, n.19, Eng. Office: Surg. Unit, U. College Hosp. Med. Sch., University St., London W.C.1, Eng.*

LUCAS, Colin C(ameron), Canadian biochemist; b. Winnipeg, Man., Can., Dec. 15, 1903; s. George Henry and Christina (Laing) L.; B.A.Sc., U. B.C., Vancouver, Can., 1925, M.A.Sc., 1926; Ph.D., U. Toronto (Ont., Can.), 1936; D.Sc., Acadia U., Wolfville, N.S., Can., 1964; m. Mary McPhedran Elliot, Apr. 30, 1942; 1 son, Colin Robert. Prof. chemistry Brandon Coll., 1926-27, 29-34; research fellow U. Toronto, 1927-29, 34-35. faculty, 1935—, prof. Banting and Best dept. med. research, 1946—. Fellow Chem. Inst. Can., Royal Soc. Can.; mem. Am. Chem. Soc., Am. Soc. Biol. Chemists, Biochem. Soc. (London, Eng.), Am. Inst. Nutrition, Nutrition Soc. Can., Canadian Physiol. Soc., Canadian Biochem. Soc. Research, numerous publs. on oceanography, silicosis, lead poisoning, amino acid fractionization, chem. composition royal jelly, chemotherapy sulfa drugs, early studies prodn. penicillin, nutrition, proteins, choline and lipotropic phenomena. Home: 43 Thorncliffe Park Dr., Toronto, 17, Ont., Can.*

LUCAS, Cyril Edward, Brit. marine biologist; b. Hull, Eng., July 30, 1909; s. Archibald and Edith (Hinch) L.; B.Sc., U. London, 1931, D.Sc., 1942; m. Sarah Agnes Rose, June 21, 1934; children—John Martin, Alison Mary, Andrew Mark. Research biologist U. Coll. of Hull, 1931-42; head dept. oceanography, 1942-48; dir. fisheries research for Scotland, dir. Marine Lab., Aberdeen, 1948—. Chmn. research and statistics com. Internat. Commn. for N.W. Atlantic Fisheries, 1953-55; chmn. cons. and liaison coms. Internat. Council for Exploration of Sea, 1962-—; mem. adv. com. marine resources research FAO, 1963-—, vice chmn., 1962-65, chmn., 1965-—. Recipient Neill medal Royal Soc. Edinburgh, 1959. Fellow Royal Soc., Inst. Biology; mem. Marine Biol. Assn., Challenger Soc., Soc. Exptl. Biology. Publs. on ecology of marine plankton, interrelationships between plankton and other organisms, aspects of fisheries research. Home: 16 Albert Terrace. Office: Marine Lab., P.O. Box 101, Aberdeen, Scotland.*

LUCAS, Edouard A., French mathematician; b. Amiens, France, 1842; ed. École normale; prof. math. Collège de France. Author: le Calcul et les machines à calculer, 1884; Formules de géométrie tricirculaire, 1877. Built (with Genaille) multiplication rod; studied primary numbers, figurative arithmetic, tricircular geometry, tetraspheric geometry. Died 1891.

LUCAS, Fred Vance, Am. physician; b. Grand Junction, Colo., Feb. 7, 1922; s. Lee Hinman and Katharine (Vance) L.; A.B., U. Cal. at Berkeley, 1942; M.D., U. Rochester, 1950; m. Rebecca Ross Dudley, Dec. 21, 1948; children—Fred Vance, Katherine Dudley. Vet. postgrad. fellow in pathology U. Rochester (N.Y.), 1950-51, asst., 1951-53, Lilly fellow, 1952-53, Gleason fellow, 1953-54, faculty, 1954-55; asso. prof. pathology Columbia, 1955-60; asso. attending pathologist Presbyn. Hosp., N.Y.C., 1955-60; prof., chmn. dept. pathology U. Mo. Sch. Medicine, Columbia, 1960-—, research asso. Space Sci. Research Center, 1964-—. Chmn. acad. sect. Coll. Am. Pathologists, 1964-65. Recipient Lederle Med. Faculty award U. Rochester, 1954-56. Mem. Am. Soc. Clin. Pathologists (mem. council on clin. chemistry), Am. Soc. Exptl. Pathology, Am. Assn. Pathologists and Bacteriologists, Internat. Acad. Pathology, Harvey Soc., Assn. Am. Med. Colls., Boone County Med. Soc., Mo. State Med. Assn., A.M.A., Am. Assn. Blood Banks, Mos. Soc. Pathology. Research, publs. on heterogeneity of hemin of hemoglobin, ultrastructure of normal and abnormal human endometrium; demonstrated activation and inactivation of human X chromosome; discovered uterine peroxidase and relation of estrogen to enzyme in normal and abnormal conditions; plasma protein studies using carbon 14. Home: 2 Ridgely Rd., Columbia, Mo. 65201.*

LUCAS, Frederic Augustus, Am. biologist; b. Plymouth, Mass., Mar. 25, 1852; s. Augustus Henry and Eliza (Oliver) L.; pub. sch. edn.; Sc.D., U. Pitts., 1909; m. Annie J. Edgar, Feb. 13, 1884; children—Jannette May, Ann Edgar. Asst., Ward's Natural Sci. Establishment, Rochester, N.Y., 1871-82; osteologist U. S. Nat. Mus., 1882-87, asst. curator div. comparative anatomy, 1887-93, curator, 1893-1904; curator-in-chief museums Bklyn. Inst. Arts and Scis., 1904-11; dir. Am. Mus. Natural History, N.Y., 1911-23, hon. dir., 1924-—. Mem. commn. to investigate condition of fur seal herd of Pribilof Islands. Fellow Am. Ornithologists' Union, N.Y. Acad. Scis.; a founder Soc. Am. Taxidermists (contbd. to higher mus. standards in U. S.). Author: Animals of the Past, 1901; Animals Before Man in North America, 1902. Died Feb. 9, 1929.

LUCAS, George Blanchard, Am. plant pathologist; b. Philipsburg, Pa., Mar. 3, 1915; s. Reuben and Rebie (Jodon) L.; B.S., Pa. State U., 1940; M.Sc., La. State U., 1942, Ph.D., 1946; m. Jennie Elizabeth Boyd, Dec. 27, 1940 (dec. Nov. 1950); children—Irvin St. John, Glenn Boyd, Guy Boyd; m. 2d, Vernelle Violet Vangher, July 17, 1955; stepchildren—Woodson Vaughan Byrd, Claude Lee Byrd III; children—Candace Vernelle, George Blanchard. Faculty, N.C. State U., Raleigh, 1946-—, prof. plant pathology, 1963-—. Mem. Am. Phytopath. Soc. (pres. So. div. 1958), Phi Beta Kappa, Sigma Xi. Author: Diseases of Tobacco, 2d edit. 1965. Asso. editor Phytopathology, 1961-63. Research, publs. on diseases of tobacco and control of tobacco. Home: 3040 Churchill Rd., Raleigh, N.C. 27607.*

LUCAS, George Herbert William, Canadian pharmacologist; b. Parkhill, Ont., Can., Aug. 25, 1895; B.A., U. Toronto, 1921, M.A., 1922, Ph.D., 1923; m., 1921; 2 children. Demonstrator chemistry U. Toronto, 1919-21, 22-24, Banting chair med. resident, 1924-26, faculty, 1926-63, prof. pharmacology, 1945-63, emeritus, 1963-—; spl. lectr., 1963-66; cons. Ont. Dept. Health, 1964-—. Chemist Ont. Fisheries Research Lab., 1928-32; sec. Canadian Com. Pharm. Standards. Mem. Soc. Pharmacology; Chem. Soc., Soc. Biol. Chemistry, Chem. Inst. Can., Canadian Soc. Forensic Sci., Canadian Physiol. Soc. Joint editor Canadian Formulary. Contbr. papers to jours. Research and publs. on anesthesia; blood and urine changes in adrenalectomized dogs; activated sludge; growth-promoting factors in yeast bios; cyclopropane; mucous secretions; physiology and pharmacology of iron; bismuth; alcohol and road traffic; toxicology; isolation and identification of alkaloids; forensic chemistry, detection of drugs in racing animals. Address: Univ. of Toronto, Toronto, Ont., Can.

LUCAS, Henry Laurence, Jr., Am. statistician, biomathematician; b. Pasadena, Cal., Jan. 8, 1916; s. Henry L. and Lilla Alice (La Spada) L.; student Chaffey Jr. Coll., Ontario, Cal., 1932-35; B.S., U. Cal. at Davis, 1937; Ph.D. in Animal Nutrition, Cornell U., 1943; m. Jane Marie Murray, Jan. 20, 1945; children—William Henry, Carole Jane, Robert Murray, Barbara Ann, Janet Louise. Supr. advanced registry cow testing, Cal., 1938; faculty, 1938-39, U. Cal. at Davis, Cornell U., 1940-45; asso. prof. exptl. statistics N.C. State Coll. at Raleigh, 1946-48, prof., 1948-—, William Neal Reynolds prof. exptl. statistics, 1957-—, dir. biomathematics tng., 1961-—. Mem. com. NRC, 1946-58, Internat. Grassland Congress, 1952; mem. Gov.'s sci. adv. com., 1961-—; mem. com. Asso. Research Councils, 1961-63; mem. President's study com. on NIH, 1964; mem. Pres.'s Study Panel Environmental Pollution, 1965. Fellow A.A.A.S., Am. Statis. Assn.; mem. Am. Soc. Range Mgmt., Math. Assn. Am., Inst. Math. Statistics, Am. Soc. Animal Prodn., Am. Dairy Sci. Assn., Biometric Soc. (pres. Eastern N.Am. region 1962), N.C. Acad. Scis. Nat. Inst. Gen. Scis., Sigma Xi, Phi Kappa Pi, Alpha Zeta, Gamma Alpha, Gamma Sigma Delta. Research and publs. on dairy and poultry nutrition; design and analysis of animal, pasture and range expts.; pasteurization of milk and milk products; prins. and practice of using stastics and maths. in biology. Home: 2612 St. Mary's St., Raleigh, N.C. 27609.*

LUCAS, Howard Johnson, Am. chemist; b. Marietta, O., Mar. 7, 1885; s. William W. and Marian (Curtis) L.; B.A., Ohio State U., 1907, M.A., 1908, D.Sc., 1953. Asst. chemistry Ohio State U., 1907-09; fellow U. Chgo., 1909-10; asst. chemist U. S. Govt., 1910-13; instr. chemistry Cal. Inst. Tech., 1913-—, prof., 1940-55, emeritus prof., 1955-—; vis. prof. chemistry U. Hawaii, 1953, Ohio State U., 1954-55. Chmn. com. nomenclature NRC, 1952-53. Mem. A.A.A.S., Am. Chem. Soc. (Sci. Apparatus Makers award 1953). Nat. Acad. Scis., Am. Assn. U. Profs., Phi Beta Kappa, Sigma Xi, Phi Kappa Sigma. Author: Organic Chemistry, 1935, rev. 1953; Principles & Practice in Organic Chemistry Laboratory (with D. Pressman), 1949. Contbr. articles Jour. Am. Chem. Soc., Indsl. and Engring. Chemistry. Deceased.

LUCAS, Keith, Brit. physiologist; b. Greenwich, Eng., Mar. 8, 1879; s. Francis Robert and Katharine (Riddle) L.; 1st class natural sci. tripos Trinity Coll., Oxford, Eng., 1848-1901; became fellow Trinity Coll., Cambridge, Eng., 1904, Sc.D., 1911; m. Alys Hubbard, 1909; 3 sons. Became lectr. sci. Trinity Coll., 1904; research staff Royal Aircraft Factory, Farnborough, Eng., 1914-16; became Croonian lectr., 1912. Fellow Royal Soc., 1913. Research and publs. on physiology of nerve and muscle; measured depths of some unsounded New Zealand Lakes, 1904; showed contraction response of ordinary muscle fiber was of an all or none variety; measured influence of duration on effectiveness of an electric stimulus, temperature coefficient of nerve-conduction was determined; recognized subnormal impulses, time-relations for recovery of nerve-fibre from refractory phase following impulse. Died Oct. 5, 1916.

LUCAS, Richard Clement, Brit. surgeon; b. Sussex, Eng., 1846; s. William Lucas; ed. U. London, Guy's Hosp.; B.S. with Honours; M.B. (Gold medal); m. Kathleen Pelly (dec. 1912); 2 sons. Successively demonstrator anatomy, lectr. operative surgery, lectr. anatomy, lectr. surgery Guy's Hosp., also sr. surgeon, after retirement cons. surgeon; became Bradshaw lectr., 1911. Fellow Royal Soc. Medicine, Royal Coll. Surgeons (mem. council, v.p.); mem. Société de Chirurgie de Paris. Recipient Coronation medal. Research and numerous publs. on medicine and surgery; 1st description of abdominal distention occurring in most cases of early rickets, 1887. Died June 30, 1915.

LUCAS-CHAMPIONNIÈRE, Just Marie Marcellin, French surgeon; b. Saint-Leonard, France, Aug. 15, 1843; M.D., Faculty of Medicine, Paris, France, 1872, agrégé, 1874; studied under Lister at Glasgow, Scotland. With Faculty of Medicine, from 1872; became surgeon hosp., Paris, 1874; founder Med. Assn. Hosps. Paris. Mem. Acad. Medicine, French Inst. Author: De la fièvre traumatique, 1872; Chirurgerie antiseptique, 1876; De la trépanation guidee par les localisations cerebrales, 1878; Cure radicale des hemies, 1886; Le massage et la mobilisation dans le traitment des fractures, 1890. Chief editor Journal du medicine et de chirurgie pratiques. Important influence in introduction and extension of antiseptic surgery methods in France; described ganglion (nerve cells serving as center of nervous influence) named after him; discovered disease pseudomembraneous bronchitis; advocated massage and passive motion after injury, especially fractures; performed 1st valvotomy for relief of chronic aortic stenosis, 1913. Died Oct. 22, 1913.

LUCE, John Sidney, Am. physicist; b. New Orleans, Apr. 11, 1909; s. Sidney Barkley and Luisanta (Bogran) L.; D.Sc. (hon.), Auburn U., 1961; m. Audrey Tolen, Apr. 30, 1932; children—Audrey Lee (Mrs. Edward Schroeder Perritt, Jr.), Joan. Partner, Webb Appliance Co., 1938-43; engring. coordinator, experimentalist U. Cal. at Berkeley, 1943-45; physicist Tenn. Eastman Corp., 1945-47, Union Carbide Nuclear Co., 1947-50; head exploratory physics group, asst. dir. thermonuclear exptl. div. Oak Ridge Nat. Lab., 1950-60; mgr. research div. Aerojet-Gen. Nucleonics, San Ramon, Cal., 1961-—. U. S. del. Internat. Conf. on Peaceful Uses Atomic Energy, Geneva, Switzerland, 1958, Conf. on Plasma Physics and Controlled Nuclear Fusion Research, Salzburg, Austria, 1959; speaker Internat. Conf. on Ionization Phenomena in Gases, Uppsala, Sweden, 1959, Munich, Germany, 1961, Paris, France, 1963; mem. research adv. com. on elec. energy systems NASA, 1959-—. Fellow Am. Phys. Soc.; mem. Am. Inst. Aeros. and Astronautics. Contbr. articles on plasma physics, arc tech. to profl. jours. Patentee ion sources, energetic gas arc, ion acceleration system, numerous others. Home: 51 Corte Encanto, Danville, Cal. 94526. Office: P.O. Box 77, San Ramon, Cal. 94583.*

LUCIA, Salvatore Pablo, Am. internist; b. San Francisco, Mar. 9, 1901; s. David and Julia (Casino) L.; A.B., U. Cal., 1926, M.D., 1930, D.Sc., 1948; NRC fellow, Naples, Italy, also London Eng., 1930-31; m. Marilyn Matys, 1959; children—Salvatore Pablo, Darryl. Teaching staff U. Cal. Med. Sch., 1931-—, prof. medicine, 1947-—, chmn. dept. preventive medicine, 1938-64, also prof. preventive medicine; cons. internal medicine, San Francisco Hosp. 1931-—. Dir. med. research unit Wine Adv. Bd. Cal. Recipient Research award Soc. Med. Friends Wine, 1964. Diplomate Nat. Bd. Medicine, Am. Bd. Internal Medicine. Fellow A.C.P.; mem. Pan-Am. Med. Assn., Med. Friends Wine (pres. 1947). Wine and Food Soc., Chevalier du Tastevin, Internat. Soc. Hematology, Am. Pub. Health Assn., Am. Geriatric Soc., Assn. Tchrs. Preventive Medicine, Am. Soc. Hematology, Phi Beta Kappa, Sigma Xi, Alpha Omega Alpha. Author: The Hemorrhagic Diseases 1949; Wine as Food and Medicine, 1954; A History of Wine as Therapy, 1963; Alcohol and Civilization, 1963; Drink to Your Health, 1968. Research and publs. on hematology, blood clot retraction, simple easy bruisability; influence of the spleen upon the liver; also nutrition, dietary sodium and potassium in Cal. wines. Home, 20 Mercedes Way, San Francisco 94127.*

LUCIANI, Luigi, Italian physiologist; b. Ascoli-Piceno, Italy, Nov. 23, 1840; s. Serafino and Aurora (Vecchi) L.; studied medicine, Naples, Italy, Bologna, Italy; grad. 1868. Worked under Ludiwig, Leipzig, Germany, 1872-74; tchr. gen. pathology, Parma, Italy, 1875-80; prof. physiology, Siena, Italy, 1880-82, Florence, Italy, 1882-93, Rome, 1893-1917; became senator, 1905; rector U. Rome, for 2 years; mem. Superior Council Edn., for 10 years. Fellow Royal Soc., 1918; mem. Med. Acad. Rome (prin. mem.), Royal Acad. Turin, Royal Acad. Naples, Leopold-Caroline Acad., Royal Med. Acad. Belgium, Neurol. Soc. London, Med. Soc. Vienna, Acad. Sci. Göttingen, Acad. Sci. Amsterdam. Author: Treatise on human physiology. Research on heart, cerebral cortex, physiology of fasting; studied cerebellum in decerebellated dogs and proved that it was concerned with tonic and static functions, 1891. Died June 23, 1919.

LUCK, Erich, German food chemist; b. Bad Ems, Germany, July 2, 1929; s. Alfons L. and Lucie (Becker) L.; M.S., U. Mainz, 1952; Ph.D., U. Frankfort/Main, 1956; m. Gisela Schiffner, Aug. 27, 1959; 1 dau., Dorothea. Food chemist, technologist Farbwerke Hoechst AG, Frankfort/Main-Hoechst, 1957-—. Mem. German Chem. Soc. Author: Dictionary of Food Technology, English-German, 1963; articles, patents, research on food preservation, preservatives, especially sorbic acid. Home: 14 Johannesallee. Office: Farbwerke Hoechst, Frankfort/Main-Hoechst, W. Germany.*

LUCK, Hans, chemist, microbiologist; b. Ziegenhain, Germany, Dec. 16, 1921; s. Alfred and Anna (Alberts) L.; Dipl.-Chem., U. Kiel, D.Sc.; m. Hanna Scheer, May 5, 1945. With Fed. Dairy Research Inst., Kiel, Germany, 1949-53, Dairy Industry Control Bd., Windhoek S.W. Africa, 1953-55, German Research Inst. Food Chemistry, Munich, 1955-64, Animal Husbandry and Dairy Research Inst., Irene, S. Africa, 1964-—. Mem. German Chem. Soc., German socs. Hygiene and Microbiology, Dairy Sci., S. African Dairy Technol. Assn. Research, publs. on food and dairy science. Home: 356 Fonteine Av., Lyttelton, Transvaal. Office: Animal Husbandry and Dairy Research Inst., Irene, Transvaal, South Africa.*

LUCKE, John Becker, Am. geologist; b. N.Y.C., Feb. 26, 1908; s. Charles E. and Ida (Becker) L.; B.S. maxima cum laude, Princeton, 1929, A.M., 1932, Ph.D., 1933; postgrad. Columbia; m. Virginia Lee Duncan, Aug. 25, 1937; children—Patricia (Mrs. William F. Morris), Helen Elizabeth, Joan Ida. Geologist, Sloan & Zook Co., Bradford, Pa., 1929-30; with Tex. Co., Pampa, Tex., 1934-35; asst. prof. geology W.Va. U., Morgantown, 1936-40; prof. geology U. Conn., Storrs, 1940-63, Grand Valley Coll., Allendale, Mich., 1964-—. Dir. Conn. Geol. and Natural History Survey, 1954-60; geol. cons. Quinn & Assos., Wyncote, Pa., 1964-—. Fellow A.A.A.S., Geol. Soc. Am.; mem. Am. Assn. Petroleum Geologists (life), Am. Geophys. Union, Am. Soc. Photogrammetry, Arctic Inst. N.Am., Nat. Assn. Geology Tchrs. (v.p. 1953), Sigma Xi (past pres. U. Conn. chpt.). Research in marine shoreland processes and devel., oceanography and related fields; publs. gen. geology manuals. Home: 1156 Gladstone St. S.E., Grand Rapids, Mich. 49506. Office: Grand Valley College, Allendale, Mich. 49401.*

LUCKE, Kurt, German physicist; b. Halberstadt, Germany, June 28, 1921; s. Heinrich and Luise (Baganz) L.; Dipl.Phys., U. Göttingen, 1947, Dr.rer.nat., 1949; m. Helen Reenstierna, 1955; children—Christian, Hanna Melanie. Wiss. asst. U. Göttingen, 1953; privatdozent Technische U. Berlin (Germany), 1953; faculty Brown U., Providence, 1954-57, prof. physics, 1956-57; prof., dir. Institut für Allgemein Metallkunde und Metallphysik, Rein.-Westf. Technische Hochschule, Aachen, Germany, 1957-—; dir. Inst. for Reactor Materials, Nuclear Research Center, Jülich, Germany, 1958-—. Mem. Deutsche Gesellschaft für Metallkunde Deutsche Physikalische Gesellschaft,

Am. Inst., Mining and Metall. Engrs., Am. Phys. Soc., Brit. Inst. Metals. Author: (with Georg Masing) Lehrbuch der Allgemeinen Metallkunde, 1950; also articles. Research in phys. metallurgy and solid state physics especially in areas of plasticity, recrystallization, diffusion, internal friction and radiation damage of metals. Home: Morillenhang, 51 Aachen, Germany.*

LUCKEY, Egbert Hugh, Am. physician; b. Jackson, Tenn., Jan. 1, 1920; s. David William and Ethel May (Freeman) L.; B.S., Union U., Jackson, 1941; Sc.D., 1954; M.D., Vanderbilt U., 1944; Dr. honoris causa, U. Bahia (Brazil), 1961; m. Betty Ann Black, Dec. 25, 1942; children—Linda Ann, James Hugh, John William, Robert Powers. Staff, Bellevue Hosp., N.Y.C., 1949-54, vis. physician, 1949—; staff N.Y. Hosp., N.Y.C., 1944—, physician-in-chief, 1957-66; pres. N.Y. Hosp.-Cornell Med. Center, 1966—; chmn. dean's com. VA Hosps., N.Y.C., 1956-57; chmn. bd. mgrs. Vincent Astor Diagnostic Service, 1957-60; faculty Cornell U. Med. Coll., N.Y.C., 1945—, dean Med. Coll., asso. dean Grad. Sch., 1954-57, prof., chmn. dept. medicine, 1957—. Mem. com. medicine W. K. Kellogg Found., 1955—; mem. exec. com. Josiah Macy, Jr. Found., 1958—; med. dir., bd. trustees, sec.-treas. Russell Sage Inst. Pathology, 1958—; cons. mem. adv. coms. to govt. Diplomate Am. Bd. Internal Medicine (ofcl. examiner 1952); Fellow A.C.P. (regent), N.Y. Acad. Medicine; mem. A.M.A. (vice chmn. sect. internal medicine 1960—), Harvey Soc., Peripatetic Club, Practitioners Soc., Am. Fedn. Clin. Research, Assn. Am. Med. Colls. (chmn. com. internships and residencies and grad. med. edn. 1955-60), Assn. Am. Physicians, Soc. Cons. to Armed Forces, Alpha Omega Alpha. Editorial bd. Jour. Med. Edn., 1961—. Home: 435 E. 70th St., N.Y.C. 10021.*

LUCKEY, Thomas Donnell, nutritional biochemist; b. Casper, Wyo., May 15, 1919; s. Frank S. and Lillian (Waggener) L.; B.S., Colo. State U., 1941; M.S., U. Wis., 1944, Ph.D., 1946; m. Pauline May Miller, Sept. 1, 1943; children—Jane, Mary, Donna. Chemistry asst. Wyo. U., 1938-39; teaching asst. Tex. A. and M. U., 1941-42; research, teaching asst. U. Wis., 1942-46; asst., asso. research prof. LOBUND, Notre Dame U., 1946-54; prof. biochemistry U. Mo. Med. Sch., 1954—, chmn. biochemistry, 1954-68. Cons. Gen. Electric Co., McDonnell Aircraft Corp., Whitehall Pharmacy; organizer, moderator Symposium on Gnotobiology IX, Internat. Congress of Microbiology, Moscow, 1966. Commonwealth fellow, 1961-62. Mem. Am. Chem. Soc., A.A.A.S., Nutrition Inst., Am. Soc. Microbiology, Soc. Exptl. Biology and Medicine, Assn. for Gnotobiotics, Am. Assn. Contamination Control, Am. Inst. Biol. Sci., Am. Assn. U. Profs., Sigma Xi. Author: Germfree Life and Gnotobiology, 1965; also numerous articles. Co-editor: Advances in Germfree Research and Gnotobiology, 1968. Co-discoverer Vitamins B10 and B11, 1st colony of germfree mammals, antibiotic growth stimulation in animals. Home: 812 S. Edgewood Av., Columbia, Mo. 65201.

LUCKHARDT, Arno Benedict, Am. physiologist; b. Chgo., Aug. 26, 1885; s. Gustav Adolph and Aurelia (Weber) L.; student Conception Coll., 1897-1903, LL.D., 1933; B.S., U. Chgo., 1906, M.S., 1908, Ph.D., 1911; M.D., Rush Med. Coll., 1912; Sc.D., Northwestern U., 1934; m. Luella Catherine LaBolle, Apr. 24, 1912; children—Hilmar Francis, Paul Gregory, Mary Aurelia. Asst. in bacteriology U. Chgo., 1908-09, in physiology, 1909-11, asso., 1911-12, from instr. to prof., 1912-23, prof., 1923-41, chmn. dept. physiology, 1941-50, Dr. William Beaumont Distinguished Service prof. physiology, 1947-50, emeritus; research cons. J. B. Roerig & Co., Pharm. House of Chas. Pfizer & Co., Inc. N.Y.C. Recipient Callahan Meml. Award medal Ohio Dental Soc. Fellow A.M.A., Soc. for Exptl. Biology and Medicine, Internat. Coll. of Anesthetists, Internat. Soc. Dental Research, Am. Coll. Dentists (hon.); hon. mem. German Med. Soc. Chgo., St. Louis Med. Soc.; mem. Internat. Anesthesia Research Soc., Am. Psychol. Soc. (past pres.), Inst. Traumatic Surgery Fedn. Am. Socs. for Exptl. Biology (pres. 1933-35), Gorgas Soc., Walter Reed Soc. (life mem., certificate of award 1952), A.A.A.S., Ill. Med. Soc., Am. Dental Assn. (council on therapeutics), Kaiserlich Deutsche Akademie der Naturforscher, Sigma Xi, Phi Beta Kappa. Discoverer (with J. Bailey Carter) of ethylene gas as an anesthetic agt. with properties superior to nitrous oxide, commonly known as laughing gas; researches in physiology of parathyroid glands, gastric and pancreatic secretion, and in history of physiology, dentistry and medicine. Died Nov. 6, 1957.

LUCRETIUS, Titus Caras, Roman philosopher; b. Rome, Italy, circa 95 B.C.; disciple of Empedocles and Epicurus. Author: De natura rerum (On the Nature of Things), presents Epicurean, mech. view of universe, treating physics, psychology, ethics in poem form, 6 books. Stated atomic theory; held that all things were composed of atoms; believed world emerged from chaos by pure chance; advocated rights of reason against superstition. Reputed to have committed suicide in fit of madness (caused by love philter given him by his wife), circa 55 B.C.

LÜDERS, Gerhart Claus Friedrich, German physicist; b. Hamburg, Germany, Feb. 25, 1920; s. Arnold C. F. and Toni (Toenniessen) L.; diploma in physics Hamburg U., 1947, Dr.rer.nat., 1950; m. Ingeborg Suhrmann, Aug. 16, 1951. Physicist, Max Planck Inst. of Physics, Göttingen, Munich, 1950-60; prof. physics Göttingen U., 1960—. Recipient Physics prize Göttingen Acad., 1959. Mem. Göttingen Acad. Scis., German (Max Planck medal 1966), Am. phys. socs. Research, publs. on symmetry properties of elementary particles, theory of superconductivity, particle orbits in accelerators. Home: 10 Hasenwinkel, 3401 Hetjershausen, Germany.*

LÜDICKE, Manfred Erich Walter, German zoologist; b. Berlin, Feb. 22, 1911; s. Walter and Elfriede L.; Ph.D., agrégé, U. Berlin; m. Lotte Hohmann, Mar. 27, 1937. Successively prof. agrégé, 1940, prof., 1950, curator, 1959, cons. sci., 1961. Mem. German Zoology Soc. Author: Gattung Quadraspidiotus, Sorauer; Serpentes. Research and publs. on anatomy of beaks and feathers of birds. Home: Rottmannstrasse 27, Heidelberg. Office: Zool. Inst., Heidelberg, W. Germany.

LÜDKE, Werner Robert Conrad Eugen, German phys. chemist; b. Stettin, Germany, Sept. 27, 1899; s. August and Martha (Lüdke) L.; Dipl.-Chem., U. Marburg/Lahn, 1923, Dr.Phil., 1927. Scientist, Kaiser Wilhelm Inst. Textile Chemistry, Berlin-Dahlem, mineral. inst. U. Leipzig; lab. mgr. VEB Carl Zeiss, Jena, Germany. Mem. Geol., Chem., Biophys. socs. East Germany, German Mineral. Soc., Leopoldina. Research, publs. in mineralogy, chemistry, biology, synthesis of amphiboles, asbestos; investigation of crystal growth, phys. properties of crystals, biol. replacability, physiol. isomorphy. Home: 27 Leibniz-Strasse, Leipzig 701, E. Germany. Office: 1 Carl-Zeiss-Strasse, Jena 69, E. Germany.*

LUDLAM, William, Brit. astronomer; b. Leicester, Eng., 1717; s. Richard and Anne (Drury) L.; ed St. John's Coll., Cambridge; B.A., 1738, M.A., 1742, B.D., 1749. Became vicar, Norton-by-Galby, 1749; jr. deacon of his coll., 1754-57; Linacre lectr. physic, 1767-69; became rector Corkfield, Eng., 1768, and remained there for the rest of his life. Author: Rudiments of Mathematics, 1785; also publs. on astronomy. Performed transformations in Euclidean postulates; postulated that 2 parallels to a given line cannot intersect (also Playfair's axiom), 1794. Died Leicestershire, Eng., Mar. 16, 1788.

LUDLUM, Seymour DeWitt, Am. psychiatrist; b. Goshen, N.Y., Aug. 1, 1876; s. John Frank and Loisa May (Minturn) L.; B.S., Rutgers U., 1897, Sc.D., 1951; M.D., Johns Hopkins, 1902; 1 son, Seymour D., by 1st marriage; m. 2d, Mabel Stewart, Oct. 9, 1920 (dec.). Chief staff, neuro-psychopathic dept. Phila. Gen. Hosp., 1910-47, cons. from 1947; med. dir., owner, Gladwyne (Pa.) Colony, pvt. sanitarium for mental and nervous disease, from 1912, also dir. Gladwyne Research Lab. for Mental Research; prof. neurology, U. Pa. Grad. Sch. Medicine, 1920, prof. psychiatry, 1922-53. Mem. Am. Psychiat, Assn., Am. Neurol. Assn., A.M.A., Am. Chem. Soc., Johns Hopkins Medical Soc., Assn. Research Nervous and Mental Diseases, Eugenic Research Assn., Am. Therapeutic Assn. Research and articles on mental and nervous diseases. Died Dec. 2, 1956.

LUDWIG, George Doring, Am. physician; b. Johnstown, Pa., Jan. 4, 1922; s. Karl D. and Kathryn (Palmer) L.; B.S., St. Vincent Coll., Latrobe, Pa., 1943; M.D., U. Pa., 1946; m. Rosemary N. Lynch, Oct. 1950 (dec. Sept. 1963); children—Rosemary, Elizabeth, Kathleen, Mary Lou, George Doring; m. 2d, Mary Ellen Whitmore, May 20, 1967. Practice medicine, specializing in internal medicine, Phila., 1952—; faculty U. Pa., 1952—, prof. medicine, 1967—; vis. scientist Med. Nobel Inst., 1956-57. Mem. exec. com. Nat. Bd. Med. Examiners. Recipient Christian Linback award for distinguished teaching U. of Pa., 1962. Markle scholar in med. scis., 1953-58. Diplomate Am. Bd. Internal Medicine. Fellow A.C.P. (gov. Eastern Pa. 1967—); mem. Am. Soc. Clin. Investigation, Am. Physiol. Soc., A.A.A.S. Research, numerous publs. in echo-ranging ultrasound for diagnosis, measurement of biophysical properties of tissues to ultra-sound, prodn. of carbon monoxide from hemin breakdown, diseases of porphyrin metabolism, metabolic and human genetic diseases. Home: 629 New Gulph Rd., Bryn Mawr, Pa. 19010. Office: Hosp. U. Pa., Maloney Bldg., 3600 Spruce St., Phila. 19104.*

LUDWIG, George Harry, Am. engring. physicist; b. Sharon, Ia., Nov. 13, 1927; s. George M. and Alice (Heim) L.; B.A. in Physics, State U. Ia., 1956, M.S., 1959, Ph.D. in Elec. Engring., 1960; m. Rosalie F. Vickers, July 21, 1950; children—Barbara Rose, Sharon Lee, George Vickers, Katy Ann. Research aide State U. Ia., Iowa City, 1953-56, research asst., 1956-58, U. S. Steel Found. research fellow, 1958-60, research asso., 1960; head energetic particles instrumentation sect., 1961-65, project sci. eccentric orbiting geophys. observatories NASA, Goddard Space Flight Center, Greenbelt, Md., 1960—, chief information processing div., 1965—. Recipient Golden Plate award Acad. of Achievement. Mem. Phi Beta Kappa, Sigma Xi, Phi Eta Sigma. Mem. editorial bd. Space Sci. Revs.; mem. adv. bd. Astrophysics and Space Sci. Library. Co-discoverer Van Allen Radiation Belts, 1958; designer sci. instrumentation for numerous satellite and space probe energetic particles expts. Home: 10522 Calumet Dr., Silver Spring, Md. 20901. Office: Goddard Space Flight Center, Greenbelt, Md. 20771.*

LUDWIG, Gerald Wilbur, Am. physicist; b. Bronx, N.Y., Jan. 7, 1930; s. Daniel and Helen (Herber) L.; A.B., Harvard, 1950, A.M., 1951, Ph.D., 1955; m. Therese Rosenvinge, June 16, 1951; children—Warren, Catherine, Arthur. Physicist, Gen. Electric Research and Devel. Center, Schenectady, 1955-63, liaison scientist, 1963-65, physicist, 1965—. Fellow Am. Phys. Soc.; mem. I.E.E.E., Sigma Xi. Research on elec. properties of semiconductors, impurity centers using electron paramagnetic resonance as a tool, microwave effects in bulk semiconductors, phosphors. Home: 112 Glenhill Dr., Schenectady 12302. Office: P.O. Box 8, Schenectady 12301.*

LUDWIG, Gunther Hermann Paul, German physicist; B. Zäckerick, Germany, Jan. 12, 1918; s. Hermann and Herta (Kurts) L.; Dr.rer.nat., U. Berlin, 1943; m. Lucie Staneczek, Aug. 19, 1944. Asst., U. Göttingen, Germany, 1946-49; prof. Free U. Berlin, 1949-63; prof. U. Marburg, Germany, 1963—. Author: Fortschritte der Projektiven Relativitäts-theorie, 1950; Grundlagen der Quantenmechanik, 1953; also articles in field. Research on projective relativity theory, ergodic theory, axiomatic foundation of quantum mechanics. Home: 11 Grüner Weg, 3554 Cappel, Germany. Office: 7 Renthof, 355 Marburg, West Germany.*

LUDWIG, Karl Friedrich Wilhelm, physiologist; b. Witzenhausen, Germany, Dec. 29, 1816; studied medicine, univs. Erlangen, Marburg, (both Germany). Apptd. to faculty U. Marburg, 1846; mem. faculties Zurich, Switzerland, Vienna, Austria; dir. physiol. inst., prof. physiology, Leipzig, Germany. Mem. French Acad. Scis. Author: Beiträge zur Lehre vom Mechanismus der Harnsecretion, 1843; Lehrbuch der Physiologie des Menschen, 1852-56; Died physiologischen Leistungen des Blutdruckes, 1865. Helped found nonvitalistic physiology in Germany; first to keep organs removed from animals alive by perfusion, 1856; used kymograph to study blood pressure; inventor (independently of Jan Dogiel) stromuhr for measuring speed of blood stream, circa 1867; inventor mercurial blood-gas pump for separation of gases from blood (led to understanding of role of gases in blood purification); studied role of nervous system in circulation, glandular secretions. Died Leipzig, Apr. 23, 1895.

LUECK, Roger Hawks, Am. chemist; b. Fox Lake, Wis., Dec. 17, 1896; s. George W. and Flora E. (Hawks) L.; B.S., Carroll Coll., 1919, D.Sc., 1943; M.S., U. Wis., 1922; m. Margaret M. McCaslin, May 31, 1924; children—Peggy E., Nancy J. (Mrs. Robert I. Dietrich). Chemist, Am. Can Co., Maywood, Ill., 1922-26, mgr. San Francisco Lab., 1926-34, mgr., Hawaiian operations, Honolulu, 1935-36, dir. research, Maywood and N.Y.C., 1937-43, sales mgr. Pacific div., San Francisco, 1944-50, gen. mgr., research and devel. dept., N.Y.C., 1951-55, v.p., N.Y.C., 1956-62, cons. in research adminstrn. and packaging tech., Saratoga, Cal., 1963—. Bd. dirs. Dole Engring. Co., San Francisco. Mem. Inst. Chemists, Inst. Food Technologists, Am. Chem. Soc., A.A.A.S., Assn. Research Dirs., Indsl. Research Inst., Can Mfrs. Inst. (chmn. tin conservation com. 1942-46), Am. Mgmt. Assn. (v.p. 1958-60), N.A.M. (dir. 1960-61), Sigma Xi, Tau Kappa Epsilon, Alpha Chi Sigma. Research on C-Enamel to prevent ferrous sulphide discoloration in canned foods, mechanism of sub-aqueous corrosion of tinplate. Address: 20016 Winter Lane, Saratoga, Cal. 95070.*

LUECKE, Richard William, Am. biochemist; b. St. Paul, July 12, 1917; s. Frederick William and Susan (Trautz) L.; B.A., Macalester Coll., 1939; M.S., U. Minn., 1941, Ph.D., 1943; m. Eleanor V. Rohrbacher, Nov. 1, 1941; children—Glenn R., Joan E., Ruth A. Research asso. dept. biochemistry U. Minn., 1941-43; research biochemist A. and M. U. Tex., 1943-45; faculty Mich. State U., East Lansing, 1945—, prof. dept. biochemistry, 1949—. Cons. animal prodn. and health div. FAO, 1959-62. Recipient Nutrition Research award Am. Feed Mfrs. Assn., 1956. Mem. Am. Chem. Soc., Am. Inst. Nutrition (com. on exptl. nutrition 1964—), Soc. for Exptl. Biology and Medicine, A.A.A.S., N.Y. Acad. Sci., Am. Soc. Animal Sci. Research, numerous publs. on B vitamin deficiencies in swine, feasibility vitamin fortification swine diets, relationship between zinc metabolism and porcine parakeratosis; discovered growth-promoting effects oral streptomycin. Home: 1026 Beech St., East Lansing, Mich. 48823.*

LUEDDE, William Henry, Am. ophthalmologist; b. Warsaw, Ill.; s. Henry J. M. and Emilie M. (Naumann) L.; M.D., Washington U. St. Louis, 1900; vol. asst. eye clinic Royal U., Kiel, Germany, 1904-05; student Laboratoire d'Ophthalmologie, Sorbonne, Paris, 1906; m. Nettie B. Shryock, Mar. 24, 1909 (dec. Nov. 1946); children—Philip S., Fullerton W., Henry W.; m. 2d, Irene E. Garbarino, Jan. 2, 1948. Asst. to Drs. Green, Post and Ewing, 1901-04; practiced in St. Louis, from 1906; asst. surgeon, Eye Clinic, Washington U., 1908-12; ophthalmic surgeon St. Louis Eye, Ear, Nose and Throat Infirmary, 1912-16; prof. ophthalmology St. Louis U., from 1921; ophthalmologist in chief Firmin Desloge Hosp., St. Mary's Hosp. and Infirmary; oculist Mo. Bapt. Sanitarium; attending ophthalmologist U. S. Marine Hosp.; cons. in ophthalmology St. Louis City, St. Louis County and St. Johns hosps. Recipient Gill prize (disease of children) Washington U., 1900. Leslie Dana medal (prevention of blindness), 1933. Dir. St. Louis Soc. for the Blind. Fellow A.M.A., A.A.A.S.; mem. many nat. internat. and fgn. med. and ophthal. assns. Alpha Omega Alpha. Died Mar. 19, 1952.

LÜERS, Heinrich, German chemist; b. Munich, Feb. 12, 1890; s. Jacob and Anna (Bauer) L.; Dr.Engring., agrégé, Munich Tech. Coll.; m. Ida Monatsberge, Dec. 18, 1918; 1 son, Heinrich. Became asst., 1913; named prof. Acad. for Agronomy, Weiheinstephan, 1914; became agrégé Munich Tech. Coll., 1930; ret., 1948. Named prof. honoris causa Gand Inst. Fermentation. Hon. mem. Soc. Master Brewers of Fedn. of German Master Malters. Author: Chemie des Brauwesens, 1929; Die Wissenschaftlich Gründlagen von Mälzerei und Brauerei, 1950; Grundruss der Bierbrauerei, 1949; Chemie und Technologie der Landwirtschaft gewerbe, 1944. Research on enzymes and biochemistry. Home: Aberhof/Rottoch-Egern, West Germany.

LUETH, Harold C(harles), Am. internist; physician; b. Chgo.; July 24, 1904; s. Charles Frederick and Matilda (Beck) L.; B.S., Northwestern U., 1927, M.S., 1928, B.M., 1929, Ph.D., 1930, M.D., 1930; m. Elizabeth A. Bullock, Sept. 3, 1932; children—Mary Matilda, Virginia (Mrs. George E. Keith, Jr.), Winfred (Mrs. Robert S. Marshall), Charles, Margaret (Mrs. Thomas Lovaas). Practice medicine specializing in internal medicine, 1932—, faculty U. Ill., 1940-46, clin. prof. medicine, 1952—; prof. medicine, dean Coll. of Medicine, supt. U. Hosp., U. Neb., 1946-52; cons. Office Civil Def., Surgeon Gen. Army, Surgeon 5th U. S. Army, 1955—. Honorable Mention Wellcome Prize. Fellow A.C.P.; Soc. Med. Consultants Armed Forces, Inst. of Medicine, Soc. Internal Med. (v.p. 1941-42), Am. Heart Assn., Central Research Club, Soc. of Exptl. Biology and Medicine, Central Soc. Clin. Research, A.M.A. (chmn. council 1960-63) Sigma Xi, Alpha Omega Alpha. Co-author: Diseases of Coronary Arteries; Chronic Myocarditis, 1932; contbr. various med. periodicals. Home: 2678 Sheridan Rd. Office: 636 Church St. Evanston, Ill. 60201.*

LUETSCHER, John Arthur, Am. Physician; b. Balt., Aug. 30, 1913; s. John A. and Elizabeth (Tumbleson) L.; A.B. magna cum laude, Princeton, 1933; M.D., Johns Hopkins, 1937; m. Genevieve Buckler, May 13, 1934; children—John Duer, R. Buckler. Research fellow Harvard, Med. Sch., 1938-40; faculty Johns Hopkins, 1940-48, asst. prof. medicine, 1945-48; faculty Stanford Med. Sch., 1948—, prof., 1955—. Spl. cons. to surgeon gen. USPHS, 1955-59, endocrinology study sect. NIH, heart Program project com., 1962-63, tng. grants com., 1965—. Recipient John Phillips Meml. award A.C.P., 1963. Mem. Am. Soc. for Clin. Investigation (pres. 1959), Assn. Am. Physicians, Endocrine Soc. (mem. council 1963-66, editorial bd. 1961—), Western Soc. for Clin. Research, Western Assn. Physicians (v.p. 1966-67). Research, numerous publs. on blood proteins, kidney disease, fluid balance, and adrenal hormones. Home: 635 Salvatierra, Stanford, Cal. 94305.*

LUFT, Rolf, Swedish endocrinologist; b. Stockholm, Sweden, June 29, 1914; s. Jankel and Ester (Milner) L.; M.D., Karolinska Inst., Stockholm, 1940, Ph.D., 1944. Asso. prof. endocrinology Serafimer Hosp., Stockholm, 1949-61; prof. endocrinology Karolinska Inst. and Hosp., 1961-66, prof. internal medicine, 1966—, dir. dept. endocrinology and metabolism Karolinska Hosp., 1958—. Author: Textbook of Endocrinology, 1962; (with Rachnud Levine) Advances in Metabolic Disorders; also numerous articles. Research on pathophysiology of Cushing's Syndrome, metabolic action of corticosteroids and ACTH, hypophysectomy in man, body composition, clin. research on subcellular level, metabolic action of human growth hormone, diabetic action of hormones, prediabetic state. Home: 78 Norr Mälarstrand, Stockholm K, Sweden.*

LUFT, Ulrich Cameron, physiologist; b. Berlin, Germany, Apr. 25, 1910; s. Friedrich J. and Mary (Wilson) L.; Friedenauer Gymnasium, Abiturium, 1929; student U. Freiburg, 1929-30, U. Munich, 1930-31; grad. U. Berlin Med. Sch., 1935, M.D., 1937; Dr. med. habil., Berlin, 1942; m. Alice Hentzelt, Apr. 17, 1941; 1 son, Friedrich C. Came to U. S., 1947, naturalized, 1955. With Aeromed. Research Inst., Berlin, 1937-45; with Himalayan expdns., 1937, 38; faculty U. Berlin, 1942-47; research physiologist, asso. prof. Air. U. Sch. of Aviation Medicine, Randolph Field, Tex., 1947-53; head physiology dept. Lovelace Found. Med. Edn. and Research, Albuquerque, 1954—. Cons. physiologist VA, bur. hearings and appeals USPHS, Surgeon Gen. U. S. Army; mem. med. adv. council NASA, 1965—. Recipient Spl. citation A.M.A., 1963. Fellow Am. Coll. Chest Physicians, Aerospace-Med. Assn., Am. Coll. Sports Medicine, A.A.A.S.; mem. Am. Physiol. Soc., Internat. Acad. Astronautics. Research, publs. on respiratory physiology in aviation and space flight, human adaptations to extreme pressure and thermal environments; phys. performance in relation to age and body composition, clin. physiology of pulmonary and cardiovascular disease. Home: 1900 Ridgecrest Dr. S.E., Albuquerque 87108. Office: 5200 Gibson Blvd. S.E., Albuquerque 87108.*

LUGEON, Jean, Swiss meteorologist; b. Lausanne, Switzerland, Aug. 4, 1898; s. Maurice and Ida (Welti) L.; ed. Lausanne, Zurich, Switzerland; Ph.D.; Dr.engring. Polytechnica Varsovie; m. Marguerite Faes, June 21, 1924. Hydraulics expert, Paris, 1923-24; meteorologist, Zurich, 1924-28; dir. Nat. Meteorol. Inst. Poland, 1929-36; vice dir. Swiss Inst. Meteorology, 1937-44, dir., 1945-63; prof. Swiss Fed. Inst. Tech., 1937-63. Author numerous books, articles. Research on climatology, aerology, hydrometeorology, radiometry, atmospheric electricity, air pollution; devel. equipment. Home: 23, rue Secrétan, Lausanne, Switzerland. Office: Kräbülstrasse 56, Zurich, Switzerland.

LUGEON, Maurice, geologist; b. Poissy, France, July 10, 1870; s. David and Adèle (Cauchois) L.; Sc.D., U. Lausanne, 1895; degree, 1893; studied Munich, Paris; hon. doctorates: U. Paris, Strasbourg, Aix-Marseille, Grenoble, Liége, Louvain, Brussels, Zurich, Lwów, Bucarest; m. Ida Welti, 1897; one son, Jean. Extraordinary prof., geology and physical geography, U. Lausanne, 1898-06; ordinary prof., 1906—; rector of U. Lausanne, 1918-20. Recipient Prestwich Medal, French Geol. Soc., 1906; Wollaston Medal, London Geol. Soc., 1938. Pres., Helvetic Soc., 1923-28; vice-pres., Swiss Geological Commn.; mem., Inst. of France; French Acad. Scis.; Royal Soc., London and Edinburgh; Acads. of Sci. of Spain, Belgium, Italy, Rumania, U.S.S.R. Author: Les grandes nappes de recouvrement des Alpes du Chablais et de la Suisse, 1901; Étude géologique sur le projet de barrage du Haut-Rhône français à Génissiat, 1912; Barrages et geologie, 1933; over 200 pub. articles. Assoc. with H. Schardt's tectonic discovery of far-traveled nature of rocks of Prealps; played leading role in development of tectonic concepts; their application, and in comprehensive interpretation of Alps; internat. consultant on dam sites. Died Lausanne, Switzerland, Oct. 23, 1953.

LUGOL, Jean Guillaume Auguste, French physician; b. Montanbau, France, Aug. 10, 1786; degree of doctor, Paris, 1812; physician Hosp. Saint-Louis. Author: Mémoires sur l'emploie de l'iode dans les maladies scrofulenses, 1826; Sui d'emploie de l'iode, suivi d'un précis de l'art de formuler les préparations iodurées, 1831; Recherches sur les causes des maladies scrofulenses, 1844. Research on skin diseases, therapeutic use of iodine; originated solution of iodine for use in tuberculous conditions (Liquor iodi fortis, strong solution of iodine, compounds solution of iodine or Lugol's solution), 1829. Died Sept. 16, 1851.

LUH, Bor Shiun, chemist; b. Shanghai, China, Jan. 13, 1916; s. Tsung F. and King (Chen) L.; B.Sc. in Chemistry, Chiao-Tung U., 1938; M.Sc. in Food Sci., U. Cal., Berkeley, 1948, Ph.D. in Agrl. Chemistry, 1952; m. Bai Tsain Liu, Nov. 23, 1940; 1 dau., Janet Shirley. Came to U. S., 1946, naturalized, 1958. Instr., Chiao Tung U., 1938-41; chemist Maling Canned Foods Co., Ltd., Shanghai, 1941-46, cons., 1965—; research asst. U. Cal. at Berkeley, 1948-51; jr. specialist U. Cal. at Davis, 1952-56, jr. food technologist, 1956-57, asst. food technologist, 1957-63, asso. food technologist, 1963—. Food tech. cons. Food Industry Research and Devel. Inst., 1965—. Recipient Tippet award Chiao-Tung U., 1938. Mem. Am. Chem. Soc., Inst. Food Technologists, Am. Oil Chem. Soc., A.A.A.S., Am. Soc. Hort. Scis., Phytochem. Soc. N.Am., Sigma Xi. Research, publs. on food chemistry and processing.*

LUHAN, Joseph Anton, Am. physician; b. Chgo., Feb. 2, 1901; s. Joseph and Bessie (Mansfield) L.; student U. Chgo., 1921-22, Crane Jr. Coll., 1922-23; M.D., Northwestern U., 1928, M.S., 1931, Ph.D., 1934; m. Wilha Stedem, Apr. 5, 1944. Faculty dept. neurology and psychiatry Northwestern U. Med. Sch., 1928-41; practice medicine specializing in neurology and psychiatry Chgo., 1930—; faculty Strich Sch. Medicine, Loyola U., Chgo., 1941—, clin. prof. neurology and psychiatry, 1945-54, prof. neurology and psychiatry, 1954—; chief dept. neurology and psychiatry Loretto Hosp., Chgo., 1962—; attending neurologist Cook County Hosp., Chgo., 1938-59, dir. neuropathology lab., 1941-61, cons. neurologist and neuropathologist, 1961—. Cons. neuropsychiatrist Mercy, St. Anthony hosps., Chgo., MacNeal Meml. Hosp., Berwyn, Ill. Mem. Am. Neurol. Assn., Am. Assn., Neuropathologists, Am. Psychiat. Assn., A.A.A.S., Ill. Psychiat. Soc. (past pres), Am. Acad. Neurology, Central Neuropsychiat. Assn., Chgo. Neurol. Soc., (past pres.). Research, publs. on pathology of nervous system, exptl. poliomyelitis, human poliomyelitis, meningitis, clin. neurol. studies, psychiat. studies particularly mental illness of curable form in older individuals. Home: 808 Wisconsin Av., Oak Park, Ill. 60304. Office: 720 Lake St., Oak Park, Ill. 60301.*

LUISADA, Aldo Augusto, Am. cardiologist; b. Florence, Italy, June 26, 1901; s. Ezio and Elisa (Rignano) L.; M.D. cum laude, Royal U., Florence, Italy, 1924; m. Anna Passigli, Apr. 12, 1931; 1 son, Claude G. Came to U. S., 1939, naturalized, 1944. Asso., Royal U. Padua (Italy) Med. Sch., 1927-30; faculty Royal U. Naples (Italy) Med. Sch., 1930-35; prof. medicine Royal U. Sassari (Italy), 1935-36, U. Ferrara (Italy), 1936-38; faculty Tufts Coll. Sch. Medicine, 1943-49; faculty Chgo. Med. Sch., 1949—, prof. medicine, 1960—, prof., dir. div. cardiovascular research, 1961—; attending cardiologist Mt. Sinai Hosp., Chgo., 1960—. Cons. medicine Hines (Ill.) VA Hosp., 1955—. Recipient Morris Parker award for research Chgo. Med. Sch. 1953. Fellow A.M.A. (Gold medal for exhibit 1954), A.A.A.S., A.C.P., Am. Heart Assn., Am. Coll. Cardiology, Am. Coll. Angiology Am. Coll. Chest Physicians, Inst. Medicine, Chgo., Am. Physiol. Soc.; mem. Sigma Xi, Alpha Omega Alpha; hon. mem. cardiol. socs. of Argentina, Brazil, Chile, Peru, Uruguay, India, Piedmont Soc., Cardiac Surgery, Japan Coll. Angiology, Philippine Heart Assn. Author: Hypertension, 1929; Constrictive Perioarditis, 1936; Cardiologia, 1938; Heart, 1949; (with C. K. Liu) Cardiac Pressures and Pulses, 1956; Intracardiac Phenomena, 1958; (with L. M. Rosa) Therapy of Cardiovascular Emergencies, 1960; (with A. Jacono) Attualitá Cardiologiche, 1965; From Auscultation to Phonocardiography, 1965; (with S. J. Slodki) Differential Diagnosis of Cardiovascular Diseases, 1965; also numerous articles. Editor-in-chief: Cardiology, 5 vols., 1955—. Research on normal function of bronchi, pulmonary edema, and treatment of it by defoaming method; phonocardiography, heart failure. Home: 5000 S. Cornell Av., Chgo. 60615.*

LUKACS, Eugene, mathematician; b. Szombathely, Hungary, Aug. 14, 1906; s. Emil and Margit (Wohl) L.; Tchrs. Exam., U. Vienna (Austria), 1929, Ph.D., 1930; m. Elizabeth C. Weisz, Jan. 20, 1935. Came to U. S., 1939, naturalized, 1945. Asst. prof. Ill. Coll., Jacksonville, 1942-44; asso. prof. Berea Coll. (Ky.), 1944-45; prof. Our Lady of Cin. Coll., 1945-53; staff U. S. Naval Ordnance Test, Sta., China Lake, Cal., 1948-50, Nat. Bur. Standards, 1950-53; head statistic br. office Naval Research, 1953-55; prof., dir. statis. lab. Cath. U., Washington, 1955—; lectr. Am. U., Washington, 1951-55, adj. prof., 1956-57; vis. prof. U. Paris (France), 1961-62, 66, Eidgenossische Technische Hochschule, Zurich, Switzerland, 1962. Fellow Inst. Math. Statists., A.A.A.S.; mem. Internat. Statis. Inst., Am. Math. Soc., Am. Statis. Assn., Biometric Soc., Société Math. de France, Math. Assn. Am., Actuarial Soc. Switzerland. Author: Characteristic Function, 1960; (with R. G. Laha) Applications of Characteristic Function, 1964; Stochastic Convergence, 1968; also numerous articles. Research on theory characteristic functions, application methods classical analysis to theory probability and math. statistics. Home: 3727 Vanness St. N.W., Washington, 20016.*

LUKASIEWICZ, Ignacy, Polish pharmacologist; b. Zadusniki, Poland, Mar. 23, 1822; Master of Pharmacology, Jagiellonian U. of Cracow. Mem., Polish Petroleum Assoc. from 1862. By a petroleum distillation process, created kerosine; made 1st kerosine lamp which he demonstrated in 1853 by illuminating hospital in Lwów; one of 1st in world to apply drilling to extraction of oil, near Krosno, Poland, 1854. Died Lwów, Poland, Jan. 7, 1882.

LUKASIEWICZ, Jan, Polish mathematician; b. Lvov, Poland, Dec. 21, 1878; s. Paul and Leopoldine (Holtzer) L.; Ph.D., U. Lvov, 1902; hon. doctorate U. Munster (Germany); m. Regina Barwinska, Apr. 30, 1929. Became lectr. philosophy U. Lvov, 1906; named lectr. philosophy U. Warsaw (Poland), 1915, prof., 1920, hon. prof., 1924, rector, 1933-1922-23, 31-32; prof. math. logic Royal Irish Acad., 1946—. Mem. Polish, Lvov, Warsaw acads. scis. Author: Philosophical Remarks on Many Valued Systems of Logic, 1930; A Proof of Completeness of the Two-Valued Propositional Calculus, 1932; On the History of Propositional Calculus, 1935; Logic and Foundations of Mathematics, 1941. Died 1956.

LUKE, Charles Daniel, Am. chem. engr.; b. Coin, Ia., Oct. 4, 1907; s. Edward and Delia (Johnson) L.; B.S. in Chem. Engring., State U. Ia., 1929, S.B., 1931; Sc.D., Mass. Inst. Tech., 1934; m. Minerva M. Eagan, May 29, 1941. Process engr. Standard Oil Co. La., Baton Rouge, 1933-35; cons. engr. Luis de Florez Engring. Co., N.Y.C., 1935-37; asst. prof. dept. chem. engring. Syracuse (N.Y.) U., 1937-39, asso. prof., dept. chmn., 1939-40, prof., dept. chmn., 1940-42, 1946-54; dir. div. classification U. S. AEC, Washington, 1954-55, asst. to dir., div. licensing and regulation, 1955-60, chief criticality evaluation br., 1960—. Research on thermodynamics, distillation, drying, nuclear safety filtration. Home: 5607 Parkston Rd., Washington 20016. Office: U. S. AEC, Washington, 20545.*

LUKE, Josephus Corbus, surgeon; b. Ottawa, Can., Apr. 18, 1907; s. Edward B. and Jane (Corbus) L.; B.A., McGill U., 1927, M.D.C.M., 1931; m. Eleanore Wallace, June 28, 1939; children—Anthony E., Roslyn Joelle. Attending staff Royal Victoria Hosp., Montreal, Que., Can., 1937—, sr. surgeon, 1954—; faculty McGill U., 1937—, asso. prof., 1954—; surgeon-in-chief Lakeshore Gen. Hosp., Pointe Claire, Que., 1963—. Fellow Royal Coll. Surgeons Eng., Royal Coll. Surgeons Can., A.C.S.; mem. Internat. Cardiovascular Soc. (v.p.), numerous other socs. Author: (with H. F. Moseley) Text Book of Surgery, 1956. Pioneer in Can. in surgery of diseases of arteries and veins. Home: Quarry Point, Como, Que. Office: Lakeshore Gen. Hosp., Pointe Claire, Que., Can.*

LUKENS, Francis Dring Wetherill, Am. physician; b. Phila., Oct. 5, 1899; s. William Weaver and Isabella (Wetherill) L.; B.A., Yale, 1921; M.D., U. Pa., 1925; m. Emma Martyn George, Oct. 10, 1933; children—Marian (Mrs. Terence M. Carney), Isabella M. Jacques Loeb fellow Johns Hopkins, 1928-30; faculty U. Pa., Phila., 1930-66, prof. medicine, 1952-66, dir. George S. Cox Med. Research Inst., 1936-66. Chief of staff VA Hosp., Pitts., 1966—. Banting lectr., Joslin lectr., 1964. Mem. Am. Diabetes Assn. (past pres. Banting medal 1960), Endocrine Soc. (past pres.), Am. Clin. and Climatol. Assn. (past pres.), New Eng. Diabetes Assn. (Elliott Proctor Joslin medal 1964), Am. Physiol. Soc., Soc. for Clin. Investigation, Assn. Am. Physicians, Soc. Exptl. Biology and Medicine. Research, numerous publs. on influence adrenal cortex on metabolism and in diabetes, exptl. diabetes, prodn. of and recovery from damage islands Langerhans, growth hormone and diabetes, insulin, protein metabolism. Home: VA Hosp., University Dr., Pitts. 15240.*

LUKER, James Allison, Am. chem. engr.; b. Yazoo City, Miss., Feb. 5, 1923; s. Moses Beauregard and Laura Jordan (Jones) L.; B.S., La. State U., 1944; S.M. in Chem. Engring., Mass. Inst. Tech., 1946; Ph.D., Northwestern U., 1950; m. Margaret Burpitt, July 10, 1948; children—Laura Lee, Linda Lorraine, James Allison, Julie Ann. Research engr. Shell Oil Co., 1950; asst. prof., asso. research engr. Denver U., 1950-52; prof. chem. engring. U. Miss., 1952-53; asso. prof. chem. engring. Syracuse (N.Y.) U., 1953-59, prof., 1959—, chmn. dept. chem. engring. and metallurgy, 1960—. Sr. designer Oak Ridge Nat. Lab., summer 1953. Mem. Am. Inst. Chem. Engrs. (exec. com. Syracuse sect., mem. nat. career guidance com.), Am. Chem. Soc., Soc. for History Tech., Am. Assn. U. Profs., Sigma Xi, Tau Beta Pi, Phi Lambda Upsilon, Kappa Mu Epsilon. Research in detonation in gaseous mixtures, programmed learning principles in teaching engring. Home: 68 Sullivan St., Cazenovia, N.Y. 13035. Office: Hinds Hall, Syracuse U., Syracuse, N.Y. 13210.*

LUKIAVON, Stepan Iurievich, Russian physicist; b. 1912; grad. Leningrad Poly., 1936; D.Physico-Math. Scis. Sect. dir. Inst. Atomic Energy, USSR Acad. Sci., 1949—; faculty Moscow U., 1954—; prof., 1949—. Recipient Lenin prize, 1958. Research in electronics, atomic physics; contributed to research on discharges in a gas to receive high-temperature plasma.

LUKIRSKY, Petr Ivanovich, Russian physicist; b. 1894; prof., Leningrad, USSR; mem. theoretical and exptl. physics br. USSR Acad. Scis., corr. mem. Physico-Math. Union. Author many sci. publs. Discovered a new kind of disintegration of atomic nucleus, 1946.

LUKOMSKY, Pavel Yevgenevich, Russian therapeutist; b. 1898; grad. Med. Faculty, 1st Moscow U., 1923; D.Med. Sci. Intern, asst., lectr. univ. clinic of hosp. therapy, until 1941; head chair propedeutics of internal medicine 1st Moscow Med. Inst., 1941-43, prof. chair therapy Health-Hygiene Faculty, 1943-44; head chair faculty therapy Chelyabinsk Med. Inst., 1944-49; head chair faculty therapy Pediatrics Faculty, 2d Moscow Med. Inst., 1949-53; chief therapeutist USSR Ministry Health, 1949—; head chair hosp. therapy Therapeutic Faculty, Pirogov 2d Moscow Med. Inst., 1953—. Decorated Order of Lenin. Mem. USSR Acad. Med. Sci., Moscow Therapeutical Soc. (chmn. cardiological sect.), All-Union (Presidium mem.), All-Russian (dep. chmn.), Moscow (bd. mem.) socs. therapeutists. Mem. editorial bd. Soviet Medicine; coeditor Therapeutics sect. Large Med. Ency., 2d edit. Research and publs. on electrocardiogram changes in diseases of cardiovascular system, coronary insufficiency, myocardial infarction, treatment and prophylaxis of atherosclerosis, effects of choline, methionine, vitamin B12 on lipoid, protein and lipoprotein metabolism, antibiotics treatment. Address: Pirogov 2d Moscow Med. Inst., Malaya Pirogovskaya ulitsa 1, Moscow, USSR.

LULIO, Raimundo, see Lullius, Raimundus.

LULL, Richard Swann, Am. paleontologist; b. Annapolis, Md., Nov. 6, 1867; s. Edward P. and Elizabeth F. (Burton) L.; B.S., Rutgers U., 1893, M.S., 1896, hon. D.Sc., 1918; Ph.D., Columbia, 1903; hon. M.A., Yale, 1911; m. Clara Coles Boggs, July 2, 1894; 1 dau., Dorothy. With div. entomology U. S. Dept. Agrl., 1893; asst. and asso. prof. zoölogy Mass. State Coll., 1894-1906; asst. prof. vertebrate paleontology Yale, 1906-11, prof., 1911-23, prof. paleontology, 1923-27, Sterling prof. paleontology, 1927-36, later emeritus. Asso. curator of vertebrate paleontology Peabody Museum, Yale, 1906-20, curator vertebrate paleontology, 1920-26, hon. curator, from 1932, dir., 1922-36, acting dir., 1937-38; asso. fellow Jonathan Edwards Coll., Yale Recipient Elliot medal Nat. Acad. Scis., 1933. Fellow Am. Acad. Arts and Scis., Geol. Soc. Am., Paleontol. Soc. (treas. 1911-24, pres. 1925); mem. Soc. Vertebrate Paleontology, Am. Museum Natural History (hon. life), Sigma Xi, Phi Beta Kappa. Author: Organic Evolution, and other books and memoirs. Editor Am. Jour. of Sci., 1933-49. Extensive research into vertebrate paleontology, organic evolution, Dinosauria. Died Apr. 22, 1957.

LULLIN DE CHATEAURIEUX, Jacob-Fréderic, Swiss agronomist; b. Geneva, Switzerland, May 10, 1772; lt. French Army; mem. French Acad. Scis. Author: Abrégé d'agriculture et d'économie domestique; observations sur les betes à laine, 1804; Des prairies artificielles, 1806. Studied sheep raising, irrigation of prairies; publ. on domestic agr. Died Geneva, Switzerland, Sept. 24, 1841.

LULLIUS, Raimundus, (or Raimundo, Lulio, Ramón Llull, also Dr. Illuminatus), physician, alchemist, philosopher; b. Palma, Mallorca, circa 1235; religious conversion, circa 1266; studied philosophy, theology, Arabic, 9 years; became mem. Franciscan order; founder Franciscan sch. for missionaries, Mallorca, 1276, tchr., 10 years; travelled, lectured in Italy, 30 years; Miramar, Mallorca. Author: Libre de contemplacio; Blanguerna (on his voyages to Muslim countries); Ars generales; Ars compendiosa inveniendi veritatem, seu ars magna et major (influenced Leibnitz; forerunner of modern logistic thought); nearly 290 other works on medicine, chemistry, math, philosophy, theology, poetry, in Catalan, Arabic, Latin. Tried to convert Muslims to Christianity and to combat Averroism, he became well versed in Arabic sci.; devised what he considered an infallible method of proving faith and reason are compatible; invented mech. device (ars magna) which combined subjects and predicates of propositions thus producing valid conclusions; many alchemical works ascribed to him. Twice imprisoned and banished for preaching to Muslims in Tunis; stoned to death in 3d attempt, 1315.

LUMHOLTZ, Carl, explorer; b. nr. Lillehammer, Norway, 1851; ed. U. Christiania (now Oslo); zool. collector Museums Christiania U., Australia, 1880-84; traveled in Sierra Madre and other parts of Mexico, 7 years, Dutch Borneo, 1914-17. Author: Among Cannibals, 1889; Unknown Mexico, 1902; New Trails in Mexico, 1912; Through Central Borneo, two years travel in the Land of the Headhunters, 1920; Symbolism of the Huichol Indians; Decorative Art of the Huichol Indians (memoir Am. Mus. Natural History). Discovered tree kangaroo, other Australian mammals. Died May 5, 1922.

LUMIÈRE, Auguste Marie Louis, French inventor, biologist; b. Besancon, France, Oct. 19, 1862; s. Antoine and Joséphine (Costille) L.; ed. U. Bern (Switzerland); m. Marguerity Winckler, Aug. 31, 1894. Adminstrv. dir. Lumière and Jougla, photog. mfg. firm. Mem. French Acad. Scis. Author: Le rôle des colloides chez les etres vivants, 1922; Les lois de la cicatrisation des plaies cutanées, 1922; Le cancer, maladie des cicatrices, 1929. Inventor (with bro. Louis Jean) motion picture camera, 1893, 1st machine to project images on screen before audience (cinematograph), 1895, color photography process, developed (with brother) theory of colloidal substances in living beings; research on vitamins, oral vaccination, cancer, tetanus treatment, improvements in photography. Died Lyons, France, Apr. 10, 1954.

LUMIÈRE, Louis Jean, French inventor, chemist; b. Besancon, France, Oct. 5, 1865; s. Antoine and Joséphine (Costille) L.; ed. École technique, La Martinière; mfr. photog. materials; mem. French Acad. Scis. Inventor (with brother Auguste Marie Louis) camera for motion pictures, 1893, 1st machine to project images on screen before audience (cinematograph), 1895, process of color photography; developed (with bro.) theory of colloidal substances in living beings; built 1st movie theater (cinema), Lyons, 1895; suggested screen (trichrome) for direct color photography, 1903; tried to develop photog. plates with silver bromide gelatin; developed plastic photography (photostereosynthesis). Died Bandol, France, June 6, 1948.

LUMIÈRE, Marey, French inventor, physiologist; b. 1830; clin. physician; became prof. natural history organisms Collège de France, 1869. Research on muscle functions, heart, including heart contraction by cutting open hearts of living animals; measured arterial pressure; developed sphygmograph; built Mareyisch capsule; one of creators of filming work. Died 1903.*

LUMLEY, Frederick Elmore, sociologist; b. Iona, Ont., Can., June 7, 1880; s. Moses Willey and Dama Edith (Williams) L.; grad. Coll. of Disciples, St. Thomas, Ont., 1901; student McMaster U., 1901-03, M.A., 1907; A.B., Hiram (Ohio) Coll., 1905; B.D., Yale, 1909, Ph.D., 1912 (DeForest scholar); m. Margaretta Sewell, Oct. 4, 1905; 1 son, Frederick Hillis (dec.). Minister, Disciples of Christ Ch., Toronto, Can., 1901-03; prin. Sinclair Coll., St. Thomas, 1905-07; minister Congl. Church, Northford, Conn., 1908-12; hon. fellow and asst. in dept. of sociology Yale, 1910-12; prof. sociology Coll. of Missions and Butler Coll., Indpls., 1912-20; prof. sociology Ohio State U., Columbus, from 1920, chmn. dept., 1932-40; vis. prof. Yale, Northwestern U., Syracuse U., Western Res. U. Mem. Sherwood Eddy Seminar for Foreign Study, 1923. Elected asso. Internat. Inst. Sociology, 1934. Mem. Am. Sociol. Soc., A.A.A.S., Fgn. Policy Assn., Ohio Valley Sociol. Soc. (pres. 1937-38), Am. Assn. Univ. Profs., Alpha Kappa Delta, Phi Beta Kappa. Author: Means of Social Control, 1925; Principles of Sociology, 1928, 2d edit., 1934; Ourselves and the World (with B. H. Bode), 1931; The Propaganda Menace, 1933; also articles. Editor The Ohio Valley Sociologist, 1930-44. Died July 26, 1954.

LUMMER, Otto R., German physicist; b. July 17, 1860, Gera, Germany; ed. German univs.; became prof. U. Breslau (now Wroclaw, Poland), 1905, named dir. Inst. Physics, 1904. Designed (with Brodhun) photometer with arrangement of prisms, 1889; built (with Arons) mercury vapour lamp for monochromatic illumination; research (with Pringsheim) on radiation energy and temperature including verification of various theoretical laws, black body radiation (which led Planck to his quantum theory); discovered interference in plane-parallel glass plate (developed into Lummer-Gehrcke plate). Died Breslau, July 5, 1925.

LUMRY, Rufus Worth II, Am. chemist; b. Bismarck, N.D., Nov. 3, 1920; s. Rufus Worth and Mabel (Will) L.; A.B., Harvard, 1942, M.S. (NRC fellow), 1948, Ph.D., 1948; m. Gayle Kelly, Mar. 27, 1943; children—Rufus Worth III, Ann E., Stephen E. Faculty, U. Minn., Mpls., 1953—, prof., 1956—, chmn. phys. chemistry div. chemistry dept., 1956, dir. lab. for biophys. chemistry, chmn. Biophysics Grad. Program, 1963—; vis. prof. Inst. For Protein Research, Osaka, Japan, 1961, Inst. For Biochemistry U. Rome, 1964. NSF Sr. Postdoctoral fellow, 1959-60. Mem. Am. Chem. Soc., Am. Phys. Soc., Am. Biol. Chem. Soc., Biophys. Soc., Soc. Plant Physiology. Author: (with Warren Reynolds) Mechanisms of Electron-transfer Reactions, 1966. Research, numerous publs. in application of phys. chemistry to biol. systems, specifically thermodynamics and kinetics of protein folding, mechanism of enzymic catalysis, oxygenation mechanisms in heme-proteins, photochemistry of biologically important molecules, mechanism of photosynthesis, fast reactions in protein systems. Home: 2808 W. River Rd., Mpls. 55406.*

LUMSDEN, Thomas William, Brit. physician; b. Dec. 20, 1874; s. Edward and Jessie (Downie) L.; M.D., Ch.B., Aberdeen U., 1897; M.D., 1901. Practiced medicine, Belgravia, 1900-25; hon. asst. path. dept. Lister Inst., 1925-30; cancer researcher Brit. Empire Cancer Campaign, 1925-42; dir. cancer research London Hosp., 1930-42; pathologist Emergency Med. Service, 1939-41. Fellow Royal Soc. Medicine, Physiol. Soc., Path. Soc. Contbr. papers on cancer research to jours. Discovered 4 centers which regulate respiration, 1922-24; research on anti-cancer serum, immunity with regard to malignant tumors. Died Nov. 24, 1953.

LUNARDI, Vincenzo, Italian aero. pioneer; b. Jan. 11, 1759; sec. to Neopolitan ambassador, Eng.; named hon. mem. Honorable Artillery Co. Author: An Account of Five Aerial Voyages in Scotland, 1786. Made several successful ascent in 33-foot hydrogen balloon from London, Edinburgh and Glasgow, 1784-85; 1st aerial journey in Britain in hydrogen balloon from London to Standon, 1784. Died 1806.

LUND, Charles Carroll, Am. surgeon; b. Boston, Apr. 15, 1895; s. Fred Bates and Zoe M. (Griffing) L.; A.B., Harvard, 1916, M.D. cum laude, 1920; m. Alice C. Marden, May 22, 1925. Intern Mass. Gen. Hosp., Boston, 1920-22, resident in surgery, 1922-23; pvt. practice surgery, Boston, 1923—; mem. faculty, med. sch., Harvard, 1923—; clin. prof. surgery, 1955-61, professor emeritus, 1961—; asso. Boston City Hosp., 1924—; surgeon-in-chief 1951-60, cons. surgery, 1961—; surgeon New Eng. Deaconess Hosp., 1926—. Pres., dir. Blood Research Found., Inc., chmn. med. adv. com., blood program A.R.C., 1963—. Fellow A.C.S.; mem. A.M.A., Am. Surg. Soc., Soc. Clin. Surgery, Am. Cancer Soc. (bd. dirs., 1947, pres. 1951-52), NRC, N.E. Surg. Soc., N.E. Cancer Soc., Research and publs. on cancer, nutrition and burns. Aesculapian, Alpha Omega Alpha. Home. 37 Randolph Rd., Chestnut Hill 67, Mass. Office: 319 Longwood Av., Boston 15.*

LUND, Curtis Joseph, Am. obstetrician, gynecologist; b. LaSita, Kan., June 8, 1907; s. Frank Julius and Ida (Lund) L.; B.S., Kan. State Coll., 1929; M.S., U. Wis., 1930, M.D., 1935; m. Costance Graham, Aug. 29, 1931; 1 son, Graham Curtis. Asst. resident U. Hosp., Madison, 1936, resident, 1938-39; asst. resident obstetrics and gynecology Chgo. Lying-In Hosp., 1937; asst. research fellows, med. sch. U. Wis., 1939-41, asso. research fellow, 1941-43; faculty U. Minn., 1943-47, La. State U., 1947-52; prof. obstetrics and gynecology, sch. medicine and dentistry U. Rochester, also obstetrician and gynecologist in chief U. Rochester Med. Center, 1952—. Bd. visitors Roswell Park Meml. Hosp. Diplomate Am. Bd. Obstetrics and Gynecology (dir., sec.). Mem. Am. Gynecol. Society, Am. Assn. Obstetricians and Gynecologists, Am. Coll. Obstetrics and Gynecology, Soc. Gynecol. Investigation, Sigma Xi, Alpha Omega Alpha. Research and publs. on physiology of reproduction with emphasis on patho-physiology of pregnancy. Home: 36 N. Country Club Dr., Rochester, N.Y. 14618.*

LUND, Ebba, Danish virologist; b. Copenhagen, Denmark, Sept. 22, 1923; d. Soren Kierkegaard and Anna (Lindberg) Lund; M.Sc., Danish Inst. Tech., 1947, Ph.D., 1950; Ph.D., U. Copenhagen, 1963; m. Soren Lovtrup, 1944, (div. 1957); children—Vita, Susanne, Anders. Engr., Lab. Ministry Fisheries, Copenhagen, 1947-50; research asst. dept. gen. zoology U. Copenhagen, 1950-51; fellow Nat. Cancer Inst., NIH, Bethesda, Md., 1951-52; research fellow Biol. Inst. Carlsberg Found., Copenhagen, 1952-54; with Municipal Virological Lab. and Virological Lab. Dept. Microbiology, Gothenburg, Sweden, 1954-66; head dept. vet. virology and immunology Royal Vet. and Agrl. Coll. of Copenhagen, 1966—. Author: Oxidative Inactivation of Poliovirus, 1953; also numerous articles. Research using tissue culture methods, virus inactivation by compounds such as chlorine, these works have practical consequences also for water pollution control. Home: 21, Ostbanegade, Copenhagen O. Office: 13, Bulowsvej, Copenhagen V., Denmark.*

LUND, Henning, Danish chemist; b. Copenhagen, Denmark, Sept. 15, 1929; s. Hakon and Bergljot (Dahl) L.; cand.polyt., Danmarks Tekniske Hojskole,

1952; Dr.phil., U. Copenhagen, 1961; m. Else Margrethe Thorup, May 26, 1953; children—Torben, Marianne, Susanne, Bettina. Research chemist LEO Pharm. Co., Copenhagen, 1953-54, 55-60; research fellow Harvard, 1954-55; reader U. Aarhus (Denmark), 1960—. Mem. Dansk Ingeniorforening, Jysk Selskab for Fysik og Kemi. Author: Elektrodereaktioner i Organisk Polarografi og Voltammetri, 1961; also articles. Research on electrolysis of organic compounds in which reaction is guided by control of potential of working electrode enabling selective reactions to be performed. Home: 8 A Vinkelvej, Risskov 8240, Denmark. Office: Langelandsgade, Aarhus C 8000, Denmark.*

LUND, Morgans Christian, Danish neurologist; b. Copenhagen, Denmark, May 25, 1913; s. Hans Peter and Bertha (Thygesen) L.; M.D., U. Copenhagen; m. Grethe Neergaard-Moller, June 23, 1935; children—Elisabeth, Trine, Hans Peter. With univ. neurol. clinics, Copenhagen, also Aarhus, Denmark; head physician Neurol. Clinic Odense, 1947-58; head physician Glostrup Hosp. Neurol. Clinic, Copenhagen, 1958—. Mem. Danish Epilepsy Assn. (pres.). Author: Epilepsy in Association with Intracranial Tumour, 1952. Office: Glostrup Hosp., Copenhagen, Denmark.

LUND, Sören Jensen, Danish botanist; b. Ölsted, Horsens, Denmark, Dec. 2, 1905; s. Jens Peter and Ane Nielsine (Jensen) L.; M.Sc., U. Copenhagen, 1932, D.Ph., 1959; m. Mimi Ethel Larvig, Sept. 14, 1960; 1 son, Flemming Ejnar. Phycologist, expdn. to East Greenland, 1933; research work with L. Kolderup Rosenvinge, 1932-39; marine biologist Danish Biol. Sta. (now Danish Inst. for Fisheries and Marine Research), Charlottenlund Slot, 1932-44, librarian, 1944—. Research, publs. on marine algae of Denmark, Greenland. Home: 14 Ved Bellahöj, Brönshöj, Copenhagen, Denmark. Office: Danmarks Fiskeri- og Havundersögelser, Charlottenlund Slot, Denmark.*

LUNDBERG, Walter Oscar Paul, Am. biochemist; b. Mpls., Dec. 15, 1910; s. Oscar Manfred and Annette (Carlson) L.; student U. Minn., 1928-30; Ph.D. (Hormel Research Found. fellow), Johns Hopkins, 1934; m. Olga B. Kalvig, Jan. 24, 1942; children—Stephen O., Stewart W., Sandra A. and Susan I. (twins). Faculty, Johns Hopkins, 1934-35; chemist U. S. Steel Corp., Kearny, N.J., 1935; postgrad. fellow U. Minn., 1941-44, faculty, 1945—, prof. biochemistry, 1947—; resident dir. Hormel Inst., Austin, Minn., 1944-49, exec. dir., 1949—. Mem. numerous adv. coms. and panels govt. agys., including NIH, NSF, NRC (liaison panel food protection com.). Recipient NSF Travel award, 1952; Normann medal German Soc. for Fat Sci., 1957; Glycerine Producers' award, 1959; named 1st Internat. Lectr. Fats and Oils Group, Soc. Chem. Industry, London, 1964. Fellow A.A.A.S.; mem. Am. Oil Chemists' Soc. (editor Lipids 1967—, pres. 1963-64; recipient Alton E. Bailey award 1967), Am. Chem. Soc., Am. Soc. Biol. Chemists, Inst. Food Technologists, Am. Heart Assn., Internat. Conf. on Biochemistry Lipids, Internat. Soc. Fat Research, Minn. Acad. Sci. (pres. 1961-62), Am. Assn. U. Profs., Soc. for Exptl. Biology and Medicine, Phi Beta Kappa, Sigma Xi, Phi Lambda Upsilon, Gamma Sigma Delta, Gamma Alpha. Editor: Autoxidation and Antioxidants, vol. 1, 1961, vol. 2, 1962; Progress in the Chemistry of Fats and Other Lipids, vols. 1 to 6, 1952-63. Chmn. editorial bd. Pergamon Press, 1964—. Publs. on processing, nutrition, chemistry, and metabolism of fats and oils. Home: 243 11th Av. N.E. Office: 801 16th Av. N.E., Austin, Minn. 55912.*

LUNDBY, Arne, Norwegian physicist; b. Oslo, Norway, June 12, 1923; s. Arne and Jensine (Jonsen) L.; ed. U. Oslo; Ph.D., U. Birmingham (Eng.); m. Mary-Ann Holmes Gwillim, Jan. 21, 1966. Asst., Uppsala Observatory, 1944-46; physicist, Defense Research, Kjeller, Norway, 1948-50; research assoc., U. Chicago, 1950-52; physicist, Institute for Atomic Energy, Norway, 1952-56; physicist, CERN, Geneva, Switzerland, 1956—; prof., U. Oslo, Norway, 1966—. Home: 40 Chemin des Chataigniers 1292 Chambésy, Geneva, Switzerland. Office: CERN, European Organization for Nuclear Research, 141 Geneva, Switzerland.*

LUNDEN, Walter A., Am. sociologist; b. Spencerbrook, Minn., Mar. 27, 1899; s. Andrew and Lydia Lunden; B.A. cum laude, Gustavus Adolphus Coll., 1922; M.A., U. Minn., 1929; Ph.D., Harvard, 1934; m. Lillian M. Chack. Asst. prof. sociology dept. U. Pitts., 1931-41; prof. sociology and anthropology Ia. State U., Ames, 1947—. Cons. Ia. correctional instns., 1947—; chmn. Ia. Gov.'s Com. on Penal Affairs, 1957-62. Fellow A.A.A.S., Am. Sociol. Assn., Am. Soc. Criminology. Author numerous books including: Facts on Crime and Criminals, 1961; The Prison Warden and the Custodial Staff, 1962; War and Delinquency, 1963; Statistics on Delinquents and Delinquency, 1964; The Iowa Parole System, 1964; Crime in Iowa, 1966; Crimes on the Prairie, 1967; Crimes and Criminals, 1967; also articles. Home: 711 Beech Av., Ames, Ia. 50010.*

LUNDIN, Carl Axel Robert, optical expert; b. Venersborg, Sweden, Jan. 13, 1851; s. C. F. and U. H. (Anderson) L.; ed. high sch., Falun, Sweden; hon. M.A., Amherst, 1905; m. Hulda M. Hansen, Apr. 3, 1875. Worked with Alvan Clark & Sons Corp., from 1874, on projects including the 30-inch objective for Pulkowa Obs., 1883 (which was transported and installed by him under spl. decree of Russian Govt.), 36-in. objective of Lick Obs., 1887, and 40-inch objective for Yerkes Obs., 1895; completed 24-inch objective for Lowell Obs., 1895; made 16-inch objective for U. Cin., 1904, and 18-inch objective for Amherst Obs., 1905; devised and put into operation several important optical tests. Died Nov. 28, 1915.

LUNDQUIST, Frank, Danish biochemist; b. Copenhagen, Denmark, May 31, 1916; s. Otto A.V. and Maria (Sort) L.; Chem.Eng., Poly. Inst., 1939; M.Sc., U. Copenhagen, 1942, Dr.Phil., 1949; m. Ida Suppli, Oct. 3, 1942; children—Lisbeth, Lars, Jesper. Staff, Ferrosan (pharm. co.), 1942-43; head biochem. div. dept. forensic medicine U. Copenhagen, 1943-62, prof. biochemistry, head dept. biochemistry, 1962—. Mem. Internat. Union Biochemistry (mem. council 1964—), European Fedn. Biochem. Socs. (mem. council 1963—), Danish Chem. Soc., Danish Biochem. Soc., Danish Biol. Soc., Biochem. Soc. (Eng.). Author: Aspects of the Biochemistry of Human Semen, 1949. Editor: Methods of Forensic Science, vol. I, 1962, vol. II, 1963. Research, publs. on biochemistry of male accessory sex glands, metabolism of alcohol, regulatory metabolic processes in liver, fructose metabolism; discovered number of enzymes and other components in human semen. Home: 46 Kastanievej, Holte, Denmark. Office: 30, Juliane Mariesvej, Copenhagen, Denmark.*

LUNDQUIST, Stig Olov, Swedish physicist; b. Gudmundra, Sweden, Aug. 9, 1925; s. Helmer and Ester (Landgren) L.; Fil.Kand., Uppsala (Sweden) U., 1949, Fil.lic., 1952, Fil.mag., 1953, Ph.D., 1955; m. Eva Hamren, June 29, 1949; children—Gunilla, Karin, and Anders (twins). Asst. prof. theoretical physics Uppsala U., 1955-61; faculty Chalmers U. Tech., Göteborg, Sweden, 1961—, prof., 1963—. Mem. Swedish Natural Sci. Research Council. Research in solid state theory and atomic physics. Home: 21 Lövviksvägen, Hovas, Sweden. Office: 5B Gibraltargatan, Göteborg, Sweden.*

LUNDY, John Silas, Am. physician; b. Inksten, N.D., July 6, 1894; s. Frederik George and Lila (Woods) L.; B.A., U. N.D., 1917, D.Sc. (hon.), 1948; M.D., Rush Med. Coll., 1920; LL.D., Hahnemann Coll., 1943; m. Lenore Mittelstadt, Sept. 5, 1925; children—Richard Allen, Joan Lenore (Mrs. Donald N. Robinson), John Charles. Practice medicine specializing in anesthesia, Seattle, 1920-24, Chgo., 1959-64, anesthetics and chronic pain, Seattle, 1964—; head anesthesia Mayo Clinic, 1924-52, sr. cons., 1952-59; faculty U. Minn., Mpls., 1924-59, emeritus prof., 1959—; faculty Northwestern U., Evanston, Ill., 1960—, asso. prof. surgery, 1960-63, emeritus asso. prof., 1963—; prof. clin. anesthetics U. Wash. Med. Sch., Seattle, 1964—. Recipient Distinguished Service award Am. Soc. Anesthesia, 1948; award of merit Horace Wells Club Conn., 1957; certificate appreciation Cal. Dental Assn., 1966; award of merit Am. Acad. Ophthalmology and Otolaryngology. Mem. A.M.A., Am. Surg. Assn., Acad. Anesthetics (Distinction award 1965). Author: Clinical Anesthesia, 1942; also numerous articles. Research on pentothal anesthesia, post anesthetics observation room, fundamental research on spinal anesthesia, new airway, peripheral nerve research, drugs for chem. use. Home: 100-W. Highland Dr., Seattle 98119. Office: 120 W. Highland Dr., Seattle 98119.*

LUNGE, Georg, chemist; b. Breslau (now Wroclaw, Poland), Sept. 15, 1839; indsl. chemist, Newcastle-upon-Tyne, Eng., 1864-76; prof., Zurich, Switzerland, 1876-1908. Author: Steinkohlenteer und Ammoniak, 2 vols., 1867; Handbuch der Soda-Industrie, 2 vols., 1879; Untersuchungsmethoden, 4 vols., 1910. Prepared 100 per cent sulfuric acid using freezing method and discovered nitrosylsulfuric acid caused the reaction in lead chamber; introduced use of methyl orange as indicator in analysis of alkali carbonates; invented Lunge nitrometer for determination of gaseous vols. liberated in reaction. Died Zurich, Switzerland, Jan. 3, 1923.

LUNING PRAK, Jacob, Dutch psychologist; b. Apel, Oct. 20, 1898; s. Johannes and Maria Christina (Speckmann) L.; Ph.D., license in law, U. Groningen; m. Jetske G. H. Veenstra, June 10, 1925; children—Niels, Anco, Job, Marianne-Dorothee, Jan-Willem. Asst. psychol. lab.; lectr. econs.; indsl. psychology; dir. Applied Psychology Inst. Author: Mensen en Mogelijkheden; Tests op school; also numerous articles. Editor intelligence tests. Address: Burgem. van Karnebeeklaan 8, s-Gravenhage, Netherlands.

LUNN, Arthur Constant, Am. mathematician; b. Racine, Wis., Feb. 19, 1877; s. John C. and Emma R. (Martin) L.; A.B., Lawrence Coll., 1898; A.M., U. Chgo., 1900, Ph.D., 1904; m. Anna J. Gowan, Sept. 27, 1900. Instr. math. and astronomy, Wesleyan U., Conn., 1901-02; asso. in applied math. U. Chgo., 1902-03, instr., 1903-10, asst. prof., 1910-17, asso. prof., 1917-23, prof. from 1923. Mem. Am. Phys. Soc., Am. Math. Soc., Am. Astron. Soc., Math. Assn. Am., A.A.A.S., Circolo Mathematico di Palermo, Sigma Xi, Phi Beta Kappa, Gamma Alpha. Died Nov. 19, 1949.

LUNZ, George Robert, Am. marine biologist; b. Charleston, S.C., Feb. 27, 1909; s. George Robert and Minnie (Lofton) L.; B.S., Coll. Charleston, 1930, M.S., 1932; D.Sc., Clemson U., 1958; m. Elsie Melchers, Sept. 2, 1933; 1 dau., Elisabeth. Asso. biol. curator Charleston Mus., 1932-48; dir. Bears Bluff Labs., Wadmalaw Island, S.C., 1946—; dir. div. comml. fisheries S.C. Mildlife Research dept., Charleston, 1959—. Mem. Atlantic States Marine Fisheries Commn., 1954-61, chmn., 1959-61. Guggenheim fellow, 1949. Fellow A.A.A.S.; mem. S.C. Acad. Sci. (Jefferson award for research 1941, past pres.), Atlantic Estuarine Soc. (past pres.), Nat. Shellfish Assn. (mem. exec. com. 1957—), Assn. Southeastern Biologists, Soc. Systematic Zoology, Am. Fisheries Soc., Charleston Edn. and Sci. Found. Contbr. numerous articles to tech. jours. Pioneered shrimp farming in U. S., cultivation oysters and other marine organisms in salt water ponds. Home: 7 Formosa Rd., Wappoo Heights, Charleston 29407. Office: Bears Bluff Labs., Route 1, Box 39, Wadmalaw Island, S.C. 29487; also Div. Comml. Fisheries, 2024 Maybank Hwy., Charleston, S.C. 29407.*

LUOMALA, Katharine, Am. anthropologist; b. Cloquet, Minn., Sep. 10, 1907; d. John Erland and Eliina (Forsness) L.; A.B., U. Cal., 1931, M.A., 1933, Ph.D., 1936. Hon. asso. anthropology Bernice P. Bishop Mus., Honolulu, 1941—; chmn. dept. anthropology U. Hawaii, 1954-57, 60, 64, prof. anthropology, 1946—. Fellow Am. Anthrop. Assn., Polynesian Soc., Internat. Soc. for Folk-Narrative Research, Bishop Mus. Assn., Am. Folklore Soc. (past v.p.), Anthrop. Soc. Hawaii (past pres.), Phi Beta Kappa, Sigma Xi. Author: Menehune and Other Little People of Oceania; Navaho Life of Yesterday and Today, 1938; Oceanic, American Indian and African Myths of Snaring the Sun, 1940; Maui-of-a-Thousand-Tricks, 1949; Ethnobotany of the Gilbert Islands, 1953; Voices of the Wind, 1955. Cross-cultural distbn. and comparison of form and style of orally transmitted narratives among nonliterate peoples; place of such narratives in life of people, especially Pacific area; description of specific cultures or aspects of them, especially in Pacific area, but also some American Indian tribes. Office: 2550 Campus Rd., Dept. Anthropology, U. Hawaii, Honolulu 96822.*

LUPTON, Arnold, Brit. civil and mining engr.; s. Arthur and Elizabeth (Wicksteed) L.; m. Jessie Ramsden. Apprenticed to mining engr.; colliery mgr.; instr., prof. coal mining Yorkshire Coll., Victoria U. (now Leeds U.); cons. mining engr. Fellow Geol. Soc. Author: Treatises on Mining, Mining Surveying, and Electricity as applied to Mining; Happy India; Vaccination and the State; also numerous leaflets. Died May 23, 1930.

LUPU, Nicolae Gh., Rumanian physician; b. 1884; specialized in internal medicine; prof. Bucharest (Rumania) U.; mem. Acad. Socialist Republic Rumania. Author: Role of the Nervous System in the Pathogeny of Pneumoconioses, 1953; Pulmonary Sclersoses, 1958. Research and contbns. in physiopathology, pulmonary sclerosis, icterus, blood diseases. Died 1966.

LURIA, Alexander Romanovich, Russian psychologist; b. Kazan, USSR, July 16, 1902; s. Roman Albertovich and Eugenia (Haskin) L.; grad. U. Kazan, 1921; M.D., U. Moscow, 1936, DEd., 1936, D.Med., 1943; m. Lana P. Lipchina, July 15, 1933; 1 dau., Helen A. With Inst. Psychology, Moscow, 1923-36; prof., head dept. Ukrainian Psychoneurol. Acad., Kharkov, 1933-36; head dept. Inst. Neurology, Inst. Neurosurgery, USSR Acad. Med. Sci., Moscow, 1936-43, 45-51; prof. dept. psychology, head dept. neuropsychology Moscow U., 1945—. Decorated Order of Lenin. Fellow Acad. Pedagogical Sci. of USSR, Brit. Psychol. Soc. (hon.); mem. French Neurol. Soc. (hon.), Swiss Psychol. Soc. (hon.), Am. Acad. Arts and Scis. (hon. fgn.), Am. Acad. Edn. (fgn. affiliate). Author: Nature of Human Conflicts, 1932; Traumatic Aphasia, 1947, (English, 1967); Restoration of Functions after Brain Trauma, 1948, (English, 1961); Higher Cortical Functions in Man, 1962, (English, 1966); The Role of Speech in Regulation of Normal and Abnormal Behavior, 1961; Human Brain and Psychological Processes, 1963 (English, 1966). Editor: The Mentally Retarded Child, 1960. Research and numerous articles on local functions of brain, neuropsychol. analysis of brain injuries, role of language in regulation of human behavior, theory of language disorders, theory of frontal lobes. Home: 13 Frunze St., Moscow G-19. Office: Dept. of Psychology, Moscow University, 20 Karl Marx Av., Moscow K-9, USSR.*

LURIA, Salvador Edward, biologist; b. Turin, Italy, Aug. 13, 1912; s. Davide and Ester (Sacerdote) L.; M.D., U. Turin, 1935; Sc.D. (hon.), U. Chgo., 1967; m. Zella Hurwitz, Apr. 18, 1945; 1 son, Daniel D. Came to U. S., 1940, naturalized, 1947. Centre National Recherche Scientifique fellow Inst. Radium, Paris, France, 1938-40; faculty Columbia, 1940-42, Ind. U., 1943-50, U. Ill., 1950-58; prof. microbiology Mass. Inst. Tech., Cambridge, 1958-65, Sedgwick prof. biology, 1965—, Salk Inst. for Biol. Studies non-resident fellow, 1965—. Recipient Lenghi prize Accademia Nazionale Lincei, Rome, Italy, 1965. Mem. Nat. Acad. Scis., Am. Philol. Soc., Am. Acad. Arts and Sci., Am. Soc. Microbiology, Genetics Soc., Am. Nat. Growth Soc., Am. Soc. Microbiology (v.p. 1966-67, pres. 1967—). Author: General Virology, 1st. edit., 1953, (with J. E. Darnell) 2d edit., 1967. Research, publs. on microbial genetics; demonstrated that bacterial variation consists of spontaneous mutations, 1943; made discovery and analysis of bacteriophage mutation, 1945; discoverer reactivation of viruses killed by radiation, 1947; studies on lysogeny, transduction, and host-controlled properties of viruses. Home: 48 Peacock Farm Rd., Lexington, Mass. 02173.*

LURIE (or LURYO), Anatolii Isakovich, Russian mech. engr.; b. 1901; grad. Faculty Physics and Mechanics, Leningrad Poly. Inst., 1925, became prof., chmn. dept. theoretical mechanics, 1935, prof. machine strength and dynamics, 1944. Corr. mem. USSR Acad. Scis. Author: (with L. G. Loitsianskii) Theoretical Mechanics, 3 vols., 1932-55; Operational Calculus with Application to Mechanical Problems, 1938; Statics of Thin-Walled Elastic Shells, 1947; Some Nonlinear Problems of the Theory of Automatic Control, 1957; Three-dimensional Problems of the Theory of Elasticity, 1955. Research on tensile strength, stability of automatic control systems, analytical mechanics. Home: Polytechnical Rd. 3, app. 90, Leningrad K-64. Office: Dept. of Machine Strength and Dynamics, Leningrad Polytechnical Inst., Leningrad, USSR.

LÜSCHER, Martin, Swiss zoologist; b. Basel, Switzerland; July 4, 1917; s. Jean-Jacques and Adèle (Simonius) L.; Ph.D., U. Basel; m. Noémie Stoecklin, Mar. 28, 1944; 1 son, Niklaus. Instr., U. Basel, 1948; Rockefeller Found. scholar, 1951-52; prof. U. Bern, Switzerland, 1954. Mem. Swiss Zoology Soc., Union for Social Insect Research. Author: Die Produktion und Elimination von Ersatzgeschlectstieren bei der Termite Kalotermes flavicollis Fabr, 1952; Hormonal Control of Caste Differentiation in Termites, 1960; Social Control of Polymorphism in Termites, 1961. Home: 10, Rainweg, Muri-Bern, Switzerland. Office: Institut de Zoologie, U. Berne, Sahlistrasse 8, Berne, Switzerland.

LUSCHKA, Hubert von, see von Luschka, Hubert.

LUSE, Sarah Amanda, Am. anatomist, pathologist; b. Emmetsburg, Ia., Dec. 12, 1918; d. David Newmyer and Mae (Ryckman) Luse; grad. Emmetsburg Jr. Coll., 1938; A.B., Rockford Coll., 1940; postgrad. Mayo Clinic; M.D., Western Res. U., 1949. Asst. neuropathologist Mayo Clinic, 1953-54; Am .Cancer Soc. fellow Washington U. Sch. Medicine, 1954-55, faculty, 1955—, prof. anatomy and pathology, 1963—, acting head dept. anatomy, 1964-65, 66-67; cons. Armed Forces Inst. Pathology, 1959. Mem. com. NIH, 1961—; sci. adviser Nat. Multiple Sclerosis Soc., 1958—. Recipient Lederle Med. Faculty award, 1956-59; Outstanding Achievement award U. Minn.'s Mayo Found., 1964. Mem. Eastern Electroencephalographic Soc., Am. Assn. Exptl. Pathology, Am. Assn. Pathologists and Bacteriologists, Am. Assn. Anatomists, Internat. Acad. Pathology, N.Y. Acad. Scis., Am. Acad. Neurology, Harvey Chsuhing Soc., Phi Beta Kappa, Alpha Omega Alpha. Research, publs. on electron microscopy of nervous system, ultrastructure of aging brain, ultrastructure of normal and neoplastic cells, diseases of nervous system and their therapy. Address: 630 W. 168th St., N.Y.C. 10032.*

LUSH, Jay Laurence, Am. animal geneticist; b. Shambaugh., Ia., Jan. 3, 1896; s. Henry and Mary Eliza (Pritchard) L.; B.S., Kansas State Agrl. Coll., Manhattan, Kan., 1916, M.S., 1918; Ph.D., U. of Wis., 1921; Dr. Agr., Agrl. Coll. Sweden, U. Giessen, 1957, Royal Danish Veterinary and Agrl. Coll., 1958; Doctor of Laws, Michigan State University, 1964; m. Adaline Lincoln, Dec. 20, 1923; children—Mary Elizabeth, David Alan. Research in animal genetics, Tex. Agrl. Exptl. Station, 1921-29, prof., animal breeding, Iowa State Coll., since 1930; temp. duty U. S. Dept. of State in Gt. Britain and with Council for Sci. and Indsl. Research in Australia, 1948; National Research Council fellow in Denmark, 1934. Received First Morrison award Am. Soc. Animal Prodn. 1946. Borden award for research dairy production, 1958, von Nathusius medal, 1960. Member of the American Society of Naturalists, Genetics Soc. of Am., Am. Soc. Animal Sci., Nat. Acad. Sci., Sigma Xi. Author: Animal Breeding Plans, 1945. Guest lecturer at Advanced Sch. of Agriculture and Veterinary Medicine, Vicosa, Brazil, 1941, FAO Center, New Delhi, India, 1954; lectr. INTA, Argentina, 1966-67. Research on statistical genetics of populations; genetics of farm animals. Home: 3226 Oakland St., Ames, Ia.

LUSITANUS, Amatus Juan Roderigo, Portuguese physician; b. Castello Branco, Portugal, 1511; grad. in medicine U. Salamanca (Spain), 1530. Prof. U. Ferrara (Italy), until 1547. Author: Curationum medicialium, 1556; Index dioscoridites; Centuria, 7 vols., 1561. First to describe case of purpura not associated with fever, as a med. entity, 1556. Died 1568.

LUSK, Graham, Am. physiologist; b. Bridgeport, Conn., Feb. 15, 1866; s. William T. and Mary Hartwell (Chittenden) L.; Ph.B., Columbia, 1887; Ph.D., Munich, 1891; hon. A.M., Yale, 1896, Sc.D., 1908; LL.D., U. Glasgow, 1923; M.D. (hon.), U. Munich, 1929; m. May W. Tiffany, Dec. 20, 1899; children—William T., Louise (Mrs. Collier Platt), Louis T. Instr. physiology Yale, 1891-92, asst. prof., 1892-95, prof., 1895-98; prof. physiology, Univ. and Bellevue Hosp. Med. Coll., 1898-1900; prof. physiology. Cornell U. Med. Coll., N.Y., 1909—; sci. dir. Russell Sage Inst. of Pathology. Fellow Royal Soc., 1932, Royal Soc. of Edinburgh, Imperial Soc. of Physicians Vienna (corr.). Author: Elements of the Science of Nutrition, 4th edit., 1928. Died July 19, 1932.

LUSSIER, Jean Paul, Canadian dentist; b. Montreal, Que., Can., Sept. 17, 1917; s. Eugene J. and Parmelia (Gauthier) L.; B.A., U. Montreal, 1938, D.D.S., 1942, M.S., 1952; Ph.D., U. Cal. at Berkeley, 1959; m. Juliette Laurin, May 4, 1943; children—Louis, Renee, Josee, Anne, Pierre, Helen, Andre. Faculty, Faculty of Dentistry, U. Montreal, 1944-52, 54—, prof., 1958—, acting dean, 1958-62, dean, 1962—. Chmn. asso. com. on dental research NRC, 1963-67. Fellow A.A.A.S.; mem. Canadian Dental Assn. (chmn. council on edn. 1960-63), U. Montreal Tchrs. Assn. (past chmn.), Am. Coll. Dentists, N.Y. Acad. Sci., Internat. Assn. for Dental Research, Fedn. dentaire internationale. Research, publs. in bone biology, influence nutritional factors. Home: 3507 Vendome St., Montreal, Que., Can.*

LUST, Reimar Heinz Fritz, German physicist; b. Barmen, Germany, Mar. 25, 1923; s. Hero and Grete (Strunck) L.; Ph.D., U. Göttingen, 1951; m. Rhea Kulka, Aug. 1, 1953; children—Dieter, Martin. Research physicist Max Planck Inst., Göttingen, 1951-55, Enrico Fermi Inst., U. Chgo., 1955-56, Princeton, 1956, Max Planck Inst. for Physics and Astrophysics, Munich, 1957-59; vis. prof. N.Y. U., 1959; head dept. Max Planck Inst. for Physics and Astrophysics, Munich, 1960-61, dir. Inst. for Extraterrestial Physics, 1963—; vis. prof. Mass. Inst. Tech., 1961, Cal. Inst. Tech., 1962. Mem. Internat. Astron. Union, Astronomische Gesellschaft, Deutsche Physikalische Gesellschaft, Wissenschaftliche Gesellschaft fur Luft und Raumfahrt. Research, numerous publs. on magnetohydrodynamics, space research. Home: Sondermeierstrasse 70, 8 Munich 45, Germany.*

LUSTED, Lee Browning, Am. physician; b. Mason City, Ia., May 22, 1922; s. George Charles and Maude (Browning) L.; B.A., Cornell Coll., Mt. Vernon, Ia., 1943, D.Sc., 1963; M.D., Harvard, 1950; m. Winifred Chamberlin, Aug. 24, 1943; children—Lee Browning, Hugh Sherbon. Faculty, U. Cal. at San Francisco, 1954-56, asst. prof., 1955-56; asst. radiologist NIH, 1956-58; faculty U. Rochester, 1958-62, prof. biomed. engring., 1961-62; prof. radiology U. Ore. Med. Sch., sr. scientist Ore. Regional Primate Research Center, Portland, 1962-67; prof., chmn. dept. radiology Loyola U. Med. Center, Hines, Ill., 1968—. Chmn. adv. com. on computers NIH, 1960-64, com. on computers in biology and medicine Nat. Acad. Scis.-NRC, 1958-59. Fellow Am. Coll. Radiology, I.E.E.E., A.A.A.S., N.Y. Acad. Scis.; mem. Am. Roentgen Ray Soc., Am. Bd. Radiology, I.R.E. (chmn. profl. group on med. electronics 1958, editor Transion Bio Med. Electronics 1958-60), Radiol. Soc. N.Am. (Cum laude award 1965, Meml. Fund lectr. 1959), Biophys. Soc. Author: (with Theodore E. Keats) Atlas of Roentgenographic Measurement, 1959; Logical Analysis in Medical Diagnosis, 1966; PRIME, An Automated Information System for Hospitals and Biomedical Research Laboratories, 1967; An Introduction to Medical Decision Making, 1968; also numerous articles. Research on use electronic principles and equipment in medicine; developed logical analysis procedures for automatic data processing. Home: 317 S. Park Av., Hinsdale, Ill. 60521. Office: Dept. Radiology, Loyola U. Med. Center, 1400 S. 1st Av., Hines, Ill. 60141.*

LUSTGARTEN, Sigmund, bacteriologist, dermatologist; b. 1858; M.D., U. Vienna, 1881; privat docent dermatology U. Vienna; cons. dermatologist Montefiore Home, also Hebrew Orphan Asylum; attending dermatologist Mt. Sinai Hosp., N.Y. Mem. Med. Soc. State of N.Y., Am., N.Y., Austrian, German dermatol. socs. Discovered syphilis bacillus. Died 1911.

LUSTIG, Alessandro, Italian bacteriologist, pathologist; b. Trieste, 1857; prof. gen. pathology, U. Florence (Italy). Introduced preventive inoculation for an Indian plague; studied leprosy, malaria, and other infectious diseases. Died 1937.

LUSTIG, Harry, physicist; b. Vienna, Austria, Sept. 23, 1925; s. Hans and Hedwig (Faltitschek) L.; came came to U. S., 1939, naturalized, 1944; B.S., Coll. City N.Y., 1948; M.S., U. Ill., 1949, Ph.D., 1953; m. Judith Hirshfield, Sept. 20, 1953; 1 son, Lawrence. Faculty, Coll. City N.Y., 1953—, asso. prof. physics, 1963—, chmn. dept. physics, 1965—. Prin. scientist Nuclear Devel. Corp. Am., 1956-61; vis. research asst. prof. U. Ill., 1959-60; Fulbright lectr. U. Coll., Dublin, Ireland, 1964-65. Fellow N.Y. Acad. Scis.; mem. Am. Phys. Soc., Am. Assn. Physics Tchrs., Fedn. Am. Scientists, Am. Assn. U. Profs., Phi Beta Kappa, Sigma Xi. Research, publs. on math. analysis exptl. data about nuclear reactions. Home: 65 Payson Av., N.Y.C. 10034.*

LUSTMAN, Seymour Leonard, Am. physician; b. Chgo., Apr. 23, 1920; s. Irving and Anna (Lee) L.; B.S., Northwestern U., 1941; Ph.D., U. Chgo., 1950, M.D., 1954; m. Katharine L. Ritman, June 10, 1941; children—Jeffrey S., Susan T. Prof. psychiatry Child Study Center, dept. psychiatry Yale, 1958—; faculty Western New Eng. Inst. for Psychoanalysis, New Haven, 1960—, sec. edn. com., 1960—. Mem. adv. com. Nat. Inst. Mental Health, USPHS, 1964—. Recipient Chandler prize U. Ill., 1954; David Rapaport award Western New Eng. Inst. for Psychoanalysis, 1962. Diplomate Am. Bd. Psychology and Neurology, Adult Psychiatry, child Psychiatry. Mem. Am. Psychoanalytic Assn., Am. Orthopsychiat. Assn., Am. Acad. Child Psychiatry, Am. Psychosomatic Soc., Am. Psychiat. Assn., Sigma Xi, Alpha Omega Alpha. Editor: The Psychoanalytic Study of the Child; editorial bd. Jour. Child Psychiatry, 1965—, Jour. Psychosomatic Medicine. Research, publs. on normal and path. personality devel., autonomic function, character devel., symptom formation, devel. impulse control, infantile autism, psychoanalysis. Home: 590 Ellsworth Av., New Haven 06511.*

LUTEMBACHER, René, French cardiologist; b. Jouy-en-Josas, France, Sept. 21, 1884; M.D., Paris, 1912; physician Centre du cardiologie de Seine-et-Oise (France). Author: Nouvelles méthodes diexamen du coeur en clinique, 1921; les Troubles fonctionnels du coeur, 1924; Étude elementaire des arythmies, 1929; les Lésions organiques du coeur, 1936. Described condition in which mitral stenosis is asso. with interatrial septal defects (Lutembacher's complex, syndrome, or disease), 1916.

LUTHE, Wolfgang, German psychologist; b. Pansdorf, Germany, Oct. 27, 1922; s. Richard and Elfriede N. Luthe; ed. U. Hamburg (Germany), U. Kiel (Germany); Ph.D., M.D.; m. Elisabeth Heberling, 1948; children—Lorenz, Matthias, Sibylle, Jan Sebastian, Corinna. Asst., Physiol. Inst., U. Hamburg (Germany); later U. Clinic Hamburg; lectr., asst. prof. psychophysiology U. Montreal. Mem. Assn. for Devel. Psychotherapy N.Y., Canadian Med. Assn. Author: (with J. Schultz) Autogenic Training, a Psychophysiologic Approach in Psychotherapy, 1959; Autogenic Training, 1964; Autogenic Abreaction, a Psychophysiologic Approach in Psychotherapy, 1965-66; Research and publs. on psychophysiology. Office: Med. Centre, 5300 côte des Neiges Rd., Montreal, 26, Que., Can.

LUTHER, Wolfgang, zoologist, radiobiologist; b. Moscow, Apr. 2, 1903; s. Arthur and Meta (Eberhardt) L.; ed. Iena, Fribourg, Gottingen; Ph.D., 1930; m. Aimée Runge, 1947; children—Eva, Irene, Helmut. Foreman hemp and flax industry, 3 years; tchr. biology until 1932; asst. U. Erlangen Inst. Zoology; radiobiologist Katherinenhospital, Stuttgart, Germany; later radiobiologist U. Marburg Inst. Radiology, also asso. prof. zoology; dir. Zoology Inst., Darmstadt Tech. Coll., 1954—. Mem. German Soc. Zoology, German Röntgen Soc. Author: (with Fiedler) Unterwasserfauna der Mittelmeerküsten, 1961; also articles. Research on radiobiology, embryology, physiology, marine life, regeneration, radiation effects. Home: Karolinenstrasse 12, Seeheim-Bergstrasse, Germany. Office: Zoologisches Institut, Darmstadt, W. Germany.

LÜTHI, Bruno Meinrad, physicist; b. Weinfelden, Switzerland, Oct. 6, 1931; Diploma in Physics, Fed. Inst. Tech., Zurich, Switzerland, 1955, Doctorate in Physics, 1959. Instr., U. Chgo., 1960-61; scientist IBM Research Lab., Zurich, 1962-63, mgr., 1964-66; asso. prof. Rutgers, The State U., New Brunswick, N.J., 1966—. Mem. Swiss, Am. phys. socs. Research, publs. on transport properties in metals, spinwave transport and interactions, magnetoacoustic effects, magnetic phase transitions.*

LUTHIN, James Nicholas, Am. soil scientist; b. Berkeley, Cal., Dec. 4, 1915; s. John Harry and Elizabeth (Van den Beukel) L.; B.S., U. Cal. at Berkeley, 1938; M.S., Mich. State Coll., 1947; Ph.D., Ia. State Coll., 1949; m. Adalyn Merrill, Apr. 25, 1946; children—James Nicholas, William Merrill. Soil scientist Soil Conservation Service, 1938-42; faculty U. Cal. at Davis, 1949—, prof. water sci. and civil engring., 1965—. Cons. to govt. and state agys., Parliament S. Australia; guest prof. Swiss Fed. Tech. Inst., Zurich, 1962-63. Fulbright Research scholar, Australia, 1958. Mem. Am. Geophys. Union, Brit., Internat. socs. soil sci., Internat. Commn. on Irrigation and Drainage, Am. Soc. Agronomy, Soil Sci. Soc. Am., Am. Soc. Agrl. Engrs., Sigma Xi. Author: Drainage Engineering, 1966; also numerous articles, monograph. Editor: Drainage of Agricultural Lands, 1957. Developed numerical, computer and elec. network methods solving problems flow through porous media; research on physics flow in capillary region above water table, velocity sound and effect air on movement water in soil; developed procedures for investigation drainage problems. Home: 609 Oeste Dr., Davis, Cal. 95616.*

LUTHRINGER, George Francis, Am. economist; b. Petersburg, Ill., Feb. 17, 1904; s. George F. and Pearl Alnutt (Sampsell) L.; student Phillips Exeter, 1921-22; B.S., Princeton, 1926, Ph.D., 1932; m. Winifred Jutten, June 18, 1930; children—Janet Irene, David George. Instr. and asst. prof. economics and finance Princeton, 1930-38; financial adv. to joint prep. com. on Philippine affairs State Dept., Washington, 1937-38; divisional asst., econ. adv. office. State Dept., 1938-41, asst. chief div. financial affairs, 1941-43, chief, 1944-46, dir. office financial and devel. policy, 1946; served as financial expert Office of High Commr. to P.I., 1943-44; attended Bretton Woods and Savannah monetary confs. as tech. adv. to U. S.; mem. U. S. del. of Allied Commn. on

Reparations, Moscow, 1945; apptd. U. S. alternate exec. dir. Internat. Monetary Fund, 1946, dep. dir. research dept., 1948, dept. dir. Far and Middle East and Latin Am. Dept., 1950, dir. Latin Am., Middle and Far Eastern dept., 1952-53, dir. western Hemisphere Dept., 1953-—; rep. Fund, London Preparatory meeting of Internat. Conf. on Trade and Employment, 1946. Mem. Am. Econ. Assn. Author: The Gold Exchange Standard in the Philippines, 1934; (with L. V. Chandler and D. C. Cline) Money, Credit and Finance, 1938; (with B. Dell). Population, Resources and Trade, 1938. Died Mar. 11, 1955.

LUTHY, Ernst Fritz, Swiss physician; b. Solothurn, Switzerland, Nov. 18, 1925; s. Ernst and Hanna (Steiner) L.; state bd. med. schs. U. Bern (Switzerland), also U. Geneva (Switzerland), 1951; m. Franziska Jaeggi, Feb. 28, 1953; children—Antonia, Philipp, Marie-Rose, Bertram. Research fellow dept. physiology U. Basel (Switzerland), 1952-53, dept. medicine Washington U., St. Louis, 1957-58; staff Med. Policlinic, U. Zurich (Switzerland), 1958-—, asst. prof., chief cardiovascular unit, 1964-—; guest prof. dept. physiology Emory U., Atlanta, intermittently, 1962-—. Author: Die Hämodynamik des suff. und insuff. rechten Herzens, 1962; (with R. Hegglin, W. Rutshauser, G. Kaufmann, H. Scheu) Kreislaufdiagnostik mit Farbstoffverdünngsmethode, 1962; also numerous articles. Metabolic and hemodynamic studies in failing and non-failing hearts, methodical studies in dye dilution and thermodilution techniques. Home: Lerchenbergstr. 21, 8703 Erlenbach, Switzerland.*

LÜTHY, Herbert F. R., Swiss biophysicist; b. St. Gallen, Switzerland, Sept. 26, 1914; s. Ferdinand and Thekla (Kessler) L.; student U. Göttingen (Germany), U. Leipzig (Germany), U. Heidelberg (Germany); Ph.D., U. Basel (Switzerland), 1942; m. Verena Stebler, Sept. 1, 1950; 1 dau., Cornelia Beatrix. Staff, Physiology Inst., Bern; staff Rötgeninstit., U. Bern, 1945-50, U. Stockholm, also U. Uppsala, 1950; with Röntgeninstitu., U. Basel, 1954-—, faculty; head ABC br. Civil Def., Basel; head isotope div. Bürgerspital, Basel. Recipient Jubiläunspreis, Schweiz. Röntgen Gesellschaft, 1961. Research, publs. in submicroscopic structure of nervous fibre by ultraviolet-dichroism, dosimetry, dosimetric reactions in biol. macromolecules. Home: 22 Waltersgraben Riehen/BS, Switzerland. Office: Bürgerspital, Basel, Switzerland.*

LUTZ, Frank Eugene, Am. biologist; b. Bloomsburg, Pa., Sept. 15, 1879; s. Martin Peter and Anna Amelia (Brockway) L.; A.B., Haverford Coll., 1900; A.M., U. Chgo., 1902, Ph.D., 1907; studied Univ. Coll., London, Eng., 1902; m. Martha Ellen Brobson, Dec. 30, 1904; children—Anna, Eleanor, Frank Brobson, Laura. Entomologist biol. lab. Bklyn. Inst., 1902; asst. in zool. dept. U. Chgo., 1903; resident investigator, Sta. for Exptl. Evolution (Carnegie Instn.), Cold Spring Harbor, 1904-09; asst. curator invertebrate zoology Am. Mus. Natural History, 1909-16, asso. curator, 1916-21, curator of entomology since 1921, also editor of tech. papers, in charge Sta. for Study of Insects, Tuxedo, N.Y., 1925-28; lectr. Columbia U., 1937. Fellow A.A.A.S., N.Y. Acad. Scis., Entomol. Soc. Am. (pres., 1927); mem. Am. Soc. Zoologists, Sigma Xi, Phi Beta Kappa. Author: Field Book of Insects, 1917; A Lot of Insects, 1941. Contbr. numerous papers on variation, heredity, assortive mating, entomology; noted for studies of insect flight and diurnal rhythms. Died Nov. 27, 1943.

LUTZ, Samuel Gross, Am. engineer; b. Lafayette, Ind., July 9, 1907; s. William J. and Eugenia (Gross) L.; student Oberlin Coll., 1925-26; B.S., Purdue U., 1929, E.E., 1933, M.S., 1934, Ph.D., 1938; m. Kathryn Cornett, Aug. 7, 1930. Tech. staff Bell Telephone Labs., N.Y. City, 1929-32; research fellow, Purdue Research Found., 1933-38; asst. prof. elec. engring., So. Methodist U., 1938-40; head measurement and direction finding sect., Naval Research Labs., Washington, 1940-46; prof. and chmn. dept. elec. engring. N.Y. U., 1945-51; mem. sr. tech. staff Hughes Aircraft Co., Culver City, Cal., 1951-—, dir. engring. Communication Systems div., 1951-59; senior scientist Hughes Research Laboratories, 1959-61, chief scientist, 1961-—. Recipient USN Distinguished Civilian Service award. Fellow of the I.E.E.E. Author: (with Ley and Rehberg) Linear Circuit Analysis, 1959. Contbr. articles on elec. engring. subjects in profl. publs. Developed (early in World War II) countermeasures against radio guided missiles; also high speed digital communication at ionespheric frequencies; pioneered in studies of satellite communication systems; frequency sharing, multiple access, and routing techniques. Home: 144 N. Woodburn Dr., Los Angeles 49. Office: Hughes Research Labs., Malibu Canyon Rd., Malibu, Cal.

LUUKKAINEN, Tapani, Jouni Valter, Finnish physician; b. Sortavala, Finland, Feb. 28, 1929; s. Johan Valter and Helmi (Korhonen) L.; M.D., U. Helsinki (Finland), 1954, D.Med.Sc., 1958; m. Inka-Taina Takki, June 22, 1952; children—Jukka, Timo, Vesa. Instr. dept. med. chemistry U. Helsinki, 1954-60, docent, 1962-65, docent obstetrics and gynecology, 1965-—; research asso. Rockefeller Inst., also NIH, 1960-61; asst. head clinic dept. obstetrics and gynecology U. Helsinki, 1965-—. Author: Studies in the Acetylation of p-Aminobenzoic Acid and Its Regulation by Sex Glands, 1958; also articles. Research on regulation of basic metabolic reactions of sex hormones and their control of onset of labor; identification of steroids in body; discovered new steroid in urine of pregnant women. Home: 12 Solnantie, Helsinki 33, Finland.*

LUXEMBURG, Wilhelmus Anthonius Josephus, mathematician; b. Delft, Holland, Apr. 11, 1929; s. Everardus H. and Digna (Van Kranendonk) L.; B.A., U. Leiden (Holland), 1950, M.A., 1953; Ph.D., Delft Inst. Tech., 1955; m. Geertruida Zappeij, Aug. 2, 1955; children—Ronald Ph., Jacqueline T. Post-doctoral fellow NRC Can., 1955-56; faculty U. Toronto, 1956-58; faculty Cal. Inst. Tech., Pasadena, 1958-—, prof. math., 1962-—. Cons. Burroughs Corp., Pasadena, 1963-64. Mem. Am., Dutch math. socs., Math. Assn. Am., Canadian Math. Congress, Soc. Indsl. and Applied Math. Research, publs. on theory of integration, spaces of measurable functions, ordinary differential equations, numerical analysis, topological linear spaces, Boolean algebras, axiomatic set theory, theory of Riesz spaces, non-standard analysis. Home: 817 S. El Molino Av., Pasadena, Cal. 91106.*

LUYS, Georges, French urologist; b. 1870; s. Jules Luys; studied medicine at Paris; physician, Paris; mem. French Acad. Medicine. Author: les Sinus de la Dure-Mère; Exploration de l'appareil urinaire; L'endoscopie de l'uretre (et) de la vessie; Traité des maladies des vésicules séminales. Originated instrument to collect urine from each kidney separately (Luys's segregator).

LUYS, Jules Bernard, French neurologist; b. Paris, 1828; became hosp. physician, 1862; agregé, 1863; studied and taught about mental illness, Salpetrière; dir. Mental Hosp., Ivry; mem. Acad. Medicine. Author publs. including: Lecons sur la structures et les maladies du systéme nerveux, 1875; Traité clinique et pratique des maladies mentales, 1881; Lecons clinique sur les principaux phénomenes de le hypnotisme, 1889; Traitement de la folie, 1894. Described degeneration of anterior horn cells in patients with progressive muscular atrophy, 1860, nucleus in hypothalamus which is a part of descending pathway from corpus striatum (Luys' nucleus, body of Luys, nucleus hypothalamicus), 1865, subthalamic nucleus whose degeneration causes hemiballism, 1865. Died Divonne-les-Bains, France, 1895.

LUYTEN, Willem Jacob, astronomer; b. Semarang, East Indies, Mar. 7, 1899; s. Jacob and Cornelia (Francken) L.; B.A., U. Amsterdam (Netherlands), 1918; M.A., U. Leiden (Netherlands), 1920, Ph.D., 1921; m. Willemina Miedema, Feb. 5, 1930; children—Mona (Mrs. Johan Coetzee), Ann (Mrs. A. Willem Dieperink), James. Came to U. S., 1921, naturalized, 1927. Fellow, Lick Obs., U. Cal. at Mt. Hamilton, 1921-23; astronomer Harvard, 1923-27, asst. prof. astronomy, 1927-30; prof. astronomy U. Minn., Mpls., 1931-—. Guggenheim Found. fellow, 1928-30, 37-38. Recipient James Craig Watson medal Nat. Acad. Sci., 1964. Fellow Am. Acad. Arts and Scis.; mem. Internat. Astron. Union, Am. Astron. Soc., A.A.A.S., Commandeur Confrerie du Tastevin, Sigma Xi. Research in stellar motions, White Dwarfs, galactic structure; discovered smallest known star, 1963. Home: 1940 E. River Terrace, Mpls. 55414.*

LWOFF, André Michael, French microbiologist, virologist; b. Ainy-le-Chateau, France, May 8, 1902; s. Salomon and Marie (Siminovitch) L.; Licencié ès Scis. Naturelles, Paris, 1921, Dr. Med., 1927, Dr. Scis. Naturelles, 1932; hon. doctorates U. Chgo., Oxford U., 1959; m. Marguerite Bourdaleix, Dec. 5, 1925. Became fellow Pasteur Inst., Paris, 1921, asst., 1925, head lab., 1929, head dept. microbiol. physiology, 1938; prof. microbiology Faculty Scis., Sorbonne, Paris, 1959-—. Recipient Nobel prize (with François Jacob and Jacques Monod) in medicine and physiology, 1965. Fellow Royal Soc. (fgn.), 1958. Mem., N.Y. Acad. Scis., Nat. Acad. Scis. Author: Problems of Morphogenesis in Ciliates: the Kinetosomes in Development, Reproduction and Evolution, 1950; Biological Order, 1962; also articles. Research on nature and function of growth factors, physiology of viruses; induction and repression of enzymes; explained phenomenon of lysogenic bacteria; demonstrated existence of latent bacterial virus; conducted studies on protozoa nutrition; identified vitamins as microbial growth factors and demonstrated that vitamins function as coenzymes. Address: Institut Pasteur, 28 rue du Docteur Roux, Paris 15, France.*

LYCAN, William H(iram), Am. chemist; b. Vermilion, Ill., July 15, 1903; s. Harry and Coral E. (Wilkin) L.; B.S., U. Ill., 1924, M.S., 1926, Ph.D., 1929; m. Janet Grace Brown, Nov. 7, 1931; 1 son, William Gregory. Instr. chemistry U. Tenn., 1925-26; research chemist E. I. du Pont de Nemours, Jackson Lab., Wilmington, Del., and Carrolville, Wis., 1929-36; sr. fellow Mellon Inst. Indsl. Research, 1937-38; with Pitts. Plate Glass Co., 1938-49; dir. research Johnson & Johnson, New Brunswick, N.J., 1949-64, v.p., 1951-64, now dir., mem. exec. com, 1964-—; vice-chmn. Johnson & Johnson Internat., 1964-—. Fellow Am. Inst. Chemists; mem. Am. Chem. Soc., A.A.A.S., Am. Oil Chemists Soc., N.Y. Acad. Scis., Soc. Chem. Industry (chmn. Am. sect. 1960-61), Sigma Xi, Phi Lambda Upsilon, Alpha Chi Sigma, Theta Xi, Gamma Alpha. Chmn. editorial bd., bd. editors Research Mgmt., 1959-62. Home: 17 M Colony House, New Brunswick, N.J. 08903. Office: Johnson & Johnson, New Brunswick, N.J.*

LYCHE, Ralph Tambs, mathematician; b. Macon, Georgia, U. S. A., Sept. 6, 1890; s. Hans Tambs and Mary (Godden) L.; ed. U. Oslo; D.Sc., U. Strasbourg, 1927; m. Else Rasmussen, Jan. 20, 1916; children: Guri, Karen, Helge. Asst., Technical U. of Norway (Trondheim), 1910; lectr., Trondheim, 1918; prof., 1951; prof., U. Oslo, 1950; emeritus, 1964; visiting prof., U. Colorado, Boulder, 1961-62. Address: Drammensveien 78; Oslo, Norway.

LYDEKKER, Richard, Brit. geologist, palaeontologist; b. 1849; s. G. W. Lydekker; B.A., Trinity Coll., Cambridge; m. Lucy Marianne Davys, 1882; 2 sons, 3 daus. Mem. staff Geol. Survey India, 1874-82. Fellow, Royal Soc., 1894, Geol. Soc., Zool. Soc. Author: Indian Tertiary Vertebrata; Geology of Kashmir; Catalogues of Fossil Mammals, Reptiles, and Birds in British Museum, 10 vols.; Phases of Animal Life; Life and Rock; Geographical History of Mammals; (with H. A. Nicholseon) A Manual of Palaeontology, 2 vols., 1889; The Deer of All Lands; Wild Oxen, Sheep, and Goats of All Lands; The Great and Small Game of Europe, N. and W. Asia, and America; Descriptions of South American Fossil Animals; Mostly Mammals; Horns and Hoofs; The Game Animals of India, Burma, and Tibet; The Game Animals of Africa; The Sportsman's British Birds; A Trip to Pilawin; A Geography of Hertfordshire; The Horse and its Relatives; The Sheep and its Cousins; The Ox and its Kindred; (with W. H. Flower) An Introduction to the Study of Mammals, 1891, The Royal Natural History, 1893-96. Died Harpenden, Eng., Apr. 16, 1915.

LYELL, Sir Charles, Brit. geologist; b. Kinnordy, nr. Kirriemuir, Forfarshire (now Angus), Scotland, Nov. 14, 1797; s. Charles and Miss Smith L.; ed. Exeter College, Oxford (Eng.) U., B.A., 1819, M.A., 1821; studied for bar at Lincoln's Inn, 1819 (but left before completing studies); resumed study of law at Gray's Inn, 1825-27; accepted for bar; D.C.L., Oxford, 1854; m. Mary Horner, July 12, 1832. Often traveled in Britain and on continent; prof. geology, King's College, London, 1831; visited America, 1841, 1845-6, 1852, 1853, Canary Islands, 1853-54. Fellow, Royal Soc., 1826 (Royal medal, 1834); mem. French Acad. Scis., 1862; mem. Linnaean Soc., Brit. Assn. Advancement Sci. (pres. 1864); Geological Soc. (sec. 1823-26; for. sec. 1826; pres. 1835-36, 1849-50; Wollaston medal, 1866); knighted, 1848; created baronet, 1865. Author: Principles of Geology: being an attempt to explain the former changes of the earth's surface, by reference to causes now in action, 3 vols., 1830-33; Elements of Geology, 1838; Travels in North America with Geological Observations, 2 vols., 1845; A Second Visit to the United States of North America, 2 vols., 1849; The Antiquity of Man, 1862; The Students' Elements of Geology, 1871; 76 scientific memoirs. Champion of uniformitarianism; vigorous opponent of catastrophist geology; insisted same causes now operating (e.g. heat, erosion) to change surface of earth have operated uniformly during vast geological periods in the past; divided tertiary period into eocene, miocene and pliocene epochs, 1829; introduced pleistocene as further division, 1839; made extensive study of paleontology and conchology; arranged for publication of views of Darwin and Wallace on origin of species; one of Darwin's earliest supporters. Died London, Eng., Feb. 22, 1875.

LYGHT, Charles Everard, physician, pharm. cons.; b. Hamilton, Ont., Can., July 26, 1901; s. George Albert and Florence (Pountney) L.; M.D., C.M., Queen's U. Faculty Medicine, 1926; m. Mona Havergal Kerruish, June 29, 1927; 1 dau., Mona Mary (Mrs. Perry E. Massey). Came to U. S., 1927, naturalized, 1934. Instr. to asso. prof. clin. medicine U. Wis., 1927-36, dir. dept. student health, 1931-36; prof. health, phys. edn., dir. coll. health service Carleton Coll., 1936-42; dir. health edn. Nat. Tb Assn., 1942-47; dir. med. communications Merck Sharp & Dohme Research Labs., Rahway, N.J., West Point, Pa., 1947-66, editor Merck Manual, 1948-66. Fellow A.C.P., Am. Coll. Chest Physicians, Am. Geriatrics Soc. (sec.), Am. Pub. Health Assn. (resolutions, internat. health coms.), Am. Med. Writers Assn. (pres. 1957-58, trustee, chmn. edn. com. 1961-62); mem. A.M.A., Royal Soc. Health, Fla., Marion County med. socs., Am. Thoracic Soc., Pharm. Mfrs. Assn. (med. sect. chmn. 1956-58, steering com. 1963-66), Am. Therapeutic Soc., A.A.A.S., Assn. Med. Dirs. (pres. 1963), Nat. Conf. Tb Workers, Nat. Vitamin Found., numerous others. Contbr. articles to profl. jours., sci. exhibits, films. Home: Coronado Pines, Route 1, Oklawaha, Fla. 32679.*

LYKOS, Peter George, Am. chemist; b. Chgo., Jan. 22, 1927; s. George Peter and Theodora (Psimoulis) L.; B.S., Northwestern U., 1950; Ph.D., Carnegie Inst. Tech., 1955; m. Marie Shumicki, July 2, 1950; children—George, Kristina, Andrew. Faculty, Carnegie Inst. Tech., 1954-55; faculty Ill. Inst. Tech., Chgo., 1955-—, prof., 1964-—, dir. Computation Center, 1964-—. Cons. Argonne Nat. Lab., 1958-—. Mem. Am. Chem. Soc., Am. Phys. Soc., Assn. Computing Machinery, Four Pi (pres.). Research, publs. on application quantum mechanics to chemistry, elucidation nature pi-electron approximation. Home: 415 N. Scoville Av., Oak Park, Ill. 60302. Office: Ill. Inst. Tech., Chgo. 60616.*

LYKOV, Aleksey Vasilevich, Russian thermal physicist; b. 1910; grad. Yaroslavl Pedagogical Inst., 1930; D.Tech. Sci. Instr., Moscow Tech. Inst. of Food Industry, 1942-52; asso. Power Engring. Inst., Belorussian Acad. Sci. (name now Inst. Heat and Mass Exchange), 1954-57, dir., 1957—. Mem. Belorussian delegation 2d Internat. Conf. on Peaceful Use Atomic Energy, Geneva, 1958. Recipient Stalin prize, 1951. Mem. Belorussian Acad. Sci. Author: The Theory of Drying, 1950; The Theory of Thermal Conductivity, 1952; Heat and Mass Exchange in Drying Processes, 1956; Theoretical Principles of Thermal Physics in Construction, 1961. Research and publs. on molecular energy and matter transfer, theory and practice of drying moist materials, thermodynamics of irreversible processes. Address: Inst. Heat and Mass Exchange, Belorussian Acad. Sci., Minsk, Belorussian SSR, USSR.

LYLE, Henry Hamilton Moore, surgeon; b. Connor, Ulster, Ireland, Nov. 15, 1874; s. Samuel and Elizabeth (Orr) L.; med. prep. edn., Cornell U., 1896; M.D., Coll. Physicians and Surgeons (Columbia), 1900; m. Clara Schlemmer, May 17, 1910 (dec. Jan. 1916); m. 2d, Jessie Benson Pickens, Apr. 16, 1919. Practiced at N.Y.C., since 1900; prof. clin. surgery Coll. Phys. and Surg., 1913-19; asst. prof. surgery, Cornell U. Med. Sch., 1919-31, prof. clin. surgery since 1931; attending surgeon St. Luke's Hosp.; dir. cancer service N.Y. Skin and Cancer Hosp.; attending surgical specialist U. S. Vets.' Bur., Dist. No. 2; cons. surgeon Elizabeth A. Horton Meml. Hosp., Middletown, N.Y., N.Y. State Reconstrn. Home, W. Haverstraw, Cornwall (N.Y.) Hosp.; cons. St. Luke's Hospital, Newburgh, N.Y. Fellow A.C.S.; mem. Am. Surg. Assn., N.Y. Surg. Soc., Am. Soc. Clin. Surgeons, Internat. Surg. Soc. of Brussels, A.M.A., Acad. Medicine (N.Y.), N.Y. State Soc. Indsl. Medicine, Nat. Inst. Social Scis. Died Mar. 11, 1947.

LYLE, Henry Samuel, Brit. surgeon; b. Bideford, Eng., Oct. 24, 1857; s. D. Lyle; ed. Liverpool (Eng.) Royal Infirmary Sch. Medicine, King's Coll., London; m. Millicent O'Brien. Asst. surgeon Liverpool Hosp. for Cancer and Skin Diseases, 1892-99, sr. surgeon, from 1899. Mem. Royal Coll. Surgeons Eng. Author: Treatment of Ringworm. Treatment of Vesicular Eczema; Case of Cancer cured by X-Rays; Relations between Cancer and Chronic Inflammation; Improved mouth gag. Died Mar. 29, 1916.

LYLE, Robert Edward, Am. educator; b. Atlanta, Jan. 26, 1926; s. Robert Edward and Adeline (Cason) L.; B.A., Emory U., 1945, M.S., 1946; Ph.D., U. Wis., 1949; m. Gloria Gilbert, Aug. 28, 1947. Asst. prof. Oberlin Coll., 1949-51; faculty U. N.H., Durham, 1951—, prof. chemistry, 1957—; cons. Arthur D. Little, Cambridge, Mass., 1963—. Guest, vis. scientist NIH, Bethesda, Md., 1958-59; USPHS fellow Oxford U., Eng., 1965—. Mem. Am. Chem. Soc., Chem. Soc. (London), A.A.A.S., Sigma Xi. Research, publs. on molecular nature of cyclic compounds with particular emphasis on pharmacol., conformation, reductions and molecular rearrangements. Home: 7 Hoitt Dr., Durham, N.H. 03824.*

LYLE, Samuel James, chemist; b. County Down, No. Ireland, July 19, 1931; s. Robert John and Margaret (Craig) L.; B.Sc. with honors, Queen's U., Belfast, Ireland, 1953, M.Sc., 1954; Ph.D., U. Birmingham (Eng.), 1957; m. Joan Margaret Lazenby, Sept. 14, 1957; children—Helen Margaret, Stephen Nicholas. Sr. analytical chemist W. Canning & Co., Birmingham, 1957-58; lectr. U. Durham (Eng.), 1958-64; lectr. U. Kent, Canterbury, Eng., 1965—. Mem. adv. bd. Talanta, 1958-62. Fellow Chem. Soc.; mem. Royal Inst. Chemistry (asso.), Soc. for Analytical Chemistry (com. mem. radiochem. group). Contbg. author: Comprehensive Analytical Chemistry, 1961; Bromine and its Compounds, 1966. Research, publs. on lanthanide and actinide elements in ionexchange, coprecipitation phenomena, liquid-liquid extraction of metal ions using chelating substances and liquid ion-exchangers, mass and charge distbn. in neutron induced fission of heavy elements, delayed neutrons in fission. Home: Roundhill, Mill Lane, Shepherdswell, Kent. Office: Chem. Lab., U. Kent, Canterbury, Kent, Eng.

LYLE, Sir Thomas Ranken, Brit. physicist; b. Coleraine, N. Ireland, Aug. 26, 1860; s. Hugh Lyle; grad. Dublin U., 1883; M.A., D.Sc.; m. Clare Millear, 1892; 1 son, 3 daus. Lectr. math. Catholic Univ. Coll., Dublin; rep. Irish Bd. Lights at expts. on lighthouse illuminants, S. Foreland, 1884-85; prof. natural philosophy Melbourne (Australia) U., 1889-1915, emeritus, from 1915. Chmn. bd. visitors Melbourne Obs.; mem. State Electricity Commn. Victoria (Australia), 1917-37. Fellow Royal Soc. Contbr. to sci. publs. Died Mar. 31, 1944.

LYMAN, Chester Smith, Am. astronomer, physicist; b. Manchester, Conn., Jan. 13, 1814; s. Chester and Mary (Smith) L.; grad. Yale, 1837; postgrad. Union Theol. Sem., 1839-40; m. Delia Williams Wood, June 20, 1850. Pastor, 1st Ch. (Congregational), New Britain, Conn., 1843-45; traveled, pursued varied occupations including surveyor in Hawaii, gold digger in Cal., 1845-50; helped prepare definitions for Webster's Dictionary, 1850-circa 1855; prof. indsl. mechanics and physics Sheffield Sci. Sch. of Yale, New Haven, 1859-71, prof. astronomy and physics, 1871-84, prof. astronomy, 1884-90. Mem. A.A.A.S. (v.p. 1874), Conn. Acad. Arts and Sci. (pres. 1859-77). Made 1st satisfactory observation of planet Venus, 1866; inventions include: 1st combined transit and zenith instrument for determining latitude; an apparatus for demonstrating wave motion; improvements in clock pendulums. June 29, 1890.

LYMAN, George Dunlap, Am. pediatrician; b. Virginia City, Nev., Dec. 12, 1882; s. Dean Briggs and Anna Louise (Dunlap) L.; A.B., Stanford U., 1905; M.D., Coll. Phys. and Surg., Columbia, 1909; postgrad. univs. Munich, Vienna and Berlin, 1912-14; m. Dorothy Quincy Van Sucklen, Dec. 28, 1911; children—Dorothy Quincy (Mrs. J. Wm. Beatty), Elizabeth Ann (Mrs. David Potter). Practiced in San Francisco, from 1914; mem. faculty of medicine, Stanford; mem. vis. staff Stanford U. Hosp. and St. Mary's Hosp. Fellow Am. Acad. of Pediatrics; mem. A.M.A., Soc. Am. Historians. Author: Care and Feeding of the Infant, 1915, 2d edit., 1922; John Marsh, Pioneer, 1930; Wierzbicki—The Book and the Doctor, 1933; Saga of the Comstock Lode, 1934 (awarded Commonwealth Club gold medal, 1934); Ralston's Ring, 1937; A Friend to Man, 1938. Owner of over 6,000 vols. of Californiana. Died July 26, 1949; buried Oakland, Cal.

LYMAN, Henry Herbert, Canadian entomologist; b. Montreal, Que., Can., 1854; B.A., McGill U., Montreal, 1876, M.A., 1880; m. a dau. of William Kirby, Mar. 1912. Joined father's firm Lyman, Clare & Co., wholesale chemists and druggists, Montreal, 1880, became pres. after his father's death; joined 5th bn. Canadian Vol. Force (now Royal Scots of Can.), 1877; ret. with rank of maj., 1891; dir. Brit. and Colonial Press Service. Fellow Royal Geog. Soc., Royal Colonial Press Service. Fellow Royal Geog. Soc., Royal Colonial Inst., Entomol. Soc. Ont. (v.p. Montreal br. 1877, pres., 1881-83, 88-99), Canadian Entomol. Soc. (became v.p. 1895, pres. 1897-99). Contbr. numerous articles on Lepidoptera to entomol. jours. Died 1914.

LYMAN, John, Am. oceanographer; b. Berkeley, Cal., Oct. 28, 1915; s. Theodore Benedict and Rowena (Wilson) L.; B.S., U. Cal. at Berkeley, 1936; M.S., Scripps Instn. Oceanography, U. Cal. at Los Angeles, 1951, Ph.D., 1958; m. Mitchell Forrest, May 4, 1946; children—John F. W., Richard D. McK. Chemist, Union Oil Co. Cal., Oleum, 1937; research asst. Scripps Instn. Oceanography, 1937-41; oceanographer USN Hydrog. Office, 1946-52, dir. div. oceanography, 1952-59; asso. program dir. oceanography NSF, Washington, 1959-62, program dir., 1963-64; oceanography coordinator Bur. Comml. Fisheries, Washington, 1964-66; cons. oceanographer, 1966—. Mem. A.A.A.S., Am. Soc. Limnology and Oceanography, Geochem. Soc., Am. Geophys. Union (editor for oceanography Transactions 1948-59), Internat. Assn. Phys. Oceanography, (past mem. exec. com.). Contbr. articles to tech. jours, Ency. Brit. Research on measurement dissociation constants carbonic acid and boric acid in sea water; established ionic composition salts in sea water; computation areas oceans and seas. Address: 5310 Rayburn Ct., Washington 20031.*

LYMAN, Richard Lee, Am. nutritionist; b. Gilroy, Cal., Apr. 7, 1927; s. Oren L. and Augusta (Young) L.; B.A., U. Cal. at Berkeley, 1949, Ph.D., 1957; M.S., U. Wis., 1951; m. Marian L. Meyer, Aug. 10, 1952; children—Ronald, Eric, Laura. Chemist, Western Utilization and Research Labs., Albany, Cal., 1957-58; faculty U. Cal., Berkeley, 1958—, prof. nutrition, 1967—. Mem. A.A.A.S., Am. Inst. Nutrition, Soc. Exptl. Biology and Medicine. Research on endocrine and diet relationships to lipid and phospholipid metabolism, physiol. effects of amino acid deficiency and imbalances with emphasis on gastrointestinal function. Home: 2612 Marquette Ct., Richmond, Cal. 90846.*

LYMAN, Theodore, Am. zoologist; b. Waltham, Mass., Aug. 23, 1833; s. Theodore and Mary Elizabeth (Henderson) L.; A.B., Harvard, 1855, B.S., 1858; m. Elizabeth Russell, Nov. 28, 1856. Studied zoology under Louis Agassiz at Lawrence Scientific Sch., joined Agassiz' sci. expdn. in Fla., 1855-58; elected as original mem. Museum of Comparative Zoology at Harvard, 1859; chmn. Fisheries Commn. of Mass., 1866; pres. Am. Fish Cultural Assn., 1884; overseer Harvard, 1868; mem. U. S. Ho. of Reps. from Mass, 1883-85. Mem. Am. Acad. Arts and Scis. Collected sci. data on ophiuridae in Europe, 1861-63. Died Nahant, Mass., Sept. 9, 1897.

LYMAN, Theodore, Am. physicist; b. Boston, Nov. 23, 1874; s. Theodore and Elizabeth (Russell) L.; A.B., Harvard, 1897, Ph.D., 1900. Instr. physics Harvard, 1902-07, asst. prof., 1907-17, dir. Jefferson Phys. Lab., 1910-47, prof. physics, 1917-26, Hollis prof. emeritus, from 1926. Recipient Rumford medal Am. Acad. Arts and Scis.; Elliott Cresson medal Am. Philos. Soc.; Frederick Ives medal Optical Soc. Am. Fellow Am. Acad. Arts and Scis. (pres.), A.A.A.S., Royal Geog. Soc.; mem. Nat. Acad. Scis., Am. Phys. Soc. (pres.); known for research on properties of light of very short wave lengths. Died Oct. 11, 1954.

LYNAM, Edward William O'Flaherty, Brit. geographer; b. London; s. J. P. D. and Anges (O'Flaherty) L.; ed. Univ. Coll., Cork, Royal U. Ireland, U. Besancon (France), U. London; m. Martha Perry; 1 son, 1 dau. Supt. map room Brit. Mus. Fellow Royal Geog. Soc., Soc. Antiquaries; mem. Internat. Geog. Union, Internat. Union of History of Scis., Viking Soc. (v.p.), Hakluyt Soc. (pres. 1945-49, v.p. 1949—). Author: Maps of the Fenlands, 1934; The First Engraved Atlas of the World, 1941; British Maps and Map Maker, 1944; The Character of England in Maps, 1945; Period Ornament, Symbols and Lettering on Maps, 1946; The Carta Marina of Alaus Magnus, 1949; also articles and revs. on history of cartography, exploration, Ireland and Scandinavia. Co-editor: Imago Mundi; editor Richard Hakluyt and his Successors, 1947. Died Jan. 29, 1950.

LYNCH, Frank W(orthington), Am. gynecologist; b. Cleve., Nov. 5, 1871; s. Frank W. and Rebecca (Nevin) L.; A.B., Western Res. U., 1895; M.D., Johns Hopkins, 1899; grad. study Vienna and Munich, 1910, 12; m. Rowena Tyng Higginson, Apr. 20, 1904; 1 son, Frank W. Asst. instr. and asso. in obstetrics Johns Hopkins U. Med. Sch., 1900-04; instr. in obstetrics Rush Med. Coll. (U. Chgo.), 1905-09, asst. prof. obstetrics and gynecology, 1909-15; prof. obstetrics and gynecology U. Cal., 1915-42, emeritus since 1942. Mem. Am. Bd. Obstetrics and Gynecology. Fellow A.C.S. (v.p. 1937-38); mem. A.M.A., Am. Gynecol. Soc. (1st v.p. 1927; pres. 1933), Pacific Coast Surg. Soc., Pacific Coast Obstet. and Gynecol. Soc. (pres. 1931), Central Assn. Obstetricians and Gynecologists (hon.), Pan-Am. Med. Assn., Chgo. Gynecol. Soc. (v.p., pres., editor), other med. socs. Author: Pelvic Neoplasms (with A. F. Maxwell), 1922. Contbr. chapters to Am. Practice of Surgery, 1911; Oxford Surgery, 1921; Nelson's Looseleaf Surgery, 1928; Davis' Obstetrics and Gynecology, 1933; Curtis' Obstetrics and Gynecology, 1933; The Treatment of Cancer, 1937. Editorial bd. Surgery, Gynecology and Obstetrics (abstract dept.), Am. Jour. Obstetrics and Gynecology, Western Jour. Surgery. Contbr. on obstet. and gynecol. subjects to Am. and German med. jours. Died Jan. 12, 1945.

LYNCH, Kenneth Merrill, Am. pathologist; b. Hamilton County, Tex., Nov. 27, 1887; s. William Warner and Martha Isabel (Miller) L.; M.D., U. Tex., 1910; LL.D., U. of S.C., 1930; Charleston Coll., 1945; D.Sc., Clemson U.; m. Lyall Wannamaker, Nov. 8, 1941; children—Kenneth Merrill, Martha Juanita, Merrill, William. Mem. faculty 1911-13; prof. pathology, Med. Coll. S.C., 1913-21, 26—, chancellor, 1960—, vice dean, 1935-43, dean, 1943-49, pres., dean faculty, 1949-60; private medicine, Dallas, Tex., 1921-26; pathologist Roper Hosp., Charleston, 1913-21, 1927-60; mem. adv. council NIH, 1957-61. also pathologist other S.C. hosps. Recipient Distinguished Service award So. Med. Assn., 1957, Distinguished Service Citation and Medal, Am. Cancer Soc., 1958. Fellow A.C.P. (bd. govs. 1925-27, 36-47), and 1936-47), A.A.A.S., A.M.A. (v.p. 1935-36); mem. Am. Soc. of Tropical Medicine (pres. 1929-30), Am. Soc. Cancer Control (past dir.), Am. Soc. Clin. Pathology (past pres.), dirs. 1939-43), Phi Beta Kappa, Alpha Omega Alpha. Revisor: (H. W. C. Vines) Green's Manual of Pathology, 1934. Author: Protozoan Parasitism of the Alimentary Tract, 1930. Research and publs. on grahuloma inguinale; asbestosis and lung cancer; first in vitro cultivation of flagellates; pneumoconiosis from exposure to kaolin dust, kaolinosis. Home: P.O. Box 811, Summerville, S.C. 24983. Office: 80 Barre S., Charleston, S.C. 29401.*

LYNCH, Richard Irwin, Brit. botanist; b. St. Germans, 1850; M.A. (hon.), Cambridge U., 1906; entered Kew, 1867; curator Botanic Garden, Cambridge, Eng., 1879-1919. Recipient Veitch Meml. medal in silver, 1901, in gold, 1924, Victoria medal in honor, 1906. Asso. Linnean Soc., Royal Botanic Soc.; cor. mem. Royal Hort. Soc. Author: Book of the Iris; also articles. Cultivated hybrids, especially Gerberas from Riviera; raised universal window plant Campanula isophylla alba, at Kew. Died Dec. 7, 1924.

LYNDON, Edward, elec. engr.; b. 1879; asso. with Simon Lake in early submarine work; became pres. co. mfg. searchlight mirrors, tech. lenses, and prisms after World War I; invented prism reflector which is used to magnify coins in N.Y. subway turnstiles; helped to develope searchlight mirrors used during World War I. Died 1940.

LYNDON, Roger Conant, Am. mathematician; b. Calais, Me., Dec. 18, 1917; s. Percy Emmons and Ann (Milliken) L.; A.B., Harvard, 1939, M.A., 1941, Ph.D., 1946; m. Freda Jones, Apr. 30, 1961. Sci. officer Office Naval Research, London, Eng., 1946-48; faculty Princeton, 1948-53; faculty U. Mich., Ann Arbor, 1953—, prof. math., 1958—. Vis. asso. prof. U. Cal. at Berkeley, 1956-57; mathematician Inst. for Def. Analyses, Princeton, 1959-60; vis. prof. Queen Mary Coll., U. London, 1960-61, 64-65. Mem. Am., London math. socs., Math. Assn. Am., Assn. for Symbolic Logic. Author: Notes on Logic, 1966; also articles. Research on analysis formal structure, its gen. regularities and spl. peculiarities, math. systems descriptive symmetries and motions geometric figures, connections between objects and linguistic entities.

Home: 1889 Superior Rd., Ypsilanti, Mich. Office: Angell Hall, U. Mich., Ann Arbor, Mich. 48104.*

LYNEN, Feodor, German biochemist; b. Munich, Germany, Apr. 6, 1911; s. Wilhelm and Frieda (Prym) L.; Dr.phil., U. Munich, 1937; Dr.med. h.c., U. Freiburg (Germany), 1960; m. Eva Wieland, May 17, 1937; children—Peter, Annemarie, Susanne, Eva-Marie, Heinrich. Lectr. chemistry U. Munich, 1942-47, prof., 1947—; dir. Max-Planck-Inst. für Zellchemie, Munich, 1954—. Recipient Neuberg medal Am. Soc. European Chemists and Pharmacists, 1954; Liebig-Denkmünze medal Gesellschaft Deutsche Chemiker, 1955; Carus medal Deutsche Acad. der Naturforscher Leopoldina, 1961; Otto Warburg medal Gesellschaft für Physiolog. Chemie, 1963; Nobel prize for medicine, 1964. Mem. Bayerische Acad. der Wiussenschaften, Deutsche Acad. der Naturforscher Leopoldina; fgn. mem. Am. Soc. Biol. Chemists, Nat. Acad. Scis., Am. Acad. Arts and Scis.; corr. mem. Austrian Acad. Scis. Research and publs. on chem. structure of active acetate, isoprene and bicarbonate and of cytohaeme; biol. oxidation of fatty acids and the formation of acetoacetate; elucidation of the biosynthesis of cysteine, terpenes, rubber and fatty acids. Home: Schiesstättstrasse 10, 813 Starnberg, Germany. Office: Max-Planck-Inst. für. Zellchemie, Karlstrasse 23, 8 Munich 2, Germany.*

LYNN, Hugh Bailey, Am. physician; b. Verona, N.J., Aug. 13, 1914; s. Hugh and Mary (Doggart) L.; A.B., Princeton, 1936; M.D., Columbia, 1940; m. Lillian Smith, June 19, 1940; children—H. Bailey, Jonathan S., Michael Anne. Practice medicine specializing in pediatric surgery, Pompton Plains, N.J., 1946-47; faculty U. Louisville Sch. Medicine, 1953-61, asso. prof., 1957-60, chief sect. on pediatric surgery, 1957-61; surgeon-in-chief Children's Hosp., Louisville, 1953-61; asso. prof. Mayo Grad. Sch. Medicine, head sect. pediatric surgery Mayo Clinic, U. Minn., Rochester, 1961—. Recipient award for exhibit N.J. State Med. Conv., 1953. Mem. Am. Acad. Pediatrics, A.C.S. (awards for med. motion pictures 1954, 56, 58, 59, 64), A.M.A., Brit. Assn. Paediatric Surgeons, Central Surg. Assn., Minn. State Med. Assn., Minn. Surg. Soc., N.W. Pediatric Soc. Contbr. numerous articles to tech. jours. Home: Cairnbrae Farm, Route 2, Rochester, Minn. 55901.*

LYNN, Melvyn Stuart, mathematician; b. London, Eng., July 7, 1937; s. Richard and Julie (Shavick) L.; B.A., Merton Coll., Oxford U., 1958, M.A., 1965; M.A., U. Cal. at Los Angeles, 1960, Ph.D., 1962; m. Barbara Berkson, Aug. 26, 1960; children—Monica Georgette, Anthea Suzanne, Matthew David. Research mathematician Cal. Research Corp., 1962; research fellow math. div. Nat. Phys. Lab., Eng., 1962-63, sr. sci. officer, 1963-64; staff mem. IBM, Los Angeles, 1964-65, mgr. math. dept., Houston, 1965-66, mgr. research dept., 1966-67, mgr. Sci. Center, 1968—. Mem. Am., Canadian math. socs., Assn. Computing Machinery (editor Sci. Applications), Soc. Indsl. and Applied Math. Research in partial differential equations, matrix theory, integral equations, and biomath. Home: 931 Magdalene St., Houston 77024. Office: IBM Sci. Center, 6900 Fannin St., Houston 77025.*

LYNN, William Gardner, Am. biologist; b. Washington, Dec. 26, 1905; s. William Lee and Ann (Gardner) L.; A.B., Johns Hopkins, 1928, Ph.D., 1931; m. Harriett Naomi Walker, Sept. 2, 1933; children—Richard Gardner, Robert Walker. With Johns Hopkins, 1931-42, asst. prof., 1938-42; with Cath. U. Am., 1942—, prof., 1946—, head dept. biology, 1958-63. Rockefeller Found. Research fellow Yale, 1939-40; Fulbright Research scholar U. Coll. W.I., Jamaica, 1952-53. Mem. A.A.A.S., Am. Soc. Zoology, Am. Soc. Ichthyology and Herpetology, Am. Soc. Naturalists, Washington Acad. Sci., Philos. Soc. Washington, Phi Beta Kappa, Sigma Xi. Author (with Chapman Grant): The Herpetology of Jamaica, 1940. Research, numerous publs. on structure and function of thyroid gland in cold-blooded vertebrates especially with reference to amphibian metamorphosis and high temperature tolerance. Home: 2935 Northampton St. N.W., Washington 20015.*

LYNTON, Ernest Albert, physicist; b. Berlin, Germany, July 17, 1926; s. Arthur J. and Martha L. (Kiefe) Lowenstein; came to U. S., 1941, naturalized, 1945; B.Sc., Carnegie Inst. Tech., 1947, M.Sc., 1948; Ph.D., Yale, 1951; m. Carla E. Kaufmann, Aug. 4, 1953; children—David Michael, Eric Daniel. AEC postdoctoral fellow Leiden (Holland) U., 1951-52; faculty Rutgers U., New Brunswick, N.J., 1952—; prof. physics, 1964—; dean Livingston Coll., 1965—; vis. prof. U. Grenoble (France), 1959-60. Research in superconductivity and low temperature physics. Author: Superconductivity, 1962; also articles. Home: 665 Snowden Lane, Princeton, N.J. 08540.*

LYON, Charles Julius, Am. plant physiologist; b. Grand Gorge, N.Y., Oct. 27, 1896; s. Frank Emory and Lizzie (Yerdon) L.; B.S., Middlebury Coll., 1918; A.M., Harvard, 1920, Ph.D., 1926; A.M. (hon.), Dartmouth, 1934; m. Gertrude Morrow Adair, June 25, 1921; 1 son, George Adair. Instr. chemistry Middlebury Coll., 1918-19; spl. instr. math. Simmons Coll., 1919-20; faculty botany Dartmouth, Hanover, N.H., 1920—, prof., 1934-63, prof. emeritus, research prof., 1963—, chmn. dept. botany, 1937-41, 53-57. Dir., botanist N.H. Nature Sch., North Woodstock, N.H., summers 1931-33. Research grantee NSF, NASA, 1959—. Mem. Am. Soc. Plant Physiologists, Am. Inst. Biol. Scis., Phi Beta Kappa. Author: Flowering Plants and Vegetation, 3d edit., 1950. Editor, translator: Plant Respiration (Kostychev), 1927; Chemical Plant Physiology (Kostychev), 1931. Research on plant respiration, tree ring records of microclimate, growth hormones in leafy plants, seedlings for 1st NASA biosatellite. Home: 4 S. Balch St., Hanover, N.H. 03755.*

LYON, David Murray, Scottish physician; b. 1888; s. William Malcolm and E. (Campbell) L.; ed. George Watson's Coll., Edinburgh, Scotland, Edinburgh U.; M.D., D.Sc., Ch.B., D.P.H.; m. Edith Dona Lloyd; 2 sons, 1 dau. Prin. med. officer Scottish Widows' Fund and Life Assurance Soc., 1936-54; lectr. pathology Edinburgh U., also Sch. Medicine, Royal Colls., Edinburgh; Christison prof. therapeutics, 1926-36; prof. clin. medicine U. Edinburgh, 1924-53; successively resident physician, clin. asst., asst. pathologist, asst. physician, physician cons. Royal Infirmary Edinburgh. Mem. Sci. Adv. Com. Stark scholar clin. medicine; Crichton scholar pathology. Fellow Royal Coll. Physicians, Royal Soc.; mem. Royal Coll. Physicians Edinburgh (pres.). Author: The Essentials of Medical Treatment, 1939. Editor, Edinburgh Med. Jour. Research and numerous publs. in medicine, therapeutics, applied physiology. Died Nov. 16, 1956.

LYON, Dorsey Alfred, Am. metallurgist; b. Bureau County, Ill., July 17, 1871; s. Walter S. and Sarah S. (McKune) L.; A.B., Stanford, 1898; A.M., Harvard, 1902; D.Sc., U. Utah, 1922. Instr. geology and mining U. Wash., 1898-99, asst. prof. mining and metallurgy, 1899-1900, prof., 1901; with U. S. Mining and Smelting Co., Midvale, Utah, 1902-03; asst. prof. metallurgy Stanford, 1903-07; mgr. Noble Electric Steel Co., Heroult, Cal., 1907-11; cons. metallurgist Electro-Metals Co., London, Eng., 1911-13; metallurgist U. S. Bur. Mines, 1913-19; chief metallurgist, 1919-27, supr. stas., 1917-27, asst. dir., 1923-26, supervising engr. Intermountain Station, 1927-31; dir. Utah Engring. Expt. Sta., 1931-35, retired. Fellow A.A.A.S., Soc. for Promotion of Engring. Edn.; mem. Am. Chem. Soc., Am. Mining Congress, Electrochem. Soc., Am. Inst. Mining Engrs., Sigma Xi. A pioneer in electric furnace work in U. S. Co-author several bulls. of U. S. Bur. Mines, dealing with use of electric furnace in metall. work; contbr. to tech. publs. on metall. subjects. Died Oct. 16, 1945.

LYON, George Marshall, Am. pediatrician; b. Union City, Pa., Feb. 8, 1895; s. Marshall Allen and Harriet Belle (Law) L.; student Marshall U., 1909-13; B.S., Denison U., 1916, D.Sc. (hon.), 1956; M.D., Johns Hopkins, 1920; m. Virginia Berkeley Sutherland, June 24, 1922 (dec. June, 1926); children—Virginia Berkeley (Mrs. Burley McCraw), Natalie Sutherland (Mrs. Eugene J. Olmi, Jr.), Elizabeth Harriet (Mrs. James E. Miller); m. 2d, Theeta Carrington Searcy, July 29, 1927; 1 son, George Marshall. Practiced medicine specializing in pediatrics, Huntington, W.Va., 1921-42; spl. asst. for atomic medicine Office of chief Med. Dir. U. S. VA, 1947-56, asst. med. dir. for research and edn., 1952-56; ret., 1965; cons. medicine VA Hosp., Miami, 1967—. Mem. adv. com. maternal and child health Children's Bur. U. S. Dept. Labor, 1936-42; mem. Joint Com. on Health Problems in Edn., A.M.A. and N.E.A., 1940-50; pediatric rep., bd. res. consultants Office Surgeon Gen., USN, 1946-65; mem. com. on atomic casualties NRC, 1946-49, mem. div. med. scis., 1952-56. Decorated Bronze star medal with Combat V; Legion of Merit with gold star; recipient Exceptional Service award U. S. VA, 1959; Alumni Citation, Denison U., 1949; Maj. Livingston Seaman prize Assn. Mil. Surgeons U. S., 1956, Pfizer award Nat. Civil Def. Council, 1959. Diplomate Am. Bd. Pediatrics. Fellow A.C.P., Am. Pediatric Soc. (emeritus), Am. Acad. Pediatrics (emeritus), Soc. for Devel. Child Growth and Devel., Radiation Research Soc., A.M.A. (cons. on nat. def. 1948-54); mem. Assn. Mil. Surgeons U. S., Soc. Med. Consultants to Armed Forces, Assn. Am. Med. Colls., So. Med. Assn., Omicron Delta Kappa, others. Research, publs. on studies of infectious and nutritional diseases of childhood, chemotherapy of bacillary dysentery, effects of radiation and use of radioisotopes in medicine, aspects of mil. medicine. Address: 1400 Dale Lane, Delray Beach, Fla. 33444.*

LYON, James Alexander, Am. cardiologist; b. Broome County, N.Y., Feb. 28, 1882; s. Henry and Catherine (Murray) L.; student Ohio U., Syracuse U.; M.D., Md. Med. Coll., 1906; postgrad. Harvard Med. Sch., also Univ. Coll. Hosp. and Nat. Hosp. for Disease of Heart, London, 1923, U. Vienna, 1924, Mass. Gen. Hosp., 1926; m. Irene Elizabeth Moore; 1 dau., Elizabeth Moore. Intern Bay View Hosp., Balt., 1906-07; asst. physician Loomis Sanatorium, Liberty, N.Y., 1907-09; asst. supt. and sr. physician Mass State Hosp. for Tb, Rutland, Mass., 1909-16; prof. clin. cardiology Georgetown U., 1929-40; cardiologist and mem. cardiac com. The Doctors Hosp., Inc.; cons. cardiologist and mem. med. bd. Children's Hosp.; cons. cardiologist Homeopathic, and Columbia hosps.; asst. physician Out-Patient Dept. Johns Hopkins' Hosp., 1925-26; attending cardiologist and chief of cardiac clinic Emergency Hosp., 1929-40; practiced medicine, Washington, 1925—. Diplomate Am. Bd. Internal. Medicine. Fellow A.M.A., A.A.A.S., N.Y. Acad. Medicine, A.C.P. (life mem.); mem. So. Med. Soc., Am. Therapeutic Soc. (pres.), Am., Washington (pres. and sec.) heart assns., Assn. Mil. Surgeons U. S., Pan-Am., Internat. med. assns., Am. Assn. History Medicine, Nat. Geog. Soc., The Hippocrates-Galen Soc. Washington, other med. socs. Died Aug. 4, 1955.

LYON, John Alexander Melvin, Am. elec. engr.; b. Saginaw, Mich., Oct. 9, 1912; s. Melvin Edgar and Mabel (Graham) L.; student Wayne State U., 1929-30; B.S., M.S., U. Mich., Ph.D., 1936; student Harvard, Mass. Inst. Tech., 1943; m. Betty Longenecker, June 24, 1938; children—William Graham, Nancy Jean. Illuminating engr. Pitts. Reflector Co., 1936; elec. engr. Utility Mgmt. Corp., Reading, Pa., 1937-38; erection engr. Cape & Vineyard Electric Co., Mass., 1939; elec. engr. Ebasco Services, Inc., N.Y.C., 1941-42; faculty engring. dept. Northwestern U., 1946-59; prof. elec. engring., U. Mich., 1959—. Dir. Nat. Electronics Conf., pres., 1952. Cited as distinguished alumnus U. Mich., 1953. Fellow A.A.A.S., I.E.E.E.; mem. Am. Soc. Engring. Edn., Sigma Xi, Phi Kappa Phi, Eta Kappa Nu (nat. dir.; nat. pres. 1962-63), Tau Beta Pi. Research and publs. on antennas, electromagnetics, plasma, electronic tubes, transistor circuits and transmission lines. Home: 2427 Shannondale Rd., Ann Arbor, Mich. 48104.*

LYON, Leverett Samuel, Am. economist; b. Will County, Ill., Dec. 14, 1885; s. Edward Payson and Charlotte (Rose) L.; student Beloit (Wis.) College, 1906-07 L.H.D., 1953; Ph.B., U. Chgo., 1910; LL.B., Kent Coll., Law, 1915; A.M., U. Chgo., 1919, Ph.D., 1921; LL.D., Ill. Inst. Tech., 1950, Northwestern U., 1951; m. Lucille Norton, June 26, 1915; children—Richard Norton, David Mansfield. Head dept. civic sci. Joliet Twp. High Sch., 1910-14, 15-16; admitted to Ill. bar, 1916; asst. in econs. U. Chgo., 1916-17, instr. 1917-19, asst. prof., 1919-23, asso. prof., 1923; dean Sch. Commerce and Finance, prof. economics and head dept. Washington U. 1923-25; lectr. Columbia, summer, 1925, U. Denver, summer 1931; prof. econs. Robert Brookings Grad. Sch. Econs. and Govt., 1925-29; mem. research staff, dir. ednl. activities and pub. relations Brookings Instn., 1929-32, exec. v.p., 1932-39; chief exec. officer Chgo. Assn. Commerce and Industry, 1939-54, chmn. exec. com., 1954—. Home Rule Commn., 1953-55; Recipient citation for pub. service U. Chgo., 1941. Mem. Chgo. Bar Assn., Am. Econ. Assn., Acad. of Polit. Sci., Am. Statis. Assn., Am. Marketing Assn. (pres. 1933), Newcomen Soc., Phi Beta Kappa. Author or co-author: Elements of Debating, 1913; Eight Lessons (Bull. of Nat. and Community Life), 1917; A Survey of Commercial Education in the Public High Schools of the United States, 1919; A Functional Approach to Social Economic Data, 1920; Our Economic Organization, 1921; Education for Business, 1922, 3d edit., 1931; Business Cases and Problems, 1925; Making a Living, 1926; Salesmen in Marketing Strategy, 1926; Vocational Readings, 1927; Hand-to-Mouth Buying, 1929; Some Trends in the Marketing of Canned Foods, 1930; Advertising Allowances, 1932; The Economics of Free Deals, 1933; The ABC of the NRA, 1934; The National Recovery Administration—An Analysis and an Appraisal, 1935; The Economics of Open Price Systems, 1936; A Preliminary Analysis for a Program of Economic Education, 1937; Government and Economic Life, Vol. 1, 1939, Vol. II, 1940; Your Business and Postwar Readjustment, 1944; Great Lakes-St. Lawrence Seaway and Power Project, 1951; Modernizing a City Government, 1954; Nothing but Nonsense (verse), 1954. Co-editor: Textbooks in the Social Studies, 11 vols.; Prospects and Problems in Aviation, 1945; Governmental Problems in the Chicago Metropolitan Area, 1957. Contbr. numerous articles to profl. jours. and encys. Died Sept. 7, 1959.

LYON, Marcus Ward, Jr., Am. zoölogist, bacteriologist, pathologist; b. Rock Island Arsenal, Ill., Feb. 5, 1875; s. Marcus Ward and Lydia Anna (Post) L.; Ph.B., Brown U., 1897; N.C. Med. Coll., 1897-98; M.S., George Washington U., 1900, M.D., 1902, Ph.D., 1913; m. Martha Maria Brewer, Dec. 31, 1902; 1 dau., Charlotte. Instr. bacteriology N.C. Med. Coll., 1897-98; aid, later asst. curator Div. of Mammals, U. S. Nat. Mus., Washington, 1898-1912; asst. prof. physiology Howard U., 1903-04, 1907-09, prof. bacteriology, 1909-15; prof. bacteriology and pathology George Washington U., 1915-17, prof. veterinary zoölogy and parasitology, 1917-18. Fellow A.M.A., A.A.A.S.; mem. St. Joseph County Med. Soc. (pres.), Soc. Am. Bacteriologists, Washington Acad. Scis., Biol. Soc. Washington (sec. 1915-19), Am. Chem. Soc., Am. Soc. Parasitologists, Ecol. Society Am., Am. Soc. Mammalogists (pres. 1931-33), Washington Biologists' Field Club, Am. Assn. Pathologists and Bacteriologists, Am. Ornithologists' Union, Am. Soc. Clin. Pathologists, Wildlife Soc., Am. Geog. Soc., N.Y. Acad. Science, Am. Soc. Tropical Medicine; Phi Beta Kappa, Sigma Xi, other socs. Author: Mammals of Indiana; also papers on biol. and med. subjects. Died May 19, 1942.

LYON, Waldo Kampeier, Am. physicist; b. Los Angeles, May 19, 1914; s. Charles R. and Anna (Kampmeier) L.; A.B., U. Cal. at Los Angeles, 1936, M.A., 1937, Ph.D., 1941; m. Virginia Louise Backus, Aug. 28, 1937; children—Lorraine Mae (Mrs. Fred Minning), Russell Roy. Physicist, USN Electronics Lab., San Diego, 1941, chief scientist arctic submarine research, 1952—; physicist submarine operations USN

Antarctic Expdn., 1946; chief scientist Joint U. S.-Can. Beaufort Sea Oceanographic Expdns., 1949-55. Sr. scientist 1st polar crossing by submarine U.S.S. Nautilus, 1958, 1st winter submarine North Pole expdn. U.S.S. Skate, 1959, arctic submarine expdn. U.S.S. Sargo, 1960, U.S.S. Seadragon, 1960, 1962, U.S.S. Skate, 1962. Recipient Navy Distinguished Civilian Service award, 1955-58; Dept. Def. Distinguished Civilian Service award, 1956; Distinguished Civilian Service award Pres. Kennedy, 1962. Fellow Am. Phys. Soc., Acoustical Soc. Am., Arctic Inst. N. Am., A.A.A.S.; mem. N.Y. Acad. Scis. (life mem.), Am. Soc. Naval Engrs. (life, Gold medal for naval engring. 1959), Sigma Xi. Design of techniques and equipment for submarine operations under ice and breaking through ice, leading to first submarine crossing of Arctic Ocean and to successful winter submarine operations in Arctic Ocean. Home: 1330 Alexandria Dr., San Diego 92107. Office: USN Electronics Lab., San Diego 92152.*

LYON, William Southern, Jr., Am. chemist; b. Pulaski, Va., Jan. 25, 1922; s. W. S. and Irene (Hunter) L.; B.S. in Chemistry, U. Va., 1943; postgrad. U. Tenn., 1947-50, 56-58; m. Carey Helen Greer, May 13, 1946; children—Victoria Carey, William Southern III. Chemist, E. I. DuPont de Nemours & Co., Belle, W.Va., 1943-44, Hanford Engring. Works, Richland, Wash., 1944-45; lab. supr. Tenn. Eastman Co., Oak Ridge, 1945-47; group leader nuclear-radiochem. analysis Oak Rdige Nat. Lab., 1947—. Mem. Am. Chem. Soc., Am. Phys. Soc., Sci. Research Soc. Am., MENSA, Am. Soc. for Testing and Materials (vice chmn. com. E10 1962—), Am. Standards Assn. (mem. com. N.5.4 1959—). Editor: Guide to Activation Analysis. Research, numerous publs. on nuclear measurements of certain decay systems, activation analysis and its application, interpretation nuclear data. Home: 7007 Rockingham Dr., Knoxville, Tenn. 37919. Office: P.O. Box X, Oak Ridge 37830.*

LYONET, Pierre, Dutch entomologist, naturalist; b. Maestricht, Holland, July 21, 1707; educated as lawyer. Fellow Royal Soc., 1748. Author: Traité anatomique de la chenille qui ronge le bois de Saule, 1760; Recherches sur l'anatomie et les métamorphoses de différentes espèces d'insectes, 1832. Distinguished more than 4,000 muscles in caterpillar which feeds on willow-wood. Died The Hague, Netherlands, Jan. 10, 1789.

LYONS, Albert Brown, chemist; b. Waimea, Hawaii, Apr. 1, 1841; s. Lorenzo and Lucia Garratt (Smith) L.; ed. Oahu Coll, 1857-63; A.B., Williams, 1865; M.D., U. Mich., 1868; m. Edith M. Eddy, Apr. 25, 1878; children—Lucia Eddy, Albert Eddy. Prof. chemistry Detroit Med. Coll., 1868-80; cons. chemist Parke, Davis & Co., Detroit, 1881-86; 1st editor Pharm. Era, 1887-88; head sci. dept. Oahu Coll., also govt. chemist for Hawaii, 1888-95; chemist Nelson, Baker & Co., Detroit, 1897—. Mem. Com. of Revision of U. S. Pharmacopoeia, 1900-20. Author, Manual of Pharmaceutical Assaying, 1887; Practical Assaying of Drugs and Galenicals, 1899; Plant Names, Scientific and Popular, 1900, enlarged edit., 1907; The Lyon Families of New England, 1905, 1907, 1908; Standardization by Chemical Assay of Organic Drugs, 1920. Died Apr. 13, 1926.

LYONS, Chalmers J., Am. oral surgeon; b. Martinsburg, O., Apr. 30, 1874; s. John P. and Manilla (White) L.; ed. Central Mich. Normal Sch., Mt. Pleasant, Mich.; D.D.S., U. Mich., 1898, D.D.Sc., 1911; m. Grace B. Driggs, 1909; 1 son, Richard Hugh. Practiced at Adrian, Mich., 1898-1907; moved to Jackson, Mich., 1907, asso. in practice with Dr. J. W. Lyons; instr. oral surgery U. Mich., 1907-15, prof. 1915—. Author: Fractures and Dislocations of the Jaws, 1919, rev., 1926. Died May 18, 1935.

LYONS, Don Chalmers, Am. oral surgeon; b. Jackson, Mich., May 5, 1899; s. James White and Estelle (Stonerode) L.; D.D.S., U. Mich., 1921, M.S. in Oral Surgery, 1932; Ph.D. in Med. Bacteriology and Biochemistry, Mich. State U., 1935; m. Gertrude Campbell Rosecrans, Sept. 28, 1922; children—Don, Rodger, Shirley (Mrs. Albert Robinson). Fellow Mayo Clinic, Mayo Found., U. Minn., Rochester, 1921-23; practice dentistry and oral surgery, Jackson; cons. oral surgery Mercy Hosp., Jackson; chief dental sect. of staff, mem. exec. bd. W. A. Foote Meml. Hosp., Jackson; vis. oral surgeon Sheldon Hosp., Albion, Mich.; mem. bd. Beth Moser Mental Clinic, Jackson. Mem. confs. on dental specialties and specialization Am. Dental Assn., 1959-60; mem. Internat. Conf. on Dental Journalism and Documentation, 1959, Conf. on Nat. Orgns. for Areas of Dental Practice, 1960, Oriental Dental Seminars, 1960; Fulbright prof. oral surgery U. San Marcus, Lima, Peru, 1961-62, U. Central and U. Cuence, Ecuador, 1963. Diplomate Am. Bd. Oral Medicine (bd. examiners 1958—). Fellow Am. Pub. Health Assn., A.A.A.S., Am. Med. Writers Assn. (bd. dirs., v.p. Mich. sect.), Am. Acad. Dental Medicine, Royal Health Soc. (Eng.), Am. Coll. Dentists; mem. Am. Dental Assn., Mich. Acad. Sci., Arts and Letters, Am. Acad. Dental Medicine (pres. 1958), Am. Soc. Bacteriologists, N.Y. Acad. Sci., Research Soc. Am., Federation Dentaire Internatioinal, Internat. Conf. in Microbiology, Am. Dental Editors Soc., Sigma Xi, Phi Sigma, Omicron Kappa Upsilon, Delta Sigma Delta. Writer dental health columns Joint Com. on Pub. Health Edn.; contbr. numerous articles to dental and popular publs., also sect. on bacteriology in Miller's Dental Diagnosis and Medicine. Research on antibiotics and antibiosis of bacterial activities; bacteriology of dental caries; growth problems of crippled children, especially related to dental conditions in children with cerebral palsy; enzyme chemistry as related to oral medicine and bacteriology. Home: 512 S. Wisner St. Office: 420 W. Michigan Av., Jackson, Mich.

LYONS, Harold, Am. physicist; b. Buffalo, Feb. 16, 1913; s. Louis and Rose (Siegel) L.; B.A. magna cum laude, U. Buffalo, 1933, postgrad., 1933-34; M.A., U. Mich., 1935, Ph.D., 1939; m. Edna Beatrice Maenick, Sept. 17, 1937; children—Glenn Alan, Sherrie Lynne. Staff, Naval Research Labs., 1939-41, Nat. Bur. Standards Radio Standards, 1941-55; sr. staff scientist Hughes Research Labs., Hughes Aircraft Co., Culver City, Cal., 1955, head atomic physics sect. Microwave Lab., 1956, head atomic physics dept. Physics Lab., 1957-60; v.p. research optical systems, corporate chief scientist, mgr. quantum physics div. Electro-Optical Systems, Inc., Pasadena, Cal., 1960-62; cons. physicist, Pacific Palisades, 1962—; research theoretical physicist in anatomy Brain Research Inst., Space Biology Lab., U. Cal. at Los Angeles, faculty, research physicist, 1966—. Recipient certificate of merit Franklin Inst., 1958; Arthur S. Flemming Govt. award Washington Jr. C. of C., 1949; Nat. Bur. Standards Superior Accomplishment award, 1949; U. S. Dept. Commerce Gold medal, 1949; Washington Acad. Scis. award, 1953; cited by U. Buffalo, 1957. Fellow Am. Phys. Soc., I.E.E.E.; mem. Research Soc. Am., Phi Beta Kappa, Sigma Xi. Invented first two atomic clocks including cesium beam atomic clock now used as world's standard of time; initiated and directed program leading to first working laser; invented first molecular chelate laser. Patentee in field. Home: 1101 El Medio Av., Pacific Palisades, Cal. 90272. Office: Brain Research Inst., Space Biology Lab., U. Cal., Los Angeles 90272.*

LYONS, Harold Aloysius, Am. physician; b. Bklyn., Sept. 14, 1913; s. Harry A. and Louise (Torreille de Castelbou) L.; B.S. cum laude, St. John's U., 1935; M.D., L.I. Coll. Medicine, 1940; postgrad. Montreal U., U. Mich., Thorndike Lab. Harvard, Tufts U.; m. Rita Mary Wood, Mar. 21, 1940; children—Harold, Frances Louise, Gail Jeanne, Robert Louis, George Christopher Bernard, Margaret Alida Marie, Jules Lawrence, Ann Marie. Faculty, State U. N.Y., Downstate Med. Center, Bklyn., 1953—, prof. medicine, also dir. cardiopulmonary lab., 1958—; dir. pulmonary diseases div. Kings County Hosp. Center, Bklyn., 1953—. Cons. medicine U. S. Naval Hosp., St. Albans, VA Hosp., Ft. Hamilton, Bklyn., Bklyn.-Cumberland Hosp., Mercy Hosp., Rockville Centre, USPHS Hosp., S.I., N.Y.; med adviser USPHS, Social Security Administrn.; ofcl. examiner Am. Bd. Internal Medicine, 1952—. Decorated Air medal. Diplomate Am. Bd. Internal Medicine, Pan Am. Med. Assn. (hon.). Fellow A.C.P., Am. Coll. Chest Physicians. Contbr. chpts. to Disease and Injury to Body, 1960. Editor: Vascular Diseases of the Lung, 1967. Contbns. to mechanics of respiration, respiratory physiology, and clin. medicine; studies in angiocardiography and cardiac disorders. Home: Harbor Rd., Harbor Acres, Sands Point, N.Y. 10050. Office: 450 Clarkson Av., Bklyn. 11203.*

LYONS, Sir Henry George, Brit. meteorologist, geologist; b. London, Oct. 11, 1864; s. T. C. Lyons; ed. Wellington Coll.; D.Sc. (hon.), Oxford U., Dublin U.; m. Helen Julia Hardwick, 1896; 1 son, 1 dau. Commd. lt. Royal Engrs., 1884, ret., 1901; dir. Gen. Geol. Survey, Egypt, 1896-98, Survey Dept., Egypt, 1898-1909; comdt. Army Meteorol. Services; dir. Meteorol. Office; dir., sec. Sci. Mus., 1920-33. Recipient Victoria Research medal Royal Geog. Soc., 1911, Symons medal Royal Meteorol. Soc., 1922. Fellow Royal Soc., 1906 (treas. 1929-39). Author: Report on the Island and Temples of Philae; Physiography of the Nile; The Cadastral Survey of Egypt; The Royal Society, 1660-1940, 1944. Died Aug. 10, 1944.

LYONS, Richard Hugh, Am. physician; b. Jackson, Mich., May 7, 1910; s. Chalmers J. and Grace (Driggs) L.; A.B., U. Mich., 1932, M.D., 1935; m. Elizabeth Kane, Aug. 14, 1937; children—Richard J., Chalmers J., Elizabeth S. Instr. medicine U. Mich., 1939-41, asst. prof. medicine, 1941-45, asso. prof. medicine, 1945-47; instr. medicine Wayne U., 1940-42; med. dir. William J. Seymour Hosp., Eloise, Mich., 1940-41; prof. medicine, chmn. dept. medicine State U. N.Y. Coll. Medicine at Syracuse, 1947—; chief of medicine Syracuse U. Hosp., 1947—, Syracuse Meml. Hosp., 1952-64; cons. Auburn (N.Y.) City Hosp., 1949—, Syracuse VA Hosp., 1953—, Potsdam (N.Y.) Hosp., 1959—, St. Joseph's Hosp., Syracuse, 1957—, Army Med. Service, Walter Reed Army Med. Center, 1953—, Crouse-Irving Hosp., Syracuse, 1961—; cons. physician Syracuse Psychiat. Hosp., 1947—; cons. staff Chenango Meml. Hosp., Norwich, N.Y., 1953—. Mem. adv. council Masonic Research Lab., Utica, N.Y., 1960—; mem. med. adv. com. Planned Parenthood Center, Syracuse, N.Y.; bd. dirs. Cerebral Palsy Clinic, Syracuse; mem. nat. adv. council Monsignor Toomey Cardiopulmonary Lab., 1964—. Diplomate Am. Bd. Internal Medicine. Fellow A.C.P.; mem. A.M.A., Am. Fedn. for Clin. Research (pres. 1947-48), Am. Soc. for Clin. Investigation, Central Soc. for Clin. Research, Am. Clin. and Climatol. Assn., Am. Heart Assn. (fellow in clin. cardiology, chmn. program com. 1961—, mem. exec. com. Council on Clin. Cardiology 1961—), A.A.A.S., Am. Coll. Cardiology, Assn. Profs. Medicine, Alpha Omega Alpha, Phi Gamma Delta, Kappa Beta Phi. Research and publs. on blood volume and cardiovascular dynamics, sulfonamide hypersensitivity, autonomic blockade, renal circulation. Home: 904 Manlius Rd., Fayetteville, N.Y. 13066. Office: 750 E. Adams St., Syracuse, N.Y. 13210.

LYONS, Robert Edward, Am. chemist; b. Bloomfield, Ind., Oct. 24, 1869; s. Mathew J. and Alice (Eveleigh) L.; A.B., Ind. U., 1889, A.M., 1890; student Frocenius' Labs., Wiesbaden, univs. of Heidelberg, Munich, and Berlin, and Joergensen's Inst. for Physiology of Fermentation, Copenhagen, 1892-95; Ph.D., U. Heidelberg, 1894; m. Eleanor Joslyn, Mar. 23, 1898; 1 son, Robert Edward. Mem. faculty chemistry Ind. U., 1889-92, 95—, prof. chemistry, head dept., 1895-38, emeritus, 1938—, also dir. biol. sta., 1900; pvt. asst. to Prof. Krafft, U. Heidelberg, 1895; chief chemist Ind. State Dept. of Geology and Natural Resources, 1900-15; prof. chemistry Central Coll. Physicians and Surgeons, Indpls., 1903-04; prof. chemistry, toxicology, and forensic medicine, dir. chem. lab. Med. Coll. of Ind., Indpls., 1904-05; became chmn. dept. chemistry Ind. U. Sch. of Medicine, Indpls., 1907; prof. in charge courses organic chemistry U. Wis., summer 1907. Fellow A.A.A.S., Am. Inst. of Chemists, Ind. Acad. Sci., Am. Chem. Soc.; mem. Deutsche Chemische Gesellschaft, Sigma Xi. Author: Qualitative Analysis Inorganic Substances, 1897, 2d edit., 1900; Manual of Toxicological Analysis, 1899; also many articles on subs. in physiol., synthetic, organic and analyt. chemistry in Am. and German publs. Inventor of processes for amalgamation of platinum and of refractory gold, for recovery of used soap from laundry suds, for rapid polymerization and oxidation of drying oils, for light and weather proof coloring of oolitic limestone, for recovery of pectin from certain fruit and vegetable waste, for silver and gold mirror decoration, for reduction of nitro compounds. Died Nov. 25, 1946.

LYONS, William James, Am. physicist; b. Duluth, Minn., July 6, 1904; s. Thomas Henry and Pauline (Schultz) L.; Ph.B., U. Chgo., 1930; M.S., St. Louis U., 1933, Ph.D., 1935; m. Polly Wells Kennedy, May 22, 1957; children—Paul Joseph, Francis Thomas. Physicist, Western Cartridge Co., 1935-37; instr. math. and physics Loyola U., New Orleans, 1937-40; physicist So. Regional Research Lab., New Orleans, 1940-45; group leader Firestone Research Labs., Akron, O., 1945-54; sect. chief U. S. Army Natick (Mass.) Labs., 1954-57; group leader Textile Research Inst., Princeton, N.J., 1957-60, asso. research dir., 1960—. Fellow Am. Phys. Soc. (founding, sec. div. high polymer physics 1944—), A.A.A.S.; mem. Fiber Soc., Soc. Rheology, Textile Research Inst., Phi Beta Kappa, Sigma Xi. Asso. editor Jour. Applied Physics, 1944-54. Author: Impact Phenomena in Textiles, 1963. Patentee spring balance, tire cord. Research and publs. on plastic deformation in metals, textile cords and fibers, viscosity of cellulose solutions; dynamic and impact properties of cords and yarns; heat-treating methods for tire cords; fatigue in textile filaments under cyclic tension and biaxial rotation; interpretation of mech. properties of polymeric materials in terms of molecular theory. Home: 52 Patton Av., Princeton, N.J. 08540.*

LYONS, William Reginald, anatomist; b. Kingston, Ont., Can., Nov. 5, 1901; s. Edward and Mary (O'Brien) L.; B.A., Queens U., 1924; M.A. U. Cal. at Berkeley, 1929, Ph.D., 1932; M.D., Duke, 1950; m. Gail Alene Teel, June 12, 1925; 1 son, William Teel. Came to U. S., 1924, naturalized, 1941. Faculty anatomy U. Cal. Sch. Medicine, San Francisco, 1927—, prof., 1949-64, prof. emeritus, 1965—; pathologist Walter Reed, Halloran Gen. hosps., 1944-46; vis. prof. histology U. Indonesia Sch. Medicine, 1954-55. Mem. Am. Assn. Anatomists, Endocrine Soc., Soc. Exptl. Biology and Medicine, A.A.A.S. Author: (with Barnes Woodhall) Atlas of Peripheral Nerve Injuries, 1949; also numerous articles. Research on breast growth and milk secretion induced by hormones, maintenance pregnancy by hormones, chem. isolation pituitary hormone responsible for milk secretion, bends at high altitudes, hormonal induction and inhibition mouse breast cancer, immunology. Office: U. Cal. Sch. Medicine, San Francisco 94122.*

LYOT, Bernard Ferdinand, French astronomer; b. Paris, Feb. 27, 1897; s. Andre Constant and Alice (Ferrand) L.; D.Sc., U. Paris, 1923; m. Madeleine Garonne, Feb. 8, 1927; 2 sons. Asst. in physics Polytech. Sch., 1918-29; with Meudon Obs., from 1918, asst. astronomer, 1925, asso. astronomer, 1930, astronomer, from 1944; titular astronomer Paris Obs. Mem. French Astron. Soc., French Acad. Scis., 1939. Recipient Gold medal Royal Astron. Soc., 1939, Bruce medal Astron. Soc. of Pacific, 1946. Author: Sur la polarisation de la lumière des planètes, 1929; A Study of the Solar Corona and Prominences without Eclipses, 1939; Le filtre monocromatique polarisant et ses applications en physique solaire, 1944.

LYRA, Gerhard, German physicist; b. Riga, Latvia, June 23, 1910; s. Julius and Elsa (Moses) L.; ed. univs. of Berlin, Tübingen, Göttingen; Ph.D.; m. Lies Erbe, Dec. 6, 1936; children—Renate, Gabriele, Justus. Research staff Max Planck Inst., 1936-39; asst. math. U. Göttingen (Germany), 1939-55, instr.,

1955-—. Mem. Wissenschaftliche Gesellschaft für Luft und Raumfahrt, Gesellschaft für angewandte Math. und Mechan. Author: Theorie der stationären Leewellenströmung in freier Atmosphäre, 1943; Über eine Modifikation der Riemannschen Geometrie, 1951. Home: Am weissen Stein 5, Göttingen, W. Germany.

LYSE, Inge Martin, Norwegian constrn. engr.; b. Forsand, Oct. 22, 1898; ed. in civil engring. Tech. U. Norway; s. Peter Kristian and Martha (Likvam) L.; m. Aasta Johanne, Aug. 16, 1930; children—Peter Knute, Arne Inge, Kari Margrethe. Constrn. engr., Cal., 1923-25; research engr. Exptl. Arch Dam, Cal., 1927-31; research engr. Portland Cement Assn., Chgo., 1927-31; prof. in charge research Lehigh U., Bethlehem, Pa., 1931-38; prof. applied physics Tech. U. Norway, Trondheim, 1938-—. Recipient Louis E. Levy medal Franklin Inst. Mem. Am. Concrete Inst. (hon.), Royal Norwegian Soc. Sci., Soc. Norwegian Engrs., Norwegian Acad. Tech. Sci., Fédération internationale de la précontrainte, Réunion des laboratoires expérimentaux matériaux, Am. Soc. C.E. (J. J. R. Croes medal), Am. Soc. for Materials and Testing, Am. Soc. for English Edn., Comité Européan du Béton; Research and numerous publs. on constrn. materials, especially uses of plain and reinforced concrete. Home: Tyholtvegen 74, Trondheim, Norway.

LYSENKO, Trofim Denisovich, Russian biologist, agronomist; b. Karlovka (now Poltava Oblast), 1898; grad. Uman Sch. Horticulture, 1921, Kiev Agrl. Inst., 1925; D.Agrl. Sci. With Belaya Tserkov Selection Sta., from 1921, Gandzha (now Kirovabad) Exptl. Selection Sta., 1925-29; sr. specialist dept. physiology Ukrainian Inst. Selection and Genetics, 1929-34; sci. dir. All-Union Selection and Genetics Inst., Odessa, 1935-36, dir., 1936-38; dir. Inst. Genetics, USSR Acad. Sci., 1940-65. Decorated Order of Lenin (8); recipient Stalin prize, 1941, 43, 49. Mem. USSR, Ukrainian acads. sci., All-Union Lenin Acad. Agrl. Sci. (pres. 1938-56, 61-62), Czechoslovak Acad. Agrl. Sci. (hon.). Author: Agrobiology, 1952; Selected Works, 1953. Attacked Mendelian genetics; promoted Michurinism; denied existence of genes and distinction between genotype and phenotype; claimed all parts of organism equally involved in heredity; during later Stalin era (1948-53) his views received official support; edn, research in Mendelian genetics virtually outlawed (some Mendelians suffered secret arrest); since death of Stalin, his programs (which were unsuccessful) have been criticized, abandoned and (since 1954) discredited. Address: USSR Acad. Sci., Leninsky prospect 14, Moscow, USSR.

LYSGAARD, Leo, Danish meteorologist; b. Vildbjerg, Denmark, Aug. 21, 1899; s. Andreas and Else N. Lysgaard; M.Sc., U. Copenhagen (Denmark); m. Enny Tegner, June 17, 1930. Became meteorologist Danish Inst. for Meteorology, 1924; apptd. dir. meteorol. service, Charlottenlund, Denmark, 1942. Mem. commn. on hydrometeorology and synoptic meteorology World Orgn. on Meteorology. Author: Lufthav, Vejr og Klima, 1943; Folia Geographica Danica, 1949. Home: L. Strnadvej 18D, Hellerup. Office: Gamlehave Alle 22, Charlottenlund, Denmark.

LYSINSKI, Edmund, German psychologist; b. Kolmar, Feb. 4, 1889; s. Anton and Marta Lysinski; ed. univs. Leipzig, Berlin (both Germany); Ph.D.; m. Johanna Nielius, Mar. 30, 1922. Became asst. Mannheim High Sch., 1918; named prof. U. Heidelberg (Germany), 1935; asso. prof. Mannheim Sch. Econs., 1947-56, prof., 1956-62, emeritus, 1962-—. Cons., Zentralausschuss der Werbewirtschaft, Berufsverband deutscher Psychologen, Deutsche Werbewissenschaftl. Gesellschaft. Author: Die Kategoriensystem der Philosophie der Gegenwart, 1913; Psychologie des Betriebes, 1923; Die Organisation der Reklame, 1924; also papers. Home: Kleinschmidtstrasse 44, Heidelberg. Office: Wirtschaftshochschule, Mannheim, W. Germany.

LYSTER, Henry Francis Le Hunte, physician; b. Sander's Court, Ireland, Nov. 8, 1837; s. William N. and Ellen Emily (Cooper) L.; brought to U. S., 1838; grad. U. Mich., 1858, M.D., 1860; m. Winifred Lee Brent, Jan. 30, 1867; 5 children. Surgeon during Civil War, 1861-65; lectr. on surgery U. Mich., 1868-70, prof. theory and practice of medicine, 1888-90; a founder Mich. Coll. Medicine; prof. practice medicine Detroit Coll. Medicine, 1885-93. Died Oct. 3, 1894.

LYSTER, Theodore Charles, Am. physician, surgeon; b. Ft. Larned, Kan., July 10, 1875; s. William John and Martha Guthrie (Doughty) L.; Ph.B., U. Mich., 1897, M.D., 1899; m. Lua Withenbury, Jan. 10, 1907; children—Russell Withenbury, Theodore Charles. Pvt. and acting hosp. steward, Hosp. Corps, U. S. Army, June 1898-Feb. 1899, served with M.C., advancing from 1st lt. to col., 1900-19; asst. surgeon Manhattan Eye and Ear Hosp., N.Y., 1901-04; chief of eye, ear, nose and throat, Ancon Hosp., C.Z., 1904-09; chief of eye service Philippine Univ., Manila, 1911; chief health officer, Vera Cruz, Mexico, during Am. occupation, 1913; chief of aviation and profl. services Surgeon-General's Office, 1917-18, in France, winter 1917-18; ret., 1919; a dir. of yellow fever elimination Rockefeller Found., 1918-22; in practice at Los Angeles, mem. firm Lyster and Jones, 1920-—. Fellow A.C.S., Am. Laryngol., Rhinol. and Otol. Soc. Died Aug. 5, 1933.

LYTE, Henry, Brit. botanist; b. Lytescary Somerset, Eng., circa 1529; s. John and Edith (Horsey) L.; ed. Oxford (Eng.) U.; m. Agnes Kelloway, Sept., 1546; 5 daus.; m. 2d, Frances Tiptoft, July 1565; 2 daus.; m. 3d, Dorothy Kelloway, Gover, 1591; 2 sons, 1 dau. Managed his father's estate, 1559-76. Served as sheriff or undersheriff of Somerset during the reigns of Queen Mary and Queen Elizabeth. Author: The Light of Britayne; a Recorde of the Honorable Originall and Antiguitie of Britaine; also works never printed: Records of the True Origin of the Noble Britons; The Mysteral Oxon of Oxenford, 1592. Translator: Cruyde-boeck (Rembert Dodoens), 1554 (one of most important of 16th century herbals). Died 1607.

LYTE, Henry, Brit. mathematician; flourished circa 1619; s. Henry and Frances (Tiptoft) L.; taught arithmetic, London. Author: The Art of Tens and Decimall Arithmeticke, 1619. One of 1st to use decimal fractions.

LYTTLETON, Raymond Arthur, Brit. astronomer; b. Warley Woods, Eng.; s. William John and Agnes (Kelly) L.; ed. Clare Coll., Cambridge; M.A., Ph.D.; m. Meave Hobden. Proctor vis. fellow Princeton U., 1935-37; research fellow St. John's Coll., Cambridge, 1937-40, also fellow; faculty asst. lectr. math. Cambridge U., 1937-45, univ. lectr. math., 1945-59. Recipient Hopkins prize, 1951, Tyson medal. Fellow Royal Soc. (council 1959-61), Royal Astron. Soc. (council 1950-61, Gold medal); mem. Nat. Com. for Astronomy, Internat. Astron. Union. Author: The Comets and their Origin, 1953; The Stability of Rotating Liquid Masses, 1953; The Modern Universe, 1956; Rival Theories of Cosmology, 1960; Man's View of the Universe, 1961. Research on theroetical astrophysics, cosmogony, physics, dynamics, geophysics. Address: St. John's Coll., Cambridge, Eng.

M

MA, Tsu Sheng, chemist; b. Canton, China, Oct. 15, 1911; s. Shao-Ching and Sze (Mai) M.; B.S., Nat. Tsing Hua U., 1931; Ph.D., U. Chgo., 1938; m. Gio-fang Dju, Aug. 27, 1942; children—Chopo, Juliana. Came to U. S., 1934, naturalized, 1956. Faculty, U. Chgo., 1938-46, Nat. Peking (China) U., 1946-49; sr. lectr. microchemistry U. Otago, Dunedin, New Zealand, 1949-51; faculty N.Y. U., 1951-54; faculty City U. N.Y., 1954-—, prof. chemistry, 1958-—; vis. prof. Nat. Tsing Hua U., 1947, Linnang U., Canton, 1949, N.Y. U., 1954-60, Nat. Taiwan U., 1961. Lectr., cons. U. S. Dept. State Bur. Ednl. and Cultural Affairs, Ceylon, Burma, Thailand, Hong Kong, P.I., 1964; mem. commn. on analytical reagts. and reactions Internat. Union Pure and Applied Chemistry, 1964-—. Fulbright lectr. Internat. Ednl. Exchange Program, 1961-62. Fellow N.Y. Acad. Scis., A.A.A.S., Am. Inst. Chemists, Chem. Soc. London, Royal Micros. Soc.; mem. Am. Chem. Soc., Soc. Applied Spectroscopy, Assn. Asian Studies, Am. Microchem. Soc., Am. Assn. U. Profs., Am. Assn. Overseas Educators, Sigma Xi. Author: Small Scale Experiments in General Chemistry, 1962; Quantitative Microchemical Analysis, 1963; Organic Functional Group Analysis, 1964; Organic Analysis: Fluorine, 1965; also articles. Editor: Mikrochimica Acta, 1965-—; editorial bd. Microchem. Jour., 1957-—, Analytical Letters, 1967-—. Research in microchemistry, organic chemistry, synthetic drugs, medicinal plants, natural products; devel. teaching methods in chemistry with simple equipment; orgn. research labs. Home: 90 La Salle St., N.Y.C. 10027. Office: Dept. Chemistry, City U. N.Y., Bklyn. 11210.*

MAAK, Ernst Adolf Wilhelm, German mathematician; b. Hamburg, Germany, Aug. 13, 1912; s. Wilhelm and Erna (Salje) M.; ed. U. Hamburg (Germany), U. Copenhagen (Denmark); Ph.D.; m. Truda Lepper, Aug. 29, 1939; children—Katharina, Johanna. Became asst., Heidelberg, Germany, 1940; named prof., Munich, Germany, 1952, Göttingen, Germany, 1958. Mem. Bavarian Acad. Sci., Göttingen Acad. Sci. Author: Fastperiodische Funktionen, 1950; Differentialune Integralrechnung, 1963; Modern Calculus, 1963. Address: Ewaldstrasse 69, Gottingen, West Germany.

MABBOX-STROMBERG, Claude, French physicist; b. Paris, Mar. 6, 1921; ed. Faculté des Sciences de Grenoble, Faculté des Sciences de Paris; m. Geneviève Tariel, Apr. 20, 1954; children—Denis, Bruno, Dominique. Asst., Faculté des Sciences de Grenoble, 1945; sci. collaborator Ministry of Air, 1946; asst. École Polytechnique, 1947; research attaché Nat. Com. for Sci. Research, 1947, mem. sect. corpuscular physics, 1957-62; acting examiner École Polytechnique, 1955; prof. Faculté des Sciences de Caen, 1962-—. Contbr. articles to profl. jours. Research in cosmic ray physics, including beta decay spectrum of mu meson, work on Kappa mesons, galactic anisotropy at 500 Gev; other work on quantum theory of radiation, Bose Einstein statistics, properties of photons at high density in phase space, Hanbury-Brown effect, nuclear emulsion, Gaz Czerenkov radiation, noise electronic correlators. Home: 38 avenue du six juin, Caen. Office: Faculté des Sciences, Caen, France.*

MACADAM, David L(ewis), Am. optics researcher; b. Phila., July 1, 1910; s. Samuel David and Louisa Anna Jane (Pearce) MacA.; B.S., Lehigh U., 1932; Ph.D., Mass. Inst. Tech., 1936; m. Muriel Snow Faulkner, Oct. 29, 1938; children—David Pearce, Keith Bradford, Lewis, Muriel. With Bartol Research Lab., 1932; teaching fellow Mass. Inst. Tech., 1932-36; with Eastman Kodak Co. Research Lab., 1936-—, sr. research asso., 1960-—; spl. research color measurement and applications to color photography and television. U. S. expert on colorimetry Internat. Commn. on Illumination, 1968-—. Recipient Adolph Lomb medal for noteworthy contbns. to optics Optical Soc. Am., 1940. Mattiello Meml. lectr. Fedn. Socs. for Paint Tech., 1965; Hurter and Driffield Meml. lectr. Royal Photog. Soc., London, 1966. Fellow Optical Soc. Am. (pres. 1962, editor Jour. 1964-—); mem. Am. Inst. Physics (bd. govs.), Intersoc. Color Council (Optical Soc. Am. del.), Phi Beta Kappa, Sigma Xi. Author: (with A. C. Hardy) Handbook of Colorimetry, 1936; The Science of Color, 1953. Home: 68 Hammond St., Rochester 14615. Office: Research Labs., Eastman Kodak Co., Rochester, N.Y. 14650.*

MACALISTER, Alexander, Brit. anatomist; b. Dublin, Ireland, Apr. 9, 1844; s. Robert and Margaret Anne (Boyle) M.; ed. Trinity Coll., Dublin; LL.D., M.D., D.Sc., M.A. Became demonstrator anatomy Coll. Surgeons, 1860; named prof. zoology Univ. Dublin, 1869; apptd. prof. anatomy and surgery, Dublin, 1877; prof. anatomy Cambridge U., from 1883. Fellow Royal Soc., 1881, Soc. Apothecaries. Author: Introduction to Animal Morphology, 1876; Morphology of Vertebrate Animals, 1878; Text-Book of Human Anatomy, 1889; Memoir of James Macartney, 1900; also textbooks on zoology and physiology, numerous papers, a work on evolution in church history. Died Sept. 2, 1919.

MACALISTER, Sir Donald, Brit. anatomist, physician; b. Perth, Scotland, May 17, 1854; s. Donald and Euphemia (Kennedy) MacA.; M.A., M.D., U. Cambridge; B.Sc., London; studied Aberdeen, Liverpool, Leipzig; D.C.L., Durham, 1905; hon. LL.D., Montreal, Toronto, Aberdeen, 1906; Glasgow, 1907; St. Andrews, 1908; Liverpool, 1909; Belfast, 1913; hon. D. Phil., Athens, 1912; non D.Sc., Bristol, 1912; hon. LL.D., Dublin, 1920; Wales, 1924, Birmingham, 1925; m. Edith F. Boyle, 1895. Math. master, Harrow, 1877; lectr., nat. philos., St. Bartholomew's Hosp., 1879; Goulstonian Lectr., Royal College Physicians, 1887; 1st Croonian Prof., Royal College Physicians, 1888; Thomson Lectr., Aberdeen, 1889; examiner, Victoria U., 1896-98; Birmingham U., 1901-2; Cambridge U. and Conjoint Board, London; principal and vice-chancellor, Glasgow U., 1907-29, chancellor, 1929-—; chmn. committee that prepared Brit. Pharmacopoeia, 1898. Pres., General Med. Council, 1904-31; hon. pres., Internat. Congress of Hygiene, Madrid, 1898; Brit. delegate, Internat. Pharmacopoeia Cong., Brussels, 1902; vice-pres., Internat. Congress of School Hygiene, London, 1907; mem. med. board, U. Wales, 1908; gov. Glasgow Royal Technical College and Commercial College. Created baronet, 1924; Knight Commander of Bath, 1908; Commander Legion of Honor, 1919; Cavalier Crown of Italy, 1919. Fellow Royal Soc., Edinburgh; Fellow Royal College Physicians, London, Edinburgh, Ireland; mem., Royal Geographical Soc. (gold medal, queen's gold medal); Royal Soc. Med.; hon. mem., Pharmaceutical Soc. of Great Britain; corr. mem., Société de Pharmacie, Paris. Editor: Ziegler's Pathological Anatomy, 1833. Author: Nature of Fever, 1887; Antipyretics, 1888; Advanced Study and Research in Cambridge, 1903; The Practicioner, 1882-94; The General Medical Council: its Powers and its Work, 1906; Echoes, 1907; Romani Versions, 1928. Died Cambridge, England, Jan. 15, 1934.

MACALLUM, Archibald Byron, Canadian physician; b. Belmont, Ontario, 1858; s. Alexander Macallum; ed. U. Toronto, Johns Hopkins; M.A.; M.D.; Ph.D., 1888; m. Minnie Isabel Bruce; 3 sons. Tutorial fellow U. Toronto, 1884-87, lectr. physiology, 1887-90, prof., 1890-1918; prof. biochemistry McGill U., 1920-29; administrv. chmn. Hon. Advisory Council Sci. and Indsl. Research, Canada, 1916-20. Fellow Royal Soc., 1906, Royal Soc. Canada; pres. Canadian Inst., 1895-97, Royal Soc., Canada, 1916-17. Contbr. numerous articles to jours. Died Apr. 5, 1934.

MACARTNEY, James, Irish anatomist; b. Armagh, Ireland, Mar. 8, 1770; s. Andrew and Mary Macartney; apprenticed as surgeon, Dublin; ed. Hunterian Sch. Medicine, London, Guy's, St. Thomas's, St. Bartholomew's hosps.; M.D., St. Andrew's U., 1813; M.D. (hon.), Cambridge U., 1833; m. Miss Ekenhead, Aug. 10, 1795. Prof. anatomy and surgery Dublin U., 1813-37. Fellow Royal Soc., 1811; mem. Royal Coll. Surgeons; Royal Coll. Physicians, Ireland (hon.). Author: Lectures on Comparative Anatomy, 1802; Observations on Curvature of the Spine, 1817; A Treatise on Inflammation, 1838; also numerous papers in Philos. Trans. Described vascular system of birds; his studies on animal luminosity formed basis of later research; gave 1st satisfactory account of rumination in herbivora; discovered fibrous texture of white substance in brain, also connection between grey matter of cerebral hemispheres and subcortical nerve fibers. Died Mar. 6, 1843.

MACBRIDE, David, Brit. physician; b. Ballymoney, Ireland, Apr. 26, 1726; s. Robert and (Boyd) M.; ed. Edinburgh, London; M.D., U. Glasgow, 1764; m. Margaret Armstrong, Nov. 20, 1753; m. 2d, Dorcas

Cumming, June 5, 1762. Surgeon's mate on hosp. ship, surgeon in navy; practiced medicine; lectr. on medicine in own home, 1776-77. Recipient Silver medal, Dublin Royal Soc., 1768; Gold medal Soc. Arts London. Mem. Medico-Philos. Soc. (sec. 1762). Author: Experimental Essays, 1764; Historical Account of the New Method of Treating the Scurvey at Sea, 1768; Introduction to the Theory and Practice of Physic, 1772. Suggested use of malt infusion to treat scurvy, 1762; advocated use of lime-water in tanning process. Died Dec. 28, 1778.

MACBRIDE, Ernest William, Brit. zoologist; b. Belfast, Ireland, Dec. 12, 1866; s. Samuel MacBride; student Queen's U., Belfast; B.Sc., London, 1889; B.A., St. John's Coll., Cambridge, 1891; m. Constance Harvey Chrysler, 1902; 2 sons. Became fellow St. John's Coll., 1893; prof. zoology McGill U., Montreal, Que., Can., 1897-1909; prof. zoology Imperial Coll. Sci., South Kensington, Eng., 1913-34, emeritus, 1934-40; Chmn. council Marine Biol. Assn. U.K.; chmn. adv. com. Devel. Commn. on Fishery Research; Fellow Royal Soc. (chmn. Bermuda com.); mem. Am. Soc. zoologists, Zool. Soc., Royal Soc. Med. and Natural Scis., Brussels, Inst. for Histology and Embryology, Lisbon. Author: Text-book of Invertebrate Embryology, 1914; Introduction to the Study of Heredity, 1924; Evolution, 1927; Embryology, 1929; Huxley, 1934; also articles in Ency. Brit., 12th edit. Co-author: A Text-Book of Zoology, 1901. Leading exponent of neo-Lamarckianism in evolutionary theory; known for his work in embryology, esp. that of invertebrates. Died Nov. 19, 1940.

MACCALLUM, William George, pathologist; b. Dunnville, Ont., Can., Apr. 18, 1874; s. George Alexander and Florence O. (Eakins) MacC.; B.A., U. Toronto, 1894; M.D., Johns Hopkins, 1897. Asso. prof. pathology Johns Hopkins, 1900-08, prof. pathol. physiology, 1908-09, prof. pathology and bacteriology, 1917-44; prof. pathology, Columbia, 1909-17. Fellow A.A.A.S., Royal Soc. Medicine (London) (hon.); mem. Assn. Am. Physicians, Nat. Acad. Scis., Soc. Medicorum Sverana (Stockholm) (hon.). Author: Textbook of Pathology, 1916, 1940; Inflammation, 1922; also articles. First to show sexual cycle of malarial parasite in blood; described (with Voegtlin) function of parathyroid glands in regulating level of calcium in bloodstream, 1909; discovered MacCallum's stain for microscopic identification of influenza organism; proved walls of lymphatics possess complete endothelial lining, without openings, comparable to lining of blood capillaries; studied pneumonias, lesions of rheumatic fever. Died Feb. 3, 1944.

MACCANON, Donald M(oore), physiologist; b. Norwood, Ia., June 17, 1924; s. G. E. and Edna (Curtis) MacC.; A.B., Drake U., 1948; M.S., State U. Ia., 1951, Ph.D., 1953; m. Nellie F. Hammond, Feb. 3, 1946; children—Julie A., Donald E. Faculty, State U. S.D., 1954-60; faculty Chgo. Med. Sch., 1960-—, asso. prof. div. cardiovascular research, asst. prof. physiology, 1961-—, asso. div. dir., 1962-—, asso. prof. physiology, 1964-—, prof., 1967-—; cons. physiology fellowship review panel NIH, 1965-—. Recipient Morris L. Parker award for meritorious research, 1964. Fellow Am. Coll. Cardiology, A.A.A.S.; mem. Am. Physiol. Soc., Am. Heart Assn., Sigma Xi. Research and publs. on relationship between breathing movements and blood pressure; adjustment of heart in the failing state; phonocardiography; effect of drugs on cardiac efficiency; inventor timer of aortic valve closure, thin flexible wall tension guage, plastic clamp for precisely controlled constriction of great vessels.*

MACCARTY, Collin Stewart, neurosurgeon; b. Rochester, Minn., Sept. 20, 1915; s. William C. and Helen M. (Collin) MacC.; A.B., Dartmouth, 1937; M.D., Johns Hopkins, 1940; M.S. in Neurosurgery, Mayo Found., U. Minn., 1944; m. Margery Deal, June 15, 1940; children—Collin Stewart, Robert Lee, Helen Carpenter. Surg. house officer Johns Hopkins Hosp., Balt., 1940-41; fellow in neurosurgery Mayo Found., 1941-44; mem. neurosurg. staff Mayo Clinic, Rochester, 1946-—, chmn. neurosurg. dept., 1963-—; instr. neurol. surgery U. Minn. Grad. Faculty, Rochester, 1947-53, asst. prof., 1953-57, asso. prof., 1957-61, prof., 1961-—. Served with M.C., USNR, 1944-46. Diplomate Am. Bd. Neurol. Surgery. Mem. Am. Coll. A.C.S., Neurosurg. Soc. Am., Harvey Cushing Soc., A.M.A., N.Y. Acad. Scis., Neurosurg. Travel Club, Dragon, Sigma Xi, Nu Sigma Nu, Alpha Delta Phi. Author: (monograph) The Surgical Treatment of Intracranial Meningiomas, 1961; (with J. L. Slooff, J. W. Kernohan) Primary Intramedullary Tumors of the Spinal Cord and Filum Terminale, 1964. Research and publs. on devel. surg. procedure for spinal cordectomy, sacral tumor treatment, use of profound hypothermia in neurosurgery. Home: 725 10th Av. S.W., Rochester 55902. Office: Mayo Clinic, Rochester, Minn. 55902.*

MACCARTY, William Carpenter, Am. pathologist; b. Louisville, Ky., June 10, 1880; B.S., U. Ky., 1900, M.S., 1909; M.D., Johns Hopkins, 1904, D.Sc., 1937; a son, Collin Stewart. Asst. pathologist Konigen Augusta Hosp., Berlin, Germany, 1904-06; prof. pathology Mayo Found., Minn., 1907-48, emeritus, 1948-—; surg. pathologist, dir. labs., clin. cons. Mayo Clinic, 1907-45, emeritus cons. physician, 1945-—, also supr. autopsy service. Mem. Soc. Clin. Pathologists, Am. Cancer Soc., Osler Med. Hist. Soc., Pan-Am. Med. Soc., Am. Assn. Pathologists and Bacteriologists, A.M.A., Gastroent. Assn., Assn. for Cancer Research, A.C.P., Internat. Coll. Surgery, Research in tumor pathology, color photography, neoplasms; described a hyperplasia of lining fold of gall bladder, in which cholesterin-fat granules are embedded; developed internat. terminology and classification for neoplasms and neoplasmata.

MACCHI, I. Alden, endocrinologist; b. Bologna, Italy, Feb. 21, 1922; s. Elio and Ida (Gamberini) M.; B.A., Clark U., 1947; M.A., 1950; Ph.D., Boston U., 1954; m. Joan Mary Shiminski, July 6, 1953; 1 dau., Deborah Joan. Research staff Worcester Found. Exptl. Biology, Shrewsbury, Mass., 1950-54; asst. prof. physiology Clark U., Worcester, Mass., 1954-56; asst. prof. biology Boston U., 1956-58, asso. prof., 1958-64, prof., 1964-—; sr. research fellow, vis. lectr. U. Sheffield (Eng.), 1962-63; exec. asst. for research Biol. Sci. Center, Boston U., 1963-—. Lalor Found. fellow, 1955; Dept. Sci. and Indsl. Research, Sr. Research fellow, 1962-63. Mem. Am. Inst. Biol. Scis., Am. Diabetes Assn., Am. Physiol. Soc., Am. Soc. Zoologists, Endocrine Soc., N.Y. Acad. Scis., Royal Soc. Medicine, Soc. Exptl. Biology and Medicine, Sigma Xi. Contbr. articles to tech. jours. Research on biochem. pathways by which adrenal cortex of mammals produces steroid hormones; transplantation of adrenal glands, assessment of their ability to function normally, evolution of regulation of adrenocortical hormone biosynthesis, transplantation of endocrine pancreas, evaluation of its capacity to produce insulin. Home: 52 Roundwood Rd., Newton, Mass. 02164. Office: 2 Cummington St., Boston 02215.*

MACCLINTOCK, Paul, Am. geologist; b. Aurora, N.Y., Feb. 2, 1891; s. William and Lucia (Lander) MacC.; B.S., U. Chgo., 1912, Ph.D., 1920; m. Sept. 4, 1925 (dec.); children—Copeland, Lucia (Mrs. R. S. Barbour). Faculty, Princeton (N.J.), 1921-58, prof. geology emeritus, 1958-—; geologist Ill. Geol. Survey, 1921-28, Vt. Geol. Survey, 1960-65. Mem. Geol. Soc. Am., A.A.A.S. Author; textbooks, also articles. Research on glacial history N.Am. Home: 37 Lake Lane, Princeton, N.J. 08540.*

MACCONKEY, Alfred Theodore, English bacteriologist; b. Liverpool, Eng., 1861; ed. Cambridge, Guy's Hosp., London; bacteriologist, Liverpool, also Leeds, Eng.; became mem. staff Lister Inst., 1901, later head antitoxid dept.; ret., 1926. Contbr. numerous articles on bacteriology to jours. Demonstrated value of VogesProskauer reaction in indentifying bacterial strains separated from feces. Died 1931.

MACCORMAC, Sir William, Brit. surgeon; b. Belfast, Ireland, Jan. 17, 1836; s. Henry and Mary (Newsham) MacC.; ed. Belfast, Dublin, Paris; B.A., Queen's U., 1855, M.A., 1858, M.D.; m. Katharine Maria Charters, 1861. Became surgeon-in-chief Anglo-Am. Ambulance, 1870; cons. (civil) surgeon S. African forces, 1899-1900; cons. surgeon, lectr. on clin. surgery St. Thomas's Hosp.; practiced medicine, specializing in surgery, Belfast, 1864-70, London, 1870-1901. Mem. Royal Coll. Surgeons, Paris Acad. Medicine. Author: Notes and Recollections of an Ambulance Surgeon, 1871; On Abdominal Section for the Treatment of Intraperitoneal Injury, 1887; Surgical Operations, 1885-89. Helped break down insularity of Brit. surgery by introducing knowledge and practices from abroad. Died 1901.

MACCORQUODALE, Kenneth, Am. psychologist; b. Olivia, Minn., June 26, 1919; s. Alexander Ross and Hallie (Chambers) MacC.; B.A., U. Minn., 1941, M.A., 1942, Ph.D., 1946. Faculty U. Minn., Mpls. 1946-—, prof., 1955-—, chmn. dept. psychology, 1960-64. Cons. to Surgeon Gen. U. S. Navy, 1959-—, Surgeon Gen. Nat. Inst. Mental Health 1960-—; mem. Minn. Bd. Examiners, 1962-—; mem. Dartmouth Conf. on Learning Theory Social Sci. Research Council, 1951. Ford Found. Fund for Advancement Edn. fellow, 1954-55. Mem. Am., Midwestern psychol. assns., A.A.A.S., Psychonomics Soc., N.Y. Acad. Scis., Sigma Xi. Author: (with others) Modern Learning Theory, 1954. Editor: Century Psychology Series, 1952-—. Contbr. articles to profl. jours. Home: Campus Club, U. Minn., Mpls. 55455.*

MACCULLAGH, James, Irish mathematician, physicist; b. Glenellie, Ireland, 1809; ed. Trinity Coll., Dublin, named scholar, 1827, fellow, 1832, prof. math., 1836, prof. natural history, 1843. Sec., council Royal Irish Acad., 1840-42, sec., 1842-46. Fellow Royal Soc., 1843. Produced geometrical work of lasting importance, phys. work of hist. interest; advanced (with Neumann MacCullagh) hypothesis that light oscillations of planes of polarization are parallel and conform to same density or thickness in different transparent media (Fresnel's theory that oscillations are perpendicular and conform to like elasticity is correct); contbd. to theory of quadric surfaces. Died 1847.

MACCULLOCH, John, Scottish geologist, physician; b. Guernsey Island, Oct. 6, 1773; s. James and Elizabeth (Lisle) M.; M.D., U. Edinburgh (Scotland), 1793; m. Miss White, 1835; served as army physician; prof. geology Woolwich (Eng.) Mil. Acad., from 1814. Fellow Royal Soc., 1840; pres. Geol. Soc., 1816-17. Author: A Description of the Western Isles of Scotland, including the Isle of Man, 1819; A Geological Classification of Rocks, 1821; Highlands and Western Isles of Scotland (letters to Walter Scott), 1824; A System of Geology, 1831. Commd. by govt. to explore geology of Scotland, 1811-21; commd. to prepare geol. map of Scotland, 1826; made extensive contbns. to structural geology and petrology. Died Poltair, Aug. 20, 1835.

MACCURDY, George Grant, anthropologist; b. Warrensburg, Mo., Apr. 17, 1863; s. William J. and Margaret (Smith) M.; grad. State Normal Sch., Warrensburg, Mo., 1887; A.B., Harvard, 1893, A.M., 1894; univs. of Vienna, Paris (Sch. of Anthropology) and Berlin, 1894-98; Ph.D., Yale, 1905; m. Glenn Bartlett, June 30, 1919. Instr. anthropology, 1898-1900, lecturer and curator anthrop. collections, 1902-10, asst. prof. prehistoric archaeology and curator anthropol. collections, 1910-23, research asso. with rank of prof. and curator anthropol. collections, 1923-31, now emeritus, all of Yale U. Dir. Am. Sch. of Prehistoric Research in Europe, 1921-22, and since 1924; hon. collaborator Smithsonian Instn., 1927-—; trustee Lab. of Anthropology, Santa Fe, N.M.; mem. bd. mgrs. Sch. of Am. Research, Santa Fe, N.M. Fellow Galton Soc. (New York), A.A.A.S. (ex-v.p.); mem. Am. Philos. Soc., Anthrop. Soc. Paris and Brussels, Archaeologist Institute America (vice pres., 1947), Am. Ethnological Society America Anthropological Assn. (sec. 1903-16; pres. 1931), Nat. Research Council since 1925, Sigma Xi; corr. mem. Inst. of Coimbra (Portugal), Sch. Anthropology (Paris), Anthrop. Soc. Washington, Mo. Hist. Soc., Soc. des Américanistes de Paris, Numismatic and Antiquarian Soc. Phila., Anthrop. Soc. Rome, British Speleological Assn. (hon.), British Prehistoric Soc. (hon.). Life mem. Navy League, U. S. Clubs; Graduate (New Haven, Conn.); Harvard of Connecticut (president 1919-20). Author: The Eolithic Problem, 1905; Some Phases of Prehistoric Archaeology, 1907; Antiquity of Man in Europe, 1910; A Study of Chiriquian Antiquities, 1911; Human Skeletal Remains from the Highlands of Peru, 1923; Human Origins—a Manual of Prehistory (in 2 vols.), 1924; Prehistoric Man, 1928; The Coming of Man, 1932; also papers on anthrop. subjects. Editor: Early Man, 1937; and Bull. of Am. School of Prehistoric Research. Research on cranial characteristics of New Britain and Bismark native islanders. Died Nov. 15, 1947.

MACDIARMID, Alan Graham, chemist; b. Masterton, New Zealand, Apr. 14, 1927; s. Archibald Campbell and Ruby (Graham) MacD.; B.S., U. New Zealand, 1948, M.S., 1950; M.S., U. Wis., 1952, Ph.D., 1953; Ph.D., U. Cambridge, 1955; m. Marian Laurene Mathieu, July 10, 1954; children—Heather Sheila, Dawn Frances, Duncan Campbell, Gail Ruby. Asst. lectr. chemistry Queens Coll. U. St. Andrews, Scotland, 1955; faculty U. Pa., Phila., 1955-—, prof. chemistry, 1964-—. Alfred P. Sloan fellow, 1959-63. Mem. Am. Chem. Soc., Chem. Soc. (London), Faraday Soc., Sigma Xi, Phi Lambda Upsilon, Alpha Chi Sigma. Mem. editorial bd. Inorganic Chemistry, 1964-—, Inorganic Syntheses, 1968-—. Contbr. numerous articles to profl. jours. Research and publs. primarily in field of synthesis and structural studies silicon hydrides, silicon fluorides, sulfur fluorides, organosilicon compounds. Home: 635 Drexel Av., Drexel Hill, Pa. 19026. Office: Dept. Chemistry, U. Pa., Phila. 19104.*

MAC DONALD, Arthur, Am. anthropologist; b. Caledonia, N.Y., July 4, 1856; s. Angus and Virginia (Dibble) M.; A.B., U. of Rochester, 1879, A.M., 1883; studied law, 1879; Princeton Theol. Sem., 1880; grad. Union Theol. Sem., 1883; post-grad. courses in philosophy, metaphysics, etc., Harvard, 1883-85; apptd. fellow in psychology, Johns Hopkins, but declined appmt. in order to pursue studies, 1885-89, at Berlin, Leipzig, Paris, Zürich and Vienna, in medicine (full course), psycho-physics, and spl. courses in insanity, hypnotism and criminology; m. Margaret J. Porterfield, Sept. 29, 1904. Docent in criminology, Clark U., 1889-91; with U. S. Bur. of Edn., 1892-1904, as specialist in edn. as related to the abnormal and weakling classes. U. S. del. to 3 Internat. Psychol. and Criminol. congresses; hon. pres. 3rd Internat. Congress of Criminal Anthropology of Europe; Author: Abnormal Man, 1893; Criminology, 1894; Education and Patho-Social Studies, 1896; Le Criminel-Type, 1895; Emile Zola, 1899; Experimental Study of Children, 1899; Hearing on the Bill to Establish a Laboratory for the Study of the Criminal Pauper and Defective Classes, 1902; Plan for the Study of Man, etc., 1902; Statistics of Crime, Suicide and Insanity, 1903; Man and Abnormal Man, 1905; Juvenile Crime and Reformation, 1908; Senate Documents; Mentality of Nations and Social Pathology, 1912, and contbns. to Am. and European publications on criminology, human abnormalities, hypnotism, etc.; also articles on criminology in Ency. Americana and Nelson's Ency., and spl. articles: A Study of the United States Senate, Scientific Political Training of President Coolidge, Old Age in Man, Death in Man, Study of Man After Death, Education and Psychoanalysis, Education and Eugenics, Legislative Anthropology, etc. Studies on Am. and European prisons, insane and inebriate asylums, slums; established standards of comparison for normal and abnormal children. Died Washington, Jan. 17, 1936.

MACDONALD, Donald Francis, geologist; b. Pictou County, N.S., Can., 1875; s. John and Mary

(MacLean) MacD.; B.S., U. Washington, 1905; M.S., George Washington U., 1906; m. Lucy Elizabeth Hagan, Nov. 7, 1918. Mem. field parties Geol. Survey, 1903, other years, asst. geologist, 1910, geologist, 1914; fellow, instr. U. Chgo., 1907-08; instr. geology and chemistry Tulane U., 1909; geologist Panama Canal Commn., 1911-14, cons., 1940-42; geologist, mining engr. Bur. Mines, 1915-17; geologist Sinclair Oil Co., 1917-24, Allied Chem. & Dye Co., 1924; prof. geology St. Francis Xavier's U., N.S., 1932-39. Research on stratigraphy and fossils of Panama; made geologic analysis which led to correction of slides at Panama Canal. Died 1942.

MACDONALD, Eleanor J., Am. epidemiologist; b. West Somerville, Mass., Mar. 4, 1909; d. Angus Alexander and Catharine (Boland) Macdonald; A.B., Radcliffe Coll., 1928; postgrad. Harvard Sch. Pub. Health, 1928-29. Mem. staff Mass. Dept. Pub. Health, 1930-40; research statistician, div. cancer research Conn. Dept. Health, 1941-48; prof. epidemiology Grad. Sch. Biomed. Scis., U. Tex., Houston, 1948—, epidemiologist, head dept. M. D. Anderson Hosp. and Tumor Inst., 1948—; asst. clin. prof. Yale U. Sch. Medicine, 1948-60. Mem. A.A.A.S., Am. Radium Soc., N.Y. Acad. Scis., Pub. Health Cancer Assn. Am., Am. Assn. Cancer Research, Am. Statis. Assn., Biometric Soc., Phi Beta Kappa, others. Research and publs. on treatment of cancer and allied diseases, tumors of the skin and the gastrointestinal tract; evolved improved methods for developing cancer programs and for measuring results of treatment for cancer by continuing evaluation of existing ones and testing results of applied findings for improvement. Home: 2107 University Blvd., Houston 77025. Office: 6723 Bertner St., Houston 77025.*

MACDONALD, Gordon Andrew, Am. volcanologist; b. Boston, Oct. 15, 1911; s. John Austin and Grace Blanche (Griffin) M.; A.B., U. Cal. at Los Angeles, 1933, M.A., 1934; Ph.D., U. Cal. at Berkeley, 1938; m. Ruth Carol Binkley, May 22, 1938; children—John Alan, James Gordon, Duncan Edwin, William Andrew. Asst. geologist Shell Oil Co., 1938-39; with with U. S. Geol. Survey, Honolulu, 1939-47, dist. geologist, 1946-47, Hawaiian Volcano Obs. geologist, 1948-56; asst. prof. U. So. Cal., 1947-48; sr. prof. geology U. Hawaii, Honolulu, 1958—; dir. Hawaiian Volcano Obs., 1951-56. Recipient Citation of Honor, A.R.C., 1956. Fellow Geol. Soc. Am., Mineral. Soc. Am., A.A.A.S.; mem. Am. Geophys. Union, Internat. Volcanological Assn. (pres. 1967—). Research and numerous publs. on behavior, structure and product of active volcanoes, composition and origin of volcanic rocks and magmas, geol. history of volcanic regions in Hawaii, Cal. Home: 326 Lanipo Dr., Kailua, Hawaii 96734.*

MACDONALD, Gordon J(ames) F(raser), geophysicist; b. Mexico City, D.F., Mexico, July 30, 1929; s. Gordon and Josephine (Bennett) MacD.; A.B. summa cum laude, Harvard, 1950, A.M., 1952, Ph.D., 1954; m. Marcelline Kuglen, August 5, 1960; children—Gordon James, Maureen Ann, Michael Andrew. Asst. prof. geology and geophysics Mass. Inst. Tech., 1954-55, asso. prof. geophysics, 1955-58; staff asso. geophysics lab. Carnegie Inst. of Washington, 1955—; cons. U. S. Geol. Survey, 1955-60; prof. geophysics U. Cal. at Los Angeles, 1958—, dir. atmospheric research lab., 1960—, asso. dir. Inst. Geophysics and Planetary Physics, 1960—, chmn. dept. planetary and space sci., 1965—; mem. Pres.'s Sci. Adv. Com., 1965—; cons. NASA, 1960—. Fellow Am. Acad. Arts and Scis., A.A.A.S.; mem. Am. Math. Soc., Nat. Acad. Scis., Royal Astron. Soc., Am. Mineral. Soc., Am. Geophys. Union Geochem. Soc. Am., Geol. Soc. Am., Am. Astron. Soc., Am. Meteorol. Soc., Am. Philos. Soc., Seismol. Soc. Am., Soc. Indsl. and Applied Math., Woods Hole Oceanographic Instn. Corp. Author: The Rotation of the Earth, 1960. Editor Revs. of Geophysics, 1962—, Jour. Atmospheric Sci., 1964—. Research in dynamics and evolution of solar system, physics of earth's interior, structure of inner planets and moon, investigation of propagation of energy in upper atmosphere, analysis of time series. Contbr. articles sci., tech. jours.*

MACDONALD, Hector Munro, Brit. math. physicist; b. Edinburgh, Scotland, Jan. 19, 1865; s. Donald and Annie (Munro) MacD.; B.Sc., U. Aberdeen (Scotland), 1886; postgrad. (Fullerton scholar) Clare Coll., Cambridge, Eng., 1886-89; LL.D., Glasgow U., 1934. Fellow Clare Coll., 1890-1908, named hon. fellow, 1914, also sr. bursar; prof. math., Aberdeen, from 1904, rep. on univ. ct., 1907. Recipient Royal medal, 1916, Adams prize, 1901. Fellow Royal Soc., 1901; pres. London Math. Soc., 1916-18. Author: Electric Waves, 1902; Electro-magnetism, 1934. Research and publs. on relations between convergent series and asymptotic expansions, zeros and addition theorem of Bessel functions, some Bessel integrals, spherical harmonics, Fourier series; gave solution to problem of diffraction at edge of totally reflecting prism, 1901; 1st to formulate explanation of mechanism of transmission of wireless signals as problem of diffraction. Died Aberdeen, May 16, 1935.

MACDONALD, Ian, Brit. clin. nutritionist; b. London, Eng., Dec. 22, 1921; s. Ronald and Elizabeth (Stutz) M.; M.B., B.S., Guy's Hosp. Med. Sch., U. London, 1944, Ph.D., 1953, M.D., 1959, D.Sc., 1966; m. Nora Cussen, Feb. 2, 1946; children—Graham, Peter, Helen. Faculty dept. physiology Guy's Hosp. Med. Sch., U. London, 1948—, reader physiology, 1963—; lectr. U. Witwatersrand, Johannesburg, S. Africa, 1958. Mem. milk fat panel Ministry Health, 1965—. Leverhulme Research scholar, 1952-53. Mem. Nutrition Soc. (programmes sec. 1962-65), Physiol. Soc., Med. Research Soc., Nutrition Soc., Am. Soc. Clin. Nutrition. Research, publs. on influence of various dietary carbohydrates on lipid metabolism and serum proteins. Home: Hillside, Fountain Dr., London, S.E. 19, Eng.*

MACDONALD, Ian (Gibbs), surgeon; b. Calgary, Alberta, Can., Apr. 9, 1903; s. Alexander D. and Gertrude (Gibbs) M.; student Dalhousie U., 1926; M.D., McGill U., 1928, M.C., 1928; m. Eleanor G. Clark, Mar. 16, 1963; children—Alexander C., Sharon (Mrs. Mathew Stratico), Bruce W., Katharine M. Intern, house surgeon Montreal Gen. Hosp., Can., 1917, 28-30; resident pathologist U. Mich. Hosp., 1930-31; chief resident surgeon, Toronto Gen. Hosp., Can., 1931-32, instr. surgery U. Toronto, 1931-32; instr. surgery, pathology U. So. Cal., 1932-34; dir. Ogden Meml. Tumor Clinic, Cornwall (N.Y.) Hosp., 1935-38; asst. clin. research A.C.S., Chgo., 1938-40; private practice specializing in cancer therapy, Los Angeles, 1941—; clin. prof. surgery U. So. Cal.; sr. attending surgeon Los Angeles County Hosp., 1941—; surgery staff St. Vincent's Hosp., Hollywood-Presbyterian Hosp., Los Angeles; cons. radiotherapy Children's Hosp., Los Angeles. Recipient medal Am. Cancer Soc., 1951. Fellow Am. Coll. Radiology; mem. A.C.S. (cancer com.), Am. Assn. Cancer Research, Am. Radium Soc., Soc. Head and Neck Surgeons (founding mem.), Pacific Coast Surg. Soc., A.A.A.S., Cal. Roentgen Soc., Los Angeles Surg. Soc., Los Angeles County Med. Assn. (sec.-treas. 1960-61, pres. 1962). Developed original theory of biol. predeterminism of cancer, 1952; research and publs. on breast cancer, mgmt. of major wounds. Office: 1930 Wilshire Blvd., Los Angeles 90057.*

MACDONALD, James Ross, Am. physicist; b. Savannah, Ga., Feb. 27, 1923; s. John Elwood and Antonina (Hansell) M.; B.A. in Physics (Tyng fellow), Williams Coll., 1944; S.B. in Elec. Engring., Mass. Inst. Tech., 1944, S.M. in Elec. Engring., 1947; Ph.D. in Physics (Rhoades scholar), Oxford (Eng.) U., 1950; m. Margaret Milward Taylor, Aug. 3, 1946; children—Antonina Hansell, James Ross IV, William Taylor. Div. Indsl. Cooperation fellow Mass. Inst. Tech., 1946-47; physicist Armour Research Found., Chgo., 1950-52; asso. physicist Argonne Nat. Lab., AEC, Lemont, Ill., 1952-53; physicist Central Research labs. Tex. Instruments, Inc., Dallas, 1953-56, dir., 1963—. Clin. asso. prof. med. electronics Southwestern Med. Sch., U. Tex., 1954—. Recipient Sr. Paper award Profl. Group on Audio, I.R.E., 1957, Achievement award, 1962. Fellow Am. Phys. Soc., I.E.E.E., A.A.A.S.; mem. Phi Beta Kappa, Sigma Xi. Contbr. articles on solid state physics, theoretical physics, phys. chemistry, electronics to tech. publs. Home: 6415 Meadow Rd., Dallas 75230. Office: Central Research Labs., Tex. Instruments, Inc., P.O. Box 5936, Dallas 75222.*

MACDONALD, John Barfoot, Canadian microbiologist; b. Toronto, Ont., Can., Feb. 23, 1918; s. Arthur A. and Gladys L. (Barfoot) M.; D.D.S. with honors, U. Toronto, 1942; M.S. in Bacteriology, U. Ill., 1948; Ph.D. in Bacteriology (Kellogg fellow, Canadian Dental Assn. research scholar), Columbia, 1953; A.M. (hon.), Harvard, 1956; LL.D., U. Man. (Can.), 1962, Simon Fraser U., 1965; D.Sc., U. B.C., 1967; m. Beatrice Kathleen Darroch, June 5, 1942; children—Kaaren C., John G., Scott Arthur. Pvt. practice dentistry, Toronto, 1942-44, 46-47; lectr. preventive dentistry U. Toronto, 1942-44, instr. bacteriology, 1946-47, asst. prof., 1949-53, asso. prof., 1953-56, chmn. div. dental research, 1953-56, prof. bacteriology, 1956; research asst. U. Ill., 1947-48; cons. dental edn. U. B.C., Vancouver, Can., 1955-56, pres. univ., 1962-67; dir. Forsyth Dental Infirmary, Boston, 1956-62, cons. bacteriology, 1962—; prof. microbiology Harvard Sch. Dental Medicine, Boston, 1956-62, dir. postdoctoral studies, 1960-62; prof. higher edn. U. Toronto, 1968—; cons. several orgns., including Canadian Pharm. Assn. Mem. com. on dental research Canadian NRC, 1950-60, chmn., 1954-57; mem. adv. bd. Mem. Dental Hygienists Assn., 1958-62; cons. dental medicine sect. Corporate Research div. Colgate-Palmolive Co., 1958-62; mem. med. adv. bd. Iran Found., 1959-68; mem. standing com. on survey dentistry Am. Assn. Dental Schs., 1959-60, spl. com. on survey dentistry, 1960-61; mem. dental study sect. NIH, 1961-66; mem. sci. commn. on dental research Federation Dentaire Internationale, 1957-58. Hon. pres. Vancouver Inst., 1962-67; hon. dir. Muscular Dystrophy Assn. Can., 1963—; hon. v.p. Canadian Red Cross Soc., 1963—. Fellow Am. Coll. Dentists, Am. Acad. Dental Sci. (hon.); mem. Canadian Dental Assn. (chmn. research com. 1951-54), Canadian Soc. Microbiologists, Internat. Assn. Dental Research, N.Y. Acad. Scis., A.A.A.S., Am. Soc. Microbiologists, Harvard Odontological Soc. (hon.), New Eng. Dental Soc. (hon.). Author: Dental Education in British Columbia, 1961. Asso. editor Jour. Dental Research, 1958-61; regional editor Archives Oral Biology, 1958-62, hon. editorial adv. bd., 1962—; editor Internat. Series on Oral Biology, 1958-63; Higher Education in British Columbia, 1962. Contbr. numerous articles to profl. jours. Research on a disease process of mucous membrane infections; other studies on anaerobic bacterial growth requirements and metabolic abilities. Address: 4 Devonshire Pl., Toronto, Ont., Can.*

MACDONALD, Sir John Denis, Brit. physician; b. Cork, Ireland, Oct. 26, 1826; s. James and Catherine (McCarthy) M.; ed. Cork Sch. Medicine, Kind's Coll. Med. Sch.; M.D.; m. Sarah Phoebe Walker, 1863; m. 2d, Erina Archer; Joined Navy as asst. surgeon, 1849; placed in charge Plymouth Hosp. Mus.; joined H.M.S. Herald, 1852; prof. naval hygiene Army Med. Sch., Netley, for nine years; became insp. gen. to Royal Naval Hosp., Plymouth, Eng., 1883; ret., 1886. Recipient Macdougal-Brisbane medal Royal Soc. Edinburgh, 1862; Sir. Gilbert Blane medal, 1871. Fellow Royal Soc., 1859. Author: Analogy of Sound and Colour, 1869; Outlines of Naval Hygiene, 1881; A Guide to the Microscopical Examination of Drinking Water, 1883; also numerous articles. Micros. research on products of sounding lead, dredge, towing net. Died Feb. 7, 1908.

MACDONALD, Roderick Patterson, Am. clin. chemist; b. Detroit, Nov. 9, 1924; s. Roderick Peter and Maude (McLellan) MacD.; B.S., Mich. State U., 1947; M.S., U. Detroit, 1949; Ph.D., Wayne State U., 1952. Grad. asst. U. Detroit, 1947-49; spl. instr. Wayne State U. Mortuary Sch., 1949-51; with Harper Hosp., Detroit, 1952—, dir. clin. chemistry, 1957—; instr. dept. pathology Wayne State U. Sch. Medicine, 1964—. Recipient Angus McLean Memorial award, 1952. Fellow Am. Assn. Clin. Chemists (nat. sec.), Am. Bd. Clin. Chemistry, Canadian Bd. Clin. Chemistry; mem. Canadian Soc. Clin. Chemistry, Am. Chem. Soc., N.Y. Acad. Sci., Sigma Xi. Author (with others) Ultramicro Methods for Clinical Laboratories, 1957, 2d ed., 1962. Pioneer studies, publs. on use, methods of ultramicro analysis; research in clin. enzymology. Home: 31356 Churchill Dr., Birmingham, Mich. 48009. Office: Harper Hosp., 3825 Brush St., Detroit 48201.*

MACDONALD, William McCullough, Am. physicist; b. Salem, O., Nov. 25, 1927; s. Robert Boles and Dorothy (Berkey) MacD.; B.S., U. Pitts., 1950; Ph.D., Princeton, 1954; m. Barbara Blakeley, July 12, 1951 (div. Jan. 1965); children—Pamela Lynne, Jeffrey Drew, Melinda Lou, Todd Dellas; m. 2d, Rosemary Anne Coldwell-Horsfall, Feb. 6, 1965; 1 son, Colin Robert Coldwell. Theoretical physicist Lawrence Radiation Lab., U. Cal. at Berkeley, 1954-55; vis. lectr. U. Wis., 1955-56; faculty U. Md., College Park, 1956—, prof. physics, 1963—. Cons. Lockheed Missiles & Space Lab., Palo Alto, Cal., 1957, 59—. Fellow Am. Phys. Soc.; mem. Am. Geophys. Union, Soc. Engring Sci., Am. Assn. U. Profs., Sigma Xi. Author: Nuclear Spectroscopy, 1960. Study of effect of coulomb forces on nuclear wave functions, (with Rosenbluth and Judd) equation for effect of collisions in an ionized gas, (with Walt) equation for effect of atmospheric scattering on geomagnetically trapped charged particles, theory unifying nuclear shell model and nuclear reaction theory. Home: 1219 Noyes Dr., Silver Spring, Md. 20910. Office: Dept. Physics, U. Md., College Park, Md. 20742.*

MACDOUGAL, Daniel Trembly, Am. botanist; b. Liberty, Ind., Mar. 16, 1865; s. Alexander and Amanda MacD.; B.S., DePauw U., 1890, A.M., 1894, LL.D., 1912; M.S., Purdue U., 1891, Ph.D., 1897; student Tübingen and Leipzig, 1895-96; LL.D., U. Ariz., 1915; m. Louise Fisher, Jan. 24, 1893. Agt., U. S. Dept. Agr. on explorations in Ariz. and Ida., 1891-92; instr. plant physiology U. Minn., 1893-95, asst. prof., 1895-99; dir. labs. and asst. dir. N.Y. Bot. Garden, 1899-1905; dir. dept. bot. research and Lab. for Plant Physiology, Carnegie Instn., Washington, 1905-33. Fellow Cal. Acad. Sci.; life mem. Bot. Soc. Am., Am. Soc. Plant Physiology, A.A.A.S. (gen. sec. 1920-25), Torrey Bot. Club; mem. Am. Philos. Soc., Am. Soc. Naturalists (pres. 1910); fgn. mem. Hollandsche Maatschappe d. Wetenschappen, Societé Nationale D'Acclimatation de France; hon. mem. Bot. Soc. of Edinburgh; corr. mem. Czechoslovak Bot. Soc. Author: Influence of Light and Darkness Upon Growth and Development, 1903; Botanical Features of North American Deserts, 1908; The Water-Balance of Succulent Plants, 1910; Conditions of Parasitism in Plants, 1910; Alterations in Heredity Induced by Ovarial Treatment, 1911; The Salton Sea, 1913; Hydration and Growth, 1919; Growth in Trees, 1921, 1924; Hydrostatic System of Trees, 1926; The Green Leaf, 1930; Pneumatic System of Plants, Especially Trees, 1933; Studies in Tree Growth by the Dendrographic Method, 1935; Tree Growth, 1938. Editor: Species and Varieties (De Vries). Research in plant physiology, heredity, cambium layer and growth phenomena, plant evolution. Died Feb. 22, 1958.

MACDOWELL, Samuel Wallace, physicist; b. Pernambuco, Brazil, Mar. 24, 1929; s. Samuel Wallace and Maria Anita (Amazonas) MacD.; B.Sc., U. Recife, 1951; Ph.D. in Math. Physics, U. Birmingham (Eng.), 1958; m. Myriam Ramos Da Silva, Feb. 2, 1953; children—Ana Myriam, Samuel, Maria Dolores. Came to U. S., 1963. Research asst. Centro Brasileiro de Pesquisas Fisicas, Rio de Janeiro, 1954-56, asso. prof., 1960-61, prof., 1962-63; instr. physics Princeton, 1959-60; vis. prof. Cath. U., Rio de Janeiro, 1962-63; mem. Inst. for Advanced Study, Princeton, N.J., 1963-65; asso. prof. Yale, 1965-67, prof., 1967—. Mem. Am. Phys. Soc., Brazilian Assn.

for Advancement Sci. Contbg. author Theoretical Physics, 1963, High Energy Physics and Elementary Particles, 1965. Contbr. articles to sci. jours. Contbns. to theory of scattering of elementary particles; dispersion relations; high energy behavior of scattering amplitude; theory of leptonic decay of K-Mesons. Home: Carriage Dr., Woodbridge, Conn. Office: Dept. Physics, Yale, New Haven.*

MACELWANE, James B(ernard), Am. geophysicist; b. nr. Port Clinton, Ottawa County, O., Sept. 28, 1833; s. Alexander and Catherine Agnes (Carr) M.; student St. Stanislaus Coll. and John Carroll U., Cleveland, until 1907; A.B., St. Louis U., 1910, A.M., 1911, M.S., 1912; Ph.D., U. of Calif., 1923; D.Sc. (honorary), Saint Norbert College, 1949; LL.D. (honorary), Washington University, 1953; D.Sc. (honorary), John Carroll University 1954. Joined Society of Jesus (Jesuits), 1903; ordained priest R.C. Ch., 1918. Instr. in mathematics, St. John's Coll. High Sch., Toledo, 1907-08, in physics, St. Louis U., 1912-13; asst. prof. physics, same univ., 1913-15, 1918-19; asst. prof. geology, U. of Calif., 1923-25; prof. geophysics and dir. dept., St. Louis Univ., since 1925, prof. geophysical engring. since 1949, dean Graduate School. 1927-33, dean Inst. Tech. 1944-—. Recipient Jackling Lecturer award, Am. Inst. Mining and Metall. Engrs. Mem. Nat. Science Board. Fellow A.A.A.S. (past v.p.), Geol. Soc. America, Am. Geog. Soc.; mem. Am. Physical Soc., Seismol. Soc. America (ex-pres.), Jesuit Seismol. Assn. (pres.), Am. Meteorol. Soc. Am. Geophys. Union (president, sect. of seismology), Am. Inst. Mining and Metall. Engrs., Mo. Acad. Sciences (ex-pres.), St. Louis Acad. Sciences (ex-pres.), Optical Soc. America (asso.), Nat. Research Council, Nat. Acad. Scis., Societa Sismologica Italiana, Societa Meteorologica Italiana, Soc. Exploration Geophysicists, Am. Assn. Petroleum Geologists, Sigma Xi. Received Bowle Medal of Am. Geophys. Union, 1948. Author: Loose Leaf Manual of Laboratory Experiments in Coll. Physics, Parts I, II, III, IV, V (with J. I. Shannon), 1914; Theoretical Seismology, Vol. I, Geodynamics, 1936, new edition, 1949; When the Earth Quakes, 1947; also more than 100 papers and articles, alone and with others, on seismology and other subjects. Editor and joint author of Bull. of Nat. Research Council on Seismology, 1933. Co-author: Internal Constitution of the Earth, 1939; Compendium of Meteorology, 1951. Specialist in seismology; developed method of tracking hurricanes at sea. Died St. Louis, Mo., Feb. 15, 1956.

MACEWEN, Sir William, Scottish surgeon; b. Rothesay, Isle of Bute, Scotland, June 22, 1848; s. John and Janet (Stevenson) M.; M.B., C.M., U. Glasgow (Scotland), 1869, LL.D., 1890; studied under Lister; hon. degrees from Oxford, Dublin, Liverpool, Durham; m. Mary Watson Allen, 1873; 3 sons, 3 daus. Housesurgeon to Sir G. H. B. Macleod, Old Infirmary, 1869; regius prof. surgery U. Glasgow, from 1892; also surgeon to Royal Infirmary; cons. surgeon to Navy, Scotland, World War 1. Knighted, 1902; Fellow Royal Soc., 1895, Royal Coll. Surgeons Eng. (hon.); mem. Brit. Med. Assn. (pres. 1922). Author: Osteotomy, 1880; Pyogenic Infective Disease of the Brain and Spinal Cord, 1893; Atlas of Head Sections, 1893. Growth of the Bone, 1912; Growth and Shedding of the Antler of the Deer, 1921. Important pioneer in the advancement of surgery of brain and spinal cord and bone grafting; 1st to deliberately operate to relieve brain disorders; laid basis for modern brain surgery; introduced method of implanting small grafts to replace missing parts of limb bones, also method for rectifying knock-knee by cutting through thigh bone; proved that fear of lung collapsing during chest operations was groundless; one of 1st to intubate the larynx in order to keep the airway open, instead of resorting to tracheotomy; recognized that disease of middle ear was the commonest cause of abscess of brain; improved methods of mastoid operation; worked on radical cure of hernia. Died Glasgow, Mar. 22, 1924.

MACFADDEN, Clifford Herbert, Am. geographer; b. Salem, Mich., Jan. 30, 1908; s. James W. and Augusta (Bigg) MacF.; A.B., U. Mich., 1937, M.A., 1939, Ph.D., 1948; m. Alyce Keller, 1934; children —Lillian Joyce, Herbert Craig. Chief, cartographic sect. M.I.S., War Dept., Washington, 1942-46; faculty U. Cal., Los Angeles, 1946-—, prof. geography, 1958-—, chmn. dept., 1961-66. Chmn. dept. geography U. Ceylon, 1950-51, 53-56; Fulbright prof. U. Delhi Sch. Econs., 1960-61. Mem. Am. Assn. Geographers, Am. Geog. Soc., and others. Author: An Atlas of Primary Productions of the World, 1938; Atlas of World Review, 1940; Bibliography of Pacific Area Maps, 1941; (with H. M. Kendall, G. F. Deasy) Atlas of World Affairs, 1946; (with Kendall, R. M. Glendinning) Introduction to Geography, 1951, 58, 62, 67; Introduction to Physical Geography, 1952, 57; also articles, chpts. in textbooks. Home: 14907 Bestor Blvd., Pacific Palisades, Cal. 90272.*

MACFADYEN, Allan, bacteriologist; b. Glasgow, Scotland, 1860; s. Archibald Macfadyen; M.D., U. Edinburgh; postgrad. under Flügge, Nencki, Pettenkofer at univs. Berne, Gottingen, Munich; M.B., Ch.M.; m. Marie Bartling, Jan. 7, 1890. Research scholar Grocers' Co.; lectr. bacteriology Coll. State Medicine; head bacteriological dept., sec. to governing body, dir. Jenner Inst. Preventive Medicine (now Lister Inst.), London; Fullerian prof. physiology Royal Instn.; examiner Edinburgh U. Author: The Cell as the Unit of Life, 1908. Contbr. to sci. publs. Studied effects of low temperatures on bacteria, technique of disintegrating bacteria, endotoxins or intracellular juices of some bacteria; showed that small injections of endotoxins provide immunity from disease. Died Mar. 1, 1907.

MACFADYEN, Amyan, Brit. biologist; b. Weald, Kent, Eng., Dec. 11, 1920; s. Eric and Violet (Champheys) M.; B.A., Oxford (Eng.) U., 1939, M.A., 1947, D.Sc., 1964; m. Ursula Margaret Hampton, Sept. 8, 1949; children—Timothy Eric, Matthew Robert, Peter Floyd, Sophie Madeleine. Mem. Jan. Mayer Expdn., Oxford U., 1947; research staff Bur. Animal Population, Oxford, 1947-56; faculty U. Coll., Swansea, Wales, 1956-—, reader zoology; guest prof. Molslaboratoriet, Aarhus U., Arhus, Denmark, 1965-67; lectr. Balliol Coll., also tutor Oxford U., 1950-56. Mem. P.T. sub com. Brit. nat. com. Internat. Biol. Programme, 1964-—. Mem. Brit. Ecol. Soc. (past mem. council), Royal Entomol. Soc., Zool. Soc. London, Inst. Biology, Ecol. Soc. Am., Soc. Exptl. Biology. Author: Animal Ecology: Aims and Methods, 1957; also articles. Research on soil including role and importance of organisms, measurement of climatic factors. Home: Reynoldston, Swansea, Wales, U.K.*

MACFARLAND, Thomas, geologist; b. Pollokshaws, Scotland, Mar. 5, 1834; s. Thomas Macfarlane; ed. Andersonian U., Glasgow, Scotland; Royal Mining Acad., Freiberg, Germany; m. Margaret Skelly. Became chief analyst Inland Revenue Dept., Ottawa, Ont., Can., 1886; with office William Hector, Procurator-Fiscal, Pollokshaws; chemist, later dir. Modums Blue Colour Works, nr. Drammen, Norway; served under Sir W. E. Logan, Geol. Survey Can.; explored mineral lands, managed mines, conducted smelting works, Que., Ont., Can., U. S., S.Am. Fellow Royal Soc. Can. Author: To the Andes; also articles, pamphlets. Research on chem. tech., geology, mineralogy, metallurgy. Died July 12, 1907.

MACFARLANE, Robert Gwyn, English physician; b. Worthing, Eng., June 26, 1907; s. Robert Gray and Eileen (Sanderson) M.; M.D., U. London (Eng.), 1938, M.B., B.S., 1933; m. Hilary Carson, Jan. 25, 1936; children—Susan, Robert Grey, John Carson, Donald Edward, Richard Oke. Demonstrator pathology St. Bartholomews Hosp., 1934-36; Sr. Halley Stewart Research fellow, 1935; asst. clin. pathologist Brit. Postgrad. Med. Sch., 1936-39; asst. bacteriologist Wellcome Research Labs., 1939-41; clin. pathologist Radcliff Infirmary, Oxford, Eng., 1941-44; with Oxford U., 1947-—, reader hematology, 1957-64, prof. clin. pathology, 1964-—; dir. blood coagulation research unit Med. Research Council, 1959-—; fellow All Souls Coll., Oxford, U., 1963-—. Named Comdr. Order Brit. Empire. Fellow Royal Soc., 1956, Royal Coll. Physicians. Author: (with R. Biggs) Human Blood Coagulation and its Disorders, 1953, 3d edit., 1962; also numerous articles. Editor: (with A. M. T. Robb-Smith) Functions of the Blood, 1961; (with R. Biggs) Treatment of Haemophilia and Other Coagulation Disorders, 1966. Research on elucidation of mechanism of blood coagulation and identification of some factors, hemostasis and its disorders. Home: Downhill Farm, Witney, Oxfordshire, Eng. Office: Dept. Hematology, Radcliffe Infirmary, Oxford, Eng.*

MACFIE, John William Scott, Brit. parasitologist; b. 1879; s. J. W. Macfie; M.A., Cambridge, Eng.; M.B., Ch.B., D.Sc., Edinburgh, Scotland; D.T.M., Liverpool, Eng. W. African med. staff, 1910-23; acting dir. Med. Research Inst., Lagos, 1913; dir. med. research Inst., Accra, 1914-23; staff War Office Malaria Investigation, Liverpool, 1917-19; lectr. protozoology Liverpool Sch. Tropical Medicine; staff Med. Research Council Investigation on Chemotherapy of Malaria, London Sch. Hygiene and Tropical Medicine, 1927-31; with Brit. Ambulance Service, Brit. Red Cross, Ethiopia, 1935-36; emergency officer St. Pancras Med. War Com., 1939-41; malariologist Numbers 3 and 8 Malaria Field Labs., Egypt, Palestine, Syria. Found. scholar Gonville and Caius Coll., Cambridge; recipient Frank Smart prize; Mary Kingsley medal Liverpool Sch. Tropical Medicine, 1919. Author: An Ethiopian Diary, 1936; John Macfie of Edinburgh and his Family, 1938. Research on tropical diseases and parasitology, especially malaria and trypanosomiasis; studied Ceratopogonidea (biting midges), mosquitoes, tse-tse flies, including descriptions of new species. Died Oct. 11, 1948.

MACGILLAVRY, Carolina Henriette, Dutch crystallographer; b. Amsterdam, Netherlands, Jan. 22, 1904; d. Donald and Ida (Matthes) MacG.; Doctor's degree with honors, U. Amsterdam, 1937. With U. Amsterdam, 1927-33, 37-—, prof. crystallography, 1950-—; asst., U. Leiden (Netherlands), 1937. Mem. Royal Netherlands Chem. Soc. (bd. officers), Royal Netherlands Acad. Scis. (gen. sec. 1960-—), Internat. Union Crystallography (exec. com. 1954-60), A.A.A.S., Am. Phys. Soc., Am. Crystall. Assn. Author: (with J. M. Bijvoet, N. H. Kolkmeijer) X-ray Analysis of Crystals, 1948; (with G. D. Rieck) International Tables for X-ray Crystallography, vol. III, 1962; Symmetry Aspects of M. C. Escher's Periodic Drawings, 1965; also articles. Editorial bd. Ned. Tijdschrift voor Natuurkunde. Research on atomic arrangement in crystals, including its relation to chem. and phys. properties of crystals and molecules; acid-forming oxides, phosphates, vitamin-A and carotenoids, alkaloids; physics of interaction of X-rays and crystals. Home: 40 Vijverhoef Amsterdam-Buitenveldert, Netherlands. Office: 126 Nieuw Prinsengracht, Amsterdam C., Netherlands.*

MACGREGOR, Gordon Scott, Brit. physiologist; b. Glasgow, Scotland, Dec. 3, 1901; s. George Scott and Jessica (Farquhar) MacG.; B.A. with honors, Worcester Coll., Oxford (Eng.) U., 1924, B.M., Ch.B., 1927, M.A., D.M., 1939; postgrad. St. George's Hosp., London; M.D. (hon.) U. Malaya, 1951; m. Ruth Adele Tobison, Oct. 18, 1934. Research asst. safety in mines Leeds (Eng.) Dept. Physiology, 1928; lectr. physiology U. Birmingham (Eng.), 1929-36; prof. physiology Coll. Medicine, Singapore, also U. Malaya, 1936-51; lectr. London Sch. Tropical Medicine and Hygiene, part-time 1953-57; in charge blood bank Dept. Physiology, Singapore, 1939-41. Mem. med. reference com. Brit. Internment Camp, Singapore, 1942-45. Fellow Royal Soc. Medicine (London); mem. Royal Coll. Surgeons, Physiol. Soc. (life). Author: Structures and Functions of the Human Body, 1966; also articles. Research on histamine, microscopy of lung circulation, influence of tropical environment on human body, leucocyte count in blood. Home: Chyverton, Mawnan Smith, Falmouth, Cornwall, Eng.*

MACGREGOR, James Gordon, natural philosopher; b. 1852; s. Peter Gordon MacGregor; M.A., Dalhousie Coll., Halifax, N.S., Can.; student Edinburgh (Scotland) U., U. Leipzig (Germany); D.Sc., London U.; LL.D.; m. Marion Miller Taylor; 1 son, 1 dau. Lectr. physics Dalhousie Coll., 1876-77, Clifton Coll. (Eng.), 1877-79; Munro prof. physics Dalhousie Coll., Halifax, 1879-1901; became prof. natural philosophy U. Edinburgh, 1901. Fellow Royal Soc., 1906, Royal Soc. Edinburgh. Author: Kinematics and Dynamics; Physical Laws and Observations; also articles, pamphlets on ednl. subjects. Died May 21, 1913.

MACH, Ernst, Austrian physicist, philosopher; b. Turas, Moravia, Feb. 13, 1838; ed. U. Vienna. Dozent, Vienna; became prof. Math., Graz, Austria, 1864, prof. physics, 1866; prof. physics, Prague, Czechoslovakia, 1867-95; prof. philosophy, Vienna, 1895-1901; elected to Austrian House of Peers, 1901. Author: Lehre von den Bewegungsempfindungen, 1875; Die Mechanik in ihrer Entwicklung, 1883; Analyse der Empfindungen, 1886; Die Prinzipien der Wärmelehre, 1896; Populärwissenschaftliche Vorlesungen; Erkenntnis und Irrtum, 1905; Die Prinzipien der physikalischen Optik, 1921. Research in supersonic projectiles, flow of gases (important in aero. design and sci. of projectiles; Mach number (in supersonic flight) named after him; studies in perception of bodily rotation; developed ideas of sci. positivism, epistemology of relationship between psychology and physics; believed sensation is data of all sciences; sometimes said to have initiated systematic study of philosophy of science. Died Feb. 19, 1916.

MACHATSCHEK, Fritz, Austrian geographer, geologist; b. Wischau/Mahren, Sept. 22, 1876; prof. univs. Prague, Zurich, Vienna, Munich. Author: Geomorphologie, 1934; Das Relief der Erde, 2 vols., 1938-40; Gletscherkunde, 1942. Leading glacial geologist and geomorphologist; supporter of theory that Alps were made more rugged and irregular by last glaciation, but had been so in pre-glacial times. Died Munich, Sept. 25, 1957.

MACHATSCHKI, Felix Karl Ludwig, Austrian mineralogist; b. Arnfels, Styria, Austria, Sept. 22, 1895; s. Felix and Christine (Schallmun) M.; Tchrs. Degree, U. Graz (Austria), 1920, Ph.D., 1922. Asst., lectr. U. Graz, 1920-27; research fellow U. Oslo (Norway), 1927-28; hon. research fellow U. Manchester (Eng.), 1928-29; guest lectr. U. Göttingen (Germany), 1929-30; prof. U. Tübingen (Germany), 1930-41; dean faculty sci., 1931-33; prof. mineralogy and petroleum U. Munich., 1941-44, U. Vienna, 1944-—. Recipient Schroedinger prize Austrian Acad. Sci., 1958; Roebling medal Mineral. Soc. Am., 1959; Austrian decoration for sci. and arts, 1961. Hon. fellow Mineral. Soc. Am.; mem. Norwegian, Yugoslav, German acads. sci., German Mineral. Soc. (hon.), Hungarian Geol. Soc.; acting mem. Austrian Acad. Sci.; fgn. mem. Royal Swedish Acad. Sci., Acc. Naz. dei Lincei (Roma); corr. mem. Acad. Sci. Munich and Göttingen, Geol. Soc. Am., Geol. Soc. Stockholm, Mineral. Soc. Italy, Internat. Union Crystallography (Austrian sec.), Assn. Natural Sci. Vienna (pres. 1945-60), Austrian Mineral. Soc. (v.p.), Mineral Soc. Gt. Britain and Ireland (hon.). Author: Grundlagen der allgemeinen Mineralogie und Kristallchemie, 1946; Vorräte und Verteilung der Mineralischen Rohstoffe, 1948; Spezielle Mineralogie auf geochemischer Grundlage, 1953. Editor Tscherm. Miner. u. Petrogr. Mitteilungen, 1944-—. Research and publs. on mineralogy, structure of inorganic compounds, crystal chemistry. Home: 4/8 Adamsstr., Vienna III, Austria. Office: 1 Luegerring, Vienna III, Austria.*

MACHIAVELLI, Niccolo, Italian statesman, polit. writer; b. Florence, Italy, May 3, 1469; s. Bernardo Machiavelli; ed. for pub. career; m. Marietta Corsini, 1502. Apptd. clk. 2d chancery of commune under Marcello Virgilio Adriani Republic of Florence, 1494, 2d chancellor, 1498-1512; diplomatic missionary to fgn. cts.; govt. rep. with Pope Julius II on expdn. to strip Cesare Borgia of powers; imprisoned on suspicion of plotting against Medici family which reascended to power in Florence, 1512, ret. from pub. life upon release, returned to pub. life, 1521-27. Author novels, songs, poems, comedies, history of Florence, treatise on war. Chiefly noted as author of The Prince, a polit. work written to explain successful techniques of rulership (book based on ancient history as well as events of Renaissance Italy); divorced politics from ethics and religion; felt that state was non-ethical and its actions should be judged by their results, not their means, also felt that leader should appear to have good qualities, but not necessarily actually possess them; believed that form of govt. should be determined by conditions in state; felt that rulers should avoid flatterers and seek to hear only truth. Died Florence, June 22, 1527.

MACHIDA, Seishi, Japanese chemist; b. Kyoto, Japan, Nov. 17, 1913; s. Hiroyuki and Nao Machida; D.Sc., Kyoto U., 1948; m. Chieko Machida, May 5, 1941; children—Hiroko, Yoriko. Prof., Kyoto U. Indsl. Arts and Textile Fibers, 1948—. Mem. Chem. Soc. Japan, Soc. Polymer Sci. Japan, Japanese Tech. Assn. Pulp and Paper Industry. Author: Fundamentals of Organic Chemistry; also others. Research, numerous publs. on water-soluble polymers, application of petroleum resin and polyethyleneimine to paper making. Home: Takanawa-cho 38, Kita-ku. Office: Sakata-machi, Kita-ku, Kyoto, Japan.*

MACHIN, John, Brit. mathematician; prof. astronomy Gresham Coll., 1713-51. Fellow Royal Soc., 1710 (sec. 1718-47). Contbr. math. papers to Philosophical Transactions; calculated pi to 100 decimal places; studied quadrature of the circle. Died 1751.

MACHLIS, Leonard, Am. plant physiologist; b. Seattle, Apr. 13, 1915; s. Samuel and Beatrice (Lapiroff) M.; B.S., State Coll. Wash., 1937; M.S., U. Hawaii, 1939; Ph.D., U. Cal. at Berkeley, 1943; m. Gertrude Therese Rafferty, June 28, 1946; children—Lee Ellen, Joan Louise, Paul Laurence, Gail Elizabeth. Instr. U. Ill., 1945-46; mem. faculty U. Cal. at Berkeley, 1946—, prof. botany, 1959—, chmn. dept., 1961—, Miller research prof., 1959-61, chmn. Biology Council, 1963-64, 65—. Cons. sci. facilities sect. NSF, 1965—; mem. cell biology study sect. NIH, 1965—. Guggenheim fellow U. Upsalla (Sweden), 1957-58. Mem. Am. Soc. Plant Physiologists, Scandinavian Soc. Plant Physiology, Bot. Soc. Am., also Phi Beta Kappa. Author: (with John G. Torrey) Plants in Action, 1956, also papers. Editor Annual Rev. of Plant Physiology, 1959—. Research on nutrition of algae and fungi; physiology and chemistry of hormones in plants. Home: 1871 Thousand Oaks Blvd., Berkeley, Cal. 14707. Office: Botany Dept., Univ. of Calif., Berkeley, Calif.*

MACHTA, Lester, Am. meteorologist; b. N.Y.C., Feb. 17, 1919; s. Nathan and Bertha (Leavitt) M.; B.A., Bklyn. Coll., 1939; M.S., N.Y. U., 1946; Sc.D., Mass. Inst. Tech., 1948; m. Phyllis M. Margaretten, Jan. 27, 1947; children—Jonathan Lee, Deborah Jo. With U. S. Weather Bur., 1948-65; dir. Air Resources Lab., Environmental Sci. Services Adminstrn., Washington, 1965—. Recipient gold medal U. S. Dept. Commerce. Mem. Am. Meteorol, Soc., Am. Geophys. Union, A.A.-A.S., Washington Acad. Scis., Royal Meteorol. Soc. Contbr. numerous articles to profl. publs. Initiator program on forecasting air pollution potential. Home: 6601 Brigadoon Dr., Bethesda, Md. 20034. Office: 8060 13th St., Silver Spring, Md. 20910.*

MACINTOSH, Charles, Scottish chemist, inventor; b. Glasgow, Dec. 29, 1766; s. George and Mary (Moore) MacI.; student of William Irvine, Glasgow, Joseph Black, Edinburgh; m. Mary Fisher, 1790. Counting house clerk, Glasgow; engaged in manufacture of sal ammoniac and other chemicals in his own plant, Glasgow, 1786, sugar of lead, acetate of aluminia, Prussian blue; founded 1st alum works in Scotland, at Hurlet, 1797, ret., 1814; with St. Rollox chem. works, until 1814; established works to manufacture waterproof fabric, Manchester, Eng., during 1820's. Fellow Royal Soc., 1823. Inventor (with Charles Tennant) bleaching powder, 1799; inventor (with Neilson) hot blast process, process of converting malleable iron to steel; inventor mackintosh (his name misspelled) waterproof fabric made by dissolving rubber in low-boiling naphtha, patented 1823; prepared sugar of lead. Died near Glasgow, July 25, 1843.

MACINTOSH, Frank Campbell, Canadian physiologist; b. Baddeck, N.S., Can., Dec. 24, 1909; s. C. C. and Beenie A. (Matheson) MacI.; B.A., Dalhousie U., N.S., 1930, M.A., 1932; Ph.D., McGill U., Que., Can., 1937; LL.D., U. Alta., Can., 1964; LL.D., Queen's U., 1965; m. Mary MacLachlan MacKay, Oct. 25, 1938; children—Christine (Mrs. A. G. Lejtenyi), Barbara (Mrs. J. N. M. Hardt), Andrew, Janet, Roderick. Research staff Med. Research Council Gt. Britain, London, Eng., 1938-49; Joseph Morley Drake prof. physiology McGill U., Montreal, Que., Can., 1949—, chmn. dept. physiology, 1949-64; vis. prof. U. Lund, 1956, Tulane U., 1962, Central U. Venezuela, 1964. Mem. Med. Research Council Can., 1961-63; mem. Sci. Council Can., 1966—; chmn. Canadian Nat. Com. for Internat. Biol. Progress, 1964—. Recipient Coronation medal, 1953. Fellow Royal Soc., 1954, Royal Soc. Can.; mem. Council Internat. Union Physiol/Scis. (treas. 1962—), Canadian (past pres.), Am. physiol. socs., Pharmacological Soc. Can., Physiol. Soc. (U.K.), N.Y. Acad. Scis., Am. Soc. for Pharmacology and Exptl. Therapeutics, Sigma Xi, Nu Sigma Nu. Asso. editor Canadian Jour. Physiology and Pharmacology, 1963—. Research, publs. on physiology of efferent nerve endings and synthesis, storage and release of pharmacologically active materials in mammalian tissues. Home: 145 Wolseley Av., Montreal West, Que. Office: Dept. Physiology, McGill U., Montreal 2, Que., Can.*

MACINTYRE, John, Brit. physician, surgeon; b. Nov. 2, 1859; ed. Glasgow U.; Vienna; M.B., C.M., LL.D., Glasgow U.; m. Agnes Jean Hardie, 1892. Various positions including dispensary surgeon, cons. surgeon for diseases of nose and throat, cons. med. electrician Glasgow Royal Infirmary; demonstrator Sch. Medicine; lectr. diseases nose and throat Glasgow U.; v.p. Med. Faculty, emeritus prof. laryngology Anderson Coll. Medicine, Glasgow. Fellow London Laryngol. Soc., Brit. Laryngol., Rhinological and Otol. Assn. (became pres. 1893-1900); mem. Brit. Med. Assn. (pres. W. of Scotland br.), Société d'Otologie, de Laryngologie, et de Rhinologie (Paris), Am. Laryngol. Assn. (corr.), Roentgen Soc. London, Brit. Electro-Therapeutic Soc. (hon.), Instn. Elec. Engrs. Fellow Royal Soc. Engrs., Royal Micros. Soc., Royal Faculty Physicians and Surgeons Glasgow. Sr. editor Brit. Jour. Laryngology. Research and numerous publs. on surgery, X-rays, applications of elec. and phys. scic. to medicine; 1st prodn. of X-ray motion pictures, 1897. Died Oct. 29, 1928.

MACINTYRE, William, physician; b. Eng.; gave 1st description of myelomatosis, or multiple myeloma (malignant tumor of bone marrow), 1850. Died 1875.

MACINTYRE, William James, Am. physicist; b. Canaan, Conn., Nov. 26, 1920; s. William M. and Helen (Hoyt) MacI.; B.S. in Physics, Western Res. U., 1943, M.A., 1947; M.S., Yale, 1948, Ph.D., 1950; m. Patricia Nelle Grossman, Sept. 16, 1947; children—Kathleen Suzanne, Steven James. Sect. chief atomic energy med. research project Western Res. U., Cleve., 1951-58, faculty, 1952—, asso. prof. biophysics, 1959—; cons. physicist Highland View Hosp., 1955-64, Huron Rd. Hosp., 1957—, Harshaw Chem. Co., 1957—(all Cleve.). Chmn. task group on scanning Internat. Commn. on Radiol. Units and Measurement, 1963—; with Found. Fund for Research in Psychiatry, 1960-66. Mem. Am. Phys. Soc., Biophys. Soc., Radiation Research Soc., Soc. Nuclear Medicine (past trustee), Central Soc. for Clin. Research, Am. Assn. Physicists in Medicine, Phi Beta Kappa, Sigma Xi. Contbg. author: Medical Physics, 1960; Progress in Radioisotope Scanning, 1963; Radioactive Isotope, vol. 5, 1963, vol. 6, 1965; also numerous articles. Research on characteristics of Scintillation Counter and application to clin. and physiol. measurements radioisotopes, electric parameters of brain tissue; developed techniques for electro-encephalographic analysis. Home: 3108 Huntington Rd., Shaker Heights, O. 44120. Office: 2040 Abington Rd., Cleve. 44106.*

MACIVER, Robert Morrison, polit. scientist, sociologist; b. Stornoway, Scotland, Apr. 17, 1882; s. Donald and Christine (Morrison) M.; M.A., Edinburgh (Scotland), 1903, Ph.D., 1915; B.A., Oxford (Eng.), 1907; Litt.D., Columbia, 1929, Harvard, 1936, Princeton, 1947, Jewish Theol. Seminary, 1950; D.Sc., New Sch. for Social Research, 1950; L.H.D., Yale, 1951; LL.D., Edinburgh, 1952, Toronto (Can.), 1957; m. Ethel Marion Peterkin, May 14, 1911; children—Ian Tennant Morrison (dec.), Christine Elizabeth (Mrs. Robert Bierstedt), Donald Gordon. Came to U. S. 1927, naturalized 1934. Lectr. polit. sci. Aberdeen (Scotland) U., 1907-10, sociology, 1911-14; prof. polit. sci. U. Toronto, Can., 1915-27, head dept. 1922-27; prof. polit. sci. Barnard Coll., 1927-36; Lieber Prof. polit. philosophy, sociology Columbia, 1929-50; dir. City N.Y. Juvenile Delinquency Evaluation Project, 1956-61; pres. New Sch. for Social Research, N.Y.C., 1963—. Vice-chmn. Dominion of Can. War Labor Bd., Ottawa, 1917-19. Fellow Royal Soc. Can., Am. Acad. Arts, Scis., Am. Philos. Soc.; mem. Am. Sociol. Soc., Inst. Internationale de Sociologie, Phi Beta Kappa. Author: Elements of Social Science, 1921; Economic Reconstruction, 1934; The Web of Government, 1947; Democracy and Economic Challenge, 1952; Life: Its Dimensions and its Bounds, 1960; Power Transformed, 1964; numerous other books. Home: Heyhoe Woods, Palisades, N.Y. Office: 66 West 12th St., N.Y.C. 10011.*

MACK, Pauline Beery, Am. chemist; b. Norborne, Mo., Dec. 19, 1891; d. John Perry and Dora (Woodford) Beery; A.B., Mo. State U., 1913; A.M., Columbia, 1919; Ph.D., Pa. State Coll., 1932; Sc.D., Moravian College for Women, 1952, Western College for Women, Oxford, Ohio, 1952; m. Warren B. Mack, Dec. 27, 1923. Science teacher, Norborne High Sch., 1913-15, Webb City (Mo.) High Sch., 1915-18, Springfield (Mo.) High Sch., 1918-19, mem. faculty Pa. State U., 1919-52, dir. Ellen H. Richards Inst., Coll. Chemistry and Physics, 1940-52; dean and dir. research Coll. Household Arts and Sciences, Texas Women's University, Denton, Tex., 1952-62, dir. Research Found., 1962—; director Phila. Mass. Studies in Human Nutrition, 1940-45, Pa. Mass. Studies, 1935-52. Distinguished Dau. Pa. Medal, 1949; Garvan Medal of American Chem. Soc., 1950. Fellow Am. Pub. Health Assn., Am. Inst. Chemists, A.A.A.S., Soc. for Research in Child Development. Am. Sch. Health Assn., Gerontol. Soc.; mem. Am. Assn. Textile Tech., Am. Assn. Phys. Anthropologists, Royal Soc. Health (Great Britain), Soc. Chem. Industry (Great Britain), Am. Council Edn., Am. Ordnance Assn., Geriatrics Soc., Fashion Group, Inc., Am. Chem. Soc., Am. Soc. for Testing Materials, Am. Assn. Textile Chemists and Colorists, Am. Home Econ. Assn., Texas Public Health Association, Tex. Soc. Aging, Textile Research Inst., Texas Acad. of Science, Am. Association University Profs., American Assn. University Women, Am. Dietetic Assn., Sigma Xi. Author: Chemistry Applied to Home and Community, 1926; Stuff, 1936. Cons. Editor of Chemistry. Research on textile chemistry, nutrition, and calcium chemistry in osteopathy; developed microphotometric technique for bone density. Home: Old McKinney Rd., R.D. 2, Denton, Texas.

MACKAY, Alexander Howard, Canadian biologist; b. North Dalhousie, N.S., Can., May 19, 1848; s. John and Barbara (MacLean) M.; ed. Dalhousie U.; B.A., B.Sc., LL.D.; m. Maude Augusta Johnstone, 1882; 1 son, 1 dau. Tchr. pub. schs., Pictou Acad., Halifax (N.S.) Acad.; lectr. biology Med. Coll., Dalhousie U.; supt. edn. Province of N.S., 1890-1926; pres. Victoria Sch. Art and Design. Mem. Geog. Bd. Can., Biol. Bd. Can., Biol. Com. for Sci. and Indsl. Research. Fellow Royal Soc. Can.; mem. Ednl. Assn. N.S. (pres.), Dominion Ednl. Assn. (pres.), N.S. Inst. Sci. (pres.). Author: The Flora of Nova Scotia; Freshwater Sponges; Phenology of Canada; also numerous articles. Died May 19, 1929.

MACKAY, Angus, Canadian agriculturist; b. 1840; supt. Dominion Exptl. Farm, Indian Head, Can.; bred (as expt. for William Saunders) Marquis wheat; discovered land must lie fallow 1 year to produce wheat the next. Died 1932.

MACKAY, Clarence Hungerford, Am. engr.; b. San Francisco, 1874; s. John W. and Marie Louise (Hungerford-Bryant) MacK.; ed. Collège Vaugirard, Paris; grad. Beaumont Coll., Windsor, Eng., 1892; m. Katherine Alexander Duer, May 17, 1898 (div.); children—Katherine (Mrs. Kenneth O'Brien), Ellin (Mrs. Irving Berlin), John William; m. 2d, Anna Case, July 18, 1931. Chmn. bd. Postal Telegraph Cable Co.; pres. Comml. Cable Co., Comml. Pacific Cable Co., Mackay Cos., Cuban All Am. Cables; dir. Internat. Tel. & Tel. Corp. Supervised laying 1st trans-Pacific cable, also cables from N.Y. to Cuba, in Azores and Ireland. Died 1938.

MACKAY, Ian Reay, Australian physician; b. Melbourne, Australia, Mar. 22, 1922; s. Eric Reay and Ethel (Carney) M.; M.B., B.S., U. Melbourne, 1945, M.D., 1952; m. Patricia Wilson, July 29, 1958; children—Colyn Erica, Charles Reay, Eric Ashfield, Indi Patricia, Ian Migdale. Commonwealth med. officer, Berlin, London, The Hague, 1948-50; house physician, research asso. Postgrad. Med. Sch., London, 1951; research fellow dept. medicine U. Wash., Seattle, 1952-54; research fellow Meml. Center and Sloan-Kettering Inst., N.Y.C., 1955; staff Walter and Eliza Hall Inst. Med. Research, Melbourne, 1956—, head clin. research unit, 1963—, acting asst. dir., 1966-67; staff Royal Melbourne Hosp., 1956—, head clin. research unit, hon. physician, 1963—. Fellow Royal Australasian Coll. Physicians (Eric Susman prize 1966); mem. Royal Coll. Physicians London. Author: (with F. M. Burnet) Autoimmune Diseases: Pathogenesis, Chemistry and Therapy, 1963; also numerous articles. Devel. of concept of autoimmunity as cause of self-perpetuating human diseases especially lupus erythematosus and chronic hepatitis; introduced concept of lupoid hepatitis as prototype of autoimmune disease of liver; recognized autoimmune hepatitis as cause of cirrhosis of liver. Home: 3 Beamsley St., Melbourne, Victoria. Office: Walter and Eliza Hall Inst., Post Office, Royal Melbourne Hosp., Victoria, Australia.*

MACKAY, Roland Parks, Am. physician, neurologist; b. Atlanta, Oct. 25, 1900; s. William Charles and Alice (Farmer) M.; B.A., Emory U., 1920; M.D., U. Toronto, 1925; m. Margaret Pomroy, July 29, 1929; children—Virginia Carol (Mrs. David Larson), Kathleen Louise (Mrs. William G. Davis), William C. Asso. prof. neurology Rush Med. Coll., Chgo., 1929-34; faculty U. Ill. Chgo., 1934-61, prof. neurology, 1942-61; prof. neurology Northwestern U. Med. Sch., Chgo., 1961—; neurologist, head dept. neurology Presbyn.-St. Luke's Hosp., Chgo., 1959-61; attending neurologist Ill. State Neuropsychiat. Inst., 1934-61; sr. attending neurologist Chgo. Wesley Meml. Hosp., 1961—; cons. VA Hosp., Hines, Ill., 1961—, VA Research Hosp., Chgo., 1964—. Dir., Am. Bd. Psychiatry and Neurology, 1945-53, pres., 1953. Mem. Am. Neurol. Assn. (pres. 1954), A.M.A. (chmn. sect. on nervous and mental disease 1947), Am. Epilepsy Soc. (pres. 1958), Soc. for Biol. Psychiatry (pres. 1952), Chgo. Neurol. Soc. (pres. 1944), Assn. for Research in Nervous and Mental Disease (v.p. 1951), A.C.P., Am. Psychiat. Assn., A.A.A.S Phi Beta Kappa, Sigma Xi, Alpha Omega Alpha. Author: Archives of Neurology. Editor sect. on neurology Year Book of Neurology, Psychiatry and Neurosurgery, 1949—; asso. editor Archives of Neurology, 1956-66, Neurology, 1954—. Research on 68 twin pairs with multiple sclerosis and their 2900 relatives. Home:

336 N. Forest Rd., Hinsdale, Ill. 60521. Office: 8 S. Michigan Av., Chgo. 60603.*

MACKENRODT, Alwin Karl, German gynecologist; b. 1859; described fibrous bands from lower portion of broad ligament to supravaginal part of cervix uteri (cardinal, uterosacral, or cervicopelvic ligament), 1895; devised operation for plastic reconstrn. of vagina, 1896. Died 1925.

MACKENZIE, Arthur Stanley, Canadian mathematician; b. Pictou, N.S., Can., Sept. 20, 1865; s. George A. and Catherine D. (Fogo) M.; B.S. (Young Gold medal) Dalhousie U., 1885; Ph.D., J.H. U., 1894; D.C.L., Vind; LL.D., Queen's U., McGill U.; m. Mary Lewis Taylor, 1895; 1 dau. Asst. master Acad., Yarmouth, N.S., 1885-87; tutor math. Dalhousie U., 1887-89; scholar physics Johns Hopkins, 1889-90; fellow physics, 1890-91; successively asso., asso. prof., prof. physics Bryn Mawr Coll., 1891-1905; prof. physics Dalhousie U., Halifax, N.S., 1905-10; prof. physics Stevens Inst. Tech., 1910-11; pres., Dalhousie U., Halifax, 1911-31, emeritus pres., 1931——. Chmn., Econ. Council N.S.; mem. Nat. Research Council Can. Mem. Am. Phys. Soc., Am. Philos. Soc. Fellow A.A.-A.S., Royal Soc. Can. Author: The Laws of Gravitation, 1900; also articles. Died Oct. 1938.

MACKENZIE, Cosmo Glenn, Am. biochemist; b. Balt., May 22, 1907; s. George Norbury III and Mary (Forwood) M.; A.B., Johns Hopkins, 1932, Sc.D., 1936; m. Julia Frances Buzz, Apr. 5, 1936; children—Thomas B., Julia A. V. Research asso. in biochemistry Johns Hopkins Sch. Hygiene and Pub. Health, 1936-38, asst. prof., 1938-42; research asso. Cornell U. Med. Coll., 1946-48, asst. prof. biochemistry, 1948-49, asso. prof., 1949-50; prof., chmn. dept. biochemistry U. Colo. Sch. Medicine, Denver, 1950——. Discovered antithyroid action of thioureas, antidystrophic activity of vitamin E, formation and oxidation of one-carbon compounds in animals. Home: 1 Bellaire St., Denver 80220.*

MACKENZIE, Donald Robertson, chemist; b. Beausejour, Man., Can., Dec. 9, 1921; s. Frederick Donald and Wilma (Robertson) MacK.; B.A., Queen's U., Kingston, Ont., 1943, M.A., 1944; Ph.D., U. Toronto, 1950; m. Susan Veronica Felstead, Oct. 15, 1949; children—Donald Hugh, Ian Robertson, Catriona Mary Jean. Asst. research officer Atomic Energy Can., Ltd., Chalk River, Ont., 1950-58; chemist Brookhaven Nat. Lab., Upton, N.Y., 1958——. Mem. Am. Chem. Soc., Am. Nuclear Soc., Fedn. Am. Scientists. Research, publs. on nuclear energy levels, effects of radiation and high temperature on fluorocarbons, noble gas compounds, developing use of cyclotron beams of high energy particles together with low temperature for synthesis of normally unstable inorganic fluorides. Home: 4 George Ct., Bellport, N.Y. 11713. Office: Upton, N.Y. 11973.*

MACKENZIE, Sir James, Brit. physician; b. Apr. 12, 1853; s. Robert Mackenzie; ed. Edinburgh (Scotland) U.; M.D.Ed.; LL.D., Aberdeen and Edinburgh, Scotland; m. Frances Jackson, 1887; 1 dau. Apptd. house physician, asst. to profs. clin. medicine Royal Infirmary, Edinburgh; practiced medicine, Burnley, 1879-1907; cons., London, 1907-18; Liver-Sharpey lectr. Royal Coll. Physicians, 1911; Schorstein lectr. London Hosp., 1911; Gibson Meml. lectr. Edinburgh Coll. Physicians, 1914. Fellow Royal Coll. Physicians, Royal Soc., 1915. Author: The Study of the Pulse, . . . , 1902; Diseases of the Heart, 1908; Symptoms and their Interpretation, 1918; Principles of Diagnosis and Treatment in Heart Affections, 1916; The Future of Medicine, 1919; also numerous articles. Demonstrated action of digitalis in auricular fibrillation, 1905; invented phlebograph to record heart action (forerunner of polygraph), 1892; studied herpes zoster (shingles). Died Jan. 26, 1925.

MACKENZIE, James Kenneth, Australian physicist; b. Melbourne, Australia, June 14, 1920; s. Robert Kenneth and Alice (Beavis) M.; B.A. with honors, (Dixon Research scholar in math.), B.Sc., U. Melbourne, 1941; Ph.D., U. Bristol (Eng.), 1950; m. Zara Lennard, Dec. 17, 1948; 1 dau., Terri Anne. Engr., A.W.A. Research Lab., 1942-45; staff Commonwealth Sci. and Indsl. Research Orgn., 1945-47, prin. research officer div. tribophysics, Melbourne, Australia, 1950-62, sr. prin. research scientist, div. chem. physics, 1964——; vis. lectr. Mass. Inst. Tech., Boston, 1963; vis. prof. U. Ill., Urbana, 1963. Vis. scientist Research Inst. for Advanced Study, hon. vis. fellow Johns Hopkins, Balt., 1963. Fellow Inst. Physics, and Phys. Soc. (past sec. Australian br.), Australian Inst. Physics; mem. Australian Math. Soc., Statis. Soc. Australia, Australia and New Zealand Assn. for Advancement Sci. Research, publs. on devel. theory of geometrical relations arising when crystalline solid undergoes change of phase from one crystal structure to another such as martensitic transformation in steel; studies in theoretical yield strength of solids, theory of sintering, geometrical probability, crystalography of surfaces. Home: 15 Ronald St., Box Hill N., Melbourne, Victoria. Office: David Rivett Lab., Box 160, Clayton, Victoria, Australia.*

MACKENZIE, John Douglas, phys. chemist; b. Hong Kong, Feb. 18, 1926; s. John and Hannah (Wong) M.; B.Sc., London (Eng.) U., 1952, Ph.D., 1954; m. Jennifer Russell, Oct. 2, 1954; children—Timothy, Andrea, Peter. Came to U. S., 1954, naturalized, 1962. Fellow, lectr. Princeton, 1954-56; I.C.I. fellow, demonstrator Cambridge (Eng.) U., 1956-57; research scientist Gen. Electric Research Lab., Schenectady, 1957-63; prof. materials sci. Rensselaer Poly. Inst., Troy, N.Y., 1963——. Cons. Pitts. Plate Glass Co., 1963——; chmn. com. on elec. properties glass Internat. Glass Commn., 1964——. Mem. Am. Chem. Soc., Am. Ceramic Soc. (Meyer award 1964), Electrochem. Soc., Soc. Glass Tech., Brit. Royal Inst. Chemistry. Author: Modern Aspects of the Vitreous State, vol. I, 1960, vol. II, 1962, vol. III, 1964; (with Bockris, White) Physical Chemical Measurements of High Temperatures, 1958; also numerous articles. Research on glass structure; discovered numerous new glasses, new crystallin compounds and ceramic materials. Home: 839 Maxwell Dr., Schenectady 12309. Office: Rensselaer Poly. Inst., Troy, N.Y. 12181.*

MACKENZIE, Kenneth Ross, Am. physicist; b. Portland, Ore., June 15, 1912; s. Thomas Purvis and Beulah (Westlake) Mack.; B.A., U. B.C., Can., 1935, M.A., 1937; Ph.D., U. Cal. at Berkeley, 1937; m. Lilian Annie Stark, Oct. 17, 1937; children—Robert, Maryann, Wallace. Faculty, U. Cal., Berkeley, 1940-45, U. B.C., Can., 1946-47; faculty U. Cal., Los Angeles 1947——, prof. physics, 1953——. Mem. Am. Phys. Soc. Developer radio frequency systems for accelerators, 1947; researcher in plasma physics, 1961——. Home: 646 Haverford Av., Pacific Palisades, Cal. 90772.*

MACKENZIE, Sir Morell, Brit. physician, laryngologist; b. Leytonstone, Eng., July 7, 1837; s. Stephen M. Mackenzie; ed. London Hosp., also Paris, Vienna, Perth; M.D., U. London, 1862; m. Margaret Bouch, 1863. Specialized in throat diseases; helped found Hosp. of Diseases of Throat, Golden Sq., London, 1863; attending physician to crown prince of Germany (later Emperor Frederick III), 1887. Knighted, 1887. Recipient Jacksonian prize Royal Coll. Surgeons, 1863. Author: Treatment of Hoarseness and Loss of Voice, 1863; On the Pathology and Treatment of Diseases of the Larynx; Use of the Laryngoscope, 1865; Manual of Diseases of the Throat and Nose, 1880; Frederick the Noble (work justifying treatment of Frederick, censored by Royal Coll. Surgeons), 1888. First English expert in larynx and throat operations. Died London, Feb. 3, 1892.

MACKENZIE, Richard James, Scottish surgeon; b. 1821; resident clerk Royal Infirmary, Edinburgh; physician in Crimean War. Author treatise on excision of knee-joint, 1853. Modified Syme's amputation at ankle by internal flap (MacKenzie's amputation); described successful operation for ligation of subclavian artery for hemorrhage from auxillary. Died 1854.

MACKENZIE, Robert Earl, Am. mathematician; b. Los Angeles, Mar. 17, 1920; s. Reginald Pinney and Julia (Kendig) MacK.; B.S., Cal. Inst. Tech., 1942; A.M., Princeton, 1947, Ph.D., 1950; m. Mildred Anne Cheatham, June 19, 1950; children—Nona Deane (Mrs. Marc Estrin), Charlotte Curtis, Margaret Delia, Anne Robin. Physicist, Naval Ordnance Lab., 1942-45; faculty Ind. U., Bloomington, 1950——, asso. prof. math., 1962——, asst. dir. Grad. Inst. for Math. and Mechanics, 1960-62, asst. chmn., 1962——. Mem. Am. Math. Soc., Sigma Xi. Author: (with Louis Auslander) Introduction to Differentiable Manifolds, 1963. Editor: Jour. Math. and Mechanics, 1956-62. Research in theory of numbers and other related algebraic structures. Home: R.F.D. 2, Box 267, Bloomington, Ind. 47401.*

MACKENZIE, Walter Campbell, Canadian surgeon; b. Glace Bay, N.S., Can., Aug. 17, 1909; s. John Kenneth and Anna (Macaulay) MacK.; B.Sc., Dalhousie U., 1929, M.D., C.M., 1933; M.S. in Surgery U. Minn., 1937; LL.D., McGill U., 1965, Dalhousie U., 1966; m. Dorothy Martin Rosier, June 25, 1938; children—Kenneth Claude, Richard Bruce, Sally Ann. Practice surgery, Edmonton, Alta., Can., 1938-40, 46——; surg. staff U. Alta. Hosp., Edmonton, 1946——, clin. prof. surgery U. Alta. Med. Sch., 1948-50, prof., 1950——, chmn. dept. surgery, 1950-60, dean faculty medicine, 1959——. Cons., Dept. Vets. Affairs No. Alta., 1948-66; adviser in surgery to dir. gen. treatment services Dept. Vets. Affairs, 1966——; hon. surgeon Her Majesty the Queen, 1966——. Recipient Malcolm Honor Soc. medal Dalhousie U., Halifax, N.S., 1933; Outstanding Achievement award U. Minn., 1964. Hon. fellow Royal Coll. Surgeons Eng., Royal Coll. Surgeons Edinburgh, Assn. Surgeons of Gt. Britain and Ireland; mem. Def. Med. Assn. Can. (past pres.), Royal Coll. Phys. and Surg. Can. (past pres.), A.C.S. (pres. 1966——), Royal Soc. Medicine, Internat. Surg. Group, James IV Assn. Surgeons, Internat. Soc. Surgery, Am., Western surg. assns., Assn. Canadian Med. Colls., Assn. Am. Med. Colls. Research, publs. on gastrointestinal diseases especially pancreatitis. Home: 10131 Clifton Pl. Office: U. Alta Hosp., Edmonton, Alta., Can.*

MACKENZIE, William, Scottish ophthalmologist; b. Glasgow, 1791; s. James M.; student U. Glasgow; M.D., Royal Infirmary, Glasgow, 1815; postgrad. St. Bartholomew's Hosp., London; 1 son. Became resident clerk Royal Infirmary, 1813; began gen. practice medicine, Glasgow, 1819; lectr. Anderson's Coll., Glasgow; founder (with Monteath) Eye Infirmary, 1824; became lectr. on eye U. Glasgow, 1828; named surgeon-oculist to Queen of Scotland, 1838. Mem. Royal Coll. Surgeons, London. Author: Practical Treatise on the Diseases of the Eye, 1830; The Physiology of Vision, 1841; Outlines of Ophthalmology, 1856. Helped elevate ophthalmology as br. of medicine; 1st to describe sympathetic ophthalmia as distinct entity, also probably made 1st assn. of increased intraocular pressure with glaucoma, introduced concept and term asthenopia, 1830. Died 1868.

MACKEY, George Whitelaw, Am. mathematician; b. St. Louis, Feb. 1, 1916; s. William Sturges and Dorothy Frances (Allison) M.; B.A., Rice Inst., 1938; A.M., Harvard U., 1939, Ph.D., 1942; M.A., Oxford, 1966; m. Alice Willard, Dec. 9, 1960; one daughter, Am. Sturges Mackey. Instr. mathematics Ill. Inst. Tec 1942-43; faculty instr. mathematics Harvard, 1943-46, asst. prof., 1946-48, asso. prof., 1948-56, prof. mathematics, 1956——; vis. prof. U. Chgo., summer 1955, U. Cal. Los Angeles, summer 1959; Walker Ames vis. prof. University of Washington, summer 1961; Eastman visiting professor at Oxford, 1966-67. Served as civilian, operational research sect. 8th Air Force, 1944, applied math. panel NDRC, 1945. Guggenheim fellow, 1949-50, 61-62. Mem. Am. Mathematical Society (vice president 1964-65), also National Acad. Scis., Société Mathematique de France, Am. Acad. Arts and Scis., Phi Beta Kappa, Sigma Xi. Author: Mathematical Foundations of Quantum Mechanics, 1963. Contbr. articles math. jours. Study of abstract analysis; infinite dimensional vector spaces; infinite dimensional representations of locally compact groups and applications to quantum mechanics. Home: 25 Coolidge Hill Rd., Cambridge 38, Mass.

MACKIN, Joseph Hoover, Am. geologist; b. Oswego, N.Y., Nov. 16, 1906; s. William David and Catherine (Hoover) M.; B.S., N.Y. U., 1930; M.A., Columbia, 1932, Ph.D., 1937; m. Esther Fisk, Sept. 16, 1930; children—Barbara Catherine, Robert Fisk. Mem. faculty U. Wash., 1934-61, prof. geology, 1946-61; Farish prof. geology, U. Tex., 1961——; part-time geologist U. S. Geol. Survey, 1943-54. Cons. engring. geology to govt., state hwy. depts., bus. firms. Chairman div. of earth scis. NRC, 1965-67. Fellow Geol. Soc. Am. (chmn. cordilleran sec. 1950, councillor 1950-53); mem. Nat. Acad. Scis., Soc. Econ. Geology, Am. Geophys. Union, Am. Assn. Petroleum Geologists (distinguished lectr. 1953), A.A.A.S., Sigma Xi (nat. lectr. 1953), A.A.A.S., Sigma Xi (nat. lectr. 1963). Research on geologic work of rivers; geomorphology; postorogenic deformation in Cordilleran region; engineering geology. Home: 300 Mt. Larson Rd., Austin, Tex. 78746.

MACKINDER, Halford John, Brit. geographer; b. Gainsborough, Feb. 15, 1861; s. Draper Mackinder; student Christ Ch., Oxford, 1892-1903, also Epson Coll.; m. Emilie Catherine Ginsburg. Barrister, Inner Temple, 1886; reader geography Oxford U., 1887-1905; prin. Univ. Coll., Reading, 1892-1903; reader, later prof. geography U. London, 1900-25; dir. London Sch. Econs. and Polit. Sci., 1903-08. Mem. Parliament, Camlachie Div., Glasgow, 1910-22; Brit. high commr. for S. Russia, 1919-20; chmn. Imperial Shipping Com., 1920-45, Imperial Econ. Com., 1926-31; mem. Royal Commns. on Income Tax, 1912, other coms. Recipient Daly Gold medal Am. Geog. Soc., 1943, Patron's Gold medal Royal Geog. Soc. London, 1945. Mem. Berlin, Hungarian, Royal (v.p. 1933-36) geog. socs., Oxford Union Soc. (pres. 1883). Author: Britain and the British Seas, 2d edit., 1907; The Rhine, Its Valley and History, 1908; Elementary Studies in Geography, 18th edit., 1930; Eight Lectures on India, 1910; Democratic Ideals and Reality, 1919; also paper The Geographical Pivot of History (evaluation of conflicts between peoples of heartland—regions accessible to horsemen, e.g., Russia—and coastland—regions accessible to shipmen). Made 1st ascent of Mt. Kenya, 1899; promoted teaching of world and regional geography. Died Parkstone, Dorset, Mar. 6, 1947.

MACKINNEY, Gordon, biochemist; b. London, Eng., May 28, 1905; s. Frederick W. and Minnie L. (Brenchley) M.; B.S.A., Ont. Agrl. Coll., Guelph, 1926; Ph.D., U. Cal. at Berkeley, 1933; m. Marion A. Esdale, May 18, 1928; 1 son, John G. Came to U. S., 1930, naturalized, 1937. Faculty, U. Cal., Berkeley, 1936——, prof. food technology, 1949——. Cons. Office Q.M. Gen., 1943-45; mem. Tech. Intelligence Com. Joint Chiefs of Staff, 1945. Mem. Am. Soc. Biol. Chemists, Am. Chem. Soc., Inst. Food Technologists, Am. Soc. Plant Physiologists, Academie d'Agriculture de France (fgn. mem.). Author: (with Angela C. Little) Color of Foods, 1962. Asso. editor Ann. Rev. Biochemistry, 1947-65. Research, publs. on chloroplast pigments, their changes in vivo and in vitro, measurement of color of foods. Office: Dept. Nutritional Scis., U. Cal., Berkeley, Cal. 94720.*

MACKINNON, Lachlan, Brit. physicist; b. Aldershot, Eng., Feb. 17, 1918; s. Lachlan and Marjory (Gordon) M.; B.A., Clare Coll., Cambridge, Eng., 1940, M.A., 1945, Ph.D., 1949; m. Kathleen Jean MacDonald, Aug. 31, 1955; children—Lachlan, Alastair, John. Research asso. dept. physics Brown U., 1953-55; lectr. U. Leeds (Eng.), 1956-61, sr. lectr., 1961-63; vis. prof., NSF Sr. Fgn. scientist Wayne State U., 1963-64; reader physics U. Essex (Eng.), 1964——. Mem. Am. Phys. Soc., Phys. Soc., Cambridge Philos. Soc., Sigma Xi. Author: Textbook on Light, 1961; Experimental Physics at Low Temperatures, 1966; also articles. Research in cryogenics especially magnetoacoustics. Home: Cumberland House, South Hill, Manningtree, Essex, Eng. Office; Dept. Physics, U. Essex, Wivenhoe Park, Colchester, Essex, Eng.*

MACKINTOSH, Allan Roy, physicist; b. Nottingham, Eng., Jan. 22, 1936; s. Malcolm Roy and Alice (Williams) M.; B.A., U. Cambridge (Eng.), 1957, Ph.D., 1960; m. Jette Stannow, Aug. 30, 1958; children—Anne Karen, Paul Erik, Ida Alys. Faculty, Ia. State U., 1960-63; asso. prof., 1964-66; vis. scientist Danish Atomic Energy Commn., 1963-64; physicist Ames (Ia.) Lab., U. S. AEC, 1964-66; prof. physics Tech. U., Copenhagen, Denmark, 1966——. Alfred P. Sloan fellow, 1964-66. Fellow Phys. Soc. (London), Am. Phys. Soc. Research, publs. on absorption of ultrasonic waves in metals at very low temperatures leading to improved understanding of electronic structure of metals, exptl. and theoretical study magnetic metals especially chromium andrare earth metals. Home: 4 Henrik Thomsensvej, Birkerod, Denmark. Office: Lab. of Electrophysics, Tech. Univ., Lyngby, Denmark.*

MACKLER, Bruce, Am. biochemist; b. Phila., May 23, 1920; s. Louis and Fanny (Albertman) M.; student Mass. Inst. Tech., 1936-38; M.D., Temple U., 1943; postgrad. U. Cin., 1948-49; m. Jeanne Underwood, June 18, 1949; children—Robert, Suzanne. Asst. prof. pediatrics U. Cin. Coll. Medicine, 1953-54; asst. prof. enzyme chemistry Inst. for Enzyme Research, U. Wis., Madison, 1953-55; research asso. Johnson Found., U. Pa., 1955; established investigator Am. Heart Assn., Madison, Seattle, 1955-60; prof. pediatrics U. Wash. Sch. Medicine, Seattle, 1961——; vis. prof. Laboratoire de Genetique Physio, Gif-sur-Yvette, France, 1962-63. Recipient Outstanding Alumnus award Temple U., 1964. Mem. Am. Soc. Biol. Chemists, Am. Soc. for Cell Biology, Soc. for Pediatric Research, Western Soc. for Pediatric Research. Research, numerous publs. on oxidative enzymes, terminal electron transport, metalloenzymes, flavin enzymes, congenital malformations, inherited biochem. diseases, carbohydrate metabolism. Home: 3811 40th N.E. St., Seattle.*

MACKLIN, Richard Lawrence, Am. physicist; b. Jamaica, N.Y., Dec. 24, 1920; s. Egbert Chalmer and Margaret (Griswold) M.; B.S., Yale, 1941, Ph.D., 1944; m. Grace Virginia Walz, Oct. 6, 1945; children—Carol Elizabeth, Roger Lawrence, Donald Dixon, Linda Joyce. Various positions Oak Ridge Gaseous Diffusion Plant, 1944-52, researcher nuclear structure physics div., 1952——. Fellow A.A.A.S., Am. Phys. Soc.; mem. Research Soc. Am., Sigma Xi, Phi Beta Kappa. Author: (with H. S. Pomerance) Progress in Nuclear Energy vol. I, 1956. Patentee nuclear physics field. Research in measurements of neutron reactions with nuclei, and nuclear gamma rays; inventor detection equipment; reactor design and astrophys. studies. Home: 225 Outer Dr. Office: Oak Ridge Nat. Lab., P.O. Box X, Oak Ridge 37830.*

MACKOWSKY, Marie Therese, German mineralogist; b. Coblenz, Germany, Dec. 7, 1913; d. Fritz and Annie (Russell) Mackowsky; student U. Freiburg (Germany), 1933-34, U. Königsberg (Germany), 1934; Dr.-rer.nat., U. Bonn (Germany), 1938. Asst., Verein für die bergbau. Interessen, 1940-44; faculty Bergakemie Clusthal, 1944-50; lectr. U. Münster (Germany), 1951, apl. prof., 1957-66, hon. prof., 1966——; dir. Petrographic-phys. lab. Steinkohlenbergbauverein, 1951-57; dir. mineral.-petrographic dept. Bergbau-Forschung, Essen, Germany, 1957——. Mem. Deutsche Mineralogische Gesellschaft, Deutsche Geologische Gesellschaft, Am. Mineral. Soc., Inst. Fuel. Author: (with others) Encyclopädie der technischen Chemie Sammelwerk des deutschen Bergbaus, 1957-65. also articles. Devel. new research methods for applied coal petrography, transition coal into coke, mineral content coal, power sta. problems especially dirty heating surface, slag refining. Home: 25 Mooren. Office: 351 Frillendorfer, 43 Essen, Germany.*

MACLANE, Saunders, Am. mathematician; b. Taftville, Conn., Aug. 4, 1909; s. Donald Bradford and Winifred (Saunders) MacL.; Ph.B., Yale, 1930; A.M., Univ. of Chicago, 1931; D.Phil., Göttingen, Germany, 1934; Doctor of Science (hon.), Purdue University, 1965; m. Dorothy M. Jones, July 21, 1933; children —Margaret Ferguson, Cynthia R. Hay. Sterling Research fellow, Yale, 1933-34; Benjamin Pierce instr., Harvard, 1934-36; instr. Cornell Univ., 1936-37, Univ. of Chicago, 1937-38; asst. prof. Harvard, 1938-41, asso. prof. 1941-46, prof. 1946-47; professor of mathematics, U. of Chgo., 1947——, chmn. dept., 1952-58, Max Mason Distinguished Service prof. of math., 1963——; mem. council Nat. Acad. Sci., 1958-61; exec. com. mem. Internat. Math. Union, 1954-58; research mathematician Applied Math. Group, Columbia, 1943-44, dir., 1944-45. John Simon Guggenheim fellow, 1947-48. Awarded Chauvenet prize for mathematical exposition by Mathematical Association of Am., 1941, Montclair Yale Cup, 1929. Mem. Am. Math. Soc. (vice pres. 1946-47, mem. council, 1939-41), Math. Assn. of Am. (v.p. 1948-49, pres. 1950-52), Nat. Acad. of Sci. (council 1958-61; chmn. editorial board Proc. 1960——), American Philos. Society (member of council 1960-63), Association for Symbolic Logic (mem. exec. committee 1945-47), Am. Acad. of Arts and Sci., Phi Beta Kappa, Sigma Xi.

Author: (with Garrett Birkhoff) Survey of Modern Algebra, 1942; Homology, 1963; (with Garrett Birkhoff) Algebra, 1967. Editor Bulletin of the American Mathematics Society, 1943-45, mng. editor, 1946-47; editor, Transactions Am. Math. Soc., 1949-54. Chmn. editl. com. Carus Math. Monographs, 1940-45. Contbr. various sci. articles on mathematics to the Bull. of Am. Math. Soc., Transactions of the Am. Math. Soc., Duke Math. Jour. Described separable class of transcendental extensions (basic result now known as MacLane's theorem), 1939; Studies on elucidation and develop. of interrelationships between algebra and geometry, algebraic topology, also systematic study of connectivity of topological space measured by homotopy groups and cohomology groups; devised MacLane's cohomology of rings; research on notion of category as organizational tool in math. Home: 5712 S. Dorchester Av. Office: Eckhart Hall, Univ. of Chicago, Chicago 60037.*

MACLAURIN, Colin, Scottish mathematician; b. Kilmodan, Scotland, Feb. 1698; s. John Maclaurin; M.A., U. Glasgow (Scotland); m. Anne Stewart, 1733; 7 children, including John, Colin. Prof. math. Marischall Coll., Aberdeen, Scotland, 1715-26; named dep. prof. Edinburgh (Scotland) U. (salary paid by Newton), 1725; organized defense of Edinburgh, 1745; fled from rebels to York, Eng. Recipient prize for work on percussion of bodies French Acad. Scis., 1724, (with Euler and Daniel Bernoulli) prize for essay on tides, 1740. Fellow Royal Soc. Author: Geometria organica, sive descriptio linearum, 1720; A Treatise of Fluxions, 1742; A Treatise of Algebra, pub. 1748; An Account of Sir Isaac Newton's Philosophy, pub. 1748; also various papers. Ranked with continental mathematicians of the day; developed Newton's method of calculus; introduced method of generating conics (named in his honor); demonstrated that homogeneous rotating fluid mass revolves in elliptic figure (Maclaurin's ellipsoids); showed that 3d and 4th degree curves can be described by the intersection of 2 movable angles; 1st to give correct theory for distinguishing maxima from minima. Died York, June 14, 1746.

MACLEAN, John, chemist; b. Glasgow, Scotland, Mar. 1, 1771; s. John and Agnes (Lang) M.; grad. U. Glasgow, 1791; M.D., U. Aberdeen, 1797; m. Phebe Brainbridge, Nov. 7, 1798, 2 daus., 4 sons including John. Came to U. S., 1795; prof. chemistry and natural history Coll. of N.J. (now Princeton), 1795-1812, also prof. math. and natural philosophy, 1797-1812; mem. Am. Philos. Soc., 1805; naturalized Am. citizen, 1807; prof. natural philosophy and chemistry Coll. William and Mary, Williamsburg, Va., 1813; returned to Princeton, 1814. Author: Two Lectures on Combustion: Supplementary to a Course of Lectures on Chemistry Read at Nassau Hall; Containing an Examination of Dr. Priestley's Considerations on the Doctrine of Phlogiston, and the Decomposition of Water, 1797. Taught erroneousness of phlogiston theory. Died Princeton, N.J., Feb. 17, 1814.

MACLEAN, Lloyd D., Canadian surgeon; b. Calgary, Alta., Can., June 15, 1924; s. Fred Hugh and Azilda (Trudelle) MacL.; B.Sc., U. Alta., 1946, M.D., 1949; Ph.D., U. Minn., 1957; m. Eleanor Colle, June 30, 1954; children—Hugh, Charles, Ian, James, Martha. Surgeon-in-chief Ancker Hosp., St. Paul, 1956-62, asso. prof. surgery, 1956-62; surgeon-in-chief Royal Victoria Hosp., Montreal, Que., Can., 1962——; prof. surgery McGill U., Montreal, 1962——. Bd. dirs. Life Ins. Med. Research Fund. Mem. Nat. Cancer Inst. Can. (dir.), Soc. U. Surgeons, Am., Central surg. assns., Am. Assn. for Thoracic Surgery, Halsted Soc., Canadian Soc. for Clin. Investigation. Research, numerous publs. on shock and transplantation. Home: 5 Renfrew Av., Montreal 6. Office: Royal Victoria Hosp., Pine Av. W., Montreal 2, Que., Can.*

MACLEAN, Paul Donald, Am. physician; b. Phelps, N.Y., May 1, 1913; s. Charles C. and Elizabeth (Dreyfus) MacL.; B.A., Yale, 1935, M.D. cum laude, 1940; postgrad. U. Edinburgh (Scotland); m. Alison Stokes, July 16, 1942; children—Paul Donald, David B., Alexander S., James T., Alison S. Clin. instr. U. Wash. Sch. Medicine, 1946-47; practice internal medicine, Seattle, 1946-47; USPHS fellow Mass. Gen. Hosp., Harvard Sch. Medicine, 1947-49; dir. electroencephalographic lab. New Haven Hosp., 1951-52; faculty Yale, 1949-57, asso. prof. physiology, 1956-57; Sr. NSF fellow Inst. Physiology, Zurich, Switzerland, 1956-57; chief sect. on limbic integration and behavior Nat. Inst. Mental Health, NIH, Bethesda, Md., 1957——. Recipient award for distinguished research Assn. for Research in Nervous and Mental Disease, 1964. Mem. Am. Physiol. Soc., Am. Neurol. Assn., Am. Electroencephalographic Soc., Am. Soc. Naturalists, Am. Assn. History Medicine, Am. Soc. for Cybernetics, Eastern Assn. Electroencephalographers, Sigma Xi, Alpha Omega Alpha. Author: (with J. A. Gergen) A Stereotaxic Atlas of the Squirrel Monkey's Brain, 1962; also articles on functions and anatomy of the limbic system. Research on brain mechanisms of emotion. Home: 9916 Logan Dr., Potomac, Md. 20854. Office: Bldg. 10, NIH, Bethesda, Md. 20014.*

MACLEAR, Sir Thomas, astronomer; b. Newtown Stewart, Ireland, Mar. 17, 1794; s. James MacLear; student medicine Guy's, St. Thomas hosps.; self-ed. astronomy, math.; m. Mary Pearse, 1825. Became

house surgeon (Eng.) Bedford Infirmary, circa 1815; partner with uncle at Biggleswade, Eng., from 1823; interest in astronomy from 1829; royal astronomer Cape of Good Hope, 1834-70. Mem. commn. weights and measures, promoter sanitary improvement in South Africa. Mem. Royal Coll. Surgeons, Eng. Recipient Laland prize, 1867, Royal medal, 1869. Fellow Royal Soc., 1831; mem. Royal Coll., Surgeons, Eng., French Acad. Sci., 1863, Astron. Soc., Acad. Sci. Palermo, Sicily, Imperial Geog. Inst. Vienna, Austria. Research, publs. on observations including meteoric showers, tides, magnetic field, many of major comets; Mars, 1862 (later used by Stone, Winnecke, and Newcomb as new determination distance of sun), all the southern stars, 1849-52, maximum of Eta Argus, 1843; remeasured, extended Lacaille's arc, 1837-47; made determinations of Alpha Centauri, 1839-40, 42-48; made set of measures of southern double stars; established elec. communication time-signals to Port Elizabeth and Simon's Town (South Africa), also lighthouses in South Africa. Died Cape Town, July 14, 1879.

MACLEOD, John James Rickard, Scottish physiologist; b. New Clunie, Scotland, Sept. 6, 1876; s. Robert Macleod; M.B., Ch.B. (Anderson scholar); student Leipzig, Germany; D.P.H., Cambridge, Eng., 1902; D.Sc., Toronto U., 1923, U. Pa., 1928, Jefferson Med. Coll., 1928; LL.D., Aberdeen, 1924, Western Res. U., 1928; m. Mary McWalter. Demonstrator physiology London Hosp., 1899-1902, became lectr. biochemistry, 1902; Mackinnon research scholar Royal Soc., 1901-03; prof. physiology Western Res. U., Cleve., 1903-18; prof. physiology Faculty Medicine, U. Toronto (Ont., Can.), 1918-28, also asso. dean; became Regius prof., Aberdeen, Scotland, 1928-35. Mem. Med. Research Council, 1929-33. Recipient: Cameron prize Edinburgh (Scotland) U., 1923; (with Banting) Nobel prize in physiology and medicine, 1923. Fellow Royal Soc., 1923, Royal Coll. Physicians, Royal Soc. Can., Royal Soc. Edinburgh, Acad. Medicine Toronto (hon.), Coll. Physicians Phila. (fgn. asso.); mem. Royal Canadian Inst. (became pres. 1925), Am. Physiol. Soc., Soc. Exptl. Biology, Soc. Biol. Chemistry, Assn. Am. Physicians, A.A.A.S., London Physiol. Soc., Biochem. Soc.; corr. mem. Med. Chirurg. Soc. Bologna, Kaiserliche Deutsche Akad. der Naturforscher zu Halle. Author: Practical Physiology, 1903; Recent Advances in Physiology (edited by Leonard Hill), 1905; Diabetes, its Physiological Pathology, 1913; Fundamentals of Physiology, 1916; Laboratory Manual in Physiology, 1919; Carbohydrate Metabolism and Insulin, 1926; The Fuel of Life, 1928; Physiology and Biochemistry in Modern Medicine, 1918; 9th ed., 1941. also numerous articles Research on phosphorous content of muscles, lactic acid metabolism, physiology of respiration, chemistry of Tb. bacillus, electric shock, purine bases, biochemistry of carbonates; discovered (with Frederick Grant Banting, Charles Herbert Best) insulin, 1922; studied its use in treating diabetes. Died Aberdeen, Mar. 16, 1935.

MACLEOD, Robert Brodie, psychologist; b. Martintown, Ont., Can., Jan. 31, 1907; s. John B. and Helena (Brodie) MacL.; B.A., McGill U., 1926, M.A., 1927; postgrad. univs. Berlin, Frankfurt (Germany), 1928-29; Ph.D., Columbia, 1932; m. Beatrice Fullerton Beach, Oct. 17, 1936; children—Ian Fullerton, Alison Stuart. Came to U. S., 1929, naturalized, 1938. Faculty, Cornell U., 1930-33, 48——, Susan Linn Sage prof. psychology, 1948——; faculty Swarthmore Coll., 1933-46, McGill U., 1946-48. Cons. Dept. Def., NSF, Ford Found., NRC, Carnegie Corp., Social Sci. Research Council; civilian with OSS, ETO, 1942-45. Mem. Am., Canadian psychol. assns., A.A.A.S., Soc. for Religion in Higher Edn., Sigma Xi. Author: Undergraduate Education in the Liberal Arts and Sciences, 1958; (with D. Wofle et al): Improving Undergraduate Instruction in Psychology, 1952. Translator: The World of Colour (David Katz), 1935. Research in psychology of perception, lang. and thinking; methodology of cross-cultural study; contbns. to history of psychology, psychol. phenomenology. Home: 957 E. State St., Ithaca, N.Y. 14850.*

MACLOSKIE, George, biologist; b. Castledawson, Ireland, Sept. 14, 1834; s. Paul and Mary (McClure) M.; A.B., Queen's U., Ireland, 1857, A.M., 1858; LL.B., U. London, 1868, LL.D., 1871; D.Sc., Queen's U., 1887; A.M. (hon.), Princeton, 1896; m. Mary C. Dunn, Aug. 18, 1863 (dec. 1907); 2d, Lila M. Campbell, June 19, 1909. Ordained Presbyn. ministry, 1861; pastor Ballygoney, Ireland, 1861-72; prof. biology Princeton 1875-1906, emeritus prof., 1906. Author: Elementary Botany, 1887; Flora of Patagonia, 1905. Research on Insect Anatomy and Physiology. Died Jan. 4, 1920.

MACLURE, Kenneth Cecil, Canadian physicist; b. Montreal, Que., Can., Oct. 14, 1914; s. Alexander M. and Annie H. (Crocker) M.; B.Sc., McGill U., 1934, M.Sc., 1950, Ph.D., in Nuclear Physics, 1952; m. Alice Margaret Blackmore, May 28, 1949; children—Richard, Malcolm, Margaret Anne, Rowena. With Sun Life Assurance Co. of Can., 1934-39; commd. flying officer RCAF, 1939, advanced through grades to group capt., 1954; officer in charge test and devel. Empire Air Nav. Sch., Shawbury, Eng., 1944-46; head arctic research Def. Research Bd., Ottawa, Ont. 1947-48; dir. armament engring. Air Force Hdqrs., Ottawa, 1954-58; attache, Warsaw,

Poland, 1958-61; project leader Def. Research Bd., Pacific Naval Lab., Esquimalt, B.C., Can., 1961-—. Lectr. dept. physics U. Victoria, part-time 1967-—. Decorated Air Force Cross (Can.); recipient U. S. Inst. Nav. Thurlow award, 1945. Fellow U.K. Inst. Nav., Arctic Inst. N.Am., Canadian Aeros. and Space Inst. (asso.); mem. Canadian Assn. Physicists, U. S. Inst. Nav., Sigma Xi. Research, publs. on Greenwich Grid nav. for polar latitudes; in charge sci. observations for N.Polar flights; developed floating buoy for rec. micropulsations at sea. Home: 3290 Woodburn Av., Victoria, B.C. Office: Pacific Naval Lab., H.M.C. Dockyard, Esquimalt, B.C., Can.*

MACMAHON, Brian, physician; b. Sheffield, Eng., Aug. 12, 1923; s. Desmond and Gladys (Nelson) MacM.; M.B., Ch.B., U. Birmingham (Eng.), 1948, M.D. with honours, 1955, D.P.H., 1949, Ph.D., 1952; S.M. in Hygiene, Harvard, 1953; m. Heidi Graber, Aug. 28, 1948; children—Michael, Kevin, Kathleen, Mary. Came to U. S., 1952, naturalized, 1962. Research fellow U. Birmingham, 1949-52, lectr. dept. social medicine, 1954-55; asso. prof. environmental medicine State U. N.Y., Bklyn., 1955-57, prof., 1957-58; prof. epidemiology Harvard Sch. Pub. Health, Boston, 1958-—. Chmn., U. S. Nat. Com. on Vital and Health Statistics, 1960-62; chmn. adv. com. on biometry and epidemiology NIH, 1964-65; cons. WHO, Am. Cancer Soc., Am. Heart Assn. Fellow Am. Pub. Health Assn. (chmn. epidemiology sect. 1963-64), Am. Epidemiologic Soc. Author: Epidemiologic Methods, 1960; Epidemiologic Findings in U. S. Mental Hospital Data, 1962. Editor: Preventive Medicine, 1967. Home: 89 Warren St., Needham, Mass. 02192. Office: 1 Shattuck St., Boston 02115.*

MACMAHON, Harold Edward, pathologist; b. Aylmer, Ont., Can. Mar. 30, 1901; s. Hugh Percival and Ethel Clive (Holmes) MacM.; student St. Johns Coll., Ridley Coll., Can.; B.A. with honors, U. Western Ont., 1922, M.D., 1925, Sc.D. (hon.), 1944; m. Marian Ross, June 30, 1934; children—Elizabeth, Hugh, D'Arcy, James. Asst. pathology U. Hamburg, 1929-30, U. Berlin, 1930-31; instr. Harvard Med. Sch., 1928-29; prof. pathology, chmn. dept. Tufts U., 1931-—; pathologist-in-chief New Eng. Med. Center, cons. pathology asso. hosps.; pathologist-in-chief Mt. Auburn Hosp.; cons. Army, USPHS, VA, and Armed Forces Inst. Pathology. Recipient Tufts U. Alumni award, 1960. Diplomate Am. Bd. Pathology. Fellow A.A.A.S., Royal Coll. Physicians London; mem. A.M.A., Mass. Med. Soc., German Path. Soc., Sigma Xi, Alpha Omega Alpha. Devel. hosp. labs. in rural areas and larger cities New Eng.; investigation, research in anatomic pathology. Home: 19 Hubbard Park Rd., Cambridge, Mass. 02138. Office: 136 Harrison Av., Boston 02111.*

MACMAHON, Percy Alexander, Brit. mathematician; b. Malta, Sept. 26, 1854; s. Patrick W. and Ellen (Curtis) MacM.; ed. Cheltenham Coll., Royal Mil. Acad., Woolwich; D.Sc. (hon.), Trinity Coll., Dublin, 1897, Cambridge, 1904; LL.D., Aberdeen, St. Andrews, 1911; m. Grace Elizabeth Howard, 1907. Instr. math. Royal Mil. Acad., 1882-88; prof. physics Ordnance Coll., 1890-97; dep. warden Standards Bd. of Trade, 1906-20. Fellow, gov. Winchester Coll.; Brit. mem. Comité Internationale des Poids et Mesures, Paris, 1920. Fellow Royal Soc., 1890 (Royal medal 1900, Sylvester medal 1919, v.p. 1917), Royal Astron. Soc. (pres. 1917); mem. Cambridge Philos. Soc. (hon.), London Math. Soc. (de Morgan medal 1923, mem. council), Royal Soc. Arts (council), Royal Irish Acad. (hon.), Brit. Assn. (pres. math. and phys. sect. Glasgow 1901, gen. sec. 1902-14, trustee), Author: Combinatory Analysis, 1915; Introduction to Combinatory Analysis; New Mathematical Pastimes, 1921; also numerous papers on pure math. Advanced abstract algebra, especially group theory and quantum theory. Died Dec. 25, 1929.

MACMILLAN, William Duncan, Am. mathematician, astronomer; b. LaCrosse, Wis., July 24, 1871; s. Duncan D. and Mary Jane (MacCrea) MacM.; ed. Lake Forest Coll., 1888-90, Sc.D., 1930; ed. U. Va., 1895; A.B., Ft. Worth U., 1898; A.M., U. Chgo., 1906, Ph.D., 1908. Research asst. in geology U. Chgo., 1907-08, asst. in math. and astronomy, 1908-09, instr. astronomy, 1909-12, asst. prof., 1912-19, asso. prof., 1919-24, prof., 1924-36, prof. emeritus, from 1936. Fellow A.A.A.S., Royal Astron. Soc.; mem. Am. Math. Soc., Math. Assn. Am., Astron. and Astrophys. Soc. Am., Société Astronomique de France. Author: Statics and the Dynamics of a Particle, 1927; Theory of the Potential, 1930; Dynamics of Rigid Bodies, 1936; and many sci. memoirs. Contbd. to modern theory of differential equations with periodic coefficients, creation of theory of automorphic functions. Died Nov. 14, 1948.

MACNAE, William, zoologist; b. Middlebie, Scotland, June 30, 1914; s. James and Janet (Steel) M.; B.Sc. with honors, U. Glasgow (Scotland), 1937; Ph.D., U. Cape Town (S. Africa), 1956; m. Maron Maeson Walgate, Jan. 16, 1949; children—James Charles, William David, Andrew Percival, Janet Marion. Research asst. dept. oceanography U. Hull (Eng.), 1937-39; tchr. Bryanston Sch., Eng., 1941-46; jr. lectr. U. Cape Town, 1947-49; lectr. Rhodes U., Grahamtown, S. Africa, 1950-56; sr. lectr. zoology U. Witwatersrand, Johannesburg, S. Africa, 1957-—. Fellow Zool. Soc. London; m. Brit., Am. ecol. socs. Author: (with M. Kalk) A Natural History of Inhaca Island Mocambique, 1958; also articles. Research on gen. ecology of plants and animals on mud flats and mangrove swamps of tropical and sub-tropical Indo Pacific region. Home: 52 Fifth Av., Partown N., Johannesburg. Office: Zoology Dept., U. Witwatersrand, Johannesburg, S. Africa.*

MACNAIR, Peter, Scottish geologist; b. Glasgow, Scotland, Sept. 12, 1868; ed. Kinnoull Sch., Perth Acad.; m. Rebecca Mackenzie, 1894; 4 sons, 3 daus. In business, 1885-99; named curator Green Br. Mus., Glasgow, 1899; became asst. fine art and hist. sect. Glasgow Internat. Exhbn., 1901; named curator natural history collections Glasgow Museums, 1902; prof. zoology Anderson's Med. Coll., Glasgow; examiner geology U. Aberdeen (Scotland), 1908-10; became examiner biology Royal Faculty Physicians and Surgeons, Glasgow, for Triple Qualification of Scottish Licensing Bodies, 1922. Became tutor geology Workers Ednl. Assn., 1921; cons. geologist mining and civil engring. Fellow Royal Soc. Edinburgh, Geol. Soc.; mem. Glasgow Geol. Soc. (pres., editor Trans.). Author: The Building of the Grampians, 1903; The Geology of Rouken Glen, 1906; The Intrusive Dolerites in the Neighbourhood of Glasgow, 1907; The Geology and Scenery of the Grampians and the Valley of Strathmore, 2 vols., 1908; Editor, contbg. author: History of the Geological Society of Glasgow, 1909; Introduction to the Study of Minerals, 1910; Introduction to the Study of Rocks, 1911; Perthshire, Argyll and Bute, 1911-13; Introduction to the Study of Fossils, 1912. Research and publs. on Scottish paleozoic geology, geol. structure of Scottish Highlands. Died Mar. 28, 1929.

MACNEAL, Ward J., Am. bacteriologist; b. Fenton, Mich., Feb. 17, 1881; s. Edward and Jane Elizabeth (Pratt), MacN.; A.B., U. Mich., 1901, Ph.D., 1904, M.D., 1905, Sc.D., 1939; m. Mabel Perry, Dec. 28, 1905; children—Edward Perry (dec.), Herbert Pratt, Perry Scott, Mabel Ruth. Asst. and fellow in bacteriology U. Mich., 1901-04, instr. histology, 1905-06; instr. anatomy and bacteriology W. Va. U., 1906-07; asst. chief in bacteriology, Ill. Agrl. Expt. Sta., 1907-11; asst. prof. bacteriology, U. Ill., 1908-11; lectr. pathology and bacteriology, 1911-12, prof., asst. dir. labs., 1912-15, prof., dir. of labs., 1915-22, prof., dir. dept. pathology and bacteriology, 1922-24, 30-39, prof. bacteriology, from 1939, prof., dir. labs., 1924-30. Mem. Ill. Pellagra Commn., 1909-12; mem. Thompson-McFadden Pellagra Commn., Am. Trench Fever Commn., France, 1918. Fellow A.A.A.S.; mem. Soc. Am. Bacteriologists, Am. Assn. Pathologists and Bacteriologists (council 1929-35, pres. 1932), Assn. for Cancer Research (council 1925-33; pres. 1934), Nat. Assn. Tb, A.M.A., Soc. Exptl. Biology and Medicine, Harvey Soc., Sigma Xi. Author: Studies in Nutrition, Vols. I to V (with H. S. Grindley), 1911-29; Pathogenic Microörganisms, 1914, 2d edit., 1920. Contbr. to Marshall's Microbiology, 1917, 20, etc. Editor Third Report Thompson Pellagra Commission, 1917. Compounded (with Frederick George Novy) nutrient blood agar for cultivation of trypanosomes (known as Novy-MacNeal agar or medium), 1904; investigated endocarditis by exptl. methods. Died Aug. 15, 1946.

MACNEILL, Arthur Edson, Am. physician; b. Waltham, Mass., July 14, 1912; s. Charles Alfred and Florence (Wright) MacN.; A.B., Harvard, 1933, M.D., 1937; m. Lydia Mae Rhoades, June 25, 1941; children—Marjorie Ann, Arthur Edson II. Physician, Dartmouth College Health Service, 1938-41, acting med. dir., 1944-45, instr. anatomy, 1944-45; physician U. Fla., 1946-47; research asso. dept. physiol. scis. Dartmouth Med. Sch., 1946-50; research cons. Fgn. Service, U. S. Dept. State, 1950; dir. labs. Chronic Disease Research Inst., U. Buffalo, 1951-55, asst. research prof. surgery, 1957-61, asso. research prof. Sch. Medicine, 1961-64; research cons. Buffalo Gen. Hosp., 1958-64, research adviser Children's Hosp., Buffalo, 1958-65; dir. Inst. for Therapeutic Engring. (name later changed to Dialysis Research Inst.), Sunapee, N.H., 1964-—; research cons. industry, edn., hosps. Mem. Instrument Soc. Am. (dir. med., biol. div. 1956-57), A.M.A., A.A.A.S., Am. Heart Assn., Assn. Mil. Surgeons, N.Y. Acad. Sci., Sigma Xi. Research, publs. on uremia, heart disease; design of synthetic urinary bladder control mechanisms, blood pumps, blood dialyzer-ultrafilter, membrane blood oxygenator, blood heat exchangers, blood rheometers, blood flowmeters, diffusion computers for biol. use; other automatic treatment systems and research on hydraulic and rheological theories of arteriosclerosis. Home: Garnet Hill Rd. Office: Dialysis Research Inst., Box 334, Sunapee, N.H. 03782.*

MACNEISH, Richard Stockton, Am. archaeologist; b. N.Y.C., Apr. 29, 1918; s. Harris Franklin and Elizabeth (Stockton) MacN.; B.A., U. Chgo., 1940, M.A., 1944, Ph.D., 1949; m. Phyllis Diana Walters, Sept. 26, 1963; 1 son, Richard Roderick. Archeol. field work Peru to Arctic Ocean, 1936-51; sr. archaeologist Nat. Mus. Can., Calgary, Alta., 1949-63, research asso. R.S. Peabody Found., 1961-66; head dept. archaeology U. Calgary, 1965-68; dir. R.S. Peabody Found. for Archaeology-Andover, Mass., 1968-—. Recipient Spinden medal, 1965; Verrill medal, 1966; Lucy Drexel Wharton medal, 1966. Guggenheim fellow, 1955. Mem. Soc. Am. Archaeologists, Am. Anthrop. Assn., Sigma Xi. Contbg. author: Archaeology of the Eastern U. S., 1952; Iroquois Pottery Types, 1952; Kincaid, 1951; Panuco, 1954; Introduction to Manitoba Archaeology, 1958; Tamaulipas, 1958; Southwest Yukon, 1964; The Prehistory of the Tehuacan Valley. Contbr. numerous articles to profl. jours. Research on origin agr. and civilization in Mexico, prehistoric peopling in New World. Home: 113 Kirkland Dr. Office: Box 71, Andover, Mass. 01810.

MACNEVEN, William James, physician; b. County Galway, Ireland, Mar. 21, 1763; s. James and Rosa (Dolphin) MacN.; studied medicine U. Prague; M.D., U. Vienna, 1784; m. Jane Riker, 1810, 3 children. Participated in Irish Revolution, 1797-98, polit. prisoner, 1798-1802; served with Irish brigade French Army, 1804-05; came to N.Y.C., 1805; elected prof. obstetrics Coll. Physicians and Surgeons, 1808, prof. chemistry, 1812, taught materia medica, 1816-20; established 1st chem. lab. in N.Y.C.; co-editor N.Y. Med. and Philos. Jour. and Rev.; elected mem. Am. Philos. Soc., 1823; with colleagues founded med. sch. affiliated with Rutgers Coll., 1826-30. Author: Rambles through Switzerland in the Summer and Autumn of 1802, 1803; Pieces of Irish History, 1807; Expositions of the Atomic Theory, 1819 (well received at time, contained no original work but was 1st book on atoms in Am.). Editor: Brandes Chemistry, 1821. Died N.Y.C., July 12, 1841.

MACNIDER, William de Berniere, Am. pharmacologist; b. Chapel Hill, N.C., June 25, 1881; s. Virginius St. Clair and Sophia Beatty (Mallett) M.; U. N.C., 1898-1903, M.D., 1903; student U. Chgo.; spl. student Western Res. U.; D.Sc., Med. Coll. Va., 1933; LL.D., Davidson Coll., 1934; m. Sarah Foard, Jan. 23, 1918; 1 dau., Sarah Foard. Kenan prof. pharmacology U. N.C., 1905-51, Kenan research prof. pharmacology, 1920, dean med. sch., 1937-40; Harvey Soc. lectr., 1928-29; Chandler Meml. lectr. Columbia, Smith-Reed-Russell lectr. George Washington U. Sch. Medicine, 1938; Brown-Sequart lectr. Med. Coll. Va., 1938; Mayo Found. lectr., 1939; former spl. lectr. pharmacology Duke Med. Sch.; physician in chief pro tem Peter Bent Brigham Hosp., 1925; cons. on gerontology NIH; mem. Nat. Bd. Med. Examiners (chmn. exam. com.); mem. NRC com. on nutritional aspects of ageing; chmn. Am. div. Internat. Club for Research on Ageing. Recipient Kober medal Assn. Am. Physicians, 1941. Fellow A.A.A.S., Am. Acad. Arts and Scis., A.C.P.; mem. Am. Soc. for Pharmacology and Exptl. Therapeutics (pres. 1932-34, council), A.M.A. (chmn. sect. of pharmacology and therapeutics 1927), Am. Physiol. Soc., Am. Assn. Pathologists and Bacteriologists, Am. Assn. Biol. Chemists, Soc. Exptl. Biology and Medicine (pres. 1941-42), Am. Soc. Exptl. Pathology, Assn. Am. Physicians, Am. Assn. Anesthetists, Nat. Anaesthesia Research Soc. (research com.), Internat. Anaesthesia Research Soc. (pres. 1934-35), Elisha Mitchell Sci. Soc. (pres.), Am. Assn. Hist. Medicine NRC (com. on cellular physiology; exec. com. med. div.), Nat. Acad. Scis., N.Y. Acad. Scis., Am. Soc. Naturalists, Am. Philos. Soc., Harvey Soc. (hon.), Path. Soc. Gt. Britain and Ireland, Brit. Physiol. Soc. (asso. fgn. mem.), Gerontol. Soc. (chmn. council), Phi Beta Kappa, Sigma Xi (pres. U. chpt.), Phi Chi, Alpha Omega Alpha. Asso. editor Proc. Soc. for Exptl. Biology and Medicine, Quar. Jour. Alcohol Study, Jour. Pharmacology and Exptl. Therapeutics, also numerous articles. Research on prodn. of acute and chronic nephritis; toxicity of general anaesthetics for kidney with methods for protection; stability of acid-base equilibrium of blood in nephritis and in animals of different age periods; toxaemias of pregnancy, toxic action of alcohol, uranium and chloroform on liver, liver regeneration, resistance of atypical regenerated liver cells to above mentioned toxic agents, influence of liver degeneration and repair on acid base equilibrium of blood, resistance of fixed tissue cells. Died May 31, 1951; buried Chapel Hill.

MACON, Nathaniel, Am. mathematician; b. Durham, N.C., Nov. 15, 1926; s. Nathaniel and Helen (Leach) M.; B.A., U. N.C., 1946, M.A., 1948, Ph.D., 1951; postgrad. U. Amsterdam (Netherlands), 1951-52; m. Ruth Vivian Muse, Mar. 21, 1953; children—Patricia, Susan Caroline, Mary Ann, Nathaniel. Instr., U. N.C., 1946-50; prof. Auburn U., 1952-65; head. numerical analysis dept. Inst. for Def. Analyses, Arlington, Va., 1965-—; research participant Oak Ridge Nat. Lab., 1953-54; mathematician Gen. Electric Co., Evendale, O., 1956-57; chief br. SHAPE Tech. Center, The Hague, Netherlands, 1962-63. Cons. govt. agys. Mem. Am. Math. Soc., Math. Assn. Am., Assn. for Computing Machinery, Soc. Indsl. and Applied Math., Sigma Xi, Pi Mu Epsilon. Author: Numerical Analysis, 1963; also articles. Research in theory numbers and numerical analysis. Home: 6104 Namakagan Rd., Washington 20016. Office: 400 Army-Navy Dr., Arlington, Va. 22202.*

MACOVEI, Gheorghe, Rumanian geologist; b. 1880; mem. Acad. Socialist Republic Rumania. Author: Rumania's Geological Evolution—The Cretaceous System, 1934; Petroleum Fields, 1938; Stratigraphic Geology, 1954. Research in So. Dobrogea, flysch of Eastern Carpathians; studied stratigraphy and tectonics of Rumania, crude oil and salt reserves.

MACOVSCHI, Eugen, Rumanian chemist; b. 1906; prof. Bucharest (Rumania) U.; mem. Acad. Socialist Republic Rumania. Author: New Trends of Research in the Field of the Permeability of Membranes, 1950; Nature and Structure of Living Matter, 1963. Investigated mechanism of formation of nitroxiderivates, also permeability, constitution and activity of live membranes.

MACPHERSON, Archibald Ian Stewart, Brit. surgeon; b. Newtonmore, Scotland, Aug. 10, 1913; s. Thomas Stewart and Helen (Cameron) M.; M.B., Ch.B., U. Edinburgh (Scotland), 1936, Ch.M., 1953. Crichton Research scholar Edinburgh U., 1939-42; Rockefeller Research fellow Presbyn. Hosp., N.Y.C., 1948-49; faculty dept. surgery Edinburgh U., 1950-—, sr. lectr., 1965-—. Cons. surgeon Royal Infirmary Edinburgh, 1954-—, Royal Edinburgh Hosp., 1955-—, Leith Hosp., 1962-—. Fellow Royal Coll. Surgeons Edinburgh, Royal Soc. Edinburgh; mem. Scottish Soc. for Exptl. Medicine, Surg. Research Soc., Assn. Surgeons Gt. Britain, Vascular Surgery Soc. Research, publs. on deficiency of vitamin K. in newborn babes, synthetic oestrogens, disorders of spleen especially congenital saherocytosis, causation, effects and treatment of portal hypertension; replacement of diseased arteries with preserved homografts and synthetic fabric protheses. Home: 26 Learmonth Terrace, Edinburgh 4. Office: Royal Infirmary, Edinburgh 3, Scotland.*

MACPHERSON, C(rawford) Brough, Canadian polit. scientist; b. Toronto, Ont., Can., Nov. 18, 1911; s. Walter Ernest and Elsie (Adams) M.; B.A., U. Toronto, 1933; M.Sc. in Econs., U. London, 1935, D.Sc., 1955; m. Kathleen Margaret Walker, Sept. 25, 1943; children—Susan, Stephen, Sheila. Faculty, U. Toronto, 1935-42, 44-—, prof. polit. sci., 1956-—; faculty U. N.B., Fredericton, Can., 1942-43; exec. officer Wartime Information Bd., Ottawa, 1943-44. Fellow Royal Soc. Can.; mem. Am., Canadian (pres. 1963-64) polit. sci. assns., Am. Soc. for Polit. and Legal Philosophy. Author: Democracy in Alberta, 1953; The Political Theory of Possessive Individualism, 1962; The Real World of Democracy, 1966. Formulated concept of quasi-party system, and demonstrated its congruence with semi-colonial societies; formulated concept of possessive individualism.*

MACPHERSON, Herbert Grenfell, Am. physicist; b. Victorville, Cal., Nov. 2, 1911; s. Duncan William and Minnie (Morrison) MacP.; student San Diego State Coll., 1928-31; A.B., U. Cal. at Berkeley, 1932, Ph.D., 1937; m. Jeanette Taylor Wolfenden, June 5, 1937; children—Candy, Robert D. With Nat. Carbon Research Labs., Cleve., 1937-56, asst. dir. research, 1950-56; with Oak Ridge Nat. Lab., 1956-—, asst. lab. dir., 1963-64, dep. dir., 1964-—. Mem. Am. Phys. Soc., A.A.A.S., Am. Soc. for Metals, Am. Nuclear Soc., Phi Beta Kappa, Sigma Xi. Author: (with J. A. Lane, Frank Maslan) Fluid Fuel Reactors, 1958; (with L. M. Currie, V. C. Hamister) The Production and Properties of Graphite for Reactors, 1955; also articles. Research on phys. properties graphite; developed theory radiation from high intensity carbon arcs, pure graphite for use in nuclear reactors, molten salt reactors for civilian power reactors. Home: 102 Orchard Circle, Oak Ridge 37830. Office: P.O. Box X, Oak Ridge 37831.*

MACQUER, Pierre Joseph, French chemist, physician; b. Paris, Oct. 9, 1718; Docteur en médecine, 1742; succeeded Bourdelin as prof. chemistry Jardin du Roi; dir. work at porcelain factory, Sèvres, France. Mem. French Acad. Scis., 1745, Acads. Stockholm and turin, Soc. Physicians (Paris). Author: Elémens de Chymie théorique, 1741; Elémens de chymie pratique, 1751; Plan d'un cours de chimie expérimentale et raisonnée, 1757; Pharmacopoea Parisiensis, 1758; Formula medicamentorum magistralium, 1763; Art de la teinture en soie, 1763; Dict. de chymie (1st chem. dictionary in France), 2 vols., 1766; Dict. portatif des arts et métiers, 2 vols., 1766. Editor sci. dept. Jour. des Savants, 1768-76. Discovered new properties in various chem. substances; witnessed proof that diamond is combustible, 1771; one of 1st to study platinum; studied arsenates of potassium and sodium; described properties of alumina, magnesia, lime sulfate; studied milk composition. Died Paris, Feb. 15, 1784.

MACRAE, Alfred Urquhart, Am. physicist; b. N.Y.C., Apr. 14, 1932; s. Fred and Eliza (Urquhart) MacR.; Ph.D. in Physics, Syracuse U., 1960. Physicist physics research dept. Bell Telephone Labs., Murray Hill, N.J., 1960-—. Fellow Am. Phys. Soc.; mem. A.A.A.S., Am. Vacuum Soc. Editorial bd. Vacuum Sci. and Tech., 1964-—. Research, publs. on infra red photoconductivity, elec. noise in solids, low energy electron diffraction studies, arrangement atoms at surfaces, optical properties solids.*

MACRAE, Duncan, Jr., Am. social scientist; b. Glen Ridge, N.J., Sept. 30, 1921; s. Duncan and Rebecca (Kyle) MacR.; A.B., Johns Hopkins, 1942; M.A., Harvard, 1943, Ph.D., 1950; m. Edith Judith Krugelis, June 24, 1950; 1 dau., Amy Frances. Staff, Mass. Inst. Tech., Radiation Lab., 1943-46; instr., lectr. sociology Princeton, 1949-51; research asso. Lab. of Social Relations, Harvard, 1951-53; asst. prof. sociology U. Cal. at Berkeley, 1953-57; faculty U. Chgo., 1957-—, prof. polit. sci. and sociology, 1964-—. Fellow A.A.A.S.; mem. Am. Sociol. Assn., Am. Econ. Assn., Am. Statis. Assn., Am. Polit. Sci. Assn. Editor: (with Greenwood, Holdam), Electronic Instruments, 1948; Dimensions of Congressional Voting, 1958; Parliament, Parties and Society in France, 1967; Research, publs. on application of sci. methods to polit. sociology including computer data processing and statis. analysis, comparative study legislative bodies. Home: 5514 S. Harper Av., Chgo. 60637.*

MACRIS, Constantin, Greek astronomer; b. Tirnavos-Larissa, Greece, Dec. 5, 1913; s. Jean and Euphemie (Papicannou) M.; ed. U. Athens (Greece), Paris Obs., Coll. France; Ph.D.; m. Helen Argyris, 1952. Asst., Nat. Obs.; prof. agrege astrophysics U. Athens; dir. Solar Inst., Nat. Obs. Mem. Internat. Astron. Union, French, Italian astron. socs. Research and publs. on solar physics. Home: 19, Odos Xanthipou, Holargos, Greece. Office: Observatoire national, Athens, Greece.

MACROBIUS, Ambrosius Theodosium, Neoplatonist; flourished 395-423; believed to have lived in Rome in 5th century. Author: Saturnaliorum conviviorum, 7 books (presents miscellany of history, philology, and mythology); also Commentarii in Somnium Scipionis, narrated by Cicero in his De republica (demonstrates math., phys. and astron. ideas of the time); De differentiis et societatibus Graeci Latinique verbi.

MACVICAR, Donald George, Jr., Am. geologist; b. Conn.; s. D.G. MacV.; B.A., Amherst Coll., 1951; postgrad. Yale, 1954; conducted geol. mapping project, Brooks Range, Alaska, summers, 1955-56. Research on supergene enrichment of copper ores; devised techniques for thermal disintegration of sedimentary rock. Drowned during geol. field work, Chandler Lake, Alaska, July 20, 1956.

MACWILLIAM, John Alexander, physician; b. Culmill, New Brunswick, July 31, 1857; s. William MacWilliam; ed. U. Edinburgh (Scotland), Leipzig (Germany) U., U. Coll., London, Eng.; M.D., U. Aberdeen (Scotland), 1882; LL.D.; m. Edith Wise, 1889; m. 2d, Florence Thomas, 1898. Regius prof. of physiology U. Aberdeen, 1886-1927, then prof. emeritus. Fellow Royal Soc., 1916 (contbr. papers to proc.). Research, publs. on physiology of heart and arteries, action of chloroform and ether, proteins, muscle sound, blood pressure; demonstrated that ventricular fibillation can be produced by injecting poisons into blood stream, 1887. Died Jan. 13, 1937.

MACY, Josiah, Jr., Am. biophysicist, mathematician; b. N.Y.C., Dec. 17, 1924; s. Josiah Noel and Mary (Martin) M.; S.B., Mass. Inst. Tech., 1949; postgrad. Harvard, 1950-51; Ph.D. in Math., Mass. Inst. Tech., 1954; m. Elizabeth Weston Murray, Sept. 12, 1945; children—Barbara Weston, Josiah Macy III, Elizabeth Martin, Ann Murray. Staff, Research Lab. Electronics, Mass. Inst. Tech., 1951-54; operations research office analyst Johns Hopkins, 1954-58; faculty Albert Einstein Coll. Medicine, Yeshiva U., N.Y.C., 1958-67, asso. prof. physiology, 1961-67; prof. biomathematics, prof. physiology and biophysics, dir. div. biophys. scis. U. Ala. Med. Center, Birmingham, 1967-—. Mem. adv. com. on computers in research and computer research study sect. NIH, 1960-—. Mem. Am. Math. Soc., Soc. Indsl. and Applied Math., Biophys. Soc. (mem. council 1964-—), Biometric Soc., N.Y. Acad. Scis., A.A.A.S., Sigma Xi. Contbg. author: Operations Research for Management, 1956; Computers in Biology and Medicine, 1965; also articles. Research on theoretical models for information transfer in human groups and in central nervous system, application of hybrid analog-digital computing methods to physiol. expts. and simulation. Home: 3301 Spring Hill Rd., Mountain Brook, Ala. 35223. Office: 1919 7th Av. S., Birmingham, Ala. 35233.*

MACY-HOOBLER, Icie Gertrude, Am. chemist; b. Gallatin, Mo., July 23, 1892; d. Perry and Ollie (Critten) Macy; B.S., U. Chgo., 1916; M.A., U. Colo., 1918; Ph.D., Yale, 1920; Sc.D., Wayne State U., 1945; m. B. Raymond Hoobler, June 11, 1938 (dec. June 1943); 1 stepson, Sibley W. Faculty, U. Colo., 1916-18; asst. biochemist Western Pa. Hosp., 1920-21; instr. U. Cal., 1921-23; dir. Nutrition Research Labs., Merrill-Palmer Sch. and Children's Hosp. Mich., 1923-30; dir. Research Lab., Children's Fund Mich., 1930-54; staff Merrill-Palmer Inst. Human Devel. and Family Life, Detroit, 1954-—. Mem. Food and Nutrition Bd., NRC, 1940-52; mem. sci. adv. com. Nutrition Found., N.Y.C., 1941-59. Recipient Norlin Achievement award U. Colo., 1938; Borden award, 1939; Modern Medicine award, 1955. Fellow Am. Inst. Nutrition (pres. 1955, recipient Osborn and Mendel award 1952); mem. Am. Chem. Soc. (Francis P. Garvan award 1952, chmn. div. biochemistry 1930-31), Am. Soc. Biochemists, Am. Bd. Nutrition, Am. Bd. Clin. Chemistry Soc. for Pediatric Research, Soc. for Research in Child Devel., Mich. Acad. Sci, Arts and Letters (past pres.), Engring. Soc. Detroit, Phi Beta Kappa, Sigma Xi, Iota Sigma Pi, Pi Beta Phi. Club: Women's Research (U. Mich.). Author: Nutrition and Chemical Growth in Childhood, vol. I, 1942, vol. II, 1946, vol. III, 1951; Hidden Hunger, 1945; Maternal and Child Health, 1950; Plasma Proteins Characteristic of Human Reproduction, 1951; Composition of Human, Cow and Goat Milk, 1953; Chemical Anthropology, 1957; also articles. Home: 502 Burson Pl., Ann Arbor, Mich. 48104. Office: 71 E. Ferry Av., Detroit 48202.*

MADANSKY, Leon, Am. physicist; b. Bklyn., Jan. 11, 1923; s. Charles and Fannie (Sobler) M.; B.S., U. Mich., 1942, M.S., 1944, Ph.D. (NRC fellow 1946-48), 1948; m. Rena Goldstein, June 8, 1947; children—Deborah, Charles. Research asst. U. Mich., 1946-48; faculty physics dept. Johns Hopkins, 1948-—, prof., 1958-—; chmn. dept., 1965-68; spl. research nuclear particle physics. NSF sr. post-doctoral fellow, 1960-61, 68-69. Mem. Am. Phys. Soc., Phi Beta Kappa, Sigma Xi. Contbr. numerous articles. Research in radiative effects in nuclei and elementary particles, particle detectors, low energy nuclear reactions and atomic properties of elementary particles. Home: 6602 Edenvale Rd., Balt. 9.*

MADDEN, John Leo, Am. physician; b. Washington, Aug. 15, 1912; s. John Joseph and Katherine (McLaughlin) M.; grad. Holy Cross Coll., 1933; M.D., George Washington U., 1937; L.H.D., Siena Coll., 1955; m. Bertha Marie Antonades, Jan. 31, 1942; children—Patricia Ann, John Joseph, Brian James, Michael Robert, Thomas Francis, Kathleen Ellen. Practice medicine specializing in surgery, N.Y.C., 1946-—; dir. surgery St. Clare's Hosp., 1949-—; clin. prof. surgery N.Y. Med. Coll., 1960-—. Recipient Hektoen gold medal A.M.A., 1952. Diplomate Am. Bd. Surgery. Fellow A.C.S. (gov.); mem. Am. Surg. Assn., Internat., N.Y. (council) surg. socs., Soc. for Vascular Surgery, Internat. Cardiovascular Soc., N.Y. Soc. for Cardiovascular Surgery (past pres.), N.Y. Acad. Medicine (pres. 1965-67). Author: Atlas of Technics in Surgery, 1958, 2d edit., 1964; also numerous articles. Research on cardiovascular and gastrointestinal surgery; 1st to suggest concept that obstruction of hepatic veins (outflow tract) is prime factor in prodn. of ascites. Inventor malleable ligature carrier, clamps for cardiovascular and gen. surgery. Address: 123 E. 69th St., N.Y.C. 10021.*

MADDEN, Richard, Am. ednl. psychologist; b. nr. Burchard, Neb., Aug. 18, 1901; s. John Kyle and Della (Stake) M.; A.B., Peru (Neb.) State Tchrs. Coll., 1926; A.M., Columbia, 1930, Ph.D., 1931; m. Fern A. Wells, July 20, 1927; 1 son, Ray Allan. Tchr., Chester, Neb., 1923-25, prin. high sch., 1923-25, supt. schs., 1925-29; asst. ednl. psychology Columbia, 1930-31; tchr. psychology, clinician, dir. lab. sch., Ind. State Tchrs. Coll., Indiana, Pa., 1931-39; dir. campus sch., San Diego State Coll., 1939-46, elementary dir. edn., 1946-47, dean, 1947-51, chmn. grad. studies, 1951-55, prof. edn., 1955-61, 64-—; prof. edn. Sonoma State Coll., Rohnert Park, Cal., 1961-64. Mem. Am. Psychol. Assn., Nat. Council Tchrs. Math., N.E.A., Phi Delta Kappa, Kappa Delta Pi. Author: School Status of the Hard of Hearing, 1931. Co-author: Stanford Achievement Tests, 1953, 64; Success in Spelling, 1955; The Scribner Arithmetic, 1955; Sound and Sense in Spelling, 1964; Stanford Diagnostic Reading and Arithmetic Tests, 1965; Analysis of Learning Potential; also elementary textbooks. Research and publs. in measurement of sch. achievement, tests for adults, diagnostic reading and arithmetic tests, modern math. tests.*

MADDEN, Robert Phyfe, Am. physicist; b. Schenectady, Dec. 20, 1928; s. Edward J. and Marion (Phyfe) M.; B.S., U. Rochester, 1950; Ph.D., Johns Hopkins, 1956; m. Ingrid Persson, Sept. 16, 1950; children—Donald, Deborah, Karl. Physicist physics research lab. Engr. Research and Devel. Labs., Ft. Belvoir, Va., 1958-60, chief spectroscopy sect., 1960-61; chief far ultraviolet physics sect., atomic physics div. Nat. Bur. Standards, Washington, 1961-—; adj. prof. Pa. State U., 1962-—. Recipient Outstanding Achievement award Engr. Research and Devel. Labs., 1961, Silver metal U. S. Dept. Commerce, 1963. Fellow Optical Soc. Am. (dir. 1966-—), Am. Phys. Soc.; mem. Washington Acad. Scis. (Sci. Achievement award 1965). Research and publs. on atomic and molecular spectroscopy, optical properties of thin films in far ultraviolet spectral region; spectroscopic instrumentation; molecular spectroscopy in infrared spectral region. Office: Physics Bldg., Nat. Bur. Standards, Washington 20234.*

MADDEN, Sidney C(larence), Am. pathologist; b. Fresno, Cal., Oct. 27, 1907; s. Edward Clarence and Imogene (Johnson) M.; A.B., Stanford U., 1930, M.D., 1934; m. Lynnette May Watt, Aug. 18, 1933; children—John Kirk, Donn Edward, Janet Lynn, Carol Lynn. Mem. faculty U. Rochester Sch. Medicine and Dentistry, 1934-45; prof. pathology, Emory U. Sch. Medicine, 1945-48; head, div. pathology, Brookhaven Nat. Lab., 1949-51; prof. and chmn. dept. pathology Sch. Medicine, U. Cal., Los Angeles, 1951-—; pathologist-in-chief U. Cal. Hosp., 1955-—; cons. V.A. Center, Los Angeles, Los Angeles Co. Harbor Gen. Hosp., 1951-—; cons. Armed Forces Inst. Pathology, 1955-—, USPHS, 1948-53, 1959-63; cons. pathology Atomic Bomb Casualty Commn., 1959-—. Chief of pathology Atomic Bomb Casualty Com., Japan, 1958-59, now mem. adv. com.; chmn. rev. com. Argonne Nat. Lab., 1962. Mem. NRC; mem. Nat. Bd. Med. Examiners. Recipient Theobold Smith Award A.A.A.S., 1943. Diplomate Am. Bd. Pathology. Fellow Coll. Am. Pathologists; mem. Am. Soc. Exptl. Pathology (pres. 1952-53), Am. Assn. Pathologists and Bacteriologists, (pres. 1962-63), A.A.A.S., Soc. Exptl. Biology, Western Soc. Clin. Research, A.M.A., Sigma Xi, Alpha Omega Alpha, Nu Sigma Nu. Contbns. to knowledge of blood and body tissue protein formation; discoverer of the ten amino acids essential for new plasma protein prodn.; found that added protein intake can counterbalance rapid protein breakdown from acute bodily injury. Home: 218 20th St., Santa Monica. Cal. 90402.*

MADDEN, Theodore R., Am. geophysicist; b. Boston, Mar. 14, 1925; s. Thomas R. and Mercedes (Johnson) M.; B.S. in Physics, Mass. Inst. Tech., 1949, Ph.D. in Geophysics, 1961; m. Sheila Murphy, Aug. 11, 1959; 1 dau., Jennifer. Faculty, Mass. Inst. Tech., Cambridge, 1956-—, prof. dept. geology and geophysics, 1967-—; vis. asso. prof. Inst. Geophysics and Planetary Physics, U. Cal. at San Diego, 1963-64. Mem. Am. Geophys. Union. Research on phys.-chem. basis of induced polarization and self-potential properties of rocks, deep elec. properties of earth using magnetotelluric and magnetic variation measurements, electromagnetic and Alfven wave resonances in earth-ionosphere cavity and magnetosphere. Home: 11 Locust Rd., Weston, Mass. 02193. Office: 77 Massachusetts Av., Cambridge, Mass.*

MADDOX, Ernest Edmund, ophthalmologist; b. Shipton, Oxon, 1860; s. J. F. Maddox; ed. Edinburgh U.; M.D.; m. Grace Rivers Monteath; 7 sons, 6 daus. Cons. opthalmic surgeon Royal Victoria and West Hants Hosp., Bournemouth; resident physician, then asst. ophthalmic surgeon, Edinburgh Royal Infirmary, also ophthalmoscopic tutor. Recipient meml. medal Oxford Ophthal. Congress, 1921. Fellow Royal Soc.; mem. Soc. Francaise d'Optalmologie; v.p. Ophthal. Soc., pres. 1931. Author: The Clinical Use of Prisms; Tests and Studies of the Ocular Muscles; Golden Rules of Refraction. Research on relation between accommodation and convergence, refraction, ocular muscles in Latham and English system of treatment; devised safety device for dangerous cataract extraction, also instrument for testing and measuring degree of heterophoria (Maddox Rod). Died Nov. 4, 1933.

MADEY, Richard, Am. physicist; b. Bklyn., Feb. 23, 1922; s. Elia D. and Dorothy (Diab) M.; B.E.E. (Conn. Alumni scholar), Rensselaer Poly. Inst., 1942; Ph.D. in Physics, U. Cal. at Berkeley, 1952; m. Mary Lou Kirch, Sept. 8, 1951; children—Doren Louise, Diane Claire, Daryl Jane, Richard Kirk, Ronald Eliot, Randall Clarke. Elec. engr. Allen B. Dupont Labs., Passaic, N.J., 1943-44; physicist Lawrence Radiation Lab., Berkeley, Cal., 1947-53; asso. physicist Brookhaven Nat. Lab., Upton, N.Y., 1953-56; scientist Republic Aviation Corp., Farmingdale, N.Y., 1956-58, sr. scientist, 1958-61, chief staff scientist for modern physics, 1961-62, chief, applied physics subdiv. research div., 1963-64; prof. physics, Clarkson Coll. Tech., Potsdam, N.Y., 1965-—, chmn. dept., 1965. Cons. to industry, 1953-56, AEC, 1965-—. Recipient Naval Ordnance Devel. award, 1946. Fellow N.Y. Acad. Scis.; asso. fellow Am. Inst. Aeros. and Astronautics; mem. Am. Phys. Soc., Am. Geophys. Union, Am. Nuclear Soc. (treas. aerospace div. 1964-65, Meritorious Service award N.Y. met. sect. 1965), Inst. Colloid and Surface Sci., Sigma Xi. Research, publs. in fields of space radiation shielding, interaction of radiation with matter, radiation dosimetry, high energy physics, radiation detection, phys. adsorption of radioactive and stable gases, nuclear test reactors, and electronic istrumentation; patentee nanosecond coincidence and pulse-shaping circuits, multi-color cathode-ray display tube, fissionable fuel power plant, energetic charged particle detector. Home: Route 1, Potsdam, N.Y. 13676.*

MÄDLER, Johann Heinrich von, see von Mädler, Johann Heinrich.

MADISON, James, Am. political theorist; b. Port Conway, Va., Mar. 16, 1751; s. James and Eleanor (Conway) M.; A.B., Coll. of N.J. (now Princeton), 1771; m. Dolly Payne Todd, Sept. 15, 1794, no children. A founder Am. Whig Soc. (debating club) at Coll. of N.J.; admitted to Va. bar; mem. Com.of Safety for Orange County (Va.), 1774; del. to Williamsburg (Va.) Conv., 1776, mem. com. which framed constn. and declaration of rights for Va.; mem. 1st Gen. Assembly of Va., 1776, Va. Exec. Council, 1778; mem. Continental Congress from Va., 1780-83, 86-88; wrote instrns. to John Jay (U. S. minister to Spain) concerning U. S. rights to navigation of Mississippi River, 1780; proposed "3/5 Compromise" before Congress to break deadlock on changing basis of state contbns. from land values to population by counting 5 slaves as 3 free people; returned to Va., 1783, began study of law and natural history of U. S.; mem. Va. Ho. of Dels. from Orange County, 1783-86, completed disestablishment of Anglican Ch. in Va. (begun by Thomas Jefferson), 1779, favored admission of Ky. to statehood, inaugurated series of surveys for improvement of transmountain communications, urged power to be granted Congress to regulate commerce, leader in effecting a series of interstate confs.; del. from Va. too Annapolis Conv., 1786; published "Vices of the Political System of the United States"; del. from Va., chief recorder U. S. Constl. Conv., 1787, kept records published in Journal of the Federal Constitution, 1840; asso. with Alexander Hamilton and John Jay in writing essays known as The Federalist (published under signature Publius), 1788, described constl. system of govtl. checks and balances, emphasized protection of pvt. property; largely responsible for ratification of U. S. Constn. by Va., 1788; mem. U. S. Ho. of Reps., from Va., 1st-4th congresses, 1789-97, participated in passage of revenue legislation, creation of exec. depts., framing of Bill of Rights; a leader of Democratic-Republic Party which opposed creation of U. S. Banks and pro-British sympathies; published series of letters under name "Helvidius" in Gazette of the United States, Aug. 24-Sept. 18, 1793, criticized George Washington's neutrality proclamation; declined mission to France and post of U. S. sec. of state, 1794; wrote Va. Resolutions against Alien and Sedition Acts, expressed opinion that states could declare acts of Congress unconstl.; U. S. sec. of state under Pres. Jefferson, 1801-09, supported Jefferson's Embargo Act of Dec. 22, 1807, repealed, Mar. 1, 1809; elected 4th Pres. U. S. (Democratic-Republic; defeated Charles Cotesworth Pinckney), 1808, inaugurated, Mar. 4, 1809, re-elected, 1812; signed bill providing for 2d Bank of U. S., 1816, Tariff Act of 1816; enrolled in soc. for encouragement of Am. mfrs., 1816; left office, Mar. 3, 1817; became rector U. Va. (succeeded Jefferson), 1826; del. Va. Constl. Conv., 1829; ret. to "Montpellier," Orange County. Made proposals incorporated in Virginia or Randolph Plan (drafted by Edward Randolph), including change in principle of representation to give larger states more influence, uniform nat. laws, a fed. veto on state legislation, extension of nat. authority to a judiciary dept., a 2 house fed. legislature with differing terms of office, a nat. exec., an article guaranteeing defense of states by fed. govt., ratification of amendments to U. S. Constn. by people as well as legislature; described as "the masterbuilder of the constitution." Died "Montpellier," June 28, 1836.

MADSEN, Thorvald Johannes Marius, Danish bacteriologist; b. Frederiksberg, Denmark, 1870; M.D., U. Copenhagen, 1893; asst., univ. lab. for med. bacteriology, Copenhagen; worked (with Ehrlich), Frankfort, Germany, also Pasteur Inst., Paris; dir. Statens Serum Inst., Copenhagen, from 1902; pres. hygienic commn. League Nations; head UNICEF mission to Italy, 1947-50; pres. 4th Internat. Congress on Microbiology, Copenhagen, 1947. Editor Communication de l'Institut Serothérapique de l'État Danois, 1906-40. Research and publs. on immunology and immunochemistry; studied diphtheria toxin and antitoxin. Died Denmark, Apr. 15, 1957.

MAEHLUM, Bernt Neeb, physicist; b. Oslo, Norway, Feb. 5, 1929; s. Eivind and Ingrid (Neeb) M.; Cand. Mag., U. Oslo, 1953, Cand. Real, 1955, Dr.Philos., 1961; m. Ellen Viggen, Nov. 20, 1954; 1 dau., Guri. State Meteorologist Norwegian Weather Forecasting Bur., Tromse, Norway, 1955-56; research physicist Norwegian Def. Research Est., Kjeller, Norway, 1956-61, sr. research physicist, 1962-66; postdoctoral scholar dept. physics and astronomy, State U. Ia., 1961-62; sr. research asso. dept. space sci. Rice U., Houston, 1966-—. Mem. Norwegian Geophys. Union (sec. 1965-—), Am. Geophys. Union, Norwegian Phys. Union. Author: (with others) Principles of High Frequency Radio Communication, 1968; also articles. Editor: Electron Density Profiles in the Ionosphere and Exosphere, 1962; High Latitude Particles and the Ionosphere, 1965. Research on irregular ionization in upper atmosphere at high geog. latitudes and sources of this ionization using observations from ground, balloons and rockets and satellites.*

MAEKAWA, Magojiro, Japanese physician; b. Kagawa Prefecture, Japan, 1902; grad. Kyoto U., 1925; D.Sc. Became asst. prof. Kyoto U., 1936, prof., 1941. Author: Interpretation of Electrogram of Action of Ventricle, 1931. Research in electrobiology, especially electrocardiogram; allergy; immunity; developed theory on devel. of consciousness, neurosis, and cause of high blood pressure.

MAENO, Ryotaku (pseudonyms Rakuzan, Ranka), Japanese physician; b. 1732; mem. family of physicians to Nakatsu clan, Oita Prefecture; reared by Zentaku Tamiya; began study of Dutch under Konyo Aoki, circa 1770, later studied Dutch medicine, Nagasaki; performed (with Gempaku Sugita) autopsy to check Kulmus' anat. information, then translated the work as Kaitai Shinsho (New Book on Anatomy), thus helping to spread Western learning in Japan. Died 1803.

MAEVSKY, Mikhail Mikhaylovich, Russian microbiologist; b. 1894; grad. Med. Faculty, St. Vladimir U., Kiev, 1916; D.Med. Sci., 1945. Asso., Kiev Health-Bacteriological Inst., 1920-40; sr. asso. All-Union Inst. Exptl. Medicine, 1940-50; head lab. exptl. biotherapy Inst. Exptl. and Clin. Oncology, USSR Acad. Med. Sci. Mem. USSR Acad. Med. Sci. (corr.), All-Union Soc. Oncologists (bd. mem.). Author: The Efficacy of Typhus Vaccine According to Clinical Data, 1944; Rickettsiosis, 1949. Mem. editorial bd. Jour. Microbiology, Epidemiology and Immunobiology, Herald of USSR Acad. Med. Scis. Research and numerous publs. on tissue cultures, rickettsiosis, synthesis of microorganic vitamins, polysaccharide lipoid antigens of bacteria, immunology of typhus, toxic substance in rickettsia (devised method to obtain this substance from lungs of mice and lice); developer prodn. of typhus vaccine, corpuscular ether vaccine against typhus; established fluctuations of virulence in rickettsia and developed method to preserve virulent strains. Address: Inst. Exptl. and Clin. Oncology, USSR Acad. Med. Sci., 3-ya Meshchanskaya 61-2, Moscow I-110, USSR.

MAFFII, G., Italian pharmacologist; b. Florence, Italy, Feb. 17, 1928; s. Ferdinando and Margherita (Nannelli) M.; M.D., U. Florence, 1951; m. Neva Agazzi, Jan. 18, 1952; children—Ferruccio, Silvia, Margherita, Ferdinando. Asst., Inst. Pharmacology, U. Florence, 1951-53; with Lepetit S. p. A., Milan, Italy, 1954-—, dir. pharmacology, 1961, joint dir. Research Labs., 1962-—. Mem. Soc. for Applied Pharmacology Scis. (pres. 1967), European Soc. for Study Drug Toxicity (mem. com. 1966-—). Research and numerous publs. on pharmacology of C.N.S., behavior and structure-activity relationships of hypnotics and respiratory stimulants, pharmacokinetics and toxicology of rifamycin antibiotics. Home: 2 Via Gioberti, Milan, 20123. Office: 38 Via Durando, Milan, 20158, Italy.*

MAGALHAES GOMES, Francisco de Assis, Brazilian physicist; b. Ouro Preto, Brazil, Jan. 16, 1906; s. Francisco and Amalia (Brandao) M. G.; civil and Mining Engr., Nat. Mining Sch., Ouro Preto, 1928; laureate D.Sc., U. Minas Gerais, 1938; m. Maria Clara Birchal, Aug. 6, 1935; children—Francisco, M. Aparecida (Mrs. Newton Matos), Clara (Mrs. Eucler Paniago), M. Conceicao (Mrs. Fernando V. Mello), M. Amália (Mrs. Santos Fagundes), Frederico, Alberto, Luis, Joana, M. Cecilia, M. Ines, Leonardo, M. Leticia. Pub. works engr. B. Hortizonte, 1928-38; prof. physics U. Minas Gerais, also Nat. Mining Sch., 1938-50; dir. Inst. for Radioactive Researches, 1951-62; dir. Central Inst. Physics, 1966-—. Mem. NRC, 1952-55, Nat. Commn. for Nuclear Energy, 1964-65; cons. mem. UNESCO Conf. Radioisotopes, Paris, France, 1950. Decorated Order of Acad. Palms (Paris). Mem. Engring. Assn. Minas Gerais (past chmn.), Brazilian Acad. Scis. Research, publs. on 2d research nuclear reactor in Brazil, B. Horizonte, 1959. Home: 198 Felipe dos Santos-B. Horizonte-M. Gerais, Brazil. Office: Cidade Universitária-Pampulha-B. Horizonte-Brazil.*

MAGALINI, Sergio Ivano, physician; b. Amatrice, Italy, July 27, 1927; s. Ivon T. and Gloria (Costabile) M.; A.B., Inst. Virgilio, 1945; M.D., U. Rome, 1951; m. Amedea DiGirolamo, Oct. 6, 1956; children—Sabina, Fabio. Practice medicine, specializing in hematology, Boston, 1955-57, Rome, 1957-58, Providence, 1965-—; research asso. Joseph Stanton Meml. Research Lab. St. Elizabeth's Hosp., Boston, 1955-56; research fellow Sch. Medicine Tufts U., 1956; clin. research fellow Childrens Med. Center, Childrens Cancer Research Found., Boston, 1956-57, research asso. pathology, 1963-65; faculty U. Rome, 1957-63; research asso. pathology Med. Sch. Harvard, 1963-65; dir. med. edn., research St. Joseph's, Our Lady Fatima hosps., Providence, 1965-—. Recipient Lepetit-2d prize for best M.D. thesis, 1951. Damon Runyon fellow Boston, 1955-56. Mem. Clin. Research Soc., Internat. Soc. Hematology, Soc. Italiana di Ematologia, N.Y. Acad. Scis., Am. Soc. Exptl. Pathology, Am. Assn. Cancer Research. Author: (with others) Ion Exchangers in Organic and Biochemistry, 1957; (with others) La Prevenzione Del Cancro, 1963; (with others) Il Tecnico di Laboratorio Clinico, 1966. Numerous publs. on clin., serum, cellular biochem. studies in hematology, cancer chemotherapy, discovery of a new transplantable leukemia in Mongolian gerbil. Home: 96 Savoy St., Providence 02906. Office: St. Joseph's Hosp., 21 Peace St., Providence 02907.*

MAGAR, Narahar Gangaram, Indian biochemist; b. Nasik, India, May 8, 1914; s. Gangaram P. and Godubai Magar; B.Sc., Inst. Sci., Bombay, India, 1936; M.Sc., Seth G.S. Med. Coll., Bombay, 1939, Ph.D., 1946; postgrad. McGill U., Montreal, Que., Can., Wis. U.; m. Anjana Kalanke, Dec. 20, 1940; 1 son, Arun. Research asst. nutrition research unit I.R.F.A., Seth G.S. Med. Coll., 1939-46; lectr. biochemistry Inst. Sci., Bombay, 1949-—, tchr. biochemistry dept., 1949-—, prof. biochemistry, 1966-—. Fellow A.A.-A.S., Am. Oil Chemist Soc. Research, numerous publs. on chemistry and nutritive value of processed fish and foodstuffs in India, component fatty acids of shark liver oils in India, biochem. changes in blood and tissues in atherosclerosis and leprosy. Home: 303 Agra Rd., Bombay-70, India.*

MAGAT, Michel, physicochemist; b. Kharkoff, Russia, Oct. 28, 1908; s. Israel and Eveline (Guerchkovitch) M.; Ph.D., U. Berlin, 1932; Sc.D., U. Paris, 1936; m. Marguerite Lamorlette, Sept. 20, 1935; 1 dau., Elizabeth (Mrs. Jean Paul Salvan); m. 2d Marguerite Lautout, Dec. 16, 1959; children—Yves, Michel, Nicolas. Mem. staff Centre National de la Recherche Scientifique, 1936-—, dir. Lab. Radiation Chemistry, 1957-—; research asso. Frick Chem. Lab., Princeton U., 1942-44; maitre de conference Faculty Scis., Paris, 1957-63, prof., 1963-67; prof. Faculty Scis., Orsay, 1967-—. Mem. French Nat. Sci. Council, 1948-50, 58-—; cons. French Atomic Energy Commn., 1958. Mem. Am. Chem. Soc., Faraday Soc., French Soc. Chem. Physics, French Soc. Physics, French Soc. Chemistry, Miller Trust. Author: (with A. Talaly) Synthetic Rubber from Alcohol, 1945; also numerous articles. Research on determination of Van der Waals radii and molecular models; Raman spectra and structure of liquid water, discovery of intermolecular bands; electrostatic theory of hydrogen bonding;

thermodynamics of polymer solutions; radiation polymerisation; radiation grafting of polymers; radiothermoluminescence. Home: 47 av. de Paris, 91-Bievres, France. Office: Lab. Chem. Physics, Faculty Scis., 91-Orsay, France.*

MÄGDEFRAU, Karl Hermann, German botanist; b. Iéna, Feb. 8, 1907; s. Otto and Elisabeth (Koch), M.; ed. U. Iéna, U. Munich; Ph.D.; m. Paul Götz, 1940; children—Meinhart, Dieter, Gerlinde, Wolfgang, Helmut. Successively became instr. Erlangen, 1936, prof. at large, Strasbourg, 1942; named asso. prof., Munich, Germany, 1951, prof., 1956; became prof., Tübingen, Germany, 1960. Mem. German Acad. Naturalists, Bavarian Bot. Soc. (hon.). Author: Palaobiologie d. Pflanzen, 1956; Vegetationsbilder der Vorzeit, 1959; also articles in bot. and geol. jours. Home: Bohnenbergerstrasse 28, Tübingen, West Germany.

MAGEE, Donal Francis, physiologist; b. Aberdeen, U.K., June 4, 1924; M.; B.A., Oxford U., 1945, M.A., B.M., B.Ch., 1948; Ph.D., U. Ill., Chgo., 1952. Research asst. Nuffield Inst., Oxford, Eng., 1946; instr. U. Ill., 1948-52; with U. Wash., 1952—, prof., 1962-65; chmn. physiology, pharmacology Creighton U., 1965—. Mem. Am. Physiol. Soc., Am. Gastroenterol. Assn., British Med. Assn., Am. Soc. Pharm. and Exptl. Therapeutics. Author: Gastro-intestinal Physiology, 1962. Contbr. numerous articles in field to sci. jours. Research in juices secreted by stomach and pancreas and movements of gall-bladder. Home: 310 S. 55th St. Office: Medical Research 507, 27th St. and California Av., Omaha 68131.*

MAGEE, John Lafayette, Am. phys. chemist; b. Franklinton, La., Oct. 28, 1914; s. John Lafayette and Edith (Jenkins) M.; A.B., Miss. Coll., 1935; M.S., Vanderbilt U., 1936; Ph.D., U. Wis., 1939; m. Priscilla Williams, Jan. 1, 1948; children—Lawrence Edward, Linda Sue, John Bradley. Research chemist B.F. Goodrich Co., Akron, O., 1941-43; staff Los Alamos Sci. Lab., 1943-45; group leader Naval Ordnance Test Sta., Cal., 1945-46; sr. scientist Argonne Nat. Lab., 1946-48; prof. chemistry U. Notre Dame (Ind.), 1948—, asso. dir. radiation lab., 1954—. Cons. AEC, Argonne Nat. Lab., Inst. Def. Analysis; bd. dirs. Argonne Univs. Assn., Mem. Am. Chem. Soc., Phys. Soc., Radiation Research Soc. (v.p., pres. elect 1966-68), Faraday Soc., Sigma Xi. Editor: (with M. Burton, J. S. Kirby-Smith) Nat. Acad. Scis. publ. Comparative Effects of Radiation, 1960. Research, publs. in photochemistry, rate reactions, radiation chemistry, chem. physics; application of statis. and quantum mechanics to chemistry. Home: 122 S. Hawthorne Dr., South Bend, Ind. 46617.*

MAGEE, Kenneth Raymond, Am. neurologist; b. Gardner, Ill., July 30, 1926; s. Raymond F. and Edna (Roager) M.; B.S., Purdue U., 1946; M.D., U. Chgo., 1948, M.S., 1949; M.A., U. Mich., 1953; m. Bettie Wendell Morris, Dec. 29, 1948; children—Robert Morris, Benjamin Rush, Kenneth Wendell, James Lyndon. Faculty, U. Mich., Ann Arbor, 1950—, prof. neurology, 1964—; clin. asst. prof. Georgetown U., 1954-56; clin. asso. Nat. Inst. Neurol. Diseases and Blindness, NIH, 1954-56; sr. asst. surgeon USPHS, 1954-56. Cons. to hosps; mem. med. adv. bd. Myasthenia Gravis Found. Am., 1963—. Diplomate Am. Bd. Psychiatry and Neurology. Fellow A.C.P.; mem. Soc. Clin. Neurologists (exec. bd. 1963—, pres. 1965—), Am. Acad. Neurology (exec. bd. 1965—), Am. Neurol. Assn. (exec. bd 1966-68), A.M.A., Assn. for Research in Nervous and Mental Diseases, Soc. Clin. Neurologists, Am. Med. Writers Assn., Am. Epilepsy Soc. Contbr. numerous articles to tech. jours., sects. to books. Research in clin. and research neurology, muscle diseases, myasthenia gravis, Parkinson's disease, headache, epilepsy. Home: 2107 Hill St. Office: 1405 E. Ann St., Ann Arbor, Mich. 48104.*

MAGEE, Robert John, chemist; b. Belfast, No. Ireland, June 10, 1922; s. Robert and Margaret (Jeffers) M.; B.Sc. with honors, Queen's U., Belfast, 1947, M.Sc., 1948; Ph.D., U. Edinburgh, 1951, D.Sc., 1965; m. Christina Dornap, July 9, 1955; children—Robert Kevin, Gary Bryan. Lectr. chemistry U. Edinburgh, 1948-56; faculty Queen's U., 1956-66, reader in chemistry, 1963-66; vis. lectr. U. Graz, Austria, 1958; vis. asso. prof. Case Inst. Tech., Cleve., 1962-63, vis. prof., 1966; Found. prof. chemistry, chmn. dept. chemistry La Trobe U., Melbourne, Australia, 1966—. Sci. adviser Ministry Home Affairs, No. Ireland, 1961-67; adviser Commonwealth of Australia Civil Def. Sch., 1967—. Fellow Royal Inst. Chemistry. Research, numerous publs. on spectra and structure of transitional metals, electrochemistry, polarographic and chronopotentiometric studies in inorganic systems, radiation chemistry, inorganic analytical chemistry. Home: 38 McArthur Rd., East Ivanhoe, Victoria. Office: La Trobe U., Bundoora, Victoria, Australia.*

MAGELLAN, Ferdinand, explorer; b. Sabrosa, Trazos-Montes, circa 1480; s. Pedro de Magalhaes Magellan m. dau. of Diogo Barbosa, circa 1517. Page to Queen Leonor; joined service of Manuel the Fortunate, 1495; volunteered for voyage of Francisco d'Almeida (1st Portuguese viceroy) to India, 1504; sent to Sofala, to build Portuguese fortress, 1506; returned to India; wounded in Battle of Diu, 1509; promoted to capt., 1510; sent to explore Spice Islands by Antonio d'Abreu, 1511; became fidalgo escudeira, 1512; joined Portuguese expdn. against Azamor, Morocco, 1513; accused of trading with Moors and fell into disfavor; left for Spain, circa 1514; began expdn. to reach East Indies under patronage of Charles V. Discovered Magellan straits; named Pacific Ocean; discovered St. Paul and Shark Islands between Cabo Deseado and Ladrones, 1521; reached Guam, 1521; discovered Philippines, 1521; his ship was 1st to circumnavigate globe, thus proving earth is round, Americas are separate from Asia and a single stretch of water circled the earth. Died Philippines, Apr. 27, 1521.

MAGELLAN, Jean Hyacinthe de, see de Magellan, Jean Hyacinthe.

MAGENDIE, François, French physiologist; b. Bordeaux, France, Oct. 15, 1783; ed. under Boyer at Hôtel-Dieu, Paris; became physician Hôtel-Dieu; named prof. medicine Coll. de France, 1831; became pres. Com. Pub. Hygiene, 1843; mem. French Acad. Scis., Acad. Medicine. Author: Précis élémentaire de physiologie, 1816-17; Anatomie comparative du cerveau, 1826; Lecons sur le choléra-morbus, 1832; Lecons sur les fonctions et les maladies du système nerveaux, 1839; Lectures on the Blood, 1839; Phénomènes physiques de la vie, 4 vols., 1842; also papers in Jour. de physiologie experimentale, 1821-31. Pioneer in exptl. physiology, exptl. pharmacology; discovered (independently but later than Bell) functions of cerebro-spinal nerve roots (Bell-Magendie law of anat. and functional discreteness of sensory and motor nerves); studied effect of air in arteries; established that blood vessels absorb fluids, 1821; showed that digested food reaches the liver by means of portal circulation, 1844; discovered that some eye diseases are caused by food deficiency; laid found. for cellular physiology; introduced iodine, bromine, strychnine and morphine compounds into gen. med. use; investigated effects and uses of various drugs; studied mechanisms of vomiting, blood flow, deglutition. Died Sannois, France, Oct. 7, 1855.

MAGGI, Bartolommeo, Italian surgeon, anatomist; b. Bologna, Italy, circa 1516; prof. anatomy and surgery, Bologna; tchr. Giulio Casare Aranzi; physician to Pope Julius III; later returned to Bologna. Author: De vulnorum sclopetorum, et bombardarum curatione tractus, posthumously 1552. First surgeon in Italy to combat prevalent theory that gunshot wounds were poisoned; treated wounds with applications of egg albumin and salt, also advocated dressings be changed once daily in winter and twice in summer. Died 1552.

MAGHIDSON, Onisim Yulevich, Russian organic chemist; b. 1890; grad. Moscow U., 1913. Faculty chemistry dept. Nat. U., 1914-19; founder, head Synthesis Research Lab. (now part All-Union Ordzhonikidze Chemopharm. Research Inst.), 1919—. Author: The Synthesis of Anti-Malarial Agents, 1935; New Means of Treating Bacterial Diseases, 1939; The Eighth Edition of the USSR State Pharmacopoeia 1944; Sulfamide Drugs, 1946; The Major Chemotherapeutic Preparations, 1948. Research and numerous publs. on synthesizing sulfanilamides, antimalarial agents (based on quinoline, acridine and other heterocyclic systems), somnifacients, anesthetics, cardiovascular drugs, vitamin preparations; developer absorption method to extract iodine from oil-well brine. Patentee in field. Address: All-Union Ordzhonikidze Chemopharm. Research Inst., Zubovskaya 7, Moscow, USSR.*

MAGINI, Giovanni Antonio, Italian mathematician, astronomer, geographer; b. Padova, Italy, June 14, 1555; s. Pasquale; diploma in math., Studio, Bologna, Italy, 1579; m. Angela de'Poggi di Gravoli; apptd. prof. math. Studio, Bologna, 1588; became pvt. math. tutor to children of Duke Vincenzio of Mantua, 1599. Author: Tabulae secundorum mobilium caelestium, 1585; Geographiae universae, 1606; Italia, 1620. Combined study of culture with geog. analysis of Italy; supported theories of Kepler, Copernicus, Galileo, with research in math. and astronomy. Died Feb. 12, 1617.

MAGITOT, Emile, French dentist; b. Paris, France, 1833; M.D., 1866. Recipient prize for research on cavities, Acad. Medicine, France, 1868. Charter mem. Société d'Anthropologie (later pres.). Author: Traité de la Carie Dentaire, 1868. A founder of modern stomatology; contbr. classic works on classification of dental anomalies, studies of morphology of dental follicle and chronology of formation of teeth; credited with description osteoperiostitis of alveloli of teeth (Magitot's Disease). Died Paris, 1897.

MAGNAC-VALETTE, Denyse Juliette, French physicist; b. Colmar, France, Jan. 14, 1924; d. Louis Augustin and Valentine (Dupuis) Vallette) Licence ès Sciences, U. Strasbourg (France), 1946, Doctorat d'Etat, 1956; m. Claude Magnac, Mar. 28, 1948; 1 child, Thierry. Research asso. Centre National de la Recherche Scientifique, 1949-56, research dir., 1956-58; faculty Strasbourg U., 1958—, titular prof., 1963—. Mem. French Physics Soc. Research and publs. on low energy physics, especially study of nuclear reactions induced by complex particles.*

MAGNÉLI, Arne, Swedish chemist; b. Stockholm, Sweden, Dec. 6, 1914; s. Agge and Valborg (Hultman) M.; M.Sc., U. Stockholm, 1940, Ph.D., 1942; D.Sc., U. Uppsala (Sweden), 1950; m. Barbro Wigh, Aug. 10, 1946; children—Christina, Lars, Peter. With U. Uppsala, 1941-53, asst. prof. chemistry, 1950-53; with U. Stockholm, 1953—, prof. inorganic chemistry, 1961—, chmn. Inst. Inorganic and Phys. Chemistry, 1964—. Mem. Swedish Nat. Com. for Crystallography, 1953—, Nat. Com. for Chemistry, 1968—, Swedish Natural Sci. Research Council, 1965—; sec. Nobel Com. for Physics and Chemistry Royal Acad. Sci., 1966—. Recipient Norblad-Ekstrand Gold medal Swedish Chem. Soc., 1954. Mem. several socs. chemistry, crystallography, metallography and physics in Sweden, U.K., U. S. A. Research, numerous publs. on chemistry of solid state (chem. crystallography and materials sci.), transition metal oxides, mechanisms of defects in crystal structures and concepts of crystallographic shear and homologous series, nonstoichiometry, order-disorder phenomena, phase transitions. Home: 5A, Odensgatan, Uppsala, Sweden. Office: 45 Kungstensgatan, Stockholm, Sweden.*

MAGNENAT, Pierre, Swiss physician; b. Renens, Switzerland, May 26, 1924; s. Alfred and Alice (Monnet) M.; ed. univs., Lausanne and Zurich; m. Marguerite Lugeon, Apr. 14, 1952; children—Anne, Lise, Eggbert, Claire. Prof., Propedeutic Med. Clinic, Lausanne, Switzerland, 1966—; chief diagnostic dept. U. Med. Clinic, Lausanne, 1965—. Mem. World Gastroenterology Soc. (mem. research com. 1964—). Research and publs. on liver diseases. Home: 51 Av. d'Ouchy. Office: Hopital Cantonal, Lausanne, Switzerland.*

MAGNER, Desmond, pathologist; b. Cork, Ireland, Jan. 5, 1913; s. William and Lily (Redmond) M.; matriculation Upper Can. Coll., 1930; M.D., U. Toronto, 1936, B.Sc. in Medicine, 1938; m. Miriam Margaret Anglin, Dec. 27, 1944; children—Eilis, Ann, Madeleine, Shelagh, Arthur, Peter, Brian, Myles. Lectr. pathology U. Toronto (Ont., Can.), 1939-45; pathologist Regina (Sask., Can.) City Hosps., 1946-47; prof., head dept. pathology U. Ottawa (Ont.), 1947—. Registrar Canadian Tumour Registry, Nat. Cancer Inst. Can., 1950—. Fellow Royal Coll. Physicians and Surgeons Can.; mem. Canadian, Ont. assns. pathologists, Am. Assn. Pathologists and Bacteriologists. Contbr. articles to med. jours. Research on tumor morphology and carcinogenesis. Home: 231 Clenow Av., Ottawa, Ont., Can.*

MAGNITSKY, Vladimir A., Russian geophysicist; b. Penza, USSR, June 12, 1915; s. Alexandre N. and Mary N. (Skvortsova) M.; candidate Sc., Moscow Geodetic Inst., 1944, engr., 1940, d.Sc., 1948; m. Catherine I. Magnitskaya, 1941; children—Boris, Vera. With Moscow (USSR) Geodetic Inst., 1942-54, prof., 1950-54; prof. geophysics Moscow State U., 1954—, head geophys. dept., 1963—. Head theoretical dept. Inst. Physics of Earth, USSR Acad. Sci., 1964—; chmn. upper mantle Nat. Com. USSR, 1962—; reporter Internat. Upper Mantle Com., 1962—. Mem. USSR Acad. Scis. (corr.). Author: Principles of Physics of the Earth, 1953; (with others) Theory of the Figure of the Earth, 1961; Interior Constitution and Physics of the Earth, 1965; also articles. Research on phys. nature of movement of the Earth's crust, physics of mantle of Earth, especially phase transitions, nature of chem. bonds, density and elasticity, degassing and zonal cleaning of the mantle. Home: Moscow B-296, Lomonosovsky Prospect, USSR. Office: Moscow B-296, Molodezhnaya 3, Soviet Geophy. Com., USSR.*

MAGNOL, Pierre, French botanist, physician; b. Montpellier, France, June 8, 1638; named prof. medicine, Montpellier, 1694; dir. bot. garden, Montpellier. Magnolia trees named in his honor by Linnaeus. Author: Botanicum monspelliense, 1676; Prodromus historiae generalis plantarum, 1689; Novus character plantarum, pub. 1720. Developed plant classification by families (1st attempt at simple method of classes). Died May 21, 1715.

MAGNUS, Heinrich Gustav, German chemist, physicist; b. Berlin, May 2, 1802; Ph.D., U. Berlin, 1827; postgrad. under Berzelius at Stockholm, 1828, under Gay-Lussac and Thenard at Paris, 1829; became pvt. docent, Berlin, 1831; named extraordinary prof. physics and tech. U. Berlin, 1834, prof., 1845-69. Fellow Royal Soc., 1863. Author: Über die Selbstenzündlichkeit des Feinzerteilten Eisens, 1825. Research in mineral. chem. analysis, agrl. and physiol. chemistry, mechanics, hydrodynamics, electrolysis; heat polarization, thermal expansion of gases, optics, magnetism, thermoelectricity; made early chem. studied of tellurium; and platinum; invented process for recovering selenium in sulfuric acid plants; discovered green salt of Magnus, 1828; made 1st quantitative analysis of blood gases, also gave 1st demonstration of tissue respiration by showing arterial blood has higher concentration of oxygen than venous blood, 1837; discovered deviation of projectile caused by its rotation (Magnus effect), 1835. Died Berlin, Apr. 4, 1870.

MAGNUS, Rudolf, physiologist; b. Germany, 1873; prof. pharmacology, Utrecht, Netherlands. Author: Körperstellung, 1924. Research in neurophysiology, study of posture and postural mechanisms; observed (in work with Sherrington) rotation of head in decerebrate animal alters muscle tone in limbs, 1908;

showed function of otoliths and semicircular canals of inner ear in regulating position and balance of body, 1926. Died 1927.

MAGNUS, Wilhelm, mathematician; b. Berlin, Germany, Feb. 5, 1907; s. Alfred and Paula (Kalkbrenner) M.; Ph.D., U. Frankfort (Germany), 1931; m. Gertrud Remy, Aug. 5, 1939; children—Jutta, Bettina, Alfred. Came to U. S., 1948, naturalized, 1955. Lectr., U. Frankfort, 1933-38; prof. math. U. Göttingen, 1947-48; research asso. Cal. Inst. Tech., 1948-50; prof. math. N.Y. U., N.Y.C., 1950—. Rockefeller fellow 1934. Mem. Am. Math. Soc., Göttingen Acad. Scis. Author: Combinatorial Group Theory, 1966; Hill's Equation, 1966. Research in group theory, diffraction of electromagnetic waves, spl. functions; linear differential equations, quadratic forms. Home: 11 Lomond Pl., New Rochelle, N.Y. 10804. Office: 251 Mercer St., N.Y.C. 10012.*

MAGNUS-LEVY, Adolf, physician, physiologist; b. Germany, 1865; prof. medicine, Berlin, Germany; also lived in U. S.; 1st to determine typical metabolism in various diseases; demonstrated that metabolic rate increases in exophthalmic goiter (previously recorded by Friedrich von Muller), also established assn. of thyroid deficiency with reduced metabolic activity and administered desiccated thyroid as treatment, pub. 1895; demonstrated (with Falk) that metabolism is high in childhood, low in old age, 1899; elucidated nature of diabetes through chem. studies; studied intermediate products of metabolism. Died 1955.

MAGNUSON, John Joseph, Am. biologist; b. Evanston, Ill., Mar. 8, 1934; s. John G. and Florence (Hellstrom) M.; student No. Ill. State Tchrs. Coll., 1952-53; B.Sc. with distinction U. Minn., 1956, M.Sc., 1958; Ph.D., U. B.C., 1961; m. Norma Edna Domian, June 14, 1959; children—Susan Florence, Jennifer Lou. Chief tuna behavior and physiology program Bur. Comml. Fisheries Biol. Lab., Honolulu, 1961-67; mem. affiliate grad. faculty dept. zoology U. Hawaii, Honolulu; asst. prof. lab. of limnology, dept. zoology U. Wis., Madison, 1968—. Mem. Animal Behavior Soc., Am. Soc. Ichthyologists and Herpetologists, Ecol. Soc. Am., Am. Fisheries Soc., A.A.A.S., Am. Inst. Biol. Scis., Sigma Xi. Contbr. articles to tech. jours. Research on role aggression in competition for food and space in lab. populations fish medaka, feeding, digestion, courtship and aggression and hydrostatic function of continuous swimming scombrid fish. Address: Lab. of Limnology, U. Wis., Madison, Wis. 53706.*

MAGOUN, Horace Winchell, Am. neuroanatomist; b. Phila., June 23, 1907; s. Roy Winchell and Minnie Sheida (Perkins) M.; B.S., R.I. State Coll., 1929; M.S., Syracuse U., 1931; Ph.D., Northwestern U. Med. Sch., 1934; D.Sc., Northwestern U., 1959; D.Sc., U. Rhode Island, 1960; H.H.D., Wayne State University, 1965; m. Jeanette Alice Jackson, June 27, 1931; children—Ann, Elizabeth, James. Instr., Inst. Neurology, Northwestern U., 1934-37, asst. prof., 1937-40, asso. prof., 1940-43, prof. microanatomy, 1943-50; Rockefeller fellow Johns Hopkins Med. Sch., 1939-40; prof., chmn. dept. anatomy U. Cal. Sch. Medicine, Los Angeles, 1950-55, prof., 1955—, faculty research lectr., 1957, lectr. med. history, 1958—, mem. Brian Research Inst., 1960—, dean of the grad. division, 1962—. Harvey lectr., N.Y., 1952; recipient Max Weinstein award United Cerebral Palsy, 1953; James Arthur lectr. Am. Mus. Natural History, 1954; George W. Jacoby award Am. Neurol. Assn., 1956; Thomas William Salmon Meml. lectr., 1957; award Modern Medicine, 1958; Borden award Am. Assn. Med. Colls., 1961; Passano award Am. Medical Assn., 1963. Member American Academy Neurology, Am. Assn. Anatomists, Am. Electroencephalographic Soc., Am. Neurol. Assn., Am. Physiol. Soc. Nat. Acad. Scis., Am. Acad. Arts and Scis. Author: The Waking Brain, 1958; The Historical Development of Physiological Thought, 1959. Contbg. author: Parkinsonism and its Treatment, 1954; Symposium on Sedative and Hypnotic Drugs, 1954; Brain Mechanism and Consciousness, 1954; Recticular Formation of the Brain, 1958. Contbg. author, editor (with John Field); Handbook, of Physiology, 1959. Study of neurophysiology. Home: 427 25th St., Santa Monica, Cal. Office: University of California School of Medicine, Los Angeles 24.

MAGROU, Joseph Emile, French physician, botanist; b. Béziers, France, Aug. 6, 1883; head dir. Pasteur Inst., Paris; mem. French Acad. Scis. Author: Des orchidées à la pomme de Terre; Maladies des vegetaux. Pioneer in study of plant tissue; perfected synthetic culture of potato, thus making possible study of tuberization; studied plant tumors, also symbiosis and its role in evolution; demonstrated that a staphylococcus (now known as Micrococcus ascoformans) causes botryomycosis, 1914. Died Paris, Feb. 10, 1951.

MAGUIRE, Bassett, Am. botanist; b. Alabama City, Ala., Aug. 4, 1904; s. Charles T. and Rose (Bassett) M.; student U. Pitts., 1926; B.S., U. Ga., 1926; Ph.D., Cornell U., 1938; m. Ruth Richards, 1926 (div. Sept. 1950); children—Bassett, Grace (Mrs. Daniel N. MacLemore, Jr.); m. 2d, Celia Kramer, Mar. 25, 1951. High sch. tchr. sci., biology, Athens, Ga., 1926-27; instr. botany U. Ga., 1927-29, Cornell U., 1929-31, 1937-38; prof. Utah State U., 1931-43; with N.Y. Bot. Garden, N.Y.C., 1943—, coordinator tropical botany, 1951—, head curator, 1958—, Nathaniel Lord Britton distinguished sr. curator, 1961—; aquatic botanist N.Y. State Conservation Dept. U. S. Bur. Fisheries; ecologist, spl. agt. U. S. Conservation Service, Dept. Agr.; adj. prof. botany Columbia, 1961—. Del. UNESCO Confs., 1947, 64-66, FAO, 1961; exec. dir. Orgn. for Flora Neotropica, 1964—; cons. various cos. Recipient Sarah Gildersleeve Fife Meml. award, 1952; David Livingstone centenary Gold medal Am. Geog. Soc., 1965. Mem. Assn. Tropical Biology (pres., mem. council), Orgn. Tropical Studies, A.A.A.S., Sociedad Venezolana de Ciencias Naturales (hon. life), Bot. Soc. Am., Internat. Soc. Plant Taxonomy, Am. Soc. Plant Taxonomists, Royal Netherlands Bot. Soc. (hon. life). Contbr. numerous articles field botany, neotropics vegetation and geography to sci. jours. Discoverer, namer astronomic geog. location (with John J. Wurdack) Serrania de la Neblina, mountain complex on Venezuelan-Brazilian frontier, 1953; bot. exploration N.Am., especially Rocky Mountains and Intermontane U. S., 1923-55; mem., leader, dir. numerous expdns. to tropical S.Am., including Roraima Formation (Lost Worlds) and Amazon Basin. Home: 2974 Perry Av., N.Y.C. 10458. Office: N.Y. Bot. Garden, Bronx Park, N.Y.C. 10458.*

MAGYAR, Imre, Hungarian physician; b. Losonc, Hungary, Oct. 14, 1910; s. László and Adele (Deutsch) M.; student U. Med. Sch., Budapest, 1928-34, D.med.-scis., 1960; m. Eve Fodor, Apr. 22, 1944; children—Mary, Anne, Elisabeth, László. staff U. Budapest, 1934—, dir. Postgrad. Sch. Medicine, 1960-65, prof., dir. 1st med. clinic, 1965—. Leading mem. Sci. Council Pub. Health, 1965—; mem. expert com. for diabetes WHO, 1962—. Mem. Hungarian Gastroent. Soc. (pres.), Hungarian Soc. Internal Medicine (pres.), Korányi Sandor Sci. Assn. (sec. 1964—), Internat. Assn. for Study Liver, Internat. Assn. for Study Diabetes, Soc. Internat. Internists. Author: Textbook of Medicine, 1947; Differential Diagnosis of Internal Diseases, 1960; Diseases of the Liver and Biliary Ducts, 1960; Diabetes mellitus, 1964; also numerous articles. Developed test for demonstrating B1 vitamin insufficiency; research on hepatitis with intrahepatic obstruction, differential diagnosis, diagnosis of pancreatic tumors without jaundice, insulin-resistance, achrestic diabetes (new form of diabetes). Home: 21.b. Sziget, Budapest, Hungary.*

MAHAN, Bruce Herbert, Am. chemist; b. New Britain, Conn., Aug. 17, 1930; s. Arthur E. and Clara (Gray) M.; A.B., Harvard, 1952, A.M., 1954, Ph.D., 1956. Faculty U. Cal. at Berkeley, 1956—, prof. chemistry. Author: Elementary Chemical Thermodynamics, 1963; University Chemistry, 1965; College Chemistry, 1966. Research, publs. on gas reaction kinetics. Home: 401 Michigan Av., Berkeley, Cal. 94707. Office: Dept. of Chemistry, Univ. of Cal., Berkeley, Cal.*

MAHAUX, Claude Charles, Belgian physicist; b. Belgium, Sept. 25, 1937; s. Marcel Valentin and Mary (van Hove) M.; Ph.D., U. Liège (Belgium), 1964; m. Eve Peeters, Mar. 28, 1961; 1 dau., Veronica. Staff, Cal. Inst. Tech., 1958-59; staff Institut Interuniversitaire des Sciences Nucléaires, Liège, 1959—; vis. prof. U. Heidelberg (Germany), 1966-Belgium, Sept. 25, 1937; s. Marcel Valentin and 67; chief confs. U. Liège, 1966-67, asso. prof., 1967—. Recipient Prize Louis Empain for Physics, 1965. Mem. Belgian Phys. Soc. Research and publs. on theory of nuclear reactions, structure of atomic nuclei, formal theory of scattering. Home: 21c, rue Justice, Boncelles, Belgium. Office: 15, avenue Tilleuls, Liège, Belgium.*

MAHAVIRA (Vardhamana) (Vardhamana Jñatiputra) (Nataputta) (Jina), Indian mathematician, theologian; b. Kollaga; flourished circa 500 B.C. Taught, preached, prophet and probably founder Jainism. Author: Ganita-Sara-Sangraha. Expounded Jaina physics which postulates existence dissimilar atoms endowed with various qualities; wrote only Hindu work dealing with ellipses (inaccurately); wrote on relations of dimensions of right triangle; related certain square roots to area quadrilateral. Died Pava.

MAHESH, Virendra Bhushan, biochemist, physiologist; b. Khanki Punjab, India, Apr. 25, 1932; s. Narinjan Prasad and SobhagyaVati (Bansal) M.; B.Sc. with honors, Patna U., India, 1951; M.Sc., Delhi (India) U., 1953, (Council Sci. and Indsl. Research fellow) Ph.D., 1955; D.Phil. (Assam Oil Co. fellow), 1958; Wellcome Found. fellow Organisch-chemische Anstalt of U. Basel, Switzerland, 1958; James Hudson Brown Meml. fellow Yale, 1958-59; m. Sushila Kumari Aggarwal, June 29, 1955; children—Anita, Vinit, Angela. Mem. faculty Med. Coll. Ga., Augusta, 1959—, prof. endocrinology, 1966—. Recipient Rubin award, 1963, Billings Silver medal A.M.A., 1965. Mem. Chem. Soc., Eng., Biochem. Soc., Eng., Endocrine Soc., Soc. Biol. Chemists, Soc. Gynecologic Investigation, N.Y. Acad. Scis. Research and publs. on biosynthesis, metabolism and secretion of various steroid hormones in health and various disease processes; steroid methodology and mechanism of hormone action; hirsutism of adrenal and ovarian origin; reported the first isolation of dehydroepiandrosterone, an androgen hitherto considered of only adrenal origin, from polycystic ovaries and ovarian venous blood. Home: 2911 Sussex Rd., Augusta, Ga. 30904.*

MAHESHWARI, Panchanan, Indian botanist; b. Jaipur, India, Nov. 9, 1904; s. Bijelal and Laxmi (Daga) M.; M.Sc., Allahbad U., 1927, D.Sc., 1931; D.Sc. (hon.) McGill U., Montreal, Can., 1959; m. Shanti Sharda, Nov. 8, 1925; children—Satish, Girish, Kamla (Mrs. Rajendra Garg), Ramesh, Sushila, Saubhagyawati. Lectr. in botany Agra Coll., 1930-37; lectr. Allahabad U., 1937-39; reader Dacca U., 1939-47, prof. biology, 1947-49; prof., head dept. botany Delhi (India) U., 1949—, dean faculty sci., 1954-56. Biol. sec. Nat. Inst. Scis. India, 1956-60, editor publs., 1962—. Recipient, Sahni Meml. medal, 1959, Hora Meml. medal, 1964. Fellow Indian Bot. Soc., Royal Soc. London, Nat. Inst. Sci. India, Nat. Acad. Sci. India, Indian Acad. Scis., Am. Acad. Arts and Scis. (hon. fgn.); mem. Am. Bot. Soc. (corr. mem.), Internat. Soc. Plant Morphologists (life Mem.), Kaiserlich Deutsche Akademie der Naturforscher, Deutsche Botanisches Gesselschaft (corr. mem.), Royal Dutch Bot. Soc. (hon. mem.). Author: An Introduction to the Embryology of Angiosperms, 1950; (with Umrao Singh). Dictionary of Economic Plants of India, 1965. Editor: Recent Advances in the Embryology of Angiosperms, 1963; Phytomorphology, 1950—. Home: 29 University Rd., Civil Lines, Delhi 7, India.*

MAHIN, Edward Garfield, Am. metallurgist; b. Lafayette, Ind., Aug. 16, 1876; s. Charles Wesley and Mary (Ogden) M.; B.S., Purdue U., 1901, M.S., 1903, D.Sc., 1950; Ph.D., Johns Hopkins, 1908; m. Margaret Parsons, June 10, 1903; children—Marjorie Felicia (Mrs. W. J. Berwanger), William Edward, Mary Elizabeth (Mrs. J. J. Dunningan), Carol Dorothea (Mrs. J. L. Ockert). Asst., instr., asso. prof., prof. analytical chemistry Purdue U., 1901-25; prof. metallurgy U. Notre Dame, 1925-32, prof., head dept. metallurgy, 1932-49; metallurgist. Mem. Am. Soc. for Metals (publs. com. 1941-43, chmn. 1943-44), Am. Inst. Mining and Metall. Engrs., Am. Foundrymen's Assn., Am. Soc. Engring. Edn., Iron and Steel Inst. (Brit.), Inst. Metals (Brit.), Phi Beta Kappa, Sigma Xi, Tau Beta Pi, Phi Lambda Upsilon. Author: Quantitative Analysis, 1914, 4th edit., 1932; Quantitative Agricultural Analysis (with R. H. Carr), 1923; Introduction to Quantitative Analysis, 1929; Elementary Physical Metallurgy, 1948, also articles. Research in hardness, carburization and inclusions in steel. Died Feb. 2, 1952.

MAHL, Hans, German physicist; b. July 17, 1909; s. Hans and Katharina (Heggenstaller) M.; student Technisch Hochschule, München, Germany, 1929-31; Diplom Ing., H.T.H. Danzig (Germany), 1934; Dr.Ing., Technisch Hochschule, Berlin, Germany, 1935; m. Franziska Küffner, Sept. 14, 1964. Sci. staff mem. AEG Forscungs Institut, Berlin, 1934-45; head dept. electron optics Fa. Carl Zeiss, Oberkochen, West Germany, 1953—. Recipient Silberne Leipniz-Medaille, Preussiche Akademie für Wissenschaften, 1943. Mem. Deutsche Gesellschaft für Elektronenmikroskopie. Author: Elektronen-Mikroskopie, 1951; also articles. Devel. electrostatic electronmicroscope, stigmator, replica method for electron microscopic investigation; research on textile fibers. Home: 38 Lenzhalde. Office: Fa. Carl Zeiss, Oberkochen, West Germany.*

MAHLER, Henry Ralph, biochemist; b. Vienna, Austria, Nov. 12, 1921; s. Hans and Elly (Schulhof) M.; came to U. S., 1938, naturalized, 1944; A.B., with honors, Swarthmore Coll., 1943; Ph.D., U. Cal. at Berkeley, 1948; m. Annemarie Ettinger, Feb. 2, 1948; children—Anthony S., Andrew M., Barbara A. Sr. chemist, Tex. Research Found., 1948-49; research asso. U. Wis., Madison, 1949, 50, asst. prof., 1950-55; faculty Ind. U., Bloomington, 1955—, prof. chemistry, 1957-66, research professor, since 1966—; Rockefeller Found. Spl. fellow, vis. prof. U. Sao Paulo (Brazil), 1957; vis. investigator Laboratoire de Génétique Physiologique, Gifsur-Yvette, France, 1962-63; vis. investigator, mem. corp. Marine Biol. Lab., Woods Hole, Mass., 1960—; Recipient Research Career award NIH, USPHS, 1962—, NSF Travel award, 1955, 61. Mem. U. S. Soc. Biol. Chemists, Biochem. Soc. (London, Eng.), A.A.A.S., Am. Chem. Soc., Soc. Cell Biology, Internat. Soc. for Neurochemistry, Sigma Xi. Author: (with E. C. Cordes) Biological Chemistry, 1966; Basic Biological Chemistry, 1968; also numerous articles. Research on isolation, modes action, biosynthesis, respiratory enzymes formation, integration in respiratory particles, control and mechanisms respiratory deficiency, nucleic acids, molecular psychology, role macromolecular synthesis in learning and memory. Home: 1880 Covenanter Dr., Bloomington, Ind. 47401.*

MAHLER, Kurt, mathematician; b. Krefeld, Germany, July 26, 1903; s. Hermann and Henriette (Stern) M.; Ph.D., U. Frankfurt (Germany) 1929; D.Sc., U. Manchester (Eng.), 1940; Faculty, U. Manchester, 1937-63, personal prof., 1952-63; prof. Inst. Advanced Studies, Australian Nat. U., Canberra, 1963—. Fellow Royal Soc., 1948, Australian Acad. Scis.; mem. Dutch Math. Soc. (hon.), London Math. Soc., Am. Math. Soc., Australian Math. Soc. Author: Diophantine Approximations, I, 1961; also numerous articles. Research on number theory, geometry of numbers, transcendental numbers. Home: University House, Canberra ACT, Australia.*

MAHON, Raymond Pierre, French physician; b. St. Martin de Seseas, France, Jan. 27, 1902; s. André and Marie (Dejean) M.; ed. Faculty Medicine, Bordeau France; m. Marie Antoinette Flous, Sept. 3, 1931; children—Maylis (Mrs. Jean Pierre Maumaut), Quitterie, Pierre, André, François. Successively became intern Hopitaux Bordeaux, 1924, head faculty clinic, 1931, hosp. obstetrician, 1934, agrégé, 1939, head obstetric service, 1944, prof., 1953, prof. obstetrics and social obstetrics, 1957; prof. at Clinique obstéricale Hopital Pellegrin, Bordeaux, 1966—. Mem. Bordeaux-Gironde Physicians Assn. (past pres.), Aquitaine Regional Council Physicians, Med. Cons. Commn. Bordeaux Regional Hosp. Center (past pres.), French Soc. Gynecology and Obstetrics, Internat. Coll. Surgeons. Contbg. author: Encyclopédie Médico-Chirurgicale, 1934—. Research and numerous publs. on physiology of uterine contraction, 1929-39, maternal mortality, 1928-59, amiotic embolism, 1958-64, toxemia in pregnancy, 1939-63, fibromes and appendicitis linked with pregnancy, 1931-47, prolonged pregnancies 1940-57, cardiopathies in course of pregnancy, 1938-66. Home: 58 rue de St. Genès, Bordeaux (Gironde). Office: Clinique obstétricale, Hôpital Pellegrin, Bordeaux, France.*

MAHONEY, Earle Barnes, Am. physician; b. Penn Yan, N.Y., July 3, 1909; s. Francis and Bessie (Barnes) M.; B.S., Hobart Coll., 1930, D.Sc., 1957; M.D. with honors, U. Rochester, 1934; m. Mary Anne Barnum, Dec. 27, 1937; children—Richard Barnum, David Barnes, Sheila Ann, Mary Elizabeth. NRC fellow Cin. Gen. Hosp., 1937-38; faculty U. Rochester (N.Y.), 1938—, prof. surgery, surgeon, 1957—. Cons. to hosps. Recipient award Rochester Acad. Medicine, 1965. Fellow A.C.S., Am. Acad. Sci.; mem. N.Y. State, Monroe County med. socs., A.M.A., Rochester Acad. Medicine (past pres.), Soc. U. Surgeons (past pres.), Am., Central, Pan Pacific surg. assns., Soc. Clin. Surgery, Soc. Vascular Surgery (past pres.), Upstate Soc. Thoracic Surgery, Internat. Cardiovascular Soc., Am. Assn. Thoracic Surgery, Am. Heart Assn., Am. Assn. U. Profs., Internat. Soc. Surgery, Phi Beta Kappa, Sigma Xi, Alpha Omega Alpha. Contbg. author: Atlas of Pediatric Surgery, 1965. Research and numerous publs. on surgery of cardiovascular disease, hypothermia, shock. Home: 101 Country Club Dr., Rochester, N.Y. 14618.*

MAHONEY, John Friend, Am. physician; b. Fond du Lac, Wis., Aug. 1, 1889; s. David and Mary Ann (Hogan) M.; M.D., Marquette U., 1914; m. Leah Ruth Arnold, Sept. 29, 1926; children—Janet Ann, John Friend. Commd. officer USPHS, 1917; dir. Venereal Disease Research Lab., U. S. Marine Hosp., S.I., 1929-49; commr. health City of N.Y., 1950-53; gen. dir. Bur. Labs., N.Y.C. Dept. Health, 1954-57. Fellow Am. Pub. Health Assn. (Lasker award 1946); mem. A.M.A., N.Y. Acad. Scis. Introduced (with R. C. Arnold and A. Harris) use of penicillin in treatment of early syphilis, 1943. Died Feb. 23, 1957.

MAI, William Frederick, Am. nematologist; b. Greenwood, Del., July 23, 1916; s. William F. and Laurana (Owens) M.; B.S., U. Del., 1939; Ph.D., Cornell U., 1945; m. Barbara Morrell, June 2, 1941; children—Virginia Austin, William Howard, Elizabeth Hardy. Faculty, Cornell U., Ithaca, N.Y., 1946—, prof. plant pathology, 1952—. Mem. Nat. Acad. Scis. (chmn. subcom. on nematodes 1965—), Phytopath. Soc. (adv. com. on nematology 1963—), Am. Phytopath. Soc. (past pres. Northeastern div.), Am. Inst. Biol. Scis., Helminthological Soc. Washington, Potato Assn. Am., Soc. European Nematologists, Sigma Xi, Phi Kappa Phi. Author: Pictorial Key to Genera of Plant Parasitic Nematodes (with H. H. Lyon, T. H. Kruk), 1960; also numerous articles. Editorial bd. Phytopathology, 1960-63. Developed control programs for golden nematode disease of potatoes, replant diseases of tree fruits, onion bloat. Home: 613 E. Shore Dr., Ithaca, N.Y. 14850.*

MAICKEL, Roger Philip, Am. pharmacologist; b. Floral Park, N.Y., Sept. 8, 1933; s. Philip V. and Margaret (Rose) M.; B.S., Manhattan Coll., 1954; postgrad. Bklyn. Poly. Inst., 1954-55; M.S., Georgetown U., 1957, Ph.D., 1960; m. Lois L. Pivonka, Sept. 8, 1956; children—Nancy E., Carolyn S. Staff, Nat. Heart Inst., NIH, Bethesda, Md., 1955-65, pharmacologist, 1962-64, sect. head, 1964-65; asso. prof. pharmacology Ind. U., Bloomington, Ind., 1965—. Mem. Am. Chem. Soc., Am. Soc. Pharmacology and Exptl. Therapeutics, Chem. Soc. (London), Am. Assn. U. Profs., A.A.A.S., Am. Inst. Chemists, Am. Therapeutic Society, Sigma Xi. Editor: Biochemical Factors in Alcoholism, 1966. Exec. editor Life Sci., 1965—; cons. editor Jour. Psychopharmacology, 1966-68. Research and publs. on psychopharmacology including effects drugs on brain function, methods of assay for drugs in biol. materials, mechanism of action of psychoactive drugs. Home: 104 Hampton Ct., Bloomington, Ind. 47401.*

MAIDEN, Joseph Henry, botanist; b. St. John's Wood, London, Apr. 25, 1859; s. Henry Maiden; m. Jeannie Hammond; 4 daus. Apptd. asst. to 1st curator Technol. Mus., New South Wales, 1881; supt. tech. edn.; cons. botanist Forest and Agr. depts.; dir. Botanic Gardens, Sydney; ret. 1924. Recipient Linnean medal, 1915. Fellow Royal Soc., 1916. Author: Useful Native Plants of Australia; Illustrated Flowering Plants and Ferns of New South Wales; Manual of the Grasses of New South Wales; Wattles and Wattle-barks; Forest Flora of New South Wales; Critical Revision of the Genus Eucalptus; Sir Joseph Banks, the Father of Australia. Died Nov. 16, 1925.

MAIER, Herbert Caille, Am. surgeon, educator; b. N.Y.C., Mar. 24, 1908; s. Otto and Dina (Caille) M.; A.B., Columbia, 1928, M.D., 1932, Med. Sc.D., 1937; M.S. in Surgery, U. Mich., 1938; m. Janet Sterling Baldwin, Jan. 10, 1948; 1 son, Donald. Practiced medicine specializing in surgery, N.Y.C., 1939—; thoracic surgeon Kings County Hosp. Bklyn., 1939-46, Meml. Hosp., N.Y.C., 1942-46; dir. surgery Triboro Hosp., N.Y.C., 1942-47; asst. prof. surgery N.Y. U., 1946-49, L.I. Coll. Medicine, 1945-55; asso. clin. prof. surgery Columbia, 1956—; clin. prof. surgery N.J. Coll. Medicine and Dentistry, Jersey City, 1963—. Cons. Lenox Hill Hosp., N.Y.C., Jersey City Med. Center, U. S. Naval Hosp., St. Albans, N.Y., Manhattan VA Hosp. N.Y., Meth. Hosp. Bklyn., U. S. VA. Named Darholt lectr., 1951, John Alexander Meml. lectr., 1958. Mem. Bd. Thoracic Surgery (chmn. emeritus), Am. Assn. for Thoracic Surgery (pres. 1965-66), Internat. Surg. Group (treas. 1960—), N.Y. Soc. for Thoracic Surgery (pres. 1947-48), N.Y. Soc. Cardiovascular Surgery (pres. 1955-57), Soc. Thoracic Surgeons Gt. Britain and Ireland (hon.), Sigma Xi, Alpha Omega Alpha. Contbg. author: Nelson's Loose-leaf Surgery, 1941, 42; Cecil and Loeb's Textbook of Medicine, 1955, 59, 63; Gibbon's Surgery of the Chest, 1962. Editor: (with Edgar Mayer) Pulmonary Carcinoma, 1956; adv. editorial bd. Jour. Thoracic and Cardiovascular Surgery, 1946—. Contbr. articles to tech. jours. Research on diagnosis and surg. treatment of diseases of chest, physiologic and path. aspects of anesthesia and surgery, tumors of chest and thoracic lymphatics. Home: 3 E. 71st St., N.Y.C. 10021.*

MAIER, Ludwig, chemist; b. Schmidhausen, Germany, May 21, 1929; s. Josef and Barbara (Linseisen) M.; Diplom Chemiker, U. Munich, 1953, Dr.rer.nat., 1955; m. Irene Aigner, Nov. 15, 1957; children—Angelika, Wilfried. Instr., U. Munich, 1955-56; research fellow Harvard, 1956-57; staff scientist Monsanto Research SA, Zürich, Switzerland, 1957-60, 61—; research chemist Monsanto Co., St. Louis, 1960-61. Mem. Swiss Chem. Soc. Contbg. author: Progress in Inorganic Chemistry, 1963; Topics in Phosphorus Chemistry, 1965; Forstschritte der chem. Forschung, 1967. Research, publs., numerous patents on devel. new syntheses of organophosphorus compounds; prepared and identified meso- and racemic-biphosphines and biphosphine disulfides for 1st time; invented direct synthesis of organo-halo-phosphines (from white phosphorus and alkylor aryl halides), tetramethylphosphonium halides (from white phosphorus and alkyl halides under pressure, phosphine oxides, phosphonic and phosphinic acids (from white phosphorus, an aldehyde and an amine). Home: 17, Tiergartenstrasse, Kilchberg, Switzerland 8802 ZH. Office: 39, Binzstr. (Monsanto) Zürich, Switzerland, 8045 ZH.*

MAIER, Norman Raymond, Am. psychologist; b. Sebewaing, Mich., Nov. 27, 1900; s. Charles and Emma (Cramer) M.; student Wayne U., 1919-21; A.B., U. Mich., 1923, M.A., 1925, Ph.D., 1928; postgrad. U. Berlin; m. Ayesha Frances Ali, Feb. 17, 1930; children—Richard A., Carl C., John L. Faculty, U. Mich., Ann Arbor, 1931—, prof. psychology, 1945—. Cons. European Productivity Agy., 1959-60; indsl. cons. to govt. agys., industry. Recipient $1000 prize A.A.A.S., 1938; Henry Russel award U. Mich., 1939. NRC fellow U. Chgo., 1929-31. Diplomate Am. Bd. Indsl. Psychology. Author: (with Renninger) A Psychological Approach to Literary Criticism, 1933; (with Schneirla) Principles of Animal Psychology, 1935; Studies of Abnormal Behavior in the Rat, 1939; Frustration, 1949; Principles of Human Relations, 1952; The Appraisal Interview, 1958, Problem Solving Discussions and Conferences, 1963; Psychology in Industry, 3d edit., 1965; (with Solem, A. Maier) Supervisory and Executive Development, 1961; (with Hayes) Creative Management, 1962; also numerous articles. Home: 1111 Fair Oaks Pkwy., Ann Arbor, Mich. 48104.*

MAIER, Rudolf Robert, German physician; b. Freiburg in Breisgau, Germany, Apr. 9, 1824. Author: Ueber den Bau der thränenorgane, insbesondere der thränenleitenden Wege; Thräneorgane des Menschen, 1859; Lehrbuch der allgemeinen pathologischen Anatomie, 1871; Arbeiten über Periarteriitis nodosa und muskuläre Pylorusstenose-Johannes Schenck von Grafenberg, 1878. Described (with Adolf Kussmaul) inflammatory disease of coats of medium-sized and small arteries, known as periarteritis nodosa, or Kussmaul's disease, Kussmaul-Maier disease, 1866. Died Freiburg, Nov. 7, 1888.

MAIGNAN, Emanuel, geometer, philosopher; b. Toulouse, France, 1601; monk; became prof. mathematics in Minim convent, Rome, 1636. Author: Perspectiva honaria; Treatise on Catoptrics, 1648; other works. Developed theory of light refraction, 1648; studied problem of vacuum; admitted void; opposed scholasticism; followed Descartes on many points; held that animals have thoughts and preception. Died 1676.

MAILLET, Benoit de, see de Maillet, Benoit.

MAILLET, Pierre L., French biologist; b. Quimper, France, Nov. 3, 1923; s. Francois and Marie (Keryvel) M.; ed. Quimper Lycee; Licence, Rennes Faculty Scis.; D.Sc., Paris Faculty Scis.; m. Claude Gayet, Nov. 26, 1949; 7 children. Asst., Paris Faculty Scis.; lectr. Rennes Faculty Scis., then full prof. gen. biology, mem. Bd. Trustees, from 1960. Recipient medals Militaire, Resistance, Evadés, Déportés, Croix de Guerre. Mem. Biol. Soc., French Soc. Electronic Microscopy, Soc. Zoology. Author: le Phylloxe de la Vigne, 1955. Research on homoptera, spermatogenesis of insects through use of electronic microscope, transmission of rickets by insects. Home: 5 bis, Bd. Burloud, Rennes, France. Office: Faculty of Sciences, Rennes, Beaulieu, France.*

MAIMONIDES (or Moses ben Maimon), philosopher, physician; b. Cordova, Spain, Mar. 30, 1135; s. Maimon ben Joseph; studied sci., art, medicine, philosophy, lit. under Arab tchrs. in Spain; his father trained him in Hebrew and Jewish scholarship; m. sister of Ibn al Mali; 1 son, Abraham. Left Cordova, 1148, went to Fez, 1160, then Egypt, 1165, and became physician to Saladin under whose auspices he wrote many works on medicine. Author: Mishneh Torah; Guide of the Perplexed; also treatises on logic, the calendar, medicine. Devoted attention to various aspects of medicine such as asthma, diet, poisons and their cures; exponent of what is now known as psychosomatic medicine; believed in healing power of nature; laid great stress on importance of hygiene; attempted to organize overwhelming mass of Jewish oral law into a reference book for both laymen and rabbis; studied astronomy, and well ahead of his time disdained astrology; attempted to reconcile teachings of O.T. with teaching of Aristotle; said to be most influential Jewish philosopher since Moses; influenced medieval medicine and many later medieval Christian philosophers. Died Cairo, Egypt, Dec. 13, 1204.

MAINE, Sir Henry James Summer, Brit. polit. theorist, jurist; b. Aug. 15, 1822; s. James and Eliza (Fell) M.; grad. U. Cambridge, 1844; m. Jane Maine, 1847; 2 sons. Became tutor Trinity Hall, U. Cambridge, then regius prof. civil law, 1847-54, elected master Trinity Hall, 1877, named Whewell prof. internat. law, 1887; called to bar, 1850; legal mem. council in India, 1862-69; also vice-chancellor U. Calcutta; apptd. corpus prof. jurisprudence U. Oxford, 1869; mem. sec. of state's council for India, from 1871. Fellow Royal Soc., 1874; mem. Am. Acad., Washington Anthrop. Soc., Academie des Sciences Morales et Politiques. Author: Ancient Law, 1861; Village Communities in the East and the West, 1871; Early History of Institutions, 1875; Early Law and Custom, 1883; Popular Government, 1885; International Law, 1888; also articles to periodicals. Evolved method for study of instns. based on analysis of law and soc. since archaic times; believed basic unit of soc. has changed from family in ancient times to the individual; challenged reality of natural rights and benefit of equality. Died Cannes, France, Feb. 3, 1888.

MAINE DE BIRAN, see de Biran, Marie-François-Pierre Gonthier.

MAIRE, Christopher, astronomer, mathematician; b. Durham, Eng., Mar. 6, 1696; s. Christopher and Frances (Ingleby) M.; ed. Jesuit Sch. St. Omer; became mem. Soc. Jesus, 1715; rector English Coll. of Rome, 1744-50; mem. French Acad. Scis., 1753. Author: Observationes cometae, 1744; Table of Longitudes and Latitudes for the Principal Towns of the World, 1747. Observed eclipse of moon, 1745; mapped Vatican; measured St. Paul's Cathedral. Died Grand, Belgium, Feb. 22, 1767.

MAIRE, René, French botanist; b. Lons-le-Saunier, France, May 29, 1878; Docteur ès scis., 1902; M.D.; became researcher Nancy (France) Faculty Scis., 1905; named lectr., Caen, France, 1908; prof. botany Faculty Algiers (Algeria), 1911-49; mem. French Acad. Scis., 1923. Author: Flore de l'Afrique du Nord. Authority in systematics; research in cytology and systematics of mushrooms,flora of N. Africa; pub. phytogeographic map of Algeria and Tunisia. Died Algiers, Nov. 24, 1949.

MAISONNEUVE, Jules-Germain-François, French urologist; Nantes, France, 1809; pupil of Dupuytren and Récamier; named surgeon Hôtel Dieu, 1862. Author: Du perioste et de ses maladies, 1834; Sur la coxalgie, 1844; Clinique chirurgicale, 2 vols., 1863-64; also articles. Devised hair cathether, 1845; credited with 1st description of gas gangrene, 1853; inventor urethrotome, 1855; one of 1st to apply theory of regeneration of bone by periosteum to surg. practice. Died Missillac, France, 1897.

MAITLIS, Peter Michael, chemist; b. Berlin, Germany, Jan. 15, 1933; s. Jacob and Judith (Ebel) M.; B.Sc., Birmingham (Eng.) U., 1953; Ph.D., London (Eng.) U., 1956; m. Marion Basco, July 19, 1959; children—Niccola Anne, Sarah Louise. Asst. lectr. London U., 1956-58, B.P. Research fellow, 1958-60; research asso. Cornell U., 1960-61; research fellow Harvard, 1961-62; faculty McMaster U., Hamilton, Ont., Can., 1962—, prof. chemistry, 1967—. Alfred P. Sloan fellow, 1967—. Mem. Am. Chem. Soc., Canadian Inst. Chemistry, Chem. Soc.

London. Research and publs. on preparation and properties of cyclobutadiene-metal complexes, ligand-transfer reactions of pi bonded organic groups, reactions of noble metal compounds with acetylenes and mechanisms of reactions. Home: 228 Governors Rd., Dundas, Ont., Can. Office: Dept. Chemistry, McMaster U., Hamilton, Ont., Can.

MAITRE-JEAN, Antoine, French ophthalmologist; b. Méry-sur-Seine, France, circa 1650; royal surgeon. Author: Traité des maladies de l'oeil, 1707. Founder of French ophthalmology; 1st to recognize cataract consists of opacity of lens rather than pellicle inside capsule of lens, 1692, pub., 1707. Died 1730.

MAJ, Jerzy, Polish pharmacologist; b. Brzesc, Poland, July 14, 1922; s. Joseph and Franciszka (Dudek) M.; Ph.D., Jagiellonian U., Cracow, Poland, 1951; m. Maria Tabeau, Jan. 30, 1950; children—Andrew, Christopher. Staff, Med. Acad., Cracow, 1949-63, docent pharmacology, 1962-63; asso. prof. pharmacology Med. Acad., Lublin, Poland, 1964—. Research and publs. on pharmacology of plants native in Poland, pharmacology of new synthetic derivatives of hydantoine and penthaerithrite, catecholamines and action of psychotropic drugs. Home: 19 Pulawska, Lublin, Poland.*

MAJER, Vladimír, Czechoslovakian chemist; b. Prague, Czechoslovakia, Mar. 29, 1903; s. Alois and Karla (Suranová) M.; Ing., Tech. U. Prague, 1925; postgrad. Charles U., Prague, 1924-26, Dr.techn., 1928, D.Sc., 1956; postgrad. U. Leipzig (Germany), U. Copenhagen (Denmark); m. Miroslava Voríšková, Aug. 25, 1948; 1 son, Vladimír. Chemist, Chem. Works Ustíí n.L., 1929-31; research worker dept. phys. chemistry Charles U., 1932-37; asst. Tech. U. Prague, 1937-57, prof. Faculty Tech. and Nuclear Physics, 1957—, head dept. nuclear chemistry, 1959—. Mem. numerous sch. and sci. councils, bds. Author books including: Radiochemistry, 1942; Radiometric Analysis, 1949; Principles of Nuclear Chemistry, 1961; also articles. Research in phys. chemistry, polarography, microchemistry, radioactivity, radiochemistry, nuclear chemistry. Home: 1 Brehová, Prague, Czechoslovakia.*

MAJEWSKI, Czeslaw, Polish pathologist; b. Lwów, Poland, Mar. 21, 1914; s. Józef and Leontyna (Cewicka) M.; Diplom of Physician, U. Jan Kazimierz, Lwów, 1938, Diplom of DM, 1949; m. Helena Laniewska, Aug. 31, 1939; children—Waclaw, Przemyslaw, Bozena. Asst. dept. obstetrics Faculty Medicine, U. Lwów, 1939-41; staff High Med. Sch., Poznan, Poland, 1947-60, adj. dept. path. anatomy, 1952-60; chmn. dept. path. anatomy J. Strus City Hosp., Poznan, 1960—; staff Inst. Balneo-Climatologic, Poznan, 1960—. Recipient Gold Cross of Merit, 1957. Mem. Polish Path. Anatomy Assn. (prize 1952), Polish Physicians Assn., Polish Dermatol. Assn., Polish Electron Microscope Assn. Research, numerous publs. on histochemistry of tumors, biophysics and biochemistry of tumors, ultrasonic interactions with living body, indsl. medicine. Home: 14 a Sadowa. Office: 9 Kozia, Poznan, Poland.*

MAJNO, Guido, pathologist; b. Milan, Italy, Feb. 9, 1922; s. Edoardo and Elda (Bernstein) M.; M.D., U. Milan, 1947; Chef des Travaux, Inst. Pathology, U. Geneva (Switzerland), 1952; M.A. (hon.), Harvard, 1961; m. Friedericke Beatrice Schuyer, June 20, 1950; children—Corinne, Lorenzo, Luca. Came to U. S., 1952. Research asso. Tufts Med. Sch., Boston, 1952-53; faculty Harvard, Boston, 1953—, asso. prof. pathology, 1961—. Guest investigator Rockefeller Inst., N.Y.C., 1958-59. Recipient Lederle Med. Faculty award, 1956-59. Sr. Research fellow USPHS, 1959-64. Mem. Am. Soc. Cell Biology, Am. Assn. Pathology and Bacteriology, Am. Soc. Exptl. Pathology, Histochem. Soc. Research, publs. on demonstration of histamine in tissue injuries, tissue death and injury and wound healing. Home: 15 Peacock Farm Rd., Lexington, Mass. 02173. Office: 25 Shattuck St., Boston 02115.*

MAJOOR, Cornelis L. H., Dutch physician; b. Bussum, Jan. 24, 1912; s. J.G.J. and Johanna M. (van Dijk) M.; M.D., U. Amsterdam (Netherlands); m. Anne Hermans, Jan. 25, 1943; children—Koosanne, Jan. Karel, Fien, Kees, Marianne. With med. clinic U. Amsterdam, 1939-42; with Berg en Bosch sanatorium, Bilthoven, Netherlands, 1941; head med. clinic Charitas Hosp., Roosendaal, 1942-55; became prof. internal medicine U. Nimègue, 1955. Mem. Dutch Assn. Clin. Studies (1st sec.). Author: Thèse et publications sur la précipitation des protéines sériques, 1942-47; Etudes clinique sur la lithiase biliare, 1947-54; Etudes sur la natriurie et la diminution du débit urinaire de l'aldostérone provoqués par l'héparine et les héparinoides, 1957-63. Home: van Slichtenhorststraat 108 Nijmegem. Office: Hôpital de l'-Université St-Radboud, Sint Annastraat Nijmegen, Netherlands.

MAJOR, Johann Daniel, German physician; b. Breslau, Germany, 1634; practiced medicine, Hamburg, also Wittenberg, Germany, until 1665; apptd. prof. medicine, Kiel, Germany, 1665; planted bot. garden, Kiel. Author dissertation on petrified crabs and snakes, 1664; Chirurgia infusoria (on infusory surgery), 1667. Credited with 1st successful injection of medicinal substance into vein of human being, recorded 1662; developed new method of transplantation; introduced a doctrine of circulation of nutritive juices of plants; anointed scalp to cure certain diseases. Died 1693.

MAJOR, John Keene, Am. physicist; b. Kansas City, Mo., Aug. 3, 1924; s. Ralph Hermon and Margaret Norman (Jackson) M.; student U. Kansas City, 1940-41; B.S., Yale, 1943, M.S., 1947; D.Sc. (Fulbright fellow), U. Paris, 1951; m. Gracemary Somers Westing, Apr. 9, 1950 (div. Aug. 1964); children—John W., Ann Somers, Richard J. Lab. asst. physics Yale, 1943-44, instr., research asst. physics, 1952-55; sci. staff spl. studies group, div. war research Columbia, 1944; instr. physics and chemistry Am. Community Sch., Paris, France, 1948-49; research fellow Centre National de la Recherche Scientifique, Laboratoire de Chimie Nucleaire, Coll. de France, Paris, 1951; Carnegie Found. fellow Laboratoire Curie, Institut du Radium, Paris, 1951; asso. prof. physics Western Res. U., 1955-57, chmn. dept., 1955-60, 61-64, Perkins prof. physics, 1957—; staff asso. div. instnl. programs NSF, 1964—; fellow at Laboratorium für Technische Physik, Technische Hochschule München, W. Germany, 1960-61. Sci. cons. Sonar Analysis Group, 1946-47. Served as lt. (j.g.) USNR, 1944-46. Mem. A.A.A.S., Am. Phys. Soc., Am. Assn. Physics Tchrs., Fedn. Am. Scientists (council del.-at-large 1954-56, 63-65), Am. Assn. U. Profs., Phys. Soc. London, Societe Francaise de Physique of Paris, Sigma Xi. Studies in K-electron capture and positron emission, nuclear instrumentation, recoil-free scattering of gamma rays. Home: 1521 27th St. N.W., Washington 20007. Office: 1951 Constitution Av. N.W., Washington 20550.

MAJUMDAR, Sudhansu Datta, Indian theoretical physicist; b. Habiganj, Sylhet, East Pakistan, Feb. 13, 1915; s. Upendra Datta and Priyatama (Biswas) M.; B.Sc. with honors in Physics, Calcutta (India) U., 1935, M.Sc., 1937, M.Sc. in Math., 1944, D.Sc. in Theoretical Physics, 1959; m. Namita Purkaystha, Mar. 8, 1942; 1 dau., Nandini. Faculty, Calcutta U., 1937-44, researcher, 1945-48, faculty, 1951-65, reader, 1960-65; Nat. Inst. Scis. Research fellow Saha Inst. Nuclear Physics, 1948-51; prof. theoretical physics Indian Inst. Tech., Kharagpur, India, 1965—. Mem. Indian Assn. for Cultivation Sci. (past mem. council), Calcutta Math. Soc. (mem. council 1965—). Research, publs. on gen. solution of Einstein's field equations for case when gravitational and electric forces balance at every point, formulation of theory of elementary particles in gen. relativity and derivation of new interaction terms from invariance considerations, solution of problem in theory of molecular spectra leading to new method of investigation of rotation and Lorentz group, Dirac equation at high energies. Office: Dept. Physics, Indian Inst. Tech., P.O. Kharagpur Tech., India.*

MAKAI, Endre, Hungarian mathematician; b. Budapest, Nov. 5, 1915; s. Endre and Elisabeth (Garay) M.; Ph.D., Budapest U., 1942; Dr. math. scis., Hungarian Acad. Scis., 1955; m. Elisabeth Koti, Aug. 3, 1946; children—Endre, Elisabeth, Joseph, John. Chemist, Chinoin Pharm. Factory, Budapest, 1941-44, Tungsram Incandescent Lamp Co., Budapest, 1945-51; sr. lectr. math. Budapest Tech. U., 1951-62; research officer Math. Research Inst., 1962—. Cons. Inst. Tech. Physics, Hungarian Acad. Scis., Budapest, 1962-64. Research, numerous publs. on differential equations, membrane eigenvalues, torsional rigidity of bars. Address: 13-15 Realtanoda, Budapest, Hungary.*

MAKARCHENKO, Aleksandr Fedorovich, Russian neurophysiologist; b. Mariupol (now Donetsk Oblast), 1903; grad. Kharkov Med. Inst., 1933, postgrad. in neurology, until 1936; Cand. Med. Sci., 1941; D.Med. Sci., 1954. Dir., Kharkov Postgrad. Med. Inst., 1937-39, Lvov Med. Inst., 1939-41; dep. peoples commissar of health of Tadzhikistan, Ukrainian and USSR dep. minister health, 1942-50; sr. asso. Kiev Postgrad. Med. Inst., 1950-53, Bogomolets Inst. Physiology, 1950-55; head dept. neurology and neurophysiology Bogomolets Inst. Physiology, Ukrainian Acad. Sci., 1950—, dir., 1955—, chmn. dept. biol. sci., 1961-62. Chmn., Coordinating Council on Physiology. Recipient Bogomolets prize, 1954. Mem. Ukrainian Acad. Sci. (v.p. 1962—), Ukrainian Soc. Neuropathologists and Psychiatrists (chmn.), Kiev Soc. Physiologists (chmn.). Author numerous works including: Atomic Energy in Biology and Medicine, 1958; The Creative Development of Physiology, 1960. Editor: Physiology Jour. of Ukrainian Acad. Sci.; mem. editorial ed. Ukrainian Soviet Ency. Address: Bogomolets Inst. Physiology, Ukrainian Acad. Sci., ulitsa Bogomoltsa 4, Kiev, USSR.

MAKAREVSKII, Aleksandr Ivanovich, Russian aero. engr.; b. Apr. 6, 1904; grad. Moscow (USSR) Tech. Sch., 1929. With Central Aero-Hydrodynamic Inst., from 1927, dir., from 1950; prof. Moscow Physico-Tech. Inst., 1952—. Recipient Stalin prize, 1943. Corr. mem. USSR Acad. Scis. Author: General Permissible Deformations in Aircraft Design, 1936; The Wing and Tail Unit Loading of a Fighter Aircraft in Flight, 1940; (with others) Handbook for Designers, 1940-42. Studied external stress and loads acting upon aircraft in flight; work resulted in establishment of domestic standards on reliability and durability (including high speed aircraft). Office: Moscow Physico-Tech. Inst., Moscow, USSR.

MAKAROV, Petr Vasilevich, Russian biologist, histologist; b. 1905; grad. biol. dept. Physico-Math. Faculty, Leningrad U., 1928; D.Biol. Sci., 1939—. Asst., Biol. Faculty, Leningrad U., 1929-35, lectr., 1936-41, prof. chair gen. and comparative physiology Biol. and Pedological Faculty, 1945-48, head cytology lab. dept. genetics and selection, 1948—, also head chair anatomy and histology; prof., 1944—; head chair gen. biology Leningrad Health-Hygiene Med. Inst., 1944—. Mem. USSR Acad. Med. Sci. (corr.). Author: Problems of General and Cellular Narcosis, 1938; The Structure of the Cellular Nucleus and Methods of Studying It, 1949; New Data on the Structure and Characteristics of the Cellular Nucleus, 1958. Mem. editorial bd. Archives of Anatomy, Histology and Embryology. Research and numerous publs. on fine morphology and physiology of cells, types of gen. narcosis (cellular, characteristic of lower animals, and neural, relating to higher organisms); developer method to preserve in vivo structure of cell nuclei in permanent microscopic specimens; devised method to diagnose gas gangrene quickly. Address: Leningrad University, Universitetskaya n. 7-9, Leningrad, USSR.

MAKEHAM, William Matthew, Brit. statistician. Formulated law of human mortality (Makeham's law) used in computing joint life contingencies, circa 1860. Died 1892.

MÄKELÄ, (Valto) (Eero) Olavi, Finnish immunologist; b. Kuusankoski, Finland, Aug. 31, 1929; s. Valto and Lyyli (Jaakkola) M.; M.D., U. Helsinki (Finland), 1957; m. Pirjo Helena Hautala, May 2, 1954; children—Matti Olli, Jukka Eerikki, Anna Helena, Tomi Pekka. Postdoctoral fellow dept. genetics Stanford Med. Sch., 1959-61; faculty U. Helsinki, 1961-65, asst. prof. med. microbiology, 1965—; dir. div. immunobiology State Serum Inst., Helsinki, 1965—. Mem. sci. council State Med. Bd., Finland, 1962—. Mem. A.A.A.S., Finnish Soc. Pathologists and Bacteriologists (past sec.), Royal Soc. Medicine (affiliate). Contbr. numerous articles to tech. jours. Discovered human blood group Bv; demonstrated that each plasma cell makes antibody to a small fraction of the antigen molecule. Home: Puuskaniemi, Helsinki 85. Office: Mannerheimintie 166, Helsinki, Finland.*

MAKEMSON, Maud Worcester (Mrs. Thomas Emmet Makemson), Am. astronomer; b. Center Harbor, N.H., Sept. 16, 1891; d. Ira Eugene and Fanny (Davisson) Worcester; student Radcliffe Coll., 1908-09; A.B., U. Cal., Berkeley, 1925, M.A., 1927, Ph.D., 1930; m. Thomas Emmet Makemson, Aug. 7, 1912; children—Lavon (dec.), Donald, Harris (dec.). Faculty U. Cal., Berkeley, 1930-31, Rollins Coll., 1931-32, Vassar Coll., 1932-57; prof. astronomy, research astronomer U. Cal., Los Angeles, 1959-64. Cons. Lockheed-Cal., 1961-63, Gen. Dynamics, Ft. Worth, 1965. Fulbright lectr., Japan, 1953-54, W. Pakistan, 1957-58; Guggenheim fellow, 1941-42. Fellow A.A.A.S.; mem. Am. Inst. Aeros. and Astronautics, Am. Astron. Soc., Astron. Soc. Pacific. Author: The Morning Star Rises, 1941; Book of the Jaguar Priest, 1951; (with Robert M. L. Baker, Jr.) Introduction to Astrodynamics, 1961, 2d edit., 1967. Editor (with Baker): 12th Astronautical Fedn. Congress, 2 vols., 1963. Research, publs. on moon, orbits of minor planets, comets, and double stars; studies in early history of astronomy. Home: 1114 N. Lacy St., Santa Ana, Cal. 92701.*

MAKENS, Royal Francis, Am. nuclear scientist, govt. ofcl.; b. Mpls., Mar. 8, 1901; s. Joseph A. and Sarah (McGinty) M.; B.S., U. Minn., 1926; M.S., Mich. Tech. U., 1930; Ph.D., State U. Ia., 1938; postgrad. Oak Ridge Inst. Nuclear Studies, Internat. Inst. Nuclear Sci. and Engring.; m. Gladys Mildred Larson, Sept. 4, 1930; children—Hugh Harvey, Neal R. Faculty phys. chemistry Mich. Tech. U., Houghton, 1930-62, prof., 1948-56, prof., chmn. dept. chemistry and chem. engring., 1956-60, dir. nuclear engring., 1960-62; prof. phys. chemistry Tex. Coll. Arts and Industries, 1947-48; chief reactor tech. br. AEC, Idaho Falls, Ida., 1962-65, AEC edn. officer and sr. tech. specialist, 1965—. Scientist-dir. USPHS, 1956—, lectr., summers 1960-61; research asso., cons. Argonne Nat. Lab., 1956-60; tchr. radiation health, participant Atomic Bomb Tests, summers 1957-58. Recipient Faculty Research award Mich. Tech. U., 1955. Mem. Am. Inst. Chem. Engrs., Am. Chem. Soc., Am. Nuclear Soc., Instn. Nuclear Engrs. (London), Sigma Xi, Phi Kappa Phi, Alpha Chi Sigma, Phi Lambda Upsilon. Author: Organic Reactor Coolants, 1964; also articles on nuclear fuels processing, chem. analyses and kinetics. Mem. editorial bd. Jour. Engring. Edn. Home: 1595 Curtis Av. Office: Idaho Operations Office, AEC, Idaho Falls, Ida. 83401.*

MAKI, Tetsuo, Japanese surgeon; b. Shichinohemachi, Aomori Perfecture, Japan, Jan. 23, 1908; s. Taizo and Hide (Morita) M.; M.D., Tohoku U., 1933, D.M.Sc., 1939; m. Shizue Kanno, Nov. 1, 1936; children—Hiroko (Mrs. Kahei Rokumoto), Michiko. Faculty, Tohoku U. Sch. Medicine, Sendai, Japan, 1939-45, prof., chmn. dept. surgery, 1961—, dean, 1967—; faculty Akita Prefectural Med. Coll., 1945-49, Hirosaki U. Sch. Medicine, 1949-61. Mem., trustee Japanese Surg. Soc., Japanese Soc. Parasitology, Japanese Gastroent. Assn.; pres. First Japanese Smooth Muscle Research Soc., 12th Japanese Electromyographic Study Soc. Research, publs. on pathogenesis of calcium bilirubinate gallstones and clarified difference of cholelithiasis between Japanese and Western people. Home: 183 Tsuchidoi, Sendai, Japan.*

MAKINO, Hiroyasu David, Japanese physician; b. Tokyo, Japan, Jan. 1, 1926; s. Tetsutaro and Sue (Takakuwa) M.; B.S., Coll. Sci. and Lit., Imperial U. Seoul, Korea; M.D., Chiba U. Sch. Medicine. Asst. prof. Chiba U. Sch. Medicine, U. Kan. Med. Center, 1960. Mem. Japan Neurol. Surgery Soc. (dir.), Japan Surg. Soc., Japan Clin. Surg. Soc., Japan Radiol. Soc., Japan Neurol. Soc., Japan Chest Surgery Soc. Author: (with others) Post Operative Management Following Brain Surgery, 1966; also numerous articles. Patentee various devises for intercranial arterial angiogram, new technique in peripheral nerve anastomosis, treatment of cancer of esophagus with Prof. Nakayama. Home: 22 Araki cho, Shinjuku-ku, Tokyo, Japan. Office: 313 Inohana-cho, Chiba, Japan.*

MAKINO, Sajiro, Japanese zoologist, cytologist; b. Chiba Prefecture, Japan, June 21, 1906; s. Rynzo and Tama (Arai) M.; grad. dept. biology Faculty of Agr., Hokkaido U., Sapporo, Japan, 1930, D.Sc., 1940; m. Sumi Aoyagi, Nov. 17, 1937; children—Yoko (Mrs. H. Y. Nakanishi), Sorako (Mrs. Takeshi Hattori), Shio, Yumiko, Ryu. Faculty, Faculty Sci., Hokkaido U., 1930-—, prof. zoology, 1940-—. Recipient prize Genetics Soc. Japan, 1944, Zool. Soc. Japan, 1953, Japan Acad., 1958, Toyo-Layon Sci. Fund, 1962, Japan Soc. Human Genetics, 1967, Japanese Med. Assn., 1967. Mem. Zool. Soc. Japan (mem. council), Genetic Soc. Japan (council), Japanese Cancer Assn. (council), Chromosome Research Assn. (council), Japan Soc. Human Genetics (council), Japan Soc. Cell Biology (council), Japan Soc. Clin. Cytology (council), Congenital Anomalies Research Assn. Japan (council), Am., Internat. socs. cell biology. Author: An Atlas of the Chromosome Numbers in Animals, 1952; A Review of the Chromosome Numbers in Animals, 1956; also numerous articles. Editor: Cytologia, Kromosomo, Cytogenetics, Caryologia, Internat. Rev. Cytology. Research and publs. on animal cytogenetics and cancer cytology, chromosomes and karyotype analyses; determination of sex in insects, fish, amphibia and mammalia; tumors, congenital disorders, human population. Home: 22 W. 5 S., Sapporo, Hokkaido, Japan.*

MAKINO, Tomitaro, Japanese botanist; b. Kochi Prefecture, 1862; Sc.D., Tokyo U., 1927; instr. Tokyo U., 1913-34; mem. Japan Acad. Author: Illustrations of Japanese Plants. Catalogued more than 1000 new plants, 1500 species. Died 1957.

MAKISHIMA, Shoji, Japanese chemist; b. Tokyo, 1907; grad. Tokyo U., 1930; D.Engring., 1945. Became instr. Tokyo U., 1932, asst. prof., 1938, prof., 1946. Recipient Imperial Acad. prize, 1945. Research and publs. on printing, attraction of light for moths, fish, fluorescent light, phosphorus phenomenon. Author: Light Chemistry.

MAKOVER, Henry Benedict, Am. physician; b. Balt., July 28, 1908; s. Bernard and Rose (Sworzyn) M.; A.B., Johns Hopkins, 1929, M.D., 1933; m. Mildred H. Weinberg, July 6, 1931; children—Richard B., Michael E. Asst. prof. Johns Hopkins Med. Sch., 1936-42; asso. med. dir. Montefiore Hosp., N.Y.C., 1946-48; research dir. study med. facilities Fedn. Jewish Charities, Phila., 1948-49; research dir. spl. Health Ins. Plan Greater N.Y., 1949-50; fellow in psychiatry Bronx Municipal Hosp., 1957-60; med. dir. Central Manhattan Med. Group, 1950-56; professor of psychiatry Albert Einstein College of Medicine, N.Y., 1956-—. Spl. cons. mental health Jud. Conf., State of N.Y., 1964-—. Fellow Am. Pub. Health Assn., N.Y. Acad. Medicine; mem. A.M.A., Am. Psychiat. Assn., N.Y. State, N.Y. County med. socs., Internat. Epidemiological Assn. Author: (with L. Rosenfeld) The Rochester Regional Hospital Council, 1956; also numerous articles. Research in mental health services, epidemiology cancer, quality med. care, group practice, med. edn., epidemiological approaches to distbn., effects, control disease through med. care. Home: 3 Country Rd., Mamaroneck, N.Y. 10543. Office: Albert Einstein Coll. Medicine, N.Y.C. 10461.*

MAKSIMENKOV, Aleksey Nikolaevich, Russian surgeon, anatomist; b. 1906; grad. Leningrad Mil. Med. Acad., 1931; D.Med. Sci., 1938. Asst. chair operative surgery Leningrad Mil. Med. Acad., 1933-36, instr., 1937-40, prof., 1941-—, head chair operative surgery, 1948-—. Recipient Stalin prize class I, 1943, Burdenko prize, 1956. Mem. USSR Acad. Med. Sci. (corr.). Co-author: Atlas of the Peripheral Nervous and Venous Systems, 2d edit., 1943; author: Surgical Anatomy of the Thorax, 1956; The Mechanism of Gunshot Wounds, 1958. Co-editor Surgery sect. Large Med. Ency., 2d edit.; dep. editor Grekov Herald of Surgery; mem. editorial council Archives of Anatomy, Histology and Embryology. Research and publs. on surg. anatomy of wounds of chest and extremities, Russian surg. history, effects of surgery of major arterial vessels and thoracic and abdominal viscera on body's physiol. functions; determined structural pattern of peripheral venous and nervous systems in connection with gen. constrn. of organs and their functional roles. Address: Leningrad Mil. Med. Acad., ulitsa Lebedeva 6, Leningrad, USSR.

MAKSUTOV, Dmitriy Dmitrievich, Russian optician; b. Apr. 23, 1896; grad. Mil. Engring. Sch., 1914, Petrograd U. With Pulkovo Obs. and Inst. Physics, USSR Acad. 'sci., until 1930, asso. Pulkovo Main Astron. Obs., 1952-—; organizer, dir. Astron. Optics Lab., Leningrad State Inst. Optics, 1930-44, prof., 1944-—. Decorated Order of Lenin (2); recipient Stalin prize, 1941, 45. Mem. USSR Acad. Sci. (corr.). Author: Shadow Methods of Investigating Optical Systems, 1934. Astronomical Optics, 1946; Production and Investigation of Astronomical Optics, 1948; Research and publs. on astron. optics, shadow and other optical methods of investigation, theory of aspherical surfaces; invented catadioptric (meniscus systems for optical devices. Address: Pulkovo Main Astron. Obs., USSR Acad. Sci., Leningrad-Pulkovo, USSR.

MALAGUTI, Faustino Jovita Marianus, chemist; b. Bologna, Italy, Feb. 15, 1802; Docteur ès scis.; worked with Guy-Lussac; became French citizen, 1840; named prof. chemistry Rennes (France) Faculty Scis., 1850; decorated officer Legion of Honor; mem. French Acad. Scis., 1855, Acad. Rennes. Author: Lecons de chimie agricole, 1848; Recherches sur l'association de l'argent aux minéraux métalliques. Research on agrl. chemistry; prepared acetamide acid, 1830, also amide from propionic acid. Died Rennes, Apr. 26, 1878.

MALAMANI, Vittorio, Italian physician; b. Poppi, Italy, May 16, 1911; s. Rullio and Teodolinda (dei baroni Valentoni) M.; ed. in medicine; m. Caterina Piras, Aug. 3, 1946; children—Anita, Tullio, Giandomenica. Dir. Semiology Inst., U. Pavia (Italy); dir. cardiology service Ronzoni Inst., Milan, Italy. Recipient Order of Merit of Italian Republic, Cross of Merit. Mem. Deutsche Gesellschaft für Kreislaufförschung, French Soc. Angiology. Research and numerous publs. in angiology, cardiology, internal medicine. Home: via Scopoli 18, Pavia, Italy.

MALAMOS, Basile, Greek physician; b. Athens, July 30, 1909; s. Constantin and Chariclé (Lambros) M.; ed. U. Hamburg (Germany); M.D., agrégé; m. Helly Pesmazoglou, Nov. 9, 1940; children—Irene, Constantin. Faculty, U. Athens, 1938-—, prof., dir. therapeutic clinic, 1953-—; staff Alexandra Hosp., Athens. Research and publs. in tropical medicine, radioisotopes, hematology, thyroid diseases. Home: Odos Strofyliou 19, Kiffisia, Greece. Office: Hopital Alexandra, Odos Vassilissis, Sophias, Athens, Greece.

MALAPRADE, Léon, French chemist; b. Dieppe, France, Aug. 27, 1903; s. Léon and Alice (Morel) M.; student Chem. Inst., Nancy, also Faculty Scis., Nancy, 1920-28; D.Sc.; m. Simone Milleur, July 19, 1932; 3 daus., 5 sons. With Faculté des Sciences, Nancy, 1923-—, prof., 1957-—. Mem. Chem. Soc. France. Author: (with P. Pascal) Traité de Chimie Minérale, also articles. Research on reaction of periodic acid on organic compounds and complex metal ions, especially silver and copper; analytical chemistry. Home: 149 rue de Mon Désert, Nancy (54), France.*

MALASSEZ, Louis Charles, French physiologist; b. Nevers, France, 1842; studied under Claude Bernard and Ranvier, also at Coll. de France; research on origin, histologic and pathogenic nature of tumors, also on syphilis, Tb; credited with making 1st hemocytometer, pub. 1874. Died 1909.

MALAURIE, Jean-Noël, ethnologist; b. Mainz, Germany, Dec. 23, 1922; s. Albert and Isabelle Regnault (Carmichaël de Baiglie) M.; Ph.D., U. Paris; m. Monique Laporte, Dec. 27, 1952; 1 son, Guillaume. Mem. Paul-Emile-Victor expdn., Greenland; research asso. Paris Nat. Research Center; prof. Sorbonne; dir. Center for Arctic Studies, Sorbonne. Pres., Foundation d'etudes nordiques, Rouen, France; Terre humaine explorations, central Sahara, 1948-49, N. Greenland, 1950-51, Central Canadian Arctic, 1961-64. Recipient gold medal Geography Soc. Paris for polar exploration. Laureate French Acad. Scis; hon. mem. Anvers Royal Geog. Soc. Research in social ethnography of boreal populations, geo-economy in N. countries. Home: 1, rue du Mail, Paris 2, France.

MALEBRANCHE, Nicolas de, French physicist, philosopher; b. Paris, Aug. 6, 1638; s. Nicolas de and Catherine (de Lauzon) M.; ed. Coll. de la Marche, Sorbonne, Paris. Joined Congregation of the Oratory, 1660; mem. French Acad. Scis. Author: Des loix de la communication des mouvements, 1692; Réflexions sur la lumière et les couleurs; Entretiens sur la metaphysique et sur la religion, 1688; Traité de la nature et de la grace, 1680; Traité de l'amour de Dieu, 1697; De la recherche de la vérité, 1674. Studied nature of light and color, psychol. conditions of vision, founds. of infinitesimal calculus; his philosophy led to doctrine of occasionalism or interference (objective thing and subjective impression are made to coincide by God's interposition); connecting link between Descartes and Spinoza; believed nothing could be demonstrated but math. and physics. Died Paris, Oct. 13, 1715.

MALEK, Zdenek, Czech. physicist; b. Pilsen, Czechoslovakia, Mar. 11, 1930; s. Frantisek and Anezka (Solová) M.; M.Sc., Charles U., Prague, 1953; Ph.D., Czechoslovak Acad. Sci., 1957; m. Zdenz Halíková, Sept. 21, 1956; 1 dau., Ivanka. Research fellow Inst. Magnetic Materials, Jena, Germany, 1958-59; asso. prof. U. Baghdad (Iraq), 1961-62; research fellow Inst. Physics, Czechoslovak Acad. Scis., Prague, 1956-58, head dept. magnetism, 1960-61, head 2d dept. dielectrics, 1963-65, head dept. nonlinear dielectrics Inst. Radio Engring. and Electronics, 1965-—; tchr. Charles U., Prague, 1959-—, Prague Tech. U., Safarík U., Kosice, 1964-—. Author: (with others, in Russian) Ferromagnetic Thin Films, 1964; also numerous papers. Research on ferromagnetics: influence of lattice defects on magnetic properties, domain structures and magnetic anisotrophy in thin ferromagnetic films; ferroelectrics: devel. temperature autostabilizing nonlinear dielectric element (tandel). Home: 11 Ul. Fr. Kadlece, Prague 8. Office: Inst. Radio Engring. and Electronics, Kobylisy, Prague 8, Czechoslovakia.*

MALENKA, Bertram Julian, Am. physicist; b. N.Y.C., June 8, 1923; s. Morris and Mollie (Wichtel) M.; A.B., Columbia U., 1947; M.A., Harvard, 1929; Ph.D., 1951; m. Ruth D. Stolper, Mar. 28, 1948; children—David Jonathan, Robert Charles. Research fellow Harvard, 1951-54; asst. prof. physics Washington U., St. Louis, 1954-56; asso. prof. Tufts U., Medford, Mass., 1956-60; faculty Northeastern U., Boston, 1960-—, prof. physics, 1962-—. Mem. sci. adv. group Harvard-Mass. Inst. Tech. Cambridge Electron Accelerator, 1956-—. Mem. Am., Italian phys. socs., N.Y. Acad. Scis., Phi Beta Kappa, Sigma Xi. Research and publs. on theory of nuclear forces and structure of nucleus, explanation polarization phenomena in high-energy scattering, gamma radiation, electric polarization deuteron, accelerator design. Home: 16 Rutledge Rd., Belmont, Mass. 02178. Office: Dept. Physics, Northeastern U., Boston 02115.*

MALFATTI, Giovanni Francesco Giuseppe, Italian mathematician; b. Ala di Trento, Tyrol, 1731; student Jesuit Coll., also Bologna, Italy (under Riccoti), 1748. Prof. math., Ferrara, Italy, 1771-1807. Proposed and solved analytically Malfatti's problem (inscribe in given triangle 3 circles, each tangent to the other 2 and to 2 sides of triangle). Died Ferrara, Oct. 9, 1807.

MALGAIGNE, Joseph-François, French surgeon; b. Charmes-sur-Moselle, France, Feb. 14, 1806; student, Nancy, France, also École Practique, Paris, 1826; m. Aglaé-Francoise Pommier, May 24, 1843. Officer of health, Nancy; became extern in hosps., 1827; apptd. surgeon St. Louis Hosp., 1845; named prof. operative medicine, 1850. Recipient 1st prize in surgery, 1829. Mem. Acad. Medicine (pres. 1865), Soc. Surgery (a founder 1843). Author: Une nouvelle théorie de la voix humaine, 1827; Une nouvelle théorie de la vision, 1830; Manuel de médecine opératoire fondée sur l'anatomie et l'anatomie pathologique, 1834; Traité d'anatomie chirurgicale et de chirurgie expérimentale, 2 vols., 1838; Oeuvres d'Ambroise Paré, 1840 (ed.); Traité des fractures et des luxations, 2 vols., 1847; Lecons sur l'orthopédie, 1862. Chief editor Med. Gazette Paris. Developed Malgaigne operation for amputating at ankle; discovered triangular depression of prin. artery of neck (Malgaigne's fossa), also fatty pads in knee joint (Malgaigne's pads). Died Oct. 17, 1865.

MALI, J. W. H., Dutch dermatologist; b. Groningen, Netherlands, June 18, 1918; s. W. P. T. and P. (Kuipers) M.; M.D.; m. Orie Mali, June 8, 1946; 5 children. Conservator dermatology Utrecht (Netherlands) State U., until 1951; conservator physiology U. Groningen, 1951-56; prof. dermatology Nijmegen (Netherlands) U., 1956-—; head dept. dermatology Cath. Mem. Concilium Dermatologicum (pres.), Netherlands Assn. Dermatology (v.p.). Research, publs. on pathogenesis of chromium and nickel dermatitis, venous stasis syndrome and atopic dermatitis. Home: Pompweg 7, Ubbergen, Netherlands. Office: Javastraat 104, Nijmegen, Netherlands.*

MALIK, S. R. K., physician; b. Rawalpindi, Pakistan, Mar. 26, 1931; s. Mali Ram. and Sarup (Kaur) M.; M.B.B.S., U. Agra (India), 1955, M.S., 1957; m. R. Malik, Dec. 12, 1957; children—Nina, Sanjiv, Anjili. Demonstrator ophthalmology Med. Coll., Agra, 1955-57; asst. prof. All India Inst. Med. Scis., New Delhi, 1959-62; asso. prof. Maulana Azad Med. Coll., New Delhi, 1952-64; prof. ophthalmology Maulana Azad Med. Coll., 1964-—; officer in charge Eye Bank, Irwin Hosp., Delhi, 1963-—. Fellow Royal Coll. Surgeons Ophthalmology, Edinburgh; mem. Delhi Ophthalmic Soc. (past sec.), Nat. Soc. for Prevention of Blindness (joint sec. 1965-—), Ophthalmic Soc. U.K. Asso. editor Oriental Archives of Ophthalmology, 1962-—; editorial bd. Indian Jour. Orthoptics, 1965-—. Research and numerous publs. on evaluation of drugs which lower intraocular pressure, evolution of clin. picture of trachoma and its therapy, keratoprosthesis and corneal stromectomies, role of cryocautery in ophthalmology, pleoptics, toxoplasmosis, electroretinography. Home: 44 Kotla Rd., New Delhi, India.*

MALING, Harriet Florence Mylander (Mrs. Henry Forbes Maling, Jr.), Am. pharmacologist; b. Balt., Oct. 2, 1919; d. Walter Conrad and Matilda (Hopf) Mylander; A.B., Goucher Coll., 1940; A.M., Radcliffe Coll., 1941, Ph.D., 1944; m. Henry Forbes Maling, Jr., Sept. 1, 1943; children—Joan, Walter, Anne, Charles. Faculty, Harvard Med. Sch., 1944-46; faculty George Washington U. Med. Sch., Washington, 1951-54; asst. research prof. in pharmacology, 1952-54; pharmacologist Lab. Chem. Pharmacology, Nat. Heart Inst, Bethesda, Md., 1954-—, head sect. on physiology, 1962-—. Mem. Am. Soc. Pharmacology and Exptl. Therapeutics, Soc. for Biology and Exptl. Medicine, N.Y. Acad. Scis., A.A.A.S., Phi Beta Kappa, Sigma Xi, Sigma Delta Epsilon, Editorial bd. Jour. Pharmacology and Exptl. Therapeutics, 1962-65. Contbr. articles to tech. jours. Research on pharmacology, cardiovascular and autonomic drugs, cardiac arrhythmias, physiology of dogs with exptl. myocardial infarcts produced by coronary artery ligation, fat moblzn., role sympathetic nervous system in fat and carbohydrate moblzn., effects drugs upon these processes. Home: 406 Taylor Av., Annapolis, Md. 21401. Office: 9000 Wisconsin Av., Bethesda, Md. 20014.*

MALINOVSKY, Mikhail Sergeevich, Russian obstetrician, gynecologist; b. Neklyudovo (now Ulyanovsk Oblast), 1880; grad. Med. Faculty, Kazan U., 1907; D.Med. Sci., 1913. Intern, asst. Prof. V. S. Gruzdev's Obstetrics and Gynecology Clinic, Kazan, 1907-19; a founder, prof. chair obstetrics and gynecology Irkutsk U., 1919-23, also a founder Med. Faculty; head chair obstetrics and gynecology 1st Moscow U. (now 1st Moscow Med. Inst.), 1923-48, Central Postgrad. Med. Inst., Moscow, 1930-34, 48-—; dir. Research Inst. Obstetrics and Gynecology, USSR Ministry Health, 1936-44, now dir. bd. obstet. services; dir. Inst. Obstetrics and Gynecology, USSR Acad. Med. Sci., 1944-48; sci. dir. Inst. Obstetrics and Gynecology, RSFSR Ministry Health, 1959-—. Decorated Order of Lenin. Mem. USSR Acad. Med. Sci. (a founder 1944, v.p. 1944-47), All-Union Soc. Obstetricians and Gynecologists (hon., chmn.). Co-author: Postnatal Infection, Its Diagnosis and Modern Treatment; Gynecology, 1957; author: Operative Obstetrics, 1955; Nonoperative Gynecology, 1955. Editor Obstetrics and Gynecology sects. Large Med. Ency., 2d edit.; mem. editorial council, past editor Obstetrics and Gynecology. Home: B. Gruzinskaya 36. Office: Inst. Obstetrics and Gynecology, Lepekhinsky tup 3, Moscow, USSR.

MALINOWSKI, Bronislaw Kasper, anthropologist; b. Cracow, Poland, Apr. 7, 1884; s. Lucyan and Józefa (Lacka) M.; Ph.D., Polish U., Cracow, 1908; student U. of Leipzig, Germany, 1908-10; D.Sc., U. of London, 1916; D.Sc. (hon.), Harvard (Tercentenary), 1936; came to U. S., 1938; m. Elsie Rosaline Masson, Mar. 6, 1919 (dec. 1935); children—Józefa Marya, Wanda, Helena; m. 2d, Valetta Hayman-Joyce, June 6, 1940. Lecturer London Sch. of Econs., 1912-13; on anthropol. expdn. to New Guinea, 1914-20; reader in social anthropology, U. of London, 1924-27, prof. anthropology since 1927, on leave since 1939; visiting prof. anthropology and fellow of Timothy Dwight Coll., Yale, since 1939. Hon. fellow Royal Soc. of New Zealand; mem. Royal Acad. of Science of Netherlands, Polish Acad. of Sciences and Arts, Phi Beta Kappa, Sigma Xi. Author: The Family Among the Australian Aborigines, 1913; Primitive Religion and Social Differentiation (in Polish), 1915; Argonauts of the Western Pacific, 1922; Crime and Custom in Savage Society, 1926; Myth in Primitive Psychology, 1926; Sex and Repression in Savage Society, 1926; The Sexual Life of Savages in N.W. Melanesia, 1929; Coral Gardens and Their Magic (2 vols.), 1935; The Foundations of Faith and Morals, 1936; posthumous publications: The Dynamics of Culture Change, 1945; Magic, Science, and Religion; also author articles in Nature, Jour. of the Royal Anthropol. Inst., Man, etc. Noted for researches among Trobriand Islanders and S. Pacific; founder of functionalism in anthropology; studied primitive socs.'s handling of problems of sex and fear of death. Died New Haven, May 16, 1942.

MALINOWSKI, Jordan Petrov, Bulgarian chemist; b. Sliven, Bulgaria, June 3, 1923; s. Peter and Ivanka (Jeliazkowa) M.; grad. U. Sofia, 1948, D.Phys. Chemistry, 1957; m. Dimka Iontcheva, June 16, 1950; 1 son, Peter. Faculty U. Sofia, 1948, 58-—, prof. phys. chemistry, 1964-—; research worker Phys. Inst. Bulgarian Acad. Scis., 1949-58; faculty Inst. Phys. Chemistry, Bulgarian Acad. Scis., 1958-—, prof. phys. chemistry, 1964-—, vice dir., sci. sec., 1962-—, head Lab. for Phys. Chemistry of Solids, 1960-—, mem. council for coordination of research br. phys. chemistry Bulgarian Acad. Scis., 1963-—. Mem. Assn. Sci. Workers in Bulgaria. Research and publs. on polarization phenomena by electrodeposition of metals, theory of crystal growth and its application to electrocrystallization theory of photog. process, electronic and ionic processes in silver bromide crystals, properties of photoexcited charge carriers, their trapping and neutralization, photochem. reactions in solids. Home: 50 Parchevich St., Sofia C. Office: Inst. Phys. Chemistry, Bulgarian Acad. Scis., Sofia 13, Bulgaria.*

MALKIEL, Saul, Am. physician; b. Boston, Dec. 28, 1912; s. Harry and Fanny (Sugarman) M.; A.B., Clark U., 1934; M.A., Boston U., 1936, Ph.D., 1942, M.D., 1944; m. Evelyn Ruth Holzman, Feb. 4, 1945; children—Alan Stephan, Charles Michael, Andrew Jay. Asst. pathology Yale, 1944-45; vis. investigator Rockefeller Inst., 1945-48; lectr., U. Pa., 1948; asst. prof. medicine Northwestern U. Med. Sch., 1948-54; research asso. Harvard Med. Sch., 1954-—; asso. in medicine Peter Bent Brigham Hosp., Boston, 1954-63, Children's Med. Center, Boston, 1963-—, Children's Cancer Research Found., Inc., Boston, 1963-—. NRC fellow in med. scis., 1945-46; Am. Cancer Soc. Sr. fellow, 1946-48. Mem. A.A.A.S., Am. Assn. Immunology, Am. Soc. for Microbiology, Am. Acad. Allergy, Soc. Exptl. Biology and Medicine, Mass. Med. Soc. Author: (with S. M. Feinberg, A. R. Feinberg) The Antihistamines, 1950; also numerous articles. Research on immunochemistry hemocyanins, hemoglobin, viruses, immunochemistry allergic reactions, effects stress on exptl. asthma, clin. and exptl. studies with antihistamines, effect cortisone on antibody prodn., quantitative precipitin reactions, hypersensitivity in white mouse, fractionation studies on Bordetella pertussis, allergens from ragweed pollen. Home: 28 Beeching St., Worcester, Mass. 01602.*

MALKUS, Willem V(an) R(ensselaer), Am. physicist; b. Bklyn., Nov. 19, 1923; s. Hubert Paul and Alida Fitzhugh (Wright) M.; student U. Mich., 1940-42, Cornell U., 1942-43; Ph.D. in Physics, U. Chgo., 1950; m. Joanne Gerould, June 13, 1948; children—David Starr, Steven Willem, Karen Elisabeth; m. 2d, Ulla Aronsson, Dec. 28, 1964. Asst. prof. natural sci. U. Chgo., 1950-51; phys. oceanographer Oceanographic Inst., Woods Hole, Mass., 1951-60; prof. oceanography Mass. Inst. Tech., 1959-60; prof. geophysics U. Cal. at Los Angeles, 1960-65, prof. planetary physics, 1965-—. Mem. Am. Acad. Arts and Scis., Am. Phys. Soc., Sigma Xi. Contbr. papers on modern particle physics, statis. physics, geophys. fluid dynamics, nonlinear stability theory, electro and magneto hydrodynamics, theory turbulence, origin of geomagnetism. Home: 24 Salt Pond Rd., Falmouth, Mass. Office: Inst. of Geophysics, Univ. of California, Los Angeles 90024.*

MALL, Franklin Paine, Am. anatomist; b. Belle Plaine, Ia., Sept. 28, 1862; s. Francis and Louise (Miller) M.; M.D., U. Mich., 1883, A.M. (hon.), 1900, Sc.D., 1908; student Heidelberg, Leipzig and Johns Hopkins; LL.D., U. Wis., 1904; LL.D., Washington U., St. Louis, 1915; m. Mabel Stanley Glover, Mar. 28, 1895. Fellow, Johns Hopkins, 1886-88, instr. pathology, 1888-89, prof. anatomy, from 1893; adj. prof. vertebrate anatomy Clark U., 1889-92; prof. anatomy U. Chgo., 1892-93. Dir. dept. embryology, founder lab. embyrology Carnegie Instn. of Washington, 1915. Mem. Commn. for Neurol. Research, Internat. Assn. Acads., 1903-17. Mem. Soc. Am. Naturalists (v.p. 1900, chmn. 1904), Assn. Am. Anatomists (pres. 1905-07), Institute International d'Embryologie. Author: Causes Underlying the Origin of Human Monsters; On The Fate of the Human Ovum in Tubal Pregnancy. Joint editor of Handbuch der Entwicklungsgeschichte des Menschen; co-editor, a founder Am. Jour. Anatomy, Anat. Record; asso. editor Jour. Morphology; editor Studies from Anat. Lab. of Johns Hopkins. Developed formula for estimating age of human embryo (known as Mall's formula), 1893. Died Nov. 17, 1917.

MALLAMS, John Thomas, Am. physician; b. Ashland, Pa., Aug. 29, 1923; s. Raymond E. and R. Elizabeth (Bevan) M.; student Lafayette Coll., 1941-43; M.D., Temple U., 1946; m. Ruth Smith, June 21, 1945; children—David, John Faith. Radiologist U. S. Naval Hosp., 1948-49, chief radiology, 1949-50; cons. radiologist Beaufort (S.C.) County Hosp., 1949-50; fellow radiology Robert Pucker Hosp. and Guthrie Clinic, Sayre, Pa., 1950-51; ACS fellow clin. radiology therapy Frances Delafield Hosp., N.Y.C., 1951-52; with Baylor U., 1952-—, dir. irradiation therapy, dir. tumor clinics, cancer registry, 1954-—, attending radiologist, 1954-—, dir. cancer research dept., 1962-—, clin. prof. radiotherapy Coll. Dentistry, 1961-—, clin. prof. radiology U. Tex. S.W. Med. Sch., prof. radiology exptl. radiotherapy, 1966-—; cons. U. S. Air Force Lackland AFB, Tex., 1964-—; cons. radiology VA Hosps., McKinney and Dallas, 1954-—. Diplomate Am. Bd. Radiology; fellow Am. Coll. Radiology; mem. Dallas—Ft. Worth Radiol. Soc. (pres. 1960), Am. Cancer Soc., Am. Radium Soc., Am. Assn. Cancer Research, Radiation Research Soc., Am. Club Therapeutic Radiologists, A.M.A. Contbr. numerous articles in field to sci. jours. Development new presurgical adjuvant techniques of irradiation in treatment of cancer. Home: 11507 Royalshire Blvd., Dallas 75230.*

MALLARD, (François) Ernest, French mineralogist; b. Chateauneuf-sur-Cher, France, Feb. 4, 1833; prof. mineralogy École des Mines, 1872-94; mem. French Acad. Scis., 1890. Author: les Groupements cristallins, 1878; Traité de cristallographie géométrique et physique, 1879-84; Examen de diverses substances cristalisées, 1887. Research in crystallography and isomorphism; his investigations in gas and mine explosions helped provide safer working conditions for miners; participated in geol. mapping of France. Died Paris, July 6, 1894.

MALLER, Ramappa Krishna, Indian chemist; b. Tellicherry, India, Jan. 31, 1918; s. Babu Ramappa and Kamala (Rao) M.; B.Sc., Presidency Coll., U. Madras (India), 1939, B.Sc. with honors, 1941, M.A., 1946, Ph.D., 1951; A.I.I.Sc., Indian Inst. Sci., 1946; m. Premala Pai, Nov. 9, 1949; Research asst. Council Sci. and Indsl. Research, India, Presidency Coll., 1946-53; postdoctoral fellow McArdle Cancer Research Lab., Madison, Wis., 1953-55, project asso., 1955-56; lectr. biochemistry Indian Inst. Sci., Bangalore, 1957-63; head radiotracer lab. CIBA Research Centre, Bombay, India, 1963-—. Recipient Wilson Gold medal Presidency Coll., Madras, 1939. Mem. Royal Inst. Chemistry London (asso.), Soc. Biol. Chemists. Research and publs. on electro-organic chemistry and drug metabolic studies using radioactive compounds, use of electric current as reagent, intermediates for dye-stuff and photo. industry, radioactive analogue of anti tumor agt. in animals. Home: D-304 Housing Colongy, CIBA Research Centre, Goregaon, Bombay-63, Maharashtra, India.*

MALLET, Robert, Brit. seismologist, engr.; b. Dublin, Ireland, June 3, 1810; s. John Mallet; B.A., Trinity Coll., 1830; M.A., Master in Engring., 1862; m. Cordelia Watson; 1 son, John William. Became partner in father's iron works, 1831; in bridge bldg., 1836; built ry. stas., 1845-48, Fastnet Rock lighthouse, 1848-49. Fellow Royal Soc., 1854; mem. Instn. Civil Engrs. (Walker premium 1841). Author: On the Physical Conditions involved in the Construction of Artillery . . . , 1856; (with son) The Earthquake Catalogue of the British Association, 1858; Greal Neapolitan Earthquake of 1857, 2 vols., 1862; The First Principles of Observational Seismology, 1862; also articles. Inventor buckled plate method of flooring, 1852; studied water action on iron and alloys of copper; catalogued over 6800 earthquakes; tried to explain earthquakes and volcano eruptions; introduced terms seismology, seismic focus, isoseismal line, meizoseismal line, angle of emergence. Died London, Nov. 5, 1881.

MALLET-GUY, Pierre Albert, French surgeon; b. Beaune, France, May 19, 1897; s. Pierre Auguste and Marguerite (Petitjean) M.-G.; Certificat, Faculté des Sciences, U. Lyon, France, 1915; Docteur en Medecine, Faculté de Médecine, 1925; agrégé U. Lyon, 1939; Doktor honoris causa, U. Giessen (Germany), 1957; m. Yvonne Depaoli, Dec. 10, 1945; children—Marguerite, Michel, Simone, Pierrette, Claude. Became head surgery clinic Faculté de Médecine, Lyon, 1927; became prof. surg. pathology, 1946, prof. surgery clinic, 1958; became dir. union surg. research Institut National de la Santé et de la Recherche Médicale, 1958, pres. commn. exptl. surgery, 1964. Laureate Institut de France. Mem. Paris Acad. Medicine, Paris Acad. Surgery, many foreign surgical societies. Author: Chirurgie biliaire sous contrôle radiographique et manométrique per-opératoire, 1947; Pancréatite chronique et récidivante, 1962. Became editor Lyon Chirurgical, 1943; cons. editor Excerpt-Medica, 1947, Revue Internationale d'Hepatologie, Surgery Gynecology and Obstetrics, 1950, Chirurgische Praxis, 1957, Voice of Medicine, 1960, Surgery, 1966. Surg. research on chronic and recurring pancreatitis, biliary surgery under manometric and radiographic controls, surg. treatment of chronic hepatitis with jaundice; studies on hepatic circulation and lymphatic circulation of liver, innervation of liver and bile ducts, pancreas, hepatic regeneration after hepatectomy, hepatic transplants; performed 1st pancreatectomy, 1933. Home: 2 Duquesne, Lyon, Rhône, 69, France, Office: Faculté de Médecine, Clinique Chirurgicale A, Lyon, France.

MALLETTE, John Michael, Am. biologist; b. Houston, Aug. 6, 1932; s. Jules L. and Lydia (Myers) M.; B.S., Xavier U., 1954; M.S., Tex. So. U., 1958; Ph.D., (Nat. Found. fellow), Pa. State U., 1962; m. Pazetta Berryman, Aug. 19, 1959; children—John Michael, Adelaide Veronica. Research technician dental br. U. Tex., Houston, 1957-58; instr. Tex. So. U., Houston, 1958-59; prof. biology Tenn. Agr. and Indsl. State U., Nashville, 1962-—. Dir. undergrad. research participation program NSF, 1964-—. Mem. A.A.A.S., Am. Zool. Soc., Am. Assn. U. Profs., Tenn. Acad. Sci. (chmn. zoology sect. 1964-—). Research, publs. on retention cholesterol in animals, growth in culture of Trypsin dissociated thyroid cells from control and altitude acclimated rats, teratogenic effects of various drugs. Home: 4011 W. Hamilton Rd., Nashville 37218.*

MALLETTE, M. Frank, biochemist; b. Leon, Ia., May 28, 1917; s. Frank G. and Isabel (Freestone) M.; B.S., Ia. State Coll., 1940; M.S. (Lydia Roberts fellow), Columbia, 1943, Ph.D., 1945; m. Ruth Elinor Heinl, Dec. 3, 1942; children—Ruth Elinor, Beth Anna. Post-doctorate fellow chemistry Cornell U., 1945-47; instr. chemistry U. Wyo., 1947-48; asst. prof. biochemistry Johns Hopkins, 1948-51; faculty Pa. State U., University Park, 1955-—, prof., 1959-—; vis. prof. U. B.C. (Can.), 1963. Fellow A.A.A.S.; mem. Am. Soc. Biol. Chemists, Am. Soc. Microbiologists, Am. Chem. Soc., Am. Assn. U. Profs., Sigma Xi. Author: (with C. Lamanna) Basic Bacteriology, 1953; (with P. M. Althouse, C. O. Clagett) Biochemistry of Plants and Animals; also numerous articles. Research multiple function enzymes, function ribonucleic acids, control enzyme formation in cells, nutritional interactions in microorganisms, occurrence and metabolism sulfur-containing compounds and metabolic inhibitors. Home: 277 E. McCormick Av., State College, Pa. 16801. Office: 203 Frear Lab., University Park, Pa. 16802.*

MALLMANN, Walter Leroy, Am. microbiologist; b. Chgo., Oct. 29, 1895; s. Joseph John and Katherine (Willson) M.; B.S., Mich. Agrl. Coll., 1917, M.S., 1924; Ph.D., U. Chgo., 1931; m. Margaret Skoll, Dec. 21, 1917; children—Roy Stoll, Mrs. John F. Bozman, Mrs. Howard Hollenbeck. Faculty, Mich. State U., East Lansing, 1918-—, prof. dept. microbiology, pub. health, 1940-66, prof. emeritus, 1966-—. Cons. U. S.

Dept. Agr., USPHS, WHO. Recipient numerous sci. awards. Fellow Am. Pub. Health Assn., A.A.A.S.; mem. Am. Acad. Microbiology, Am. Soc. Microbiology, Soc. Exptl. Biology and Medicine, Inst. Food Technologists, Water Pollution Control Fedn., Am. Water Works Assn., Internat. Assn. Milk and Food Technologists, Nat. Tb and Respiratory Disease Assn., Sigma Xi, Phi Kappa Phi, Phi Sigma, Alpha Zeta, Phi Zeta. Author: (with MacMann, Eldredge) Analysis of Water and Sewage; Research, numerous publs. on pathogenesis of typical mycobacteria for cattle and swine, devel. methods for isolation of coliforms from water, isolation of streptococci from water, pub. health standard for swimming pools. Home: 1706 Germany Rd., Williamston, Mich. 48895. Office: Dept. Microbiology, Mich. State U., East Lansing, Mich. 48823.*

MALLORY, Edith Brandt, Am. psychologist; b. Phila., 1901; d. Frank Burke and Lida (Roberts) Brandt; A.B., Wellesley Coll., 1923; M.A., Columbia, 1924, Ph.D., 1925; m. Tracy Burr Mallory, June 6, 1925; children—Kenneth Brandt, Jean Roberts (Mrs. William Jeffries Childs). Lectr. psychology Brown U., 1943-45; asst. dept. psychology Wellesley Coll., 1927-28, instr., 1928-33, asst. prof., 1933-41, asso. prof., 1941-56, prof., 1958-64, chmn. dept., 1954-64, prof. emeritus, 1964—. Past Bd. dirs., v.p. New Eng. Home for Little Wanderers; past bd. mgrs. Orchard Home. Mem. Am. Psychol. Assn., A.A.A.S., Internat. Council Psychologists, Eastern, Mass. psychol. assns., Phi Beta Kappa, Sigma Xi. Author: Wellesley Spelling Test, also jour. articles. Studies on voice and personality, test construction. Home: Plymouth Harbor, John Ringling Blvd., Sarasota, Fla. 33577.*

MALLORY, Frank Burr, Am. pathologist; b. Cleve., Nov. 12, 1862; s. George Burr and Anna (Faragher) M.; A.B., Harvard, 1886, A.M., M.D., 1890; Sc.D., Tufts, 1928, Boston U., 1932; m. Persis McClain Tracy, Aug. 31, 1893; children—Tracy B(urr), G(eorge) Kenneth. Asst. in histology Harvard, 1890-91, pathol. anatomy, 1891-92, instr. pathology, 1894-96, asst. prof., 1896-1901, asso. prof., 1901-19, prof. pathology Med. Sch., 1928-32; pathologist Boston City Hosp., 1897-1932; cons. pathologist since 1932. Fellow Am. Acad. Arts and Scis.; mem. Assn. Am. Physicians, Am. Assn. Pathologists and Bacteriologists, Am. Assn. for Cancer Research, A.M.A., Mass. Med. Soc. Am. Social Science Assn., Internat. Assn. Med. Museums. Author: The Principles of Pathologic Histology, 1914; (with James H. Wright) Pathological Technique, 1897, 8th edit., 1923. Contbr. to med. jours. Introduced term endothelian leukocytes for large wandering cells of tissues and circulating blood with marked phagocytic power, 1898; demonstrated (with A. A. Homer and F. F. Henderson) causal relation of Bordet-Gengou bacillus (Hemophilus pertussis) to whooping cough, 1913. Died Sept. 27, 1941.

MALLOWAN, Max Edgar Lucien, English archeologist; b. London, England, May 6, 1904; s. Frederick and Marguerite (Duvivier) M.; student Lancing Coll., 1917; M.A., New Coll. Oxford U., 1925; D.Lit., London U., 1950; m. Agatha Mary Clarissa Miller, Sept. 11, 1930. Prof., Western Asiatic archaeology U. London, Eng., 1947-62; fellow All Souls Coll., Oxford, Eng., 1962—, emeritus prof., 1962—; dir. Brit. Sch. of Archaeology in Iraq, 1947-61, chmn., 1966—. Pres., Brit. Inst. Persian Studies, 1961—. Decorated Commander Brit. Empire; recipient Lucy Wharton Drexel Gold medal, 1957. Mem. German Archaeol. Inst., Académie des Inscriptions et Belles-Lettres. Author: Early Mesopotamia and Iran, 1965; Nimrud and Its Remains, 2 vols., 1966; also numerous articles. Editor: Nr. Eastern, Western Asiatic series Penquin Books, 1950-63. Archaeol. excavations principally in Iraq and Syria on sites ranging from pre-historic periods down to 4th century B.C.; discoveries in ruins of an Assyrian city at site of Nimrud where excavations revealed largest ancient mil. establishment yet discovered in Western Asia, namely Fort Shalmaneser, containing many thousands of fragments of beautifully carved ancient ivories manufactured principally between 840 and 700 B.C. Home: Winterbrook House, Wallingford, Berkshire, Eng. Office: All Souls Coll., Oxford, England.*

MALM, Carl Pehr Hakan, Finnish surgeon; b. Helsinki, Finland, Sept. 21, 1906; s. Carl Wilhelm and Gundel (Wasenius) M.; M.D., U. Helsinki; m. Gretel Fabritius, June 6, 1936; children—Tua, Gundel, Carl. Attache, U. Hosp. Helsinki, 1937-38; staff Maria Hosp. Helsinki, 1939-45, Orthopedic Hosp. Found. for Invalides Helsinki; practice medicine, specializing in orthopedic surgery; dir., dr. in chief Eira Hosp., Helsinki. Mem. Finnish Surgery Soc., Scandinavian Soc. Orthopedic Surgery, Scandinavian Soc. Surgery, Swedish Soc. Surgery, Soc. Dentists. Author: The Danger of Malignancy in Goitre, 1953; The Role of Surgery in Modern Treatment of Thyroid Diseases, 1958; Surgical Diseases of Salivary Gland, 1960. Research and publs. in treatment of thyrohypophysical syndrome. Home: Skepparegatan 43. Office: Eira Hosp., Helsinki, Finland.

MALM, James Royal, Am. surgeon; b. Cleve., Sept. 7, 1925; s. Royal D. and Theodora (Dumont) M.; A.B., Princeton, 1949; M.D., Columbia, 1949; m. Constance Brooks, July 8, 1950; children—Martha, Melissa, Karen, Sarah. Resident Presbyn. Hosp., N.Y.C., 1953-59, dir. open heart surg. program, 1960—; prof. clin. surgery Columbia, 1967—. Mem. Soc. U. Surgeons, A.C.S., Am. Assn. Thoracic Surgery, Am. Heart Assn. Research and publs. on advances in cardiovascular surgery; devel. of the aortic homograft; improved surgery of cyanotic congenital heart disease; mgmt. of and base balance during cardiac surgery. Home: Hudson Rd. E., Irvington, N.Y. 10533. Office: 180 Ft. Washington Av., N.Y.C. 10032.*

MALMESTEN, Pehr Henrik, physician; b. Sweden, 1811; discovered Balantidium coli (1st parasitic protozoan identified as such), 1857. Died 1883.

MALMGREN, Richard Axel, Am. physician; b. St. Paul, Dec. 31, 1921; s. Axel Arthur and Marjorie (Root) M.; student Cornell U., 1939-40, M.D., 1945; B.S., Wagner Coll., 1942; m. Elizabeth Olivia Hanson, Aug. 17, 1946; children—Elizabeth Oliva, Richard Axel. Commd. sr. asst. surgeon USPHS, 1948, advanced through grades to med. dir.; instr. pathology U. Tenn., 1952-56; head cytodiagnosis service Nat. Cancer Inst., Bethesda, Md., 1957—. Mem. Am. Assn. Cancer Research, Am. Soc. Exptl. Pathology, Am. Soc. Cytology. Fellow Am. Coll. Pathology. Contbr. numerous articles to tech. jours. Research on effect carcinogens and cancer chemotherapeutic agts. on immunity, intracellular localization tumor producing virus, exfoliate cytology studies cervical cancer and cancer cells in blood. Home: 4622 Edgefield Rd. Office: Nat. Cancer Inst., Bethesda, Md. 20014.*

MALMSTADT, Howard Vincent, Am. chemist; b. Marinette, Wis., Feb. 17, 1922; s. Guy A. and Nellie (Rusch) M.; B.S., U. Wis., 1943, M.S., 1948, Ph.D., 1950; m. Carolyn Gay Hart, Aug. 3, 1947; children—Cynthia Sue, Alice Ann, Jonathan Howard. Faculty, U. Wis., 1947-51; faculty U. Ill., Urbana, 1951—, prof., 1961—; mem. analytical adv. bd. Oak Ridge Nat. Labs., 1963—. Dir. NSF Confs., 1963—. Mem. Am. Chem. Soc. (award in chem. instrumentation 1963, div. chmn.), Soc. Applied Spectroscopy, Sigma Xi, Phi Lambda Upsilon, Alpha Chi Sigma. Author: (with C. G. Enke) Electronics for Scientists, 1962. Research, numerous publs. on emission and absorption spectroscopy, analytical reaction rate and titration procedures and automation, pioneer automatic derivative titrators and equipment for precision null-point potentiometry and spectrophotometry, complete electronics instrumentation sta. for tng. scientists for lab. research and devel. work. Home: 305 Sunnycrest Ct., Urbana, Ill. 61801.*

MALONE, Thomas Francis, Am. meteorologist, ins. co. exec.; b. Sioux City, Ia., May 3, 1917; s. John and Mary (Hourigan) M.; B.S., S.D. Sch. Mines and Tech., 1940, D.Engring., 1962; Sc.D., Mass. Inst. Tech., 1946, D.H.L., 1965; m. Rosalie A. Doran, Dec. 30, 1942; children—John H., Thomas F., Mary E., James K., Richard K., Dennis P. Instr. to asso. prof. Mass. Inst. Tech., 1941-55; dir. research Travelers Ins. Co., Hartford, Conn., 1955-64, 2d v.p., 1964-67, vice president, director of research, 1967—, chmn. bd. Travelers Research Center, Inc., 1961—. Mem. adv. panel on sci. and tech. Com. on Sci. and Astronautics, U. S. Ho. of Reps., 1960—; mem. nat. adv. com. on community air pollution Dept. Health, Edn. and Welfare, 1963-66; mem. sci. information council NSF, 1963-66; mem. adv. com. on internat. orgns. and programs Office of Fgn. Sec., 1964—. Recipient Losey award Inst. Aerospace Scis., 1960. Fellow A.A.A.S., N.Y. Acad. Scis.; mem. Am. Meteorol. Soc. (Charles Franklin Brooks award 1964, past pres.), Am. Geophys. Union (past pres., sec. for internat. participation), Nat. Acad. Scis. (chmn. com. on atmospheric scis.), Econometric Soc., Internat. Union Geodesy and Geophysics (sec.-gen. com. atmospheric sciences), Sci. Research Soc. Am., Sigma Xi. Editor: Compendium of Meteorology, 1951. Contbr. articles to profl. publs., chpts. to books. Research in application of statis. methods to weather prediction, implications of weather modification. Home: 319 N. Quaker Lane, Hartford 06119. Office: Travelers Ins. Co., 1 Tower Sq., Hartford, Conn. 06115.*

MALONEY, James O'Hara, Am. chem. engr.; b. St. Joseph, Mo., Apr. 29, 1915; s. John C. and Jeannette (Clay) M.; Asso. in Sci., Kansas City Jr. Coll., 1931-34; B.S., U. Ill., 1936; M.S., Pa. State Coll., Ph.D., 1941; m. Dorothy Burkholder, Sept. 10, 1940; children—John C., Nancy Jean, Kathleen Sue. Grad. asst. chem. engring. Pa. State Coll., 1937-41; chem. engr. unit operations research and design E. I. du Pont de Nemours & Co., 1941-45; on loan to assist in heavy water program, Columbia, 1942, 45, to assist in plutonium project, U. Chgo., 1943-45; prof. chem. engring. U. Kan., 1945—; lectr. chem. engring. U. Naples, Italy, 1956-57, UAR, 1960. Awarded Frederick Gardner Cottrell grant from Research Corp., 2 research grants AEC, 1949-50, 50-51, research grant from Office Naval Research. Mem. subcom. mil. petroleum adv. bd. aromatics, 1950-51. Mem. Am. Inst. Chem. Engrs., Sigma Xi, Phi Lambda Upsilon, Tau Beta Pi. Contbr. articles chemical jours. Home: 808 Broadview Dr., Lawrence, Kan.*

MALORNY, Guenther Hermann, pharmacologist; b. Borislawitz, Silesia, Aug. 11, 1912; s. Franz Constantin and Vally (Pantke) M.; student U. Graz, 1932-33; M.D., U. Kiel, 1938; m. Zita Dorothea Rehde, Sept. 22, 1942; children—Hans-Dieter, Ursula, Anita. Asst. prof. pharmacology U. Kiel, 1943-56; prof. pharmacology, dir. Inst. Pharmacology, U. Hamburg (Germany), 1957—, dean Faculty Medicine, 1961-62. Med. adviser European Soc. Coal and Steel, 1961—. Mem. German Pharm. Assn. Research, numerous publs. on pharmacology of autonomic nervous system and potentiated anaesthesia, pharm. aspects of coffee and caffeine, shiftings in water and electrolytes between blood and tissues, biochem.-pharm. studies on food preservation, oxygen poisoning, chronic carbon monoxide poisoning, toxicology of fire-resistant hydraulic fluids. Home: 23, Hoheneichen, Hamburg 64. Office: 52, Martinistrasse, Hamburg 20, Germany 2000.*

MALOTT, Clyde A(rnett), Am. geologist; b. Atlanta, Ind., Sept. 10, 1887; s. John Franklin and Alice (Fippen) M.; A.B., Ind. U., 1913, A.M., 1915, Ph.D., 1919; m. Mary Orda Clayton, July 30, 1911; children—Alice, Roland Floyd. Tchr. pub. schs., Ind., 1909-15; staff, dept. geology, Ind. U., 1916-24, prof. geology 1924-47, acting head dept. geol. and geography, 1941-45; engaged in geol. research in Ind., 1947—; mem. staff Okla. Geol. Survey, summer 1916, Ind., 1919-21; geologist Empire Gas & Fuel Co., summers 1918, 23; geologist Pure Oil Co., 1924; acting prof. geology Williams Coll., 1st semester 1929-30; mem. staff Ill. Geol., summer 1930; cons. Sun Oil Co., 1938-40. Recipient award for paper Proc. Ind. Acad. Sci., 1948. Mem. A.A.A.S., Geol. Soc. Am., Ind. Acad. Sci. (v.p. 1937, pres. 1944), Nat. Speleological Soc., Phi Beta Kappa, Sigma Xi. Author: Physiography of Indiana, 1922; The Geology of Dicksburg Hills, Knox County, Indiana, 1948; also sci. papers and many pvt. reports chiefly on Indiana geology; studies in Indiana Caverns a specialty; authority on stratigraphy of Ind. Mississippian and Pennsylvanian. Died Aug. 26, 1950.

MALOUIN, Paul-Jacques, French chemist; b. Caen, France, Feb. 8, 1742; prof. chemistry Royal Bot. Gardens; prof. medicine Coll. de France; physician in ordinary to Queen; in charge control of plague in Paris, 1753; friend of Voltaire. Fellow Royal Soc., 1753; mem. French Acad. Scis., 1744. Author: Traité de chimie, 1734; Sur les maladies épidémiques observées à Paris. Made chem. expts. with zinc and tin. Died Dec. 31, 1777.

MALPIGHI, Marcello, Italian biologist, physician; b. Crevalcore, Italy, Mar. 10, 1628; med. degree, Bologna, Italy. Became lectr., Bologna, 1656; prof. theoretical medicine, Pisa, Italy, 1656-59, Bologna, 1660-62; became prof. primarius, Messina, Italy, 1662, Bologna, 1666-91; named physician to Pope Innocent XIII, 1692; prof. medicine Papal Med. Coll., Rome. Fellow Royal Soc., 1669. Author: De pulmonibus observationes anatomicae, 1661; De formatione pulli in ovi, 1673; Anatomia plantarum, 1675-79; De ovo incubato observationes, 1673. Pioneered study of chick embryology; numerous discoveries including capillary circulation in lung and bladder of frog, 1661, deeper layer of epidermis (Malpigian layer), adenoid tissue in spleen (Malpighian corpuscles), loops of capillaries in kidney; described secreting glands, metamorphosis of silkworm (wrote 1st monograph on invertebrate), stomata in plants, respiratory vessels in insects, structure of human lung, brain and spinal cord; one of 1st to use microscope in study of animal and plant structure. Died Rome, Nov. 30, 1694.

MALQUORI, Giovanni, Italian chemist; b. Florence, Italy, July 15, 1900; s. Enrico and Domenica (Camaiti) M.; Ph.D. in Chemistry, U. Naples (Italy); m. Amelia Bastaglini, Aug. 10, 1929; children—Anna, Enrico, Gianna, Pietro. Prof. indsl. chemistry U. Naples; pres. Cementerie del Tirrena. Mem. Cellulosa d'Italia Soc. (pres.), Lincei Acad. 200 sci. publications. Home: largo San Marcellino 10, Naples, Italy.

MALSKY, Stanley Joseph, Am. physicist; b. N.Y.C., July 15, 1925; s. Joseph and Nellie (Karpinski) M.; B.Sc, N.Y. U., 1949, M.A., 1950, M.S., 1953, Ph.D., 1964; m. Gloria E. Gagliardi, Oct. 15, 1965. Nuclear physicist U. S. Naval Nat. Lab., Bklyn., 1950-54; chief physicist U. S. VA Hosp., Bronx, N.Y., 1954—; research asso. prof. physics dept. Manhattan Coll., Riverdale, N.Y., 1962—; asst. prof. N.Y., 1960-64; research collaborator med. div. Brookhaven Nat. Lab., Upton, L.I., N.Y., 1961—; co-dir. radiol. inst. Manhattan Coll., 1964—. Recipient James Picker Found. award Nat. Acad. Sci.-NRC, 1964-66; Founders Day award N.Y. U., 1964. Mem. N.Y. Acad. Sci., Health Physics Soc., Soc. Nuclear Medicine, Am. Nuclear Soc., Sigma Xi, Sigma Pi Sigma, Phi Delta Kappa. Contbg. author: Progress in Clinical Cancer, 1965; also articles. Research in solid state miniature radiation dosimetry as applied to human and biol. situation, health physics, radiation protection. Home: 809 W. Hartsdale Rd., White Plains, N.Y. 10607. Office: VA Hosp., 130 W. Kingbridge Rd., Bronx, N.Y. 10468.*

MALT, Ronald A., Am. surgeon; b. Pitts., Nov. 12, 1931; A.B., Washington U., St. Louis, 1951; M.D., Harvard, 1955; 3 children. Research asso. USN Sch. Aviation Medicine, Pensacola, Fla., 1956-58; research asso. dept. biology Mass. Inst. Tech., 1962-64, fellow Sch. for Advanced Study, 1963; staff Harvard, 1961—, asso. surgery, 1967—; resident staff Mass. Gen. Hosp., Boston, 1955-56, 58-62, asst. surgeon, 1962—; asso. surgeon Shriners Burns Inst., Boston, 1965—; established investigator Am. Heart Assn., 1963—. Hon. fellow U. Leeds (Eng.), 1964. Mem. Med. Fedn. Ecuador (hon.), Soc. U. Surgeons, A.C.S., A.M.A., Biophys. Soc., Soc. for Developmental Biology, Boston Surg. Soc., Am. Soc. Nephrology. Research, numerous publs. on replantation of severed limbs, mechanism of regeneration especially compensatory growth of kidney, molecular pathology of renal injury. Asso. editor New Eng. Jour. Medicine, 1965—. Office: Mass. Gen. Hosp., Boston 02114.*

MALTBY, Per, Norwegian astrophysicist; b. Oslo, Cand.Real, U. Oslo, 1957, Ph.D., 1964; m. Elisabet Norway, Nov. 3, 1933; s. Olaf and Else (Raastad) M.; Ruud, July 28, 1956; children—Bente, Lars Petter. Research fellow Cal. Inst. Tech., 1960-61, sr. research fellow, 1964-65; amanuensis U. Bergen (Norway), 1961-63; lectr. U. Oslo, 1963, forste-amanuensis, 1965—, dozent, 1967—. Fellow A.A.A.S.; mem. Am. Astron. Soc. Research, publs. on angular sizes of radio sources, spectra of intensity variations in quasistellar objects, solar physics especially velocity fields in sunspots. Home: Lysehagan 71I, Oslo, Norway.*

MALTE-BRUN, Conrad, geographer; b. Thister, Denmark, Aug. 12, 1775; at least 1 son, Victor-Adolphe; devoted himself to politics and literature in Copenhagen; his favoring of enfranchisement of peasants, of liberty of press, and pro-French Revolution writings forced him to flee to Sweden; exiled from Denmark, 1800; went to Paris, where he became known as geographer. First sec. gen. Soc. Geography. Author: (with Mentelle and Herbin) Geographie mathématique, physique et politique de toutes les partes du monde, 1803-07; Précis de geographie universelle, 8 vols., 1810-29; Annales des voyages, de la géographie et de l'histoire, 24 vols., 1808-14; (begun 1810, completed posthumously by Huot); also articles, non-sci. books; edited fgn. political dept. of Journal des debats. Died Paris, Dec. 14, 1826.

MALTHUS, Thomas Robert, English polit. economist; b. nr. Guildford, Eng., Feb. 17, 1766; s. Daniel Malthus; ed. by father, also Gilbert Wakefield, Richard Graves; M.A. (fellow), Jesus Coll., Cambridge (Eng.) U., 1791; m. Harriet Eckersall, 1804. Took holy orders, 1797; curate Albury, Surrey, Eng., by 1798; researcher on population in Germany, Sweden, Norway, Finland, Russia, 1799, France, Switzerland, 1800; prof. history, polit. economy Coll. Haileybury (Eng.), from 1805. Fellow Royal Soc., 1818, Statis. Soc.; mem. Polit. Economy Club, Royal Soc. Lit. (royal asso.). Author: An Essay on the Principle of Population as it Affects the Future Improvement of Society, 1798, rev. 1803; An Inquiry into the Nature and Progress of Rent, 1815; Principles of Political Economy, 1820; Definitions in Political Economy, 1827. Pioneer modern population study; doctrines influenced neo-Malthuseans, classical economists; helped form evolutionary ideas of Charles Darwin and Alfred Wallace by stressing struggle for existence; most famous principle is that unchecked human breeding causes population to grow geometrically while food supply grows arithmetically (felt that only natural checks on population growth were war, famine, disease, continence); felt that real wages could not rise much above subsistence level because an increase in well-being only led to larger supply of workers (was found. of law of diminishing returns). Died St. Catherine's, nr. Bath, Eng., Dec. 23, 1834.

MALTSEV, Anatilii Ivanovich, Russian mathematician; b. Moscow, USSR, Nov. 27, 1909; grad. Moscow U., 1931; D. in Physico-Math. Sci., 1941. Mem. staff Ivanovo Pedagogical Inst., 1932—, prof., 1943—; staff Math. Inst., USSR Acad. Scis., 1942—. Recipient Stalin prize, 1946; named Honored Scientist, 1956; dep. 4th and 5th sessions Supreme Soviet USSR. Mem. USSR Acad. Scis. Author: On the Immersion of an Algebraic Ring into a Field, 1937; On the Inclusion of Associated Systems in Groups, 1939; A General Method of Obtaining Local Group Theory Theorems, 1941; The Semi-simple Sub-groups of the Lie Groups, 1944; The Theory of the Lie Groups in the Large, 1945; A Class of Uniform Spaces, 1949; The General Theory of Algebraic Systems, 1954; Fundamentals of Linear Algebra, 1956; Classes of Models with a Generative Operation, 1957. Specialist in algebra, theory of algebraic systems, theory of continuous groups, math. logic. Office: V.A. Steklov Math. Inst., 1-y Akademischiskii Proyezd 28, Moscow, USSR.

MALUS, Étienne Stephen Louis, French physicist; b. Paris, June 23, 1775; s. Anne-Louis Malus; ed. Mezières mil. sch., École Polytechnique; Served as engr. captain in army, Germany, Egypt; returned to France, 1801; supr. constrn. of supr. fortifications, Antwerp, Belgium, Strasbourg (now France), 1804; became entrance examiner, dir. studies École Polytechnique, 1820. Mem. French Acad. Scis. (prize 1810). Author: Traité d'optique analytique, 1810; Théorie de la double réfraction de la lumière dans les substances cristallines, 1811; Sur les phenoménes qui dépendent des formes des molécules de la lumière, 1811; also papers. Discovered polarization of light, 1808, also Malus' law that sum of intensities of transmitted rays equals intensity of incident ray; explained theory of double refraction. Died Paris, Feb. 23, 1812.

MALZBERG, Benjamin, Am. statistician; b. N.Y.C., Dec. 2, 1893; s. Nathan and Anna (Elson) M.; B.S., Coll. City N.Y., 1915; M.A., Columbia, 1917, Ph.D., 1934; m. RoseHershberg, Aug. 25, 1935; children—Judith Ann, Amy Susan, and Ruth Ellen (twins). With N.Y. State Dept. Social Welfare, 1923-28; asst. dir. Bur. Statistics, N.Y. State Dept. Mental Hygiene, Albany, 1928-44, dir., 1944-56; prin. research scientist Research Found. for Mental Hygiene, Albany, 1956—. Cons. to insts., schools. Am. Field Service fellow U. Paris (France), 1919-21; Nat. Inst. Mental Health grantee, 1949, 56-64. Fellow A.A.A.S., N.Y. Acad. Sci., Am. Sociol. Assn. N.Y. Acad. Medicine (asso.); mem. Sigma Xi. Author numerous books including: Social and Biological Aspects of Mental Disease, 1940; (with E. S. Lee, Dorothy Swaine Thomas) Migration and Mental Disease, 1956; Cohort Studies of Mental Disease in New York State, 1956; Mental Disease among Jews in New York State, 1959; Alcoholic Psycoses at Mid-Century, 1959; Mental Health of the Negro, 1963; also numerous articles. Research on statis. aspects mental diseases, especially mortality among mental patients, ethnic and racial variations in incidence mental disease, expectation of mental disease. Home: 33 Bancker St., Albany, 12208. Office: 119 Washington Av., Albany,N.Y. 12225.*

MAN, Evelyn Brower, Am. biochemist; b. Lawrence, L.I., N.Y., Oct. 7, 1904; d. Edward and Mary (Hewitt) Man; B.A., Wellesley Coll., 1925; Ph.D., Yale, 1932. Faculty, Yale, 1934-61; asso. mem. Inst. Health Scis., Brown U., Providence, 1961—; dir. thyroid lab. Providence Lying-In Hosp. Mem. Am. Soc. Biol. Chemists, A.A.A.S., Am. Chem. Soc., N.Y. Acad. Scis., Am. Thyroid Assn., Endocrine Soc., Assn. Clin. Scientists, Am. Assn. Clin. Chemists, Royal Soc. Medicine, Am. Assn. U. Women, Phi Beta Kappa, Sigma Xi. Research, numerous publs. on lipidemia and iodinemia in health, disease, infancy and pregnancy. Home: Rural Route 2, Box 370A, North Stonington, Conn. 06339. Office: Lying-In Hosp., 50 Maude St., Providence 02908.*

MANCALL, Elliott Lee, Am. physician; b. Hartford, Conn., July 31, 1927; s. Nicholas and Bess (Tuch) M.; B.S., Trinity Coll., 1948; M.D., U. Pa., 1952; m. Jacqueline Cooper, Dec. 27, 1953; children—Andrew Cooper, Peter Cooper. Asst. resident Neurol. Inst. N.Y., 1955-56; resident neuropathology Mass. Gen. Hosp., 1956-57, clin., research fellow neurology, 1957-59; teaching fellow neuropathology Harvard, 1956-57; practice medicine, specializing in neurology, Phila., 1958—; faculty Jefferson Med. Coll., 1958-65; prof. medicine Hahnemann Med. Coll., 1965—; sr. attending physician neurology, head sect. Hahnemann Hosp., 1965—; chief sect. dept. neurology Phila. Gen. Hosp., 1959—; cons. neurology N.J. State Hosp., Ancora, 1959-61, Phila. Home for Incurables, 1964—. Vis. lectr. U. S. Naval Hosp., Phila., 1964—; mem. med. adv. bd. Delaware Valley chpt. Myasthenia Gravis Found. Diplomate Am. Bd. Psychiatry and Neurology. Fellow Am. Acad. Neurology; mem. Am. Neurol. Assn., Am. Assn. Neuropathologists, Assn. Research in Nervous and Mental Diseases, Am. Assn. U. Profs. Author: (with J. B. Angevine and P. Yakovlev) The Human Cerebellum: A Topographical Atlas, 1961. Research, publs. on clin. and pathol. features of metabolic diseases of nervous system, features related to remote effects of malignancy upon nervous system, neuropathologic studies of other metabolic diseases such as Pompe's Disease, anatomic basis for memory. Home: Harts Lane, Miquon, Pa. 19452. Office: 230 N. Broad St., Phila. 19102.*

MANCILL, Julian Dossy, Am. mathematician; b. Dixie, Ala., Jan. 21, 1904; s. Battle Calvin and Alice (Clements) M.; A.B., U. Ala., 1926, M.A., 1927; Ph.D., U. Chgo., 1934; m. Dora Mae DaLee, Sept. 2, 1929; children—Julianne (Mrs. Ray Vaugh Allen), Alice Louise (Mrs. William Curtis Nichols), Marilyn (Mrs. Jack Milton Dollar). Faculty, U. Ala., Tuscaloosa, 1925—, prof., head dept., 1955—; vis. prof. summers U. Fla., 1947, U. S.C., 1958. Math. cons., grad. tng. dir. Redstone Arsenal, 1951-54; Math. Assn. lectr., 1961—. Recipient Comer medal U. Ala., 1927. Mem. Am. Math. Soc., Math. Assn. Am. (award for expository writing of high quality 1964), Assn. Philosophy Math., Nat. Council Tchrs. Math., Am. Assn. U. Profs., Scabbard and Blade, Phi Beta Kappa, Sigma Xi, Pi Mu Epsilon, Kappa Delta Phi. Author: (with M. O. Gonzalez) Algebra Elemental Moderna, 1956, Introduction to College Mathematics, 1958, rev., 1959, Modern College Algebra, 1960, Basic College Algebra, 1962, Algebra Elemental Moderna, 1962, Introductory College Mathematics, A Contemporary Approach, 1966; Modern Analytic Trigonometry, 1960. Contbns. to calculus of variations, founds. of elementary math, philosophy of math. Home: 1021 Myrtlewood Dr., Tuscaloosa, Ala. 35401.*

MANCINI, Roberto E(usebio), Argentinian physician; b. Buenos Aires, Argentina, Sept. 26, 1916; s. Antonio and Adelina (Doria) M.; student Med. Sch. Buenos Aires, 1938-46, M.D., 1948; m. Nelly Maglio, Jan. 8, 1949; children—María Susana, Roberto Claudio. Instr. histology Med. Sch. Buenos Aires, 1940-46; head histology Nat. Inst. Endocrinology, Buenos Aires, 1948-53; mem. dept. biology Nat. Atomic Commn., Buenos Aires, 1953-55; prof. histology U. Buenos Aires Med. Sch., 1956—, vice dean, 1958-59. Mem. sci. career NRC, 1963—. Recipient Gold medal Endocrine Soc. Buenos Aires, 1961; award in biology NRC, 1962. Mem. Endocrine Soc., Am. Assn. Anatomists, Histochem. Soc., Internat. Cell Biologists Belgium, Soc. Argentine de Biologia. Numerous publs. on histochemistry, histoimmunology, chemistry of animal and human connective tissue; research on male gonad. Home: 1318th J.E. Uriburu, Buenos Aires, Argentina.*

MANDEL, H(arold) George, pharmacologist; born Berlin, Germany, June 6, 1924; s. Ernest A. and Else (Crail) M.; came to U. S., 1937, naturalized, 1944; B.S., Yale, 1944, Ph.D., Organic Chemistry, 1949; m. Marianne Klein, July 25, 1953; children—Marcia Vivian, Audrey Lynn. Faculty George Washington U. Sch. Medicine, 1949—, prof., 1958—, chmn. dept. pharmacology, 1960—. Cons. various brs. govt., instns. and bus. firms; also lectr.; mem. coms. Therapeutic Research Found., NRC, USPHS. Recipient John J. Abel award Eli Lilly & Co., 1958. Advanced Commonwealth Fund fellow Molteno Inst., Cambridge U., Eng., 1956, Pasteur Inst., Paris, France, 1957, U. Auckland, New Zealand, U. Med. Scis., Bangkok, Thailand, 1964. Mem. Am. Chem. Soc., N.Y. Acad. Sci., Am. Soc. Biol. Chemists, Am. Soc. Pharmacology and Exptl. Therapeutics (sec. 1961-63), Am. Assn. Cancer Research, A.A.A.S., Am. Assn. U. Profs., Washington Acad. Scis. (Distinguished Achievement award 1958), Assn. Am. Med. Colls., Sigma Xi. Mem. editorial bd. Jour. Pharmacology and Exptl. Therapeutics, 1960-65, Molecular Pharmacology, 1965—. Research, publs. on biochem. mechanisms of drug action, effects of anti-metabolites on cancer, growth-inhibitory drugs, metabolic fate of analgesics. Home: 5500 Christy Dr., Washington 20016. Office: 1339 H St. N.W., Washington 20005.

MANDEL, Irwin Daniel, Am. dentist; b. N.Y.C., Apr. 9, 1922; s. Samuel and Shirley (Blankstein) M.; B.S., Coll. City N.Y., 1942; D.D.S., Columbia, 1945; m. Charlotte Lifschutz, Apr. 1, 1944; children—Carol, Nora, Richard. With Columbia Dental Sch., 1946—, asso. clin. prof. dentistry, 1960—, dir. Lab. for Clin. Research, 1959—; practice dentistry, N.Y.C., 1946—. Fellow A.A.A.S.; mem. Am. Dental Assn., N.Y. Acad. Sci., Sigma Xi. Research, articles on salivary calculus and mechanism by which it is deposited around teeth, characterization of salivary proteins, alterations in composition of saliva in various oral and systemic diseases. Home: 60 Pine Dr., Cedar Grove, N.J. 07009. Office: 630 W. 168th St., N.Y.C. 10032; also 119 W. 59th St., N.Y.C. 10019.*

MANDELBAUM, David Goodman, Am. anthropologist; b. Chgo., Aug. 22, 1911; s. Samuel and Lena (Goodman) M.; B.A., Northwestern, 1932; Ph.D., Yale, 1936; m. Ruth Weiss, May 23, 1943; children—Michael, Susan, Jonathan. Research asso., Am. Mus. Natural History, 1936; fellow NRC, 1937-38; instr., asst. prof. U. Minn., 1938-41; fellow Carnegie Corp., 1941-42; asso. prof. U. Cal. at Berkeley, 1946, prof., 1948—, fellow Guggenheim Found., 1949-50, dept. chmn., 1955-57, chmn. Center for S. Asia Studies, 1964—. Fulbright research prof. U. Cambridge (Eng.), 1953; fellow Center for Advanced Study in Behavioral Scis., Stanford, 1957-58; sr. fellow Am. Inst. Indian Studies, New Delhi, 1963-64. Dir. ednl. resourses in anthropology project NSF, U. Cal. at Berkeley, 1959-63; mem. U. S. nat. commn. UNESCO, 1957-62, chmn. soc. sci. com., 1960-62; cons. Dept. State, 1946-48, NSF, 1958—. Fellow Royal Anthropol. Inst. Great Britain. Author: The Plains Cree, 1940; Soldier Groups and Negro Soldiers, 1952. Editor: (with E. Albert, G. Lasker) The Teaching of Anthropology, 1963. Anthrop. field work with Apache, Plains Cree, in South India, Burma, N. India; Research on cultural use of alcohol, nature of caste in India. Home: 911 Mendocino Av., Berkeley, 7, Cal.*

MANDELBAUM, Hugo, geophysicist; b. Sommerhausen, Bavaria, Germany, Oct. 18, 1901; s. Philip and Rachel (Berlinger) M.; M.A., Hamburg (Germany) U., 1930, Dr.rer.nat., 1934; m. Sophie Fraenkel, May 6, 1931; children—Noa (Mrs. Israel Flam), Uriel, Shoshana (Mrs. Alexander Freemann), Yehuda N., Rachel M. Came to U. S., 1940, naturalized, 1945. Faculty, Tchr.'s Prep. Sch., Burgpreppach, Germany, 1920-24, T.T. High Sch., Hamburg, 1925-39; prin. Beth Yehudah Day Sch., Detroit, 1940-48; prof. geology Wayne State U., Detroit, 1948—. Mem. Am. Geophys. Union, Seismol. Soc. Am., Soc. Exploration Geophysicists, A.A.A.S., Mich. Acad. Sci., Arts and Letters, Assn. Orthodox Jewish Scientists, Mich. Basin Geol. Soc. Author: (with Samuel Conte) Solid Geometry, 1957; Gezeitenstroeme and Reststroeme bei Borkum-Riff, 1934; also articles. Research on influence of meteorol. elements on ocean currents and tidal currents, statis. analysis problems sedimentation, harmonic analysis. Home: 18460 Griggs St., Detroit, 48221.*

MANDELBROJT, Szolem, mathematician; b. Warsaw, Poland, Jan. 20, 1899; s. Saloman and Miriam (Rabinowicz) M.; Docteur ès Sciences, Mathématiques, U. Paris (France), 1923; m. Gladys Grunwald, May 25, 1926; children—Jacques. Fellow, Rockefeller Found., 1924-26; lectr. Rice Inst., 1926-27; maitre de conférences U. Lille (France), 1928-29; prof. Faculté des Sciences, Clermont-Ferrand, France, 1929-38; prof. math. Collège de France, Paris, 1938—. Dir. Monographies internationales de Mathématiques Modernes. Decorated officier la Légion d'Honneur; recipient 5 prices l'Académie des Sciences, Paris, including Grand Prix des sciences mathematiques et physiques, 1960, Médaille Emile Picard, 1965. Mem. Comité National de la Recherche Scientefiques, 1960—. Mem. Société Mathimatique de France (past pres.), Am. Math. Soc. Author or co-author several books; numerous articles. Analysis, including theory of functions, especially analytic functions, Dirichlet series, functionals, number theory. Home: 20 Levepri-er, Paris, France.*

MANDELBROT, Benoit, mathematician; b. Warsaw, Poland, Nov. 20, 1924; s. Charles and Belle (Lourie) M.; Engr., Ecole Polytechnique, Paris, France, 1947; M.S., Cal. Inst. Tech., 1948, Engr., 1949; Ph.D., U. Paris, Sorbonne, 1952; m. Aliette Kagan, Nov. 5, 1955; children—Laurent Mathieu, Didier Alain Jerome. Research asso. Mass. Inst. Tech., 1953; mem. Inst. for Advanced Study, Princeton, N.J., 1953-54; research asso. Institut Henri Poincaré, Paris, 1954-55; jr. prof. U. Geneva (Switzerland), 1955-57, U. Lille (France), Ecole Polytechnique, 1957-58; vis. prof. Harvard, 1962-64; research staff mem. IBM Research Center, Yorktown Heights, N.Y., 1958—; Inst. lectr. Mass. Inst. Tech., Cambridge, 1964—. Author: Logique, Langage et théorie de l'Information (with Leo Apostel, A. Morf), 1957; also numerous articles. Provided math. theories for erratic chance phenomena and self-similarity methods in probability, research on sporadic processes. Home: 7 Valley View Rd., Chappaqua, N.Y. 10514. Office: Box 218 Yorktown Heights, N.Y. 10598; also Mass Inst. Tech., Cambridge, Mass. 02139.*

MANDELSTAM, Stanley, physicist; b. Johannesburg, S. Africa, Dec. 12, 1928; s. Boris and Beatrice (Liknaitzky) M.; B.Sc., U. Witwatersrand, Johannesburg, S. Africa, 1951; B.A., U. Cambridge (Eng.), 1954; Ph.D., U. Birmingham, 1956. Research asst. U. Birmingham, 1956-57; Boese fellow Columbia, 1957-58; asst. research physicist U. Cal. at Berkeley, 1958-60; prof. math. physics U. Birmingham, 1960-63; prof. physics U. Cal. at Berkeley, 1963—. Fellow Royal Soc., 1962; mem. Am. Phys. Soc. Research in theoretical physics elementary particles. Home: 2534 Benvenue Av., Berkeley, Cal. 94704.*

MANDIL, I. Harry, elec. engr.; b. Istanbul, Turkey, Dec. 11, 1919 (parents Am. citizens); s. Harry R. and Bertha (Presente) M.; B.S., U. London, 1939; M.S., Mass. Inst. Tech., 1941; grad. Oak Ridge Sch. Reactor Tech., 1950; D.Sc., Thiel Coll. (Pa.), 1960; m. Beverly Ericson, June 22, 1946; children—Jean Dale, Eric Robert. Dir. reactor engring. div. Bur. Ships USN, chief Reactor Engring. br. naval reactors U. S. AEC, project mgr. Shippingport Atomic Power Sta., 1950-64; prin. officer, dir. MPR Assos., Inc., Washington, 1964—. Recipient Meritorious Civilian Service award USN, 1952, Distinguished Civilian Service award, 1959. Registered profl. engr., D.C. Devel. of nuclear plants for naval propulsion and generation of electricity. Home: 6900 Forest Hill Dr., University Park, Md. 20782. Office: 1140 Connecticut Av. N.W., Washington 20036.*

MANDL, Ines, biochemist; b. Vienna, Austria, Apr. 19, 1917; d. Ernst and Ida (Bassan) Hochmuth; M.S., Poly. Inst. Bklyn., 1947, Ph.D., 1949; m. Hans Alexander Mandl, May 31, 1936. Asst., Interchem. Corp. Research Labs., N.Y. U., 1945-49; research asso. Columbia Coll. Physicians and Surgeons, N.Y.C., 1949-55, asso., 1955-56, asst. prof. biochemistry, 1956—; dir. gynecol. research labs. Francis Delafield Hosp., N.Y.C., 1959—. Mem. Am. Chem. Soc., Am. Soc. Biol. Chemists, Biochem. Soc. (Eng.), Am. Soc. for Cancer Research, Soc. for Exptl. Biology and Medicine, A.A.A.S., Gerontological Soc., Am. Heart Assn., Basic Sci. Council, Sigma Xi. Contbg. author: Collagenases and Elastases, 1961; also numerous articles. Research on mode action Clostridium Histolyticum collagenase, elastolytic enzymes microbial origin, microstructure collagen and elastin, proteolytic enzymes and gynecol. tumors. Home: 166 W. 72d St., N.Y.C. 10023.*

MANDLER, George, psychologist; b. Vienna, Austria, June 11, 1924; s. Richard and Hede (Goldschmied) M.; came to U. S., 1940, naturalized, 1943; student U. Basel, Switzerland, 1947-48; B.A. cum laude, N.Y. U., 1949; M.S., Yale, 1950, Ph.D., 1953; m. Jean Matter, Jan. 19, 1957; children—Peter Clark, Michael Allen. Psychologist, C. W. Beers Guidance Clinic, New Haven, 1952-53; asst. prof. Harvard, 1953-56; vis. research fellow Yale, 1956; lectr., research asso. Harvard, 1956-60; fellow Center for Advanced Study in Behavioral Scis., Stanford, 1959-60; prof. U. Toronto, 1960-65; prof., chmn. dept. U. Cal. at San Diego, 1965—, dir. Center for Human Information Processing, 1967—. Mem. com. on exptl. psychology NRC, Can., 1962-65. Recipient Social Sci. Research Council Research award, 1958. Fellow Am. Psychol. Assn., A.A.A.S.; mem. Psychonomic Soc., Canadian Psychol. Assn., Am. Assn. U. Profs., Philosophy Sci. Assn., Soc. for Psychophysiol. Research, Phi Beta Kappa, Sigma Xi. Author: (with W. Kessen) The Language of Psychology, 1959; (with R. W. Brown, E. Galanter, E. Hess) New Directions in Psychology, 1962; (with Jean M. Mandler) Thinking: From Association to Gestalt, 1964. Contbr. numerous articles to sci. jours. Theoretical, exptl. contbns. to theory of anxiety and emotion; research in human learning particularly verbal behavior and devel. organizational principles underlying human memory; philosophy of sci. in psychology. Address: Dept. Psychology, U. Cal. San Diego, La Jolla, Cal. 92037.*

MANDROUX, Louis-Victor, French physicist; b. Jouy-en-Josas, France, 1843; insp. French Postal and Telegraph System. Author: Étude sur les commutateurs centraux, 1898. Invented automatic starter of Hughes' telegraph, 1875, polarized relay built for undersea cables, 1877; applied Hughes telegraph to undersea use, 1879; adapted Marseilles-Algeria cables, 1882. Died 1913.

MANDUCHI, Claudio, Italian physicist; b. Trento, Italy, May 10, 1924; s. Salvatore and Maria (Vielmetti) M.; grad. Politecnico di Milano (Italy), 1942; physicist Università di Padova (Italy), 1951; m. Maria Teresa Russo, July 15, 1959; children—Gabriele, Roberto. Mem. Istituto Nazionale di Fisica Nucleare, Roma, 1951; head mass spectrometer dept. Soc. Montecatini, Milan, 1954-57; asst. prof. U. Padua, 1957-60, prof. in charge radioactivity, 1961—. Mem. Societa Italiana di Fisica. Research and publs. on cosmic rays, including positive mesons excess, extensive showers, nucleon cascade; beta deay; electron capture including 1st exptl. data in M//L capture ratio; nuclear reactions including polarization phenomena of nucleons. Home: 9 Barbo, Padua, Italy.*

MANDUIT, John, see Mauduith, John.

MANENKOV, Alexander Alexeevich, Russian physicist; b. Tatarskaja, USSR, Jan. 2, 1930; s. Alexey N. and Marija V. Manenkov; diploma of physicist Kasan U., 1952; candidate phys.-math. scis. Phys. Inst., Acad. Scis., Moscow, USSR, 1955, D.Phys.-Math. Scis., 1965; m. Alevtina B. Gagarina, Apr. 2, 1957; children—Margarita, Yulya. Sci. worker, head group Oscillation Lab., P. N. Lebedev Phys. Inst., Moscow, 1955-65, head div. Oscillation Lab., 1965—, mem. sci. council, 1962—, tchr., 1955—. Author: (with N. V. Karlov) Quantum Amplifiers, 1966; also articles. Editor: (with R. Orbach) Spin-Lattice Relaxation in Ionic Solids, 1966. Research on electron paramagnetic resonance, materials and proposal of ruby for masers, investigation of dinamic nuclear orientation; discovered new phenomena in relaxation; devel. theory of transient processes in masers; 1st determination of momenta of some nuclei. Home: 7/43 Krasikov St. Office: 53, Leninsky prospect, Moscow, USSR.*

MANESCHI, Mainaldo, Italian physician; b. Foligno, Italy, June 24, 1926; s. Antonio and Anita (Zopegni) M.; M.D., U. Parma (Italy) 1948; m. Maria Parodi Salvo, Aug. 29, 1956; children—Francesco, Marco, Massimiliano. Vol. asst. in obstetrics and gynecology clinic U. Parma and Genoa, 1948-53; effective asst. Civil Hosp. S. Martino, Genoa, 1954-55; asst. in obstetrics and gynecology clinic U. Modena, 1956—; asst. in obstetrics and gynecology U. Palermo, 1956—, 1st asst., 1960—. Mem. Italian Obstetrics and Gynecology Soc., Exptl. Biol. Soc., Italian Soc. for Study Sterility. Author: Pneumogynecography, 1955; Pelvin Lymphatic Pattern in Gynecological Oncology, 1965. Research, numerous publs. on cortico-suprarenal function in newborn child, radiologic techniques in gynecology, pneumogynecography and its relation to retropneumoperitoneum and laparoscopy, pelvic lymphadenography metabolism of P32 and I131 at level of genital apparatus, esp. in ovary, localization of P32 in cancer of cervix and endometrium; evaluation of thyroid function in gynecol. conditions especially puberty metrorrhagia, therapeutic effectiveness of pseudo-pregnancy in uterine hypoevolution and evaluation of endometrial modifications due to some synthetic progestational substances. Home: 59 Giovanni Bonanno, Palermo, Italy.*

MANFREDI, Angelo, Italian physicist; b. Rome, Italy, Apr. 6, 1913; s. Manfredo and Albertina (Gardella) M.; grad. physics and math. U. Rome, 1936; m. Maria Positano, Nov. 15, 1943; Tchr., Dante Alighteri, 1932-36; head of research l'Instituto Elettroacutica di Roma, 1936—; dir. electrophysiol. lab. hearing and audiology San Giovanni Hosp., Rome, 1958—. Lectr., U. Milano, 1951, United Hosp. Rome, 1961. Recipient Italian gold medal Consiglio Nationale Ricerche, 1962. Mem. Collegium Othologicum Amicitiae Sacrum, Internat. Soc. Audiology, Italian Soc. E.N.T. Author: (with Fiori-Ratti) Electrophysiology of Hearing, 1947; Corso di Aggiornamento in Audiologia, 1961. Research, publs. on electrophonic effect and its application to hearing; contbns. to understanding of functional role of ear lobe in man, psychovoltaic reflex and its applications to objective audiometry; electro optic rotatory effect (rotatory phosphenes), radioacustic effect which derived most important in therapy, radio acustical treatment of deafness. Home: Viale Bruno Buozzi 14. Office: San Giovanni Hosp., Rome, Italy.*

MANFREDI, Eustachio, Italian geometer, astronomer; b. Bologna, Italy, Sept. 20, 1674; s. Adolphe Manfredi. Prof. math. U. Bologna, from 1698; astronomer Bologna Inst., from 1711; supt. of waters. Fellow Royal Soc., 1728; fgn. asso. mem. French Acad. Scis., 1726. Author: Instituzioni astronomiche; pub. Ephemerides, 1715-50. Numerous astron. observations. Died Bologna, Feb. 15, 1739.

MANFREDI, Gabriel, Italian mathematician; b. Bologna, Italy, 1681; s. Adolphe Manfredi. Prof. math. U. Bologna, from 1720; supt. of waters, from 1739. Author: De Constructione aequationum differentialium primi gradus, 1707. Investigated equations of 1st degree. Died 1761.

MANGANIELLO, Eugene Joseph, Am. research dir.; b. N.Y.C., June 8, 1914; s. Joseph and Mary (Pascuso) M.; B.S. in Engring., Coll. City N.Y., 1934, E.E., 1935; m. Helen Gosney, Nov. 25, 1945; children—Eugene Joseph, Gail Lynn, Guy, Gwen. Research engr. Langley Aero. Lab., NACA (now NASA), 1936-43, head heat transfer sect. Lewis Flight Propulsion Lab., 1943-45, chief thermodynamics br., 1945-49, asst. dir., 1949-58, asso. dir. Lewis Research Center, 1958-61, dep. dir., 1961—. Fellow A.A.A.S., Am. Inst. Aeros. and Astronautics (chmn. Cleve. 1954-55, regional dir. 1963); mem. Soc. Automotive Engrs. (chmn. Cleve. 1957-58, bd. dirs. 1961, 62), Pi Tau Sigma. Research, publs. in field of aircraft propulsion; exptl. and analytical investigations jet engines; pioneer in application of nuclear energy to aircraft propulsion. Home: 4045 Diane Dr., Fairview Park O. 44126. Office: 21000 Brookpark Rd., Cleve. 44135.*

MANGELSDORF, Paul Christoph, Am. biologist; b. Atchison, Kan., July 20, 1899; s. August and Mary (Brune) M.; B.S., Kan. State Coll., 1921; S.M., Sc.D., Harvard, 1925; D.Sc. (honorary), Park Coll., 1960, St. Benedict's College, 1965; LL.D., Kan. State U., 1961; m. Helen Parker, June 27, 1923; children—Paul Christoph, Clark Parker. Asst. geneticist, Conn. Agrl. Expt. Sta., 1921-27; agronomist Tex. Agrl. Expt. Sta., 1927-40, asst. dir., 1936-40; vice-dir., 1940; prof. botany and asst. dir. Bot. Mus., Harvard, 1940-45, dir., 1945-67; Fisher prof. natural history Harvard, 1962-67; faculty botany dept. U. N.C., Chapel Hill, 1967—; cons. agr. Rockefeller Found., 1949-52; hon. prof. U. San Carlos, Guatemala, 1956, Nat. Sch. Agr., Peru, 1959. Mem. Rockefeller Found. Agrl. Commn. Mexico, 1941; chmn. bd. cons. for agr. Rockefeller Found., 1956. Fellow Am. Soc. Agronomy; mem. Am. Soc. Naturalists (pres. 1951), A.A.A.S., Am. and N.E. bot. socs., Genetics Soc. America (pres. 1955), Am. Philos. Soc., Am. Acad. Arts and Scis., Nat. Acad. Scis., Linnean Soc. London, Soc. Econ. Botany (pres. 1962), Sigma Xi. Co-author (with R. G. Reeves): The Origin of Indian Corn and its Relatives, 1939; (with others) Races of Maize in Mexico, 1952; Race of Maize in Colombia, 1957; Race of Maize in Central America, 1957; The Origin of Corn, 1959; Races of Maize of Peru, 1962; Campaigns Against Hunger, 1967. Genetic and cytological studies of origin, evolution and improvement of cultivated plants, especially corn. Home: 510 Caswell Rd., Chapel Hill, N.C.*

MANGER, William Muir, Am. internist; b. Greenwich, Conn., Aug. 13, 1920; s. Julius and Lilian (Weissinger) M.; B.S., Yale, 1944; M.D., Columbia, 1946; Ph.D. (fellow internal medicine), Mayo Found., 1957; m. Lynn Seymour Sheppard, May 30, 1964; children—William Muir, Lilian Wade. Instr. medicine Columbia, Coll. Phys. and Surg., 1957-66, asso. in medicine, 1966—, clin. asst. vis. physician, med. div. Bellevue Hosp., 1964—; staff Presbyn. Hosp., 1957—, asst. attending physician, 1966—; dir. Wm. M. Manger Research Found., 1958—; vice chmn. bd. Manger Corp., 1957—. Recipient Mayo Found. Alumni award for meritorious research, 1955. Author: (with K. G. Wakim, J. L. Bollman) Chemical Quantitation of Epinephrine and Norepinephrine in Plasma, 1959. Editor: Hormones and Hypertension, 1966. Research, publs. primarily on studies on plasma catecholamines in hypertension, hemorrhagic shock and metabolic disturbances of acid-base balance, also studies on mechanisms of release of catecholamines. Home: 8 E. 81st., N.Y.C. 10028. Office: 100 W. 58th St., N.Y.C. 10019.*

MANGERON, Démetre, Rumanian engr., mathematician; ed. Faculty Scis., U. Jassy (Rumania); postgrad. Naples, Rome (both Italy), Göttingen (Germany) univs.; D.Sc., Ph.D. Prof. math. for engrs., theoretical mechanics, mechanics of vibrations Poly. Inst. Jassy, editor-in-chief Bull. Poly. Inst. Jassy. Mem. hon. editorial bd. Jour. Mechanisms. Research, publs. on functional equations, mechanisms of vibrations, symbolic calculus, calculus of variations, theory of mechanisms and machines; introduced Mangeron's polyvibrating equations, Mangeron's tensorial method, Mangeron's functions, Mangeron's tangential and reduced accelerations methods, Mangeron's generalized equations of analytical dynamics, Mangeron's variational methods. Address: Jassy, Rumania, Allea Grigore Ghica, 25.*

MANGET, Jean-Jacques, path. anatomist; b. Geneva, Switzerland, June 19, 1652; M.D., U. Valence (France), 1678. Author: Bibliotheca anatomica, 2 vols., 1685; Bibliotheca medico-practica, 4 vols., 1695-1739; Bibliotheca chemica curiosa, 2 vols., 1702; Theatrum anatomicum, 2 vols., 1717; Bibliotheca scriptorum medicorum, 4 vols., 1731. Made early study of miliary Tb and caverns in lung; tried to summarize most important med., surg., pharmacological findings of earlier and contemporary sci. Died Aug. 15, 1742.

MANGIN, Jean Philippe Maurice, French geologist; b. Escarmain, France, Apr. 26, 1926; s. Maurice Raoul and Madeleine (Leroy) M.; student Faculté des Sciences, Bordeaux, France, 1945-51; Licencié es sciences, U. Dijon (France), 1953, Docteur es Sciences, 1958; m. Michèle Mangin, July 24, 1952; children—Claire, Cécile, Laure, Rémi, Frédérique, Antoine. Asst. geologist ESSO-REP, Bordeaux, France, 1951; with Faculté des Sciences, Dijon, France, 1952-

66, prof. geology, 1963-66; prof. geology U. Nice (France), 1966—. Cons. in geoscis. to French govt. offices, pvt. cos. Decorated chevalier du Mérite Agricole et des Palmes Académiques. Mem. Internat. Assn. Sedimentologists, Sté Géologique de France, Internat. Union Paleontologists, Marine Technol. Soc., Comité Internat. d'Etudes Pyrenéennes. Author: Le Nummulituque sud Pyrenéen à l'Quest d l'Aragon, 1959; also articles. Research on chronology of tertiary bldg. of Pyreneen ridge, mechanisms of deposition in various rock faces, mechanisms and results of erosion and sedimentation in various hydrographic systems under intertropical and Mediteranean climates, relations between coastal and lagoonal environments and Foraminifera, discovered various underground water supplies in Burgundy; invented gen. purpose sampler of sediments in rivers and oceans. Home: 48 Av. J. Lorrain, Nice 06, France.*

MANGIN, Louis Alexandre, French botanist; b. Paris, Sept. 8, 1852; docteur ès sciences naturelles; became prof. Lycée de Nancy, France, 1873, Lycée Louis le Grand, Paris, 1881; named prof. Muséum d' histoire naturelle, 1904, dir., 1920. Mem. French Acad. Sci. (became v.p. 1928, pres. 1929), Acad. Agr. Author: Botanique elementaire, 1883; Eléments de botanique, 1884; Cours elementaire de botanique, 1885; Anatomie et Physiologie vegétales, 1895. Research in gen. physiology and anatomy, vegetable pathology, color reactants for microscope use, phytoplankton of Antarctic (from Charcot expdn. of 1915), diseases of chestnut trees, phtheriose of vines. Died June 27, 1937.

MANGOLDT, Hans von, see von Mangoldt, Hans.

MANGON, (Charles-François) Hervé, French engr.; b. Paris, July 31, 1821; chief engr. Hwys. Dept.; dir. Conservatory Arts and Crafts; mem. French Acad. Scis. 1872 (became v.p. 1887), Soc. Agr. Author: Études sur les irrigations de la campine, 1850; Études sur le drainage, 1853. Developed methods of drainage and irrigation; invented various farming machines. Died May 15, 1888.

MANGULIS, Visvaldis, physicist; b. Tukums, Latvia, Nov. 25, 1930; s. Oskars Valdemars and Ieva (Sunins) M.; came to U. S., 1950, naturalized, 1953; B.S., Bklyn. Coll., 1956; M.S., N.Y. U., 1958; m. Vija Blumfelds, Apr. 12, 1953; children—Mara, Antra. With Techn. Research Group (Name later changed to TRG, Inc.), Melville, N.Y., 1956-68, head applied math. sect. systems dept., 1962-68; with Gen. Telephone & Electronics Labs., Inc., Bayside, N.Y., 1968—. Mem. Am. Phys. Soc., Acoustical Soc. Am., Soc. Engring. Sci., Phi Beta Kappa, Sigma Xi. Author: Handbook of Series for Scientists and Engineers, 1965; also articles. Research on nuclear reactor shielding, hydrodynamics and underwater acoustics; devel. advanced sonar systems. Invented dielectric reflector of electromagnetic waves. Home: 11 Abby Dr., East Northport, N.Y. 11731. Office: Gen. Telephone & Electronics Labs., Inc., Bayside, N.Y.*

MANHEIM, Frank Tibor, geochemist; b. Leipzig, Germany, Oct. 14, 1930; s. Ernest and Anna S. (Vitters) M.; came to U. S., 1936, naturalized, 1944; A.B., Harvard, 1951; M.Sc., U. Minn., 1953; Ph.D. in Geochemistry, Stockholm U., 1961; m. Ose Landergren, Aug. 12, 1960; children—Ose, Leif. Analytic chemist Geol. Survey Sweden, 1960-62; geochemist, oceanographer U. S. Geol. Survey, Woods Hole, Mass., 1964—; staff dept. geology Yale, 1963. Mem. Am. Geophys. Union (life), A.A.A.S., Geochem. Soc., Soc. Applied Spectroscopy, Geologiska Föreningen i Stockholm, Sigma Xi. Research, publs. on chemistry of recent bottom deposits Baltic Sea, Swedish Lakes, chem. composition of sediments, pore waters of Atlantic continental shelf and slope, E. coast U. S., hot brines and iron deposits on bottom Red Sea. Home: 15 Taylor Ct., Falmouth, Mass. 02540. Office: Woods Hole Oceanographic Instn., Woods Hole, Mass. 02543.*

MANHOLD, John Henry, Am. pathologist; b. Rochester, N.Y., Aug. 20, 1919; s. John Henry and Helen (Schulz) M.; B.A., U. Rochester, 1940; D.M.D., Harvard, 1944; M.A. in Psychology, Washington U., St. Louis, 1956; m. Beverly Schecter, May 8, 1953; 1 son, Gordon. Instr. pathology, dir. cancer teaching program Tufts U. Coll. Medicine and Dentistry, 1947-50; asst. prof., acting chmn. dept. pathology and oral pathology Washington U. Coll. Dentistry, 1954-56; faculty N.J. Coll. Medicine and Dentistry, Jersey City, 1957—, prof., 1958—, dir. dept. pathology and oral diagnosis, 1957—; dir. cancer teaching program, attending pathologist Jersey City Med. Center, 1958—. Cons. to pvt. cos. Carnegie fellow in pathology U. Rochester, 1944. Fellow Internat., Am. colls. dentists, Acad. Psychosomatic Medicine, A.A.A.S.; mem. A.M.A., Am. Psychol. Assn., Am. Dental Assn., Internat. Assn. for Dental Research, Sigma Xi. Author: Introductory Psychosomatic Dentistry, 1956; (with T. Bolden) Outline of Pathology, 1960; Clinical Oral Diagnosis, 1965; also numerous articles; contbr. to Orban's Periodontics, 1963; Pharmacotherapeutics of Oral Disease, 1964; Dental Clinics of North America, 1962. Research on application psycho-physiologic research methods to study dental disease, tissue metabolism studies. Home: 6600 Kennedy Blvd. E., West New York, N.J. 07093. Office: N.J. Coll. Medicine and Dentistry, Jersey City 07304.*

MANI, Gopalakvishnan Subramanian, physicist; b. Rangoon, Burma, Jan. 16, 1929; s. G. K. and Sarada (Narayan) Iyer; B.Sc. with honors in Physics, Madras (India) U., 1949; m. Vijaya S. Mani, Apr. 29, 1953; children—Gopinath, Maya, Prema. Research officer Indian Atomic Energy Establishment, 1949-57; vis. physicist Atomic Energy Research Establishment, Harwell, Eng., 1957-59; research asso. Rice U., Houston, 1959-61; group leader SPNBE, CEA, Saclay, Paris, France, 1961-64; lectr. physics dept. Manchester (Eng.) U., 1964-67, sr. lectr., 1967—. Cons. physicist Commissariat a l'énergie Atomique, Saclay, Paris, 1964-65. Research, publs. on interactions between nuclei by studying products produced in reaction induced by particles accelerated by suitable accelerating machines. Home: 31, Vicarage Av., Cheadle Holme, Cheshire, U.K. Office: Physics Dept., Manchester, U., Manchester, U.K.*

MANILIUS, Marcus, Roman astrologer; flourished Rome, end of Augustine Age. Author: Astronomica, an Astrological Poem, 5 books (describes creation of heavens, signs of zodiac). Used simple arithmetic progressions (Neugebauer showed these methods are related to Babylonian lunar calculations); believed in fixity of laws of nature and absence of miracles.

MANIRE, George Philip, Am. bacteriologist; b. Roanoke, Tex., Mar. 25, 1919; s. Ernest L. and Zera (Ballew) M.; B.S., N. Tex. State Coll., 1940, M.S., 1941; Ph.D., U. Cal. at Berkeley, 1949; m. Ruth Jacobs, Apr. 10, 1943; children—Sarah, Philip. Instr., Southwestern Med. Sch., U. Tex., Dallas, 1949-50; faculty U. N.C. Med. Sch., Chapel Hill, 1950—, prof. dept. bacteriology and immunology, 1959—, asst. vice chancellor health affairs, 1965-66, chmn. dept., 1966—. Fulbright Research scholar Statens Seruminstut, Copenhagen, Denmark, 1956; Alan Gregg Travel fellow in med. edn. China Med. Bd., 1963-64; vis. prof. Inst. for Virus Research, Kyoto (Japan) U., 1963-64. Research, publs. on chem., biol. and structural characteristics psittacosis organisms, methods for large scale cultivation; devised classification method based on toxin neutralization tests; prepared purified cell wall suspensions. Home: 710 Coker Dr., Chapel Hill, N.C. 27514.*

MANIS, Jerome G., Am. sociologist; b. Albany, N.Y., Mar. 18, 1917; s. Joseph and Sarah (Zubris) M.; B.A., Wayne State U., 1947; M.A., U. Chgo., 1949; Ph.D., Columbia, 1952; m. Laura May Glance, May 31, 1949; children—Robert Ethan, Lisa Ann. Lectr., Roosevelt U., Chgo., 1948; asst. prof. Central Mich. U., Mt. Pleasant, Mich., 1952; faculty Western Mich. U., Kalamazoo, 1952—, prof. sociology, 1960—, dir. Center for Sociol. Research, 1956—; Fulbright prof. Silliman U., Philippines, 1959-60. U. S. Office Edn. grantee, 1956-58; USPHS grantee, 1959-61. Mem. Mich., Am. sociol. assns., Soc. for Study Social Problems. Author: (with S. I. Clark) Man and Society, 1960; also articles. Research on effects acad. situation upon publ. productivity social scientists, survey techniques for estimating prevalence alcoholism and mental illness, validation mental health scales, comparison pub. and psychiat. conceptions mental illness, influences surburban residence upon polit. behavior. Home: 3607 Middlebury Rd., Kalamazoo 49007.*

MANITIUS, S. B., naturalist; flourished 17th century. Author: Chymica Formicarum Analysis, 1689. Pub. work on chem. analysis of ants includes definition generation, microscopic observations, differences in ants, internal principles, gen. and medicinal uses, and methods of extermination.

MANKOVSKII, Boris Nikitich, Russian neuropathologist; b. Kozelets, USSR, Mar. 23, 1883; grad. Kiev (USSR), 1910; M.D.; student U. Leipzig (Germany). Successively intern, asst., lectr. nervous diseases Kiev U. until 1922; head nervous diseases Kiev Postgrad. Med. Inst., 1922-41; founder, sci. dir. Kiev Psychoneurol. Research Inst., 1931-50, also founder blastomatosis clinic; head clinic nervous diseases Kiev City Hosp.; prof. Kiev Inst. for Improvement Physicians, 1923-41, 45-58. Named hon. sci. worker Ukraine, 1942—; recipient Order of Lenin, twice; Order Red Banner of Labor; Badge of Honor. Author: (monographs) Encephalomyelitis disseminata sclerotican periaxillis, 1941, The Pathogenesis of Neuroinfectious Diseases and the Body's Response, 1950; also numerous articles. Founder, dir. Modern Psychoneurology; mem. editorial council Korsakov Jour. Neuropathology and Psychiatry, Clin. Medicine. First description of liquorrhea nasalis in cerebral tumors; research in metastatic cancer of central nervous system, spinal cord diseases, secondary neural infections in measles and typhoid. Office: A. USSR, Kiev, Ukraine SSR, ul. Shevchenko 13, Meditsinsky institut.

MANLY, Charles Matthews, Am. mech. engr.; b. Staunton, Va., Apr. 24, 1876; s. Charles and Mary Esther Hellen (Matthews) M.; grad. Master Math. and Mech. Philosophy, Furman U., 1896; M.E., Cornell, 1898; m. Grace Agnes Wishart, June 9, 1904; children—Charles Wishart, John Frederick. Chief asst. to Dr. Samuel P. Langley in aviation devel. work Smithsonian Instn., 1898-1905; v.p. and chief engr. Manly Drive Co., N.Y., 1905—; mem. Manly & Veal, cons. engrs., N.Y.C., cons. engr. British War Office, on devel. of large airplanes in Am., 1915; cons. engr. Curtis Aeroplane & Motor Corp., Buffalo, 1915-19, asst. gen. mgr. N.Y.C., 1919-20. Mem. Soc. Automotive Engrs. (pres. 1919). Author: (with S. P. Langley) Langley Memoirs on Mechanical Flight (Smithsonian Instn.), 1911. Built and piloted the historic Langley aeroplane, in its tests, 1903, the work being stopped by lack of funds before complete tests could be made; same machine flown by G. H. Curtiss, 1914; holder about 50 patents on automotive transp., power generation and transmission. Died Oct. 17, 1927.

MANLY, Richard S(amuel), Am. biochemist; b. Malta, O., May 31, 1911; s. Fred B. and Ella (Betts) M.; B.S., Antioch Coll., 1933; M.A., Oberlin Coll., 1934; Ph.D., U. Rochester, 1938; m. Marian L. Lefevre, Apr. 15, 1939; children—Kenneth, Philip, Linda. Sr. fellow U. Rochester (N.Y.), 1938-39; research chemist Proctor & Gamble Co., Cin., 1939-45; faculty Tufts U.: Sch. Dental Medicine, Boston, 1945—, prof. biochemistry, 1947-65, research prof., 1965—; dir. Westwood Research Lab. (Mass.), 1953-65, pres., 1965—. Mem. biochem. subcom. Dental Research Council NRC, 1952-54; mem. dental study sect. NIH, 1955-59. Mem. A.A.A.S. (sec. sect. ND 1967—), Internat. Assn. Dental Research (pres. 1966—), Am. Chem. Soc., Inst. Chemists, Omicron Kappa Upsilon (hon.). Research, numerous publs. on methods of study of tooth abrasion by dentifrices, decalcification resistance of tooth enamel, carbohydrate polymerization by oral organisms, diffusion in gels, glycolysis inhibitors. Home: 20 Lanark Dr., Westwood, Mass. 02090. Office: Tufts U. Sch. Dental Medicine, 136 Harrison Av., Boston 02111; also Westwood Research Lab., Inc., 543 High St., Westwood, Mass. 02090.*

MANN, Alfred Kenneth, Am. physicist; b. N.Y.C., Sept. 4, 1920; s. David and Belle (Mann) M.; A.B., U. Va., 1942, M.S., 1946, Ph.D., 1947; m. Jayne Bowers, June 29, 1946; children—Stephen, Cecile, David, Brian. Civilian with OSRD, 1942-43, USN, 1944; instr. Columbia, 1947-49; faculty U. Pa., Phila., 1949—, prof. physics, 1957—. Fulbright fellow Australian Nat. U., 1955-56; NSF fellow CERN, Geneva, Switzerland, 1962-63. Fellow Am. Phys. Soc., Royal Soc. Arts, N.Y. Acad. Sci.; mem. Phi Beta Kappa, Sigma Xi. Research, numerous publs. on interactions of elementary particles and forces that lead to their spontaneous disintegration. Home: 18 Penarth Rd., Bala-Cynwyd, Pa. 19004. Office: 209 S. 33d St., Phila. 19104.*

MANN, Douglas Bradwell, Am. engr.; b. Chgo., Aug. 25, 1928; s. Lester B. and Helen (Moore) M.; B.S. in Mech. Engring., 1952, M.S. in Mech. Engring., 1959; m. Jeanne Marie Webber, Aug. 29, 1952; children—Douglas Brian, Elizabeth Joanne. With cryogenic div. Nat. Bur. Standards, Inst. for Materials Research, Boulder, Colo., 1952—, chief cryogenic metrology sect., 1967—. Mem. Research Soc. Am. Research, publs. on prodn., transport and storage of cryogenic fluids; thermophys. properties correlations, liquid helium tech., liquid hydrogen bubble chambers, refrigeration processes. Home: 3110 23d St. Office: 325 Broadway, Boulder, Colo. 80302.*

MANN, Frederick George, English organic chemist; b. West Norwood, London, Eng., June 29, 1897; s. William Clarence and Elizabeth (Casswell) M.; B.Sc. with honors, Battersea Poly., U. London, 1919, D.Sc., 1929; Sc.D., Cambridge (Eng.) U., 1932; m. Margaret Reid Shackleton, Aug. 16, 1930 (dec.); children—Shirley Margaret Shackleton (Mrs. Desmond John O'Donnell-Bourke), Carola Mary Shackleton (Mrs. Kenneth Garfield Carr-Brion); m. 2d, Barbara Thornber, Sept. 18, 1951; 1 dau., Elizabeth Hilary Frances. Asst., Cambridge U., 1922-30, lectr. chemistry, 1930-46; reader organic chemistry, 1946-66, now emeritus reader; fellow Trinity Coll., Cambridge, 1931—, life fellow, 1952, lectr., 1931-64; vis. sr. prof. U. Hawaii, 1946-47. Fellow Royal Soc., Royal Inst. Chemistry (past mem. council), Chem. Soc. (past mem. council, past v.p.); mem. Soc. Chem. Industry. Author: Heterocyclic Derivatives of Phosphorus, Arsenic, Antimony, Bismuth and Silicon, 1950, 2d edit., 1968; (with B. C. Saunders) Practical Organic Chemistry, 1936; Introduction to Practical Organic Chemistry, 1939; also numerous articles. Research on organic, coordination, and inorganic chemistry with particular studies on cyclic nitrogenous derivatives, cyanin dyes, synthesis and use of tertiary phosphines and arsines, coordination compounds. Home: 24 Porson Rd., Cambridge, Eng.*

MANN, Henry Berthold, mathematician; b. Vienna Austria, Oct. 27, 1905; s. Oscar and Friederike (Schoenhof) M.; Ph.D., U. Vienna, 1935; m. Anna Loeffler, July 19, 1935; 1 son, Michael. Came to U. S., 1938, naturalized, 1945. Carnegie fellow, 1942-43; instr. Bard Coll., 1943-44; research asso. Ohio State U., 1944-45, Brown U., 1945; prof. math. Ohio State U., 1946-64; prof. U. Wis., Madison, 1964—, mem. Math. Research Center, 1964—; vis. prof. U. Cal. at Berkeley, 1949-50. Fellow Inst. Math. Statistics; mem. Am. Math. Soc. (Frank Nelson Cole prize in theory numbers 1946). Author: Analysis and Design of Experiments, 1949; Theory of Stochastic Processes Depending on a Continuous Parameter, 1953; Introduction to the Theory of Algebraic Numbers, 1955; Additions Theorems, 1965; also numerous articles. Developed metric theory of addition in groups and semi-groups; research on combinatorial analysis, number-theory, math. statistics. Home: 5612 Lake Mendota Dr., Madison, Wis. 55705.*

MANN, James, Am. physician; b. Wrentham, Mass., July 22, 1759; s. David and Anna Mann; grad. Harvard, 1776; M.D. (hon.), Brown U., 1815; studied with Dr. S. Danforth; m. Martha Tyler, Dec. 12, 1788; 5 children. Served as surgeon with Col. W. Shepard's 4th Mass. Regt., 1779-82, imprisoned by British, 2 months, 1781; practiced medicine, N.Y.C.; head of med. dept. U. S. Army on No. frontier during War of 1812; sr. hosp. surgeon, Detroit, 1816-18; post surgeon, 1818-21, asst. surgeon, 1821-32. Author: Medical Sketches of the Campaigns of 1812, 13, 14, to Which are Added Surgical Cases, Observations on Military Hospitals; and Flying Hospitals Attached to a Moving Army, Also an Appendix . . . , 1816. Credited with 1st excision of elbow joint in U. S., circa 1822. Died Nov. 7, 1832.

MANN, Lloyd Godfrey, Am. physicist; b. Sterling, Mass., July 2, 1922; s. Walter and Ethel (Godfrey) M.; B.S., Worcester Poly. Inst., 1944; Ph.D. in Physics, U. Ill., 1950; m. Patricia Ann Barraclough, Mar. 7, 1959; children—Martha Louise, Lorraine Frances, Douglas Lloyd. Staff, Radiation Lab., Mass. Inst. Tech., Cambridge, 1944-45; instr. physics Stanford, 1950-53; physicist Lawrence Radiation Lab., Livermore, Cal., 1953——. Mem. Am. Phys. Soc. Research on properties low-lying nuclear energy levels, measurements circular polarization gamma rays. Home: 4345 Baylor Way. Office: Box 808 Livermore, Cal. 94550.*

MANN, Thaddeus Robert Rudolph, biochemist; b. Lwow, Poland, Dec. 4, 1908; s. William and Emilia (Quest) M.; M.D., U. Lwow, 1934; Ph.D., Cambridge U., 1937, Sc.D., 1950; m. Cecilia Lutwak, Apr. 2, 1934. Lectr., docent U. Lwow, 1934-35; with Molteno Inst., Cambridge, Eng., 1935——, Rockefeller Found. research fellow, 1935-37, Beit Meml. research fellow, 1937-44, prin. sci. officer Agrl. Research Council Gt. Britain, 1944-52, univ. reader in physiology of animal reprodn., dir. unit reproductive physiology and biochemistry, 1953——; vis. prof. biology Fla. State U., 1962; sr. Lalor fellow Woods Hole, Mass., 1960. Fellow Royal Soc., 1951; mem. Biochem. Soc., Physiol. Soc., Inst. Biology, Soc. Endocrinology, Nutrition Soc., Soc. Exptl. Biology. Author: Biochemistry of Semen, 1954, The Biochemistry of Semen and of the Male Reproductive Tract, 1964. Publs. on muscle biochemistry, chem. mycology, discovery of prosthetic metallic groups in certain enzymes; composition of seminal fluid and identification of fructose as sugar present. Home: 1 Courtney Way. Office: Molteno Inst., Downing St., Cambridge, Eng.*

MANN, William Alfred, Am. ophthalmologist; b. Chgo., Mar. 21, 1898; s. William A. and Anna D. (Cram) M.; B.S., U. Ill., 1921, M.D., 1923; M.S., Northwestern U., 1939; m. Maud Lucille Davison, May 30, 1931; children—William A. III, Nancy D. (Mrs. Hugh Germanetti), David L. Practice medicine, specializing in ophthalmology, Chgo., 1926——; faculty Northwestern U. Med. Sch., Chgo., 1927——, prof. ophthalmology, 1949-66, emeritus prof., 1966——; chmn. dept. ophthalmology Chgo. Wesley Meml. Hosp., 1940-66, emeritus, 1966——. Cons.-chief VA Hosp., Hines, Ill., 1945——; mem. revision com. U. S. Pharmacopia, 1950-60; sec.-treas. Ophthalmic Pub. Co., 1948——. Mem. Chgo. Ophthal. Soc. (pres. 1946-47), Sigma Xi, Lambda Chi Alpha, Omega Beta Pi, Alpha Kappa Kappa (primarius 1949-53, 63-65). Contbr. articles to profl. jours., med. textbooks. Home: 229 E. Lake Shore Dr., Chgo. 60611. Office: 251 E. Chicago Av., Chgo. 60611.*

MANNARI, Isao, Japanese physicist; b. Soja City, Okayama, Japan, Sept. 3, 1929; s. Kunihei and Kiyoko (Fujii) M.; B.S., Kyoto (Japan), 1953, D.Sc., 1962; m. Teruko Saito, Jan. 8, 1955; children—Hiroyuki, Koji. Research asst. Kyoto U., 1953-61; asst. prof. physics Shizuoka U., 1961-64; asst. prof. physics Okayama U., 1964-65, prof. physics, 1965——, chief div. digital computer Electronic Computer Center, Okayama U., 1953——. Mem. Phys. Soc. Japan, Sanyo Assn. for Advancement Sci. and Technics. Research, publs. transport phenomena, phase transitions, theory of elec. conductivity of ferromagnets, phase transition in ferromagnets; table of extended Watson integral useful in theory of phase transitions. Address: Tsushima, Okayama, Okayama, Japan.*

MANNERHEIM, Carl Gustav von, see von Mannerheim, Carl Gustav.

MANNERING, Gilbert J., Am. biochemist; b. Racine, Wis., Mar. 9, 1917; s. John Knight and Louise (Holm) M.; B.S., U. Wis., 1940, M.S., 1943, Ph.D., 1944; m. Virginia Agnes Krahn, Feb. 1, 1939; children—Michael John, Gail Phyllis, Barbara Lynn. Sr. research biochemist Parke, Davis & Co., Detroit, 1944-50; cons. in chemistry 406th Med. Gen. Lab., Tokyo, Japan, 1950-54; with U. Wis., 1954-62; prof. pharmacology U. Minn., Mpls., 1962——. Cons. toxicology Wis. Crime Lab., 1954-62, Wis. Dept. Agr., 1954-57; spl. cons. interdepartmental com. on nutrition for nat. def. NIH, Ethiopia, 1958; mem. toxicology study sect. USPHS, 1962-65, mem. pharmacology-toxicology rev. com., 1965——; mem. com. applications of biochem. studies in evaluating drug toxicity NRC, 1965——. Mem. Am. Soc. Pharmacology and Exptl. Therapeutics, Am. Acad. Forensic Scis., Soc. Toxicology, Sigma Xi. Studies, publs. drug metabolism; metabolism of narcotics, alcohols, kinetics of drug metabolizing enzymes, inhibition of drug metabolism; devel. chem. analytical procedures used in drug metabolism studies and in toxicology. Home: 1865 N. Fairview St., St. Paul 55113. Office: Dept. Pharmacology, U. Minn., Mpls. 55435.*

MANNERS, David John, Brit. biochemist; b. Castleford, U.K., Mar. 31, 1928; s. George and Gwendoline (Shortridge) M.; B.A., Cambridge (Eng.) U., 1949, Ph.D., 1952; D.Sc., Edinburgh U., 1960; m. Gweneth Mary Chubbock, Aug. 23, 1952; children—Roger David, Stephen John, Judith Rosemary. Faculty, U. Edinburgh (Scotland), 1952-65, reader chemistry, 1964-65; prof. brewing and applied biochemistry Heriot-Watt U., Edinburgh, 1965——. Fellow Royal Inst. Chemistry (Meldola medal 1957), Royal Soc. Edinburgh. Research, numerous publs. on structure and metabolism of complex carbohydrates in plant, mammalian and protozoal tissues. Home: 165 Mayfield Rd., Edinburgh 9, U.K.*

MANNERS, Robert A., Am. anthropologist; b. N.Y.C., Aug. 21, 1913; s. Abraham and Dora (Kniaz) M.; B.S., Columbia U., 1935, M.A., 1939; Ph.D., 1950; m. Margaret Daviss Hall; children—Karen Elizabeth, John Hall; m. 2d, Jean Ingalls Hall, Sept. 12, 1955; children—Stephen David, Katherine Dora. Instr., U. Rochester, 1950-52; faculty Brandeis U., Waltham, Mass., 1952——, prof., chmn. dept. anthropology. Fellow Am. Anthrop. Assn., African Studies Assn.; mem. Am. Ethnol. Soc. Author: (with others) Peoples of Puerto Rico, 1956; (with Duffy) Africa Speaks, 1961; Process and Pattern in Culture, 1964; also articles. Contbg. author: Contemporary Change in Traditional Societies, 1967. Study of culture change in developing areas of Africa, Caribbean and among Am. Indians of S.W., problems of these areas and methods for study and analysis. Home: 134 Sumner St., Newton Centre, Mass. 02159. Office: Brandeis U., Waltham, Mass. 02154.*

MANNHEIM, Amédée, French mathematician; b. Paris, France, July 17, 1831; ed. École Polytechnique, France. Prof. descriptive geometry École Polytechnique. Author: Théorie, géométrique de l'hyperboloide; Mémoire d'optique géométrique; Cours de géométrie descriptive de l'École polytechnique, 1886; Principes et developpements de géométrie cinématique, 1894. Developed Mannheim slide rule circa 1850. Died Paris, 1906.

MANNHEIM, Karl, sociologist; b. Budapest, Mar. 27, 1893; s. Gustav and Rosa (Eylenburg) M.; ed. univs. of Budapest, Berlin, Paris, Freiburg i/Br., Heidelberg; m. Julia Lang, 1921. Privatdozent in sociology U. Heidelberg (Germany), 1926-30; prof. sociology, head dept. U. Frankfort/Main (Germany), 1930-33; lectr. sociology London Sch. Econs., U. London, 1933-45; prof. edn. U. London, 1945-47. Author: Strukturanalyse der Erkenntnis Theorie, 1922; Das konservative Denken, 1927; Die Gegenwartsaufgaben der Soziologie, 1932; Ideology and Utopia, 1929, enlarged English edit., 1936; Man and Society in an Age of Reconstruction, 1940; Diagnosis of Our Time, 1943; Essays on Sociology and Social Psychology, posthumously 1953. Editor, Internat. Library of Sociology and Social Reconstruction. Pioneered sociology of knowledge; studied interconnections between interest groups in society, their ideas and modes of thought. Died London, Jan. 9, 1947.

MANNHEIMS, Bernhard Josef, German zoologist; b. Burg Irnich, July 12, 1909; s. Josef and Margarete (Biermanns) M.; ed. U. Cologne (Germany), U. Frankfort, U. Innsbruck (Austria); Ph.D.; m. Helen Hieronimi, June 8, 1937; children—Hans-Heiner, Brigitte, Wolfgang, Gretel, Elisabeth, Hildegard. Phytopathologist, Office for Protection Plants, Bonne, Germany, 1937-40; army entomologist, Greece, 1942-45; sect. chief, sci. asst. Koenig Mus., Bonn. Mem. German Entomol. Soc. Author: Tipulidae in Lindner; Die Fliegen der pal. Region; Das kleine Fliegenbuch; also articles. Home: Koblenzer Strasse 150, 53 Bonn. Office: Zoologisches Forschungsinstitut and Mus. Alex. Kcenig, Bonn, West Germany.

MANNICH, Carl, German pharm. chemist; b. Breslau (now Wroclaw, Poland), Mar. 8, 1877; prof., Berlin; developed Mannich reaction for isolating tertiary bases by reaction of formalin and secondary amine on compounds with active hydrogen. Died Karlsruhe, Germany, Mar. 5, 1947.

MANNICK, John Anthony, Am. surgeon; b. Deadwood, S.D., Mar. 24, 1928; s. Alfred and Catherine (Schuster) M.; A.B., Harvard, 1949, M.D., 1953; m. Alice Virginia Gossard, June 9, 1952; children—Catherine Virginia, Elizabeth Eleanor, Joan Barbara. USPHS research fellow Mary Imogene Bassett Hosp., Cooperstown, N.Y., 1958-59; faculty Med. Coll. Va., 1960-64; prof. surgery Boston U. Med. Sch., 1966——; vis. surgeon U. Hosp., 1966——; clin. asso. surgery Boston City Hosp., 1964——; cons. surgery Providence Vets. Hosp., 1964——; attending surgeon Boston Vets. Hosp., 1964——. John and Mary R. Markle scholar in med. scis., 1961——. Mem. Am. Fedn. Clin. Research, Am. Soc. Exptl. Pathology, N.Y. Acad. Scis., A.M.A., Soc. U. Surgeons, Am. Surg. Assn., Am., So. socs. clin. investigation, A.C.S., Am. Soc. Vascular Surgery, Transplantation Soc., Phi Beta Kappa. Contbr. articles to med. jours. Investigation of transfer of transplantation immunity by means of lymphoid RNA; induction of immunological tolerance with subcellular transplantation antigenic preparations; application of electromagnetic flowmeter to clin. vascular surgery. Home: 81 Bogle St., Weston, Mass. 02193. Office: 750 Harrison Av., Boston 02118.*

MANNING, George William, Canadian physician; b. Toronto, Ont., Can., Aug. 4, 1911; s. George Arthur and May (Wright) M.; Malvern Collegiate, 1927-31. Trinity Coll., Toronto, 1931-35; B.A., U. Toronto, 1935, M.D., 1940; m. Marie Lillian Collins, Dec. 14, 1940; children—Cheryl, Helen, Christopher. Aviation med. research RCAF, 1940-45; Nuffield fellow cardiac unit London (Ont.) Hosp., 1946-47; faculty U. Western Ont., London, 1947——, prof. medicine, dept. cardiology, 1960——, dir. heart unit U., Victoria Hosp., London, 1947——; cardiac cons. Canadian Armed Forces, 1948——, dept. Vets. affairs Westminster Hosp., London, 1949——. Chmn. med. com. Ont. and Canadian Heart Found., 1952——, Can. Heart Found., Can. Cardiovascular Soc. Mem. Can., Am. phys. socs., Am. Coll. Cardiology, A.C.P., Can. Soc. Clin. Investigation, others. Research, publs. on exptl. and clin. cardiovascular research with spl. reference to coronary artery diseases and electro and vectorcardiography; recently developed a new vector rec. apparatus. Home: 9 Harrison Crescent. Office: Victoria Hosp., London, Ont., Can.*

MANNING, Melvin Lane, Am. elec. engr.; b. Miller, S.D., Nov. 26, 1900; s. James David and Linnie (Durrin) M.; B.S. in Elec. Engring., S.D. State Coll., 1927; M.S. in Elec. Engring., U. Pitts., 1932; coll. tchr. certificate, 1936; m. Elizabeth Harrington, Sept. 30, 1941; children—Peggy Jean, James David. With Westinghouse Elec. Corp., 1929-32, supr. high voltage labs. and edn. program, transformer div., 1936-42; instr. math. U. Pitts., 1932-36; asso. prof. elec. engring. Ill. Inst. Tech., 1943-44. Cornell U., 1944-45; chief engr. Kuhlman Elec. Co., Bay City, Mich., 1945-49, devel. engr. McGraw-Edison Co., Canonsburg, Pa., 1950-59; dean engring., dir. engring. expt. sta. S.D. State U., 1959-66, prof. elec. engring. Sec., S.D. Bd. Engring. and Archtl. Examiners; chmn. uniform exam. com. Nat. Council Engrs. Examiners. Recipient prize for paper Am. Inst. E.E.-Nat. Elec. Mfrs. Assn., 1960. Registered profl. engr., Pa., S.D. Fellow I.E.E.E. (chmn. gaseous insulation com. 1962——); mem. Jr. Engring. Tech. Soc. (dir.), Engrs. Council Profl. Devel. (chmn. reguidance com. 1960——), Nat., S.D. (chmn. edn. com.) socs. profl. engrs. Am. Soc. Engring. Edn, Am. Soc. Testing Materials, Nat. Acad. Sci., A.A.A.S., Sigma Xi, Sigma Tau, Phi Kappa Phi, Eta Kappa Nu. Contbr. profl. papers. Patentee application silicone insulation dry-type transformers, perfluoro-carbon gases in transformers. Home: 405 State Av., Brookings, S.D. 57006. Office: S.D. State U., Brookings, S.D.*

MANNING, Raymond Brendan, Am. zoologist; b. Bklyn., Oct. 11, 1934; s. F. B. and Elizabeth (Smith) M.; B.S. in Zoology, U. Miami, 1956, M.S. in Marine Sci., 1959, Ph.D., 1963; m. Lilly D. King, Aug. 1, 1957; children—Marian, Barbara Ann, Elaine. Research instr. Inst. Marine Scis., U. Miami, Coral Gables, Fla., 1959-63; asso. curator div. marine invertebrates Smithsonian Instn., Washington, 1963-65, curator-in-charge div. Crustacea, 1965-67, chmn. dept. invertebrate zoology, 1967——. Mem. Biol. Soc. Washington, Marine Biol. Assn., (U.K.), Marine Biol. Assn. India. Research, publs. on distbn. and relationships of decapod and stomatopod crustaceans of tropical waters. Home: 2401 Jackson Pkwy., Vienna, Va. 22180. Office: Dept. Invertebrate Zoology, Smithsonian Instn., Washington 20560.*

MANNING, Winston Marvel, Am. chemist; b. Washington, Mar. 26, 1909; s. William Ray and Mabel (Marvel) M.; B.A., Am. U., 1930; M.S., Brown U., 1931, Ph.D., 1933; m. Dorothy Bauch, June 25, 1935; children—Stephen D., Joan E. Research fellow Brown U., 1933-34; research asst. U. Wis., 1934-36; research asso. U. Wis., 1936-41; phys. chemist Carnegie Inst. Washington, Stanford, Cal., 1941-43; dir. chem. div. Argonne (Ill.) Nat. Lab., 1946-65, asso. lab. dir., 1965——, acting dir., 1967. Mem. Am. Nuclear Soc., Am. Chem. Soc., Sigma Xi. Author: (with J. J. Katz, G. T. Seaborg) The Transuranium Elements, 1949; also numerous articles. Research on photochemistry in green plants, role yellow pigments as well as green pigments in photosynthesis, chem. properties plutonium, new isotopes heavy elements and their properties, application chem. properties transuranium elements to devel. processes for their recovery and purification, Home: 5524 Carpenter St., Downers Grove, Ill. 60515. Office: 9700 S. Cass Av., Argonne, Ill. 60439.*

MANOUVRIER, Léonce-Pierre, French physician, anthropologist; b. Guéret, France, 1850. Dir. anthropology lab. École des Hautes Études; formerly asst., later prof. École d'Anthropologie; dir. physiol. sta. Collège de France. Research on brain and its functions, also work on philosophy of sci. Died Paris, France, 1927.

MANOWITZ, Bernard, Am. chem. engr.; b. Jersey City, Mar. 6, 1922; s. David and Mary (Segall) M.; B.S. in Chem. Engring., Newark Coll. Engring., 1943;

M.S. in Chem. Engring., Columbia, 1947; m. Adele Zena Sklar, July 14, 1951; children—Douglas C., Gary E. Chem. engr. Oak Ridge Nat. Lab. (formerly Clinton Labs.), 1944-46; sr. chem. engr., head radiation div. nuclear engring. dept. Brookhaven Nat. Lab., Upton, L.I., N.Y., 1947—. Fellow Am. Nuclear Soc. (past dir. isotope and radiation div.); mem. Am. Inst. Chem. Engrs. (past chmn. nuclear engring. div.), Phi Lambda Upsilon. Author: (with John H. Lawrence, Benjamin S. Loeb) Radioisotopes and Radiation, 1964; also numerous articles. Editor: Internat. Jour. Applied Radiation and Isotopes, 1955—. Design of systems for disposal of radioactive wastes; research on behavior of entrainment in evaporators; devel. radioisotope radiation sources, methods of irradiator design, radiation induced chem. processes. Home: 216 Lakeview Av. E., Brightwaters, N.Y. 11718. Office: Brookhaven Nat. Lab., Upton, N.Y. 11973.*

MANRIQUE, Jorge, Argentinian physician; b. Mendoza, Argentina, Sept. 12, 1921; s. Francisco and Elvira (Romero) M.; M.D., U. Buenos Aires Med. Sch., 1946; m. Nivea Esther Sobredo, Sept. 11, 1946; children—Jorge Luis, Maria Graciela. Asst. surgery Hosp. de Clinicas, Buenos Aires, Argentina, 1946-56; asst. in medicine Peter Bent Brigham Hosp., fellow Harvard Med. Sch., Boston, 1956-57; 2d chief surg. service Policlinico San Martin, Buenos Aires, 1958—; faculty Buenos Aires Med. Sch., 1961—, asst. prof. surgery, 1961—. Pres. com. control and evaluation residency Nat. Pub. Health Sec. of State, 1966—; mem. Nat. Commn. Med. Residencies, 1967—. Recipient Gold medal Buenos Aires Med. Sch., 1945; Arce prize for best surg. paper, 1954. Fellow A.C.S.; mem. Argentine Med. Assn. (pres. com. residency control and evaluation 1966—), Argentine Coll. Surgeons (gen. sec. 1964), Argentine Surg. Assn., Argentine Soc. Surgeons, Argentine Soc. Gastroenterology. Author: Internal Medium on Surgical Practice, 1956; Upper Gastrointestinal Hemorrhages, 1967; also numerous articles. Research on surg. shock, diagnosis, ethiology, mgmt. upper gastrointestinal hemorrhage, ethical and tech. aspects of profl. surg. activities. Home: 1330 11° B/Uruguay, Buenos Aires. Office: 1789 Charcas, Buenos Aires, Argentina.*

MANSA, Johan Ludvig, Danish mech. engr.; b. Frederiksberg, Denmark, June 17, 1901; s. Harald Hoff and Antonia (Jessen) M.; Mech. Engr., Tech. U., Copenhagen, Denmark, 1925; Dr.-Ing., Technische Hochschule, Karlsruhe, Germany, 1929; m. Rigmor Charlotte Jensen, Jan. 16, 1934; children—Anette Juul, Soren, Birgitte Brorsen. Designer diesel engine dept. Burmeister & Wain, Ltd., Copenhagen, 1925-27; research engr. Technische Hochschule Karlsruhe, 1927-29; designer diesel engring. dept. Ingersoll-Rand Co., Phillipsburg, N.J., 1929-32, Elsinor Shipyard (Denmark), 1932-33; prof. mech. engring. Tech. U., Trondheim, Norway, 1932-40, Tech. U., Copenhagen, 1941-53, 66—; dir. Titan Co., Copenhagen, 1954-66. Decorated Ridder of Dannebrog. Mem. Danish Engring. Assn., Instn. Mech. Engrs. (London). Author: Maskinteknisk Varmekere, 1942; Maskinteknisk Termodynamik, 1951; also articles. Research on damping of vibration in machinery; improvement of efficiency of steam engines and diesel engines; art of directing research. Home: 20 Ordrupdalvej, Charlottenlund, Denmark. Office: Bldg. 424, DTH, Lyngby, Denmark.*

MANSCHOT, Willem Arnold, Dutch ophthalmologist; b. Winterswijk, Netherlands, Jan. 3, 1915; s. Gerrit Willem and Sophia J. (Hogeweg) M.; M.D., U. Amsterdam, 1940; m. Wilhemina Gertrude Leupen, Aug. 22, 1940; children—Wilhelmina G., Dirk, Hannie, Jan. Practice medicine specializing in ophthalmology, Rotterdam, Netherlands, 1945—; head ophthalmic pathology dept. Municipal Hosps., Rotterdam, 1960—. Recipient Treacher Collins prize Ophthal. Soc. U.K., 1960. Mem. European Ophthalmic Pathology Soc. (pres.). Research, numerous publs. on collagen diseases, subarachnoid haemorrhage, expulsive haemorrhage, retrolental fibroplasia, hyperplastic primary vitreous, toxoplasmosis, 13-15 trisomy, Coat's disease, uveal melanoma. Address: 157 Nieuwe Binnenweg, Rotterdam 3, Netherlands.*

MANSFIELD, George Rogers, Am. geologist; b. Gloucester, Mass., Aug. 30, 1875; s. Alfred and Sarah Jane (Hubbard) M.; student Mass. Agr. Coll. (now U. Mass.), 1893-94, B.S., Amherst Coll., 1897, M.A., 1901; A.M., Harvard, 1904, Ph.D., 1904; m. Adelaide Claflin, Aug. 18, 1903; children—Harvey Scott, James Scott, Robert Hubbard, Marion Claflin (Mrs. George Patterson III), Helen Rosero (Mrs. John Carroll). Research and publs. on phosphates in Ida., Wyo. and Fla., nitrates in Ida. and Cal., potash in green sands of N.J., potash in salt beds of Tex. and N.M., coal in Eastern Ida., sulphur in N.M., also stratigraphy and structure Rocky mountains S.E. Ida. Died 1917.

MANSFIELD, Wendell Clay, Am. geologist; b. N.Y., June 9, 1874; s. Curtis and Huldah (Phillips) M.; grad. Fredonia Normal Sch., 1898; B.S., Syracuse U., 1908; M.A., George Washington U., 1913, Ph.D., 1927; m. Katherine Gibson, 1919 (dec. 1935). Instr., Dansville (N.Y.) High Sch., 1908-10; joined staff Geol. Survey, 1910, became asst. geologist, 1917, then geologist; field work Tullock coal field, Atlantic Coastal Plain. Fellow Geol. Soc.; mem. Paleont. Soc., Washington Acad. Scis. Research and publs. on marine deposits, fauna, stratigraphy; 1st to show extension of Yorktown formation inland to Petersburg, Va.; described faunas of upper Miocene Choctowhatclee formation, Fla.; studied relations between Duplin formation of Carolinas and Yorktown formation; Tampa and Suwannee limestones. Died 1939.

MANSKE, Richard Helmuth Fred, chemist; b. Berlin, Germany, Sept. 14, 1901; s. John August and Augusta (Wruck) M.; B.Sc., Queens U., Kingston, Ont., Can., 1923, M.Sc., 1924; Ph.D., Manchester (Eng.) U., 1926, D.Sc., 1937; D.Sc., McMaster U., Hamilton, Ont., 1960; m. Jean B. Gray, Sept. 12, 1924; (dec.); children—Barbara Jean (Mrs. Hugh R. MacCallum), Corydalis C. (Mrs. Richard J. C. Burgener). Research fellow Yale, 1927-30; research asso. NRC Can., 1930-43; dir. research Dominion Rubber Co., Ltd., Guelph, Ont., 1943-67; sr. lectr. U. Waterloo (Ont.), 1968—. Fellow Chem. Inst. Can. (medal 1959, past pres.); mem. Am. Chem. Soc., London Chem. Soc., Royal Soc. Can. Editor: The Alkaloids, 10 vols., 1950-65; also numerous articles. Research on chemistry natural products, isolation structure, synthesis, mechanisms reactions. Patentee chems. Home: 27 Chadwick St. Office: 120 Huron St., Guelph, Ont., Can.*

MANSKI, Wladyslaw Julian, immuno-chemist; b. Lwow, Poland, May 15, 1915; s. Julian and Helena (Lewicka) M.; M.Philosophy in Chemistry, U. Warsaw (Poland), 1939; D.Sc., U. Wroclaw, 1951; m. Anna Z. Artymowicz, June 20, 1941; children—Chris, Louis. Came to U. S., 1958, naturalized, 1963. Instr., U. Warsaw, 1936-39; research asso. Inst. Lwow, 1940-41, Inst. Agr., Pulawy, Poland, 1943-44; instr. U. Lublin (Poland), 1944-45; docent Inst. Immunology, Wroclen, Poland, 1945, Inst. Biochemistry, Polish Acad. Sci., Warsaw, 1954; Rockefeller fellow, U.S.A., 1949-50; head immunochemistry Inst. Immunology and Exptl. Therapy, Polish Acad. Sci., Wroclaw, 1951-55, head macromolecular biochemistry, Polish Acad. Sci., Warsaw, 1955-57; head biochemistry lab. State Inst. Hygiene, Warsaw, 1955-57; asst. prof. microbiology, faculty Columbia Coll. Phys. and Surg., 1958—. Mem. Assn. Am. Immunologists, Brit. Biochem. Soc., Harvey Soc., Transplantation Soc., A.A.A.S., Am. Chem. Soc. Research, numerous publs. on immunochemistry as applied to blood group substances, molecular evolution of lens proteins comparative serology, organ specificity and autoimmunity, hypersensitivity reaction by particulate and soluble forms of same antigen. Home: 215 W. 90th St., N.Y.C. 10024.*

MANSON, Numa, physicist; b. Ekaterinoslav, Russia, Oct. 26, 1913; s. Peter and Maria (Levin) M.; Bachelor, U. Paris, 1931, Licencié es Sciences, 1939, Docteur es Sciences, 1946; m. Madeleine Weber, Sept. 27, 1938; 1 dau., Dominique. Technician, Institut de Sondure Autogène, Paris, 1935-36, 41-43; research engr. French Petroleum Inst., Sta. Cl. Bonnier, Belvue, France, 1943-54; prof. U. Poitiers (France), 1954—, dean Faculty Scis., 1963—; prof. Faculty Scis. and Ecole Nationale Supérieur de Mecanique et d'Aerotechnique, 1954—, dir., 1962-64. Mem. Nat. Com. for Sci. Research, 1960-66. Recipient Prix des Laboratories de la French Acad. Scis., 1959; decorated chevalier la Legion d'Honneur, 1967. Mem. French Astronautical Soc. (pres., founder Poitiers sect.), French Soc. Physics, French Soc. Chemistry, Combustion Inst. Author: Propagation des detonations et des deflagrations, 1947; Equilibres physico-dubuques dans les gaj a haute températures; Applications, 1954, (with G. Ribaud), 1958; also numerous articles. First quantitative theory of spinning detonations, 1946; complete theory of inverse method for stable self-sustained detonations, 1958; research on instability of self-sustained detonations in gases, thermodynamic properties of gases at high temperatures. Home: Les Marronniers, 86-Migné-Auxances, France. Office: Faculté des Sciences 86-Poitiers, France.*

MANSON, Sir Patrick, physician, parasitologist; b. Aberdeenshire, Scotland, Oct. 3, 1844; s. John and Elizabeth (Blaihie) M.; M.B., Aberdeen U., 1865, C.M., 1866, M.D., LL.D., 1886; D.Sc., Oxford (Eng.) U., 1904; m. Henrietta Isabella Thurburn, 1875; 3 sons, 1 dau. Med. officer for Formosa to Chinese Imperial Maritime Customs, from 1866; pvt. practice medicine Hong Kong, 1883-89, ret., 1889; physician Seaman's Hosp. Soc., Scotland, from 1892; physician, med. adviser Colonial Office, from 1892. Founder med. sch. (became U. Hong Kong); aided in found. London (Eng.) Sch. Tropical Medicine, 1899; lectr. tropical diseases. Fellow Royal Soc., 1900; mem. French Acad. Scis. Author: Tropical Diseases: a Manual of the Diseases of Warm Climates, 1898; Lectures on Tropical Diseases, 1905; (with C. W. Daniels) Diet in the Diseases of Hot Climates, 1908. Research in removal of massive tumors of elephantiasis, causal relationship of filaria worms to elephantoid diseases, 1879; discovered organism filaria homines, 1st to describe tinea nigra (both 1898); 1st to give hypothesis (later confirmed by others) that mosquito was agent in spreading malaria; called father of tropical medicine. Died London, Apr. 9, 1922.

MANTEGAZZA, Paolo, Italian physician, anthropologist; b. Monza, Italy, Oct. 31, 1831; ed. univs. Pisa, Milan (Italy); M.D., U. Pavia (Italy), 1854. Traveled Europe, India, Am.; practiced medicine, Argentina, Paraguay; surgeon Milan (Italy) Hosp., from 1858; prof. gen. pathology at Pavia, from 1860; prof. anthropology Instituto di Studii Superiori, Florence, Italy, from 1870; founder Archivio per l'Antropologia e la Etnologia. Author: Fisiologia del Dolore, 1850; Elementi d'Igiene, 1875; Fisonomia e Mimica, 1883; Le Estasi Umane, 1887; Fisiologia dell' Amore, 1896. Founded 1st mus. anthropology and ethnology in Italy, Florence, also 1st lab. gen. pathology in Europe. Died San Terenzo, Italy, Oct. 28, 1910.

MANTELL, Gideon Algernon, English geologist, paleontologist; b. Lewes, Eng., Feb. 3, 1790; ed. pvt. sch., Wiltshire, Eng.; children—Walter, Joshua, 1 dau. Articled to surgeon, Lewes, later became partner; became lectr., Brighton, Eng., 1835; practiced medicine at Clapham, Eng. Recipient Wollaston medal. Fellow Royal Soc., 1825, Royal Coll. Surgeons (hon.); mem. Linnean Soc., Geol. Soc. (a sec. 1841-42, v.p. 1848-49). Author of 67 books, including The Fossils of the South Downs, 1822; The Geology of the South-East of England, 1833; The Wonders of Geology, 2 vols., 1838; also numerous articles. Research in geology and paleontology, particularly study of Waelden formations in Sussex, 1st to show freshwater origin of Wealden strata; discovered 4 out of the five genera of dinosaurs known at time of his death; discovered remains of Iguanodon, 1825. Died London, Nov. 10, 1852.

MANTEN, Adriaan, Dutch chemotherapist, microbiologist; b. Hilversum, Netherlands, June 22, 1918; s. D. and T. (Griffioen) M.; student U. Utrecht (Netherlands), 1937-43; Cand.ex., U. Delft (Netherlands), 1940, Drs.ex. cum laude, 1946, D.Sc. cum laude, 1948; m. Willemina J. J. Wever, Aug. 29, 1951; children—Fredrike J. C., Joanne W. Asst., U. Utrecht, 1944-47; microbiologist to exptl. sta., Groningen, Utrecht, 1947-48; research officer Nat. Inst. Pub. Health, Utrecht, 1950-57, head dept. chemotherapy, 1957—. Cons. med. microbiology to hosps. Internat. Soc. Chemotherapie, 1963; mem. expert com. WHO, 1962—. Recipient Dr. Saal van Zwanenberg award, 1965. Fellow Am. Coll. Chest Physicians. Editor: (with P. de Somer) Nieuwe Antibiotica, 1966. Research, publs. on genetics aspects of drug resistance in microorganisms, interaction between antibiotics, typing of anonymous mycobacteria by their susceptibilities to anti-Tb drugs. Home: 11 Julianalaan, Hilversum, Netherlands. Office: 1 Sterrenbos, Utrecht, Netherlands.*

MANTER, Harold Winfred, Am. parasitologist; b. Anson, Me., June 18, 1898; s. Fred Augustus and Augusta (Tinkham) M.; B.A., Bates Coll., 1922; M.A., U. Ill., 1923, Ph.D., 1925; m. Esther Ruby Welch, Aug. 16, 1927. Instr., La. State U., Baton Rouge, 1925-26; faculty U. Neb., Lincoln, 1926—, prof. parasitology, 1935—, chmn. dept. zoology, 1953-61; guest investigator Tortugas Lab., Carnegie Instn., 1930-32. Cons. NSF, 1960-63, NIH, 1956-60. Fulbright scholar, New Zealand, 1951. Fellow A.A.A.S.; mem. Am. Soc. Parasitologists (past pres.), Am. Micros. Soc. (past pres.), Am. Soc. Zoologists, Soc. Systematic Zoology, Phi Beta Kappa, Sigma Xi. Author: (with Dwight D. Miller) Introduction to Zoology, 1959; also numerous articles, monographs. Research on collection, description, and classification many parasites marine fishes, geog. distbn. parasites fishes and its significance. Home: 1300 N. 41st St., Lincoln, Neb. 68503.*

MANTOUX, Charles, French physician; b. 1877; ed. Paris, France; lived nr. Cannes, France. Research on Tb; developed intracutaneous tuberculin test (Mantoux test), 1908; prodn. artificial pneumothorax, pictures of cavities in lungs. Died 1947.

MANUILA, Alexandre, physician; b. Cluj, Rumania, June 14, 1921; s. Sabin and Veturia (Leucutia) M.; ed. U. Lausanne (Switzerland), U. Geneva; M.D., Ph.D.; m. Ludmila Psenit, Sept. 24, 1944; children —Radu, Ruxandra. Research in biology and anthropology, Switzerland, Rumania, 1947; editor in chief tech. bull., staff chronicles, reports, bulls. WHO. Recipient Claparède prize U. Geneva, 1957. Mem. Interat. Inst. Sociology, Swiss Soc. Natural Sci. Author: Dictionnaire francais de médecine et de biologie, 1966; Enzyklopedie der medizinischen Fachausdrucke, 1966. Research and publs. in phys. anthropology, data processing and information sci., blood groups, med. lexicography. Home: 8 ch. Falletti. Office: Palais des Nations, Geneva, Switzerland.

MANWELL, Reginald Dickinson, Am. zoologist; b. Harford, Pa., Dec. 24, 1897; s. John Parker and Stella (Dickinson) M.; grad. Deerfield Acad., 1914; A.B., Amherst Coll., 1919; M.A., Johns Hopkins, 1926, Sc.D., 1928; Sc.D. (hon.), Syracuse U., 1963; m. Elizabeth Skelding Moore, Aug. 6, 1930 (dec. Oct. 1964); children—John Parker, Henry Dickinson. Tchr., St. Charles (Mich.) High Sch., 1920-21; prin. Waterport (N.Y.) Union Sch., 1921-24; prof. biology W.Va. Wesleyan Coll., 1928-29; instr. protozoology John Hopkins Sch. Hygiene and Pub. Health, 1929-30; prof. zoology Syracuse U., 1930-63, prof. emeritus, research asso., 1963—. Summer faculty Rocky Mountain Biol. Lab., Crested Butte, Colo., 1929, 31, 48, 50, 52, 54, trustee, 1956—, v.p., 1960-62, pres., 1962—; vis. prof. zoology Mountain Lake Biol. Sta., U. Va., summer 1962; cons. blood parasites USN, 1961—. Mem. Soc. Protozoologists (treas. 1951-54, pres. 1962-63), Soc. for Exptl. Biology and Medicine, Marine Biol. Lab. (Woods Hole,

Mass.), Am. Soc. Zoologists, Am. Soc. Parasitologists, Am. Soc. Naturalists, Am. Soc. Tropical Medicine and Hygiene, Sigma Xi (Annual Research award Syracuse chpt. 1959), Delta Tau Delta, Kappa Theta. Author: (with Sophia L. Fahs) The Church Across the Street, 1946, rev., 1963; (with P. F. Russell, L. S. West, G. MacDonald) Practical Malariology, 1946, rev., 1963; Introduction to Protozoology, 1961. Bd. editors Jour. Parasitology, 1938-42, 63-—; Jour. Protozoology, 1954-57, Rivista di Malariologia (hon.). Research, publs. on cause and disease process of malaria in birds, bats, other parasites, bird homing habits. Home; Hoag Lane, Fayetteville, N.Y. Office: Lyman Hall, Syracuse U., Syracuse, N.Y. 13210.*

MANZO, see Ajima, Naonobu.

MANZON, Luigi, Italian agronomist; b. Agordo, July 29, 1888; s. Francesco and Anna (Coranlo) M.; Ph.D. in Agrarian Sci. With Agr. Tech. Inst., Conegliano, Italy, 1912-—, prof., later dean; prof. plant pathology U. Padua. Mem. Italian Acad. 'della vite e del vino, Turin Acad. Agr. Research and publs. on transpiration and absorption of water, anatomy and physiology of plants. Address: Istituto Tecnico Agrairio, Conigliano (Treviso), Italy.

MANZULLO, Alfredo, Argentinian immunologist; b. Buenos Aires, Argentina, Feb. 9, 1909; s. Pascual and Rose Maria (Santoianni) M.; B.A., Brit. Sch. Buenos Aires, 1927; Ph.D. in Vet. Sci., Nat. U. La Plata (Argentina), 1931, Ph.D. in Bacteriology, 1933; m. Damasia Cenzano Hagem, Sept. 12, 1936. Bacteriologist, Malbran Inst., Ministry Publ. Health, 1933-40, investigations chief, 1940-48; prof. microbiology Nat. U. La Plata, 1948-49, prof. immunology, 1949-63, hon. prof., 1963-—, dir. Inst. Applied Immunology, 1963-—; dir. tech. splty. Ministry Pub. Health and Social Assistance, Buenos Aires, 1949-52. Cons. Med. Sch. Contagious Diseases, Buenos Aires, 1959. Recipient award for best work Sch. Medicine Buenos Aires, 1949-50. Mem. Argentine Soc. Transmissable Diseases, Argentine Sci. Soc., Argentine, Latin Am. microbiol. socs., Argentine Soc. Immunology, Latin Am. Parasitological Soc., Argentine Soc. Leprosy. Author: (with J. Herran) Diphtheria, 1952; also numerous articles. Research on diagnosis in vitro and in situ of diphtheria, method of rapid diphtheria diagnosis, blood types in Indians of Tierra del Fuego, Brucellosis. Home: 410 Brasil, Buenos Aires, Argentina.*

MAPLE, Clair George, Am. mathematician; b. Glenwood, Ind., Mar. 17, 1916; s. Clair C. and Ora E. (Dunn) M.; B.A., Earlham Coll., 1939; M.A., U. Cin., 1940; D.Sc., Carnegie Inst. Tech., 1948; m. Louise Catron, Sept. 20, 1942. Instr. math. W.Va. Inst. Tech., 1940-41, Ohio State U., 1941-44; research asso. Carnegie Inst. Tech., 1946-48; faculty N. Tex. State U., 1948-49; faculty Ia. State U., 1949-—, prof. math., 1955-—, dir. computation center, 1963-—, chief math and computer sci. div. Ames Lab., 1966-—; cons. U. S. Govt., 1953-60, Zenith Radio, 1954-66. Mem. Am. Math. Soc., Math. Assn. Am., Am. Assn. Computer Machines Soc. Indsl. and Applied Math., A.A.A.S., Sigma Xi, Phi Kappa Phi, Pi Mu Epsilon. Author: (with D. L. Holl and B. Vinograde) Introduction to the Laplace Transform, 1959. Research and publs. primarily on partial differential equations. Numerical math., subscription TV, dynamics of potentially colliding aircraft, topics of interest to the Nat. Security Agy. Home: 333 Pearson Av., Ames, Ia. 50010.*

MAPLETON, Robert Allan, Am. physicist; b. San Francisco, June 25, 1910; B.S., Purdue U., 1948; M.A., Harvard, 1949, M.E., 1955; Ph.D. in Applied Math. Queen's U., Belfast, N. Ireland, 1966; m. Hazel Marie Powers, May 23, 1943; children—Thomas Robert, Georgiana Elizabeth. Research engr. Lab. for Electronics, Boston, 1950-52; research physicist AF Cambridge Research Labs., Bedford, Mass., 1952—. Indsl. cons. on solid and liquid ultrasonic delay lines, Boston, 1952-58. Guenter Loeser Meml. lectr., 1962. Mem. Sci. Research Soc. Am., Am. Inst. Physics. Analysis of ultrasonic wave propagation in solid delay lines; calculations on ionization of atomic gases by collisions; calculations on electron capture from atomic gases by protons; math. analysis of problems in quantum and classical theories of electron capture. Home: 31 Fairview Av., Reading, Mass. 01867. Office: 259 (CRUS), AFCRL, L. G. Hanscom Field, Mass. 01730.*

MAPOTHER, Dillon Edward, Am. physicist; b. Louisville, Aug. 22, 1921; s. Dillon Edward and Edith (Rubel) M.; B.S. in Mech. Engring., U. Louisville, 1943; D.Sc. in Physics, Carnegie Inst. Tech., 1949; m. Elizabeth Beck, June 29, 1946; children—Ellen Susan, Anne. Engr., Westinghouse Research Labs., East Pittsburgh, Pa., 1943-46; instr. Carnegie Inst. Tech., Pitts., 1946; faculty U. Ill., Urbana, 1949-—, prof. physics, 1959-—. Cons. AEC labs., Los Alamos, Oak Ridge, Argonne, Ill. Alfred P. Sloan fellow, 1958-61, Guggenheim fellow, 1960-61. Fellow Am. Phys. Soc.; mem. Sigma Xi. Research on ionic mobility in alkali halides, thermodynamic properties of superconductors, calorimetric study of critical points. Home: 808 S. Foley Av., Champaign, Ill. 61820. Office: Physics Dept., U. Ill., Urbana, Ill. 61801.*

MAPOTHER, Edward, Brit. psychiatrist; b. July 12, 1881; s. Edward Dillon Mapothei; ed. Univ. Coll. Sch. and Hosp.; M.D., U. London; m. Barbara Mary Reynolds. Prof. psychiatry U. London; physician, lectr. psychol. medicine, King's Coll. Hosp., London; cons. psychol. medicine Queen Alexandria Mil. Hosp., Milbank; chief cons. Ex-Services Welfare Soc.; late hon. psychiatrist Hosp. Epilepsy and Paralysis, Maida Vale; med. officer Long Grove Mental Hosp., Epsom, 1908-14; med. supt. Maudsley Hosp., of Ministry of Pensions, 1914-20; Norman Kerr lectr. Soc. Study Inebriety, 1938. Examiner psychol. medicine U. London, also Royal Coll. Physicians and Surgeons. Fellow Royal Coll. Physicians (mem. council; Bradshaw lectr. 1936), Royal Coll. Surgeons; mem. Royal Coll. Medicine (pres. sect. psychiatry 1933), Brit. Med. Assn. (v.p. sect. neurology and psychiatry 1934); corr. mem. Société Suisse de Psychiatrie, Sociedad de Neurologia y Psiquiatria. Author: The Schizophrenic Paranoid Series, vol. 1, Early Disorders, 1926; Assessment of Alcoholic Morbidity, 1929; Mental Symptoms Associated with Head Injury, 1937; The Physical Basis of Alcoholic Mental Disorders, 1939. Died Mar. 20, 1940.

MAQUENNE, Léon-Gervais-Marie, French chemist; b. Paris, Dec. 2, 1853; D.Sc., 1880; tchr. Chaptal Coll.; became asst. naturalist Mus. Natural History, 1880, later became prof. plant physiology. Mem. French Acad. Scis., 1904, Acad. Agr. Author: L'Azote atmosphérique et la végetation, 1891. Research on essential functions of plant life, especially lineage and determinants of principles, evaporation of water from plants, absorption of carbon dioxide by leaves, constitution of inositol and pinitol, synthesis of erythrose, transformation of farina. Died Paris, Jan. 19, 1925.

MARAINI, Giovanni, Italian physician; b. Tripoli, Lybia, Dec. 5, 1933; s. Giovanni Giocondo and Lidia (Frojo) M.; Laurea in Medicine, U. Torino (Italy), 1956, Specialization in cardiology, 1958; Specialization in Ophthalmology, U. Parma (Italy), 1962; m. Bruna Rossi, May 8, 1965; 1 son, Giovanni. Asst. med. clinic Torino U., 1956-60; asst. ophthal. clinic Parma (Italy) U., 1960-—, asst. prof. Orthoptic Sch., 1964-—, Postgrad. Sch. Ophthalmology, 1965-—. Mem. Italian Soc. Ophthalmology, Internat. Strabismus. Soc. Author: (with L. Pasino) La Visione Binoculare Anomala, 1965; also articles. Research on hematology including proliferative and maturation defect in leukemic cells, ophthalmology including physiopathology of sensorial anomalies in strabismus, biochem. studies on lens and retina. Home: 13 Duca Alessandro, Parma, Italy.*

MARAIS DE BEAUCHAMP, Charles Alfred Paul, French zoologist; b. Paris, Mar. 3, 1883; s. Arthur and Elisabeth (Nicard) Marais De B.; ed. Faculty Paris; M.D., Ph.D. Became asst. Faculty Sci., Paris, 1908; named instr. Faculty Sci., Dijon, France, 1919; became lectr. Strasbourg (France) Faculty Sci., 1922, prof. at large, 1927, titular prof. zoology, 1933, ret., 1945, now hon. prof. Recipient Order of Merit in Agr. Corr. mem. French Acad. Scis. Author: Les grèves de Roscoff, 1914; Lecons préliminaires de zoologie, 1925; also numerous articles in zoology. Home: 5, rue Stanislas, Paris 6. Office: Laboratoire d'Evolution des Etres organisés, 105, bd Raspail, Paris 6, France.

MARALDI, Giacomo Filippo, astronomer; b. Perinaldo, Italy, Aug. 21, 1665; lived in Paris, beginning 1687; 1st geographer to king. Mem. French Acad. Scis., 1694. Participated in survey for lengthening meridian nr. Bourges, France, also from Amiens to Dunkirk; discovered dark division (observed by Cassini) is line of demarcation between two of Saturn's rings; pointed out variability of a star in Hydra constellation, 1704; wrote star catalogue. Died Paris, Dec. 1, 1729.

MARALDI, Giovanni Domenico, astronomer; b. Perinaldo, Italy, Apr. 17, 1709; nephew of Giacomo Filippo Maraldi; astronomer Obs. of Paris; mem. French Acad. Scis., 1733; pioneer in calculation of orbits of comets; studied Jupiter and its satellites. Died Paris, Nov. 14, 1788.

MARAMOROSCH, Karl, virologist; b. Vienna, Austria, Jan. 16, 1915; s. Jacob and Stefanie Olga (Schlesinger) M.; M.S. magna cum laude in Entomology, Agrl. U., Warsaw, Poland, 1938; postgrad. Poly. U. Bucharest (Rumania), 1944-46; fellow Bklyn. Bot. Garden, 1947-48; Ph.D. (predoctoral fellow Am. Cancer Soc. 1948-49), Columbia, 1949; m. Irene Ludwinowska, Nov. 15, 1938; 1 dau., Lydia Ann. Came to U. S., 1947, naturalized, 1952. Civilian internee in Rumania, 1939-46; asst., then asso. Rockefeller Inst., N.Y.C., 1949-61; sr. entomologist Boyce Thompson Inst., Yonkers, N.Y., 1961-—, program dir. virology and insect physiology, 1962-—; vis. prof. agr. U. Wageningen (Netherlands), 1953, Cornell U., 1957; Mendel lectr. St. Peters Coll., Jersey City, 1963. Virologist, FAO to Philippines, 1960, cons. world-wide survey, 1963; chmn. U. S.-Japan Coop. Seminar, 1965. Recipient Sr. Research award Lalor Found., 1957. Fellow N.Y. Acad. Scis. (A. Cressy Morrison prize natural sci. 1951; chmn. div. microbiology 1956-60, recording sec. 1960-61, v.p. 1962-63), A.A.A.S. (Campbell award 1958); mem. Harvey Soc., Growth Soc., Phytopath. Soc. (councillor), Entomol. Soc. Am., Entomol. Soc. Japan, Japan, Indian, Canadian phytopath. socs., Internat. Com. Virus Nomenclature, Electron Microscope Soc., Sigma Xi. Author: Comparative Symptomatology of Coconut Diseases of Unknown Etiology, 1964. Editor: Biological Transmission of Disease Agents, 1962. Editor Methods of Virology, 1964-—; asso. editor Virology, 1964-—. Studied virus transmission by insects; electron microscopy of viruses in cells of animals and plants; plant virus tumors; virus diseases of plants; effects of viruses on hosts; tropical plant diseases of unknown causes. Home: 17 Black Birch Lane, Scarsdale, N.Y. 10585. Office: Boyce Thompson Inst., Yonkers, N.Y. 10701.*

MARAN, Stephen Paul, Am. astrophysicist; b. Bklyn., Dec. 25, 1938; s. Alexander P. and Clara (Schoenfeld) M.; B.S., Bklyn. Coll., 1959; M.A., U. Mich., 1961, Ph.D., 1964. Physicist, Inst. for Space Studies, N.Y.C., 1961-64; astronomer in charge, remotely controlled telescope Kitt Peak Nat. Obs., Tucson, Ariz., 1964-—. Bd. dirs. Jr. Astronomy Corp., N.Y.C., 1964-—. Fellow Royal Astron. Soc.; mem. Am. Astron. Soc., Am. Phys. Soc., International Astronomical Union. Editor: (with A. G. W. Cameron) Physics of Nonthermal Radio Sources, 1964; also articles. Developed 1st computer-controlled, remotely operated astron. telescope; research in measurement and explanation of phys. conditions in an exploding star. Home: Route 9, Box 998, Tucson 85705. Office: 950 N. Cherry Av., Tucson, Ariz. 85717.*

MARANON, Gregorio, Spanish pathologist; b. Madrid, Spain, May 19, 1888; s. Manuel and Carmen (Posadillo) M.; ed. U. Madrid; m. Dolores Moya, July 19, 1911; 5 children. Asst. to Ehrlich; prof., Madrid; founder, dir. Instituto de Patología. Author: La edad crítica, 1919; La evolución de la sexualidad y los estados intersexuales, 1930; Ginecología endocrina, 1935; Pathologie de l'hypophyse, 1943; Manual de diagnóstico etiológico, 1948; La enfermedad de Addison, 1949; El crecimiento y sus trastornes, 1953. Founder (with Falta, Pende) endocrinology; historian, biographer. Died Madrid, Mar. 27, 1960.

MARANTA, Bartolomeo, Italian botanist, physician; b. Venosa, Kingdom of Naples; student botany, mineralogy (with L. Ghini), Pisa, Italy. With bot. garden, Naples, 1554-56; attending physician to Vespusian Gonzaga, 1556, later to Cardinal Castiglioni of Trinità; developed garden in Rome, Italy, 1568. Author: De Aguae, 1559; Methodis Cognoscendorum Medicamentorum Simplicium, 3 vols., 1559; (with Ferrante Imperato) Natural History. Developed rare and esoteric herbs and plants, 1554-56, also original method of identifying location, phys. attributes, medicinal uses of herbs; worked on analysis of type of mineral water in Naples, 1558. Died Mar. 24, 1571.

MARBE, Karl, psychologist, philosopher; b. Paris, France, Aug. 31, 1869; s. August and Wilhelmine (Wagner) M.; ed. univs. Freiburg, Bonn, Berlin, Leipzig (all Germany); m. Milly Fries, Aug. 6, 1908. Docent, U. Würzburg (Germany), 1896-1901; prof. Psychol. Inst., from 1909; prof. philosophy, dir. Psychol. Inst. Frankfort (Germany) Acad., 1905-09. Author: Natur-Philosophie Untersuchungen zur Wahrscheinlichkeitslung 1899; Experimentell-Psychologische Untersuchungen über den Urteil, eine Einleitung in der Logik, 1901; Experimentische Uuntersuchungen über den Grundlage der Sprachlische Analogiebildung, 1901; Über den Rhythmus der Prosa, 1904; Die Gleichförmigkeit in der Welt, 1916; Theorie der kinematographischen Projektionen, 1910. Mem. Würzburg Sch. of imageless thought; made exptl. study of judgment. Died 1953.

MARBLE, Alexander, Am. physician; b. Troy, Kan., Feb. 2, 1902; s. Charles and Minnie I. (Dittemore) M.; A.B., U. Kan., 1922, A.M., 1924; M.D., Harvard, 1927; m. Beula Frances Becker, Sept. 20, 1930; children—Elizabeth Myers (Mrs. Richard E. Hartwell), Moseley Traveling fellow Harvard Med. Sch., Austria, Germany, Eng., 1931-32; practice internal medicine, Boston, 1932-—; clin. prof. medicine Harvard Med. Sch., 1966-—; med. dir. Joslin Clinic; staff New Eng. Deaconess Hosp., Boston; sr. asso. in medicine Peter Bent Brigham Hosp., Boston. Cons., VA, 1946-—, USPHS, 1946-—. Pres., Diabetes Found., Inc. Recipient award for Distinguished Service, U. Kan., 1959. Fellow A.C.P.; mem. Am. Soc. Clin. Investigation, Assn. Am. Physicians, Am. Diabetes Assn. (past pres.; Banting Meml. lectr. 1967), Endocrine Soc., A.M.A., A.A.A.S., Assn. Mil. Surgeons. Author: Treatment of Diabetes Mellitus (with Joslin, Root, White), 10th edit. 1959; (with Cahill) Chemistry and Chemotherapy of Diabetes, 1962; also numerous articles. Research on non-diabetic meliturias, action various types insulin and oral hypoglycemic agts., origins and early stages diabetic state, relations control diabetes to nature and extent vascular complications. Home: 131 Laurel Rd., Chestnut Hill, Mass. 02167. Office: 15 Joslin Rd., Boston 02215.*

MARBRES, see Canonicus, Joannes.

MARBURG, Otto, physician; b. Roemerstadt, Austria, May 25, 1874; s. Max and Adele (Berg) M.; came to U. S., 1838, naturalized; M.D., U. Vienna (Austria), 1899; m. Malvine Knoepflmacher, Sept. 5, 1916. Pvt. docent U. Vienna, 1905-12, title prof., 1912-16, real prof., 1916-38, chief of neurol. inst.,

1919-38; clin. prof. neurology Columbia, since 1938. Hon. mem. Am. N.Y., Phila. neurol. assns., A.M.A. Author: Mikroskopisch-topographischer Atlas des menschlichen Zentralnerven-systems, 1904 (3d edit. 1927); Die sogenannte akute multiple Sklerose, 1906; Syphilis des Nervensystems (with I. A. Hirschl), 1914; Handbuch der Neurologie des Ohres (with Alexander and Brunner), 1924-26; Die Roentgenbehandlung der Nervenkrankheiten (with Sgalitzer), 1931; Unfall & Hirngeschwulst, 1934; Injuries of the Nervous System, 1939; Hydrocephalus, Its Symptomology, Pathology, Pathogenesis and Treatment, 1940. Editor "Arbeiten aus Dem Neurologischen Institut der Universitaet Wien" from 1919. Contbr. numerous med. articles. Described leukoencephalitis concentrica (encephalomyelitis periaxialis schleroticans), 1906. Died June 13, 1948.

MARBUT, Curtis Fletcher, Am. agriculturist, geologist; b. Lawrence County, Mo., July 19, 1863; s. Nathan T. and Jane (Browning) M.; B.S., U. of Mo., 1889; A.M., Harvard, 1895; studied in Europe, 1899-1900; LL.D., U. of Mo., 1916; m. Florence Martin, Dec. 17, 1891 (dec.); children—Louise, Thomas Fiske, William Martin, Helen, Frederick Browning. Inst. geology and mineralogy, 1895-97, asst. professor, 1897-99, prof. and curator Geol. Mus., 1899-1913, U. of Mo.; dir. Soil Survey of Mo., 1905-13; spl. agt. Bur. of Soils, U. S. Dept. Agr., 1909-10; in charge Soil Survey, U. S. Dept. Agr., 1910——. Author: Soils of the United States, 1935; also numerous articles in Atlas of Am. Agr., others. Directed mapping of soils of U. S., studied soils of Africa; joint-work on an internat. system of soil study, pedology. Died Aug. 25, 1935.

MARCELLI, Francesco, Italian gynecologist; b. Rome, Italy, July 14, 1930; s. Umberto and Caterina (Pargani) M.; ed. in medicine and surgery U. Rome, 1958-59, obstetrics and gynecology, 1963; m. Blasi Michela, Oct. 31, 1953; children—Vittoria, Flavio. Extraordinary asst. Obstet. and Gynecol. Clinic, U. Rome. Mem. Theatine Acad. Sci. Research, publs. on conjugal sterility. Home: 30, via Pittaco (Casalpalocco), Rome. Office: 2, via Pomponio Leto, Rome, Italy.*

MARCET, Alexander (John Gaspard), physician, chemist; b. Geneva, Switzerland, 1770; M.D., Edinburgh, 1797; m. Jane Haldimand, Dec. 4, 1799; children include Mrs. Edward Romilly. Physician, chem. lectr. Guy's Hosp., London, Eng., 1804; placed in charge temporary mil. hosp., Portsmouth, Eng., 1809; named hon. prof. chemistry, Geneva, 1819. Fellow Royal Soc., 1808. Author: An Essay on the Chemical History and Medical Treatment of Calculous Disorders, 1817; also essays on chemistry. Probably 1st to suggest pain of renal calculus is usually caused by its passage down a ureter, also that it may grow in kidney without acute pain; research in chemistry. Died London, Eng., Oct. 19, 1822.

MARCET, Jane Haldimand, physician, chemist; b. Geneva, Switzerland, 1785; d. Anthony Francis Haldimand; m. Alexander Marcet, Dec. 4, 1799; children include Mrs. Edward Romilly. Author: Conversations on Chemistry, 1805; Conversations in Political Economy, 1816; Conversations on Natural Philosophy, 1820. Wrote popular textbooks at a time when simple sci. lit. was almost unknown. Died June 28, 1858.

MARCET, William, physician; b. Geneva, Switzerland, 1829; s. Francois Marcet; student Geneva U.; M.D., Edinburgh U., 1850; asst. physician Westminster Hosp., Hosp. for Consumption and Diseases of the Chest, Brompton, Eng.; fellow Royal Soc., Royal Coll. Physicians; pres. Royal Meteorol. Soc., 1857. Author: Southern and Swiss Health Resorts, 1883; A Contribution to the History of the Respiration of Man, 1897. Described presence of coprostanol (reduced form of cholesterol, in normal feces, 1858. Died Mar. 4, 1900.

MARCGRAF, Georg, naturalist; b. Liebstadt, Saxony, 1610. Author: (with W. Piso) Historia naturalis Braziliae, 1648. Introduced ipecacuanha from Brazil; made important topog., meteorol., astron. studies; observed eclipse of 1640. Died Guinea, 1644.

MARCH, Arthur, physicist; b. Brixen, Tyrol, Feb. 23, 1891; prof., Innsbruck, Austria, also Oxford, Eng. Author: Die Grundlagen der Quantenmechanik, 1930; Einführung in die moderne Atomphysik, 1933; Der Weg des Universums, 1948. Introduced into quantum theory concept of minimum span as constant which exists in nature and limits measurement possibility. Died Innsbruck, 1957.

MARCH, Hans, German neurologist, psychiatrist; b. Bin-Charlottenburg, Germany, June 14, 1895; s. Otto and Maria (Vorster) M.; ed. U. Berlin, U. Marbourg (Germany); M.D.; m. Katharina Plinzner, Oct. 31, 1926; children—Brigitte, Wolfgang. Asst., U. Greifswald Neurol. Clinic; prof. medicine, Landesversicherungsanstalt Berlin; chief physician various depts. neurology Municipal Hosp.; ind. practice medicine. Recipient Red Cross medal. Mem. Internat. Soc. Psychoanalysis. Author: Der religiöse Sinn der sexuellen Krise, 1930; Psychologische Seelsorge, 1930; Lebensschicksale in psychiatrischen Gutachten, 1959; Kernfragen des Leben, 1953; Verfolgung und Angst, 1960;

Research in psychosomatic illness. Address: Hohenzollerndamm 83, Berlin 33, West Germany.

MARCH, Norman Henry, Brit. physicist; b. Coalville, Eng., July 9, 1927; s. William Henry and Elsie (Brown) M.; B.Sc. with Spl. Physics Class I, King's Coll., London U., 1948, Ph.D., 1951; m. Margaret Joan Hoyle, Apr. 23, 1949; children—Peter Henry, Anthony John. Successively lectr., reader U. Sheffield (Eng.), now prof. physics, joint head physics dept.; vis. prof. lab. atomic and solid state physics Cornell U., 1965-66. Fellow Phys. Soc. London; mem. Am. Phys. Soc. Author: (with W. H. Young, S. Sampanthar) The Many-Body Problem in Quantum Mechanics, 1967; also numerous articles. Research in many-body problem, especially electrons in solid and liquid metals; behaviour of electrons in molecules; discovered new features in forces between ions in liquid metals and originated theory of melting of metals. Home: 25 Clarendon Rd., Sheffield, Yorkshire, Eng.*

MARCH, Ralph Burton, Am. entomologist; b. Oshkosh, Wis., Aug. 5, 1919; s. Albert Harold and Vanita (Siewert) M.; student Wis. State Coll., Oshkosh, 1937-38; A.B., U. Ill., 1941, A.M., 1946, Ph.D., 1948; m. Robinetta Tompkin, Dec. 26, 1942; children—John Stephen, Janice Allyn, Susan Edith. Mem. faculty U. Cal. at Riverside, 1948—, dean grad. div., 1961-68, prof. dept. entomology Coll. Biol. and Agrl. Scis., 1961——. Mem. effects of atomic radiation com. Nat. Acad. Scis.-NRC, 1955-62; mem. study sect. toxicology USPHS-NIH, 1958-62. Mem. Entomol. Soc. Am., Am. Chem. Soc., A.A.A.S., Sigma Xi, Phi Beta Kappa, Phi Kappa Phi. Research and publs. on the physiology, biochemistry and toxicology of insecticides and the devel. of resistance by insects to insecticides. Home: 3522 Mt. Vernon Av., Riverside, Cal. 92507.*

MARCHAL, Émile-Jules-Joseph, Belgian cytologist; b. Maeseyck, Belgium, Apr. 10, 1871; s. Éliè and Juliette Hoton Marchal; ed. Grembloux Agrl. Inst. 1888-91; m. Sophie Bergh, 1897. From asst. to prof. botany Gembloux Agrl. Inst., 1904; named dir. Phytopathology Inst., Gembloux; Mem. French Acad. Scis, 1933. Author: Éléments de pathologie végetale, 1925; Éléments de physiologie végétale, 1926. Studied prodn. ammoniac in soil; publs. on vegetable genetics, plant diseases and plant physiology. Died Brussels, Belgium, Nov. 17, 1954.

MARCHAL, Paul-Alfred, French biologist; b. Paris, Sept. 27, 1862; M.D.; docteur-ès-sciences; prof. zoology Nat. Agronomic Inst.; dir. Paris Entomol. Lab. Mem. French Acad. Scis., Scis., 1912, Acad. Agr. Author: l'Entomologie appliquée en Europe, 1897; les Coccinellides nuisibles, 1895. Research in marine zoology, reprodn. wasps, including discovery of phenomenon of nutritional castration and polyembryology, agrl. parasites; advocated use of antagonism between species for control agrl. pests; established 5 entomol. labs., France. Died Paris, Mar. 2, 1942.

MARCHAL, Raymond, French aero. engr.; b. Paris, Mar. 11, 1910; s. Hippolyte and Lucie (Guyou) M.; ed. École Polytechnique; Nat. Sch. Advanced Aeros.; Inst. Advanced Studies for Nat. Defense; m. Denise Massenet, Jan. 7, 1935; children—Claude, Philippe, Francoise, Bruno, Claire, Marie-Henriette. Tech. dir. Nat. Soc. for Study and Constrn. of Aeromotors, 1945-54, dir. atomic div., 1956-64, sci. dir., from 1964; hon. prof. Nat. Sch. Advanced Aeros.; prof. Nat. Inst. Nuclear Scis.; prof. thermodynamics Nat. Sch. Rural Engring. Recipient Aeros. medal. Author: Moteurs d'Avions; La Thermodynamique; also articles. Research on perfection of numerous aeromotors, pulse jet engines and jet deflectors, also on core of exptl. reactor at Cadarache Testing Center, variable pitch nuclear blowing engines, turbomolecular vacuum pump. Address: S.N.E.C.M.A., 22 quai Gallieni, Suresnes, France.*

MARCHAND, Felix Jacob, German pathologist; b. 1846; Ph.D., Marburg, Germany; prof. gen. pathology and path. anatomy, dir. Path. Inst., Leipzig (Germany) U. Probably 1st to describe pathology of carotid body tumors, 1891. Died 1928.

MARCHAND, Marcel Pascal, French physician, health officer; b. Villereau, France, Nov. 19, 1901; s. Pierre and Marie-Louise (Laine) M.; ed. U. Lille (France); M.D.; m. Amélie Alphant, Apr. 14, 1928. Prof. legal medicine Faculty Medicine, Lille; med. insp. for laborers in No. region; med. expert Communauté économique européenne. Author 5 books on hygiene, security of work; also numerous reports. Decorated laureate Acad. Medicine; Order of Palms Acad.; Order Pub. Health. Mem. Sci. Soc. for Legal and Work Medicine. Home: 12, rue de Tenremonde, Lille (Nord), France.

MARCHANT, Jean, French botanist; b. 1650; s. Nicolas Marchant; dir. cultivation Royal Bot. Gardens; mem. French Acad. Scis. Author: Memoires; Dissertation sur la préférence que nous devons attacher aux plantes de notre pays, par-dessus les plantes étrangères 1701; also papers on mosses. Developed theory of formation of species; named (in honor of his father) common liverwort Marchantia. Died Paris, Nov. 11, 1738.

MARCHANT, Nicolas, French botanist; flourished Paris; M.D., Padua, Italy; 1 son, Jean. Dir. bot. garden Gaston d'Orleans; mem. French Acad. Scis. (a founder 1666). Marchantia, genus of liverworts, named in his honor. Contbg. author: Description des plantes, 1676. Died Paris, 1678.

MARCHETTI, Andrew A., Am. obstetrician, gynecologist; b. Richmond, Va., July 2, 1901; s. Louis and Bianca (Iaccheri) M.; student Johns Hopkins, 1920-22, M.D., 1928; A.B., U. Richmond, 1924; m. Catherine E. Fopeano, Jan. 2, 1935; children—Marco Anthony, Peter Luigi, Michael Joseph, John Philip. Instr., later asso. prof. obstetrics and gynecology Med. Coll. Cornell U., 1933-47; prof. obstetrics and gynecology Georgetown U. Sch. Medicine, 1947——; civilian cons. Army Med. Center, Walter Reed Hosp. Diplomate Am. Bd. Obstetrics and Gynecology (dir., pres.). Fellow Am. Coll. Obstetricians and Gynecologists (1st v.p. 1958-59), Am. Assn. Obstetricians and Gynecologists, Am. (sec. 1957-61; pres. 1966-67), Washington (pres. 1952-53) gynecol. socs., A.C.S.; mem. Soc. Pelvic Surgeons, A.M.A., N.Y. Obstet. Society, Med. Soc. D.C., Sigma Xi, Alpha Omega Alpha, Kappa Sigma, Alpha Kappa Kappa. Co-author: The Epithelia of Woman's Reproductive Organs, 1948. Research in pathology of ovarian tumors; coworker in vaginal cytology; collaborator in new approach to surg. correction of urinary stress incontinence in the female. Home: 3495 S. Leisure World Blvd., Silver Spring, Md. 20906.* Office: Georgetown U. Hospital, Washington 20007.*

MARCHETTI, Mario, Italian physicist; b. Massarosa, Italy, July 29, 1898; s. Vincenzo and Alduina (Gori) M.; Dr.engring.; m. Paolina Martinelli, Oct. 16, 1930; children—Marco, Anna. Dir., Hydraulics Inst., Milan (Italy) Poly. Coll.; prof. hydraulic constrns. Cons., Pub. Works Com., Fire Fighting Service. Recipient Medal of Merit for teaching, culture and arts. Mem. Lombardy Inst. Sci. and Letters, Italian Acad., Tessin Soc. (pres.). Author 3 vols. on hydraulics; also articles. Home: via A. Grossich 17, Milan, Italy. Office: piazza Leonardo da Vinci 32, Milan, Italy.

MARCHIONINI, Alfred, German dermatologist; b. Königsburg (now Kaliningrad, USSR), Jan. 12, 1899; s. Karl and Auguste (Domnick) M.; ed. U. Fribourg, U. Königsberg, U. Leipzig; M.D., 1922, agrégé in dermatology, 1928; Dr.honoris causa U. Strasbourg (France); m. Mathilde Soetbeer, Feb. 19, 1931. Became prof. dermatology, 1934, polit. emigrant, 1938; exiled to Turkey; head dermatology clinic Ankara (Turkey) Hosp.; became titular prof. U. Ankara, 1945; named titular prof. dermatology, dir. clinic U. Hamburg (Germany), 1948, U. Munich, 1950. Recipient Bronze medal City of Paris; Keizo-Dohi medal Japanese Dermatol. Soc.; Order Public Health France. Mem. dermatol. socs. of Argentina, Australia, Belgium, Brazil, Gt. Britain, Denmark, Finland, France, Greece, Am. Dermatol. Soc., French Allergy Soc., Barcelona Acad. Medicine, Spanish Acad. Med. Sci. Founder, editor Hautarzt; coeditor Archiv für klinische und experim. Dermatologie; Author: Hauttuberkulose; Asthma, Ekzem und Neurodermitis; Orientbeule; Penicillinbehandlung der Hautkrankheiten; also numerous articles. Address: Frauenlobstrasse 11, Munich 15, West Germany.

MARCHIONNA, Ermanno, Italian mathematician; b. Castel di Sangro, Italy, Nov. 16, 1921; s. Mario and Domenica (Di Biase) M.; Ph.D. in Math. Sci.; m. Cesarina Tibiletti, Oct. 31, 1953; children—Clelia, Mario-Vito, Emilio. Prof. geometry, U. Turin (Italy). Corr. mem. Accademia delle Scienze di Torino. Research and publs. on geometry, algebra. Home: viale Abruzzi 44, Milan, Italy. Office: via Principe Amedeo 8, Turin, Italy.

MARCHIONNA-TIBILETTI, Cesarina, Italian mathematician; b. Milan, Italy, Nov. 17, 1920; d. Pasquale and Anna Maria (Cambianco) Tibiletti; Ph.D. in Math.; m. Ermanno Marchionna, Oct. 31, 1953; children—Clelia, Mario Vito, Emilio. Prof. geometry U. Ferrara, until 1962; now prof. math. U. Milan. Research and publs. on algebra, algebraic geometry. Home: viale Abruzzi 44, Milan, Italy.

MARCHIORO, Thomas Louis, Am. surgeon; b. Spokane, Wash., Aug. 1, 1928; s. Americo and Gertrude (Dennehy) M.; student Colo. Coll., 1945-46, U. Cal. at Berkeley, 1949; B.S., Gonzaga U., 1951; M.D., St. Louis U., 1955; m. Karen Byus, Apr. 2, 1956; children—Thomas, John, Elizabeth, Stephen, Joan, Katherine, Robert. Intern, St. Mary's Group Hosps., St. Louis, 1955-56; resident Henry Ford Hosp., Detroit, 1956-57, U. Colo. Med. Center, 1957-62; faculty U. Colo. Med. Center, Denver, 1959——, asso. prof. surgery, 1966——, Markle scholar, 1965——; asst. chief surgery, attending surgeon Denver VA Hosp., 1962——; mem. surg. staff Colo. Gen. Hosp. 1962——. Diplomate Am. Bd. Surgery, Am. Bd. Thoracic Surgery. Fellow A.C.S.; mem. A.M.A., Am. Thoracic Soc., Soc. U. Surgeons, Internat. Cardiovascular Soc., Soc. Vascular Surgery, Alpha Sigma Nu, Alpha Omega Alpha. Research, publs. primarily on renal transplantation. Home: 942 Jamaica Ct., Denver 80010. Office: 1055 Clermont St., Denver 80220.*

MARCHUK, Gurii Ivanovich, Russian physicist; mem. USSR Council Ministers' Main Adminstrn. for Use of Atomic Energy, Moscow, 1961—. Corr. mem. USSR Acad. Scis. Research, publs. on atomic energy, nuclear reactors. Office: USSR Council Ministers' Main Administration for Use of Atomic Energy, Moscow, USSR.

MARCI VON KRONLAND, Jan (or Johannes) Marcus, Bohemian physician; b. Landskron, Bohemia, June 13, 1595; ed. Prague, Czechoslovakia. Prof. medicine Prague U., 1620-60. Author: Idearum operatricium idea, 1635; De proportione motus seu regula sphygmica ad celeritatem et tarditatem pulsuum, 1639; Liturgia mentus seu dissertatio . . . de natura epilepsiae illius ortu et causis . . . ; Thaumantias liber de arcu coelesti deque colorum apparentium natura . . . , 1648; Dissertatio de natura iridis, 1650. Excelled in mechanics; defined kinetic motion of light as result of compressions, expansions. Died Prague, Apr. 10, 1667.

MARCINKOWSKI, Marion John, Am. metallurgist; b. Balt., Feb. 27, 1931; s. Marion Theodore and Katharine (Printki) M.; B.S., U. Md., 1953; M.S., U. Pa., 1955, Ph.D., 1959; m. Carlita Sue Robertson, Oct. 2, 1966. Supervising scientist Edgar C. Bain Lab. for Fundamental Research, U. S. Steel Research Labs., Monroeville, Pa., 1956-63; metallurgist, asso. prof. metallurgy Inst. for Atomic Research and Dept. Metallurgy, Ia. State U., Ames, 1963—. Mem. Am. Inst. Metall Engrs., Am. Inst. Physics, Am. Soc. for Metals, Deutsche Gesellschaft für Metallkunde, Inst. Metals, Electron Microscopy Soc. Am. Studies on arrangement of atoms on strength and deformation behavior of alloys with the aid of electron microscopy techniques; effect of atomic order on magnetic and elec. properties of alloys; contbns. toward understanding of strengthening mechanisms in gen. occurring in pure metals and disordered alloys. Home: 1822 N. Douglas Av., Ames, Ia. 50010.*

MARC OF TOLEDO (Marcus Toledanus), Spanish physician; probably flourished in later 12th century; canon of Toledo, Spain. Translator: The Qur'ran; De Aëre aquis locis (Hippocrates), De tactu pulsus, De utilitate pulsus, De motu membrorum (earliest Greek treatise on biol. subject available in Christian West), De Motibus liquidis; Isagoge ad Tegni Galeni (Hunain ibn Ishaq).

MARC THE GREEK, alchemist; flourished 2d half 13th century. Author: Liber Ignium ad Comburendos Hostes (collection recipes dealing with incendiary and pyrotechnic substances, so-called Greek fire, phosphorescent substances, and explosives containing saltpeter). Furnished probably earliest recipe of its kind for gunpowder.

MARCONI, (Guglleimo) William, Italian inventor, elec. engr.; b. Bologna, Italy, Apr. 25, 1874; s. Giuseppe and Arnie (Jameson) M.; ed. Leghorn and U. of Bologna; LL.D., Glasgow, Aberdeen, U. of Pa.; D.Sc., Oxford; m. Beatrice O'Brien, Mar. 16, 1905, 1 son, 2 daus.; m. 2d, Contessa Maria Criltina Bezzi-Scali, 1927, 1 dau. Began, 1890, on his father's estate, experiments to test the theory that the electric current is capable of passing through any substance, and, if started, in any given direction, of following an undeviating course without need for a wire or other conductor. He invented an apparatus for wireless telegraphy which attracted attention of Sir William Henry Preece, engr. and electrician-in-chief English Postal Telegraph, who tested the apparatus, with success, in England; soon afterward succeeded in sending messages from Spezia to a steamer 15 kilometers distant; also sent messages from Queen Victoria ashore to Prince of Wales on royal yacht, 1897; founded Marconi's Wireless Telegraph Co., Ltd., London, 1897; came to U. S., 1899; used his method in reporting election, 1900; sent message from Cornwall to Newfoundland, 1901; succeeded in establishing wireless telegraphic communication across Atlantic Ocean, 1902; daily ocean news service by wireless telegraphy inaugurated by him on trans-Atlantic liners, 1904. Invented directive method of wireless telegraphy, 1905, and continuous-wave system, 1906. Recipient (with Braun) Nobel Prize, for physics, 1909; Albert medal Royal Soc. Arts; numerous other decorations and awards. Producer of 1st apparatus to succeed in sending messages through space on comml. scale; worked on short-wave, ultra-short-wave transmission, multiplex transmission of voice and code. Died Rome, Italy, July 19, 1937.

MARCOU, Jules, geologist; b. Salins, France, Apr. 20, 1824; attended Coll. of St. Louis 1842-44; m. Jane Belknap, 1850; 2 sons. Prof. mineralogy Sorbonne, 1846-48; traveling geologist Jardin des Plantes, 1848-50; collected for Paris museums in North Am., especially in Lake Superior region, until 1854; sent by Am. govt. on expdn. to Cal. deserts, 1853-54; prof., Zurich, 1855; prof. paleontology École Polytechnique, Zurich, 1856-60; returned to Am., 1860; geologist Mus. of Comparative Zoology, Harvard, 1862-64, made several field trips to Western U. S. Author: Lettres sur les Roches du Jura, 1857-60; Geology of North American, 1858; also produced Geological Map of the World, 1862. Made 1st systematic geol. survey of N.Am. continent; 1st to recognize existence of Jurassic fossils in New World (in Cal. deserts). Died Apr. 17, 1898.

MARCOVICI, Graziella, Rumanian neurohistopathologist; b. Bucharest, Rumania, June 3, 1920; s. Victor and Zoe (Gerbea) M.; student Inst. de hautes etudes françaises 1940-43; ed. Faculty of Medicine in Bucharest, 1944-50; m. Marcovici Neumann, Dec. 15, 1945; children—Marcovici Alexandru, Marcovici Natalia. With Inst. Neurology, RSR Acad., Bucharest, 1950—, physician specialist, 1960—. Research and publs. on traumatic injuries of head, epileptic and non-epileptic cerebral scars, exptl. cerebral hypoxia, microvasculature changes in cerebral artheroslerosis. relationship between tumors and head trauma. Home: Bucharest, RSR Bd. Ilie Pintilie 37, R 30 Dec., Romania. Office: 42 Povernei, Bucharest, Rumania.*

MARCUS, Froim, Rumanian mathematician; b. Bucecea-Roumania, Sept. 8, 1904; s. Inghel and Haia (Argintaru) M.; Dr. Ing., R. Scuola D'Ingegneria, Di Torino, Italy, 1929; m. Dora Iwanier, Sept. 10, 1935. Chief engr. tech. service, Region BotoSani, 1944-52; prof. math. Politech. Inst., Jasi, R.S. Romania, 1953—; prin. researcher Math. Inst. Roumanian Acad. Jassy. Research, numerous publs. on projective differential geometry of surfaces, conjugates, nets, rectilinear congruences, infinitesimal projective transformation of surfaces, converse of Bianchi's permutability theorem, surfaces which are simultaneously minimal in projective, affine and metrical geometry. Home: BotoSani Calea Nationala 363. Office: Institutul Politehnic, Jasi, R.S., Romania.*

MARCUS, Irwin M., Am. psychiatrist; b. Chgo., Mar. 18, 1919; s. Max U. and Belle (Rothbaum) M.; B.S., U. Ill., 1939, M.D., 1943; certificate in psychoanalysis Columbia, 1949; m. Dorothy Mann, June 29, 1948; children—Randall, Sherry, Melinda. Fellow child psychiatry Hawthorne Cedar-Knolls Inst., White Plains, N.Y., 1949-51, Jewish Bd. Guardians, Child Guidance Inst., N.Y.C., 1949-51, Council Child Devel. Center, N.Y.C., 1950-51; practice medicine, specializing in child psychiatry and psychoanalysis, New Orleans, 1951—; sr. vis. physician Charity Hosp., New Orleans, 1951—; sr. Touro Infirmary, 1956—; supervisory and tng. psychoanalyst New Orleans Psychoanalytic Inst., 1956—; lectr. Sch. Social Work, Tulane, 1951—, faculty Center Tchr. Edn., 1966—; clin. prof. psychiatry La. State U., 1956—; chief cons. Orleans Parish Dept. Pub. Welfare; cons. Family Service Soc., New Orleans. Fellow Am. Orthopsychiat. Assn.; mem. Am. Psychoanalytic Assn. (bd. profl. standards 1965—), Am. Psychiat. Assn., Am. Acad. Child Psychology, Internat. Assn. Child Psychiatry, Am. Group Psychotherapy Assn., Am. Assn. Child Psychoanalysis, Am. Assn. Adolescent Psychiatry, Sigma Xi. Author: (with Niles Newton, Paul György, Thaddeus L. Montgomery) Family Book of Child Care, 1957; also articles. Cons. editor Child Devel. Publs., 1957-60. Originator costume play therapy, group treatment of fathers of disturbed children; research on action prone child, learning problems in adolescence. Home: 4231 Vendome PL., New Orleans 70125. Office: 3619 Prytanna St., New Orleans 70115.*

MARCUS, Jules Alexander, Am. physicist, educator; b. Coytesville, N.J., May 10, 1919; s. Alexander and Julia (Parks) M.; B.S., Yale, 1940, Ph.D., 1947; m. Ruth C. Barcan, Aug. 28, 1942; children—James S., Peter W., Katherine H., Elizabeth P. Instr., Yale, 1943-44, research asst., 1946-47; physicist Johns Hopkins, 1944-46; postdoctoral fellow U. Chgo., 1947-49; faculty physics Northwestern U., Evanston, Ill., 1949—, prof., 1961—. Fellow Am. Phys. Soc.; mem. Am. Assn. Physics Tchrs., Sigma Xi. Research, publs. on low temperature properties of metals especially magnetic properties of zinc and chromium and "Quadratic Hall effect"; discovered de Haas-van Alphen effect in zinc, 1947, also symmetry of chromium can be reduced from cubic to tetragonal by cooling in magnetic field, 1964. Home: 2801 Girard Av., Evanston, Ill. 60201.*

MARCUS, Marvin, Am. mathematician; b. Albuquerque, July 31, 1927; s. David Clarence and Esther (Rosenthal) M.; B.A., U. Cal. at Berkeley, 1950, Ph.D., 1954; m. Arlen Ingrid Sahlman, Sept. 14, 1951; children—Jeffrey Thomas, Karen Melissa; m. 2d, Rebecca Elizabeth Michael, Oct. 12, 1965. Instr., then asst. prof. U. B.C., 1954-56, asso. prof., 1957-62; postdoctoral research fellow Nat. Bur. Standards, Washington, 1956-57; prof. U. Cal. at Santa Barbara, 1962—, chmn. dept. math., 1963—. Cons. Bur. Naval ordnance, Pasadena, Cal. Served with USNR, 1945-46. Mem. Am. Math. Soc., Math. Assn. Am., Soc. Indsl. and Applied Mathematics, Sigma Xi. Author articles in field. Research in linear and multilinear algebra; theory of inequalities; differential equations. Home: 2937 Kenmore Pl., Santa Barbara, Cal.*

MARCUS, Solomon Alter, Rumanian mathematician, linguist; b. Bacau, Rumania, Mar. 1, 1925; s. Alter and Sima (Herscovici) M.; B.Sc., Faculty Math., Bucharest, Rumania, 1950, D.Math., 1956; m. Paula Floarea Diaconescu, Apr. 28, 1966. Asst. lectr. Bucharest Poly. Inst., also Bucharest U., 1950-55; faculty U. Bucharest, 1955—, prof. dept. math., 1966—; chief sector set theory and real functions Acad. R.S. Rumania, Inst. Math., 1964—. Recipient Laureate of Prize Timotei Cipariu, Rumanian Acad., 1964. Mem. Rumanian, Am. math. socs. Author: (with M. Nicolescu, N. Dinculeanu) Manual of Mathematical Analysis, vol. 1, 1962, vol. 2, 1964; Mathematical Linguistics, 1963; Grammars and Finite Automata, 1964; Algebraic Linguistics, 1966; also other books, numerous articles. Establishment of structure of various types of real functions of real variables especially arbitrary functions, Darboux functions, derivatives, differential structure of discontinuous functions; constrn. of algebraic and set theoretic models for various linguistic notions. Home: 47 M.Eminescu, Bucharest g, Rumania.*

MARCUS, Stanley, Am. microbiologist; b. N.Y.C., Jan. 20, 1916; s. Israel L. and Raechel (Tow) M.; B.A., City Coll. N.Y., 1937; Ph.D., U. Mich., 1942; m. Harriet Lynn, Aug. 20, 1939; children—Charles J., Rand T. Mem. staff Rackham Arthritis Unit, U. Mich., 1947-49; prof. microbiology U. Utah, Salt Lake City, 1949—; cons. Am. Sterilizer Co., 1956, USPHS, 1955-60, Latter-day Saints Hosp., 1950-67. Recipient Research Career award Nat. Inst. Allergy and Infectious Diseases, USPHS, 1962. Mem. Am. Soc. Microbiology, Am. Assn. Immunologists, Soc. Exptl. Biology and Medicine, Am. Trudeau Soc., Am. Pub. Health Assn., A.A.A.S., Sigma Xi. Research and publs. on nature of resistance, both enhanced and depressed, to infectious disease and to cancer; contbd. to theory of cellular resistance; rapid preparation and standardization of skin test reagents for fungal disease; microbiological test for pyrogens in freshly prepared solutions; test for malignant melanoma. Home: 1400 Federal Way, Salt Lake City 84102.*

MARCUS, Yizhak, chemist; b. Kolberg, Germany, Mar. 17, 1931; s. Fritz and Rosa (Nelken) M.; M.Sc., Hebrew U., Jerusalem, Israel, 1952, Ph.D., 1956; m. Tova Semmel, Oct. 10, 1954; children—Tamar, Ruth. Research chemist Israel Atomic Energy Commn., Rehovoth, 1952-57, head radiochemistry dept., 1958-65, dir. chemistry div., 1964-65; research asso. Mass. Inst. Tech., Cambridge, 1957; prof. inorganic chemistry, head dept. inorganic and analytical chemistry Hebrew U., Jerusalem, 1965—. Cons., Israel Atomic Energy Commn., 1965—. Mem. Israel, Am. chem. Socs. Author: (with A. S. Kertes) Ion Exchange and Solvent Extraction of metal Complexes, 1968; also articles. Research on theory of sorption of metal complexes on anion exchangers, theory for stability of mixed ligand metal complexes, exptl. studies on solvent extraction from molten salts, solution chemistry of lanthanide and actinide elements. Home: 36 Hapalmach, Jerusalem, Israel.*

MARCUZZI, Giorgio, Italian zoologist; b. Trieste, Italy, Aug. 16, 1919; s. Giorgio and Pina (Jurza) M.; Dr.Nat.Sci., U. Padova (Italy), 1946, Dr.Biol.Sci., 1947; m. Maria Grazia Gregorutti, Dec. 1, 1949; 1 dau., Fabia. Asst., Inst. Zoology, U. Padova, 1946-47, lectr. zoology, 1950—, lectr. animal ecology, 1963—; prof. Caracas (Venezuela) U., 1948-50; lectr. zoology U. Trieste, 1957—, dir. Inst. Zoology, 1961—. In charge program for secondary productivity I.B.P., Italy, 1965—. Recipient Garbini prize for Biogeography, Verona, Italy, K2 prize for biogeography, Udine; grantee Cambridge (Eng.) U., U. Birmingham (Eng.). Mem. Ecol. Soc. Am., Zool. Union Italy, Entomol. Soc. (Italy), Soc. Natural Scis., Soc. Ciencias Naturales La Salle. Author: Fauna delle Dolomiti, 1956; also numerous articles. Research on animal ecology of arid zones of Venezuela, soil fauna of Italy, zoogeography of Venezuela and Antilles and Italy, insect physiology including water relations in Tenebrionid beetles, function and structure of Malpighian tubes, systematics of Tenebrionid beetles. Home: 42 Palermo, Padova, Italy.*

MARCY, Étienne-Jules, French physiologist; b. Beaune, France, Mar. 5, 1830. Became prof. natural history Coll. de France, 1867; Mem. French Acad. Scis., 1878. Author: Recherches sur le pouls au moyen d'un nouvel appareil enregisteur, le sphygmographe, 1860; Physiologie médicale de la circulation du sang, 1863; Études physiologiques sur les caractères graphiques des battements du coeur et des mouvements respiratoires, 1865; Physiologie expérimentale, 1875; La méthode graphique dans les sciences expérimentales et particulièrement en physiologie et en médecine, 1878; Étude da la locomotion animale par la chronophotographie, 1887; Physiologie du mouvement: le vol des oiseaux, 1890; Le mouvement, 1894; Travaux du laboratoire, pub. annually. Studied physiology of heart and circulation, elec. phenomena in animals, action of poisons on nerves and muscles, flight of insects and birds; inventor sphygmograph for recording pulse, 1863; made 1st series of photographs of successive phases of heart movement; 1st to recognize relationship between blood pressure and heart rate. Died Paris, May 15, 1904.

MARDASHEV, Sergey Rufovich, Russian biochemist; b. 1906; grad. 2d Leningrad Med. Inst., 1930; postgrad. 1st Moscow Med. Inst., until 1935; D.Biol. Sci., 1940. Head biochem. lab. All-Union Inst. Exptl. Medicine, 1935-39; lectr. dept. biol. chemistry 1st Moscow Med. Inst., 1937-43; former dir. research lab. Lenin's Mausoleum; prof. Sechenov 1st Moscow Med. Inst., 1943—, head chair biol. and organic chemistry, 1952—. Decorated Order of Lenin (2); recipient Stalin prize, 1948. Mem. USSR Acad. Med. Sci. (v.p. 1963—). Author: The Enzymology of Tumors, 1948. Mem. editorial bd., former chief editor Problems of Med. Chemistry. Research and numerous publs. on enzyme processes, protein metabolism in animals; dir. work for embalming Stalin's and Gottwald's bodies. Address: Sechenovs 1st Moscow Med. Inst., B. Pirogovskaya ulitsa 2-6, Moscow, USSR.*

MARDEN, Morris, Am. mathematician; b. Boston, Feb. 12, 1905; s. Abraham and Fannie B. (Lieberman) M.; A.B., Harvard, 1925, A.M., 1927, Ph.D., 1928; postgrad. Princeton, Swiss Fed. Tech. U., U. Paris; m. Miriam L. Goldman, Sept. 11, 1932; children—Albert, Philip. NRC fellow U. Wis., Princeton, U. Zurich (Switzerland), U. Paris, 1928-30; faculty U. Wis., Milw., 1930—, chmn. math. dept., 1957-61, 63-64, distinguished prof. math., 1964—; vis. lectr. Math. Assn. Am. NSF grantee, 1961—. Fellow A.A.A.S.; mem. Am. Math. Soc., Math. Assn. Am., Am. Assn. U. Profs., French, Indian math. socs., Phi Beta Kappa, Sigma Xi. Author: Geometry of Zeros of a Polynomial in a Complex Variable, 1949; Geometry of Polynomials, 1966. Research, numerous publs. on zeros of polynomials and allied functions of a complex variable. Home: 7100 N. Barnett Lane, Milw. 53217.*

MARDEN, Philip Ayer, Am. physician; b. Newport, N.H., Oct. 31, 1911; s. Albion Sullivan and Laura Isobel (McEchern) M.; A.B., Dartmouth, Coll., 1933, postgrad. Med. Sch., 1933-34; M.D., U. Pa., 1936; m. Magdalen Rekus, Aug. 5, 1950. Fellow otolaryngology U. Pa. Hosp., 1939-40, chief otolaryngology, 1959—; mem. faculty U. Pa. Med. Sch., 1940—, prof., chmn. dept. otolaryngology, 1959—; chief otolaryngology, div. A, Phila. Gen. Hosp., 1960—; cons. Presbyn. Hosp., Phila., VA Hosp., Phila. Mem. A.M.A., Am. Acad. Ophthalmology and Otolaryngology, A.C.S., Coll. Physicians Phila., Phi Beta Kappa, Sigma Xi. Home: 163 Vassar Rd., Bala-Cynwyd, Pa. 19004. Office: 3400 Spruce St., Phila. 19104.*

MAREN, Thomas Hartley, Am. pharmacologist; b. N.Y.C., May 26, 1918; s. James M. and Roxane (Holstein) M.; A.B., Princeton, 1938; M.D., Johns Hopkins, 1951; m. Ruth Eleanor Hendricks, Sept. 5, 1941; children—Peter, James, David. Chemist, Wallace Labs., New Brunswick, N.J., 1938-40, 41-43; chemist Office of Naval Research, Johns Hopkins, 1943-46, instr. pharmacology, 1946-51; group leader in pharmacology Am. Cyanamid Co., Stamford, Conn., 1951-55; prof. pharmacology, chmn. dept. U. Fla. Coll. Medicine, Gainesville, 1955—. Mem. Am. Soc. Pharmacology and Exptl. Therapeutics, Alpha Omega Alpha. Research and introduction of new drugs for treatment of filariasis, bacterial disease, fluid imbalance, action of drugs on enzymes. Home: 1228 S.W. 14th Av., Gainesville, Fla.*

MARÈS, Henri (Pierre-Louis), French chemist, agronomist; b. Caslons-sur-Saône, France, Jan. 18, 1820; mem. French Acad. Scis. 1886, Soc. Agr. Author: Action du soufre en poudre sur l'ordium de la vigne, 1856; Note sur la pourriture des racines des vines, 1868. Research on various vine diseases; developed method sulphuring grape vines to protect them from vine mildew. Died Montpellier, France, May 9, 1901.

MARET, Hughes, French physician; b. Dijon, France, Oct. 6, 1726; M.D., Montpellier, France; practiced medicine, Dijon; mem. French Acad. Scis. Author: Éléments de chimie; Sur le traitement d'une fièbre épidémique, 1775; Traité de la fièvre pétéchiale maligne, 1762; Mémoire sur l'influence des moeurs des Français sur leur santé, 1772. Research on contagious diseases; studied history of epidemics in 18th century France. Died June 11, 1786.

MARETT, Robert Ranulph, English philosopher, anthropologist; b. Isle of Jersey, Eng., June 13, 1866; s. Robert Pipon and Julia Anne Marett; ed. Victoria Coll., Jersey; prizeman Balliol Coll. Oxford (Eng.) U., 1884, 86, 87; M.A., Sc.D., 1913, D. Litt. (hon.); LL.D., U. Berlin (Germany), 1888, U. St. Andrews, Scotland, 1929; admitted to Jersey Bar, 1891; m. Nora Kirk, 1898; 4 children. Tutor, Lord Basil Blackwood, also gen. coach Balliol Coll. prior to 1891; fellow Exeter (Eng.) Coll., 1891, tutor philosophy from 1893, sub-rector, 1893-98, examiner Literae Humaniores, 1905-08, proctor, 1918-19, rector, 1928-43; reader, acting prof. social anthropology Oxford U., 1910-36; expdns. caves France, Spain; conducted excavations (successful) La Cotte de St. Brelade, 1912-15; traveled to Australia, 1914. Recipient Green prize, 1893, Huxley medal Royal Anthrop. Inst., 1939; Frazer lectr. Cambridge (Eng.) U., 1927, Andrew Lang lectr. St. Andrews U., 1928, Gifford lectr., 1931-33, Lowell lectr. Boston U., 1930, Thomson lectr. Aberdeen (Scotland) U., Donelian lectr. Trinity Coll. Dublin (Ireland), 1933. Fellow Royal Anthrop. Inst., Brit. Acad., Royal Soc. Letters; mem. Sociol. Inst. (pres. 1932-35), Brit. Assn. (pres., sec. 1916), Speleological Assn. (pres.), Folklore Soc. (pres. 1913-18), École d'Anthropologie; founder Anthrop., Dialectical socs. Oxford U. Author: Essay on Origin and Validity in Ethics in Personal Idealism, 1902; The Threshold of Religion, 1909; The Birth of Humility, 1910; Anthropology, 1912; Psychology and Folklore, 1920; The Diffusion of Culture, 1927; Man in the Making, 1928; The Raw Material of Religion, 1929; Faith, Hope, and Charity in Primitive Religion, 1932; Sacraments of Simple Folk, 1933; Head, Heart and Hands in Human Evolution, 1935; Tylor, 1936. Editor: Anthropological Essays in Honour of E. B. Tylor, 1908; Anthropology and the Classics, 1908. Contbr. articles literary mags., research, publs. on philosophy, excavations, parallel between primitive religion and his own conception of Anglicanism, developed concept Präanimismus; as contbn. to Oxford sch. anthropology opposed rational anthrop. categories of Tylor and Frazer with his own, based on primitive logic of heart, so-termed. Died Feb. 18, 1943.

MARFAN, Antonin Bernard, French pediatrician; b. Castlenaudary, France, June 23, 1858; ed. Toulouse and Paris, France; M.D., 1887. Practiced medicine specializing in diseases of children; prof. therapeutics Faculty of Paris, 1911, later prof. hygiene and pediatrics. Hon. fellow Royal Soc. Medicine; mem. Acad. Medicine. Author: (with Grancher and Comby) Traité des maladies de l'enfance, 1896-98; Clinique des maladies de la première enfance and Traité de l'allaitement, 1899-1903. Founder-editor: Le Nourrisson jour.; co-editor: (with Charcot, Bouchard, Brissord) System of Medicine. Research, publs. on periodic vomiting with acetonemia, puncture of pericardium by epigastric routes, prevention, diagnosis, treatment of children's diseases; described progressive spastic paraplegia in children with hereditary syphilis (Marfan's disease), 1892. Died 1942.

MARG, Elwin, Am. visual scientist; b. San Francisco, Mar. 23, 1918; s. Sigmund and Fannie (Sockolov) M.; A.B., U. Cal. at Berkeley, 1940, Ph.D., 1950; m. Helen Eugenia Kelly, Apr. 1, 1942; 1 dau., Tamia. Faculty, U. Cal. at Berkeley, 1950—, prof. physiol. optics, 1962—. NSF Sr. Postdoctoral fellow, 1957; Guggenheim fellow, 1964; recipient Apollo award Am. Optometric Assn., 1962. Mem. Am. Physiol. Soc., Am. Psychol. Assn., Optical Soc. Am., I.E.E.E. (senior mem.), Am. Acad. Optometry. Contbr. numerous articles to tech. jours. Co-inventor tonometer used to test eye pressure to prevent blindness due to glaucoma. Home: 4849 Grizzly Peak Blvd., Berkeley, Cal. 94705.*

MARGALEF, Ramón, Spanish biologist; b. Barcelona, Spain, May 16, 1919; s. Ramón and Vicenta (López) M.; Lic. Cienc, U. Barcelona, 1950; Dr.Sc., U. Madrid (Spain), 1952; m. María Mir, July 4, 1952; children—Neus, Nuria, Ramón, Bartolomé. Staff, Instituto de Biología Aplicada, 1946-51; biologist Instituto de Investigaciones Pesqueras, Barcelona, 1951—. Lectr. gen. ecology U. Barcelona, 1955—. Mem. Acad. Scis. Barcelona, Research and numerous publs. on methods of assessing orgn. of communities of organisms, relations between such orgns. and productivity, exploitability. Home: 31 Ronda Guinardó, Barcelona. Office: s/n Paseo Nacional, Barcelona, Spain.*

MARGARIA, Rodolfo, Italian physiologist; b. Chatillon, Italy, Nov. 15, 1901; s. Giovanni Margaria; M.D., U. Turin (Italy), 1924; m. Maria Maria Giovanna De Luca; children—Riccardo, Paola, Elsa (Mrs. Roberto Mutani), Cosetta, Roberto. Lectr. biochemistry, from 1928; prof. physiology Ferrara, Parma, Pavia (all Italy) univs., 1933—, U. Milan, 1938—. Rockefeller fellow U. Coll., London, Eng., 1930; with Marine Biol. Lab., London, 1930, Physiology Lab., Cambridge (Eng.) U., 1931, Fatigue Lab., Harvard, 1932, Rockefeller Hosp., N.Y.C., 1933, Johnson Found. for Med. Physics, Phila., 1933; vis. prof. physiology Yale, 1948-50; dir. Sch. Sports Medicine, 1957—, Sch. Phys. Fitness, 1965—. Recipient Gold medals for contbn. in fields of edn., pub. health; Einaudi prize for physiology and pathology, 1953. Mem. Nat. Acad. Lincei, Société Philomatique (Paris), Internat. Acad. Aviation Medicine, Internat. Acad. Astronautics. Fellow Am. Coll. Sports Medicine, Physiol. Soc. Eng., Schweizersche Verein der Physiologen and Pharmacologen (Switzerland), Internat. Union of Physiol. Scis. (council); Author: Principi di Fisiologia, 2 vols., 3d edit.; Principi di Chimica e Fisicochimica fisiologica, 9th edit., 1966; Fisiologia, 2d edit., 1965. Research, publs. on space physiology (avoidance of acceleration forces by immersion in water, also orientation in space and human locomotion in subgravity), physiology of muscular exercise (energy cost of walking and running, energy equivalent of glycogen to lactic acid process, computation of energy cost of exercise performed in anaerobic conditions, analysis of mech. work in walking and running, mechanics of sound prodn. and its efficiency), respiratory physiology (energy absorbed in respiration in muscular exercise). Home: 9 Piazza Grandi, Milan. Office: 32 Via Mangiagalli, Milan, Italy.*

MARGENAU, Henry, physicist; b. Bielefeld, Germany, Apr. 30, 1901; s. Frederick and Karoline (Wagemann) M.; Tchr.'s Diploma, Tchr.'s Coll., Herford, Germany, 1921; A.B., Midland Coll., 1924, Dr. Pub. Service, (hon.), 1965; M.Sc., U. Neb., 1927, D.Sc. (hon.), 1957; Ph.D., Yale, 1929; L.H.D. (hon.), Carleton Coll., 1954; LL.D., Dalhousie U., 1960, R.I. Coll., 1962; D.Sc. (hon.), Hartwick Coll., 1964; m. Louise M. Noe, May 28, 1932; children—Rolf Carol, Annemarie Louise (Mrs. Carl Lindskog), Henry F. Came to U. S., 1923, naturalized, 1930. Faculty, U. Neb., 1925-27; faculty Yale, New Haven, 1928-29, 30—, Eugene Higgins prof. physics and natural philosophy, 1950—. Mem. Inst. for Advanced Study, Princeton, N.J., 1939; staff mem. Radiation Lab., Cambridge, Mass., 1945; cons. to govt. and industry; chmn., research dir. Found. for Integrative Edn., 1950—. Trustee Conn. Coll. for Women, New London. Sterling fellow, 1930-31, Guggenheim fellow, Fulbright lectr., 1961. Fellow Am. Acad. Arts and Scis., Am. Phys. Soc. (pres. N.E. sect. 1943-44), Philosophy Sci. Assn. (pres. 1950-58), Conn. Acad. Arts and Scis. (v.p. 1950-52), Academie Internationale de Phil. de Sci. (hon.), Phi Beta Kappa. Author numerous books, including Open Vistas, 1961, Ethics and Science, 1964, The Scientist, 1965. Editor Am. Jour. Sci., 1935, Philosophy of Sci., 1936—, Rev. of Modern Physics, 1955-56, Jour. Chem. Physics, 1956-58, Jour. Quantitative Spectroscopy and Radiative Transfer, 1961-65. Cons. editor Time and Life Sci. series, 1964—. Research contbns. to theory of forces between molecules; structure of spectral lines; absorption of radar waves in ionized media; philos. founds. physics. Home: 173 Westwood Rd., New Haven 06515.*

MARGERIE, Emmanuel-Marie-Pierre-Martin Joaquin de, see de Margerie.

MARGGRAF, Andreas Sigismund, German chemist; b. Berlin, Germany, Mar. 3, 1709; studied under Neumann, Berlin, circa 1725, at Frankfort on the Main, Germany; Strasbourg; studied at Halle, Germany, under Hoffmann and Juncker; then at Freiberg, Germany, under Henckel, 1734. Entered service Royal Court Pharmacy, Prussia, 1735; mem. Berlin Acad. Scis., dir. chem. lab., from 1754, became dir. phys.-chem. sect., 1760-61, dir. acad., 1767; corr. mem. Paris (France) Acad. Author: Chemische Versuche zur Gewinnung des Zuckers, 1747; Chemischer Schriften, 2 vols., 1761-67; also many articles. Pioneer in chem. analysis; introduced use of microscope in chem. studies; discovered beet sugar in beetroot, 1747; called father of the sugar industry; recognized alumina in lime, 1754; used flame test to distinguish potash from soda; isolated zinc from calamine; classified silver as silver chloride; obtained phosphides of zinc and copper; knew of formation of phosphorus pentoxide on burning phosphorus; gave many reactions of microcosmic salt; demonstrated that burning magnesium salts produce green flame; obtained many metallic formates; discovered refined camphor, magnesia, several acid anhydrides; investigated (for Frederick the Great) plant growth and fertilizers; studied distillation of water, platinum; last eminent adherent of phlogiston theory in Germany. Died Berlin, Aug. 7, 1782.

MARGGRAF, Georg, see Marcgraf, Georg.

MARGOLIN, Sydney G(erald), Am. physician; b. N.Y.C., Apr. 25, 1909; B.S., Columbia, 1930, M.A. (Columbia fellow), 1931; M.D. (univ. fellow), N.Y. State U., 1936. Abrahamson fellow neurology Mt. Sinai Hosp., 1939-40; Josiah Macy, Jr. Found. grantee Columbia, 1941-42; psychiatrist in charge male med. service Worcester (Mass.) State Hosp., 1941; asso. attending psychiatrist Mt. Sinai Hosp., 1946-55; prof. psychiatry U. Colo. Sch. Medicine, 1955—, dir. human behavior lab., 1956—, dir. Ute Indian project, 1956—. Faculty, Columbia, 1930-32, N.Y. Psychoanalytic Inst., 1946-55; chief, div. psychosomatic medicine U. Colo. Med. Center, Denver, 1955-58. Diplomate Am. Bd. Psychiatry and Neurology, Nat. Bd. Med. Examiners. Fellow A.C.P., Am. Psychiat. Assn., Am. Orthopsychiat. Assn., A.A.A.S.; mem. Am. Psychoanalytic Assn., Am. Psychosomatic Soc. (pres. 1952-53), Group for Advancement Psychiatry (past chmn. com. therapy), Assn. Research in Nervous and Mental Disease, Soc. Psychophysiol. Research (dir.), Bioengring. Symposium (past v.p.), I.E.E.E., also others. Research in states of consciousness, psychosensory perception, electro-physiol. epiphenomena of brain activity. Home: 4375 S. Lafayette St., Englewood, Colo. 80110.*

MARGOLIS, Bernard, Canadian physicist; b. Montreal, Que., Can., Aug. 15, 1926; s. John and Esther (Flisfeder) M.; B.Sc., McGill U., 1947, M.Sc., 1949; Ph.D., Mass. Inst. Tech., 1952; m. Emelia Barbara Levin, Aug. 29, 1954; children—Jared Richard, Julie Anne. Research asso., instr. Mass. Inst. Tech., 1952-53; instr. Columbia U., 1954-57, research physicist, 1957-59; asso. prof. Ohio State U., 1959-61; faculty McGill U., Montreal, 1961—, prof. physics, 1963—; vis. scientist CERN, Geneva, Switzerland, 1967-68. Cons., Brookhaven Nat. Lab., 1963-67. Mem. Am. Phys. Soc., Canadian Assn. Physicists (past vice chmn. theoretical div.). Editor: (with C. S. Lam) Proceedings of McGill Summer Institute in Nuclear and Particle Physics, 1967; Research and publs. on theory of nuclear reactions, nuclear structure, particle physics.*

MARGRAVE, John Lee, Am. chemist; b. Kansas City, Kan., Apr. 13, 1924; s. Orville Frank and Bernice June (Hamilton) M.; B.S. in Engring., Physics, U. Kan., 1948, Ph.D., in Chemistry, 1951; m. Mary Lou Davis, June 11, 1950; children—David Russell, Karen Sue. AEC Postdoctoral fellow U. Cal. at Berkeley, 1951-52; faculty U. Wis., Madison, 1952-63, prof. chemistry, 1960-63; prof. chemistry Rice U., Houston, 1963—, chairman department of chemistry, 1967—. Cons. to pvt. cos., nat. labs.; chmn. com. on high temperature phenomena NRC, 1962-67. Recipient Teaching award U. Wis., 1957. Mem. Am. Chem. Soc. (inorganic chemistry award 1967), Am. Phys. Soc., Am. Ceramic Soc., A.A.A.S. Editor: Characterization of High Temperature Vapors, 1967. Research and publs. in high temperature chemistry, thermodynamics with mass and optical spectrometers, calorimeters, plasma torches, fluorine chemistry, pioneered demonstration utility silicon difluoride for syntheses organosilicon compounds, borosilanes, silicon oxyfluorides. Home: 5012 Tangle Lane, Houston 77027.*

MARGUET, Frédéric-Philippe, astronomer; b. Algiers, Algeria, June 11, 1874; capt. French Navy; prof. astronomy Brest Naval Sch.; mem. French Acad. Scis., 1937. Author: Cours d'astronomie; Cours de navigation. Studied satellites of Jupiter, theory of curves, height of spheric segments. Died Villeneuve, France, June 2, 1951.

MARGULIS, Alexander Rafailo, physician; born Belgrade, Yugoslavia, Mar. 31, 1921; s. Rafailo and Olga (Weiss-Belic) M.; came to U. S., 1946, naturalized, 1955; student Faculty Medicine, U. Belgrade, 1939-41, 45-46; M.D., Harvard, 1950; m. Renee Reisner, Mar. 23, 1946. Jr. clin. instr. U. Mich., 1953-54; faculty U. Minn., 1954-59, asst. prof., 1956-59; asso. Duke Sch. Medicine, 1957-59; asst. prof. Washington U. Sch. Medicine, St. Louis, 1959-60, asso. prof., 1960-61, prof. radiology, 1961-63; prof., chmn. dept. radiology U. Cal. Sch. Medicine, San Francisco, 1963—. Cons. to hosps.; mem. com. to develop tech. guide on radiology lab. animals VA, 1963—, com. on radiology Nat. Acad. Scis.-NRC, 1964—. Diplomate Am. Bd. Radiology. Mem. A.M.A. (cons. on drugs 1961—), Radiol. Soc. N.Am., Am. Coll. Radiology, Am. Roentgen Ray Soc., Assn. U. Radiologists pres. 1966-67), Cal. Med. Assn., San Francisco Med. Soc., Cal. Radiol. Soc., Bay Area Gut Club, San Francisco Radiol. Soc. Author: (with C. M. Nice, L. G. Rigler) Roentgen Diagnosis of Abdominal Tumors in Childhood, 1957; also numerous articles. Co-editor: Alimentary Tract Roentgenology, 1967. Research on vascular patterns of tumors in exptl. animals and humans; developed foam cast exam. lower colon, diagnostic signs for abdominal tumors. Home: 8 Tara Hill Rd., Tiburon, Cal. 94920. Office: Dept. Radiology, U. Cal., San Francisco Med. Center, San Francisco 94122.*

MARHERR, Philipp Ambrosius, Austrian physician; b. Vienna, Austria, 1738; M.D., Vienna, 1762. Prof., Insts. Medicine, Prague, 1766. Author: Dissertatio chymica de affinitate corporum, 1762; Programma de electricitatis aëreae in corpus humanum actione, 1766. Stated that course of blood through lungs was quicker than through rest of body. Died Mar. 28, 1771.

MARIAT, François, French med. mycologist; b. Paris, France, Apr. 28, 1921; s. Emile and Marie (Billault) M.; Dr. ès Sciences, U. Paris, 1951; m. Denise Cottet, Oct. 19, 1946. With Mus. Natural History, Paris, 1940-42; with Inst. Pasteur, Paris, 1942—, chief lab., dept. med. mycology, 1958—, prof. med. mycology, 1953—; lectr. med. mycology Sch. Medicine, Paris, 1964—. Vis. prof. med. mycology U. Mexico, 1960, 65, 67, Guatemala City, 1965, San Jose, Costa Rica, 1965. Mem. Internat. Soc. Human and Animal Mycology (gen. sec. 1967), French Soc. Microbiology, others. Author: (with G. Segretain and E. Drouhet) Diagnostic de Laboratoire en Mycologie Medicale, 2d edit., 1964. Contbr. over 125 articles to med. jours. Research on symbiosis and parasitism, physiology of symbiosis in orchids, pathogenic fungi and actinomycetes and its application to epidemiology, diagnosis and therapy of mycoses. Home: 72 Av. Henri Barbusse, Clamart, 92, France. Office: Institut Pasteur, 28 rue du Dr. Roux, Paris 15, 75 France.

MARIE, Charles François Maximilien, French mathematician; b. Paris, France, 1819; ed. l'École polytechnique, Paris, 1838, l'École de Metz. Left mil. career to become asst. master l'École polytechnique, later admissions examiner. Author: Théorie des fonctions de variables imaginaires, 1874-75; Historie des sciences mathématiques et physiques, 1883-88. Graphic representation of imaginaries of analytic geometry in different way than that of Von Staudt. Died Paris, 1891.

MARIE, Pierre, French neurologist; b. 1853; M.D., U. Paris, 1883; student law; m.; became physician under Charcot, Salpetrière, 1885; became prof. neurology Paris Faculty, 1907; ret., 1925. Mem. French Acad. Medicine. Author: Graves disease, 1883; Leçon sur les maladies de la moelle épinière, 1892. Research on diseases of nervous system; 1st description of acromegaly, amyotrophy of Charcot-Marie, rhizometic spondylosis; also described pulmonary osteoarthropathy (Strumpell-Marie disease), hereditary cerebellar ataxis; studied aphasia. Died 1940. Paris, Apr. 10, 1940.

MARIÉ-DAVY, Edme Hippolyte, French physicist; b. Clamecy, France, 1820; ed. l'École normale supérieure; became prof. physics, Montpellier, 1845; named titular astronomer Obs. Paris, 1862; organizer Internat. Meteorology Service, France, 1863; dir. Obs. Montsouris, Paris, 1873-87. Author: Atlas des mouvements généraux de l'atmosphère; Atlas des orages. Research on terrestrial magnetism, gen. movements of atmosphere; application of meteorology to hygiene and agr.; invented cell using bisulfate of mercury. Died Dornecy, Nièvre, France, 1893.

MARIETTE, Auguste Édouard, French Egyptologist; b. Boulogne, Feb. 11, 1821; ed. Boulogne Municiple Coll.; grad. Douai, 1841. Prof. French and drawing at boy's sch., Stratford-on-Avon, 1839; prof. at Boulogne, 1841; asst. Louvre, 1848, asst. conservator, 1854-58; sent to Cairo to buy papyri, 1850; excavated in Egypt, 1850-54; conservator monuments and antiquities of Egypt, 1858-81 Mem. Acad. Inscriptions (prize). Author: Mémoire sur la mère d'Apis, 1856; Le Sérapéum de Memphis, 1857; Fouilles exécutées en Égypte, en Nubia, et au Soudan d'après les ordres du vice-roi d'Égypte, 1867; Dendérah, 5 vols., 1873-75; Abydos, 2 vols., 1870-80; Karnak, 1875; Deir el-Bahari, 1877; Les Mastabas de l'Ancien Empire, 1881. Established world's largest museum of Egyptian antiquities at Bulak, 1863; excavations include discovery of Serapeum (or alley of Sphinxes), catacombs of Apisbulls, Karnak, Medinet-Habu, Deir-el-Bahri, Tanis; although his discoveries were important, was especially noted for work in conserving ancient treasures in Egypt. Died Cairo, Jan. 19, 1881.

MARIGNAC, Jean Charles Galissard de, see de Marignac, **Jean Charles Galissard.**

MARINELLI, Leonidas D., physicist; b. Buenos Aires, Argentina, Nov. 28, 1906; s. Vincent and Amelia (Sammartino) M.; came to U. S., 1926, naturalized, 1932; student Institute A. Volta, Naples, Italy, 1925; B.S. in Elec. Engring., Cooper Union Sch. Engring., 1931; M.A. in Physics, Columbia U., 1936; m. Helen Acampora, Apr. 14, 1934; children—Linda (Mrs. John N. Landor), Judith (Mrs. Thomas J. Godfrey). Asst. physicist Meml. Hosp., N.Y.C., 1935-43, physicist in charge, 1943-45; mem., head div. physics and biophysics, 1945-49, sr. biophysicist, asso. dir. radiol. physics div. Argonne (Ill.) Nat. Lab., 1948-63, dir. radiol. physics div., 1963-67; asst. prof. Cornell U. Med. Coll., 1944-49; research asso. U. Chgo., 1948—, faculty, 1948—, asso. prof. radiology, 1955—. Mem. adv. com. for biology and medicine AEC, 1957-63; mem. expert adv. panel on radiation WHO, 1960-64; mem. radiol. health adv. com. on biol. effects radiation USPHS, 1965—. Recipient Janeway medal Am. Radium Soc., 1958. Contbr. numerous articles to tech. jours. Research on dosimetry and uses radioactive isotopes in industry, clin. therapy, med. research, radiation protection internal emitters, spectrometry low-level gamma radiation emitted by human body, epidemiological investigations concerning effects chronic radiation. Home: 320 E. Claymoor Dr., Hinsdale, Ill. 60521. Office: Argonne Nat. Lab., Argonne, Ill. 60439.*

MARINESCO, Georges, Rumanian virologist; b. Izvor, Rumania, Mar. 23, 1922; s. Voicu N. and Ioana (Tudose) M.; grad. Faculty Law, Bucharest, Rumania, 1946; dr. Med. Scis., Faculty Medicine, Bucharest, Rumania, 1947; m. Speranta Hallunga, Jan. 19, 1948; children—Michaela, Alexandru-Octavian, Mugurel-Ionut. Reader, Faculty Medicine, Clinic Infectious Diseases, Bucharest, 1950-62; staff Inst. Inframicrobiology, Rumanian Acad., Bucharest, 1950-58, head lab. immunology, 1958—; sr. physician children's dept. Dr. V. Babes Hosp. for Infectious Diseases, Bucharest. Recipient the Hadot, French Acad. Medicine, 1963, Jansen, 1964; Medalia Muncii, Rumania, Ordinul Muncii. Mem. Soc. Microbiology (Paris), French Soc. Exotic Pathology, N.Y., Lombard (Italy) acads. sci. Author: Infectious Mononucleosis, 1960; (with St. Nicolau, N. Constantinescu, N. Cajal) Poliomyelitis, 1963; Infectious Lymphocytosis, 1965; also numerous articles. Research on infectious diseases caused by viruses, pneumonia in poliomyelitis and measles myocarditis (prize from Acad. Medicine in Paris). Home: 211 sos. Mihai Bravu. Office: Institutul de Inframicrobiologie, 285 sos. Mihai Bravu, Bucharest, Rumania.*

MARINESCO, Georges, Rumanian pathologist; b. Rumania, 1864; s. Mariu Marinesco; M.D., Bucharest, Rumania. Prof. medicine Faculty Bucharest; med. research on lethargic cephalagy, London, during World War 1. Mem. Rumanian Acad., Med. Acad. Paris. Author: La célule nerveuse, 2 vols.; also numerous publs. on pathology, chronic rheumatism, myopathy, amputation, tropical diseases. Credited with introduction of term chromatolysis, 1909; studied biology of nerve cell under normal and path. conditions. Died 1938.

MARINETTI, Guido V., Am. biochemist; b. Rochester, N.Y., June 26, 1918; s. Michael and Nancy (Lippa) M.; B.S. with high distinction in chemistry, U. Rochester, 1950, Ph.D. in Biochemistry, 1953; m. Antoinette Francione, Sept. 19, 1942; children—Timothy D., Hope L. Research biochemist Western Regional Lab., Albany, Cal., 1953-54; faculty U. Rochester Sch. Medicine and Dentistry, 1954—, prof. biochemistry, 1966—; vis. prof. Cornell U., 1964. Mem. sub-com. NRC, 1961-63. Recipient Lederle Med. Faculty award Am. Cyanamid Co., 1955-57; NSF fellow, 1952-53. Mem. Am. Soc. Biol. Chemists, Am. Chem. Soc., N.Y. Acad. Scis., A.A.A.S., Am. Assn. U. Profs., Sigma Xi, Phi Beta Kappa (Glycerine Research award, 1957). Editor: Lipid Chromatographic Analysis, vol 1, 1967. Research and publs. on devel. micro-paper chromatographic techniques for the separation and analysis of phospholipids; studied the mechanism of biosynthesis of phospholipids and glycerides in mammalian systems; elucidated structure of sphirgomyelin, cerebrosides and plasmologens; also the role of lipids in blood clotting; analyzed lipid content of biol. preparations; studied the action of snake venom phospholipase A.*

MARINI-BETTOLO, Giovanni Battista, Italian chemist; b. Rome, Italy, June 27, 1915; s. Rinaldo and Evelina (Bettolo) Marini; Doctor's degree, U. Rome, 1937; Dr.h.c., U. Cat. Chile, 1948; m. Luisa Piva, Sept. 12, 1945; children—Rinaldo, Priscilla, Umberto, Maria Vittoria. Staff, Rome U., 1938-46, asst. prof., 1941-46; prof. chemistry U. Cat. Santiago (Chile), 1947-48; prof. Faculty Chemistry, Montevideo, Uruguay, 1948-49; research prof. Istituto Superiore Sanità, Rome, 1950-60, head dept. biochemistry, 1960—, acting dir., 1964—; prof. chemistry, U. Catolica, Rome, 1961—. Chmn., European Pharmacopoeia Commn., 1964-68. Fellow Accademia Nazionale dei XL, Casa de la Cultura Ecuadoriana, Am. Chem. Soc., Société Suisse di Chimie, Chem. Soc., Società Chimica Italiana. Author: Reazioni Organiche, 1951; (with Bovet, Bovet-Nitti) Curare and Curare like Agents, 1959; also numerous articles. Developed new methods for preparation of flavones; elucidated structure of new flavones; research on strychnos alkaloids, new quinones, relation between activity and structure of synthetic sympatholytics on model of ergotamine; demonstrated new alkaloids in Apocynaceae. Home: 58, Via Crecenzio, Rome, Italy.*

MARINONI, Hippolyte, inventor; b. nr. Melun, France, 1823; apprentice to printer at age 12; laborer in hand printing press factory; joined printer Emile de Girardin as inventor, 1848; later dir. Petit Jour. Invented a cotton gin, 1845; later a printing press which quadrupled printing speed; rotation press, 1866, improved it 1872, and used it in color printing, 1899. Died 1904.

MARINUS, Roman physician, anatomist; possibly lived in Alexandria; mem. Eclectic Sch.; flourished during reign of Nero, 54-68 A.D.; tchr. of Quintus, Numisianos of Corinth. Author: of an anatomy in 20 books; book on roots of nerves; an anatomy of muscles; a commentary on aphorisms. Discovered inferior laryngeal nerves, mesenteric glands, and probably vagus nerve (referred to as 6th nerve); gave an accurate description of muscles; discussed many physiol. problems; Galen spoke of him as the reviver of anat. knowledge.

MARINUS OF TYRE, Greek geographer, mathematician; flourished 2d century; founder math. geography; located places by latitude and longitude, established the prime meridian through Fortunatae Insulae (adopted by Ptolemy).

MARIOLOPOULOS, Elias, Greek meteorologist; b. Athens, Greece, Aug. 19, 1900; s. George and Hellene M.; ed. U. Athens, Cambridge U., U. London, U. Paris; Ph.D.; m. Catherine Kanaguina, 1938. Head meteorol. service Athens Obs.; rector U. Athens, 1959-61. Mem. Royal Meteorol. Soc. London; hon. mem. Budapest Meteorol. Soc., Belgrade Geog. Soc. Home: Odos Vizinou 7, Athens. Office: Nat. Obs., Athens, Greece.

MARION, (Antoine) Fortuné, French biologist; b. Aix, France, Oct. 10, 1846; prof. zoology Marseille (France) Faculty Scis.; dir. Marseille Marine Zoology Lab.; mem. French Acad. Scis., 1887. Author: Essai sur l'état de la végétation à l'époque des marnes heersiennes de Gelinden, 1873; Recherches sur les végétaux fossiles, 1876; L'évolution du régne végétale, 1881. Specialist in marine fauna; supported neo-Lamarkian belief in influence of environment on heredity; research on anti-phylloxera measures. Died Marseille, Jan. 23, 1900.

MARION, Jerry Baskerville, Am. physicist; b. Mobile, Ala., Dec. 10, 1929; s. Lester L. and Virginia (Hensel) M.; B.A., Reed Coll., 1952; M.A., Rice U., 1953, Ph.D., 1955; m. Adelia I. Macnab, May 31, 1952; children—Forrest Lee, Kathryn Ann. Research asso. Rice U., 1955; NSF fellow Cal. Inst. Tech., 1955-56, vis. asso. in physics, Guggenheim fellow, 1965-66; faculty U. Rochester, 1956-57; physicist Los Alamos Sci. Lab., 1957; faculty U. Md., College Park, 1957—, prof. physics, 1962—. Sr. staff scientist Convair/San Diego, 1960-61; com. mem. Nat. Acad. Sci.-NRC; cons. govt. agys., indsl. firms. Mem. Am. Phys. Soc., Am. Inst. Physics. Author: Nuclear Reaction Graphs, 1960; Classical Dynamics of Particles and Systems, 1965; Classical Electromagnetic Radiation, 1965; Principles of Vector Analysis, 1965. Editor: (with J. L. Fowler) Fast Neutron Physics, 1960; (with G. C. Phillips, J. R. Risser) Progress in Fast Neutron Physics, 1963. Mem. internat. editorial bd. Nuclear Data, 1965—. Research, numerous publs. on structure of light nuclei, especially effects relative to isoboric spin, single particle states, configuration mixing. Home: 705 Hobbs Dr., Silver Spring, Md. Office: Dept. Physics and Astronomy, U. Md., College Park, Md. 20740.*

MARION, Leo (Edmund), Canadian chemist; b. Ottawa, Ont., Can., Mar. 22, 1899; s. Joseph and Emma (Vezina) M.; B.Sc., Queen's U., Kingston, Ont., 1926, M.Sc., 1927; Ph.D., McGill U. Montreal, Que., Can., 1929; postgrad. U. Vienna, 1934-35; hon. degrees Laval U., 1954, U. Ottawa, 1958, Queen's U., 1961, Montreal U., 1961, U. Toronto, 1962, Sorbonne, 1962, U. B.C., 1963, Royal Mil. Coll., 1965, Carleton Coll., 1965, McGill, Bishop's U., 1966, Poznan, 1967; m. Marie-Paule Lefort, Oct. 3, 1933. Chemist, NRC, 1929-34, 34-43, head chem. sect., 1943-65, sr. dir., 1960-63, v.p. sci., 1962-65; dean Faculty Pure and Applied Sci., U. Ottawa, 1965—; editor-in-chief Can. Jours. Research, 1947-65. Decorated Order Brit. Empire. Recipient medal Assn. Can. pour l'Advancement des Sci., 1948; Chem. Inst. Can. 156; Profl. Inst. Can. medal, 1959; Jecker prize Acad. des Sci. Paris, 1963. Fellow Royal Soc. Can., Royal Soc., 1961; mem. Soc. Chem. France (hon.), Chem. Inst. Can., Am. Chem. Soc., Chem. Soc. Britain. Contbg. author The Alkaloids; also author numerous articles. Determined structure lupine alkaloids aconite alkaloids, biosynthesis alkaloids.*

MARIOTTE, Edme, French physicist; b. Dijon, France, 1620; ordained priest Roman Catholic ch., prior St. Martin-sous-Beaune, nr. Dijon. Mem. French Acad. Scis. (one of 1st). Author: Traité de la percussion ou choc des corps; Essai sur la nature de l'air; Essai du chaud et du froid; Essai sur la nature des couleurs; Traité du mouvement des eaux et des autres corps fluides. Performed many expts. to elucidate nature of heat, cold, light, sight, color, water; papers on hydrostatics and hydraulics confirmed Torricelli's work on outflow speed of fluids and demonstrated flow resistance in supply pipes; noted that frozen water occupied greater vol. than liquid water; performed barometric expts., suggested that barometer might be used to measure altitudes of mountains; discovered (independently of Boyle) that vol. of a gas varies inversely as the pressure (went further than Boyle in noting that this inverse relation holds only for constant temperature); investigated sap pressure in plants; demonstrated absorption of heat waves in glass; discovered punctum caecum in eye; described macula lutea; investigated laws of impact between elastic bodies, also invented apparatus to verify them; worked on theory of coronas, also discussed rainbow, halos, parhelia, diffraction; studied notes of the trumpet; investigated recoil of guns and motion of pendulum; produced numerous pieces of lab. equipment, such as Mariotte's tube. Died Paris, May 12, 1684.

MARISCOTI, Galeotto, Italian physicist; flourished circa 1617. Jesuit. Attempted to refute Aristotelian color doctrine in 1617; showed that colors not produced by prism itself but by refraction of light rays.

MARIUS, Simon (also: Mayr, Mayer), German astronomer; b. Gunzenhausen, Germany, 1570; studied astronomy under Tycho Brahe; studied medicine in Italy. Court astronomer in Ansbach to Elector of Brandenburg. Author: Tabulae directionum novae universae Europae inservientes, 1599; Frankischer Kalender oder practica, 1610; Mundus Iovialis anno, 1609 detectus. Observed faint nebula in Andromeda; on Mayr's instigation, one of his students obtained copy of Galileo's handbook for users of Galileo's proportional compass, and with his father rewrote it and published it as their own; later claimed to have discovered prior to Galileo 4 satellites of Jupiter, which he named Io, Europa, Ganymede, and Callisto; Galileo showed this priority claim unwarranted, a judgment sustained by later investigators. Died Ansbach, Dec. 26, 1624.

MARJOLIN, Jean-Nicolas, French surgeon; b. Ray-sur-Saône, France, Dec. 6, 1780; became intern, Paris, France, 1803; 1 son, René. Became candidate for chair operative surgery after death of Sabatier, 1812; named 2d surgeon Hôtel Dieu, Paris, 1818, prof. surg. pathology, 1819; practiced in Paris; physician to King; asso. with Roux, Cloquet, Mural. Mem. French Acad. Medicine. Author: Manuel d'anatomie, 1815; Cours de pathologie chirurgicale, 1837. Research on surgery for strangulated hernia. Died Paris, 1850.

MARK, Hans Michael, physicist, nuclear engr.; b. Mannheim, Germany, June 17, 1929; s. Herman Francis and Maria (Schramek) M.; came to U. S., 1940, naturalized, 1945; A.B., in Physics, U. Cal. at Berkeley, 1951; Ph.D., Mass. Inst. Tech., 1954; m. Marion G. Thorpe, Jan. 28, 1951; children—Jane H., James P. Research asso. Mass. Inst. Tech., 1954-55, asst. prof., 1958-60; research physicist Lawrence Radiation Lab., U. Cal., Livermore, 1955-58, 60—, exptl. physics div. leader, 1960-64; asso. prof. nuclear engring. U. Cal. Berkeley, 1960—, dept. chmn. nuclear engring., 1964—. Fellow Am. Phys. Soc., Am. Nuclear Soc., Am. Geophys. Union. Author: (with N. T. Olson) Experiments in Modern Physics, 1966; also numerous articles. Research on nuclear energy levels, nuclear reactions, applications nuclear energy for practical purposes, atomic fluorescence yields, measurement X-rays above atmosphere. Home: 90 Avenida Dr., Berkeley, Cal. 94720.*

MARK, Herman F(rancis), chemist; b. Vienna, Austria, May 3, 1895; s. Herman Carl and Lili (Mueller) M.; Ph.D., U. Vienna, 1921; E.D. (hon.) U. Liége, Belgium; (hon.) Ph.D., U. Berlin, U. Upsala, Karl-Franzens U., Graz; D.Sci. (hon.), Johannes Gutenberg U., Mainz, Lowell Inst.; Eng. D. Technical University Munich; m. Mary Schramek, Aug. 19, 1922; children—Hans Michael, Peter Herman. Came to U. S., 1940, naturalized, 1945. Research asso. Kaiser Wilhelm Inst., Berlin, 1922-26, I. G. Farben at Ludwigshafen, 1927-32; prof. chem., U. Vienna, 1932-38; research mgr. Canadian Internat. Paper Co., Hawkesbury, Can., 1938-40; prof. organic chemistry Polytechnic Inst., Brooklyn, 1940-44; dir. Inst. Polymer Research, 1944-64, dean of the faculty, 1961—. Recipient numerous awards. Fellow N.Y. Acad. Scis., Am. Phys. Soc., Textile Inst. Gt. Britain; mem. Am. Chem. Soc. (past chmn. N.Y. and Met. L.I. sects.), A.A.A.S., Nat. N.Y. acads. scis., Am. Acad. Arts and Scis., Dutch, Hungarian, Roumanian, Austrian, Spanish acads. sci., Max Planck Soc.; hon. mem. Indian Acad. Sci., Weizmann Inst. Sci., Istituto Politecnico Milan. Author 14 books on fibers, rubbers and plastics. Contbr. articles to numerous publs. Synthesized several organic free radicals that remained uncombined in solid crystalline state; determined structure of several complex compounds (including urea and pentaerythritol); gave first quantitative measurement of anamalous dispersion and polarization of x-rays; determined structure of cellulose, silk, rubber, chitin and starch, and developed new polymers (including polyvinyls and polyacrylics). Home: 325 Ocean Av., Brooklyn 25. Office: 333 Jay St., Bklyn. 1.*

MARK, Lester Charles, Am. anesthesiologist; b. Boston, July 16, 1918; s. Harry and Jean (Akabas) M.; M.D., U. Toronto (Ont., Can.), 1941; postgrad. anesthesiology N.Y. U.; m. Muriel Harriet Widman, Nov. 17, 1946; children—Dana F., Laurence Peter. Practice medicine specializing in anesthesiology, N.Y.C.; staff Presbyn. Hosp., N.Y.C.; research fellow N.Y. U. Research Service, Goldwater Meml. Hosp., 1948-51; clin. instr. State U. N.Y., Bklyn., 1952-53; with Columbia, 1953—, prof. anesthesiology, 1965—; attending anesthesiologist Presbyn. Hosp., N.Y.C., 1965—; Fulbright prof., Denmark, 1960-61; China Med. Bd. prof. Sapporo Med. Coll., Japan, 1967. Cons. Anaesthesiology Centre, Copenhagen, Denmark, 1960-61, A.M.A. Council on Drugs, 1963—; mem. adv. com. on respiratory and anesthetic drugs Food and Drug Administrn., 1967—; pharmacology-toxicology rev. com. Nat. Inst. Gen. Med. Scis., 1968. Am. Heart Assn. Research fellow, 1949-51; John Simon Guggenheim Meml. fellow, 1960-61. Recipient Distinguished Service award N.Y. State Jour. Medicine, 1967; Hiroshima U. medal, 1967. Diplomate Am. Bd. Anesthesiology. Fellow Am. Coll. Anesthesiologists, N.Y. Acad. Medicine, A.A.A.S.; mem. A.M.A., Am., N.Y. Socs anesthesiologists, Pan Am. Med. Assn. (Med. Ambassador of Goodwill bronze plaque 1962, past pres.), Am. Soc. for Pharmacology and Exptl. Therapeutics (exec. com. clin. pharmacology div.), Am. Fedn. for Clin. Research, Sigma Xi. Editor: Clinical Anesthesia Conferences; co-editor Advances in Anesthesiology: Muscle Relaxants; editorial cons. Jour. Pharmacology and Exptl. Therapeutics 1959—; collaborator Medicina et Pharmacologia Experimentalis, 1959—; editorial bd. Jour. Oral Therapeutics and Pharmacology, 1964—. Research, numerous publs. on mechanisms of drug action in man, blood-brain, placental and peritoneal barriers, barbiturate studies; co-discoverer of procainamide. Home: 195 Chestnut Dr., Roslyn, N.Y. 11576. Office: 622 W. 168th St., N.Y.C. 10032.*

MARKALOUS, Eugen, Czech. neurologist, endocrinologist; b. Prague, Czechoslovakia, Apr. 19, 1906; s. Bohumil and Gisela M.; M.D., Charles U., Prague, Czechoslovakia. Asst. 1st internal clinic Charles U., Prague. Mem. Assn. Czechoslovak Physicians. Research in internal secretion, ovarian follicles, hyperpsychosis. Home: Prague 2, Trojická 1, CSSR.*

MARKARYAN, Partev Ambartsumovich, Russian obstetrician, gynecologist; b. Arazap, Armenia, 1898; grad. Med. Faculty, Yerevan U., 1927; D.Med. Sci., 1952. Asst. dept. obstetrics and gynecology Yerevan Med. Inst., 1932-35, lectr., 1936-50, dean Therapy Faculty, 1939-43, head chair obstetrics and gynecology Postgrad. Med. Faculty, Health-Hygiene Faculty, 1958—; Armenian dep. peoples commissar of health, 1943-45; dir. Research Inst. Obstetrics and Gynecology, Armenian Ministry Health, 1949—, also chief obstetrician and gynecologist, chmn. obstetrics council. Plenum mem. council obstetrics and gynecol. service USSR and RSFSR Ministry Health. Author: Manual of Obstetrics for Medical Colleges, 3d edit., 1960; Prescriptions Manual for Obstetrics and Gynecology, 1960. Mem. editorial council Obstetrics and Gynecology. Research and publs. on malaria and pregnancy, metabolism, birth rate, child mortality. Address: Research Inst. Obstetrics and Gynecology, Yerevan, Armenian SSR, USSR.

MARKEE, Joseph Eldridge, Am. anatomist, endocrinologist; b. Neponset, Ill., May 22, 1903; s. Joshua W. and Josephine (Eldridge) M.; student Knox Coll., 1921-24; B.S., U. Chicago, 1924, Ph.D., 1929; m. Myrtle Clapp, July 2, 1927; children—Shirley J., Joseph Eldridge. Mem. faculty Stanford, 1929-43, prof. anatomy, 1943; prof., chmn. dept. anatomy Duke U. Med. Sch., 1943-53, James B. Duke prof., 1953—, chmn. dep. anatomy, 1953-63, asst. dean medical admissions; visiting prof. anatomy U. Tenn. Med. Sch., summer 1942. Recipient Golden Apple award Student Am. Med. Assn., 1963. Hon. fellow Am. Soc. Orthopedic Surgeons, Am. Orthopedic Assn., mem. Assn. Medical Colleges (audio-visual com.), Am. Assn. Anatomists (adv. com. med. Film Inst.; exec. com.), Am. Physiol. Soc., Am. Zool. Soc., N.Y. Acad. Sci., Phi Beta Kappa, Sigma Xi. Asso. editor Journal Morphology. Clin. studies of skeletal muscle activity, relation of nervous system to hypophysis, and menstruation; produced visual aids for teaching of medicine. Home: 1015 Demerius St., Durham, N.C.

MARKERT, Clement Lawrence, Am. biologist; b. Las Animas, Colo., Apr. 11, 1917; s. Edwin John and Sarah (Norman) M.; B.A. summa cum laude, U. Colo., 1940; M.A., U. Cal. at Los Angeles, 1942; Ph.D., Johns Hopkins, 1948; m. Margaret Rempfer, July 29, 1940; children—Alan Ray, Robert Edwin, Betsy Jean. Merck-NRC fellow Cal. Inst. Tech., 1948-50; faculty U. Mich., 1950-57, asso. prof., 1956-57; prof. biology Johns Hopkins, 1957-65; prof., chmn. dept. biology, Yale, 1965—. Trustee Bermuda Biol. Sta. for Research, Inc.; panelist NSF, 1959-63. Mem. Nat. Acad. Scis., Am. Inst. Biol. Scis. (pres. 1966), Soc. for Devel. Biology (past pres.), Am. Soc. Biol. Chemists, Genetic Soc. Am., Am. Soc. Naturalists (v.p. 1968), Am. Acad. Arts and Scis., Am. Soc. Zoologists (pres. 1969), Internat. Inst. Embryology, Internat. Soc. Cell Biology. Editor: Archives of Biochemistry, 1963—; mng. editor Jour. Exptl. Zoology, 1963—. Contbr. numerous articles to tech. jours. Research on multiple gene control enzyme synthesis, role genes in cell differentiation, isozymic forms enzymes, their structure, genetic control and synthesis during cell differentiation. Home: 64 Hartford Turnpike, Hamden, Conn. 06517. Office: Dept. Biology, Yale U., New Haven 06520.*

MARKGRAF, Friedrich, botanist; b. Berlin, Germany, Feb. 1, 1897; s. Paul and Maria (Zillmann) M.; Dr.Phil., U. Berlin, 1921, Privatdozent, 1927; m. Ingeborg Dannenberg, Nov. 7, 1939; 1 dau., Vera. With Bot. Garden and Mus., Berlin-Dahlem, 1920-48, prof., 1941-48; conservator Bot. Garten, Munich, 1948-56, dir., 1956; asso. prof. U. Munich, 1948-58; prof. U. Zürich, Switzerland, 1958-67, dir. Inst. Systematic Bot. and Bot. Garden, until 1967. Recipient Medalha de Curo Dom Goas VI (Brazil). Hon. mem. Bayer. Bot. Gesellschaft; corr. mem. Svenska Värtgeog. Sellsk.; mem. Internat. Assn. Plant Taxonomy, Deutsch Bot. Gesellschaft, Schweiz. Naturf. Gesellschaft, Schweiz Bot. Gesellschaft. Author: Ökologie der Bredower Forst, 1921; Gnetales, 1926; Praktikum der Vegetationskunde, 1926; Hayeks Prodromus Florae Penins Balcan, 1928-33; In Albaniens Bergen, 1930; Pflanzengeog. v. Albanien, 1932. also numerous articles. Research in phytogeography, phytomorphology, systematics of higher plants; phytogeographic studies in Albania, Italy, Greece, Alps, Brazil, Turkey. Home: 2 Bruderwiesweg, Zürich, Switzerland CH 8041.*

MARKHAM, Sir Clements Robert, Brit. geographer; b. Stillingfleet, County York, July 20, 1830; s. David F. and Catherine (Milner) M.; m. Minna Chishester. Served in navy, 1844-52, on artic expdn., 1850-51; traveled in Peru, 1852-54; geographer Abyssinian expdn.; asst. sec. India Office, 1867-77. Fellow Royal Soc., 1873; mem. Hakluyt Soc.; sec. Royal Geog. Soc., 1863-88, pres. Author: Franklin's Footsteps, 1853; Cuzco and Lima, 1856; Travels in Peru and India, 1862; History of the Abyssinian Expedition, 1869; The Threshold of the Unknown Region, 1874; Life of John Davis, 1889; Christopher Columbus, 1892; Quichua Dictionary, 1908; Life of Sir Leopold M'Clintock, 1909; The Incas of Peru, 1910; also others. Authority on Inca civilization; introduced cultivation of quinine-yielding chinchona trees from Peru into Brit. India, 1859-62. Died Jan. 30, 1916.

MARKHAM, Edwin C(arlyle), Am. chemist; b. Durham, N.C., Dec. 24, 1902; s. James William and Anna (Leigh) M.; A.B., Trinity Coll., 1923; Ph.D., U. Va., 1927; m. Anne Janet Whitlock, May 27, 1928 (dec.); children—Carlyle, Allan Whitlock. Asst. prof. chemistry, U. Va., 1927-30, research asso., 1930-33; instr. chemistry, U. Del., 1933-34; asst. prof. chemistry, U. N.C., 1934-37, asso. prof., 1937-41, prof., 1941-51, Smith prof. chemistry 1951—. Mem. Am. Chem. Soc., N.C. Acad. Sci. Sigma Xi, Alpha Chi Sigma. Research in heterogeneous reaction rates, adsorption of gases mixtures by metals and metallic oxides, adsorption of gases at high pressures, use of organic reagents in analytical chemistry, indicator constants in organic media, polarography. Home: Chapel Hill, N.C.*

MARKHAM, John Raymond, Am. aero-engr.; b. Cambridge, Mass., July 23, 1895; s. John Henry and Mary (Williams) M.; M.E., Mass. Inst. Tech., 1918; m. Genevieve Triquera, June 5, 1921; 1 son, James Paul (killed in action, Mar. 8, 1945). Began as research asso. aero. engring. dept., Mass. Inst. Tech., 1922, prof. aero. engring. 1946—, dir. supersonic lab. 1947—; dir. Wright Bros. Wind Trunnel, Mass. Inst. Tech.; cons. engr. Argentine and Brazilian govts. in design of wind tunnels and equipment, also USAAF, Boeing Aircraft, United Aircraft Corp. and Gen. Motors Corp.; mem. sci. adv. bd. USAAF, 1945—; chmn. industry and ednl. adv. bd. USAF sci. com. and subcom. Nat. Adv. Council for Aeros. Chmn. bd. Mithras, Inc. (Cambridge, Mass.). Fellow Inst. Aero. Sci.; mem. Sigma Xi. Contbr. articles to engring. and sci. publs. Home: 73 Somerset St., Belmont, Mass. Office: Mass. Inst. of Tech., Cambridge, Mass.*

MARKLEY, Klare Stephen, Am. chemist; b. Phila., Dec. 16, 1895; s. Jonah Jacob and Mabel Eileen (Montague) M.; B.S. in Chem. Engring., George Washington U., 1924, M.S., 1925; Ph.D., Johns Hopkins, 1929; m. Calla Inez Lepper, Dec. 24, 1921 (dec. Dec. 1954); m. 2d, Carmen Nogueira de Mello, Mar. 14, 1955. From asst. to asso. biochemist U. S. Dept. Agr., 1927-37, chief oil sect. Regional Soybean Lab., Urbana, Ill., 1937-39; chief oilseed div. So. Regional Research Lab., New Orleans, 1939-52, cons. Brazil, Uraguay, Columbia, 1960; with ICA, Paraguay, 1952-54, mem. staff U. S. operations mission, Brazil, 1954-60; mem. missions FAO, Venezuela, Guatemala, 1948; indsl. cons. Rio de Janeiro, 1961—. Sci. cons. tech. intelligence com. Joints Chiefs Staff, 1945. Recipient Superior Service award U. S. Dept. Agr., 1950. Fellow A.A.A.S.; mem. Am. Chem. Soc. (past sec. Washington sect., S.W. regional award 1951), Am. Oil Chemist Soc. (past pres.), Am. Soybean Assn., Palm Soc., Fgn. Service Assn., Sigma Xi, Phi Lambda Upsilon, Alpha Chi Sigma. Author: (with W. H. Goss) Soybean Chemistry and Technology, 1944; Fatty Acids—Their Chemistry and Physical Properties, 1947; Fatty Acids—Their Chemistry, Properties, Production and Uses, 4 vols., 1960-66. Editor: Soybeans and Soybean Products, 2 vols., 1950-51. Patentee in field. Research, publs on chemistry and phys. properties of soybeans and fatty acids. Home: 520 Anderson Av., Rockville, Md. Office: Av. Copacabana 455, Rio de Janeiro, ZC-07 Brazil.*

MARKOSYAN, Akop Artashesovich, Russian physiologist; b. Leninakan, Armenia, Sept. 20, 1904; s. Artashes A. and Vardanush A. (Vantsyan) M.; grad. Moscow State Pedagogical Inst., 1934; completed postgrad. course in physiology, 1937; m. Valentina A. Mokeicheva, June 29, 1934; children—Ruben A., Alla A. Asst. prof. physiology Pedagogogical Inst., also Moscow State Med. Inst., until 1949; dir. age physiology inst., Moscow, also head lab. vegetative functions, 1949—; head chair physiology Moscow City Pedagogical Inst. Phys. Edn., until 1953; dir. Med. Lit. Pub. House, Moscow, 1941-48. Decorated Badge of Honour, Red Labour Banner, other medals. Corr. mem. USSR Acad. Pedagogics, 1953, mem., 1965, chmn. problematic commn. on age physiology USSR Acad. Sci. Mem. Moscow Physiology Soc. (v.p.), Physiology Soc. USSR. Author: An appliance on physiology for corresponding students, 1943; A text-book on normal physiology, 1949-65; (monographs) Pavlov's Works in Blood Circulation and Digestion, 1953, Nervous Regulation of Blood Coagulation, 1960, Physiology of Blood Coagulation, 1966; also numerous articles. Conducted studies of physiol. functions in children and teen-agers, studies devels. in physiol. system, beginning with embry ogenesis till old age, studies of nervous and humoral mechanisms to regulate blood coagulation. Home: 24, flat 21 Novo-pestchanaja, Moscow A-252. Office: 8 Pogodinskaja, Moscow, Y-117, USSR.*

MARKOV, Andrey Andreevich, Russian mathematician; b. Leningrad, Sept. 22, 1903; grad. Leningrad U., 1924; D.Physico-Math. Sci., 1935. With Leningrad U., 1935-42, 44-53; prof., 1935—; sr. asso. Math. Inst., USSR Acad. Sci., 1939—, bur. mem. dept. physico-math. sci., 1960-63. Mem. USSR Acad. Sci. (corr.). Author: Free Topological Groups, 1945; Algorithm Theory, 1951; Insoluble Algorithmic Problems, 1952; Alphabet Coding, 1960. Research and publs. on topology, topological algebra, dynamic systems, algorithm theory; demonstrated impossibility of algorithmic solution of problems of number matrices and asso. systems. Address: Math. Inst., USSR Acad. Sci., 1-y Akademichesky prospect 28, Moscow, USSR.

MARKOV, Daniil Aleksandrovich, Russian neuropathologist, physiotherapeutist; b. 1895; grad. Med. Faculty, Saratov U., 1919; D.Med. Sci., 1936. Asst., lectr. clinic nervous diseases Belorussian U., 1920-33; dir. Belorussian Research Inst. Physiatrics, Orthopedics and Neurology, 1929-41; prof., head chair nervous diseases Minsk Med. Inst., 1931-41; prof., 1931—; sci. dir. Belorussian Research Inst. Neurology, Neurosurgery and Physiotherapy, 1945—, Belorussian Respiratory Center, Minsk Isolation Hosp.; head chair nervous diseases Belorussian Postgrad. Med. Inst., 1947—. Mem. Belorussian Acad. Sci. Coauthor: Epilepsy and Its Treatment, 1954; author: Clinical Chronaximetry, 1956. Co-editor Neurology sect. Large Med. Ency., 2d edit.; mem. editorial council Korsakov Jour. Neuropathology and Psychiatry, Problems of Balneology, Physiotherapy and Kinesiatrics, Pub. Health Care in Belorussia. Research and publs. on functional diagnosis of nervous diseases using clin. and physiol. methods of exam. and their therapy, revision of function of cerebellum, diagnosis and treatment of epilepsy in intermittent period, exptl. treatment of extrapyramidal diseases and methyl alcohol poisoning, tick-borne encephalitis; described tibiodigital reflex, 1921, tabetic arthropathy of spine, 1929, scleromous polyneuritis, 1931, also numerous chronaximetric syndromes in clin. treatment; pioneer clin. use of plethysmographic methods, 1929; 1st in USSR to use chronaximetry in neurol. clin. practice. Address: Belorussian Research Inst. Neurology, Neurosurgery and Physiotherapy, Podlesnaya ulitsa 9, Minsk, Belorussian SSR, USSR.

MARKOV, Konstantin Konstantinovitch, Russian geographer; b. Viborg, USSR, May 20, 1905; s. Konstantin Vasilievitch and Maria (Doss) M.; student Leningrad (USSR) U., 1921-25, D.Geog. Scis., 1937; D.Sc., U. Lodz (Poland), 1961; m. Anastasia Youze, Apr. 15, 1932; 1 dau., Anastasia. Faculty, Leningrad U., 1926-37; head dept. geography Moscow (USSR) U., 1946—, dean Geography Faculty, 1945-55, Recipient Gold medal Geog. Soc. USSR, 1959; medal Hungarian Acad. Scis., 1961. Mem. Orders Soviet Union, Croatian Geog. Soc. (hon.), Polish Geog. Soc. (hon.). Author numerous books including: The Quaternary Period, Vols. I, II, 1955; (with V. I. Bardin, V. L. Leldev, A. J. Orlov, J. A. Suyetova) The Geography of the Antarctica, 1966; also numerous articles. Research on gen. geography, geomorphology and Quarternary geology in various regions of USSR and Antarctic. Home: U. Lodging 56, Moscow B-234, U.S.S.R.*

MARKOV, Mikhail Vasilevich, Russian geobotanist; b. 1900; grad. Kazan U., Kazan Inst. Agr., 1924; postgrad. Kazan Agrl. Inst., from 1926; D.Biol. Sci., 1938. Head chair botany Kazan Agrl. Inst., 1930-32; head chair agrobotany and meadow cultivation Inst. Dairy Farming, 1930-32; head botany dept. Kazan U., 1932-44, prof., 1940—, head chair geobotany 1944—, former dean Biol. Faculty, pro-rector for sci. work. Decorated Order of Lenin. Author: The Study of the Meadows and Marshes of the Tatar ASSR, 1930; Fifteen Years of Geobotanical Research in the Tatar Republic, 1935; Darwinism and Geobotany, 1941; Wild Medicinal Plants of the Tatar ASSR, 1946. Research and publs. on meadows, weed and forest vegetation, theoretical aspects of geobotany. Address: Kazan University, ulitsa Chernyshevskogo 18, Kazan, RSFSR, USSR.

MARKOV, Moisey Aleksandrovich, Russian theoretical physicist; b. May 13, 1908; grad. Moscow U., 1930. Asso., Physics Inst., USSR Acad. Sci., 1934—, dir. Inst. Semicondrs., until 1962. Mem. USSR Acad. Sci. (corr.). Author: Hyperons and K-Mesons, 1958; The Present Forum of Atomic Theory, 1960; The Future Theory of Elementary Particles, 1960; Neutron Physics of High Energies and Cosmic Rays, 1962. Research and publs. on quantum electrodynamics, theory of elementary particles. Address: Physics Inst., USSR Acad. Sci., Miusskaya ulitsa 3, Moscow, USSR.

MARKOVIC, Tihomil, Yugoslavian chemist; b. Zemun, Yugoslavia, Aug. 6, 1919; s. Jakob and Ruza (Visnjevski) M.; grad. Faculty Tech., U. Zagreb (Yugoslavia), 1947, grad. Faculty Natural Scis., 1951, Ph.D., 1954, D.Sc., 1956; m. Vera Fordren, Sept. 15, 1946; 1 son, Damir. Staff, Faculty Tech., U. Zagreb, 1950—, prof., 1960—; dean Faculty Tech., Tuzla, Yugoslavia, 1962—. Mem. Bunsengesellschaft für Physikalische Chemie, CEOCOR. Author: Bodenkorrosion und ihre Verhütung, 1963; Thermodynamische Korrosion, Passivität und Oberflächenschutz, 1967; also numerous articles. Research on thermodynamic and kinetic formulation of soil aggressivity; system metal-solution as mixed polyelectrode; thermodynamic electrochem. affinity-overpotential; diagram for active, passive, transpassive and immune behavior of metals; system of passive metals. Home: 25 Irceva, Tuzla, Yugoslavia.*

MARKOWITZ, J(acob), Canadian physiologist; b. Toronto, Ont., Can., Sept. 17, 1901; s. Harry and Jeanette (Marcus) M.; M.B., U. Toronto, 1923, Ph.D., 1926; M.S. in Exptl. Surgery, U. Minn., 1930; m. Ruth McCullough, Jan. 12, 1946; children—Jane Catherine, Thomas Henry. Prof. physiology Georgetown, U. Sch. Medicine, 1930-32; prof. physiology Toronto U., 1959—. Pres., Canadian Forum, 1948—. Decorated Order Brit. Empire. Mem. Canadian Physiol. Soc., Am. Physiol. Soc. Author: (with J. Archibald, A. Downie) Experimental Surgery, 1937; also numerous articles. Research on exptl. surgery and transplantation of organs. Home and office: 46 Forest Hill Rd., Toronto 7, Ont., Can.*

MARKOWSKI, Adam, Polish plant physiologist; b. Lvov, Poland, June 23, 1921; s. Zdzislaw and Zuzanna (Glogowska) M.; M.Sc.Agr., Jagellonian U., 1946, doctor's degree Agr., 1952; m. Danuta Czepiel, 1952; 1 dau., Elizabeth. Head plant breeding research sta. Grodkowice, nr. Cracow, Poland, 1946-50; lectr. Jagellonian U., 1950-51, head dept. plant physiology, 1951-54; head dept. plant physiology Agr. U., Cracow, 1954—; head lab. plant physiology Polish Acad. Sci., 1956—; vice dean Agr. Faculty, Agr. U., 1960-66. Decorated Krzyz Kawalerski Orderu Odrodzenia Polski. Mem. Polish Bot. Soc., Polish Biochem. Soc., Soc. Cryobiology. Research and publs. on influence of climatic condition on generative devel. cultivated plants; formulated new hypotheses on interaction between temperature and light conditions during devel. plants; studies on physiol. bases of frost resistance of plants; biochem. metabolism during vegetation plants. Home: 11 Rydla, Cracow. Office: 21 Mickiewicza, Cracow, Poland.*

MARKS, Paul Alan, Am. physician, biochemist; b. N.Y.C., Aug. 16, 1926; s. Robert R. and Sarah (Boharad) M.; A.B. with gen. honors, Columbia, 1945, M.D., 1949; m. Joan Harriet Rosen, Nov. 28, 1953; children—Andrew Robert, Elizabeth Susan, Matthew Stuart. Fellow, Columbia Coll. Phys. and Surg., 1952-53, asso., 1955-56, faculty, 1956—, dir. hematology tng., 1961—, prof. medicine, 1967—; asso. investigator Nat. Inst. Arthritis and Metabolic Diseases, NIH, Bethesda, Md., 1953-55; instr. George Washington U. Sch. Medicine, 1954-55; vis. scientist Lab. Cellular Biochemistry, Pasteur Inst., 1961-62. Mem. NSF adv. panel on developmental biology, 1964-67; cons. VA Hosp., N.Y.C., 1962-66; vis. prof. Ist. di Chimica Biologica, U. Genova (Italy), 1963. Recipient Charles Janeway prize Columbia, 1949, Joseph Mather Smith prize, 1959, Stevens Triennial prize, 1960; Commonwealth Fund fellow, 1961-62; recipient Swiss-Am. Found. award in Med. Research, 1965. Mem. Red Cell Club (past chmn.), Am. Fedn. for Clin. Research (past Eastern councillor), Am. Soc. for Clin. Investigation, Am. Soc. Biol. Chemists, Am. Soc. Human Genetics (past program Com.). Editor: Monographs in Human Biology, 1963; editorial bd. Blood, 1964—; editor-in-chief Jour. Clin. Investigation, 1967—. Research, numerous publs. on biochemistry erythrocyte aging, predisposing factors to anemia, messenger RNA for hemoglobin, polyribosomes, molecular defect in thalassemia. Home: 25 Claremont Av., N.Y.C. 10027. Office: 630 W. 168th St., N.Y.C. 10032.*

MARKS, Vincent, English physician; b. London, Eng., June 10, 1930; s. Lewis Myer and Rose (Goldbaum), M.; Hume Exhibitioner, Brasenose Coll., Oxford (Eng.) U., 1948-52, M.A., 1954, B.M.,B.Ch., 1954; m. Averil Rosalie Sherrard, Feb. 10, 1957; children—Alexandra Louise, Lewis Adam. Faculty, Inst. Neurology, London, 1957-61, sr. lectr., 1961-62; Med. Research Council fellow King's Coll. Hosp., 1959-61; dir. labs., cons. chem. pathologist Area Lab., West Park Hosp., Epsom, Eng., 1962—; clin. tutor Epsom Dist. Hosp., 1966—. Mem. Med. Research Soc., Brit. Diabetic Assn., Royal Coll. Physicians, Coll. Pathologists, Assn. Clin. Biochemists, Com. for Study Drug Addiction. Author: (with F. C. Rose) Hypoglycaemia, 1965; also numerous articles. Research on interrelationship between hormones especially glucagon, insulin and growth hormone, regulating glucose metabolism, diagnosis and treatment of diseases asso. with hypoglycemia; elucidation of cellular composition of cerebrospinal fluid in man. Home: 8 The Downsway, Sutton, Surrey. Office: The Area Lab., West Park Hosp., Epsom, Surrey, Eng.*

MARKSTEIN, George Henry, physicist; b. Vienna, Austria, June 22, 1911; s. Frederick and Martha (Persicaner) M.; Ingenieur, Technische Hochschule, Vienna, 1935, Dr. Tech. Sci., 1937; m. Hedil M. Wolf, Sept. 29, 1937; 1 dau., Eva K. Came to U. S., 1946, naturalized, 1951. Research physicist Allgemeine Gluehlampen Fabriks A.G., Vienna, 1937-38; seismologist Cia. de Petroleo Shell De Colombia, Bogota, Colombia, 1939-40; various indsl. and engrg. positions in Bogota, 1941-46; with Cornell Aero. Lab., Inc., Buffalo, 1946—, prin. physicist aerodynamic research dept., 1956—. Mem. Am. Phys. Soc., Am. Inst. Aeros. and Astronautics, A.A.A.S., Combustion Inst., N.Y. Acad. Sci., Sigma Xi. Editor, Author: (with H. Guenoche, A. A. Putnam) Nonsteady Flame Propagation, 1964. Research and publs. on interaction between flames and flow disturbances, including spontaneous cell structure of flames, vibratory flame motion, interaction between flames and shock waves; combustion of metals, particularly kinetics of heterogeneous reactions in metal combustion. Home: 78 Hillside Dr., Williamsville, N.Y. 14221. Office: P.O. Box 235, Buffalo 14221.*

MARKUS, Lawrence, Am. mathematician; b. Hibbing, Minn., Oct. 13, 1922; s. Benjamin and Ruby (Friedman) M.; B.S. in Math., U. Chgo., 1942, M.S. in Meteorology, 1946; Ph.D. in Math., Harvard, 1951; m. Lois Shoemaker, Dec. 9, 1950; children—Sylvia, Andrew. Instr., U. Chgo., 1942-44, Harvard, 1951-52, Yale, 1952-55; lectr. Princeton, 1955-57; faculty U. Minn., Mpls., 1957—, prof. math., 1960—, asso. head math. dept., 1961, dir. Center for Control Scis., 1964. Adviser to govt. agys.; vis. prof. Yale, 1960, U. Cal. at Berkeley, 1961, Columbia, 1963. Fulbright fellow, 1950; Guggenheim fellow, 1963. Mem. Am. Math. Soc. (past mem. council), Math. Assn. Am. (exec. officer 1962), Soc. for Indsl. and Applied Math. (nat. lectr. 1965), Phi Beta Kappa, Sigma Xi. Author: (with L. Auslander) Flat Lorentz 3-Manifolds, 1958; (with L. Auslander, F. Hahn, L. Green) Flows on Homogeneous Spaces, 1963; (with E. B. Lee) Mathematical Foundations of Optimal Control Theory, 1967; also numerous articles. Editor math. jours. Contr. to Differential Equations, SIAM Control Jour., Internat. Jour. Nonlinear Mech. Research in qualitative theory dynamical system within math. framework of differential equations and global differential geometry, applications to automatic control theory and nonlinear oscillations and stability, geometry relativistic cosmology. Home: 5004 Belmont Av., Mpls. 55419.*

MARKUSHEVICH, Aleksey Ivanovich, Russian mathematician; b. Petrozavodsk, 1908; grad. Central Asian U., Tashkent, 1930; postgrad. Moscow U., until 1934; D.Physico-Math. Sci., 1944. Prof., Moscow U., 1944—; RSFSR 1st dep. minister edn., 1958—. Mem. RSFSR Acad. Pedagogical Sci. (v.p. 1950-58, Presidium mem. 1959—). Author: Elements of Analytic Functions Theory, 1944; Real Numbers and the Basic Principles of Limit Theory, 1948; The Theory of Analytic Functions, 1950; Outline of the History of the Theory of Analytic Functions, 1951; A Short Course of Analytic Functions, 1957. Chief editor Ency. of Elementary Mathematics, Children's Ency. Address: RSFSR Acad. Pedagogical Sci., B. Polianka 58, Moscow, USSR.*

MARLATT, Charles Lester, Am. entomologist; b. Atchison, Kan., Sept. 26, 1863; s. Washington M. and J. A. (Bailey) M.; B.S., Kan. State Coll., 1884, M.S., 1886, D.Sc., 1921; m. Florence L. Brown, Dec. 1, 1896 (dec. Oct. 1903); m. 2d, Helen Stuart Mackay

Smith, July 5, 1906; children—Florence, Virginia, Charles Lester, Helen, Dorothy, Constance. Asst. prof. Agrl. Coll., Manhattan, Kan., 2 yrs.; asst. entomologist U. S. Dept. Agr., 1889-94, 1st asst. and asst. chief entomologist, 1894-1922, asso. chief, 1922-27, chief Bur. Entomology, 1927-33; engaged in hist. and geneal. work, 1933-—. Directed effort to secure nat. law to prevent importation of infested and diseased plants into U. S., 1909-12, resulting in Plant Quarantine Act of Aug. 20, 1912; chmn. Federal Hort. Bd. to supervise enforcement of this act, 1912-28; responsible for reorgn. and assembling from other burs. of Dept. Agr., of all plant quarantine and regulatory work, under new office of Plant Quarantine and Control Adminstrn., chief of office, 1928-29. Fellow A.A.A.S.; pres. Entomol. Soc. Washington, 1897-98, Assn. Econ. Entomologists, 1899; mem. Washington Acad. Scis. Biol., Archaeol. and Geog. socs. Author: An Entomologist's Quest, 1953; also many papers and bulls. on systematic and econ. entomology and on plant quarantine; also 16 vols. of service and regulatory announcements, recording 68 foreign and domestic plant quarantines, the regulations thereunder and explanatory papers, 1914-1929. Editorial com. Jour. Agrl. Research, 1919-26. Died Mar. 3, 1954.

MARLIER, George Juste Leon, biologist; b. Bordeaux, France, Jan. 28, 1917; s. Juste Edouard and Amelie (Samain) M.; D.Zoology, Free U. Brussels (Belgium), 1940; m. Mathilde Decoen, Aug. 12, 1958. Asst., U. Brussels Zool. Lab., 1935-48, suppletive prof., 1951; head Research Center Tanganyika, Uvira, 1949-60; charge de recherche Institut Belge Recherche Scientifique Outre-Mer, Brussels, 1961-63; staff Mission for Limnol. Research in Amazonia, 1963-64; dir. lab. Inst. Royal des Sciences Naturelles, Brussels, 1964-—. FAO André Mayer fellow, 1962. Pres. nat. subcom. P.F. sect. Internat. Biol. Programme, 1965-—. Author: Genera des Trichoptères de l'Afrique, 1962; also numerous articles. Co-editor Hydrobiologia, 1954-—. Research on biology and taxonomy of African trichoptera, biology of Lake Tanganyika fishes, ecology of some trop. freshwater environments. Home: 229 Av. Montjoie, Brussels 18. Office: 31 Rue Vautier, Brussels 4, Belgium.*

MARLOWE, Donald Edwards, Am. mech. engr.; b. Worcester, Mass., Mar. 27, 1916; s. Daniel James and Helen (Gillogly) M.; B.C.E., U. Detroit, 1938; M.S. in Engring., U. Mich., 1939; Sc.D., Merrimack Coll., 1962; m. Corinne Mildred Ingersoll, Nov. 23, 1939; children—Donald Edward, Christopher James. Research asso. dept. engring. research U. Mich., Ann Arbor, 1939-41; asso. dir. U. S. Naval Ordnance Lab., Whiteoak, Md., 1941-55; dean Sch. Engring. and Architecture, Catholic U. Am., Washington, 1955-—. Dir. Jitco Engring. Co., Washington. Cons. to govt. agys., pvt. industry; chmn. Bd. Registration for Profl. Engrs., D.C., 1958-66. Recipient Distinguished Civilian Service award U. S. Navy, 1946; Distinguished Alumnus award U. Mich., 1953. Registered profl. engr., D.C. Fellow Am. Soc. M.E. (Wright lectr. 1963, Distinguished Service award 1965), A.A.A.S.; mem. I.E.E.E. (sr.), Nat. Soc. Prof. Engrs. (Distinguished Service award 1962), Nat. Council State Bds. Engring. Examiners (pres. 1966-67), Distinguished Service award, 1965), Blue Key, Sigma Xi, Tau Beta Pi, Alpha Sigma Nu. Pioneered underwater ballistics. Home: 9332 Wilmer St., Silver Spring, Md. 20901. Office: Sch. Engring. and Architecture, Catholic U. Am., Washington, 20017.*

MARLOWE, Thomas Johnson, Am. geneticist; b. Fairview, N.C., Sept. 15, 1917; s. Alonzo Garland and Nellie (Bass) M.; A.A., Mars Hill Jr. Coll., 1938; B.S., N.C. State U., 1940, M.S., 1949; Ph.D., Okla. State U., 1954; m. Christine Odelle Williamson, Mar. 3, 1945; children—Michael Thomas, Ronald Jay, Donald Ray, Melody Ann. Tchr., Oxford High Sch., Catawba County, N.C., 1940-42; asst. county agt., livestock specialist Mecklenburg County, Va., 1949-50; instr. animal husbandry Miss. State U., 1951-52; grad. asst. Okla. State U., 1952-54; faculty Va. Poly. Inst., Blacksburg, 1954-—, prof. animal sci., 1964-—. Tech. advisor Va. Beef Cattle Improvement Assn. Mem. Va. Acad. Sci., Am. Soc. Animal Sci., Am. Assn. U. Profs., Am. Genetic Assn., A.A.A.S., Res. Officers Assn., Assn. So. Agrl. Workers, Sigma Xi, Phi Kappa Phi. Editorial bd. Va. Jour. Sci. Research, numerous publs. in field animal genetics, pathology of dwarfism in beef cattle; discovered growth hormone deficiency in snorter dwarf; conducted research, organized first state beef cattle performance testing program; established adjustment factors for nongenetic influences on growth and conformation. Home: Melody Acres, Route 1, Blacksburg, Va. 24060.*

MARMIER, Pierre Edouard, Swiss physicist; b. Neuhausen, Switzerland, Jan. 8, 1922; s. Edouard and Berthe (Gottrau) M.; diploma Fed. Inst. Tech., Zurich, Switzerland, 1946, Ph.D., 1951. Faculty, Fed. Inst. Tech., Zurich, 1946-51, 55-—, prof. nuclear physics, 1958-—, dir. nuclear physics lab., 1960-—, chmn. dept. math. and physics, 1964-66. Sr. research fellow Cal. Inst. Tech., 1952-55. Mem. Swiss, Am. phys. socs. Author: (with Eric Sheldon) Physics of Nuclei and Particles, 1968; also articles. Research in nuclear spectroscopy and nuclear reactions, use of nuclear methods on biol. research. Home: 18 Steinbrüchelstr., Zurich 8053. Office: Laboratorium für Kernphysik E.T.H., Zurich 8049, Switzerland.*

MARMION, Barrie Patrick, Brit. microbiologist; b. Alverstoke, Hants., Eng., May 19, 1920; s. J. P. and M. H. (Bryan) M.; M.B., B.S., London (Eng.) U., 1943, M.D., 1947, D.Sc., 1963; m. Diana Ray Newling, Sept. 24, 1953; 1 dau., Jane Louise. Rockefeller fellow virology Walter and Eliza Hall Inst., Melbourne, Australia, 1951-52; cons. bacteriologist Pub. Health Lab. Service, Leeds, Eng., 1955-62; Found. prof. microbiology, chmn. microbiology dept. Monash U., Melbourne, 1963-67; prof bacteriology U. Edinburgh (Scotland), 1968-—. Fellow Coll. Pathologists; mem. Coll. Pathologists Australia, Path. Soc. (Gt. Britain), Soc. Gen. Microbiology, Soc. Immunology, Contbg. author: Recent Advances in Medical Microbiology, 1967. Research and publs. on biology of Q fever organism and its epidemiology, virus diseases, properties of mycoplasmas. Office: Dept. Bacteriology, Med. Sch., U. Edinburgh, Edinburgh, U.K.*

MARMO, Frederick Francis, Am. chem. physicist; b. East Boston, Mass., Oct. 25, 1920; s. Nicholas and Jane (LaBelle) M.; A.B. in Chemistry summa cum laude, Boston U., 1949; M.S., Harvard, 1951, Ph.D. in Chem. Physics, 1953; m. Maryann LaRosa, Dec. 12, 1943; children—Frederick Joseph, Joanne, Malia A. Staff scientist Air Force Cambridge Research Center, Bedford, Mass., 1951-56, chief, chem. physics br., LG Hanscom Field, 1956-58; mgr. chem. physics dept. Geophysics Corp. Am. (now GCA Corp.), Bedford, 1958-61, dir. space scis. lab. GCA Tech. Div., 1961-66, dir. space scis. operations, 1966-67, v.p., tech. dir., 1967-—. Mem. reaction rate working group Def. Atomic Support Agy., 1963-—. Mem. Am. Chem. Soc., Am. Rocket Soc., Am. Optical Soc., Am. Inst. Aeros. and Astronautics, Am. Geophys. Union, Am. Meteorol. Soc., Am. Phys. Soc., Phi Beta Kappa, Sigma Xi. Theoretical and exptl. research studies on rockets, atmospheric probing. Home: 138 Main St., Wakefield, Mass. 01880. Office: Burlington Road, Bedford, Mass. 01730.*

MARMON, Howard C., Am. inventor; b. Richmond, Ind., May 24, 1876; s. Daniel W. and Elizabeth (Carpenter) M.; student Earlham Coll., 1892-94; degree in mech. engring., U. Cal. at Berkeley; m. Florence Myers, 1901; 1 dau., Carol Carpenter (wife of Prince Nicolas Tchkotoua); m. 2d, Martha Foster, 1911. Began as asso. with father in flour mill machinery bus., absorbed by automobile industry; became v.p. in charge engring. Marmon Motor Car Co., 1902; invented Marmon automobile, pioneer in designing and producing racing cars; designed Marmon Wasp, which won 1st 500-mile internat. sweepstakes on Indpls. Speedway, 1911 (average speed of 74.61 miles per hour for the course); invented duplex downdraft manifold, widely used in building straight eights; reduced weight of 16 cylinder engine by use of aluminum parts, thus making engine commercially practical; a developer of Liberty airplane motor during World War I. Pres. Am. Soc. Automotive Engrs., 1913, 14 (awarded medal by Met. sect., 1931, for year's outstanding automotive design, the Marmon Sixteen); only Am. hon. mem. English Soc. Automotive Engrs., 1913. Died Apr. 4, 1943.

MARMOR, Leonard, Am. physician; b. Passaic, N.J., Nov. 17, 1926; s. Max and Vilma (Fruchter) M.; student U. Nev., 1944-45, 46-47, Syracuse U., 1947-48; M.D., State U. N.Y. at Syracuse, 1952; m. Caryl Tonn, June 11, 1960. Instr. surgery, orthopedics U. Cal. Los Angeles Sch. Medicine, 1958-60, asst. prof., 1960-64, asso. prof., 1964-—. Recipient Nicolas Andry award, 1963. Mem. Western, Brit. orthopedic assns., Am. Acad. Orthopedic Surgeons, A.M.A., World Med. Assn., Assn. Bone and Joint Surgeons. Author: (with others) Upper Extremith Orthotics; Surgery of Rheumatoid Arthritis, 1967; Peripheral Nerve Regeneration, 1967; also numerous articles. Research on use of irradiation to prepare nerves for grafting in peripheral lesions, use of homo and hetero grafts, reconstructive surgery of patient with severe deformities due to rheumatoid arthritis. Home: 1401 Roscomare Rd., Los Angeles 90024.*

MARMOREK, Alexander, French physician, bacteriologist; b. Mielnica, France, Feb. 19, 1865; ed. U. Vienna (Austria), 1889; chief lab. Pasteur Inst., Paris, France. Pres., Zionist Fedn. in France. Author: Versuch einer Theorie der seplischen Krankheiten, 1894. Discovered serum antitoxin for streptococcus pyogenes, 1896, to tubercle bacillus, 1903. Died July 12, 1923.

MARNECK, F. H., see Bolza, Oskar.

MARON, Samuel Herbert, phys. chemist; b. Warsaw, Poland, May 28, 1908; s. Harry and Bertha (Zellin) M.; came to U. S., 1922, naturalized, 1932; B.S., Case Inst. Tech., 1931, M.S., 1933; Ph.D., Columbia, 1938; m. Pearl Weinstein, June 28, 1936; 1 dau., Linda Anne (Mrs. Ronald E. Posner). Faculty, Case Inst. Tech., Cleve., 1931-—, prof. phys. chemistry and polymer sci., 1965-—, prof. phys. chemistry, 1945-65; dir. research project Office Synthetic Rubber, 1943-53, mem. latex adv. com., 1945-50. Cons. to various pvt. cos. Recipient Certificate of Merit, Chem. Profession Cleve., 1960; Research award Case chpt. Sigma Xi, 1962. Mem. Am. Chem. Soc., A.A.A.S., Am. Inst. Chem. Engrs., Soc. Rheology, Am. Assn. U. Profs., Sigma Xi, Tau Beta Pi, Alpha Chi Sigma. Author: (with Prutton) Principles of Physical Chemistry, 1944; also numerous articles. Research on thermodynamics of gases and non-electrolyte solutions of both low and high molecular weight substances, phys. chemistry synthetic rubber latices, polymerization, solution kinetics, light scattering by colloidal systems, rheology colloids and polymer solutions. Home: 2632 Milton Rd., University Heights, O. 44118. Office: Case Western Res. U., University Circle, Cleve., 44106.*

MAROS, Tiberius, Rumanian physician; b. Hateg, Rumania, Nov. 9, 1923; s. Eugene and Ibolya (Kardos) M.; M.D., U. Bolyai, Cluj, Rumania, 1947; m. Ilona Nagy, Oct. 8, 1949; children—Susana Maria, Tibor Gábor. Faculty, Med. Faculty U. Tirgu-Mures (Rumania), 1945-—, prof. anatomy, 1958-—, dean Faculty Gen. Medicine, 1953-56, 64-—. Recipient Ordinul Muncii, Govt. of Rumania, 1965; Laureate of Babes prize Rumanian Acad. Scis., 1966. Mem. Balkan Med. Assn. Author: The Viscero-Visceral and Neuro-visceral interrelations of the liver, 1964; also numerous articles. Research in inter-relations existing between liver and nervous system; demonstrated that there is an area in diencephalon directing function of liver; cellular mechanism of hepatic regeneration, morphology of exptl. allergic encephalomyelitis in dogs especially vascular reaction, mechanism of demyelinization and problem of glia-reaction. Home and office: No. 38, Gh. Marinescu, Tirgu-Mures, Rumania.*

MAROTI, Mihály, Hungarian biologist; b. Lovasberény, Hungary, Jan. 7, 1917; s. Mihály and Erzsébet (Mong) M.; grad. U. P. Pázmány, Budapest, Hungary, 1944, tchr.'s diploma in biology and geography, 1946, Ph.D. in Biology, Zoology, Mineralogy, 1948; m. Mária Pill, Aug. 9, 1952; 1 dau., Éva. Asst., U. P. Pázmány, Budapest, 1943-52; 1st asst. U. L. Eötvös, Budapest, 1952-53, asst. prof. plant physiology, physiology of cells, 1953-—, dir. Biol. Sta., Alsógöd, Hungary, 1954-—. Recipient ministerial praise and decoration for popularizing biology, 1966. Contbg. author: A kisérleti orvostudomány vizsgáló módszerei. IV, 1958. Research and numerous publs. on devel. of plant cells, cell cultures and organ; demonstrated nucleic acids, proteins, auxins, inhibitors are all active in cell. devel.; built some tech. instruments including root segment cutter. Home: 10/I.II. Lobogó U., Budapest, Hungary.*

MARQUAND, Allan, Am. archeologist; b. N.Y.C., Dec 10, 1853; s. Henry G. and Elizabeth L. (Allen) M.; A.B., Princeton, 1874; postgrad. U. Berlin, 1877-78; Ph.D., Johns Hopkins, 1880; L.H.D., Hobart, 1888; m. Eleanor Cross, June 18, 1896. Tutor and lectr. Princeton, 1881-83, prof. archeology and history of art, 1883-1905, prof. art and archeology, 1905-—, dir. Mus. of Historic Art, 1890-1921. Prof. archeology Am. Sch. Classical Studies, Rome, 1896-97. Fellow Am. Acad. Arts and Scis. Joint author and editor: Iconographic Encyclopedia, Vol. III, 1896. Joint author: History of Sculpture, 1896-99. Author: Greek Architecture, 1909; Della Robbias in America, 1912; Luca Della Robbia, 1914; Robbia Heraldry, 1919; Giovanni Della Robbia, 1920; Benedetto and Santi Buglioni, 1921; Andrea dell Robbia, 2 vols., 1922. Asso. editor Am. Jour. Archeology, 1885-—. Died Sept. 24, 1924.

MARQUARDT, Ernst Günter, German orthopedist; b. Meiningen, Thüringen, Germany, Feb. 2, 1924; s. Erich and Elsa (Beck) M.; student U. Tübingen, U. Strasbourg; M.D., U. Kiel, 1951; m. Wiebke Jebens, Oct. 14, 1950; children—Regina Susanna, Anna Elsabe. Staff, Orthopedic Hosp. Heidelberg (Germany), 1955-—, head dysmelia dept. and research lab. orthopedic technics, 1963-—, tchr. univ.; also lectr. various countries. Recipient regional prize for Europe, Readers Digest Internat. Rehab. award, 1963. Mem. German Orthopedic Soc., German Soc. Rehab. of Disabled, others. Author: Rehabilitation bilateral armamputees, 1962; Malformations of upper extremities, 1963; (with others) Treatment of limb malformations and amputees of upper extremities. Research, publs. on constrn. and adaption of external powered prosthesis for needs of armless and legless adults and children; coordinator med. edn. and rehab. of amputees, also parents edn. in care of handicapped children. Home: 20 Gutleuthofweg, Heidelberg, 69, Germany.*

MARQUARDT, Hans Ferdinand, German biologist; b. Oehringen, Germany, Oct. 1, 1910; s. Emil and Eugenie (Luppold) M.; student natural sci. and musicology U. Tübingen, 1929-34, Ph.D. in Musicology, 1937, Ph.D.N.-Sci., 1937; m. Charlotte Jung, Aug. 26, 1939. Faculty, U. Freiburg (Germany) 1940-—, prof. biology, 1963-—, dir. Forstbotanischen Inst., 1952-—. Mem. Deutsche Botanische Gesellschaft, Internat. Soc. for Cell Biology, Verband Deutscher Biologen. Contbg. author: Rowholts Deutsche Enzyklopädie. Research, numerous publs. on genetics of microorganisms, cytogenetics of plants, forest botany. Home: 2, Marzeller Weg, Badenweiler, Germany. Office: 17, Bertoldstrasse, Freiburg i.Br., Germany.*

MARQUARDT, Peter, German biochemist; b. Berlin, Germany, Oct. 8, 1910; s. Hans and Margarete (Göppert) M.; M.D.; diploma in chemistry; m. Victoria Quilitz; 3 children. Dir. dept. exptl. therapy U. Freiburg, Germany. Research on biochemistry of autonomous nervous system, especially acetylcholine and adrenaline. Office: U. Freiburg, Freiburg i Br., Germany.

MARQUARDT, William Charles, Am. zoologist; b. Ft. Wayne, Ind., Oct. 9, 1924; s. William Charles and Lucy E. (Pluess) M.; B.S., Northwestern U., 1948; M.S., U. Ill., 1950, Ph.D., 1954; m. Barbara Ann Schucker, June 19, 1948; children—Katherine, William Charles IV, Joan. Research asst. Coll. Vet. Medicine U. Ill., 1952-54, asso. prof. parasitology, 1962-66, also sr. mem. Center Zoonoses Research; faculty Mont. State Coll., 1954-61; vis. investigator So. Regional Animal Disease Lab. U. S. Dept. Agr., Auburn, Ala., 1959; asso. prof. biology De Paul U., Chgo., 1961-62; prof. zoology Colo. State U., Ft. Collins, 1966——. Mem. Am. Soc. Parasitologists, Soc. Protozoologists, Am. Soc. Zoologists, Wildlife Disease Associaton, Am. Soc. for Trop. Medicine and Hygiene, Sigma Xi. Research and publs. on parasitic diseases domestic and wild animals, coccidiosis of cattle, nematodes of sheep, heartworms of dogs, chemotherapy parasitic infections, preservation ciliated and flagellated protozoa by freezing, cross transmission parasitic protozoa. Home: 1112 E. Lake Pl., Ft. Collins, Colo. 80521.*

MARQUES DE ALMEIDA, Fernando Flavio, Brazilian geologist; b. Rio de Janeiro, Brazil, Feb. 18, 1916; s. Gerson and Nair (Marques) de Almeida; C.E., Poly. Sch. U. Sao Paulo, 1939; m. Beatriz Eugenia Bastos, Mar. 15, 1940; children Joao Paulo, Fernando Luiz, Carlos Alberto. Faculty, Poly. Sch., U. Sao Paulo, 1940——, prof. geology, 1962——; chief mining engring. dept., 1965——; mining engr. Departamento Nacional Producao Mineral, Ministerio das Minas e Energia, 1945——. Recipient Jose Bonifacio medal Sociedade Brasileira de Geologia. Mem. Brazilian Acad. Scis. Author: Brasil—A Terra e o Homem, Vol. I, 1964; also numerous articles. Research in Brazilian geology and geomorphology especially on sedimentary and basaltic rocks of Parana basin, Precambrian geology Mato Grosso, petrology oceanic islands of Trindade and Fernando de Noronha. Home: Rus Paula Ney n° 509, Sao Paulo, Brazil.*

MARQUEZ-MONTER, Hector, Mexican pathologist; b. Mexico City, Mexico, Oct. 4, 1929; s. Juan B. and Ciria (Monter) M.; B.S., Colegio Franco-Español, Mexico, 1948; M.D., U. Nacional Autonoma Mexico, 1954; m. Irma Lugo, June 11, 1954; children—Angeles, Hector, Patricia, Irma Marquez. Fellow pathology U. Va. Hosp., 1955-56, 58-59, Mayo Clinic, Rochester, Minn., 1956-58; instr. pathology U. Va. Med. Sch., 1958-59; prof. pathology Escuela de Medicine, U. Autónoma de Guadalajara, Mexico, 1959-61; prof. pathology, 1961——. Diplomate Am. Bd. Pathology. Mem. Internat. Acad. Pathology, Asociacion Mexicana de Patologos (award 1963), Sociedad Latinoamericana de Patologos. Contbg. author: Metodos de Autopsias en Anatomia Patologica, 1965; also numerous articles. Research on biology of chorionic tumors of women, epidemiology, biochemistry, electron microscopic studies, gynecol. pathology, malignant tumors in women, cytogenetics, sex chromatin surveys in Mexican population. Home: 66 comision Monetaria, Mexico 12. Office: Unidad de Patología, Hosp. General, Mexico 7, D.F., Mexico.*

MARRAZZI, Amedeo S., Am. pharmacologist; b. N.Y.C., Feb. 6, 1905; s. Raffaele and Felicia (Sorrentino) M.; student City Coll. N.Y., 1922-24; M.D., N.Y. U., 1928; m. Rose Florence Netter, June 15, 1926; children—Robert, Mary Ann. Instr., City Coll. N.Y., 1928-29; practice medicine, N.Y.C., 1930-35; with N.Y. U. Coll. Medicine, 1931-43, asst. prof. pharmacology, 1939-43; prof., head pharmacology and therapeutics Loyola U. Sch. Medicine, Chgo., 1943-44; prof., head pharmacology and therapeutics Wayne U. Coll. Medicine, 1944-48; chief toxicology br. Chem. Corps Med. Labs., Army Chem. Center, Md., 1948-51, chief research physician, chief clin. research div., 1951-56, asst. sci. dir., med. directorate Chem. Warfare Labs., 1956; dir. VA Research Labs. in Neuropsychiatry, Pitts., 1956-64; prof. physiology, pharmacology U. Pitts. Sch. Medicine, 1956-64; Hill prof. neuropharmacology U. Minn., 1964——. Cons., NASA, 1964——, VA Mpls., 1965——. Mem. A.M.A., A.A.A.S., Soc. Exptl. Biology and Medicine, Sigma Xi, numerous others. Author: (with M. H. Aprison) Studies of Function in Health and Disease, 1960. Editorial bd. Arch. Intern. Pharmacodyn. et Therap. Publs. on research on mechanisms, measurement of function of nervous system; pioneered use of direct elec. recording in analysis of drug action on nervous system. Home: 2135 W. Hoyt St., St. Paul, 55108. Office: U. Minn., Millard Hall, Mpls. 55455.*

MARRIOTT, McKim, Am. social anthropologist; b. St. Louis, Feb. 1, 1924; s. Williams McKim and Elizabeth (Robinson) M.; student Harvard, 1941-43, Stanford U., 1944; A.M. in Anthropology, U. Chgo., 1949, Ph.D., 1955; postgrad. U. Pa., 1949; m. Jacqueline Fay, June 25, 1946; children—Diana, Robinson, Elizabeth, Lucretia. Jr. research anthropologist Inst. E. Asiatic Studies, U. Cal., Berkeley, 1953-55; mem. faculty U. Chgo., 1957——; prof. anthropology and social sci., 1964——. Recipient Aux. Research award Social Scis. Research Council, 1958, Quantrell award for excellence in undergrad. teaching U. Chgo., 1963; Lewis Henry Morgan lectr. U. Rochester, 1967. Social Sci. Research Council fellow, India, 1950-52, Ford Found. fellow, India, 1955-57, NIH and Center for Advanced Study in Behavioral Scis. fellows, 1961-62.

Fellow Am. Anthrop. Assn., Royal Anthrop. Soc.; mem. Assn. Asian Studies (adv. editorial bd. 1957-58), Current Anthropology (asso.). Author: Caste Ranking and Community Structure in Five Regions of India and Pakistan, 1960; (with Albert Mayer and Richard L. Park) Pilot Project, India, 1958. Editor: Village India, 1955. Home: 5800 S. Harper Av., Chgo. 60637.*

MARRIOTT, Ross W., Am. mathematician; b. Paxton, Ill., Dec. 30, 1882; s. Joshua H. and Elizabeth (Kelley) M.; B.S., Valparaiso U., 1904; A.B., Ind. U., 1906; A.M., Swarthmore, 1907; Ph.D., U. Pa., 1911; m. Marian Redfield Stearne, Sept. 8, 1915; 1 dau., Alice Elizabeth. Instr. math. Swarthmore, 1907-10, asst. prof., 1910-22, asso. prof., 1922-27, prof., 1927——, from 1927, mem. eclipse expdns., Mexico, 1923, New Eng., 1925, 32, Sumatra, 1926, 29; mem. U. S. Naval Obs. Eclipse Expdn., 1930. Research ballistician on spl. aircraft ammunition, E. I. du Pont de Nemours & Co., 1918. Fellow A.A.A.S., Royal Astron. Soc.; mem. Am. Astron. Soc., Am. Math. Soc., Math. Assn. Am., Sigma Xi. Contbr. research papers on astron. subjects. Died Oct. 19, 1955.

MARSAGLIA, George, Am. mathematician; b. Denver, Mar. 12, 1925; s. John and Mabel (Tappan) M.; B.Sc., Colo. State U., 1947; M.A., Ph.D., Ohio State U., 1950; m. Lee Ann Stewart, Sept. 7, 1954; 1 son, John Winston. Inst. math. Ohio State U., 1948; Fulbright scholar U. Manchester (Eng.), 1949-50; asst. prof. math. and astronomy Mont. State U., 1950-53, dir. statis. lab., 1952-53; vis. lectr. U. N.C., 1953-54, Okla. State U., 1954-55; Fulbright prof. math. statis. U. Rangoon (Burma), also adviser Burma Dept. Commerce, Census and Agr., 1955-56; cons. Westinghouse Electric Corp., 1956; lectr. U. Wash., Seattle, 1957——, lectr. dept. medicine, 1962——; staff Boeing Sci. Research Lab., Seattle, 1957-——. Cons., RAND, 1963——, Philco, 1966——. Mem. Am. Math. Soc., Inst. Math. Statis., Soc. for Indsl. and Applied Math. Research, publs. in probability and statistics; developed methods for Monte Carlo theory; applied math. to problems in medicine, iron metabolism; treatment of leukemia by irradiation; mathematics of radioactive tracers. Home: 4169 133d St. S.E., Bellevue, Wash. 98004. Office: 1422 S. Trenton St., Seattle 98108.*

MARSCHAK, Jacob, economist; b. Kiev, Russia, July 23, 1899; s. Israel and Sophie (Gorlov) M.; Ph.D., U. Heidelberg, 1922; M.A., U. Oxford, 1935; hon. doctorate U. Bonn; m. Marianne Kamnitzer, Feb. 17, 1927; children—Ann Irene, Thomas Andrew. Came to U. S., 1939, naturalized 1945. Instr. econs. and statistics, U. Heidelberg, 1930-33; dir. Inst. of Statistics, U. Oxford, 1933-39; prof. econs. with grad. faculty, New Sch. for Social Research, 1939-42; prof. econs. U. Chgo., 1943-55, research dir., Cowles Commn. for Research in Econs., 1943-48; prof. econs. Yale, 1955-60; prof. econs. and bus. adminstrn. U. Cal. at Los Angeles, 1960——; Ford distinguished research prof. Carnegie Inst. Tech., 1958-59; cons. RAND Corp. Center for Behavioral Studies, fellow, 1955-56, Ford sr. scholar, 1963——; Distinguished fellow Am. Econ. Assn. Fellow Inst. Math. Statistics, Am. Statis. Assn. (v.p. 1947, dir.), Econometric Soc. (pres. 1946), Am. Acad. Arts and Scis., Royal Statis. Soc. (London) (hon.); mem. Internat. Statis. Inst. Joint editor; Behaviorial Science; Management Science. Application of math. and statistics to econ. problems, especially problems in decision-making, information, and orgn. Address: 968 Stone Hill Lane, Los Angeles 90049.*

MARSDEN, Philip Law, English physicist; b. Brighouse, Eng., Feb. 14, 1924; s. Arthur Beeton and Dorothy (Whiteley) M.; B.Sc. with honors, U. Birmingham, 1948; Ph.D., U. Leeds, 1951; m. Doreen I. Mounty, June 22, 1946; children—D. Jane, Simon L., Andrew P. Jr. sci. officer Telecommunications Research Establishment, Malvern, Eng., 1944-46, Nat. Phys. Lab., Teddington, Eng., 1946-47; faculty physics dept. U. Leeds (Eng.), 1948——, reader, 1965——. Mem. Am. Geophys. Union. Research and numerous publs. on cosmic ray physics, solar-terrestrial relationships, space physics. Home: 18 Park Crescent, Leeds 8, Eng.*

MARSH, C(harles) Dwight, Am. physiologist; b. Hadley, Mass., Dec. 20, 1855; s. J. Dwight and Sarah (Ingram) M.; A.B., Amherst Coll., 1877, A.M., 1880, Sc.D., 1927; Ph.D. U. Chgo., 1904; m. Florence Lee Wilder, Dec. 27, 1883; children—Hadleigh, Charles Wilder. Prof. of chemistry and biology Ripon (Wis.) Coll., 1883-1889, biology, 1889-1904, dean of faculty, 1900-04; prof. biology Earlham Coll., 1904-05; physiologist in charge field investigations of poisonous plants Bur. Plant Industry, Dept. Agr., 1905-15, transferred to Bur. Animal Industry, 1915, in charge poisonous plant investigations, ret., 1931. Hon. curator fresh water copepoda, U. S. Nat. Mus.; ex-officio mem. commrs. Wis. Geol. and Natural History Survey, sec. bd., 1897-99; biologist on Geol. and Natural History Survey, Wis.; lectr. biology Milw. Med. Coll., 1903-04. Author: Limnetic Crustacea of Green Lake; The Plankton of Lake Winnebago and Green Lake; A Revision of the North American Species of Diaptomus; A Revision of North American Species of Cyclops; The Loco Weed Disease of the Plains; Stock-Poisoning Plants of the Range; sr. joint author of Zygadenus or

Death Camas; Larkspur Poisoning of Live Stock; Lupines as Poisonous Plants. Died 1932.

MARSH, James, English chemist; b. London, Sept. 2, 1794; practical chemist Woolwich Arsenal; became asst. to Faraday, Royal Mil. Acad., 1829, remained there until his death. Recipient Silver medal Soc. Arts, 1823, Gold medal, 1836. Research and publs. on chemistry, especially poisons and their effects; invented electro-magnetic apparatus, Marsh arsenic test, 1836. Died June 21, 1846.

MARSH, James Alton, Am. elec. engr.; b. Youngstown, O., Jan. 7, 1922; s. Harold E. and Mary (Peck) M.; A.B. in Math. and Physics, Ohio Wesleyan U., 1946; M.Sc. in Elec. Engring., Ohio State U., 1947, Ph.D., 1949; m. Dolores Mae Wisler, Dec. 24, 1943; children—Richard A., James R. Instr. elec. engring. Ohio State U., 1947-48, research fellow Antenna Lab., 1948-50; group leader Radar Autonetics div. N.Am. Aviation, 1950-56; pres. Systems Labs. Corp., Sherman Oaks, Cal., 1956-59; v.p. engring. Electronic Splty. Co., Glendale, Cal., 1959-61; group dir. satellite control facility directorate Aerospace Corp., El Segundo, Cal., 1961——. Mem. I.E.E.E. (sr.), Phi Beta Kappa, Sigma Xi, Eta Kappa Nu, Pi Mu Epsilon. Designer monopulse radar systems, mil. antennas, electronic ground systems. Home: 4655 Libbit Av., Encino, Cal. 91316. Office: 2400 E. El Segundo Blvd., El Segundo, Cal. 90045.*

MARSH, Julian Bunsick, Am. biochemist; b. N.Y.C., Jan. 21, 1926; s. Arin and Natalie (Bunsick) M.; student U. Pa., 1941-43, M.D., 1947; m. Priscilla Kent, June 18, 1948; children—Constance, Gail. Post-doctoral fellow in biochemistry dept. research medicine U. Pa., 1948-50, asso. biochemistry Grad. Sch. Medicine, 1952-54, faculty, 1954——, acting chmn. dept. biochemistry, 1962-63, prof. biochemistry, 1963-65, prof., chmn. dept. bio-chemistry Sch. Dental Medicine, 1965——. Cons., VA Hosp., 1962——. Recipient M.E. Bell prize Sch. Medicine, U. Pa., 1946, Borden award, 1947; J. S. Guggenheim fellow Nat. Inst. for Med. Research, London, Eng., 1960-61; Lalor fellow Marine Biol. Labs., Woods Hole, Mass., 1959. Mem. Am. Soc. Biol. Chemists, Biochem. Soc. (Gt. Britain), Am. Chem. Soc. Author: (with Haugaard) The Action of Insulin, 1953; also numerous articles. Research leading to discovery of binding insulin to tissue cells, biosynthesis of cytochrome by isolated tissues, mechanism of elevation of plasma lipid in lipid nephrosis, synthesis of plasma lipoproteins by isolated liver cells and cell-free enzyme systems. Home: 4054 Irving St., Phila. 19104.*

MARSH, Othniel Charles, Am. paleontologist; b. Lockport, N.Y., Oct. 29, 1831; s. Caleb and Mary (Peabody) M.; grad. Phillips Andover Acad., 1856, Yale, 1860; postgrad. Heidelberg, Berlin, Breslau (all Germany) univs., 1862-65. Prof. paleontology Yale, 1866-99; numerous trips to Western U. S. to collect materials, from 1866; explored Pliocene deposits Neb., Miocene deposits No. Colo.; apptd. vertebrate paleontologist to U. S. Geol. Survey, 1882. Mem. Nat. Acad. Scis. (pres. 1883-95). Author numerous publs. including: Fossil Horses in America, 1874; Introduction and Succession of Vertebrate Life in America, 1877; Odontornithes: A Monograph on the Extinct Toothed Birds of North America, 1880; Dinocerata: A Monograph on an Extinct Order of Gigantic Mammals, 1884; The Dinosaurs of North America, 1896. Discovered over 1000 new fossil vertebrates and described more than 500; found 1st pterodactyl in Am., 1871; findings helped substantiate theory of evolution. Died New Haven, Mar. 18, 1899.

MARSH, Richard Edward, Am. structural chemist; b. Jackson, Mich., Mar. 6, 1922; s. Howard Roland and Dorothy (Robinson) M.; B.S., Cal. Inst. Tech., 1943; Ph.D. U. Cal., Los Angeles, 1950; m. Helena Joan Laterriere, Aug. 9, 1947; children—Susan Elizabeth, Richard Charles, Kirby Joan, Stephen Hayden. Civilian physicist, electronics technician USN, San Francisco, New Orleans, 1943-46; mem. faculty U. Cal., Los Angeles, 1953; research fellow Cal. Inst. Tech., Pasadena, 1950-55, sr. research fellow, 1955-——. Mem. U. S. Nat. Com., Internat. Union Crystallography, 1966——, mem. Am. Crystallographic Assn., A.A.A.S., Am. Soc. Biologic Chemists, Sigma Xi. Co-editor Acta Crystallographica, 1964——. Research, publs. on crystal structure analysis; molecular structure; structure of biologic molecules; bridged di-cobalt compounds. Home: 1947 Sherwood Rd., San Marino, Cal. 91108.*

MARSH, Robert Mortimer, Am. sociologist; b. Everett, Mass., Jan. 22, 1931; s. Henry Warren and Ruth (Dunbar) M.; student Boston U., 1948-50; A.B., U. Chgo., 1952; M.A., Columbia U., 1953, Ph.D., 1959; m. Susan S. Han, Oct. 12, 1957; children—Eleanor L., Christopher S. H. Fellow Ford Found., Japan, Taiwan, Hong Kong, 1956-58; instr. sociology U. Mich., 1958-61; asst. prof. sociology Cornell U., 1961-65; asso. prof. sociology Duke, 1965-67; asso. prof. sociology Brown U., 1967——; East Asian Inst. Summer fellow in Chinese, Columbia U., 1955. Mem. Am., Eastern sociol. assns. Author: The Mandarins: The Circulation of Elites in China, 1961; Comparative Sociology, 1967. Asso. editor Admstrv. Sci. Quar., 1963-67. Contbr. articles to sci. jours.

Studies of influence of family origin and social stratification on career path in Chinese Imperial bureaucracy, 1600-1900; codification of recent comparative cross-societal research in sociology and social anthropology in terms of theory of social differentiation; analysis of relationship between modernization of societies and strength of Communist Party; study of social life in Taipei, Taiwan.

MARSH, Sylvester, Am. inventor; b. Compton, N.H., Sept. 30, 1803; s. John and Mehitable (Percival) M. Owner meat packing business, Ashtabula, O., 1828-33; operated beef marketing firm, Chgo., 1833-37; in grain bus., Chgo., 1837-55, invented more efficient grain dryers and mfg. process for meal; used process to produce Marsh's Caloric Dry Meal, exported product, largely to W.I.; originated plan to build railroad up Mt. Washington, N.H. (railroad completed 1869); lived in Littleton, N.H., 1865-79; patented locomotive engines designed to ascend grades, cog rail for railroads, atmospheric brake for railroad cars; lived in Concord, N.H., 1879-84. Died Concord, Dec. 30, 1884; buried Concord.

MARSHAK, Alfred George, Am. biologist; b. N.Y.C.; June 2, 1907; s. William and Anna (Sokolow) M.; B.S., Cornell U., 1930; Ph.D., Harvard, 1934; m. Celia Levine, Feb. 1, 1952; children—David, Daniel. Research asso. N.E. Deaconess Hosp., Boston, 1935-38, Radiation Lab., U. Cal., 1938-44; biochemist Tb control div. USPHS, 1945-52; adj. asso. prof. dept. chemistry N.Y. U. Coll. Medicine, 1947-52; prof. physiology U. Notre Dame, 1958-59; prof. radiobiology Jefferson Med. Coll., Phila., 1956-64; prof. exptl. pathology Tulane U. Sch. Medicine, New Orleans, 1964—. Guggenheim Found. fellow, 1938-40. Fellow A.A.A.S.; mem. Am. Physiol. Soc., Am. Chem. Soc., Biophys. Soc., Radiation Research Soc., Am. Assn. for Cancer Research, Genetics Soc. Am., Sigma Xi. Research and numerous publs. on effects of X-rays and neutrons on chromosomes, nucleic acids of nuclei and chromosomes, and of microorganisms, nucleic acids in cancer, inheritance in mammals, chromosome structure in plants, animals and bacteria. Home: 6323 Barrett St., New Orleans 70118.*

MARSHAK, Moisey Yefimovich, Russian physiologist; b. 1894; grad. Med. Faculty, Moscow U., 1919; D.Med. Sci. Head lab. labor hygiene and physiology Moscow Inst. Labor Protection, 1924-33; head lab. climatophysiology, later lab. respiration and blood circulation All-Union Inst. Exptl. Medicine, 1933-44; prof. chair physiology Inst. Phys. Tng., 1938-49; head lab. physiology and pathology of respiration and blood circulation Inst. Normal and Path. Physiology, USSR Acad. Med. Sci., 1945—. Mem. USSR Acad. Med. Sci. (corr.). Author: The Physiology of Man, 1946. Co-editor Physiology sect. Large Med. Ency., 2d edit.; mem. editorial council Sechenov Jour. Physiology of USSR. Research and numerous publs. on regulation of respiration and blood circulation under conditions of oxygen starvation, regulation of regional blood circulation; developer automatic device to take samples of alveolar air in man, thermoelectrode to study blood circulation in organs, method to record continuously temperature of organs in chronic expts.; analyzed role of cerebral cortex in regulation of human respiration under conditions of vital activity. Address: Inst. Normal and Path. Physiology, USSR Acad. Med. Sci., Solyanka 14, Moscow, USSR.

MARSHAK, Robert E(ugene), Am. physicist, astrophysicist; b. N.Y.C., Oct. 11, 1916; s. Harry and Rose (Shapiro) M.; A.B., Columbia, 1936, student 1936-37; Ph.D., Cornell, 1939; m. Ruth Florence Gup, Apr. 18, 1943; children—Rachel Ann, Robert Stephen. Instr., dept. physics, U. Rochester, 1939-43, asst. prof., 1943-46, asso. prof., 1946-49, prof., 1949—, chmn. dept. physics and astronomy, 1950-64, Harris professor, 1952-64, Distinguished University profssor, 1964—; lectr. Harvard Obs., summer 1940; professeur d'Echange (Guggenheim fellow) at Sorbonne, 1953-54; vis. prof. Columbia summer 1950, Mich., 1952, Tata Inst. Bombay, 1953. French Sch. for Theoretical Physics, 1954; guest prof. at Cern, Geneva, Switzerland under Ford Found. and Guggenheim fellow; mem. Inst. Advanced Study, Princeton, spring 1948; physicist, radiation laboratory, Massachusetts Institute Technology, 1942-43, Montreal Atomic Energy project under Department Scientific and Indsl. Research of Great Britain, 1943-44; deputy group leader in theoretical physics, Los Alamos Scientific Lab., 1944-46. Vice chmn. N.Y. State Adv. Com. on Atomic Energy, 1958; cons. Eastman Kodak Co.; chmn. vis. physics com. Brookhaven Nat. Lab., 1964-65; Niels Bohr vis. prof. Inst. Math. Sci., Madras, India, 1963; chmn. adv. com. on Soviet Union and Eastern Europe, Nat. Acad. Scis.; trustee Atoms for Peace Award. Recipient A. Cressy Morrison astron. prize N.Y. Acad. Scis., 1940. Fellow Am. Phys. Soc. (council), A.A.A.S.; mem. Am. Association University Professors, Federation American Scientists (chairman 1947-48), National Acad. Scis. (chmn. adv. com. on sci. exchanges with USSR and Eastern Europe), American Acad. Arts and Scis., Internat. Union Pure and Applied Physics (past sec. commn. on high energy physics), Sigma Xi, Phi Beta Kappa, Phi Kappa Phi. Author: Meson Physics, pub. in 1952; (with L. I. Schiff and E. C. Nelson) Our Atomic World, 1946; (with E. C. G. Sudarshan) Elementary Particles, 1961. Associate editor Phys. Rev., 1953-55. Editor: Interscience Tracts and Monographs in Physics and Astronomy; mem. internat. editorial bd. Il Nuovo Cimento. Research on energy sources of stars, atomic nuclei, neutron diffusion, elementary particles, white dwarf stars, and theory of universal weak interaction. Home: 1592 Highland Av., Rochester, N.Y. 14618.

MARSHAK, Robert Reuben, Am. veterinarian; b. N.Y.C., Feb. 23, 1923; s. David and Edith (Youselefsky) M.; student U. Wis., 1940-41; D.V.M., Cornell U., 1945; m. Ruth Lyons, Dec. 4, 1948; children—William L., John B., Richard B. Practice vet. medicine, Springfield, Vt., 1946-56, Phila., 1956—; prof. medicine, chmn. dept. clin. studies Sch. Vet. Medicine, U. Pa., 1956—; cons. Brookhaven Nat. Lab., 1963—. Fellow N.Y. Acad. Sci.; mem. A.A.-A.S., Am. Assn. Cancer Research, Am. Vet. Med. Assn., Am. Pub. Health Assn., Sigma Xi. Contbr. numerous articles to profl. jours. Studies on metabolic diseases of cattle, leptospirosis, bovine leukemia.

MARSHALL, Alfred, English economist; b. London, July 26, 1842; s. William and Rebecca (Oliver) M.; ed. St. John's Coll., B.S., 1865; M.A., hon. D.Sc., Oxford, Cambridge; hon. LL.D., Edinburgh, Bristol; m. Mary Paley, 1877. Prin., U. Coll., Bristol, 1877-82; lectr., fellow Balliol Coll., Oxford, 1883-84; prof. polit. economy Cambridge, 1885-1908; emeritus Cambridge, 1918. Fellow Brit. Acad.; fgn. mem. insts. France, Sweden, Rome, Milan, Turan, v.p. Royal Econ. Soc., mem. Royal Commn. on Labour, 1891. Author: (with Mary Marshall) Economics of Industry, 1879; Principles of Economics, vol. 1, 1890; Elements of Economics, vol. 1, 1891; Memorandum on the Fiscal Policy of International Trade, 1903; Industry and Trade, 1919; Money, Credit and Commerce, 1923. Gave utility theory of value and concept of elasticity of demand; leader in devel. of Brit. classical econ. theory in 20th century. Died Cambridge, July 13, 1924.

MARSHALL, Eli Kennerly, Jr., Am. physiologist, pharmacologist; b. Charleston, S.C., May 2, 1889; B.S., Coll. Charleston, 1908, Ph.D. in Chemistry, 1911, M.D., 1917, LL.D., 1941. Asst. physiol. chemistry Johns Hopkins, 1911-13, asso., 1913-14, pharmacologist, 1914-17, asso. prof., 1917-19; prof. Washington U., St. Louis, 1919-21; physiologist Sch. Medicine, Johns Hopkins, Balt., 1921-32 prof. pharmacology and exptl. therapeutics, 1932-55, emeritus prof., 1955—. Mem. Nat. Acad. Scis., A.A.A.S., Am. Soc. Pharmacology (pres. 1942), Am. Physiol. Soc., Soc. Biol. Chemists, Am. Philos. Soc. Editor, Jour. Pharmacology and Exptl. Therapeutics, 1932-37. Research in physiol. chemistry, kidney function, urea determination, urine secretion, chemotherapy, antibiotics, respiratory stimulants, alcohol metabolism; introduced sulfaguanidine (with A. C. Bratton, H. J. White, J. T. Litchfield) for treatment intestinal infections, 1940, (with A. C. Bratton, L. B. Edwards, E. L. Walker) for treatment bacillary dysentery, 1941. Office: Johns Hopkins Hosp., Balt.

MARSHALL, Frederick Joseph, Am. chemist; b. Detroit, Aug. 14, 1920; s. Frederick Joseph and Nora (Orleman) M.; B.S., U. Detroit, 1941, M.S., 1943; Ph.D., Ia. State U., 1948; m. Marcella E. Campbell, Dec. 28, 1946; children—Mary Margaret, Suzanne L., Frederick Joseph, III, Rita J., Jane F., Timothy G., Maureen A. Sr. research chemist Eli Lilly & Co., Indpls., 1948—. Mem. Am. Chem. Soc., N.Y. Acad. Scis. Research, publs., and patent in synthetic methods in organic chemistry, synthesis of sulfonylureas for studies in blood sugar lowering, structure studies on high molecular weight antibiotic, synthetic methods for preparation of carbon14 labeled drugs. Home: 3120 Shady Grove Ct., Indpls. 46222. Office: Lilly Research Labs., Indpls. 46206.*

MARSHALL, Hugh, Scottish chemist; b. Edinburgh, Scotland, Jan. 7, 1868; ed. Moray House Normal Sch., univs. Edinburgh, Munich (Germany), Ghent (Belgium); D.Sc. Became asst. in chemistry U. Edinburgh, 1887, lectr. mineralogy and crystallography, 1894, lectr. chemistry, 1902; prof. chemistry U. Coll., U. St. Andrews, Dundee, Scotland, 1908—. Fellow Royal Soc., Royal Soc. Edinburgh, Chem. Soc. Research and publs. on chemistry and crystallography. Died Sept. 6, 1913.

MARSHALL, Humphrey, Am. botanist; b. Chester County, Pa., Oct. 10, 1722; s. Abraham and Mary (Hunt) M.; m. Sarah Pennock, Sept. 16, 1748; m. 2d, Margaret Minshall, Jan. 10, 1788. Engaged in farming, Chester County, Pa., from 1748; built 1st conservatory for plants in area, circa 1768; constructed hot house and bot. garden with collection of fgn. and domestic plants, at his home, Marshallton, Pa.; corresponded with Dr. John Fothergill and Peter Collinson in Eng.; mem. Am. Philos. Soc. Author: Arbustrum Americanum, the American Grove (list of native forest trees and shrubs), 1785. Died Marshallton, Nov. 5, 1801.

MARSHALL, John, English anatomist, surgeon; b. Ely, Eng., Sept. 11, 1818; s. William Marshall; entered U. Coll., London, Eng., 1838; LL.D., Edinburgh, Scotland; M.D. (hon.), Dublin, Ireland, 1890; 1 son, 2 daus. Asst. to Robert Liston; became demonstrator anatomy U. Coll., London, 1845, prof. surgery, 1866, later prof. clin. surgery; became cons. surgeon U. Coll. Hosp., 1884; prof. anatomy Royal Acad., 1873-91; Fullerian prof. physiology Royal Instn. Hunterian lectr., 1885; Morton lectr., 1889. Fellow Royal Soc., 1857, Royal Coll. Surgeons Eng.; mem. Royal Med. and Chirurg. Soc. London (pres. 1882-83), Gen. Med. Council (became pres. 1887). Author: The Outlines of Physiology, 1867; A Description of the Human Body, its Structure and Functions, 1860; Anatomy for Artists, 1878; also articles. Introduced galvanocautery and excision of varicose veins; one of 1st to demonstrate that cholera is spread by potable water. Died Jan. 1, 1891.

MARSHALL, John Albert, Am. biochemist, dental pathologist; b. Chgo., Aug. 30, 1884; s. John Sayre and Isabelle M. (Carter) M.; B.S., U. Cal., 1907, M.S., 1914, D.D.S., 1916, Ph.D., 1917; post-grad. U. Berlin, Tech. U., Charlottenburg (Germany), 1909-10; m. Hazel C. Knowles, May 18, 1907; children—John A., Muriel, Shirley; m. 2d, Irene Byram Kuechler, Dec. 28, 1932. Prof. biochemistry and dental pathology U. Cal. Mem. A.M.A., Am. Dental Assn., Internat. Assn. Dental Research, Soc. Exptl. Biology and Medicine, Pacific Coast Soc. Orthodontists (hon.), Am. Soc. Orthodontists (hon.), Sigma Xi. Author: Military Explosives (pub. by U. S. War Dept.), 1919; Manufacturing and Testing of Military Explosives, 1919; Diseases of the Teeth, 1926; (with C. N. Johnson) Operative Dentistry, 1923; Anatomy of the Rhesus Monkey (with Hartman and Straus), 1933. Contbr. research papers to jours. Asso. editor Jour. Dental Research. Died May 7, 1941.

MARSHALL, John Hart, Am. biophysicist; b. Chgo., Feb. 14, 1925; s. Thomas Linder and Elizabeth (Carpenter) M.; A.B. magna cum laude, Harvard, 1945; Ph.D., Mass. Inst. Tech., 1952; m. Constance E. Leighton, May 21, 1955; children—Deborah L. (Mrs. James L. Woodward), Nancy H. Research asso. Mass. Inst. Tech. Radioactivity Center, 1952-55; asso. physicist radiol. physics div. Argonne (Ill.) Nat. Lab., 1955-67, sr. biophysicist, 1967—, Mem. Am. Phys. Soc., Radiation Research Soc., Orthopaedic Research Soc. Publs. on work on liquid argon ionization chambers; measurement of beta ray spectra; microscopic metabolism of calcium in bone; math. theory of alkaline earth metabolism. Home: 836 S. Park Av., Hinsdale, Ill. 60521. Office: 203-A-137 Argonne Nat. Lab., Argonne, Ill. 60439.*

MARSHALL, John Stewart, Canadian meteorologist; b. Welland, Ont., Can., July 18, 1911; s. John Wells and Catherine (Stewart) M.; B.A., Queen's U., Kingston, Can., 1931, M.A., 1933; Ph.D., Cambridge U., 1940; m. Helen Elizabeth Reburn Scott, Dec. 19, 1940; children—Claire Elizabeth, Heather. Research physicist Nat. Research Council Labs. Can., Ottawa, 1939-42; operational research scientist Canadian Army Operational Research Group, 1943-45; faculty McGill U., Montreal, 1945—, Macdonald prof. physics, 1959—, chmn. dept. meteorology, 1961-64. Chmn., Inter-Union Com. on Radio Meteorology. Recipient Patterson medal for Canadian meteorology, 1961; Hugh Robert Mill Brit. Rainfall medal, 1962. Mem. Canadian Assn. Physicists (pres. 1950), Phys. Soc. London, Royal Meteorol. Soc., Am. Phys. Soc., Am. Meteorol. Soc. (councillor 1965—), Royal Soc. Can. (pres. sect. 1966). Author: (with E. R. Pounder) Physics, 1957, rev. (with E. R. Pounder, R. W. Stewart), 1967; Why the Weather, 1964; also articles. Developed use of radar to study rain, snow, hail. Home: 20095 Lakeshore Rd., Baie d'Urfe, Que., Can. Office: Physics Bldg., McGill U., Montreal 2, Que., Can.*

MARSHALL, Lauriston Calvert, Am. physicist; b. Canton, China, June 27, 1902 (parents Am. citizens); s. George Washington and Edmonia Bell (Sale) M.; A.B., Park Coll., 1923; Ph.D. (Whiting fellow), U. Cal. at Berkeley, 1929; postgrad. (NRC fellow) Princeton; m. Lucie Welch Sewell, Aug. 20, 1949; children—Clarice Sewell, Katherine Stow, Lauriston Calvert. With U. S. Dept. Agr., 1928-37; faculty U. Cal. at Berkeley, 1937-54, prof. elec. engring., 1945-54, dir. microwave power lab., 1946-52; dir. phys. research lab. Link-Belt Co., Indpls., 1952-59; asso. tech. dir. microwave power lab. Varo, Inc., Garland, Tex., 1959-61; prof., chief office sci. personnel, dir. materials research div. S.W. Center for Advanced Studies, Dallas, 1961-67; vis. prof. U. So. Ill., Carbondale, 1967—. Mem. staff, div. head Radiation Lab., Mass. Inst. Tech., 1941-46, OSRD, 1944-46; staff Lawrence Radiation Lab., U. Cal. at Berkeley, 1946-54, cons., 1956. Recipient citations, War Theater, 1945, OSRD, 1946, Presdl., 1948; Distinguished Alumnus award Park Coll., 1952. Guggenheim fellow, 1950-51. Fellow Am. Inst. Physics, I.E.E.E., Inst. of Physics, Am. Phys. Soc.; mem. Audio Engring. Soc., Am. Soc. for Testing and Materials, Am. Geophys. Soc., Am. Chem. Soc., Am. Astronautical Soc., Am. Optical Soc., Sigma Xi, Eta Kappa Nu. Research, publs. in Gaseous Conduction phenomena, biophys. investigations, high voltage research, prodn. of very high power at microwave frequencies; devel. radar systems, including 1st microwave radar systems for aircraft interception, anti submarine warfare and GCA aircraft blind landing; work in high energy physics and particle accelerators; study of origins and history of atmospheres of Earth and planets. Office: U. So. Ill., Carbondale, Ill.*

MARSHALL, Lawrence Marcellus, Am. biochemist; b. Pitts., Mar. 31, 1910; s. James Henry and Mary (Mosby) M.; B.S., Duquesne U., 1932, M.S., 1940; Ph.D. (USPHS fellow 1945-48), Wayne State U., 1949; m. Mary C. Williams, June 6, 1939; children—Lawrence Marcellus, Gwendolyn, Judith. Asst. prof. chemistry Clark Coll., Atlanta, 1937-39; prof. chemistry and biology Ark. State Coll., Pine Bluff, 1940-44; asst. chemist Pine Bluff Arsenal, 1942-44; asso. chemist Jeffersonville Q.M. Depot, 1944-45; mem. faculty Howard U. Med. Sch., 1948—, prof. biochemistry, head dept., 1959—. Recipient Lederle Med. Faculty award Lederle Lab. div. Am. Cyanamid Co., 1954. Mem. Am. Soc. Biol. Chemists, Am. Chem. Soc., Sigma Xi. Design and study of procedures for separation and identification of compounds in mixtures of biol. interest through such methods as chromatography; application of chromatography to carbon dioxide utilization in animal tissues. Home: 1719 Otis St. N.E., Washington 20018.*

MARSHALL, Leona Woods, Am. physicist; b. LaGrange, Ill., Aug. 9, 1919; d. Weightstill and Mary (Holderness) Woods; B.S., U. Chgo., 1938, Ph.D., 1943; m. John Marshall, July 3, 1943. Research fellow U. Chgo., 1939-42, 46-48, research asso. Metall. Lab., 1942-44; research asso. Inst. Nuclear Studies, 1948-53, asst. prof., 1953-60; asso. prof. N.Y. U., 1960-62, prof., 1962-63; professor of physics Univ. Colorado, Boulder, 1963—; physicist Hanford Engr. Works, 1944-46; vis. scientist Brookhaven Nat. Laboratory, 1958—; cons. Los Alamos Nat. Lab., vis. staff mem., 1964—; fellow Inst. for Advanced Study, Princeton, N.J., 1957-58. Fellow Am. Phys. Soc., Royal Geog. Soc. Asso. editor Phys. Review, 1960-62. Home: 727 13th St., Boulder, Colo.

MARSHALL, Max Skidmore, Am. microbiologist; b. Lansing, Mich., Dec. 20, 1897; s. Charles Edward and Maud Alice (Skidmore) M.; B.S., U. Mass., 1920; M.A., U. Mich., 1922, Ph.D., 1925; m. Barbara Goldsmith Trask, June 22, 1957. Research bacteriologist Mich. Dept. Health, Lansing, 1923-27; prof. microbiology U. Cal. Med. Center, San Francisco, 1927-65, chmn. dept., 1948-62; cons. San Francisco Dept. Pub. Health, 1936-52. Distinguished vis. prof. Mich. State U., 1958. Author: Applied Medical Bacteriology, 1947; Two Sides to a Teacher's Desk, 1951; Crusader Undaunted, 1958; Teaching without Grading, 1968; (with J. C. Geiger) Handbook of Public Health Bacteriology, 1939. Editorial bd. Ore. State U. Publ. Improving College and University Teaching, 1954—. Research and publs. on bacterial vaccines, toxic products; bacteriophage, immunity, edn. Home: 405 Davis St., San Francisco 94111.*

MARSHALL, Nelson, Am. oceanographer; b. Yonkers, N.Y., Dec. 16, 1914; s. E. W. and Mabel (Austin) M.; B.S., Rollins Coll., 1937; M.S., Ohio State U., 1938; Ph.D., U. Fla., 1941; m. Grace Terry, Feb. 3, 1940; children—Terry Sue, John Murray, Catherine (Mrs. Stephen Wilson), Richard Nelson. Asst. prof. U. Conn., 1941-45, U. Miami, 1945-46; asso. prof. U. N.C., 1946-47; dir. Va. Inst. Marine Scis., 1947-51; dean Coll. William and Mary, 1949-51; asso. dir. Oceanographic Inst., Fla. State U., 1952-54; dean Coll. Liberal Arts, Alfred U., 1955-59; prof. oceanography Grad. Sch. Oceanography, U. R.I., Kingston, 1959—; vis. prof. Bingham Oceanographic Lab., Yale, 1944-55; exec. com. Chesapeake Bay Inst., Johns Hopkins, 1948-51. Cons. marine sci. So. Regional Edn. Bd., 1952-54; cons. in ecology Millstone Point Corp., 1966—. Trustee Rollins Coll. Fellow A.A.A.S.; mem. Atlantic Estuarine Research Soc. (hon.), Am. Soc. Limnology and Oceanography, Ecol. Soc. Am., Nat. Shellfish Assn., Sigma Xi. Contbr. numerous articles to tech. jours. Research on trophic processes shoal estuarine environments, coral reefs. Home: 2 Locust Dr., Kingston, R.I. 02881.*

MARSHALL, Robert J., physician; b. Ballymena, North Ireland, May 5, 1926; s. Robert James and Margaret (Robinson) M.; M.B., B.Ch., B.A.O., Queen's U., Belfast, Ireland, 1948, M.D., 1952; m. Mabel Margaret Stevenson, Feb. 12, 1957; children—Stephen Robert, Deirdre Margaret, Ian William. Mem. faculty Queen's U., 1950-57; research fellow Alfred Hosp., Melbourne, Australia, 1957-58, Mayo Clinic, Rochester, Minn., 1958-61; faculty W.Va. U., Morgantown, 1961—, prof. medicine, 1963—. Mem. Am. Soc. Clin. Investigation, Am. Physiol. Soc., Assn. U. Cardiologists, Am. Fedn. Clin. Research, So. Soc. for Clin. Research, Soc. Exptl. Biology and Medicine, Brit. Med. Assn., others. Author: (with T. Darbey) Shock: Pharmacological Principles in Treatment, 1966; (with J. T. Shepherd) Cardiac Function in Health and Disease, 1967. Research and publs. into dynamics of blood flow through lungs, peripheral circulation, and other areas in health and disease. Home: 664 Villa Place, Morgantown, W.Va. 26506.*

MARSHALL, Royal Richard, Am. geochemist; b. St. Paul, Sept. 18, 1926; s. M. Royal and Victoria (Woodle) M.; B.Sc., U. Minn., 1950; Ph.D., Cal. Inst. Tech., 1955; m. Cynthia C. Gilbert, Jan. 2, 1959. Research fellow Cal. Inst. Tech., Pasadena, 1954-55; research asso. Enrico Fermi Inst. Nuclear Studies, U. Chgo., 1955-58; NSF grantee Physikalisches Institut, U. Bern (Switzerland), 1959-60; sr. scientist Jet. Propulsion Lab., Cal. Inst. Tech., Pasadena, 1961—. Mem. Am. Geophys. Union, Am. Mineral. Soc., Phi Beta Kappa, Sigma Xi. Research and publs. on chem. kinetics of recombination of iodine atoms in gases and liquids, alteration of volcanic glass and crystal habit of minerals; determination of isotopic composition and concentration of minute amounts of lead in meteorites and rocks, relations of this to origin of elements and formation of meteorites; studies on transfer of heavy elements to surface of earth. Home: 1582 Poppy Peak Dr., Pasadena, Cal. 91105.*

MARSHALL, Sheina Macalister, Scottish marine biologist; b. Glasgow, Scotland, Apr. 20, 1896; d. John Nairn and Jean (Binnie) M.; B.Sc., U. Glasgow, 1919, D.Sc., 1934. Staff, Marine Sta., Millport, Isle of Cumbrae, Scotland, 1922-64, dep. dir., 1962-63; mem. Gt. Barrier Reef Expdn., 1928-29. Fellow Royal Soc., Royal Soc. Edinburgh, Inst. Biology, Challenger Soc. Author: (with L. Newton, A. P. Orr) British Seaweed and the Preparation of Agar, 1949; (with A. P. Orr) The Biology of a Marine Copepod, Calanus finmarchicus, 1955, Seashores, 1965; also articles. Research on productivity in sea, relation between phytoplankton and zooplankton with special reference to physiology of copepods. Home: Bellevue, Kames Bay, Millport. Office: Marine Sta., Millport, Isle of Cumbrae, Scotland.*

MARSHALL, Thomas Robert, Am. physician; b. Charlestown, Ind., Feb. 5, 1930; s. Edmond W. and Frances (Wessel) M.; A.B., Ind. U., 1951, postgrad. Med. Center, 1956-59; M.D., U. Louisville, 1955; postgrad. Boston Coll. Law, 1960-61; m. Ann Elizabeth Glass, June 11, 1955; children—Thomas Robert, Diane, Kathy, Linda, Cheryl. Practice medicine, specializing in radiology and nuclear medicine Louisville, 1961—; mem. staffs Washington County Hosp., Salem, Ind.; faculty U. Louisville Sch. Medicine, 1961—, asso. prof. radiology, 1964—, chief sect. diagnostic radiology 1964—. Diplomate Am. Bd. Radiology. Mem. A.M.A., Ky. Med. Assn., Am. Coll. Angiology, Am. Med. Writers Assn. Internat. Coll. Angiology, Jefferson County Med. Soc., Radiol. Soc. N.Am., Assn. U. Radiologists, Am. Coll. Radiology (pres. Ky. 1965—). Editor Current Med. Digest, 1962—. Publs. on devel. non-catheter brachial angiography, Marshall automatic pressure injector; numerous vascular needles and catheters. Home: 25 S. Sheridan Rd., Louisville 40220. Office: Med. Towers South, 633 S. Floyd St., Louisville 40202.*

MARSHALL, Wade Hampton, Am. neurophysiologist; b. Pitts., Dec. 17, 1907; s. Francis James and Ann (Miller) M.; B.S., Beloit Coll., 1930; M.S., U. Chgo., 1931, Ph.D., 1934; m. Louise Hanson, Dec. 31, 1934; children—Thomas H., Alice. Asst. physiologist U. Chgo., 1931-34; physiol. instr. George Washington U. Med. Sch., 1934-36; NRC fellow Med. Sch., Johns Hopkins, 1936-38, instr. Wilmer Ophthalmol. Inst., 1938-40, asso. Lab. Physiol. Optics, Wilmer Ophthal. Inst., 1940-43, sr. physicist Applied Physics Lab., 1946-47; research engr. Bowen & Co., 1944-46; spl. research fellow NIH, Bethesda, Md., 1947-49, physiologist Nat. Inst. Mental Health, 1949-53, chief lab. neurophysiology Nat. Inst. Mental Health—Nat. Inst. Neurol. Diseases and Blindness, 1954—. Mem., Am. Electroencephalographic Soc., Am. Neurol. Assn., Am. Physiol. Soc., Assn. for Research in Nervous and Mental Diseases, Biophys. Soc., Brazilian Acad. Sci. (fgn. mem.), Eastern Assn. Electroencephalographers, Internat. Fedn. for Med. Electronics, Washington Acad. Scis., Washington Philos. Soc., N.Y. Acad. Scis., Sigma Xi. Research in physiology of central nervous system, elec. examination of nervous system, pathways of sensory systems, synaptic transmission and spreading depression. Home: 4209 Everett St., Kensington, Md. 20795. Office: NIH, Bethesda, Md. 20014.*

MARSHALL, William Leitch, Am. chemist; b. Columbia, S.C., Dec. 3, 1925; s. William Leitch and Georgia (Kittrell) M.; B.S., Clemson U., 1945; Ph.D., Ohio State U., 1949; m. Phyllis Joanne Fox, Apr. 16, 1949; children—Nancy Diane, William Fox. Asst. chemistry Ohio State U., 1945-46, Naval Research fellow, 1947-49; Guggenheim fellow U. Amsterdam (Holland) Van der Waals Lab., 1956-57; sr. research chemist Oak Ridge Nat. Lab., 1949—, research group leader, 1957—. Mem. Am. Chem. Soc., Research Engring. Soc. Am. Research and publs. on solubilities, conductance, chemistry of water electrolyte systems; co-discovered double oxides; nature of Grignard-type reagents. Home: 101 Oak Lane. Office: Oak Ridge Nat. Lab., P.O. Box X, Oak Ridge 37830.*

MARSIGLI, Count Luigi Ferdinando, Italian naturalist, geographer; b. Bologna, Italy, July 10, 1658; s. Count and Countess Marsigli; ed. Bologna (under Malpighi, Triofetti, Montarnari), also Padova, Venice, Italy. Visited Rome, 1676, Naples, Pozzuoli, Mt. Vesuvius, 1677. Sent. to Turkey by Republic of Venice to study natural history and mil. orgn.; visited Constantinople (with Pietro Civrano), 1679; Andrianople, Sofia, Belgrade, Bosnia; served with Venetian and Austrian armies; captured by Turks, 1683, ransomed, 1684; participated in battles at Neuhausel, 1684, Buda, 1686, Belgrade, 1693; engaged in army maneuvers, road, bridge bldg., to 1699; expelled from army after surrender of Breisach, 1703; research in Cassis, France, 1705-06; became acquainted with Newton and Haley on visit to Eng.; also visited Holland; returned to Maderno sul Garda, 1725; founder Bologna Inst. Arts and Scis., 1712, also press (later run by Dominicans). Fellow Royal Soc., 1692; mem. French Acad. Scis. Author: Osservazioni intorno al Bosforo Tracio, 1681; Breve restritto del Soggio fisico intorno, 1711; Danubius ponnonico-mysicus, 6 vols., 1725; Histoire physique de la mer, 1725; Danubis Pannonico-Mysicus, 1725; Stato militare dell' impero ottomano, 1732. Worked in areas of geography, ethnography, astronomy, hydrography; inspired by exptl. work of Galileo; made 1st study of birds in Danube; studied vegetation of Mediterranean Sea floor and its effect on fish life; systematically investigated coral polyps; realized that biology and physics are related, that chem. and phys. investigations are needed to study biology of sea; discovered and explained density currents of sea; saw structural relationship between ocean basins and surrounding and surrounding mountains. Died Bologna, Nov. 1, 1730.

MARSILIUS OF PADUA (Marsiglio Mainardino), Italian polit. theorist; b. Padua, Italy, 1270; studied medicine; after practising several professions, went to Paris, circa 1311; became rector U. Paris, 1313; collaborated with philosopher John of Jandun, and together sided with Louis of Bavaria, king of the Romans, during the struggle with Pope John XXII; went to Italy with Louis of Bavaria to support supremacy of Holy Roman Empire; with John of Jandun played an active role in Roman Revolution, 1328; apptd. imperial vicar, later probably archbishop of Milan. Author: (with John of Jandun) Defensor pacis (which held that polit. authority was vested in adult male citizens of state who in turn chose exec. council or prince, also that church should be spiritual power without worldly wealth and subject to the state), 1324; De translations imperii romani; De jurisdictione imperatoris in causa matrimoniale; Defensor minor. Died circa, 1342.

MARSLAND, Douglas Alfred, Am. physiologist; b. Bklyn., Feb. 17, 1899; s. Albert Edward and Mebel (Douglas) M.; B.S., N.Y. U., 1922, Ph.D., 1934; M.A., Columbia U., 1928; m. Alice Sansom, June 17, 1924. Tchr., Silver Bay Sch., N.Y., 1922-24; instr. biology dept. Washington Sq. Coll., N.Y. U., 1924-27, asst. prof., 1927-42, asso. prof., 1942-47, prof., 1947-62, research prof., 1962—. Trustee Marine Biol. Lab., Woods Hole, Mass., 1945—, exec. com., 1948-56, sec. bd., 1964—. Trustee Bermuda Biol. Sta. Recipient Newcomb Cleveland prize A.A.A.S., 1941; Guggenheim fellow, 1951-52, 59-60; Fulbright sr. research fellow, 1959-60. Fellow A.A.A.S., N.Y. Acad. Sci.; mem. Am. Naturalists, Harvey Soc., Am. Soc. Zoology. Author: Principles of Modern Biology, 1964. Research and numerous publs. on cell div., ameboid movement, pigmentary responses, cell membranes, narcosis, protoplasmic streaming, physiol. effects of high pressure. Home: 48 Church St., Woods Hole 02543. Office: Marine Biol. Lab., Woods Hole, Mass.*

MARSTON, Anson, Am. civil engr.; b. Seward, Ill., May 31, 1864; s. George W. and Sarah (Scott) M.; studied Berea Coll., 1884; C.E., Cornell, 1889; Eng.-D., U. Neb., 1925, Mich. State Coll., 1927; m. Alice Day, Dec. 14, 1892; children—Morrill Watson, Anson Day. Engr. Mo.P. Ry. on location and constrn., 1889-92; in charge constrn. Ouachita River Bridge, 1891-92; prof. civil engring. Ia. State Coll., 1892-1920, dean and dir. engring. div., 1904-32, sr. dean, 1932-37, emeritus, 1937—. Mem. Ia. Hwy. Commn., 1904-27, chmn. 1913-15; mem. Engring. Bd. Review, Sanitary Dist. Chgo., 1924, 25; cons. engr. Miami, Fla., sewerage, 1925-27; mem. (Fla.) Everglades Engring. Bd. Rev., 1927; mem. Interoceanic Canal Bd., to advise on Nicaragua Canal and enlargement Panama Canal, 1929-32; mem. Mississippi River Engring. Bd. Review, 1932, 33; mem. NRC (rep. Am. Soc. C.E.), 1919. Recipient Chanute medal Western Soc. Engrs., 1903; Fuertes medal Cornell U., 1904; Lamme medal Soc. for Promotion of Engring. Edn., 1941. Mem. Am. Soc. C.E. (pres. 1929), Am. Soc. for Testing Materials, Ia. Engring. Soc. (pres. 1900), Soc. Promotion Engring. Edn. (pres. 1914-15), Land Grant Coll. Engring. Assn. (pres. 1913-14), Am. Assn. of Land Grant Colls. and Univs. (pres. 1929). Author: Sewers and Drains, 1907; Engineering Valuation (with T. R. Agg), 1936. Contbr. engring. jours. and trans. Died Oct. 21, 1949.

MARSTON, Jeffery Allen, English physician; b. Eng., 1831; s. Thomas Marston; M.D.; m. Annie Webb; with Army Med. Dept., from 1854, served in Egypt, 1882, named surgeon-gen., 1889. Fellow Royal Coll. Surgeons. Described leptospiral jaundice, 1863 (23 years before Adolf Weil); credited with distinguishing Malta fever from other fevers, 1863. Died 1911.

MARSTON, William Moulton, Am. psychologist; b. Cliftondale, Mass., May 9, 1893; s. Frederick William and Annie Dalton (Moulton) M.; A.B., Harvard, 1915, LL.B., 1918, Ph.D., 1921; m. Elizabeth Holloway, Sept. 16, 1915; children—Fredericka (dec.), Moulton, Olive Ann, Byrne Holloway, Donn Richard. Asst. in psychology, Radcliffe Coll., 1915; admitted to Mass. bar, 1918; atty. for Boston Legal Aid, 1918; prof. legal psychology, Am. U., Washington, D.C., 1922-23; psychologist Nat. Com. Mental Hygiene, Staten Island sch. survey, Tex. penitentiary survey, 1924; asst. prof. psychology and dir. student clinic, Tufts Coll., 1925-26; cons. psychologist since 1925; lecturer in psychology, Columbia and New York U., 1926-29; dir. pub. service, Universal Pictures Corp., 1929; lecturer in psychology, U. of Southern Calif.,

1929-30; prof. psychology, Long Island U., 1931-32; v.p. Hampton, Weeks & Marston (advertising agts.), 1931-32; lecturer psychology, New Sch. Social Science and Rand School, 1933, 34; public lecturer psychological topics since 1935. Dir. and v.p. Brunswick Sch. since 1944. Served as 2d lt., United States Army, 1918-19. Fellow A.A.A.S.; mem. Psychol. Assn., Orthological Inst. (assessor), Am. Assn. Criminal Law and Criminology, Phi Beta Kappa. Author: Emotions of Normal People, 1928; The Art of Sound Pictures (with W. B. Pitkin), 1930; Integrative Psychology (with wife and C. Daly King), 1930; Venus with Us, 1932; Try Living, 1937; The Lie Detector Test, 1938; March On, 1941; F. F. Proctor, Vaudeville Pioneer (with J. H. Feller), 1943. Contributor to Am. Jour. Psychology, Abnormal and Social Psychology, Psyche, Exptl. Psychology, Encyclopedia of Psychology; Ency. Britannica, popular psychol. articles to various mags., many syndicated articles on psychology to newspapers. Discoverer of systolic blood pressure deception test (popularly called the "lie detector"), 1915. Died May 2, 1947.

MARTEL, Édouard-Alfred, French geographer, speleologist; b. Pontoise, France, 1859; solicitor, tribunal of commerce, Paris, 1886-98; became prof. subterranean geography Sorbonne, 1899; named mem. staff Dept. Geol. Maps of France, 1901. Mem. Soc. Speleology (founder, sec.-gen. 1895), Soc. Geography (pres. from 1928). Author: Les Cévennes et la région des causses, 1889; Les abimes, 1894; L'Irlande et les cavernes anglaises, 1897; La spéléologie, 1900; Padirac, 1901; La France ignorée, 1928-30. A founder of speleology; explored limestone caves of Cévennes, 1883-87; made descents (with Gaupillat and de Launay) into previously unknown chasms in France, Ireland, Austria, Greece, Majorca, after 1888. Died Montbrison, France, 1938.

MARTELL, Arthur Earl, Am. chemist; b. Natick, Mass., Oct. 18, 1916; s. Ambrose R. and Dorina (Lamoureux) M.; B.S., Worcester Poly. Inst., 1938, D.Sc., 1962; Ph.D. (Univ. fellow), N.Y. U., 1941; m. Norma June Saunders, 1944; children—Stuart A., Edward S., Janet E., Judith S., Jon V., Elaine C.; m. Mary Austin, 1965; 1 dau., Helen E. Teaching fellow N.Y. U., 1938-40; instr. Worcester (Mass.) Poly. Inst., 1941-42; asst. prof. chemistry Clark U., Worcester, 1942-46, asso. prof., 1946-51, prof., chmn. dept. chemistry, 1951-61; prof., chmn. dept. chemistry Ill. Inst. Tech., Chgo., 1961-66; prof., head dept. chemistry Tex. A. and M. U., College Station, 1966—; research fellow chemistry U. Cal., 1949-50. Guggenheim fellow U. Zurich (Switzerland), 1954-55; NSF sr. postdoctoral fellow, also fellow Sch. for Advanced Studies, Mass. Inst. Tech., 1958-59; NIH postdoctoral fellow Lab. Chem. Biodynamics, U. Cal. at Berkeley, 1964-65. Fellow A.A.A.S., Am. Acad. Arts and Scis.; mem. Am. Chem. Soc., Internat. Union Pure and Applied Chemistry (mem. commn. on equilibrium data, chmn. 1963-67), Sigma Xi. Author: (with Melvin Calvin) Chemistry of the Metal Chelate Compounds, 1952; (with Stanley Chaberek) Organic Sequestering Agents, 1958. Editor: (with L. G. Sillen) Stability Constants, 2d edit., 1964. Research and numerous publs. on equilibrium, kinetics and mechanism of formation of metal chelate compounds. Home: 1211 Orr St., College Station, Tex. 77840.*

MARTIN, Adolphe, French physicist; b. Paris, 1824; D.Sc.; asst. master physics Inst. Agronomy; worked with Foucault at Paris obs. Author: Sur une méthode d'autocollimation direct des objectifs astronomiques, 1857; Méthode directe pour la détermination des courbures des objectifs de photographie, 1894. Inventor ferrotyping, 1852, also process for silverplating glass; helped build reflecting mirror of great telescope in Paris obs., 1874. Died 1896.

MARTIN, Archer John Porter, English chemist, engr.; b. London, Eng., Mar. 1, 1910; s. William Archer Porter and Lilian Kate (Brown) M.; student Peterhouse, Cambridge, Eng., 1929-32, Ph.D., 1936; m. Judith Bagenal, Jan. 9, 1943. Nutritional Lab., Cambridge, 1933-38; chemist Wool Industries Research Assn., Leeds, Eng., 1938-46; research dept. Books Pure Drug Co., Nottingham, 1946-48; staff Med. Research Council since 1948; head phys. chemistry div. Nat. Inst. Med. Research, Mill Hill, 1952-56; chem. cons., 1956-59; Extraordinary professor at Technological University, Eindhoven, The Netherlands, 1964. Recipient Berzelius gold medal Swedish Med. Soc., 1951; Nobel prize chemistry (with R. L. M. Synge), 1952; John Scott award, 1958; John Price Wetherill medal, 1959. Mem. Biochem. Soc.; Fellow Royal Soc., 1950 (Leverbulme medal 1964). Developed partition chromatography in organic analysis (technique applicable to all areas of biochem. research, led to discovery of new antibiotics and amino acids); developed techniques of gas-liquid chromatography; research on micro-manipulators and prodn. of small accurate parts. Home: Abbotsbury, Barnet Lane, Elstree Herts, Eng.

MARTIN, Arlene Patricia, Am. biochemist; b. Binghamton, N.Y., June 30, 1926; d. Edward Joseph and Helena F. (Hogan) Martin; A.B., Cornell U., 1948, M.Nutritional Sci., 1952; Ph.D., U. Rochester, 1957. Postdoctoral fellow dept. biochemistry U. Rochester Sch. Medicine and Dentistry, 1957-58, instr., 1958-65; asst. prof. dept. radiology Jefferson Med. Coll., Phila., 1965-68; asso. prof. biochemistry in pathology U. Mo., Columbia, 1968—. Fellow A.A.A.S., N.Y. Acad. Scis.; mem. Am. Chem. Soc., Sigma Xi, Sigma Delta Epsilon. Contbr. articles to tech. jours. Research on isolation and characterization of various mammalian enzymes involved in oxidative metabolism. Address: Dept. Pathology, U. Mo. Sch. Medicine, Columbia, Mo. 65201.*

MARTIN, Arthur Wesley, Am. zoologist; b. Nanking, China, Dec. 13, 1910 (parents Am. citizens); s. Arthur W. and Alice (Bull) M.; B.S., U. Puget Sound, 1931; Ph.D., Stanford, 1936; m. Mary Roberta Dubois, Aug. 23, 1931 (div. Mar. 1955); children—Arthur Wesley, III, Linda Gail (Mrs. David Kimball); m. 2d, Effie Mae Crowther, Aug. 29, 1958. Instr., Stanford, 1936-37; faculty, U. Wash., Seattle, 1937—, prof. physiology, 1949—, exec. officer dept. zoology, 1948-63; program dir. for regulatory biology NSF, Washington, 1958-59. Mem. A.A.A.S., Am. Physiol. Soc., Soc. Exptl. Biology and Medicine, Am. Soc. Zoologists (chmn. div. comparative physiology 1966—), Soc. Gen. Physiologists, Western Soc. Naturalists (past pres.). Research and numerous publs. on metabolic rate, trophic control of mammalian skeletal muscle, comparative physiology of excretion, tissue respiration, cephalopod circulation and excretion, carbohydrate metabolism. Home: 17737 15th Av. N.W., Seattle 98177.

MARTIN, Benjamin, English encyclopedist, mathematician, instrumentmaker; b. Worplesdon, Eng., 1704; self-taught. Tchr. Guildford, Eng.; traveler, lectr. natural philosophy, Eng; founder sch., also invented, prod., sold optical instruments, spectacles Chichester, Eng.; at London, Eng., from 1740. Author: Elements of Geometry, 1733; Spelling Book of Arts and Sciences; Philosophical Grammar in Four Parts; The Young Student's Memorial Book, 1735; A New System of Decimal Arithmetic, 1735; Trigonometer's Complete Guide, 2 vols., 1736; Description and Use of Both the Globes, 1736; Elements of All Geometry, 8 vols., 1739; Description and Use of a Newly Invented Pocket Microscope, 1740; Logarithmologia, 1740; Micrographia Nova, 1742; Description and Use of a Case of Mathematical Instruments, 1745; An Essay on Electricity, 1746; Supplement Containing Remarks on a Rhapsody of Adventures of a Modern Knight-Errant in Philosophy, 1746; Philosophia Britannica, 2 vols., 1747; An English Dictionary, 1749; Panegyric of the Newtonian Philosophy, 1749; On the New Construction of the Globes, 1755; Essay on Visual Glasses, 1756; Philological Library of Literary Arts and Sciences, 1757 Essay on the Use of Globes, 1758; New Elements of Optics, 1759; A Sure Guide to Distillers, 1759; Venus in the Sun, 1761; A Plain and Familiar Introduction to the Newtonian Philosophy, 5th edit., 1765; Institutions of Astronomical Calculations, 1765; The Mariner's Mirror, Part 2, 1769; Description and Use of a Table Clock upon a New Construction, 1770; Description and Use of an Orrery, 1771; Description . . . of a Graphic Perspective and Microscope, 1771; Optical Essays, 1770; Logarithmologia Nova, London, 1772; The Young Gentleman and Lady's Philosophy, 3d edit., 1781; Pub. Martin's Magazine, 1755-64. Renown as maker sci., optical instruments in Eng.; collected fossils, curiosities; writings included most aspects contemporary knowledge comprising scis., math., arts, geographies, biographies, customs. Died Feb. 9, 1782.

MARTIN, Charles F., physician; b. Montreal, Can., 1868; B.A., M.D., C.M., McGill U.; student Pasteur Inst., Paris, France; LL.D., univs. Queens, McGill and Harvard; D.C.L., U. Lennoxville; married. Hon. pres. Alexandra Hosp., Montreal; hon. v.p. Royal Edward Laurentian Hosp.; emeritus dean med. faculty, also emeritus prof. McGill U.; cons. physician Royal Victoria Hosp. Hon. pres. Montreal Museum Fine Arts; trustee Children's Meml. Hosp. Mem. Canadian Med. Assn. (Starr medal 1953), Assn. Am. Physicians (past pres.), A.C.P. (past pres.), Assn. Am. Med. Colls. (past pres.). Publns. on blood, stomach, heart, metabolism, med. edn. Died Oct. 28, 1952.

MARTIN, Sir Charles James, Brit. physiologist; b. London, 1866; s. Josiah Martin; ed. King's Coll., London, St. Thomas's Hosp., U. Leipzig (Germany); M.B., D.Sc., London; D.Sc., Melbourne, Australia; D.Sc. honoris causa, Sheffield, Eng., Dublin, Ireland; LL.D., Edinburgh; D.C.L. (hon.), Durham, Eng.; M.A. (hon.), Cambridge; m. Edythe Cross; 1 dau. Exhibitioner, scholar in physiology, U. London; demonstrator biology and physiology, evening lectr. comparative anatomy King's Coll., London, 1887; became demonstrator physiology U. Sydney, Australia, 1891; named lectr. physiology U. Melbourne, 1897, prof. physiology, 1901; dir. Lister Inst. Preventive Medicine, 1903-30; prof. exptl. pathology U. London, now prof. emeritus. Chief div. animal nutrition Australian Council Sci. and Indsl. Research, 1931-33; prof. biochemistry and gen. physiology Adelaide U., Australia, 1931-33. Mem. adv. com. for investigation plague in India, India Office, Royal Soc. and Lister Inst., 1905-13. Recipient Gold medal, London. Became fellow Royal Soc., 1901, Royal Coll. Physicians. Research and numerous publs. in biology, pathology, physiology. Died Feb. 15, 1955.

MARTIN, Charles Louis, Am. physician; b. Massey, Tex., Dec. 2, 1893; s. James Madison and Emma (Auerbach) M.; E.E., U. Tex., 1914; M.D., Harvard, 1919; m. Maybelle Ober, July 17, 1920; 1 son, James Addison. Prof. radiology Baylor Med. Sch., Dallas, 1920-40; dir. Martin X-Ray and Radium Clinic, Dallas, 1940—, cons., 1966—; prof. radiology U. Tex. Med. Sch., Dallas, 1950—, emeritus prof., 1966—. Radiol. cons. Gaston, Baylor, Parkland, VA hosps., Dallas, 1943—. Mem. Dallas County Med. Soc. (pres. 1951), Dallas So. Clin. Soc. (Marchman award 1953, pres. 1953), Am. Radium Soc. (pres. 1947, Janeway medal 1949), Am. Roentgen Ray Soc. (pres. 1952), Am. Coll. Radiology (bd. chancellors 1957-61), Dallas Tech. Club, (past pres.), Am. Club Therapeutic Radiologists, Alpha Omega Alpha, Delta Kappa Epsilon. Author (with James A. Martin) Low Intensity Radium Therapy, 1959. Contbr. sci. articles to med. jours., textbooks. Home: 4605 Watauga Rd., Dallas 75209. Office: 3501 Gaston Av., Dallas 75246.*

MARTIN, Charles-Noël, French physicist; b. Paris, France, Dec. 25, 1923; s. Charles Antoine and Jeanne (Saupin) M.; student U. Tunis (Tunisia), 1942, U. Algier (Algeria); Ph.D., Sorbonne, Paris, 1949; m. Huguette Oddo, June 20, 1963. Staff, Institut du Radium, 1950-51; research attaché Institut Henri Poincaré, 1952-57; independent researcher, 1958—. Sci. cons. French radio-TV, Swiss radio, Sottens. Decorated commander Nicham Iftikhar (Tunisia), 1955. Mem. Société Mathématique de France, Am. Phys. Soc., A.A.A.S., Società Italiana di Fisica, Phys. Soc. Japan, Société Suisse de Physique. Author numerous books including: (with Franklin Watts) The 13 Steps to the Atom; (with Hutchinson) The Role of Perception in Science; (with F. Watts) The Atom Friend of Foe?; (with Hill, Wang) The Universe of Science; Nuclear Physics Tables (contains calculated masses and binding energies of more than 3000 nuclides), 1954; also numerous popular and sci. articles. Sci. cons., sci. editor French newspaper Le Figaro. Research on alpha radio-activity, nuclear structure and theory of elementary particles; classification and analysis of 32 transformations of energy. Home: 6 Rue des Saints Sauveurs, Fontenay aux Roses (92) France. Office: Institut Henri Poincaré, II Rue Pierre Curie, Paris (5) France.*

MARTIN, Daniel S., Am. physician; b. Bklyn., Oct. 29, 1921; s. Jacob and Rose (Abelson) Shapiro; student Cornell U., 1938-41; M.D., N.Y. U., 1944; m. Bette Justine Finch, Feb. 15, 1947; children—Scott Vincent, Gail Justine, Dee Marie, Lee Ann. With U. Miami, 1958—, asso. prof. surgery; chief surg. service Jackson Meml. Hosp., Miami, Fla., 1958—; cons. surgery VA Hosp., Coral Gables, Fla. Recipient Mead Johnson award, 1955; Dazian Found. for Med. Research fellow, 1951. Damon Runyon Cancer Research fellow, 1949-50. Mem. Am. Assn. Cancer Research, Inc., Am. Soc. Artificial Internal Organs, Am. Soc. Clin. Oncology, Harvey Soc., Soc. Exptl. Biology and Medicine. Author: Progress in Clinical Cancer, Cancer Chemotherapy by Regional Perfusion, 1965. Basic research, publs. in cancer chemotherapy, treatment of shock, devel. membrane oxygenator, studies in clin. perfusion and gastric physiology. Home: 3939 Leafy Way, Coconut Grove, Fla. 33133. Office: 1700 N.W. 10th Av., Miami, Fla. 33136.*

MARTIN, Daniel William, Am. acoustical physicist; b. Georgetown, Ky., Nov. 18, 1918; s. Dean William and Ethel (Weigle) M.; A.B., Georgetown Coll., 1937; M.S., U. Ill., 1939, Ph.D., 1941; m. Martha Elizabeth Parker, June 9, 1941; children—Mary Elizabeth, David William, Nancy Jane, Donald Warren. Asst. physics U. Ill. at Urbana, 1937-41; with RCA, 1941-49, tech. coordinator, 1946-49; acoustical research supr. Baldwin Piano Co., Cin., 1949-57; research dir. D. H. Baldwin Co., Cin., 1957—; extension instr. math. Purdue U., 1941-46; asst. prof. musical acoustics U. Cin., 1964—. Fellow Acoustical Soc. Am. (exec. council 1957-60), Audio Engring. Soc. (pres. 1964-65), I.E.E.E. (nat. chmn. profl. group audio I.R.E. 1956-57); mem. Engring. Soc. Cin. (bd. dirs. 1964—), Tech. and Sci. Socs. Council Cin. Author, patentee in field. Editor I.R.E. Trans. on Audio, 1953-55; asso. editor Sound, 1961-63. Research in physics of brass instruments and piano, tone of piano, organ, orchestral instruments and voice; invented throat microphones, ear cushions, loud-speaker cabinets, electronic mus. instruments, reverberation devices; developed speech communication systems for mil. aircraft in intense noises. Home: 7349 Clough Pike, Cin. 45244. Office: 1801 Gilbert Av., Cin. 45202.*

MARTIN, Douglas Leonard, physicist; b. London, Eng., Nov. 11, 1930; s. Leonard William and Jenny (Anderson) M.; B.Sc., U. London, 1951, Ph.D., 1954. Postdoctorate fellow Nat. Research Council, Ottawa, Ont., Can., 1954-55, physicist, 1957—, now sr. research officer; sci. officer Royal Aircraft Establishment, Farnborough, Eng., 1955-56. Fellow Phys. Soc. (London). Research and publs. on calorimetry, cryogenics, solid state physics, especially metals and alloys. Office: Nat. Research Council, Ottawa 2, Ont., Can.*

MARTIN, Ernest Gale, Am. physiologist; b. Mpls., Nov. 16, 1876; s. John Wesley and Mary Esther (Bullard) M.; Ph.B., Hamline U., 1897; Ph.D., Johns Hopkins, 1904; m. Ruby A. Ticknor, Aug. 31, 1904; 1 dau., Lois Ticknor. Fellow and asst. in physiology, Johns Hopkins, 1902-04; instr. physiology Purdue U., 1904-06; instr. physiology Harvard, 1906-10, asst. prof., 1910-16, also lectr. Sargent Sch. for Phys. Edn., 1906-14; asst. prof. physiology Radcliffe Coll.,

1914-16; physiologist Vt. Bd. Health, 1915-16; prof. physiology Stanford, 1916—. Fellow Am. Acad. Arts and Scis., A.A.A.S. Author: The Measurement of Induction Shocks, 1912. Revised 9th, 10th and 11th edits. of The Human Body (by Henry N. Martin), 1910; vol. on physiology (Collier's Popular Science Library), 1921. Joint author: General Biology (with Burlingame, Heath, and Peirce); Elements of Physiology (with Weymouth), 1928. Died Oct. 17, 1934.

MARTIN, Ettore Leonida, Italian astronomer; b. Latisan, Italy, Nov. 21, 1890; s. Giovanni and Angela (Moro) M.; Ph.D.; m. Carlina Donati, Apr. 15, 1925; children—Gianfranco, Giuliano. Dir., Astronomic Obs., Trieste, Italy. Author: Stelle doppie, 1932-50; Astronomia sferica e teorica, 1956-57. Mem. Internat. Astron. Union, Veneto Inst., Acad. Sci. Letters and Arts of Udina and Padua, Veneto Inst., Adriatic Acad. of Sci. Trieste. Research in celestial mechanics. Home: via G. B. Tiepolo 11. Died Venice, Aug. 9, 1966.

MARTIN, Everett Dean, Am. sociologist; b. Jacksonville, Ill., July 5, 1880; s. Buker E. and Mollie (Field) M.; B.A., Ill. Coll., Jacksonville, 1904, Litt.D., 1929; grad. McCormick Theol. Seminary, 1907; m. Esther W. Kirk, 1907 (divorced 1915); children—Mary, Margaret, Elizabeth; m. 2d, Persis E. Rowell, 1915 (divorced); 1 son, Everett Eastman; m. 3d, Daphne Crane Drake, 1931. Ordained Congl. ministry, 1907; pastor 1st Ch., Lombard, Ill., 1906-08, Peoples Ch., Dixon, Ill., 1908-10, Unitarian Ch., Des Moines, Ia., 1910-14; writer on philos. subjects, 1914-16; lecturer on social philosophy, Peoples Inst., N.Y., 1916-36; asst. dir. and sec., 1917-22, and dir., 1922-38; dir. Cooper Union Forum, the largest center for free discussion of polit. and edni. subjects in America, head of its dept. social philosophy, 1918-38; prof. of social philosophy, Cooper Union, 1934-38, Graduate School, Claremont (Calif.) Colleges, 1936—; mem. of Faculty Graduate School of Banking, 1938—; lecturer social psychology, New School for Social Research, 1922; instructor in social psychology, Brookwood Workers Coll., Katonah, N.Y., 1922-23; traveling lecturer and prof. of social philosophy, Assn. of Am. Colls. 1939. Chmn. Nat. Bd. of Review of Motion Pictures, 1919-22; Culver lecturer, Brown University, 1930. Author: The Behavior of Crowds, 1920; The Mystery of Religion; Psychology; Psychology and Its Uses, 1926; The Meaning of a Liberal Education, 1926; Liberty, 1926; The Conflict of the Individual and the Mass; Civilizing Ourselves, 1932; Farewell to Revolution; Philosophical Backgrounds of Current Economic and Social Problems; Some Principles of Political Association; The Nature of the State. Best known for his sociology of crowds and description of how a crowd is formed from an aggregation of individuals. Died Claremont, May 10, 1941.

MARTIN, Fernando Wood, Am. chemist; b. Volga, W.Va., May 5, 1863; s. Washington and Matilda (Cool) M.; B.S., Chaddock Coll., 1886; Ph.D., Syracuse U., 1893; postgrad. U. Leipzig, 1897, 97-98; m. Emma Herron, June 26, 1889. Prof. natural sci. Chaddock Coll., 1886-90; lectr. chemistry and toxicology Quincy (Ill.) Med. Coll., 1889-90; prof. natural sci., v.p. Ft. Worth U., 1890-92; prof. chemistry Randolph-Macon Women's Coll., 1893-1929, v.p., 1894-1907, prof. emeritus, 1929. Author: Qualitative Analysis with the Blow Pipe, 1903; Text-Book on Inorganic Chemistry, 1904; Qualitative Analysis, 1907; Introduction to Anthropology, 1913; Essentials of Organic Chemistry, 1915. Died Mar. 22, 1933.

MARTIN, Franklin H., Am. surgeon; b. Ixonia, Wis., July 13, 1857; s. Edmond and Josephine (Carlin) M.; M.D., Chgo. Med. Coll. (now med. dept. Northwestern U.), 1880; LL.D., Queen's Univ., Belfast, Ireland, U. Wales, U. Pitts.; D.P.H., Detroit Coll. of Medicine and Surgery; D.Sc., Northwestern U.; m. Isabelle Hollister, 1886. Prof. gynecology Polyclinic, Chgo., 1886-88; organized with Dr. W. F. Coleman, Post-Grad. Med. Sch., Chgo., 1888; gynecologist Woman's Hosp. many yrs.; organized Charity Hosp.; founded Surgery, Gynecology and Obstetrics (med. jour.), 1905, editor in chief, added Internat. Abstract of Surgery, 1913; organized clin. Congress Surgeons of N. Am. (now Clin. Congress of A.C.S.), 1910, dir. gen., organized A.C.S., 1913, dir. gen. and mem. bd. regents, also pres., 1928-29; asso. editor Am. Jour. Obstetrics and Gynecology. Author: Treatment of Fibroid Tumors of the Uterus, 1897; Treatise on Gynecology, 1903; South America from a Surgeon's Point of View, 1923; (monograph) Australia and New Zealand, 1924; The Joy of Living—An Autobiography, 1933. Died Mar. 7, 1935.

MARTIN, Franklin W., Am. geneticist; b. Salt Lake City, Apr. 14, 1928; s. Herman and Anna (Jewett) M.; B.S., Okla. Baptist U., 1948; postgrad. Wash. State Coll., 1948-49, Coll. of Pacific, 1949-51; Ph.D., U. Cal., Davis, 1960; m. Madelain Nancy Dawson, Sept. 28, 1956; children—Jonathan Dean, Roderic Lee, Alison Gale, Cecilia Ann. Research technician U. Cal., Riverside, 1954-56, Davis, 1956-60; asst. horticulturist Western Wash. Expt. Sta., Puyallup, 1960-61; geneticist Fed. Expt. Sta., Mayaquez, P.R., 1961—. Mem. Genetics Soc. Am., Bot. Soc. Am., Am. Genetic Assn. Studies and numerous publs. of incompatibilities and sterilities in flowering plants, including tomato, broccoli, black pepper, sweet potato and Tephrosia vogellii; identification of their physiol. mechanisms and genetic control; genetics and breeding of sapogenin-bearing yams. Address: Fed. Expt. Sta., Mayaquez, P.R. 00708.*

MARTIN, F(redrick) O(skar), engr., geologist; b. Middweida, Saxony, Germany, Aug. 20, 1871; s. Frederick August and Anna Emmeline (Heyne) M.; ed. Bealschule, Mittweida, Saxony; student Columbian Univ., D.C., 1900-02, Harvard and Catholic U.; m. Agnes Elizabeth Riese, Aug. 13, 1908; children—Anna Elisabeth, Agnes Fritzi, Mrs. Margareth Martin Hamer. Engaged in mining and prospecting in Alaska, Ida., Cal., Wash., Mont., 1894-1900; asst. in soil survey and scientist Bur. of Soils, U. S. Dept. Agr., 1901-05; asst. engr. Panama Canal, div. of meteorology and river hydraulics, 1905-06; engring. work and ry. contractor, 1906-09; mineral insp. U. S. Dept. Interior, 1909-19, principally in Cal. oil fields; geologist, Union Oil Co. of Cal., 1919-30, principally in Colombia; later in pvt. practice; mining engr. Div. Investigations, U. S. Dept. Interior, 1933-41; then pvt. practice. Fellow Royal Geog. Soc., Am. Geog. Soc., Pacific Geog. Soc.; mem. Am. Inst. Mining and Metall. Engrs., Am. Assn. Petroleum Geologists. Author: Explorations in Colombia, South America. Died June 30, 1951.

MARTIN, George Willard, Am. mycologist; b. Bklyn., Oct. 27, 1886; s. George Augustus and Sarah Ann (Harned) M.; Litt.B., Rutgers U., 1912, M.Sc., 1915, D.Sc., 1963; Ph.D., U. Chgo., 1922; m. Mary Gillespie, Dec. 23, 1916; children—Elizabeth Gillespie (Mrs. Edwin Stephen McCollister), Ann Harned (Mrs. Richard Joseph Edelman). Faculty, Mass. Agrl. Coll., 1916-17, Rutgers U., 1919-23; faculty U. Ia., Iowa City, 1919-55, prof., head, dept. botany, 1952-55, emeritus prof., 1955—. Vis. prof. U. Ill., 1956-57; chief, biol. lab. U. S. Army Q.M. Depot, Jeffersonville, Ind., 1944-45. Mem. Mycol. Soc. Am. (past pres.), Am. Assn. U. Profs., Bot. Soc. Am., Am. Soc. Plant Taxonomy, Soc. Indsl. Biology, Wash. Acad. Sci., Phi Beta Kappa, Sigma Xi. Author: (with Macbride) The Myxomycetes, 1934, also articles. Research in plant pathology, marine ecology; study of fungi, particularly in gen. classification of these organisms and in spl. taxonomy of slime molds and jelly fungi. Home: 1685 Ridge Rd., Iowa City 52240.*

MARTIN, Gustav Julius, Am. biochemist; b. Hartline, Wash., Dec. 5, 1910; s. Charles and Pauline Christine (Haas) M.; B.S. in Chemistry, U. Wash., 1932; Sc.D. in Biochemistry, Johns Hopkins, 1935; Sc.D. (hon.), Phila. Coll. Pharmacy and Sci., 1958, Upsalla Coll., 1964; m. Dorothy Patricia Rogers, July 5, 1936. Faculty chemistry Coll. St. Teresa, Winona, Minn., 1936, pharmacology Middlesex Coll., Waltham, Mass., 1938; asso. research dir. Warner Inst., N.Y.C., 1940-44; dir. research Nat. Drug, Phila., 1944-60, William H. Rorer, Ft. Washington, Pa., 1960; faculty research biology Gwynedd Mercy Coll., Ambler, Pa., 1963; dir. biochem. research Camden (N.J.) Gen. Hosp., 1962—; vis. research scientist Eastern Pa. Sch. and Hosp., Ft. Washington, 1964—. Vice pres. Alcoholism Research Found., Phila. 1965—. Fellow 8 sci. socs.; mem. Carl Neuberg Soc. for Internat. Sci. Relations (pres.) and 22 other sci. socs., Chemists Club, N.Y.C. Author: Biological Antagonism, 1951; Ion Exchange and Absorption Agents in Medicine, 1955; Clinical Enzymology, 1958; Biological Fourth Dimension, 1964. Creator over 100 drug products, including antacids, 1942, bioflavonoids, 1946, acridines, 1947, enzymes, 1952. Home: 372 W. Johnson St., Phila. 19144. Office: 500 Virginia Dr., Ft. Washington, Pa. 19034.*

MARTIN, Hans, German chemist; b. Ludwigshafen, Germany, Apr. 21, 1908; s. Wilhelm and Henriette M.; ed. U. Heidelberg (Germany); Ph.D.; m. Marie-Theresa Nobisch, Feb. 14, 1942; children—Joachim, Wolfgang. Asst., Karlsruhe (Germany) Tech. Coll., 1933-36; asst. U. Kiel (Germany), 1937-41, agrege, 1941, became asso. prof., dir. Inst. Physics and Chemistry, 1949, prof., 1957. Mem. Bunsen Soc., German Soc. Physicist, Assn. German Chemists, Electrochem. Soc., Faraday Soc. Research and publs. on photochem. separation of isotopes, molecular reactions of radiation, gas and solvents. Home: Moltkestrasse 11, 23 Kiel, West Germany.

MARTIN, Henry Austin, surgeon; b. London, Eng., July 23, 1824; s. Henry James Martin; grad. Harvard Med. Sch., 1845; m. Frances Coffin Crosby, 1848; 5 children, including Stephen Crosby, Francis Coffin. Practiced medicine, Roxbury, Mass.; apptd. surgeon U. S. Army, 1861; became med. dir. at Norfolk, Portsmouth and Newbern, Va.; apptd. surgeon-in-chief 1st Div., II Corp., Army of Potomac, circa 1864; ret., 1865; introduced smallpox vaccine produced from cowpox (together with innoculation method) to U. S., 1870; developed use of rubber bandage for ulcers of leg, tracheotomy operation without tube. Died Boston, Dec. 7, 1884.

MARTIN, Henry Newell, physiologist; b. Newry, Ireland, July 1, 1848; B.Sc., Cambridge (Eng.) U.; M.B., U. London (Eng.); m. Hetty Pegram, 1878. Fellow, Trinity Coll., Cambridge U., 1874-76; came to U. S., 1876; prof. biology Johns Hopkins U., 1876-93, specialized in study of cardiac physiology. returned permanently to Eng., 1893. Fellow Royal Soc., 1885 (Croonian lectr. 1883). Author: (with Julian Huxley) A Course of Practical Instruction in Elementary Biology, 1875; The Human Body, 1881; Physiological Papers, 1895. Founder, editor Johns Hopkins U. Studies from Biol. Lab., 1877-93; Established (with E. M. Hartwell) function of intercostal muscles in respiration, circa 1879-80; 1st to isolate mammalian heart in a perfusion chamber, 1881. Died Burley-in-Wharfedale, Yorkshire, Eng., Oct. 27, 1896.

MARTIN, James Paxman, Am. soil microbiologist; b. Cowley, Wyo., Sept. 22, 1914; s. Thomas Lysons and Hattie (Paxman) M.; B.A., Brigham Young U., 1938; Ph.D., Rutgers U., 1941; m. Cleo Mary Long, Sept. 15, 1937; children—Karen (Mrs. Maurice L. Campbell), Janet (Mrs. Francis E. Burk). Research asst. N.J. Agr. Exptl. Sta., New Brunswick, N.J., 1938-41, asst. soil microbiologist, 1941-43; asst. bacteriologist U. Ida., Moscow, 1943-45; with Citrus Research Center and Agr. Expt. Sta., U. Cal. at Riverside, 1945—, chemist, 1958—, prof. soil sci., 1963—; researcher S. African Dept. Agr. and Citrus Exchange, 1959; vis. staff Inst. for Biochemistry of Soil, Branschweig, Germany, 1966. Fellow A.A.A.S., Am. Soc. Agronomy; mem. Western Soc. Soil Sci., Am. Soc. Microbiologists, Soil Sci. Soc. Am. Research and numerous publs. on soil microorganisms in relation to soil tilth, synthesis of resistant organic humus compounds by soil organisms, factors causing reduced growth of citrus replants, influence of pesticides on plant growth and soil properties. Home: 1180 La Subida Ct., Riverside, Cal. 92507.*

MARTIN, Jean Charles, French biologist; b. Reims, France, Mar. 6, 1929; s. Charles and Louise (Chambon) M.; m. Michele Bon, Sept. 6, 1958; children—Jean-Remi, Clotilde. With Internat. des hôpitaux, Reims, France, 1954-60; researcher biology Centre d'Etudes et Recherches des Charbonnages de France, Verneuil-en-Halatte, 1960—. Mem. Société francaise de Pathologie Respiratoire, Société Francaise de Microscopie Electronique. Research, publs. on lung pathology, environmental diseases and pneumoconiosis, histophysiology and lung cytology. Office: CERCHAR-B.P. 27, Creil, (60), France.*

MARTIN, John Hume, Scottish physicist; b. Dundee, Scotland, Sept. 26, 1919; s. William Dalrymple and Mary (Crammond) M.; B.Sc. with honors, U. St. Andrews (Scotland), 1940; Ph.D., U. London (Eng.), 1950; m. Mary N. Hall, July 24, 1948; children—Linda Jean, John Graham, Stephen William. Jr. sci. officer Royal Aircraft Establishment, Farnborough, Eng., 1940-44; physicist Royal Cancer Hosp., 1944-50; head physics dept. Cancer Inst., Melbourne, Australia, 1950-60; mgr. health and safety U.K. Atomic Energy Authority, Chapelcross, Annan, Scotland, 1960—. Mem. various internat. adv. panels on radiation dosimetry and radiobiology. Fellow Inst. Physics; mem. Brit. Inst. Radiology, Hosp. Physicist Assn., Soc. for Radiol. Protection. Editor: Proceedings of 2d Australasian Conference on Radiation Biology, 1958; (with J. M. A. Lenihan, J. F. Loutit) Proceedings of Symposium on Some Aspects of Strontium Metabolism, 1967; also articles. Research on radiation dosimetry especially of high energy X-rays and particle radiation; phys. and biol. studies relating energy deposition mechanisms and biol. effects. Home: Mt. Pleasant, Hill St., Dumfries, Dumfriesshire. Office: Chapelcross Works, Annan, Dumfriesshire, Scotland.*

MARTIN, John Joseph, Am. mech. engr.; b. Detroit, Oct. 19, 1922; s. Anthony L. and Helen (Jason) M.; B.S. in Mech. Engring., U. Notre Dame, 1943, M.S., 1950; Ph.D., Purdue U., 1951; m. Carol C. Kline, June 10, 1948; children—Peter, Anne, Stephen. Instr., U. Notre Dame (Ind.), 1946-47, 49-50; research engr. N.Am. Aviation, Inc., Downey, Cal., 1951-53; chief engr. Bendix Corp., South Bend-Mishawaka, Ind., 1953-60; resident cons. Royal Aircraft Establishment, Farnborough, Eng., 1963-64; staff Inst. Def. Analyses, Arlington, Va., 1960-63, 64—, spl. asst. to v.p. for research, 1966-67, asst. to pres., 1967—. Mem. Acoustical Soc. Am., Am. Inst. Aero. and Astronautics, Philos. Soc. Washington, Am. Soc. M.E. (past sec. Los Angeles), Sigma Xi, Tau Beta Pi, Pi Tau Sigma. Author: Atmospheric Reentry, 1966; also numerous articles. Research on atmospheric reentry ballistic missiles and manned space craft; description of reflection characteristics of ocean surfaces and bottoms. Home: 7818 Fulbright Ct., Bethesda, Md. 20034. Office: Inst. Def. Analyses, 400 Army Navy Dr., Arlington, Va. 22202.*

MARTIN, Lay, Am. physician; b. Balt., Nov. 12, 1892; s. Augustus Warfield and Anne Carberry (Lay) M.; B.S., Princeton, 1915; M.D., Johns Hopkins, 1920; m. Lettice Lee Coulling Streett, Aug. 6, 1943; children—Lay, Warfield, Sarah Lee, William B. Faculty, Johns Hopkins, Balt., 1925—, now asso. prof. medicine; physician Johns Hopkins Hosp., Balt., 1925—; practice internal medicine, Balt., 1923—. Mem. Am. clin. and climatological Assn., A.M.A., Am. Gastroent. Assn. Research and numerous publs. on biochemistry gastric secretion and mucosa, fgn. protein reactions in immunology. Home: 220 Wendover Rd., Balt. 21218. Office: 1201 Calvert St., Balt. 21202.*

MARTIN, Lillien Jane, Am. psychologist; b. Olean, N.Y., July 7, 1851; d. Russell and Lydia (Hawes) Martin; A.B., Vassar Coll., 1880; postgrad. U. Göttingen, 1894-98; hon. Ph.D., U. Bonn, 1913. Science tchr. Indpls. High Sch., 1880-89; vice prin. and head dept. science Girls' High Sch., San Francisco, 1889-94; asst. prof. psychology Stanford U., 1899-1909, asso. prof., 1909-11, prof., 1911-16, prof. emeritus since 1916; cons. psychologist, San Francisco; psychopathologist and chief of mental hygiene clinic San Francisco Polyclinic and Mt. Zion Hosp. Founder and dir. Old Age Counselling Center, San Francisco; pres. Cal. Soc. Mental Hygiene, 1917-21; mem. Kongress für experimentelle Psychologie. Fellow A.A.A.S.; mem. Am. Psychol. Assn., Sigma Xi. Author: Zur Analyse der Untershiedsempfindlichkeit, 1899; Über Asthetische Synästhesic, 1909; Zur Lehre von der Bewegungsvorstellungen, 1910; Die Projektions Methode, 1912; Ein experimenteller Beitrag zur Erforschung des Unterbewussten, 1915; Personality as Revealed by the Content of Images, 1917; Mental Hygiene and the Importance of Investigating It, 1917; Two Years' Experience as a Clinical Psychologist, 1920; Mental Training of the Pre-School Age Child, 1923; Round the World with a Psychologist, 1927; Salvaging Old Age, 1930; Sweeping the Cobwebs, 1933; The Home in a Democracy, 1937; also various articles in psychol. and other jours. Died Mar. 26, 1943.

MARTIN, Louis, French physician, bacteriologist; b. Puy, France, Sept. 20, 1864; asst. dir., dir. Pasteur Inst., Paris; mem. Acad. Medicine, French Acad. Scis., 1937. Introduced culture media for cultivation of typhoid bacillus, 1915, for isolation of diphtheria bacillus, 1916, for spirochaeta icterohemorrhagiae, 1917; described bacillus of pseudo-diphtheria, also differentiated between diphtheria and pseudo-diphtheria, 1919; co-discoverer of antidiphtheric serum; dispensed (with Roux and Chaillou) 1st injections of anti-diphtheria serum. Died Paris, June 13, 1946.

MARTIN, Monroe H(arnish), Am. mathematician; born Lancaster, Pa., Feb. 7, 1907; s. Amos Z. and Mary (Harnish) Martin; B.S., Lebanon Valley Coll., 1928, Sc.D., 1958; Ph.D., Johns Hopkins, 1932; m. Virginia Parker, June 18, 1932; 1 dau., Mary Helen. Nat. Research fellow Harvard, 1932-33; instr., math. Trinity Coll., 1933-36; asst. prof. math., U. Md., 1936-38, asso. prof., 1938-42, prof. and acting head dept., 1942-43, prof. and head 1943-54; mem. part time Inst. for Fluid Dynamics and Applied Math., 1949—— (chmn. Conf. on Differential Equations 1950), acting dir., 1952-54, dir., 1954——; exec. sec. div. math. Nat. Acad. Sci. NRC, 1955-57, 58-59, 61-63, chmn. com. on application of math., 1958-59; Guggenheim fellow, and vis. lectr. St. Andrews, 1960. Mem. Am. Math. Soc. (chmn. com. applied math. 1951-54, rep. U. S. Nat. Commn. on Theory and Applied Mechanics 1953-56), Am. Math. Assn., Sigma Xi, Phi Beta Kappa, Kappa Sigma, Gamma Alpha. Contbr. articles fields of celestial mechanics, dynamics, analysis and gas dynamics in U. S. and European math. jours. Home: College Park, Md.*

MARTIN, Paul Cecil, Am. physicist; b. Bklyn., Jan. 31, 1931; s. Harry and Helen (Salzberger) M.; A.B., Harvard, 1952, Ph.D., 1954; m. Ann Wallace Bradley, Aug. 7, 1957; children—Peter, Stephanie. Faculty, Harvard, 1957——, prof. physics, 1964——; vis. prof. École Normale Supérieure, Paris, France, 1963, 66. NSF postdoctoral fellow, 1955; Sloan Found. fellow, 1959; Guggenheim fellow, 1966. Mem. Am. Acad. Arts and Scis., Am. Phys. Soc. Bd. editor Jour. Math. Physics, 1965——. Contbns. to theoretical physics in areas of quantum electrodynamics, nuclear theory, statis. mechanics, solid state and low temperature physics. Home: 27 Stone Rd., Belmont, Mass. 02178. Office: Dept. Physics, Harvard Cambridge, Mass. 02138.*

MARTIN, Pierre-Emile, French engr.; b. Bourges, France, Aug. 18, 1824; student Sch. Mines; worked in forges, Fourchambault, France; became dir. Sireul forges, 1854-83. Invented open-hearth process for prodn. steel by incomplete decarbonization in smelting (basis for present iron and steel industry and now known as Siemens-Martin process), 1865. Died Fourchambault, May 24, 1915.

MARTIN, R(obert) Bruce, Am. chemist; b. Chgo., Apr. 29, 1929; s. Robert Frank and Helen (Woelffer) M.; B.S., in Chemistry, Northwestern U., 1950; Ph.D. in Phys. Chemistry, U. Rochester, 1953; m. Frances May Young, June 7, 1953. Asst. prof. chemistry Am. U., Beirut, Lebanon, 1953-56; research fellow Cal. Inst. Tech., Pasadena, 1956-57, Harvard, 1957-59; faculty U. Va., Charlottesville, 1959——, prof., 1965——; program dir. molecular biology NSF, Washington, 1965-66. Spl. fellow NIH, Oxford (Eng.) U., 1961-62. Mem. Am. Soc. Biol. Chemists, A.A.A.S. Author: Introduction to Biophysical Chemistry, 1964; also numerous articles. Research on rates and equilibria in chem. reactions of biol. interest; structure analysis by circular dichroism and nuclear magnetic resonance spectroscopy. Address: Chemistry Dept., U. Va., Charlottesville, Va. 22901.*

MARTIN, Samuel Preston, Am. physician; b. East Prairie, Mo., May 2, 1916; s. Samuel Preston and Lucy (Simmons) M.; M.D. cum laude, Washington U., St. Louis, 1941; m. Ruth Campbell, July 2, 1939; children—Samuel P., William B., Celia S. Vis. investigator Rockefeller Inst., N.Y.C., 1948; with Duke, 1949-56, asso. prof., 1953-56; prof., head dept. medicine U. Fla., Gainesville, 1956-62, provost, 1963——. Con., mem. nat. adv. council med., dental, optometric and podiatric edn. USPHS, 1966——; mem. com. on chemotherapy NRC, 1956-58. Decorated Order of Leopold (Belgium). Mem. Am. Assn. Physicians, Am. Assn. Immunologists, Am. Soc. Clin. Investigation, A.C.P., So. Soc. for Clin. Research. Research and numerous publs. in mechanisms of natural immunity and host metabolic response to infection, metabolic response of human leukocytes to disease. Home: 1900 S.W. 8th Dr., Gainesville, Fla. 32601.*

MARTIN, Stevens John, Am. physician; b. N.Y.C., July 4, 1906; s. John Timothy and Annette (La Pane) M.; B.A., U. Wis., 1927, M.A., 1928, Ph.D., 1930, M.D., 1935; m. Louise Jane Minshall, Aug. 7, 1933; children—Sandra Louise (Mrs. Martin Sereque), Stevens John. Asso. prof. Albany Med. Sch., 1936-39; dir. sch. and dept. anesthesiology St. Francis Hosp., Hartford, Conn., 1946——; practice medicine specializing in anesthesiology, Hartford, 1946——. Mem. Am. Soc. Anesthesiologists (past dir., pres.), Am. Phys. Soc., A.M.A., Am. (past pres., past mem. ho. of dels.), Conn. State, (past pres.), New Eng. (past pres.) socs. anesthesiologists, Acad. Anesthesiology (past pres.), Soc. Exptl. Biology and Medicine, Conn. State Med. Soc. (mem. ho. of dels. 1954——, chmn. council 1961——), Hartford County Med. Assn. (dir. 1954——, past pres.), Hartford Med. Soc. (past pres.). Research and numerous publs. on physiology of circulation, pharmacology of anesthetic drugs, anesthetic considerations. Home: 115 Ferncliff Dr., West Hartford, Conn. 06115. Office: St. Francis Hosp., 114 Woodland St., Hartford, Conn. 06105.*

MARTIN, Walter Edwin, Am. parasitologist; embryologist, educator; b. DeKalb, Ill., Jan. 14, 1908; s. Walter Sylvester and Tillie Lula (Secora) M.; B.E., No. Ill. State Tchrs. Coll., 1930; M.S., Purdue U., 1932, Ph.D., 1937; m. Ruth Virginia Butler, June 16, 1934; children—Carol Ann (Mrs. Richard Fallis), John Walter, Judith Kathryn, David Butler. Asst., Purdue U., 1930-34, instr., 1934-37; faculty DePauw U., 1937-47, prof., 1946-47; asso. prof. U. So. Cal., Los Angeles, 1947-48, prof., 1948——, head zoology dept., 1948-54, head biology dept., 1954-58. Mem. sci. expdn., Honduras, C.Am., summer 1934; staff Marine Biol. Lab., summers 1935-42; naval technician, Egypt, summers 1953, 55, Japan, Taiwan, P.I., 1957; sabbatical leave U. Hawaii Marine Lab., 1956-57; research U. Newchatel (Switzerland), 1963-64. Fellow Ind., So. Cal. acads. sci., A.A.A.S.; mem. Am. Soc. Zoologists, Am. Soc. Parasitologists, Am. Micros. Soc., Am. Assn. U. Profs., Western Soc. Naturalists, Sigma Xi, Phi Sigma, Alpha Epsilon Delta. Studies, publs. on taxonomy, life cycles, electron microscopy of helminths; exptl. embryology. Home: 2185 Warmouth St., San Pedro, Cal. 90732. Office: Sci. Bldg., U. So. Cal., Los Angeles 7.*

MARTIN, Walter Tilford, Am. sociologist; b. Sherwood, Ore., Aug. 26, 1917; s. James T. and Clara (Brown) M.; B.A., U. Wash., 1943, M.A., 1947, Ph.D., 1949; m. Rena Elizabeth Buckley, Jan. 7, 1939; children—Susan Elizabeth, Kathleen Ann, Lawrence Arthur, David Tilford. Faculty, U. Ore., Eugene, 1947-49, 49——, prof. sociology, 1959——, chmn. dept., 1957——. Dep. warden care and treatment Ore. State Penitentiary, 1952-53. Mem. Am. (exec. council 1960-63), Pacific (pres. 1964-65) sociol. assns., Population Assn. Am. (dir. 1963-66), The Sociological Research Association, The Internat. Union Scientific Study Population, Am. Assn. U. Profs. Author: The Rural Urban Fringe, 1953; (with R. W. O'Brien, Clarence Schrag) Readings in General Sociology, 1951; (with Jack P. Gibbs) Status Integration and Suicide, 1964; also articles. Research on understanding of adjustment problems of residents in rural-urban fringe, problems of urbanization; theories on variations in suicide rates of populations; social stress and its relationship to chronic disease rates of populations. Home: 2730 Emerald St., Eugene, Ore. 97403.*

MARTIN, W(eston) J(oseph), Am. plant pathologist; b. Church Point, La., Jan. 15, 1917; s. J. Camile and Ada (Daigle) M.; B.S., Southwestern La. Inst., 1937; M.S., La. State U., 1939; Ph.D., U. Minn., 1942; m. Aimee Norma Miller, Feb. 9, 1948; children—Jeanne Weslyn, Wayne Mark, Carol Ann, Grace Ellen, Blair James. Asso. pathologist U. S. Dept. Agr. Rubber Plant Investigations, Mexico, Central Am., 1942-45, pathologist, 1945-47; staff La. State U., Baton Rouge, 1947——, plant pathologist, 1954-58, prof. plant pathology, 1958——. Mem. tech. com. So. Regional Project on Plant Parasitic Nematodes, 1953-60. Fellow A.A.A.S.; mem. Am. Phytopath. Soc (past pres. So. div.), Helminthological Soc. Washington, La. Acad. Sci., Sigma Xi, Gamma Sigma Delta. Research and numerous publs. on biology of smut fungi, diseases of para rubber tree in Mexico and Central Am., role of plant parasitic nematodes in agr. So. U. S., control of econ. plant diseases caused by plant parasitic nematodes, control of sweet potato diseases. Home: 1944 Richland Av., Baton Rouge 70808.*

MARTIN, William, English geologist; b. Marsfield, Eng., 1767; student of James Bolton, Halifax, 1779; m. Mrs. Adams, 1797; 6 children; became drawing master Macclesfield (Eng.) grammar sch., 1805; fellow Linnean Soc. Author: Figures and Descriptions of Petrifications collected in Derbyshire, 1793, rev. and reissued as Petrificata Derbiensia, 1809; Outlines of an Attempt to establish a Knowledge of extraneous Fossils on Scientific Principles, 1809; also papers. Died Macclesfield, May 31, 1810.

MARTIN, William Butler, Jr., Am. chemist; b. Winchendon, Mass., Aug. 31, 1923; s. William Butler and Elizabeth (Ela) M.; A.B., Clark U., 1948, A.M., 1949; Ph.D., Yale, 1953; m. Nancy Coffin, June 25, 1950; children—Timothy George, Pamela Jean, Cynthia Coffin. Postdoctoral fellow Hickrill Chem. Research Found., Katonah, N.Y., 1952-53; faculty Union Coll., Schenectady 1953——, prof. chemistry, 1963——. Cons., research asso. Schenectady Chems., Inc., 1961——. Fellow Sch. for Advanced Studies, Mass. Inst. Tech., Cambridge, 1959-61; DuPont fellow, Yale, 1951-52. Mem. Am. Chem. Soc. (councillor E. N.Y. sect. 1965——), Mohawk Assn. Scientists and Engrs. (pres. 1965——), Fedn. Am. Scientists, Am. Assn. U. Profs., A.A.A.S., Sigma Xi. (past pres. local chpt.). Contbr. articles to tech. jours. Research on syntheses of organic chemicals, dielectric properties of chlorinated diphenyl ethers, quantitative studies of photochemistry of aqueous ferrous-ferric systems, biochem. studies of paramecium aurelia. Home: 3 Dexter St., Schenectady 12309.*

MARTIN, William Clyde, Am. physicist; b. Cullman, Ala., Nov. 27, 1929; s. William Clyde and Bertha (Roberts) M.; student Coll. Charleston, 1947-49; B.S., U. Richmond, 1951; Ph.D., Princeton, 1956; m. Dolores Moyano, June 7, 1959; children—Eric Barraud, Christian Bryan. Instr. physics Princeton, 1955-57; physicist Nat. Bur. Standards, Washington, 1957——, chief spectroscopy sect., 1962——; mem. com. on line spectra NRC-Nat. Acad. Scis. Fellow Optical Soc. Am., A.A.A.S.; mem. Am. Phys. Soc., Am. Astron. Soc., Sigma Xi, Sigma Pi Sigma. Research on measurement of wavelengths in optical region for several atomic spectra, analyses of these spectra to yield atomic energy levels with their quantum characteristics. Home: 5116 Duvall Dr., Washington 20016. Office: Nat. Bur. Standards, Washington 20234.*

MARTIN, William Joseph, Am. physician; b. Freehold, N.J., Mar. 19, 1918; s. William Redmond and Julia (Conway) M.; B.S., Mt. St. Marys Coll., 1940; M.D., Georgetown U., 1943; M.S., U. Minn., 1952; m. Mary Adams, Apr. 22, 1944; children—Mary Jo, Julie Ann, William Joseph. Fellow in medicine Georgetown U. Sch. Medicine, 1944-46, Mayo Grad. Sch. Medicine, 1949-53; cons. in medicine Mayo Clinic, 1953——; now asso. prof. medicine Mayo Grad. Sch., U. Minn. Diplomate Am. Bd. Internal Medicine. Fellow A.C.P.; mem. N.Y. Acad. Scis., A.A.A.S., Sigma Xi, Alpha Omega Alpha. Contbr. chpts. Conn's Current Therapy, Brenneman's Pediatrics, Wohl's Chronic Brucellosis; also numerous papers and articles. Bd. editors Minn. Medicine. Investigation of antimicrobial agts., especially penicillin V, novobiocin, amphotericin B, nitrofurans, phenethecillin; investigations of brucellosis, skeletal infections, endocarditis, bacteremias, other diseases due to infectious agts.; investigations of certain esoteric diseases including porphyria, alkaptonuria, pyroglobulinemia and pseudopseudohypoparathyroidism. Home: 606 Memorial Pkwy. Office: Mayo Clinic, 200 1st St. S.W., Rochester, Minn.*

MARTIN, William Randolph, Am. microbiologist; b. Knoxville, Tenn., Apr. 19, 1922; B.A. 1947, M.S. 1950, U. Tennessee, Ph.D. U. Texas, 1955; m. 1949; 2 children. Research asst., Oak Ridge Nat. Lab., 1947-49; research asst., Texas, 1951-52; research scientist, 1952-55, assoc. bacteriologist, Am. Meat Inst. Foundation, 1955-57; instr. microbiology, U. Chicago, 1957-58, asst. prof. 1958-64, assoc. prof. 1964——. Mem. Soc. Microbiologists; N.Y. Acad. Sci., Brit. Soc. General Microbiology. Research on mechanisms of cellular resistance to antitumor drugs; microbial metabolism; filamentous fungi. Office: Dept. of Microbiology, University of Chicago, Chicago, Ill. 60637.

MARTIN, William Ted, Am. mathematician; b. Springdale, Ark., June 4, 1911; s. James Ellsworth and Dora (Smyer) M.; B.A., U. Ark., 1930; M.A., U. Ill., 1931, Ph.D., 1934; m. Lucy Dodge Gray, July 26, 1938; children—Seelye, James Ellsworth, William Gray, Thomas Edward. Faculty, Mass. Inst. Tech., Cambridge, 1936-43, 46——, prof., 1946——, head dept., 1947——; chmn. steering com. African Math. Program, 1961——; co-chmn. Cambridge Conf. on Sch. Math., 1962——. Fellow A.A.A.S., Am. Acad. Arts and Scis.; mem. Am. Math. Soc. (trustee, treas.) Math. Assn. Am.; Phi Beta Kappa, Sigma Xi. Author: (with S. Bochner) Several Complex Variables, 1948; (with E. Reissner) Elementary Differential Equations, 1956; Foreword to Selected Papers of Norbert Wiener, 1964; (with N. Wiener, A. Siegel, B. Rankin) Differential Space, Quantum Systems and Prediction, 1966. Editor: (with I. E. Segal) Analysis in Function Space, 1964; (with D. C. Pinck) Curriculum Improvement and Innovation: A Partnership of Students, School teachers and Research Scholars, 1966. Research in analytic functions of several complex variables, theory of integration in function space. Home: 16 Swan Lane, Lexington, Mass. 02173. Office: 77 Massachusetts Av., Cambridge, Mass. 02139.*

MARTINDELL, Jackson, Am. mgmt. scientist; b. Amarillo, Tex., July 26, 1900; U. Colo., Columbia U.; LL.D., Whitman Coll.; D.Sc., Pa. Mil. Coll.; m. Anne Clark. Chmn. bd., chief exec. officer Marquis-Who's Who, Inc., Chgo., Am. Inst. Mgmt., Inc., N.Y.C.

(founder); chmn. A. N. Marquis Co., 1964-67, now dir.; trustee Power Reactor Devel. Co. (not-for-profit). Author: Scientific Appraisal of Management, 1950; The Appraisal of Management, 1962. Originator of the only systematic critique for evaluation of mgmt. (assigning of points in evaluation of each of the 12 maj. phases of mgmt.). Home: 132 Elm Rd., Princeton, N.J. Office: 125 E. 38th St., N.Y.C.

MARTINE, George the Younger, Scottish physician; b. Scotland, 1702; s. George Martine the elder; ed. St. Andrews, Edinburgh, Leiden, Netherlands; M.D., 1725; practiced medicine, St. Andrews, Scotland; accompanied Charles Cathcart as physician to forces on Am. expdn., 1740; 1st physician to expdn. against Carthagena (under Adm. Veinon), 1741. Fellow Royal Soc., 1740. Author: Desimilibus animalibus et de animalibus calore libri duo, 1740; Essays Medical and Philosophical (1st significant work on clin. thermometry), 1740; also articles. Credited with performing 1st tracheotomy for diphtheria in Britain, circa 1730. Died of fever contracted at Carthagena, 1741.

MARTINEAU, André, French mathematician; b. Aubigny, France, May 14, 1930; s. Jean and Blanche (Largeaud) M.; Agregation de Math., Ecole Normale Supérieure, Paris, 1952; m. Christine Genieys, July 10, 1954; children—Anne, Jean, Pierre, Jacques. Staff, Centre National Recherche Scientifique, 1953-55; with Sorbonne, 1958-61; with U. Montpellier (France), 1961-67, prof., 1964-67; prof. U. Nice (France), 1967—. Mem. Société Mathématique de France (mem. council). Author: (with Treves) Éléments de la theorie des distributions . . . , 1955; also articles. Research on classic analytic functions, including fundamental results on entire function of exponential type in several complex variables; devel. of hyperfunction theory from Sato; several complex variables of boundary-value theory for distbns. Home: 21 Bd F. Grosso, Nice, France.*

MARTINELLI, Enzo Carlo, Italian mathematician; b. Pescia, Italy, Nov. 11, 1911; s. Alfredo and Maria (Gulaco) M.; ed. U. Rome; Ph.D.; m. Luigia Panella, Apr. 29, 1946; children—Roberto, Maria Renata. Asst., prof. agrégé math. analysis U. Rome, 1935-46; became asso. prof. U. Genes, 1947, later prof. geometry; prof. geometry U. Rome, 1955—. Mem. dei Lincei Acad., Austrian Math. Soc., Italian Math. Union, Acad. Sci. and Letters. Author: Lezioni di Geometria; Lezioni di Topologia; also articles. Research in theory of analytic function, differential geometry, topology. Home: via Aladino Govoni 24, Rome, Italy.

MARTINI, Paul, German physician; b. Frankenthal, Germany, Jan. 25, 1889; s. Paul M. and (Bauer) M.; ed. U. Munich, U. Kiel. Became lectr. U. Munich, 1922; named med. supt. Hedwigs Hosp., Berlin, 1928; dir. Med. Clinic, U. Bonn (Germany), 1932—. Mem. Sci. Council, Fed. Council Health, med. commn. German Physicians' Chamber. Recipient Paracelsus medal. Author: Die unmittelbare Krankenuntersuchung; Methodenlehre der therapeutische klinischen Forschung, 3d edit., 1953. Home: 38 Haagerweg, Bonn, Germany.

MARTINOVITCH, Petar Nichola, Yugoslavian biologist; b. Titograd, Yugoslavia, July 11, 1897; s. Nichola and Plana (Popovic) M.; B.S., Syracuse U., 1927, M.A., 1928; Ph.D., U. Belgrade (Yugoslavia), 1932; m. Militza Mila, Apr. 19, 1947; children—Maria, Maya. Research asso. dept. biology Central Inst. Hygiene, Belgrade, 1930-45; dir. lab. for Exptl. Biology Medicine, Novi Sad, Yugoslavia, 1946-49; head dept. exptl. histology and tissue culture Radiobiol. Lab., Inst. Nuclear Scis. Boris Kidric, Belgrade, 1949—. Decorated Order Labor. N°1, Order Merit N°1. Mem. Internat. Soc. Cell Biology, Am. Tissue Culture Assn., European Tissue Culture Club, Internat. Inst. Embryology, Serbian and Yugoslav Acad. Scis. and Arts, Sigma Xi. Research and publs. on cultivation in vitro and transplantation of rat endocrine glands, exptl. chimerae in domestic fowl and duck, hereditary microcytic-hypochromic anemia in lab. rat. Home: 6 Pere Todroviica, Belgrade. Office: P.O. Box 522, Inst. Nuclear Sci., B.Kidric, Belgrade, Yugoslavia.*

MARTINS, Charles-Frédéric, French botanist; b. Paris, Feb. 6, 1806; asst. naturalist Faculty Medicine, U. Paris; became prof. botany Faculty Scis., Montpellier, France, 1846; mem. sci. expdns. to North, 1838-40; mem. French Acad. Scis., 1863. Author: Voyage botanique en Norvége, 1848; Essai sur l'ancien glacier dela valée d'Argelès, 1868; l'Origine glaciére des tourbiéres du Jura, 1871. Founder (with Berigny, Hoegheus), Annuaire météorologique, 1848. Research and publs. on flora of Asia Minor, valleys formed by glaciers. Died Paris, Mar. 7, 1889.

MARTINS, Thales (Cesar), Brazilian physiologist; b. Rio de Janeiro, Brazil, Sept. 29, 1896; s. José and Isaura (de Padua) M.; B.S., Bapt. Coll., 1913; M.D., U. Rio de Janeiro, 1918; m. Ara Pederneiras, Nov. 10, 1920. Staff, Faculdade de Medicina, U. Rio de Janeiro, 1926—, prof. physiology, 1952—, head dept. endocrinology, 1942—; head dept. physiology Instituto Oswaldo Cruz, Rio de Janeiro, 1952—; prof. physiology Escola Paulista de Medicina, Sao Paulo, Brazil, 1934-40. Decorated Brazilian Order Med. Merit. Guggenheim fellow Yale, 1948. Fellow A.A.A.S.; mem. Internat. Congress Pharmacology (hon. pres. 1966), Acad. Sci., Brazilian, Lisbon, N.Y. acad. scis., Nat. Acad. Med. Brazilian, Soc. Biology Sao Paulo (past pres.), Deutsche endokrin. Gesellschaft. Author: Glandulas sexuais e Hypophyse anterior, 1936; also numerous articles. Research on tests for androgens, feedbacks in regulation of pituitary by gonads, sexual differences of pituitary function, parabiosis, endocrine control of motility of male accessory genitals, maternal behavior in dogs, endocrine control of micturition behavior, manipulation of biol. cycles. Home: 216 Av. Beira Mar, Rio de Janeiro, G.B., ZC39 Brazil.*

MARTIUS, Heinrich Emil Fedor; b. Berlin, Germany, Jan. 2, 1885; s. Friedrich M. and Martha (Leonard) M.; ed. U. Freiburg, U. Leipzig, U. Rostock; m. Berta Weinlig, 1919. Staff, Eppendorfer Hosp., Hamburg, Germany, Women's Clinic, Bonn, Germany; became prof. U. Bonn, 1922, U. Göttingen (Germany), 1926; dir. U. Womens Clinic, Gottingen, until 1954. Pres., German Central Com. for Combating Cancer and Cancer Research. Mem. Fédération internationale de gynécologie et d'obstétrique. Author: Geburtshilfliche Operationen, 1934; Gynäkologische Operationen, 1936; Lehrbuch der Geburtshilfe, 1943-48, 2 vols.; Lehrbuch der Gynäkologie, 1946. Research in X-ray therapy and diagnostics. Home: 4 Bismarckckstr., Göttingen, Germany.

MARTIUS, Karl Friedrich Philipp, see von Martius, Karl Friedrich Philipp.

MARTLAND, Harrison Stanford, Am. physician, pathologist; b. 1883; B.A., Western Md. Coll., 1901; M.D., Columbia U., 1905; m. Myra C. Ferdon; 2 children. Intern, N.Y.C. Hosp., Welfare Island, 1905-06; staff pathology and bacteriology Russell Sage Lab., 1906-08; pathologist Newark City Hosp., 1908—; chief med. examiner Essex County, 1925—; asst. prof., then prof. forensic medicine N.Y. U. Coll. Medicine, 1935—. Mem. N.Y. Acad. Medicine. Author: Punch Drunk, 1928; Dr. Watson and Mr. Sherlock Holmes, 1939; also numerous articles. Research on radium poisoning in watch dial industry; studied adrenal changes caused by acute infectious disease including Waterhouse-Friderichson syndrome. Died 1954.

MARTON, Ladislaus Laszio, physicist; b. Budapest, Hungary, Aug. 15, 1901; s. James L. and Helena (Ring) M.; Ph.D. in Physics, U. Zurich (Switzerland), 1924; m. Claire Perl De Kisker, Aug. 25, 1933. Came to U. S., 1938, naturalized, 1944. Research asst. U. Zurich, 1924-25; research physicist Tungsram Lamp Co., Hungary, 1925-28; faculty U. Brussels (Belgium), 1928-38, asst. prof., 1933-38; lectr. U. Pa., 1938-39; research physicist RCA Mfg. Co., 1938-41; asso. prof. electron optics, head div. Stanford, 1941-46; physicist Nat. Bur. Standards, Washington, 1946—, chief electron physics sect., 1948-62, chief internat. relations, 1962—, coordinator spl. internat. programs, 1964—; research fellow Internat. Inst. Physics Solvay, Brussels, Belgium, 1936-38; vis. prof. Tex. A. and M. U., 1946, Cambridge (Eng.) U., 1948, Linfield Coll., 1956, U. Paris (France), 1962-63. Recipient medal Internat. Union against Cancer, 1938, Verhagen medal U. Brussels, 1947, Exceptional Service award, Gold medal U. S. Dept. Commerce, 1955, U. Brussels medal, 1963. Fellow Am. Phys. Soc. (past chmn. div. electron physics), A.A.A.S. (mem. council 1960—), Phys. Soc. London; mem. Acad. Belgium (fgn. mem.), Electron Microscope Soc. France (hon.), Philos. Soc. Washington (past pres.), Washington Acad. Scis. (past v.p.). Editor-in-chief, Advances in Electronics and Electron Physics, 1947—, Methods of Experimental Physics, 1957—. Patentee in electron physics, electron optics, electron scattering, electron interferences. Pioneered electron microscopy. Home: 4515 Linnean Av. N.W., Washington 20008. Office: Nat. Bur. Standards, Washington 20234.*

MARTORELL, Fernando, Spanish physician; b. Barcelona, Spain, June 4, 1906; s. Vicente and Emilia (Otzet) M.; student Barcelona Sch. Medicine; M.Surgery honoris causa, U. Ireland, 1961; m. Carmen Oliveras, July 1, 1937; children—Maria Teresa (Mrs. José M. Piera), Guillermo, Fernando, Carlos, Jorge, Carmina, Eduardo. Asst. prof. Sch. Medicine, Barcelona, 1934-40; chief Vascular Clinic, Instituto Policlínico, Barcelona, since 1940—. Decorated Gran Cruz de la Orden Civil de Sanidad; Gran Cruz de la Orden de Isabel la Católica; Cavaliere Ufficiale dell Ordine al Merito della Republica Italiana. Hon. mem. Soc. Angiology, Soc. Vascular Surgery. Author several books on vascular surgery, also numerous articles. Founder, Jour. Angiologia, 1949, dir., 1949—. Pioneered in vascular diseases, Martorell used for aortic arch syndrome and hypertensive-isquemic ulcer. Home: 29, Travesera de Gracia, Barcelona, Spain.*

MARTYN, David Forbes, physicist; b. Cambuslang, Scotland, June 27, 1906; s. Harry Somerville and Elizabeth (Thom) M.; B.Sc., Imperial Coll., London, Eng., 1926, Ph.D., 1928, D.Sc., 1936; m. Margot Adams, May 12, 1944. Chief radiophysics lab. Commonwealth Sci. and Indsl. Research Orgn., Camden, Australia, 1939-42, chief officer-in-charge upper atmosphere sect., 1944—; staff Radio Research Bd. Australia, 1930-39. Chmn., Australian Nat. Com. Space Research, 1958—, Australian Nat. Com. Antarctic Research, 1958-66, sci. and tech. com. for peaceful uses of outer space UN, 1962—. Fellow Australian Acad. Sci. (found.; 1st sec. 1954-55), Royal Soc., 1950, Inst. Physics; mem. Phys. Soc., Royal Meteorol. Soc., I.R.E., Royal Coll. Sci. (asso.). Research and numerous publs. of temperature distbn. of upper atmosphere; devel. (with Bailey) of theory of interaction of radio waves; devel. (with Baker) of modern theory geomagnetic variations; devel. theory of equatorial ionospheric anomaly. Address: Coresearch, Camden, N.S.W., Australia.*

MARTYN, John, English botanist; b. London, Sept. 12, 1699; s. Thomas and Katherine (Weedon) M.; student Emmanuel Coll., Cambridge, Eng.; m. Eulalia King, 1732 (dec. 1749); 3 sons, including Thomas, 5 daus.; m. 2d, Mary Anne Fonnereau, 1750; a son, Claudius. Entered father's counting house, 1727; practiced as apothecary and physician; lectr. botany and materia medica; prof. botany Cambridge, 1732-62; dir. Bot. Gardens, Cambridge. Fellow Royal Soc., 1727. Corresponded with C. Linnaeus. Author: Historia plantarum rariorum, 1728-37; also articles; contbg. author: Dictionary (Bailey), 1725. Translator: Georgics (Vergil), 1741, Bucolics with agrl. and bot. notes, 1749. Introduced valerian, pepper mint water, black currants to pharmacy; collected fgn. plant specimens; explained tech. terms in botany; studied history and modern langs. Died Chelsea, Eng., Jan. 29, 1768.

MARTYN, Thomas, English botanist; b. Chelsea, England, 1735; s. John Martyn; prof. botany, Cambridge, Eng., 1762-1825. Fellow Royal Soc., 1786. Author: Flora rustica, 1792-94; The Language of Botany, 1793; The Universal Conchyliologist. Translated, continued Letters on the Elements of Botany (Rousseau), 1785. Introduced Linnaeus' system to Cambridge; studied molluscs and entomology. Died Bedford, Eng., 1825.

MARTYNOV, Dmitriy Yakovlevich, Russian astronomer; b. 1906; grad. Crimean U., 1924, Kazan U., 1926; Dir., Engelhardt Astron. Obs., nr. Kazan, 1931-54; prof. Kazan U., 1935-54, rector, 1951-54; prof. Moscow U., 1954—; dir. Shternberg Astron. Inst., 1956—. Mem. All-Union Astron. and Geodetical Soc. (pres.). Author: Eclipsing Variable Stars, 1939 Practical Astrophysics, 1960, 67; General Astrophysics, 1965. Discovered and explained relationship between period and spectral class of eclipsed variable stars, 1937, also periodic disturbances in motion of these stars; established connection between stellar chains and dark interplanetary matter. Address: Moscow University, Leninskie gory, Moscow, USSR.*

MARUASHVILI, Georgiy Minaevich, Russian parasitologist; b. 1910; grad. Tiflis Med. Inst., 1933; D.Med. Sci. With I. S. Bertashvili's Physiol. Lab., 1933-41; former asst. dept. parasitology Tbilisi Med. Inst., lectr., 1942-54, prof., 1955-59; dir., sci. dir. Virsaladze Inst. Malaria, Parasitology and Tropical Medicine, 1944-59; head chair med. parasitology and tropical medicine Tbilisi Postgrad. Med. Inst., 1959-60; dir. Virsaladze Inst. Med. Parasitology and Tropical Medicine, Georgian Ministry Health, 1960—. Mem. USSR Acad. Med. Sci. (corr.). Author: The Clinical Aspects and Therapy of Amebiosis, 1948. Mem. editorial council Med. Parasitology and Parasitic Diseases. Research and numerous publs. on med. parasitology and tropical medicine, leishmaniasis; 1st to determine epidemiological and clin. features of leishmaniasis in different geog. zones of Georgia. Address: Virsaladze Inst. Med. Parasitology and Tropical Medicine, Georgian Ministry Health, Tbilisi, Gruz. SSR, USSR.

MARUHN, Karl, German mathematician; b. Chemnitz, Dec. 5, 1904; s. Karl and Charlotte (Schmitz) M.; ed. U. Leipzig; Ph.D.; agrégé; m. Eva Feldt, June 27, 1934; children—Jurgen, Rainer. Collaborator, Deutsche Versuchanstalt für Luftfahrt, 1935-44; instr. Berlin, Germany, also Prague (Czechoslovakia) Tech. Coll., 1938-45; prof. math. Iena, Dresden, Germany, 1946-59; became prof. Giessen, Germany, 1959. Mem. Deutsche Mathematikervereinigung, Gesellschaft für Angewandt Mathematik und Mechanik, Österreichische mathematische Gesellschaft. Research and publs. on potential theory, hydromechanics. Home: Am. Unteren Rain 10, 63 Giessen, West Germany.

MARUM, Martin van, see van Marum, Martin.

MARUMORI, Toshio, Japanese physicist; b. Sendai-City, Japan, Oct. 3, 1929; s. Iwao and Masako (Yagi) N.; grad. Nagoya (Japan) U., 1952, D.Sc., 1957; m. Fumiko Aihara, Oct. 20, 1958; 1 son, Shinichi. Research asst. Research Inst. for Fundamental Physics, Kyoto (Japan) U., 1956-57, asst. prof. dept. physics, 1957-66; prof. dept. physics Kyushu (Japan) U., 1966—. Research fellow Niels Bohr Inst., U. Copenhagen (Denmark), 1964-66. Mem. Phys. Soc. Japan. Asso. editor Progress Theoretical Physics, 1957-64. Research, publs. on theory on structure of atomic nuclei from standpoint of nuclear many body problem. Home: 6-21, Kokusetsu-Tataya Jutaku, Oaza-Matsuzaki, Fukuoka, Japan.*

MARUSSI, Antonio, Italian geodesist; b. Trieste, Italy, Oct. 12, 1908; s. Gustavo and Maria (Leutheuser) M.; D.degree in Math., U. Bologna (Italy), 1931; m. Dolores de Finetti, May 8, 1943. Geodesist, Istituto Geografico Militare, Florence, Italy, 1932-52; prof. geodesy and surveying U. Trieste, 1952—, dir. Inst. Geodesy and Geophysics, 1946; sci. mem. expds. to Karakorum-Hindu Kush; sci. sec. Conf. for Applications of Sci. and Tech., UN, Geneva, Switzerland, 1963. Decorated Commendatore al merito della Repubblica Italiana, award of order of George I, Greece. Corr. mem. Accademia Nazionale dei Lincei, Roma, Italy; asso. Royal Astron. Soc. London. Author: Geophysics of the Karakorum, 1964; also numerous articles. Originated concept of intrinsic geodesy; research on gravimetric surveys of Karakorum and Hindu Kush ranges, tides and free oscillation of earth.*

MARVEL, Carl Shipp, Am. chemist; b. Waynesville, Ill., Sept. 11, 1894; s. John Thomas and Mary Lucy (Wasson) M.; A.B., Ill. Wesleyan U., 1915, M.S., 1915, D.Sc., 1946; A.M., U. Ill., 1916, Ph.D., 1920, D.Sc., 1963; m. Alberta Hughes, Dec. 26, 1933; children—Mary Catharine, John Thomas. Faculty, U. Ill., Urbana, 1920—, research prof. organic chemistry, 1953-61, research prof. organic chemistry emeritus, 1961—; prof. chemistry U. Ariz., Tucson, 1961—. Mem. various coms., panels, bds. NRC, NDRC, NSF, Nat. Bur. Standards. Recipient Internat. award in plastic sci. and engring. Soc. Plastics Engrs., 1964, Perkin medal Am. sect. Soc. Chem. Industry, 1965. Fellow N.Y. Acad. Scis.; mem. Am. Chem. Soc. (Nichols medal N.Y. sect. 1944, Willard Gibbs medal Chgo. sect. 1950, Priestly medal 1956, Witco award in Polymer chemistry 1964), Am. Inst. Chemists (Gold medal 1955), Nat. Acad. Scis., Am. Philos. Soc., Am. Acad. Arts and Scis., Soc. Suisse de Chemie, A.A.A.S., Phi Beta Kappa, Phi Lambda Upsilon (hon.), Sigma Xi. Author: Introduction to the Organic Chemistry of High Polymers, 1959; also numerous articles. Editor-in-chief. Organic Syntheses, Vol. V, 1925, Vol., XI, 1931. Research in organic chemistry; synthesis of high polymers, amino acids, organo-metallic compounds; developed polymer of benzimidazole units having unusual high temperature properties; established head to tail structure of recurring units in vinyl polymers; developed polyaromatic heterocyclic polymers and thermally stable polymers. Home: 2332 E. 9th St., Tucson 85719.

MARVIN, Charles Frederick, Am. meteorologist; b. Putnam, O., Oct. 7, 1858; s. George F. and Sarah A. (Speck) M.; ed. pub. schs., Columbus, O., grad. (in mech. engring.), Ohio State U., 1883, Sc.D., 1932; m. Nellie Limeburner, June 27, 1894 (dec. Feb. 1905); children—Charles Frederick, Cornelia Theresa, Helen Elizabeth; m. 2d, Mabel Bartholow, Nov. 8, 1911 (dec. 1932); m. 3d, Sophia A. Beuter, Nov. 12, 1932. Instr. mech. drawing, Ohio State U., 1879-83; apptd. on civilian corps of signal service, 1884, also prof. meteorology, U. S. Weather Bur.; chief U. S. Weather Bur., 1913-34. Co-author: Moses, the Greatest of Calendar Reformers. Conducted expts. upon which are based tables used by Weather Bur. for deducing moisture in air; made important investigations of anemometers for measurement of wind velocities and pressures; invented instruments for measuring and automatically recording rainfall, snowfall, sunshine, atmospheric pressure; also clinometer; made extensive studies and wrote on use of kites for ascertaining meteorol. conditions in free air, registration of earthquakes, measurement of evaporation, solar radiation, temperature with elec. resistance thermometers; contbr. papers on meteorology and simplification of calendar, including proposal to improve Gregorian rule for leap years by omitting 4 leap years in 500 yrs. which will keep reckoning accurate for more than 10,000 yrs. Died June 5, 1943.

MARVIN, Harry, Am. inventor; b. Jordan, N.Y., 1863; s. Daniel W. and Ellen (Weed) M.; grad. U. Syracuse, 1883; children—Robert S., Marguerite (Mrs. Smith). Asso., Thomas Edison; became pres. Edison Office, N.Y., 1888; staff Gen. Electric Co., Schenectady; founder, v.p. Biograph Co.; pres. Motion Pictures Patent Co.; ret., 1912. Invented electric rock drill, biograph (1st true movie machine), machine for stamping names on metal plates, automatic tuning device for radio sets, automatic palm reading device for penny arcades, (with Herman Casler) mutoscope (forerunner motion picture machine). Died 1940.

MARVIN, Horace Newell, Am. educator; b. Camden, Del., Apr. 20, 1915; s. Charles Howard and Emma Lavinia (Greene) M.; B.A., Morningside Coll., 1936; M.A., U. Wis., 1938, Ph.D., 1941; m. Leah Margaret Banta, Aug. 28, 1940; children—Horace Newell II, Peter, Timothy, Tamara, Tad. Spl. investigator Carnegie Inst. Washington, 1941-42; instr. U. Ark., Little Rock, 1942-45, asst. prof., 1945-48, asso. prof., 1949-58, prof., chmn. dept. anatomy, 1958—, asso. dean Sch. Medicine 1965—; head dept. biology M.D. Anderson Hosp. for Cancer Research, 1948-49; cons. radiobiology health research div. Los Alamos Sci. Lab., AEC; vis. prof. U. Lagos Med. Sch., Nigeria, 1963; dir. Human Whole Body Radioisotope Detector Lab. Fulbright fellow, 1963; Commonwealth Fund Travel fellow 1963, 65; Wis. Alumni Research fellow, 1936-37. Mem. Am. Assn. Anatomists, Soc. for Exptl. Biology and Medicine, Soc. for Study Internal Secretion, Sigma Xi. Author (with W. C. Langston) Laboratory Manual of Medical Microanatomy, 1952, 2d edn., 1961. Research and numerous publs. on dependence of testicular devel. on pituitary activity; endocrine factors responsible for cystic ovaries; use radioisotope methods to study role of tocopherol and other vitamins in red cell production and survival. Home: 4015 N. Lookout St., Little Rock 72205. Office: 4301 W. Markham St., Little Rock 72205.*

MARVIN, James W., Am. plant physiologist; b. E. Norwalk, Conn., Apr. 22, 1909; s. James W. and Edith A. (Cousins) M.; B.S. in Agr., U. Vt., 1932, M.S., 1933; Ph.D., Columbia, 1939; m. Alice Lee McConnell, Aug. 28, 1934; children—Nancy Lee (Mrs. William G. M. Hardison), Mary Wallace (Mrs. Douglas H. Philpsen), James Wallace III, David Raymond. Lab. asst. Columbia, 1933-39; faculty U. Vt., 1939—, prof. botany, 1949—, chmn. dept., 1944-64, asst. botanist Agr. Expt. Sta., 1939-45, botanist, 1945-47, plant physiologist, 1947—. Trustee Conservation and Research Found. Guggenheim fellow, 1953-54. Contbr. numerous articles in field. Studies in tree growth and movement of materials in woody plants. Home: 303 Swift St., South Burlington 05403. Office: Univ. of Vermont, Burlington, Vt. 05401.*

MARX, George, entomologist; b. Laubach, Germany, 1838; student pharmacy, Giessen, Germany; M.D., Columbia U., 1885. Came to U. S., 1860. Vol., Am. Civil War; pharmacist, N.Y., 1862-65; in business, Phila., 1865-78; natural history draftsman div. entomology U. S. Dept. Agr., 1878-79, chief div. illustrations, 1889 until retirement. Mem. Entomol. Soc. Washington (became pres. 1891). Completed E. Keyserling's Die Spinnen Amerikas, 1891. Research, publs. on spiders and ticks; named some Western spiders and a few Eastern species; contributed to knowledge of Arachnida; known for excellency of his illustrations. Died 1895.

MARX, George, Hungarian physicist; b. Budapest, Hungary, May 25, 1927; s. Steve and Julie (Laszlo) M.; Ph.D., Eötvös U., Budapest, 1950; m. Edith Koczkas, July 31, 1953; children—Leslie, Thomas, Suzie. Asst., Inst. Theoretical Physics, Eötvös U., 1948-61, prof. physics, 1961—; vis. prof. various European univs., also Stanford, 1964-65. Recipient Kossuth prize, 1955. Mem. Internat. Astron. Union, Hungarian Phys. Soc. (mem. presidency 1952—). Author: Introduction to Quantum Mechanics, 1957; Beyond Atomic Physics, 1962; Statistical Mechanics, 1966; Problems for Theoretical Physics, 1954; also numerous articles. Editor in chief Hungarian periodical Fizikai Szemle, 1959—. Discovered conservation of leptonic charge; research in relativistic mechanics and electrodynamics, prediction of temperature dependence of solar neutrino spectrum, particle theory, possibilities of interstellar space travel. Home: 20 Lagymanyosi, Budapest XI. Office: 5 Pushkin, Budapest VIII, Hungary.*

MARX, Karl (Heinrich Karl), German social philosopher, economist; b. Trèves, Rhenish Prussia, Germany, May 5, 1818; s. Heinrich (Hirschel) M., a lawyer; studied law at U. Bonn, history and philosophy at U. Berlin; Ph.D., U. Jena, 1841; m. Jenny von Westphalen, 1843; 1 son, Edgar, five daus. Editor newspaper, Rheinische Zeitung, Cologne, 1842-43; went to Paris, 1843; there briefly co-edited a review Deutsch-Französische Jahrbücher studied reformist writings, met Friedrich Engels; with Engels, acquired a workers' weekly in Brussels, Belgium; joined "League of the Just," wrote (with Engels) Communist Manifesto, pub. 1848, presenting view of history as class struggle, and exhorting working class to unite and take over means of production; expelled from Paris and Germany; lived in London, Eng., 1849-83; correspondent N.Y. Tribune (staff of which included Fourierists), wrote on Crimean War, Eastern Question, Am. Civil War; most prominent figure in Internat. Workingmen's Assn., 1864-76. Author: Die heilige Familie oder Kritik der kritschen Kritik, (with Engels) 1845; La Misère de la philosophie, 1847 (Eng. trans: The Poverty of Philosophy, 1900); Manifest der kommunistichen Partei, (with Engels), 1848 (Eng. trans: The Communist Manifesto, 1848); Lohnarbeit und Kapital, 1848 (Eng. trans: Wage, Labour and Capital, 1900); Der Achtzehnte Brumaire des Louis Napoleon Bonaparte, 1852; Enthüllunger über den Kölner Kommunistenprozess, 1852; Zur Kritik der politischen Ökonomie, 1859; Herr Vogt, 1860; Inaugural Address of the International Working Men's Association, 1864; Value, Price and Profit, 1865; Das Kapital, Vol. I, 1867, Vols. II and III ed. by Engels, 1885-94; The Civil War in France, 1871; L'Alliance de la démocratie socialiste, 1873; Critique of the Gotha Program, 1875; Die Klassenkämpfe in Frankreich, 1895; European Revolutions and Counter-Revolutions, 1897; The Eastern Question, 1898. Father of modern "scientific" socialism; propounded doctrine of dialectical materialism; Theory of Surplus Value held that capitalism causes surplus of goods and unemployment crises; predicted that after a final crisis state socialism would replace capitalism, with the state eventually withering away. Died London, England, Mar. 14, 1883.

MARX, Walter, biochemist; b. Karlsruhe, Germany, June 26, 1907; s. August and Lise (Gutmann) M.; Dipl.Ing., Tech. Hochsch, Karlsruhe, 1931, D. Ing., 1933; m. Marianne Wormser, Sept. 24, 1954; 1 son, Thomas. Came to U. S., 1934, naturalized, 1941. Instr., Inst. Phys. Chemistry, Tch. Hochsch., Karlsruhe, 1930-33; research fellow Kaiser Wilhelm Inst., Heidelberg, 1933-34, Mt. Sinai Hosp., N.Y.C., 1934-37; research asso. Duke, 1937-39, Inst. Exptl. Biology, U. Cal. at Berkeley, 1939-44; research fellow Cal. Inst. Tech., 1945-46; mem. faculty Sch. Medicine, U. So. Cal., 1946—, prof. biochemistry, 1954—, acting chmn. biochemistry, 1962-63. Mem. A.A.A.S., Soc. Exptl. Biology and Medicine, Am. Soc. Biol. Chemists. Research on mucopolysaccharides; heparin biosynthesis; sulfate metabolism; thyroxine and yeast metabolism; cholesterol metabolism; pituitary hormones. Home: 144 N. Gower St., Los Angeles 90004.*

MARY, Louis-Charles, French engr.; b. Metz, France, 1791; student École polytechnique; built roads and bridges; became divisional insp., 1848, 1st class insp. gen., 1855; prof. navigation Sch. Rds. and Bridges. Author: Fondation de l'écluse de Froissy, 1831; De l'emploi du béton dans la fondation des écluses, 1832. Research on use of concrete in canal and lock construction. Died Paris, 1870.

MASAITIS, Ceslovas, mathematician; b. Kaunas, Lithuania, Mar. 2, 1912; s. Juozas and Jadvyga (Butkevicius) M.; Univ. Diploma, U. Kaunas, 1937; postgrad. U. Ky., 1952-53; Ph.D., U. Tenn., 1956; m. Elena Balciunas, Aug. 3, 1940; 1 dau., Elena Nijole. Came to U. S., 1949, naturalized, 1955. Asst., U. Kaunas, 1937-40, U. Vilnius (Lithuania), 1940-44; instr. Nazareth (Ky.) Coll., 1950-52, U. Ky., Lexington, 1952-53, U. Tenn., Knoxville, 1953-56; mathematician Ballistic Research Labs., Aberdeen Proving Ground, Md., 1956-63, research mathematician, 1963-64, supervisory research mathematician, 1964—; lectr. U. Del., Newark, 1957—; research asso. Sch. Medicine, U. Md., 1964—. Recipient Sustained Superior Performance award, U. S. Army, 1962. Mem. Am. Math. Soc., Math. Assn. Am., Sigma Xi, Pi Mu Epsilon, Phi Kappa Phi. Contbg. author: Recent Contributions to Soviet Mathematics, 1962; also articles. Contbg. author, editor astronomy sect.: Lithuanian Ency., 36 vols., 1966. Research on analytic representation aiming data for guided missles, math. analysis of inertial fuzing systems, math. model of wound ballistics physiology, math. models in physiol. and med. research. Home: 951 Nena Av., Havre de Grace, Md. 21078. Office: Aberdeen Proving Ground, Md. 21005.*

MASAMUNE, Satoru, organic chemist; b. Fukuoka, Japan, July 24, 1928; s. Hajime and Chikako (Kondo) M.; A.B., Tohoku U., 1952; Ph.D., U. Cal. at Berkeley; m. Takako Nozoe, May 1, 1956; children—Hiroko, Tohoru. Project asso., research chemist U. Wis., Madison, 1956-59, lectr., 1959-61; fellow Mellon Inst., Pitts., 1961-64; asso. prof. dept. chemistry U. Alta. (Can.), Edmonton, 1964-67, prof., 1967—. Mem. Am. Chem. Soc., Chem. Soc. (London), Chem. Inst. Can. Contbr. articles to tech. jours. Research on structures and stereochemistry of senecio alkaloid, total syntheses of diterpenes and diterpene alkaloids, cyclic pi electron system, chemistry of extremely strained compounds. Home: 7223 119th St., Edmonton, Alta., Can.*

MASCAGNI, Paolo, Italian anatomist; b. Castelleto, nr. Siena, Tuscany, Feb. 5, 1752; studied medicine U. Siena, anatomy under Tabarrani. Prof. anatomy U. Siena, 1774, Pisa, 1800; prof. anatomy, physiology and chemistry Hosp. Santa Maria Nuova, Florence, 1801. Mem. French Acad. Scis. (corr.). Author: Vasorum lymphaticorum corporis humani historia et ichnographia, 1787; Anatomia universa. His book on lymphatics a classic, included notable passage on diapedesis of blood corpuscles through blood vessels in areas of inflammation. Died Florence, Tuscany, Oct. 19, 1815.

MASCARENHAS, S., Brazilian physicist; b. Rio de Janeiro, Brazil, May 2, 1928; s. Sebastian and Bartyra (Oliveira) M.; B.Sc. in Physics, U. Brazil, 1952; Ph.D. cum laude, U. Sao Paulo, 1956; m. Yvonne Primerano, June 26, 1954; children—Sergio Roberto, Yvonne Maria, Helena. Faculty, U. Brazil, 1952-56, Catholic U., Rio de Janeiro, 1953-56; prof. physics, head physics dept. U. Sao Paulo, 1956—. Research scientist Carnegie Inst. Tech., 1959-60; vis. prof. Princeton, 1962; guest scientist Brookhaven Nat. Lab., 1962. Fulbright fellow, 1959; Guggenheim award, 1960. Mem. Am., Brazilian phys. socs., Brazilian Soc. Advancement Sci. (council), Brazilian Acad. Scis. Author: University Physics, vols. I. and II, 1958. Research, numerous publs. on elec. phenomena during phase changes, color centers in solids caused by radiation damage, developed photoelastic technique to very low temperatures, showed that thermal conductivity may be changed by electric fields. Office: Physics Dept., Sch. Engring., U. Sao Paulo, Sao Paulo, Brazil.*

MASCART, Éleuthère-Élie-Nicolas, French physicist; b. Quarouble, France, Feb. 20, 1837; entered Ecole normale, 1858; D.Sc., 1864. Became dir. Central Meteorol. Bur., 1871; became prof. regent Collège de France, 1872. Mem. Bur. Weights and Measures. Recipient Legion d'Honneur, 1889, Grand Prix des Sciences Mathèmatiques, 1874. Fellow Royal Soc., 1892; mem. French Acad. Scis. (became pres. 1904, also perpetual sec.). Author: Élements de mécanique, 1866; Traité d'électricité statique, 1876; la Météorological appliquée à la prévision du temps,

1881; Leçons sur l'électricité et magnétisme, 1882-86; Traité d'optique, 4 vols., 1889-93; Traité du magnetisme terrestre, 1900. Studied optics, atmospheric electricity, terrestrial magnetism; originated theory of currents of electrodynamic induction; perfected Thomson's electrometer-in-quadrants, 1880; built spectrograph; determined standard wave lengths, electro-chem. equivalent of silver; applied photography to mapping of spectra. Died Paris, Aug. 26, 1908.

MASCART, Henri, French mathematician; b. Ville-d'Avray, France, July 30, 1928; s. Georges and Mathilde (Decoudu) M.; Licence ès Sci., 1949, Diplôme d'Etudes Supérieures, 1950, Agrégation de math., 1951; Doctorat ès Sci., 1960, Ecole Normale Supérieure; m. Marie-Madeleine Delbouis, July 1, 1955; 1 dau., Marie-Hélène. Prof. agrégé de Lycée, Chaumont, Cannes, Aire-sur-l'Adour, Paris, 1951-56; chef de travaux, 1956; lectr., 1961; prof. extraordinary, 1965; titular prof., Faculty Scis., Toulouse, 1967. Reviewer, Zentralblatt für Mathematik und ihre Grenzgebiete. Mem., French Math. Soc.; pres., Toulouse Soc. Math.; mem., Austrian Math. Soc. Author several published articles. Investigations of linear differential operators of infinite order; topological modules. Home: 4 Place des Avions, 31-Toulouse, France. Office: 118 Route de Narbonne, 31-Toulouse, France.*

MASCHERONI, Lorenzo, Italian mathematician; b. Castagnetto, Italy, May 14, 1750; ed. sem., Bergamo, Italy; took Holy Orders; tchr. Greek and poetry, Pavia, Italy; fought for Cisalpine Republic; later studied geometry; became prof. math., Pavia, 1786. Author: Geometria del compasso, 1797; Nuove ricerche sei l'equilibrio delle volte, 1785; Adnotationes ad calculum integrali Euleri, 1790-92. Proved that all constructions in elementary geometry, possible with a ruler and compass, are possible using only a compass. Died July 30, 1800.

MASÈRES, Francis, Brit. mathematician, lawyer; b. London, Dec. 15, 1731; B.A., Cambridge, Eng., 1752, M.A., 1955. Fellow Clave Coll., Cambridge, 1756-59; atty. gen. of Quebec, 1766-69; curistor baron of Exchequer, 1773-1824; became sr. judge of London Sheriff's Ct., 1780. Fellow Royal Soc., 1771. Author: Dissertation on the Use of the Negative Sign in Algebra, 1759; Tract on the Resolution of Cubic Equations, 1800; A View of the English Constitution, 1781; Principles of the Doctrine of Lifetime Annuities, 1783; Scriptores optici, 1823; also treatise on permutations and combinations, 1795. Died May 19, 1824.

MA'SHAR JA'FAR IBN MUHAMMAD, see Albumasar.

MASHIMA, Riko (Toshiyuki), Japanese chemist; b. Kyoto, Japan; grad. Tokyo U., 1899; student in Europe, 1907-09; Sc.D. Prof., Tohoku U.; dean Hokkaido U. Faculty Sci.; dean Osaka U. Faculty Sci., now emeritus prof.; became prof. Tokyo Coll. Tech., 1932-39; pres. Osaka U., 1942-46, now emeritus prof. Recipient Imperial prize, 1917, Order Cultural Merit, 1949. Mem. Acad., Physico-Chem. Research Inst. Author: Complete Research in Japanese Chemistry.

MASIAK, Michal Antoni, Polish biochemist; b. Poznan, Poland, Jan. 15, 1924; s. Pawel and Wanda (Idaszewska) M.; magister pharm. Med. Acad., Wroclaw, Poland, 1950, Pharm.Doctor, 1960; m. Krystyna Roszek, June 28, 1952. Asst. chair applied pharmacy Med. Acad. Wroclaw, 1947-48; head lab. Dermatol. Clinic, Med. Acad. Wroclaw, 1948-50, head biochem. lab. 2d Surg. Clinic, 1957—; with State Med. Service, 1950-57. Mem. Polish Biochem. Soc., Polish Physiol. Soc., Polish Chem. Soc., Polish Pharm. Soc. Research and publs. on new aspects of blood hemolysis; co-discoverer metalo-lipids; invented linomag (a medicine with wide application in medicine). Home: 27 Abramowskiego. Office: 66 Curie-Sklodowskiej, Wroclaw, Poland.*

MASING, Georg, metallurgist; b. St. Petersburg (now Leningrad) Russia, Nov. 2, 1885; s. Heinrich M. Masing) student chemistry, St. Petersburg U., Göttingen (Germany) U.; m. Marthe Tischer, 1915. In tech. seminar, 1916-37; prof. metallurgy Göttingen U.; ret., 1953; prof. Technische Hochschule, Berlin. Recipient Heyn Gold medal; Platinum medal Inst. Metals, 1953. Author: Ternäre Systeme, 1938; Grundlage der Metallkunde, 1940; Lehrbuch der Metallkunde, 1951. Died 1956.

MASIUS, Morton, Am. physicist; b. Egg Harbor City, N.J., Oct. 6, 1883; s. Alfred and Edith (Bailey) M.; A.M., Ph.D., U. Leipzig, 1908; D.Sc., Worcester Poly. Inst., 1946; m. Paula Maria Wagner, Aug. 5, 1910; children—Marguerite (Mrs. Robert M. Hammond), Vera Mildred (Mrs. Alan D. Ferguson). Whiting fellow Harvard, 1908-09; mem. faculty Worcester (Mass.) Poly. Inst., 1909—, prof. physics, 1919-54, prof. emeritus, 1954. Fellow Am. Phys. Soc., A.A.A.S.; mem. Am. Physics Tchrs. Assn., Soc. for Freedom in Sci. Author: Problems in General Physics, 1921; College Physics, 1941. Research and publs. on adsorption, electricity, optics, thermodynamics, kinetic theory. Home: 74 Cold Spring St., New Haven 06511.*

MASKELL, William Miles, entomologist; b. Hampshire, Eng., 1840; ed. Catholic Coll. St. Mary, Oscott, Eng., also Paris, France. Served in English Army; farmer, New Zealand and Eng., 1864-72; mem. Privy Council, 1865-75; became provincial sec., treas., Canterbury, 1874; registrar New Zealand U. Mem. Philos. Inst. Author: An Account of New Zealand Scale Insects, 1887; also articles. Pioneered in coccidology of New Zealand, Australia, and neighboring islands; described many species of scale insects. Died 1898.

MASKELYNE, Mervyn Herbert Nevil Story, English mineralogist; b. Sept. 3, 1823; s. A. M. Story-Maskelyne; ed. Wadham Coll., Oxford (Eng.) U.; M.A.; hon. D.Sc., Oxford U.; m. Theresa Dillwyn-Llewelyn, 1858; 3 daus. Keeper minerals Brit. Mus. until retirement; M.P., Cricklade, 1880-85, N. Wilts, 1885-86; prof. mineralogy Oxford U., 1856-95; magistrate County of Wilts. Fellow Royal Soc., 1870. Author: Catalogue of the Intaglio and Cameos known as the Marlborough Gems. Research and publs. on morphology of crystals, crystallography, mineralogy, petrology, meteorites. Died May 20, 1911.

MASKELYNE, Nevil, English astronomer; b. London, Oct. 6, 1732; s. Edmund and Elizabeth (Booth) M.; ed. Trinity Coll., Cambridge, Eng.; became fellow, 1757; M.A., 1757, D.D., 1777; m. Sophia Rose, circa 1785; 1 dau., Margaret (Mrs. Anthony Mervyn Story). Became clergyman in Shrawardine, 1775, North Runcton, 1782; asst. to Bradley; astronomer, dir. Greenwich (Eng.) Obs., 1765-1811. Fellow Royal Soc., 1758 (Copley medal, 1775); mem. French Acad. Scis., 1802. Author: British Mariner's Guide, 1763. Founder, Nautical Almanac, 1766. Editor: Lunar Tables (by Mayer, corrected by Mason), 1787. Contbr. papers to Philos. Trans. During unsuccessful expdn. to observe transit of Venus at St. Helena, developed method of determining longitude by lunars, 1761; perfected method of observing transits, 1772; improved method of dealing with effects of parallax; made 1st time measurements at Greenwich to tenths of second; used deviations of plumb line to measure earth's density, nr. Schiehallion (mountain), Scotland, 1774; made some 90,000 observations, pub. 1776-1811. Died Greenwich, Feb. 9, 1811.

MASLAMA IBN AHMAD AL-MAJRITI, see al-Majriti.

MASLAND, Richard Lambert, Am. physician; b. Phila., Mar. 24, 1910; s. Charles William and Mary (Gillinder) M.; B.A., Haverford Coll., 1931; M.D., U. Pa., 1935; m. Mary Wootton, Nov. 2, 1940; children—Richard H., Frances W., S. Eleanor, Thomas W. Fellow neurology U. Pa., 1937-40, asso. neurology, 1940-45; asst. prof. psychology and neurology Bowman Gray Sch. Medicine, 1946-47, asso. prof., 1947-55, prof., 1955-57; asst. dir. Nat. Inst. Neurol. Diseases and Blindness, NIH, Bethesda, Md., 1957-59, dir., 1959—. Recipient Myopia award Myopia Research Found., 1964. Mem. Am. Neurol. Assn., Am. Acad. Neurology, Am. Electroencephalographic Soc., Am. Epilepsy Soc. (pres. 1954), Phi Beta Kappa, Alpha Omega Alpha. Home: 4700 Jamestown Rd., Washington 16. Office: National Inst. Neurological Diseases and Blindness, Bethesda, Md.*

MASLOW, Abraham Harold, Am. psychologist; b. Bklyn., Apr. 1, 1908; s. Samuel and Rose (Schilofsky) M.; B.A., U. Wis., 1930, M.A., 1931, Ph.D., 1934; LL.B. (hon.), Xavier U., 1965; m. Bertha Goodman, Dec. 31, 1928; children—Ann (Mrs. Jerome Kaplan), Ellen. Asst. instr. U. Wis., 1930-35; Carnegie fellow Columbia Tchrs. Coll., Columbia U., 1935-37; from instr. to asso. prof. psychology Bklyn. Coll., 1937-51; plant mgr. Maslow Cooperage Corp., 1947-49; faculty Brandeis U., Waltham, Mass., 1951—, prof. psychology, 1962—. Named Humanist of Year, Am. Humanist Assn., 1967. Mem. Am. (pres. 1967—), New Eng. (past pres.), Mass. (past pres.) psychol. assns., Soc. for Psychol. Study Social Issues (past mem. council), Am. Assn. Humanistic Psychology. Author: New Knowledge in Human Values, 1959; Religions, Values and Peak Experiences, 1964; Toward a Psychology of Being, 1962; Eupsychian Management, 1965; Psychology of Science, 1966; also numerous articles. One of originators of humanistic psychology; pioneered research on psychologically healthy persons. Office: Brandeis U., Waltham, Mass. 02154.*

MASON, Brian Harold, geologist; b. New Zealand, Apr. 18, 1917; s. George Harold and Catherine (Fairweather) M.; M.Sc., U. New Zealand, 1938; Ph.D., U. Stockholm, 1943. Came to U. S., 1947, naturalized, 1953. Lectr. geology Canterbury Coll., New Zealand, 1944-47; prof. mineralogy Ind. U., 1947-53; chmn. dept. mineralogy Am. Mus. Natural History, N.Y.C., 1953-65; research curator div. meteorites U. S. Nat. Mus., Washington, 1965—. Sec. Internat. Commn. Meteorites, 1960—. Fellow Mineral. Soc. Am., Geol. So. Am.; mem. Geochem. Soc., Royal Soc. New Zealand, Swedish Geol. Soc. Author: Principles of Geochemistry, 3d edit., 1966; Meteorites, 1962; The Literature of Geology, 1958; (with L. G. Berry) Mineralogy, 1959. Research on problems of mineral genesis, especially in meteorites. Home: 809 E. Capitol St., Washington 20003. Office: U. S. Nat. Museum, Washington 20560.*

MASON, Charles, astronomer; b. England, 1730; astronomer Greenwich Obs., 1756-60; employed (with Jeremiah Dixon) by Lord Baltimore and William Penn to survey boundary line between Md. and Pa. (Mason-Dixon line), 1763-67; measured (with Dixon) arc of meridian in Am.; returned to Eng., 1769; later returned to America. Author: Mayer's Lunar Tables Improved by Charles Mason, 1787. Died Phila., February 1787.

MASON, Clarence Tyler, Am. chemist; b. Chgo., Apr. 21, 1908; s. Clarence L. and Louise (Tyler) M.; B.S., Northwestern U., 1931; M.S., McGill U., 1933, Ph.D., 1935; m. Lois L. Lucas, June 12, 1936; children—Clarence, Karen (Mrs. Gatewood). Faculty Dillard U., New Orleans, 1935-44; faculty Tuskegee Inst., 1944—, dir., chmn. div. sci. Carver Research Found., prof. chemistry, 1944—. Instl. rep. Emergency Sci. War Tng. 1942-44; Rosenwald fellow, 1936. Fellow Am. Inst. Chemists, Am. Chem. Soc.; mem. Nat. Inst. Sci. (past pres.), A.A.A.S., N.E.A., Sigma Xi, Scabbard and Blade, Beta Kappa Chi (past pres., editorial bd. Bull. 1960—). Contbr. articles to profl. jours. Research on halogenated ethers, thioethers, thioxane, dioxane. Home: 1005 Montgomery Rd., Tuskegee Inst., Ala. 36088.*

MASON, Clyde Walter, Am. chemist; b. Watertown, S.D., June 17, 1898; s. George Walter and Cora (Pitt) M.; A.B., U. Ore., 1920; Ph.D., Cornell U., 1924; m. Elizabeth Mandana Peterson, Aug. 2, 1920; children—George William, Phoebe Jane. Asst. in chemistry U. Ore., 1917-20; faculty dept. chemistry Cornell U., Ithaca, N.Y., 1920—, asst. prof., 1927-33, prof., 1933-42, prof. chem. microscopy and metallography Sch. Chem. Engring., 1942-67, emeritus, 1967—. Tech. rep. OSRD, 1944; cons. on chem. microscopy to various industries, 1920—. Fellow N.Y. Micros. Soc.; mem. Am. Chemical Society, American Soc. Metals, N.Y. Acad. Scis., Sigma Xi, Tau Beta Pi, Phi Kappa Phi, Alpha Chi Sigma. Author: (with E. M. Chamot) Handbook of Chemical Microscopy, 2 vols., 1931, 58; Introductory Physical Metallurgy, 1947. Research in chem. microscopy; qualitative microscopical analysis; structural colors; metallography. Home: 411 Hanshaw Rd., Ithaca, N.Y. 14850.

MASON, Dean Towle, Am. physician; b. Berkeley, Cal., Sept. 20, 1932; s. Ira Jenckes and Florence (Towle) M.; B.A., Duke, 1954, M.D., 1958; m. Maureen O'Brien, June 22, 1957; children—Kathleen, Alison. Clin. asso. Nat. Heart Inst., Bethesda, Md., 1961-64, sr. investigator cardiology br., 1964—, attending physician, cons. cardiologist to surgery br., 1964—, asst. chief sect. cardiovascular diagnosis, 1964—; clin. asst. prof. medicine Georgetown U. Sch. Medicine, 1965—. Recipient Appleton-Century-Crofts Book award, 1956; C. V. Mosby Scholarship award, 1958, Am. Therapeutic Soc. prize essay award, 1965. Diplomate Nat. Bd. Med. Examiners, Am. Bd. Internal Medicine, Subsplty. Bd. Cardiovascular Diseases. Mem. A.M.A., Am. Heart Assn., A.C.P., Am., Internat. colls. angiology, A.A.A.S., Am. Coll. Chest Physicians, Internat. Soc. Nephrology, Johns Hopkins Med. and Surg. Assn., Am. Physiol. Soc., Am. Coll. Cardiology, Am. Fedn. for Clin. Research, Phi Beta Kappa, Alpha Omega Alpha. Author: (with A. Brest, J. Moyer) Cardiovascular Drug Therapy, 1965; (with W. H. Blahd) Nuclear Medicine, 1965; (with E. Braunwald) Digitalis and Related Preparations, 1966; (with W. F. M. Fulton) Mechanisms of Action and Therapeutic Uses of Cardiac Drugs, 1966. Research, numerous publs. in field cardiovascular physiology and pharmacology, cardiovascular diagnostic techniques. Home: 12009 Hitching Post Lane, Rockville, Md. 20852. Office: Cardiology Br., Nat. Heart Inst., NIH, Bethesda, Md. 20014.*

MASON, Edward Allen, Am. chem. physicist; b. Atlantic City, Sept. 2, 1926; s. Edward Paul and Olive Margaret (Lorah) M.; B.S., Va. Polytech. Inst., 1947; Ph.D., Mass. Inst. Tech., 1951; m. Ann Courtenay Laufman, July 6, 1952; children—Catherine Hubbard, Stephen Edward, Elizabeth Margaret, Sarah Lois. Research asso. Mass. Inst. Tech., 1950-52; NRC fellow U. Wis., 1952-53; mem. faculty Pa. State U., 1953-55, U. Md., 1955-67; prof. chemistry and engring. Brown U., Providence 1967—. Fellow Am. Phys. Soc., Washington Acad. Scis. (Phys. Scis. award 1962), Random Soc.; mem. A.A.A.S., Am. Assn. Physics Tchrs., Am. Assn. U. Profs., Am. Chem. Soc., Sigma Xi, Phi Kappa Phi, Phi Lambda Upsilon. Author: (with J. T. Vanderslice and H. W. Schamp) Thermodynamics, 1966. Research, publs. on interactions and collisions among atoms, molecules and ions; their effects on properties and behavior of gases. Home: 26 Nayatt Rd., West Barrington, R.I. 02890. Office: Dept. Chemistry, Brown U., Providence 02912.*

MASON, Harold Lawrence, Am. biochemist; b. Compton, Cal., Apr. 17, 1901; s. Charles Isaac and Mary (Whaley) M.; B.A., U. So. Cal., 1923, M.A., 1924; Ph.D., U. Chgo., 1927; m. Maude McKenzie, Sept. 7, 1927; children—Lawrence Harold, Norman Ronald, Kathleen (Mrs. Ronald Ward Hedlund). Instr., Northwestern U., 1927-28; instr. Mayo Grad. Sch. Medicine (formerly Mayo Found.), Rochester, Minn., 1929-33, asst. prof., 1933-40, asso. prof., 1940-48, prof., 1948-66, emeritus, 1966—, asst. Mayo Clinic, Rochester, 1928-29, cons. 1929-66; sr. staff Lovelace Found., 1966—. Diplomate Am. Bd. Clin. Chemistry; mem. Am. Chem. Soc., Am. Soc. Biol. Chemists, Minn. Acad. Sci. (dir. 1959-63), Endocrine Soc. (pres. 1963-64). Research on isolation and structure of adrenal cortical hormones; structure of glutathione and method for its determination; porphyrins; vitamins; steroids.*

MASON, Herman Charles, Am. pub. health adminstr.; b. Chgo., Sept. 26, 1910; s. Nissl and Minna (Bordonsky) Fischer-Mason; B.Sc., U. Chgo., 1932; M.S., U. Ill. Coll. Medicine, 1936, Ph.D., 1938; grad. Indsl. Coll. Armed Forces, 1964; m. Beryl Imogene Troxell, Dec. 26, 1935; children—Jane Addams, Max William. Courtesy fellow Johns Hopkins Sch. Hygiene, 1939-40; faculty Wash. State Coll., 1941; asso. prof. U. N.C. Schs. Medicine and Pub. Health, 1942-44; cons., 1944-48; adviser labs. Korean NIH, 1948; dir. labs., Great Lakes, Ill., 1949-50; cons. pub. welfare Ill. Neuropsychiat. Inst., Chgo., 1950-57; dir. pub. health labs. Wash. Dept. Health, 1957-59, chief, psychophysiology and sr. pathologist Boeing Airplane Co., Seattle, 1959-61; adminstr. labs. and research biochemist VA Hosp., Topeka, 1961-64; dir. research labs., cons. Ark. State Hosp., Little Rock, Benton, 1964—. Research prof. U. Ill., 1955-56, U. Wash., 1958; cons. USPHS, Kans. Heart Assn.; chmn. subcom. on mental retardation planning, Ark., 1966. Fellow Am. Pub. Health Assn. (life), Royal Soc. Health, Royal Soc. Tropical Medicine and Hygiene; mem. Am. Assn. Immunologists, Am. Assn. Bacteriologists, Pacific N.W. Soc. Pathologists, Am. Acad. Polit. and Social Sci., Sigma Xi. Contbr. articles to sci. jours. Research in virus diseases, epidemiology of human and animal diseases, tissue culture, radiation physiol. pathology, lab. methods, biochemistry and physiology of behavioral disorders, patho-physiology in space medicine, instrumentation for space flight, histochemistry, chemotherapy. Home: 1909 Schiller St., Little Rock 72202. Office: Ark. State Hosp., Markham and Elm Streets, Little Rock 72202.*

MASON, John Alden, Am. anthropologist; b. Phila., Jan. 14, 1885; s. William Albert and Ellen Louise (Shaw) M.; A.B., U. Pa., 1907; Ph.D., U. Cal. at Berkeley, 1911; Litt.D., Franklin and Marshall Coll., 1958; m. Florence Roberts, Dec. 23, 1921; 1 son, John Alden. Asst. curator Field Mus. Natural History, Chgo., 1917-24; asst. curator Mexican archeology Am. Mus. Natural History, N.Y.C., 1924-25; curator Am. sect. U. Mus., U. Pa., Phila., 1926-55, curator emeritus, 1955—, editor New World Archeol. Found., Provo, Utah, 1958—. Mem. Soc. Am. Archaeology (pres. 1944), Eastern States Archeol. Fedn. (pres. 1942-46), Soc. Pa. Archaeology (pres. 1929-30), A.A.A.S. (v.p. 1944), Am. Anthrop. Assn. (v.p. 1944), Phila. Anthrop. Soc. (pres. 1927, 54), Sigma Xi. Unitarian. Editor, Am. Anthropologist, 1945-48. Author: The Ethnology of the Salinan Indians, 1912; The Language of the Salinan Indians, 1917; The Mutsun Dialect of Costanoan, 1916; Tepecano, A Language of Western Mexico, 1918; Puerto Rican Folklore, 1918; Archeology of Santa Marta, Colombia, 1931, 36, 40; Language of the Papago of Arizona, 1950; The Languages of South American Indians, 1950; Ancient Civilizations of Peru, 1957. Research and publs on Am. Indian archeology, linguistics, ethnology, and folklore; archeol. study of Santa Marta, Colombia; collection of Puerto Rican folklore. Home: 725 Conestoga Rd., Berwyn, Pa. 19312. Office: University Mus., 33rd and Spruce Sts., Phila. 19104.*

MASON, Leonard Edward, Am. anthropologist; b. Seattle, June 26, 1913; s. Roy Edward and Mattie (Reed) M.; B.A., U. Minn., 1935, M.A., 1941; Ph.D., Yale, 1955; m. Hazel Lillian Cates, June 23, 1939; children—William, Jacqueline, Nancy. Research asst. St. Paul Inst. (Sci. Mus.), 1935-41; research asst. cross cultural survey, Yale, 1941-43, faculty, 1943-44; research analyst OSS, Washington, 1944-45, U. S. Dept. State, Washington, 1945-46; faculty U. Hawaii, Honolulu, 1947—, prof. anthropology, 1955—. Anthropologist, U. S. Comml. Co., Micronesia, 1946; cons. USN Trust Ter., Micronesia, Bikini Marshallese, 1948; mem. exec. com. Tri-Instl. Pacific Program, 1953-63. NSF fellow, 1956-57. Mem. Am. Anthrop. Assn., Am. Ethnol. Soc., Soc. Applied Anthropology, A.A.A.S. Author: Economic Organization of Marshall Islanders, 1947; also articles. Research in ethnology of Micronesia; adaptation of resettled Marshallese communities (Bikini, Eniwetok). Home: 5234 Keakealani, Honolulu 96821.*

MASON, Max, Am. mathematician; b. Madison, Wis., Oct. 26, 1877; s. Edwin Cole and Josephine (Vroman) M.; B.Litt., U. Wis., 1898, LL.D., 1926; Ph.D., U. Göttingen, 1903; D.Sc., Columbia, 1926; LL.D., Yale, 1926, Dartmouth, 1927, Pomona Coll., 1937; m. Mary Louise Freeman, June 16, 1904 (dec. 1928); children—William Vroman, Maxwell, Molly; m. 2d, Helen Schermerborn Young, Aug. 5, 1939 (dec. 1944); m. 3d, Daphine Crane Martin, Nov. 6, 1945. Instr. math. Mass. Inst. Tech., 1903-04, asst. prof. math. Yale, 1904-08; prof. math. physics U. Wis., 1908-25; pres. U. Chgo., 1925-28; dir. natural scis. Rockefeller Found., N.Y.C., 1928-29, pres., 1929-36; named chmn. council to direct. constrn. Mt. Palomar (completed 1948), also mem. exec. council Cal. Inst. Tech., 1936; ret., 1949. Lectr. math. physics 2d semester Harvard, 1911-12; staff Naval Exptl. Sta., New London, Conn.; mem. submarine com. NRC, 1917-19. Fellow A.A.A.S.; mem. Nat. Acad. Sciences, Am. Math. Soc. (colloquium lectr. 1906), Am. Phys. Soc., Deutsche Mathematiker Vereinigung, Phi Beta Kappa, Sigma Xi, Gamma Alpha, Psi Upsilon, Phi Kappa Phi. Author: The New Haven Mathematical Colloquium, 1910. Co-author: The Electromagnetic Field. Asso. editor Trans. Am. Math. Soc., 1911-17. Contbr. papers on math. research to sci. jours. Developed relation between algebra of matrices and integral equations, as well as related boundary problems; studied electromagnetic theory, differential equations, calculus of variations, existence theorems, oscillation properties, asymptotic expressions, devel. theorems; inventor acoustical compensators, submarine detection devices. Died Claremont, Cal., Mar. 23, 1961.

MASON, Otis Tufton, Am. ethnologist; b. Eastport, Me., Apr. 10, 1838; s. John and Rachel Thompson (Lincoln) M.; grad. Columbian, 1861 (A.M., 1862; Ph.D., 1879; LL.D., 1898); in charge prep. sch. Columbian Univ., 1861-84; from 1884 curator ethnology Nat. Museum, and Head curator, dept. anthropology, 1902—. Hon. and corr. mem. many Am. and European scientific socs. Author: The Hupa Indian Industries; Woman's Share in Primitive Culture, 1894; Origin of Inventions, 1895 (London); Primitive Transportation; The Land Problem; Cradles of the North American Indians; The Antiquities of Guadeloupe; Aboriginal American Basketry (2 vols.), 1904. Died 1908.

MASON, Silas Cheever, Am. horticulturist; b. East Greensboro, Vt., Apr. 19, 1857; s. Elkanah Phillips and Adaline (Cheever) M.; B.S., Kan. State Agrl. Coll., 1890, M.S., 1893, D.Sc. (hon.), 1928; m. May V. Quinby, Jan. 1, 1884. Tchr. pub. schs. 3 yrs.; asst. prof. horticulture and forestry Kan. State Agrl. Coll., 1890, prof., 1894-97; prof. horticulture and forestry Berea (Ky.) Coll., 1897-1906; arboriculturist (under new classification horticulturist) U. S. Dept. Agr., 1907-31, detailed by sec. of agr. to study date palm culture in Egypt and Sudan, and secure offshoots of valuable varieties, 1913-14; again in Egypt, 1920, in Algeria, 1921-22; leave of absence, 1924-25, as cons. expert on date culture to Sudan Govt. Discovered identity, history and geog. range of Saidy date of oases of Libyan Desert, secured 7,000 offshoots for U. S. Dept. Agr., for planting in So. Cal. Died Oct. 19, 1935.

MASON, Stephen Finney, English chemist; b. Leicester, Eng., July 6, 1923; s. Leonard Stephen and Christine (Finney) M.; B.A., Oxford, 1945, M.A., D.Phil., 1947, D.Sc., 1967; m. Joan Banus, Oct. 5, 1955; children—Oliver Neil, Andrew Lawrence, Lionel Jeremy. Lectr. history of sci. Oxford U., 1947-53; research fellow in med. chemistry Australian Nat. U., 1953-56; lectr., sr. lectr., reader chemistry Exeter U., 1956-64; prof. chemistry U. East Anglia Sch. Chem. Scis., Norwich, Eng., 1964—. Mem. Brit. Am. chem. socs., Optical Soc. Am., Faraday Soc. Author: A History of the Sciences, 1953. Research, numerous publs. on modes of drug action, mechanisms of chem. reactions, interactions of molecules with radiation, theoretical chemistry. Home: 168 Unthank Rd., Norwich, Norfolk, Eng.*

MASON, Thomas Godfrey, Brit. physiologist; b. July 24, 1890; ed. Cheltenham Coll., Trinity Coll., Dublin, Ireland. Lectr. botany, Alta., Can., 1919-20; botanist, Barbados, 1920-22; prof. botany Imperial Coll. Tropical Agr., Trinidad, 1922-23; sr. botanist Nigeria, 1924-25; physiologist Cotton Research Sta., Trinidad, W.I., 1926—. Fellow Royal Soc., 1937. Research and publs. on movement of food materials in plants. Died Oct. 22, 1959.

MASON, Warren Perry, Am. physicist; b. Colorado Springs, Colo., Sept. 28, 1900; s. Edward Luther and Kate (Sagendorph) M.; B.S. in Elec. Engring., U. Kan., 1921; M.A. in Physics, Columbia, 1924, Ph.D. in Physics, 1928; m. Evelyn Stuart McNally, May 10, 1929 (dec. 1953); 1 dau., Penelope E. (Mrs. Lynn C. Scall); m. 2d, Edith Ewing Aylesworth, Apr. 14, 1956. Mem. tech. staff Bell Telephone Labs., Murray Hill, N.J., 1921-35, head piezoelectric research, 1935-48, head mechanics research, 1948—. Adj. prof. Columbia, N.Y.C., 1964—. Fellow Am. Phys. Soc., Acoustical Soc. Am. (pres. 1955-56), I.E.E.E., Audio Engring. Soc. (hon.); mem. A.A.A.S., Rheological Soc., Instrument Soc. Am. (Arnold O. Beckman award 1964), Sigma Xi, Tau Beta Pi. Author: Electromechanical Transducers and Wave Filters, 1942; Piezoelectric Crystals and Their Application to Ultrasonics, 1950; Physical Acoustics and the Properties of Solids, 1958; also numerous articles. Editor: Physical Acoustics, 8 vols., 1964-66. Holder 200 patents on telephone equipment. Discoveries include: quartz crystal filters, 1929, 1st application coaxial transmission lines to filters and transformers, 1937, 1st measurement of shear elasticity in liquids, 1946, theory of ferroelectric effect in Rockelle salt, 1947, 1st thermodynamic derivation of electro optic and piezo optic effects in crystals, 1950, 1st derivation of attenuation of acoustic waves damped by electrons, 1955, internal friction and fatigue of metals at large strain amplitudes, 1956, damping of dislocations by phoron viscosity, 1960, and by electrons, 1965, 1st determination of 3d order elastic moduli in solids, 1961, attenuation and velocity changes of sound waves due to doping of semiconductors, 1963. Home: 50 Gilbert Pl., West Orange, N.J. 07052. Office: Bell Telephone Labs., Murray Hill, N.J. 07971.

MASOUREDIS, Serafeim P(anagiotis), Am. physician; b. Detroit, Nov. 14, 1922; s. Peter George and Lemonia (Moniodis) M.; A.B., U. Mich., 1944, M.D., 1948; Ph.D. in Med. Physics, U. Cal. at Berkeley, 1952; m. Marion Helen Mykytew, Oct. 2, 1943; children—Claudia, Linus. Faculty, U. Cal. at San Francisco., 1950-52, asso. prof. preventive medicine, 1959-62, research asso. Cancer Research Inst., 1959-67, asso. prof. medicine, staff specialist, chief blood bank, 1962-67; exec. dir. Milw. Blood Center, 1967—; prof. medicine Marquette Sch. Medicine, Milw., 1967—. research asso. med. physics Donner Lab., U. Cal. at Berkeley, 1953-54; faculty U. Pitts. Med. Sch., 1955-59, asso. prof. pathology, 1958-59, asst. dir. Central Blood Bank, 1955-59. Cons., VA Hosp., San Francisco, 1964-65; reviewer Cancer Virology Alerting Service, NIH, 1964—. AEC fellow U. Cal., 1948-50; Nat. Cancer Inst. fellow U. Cal. at Berkeley, 1950-52; recipient 1st prize for exhibit Am. Assn. Blood Banks, 1960; Nat. Heart Inst. Spl. fellow U. Lausanne (Switzerland), 1965-66; Britton Fund grantee, 1959-60; USPHS grantee, 1955-59, Nat. Cancer Inst. grantee, 1956-65, Nat. Heart Inst. grantee, 1957—. Mem. Western Assn. Immunologists (councilor 1960-—), Am. Assn. Blood Banks (mem. sci. program com. 1964-—), Am. Assn. for Cancer Research, Am. Assn. Immunologists, Am. Fedn. for Clin. Research, Am. Hellenic Inst. for Med. Research, Am. Inst. Biol. Scis., Am. Physiol. Soc., Am. Soc. for Clin. Investigation, Am. Soc. Hematology, Am. Soc. Human Genetics, Brit. Soc. Immunology, Internat. Soc. Blood Transfusion, Internat. Soc. Hematology, Soc. Nuclear Medicine, Soc. for Exptl. Biology and Medicine, Western Assn. Physicians, Western Soc. for Clin. Research. Cons. editor Transfusion, 1960—. Research and publs. on application of radioactive tracers to immunology, immunohematology and cancer research. Home: 2927 E. Newberry Blvd., Milw. 53211. Office: 763 N. 18th St., Milw. 53233.*

MASPERO, Gaston Camille Charles, French Egyptologist; b. Paris, June 23, 1846; ed. Ecole Normale, 1868-73; self-educated in Egyptology. Prof. Egyptology, College de France, 1874-80; 86-99; led archeol. expdn. in Egypt, 1880; dir. gen. excavations and antiquities for Egyptian Govt., dir. mus. at Bulak (later at Cairo), 1881; opened pyramid of 5th dynasty king Unis at Saqqara; initiated discovery of royal mummies at Deir el-Bahri, 1881; cleared great Sphinx at Giza and temple at Luxor; dir. excavations in Egypt, especially at Karnak, 1899-1914. Mem. Academie des Inscriptions (permanent sec. 1914-16). Author: Les contes populaires de l'Egypte ancienne, 1882; L'archéologie égyptienne, 1887; Etudes de mythologie et d'archéologie égyptiennes, 8 vols., 1893-1916; Histoire ancienne des peuples de l'Orient classique, 3 vols., 1895-1908. Set up French Sch. Oriental Archaeology, Cairo; early leader archaeology in Egypt, most important excavations at Karnak and Luxor. Died Paris, June 30, 1916.

MASSA, Niccolo, Italian physician; b. Padua, or Venice, Italy, 1499; student medicine, Padua; practiced medicine, specializing in syphilogy, Padua. Author: De morbo gallico, 1532; Anatomiae liber introductorius, 1536; De febre pestilentiali, 1540; Epistolarum medicinalium, 1542; Examen de venae sectione, 1560. 1st to describe gummata, 1532; one of earliest to describe function of ossicles, 1563; 1st to point out syphilis can cause mental disease; studied cholera. Died 1569.

MASSALSKI, Tadeusz Bronislaw, phys. metallurgist; b. Warsaw, Poland, June 29, 1926; s. Piotr and Stanislawa (Andrukaniec) M.; student London (Eng.) U., 1947-49; B.Sc., Birmingham (Eng.) U., 1952, Ph.D., 1954, D.Sc., 1964; fellow Institute Study Metals, University of Chicago, 1954-56; m. Sheila Joan Harris, September 19, 1953; children—Irena, Peter, Christopher. Came to U. S., 1959, naturalized, 1965. Lectr. Birmingham U., 1956-59; head metal physics group Mellon Inst., Pitts., 1959—, staff fellow, 1961—, mem. adv. bd. to pres., 1961—; cons. Batelle Meml. Inst., 1957; vis. prof. U. Buenos Aires, 1962, Cal. Inst. Tech., 1962, Stanford, 1963; Guggenheim fellowship Oxford University, 1965-66. Fellow Brit. Inst. Physics; mem. British Inst. Metals, Am. Inst. Mining, Metall. and Petroleum Engrs., Phys. Soc., Geophys. Union, Internat. Soc. Stereology, Polish Inst. Arts and Scis. in Am. Author: (with C. S. Barrett) The Structure of Metals, 3d edit., 1966; also papers and articles on alloy theory, crystellography, metals physics, meteorites. Editor: Alloying Behaviour and Effects in Concentrated Solid Solutions, 1965. Research on theories of alloy phases, solid state transformations, structure of metals, crystalline imperfections. Home: 900 Field Club Rd., Pitts. 15238. Office: 4400 5th Av., Pitts. 15213.*

MASSART, Jean, Belgian botanist; b. Etterbeck, Belgium, Mar. 7, 1865; M.D.; Docteur ès sciences; studied under Heger; became prof. botany U. Brussels (Belgium), 1895; research staff Paris (France) Pasteur Inst.; dir. Brussels Bot. Garden, 1902-06; made voyages to Brazil, Sahara, Mexico, Am. Head campaign to preserve Belgian natural sites. Mem. French Acad. Scis., 1921. Author: Nos arbres, 1911; Pour la protection de la nature en Belgique, 1912; Éléments de biologie générale et de botanique, 1918. Research on geog. botany of Brussels, plants with respect to their environment. Died Houx, France, Aug. 16, 1925.

MASSAZZA, Serafino Mario, Italian gynecologist; b. Pavia, Italy, Apr. 16, 1894. Became prof., 1948, full prof., 1951; past prof., dir. obstetrics inst. U.

Pavia; also dir. Sch. for Specialization in Obstetrics and Gynecology; prof. obstetrics and gynecology U. Milan (Italy). Mem. Soc. Allergology, Mediterranean Med. Union (v.p.), Lombard Soc. Obstetrics and Gynecology, Soc. Exptl. Biology, Lombard and Venetian Soc. Obstetrics and Gynecology. Author: Manuale di ostetrica' e ginecologia, 1955. Research in hysteriotemy, allergies in obstetrics and gynecology, vaginal tumors. Address: Università degli Studi, Milan, Italy.

MASSÉ, Henri Claude, French surgeon; b. Bordeaux, France, June 6, 1923; s. Lucien and Anne (Bastit) M.; D.M., Faculté de Médecine de Bordeaux, 1953; m. Colette Desmettre, Dec. 28, 1946; children—Nicole, Jean-François, Michel, Caroline, Arnaud. Head surg. clinic; successively became hosp. asst., 1956, head lab. operative medicine, prof. agrégé, 1958, hosp. surgeon, 1963, prof. without chair gen. surgery, 1965. Decorated Chevalier des Palms académiques; laureat French Acad. Medicine. Mem. Société internationale de Chirurgie. Fellow French Coll. Vascular Pathology (founding). Research and numerous publs. on arterial grafts, including methods of preservation and results, histological, physiol. and biol. features; bacterial contamination of blood, effects and chem. reactions, bacteriological problems, prevention; response of lymph node to skin homografts; lymph nodes structure and prognosis of cancer; arterial pathology and surgery. Home: 44 Place Gambetta. Office: Hopital Tatet-girard, Pellogron, Bordeaux, Gironde, 33, France.*

MASSERMAN, Jules H., psychiatrist; b. Chudnov, Poland, Mar. 10, 1905; s. Abraham and Czerna (Baker) M.; Ph.G., Sandusky Coll. Pharmacy, 1924; B.S., Wayne U., 1926, M.D., 1930; m. Christine McGuire, Feb. 20, 1943. Instr., asst. prof. psychiatry U. Chgo., 1935-46; asso. prof. Northwestern U., 1946-49, prof., 1952—, co-chmn. psychiatry, 1956—; chmn. Ill. Tng. and Research Authority, 1965—. Recipient Lasker award in psychiatry, 1946. Mem. Am. Psychoanalytic Assn., A.M.A., Am. Psychiat. Assn. (councillor 1964—), Am. Soc. Biol. Psychiatry, (pres. 1956), Am. Acad. Psychoanalysis (pres. 1957), Am. Soc. Group Therapy (pres. 1957), Phi Lambda Kappa. Author: Behavior and Neuroses, 1943; Principles of Dynamic Psychiatry, 1946; Practice of Dynamic Psychiatry, 1955; Current Psychiatric Therapies vol. I-VII, 1960-67; Science and Psychoanalysis vol. I-XII, 1955-67; also numerous articles. Clarified causes and advancing treatment of deviations of human behavior through animal exptl., psychoanalytic and clin. research. Home: 2231 E. 67th St., Chgo. 60649. Office: 8 S. Michigan Av., Chgo. 60603.*

MASSEVITCH, Alla Genrikhovna (Mrs. Joseph N. Friedlander), Russian astrophysicist; b. Tbilisi, USSR, Oct. 9, 1918; d. Genrikh C. and Natalie A. (Zhgenti) Massevitch; grad. Moscow (USSR) U., 1941, D. Phys. and Math. Scis., 1946; m. Joseph N. Friedlander, June 1942; 1 dau., Natasha. Prof. astrophysics Moscow U., 1946—; chief Satellite Optical Tracking Stas. Net USSR, 1957—. Pres., Soviet Commn. for Stellar Structure and Evolution of Stars, 1961, Com. E. European Countries for Sci. Use Satellite Observation Results, 1960. Recipient govtl. decoration, 1947, 50, 61, Internat. Astronautical award Galambert, 1964. Mem. Internat. Astron. Union (v.p. commn. extraterrestrial astronomy 1961—), Soviet Assn. Sci. (mem. bd.). Research, numerous publs. on evolutionary tracts for massive stars, geophy. application of satellite tracking results. Home: I Vosstania Sq., Moscow, USSR.*

MASSEY, Sir Harrie Stewart Wilson, physicist; b. Melbourne, Australia, 1908; s. Harrie and Eleanor Massey; M.; B.A., M.Sc., Melbourne U., 1929, LL.D. (hon.); Ph.D., Cambridge U., 1932; LL.D. (hon.), U. Glasgow; D.Sc. (hon.), U. Belfast, U. Leicester; m. Jessica Barton Bruce; 1 dau., Mrs. L. A. Duncanson. Ind. lectr. math. physics Queens U., Belfast, 1933-38; Goldsmid prof. math. U. London, 1938-50, Quain prof., head dept. physics, 1950—; With Admiralty Research Lab., 1940-45; mem. gov. bd. Nat. Inst. Research in Nuclear Sci., 1957-65; chmn. Brit. Nat. Com. for Space Research, 1959—; chmn. Council for Sci. Policy, 1965—. Knighted, 1960. Fellow Royal Soc., 1940 (mem. council 1949-51, 59-60, Hughes medal 1955, Royal medal 1958); mem. Atomic Sci. Assn. (pres. 1953-57), Royal Astron. Soc., Phys. Soc. (pres. 1954-56). Author 14 books including Space Physics, 1964; (with others) Scientific Research in Space, 1964; (with H. Kestleman) Ancillary Mathematics, 1964; L'Ere Nouvelle Des Sciences Physiques, 1964; Het Niewe Tijdperk der Fysica, 1964; (with N. F. Mott) The Theory of Atomic Collisions, 1965; also numerous articles. Research in atomic, nuclear and space physics, especially theory of atomic collisions. Home: Kalamunnda, Pelham's Walk, Esher, Surrey, Eng. Office: U. Coll. London, Gower St., London, Eng.*

MASSEY, Vincent, biochemist; b. Berkeley, Australia, Nov. 28, 1926; s. Walter and Mary (Mark) M.; B.Sc. with honours, U. Sydney (Australia), 1947; Ph.D., U. Cambridge (Eng.), 1953; m. Margot Eva Gruenwald, Mar. 4, 1950; (div. 1966); children—Charlotte, Andrew, Rachel. Research officer Commonwealth Sci. and Indsl. Research Orgn., Australia, 1947-50; Imperial Chem. Research fellow, U. Cambridge, 1953-55; research staff Edsel Ford Research Inst., Detroit, 1955-57; sr. lectr. U. Sheffield, Eng., 1957-63; prof. U. Mich., Ann Arbor, 1963—. Mem. Biochem. Soc., Am. Soc. Biol. Chemists. Contbr. numerous articles to tech. jours. Research on effect temperature on enzyme action, lipoyl dehydrogenase, ketoglutarate dehydrogenase complex, flavoproteins, lipoyl dehydrogenase, glutathione reductase, glucose oxidase, dextro and levo amino acid succinic and dihydroorotic dehydrogenases, xanthine oxidase; discovered labile sulfur in metalloflavoproteins, oxidation-reduction role protein disulfide in lipoyl dehydrogenase, glutathione reductase. Home: 2536 Bedford, Ann Arbor, Mich. 48104.*

MASSEY, William S., Am. mathematician; b. Granville, Ill., Aug. 23, 1920; s. Robert R. and Alma (Schumacher) M.; student Bradley U., 1937-39; B.Sc., U. Chgo., 1941, M.Sc., 1942; Ph.D., Princeton, 1948; m. Ethel Heap, Mar. 14, 1953; children—Eleanor, Alexander, Joan. Mem. research dept. Princeton, 1948-50; from asst. prof. to prof. Brown U., 1950-60; prof. math. Yale, 1960—. Fellow Am. Acad. Arts and Scis.; mem. Am. Math. Soc. Mem. editorial staff math. jours. Contbns. to field of algebraic topology, especially homotopy theory, theory of fibre spaces, cohomology theory, and differential manifolds. Home: 101 Haverford St., Hamden, Conn. Office: Math. Dept., Yale Univ., New Haven.*

MASSIEU, François, French geologist, mineralogist; b. France, 1832; prof. geology and mineralogy Faculté de Rennes; insp. gen. mines. Proved existence of function which is characteristic for adiabatic change in thermodynamics. Died 1896.

MASSION, Walter Herbert, physiologist; b. Eitorf, Germany, June 4, 1923; s. Rudolf and Margarthe (Polch) M.; student U. Bonn, 1943-44; B.S., U. Cologne, 1947; M.D., U. Heidelberg, 1951; postgrad. U. Berne, 1949-50, U. Zurich, 1951, U. Copenhagen, 1952-53; m. Rose Marie Agnes Kuemin, July 26, 1956; children—Birgit E., Stuart P., Iris C. Came to U. S., 1957, naturalized, 1962. Asst. prof. physiology U. Basle (Switzerland), 1953-54; postdoctoral fellow physiology U. Rochester (N.Y.), 1954-56; research asso. anesthesiology U. Okla., Oklahoma City, 1957-58, faculty, 1958-59, 61—, asso. prof., 1961-66, dir. anesthesia research, 1961—, research asso. prof. surgery, 1966—; postdoctoral fellow U. Cal. at San Francisco, 1959-60. Mem. Am. Physiol. Soc., Am. Soc. Anesthesiologists, A.A.A.S., Sigma Xi, Sigma Delta Chi (Newsmaker of 1964). Contbr. articles to tech. jours. Research on respiration, circulation, anesthesia, role high energy compounds in shock and reflex vasospasm in pulmonary embolism; codiscoverer chemosensitive receptor area in brain regulating respiration. Home: 4700 Willard Av., Oklahoma City 73105.*

MASSON, Antoine-Philibert, French physicist; b. Auxonne, France, 1806; ed. École normale; agrégé, 1835; prof. Lycée Louis-le-Grand; became prof. École centrale, 1845. Author: Théorie physique et mathématique des phénomène électro-dynamiques, 1838; Études de photométrie électrique, 1845. Produced high tension induction current (using geared wheel to rapidly store current of battery in inductor), 1836; built (with Bréguet) electric telegraph, 1838; research in photometry, 1845-55, motion of elastic liquids and gases; proved (with Jamin) proportionality of dilution of light and heat by various absorbant substances, 1850. Died 1880.

MASSON, Henry James, Am. chem. engr.; b. N.Y.C., Dec. 24, 1891; s. Henry James and Sarah (Pancoast) M.; Ch.E., Columbia, 1914, M.A., 1916; M.S., N.Y. U., 1915, Ph.D., 1918; m. Edith Virginia Hewitt, Nov. 28, 1917; children—Robert Hewitt, Helen Virginia. Mem. faculty Columbia, 1917-18; faculty N.Y. U., 1918—, prof. chem. engring., 1935—, dir. engring. div., 1942-45, asst. dean grad. div. Coll. Engring., 1942-58, prof. emeritus, 1964—. Registered prof. engr. Mem. Phi Lambda Upsilon, Tau Beta Pi, Sigma Xi. Research, publs. and patents on petroleum tech. Home: 27 Oakwood Dr., Delaware, O. 43015.*

MAST, Samuel Ottmar, Am. biologist; b. Ann Arbor Twp., Mich., October 5, 1871; s. G. F. and Beata (Staebler) M.; certificate, State Normal Coll., Ypsilanti, Mich., 1897, M.Pd., 1912; B.S., U. Mich., 1899, Sc.D., 1941; Ph.D., Harvard, 1906; Johnston scholar, Johns Hopkins, 1907-08; m. Grace Rebecca Tennent, 1908; children—Louise Rebecca, Elisabeth Tennent, Margaret Tennent. Prof. biology and botany Hope College, 1899-1908; asso. prof. biology and prof. botany Goucher Coll., Balt., 1908-11; asso. prof. and prof. zoölogy Johns Hopkins, since 1911, head of dept. zoölogy, dir. zoöl. lab., since 1938, emeritus, 1942. Recipient Cartwright prize Columbia, 1909. Fellow A.A.A.S.; mem. Am. Soc. Zoölogists, Am. Physiol. Soc., Am. Soc. Naturalists Soc. for Study of Growth and Devel., Acad. Sci., Phila., Phi Beta Kappa, Sigma Xi. Author: The Structure and Physiology of Flowering Plants, 1907; Light and the Behavior of Organisms, 1911; Motor Response to Light in the Invertebrate Animals, 1936; Factors Involved in the Process of Orientation of Lower Organisms in Light. 1938; Motor Response in Unicellular Animals, 1941; also papers on responses in organisms and on growth in protozoa. Died Feb. 3, 1947.

MASTERS, Colin John, biochemist; b. Taihape, New Zealand, June 6, 1930; s. Clarence and Jane (Maclean) M.; M.Sc., U. New Zealand, 1952; Ph.D. U. Queensland, 1960; m. Josephine Horman, Oct. 3, 1955; children—Gregory, Alistair, Scott. Sr. lectr. biochemistry U. Queensland, Brisbane, Australia, 1960-66; vis. prof. biochemistry Brandeis U., Waltham, Mass., 1966—. Mem. Australian Biochem. Soc. (past mem. council). Research and publs. on molecular basis of tissue differentiation, nature multiple enzyme forms, isoenzyme status of lactate dehydrogenase, esterases and aldolase. Home: 23 Northgate St., Wellesley, Mass. 02181. Office: Grad. Dept. Biochemistry, Brandeis U., Waltham, Mass. 02154.*

MASTERS, Maxwell Tylden, English botanist; b. Canterbury, Eng., Apr. 15, 1833; s. William Masters; ed. King's Coll., London, Eng.; M.D., St. Andrews, Scotland, 1862; m. Ellen Tress, 1858; 4 children. Sub curator, Oxford, Eng.; lectr. botany St. George's Hosp. Med. Sch., 1855-68; began practice gen. medicine, Peckham, Eng., 1856. Sec., Internat. Hort. Congress of 1866, editor Proc. Mem. Royal Hort. Soc. (chmn. sci. com.), French Acad. Scis. (corr.), 1888. Royal Soc., 1870. Linnean Soc. Author: Vegetable Teratology, 1869; also numerous articles; contbg. author: Treasury of Botany (Lindley and Moore), 1866, 73. Became editor Gardeners Chronicle, 1865. Research on many plant groups, plant anomalies and monstrosities. Died Ealing, May 30, 1907.

MASTERS, William Howell, Am. physician; b. Cleve., Dec. 27, 1915; s. Francis Wynne and Estabrooks (Taylor) M.; grad. Lawrenceville Sch., 1934; B.S., Hamilton Coll., 1938; M.D., U. Rochester, 1943; m. Elisabeth Ellis, June 13, 1942; children—Sarah Worthington, William Howell III. Practice medicine, specializing in obstetrics, gynecology, St. Louis, 1943—; faculty Sch. Medicine Washington U., St. Louis, 1943—, asso. prof. clin. obstetrics, gynecology, 1964—, dir. Reproductive Biology Research Found., 1964—; asso. obstetrician, gynecologist St. Louis Maternity and Barnes Hosps, Washington U. Clinics; asso. gynecologist St. Louis Childrens Hosp.; cons. gynecologist St. Louis City Infirmary, Salem (Ill.) Meml. Hosp. Diplomate Am. Bd. Obstetricians and Gynecologists. Mem. A.M.A., Gerontol. Soc., Am. Fertility Soc., Internat. Fertility Assn., Endocrine Soc., Am. Coll. Obstetrics and Gynecology, N.Y. Acad. Scis., Soc. for Sci. Study Sex, A.A.A.S., Internat. Soc. Comprehensive Medicine, Pan-Am. Med. Assn., Sigma Xi, Alpha Omega Alpha, Alpha Delta Phi. Author: (with Mrs. Virginia E. Johnson) Human Sexual Response, 1966. Studies and numerous publs. on steroid replacement in aged; physiology of human sexual response. Home: 34 Oakleigh Lane, Ladue, Mo. 63124. Office: 4910 Forest Park Blvd., St. Louis 63108.*

MASTERSON, John Joseph, Am. physician; b. Bklyn., Aug. 16, 1881; s. William Henry and Margaret Ann (Donohue) M.; student St. Patrick's Acad., Bklyn., 1887-96, Heffley Inst., 1902-04; M.D., L. I. Coll. Hosp., 1908; m. Lucille McGuire, Jan. 26, 1910; children—William F., Lucille (Mrs. Edward A. Harvey), John G. Individual practice medicine, Bklyn., 1910—, specialist in roentgenology 1919—; attending roentgenologist Norwegian Hosp., 1919-53; cons. roentgenologist Victory Meml., Bay Ridge and Norwegian hosps. Dir. United Med. Service N.Y., 1943-60, Asso. Hosp. Service N.Y., 1937-59. Served as mem. Med. Adv. Bd., World War I and II. Mem. med. adv. com. N.Y. World's Fair, 1939; mem. N.Y. State com., 1950 White House Conf. on Children and Youth; mem. med. adv. com. on civilian def. N.Y. State Health Dept., 1950-53; dir. Physicians Home, Inc., N.Y.C., 1955—; med. cons. adviser on Selective Service, N.Y.C., 1948-59. Bd. dirs. Cath. Med. Mission Bd. Named Cath. Physician of Year, Nat. Fedn. Cath. Physicians Guilds, 1959; recipient certificate of distinction in field of medicine State U. N.Y. Downstate, Med. Center, 1960. Diplomate Am. Bd. Radiology. Fellow Am. Coll. Radiology, N.Y. Acad. Medicine; mem. A.M.A. (del. from N.Y. to House of Dels. since 1935), Radiol. Soc. N. Am., Nat. Fedn. Cath. Physician Guilds (past pres.), Pan Am. Med. Assn., Med. Soc. State N.Y. (past pres., trustee; chmn. emergency med. preparedness com. 1950-52; chmn. publ. com. jour., 1952-58; chmn. bd. trustees), Alpha Kappa Kappa, Alpha Omega Alpha. Home: 9425 Shore Rd., Bklyn.*

MASTIN, Claudius Henry, Am. surgeon; b. Huntsville, Ala., June 4, 1826; s. Francis Turner and Ann Elizabeth Caroline (Levert) M.; attended U. Va.; M.D., U. Pa., 1849; studied medicine with Dr. J. Y. Bassett, Huntsville; attended U. Edinburgh, Royal Coll. Surgeons, U. Paris; m. Mary Eliza McDowell, Sept. 20, 1848; at least 2 sons. Practiced medicine briefly, then went to Europe for further edn.; began practice medicine specializing in genito-urinary surgery, Mobile, Ala., 1854; asso. with U. S. Marine Hosp. Service, Mobile, 1854-57; surgeon Mobile Hosp., 1855; med. dir. on staffs of gens. Polk, Bragg, G. T. Beauregard, 1861-65; a founder Congress Am. Physicians and Surgeons; fellow Am. Surg. Assn., pres., 1890-91; mem. So. Surg. and Gynecol. Assn.; founder Am. Genito-Urinary Assn., pres., 1895-96; wrote many articles pub. by orgns. to which he belonged. Died Oct. 3, 1898.

MASTLIN, Michael, German astronomer; b. Göppingen, Württemberg, Sept. 30, 1550; ed. U. Tübingen (Württemburg). Prof. math. univs. Heidelberg (Baden); tutor of astronomer Kepler; asso. with Galileo during stay in Italy. Author: Thesis de Eclipsibus; Epitome Astronomiae, 1597. Pub. Mysterium Cosmographicum (Kepler), 1596. Early Advocate Copernican system and said to have converted Galileo to its acceptance; oriented Kepler toward study astronomy; explained reason for ashen light of moon; systematically studied Great Comet of 1577; developed method alignments. Died Tübingen, Dec. 20, 1631.

MATARAZZO, Joseph Dominic, Am. psychologist; b. Caiazzo, Italy, Nov. 12, 1925 (parents Am. citizens); s. Nicholas and Adeline (Mastroianni) M.; student Columbia, 1944; B.A., Brown U., 1946; M.S., Northwestern U., 1950, Ph.D., 1952; m. Ruth Wood Gadbois, Mar. 26, 1949; children—Harris, Elizabeth, Sara. Fellow med. psychology, faculty Washington U., St. Louis, 1950-55; research asso. Harvard Med. Sch., also asso. psychologist Mass. Gen. Hosp. Boston, 1955-57; head med. psychology dept. U. Ore. Med. Sch., Portland, 1957——. Dir. Psychology/ Press, Brandon, Vt. Mem. nursing research study sect. NIH, 1963——. Recipient Hofheimer Research award Am. Psychiat. Assn., 1962. Fellow Am. Psychol. Assn., A.A.A.S.; mem. Nat. Assn. for Mental Health (dir. 1963——), Ore. Mental Health Assn. (dir. 1958-68, pres. 1962-63); Am. Assn. State Psychology Bds. (pres. 1963-64). Asso. editor Human Orgn., 1954——; editorial bd. Jour. Clin. Psychology, 1962——; Psychotherapy: Theory, Research and Practice, 1964——; cons. editor Contemporary Psychology, 1962——; psychology series editor Aldine Pub. Co., 1964——. Research, publs. on behaviour and methods of, influences on interview. Home: 1934 S.W. Vista Av., Portland, Ore. 97201.*

MATARÉ, Herbert Franz, physician; b. Aachen, Germany, Sept. 22, 1912; s. Josef P. and Paula (Broicher) M.; Abiturium, U. Geneva (Switzerland), 1934; Dipl. Ing., U. Aachen; Dr.Engring., Tech. U. Berlin (Germany), 1942; Ph.D., Ecole Normale Super, Paris, France, 1950; m. Ursula Krenzien, Dec. 9, 1939; children—Felicitas, Vitus. Came to U. S., 1953, naturalized, 1966. Head microwave receiver lab. Telefunken, Berlin, 1939-46; head semiconductor lab. Westinghouse, Paris, 1946-51; pres. Intermetall. Corp., Dusseldorf, Germany, 1951-53; asst. dir. research Tung-sol Electric, Bloomfield, N.J., 1953-55; head semiconductor lab. Sylvania Co., N.Y.C., 1957-59; head quantum electronics lab. Bendix Research Lab., Southfield, Mich., 1960-62; tech. dir. Lear-Siegler Corporate Research Lab., Santa Monica, Cal., 1962-63; head electronics research dept. Douglas Corp., Santa Monica, 1963-66; sci. adviser autonetics div. N.Am. Aviation, Inc., Fullerton, Cal., 1966——. Lectr. engring. U. Cals., Los Angeles, 1966——. Mem. I.E.E.E. (sr.), Am. Phys. Soc., Electrochem. Soc. Am., Vacuum Soc. Author: Receiver Problems in the U.H.F. Region, 1951; also numerous articles. Patentee in solid state electronics field including transistor, crystal growth, cryotron; Research in deduction signal/noise radio for diode mixers, derivation of diode-noise law, 1st low temperature transistor based on bicrystals, bicrystal photoresponse. Home: 17408 Revello Dr., Pacific Palisades, Cal. 90272. Office: 1621 S. Pomona Av., Fullerton, Cal. 92632.*

MATAS, Rudolph, Am. surgeon; b. Bonnet Carre, nr. New Orleans, La., Sept. 12, 1860; s. Dr. N. Hereu and Teresa (Jorda) M.; ed. Paris, Barcelona, Brownsville (Tex.), Soule's Coll. (New Orleans); grad. Lit. Inst. of St. John, Matamoros, Mexico, 1876; M.D., Tulane, 1880; LL.D., Washington U., 1915, U. of Ala., 1926, Tulane, 1928; Sc.D., U. of Pa., 1925, Princeton, 1928; M.D., honoris causa, Nat. U. of Guatemala, 1934; m. Adrienne Goslee Landry, Jan. 20, 1895. Began practice at New Orleans, 1880, specializing in surgery from 1895; prof. surgery, Tulane Med. Dept., 1895-1927; sr. surgeon Charity Hosp., 1894-1928, cons., from 1928; chief sr. surgeon Touro Infirmary, 1905-35, hon. chief surgeon, 1935——; cons. surgeon Eye, Ear, Nose and Throat Hosp. Fellow Am. Coll. Surgeons (v.p. 1913, 20; pres. 1924-25), A.A.A.S., mem. A.M.A. (chmn. sect. surg. 1908; v.p. 1920, 32-33), Am. Surg. Assn. (pres. 1909), Am. Assn. Thoracic Surgeons (pres. 1920), Am. Soc. Clin. Surgery (v.p. 1908-10), Am. Assn. Cancer Research, hon. fellow Royal Coll. of Surgeons, Eng., 1927; pres. internat. Soc. of Surgery, 1936-38; mem. Assn. Française de Chirurgie (rapporteur by invitation, and hon. pres. 1922); mem. Soc. Internat. de Chirurgie (rapporteur by invitation); hon. mem. Royal Acad. Medicine (Rome), or hon. mem. numerous fgn. socs.; Recipient first distinguished service medal A.M.A., 1938, other awards. Editor New Orleans Med. and Surg. Jour., 1883-85. Author of many treatises and monographs on surg. subjects, specially vascular surgery, contbr. to med. jours. and text books. Studied and helped suppress epidemics of yellow fever, 1878-82; worked on new methods and appliances for anaesthesia, 1886-1902; began intravenous use of saline solutions and other serums for treatment hemorrhage and shock, 1891; developed continuous intravenous drip and continuous syphonage of duodenum; developed operation to treat aneurysm by stitching orifices of vessels supplying the aneurysm from within the aneurysmal sac (Matas operation); 1st recorded cure of aortal aneurysm, 1923. Died New Orleans, Sept. 23, 1957.

MATEEV, Evgeni Georgiev, Bulgarian economist; b. Targorishte, Bulgaria, Apr. 1, 1920; s. Georgi Ivahov and Christina (Stolkova) M.; grad. in state and law scis. U. Sofia (Bulgaria), 1943; m. Izvetana Izvetkova Mateeva, Sept. 30, 1944; 1 dau., Zoya. Faculty, U. Sofia, 1948——, prof. econ. planning, 1951——; v.p. State Planning Com., 1951, 60-63, pres., 1952; pres. Central Statis. Bd., 1960; minister without portfolio, 1963-65. Recipient Nat. Dimitrov prize for sci.; several decorations. Mem. Bulgarian Acad. Scis., Bulgarian Assn. Scientists. Author: Labour Productivity and Reproduction, 1961; National Accounts, 1960; Long Term Planning, 1963; International Division of Labour in the Socialist System, 1965; Long term Planning and Economical Cybernetics, 1966; also articles. Research on math. instruments for macroecon. analysis, problems of econ. efficiency and prices, math. instruments and methods of macroecon. long-term planning. Home: 11 Oborishte. Office: 3 Aksakov, Sofia, Bulgaria.*

MATEYKO, Gladys Mary, Am. biologist; b. N.Y.C., Aug. 22, 1921; d. William and Jennie (Kostur) Mateyko; A.B., Hunter Coll., 1942; A.M., Mt. Holyoke Coll., 1944; Ph.D., N.Y. U., 1952. Microbiologist, Lederle Labs., 1944-45; instr. Hunter Coll., 1946-48, Bklyn. Coll., 1947-48; research asso. Brookhaven Nat. Lab., 1950-51; with N.Y. U., N.Y.C., 1952——, prof. biology, 1965——. Damon Runyon Cancer Research fellow, 1952-55. Fellow N.Y. Acad. Scis.; mem. A.A.A.S., Am. Micros. Soc., Am. Soc. Cell Biology, Am. Assn. Cancer Research, Soc. Gen. Physiologists, Growth Soc., Sigma Xi, Phi Beta Kappa, Phi Sigma. Research and numerous publs. on biology cells, cytophysiology tumor cells, ultracentrifugation, pycnotic cushioning, microsurgery, cytochemistry, cytochemistry and cytophysiology frog renal tumors.*

MATHER, Cotton, Am. natural philosopher; b. Boston Feb. 23, 1663; s. Increase and Maria (Cotton) M.; A.B., Harvard, 1678, M.A., 1681; D.D. (hon.), U. Glasgow (Scotland), 1710; m. Abigail Phillips, 1686; m. 2d, Elizabeth (Clark) Hubbard; m. 3d, Lydia George; 15 children, including Samuel. Ordained to ministry Congl. Ch., 1684; tchr. 2d Congl. Ch., Boston, 1685-1723, 23-28; fellow Harvard, 1690-1703; wrote statement on witch trial evidence condemning reliance on spectral evidence, 1692; defended Salem witch trials in book Wonders of the Invisible World, 1693 (later attacked by Robert Calef in his book More Wonders of the Invisible World); a founder Yale Coll., refused to become its pres.; became 1st Am.-born mem. Royal Soc., 1713, contbd. many articles to its Proc., including one on mastadons in Albany, 1714; informed Dr. Zabdiel Boyleston of process of inoculation for smallpox, supported the radical idea financially and by his influence during smallpox epidemic of 1721; described expts. on plant hybridization; a leader in Boston charities, ednl. and social improvement plans; campaigned against intemperance, mistreatment of slaves. Author over 450 books including: A Family Well-Ordered, 1699; Magnalia Christi Americana: or the Ecclesiastical History of New England from its First Planting (most famous work), 1702; Some Few Remarks upon a Scandalous Book . . . by one Robert Calef, 1704; The Good Education of Children, 1708; Essays to do Good, 1710; Christian Philosopher (1st gen. work on science pub. in N.Am.), 1721; Sentiments on the Smallpox Inoculated, 1721; An Account . . . of Inoculating the Smallpox, 1722. Died Boston, Feb. 24, 1728.

MATHER, John Russell, Am. climatologist; b. Boston, Oct. 9, 1923; s. John and Mabelle (Russell) M.; B.A., Williams Coll., 1946; B.S. in Meteorology, Mass. Inst. Tech., 1947, M.S., 1948; Ph.D. in Geography-Climatology, Johns Hopkins, 1951; m. Amy L. Nelson, June 21, 1946; children—Susan, Thomas, Ellen. Research asso., climatologist Lab. Climatology, Seabrook, N.J., 1948-54, prin. research scientist, Centerton, N.J., 1954-63; pres. Lab. Climatology, C. W. Thornthwaite Assos., Centerton, 1963——; asst. prof. Johns Hopkins, 1951-53; asso. prof. climatology Drexel Inst. Tech., Phila., 1957-60; prof. geography U. Del., Newark, 1961——; vis. lectr. geography U. Chgo., 1957-61. Mem. Am. Meteorol. Soc., Am. Geog. Soc., Am. Geophys. Union, Tau Beta Pi. Contbr. numerous articles to tech. jours. Contributed to concept of potential evapotranspiration, its measurement, use in climatic water balance; application of climatic water balance to studies in agr., hydrology, climatology, geography. Home: R.D. 3. Office: R.D. 1, Elmer, N.J. 08318.*

MATHER, Keith Benson, physicist; b. Adelaide, Australia, Jan. 6, 1922; s. Augustus Benson and Minnie Rabone (Rooney) M.; B.Sc. (Eng.), U. Adelaide, 1942, M.Sc., 1944; m. Betty Tossell Ralph, July 18, 1946; children—Elizabeth Sharman, Bernita Kaye. Came to U. S., 1961. Demonstrator in physics U. Adelaide, 1943-45, lectr., 1946; research asst. Washington U., St. Louis, 1946-49; lectr. physics U. Ceylon, 1950-51, Melbourne U., 1958-61; research officer Commonwealth Sci. and Indsl. Research Orgn., Australia, 1952-54; sr. research officer Australian AEC, 1954-56; physicist-in-charge Mawson Sta., Antarctic, 1957; faculty U. Alaska, College, 1961——, dir. Geophys. Inst., prof. physics, 1963——. Mem. council U. Corp. for Atmospheric Research, 1964——. Recipient Polar medal Brit. Commonwealth, 1959; Imperial Chem. Industries fellow, 1949-50; Fulbright grantee, 1961-62. Mem. Am. Geophys. Union, Arctic Inst. N.Am. (auroral com. 1963-66), Australian Inst. Physics, Nat. Council U. Research Adminstrs. Author: Experiments in Physics for First Year Students, 1951; (with P. Swan) Nuclear Scattering, 1958. Research on forces between nucleons; phenomena at geomagnetically conjugate points; katabatic winds or downslope winds on icecaps. Home: 703 Gradelle Av., McKinley Acres, College, Alaska 99701.*

MATHER, Kirtley Fletcher, Am. geologist; b. Chgo., Feb. 13, 1888; s. William Green and Julia (King) M.; B.Sc., Denison U., 1909, D.Sc. (hon.), 1919; Ph.D., U. Chgo., 1915; A.M. (hon.), Harvard, 1926; Sc.D. (hon.), Colby Coll., 1936; Litt.D. (hon.), Union Coll., 1942; L.H.D. (hon.), Bates Coll., 1943; LL.D., Beloit Coll., 1949; m. Marie Porter, June 12, 1912; children—Florence Margaret (Mrs. Sherman A. Wengerd), Julia Carolyn (Mrs. Leroy G. Seils), Jean Marie (Mrs. Dean W. Seibel). Faculty, U. Ark., 1911-14, Queen's U., Kingston, Ont., Can., 1915-18, Denison U., Granville, O., 1918-24; faculty Harvard, 1924——, prof. geology, 1927-54, prof. emeritus, 1954——. Jr. to sr. geologist U. S. Geol. Survey, 1911-45; geologist and cons. to industry; Danforth vis. lectr. Assn. Am. Colls., 1961-62; pres. Ednl. Research Corp., 1943-51. Recipient Bradford Washburn medal Boston Mus. Sci., 1964; Cullum medal Am. Geog. Soc., 1965; Thomas Alva Edison Found. award for best book on sci. for young people, The Earth Beneath Us, 1965. Fellow Geol. Soc. Am., Royal Geog. Soc., Seismol. Soc. Am., A.A.A.S. (pres. 1951), Am. Acad. Arts and Scis. (pres. 1957-61); mem. Am. Geophys. Union, Am. Inst. Mining and Metall. Engrs., Am. Assn. Petroleum Geologists, Phi Beta Kappa (vis. scholar 1960-61, 64), Sigma Xi, Tau Kappa Alpha. Author: Old Mother Earth, 1928; Science in Search of God, 1928; Sons of the Earth, 1930; Enough and to Spare, 1944; Crusade for Life, 1949; The World in Which We Live, 1961; (with Mason) Source Book in Geology, 1939. Editor: The Earth Science Series, 1932——. Research, publs. in geology and petroleum resources of Bolivia, N.S., Alaska, Cal., Okla., Ky., Tenn.; glacial geology of Southwestern Colo., Central O., Eastern Mass.; use of telecopic alidade in geologic mapping. Home: 3 Concord Av., Cambridge, Mass. 02138.*

MATHER, Thomas, elec. engr.; Sr. Wh. Sch., 1878. Prof. elec. engring. City and Guilds (Engring.) Coll., Imperial Coll. Sci. and Tech., South Kensington, Eng., 1909-22, emeritus prof., 1922——; examiner elec. engring., mem. bd. studies elec. engring. U. London. Became fellow Royal Soc., 1902. Author: (with Ayrton) Practical Electricity, 1911; (with Howe) Exercises in Electrical Engineering, 1910; also articles. Research on constrn. elec. instruments, especially galvanometers; determination fundamental elec. units; improvements in indsl. elec. measuring instruments. Died June 23, 1937.

MATHER, William Williams, Am. geologist; b. Brooklyn, Conn., May 24, 1804; s. Eleazar and Fanny (Williams) M.; grad. U. S. Mil. Acad., 1828; m. Emily Maria Baker, 1830; 6 children; m. 2d, Mary Harry, Aug. 1857; 1 son. Brevetted 2d lt. U. S. Army, 1828; prof. of chemistry and mineralogy at U. S. Mil. Acad., West Point, N.Y., 1829-35; also prof. chemistry, mineralogy and geology Wesleyan U.; promoted 1st lt., 1834; aided G. W. Featherstonhaugh in survey of Green Bay (Wis.) region, 1835; resigned commn., 1836; prof. chemistry U. La., 1836; geologist for 1st Dist., N.Y. State, 1836-44; dir. Ohio Geologic Survey, 1837-38; state geologist Ky., 1838-39; prof. natural science Ohio U., Athens, 1842-45, acting pres., 1845-47; acting prof. Marietta Coll. Author: Elements of Geology for the Use of Schools, 1833; also reports Geol. Survey N.Y. State, circa, 1836. Editor Western Agriculturist. Measured temperature of water at bottom of Hudson River in winter; discovered that bromide could be extracted at low cost from salt spring waters nr. Athens, O., 1845. Died Columbus, O., Feb. 26, 1859.

MATHERON, Philippe, French civil engr., geologist; b. Marseilles, France, Oct. 18, 1807; mem. French Acad. Scis., 1895. Author: Catalogue méthodique et descriptif des corps organisés fossiles du département des Bouches-du-Rhône, 1842; Recherches comparatives sur les dépôts fluvio-lacustres tertiaires des environs de Montpellier, de l'Aude et de la Provence, 1862; also publs. on geol. structure of Provence. Died Marseilles, Dec. 31, 1900.

MATHESIUS, Johann (Matthesius), German theologian; b. Rochlitz, June 24, 1504; Lutheran pastor, Joachimsthal, Germany. Author: Sarepta . . . , 1571. Wrote on bismuth, pyrites, magnet and compass in his book of sermons; disagreed with Pliny's statement that only gold is found native. Died circa 1565.

MATHESON, Max Smith, Am. chemist; b. McGill, Nev., May 24, 1913; s. Henry Thompson and Ethel (Smith) M.; B.A. U. Utah, 1936; M.S., Brown U., 1938; Ph.D., U. Rochester, 1940; m. Gladys Hamilton Mulchahey, Sept. 30, 1939; children—Linda Jean (Mrs. Kenneth R. Boland), Marc Edward; m. Georgine M. Sims, Apr. 29, 1967. Research chemist Gen. Labs., U. S. Rubber Co., Passaic, N.J., 1940-50; staff Argonne (Ill.) Nat. Lab., 1950——, sr. chemist, 1952——, dir. chemistry div., 1965——. Guggenheim fellow

Institut du Radium and Laboratoire de Chimie Physique, Paris, France, 1960-61. Mem. Am. Chem. Soc., Am. Phys. Soc., Chem. Soc. (Eng.), A.A.A.S., Research Soc. Am., Phi Beta Kappa, Sigma Xi, Phi Kappa Phi. Research and publs. on application of rotating sector to measurement rate constants in free radical polymerization; among 1st users of pulse radiolysis for detection of short-lived species, use of pulse radiolysis to determine rate constants for reactions of short-lived species in radiolysis of water. Home: 22 W. 60th St., Westmont, Ill. 60559. Office: 9700 S. Cass Av., Argonne, Ill. 60439.*

MATHESON, Norman Alistair, Scottish surgeon; b. Inverness, Scotland, Mar. 26, 1932; s. Allan John and Margaret (Mutch) M.; M.B., Ch.B., U. Aberdeen (Scotland), 1956, Ch.M., 1965; m. Helen Grant, Sept. 10, 1958; children—Malcolm Grant, Fiona Anne, Catriona Margaret. Faculty, U. Aberdeen, 1957-59, 61—, sr. lectr. surgery, 1964—; registrar in surgery Aberdeen Gen. Hosp., 1959-61. Cons. surgeon Aberdeen Gen. Hosps., 1964—, Royal Aberdeen Hosp. for Sick Children, 1964—. Fellow Royal Coll. Surgeons; mem. Surg. Research Soc., Scottish Soc. for Exptl. Medicine, Brit. Microcirculation Soc., European Microcirculation Soc., European Transplant and Dialysis Soc., Brit. Assn. Urol. Surgeons. Research and publs. on various aspects of surg. care, renal function, blood platelet function, blood flow characteristics. Home: 36 Airyhall Dr. Office: Dept. Surgery, U. Aberdeen, Aberdeen, Scotland.*

MATHESON, Robert, entomologist; b. West River, N.S., Can., Dec. 20, 1881; s. Walter Alexander and Mary (Anderson) M.; student N.S. Sch. Science; B.S., Cornell U., 1906, M.S., 1907, Ph.D., 1911; m. Margaret Katherine Macpherson, Aug. 25, 1911; 1 son, Robert Macpherson. Prof. entomology S.D. State Coll., 1907-09; prof. zoology N.S. Coll. Agr., 1912-13; prof. entomology (med. entomology and parasitology) Cornell U., 1914-49; cons. health sect. TVA. Fellow A.A.A.S., Entomol. Soc. Am.; mem. Am. Soc. Parasitologists (v.p. 1940; mem. editorial bd.), Am. Assn. Econ. Entomologists, N.S. Inst. Science, Am. Soc. Tropical Medicine, Washington Acad. Scis., Sigma Xi; corr. mem. Venezuelan Soc. Natural Scis., Acad. Natural Chile, of Chile, Phila. Acad. Natural Scis. Author: Handbook of the Mosquitos of North America, 1929, rev. edit. 1945; Medical Entomology, 1932; Laboratory Guide in Entomology, 1939; Entomology, 1944, rev. edit. 1951; also many articles in jours. and expt. sta. publs. Died Dec. 14, 1958.

MATHEWS, Albert Prescott, Am. physiol. chemist; b. Chgo., Nov. 26, 1871; s. William Smith Babcock and Flora E. (Swain) M.; S.B., Mass. Inst. Tech., 1892; studied biology, Cambridge, Eng., Naples, Italy, Marburg, Germany, 1895-97; Ph.D., Columbia, 1898; D.Sc., Institutum Divi Thomae, 1940; m. Jessie Glyde Macrum, Feb. 7, 1895; 1 dau., Mrs. Noreen Macrum Koller. Asst. in biology Mass. Inst. Tech., 1892-93; asst. prof. physiology Tufts Coll. Med. Sch., 1899-1900; instr. physiology Harvard Med. Sch., 1900-01; asst. prof. physiol. chemistry U. Chgo., 1901-04, asso. prof., 1904-05. prof., 1905-18, chmn. dept. physiology, 1909-16; prof. biochemistry, U. Cin., 1918—, emeritus, 1940—. Fellow A.A.A.S.; mem. Am. Chem. Soc., Am. Physiol. Soc., Biochem. Soc., Soc. de chimie biologique, Bio-chem. Soc. (British), Soc. Exptl. Biology Gt. Britain, Academia Nationale dei Lincei (fgn., Rome). Author: Text Book of Physiological Chemistry (6th edit.); The Nature of Matter, Gravitation and Light; Gravitation, Space-Time and Matter, 1934; Principles of Biochemistry. Contbr. to sci. jours. Noted for original research on parthogenesis, nature of nerve impulse, pharmacology, chem. biology. Died Sept. 21, 1957.

MATHEWS, Chester Ora, Am. psychologist; b. Crab Orchard, Neb., Apr. 25, 1895; s. Urias Grant and Rosa Allie (Myers) M.; A.B., Kans. Wesleyan U., 1919; postgrad. Northwestern U.; M.A., Columbia, 1924, Ph.D., 1927; m. Martha Selma Rader, Aug. 20, 1920 (dec. June 1938); 1 son, Warren Edward; m. 2d., Vera Elizabeth Nichols, Aug. 4, 1939. Prin. high schs., Ada, Covert, Kans., 1919-23; asst. Inst. Ednl. Research Tchrs. Coll. Columbia, N.Y.C., 1925-27; faculty Ohio Wesleyan U., Delaware, 1927-63, prof. edn., dir. ednl. research and evaluation service, 1934-63, prof., dir. emeritus, 1963—; psychometrist U. Cal. Counseling Center, Santa Barbara, 1963—. Ednl. cons. Bennett Coll., Greensboro, N.C., 1953, 63. Mem. Am. Ednl. Research Assn., Am. Psychol. Assn., A.A.A.S., Am. Personnel and Guidance Assn., Ohio Statis. Assn. (chmn. ednl. div. 1963), Ohio Ednl. Assn. (pres. research dept. 1936), Ohio Coll. Assn. (pres. psychology sect. 1940), Ohio Acad. Sci. (v.p., chmn. psychology sect. 1950-51), Phi Delta Kappa, Kappa Delta Pi. Author: The Grade Placement of Curriculum Materials in the Social Studies, 1926; The Development of Attitude Scales in Practical Politics, 1955; (with Ruth Larson) Changes in Student Attitudes toward Practical Politics During Their First Two Years, 1957, Over a Four-Year Period, 1958; also articles. Home: 5950 Encina Rd., Goleta, Cal. 93017. Office: Counseling Center, U. Cal., Santa Barbara, Cal. 93018.*

MATHEWS, Gregory Macalister, Australian ornithologist; b. Biamble, Australia, Sept. 10, 1876; s. Robert H. Mathews; ed. King's Sch., Parramatta, Australia; m. Marian White; 1 son. Pastoral duties, N. Queensland, Australia; travelled around world, studying ornithology. Mem. Internat. Ornithol. Com.; presented his Australian Ornithol. Library to Commonwealth of Australia, 1939. Mem. Brit. Ornithol. Union (became v.p. 1924, asst. editor, 1931-41), Brit. Ornithologists Club (chmn. 1935-38). Fellow Royal Soc. Entomologists. Author: Hand List of the Birds of Australia, 1908; Birds of Australia, 12 folio vols., 1910-27; A Reference List of the Birds of Australia, 1912; A Reference List of the Birds of New Zealand, 1913; A List of Birds of Australia, 1913; Manual of Birds of Australia, 1921; Birds of Norfolk and Lord Howe Islands, 1928; Systema Avium Australasianarum, Vol. I, 1927, Vol. II, 1930; Supplement to the Birds of Australia and of New Zealand, 1936; Books and Birds, 1942; Notes on Order of Petrels, 1943; Working List of Australian Birds, 1946. Editor, Austral Avian Record, 1912—. Died Mar. 27, 1949.

MATHEWS, John Alexander, Am. metallurgist; b. Washington, Pa., May 20, 1872; s. William Johnston and Frances (Pelletreau) M.; B.S., Washington and Jefferson Coll., 1893, M.S., 1896, Sc.D., 1902; A.M., Columbia, 1895, Ph.D., 1898; postgrad. Royal Sch. Mines, London, 1900-01; m. Florence Hosmer King, Jan. 29, 1903; children—Margaret King, John Alexander. Asst. in assaying Columbia, 1896-97, tutor chemistry, 1898-1900; Barnard fellow, 1900, 01, 02; Andrew Carnegie Research scholar Iron and Steel Inst. of Gt. Britain, 1901; mem. U. S. Assay Commn., 1900, 05, 11; metallurgist and asst. mgr. Sanderson works, Crucible Steel Co. of Am., 1902-08, later v.p., dir.; gen. mgr. Halcomb Steel Co., Syracuse, N.Y., 1908-15, pres., 1915-20. Recipient 1st Andrew Carnegie Gold medal for research Iron and Steel Inst., 1902, Robert W. Hunt Gold medal Am. Inst. Mining and Metall. Engrs., 1928. Mem. Am. Chem. Soc., Am., Internat. socs. for testing materials, Am. Inst. Mining Engrs., Am. Electrochem. Soc. Contbr. to tech. jours. Pioneer in Am. in experimentation with vanadium and its effect on steel; patentee use of vanadium in high speed steel, 1905. Died Jan. 11, 1935.

MATHEWS, Jon, Am. physicist; b. Los Angeles, Feb. 10, 1932; s. John H. and Grace (Logie) M.; B.A., Pomona Coll., 1952; Ph.D., Cal. Inst. Tech., 1957; m. Charlotte Dallett, June 10, 1952; children—Valerie, Jancis, Richard, Hays, William Salisbury. Faculty, Cal. Inst. Tech., Pasadena, 1957—, prof. physics, 1966—. Mem. Am. Phys. Soc., A.A.A.S. Author: (with R. L. Walker) Mathematical Methods of Physics, 1964; also articles. Research in elementary particle physics and gravity. Home: 459 W. Loma Alta Dr., Altadena, Cal. 91002. Office: Physics Dept., Cal. Inst. Tech., Pasadena, Cal. 91109.*

MATHEWS, Joseph Howard, Am. chemist; b. Auroraville, Wis., Oct. 15, 1881; s. Joseph and Lydia (Cate) M.; B.S., U. Wis., 1903; M.S., 1905; M.A., Harvard, 1906, Ph.D., 1908; m. Ella Barbara Gilfillan, June 26, 1909; children—Marion Z. (Mrs. N. H. Withey), Jean Barbara (Mrs. C. C. Watson). Instr. chemistry Case Inst. Applied Sci., 1906-07; faculty U. Wis., Madison, 1908-52, prof. chemistry, chmn. dept., 1919-52. Expert in criminology, 1925—; founder Nat. Symposium in Colloid Chemistry, 1922. Fellow A.A.A.S.; mem. Am. Chem. Soc., Alpha Chi Sigma (founder). Author: (with F. Daniels, J. W. Williams) Experimental Physical Chemistry, 1929; Identification of Firearms, 2 vols., 1964. Research, publs. on devel. of instruments and techniques used in identification of firearms. Home: 128 Lathrop St., Madison, Wis. 53705.*

MATHEWS, Piravonu Mathews, Indian physicist; b. Maramon, Kerala, India, May 12, 1932; s. Piravonu Cherian and Rachel (Abraham) M.; B.Sc. (Hons.), Madras Christian Coll., U. Madras (India), 1952; M.Sc., U. Madras, 1953, Ph.D., 1956; m. Mary Kuruvila, Sept. 14, 1959; children—Ranjit, Prasad Kuruvila, Pradeep Cherian. NRC Can. Research fellow, Ottawa, Ont., 1956-58; research fellow Brandeis U., Waltham, Mass., 1958-59, vis. sr. research asso., 1964-65; faculty U. Madras, 1959—, prof. physics, 1964—. Bd. govs. Inst. Math. Scis., Madras. Research, publs. on fluctuation in numbers of photons and electrons in cosmic ray showers, theory of random processes in physics, dynamics of crystal lattices with impurities, statis. mech. theory of irreversible behavior of phys. systems; developed new relativistic theory of elementary particles of arbitrarily high spin. Home: 2 Lady Madhavan Nair Rd., Madras 34, India.*

MATHIAS, Émile-Ovide-Joseph, French physicist; b. Paris, Aug. 15, 1861; prof. physics, U. Toulouse (France), U. Clermont-Ferrand (France); dir. geophys. obs. on Puy-de-Dôme; mem. French Acad. Scis. Author: Sur la chaleur de vaporisation des gaz liquidéfiè, 1890; le Poids moléculaire des liquides, 1904; le Point critique des corps purs, 1904. Determined (with Cailleteto) latent heat of vaporization and specific heat of liquids and gases near the critical point, 1890-94; studied molecular weights of liquids. Died Clermont-Ferrand, Mar. 7, 1942.

MATHIAS, Mildred Esther, Am. botanist; b. Sappington, Mo., Sept. 19, 1906; d. John Oliver and Julia Hannah (Fawcett) Mathias; A.B., Washington U., St. Louis, 1926, M.S., 1927, Ph.D., 1929; m. Gerald L. Hassler, Aug. 30, 1930; children—Frances Jane (Mrs. Kenneth C. Hill), John Mathias, Julia Ellen (Mrs. Michael Taylor), James Barker. Research asst. Mo. Bot. Garden, 1929-30; research asso. N.Y. Bot. Garden, 1932-36, U. Cal. at Berkeley, 1937-42; faculty U. Cal. at Los Angeles, 1947—, prof. botany, 1962—, dir. Bot. Garden, 1956—. Mem. adv. com. Hunt Bot. Library, 1961—; adv. council Santa Barbara Bot. Garden, 1951—, Saratoga Hort. Found., 1962—, So. Cal. Hort, Inst., 1959—; adv. com. Huntington Bot. Gardens, 1964—. Bd. dirs. Orgn. for Tropical Studies; trustee Cal. Arboretum Found. Recipient merit award Cal. Conservation Council, 1962; Woman of Sci. award U. Cal. at Los Angeles Med. Aux., 1963; Nature Conservancy Nat. award, 1964; Woman of Achievement award Los Angeles Times, 1964; Washington U. Alumni Assn. award, 1966. Mem. Am. Soc. Plant Taxonomists, Western Soc. Naturalists, Internat. Assn. for Plant Taxonomy, Internat. Soc. for Hort. Sci., Nature Conservancy (bd. govs.), A.A.A.S., Am. Inst. Biol. Scis., Am. Soc. Naturalists, Bot. Soc. Am. Research and numerous publs. on taxonomy of flowering plant family Umbelliferae particularly for Western hemisphere, treatments of this plant family for N.Am. flora, local floras, floras of S.Am. Home: 446 S. Bentley Av., Los Angeles 90049.*

MATHIASSEN, Therkel, Danish archeologist; b. Faurbo, Sept. 5, 1892; s. Mathias Jens and Nicoline (Nielsen) M.; ed. U. Copenhagen (Denmark); M.A., Ph.D.; m. Asta Gammeltoft, May 7, 1919; children—Inger, Agneta, Kirsten, Henning. Mem. Knud Rasmussen expdn., N.Am., 1921-23; archeol. research in Greenland, 1929-34; became cons. dept. prehistory Nat. Mus. Copenhagen, 1933, chief cons., 1941. Decorated Danebrog, French Legion Honor; recipient Hans Egede medal Danish Geography Soc. Mem. Royal Soc. Nordic Antiquities (sec.), Soc. for Greenland (pres.). Research and publs. on Danish and Eskimo archeology. Home: Grönnemose Allé 101, Söborg, Denmark. Office: Vestervoldgade 115 4, Copenhagen, Denmark.

MATHIEU, Claude Louis, French astronomer; b. Macon, France, Nov. 25, 1783; entered L'École Polytechnique, Paris, 1803, L'École des Ponts, Chaussées, 1805; m. sister of Francois Arago, 1834. Worked briefly as engr.; became astronomer Paris Obs., also Bur. of Longitudes (titular mem. from 1862, later pres.); prof. astronomy Coll. de France; prof. analysis L'École Polytechnique, from 1829, later examiner; dep. from Macon, Chamber of Deputies, 1834-48. Recipient Lalande astron. prize, 1809, 16. Mem. French Acad. Scis., 1817. Contbr. articles to La Connaissance des Temps; also editor population statistics L' Annuaire du Bureau des Longitudes; published l'Histoire de l'astronomie au XVIIIe siècle (Delambre), 1827. Research concentrated on determining distances of stars from earth. Died Paris, Mar. 5, 1875.

MATHOV, Enrique, Argentinian physician; b. Buenos Aires, Argentina, June 29, 1914; s. Jose and Ester (Spivakosky) M.; student U. Buenos Aires Sch. Med. Sci., 1932-39; m. Estrella Mazzolli, Dec. 21, 1940; 1 dau., Ada Silvia. Clin. physician Rawson Hosp., Buenos Aires, 1940-48; chief allergy services Lanus Polyclinic, 1956-60, chief allergy services Avellaneda Polyclinic, 1960-66, asso. prof. U. Buenos Aires Sch. Medicine, 1963—; dir. Centro Privado de Enfermedades Alergicas, Buenos Aires, 1950—, Revista Argentina de Algeria, 1954-65. Mem. Argentine Assn. Allergy and Immunology (pres. 1966—), Peruvian Allergy Acad. (hon. academician), Colombian Allergy Soc. (corr.). Author: Alergia a drogas, 1967; also articles. Modification of method of dosage of hipuric acid in urine; research on L.E. cells in blood of asthmatics, method of controlled gradual inducement for diagnosis of allergy to drugs, cause of rhinitis in patients taking drugs with reserpin, lack of action of histamine in Arthus phenomenon, lack of effect of triple vaccination in allergic children, respiratory allergy toward cold in cold storage plant workers. Home: 3431 Cordoba, Buenos Aires, Argentina.*

MATHUR, Ram Nath, Indian entomologist; b. Lahore, Pakistan, July 19, 1903; s. Kanhiya Lal and Mor Bibi; M.Sc., Govt. Coll. at Lahore, 1925; Ph.D., Panjab U., 1945; m. Jagmohini Devi, June 8, 1922; 8 children. With Forest Research Inst. and Colls., Dehra Dun, India, 1927-61, chief research officer, head div. forest protectuon, 1955-61, ret., 1961. Tchr., Indian Forest Coll., Indian Forest Ranger Coll. Fellow Royal Entomol. Soc. London; mem. Entomol. Soc. India (life), Current Sci. Assn. Research, numerous publs. on biology and ecology of forest insect pests, control of forest and timber pests, virus disease entomology, termitology. Home: 49 Subash Rd., Dehradun, India.*

MATIGNON, Camille, French chemist; b. Saint-Maurice, France, Jan. 3, 1867; sr. student École normale supérieure, 1886-89; passed compet. exam in phys. scis.; D.phys. scis., 1892. Became tutor Sch. Higher Studies, Paris, 1889; later became mem. staff lab. of M. Berthelot, Collège de France; became master lectures Faculty Scis., Lille, France, 1893, asst. prof., 1897; named lectr. Sorbonne, Paris, France, 1898; apptd. prof. mineral chemistry Collège de France, 1908. Mem. French Acad. Scis. Author: l'Électrométallurgie des fontes, 1906; Contribution à l'histoire de la synthèse de l'acide nitrique, 1913; Note sur l'évaporation des vinasses; also monographs on rare metals, preparation of anhydrous chlorides and catalysts of synthesis of ammomia. Research in thermochemistry. Died Paris, Mar. 18, 1934.

MATKOVICS, Béla, Hungarian chemist; b. Csongrád, Hungary, July 12, 1927; s. Béla and Mária (Kiss) M.; M.D., Med. U., 1951; Candidate Chem. Scis., U. Natural Scis., Szeged, Hungary, 1964, Ph.D., 1964; divorced; children—Éva, Ilona. Asst., Med. Chem. Inst., Med. U. Szeged, 1950-55; sr. asst., docent Organic Chem. Inst., JATE U., Szeged, 1955—. Mem. Am., Royal, Polish, Hungarian chem. socs., Am. Nuclear Soc. Research, numerous publs. on measurement of oxidation reduction potential of different microorganismus, influential factors of potential, 1950-55, catalase and perioxidase enzymes, stereochemistry of different steroid-oximes, metabolism of bile acids. Home: 3, Arpád, Szeged, Hungary.*

MATOLTSY, Alexander Gedeon, cell biologist; b. Kaposvar, Hungary, Feb. 27, 1920; s. Alexander and Vilma (Jamrich) M.; M.D., U. Budapest, 1944; m. Margit Nagy, Dec. 22, 1945. Came to U. S., 1949, naturalized, 1955. Mem. faculty U. Budapest, Hungary, 1943-45, Hungarian Biol. Research Inst. Tihany., 1946-47; research fellow Karolinska Inst., Stockholm, Sweden, 1947-49; faculty Inst. Muscle Research, Woods Hole, Mass., 1949, Harvard, 1949-56, Rockefeller Inst., N.Y.C., 1956-59; research prof. dermatology U. Miami (Fla.), 1959-61; research prof. anatomy and dermatology Boston U., 1961—. Mem. dermatology tng. grant com. NIH, 1965—; vis. prof. Australian Nat. U., Canberra, 1965. Mem. Am. Soc. Cell Biology, Internat. Soc. Cell Biology, Electron Microscope Soc. Am., Soc. Investigative Dermatology, Am. Acad. Dermatology, Sigma Xi. Research and publs. on structure and function of skin epidermis; isolated and characterized in detail various protein constituents of epidermis. Home: 9 Chilton St., Belmont, Mass. 02178. Office: 15 Stoughton St., Boston 02118.*

MATOSSI, Frank, German physicist; b. Frankfurt Main, Germany, Feb. 24, 1902; s. Lorenz and Lilli (Armbruster) M.; student U. Frankfurt, 1921-22, U. Göttingen (Germany), 1922-23; Dr.phil., U. Marburg (Germany), 1926; m. Ottilie Riechemeier, July 7, 1937. Asst. Physics Inst., U. Giessen (Germany), 1926-28; asst. Physics Inst., U. Breslau (Germany), 1928-35, lectr. 1935-39; asso. prof. U. Graz (Austria), 1940-45; physicist U. S. Naval Ordnance Lab., White Oak, Md., 1947-58; prof. phys. chemistry U. Freiburg (Germany), 1958—. Recipient Meritorious Civilian Service award USN, 1957. Fellow Washington Acad. Sci., Am. Phys. Soc.; mem. German, Swiss phys. socs., Soc. German Chemists. Author: Das Ultrarote Spektrum, 1930; (with C. Schaefer) Der Ramaneffekt, 1944; Elektrolumineszenz und Elektrophotolumineszenz, 1957; Gruppentheorie der Eigenschwingungen von Punktsystemen, 1961; Optik, 1966; also numerous articles. Research on infrared and Raman spectroscopy especially crystals, Rayleigh scattering in crystals, electroluminescence and related phenomena in ZnS semiconductors. Home: 72 Pfaffendobel, 7801 Buchenbach, Baden. Office: 38 Hebelstr., 7800 Freiburg, Baden, West Germany.*

MATRONE, Gennard, Am. biochemist; b. Batavia, N.Y., Jan. 22, 1914; s. Joseph and Margaret (Sogia) M.; B.S. in Dairy Chemistry, Cornell U., 1938, M.S. in Nutrition and Biochemistry, 1944; Ph.D., N.C. State Coll., 1950; m. Alma Cox, Nov. 14, 1939; children—Margaret, Kenneth. Chemist, U. S. Plant, Soil and Nutrition Lab., Ithaca, N.Y., 1941-52; faculty N.C. State U., Raleigh, 1952—, William Neal Reynolds distinguished prof., 1962—, acting head biochemistry dept., 1965-67, head, 1967—. Cons. animal industry, P.R., 1955, agr., Peru, 1962; mem. govt., NIH coms. Recipient Research award of merit Gamma Sigma Delta, N.C., 1963, Nutrition Research award Am. Feed Mfrs. Assn., 1964, Moorman Travel award, 1966. Fellow A.A.A.S.; mem. Am. Inst. Nutrition, Biometrics Soc., Am. Chem. Soc., Soc. for Exptl. Biology and Medicine, Am. Soc. Animal Sci., Am. Dairy Sci. Assn., Nat. Acad. Sci., Am. Soc. Biol. Chemists, N.Y. Acad. Scis., Sigma Xi, Gamma Sigma Delta, Phil Kappa Phi. Editorial bd. Jour. Nutrition, 1964. Research and publs. in biochem. field. Home: 1333 Duplin Rd., Raleigh, N.C. 27607.*

MATSON, Gustave Albin, Am. microbiologist; b. Rexburg, Ida., June 24, 1899; s. John Edward and Anna Mathilda (Kjelin) M.; A.B., U. Utah, 1927; M.A., U. Kans., 1929; Ph.D., Washington U., St. Louis, 1935; m. Vadna Evie Cude, Oct. 10, 1924; children—Jean Athalee (Mrs. Lee B. Nielsen), Lavon (Mrs. Richard W. Christensen), Gustave Albin. Faculty, U. Mont., 1930-32, 34-37; faculty U. Utah, 1937-49, research prof., 1964—; asst. prof. bacteriology U. Minn., 1950-62; dir. Mpls. War Meml. Bank, 1949-64. Fellow Internat. Soc. Hematology, Internat. Soc. Blood Transfusion; mem. Soc. Am. Bacteriologists, Am. Assn. Phys. Anthropologists, Am. Assn. Immunologists, Am. Assn. Blood Banks (pres. 1947-48, dir. 1948-50, John Elliott Meml. award 1961), A.A.A.S., Am. Acad. Microbiology, Am. Soc. Human Genetics, Sigma Xi, Phi Kappa Phi. Studies, publs. on blood groups among Am. Indians; determination of blood groups in mummies; new blood groups; antibiotics in plants. Home: 3169 Kaibab Way, Salt Lake City 84109.*

MATSUDA, Kojiro, Japanese physiologist; b. Osaka, Japan, Sept. 10, 1908; s. Takejiro and Mitsu (Nakayama) M.; M.D., U. Tokyo (Japan), 1932, Ph.D., 1941; m. Miye Otsuka, Feb. 3, 1937; children—Shinobu, Tasuku, Yasuko. Research asst. U. Tokyo, 1932-41, prof. physiology, chmn. dept., 1957—; research cons. physiology Naval Research Inst., 1941-42; faculty Tohoku U., Sendai, Japan, 1942-57, prof. applied physiology, 1944-57; vis. prof. Downstate Med. Center, State U. N.Y., 1966. Recipient Asahi Sci. award for research, 1957. Mem. Physiol. Soc. Japan (chmn. exec. com. 1957—), Japanese Soc. Med. Electronics and Human Engring. Author: Textbook of Physiology, 1958; also articles. Research on electrophysiology of heart, heart rate and rhythm and its regulation, physiology of blood circulation. Home: Oizumi-gakuen-cho, Merima-ku, Tokyo, Japan.*

MATSUDA, Ryuichi, biologist; b. Kagoshima City, Japan, July 8, 1920; s. Takemaro and Mise (Hirayama) T.; B.A. in Agronomy, Coll. Agr., Kyushu U., 1950; D.Sc. in Agronomy, 1961; Ph.D., in Biology, Stanford, 1956. Asst. dept. entomology U. Kan., Lawrence, 1955-56, research asso., 1956-63; research asso. Mus. Zoology, U. Mich., Ann Arbor, 1963-68; research scientist Entomology Research Inst., Ottawa, Ont., Can., 1968—. Mem. Entomol. Soc. Japan, Entomol. Soc. Am., Soc. for Study Evolution, Soc. Systematic Zoology. Author: (with R. L. Usingen) Classification Aradidae, 1959; Evolution and Classification of the Gerridae, 1960; Morphology and Evolution of the Insect Head, 1965; also numerous articles. Research on Hemiptera, Insecta, evolution of insect head; postulated theory regarding evolutionary trends of relative growth in animals; described new subfamily, genera and species of Hemiptera. Home: 2045 Carling Av. Office: Entomology Research Inst., Central Exptl. Farm., Ottawa, Ont., Can.*

MATSUMOTO, Junji, Japanese neurophysiologist; b. Ise City, Japan, Mar. 30, 1917; s. Sohichi and Tane (Ikeda) M.; M.D., Osaka U., 1941; m. Hiroko Inove, Apr. 6, 1941. With Med. Sch., Osaka (Japan) U., 1941—, asso. prof. Lab. Neurophysiology, 1961—; staff Lab. Exptl. Pathology, Lyon (France) U., 1963-64; prof. 2d dept. physiology Med. Sch., Tokushima (Japan) U., 1964—. Mem. Physiol. Soc., Electroencephalographic Soc., Assn. Psychophysiol. Study Sleep, Soc. Conditioned Reflex Japan. Research and publs. on factors in induction paradoxical sleep; electroencephalographic study of conditioned reflex; observations on phenol induced convulsions. Home: 4-146, Nakashimada, Tokushima, Japan.*

MATSUMOTO, Seiichi, Japanese obstetrician, gynecologist; b. Kamakura, Japan, Nov. 8, 1916; s. Kiyoharu and Teruko (Mochizuki) M.; grad. Tokyo U. Sch. Medicine, 1941, Dr.Med.Sci., 1947; m. Sadako Okochi, Nov. 8, 1948; children—Kiyoshi, Rumiko. Faculty, Showa Med. Coll., 1945-57, lectr. obstetrics and gynecology, 1950-57; lectr. Tokyo (Japan) U. Sch. Medicine, 1957-58; prof., chmn. obstetrics and gynecology Gunma U. Sch. Medicine, Maebashi, Japan, 1958—; chief dept. maternal health Aiiku Inst., 1950-54; chief dept. obstetrics and gynecology Kanto Teishin Hosp., 1954-58. Dir. Sch. Midwives, Gunma U. Recipient award for paper Japaenese Soc. Obstetrics and Gynecology, 1964. Mem. Gunma Obstet. and Gynecol. Soc. (chmn.), Gunma Soc. Maternal Health (chmn.), Internat. Fedn. Gynecology and Obstetrics, Internat. Fertility Soc., Internat. Planned Parenthood Fedn. Author: (with Igakushoin) Menstrual Disorders, 1956; (with Igakuno sekaisha) Menstruation and Its Disorders, 1962; Physiology of Menstruation, 1964; Maternal Care, 1966; also numerous articles. Research on classification, treatment and pathophysiology of menstrual disorders, statis. study of basal body temperature records and contraceptive methods using oral pills and intrauterine devices; maternal care. Home: 3-11-4 Showamachi, Maebashi, Gunma, Japan.*

MATSUMOTO, Shin-ichi, Japanese pathologist, physician; b. Wakamatsu, Japan, 1884; grad. Kyoto U., 1909; M.D.; Hon. Dr., Athens U.; m. Masako Yamamoto. Became asst. prof. Kyoto U., 1913, prof., 1919-45, chief med. dept., 1938-40, emeritus prof., 1945—; pres. Osaka Med. U. Recipient 2d Hideyo Noguchi Med award, Tokyo, 1958. Mem. Acad. Japan, Japan Cutaneous Acad. (hon. pres.), German Natural Sci. Acad., Cutaneous Sci. Acad. (sr.). Research and publs. on pathology and immunity of syphilis and franbeshia.

MATSUMURA, Seiji, Japanese geneticist; b. Kyoto, Japan, Feb. 28, 1909; s. Seisuke and Yae (Soza) M.; grad. Faculty Agr., Kyoto U., 1935, Dr.Agr., 1945; m. Miyoko Imai, Jan. 8, 1939; children—Eiko, Shosaku, Shinsaku. Prof. dept. textile fibres Kyoto Tech. U. (formerly Kyoto Coll. for Textile Fibres), 1945-49; subhead 1st dept. Nat. Inst. Genetics, Misima, Japan, 1949-55, head dept. induced mutation, 1955—; staff Kihara Inst. for Biol. Research, Yokohama, Japan, 1942—; faculty Nihon U., 1952-56, Yokohama Municipal U., 1957-60. Mem. Genetics Soc. Japan (medal 1951), Japanese Soc. Breeding, Japanese Soc. Environmental Control in Biology, Bot. Soc. Japan, Japan Radiation Research Soc. Author: Cytogenetics and Breeding in Wheat, 1950; Introduction to Cytogenetics, 1951; also numerous articles. Editor: (with others) Environmental Control in Biology, 1962; (with Tazima) Radiation Genetics, 1964. Research on cytogenetics of wheat and related species, triploid sugar beets, radiation genetics, environmental control in biology especially growth cabinet, phytotron. Home: 151-2 Yata. Office: 1,111 Yata, Misima, Sizuokaken, Japan. Died Feb. 19, 1967.*

MATSUNAGA, Fujio, Japanese physician; b. Kamikuya, Numata-shi, Japan, May 12, 1911; s. Tasuku and Teru (Shimada) M.; grad. Tohoku U. Sch. Medicine, 1936, Ph.D., 1942; m. Yoko Kasai, May 14, 1953; children—Muneo, Masato, Keiko, Manabu. Faculty, Tohoku U. Sch. Medicine, 1941-46, prof. internal medicine, 1944-46; prof., dir. internal medicine Aomori Med. Coll., 1946-48; prof. dir. med. dept. Hirosaki Med. Coll., 1948-51, prof., dir. 1st med. dept., 1951—, dir. Hirosaki U. Hosp., 1959-63. Mem. Internat. Congress Internal Medicine, Internat. Congress Hematology, Japanese Soc. Chemotherapy, numerous others. Author: Radiography of Large Intestine, 1958; numerous other books, articles. Research on gastroenterology; devised sigmoidocamera for taking pictures of mucous membrane of colon; diagnostics and treatment of gastrointestinal tract especially of large intestine disorders; discovered airpocket phenomenon; massive intermittent adminstrn. method in cancer chemotherapy; vitamin B2 and B1. Home: 203 Kikyono, Hirosaki, Japan.*

MATSUNAGA, Yoshisuke, Japanese mathematician; b. 1693. Rearranged teachings of Takakazu Seki in order, laid found. for study Seki sch. math.; a master of Seki sch.; instr. math. to Masaki Naito (lord of Nobeoka clan in Miyazaki Prefecture). Obtained results down to 49 decimal points of math. symbol pi (never before achieved by Japanese mathematician). Died 1744.

MATSUSAKA, Teruhisa, mathematician; b. Kyoto, Japan, May 4, 1926; s. Haruhisa and Reiko (Yagi) M.; B.S., Kyoto U., 1948, M.S., 1950, D.Sc., 1952; m. Mioko Yamamoto, Nov. 12, 1950; children—Yoshihisa, Mutsuo, Junko, Kazuko. Faculty, Ochanomizu U., Tokyo, 1952-54; research asso. U. Chgo., 1954-57; asso. prof. Northwestern U., 1957-60, prof., 1960-61; prof. Brandeis U., Waltham, Mass., 1961—. Guggenheim Meml. Found. fellow, 1959. Mem. Am., Japan math. socs., Am. Acad. Arts and Scis. Author: Theory of Q-varieties, 1965. Research and publs. primarily in field of geometric objects defined as set of common zeros of a finite set of polynomials. Home: 83 Agawam Rd., Newton, Mass. 02168. Office: Dept. Math., Brandeis U., Waltham, Mass. 02154.*

MATSUSHITA, Sadami, physicist; b. Ehime, Japan, Feb. 12, 1920; s. Kiyomi and Taka (Taniguchi) M.; M.Sc. Kyoto (Japan) U., 1944, Dr.Sc., 1951; m. Kyoko Nakajima, Mar. 23, 1946; children—Hiromi, Hidemi. Came to U. S., 1955, naturalized, 1962. Lectr. geophysics dept. Kyoto U., 1945-59; research staff physics dept. U. Coll., London, Eng., 1954-55; sr. research staff High Altitude Obs., Colo., 1955—. Guest cons. Nat. Bur. Standards, Boulder, 1955—; prof. adj. astro-geophysics dept. U. Colo., Boulder, 1962—. Recipient Tanakadate award Soc. Terrestrial Magnetism and Electricity, 1950; Scientist award Sci. Research Soc. Am., 1963. Fellow A.A.A.S.; mem. Am. Geophys. Union, Am. Meterol. Soc., Sigma Xi. Author: (with E. K. Smith) Ionospheric Sporadic E, 1962; (with W. H. Campbell) Physics of Geomagnetic Phenomena, 1966; also numerous articles; Editor: Jours. Geomagnetism and Geoelectricity, 1949-56. Research on solar terrestrial relationship and phys. behavior ionosphere, magnetosphere, and earth's magnetic field; discovered lunar effects on lower ionosphere, special ionospheric zone of sporadic E over magnetic equator. Home: 3984 Fuller Ct., Boulder, 80302. Office: High Altitude Obs., Boulder, Colo. 80302.*

MATTAUCH, Josef Heinrich Elizabeth, Austrian physicist; b. Mährisch-Ostrau, Austria, Nov. 21, 1895; s. Josef and Karoline (Wähner) M.; Ph.D., U. Vienna, 1920; hon. dr., Vienna Inst. Tech., 1965; D.Sc., U. Man. (Can.), Winnipeg, 1967; m. Esther-Maria Amren, Oct. 22, 1949. Asst., U. Vienna, 1920-39; Rockefeller Found. fellow Cal. Inst. Tech., Pasadena, 1926-27; with Max Planck (formerly Kaiser Wilhem) Inst. for Chemistry, Mainz, 1939-52, asst. dir., 1943-47, dir., 1947-65, dir. emeritus, 1965—; hon. prof. U. Mainz, 1952—. Recipient Haitinger prize Vienna Acad. Scis., 1935; Wilhelm Exner medal Austrian Indsl. Assn., 1958; Austrian Medal Honor for Science and Art, 1965. Corr. mem. Austrian Acad. Scis. Author: (with S. Fluegge) Kernphysikalische Tabellen, 1942; (with A. Flammersfeld) Isotopenbericht, 1949; numerous articles. Defined Mattauch's rule (regularities of binding energies dependent on even or odd nuclear numbers); invented Mattauch-Herzog mass spectograph; compilation of nuclear binding and reaction energies. Address: 23 Saarstrasse, Mainz, West Germany.*

MATTE-LAFAVEUR, Jean, French chemist; b. Montpellier, France, Feb. 1, 1660; 1 son, Jean; became instr. chemistry, Montpellier, 1675; prof. chemistry U. Paris (France); mem. French Acad. Scis., 1699. Author: Pratique de chimie, 1671. Discovered styptic water. Died Montpellier, Aug. 7, 1742.

MATTERSON, Lloyd Daniel, Am. animal scientist; b. Filer, Ida., Jan. 3, 1907; s. Adelbert and Mary

Ellen (Struppler) M.; B.S., Wash. State U., 1929, M.S., 1936; Ph.D., U. Minn., 1942; m. Helene Pease, Apr. 16, 1938; 1 son, Stephen Lloyd. Asst. prof. poultry husbandry U. Conn., Storrs, 1942-47, asso. prof., 1947-53, prof., 1953—. Chmn. exec. com. Inst. Nutrition and Food Sci., 1961—; exec. com. Animal Nutrition Research Council, 1964-65. Recipient Am. Feed Mfrs. Nutrition Council award, 1957, Distinguished Nutritionist award Distillers Feed Research Council, 1964. Mem. Am. Chem. Soc., Am. Inst. Nutrition, Poultry Sci. Assn., Sigma Xi, Gamma Sigma Delta, Alpha Gamma Rho. Research and numerous publs. on synthetic antioxidants in prevention of nutritional encephalomalacia in chick. Home: Moulton Rd., Storrs, Conn. 06268.*

MATTEUCCI, Carlo, Italian physicist, physician; b. Forli, Italy, June 21, 1811; doctorate, Bologna, Italy; student L'École polytechnique, Paris. Prof. physics, U. Bologna, 1832; joined faculty physics coll. in Ravenna, Italy, 1838; became prof. physics, Pisa, Italy, 1841; named commr. from Tuscany to Charles Albert, 1848; became dir. Tuscany Telegraph System, 1849; represented Tuscany at Turin, Italy, 1859; became insp. gen. Italian Telegraph System, 1860; mem. Italian Senate; named minister pub. instrn., 1862. Mem. French Acad. Scis. (corr.), 1857. Author: Lezioni di fisica, 2 vols., 1841; Lezioni sui fenomeni fisicochimici dei corpi viventi, 1844; Manuale di telegrafia elettrica, 1850; Cours spécial sur l'induction, le magnétisme de rotation, et le diamagnétisme, 1854. Research in chemistry, optics, geology, telegraphy, and politics; studied physiol. effects of electricity, polarization of electrodes, action of gas batteries, magnetic rotation of light, torsion effects on magnetism; proved there was a difference in electric potential between an injured nerve and its muscle, 1838; demonstrated passage of electric current from contracting muscle, 1843. Died Ardenza, Italy, June 25, 1868.

MATTHES, François Emile, geologist; b. Amsterdam, Holland, Mar. 16, 1874; s. Willem Ernst and Johanna Suzanna (van der Does de Bije) M.; came to U. S., 1891, naturalized, 1896; B.S., Mass. Inst. Tech., 1895; LL.D., U. Cal., 1947; m. Edith Lovell Coyle, June 7, 1911. With topographic br. Geol. Survey, 1896-1914, with geologic br., 1914-47; in charge of topographic surveys in Big Horn Mountains, Wyo., Glacier Nat. Park, Grand Canyon of Colo. River, Yosemite Valley, Mt. Rainier Nat. Park; geol. studies in Yosemite and Sequoia nat. parks, Central Sierra, Nevada, Cal., central Mississippi Valley. Fellow Geol. Soc. Am., A.A.A.S.; mem. Assn. Am. Geographers (pres. 1933), Washington Acad. Scis. (v.p. 1933), Am. Geophys. Union, Internat. Commn. Snow and Glaciers (sec.), Internat. Assn. Sci. Hydrology (asst. sec.), Geol. Soc. Washington (pres. 1932), Am. Soc. Chem. Engrs. (life), Brit. Glaciological Soc. Author: Glacial Sculpture of Bighorn Mountains, Wyo., 1900; Mt. Rainier and Its Glaciers, 1914; Geologic History of the Yosemite Valley, 1930; Geologic History of Mt. Whitney, 1937; various sci. reports and articles. Research on alpine glaciation, post-Pleistocene glaciation, on geomorphology and glaciology of Yosemite Valley region and central Sierra Nevada of Cal., also on topographic delineation; studied geomorphology of central Mississippi Valley. Died June 21, 1948.

MATTHES, Gerard Hendrik, civil and hydraulic engr.; b. Amsterdam, Holland, Mar. 16, 1874; s. Willem Ernst and Johanna (van der Does) M.; B.S. in Civil Engring., Mass. Inst. Tech., 1895; m. Mary M. Bewick, Mar. 3, 1904; 1 dau., Florence B. (Mrs. H. E. Stephens). Came to U. S., 1891, naturalized, 1896. Instrument man and draftsman, 1895-97; asst. hydrographer U. S. Geol. Survey, 1897-1902; engr. asst. supervising engr. U. S. Reclamation Service, 1902-07; designing engr., resident engr., and supt. constrn. Colo. Power Co., 1907-11; prin. engr. hydroelectric dept. Am. Water Works and Guaranty Co., on constrn. of power devel. in W.Va., 1911-13; div. engr. State Water Supply Commn. of Pa., in charge of flood problems, 1913-15; hydraulic engr. Miami Conservancy Dist. on flood control, 1915-20; U. S. asst. engr. War Dept., Chattanooga, charge survey of Tenn. River and tributaries which include 1st aerial photog. survey of rivers undertaken by war dept., 1920-23; con. engr., N.Y., specializing aerial surveys and hydro-elec. power projects, 1923-28; with War Dept. as sr. hydro-elec. engr. and later prin. engr. charge comprehensive studies relating to water power, flood control, and nav. improvements in southeastern states, 1929-32; prin. engr., head engr. and cons. to pres. Miss. River Commn. on flood control of lower Miss. River, 1932-45; also head engr. and dir. U. S. Waterways Expt. Sta., 1942-45; pvt. cons. practice specializing river, harbor and irrigation projects, N.Y.; cons. to sec. hydraulic resources, Mexico, 1948-50; cons. Asso. Nav. Cos., Colombia, S.Am., on improvement Magdalena River, 1952. Member astronomic expdn. to Sumatra sent by Mass. Inst. Tech., 1901; charge geol. investigation of alluvial valley of Miss. River for Miss. River Commn., 1941-45. Recipient citation and award for exceptional civilian service War Dept., 1944. Mem. Am. Soc. C.E. (Norman medalist 1949), Am. Geophys. Union, Engineers Club. Inventor tetrahedral block revetments for river banks; designer of topographic slide rule for use in planetable surveying. Author and illustrator: River Engineering, pub. in American Civil Engineering Practice, Vol. II, 1956; also articles and reports. Died Apr. 8, 1959.

MATTHEWS, Donald Rowe, Am. polit. scientist; b. Cin., Sept. 14, 1925; s. William P. and Janet (Williams) M.; student Kenyon Coll., 1943, Purdue U., 1944-45; A.B., Princeton, 1948, A.M., 1951, Ph.D., 1953; m. Margie Craig Richmond, June 28, 1947; children—Mary Rowe, Jonathan Wingate. Asst. instr. Princeton, 1949-51;, instr. Smith Coll., 1951-53, asst. prof., 1953-57; Falk lectr. U. N.C., 1957-58, asso. prof., 1958-63, prof. polit. sci., 1963—, research prof. Inst. for Research in Social Sci., 1963—, dir. polit. studies program, 1957—, dir. N.C. Center for Edn. in Politics, 1963—. Cons., U. S. Civil Rights Commn. Center for Advanced Study in Behavioral Scis. fellow, 1964-65. Mem. Am., So. polit. sci. assns., Phi Beta Kappa. Author: Social Background of Political Decision Makers, 1954; U. S. Senators and their World, 1960; Negroes and the New Southern Politics, 1966. Contbr. articles to sci. jours. Description and analysis of process by which polit. elites are recruited, behavior of legislators; variables affecting Negro polit. participation in contemporary Am. South. Home: N. Lakeshore Dr., Chapel Hill, N.C. 27514.*

MATTHEWS, James Lester, Am. biologist; b. Denton, Tex., July 3, 1926; s. James Carl and Rena Mae (Waggoner) M.; student N. Tex. State U., 1943-44; B.S., N. Tex. State U., 1948; M.S., 1949; Ph.D. in physiology, U. Ill., 1955; m. Betty Sue Sharp, June 2, 1950; children—Janet Lynn, Joan Leigh, Leslie Sue. Faculty, Baylor U. Dental Coll., 1955—, now prof. anatomy and physiology, chmn. dept. anatomy. Edn. cons., instruction liaison officer Army Grad. Tng. Program, 1960-65. Mem. Am. Physiol. Soc., Internat. Assn. for Dental Research, Electron Microscope Soc., Sigma Xi, Phi Delta Kappa. Contbr. articles to tech. jours. Research on effects of radiation on bones and teeth; circulation, oxygen, tension and hormone effects in bones and teeth; fate of intramembranous bone grafts, particularly mineral exchange and circulation changes. Home: 6944 Lakewood Blvd., Dallas 75214.*

MATTHEWS, Leonard Harrison, Brit. zoologist; b. Bristol, Eng., June 12, 1901; s. Harold Evan and Sarah (Harrison) M.; B.A., King's Coll., Cambridge, Eng., 1922, M.A., 1930, Sc.D., 1939; m. Dorothy Helène Harris, Nov. 7, 1924; children—Jean Dorothy (Mrs. Christopher Trewhella), John Michael. Sci. staff Discovery expdn., 1924-28; spl. lectr. zoology U. Bristol (Eng.), 1935-41, research fellow, 1945-51; sr. sci. officer Telecommunications Research Establishment, 1941-45; sci. dir. Zool. Soc. London (Eng.), 1951-66; sci. expdns. to S.Am., Arctic, E. Africa. Chmn., World hist. Sci. Periodicals, 1957-66. Fellow Royal Soc., 1954; mem. Brit. Assn. for Advancement Sci. (pres. sect. D 1959), Linnean Soc. London (past mem. council), Zool. Soc. London, Assn. Brit. Zoologists (sec. 1948-49), Ray Soc. (pres. 1966—), Mammal Soc. Brit. Isles (chmn. 1956-63), Brit. Acad. Forensic Scis. (pres. 1962). Author: South Georgia, 1929; Wandering Albatross, 1952; British Mammals, 1952; Sea Elephant, 1953; Beasts of the Field, 1954; Animals in Colour, 1959; (with M. Knight) Senses of Animals, 1962; also numerous articles. Research on reproductive cycles in many mammals, anatomy of basking shark, birds of S. Georgia, seals both Antarctic and Northern. Home: The Old Rectory, Stansfield, Sudbury, Suffolk, Eng.*

MATTHEWS, Samuel Arthur, Am. zoologist; b. Grand Lake Stream, Me., Aug. 4, 1902; s. Thomas Murray and Lottie (Hoar) M.; S.B., Boston U., 1923, A.M., 1924; M.A., Harvard, 1925, Ph.D., 1928; Sc.D., Williams Coll., 1964; m. Dorothy Jean Stuart Robertson, Aug. 16, 1927; 1 dau., Jean Stuart (Mrs. William Everett). Instr. biology Boston U., 1923-28; instr., tutor Harvard, 1927-28; instr. anatomy Med. Sch., U. Pa., Phila., 1928-32; asso. Marine Biol. Lab., Woods Hole, Mass., 1932-37, instr., 1932-40; asst. prof. biology Williams Coll., Williamstown, Mass., 1937-42, prof., 1943—, Samuel Fessendeu Clarke prof., 1957—, chmn. faculty, 1951-66. Mem. N.Y. Acad. Sci., Phi Beta Kappa, Sigma Xi. Research and publs. in pigmentation, pituitary gland and sex cycles; respiration of fishes; carbohydrate metabolism of amphibians. Home: Woodcock Rd., Williamstown, Mass. 01267.*

MATTHEY, Robert, Swiss cytologist; b. Avency, Switzerland, July 21, 1900; s. Robert Marc and Alice (Vannod) M.; Lic.ès sci., U. Lausanne (Switzerland), 1921, Dr. ès Sc., 1924; Dr.honoris causa, U. Geneva, U. Rennes; m. Suzanne Renaud, Oct. 10, 1925. Faculty, U. Lausanne, 1929—, prof., 1938—. Pres. numerous sci. socs. including Swiss Soc. Zoology, Swiss Genetic Soc., Vaudois Soc. Natural Sci. Author: Les chromosomes des vertébrés, 1948; also numerous articles. Research in cytology of arthropods, cytogenetics of reptiles and mammals. Home: 121 rue General Guilan, Lausanne, Switzerland.*

MATTHIAS, Bernd Teo, physicist; b. Frankfurt on Main, Germany, June 8, 1918; s. Ludwig and Marta (Lippman) M.; Ph.D. in Physics, Fed. Inst. Tech., 1943; m. Joan Trapp, Aug. 5, 1950. Came to U. S., 1947, naturalized, 1951. Sci. collaborator Fed. Inst. Tech., Zurich, Switzerland, 1942-47; staff div. indsl. coop. Mass. Inst. Tech., 1947-48; asst. prof. U. Chgo., 1949-51; mem. tech. staff Bell Telephone Labs., Inc., Murray Hill, N.J., 1948—; prof. physics U. Cal., San Diego, 1961—. Recipient award Research Corp., 1962; Wetherill medal Franklin Inst., 1963. Fellow A.A.A.S., Am. Phys. Soc.; mem. Nat. Acad. Scis., Am. Acad. Arts and Scis., Swiss Phys. Soc. Discoverer critical conditions for occurrence of ferroelectricity and superconductivity; conduction electron ferromagnets; established that ferroelectricity in dielec. transitions, and superconductivity in metals at low temperatures are gen. phenomena. Home: 7931 Princess St., La Jolla, Cal. 92037.*

MATTHIAS, Eckart, physicist; b. Quedlinburg, Germany, Sept. 30, 1932; s. Wilhelm F. and Marie (Huch) M.; student Friedrich-Schiller U. Jena, Germany, 1950-52; Vordiplom, Georg-August U. Göttingen, Germany, 1955; Diplom-Hauptprufung, U. Hamburg, Germany, 1959; Filosofie Licentiatexamen, Uppsala (Sweden) U., 1959, Filosofie Doktor, 1963, Docent in Physics, 1963; m. Barbro E. Jernberg, Nov. 29, 1958; children—Bjorn E., Katarina M. Came to U. S., 1963. Miller fellow Miller Inst. for Basic Research in Sci., U. Cal., Berkeley, 1963-65, research asso. Lawrence Radiation Lab., 1965-66, sr. staff mem., 1966—. Mem. Deutsche Physikalische Gesellschaft, Am. Phys. Soc. Editor: (with E. Karlsson and K. Siegbahn) Perturbed Angular Correlations, 1964. Contbns. to refinement of perturbed angular correlation method and its application to measuring nuclear moments and hyperfine interactions; participated in devel. of radiative detection methods for nuclear magnetic resonance in radioactive states. Home: 2213 Rose St., Berkeley, Cal. 94709.*

MATTHIESSEN, Augustus, Brit. chemist, physicist; b. London, Eng., Jan. 2, 1831; Ph.D., Giessen, Germany, 1852; student Heidelberg, Germany, 1853; studied under Hoffmann, London. Built lab. where he studied phys. properties of pure metals and alloys; lectr. chemistry St. Mary's Hosp., London, 1862-68; joined staff St. Bartholomew's Hosp., London, 1868; cons. chemist; apptd. examiner U. London, 1870. Fellow Royal Soc., 1861 (Royal medal; mem. council). Research and publs. on constitution of alloys and opium alkaloids; demonstrated tin, lead, zinc and cadmium behave differently in alloys than other metals; suggested small amounts of impurities cause metal with which they are alloyed to assume an allotropic form, that alloys must usually be considered a solidified solution. An editor Philos. Mag., 1869-70. Died Oct. 6, 1870.

MATTIG, Wolfgang, German astrophysicist; b. Brandenburg, Germany, Nov. 22, 1927; s. Max and Hedwig (Hilsky) M.; Diplom-Astronom, U. Berlin, 1952, dr. degree, 1957; m. Ingrid Schaack, Feb. 21, 1953; children—Claudia, Bettina. Asst., Astrophys. Obs., Potsdam, Germany, 1952-61; sci. collaborator Fraunhofer Inst., Freiburg, Germany, 1961—. Mem. Astronomische Gesellschaft, Internat. Astron. Union. Research, publs. on solar physics, especially structure and magnetic fields in sunspots and dynamic phenomena in solar surface, cosmology. Home: 63 Scherkhofenstr., Ihringen, Germany. Office: 6 Schöneckstr., Freiburg Br., Germany.*

MATTING, Alexander, German engr.; b. Berlin, Germany, Nov. 21, 1897; s. Paul and Bessie (White) M.; Diplom Hauptprüfung, Technische Hochscule Breslau (Germany), 1924; postgrad. U. London (Eng.), 1925; Dr.Ing., 1927; m. Irmgard Zschech, May 21, 1926; children—Harald, Bessie, Arnulf, Linde. Dir. Exptl. Inst. for Welding Tech., Wittenberge, Germany, 1930-35; faculty, Technische Hochschule Hannover (Germany), 1935—, prof., dir. Instituts für Werkstoff kunde, 1935—, rector, 1940-43; dir. Niedersächsischen Materialprüfamtes, 1957-60. Hon. mem. Instituto de la Soldadura, Centro Nacional de Investigaciones Metalúrgicas, Madrid, Deutscher Verband für Schweisstechnik Düsseldorf; mem. Verein Deutscher Ingenieure, Verein deutscher Eisenhüttenleute, Deutsche Gesellschaft für Metallkunde, Deutscher Verband für Schweisstechnik. Author: Anleitungsblätter für des Schweissen von Leichtmetallen, 1940; Das Schweissen der Leichtmetalle und seine Rundgebiete, 1959; also numerous articles. Fusion and pressure welding, metal and chalcedony spraying, heat cutting, destructive and non-destructive materials testing, metallurgy, adhering of metals together, plastics. Home: 7 Mainzer Strasse. Office: Im Moore 11, 3 Hannover, Germany.*

MATTINGLY, Richard Francis, Am. physician; b. Zanesville, O., Oct. 25, 1925; s. James J. and Frances K. (Shaw) M.; B.A., Ohio State U., 1949; M.D., Cornell U., 1953; m. Mary Elizabeth Kohlman, Sept. 10, 1948; children—Kevin Lewis, Kerry George, Kent Victor, Kathleen Frances, Keith Richard, Kelly James, Kristen David. Faculty, Johns Hopkins Hosp., 1958-61, dir. gynecology out patient dept., 1958-61; prof., chmn. dept. gynecology and obstetrics Sch. Medicine, Marquette U., Milw., 1961—; practice medicine specializing in obstetrics and gynecology, Balt., 1958-61, Milw., 1965—; dir. dept. gynecology and obstetrics Milwaukee County Hosp., 1961—. Cons. staff pvt. hosps., Milw.; med. adviser Catholic Social Welfare Bur., Milw., 1961—. Fellow Am. Coll. Obstetrics, Gynecology; mem. A.M.A., Soc. Pelvic Surgeons, Am. Soc. Cytology, Am. Soc. for Study Sterility, Am. Assn. U. Profs., Assn. Profs. Gynecology and Obstetrics, Central Assn. Obstetrics-Gynecology, Continental Gynecol. Soc., Med. Soc. Milwaukee

County, Wis. State Med. Soc., Wis. State Obstetrics-Gynecology Soc., Howard A. Kelley Soc., Milw. Gynecol. Soc. Co-author: (with R. W. Telinde) Operative Gynecology. Editorial bd. Obstetrics and Gynecology Med. Jour., 1964—; guest editor Symposium on Radical Pelvic Surgery, 1965. Research on gynecol. pathology, reproductive physiology and endocrinology, pathogenesis of pyelonephritis of pregnancy. Home: 3340 N. Windermere Ct., Milw. 53211. Office: 8700 W. Wisconsin Av., Milw. 53226.*

MATTIOLI, Pietro Andrea Gregorio, Italian physician, botanist; b. Siena, Italy, Mar. 23, 1501; s. Francesco and Lucrezio (Buoninsegni) M.; M.D, Padua, Italy. Invited to Prague, Czechoslovakia by Emperor Ferdinand; made aulic councillor; later apptd. 1st physician to Maximilian II. Author: Commentarii in sex libros Pedacii Dioscoridis, 1544; Papyrus, Hoc est Commentarius, 1572. Made bot. studies, genus Matthiola (stock-gilliflower) named in his honor; believed that halosanthos was flos salis and that sperm of whale had nothing in common with amber; ascribed origin of fossils found by Palissy to legendary and occult causes. Died Trento, Italy, 1577.

MATTU, Flavio, Italian pharm. chemist; b. Cagliari, Italy, Apr. 8, 1919; s. Francesco and Assuntina (Congiu) M.; grad. in chemistry, U. Cagliari, 1942, in pharmacy, 1950, Doc.L., 1958, Qualif.profession, 1961; Qualif.profession, U. Pisa (Italy), 1943; m. Regina Sala, Apr. 17, 1950; 1 son, Guido. Staff, Chem.-Pharm. Inst., U. Cagliari, 1941—, dir., 1961-65, docent, 1945—. Adviser galen industry, Cagliari, 1951—, Milan, 1960—; collaborator Smith, Kline & French Labs., Phila., 1962—. Mem. Profl. Assn. Chemists and Pharm. Chemists. Author books, articles. Research on pharm. organic substances, including synthesis of new compounds; 1st to use differential-thermal analysis in study of pharm. and organic materials; studied symmetrical and asymmetrical ethylenic cycloketones, anti-tubercular substances, structure and synthesis of alkaloidic substances and N-phenacylamine derivatives. Home: 18 Tommaseo St., Cagliari, Sardegna, Italia.*

MATTUCK, Arthur Paul, Am. mathematician; b. Bklyn., June 11, 1930; s. Jacob and Rae (Bolnik) M.; A.B., Swarthmore Coll., 1951; Ph.D, Princeton, 1954; m. Joan B. Berkowitz, Sept. 7, 1959; 1 dau., Rosemary. Research fellow Harvard, 1954-55; faculty Mass. Inst. Tech., 1955—, prof. math., 1965—. Mem. Am. Math. Soc., Math. Assn. Am., Sigma Xi. Research and publs. in algebraic geometry. Home: 35 Crowninshield Rd., Brookline, Mass. 02146. Office: Mass. Inst. Tech., Cambridge, Mass. 02139.*

MATUDA, Eizi, taxonomist; b. Nagasaki, Japan, Apr. 20, 1894; s. Hanziro and Otuna Matuda; student Taihoku U.; D.Sc., Tokyo U., 1960; m. Miduho Kaneko, Aug. 15, 1930; 1 son, Julio. Educator, botanist, Formosa, 1914-22; owner agrl. estate, Escuintla, Chiapas, Mexico, 1922—; founder, dir. Matuda Herbarium Inst., Escuintla, 1932—; prof., asso. curator Nat. U. Mexico, Mexico City, 1951—. Chief botanist Nat. Inst. Forest Research, 1954-59, Bot. Dept., 1955-65. Recipient diploma of honor Forest Soc. Mexico, 1953; Bot. Soc. Mexico, 1959; Congress Mexican Botany, 1963. Author: Mexican Aroids, 1954; Mexican Dioscoreae, 1953; Mexican Commelinae, 1955; also numerous articles. Collections and studies of Mexican plants, birds, reptiles. Home: 8 Madrono, Xotepingo, Mexico 21 D.F., Mexico.*

MATULIS, Yuozas Yuozasovich, Lithuanian chemist; b. Mar. 31, 1899; grad. Kaunas U.; D.Chem. Sci. Former sr. lab. asst., asst., lectr., head chair at univ.; dean Natural Sci. and Math. Faculty, Vilnius U., 1940—, prorector, 1944—. Decorated Order of Lenin. Mem. USSR (corr.), Lithuanian (pres. 1946—) acads. sci., Lithuanian Znanie Soc. (chmn.). Author: Colloidal Chemistry, 1947; (manual) Practical Physical Chemistry, 1948. Research and publs. on photochemistry, electrochemistry, kinetics of reactions in solutions. Address: Lithuanian Acad. Sci., Vilnius, Lithuanian SSR, USSR.

MATUMOTO, Minoru, Japanese virologist; b. Urawa, Saitama, Japan, Sept. 14, 1915; s. Yosaburo and Yoshi (Aoyama) M.; Dr. Med. Sci., U. Tokyo Sch. Medicine, 1949; m. Michiko Kubo, Nov. 18, 1949; children—Yaeko, Emiko. Asst. prof. Inst. for Infectious Diseases, U. Tokyo, 1939-48, asso. prof., 1948-57, prof., chmn. dept. virology, 1957—. Recipient Tech. Ministry award for devel. measles vaccine. Mem. Japan Virologist Soc. Research, numerous publs. on viral diseases of man and domestic animals, Japanese encephalitis, psittacosis, measles; devel. live attenuated measles-virus vaccine. Home: 7-25, 2-Chome, Kamiogi, Suginami-ku, Tokyo, Japan.*

MATZKE, Edwin Bernard, Am. botanist; b. New York City, Aug. 2, 1902; s. Conrad Joseph and Emilie (Frieling) M.; A.B., Columbia, 1924, Ph.D., 1930; student Kaiser Wilhelm Institut for Biology, Berlin-Dahlem and Univ. of Berlin (Cutting traveling fellow), 1928-29; unmarried. Asst. in botany, Columbia, 1924-28, instr. in botany, 1929-33, asst. prof., 1933-43, asso. prof., 1943-47, prof. of botany, 1947—. chairman of department of botany, 1958-66, chairman of dept. of biological scis., 1966-67, asst. to dean Columbia Coll., 1944-60; lectr. Fordham U., 1942-44; research investigator Marine Biology Lab., summers 1949-56; bd. mgrs.; exec. com. of bd. N.Y. Bot. Garden, 1958—. Fellow A.A.A.S., N.Y. Acad. Sciences. Member Botanical Soc. of Am., Am. Bryological Soc., Soc. for Study of Evolution, Torrey Botanical Club (treas., 1936, v.p., 1941, sec. 1942-44, pres. 1949, asso. editor 1950-59), Phi Beta Kappa, Sigma Xi. Contributor numerous articles on cellular structure of plants, plant growth, floral morphology, coloration in plants, in numerous scientific publications. Researcher on three dimensional shapes of cells, cell division, floral anatomy and pigmentation, developmental morphology of liverworts. Home: 3190 Perry Av., Bronx, N.Y. 10467.*

MAU, August, archaeologist; b. Kiel, Holstein, Oct. 15, 1840. Secondary sch. tchr., Glückstadt; settled in Rome, 1872; attached to German Archaeol. Inst., Rome. Author: Pompejanische Beiträge, 1879; Geschichte der dekorativen Wandmalerei in Pompeji, 1882; Pompeji in Leben und Kunst, 1900. Concerned with excavations at Pompeii, studied mural painting and other art of Pompeii. Died Rome, Italy, Mar. 6, 1909.

MAUCHART, Burkhard David, German anatomist; b. Marbach, Germany, 1696; s. Johann David Mauchart; student, Tübingen, Altdorf, Strasbourg, Paris, 1712-19; began med. practice, Tübingen, Germany, 1722; became prof. anatomy and surgery, Tübingen, 1726. Described lateral odontoid ligaments (check, or Mauchart's, ligaments), circa 1738. Died 1751.

MAUCHER, Albert, German mineralogist; b. Freiburg, Dec. 22, 1907; s. Wilhelm and Frieda (Spiess) M.; ed. Aix-la-Chapelle, Munich Tech. Coll.; Dr.engring.; m. Nadina Ferreri, Nov. 3, 1945. agrégé, asst. Munich and Berlin tech. colls.; mineralogist Ankara (Turkey) Inst.; prof. agrégé U. Göttingen (Germany); prof. U. Munich. Mem. Bavarian Acad. Sci. Author: Die Lagerstatten des Urans. Research in ore deposits, relationship between volcanism and ore genesis, geochemistry. Address: Luisenstrasse 37, Munich, West Germany.

MAUCHLY, John W(illiam), Am. physicist, research engr.; b. Cin., Aug. 30, 1907; s. Sebastian Jacob and Rachel Elizabeth (Scheidemantel) M.; student Johns Hopkins Sch. Engring., 1925-27, Ph.D. in Physics, 1932; m. Mary Augusta Walzl, Dec. 30, 1930 (dec. Sept. 1946); children—James, Sidney; m. 2d, Kathleen Rita McNulty, Feb. 7, 1948; children—Sara Elizabeth, Kathleen Ann, John William, Virginia, Eva. Research asst. Johns Hopkins, 1932-33; head dept. physics Ursinus Coll., Collegeville, Pa., 1933-41; asso. dept. terrestrial magnetism Carnegie Inst., Washington, summers 1936-38; instr. elec. engring. U. Pa., 1941-43, asst. prof. elec. engring., 1943-46; cons. Naval Ordnance Lab., White Oaks, Md., 1944-45; established (with J. P. Eckert Jr.) Electronic Control Co., for design and mfr. electronic digital computing equipment, comml. and sci. applications, 1946; pres. Eckert-Mauchly Computer Corp., 1947-51; dir. Remington Rand Weather Project, 1951-53, dir. systems studies Eckert-Mauchly div. Remington Rand, 1953, dir. UNIVAC Applications Research Center, Remington Road, UNIVAC div. Sperry Rand Corp., 1954-59; pres. Mauchly Assos., Inc., 1959-65, chmn. bd., 1959—; dir. Pyrometer Corp., Jonkers Corp., Data Systems Corp, Package Devices, Inc., Duovent, Inc.; vis. prof. system engring. Carnegie Inst. Tech., Pitts., 1959-60. Recipient Howard N. Potts medal (with J. P. Eckert, Jr.), Franklin Inst., 1949, John Scott award, 1961, Modern Pioneer award N.A.M., 1965. Fellow I.E.E.E., Am. Statis. Assn.; mem. Am. Phys. Soc., Franklin Inst., Am. Geophys. Union, Inst. Math. Statistics, Am. Meteorol. Soc., Assn. for Computing Machinery (v.p. 1947-48, pres. 1948-50), Soc. Indsl. and Applied Math. (trustee 1953-54, pres. 1955-56), Am. Astron. Soc., Phi Beta Kappa, Sigma Xi. Pioneered with J. P. Eckert Jr. in devel. large electronic integrator and computer; co-inventor (with J. P. Eckert Jr.) of Eniac (1st large electronic digital computer capable of modifying a stored program), 1945, Univac (1st comml. data processor using magnetic tape), 1951, also Edvac, Binac electronic computers; 1st statis. corroboration that moon affects precipitation, 1954; directed devel. of critical-path scheduling, 1957, resource allocation, 1960. Contbr. articles sci. jours., also to Industrial Engineering Handbook, 1963. Home: Little Linden Farm, 1230 Cedar Rd., Ambler, Pa. 19002. Office: Mauchly Assos., Inc., Fort Washington Indsl. Park, Fort Washington, Pa. 19034.*

MAUCHLY, Sebastian Jacob, Am. physicist; b. Swanton, O., July 9, 1878; s. John William and Mary Jane (Ziegler) M.; A.B., U. Cin., 1911, Ph.D., 1913; postgrad. U. Chgo., m. Rachel Elizabeth Scheidemantel, Dec. 27, 1905; children—John William, Helen Elizabeth. Prin. and instr. physics, Hartwell High Sch., Cin., 1905-11; Hanna research fellow, dept. physics U. Cin., 1911-13; head dept. physics Woodward High Sch., Cin., 1913-14; apptd. asso. physicist, dept. terrestrial magnetism Carnegie Instn. of Washington, 1914, asst. chief obs. div., 1917-18, chief of sect. terrestrial electricity, 1919—. Fellow Am. Phys. Soc., A.A.A.S. Co-author of Vol. V, Researches Dept. of Terrestrial Magnetism (Carnegie Instn., Washington), 1926. Research in earth currents and in elec. conduction in gases, including atmospheric electricity. Died Dec. 24, 1928.

MAUDSLAY, Henry, English engr.; b. Woolwich, Eng., Aug. 22, 1771; m. Sarah (dec. 1828); children include: Thomas Henry, Joseph. Joined arsenal where his father worked; employed by Bramah in manufacture of locks, 1789; founded his own engring. bus., 1798. Patented method for printing calico, 1805, (with B. Donkin) differential motion of raising weights, 1806; patented steam engine, 1807; devised (with Dickinson) method for purifying water aboard ship by blowing air through it, (with Field) method of regulating water supply to boilers at sea and method of preventing formation of brine in boilers; improved lathe; invented measuring machine which was accurate to 1/10,000 of an inch. Died Lambeth, Eng., Feb. 14, 1831.

MAUDUITH (or MANDUIT), John, Brit. mathematician, astronomer; flourished 1310; became fellow Merton Coll., circa 1305; joined faculty, Oxford, 1340; practice medicine. Author: Parvus tractatus editus a magistro Johanne Mauduth super quattuor tabulis mirabiliter inventis in civitate, 1310 (source of Wallingford's Quadripartitum). A founder of Western trigonometry.

MAUER, Paul Herbert, Am. immunologist; b. N.Y.C., June 29, 1923; s. Joseph H. and Clara (Vogel) M.; B.S., Coll. City N.Y., 1944; Ph.D., Columbia, 1950; m. Miriam E. Merdinger, June 27, 1948; children—Susan G., David M., Philip M. Research biochemist Gen. Foods Corp., 1944-46; instr. Coll. City N.Y., 1946-51; research asso. depts. medicine surgery, pediatrics Columbia, 1950-51; asst. research prof. U. Pitts. Sch. Medicine, 1951-54, asso. prof. immunochemistry, 1954-60; faculty Seton Hall Coll. Medicine, Jersey City, 1960—; prof. microbiology N.J. Coll. Medicine, Jersey City, 1962—. Fellow N.Y. Acad. Scis.; mem. Am. Chem. Soc., Biochem. Soc. (London), Harvey Soc., Am. Assn. Immunologists, A.A.A.S., Sigma Xi. Research, numerous publs. on chem. modification proteins, effect on structure and immunological properties, effect complement on antigen-antibody reactions, antigenicity of proposed plasma expanders, antigenicity of synthetic polymers of amino acids nature of immunogenicity. Home: 403 Elmwood Av., Maplewood, N.J. Office: 24 Baldwin Av., Jersey City 07304.*

MAUGUIN, Charles-(Victor), French crystallographer; b. Provins, France, Sept. 19, 1878; prof. mineralogy and crystallology Nancy (France) Faculty Scis., then at Paris, France; mem. French Acad. Scis. Author: Les amides bromées-sodées, 1910; la Structure des cristaux déterminée au moyen des rayons-X, 1924. Research on liquid crystals, combat gas, organic molecules, crystaline networks; used X-ray in determining crystaline formation. Died Mar. 1, 1937.

M'AULAY, Alexander, mathematician; b. Luton, Eng., Dec. 9, 1863; ed. Gonvill and Caius Coll., Cambridge (Eng.) U.; M.A.; m. I. M. Butler, 1895; 1 son, 2 daus. Tutor, lectr. math. and physics Ormond Coll., Melbourne, Australia, 1888-92; research prof. Tasmania U.; ret., 1929. Author: Utility of Quaternions in Physics; Octonions, a Development of Clifford's Bi-Quaternions; Five-figure Logarithms and Other Tables; also articles math. physics. Made (with Prof. Hogg) magnetic survey of Tasmania, 1900, 01. Died July 5, 1931.

MAUNDER, Annie Scott Dill Russell, Brit. astronomer; b. Strabane County Tyrone, Apr. 14, 1868; d. W. A. Russell; ed. Victoria Coll., Belfast, Ireland; Girton Coll., Cambridge; Pfeiffer student for research Girton Coll., 1897-98; m. E. Walter Maunder, 1895 (dec. 1928). Computer, Royal Obs., Greenwich, Eng., 1891-95, 15-20. Rep. astronomy Women's Internat. Congress, London, 1899; mem. eclipse expdns., Lapland, 1896, India, 1898, Algiers, 1900, Mauritius, 1901, Labrador, 1905. Mem. Brit. Astron. Assn. (editor jour. 1894-96, 1917-30, v.p. 1896-99, 1900-08, 42). Author: Catalogue of Recurrent Groups of Sunspots; Astronomical Allusions in the Sacred Books of the East; Origin of the Planetary Symbols. Research and publs. in astronomy; obtained longest coronal extension then photographed, India, 1898. Died Sept. 15, 1947.

MAUNDER, Edward Walter, Brit. astronomer; b. Apr. 12, 1851; s. George Maunder; ed. U. Coll. Sch., also King's Coll., London; m. Edith Hannah Bustin, 1875; m. 2d, Annie Scott Dill Russell, 1895; 3 sons, 2 daus. Supt. solar dept. Royal Obs., Greenwich, Eng., 1873-1913, 16-19; joined as asst. Royal Obs., Greenwich, Eng., 1873; mem. eclipse expdns. to W.I., 1886, Lapland, 1896, India, 1898, Algiers, 1900, Mauritius, 1901, Labrador, 1905. Fellow Royal Aston. Soc. (mem. council 1885-89, 91-92, sec., 1892-97, became v.p., 1897); mem. Brit. Astron. Assn. (founder 1890, pres., 1894-96), Victoria Inst. (sec. 1913-18). Author: Royal Observatory, Greenwich, its History and Work; Astronomy without a Telescope, 1902; The Astronomy of the Bible, 1908; Science of the Stars, 1912; Are the Planets Inhabited?, 1913; Sir W. Huggins, and Spectroscopic Astronomy, 1913; (with Annie Scott Dill Russel Maunder) The Heavens and their Story, 1908. Editor, The Obs., 1881-87; Jour. and Memoirs Brit. Astron. Assn., 1890-94, 1896-1900; Astron. Dept. Knowledge, 1895-1904; Indian Eclipse of 1898; Total Solar Eclipse of 1900. Died Mar. 21, 1928.

MAUNOIR, Jean Pierre, Swiss surgeon; b. Geneva, Switzerland, Oct. 30, 1768; studied under Desault, Paris, France; doctorate in surgery, 1792; prof. anatomy Acad. Geneva; mem. French Acad. Scis. Research on surgery for cystic tumours, ophthalmology, including construction of artificial pupil; described congenital lymphatic cyst of neck (cervical hydrocele, also Maunoir's hydrocele), 1825. Died Geneva, Jan. 16, 1861.

MAUPAS, Emile, French biologist; b. Vaudry, France, July 2, 1842; curator Algiers (Algeria) Nat. Library; mem. French Acad. Scis., 1901. Research on reprodn. of threadworms; 1st to analyze sexual cycle of ciliated infusoria. Died Algiers, Oct. 18, 1916.

MAUPERTUIS, Pierre Louis Moreau de, mathematician, astronomer; b. St. Malo, Ille-et-Vilaine, France, July 17, 1698. Musketeer, capt. French dragoons, 1718-23; became instr. math. French Acad. Scis. 1723; head expdn. sent by Louis XV to measure length of meridian degree, Lapland (results confirmed Newton's theory of flattening of globe at Poles), 1736-37; taken prisoner by Austrians at battle of Mollwitz, 1741. Fellow Royal Soc., 1728; mem. French Acad. Scis., 1728, Berlin Acad. Scis. (dir. 1742, pres. 1746-59). Author: Balistique arithmétique; Commentaires de la douzième section du premier livre des principes de Newton, 1732; Discours sur la figure des astres, 1732; Sur la figure de la terre, 1738; Discours sur la parallaxe de la lune, 1741; Lettre sur la comète de 1742, 1742; Astronomie nautique, 1745; Essai de cosmologie (gives mechanistic view of universe), 1750; Essai de philosphie morale, 1751; also numerous articles. Contributed to introduction and defence of Newtonian sci. and philosophy in France and Prussia; discovered form of principle of least action; combined Snell's law of refraction and Newton's law of propagation of light; in biology foreshadowed idea of mutation in his study of generation. Died Basel, Switzerland, July 27, 1759.

MAURAIN, Charles, French physicist; b. Orleans, France, Feb. 27, 1871; ed. Normal Superior Sch.; m. Jeanne Janoty, 1902. Successively became tuor College de France, 1894, prof. Lycée Lorient, 1897, lectr. Faculty Scis., Rennes, France, 1899, prof. U. Caen (France), 1905; became instr. Faculty Scis., Paris, 1910, titular prof., 1921-41, dean, 1926-41, hon. dean, 1941——; in charge orgn. became dir. Aerotech. Inst., Paris, 1910; dir. Inst. Global Physics, U. Paris. Decorated grand officer Legion Honor. Became mem. French Acad. Scis., 1930, v.p. astronomy sect., 1930, pres., 1944; mem. Bur. Longitudes. Author: Le magnétisme du fer, 1899; Les états de la matière, 1910; Répertoire du laboratoire francais, 1922; Physique du globe, 1923; (with Painlene, E. Borel) L'Aviation, 1923; (with Eble) Atlas du magnétisme terrestre; Royonnements solaire, atmosphérique et terrestre; L'Étude physique de la terre. Research on magnetism and terrestrial electricity, propagation of sound waves in atmosphere, seismology, meteorology, effect of air on airplanes in flight. Home: 79, rue du Faubourg-Saint-Jacques, Paris 14, France.

MAURER, Fred Dry, Am. veterinarian; born Moscow, Ida., May 4, 1909; s. Oran J. and Lavina (Stone) M.; B.S., U. Ida., 1934; D.V.M., Wash. State Coll., 1937; Ph.D., Cornell U., 1948; m. Irene L. Haworth, Aug. 25, 1935; children—Allen D., Linda J. Commd. lt. Vet. Corps U. S. Army, 1941, advanced through grades to col., 1959; research on equine diseases, Front Royal, Va., 1941-43; virus research War Disease Control Sta., Grosse Isle, Can., Africa, 1943-46, Walter Reed Army Inst. Research, Washington, 1946-47, 48-51; research on fgn. animal diseases, Africa, 1951-54; research in virology and pathology Armed Forces Inst. Pathology, Washington, 1954-61; dir. pathology div. Army Med. Research Lab., Ft. Knox, Ky., 1961-64; ret., 1964; asso. dean, distinguished prof. pathology Coll. Vet. Medicine, Tex. A. and M. U., College Station, 1964-——. Cons. U. S. Dept. Agr., 1951——, Morris Animal Found., 1963——; Ill. Zoonoses Center, 1963-66, Animal Health Com., NRC, 1965——. Decorated Legion of Merit; recipient A rating for research Surgeon Gen. U. S. Army, 1961. Mem. U. S. Livestock San. Assn. (chmn. fgn. animal disease com. 1965-66), Am. Vet. Med. Assn. (research council 1965——), Am. Coll. Vet. Pathologists (pres. 1964). Contbg. author: Diseases of Cattle, 1956; Diseases of Swine, 1964; Virus Diseases, 1965. Research in diagnosis and control of infectious diseases in animals with emphasis on virology and pathology of fgn. diseases of food producing animals; devel. of tng. aids; research adminstrn. Home: 2408 Morris Lane, Bryan, Tex. 77801. Office: Coll. Vet. Medicine, Tex. A. and M. U., College Station, Tex. 77843.*

MAURER, Georg, German surgeon; b. Munich, Germany, May 29, 1909; ed. U. Munich; MD. Became instr., Munich, 1941, asso. prof., 1948; chief physician Munich-Perlach Area Hosp., 1946-53; head physician surg. div. Munich Municipal Hosp., 1953-——, med. dir., 1959-——; municipal adviser, 1952-——. Mem. Munich Doctors Union, Bavarian Assn. Surgeons, Anglo-German Assn. Physicians, Profl. Fedn. German Surgeons, German Anesthesiology Soc., German Blood Transfusion Soc., German Surgery Soc., Munich Doctor's Assn., German Soc. for Med. Sci. on Accidents and Ins., Internat. Soc. Doctors Diseases of Blood and Tumors, Union Med. Dirs. Hosps. Germany, Internat. Cardiovascular Soc., Internat. Surgery Soc., Austrian Surgery and Traumatology Soc. Author: Schematliche Darstellung der Gefässe und Nerven; Wetter und Jahreszeit in der Chirurgie, 1938; Umbau, Dystrophie und Atrophie an der Gliedmassen, 1940; Das Gasödem und sein Behandlung, 1944; Zur operativen Behandlung und deren Erfolgsaussichten bei peripheren Nervenverletzungen, 1958. Research in gen. surgery, abdominal surgery, accident surgery, surgery of peripheral nerves and facial nerve. Home: Gabrielvon-Seidel-Strasse 32, Munich-Grunwald, Germany Office: Ismaninger 22, Munich, West Germany.

MAURER, Paul Herbert, Am. immunochemist; b. N.Y.C., June 29, 1923; s. Joseph and Clara (Vogel) M.; B.S., Coll. City N.Y., 1944; Ph.D., Columbia U., 1950; m. Miriam E. Merdinger, June 27, 1948; children—Susan Gail, David Mark, Philip Mitchell. Asso. prof. immunochemistry U. Pitts., 1954-60; asso. prof. microbiology Seton Hall Coll. Medicine, 1960-62; prof. microbiology N.J. Coll. Medicine, 1962-66; prof., chmn. dept. biochemistry Jefferson Med. Coll., Phila. 1966-——. Recipient Research Career award NIH, 1962-66, Chemistry medal Coll. City N.Y., 1944. Mem. Am. Chem. Soc., N.Y. Acad. Scis., Biochem. Soc. (London), Harvey Soc., Am. Assn. Immunologists, A.A.A.S., Sigma Xi. Contbr. numerous articles to profl. jours. Research in immunochemistry and immunology relating to chem. basis for immunogenicity, chem. composition of antigen as it relates to its property of inducing immune response in appropriate host, synthetic and natural plasma expanders. Home: 1001 Dell Lane, Wyncote, Pa. 19095. Office: Jefferson Med. Coll., 1025 Walnut St., Phila. 19107.*

MAURER, Robert Joseph, Am. physicist; b. Rochester, N.Y., Mar. 26, 1913; s. Armand Augustine and Louise (Ruebsam) M.; B.S., U. Rochester, 1934, Ph.D., 1939; m. Dorothy Whelan, Aug. 17, 1940; children—Christopher Armand, Kathryn Rebecca. Research asso. Mass. Inst. Tech., Cambridge, 1939-42; asst. prof. U. Pa., Phila., 1942; from asst. prof. to asso. prof. Carnegie Inst. Tech., Pitts., 1942-49; faculty U. Ill., Urbana, 1949-——, prof., 1951-——, dir. materials research lab., 1963-——; physicist Manhattan Project, 1944-45; head physics br. Office Naval Research, USN, 1948. Research on photoelectric effect and optical constants of metals, elec. conductivity of ionic crystals; diffusion of ions in ionic crystals, elec. properties of semiconductors. Home: 610 W. Delaware St., Urbana, Ill. 61801.*

MAURER, Werner, German biophysicist; b. Solingen, Germany, Mar. 22, 1906; s. Peter and Elisabeth (Becher) M.; doctorate Darmstadt U., 1933. Mem. staff Phys. Inst., U. Darmstadt, 1930-45; faculty Med. Clinik, U. Cologne, 1947-57, Inst. for Med. Isotope Research, 1957-65, Inst. for Med. Research, U. Würzburg, Germany, 1965-——. Research, numerous publs. on physics of atom, biophysics, autoradiography, radiation physics. Address: Inst. für Med. Strahlenkunde der Univ. Würzburg, Versbacher Landstr., Würzberg, Germany.*

MAURIAC, Charles, French physician; b. Saint-Aquilin, France, 1832; staff Hosp. Midi, 1868-97; recipient Gold medal internship Hosp. Paris, 1858. Author: Premières phases de la syphilis, 1883; Syphilis tertiare et syphilis héréditaire, 1890; Traitment de la syphilis, 1896. Research on diseases of circulatory system, venereal diseases. Died Pontours, France, 1905.

MAURICE, Baron Jean-Frédéric-Théodore, mathematician; b. Geneva, Switzerland, Oct. 13, 1775; s. Frédéric-Guillaume Maurice; became prof. math. Acad. Geneva, 1798; named examiner École polytechnique, 1801; mem. French Acad. Scis. Died Geneva, Apr. 16, 1851.

MAURICEAU, François, French surgeon; b. Paris, 1637; provost Coll. Surgery; 1st doctor lying-in sect. Hôtel-Dieu. Author: Traité des maladies des femmes grosses et de celles qui sont nouvellement accouchés, 1668; Observations sur la grossesse et l'accouchement des femmes et sur leurs maladies et celles des enfants nouveau-nés, 1695. Invented obstet. method named after him; described tubal pregnancy, strangulation by umbilical cord, fetal brow presentation, rupture of fetal membrane to induce labor, 1668; introduced terms pudendal, fossa navicularis. Died Paris, 1709.

MAURO, Jack Anthony, Am. optical physicist; b. Bklyn., Feb. 21, 1916; s. Francis and Lucia (Buono) M.; student N.Y. U., 1936-42; B.S. in Physics, certificate optics and optometry, Columbia, 1947, grad. student physiol. optics, 1947-48; D.Optometry, Phila. Optical Coll., 1951; m. Camille Dolores Montaperto, June 19, 1937; children—Richard F., George E., Barbara L. Optician-mgr. Equitable Optical Co., N.Y.C., 1934-43; chief instr. theoretical optics N.Y. Inst. Optics, also Applied Optics Lab., 1947-53; dir. engring. Saratoga (N.Y.) div. Espey Mfg. Co., 1950-55; cons. optics engr. Gen. Elec. Co., 1955-——; mem. Polaris missile design team, also design and devel. Atlas and Thor missiles; optical cons. on 1st 151 inch diameter quartz primary mirror for Kitts Peak Obs. telescope, 1965. Bd. dirs. Columbia; trustee N.Y. Inst. Optics. Recipient Letter of Achievement for Polaris, Navy Dept., 1960. Letter of Achievement for Thor and Atlas, USAF, 1959. Fellow Am. Acad. Optometry; mem. Am. Phys. Soc., Mohawk Assn. Scientists and Engrs., Soc. Photog. Instrumentation Engrs., Am. Inst. Physics, Optical Soc. Am., Omega Epsilon Phi. Editor, co-author: Optical Engineering Handbook, 1957, optical sect. Industrial Electronics Handbook, 1958. Patentee radiometric flaw detector, toroidal lens; designed free-fall and recoverable cine camera. 1958, which obtained 1st photographs of earth and booster separation from outer space; designed and built ultra-high speed drum camera for studying missile models in free-flight up to meteorite speeds, 1955; also optical alignment systems for interplanetary navigation. Home: 615 Keebler Rd., King of Prussia, Pa. Office: Gen. Electric Co., Spacecraft Dept., Valley Forge, Pa.*

MAUROLICO (or Maurolycus), Francesco (or Franciscus), Sicilian geometer, optician; b. Messina, Sicily, Sept. 16, 1494. Ordained priest Roman Catholic Ch., became abbot; tchr. math. Messina. Author: Cosmographia de forma, 1543; Gnomonica, 1553; Opuscula Mathematica, 1575; Arithmeticorum libri duo, 1575; Photismi de lumine et umbra, 1575; Problematica mechanica, pub., 1613. One of gt. geometers of 16th century; work on conic sects.; introduced secants into trigonometrical calculations; discovery of inference by math. induction has been ascribed to him; translated works of Theodosios and Menelaos into Latin, also treatise on sphere (Autolycus), 1558, Phaenomena (Euclid); pub. works on Appollonios, Archimedes; systematically studied prisms, spherical mirrors, human eye; anticipated many of Kepler's findings (his work, while done before Kepler's, was published later); observed spherical abberation; gave approximate value of refraction of light and explanation of rainbow; explained operation of camera obscura; gave 1st hint of combustion lines, also 1st clarification of near and far sightedness. Died Messina, July 21, 1575.

MAURY, Carlotta Joaquina, Am. paleontologist; b. Hastings-on-Hudson, N.Y., Jan. 6, 1874; d. Mytton and Virginia (Draper) M.; studied at Radcliffe Coll., Columbia, U. Paris; Ph.B., Cornell U., 1896, Ph.D., 1902, Sarah Berliner fellow, 1916. Asst. in dept. of paleontology Columbia, 1904-06, La. Geol. Survey, 1907-09; lectr. geology Barnard Coll. (Columbia), 1909-12; paleontologist Venezuelan Geol. Expdn., 1910-11; prof. geology and zoölogy Huguenot Coll. (U. Cape of Good Hope), S. Africa, 1912-15; organized Maury expdn. to Dominican Republic, 1916; cons. paleontologist Royal Dutch Shell Petroleum Co.; paleontologist Brazilian Govt.; specialist in Antillean, Venezuelan and Brazilian stratigraphy and fossil faunas. Fellow Geol. Soc. Am., A.A.A.S., Am. Geog. Soc. Author: Eocene of Trinidad, 1912; Dominican Type Sections and Fossils, 1917; Mollusca Gulf of Mexico, 1922; Miocene of Trinidad, 1925; Tertiary and Cretaceous of Brazil, 1925; Silurian of Santa Catharine, 1927; Silurian of Para, 1929; Puerto Rican, Dominican, and Soldado Stratigraphy and New Formational Names, 1929-31; Cretaceous of Parahyba do Norte, 1930; Bartonian and Ludian Upper Eocene in the Western Hemisphere, 1931; Cretaceous of Sergipe, Brazil; Triassic and Cretaceous of Northeastern Brazil, 1934; Lovenilampas, A New Echinoidean Genus from the Brazilian Cretaceous, 1934; New Genera and New Species of Fossil Terrestrial Mollusca from the States of Rio de Janeiro and Sao Paulo, Brazil, 1935. Died Jan. 3, 1938.

MAURY, Francis Fontaine, Am. surgeon; b. Danville, Ky., Aug. 9, 1840; grad. Center Coll., 1860; attended U. Va., 1860-61; M.D., Jefferson Med. Coll., 1862. Resident intern, Phila., 1862-63; asst. surgeon U. S. Army at South St. Gen. Hosp., Phila., 1863-65; clin. asst. to Prof. S. D. Gross, Jefferson Med. Coll., 1863-64, chief of surgeons, 1865; mem. Am. Dermatol. Assn., Path. Assn. Phila., Coll. of Physicians, Acad. of Natural Scis., Phila. County Med. Soc. Coeditor Photog. Rev. of Medicine and Surgery, 1870-72. Credited with 1st operation of gastrostomy in U. S., 1870; performed ligation of common cartoid and subclavian arteries for aortic aneurism; performed 1st resection of bronchial plexus to relieve pain in neuroma of skin of upper extremity, 1874; removed cystic goiters; used flap from perineum and scrotum in a spl. plastic operation for exstrophy of bladder; performed 1st amputation of hip joint in patient who survived in Am. Died Phila., June 4, 1879.

MAURY, Matthew Fontaine, Am. oceanographer; b. Fredericksburg, Va., Jan. 14, 1806; s. Richard and Diana (Minor) M.; LL.D. (hon.), Cambridge; m. Ann Hall Herndon, July 15, 1834, 5 daus., 3 sons. Became midshipman U. S. Navy, 1825, promoted lt., 1836; on surveying duty in harbors of S.E. U. S.; contbd. many articles on U. S. Navy to newspapers and periodicals, 1838-42; supt. depot of charts and instruments Navy Dept., 1842-55, 58-61, developed wind and current charts; promoted comdr., 1858; served as comdr. Confederate States Navy, 1861-65; spl. agt. of Confederacy in Eng.; imperial commr. of immigration to emperor of Mexico, 1865-66; lived in Eng., 1866-68; prof. meteorology Va. Mil. Inst., Lexington, 1868. Author: Wind and Current Chart of the North Atlantic, 1847; Abstract Log for the Use of American Navigators, 1848; The Physical Geography of the Sea, 1855; First Lessons in Geography, 1868. Founder of modern oceanography; developed internat. system of data notation; prepared chart of Atlantic bottom for cable route. Died Lexington, Va., Feb. 1, 1873.

MAUSER, Heinz, German chemist; b. Plochingen, Germany, Feb. 27, 1919; s. Ludwig Christian and Lisa (Röllig) M.; Diplom Chem., U. Tübingen, 1955, Dr.rer.-nat., 1956; m. Inge Bevernick, Aug. 9, 1952; children—Werner Wolfram, Rainer Roland Peter, Ute Beate, Gernot Gilbert. Research fellow U. Pa., 1956-57; asst. Phys. Chem. Inst., U. Tübingen, 1957-61, docent, 1961-65, unscheduled prof., 1965—. Mem. Soc. German Chemists, German Bunsen Soc. for Phys. Chemistry. Research, publs. on thermodynamics of optical antipodes, derivation of thermodynamic relations with jacobians, devel. pure calorimetric method for determination of free enthalpy of mixing, Becquerel effect, kinetics of photochem. reactions. Home: 27 Jahnstrasse, 741 Reutlingen, Germany. Office: 56 Wilhelmstrasse, 74 Tübingen, West Germany.*

MAUTHNER, Ludwig, ophthalmologist; b. Prague, Czechoslovakia, Apr. 13, 1840; M.D., U. Vienna, 1861; became privat-docent ophthalmology U. Vienna, 1864; became prof. U. Innsbruck (Austria), 1877; named asst. chief prof. eye dispensary U. Vienna, 1890, prof., 1894. Monument erected in his honor Arcaden of Vienna U. Contbr. articles on neuropathology and ophthalmology to med. jours. Described the sheath of an axon (axolemma, axilemma of Mauthner, or Mauthner's sheath), 1882. Died Oct. 20, 1894.

MAUTNER, Henry George, medicinal chemist; b. Prague, Czechoslovakia, Mar. 30, 1925; s. Frank Thomas and Maria (Neumann) M.; brought to U. S., 1941, naturalized, 1946; B.S., U. Cal., Los Angeles, 1946, Ph.D., 1955; M.S., U. So. Cal., 1950; m. Dorothea Johanna Baukemeyer, Nov. 21, 1967; children—Monica Ann, Matthew Erich. Teaching asst. U. Cal., 1951-55; research chemist Productol Co., Santa Fe Springs, Cal., 1950; with Yale, 1955—, prof. dept. pharmacology, 1967—, head sect. medicinal chemistry, 1962—. Mem. Am. Chem. Soc., Chem. Soc. (London), Am. Assn. Pharmacology, Am. Assn. Cancer Research, Biophys. Soc. Publs. on comparison of binding capacities and reactivities of analogous oxygen, sulfur and selenium compounds; molecular basis of nerve conduction; mechanisms of coenzyme action. Home: 111 Park St., New Haven.*

MAUTZ, Frederick Robert, Am. surgeon; b. nr. Marian, O., Oct. 26, 1907; s. John Frederick and Katherine (Peuser) M.; B.S. in Physics, Case Sch. Applied Sci. (now Case Inst. Tech.), 1929; M.D., Western Res. U., 1933; m. Bertha Specht, Sept. 29, 1934; children—Elsa Catherine (Mrs. John Quincy Adamson), John Frederick, Margaret Louise, Katherine Peuser. Practice medicine specializing in surgery, Cleve. 1938—, Chardon, O., 1959—; faculty Western Res. U., 1938-55, asst. clin. prof. surgery, 1950-55; dir. dept. surgery Marymount Hosp. Cleve., 1959-63, Geauga Community Hosp., Chardon, O., 1966—; cons. USPHS. Diplomate Am. Bd. Surgery, Am. Bd. Thoracic Surgery. Fellow A.C.S., Am. Coll. Chest Physicians; mem. Soc. Exptl. Biology and Medicine, Am. Physiol. Soc., Am. Assn. Thoracic and Cardiovascular Surgery, A.M.A., Sigma Xi, Alpha Omega Alpha, Tau Beta Pi, others. Asso. editor (Glasser) Medical Physics, 1943. Research, publs. on measurement of gamma ray dosage in R units, collateral circulation after coronary arterial occlusion, antiarrhythmic action of procaine hydrochloride, devel. of a mech. respirator, a hosp. resuscitation program. Home: 1204 Brentwood Rd., Cleveland Heights, O. 44121. Office: 13241 Ravenna Rd., Chardon, O. 44024; also 10515 Carnegie Av., Cleve. 44106.*

MAUVAIS, Félix-Victor, French astronomer; b. Maiche, France, Mar. 7, 1809; student Besançon, France; became astronomer Obs. Paris, 1836; became asst. astronomer Bur. Longitudes, 1843, lost his position, 1858. Mem. French Acad. Scis., 1843. Discovered 4 periodic comets; wrote unfinished work on determination of fundamental stars. Died Paris, Mar. 22, 1854.

MAUZEY, Armand Jean, Am. physician; b. Findlay, Ill., Apr. 18, 1907; s. George Washington and Catherine Elizabeth (Cloos) M.; A.B., Eureka Coll., 1928; B.S., M.D., U. Ill. Coll. Medicine, 1933; M.Sc., U. Pa. Grad. Sch. Medicine, 1940, D.Sc., 1948; m. Virginia Elizabeth Tompkins, May 25, 1945; children—Katherine, Elizabeth, John Michael, Suzanne Rachel. Practice gen. medicine, Shelbyville, Ill., 1933-36; faculty U. Ill. Coll. Medicine, Chgo., 1947—, asso. prof., 1961—; asso. attending obstetrician-gynecologists Research and Edn. Hosp., U. Ill., Chgo., 1948—. Recipient 25 Year Service award U. Ill. Coll. Medicine, Chgo., 1962. Diplomate Am. Bd. Obstetrics and Gynecology. Mem. Chgo. Gynecol. Soc., A.C.S., Am. Coll. Obstetricians and Gynecologists, Internat. Coll. Surgeons. Research and publs. on premature infants, physiology of ovary, pregnandiol excretion time in labor, infertility, ovarian endocrine tumors, pelvic surgery in elderly female. Home: 310 Prospect Av. Office: 135 S. Kenilworth St., Elmhurst, Ill. 60126.*

MAWSON, Colin Ashley, Canadian biochemist; b. Sheffield, Eng., Oct. 22, 1908; s. James and Florence (Anders) M.; B.Sc. with honors in Chemistry, Victoria U., Manchester, Eng., 1929, M.Sc., 1930, Ph.D. in Physiology, 1933; m. Mary Elinor Huntsman, June 6, 1935; 1 dau., Janet Beatrice. Staff, Royal Cancer Hosp., London, Eng., 1933-35, Westminster Hosp., London, 1935-37, Royal Berkshire Hosp., Reading, Eng., 1937-49; staff Atomic Energy Can. Ltd., Chalk River, Ont., Can., 1949—. Vis. prof. India Atomic Energy Establishment, Trombay, 1966; cons., IAEA, 1960—, Internat. Commn. on Radiol. Protection, 1956—. Mem. Physiol. Soc. (London), Biochem. Soc. (London), Canadian Physiol. Soc., Canadian Biochem. Soc., Health Physics Soc. Author: Management of Radioactive Wastes, 1965; also articles. Research on glycolysis in muscle and tumour tissues, Rous fowl sarcoma, metabolism of vitamins, clin. biochemistry, trace metal metabolism, radioactive waste mgmt. and environmental monitoring. Home: P.O. Box 15, Deep River, Ont., Can. Office: Atomic Energy Can. Ltd., Chalk River, Ont., Can.*

MAWSON, Sir Douglas, geologist; b. Bradford, Eng., May 5, 1882; s. R. E. and M. A. (Moore) M.; grad. B.Mining Engr., Sydney U., 1901; B.Sc., 1904; D.Sc., 1909; m. Paquita Delprat, 1914; 2 daus. Became demonstrator chemistry Sydney (Australia) U., 1902; geol. exploration New Hebrides Islands, 1903; became lectr. Adelaide U., 1905; prof. geology U. Adelaide (Australia); joined sci. staff Sir Ernest Shackleton's Antarctic Expdn., 1907, ascent of Mt. Erebus, magnetic pole journey, 1908; leader Australasian Antarctic Expdn., 1911-14, Brit., Australian and New Zealand Antarctic Expdn., 1929-31. Became mem. Commonwealth Govt.'s Antarctic Adv. Com., 1946. Recipient gold medals Am., Chgo., Paris geog. socs.; Bigsby medal Geol. Soc. London, 1919; Nachtigal Gold medal Gesellschaft für Erdkunde, Berlin, 1928; Von Mueller Meml. medal, 1930; Sir Joseph Verco medal S. Australian Royal Soc., 1931. Fellow Royal Soc., 1923; mem. Royal Geol. Soc. (Antarctic medal 1909, Founder's medal 1915). Author: The Home of the Blizzard; also articles. Editor, contbg. author: Scientific Reports of Australasian Antarctic. Expedition, 22 vols; editor: Reports of Banzar Expedition. Discoveries and exploration of uncharted areas of Antarctica. Died Oct. 14, 1958.

MAXCY, Kenneth Fuller, Am. epidemiologist; b. Saco, Me., July 27, 1889; s. Frederick Edward and Estelle Abbey (Gilpatric) M.; A.B., George Washington U., 1911; M.D., Johns Hopkins, 1915, D.P.H., 1921; m. Gertrud Helene McClellan, June 22, 1918; children—Kenneth Fuller, Frederic Reynolds, Selina Gilpatric. Resident house officer, Johns Hopkins Hosp., 1915-16, asst. resident pediatrician, 1916-17; asst. in medicine, Henry Ford Hosp., Detroit, Mich., 1917; fellow Johns Hopkins Sch. of Hygiene and Pub. Health, 1919-21; asst. surgeon, passed asst. surgeon and surgeon U. S. Pub. Health Service, 1921-29; prof. of bacteriology and preventive medicine, U. of Va., 1929-36; prof. of pub. health and preventive medicine, U. of Minn., 1936-37; prof. of bacteriology, Sch. of Hygiene and Pub. Health, Johns Hopkins, 1937-38, prof. epidemiology 1938-54, emeritus, 1954; cons. internat. Health Div., Rockefeller Found., 1937-40, 42-45, 48-52; cons. Secretary of War, Army Epidemiological Board, 1941-49; member Nat. Adv. Health Council, 1942-46; Consultant Research and Development Bd. of Nat. Mil. Establishment since 1946; mem. exec. com. Advisory Bd. on Health Services, American Red Cross, 1945-48; mem. Med. Advisory Com., National Foundation for Infantile Paralysis 1940-48; mem. bd. trustees Internat. Polio Congress; chmn. com. on research and standards, Am. Public Health Assn., 1939-46. Fellow Am. Public Health Assn. (Sedgwick Meml. medalist 1952); mem. Nat. Acad. Scis., American Society Epidemiologists, Assn. Am. Physicians, Pithotomy Club, Phi Beta Kappa Sigma Xi. Contbr. to knowledge of typhus, awarded U.S.A. Typhus Commn. medal, 1946. Editor: Papers of Wade Hampton Frost, 1941; Rosenaus' Preventive Medicine and Hygiene, 8th edit., 1956. Studies of infectious diseases and public health, epidemiology of malaria and typhus, use of quinine in treatment of malaria, and poliomyelitis; described a form of typhus endemic to southern U. S. Died Dec. 12, 1966.

MAXEY, Edward Ernest, Am. physician; b. Irvington, Ill., Aug. 21, 1867; s. W. C. and Sarah G. (Lane) M.; M.D., Coll. Physicians and Surgeons, Chgo., 1891; post-grad., Balt., Phila., N.Y., and 1 yr. in European clinics; m. Edna Horn, Dec. 19, 1900; children—Marie, Edward. Specialized in eye, ear, nose and throat. Fellow A.C.S., A.M.A. Credited with 1st description of Rocky Mountain spotted fever (under name spotted fever of Idaho), 1899. Died Aug. 31, 1934.

MAXEY, George Burke, Am. hydrogeologist; b. Bozeman, Mont., Apr. 4, 1917; s. David and Opal (Grigsby) M.; A.B., Mont. State U., 1939; M.S., Utah State U., 1941; A.M. Princeton, 1950, Ph.D., 1951; m. Jane Judith Clow, Sept. 11, 1941; children—Lee Burke, Nelle Jane (Mrs. Perry Olson), Dan Howell, Ann Dee, Permilla Jean. Geologist, U. S. Geol. Survey, Utah, Ky., Nev., Conn., 1941-48; faculty U. Conn. Storrs, 1949-54; geologist FOA, Libya, 1952-54; geologist, head div. ground water Ill. Geol. Survey, also prof. geology U. Ill., Urbana, 1955-62; prof. hydrology and geology U. Nev., Reno, 1962-67; head Center for Water Resources Research Nev., 1967—. Cons., AEC, 1957—, State of Nev., 1950-62. Fellow Geol. Soc. Am., A.A.A.S.; mem. Am. Assn. Petroleum Geologists, Soc. Econ. Geologists, Am. Geophys. Union. Contbr. articles to profl. lit. Studies of occurrence, motion and storage of ground water in Gt. Basin, central interior and E. coast U. S., N. Africa; Introduced concept of hydrostratigraphic units in geology. Home: 380 Morningside Dr., Reno 89507.*

MAXFIELD, James Robert, Jr., Am. radiologist; b. Grand Saline, Tex., Nov. 11, 1910; s. James Robert and Marie (Streeter) M.; A.B., M.D., Baylor U., 1935; m. Kathryn Morgan Jester, Aug. 21, 1936; children—James Robert III, Morgan Jester, Jordan Henly Work Streeter, Streeter Workman. Clin. asst. prof. radiology Southwestern Med. Sch., U. Tex., 1945—; cons. Los Alamos Sci. Labs., 1949-57; dir. Tex. Radiation and Tumor Inst., Dallas, 1946—; lectr. Blackford Meml. Cancer Lectures, Grayson County Med. Soc., 1957; Equen Meml. lectr. Ga. Med. Soc., 1958. Mem. regional adv. council nuclear energy So. Gov.'s Conf.; mem. Tex. Gov.'s Radiation Study Commn., AEC, also mem. adv. com. div. isotope devel.; chmn. Tex. Radiation Adv. Bd.; mem. permanent adv. com. So. Interstate Nuclear Bd. Mem. bd. trustees, past pres. Dallas Health and Sci. Museum. Fellow Internat. Coll. Surgeons; asso. fellow Am. Inst. Aeros. and Astronautics; mem. Am., So. med. assns., Am. Coll. Radiology, Radiol. Soc. N.Am., Inter-Am. Congress Radiology, Am. Radium Soc., Brit. Inst. Radiology, Tex., Dallas-Fort Worth, Rocky Mountain radiol. socs., Soc. Nuclear Medicine (pres. 1963-64) Southwestern Soc. Nuclear Medicine (pres. 1966-67), Flying Physicians Assn., Contbr. papers to tech. lit. Developed cliniscanner for scanning radioisotope deposition in the human body. Home: 5918 Loma Alto, Dallas 75205. Office: 2711 Oak Lawn Av., Dallas 75219.*

MAXFIELD, John Edward, Am. mathematician; b. Los Angeles, Mar. 17, 1927; s. Chauncey George and Rena Lucile (Cain) M.; B.S., Mass. Inst. Tech., 1947; M.S., U. Wis., 1949; Ph.D., U. Ore., 1951; m. Margaret Alice Waugh, Nov. 24, 1948; children—Frederick George (dec.), David Glen, Elaine Rebecca, Nancy, Daniel. Instr., U. Ore., 1950-51; mathematician U. S. Naval Ordnance Test Sta., China Lake, Cal., 1949-56, head computing br., 1956-57, head math. div., 1957-60; lectr. U. Cal. at Los Angeles, 1951-60; head prof. dept. math. U. Fla., 1960—. Mem. Am. Math. Soc., Math. Assn. Am., Soc. Indsl. and Applied Math., Sigma Xi. Research in the theory of numbers and numerical analysis. Home: Route 3, Box 188C, Gainesville, Fla.*

MAXFIELD, J(oseph) P(ease), Am. physicist, electronics engr.; b. San Francisco, Dec. 28, 1887; s. Joseph Elwyn and Harriett (Mansfield) M.; B.S. in Electro Chemistry, Mass. Inst. Tech., 1910; m. Milicent Arnold Harrison, June 20, 1914; children—Katherine Hayward (Mrs. Robert B. Gibney), Eleanor Taylor (Mrs. Charles W. Weis). Instr., Mass. Inst. Tech., 1910-14; research physicist Western Electric Co. (now Bell Telephone Labs.), 1914-26; mgr. engring. and research Victor Talk Machine Co., 1926-29; cons. and staff engr. Electrical Research Products, Inc. subsidiary Western Electric Co., 1929-36, dir. comml. engring., 1936-42, supt. comml. engring. Western Electric Co., 1936-42; prof., dir. div. physics war research Duke, Durham, N.C., 1942-46; supt. scientist, tech. dir. USN Electronics Lab., 1948-53; cons. Navy Air Missile Test Center, Point Mugu, Cal., 1956-59, acting chief scientist, 1959. Cons. on auditorium acoustics and sound rec., transmission and reprodn. Recipient Meritorious Civilian Service award U. S. Army, 1947, USN, 1953; John Potts Meml. award for achievement in audio engring., 1953. Fellow I.E.E.E., Am. Phys. Soc., Acoustical Soc. Am., A.A.A.S.; mem. Sigma Xi. Author: (with others) Recording Sound for Motion Pictures, 1931; (with Douglas Stanley) The Voice, Its Production and Reproduction, 1933. Research on elec. rec., transmission and reprodn. of sound for phonograph, radio broadcast and talking pictures, acoustic control of rec. conditions to match sound perspective with camera perspective in talking pictures, auditorium acoustics, electronics of sound detecting and measuring equipment. Patentee in field, including negative feedback through amplifiers, stereophonic sound (resulting in 1st comml. elec. rec. and talking pictures in U. S.). Home: 1120 E. Washington Av., Escondido, Cal. 92025.*

MAXFIELD, Myles, Am. biophysicist, physician; b. Portland, Me., Aug. 5, 1921; s. Horatio W. and Inez (Van Blarcom) M.; A.B. in Physics, Harvard, 1942, M.D., 1945; Ph.D. in Biophysics, Mass. Inst. Tech., 1950; m. Ruth Ford Brown, Apr. 2, 1945; children—Myles, Mac Rae, Nancy, Dana. Mem. faculty Mass. Inst. Tech., 1950-56; asso. scientist, asst. physician med. dept. Brookhaven Nat. Lab., Upton, N.Y., 1956-59; chief phys. scis. div. U. S. Army Biol. Lab., Ft. Detrick, Md., 1959-61; dir. biophysics program, prof. U. So. Cal., Los Angeles, 1961—. Mem. vis. com. med. dept. Mass. Inst. Tech., 1957-61; mem. biophysics and biophys. chemistry study sect. NIH, 1964—; cons. U. S. Army, 1962—. Mem. Am. Physiol. Soc., A.A.A.S., Electron Microscope Soc. Am., Sigma Xi. Editor: (with A. Callahan, L. J. Fogel) Biophysics and Cybernetic Systems, 1965. Studies and publs. in mechanism of devel. of cystic fibrosis as reflected by glycoproteins; proteins of nerve axoplasm; air pollution, smog effects on animals. Home: 722 S. Euclid Av., Pasadena, Cal.*

MAXIA, Carlo, Italian anthropologist; b. Rome, Italy, Feb. 18, 1907; s. Francesco and Teresa (Arangino) M.; M.D.; m. Maria Olivieri, Dec. 8, 1938; children—Teresa, Francesco. Dir. Anthropol. Inst., U. Cagliari (Italy); pres. Internat. Center for Sardinian Studies. Decorated Order of Merit, Order of Alphonse X the Wise. Mem. French Prehist. Soc., Italian Soc. 'di antropologia et etnografia, Royal Anthrop. Inst. Gt. Britain and Ireland, Colegie Anatomico Brasileiro. Research and numerous publs. in morphology, radiobiology, ethnology, anthropology. Home: via Calabria 15, Cagliari, Italy.

MAXIM, Hiram Percy, Am. inventor, mech. engr.; b. Bklyn., Sept. 2, 1869; s. Sir Hiram Stevens and Louisa Jane (Budden) M.; grad. Sch. Mechanic Arts of Mass. Inst. Tech., 1886; D.Sc., Colgate; m. Josephine Hamilton, Dec. 21, 1898; children—Hiram Hamilton, Percy (dau.). Elec. engr. Ft. Wayne Jenney Electric Co. (Ind.), 1886-87, W. S. Hill Electric Co., Boston, 1887-88, Thomson Elec. Welding Co., Lynn, Mass., 1888-90; supt. Am. Projectible Co., Lynn, 1890-95; chief engr. Electric Vehicle Co., Hartford, Conn., 1895-1907; pres. Maxim Silencer Co., Hartford. Pres. Am. Radio Relay League, Internat. Amateur Radio Union. Author: Life's Place in the Cosmos, 1933. Inventor elec. devices and ordnance, Columbia automobiles, the Maxim silencer for firearms and silencer for automobile engines. Died Feb. 17, 1936.

MAXIM, Sir Hiram Stevens, inventor; b. Sangerville, Me., Feb. 5, 1840; s. Isaac Wetson and Harriet M.; common sch. edn.; m. Louisa Jane Budden; 2d, Sarah Haynes; children—Florence and Hiram Percy M. Coach-bldg. apprentice 4 years, with various iron works; patentee numerous inventions in U. S.; went to Eng., 1881; Brit. subject, 1900; cons. engr. Vicker's Ltd. 1881; patented many elec. inventions, including incandescent Lamps, self-regulating current machines; engine governors; inventor Maxim gun, automatic system of firearms, which makes recoil of the gun serve as power for reloading, delayed action fuse, also other ordnance inventions, Cordite, a smokeless powder; devoted much time and invention to aerial nav. Author: My Life, 1915. Died Nov. 24, 1916.

MAXIM, Hudson, Am. inventor, mech. engr.; b. Orneville County, Me., Feb. 3, 1853; s. Isaac and Harriet Boston (Stevens) M.; ed. Me. Wesleyan Sem.; D.Sc., Heidelberg, 1913; LL.D., St. Peters, 1918; m. Jane Morrow; m. 2d, Lilian Durban, Mar. 26, 1896. In printing and subscription pub. business, Pittsfield, Mass., 1883; took up bus. of ordnance and explosives, 1888; 1st to make smokeless powder in U. S., 1st to submit samples to U. S. Govt. for trial; built (named for him) dynamite factory and smokeless powder mill, Maxim, N.J., 1890; sold smokeless powder inventions to E. I. du Pont de Nemours & Co., Wilmington, Del., 1897, became its cons. engr., expert in devel. dept. Pres. Aero. Soc. N.Y. Author: The Science of Poetry and the Philosophy of Language, 1910; Defenseless America, 1915; Dynamite Stories, 1916. Formulated hypothesis of so-called atoms, 1875; inventor process for printing newspapers in color, 1882; U. S. Govt. adopted his smokeless powder; sold to U. S. Govt. formula of Maximite, (1st high explosive to be fired through heavy armor plate), 1901; perfected Stabillite, a smokeless powder producing much better ballistic results than any other; inventor U. S. service detonating fuse for high explosive armor-piercing projectiles also of motorite, a new self-combustive material for driving automobile torpedoes, of process and apparatus for mfg. multi-perforated powder grains, of improvements in smokeless powder grains. Died May 6, 1927.

MAXIMILIAN, Constantin, Rumanian physician; b. Bucharest, Rumania, Aug. 1, 1928; s. Niculae and Maria (Chiritescu) M.; M.D., Med. Sch. Bucharest, 1953. Staff, Anthropol. Inst., Bucharest, 1953-59; staff Inst. Endocrinology, Bucharest, 1959—, chief genetics lab., 1960—. Recipient Acad.'s prize for medicine, 1962. Mem. Anthropol. Soc. Paris. Author: (with others) Sarata Monteoru, 1962; (with St. Milcu) Human Genetics, 1966; (with Milcu) Introduction to Physical Anthropology, 1966; also numerous articles. Research on devel. of anthropology and human genetics in Rumania, role of chromosomal aberrations in human pathology. Home: Bucaresti Str. Bujoreni nr.19 B, Bloc T 9 et.IV. ap. 54. Office: Inst. Endocrinology, Bd. Aviatori 34, Bucuresti, Rumania.*

MAXON, Joseph, English math. instrument-maker; b. Wakefield, Eng., 1627; student Hooke's pub. lectures, Gresham Coll., 1672; globe and instrument maker, London; apptd. hydrographer to Charles II, 1670. Fellow Royal Soc., 1678. Author various works on practical math. and math. instruments including: A Tutor to Astronomie and Geographie, 1659; also earliest math. dictionary, 1679. Died circa 1700.

MAXON, William Ralph, Am. botanist; b. Oneida, N.Y., Feb. 27, 1877; s. Samuel Albert and Sylvia Louisa (Stringer) M.; Ph.B., Syracuse U., 1898, Sc.D., 1922; m. Edith Hinckley Merrill, June 2, 1908; 1 dau., Mary, Aid, U. S. Nat. Mus., Washington, 1899, later asso. in botany; asst. curator, asso. curator and curator U. S. Nat. Herbarium, 1937-46. Fellow Am. A.A.A.S., Am. Acad. Arts and Scis.; mem. Bot. Soc. Washington, Biol. Soc. of Washington, Washington Acad. Scis., Sigma Xi. Contbr. series of papers entitled Studies of Tropical American Ferns to Contbns. U. S. National Herbarium; also Ferns as a Hobby Pteridophyta of Porto Rico, and numerous other articles and reports upon ferns. Specialist in study of Pteridophyta of tropical Am. Died Feb. 25, 1948.

MAXWELL, Arthur Eugene, Am. oceanographer; b. Maywood, Cal., Apr. 11, 1925; s. John H. and Nelle (Arnold) M.; B.S. in Physics with honors, N.M. State U., 1949; M.S. in Oceanography, U. Cal., 1952, Ph.D., 1959; m. Rita Louise Johnson, Nov. 23, 1963; children—Delle Rae, Eric Arnold, Lynn Marie, Brett Alan, Gregory James. Head oceanographer geophysics br. Office Naval Research, Washington, 1955-59, head geophysics br., 1959-65; asso. dir. Woods Hole (Mass.) Oceanographic Inst., 1965—. Mem. adv. bd. Nat. Oceanographic Data Center, Washington, 1961-65. Recipient Meritorious Pub. Service, U. S. Navy, 1966. Distinguished Pub. Service, 1963. Fellow Am. Geophys. Union (sec. oceanography sect.); mem. A.A.A.S., Marine Tech. Soc. (v.p. 1964-65). Contbr. articles to tech. jours. Research on heat flow through ocean floor. Home: 512 Woods Hole Rd. Office: Woods Hole Oceanographic Instn., Woods Hole, Mass. 02543.*

MAXWELL, James Clerk, Brit. physicist; b. Edinburgh, Scotland, Nov. 13, 1831; s. John Clerk and Frances (Cay) M.; ed. Edinburgh U., 1847-50; grad. Trinity Coll., Cambridge (Eng.) U., 1854, postgrad., 1854-56; m. Katherine Mary Dewar, 1858. Chair natural philosophy Marischal Coll., Aberdeen, Scotland, 1856-60; prof. physics and astronomy King's Coll., London, Eng., 1860-65; prof. exptl. physics Cambridge U., 1871-79, supervised bldg. Cavendish Lab., recipient Adams prize. Fellow Royal Soc. (Rumford medal 1860), 1861. Author: Theory of Heat, 1871; Treatise on Electricity and Magnetism, 1873; Matter and Motion, 1876. Editor: Electrical Researches (Henry Cavendish), 1879. Research, publs. on mech. method for tracing perfect oval curves, theory of rolling curves, equilibrium of elastic solids (found. for his later discovery of temporary double refraction produced in viscous liquids by shearing stress); demonstrated from theoretical considerations that rings of Saturn could not be solid and continuous, explained motion and permanence of Saturn's rings in terms of numerous small particles dynamically stable, 1857-59; investigated kinetic theory of gases (behavior of assemblages of colliding molecules, especially distbn. of velocities among molecules of a gas), 1860-79; studied color blindness and color perception; translated Faraday's theories on magnetic lines of force into math. notation, 1855-56; held that elec. and magnetic effects are means by which changes in an ether become known to us (ether being a material medium pervading all space); devised model having properties analogous to those necessarily possessed by electro-magnetic medium (ether), used it to provide mech. explanation in terms of stresses and motions in ether of inter-relations and effects of electricity and magnetism, 1861-62; applied dynamical equations in generalized Lagrangian form and showed that electromagnetic action travels through space in transverse waves, as does light, and with same velocity as light; hypothesized that light and electro-magnetism are of same ultimate nature, both being electro-magnetic radiations; expanded these investigations into Treatise on Electricity and Magnetism, 1873, in which he implicitly abandons necessity of using ether or any other mech. model and rests his theory on two pair of symmetrical equations (Maxwell's equations), 1st of which expresses continuous nature of electric and magnetic fields, 2d of which expresses principle by which changes in one field produce changes in other; equations are today foundation of physics of electricity, magnetism, light and radio waves (as extension of Maxwell's work, Hertz discovered radio waves, 1888). Died Cambridge, Nov. 5, 1879.

MAXWELL, Louis Rigby, Am. physicist; b. Waterloo, Ia., May 28, 1900; s. Ezra A. and Elsie (Rigby) M.; B.A., Cornell Coll., Ia., 1923; Ph.D., U. Minn., 1927; m. Marie Ellis, Sept. 12, 1925 (dec. Sept. 1952); children—Jean (Mrs. Gale Stringham), Louis Rigby; m. 2d, Marion O'Dell, Aug. 31, 1957. Nat. Research fellow, Bartol Research Found., Phila., 1927-29, Bartol Research fellow, 1929-31; physicist U. S. Dept. Agr., 1931-41; research physicist mine counter measures sect. Bur. Ships, USN, 1941-47; chief applied physics dept. U. S. Naval Ordnance Lab., 1947—. Recipient Distinguished Service award Bur. Ships, also Bur. Ordnance, 1930. Fellow Am. Phys. Soc.; mem. Philos. Soc. (pres. 1960). Research in excited atomic state, diffraction of gases by electrons. magnetic structures and magnetism of spinels. Home: 3506 Leland St., Chevy Chase, Md. Office: U. S. Naval Ordnance Lab., Silver Springs, Md.*

MAXWELL, Morton Harrison, Am., physician; b. N.Y.C., Feb. 23, 1924; s. William and Esther (Levy) M.; B.A., Columbia, 1943, M.D., 1946; m. Joanne Wilson, Dec. 25, 1960; children—Thomas, Susan, Jennifer, Lisa. Faculty, N.Y.U. Coll. Medicine, 1948-49, Cornell U. Sch. Medicine, 1950-52; clin. prof. medicine U. Cal. at Los Angeles Sch. Medicine, 1952—; practice medicine specializing in internal medicine, Los Angeles, 1950—; affiliated with U. Cal. at Los Angeles Med. Center, Cedars of Lebanon, Harbor Gen., Good Samaritan, St. John's hosps., VA Med. Center, Los Angeles; chief renal hypertension service Mt. Sinai Hosp. Mem. com. on urology NRC, 1965—; mem. nat. sci. adv. bd. Nat. Kidney Disease Found., 1959—; mem. project rev. com., high blood pressure research council Am. Hosp. Assn. Diplomate Am. Bd. Internal Medicine, Nat. Bd. Med. Examiners. Fellow A.C.P., Am.. Coll. Cardiology, Am. Physiol. Soc., Am. Soc. for Clin. Investigation, Harvey Soc., numerous other profl. assns., Sigma Xi. Author: textbooks, numerous articles. Maj. contbns. to pathophysiology, diagnosis, and treatment of kidney diseases, fluid and electrolyte disorders and high blood pressure; developed technique for treatment of uremia (peritoneal dialysis) which is widely used throughout world. Office: 10921 Wilshire Blvd., Los Angeles 90024.

MAY, Charles Henry, Am. ophthalmologist; b. Balt., Aug. 7, 1861; s. Henry and Henrietta M.; student Coll. City of N.Y., 1875-77; spl. studies chemistry, 1877-79; M.D., Coll. Phys. and Surg. (Columbia), 1883; study in eye abroad, 1887; m. Rosalie Allen, Nov. 7, 1893. Lectr. diseases of eye N.Y. Polyclinic, 1887-90; instr. ophthalmology and chief of clinic Columbia, 1890-1903; dir. and vis. surgeon, eye dept. Bellevue Hosp., 1915-25; ophthalmic surgeon, Mt. Sinai. Hosp.; cons. ophthalmic surgeon to Bellevue French and Monmouth hosps. Fellow A.C.S.; mem. Am. Acad. Ophthalmology, Am. Ophthalmol. Soc., N.Y. Acad. Medicine, A.M.A., N.Y. State and County med. socs. Author: Manual of Diseases of the Eye 1900, 17th edit., 1941; also 10 transls. into foreign langs. Contbr. to New Internat. Ency., and Reference Handbook of Med. Scis. Introduced improved system of illumination in electric ophthalmoscopes in 1914, now universally adopted. Died Dec. 7, 1943.

MAY, Everette Lee, Am. chemist; b. Timberville, Va., Aug. 1, 1914; s. Lee and Ida (Hinegardner) M.; B.A., Bridgewater Coll., 1935, Ph.D., U. Va., 1939; m. Lois Marie Lee, June 21, 1940 (dec. July 1964); children—Everette Lee, Philip Alan; m. 2d, Helen V. Sheehy, Dec. 1, 1965; 1 adopted dau., Victoria Lee. Staff, Nat. Oil Products Co., Harrison, N.J., 1939-41; with NIH, Bethesda, Md., 1941—, scientist dir., chief sec. on medicinal chemistry, 1960—. Mem. panel drugs liable to cause addiction WHO, 1959—; mem. com. on drug addiction and narcotics NRC, 1960—. Recipient Alumnus of Year award Bridgewater Coll., 1959. Mem. Am. Chem. Soc., Sigma Xi, Alpha Chi Sigma. Author: (with Nathan Eddy) Synthetic Analgesics, Part II, 1964. Research in organic and medicinal chemistry and analgesics; discovered phenazocine, a potent analgesic. Home: 6509 Westland Rd., Bethesda 20034. Office: 9000 Wisconsin Av., Bethesda, Md. 20014.*

MAY, Jack Truett, Am. forester, soil scientist; b. nr. McComb, Miss., Dec. 24, 1909; s. Hua Joe and Victoria E. (Boyd) M.; student Miss. Coll., 1926-27; B.S.F., La. State U., 1932; M.S.F., U. Ga., 1937; Ph.D., Mich. State U., 1957; m. Martha Louise Love, Nov. 18, 1942; children—Jack, James, Martha, Nadine, Marianne, Joe, Thomas; 1 foster dau., Mary Love Henderson (Mrs. James Helms). With Forest Service, U. S. Dept. Agr., Southwestern Region, 1929-49, forester, 1933-49; prof. silviculture Auburn U., 1949-58; prof. silviculture and forest soils U. Ga., Athens, 1958—. Cons. govt. agys., state forestry orgns., 1955-57, paper and pulp cos. in So. region, 1955—. Fellow A.A.A.S.; mem. Soc. Am. Foresters, Ga. Acad. Sci. (past sec.), Ga. Geol. Soc., Soc. Am. Foresters (past chmn. Southeastern sect.), Soil Sci. Soc. Am. Contbg. author: Direct Seeding in the South, 1959; Forest-Soil Relationships in North America, 1963; A Guide to Loblolly and Slash Pine Plantation Management in Southeastern U. S. A., 1965. Research, publs. on soil-site relationships, regeneration methods and mgmt. of young forest stands; devel. sci. techniques for mgmt. forest tree nurseries, including devel. of equipment and soil-plant nutrition relationships. Home: 180 Cloverhurst Terrace, Athens, Ga. 30601.*

MAY, Jacques Meyer, physician, ecologist; b. Paris, France, Jan. 27, 1896; s. Paul M. and Yvonne D. (Cardozo) M.; M.D., U. Paris, 1925, Agregation, 1936; m. Marie Anne Legrand, Jan. 26, 1946; children—James Patrick and Xavier Paul (twins). Came to U. S., 1940, naturalized, 1954. Prof. medicine Med. Coll. Hanoi (Indochina), 1935-40; dir. med. geography program Am. Geog. Soc., N.Y.C., 1948-60; chief med. edn. adviser USOM, Saigon, Vietnam, also prof. preventive medicine U. Saigon, 1960-63; acting chief pub. health adviser Latin Am. Bur., AID, Washington, 1964-65. Lectr., Columbia, 1955—. Served with French Army, 1915-18, 40-45. Decorated knight French Legion of Honor; recipient award Am. Assn. Geographers, 1953, Walter Reed Army Med. Center, 1958. Mem. Am. Pub. Health Assn., Am. Assn. Human Genetics. Author: Atlas of Disease, 1948, rev., 1955; Ecology of Human Disease, Vol. 1, 1958, Vols. 2-5, 1961, 63-65. Research and numerous publs. on ecology of human diseases, especially that of malnutrition. Home: 185 Sea View St., Chatham, Mass. 02633. Office: Nat. Insts. Health, Bethesda, Md.*

MAY, Kenneth Nathaniel, Am. food scientist; b. Livingston, La., Dec. 24, 1930; s. Robert William and Mary (Caraway) M.; B.S., La. State U., 1952, M.S., 1955; Ph.D., Purdue U., 1959; m. Patsy Jean Farr, Aug. 4, 1953; children—Sherry Alison, Nathan Elliott. Asst. state poultry supr. La. State Livestock San. Bd., 1954; research asso. poultry dept. La. State U.,

1954-56; faculty U. Ga., Athens, 1958—, asso. prof. food sci. and poultry sci., 1962—; cooperative agt. U. S. Dept. Agr., 1958-64. Recipient Ga. Egg Commn. award for meritorious service in research, 1964. Mem. Inst. Am. Poultry Industries (mem. research council 1962—, Research award 1963), Poultry Sci. Assn. (dir. 1965—), Soc. Heating, Refrigeration and Air Conditioning Engrs. (mem. com. for poultry and dairy products 1963—), Inst. Food Technologists, Poultry Sci. Assn., World Poultry Sci. Assn., Sigma Xi, Alpha Zeta, Gamma Sigma Delta (Ga. Young Scientist Research award 1963). Research and publs. on microbial chem., biochem., phys. and organoleptic factors affecting quality of poultry products. Home: 175 Lullwater Rd., Athens, Ga. 30601.*

MAY, Kenneth Ownsworth, Am. mathematician; b. Portland, Ore., July 8, 1915; s. Samuel Chester and Eleanor (Parkin) M.; A.B., U. Cal. at Berkeley, 1936, M.A., 1937, Ph.D., 1946; postgrad. U. London (Eng.), U. Paris (France); m. Miriam Dyer-Bennet, Nov. 2, 1963. Teaching asst. U. Cal. at Berkeley, 1936-37, 39-40; instr. U. Study Center, Florence, Italy, 1945; faculty math. Carleton Coll., Northfield, Minn., 1946-66, prof., 1952-66, chmn. dept., 1951-61; prof. math. U. Toronto (Can.) Coll. Edn., 1966—; lectr. U. Minn., 1947, 61, Okla. A. and M. U., 1948, U. Mich., 1956, U. Cal. at Berkeley, 1965. Dir. NSF Undergrad. Research Projects, 1956-60, 61—; research cons. Cowles Commn., 1946-48; cons. Ed Holden-Day, Inc., 1962—; regional cons. sci. teaching improvement program A.A.A.S., 1956-58; math. adv. com. Ednl. Testing Service, 1958-60; adv. com. Sch. Math. Testing Group, 1959-61. Fellow Inst. Current World Affairs, 1937-38, Fund for Advancement Edn., 1953-54; sci. faculty fellow NSF, 1962-64. Fellow A.A.A.S.; mem. Math. Assn. Am. (bd. govs. 1953-55, 64-66, mem. com. ednl. media, 1962-64, vis. lectr. 1962-66), Phi Beta Kappa, Sigma Xi, Pi Mu Epsilon (mem. nat. council 1963—). Author: Elementary Analysis, 1952; Elements of Modern Mathematics, 1959; Undergraduate Research in Mathematics, 1961; Programed Learning and Mathematical Education, 1965; also articles. Asso. editor Econometrica, 1951-53, Math. Monthly, 1963-65. Home: 16 Douglas Crescent, Toronto, Ont., Can.*

MAY, Paul S., Am. microbiologist; b. N.Y.C., July 12, 1931; s. Albert and Laura (Pinchuck) M.; B.S., Coll. City N.Y., 1951; M.S., Syracuse U., 1952; D.Sc., Phila. Coll. Pharmacy and Sci., 1955; m. Claire Jean Shereshefsky, Nov. 4, 1956; children—Melissa Naomi, Heather Jae. Lectr., Phila. Coll. Pharmacy, 1952-55; research microbiologist S. B. Penick & Co., Jersey City, 1955-58; asst. chief microbiologist Beth Israel Hosp., N.Y.C., 1958-62; sr. scientist Melpar, Inc., Falls Church, Va., 1962-64; asst. dir. N.Y.C. Dept. Health, Bur. Labs., 1964—. Lectr. epidemiology Columbia U. Sch. Pub. Health and Administrative Medicine. Mem. Am. Soc. Microbiologists, Am. Pub. Health Assn., Soc. for Indsl. Microbiology. Contbr. articles to tech. jours. Research on antibiotic fermentations, enteric microbiology, biochem. fuel cells, life support systems, disinfection. Home: 23 Fairview Lane, Orangeburg, N.Y. 10962.*

MAY, Raoul Michel, biologist; b. Hermosillo, Mexico, Sept. 1, 1900; s. Alfred and Henriette (Frey) M.; A.B., Stanford, 1921, A.M., 1923; Ph.D., Harvard, 1924; Dr.Sci., U. Paris, 1927, Dr. Medicine, 1939; m. Henriette Ducreux, May 28, 1927. Instr., N.Y. U., 1927-28; asst. Institut Pasteur, Paris, 1928-35; charge de recherches Centre National de la Recherche, Paris, 1935-37; faculty U. Paris, 1937—, prof. zoology, 1960—. Pres., gen. sec. various internat. biol. meetings, congresses. Decorated Officier de l' Instruction Publique, 1946, Officier de l'Ordre du Merite pour la Recherche et l'Invention, 1962; Chevalier de la Légion d'Honneur, 1965; Laureate, Academie des Sciences, Soc. Biologie, Académie de Médecine. Author: La transplantation animale, 1932; Les cellules embryonnaires, 1938; La formation du systeme nerveux, 1945; La vie des Tardigrades, 1948; La greffe, 1952; La culture des tissues, 1956; Culture et greffe des cellules nerveuses chez les Vertébres supérieurs, 1966; also numerous articles. Research in transplantation of embryonic tissues in adult hosts, cytology of dormant tissues and organisms, culture and transplantation of nerve cells, cytopathology, healing of tissues, history of biology. Home: 2 Rue Alexandre Parodi, Paris 10, France. Office: Laboratoire d'Anatomie Comparée et de Morphologie Expérimentale, Faculté des Sciences d'Orsay (Essonne) de l'Université de Paris, France.*

MAY, Robert McCredie, Australian physicist; b. Sydney, Australia, Jan. 8, 1936; s. Henry W. and Kathleen (McCredie) M.; B.Sc. with Class I Honors, Sydney U., 1956, Ph.D. (GMH Research fellow), 1959; m. Judith Feiner, Aug. 3, 1962; 1 dau., Naomi Felicity. Research fellow Harvard, 1960-61, 66, lectr., 1961, 66; faculty Sydney U., 1962—, reader, 1965—. Research and publs. on basic theory superconductivity and formal aspects of statis. mechanics, plasma physics especially large amplitude waves, calculation of various collision cross-sects., low energy nuclear physics. Home: 8 Karingal Rd., Lane Cove, Sydney, NSW, Australia.*

MAYALL, Nicholas Ulrich, Am. astronomer; b. Moline, Ill., May 9, 1906; s. Edwin Lyman and Olive (Ulrich) M.; A.B., U. Cal. at Berkeley, 1928, Ph.D., 1934; m. Kathleen Czarina Boxall, June 30, 1934; children—Pamela Ann, Bruce Ian. Astronomer, Lick Obs., Mt. Hamilton, Cal., 1934-42, 45-60; staff mem. Radiation lab. Mass. Inst. Tech., Cambridge, 1942-43, Kellogg Lab., Cal. Inst. Tech., Pasadena, 1943-45; dir. Kitt Peak Nat. Obs., Tucson, 1960—. Mem. Nat. Acad. Sci. (mem. space sci. bd. 1964—), Am. Philos. Soc., Am. Acad. Arts and Scis., Am., Royal astron. socs., Astron. Soc. Pacific, Internat. Astron. Union. Research and publs. on redshifts of galaxies, internal-motions of galaxies, velocities of globular clusters, velocities of galactic nebulae and supernovae remnants. Home: 5945 Mina Vista, Tucson 85718. Office: 950 N. Cherry Av., Tucson 85717.*

MAYAT, Valentin Sergeevich, Russian surgeon; b. 1903; grad. Med. Faculty, 2d Moscow U., 1925; D.Med. Sci., 1945. Asst. dept. anatomy 2d Moscow U., 1922-25; intern, asst., lectr. chair hosp. surgery 2d Moscow Med. Inst., 1925-53, prof., head chair hosp. surgery Med. Faculty, 1953—, former dean Med. Faculty. Chmn. surgery problems commn., learned council RSFSR Ministry Health. Decorated Order of Lenin. Mem. All-Union Cardiological Soc. (bd. mem.). Mem. editorial bd. Surgery. Research and numerous publs. on abdominal and cardiovascular surgery, orthopedics, traumatology, field surgery. Address: 2d Moscow Med. Inst., M. Pirogovskaya ulitsa 1, Moscow, USSR.*

MAYBACH, Wilhelm, German engr.; b. Heilbronn, Germany, Feb. 9, 1846. With G. Daimler established workshop at Bad Cannstatt, 1883; technical dir., Daimler Motor Co., from 1895; with Zeppelin established plant at Friedrichshafen to build motorcars and airship motors, 1909. Pioneer in engineering and construction of automobiles; Produced (with G. Daimler) early gasoline motor, Cannstadt, 1882; credited with inventing spray-nozzle carburetor; built 1st Mercedes with chassis of pressed sheet metal, 1901, also honeycombed radiator, float-feed carburetor, and shifting gear mechanism; worked with Levassor and Brit. engr. Simms. Died Dec. 29, 1929.

MAYDL, Karel, surgeon; b. Bohemia, 1853; devised operation transplanting ureters into rectum, for extrophy of bladder; credited with performing 1st successful colostomy, circa 1888; devised method by which opened colon is secured by glass rod until adhesions form. Died 1903.

MAYER, Alfred Marshall, Am. physicist; b. Balt., Nov. 13, 1836; s. Charles F. and Eliza (Blackwell) M.; attended St. Mary's Coll.; attended U. Paris, 1863-65; Ph.D., U. Pa., 1865; m. Katherine Duckett Goldsborough, 1865; m. 2d, Louisa Snowden, 1869; 3 children, including Alfred. Asst. prof. chemistry and physics U. Md., 1856-58, Westminster Coll., Fulton, Mo., 1858-63; prof. phys. science Pa. Coll., Gettysburg, 1865-67, prof. physics and astronomy, 1867-71; prof. Stevens Inst. Tech., 1871-97; mem. Nat. Acad. Scis. Author: The Earth a Great Magnet, 1872, Sound, 1878. Co-author; Light, 1877. Contbr. many articles to sci. jours. Research in acoustics, electricity, gravity, heat and light. Died July 13, 1897.

MAYER, Alfred Max, plant physiologist; b. Halberstadt, Germany, Oct. 5, 1926; s. Rudolf and Paula (Joseph) M.; B.Sc. with honors in Chemistry, U. London, 1945, Ph.D., 1950. Faculty, Hebrew U., Jerusalem, Israel, 1950—, asso. prof., 1964—; staff dept. biochemistry Cambridge & Low Temperature Research Sta., Cambridge, Eng., 1958-59; co-dir. tng. course in plant physiology UNESCO, Delhi, India, 1965; vis. prof. Mich. State U., 1966. Chmn. physiol. studies, Faculty Scis., Hebrew U., 1960-63, chmn. bot. studies, 1963-65. Mem. Biochem. Soc. (Eng.), Biochem. Soc. (Israel), Am. Soc. Plant Physiology, Scandinavian Soc. Plant Physiologists, Phytochem. Group (Eng.). Author: (with A. Poljakoff-Mayber, D. Koller) Plant Physiology, 1959; (with A. Poljakoff-Mayber) Germination of Seeds, 1963; also numerous articles. Research on biochemistry of seed germination, blooming in fruits, mass culture of algae for food and feedstuff. Home: 5 Palmach, Jerusalem, Israel.*

MAYER, André, French physiologist; b. Paris, France, Nov. 9, 1875; M.D., 1900; studied under Lapique; became prof. physiology Collège de France, Strasburg (France), then Faculty Medicine, 1919; work in Algeria, 1941, U. S., 1942. Became mem. French Acad. Scis., 1950, French Acad. Medicine. Author: Métabolisme de l'eau et métabolisme minimum, 1929; Sur les variations du métabolisme du lapin, 1929. Research on structure of protoplasm, physicochem. equilibrium of organic cells, thermal regulations of organisms, nutrition control, viscosity of liquids of organisms, colloidal material of protoplasm, Died June 26, 1950.

MAYER, Christian, astronomer; b. Mederizenhi, Moravia (now Czechoslovakia), Aug. 20, 1719; joined Soc. Jesus, 1745; tchr. Aschaffenburg, Germany; later prof. math. and physics, Heidelberg, Germany; built astron. obs., Schwetzingen; built and became dir. Mannheim (Germany) Obs., 1755. Fellow Royal Soc., 1765. Author: Gründliche verteidigung neuer Beobachtungen von Fixsterntrabanten, 1778; Expositio de transitu Veneris; Observations de la comète de 1781. Died Heidelberg, Apr. 16, 1783.

MAYER, Dennis Thomas, Am. biochemist; b. Lexington, Mo., Jan. 3, 1901; s. August John and Agatha (Gavin) M.; student U. Ill., 1923-25; A.B., U. Mo., 1931, M.A., 1933, Ph.D., 1938; m. Virginia Louise Miller, May 8, 1937; children—Dennis T., David R., Michael J. Agt., U. S. Dept. Agr., 1937-42; faculty U. Mo., Columbia, 1937—, prof. agrl. biochemistry, 1952—, dir. research in physiology and biochemistry reprodn., 1942—, dir. interdisciplinary reproductive biology tng. program, 1966. Mem. Am. Soc. Biol. Chemists, Soc. for Exptl. Biology and Medicine, Am. Soc. Animal Sci., N.Y. Acad. Scis., Mo. Acad. Sci., Phi Beta Kappa, Sigma Xi, Gamma Sigma Delta. Contbr. numerous articles to tech. jours. Analyzed spermatozoa chemically; pioneered isolation of DNA, lipid complexes; research on biochem. and enzymatic components of uterus of importance to embryo survival; isolated estriol and isomers from sow urine. Home: 319 Stewart Rd., Columbia, Mo. 65201.*

MAYER, Edmund, exptl. pathologist; b. Berlin, Germany, Dec. 20, 1889; s. Joseph and Bertha (Wilder) M.; student U. Berlin, 1908-14, U. Freiburg (Germany), 1912; M.D., U. Berlin, 1919; m. Hildegard Wertheim, Mar. 22, 1917; children—Ursula, Maria. Came to U. S., 1941, naturalized, 1947. First asst. dept. pathology Virchow Hosp., Berlin, 1919-25; dir. dept. pathology Hosp. Berlin-Lankwitz, 1925-26, Urban Hosp., Berlin, 1927-33; Rockefeller Found. fellow, Michaelsen Fund fellow Biol. Inst. Carlsberg Found., Copenhagen, Denmark, 1933-36; asso. prof. pathology, head dept. Am. U., Beirut, Lebanon, 1936-41; guest dept. anatomy Harvard Med. Sch., 1941-43; exptl. pathologist Am. Cyanamid Co., 1943-61, head dept. Stamford (Conn.) Labs., 1943-55, Lederle Labs., Pearl River, N.Y., 1960-61. Fellow A.A.A.S.; mem. Am. Assn. Anatomists, Am. Assn. Pathologists and Bacteriologists, Tissue Culture Assn., N.Y. Acad. Scis. Author: Introduction to Dynamic Morphology, 1963; also numerous articles. Research on human pathology, validity of bacteriological analysis in human corpses, use of cell colonies in vitro for studies of growth and form and for determination ultraviolet sensitivity; discovered important variations in histology of dog kidneys, low demand for thyroid activity in dogs. Home: 15 Pacific Av., Nanuet, N.Y. 10954. Office: Lederle Labs., Pearl River, N.Y. 10965.*

MAYER, Herbert, physicist; b. Czernowitz, Austria, Jan. 13, 1900; s. Ernst and Ottilie (Schmidt) M.; ed. U. Czernowitz; Ph.D.; 1926; agrégé, 1929; m. Anny Lugert, 1928; 1 child, Ingolf. Became scholar Rockefeller Inst., 1928; instr. U. Czernowitz, became asso. prof. physics 1933; named prof. Physics Inst., U. Posen, 1941; became asso. prof. physics U. Munich, 1950; prof., dir. physics Inst., Clausthal (W. Germany), Sch. Mines, 1951—. Rockefeller grand winner for study Physics Inst., U. Bristol, Eng., 1932-33; Author: Aktuelle Forschungsprobleme an der Physik dunner Schichten; Physik dünner Schichten. Research in solid state physics, especially physics of thin films. Home: Sorgestrasse 27, Clausthal, West Germany.

MAYER, Jean, nutritionist; b. Paris, France, Feb. 19, 1920; s. Andre and Jeanne Eugénie (Mayer) M.; B.Litt. summa cum laude, U. Paris, 1937, B.Sc. magna cum laude, 1938, M.Sc., 1939; Ph.D. (Rockefeller Found. fellow), Yale, 1948; D.Sc. in Physiology with highest honors, Sorbonne, 1950; A.M. (hon.), Harvard, 1965; m. Elizabeth Van Huysen, Mar. 16, 1942; children—Andre VanH., Laura D. VanH., John-Paul VanH., Theodore VanH., Pierre VanH. Came to U. S., 1939, naturalized, 1955. Research asso. George Washington U., 1948-49; faculty Harvard, Boston, 1950—, prof. nutrition, 1965—. Mem. adv. missions, coms. FAO-WHO; cons. to surgeon gen. U. S. Army, 1964—; mem. com. U. S. Pharmacopeia, 1960—, U. S. Army Research Center, 1962—; mem. adv. bd. Sargent Coll., 1957—; lectr. various instns. Ecole Normale Superieure fellow, 1939-40. Decorated knight Legion of Honor; Croix de Guerre with Stars and palms; Resistance medal. Fellow A.A.A.S., Am. Coll. Sport Medicine, Am. Acad. Arts and Scis.; mem. Am. Inst. Nutrition, Am. Physiol. Soc., Endocrine Soc., N.Y. Acad. Scis., Brit. Nutrition Soc., Am. Fedn. Clin. Research, Am. Soc. Clin. Nutrition, Am. Bd. Nutrition, Sigma Xi, Delta Omega, others. Nutrition editor Postgrad. Medicine, 1959—; bd. editors Am. Jour. Physiology, Jour. Applied Physiology, 1960—. Research and numerous publs. on regulation of appetite, mechanism of hunger, various types of exptl. and human obesities, gen. nutrition. Home: 23 W. Cedar St., Boston 02108. Office: 665 Huntington Av., Boston 02115.*

MAYER, Johann Tobias, German astronomer, mathematician; b. Marbach, Germany, Feb. 17, 1723; studied math. and art under his father; became prof. econs. and math. U. Göttingen (Germany), 1751; named dir. Göttingen Obs., 1754. Author: Theoria lunae juxta systema Newtonianum, 1767; Tabulae motuum solis et lunae, 1770; star catalog with 998 zodiacal stars, 1756-61; also map of moon, 1775. First to describe longitude and latitude of moon with accurate algebraic formulae; developed method of repetition of angles (usually attributed to Borda); invented repeating circle (later used in measuring arc of meridian); improved map-making techniques; compiled lunar tables, 1752. Died Feb. 20, 1762.

MAYER, Joseph Edward, Am. chem. physicist; b. N.Y.C., Feb. 5, 1904; s. Joseph and Catherine (Proescher) M.; B.S., Cal. Inst. Tech., 1924; Ph.D., U. Cal. at Berkeley, 1927; Ph.D., U. Libre de Bruxelles, 1962; m. Maria Goeppert, Jan. 12, 1930; children—Maria Ann (Mrs. Donat G. Wentzel), Peter Conrad. Internat. Edn. Bd. fellow Göttingen, Germany, 1929-30; asso. Johns Hopkins, 1930-39; asso. prof. Columbia, 1939-45; prof. U. Chgo., 1946-60, U. Cal., San Diego, 1960——. Recipient Gilbert N. Lewis medal Am. Chem. Soc., 1958; Chandler medal Columbia, 1966. Mem., Nat. Acad. Scis., Am. Acad. Arts and Scis., Heidelberg Akademie der Wissenschaften (corr.). Author: (with M. G. Mayer) Statistical Mechanics, 1940. Editor: Jour. Chem. Physics, 1940-52. Research, publs. on statis. and quantum mechanics. Home: 2345 Via Siona, La Jolla, Cal. 92037.*

MAYER, Julius Robert von, see von Mayer.

MAYER, Kurt Bernd, Swiss sociologist; b. Zurich, Switzerland, Sept. 6, 1916; s. Salomon and Anna (Hirsch) M.; student U. Zurich, 1935-36, 1937-39, London Sch. Econs. 1936-37; Ph.D., Columbia U., 1951; m. Elizabeth Meyer, Mar. 28, 1942; children—Charles, Eva, David. Lectr. sociology New Sch. Social Sci., N.Y.C., 1947-50; instr. sociology Rutgers U., 1948-50; with Brown U., 1950-66, prof. sociology, 1956-66, chmn. dept. sociology, anthropology, 1956-63; prof. sociology, dir. sociology inst. U. Berne, Switzerland, 1966——; vis. prof. sociology Australian Nat. U., 1963. John Simon Guggenheim Meml. fellow, 1964. Author: The Population of Switzerland, 1952; Economic Development and Population Growth in Rhode Island, 1953; Class and Society, 1955; Migration and Economic Development in Rhode Island, 1958; The First Two Years: Problems of Small Business Growth and Survival, 1961; Social Stratification in Two Equalitarian Societies; Australia and the United States, 1965; The Impact of Postwar Immigration on the Demographic and Social Structure of Switzerland, 1966. Studies on nature of social classes in modern societies and research about population problems. Home: Stegmatt 704 D, 3032 Hinterkappeln. Office: Brückfeldstrasse 14, 3012 Berne, Switzerland.*

MAYER, Manfred Martin, immunologist; b. Frankfort, Germany, June 15, 1916; s. Gustav and Julie (Sommer) M.; came to U.S., 1933, naturalized, 1940; B.S., Coll. City N.Y., 1938; Ph.D., Columbia U., 1946; m. Elinor S. Indenbaum, Dec. 6, 1942; children—Jonathan Marnin, David Michael, Dan Ellis, Matthew Jared. Instr. biochemistry Columbia U., N.Y.C., 1946; asst. prof. bacteriology Johns Hopkins, Balt., 1946-48, asso. prof., 1948-60, prof. microbiology, 1960——. Cons. USPHS, 1957-6 , Office Naval Research, 1953-57, U. S. Dept. Agr., 1960——. Recipient citation for work in Div. War Research, Columbia, 1945, Kimble award for methodology, 1953, medal of Theobald Smith Soc., 1957. Mem. A.A.A.S., Am. Assn. Immunologist, Soc. Exptl. Biology and Medicine. Author: (with E. A. Kabat) Experimental Immunochemistry, 1961. Asso. editor Biol. Abstracts, Jour. Immunology, Analytical Biochemistry, Immunochemistry. Research and numerous publs. on allergy and immunity, especially on complement (a group of immunity factors in blood). Home: 562 Sudbrook Lane, Balt. 21208.*

MAYER, Marie Goeppert, see Goeppert-Mayer, Marie.

MAYER, Simon, see Marius, Simon.

MAYER, William Vernon, biologist; b. Vancouver, B.C., Can., Mar. 25, 1920; s. Carlos Sieber and Esther (Vernon) M.; A.B. in Zoology, U. Cal., Berkeley, 1941; Ph.D. in Biology, Stanford, 1949; m. Margaret Laird, Dec. 23, 1941; children—Ann Elizabeth, William Laird. Grad. asst. Mont. State Coll., Bozeman, 1941-42; faculty U. So. Cal., Los Angeles, 1948-57; prof. biology Wayne State U., Detroit, 1957——, asso. dean Coll. Liberal Arts, 1960-65. Research investigator Arctic Research Lab., Point Barrow, Alaska, 1952-53, Arctic Aeromed. Lab., Fairbanks, Alaska, 1954-56; with Biol. Scis. Curriculum Study, 1960—, dir., 1965——; mem. exec. com. Mich. Commn. on Coll. Accreditation, 1964-65. Fellow A.A.A.S.; mem. Am. Soc. Mammalogists, Am. Inst. Biol. Scis., Nat. Sci. Tchrs. Assn., Nat. Assn. Biology Tchrs. (pres. 1967), Sigma Xi, Phi Sigma, Alpha Epsilon Delta. Author: Biological Science: An Inquiry into Life, 1963; Human Anatomy, 1955. Co-editor: Physiological Mammalogy, vol. I, 1963, vol. II, 1964. Research, publs. on effects of low temperatures; ecology of arctic mammals; improvement of biol. edn. Home: 304 W. Maplehurst, Ferndale, Mich. 48220.*

MAYERNE, Théodore Turquet, see Turquet de Mayerne, Théodore.

MAYERSON, Hymen Samuel, Am. physiologist; b. Providence, Sept. 10, 1900; s. Moses and Frances (Shepper) M.; A.B., Brown U., 1922, D.Sc., 1962; Ph.D., Yale, 1925; m. Caroline Wolf, June 10, 1930; children—Peter, Mary. Asst. in biology Brown U.,

1921-22; with Yale, 1922-26, instr. physiology, 1925-26; with Tulane U., 1926—, prof., chmn. dept. physiology, 1945-65, prof. emeritus, 1965——; research cons. Touro Infirmary, 1949-65, asso. dir. 1965——. Research cons. VA Hosp., 1953——. Mem. fellowship panel for anatomy and physiology USPHS, 1960-63; Mary Scott Newbold lectr. Phila. Coll. Physicians 1960; Baxter lectr. A.C.S., 1962; mem. nat. com. for internat. union physiol. scis. NRC, 1961-65. Mem. Am. Physiol. Soc. (exec. council, pres. 1962), Soc. Exptl. Biology and Medicine, Am. Assn. U. Profs. (pres. Tulane chpt. 1941-42), New Orleans Acad. Scis. (v.p. 1950-52), La. Heart Assn. (dir. 1955-65), Am. Heart Assn. (dir. 1954-60), Nat. Acad. Scis. (mem. fellowship panel NSF grad. fellowship program 1963-66), Nat. Bd. Med. Examiners (chmn. physiology test com. 1958-60), Fedn. Am. Socs. for Exptl. Biology (chmn. adv. com. 1963-64). Research on cardiovascular changes with posture, blood volume, capillary permeability and lymphatic system. Home: 1140 7th St., New Orleans 70115. Office: 1400 Foucher St., New Orleans 70115.*

MAYNARD, Charles Johnson, Am. naturalist; b. West Newton, Mass., May 6, 1845; s. Samuel and Emeline M.; ed. common schs.; m. 1 dau.; worked on his mother's farm, and studied nature from earliest youth; instr. econ. bird study Mass. Agrl. Coll., Amherst, 1910-19; v.p. Nuttall Ornith. Club, Cambridge, Mass., 1875; an original mem. Newton Natural History Soc., pres., 1891. Author: Contributions to Science (3 vols.); Nature Studies No. 2—"Sponges"; Manual of Taxidermy; Methods in Moss Study; Field Directory to the Birds of Eastern North America; Atlas to the Directory of the Birds of Eastern North America; Records of Walks and Talks with Nature, 12 vols.; Field Ornithology; Plates to Field Ornithology; Migration of Birds and Other Animals; Vocal Organs of Talking Birds and of Other Species, 1918, 2d edit., 1922; Notes on Life History of Cerions, 1919. Founder, editor Nuttall Bull. Research on vocal organs of birds; discovered vocal organs of Am. bittern; studied land shells of W. Indian genus Strophia (now Cerion). Died Oct. 15, 1929.

MAYNARD, Donald More, Jr., Am. zoologist; b. Avon Park, Fla., Apr. 6, 1929; s. Donald More and Hester (Douglas) M.; A.B., Harvard, 1947-51; Ph.D., U. Cal., 1955; m. Edith A. Holst, Sept. 25, 1954; children—Katharine A, Elizabeth T., Donald More III. NSF fellow Cambridge (Eng.) U., 1955; research asso. Harvard Biol. Lab., 1956; faculty U. Mich., Ann Arbor, 1956—, prof. zoology, 1965—, research asso. Mental Health Research Inst., 1956——. Mem. Am. Physiol. Soc., Am. Soc. Zoology, Soc. Exptl. Biology, A.A.A.S., Sigma Xi, Phi Beta Kappa, Biophys. Soc. Research, publs. on patterned activity in small nervous systems and explorations of properties of neurons in arthropod brains. Home: 1706 Cambridge St., Ann Arbor, Mich. 48104.*

MAYNARD, Edward, Am. dentist, inventor; b. Madison, N.Y., Apr. 26, 1813; s. Moses and Chloe (Butler) M.; attended U. S. Mil. Acad.; m. Ellen Sophia Doty, 1839; m. 2d, Nellie Long, 1869; 8 children, including George W. Practiced dentistry, Washington, 1836-91; 1st dentist to fill teeth with gold foil, 1838; patented firearm priming system, 1845; prof. theory and practice Balt. Coll. of Dental Surgery, 1857-91; developed and patented Maynard carbine (one of 1st breech-loading rifles in Am.); court dentist to Emperor Nicholas I of Russia; prof. dental theory and practice Nat. U., Washington, 1887-91; hon. mem. Am. Acad. Dental Scis. Co-editor Am. Jour. Dental Sci., 1843-46. Died Washington, May 4, 1891.

MAYNARD, Poole, Am. geologist; b. Balt., Feb. 15, 1883; s. Albert and Emma Dorsey (Poole) M.; A.B., Johns Hopkins Univ., 1905, Ph.D., 1909; m.; children—Mary Cary, Albert. Student asst. in econ. geology Johns Hopkins, 1905-09; mem. Md. Geol. Survey and Va. Geol. Survey, 1905-09; spl. employment U. S. Geol. Survey, 1907; asst. state geologist Ga. Geol. Survey, 1909-12; cons. geologist Central of Ga. Ry., 1912-24; indsl. geologist A., B. & C. R.R. since 1925, A.C.L. R.R. since 1938; v.p. Maynard Furniture Co., Belton, S.C.; adviser in nor-metallics U. S. Bur. of Mines since 1939, also cons. engr.; established bus. as cons. geologist, 1912, particular reference to indsl. processes in non-metallics Fellow Geol. Soc. Am., A.A.A.S.; mem. Am. Inst. M.E., Soc. Econ. Geologists, other profl. socs. Author bulls., papers. Originator patents for manufacture heavy clay vitrified products from slag and clay; coloring. of burned clay granules by precipitation of colors; collaborator in patents and processes for concentration of bauxites, for mfr. of magnesia and other chem. products from dolomite; 1st. to recover potash commercially from shales and to locate in Ga. bentonites, roofing slates. Died Aug. 22, 1952; buried Atlanta.

MAYNERT, Everett William, Am. pharmacologist; b. Providence, Mar. 18, 1920; s. William W. and Anna (Erler) M.; Sc.B. in Chemistry, Brown U., 1941; Ph.D. in Chemistry, U. Ill., 1945; M.D., John Hopkins, 1957. Research chemist Interchem. Corp., N.Y.C., 1945-47; faculty Columbia U., 1947-52; Am. Cyanamid fellow, asso. prof. pharmacology John Hopkins, 1957-65; prof. U. Ill., Chgo., 1965——. Cons. USPHS, 1962——. Recipient Lederle Med. Faculty

award, 1957-60. Fellow A.A.A.S.; mem. Am. Chem. Soc., Am. Soc. Pharmacology, Harvey Soc. Contbg. author: Drill's Pharmacology in Medicine, 1965. Research and numerous publs. on absorption, distbn., metabolism, and excretion of drugs; mechanisms of tolerance to hypnotic and narcotic drugs, drug addiction, neurohormones, nerve ending particles, synaptic vesicles. Home: 339 Barry St., Chgo. 60657.*

MAYO, Charles Horace, Am. surgeon; b. Rochester, Minn., July 19, 1865; s. William Worrall (M.D.), and Louise Abigail (Wright) M.; M.D., Northwestern U., 1888, M.A., 1904; post-grad. N.Y. Polyclinic, N.Y. Post-Grad. Med. Sch.; numerous hon. degrees; m. Edith Graham, 1893; children—Margaret, Dorothy, Charles William, Edith (Mrs. Fred W. Rankin), Joseph Graham, Louise (Mrs. George T. Trenholm), Rachel, Esther (Mrs. John B. Hartzell). Practiced surgery at Rochester, 1888——; established (with brother) Mayo Found. for Med. Edn. and Research at Rochester, in affiliation with U. of Minn.; founded (with brother) Mayo Properties Assn., 1919; surgeon and asso. chief of staff Mayo Clinic; surgeon to St. Mary's, Colonial and Worrall hosps.; prof. surgery Med. Sch., U. Minn., 1919-36, prof. surgery Grad. Sch. (Mayo Found.), 1915-36, emeritus. Recipient letter of commendation Minn. State Med. Assn., 1934; commemorative plaque presented by Pres. of United States in person, 1934; certificate in recognition of service to U. of Minn. and to State as prof. of surgery, 1936; Bronze medal Interstate Post Grad. Med. Assn. of N.Am. for contbns. to sci. medicine, 1936; other awards. Fellow or mem. numerous sci. and non-sci. orgns. in U. S. and fgn. countries. Services to sci. and non-sci. periodicals, including Anales de Cirugia la Habana, Cuba (fgn. collaborator), Archives of Clin. Cancer Research (editorial bd. 1924-32), Gaceta Medica Española (del. in U. S. 1926-30; internat. patron in U. S., 1931—), Internat. Clinics (collaborating editor 1907-33), Narkose Und Anaesthesie (contbr. 1928), Nosokomien (editorial bd.), Ency. Brit. (adv. bd.). Skilled in goiter surgery; reduced death rate for goiter by 50 per cent; also studies in ophthalmology. Died May 26, 1939.

MAYO, Charles William, Am. surgeon; b. Rochester, Minn., July 28, 1898; s. Charles Horace and Edith (Graham) M.; prep. edn. The Hill Sch., Pottstown, Pa., 1912-17; A.B., Princeton, 1921; M.D., U. Pa., 1926; M.S. in Surgery, U. Minn., 1931; LL.D., St. Lawrence U., Canton, N.Y., 1947, Drake U., Des Moines, 1957, Gonzaga U., 1962; D.Sc. (hon.), Franklin and Marshall Coll., 1949; L.H.D., Neb. Wesleyan U., 1962; m. Alice Varney Plank, June 24, 1927; children—Mildred (Mrs. Hugo Torres), Charles Horace II, Edward M., Joseph G., Edith Maria (Mrs. Donald Sones) and Alexander Steward; (wards) David Graham Mayo (dec. July, 1958), William James Mayo. Fellow in surgery, Mayo Found., Rochester, Minn., 1927-31, surgeon 1931-63; instr. surgery Mayo Found. Grad. Sch., U. Minn., 1933-35, asst. prof. surgery, 1935-40, asso. prof. surgery, 1940-47, prof., 1947-63; gov. Mayo Clinic, Med. adviser N.W. Airlines, 1948-63; bd. dirs. Mut. Benefit Health and Accident Assn. Trustee Carleton College; chmn. Mayo Assn.; chmn. bd. regents U. Minn., 1961——. Recipient Honor award, Am. Med. Writers Assn. Fellow. Royal Coll. of Surgeons of Eng. (hon.), Royal Australasian Coll. Surgeons (hon.), A.C.S.; mem. A.M.A., Am., So., Western surg. assns., Seaboard Med. Assn. of Va. and N.C. (hon. fellowship), The Surgeons' Club, Am. Bd. of Surgery (Founders Group), Minn. Acad. Medicine, Internat. Soc. Surgeons, Soc. U. S. Med. Cons. of World War II, Societa Italiana di Chirurgia, Rome, Italy (hon.), Sigma Xi, Nu Sigma Nu. Author: Surgery of the Small and Large Intestine, 1955. Contbr. med. and surg. jours. Editor-in-chief Postgraduate Medicine. Died Rochester, Minn., July 28, 1968.

MAYO, George Elton, management scientist; b. Adelaide, Australia, Dec. 26, 1880; s. George Gibbs and Hetty Mary (Donaldson) M.; Westminster classical scholar, St. Peter's Coll., Adelaide, 1894-95; Adelaide Univ., 1898-99, B.A., 1910, M.A., 1919; hon. M.A., Harvard, 1942; m. Dorothea McConnel, Apr. 18, 1913; children—Patricia Elton (Mrs. Walter Goetz), Ruth Elton (Mrs. Guy Vincent). Lecturer in logic and psychology, Queensland U., 1911, prof., 1919-23; research associate, U. of Pa., 1923-26; asso. prof. industrial research, Harvard, 1926, professor, 1929-47, professor emeritus since September 1947. Roby Fletcher prizeman (Adelaide), 1906, David Murray research scholar, 1911. Fellow American Academy Arts and Sciences. Author: Democracy and Freedom, 1919; The Human Problems of an Industrial Civilization, 1933, 2d edition 1946; The Social Problems of An Industrial Civilization, 1945. Applied sociopsychological techniques to management; developed theory of human behavior in organizations stressing feeling of group membership. Died Sept. 7, 1949.

MAYO, Herbert, English physiologist, anatomist; b. London, Apr. 3, 1796; s. John Mayo; studied under Sir Charles Bell, 1812-15; M.D., Leiden, Netherlands; m. Jessica Matilda Arnold; 1 son, 2 daus. Surgeon, Middlesex Hosp., 1827-42, founder med. sch., 1836; prof. anatomy and surgery Royal Coll. Surgeons, 1828-29; prof. anatomy King's Coll., London, 1830-36; became physician hydropathic establishment, Boppart, 1843, later Bad Weilbach. Fellow Royal Soc., 1828, Geol. Soc.; mem. Royal Coll. Surgeons. Author: Anatomical and Physiological Commentaries, 1822-23; Course of Dissections for Students, 1825; Outlines of Human Physiology, 1827; A Series of Engravings of Brain and Spinal Cord in Man, 1827. Discovered real function of facial nerves, 1822. Died May 15, 1852.

MAYO, William James, Am. surgeon; b. Le Sueur, Minn., June 29, 1861; s. William Worrall (M.D.) and Louise Abigail (Wright) M.; M.D., U. Mich., 1883, A.M. (hon.), 1890; certificate N.Y. Post Grad. Med. Sch., 1884; M.D., N.Y. Polyclinic, 1885; numerous hon. degrees; m. Hattie M. Damon, Nov. 20, 1884; children—Carrie L. (Mrs. D. C. Balfour), Phoebe G. (Mrs. H. Waltman Walters). Engaged in practice of surgery, Rochester, Minn., 1883-—; specialist in stomach surgery; surgeon Mayo Clinic (St. Mary's Hosp.), 1889-—, also asso. chief of staff established (with brother) Mayo Found. for Med. Edn. and Research, Rochester, in affiliation with U. Minn. Recipient Gold medal Nat. Inst. Social Scis., 1918; Henry Jacob Bigelow Gold medal Boston Surg. Soc., 1921; Gold medal A.M.A., 1930; spl. award for distinguished service to science Sigma Xi (U. Minn. chpt.), 1933; letter of commendation Minn. State Med. Assn., 1934; other awards. Mem. numerous Am. and fgn. sci. socs. Died July 28, 1939.

MAYO, William Worrall, physician; b. Manchester, Eng., 1819; s. James and Ann (Bousal) M.; studied physics under John Dalton, Owens Coll., Manchester; M.D., U. Mo., 1854; m. Abigail Wright, 1851; children—William James, Charles Horace, Gertrude (Mrs. David M. Berkman). Came to U. S., 1845. Became provost surgeon for So. Minn., hdgrs. at Rochester, 1863; headed St. Mary's Hosp., Rochester, from 1889; St. Mary's became nucleus of Mayo Clinic, developed later by his sons. Contbr. numerous med. articles to tech. jours. Research on surg. removal of ovarian tumors; one of earliest physicians in West to use microscope for diagnostic work. Died Rochester, Mar. 6, 1911.

MAYOR, Heather Donald, med. scientist; b. Melbourne, Australia, July 6, 1930; d. Joseph Arthur and Elizabeth (Boyd) Donald; B.S., U. Melbourne, 1948, M.S., 1950; Ph.D., U. London, Eng., 1954; m. Richard Blair Mayor, May 28, 1956; children—Diana Boyd, Philip Hastings. Came to U. S., 1956, naturalized, 1962. Research asso. in virology Walter and Eliza Hall Inst. Med. Research, Melbourne, 1955-56; research asso. in bacteriology and immunology Harvard Med. Sch., Boston, 1956-60; asst. prof. virology Coll. Medicine, Baylor U., Houston, 1960-63, asso. prof., 1963-—; vis. prof. U. Melbourne, 1965. Mem. Biophys. Soc., Am. Assn. Immunologists, Am. Assn. for Cancer Research, Electron Microscope Soc. Am., Am. Soc. for Cell Biology, Sigma Xi. Research and publs. in structure and symmetry of viruses and biophys. properties of viruses, viruses and cancer, virus-host cell interactions using electron and fluorescent microscopy, nature of viral nucleic acids. Home: 4040 San Felipe, Houston, 77027.*

MAYOR, Mathias, Swiss physician; b. Cudrefin, Switzerland, Apr. 21, 1775; student medicine, Montpellier, France; student Zurich, Switzerland, Mailand, Pavia, Italy, Paris, France; physician, Morcel. Author: Nouveau Système de déligation chirurgicale, 1832; Essai sur l'anthropotaxidermie, 1838; la Chirurgie simplifiée, 1841; la Chirurgie populaire manuel du baigneur; Excentricités chirurgicales. Improved treatment of wounds; 1st to use hydrophilic cotton for dressings; invented Mayor's hammer which was used as a counter-irritant in skin operations. Died 1847.

MAYOW, John, English chemist and physiologist; b. Morval, nr. Looe, Eng., May 24, 1640; s. William and Elizabeth M.; student Wadham Coll., Oxford (Eng.) U., from 1658, B.C.L., All Soul's Coll., 1665, D.C.L., 1670; student medicine. Physician, London, Bath (both Eng.), from 1670. Fellow Royal Soc., 1678. Author: Tractatus duo, de respiratione et de rachitide, 1668; Tractatus quinque medico-physici, 1674. Made chem. analysis of salt springs at Bath; described action of intercostal muscles, ribs and diaphragm; noted and studied similarity between chem. process of combustion and physiol. function of respiration; held (with others) that there is a substance in atmosphere (nitro-aerial particles) which take part in calcination of metals and in respiration; showed that only part of air is used in combustion and respiration and that remainder is insoluble in water and does not support combustion or respiration; held that breathing is means of converting venous into arterial blood and of heating blood, also that inspired air communicates subtle nitro-aerial particles to blood by way of lungs, helps produce fermentation in blood which is cause of animal heat and hence necessary to life; demonstrated that fetal respiration is effected through placenta. Died London, Oct. 10, 1679.

MAYR, Ernst (Walter), biologist; b. Kempten, Germany, July 5, 1904; s. Otto and Helene (Pusinelli) M.; student U. Greifswald, Berlin, 1923-26; Ph.D., U. Berlin, 1926; Ph.D. (hon.), Uppsala U. (Sweden), 1957; D.Sc. (hon.), Yale, 1959, U. Melbourne, 1959, Oxford U., 1966; m. Margarete Simon, May 4, 1935; children—Christa Elizabeth, Susanne. Came to U. S., 1931, naturalized. Asst. curator Zool. mus. U. Berlin, 1926-32; mem. Rothschild expdn. to Dutch New Guinea, 1928, expdn. to Mandated Ty. of New Guinea, 1928-29, Whitney South Sea expdn. to Solomon Islands, 1929-30; Whitney research asso. Am. Mus. Natural History, N.Y.C., 1931-32, asso. curator, 1932-44, curator, 1944-53; Jesup lectr. Columbia, 1941; Alexander Agassiz prof. zoology Harvard, 1953-—, dir. Mus. Comparative Zoology, 1961-—. Pres. XIII Internat. Ornithol. Congress, 1962. Recipient Leidy Medal, 1946, Wallace Darwin medal, 1958, Brewster Gold medal, 1965; Verrill medal, 1966; Daniel Giraud Elliot medal, 1967. Fellow Linnean Soc. N.Y. (past sec. editor), Am. Ornithol. Union (pres. 1956-59), N.Y. Zool. Soc.; mem. Am. Philos. Soc., Nat. Acad. Sci., Am. Acad. Arts and Scis., Am. Soc. Zoologists, Soc. Systematic Zoology (pres. 1966), Soc. Study Evolution (sec. 1946, pres. 1950); hon. or corr. mem. Royal Australian, Brit. ornithol. unions, Zool. Soc. London, Soc. Ornithologists France, Royal Soc. New Zealand, Bot. Gardens Indonesia, S. Africa Ornithol. Soc., Linnean Soc. London. Author: List of New Guinea Birds, 1941; Systematics and the Origin of Species, 1942; Birds of the Southwest Pacific, 1945; Birds of the Philippines (with Jean Delacour), 1946; Methods and Principles of Systematic Zoology (with E. G. Linsley and R. L. Usinger), 1953; Animal Species and Evolution, 1963; Principles of Systematic Zoology, 1968. Editor: Evolution, 1947-49. Home: 11 Chauncy St. Office: Mus. Comparative Zoology, Harvard U., Cambridge, Mass. 02138.*

MAYR, Franz Xaver, Austrian physician; b. Gröbning, Nov. 28, 1875; s. Anton and Serafine (Leypold) M.; ed. U. Graz (Austria), Paris, London, Berlin; M.D.; m. Alice Rougon, June 14, 1956; Began practice medicine, Carlsbad, 1903. Mem. Med. Soc. Vienna. Author: Darmträgheit (Stuhlverstopfung und ihre radikale Bekämpfung), 1912; Schönheit und Verdauung, 1920; Fundamente zur Diagnostik der Verdauungskrankheiten, 1921; Ein Weg aus der Not und Friedosigkeit der Menschen und Völker zum internationalem Wirtschaftsstaat, 1927; Wann ist unser Verdauungsapparat in Ordnung? Die verhängnisvollste Frage, 1949. Address: Heiligenstaedterstrasse 11, Vienna 9, Austria.

MAYR, Franz Xaver Josef, German chemist; b. Pfronten, Feb. 21, 1887; s. Jakob and Karoline (Freiin von Köppelle) M.; ed. Ratisbonne, Munich, Kiel, Erlangen, Innsbruck, Freising; student Catholic theology, 1921-23; Ph.D. Prof. chemistry, biology geography Insts. of Ratisbonne, Landshut and Aschaffenberg, 1914-21; ordained priest, 1923; prof. chemistry, biology and geology U. Eichstätt (Germany); ret., 1958; keeper sci. collections U. Eichstätt, spiritual adviser, 1939-—. Mem. Bavarian Bot. Soc. Ratisbonne, German Geol. Soc., Soc. Paleontology, Soc. for Frankish History, Ratisbonne Soc. Natural Sci. (hon.). Author: Hydropoten an Wasser- und Sumpfpflanzen, 1915; Beiträge zur Anatomie der Alismataccen, 1932-43; Woher der Mensch?, 1954; Arbeiten zur Geologie und Paläontologie des Jura. Died Sept. 21, 1965.

MAYR, Hubert Hermann, Austrian biochemist; b. Wels, June 19, 1928; s. Hubert and Maria M.; ed. Gratz (Austria) Tech. Coll.; m. Johanna Repp, May 27, 1953; children—Ursula, Wolfgang, Helmut. Biochemist, Stickstoffwerke Labs., Linz, 1951-60; civil engring. and tech. chemistry, beginning 1959; named dir. biol. research Stickstoffwerke Soc. Labs., 1961. Mem. Union Austrian Surgeons, Austrian Soc. Pure and Applied Biophysics, Austrian Biochem. Soc., Austrian Agr. Soc., Munich Alimentary Biology Soc. Research and publs. in biochemistry, especially of plants; plant growth regulators, agrl. chemistry; use of radioactive isotopes; soil sci. Home: Leonding/H.A. Office: Haagstrasse 19, Linz/D., Austria.

MAYR, Simon, see Marius, Simon.

MAYS, Charles William, Am. physicist; b. Corsicana, Tex., Mar. 17, 1930; s. Charles W. and Fay (Lockhart) M.; student U. Mo., 1947-48; B.S., U. Utah, 1951, Ph.D., 1958; m. Evelyn Ekker, June 24, 1951; children—Shelby Marie, Sharon Eve, Susan Fay. Physics group leader radiobiology div. U. Utah, Salt Lake City, 1957-—, asso. research prof. anatomy, 1966-—. Mem. Nat. Council on Radiation Protection and Measurements, Radiation Research, Health Physics, Sigma Xi, Phi Kappa Phi, Tau Beta Pi, Theta Tau. Research, publs. on measurement of radioactivity in living animals; dosimetry of radionuclides in the skeleton; radiation-induced cancer. Home: 3126 S. 2750 E., Salt Lake City 84109.*

MAYSUYAMA, Motonori, Japanese geophysicist; b. Oita Prefecture, Japan, 1884; grad. Kyoto (Japan) U., 1911; D.Sc.; eldest dau., Tazuko (Mrs. Susumu Matsushita). Successively instr., asst. prof., prof., also dean sci. dept. Kyoto U.; ret.; became pres. Yamaguchi U., 1949. Recipient Imperial Acad. prize, 1932. Mem. Japan Acad. Research and numerous publs. on lithosphere, interior of earth, phys. methods of locating underground resources. Author: Physics of Lithospheres and Interior of the Earth. Died Jan. 27, 1958.

MAZEIKA, Paul Almis, oceanographer; b. Vainutas, Lithuania, Jan. 9, 1915; s. Juozas and Maria (Gecaitel) M. Tchrs. diploma, Tchrs. Coll., Lithuania, 1935; D. Marine Scis., Instituto Universitaria Navale, Napoli, Italy, 1943; m. Elena Janniskis, Sept. 1, 1945; children—Jurate, Gediminas. Came to U. S., 1949, naturalized, 1955. Lectr. DP Navigation and Sea-Engring. Sch., Feinsburg, Germany, 1946-48; tchr. math. Oldenburg, Germany, 1948-49; with Atlantic Fisheries, New Bedford, Mass., 1950-56; phys. oceanographer U. S. Naval Oceanographic Office, Washington, 1956-63, 66-—, Bur. Comml. Fisheries, Washington, 1963-66. Mem. Am. Geophys. Union. Research in circulation, thermal structure and other phys. properties distbn. in upper layers of north and tropical Atlantic. Home: 4718 Park Lane, Washington 20023. Office: U. S. Naval Oceanographic Office, Code 8500, Naval Research Lab., 29390.*

MAZET, Robert Jean Sosthène, mathematician, physicist; b. Paris, Feb. 7, 1903; s. Gaston and Julia (Houbert) M.; Agrégation de Mathématiques, Ecole Normale Supérieure, 1924, D.Sc., 1929; m. Huguette Colin, May 24, 1939; children—Françoise (Mrs. Jacques Weber), Jean-Luc, Pascale, Dominique, Bruno. Asst. prof. Faculty Sci., Lille, 1929-32, asso. prof., 1932-36, prof.; recteur d'académie, Caen, 1944-46, Poitiers, 1946-47; prof. Faculty Sci., Poitiers 1848-60; asso. prof. Faculty Sci., Orsay-Paris 1960-61, prof., 1961-—; sci. dir. structures Nat. Office Aerospace Research 1947-—; prof. Ecole Nationale Supérieure de l'Aéronautique 1950-—. Recipient Médaille de l'Aéronautique; officer, Légion d'honneur; commdr., Ordre des Palmes Académiques. Mem. French Acad. Sci. (corr.), Nat. Com. on Rheology (chmn.), French Mech. Soc. (dep. chmn.). Author: Mécanique Vibratoire, 2d edit., 1966, also numerous articles. Research on theoretical mechanics, friction between solid bodies, airplane wing vibrations (flutter), deformation of structures beyond elastic range (plasticity, creep). Home: 38 rue Madeleine Crenon, Sceaux 92, France. Office: Faculté des Science, Orsay 91, France.

MAZIA, Daniel, Am. biologist; b. Scranton, Pa., Dec. 18, 1912; s. Aaron and Bertha (Kurtz) M.; A.B., University of Pennsylvania, 1933, Ph.D., 1937; married Gertrude Greenblatt, June 19, 1938; children—Judith Ann, Rebecca Ruth. National Research Council fellow in biology science, Princeton, 1937-38; asst. prof. zoology, U. of Mo., 1938-41, asso. prof., 1942-47, prof. since 1947; Gosney fellow, Calif. Inst. Tech., spring 1950; asso. prof. zoology U. Cal., Berkeley, 1951-53, now prof. Mem. Corp., Marine Biol. Lab., Woods Hole, mem. bd. trustees and physiology teaching staff, 1950-—. Fellow Am. Acad. Arts and Scis.; mem. Internat. Soc. Cell Biology, Am. Soc. Zoologists, Am. Soc. Naturalists, Soc. Gen. Physiologists, National Academy of Sciences. Member of the editorial bd. Exptl. Cell Research, Jour. Biophys. and Biochem. Cytology, Quarterly Rev. of Biology. Research on chemistry of chromosomes; nuclear and cellular physiology; ionic changes in stimulation; ion accumulation and exchange; biochemistry of mitosis; chemistry of enzymes. Home: 900 Shattuck, Berkeley, Cal.

MAZUR, Jacob, physicist; b. Lodz, Poland, Dec. 17, 1921; s. Elieser Haim and Itta (Cohen) M.; M.Sc., Hebrew U., Jerusalem, Israel, 1945, Ph.D., 1948; m. Adeline June Garfield, Feb. 24, 1951; children—David, Jonathan. Came to U. S., 1955, naturalized, 1959. Research fellow Cal. Inst. Tech., Pasadena, 1948-50; vis. fellow U. Chgo., 1950-51; sr. scientist Weizmann Inst. Sci., Rehovoth, Israel, 1951-55; research asso. U. Ill., Urbana, 1955-57; research scientist Dow Chem., Midland, Mich., 1957-60; research chemist Nat. Bur. Sci., Washington, 1960-—. Recipient A. Cressy Morrison award in natural scis. N.Y. Acad. Scis., 1959. Fellow Am. Phys. Soc.; mem. Washington Acad. Sci., Washington Philos. Soc., Sigma Xi. Research and publs. on phase transitions in macromolecular systems, theory of reaction rates, theory of electrically charged polymers in solution, quantum-mech. theory of reactions of atomic exchanges, diffusion in saturated binary systems, statis. thermodynamics of lattice models of polymer systems. Home: 8305 Donnybrook St., Chevy Chase, Md. 20015. Office: Nat. Bur. Standards, Connecticut and Vanness Sts., Washington 20234.*

MAZUR, Peter, Am. biologist; b. N.Y.C., Mar. 3, 1928; s. Paul M. and Adolphia (Kaske) M.; A.B. magna cum laude, Harvard, 1949; Ph.D., 1953; m. Drusilla Stevens, May 28, 1953; 1 son, Timothy. NSF fellow dept. biology Princeton, 1957-59; staff biology div. Oak Ridge Nat. Lab., 1959-—. Mem. Am. Inst. Biol. Scis. adv. bd. to medicine and dentistry br. Office Naval Research, 1963-65, to environmental biology NASA, 1966-—. Fellow A.A.A.S.; mem. Soc. Gen. Physiologists, Soc. for Cryobiology, Am. Soc. for Microbiology, Biophys. Soc., Am. Inst. Biol. Scis., Bot. Soc. Am., Phi Beta Kappa, Sigma Xi. Editorial bd. Cryobiology, 1966-—. Research, publs. on basic mechanisms causing injury in living cells subjected to freezing and low temperature storage; determined that one of chief causes of death is formation of ice crystals in cells; that with proper cooling, cells could be kept alive indefinitely. Home: 125 Westlook Circle. Office: Biology Div., Oak Ridge Nat. Lab., P.O. Box Y, Oak Ridge 37830.*

MAZZOTTI, Luis, Mexican med. parasitologist; b. Lerdo, Mexico, Jan. 12, 1900; s. Juan and Manuela (Galindo) M.; B.S., U. Mexico, 1919, M.D., 1925; M.P.H., Johns Hopkins, 1935; D.T.M., U. Liverpool

(Eng.), 1937; m. Margarita Palomo, May 21, 1932; children—Margarita, Luis. Health officer various towns, Mexico, 1925-30; supr. Health Services, 1931-32; dir. rural health, 1933-35; dir. supervision Health Services, 1936-38; chief lab. helminthology Inst. Tropical Diseases, Mexico City, Mexico, 1938-66; cons. parasitology Mexican Inst. Social Security, Mexico City, 1967—. Expert parasitic diseases WHO, 1950—. Recipient honor diplomas Soc. Mexican Hig., 1959, Pres. of Mexico, 1960, Ministry of Health, 1965; prix Monbinne, Acad. Medicine Paris, 1944. Mem. Acad. Medicine (Mexico), Soc. Path. Exot. (France), Soc. Belge Med. Trop., Am. Acad. Microbiology. Research, numerous publs. on tropical diseases in Mexico, med. parasitology and entomology; developed Mazzotti test for diagnosis of Onchocerciasis and other filariasis; demonstrated high death rate from scorpion stings in Mexico and devised method of protection by using a strip of tile around walls. Address: Calle 19 N. 109, Mexico 18, D.F., Mexico.*

McADAM, John Loudon, engr.; b. Ayr, Scotland, Sept. 21, 1756; s. James and Suzannah (Cochrane) McA.; m. Miss Nichol; 4 sons, 3 daus.; m. 2d, Miss DeLancy. Came to N.Y., 1770 and remained until end of Revolutionary War; partner in John McAdam and Co., 1777-80; McAdam, Watson, and Co., 1781-83; returned to Scotland, 1783; coal tar manufacturer in Ayrshire, 1783-98; magistrate, dep. lt., for County of Ayrshire, Scotland; became agt. for revictualling navy, 1798; became paving commr. Bristol, Eng., 1806, named surveyor-gen. Bristol (Eng.) rds., 1816; named gen. surveyor of rds., 1827. Parliament voted him £2,000 for significant public service, 1825. Author: A Practical Essay on the Scientific Repair and Preservation of Roads, 1819; Present State of Roadmaking, 1820. Originated method of road making using thin layers of crushed rock; insisted on proper drainage procedure for roads. Died Moffat, Scotland, Nov. 26, 1836.

McADIE, Alexander George, Am. meteorologist; b. N.Y.C., Aug. 4, 1863; s. John and Anne (Sinclair) M.; A.B., Coll. City N.Y., 1881, A.M., 1884; A.M., Harvard, 1885; M.S. (hon.), Santa Clara Coll.; m. Mary Randolph Browne, Oct. 7, 1893. In phys. lab. U. S. Signal Office, 1886-87; fellow in physics and lectr. in meteorology Clark U., Worcester, Mass., 1889-90; with Weather Bur., Washington, 1891-95; local forecast ofcl., New Orleans, 1898; forecast ofcl., San Francisco, 1899; prof. meteorology U. S. Weather Bur., 1903-13; A. Lawrence Rotch prof. meteorology Harvard, dir. Blue Hill Obs. since 1913. Fellow Am. Acad. Arts and Scis.; mem. Astron. Soc. Pacific (pres. 1912), Seismol. Soc. Am. (pres. 1914), Am. Antiquarian Soc. Author: Principles of Aerography; Climatology of California; Rainfall of California; The Fogs and Clouds of San Francisco; The Winds of Boston; The Ephebic Oath; Wind and Weather; Cloud Atlas; Making the Weather; War Weather Vignettes; Man and Weather; Clouds, Airgraphics, Fog; also bulls. and pamphlets on meteorol. subjects, especially lightning, frost, fog. and sci. units. Pioneer in study of relation of weather conditions to history of man; divided atmosphere above and around the earth into distinct classifications. Died Nov. 1, 1943.

McAFEE, John G., physician; b. Toronto, Ont., Can., June 11, 1926; s. Robert Duncan and Susan Jane (Damery) McA.; M.D., U. Toronto, 1948; m. Joan Margaret Weber, Feb. 9, 1952; children—Paul Clifton, Carol Joan, David Robert. Came to U. S., 1951, naturalized, 1960. Practice medicine, specializing in radiology, Balt., 1951-65, Syracuse, N.Y., 1966—; faculty Johns Hopkins Hosp., 1953-65; prof., chmn. radiology State U. N.Y. Upstate Med. Center, Syracuse, 1965—. Mem. Soc. Nuclear Medicine, Assn. U. Radiologists, Radiol. Soc. N.Am. Contbr. numerous articles to profl. jours. Developed radioactive agts. for detection brain tumors and kidney lesions in man, method for visualization of human placenta. Home: 5 Lynacres Blvd., Fayetteville, N.Y. 13066. Office: 750 E. Adams St., Syracuse, N.Y. 13210.*

McALESTER, A(rcie) Lee, Am. geologist; b. Dallas, Feb. 3, 1933; s. Arcie Lee and Alverta (Funderburk) McA.; B.A., So. Methodist U., 1954, B.B.A., 1954; M.S., Yale, 1957, Ph.D., 1960. Faculty, Yale, New Haven, 1959—, prof. geology, 1966—; research asso. Am. Mus. Natural History, N.Y.C., 1963—, U. Rochester (N.Y.), 1965—. Guggenheim Found. fellow, Glasgow (Scotland) U., 1965. Mem. Geol. Soc. Am., Paleontol. Soc., Soc. Systematic Zoology, A.A.A.S. Author: The History of Life, 1967; also articles, monographs. Application of principles of evolutionary and ecol. biology to study of fossil marine invertebrate life. Home: 123 York St., New Haven 06511.*

McALLISTER, Raymond Francis, Am. oceanographer; b. Ithaca, N.Y., June 26, 1923; s. Raymond Francis and Jessie (Pinder) McA.; student Coll. City N.Y., 1941-42; B.S. in Agr., Cornell U., 1950; M.S. in Geology, U. Ill., 1950-51; postgrad. Scripps Instn. Oceanography, 1951-54; Ph.D. in Oceanography, A. and M. Coll., Tex., 1958; m. Joan Margaret Simonson, Sept. 1, 1951; children—Keith Raymond, Karen Kathleen, Kevin Michael. Research oceanographer Scripps Instn. Oceanography, 1951-54; research scientist Tex. A. and M. Research Found., 1954-58; sr. oceanographer Geophys. Field Sta., Columbia U., St. David Island, Bermuda, 1958-63, research scientist Bermuda Field Lab., Hudson Labs., 1963; asst. dir. N.Am. Aviation-Marine Tech. Group, 1964-65; prof. oceanography, Fla. Atlantic U., Boca Raton. Marine geol. cons. Gulf Cons., Inc., College Station, Tex., 1952-53. Mem. Marine Tech. Soc. Research and publs. on nature of marine clay minerals nr. Mississippi River Delta, marine geologic and phys. oceanographic instrumentation, ocean engring. Home: 987 S.W. 13th Pl., Boca Raton, Fla. 33432.

McBAIN, James William, chemist; b. Chatham, N.B., Can., Mar. 22, 1882; s. James Afleck Fraser (D.D.) and Mary Morrison (Quin) McB.; B.A., U. Toronto, 1903, M.A., 1904; postgrad. U. Leipzig, 1904-05; Ph.D., U. Heidelberg, 1906; D.Sc. (hon.), Brown U., 1923, U. Bristol (Eng.), 1928; m. Mary Evelyn Laing, Jan. 1, 1929; children—Janet Quin, John Keith. Lectr. phys. chemistry U. Bristol, 1906-19, Leverhulme prof. in phys. chemistry, 1919-26; vis. prof. U. Cal., 1926; prof. chemistry Stanford, 1927-47, emeritus, 1947—. Dir. Nat. Chem. Lab. Poona, India, 1949-52. Fellow Royal Soc., 1923 (Davy medal 1939), Royal Inst. Chemistry; mem. Am. Chem. Soc., Nat. Inst. Social Scis., Chem. Soc. (British), Bunsen Gesellschaft, Faraday Soc. (hon. life mem., v.p. 26-29), Soc. Rheology (v.p. 1946), Sigma Xi, other socs. Author: The Sorption of Gases and Vapours by Solids, 1931; Textbook of Colloid Science, 1949; also papers. Editor Jour. Colloid Science. Died Mar. 12, 1953.

McBEE, Earl Thruston, Am. chemist; b. Braymer, Mo., July 6, 1906; s. William and Lydis (Post) McB.; A.B. in Sci. and Edn., William Jewell Coll., 1929; M.S., Purdue U., 1931; Ph.D., 1936; m. Viola Lowman, Feb. 15, 1962; children (by previous marriage)—Beverly (Mrs. Thomas Herbig), Robert; stepchildren—David Lowman, Donald W. Lowman, Shirley (Mrs. Joseph Toscano). Mem. faculty Purdue U., Lafayette, Ind., 1929—, prof. chemistry, 1943-67, 1st Shreve prof. indsl. chemistry, 1967—, head dept. chemistry, 1949-67. Mng. dir. Gt. Lakes Chem. Corp., Lafayette, 1957—; pres. Ark. Chems., Inc., Lafayette, 1960—; dir. Chapman Chem. Corp., Memphis. Cons. to govt. and industry; chmn. adv. bd. U. S. Naval Propellent Plant, Indian Head, Md., 1953-62. Recipient Modern Pioneer award N.A.M., 1940, Certificate of Effective Service in Prodn. Atomic Bomb. U. S. War Dept. Manhattan Project, 1945, Certificate Effective Service in Prosecution 2d World War OSRD, 1945, Achievement award William Jewell Coll., 1949. Fellow Ind. Acad. Scis.; mem. Am. Chem. Soc., A.A.A.S., Am. Inst. Chemists, N.Y. Acad. Sci., Sigma Xi (Ann. Research award Purdue chpt. 1946), Phi Lambda Upsilon. Patentee chem. field. Research and numerous publs. in chlorination, fluorination, oxidation, nitration, silanes, silicones, various polymerization reactions. Home: 420 Forest Hill Dr., West Lafayette, Ind. 47906.*

McBEE, Richard Harding, Am. microbiologist; b. Eugene, Ore., May 15, 1916; s. Elmer Francis and Cora (Clow) McB.; B.S., Ore. State U., 1938, M.S., 1940; Ph.D., Wash. State U., 1948; m. Virginia Helen Brown, June 15, 1940; children—Gail Elizabeth (Mrs. James O. Smith), Richard Harding, Christopher Alan, Anne Katherine. Faculty, Mont. State U., Bozeman, 1949—, prof. microbiology, 1955—, head dept., 1964—, dean Coll. Letters and Sci., 1968—; dir. McBee Lab., 1952—; cons. Anaconda Co., 1962—. Diplomate Am. Acad. Microbiology. Fellow A.A.A.S., Am. Acad. Microbiology; mem. Am. Soc. Microbiology, Am. Chem. Soc., Am. Soc. Animal Sci., Am. Soc. Profl. Biologists. Author: (with W. G. Walter) General Microbiology, 1955. Research in thermophilic fermentations of cellulose, rumen fermentations in wild animals, cecal fermentation in birds and rodents, anaerobic culture methods for clin. labs. Home: Sourdough Rd., Bozeman, Mont. 59715.*

McBRIDE, Earl Duwain, Am. orthopedic surgeon; b. Servery, Kan., June 16, 1891; s. Aaron and Almeda (Tucker) McB.; B.S., U. Okla., Med. Sch., 1912; M.D., Columbia, 1914; m. Pauline Mary Wahl, Sept. 2, 1913; children—Pollyana (Mrs. William K. Ishmael), Mary Frances (Mrs. Richard Tullins), Dorothy Lu (Mrs. Patrick Malloy). Practice gen. medicine, Ralston, Okla., 1915-17; practice orthopedic surgery, Oklahoma City, 1919—; founder Crippled Children's Clinics Okla., 1920, McBride Clinic, Oklahoma City, Bone and Joint Hosp., Oklahoma City, 1926; clin. prof. emeritus Okla. U. Sch. Medicine. Mem. A.M.A., Am. Orthopedic Assn., Am. Acad. Orthopedic Surgery, Assn. Bone and Joint Surgeons. Author: Crippled Children, Their Nursing Treatment; Disability Evaluation; also numerous articles. Invented operation for correction of bunion deformity of foot, back fusion operation, one of 1st 3 hip replacement prostheses. Home: 1705 Dorchester Pl., Oklahoma City 73120. Office: 605 N.W. 10th St., Oklahoma City 73103.*

McBURNEY, Charles, Am. surgeon; b. Roxbury, Mass., Feb. 19, 1845; s. Charles and Rosine (Horton) McB.; A.B., Harvard, 1866, A.M., 1869; M.D., Columbia U., 1870; m. Margaret W. Weston, Oct. 8, 1874. Asst., demonstrator anatomy Coll. Phys. and Surg., Columbia U., 1872-89, prof. surgery, 1889-92, prof. clin. surgery, 1892-1907, became emeritus, 1907; vis. surgeon St. Luke's Hosp., 1875-79, Bellevue Hosp., 1888-1900; cons. surgeon N.Y. Presbyn., St. Luke's, St. Mark's, Orthopedic hosps., Hosp. for Ruptured and Crippled, N.Y.C. Hon. fellow Royal Coll. Surgeons Edinburgh; mem. Coll. Physicians Phila. Discovered McBurney's point, a pressure point used in diagnosis of appendicitis; developed technique for appendectomy which avoided cutting across abdominal muscle fibers (McBurney's incision). Died Brookline, Mass., Nov. 7, 1913.

McCABE, Louis C(ordell), Am. geologist; b. Graphic, Ark., Feb. 5, 1904; s. Mark L. and Hattie Nolan (Stokes) McC.; student U. Ark., 1923-25, Northwestern, 1926-28; B.S., U. Ill., 1931, M.S., 1933, Ph.D., 1937; m. Catherine Hesselschwerdt, Dec. 20, 1936; children—Michael, Dorothy, John, Jean. Asst. Ill. Geol. Survey, 1927, asst., asso. geologist, later geologist, 1930-41; geologist Miss. Coal Corp., 1928-30; mining engr. Am. Gas & Elec. Co., 1945-46; chief coal br. U. S. Bur. Mines, 1946-47; dir. Los Angeles Co. Air Pollution Control Dist., 1947-49; chief, Office Air and Stream Pollution, U. S. Bur. Mines, 1949-51, chief fuels and explosives div., 1951-55, sci. dir. USPHS, 1955; pres. Resources Research, Inc., 1956—; chmn. bd. Hazleton Labs., Inc., Falls Church, Va., 1963—; cons. engring. scientist, N.Y. U., 1956-59. Dept. State rep. coal mining com. Internat. Labor Orgn., Geneva, Switzerland, 1947, 51. Decorated Legion of Merit; Order of Brit. Empire; Order of Crown (Belgium). Fellow Geol. Soc. Am., A.A.A.S.; mem. Am. Inst. Mining and Metall. Engrs., Soc. Econ. Geologists, Air Pollution Control Assn. (v.p.), Am. Soc. for Testing and Materials, Washington Acad. Sci., Ill. Mining Inst., Am. Chem. Soc., Sigma Xi. Contbr. articles profl. jours. Editor: Air Pollution, 1952. Contbg. editor profl. jours. Research in environmental and phys. scis. Home: 7304 Brennon Lane, Chevy Chase, Md. 20015. Office: Box 90, Falls Church, Va. 22046.*

McCABE, Warren Lee, Am. chem. engr.; b. Bay City, Mich., Aug. 7, 1899; s. James C. and Frances (Cooke) McC.; B.S., U. Mich., 1922, M.S., 1923, Ph.D., 1928; m. Lillian F. Hoag, July 29, 1925; children—Warren Lee, Barbara (Mrs. Paul C. Moran, Jr.). Instr., Mass. Inst. Tech., Cambridge, 1923-25; faculty U. Mich., Ann Arbor, 1925-36; prof., head dept. chem. engring. Carnegie Inst. Tech., 1936-47; adminstrv. dean Polytech. Inst. Bklyn., 1953-64, acting head dept. engring., 1961-64; dir. research Flintkote Co., Whippany, N.J., 1947-53, v.p., 1949-53; faculty N.C. State U., Raleigh, 1964—, R. J. Reynolds prof. chem. engring. 1965—. Vice pres. Engrs. Council Profl. Devel., 1957-59; mem., cons. Engring. Manpower Commn., 1953-64; cons. to industry, 1925—; Fulbright lectr., 1965. Co-recipient (with Adm. Arleigh Burke) Golden Key award N.E.A., 1961. Mem. Am. Inst. Chem. Engrs. (pres. 1950, William H. Walker award 1938), Am. Chem. Soc., Am. Soc. M.E., Am. Soc. for Engring. Edn., Sigma Xi, Tau Beta Pi, Phi Lambda Upsilon. Author: (with W. L. Badger) Elements of Chemical Engineering, 1930; (with J. C. Smith) Unit Operations of Chemical Engineering, 1956, 67. Co-developer McCabe-Thiele graphical method for design indsl. separation equipment for complex mixtures; originator of McCabe law of crystal growth; introducer of thermodynamic diagrams for design of process equipment. Home: 222 Ransom St., Chapel Hill, N.C. 27514.*

McCAFFERY, Edward Lawrence, Am. chemist; b. Bklyn., Mar. 13, 1929; s. Edward L. and Madeline (Quinn) McC.; B.S., St. John's U., 1951; M.S., U. Md., 1954, Ph.D., 1955; m. Marie Burke, June 22, 1957; children—Edward, Marianne, Eileen, Jean. Research chemist E. I. du Pont de Nemours, Phila., 1955-58; cons. Avco Research and Advanced Devel., Wilmington, Mass., 1959-63; prof. chemistry Lowell Tech. Inst., Lowell, Mass., 1958—. Recipient Outstanding Contribution to Devel. of Youth award Columbian Squires, 1965. Mem. Am. Chem. Soc. (regional rep. for polymer sci. sect. 1964—), Chem. Soc. London, Sigma Xi. Publs. on determination of magnitude of steric effects in some displacement reactions, formulated a mechanism for ablation of polymers, synthesized a number of new polymer systems; research in chemistry of polymetallophilic organic compounds. Home: Russell Rd., R.F.D. 1, Greenville, N.H. 03048.*

McCAIN, Arthur Hamilton, Am. plant pathologist; b. San Francisco, Aug. 31, 1925; s. Arthur Hamilton and Eleonore (Fentzling) McC.; B.S., U. Cal. at Davis, 1949, Ph.D., 1959; m. Elma Doris Friesen, May 30, 1959; 1 son, Arthur H. Agriculturist, Holly Sugar Corp., Wasco, Cal., 1953-55; research asst. dept. plant pathology U. Cal. at Davis, 1955-59; plant pathologist Agrl. Extension Service, U. Cal. at Berkeley, 1959—. Cons., U. S. Forest Service, 1962—. Mem. Am. Phytopath. Soc., A.A.A.S., Sigma Xi. Author: Chemicals for Plant Disease Control, 1965; also numerous articles. Research on inheritance of pathogenicity of rust of safflower, chem. control of root rots of container-grown plants; discovered leaf spot of ti, root rot of ivy and pines; concentration and isolation of Phytophthra cinnamoni from soil. Home: 16 Edwin St., Kensington, Cal. 94707. Office: Hilgard Hall, U. Cal., Berkeley, Cal. 94720.*

McCALL, Raymond Joseph, Am. psychologist; b. N.Y.C., Oct. 16, 1913; s. John Joseph and Helen G. (Whalen) McC., A.B., Fordham Coll., 1934; M.A., Cath. U., 1936; Ph.D, Fordham U., 1941; M.A., Columbia U., 1949, Ph.D, 1951; m. Mary Elizabeth McDonald, June 19, 1937; children—Raymond Joseph, Anthony L., Timothy B. Faculty, St. John's U., Bklyn, 1936-51, prof., 1946-51, chmn. dept. philosophy and psychology, 1946-51; Rorschach examiner depts. research psychology and psychiatry N.Y. Psychiat. Inst., 1949-51; prof. chmn. dept. psychology De Paul U., Chgo., 1951-56; prof. Marquette U, Milw., 1956—, chmn. dept. psychology, dir. Guidance Center, 1956-61, dir. clin. tng., 1961—; Ford Found. Faculty fellow Harvard, 1954-55. Cons. psychologist Grace Clinic, Bklyn., 1950-51, Chgo. Municipal Cts. Psychiat. Inst, 1951-52. Fellow Am. Psychol. Assn.; mem. Wis., (sec. 1958-61, 62—, pres. 1965-66), Am. Cath. (past pres.), Milw. County (past pres.) psychol. assns., Am. Assn. U. Profs. (past pres. Marquette U. chpt.), Sigma Xi. Author: Basic Logic, 1952; A Preface to Scientific Psychology, 1959. Contbr. articles to profl. jours. Home: 2814 E. Newberry Blvd. Milw. 53211.*

McCALLUM, Graham John, New Zealand physicist; b. Christchurch, New Zealand, Feb. 11, 1928; s. John Hay and Estella (Clarkson) McC.; B.Sc., Victoria U., Wellington, New Zealand, 1949, M.Sc. with honors, 1951; m. Ann Allan, Mar. 3, 1956; children—John, Bruce, Katrina. Staff div. nuclear scis. New Zealand Dept. Sci. and Indsl. Research, Lower Hutt, 1952-56; physicist nuclear physics div. A.E.R.E., Harwell, Eng., 1956-59, Chalk River (Ont. Can.) Labs., A.E.C.L., 1959-61; scientist Inst. Nuclear Scis., Lower Hutt, 1961-65, head accelator physics group, 1966—; vis. fellow dept. nuclear physics Australian Nat. U., Canberra, Australia, 1965-66. Mem. Am. Phys. Soc. Research and articles on natural radioactivity in soils, Carbon 14 age determination measurements, low energy nuclear physics research principally in light nuclei. Home: 78 Bell Rd. Office: Inst. Nuclear Scis., Pvt. Bag, Lower Hutt, New Zealand.*

McCALLUM, K(enneth) J(ames), Canadian chemist; b. Scott, Sask., Can., Apr. 25, 1918; s. James Alexander and Alice (Fines) McC.; B.Sc., U. Sask., Saskatoon, 1936, M.Sc., 1939; Ph.D., Columbia U., 1942; m. Christine Chorneyko, Sept. 2, 1950; children—Patricia Jean, Douglas James. Jr. research officer Nat. Research Council Can., 1942-43; faculty U. Sask., 1943—, prof. chemistry, 1953—, head dept. chemistry and chem. engring., 1959—; dir. sci. affairs Chem. Inst. Can., 1961-64. Fellow Royal Soc. Can., A.A.A.S., Chem. Inst. Can. Asso. editor Jour. Chem. Physics, 1947-50. Research and publs. in cement chemistry, electron affinities, radioisotope exchange reactions, chem. effects nuclear transformations, radiation chemistry. Home: 1622 Park Av., Saskatoon, Sask., Can.*

McCAMAN, Richard Eugene, Am. biochemist; b. Barberton, O., Mar. 28, 1930; s. Orven L. and Iva (Stoll) McC.; A.B., Wabash Coll., 1952; Ph.D., Washington U., St. Louis, 1957; m. Marilyn Ruth Wales, Feb. 8, 1953; children—Michael Thomas, David Patrick, Judith Suzanne, Sandra Jean, Christopher Andrew. Mem. faculty Sch. Medicine Ind. U., Indpls., 1957—, asso. prof. pharmacology and psychiatry, 1964—, research asso. Inst. Psychiat. Research, 1957-62, prin. investigator in neurochemistry, 1963—. Mem. A.A.A.S., Am. Oil Chemists Soc., Histochem. Soc., Am. Soc. Biol. Chemists, Assn. for Research in Nervous and Mental Diseases. Research and publs. concerned with identification of significant chem. processes asso. with specific functional changes in normal and pathologic brain, especially study of neurohumoral metabolism and intermediary metabolism of phospholipids. Home: 1750 E. 80th St., Indpls. 46240.*

McCANCE, Andrew, British metallurgist; b. Cadder, Scotland, Mar. 30, 1889; s. John and Jessie (Ferguson) McC.; D.Sc., Royal Sch. Mines, London, Eng., 1910, LL.D., 1948; m. Joya Harriet Gladys Burford, Mar. 6, 1936; children—Jean Nathalie, Margaret Charmian. With William Beardmore & Co., Glasgow, Scotland, 1910-19; founder Clyde Alloy Steel Co. Ltd., Motherwell, Eng., 1919; adv. metallurgist David Colvill & Sons, 1919-30, dir., 1930—; dir. Colvilles Ltd.-David Colville & Sons Ltd., also James Dunlop & Co. Ltd., Glasgow, 1931—, chmn., mng. dir., 1956-65, hon. pres., 1965—; dir. various cos. including Nat. Comml. Bank Scotland Ltd. Fellow Royal Soc., 1943; mem. Iron and Steel Inst., West of Scotland Iron and Steel Inst., Instn. Engrs. and Shipbuilders in Scotland, Brit. Iron and Steel Research Assn., Inst. Welding. Research and publs. on sci. steel mfr. and properties of metals. Home: The Craigs, Carmunnock, Glasgow. Office: 195 W. George St., Glasgow, C.2., Scotland.*

McCANN, Samuel McDonald, Am. physiologist; b. Houston, Sept. 8, 1925; s. Samuel Glenn and Margaret (Brokawa) McC.; student Rice U., Houston, 1942-44; M.D., U. Pa., 1948; m. Barbara Lorraine Richardson; children—Samuel Donald, Margaret, Karen Elizabeth. Faculty, U. Pa., 1952-65, prof. physiology Sch. Medicine, 1964-65; prof., chmn. physiology U. Tex. Southwestern Med. Sch., Dallas, 1965—. Cons., Schering Corp., Bloomfield, N.J., 1958; mem. gen. medicine B study sect. NIH, 1965—. Recipient Spencer Morris prize U. Pa. Med. Sch., 1948, Lindback award for distinguished teaching, 1965, Oppenheimer award for research in endocrinology, 1966. Mem. Am. Physiol. Soc., Endocrine Soc., Soc. for Exptl. Biology and Medicine, N.Y. Acad. Scis., Neuroendocrine Discussion Group (chmn. 1965). Contbr. numerous articles to tech. jours, chpts. to books. Editorial bd. Endocrinology, 1963—. Research on hypothalamus; demonstrated that hypothalamus controls urge to drink water, output pituitary hormones by secreting polypeptide hormones.*

McCANN, William Sharp, Am. physician; b. Cadiz, O., July 6, 1889; s. Charles F. and Carolyn (Sharp) McC.; B.A., Ohio State U., 1911, D.Sc., 1934; M.D., Cornell U., 1915; LL.D., Hobart- William Smith Coll., 1954; m. Gertrude Fisher, Dec. 29, 1916 (dec.); children—Elizabeth M. (Mrs. Adams), William P. Research fellow Russell Sage Inst. Pathology, Cornell Med. Sch., 1919-21; asso. prof. Johns Hopkins, 1921-24; prof. medicine, physician in chief U. Rochester Sch. Medicine, 1925-57, emeritus prof., 1957—. Dep. chmn. comm. on med. scis. research and devel. bd. U. S. Dept. Def. Diplomate Am. Bd. Internal Medicine (chmn.). Fellow A.C.P. (hon. master 1960, regent), mem. A.M.A., Am. Soc. for Clin. Investigation (v.p.), Assn. Am. Physicians (past pres.). Author: Calorimetry in Medicine, 1923; also numerous articles. Research in metabolism fever, calorimeter metabolism in diabetes, nephritis, pernicious anemia, pulmonary function during exercise, pneumono-koniosis, and emphysema, relation blood flow and ventilation in lungs in various disorders lungs and heart. Home: 80 Rossiter Rd., Rochester 14620. Office: Strong Meml. Hosp., Rochester, N.Y. 14620.*

McCARROLL, Russell Hudson, Am. chem. and metall. engr.; b. Detroit, Feb. 20, 1890; s. John and Emily (Roberts) McC.; B.S. in Chem. Engring., U. Mich., 1914, M.S. in Engring. (hon.), 1937; m. Muriel C. Channer, Sept. 30, 1916; children—Charlotte Jane (Mrs. Charles R. Vincent, Jr.), Marjorie. With Solvay Process Co., Detroit, 1914-15; with dept. of chem. and metall. engring. Ford Motor Co., Dearborn, Mich., since 1915, in charge dept. since 1921, became co. exec. engr., in charge of research, metall. engring., chem. engring. and automotive engring., 1944, later dir. chem. and metall. engring. director of chemical and metallurgical engineering and research. Mem. Am. Soc. for Metals, Am. Chem. Soc., Iron and Steel Inst. (London), Soc. of Automotive Engrs., Engring. Soc. Detroit (sec.), Am. Foundrymen's Assn., Soc. for Promotion of Engring. Edn., Contbr. articles to profl. jours. Granted more than 20 metall. and engring. patents; engaged in expts. designed to widen indsl. uses of farm produced materials. Died Mar. 31, 1948.

McCARTHY, Paul Joseph, Am. mathematician; b. Chgo., Oct. 23, 1928; s. Francis Anthony and Elizabeth (Loosen) McC.; B.S. in Elec. Engring., U. Notre Dame, 1950, M.S., 1952, Ph.D., 1955; m. Lois Jean White, June 11, 1952; children—Margo Jean, Mark Joseph. Faculty, U. Notre Dame, 1954-55, Coll. of Holy Cross, Worcester, Mass., 1955-56, Fla. State U., 1956-61; faculty U. Kans., Lawrence, 1961—, prof. math., 1965—. Mem. Am., London, Indian math. socs., Math. Assn. Am., Am. Assn. U. Profs., Sigma Xi, Pi Mu Epsilon. Contbns. to number theory, algebra, and spl. functions. Home: 323 Dakota St., Lawrence, Kans. 66044.*

McCARTNEY, James Lincoln, Am. neuropsychiatrist; b. Chungking, China (parents Am. citizens), July 24, 1898; s. James Henry and Saddie (Kissack) McC.; student O. Wesleyan U., Delaware, 1914-17; B.S., U. Chgo., 1921; postgrad. Peking Union Med. Coll., China, 1920-21; M.D., Rush Med. Coll., 1923; postgrad. Columbia, 1928, 1942, N.Y. U., 1946; m. Edith Tufts, Dec. 27, 1924; children—Helen E. (dec.), Helen C. (Mrs. James N. Gettemy), Joan E. (Mrs. Frederick J. Cusick), James R. Asst. physiology U. Chgo., 1918-20, 1921-23; instr. neurophysiology Peking Union Med. Coll., China, 1920-21; intern Bklyn. Hosp., 1923-24; practice medicine, China, 1924-27; psychiatrist Inst. Child Guidance, N.Y.C., 1928-29; dir. Bur. Mental Hygiene, Conn. Dept. Health, 1929-31; dir. Classification Clinic, Elmira Reformatory, N.Y. Dept. Correction, 1931-34, Boy's Indsl. Sch., 1935-37, dir. N.W. Retreat, 1934; dir. psychiat. dept. Battle Creek Sanitarium, 1935; asso. med. dir. Sharpe & Dohme Co., 1937-40; med. dir. William R. Warner, 1940-42; practice neuropsychiatry, Garden City, N.Y., 1945-65; asso. prof. Southampton Coll., L.I. U., 1965—. Dir. Transinternat. Psychosomatic Seminars, 1960—. Recipient, John Freer medal, Rush medal, 1923, Fisk prize, R.I. Med. Soc., 1924, Salmon award, N.Y. Acad. Medicine, 1933. Fellow A.C.P., Am. Psychiat. Assn., Med. Writer's Assn., Nassau Acad. Medicine, N.Y. Acad. Medicine, Am. Soc. Clin. Hypnosis, Nassau Neuropsychiat. Soc. (organizer, pres. 1947), Acad. Psychosomatic Medicine; mem. A.M.A., N.Y. State Med. Soc., Nassau Med. Assn., Soc. Clin. and Exptl. Hypnosis, A.A.A.S., Assn. for Advancement Psychotherapy, Soc. Psychoanalytic Physicians. Author: Classification of Prisoners, 1935; Understanding Human Behavior, 1956; Frustrated Martyr, 1953; Drama of Sex, 1946. Research and numerous publs. in female sterilization, practice of psychiatry throughout the world, climate and mental illness. Office: 45 Oneck Lane (P.O. Box 1309), Westhampton Beach, N.Y. 11978.*

McCARTY, Daniel John, Jr., Am. physician; b. Phila., Oct. 31, 1928; s. Daniel John and Margaret (Gallagher) McC.; B.S. in Biology, Villanova Coll., 1950; M.D., U. Pa., 1954; m. Constance Helen Paakh, Aug. 28, 1954; children—Constance Elizabeth, Claire Louise, Margaret Helen, Daniel John, Brian Albert. Faculty, Hahnemann Med. Coll. and Hosp., Phila., 1960-67, asso. prof. medicine, 1963-67, head sect. rheumatology, sr. attending physician, 1960-67; chief arthritis clinic div. B. Phila. Gen. Hosp., 1962-67; prof. medicine, head sect. arthritis and metabolic diseases, div. biol. scis., dept. medicine, U. Chgo., 1967—. Cons., VA Hosp., Phila., Cons., Surgeon Gen., mem. tng. grants com. USPHS. Recipient Russell Cecil Nat. award Arthritis Found., 1965, Gairdner Found. Internat. Ann. award, 1965; Markle scholar in acad. medicine, Fellow A.C.P.; mem. Am. Fedn. Clin. Research, Am. Soc. Clin. Investigation, A.M.A. (Hektoen silver medal, 1963), Am. Rheumatism Assn., Am. Therapeutic Soc., Arthritis and Rheumatism Found. (med. adv. council 1960—, research com., 1964—), N.Y. Acad. Scis., A.A.A.S. Editor-in-chief: Arthritis and Rheumatism. Contbr. numerous articles to tech. jours. Developed technique for specific identification of sodium urate crystals in gouty joint fluid; discovered calcium pyrophosphate crystals in joint fluid of other patients; characterized signs and symptoms of pseudogout; research on mechanisms crystal-induced joint inflammation. Home: 6918 S. Euclid, Chgo. 60649. Office: 950 E. 59th St., Chgo. 60637.

McCARTY, Harold Hull, Am. geographer; b. Hiteman, Ia., Sept. 10, 1901; s. John Harvey and Maude (Hull) McC.; B.S., U. Ia., 1923, Ph.D., 1929; m. Vivian E. McClenahan, June 6, 1924; children—John E., Robert W. Faculty, U. Ia., Iowa City, 1923—, prof. geography, 1946—; vis. prof. Canterbury U., Christchurch, New Zealand, 1962. Mem. Assn. Am. Geographers (Outstanding Achievement award 1965), Am. Geog. Soc., Nat. Council Geog. Edn., A.A.A.S. Author: Geographic Basis of American Economic Life, 1940; (with James B. Lindberg) Preface to Economic Geography, 1966; also articles. Research on application of sci. methods to problems in econ. geography especially those related to mfg. and agr. Home: 441 Magowan Av., Iowa City 52240.*

McCARY, James Leslie, Am. psychologist; b. Winkler, Tex., Apr. 5, 1919; s. F. Leslie and Lucille (Ferguson) McC.; B.S., N. Tex. State Coll., 1940; M.S., U. Houston, 1942; postgrad. U. Pitts., 1946-47; Ph.D., U. Tex., 1949; m. LaVirle Precise, Aug. 15, 1938; children—Mary Lesley (Mrs. Pierre Marcel Schlumberger), Stephen Paul. Psychologist, Austin (Tex.) State Hosp., 1947-48; instr. U. Tex., 1947-48; prof. psychology U. Houston, 1948—; chief psychologist Almeda Clinic, Houston, 1948—, Oak Pl. Hosp., Houston, 1949-66; lectr. U. Tex. Med. Sch., Galveston, 1958—. Diplomate Am. Bd. Examiners in Profl. Psychology. Mem. Am., S.W., Tex. psychol. assns., Am. Assn. Marriage Counselors, Soc. for Projective Techniques. Author: (with D. E. Sheer) Six Approaches to Psychotherapy, 1955; The Psychology of Personality, 1956; Introduction to Sexology, 1966; Human Sexuality, 1967; also articles. Research on frustration and aggression. Home: 8730 Meml. Dr., Houston 77024.*

McCASKEY, Hiram Dryer, Am. geologist; b. Ft. Totten, Dakota Ty. (now N.D.), Apr. 10, 1871; s. William Spencer and Eleanor Forsythe (Garrison) M.; B.S., Lehigh U., 1893, M.S., 1907; m. Mary Louise Fuller, June 7, 1913. Chemist, Boston & Mont. Smelter, Great Falls, Mont., 1893-95; instr. math. St. Thomas Hall, Miss., 1895-96, headmaster and instr. English, 1896-98; instr. math. and English St. Matthew's Sch., Cal., 1898-1900; mining engr. Mining Bur., Manila, P.I., 1900-03; chief of Mining Bur. and Div. of Mines, Bur. Science, Manila, 1903-06; asst. geologist U. S. Geol. Survey, 1907-10, geologist, 1911—, chief sect. of metal resources, 1912-19, geologist in charge Div. Mineral Resources, 1915-19. Del. from P.I. to 10th Internat. Geol. Congress, Mexico City, 1906; exploratory field work in P.I.; reorganized Mining Bur. in P.I.; asst. in reorgn. of work of U. S. Geol. Survey in multiple mineral resources of U. S. and in charge 1912-19; U. S. del. Internat. Engring. Congress, San Francisco, 1915, Pan-Am. Sci. Congress, 1917. Died Apr. 26, 1936.

McCASLIN, Murray Frew, Am. ophthalmologist; b. Lawrence County, Pa., Feb. 21, 1904; s. Scott and Elizabeth (Frew) McC.; B.S., U. Pitts., 1926, M.D., 1929; postgrad. Mayo Clinic, U. Minn., 1930-31; m. Harriet Wilson, Oct. 3, 1936; children—Janet (Mrs. Christopher Larsen), Ellen (Mrs. Chas. Srodes), W. Scott. Practice medicine, specializing ophthalmology, Pitts., 1931—; prof., chmn. dept. ophthalmology U. Pitts. Sch. Medicine, 1952—; prof. Sch. Spl. Edn. and Rehab.; chief, eye service Eye and Ear Hosp., Children's Hosp., Pitts.; cons. ophthalmology VA Hosp., Pitts., also Allegheny, Columbia, Elizabeth Steel Magee, Howard Marcy hosps., Western Pa. Sch. and Hosp., Canonsburg, Western Pa. Sch. for Blind Children. Mem. bd. United Fund, Health, Service and Research Found. Mem. Am. Ophthal. Soc., A.C.S., Am. Acad. Ophthalmology and Otolaryngology, Am. Assn. Ry. Surgeons. Home: 2731 Oak Hill Farms, Allison Park, Pa. 15101. Office: 550 Grant St., Pitts. 15219.*

McCAULEY, Roy Barnard, Jr., Am. phys. chemist; b. Chgo., Feb. 9, 1919; s. Roy Barnard and Irene E. (Morris) McC.; B.A., Cornell Coll., 1940; M.S., Ill. Inst. Tech., 1943; m. Audrey Paulsen, Oct. 10, 1941; children—Roy Barnard III, Paul Thomas, Robert William, Andrew John. Faculty, Ill. Inst. Tech., 1940-50; prof., chmn. dept. welding engring., dir. welding research Ohio State U., Columbus, 1950—. Cons. Armour Research Found. Recipient Meritorious Certificate award Am. Welding Soc., 1959, Adams Meml. mem. Am. Welding Soc., 1960. Mem. Am. Welding Soc. (pres.), Internat. Inst. Welding (chmn., pres. commn. edn.), Am. Soc. Engring. Edn., Am. Soc. Metals, Soc. Nondestructive Testing, Sigma Xi, Tau Beta Pi, Phi Lambda Upsilon, Pi Tau Sigma, Sigma Gamma Epsilon. Research on effect of defects on welds and organized information concerning discontinuities detectable by nondestructive testing methods, states of stress, bond mechanisms, discontinuity evaluations of manufactured metals and alloys. Home: 7265 Linworth Rd., Worthington, O. 43085. Office: 190 W. 19th Av., Columbus, O. 43210.*

McCAULEY, William John, Am. zoologist; b. Ray, Ariz., Aug. 11, 1920; s. Willard Jay and Evelyn (Boyer) McC.; B.S., U. Ariz., 1947; Ph.D., U. So. Cal., 1955; m. Mary Apman Michaels, Sept. 27, 1949; children—Lorna Ann, Heather Robin. Research asso. Lederle Labs. div. Am. Cyanamid Co., 1947-49; NIH fellow U. So. Cal., 1950-52, instr., 1952-55; faculty U. Ariz., Tucson, 1955—, prof., 1964—. Mem. Ariz. State Bd. Examiners in Basic Scis., 1956-67, pres., 1963-67. Mem. A.A.A.S., Am. Soc. Zoologists, Western Soc. Naturalists, Mexican Soc. Natural History, Sigma Xi. Author: (with E. L. Cockrum) Zoology, 1965; Human Anatomy, 1962; (with Cockrum, N. Younggren) Biology, 1966; also articles. Research in pharmacology of aureomycin, lymphatic system of reptiles. Home: 1714 N. Forgeus St., Tucson 85716.*

McCAWLEY, Elton Leeman, Am. pharmacologist; b. Long Beach, Cal., Nov. 1, 1915; Arthur Temple and Lavernia (Leeman) McC.; A.B., U. Cal. at Berkeley, 1938; Ph.D., U. Cal. Med. Sch., San Francisco, 1942; m. Leila Annette Calongé, Nov. 16, 1941; children—Melindé Devon, Douglass Floyd. Postdoctoral fellow U. Cal. Med. Sch., San Francisco, 1942-43, instr., 1942-43; faculty Yale Sch. Medicine, 1943-45, asst. prof. pharmacology, 1945-49; prof. pharmacology U. Ore. Med. Sch., Portland, 1949—; asst. toxicologist City of San Francisco. Recipient Honors Achievement award Angiology Research Found. Mem. Am. Soc. Pharmacology and Exptl. Therapeutics, Ore. Med. Assn., Western Pharmacology Soc. (past pres.). Contbg. author: Cardioactive Alkaloids in the Alkaloids, 1955. Research, publs. on synthesis and biol. evaluation of 1st narcotic antagonist N-allylnormorphine, prodn. of cardiac disease in exptl. animals (atrial fibrillation), changing sensitivities of drugs in transplanted cadaver hearts (canine). Home: 210 N.E. Laurelhurst Pl., Portland, Ore. 97202.*

McCAY, Leroy Wiley, Am. chemist; b. Rome, Ga., Aug. 9, 1857; s. Robert T. and Susan L. (Wiley) M.; A.B., Princeton, 1878, A.M., 1881, D.Sc., 1883; student Freiberg Sch. of Mines, U. Heidelberg. Asst. in analytical chemistry Princeton, 1883-86, instr., 1886-89, asst. prof., 1889-92, prof. inorganic chemistry, 1892-1928, emeritus prof., 1928—. Research on cobalt, nickel and iron pyrites, methods for determining arsenic, nonexistence of sulpharsenic acid, sulphoxyarsenic acids and their salts, separation of the metals of the tin group, use of hydrofluoric acid in ordinary and electro-chem. analysis, use of mercury as a reducing agt. Died Apr. 13, 1937.

McCHESNEY, Evan William, Am. chemist; b. Oconto Falls, Wis., Dec. 11, 1904; s. William Henry and Lydia (Glise) McC.; S.B., U. Chgo., 1926, M.S., 1928; Ph.D., Northwestern U. Med. Sch., 1931; m. Arline M. Feltham, Mar. 26, 1933; children—Richard W., Ruth E. (Mrs. Robert S. Becker), Margaret E. Asso. prof. U. N.C., Chapel Hill, 1931-37, Baylor U., Dallas, 1937-38; sr. biochemist Sterling-Winthrop Research Inst., Rensselaer, N.Y., 1938-68; asso. prof. toxicology Inst. Exptl. Pathology and Toxicology, Albany (N.Y.) Med. Coll., 1968—. Lectr. pharmacology Albany (N.Y.) Med. Coll., 1956-68. Mem. Am. Chem. Soc., Am. Physiol. Soc., Soc. Toxicology, Soc. for Exptl. Biology and Medicine. Contbr. numerous articles to tech. jours. Research on absorption, excretion and metabolism drugs. Home: 14 Alden Ct., Delmar, N.Y. 12054. Office: Inst. Exptl. Pathology and Toxicology, Albany Med Coll., Albany, N.Y. 12208.*

McCLEARY, Robert Altwig, Am. physiol. psychologist; b. Dayton, O., Jan. 9, 1923; s. Charles H. and Mae (Altwig) Mc.; B.A., Harvard, 1944; M.D., Johns Hopkins, 1947, Ph.D., 1951; m. Nan S. Brown, Feb. 3, 1945; children—Robert E., Beverly N., Susan E. Asst. prof. U. Mich., Ann Arbor, 1953-57, asso. prof., 1957-61; prof. depts. psychology, physiology U. Chgo., 1961—. Cons. to NIH, 1961—; cons. on aerospace medicine U. S. Air Force, 1960—. Mem. Am. Psychol. Assn., Psychonomic Soc., Soc. Exptl. Psychologists. Author: (with R. Y. Moore) Subcortical Mechanisms of Behavior, 1965. Research and publs. on behavioral functions of brain. Home: 5715 S. Kenwood Av., Chgo. 60637.*

McCLELLAN, George, Am. anatomist, surgeon; b. Woodstock, Conn., Dec. 23, 1796; s. James and Eunice (Eldredge) McC.; grad. Yale, 1816; grad. in medicine U. Pa., 1819; m. Elizabeth Brinton, 1829; 3 children, including George B., John Hill Brinton. Founder Jefferson Med. Coll. (series of pvt. lectures on anatomy and surgery resulted in charter), 1825, prof. surgery, 1825-39, prof. anatomy, 1827-30; a leading ophthalmic surgeon, 1st to remove lens of an eye; obtained charter for Pa. Coll. Med. Sch., 1838, lectr., 1839-43. Author: Principles and Practice of Surgery (completed by son, John Hill Brinton), 1848. Editor: Theory and Practice of Physic (Eberle), 1840. Died Phila., May 9, 1847.

McCLELLAN, Roger Orville, Am. radiobiologist; b. Tracy, Minn., Jan. 5, 1937; s. Orville and Gladys (Paulson) McC.; D.V.M., Wash. State U., 1960; m. Kathleen Mary Dunagan, June 23, 1962; children—Eric John, Elizabeth Christine. Jr. to sr. scientist biology lab. Hanford Labs., Gen. Elec. Co., Richland, Wash., 1957-64; sr. scientist, biology dept. Battelle-N.W., Richland, 1964-65; scientist div. biology and medicine AEC, Washington, 1965-66; dir. fission product inhalation program Lovelace Found., Albuquerque, 1966—; asst. dir. research Lovelace Found., Albuquerque, 1967—. Lectr., Wash. State U., 1963-64. Diplomate Am. Bd. Vet. Toxicology. Mem. Radiation Research Soc., Am. Vet. Med. Assn., A.A.A.S., Health Physics Soc., Soc. for Exptl. Biology and Medicine. Contbns. to improved understanding of metabolism and toxicity of significant radionuclides in exptl. animals; pioneered in devels. leading to increased use of swine and especially miniature swine in biomed. research. Home: 3505 Smith St. S.E., Albuquerque 87106. Office: 5200 Gibson Blvd. S.E., Albuquerque 87108.*

McCLENDON, Jesse Francis, Am. physiologist; b. Lanette, Ala., Dec. 21, 1880; s. James Wooten and Annie (Thompson) McC.; B.S., U. Tex., 1903, M.S., 1904; Ph.D., U. Pa., 1906; m. Maren Petersen, Oct. 29, 1936; children—James Stewart, John Haddaway. Faculty, Randolph-Macon Coll., Ashland, Va., 1906-07, Cornell U. Med. Coll., N.Y.C., 1909-14; faculty U. Minn. Med. Sch., 1914-39, prof. physiol. chemistry, to 1939; prof. physiology Hahnemann Med. Coll., Phila., 1939-49, prof. emeritus, 1949—; investigator radiology Albert Einstein Med. Center, Phila., 1950-62; vis. prof. biology Tohoku U., Sendai, Japan, 1932-33. Mem. Am. Physiol. Soc., A.M.A., Am. Soc. Biol. Chemistry, Am. Inst. Nutrition, Am. Thyroid Assn. Author: Physical Chemistry of Vital Phenomena, 1917; Physical Chemistry in Biology and Medicine (with Medes), 1925; Iodine and Incidence of Goiter, 1939; also numerous articles. Research on application of phys. chemistry to biology and medicine especially to living cell. Home: Route 1, Box 393, Rural Route 1, Norristown, Pa. 19401.*

McCLUER, Robert Hampton, Am. biochemist; b. San Angelo, Tex., Apr. 13, 1928; s. Robert Dabney and Anne (Forrest) McC.; B.A., Rice Inst., 1949; Ph.D., Vanderbilt U., 1955; m. Carol Olsen, June 4, 1949; children—Robert Forrest, Christopher Hampton, Daniel Morgan, Megan Dianna. Research asso. U. Ill., Urbana, 1955-57; faculty Ohio State U, Columbus, 1958—, asso. prof., 1963—. Mem. Am. Chem. Soc., A.A.A.S., N.Y. Acad. Sci., Sigma Xi. Research and publs on chem. constituents of brain, specifically gangliosides. Home: 2775 Kensington St., Columbus, O. 43202.*

McCLUNG, Clarence Erwin, Am. zoologist; b. Clayton, Cal., Apr. 5, 1870; s. Charles Livingston and Annie Howard (Mackey) McC.; Ph.G., U. Kan., 1892, A.B., 1896, A.M., 1898, Ph.D., 1902; postgrad. Columbia, 1897, U. Chgo., 1899; Sc.D., U. Pa., 1940; Franklin and Marshall College, 1941; m. Anna Adelia Drake, Aug. 31, 1899; children—Ruth, Cromwell and Della Elizabeth. Asst. prof. zoology U. Kan., 1897-1900, asso. prof., 1900-06, prof., 1906-12, head dept. and curator vertebrate paleontol. collections, 1902-12, acting dean, Sch. Medicine, 1902-06; prof. zoology and dir. zool. lab. U. Pa., 1912-40, emeritus prof. since 1940; vis. prof. Keio U., Tokyo, 1933-34, U. Ill., 1940-41; acting head zoology dept. Swarthmore Coll., 1943. Mem. embryological staff, Woods Hole Mass., 1893, since 1914 (trustee); head of sci. expdns. to Ore., Wash., Western Kan., Japan, China, Java, Ceylon, S. Am., S. Africa. Chmn. div. biology and agr. NRC, 1919-21. Fellow A.A.A.S.; mem of Am. Zool. Soc. (pres. 1910, 14), Am. Philos. Soc., Am. Soc. Naturalists (pres. 1927), Acad. Natural Scis. (research asso.), Phila., Nat., Washington, Kan. acads. sci., Am. Assn. Anatomists, Union of Am. Biol. Socs. (pres. 1922-30), Sigma Xi (pres. 1919-21); fgn. mem. La Sociedad de Biologia de Montevideo. Author: Microscopical Technique, Chromosome Theory of Heredity (General Cytology); The Spermatocyte Divisions of the Acrididae, 1900; The Spermatocyte Divisions of the Locustidae, 1902; Restoration of the Skeleton of Bison Occidentalis, 1908; The Spermatogenesis of Xiphidium Fasciatum, 1908; Ichthyological Notes of the Kansas Cretaceous, 1908; also tech. papers on cytology, sex-determination, paleontology. Editor, Handbook of Microscopical Technique; editor, mng. editor Jour. Morphology; mem. bd. editors Acta Zoologica, Cytologia; editor Biol. Abstracts. Extensive research on mechanisms of heredity and sex determinations; formulated theory that accessory chromosomes are sex determinants (sex chromosomes, X and Y chromosomes, or McClung's chromosomes), 1902; studied chromosomes of orthopteran and other insects, with regard to heredity, taxonomy, sex determination. Died Jan. 17, 1946.

McCLUNG, L(eland) S(wint), Am. microbiologist; b. Atlanta, Tex., Aug. 4, 1910; s. Joe Baker and Roxie Buelah (Swint) McC.; A.B., U. Tex., 1931, A.M., 1932; Ph.D., U. Wis., 1934; m. Ruth Wilhelmein Exner, Dec. 25, 1944. Research bacteriologist, Am. Can Co., Maywood, Ill., 1934-36; instr. in fruit products and jr. bacteriologist, Coll. Agr., U. Calif., 1936-37; instr. research medicine, George Williams Hooper Found. for Med. Research, U. Calif., 1937-39; asst. prof., Ind. U., Bloomington, Ind., 1940-44; asso. prof., 1944-48, prof. and chmn. dept. bacteriology, 1948-66, assistant director division of biol. sciences 1965—). Fell., John Simon Guggenheim Meml. Found. and research fellow in bacteriology, Harvard U. Med. Sch., 1939-40. Fellow A.A.A.S., Am. Pub. Health Assn., Am. Acad. Microbiology (gov. 1961-66), Indiana Acad. Sci.; mem. of American Society Microbiology (archivist 1960—), American Association of Immunologists, Society American Bacteriologists (council 1943-47, editorial bd. of Journal Bacteriology, 1953-57, chairman com. on edn., 1958-61, archivist, 1953-60), Am. Inst. Biol. Scis. (governing bd. 1967—), Inst. Food Technology (mem. editorial bd., 1942-45, 1947-50), Soc. Gen. Microbiology, Soc. Exptl. Biology and Medicine (council 1949-50), National Assn. Biology Teachers (1st vice president 1963, president 1965), Soc. Am. Archivists, Soc. Indsl. Microbiology (v.p., 1958, dir. 1960-63), Sigma Xi, Phi Sigma, Gamma Alpha, Alpha Epsilon Delta, Omega Beta Pi. Author: The Anaerobic Bacteria and Their Activities in Nature and Disease: A Subject Bibliography (with Elizabeth McCoy), 2 vols., 1939, supplement 1941. Contbr. articles sci. jours. Prin. research concerns anaerobic bacteria (living only in absence of atmospheric oxygen); taxonomic studies include description of new species; reported new type of food poisoning, enterobacteriaceae, bacterial viruses, sci. edn. Office: Ind. U., Bloomington, Ind. 47401.*

McCLURE, Charles Freeman Williams, Am. anatomist; b. Cambridge, Mass., Mar. 6, 1865; s. Charles Franklin and Joan Elizabeth (Blake) McC.; A.B., Princeton, 1888, A.M., 1892; postgrad. College Phys. and Surg. (Columbia), 1890-91, univs. of Berlin, 1893, Kiel, 1895, Würzburg, 1897; Sc.D. (hon.), Columbia, 1908; m. Grace Latimer Jones, Aug. 25, 1921. Instr. biology Princeton, 1891-95, asst. prof., 1895-1901, prof. comparative anatomy, 1901-34; prof. emeritus. Mem. Peary Relief Expdn., 1899. Fellow A.A.A.S.; mem. Am. Soc. Naturalists, Am. Zool. Soc., Assn. Am. Anatomists (v.p., 1910-11, pres., 1920-21), Am. Philos, Soc., Anatomische Gesellschaft, Phi Beta Kappa, Sigma Xi. Contbr. of numerous papers on anatomy and devel. of vascular system, edema. Editor Anat. Record. Died July 23, 1955.

McCLURE, Donald Stuart, Am. phys. chemist; b. Yonkers, N.Y., Aug. 20, 1920; s. Robert Hirt and Helen (Campbell) McC.; B.Chem., U. Minn., 1942; Ph.D., U. Cal. at Berkeley, 1948; m. Laura Lee Thompson, July 9, 1949; children—Edward, Katherine, Kevin. With War Research Div., Columbia U., 1942-46; faculty U. Cal. at Berkeley, 1948-55; group leader, mem. profl. staff RCA Lab., Princeton, N.J., 1955-62; prof. chemistry U. Chgo., 1962-67; prof. chemistry Princeton U., 1967—. Vis. lectr. U. Ill., Fla. State U., U. So. Cal.; cons. Dept. Def., RCA Labs. Mem. Am. Chem. Soc., Am. Phys. Soc. Author: Electronic Spectra of Molecules and Ions in Crystals, 1950; Some Aspects of Crystal Field Theory, 1964; also numerous articles. Research on electronic spectroscopy, lifetime of phosphorescent states of organic molecules, assignment of electronic state symmetry types, photo-chemistry, application of crystal field theory to inorganic ion spectra. Home: 23 Hemlock Circle, Princeton, N.J. 08540.*

McCLURE, Frank T(relford), phys. chemist; b. Edmonton, Alta., Can., Aug. 21, 1916; s. Frank Austin and Mary (Chapman) McC.; came to U. S., 1942, naturalized, 1945; B.Sc., with first class honors, U. Alta., Edmonton, Can., 1938; Ph.D., U. Wis., 1942; m. Mary Soffa, 1939; children—Charles Frederick, Michael David. Instr., U. Rochester (N.Y.), 1942-43; chief ballistics design div. Allegany Ballistics Lab., George Washington U., Washington, 1943-45; physicist Applied Physics Lab., Johns Hopkins, Silver Spring, Md., 1946-48, acting chmn. Research Center, 1948-49, chmn. research Center, 1949—, asso. dir., 1967—. Mem. space tech. panel Pres.'s Sci. Adv. Com., 1964-67. Carnegie fellow in biophysics, 1959; Overseas fellow Churchill Coll., Cambridge (Eng.) U., 1964-65; Recipient Bur. Naval Ordnance Devel. award, 1945, Presdl. certificate of merit, 1948, Hillebrand prize Chem. Soc. Washington, 1960, NASA Contbns. award, 1961, U. S. Dept. Def. certificate of appreciation, 1964, John Scott award City Phila., 1965. Fellow Washington Acad. Sci.; mem. Am. Chem. Soc., Am. Phys. Soc., Combustion Inst., A.A.A.S., Philos. Soc. Washington, Sigma Xi, Phi Lambda Upsilon. Research and numerous articles on thermodynamics, interior ballistics and acoustic instability in rocket engines, satellite doppler navigational system for ships, concentrating mechanism in bacterial cells. Home: 810 Copley Lane, Silver Spring 20904. Office: 8621 Georgia Av., Silver Spring, Md. 20910.*

McCLURE, Howe Elliott, Am. ecologist; b. Chgo., Apr. 29, 1910; s. Howe Alexander and Clara (Phillips) McC.; B.S., U. Ill., 1933, M.S., 1936; Ph.D., Ia. State Coll., 1941; m. Lucy Esther Lou Fairchild, Oct.

1, 1933; children—Jeannette (Mrs. Sterling Davis), Clara Ann (Mrs. Harry Miles). Biologist, Neb. Game Forestation and Park Commn., 1941-44; ornithologist Hooper Med. Found., Bakersfield, Cal., 1946-50; med. ornithologist 406th Med. Gen. Lab., Dept. Army, Tokyo, Japan, 1950-58, U. S. Army Med. Research Unit, Kuala Lumpur, 1958-63; ornithologist Walter Reed Army Inst. Research Migratory Animal Path. Survey, Tokyo, 1963-66, Bangkok, Thailand, 1966—. Mem. Wildlife Soc., Am. Ornithologists Union, Wilderness Soc., Entomol. Soc. Am., Cooper, Japan ornithol. socs., Malayan Nature Soc., Hawk Mountain Assn. Research and numerous publs. on life history and mgmt. of mourning dove in western U. S., epidemiology of Japanese encephalitis and its relationship to birds, bird migration in Asia. Address: Migratory Animal Pathological Survey, APO, San Francisco 96346.*

McCLURE, Sir Robert John Le Mesurier, Brit. sci. explorer; b. Wexford, Ireland, Jan. 28, 1807; s. Robert and Jane (Elgee) McC.; ed. Eton and Sandhurst; m. Constance Ada Tudor, 1869. Joined Navy, 1824; mate under Capt. G. Bock on board Terror, Arctic voyage, 1836-37; promoted to lt., 1837; stationed on board Niagara, Can., 1838-39, Pilot, W.I., 1839-42; comdr. Romney, Havana, Cuba, 1842-46; apptd. to Coast Guard, 1846; became 1st lt. with Capt. Bird on board Investigator in Arctic expdn. of Sir James C. Ross, 1848; named comdr. Investigator, 1849; comdr. expdn. to search for Sir. John Franklin, 1850-54; promoted to capt., 1854; apptd. comdr. Esk, in Pacific, 1856; returned to Eng., 1861; named rear adm., 1867, vice adm., 1873. Knighted, 1854. Discovered Northwest passage, 1850-54. Died Eng., Oct. 17, 1873.

McCOLL, John Duncan, Canadian pharmacologist; b. London, Ont., Can., Nov. 11, 1925; s. Gordon and Mary (Clunis) McC.; B.A., U. Western Ont., 1946, M.Sc., 1950; Ph.D., U. Toronto, 1953; m. Patricia Amy Ridout, May 29, 1954; children—Pamela Patricia Amy, Susan Carol, John Duncan Gordon Ridout. Asst. research chemist Parke, Davis & Co., Detroit, 1950-51; pharmacologist Frank W. Horner Ltd., Montreal, Que., Can., 1953-57, dir. pharm. research, 1957-61, asst. dir. research, 1961—. Mem. Pharm. Soc. Can. (sec. 1961-64), Can. Assn. for Research on Drug Safety (sec.-treas. 1965—), Am. Soc. Pharmacology and Exptl. Therapeutics, Am. Coll. Clin. Pharmacologists, Biochem. Soc. Great Britain, Soc. Toxicologists. Research and publs. in pharm. evaluation of new synthetic compounds to be used as therapeutic agts. in man; research and patentee in diuretic, hypoglycemic hypotensive and central nervous system agts. Home: 722 Church Av., Dorval, Que. Office: 5485 Ferrier St., Montreal, Que., Can.*

McCOLLOCH, J. W., Am. entomologist; b. Anthony, Kan., Apr. 14, 1889; s. Robert Patterson and Sarah Bell (Walker) McC.; B.S., Kan. State Agr. Coll., 1912, M.S., 1923; m. Maude Eveline Nonamaker, Dec. 11, 1911; children—Marjorie Mabel, Robert James. Research and publs. on chinchbug egg parasite, dust sprays for control of corn earthworm, wind dispersal of Hessian fly, soil-inhabiting insects, insect ecology, plant resistance to insect attack. Died 1929.

McCOLLUM, Elmer Verner, Am. physiol. chemist; b. nr. Ft. Scott, Kan., Mar. 3, 1879; s. Cornelius Armstrong and Martha Catherine (Kidwell) M.; B.A., U. Kan., 1903; M.A., 1904; Ph.D., Yale, 1906; Sc.D., U. of Cincinnati, 1920; LL.D., U. Manitoba, 1938, Johns Hopkins, 1951, L.H.D. (hon.), Brandeis University, 1959. Instr. agricultural chemistry, 1907-08; assistant professor, 1908-11, asso. prof., 1911-13, prof., 1913-17, U. of Wis.; prof. bio-chemistry, Sch. Hygiene and Public Health, Johns Hopkins, 1917-44, Emeritus professor since 1945. Member international committee on vitamin standards of League of Nations, 1931, and international and mixed commns. on nutrition, 1935. Del. 10th Pan-Am. Sanitary Conf., Bogota, 1938; chmn. nutrition sect. Pan-Am. Sanitary Bureau, 1939; member food and nutrition bd. Nat. Research Council since 1942; cons. Lend-Lease adminstrn., 1943; cons. to Indsl. Hygienic Div., U. S. Army, 1943. Member Sci. Advisory Committee Nutrition Found.; mem. adv. com. McCollum-Pratt Inst. Johns Hopkins University. Recipient of Osborne and Mendel Award, American Institute Nutrition, 1955; Borden Fedn. Centenary award, 1958; Chas. F. Spencer award, American Chemical Society, 1958. Fellow Royal Soc., 1961, Royal Society Arts (London); member British Nutrition Society (honorary), Am. Acad. Dental Medicine (hon.), A.A.A.S., Am. Soc. Biol. Chemists (pres. 1927-29), Am. Chem. Soc., Am. Public Health Association, Am. Assn. Univ. Profs., Harvey Soc. (hon.), Am. Home Econ. Assn. (hon.), American Philosophical Society, National Academy of Sciences, Phi Beta Kappa. Des Moines Academy Medicine (hon.), Kaiserlich Deutsche Academie der Naturforscher zu Halle, Royal Acad. of Medicine (Belglum). Foreign member Swedish Academy Sciences. Received Howard N. Potts gold medal, Franklin Institute, 1921, John Scott medal from the City of Phila., 1924, Newell Sill Jenkins medal from Conn. State Dental Soc., 1927, gold medal of Am. Inst. of N.Y., 1934, Callahan medal of Ohio State Dental Society, 1935. The Borden Award in Nutrition; Award from Modern Medicine for 1960; Medal and Citation, New York Academy of Medicine, 1961. Author: Text Book of Organic Chemistry for Medical Students, 1916; The Newer Knowledge of Nutrition, 1918, 5th edit., 1939; The American Home Diet, 1919; Foods, Nutrition and Health, 1933. Authorship: History of Nutrition, 1957; From Kansas Farm Boy to Scientist (autobiography), 1964. McCollum award established by Am. Soc. Clin. Nutrition, 1965. Discovered vitamins A and D; authority on relation of nutrition and diet to disease; developed method of biol. analysis of food-stuff; synthesized several pyrimidines. Home: 2402 Talbot Rd., Balt. 16.

McCONNELL, Duncan, Am. mineralogist; b. Chgo., Jan. 30, 1909; s. John Lorenzo and Anne (Duncan) McC.; B.S. in Chemistry, Washington and Lee U., 1931; M.S. (Goldwin-Smith fellow), Cornell U., 1932; postgrad. U. Chgo., 1932-33, Stanford, 1934-35; Ph.D., U. Minn., 1937; m. Jane Washington Willis, Sept. 5, 1934; children—Joanne (Mrs. Joseph J. Moldenhauer), Thomas Duncan, Charlotte Washington (Mrs. Thomas B. Blann). Instr., U. Tex., 1937-40; mineral economist U. S. Bur. Mines, Washington, 1941; chemist-petrographer U. S. Bur. Reclamation, Denver, 1941-47, head petrographic lab., 1944-47; acting asst. div. dir. Gulf Research & Devel. Co., Pitts., 1947-50; prof. mineralogy Ohio State U., Columbus, 1950-56, 64—, chmn. dept., 1952-56, asst. dean Grad. Sch., 1954-55, research prof. dentistry, 1957—. Cons., U. S. Army, 1944, Owens-Ill. Glass Co., 1955-57, Sylvania Electric Products, 1965—; research asso. Argonne Nat. Lab., 1954-56, Ohio State U. Research Found., 1955—; sr. scientist, comdr. USPHS Res., 1956—. Spl. research fellow USPHS, 1957-61. Fellow Mineral. Soc. Am., A.A.A.S. (past mem. council), Ohio Acad. Sci.; mem. Internat. Assn. Dental Research (council 1963—), Geochem. Soc., Am. Crystallographic Assn., Mineral. Assn. Can., Mineral. Soc. (Eng.), Soc. Mineralogy and Crystallography (France), Am. Chem. Soc., Soc. Econ. Geologists, Electron Microscopy Soc. Am., Phi Beta Kappa, Sigma Xi (grant-in-aid 1939), Phi Lambda Upsilon, Gamma Sigma Epsilon, Sigma Gamma Epsilon, Tau Kappa Iota, Omicron Kappa Upsilon. Contbr. chpts. to books, encys. Discovered mineral ellestadite; introduced several fundamental principles on crystal chemistry; structure of minerals and bone, phys. chemistry of concrete deterioration, nature of biol. precipitates. Home: 2154 Fairfax Rd., Columbus, O. 43221.*

McCONNELL, James Robert, Irish physicist; b. Dublin, Ireland, Feb. 25, 1915; s. Robert John and Frances (Lennon) McC.; M.A., U. Coll., Dublin, 1936; S.T.L., Lateran U., Rome, Italy, 1940, B.C.L., 1941; D.Sc.Mat., Royal U. Rome, 1941; D.Sc., Nat. U. Ireland, 1949. Ordained priest Roman Catholic Ch., 1939; scholar Dublin Inst. for Advanced Studies, 1942-45; prof. math. physics St. Patrick's Coll., Maynooth U., Kildare, Ireland, 1945—; vis. prof. Fordham U., N.Y.C., 1959-60, Laval U., Quebec, Que., Can., 1964. Chmn., Irish Nat. Com. for Math., 1960-64, Irish Nat. Com. for Physics, 1965—; Ireland rep. Internat. Union Pure and Applied Physics. Mem. Royal Irish Acad. Author: Selected Papers of Arthur William Conway, 1953; Quantum Particle Dynamics, 1958; Introduction to the Group Theory of Elementary Particles, 1965; also articles. Editorial bd. Nuclear Physics, 1956—. Research on unification gravitationation and electromagnetics in gen. relativity, antiproton, quantum electrodynamics, statis. theory high-energy events, application group theory to conservation laws for elementary particles. Home: Lorelei, Avondale Park, Killiney, Co. Dublin, Ireland. Office: St. Patrick's Coll., Maynooth, Co. Kildare, Ireland.*

McCONNELL, James Vernon, Am. psychologist; b. Okmulgee, Okla., Oct. 26, 1925; s. James Vernon and Helen (Stokes) McC.; B.A., La. State U., 1947; M.A., U. Tex., 1954, Ph.D., 1956; postgrad. U. Oslo, 1954-55. Faculty, U. Mich., Ann Arbor, 1956—, prof., research psychologist, 1963—. Recipient Career Research Devel. award Nat. Inst. Mental Health, 1963. Fellow A.A.A.S., Am. Psychol. Assn., N.W. Pacific Soc. Neurology and Psychiatry; mem. Psychonomic Soc., Midwestern Psychol. Assn., Animal Behavior Soc., Sci. Fiction Writers Am. Author: (with others) Psychology, 1961. Editor: The Worm Returns, 1965. Editor, pub. The Worm Runner's Digest, 1959—. Research, publs. on biochemistry of memory in several species animals. Home: 4101 Thornoaks Rd., Ann Arbor, Mich. 48104.*

McCONNELL, Wallace Beverly, Canadian chemist; b. Edmonton, Alta., Can., Sept. 24, 1916; s. William John and Mary (Wallace) McC.; B.Sc. with honors in Chemistry, U. Alta., 1947; Ph.D., McGill U., Montreal, Que., Can., 1949; m. Viola Agnes Morden, July 10, 1940; children—William George, Carol Jean. Staff, Prairie Regional Lab., NRC Can., Saskatoon, Sask., 1949-66, sr. research officer, 1960-66; prof. chemistry U. Sask., Regina, 1966—, chmn. chemistry dept., 1967—. Fellow Chem. Inst. Can.; mem. Canadian Soc. Plant Physiologists. Asso. editor Canadian Jour. Biochemistry, 1965—. Research and publs. on chem. characterization of wheat flour proteins, biosynthesis of wheat kernel amino acids and protein during plant maturation, biochem. processes in wheat stem rust spores, pathways of glutamic metabolism by anaerobic organisms. Home: 75 Massey Rd., Regina, Sask., Can.*

McCORD, William Mellen, physician, chemist; b. Durban, Natal, Union S. Africa, Jan. 24, 1907; s. James Bennett and Margaret (Mellen) McC.; A.B., Oberlin Coll., 1928; Ph.D., Yale, 1931; M.D., La. State U., 1939; m. Evangeline Andrews, June 14, 1930; children—Marilyn (Mrs. Harry S. Sloane), James Andrews. Faculty La. State U., New Orleans, 1931-45, asso. prof., 1942-45, prof. Med. Coll. of S.C., Charleston, 1945-64, pres., 1964—. Mem. Charleston County Med. Soc. (pres. 1960-61), Am. Chem. Soc. (pres. S.C. 1959-60), Soc. Exptl. Biology and Medicine (chmn. S.E. sect. 1958-59), Med. Soc. S.C., A.M.A., A.A.A.S., Am. Fedn. Clin. Research, S.C. Acad. Sci., Soc. Nuclear Medicine, Sigma Xi, Alpha Omega Alpha. Research and many publs. in bichem. methods, protein chemistry, toxicological problems. Home: Route 1, Box 348, Ft. Johnson Rd., Charleston, S.C. 29407.*

McCORKLE, Willar Homer, Am. physicist; b. Palo, Ia., Apr. 21, 1902; s. John Willard and Magnolia (Woods) McC.; B.A., U. Ia., 1924, M.S., 1928, Ph.D., 1935; m. Alta K(athryn) Gayle, June 6, 1928; children—Alice K(athryn) (Mrs. Jack M. Stone), Jerry Willard, Alta Wynnell (Mrs. Donald E. Busch). Faculty, Agrl. and Mech. Coll. Tex., College Station, 1924-43, prof., 1939-43; physicist, asso. sect. chief U. Chgo. Metall. Lab., 1944-45; physicist Phillips Petroleum Co., Bartlesville, Okla., 1945-46; sr. physicist, div. dir. research reactor operations div. Argonne (Ill.) Nat. Lab., 1946-61; sr. physicist, div. chief reactor div. Ames Lab., Ia. State U., Ames, Ia., 1961—. Mem. Am. Phys. Soc., Acoustical Soc. Am., Am. Assn. Physics Tchrs., Am. Nuclear Soc., Sigma Xi. Author: (with E. E. Vezey) Laboratory Manual of Physics for Engineering Students, 1938; (with G. S. Monk) Optical Instrumentation, 1954; also articles. Research in atomic and nuclear physics. Patentee nuclear reactor designs, reactor safety mechanisms. Home: Meadow Glen Rd., Route 3, Ames, Ia. 50010.*

McCORMAC, Billy Murray, Am. physicist; b. Zanesville, O., Sept. 8, 1920; s. Samuel Dennis and Phyllis (Murray) McC.; B.S., Ohio State U., 1943; M.S., U. Va., 1956, Ph.D. in Nuclear Physics, 1957; m. Dorothy Mary Boros, Nov. 21, 1948; children—Norene Leslie, Candace Elisabeth, Lisbeth Phyllis. Physicist, Office of Spl. Weapons Devel., U. S. CONARC, 1957-60; mil. scientist Office Chief of Staff, U. S. Army, 1960-61; physicist to chief of electromagnetic br. Def. Atomic Support Agy., 1961-63; dir. geophysics div. Research Inst., Ill. Inst. Tech., Chgo., 1967—. Chmn. advanced study insts. NATO, 1965-67. Recipient Legion of Merit, Def. Atomic Support Agy., 1963. Fellow Am. Inst. Aeros. and Astronautics (asso.); mem. Am. Phys. Soc., Am. Geophys. Union, Marine Tech. Soc., Am. Ordnance Assn., A.A.A.S., Marine Tech. Soc. (sr.), Sigma Xi. Editor: Radiation Trapped in the Earth's Magnetic Field, 1966; Aurora and Airglow, 1967. Research, publs. on multidisciplinary approach to complex mil. and environmental sci. problems, including nuclear warfare, especially high altitude nuclear burst, trapped radiation, aurora and airglow, ionosphere. Home: 1123 Birch Lane, Western Springs, Ill. 60558. Office: 10 W. 35th St., Chgo. 60616.*

McCORMACK, Arthur Thomas, Am. physician; b. nr. Howardstown, Ky., Aug. 21, 1872; s. Joseph N. and Corinne (Crenshaw) McC.; B.A., Ogden Coll., 1892; M.D., Columbia, 1896; student U. Va.; M.A. (hon.), Bethel Coll., Russellville, Ky., 1900; D.P.H., Detroit Coll. Medicine and Surgery, 1925; D.Sc., Berea Coll., 1926; LL.D., Transylvania U., 1930; m. Mary Moore Tyler, Dec. 15, 1897; children—Joseph Nathaniel, Lucy Norton, Arthur Thomas, Mary Wilbur; m. 2d, Mrs. Jane Teaze Dahlman, Oct. 16, 1924. Began practice at Bowling Green, 1897; health officer, Warren County, Ky., 1897-1900; asst. state health officer, 1898-1912, state health commr. since 1912, also mem. Bd. of Health, Ky.; surg. gen. Ky. N.G., 1900-08; lt. Med. Res. Corps, U. S. Army, 1911-17; maj., 1917; organized Base Hosp. No. 59, and sent by Gen. Gorgas to Panama to succeed him as chief health officer; completed building of Ancon Hosp.; controlled epidemic of cerebro-spinal meningitis on Japanese S.S. Anyo Maru and was thanked by Mikado; spl. cons. USPHS State flood relief dir. for 1937 flood. Organizer, dean Sch. of Pub. Health, Ky. Bd. Health; mem. Gorgas Meml. Inst.; mem. Nat. Health Council. Fellow A.C.S.; mem. and officer Am., Ky. and Jefferson County med. assns., Med. Vets. of World War (sec., pres.), Assn. Mil. Surgeons, Nat. Tb Assn., Conf. of State and Provincial Health Authorities of N.A. (pres.), Ky. Conf. Social Work (pres.), Am. Pub. Health Assn. (pres.), So. Med. Assn. (pres.). Author: Course in Physical Education for the Common Schools of Kentucky, 1920. Founder Ky. Med. Jour., 1901, also editor. Died Aug. 7, 1943.

McCORMICK, Cyrus Hall, Am. inventor; b. Walnut Grove, Va., Feb. 15, 1809; s. Robert and Mary (Hall) McC.; m. Nancy Fowler, Jan. 26, 1858; 7 children. Invented and patented hillside plough, 1831; patented reaping machine, 1834, began mfg. machine commercially, 1837; erected factory in Chgo., 1847; built up nat. business for McCormick Harvesting Machine Co., by 1850; added mowing attachment to reaper, 1850's, also developed self-raking device, hand-binding harvester, wire-binder, twine-binder, 1860's; introduced reaper in London, Eng., 1851, then Europe; awarded Council medal London World's Fair, 1851; won major prizes at world fairs, Paris, London, Hamburg, Lille, Vienna, Phila., Melbourne, 1855-80; elected mem. French Acad. Scis., 1879; owner Presbyn. Expositor, newspaper, 1860; The Interior (Presbyn. newspaper, later named Continent), 1872-84; became owner Chgo. Times, 1860, publisher, 1860-61; dir. U.P. R.R. Died Chgo., May 13, 1884.

McCORMICK, Donald Bruce, Am. biochemist; b. Front Royal, Va., July 15, 1932; s. Jesse Allen and Elizabeth (Hord) McC.; B.A., Vanderbilt U., 1953, Ph.D, 1958; postdoctoral fellow U. Cal. at Berkeley, 1959-60; m. Norma Jean Dunn, June 6, 1955; children—Susan Lynn, Donald Bruce, Michael Allen. Asst. prof. Cornell U., Ithaca, N.Y., 1960-63, asso. prof., 1963—; vis. lectr. U. Ill., 1963. Cons. biochemist Interdepartmental Com. on Nutrition for Nat. Def., Spain, 1958. Westinghouse Sci. scholar, 1950; award Bausch and Lomb, 1950; fellow NIH, 1958-60. Mem. Am. Soc. Biol. Chemists, Am. Inst. Nutrition, Soc. Exptl. Biology and Medicine, Am. Chem. Soc., Am. Inst. Biol. Sci., N.Y Acad. Sci., A.A.A.S. Sigma Xi. Author: (with others) Spain: Nutrition Survey of the Armed Forces; also many articles. Aided in elucidating the mechanisms for interconversions and functions of certain vitamins and coenzymes, particularly flavin systems. Home: 115 Cayuga Park Circle, Ithaca, N.Y. 14850.*

McCORMICK, Jerry Robert Daniel, Am. chemist; b. St. Albans, W.Va., Feb. 24, 1921; s. James Moody and Georgia (Fowler) McC.; B.S. in Chemistry, Rensselaer Poly. Inst., 1942; Ph.D. in Chemistry, U. Cal. at Los Angeles, 1949; m. Catherine Mary Rello, Oct. 24, 1953; 1 son, Joshua. Chemist, Winthrop Chem. Co., Rensselaer, N.Y., 1942-46; staff Am. Cyanamid, Lederle Labs., Pearl River, N.Y., 1949—, head fermentation biochemistry dept., 1958-62, research asso., 1962-65, research fellow, 1965—. Fellow A.A.A.S.; mem. Am. Chem. Soc., N.Y. Acad. Scis. Patentee, antibiotics, chem. processes, chem. intermediates. Discoverer of demethyltetracycline antibiotics; research on elucidation of mechanism of biosynthesis of tetracycline antibiotics. Home: Pomona Country Club, Spring Valley, N.Y. 10977. Office: Lederle Labs., Pearl River, N.Y. 10965.*

McCORMICK, Robert, Am. inventor; b. Rockbridge County, Va., June 8, 1780; s. Robert and Martha (Sanderson) McC.; m. Mary Ann Hall, Feb. 11, 1808; several children including Cyrus Hall, William S., Leander James. Inventor several farm implements (none were practical or commercially valuable); inventor and patentee hempbrake, gristmill, hydraulic machine and blacksmith's bellows, 1830-31; inventor threshing machine, 1834; experimenter with grain reapers, 1809-31; built an iron furnace, 1836; mfr. reaper invented by son Cyrus, 1837-45. Died Rockbridge County, July 4, 1846.

McCORMICK, Thomas Carson, Am. agronomist, sociologist; b. Tuscaloosa, Ala., Jan. 11, 1892; s. William Thomas and Virginia Marr (Carson) McC.; A.B., U. Ala., 1911; A.M., George Peabody Coll., 1918; Ph.D., U. Chgo., 1929; m. Lillie Anderson Griffith, Aug. 15, 1918; children—Virginia Marr (Mrs. Greggar P. Sletteland), Mary Luttrell (Mrs. Rolf N. Olsen), Lillie Griffith (Mrs. Roy G. Francis). High sch. tchr., 1912-21; prof. agr. E. Central State Tchrs. Coll., Okla., 1921-27, prof. sociology, 1929-31; vis. prof. sociology U. Chgo., summer 1930; asst. prof. rural sociology U. Ark., 1931-34; research supr., acting coordinator of rural research WPA, Washington, 1934-35; prof. sociology U. Wis., 1935—, chmn. dept. sociology and anthropology, 1941-42; chmn. div. social studies, 1947-50. Chief statistician Wis. Citizens Com. on Public Welfare, 1936; research asso. Negro in America study, Carnegie Corp. of N.Y., 1939; mem. adv. com. population U. S. Census Bur. Mem. Am. Statis. Assn., Population Assn. Am., Phi Beta Kappa. Editor: (with A. S. Barr and H. L. Ewbank) Radio in the Classroom, 1942; Problems of the Post-war World, 1945. Author: Agriculture for Rural Teachers, 1929; Rural Relief and Non-Relief Households, 1935; Elementary Social Statistics, 1941; Sociology, 1952; Methods of Research in the Behavioral Sciences, 1958. Contbr. to sociol. and statis. jours. Died Nov. 9, 1954.

McCOSH, Andrew J., surgeon; b. Belfast, Ireland, 1858; s. James and Isabella (Guthrie) M.; A.B., Princeton, 1877, A.M., LL.D., 1906; M.D., Coll. Physicians and Surgeons, N.Y., 1880; postgrad. U. Vienna, 1882; LL.D., Columbia, 1905. Practiced at N.Y., 1883—; surgeon to Presbyn. Hosp., 1889—; prof. clin. surgery Columbia. Fellow Am. Surg. Assn. Author: (with Starr) A Contribution to the Localization of the Muscular Sense, 1894; also papers. Died 1908.

McCOWN, Theodore Doney, Am. anthropologist; b. Macomb, Ill., June 18, 1908; s. Chester Charlton and Harriet (Doney) McC.; A.B., U. Cal. at Berkeley, 1929, Ph.D., 1939; m. Elizabeth Ann Richards, Aug. 30, 1946; children—Ann Elizabeth, Jean Keith, Faith Carol. Faculty, U. Cal. at Berkeley, 1938—, prof. curator dept. anthropology, 1957—, chmn., 1950-55, asso. dean Coll. Letters and Sci., 1956-61. Fellow A.A.A.S., Am. Anthrop. Assn., Royal Anthrop. Inst. Gt. Britain and Ireland; mem. Prehistory Soc., Am. Assn. Phys. Anthropologists, Soc. for Am. Archaelogy. Author: (with Arthur Keith) The Stone Age of Mount Carmel, vol. II, 1939; Pre-Incaic Huamachuco, 1945; also articles. Excavated and described prehistoric Neanderthal cemetery in Palaeolithic caves, Mt. Carmel, Israel (improtant in destroying dogma that Neanderthal man and modern man are separate species). Home: 1114 Oxford St., Berkeley, Cal. 94707.*

McCOY, Elizabeth, microbiologist; b. Madison, Wis., Feb. 1, 1903; d. Cassius James and Esther (Williamson) McCoy; B.S., U. Wis., 1925, M.S., 1926, Ph.D., 1929; postgrad. U. London. Faculty, U. Wis., Madison, now prof. bacteriology. Asso. research medicine Hooper Found., U. Cal., 1938. NRC fellow Rothamsted Exptl. Sta., Eng., also Karlova U., Czechoslovakia. Fellow Am. Soc. Microbiology, Am. Acad. Microbiology; mem. A.A.A.S., Soc. Exptl. Biology and Medicine, USPHS Assn. Author: (with Fred and Baldwin) Root Nodule Bacteria, 1930; (with McClung) Anaerobic Bacteria, 2 vols., 1939. Research in indsl. microbiology, taxonomy of bacteria, fresh water bacteria and pollution. Home: Route 3, Box 57, Madison, Wis. 53711.*

McCOY, Herbert Newby, Am. chemist; b. Richmond, Ind., June 29, 1870; s. James W. and Sarah N. McC.; B.S., Purdue U., 1892; M.S., 1893, D.Sc., 1938; Ph.D., 1898; m. Ethel M. Terry, June 13, 1922. Tech. chemist Chgo., 1893-94; prof. chemistry and physics, Fargo (N.D.) Coll., 1894-96; asst. in chemistry, U. Chgo., 1898-99, instr., 1901-03, asst. prof. chemistry, 1903-07, asso. prof., 1907-11, prof., 1911-17; asst. prof. U. Utah, 1899-1901; sec. Carnotite Reduction Co., Chgo., 1915-17, pres., 1917-23; v.p. Lindsay Light & Chem. Co. since 1919. Recipient Willard Gibbs medal, 1937. Mem. Am. Chem. Soc., Am. Electrochem. Soc., Am. Phys. Soc., Inst. Chem. Engrs. Author: (with Ethel M. Terry) Introduction to General Chemistry, 1919; A Laboratory Outline of General Chemistry, 1919. Contbr. numerous papers on phys. chemistry, radioactivity and rare earths. Determined ionization constant of an indicator as measure of its sensitivity, 1904; inventor method for obtaining minute amounts of carbon dioxide from an activated nerve; extensive research on radioactive elements; gave early exptl. demonstration that radium is product of spontaneous transmutation of uranium, 1904; gave (with W. H. Ross) 1st description of isotopes, 1907; gave 1st quantitative proof that alpha ray activity of uranium compounds is directly proportional to their uranium content, also proved that radioactivity is an atomic property of elements; discovered new method for isolation of europium, 1935; Died May 7, 1945.

McCOY, Oliver Rufus, Am. biologist, physician; b. St. Louis, Aug. 1, 1905; s. William Carson and Sarah E. (Wilson) McC.; A.B., Washington U., St. Louis, 1926, M.S., 1927; Sc.D, Johns Hopkins, 1930; M.D., U. Rochester, 1942; LL.D., Keio (Japan) U., 1956; m. Julia Louise Large, Oct. 23, 1937; children—Caroline L. (Mrs. Paul E. White), William L., Charles P. Instr. helminthology Johns Hopkins, 1929-30; asst. prof. parasitology U. Rochester, 1930-42; field staff Rockefeller Found., France, 1946-48, Japan, 1948-56; asso. dir. China Med. Bd. N.Y., Inc., N.Y.C., 1956-59, dir., 1959—. Cons. Gorgas Meml. Lab., Panama, 1934; vis. prof. Nat. Med. Coll., Shanghai, China, 1936; mem. N.Y. State Trichinosis Commn., 1939-41. Decorated Order of Sacred Treasure (Japan). Fellow A.A.A.S., Am. Pub. Health Assn.; mem. Am. Soc. Tropical Medicine and Hygiene, Am. Soc. Parasitologists, Royal Soc. Tropical Medicine and Hygiene. Home: 47 Lyncroft Rd., New Rochelle, N.Y. 10804. Office: 420 Lexington Av., N.Y.C. 10017.*

McCRACKEN, Kenneth Gordon, Australian physicist; b. Brisbane, Queensland, Australia, Aug. 7, 1933; s. Richard and Elizabeth (Mensies) McC.; B.S., U. Tasmania, 1954, B.S. with honors, 1956, Ph.D., 1959; m. Gillian Mabel Filby, May 2, 1959; children—Ruth Jeanette, Helen Elizabeth. Research asso. Mass. Inst. Tech., Cambridge, Mass., 1959-61, asst. prof., 1962; prof. Grad. Research Center of Southwest, Dallas, 1962-66, U. Adelaide, (S. Australia), 1966—. Research and publs. on cosmic radiation within solar system, interplanetary magnetic fields. Home: 14 Reece Av., Klemzig, Adelaide, S. Australia.*

McCRACKEN, Ralph Joseph, Am. soil scientist; b. Guantanamo, Cuba, July 31, 1921 (parents Am. citizens); s. John Raymond and Ina (Ratliff) McC.; A.B., Earlham Coll., 1942; M.S., Cornell U., 1951; Ph.D., Ia. State U., 1956; m. Virginia Johnson, Aug. 27, 1949; children—Douglas Raymond, Jo Ellen. Soil scientist U. S. Dept. Agr., Cal., N.Y., Ia., 1947-54; asso. agronomist U. Tenn., Knoxville, 1954-56; faculty N.C. State U., Raleigh, 1956—, prof., 1962—, head dept., 1964—. Fellow A.A.A.S., Am. Soc. Agronomy; mem. Soil Sci. Soc. Am., Internat. Soc. Soil Sci., Am. Soc. Agronomy, Mineral. Soc. Am., Clay Minerals Soc., Soil Conservation Soc. Am. Research and publs. on genesis and classification of soils, soil mineralogy. Home: 1025 Warren Av., Cary, N.C. 27621. Office: Dept. Soil Sci., N.C. State U., Raleigh, N.C. 27607.*

McCRADY, Edward, Am. biophysicist; b. Canton, Miss., Sept. 19, 1906; s. Edward and Mary Ormond (Tucker) McC.; A.B., Coll. Charleston, 1927, LL.D., 1952; M.S., U. Pitts., 1930; Ph.D., U. Pa., 1933; LL.D., U. Chattanooga, 1958; D.Sc., Southwestern U. at Memphis, 1959; L.H.D., Concord Coll., 1962; m. Edith May Dowling, Aug. 15, 1930; children—Edward, John, James Waring, Sarah Heath (Mrs. E. Hayne Shumate, Jr.). Fellow U. Pa. Wistar Inst. Anatomy, 1930-34, research scientist, 1934-37; prof., head biology dept. U. South, Sewanee, Tenn., 1937-50, vice-chancellor, pres., 1951—. Biologist, Office Research and Medicine, AEC, 1948-49, chief biology div., 1949-51; mem. Fulbright Selection Bd., 1950, Tenn. Rhodes Scholarship Selection Com., 1957-59, Danforth Found. Council, 1958-60; pres. So. U. Conf., 1960. Danforth scholar, 1958-60. Fellow A.A.A.S.; mem. Am. Assn. Anatomists, Am. Soc. Zoologists, Acoustical Soc. Am., Am. Geophys. Union, Am. Inst. Physics, Assn. Southeastern Biologists, Tenn. Acad. Sci., Victoria Inst. Gt. Britain, Nat. Speleological Soc., Blue Key, Sigma Xi, Sigma Pi Sigma, Alpha Tau Omega, Omicron Delta Kappa, Sigma Upsilon. Author: The Embryology of the Opossum, 1938; (with others) Religious Perspectives in College Teaching, 1952; (with others) Research Policies of Colleges and Universities, 1954. Research on mammalian embryology, physiology of hearing, fossil jaguars, cave salamanders, age of earth, origin of matter. Home: Address: U. of South, Sewanee, Tenn. 37375.*

McCRAE, Thomas, physician; b. Guelph, Ont., Can., Dec. 16, 1870; s. David and Janet (Eckford) M.; A.B., U. Toronto, 1891, M.B., 1895, M.D., 1903; D.Sc., 1927; U. Göttingen, 1899; m. Amy Gwyn, Sept. 16, 1908. Fellow biology U. Toronto, 1892-94; asso. in medicine Johns Hopkins Hosp., 1904-12; asso. prof. medicine Johns Hopkins U., 1906-12; prof. medicine Jefferson Med. Coll., Phila., 1912—; physician to Jefferson and Pa. hosps., 1912—. Fellow Royal Coll. Physicians London (Lumleian lectr. 1924). Author: Cancer of the Stomach (with Sir William Osler), 1900. Editor: (with Osler) Modern Medicine (included McCrae's articles on typhoid fever and arthritis defor mans), 7 vols., 1907-10; Osler's System of Medicine (3d edit.); Osler's Practice of Medicine, 11th edit., 1930. Died June 30, 1935.

McCREA, William Hunter, mathematician, astronomer; b. Dublin, Ireland, Dec. 13, 1904; s. Robert Hunter and Margaret (Hutton) McC.; B.A., Cambridge (Eng.) U., 1926, Ph.D., 1929, M.A., 1931, Sc.D., 1958; student Göttingen (Germany) U., 1928-29; D.Sc., Nat. U. Ireland, 1954; m. Marian Nicol Core Webster, July 28, 1933; children—Isabella Nicol Core, Sheila Hunter (Mrs. John Rousseau), Roderick Hunter. Lectr. math. Edinburgh (Scotland) U., 1930-32; reader in math. Imperial Coll., London (Eng.) U., 1932-36; prof. math. Belfast (Ireland) U., 1936-44; prof. math. Royal Holloway Coll., London U., 1944-66; research prof. theoretical astronomy Sussex (Eng.) U., 1966—. Operational research Admiralty, London, 1943-45; Bye fellow Gonville and Caius colls., Cambridge U., 1952-53; vis. prof. U. Cal., 1956, Case Inst. Tech., 1964, Kitt Peak Nat. Obs., 1965. Recipient Keith medal Royal Soc. Edinburgh, 1941. Fellow Royal Soc., Royal Soc. Edinburgh; mem. Royal Astron. Soc. (pres. 1961-63), Brit. Assn. Advancement Sci. (pres. sect. A 1966), Royal Irish Acad. Author: Relativity Physics, 1935; Analytical Geometry of Three Dimensions, 1942; Physics of Sun and Stars, 1950; also numerous articles. Research in quantum mechanics; pioneer in quadrupole radiation, astrophys. application of turbulence, model stellar atmospheres; co-discoverer of Newtonian Cosmology; developed theory of uncertainty in cosmology, new theory of solar system. Home: 87 Houndean Rise, Lewes, Sussex, Eng.*

McCRIMMON, Hugh Ross, Canadian aquatic biologist; b. Hamilton, Ont., Can., Apr. 10, 1923; s. Leon Ross and Catherine (Clarke) McC.; B.A., McMaster U., 1946; Ph.D., U. Toronto, 1949; m. Doris Irene Davison, June 3, 1948; children—Katherine, Ian, Carol. Instr. zoology U. Toronto, 1946-48; aquatic biologist, fish and wildlife supr. Ont. Dept. Lands and Forests, Toronto, 1949-57; asso. prof. Ont. Agrl. Coll., 1957-65; prof. U. Guelph, 1966—; cons. fisheries, pollution, water mgmt., aquatic zoology. Mem. Am. Fisheries Soc., Wildlife Soc., Canadian Soc. Zoologists, Chem. Inst. Canada. Contbr. numerous articles in field to sci. jours. Research in fisheries and limnology which is of practical value in mgmt. of renewable aquatic resources for industry, food and recreation. Home: 83 University Av., Guelph, Ont., Can.*

McCROSKY, Theodore Tremain, Am. city planning engr.; b. Tecumseh, Neb., June 12, 1902; s. James Warren and Josephine (Tremain) McC.; B.S. in Civil Engring., Yale, 1923; Ingénieur Constructeur, U. Louvain (Belgium), 1924; m. Agnes Herriott James, Sept. 2, 1925; children—Marion Currie (dec.), John Warren James. Dir. planning N.Y.C. Dept. City Planning, 1938-40; exec. dir. Chgo. Plan Commn., 1941-42; exec. dir. Greater Boston Devel. Com., 1946-48; cons. engr. city and regional planning, N.Y.C., 1948-62; partner McCroskey-Reuter, cons. in city and regional planning, 1962-66, cons., 1966—. Instr., Yale 1923-24, 25-57. Mem. Am. Inst. Planners (past v.p.), Am. Inst. Cons. Engrs. (sec.), Inst. Traffic Engrs., Sigma Xi, Tau Beta Pi, Lambda Alpha, Theta Xi. Fellow Am. Soc. Civil Engrs. Author: (with Charles A. Blessing, J. Ross McKeever) Surging Cities, 1948; also articles; contbg. author: American Civil Engineering Practice, 1956. Developed statis. techniques for municipal capital budgeting and effect capital improvement programs on future tax rates and borrowing capacity.

Home: 370 1st Av., N.Y.C. 10010. Office: Am. Inst. Cons. Engrs., 345 E. 47th St., N.Y.C. 10017.*

McCUE, John Joseph Gerald, Am. physicist; b. South Orange, N.J., Dec. 25, 1913; s. John Joseph and Norita (Kerons) McC.; A.B. cum laude, Harvard, 1936; Ph.D. Cornell U., 1942; m. Mariam Eugenia Crowley, Dec. 19, 1949; 1 son, Brian. Tech. asst. Bell Telephone Labs., N.Y.C., 1930-32; asst. in physics Cornell U., 1936-41; instr., asst. prof. Hamilton Coll., Clinton, N.Y., 1941-44; staff mem. radiation lab. Mass. Inst. Tech., 1944-45, lab. nuclear sci., 1949-51, Lincoln Lab., 1951—; asso. prof. physics Smith Coll., 1946-49. Vis. prof. physics Boston U., 1956-57; faculty Harvard Summer Sch., 1959, 60. Mem. Am. Phys. Soc., Am. Assn. Physics Tchrs., Royal Astron. Soc. Can., I.E.E.E. (chmn. group on ultrasonics 1963-65, editorial bd. Proc. 1966—, editor Spectrum 1968—), Sigma Xi. Author: The World of Atoms, 1956, 2d edit., 1963; contbg. author: Radar Beacons, 1947; also articles. Measured energy levels of several nuclear species; developed techniques for suppressing radar returns from nonmoving objects; devised radars resistant to jamming; studied methods used by bats for echolocating. Home: 20 N. Hancock St. Office: Lincoln Laboratory, Mass. Inst. of Technology, Lexington, Mass., 02173.

McCULLAGH, E. Perry, physician; b. Douglas, Man., Can., May 27, 1901; s. Robert E. and Violet (Perry) McC.; M.D., U. Man., 1924; m. Lucile Florence Myers; children—Robert E., Jane (Mrs. Thomas E. Dustin), Mary (Mrs. Thomas R. Wigglesworth). Came to U. S., 1925, naturalized, 1941. With The Cleve. Clinic Found., 1925—, head dept. endocrinology, metabolism, 1934-65, now emeritus cons.; med.dir. Camp Ho Mita Koda, 1953-64. Mem. A.C.P., Diabetes Assn. Greater Cleve. (pres. 1962-64), Endocrine Soc. (pres. 1958-59), Cleve. Philosoph. Club (pres. 1960), Am. Fedn. Clin. Research, Am. Thyroid Assn., A.M.A., Am. Diabetes Assn., Central Soc. Clin. Research. Contbr. numerous articles in field to sci. jours. Research, publs. in endocrinology. Home: 2194 Chatfield Dr., Cleve. 44106. Office: 2020 E. 93d St., Cleve. 44106.*

McCULLAGH, James, Irish mathematician, physicist; b. County Tyrone, Ireland, 1809; ed. Trinity Coll., Dublin, Ireland, became scholar, 1827, fellow, 1832. Named prof. math. U. Dublin, 1836, prof. natural philosophy, 1843, Mem. Royal Irish Acad. (sec. to council 1840-42, sec. 1842-46). Research and publs. in wave theory of light; attempted to construct dynamical theory of luminiferous ether; introduced studies in electricity, heat, terrestrial magnetism to fellowship course at U. Dublin. Died Oct. 1847.

McCULLOCH, Allan Riverstone, ichthyologist; b. June 20, 1885; s. Herbert Riverstone and Ella Maud (Backhouse) McC.; ed. U. Sydney (Australia). With Australian Mus., Sydney, from 1898, zoologist, from 1906. Mem. council Linnean, zool. socs. N.S. Wales, Australia, Gt. Barrier Reef Com. Catalogued Australian fishes; made natural history, ethnol. collecting expdns. to western Papua, New Hebrides, along Australian coast from south of Tasmania to Torres Straits; publs. dealing with fish, other mammals, reptiles, crustaceans of Australia; writer popular articles in semi-sci. jours. Died Sept. 1, 1925.

McCULLOCH, Hugh, Am. physician; b. Marianna, Ark., Aug. 20, 1888; s. Edgar Allen and Harriet (Hassell) McC.; A.B., U. Ark., 1908; M.D., Johns Hopkins, 1912; m. Ida Louise Haardt, July 9, 1935; 1 son, Hugh II. With Wash. U., 1912-48, asso. prof. clin. pediatrics, to 1948; prof. clin. pediatrics U. Ill., Chgo., 1949-58, prof. emeritus, 1958—; from asst. to physician St. Louis Children's Hosp., 1915-48; dir. La Rabida Sanitarium, Chgo., 1949-58; dir. Center for Retarded Children, Montgomery, Ala., 1958-63; pvt. practice medicine, specializing in pediatrics, St. Louis, 1919-48; asso. editor Am. Heart Jour., 1925-38; co-editor Jour. Pediatrics, 1932-47, editor, 1947-53. Mem. Am. Heart Assn. (Gold Heart award 1950, Founders award 1964), Am. Acad. Pediatrics (Clifford G. Grulee award 1954), A.M.A., Am. Pediatric Soc. (pres. 1951-52), Phi Beta Kappa. Home: 1625 S. Hull St., Montgomery, Ala. 36104.*

McCULLOCH, Warren Sturgis, Am. neurophysiologist; b. Orange, N.J., Nov. 16, 1898; s. James W. and Mary H. (Bradley) McC.; B.A., Yale, 1921; M.A., Columbia, 1923, M.D., 1927; m. Rook Metzger, May 29, 1924; children—David S., Katherine (Mrs. Alfred Holland), Jean (Mrs. Alexander Vasiloff). Practice medicine, specializing in psychiatry, Cambridge, Mass., 1952—; staff mem. div. sponsored research. Research Labs. Electronics Mass. Inst. Tech., 1952—. Diplomate Nat. Bd. Medicine. Fellow A.A.A.S., I.E.E.E.; mem. Am. Acad. Arts and Scis., Am. Acad. Neurology, Am. Assn. Anatomists, EEG Soc., N.Y. Acad. Scis., Soc. Biol. Psychiatry, Aerospace Med. Assn., Sigma Xi. Author: (with others) The Isocortex of the Chimpanzee, 1950; Finality and Form, 1952; Embodiments of Mind, 1965; also numerous articles. Research on functional orgn. of brain, facilitation, extinction and functional orgn. of cerebral cortex. Home: Whippoorwill Rd., Old Lyme, Conn. Office: 77 Massachusetts Av., Cambridge, Mass. 02139.*

McCULLOUGH, Campbell Rogers, Am. nuclear and chem. engr.; b. Washington, Apr. 12, 1900; s. Charles Edmund and Emma (Rogers) McC.; A.B., Swarthmore Coll., 1921; M.S., Mass. Inst. Tech., 1922, Ph.D. (DuPont fellow 1927-28), 1928; m. Exia Drummond, Oct. 16, 1936; children—David Rogers, Diane Exia. Chemist, Hygrade Lamp Co., 1922-26; chemist, chem. engr. prodn. and devel. organic and inorganic chems. Monsanto Chem. Co., 1928-60; project mgr. NDRC Rocket Propellant Pilot Plant, 1944-46; dir. Power Pile div. Clinton Labs., 1946-47; v.p. Nuclear Utility Services, Inc., 1960-65, dir. Atomic Indsl. Forum, Nuclear Utility Services; tech. dir. So. Nuclear Engring., 1965—. cons. CNEN of Italy. Chmn., mem. adv. com. on reactor safeguards AEC, 1951-59, vice chmn., 1960-61; mem. regulatory rev. panel, 1965; sci. adviser U. S. delegation Internat. Conf. for Peaceful Uses of Atomic Energy, Geneva, 1955, 58. Recipient Presdl. Certificate of Merit, 1949. Mem. Am. Ordnance Assn., Am. Nuclear Soc. (pres. 1956-57), Am. Chem. Soc., Am. Inst. Chem. Engrs., Phi Beta Kappa, Sigma Xi. Editor: Safety Aspects of Nuclear Reactors, 1957. Contbr. to Handbook on Nuclear Engineering, Modern Nuclear Technology. Home and office: 15201 Rosecroft Rd., Rockville, Md. 20853.*

McCULLOUGH, Norman B., Am. physician, microbiologist; b. Milford, Mich., Feb. 23, 1909; s. George W. and Florence W. (Bird) McC.; B.S. with high honor, Mich. State Coll., 1932, M.S., 1933; Ph.D., U. Chgo., 1937, M.D., 1944; m. Mary Louise Young, July 22, 1939; children—Marilyn Kay, Robert Norman. Staff, Parke Davis & Co., Detroit, 1934-36, research bacteriologist, 1938-40; research bacteriologist U. Tex., 1940-42; commd. sr. asst. surgeon USPHS, 1947, advanced through grades to med. dir., 1952; chief brucellosis activities NIH, U. Chgo., 1947-51, asst. clin. prof. medicine, 1947-50; chief brucellosis unit Lab. Infectious Diseases, Nat. Inst. Allergy and Infectious Diseases, NIH, Bethesda, Md., 1951-56, chief Lab. Clin. Investigation, clin. dir., 1952-58, chief Lab. Bacterial Diseases, 1958—, mem. med. bd. Clin. Center, NIH, 1952-58, instr. Found. for Advanced Edn. in Scis., 1961—, faculty chmn. microbiology and immunology, 1963—. Lectr. Infectious diseases Georgetown U., Washington, 1952—; panelist WHO/FAO, 1952—; mem. Nat. Brucellosis Com., 1961—, bd. dirs., 1963—. Diplomate Am. Bd. Microbiology. Fellow Am. Acad. Microbiology, Washington Acad. Sci.; mem. Am. Soc. for Microbiology, A.A.A.S., N.Y. Acad. Scis., A.M.A., Sigma Xi, Phi Kappa Phi, Phi Sigma, Alpha Omega Alpha, Gamma Alpha, Phi Chi. Research and numerous publs. on characteristics of agt. of brucellosis and diagnosis; treatment of disease; other infectious parasitic diseases of man. Home: 2403 Eugene St., Silver Spring, Md. 20902. Office: Nat. Insts. Health, Bethesda, Md. 20014.*

McCULLOUGH, Timothy Pendleton, Am. physicist; b. Vardaman, Miss., Dec. 9, 1910; s. Timothy P. and Mary Anne (Wells) McC.; A.B., U. Miss., 1936; M.S., N.C. State U., 1937; m. Virginia Ball, Sept. 7, 1941; children—Robert Eugene, Jane Ball, Charles Eric. Faculty, Miss. State U., 1937-42, U. Miss., 1942-43; research physicist U. S. Naval Research Lab., Washington, 1946-51, 54—, researcher in radio astronomy, 1954—. Mem. Am. Astron. Soc., Sci. Research Soc. Am. (Pure Sci. award for significant basic contbns. to knowledge of planetary radiation 1961), Internat. Sci. Radio Union (U. S. A. nat. com.). Contbr. articles to profl. jours. First to detect and measure radiation from planets Venus and Mars at radio wave lengths, showed unexpected high temperature of Venus, 1st to measure thermal radiation from Jupiter at radio wave lengths; measurements of polarized component of radiation from celestial radio sources. Home: 6002 Loretto St., Springfield, Va. 22150. Office: Code 7133B, U. S. Naval Research Lab., Washington 20390.*

McCUMBER, Dean Everett, Am. physicist; b. Rochester, N.Y., Nov. 25, 1930; s. Ralph Henry and Anna (Bauman) McC.; B.E.E., Yale, 1952, M.E.E., 1955; A.M., Harvard, 1956, Ph.D., 1960; m. Nancy Snyder, June 22, 1957; children—David Everett, Katherine Wilson. NSF postdoctoral fellow Ecole Normale Superieure, Paris, France, 1959-60, Inst. for Theoretical Physics, Copenhagen, Denmark, 1960-61; with Bell Telephone Labs., Murray Hill, N.J., 1961—, head crystal electronics research dept., 1965—. Fellow Am. Phys. Soc. Research on analysis of the optical spectra of impurities in solids; theory of lasers, also of solid-state active plasma devices, especially negative-resistance Gunn-effect devices, superconductor devices. Office: Mountain Av., Murray Hill, N.J. 07971.*

McCURDY, Harold Grier, Am. psychologist; b. Salisbury, N.C., May 30, 1909; s. McKinnon Grier and Nellie (Curd) McC.; A.B. magna cum laude in Greek, Duke, 1930, postgrad.; Ph.D. in Psychology, 1938; m. Mary Burton Derrickson, Sept. 15, 1937; children—John Derrickson, Ann Lewis (dec. 1958). Faculty, High Point (N.C.) Coll., 1931-32, Milligan Coll. (Tenn.), 1938-41, Meredith (N.C.) Coll., 1941-48; faculty U. N.C., Chapel Hill, 1948—, Kenan prof. psychology, 1963—. Lectr., Summer Inst. in Contemporary Sci. Psychology, Beloit Coll., 1962. Mem. Am. Psychol. Assn., A.A.A.S., Elisha Mitchell Sci. Soc. (pres. 1962-63), Sigma Xi, Phi Beta Kappa. Author: Straw Flute, 1946; Personality of Shakespeare, 1953; Personal World, 1961; Personality and Science, 1965; Barbara, 1966; also articles. Research in personality devel. and expression, empirical and exptl. studies of learning, problem-solving by groups and individuals. Home: Gooseneck Rd., Chapel Hill, N.C. 27514.*

McCUSKER, Charles Brian Anthony, physicist; b. Stockton-on-Tees, Eng., Feb. 24, 1919; s. Charles Bowser and Elizabeth (Merrick) McC.; B.Sc. with 1st class honors in Physics, U. Manchester (Eng.), 1940, M.Sc., 1942, D.Sc., 1962; m. Emily Mary Maltby, Apr. 11, 1944; children—Stephen Charles, Alison Mary, Eric Maltby. Lectr., Wigan Mining and Tech. Coll., 1942-46, Liverpool (Eng.) U., 1946-48; prof. Dublin (Ireland) Inst. for Advanced Studies, 1949-59; vis. prof. U. Sydney (Australia), 1958-59, faculty, 1959—, prof. high energy nuclear physics, 1961—; prin. investigator various USAF research contracts and grants, 1956—. Mem. Royal Irish Acad. Research and publs. in cosmic radiation in high energy region, determined composition and energy spectrum; used proton component to study characteristics of nuclear interactions in this range; developed theory collisions with complex nuclei, method of extending observations to 10 to the 21st power ev. Home: 105 Hewlett St., Bronte, Sydney N.S.W., Australia.*

McCUSKEY, Sidney Wilcox, Am. astronomer; b. Cuyahoga Falls, O., Feb. 28, 1907; s. Charles and Lottie (Wilcox) McC.; B.S., Case Inst. Tech., 1929; M.S., Mass. Inst. Tech., 1930; Ph.D., Harvard, 1936; m. Jeannette Scott, Dec. 24, 1932; children—Robert S., William A. Mem. faculty Case Inst. Tech. since 1930, professor of math. and astronomy since 1936, chairman of department of mathematics, 1945-59, chmn. dept. astronomy, dir. observatories, 1959—; asst., astronomy, Harvard Obs., 1934-36. Mem. Am. Astron. Soc., Am. Math. Soc., Math. Assn. Am., A.A.A.S., Sigma Xi. Contbr. articles to Astrophys. Jour. Studies of the space distribution of stars in the galaxy; evaluation of the frequency distribution of intrinsic stellar luminosities in the solar region of the galaxy. Home: 968 Brunswick Rd., Cleveland Heights 12, O. Office: Case Institute of Technology, Cleveland 6.

McCUTCHEON, Rob S., Am. pharmacologist; b. Idaho Falls, Ida., May 10, 1908; s. Rob B. and Amy (Stewart) McC.; B.S., Ida. State U., 1933; M.S., U. Wash., 1946, Ph.D., 1948; m. Julia H. Hurt, Nov. 25, 1929; children—Laurie Ann (Mrs. R. D. Drews), Judith Mary (Mrs. J. E. Oglesby), Rob A. High sch. sci. tchr., 1933-44; asso. prof. Ore. State U., 1948-55, prof. pharmacology, 1955-64; scientist adminstr. NIH, USPHS, Bethesda, Md., 1964—. Cons., Ore. Bd. Pharmacy, 1950—, state of Ore., 1952—. Am. Found. for Pharm. Edn. fellow, 1946-48; NIH spl. fellow, 1963. Fellow A.A.A.S.; mem. Am. Pharm. Assn., Fedn. Am. Socs. for Exptl. Biology, Sigma Xi. Writer, PharmIndex, 1957—. Contbr. numerous articles to tech. jours. Research on effects drugs stimulating autonomic nervous system, agts. which prevent this stimulation and metabolic changes involved when these drugs act in the system, circulation. Home: 5300 Westbard Av., Washington 20016. Office: NIH, Bethesda, Md. 20014.*

McDANIEL, Earl Wadsworth, Am. physicist; b. Macon, Ga., Apr. 15, 1926; s. James Bruce and Jewell (Wadsworth) McD.; B.S., Ga. Inst. Tech., 1948; M.S., U. Mich., 1950, Ph.D., 1954; m. Marjorie Frances Scarratt, 1948; children—Keith Bruce, Linda Frances. Faculty, Ga. Inst. Tech., Atlanta, 1948—, prof., 1960—. Cons. Oak Ridge Nat. Lab., 1959—; sec. Gaseous Electronics Conf., 1965-66; vis. scholar U. Durham (Eng.), 1966-67. Guggenheim fellow, 1966-67; Fulbright Sr. Research scholar, 1966-67. Fellow Am. Phys. Soc. Author: Collision Phenomena in Ionized Gases, 1964. Research on basic aspects of atomic collisions and gaseous electronics, natural radioactivity. Home: 4565 Meadow Valley Dr., Atlanta.*

McDERMOTT, Walsh, Am. physician; b. Conn., 1909; s. Terence and Rosella (Walsh) McD.; B.A., Princeton, 1930; M.D., Columbia, 1934; m. Marian MacPhail, Nov. 11, 1942. With N.Y. Hosp., N.Y.C., 1934—, attending physician, to present; with Cornell U. Med. Coll., 1937—, Livingston Farrand prof. pub. health, chmn. dept., 1955—; editor Am. Review of Respiratory Diseases; co-editor Cecil-Loeb Textbook of Medicine. Recipient Lasker award, 1955, Nat. Inst. Health Dyer Lectureship award, 1958, Trudeau medal, 1963; Bruce Meml. award A.C.P., 1968. Mem. Nat. Acad. Scis., Assn. Am. Physicians, Am. Soc. Clin. Investigation, numerous others. Studies, publs. in the field of the drug therapy of microbial infections, especially on microbial presistence, biomed. aspects of socio-econ. devel. Office: 1300 York Av., N.Y.C. 10021.*

McDERMOTT, William Vincent, Jr., Am. physician; b. Salem, Mass., Mar. 7, 1917; s. William Vincent and Mary A. (Feenan) McD.; grad. Phillips Exeter Acad., 1934; A.B., Harvard, 1938, M.D., 1942; m. Blanche O'Riorden, May 15, 1943; children—Blanche Anne, William Shaw, Jane Travers. Intern Massachusetts Gen. Hosp., Boston, 1942, asst. resident surgeon, 1946-49, chief resident surgeon, 1950, mem. staff, 1951—; practice medicine, specializing in surgery, Boston, 1951—; USPHS fellow dept. biochemistry Sch. Medicine Yale, 1949-50; prof. surgery Med. Sch. Harvard, 1951—; vis. prof. pro tem Kings Coll. Hosp. Med. Sch., London, Eng., 1960; dir. Harvard surg. unit, Sears surg. labs. Boston City Hosp. Chmn. research com. Med. Found. Boston. Served to maj.,

M.C., AUS, 1943-46; ETO. Diplomate Nat. Bd. Med. Examiners (chmn. surgery test com.). Mem. A.C.S., Soc. U. Surgeons, Am. Surg. Assn., New Eng., Boston surg. socs., Harvard Med. Alumni Assn. (treas.). Mem. editorial bd.: Jour. Surg. Research, 1960—. Research on the endocrine system; gastrointestinal physiology; diseases of the liver and portal circulation. Home: 570 Bridge St., Dedham, Mass. Office: Med. Sch., Harvard, Boston 02115.

McDIARMID, Ian Bertrand, Canadian physicist; b. Carleton Place, Can., Oct. 1, 1928; s. John and Lillian (Campbell) McD.; B.A., Queen's U., Kingston, Ont., 1950, M.A., 1951; Ph.D., Manchester (Eng.) U., 1954; m. Dorothy May Folger, Aug. 18, 1951; children—John, Leslie. Postdoctoral fellow NRC Can., 1954-55, staff, 1955—, prin. research officer, head cosmic ray and high energy particle sect., Ottawa, Ont., 1966—. Lectr., U. Ottawa, 1954-57. Research, publs. on electromagnetic interactions of high energy mu-mesons; satellite studies of radiation belts; rocket studies of charged particles asso. with aurora. Home: 1981 Lynda Lane, Ottawa. Office: NRC, Sussex Dr., Ottawa, Ont., Can.*

McDONAGH, Edward Charles, sociologist; b. Edmonton, Alta., Can., Jan. 23, 1915; s. Harry Fry and Aletta (Bowles) McD.; came to U. S., 1922, naturalized, 1936; A.B. magna cum laude, U. So. Cal., 1937, A.M., 1938, Ph.D., 1942; m. Louise Lucille Lorenzi, Aug. 14, 1940; children—Eileen Louise, Patricia Ann. Faculty, U. So. Cal., Los Angeles, 1947—, prof., 1956—, head dept., 1958-62, chmn. div. social scis., 1960-63, asso. dean div. social scis. and communications, 1963-66; Smith-Mundt prof., Sweden, 1956-57. Cons. Los Angeles Sch. Dist. Fellow Am. Sociol. Assn.; mem. Am. Assn. U. Profs., A.A.A.S., Am. Assn. Pub. Opinion Research, Pacific Sociol. Soc., Skull and Dagger, Blue Key, Phi Beta Kappa, Alpha Kappa Delta (nat. pres.). Author: (with E. S. Richards) Ethnic Relations in the U. S., 1952; (with J. E. Nordskog, M. J. Vincent) Analyzing Social Problems, 1956; (with J. Simpson) Social Problems; Persistent Challenges, 1963; also numerous articles. Asso. editor: Sociology and Social Research, 1960-67; cons. editor: Sociometry, 1963-66. Research on status levels of maj. ethnic minorities, evaluation of factor of prestige in evaluation of maj. professions, suggested hypothesis that man reacts toward status goals and values with unconscious motivation of a tropism. Home: 1513 Via Montemar, Palos Verdes Estates, Cal. 90274. Office: U. So. Cal., Los Angeles 90007.*

McDONALD, Donald Fiedler, Am. physician; b. Chicago Heights, Ill., Aug. 13, 1919; s. Guy Arthur and Carrie (Fiedler) McD.; M.D., U. Chgo.; m. Virginia Caroline Vail, Dec. 23, 1942; children—Bruce Bradford, Stuart Vail, Nancy Caroline. Instr. urology U. Chgo. Clinics, 1946-49; asso. prof. urology U. Wash., Seattle, 1949-58; prof. urology U. Rochester (N.Y.), 1958—, chmn. div. urology, 1958—. Markle scholar in med. scis., 1951-56. Fellow A.C.S.; mem. Am. Urol. Assn., Internat. Soc. Urology, Clin. Soc. Genito-Urinary Surgeons, A.M.A., Research, publs. in cancer of genitourinary organs, urinary calculi, renal hypertension. Home: 239 Sandringham Rd., Rochester, N.Y. 14610.*

McDONALD, Ellice, biochemist, pathologist; b. Ft. Ellice, Manitoba, Can., Oct. 27, 1876; s. Archibald and Ellen (Inkster) McD.; ed. St. John's Coll., Winnipeg, Can., also McGill U.; (M.D. 1901); m. Ann Heebner, Oct. 15, 1907; children—Vicomtess Diane de Branges de Bourica, Ellice. Successively resident surgeon Kensington Hosp. and N.Y. Lying-in Hosp., asst. in pathology Albany Med. Sch., instr. Columbia U. Coll. Pharmacy and Sci., 1901-07; instr. surgery N.Y. Post Grad. Med. Sch. and Hosp., 1907-16; asst. prof. of gynecology Grad. Sch. Medicine, U. Pa., 1922-35; dir. Cancer Research Fund, 1928-35; dir. Biochemical Research Found. of Franklin Inst. since 1935. Recipient Gold medal Internat. Faculty Scis., London, 1937. Fellow A.C.S.; mem. Am. Inst. of City of N.Y. Biochem. Soc., Faraday Soc., Franklin Institute of Pa., Am. Assn. Cancer Research, American Phys. Soc., Am. Chem. Soc., A.A.A.S., Internat. Soc. Exptl. Cytology, N.Y. Pathol. Soc. (life), Pa. Hort. Soc. Author: Studies in Gynecology and Obstetrics, 1914; Ectopic Pregnancy, 1919. Editor: Reports of the Cancer Research Laboratories of the University of Pa., Vol. 1, 1930-31, Vol. 2, 1932-33; Reports of the Biochemical Research Foundation of Franklin Inst., Vol. 3, 1934-35, Vol. 11, 1950-51, Vol. 12, 1952 53; Neutron Effects on Animals, 1947. Contbg. author on biol. effects of external radiation, 1954. Introduced McDonald's solution of orthophenylphenate, sodium oleate, acetone and alcohol for sterilization of skin of hands and abdomen, 1915. Died Jan. 31, 1955.

McDONALD, Frederick Honour, Am. civil engr.; b. Charleston, S.C., Aug. 16, 1892; s. William Ogier and Katie (St. Clair) McD.; B.S. in Elec. and Mech. Engring., Clemson U., 1914; postgrad. extension work U. Pitts., 1915-16; m. Katharine Steed Everett, Dec. 30, 1919; children—Mary Fay (Mrs. Lester MacLean), Katharine Everett (Mrs. John T. Jeter), Jane Honour (Mrs. William E. Craver, Jr.), Anne Ewing (Mrs. James E. Bell, Jr.). With Westinghouse Electric & Mfg. Co., 1914-16; field engr. Hope Engring. and Supply Co., Tulsa, 1916-17; indsl. engr. Lockwood Greene & Co., Atlanta, 1921-23; dir., chief engr. Ga. Indsl. Bur., Atlanta, 1923-24; pres. McDonald & Co., 1924-32; pvt. cons. engr., Atlanta, 1932-39; dir. Ga. Geodetic Control Surveys, 1934-39; founder, dir. Community Research Inst., Charleston, S.C., 1939—; pres. Mgmt. Research Inst., Charleston, 1950—; devel. and indsl. engr. S.C. Pub. Service Authority, 1939-41; mgmt. cons. indsl. engr., Charleston, 1941—. Pub., editor Dixie Mag., 1947-49; mem. bd. archtl. rev. for Old and Historic Charleston, 1941—. Mem. Am. Inst. Cons. Engrs., Am. Soc. C.E. (dir. 1934-36, organizer Engring. Econs. div. 1931, chmn. 1938-39), S.C. Soc. Engrs., Huguenot Soc. S.C. Author: How to Promote Community and Industrial Development, 1938; Geodetic Survey of Ga., 1939; Manual for the Business Aid Clinic, 1940; Manual on Manpower and Incentive Principles, 1951; The Citadel of Business, 1952; Education and Race Relations, 1954; Creative Management, 1956; The New Art of Fabrication Engineering, 1958. Contbr. tech. and econ. articles to profl. and nat. publs. Designer automotive, self-driven elevator system for multiple cars in single shafts, clover leaf gear assemblies, 1959—. Home: 33 New St., Charleston 29401. Office: P. O. Box 235, Charleston, S.C. 29402.*

McDONALD, Hugh Joseph, chemist; b. Glen Nevis, Ont., Can., July 27, 1913; s. Roderick J. and Annie Sarah (McDonell) McD.; student Queen's U., Kingston, Ont., 1930-32; B.Sc. with 1st class honors, McGill U., Montreal, 1935; M.S., Carnegie Inst. Tech., 1936, D.Sc., 1939; Hon. grad. U. Catolica de Rio de Janeiro, 1962; m. Margaret Bilsland Taylor, Feb. 14, 1942 (dec. Jan. 1963); children—George Gordon, Jean Margaret, Gail Margaret; m. 2d, Avis Eugenia Nieman, Aug. 8, 1964. Came to U. S., 1935, naturalized, 1943. Teaching asst., instr. Carnegie Inst. Tech., Pitts., 1936-39; faculty chemistry Ill. Inst. Tech., Chgo., 1939-48, prof., 1946-48; prof., chmn. dept. biochemistry and biophysics Stritch Sch. Medicine, Loyola U., Chgo., 1948—. Cons. Argonne Nat. Lab., 1946—. Diplomate Am. Bd. Clin. Chemistry. Fellow A.A.A.S.; mem. Am. Chem. Soc., Am. Assn. Clin. Chemists (chmn. com. on edn. 1949-53, pres. 1953-54), Soc. Exptl. Biology and Medicine, Am. Soc. Biol. Chemists, Biophys. Soc. (charter), Inst. Medicine, Sigma Xi, Phi Lambda Upsilon, Alpha Chi Sigma. Author: Ionography; Electrophoresis in Stabilized Media, 1955; also articles. Asso. editor Clin. Chemistry, 1955-60; editorial bd. Analytical Biochemistry, 1960—. Research on chem. thermodynamics and kinetics, electrochemistry, chromatography, electrophoresis, inhibition of certain hepatic syntheses by oral hypoglycemics, artherosclerosis, diabetes mellitus. Home: 5344 Cleveland St., Skokie, Ill. 60076. Office: Loyola U. Med. Center, P.O. Box 1336, Hines, Ill. 60141.

McDONALD, Janet, Am. mathematician; b. Wesson, Miss., Sept. 3, 1905; d. Joseph and Bessie (Walden) McDonald; A.B., Belhaven Coll., 1925; M.A., Tulane U., 1929; Ph.D., U. Chgo., 1943. Tchr., Prentiss (Miss.) High Sch., 1925-28; head math. dept. Miss. Synodical Coll., 1929-32, Hinds Jr. Coll., 1932-41; instr. U. Chgo., 1943-44; mem. faculty Vassar Coll., 1944—, prof. math., 1959—, chmn. dept., 1962—; spl. research projective differential geometry. Mem. Math. Assn. Am., Am. Math. Soc., Am. Assn. U. Profs., Sigma Xi. Contbr. articles in field. Address: Vassar Coll., Poughkeepsie, N.Y. 12601.*

McDONALD, John Roland, pathologist; b. Roland, Man., Can., Jan. 23, 1910; s. John R. and Alice (Carr) McD.; M.D., U. Man., 1933; M.S., 1936; m. Ann Mildred McKay, Oct. 11, 1934; children—Mrs. David B. Hill, Walter, Jay. Practice medicine, specializing in pathology, Rochester, Minn., 1934-58, Detroit, 1958—; prof. pathology Mayo Found., 1949-58, head sect. surg. pathology, 1944-58; pathologist, dir. labs. Harper Hosp., 1958—, chief of staff; faculty Wayne State U., 1958—, prof. surg. pathology, 1959—; mem. Adv. Cancer Control Program USPHS, 1964. Recipient Gold medals for exhibits Am. Soc. Clin. Pathologists, 1948, 49, Am. Cancer Soc. Minn. div., 1955. Diplomate Am. Bd. Pathology. Fellow Am. Assn. Thoracic Surgery, Am. Coll. Chest Physicians, Coll. Am. Pathologists; mem. Am. Cancer Soc., Am. Soc. Clin. Pathologists, Am. Soc. Cytology (past pres.), A.M.A., Inter-Soc. Cytology Council (past pres.), Internat. Acad. Pathology, Am. Assn. Pathologists and Bacteriologist, Zumbro Valley Med. Soc. Contbr. numerous articles to profl. jours. Research, publs. on surg. pathology and diagnostic cytology with spl. emphasis on lung and cytology. Home: 5671 Shadow Lane, Bloomfield Hills, Mich. 48013. Office: 3825 Brush St., Detroit 48201.*

McDONALD, Joseph John, physician; born Seattle, Feb. 25, 1913; s. Joseph and Nellie (Nicholson) McD.; B.S., U. Wash., 1935; M.S., Northwestern U., 1939, M.D., 1940; m. JoJanette Gilbert, Sept. 9, 1943; children—Ann Laura, Joseph Gilbert. Faculty surgery Columbia Coll. Phys. and Surg., N.Y.C., 1945—, prof., 1948—; prof. surgery Am. U. Beirut (Lebanon), 1946—, dean Med. Sch., 1953—. Mem. Soc. U. Surgeons, Soc. for Surgery of Hand, Am. Soc. Plastic and Reconstructive Surgery, Allen O. Whipple Soc., Arthur Purdy Stout Soc. for Surg. Pathology. Author: (with J. L. Wilson) Handbook of Surgery, 1963; (with J. C. Chusid) Correlative Neuroanatomy and Functional Neurology, 1964; also articles. Research on neuroanatomy; developmental anatomy; plastic and reconstructive surgery; head and neck cancer surgery. Home: American University of Beirut, Beirut, Lebanon.*

McDONALD, Lawson, physician; b. Belfast, Ireland, Feb. 8, 1918; s. Charles Seaver and Mabel Deborah (Osborne) McD.; B.A., Cambridge (Eng.) U., 1943, M.B., B.Chir., 1946, M.A., 1946, M.D., 1952; m. Ellen Greig Rattray, July 31, 1953; 1 son, James Torquil Osborne. Staff, Nat. Heart Hosp., also Middlesex Hosp., 1942-55; Rockefeller Travelling fellow medicine, asst. medicine Peter Bent Brigham Hosp., Boston, also research fellow medicine Harvard, Boston, 1952-53; asst. dir. Inst. Cardiology U. London, 1955-61, lectr., 1961—. Cons. physician London Hosp., 1960—, Nat. Heart Hosp., 1961—. Fellow Royal Coll. Physicians; mem. Assn. Physicians Gt. Britain and Ireland, Brit. Cardiac Soc., other cardiac socs. Editor: Pathogenesis and Treatment of Occlusive Arterial Disease, 1960. Research and numerous publs. on cause and mgmt. of various types of heart disease. Home: 9 Bentinck Mansions, Bentinck St., London, W.1., Eng.*

McDONALD, Leslie Ernest, Am. physiologist; b. Middletown, Mo., Oct. 14, 1923; s. Leslie Earnest and Velma (Tagg) McD.; student U. Mo., 1941-43; B.S., Mich. State U., 1948, D.V.M., 1949; M.S., U. Wis., 1951, Ph.D., 1952; m. Evelyn Frances, June 11, 1946; children—Patricia Jo, Dennis Allen, Dale Brian. Practice vet. medicine, Berlin, Wis., 1949-50; research fellow U. Wis., 1950-52; faculty U. Ill., 1952-54; prof., head dept. physiology pharmacology Okla. State U., Stillwater, 1954—; pres. Okla. Bd. Examiners Basic Med. Scis., 1961-63; cons. NIH, USPHS, 1964—; Com. on drug efficacy Nat. Acad. Sci., 1966—. Mem. Am. Physiol. Soc., Soc. Exptl. Biology and Medicine, Soc. for Study Fertility (Brit.), Am. Vet. Med. Assn., Sigma Xi. Author: (with others) Textbook of Veterinary Pharmacology and Therapeutics, 1965. Publs. on devel. of a technique for bovine testis biopsy; use of a potent steroid for treatment of a metabolic disease of cattle; use of hormones to reduce embryonic mortality of swine. Home: 2110 W. 3d St., Stillwater, Okla. 74074.*

McDONALD, Margaret Ritchie (Mrs. Bo Prytz), biochemist; b. Govan, Scotland, May 24, 1910; d. John Cunningham and Grace (Green) McDonald; came to U. S., 1924, naturalized, 1930; B.Sc., Douglass Coll., 1930; Ph.D., Rutgers U., 1940; m. Bo Prytz, Jan. 1, 1942; 1 son, John McDonald. Staff, Rockefeller Inst. for Med. Research, Princeton, N.J., 1930-43; research asso. dept. genetics Carnegie Instn. Washington, Cold Spring Harbor, N.Y., 1943-45, investigator, 1945-63; chief biochemist Waldemar Med. Research Found., Inc., Woodbury, N.Y., 1963—; civilian OSRD, 1941-43. Mem. Phi Beta Kappa, Sigma Xi. Research, numerous publs. on crystallization and characterization enzymes, kinetics and mechanism conversion trypsinogen to trypsin, effect X-radiation on enzymes, analysis chem. orgn. chromosomes showing they contain RNA as well as DNA, preferential destruction malignant tissues by nucleases. Home: 21 McCouns Lane, Oyster Bay, N.Y. 11771. Office: Sunnyside Blvd., Woodbury, N.Y. 11797.*

McDOUGALL, Ian, Australian geologist; b. Hobart, Tasmania, Australia, May 24, 1935; s. Dugald and Caroline Jane (Grant) McD.; B.Sc. with 1st class honors, U. Tasmania, 1956; Ph.D., Australian Nat. U., 1960; m. Pamela May Hodeson, May 21, 1960; children—Scott, Geoffrey David, Sandra. Postdoctoral fellow U. Cal. at Berkeley, 1960-61; research fellow Australian Nat. U., Canberra, 1961-64, fellow, 1964—; vis. prof. U. Sao Paulo (Brazil), 1966. Mem. Geol. Soc. Australia. Research and publs. on isotopic dating of young rocks; petrological, geochem. and dating studies on large intrusions of Mesozoic basaltic rocks in Tasmania, Antarctica, S. Africa, S. Am.; showed that earth's magnetic field has reversed its polarity at infrequent intervals in geol. past. Home: Rivett, Canberra, A.C.T., Australia.*

McDOUGALL, Walter Byron, Am. botanist; b. nr. Ypsilanti, Mich., Dec. 10, 1883; s. John A. and Delphine (Fowler) McD.; A.B., U. Mich., 1911, Ph.D., 1913; m. Myrtle Dolby, Oct. 24, 1908; children—Onnolee Genee (Mrs. James F. Dippell), Jocelyn Eileen (Mrs. Joe S. Farrell), Arlene Wilma (Mrs. Walter A. Sturtevant). Faculty, U. Ill., 1913-29; faculty U. So. Cal., 1930-31; biologist Nat. Park Service, Washington, 1932-55; curator botany Mus. No. Ariz., Flagstaff, 1956—. Recipient certificate for meritorious service U. S. Dept. Interior, 1953. Mem. A.A.A.S., Bot. Soc. Am., Ecol. Soc. Am., Ill. Acad. Sci., Sigma Xi. Author: Mushrooms, 1925; Plant Ecology, 1927; (with Herma A. Baggley) Plants of Yellowstone National Park, 1936; (with Omer E. Sperry) Plants of Big Bend National Park, 1951; Grand Canyon Wildflowers, 1964. Contbr. numerous articles to profl. jours. Pioneer work on mycorrhizas of forest tree roots; formulated 1st usable classification symbiotic phenomena. Home: P.O. Box 427. Office: P.O. Box 1389, Flagstaff, Ariz. 86001.*

McDOUGALL, William, psychologist; b. Lancashire, Eng., 1871; s. I. S. and R. (Smalley) McD.; ed. Owens Coll., Manchester; St. Thomas Hosp., London; M.B., Cambridge U.; M.A., Oxford U.; studied Göttingen; D.Sc., British Soc. Psychical Research (pres.);

m. A. A. Hickmore, of Brighton, Eng., 1899; children—Mrs. Paul Brown, Duncan Shimwell, Angus Dougal, Kenneth Dougal, Janet Aline (dec.). Fellow St. John's Coll., Cambridge, 1898; reader, University Coll., London; reader in mental philosophy and fellow, Corpus Christi Coll., Oxford; prof. psychology, Harvard, 1920-27; professor of psychology Duke University, 1927——. Maj. Royal Army M.C., 1914-19. Author: Physiological Psychology, 1905; Social Psychology, 1908; Pagan Tribes of Borneo, 1911; Psychology, 1912; Body and Mind, 1912; Group Mind, 1920; Is America Safe for Democracy?, 1921; Outline of Psychology, 1923; Ethics and Some Modern World Problems, 1924; Outline of Abnormal Psychology, 1926; Janus, 1927; Character and the Conduct of Life, 1927; Modern Materialism and Emergent Evolution, 1929; World Chaos—the Responsibility of Science, 1931; Energies of Men, 1933. Pioneer in physiological and social psychology; noted for his biological approach to psychological problems; with Freud known as an early leader in psychology; studied the inheritance of acquired characteristics; psychical problems; surveyed and systematized various concepts of different schools of psychology; developed system of psychology known as Hormic Psychology. Died Nov. 28, 1938.

McDOWELL, C(harles) A(lexander), phys. chemist; b. Belfast, Ireland, Aug. 29, 1918; s. Charles and Mabel (McGregor) McD.; B.Sc., M.Sc., Queens U., Belfast, 1942; D.Sc., U. Liverpool (Eng.), 1955; m. Christine Joan Staddart, Aug. 10, 1945; children—Karen Mary Anne, Christina Anne, Avril Jeanne. Sr. asst. lectr. Queen's U., 1941-42; sci. officer U.K. Civil Def., 1942-45; lectr. U. Liverpool, 1945-55; prof., head dept. chemistry U. B.C. (Can.), Vancouver, 1955——; vis. prof. Kyoto (Japan) U., 1965. NRC Sr. Research fellow Cambridge (Eng.) U., 1963-64. Recipient Letts Gold medal in theoretical chemistry Queen's U., 1941; Sci. medal Université de Liege (Belgium), 1955. Mem. Chem. Soc. London, Chem. Inst. Can., Royal Inst. Chemistry (U.K.), Royal Soc. Can., Am. Chem. Soc., Am. Phys. Soc. Editor: Mass Spectrometry, 1963. Research, numerous publs. in fields electron impact phenomena with molecules, electron spin resonance spectroscopy, mass spectrometry, chem. kinetics and chem. physics studies. Home: 5612 McMaster Rd., Vancouver 8, B.C., Can.*

McDOWELL, Ephraim, Am. physician; b. Rockbridge County, Va., Nov. 11, 1771; s. Samuel and Mary (McClung) McD.; attended med. lectures Med. Sch., U. Edinburgh (Scotland), 1793-94; M.D., U. Md., 1825; m. Sarah Shelby, 1802; 6 children. Most noted surgeon west of Phila.; performed considerable work for charity; a founder, 1st trustee Centre Coll., Danville; received diploma of membership Med. Soc. Pa., 1817. Pioneer in abdominal surgery; performed 1st ovariotomy in U. S., 1809, had performed 12 with only one death by 1824; repeatedly performed radical operative cures for nonstrangulated hernia, at least 32 operations for stones in bladder, without a death; used lateral perineal incision. Died Danville, June 25, 1830.

McDOWELL, Frank, Am. surgeon; b. Marshfield, Mo., Jan. 30, 1911; s. Hollie A. and Louise (North) McD.; A.B., Drury Coll., 1932; M.D., Washington U., St. Louis, 1936; m. Mary Elizabeth Neal, June 10, 1934; children—Robert Lawrence, George Edward, Carole Louise. Practice medicine, specializing in plastic surgery, St. Louis, Honolulu, 1942——; mem. staffs Barnes, St. Louis Children's, DePaul hosps.; pres. med. and surg. staff Barnes, Allied hosps. Med. Center, 1958-59; cons. surgeon Shriners, Frisco, Mo. Pacific, St. Louis City hosps. faculty Washington U. Sch. Medicine, 1939-67, asso. prof. clin. surgery, 1954-67, asso. prof. maxillo-facial surgery Sch. Dentistry, 1940-67; exec. com. dept. surgery U. Hawaii Sch. Medicine, 1967——; surg. cons. Tripler Gen. Hosp., Honolulu, 1967——; surg. staff Queen's, Children's, Castle hosps., Honolulu, 1967——. Lectr. various univs.; mem. Adv. Bd. for Med. Specialists, 1955-62; v.p. 3d Internat. Congress Plastic Surgery, Washington, 1963; exec. com. Adv. Bd. for Med. Splty s., 1966——. Diplomate Am. Bd. Surgery, Am. Bd. Plastic Surgery (chmn. 1961-62). Fellow A.C.S.; mem. Am. Assn. Plastic Surgeons (pres. 1962-63), Am. Soc. Plastic and Reconstructive Surgeons, Soc. Head and Neck Surgeons (founder), Am., Western surg. assns., Am. Assn. Surg. Trauma, Israel Assn. Plastic Surgeons (hon.), Societe francaise de chirurgie reconstructive et plastique (asso.). Author: Surgery of Face, Mouth and Jaws, 1954; Neck Dissections, 1957; Skin Grafting, 1957; Plastic Surgery of the Nose, 1966; also articles. Asso. editor Directory Med. Specialists, 1955-62, Jour. Plastic and Reconstructive Surgery, 1959-65, Lawyer's Med. Cyclopedia, 1962——; editor-in-chief Jour. Plastic and Reconstructive Surgery, 1967——. Devel. neck and jaw surgery, including internal pin fixation for fracture jaws, cleft lip operations, muscle resect.; early work on homograft acceptance and rejection. Home: 100 F. N. Kalaheo, Kailua, Hawaii 96734. Office: Alexander Young Bldg., Honolulu 96813.*

McDOWELL, M. C. R., Brit. physicist; b. Belfast, N. Ireland, Jan. 30, 1932; s. Richard W. C. and Eveline (Fieldman) McD.; B.Sc., Queen's U. Belfast, 1953, Ph.D., 1957; M.A., Columbia U., 1954; m. Brenda G. Blair, June 7, 1956; children—Jonathan Christopher, David Blair. Gassiot Research fellow Queen's U., Belfast, 1956-57; lectr. math. Royal Holloway Coll., U. London, 1957-64; reader applied math. U. Durham (Eng.), 1964——; vis. scholar Ga. Inst. Tech., 1959-60, U. Colo., 1963. Cons., Oak Ridge Nat. Lab., 1959-64, Kaman Nuclear Corp., 1962-63; mem. organizing com. internat. confs. on physics and electronic and ionic collisions. Fellow Phys. Faculty, Royal Acad. Scis.; mem. Am. Phys. Soc. Editor: Atomic Collision Processes, 1964. Research and publs. in atomic collision physics, upper atmosphere, astrophysics; deduced (with D. R. Bates) temperature of thermophere before 1st satellites; studies on role of negative ion of hydrogen in catalysing molecule formation and hence star formation, ion-atom (especially proton-hydrogen atom) charge exchange. Home: 20 St. Oswald's Dr., Durham, Eng.*

McDOWELL, Sister Margaret Ann, Am. biologist; coll. adminstr.; b. Coshocton, O., Oct. 22, 1912; d. Charles S. and Gertrude (Divan) McDowell; B.A., Coll. St. Mary Springs, 1935; M.A., Ohio State U., 1942; M.S., Institutum divi Thomae, Cin., 1943, Ph.D., 1953. Joined Dominican Order, 1933; high sch. tchr., Zanesville, O., Steubenville, O., 1935-42; faculty Coll. St. Mary Springs, Columbus, O., 1945——, biology prof., 1953-62, head biology dept., dir. research, 1953-57, chmn. div. natural sci., 1957-62, exec. v.p., acad. dean, 1962-66. Fellow Ohio Acad. Sci.; mem. A.A.A.S., N.Y. Acad. Scis., Soc. Am. Bacteriologists, Ohio Biology Tchrs. Conf., Sigma Xi, Delta Epsilon Sigma. Author: Dignity of Science, 1961; also articles. Research on cancer; discovered anti-cancer factor from staphylococcus. Address: Coll. St. Mary Springs, Columbus, O. 43219.*

McELHINNEY, John, Am. physicist; b. Phila., Mar. 25, 1921; s. Joseph and Mary (Kearney) McE.; B.S., Ursinus Coll., 1942; M.S., U. Ill., 1943, Ph.D., 1947; m. Geraldine E. Walters, Dec. 28, 1942; children—Ruth Elaine, Barbara Jill. Spl. research asso. U. Ill., Champaign, 1947-48; asso. scientist Los Alamos Sci. Lab., 1948-49; staff Nat. Bur. Standards, Washington, 1949-55; head nuclear interactions br. Naval Research Lab., Washington, 1955-63, Linacs br., 1963-66, asso. supt. nucleonics div., 1957-66, supt., nuclear physics div., 1966——. Recipient Presdl. citation, 1964. Mem. Am. Phys. Soc., Research Soc. Am., A.A.A.S., Philos. Soc. Washington. Research and publs. on photofission and photonuclear reactions, radiation dosimetry of high energy X rays. Home: 11601 Stephen Rd., Silver Spring, Md. 20904. Office: Code 7600, Naval Research Lab., Washington 20390.*

McELHINNY, Michael William, geophysicist; b. Windsor, Eng., Dec. 9, 1933; s. Geoffrey William and Doris (Burrows) McE.; B.Sc. with honors in Physics, Rhodes U., Grahamstown, S. Africa, 1953, Ph.D., 1958; m. Margaret Ann Malton, May 4, 1957; children—Christopher Michael, Lindsay Anne, Julia Jane. Lectr. physics Univ. Coll. of Rhodesia, Salisbury, 1956-64, lectr. 1965-66; sr. fellow in geophysics Australian Nat. U., Canberra, 1967——. Fellow Inst. Physics (London), Royal Astron. Soc.; mem. Am. Geophys. Union. Contbr. articles to profl. jours. Research on effect of solar eclipses on ionosphere, paleomagnetic studies of rock systems in Central and So. Africa contbg. to theory of continental drift and ancient magnetic field. Address: Dept. of Geophysics, Australian National University, Box 4, P.O., Canberra, Australia.*

McELROY, Dennis Lee, Am. mining engr.; b. Moundsville, W.Va., Oct. 19, 1904; s. Arch Jones and Ella Mae (Dorsey) McE.; B.S. in Mining Engring., W.Va. U., 1927, M.S., 1930; m. Opal Lucille Hall, June 2, 1926; 1 son, Arch Leslie. Mine foreman Penn-Pitt Coal & Coke Co., Greensboro, Pa., 1927-28; research fellow W.Va. U., Morgantown, 1928-30, asst. dir. mining extension, 1930-38, dir. Sch. Mines, indsl. extension, prof. mining engring., 1939-43; head dept. mining engring. Va. Poly. Inst., Blacksburg, 1938-39; chief coal sect. WPB, 1941-43; chief engr. Consol. Coal Co., Pitts., 1946-47, v.p. operations, 1952-60, exec. v.p., 1960-65, also dir.; dir. N.W. Hanna Fuel Co. Mem. survey team U. S. Bur. Mines, 1953; mem. U. S. Dept. Commerce Indsl. Trade Commn. to Philippines, 1965. Mem. W.Va. U. Alumni Assn. (pres. 1963-64), Engrs. Soc. W. Pa. (pres. 1963-64), Soc. Mining Engrs. (dir.), Sigma Xi, Sigma Gamma Epsilon, Kappa Kappa Psi, Tau Beta Pi. Asso. editor Mineral Industry, 1939-41. Contbr. papers on mining and mining edn. to profl. jours, chpt. to Coal Mining, 1960. Home: 430 Morrison Dr., Pitts. 15216.*

McELROY, William David, Am. biochemist; b. Rogers, Tex., Jan. 11, 1917; s. William D. and Ora (Shipley) McE.; B.A., Stanford, 1939; M.A., Reed Coll., 1941; Ph.D., Princeton, 1943; LL.D., Buffalo U., 1962; m. Nella A. Winch, Dec. 23, 1940; children—Mary, Ann, Thomas, William. Faculty biology Johns Hopkins, Balt., 1946——, prof., 1952——, dir. McCollum Pratt Inst., 1948. Mem. Sch. Bd., Balt., 1958——. Recipient Barnett Cohen award in bacteriology Am. Soc. Microbiology-Md., 1950: Rumford award Am. Acad. Arts and Sci., 1965. Mem. Nat. Acad. Scis. Author: Cell Physiology and Biochemistry, 1960. Study of biochem. genetics; bacterial mutations; bacterial and mold metabolism; mechanism of inhibitor action. Home: 220 Ridgewood Rd., Balt. 21210.*

McENTIRE, Davis, Am. sociologist; b. Harrisville, Utah, Oct. 15, 1912; s. Wells and Ida (Davis) McE.; B.S., Utah State Coll., 1932; M.A., Duke, 1933; M.P.A., Harvard, 1941, Ph.D., 1947; m. Iras Leavitt, July 16, 1932; children—Marian (Mrs. Alberto Garcia Alvarez), Mark. Economist, U. S. Dept. Agr., 1934-41; asst. dir. employment U. S. War Relocation Authority, 1942-43; research dir. Commonwealth Club Cal., 1944-45; dept. chief Relief and Rehab. Mission, UN, Byelorussian S.S.R., 1946-47; prof. social welfare, lectr. agrl. econs. U. Cal. at Berkeley, 1947——; exchange prof. univs. of Rome, Padua, Urbino, Catania (all Italy), 1959-60, 65, 67-68. Research dir. Commn. on Race and Housing, N.Y.C., 1955-60; cons. U. S. Bur. Census, 1955-60; mem. com. on agrl. scis. U. S. Dept. Agr., 1965——. Guggenheim fellow, 1953-54; Fulbright Research scholar, Italy, 1959-60, 67-68; recipient Sidney Hillman Found. prize award, 1960. Mem. Am. Sociol. Assn., Rural Sociol. Soc., Am. Soc. Pub. Adminstrn., Population Assn. Am. Author: Population of California, 1946; Labor Force of California, 1952; Residence and Race, 1960; (with Nathan Glazer) Housing and Minority Groups, 1960; also articles. Research on social policy especially on rural populations, agrarian reform movements, and rural devel. in U. S. and fgn. countries; study of pub. policy toward racial discrimination in housing. Home: 1307 Bay View Pl., Berkeley, Cal. 94708.*

McEWEN, Osceola Currier, Am. physician; b. Newark, N.J., Apr. 1, 1902; s. George Floy and Antoinette (Currier) McE.; B.S., Conn. Wesleyan U., 1923, D.Sc., 1950; M.D., N.Y. U., 1926; D.Sc., Marietta Coll., 1952; m. Katherine Morgan Cogswell, June 17, 1930; children—Ann Cogswell (Mrs. Maurice Castonguay), Matilda Currier (Mrs. Ruben Mendez), Katherine Thompson (Mrs. Hubbard C. Goodrich), Ewen Currier Floy. Asst. in medicine Rockefeller Inst., 1928-30, asso. in medicine, 1930-32; instr. medicine N.Y. U., 1932-33, asst. dean, 1932-37, asst. prof. medicine, 1933-39, dean, 1937-55, asso. prof. medicine, 1939-59, prof. medicine, 1959——, chmn. rheumatic disease study group, 1948-67; cons. Surgeon Gen. Army, 1947-53; cons. VA Central Office, 1959, mem. spl. adv. group, 1959-65; mem. nat. adv. arthritis and met. disease council USPHS, 1954-58; bd. dirs. Nat. Arthritis Found., 1953-57, N.Y. chpt., 1948. Mem. Assn. Am. Physicians, Am. Soc. Clin. Investigation, Am. Rheumatic Assn. (pres. 1952), Interurban Clin. Club, A.M.A., Am. Coll. Physicians, Harvey Soc. Contbr. numerous articles in field to sci. jours. Clin., pathologic, immunologic studies of rheumatic fever, systemic lupus erythematosus and various types of arthritis. Home: 5441 Palisades Av., N.Y.C. 10471. Office: N.Y. U. Med. Center, 550 First Av., N.Y.C. 10016.*

McEWEN, William Hayward, Canadian mathematician; b. Owen Sound, Ont., Can., Feb. 7, 1902; s. Robert Cumming and Matilda (Marshall) McE.; B.S., U. Sask. (Can.), 1921, M.S., 1922; M.A., U. Minn., 1924, Ph.D., 1930; m. Olive Kathleen Leitch, Nov. 18, 1926. Tchr., Regina (Sask.) Coll., 1925-29; faculty Mt. Allison U., Sackville, N.B., Can., 1930-46, prof., head dept., 1933-46; faculty U. Man., Winnipeg, 1946——, dean grad. studies and research, 1949-63, prof., 1963——. Mem. exec. bd., v.p. Can. Math. Congress, 1946-57. Mem. Am. Math. Soc., Math. Assn. Am. Research and publs. in approximations asso. with linear differential systems, degree of convergence of derived Birkhoff series spectral theory and applications. Home: 7-828 Preston Av., Winnipeg 10, Man., Can.*

McFADDEN, William H., chemist; b. Vancouver, B.C., Can., Jan. 25, 1927; s. William J. and Effie (Hamilton) McF.; B.S., U. B.C., Vancouver, 1949, M.A., 1951; Ph.D., U. Utah, 1954; m. Mary Louise Muirhead, June 13, 1949; children—James Cameron, Heather Ruth, Shauna Louise, Roderick Bruce. Came to U. S., 1958. Research asso. Cornell U., Ithaca, N.Y., 1954-56; asst. research officer Atomic Energy Can., Chalk River, Ont., Can., 1956-58; chemist Shell Devel. Co., Emeryville, Cal., 1958-59; chemist Western Regional Labs., Albany, Cal., 1959-67; dir. Internat. Flavors & Fragrances, Union Beach, N.J., 1967——. Mem. Am. Chem. Soc., Sigma Xi. Research and publs. on devel. techniques in mass spectrometry and gas chromatography for analysis of microgram quantities of complex organic mixtures; determined important reaction mechanisms in mass spectral decompositions of complex molecules. Home: 30 Sunnylands Ct., Little Silver, N.J. 07739. Office: 1515 Hwy. 36, Union Beach, N.J. 07735.*

McFARLAN, Ronald Lyman, Am. electronic engr.; b. Cin., Mar. 8, 1905; s. Frank G. and Mary (Henninger) McF.; A.B., U. Cin., 1926; Ph.D., U. Chgo., 1930; m. Ethel Warren White, Sept. 6, 1933; children—Franklin Warren, Ethel Louise (Mrs. Charles Martin Hamann). Nat. Research fellow Harvard, 1930-32, instr. physics, 1932-35; chief physicist United Drug Co., Boston, 1935-40; chief physicist United Shoe Machinery Corp., Cambridge, Mass., 1940-43; dir. research Bulova Watch Co., N.Y.C., 1943-46; asst. dir. engring. Raytheon Co., Waltham, Mass., 1946-56; electronic cons., Chestnut Hill, Mass., 1956——. Mem. Gamma Alpha. X-ray research in structure of liquids and solids, 1930-35, electronic instrumentation to measure vitamin concentrations, 1935-40, electronic equipment for guidance, control and digital information processing, 1940——; crystal structure of high pressure ice forms, miniature aircraft instruments. Address: 20 Circuit Rd., Chestnut Hill, Mass. 02167.*

McFARLAND, David Ford, Am. chemist, metallurgist; b. Mansfield, O., Aug. 1, 1878; s. Robert S. and Mary J. (McBride) McF.; A.B., U. Kan., 1900, A.M., 1901; fellow Yale U., 1902-03, M.S., 1903, Ph.D., 1909; m. Martha Elizabeth Pittenger, June 23, 1909; children—George Robert, Mary Louise, Elizabeth Jean, David Ford. Instr. in chemistry U. Kan., 1900-02, asst. prof., 1903-10; asst. in chemistry Yale, 1908-09; asst. prof. applied chemistry U. Ill., 1910-14, asso. prof., 1914-20; prof. metallurgy and head of dept. Pa. State Coll., 1920-45, acting dean Sch. Mine and Metallurgy, 1922, 27-28, prof. emeritus, 1945; asst. chem. Kansas Geol. Survey, summers 1899-1907; asst. chem. and metallographist Engring. Expt. Sta., U. Ill., summers 1910-20. Mem. Am. Soc. for Metals Am. Chem. Soc., Phi Beta Kappa, Sigma Xi. Contbr. papers and bulls. on organic and indsl. chem. and metall. subjects. Discovered (with H. P. Cady) helium in natural gas, also methods of extraction on lab. scale. Died Feb. 5, 1955.

McFARLAND, Joseph, Am. pathologist; b. Phila., Feb. 9, 1868; s. Joseph and Susan E. (Grim) M.; M.D., U. Pa., 1889, Medico-Chirurg. Coll., 1898; postgrad. Heidelberg and Vienna, 1890, Berlin and Halle, summer 1895, Pasteur Inst., Paris, summer 1903; Sc.D., Ursinus Coll., Pa., 1913; m. Virginia E. Kinsey, Sept. 14, 1892; children—Helen Josephine, Katharine A., Ruth, Joseph. Prof. pathology and bacteriology Medico-Chirurg. Coll., 1896-1916; prof. pathology Woman's Med. Coll. Pa., 1911-13; prof. pathology U. Pa., 1916-40; prof. gen. pathology Temple U. Dental Sch. since 1940; vis. prof. pathology Jefferson Med. Coll., 1943. Fellow A.C.P., Coll. Physicians Phila., A.M.A., Acad. Natural Sci. Phila., Acad. Stomatology; mem. Med. Soc. State Pa., Phila. County Med. Soc., Am. Assn. Pathologists and Bacteriologists, Soc. Clin Pathologists. Author: Pathogenic Bacteria, 9 edits., 1896-1910; Text-book of Pathology, 2 edits., 1904, 09; Biology, General and Medical, 5 edits., 1910-26; The Breast (with Dr. John B. Deaver), 1917; Fighting Foes Too Small to See, 1923; Surgical Pathology, 1924; also many contbns. to med. lit. in English and German. Died Sept. 22, 1945.

McFARLAND, Robert White, Am. mathematician; b. Champain County, O., June 16, 1825; s. Robert and Eunice (Dorsey) M.; A.B., Ohio Wesleyan U., 1847, A.M., 1850, LL.D., 1884; m. Mary A. Smart, Mar. 19, 1851. Tchr. math. Greenfield Sem., 1848-51; supt. pub. schs., Chillicothe, O., 1851-53; prof. math. Madison Coll., O., 1853-56; prof. Miami U., 1856-73, pres., 1885-88; prof. Ohio State U., 1878-85, emeritus prof. civil engring., 1902—. Editor 6 books of Virgil, 1849. Computed eccentricity of the earth's orbit and longitude of perihelion for 4,500,000 yrs., at intervals of 10,000 yrs. Died 1910.

McFARLAND, Ross Armstrong, Am. aerospace scientist; b. Denver, July 18, 1901; s. James and Helen (Russell) McF.; A.B., U. Mich., 1923; Ph.D., Harvard, 1928; Sc.D. (hon.) Park Coll., Rutgers U., Trinity Coll.; m. Emily Frelinghuysen Bilkey, Oct. 14, 1950. Faculty, Columbia, 1928-37; faculty Harvard, 1937—, Daniel and Florence Guggenheim prof., 1962—, dir. Guggenheim Center Aerospace Health and Safety, 1957—. Cons. govt. agys. including NASA, surgeon gen. U. S. Army, USPHS. Recipient John Jeffries award Inst. Aero. Scis., 1956; Arthur William Meml. award Am. Museum Safety, others. Fellow Aerospace Med. Assn. (Longacre award 1947, Walter Boothby award 1962), Am. Acad. Arts and Scis., Am. Geog. Soc., Am. Psychol. Assn., Gerontological Soc., Am. Pub. Health Assn., N.Y. Acad. Scis. (v.p. 1935-56), Royal Aero. Soc.; mem. A.A.A.S., Am. Assn. Automotive Medicine (hon.), Am. Assn. Phys. Anthropologists, Am. Physiol. Soc., Am. Inst. Aeros. and Astronautics, Am. Psychosomatic Soc., Am. Rocket Soc. (sr.), Ergonomics Research Soc. (ann. lectureship Gt. Britain), Human Factors Soc., Internat. Acad. Aviation Medicine, Soc. Automotive Engrs. Author: Human Factors in Air Transport Design, 1946; Human Factors in Air Transportation-Occupational Health and Safety, 1953; The Human Body in Equipment Design, 1966. Research, publs. on effects of high altitude on man, air and hwy. safety and aerospace medicine, one of founders of field of human factors, engring. and equipment design. Home: 17 Fresh Pond Pkwy., Cambridge, Mass. 02138. Office: 665 Huntington Av., Boston 02115.*

McFARLAND, William N., zoologist; b. Toronto, Ont., Can., Sept. 11, 1925; B.A. in Zoology, U. Cal. at Los Angeles, 1951, M.A., 1953, Ph.D., 1958; m. Anna K. McIntire, Apr. 21, 1950; 3 children. Physiologist, Inst. Marine Sci., U. Tex., 1958-61; asso. prof. zoology Cornell U., Ithaca, N.Y., 1961—, chmn. sect. ecology and systematics div. biol. scis. Aquatic cons. Mem. A.A.A.S., Am. Soc. Ichtyologists and Herpetologists, Am. Soc. Zoologists, Sigma Xi. Research, publs. on physiol. adaptations of organisms to their environment, especially their metabolic potential, ability to regulate salt and water balance and their visual adaptations. Home: 117 E. Remington Rd., Ithaca, N.Y. 14850.*

McFEAT, T. F. S., Canadian cultural anthropologist; b. Montreal, Que., Can., Feb. 5, 1919; s. William Pearse and Cecil (Scott) McF.; B.A., McGill U., Montreal; A.M., Ph.D., Harvard; m. Mary Helen Davis, Sept. 5, 1947; children—Thomas Charles, Elaine Barbara. Asst. prof. U. N.B., 1954-59; chief ethnologist Nat. Mus. Can., 1959-63; asso. prof. anthropology Carlton U., 1963-64; chmn. dept. anthropology U. Toronto (Ont., Can.), 1964—. Fellow Am. Anthrop. Assn. Author: Indians of the North Pacific Coast, 1965; also articles. Field study of Canadian Army inf. tng., at Zuni, N.M., Tobique Point, N.B.; lab. studies in mass media, small group culture; pioneered research on community orgn. of Indians. Home: 34 Wilgar Rd., Toronto 18, Ont., Can.*

McGAVACK, Thomas Hodge, Am. internist; b. Waterford, Va., Apr. 7, 1898; s. Thomas and Esther Hodge (Clapham) McG.; A.B., Hampden-Sydney Coll., 1917; M.D., Hahnemann Med. Coll., 1923; m. Freda Nichols Wicks, May 2, 1930; 1 stepdau., Daphne Cather Durant. Mem. faculty U. Cal. at San Francisco, 1925-35, N.Y. Med. Coll., N.Y.C., 1936-57; asso. chief of staff research and edn., chief intermediate service, dir. geriatrics research lab., VA Center, Martinsburg, W. Va., 1957—; professorial lectr. George Washington U. Sch. Medicine., Washington 1957—. Conferee, Nat. Conf. World Health for World Peace, 1959, W. Va. Gov. Conf. on Aging, 1959, 60, Commn. on Aging, 1967—; cons., on endocrinology and metabolism to numerous hosps. Diplomate Am. Bd. Internal Medicine. Fellow A.C.P.; mem. A.M.A., Internat. Assn. Gerontology (mem. council), Am. Geriatrics Soc. (1st Edward Henderson lectureship award, 1963, dir., past pres., asso. editor jour.), Assn. Study Internal Secretions (mem. council), Soc. Exptl. Biology and Medicine, Harvey Soc., A.A.A.S., Gerontology Soc., N.Y. Acad. Scis., Am. Diabetes Assn., Am. Thyroid Assn., N.Y. Acad. Medicine, Sigma Upsilon, Phi Beta Kappa, Alpha Omega Alpha, others. Author: The Thyroid, 1951; (with E. P. Gelvin, T. H. McGavack) Obesity, Its Causes, Classification and Care, 1957. (with E. Simonson, T. H. McGavack) Cerebral Ischemia, 1964. Research and publs. on subjects related to endocrine and metabolic diseases, particularly as they are influenced by aging. Home: 12 Showers Lane, Martinsburg, W. Va. 25401. Office: VA Center, Martinsburg, W.Va. 25401.*

McGEE, Lemuel Clyde, Am. med. adminstr.; b. New Boston, Tex., Aug. 2, 1904; s. Thomas and Elizabeth (Murrell) McG.; A.B., Baylor U., 1924; Ph.D. in Biochemistry and Physiology, U. Chgo., 1927; M.D., Rush Med. Coll., 1929. Practice medicine, specializing in internal medicine, Dallas, 1932-34; faculty Baylor U. Coll. Medicine, part time 1932-34; research fellow medicine and biochemistry Harvard Sch. Medicine and Peter Bent Brigham Hosp., Boston, 1939-40; med. dir. Hercules, Inc., Wilmington, Del., 1940—; mem. staff Del. Hosp., Wilmington, 1944—, attending chief in medicine, 1948—; vis. prof. indsl. medicine U. Pa., 1949—. Mem. med. adv. com. Bur. Old-Age and Survivors Ins., Social Security Adminstrn., 1957-66, chmn., 1960; mem. com. fellowships in indsl. medicine AEC, 1955-63; mem. adv. com. occupational health Surgeon Gen. USPHS, 1957-65; mem. adv. com. health edn. United Health Found., Inc., 1963—. Diplomate Nat. Bd. Med. Examiners, Am. Bd. Internal Medicine, Am. Bd. Preventive Medicine in Indsl. Medicine (vice chmn. occupational medicine 1961—). Fellow Am. Acad. Occupational Medicine (award of Honor 1963), Am. Gastroent. Assn., A.C.P. (3d v.p., 1957-58, trustee life ins. plan 1960—), Am. Pub. Health Assn., Indsl. Med. Assn. (Knudsen award 1958-59), A.M.A. sec. sect. preventive medicine 1959-65), Am. Coll. Preventive Medicine; mem. Am. Heart Assn., Am. Indsl. Hygiene Assn., Am. Clin. and Climatol. Assn., N.Y. Acad. Scis., Aerospace Med. Assn., Sigma Xi, Alpha Omega Alpha. Author: Student Manual of Industrial Medicine, 1956, also articles. Editorial bd. Annals Internal Medicine, 1960-65. Home: 4301 Pyles Ford Rd., Greenville, Del. 19807. Office: 910 Market St., Wilmington, Del. 19899.*

McGEE, W. J., Am. anthropologist, geologist. hydrologist; b. Dubuque Co., Ia., Apr. 17, 1853; s. James and Martha (Anderson) M.; self ed.; (LL.D., Cornell Coll., Ia., 1901); m. Anita, Newcomb, 1888; one dau. In land surveying and justice court practice, 1873-75; invented, patented and mfd. agrl. implements, working at forge and bench, 1874-76; studied geology and archaeology, 1875-77; made geologic and topographic survey of Northeastern Iowa, 1877-81; examined and reported upon building stones of Iowa for 19th Census, 1881-82; became attached to U. S. Geol. Survey, and in 1885 assumed charge of important div.; surveyed and mapped 300,000 sq. miles in Southeastern U. S.; compiled geologic maps of U. S. and of New York; investigated Charleston earthquake, 1886; explored, 1894-95, Tiburon Island. Ethnologist in charge Bur. of Am. Ethnology, 1893-1903; resigned July 1903, to become chief dept of anthropology. St. Louis Expn., 1904; dir. St. Louis Pub. Mus., 1905-07; U. S. commr. Inland Waterways Commn., from 1907; expert U. S. Dept. of Agr., 1907. Lecturer; U. S. commr. Am. Internat. Commn. of Archaeology and Ethnology, from 1902; chmn. organizing com. for Internat. Geographic Congress, 1904; senior speaker dept. of anthropology, World's Congress of Arts and Sciences, 1904; sec. Conf. of Governors in White House, 1908. Leading founder Columbian Hist. Soc.; pres. Am. Anthrop. Assn.; pres. Anthrop. Soc. Washington; acting pres. A.A.A.S., 1897-98; pres. Nat. Geog. Soc., 1904-05; v.p. Archaeol. Inst. America, 1902-05. Author: Pleistocene History of Northeastern Iowa, 1891; Geology of Chesapeake Bay, 1888; The Lafayette Formation, 1892; The Potable Waters of Eastern U. S., 1894; The Siouan Indians, 1897; Primitive Trephining in Peru, 1898; The Seri Indians, 1900; Primitive Numbers, 1901; Outlines of Hydrology, Bull. Geol. Soc., America, 1908; Soil Erosion, 1911; The Agricultural Duty of Water, 1911; Wells and Subsoil Water, 1913. Editor dept. anthropology, Internat. Encyclopedia; asso. editor Nat. Geo. Magazine. Developed method of correlating geologic formations by identity of origin; tried to classify anthropology; studied previously unknown tribe Am. Indians. Died Washington, Sept. 4, 1912.

McGEER, Edith Graef, chemist; b. N.Y.C., Nov. 18, 1923; d. Charles and Charlotte (Ruhl) Graef; B.A., Swarthmore Coll., 1944; Ph.D., U. Va., 1946; m. Patrick L. McGeer, Apr. 15, 1954; children—Patrick Charles, Brian T., Victoria L. Research chemist Exptl. Sta. DuPont & Co., Wilmington, Del., 1946-54; research asso. Kinsmen Lab. Nuerol. Research U. B.C., Vancouver, Can., 1956—. Merck fellow, 1944-45, DuPont fellow, 1945-46. Mem. N.Y. Acad. Sci., Canadian Biochem. Soc., Phi Beta Kappa, Sigma Xi. Contbr. articles in field to sci. jours. Studies on aromatic metabolism, brain chemistry; new organic syntheses; patentee chem. tetracyanoethylene. Home: 4727 W. 2d Av., Vancouver 8, B.C., Can.*

McGEER, P(atrick) L., Canadian physician; b. Vancouver, B.C., Can., June 29, 1927; s. James Arthur and Ada (Schwengers) McG.; B.A. with honors, U. B.C., Vancouver, 1948, M.D., 1958; Ph.D., Princeton, 1951; m. Edith Graef, Apr. 15, 1954; children—Patrick, Brian, Victoria. Research chemist E. I. DuPont, Wilmington, Del., 1951-54; research asso., U. B.C., 1956-58, faculty, 1959—, asso. prof. medicine, 1962—; mem. Legislative Assembly of B.C., 1962—. Research and publs. on identification of new substances excreted by humans, identification of enzymes concerned with brain metabolism, relationships between brain metabolism drugs and behavior. Home: 4727 W. 2d St., Vancouver, B.C., Can.*

McGHEE, Robert Barclay, Am. zoologist, parasitologist; b. Cleveland, Tenn., Feb. 22, 1918; s. Charles McClung and Marian (Snavely) McG.; student King Coll., 1936-37; A.B., Berea Coll., 1940; M.S., U. Ga., 1942; Ph.D., U. Chgo., 1948; m. Ann Lewis Hinkle, Mar. 30, 1946; children—Nancy Stuart, Terence Barclay, Michael Bruce. Faculty Rockefeller Inst. for Med. Research, 1948-53; faculty U. Ga., Athens, 1953—, Alumni Found. distinguished prof. zoology, 1964—, head dept., 1955-64. Rapporteur 7th Internat. Congresses for Tropical Medicine and Malaria, 1963. Recipient Michael award for research U. Ga., 1957. Fellow A.A.A.S.; mem. Am. Soc. Parasitologists (council 1961-64), Soc. Protozoologists, Am. Soc. Naturalists, Am. Soc. Tropical Medicine and Hygiene, Am. Acad. Microbiology. Editorial bd. Jour. Parasitology, 1956-62, Exptl. Parasitology, 1965—. Research, publs. on malaria. Home: Barnett Shoals Rd., Athens, Ga. 30601.*

McGHIE, Andrew, Scottish psychologist; b. Glasgow, Scotland, Feb. 14, 1926; s. Andrew and Mary (McNaught) McG.; M.A. with 1st class honors, U. Glasgow, 1951, Ph.D., 1959; m. Olivea Ena Allison, Apr. 12, 1952; children—Austin Lindsay, Saunders. Research asst., mental insts., 1952-57; sr. clin. psychologist Gartnavel Royal Mental Hosp., Glasgow, 1957-59; prin. psychologist Dundee (Scotland) Royal Mental Hosp., 1959-63; dir. dept. clin. psychology Royal Dundee Liff Hosp., 1963—; lectr. dept. psychiatry Queen's Coll., U. St. Andrew's, Dundee, 1963—. Examiner psychology Scottish Diploma Psychol. Medicine and Royal Coll. Fellow Brit. Psychol. Soc. Author: (with T. Freeman, J. L. Cameron) Chronic Schizophrenia, 1958, Studies in Psychosis, 1966; Psychology as Applied to Nursing, 1959; also numerous articles. Research on selective attention in schizophrenic and other psychiat. patients, relationship between sleep patterns and psychiat. illness, mother-child relationship in schizophrenia. Home: 5 Deanbank, Dundee. Office: Royal Dundee Liff Hosp., Liff, by Dundee, Scotland.*

McGILL, Thomas Emerson, Am. psychologist; b. Sharon, Pa., Sept. 26, 1930; s. Emerson Dickson and Margaret (McCallen) McG.; B.A., Youngstown U., 1954; M.A., Princeton, 1957, Ph.D., 1958; m. Nancy Cipperley Welch, June 14, 1955; children—Michael Howard, Steven Emerson. Faculty dept. psychology Williams Coll., Williamstown, Mass., 1958-59, 1960—, asso. prof., 1965—, dir. NSF undergrad. traineeship program, 1962—; USPHS fellow U. Cal. at Berkeley, 1959-60; NAS-NRC fellow U. Edinburgh (Scotland), 1964-65. Mem. Am. Psychol. Assn., A.A.A.S., Am. Assn. U. Profs., Animal Behavior Soc., Soc. Study of Evolution, Sigma Xi. Author: Readings in Animal Behavior, 1965. Research, publs. on effects of sensory deprivation on human behavior; physiology of hearing; the genetics and physiology of differences in sexual behavior in mice. Home: Green River Rd., Williamstown, Mass. 01267.*

McGILVERY, Robert Warren, Am. biochemist; Coquille, Ore., Aug. 25, 1920; s. Neil and Mary (Berry) McG.; B.S. in Chemistry, Ore. State Coll., 1941;

Ph.D. in Physiol. Chemistry, U. Wis., 1947; m. Alice M. Lusby, Nov. 1, 1943; children—Jeannie, Malcolm, Laurel, Elizabeth. USPHS Sr. Research fellow U. Wis., 1947-48, faculty, 1948-57, asso. prof., 1951-57; faculty U. Va., Charlottesville, 1957—, prof. biochemistry, 1962—. Editor: (with B. M. Pogell) Fructose-1,6-diphosphatase and Role in Gluconeogenesis, 1964. Research and publs. on methods cell synthesis; discovery role ATP in formation organic sulfates, adaptability liver glucose-synthesizing enzyme. Home: Route 2, W. Leigh St., Charlottesville, Va. 22901.*

McGINNIS, Robert, Am. sociologist; b. Nacogdoches, Tex., Oct. 19, 1927; s. Brian and Dorothy (Abercrombie) McG.; B.A. with high honor, San Francisco State Coll., 1950; M.A., Stanford, 1951; Ph.D., Northwestern U., 1955; m. Mary McCulloch, Aug. 2, 1965; children—Kevin, Brian, Meaghan. Research asst. Northwestern U., 1952-53; acting asst. prof. sociology Fla. State U., 1953-55; asst. prof. U. Wis., Madison, 1955-57, asso. prof., 1957-61; prof. sociology Cornell U., Ithaca, N.Y., 1961—, chmn., div. regional studies adv. com., bd. govs. Center for Environmental Quality Mgmt. Editorial cons. Bobbs Merrill Pub. Co., 1960—; cons. Nat. Register Sci. and Tech. Personnel, NSF, 1965—. Mem. Soc. for Study Communication (asso. editor Jour. Communication 1965—), Am. Sociol. Assn., Am. Statis. Assn., Eastern Sociol. Soc., Population Assn. Am. Author: Mathematical Foundations for Social Analysis, 1965; also articles. Editor: (with Robert F. Winch) Selected Studies in Marriage and the Family, 1953, rev. edit. (with Herbert R. Barringer), 1962. Research on methods of sociol. research, math. applications in sociology; constructed Cornell Mobility Model. Home: 3 Strawberry Lane, Ithaca, N.Y. 14850.*

McGLYNN, Sean Patrick, chem. physicist, chemist; b. Dungloe, Ireland, Mar. 8, 1931; s. Daniel and Catherine (Brennan) McG.; B.Sc., Nat. U. Ireland, 1951, M.Sc., 1952; Ph.D., Fla. State U., 1956; m. Helen Magdalena Salacz-von Dohnanyi, Apr. 11, 1955; children—Sean Ernst, Daniel Julian, Brian Charles, Sheila Ann, Alan Patrick. Came to U. S., 1952, naturalized, 1957. Fellow, Fla. State U., 1956, U. Wash., Seattle, 1956-57; asst. prof. La. State U., Baton Rouge, 1957-60, prof. chemistry, 1964—; asso. prof. biophysics, 1961. Cons. to pvt. cos. Recipient award Baton Rouge Council Engring. and Sci. Socs., 1962-63; Mem. Am. Chem. Soc., A.A.A.S., Am. Phys. Soc. Author: (with T. Azumi) The Triplet State, 1966; also numerous articles. Research on anthracene, organic molecular complexes, molecular spectra, glassy solution. Home: 1503 Kenmore Av., Baton Rouge 70808.*

McGOON, Dwight Charles, Am. physician; b. Marengo, Ia., Mar. 24, 1925; s. Charles D. and Ada (Buhlman) McG.; student Ia. State U., 1942-43, St. Ambrose Coll., Davenport, Ia., 1943-44; M.D., Johns Hopkins, 1948; m. Betty Hall, Apr. 2, 1948; children—Michael, Susan, Betsy, Sarah. Practice medicine, specializing in surgery, Rochester, Minn.; asst. to staff Mayo Clinic, 1956-57, cons. physician in surgery, 1957—, also chmn. dept. cardiovascular surgery; asst. prof. surgery Mayo Grad. Sch., U. Minn. Mem. Soc. U. Surgeons, Am., Western surg. assns., Am. Assn. Thoracic Surgery, Halsted Soc., Soc. Clin. Surgery, Phi Beta Kappa, Alpha Omega Alpha. Contbr. numerous articles to profl. jours. Research in surg. treatment of lesions within the heart of a congenital and acquired nature; developed one of early types of artificial aortic valves; contbd. to devel. techniques of open heart surgery which resulted in progressive reduction in risks of this type of surgery. Home: East Meadowridge Dr., Rochester 55901. Office: 200 First St. S.W., Rochester, Minn. 55902.*

McGOVERN, John Phillip, Am. allergist; b. Washington, D.C., June 2, 1921; s. Francis Xavier and Charlotte (Brown) McG.; M.D., B.S., Duke, 1945; m. Kathrine Dunbar Galbreath, Dec. 20, 1961. Faculty, George Washington U., 1951-54, Tulane U., 1954-56, Baylor U., 1956; practice medicine specializing in allergies, Houston, 1956—; asso. prof. pediatrics and microbiology Baylor U. Coll. Medicine, 1960—; prof. allergy Grad. Sch. Biomed. Scis., U. Tex., Houston, 1963—; chief allergy service, dir. allergy clinic Tex. Children's Hosp., Houston, 1957—. Cons. Harmann Hosp., Wilford Hall USAF Hosp.; regional cons. Nat. Found. Asthmatic Children, Denver, also Tucson; chmn. bd. dirs. Tex. Allergy Research Found., 1962—; bd. dirs. Allergy Found. Am. Markle scholar, 1950-55. Fellow Acad. Psychosomatic Medicine (publs. com., contbg. editor Psychosomatics), Royal Soc. Health, Am. Coll. Chest Physicians, Internat. Soc. Tropical Dermatology, Internat. Congress Allergology, Am. Coll. Allergists (asso. editor Annals of Allergy, pres. 1968-69), Am. Acad. Allergy, mem. Internat. Corr. Soc. Allergists (editorial bd. Letters), Am. Assn. for Study of Headache (pres. 1963-64, asso. editor Headache), Am. Assn. Immunologists, Soc. Exptl. Biology and Medicine, Pan. Am. Med. Assn., So. Soc. Pediatric Research, Smith-Reed-Russell Hon. Soc., Sigma Xi, Sigma Pi Sigma, Alpha Epsilon Delta, others. Author: (with James Knight) Allergy and Human Emotions, 1967. Editor: Lecturers in Allergy, 1964; contbr. editor Rev. Allergy and Applied Immunology, 1959-66; asso. editor Jour. Asthma Research, 1963—. Research, publs. on various aspects of allergy and immunology. Home: 2349 Bellefontaine St., Houston 77025. Office: 6655 Travis St., Houston 77025.*

McGRATH, James Williamson, Am. physicist; b. Kirkland, Ill., May 17, 1912; s. Robert Timothy and Nettie (Williamson) McG.; B.S., Fort Hays Kan. State Coll., 1933; M.A., U. Kan., 1934; Ph.D., State U. Ia., 1939; m. Ruth Rigby McChesney, July 19, 1934; 1 son, Robert Leslie. From instr. to asso. prof. physics Mich. State U., East Lansing, 1939-46; faculty Kent (Ohio) State U., 1946—, asso. prof., 1946-49, prof. physics, 1949—. Fellow Am. Phys. Soc., Ohio Acad. Sci. (v.p. 1956); mem. Am. Assn. Physics Tchrs. Asso. Editor, Am. Jour. Physics, 1950-52. Research on X-ray absorption edges, missile guidance, structure of water molecules and their motion and proton relaxation in hydrates. Home: 347 Miller Av., Kent, O. 44240.*

McGREGOR, Ian A., physician; b. Cambuslang, Scotland, Aug. 26, 1922; s. John and Isabella (Taylor) McG.; student St. Mary's Coll. Medicine, Glasgow, Scotland, 1940-45; diploma in Tropical Medicine and Hygiene, London Sch. Tropical Medicine, 1949; m. Nancy Joan Small, Jan. 30, 1954; children—Ian Alistair, Lesley Joan. Med. officer human nutrition research unit Med. Research Council, 1949-53; dir. Med. Research Council Labs., Fajara, Gambia, W. Africa, 1954—. Mem. adv. panel on malaria WHO, 1961—; mem. malaria com. Med. Research Council, 1962—. Decorated Order Brit. Empire, 1959. Licentiate Royal Coll. Physicians, Royal Coll. Surgeons, Royal Coll. Physicians and Surgeons Glasgow. Fellow Royal Soc. Tropical Medicine and Hygiene (Chalmers medal); mem. Brit. Union Ornithologists. Research, and publs. on determination of patterns of growth, survival and health in African rural communities, identification of factors responsible for high mortality statistics in children, evolution and study of methods of disease prevention in such communitites, manner in which immunity to malaria is acquired, identification of serum protein fraction with antibody activity, antigenic composition of malarial parasite. Addresses: Med. Research Council Labs., Fajara, Gambia, W.Africa.*

McGREGOR, James Howard, Am. zoologist; b. Bellaire, O., July 23, 1872; s. Robert Alexander and Lucy (Watterson) McG.; B.S., Ohio State U., 1894; M.A., Columbia, 1896, Ph.D., 1899; Sc.D., 1954. Mem. zoöl. staff Columbia, 1897—, prof. zoölogy, 1924-42, emeritus since 1942; mem. staff Marine Biol. Lab., Woods Hole, Mass., 1899-1906; asso. in human anatomy Am. Mus. Natural History, since 1916. Fellow A.A.A.S., N.Y. Zoöl. Soc.; mem. Am. Soc. Zoölogists, Am. Soc. Naturalists, Am. Soc. Mammalogists, Am. Assn. Phys. Anthropologists, Soc. Vertebrate Paleontologists, Soc. for Study of Evolution, Am. Philos. Soc., Sigma Xi, Phi Beta Kappa; mem. Associé Étranger, Soc. d'Anthropologie de Paris. Contbr. zool. papers on reptilian and primate paleontology. Specialist in study of primates and fossil races of man. Died Nov. 14, 1954.

McGREGOR, Maurice, cardiologist; b. So. Africa, Mar. 24, 1920; s. Frank and Elleanor (Roechling) McG.; MB. U. Witwatersrand, Johannesburg, S. Africa, 1942, B.Ch., 1942, M.D., 1947; m. Margaret Rigsby Becklake, Mar. 20, 1948; children—James Andrew, Margaret Jane. Lectr., U. Witwatersrand, 1950-57; asso. prof. medicine, McGill U., Montreal, Que., Can., 1958—, prof. exptl. medicine, 1966—, dean of medicine, 1967—; asso. dir., joint cardiorespiratory service Royal Victoria Hosp. and Montreal Children's Hosp., 1958—. Fellow Royal Coll. Physicians (London, Eng.), Royal Coll. Physicians and Surgeons Can.; mem. Royal Coll. Physicians (London, Eng.). Contbr. numerous articles to tech. jours. Research in physiology and physiopathology function circulation and lung in health and disease. Home: 7430 Bayard St., Montreal 16. Office: 1200 Pine Av. W., Montreal, Que., Can.*

McGUIGAN, Frank Joseph, Am. psychologist; b. Oklahoma City, Dec. 12, 1924; s. Francis Leo and Edith Louise (Whiting) McG.; B.A., U. Cal. at Los Angeles, 1945, M.A., 1949; Ph.D., U. So. Cal., 1950; m. Lillian Carol Vidovich, July 29, 1950; children—Joan April, Constance, Richard Patrick. Faculty, Pepperdine Coll., 1949-50, U. Nev., 1950-51, George Washington U., 1951-55; prof. psychology Hollins Coll., Va., 1955—, chmn. dept., 1955-65. Cons., Ency. Brit., 1961—. Fellow Am. Psychol. Assn.; mem. Psychonomic Soc., Soc. Psychophysiol. Research, Sigma Xi. Author: Biological Basis of Behavior, 1963, Experimental Psychology, 1960, Thinking, 1966; (with A. D. Calvin): Current Studies in Psychology, 1958; (with Calvin, Scriven, Gallagher, McConnell & Hanley): Psychology, 1961. Contbr. articles to profl. lit. Research in covert behavior and its relation to lang. stimuli. Home: 102 Crittenden Av. N.E., Roanoke, Va. 24020. Office: Hollins Coll., Va. 24020.*

McGUIRE, Austin Dole, Am. physicist; b. Malden, Mass., Oct. 13, 1924; s. Lee Wesley and Ethel (Wines) M.; B.S. in Physics, Mass., Inst. Tech., 1949; Ph.D. in Physics, U. Rochester, 1954; m. Irene Josephine Krofsky, June 4, 1950; 1 dau., Tracy K. Staff mem. Los Alamos Scientific Lab., Los Alamos, 1954-65, Electro-optical Systems, Inc., Pasadena, Cal., 1965—. Fellow Am. Phys. Soc. (pres. 1956); mem. Am. Inst. Physics (pres. 1952). Study of Pi-mesonic X-rays. First observation of interaction of free neutrino with large scint detector at reactor, study of Mossbauer effect, measurements of high altitude radiation, satellite design, weapons testing, accelerator design. testing of nuclear rocket engines. Home: 1041 Mission Ridge Rd., Santa Barbara, Cal. 93103. Office: EG&G Inc., 130 Robin Hill Rd., Golita, Cal. 93017.*

McGUIRE, Johnson, Am. physician; b. Alexandria, Va., Oct. 21, 1899; s. Hugh Holmes and Sara Elizabeth (Johnson) McG.; B.S., U. Va., 1920; M.D., Johns Hopkins, 1924; M.A., U. Cin., 1930; m. Elizabeth Treon Livingood, Aug. 5, 1932; children—Treon (Mrs. Chadwick W. Christine), Mary (Mrs. Victor Morris Tyler II), Sara (Mrs. James Alexander Muspratt), Ann (Mrs. John C. Breckenridge). Jacques Loeb fellow internal medicine Johns Hopkins, 1926; fellow cardiac research from U. Cin. in Vienna, Austria, 1929-30; dir. cardiac lab. U. Cin. Coll. Medicine, 1935—, prof. medicine, 1958—; staff Christian R. Holmes Hosp., 1932—. Cons. to hosps. Dir. Union Central Life Ins. Co. Diplomate Am. Bd. Internal Medicine. Fellow A.C.P., Council on Clin. Cardiology; mem. Am. Clin. and Climatol. Assn., Central Soc. for Clin. Research, Am. Heart Assn. Research, numerous publs. on visualization of coronary arteries in man, measurement of output of heart by Fick principle in man, demonstrated output of heart cannot be increased to normal amount by exercise in patients with enlarged hearts. Home: 2583 Grandin Rd., Cin. 45208. Office: Cardiac Lab., Cin. Gen. Hosp., Cin. 45229.*

McHALE, Kathryn, Am. psychologist; b. Logansport, Ind.; d. Martin and Margaret (Farrell) McHale; B.S., Columbia, 1919, A.M., 1920. Ph.D., 1926; L.H.D. (hon.) Brown University, 1941, Russell Sage College, Troy, N.Y., 1942, MacMurray College, 1946. Instructor in edn. Goucher Coll., Baltimore, Md., 1920-22, asst. prof., 1922-26, asso. prof. 1926-27, prof. 1927-35, non-resident prof., 1935—; prof. edn., summers Columbia, 1918-26, U. of Minn., 1928; gen. dir. Am. Assn. University Women, 1929-50 (dir.); apptd. mem. Subversive Activities Control Bd., 1950, reapptd., 1952. Trustee Purdue, 1936-46. Mem. Alfred I. du Pont Radio Awards com., 1944-46. Chairman citizen's federal committee on education U. S. Office of Education, 1947-50; member bd. educational advisers Nat. Found. for Education in American citizenship; member American Assn. for Adult Edn.; mem. U. S. Nat. Commn. for UNESCO, 1946-50, exec. com. mem., 1946-48; mem. adv. bd. Home Library Found.; mem. Jane Adams Achievement Award Committee. Mem. hon. committee, Am. Women's Hospitals. Fellow A.A.A.S.; mem. Am. Assn. Univ. Profs., American Psychol. Assn., Nat. Society for Scientific Study of Edn., National Society Coll. Tchrs. of Edn., Kappa Delta Pi, Delta Kappa Gamma (hon.). Author: Comparative Psychology and Hygiene of the Overweight Child, 1926; Current Changes and Experiments in Liberal Arts Education, 1932; Housing Coll. Students, 1934; also 43 brochures, pamphlets and articles on psychol. and ednl. subjects. Formulated McHale Vocational Test; influenced growth of Am. Assn. U. Women; contbd. to devel. Am. edn. Died Oct. 8, 1956.

McHENRY, J(ohn) Roger, Am. chemist; b. Gering, Neb., Nov. 3, 1916; s. Matthew Henry and Amanda (Sappington) McH.; student Colo. State Coll. Edn., Greeley, 1937; B.Sc. magna cum laude, Neb. State Tchrs. Coll., Chadron, 1939; M.Sc., U. Neb., 1941; Ph.D., Ia. State U., 1944; m. Dora Gladys Hayward, Dec. 27, 1942; children—Llyn Richard, Rolf Wesley, Sally Winifred, Sherry Doris. Asst. prof. agronomy U. Neb., 1944-47; asst. prof. soil physics Wash. State U., 1947-49; soil scientist Dept. Agr., Mandan, N.D., 1949-52, research soil scientist, Oxford, Miss., 1957—; sr. scientist Hanford Atomic Products Operation Gen. Elec. Co., Richland, Wash., 1952-57; prof. chemistry and chem. engring. U. Miss., 1959—. Mem. Am. Chem. Soc., Am. Nuclear Soc., Am. Soc. Agronomy, Soil Sci. Soc. Am., Am. Geophys. Union, Geochemistry Soc., Soil Conservation Soc. Am., Sigma Xi. Contbr. articles soil physics, radiochemistry areas; devel. of nuclear instrumentation and techniques for hydrological measurements. Home: Route 3, Box 160, Oxford, Miss. 38655.*

McHUGH, John Laurence, marine biologist, oceanographer; b. Vancouver, B.C., Can., Nov. 24, 1911; s. John and Annie (Woodward) McH.; B.A., U. B.C., 1936, M.A., 1938; Ph.D., U. Cal. at Los Angeles, 1950; m. Eileen Francesca Smallwood, July 30, 1941; children—Peter Chadwick, Heather, Jan Margaret. Came to U. S., 1946, naturalized, 1959. Biologist, Fisheries Research Bd. Can., summers 1929-37, full time, 1938-41; research asso. Scripps Instn. Oceanography, 1948-51; dir. Va. Fisheries Lab., 1951-59; prof. marine biology Coll. William and Mary, Williamsburg, Va., 1951-59; chief div. biol. research Bur. Comml. Fisheries, U. S. Dept. Interior, Washington, 1959-63, asst. dir. for biol. research, 1963-67, dep. dir., 1967—. U. S. commr. Inter-Am. Tropical Tuna Commn., 1960—; dep. U. S. commr. Internat. Whaling Commn., 1961-67, commr., 1967—. mem. NRC, 1963—; adv. com. on marine resources to dir. gen. FAO, UN, 1966—; mem., head U. S. delegations to numerous internat. fishery confs. Trustee Internat. Oceanographic Found. Mem. A.A.A.S., Am. Fisheries Soc., Am. Inst. Fishery Research Biologists, Nat. Shellfisheries Assn., Atlantic Estuarine Research Soc., Sigma Xi. Contbr. numerous articles to sci. publs.

Studies life histories and subpopulation fishes, oyster biology, effects of fishing and natural forces upon fish populations, oceanography, fishery mgmt. Home: 3677 N. Harrison St., Arlington, Va. 22207. Office: Bureau of Commercial Fisheries, Dept. of Interior, Washington 20240.*

McILWAIN, Carl Edwin, Am. physicist; b. Houston, Mar. 26, 1931; s. Glenn William and Alma Ora (Miller) McI.; B.A., N.Tex. State Coll., 1953; M.S., State U. Ia., 1956, Ph.D., 1960; m. Mary Louise Hocker, Dec. 30, 1952; children—Janet Louise, Kreg David. Faculty, State U. Ia., 1957-62; asso. prof. U. Cal. at San Diego, La Jolla, 1962-66, prof., 1966—; dir. space scis. lab., 1963—. Mem. NASA Space Sci. Steering Com., 1962-66; mem. fields and particles subcom., anti submarine warfare Pres.' Sci. Adv. Com., 1964-67. Recipient on Guggenheim fellowship, 1967. Member of Am. Inst. Physics, Am. Geophys. Union, Am. Astron. Soc. Research, publs. on measurements of charged particles producing bright auroral display, Van Allen radiation. Home: 8696 Glenwick Lane, La Jolla, Cal. 92037.*

McILWAIN, Henry, English biochemist; b. Newcastle-on-Tyne, Eng., Dec. 20, 1912; s. John and Louisa (Widdowson) McI.; B.Sc., King's Coll., Newcastle, 1934, M.Sc., 1936, Ph.D., 1937, D.Sc., 1943; postgrad. Queen's Coll., Oxford (Eng.) U.; m. Valerie Durston, Apr. 4, 1941; children—Jean, Margaret Clare. Leverhulme research fellow dept. bacterial chemistry Med. Research Council, Middlesex Hosp., London, 1937-41, Sheffield, 1941-44, mem. sci. staff unit for research in cell metabolism U. Sheffield (Eng.), 1944-47; lectr. biochemistry U. Sheffield, 1944-47; faculty Inst. Psychiatry, Brit. Postgrad. Med. Fedn., U. London, 1948-54, reader, 1952-54, prof. biochemistry, 1954—; hon. biochemist Bethlem Royal Hosp. and Maudsley Hosp.; research asso. U. Chgo., 1951; vis. lectr. U. Otago (New Zealand), 1954. Fellow Chem. Soc.; mem. Biochem. Soc., Physiol. Soc., Assn. U. Tchrs., Soc. Authors, Internat. Brain Research Orgn., Internat. Neurochem. Soc. (founding). Author: Biochemistry and the Central Nervous System, 1955; Chemotherapy and the Central Nervous System, 1957; (with R. Rodnight) Practical Neurochemistry, 1962; Chemical Exploration of the Brain, 1963; also numerous articles. Mem. adv. bd. Jour. Neurochemistry, 1956—, Jour. Psychomatic Research, 1956—, Internat. Jour. Neuropharmacology, 1962—. Discovered resynthesis and maintenance in isolated tissues from mammalian brain, their main energy-rich metabolites; demonstrated such tissues are electrically excitable; discovered restoration of membrane potentials in isolated tissues; developed techniques of tissue metabolism and means of observing post-synaptic responses in mammalian tissues; research on nutrition and metabolism of microorganisms; initiated synthesis of antimetabolites; discovered new class of free radicals. Home: 73 Court Lane, London S.E. 21. Office: Dept. Biochemistry Inst. Psychiatry, Maudsley Hosp., Denmark Hill, London S.E. 5, Eng.*

McINTOSH, Henry Deane, Am. physician; b. Gainesville, Fla., July 19, 1921; s. Thomas Rivin and Nelle (Deane) McI.; B.S., Davidson Coll., 1943; M.D., U. Pa., 1950; m. Harriet Lee Owens, Nov. 6, 1945; children—Thomas Irvin, James Owens, Willia Elizabeth. Fellow, Duke Med. Center, Durham, N.C., 1952-55, sr. staff dept. medicine, 1954—; faculty Duke, 1957—, prof. medicine, 1962—, chief cardiology div., 1966—. Mem. cardiovascular B study sect. Nat. Heart Inst., NIH, 1965 ; cons. to hosps. Diplomate Am. Bd. Internal Medicine in Cardiovascular Diseases. Mem. Am. Fedn. for Clin. Research, Am. Heart Assn., A.M.A., So. Soc. for Clin. Investigation, A.M.A., So. Soc. for Clin. Investigation, Assn. U. Cardiologists, Internat. Cardiovascular Soc., Am. Clin. and Climatologic Soc., A.C.P., Am. Coll. Cardiology. Asst. editor Modern Concepts in Cardiovascular Disease, 1965—. Research and numerous publs. on devel. of techniques used in cardiac catheterizations and cineangiocardiography, myocardial function in health and disease. Home: 2705 Stuart Dr., Durham, N.C. 27707.*

McINTYRE, Archibald Ross, physiologist; b. London, Eng., Jan. 19, 1902; s. Oswald and Matilda (Smith) McI.; student U. Manchester (Eng.), 1923; B.S., U. Chgo., 1927, Ph.D., 1930, M.D., 1931; m. Margaret Day, Sept. 8, 1928; children—Ross, David, Donald, John. Naturalized Am. citizen, 1938. Instr., U. Chgo., 1929-30, U. Mich., 1930-31; asso. prof. dept. physiology and pharmacology U. Neb., Omaha, 1932-34, prof., chmn. dept., 1934—. Cons. Squibb Inst., 1940-52, VA Hosp., 1947—, Cudahy Packing Co., 1953-63; lectr. Yale, U. Coll., London, Eng. Recipient Acad. Lincei award, Rome, Italy, 1954. Mem. A.M.A., Am. Soc. Pharmacology and Exptl. Therapeutics, Am. Physiol. Soc., Sigma Xi, Alpha Omega Alpha. Contbg. author: Curare, 1957. Research, numerous articles on curare, neuromuscular physiology, pathology muscle disease. Home: 924 S. 36th St., Omaha.*

McINTYRE, Donald, Am. chemist; b. Detroit, Sept. 8, 1928; s. James and Annie (Thomson) McI.; student Wayne U., 1945-47; A.B., Lafayette Coll., 1949; Ph.D., Cornell U., 1953; m. Alison MacArthur, Sept. 21, 1957; children—Kerry, Ann, Ian. Research chemist Monsanto Chem. Co., 1953-54; research chemist Nat. Bur. Standards, Washington, 1956-66, sect. chief, 1962-66; prof. chemistry U. Akron (O.), 1966—; vis. scientist Centre de Recherche sur les Macromolecules, Strasbourg, France, 1963-64. Mem. com. on macromolecular chemistry Nat. Acad. Scis.-NRC, 1965—. Mem. Am. Chem. Soc., Am. Phys. Soc., Sigma Xi. Author: (with Gornick) Light Scattering in Polymer Solution, 1964; also articles. Research on characterization of macromolecular structure, molecular weights, sizes and thermodynamic interaction of large molecules, critical phenomena in solution. Home: 1365 Hillandale Dr., Akron, O. 44313.*

McINTYRE, John Armin, Am. physicist; b. Seattle, June 2, 1920; s. Harry John and Florence (Armin) McI.; B.S., U. Wash., 1943; M.A., Princeton, 1948, Ph.D., 1950; m. Madeleine Forsman, June 15, 1947; 1 son, John Forsman. Faculty elec. engring. Carnegie Inst. Tech., Pitts., 1943; radio engr. Westinghouse Elec. Co., Balt., 1944; research asso. Stanford, 1950-57; faculty Yale, 1957-63; prof. physics Tex. A. and M. U., College Station, 1963—, asso. dir. Cyclotron Inst., 1965—. Mem. council Oak Ridge Asso. Univs., 1964—. Mem. Am. Phys. Soc., A.A.A.S., Am. Assn. Physics Tchrs., Am. Sci. Affiliation. Research, publs. on scintillation counter for gamma ray spectroscopy; determination of nuclear charge distbns. by electron scattering; study of nuclear structure by neutron transfer reactions; devel. variable energy gamma ray beams. Home: 2316 Bristol St., Bryan, Tex. 77801. Office: Dept. Physics, Tex. A. and M. U., College Station, Tex. 77843.*

McIVER, Monroe Anderson, Am. physician; b. Gulf, N.C., Oct. 15, 1890; s. John McMillan and Lois Anderson) McI.; A.B., U. N.C., 1912; student U. Va. Med. Sch., 1913-15; M.D., Harvard, 1917; m. Elizabeth C. Putnam, Aug. 21, 1923; children—Elizabeth C. (Mrs. John T. Flavin), Marian C. (Mrs. Martin Prochnik). Fellow in physiology NRC, 1922-23; asso. in surgery Mass. Gen. Hosp., Boston, 1923-27, asst. vis. surgeon, 1927-30; asst. vis. surgeon Mass. Eye and Ear Infirmary, Boston, 1927-30; surgeon in chief Mary Imogene Bassett Hosp., Cooperstown, N.Y., 1930-56, emeritus, trustee, 1956—; asso. attending surgeon Albany (N.Y.) City Hosp., 1930-56; chief surgeon Nemazee Hosp., Shiraz, Iran, 1956-58, vis. prof. surgery Shiraz U. Med. Sch., 1956-58; with Harvard U. Med. Sch., 1920-30, asst. prof., 1928-30; asso. prof. surg. Albany Med. Sch., 1930-56; with Columbia, 1947-56, clin. prof. surgery, 1948-56; mem. Harvard Typhus Research Unit to Poland, 1920; Diplomate Am. Bd. Surgeons. Mem. A.M.A., Allen O. Whipple, Excelsior, Halsted surg. socs., soc. Clin. Investigation, A.C.S. (bd. govs., 1955-56). Author: Acute Intestinal Obstruction, 1934. Research, publs. on glands of internal secretion, particularly thyroid: devised choledochoscope, an instrument for visualizing interior of common duct at operation; research on gastro-intestinal physiology and pathology. Home: 12 Main St., Cooperstown, N.Y. 13326.*

McKAY, Kenneth Gardiner, physicist; b. Montreal, Que., Can., Apr. 8, 1917; s. James Gardiner and Margaret (Nicholas) McK.; B.Sc., McGill U., 1938, M.Sc., 1939; Sc.D., Mass. Inst. Tech., 1941; m. Irene C. Smith, July 25, 1942; children—Margaret Craig, Kenneth Gardiner. Came to U. S., 1946, naturalized, 1953. Research engr. NRC, Ottawa, Ont., Can., 1941-46; research in phys. electronics and solid state physics Bell Telephone Labs., Murray Hill, N.J., 1946-52, head research phys. electronics, 1952-54, head research solid state physics, 1954-57, dir. solid state device devel., 1957-59, v.p. systems engring., 1959-62, exec. v.p., 1962-66; v.p. engring. Am. Tel. & Tel. Co., 1966—. Fellow Am. Phys. Soc., I.E.E.E.; mem. Nat. Acad. Engring. Bd. editors: Phys. Rev. Research, publs. on avalanche breakdown in semicondrs., interactions between electrons and solids. Home: 100 Wildwood Lane, Summit, N.J. Office: Am. Tel. & Tel. Co., 195 Broadway, N.Y.C. 10007.

McKEACHIE, Wilbert James, Am. psychologist; b. Clarkston, Mich., Aug. 24, 1921; s. Bert A. and Edith (Welberry) McK.; B.A., Mich. State Normal Coll., 1942; M.A., U. Mich., 1946, Ph.D., 1949; LL.D Eastern Mich. U., 1957; m. Virginia Mack, Oct. 30, 1942; children—Linda, Karen. Faculty, U. Mich., Ann Arbor, 1946—, prof. psychology, 1959—, chmn. dept., 1961—. Mem. com. in-service tng. NIH, 1963-66; cons. biol. sci. facilities U. S. Office Edn., 1962—. Recipient $1,000 award for Outstanding Teaching, U. Mich. Class of 1923, 1955. Mem. Am Psychol. Assn. (dir. 1964—), Am. Council Edn., Am. Assn. U. Profs., A.A.A.S., Sigma Xi. Author: Teaching Tips, 1965; (with Doyle) Psychology, 1966. Established importance of interactions between teaching method, student characteristics and goals in determining teaching effectiveness. Home: 4660 W. Joy Rd., Rural Route 1, Dexter, Mich. 48130.*

McKEAN, Harlley Ellsworth, Am. statistician; b. Chgo., June 23, 1931; s. Elliott Ellsworth and Bonnibell (Child) McK.; B.A., Cornell Coll., Mt. Vernon, Ia, 1952; M.S., Purdue U., 1954, Ph.D., 1958; m. Ellen Bowden Pengilly, 1967; children—Gregory Chase, Kerry Louise, Jennifer Jeanne, Michael Elliott, John Andrew. Faculty, Purdue U., 1956-67, asso. prof., 1962-67; dir. statis. lab. N.M. State U., Las Cruces, 1967—; research asso. Ohio State U., 1960-61; asso. Bayer & McElrath, Inc., mgmt. consultants, 1963—; vis. asso. prof., research asso. U. Cal. at Berkeley, 1965-66. Cons. to pvt. cos. Mem. Inst. Math. Statistics, Am. Statis. Assn. (past mem. council), Biometric Soc. (past treas. Eastern N.Am. region), Am. Assn. U. Profs., Genetics Soc. Am., Sigma Xi. Research, publs. on theory and application of population genetics, plant and animal breeding, epidemiology of coronary heart disease. Home: 1900 Myrtle Av., Las Cruces, N.M. 88001.*

McKEE, Edwin Dinwiddie, Am. geologist; b. Washington, D.C., Sept. 24, 1906; s. Edwin Jones and Ethel (Swope) McK.; A.B., Cornell U., 1928; postgrad. U. Ariz., U. Cal. at Berkeley, Yale; D.Sc., Ariz. State U., 1957; m. Barbara Hastings, Dec. 31, 1929; children—William D., Barbara (Mrs. John Lajoie), Edwin H. Park naturalist Grand Canyon Nat. Park, 1929-40; asst. dir. charge research Mus. No. Ariz., 1941-42, summers 1942-53; faculty U. Ariz., 1942-53; prof., head dept., 1951-53; chief paleotectonic map sect. U. S. Geol. Survey, Denver, 1953-61, research geologist, 1962—; collaborator Nat. Park Service, 1941—. Cons. AEC, 1951-53; dir. mineral resources survey Navaho Indian Reservation, 1952-53. Recipient Distinguished Service award Dept. Interior, 1962. Mem. Geol. Soc. Am., Paleontol. Soc., Soc. Econ. Paleontology and Mineralogy, Am. Assn. Petroleum Geologists, A.A.A.S., Sigma Xi. Author: Environment and History of the Toroweap and Kabab Formations, 1938; (with C. E. Resser) Cambrian History of the Grand Canyon Region, 1945; (with others) Paleotectonic Maps of Jurassic System, 1956, Paleotectonic Maps of Triassic System, 1960; also numerous articles. Research on primary structures in sediments, paleotectonic map devel. Home: 4845 Redwood Dr., Littleton, Colo. 80120. Office: U. S. Geol. Survey, Fed. Center, Denver 80225.*

McKEE, Jack Edward, Am. san. engr.; b. Pitts., Nov. 9, 1914; s. George Edward and Maria (Parsons) McK.; B.S. in Civil Engring., Carnegie Inst. Tech., 1936; M.S., Harvard, 1939, Sc.D., 1941; m. Ruth Yeaton, Dec. 31, 1941; children—Douglas Edward, Richard Carle, Katherine Alice. Engring. aide hydraulic data div. TVA, 1936-37; teaching fellow Harvard U. Grad. Sch. Engring., 1937-41; asso. pub. health engr. USPHS, 1941-42; partner camp, Drexser & McKee, cons. engrs., 1946-49, 65—; asso. prof., prof. environmental health engring. Cal Inst. Tech., 1949—. Cons. NIH, Cal. Water Pollution Control Bd., Nat. Adv. Health Council, Wash. Pollution Control Bd.; mem. adv. com. on reactor safeguards AEC. Recipient Rudolf Hering medal Am. Soc. C.E., Desmond Fitzgerald medal Boston Soc. C.E. Registered profl. engr., Cal., N.Y., Mass., Conn. Diplomate Am. Acad. San. Engrs. Mem. Water Pollution Control Fedn. (pres. 1962-63), Am. Soc. C.E. (pres. Los Angeles sect. 1960, nat. dir. 1965—), Am. Water Works Association, American Society for Engineering Education, Sigma Xi, Tau Beta Pi, Phi Kappa Phi, Chi Epsilon, Delta Omega, Pi Kappa Alpha. Author: Water Quality Criteria, rev. edit., 1963; also numerous tech., profl. papers. Research, publs. on disinfection of wastewater, marine ecology, wastewater reclamation, engring. edn., waste mgmt. in space. Home: 635 E. Orange Grove Av., Sierra Madre, Cal. 91024.*

McKEE, Ralph Wendell, Am. chemist; b. Boynton, Okla., Nov. 13, 1912; s. John Warren and Marguerite (Galbraith) McK.; B.A., Kalamazoo Coll., 1934, M.S., 1935; Ph.D., St. Louis U., 1940; m. Jeriene Ward, Aug. 6, 1938; children—Robert, Jean, James. With Harvard, 1940-52, asst. prof. Med. Sch., 1947-52; head biochemistry Cancer Research Inst., New Eng. Deaconess Hosp., Boston, 1950-52; prof. biol. chemistry U. Cal. at Los Angeles, 1952—, asst. dean Sch. Medicine. Co-recipient award A.A.A.S., 1946. Mem. Am. Assn. Biol. Chemists, Am. Chem. Soc., A.A.A.S., Soc. for Exptl. Biology, Sigma Xi. Research, numerous publs. on isolation and chem. structure identification of vitamin K, physiology and toxicology of carbon disulfide; biochem. studies of lewisite and sulfur mustard gas; growth, metabolism and nutrition of malarial parasites; chemistry, metabolism and culturing of cancer cells. Home: 858 Oreo Pl., Pacific Palisades, Cal. 90292. Office: 405 Hilgard Av., Los Angeles 90024.*

McKEE, Samuel Hanford, Canadian ophthalmologist; b. Fredericton, N.B., Can., Oct. 11, 1875; s. Samuel H. and Jean (Armour) McK.; B.A., U. N.B., 1896; M.D., McGill U., Montreal, Ont., Can., 1900; postgrad. Freiburg, Baden (both Germany) univs.; C.M.; m. Shirley Britton Cowan, 1906. Oculist Montreal Gen. Hosp., Montreal Maternity Hosp., from 1907, Alexandra Hosp.; house surgeon Royal Victoria Hosp., 1900-03; demonstrator, lectr. bacteriology, later prof. ophthalmology McGill U., from 1939. Fellow Royal Coll. Surgeons; mem. Montreal Medico-Chirurg. Soc. (sec., pres.), Am. Assn. Pathologists, Am. Acad. Ophthalmology and Oto-Laryngology (pres. 1932), Am. Ophthal. Soc., Oxford Ophthal. Congress. Research, publs. on conjunctivitis, menigococcus, bacteriology of eye. Died Nov. 25, 1942.

McKEEHAN, Louis Williams, Am. physicist; b. Mpls., Mar. 31, 1887; s. Alfred Espy and Catherine (Williams) McK.; student U. S. Naval Acad., 1903-05; B.S. in Engring., U. Minn., 1908, M.S., 1909, Ph.D., 1911; M.A., Yale, 1927; m. Grace Badger, Aug. 14, 1912 (dec. Oct. 1958); children—John Badger, Mary Kate (Mrs. Arthur E. Ronchie); m. 2d, Ada M. Wilson, Apr. 24, 1960. Faculty, U. Minn., 1906-19; physicist Western Electric Co., Bell Telephone Labs., 1921-27;

prof. physics Yale, 1927-55, dir. Sloane Physics lab., 1927-40, 46-54, prof. emeritus, 1955. Cons. physics and engring. Decorated Legion of Merit; recipient Meritorious Pub. Service award USN, 1965. Author: Yale Science, 1701-1801, 1947; Magnets, 1965; also articles. Research in fall of small spheres in air, 1908-11, electrons in gases, 1911-12, radioactivity, 1912-17, structure and properties of ferromagnetic alloys, 1921-40, 46-50. Address: Coronado St., Jamestown, R.I. 02835.*

McKELVEY, John Leyland, physician; b. Kingston, Ont., Can., Apr. 1, 1901; s. Robert John and Evelyn (Johnston) McK.; B.A., Queens U., Kingston, 1923, M.D., 1926, LL.D., 1957; m. Ruth MacKinnon, Aug. 7, 1930; children—John M., Herbert L., Robert J. Came to U. S., 1938, naturalized, 1942. Halsted fellow in Germany, 1930-32; fellow, instr. Johns Hopkins Hosp., Balt., 1932-34; faculty Peiping (China) Union Med. Coll., 1934-38; prof., head dept. obstetrics and gynecology U. Minn. Med. Sch., Mpls., 1938—; with U. Singapore Med. Sch., 1962-63. Mem. Sigma Xi. Research, numerous publs. on obstet.-gynecol. histology and cancer. Home: 30 Barton Av. S.E., Mpls. 55414.*

McKELVEY, Vincent Ellis, Am. geologist; b. Huntington, Pa., Apr. 6, 1916; s. Ellis Elmer and Eva (Faus) McK.; B.A. with honors in Geology, Syracuse U., 1937; M.A., U. Wis., 1939, Ph.D., 1947; m. Genevieve Patricia Bowman, June 5, 1937; 1 son, Gregory E. With U. S. Geol. Survey, 1941—, asst. chief geologist, 1960-65, research geologist, 1965—; dir. Econ. Geology Pub. Co., Washington. Am. Inst. Mining Engrs. Henry Krumb lectr., 1968. Recipient Distinguished Service award U. S. Dept. Interior, 1962. Fellow Am. Geophys. Union, Geol. Soc. Am.; mem. Soc. Econ. Geologists (mem. council 1967—), A.A.-A.S., Washington Acad. Sci., Geochem. Soc. Research, publs. on geology of phosphate and uranium, stratigraphy, mineral resources. Home: 6601 Broxburn Dr., Bethesda, Md. 20034. Office: U. S. Geol. Survey, Washington 20242.*

McKENNA, Malcolm Carnegie, Am. vertebrate paleontologist; b. Pomona, Cal., July 21, 1930; s. Donald Carnegie and Bernice (Waller) McK.; student Cal. Inst. Tech., 1948-50, Pomona Coll., 1952; B.A., U. Cal. at Berkeley, 1954, Ph.D., 1958; m. Priscilla Coffe, June 27, 1952; children—Douglas Mowry, Katharine Louise, Andrew Morrison, Bruce Carnegie. Instr., U. Cal. at Berkeley, 1958-59; asst. curator dept. vertebrate paleontology Am. Mus. Natural History, N.Y.C., 1960-65, asso. curator, Frick asso. curator, 1965—; asst. prof. dept. geology Columbia, 1960-65, asso. prof., 1965—; research asso. U. Colo. Mus., 1961—. Nat. Acad. Scis. Exchange scientist, 1965. Mem. Soc. Vertebrate Paleontology (editor 1952—), Geol. Soc. Am., Am. Soc. Mammalogists, A.A.A.S., Soc. for Study Evolution Research, articles on mammalian devel. Cretaceous, Paleocene and Eocene faunas, especially cranial anatomy and classification of fossil and recent insectivora. Home: 320 Walnut St., Englewood, N.J. 07631. Office: Am. Mus. Natural History, Central Park W. at 79th St., N.Y.C. 10024.*

McKENNIS, Herbert, Jr., Am. biochemist; b. N.Y.C., Jan. 29, 1916; s. Herbert and Mary (Anderson) McK.; B.S., Harvard, 1938; Ph.D., Cornell U., 1945; m. Evelyn Porter Sanborn, June 20, 1938; children—Quentin S., Claudia S., Jeffrey S. Chemist, Nuodex Products Co., Elizabeth, N.J., 1938-39, Ciba Pharm. Products Co., Summit, N.J., 1940-42; asst. biochemistry Cornell U. Med. Coll., 1942-45; asst. prof. chemistry Med. Coll. Va., Richmond, 1945, asso. prof. biochemistry, 1948, faculty, 1953—, prof. pharmacology, 1955—; instr. physiol. chemistry Johns Hopkins, 1946-48; head basic scis. research dept. Naval C.E. Lab., 1949; vis. prof. U. Chile, 1960, Royal Vet. Coll., Stockholm, Sweden, 1966. Cons. indsl., acad., govt. orgns. Mem. Am. Chem. Soc., Am. Soc. Biol. Chemists, Soc. Exptl. Biology and Medicine, Internat. Oceanographic Found., N.Y. Acad. Scis., Am. Soc. Pharmacology and Exptl. Therapeutics, A.A.A.S., Soc. Am. Mil. Engrs., Am. Inst. Chemists, Va. Acad. Sci., Soc. Toxicology, Sigma Xi, Phi Lambda Upsilon. Co-author: Tobacco Alkaloids and Related Compounds, 1965. Editor: Va. Jour. Sci. Publs. on chemistry and metabolism of hydrazine compounds and alkaloids, soil stblzn., relationship of chem. structure to biol. activity. Home: 1421 W. Laburnum Av., Richmond, Va. 23227.*

McKENZIE, John, Brit. embryologist, anatomist; b. Fraserburgh, Scotland, Oct. 18, 1922; s. John and Gertrude (Anderson) McK.; M.B., Ch.B., U. Aberdeen (Scotland), 1944, M.D., 1957; m. Christina MacFarlane, Dec. 24, 1945; children—John MacFarlane, Kenneth Gordon. Faculty, U. Aberdeen, 1945—, reader in embryology, 1963—. Mem. Anat. Soc. Gt. Britain (council), Soc. Developmental Biology (com.), Brit. Soc. Cell Biology. Editor: Faber's Anatomical Atlas, 4th edit. 1962. Research, publs. on congenital abnormalities particularly of face and head; first to describe 1st arch syndrome, actions of teratogenic drugs, fundamentals of tissue interactions during devel., tissue extracts and RNA. Home: 6 Bayview Rd. Office: Marischal Coll., U. Aberdeen, Aberdeen, Scotland.*

McKENZIE, Robert Tait, surgeon; b. Almonte, Ont., Can., May 26, 1867; s. William and Catherine (Shiells) McK.; A.B., McGill U., Montreal, Que., Can., 1884, M.D., 1892, LL.D., 1921; diploma Harvard Summer Sch. Phys. Edn., 1891; M.Phys. Edn., Springfield (Mass.) Tng. Coll., 1913; m. Ethel O'Neil, Aug. 18, 1907. Dir. phys. tng. McGill U., 1896-1904; dir. phys. edn. U. Pa., 1904-30; house physician Montreal Gen. Hosp., 1893; became surgeon Beane Line Steamer, Liverpool to Montreal, 1893; named house physician to gov. gen., Can., 1895; demonstrator, then lectr. anatomy McGill U., 1894-95, med. dir. phys. tng., 1896-1904; became lectr. artistic anatomy Montreal Art Assn., 1901; prof., dir. dept. phys. edn. U. Pa., 1904-30, research prof. phys. edn., 1931—. Sculptor of Sprinter Athlete, General Wolf Statue, Greenwich, 1930, Delano Nurses Memorial, Washington, D.C., 1933. Fellow Royal Canadian Acad. Author: Therapeutic Uses of Exercise in Education, 1894; The Fit of Breathing, 1897; Exercise in Education and Medicine, 1909; The City and Fresh Air, 1910; Influence of Exercise on the Heart, 1913; Treatment of Convalescent Soldiers by Physical Means; Reclaiming the Maimed (used by surgeon gen. of U. S. Army for reconstrn. hosps.), 1918. Studied positive medicine and phys. edn. Died Phila., Apr. 28, 1938.

McKEOWN, James Edward, Am. sociologist; b. Detroit, Sept. 3, 1919; s. Francis Joseph and Grace Margaret (Ruddon) McK.; B.A., Wayne U., 1941, M.A., 1945; Ph.D., U. Chgo., 1949; m. Mary Elizabeth McNamara, Aug. 6, 1955. Instr. social sci. St. Xavier Coll., Chgo, 1945-48; asst. prof. sociology N.M. Highlands U., Las Vegas, 1948-52; asst. prof. sociology DePaul U., Chgo., 1952-55, asso. prof. 1955-57, prof. sociology, 1957—, chmn. dept., 1962—; vis. prof. Northwestern U., Concordia Coll., 1965; cons. sociologist for delinquency prevention programs N.M., Bolivia, Ill.; fellow Fund Advancement Edn., summer 1954; Smith-Mundt lectr., Bolivia, 1958. Mem. Am., Am. Cath. sociol. socs., A.A.A.S., Am. Acad. Polit. and Social Sci., Am. Assn. U. Profs., Pi Gamma Mu, Psi Chi, Phi Sigma Iota. Author: Study Guide for Economics, Chicago, 1958, 60, 63, 64; co-editor: The Changing Metropolis, Boston, 1964. Contbr. articles profl. jours. Research on influence of econ. conditions on crime rates in large cities; parent-patient relationships of schizophrenic and neurotic patients. Home: 1269 E. 85th St., Chgo. 60619.*

McKERCHER, Delbert Grant, bacteriologist; b. Maxville, Ont., Can., July 3, 1914; s. Henry George and Mary (Grant) McK.; D.V.M., Ont. Vet. Coll., 1938; postgrad. St. Patricks Coll., 1938-40; M.A., Queens U., 1942; Ph.D., Cornell U., 1949; m. Georgina Pearl Stevens, Aug. 21, 1943; children—Ann-Louise, Mary Adrienne, John Andrew. Asst. scientist Animal Diseases Research Inst., Hull, Que., Can., 1938-40; bacteriologist B.C. Dept. Pub. Health, 1946; faculty U. Cal. at Davis, 1949—, prof. bacteriology, 1957—. Mem. viral and rickettsial study sect. NIH, 1959-64; cons. VA Hosp., Livermore, Cal., 1964—; adviser AID program, Peru, 1965. NIH Sr. research fellow, 1960-61. Diplomate Am. Coll. Vet. Microbiologists (charter). Mem. Am. Vet. Med. Assn., Am. Coll. Vet. Microbiologists, Research Workers in Animal Diseases, U. S. Livestock San. Assn. Contbg. author: U. S. Department of Agriculture Yearbook, 1956; Advances in Veterinary Science, 1959; Basic Medical Virology, 1966. Research and publs. on viruses causing respiratory infections in cattle, sheep and cats, abortion in sheep and cattle, blue-tongue in sheep, myxomatosis in rabbits and methods of preventing and controlling these diseases. Home: 749 Anderson Rd., Davis, Cal. 95616.*

McKERNS, Kenneth Wilshire, biochemist; b. Hong Kong, Mar. 5, 1919; s. Frederick William and Daisy (Peel) McK.; B.Sc., U. Alta., 1942; M.Sc., 1946; Ph.D., McGill U., Montreal, Que., Can., 1950; m. Dorothy Vivian McDuffe, Feb. 12, 1943; children—Maureen Kendra, Leslie Allison. Came to U. S. 1956, naturalized, 1963. Chief demonstrator, lectr. McGill U., 1947-50; postdoctoral fellow McGill-Montreal Gen. Hosp. Research Inst., 1951; chief biochemist Can. Packers Ltd., Toronto, Ont., 1951-55; asst. lectr. U. St. Andrews (Scotland), 1955-56; sr. research scientist, group leader Lederle Labs., Pearl River, N.Y., 1956-60; asso. prof. U. Fla. Coll. Medicine, Gainesville, 1960-65, prof. obstetrics and gynecology, 1965—. Mem. Am. Soc. Biol. Chemists, Endocrine Soc., Soc. for Gynocologic Investigation, Soc. for Exptl. Biology and Medicine, A.A.A.S. Author: Molecular Biology of the Uterus, 1966; also numerous articles. Research on hormonal stimulation of pentose phosphate shunt for glucose metabolism by enzyme combination. Home: 2025 N.W. 18 Lane, Gainesville, Fla. 32601.*

McKETTA, John J., Jr., Am. chem. engr.; b. Wyano, Pa., Oct. 17, 1915; s. John and Mary (Golet) McK.; B.S., Tri-State Coll., 1937, D.Engring. (hon.), 1967; B.S.E., U. Mich., 1943, M.S., 1944, Ph.D., 1946; m Helen Elisabeth Smith, Oct. 17, 1943; children—Charles William, John J. III, Robert Andrew, Mary Anne. Wyandotte Chems. Corp. (Mich.), 1937-41; chem. dir. C. B. Schneible Co., Detroit, 1941-42; editorial dir. Gulf Pub. Co., Houston, 1952-54; faculty U. Tex., Austin, 1946—, prof., 1951-52, asst. dir. Tex. Petroleum Research com., 1951-52, 54-55, chmn. chem. engring. dept., 1950-52, 55-63, grad. prof. 1954—, dean Coll. Engring., 1963—. Dir. Gulf Pub. Co., Inc., Houston, Chemoil Cons., Inc., Austin, Vulcan Materials Co. Chmn. adv. bd. Tex. Atomic Energy Com., exec. com. So. Interstate Nuclear Bd. Named Distinguished Alumnus, U. Mich. Coll. Engring., 1953, Tri-State Coll., 1956. Allied Chem. and Dye fellow, 1945-46. Mem. Am. Inst. Chem. Engrs. (plaque for best paper 1952, Most Distinguished Service award 1963, pres. 1962, adv. bd. Chem. Engring. Progress 1954-56), Am. Chem. Soc. (chmn. Grady award com. 1956-57), Am. Inst. Mining, Metallurgy and Petroleum Engrs., Am. Petroleum Inst., Am. Soc. for Engring. Edn., Engrs. Joint Council (dir. 1967—), Sigma Xi, Omicron Delta Kappa. Contbg. author: Unit Processes in Organic Synthesis, 1948; Ency. for Sci. and Industry, 1960; A.G.A. Handbook, 1962; Petroleum Engrs., Handbook, 1962. Editor: (with K. A. Kobe) Recent Advances in Petroleum Chemistry and Refining, 10 vols., 1958-65; Interscience Library of Chem. Engring., 1962—; chmn. editorial com. Petroleum Refiner, 1954—; editorial bd. Ency. Chem. Tech., 1960—. Research, publs. in chem. thermodynamics especially as applied to petrochems. Home: 5227 Tortuga Trail, Austin, Tex. 78731.*

McKIEL, John Albert, Canadian microbiologist; b. Fredericton, N.B., Can., Nov. 13, 1918; s. Edward Albert and Emma (Woodworth) McK.; B.S., U. N.B., 1949, M.S., 1951; Ph.D., Queen's U., Kingston, Ont., 1955; m. Marie Rosaline MacDonald, May 12, 1943; children—David, Donald, Gregory. Sci. service officer Def. Research Bd., Kingston, 1951-55, Ottawa, Ont., 1955-56; bacteriologist lab. hygiene Dept. Nat. Health and Welfare, Ottawa, 1956-57, chief zoonoses lab., 1958—. Mem. Canadian Pub. Health Assn., Canadian Soc. Microbiologists, Am. Soc. Tropical Medicine and Hygiene. Research, numerous publs. on reactions to mosquito bites, leptospirosis in rodents, existence in Ont. of Rocky Mountain spotted fever, Cal. encephalitis and Colo. tick fever; developed technique for continuous mass-rearing of mosquitoes. Home: Box 162, Richmond, Ont. Office: Lab. Hygiene, Dept. Nat. Health and Welfare, Ottawa 3, Ont., Can.*

McKINLEY, Donald William Robert, physicist; b. Shanghai, China, Sept. 22, 1912; s. David Fuller and Mabel (Burns) McK.; B.A., U. Toronto, 1934, M.A., 1935, Ph.D., 1938; m. Barbara Mabel Girdwood, Mar. 25, 1950; children—Alan Duncan, Kathryn Jean. With NRC Can., Ottawa, Ont., 1938—, dir radio and elec. engring. div., 1963—. Decorated officer Order Brit. Empire. Fellow Royal Soc. Can., I.E.E.E., Am. Phys. Soc. Author: Meteor Science and Engineering, 1961. Research, publs. on meteors, giving proof that meteors are members of our solar system. Home: 38 Dunvegan Rd., Ottawa 7. Office: Nat. Research Council, Ottawa, Ont., Can.*

McKINLEY, Earl Baldwin, Am. bacteriologist; b. Emporia, Kan., Sept. 28, 1894; s. Joseph Baldwin and Mary Elizabeth (Griffith) McK.; A.B., U. of Mich., 1916, M.D., 1922; fellow NRC, Pasteur Inst., U. Brussels, 1924-25; m. Leola Edna Royce, June 23, 1917; children—Elsbeth Janet, Royce Baldwin. Instr. in bacteriology and biochemistry U. Mich. Med. Sch., 1919-22; asst. prof. medicine Coll. Medicine, Baylor U., 1922-23, prof. hygiene and bacteriology and chmn. dept., 1923-24; asst. prof. bacteriology, Coll. Phys. and Surg., Columbia, 1925-26, asso. prof., 1926-27, prof., 1928-31; field dir., Manila, P.I., Rockefeller Found., 1927-28; mem. adv. com. to gov. gen. for control of leprosy, also lectr. U. Philippines, 1927-28; dir. Sch. Tropical Medicine, U. Puerto Rico (under auspices of Columbia), 1928-31; dean Coll. of Medicine, prof. bacteriology George Washington U., 1931—. Fellow A.M.A., A.C.P., Royal Soc. Tropical Medicine and Hygiene (London); mem. numerous sci. socs. Author: Filterable Virus and Rickettsia Diseases, 1929; a Geography of Disease, 1935; Agents of Disease and Host Resistance (with others), 1935. Died July 29, 1938.

McKINLEY, J(ohn) Charnley, Am. neuropsychiatrist; b. Duluth, Minn., Nov. 8, 1891; s. John and Alice Salome (Frizzell) McK.; B.S., U. Minn., 1915. A.M., 1917, M.D., 1919, Ph.D., 1921; postgrad. in psychiatry, Psychiatric Inst., N.Y.C., 1919; m. Doris I. Swedien, Apr. 29, 1944; children—(1st marriage) Marian Louise (Mrs. Leland Phelps), Helen Alice (Mrs. George W. Miners), Ruth Elizabeth (Mrs. Roy Pistore), John Charnley. Instr. pathology U. Minn., 1917-18, teaching fellow in nervous and mental diseases, 1918-21, asst. prof. of neuro-pathology, 1921-25, asso. prof. of neurology, 1925-29, prof. neuropsychiatry, 1929-46, acting head dept. medicine, 1932-34, head dept. medicine, 1934-43, head dept. neuropsychiatry, 1943-46, prof. emeritus 1946—. Sec.-treas. Minn. Bd. Examiners in Basic Scis., 1934-45. Guggenheim fellow (studies at univs. Breslau and Munich, Germany), 1928-29. Mem. bd. dirs. Am. Bd. of Psychiatry and Neurology, Fellow A.A.A.S., A.M.A.; mem. Minn. Acad. Medicine, Central Neuropsychiatric Assn., Am. Neurol. Assn., Sigma Xi. Contbr. to met. jours. Editor: Outline of Neuropsychiatry. Co-author: Minnesota Multiphasic Personality Inventory. Died Jan. 3, 1950.

McKINSEY, James O., Am. management scientist; b. Gamma, Mo., June 4, 1889; s. James Madison and Mary Elizabeth (Logan) McK.; Pd.B., State Teachers Coll., Warrensburg, Mo., 1912; LL.B., U. of Ark., 1913; Ph.B., U. of Chicago, 1916, A.M., 1919; m. Alice Louise Anderson, June 12, 1920; children—Robert, Richard (twins). Mem. faculty U. of

Chicago, 1917—, prof. business policies, 1926-35, lecturer on accounting, Columbia, 1920-21; C.P.A., Ill., 1919; sr. partner James O. McKinsey & Co., 1925-35; chmn. bd. Marshall Field & Co., 1935—; dir. Chicago Corp., Kroger Grocery & Baking Co., Selected Shares Corp. Dir. Woodlawn Hosp. Mem. bd. trustees Armour Inst. Tech.; mem. bd. mgrs. Chicago Y.M.C.A., mem. bd. dirs. Central Y.M.C.A. Coll.; mem. transportation com. Chicago Assn. of Commerce. Author: Bookkeeping and Accounting, 1920; Budgetary Control, 1922; Managerial Accounting, 1924; Business Administration, 1925; Accounting Principles, 1929. Stressed budgeting in management and organization planning. Died Nov. 30, 1937.

McKUSICK, Victor Almon, Am. physician, geneticist; b. Parkman, Me., Oct. 21, 1921; s. Carroll L(ee) and Ethel (Buzzell) McK.; student Tufts Coll., 1940-43; M.D., Johns Hopkins, 1946; m. Anne Bishop, June 11, 1949; children—Carol Anne, Kenneth Andrew, Victor Wayne. Staff, Marine Hosp., Balt., 1948-50; faculty Johns Hopkins Sch. Medicine, 1947—, chief div. med. genetics dept. medicine, 1957—, prof. medicine, 1960—; co-dir. short course in med. genetics, Bar Harbor, Me., 1960—. Recipient Modern Medicine Distinguished Achievement award, 1966. Author: Heritable Disorders of Connective Tissue, 1956; Cardiovascular Sound in Health and Disease, 1958; Medical Genetics 1958-60, 1961; On the X Chromosome of Man, 1964; Human Genetics, 1964; Medical Genetics, 1961-63, 1965. Research on analysis of heart sounds and murmurs by sound spectrography, delineation of heritable disorders of connective tissue, mapping human X chromosome, religious isolates (Amish). Home: 221 Northway, Balt. 21218. Office: Johns Hopkins Hosp., Balt. 21205.*

McLACHLAN, Andrew David, English chemist; b. London, Eng., Jan. 25, 1935; s. Donald H. and Katherine (Harman) McL.; B.A., Cambridge (Eng.) U., 1956, M.A., Ph.D., 1961; m. Jennifer Kerr, June 20, 1959; children—Charles, Hugh. Research fellow Trinity Coll., Cambridge, 1958-61; Harkness fellow Cal. Inst. Tech., Harvard, 1959-61; vis. prof. Cal. Inst. Tech., 1964-65; lectr. chemistry Cambridge U., 1965—. Author: (with A. Carrington) Introduction to Magnetic Resonance, 1966; also articles. Research on theoretical chemistry and electron spin resonance in organic radicals. Address: Trinity Coll., Cambridge, Eng.*

McLACHLAN, Dan, Jr., Am. phys. chemist; b. Arcola, Sask., Can., Dec. 5, 1905 (parents Am. citizens); s. Daniel and Laura (Carter) McL.; B.S., Kan. State U., 1930; M.S., Pa. State U., 19—, Ph.D., 1936; m. Fredolyn May Walker, June 11, 1934; children—Edwin Dale, Dan Houston, Wayne Carter. Phys. chemist Corning (N.Y.) Glass Works, 1936-40; molecular structure group leader Am. Cyanamid Co., Stanford, Conn., 1940-47; prof. metallurgy, physics U. Utah, Salt Lake City, 1947-53; fundamental research scientist Stanford Research Inst., Menlo Park, Cal., 1953-61; prof. metallurgy U. Denver, 1961-63; Battelle vis. prof. metallurgy Ohio State U., Columbus, 1963, prof. mineralogy, 1964—. Fellow Am. Mineral. Soc., Am. Phys. Soc., N.Y. Acad. Soc.; mem. Am. Crystallographic Assn. (pres. 1959), Am. Chem. Soc., Geochem. Soc., A.A.A.S., Am. Optical Soc., Sigma Xi, Phi Lambda Upsilon, Sigma Pi Sigma. Author: X-ray Crystal Structure, 1957; also numerous publs. Devised machines for computation of results in X-ray diffraction; research on phase problem in crystallography, statis. properties of metals. Patentee in field including a deep field microscope. Home: 1934 Langham Rd., Columbus, O. 43221.*

McLAREN, Arthur Douglas, Am. biophysicist; b. Ipava, Ill., Sept. 27, 1917; s. Homer D. and Matilda (Rose) McL.; B.A., Park Coll., 1939; Ph.D., U. Mo., 1943; m. Alice Gray Jones, Feb. 3, 1940; 1 son, Gordon Douglas; m. 2d, Helen Rodina, Sept. 21, 1965. Research chemist E. I. du Pont de Nemours & Co., Buffalo, 1943; asst. prof. Poly. Inst. Bklyn., 1946; prof. U. Cal. at Berkeley, 1951—; vis. prof. Grad. Research Center, Dallas, 1966. Mem. com. on photobiology NRC, 1960—. Fellow Rockefeller Found. fellow, 1949-50; Fulbright fellow, Australia, 1959; Guggenheim Meml. Found. fellow, 1966-67. Mem. Am. Soc. Biol. Chemists, Am. Biophys. Soc., Scandinavian Soc. Plant Physiologists. Author: (with David Shugar) Photochemistry of Proteins and Nucleic Acids, 1964; (with George Petersen) Soil Biochemistry, 1966; also numerous articles. Exec. editor Photochemistry and Photobiology, 1961—. Research on action ultraviolet radiation on viruses, translocation large molecules in plants, actions enzymes on surfaces proteins, starch and fats, mechanism action adhesives in bonding cellulose materials. Home: 37 Vicente Pl., Berkeley, Cal.*

McLAREN, Donald Stewart, med. nutritionist; b. London, Eng., Feb. 4, 1924; s. James and Gertrude (Eaton) McL.; M.B.Ch.B., U. Edinburgh, 1947, M.D., 1955; Dr. Tropical Medicine and Hygiene, London Sch. Hygiene and Tropical Medicine, 1949; Ph.D., U. London, 1957; m. Olga Mair Evans, Oct. 23, 1948; children—Gavin Stewart, Jillian Mary. Med. officer Moorshead Meml. Hosp., India, 1949-54; sr. med. research officer E. African Inst. Med. Research, Mwanza, Tanganyika, 1957-62; prof. clin. nutrition Am. U. Beirut Sch. Medicine, Lebanon, adj. prof. nutrition Columbia, N.Y.C., 1962—, dir. Am. U.-Columbia nutrition research program, 1962—. Cons. WHO, Inter-Departmental Com. on Nutrition for Nat. Devel.; mem. WHO-FAO Expert Group on Vitamin Requirements, 1956. Recipient Ciba Found. award, 1957; Barraquer Inst. award, 1960. Fellow Royal Med. Soc. Edinburgh (past pres.), Royal Soc. Tropical Medicine and Hygiene; mem. Brit. Med. Assn. (Oliver Hawthorne award 1955), Nutrition Soc. U.K., Inst. Nutrition, Am. Soc. Clin. Nutrition. Author: Malnutrition and the Eye, 1963; also numerous articles. Research on role of vitamin and protein deficiency in blindness and other visual defects especially among children, causes, nature of biochem. derangements, treatment and prevention of clin. conditions marasmus and kwasiorkor due to protein-calorie malnutrition in young children. Address: Am. U., Beirut, Lebanon.*

McLAREN, Leroy Clarence, Am. microbiologist; b. Bishop, Cal., Jan. 18, 1924; s. Leroy Clarence and Amelia E. (Barlow) McL.; A.A., San Bernardino Valley Coll., 1947; A.B., San Jose State Coll., 1949; M.A., U. Cal. at Los Angeles, 1951, Ph.D., 1953; m. Dorothy L. Appleton, Aug. 14, 1945; children—Mark D., David L., Deborrah E. Instr. infectious diseases U. Cal. at Los Angeles Sch. Medicine, 1953-55; asst. prof., asso. prof. dept. microbiology U. Minn., 1955-64; prof., chmn. dept. microbiology, U. N.M. Sch. Medicine, Albuquerque, 1964—. Research career award USPHS, 1962-64. Mem. Am. Soc. Microbiology, Am. Assn. Immunologists, Soc. Exptl. Biology and Medicine, Am. Pub. Health Assn., A.A.A.S., Am. Assn. U. Profs., Sigma Xi. Research, publs. on biochemistry of cellular infection by various animal and human viruses. Home: 7705 Palo Duro St. N.E., Albuquerque 87110.*

McLAUGHLIN, Dean Benjamin, Am. astronomer; b. Bklyn., Oct. 25, 1901; s. Michael Leo and Celia (Benjamin) McL.; A.B., U. Mich., 1923, M.S., 1924, Ph.D., 1927; m. Laura Elizabeth Hill, Dec. 27, 1927; children—Elizabeth (Mrs. Peter F. Schick), Laura Alberta (Mrs. Juan Carlos Dawson), Dean B. M., Sarah Jeanette, Margaret Louise (Mrs. Lawrence Ira Farley). Instr. math. and astronomy Swarthmore Coll., 1924-27; faculty U. Mich., Ann Arbor, 1927—, prof. astronomy, 1941—; mem. staff Swarthmore Solar Eclipse Expdn., Sumatra, 1926, U. Mich. Solar Eclipse Expdn., Me., 1932, Radiation Lab, Mass. Inst. Tech., 1943-45; cooperating geologist Pa. Geol. Survey, 1951—. Fellow A.A.A.S., Geol. Soc. Am.; mem. Am. Astron. Soc. (past sec.), Astron. Soc. Pacific, Pa., Mich. acads. sci., Phi Beta Kappa, Sigma Xi, Sigma Gamma Epsilon. Author: Introduction to Astronomy, 1961; also numerous articles. Research on novae, Be stars, Mars, Triassic in Eastern U. S., radar. Home: 1214 W. Washington St., Ann Arbor, Mich. 48103. Died Dec. 8, 1965.

McLAUGHLIN, Donald Hamilton, Am. mining geologist, engr.; b. San Francisco, Dec. 15, 1891; s. William Henry and Katherine (Hamilton) McL.; B.S., U. Cal. at Berkeley, LL.D., 1966; A.M., Harvard, 1915, Ph.D., 1917; D.Engring., S.D. Sch. Mines and Tech., Mont. Sch. Mines, Mich. Inst. Mining and Tech., 1950, Colo. Sch. Mines, 1955; m. Eleanor Eckhart, Sept. 12, 1925 (div. 1941); children—Donald Hamilton, Charles Capen; m. 2d, Sylvia Cranmer, Dec. 29, 1948; children—Jean Katherine, George Cranmer. With Cerro de Pasco Copper Corp., Peru, N.Y., 1919-25, v.p. gen. mgr., 1943-45; prof. mining geology, chmn. div. geol. scis. Harvard, 1925-51; dean engring. U. Cal., 1941-43; pres. Homestake Mining Co., San Francisco, 1944-60, chmn. bd. dirs., 1960; dir. Internat. Nickel Co. Can., N.Y.C., San Luis Mining Co., San Francisco; adv. dir. Wells Fargo Bank, San Francisco. Mem., chmn. adv. com. on raw materials AEC, 1947-52, plowshare adv. com., 1959—; nat. sci. bd. NSF, 1950-60. Coordinating Council for Higher Edn. Cal., 1958-65; regent U. Cal., 1951-66. Recipient Monell award and medal Columbia, 1964. Fellow Geol. Soc. Am.; mem. Am. Acad. Arts and Scis., Am. Inst. Mining and Metall. Engring. (hon., Rand medal 1961). Research, publs. on geol. factors in valuation mines, econ. geology. Home: 1450 Hawthorne Terrace, Berkeley, Cal. 94708. Office: 650 California St., San Francisco 94108.*

McLEAN, Donald Millis, physician; b. Melbourne, Australia, July 26, 1926; s. Donald and Nellie (Millis) McL.; B.Sc., U. Melbourne, 1947, M.B., 1950, M.D., 1954. Fellow, Rockefeller Found., N.Y.C., Hamilton, Mont., 1955; vis. instr. in bacteriology U. Minn., Mpls., 1957; med. officer Commonwealth Serum Labs., Melbourne, 1957; virologist Research Inst., Hosp. for Sick Children, Toronto, Ont., Can., 1958-67; asso. prof. microbiology, asso. in pediatrics U. Toronto, 1962-67; prof. microbiology U. B.C., Vancouver, Can., 1967—. Mem. Am. Assn. Immunologists, Am., Canadian pub. health assns., Am. Soc. Tropical Medicine, Can. Med. Assn., Pathol. Soc. Great Britain and Ireland, Canadian Soc. Microbiologists. Research, numerous publs. on ecology of arboviruses, diagnostic tests for human infections with enteroviruses, influenza, measles, rubella viruses, transmission of viruses by water; discovered Powassan and Silverwater viruses. Office: Dept. Microbiology, U. B.C., Vancouver 8, B.C., Can.*

McLEAN, Eugene Otis, Am. chemist; b. Nixa, Mo., Feb. 14, 1919; s. Alva Otis and Ethel (Buxton) McL.; B.S., U. Mo., 1942, M.A., 1943, Ph.D., 1948; m. Marjorie Inez Russell, Oct. 5, 1943; children—Eugene Otis, Stephen R. Faculty, U. Ark., 1950-56; asso. prof., prof. agronomy Ohio State U., Columbus, 1956—, chmn. grad. com. agronomy, 1958—, grad. council, 1965—. Mem. Soil Sci. Soc. Am., Internat. Soc. Soil Sci., Am. Chem. Soc., A.A.A.S., Sigma Xi, Gamma Sigma Delta. Author: (with Frank L. Himes) Chemistry of the Soil, 1967. Research, numerous publs. on ionic activities and interrelationships in colloidal systems and their effects on plant uptake, soluble aluminum in acid soils as a factor in soil acidity, cation exchange capacity and determination of lime requirements of soils, partially acidulated rock phosphate as a source of P to plants in soils of high P fixing capacity, factors affecting fixation and release of potassium from soils. Home: 1390 London Dr., Columbus, O. 43221.*

M'CLEAN, Frank, engr., astronomer; b. 1837; s. J. R. M'Clean; ed. Glasgow (Scotland) Coll., Trinity Coll., Cambridge (Eng.) U.; B.A., 1958; M.A.; LL.D.; m. Ellen Greg, 1865. Apprenticed civil engr., 1859; partner engring. firm M'Clean & Stileman, 1862-70, ret., 1870; established astron. obs., Tunbridge Wells, Eng., 1874. Fellow Royal Soc., Royal Astron. Soc. (Gold medal for photographic survey of star spectra in both hemispheres, other contbns. to advancement astronomy 1899); mem. Inst. C.E.'s. Early dock, ry. works; research, publs. on solar, spectroscopic work; discovered presence of oxygen in helium class of stars, 1897; founded Isaac Newton studentships, Cambridge U., 1890; presented Victoria Photographic Telescope (24 inch aperture) to Royal Obs., Cape Good Hope, 1894. Died Nov. 8, 1904.

McLEAN, Franklin Chambers, Am. physiologist; b. Maroa, Ill., Feb. 29, 1888; s. William Thomas and Margaret Philbrook (Crocker) McL.; B.S., U. Chgo., 1907, M.S., 1912, Ph.D., 1915; M.D., Rush Med. Coll., 1910; M.D. (hon.), U. Lund (Sweden), 1957; Docteur honoris causa U. of Bordeaux (France), 1967; m. Helen Vincent, June 11, 1923; 1 son, Franklin Vincent (dec.), Prof. pharmacology U. Ore., 1911-14; dir. Peking (China) Union Med. Coll., 1916-20, prof. medicine, 1916-23; prof. medicine U. Chgo., 1923-32, prof. pathol. physiology, 1933-53, prof. emeritus, 1953-65; vis. prof. U. Ill. Coll. Dentistry, 1966-68. Cons. C.W.S., U. S. Army, 1946-59; dep. chmn. joint panel on med. aspects atomic warfare Dept. Def., 1949-53, mem. com. on med. sci., 1950-53, mem. tech. adv. panel on biol. and chem. warfare, 1955-60; mem. subcom. on skeletal system NRC, 1952-58; mem. com. on Army Med. Edn., 1954-56; dir., sec.-treas. Nat. Med. Fellowships, Inc. Recipient award in human relations Commn. On Human Relations, Chgo., 1953; scroll of honor Nat. Med. Assn., 1964. Author: (with M. R. Urist) Bone, An Introduction to the Physiology of Skeletal Tissue, 1955, 61, 68; (with A. M. Budy) Radiation, Isotopes and Bone, 1964. Research on calcium metabolism and physiology of bone; devised formula for determining urea index which is modification of Ambard's formula, known as McLean's formula or index. Home: 5825 Dorchester Av., Chgo. 60637.

McLEAN, Helen Vincent, Am. psychoanalyst; b. Sandusky, O., May 27, 1894; d. Clarence A. and Lucy (Hall) Vincent; B.S., Mt. Holyoke Coll., 1915, D.Sc. (hon.), 1965; Ellen Richards fellow Mass. Inst. Tech., 1917-18; M.D., Johns Hopkins, 1921; m. Franklin C. McLean, June 11, 1923; 1 son, Vincent (dec.). Resident physician obstetrics gynecology Peking Union Med. Coll., 1922-23; lectr. Social Hygiene Council, Chgo., 1923-30; clinic physician Infant Welfare Soc. and Birth Control Clinic, Chgo., 1925-30; staff physician Pub. Health Inst., Chgo., 1925-30; lectr. Sch. Social Service Adminstrn. U. Chgo., 1926-30; mem. staff Inst. for Psychoanalysis, Chgo., 1932—; cons. psychiatrist Michael Reese Hosp., Chgo., 1944-48. Recipient Elizabeth Blackwell Centennial citation, 1949. Mem. Chgo. Psychoanalytic Soc. (pres., 1938-40), Am. Psychoanalytic Assn., Ill. Soc. Mental Health (dir., 1954-55), Planned Parenthood Assn. (dir., 1954-55). Research, publs. on psychoanalytic edn., interracial relations, Greek drama in context of Greek culture and psychoanalysis. Home: 5545 University Av., Chgo. 60637. Office: 180 N. Michigan Av., Chgo. 60601.*

McLEAN, I(saac) William, Jr., Am. microbiologist; b. Panama, C.Z., Oct. 27, 1917 (parents Am. citizens); s. Isaac William and Mary Lee (Rudisill) McL.; B.S. in Chemistry, Davison Coll., 1938; M.D., Duke, 1942; m. Brita Rosenquist, June 22, 1940; children—Ian William, Mary Ellen. With Parke, Davis & Co., Detroit, 1946—, dir. microbiol. research, 1963—. Diplomate Nat. Bd. Med. Examiners, Am. Bd. Microbiology, Am. Bd. Virology. Mem. Am. Acad. Microbiology, Am. Soc. Microbiology, Am. Assn. Immunologists, Am. Pub. Health Assn., Infectious Diseases Soc. Am., N.Y. Acad. Sci. Research, numerous publs. on chemotherapy and vaccines, influenza, adenovirus, poliovirus and hepatitis studies. Home: 781 Lake Shore St., Grosse Pointe Shores, Mich. 48236. Office: Joseph Campau St. at River, Detroit 48232.*

McLEAN, John M(ilton), Am. ophthalmic surgeon; b. N.Y.C., Oct. 24, 1909; s. William and Ella Louise (Powel) McL.; M.E., Stevens Inst. Tech., 1930, Ed.D. (hon.), 1965; M.D., Cornell U., 1934; m. Mary Lou Carlon, June 14, 1941; children—Ann Powel, Mary

Margaret, John Brandon, Ellen Steele. Asst. ophthalmology Johns Hopkins Med. Sch., 1935-38, Mellon fellow, 1936-37, asso., 1939-41; asso. prof. ophthalmology Cornell U. Med. Coll., 1941-42, prof., 1942——, prof. clin. surgery, 1942——; dir. dept. ophthalmology N.Y. Hosp., attending surgeon ophthalmology, 1941——; cons. ophthalmologist N.Y. Eye and Ear Infirmary, U. S. Naval Hosp., St. Albans, Hosp. Spl. Surgery, Phelps Meml. Hosp., Meml. Hosp. Center, Manhattan Eye, Ear and Throat Hosp. Mem. adv. bd. N.Y. State Athletic Commn. Diplomate Am. Bd. Ophthalmology. Fellow A.C.S.; mem. Am. Ophthal. Soc., Assn. Research Ophthalmology, A.A.A.S., A.M.A., N.Y. Acad. Med., Am. Acad. Ophthalmology and Otolaryngology, Internat. Congress Ophthalmology, ophthal. socs. Peru, Brazil, Mexico, Chile, Pan Am., N.Y. ophthal. socs., Harvey Soc., Alpha Omega Alpha, Chi Phi, Tau Beta Pi, Nu Sigma Nu. Author textbooks on ophthalmology, also articles. Research in ocular fluid balance and pressures, ocular surgery, cryosurgery, eye pathology. Died May 2, 1968.

McLEAN, Stewart, chemist; b. Moascar, Egypt, Nov. 19, 1931; s. Arthur and Elizabeth (Jackson) McL.; B.Sc., Glasgow (Scotland), 1954; Ph.D., Cornell U., 1958; m. Carrie Anne Holden, Aug. 14, 1957; children—Helen, Laurie, Catherine. Postdoctoral fellow U. Wis., Madison, 1957-58, NRC Can., Ottawa, Ont., 1958-60; faculty U. Toronto (Ont.), 1960——, asso. prof. chemistry, 1964——; hon. research asso. Harvard, 1967. Mem. Chem. Inst. Can., Chem. Soc. (London), Am. Chem. Soc. Research, publs. on applications of modern methods (especially spectroscopic) to elucidation of molecular structure; syntheses of complex molecules; studies of thermal rearrangements. Home: 281 St. Germain Av., Toronto 12, Ont., Can.*

McLELLAN, Alister George, New Zealand physicist; b. Christchurch, New Zealand, June 4, 1919; s. Hugh Carrick and Constance (Williscroft) McL.; B.Sc., U. Otago, 1941, M.Sc., 1942; Ph.D., U. Edinburgh, 1948; m. Pamela Mary Marsden, Oct. 31, 1949; children—John David, Daniel Hugh, Alexander Donald. Physicist, Dept. Sci. and Indsl. Research New Zealand, 1942-45; lectr. U. Otago, 1949-55; prof., head dept. physics U. Canterbury, Christchurch, 1955——. Nuffield fellow Cambridge U., 1955. Carnegie Travel grantee, 1961. Fellow Royal Soc. New Zealand (recipient Hector medal 1958). Publs. on research in statis. mechanics, thermodynamics, theory of solid state. Home: 7 Garreg Rd., Christchurch, New Zealand.*

McLELLAN, Crawford Reid, Am. chemist; b. Lexington, Miss., Oct. 18, 1906; s. James I. and Elizabeth (McNeer) McL.; B.S., Miss. Coll.; M.A., U. N.C.; Ph.D., La. State U.; m. Julia K. Caylor, Aug. 26, 1942; children—Judy D. (Mrs. David M. Dacus), Crawford Reid, Richard C., Kay Rinn, Rebecca A., Ronald G. Prof. chemistry Ark. A. and M. Coll., Monticello, Ark., 1930-39, La. Coll., Pinesville, 1941-42; chemist Pinebluff (Ark.) Arsenal, 1942-45; prof. chemistry La. State U., Baton Rouge, 1945——, dir. 1st year chemistry program, 1946——. Recipient Sci. Tchr. Merit award La. Sci. Tchrs. Assn. Mem. Am. Chem. Soc., La. Acad. Sci. Author: Introductory Experiments in Inorganic Chemistry, 1950; (with Tucker) Experiments in General Chemistry, 1961; (with Day, Clark) Concepts of General Chemistry, 1966; also articles. Research in organic chemistry, chem. edn. Home: Route 2, Box 149, Baton Rouge 70815.*

McLENNAN, Charles Ewart, Am. obstetrician and gynecologist; b. Duluth, Minn., Dec. 26, 1909; s. Archibald James and Grace Jane (McLean) McL.; A.B., U. of Minn., 1930, A.M., 1932, M.D., 1934, Ph.D., 1942; m. Margaret Jane Thomas, June 26, 1937; children—James Edward, Nancy Ann, Jane, Thomas. Teaching fellow in med., U. Minn., 1936-38, instr. obstetrics and gynecol., 1938-40; Commonwealth Fund fellow, dept. of medicine, U. Va., 1940-41; asst. prof. obstetrics and gynecology U. Minn., 1941-43, asso. prof., 1943-44; prof. and head dept. obstetrics and gynecology, school of medicine, U. Utah, 1944-47; prof. obstetrics and gynecol., head dept. Stanford, 1947——; chief obstet. and gynecol. service Stanford Hosp. Diplomate, Am. Bd. Obstetrics and Gynecology (dir.). Mem. Am. Coll. Obstetricians and Gynecologists, A.M.A., Am. Gynecol. Society, A.A.-A.S., Am. Fedn. Clin. Research, Cal. Med. Assn., San Francisco Gynecological Society (past president), Society Gynecologic Investigation (past president), Phi Beta Kappa, Sigma Xi. Author: Synopsis of Obstetrics, 1966. Editorial bd Cal. Medicine, Pacific Medicine. Research on venous pressure and blood volume in pregnancy; dysfunctional uterine bleeding; endometrial cancer. Home: 701 Tennyson Av., Palo Alto, Cal. 94303.*

McLENNAN, Hugh, Canadian physiologist; b. Montreal, Que., Can., Oct. 22, 1927; s. William Durie and Gyneth (Wanklyn) McL.; B.Sc. with honors, McGill U., Montreal, 1947, M.Sc., 1949, Ph.D., 1951; m. Hilda Jean Connell, June 2, 1949; children—Catriona Isabelle, Neil Stewart. Asst. lectr. U. Coll., U. London (Eng.), 1952-53; research fellow Montreal Neurol. Inst., 1953-55; asst. prof. Dalhousie, U., Halifax, N.S., Can., 1955-57; faculty U. B.C., Vancouver, Can., 1957——, prof. physiology, 1965——. Mem. Am., Canadian (sec. 1965——) physiol. socs., Physiol. Soc. Gt. Britain. Author: Synaptic Transmission, 1963; also numerous articles. Research on chem. transmission synaptic action within central nervous system. Home: 2961 W. 49th Av., Vancouver 13, B.C., Can.*

McLENNAN, James Alan, Jr., Am. physicist; b. Atlanta, Nov. 24, 1924; s. James Alan and Edith (Camp) McL.; A.B., Harvard, 1948; M.S., Lehigh U., 1950, Ph.D., 1952; m. Elizabeth Jane Rohs, Aug. 16, 1952; children—William Ross, Marie Camp. Tech. engr. Gen. Electric Co., Cin., 1952-53; faculty Lehigh U., Bethlehem, Pa., 1953——, prof. physics, 1962——; vis. prof. U. Zurich (Switzerland), 1966-67. Cons. Frankford Arsenal, 1958-60, Los Alamos Sci. Lab., 1964——. NSF fellow U. Brussels (Belgium), 1960-61. Fellow Am. Phys. Soc.; mem. Sigma Xi. Research, publs. on theory elementary particles especially symmetry properties, statis. mechanics of nonequilibrium processes. Home: 2005 Edgehill Rd., Bethlehem, Pa. 18017.*

McLENNAN, Sir John Cunningham, Canadian physicist; b. Ingersoll, Ont., Can., Apr. 14, 1867; s. David and Barbara (Cunningham) McL.; B.S., U. Toronto, 1892; studied under J. J. Thomson, Cavendish Lab., 1898-99; m. Elsie Monro Ramsay, 1910. Asst. demonstrator physics, Toronto, Ont., 1892-98, became demonstrator, 1899, asso. prof., 1902, dir. physics lab., 1904, prof. physics, 1907, dean grad. studies, 1930; ret., 1932; went to Eng. where he studied radium treatment of cancer, 1932. Mem. several war time coms. 1914-18; became mem. (with Rutherford) Admiralty Com. on Antisubmarine Warfare, 1917; Bakerian lectr., 1928; mem. Royal Commn. on Cancer Treatment, Ont. Fellow Royal Soc. Can. (pres.), Royal Soc., 1915 (Royal medal 1927); mem. Royal Canadian Inst. (pres. 1933-34). Research on cryogenics; liquified helium, 1923; reproduced (with G. M. Shram) auroral green line, 1925; (with Rutherford) developed magnetic devices for detection of submarines; numerous publs. on radio-activity, elec. conduction in gases, spectroscopy, liquification of gases. Died nr. Abbeville, France, Oct. 9, 1935.

McLESTER, James Somerville, Am. physician; b. Tuscaloosa, Ala., Jan. 25, 1877; s. Joseph and Nannie (Somerville) M.; A.B., U. Ala., 1896, LL.D.; M.D., U. Va., 1899; post-grad. Göttingen, Freiburg, 1901-02, Berlin and Munich, 1907-08; m. Ada Bowron, 1903; children—Anna, James B., Jane. Prof. medicine U. Ala., 1919-50. Chmn. subcom. on med. nutrition NRC. Fellow A.C.P.; mem. A.M.A. (Goldberger award, pres. 1935-36), Assn. Am. Physicians, Am. Soc. Clin. Investigation, Am. Climatological and Clin. Assn., Soc., Ala. (pres. 1920) med. assns. Author (textbooks): Nutrition and Diet in Health and Disease; the Diagnosis and Treatment of Disorder of Metabolism. Research papers on diseases of nutrition and metabolism; relationship between health and diet. Died Feb. 8, 1954.

McLIMANS, William Fletcher, Am. bacteriologist; b. Duluth, Minn., Aug. 22, 1916; s. John Pusey and Emma McL.; B.A., U. Minn., 1939, Ph.D., 1946; m. Helen Eleanor Besser, Mar. 1940; children—Judith Ann (Mrs. Lyle Henry Everard), Jeffrey Paul, William Fletcher IV. Teaching asst. U. Minn., Mpls., 1939-40, instr., 1947-49; night bacteriologist Minn. Dept. Health, Mpls., 1939-40, research virologist, 1941-42; research virologist Rocky Mountain Lab., USPHS, Hamilton, Mont., 1949-50; head dept. bacteriology, research div. Upjohn Co., Kalamazoo, 1950-54; research asso. prof. microbiology in medicine Sch. Medicine, Sch. Vet. Medicine, U. Pa., also mem. Wistar Inst., head microbiology in medicine U. Pa., 1955-59; chief tissue culture unit Communicable Disease Center, 1959-62; cancer research specialist Roswell Park Meml. Inst., Buffalo, 1962——. Chmn., cell culture com. NRC. Recipient Presdl. award 4th Internat. Polio Conf., Geneva. Fellow N.Y. Acad. Scis.; mem. Am. Assn. Immunologists, Am. Soc. Bacteriologists, Tissue Culture Assn., A.A.A.S., Am. Acad. Microbiology. Research and publs. regarding scrub typhus, poliomyelitis virus, tumor cells. Home: 7858 Lake Shore Rd., Lakeview, N.Y. Office: 666 Elm St., Buffalo, N.Y. 14203.*

McMAHON, Howard Oldford, phys. chemist; b. Killam, Alta., Can., Sept. 16, 1914; s. Thomas Alexander and Tryphina (Oldford) McM.; B.A., U. B.C., 1935, M.A., 1937; Ph.D., Mass. Inst. Tech., 1941; m. Edna Lucile Nelson, July 2, 1941; children—Thomas Arthur, Elizabeth Jean, Lucile Nancy. Came to U. S., 1941, naturalized, 1948. Research asso. Mass. Inst. Tech., 1941-43; research chemist Arthur D. Little, Inc., Cambridge, Mass., 1943-52, sci. dir., 1952——, v.p., 1956-60, sr. v.p., 1960-63, head research and devel. div., 1962-63, exec. v.p., 1963-64, pres., 1964——, also dir. Recipient Edward Longstreth medal Franklin Inst. 1951; Frank Forrest award Am. Ceramics Soc., 1952. Fellow Am. Acad. Arts and Scis., A.A.A.S.; mem. Am. Physics Soc., Am. Chem. Soc., N.Y. Acad. Scis., Sigma Xi. Contbr. articles to tech. jours. Patentee in field of very low temperatures. Home: 72 Shade St., Lexington, Mass. 02173. Office: 25 Acorn Park, Cambridge, Mass., 02140.*

McMASTER, Philip Duryee, Am. research physician; b. Phila., Sept. 14, 1891; s. John Bach and Gertrude (Stevenson) McM.; student pvt. schs., Phila.; B.S., Princeton, 1914; M.D., U. Pa., 1918; spl. student U. Freiburg, Germany, 1914, Columbia, 1921; m. Elizabeth Parsons Dwight, Oct. 13, 1923; children—Gail Parsons (Mrs. Charles Booth Alling, Jr.), Philip Robert Bache. Fellow Rockefeller Inst. Med. Research, 1919-20, asst., 1920-22, asso., 1922-26, asso. mem., 1926-51, mem., prof., 1951-62, prof. emeritus, 1962——; fellow research psychology Harvard, 1929-30. Fellow A.A.A.S.; mem. Nat., N.Y. acads. scis., N.Y. Acad. Medicine, Harvey Soc. (sec. 1927-28), Am. Assn. Immunologists, Am. Soc. Exptl. Pathology, Am. Assn. Pathologists and Bacteriologists, Am. Soc. Exptl. Biology and Medicine, Sigma Xi, Alpha Omega Alpha. Research, publs. on liver and gall bladder physiology, liver pigment formation, blood circulation, antibody formation in lymph glands. Home: River Rd., Cos Cob, Conn. 06807. Office: Rockefeller Inst. 66th St. and York Av., N.Y.C. 10021.*

McMICHAEL, Sir John, Brit. physician; b. Gatehouse, Scotland, July 25, 1904; s. James and Margaret (Sproat) McM.; M.B., Edinburgh (Scotland) U., 1927, M.D., 1933, LL.D., 1962; M.D., Melbourne (Australia), 1965; D.Sc., U. Newcastle (Eng.), U. Sheffield (Eng.), 1965; m. Sheila Howarth; stepchildren—Sarah, Judith Sharpey Schafer; children—Ian, Hugh, Andrew, Peter. Beit Research fellow, 1930-34; faculty Postgrad. Med. Sch. London (Eng.), 1939——, prof. medicine, 1946-66. Wellcome trustee, 1960——. Recipient Gairdner award, Toronto, Ont., Can., 1960; created knight, 1965. Fellow Royal Soc., 1957; mem. Physiol. Soc., Assn. Am. Physics. Research, numerous publs. on methods measuring lung function, cardiac output and liver biopsy. Home: 2 N. Sq., London N.W. 11., Eng.*

McMILLAN, Edwin Mattison, Am. physicist; b. Redondo Beach, Cal., Sept. 18, 1907; s. Edwin H. and Anna Marie (Mattison) McM.; B.S., Cal. Inst. Tech., 1928, M.S., 1929; Ph.D. in Physics, Princeton, 1932; D.Sc. (hon.), Rensselaer Poly. Inst., 1961, Gustavus Augustus Coll., 1963; m. Elsie Walford Blumer, June 7, 1941; children—Ann Bradford, David Mattison, Stephen Walker. Faculty physics U. Cal., Berkeley, 1935——, prof., 1946——, asso. dir. Lawrence Radiation Lab., 1954-58, dir., 1958——; researcher, Los Alamos Sci. Lab., 1942-45. Mem. gen. adv. com. U. S. AEC, 1954-58; mem. commn. on high energy physics Internat. Union Pure and Applied Physics, 1960-66. Recipient Sci. award Research Corp., 1951; Alumni Distinguished Service award Cal. Inst. Tech., 1966; (with G. T. Seaborg) Nobel prize in Chemistry, 1951, (with V. I. Veksler) Atoms for Peace prize, 1963. Fellow Am. Phys. Soc., Am. Acad. Arts and Scis.; mem. Nat. Acad. Scis., Am. Philos. Soc. Research in nuclear physics and particle accelerator devel., microwave radar, and sonar; discoverer (with Abelson) of neptunium; co-discoverer of plutonium; origination of principle of phase stability.

McMINN, Robert Matthew Hay, Brit. anatomist, histologist; b. Auchinleck, Scotland, Sept. 20, 1923; s. Robert Martin and Elsie (Kent) McM.; M.B., Ch.B., U. Glasgow (Scotland) U., 1947, M.D., 1958; Ph.D., U. Sheffield (Eng.), 1956; m. Margaret Grieve Kirkwood, Aug. 27, 1948; children—Marion Elizabeth, Robert George Allan. Demonstrator anatomy U. Glasgow, 1950-52; lectr. U. Sheffield, 1952-60; faculty King's Coll., U. London, 1960——, prof. anatomy, 1966——. Fellow Zool. Soc. London, Royal Micros. Soc., Royal Soc. Medicine; mem. Anat. Soc. Gt. Britain and Ireland (program sec. 1965-67), Am. Assn., Anatomists, Brit. Soc. Gastroenterology. Research, publs. on cytology of alimentary tract, mechanisms of repair in mucous membranes and in tympanic membranes. Home: 74 Dorling Dr., Ewell, Epsom, Surrey, Eng. Office: King's Coll., Strand, London W.C.2., Eng.*

McMURRICH, James Playfair, anatomist; b. Toronto, Ont., Can., Oct. 16, 1859; s. John and Janet (Dickson) McM.; B.A., U. Toronto, 1879, M.A., 1881, LL.D., 1931; Ph.D., Johns Hopkins, 1885; LL.D., U. Mich., 1912, U. Cin., 1923; m. Katie Moodie Vickers, 1882; 1 son, 1 dau. Became prof. biology Ont. Agrl. Coll., 1882; named instr. osteology and mammalian anatomy Johns Hopkins, 1884; prof. biology Haverford (Pa.) Coll., 1886-89; prof. animal morphology Clark U., Worcester, Mass., 1889-92; prof. biology U. Cin., 1892-94; prof. anatomy U. Mich., 1894-1907; prof. anatomy U. Toronto, 1907-30, dean Sch. Grad. Studies, 1922-30, became prof. emeritus, 1930. Mem. N.Am. Commn. on Fisheries Investigations, 1921-39; chmn. Biol. Bd. Can., 1926-34. Fellow Royal Soc. Can. (pres.); mem. Am. Assn. Anatomists (became pres. 1908), A.A.A.S. (pres.). Recipient Flavelle medal, posthumously 1939. Author: A Textbook of Invertebrate Morphology, 1894; The Development of the Human Body, 1902; Leonardo da Vinci, the Anatomist (1452-1519), 1930. Editor: Human Anatomy (Henry Morris), 4th edit., 1907. Contbg. author: Human Anatomy (G. A. Piersol), 1907. Translator: Atlas and Textbook of Human Anatomy, 1906-07. Research and numerous publs. on actiniaria or sea-anemones, morphology and embryology, history of sci. Died Toronto, Feb. 9, 1939.

McMURTIE, William, Am. chemist; b. Belvidere, N.Y., Mar. 10, 1851; s. Abram and Almira M.; E.M., Lafayette Coll., 1871, M.S., 1874, Ph.D., 1875; m. Helen M. Douglass, Apr. 5, 1876; 2 children. Asst. and chief chemist U. S. Dept. Agr., 1872-79, spl. agt. in agrl. tech., 1879-82; prof. chemistry U. Ill., 1882-

88; chemist N.Y. Tartar Co., 1888-—; cons. chemist Royal Baking Powder Co., 1899-—, 2d v.p., 1908-—; cons. prof. gen. tech. chemistry, Poly. Inst., Bklyn., 1905-—. Chemist Ill. Bd. Agr., 1884-88, Ill. Agrl. Expt. Sta., 1886-88. Author: Culture of the Beet and Manufacture of Sugar Therefrom, 1880; The Culture of Sumac, 1880; Grape Culture in the United States, 1883; Wools and Other Animal Fibres, 1886, 1901. Made study of fgn. beet-sugar industry which was instrumental in founding of beet root sugar manufacture in U. S.; found an improved, more econ. method of baking powder prodn. Died May 24, 1913.

McNAIR, Andrew Hamilton, Am. geologist; b. Victor, Mont., May 29, 1909; s. Andrew H. and May (Rohrbach) McN.; A.B., Mont. U., 1931, M.A., 1933; Ph.D., U. Mich., 1935; m. Evelyn Lyford, Jan. 28, 1939; children—Ann (Mrs. Norman J. Page), Jane (Mrs. John C. Stormer, Jr.), Peter H. Faculty geology Dartmouth, 1935-—, prof., 1946-—, chmn. dept., 1939-43, 51-53, head geol. expdn. to Victoria Island, 1964, 65. Geologist strategic minerals investigation U. S. Geol. Survey, 1942-45; cons. geologist indsl. materials, 1946-61; part-time geologist Phillips Petroleum and Gulf Oil Cos., 1949-52; charge geol. field work Dominion Explorers Ltd., Canadian Arctic Islands, 1959-60; mem. Internat. Geol. Field Inst., Gt. Britain, 1961. Fellow Geol. Soc. Am., Paleontol. Soc. Am.; mem. Am. Assn. Petroleum Geologists, Sigma Xi, Gamma Alpha. Author articles on paleontology, stratigraphy, econ. geology, arctic geology, tectonics; described (with Y. O. Fortier, R. Thorsteinsson) Innuitian orogenic system, 1954. Home: 6 Mitchell Lane, Hanover, N.H. 03755.*

McNAIR, James Birtley, Am. chemist, botanist; b. Hazleton, Pa., Mar. 18, 1889; s. Thomas Speer and Mary (Stevens) McN.; student Pomona Coll., 1912-13; A.B., U. Cal., 1916, A.M., 1917; postgrad. U. Pa., U. Chgo., U. So. Cal., Chem. Warfare Sch. Chemist various govt. burs., 1917-26; with Field Mus. Natural History, Chgo., 1923-32; cons. in ethnobotany S.W. Mus., Los Angeles, 1929-—; chemist in charge Lac Chems., Inc., Culver City, Cal., 1944-45; asst. wine maker and chemist Pacific Wines, Inc., Los Angeles, 1945-46. Recipient certificate of merit Inst. Am. Genealogy, 1939. Fellow A.A.A.S., Linnean Soc. London, Soc. Antiquaries Scotland; mem. Am. Bot. Soc., Am. Chem. Soc., Soc. Genealogists (London), Sigma Xi. Author: McNair, McNear and McNeir Genealogies, 1923; Rhus Dermatitis, Its Pathology and Chemotherapy, 1923; Citrus Products, part I, 1926, part II, 1927; The Analysis of Fermentation Acids, 1947; Simon Cameron's Adventure in Iron, 1949; With Rod and Transit; the Engineering Career of Thomas S. McNair, 1951; Chemical Plant Phylogeny, 1965; Studies in Plant Chemistry, 1965. Study of chem. products in relation to plant evolution; essential oils and resins; citrus products; plant poisons; soil acidity; spices. Address: 818 S. Ardmore Av., Los Angeles 90005.*

McNALLY, James Rand, Jr., Am. physicist; b. Boston, Nov. 10, 1917; s. James Rand and Margaret (Turley) McN.; B.S. in Physics magna cum laude, Boston Coll., 1939; S.M., Mass. Inst. Tech., 1941, Ph.D., 1943; m. Margaret Anne McKenna, Nov. 26, 1942; children—James Rand III, Peter Joseph, Mary Ellen, Francis Edward, Anne Therese, Michael Stephen, Margaret Rose. Instr., Mass. Inst. Tech., 1944-48; physicist Oak Ridge Nat. Lab., 1948-49, sr. physicist, 1949-62, research staff mem., 1962-—; instr. U. Tenn. Extension, Oak Ridge, 1952-53. Recipient Manhattan Dist. Civilian award, 1946. Fellow Optical Soc. Am., A.A.A.S.; mem. Am. Phys. Soc., Soc. Applied Spectroscopy, Am. Assn. Physics Tchrs. Research, publs. on atomic physics, energy levels, Zeeman effects, hollow cathode sources, spectro-isotopic analysis, hyperfine structure, isotope shifts, Doppler effects, electron temperature studies, excitation-heating of ions to high temperatures, fusion chain reactions. Home: 103 Norman Lane, Oak Ridge 37830. Office: P.O. Box Y, Bldg. 9201-2, Oak Ridge 37831.*

McNALLY, William James, Canadian otolaryngologist; b. Bryson, Que., Dec. 27, 1897; s. Richard and Elizabeth (Ryan) McN.; B.A. St. Francis Xavier U., LL.D., 1950; M.D.C.M., Dalhousie U.; M.Sc. McGill U., 1925, D.Sc., 1934; m. Harriet Purcell, Sept. 27, 1927; children—Herbert, Ann Elizabeth (Mrs. Richard Darling). Mem. attending staff in otolaryngology Royal Victoria Hosp., Montreal, Que., 1926-—; faculty McGill U., Montreal, 1926-—, prof., chmn. dept. otolaryngology, 1950-64, founder, dir. Inst. Otolaryngology, 1961-64. Recipient Dalby Meml. prize Royal Soc. Medicine, 1938; Cardinal Newman award, 1963. Mem. Am. Laryngol. Assn. (Newcomb award 1952, pres. 1960), Am. Otol. Soc. (pres. 1956, award of merit 1962), Canadian Otolaryn. Soc. (pres. 1951), Scottish Otol. and Laryngol. Soc. (hon.). Author: Examination of the Labyrinth, in Relation to Its Physiology and Non Suppurative Diseases, 1953, 67; also articles. Home: 25 Redpath Pl., Montreal 25, Que., Can.*

McNAMARA, Delbert Harold, Am. astrophysicist; b. Salt Lake City, June 28, 1923; s. Delbert H. and Florence (Williams) McN.; B.S., U. Cal. at Berkeley, 1947, Ph.D., 1950; m. Elmeda Louise Robison, Feb. 21, 1945; children—Marilyn, Susan, Jay. Research astronomer U. Cal., Berkeley, 1950-55; faculty Brigham Young U., Provo, Utah, 1955-—, prof. physics, 1962-—, 4th Ann. Faculty lectr., 1967; prin. scientist N.Am. Aviation, Los Angeles, 1961-62. Research, numerous publs. on photoelectric and spectrographic studies of eclipsing binaries and pulsating variable stars. Home: 1880 N. 1500 East, Provo, Utah 84601.*

McNAMEE, Raymond Wilson, Am. chemist; b. Belaire, O., Nov. 27, 1907; s. James Leo and Anna (Weisner) McN.; B.S., U. Akron, 1929; M.S., Northwestern U., 1931, Ph.D., 1933; m. Edith Lucille Irwin, Nov. 28, 1930; children—Raymond Wilson, Philip Irwin. Research chemist Carbide & Carbon Chems. Co., South Charleston, W.Va., 1933-49; supt. research and devel. dept. Union Carbide Corp., N.Y.C., 1949-53, mgr. research adminstrn., 1953-59, mgr. research, 1959-65, v.p., dir. Union Carbide & Carbon Research Labs., Inc., N.Y.C., 1957-65, dir. Union Carbide European Research Assos. Mem. Am. Chem. Soc. (Herty medal 1953), Am. Inst. Chem. Engrs., Am. Ordnance Assn., Indsl. Research Inst., N.Y. Acad. Scis., N.A.M., Soc. Chem. Industry, Dirs. Indsl. Research, Chemists Club N.Y., Sigma Xi, Phi Delta Theta, Alpha Chi Sigma, Phi Lambda Upsilon. Research, publs. on synthetic organic chems., catalysis; patentee in field. Home: 2775 Crayton Rd., Naples, Fla. 33940.*

McNEAL, Francis Harrison, Am. agronomist; b. Barlett, Ore., Dec. 9, 1920; s. Francis J. and Tillie (Zell) McN.; B.S., Ore. State U., 1943, M.S., 1948; Ph.D., U. Minn., 1953; m. Ora Helen Veenker, Feb. 28, 1947; children—Linda Rae, Allen Francis. Agronomist, Agrl. Research Service, U. S. Dept. Agr., Pendleton (Ore.) Br. Expt. Sta., 1947-48, Mont. State U., Bozeman, 1948-—. Mem. Nat. Wheat Improvement Com., 1960-—; sec. Western Wheat Improvement Com., 1962-—. Fellow Am. Soc. Agronomy; mem. Sigma Xi. Research, publs. on variety devel. wheat, wheat stem sawfly resistance, rust resistance, virus diseases, fertility and quality relationships, genetics of stem solidness in wheat. Home: Route 1, Box 323, Bozeman, Mont. 59715.*

McNEE, Robert Bruce, Am. geographer; b. Big Timber, Mont., Aug. 20, 1922; s. William Wooldridge and Nina Lucy (Gates) McN.; B.A. with high honors, Wayne U., 1949; M.A., Syracuse U., 1950, Ph.D., 1953; m. Gilberta Lyons, Sept. 20, 1949 (annulled Sept. 1954); 1 son, Robert Andrew; m. 2d, Doris Smithers, Dec. 11, 1954; children—Margaret Ann, William Smithers; step-children: Douglas, George. From lectr. to asso. prof. Coll. City N.Y., 1952-63; prof. geography, chmn. dept. U. Cin., 1963-—; cons. Ford Found., 1962-63. Mem. Assn. Am. Geographers (chmn. N.Y.-N.J. div. 1960-62), Am. Geog. Soc., Nat. Council Geog. Edn., Nat. Planning Assn., A.A.A.S., Royal Geog. Soc. Contbr. articles in field. Research on econ. devel.; location theory; urban planning; curricular reform; systems analysis. Home: 194 Lafayette Circle, Cin. 45220.*

McNEILL, K(enneth) G(ordon), physicist; b. Cheshire, Eng., Dec. 21, 1926; s. Ferguson and Elizabeth (Stevenson) McN.; B.A., Oxford U., 1947, M.A., 1950, D.Phil., 1950; m. J. Ruth S. Robertson, Nov. 6, 1959; 1 dau., Diane E.S. Postdoctoral fellow Yale, 1950-51; research fellow Glasgow U., 1951-52, lectr., 1952-57; faculty U. Toronto (Ont., Can.), 1957-—, prof. physics, 1963-—. Mem. Canadian Assn. Physicists, Am. Phys. Soc., Instn. Nuclear Engrs. (U.K.). Author: (with J. Maclachlan, J. Bell) Matter and Energy, 1963; also numerous articles. Research on low energy nuclear physics and med. physics. Home: 227 St. Leonards Av., Toronto 12, Ont., Can.*

McNEMAR, Quinn, Am. psychologist, statistician; b. Greenland, W.Va., Feb. 20, 1900; s. Dolly and Bessie (Michael) McN.; A.B., Juniata Coll., Huntingdon, Pa., 1925; Ph.D., Stanford, 1932; postdoctoral fellow, Social Sci. Research Council, Columbia, 1933-34; m. Olga Williamson, Sept. 12, 1931. Tchr. Charleston (W.Va.) schs., 1925-27; mem. faculty Stanford U., 1932-—, prof. psychology, statistics and edn., 1948-65, prof. emeritus, 1965-—; prof. psychology and edn. U. Tex., 1965-—; asso. prof. Fordham U., 1937-38. Staff, fellowship sec. Social Sci. Research Council, 1941-43; cons. Dept. War, 1941-44, VA 1948-62, Office Naval Research, 1949-52, NSF, 1952-57, NIH, 1963-66, U. S. Office Edn., 1966-67. Mem. Commn. Human Resources and Advanced Tng., 1949-53; Bd. dirs. Psychol. Corp. Mem. A.A.A.S., Am. Psychol. Assn. (pres. 1964), Social Science Research Council (bd. dirs. 1964-67), Psychometric Soc. (pres. 1951), Western Psychol. Assn. (pres. 1959), Sigma Xi. Author: The Revision of the Stanford-Binet Scale, 1942; Psychological Statistics, 3d edit., 1962. Asso. editor Annual Rev. Psychology, 1954-66. Contbr. articles profl. jours., mem. editorial bds. jours. Research on statis. techniques in psychology, psychol. measurement Stanford-Binet IQ sclae. Home: 1801 Lavaca St., Austin, Tex. 78701.*

McNESBY, James R., Am. chemist; b. Bayonne, N.J., Apr. 16, 1922; s. James Aloysius and Margaret (O'Connor) McN.; B.S., Ohio U., 1943; Ph.D., N.Y. U., 1951; m. Helen Louise Rittenhouse, Dec. 27, 1949; children—Kevin L., James R., Shawn. Phys. chemist U. S. Naval Ordnance Test Sta., China Lake, Cal., 1951-56; phys. chemist Nat. Bur. Standards, Washington, 1957-—. Recipient Rockefeller Pub. Service award, 1958-59. Mem. Am. Chem. Soc., Faraday Soc. Research, publs. on kinetics of methyl radical reactions, isomerization of alkyl radicals, vacuum ultraviolet photochemistry; discovered molecular elimination of hydrogen from hydrocarbons, non-isomerization of alkyl radicals. Home: 13308 Valley Dr., Rockville, Md. 20850. Office: Nat. Bur. Standards 223.53, Washington 20234.*

McNEW, George Lee, Am. plant pathologist; b. Alamogordo, N.M., Aug. 22, 1908; s. William Henry and Nettie (Fry) McN.; B.S., N.M. State U., 1930, D.Sc., 1954; M.S., Ia. State U., 1932, Ph.D., 1935; m. Elizabeth Anne Mehlhop, May 22, 1932; 1 dau., Freda Louise. Fellow, Ia. Agr. Expt. Sta., 1930-35; asst. mem. Rockefeller Inst., 1935-39; asso. mem. N.Y. Agr. Expt. Sta., 1939-43; mgr. research and devel. agrl. chems. U. S. Rubber Co., 1943-47; prof., head botany dept. Ia. State U., 1947-49; mng. dir. Boyce Thompson Inst., Yonkers, N.Y., 1949-—, dir., trustee, 1949-64, pres. Research Found., 1950-—. Recipient Alumni Achievement award Ia. State U., 1957. Mem. Am. Phytopath. Soc. (pres. 1952), Bot. Soc. N.Am., Am. Soc. Plant Physiologists, N.Y. Acad. Sci., Torrey Bot. Club (pres. 1959), Am. Chem. Soc. (chmn. N.Y. sect. 1960), Yonkers Tb and Pub. Health Assn. (pres. 1953-54), N.Y. Tb Soc. (pres. 1958-59), Sigma Xi, Alpha Zeta, Sigma Epsilon, Phi Kappa Phi. Contbr. articles on plant diseases, parasitism, agrl. chems., fungicides to profl. publs. Home: 255 Broadway, Hastings-on-Hudson, N.Y. Office: 1086 N. Broadway, Yonkers, N.Y. 10701.*

McNIVEN, Neal Lindsay, chemist; b. Kingston, Ont., Can., Sept. 15, 1914; s. John Lindsay and Margaret (Johnstone) McN.; B.Sc., McGill U., 1936, M.Sc, 1939; Ph.D., St. Andrews U., 1950; m. Edith M. Havlin, Dec. 15, 1945; children—Heather Anne Lynn, Andrew Ranald. Chemist, Canadian Industries Ltd., Shawinigan, Que., Can., 1936-38, research chemist, McMasterville, Que., 1939-45; research chemist Ont. Research Found., 1945-47; organic chemist Worcester Found. Exptl. Biology, Shrewsbury, Mass., 1951-—. Mem. Am. Chem. Soc. (chmn. Central Mass. sect.), Chem. Soc., Inst. Chem. Engrs. (London), Soc. Applied Spectroscopy, Am. Soc. for Testing Materials, N.Y. Acad. Sci., Chem. Inst. Can. Research, numerous publs. on steroid synthesis, gas chromatography of steroids, automatic infrared spectroscopy. Home: 14 Water St. Office: 222 Maple Av., Shrewsbury, Mass. 01545.*

McNOWN, John Stephenson, Am. civil engr., coll. dean; b. Kansas City, Kan., Jan. 15, 1916; s. William Coleman and Florence Marie (Klahr) McN.; B.S., U. Kan., 1936; M.S., U. Ia., 1937; Ph.D., U. Minn., 1942; D.Sc. (Fulbright Research scholar), U. Grenoble (France), 1951; m. Miriam Leigh Ellis, Sept. 6, 1938; children—Stephen Ellis, Robert Neville, Cynthia Leigh, Mark William. Research asso. oceanography U. Cal., Navy Electronics Lab., San Diego, 1942-43; faculty State U. Ia., Iowa City, 1943-54, U. Mich., 1954-57; dean engring. and architecture U. Kan. Lawrence, 1957-65, Alfred P. Learned prof. engring. mechanics, 1965-—; dir. overseas liaison com. Am. Council on Edn., Washington, 1967-—. Cons. edn. NSF; 1960-62, Ford Found., 1963-—, Bell Telephone Labs., 1964. Fellow African Studies Assn.; mem. A.A.A.S., Am. Soc. C. E. (J. C. Stevens award 1946, Research Program prize 1949, James R. Croes medal 1955), Am. Soc. Engring. Edn., Commn. Engring. Edn., Internat. Assn. Hydraulic Research, Sigma Xi, Phi Delta Theta, Tau Beta Pi, Theta Tau, Phi Kappa Phi. Contbg. author Engineering Hydraulics, 1950; Advanced Mechanics of Fluids, 1959; Contbr. articles to tech. jours. Research on sedimentation, analytical and exptl. fluid mechanics and hydraulics as applied to vapor cavity formation on underwater bodies, energy losses and pressure changes in manifold flow, mass oscillations in harbors, and flow systems in navigation locks. Home: 1135 Highland Dr., Lawrence, Kan. 66044.*

McNUTT, Clarence Wallace, Am. physiologist; b. Ozan, Ark., Aug. 5, 1913; s. Walter Scott and Mary Elizabeth (Wallace) McN.; A.B., Henderson State Coll., 1935; M.S., La. State U., 1938; Ph.D., Brown U., 1941; m. Alice Catherine Walsh, June 19, 1939; children—Lilla Marian, Mary Katherine, Margaret Ellen, Edith Louise. Muellhaupt fellow Ohio State U., 1941-42; aviation physiologist U. S. Army and Air Force, 1942-46; instr. anatomy U. Wis., 1946-50; faculty U. Tex., 1950-—, prof. anatomy South Tex. Med. Sch., 1967-—; cons. med. genetics U. Hosps. Counselor human genetics. Fellow A.A.A.S.; mem. Soc. Exptl. Biology and Medicine, Am. Assn. Anatomy, Genetics Soc. Am., Am. Soc. Human Genetics, Genetics Assn., Tex. Acad. Sci., Am. Inst. Biol. Sci., Sigma Xi. Research, publs. on developmental mammalian genetics with spl. reference to neuromuscular disorders in man and mutant mice. Home: 1019 Church St., Galveston, Tex. 77550.*

McPHAIL, M(urchie) K(ilburn), Canadian physiologist; b. Kilburn, N.B., Can., Jan. 26, 1907; s. Alexander Marshall and Mary Jane (Robertson) McP.; B.A. with First-Class Honours in Zoology, U. B.C., 1929; Ph.D., McGill U., 1932; m. Irene Porter Ross, May 5, 1937; children—Michael B. R., John D. R., Moya J. R. (Mrs. F. B. M. Cowan). Faculty, U. Alta., 1934-38, Dalhousie U., 1938-45; chief pharmacologist Vick Chem. Co., N.Y.C., 1945-48; head physiology sect. Suffield Exptl. Sta., Ralston, Alta., Can., 1948-63; dir. bioscis. research Def. Research Bd. Hdqrs., Ottawa, Can., 1963-67; dir. Band C div. Def. Research Bd., Def. Chem. Biol. and Radiation Establishment, Shirley's Bay, Ottawa, 1967—. Mem. Canadian Phys. Soc., Pharmacological Soc. Can., N.S. Inst. Sci., Brit. Physiol. Soc., Brit. Biochem. Soc., Brit. Soc. Endocrinology, Am. Soc. Pharmacology and Exptl. Therapeutics. Research, publs. on physiology of reprodn. and related fields of physiology. Home: 1889 Broadmoor Av., Ottawa 8, Ont. Office: Def. Research Bd., Def. Chem. Biol. and Radiation Establishment, Shirley's Bay, Ottawa, Ont., Can.*

McPHERSON, Archibald Turner, Am. chemist; b. Marceline, Mo., Feb. 22, 1895; s. Samuel A. and Adeline (Brewer) McP.; A.B., Trinity U., 1914; M.A., U. Tex., 1916; Ph.D., U. Chgo., 1923; m. Margaret F. Wilcox, Dec. 23, 1923; children—Frances (Mrs. Richard Greschel), Jean (Mrs. Harold E. Bennett). Instr. chemistry Ewing Christian Coll., Allahabad, India, 1919-21; with Nat. Bur. Standards, Washington, 1918-19, 23-61, asso. dir., 1951-61, spl. asst. to dir. for internat. standards, 1963-65; asso. dir. office tech. Services, U. S. Dept. Commerce, 1962-63; free lance writer, speaker on chem. and biochem. synthesis of food, 1965—. Recipient Exceptional Service award Dept. Commerce, 1959. Fellow Am. Phys. Soc., A.A.A.S.; mem. Am. Soc. Testing and Materials, Standards Engrs. Soc. (award 1962), Am. Chem. Soc., Soc. Rheology, Washington Acad. Sci., Philos. Soc. Washington. (Washington). Editor: Engineering Uses of Rubber, 1956. Research, numerous publs. on rubber, other polymers, food synthesis. Home: 4005 Cleveland St., Kensington, Md. 20795.*

McPHERSON, William, Am. chemist; b. Xenia, O., July 2, 1864; s. William and Mary (Rader) M.; B.Sc., Ohio State U., 1887, M.Sc., 1890, D.Sc., 1895, LL.D., 1940; Ph.D., U. Chgo., 1899; LL.D., Wittenberg Coll., 1927; m. Lucretia Heston, June 21, 1893; children—William Heston, Gertrude May; m. 2d, Mary B. Henderson, Apr. 18, 1925. Instr. chemistry and physics Toledo High Sch. and Manual Tng. Sch., 1887-89, chemistry and Latin, 1889-92; asst. in chemistry Ohio State U., 1892-93, asst. prof., professor, 1893-95, asso. prof. gen. chemistry, 1895-97, prof. gen. chemistry, 1895-97, prof. chemistry, 1897-1937, dean grad. sch., 1911-37, emeritus dean and prof., 1937-51, acting pres., 1924, 38, pres. emeritus, 1938-51. Fellow A.A.A.S.; mem. Am. Chem. Soc. (pres. 1929-30), Am. Inst. Chemists (hon.), Deutsche Chemische Gesellschaft, Phi Beta Kappa, Sigma Xi. Coauthor (with William E. Henderson) of series of text books in chemistry; contbr. to chem. jours. Died Oct. 2, 1951.

McQUILLEN, John Hugh, Am. dentist; b. Phila., Feb. 12, 1826; s. Hugh and Martha (Scattergood) McQ.; M.D., Jefferson Med. Coll., 1852; D.D.S., Phila. Coll. Dental Surgery, 1853; m. Amelia D. Schellenger, 1852; 5 children. Mem. Pa. Assn. Dental Surgeons, later pres.; prof. operative dentistry and dental pathology Pa. Coll. Dental Surgery, 1857-62; founder Phila. Dental Coll., 1863, dean and prof. anatomy, physiology and hygiene, 1863-79; an organizer Am. Dental Assn., 1859, pres., 1865; an organizer, 1st corr. sec. Odontographic Soc. Phila., 1863, pres., 1868-70; 1st corr. sec. Assn. Coll. Dentistry, 1866; mem. Acad. Natural Science at Phila., founder biol. and micros. sect. Editor Dental Cosmos, 1859, editor-in-chief, 1865-72. One of 1st in Am. to demonstrate importance of micros. knowledge of human teeth in health and disease. Died Phila., Mar. 3, 1879.

McREYNOLDS, Paul Wyatt, Am. psychologist; b. Adrian, Mo., June 18, 1919; s. William W. and Ella (McCune) McR.; B.S., Central Mo. State Coll., 1940; M.A., U. Mo., 1946; Ph.D., Stanford, 1949; m. Billie Huffsmith, Aug. 14, 1955; 1 son, David. Chief psychology research Palo Alto (Cal.) VA Hosp., 1947-60, chief Behavioral Research Lab., 1960—; instr. U. Mo., 1946; asst. prof. San Francisco State Coll., 1950-52; vis. prof. U. Ore., 1961; lectr. U. Cal. at Berkeley, 1961-64, 66; cons. asso. prof. Stanford, 1956—. Mem. Am. Psychol. Assn., A.A.A.S., Sigma Xi. Asso. editor Psychol Record, 1959—, Psychol. Reports, 1963—, Perceptual and Motor Skills, 1963—. Research, numerous publs. on motivation, psychopathology, test constrn. Home: 3502 Arbutus Dr., Palo Alto 94303. Office: VA Hosp., Palo Alto, Cal. 94304.*

McSHANE, Edward James, Am. mathematician; b. New Orleans, May 10, 1904; s. Augustus and Harriet (Butler) McS.; B.E., Tulane U., 1925, B.S., 1925, M.S., 1927, Sc.D. (hon.), 1947; Ph.D., U. Chgo., 1930; m. Virginia Haun, Sept. 10, 1931; children—Neill (dec.), Jennifer (Mrs. H. N. Ward), Virginia Patricia (Mrs. Robert B. Warfield). Faculty, U. Wichita, 1928-29, Universität, Göttingen, 1932-33, Princeton, 1933-35; prof. math. U. Va., Charlottesville, 1935—, Alumni prof., 1957—; vis. prof. Rockefeller U., 1963—. Chief mathematician Ballistic Research Lab., Aberdeen Proving Ground, Md., 1942-45; visitor Inst. for Advanced Study, Princeton, N.J., 1949-50; Fulbright prof. U. Utrecht, 1955-56. NRC fellow, 1930-32. Mem. Am. Math. Soc. (past pres.), Math. Assn. Am. (Distinguished Service award 1964, past pres.), Nat. Acad. Sci., Am. Philos. Soc., Phi Beta Kappa, Sigma Xi, others. Author: Integration, 1944; Order-preserving Maps and Integration Processes, 1953; (with J. L. Kelley and F. V. Reno) Exterior Ballistics, 1953; (with T. A. Botts) Real Analysis, 1959; also research articles. Study of existence theorems in calculus of variations; problem of Bolza; stochastic processes; integration; exterior ballistics. Home: 209 Maury Av., Charlottesville, Va. 22903.*

McTAGGART-COWAN, Patrick Duncan, meteorologist; b. Edinburgh, Scotland, May 31, 1912; s. Garry and Laura Alice (Mackenzie) McT.-C.; B.A. with honors, U. B.C., Vancouver, Can., 1933, D.Sc., (hon.), 1961; B.A. with honors, Oxford (Eng.) U., 1936; m. Margaret Lawson Palmer, Oct. 17, 1939; children—Gillian Hope, James Duncan. Instr. physics U. B.C., 1934; with Brit. Meteorol. Office. London (Eng.) Airport, 1936; officer in charge Meteorol. Service, Nfld., 1936-42; sec. for air navigation Provisional Internat. Civil Aviation Orgn., 1945; asst. dir., chief forecast div. Meteorol. Service Can., Toronto, 1946-57, asso. dir., 1958-59, dir., 1959-64; pres. Simon Fraser U., Burnaby, B.C., 1963—. Recipient Coronation medal, 1953; Order Brit. Empire, 1944; Rob. M. Losey award Inst. Aero. Scis., 1961; Patterson medal Meteorol. Service of Can., 1965; Centennial medal, 1967. Fellow Arctic Inst. N.Am. (gov.), Royal Meteorol. Soc. (pres. Canadian br. 1950-51); mem. Am. Meteorol. Soc. (Charles Franklin Brooks award 1965, past councillor), Arctic Inst. N.Am. (past gov.), Am. Geophys. Union, Canadian Assn. Physicists, World Meteorol. Orgn. (exec. com. 1960-63, regional pres. 1963-64). Research, numerous publs. on meteorology and air nav. Home: Pres.'s Residence, Simon Fraser U., Burnaby 2, B.C., Can.*

McTAMMANY, John, inventor; b. nr. Glasgow, Scotland, June 26, 1845; s. John and Agnes (McLean) McT. Came to U. S., 1862; served with 115th Ohio Vol. Inf., 1863-65, critically wounded nr. Chattanooga; while convalescing at Nashville, repaired music box, which gave him idea for an instrument operated by depressions; developed player-piano, 1866; built 3 models of player-piano, also 2 machines to prepare perforated sheets, 1866-76; gave public exhbn. of piano, St. Louis, 1876; prevented by circumstances from getting patent on his invention within prescribed time limit; declared to be original inventor of player-piano, after long and costly litigation against competitors, 1880; received 3 patents on invention, 1881; patented 1st voting machine, which pneumatically registered votes using perforated roll (1st machine ever used in an election), 1892. Died Stamford, Conn., Mar. 26, 1915.

McVEIGH, Ilda, Am. biologist; b. Fulton, Mo., Feb. 12, 1905; d. Joseph F. and Emma (Guerrant) McVeigh; A.A., Synodical Coll., 1925; B.S. in Edn., U. Mo., 1931, M.A., 1933, Ph.D., 1937. Faculty, U. Mo., 1937-40, N.W. State Coll., Alva, Okla., 1940-41, Conn. Coll., 1942-43; research asst. Yale, 1941-42, 43-45, N.Y. Bot. Garden, 1945-48; asso. prof. biology Vanderbilt U., Nashville, 1948-66, prof., 1966—. Fellow A.A.A.S.; mem. Bot. Soc. Am., Mycol. Soc. Am Soc. Am. Microbiologists, Am. Inst. Biol. Scis., Assn. Southeastern Biologists, Tenn.-Ky. Soc. Am. Microbiologists, Torrey Bot. Club, Sigma Xi, Sigma Delta Epsilon. Research, publs. on vegetative reprodn., nutrition and growth factors of microorganisms, antibiotics. Home: 2305 Elliston Pl., Nashville 37203.*

McVITTIE, George Cunliffe, astronomer, educator; b. Izmir, Turkey, June 5, 1904; s. Francis Skinner and Emily Caroline (Weber) McV.; M.A., U. Edinburgh (Scotland), 1927; Ph.D., U. Cambridge (Eng.), 1930; m. Mildred Bond Strong, Sept. 3, 1934. Asst. lectr. math. Leeds (Eng.) U., 1930-33; lectr. Edinburgh U., 1933-34, Liverpool (Eng.) U., 1934-36; reader math. Kings Coll., U. London (Eng.), 1936-48, prof. math. Queen Mary Coll., 1948-52; prof., head dept. astronomy U. Ill., Urbana, 1952—. Mem. staff Meteorol. Office, Air Ministry and Fgn. Office, 1939-45; mem. Brit. Meteorol. Research Com., 1948-52. Decorated officer Order Brit. Empire, 1946. Fellow Royal Soc. Edinburgh, Royal Astron. Soc.; mem. Am. Astron. Soc. (sec. 1961—), Internat. Astron. Union (vice president 1964-67, president 1967—. Author: Cosmological Theory, 1937; General Relativity and Cosmology, 1956, 2d edit., 1965; Fact and Theory in Cosmology, 1961. Editor: Problems of Extragalactic Research, 1962; joint editor Obs., 1938-48, Quar. Jour. Mechanics and Applied Math., 1946-52. Contbns. to analysis of gen. relativity to astronomy, especially structure of universe, theory of gravitational collapse and nature of quasars. Office: U. Ill. Obs., Urbana, Ill. 61801.*

McVOY, Kirk Warren, Am. physicist; b. Mpls., Feb. 22, 1928; s. Kirk Warren and Phyllis (Farmer) McV.; B.A., Carleton Coll., 1950; B.A., Oxford U., Eng., 1952; Dipl., U. Gottingen, Germany, 1953; Ph.D., Cornell U., 1956; m. Hilda A. Van Der Laan, Aug. 15, 1953; children—Christopher, Lawrence, Annelies. Research asso. Brookhaven Nat. Lab., Upton, N.Y., 1956-58; asst. prof. Brandeis U., 1958-62; asso. prof. U. Wis., 1963-67, prof., 1967—. Mem. Am. Phys. Soc. Research, publs. on nuclear reaction theory. Address: Dept. Physics, U. Wis. Madison, Wis. 53706.*

McWEENEY, Edmond J., Irish physician; b. Dublin, Ireland, Mar. 9, 1864; s. Theophilus and Margaret (Kendellen) McW.; ed. colls. St. Bertin, St. Omer (both France), Catholic U. Dublin; B.A., 1884; M.A., 1885; M.B., 1887; M.D., 1891; M.Ch., M.A.O., Royal U. Ireland; diploma in pub. health Conjoint Royal Colls. Ireland; m. Emilie Brazil, 1891. Prof. pathology and bacteriology U. Coll. Dublin; work pathology, bacteriology under Kolisko, Vienna, Austria, 1888, under C. Fraenkel and von Esmarch, Berlin, Germany, 1889; pathologist Mater Misericordiae Hosp., Dublin, from 1889; bacteriologist Irish Local Govt. Bd., from 1900. Fellow Royal Coll. Physicians Ireland; mem. Royal Acad. Medicine Ireland (pres. path. sect.). Translator, co-author: Chemical Methods of Clinical Diagnosis (Tappeiner). Research, publs. on human and vegetable pathology, bacteriology. Died June 20, 1925.

McWEENY, R(oy), Brit. physicist, theoretical chemist; b. Bradford, Eng., May 19, 1924; s. Maurice and Vera (Myers) McW.; B.Sc. with 1st class honors in Physics, U. Leeds (Eng.), 1945; D.Phil., Oxford (Eng.) U., 1948; m. Patricia Mary Healey, Dec. 6, 1947; children—Cherry Jane, Bruce Marlowe. Lectr. phys. chemistry King's Coll., Durham (Eng.) U., 1948-57; lectr. theoretical chemistry U. Coll. N. Staffordshire, 1957-60; asso. dir. Quantum Chemistry Inst., Uppsala (Sweden) U., 1960-61; faculty U. Keele (Eng.), 1961-66, prof. theoretical physics and theoretical chemistry, 1964-66; prof. theoretical chemistry U. Sheffield (Eng.), 1966—, dir. quantum theory research group, also computing lab., 1962—. Mem. Phys. Soc. and Inst. Physics (London), Chem. Soc., Am. Phys. Soc. Author: Symmetry and Introduction to Group Theory and its Applications, 1963. Research, publs. on quantum theory of electronic structure of atoms, molecules and crystals as means of understanding and predicting phys. and chem. properties; discoveries in math. methods used in quantum theory. Office: Chemistry Dept., Univ., Sheffield 10, Eng.*

MEAD, Albert Raymond, Am. zoologist; b. San Jose, Cal., July 17, 1915; s. Lester Albert and Jennie (Fiske) M.; B.S., U. Cal. at Berkeley, 1938; postgrad. U. Cal. at Davis; Ph.D., Cornell U., 1942; m. Eleanor Comfort Morrow, Feb. 8, 1942; children—Ruth Evelyn, James Irving. Instr., Army Coll., Gold Coast, West Africa, 1944-45; research fellow U. Cal. at Berkeley, 1946; faculty U. Ariz., Tucson, 1946—, prof. zoology, 1952-56, head zoology dept., 1956—. Research asso. Pacific Sci. Bd., Nat. Acad. Sci., Harvard 1948, Pacific Islands, 1949; investigator NSF, Ceylon, 1953-54, Hawaii, 1963—, NIH, 1956-63; cons. NRS-Nat. Acad. Sci., 1963—. Fellow A.A.-A.S. (past pres. Southwestern and Rocky Mountain div., mem. nat. council 1963—); mem. Am. Inst. Biol. Scis., Am. Malacological Union (nat. pres. 1963), Am. Soc. Tropical Medicine and Hygiene, Am. Soc. Zoologists, Malacological Soc. London (Eng.), Soc. for Study Evolution, Soc. for Systematic Zoology, Sigma Xi, Alpha Zeta, Beta Beta Beta, Phi Kappa Phi, Phi Sigma. Author: The Great African Snail, A Problem in Economic Malacology, 1961; also articles. Research on econs., dispersal, biology, control and diseases noxious terrestrial snails and slugs; identified discipline econ. malacology. Home: 401 Sierra Vista Dr., Tucson 85719.*

MEAD, George Herbert, Am. social psychologist; b. S. Hadley, Mass., Feb. 27, 1863; s. Rev. Hiram and Elizabeth Storrs (Billings) Mead; A.B., Oberlin Coll., 1883; pvt. tutor, 1883-87, A.B., Harvard, 1888; univs. Leipzig and Berlin, 1888-91; m. Helen Kingsbury Castle, Oct. 1, 1891 (dec. 1929); 1 son, Henry C. A. Instr. philosophy, 1891-93, asst. prof., 1893-94, U. of Mich.; asst. prof. philosophy, 1894-1902, asso. prof., 1902-07, prof., 1907-31, U. of Chicago. Mem. Am. Psychol. Assn.; Am. Philos. Assn. Author: The Philosophy of the Present, 1932; Mind, Self and Society, 1934; Movements of Thought in the Nineteenth Century, 1936; The Philosophy of the Act, 1938; The Social Philosophy of George Herbert Mead, 1956. Developer of a pragmatic social-behaviorism; tried to establish sci. basis for social psychology; stressed theory that individual mind and soc. cannot be separated. Died Chgo., Apr. 26, 1931.

MEAD, James Franklyn, Am. biochemist; b. Evanston, Ill., Oct. 24, 1916; s. James E. and Maxine (Cole) M.; A.B., Princeton, 1938; Ph.D., Cal. Inst. Tech., 1942; m. Marilyn Denney MacLennan, May 22, 1942; children—Victoria (Mrs. Robert A. Thomas), Robin Christine, James Maclennan. Research asst. Cal. Inst. Tech., 1942-45; asst. prof. chemistry Occidental Coll., 1945-47; research coordinator Office Naval Research, 1947-48; research biochemist, div. chief atomic energy project Lab. Nuclear Medicine and Radiation Biology, U. Cal., Los Angeles, 1951—, asso. prof., 1954-57, prof. depts. biol. chemistry and biophysics and nuclear medicine, 1957—. Recipient Career Research award NIH, 1963—. Mem. Am. Soc. Biol. Chemistry, Am. Chem. Soc., Am. Oil Chem. Soc., Soc. Exptl. Biol. Medicine. Author: (with D. R. Howton) Radioisotope Studies of Fatty Acid Metabolism, 1960; also chpts. in other books, numerous articles. Research on column chromatography lipids, metabolism essential fatty acids, metabolism brain fatty acids. Home: 1210 Las Lomas Pl., Pacific Palisades, Cal. Office: Dept. Biophysics and Nuclear Medicine, U. Cal., Los Angeles 90024.*

MEAD, Margaret, Am. anthropologist; b. Phila., Pa., Dec. 16, 1901; d. Edward Sherwood and Emily (Fogg) Mead; student De Pauw U., 1919-20; B.A., Barnard College, 1923; M.A., Columbia University,

1924, Ph.D., 1929; 1 daughter, Catherine Bateson. Fellow Nat. Research Council, in Samoa, 1925-26; assistant curator ethnology Am. Mus. Natural Hist., 1926-42, asso. curator, 1942-64, curator ethnology, 1964—; exec. sec., com. on food habits, Nat. Research Council, 1942-45; fellow Social Sci. Research Council, in Admiralty Islands, 1928-29; expdn. to New Guinea, 1931-33, 38, 53, 64, 65; expeditions to Bali, 1936-38 and 1957-58; director of Columbia University. Research in Contemporary Cultures, 1948-50; adj. professor anthropology Columbia, 1954—; vis. professor dept. psychiatry U. Cin., 1957—, Menninger Foundation, 1959—. Member American Anthropol. Assn. (pres. 1960), Am. Ethnol. Soc. N.Y. Acad. Scis., A.A.A.S. American Association University Women, American Orthopsychiatry Assn., Soc. Applied Anthropology, Society of Women Geographers, Institute for Intercultural Studies, World Federation for Mental Health (pres. 1956-57), Phi Beta Kappa. Author: Coming of Age in Samoa, 1928; An Inquiry into the Question of Cultural Stability in Polynesia, 1928; Growing up in New Guinea, 1930; The Changing Culture of an Indian Tribe, 1932; Sex and Temperament in Three Primitive Societies, 1935; And Keep Your Powder Dry, 1942; Balinese Character: A Photographic Analysis (with Gregory Bateson), 1942; Male and Female, 1949, Soviet Attitudes Toward Authority, 1951; Growth and Culture: A Photographic Study of Balinese Childhood (with Frances Macgregor, 1951; Themes in a French Culture (with Rhoda Metraux), 1954; An Anthropologist at Work, 1959; Family (with Ken Heyman), 1965; Anthropologists and What They Do, 1965. Editor: Cooperation and Competition Among Primitive Peoples, 1937; Primitive Heritage (with Nicholas Calas), 1953; The Study of Culture at a Distance (with Rhoda Metraux), 1953; Childhood in Contemporary Cultures (with Martha Wolfenstein), 1955; New Lives for Old 1956; People and Places, 1959; (with Ruth Bunzel) The Golden Age of American Anthropology, 1960; Continuities in Cultural Evolution, 1964; American Women, 1966. Research on personality and culture; studies of child development; application of psychoanalytic theory, learning theory, ethnology, and cybernetics in studies of seven Oceanic cultures; primary research in the native languages and applications to the fields of national character, mental health, and education. Home: 211 Central Park W., N.Y.C. 10024. Office: Am. Mus. Natural History, N.Y.C.*

MEAD, Richard, English physician; b. Stepney, Eng., Aug. 11, 1673; s. Matthew Mead; student under Graevius, Utrecht, 1689-92, under Paul Herman and Archibald Pitcairne, Leyden, 1692 (both Netherlands); M.D., U. Padua (Italy), 1695; M.D., U. Oxford (Eng.), 1707; m. Ruth Marsh, July, 1699; 8 children; m. 2d, Anne Alston, Aug. 14, 1724. Travelled in Italy, 1695; began med. practice, Stepney, 1696; physician St. Thomas' Hosp., 1703-15; anatomy lectr. Coll. Barbers and Surgeons, 1711-15; physician to Sir Isaac Newton, Bishop Burnett, George 1, Sir Robert Walpole; Harveian orator, 1723. Fellow Coll. Physicians, Royal Soc., 1703 (v.p. 1717). Author: Mechanical Account of Poisons, 1702; Short Discourse Concerning Pestilential Contagion, 1720; Monita et praecepta medica, 1751. Important contbr. to introduction of inoculation for smallpox; discovered itch-mite, 1703; wrote 1st book of epidemiological advice produced by med. practitioner at request of state. Died London, Feb. 16, 1754.

MEAD, Warren Judson, Am. geologist; b. Plymouth, Wis., Aug. 5, 1883; B.S., U. Wis., 1906, M.A., 1908, Ph.D., 1926; m. 1909; 3 children. From asst. to prof. geology U. Wis., 1906-34; prof., head dept. geology Mass. Inst. Tech., 1935-49. Cons. mining and engring. geology, 1906—; with OSRD, U. S. Bur. Reclamation, 1944. Mem. Nat. Acad. Scis., Geol. Soc. Am., Soc. Econ. Geologists (pres. 1942), Am. Inst. Mining, Metall. and Petroleum Engrs., Research on percentages of various stone in sediments, structural, econ. and engring. geology. Address: 88 Rutledge Rd., Belmont, Mass.

MEADE, J(ames) E(dward), Brit. economist; b. Swanage, Dorset, U.K., June 23, 1907; s. Charles Hippisley and Kathleen (Cotton-Stapleton) M.; B.A., Oriel Coll., Oxford U., 1930, M.A., 1933; B.A., Trinity Coll., Cambridge U., 1931, M.A., 1957; Hon. Dr., U. Basel (Switzerland), 1961, Hull (Eng.) U., 1965; m. Elizabeth Margaret Wilson, Mar. 14, 1933; children—Thomas Wilson, Charlotte Elizabeth (Mrs. Gordon Lewis), Bridget Ariane (Mrs. Edward Dommen), Carol Margaret. Fellow, lectr. Hertford Coll., Oxford U., 1930-37, bursar, 1934-37; mem. econ. sect. League Nations, Geneva, Switzerland, 1937-40; mem. econ. sect. U.K. Cabinet Secretariat, 1940-45, dir., 1945-47; prof. commerce London Sch. Econs. 1947-57; prof. polit. economy Cambridge U., 1957—, fellow Christ's Coll., 1957—. Trustee Urwick, Orr & Partners Ltd. Decorated Companion of Bath. Fellow Brit. Acad.; mem. Royal Econ. Soc. (past pres.), Eugenics Soc. (treas. 1963-67), Brit. Assn. for Advancement Sci. (past sect. pres.), Am. Acad. Arts and Scis. (fgn. hon.). Author numerous books including: The Balance of Payments, 1951; A Geometry of International Trade, 1952; Problems of Economic Union, 1953; The Theory of Customs Unions, 1955; Trade and Welfare, 1955; The Control of Inflation, 1958; A Neo-Classical Theory of Economic Growth, 1960; Efficiency, Equality and the Ownership of Property, 1964; The Stationary Economy, 1965; also numerous articles. Research on devel. of application of econ. theory and analysis to problems of econ. policy. Home: Low Brooms, Little Shelford, Cambridge, U.K.*

MEADOW, Jacob R(obert), Am. chemist; born Shaw, Miss., Dec. 11, 1903; s. Jacob Kinchen and Letitia Fowlkes (Box) M.; A.B., Ark. Coll., 1925; A.M., U. Ark., 1927; Ph.D., Johns Hopkins, 1933; sr. fellowship microchemistry, N.Y. U., 1929-30; m. Margaret Cobb, Aug. 1, 1929; children—Margaret Joan, Barbara Fowlkes. Instr. chemistry U. Ark., 1925-26, 1926-27, summers 1928, 29, 31; asst. prof. chemistry, Ark. Coll., 1927-28, asso. prof., 1928-29, prof. and head chem. dept., 1930-31; research chemist, expt. sta., E. I. du Pont de Nemours Co., Wilmington, Del., 1933-35; prof. chemistry and head dept., Southwestern-at-Memphis, 1935-42, 1944-45; group and sect. leader, synthetic hydrocarbons, aviation fuels, and desulfurization processes, research and devel. div. Socony-Vacuum Oil Co., Paulsboro, N.J., 1942-44; asso. prof. chemistry U. Ky., 1945-46; prof. indsl. chemistry and dir. freshmen instrn. 1946—; asst. dean coll. arts and scis., 1958—; chem. cons., 1935—; vis. prof. Georgetown U. Med. Sch., 1952-53; prof. chemistry U. Indonesia, Bandung, auspices ICA, 1956-57; acad. visitor Imperial Coll. Sci. and Tech., London, also Eidgenossische Technishe Hochschule, Zurich, 1964-65. Mem. Am. Chem. Soc. (chmn. Lexington 1948-49, Phi Beta Kappa, Sigma Xi. Contbr. articles to profl. jours. Patentee on Di-Tertiary Alkyl Sulfides, Alkylation process, and Simultaneous Alkylation and Desulfurization process, others in field medicinal chemistry; patent applications in petroleum field; co-author patents in field of chemotherapy Home: 510 McCubbing Dr., Lexington, Ky. 40503.*

MEANS, James Howard, Am. physician; b. Dorchester, Mass., June 24, 1885; s. James and Helen Goodell (Farnsworth) M.; prep. edn., Noble and Greenough's Sch., Boston; spl. student in biology and chemistry, Mass. Inst. Tech., 1902-03; A.B., Harvard, 1907, M.D., 1911, H. P. Walcott fellow 1913-16; m. Marian Jeffries, Jan. 11, 1915 (dec. Feb. 1950); 1 son, James; m. 2d Carol Lord Butler, Feb. 17, 1951. Teaching fellow medicine Harvard Med. Sch., 1916-18, instr. in medicine, 1919-21, asst. prof. medicine, 1921-24, Jackson prof. clin. medicine, 1924-51, Jackson prof. clin. medicine emeritus 1951—; asso. in medicine, Mass. Gen. Hosp., 1917-24, chief of med. services, 1924-51; physician Mass. Inst. Tech., 1951-57; cons. social medicine Moniefiore Hosp., N.Y.C. Mem. nat. adv. health council USPHS, 1952-56; hon. mem. faculty U. Cuyo (Argentina); hon. physician Mass. Gen. Hosp.; hon. perpetual student Med. Coll. St. Bartholomew's Hosp., London. Recipient Sidney Hillman Award, 1951, Squibb award, Endocrine Soc., 1952; George M. Kober medal Assn. Am. Physicians, 1964. Fellow Royal Soc. Medicine (London) (hon.), A.C.P. (pres. 1937-38); fellow Am. Acad. Arts and Scis., A.A.A.S., Med. Soc. Finland (hon.), Am. Soc. Clin. Investigation, Assn. Am. Physicians (pres. 1942), Am. Assn. for the Study of Goiter (pres. 1947-48), Alpha Omega Alpha; hon. mem. Argentine Assn. for Endocrinology and Nutrition; corr. mem. Nat. Acad. Medicine of Buenos Aires, Gorgas Meml. Inst., Tropical and Preventive Medicine. Author: Dyspnoea, 1924; The Diagnosis and Treatment of Diseases of the Thyroid Gland (with E. P. Richardson, M.D.), 1929; The Thyroid and Its Diseases, 1937, 2d ed. 1948; Doctors, People and Government, 1953; Lectures on Thyroid, 1954; Ward 4, 1958; The Association of American Physicians: Its First Seventy-Five Years, 1961; James Means and His Problem of Manflight, 1964. Contbr. papers to jours. Died Sept. 3, 1967.

MEARS, Eleanor Cowie Loudon, Brit. gynecologist; b. Cleland, Scotland, Dec. 9, 1917; d. William and Helen (Robertson) Loudon; M.B., Ch.B., Edinburgh (Scotland) U., 1940; m. Kenneth Patrick Geddes Mears, Apr. 3, 1940 (div.); children—Elspeth, Roger, Monica. Practice gen. medicine, London, Eng., 1941-46, specializing in gynecology, Christchurch, New Zealand, 1946-56, London, 1956—; med. sec. Council for Investigation of Fertility Control, London, also Family Planning Assn., U.K., part-time 1958-65; clin. adviser Family Planning Assn. U.K., 1965—. Lectr. on fertility and psychosexual problems Westminster Med. Sch., 1958—, also colls. in U.K.; cons. to pharm. cos., med. jours. Fellow Eugenics Soc.; mem. Soc. for Endocrinology, Soc. for Study Fertility, Council for Investigation of Fertility Control. Author: (with Mears, Guttmacher) Babies—By Choice or By Chance, 1960; Marriage—A Continuing Relationship, 1960; also numerous articles. Editor: Handbook of Oral Contraception, 1965. Research on fertility, procedures for evaluation of new products and tablet formulations, large scale trials of systemic agts. for fertility control. Address: 2, Kent Terrace, Park Rd., London N.W. 1, Eng.*

MEARS, Robert Bruce, Am. electrochemist; b. Scranton, Pa., Jan. 28, 1907; s. Joseph A. and Elizabeth (Milnes) M.; B.S. in Electrochem. Engring., Pa. State U., 1928; Ph.D. in Metallurgy, Cambridge (Eng.) U., 1935; m. Margaret Hart, July 4, 1929; children—Diana Elizabeth (Mrs. Robert E. Marquis), Dana Christopher. Mem. tech. staff Bell Telephone Labs., N.Y.C., 1928-32; Carnegie scholar Brit. Iron and Steel Inst., 1933-34; chief chem. metallurgy div. Alcoa Research Lab., 1935-46; dir. Applied Research Lab., U. S. Steel Corp., Pitts., 1946-60, asst. v.p. applied research, 1960-64, v.p. new product devel., 1964—. Mem. Nat. Assn. Corrosion Engrs. (pres. 1950, Whitney award 1949), Electrochem. Soc. (v.p. 1951). Research on electrochem. mechanisms of cathodic protection, inhibition, passivity and stress corrosion cracking. Home: 628 California Av., Oakmont, Pa. Office: 525 William Penn Pl., Pitts. 15230.*

MÉCHAIN, Pierre-François-André, astronomer, hydrographer; b. Laon, France, Aug. 16, 1744; hydrographer marine depot. Fellow Royal Soc., 1789; mem. French Acad. Scis., 1782, Bur. Longitudes (dir., 1798). Editor, Connaissance, 1784-94. Measured (with Delambre) meridian arc from Dunkirk to Barcelona, Spain, as basis for establishing metric system, 1791; discovered eleven comets and calculated their orbits; observed eclipses of sun, moon, stars, planets. Died Castellon de la Plana, Spain, Sept. 28, 1804.

MECHANIC, David, Am. sociologist; b. N.Y.C., Feb. 21, 1936; s. Louis and Tillie (Penn) M.; B.A., magna cum laude, Coll. City N.Y., 1956; M.A., Stanford, 1957, Ph.D., 1959; m. Margaret Newton, July 26, 1960; children—Robert Edmund, Michael Alexander. USPHS fellow U. N.C., 1959-60; faculty U. Wis., Madison, 1960—, prof. sociology, dir. grad. tng. in med. sociology and mental health, 1965—, chmn. dept. sociology, 1968—. Ednl. cons. Sociometry, 1963-65; cons. Assn. Am. Med. Colls., govt. agys. USPHS fellow Brit. Med. Research Council Inst. Psychiatry Maudsley Hosp., London, Eng., 1965-66. Mem. Am. Sociol. Assn., Am. Assn. U. Profs., A.A.-A.S., Phi Beta Kappa. Author: Students Under Stress, 1962; (with others) Social Science of Organizations, 1963; Medical Sociology, 1968. Asso. editor Jour. Health and Social Behavior. Studies, publs. on behavior in illness, factors influencing recognition and alleviation of symptoms, social factors in med. and psychiat. rehab., comparative med. systems. Home: 1821 Vilas Av., Madison, Wis. 53711.*

MECHNIKOV, Ilia Ilich, see Metchikoff, Elie.

MECKEL, Johann Friedrich, German anatomist; b. Wetzlar, Germany, July 31, 1724; prof. anatomy and surgery, Berlin, Germany. Mem. French Acad. Scis., 1752. Author: Tractatus de quinto pare nervorum cerebri, 1748; De ganglio secondi rami quinti paris nervorum nuper detecto, 1749. Discovered sphenopalatine ganglion on 5th cranial nerve (Meckel's ganglion). Died Berlin, Sept. 18, 1774.

MECKEL, Johann Friedrich, German anatomist; b. Halle, Germany, Oct. 17, 1781. Prof. physiology, anatomy U. Halle. Corr. mem. French Acad. Scis., 1829. Author: Abhandlung aus der vergleichenden und menschlichen Anatomie, 1805; Beiträge zur vergleichenden Anatomie, 2 vols., 1808-11; Handbuch der menschlichen Anatomie, 4 vols., 1815-20; Tabulae anatomicae, 1817-26; System der vergleichenden Anatomie, 5 vols., 1821-33. Translated into German: Comparative Anatomy (Curier), 4 vols., 1809-10. Work in comparative anatomy; discovered diverticullum of small intestine (named for him), 1809; described cartilage of mandibular arch (Meckel's cartilage), 1805. Died Halle, Oct. 31, 1833.

MEDAWAR, Peter Brian, biol. scientist; b. Rio de Janeiro, Brazil, Feb. 28, 1915; s. Nicholas and Muriel (Dowling) M.; M.A., Oxford (Eng.) U., 1939, D.Sc., 1945; hon. doctorates Cambridge (Eng.), U. Birmingham (Eng.), U. Liège (Belgium), U. Brazil, U. Brussels (Belgium), U. Alta. (Can.), U. Hull (Eng.), U. Glasgow (Scotland); m. Jean Taylor, Feb. 27, 1937; children—Caroline (Mrs. Peter Loizos), Charles, Louise (Mrs. Leonard Skerker), Alexander. Lectr., Oxford U., 1938-47; fellow Magdalen Coll., Oxford, 1938-47; head zoology dept. Birmingham U., 1947-51; head zoology dept. U. Coll., London, Eng., 1951-62; dir. Nat. Inst. for Med. Research, London, 1962—; prof. at large, Cornell U., Ithaca, N.Y., 1966. Recipient Nobel prize in medicine and physiology (with Burnet), 1960. Fellow Royal Soc., 1949 (Royal medal 1959), Royal Coll. Surgeons, Royal Canadian Coll. Physicians and Surgeons; fgn. fellow Nat. Acad. Sci., Am. Philos. Soc., Am. Acad. Arts and Scis., N.Y. Acad. Scis., A.C.P. Author: The Uniqueness of the Individual, 1956; The Future of Man, 1960; The Art of the Soluble, 1967; also articles. Research on growth and aging, reactions to tissue transplantation especially immunity reactions prohibiting transplantation between individuals; discovered acquired immunological tolerance. Home: Mt. Vernon House, Holly Hill, London N.W.3., Office: Nat. Inst. for Med. Research, London N.W.7., Eng.*

MEDCOF, John Carl, Canadian biologist; b. Ruthven, Ont., Can., Jan. 7, 1911; s. John Dowker and Mabel (Crick) M.; B.A., U. Toronto (Ont.), 1934; M.A., U. Western Ont., 1936; Ph.D., U. Ill., 1938; m. Bessie Wren, July 3, 1942; children—Sue M., John Wren, Carl Ranby. With Fisheries Research Bd. Can., 1938—, with Biol. Sta., St. Andrew, N.B., 1943—, prin. scientist off-shore investigations, 1963-65, asst. dir., 1965-67, sr. molluscan shellfish investigator, 1967—. Biologist attached to Canadian Colombo Plan Fisheries Project, Colombo, Ceylon, 1953-55; sec. Canadian Govt.'s Interdepartmental Shellfish Com., 1948-52, 56-60, chmn., 1961-63; sci. adviser Canadian Commn. to Internat. Commn. for N.W. Atlantic Fisheries, 1964-65. Recipient Jean Balmer prize in biology U. Toronto, 1931. Mem. Atlantic Fisheries Biologists (past pres.), Nat. Shellfisheries Assn. (asso. editor Proc. 1965—), Profl. Inst. Pub. Service Can., Am. Inst. Fisheries Research Biologists. Research, publs. on comml. fisheries; described and explained fluctuations in populations of fishes in terms of natural factors and indsl. practices; devised better methods of harvesting and culturing to produce more fish of better quality; research on san. aspects of shellfish, especially paralytic shellfish poisoning. Home: 327 Water St. Office: Fisheries Research Bd. Biol. Sta., St. Andrews, N.B., Can.*

MEDEA, Eugenio, Italian neurologist; b. Varesia, Italy, Oct. 4, 1873; s. Tranquillo and Luigia (de Vincenti) M.; ed. in neurology and psychiatry; m. Bianca Pisani, May 23, 1903; 1 child., Alba. Prof. neurology and psychiatry U. Milan (Italy), prof. nervous semiology. Recipient Gold medal Commune of Milan, Gold medal Province of Milan; Order of Italian Crown. Mem. Consiglio Consulenza Lega (pres.), Inst. Mental Hygiene, Lombard Inst. Sci. and Letters, Lombard Criminology Soc. (pres.), Pro infanze anormale (pres.). Research and publs. in criminology neurology, psychiatry. Address: via Guastalla 3, Milan, Italy.

MEDEARIS, Donald Norman, Jr., Am. pediatrician, microbiologist; b. Kansas City, Kan., Aug. 22, 1927; s. Donald N. and Gladys (Sandford) M.; A.B., U. Kan., 1949; M.D., Harvard, 1953; m. Mary Marble, Aug. 25, 1956; children—Donald H., Ellen S., John N. Research fellow pediatrics Harvard Med. Sch., research div. infectious diseases Children's Med. Center, Boston, 1956-58; faculty Johns Hopkins, Balt., 1958-65, asso. prof. pediatrics and microbiology, 1963-65; prof. pediatrics, chmn. dept. U. Pitts. Sch. Medicine, med. dir. Children's Hosp. Pitts., 1965—. Mem. Am. Assn. Immunologists, Soc. for Pediatric Research, Infectious Disease Soc. Am., Soc. for Exptl. Biology and Medicine, Am. Pediatric Soc., Am. Acad. Pediatrics. Research on pathogenesis of infections in immature animals. Home: 506 S. Linden Av., Pitts. 15208. Office: 125 DeSoto St., Pitts. 15213.*

MEDER, Albert Eugene, Jr., Am. mathematician; b. N.Y.C., Mar. 19, 1903; s. Albert Eugene and Anna (Sommer) M.; A.B., Columbia, 1922, A.M., 1923; LL.D., Fairleigh Dickinson U., 1956; L.H.D., Bloomfield Coll., 1961; m. Janet B. Davis, Oct. 14, 1956. Faculty, Columbia, N.Y.C., 1922-26; faculty Rutgers U., New Brunswick, N.J., 1926—, dean univ., prof. math., 1948—. Exec. dir. Commn. on Math., Coll. Entrance Exam. Bd., 1957-58. Mem. Am. Math. Soc. (treas. 1949-64), Assn. Symbolic Logic. Research on founds. of math. and math. edn. Home: 508 Salter Pl., Westfield, N.J. 07090.*

MEDICUS, Heinrich Adolf, physicist; b. Zurich, Switzerland, Dec. 24, 1918; s. Fritz G. and Clara (Frey) M.; dipl.naturwiss., Swiss Fed. Inst. Tech., 1943, D.Sc., 1949; m. Hildegrard Schmelz, June 25, 1961. Research asso. Swiss Fed. Inst. Tech., 1943-50; vis. scientist Radiation Lab., U. Cal., at Berkeley, 1950-51; with Mass. Inst. Tech., 1951-55; asso. prof. physics Rensselaer Poly. Inst., Troy, N.Y., 1955—. Fellow Swiss Council for Fellowships in Math. and Physics, 1950-52. Mem. Am. Phys. Soc., Swiss Phys. Soc., Am. Assn. U. Profs. Author: (with Francis Bitter), Fields and Particles, 1968; also articles. Research on radioactivity, meson physics, nuclear structure, nuclear reactions, especially photonuclear reactions. Home: East Acres, Troy, N.Y. 12180.

MEDIN, Oscar, Swedish pediatrician; b. Axberg, Sweden, 1847; prof., Stockholm, Sweden. Author: En epidemi af infantil paralysi, 1890. First to point out epidemic character of acute anterior poliomyelitis, 1890. Died Stockholm, Dec. 24, 1927.

MEDLER, John Thomas, Am. entomologist; b. Las Cruces, N.M., May 28, 1914; s. Edward Lewis and Lillian (Thomas) M.; B.S., N.M. State Coll., 1936, M.S., 1937; Ph.D., U. Minn., 1940; m. Priscilla Alden Hobbs, June 17, 1939; children—Meredith, Michael, Marcia, Marcus, Melvin; m. 2d, Jane Ann Kurtenacker Curtis, Dec. 11, 1964. Faculty, U. Wis., 1946—, now prof. John Simon Guggenheim Meml. Found. fellow, 1942. Mem. Entomol. Soc. Am., Washington, Kan., Canadian entomol. socs., Wis. Acad. Sci., Arts and Letters. Research, numerous publs. on taxonomy and ecology of Heteroptera and Homoptera, especially Aphidae and Cicadellidae, biology and physiology of plant sucking insects, including nature of salivary toxins, insect ecology with emphasis on bumble bees and other pollinators of legumes, control of legume insects. Home: 4102 Yuma Dr., Madison, Wis. 53711.*

MEDLICOTT, Henry Benedict, Brit. geologist; b. Galway, Ireland, Aug. 3, 1829; s. Samuel and Charlotte (Dolphin) M.; ed. Trinity Coll., Dublin, Ireland; m. Louisa Maunsell. Apptd. to Geol. Survey of Ireland, 1851, Eng., 1853, India, 1854; became prof. geology Thomason Coll., Roorkee, 1854; named dir. Geol. Survey of India, 1876; ret., 1887. Recipient Wollaston medal, 1888. Fellow Royal Soc., 1877. Author: A Manual of the Geology of India; (pamphlets) Agnosticism and Faith, 1888, Evolution of Mind in Man, 1892. Died Apr. 6, 1905.

MEDUNA, Ladislas Joseph, physician, psychiatrist, b. Budapest, Hungary, Mar. 27, 1896; s. Francis M. and Gisela (Eissler) M.; M.D., Royal U. of Sci., Budapest, Hungary, 1921; m. Clara Varga, 1934. Asst. prof. Budapest Interacad. Inst. Brain Research, 1924-27; asso. prof. Univ. Clinic for Mental and Nervous Diseases, 1927-33; asso. prof. Royal Univ. Clinic; dir. neuro-histology lab., Leopold Field State Hosp. for Insane, Budapest, 1933-34, head dept. male services, 1934-38; head dept. male services, Nagel Field State Hosp., 1938-39; asso. prof. psychiatry and neurol. Loyola U. of Chicago, 1939-43; asso. prof. psychiatry, U. of Ill., Neuropsychiatric Inst., Chgo., 1943-50, prof., 1950—; cons. psychiatry state Department of Public Welfare, Illinois, 1954—; honorary staff mem. Ridgeway Hospital, Chgo. Recipient Guggenheim Foundation fellowship, 1955. Honorary member Royal Medico-Psychol. Association (London), Sociedad Medicina de Pernambuco, Sociedad Brazileira, Associacao Paulista Roman Acad. Medicine, Sociedad Cubana de Neurol. y Psiquiatria, Electro-Shock Research Assn., Blue Key; mem. Chgo. Neurol. Soc., American Medieval Acad., Ill., Chgo. med. socs., Milwaukee Neuropsychiatric Society, American Psychiatric Assn., N.Y. Acad. Sci., A.M.A., American Society Med. Psychiatry (past pres.), Soc. Biol. Psychiatry (past pres.), A.A.-A.S. Sigma Xi. Die Entwicklung der Zilberdruese in Saeuglingalter, 1925; Histopathology of Microglia, 1927; On Histopathology of Epilepsy, 1932; Convulsive Treatment of Schizophrenia, 1938. Editor-in-chief: Jour. Neuropsychiatry. Discovered Metrazol convulsive treatment, application of carbon dioxide therapy to neuroses; established presence of anti-insulinic factor in blood of schizophrenics. Home: 537 W. Arlington Pl., Chgo. 14. Office: 8 S. Michigan, Chgo. 60603.

MEDVEDEV, Sergey Sergeevich, Russian chemist; b. May 17, 1891; grad. Moscow U., 1919. With Karpov Physicochem. Inst., USSR Acad. Sci., 1922—, now head lab. polymerization processes; instr. Moscow Inst. Fine Chem. Tech., 1922—. Mem. Soviet delegation 3d Internat. Conf. on High-Molecular Compounds, London, 1958. Decorated Order of Lenin; recipient Stalin prize, 1946. Mem. USSR Acad. Sci. Author: Research on the Laws of Emulsion Polymerization, 1963; co-author: Reactions of Oxygen-Containing Radicals of Type RO, 1959; Mechanism of Initiation of Cationic Polymerization in the Presence of Metal Halides, 1959; Effect of Side Chains on Rate of Oxidation of Carbon Chain Polymers, 1959; Kinetics of the Emulsion Polymerization of Styrene, 1960; Crossing Polymer Chains with Gamma-Radiation, 1960. Research and publs. on high-molecular compounds, polymerization processes used in prodn. of synthetic rubber, plastics, effects of nuclear radiation on polymerization processes. Address: Karpov Physicochemical Inst., Leninsky prospect 31, Moscow, USSR.

MEDVEDEVA, Nina Borisovna, Russian pathophysiologist; b. Saratov, 1899; grad. Med. Faculty, Saratov U., 1921; D.Med. Sci. Asst. dept. gen. pathology Saratov U., 1921-25; asst., lectr. dept. path. physiology 2d Moscow Med. U., 1925-31; sr. asso. Inst. Exptl. Biology and Pathology, Ukrainian Acad. Sci., 1931-52, dept. head Bogomolets Inst. Physiology, 1953—; prof., 1932—. Decorated Order of Lenin. Mem. Ukrainian Acad. Sci. (corr.). Mem. editorial bd. Med. Jour. Research and numerous publs. on path. physiology, effect of tiredness on tissue metabolism under normal conditions and with disturbance of functional state of some organs, blood transfusion, autocatalytic function of lungs, liver, lymph nodes, spleen, changes in tissue proteins in old age; established limit of permissible muscle loading in liver and kidney deficiency; proved experimentally that antireticular cytotoxic serum affects change in tissues' nitrogen and protein composition and has rejuvenating effect on aged; isolated cortical hormone from adrenal cortex; suggested that presence of desoxydative carbonuria be regarded as early symptom in diagnosis of cancer; compiler bibliography on path. physiology and endocrinology. Address: Bogomolets Inst. Physiology, Ukrainian Acad. Sci., Vladimirskaya 55, Kiev, USSR.

MEECHAM, William Correll, Am. physicist; b. Detroit, June 17, 1925; s. William Edward and Mabel (Wilcox) M.; B.S., U. Mich., 1948, M.S., 1948, Ph.D., 1954; m. Della F. Carson, Sept. 11, 1964; children—Janice Lynn, William James. Teaching fellow U. Mich., Ann Arbor, 1948-53, research asst., 1951-53, asso. research physicist, 1954-56, faculty, 1956-60, asst. prof., dept. physics, 1957-60, research physicist, head fluid and solid mechanics lab., 1959-60; prof. dept. aero. and engring. mechanics U. Minn., Mpls., 1960—. research asst. Brown U., Providence, 1953-54; research asso. U. Cal. at La Jolla, 1963. Cons. U. Mich., Inst. Sci. and Tech., 1960—, Space Tech. Labs., Los Angeles, 1959-63, Rand Corp., Los Angeles, 1964—. Fellow Acoustical Soc. Am.; mem. Am. Phys. Soc., Am. Phys. Soc. Research, publs. on reflection radiation from rough surfaces, generation sound by turbulent air and water, theory turbulence. Home: 1732 Lydia St., St. Paul 13. Office: Dept. Aero. and Engring. Mechanics, U. Minn., Mpls. 55455.*

MEEHL, Paul Everett, Am. psychologist; b. Mpls., Jan. 3, 1920; s. Otto John and Blanche (Duncan) M.; B.A. summa cum laude, U. Minn., 1941, Ph.D., 1945; m. Alyce Roworth, Sept. 6, 1941; children—Karen, Erik. Faculty, U. Minn., 1944—, prof. clin. psychology, 1952—, chmn. dept., 1951-57; individual practice psychotherapy, Mpls., 1951—. Recipient Distinguished Sci. Contbr. award Am. Psychol. Assn., 1958. Author: Clinical Versus Statistical Prediction, 1954. Home: 1544 E. River Terrace, Mpls. 55414.*

MEEK, Alexander, Brit. zoologist; b. Boughty Ferry, Apr. 7, 1865; ed. Royal Coll. Scis., U. St. Andrews (Scotland), U. Freiburg (Germany); B.Sc., U. St. Andrews, 1889; M.Sc., U. Durham (Eng.), 1896, hon.D.Sc., 1921. Worked at St. Andrews Marine Lab., 1889-91; lectr. agrl. zoology Aberdeen County, 1891-94; mem. faculty Armstrong Coll. (now King's Coll.), U. Durham, from 1894, then prof. emeritus; dir. Dove Marine Lab., Cullercoats, Northumberland, 1897-1932. Fellow Zool. Soc. Author: Migrations of Fish, 1916; Fishes of Northumberland; Fishes of Durham (in Victoria County Histories); History of Fisheries (in Northumberland County History) Vol. 11; Essentials of Zoology; The Progress of life; also numerous papers on growth, embryology, gen. and marine zoology. Editor Reports of Dove Marine Lab. to 1932. Died Nov. 2, 1949.

MEEK, Fielding Bradford, Am. paleontologist; b. Madison, Ind., Dec. 10, 1817. Asst. to David Dale Owen, head of U. S. Geol. Survey of Ia., Wis. and Minn., 1848-49; asst. to James Hall, paleontologist, Albany, N.Y., 1852-58; accompanied F. V. Hayden on geol. expdn. to Dakota, 1853; residence in Smithsonian Instn., Washington, 1858-76. Author: Check List of the Invertebrate Fossils of North America, 1864; (with Hayden) Paleontology of the Upper Missouri, 1865; Report on the Invertebrate Cretaceous and Tertiary Fossils of the Upper Missouri Country, 1876; other publs. Identified and defined new invertebrate fossil species; contbd. to advancement of paleontology in U. S. Died Washington, Dec. 21, 1876.

MEEK, John Millar, English physicist; b. Wallasey, Eng., Dec. 21, 1912; s. Alexander and Edith (Montgomery) M.; D. Engring., U. Liverpool (Eng.); m. Marjorie Inglesby, July 18, 1942; children—Rosalind, Sara. With research dept. Asso. Elec. Industries, Ltd., 1934-38, 40-46; faculty, physics dept. U. Cal., 1938-40; prof. tech. electricity U. Liverpool, 1946—. Fellow Inst. Physics; mem. Inst. Elec. Engrs. Author: The Mechanism of the Electric Spark; Electrical Breakdown of Gases; High Voltage Laboratory Techniques. Home: 190 Meols Parade, Hoylake, Eng.

MEEK, John Sawyers, Am. chemist; b. Madison, Wis., Aug. 12, 1918; s. Walter Joseph and Crescence (Eberle) M.; B.A., U. Wis., 1941; M.S., U. Ill., 1944, Ph.D., 1945; m. Mary Margaret McPherson, Feb. 15, 1945; children—David Sawyers, Crescence Jeanette. Research asst. U. Wis., 1943; teaching asst. U. Ill., 1941-45; faculty chemistry U. Colo., Boulder, 1945—, prof., 1959—. Mem. Am. Chem. Soc., Sigma Xi, Alpha Chi Sigma. Research, publs. in elucidation of positional and stereo isomerism in Diels-Alder reactions, preparations and reactions of bridghead compounds, synthesis of fibrinolytic and antiviral compounds, amino acids and their antagonists. Home: 1911 Columbine St., Boulder, Colo. 80302.*

MEEKS, Marion Littleton, Am. physicist; b. Gainesville, Ga., Oct. 1, 1923; s. Jesse L. and Ione (Tumlin) M.; B.S., Ga. Inst. Tech., 1943, M.S., 1947; Ph.D., Duke, 1950; m. Bennie Stone, May 25, 1944; children—Marshall Stone, Fleming Littleton, Marion Littleton. Asst. prof., asso. prof. physics dept. Ga. Inst. Tech., 1950-59; research asso. Harvard Coll. Obs., 1959-60; staff mem. Mass. Inst. Tech. Lincoln Lab., Lexington, 1961—. Mem. Am. Astron. Soc., Internat. Astron. Union, Am. Geophys. Union, Union Radio Sci. Internat. Proposed method of sounding temperature in earth's atmosphere by satellite observations of 5 mm wave length oxygen; co-discoverer of OH in interstellar medium; contbr. to theory of OH emission in radio astronomy.*

MEEM, James Lawrence, Jr., Am. nuclear scientist; b. N.Y.C., Dec. 24, 1915; s. James Lawrence and Phyllis (Deaderick) M.; B.S., Va. Mil. Inst., 1939; M.S., Ind. U., 1947, Ph.D., 1949; m. Buena Vista Speake, Sept. 5, 1940; children—James, John. Aero. research scientist NACA, 1940-46; dir. bulk shielding reactor Oak Ridge Nat. Lab., 1950-53; head nuclear operation Aircraft Reactor Expt., 1954-55; chief reactor scientist Alco Products, Inc., Schenectady, 1955-57; chmn. dept. nuclear engring., dir. reactor facility U. Va., Charlottesville 1957—. Cons. in nuclear safety to govt. agys., pvt. cos. Mem. Am. Nuclear Soc., Am. Phys. Soc., Am. Soc. for Engring. Edn., Va. Acad. Sci. Author: Two Group Reactor Theory, 1964. Research with nuclear reactors. Home: Mt. Airy, Route 2, Charlottesville, Va. 22901.*

MEEN, V(ictor) B(en), Canadian mineralogist; b. Toronto, Ont., Can., July 1, 1910; s. Benjamin and Mary Gertrude (Tidy) M.; B.A., U. Toronto, 1932, M.A., 1933, Ph.D. in Mineralogy, 1936; postgrad. U. Minn., 1937, U. S. Nat. Mus., 1938; m. Thelma Irene Stables, Mar. 27, 1937; children—Heather Elizabeth (Mrs. David Crampton), Beverley Victoria (Mrs. Peter Casson), Sharon Patricia. Asst. provincial assayer Dept. Mines, Ont., 1942-43, acting provincial assayer, 1944; asst. dept. mineralogy and petrography U. Toronto, 1932-36, lectr., 1936-44, asst. prof. mineralogy, dept. geol. scis., 1944-51, asso. prof. geology, 1951-56, prof., 1956-57, spl. lectr., 1957-59; with Royal Ont. Mus. and predecessor instns., U. Toronto, 1936-—, head div. geology and mineralogy, 1955-59, head earth scis. div., 1959-64, chief mineralogist mus., 1964-—. Canadian rep. Commn. on Meteorites, Internat. Geol. Congress, 1956-64, Commn. on Museums, Internat. Mineral. Assn., 1959-—, Commn. on Meteorites, Internat. Union Geol. Scis., 1964-—. Fellow Mineral. Soc. Am., Geol. Soc. Am., Geol. Assn. Can.; mem. Mineral. Assn. Can., Mineral. Soc. Gt. Britain and Ireland, Canadian Inst. Mining and Metallurgy, Royal Canadian Inst. (pres. Toronto br. 1949-50, mem. council 1964), NRC of Can. (asso. com. on meteorites 1960-—), Royal Astron. Soc. Can., Gemmological Assn. Can. (hon. dir.), Royal Ont. Mus. (hon. life). Author: (with D. H. Gorman) Mineral Occurrences of Wilberforce, Bancroft and Craigmont—Lake Clear Areas, Southeastern Ontario, 1953; Quetico Geology, 1959; (with J. A. Mandarino) Introductory Gemmology, 1961; Gem Hunting in Burma, 1963; (with A. D. Tushingham) The Crown Jewels of Iran, 1968; also articles. Pioneered in investigation of meteor craters in exploring Chubb Crater and Merewether Crater, Eastern Canadian Arctic; investigated occurrence of gemstones (rubies, sapphires, jade) in Burma, Cambodia, Thailand, Ceylon. Home: 34 Birchview Blvd., Toronto 18. Office: 100 Queen's Park, Toronto 5, Ont., Can.*

MEERSON, Felix Z., Russian physiologist; b. Moscow, USSR, Aug. 5, 1926; s. Zalman and Mina (Asarkh) M.; grad. 1st Med. Inst., 1949, med. doctor's degree, 1958; m. Shokhova Lya, Jan. 8, 1954; children—Helen, Natalie. Sci. worker, dept. pathophysiology Sci. Research Inst. of Phys. Methods of Treatment, Yalta, until 1955; sci. worker Inst. Higher Nervous Activity of USSR Acad. Scis., Moscow, 1955-56; reader, chair clin. and exptl. physiology Central Inst. Improvement of Physicians, Moscow, 1957-60; sci. worker Lab. of Physiology and Pathology of Myocardium, Inst. Normal and Pathol. Physiology, Acad. Med. Scis., USSR, Moscow, 1960-63, chief Lab. Exptl. Cardiology, 1963-—. Sci. sec., editorial dept. pathophysiology Gt. Med. Ency. Mem. All-Union Soc. Cardiologists (mem. mng. com.), also other sci. socs. of physiologists and pathophysiologists. Author: Compensatory hyperfunction of the heart and cardiac insufficiency, 1960; Problems of pathophysiology and cardiac insufficiency, 1962; On the interrelation of the physiological function and genetic apparatus of the cell, 1963; Myocardium in hyperfunction, hypertrophy and cardiac insufficiency, 1965; (with V. V. Parin) Contribution to clinical physiology of circulation, 1960, 2d edit., 1965; also numerous articles. Research regarding physiology of heart in compensatory enlargement due to prolonged stress; attempts to relate genetic apparatus of heart cells to physiol. changes found. Home: 80, 6, Volokolamskoye shosse. Office: 8, Baltiskaya ul., Moscow A-315, USSR.*

MEES, C(harles) E(dward) Kenneth, photographer; b. May 26, 1882; s. Charles Edward and Ellen (Jordan) M.; ed. Harrogate Coll., St. Dunstan's Coll., Catford; B.Sc., Univ. Coll., London, Eng., 1903, D.Sc., 1906; D.Sc. (hon.), U. Rochester (N.Y.), Alfred (N.Y.) U., 1950; m. Alice Crisp, 1909; one son, one dau. Mng. dir. Wratten and Wainwright, Ltd., Croydon, Eng., mfrs. photog. plates, 1906-12; joined Eastman Kodak Co., 1912, organized research lab., 1912, established dept. manufacture and supply synthetic organic chems. for research, 1918, retired, 1955; dir Kodak Ltd., London, 1948-59; fellow Univ. Coll., London, 1950. Recipient Silver medal Royal Soc. Arts, 1908, 34, John Scott medal, Phila., 1921, Adelsköld medal Swedish Photog. Soc., Franklin medal, 1954. Hon. fellow Royal Photog. Soc. (Progress medal 1913, 53, Hurter and Driffield medal 1924), Photog. Soc. Am. (Progress medal 1948). Fellow Royal Soc., 1939, Chem. Soc., Royal Astron. Soc, hon. mem. Soc. Franc. de Phot. (Janssen medal 1923), Franklin Inst.; fellow Soc. Motion Picture and TV Engrs., A.A.A.S., Am. Philos. Soc., Nat. Acad. Scis., 1950, Am. Acad. Arts and Scis., 1950 (Rumford medal 1943); mem. Am. Astron. Soc. (patron 1950), Med. Soc. Photographers Eng. Author: (with S. E. Sheppard), 1907; Organization Industrial Scientific Research, 1920, rev. edit. (with J. A. Leermakers), 1950; Photography, 1936; The Theory of the Photographic Process, 1942; The Path of Science, 1946; From Dry Plates to Ektachrome Film (posthumous), 1961; also numerous sci. papers on photography and related subjects; made available comml. panchromatic and process panchromatic plates for photography, 1906-07; developed light filters for photography. Died Aug. 1960.

MEEUSE, Bastiaan Jacob Dirk, plant physiologist; b. Sukabumi, Indonesia, May 9, 1916; s. Adriaan Dirkszoon and Jannigje (Kruithof) M.; B.Sc., U. Leiden (Netherlands), 1936, doctorandus, 1939; D.Tech. Sci., U. Delft (Netherlands), 1943; m. Johanne Jacoba ten Have, Aug. 28, 1942; children—Karen Barbara, Peter Nicholas. Came to U. S., 1952, naturalized, 1962. Tchr., Hort. Sch., Boskoop, Netherlands, 1939-42; with Lab. voor Technische Botanie, U. Delft, 1942-52, lector, 1949-52; faculty U. Wash., Seattle, 1952-—, prof. plant physiology, 1960-—. Rockefeller Found. fellow, 1947-49; NSF Sr. fellow, 1962-63. Mem. Royal Dutch Acad. Scis. (corr.), A.A.A.S., Am. Inst. Biol. Scis., Am., Royal Dutch bot. socs., Soc. for Exptl. Botany (Eng.), Am., Internat. phycological socs., Am. Soc. Protozoology, Western Soc. Naturalists, Am., Japanese socs. plant physiology. Author: Oriënterende onderzoekingen over de vorming van rietsuiker uit zetmeel in planten by lage temperatuur, 1943; The Story of Pollination, 1961; also articles. Research in plant physiology and biochemistry especially plant respiration and carbohydrates, also phycology, protozoology, animal behavior, flower pollination. Home: 10442 N.E. 124th St., Kirkland, Wash. 98033. Office: Botany Dept., Johnson Hall, U. Wash., Seattle 98105.*

MEFFERD, Roy Balfour, Jr., Am. med. researcher; b. Hico, Tex., Sept. 22, 1920; s. Roy B. and Delfa (Russell) M.; Asso. Sci., Tarleton State Coll., 1938; B.S., Tex. A. and M. U., 1940, M.S., 1940; Ph.D., U. Tex., 1951; m. Mary Louise Key, Aug. 25, 1940; children—Marsha Ellen (Mrs. Don L. Lambert), Roy Scott. Tchr. vocational agr., Rockwood, Tex., 1940, Carleton, Tex., 1941; soils technologist U. S. Bur. Reclamation, Holbrook, Ariz., 1941-42; indsl. crops specialist nat. resources sect. G.H.Q., SCAP, Tokyo, Japan, 1946-47; research scientist gene research-bacteriology U. Tex., 1948-51; dir. metabolism lab. S.W. Found. for Research and Edn., San Antonio, also prof. biology and chemistry Trinity U., San Antonio, 1951-56; research scientist V, Clayton Found. Biochemistry Inst., U. Tex., 1956-58; dir. psychiat. and psychosomatic research lab. Houston VA Hosp., also asso. prof. physiology and biochemistry Baylor U. Coll. Medicine, Houston, 1958-—. Cons. to pvt. cos. Rosalie B. Hite fellow in cancer research, 1949-51; Damon Runyon fellow in cancer research, 1952-53. Fellow A.A.A.S.; mem. Am. Physiol. Soc., Soc. Psychophysiol. Research, Am. Chem. Soc., Sigma Xi. Contbg. author: The Physiological Effects of High Altitude, 1964; Simposio Internacional de Aclimatacion al Frio, 1964; Multivariate Methods in Psychology, 1965; Automation in der Analytischen Chemie, 1965; also numerous articles. Research on radiation biology and mutagenesis, study interacting functional process at biol.-psychol. interface in humans under normal and stressful conditions. Home: 4418 Waynesboro St., Houston 77035. Office: Houston VA Hosp., 2002 Holcombe St., Houston 77020.*

MEGGERS, William Frederick, Am. physicist; b. Clintonville, Wis., July 13, 1888; s. John and Bertha (Bork) M.; B.A., Ripon Coll., 1910, D.Sc., 1951; M.A., U. Wis., 1916; Ph.D., Johns Hopkins, 1917; m. Edith Marie Raddant, July 13, 1920; children—Betty Jane (Mrs. Clifford Evans), William Frederick, John Charles. Lab. asst. Ripon Coll., 1910-11, U. Wis., 1911-12; instr. physics Carnegie Inst. Tech., 1912-14; lab. asst. Nat. Bur. Standards, Washington, 1914-16, asst. physicist, 1916-19, physicist, 1919-—, chief spectroscopy sect., 1920-58. physicist Welch Sci. Co., Chgo., 1960-—. Cons. AEC; chmn. com. on line spectra NRC, 1946-61; pres. Internat. Joint Commn. for Spectroscopy, 1952-58; pres. Rydberg Centennial Conf., 1954, Internat. Astron. Union Commn. on Standard Wavelengths and Spectral Tables, 1935-52. Recipient medals U. S. Assay Commn., 1928, 32; Alumni citation Ripon Coll., 1947; Gold medal U. S. Dept. Commerce, 1949; medal U. Liege, 1950; Elliot Cresson Gold medal Franklin Inst., 1953; award Pitts. Spectroscopy Soc., 1963. Mem. Nat. Acad. Scis., Am. Phys. Soc., Am. Astron. Soc. (medal 1952), Optical Soc. Am. (Ives medal 1947, Mees medal 1964, pres. 1950-51), Soc. for Applied Spectroscopy, Washington Acad. Scis., Philos. Soc. Washington, Am. Inst. Physics, Phi Beta Kappa, Sigma Xi. Author: Index to Literature on Spectrochemical Analysis, 1920-60; Tables of Wavenumbers, 1960; Tables of Spectral Line Intensities, 1961; also numerous articles; contbr. article on spectroscopy to Ency. Brit. Home: 2904 Brandywine St., Washington 20008. Office: Nat. Bur. Standards, Washington 20025.*

MEGHNAD, Saha, Indian physicist; b. 1893; s. Jagannath Saha; ed. Dacca Coll., Calcutta Presidency Coll.; D.Sc., U. Calcutta; D.Sc. (hon.), univs. Allahabad and Lucknow; m. 1917; three sons, four daus. Prof. physics Allahabad U., 1923-38; Palit prof. physics, 1938-55; dean sci. faculty U. Calcutta, 1951-56, pres. postgrad. council sci., 1947-49; dir. Inst. Nuclear Physics. Mem. Univ. Commn. Govt. India, 1948-49; fellow, founder-pres. Acad. Scis. United Provinces Agra and Oudhi, India; organizing sec. Nat. Inst. Sci., India, 1935, pres., 1937; gen. pres. 21st Indian Sci. Congress, 1934; mem. nat. planning com. Indian Nat. Congress; mem. Govt. India Council Sci. and Indsl. Research. Mem. of Parliament, Govt. India, 1952. Carnegie scholar of British Empire, 1936. Fellow Royal Soc., 1927, Royal Asiastic Soc. Bengal; mem. Indian Assn. Cultivation Sci., Indian Phys. Soc. (pres. 1937); hon. fellow Am. Acad. Arts and Scis.; life mem. Astron. Soc. France. Author: Theory of Thermal Ionisation of Gases and Theory of Selective Radiation Pressure, also many papers; six lectures on atomic physics, a treatise on heat and thermodynamics; Outlines of Modern Knowledge on Atoms, Molecules and Nuclei; founder sci. jour. Sci. and Culture. Died Feb. 1956.

MEGLIN, Jean Antoine, French physician; b. Sultz, Alsace, France, 1756; corr. Athenaeum Medicine of Paris. Author: Recherches et observations sur la neuralgie faciale, 1816; Notice historique sur l'état ancien de la ville de Sultz, 1817; Mémoire sur l'usage des bains dans le tétanos, 1822. Described Meglin's point of emergence of anterior palatine nerve from greater palatine foramen; analyzed waters of Sultzmatt, outside Alsace, 1779; invented pills for facial neuralgia, 1816. Died Colmar, France, Mar. 13, 1824.

MEHEDINTI, Simion, Rumanian geographer; b. 1869; prof. geography (1st in Rumania) Bucharest U.; mem. Rumanian Acad. Author: Terra—Introduction to Geography as a Science (an original system of geography, with materialistic elements), 2 vols., 1931. Died 1962.

MEHL, Robert Franklin, Am. metallurgist; b. Lancaster, Pa., Mar. 30, 1898; s. George H. and Sarah (Ward) M.; B.S., Franklin and Marshall Coll., 1919, Sc.D., 1938; Ph.D., Princeton, 1924; Doutor Honoris Causa, Universidade de Sao Paulo (Brazil), 1944; Eng.D., Stevens Inst. Tech., 1944, Colo. Sch. Mines, 1952, Case Inst. Tech., 1959; Sc.D., U. Pa., 1959; m. Helen Charles, Dec. 27, 1923; children—Robert Franklin, Marjorie (Mrs. Clarence B. Nixon), Gretchen (Mrs. Robert A. Deans). Proctor fellow in chemistry Princeton, 1922-23; head dept. chemistry Juniata Coll., 1923-25; NRC fellow Harvard, 1925-27; supt. div. phys. metallurgy Naval Research Lab., 1927-31; asst. dir. research labs. Am. Rolling Mill Co., 1931-32; faculty Carnegie Inst. Tech., Pitts., 1932-65, dir. metals research lab., head dept. metallurgy, 1935-59, dean grad. studies, 1953-60, on leave as cons. in sci. and tech. to U. S. Steel Corp., Zurich, Switzerland, 1960-65; prof. U. Del. Coll. Engring., Newark, 1966-67; prof. U. Syracuse (N.Y.), 1967-—. Mem. adv. com. Nat. Bur. Standards; attache U. S. Embassy, London, Eng., 1945; chmn. materials adv. bd. Nat. Acad. Scis.-NRC, 1950; cons. Inst. Tech. Research, Sao Paulo, 1949-53; U. S. State Dept. Point Four Program, Brazil. Recipient certificate for best sci. papers Am. Inst. Mining and Metall. Engrs., 1934, 39, 43, 44, James Douglas Gold medal, 1945; John Scott medal City of Phila., 1934; Howe medal Am. Soc. Metals, 1939, Sauveur achievement medal, 1951, Gold medal, 1952; medal Am. Indsl. Radium and X-ray Soc., 1943; medal Associacao Brasileira de Matais, 1944; Clamer medal Franklin Inst., 1953; Le Chatelier medal Société Francaise de Metallurgie, 1956; Platinum medal Inst. Metals Great Britain, 1960. Translator: The States of Aggregation (Gustav Tammann), 1925. Author: Metallurgy of Iron and Steel, 1944; Brief History of the Science of Metals, 1947; also numerous research papers. Home: 222 Melwood Av., Pitts.*

MEHLER, Alan Haskell, Am. biochemist; b. St. Louis, May 24, 1922; s. Louis A. and Estelle (Caplan) M.; A.B., Washington U., St. Louis, 1942, postgrad. Med. Sch., 1942-44; postgrad. Columbia, 1944-45; Ph.D. in Biochemistry, N.Y. U., 1948; m. Anne Feinstein, Sept. 19, 1943; children—Louise N., Edith C., Ronald W., David E. Chemist, Fleishmann Labs., 1945; mem. Weizmann Inst. Sci., Rehovoth, Israel, 1948; asst. prof. Northwestern U., 1949; research asso. U. Chgo., 1949-51; vis. scientist NIH, Bethesda, Md., 1951-52, chemist, 1952-65, chief sect. on enzyme chemistry Nat. Inst. Dental Research, 1960-65; prof., chmn. dept. biochemistry Marquette U. Sch. Medicine, Milw., 1965-—. Mem. Am. Soc. Biol. Chemistry, Am. Chem. Soc., A.A.A.S., Biochem. Soc. (Eng.), Soc. Gen. Physiologist, Soc. for Exptl. Biology and Medicine. NSF Sr. Postdoctoral fellow NSF, U. Paris, 1958-59. Author: Advances in Enzymology, 1957; also articles. Identification and characterization of enzymes, including malic enzyme which fixes carbon dioxide, tryptophan pyrrolase, important in enzyme induction in mammals, picolinic carboxylase which is controlled by hormones which control niacin biosynthesis, enzymes of histidine metabolism, amino acid activating enzymes; structure and mechanism of action of enzymes. Home: 2423 E. Beverly Rd., Milw. 53211.*

MEHROTRA, Brahma Swarup, Indian microbiologist; b. Agra, India, Oct. 10, 1926; s. Shri A. Swarup and Smt. B. Devi (Mehra) M.; B.Sc., U. Allahabad, 1945, M.Sc., 1947, Ph.D., 1949; m. Raj Rani Mehrotra, May 19, 1950; children—Rita, Ajai, Renu, Rekha, Vinay Vijai. Lectr. botany U. Allahabad, India, 1949-—, reader in botany, 1966-—. Recipient Empress Victoria Readership award, 1948. Fellow Nat. Acad. Scis., Indian Phytopath. Soc.; mem. Brit. Mycol. Soc., Mycol. Soc. Am. Research, numerous publs. on soil fungi especially with reference to taxonomical and nutritional studies of parasitic and saprophytic molds; microbiol. studies of industrially important molds. Home: 23 C.Y. Chintamani Rd., Allahabad-2, India. Office: Botany Dept., U. Allahabad, India.*

MEHROTRA, Ram C., Indian chemist; b. Kanpur, India, Feb. 16, 1922; s. Ram B. and C. (Tandon) M.; B.Sc., Allahabad U., 1941, M.Sc., 1943, D.Phil., 1948; Brit. council scholar Birkbeck Coll., London, 1950-52; Ph.D., London U., 1952, D.Sc., 1964; m. Suman Khanna, Dec. 11, 1944; children—Rashmi, Puyush, Shalini. Lectr., Allahabad U., 1944-54; reader Lucknow U., 1954-58; prof., dean Gorakhpur U., 1958-62; dean Faculty Sci., Rajasthan U., Jaipur, India, 1962-65, prof. math., 1962-—, chief rector, 1965-—. Pres. chem. sect. Indian Sci. Congress, 1966-67. Recipient E. G. Hill meml. prize, 1949. Fellow Indian Chem. Soc. (mem. council 1964-—), Chem.

Soc. London (Eng.), Nat. Acad. Scis., Nat. Inst. Scis., Royal Inst. Chemistry. Author: several text books. Editor: Jour. Indian Chem. Soc., 1965—. Research, publs. on synthesized and identified aluminum soaps; synthesized complex polymetaphosphates; extensive contbns. on alkoxide chemistry and organometallic chemistry. Home: P 4, University Campus, U. Rajasthan, Jaipur, India.*

MEHTA, Bansidhar Vithaldas, Indian soil scientist; b. Baroda, India, Aug. 13, 1913; s. Vithaldas Chimanlal and Maniben (Vithaldas) M.; student Baroda Coll., 1929-35; B.A., Coll. Agr., Anand, India, B.Sc., 1933, M.Sc., 1935, Ph.D., 1952; m. Bhanumati Bansidhar, May 10, 1933; children—Vijay, Jay, Usha (Mrs. Kishore Sanghvi), Nina, Digvijay. Lectr. chemistry Kalbhavan Tech. Inst., Baroda, 1935-47; faculty B.A. Coll. Agr., Anand, India, 1947—, prof. agrl. chemistry and soil sci., 1956—. Mem. bd. studies in agr. Gujarat U., Ahmedabad, India, 1950-55; chmn. bd. studies in agr. Sardar Patel U., Anand, 1956—; mem. agr. chemistry and soil sci. com. Indian Council Agr. Research, New Delhi, India, 1958—. Mem. Indian Soc. Agronomy, Soil Sci. Soc. India (v.p. 1966—), Soil Conservation Soc. India. Research, numerous publs. on micronutrient deficiency symptoms of crops, nutrient requirement of crops, availability of maj. and micronutrients in soils, quality of tobacco, characteristics of tobacco soils. Address: Inst. Agr., Anand, Kaira Dist., India.*

MEIBOM, Heinrich, German physician, anatomist; b. Lübeck, June 29, 1638; s. Johann Heinrich Meibom. Prof. medicine U. Helmstedt, 1664, prof. history and poetry, from 1678; also taught philosophy, philology, geometry, antiquities at Groningen, Leyden. Author: Exercitatio de incubatione . . . , 1659; Dissertatione de longaevis, 1664; De vasis palpebrarum, 1666; Pathologicae dissertationes . . . , 1669; De medicorum historia . . . , 1669; Exercitatio anatomico-medica . . . , 1682; History of Germany (in Latin), 1700, also numerous sci. papers. First to describe oscular tarsal glands (known as Meibom's glands), 1966. Died Heimstedt, Mar. 26, 1700.

MEIENHOFER, Johannes (Arnold), chemist; b. Dresden, Germany, Mar. 3, 1929; s. Emil August and Katharine (Gabriel) M.; diplom-chemiker U. Heidelberg, Germany, 1954, Dr. Rer. Nat., 1956; m. Katharina Bredol, Aug. 10, 1963; 1 son, Johannes. Teaching asst. Inst. Organic Chemistry, U. Heidelberg, 1957; research asso. biochemistry Cornell U. Med. Coll., N.Y.C., 1957-59, hormone research lab. U. Cal., Berkeley, 1959-60, Deutsches Wollforschungsinstitut, Technische Hochschule, Aachen, Germany, 1961-64, Farbenfabriken Bayer Ag. Wuppertal-Elberfeld, Germany, 1961-64; research asso. Children's Cancer Research Found., Boston, 1965—, Harvard Med. Sch., Boston, 1966—. Mem. Gesellschaft Deutscher Chemiker, Gesellschaft fur Biologische Chemie, Am. Chem. Soc., A.A.A.S. Research, publs. on crosslinking of proteins with dinitrofluorobenzene applied to insulin; synthesis of lysine vasopressin and analogs; synthesis of n-terminal nonadekapeptide of ACTH; synthesis of insulin; cancer research in peptide and protein field. Home: 59 Summit St., Newton, Mass. 02158. Office: 35 Binney St., Boston 02115.*

MEIER, Mark Frederick, Am. glaciologist; b. Iowa City, Dec. 19, 1925; s. Norman Charles and Clea (Grimes) M.; B.S., State U. Ia., 1949, M.S., 1951; postgrad. U. Innsbruck (Austria); Ph.D., Cal. Inst. Tech., 1957; m. Barbara June McKinley, Sept. 16, 1955; children—Lauren Gale, Mark Stephen, Gretchen Ann. Leader mechanics crevasse formation expdn. N.W. Greenland, Occidental Coll. and U. S. Army, 1955; geologist U. S. Geol. Survey, Tacoma, 1956—; vis. prof. dept. geology Dartmouth, 1964; research prof. geophysics U. Wash., 1964—. Vice pres. Internat. Commn. Snow and Ice, 1963—; mem. U. S. nat. com. Internat. Hydrological Decade, 1964—. Recipient Sigma Xi award for outstanding research at Cal. Inst. Tech., 1957. Fellow Am. Geophys. Union (mem. vis. scientist program 1964-65), Geol. Soc. Am., Arctic Inst. N.Am.; mem. Glaciological Soc. (v.p. 1966—), Geol. Soc. Am., A.A.A.S. Research, publs. on mechanics of glacier flow including new information on flow law of ice, first measurement of stress and strain rates on a flowing glacier, formation of geologic structures in ice, mass exchange at glacier surfaces, distbn. of glaciers, hydrology of glacier melt and stream runoff. Home: 7409 22d St. W., Tacoma 98466. Office: 1305 Tacoma Av. S., Tacoma 98402.*

MEIER, Richard Louis, Am. behavioral scientist; b. Kendallville, Ind., May 16, 1920; s. Walter A. and Mary (Lottman) M.; student No. Ill. State Tchrs. Coll., 1936-39; B.S., U. Ill., 1940; M.A., U. Cal. at Los Angeles, 1942, Ph.D., 1944; m. Gitta Unger, May 20, 1944; children—Karen, Andrea, Alan. With Cal. Research Corp., 1943-47; exec. sec. Fedn. Am. Scis., 1947-48; with Petrocarbon Ltd., 1949-50; Fulbright scholar Manchester U., Eng., 1949-50; asst. prof., program of edn. and research in planning U. Chgo., 1950-56; research social scientist Mental Health Research Inst., U. Mich., Ann Arbor, 1957—, asso. prof. conservation, 1960-65, prof., 1965-67; prof. environmental design U. Cal., Berkeley, 1967—. vis. lectr. Harvard, 1959-60; vis. prof. Grad. Sch. Ekistics, Athens, 1962, U. Cal., Berkeley, 1966. Indsl. planner P.R. Planning Bd., 1952; cons. on social planning and resources planning Joint Center for Urban Studies, Mass. Inst. Tech., Harvard, in Venezuela, 1963-65. Mem. Am. Chem. Soc., Am. Sociol. Assn., Am. Geog. Assn., Soc. for Gen. Systems Research, Regional Sci. Assn., Marine Tech. Soc., Fedn. Am. Scis., A.A.A.S., Soc. for Internat. Devel. Author: Science and Economic Development, 1956; Modern Science and the Human Fertility Problem, 1959; A Communications Theory of Urban Growth, 1962; Developmental Planning, 1965; also numerous articles. Research in acid-catalyzed polymerization of olefins, industrialization of photosynthesis, tech. systems for large scale urbanization, measurement of social and cultural growth, communications overload in instns., planning theory and technique, gaming simulation of concept of community, resource-conserving urban design. Home: 7 San Mateo Rd., Berkeley, Cal. 94707.*

MEIER-RUGE, William Alfred, pathologist; b. Rudolstadt, Germany, July 28, 1930; s. Arthur Robert and Herta (Krüger) M.-R.; German med. licensing examination U. Berlin, 1954, M.D. Diploma, 1957; m. Jutta Henrietta Ruge, May 28, 1955; children—Peer Michael, Cora Waltraut, Tilman Boris. Clin. asst. Potsdam-Babelsberg Gen. Hosp., 1954-56; research asst. pathology Pathol. Inst., U. Berlin, 1956-61, asst. forensic medicine Inst. Forensic Medicine 1958-59, research asso. pathologic chemistry and histochemistry Path. Inst., 1959-61; research asso. Path. Inst., U. Basel (Switzerland), 1963—, lectr., 1965—, head histochem. lab.,1965-67; research asso. dept. biol. and med. research div. Sandoz Ltd., Basle, 1962—, head lab. exptl. pathology and histochemistry, 1962—. Recipient Rudolf-Virchow Prize, 1960. Mem. Deutsche Gesellschaft für Pathologie, Internat. Acad. Pathology, Freie Vereinigung Schweizer Pathologen., Royal Soc. Medicine (London) Deutsche Gesellschaft für Sozialhygiene, Assoc. Suisse de Jeune Chercheurs pour le Development de la Récherche Scientifique, A.A.A.S., Swiss Soc. Natural Sci. Author: Medikamentöse Retinopathie, 1967; also articles. Research on biochemistry of iron and lipid metabolisms of human liver; histochemistry including enzyme histochemistry; pathology of drug side effects; drug induced damages on retina by psychopharmaceuticals and antimalarial drugs; enzymatic investigations in megacolon especial megacolon Hirschsprung; comparative cytology and enzyme histochem. investigation in cancer research. Home: 12 Oberwilerstrasse, Bottmingen, Switzerland CH 4103. Office: Sandoz Ltd., Lichtstrasse, Basel, Switzerland CH 4000.*

MEIGE, Henri, French physician; b. France, 1866. Author: Tics et spasmes cloniques de la face, 1893; Gigantisme et acromégalie, 1895; L'infantilisme, le féminisme, et les hermaphrodites, 1895. Authority on diseases characterized by enlarged bones, caused by malfunction of pituitary gland; described chronic hereditary trophedema of legs (Meige's or Milroy's disease), 1899. Died 1940.

MEIGS, Arthur Vincent, Am. physician; b. Phila., Pa., Nov. 1, 1850; s. John Forsyth and Ann Wilcocks (Ingersoll) M.; M.D., U. Pa., 1871; m. Mary R. Browning, Oct. 16, 1878. Physician Pa. Hosp. Pres. Coll. Physicians, Phila., 1904-06. Author: Milk Analysis and Infant Feeding, 1885; The Origin of Disease, 1899; A Study of the Human Blood-Vessels in Health and Disease, 1907. Modified cow's milk so that it was nearly as satisfactory for babies as their own mother's milk; described Meig's capillaries between muscle fibers of heart, 1891. Died Jan. 1, 1912.

MEIGS, Charles Delucena, physician; b. St. George, Bermuda, Feb. 19, 1792; s. Josiah and Clara (Benjamin) M.; grad. U. Ga., 1809, U. Pa. Sch. Medicine, 1817; m. Mary Montgomery, Mar. 15, 1815; 10 children. Came with family to New Haven, Conn., 1796; moved to Athens, Ga., 1801; practiced medicine, Augusta, Ga., 1814-17, Phila., 1817-61; prof. obstetrics and diseases of women Jefferson Med. Coll., Pa., 1841-61; ret., Hamanassett County, Pa., 1861-69. Author: Elementary Treatise on Midwifery, 1838; A Treatise on the Diseases and Special Hygiene of Females, 1845; Woman, Her Diseases and Remedies, 1847; Obstetrics, the Science and Art, 1849; Treatise on Acute and Chronic Diseases in the Neck of the Uterus, 1850; Childhood Fevers, 1854. Invented a ring pessary; advocated that forceps be applied to aftercoming head in childbirth; drew attention to embolism (cardiac thrombosis) as cause of sudden death in childbed fever, 1849. Died Phila., June 22, 1869.

MEIGS, James Aitken, Am. physician; b. Phila., July 31, 1829; s. John G. and Mary A. Meigs; grad. Jefferson Med. Coll., 1851; never married. Practiced medicine, Phila., 1851-54; prof. climatology and physiology Franklin Inst., 1854-62; physician of pulmonary diseases Howard Hosp. and Infirmary for Incurables, 1855-68; librarian Aca. Natural Scis. of Phila., 1856-59; prof. insts. of medicine, 1857-59; prof. Pa. Med. Coll., 1859-60; prof. medicine and med. jurisprudence Phila. County Med. Soc., 1868-79; trustee Poly. Coll. State of Pa., several years. Studied pulmonary diseases; in anthropology did research of cranial characteristics in man. Died Nov. 9, 1879.

MEIGS, Joe Vincent, Am. gynecologist, physician; b. Lowell, Mass., Oct. 24, 1892; s. Vincent and Sarah (Parker) M.; A.B., Princeton, 1915; M.D., Harvard, 1919; Hon. D.Sc., Northwestern, 1959, Lowell Tech., 1960; m. Elizabeth Wallace, Apr. 2, 1921; children—Wallace, Sarah, Elizabeth. Intern Mass. General Hosp., Boston, 1919-21; asst. resident, resident The Free Hosp. for Women, Brookline, Mass., 1921; cons. vis. surgeon Mass. Gen. Hosp., Boston; became chief gynecol. service Vincent Memorial Hosp., 1941, cons. vis. gynecologist, 1955; gynecologist Pondville Hosp., 1927—; instr. surgery Harvard Med. Sch., 1932-42, clin. prof. gynecology, 1942-59; Emeritus prof., 1959—. Fellow Royal Coll. Surgeons Edinboro (hon.), Royal Coll. Obstetricians and Gynecologists, American Academy of Arts and Sciences; member Swedish Obstetrical and Gynecological Society (hon.), Am. Acad. Arts and Scis., Am., So. surg. assns., Am. Assn. Obstetricians and Gynecologists, Am. Gynec. Soc. Author: Tumors of the Female Pelvic Organs, 1934; Progress in Gynecology (with Dr. Somers Sturgis), 1946, 57. Author and editor: The Surgical Treatment of Cancer of the Cervix, 1954. Specialist on uterine carcinoma; described syndrome asso. with ovarian fibroma (now called Meigs' syndrome), 1934. Died 1963.

MEIJER, Paul Herman Ernst, physicist; b. The Hague, Netherlands, Nov. 14, 1921; s. Herman Willem and Elisabet (Kossmann) M.; Ph.D., U. Leiden, 1951; m. Marianne Anita Schwarz, Feb. 17, 1949; children—Onko Frans, Miriam, Daniel, Mark, Corinne. Came to U. S., 1953, naturalized, 1959. Research asso. U. Leiden, 1952-53; vis. lectr. Case Inst. Tech., 1953-54; research asso. Duke, 1954-55; asst. prof. U. Del., 1955-56; asso. prof. Cath. U., Washington, 1956-60, prof., 1960—. Cons. Nat. Bur. Standards, Naval Ordnance Lab. Fulbright grantee, 1953-55; Guggenheim grantee, 1964-65. Mem. Am., Washington phys. socs., Phys. Soc. Netherlands, Sigma Chi. Author: (with E. Bauer) Group Theory, 1962. Editor: Group theory and Solid State Physics, 1964. Research, publs. primarily on statis. mechanics and solid state theory. Home: 1438 Geranium St., N.W., Washington 20012.*

MEIKLE, George Stanley, Am. chemist; b. Milton Mills, N.H., May 30, 1886; s. George Douglas and Emma Etta (Fox) M.; B.Engring., M. Civil Engring., Union Coll., Schenectady, 1913; m. Louise Juliet Zimmerman, Sept. 6, 1910. Chief safety engr., asst. dist. mech. engr. U. S. Steel Corp., 1909-11; sci. research Gen. Electric Co. Labs., Schenectady, 1912-17; pres. G. S. Meikle Co., cons. scientists and engrs., N.Y.C., 1919-24; research and engring. exec., 1924-28; mem. adminstrv. staff, dir. research relations with industry Purdue U., from 1928. bd. dirs. and research dir. (officer) Purdue Research Found., from 1930; v.p. Better Homes in Am., Inc.; v.p. Purdue Aeronautics Corp. In charge devel. submarine dir. for USN and model 1919 gas mask for U. S. Army (as civilian), World War I. Fellow A.A.A.S., Internat. Anesthesia Research Soc.; mem. Tippecanoe County Med. Assn. (hon.), Sigma Xi, Tau Beta Pi. Research in phys. chemistry, discovering hot cathode gas filled rectifiers Tungar, and allied devices; research in heat transfer resulting in new formula and discovery of methods and devices for heating houses with liquid and gaseous fuels. Died Mar. 30, 1960.

MEIKLEJOHN, Gordon, Am. physician; b. Providence, Apr. 8, 1911; s. Alexander and Nannine (La-Villa) M.; grad. Taft Sch., Watertown, Conn., 1927; student U. Wis., 1927-29, 30-32, Yenching U., Peking, China, 1929-30; M.D., C.M., McGill U., 1937; m. Greta Louise Hiltz, July 4, 1940; children—Robin, James, Nancy. Rockefeller Found. fellow U. Cal. Med. Sch., San Francisco, 1941-42, faculty, 1942—, asso. prof. medicine, 1951; research asso. Cal. Dept. Pub. Health, Berkeley, 1942-48, cons. in virology Viral and Rickettsial Disease Lab., 1948-51; prof. chmn. dept. medicine U. Colo. Sch. Medicine, Denver, 1951—. Cons. to govt.; mem. commn. influenza Armed Forces Epidemiological Bd., 1948—; com. vaccine devel. Nat. Insts. Allergy and Infectious Diseases, 1962—. Diplomate Am. Bd. Internal Medicine. Fellow A.C.P.; mem. Soc. Exptl. Biology and Medicine, Central (past pres.), Western socs. clin. research, Am. Soc. Tropical Medicine, Am. Assn. Immunologists, Am. Pub. Health Assn., A.M.A., Am. Soc. Clin. Investigation, Western (past pres.), Am. assns. physicians, Am. Assn. U. Profs., Arthritis and Rheumatism Found., Assn. Profs. Medicine, Am. Acad. Microbiology, Infectious Disease Soc., Sigma Xi, Alpha Omega Alpha. Home: 145 Ivanhoe St., Denver 80220.*

MEIMBERG, Paul, German agronomist; b. Düren, Germany, June 29, 1916; s. Alfred and Maria (Riicken) M.; ed. in agronomy Berlin, Munich, Göttingen (all Germany) univs.; Ph.D. agrégé; m. Inge Rolfes, May 8, 1948; children—Marianne, Margrit, Sibylle. Instr. agronomy U. Giessen; head study bur. for Zuckermarket, Bonn, Germany, from 1961; dir. Agronomy Inst., Justus Liebig U., Giessen. Author: Die Landbaugebiete Hessens; Die Landwirtschaftlich Betriebslehre in Westdeutschland; Entwicklung und Ergebnisse; Kostenrechnung in der Landwirschaft; Die Zuckerwirtschaft in Belgien, Frankreich, Italien und Niederlanden. Home: Am Riegelpfad 108, 63 Giessen. Office: Landgraf-Philipplatz 4, 63 Giessen, W. Germany.

MEINEL, Aden Baker, Am. astronomer; b. Pasadena, Cal., Nov. 25, 1922; s. John G. and Gertrude (Baker) M.; student Cal. Inst. Tech., 1941-42; A.B.,

U. Cal. at Berkeley, 1947, Ph.D., 1949; m. Marjorie Pettit, Sept. 5, 1944; children—Carolyn P., Walter B., Barbara M., Edward S., Elaine J., Mary L., David L. Research asst. Cal. Inst. Tech., 1942-44; research asso. U. Cal. at Berkeley, 1949-50, faculty, 1951-57, asso. prof., 1954-57, asso. dir. Yerkes and McDonald observatories, 1954-57; instr. U. Chgo., 1950; dir. Kitt Peak Nat. Obs., Tucson, 1958-61; prof. U. Ariz., Tucson, 1961-—, dir. Steward Obs., 1963-66, dir. Optical Sci. Center, 1966-—. Exec. sec. Nat. Obs. Commn., NSF, 1955-58; cons. sec. Air Force, 1963-—, Perkin-Elm Corp., 1955-—. Mem. Am. (past mem. council, Warner prize 1954), Royal astron. socs., A.A.A.S. (past div. pres.), Optical Soc. Am. (past dir., Lomb medal 1952), Am. Acad. Arts and Sci., Internat. Astron. Union. Research, publs. on optical instrument design, spectroscopy of airglow and aurora, accelerator induced spectra, stellar spectroscopy, space telescopes, high altitude volcanic dust, noctilucent clouds. Home: Route 2, Box 732B, Tucson 85715.*

MEINKE, W(illiam) Wayne, Am. chemist; b. Elyria, O., June 27, 1924; s. William Carl and Marian Ella (McRoberts) M.; A.B., Oberlin Coll., 1947; Ph.D., U. Cal. at Berkeley, 1950; m. Marilynn Hope Hayward, July 12, 1947; children—Sue Anne, David William. Faculty dept. chemistry U. Mich., Ann Arbor., 1950-63, prof. chemistry, 1962-63; chief analytical chemistry div. Nat. Bur. Standards, Washington, 1963-—, chief Office of Standard Reference Materials, 1964-—. Mem. subcom. physicochem. standards Nat. Acad. Scis.-NRC, 1963-—; mem. geochemistry panel Apollo Adv. Com. to NASA, 1963-—; cons. to pvt. cos., govt. agys. Fellow A.A.A.S.; mem. Am. Chem. Soc., Am. Phys. Soc., Am. Nuclear Soc. (exec. com. isotopes and radiation div. 1964-66), Internat. Union Pure and Applied Chemistry (mem. coms. 1965-—), Phi Beta Kappa, Sigma Xi, Gamma Alpha, Phi Lambda Upsilon, Alpha Chi Sigma. Co-editor Trace Characterization, Chemical and Physical; editor series monographs Radiochemistry of Elements, 1960-62. Research, numerous publs. on rapid chem. analysis by nuclear methods, fast radiochem. separations, trace element determinations; standard reference materials for measurement. Home: 8405 Peck Pl., Bethesda, Md. 20034. Office: Nat. Bur. Standards, Washington 20234.*

MEINKOTH, Norman August, Am. zoologist; b. New Baden, Ill., Jan. 29, 1913; s. Christian Herman and Elizabeth (Monken) M.; B.Ed., So. Ill. Normal U., 1938; M.S., U. Ill., 1944, Ph.D., 1947; m. Marian Catherine Richards, Dec. 26, 1938; 1 dau., Pantip. Biology tchr. Herrin Twp. High Sch., 1938-41; faculty Swarthmore (Pa.) Coll., 1947-—, asso. prof. zoology, 1951-66, prof. zoology, chmn. dept. biology, 1966-—. Fulbright lectr. biology Chulalongkorn U., Bangkok, Thailand, 1957-58. Mem. A.A.A.S., Am. Inst. Biol. Sci., Am. Soc. Zoologists, Am. Soc. Parasitologists, Am. Microbiology Soc., Pa. Acad. Sci., Sigma Xi. Research, publs. on host-parasite relations of tapeworms of fishes and chickens, trichinosis, mesozoan parasites of marine invertebrates, marine invertebrate fauna of Thailand. Home: 431 W. Woodland Av., Springfield, Pa. 19064. Office: Dept. Biology, Swarthmore Coll., Swarthmore, Pa. 19081.*

MEINSCHEIN, Warren G., Am. chemist; b. Slaughters, Ky., Nov. 12, 1920; s. Tim and Carrie (Poole) M.; student Ind. U., 1938-40; B.S. with distinction, U. Mich., 1948; Ph.D., U. Tex., 1951; m. Mary Elizabeth Williams, Feb. 5, 1944; children—Sherryl Elizabeth, Warren G., Tim Allan. Research asso. Magnolia (now Mobil) Labs., Dallas, 1951-58; research asso. Esso Research & Engring. Co., Linden, N.J., 1958-66; prof. geochemistry Ind. U., Bloomington, 1966-—. Mem. Geochem. Soc. (sec. organic geochemistry div., sec. group for analyses of carbon compounds in carbonaceous chondrites), A.A.A.S., Am. Chem. Soc., Ind. Acad. Scis., Am. Geophys. Union, N.Y. Acad. Scis. Research, publs. on theory of origin of petroleum, use of organic compounds as molecular or chem. fossils; co-discoverer chem. evidence for Precambrian life on Earth, Carbon 14 dating of fossil organic matter by methane-proportional counts, chem. evidence of life in meteorites. Home: 4418 Sheffield Dr., Bloomington, Ind. 47401.*

MEINSMA, Lenze, Dutch physician; b. Makkum, Holland, Nov. 8, 1923; s. Tiemen and Jetske (De Vries) M.; M.D., U. Leiden, 1963. Dir., Bur. Cancer Registry, Netherlands, 1953-—; sec. Sci. Bd. Cancer Soc., 1958-67; dir. Bur. Cancer Edn., 1963-—; med. dir. Netherlands Cancer Soc., 1967-—. Mem. com. on pub. edn. on cancer Internat. Union against Cancer, 1962-—. Mem. Royal Soc. Medicine (affiliate Gt. Britain). Author: Vijfjaarsoverlevingscijfers na Kankerbehandeling, 1963; Resultaten Behandeling Kankerpatienten, 1966; also articles. Research on pub. health aspects of cancer, epidemiological research. Home: 8 Valkenstein. Office: 33 De Lairessetraat, Amsterdam, Holland.*

MEINWALD, Jerrold, Am. chemist; b. Bklyn., Jan. 16, 1927; s. Herman and Sophie (Baskind) M.; Ph.B., U. Chgo., 1947, B.S., 1948; M.A., Harvard, 1950, Ph.D., 1952; m. Yun-wen Chu, June 25, 1955; children—Constance Chu, Pamela Joan. Faculty, Cornell U., 1952-—, prof. chemistry, 1961-—. Chmn. medical chemistry study sect. NIH; chem. cons. Schering Corp., Norwich Pharmacal Co., Allied Chem. Corp. Alfred P. Sloan Found. fellow, 1958-62; John Simon Guggenheim fellow, 1960-61. Mem. Am., (chmn. organic div.), Swiss chem. socs., Chem. Soc. (London), Phi Beta Kappa, Sigma Xi. Bd. editors Jour. Organic Chemistry. Research, publs. on molecular rearrangements; synthesis of strained ring systems, modified steroids; photochem. reactions, n.m.r. spectroscopy, natural products from arthropods. Home: 333 N. Sunset Dr., Ithaca, N.Y. 14850.*

MEINZER, Oscar Edward, Am. geologist; b. nr. Davis, Ill., Nov. 28, 1876; s. William and Mary Julia (Meinzer) M.; A.B. magna cum laude, Beloit Coll., 1901, D.Sc., 1946; studied U. Chgo., 1905-07; Ph.D., 1922; m. Alice Breckenridge Crawford, Oct. 3, 1906; children—Robert William (adopted), Roy Crawford. Prin. pub. schs., Frankfort, S.D., 1901-03; prof. phys. scis. Lenox Coll., Hopkinton, Ia., 1903-05; instr. geology Corr. Sch. U. Chgo., 1906-08; with U. S. Geol. Survey, 1906-46, devoting time chiefly to investigations of underground water; geologist in charge div. of ground water, 1912-46; in charge desert wateringplace survey, 1917-18. Fellow Geol. Soc. Am., A.A.-A.S.; mem. Wash. Acad. Scis. (pres. 1936-37), Geol. Soc. Wash. (pres. 1930-31), Soc. of Econ. Geologists (pres. 1945), NRC, Am. Geophys. (pres. 1947-—, Bowie medal 1943), Sigma Xi. Author: Outline of Ground-Water Hydrology; Occurrence of Ground-Water in the United States; Large Springs in the United States; Plants as Indicators of Ground-Water; Compressibility and Elasticity of Artesian Aquifiers; Outline of Methods for Estimating Ground-Water Supplies; History and Development of Gound-Water Hydrology; Our Water Supply; Hydrology (with others); Hydrology in Relation to Economic Geology; asso. editor Economic Geol. Died June 14, 1948.

MEISEL, Maksin Nikolaevich, Russian microbiologist; b. 1901; grad. Leningrad (USSR) Med. Inst., 1926, postgrad. histology, 1929; postgrad. microbiology and cytology USSR Acad. Scis., from 1932; D. Biol. Scis., Moscow State U., 1947. An organizer Far-Eastern br. USSR Acad. Scis., 1932, mem. presidium, acad. sec., 1932-34, staff Inst. Microbiology, from 1934; faculty Moscow State U., from 1946, lab. chief Inst. Radiation and Physico-Chem. Biology, 1959-—. Visitor, U. S.-Internat. Radiation Research Congress, Buckington, Vt., 1958. Corr. mem. USSR Acad. Scis. (mem. Inst. Biol. Physics). Research on microbiology of nucleic acids, polyphosphates, radiation effects. Office: Inst. Radiation and Physico-Chem. Biology, USSR Acad. Scis., Moscow, USSR.

MEISENHEIMER, Johannes, German biologist; b. Griesheim, Germany, June 30, 1873; prof., Jena, also Leipzig, Germany. Author: Entwicklungsgeschichte der Tiere, 2 vols., 1908; Studien zur Soma-und Geschlechtsdifferenzierung, 3 vols., 1909-24; Die Weinbergschnecke, 1921; Geschlecht und Geschlechter im Tierreich, 2 vols., 1921-30. Research on heredity and evolution. Died Leipzig, Feb. 24, 1933.

MEISSNER, Alexander, radio engr.; b. Vienna, Austria, Sept. 14, 1883; became technician Telefunken, 1907, Berlin. Decorated by German Fed. Republic; named hon. senator Tech. U., Berlin, 1958; recipient Heinrich Herz Gold medal, 1925, Abbe-Medaille, 1929, Gold Dresel-Medaille, 1955. Mem. I.R.E. (became v.p. 1929). Research and publs. on improvement in elec. insulators; assisted in devel. continuous wave transmission; studied heterodyne reception; 1st to use self-excited vacuum tube generator with feedback for prodn. of high frequency radio signals. Died Berlin, Germany, Jan. 4, 1958.

MEISSNER, K(arl) W(ilheim), physicist; b. Reutlingen, Germany, Dec. 15, 1891; s. Karl Emil and Ottilie (Plankenhorn) M.; student U. Munich, 1912; Dr. rer. nat. U. Tuebingen, 1915; m. Ita B. Kohn, Sept. 27, 1919 (dec. July 1939); m. 2d, Hanna Hellinger, May 22, 1942. Came to U. S., 1938, naturalized, 1943. Pvt. docent U. Zurich, 1919-25; prof. physics U. Frankfort/Main (Germany), 1925-37, head dept. physics, 1931-37; asst. prof. physics Worcester Poly. Inst., 1938-41; prof. physics Purdue U., from 1941. Mem. Am. Phys. Soc., Optical Soc. Am., Sigma Xi. Researcher in spectroscopy and atomic physics. Died Apr. 13, 1959.

MEISSNER, William Avison, Am. pathologist; b. Oregon City, Ore., May 20, 1913; s. Carl Herbert and Laura (Avison) M.; A.B., U. Ore., 1935, M.D., 1938; m. Bernice Baynard, Aug. 26, 1936; children—Janice, William. Pathologist, New Eng. Deaconess Hosp., Boston, 1942-63, pathologist-in-chief, 1963-—; pathologist New Eng. Bapt. Hosp., Boston, 1942-63, pathologist-in-chief, 1963-—; clin. prof. pathology Harvard Med. Sch., Boston, 1963-—. Cons. to hosps. Mem. Coll. Am. Pathologists, Am. Soc. Clin. Pathologists, Am. Assn. for Cancer Research, Am. Assn. Pathologists and Bacteriologists, Am. Thyroid Assn., A.M.A., Am. Soc. Exptl. Pathology, Internat. Acad. Pathology, American (president of the Massachusetts div. 1964-66) N.E. cancer socs. Author: (with S. Warren) Tumors of the Thyroid Gland, 1953; also numerous articles. Research in cancer, thyroid and other endocrine diseases. Home: 26 Park Lane, Newton Centre, Mass. 02159. Office: 185 Pilgrim Rd., Boston 02215.*

MEISTER, Arnold George, Am. physicist; b. Chgo., May 30, 1912; s. Arnold Paul and Hulda (Grossman) M.; B.S. in Math., Central YMCA Coll., 1936; Ph.D. in Physics, Ill. Inst. Tech., 1947; m. Bernice May Wischoeffer, June 17, 1939; 1 son, Cary Walter. Cost analyst Commonwealth Edison Co., Chgo., 1930-42; faculty physics Ill. Inst. Tech., Chgo., 1942-57, asso. prof., 1950-57; prof. physics Ariz. State U., Tempe, 1957-—. Cons in molecular spectroscopy and optics, 1950-—; vis. scientist for high schs. Ariz. Acad. Sci., 1960-—. Fellow Am. Phys. Soc., Phys. Soc. London; mem. Optical Soc. Am., A.A.A.S., Am. Assn. Physics Tchrs., Sigma Xi. Research on calculation of potential constants for polystomic molecules, analysis of rotation-vibration bands of symmetric rotor and asymmetric rotor molecules. Home: 33 E. Balboa Dr., Tempe, Ariz. 85281.*

MEITES, Joseph, physiologist; b. Kishinev, Russia, Dec. 22, 1913; s. Ben and Frieda (Kaminetzky) M.; came to U. S., 1920, naturalized, 1940; B.S., U. Mo., 1938, M.A., 1940, Ph.D., 1947; m. Mable E. Rumbrug, Feb. 3, 1943; Faculty, Mich. State U., East Lansing, 1947-—, prof. physiology, 1953-—. Mem. endocrinology study sect. NIH, 1966-—; mem. subcom. on use hormones in domestic animal Nat. Acad. Scis.-NRC, 1960-—. Weizmann fellow, 1955-56. Recipient Mich. State U. Sigma Xi Jr. award, 1953, Sr. award, 1966; recipient travel awards Internat. Physiol Congress, 1959, Internat. Nutrition Congress, 1963, Pharm. Congress, 1962, WHO Conf. on Location, 1963, Internat. Cancer Congress, 1966. Mem. Soc. for Exptl. Biology and Medicine (pres. Mich. sect. 1959), Endocrine Soc., Am. Physiol. Soc., Am. Inst. Nutrition, Am. Assn. for Cancer Research, A.A.A.S., Am. Soc. Animal Sci., Sigma Xi. Contbg. author: Vitamines and Hormones, 1960; author: Advances in Neuroendocrinology, 1963; (with C. E. Nicoll) Annual Review of Physiology, 1966; also numerous articles. Editorial bd. Endocrinology, 1953-56, 66-—; mem. adv. council Neuroendocrinology, 1966-—. Research on neuroendocrine control anterior pituitary, control mammary growth and lactation, hormones and cancer, hormones and nutrition, hormones and reprodn. Home: 4715 Nakoma Dr., Okemas, Mich. 48864. Office: Dept. Physiology, Mich. State U., East Lansing, Mich. 48823.*

MEITES, Louis, Am. chemist; b. Balt., Dec. 6, 1926; s. Louis and Gertrude (Harand) M.; B.A., Middlebury Coll., 1945; M.A., Harvard, 1946, Ph.D., 1947; m. Thelma Steinberg, June 10, 1947; children—Judith Ann, Norman Louis, Robin Leslie. Instr., Princeton, 1947-48; faculty Yale, 1948-55, asst. prof., 1952-55; faculty Poly. Inst. Bklyn., 1955-—, prof. chemistry, 1962-—. Mem. Am. Chem. Soc. (alternate councilor N.Y. sect. 1963-65), Internat. Union Pure and Applied Chemistry (asso.), Am. Assn. U. Profs. Author: Polarographic Techniques, 1955, 2d edit., 1965; (with Henry C. Thomas) Advanced Analytical Chemistry, 1958; also numerous articles; contbg. author, editor Handbook of Analytical Chemistry, 1963. Editorial adv. bd. Chem. Analysis Series, 1965-—, Talanta, 1966-—. Research in electroanalytical chemistry, polarography, electrolytic and coulometric techniques chem. analysis, rates and mechanisms chem. reactions in solutions, electrochem. behaviors inorganic and organic substances, kinetics redox reactions, theory potentiometric and other titration curves. Home: 26 Knowles St., Plainview, L.I., N.Y. 11803. Office: 333 Jay St., Bklyn. 11201.*

MEITNER, Lise, physicist; b. Vienna, Austria, Nov. 7, 1878; d. Philipp M. and Hedwig (Skooren) Meitner; Ph.D., U. Vienna, 1906; studied with Dr. Max Planck, Berlin, Germany, also Dr. Otto Hahn; hon. degrees univs. Rochester, Rutgers, Stockholm, Berlin, also Adelphi Coll., Smith Coll. Swedish citizen, 1949. Asst., Inst. Theoretical Physics, U. Berlin, 1912-15, lectr. U. Berlin, 1922, extr. prof., 1926; head phys. dept. Kaiser Wilhelm Inst. for Chemistry, 1917-38; staff Nobel Inst., Stockholm, Sweden, 1938-—; Royal Swedish Academy of Engineering Sciences, 1947-—. Recipient Leibnitz medal Berlin Acad. Scis., 1924, Leiben prize Austrian Acad. Scis., 1925; prize in sciences City of Vienna, 1947; Planck medal, 1949. Fgn. mem. Swedish, Copenhagen, Göteborg, Vienna, Berlin, Göttingen, Stockholm acads. scis., Royal Soc., 1955. Author: Beiträge zur Physik der Atomkerne, Atomvorgange und ihre Sichtbarmachung, 1926; (with Max Delbruck) Der Aufbau der Atomkerne, 1935; Radioaktivität und Kernphysik; Beta und Gammastrahlenspektren und ihre Deutung; (with Otto Hahn) Entdeckung einiger radioaktiver Substanzen; Erste Deutung der Uranspaltung. With Hahn discovered Thorium-C, 1908, and protoactinium, 1917; studied nuclear isomerism; found four radioactive elements resulting from neutron bombardment of uranium; with Frisch split uranium nucleus, 1939; predicted chain reaction which contributed to atomic bomb devel. Address: 16 Highsett, Hills Rd., Cambridge, England.

MEITZEN, August, German statistician, historian; b. Breslau, Germany, Dec. 16, 1822; ed. Heidelberg, Tübingen (both Germany); mem. Statis. Bur.; named hon. prof. statistics and polit. economy U. Berlin, 1892. Author: Der Boden und die landwirtschaftlichen Verhältnisse der preussichen Staates, 8 vols., 1868-1908; Die internationale land- und forstwirtschaftliche Statistik, 1873; Die Mitverontwortlichkeit der Gebildeten für das Wohl der arbeitenden Klassen, 1876; Geschichte, Theorie und Technik der Statistik, 1886; Siedlung und Agrarwesen der Westgermanen und Ostgermanen, der Kelten, Römer, Finnen und Slawen, 3 vols., 1896. Fundamental work on statistics; studied German agrarian history; introduced field maps as source in history of settlements. Died Berlin, Jan. 19, 1910.

MEJIA, Raul Hector, Argentinian physiologist; b. Buenos Aires, Argentine, Feb. 21, 1927; s. Raul Alberto and Florencia (Quesada) M.; Bachiller, Colegio del Salvador, Buenos Aires, 1943; M.D., U. Buenos Aires, 1950; m. Maria Isabel Cullen, Nov. 17, 1956; children—Raul Mariano, Ignacio Lucas, Miguel Eduardo, Andres Federico, Pablo Francisco. Asst. cardiologist Hosp. Alvear, Buenos Aires, 1951-56; faculty U. Buenos Aires, 1957—, asst. prof. physiology Facultad de Medicina, 1965—; established investigator Consejo Nacional de Investigaciones Cientificas y Tecnicas de Argentina, 1961—; research fellow dept. physiology Down State Med. Center, State U. N.Y., Bklyn., 1960-61. Sec., Consorcio de Medico Catolicos de Buenos Aires, 1956-59. Recipient Premio, Facultad de Ciencias Medicas, 1959. Mem. Asociacion Medica Argentina (council on med. edn. 1966—), Sociedad Argentina de Biologia, Sociedad Argentina de Cardiologia, Sociedad Argentina de Fisiologia. Author: (with others) Guia de Trabajos Practicos de Fisiologia, 1966; also articles. Research on diagnostic methods in cardiology especially ballistocardiography and electrocardiography, irreversibility of hemorrhagic shock, body composition. Home: 2047 Arenales, Buenos Aires, Argentina.*

MEL, Howard Charles, Am. biophysicist; b. Oakland, Cal., Jan. 14, 1926; s. Charles and Florence (Nachtrieb) M.; B.S., U. Cal. at Berkeley, 1948, Ph.D., 1953; m. Nancy Shenon, June 18, 1949; children—Amelie C., Stephanie F., Bartlett W. Faculty, U. Cal., Berkeley, 1955—, asso. prof. biophysics, 1966—, staff mem. Donner Lab. of Lawrence Radiation Lab., 1960—. Fulbright fellow U. Brussels (Belgium), 1953-54; NSF sr. postdoctoral fellow, Paris, 1965-66. Fellow A.A.A.S.; mem. Biophys. Soc. (mem. council, exec. com.), Am. Chem. Soc., Am. Assn. Physics Tchrs., Am. Inst. Physics, Soc. For Gen. Systems Research, Phi Beta Kappa, Sigma Xi. Editorial bd. Rev. Sci. Instruments, 1965—. Research, publs. on cellular and subcellular biophysics, components in complex biol. mixtures, blood cell formation and cellular devel., fertility, biol. separations, including electrophoresis, thermodynamics of open and closed systems; inventor stable-flow free-boundary method. Home: 1320 Arch St., Berkeley, Cal. 94708.*

MELA, Pomponius, Roman geographer; b. Tingentera, Spain; flourished 40 A.D. Author: De situ orbis, 3 books (earliest existing Latin work on geography). Largely followed previous geographers, especially Strabo; divided earth into 5 zones, N. Frigid, N. Temperate, Torrid, S. Temperate, S. Frigid; believed only temperate zones are inhabitable; provided account of phys. features, customs of certain countries, but omitted math. details about distances.

MELANCHTHON, Philip, German mathematician, theologian, philosopher; b. Bretten, Germany, Feb. 16, 1497; s. George and Barbara (Reuter) Schwarzerd; B.A., U. Heidelberg (Germany), 1511; M.A.; U. Tübingen (Germany), 1514; m. Katharina Krapp, Nov. 25, 1520; children—Anna, Philip, George (dec. in infancy), Magdalen. Lectr. classics U. Tübingen, 1512-18; became prof. Greek, Wittenberg, Germany, 1518; leader during Reformation, Germany; established chairs of math., Nuremburg, Germany, 1526, math. and astronomy, Wittenberg, 1536. Called "Preceptor of Germany" for endeavors in founding prep. schs. and reorganizing area's university systems; helped establish universities of Marburg, Jena, and Königsberg; instituted reforms in Wittenberg, Leipzig, Tübingen, Frankfort/Oder, and Rostock. Author: Institutiones grammaticae Graecae, 1518; Unterricht der Visitatoren, 1528; Mathematicarum disciplinarum, tum etiam astrologiae encomia, 1540. Editor: Phaenomena (Aratus), 1521; Theoricae novae planetarum (Peurbach); Tabulae astronomicae (Schöner), 1536; Rudimenta astronomica (al-Fargani), 1537; De praedictionibus astronomica (Ptolemy), 1553; Libellus de sphaera (Sacrobosco), 1560. Influenced church in teaching and acceptance of exact sciences; condemned Copernican system because it repudiated Scriptures; questioned ultimacy of man's rational powers; founder Aristotelian sch. philosophy of 16th, 17th and 18th centuries in Germanic Europe; his philosophic system dominated in area now known as Germany until it was surpassed by Wolffian rationalistic philosophy of 18th century. Died Wittenberg, Apr. 19, 1560.

MELANDER, Axel Leonard, Am. biologist; b. Chgo., June 3, 1878; s. Silas Peter and Mathilda (Bjork) M.; B.S., U. Tex., 1901, M.A., 1902; postgrad. (fellow) U. Chgo., 1902-03; Sc.D., Harvard, 1914; m. Mabel Evans, Mar. 5, 1903; 2 sons. Became instr. entomology Wash. State Coll., 1904, prof. entomology, 1906, head dept. zoology, 1907-26; prof. biology, head dept. Coll. City N.Y., 1926-43, emeritus, 1943—; research asso. U. Cal., 1943—. Fellow Am. Acad. Arts and Scis., A.A.A.S., Entomol. Soc. Am. (became pres. 1938), N.Y. Acad. Scis.; mem. N.Y. Acad. Entomology, Am. Naturalists, Am. Soc. Zoologists, Am. Assn. Econ. Entomologists, Bklyn. Entomol. Soc. Author: Empididae of the World, 1926; (with C. T. Brues) Classification of Insects, 1932; Sourcebook of Biological Terms, 1937; Key to Families of North American Insects, 1915. Contributed to devel. of insecticides; developed herbicides; studies in control of fruit insect pests, classification of Diptera. Home: 4670 Ladera Lane, Riverside, Cal.

MELANDER, Lars Conrad Samuel, Swedish chemist; b. Stockholm, Sweden, May 11, 1919; s. Nils A. and Jenny (Krogh) M.; fil.mag. U. Stockholm, 1943, fil.lic., 1947, fil.dr., 1950; m. Louise Brackett, May 18, 1957; children—Nils V., Hans A. Chemist, Swedish Research Inst. Nat. Def., 1943-46; head nuclear chemistry dept. Nobel Inst. Physics, 1946-51; acting head Nobel Inst. Chemistry, 1952-63 (all Stockholm); prof. organic chemistry U. Göteborg (Sweden), 1963—. Mem. Swedish Atomic Research Council, 1964—; mem. pub. com. Swedish Natural Sci. Research Council, 1964—; mem. Swedish Nat. Com. for Chemistry, 1965—. Mem. Swedish Chem. Soc. (Norblad-Ekstrand gold medal 1961). Author: Isotope Effects on Reaction Rates, 1960; also articles. Research in nuclear chemistry, kinetic isotope effects especially applied to organic chemistry; phys. organic chemistry especially reaction mechanisms.

MELCHERS, Johann Georg Friedrich, German biochemist, plant physiologist; b. Cordingen, Germany, Jan. 7, 1906; s. Georg Friedrich and Betty (Voss) M.; ed. Fribourgh, Kiel, Gottingen (all Germany) univs.; Ph.D.; m. Eleonore Drexler, Apr. 4, 1931; children—Friedrich, Christoph, Konrad. Asst., U. Munich (Germany), from 1931, also U. Berlin, from 1934; dir. virus research sect. Berlin, Tübingen univs., from 1941; chief Melchers sect. Max Planck Inst. for Biology, Tübingen, dir., 1947—; hon. prof. U. Tübingen, 1947—. Mem. Max-Planck Soc., German, Am. botany socs., Am. Soc. Phyto-physiologists. Author: Handbuch der Pflanzenphysiologie; Zeitschrift für Vererburgslehre; others. Research, publs. on physiol. devel. plants, genetics, virology. Home: Correnstrasse 45, Tübingen. Office: Correnstrasse 41, Tübingen, W. Germany.

MELCHIOR, Jacklyn Butler (Mrs. Norten C. Melchior), Am. biochemist; b. Sacramento; Drury Dewolf and Lalita (Jodon) Butler; B.S., U. Cal. at Berkeley, 1940, Ph.D., 1946; m. Norten C. Melchior, Dec. 21, 1939; children—Ernst Drury, June Ann. Research asso. Northwestern U., 1946-49; asso. prof. Loyola U. Stritch Sch. Medicine, Chgo., 1949-57; asso. prof. pharmacology Loyola Med. Sch., 1957-60; prof., chmn. biochemistry Chgo. Coll. Osteopathy, 1960—. Recipient Lederle Med. Faculty award, 1957-60. Fellow A.A.A.S.; mem. Am. Soc. Biol. Chemists, Sigma Xi, Iota Sigma Pi. Research, publs. on enzyme mechanisms, molecular interactions. Home: 209 S. Elmwood St., Oak Park, Ill. 60302. Office: 1122 E. 53d St., Chgo. 60615.*

MELCHIOR, Paul Jacques Léon, Belgian astronomer; b. Montsur-Machienne, Belgium, Sept. 30, 1925; s. Léon and Claire (Dupont) M.; Ph.D., U. Brussels (Belgium); m. Anne-Marie Bary, Sept. 16, 1950; children—Philippe, Jean-Paul, Pierre. Astronomer, Royal Obs. Belgium; dir. Internat. Center for Earth's Tides; lectr. U. Louvain (Belgium); v.p. Internat. Astron. Union on Commn. for Movement of the Pole. Mem. Belgian Astron. Soc., Belgian Soc. of Astronomy, Meteorology and Global Physics. Author: The Earth Tides; Latitude Variation; Earth Tides; La rotation de la terre, mouvement des pôles. Research, publs. on geodynamics, rotation and tidal deformations of earth. Home: 190 Groeselenberg, Brussels 18. Office: 3, avenue Circulaire, Brussels 18, Belgium.

MELDE, Franz Emil, German physicist; b. nr. Fulda, Germany, 1832; prof. physics and astronomy, dir. Math.-Phys. Inst., U. Marburg (Germany). Made vibrations of strings visible (each string showed a different number of vibratory loops according to its tension). Died 1901.

MELDRUM, Charles, mathematician, meteorologist; b. Kirkmichael, Scotland, 1821; ed. Marischal Coll., M.A., 1844; also U. Aberdeen (Scotland); LL.D., 1876; m. Charlotte Fitz-Patrick, 1870, Joined Bombay (India) Ednl. Dept., 1846; became prof. math. Royal Coll. Mauritius, 1848; named govt. meteorol. observer, 1862; named dir. Royal Alfred Obs., 1875. Fellow Royal Soc., 1876; mem. council Govt. of Mauritius, 1886-96, Meteorol. Soc. Mauritius (a founder). Worked out laws of cyclones in Indian Ocean. Died Edinburgh, Aug. 28, 1901.

MELEKHOV, Ivan Stapanovich, Russian sylviculturist; b. 1905; grad. Leningrad Timber Tech. Acad., 1930; D.Agrl. Sci. Asst., Arkhangelsk Timber Tech. Inst., 1930-44, prof., 1944-50, head chair sylviculture, 1952—; head forestry sect. No. Base and Arkhangelsk Sta., USSR Acad. Scis., 1938-57, dir. No. dept. Inst. Forestry, 1957-59, dir. Arkhangelsk Inst. Timber and Wood Pulp Chemistry, 1959—; prof. Leningrad Timber Tech. Acad., 1951-52; dep. chmn. State Com. for Timber, Paper and Wood Processing Industry and Forestry, USSR Gosplan. Decorated Order of Lenin. Mem. All-Union Lenin Acad. Agrl. Sci. Author: Fir Wood, 1934; The Effects of Fires on Forests, 1948; Outline Development of Forestry in Russia, 1957. Research and numerous publs. on Taiga forestry, No. forests, history of forestry in USSR. Address: Gosplan USSR, prospect Marksa 12, Moscow, USSR.

MELENTEV, Lev Aleksandrovich, Russian engr.; b. 1908; grad. Leningrad (USSR) Poly. Inst., 1930. Staff, Leningrad Energetics Inst., 1929-33; bur. chief Leningrad Commn. Energetics, 1933-35; from sr. instr. to prof. Leningrad Engring.-Econs. Inst., 1936-42, chmn. thermo-energetics dept., prof., 1945-60; sr. sci. worker Inst. Energetics, USSR Acad. Scis., 1942-60, dir. Inst. Energetics, Siberian br., 1960—. Corr. mem. USSR Acad. Scis. (chmn. presidium E. Siberian br. Siberian dept.). Author: 40 Years of Soviet Power Engineering, 1958; (with G. B. Levental) Correlation between the Thermodynamic and Power Indices of Heat-Power Plant Efficiency, 1958. Work on devel. power plants USSR. Office: Inst. Energetics, Siberian branch USSR Acad. Sciences, Irkutsk, Siberia, USSR.

MELETTI, Paolo, Italian botanist; b. Montemarciano, Italy, June 27, 1927; s. Michele and Lina (Leonori) M.; Natural Sci. Dr., Scuola Normale Superiore, U. Pisa (Italy), 1950; m. Mariapia Picciau, July 24, 1954; children—Roberta, Giorgio, Gabriele, Carlo, Anna. Faculty, U. Pisa, prof. botany, 1964—, dir. Institut Botany. Mem. Italian Soc. Botany, Italian Soc. Plant Physiology, Italian Assn. Genetics. Research, publs. on embryo transplantation technique, embryo-endosperm relations in irradiated seeds, dormancy in wheat seeds, exptl. cytology. Home: 55 S. Maria, Pisa, Italy.*

MELHORN, Wilton Newton, Am. geologist; b. Sisterville, W.Va., July 8, 1920; s. Ralph Wilton and Velma P. (Jones) M.; B.S., Mich. State U., 1942, M.S., 1951; M.S., N.Y. U., 1943; Ph.D., U. Mich., 1955; m. Agnes Leigh Ogg, Aug. 23, 1961. Hydrologist, Mich. Geol. Survey, 1946-49, U. S. Weather Bur., 1949-50; asst. prof. engring. geology Purdue U., 1955-57, asso. prof., 1957—, head dept. geoscis., 1967—. Fellow Geol. Soc. Am., A.A.A.S.; mem. Am Meteorol. Soc., Am. Assn. Petroleum Geologists, Soc. Econ. Paleontologists and Mineralogists, Nat. Speleological Soc., Clay Minerals Soc., Ind. Acad. Sci., Mich. Acad., Arts, Sci. and Lit. Research and publs. in Silurian stratigraphy, paleobotany, engring. geology, clay mineralogy, history of geology, storage and geologic environment of low-level radioactive solid waste disposal, role of geology in urban planning. Home: 2065 S. 9th St., Lafayette, Ind. 47905.*

MELICOW, Meyer Morton, physician; b. Bielystok, Russia, Dec. 25, 1894; s. Jacob and Rebecca (Todes) M.; came to U. S., 1905, naturalized, 1920; B.S., Coll. City N.Y., 1917; M.D., Columbia, 1920; m. Sylvia Friedlander, Nov. 30, 1932; 1 son, Daniel Malcolm. Practice medicine specializing in uropathology, N.Y.C., 1930—; faculty Columbia Coll. Phys. and Surg. 1928—, Given prof. uropathology, 1962, Given prof. emeritus, spl. lectr., 1962—; hon. prof. urology U. Madrid (Spain) Med. Sch.; cons. St. Albans U. S. Naval, Francis Delafield, Harlem hosps. Recipient Silver medal Columbia, 1967. Fellow N.Y. Acad. Medicine; mem. A.M.A. (Billings Gold medal 1947), Assn. Clin. Scientists, Am. Physicians Art Assn. (past pres.), Am. Japanese urol. assns., Alpha Omega Alpha. Research, numerous publs. on tumors of testis, kidney and bladder; 1st to present "Herald" lesion of urinary bladder, "carcinoma in situ" of bladder, pelvis; periodic table of sexual anomalies, tumors of dysgenetic gonads in intersexes, significance of prostatic nodule. Home: 55 Central Park W., N.Y.C. 10023. Office: 630 W. 168th St., N.Y.C. 10032.*

MELIN, Johannes Botvid Elias, Swedish botanist; b. Valstad, Sweden, July 28, 1889; s. Samuel and Hilda (Stenborg) M.; Ph.D., U. Uppsala (Sweden); dr. honoris causa in agr., Norwegian Agrl. Coll.; m. Margit Valley, Aug. 16, 1921; children—Hans Samuel, Berit Garda Margareta. Instr. plant biology U. Uppsala, 1918, prof. botany, 1930-56, now emeritus; instr. botany Stockholm (Sweden) Forestry Sch., 1919-30. Laureate Bjorken prize U. Uppsala, 1925; recipient Henri de Jouvenel prize Ministry Edn. France, 1938, Emil Hansen prize Copenhagen, 1956. Mem. Swedish Bot. Soc. (pres. 1948-59), Scandinavian Soc. Plant Physiology (pres. 1947-50, 58-61), Royal Acad. Sci. Sweden, Royal Acad. Agr. Sweden, Danish Acad. Sci. and Letters, Norwegian, Finnish acads., Royal Physiographic Soc. Lund (Lineus prize 1956), London Lineus Soc., Paris Acad. Sci. (corr.) Research, publs. on embryology and cytology of mosses, physiology of mushrooms, ecology and physiology of forest trees. Address: Institut de botanique physiologique, Université, Uppsala, Sweden.

MELISSINOS, Adrian Constantin, physicist; b. Thessaloniki, Greece, July 28, 1929; s. Constantin J. and Olympia (Abbott) M.; student Royal Hellenic Naval Acad., 1945-48; M.S., Mass. Inst. Tech., 19—, Ph.D., 1958; m. Joyce Mitchell, June 7, 1960; 1 son, Constantin John. Asst. prof. U. Rochester (N.Y.), 1960-63, asso. prof., high energy physics, 1963—. Mem. Am. Phys. Soc. Author: Elements of Vector Calculus, 1955; Experiments in Modern Physics, 1966; also articles. Research on measurement isomeric shift in atomic spectra, prodn. pi mesons and K-mesons in high energy proton-proton collisions, muon-proton scattering, pi and k-meson interactions in bubble chambers. Home: 12 Birmingham St., Rochester, N.Y. 14618.*

MELLANBY, Sir Edward, Brit. pharmacologist; b. W. Hartlepool, Eng., 1884; s. John Mellanby; ed. St. Thomas's Hosp.; research student Emmanuel Coll., Cambridge U., 1907, also M.A., M.D.; D.Sc. (hon.), univs. Sheffield, Oxford, Belfast, Chgo., Oslo; LL.D. (hon.), univs. Birmingham, Glasgow, St. Andrews, Melbourne; M.D. (hon.) univs. Witwatersrand, Adelaide; m. May Tweedy, 1914. Demonstrator physiology St. Thomas's Hosp., 1909-11; lectr., later prof. physiology King's Coll. for Women, London U., 1913-20; formerly Fullerton prof. physiology Royal Instn.; formerly hon. physician Sheffield Royal Infirmary; prof. emeritus pharmacology U. Sheffield; Oliver-Sharpey lectr. Royal Coll. Physicians, 1922, recipient Bissett-Hawkins medal, 1929, Croonian lectr., 1933, recipient Moxon medal, 1938, Harveian orator, 1938, Baly medal, 1949; Kinacre lectr. Cambridge U., 1933, Rade lectr., 1939; Charles Mickle fellow Toronto U., 1935; Croonian lectr. Royal Soc., 1943; Abraham Flexner lectr. Vanderbilt U., Nashville, 1947; Withering lectr. U. Birmingham. Chmn. Internat. Confs. Standardications of Vitamins, 1931, 34, 49; chmn. Internat. Tech. Commn. Nutrition; mem. Med. Research Council, 1931-33, sec., 1933-49. Decorated Comdr. British Empire; officer Legion of Honor (France); recipient Am. medal of freedom with silver palm, Walsingham medal, 1907, Gedge prize, 1908, Raymond Horton Smith prize; Stewart prize for med. research Brit. Med. Assn., 1924, Cameron prize U. Edinburgh, 1932; hon. fellow Emmanuel Coll., 1946. Hon. fellow Royal Coll. Surgeons Edinburgh, Danish Royal Acad. Sci., 1947, Fellow Royal Soc., 1925 (Royal medal, 1932, 47, Buchanan medal 1947), Royal Coll. Physicians. Author: Nutrition and Disease, 1934; A Story of Nutritional Research; also publs. on physiology, biochems. and medicine. First to demonstrate rickets is a diet deficiency disease with his recognition of anti-rachtis factor of vitamin D, 1918; discovered that nitrogen trichloride (bleaching agent for bread flour) could be injurious; it was later banned by F.D.A. Died Jan. 30, 1955.

MELLANBY, John, English physiologist; b. West Hartlepool, Eng., June 12, 1878; s. John and Mary (Lawson) M.; B.S., Emmanuel Coll., Cambridge, 1900, research student, Cambridge, 1907-09; postgrad., 1901; M.D., Manchester, Eng., 1907; m. Alice Mary Watson, 1911; 1 dau. In charge research labs Burroughs, Wellcome and Co., Herne Hill, Eng., 1902-05; apptd. lectr. in charge physiol. dept. St. Thomas's Hosp. Med. Sch., 1909, prof., 1920; became Waynflete prof. physiology Oxford, also fellow Magdalen Coll., 1936. Mem. Med. Research Council, 1936-39. Fellow Royal Soc., 1929. Editor, Physiol. Abstracts. Research and publs. on proteins of blood, coagulation, secretions of pancreas, globulin; developed method for isolating prothrombose as dry powder, 1930; proved curdling of milk in stomach was caused by pepsin, and curdling of pancreatic juice by trypsin. Died Oxford, Eng., July 15, 1939.

MELLINKOFF, Sherman Mussoff, Am. physician; b. McKeesport, Pa., Mar. 23, 1920; s. Albert and Helen (Mussoff) M.; A.B., Stanford U., 1941, M.D., 1944; m. June O'Connell, Nov. 18, 1944; children—Sandra H., Sherrill M., Albert J. Instr. medicine Johns Hopkins, 1951-53; fellow gastroenterology U. Pa., 1949-50; asst., then asso. prof. medicine U. Cal. at Los Angeles Med. Sch., 1953-62, prof. medicine, dean Med. Sch., 1962—; attending cons. internal medicine Wadsworth Gen. Hosp., VA Center, Los Angeles; sr. attending physician Harbor Gen. Hosp., Torrance, Cal. Diplomate Am. Bd. Internal Medicine. Mem. Am. Gastroent. Assn., A.C.P., Assn. Am. Physicians, Western Assn. Physicians, Am. (adviser council drugs 1963-65), Pan Am. med. assns., Am. Soc. Clin. Nutrition, Western Soc. Clin. Research, Am. Fedn. Clin. Research, A.A.A.S. Contbr. numerous articles med. jours. Editor: Differential Diagnosis of Abdominal Pain, 1959; Differential Diagnosis of Diarrhea, 1964. Office: Univ. Cal. at Los Angeles Med. Center, Los Angeles 90024.*

MELLIS, Otto, mineralogist; b. St. Petersburg, Russia, Mar. 10, 1906; s. Peter and Alvine (Arklins) M.; Mag.rer.nat., Latvian State U. at Riga, 1933, Dr.rer.nat., 1943; m. Irene Putnins, May 20, 1939; 1 dau., Inga. Fellow, Mineral. Mus., Oslo, Norway, 1936, 39; faculty U. Riga, 1937-46; prof. Baltic U., Hamburg, Germany, 1946; lectr. U. Uppsala, Sweden, 1947-50; asso. prof. U. Stockholm, Sweden, 1949—; research staff Swedish Research Council for Natural Scis., 1961—. Swedish del. Internat. Mineral. Assn., Internat. Gemmological Conf. Decorated Nobel Cross, Order of White Rose of Finland, 1939; Latvian Order of Merit, 1939. Mem. Mineral. Soc. Am., Soc. Econ. Paleontologists and Mineralogists, Geol. Soc. Sweden, German Mineral. Soc., numerous others. Research, publs. on minerals, especially optical properties; fibrous mineral aggregates, mineral inclusions, deep sea sediments, cosmic dust. Home: Rosenhill, Djurgarden, Stockholm, Sweden.*

MELLON, Ralph Robertson, Am. physician, med. researcher; b. Springdale, Pa., Feb. 1, 1883; s. Thomas Donnelly and Angie (Robertson) M.; B.Sc., Grove City (Pa.) Coll., 1901, Sc.D. (hon.), 1936; M.D., U. Mich., 1909, M.Sc., 1912; Dr. P.H., Harvard, 1916; m. Arda J. Esten, Sept. 18, 1912; children—Miriam Hinsdale (Mrs. James H. Pennoyer), Janet Robertson (Mrs. Dexter Farnsworth). Faculty, U. Mich., 1909-15; dir. Highland Hosp. Lab., Rochester, N.Y., 1916-27; dir. Inst. Pathology, Western Pa. Hosp., Pitts., 1927-53, cons., 1953—; dir. pneumonia research fund Mellon Inst. Indsl. Research, Pitts., 1930-40. Mem. Am. Assn. Pathologists and Bacteriologists, Soc. Am. Bacteriologists, Soc. Am. Immunologists, Am. Soc. Clin. Pathology (bd. mgrs.), Soc. Exptl. Biology and Medicine, A.M.A. Author: (with Frank Cooper, Paul Gross) Sulphanilamide Therapy of Bacterial Infections, 1938. Co-discoverer sulphathiazole for pneumonia, 1935; discovered sulphur solution for treatment of burns, especially indsl. and eye burns; research on microbic heredity. Address: Star Route 1, Box 550, Twenty-Nine Palms, Cal. 92277.

MELLONI, Macedonio, Italian physicist; b. Parma, Italy, Apr. 11, 1798; ed. U. Parma, also Paris, France. Prof. physics U. Parma, 1824-31; exiled to France; prof. Dept. Jura, France, also U. Montpellier (France); returned to Italy, 1839; dir. Cabinet of Arts and Trades, Naples, Italy, from 1839; Vesuvius Meteorol. Obs., Naples, 1847-49. Fellow Royal Soc. (Rumford medal 1834), 1839. Author: La thermochrose, ou la coloration calorifique, 1850. Noted for work on radiant heat (contbd. to knowledge of infrared portion of solar spectrum); invented term diathermancy for capacity of transmitting infrared radiations, also studied effect of various substances on transmitted rays; maintained light and radiant heat are ultimately of same nature; invented galvonometer and thermopile arrangement for measurement of radiant heat. Died Portici, nr. Naples, Aug. 11, 1854.

MELLOR, Joseph William, Brit. chemist; b. Huddersfield, Eng., 1869; ed. U. Otago, U. Manchester; D.Sc. First prin. pottery dept. N. Staffordshire Tech. Coll.; dir. research Brit. Refractories Research Assn., under Dept. Sci. and Indsl. Research. Fellow Royal Soc., 1927. Author: Higher Mathematics for Student of Chemistry and Physics, 1902; Chemical Statics and Dynamics, 1904; Treatise on Inorganic and Theoretical Chemistry, 16 vols.; Collected Papers in Ceramic Industries; also numerous articles. Made various innovations in ceramics. Died May 24, 1938.

MELNICK, Joseph Louis, Am. virologist; b. Boston, Oct. 9, 1914; s. Samuel and Esther (Melny) M.; B.A., Wesleyan U., 1936; Ph.D., Yale, 1939; m. Matilda Benyesh, 1958; 1 dau., Nancy (Mrs. John Livingston). With Yale, 1939-58, prof. epidemiology, 1954-58; prof., chmn. dept. virology and epidemiology Baylor U. Coll. Medicine, 1958—. Mem. human cancer virus task force Nat. Cancer Inst., NIH 1962—, mem. bd. virus reference reagts. Nat. Inst. Allergy and Infectious Diseases, 1962-65, chmn. panel for picornaviruses, 1963-65, mem. allergy and infectious diseases tng. grant com., 1962-65; Editorial adv. bd. Inst. for Sci. Information, Phila.; chmn. coms. on ECHO, enteroviruses Nat. Found. For Infantile Paralysis-NIH, 1955-63; mem. com. on live poliovirus vaccine USPHS, 1958-61; mem. nat. adv. bd., 1965—; mem. expert adv. panel on virus diseases WHO, 1957—, dir. Internat. Reference Center for Enteroviruses, 1961—. Named to Polio Hall of Fame, 1958; recipient Argentinian Found. Against Infantile Paralysis Gold medal award for poliomyelitis research, 1949; Modern Medicine Distinguished Achievement award, 1965; others. Diplomate Am. Bd. Microbiology. Fellow A.A.A.S., Am. Pub. Health Assn., Am. Acad. Micribiology, N.Y. Acad. Scis.; mem. Microbiol. Soc. Israel (hon.), Am. Epidemiol. Soc., Soc. Exptl. Biology and Medicine, Am. Assn. Immunologists, Am. Soc. for Microbiology, Am. Assn. for Cancer Research (pres. S.W. sect. 1965—), Alpha Omega Alpha (hon.). Editor: Progress in Med. Virology, 1958—; asso. editor Cancer Research 1965—, Am. Jour. Epidemiology, 1964—, Jour. Virology, 1967—, Exptl. and Molecular Pathology, 1965—. Research, publs. field virology, particularly picornaviruses, enteroviruses, poliovirus, echovirus, coxsackievirus, and cancer viruses including the papovaviruses, adenovirus hybrids, and the detective satellite viruses. Home: 8838 Chatsworth St., Houston 77024. Office: Baylor Med. Coll., Houston 77025.*

MELNICK, Matilda Benyesh (Mrs. Joseph L. Melnick), virologist; b. Russe, Bulgaria, Feb. 7, 1926; d. Sinto and Rachel (Farchi) Benyesh; M.B., U. Sofia, Bulgaria, 1944-49; M.D., Hebrew U., Jerusalem, 1952; m. Joseph L. Melnick, June 12, 1958. Came to U. S., 1955, naturalized, 1960. Research asst. Tel Hashomer Hosp., Israel, 1953-55; research fellow Yale Sch. Medicine, 1955-57; vis. scientist NIH, Bethesda, Md., 1957-58; faculty Baylor U. Coll. Medicine, Houston, 1958—, prof. virology, 1967—. Recipient Humanitarian award Jewish Inst. Med. Research, 1964. Mem. Am. Soc. for Microbiology, Am. Assn. for Cancer Research, A.M.A. (affiliate), Am. Assn. Immunologists. Contbd. to genetic studies on polio virus defining conditions for using live polio vaccine; devised improved methods for identifying viruses isolated from man; devised techniques for isolating and studying lines of cells from children with infectious diseases; studying role of viruses as causative agts. of human leukemia, infectious mononucleosis, congenital disease. Home: 8838 Chatsworth St., Houston 77024.*

MELNICK, P(erry) J(ulius), pathologist; b. Poland, May 6, 1901; s. Herman M. and Tillie H. Melnick; came to U. S., 1903, naturalized, 1909; B.S., U. Ill., 1927, M.S. in Pathology, 1929, M.D., 1930, Ph.D. in Pathology, 1936; m. Esther Altabe, Aug. 15, 1926; children—Philip, David, Daniel. Asso. clin. prof. pathology U. So. Cal. Med. Sch., Los Angeles, 1946-53, Wayne State U., Detroit, 1954-56, U. Cal. Med. Sch., Los Angeles, 1956-61, San Francisco, 1962—; chief lab. service VA Adminstrn. Hosp., Martinez, Cal., 1962—. Recipient 3d award Internat. Congress Radiology, 1937; Class I Certificate of merit Ill. State Med. Soc., 1937, 39. Diplomate Am. Bd. Pathology. Fellow A.C.P., Am. Coll. Pathologists, Am. Soc. Clin. Pathologists, Am. Assn. Pathology and Bacteriology, A.A.A.S., Internat. Acad. Pathology, A.M.A., N.Y. Acad. Scis.; mem. Pan Am. Med. Assn., Histochemistry Soc., Soc. Cryobiology (bd. govs.), Sigma Xi, Alpha Omega Alpha. Bd. editors, book rev. editor Am. Jour. Clin. Pathology, 1954—; bd. editors Cryobiology, 1964—. Research, publs. on pathology of tumors, radiation diseases; histochemistry of granulomas and cancer; cryobiology. Home: 2961 Benvenue Av., Berkeley, Cal. 94705. Office: VA Adminstrn. Hosp., Martinez, Cal. 94553.*

MELNIKOV, Oleg Aleksandrovich, Russian astronomer; b. 1912; grad. Kharkov State U., 1933; D.Physico-Math. Scis., 1945. With Main Astron. Obs., USSR Acad. Scis., 1933, asst. dir., 1961—; prof. dept. astrophysics Leningrad State U., 1946. Mem. USSR Acad. Scis. (corr.). Research and publs. on stellar and solar physics, interstellar matter, building astron. instruments, history of astrophysics and astron. equipment making. Address: Astronomical Observatory, USSR Acad. Scis., Pulkovo, USSR.

MELOCHE, Villiers Willson, Am. chemist; b. Pt. Huron, Mich., Dec. 31, 1895; s. John B. and Mary Ann (White) M.; B.S., U. Wis., 1925, M.S., 1923, Ph.D., 1925; m. Alice Van Patten King, Dec. 29, 1936. Chemist, Carnegie-Ill. Steel Co., Chgo., 1917-20; asst. in chemistry U. Wis., Madison, 1921-25, faculty, 1925—, prof. analytical chemistry, 1928—, prof. chemistry, 1938—. Cons., U. S. Forest Products Lab., Madison, 1951-62; coop. researcher Monsanto Chem. Co., OSRD, 1944-45. Mem. Am. Chem. Soc. (chmn. local sect. 1936), A.A.A.S., Am. Soc. Testing Materials, Sigma Xi, Alpha Chi Sigma (nat. pres. 1936-38), Gamma Alpha, Phi Lambda Upsilon. Research, publs. on selenium, tellurium, rhenium and trace elements, hydro chemistry, devel. analytical and instrumental methods. Home: 2146 Fox Av., Madison, Wis. 53711.*

MELTON, Arthur Weever, Am. psychologist, educator; b. Fayetteville, Ark., Aug. 13, 1906; s. Henry Arthur and Josephine (Curry) M.; A.B., Washington U., St. Louis, 1928; M.A., Yale, 1929, Ph.D., 1932; m. Dorothy Pennell, Sept. 23, 1929 (div. Aug. 1944); children—Walter Curry, Deborah (Mrs. Jon Mac Anderson); m. 2d., Agnes Elizabeth Gohmert, June 29, 1946. Research asso. Am. Assn. Mus., Phila., 1930-32; faculty Yale, 1932-35, U. Mo., Columbia, 1935-41; prof. psychology Ohio State U., 1946-49; tech. dir. Air Tng. Command Human Resources Research Center, San Antonio, 1949-54, USAF Personnel and Tng. Research Center, San Antonio, 1954-57; prof. psychology U. Mich., Ann Arbor, 1957—. Cons. to Dept. Def., 1947-49, 57-62. Recipient Commendation for Meritorious Civilian Service, Dept. Air Force, 1951, 59. Fellow Am. Psychol. Assn., N.Y. Acad. Sci., A.A.A.S.; mem. Soc. Exptl. Psychologists, Sigma Xi, Kappa Alpha, Gamma Alpha. Editor: Jour. Exptl. Psychology, 1951-62; editor, contbr. Apparatus Tests, 1947, Categories of Human Learning, 1964. Contbr. articles to profl. jours. Home: 2541 Washtenaw Av., Ann Arbor, Mich. 48104.*

MELTZER, Samuel James, physiologist; b. Russia, Mar. 22, 1851; s. Simon Meltzer; gen. edn. Königsberg, Prussia; M.D., U. Berlin, 1882; several hon. degrees; m. Olga T. Levitt, 1871; 2 children. Came U. S., 1883. Practiced medicine, N.Y.; head dept. physiology and pharmacology Rockefeller Inst. for Med. Research, 1906—; cons. physician Harlem Hosp. Fellow A.A.A.S., N.Y. Acad. Scis., N.Y. Acad. Medicine. Showed (with Welch) that erythrocytes can be destroyed mechanically, 1884; demonstrated (with Auer) the anesthetic properties of magnesium, 1906, also simplified procedures for tracheotomy, 1909; 1st (with Lyon) to introduce non-surg. drainage of gall bladder, 1917. Died Nov. 7, 1920.

MELVILL, Thomas, scientist; b. 1726; student Glasgow, Scotland, 1748-49. Research on optics; proposed theories of refrangibility and aberration of light; 1st to study spectra of luminous gases, 1752. Died Geneva, Switzerland, Dec. 1753.

MELVILLE, David, Am. inventor; b. Newport, R.I., Mar. 21, 1773; s. David and Elizabeth (Thurston) M.; apprenticed to a pewterer; m. Patience S. Sherman, Mar. 4, 1812; 7 children. Established as pewterer, Newport, by 1803; developed method for producing illuminating gas, succeeded in lighting his own house with coal gas, 1806, obtained 1st U. S. patent for apparatus for making coal gas, 1813; unsuccessfully attempted (with Winslow Lewis) to influence U. S. Govt. to use coal gas for light houses. Author: An Exposé of Facts Respectfully Submitted to the Government of the United States Relating to the Conduct of Winslow Lewis, 1819. Died Newport, Sept. 3, 1856.

MELVILLE, Donald Burton, biochemist; b. Netherton, Eng., Jan. 30, 1914; s. Donald and Hilda (Burton) M.; came to U. S., 1925, naturalized, 1932; B.S., U. Ill., 1936, M.S., 1937, Ph.D., 1939; m. Dorothy Robbin, July 20, 1940; children—Joan (Mrs. Walter Corcoran), Laura. Faculty, Cornell U. Med. Coll., N.Y.C., 1939-60; prof., chmn. dept. biochemistry U. Vt. Coll. Medicine, Burlington, 1960——. Mem. Am. Soc. Biol. Chemists, Am. Chem. Soc., Sigma Xi, Phi Kappa Phi, Phi Lambda Upsilon. Research, numerous publs. on determination of chem. structure of vitamin, biotin; studies on structure of penicillin; syntheses of biochems., labeled with radioisotopes; origin and function of blood ergothioneine; intermediary metabolism of amino acids. Home: 1 Fern St., Burlington, Vt. 05401.*

MELVILLE, Sir Harry (Work), Brit. chemist; b. Edinburgh, Scotland, Apr. 27, 1908; B.Sc. in Chemistry, U. Edinburgh, 1930, later Ph.D., D.Sc. Fellow Trinity Coll., Cambridge, 1933-43; asst. dir. colloid sci. lab. Cambridge U., 1938-40; prof. Aberdeen (Scotland) U., 1940-48; Mason prof., dir. chemistry dept. Birmingham (Eng.) U., 1948-56; sec. Dept. Sci. and Indsl. Research, 1956-65; Sci. adviser to chief supt. chem. def. Ministry Supply, 1940-43; supt. radar research sta., 1943-45; mem. Adv. Council for Sci. and Indsl. Research, 1946-51, sci. adv. council Ministry Supply, 1949-51, 53-56, Adv. Council on Sci. Research and Devel., 1953-56, sci. adv. council Brit. Electricity Authority, 1949-56; mem. sci. adv. council Ministry Power, 1954-60, mem. adv. council on research and devel., 1960——. Fellow Royal Soc., 1941 (Davy medal 1955). Author: Experimental Methods in Gas Reactions, 1938. Devised techniques for study of mechanism and chemistry of radical chain reactions, especially in formation of polymers; elementary photochem. decomposition of simple molecules. Home: Norwood, Dodds Lane, Chalfont St., Giles, Buckshire, Eng.

MELVILLE, Herman, Am. writer, naturalist; b. N.Y.C., Aug. 1, 1819; s. Allan and Maria (Gansevoort) M.; ed. Albany Acad.; m. Elizabeth Shaw, Aug. 4, 1847; children—Malcolm, Stanwix, Elizabeth, Frances (Mrs. Henry B. Thomas). Clk., N.Y. State Bank, 1834; decided to go to sea, 1837, became cabin boy in ship Highlander (described in book Redburn, 1849); mem. crew of whaling ship Acushnet, 1841-42 (recreated in Moby Dick, 1851); jumped ship at Marquesas Island, July 1842, lived among cannibals (described in Typee, 1846); escaped island aboard Australian ship Lucy Ann, then jumped ship in Tahiti, Sept. 1842 (described in Omoo, 1847); returned to Boston in U. S. frigate United States, Oct. 14, 1844 (described in White-Jacket or The World in a Man-of-War, 1850); his books Typee and Omoo were successful as travel books, created controversy in their description of effects of Christian missionaries on natives; went to Paris, 1849, later used city as setting for part of book Israel Potter (1855); moved to Arrowhead, Pittsfield, Mass.; pub. Moby Dick (chpts. on cetology unparalleled sci. study on the subject), 1851; wrote The Piazza Tales, 1856, The Confidence Man, 1857; moved to N.Y.C., 1863; outdoor customs insp., 1866-85; wrote his last novel Billy Budd, 1891, also wrote poetry. Died in poverty and obscurity, N.Y.C., Sept. 28, 1892.

MELVILLE, K(enneth) I(van), pharmacologist; b. Jamaica, B.W.I., July 5, 1902; s. Nathan Josiah and Rose (Smith) M.; B.Sc., McGill U., Montreal, Que., Can., 1926, M.D., C.M., 1926, M.Sc., 1931; m. Gladys Vivian Brodber, Apr. 14, 1933; children—Enid Lorraine (Mrs. Virgil Wright), David Louis. Faculty, McGill U., Montreal, 1930——, prof., chmn. dept. pharmacology, 1953——. Cons. Royal Victoria, Montreal Gen., Montreal Children's hosps. Recipient medal for research in anesthesia Internat. Coll. Anesthetists, 1952. Mem. Pharmacological Soc. Can. (past pres.), Am. Soc. Pharmacology and Exptl. Therapeutics. Contbg. author: Pharmacology in Medicine, 1965; also numerous articles. Research on actions drugs on coronary circulation, drugs affecting blood pressure and heart function. Home: 4937 Circle Rd., Montreal, Que., Can.*

MELVIN, Mael Avramy, Am. physicist; b. Israel, Mar. 27, 1913 (parents Am. citizens); s. Abraham Elijah and Norma (Gottfried) M.; B.S., U. Chgo., 1933, M.S., 1935, Ph.D., 1938; m. Sophia Jean Brown, Sept. 18, 1946; children—Jonathan David, Paul Michael. Faculty, Columbia, 1938-47; prof. physics Fla. State U., 1952-66; prof. physics Temple U., Phila., 1966——; vis. prof. U. Upsala, Sweden, 1957-58, Instituto di Fisica, Bariloche, Argentina, 1960. Guggenheim Found. fellow, 1951-52, 57-58. Fellow Am. Phys. Soc.; mem. Phi Beta Kappa, Sigma Xi. Research, numerous publs. in algebraic method for determining arrangement of atoms in crystals directly from x-ray data, theory of rates of transformation of phases of matter, method of applying symmetry to analyze phys. phenomena, proof that a magnetic field can exist stably all by itself without currents to cause it. Office: Physics Dept., Temple U., Phila. 19122.*

MENAECHMOS, Greek mathematician; flourished circa 350 B.C.; pupil of Eudoxos; asso. Plato; discovered conic sects. and used them to solve problem of doubling cube (earliest solution of cubic equation); defined meaning of term elements in geometry; discussed meaning of terms theorem, problem; no works survive.

MENAKER, Michael, Am. biologist; b. Vienna, Austria, May 19, 1934 (parents Am. citizens); s. William and Esther (Astin) M.; B.A., Swarthmore Coll., 1955; M.A., Princeton, 1959, Ph.D., 1960; m. Shirley Ann Lasch, June 4, 1955; children—Ellen Margaret, Nicholas. NSF postdoctoral fellow Harvard, 1959-61, NIH fellow, 1961-62; faculty U. Tex., Austin, 1962——, asso. prof. zoology, 1967——. Mem. Am. Inst. Biol. Scis. (mem. coms.), Am. Soc. Zoologists, Soc. Comparative Physiologists, Hibernation Information Exchange, A.A.A.S., Sigma Xi. Asso. editor Comparative Physiology, Midland Naturalist, 1965——. Research, publs. on physiology of seasonal preparation for hibernation in bats and ground squirrels; persistance of endogenous daily rhythms during hibernation, mechanisms of time measurement in photoperiodic control of reprodn. in birds, synchronization of avian biol. clocks by light via an extra-retinal light receptor. Home: 3302 Cherry Tree Circle, Austin, Tex. 78731.*

MENARD, Henry William, Jr., Am. geologist, educator; b. Fresno, Cal., Dec. 10, 1920; s. Henry William and Blanche (Hodges) M.; B.S., Cal. Inst. Tech., 1942, M.S., 1947; Ph.D., Harvard, 1949; m. Gifford Merrill, Sept. 21, 1946; children—Andrew Ogden, Elizabeth Merrill, Dorothy Merrill. Oceanographer, Navy Electronics Lab., San Diego, 1949-55; asso. prof. geology Inst. Marine Resources and Scripps Inst. Oceanography, U. Cal. at San Diego, La Jolla, 1955-61, prof., 1961——. Cons. Am. Tel.&Tel. Co., 1956——; vis. prof. Cal. Inst. Tech., 1959. Decorated Bronze Star medal. Guggenheim fellow Cambridge (Eng.) U., 1962-63. Mem. Nat. Acad. Scis., Geol. Soc. Am., Am. Geophys. Union, A.A.A.S., Sigma Xi. Author: Marine Geology of the Pacific, 1964. Geologist or chief scientist numerous oceanographic expdns.; discoverer numerous undersea mountains and fault scarps. Home: 2615 Ellentown Rd., La Jolla, Cal.*

MENCHIKOFF, Nicolas, geologist; b. Moscow, USSR, Nov. 22, 1900; s. Nicolas and Helen (Ponomareva) M.; Ph.D., U. Paris (France); m. Catherine Serikoff, July 16, 1927; Geol. explorer Sahara Desert, 1924——; with Nat. Center for Sci. Research, from 1934; dir., founder Center for Research in the Sahara, 1943——. Recipient prize in sci. research and tech. progress, 1957. Mem. Nat. Com. for Sci. Research, French Geol. Soc., Overseas Acad. Sci., Soc. Naturalists of Moscow. Research, publs. in geology and morphology in north of Occidental Sahara; geol. maps of Sahara. Home: 42, quai Louis-Blériot, Paris 16. Office: 16, rue Pierre-Curie, Paris 5, France.

MENCKE(N), Otto, German philosopher; b. Oldenburg, Germany, Mar. 22, 1644; M.A., Leipzig, Germany, 1664; licentiate in divinity, 1671; prof. morality Leipzig U., 1668-1707, rector, 5 times, dean faculty philosophy, 7 times; founder Acta Eruditorum (1st sci. jour. in Germany). Died Leipzig, Jan. 29, 1707.

MENDEL, Gregor Johann, Austrian biologist; b. Johann Mendel in Heinzendorf, Silesia, Austria, (now Hyncica, Czechoslovakia;, July 22, 1822; ed. Philosophical Institute, Olmütz. Joined Augustinian cloister (taking name Gregor), Brünn, Bohemia (now Brno, Czechoslovakia), 1843; ordained priest, 1847; reserve teacher of Greek and math. in Gymnasium at Znaim, 1849-51 (failed examination for certification as regular teacher, 1850); studied physics, chemistry, math., zoology, and botany, U. Vienna, 1851-53; sci. tchr. Brünn Realschule, 1854-68; bot. research with peas in monastery garden, 1856-65; elected abbot of monastery, 1868. Discovered dominant and recessive characteristics in pea plants; developed principle of factorial inheritance by which paired elementary units of heredity combine and sort themselves in different generations according to fixed statistical rules while maintaining their identity; 1st explanation of hybridization using controlled pollination and statistical analysis; his work although unnoticed until after his death became basis of modern genetics when it was discovered in 1900. Died Brünn, Jan. 6, 1884.

MENDEL, Kurt, German neurologist; b. 1874; demonstrated that in healthy subjects percussion of dorsum of foot causes dorsal flexion of 2d to 5th toes, whereas in cases of pyramidal tract and other diseases, flexion of toes is plantar, 1904. Died 1946.

MENDEL, Lafayette Benedict, Am. physiol. chemist; b. Delhi, N.Y., Feb. 5, 1872; s. Benedict and Pauline (Ullman) M.; A.B., Yale, 1891, Ph.D., 1893, Larned fellow, 1891-94; research student, univs. of Breslau and Freiburg, 1895-96; Sc.D. (hon.), U. Mich., 1913, Rutgers Coll., 1930; LL.D., Western Res. U., 1932; m. Alice R. Friend. Tchr., Yale, 1892, asst. prof. Sheffield Sci. Sch., 1897-1903, prof. physiol. chemistry, 1903-21, Sterling prof. physiol. chemistry Yale, 1921-. Dir. Russell Sage Inst. Pathology; research asso. Carnegie Instn. of Washington. Recipient Gold medal Am. Inst. Chemists, 1927. Mem. numerous Am. and fgn. Sci. Socs. Author: Nutrition—The Chemistry of Life; Changes in the Food Supply and Their Relation to Nutrition. Editor Jour. Biol. Chemistry, Jour. of Nutrition, chem. monographs of Am. Chem. Soc. Research on vitamins, protein metabolism, physiology of growth, digestion; discovered vitamin A, 1913, also function of vitamin C. Died Dec. 9, 1935.

MENDELEEV, Dmitri Ivanovich, Russian chemist; b. Tobolsk, Siberia, Russia, Feb. 7, 1834; s. Ivan Pavlovich and Marya (Dimitrievna) M.; student Pedagogical Inst., St. Petersburg (now Leningrad, USSR), from 1850; M. in Chemistry, U. St. Petersburg, 1856, D. in Chemistry, 1865; postgrad. U. Heidelberg (Germany), also France, 1859-61; m. Feozva Nikitichna Lescheva (div. 1876); children—Vladimir, Olga; m. 2d, Anna Ivanovna Popov, 1876; children—Lyubova, Ivan, Marya, Vasilii. Prof. chemistry St. Petersburg Inst. Tech., 1863; prof. gen chemistry U. St. Petersburg, 1866-90; dir. Bur. Weights and Measures, 1890-93. New element, mendelevium, named in his honor, 1955. Fellow Royal Soc. (Davy medal 1882, Copley medal 1905), 1892; mem. U. S. Nat. Acad. Scis. (asso.), French Acad. Scis., Russian Chem. Soc. (a founder). Author: The Principles of Chemistry, 2 vols., 1868-70; contbr. articles to sci. publs. Developed periodic classification of chem. elements (using only 63 elements) according to atomic weights, 1869-71 (thereby bringing order to list of elements); discovered periodicity of phys. and chem. properties of elements, used this discovery to predict successfully properties of several as then undiscovered elements, named by him (ekaboron, ekaaluminum, ekasilicon), later discovered and named gallium, 1875, scandium, 1879, germanium, 1886; corrected atomic weights of other elements, such as beryllium (later verified); investigated thermal expansion of liquids; formulated concept of critical temperature in studies of phys. properties of gases and liquids; investigated character of petroleum. Died St. Petersburg, Feb. 2, 1907.

MENDELEJEFF, Dmitri Ivanovich, see Mendeleev, Dmitri Ivanovich.

MENDELEYEV, Dmitri Ivanovich, see Mendeleev, Dmitri Ivanovich.

MENDELOFF, Albert Irwin, Am. physician; b. Charleston, W.Va., Jan. 29, 1918; s. Morris Israel and Esther (Cohen) M.; A.B., Princeton, 1938; M.D., Harvard, 1942, M.P.H., 1944; m. Natalie Lavenstein, Dec. 19, 1943; children—Henry, John, Katherine. Fellow in nutrition Rockefeller Found., 1943-44; nutrition cons. UNRRA mission to Greece, 1944-46; asst. prof. medicine and preventive medicine Washington U., 1949-54, asso. prof., 1955; gastroenterology cons. Barnes Hosp., St. Louis, 1952-55; asso. prof. medicine Johns Hopkins, 1955——; physician-in-chief Sinai Hosp., Balt., 1955——. Sr. surgeon Res. USPHS. Mem. Assn. Am. Physicians, Am. Soc. Clin. Investigation, Central Soc. Clin. Research, Am. Fedn. Clin. Research, Am. Gastroente. Assn., Phi Beta Kappa, Alpha Omega Alpha. Research in gastrointestinal physiology; liver disease; absorption from the small intestine; colonic physiology and disease. Home: 2109 Northcliff Dr., Balt. 9. Office: Sinai Hosp., Balt.*

MENDELSON, Elliott, Am. mathematician; b. N.Y.C., May 24, 1931; s. Joseph and Helen (Bienstock) M.; A.B., Columbia, 1952; M.A., Cornell U., 1954, Ph.D., 1955; m. Arlene Zimmerman, Jan. 25, 1959; children—Julia, Hilary, Peter. Instr., U. Chgo., 1955-56; jr. fellow Soc. Fellows, Harvard, 1956-58; Ritt instr. Columbia, 1958-61; faculty Queens Coll., City U., N.Y., 1961——, prof. math., 1965——, dir. instr. NSF math. program for high sch. students, 1964——. Mem. Am. Math. Soc., Math. Assn. Am., Assn. for Symbolic Logic, Am. Assn. U. Profs., Sigma Xi. Author: Introduction to Mathematical Logic, 1964; also articles. Research in axiomatic set theory and math. logic especially ind. various important propositions of axiomatic set theory: axiom of choice, axiom of restriction. Home: 10 Pinewood Rd., East Hills, Roslyn, N.Y. 11576. Office: Queens Coll., Flushing, N.Y. 11367.*

MENDELSSOHN, Kurt Alfred George, physicist; b. Berlin, Germany, Jan. 7, 1906; s. Ernst Moritz and Elizabeth (Ruprecht) M.; Dr. phil., M.A., Berlin U., 1930; M.A., Oxford (Eng.) U., 1944; m. Jutta Lina Charlotte Zarniko, Dec. 19, 1932; children—Corinna (Mrs. Joseph Winston Welch), Ursula (Mrs. Keith George Meadows), Monica, Diana, James. Faculty, U. Berlin, 1929-31, U. Breslau (Germany), 1931-33; faculty Oxford U., 1933——, reader physics, 1955——; vis. prof. Rice U., 1952, Purdue U., 1956, Tokyo (Japan), 1960, Kumasi U., Ghana, 1964. Fellow Royal Soc., 1951 (Hughes medal 1967), Inst. Physics; mem. Phys. Soc. London (past vice chmn.). Author: What is Atomic Energy, 1946; Cryophysics, 1960; The Quest for Absolute Zero, 1966; also numerous articles. Editor: Cryogenics, 1960——. Research on low temperature physics especially superconductivity and superfluidity; discovered magneto-caloric effect in superconductors, mobility of helium film, mechano-caloric effect in helium; med. physics; transuranic metals. Home: 235 Iffley Rd., Oxford, Eng. Office: Clarendon Lab., Parks Rd., Oxford, Eng.

MENDENHALL, Thomas Corwin, Am. physicist; b. Hanoverton, O., Oct. 4, 1841; s. Stephen and Mary

(Thomas) M.; pub. sch. edn.; Ph.D. (hon.), Ohio State U., 1878; Sc.D. (hon.), Rose Poly. Inst.; LL.D., U. Mich., 1887, Western Res. U., 1912; m. Susan Allen Marple, July 12, 1870; 1 son, Charles Elwood M. Prof. physics and mechanics Ohio State U., 1873-78, prof. physics, 1881-84, emeritus prof., 1884; prof. physics Imperial U. Japan (work contbd. to govt. meteorol. system), 1876-81; prof. U. S. Signal Corps, 1884-86; pres. Rose Poly. Inst., 1886-89; supt. U. S. Coast and Geod. Survey, 1889-94; pres. Worcester Poly. Inst., 1894-1901; in Europe, 1901-12. Supt. U. S. weights and measures, 1889-94; mem. U. S. Light House Bd., 1889-94; mem. 1st Bering Sea Commn., 1891; mem. U. S. and Gt. Britain Boundary Line Survey Commn., 1892-94. Recipient medal Paris Expn., 1900. Hon. fellow Am. (Gold medal 1901), Nat. geog. socs., Franklin Inst. (Gold medal, 1918); fellow Am. Acad. Arts and Scis. Author: A Century of Electricity. Developed state weather service of Ohio; 1st to propose ring pendulum for measurement of absolute force of gravity; conducted studies in electricity, seismology, atmospheric electricity. Died Mar. 23, 1924.

MENEELY, George Rodney, Am. physician; b. Hempstead, N.Y., Sept. 30, 1911; s. Charles Dickinson and Emily Frances (Gahn) M.; B.S., Princeton, 1933; M.D., Cornell U., 1937; children—Judith (Mrs. Jerome Wichelns), Denny (Mrs. Girard A. Chapnick), George Rodney III. Instr. medicine La. State U. Med. Sch., New Orleans, 1941-43; from instr. to asso. prof. dept. medicine Vanderbilt U., Nashville, 1943-61; asso. prof. medicine Northwestern Med. Sch., Chgo., also dir. dept. sci. assembly A.M.A., 1962-63; prof. nuclear medicine, internist U. Tex., Houston, 1963-65; prof. medicine, asso. dean La. State U. Med. Center, Shreveport Sch. Medicine, 1966—. Cons., VA, Social Security Adminstrn., Bur. Employees Compensation U. S. Dept. Labor, U. Tex. M.D. Anderson Hosp., Tex. Inst. Rehab. and Research, IBM Sci. Center, Houston. Diplomate Am. Bd. Internal Medicine. Fellow A.C.P., Am. Coll. Cardiology (pres. 1957-58), Am. Coll. Chest Physicians, A.A.A.S.; mem. Am. Soc. Exptl. Pathologists, Am. Physiol. Soc., Biophys. Soc., Soc. Nuclear Medicine, So. Soc. for Clin. Research, Soc. for Exptl. Biology and Medicine, Am. Fedn. for Clin. Research (founding mem., v.p. 1953), Am. Thoracic Soc., N.Y. Acad. Scis., Assn. for Computing Machinery, Assn. Am. Med. Colls., Am. Assn. U. Profs., Am. Therapeutic Soc., Am. Soc. Internal Medicine, World (U. S. com.), So. (vice chmn. sect. on medicine 1965-66), Pan Am. med. assns., Am. Cancer Soc., Am. Heart Assn. (fellow clin. cardiology, fellow council on epidemiology). Editorial bd. Am. Jour. Cardiology, 1957—; asso. editor Medicina Thoracalis, 1961—. Contbr. numerous articles profl. jours. Research in field of human physiology, especially heart and lung function and body composition; nuclear medicine in applied research. Home: 260 Normandy Village, Shreveport 71104. Office: 1541 Kings Hwy., Shreveport, La. 71103.*

MENEES, James H., Am. entomologist; b. Checotah, Okla., Nov. 24, 1929; s. James Matthew and Gladys (Holt) M.; A.A., Hartnell Coll., 1951; B.A., San Jose State Coll., 1953, M.S., 1957, Ph.D., 1959; m. Carolyn Anne Gabbert, June 20, 1953; children—Charolette Jean, James Jeffrey. Research asst. Cornell U., 1955-59; asso. prof. entomology Cal. State Coll., Long Beach, 1959—. NSF Research grantee, 1960-62. Mem. Entomol. Soc. Am., Am. Soc. Zoologists, A.A.A.S., Am. Inst. Biol. Sci. Author: (with Shipley, Kroman, Carpenter) Laboratory Experiments in Biology, 1963; Dissection Guide to Lubber Grasshopper, Crayfish, 1965; also articles. Research on anatomy, histology, embryology and histochemistry insects. Home: 15141 Newcastle Lane, Huntington Beach, Cal. Office: Dept. Biology, Cal. State Coll., Long Beach, 1 Cal.*

MENELAOS OF ALEXANDRIA, Greek mathematician; flourished 98 A.D.; author: Sphaerica, 3 books. First to separate trigonometry from stereometry and astronomy; 1st definition of a spherical triangle; used relation between spherical and plane triangles (Menelaos' theorem) for solving spherical triangles.

MENESTOR, Greek writer on botany; flourished circa 4th century B.C. 1st Greek to study plants inductively; writings were often quoted by Theophrastus; used Pythagorean theory of opposition of warm and cold to divide plants into those which because of their warm nature could grow in water or cold parts of earth and those with a cold nature which need a warm climate.

MÉNÉTRIÉS, Edouard, entomologist; b. Paris, France, Oct. 2, 1802; conserver of rarities, zool. mus. Imperial Acad. Scis., St. Petersburg, Russia; contbr. papers on insects from Russia and Siberia to sci. publs.; named species of Lepidoptera and Coleoptera from many parts of world; described Coleoptera collected by Vosnesensky in Cal. Died Apr. 10, 1861.

MENGEL, John Thomas, Am. electronic engr., govt. ofcl.; b. Ballston Lake, N.Y., Apr. 16, 1918; s. Arthur Clayton and Maude (Thomas) M.; B.Sc. in Physics, Union Coll., 1939; postgrad. Lafayette Coll.; m. Joan Wilson Loveland, Aug. 10, 1940; children—Nancy Ann, Judith Loveland, John Thomas. Test engr. Gen. Electric Co., Schenectady, 1940-42; radio engr. detection system devel. Bur. Ships, USN, Washington, 1942-46; electronic scientist Naval Research Lab., Washington, 1946-58; asst. dir. tracking and data systems Goddard Space Flight Center, NASA, Greenbelt, Md., 1958—. Mem. I.E.E.E. (sr.), A.A.A.S. Research and devel. on satellite tracking, command, and data acquisition and processing. Home: 700 Hobbs Dr., Silver Spring, Md. Office: Goddard Space Flight Center, Greenbelt, Md.*

MENGER, Wolfgang, German pediatrician; b. Berlin, Germany, July 19, 1919; s. Ludwig and Erika (Ruppert) M.; ed. Berlin, Danzig (Poland), Vienna (Austria) univs.; M.D.; m. Hildegard Schall, May 20, 1948; children—Irmgard, Waltraud, Dietmar, Hartmut. Med. asst. pediatric clinics Bremen, Germany, U. Mainz (Germany); head physician Norderney, Nordseebad (Germany) Pediatric Clinic. Mem. Meeresheilkunde im Deutschen Bäderverband; Landesgesundheistrat Niedersachsen, Internat. Soc. Biometeorology. Author: Häufigkeit und Art meteorotroper Erscheinungen im Kindesalter; Leitfaden der Meeresheilkunde, 1964; Meteoropathologie und Klmatherapie, Thalassotherapie. Research, publs. on thalassotherapy, bioclimatology, pediatric medicine, meteoropathology. Home: Tannenstrasse 14, Norderney. Office: Kinderkrankenhaus Seehospiz Kaiserin Friedrich, Norderney, W. Germany.

MENGERT, William Felix, Am. obstetrician, gynecologist; b. Washington, Nov. 13, 1899; s. Ulric T. and Margaret (Johnson) M.; S.B., Haverford (Pa.) Coll., 1921; M.D., Johns Hopkins, 1927; m. Ida Gaarder, June 14, 1924; children—Eric John, Ann Katharine. Univ. Ia. Research fellow Gynecean Hosp. Inst. U. Pa., 1932-34; asst. prof. obstetrics and gynecology, U. Ia., 1934-38, asso. prof., 1938-43; prof. and chmn. dept. obstetrics and gynecology. Southwestern Med. Sch. U. Tex. and Parkland Meml. Hosp., Dallas, 1943-55; prof. and head dept. obstetrics and gynecology Coll. Medicine, U. Ill., 1955—. Med. adv. bd. Planned Parenthood Chgo., 1956—, chmn., dir., 1956-61. Gen. program sec. 2d Am. Cong. Obstetrics, Gynecology, St. Louis, 1941, gen. program chmn. 3d Congress, 1947. Fellow A.C.S.; mem. A.M.A. (sec. sect. obstetrics and gynecology 1944-47, chmn. 1947-48), Am. (pres. 1957-58), Central (pres. 1947-48), assns. obstetricians and gynecologists, Am. Coll. Obstetricians and Gynecologists (pres. 1955), Am., Chgo. (pres. 1962-63) gynecol. socs., Sigma Xi, Alpha Omega Alpha, Nu Sigma Nu. Author: Postgraduate Obstetrics, 1947. Contbr. med. jours. Research in toxemia of pregnancy; devised method by X-ray of estimating pelvic capacity. Home: 145 Scottswood Rd., Riverside, Ill. 60546. Office: 840 S. Wood St., Chgo. 60680.*

MENGOLI, Pietro, Italian mathematician; b. Bologna, 1625. Author: Norae Quadraturae Arithmeticae, 1650. Did work on infinite series, proved divergence of harmonic series; demonstrated convergence of reciprocals of triangular numbers. Died 1686.

MENGUY, Rene, physician; b. Prague, Czechoslovakia, Feb. 4, 1926; s. Auguste and Beatrice (Adam) M.; B.A., U. Hanoi (Indochina), 1944; M.D., U. Paris (France), 1951; Ph.D., U. Minn., 1957; m. Emilie Rigacci, Aug. 10, 1950; children—Ghislaine, Jean. Came to U. S., 1951, naturalized, 1956. Faculty, U. Okla., 1957-61, U. Ky. Med. Center, 1961-65; prof. physiology, also prof., chmn. dept. surgery U. Chgo., 1965—. Recipient Mayo Found. Alumni award, 1956. John and Mary R. Markle scholar, 1958-63; Fulbright grantee, 1951-52. Mem. A.C.S., A.M.A., Am. Fedn. Clin. Research, N.Y. Acad. Scis., Soc. Exptl. Biology and Medicine (editorial bd.), Am. Assn. Cancer Research, Am. Gastroent. Assn., Am. Physiol. Soc., Internat. Soc. Surgery, Am. Surg. Assn., Sigma Xi, others. Editorial bd. Am. Jour. Digestive Diseases. Research, publs. on gastric function emphasizing role of gastric mucus in ulcers and influence of cortisone and aspirin on the stomach. Home: 5811 Dorchester St., Chgo. 60637.*

MÉNIÈRE, Prosper, French otologist; b. France, 1801; M.D., Paris, 1828; at least 1 son, Emile Antoine. Author: Traité des maladies de l'oreille, 1853; Cicéron médecin (med. discoveries of Cicero), 1862. 1st to describe syndrome characterized by deafness, dizziness, tinnitus, nausea and vomiting, occurring in some diseases of internal ear (Ménière's syndrome), 1861. Died 1862.

MENK, Walther, German hygientist; b. Arolsen, Feb. 18, 1892; s. Adolf and Emma (Ebersbach) M.; ed. U. Lausanne, (Switzerland), U. Marburg (Germany), U. Leipzig (Germany); M.D.; agrégé, 1938; m. Clara Scholz, Feb. 5, 1957; 1 dau., Gisela. Med. asst. Hamburg Tropical Inst., 1920-23; physician med. dept. United Fruit Co., 1923-28; prof. hygiene and tropical medicine Colombia & Baines Div.; staff bacteriology dept. Berlin Health Office, 1928-31; asst. Research Offices for Contagious Diseases, Fribourg, 1931-32; sci. dir. Inst. Behring, Rio de Janeiro, Brazil, Buenos Aires, Argentina, 1932-37; later head clin. dept.; dir. dept. Tropical Inst., Hamburg, 1940-45; became prof. hygiene tropical medicine U. Hamburg, 1939; hygienist 14th army, 1944-45; scientist Prof. von Bormann's Medico-Diagnostic Inst., Bad Nauheim, Germany, 1948-49; sci. dir. Biologische Arbeitsgemainschaft, 1949-62; staff Hamburg Pub. Health Office, 1962—. Recipient Bernhard Nocht medal for tropical med. services. Mem. Sci. Council German Green Cross (pres.), German Africa Soc., German Soc. for Hygiene and Microbiology, German Soc. for Tropical Medicine. Author: Uber Behandelungsversuche der chron. Amöbenruhr mit Yatren, 1921; Behandlung der afrikan, Schlafkrankheit mit Bayer 205; Chemotherapie der Malaria, 1943; Sontochinbehandlung der Malaria, 1951; Neuere chemische physiologische und serologische Untersuchungen über die Wirkungsweise antiretikulärer and anderer Zytotoxischer Sera, 1957. Address: Lich (Oberhess), Schloss, West Germany.

MENNE, Fritz Adolf, German biochemist; b. Allenstein, Germany, Nov. 5, 1910; s. Paul and Marie (Hellenbach) M.; M.D., U. Gottingen (Germany); m. Ingrid Kriegeskorte, Dec. 8, 1945; children—Claudia, Cornelia. Prof. biochemistry, from 1951; dir. Inst. Biochemistry, U. Rostock (Germany), 1954-56; now with U. Münster. Mem. Physiol.-Chem. Soc., Biochem. Soc. Author: Einführung in die Chemie und klinische Chemie. Research, publs. on medicine, chemistry. Home: Eichenweg 1, Münster/Westphalia. Office: Physiologisch-chemisches institut der Universität, Waldeyerstrasse 15, Münster/Westphalia, W. Germany.

MENNINGER, Karl Augustus, Am. psychiatrist; b. Topeka, Kan., July 22, 1893; s. Charles Frederick (M.D.) and Flora (Knisely) M.; student Washburn Coll., 1910-12; Indiana U., summer 1910; A.B., U. of Wis., 1914, M.S., 1915; M.D. cum laude, Harvard, 1917; L.H.D., Park Coll., 1955; LL.D., Jefferson Med. Coll. Phila., 1956, Parsons Coll., 1960; D.Sc., Washburn U., Topeka, 1949; m. Grace Gaines, Sept. 9, 1916 (div. Feb. 1941); children—Julia (Mrs. A. H. Gottesman), Robert Gaines, Martha (Mrs. William Nichols); m. 2d, Jeanetta Lyle, September 8, 1941; one daughter, Rosemary Karla Jeanetta. Chairman of the board of trustees, dir. ednl. dept. Menninger Found.; clin. prof. psychiatry U. Kan.; neuropsychiatrist Stormont-Vail Hosp., Topeka; chief staff Menninger Clinic; dean Menninger Sch. Psychiatry; mem. edn. com., sr. cons. Topeka VA Hosp., Topeka State Hosp.; cons. Fed. Bur. Prisons, Forbes AFB Hosp.; cons. Office Vocational Rehabilitation, Dept. Health, Education and Welfare. Mem. spl. commn. on psychiatry, OSRD, ETO, adv. to surgeon gen. U. S. Army, 1945. Served as lt. USNRF, 1918-21. Dir. Topeka Inst. Psychoanalysis; v.p. Am. League to Abolish Capital Punishment; cons. research staff Com. on Rights of Mentally Ill of Am. Bar Found.; adv. com. Internat. Survey Correctional Research and practice, Sacramento; adv. commn. on instnl. mgmt. Bd. Social Welfare of State of Kan.; adv. bd. Am. Assn. Phys. and Mental Rehab.; med. adv. council Am. Assn. Rehab. Therapists; adv. council Law-Medicine Research Inst., Boston U. Sch. Law; adv. council European Ednl. Center, Assn. Migros Schs., Zurich, Switzerland. Life fellow Am. Psychiat. Assn., A.C.P.; fellow A.M.A.; life mem. Am. Med. Writers Assn., Am. Orthopsychiat. Assn., Am. Psychoanalytic Assn.; charter mem. Central Neuropsychiat. Assn., World Med. Assn., Assn. for Research Nervous and Mental Diseases; mem. Nat. Assn. Music Therapy (hon. adv. bd.), Nat. Congress Am. Indians, Am. Hort. Council, A.A.A.S. Author: Why Men Fail (with others), 1918; The Human Mind, 1930 (rev. 1945); The Healthy-Minded Child (with others), 1930; Man Against Himself, 1938; America Now (with others), 1938; (with Mrs. Menninger) Love Against Hate, 1942; A Guide to Psychiatric Books (with Devereux), 1950; Manual for Psychiatric Case Study, 1952; Theory of Psychoanalytic Technique, 1958; (selected papers) A Psychiatrist's World. 1959; (with Martin Mayman and Paul Pruyser) The Vital Balance, 1963; also articles relating to field. Editor-in-chief Bull. of Menninger Clinic; mem. editorial bd. Archives of Criminal Psychodynamics, Jour. of Diseases of the Nervous System; Psychoanalytic Quar. Mem. adv. com. numerous civic and govtl. orgns. Contbn. to study of psychol. factors in somatic illnesses, criminology, penology, suicide, mil. and indsl. psychology, influenza and mental health. Home: 1819 Westwood Circle, Topeka. Office: Menninger Clinic Bldg., also Vets. Adminstrn. Hosp., Topeka, Kan.*

MENNINGER-LERCHENTHAL, Erich Theodor, neurologist; b. Hermagor-Kärnten, Germany, May 8, 1898; s. Albert and Margarete (Holzmann) M.-L.; M.D., U. Graz (Austria); m. Corinna Calvagni, Sept. 23, 1933. Physician, clinic, Graz; psychiatrist Tulln-Vienna (Austria) Sanitarium, Steinhof-Vienna Hosp.; specialist, Vienna. Mem. Vienna Soc. Doctors, Psychiatry Soc. Author: Das Truggebild der eigenen Gestalt, 1935; Der eigene Doppelgänger, 1946; Das europäische Selbstmordprobelm, 1947; Periodizität in der Psychopathologie, 1960; Forschergeist, 1964. Contbr. articles to profl. publs. Address: Peter Frankgasse 1, Vienna 9, Austria.

MENODOTOS OF NICOMEDIA, Greek physician; flourished probably circa 150; student of Antiochus of Ascalon; follower of Pyrrhon; leader empirical sch. medicine. Numerous writings including med. work which Erasmus translated into Latin; Galen wrote against Menodotos. One of 1st to synthesis empirical and sceptical medicine.

MENON, Thuppalay Kochugovinda, astronomer; b. Tattamangalam, India, Dec. 19, 1926; s. Cherucat Achuthan and Thankam (Menon) Nair; B.S., Annamalai U., India, 1947; M.S. (Gordon Mackay scholar), Harvard, 1953, Ph.D., 1956; m. Rama Chintakindi, June 10, 1962; children—Ravi Shanker, Kusum.

Lectr., Agassiz radio astronomer Harvard, 1956-58; asst. prof. astronomy, elec. engring. U. Pa., 1958-60; asso. scientist Nat. Radio Astronomy Obs., Green Bank, W.Va., 1960-62, 63-65, scientist, 1965—; asso. prof. physics, elec. engring. Ohio State U., 1962-63. Mem. Am., Royal astron. socs., Internat. Astron. Union, Internat. Sci. Radio Union. Research, publs. on radio-astron. techniques, star formation, dynamics of interstellar medium. Address: P.O. Box 2, Green Bank, W.Va. 24944.*

MENSBRUGGHE, Gustav van der, see van der Mensbrugghe, Gustav.

MENSHOV, Dmitrii Evgenevich, Russian mathematician; b. Moscow, Apr. 18, 1892; grad. Moscow U., 1916; D.Phys.-Math. Scis. Faculty, Moscow U., 1922—, prof., 1935—. Recipient Stalin prize, 1951. Mem. USSR Acad. Scis. (corr.). Author: The Limits of Indefiniteness of Fourier Series, 1952; The Limits of Indefiniteness of Partial Sums of Universal Trigonometric Series, 1954; Almost Converging Trigonometric Series, 1955; The Summation of Orthogonal Series by Linear Methods, 1960. Research on orthogonal functions and trigonometric series; obtained result in uniqueness of representing functions by trigonometric series, 1916; solved problem of representation of functions by trigonometric series, 1940. Address: Mathematics Dept., Moscow University, Moscow, USSR.

MENSUTKIN, Boris Nikolaevich, Russian chemist; b. Apr. 29, 1874; prof., St. Petersburg; author widely used textbooks; editor works of Lomonosov; research in phys.-chem. analysis. Died Sept. 15, 1938.

MENTZER, Charles, French chemist; b. Lampertheim, France, May 27, 1911; s. Michel and Salomé (Heimburger) M.; student Faculté de Pharmacie et des Scis., Strasbourg, France, then Paris, 1929-34; m. Marie Jesel, Sept. 7, 1935; 1 child, Michel. Head research lab. Establissements Roussel-Uclaf, Paris, 1939-47; dir. research Centre National de la Recherche Scientifique, 1948-50; research chief Faculté des Sciences, Lyon, France, 1950-52, prof., 1953-58; prof. chemistry Musée national d'Histoire naturelle, Paris, 1958—. Mem. Comité National de la Recherche scientifique. Mem. French, Am. chem. socs., French Biochemistry Soc. Author: Actualités de phytochimie fondamentale, 2 vols., 1964, 66; also numerous articles. Discovered (with P. Mennier, D. Molho) anticoagulant action of phenyl-indane-dione; discovered method for synthesis of plant substances (coumarins, chromones, xanthones, ratenoids) by non-catalized thermic condensation. Died April 1967.

MENZEL, David W., Am. oceanographer; b. India, Feb. 22, 1928 (parents Am. citizens); s. Emil W. and Ida (Thrun) M.; B.S., Elmhurst Coll., 1949; M.S., U. Ill., 1952; Ph.D., U. Mich., 1957; m. Dorothy A. Adamy, Sept. 7, 1951. Research biologist Bermuda Biol. Sta., 1957-62; asso. scientist Woods Hole (Mass.) Oceanographic Inst., 1962—. Mem. Am. Chem. Soc. Research, publs. on distbn., prodn. and decomposition of organic matter in oceans. Home: 75 Elm Rd., Falmouth, Mass. 02540. Office: Woods Hole Oceanographic Inst., Woods Hole, Mass. 02543.*

MENZEL, Donald Howard, Am. astrophysicist; b. Florence, Colo., Apr. 11, 1901; A.B., U. Denver, 1920, A.M., 1921, Sc.D. (hon.), 1954; A.M. Princeton, 1923, Ph.D., 1924; A.M. (hon.), Harvard, 1942; m. Florence E. Kreager, June 17, 1926; children—Suzanne Kay (Mrs. James E. Lindeman), Elizabeth Ina (Mrs. Bernard Davis). Faculty, U. Ia., 1924-25, Ohio State U., 1925-26, U. Cal., Berkeley, 1926-32; faculty Harvard, Cambridge, Mass., 1932—, prof., 1938—, chmn. dept. astronomy, 1946-49, dir. Harvard Coll. Obs., 1954-66, Paine prof. practical astronomy, 1956—; research scientist Smithsonian Astrophys. Obs., 1966—. Chmn., Boyden Obs. Council, 1954—; chief scientist GCA, 1959—; mem. numerous expdns.; lectr. various univs. Recipient A. Cressy Morrison prizes N.Y. Acad. Scis., 1926, 28, 47; Thomas Alva Edison Found. award, 1957. Fellow Am. Astron. Soc., Am. Geophys. Union, Am. Astronautical Soc.; mem. Am. Acad. Arts and Scis., A.A.A.S., Am. Assn. Variable Star Observers, Am. Math. Soc., Am. Philos. Soc., Am. Phys. Soc., Astron. Soc. Pacific, I.E.E.E. (sr.), Internat. Acad. Astronautics of Internat. Astronautical Fedn. Internat. Astron. Union (pres. commn. 17 1964-67), Internat. Geophys. Union, Internat. Radio Sci. Union, Nat. Acad. Scis., Nat. Acad. Scis. India (hon. fellow), Royal Astron. Soc., Royal Soc. Scis. Liege (corr.), Phi Beta Kappa, Sigma Xi. Author: Our Sun, 1949; Flying Saucers, 1953; Fundamental Formulas of Physics, 1955; Mathematical Physics, 1953; The Universe in Action, 1957; The Radio Noise Spectrum, 1960; Writing A Technical Paper, 1961; The Friendly Stars, 1963; The World of Flying Saucers, 1963; A Field Guide to the Stars and Planets, 1964; Principles of Atomic Spectra, 1968. Authority on sun's chromosphere; with J. C. Boyce discovered that sun's corona consists in part of oxygen, 1933; research on interpretation of stellar and nebular spectra, planetary atmospheres, surface of the moon, wave mechanics and atomic spectra, theory of reactions and equilibria at high temperatures; advanced theory that flying saucers are merely natural optical phenomena. Home: 32 Hubbard St., Cambridge, Mass. 02138.*

MENZEL, Erich Harald, physicist; b. Danzig, Poland, Aug. 13, 1918; s. Karl and Martha (Allert) M.; Ph.D., U. Tübingen (Germany); m. Christel Kopp, Sept. 8, 1947. Faculty, Phys. Inst., Technische Hochschule, Braunschweig, W. Germany. Author: Studies in the Preparation and Behavior of Nearly Perfect Metal Surfaces, 1963. Research, publs. in theory of transfer in optics, physics of surfaces of metals, surface energy of crystals, Fourier optics and ion-linear information transfer, surface physics of single crystals, electron diffraction, epitaxy. Home: Kasernenstrasse 32, Braunschweig. Office: Physikal. Institut Technische Hochschule, Braunschweig, W. Germany.

MENZEL, Heinz, German geophysicist; b. Wehlau, Germany, June 15, 1910; s. Julius and Hedwig (Kruger) M.; Ph.D., U. Königsberg (Germany); m. Helga Hansen, Oct. 14, 1950; children—Dagmar, Helmut. From instr. to prof. geophysics Institut für die Physik des Erdkörpers. Corr. mem. Colombian Acad. Scis. (Bogota), German Geophysics Soc., European Asso. Exploration Geophysics, Seismology Soc. Am. Co-editor: Geophysikalische Probleme series. Contbr. chpts. sci. publs. Home: Harksheide bei Hamburg, Am. Hochsitz 4. Office: Institut für die Physik des Erdkörpers, 2 Hamburg 13, Binderstrasse 22, W. Germany.

MENZEL, Robert Winston, Am. biologist; b. nr. Williamsburg, Va., Jan. 29, 1920; s. George Earnest and Susan (Gary) M.; B.S., Coll. William and Mary, 1940, M.A., 1943; Ph.D., Tex. A. and M. U., 1954; m. Margaret Young, Apr. 9, 1949; children—Robert Winston, Gary Patterson, Mary Linda. Research asst. Va. Fisheries Lab., Yorktown, Va., 1940-42, asst. biologist, 1940-46; biologist Tex. A. and M. Research Found., Houma, La., 1947-49; instr. Tex. A. and M. U., College Station, 1953-54; faculty Fla. State U., Tallahassee, 1954—, asso. prof. biology, 1959—. Recipient Fla. Wildlife Fedn. Conservation award, 1962. Mem. A.A.A.S., Am. Fisheries Soc., Am., Am. Soc. Ichthyologists and Herpetologists, Am. Soc. Limnology and Oceanography, Am. Soc. Zoologists, Assn. Southeastern Biologists, Gulf Caribbean Fisheries Inst., Nat. Shellfish Assn., Sigma Xi, Phi Sigma. Research, publs. on ecology, growth, feeding, predators, diseases, anatomy, physiology oysters, research on rearing and methods farming, growth rate, hybridization, cytotaxonomy, chromosomal behavior in meiosis and mitosis two species quahog clams. Home: 1605 Kolopakin Nene, Tallahassee, Fla. 32301.*

MENZEL, Ronald George, Am. soil scientist; b. Independence, Ia., Jan. 23, 1924; s. Raymond Gerald and Bonnie (Dillard) M.; B.S., Ia. State U., 1947; Ph.D., U. Wis., 1950; m. Elsie Gray Burke, Feb. 23, 1952; children—Martha Jean, Robert Gray. Soil Scientist U. S. Dept. Agr., Beltsville, Md., 1950—. Mem. Am. Soc. Agronomy, Am. Chem. Soc., A.A.A.S. Soil Conservation Soc. Am. Research, publs. on accumulation and movement of radioactive fission products in soils and plants; one of 1st to discover importance of radiostrontium among fission products taken up by plants from soils; showed liming and other soil mgmt. practices may depress uptake of radiostrontium; demonstrated importance of direct contamination of crop plants with airborne radionuclides. Home: 3420 Chatham Rd., Adelphi, Md. 20783. Office: U. S. Dept. Agr., Beltsville, Md. 20705.*

MENZIES, Robert J., Am. biol. oceanographer; b. Denver, Dec. 2, 1923; s. Walter J. and Florence (Lundberg) M.; A.B., Coll. Pacific, 1945; M.A., Pacific Marine Sta., 1949; Ph.D. (Hancock Found. fellow), U. So. Cal., 1951; m. Lucille T. Brinkley, Dec. 3, 1956; children—Diana, Kathy, Stephen, Christopher. With San Diego Natural History Mus., 1943; research asst. Hancock Found., So. Cal., 1946-47; instr. Pacific Marine Sta., 1947-49; lectr. marine biology Modesto Jr. Coll., 1948; asso. zoology U. Cal. at Davis, 1951-52; dir. biology program Lamont Geol. Obs., Columbia, 1955-60; dir. Cientifico Estacion Inv. Marinas de Margarita Venezuela Fund LaSalle de Ciencias Natural Caracas, 1960; research asso. U. So. Cal., 1960-61, asso. prof., 1961-62; prof. zoology, dir oceanographic program Duke, Durham, N.C., 1961-67; prof. oceanography, Fla. State U., 1967—. Ellsworth fellow. Recipient William F. Clapp Meml. award Beta Beta Beta; Phi Sigma award. Mem. A.A.A.S. (asso.), Phi Beta Kappa, Sigma Xi, Phi Gamma Mu. Co-author: 2 books; also numerous articles. Discovered living fossil Neopilina, numerous new species of isopod crustaceans; chief scientist of sci. cruises to Central Am., S.Am., Africa, Mediterranean; developed (with K. O. Emery, L. Smith) deep sea grab camera. Home: 217 Rochelle Dr., Morehead City, N.C. Office: Duke Marine Lab., Beaufort, N.C. 28516.*

MERANZE, David Raymond, Am. pathologist; b. Phila., Dec. 25, 1900; s. Samuel and Yetta (Rotman) M.; B.S., U. Pa., 1921, M.A., 1930; M.D., Jefferson Med. Coll., 1927; m. Yetta Kaplan, July 18, 1928; children—David, Walter. Pathologist, dir. lab., dir. research Mt. Sinai Hosp. Phila., 1930-51, acting adminstr., 1950-51; dep. pathologist Phila. Gen. Hosp., 1938-45; with So. div. Albert Einstein Med. Center, Phila., 1951—, dir. labs., pathologist, dir. research, 1955-66; head research pathology Korman Research, 1966—; cons. pathologist Phila. Psychiat. Hosp., 1942-64; prof. pathology Hahnemann Med. Coll., 1947-66. Diplomate Am. Bd. Pathology. Mem. A.M.A., Am. Soc. Clin. Pathology, A.A.A.S., Phila. Physiol. Soc., Alpha Omega Alpha. Research, publs. on prostatic alkaline phosphatase, histologic diagnosis of male sterility, isolation of a coagulation-inhibiting enzyme from bacteria, schistosomiasis, exptl. carcinogenesis. Home: 229 W. Upsal St., Phila. 19119. Office: York and Tabor Rds., Phila. 19141.*

MÉRAY, (Hughes) Charles Robert, French mathematician; b. Chalon-sur-Saône, France, Nov. 12, 1835; prof. U. Dijon (France); mem. French Acad. Scis. Author: Exposition nouvelle de la théorie des formes linéaires, 1884; Nouveaux éléments de géométrie, 1903. Contbd. to modern attack on real number system, during 1870's; one of 1st to establish firmly continuum of real numbers; defined irrational number arithmetically without reference to limits. Died Dijon, Feb. 2, 1911.

MERCADIER, Ernest, physicist, inventor; b. Montauban, France, 1836; ed. École polytechnique, became dir., 1881; named prof. physics Nat. Sch. Telegraphy, 1878; inventor multiple tone frequency telegraph, 1897, also thermaphone, electric recording tuning fork with changeable length of vibration for studying tones; studied (with Cornu) melodic scales, problems of wireless telegraphy. Died 1911.

MERCATI, Michele, Italian geologist; b. San Miniato, Tuscany, Apr. 8, 1541; M.D., Ph.D.; pupil of Cesalpino. Dir., Vatican Bot. Garden; 1st physician to Clement VIII, from 1592. Author: Istruzioni sopra la peste, podagra e paralisi, 1576; Degli obelische di Roma, 1589; Considerazioni, 1590; Metallotheca, pub. 1717-19. Catalogued Vatican geol. collection, 1574; created (in Vatican) 1st mineral. mus. in Europe 1585; in writings mentioned use of manganese for coloring glazes, also whitening green and yellow glass; described manufacture of alum. Died Rome, Italy, June 25, 1593.

MERCATOR, Gerhardus (Kremer, Gerhard), mathematician, geographer, map maker; b. Rupelmonde, Flanders, Mar. 5, 1512; grad., Louvain, Belgium, 1530, licentiate, 1532. Founder geog. establishment, Louvain, 1534; arrested and prosecuted for heresy, 1544; became prof. cosmography U. Duisburg (Germany), 1552; cosmographer to Duke of Juliers. Published his earliest known map, 1537, survey and map of Flanders, 1737-40, map of world in N. and S. hemispheres, 1538, terrestial globe, 1541, celestial globe, 1551; 1st map on Mercator's projection with parallels and meridians at right angles, 1568; began atlas, 1585, completed by son, 1594; numerous other maps. Built complete set of observation instruments for Charles V; constructed planisphere for use in navigation. Died Duisburg, Dec. 5, 1594.

MERCATOR, Nicolaus (born Kaufmann), mathematician, engineer; b. Cismar, Holstein, Germany, circa 1620; studied math. and astronomy, U. of Copenhagen and Rostock. Became engineer; went to London, circa 1660, where he became active in formation of Royal Soc.; later went to France as architect and engineer; spent most of his time after 1660 in London. Fellow Royal Soc. (charter member), 1663. Author: Cosmographia, 1651; Astronomia Sphaerica, 1651; Rationes Mathematicae Subductae, 1653; Logarithmotechnia, 1668-74; Institutionum astronomicarum, 1876. Discovered an infinite series for computing log $(1 + x)$; developed procedure for computation of logarithmns; designed and built fountains at Versailles, France, after 1683. Died Paris, France, 1687.

MERCER, Henry Chapman, Am. anthropologist; b. Doylestown, Pa., June 24, 1856; s. William Robert and Mary Rebecca (Chapman) M.; A.B., Harvard U., 1879; Sc.D., Franklin and Marshall Coll., 1916. Curator of Am. and prehistoric archaeology U. Pa., 1894-97. Found remains of extinct animals, tapir, mylodon, peccary and fossil sloth in Am. caves; compared remains of ancient man in drift gravels and flint workings of America and Europe; explored caverns of Yucatan, fixing geol. date for peninsular ruins; found several new species of extinct animals in bone cave, Port Kennedy, Pa.; examined artistic remains of Pa. German settlers, and experimented upon and developed their processes of making and decorating pottery, inventing, 1899, a new method of mfg. tiles for mural decoration, and 1902, a new process of making mosaics; invented process of printing large designs in color on fabrics and paper, 1904. Recipient Grand prize. St. Louis Expn., 1904, Craftsmanship medal Am. Inst. Architects, 1921. Author: Lenape Stone, 1885; Hill Caves of Yucatan, 1896; Researches Upon the Antiquity of Man in the Delaware Valley and the Eastern United States, 1897; Tools of the Nation Maker, 1897; Bible in Iron, or the Pictured Stoves and Stoveplates of the Colonial United States, 1915; Ancient Carpenters' Tools, 1925. Archeol. editor Am. Naturalist, 1893-97. Made a collection of utensils and implements illustrating indsl. history of colonial U. S., to preserve and exhibit which he built and endowed a museum at Doylestown, Pa., 1916. Died 1930.

MERCER, James, English mathematician; b. Liverpool, Eng., Jan. 15, 1883; ed. Univ. Coll., Liverpool, Trinity Coll., Cambridge; M.A., D.Sc.; m. Annie Barnes; 1 son. Fellow Trinity Coll., Cambridge, also asst. lectr., Liverpool, 1909-12; fellow, math. lectr. Christ's Coll., Cambridge, 1912-14; naval instr., 1914-18. Fellow Royal Soc., 1922. Contbr. papers to sci. jours. Pioneer in work on theory of integral equations and theory of orthogonal series in Eng.; originator theorem concerning kernels with positive eigen values, also theorem on divergent series. Died London, Feb. 21, 1932.

MERCER, John, English chemist; b. Dean, Eng., Feb. 21, 1791; s. Robert Mercer; self-educated; m. Mary Wolstenholme, Apr. 17, 1814 (dec. 1859); 2 sons, 2 daus. Became partner dyeing bus., 1807; became apprentice color shop Oakenshaw Print Works, 1809; became hand loom weaver, chemist in color shop, 1818; partner in Fort Bros. print works, 1825-48; joined cotton trade, 1850. Named juror Internat. Exhbn., London, 1851, 62; became mem. Commn. Peace for County of Lancaster, 1861. Recipient Council medal Internat. Exhbn., London, 1851. Fellow Royal Soc., 1852; mem. Chem. Soc. Invented treatment of cotton cloth with caustic soda which causes cloth to become semi-transparent, to take dye more rapidly and increases its strength (mercerising), 1841; discovered several calico dyes; in theoretical chemistry, suggested 1st rational theory of catalytic action. Died Nov. 30, 1866.

MERCER, Verdun Frank, Australian botanist; b. Adelaide, Australia, Apr. 15, 1918; s. Joseph Howard and Elsie (Edgar) M.; B.Sc.Hons. I, U. Adelaide, 1944; Ph.D., Cambridge (Eng.) U., 1950; m. Marie Jeannette Peebles, Feb. 9, 1955; children—Melissa Jane, Juliet Miranda. Demonstrator, U. Adelaide, 1944; faculty U. Sydney (Australia), 1945-65, prof. cell physiology, 1963-65; head Sch. Biol. Scis., prof. biology Macquarie, Eastwood, Australia, 1966—; co-dir. plant physiology research unit Commonwealth Council for Indsl. Research Orgn. and U. Sydney, 1955-65; vis. prof. biology U. Pa., 1964. Nuffield fellow, 1949; Carnegie fellow, 1956; NSF Sr. Fgn. Scientist, 1964. Mem. Australian Soc. Plant Physiologists, Am. Soc. Plant Physiology, Am. Soc. Microbiology, Linnean Soc. Contbg. author: Science for High Schools, 1963. Research, publs. on structure and submicroscopic structure of plants cells and tissues, functions of cells and tissues. Home: 24 Trafalgar Av., Roseville, N.S.W. Office: Macquarie U., Eastwood, N.S.W., Australia.*

MERCHANT, Donald Joseph, Am. microbiologist; b. Biltmore, N.C., Sept. 7, 1921; s. Oscar Lowell and Bess (Clark) M.; A.B., Berea Coll., 1942; M.S., U. Mich., 1947, Ph.D., 1950; m. Marian Adelaide Yeager, May 31, 1943; children—Nancy Adele, Barry Scott, Karen Ruth. Instr., U. S. Armed Forces Inst., Manila, P.I., 1945; faculty U. Mich., Ann Arbor, 1948—, prof. microbiology, 1964—. Mem. Am. Soc. for Microbiology, Tissue Culture Assn. (pres. 1964—), N.Y. Acad. Sci., Am. Soc. Cell Biology, Internat. Soc. Cell Biology, Royal Soc. Medicine (London, affiliate), Soc. for Exptl. Biology and Medicine, A.A.A.S., Sigma Xi. Author: (with R. H. Kahn, W. H. Murphy), Handbook of Cell and Organ Culture, 1960, 2d edit., 1964; also articles. Research on physiol. and biochem. characterization growth established lines animal cells in culture, definition genetic markers for study animal cell populations, large-scale growth animal cells, demonstration specific function by long-term animal cell lines, study antigens animal cells in culture. Home: 1424 Arlington Blvd., Ann Arbor, Mich. 48104.*

MERCIER, André P. H., Swiss physicist; b. Geneva, Switzerland, Apr. 15, 1913; s. Paul A. and Jeanne (Golay) M.; B.Sc., Geneva Coll.; Lic.ès.sci.math., Lic.ès. sci.phys., Ph.D., U. Geneva; postgrad. Inst. Henri Poincaré, U. Paris, Inst. Theoretical Physics, U. Copenhagen; m. Ruth Marie Fossum, July 24, 1938; children—Geneviève Aase-Mercier, Inga Costacurta. Former asst. depts. geology and physics, privat docent U. Geneva; asst. dept. pure and applied mechanics Fed. Inst. Tech., Zurich; asso., then full prof., head dept. theoretical physics U. Berne, also prof. philosophy, former dean Faculty Scis., now rector. Vis. fellow Yale; fgn. fellow Silliman Coll.; sec.-gen. Internat. Com. on Gen. Relativity and Gravitation. Recipient A. de Claparède prize, Plantamour-Prevost award. Mem. Swiss Phys. Soc. (past pres.), Swiss Philos. Soc. (past pres.), Internat. Inst. Philosophy (mem. bd.), Swiss Acad. Scis. (senate), Swiss Acad. Humanities (del.). Author: Leçons sur les principes de l'électrodynamique classique, 1952; Principes de Mécanique analytique, 1955; Analytical and Canonical Formalism in Physics, 1959, 63; Thought and Being, An Investigation into the Nature of Knowledge, 1959; De l'Amour et de l'Etre, 1960; Antikes und modernes Denken in Physik und Mathematik, 1964, also others. Research and numerous publs. on spl. math. methods of theoretical physics, theories on origin of earth, problem of time from phys. and philos. viewpoint, theory of knowledge. Home: 43, Gryphenhübeliweg, Berne BE 3006. Office: 5, Sidlerstrasse, Berne BE 3000, Switzerland.*

MERCIER, (Louis) Auguste, French urologist; b. Arras, France, 1811; M.D., 1839; practiced in Paris. Author: Recherches sur les maladies des organes urinaires des vieillards, 2 vols., 1844. Described transverse ridge extending between openings of ureters on inner surface of bladder which forms posterior boundary of trigone (Mercier's bar), 1854; research on hypertrophied prostrate; improved instruments for urology and lithotripsy, 1836. Died Paris, 1882.

MERCIER, Charles Anthony, physician; b. 1852; s. L. P. Mercier; ed. Merchant Taylors' Sch., London Hosp.; m. 2d, Mary MacDougall, 1913. Med. officer of various asylums; physician mental diseases Charing Cross Hosp.; examiner mental diseases London U. Fellow Royal Coll. Physicians, Royal Coll. Surgeons. Author: Sanity and Insanity; Nervous System and the Mind; Lunacy Law for Medical Men; Psychology, Normal and Morbid; Text-book of Insanity; Criminal Responsibility. Died Sept. 2, 1919.

MERCIER, Robert, Swiss physicist; b. Geneva, Switzerland, Oct. 6, 1904; s. Henri and Marguerite (Roess) M.; B.Sc., College de Geneva; Dipl.Ing.Elctr., U. Lausanne (Switzerland) 1927; Dr.Sc., Ecole Polytechnique fédérale, Zurich, 1935; m. Line Tschumi, Oct. 13, 1928; 1 dau., Michèle-Claire. Tchr. physics Technicum cantonal de Bienne, 1934-36; prof. physics U. Lausanne, also Ecole Polytechnique de l'U. Lausanne, 1936—; prof. mechanics U. Neuchâtel, 1947-56. Head sect. for sci. research OCDE, Paris, 1960-62. Mem. Swiss, French socs. for physics, Swiss Soc. for Vacuum Physics, Nat. Com. for Optics, Internat. Union for Vacuum Sci., Technique and Application (treas.). Author: A Course of Mechanics, 3 vols.; A Course of Physics, 3 vols.; also articles. Research in ultrasonics, vacuum, optics, magnetic resonance, solid state physics. Home: 8 Primerose, Lausanne, Switzerland.*

MERCK, George Wilhelm, Am. chemist; b. N.Y.C., Mar. 29, 1894; s. George and Friedrike (Schenck) Merck; A.B., Harvard, 1915; many hon. degrees; m. Serena Stevens, Nov. 24, 1926; children—George W., Jr., Albert W. (by previous marriage); Serena M. (Mrs. Francis W. Hatch, Jr.), John H. C., Judith F. With Merck & Co., Inc., mfg. chemists, Rahway, N.J., 1914—, pres., dir., 1925-50, chmn. bd., 1949—; chmn. bd., dir. subsidiaries, other cos. Mem. com. on drugs and med. supplies NRC, 1942-45. Recipient Chem. Industry medal Am. sect. Soc. Chem. Industry, 1947. Fellow Am. Geog. Soc.; mem. Am. Cancer Soc., Mfg. Chemists' Assn. (pres. 1949-51), Am. Forestry Assn. Merck labs. were 1st to synthesize vitamin B, 1936, vitamin E, 1938, vitamin B6, 1939; introduced sulfonamide derivatives, sulfanilimide, sulfapyridine, sulfathiazole. Died Nov. 9, 1957.

MERCURIO, Scipione, Italian obstetrician; b. Venice, 1568. Author: La Commare o Riccoglitrice, 1601; De gli errori popolari d'Italia, 1603. Wrote 1st book on obstetrics in Italian, indicates for 1st time that contracted pelvis is indication for cesarean operation.

MEREDITH, Howard Voas, child somatologist; b. Birmingham, Eng., Nov. 5, 1903; s. Howard J. P. and Ada (Barker) M.; A.A., Graceland Coll., 1929; B.A., U. Ia., 1931, M.A., 1932, Ph.D., 1935; m. Matilda Johnson, July 27, 1926; children—Leslie Hugh, Mavis Rogene (Mrs. Robert L. Stewart). Came to U. S., 1923, naturalized, 1938. Research asst., Ia. Child Welfare Research Sta., U. Ia., Iowa City, 1931-35, research asso., 1935-39, faculty 1939-56, prof., cons., 1952-56, prof., cons. Coll. Dentistry, 1952-56, prof. Inst. Child Behavior and Devel., 1956—. Exchange appointee Harvard summer 1935; vis. lectr. U. So. Cal., summer 1948; prof. U. Ore. Sch. Health and Phys. Edn., Eugene, 1949-52; vis. prof. U. So. Cal., 1951. Mem. Soc. for Research in Child Devel. (pres. 1953-55), Am. Assn. Phys. Anthropologists, Am. Assn. U. Profs., Am. Assn. Orthodontists (hon. mem.), Sigma Xi, Phi Epsilon Kappa, Lambda Delta Sigma. Author: The Rhythm of Physical Growth, 1935; The Physical Growth of White Children, 1936; Physical Growth from Birth to Two Years: Stature, 1943; also chpts. in books. Home: 1205 Pickard St., Iowa City, Ia. 52240.*

MERENDINO, K. Alvin Aurelius, Am. surgeon; b. Clarksburg, W.Va., Dec. 3, 1914; s. Biagio and Cira (Bivona) M.; B.A. summa cum laude, Ohio U., 1936; M.D. (Verdi scholar), Yale, 1940; Ph.D., U. Minn., 1946; LL.D., Ohio State U., 1967; m. Shirley E. Hill, July 6, 1943; children—Cira, Nancy, Susan, Nina, Maria. Faculty, U. Minn. Med. Sch., 1944-48; dir. program postgrad. med. edn. in surgery Ancker Hosp., St. Paul, 1946-48; faculty U. Wash., Seattle, 1949—, dir. exptl. surgery labs., 1949—, prof. surgery, 1955—, adminstrv. officer dept. surgery, 1957-64, chmn. dept. surgery, 1964—; surgeon-in-chief U. Wash. Hosp., 1964—. Mem. surg. study sect. NIH, 1958-62, tng. com., 1965—; mem. adv. com. on hosps. and clinics Pub. Health Service, 1963-66; cons. Battelle-N.W. Meml. Inst. Research Lab., 1967—. Recipient John Baird Thomas award for outstanding work in premed. scis. Ohio U., 1936. Diplomate Am. Bd. Surgery (past chmn.), Nat. Bd. Med. Examiners. Mem. A.C.S. (mem. surgery forum com. 1961-66, gov. 1966—), Soc. U. Surgeons, Halstead Soc., Surg. Biology Club, A.M.A., Societe Internat. de Chirurgie, Pacific Coast, Seattle surg. socs., Am., N. Pacific surg. assns., Am., Wash. State heart assns., Am. Assn. Thoracic Surgeons, Wash. State Med. Assn., Allen O. Whipple Surg. Soc., Phi Beta Kappa, Sigma Xi, Phi Beta, Pi. Editor: Prosthetic Valves for Cardiac Surgery, 1961; editorial bd. Jour. Surg. Research, Am. Jour. Surgery. Contbr. numerous articles to tech. jours, chpts. in books. Research on intestinal sensitivity intestines to acid-peptic gastric secretions, jejunal interposition operation for esophageal stricture; introduced fabric teflon to vascular surgery; described inter-atrial venous transposition operation for certain types congenital heart disease; developed a fatigue machine for testing artificial heart valves, prosthetic heart valve devel.; initiated studies and clin. application in U. S. of profound hypothermia and cardiac arrest in small infants with congenital heart defects. Home: Spring Dr., The Highlands, Seattle, 98177.*

MERETOJA, Atte Kalevi, Finnish chemist; b. Salo, Finland, May 30, 1912; s. Edvard and Helga (Kavila) M.; M.Sc., U. Helsinki, 1943, D.Ph., 1945; m. Ilona Aalto, July 7, 1935; children—Jouko, Kristina, Annika, Marjaana, Olli, Päivi. Asst. prof. U. Helsinki, 1947-49; prof. inorganic and analytical chemistry U. Turku (Finland), 1949—, dean Faculty Sci., 1956—. Mem. Nat. Bd. Finnish High Schs., 1967—; sci. adviser Minsterium of Health, 1962—. Mem. Finnish Chem. Soc., Finnish Phys. Soc. Author: Calculation in Chemistry, 5th edit., 1966. Research and publs. in inorganic reactions kinetics and equilibrium, theory of analysis. Home: Eerikink.2 0, Turku, Finland.*

MEREWETHER, Arthur Francis, Am. meteorologist; b. East Providence, R.I., July 7, 1902; s. Francis G. and Margaret Jane (Murphy) M.; Ph.B., Brown U., 1922; M.S., Mass. Inst. Tech., 1925, postgrad.; m. Genevieve Evans, July 7, 1937; children—Diana Genevieve, Catherine Evans, Francis Charles, James William. With Pitts. Pirates Baseball Team, 1922; athletic coach Phillips Acad., Andover, Mass., 1926-27; chemist Heywood-Wakefield Co., 1925-26, E. R. Squibb & Son, 1928-29, Carborundum Co., 1929; served to col. A.C., U. S. Army, 1929-46, chief of weather service, 1939-42, comdg. officer 8th weather region, 1942-46; mgr. weather service Am. Airlines, Jamaica, N.Y., 1946-66. Mem. subcom. on met. problems NACA, 1940-42; mem. Joint Control Bd. for Atlantic Air Routes, 1942-45; mem. panel on atmosphere, research and devel. bd. Dept. Def., 1947-53; mem. vis. com., meteorology dept. Mass. Inst. Tech., 1941-42, 46-56; chmn. met. com. Air Transport Assn., 1948-49, 54-55; mem. weather adv. commn. Dept. Commerce, 1953-54. Decorated Legion of Merit, comdr. Order Brit. Empire; recipient Losey award Inst. Aero. Scis., 1961. Fellow N.Y. Acad. Scis., Am. Inst. Aeros. and Astronautics (asso.); mem. Am. Meteorol. Soc. (councilor 1948-50, 53-54, pres. 1954-56; Cleveland Abbe award 1967), Am. Geophys. Union, Royal Meteorol. Soc., A.A.A.S., Phi Kappa Theta. Discovered meteoritic crater and contained lake in No. Labrador (named for him), 1943. Home: 37-02 222d St., Bayside, N.Y. 11361.*

MERGELIAN, Sergei Nikitovich, Russian mathematician; b. Simferopol, May 19, 1928; grad. Yerevan U., 1947. Head chair math. Yerevan U., 1949-53; prof. Moscow U., 1953—; establisher, dir. Computer Research Inst., Armenia, 1957—. Recipient Stalin prize, 1952. Mem. Armenian, USSR (corr., mem. delegation to U. S. 1959) acads. scis. Author: Some Problems of the Constructive Theory of Functions, 1951; The Approximation of Continuous Functions by Means of Polynomials on Arbitrary Closed Sets, 1952; Uniform Approximations of the Functions of a complex Variable, 1952; The Completeness of Systems of Analytic Functions, 1953. Directed devel. of general-purpose computors; research on theory of functions of complex variables by polynomials. Address: Armenian Acad. Scis., Yerevan, Armenian SSR, USSR.

MERGEN, François; forester; b. Grand Duchy of Luxembourg, May 1, 1925; s. Aloyse and Marie (Arens) M.; certificat de fin d'etudes Luxembourg Coll., 1946; student Laval U., 1947; B.S., U. N.B., 1950; M.F. cum laude, Yale, 1951, Ph.D., 1954; m. Andree Bodson, Aug. 2, 1947; 1 son, John Francis. Faculty, Yale, 1954—, prof. forestry, dean Sch. Forestry, 1965—. research collaborator biology dept. Brookhaven Nat. Lab. Mem. Soc. Am. Foresters participant Vis. Scientist Program 1961-65, award for outstanding achievement in biol. research 1966), Radiation Research Soc., Sigma Xi. Research, numerous publs. on vegetative propagation So. pines, flowering and sex expression pines, mutation induction, genotype-environment interaction, hybridization in forest trees, effect of ionizing radiation on reproductive capacity of pine-oak forests. Home: 86 Hill St., Hamden, Conn. 06514. Office: 205 Prospect St., New Haven 06511.*

MERGENTHALER, Ottmar, inventor; b. Hachtel, Germany, May 11, 1854; s. Johann George and Rosina (Achermann) M.; m. Emma Frederica Lachenmayer, Sept. 11, 1881; at least 4 children. Watch-maker's apprentice, Bietigheim, Württemberg, Germany, 1868-72; came to Balt., 1872; with August Hahl's sci. instrument shop, Washington, 1872-76; moved to Balt., 1876, formed partnership with Hahl, Balt., 1880; invented linotype machine, 1884. Died Balt., Oct. 28, 1899.

MERIAN, Maria Sibylla, naturalist; b. Frankfort/Main, Germany, 1647; d. Matthäus Merian; m. Johann Andreas Graff, 1665; visited Surinam (Dutch Guiana), 1699-1701. Author: Metamorphoses of Insects of Surinam, 1705; Neues Blumen Buch; also work on insects of Europe pub. posthumously. Known for accurate observations; among best early zool. drawings; published 1st book on insects, with plates she had engraved and colored, 1699; research in natural history; prepared pictorial studies, mainly in water color. Died Amsterdam, Netherlands, 1717.

MERIAN, Peter, Swiss physicist, chemist, geologist, seismologist; b. Basel, Switzerland, 1795; ed. Geneva, Switzerland, Göttingen, Paris; prof. zoology and paleontology, Basel, 1820-28, pres. Coll. Edn., 1847-65. Merian Found. named in his honor. Author: Beiträge zur Geognosie, 1821; Uber die Theorie der Gletscher, 1843; Die Mathematiker Bernoulli, 1860; also numerous others. Research on geologic formation of Jura. Died 1883.

MERICA, Paul Dyer, Am. metallurgist; b. Warsaw, Ind., Mar. 17, 1889; s. Charles Oliver and Alice (White) M.; student DePauw U., 1904-07, D.Sc., 1934; A.B., U. Wis., 1908; Ph.D., U. Berlin, 1914; D.Sc., Lehigh U., 1938, Stevens Inst., 1942; m. Florence Young, Sept. 22, 1917. Prof. chemistry Chekiang Provincial Coll., Hangchow, China, 1909-11; research physicist U. S. Bur. Standards, 1914-19; dir. research Internat. Nickel Co., 1919—; tech. asst. to pres. Internat. Nickel Co. of Can., 1929, successively asst. to pres., exec. v.p., pres., dir., cons. research and devel. work on metals and alloys—their metallurgy and metallography; spl. investigator on caustic embrittlement on steel in boilers, U. Ill. Recipient James Douglas medal, 1929; John Fritz medal, 1938; medal Inst. Metals, 1941; Franklin Inst. medal, 1942; Gold medal Am. Soc. for Metals, 1951. Fellow A.A.A.S.; mem. Am. Iron and Steel Inst., Am. Soc. for Testing Materials, Am. Phys. Soc., Am. Inst. Mining and Metall. Engrs., Am. Inst. C.E., Nat. Acad. Scis., Inst. Metals and Iron and Steel Inst. (both Brit.), Canadian Inst. Mining and Metallurgy, Mining and Metall. Soc. Am., Am. Soc. for Metals. Contbr. articles and monographs in tech. publs. Developed theory of precipitation or dispersion hardening in metals, while studying hardening changes in duralumin. Died Oct. 20, 1957.

MERIFIELD, Paul Milton, Am. geologist; b. Santa Monica, Cal., Mar. 17, 1932; s. J. Earl and Dorothy (Allard) M.; A.B. with honors, U. Cal. at Los Angeles, 1954, M.A., 1958; postgrad. U. Munich (Germany); Ph.D., U. Colo., 1963. Research scientist Lockheed-Cal. Co., Burbank, 1962-64; sr. scientist, dir. Earth Sci. Research Corp., Santa Monica, Cal., 1964—. Cons. in engring. geology and astrogeology to aerospace cos., 1961—. Fulbright scholar, 1958-59; Boettcher scholar, 1961. Mem. Am. Soc. Photogrammetry, Marine Tech. Soc., Assn. Engring. Geologists, Sigma Xi. Contbr. articles to tech. jours. Pioneered utilization satellite photography for exploration earth and planets, dating origin Earth-Moon system by geol. methods; helped develop first techniques for photographing ocean bottom from submarine. Home: 1333 N. Kenter Av., Los Angeles 90049. Office: P.O. Box 2427, Santa Monica, Cal. 90405.*

MERING, Baron Joseph von, see von Mering, Baron Joseph.

MERITT, Lucy Taxis Shoe, Am. archaeologist; b. Camden, N.J., Aug. 7, 1906; d. William Bonaparte and Mary Esther (Dunning) Shoe; A.B., Bryn Mawr Coll., 1927, M.A., 1928, Ph.D., 1935; fellow Am. Sch. Classical Studies, Athens, Greece, 1929-32; fellow Am. Acad. in Rome, 1936-37, research fellow, 1949-50; m. Benjamin Dean Meritt, Nov. 7, 1964. Asst. prof., then asso. prof. art and archaeology and Greek, Mt. Holyoke Coll., 1937-50, counselor, then chief counselor students, 1943-47; mem. Inst. Advanced Study, Princeton, 1948-49, 50—; editor publs. Am. Sch. Classical Studies, Athens, 1950—; vis. prof. Washington U., St. Louis, 1958, 60; vis. lectr. Princeton, 1959; mem. excavation staff at Cosa, 1950, at Morgantina, 1957. Mem. mng. com. Am. Sch. Classical Studies, 1937—, exec. com. 1948-52, chmn. publs. com. 1950—. Mem. Archaeol. Inst. Am. (acting gen. sec. 1962, recorder 1960—), Soc. Archtl. Historians, Hist. Soc. Princeton, Alumni Assn. Am. Sch. Classical Studies Athens (sec.-treas. 1940—), Classical Soc. Am. Acad. Rome (treas. 1942-46, v.p. 1951, pres. 1952), Princeton Soc. Archaeol. Inst. Am. (sec. 1953-56, pres. 1963-67); corr. mem. German Archaeol. Inst., Internat. Assn. Classical Archaeology. Author: Profiles of Greek Mouldings, 1936; Profiles of Western Greek Mouldings, 1952; Etruscan and Republican Roman Mouldings, 1965; also articles. Study of full size profiles of Greek archtl. mouldings, which gave new criterion for dating Greek architecture and led to new hist. observations, especially relations of Greece to Western colonies; study of Etruscan and early Roman mouldings which established independence of Etruscan architecture from Greek and emphasized strong dependence of Republican Roman on Etruscan. Home: 68 Westerly Rd., Princeton, N.J. 08540. Office: Inst. for Advanced Study, Princeton, N.J. 08540.*

MERKEL, Friedrich Siegmund, German anatomist; b. 1845; became prof., Rostock, Germany, 1872, Königsberg, 1883, Göttingen, 1885. Author: (with R. Bonnet) Ergebnisse der Anatomie und Entwicklungsgeschichte, 1892-1914. Described tactile corpuscles in submucous layer of mouth and tongue (Merkel's corpuscles or disks, also tactile disks), 1880; research on connective tissue. Died 1919.

MERKER, Hans Hermann, German physician; b. Bremerhaven, Germany, Nov. 23, 1921; s. Karl Ernst and Ida (Schuler) M.; Dr.med., U. Freiburg (Germany) Brg., 1951; m. Irmgard Sacksen, Dec. 23, 1944; children—Christa, Maria, Gudrun Katharina, Sylvia Angelika. Mem. staff dept. medicine U. Freiburg Brg., 1952—, head lab. for cytochemistry in haematology, 1960—. Sec. expert panel cytochemistry in haematology European Com. for Standardization in Haematology, 1965—. Mem. Deutsche Gesellschaft für Innere Medizine, Deutsche Gesellschaft fur Hämatologie, U. S. Soc. Histochemistry, Swiss Soc. Haematology (corr.). Author: (with O. Fresen, H. Begemann) Zur Begutachtung der Blutkrankheiten, 1959; also numerous articles, chpts. in books. Editor: Zyto- und Histochemie in der Hämatologie, 1963. Research on enzyme cytochemistry in hematology, significance of cytochem. and cytogenetic findings in DiGuglielmo's syndrome, refractory sideroblastic anemias, and chronic myeloproliferative diseases. Home: 6 Weddigenstrasse, 78 Freiburg, West Germany.*

MERKER, H. J., German anatomist; b. Merseburg, Germany, July 10, 1929; s. Rudolph and Hertha (Stelling) M.; student Med. Sch., Free U., Berlin; M.D., 1957; student Med. Sch., U. Giessen (Germany); m. Antje Hellenschmied, Feb. 25, 1966. Staff dept. electron microscopy Free U. Berlin, privat-dozent for anatomy, 1964-66, privat-dozent 2d dept. anatomy, 1966—. Mem. Assn. German Anatomists, Assn. German Electron microscopists. Contbg. author: Handbook of Pharmacology; also articles. Research on fine structure of connective tissue, including amyloid, morphological effect of hormones, effect of drugs on liver. Home: 41 Pacelli-Allee, Berlin, West Germany.*

MERKER, Philip Charles, Am. pharmacologist; b. Bklyn., July 23, 1922; s. Harvey and Tillie (Lehrer) M.; B.A., Bklyn. Coll., 1946; B.Sc. in Pharmacy, L.I. U., 1951; M.S., Purdue U., 1953, Ph.D., 1954; m. Alta Irene Hirshfeld, Aug. 20, 1953; children—Marilyn Paula, Edward Lawerence. Mem. research and teaching faculties Sloan-Kettering Inst., Cornell U., 1954-62; faculty U. Tenn. Coll. Pharmacy, 1963-65; prof. pharmacology, chmn. div. biol. sci. and pharmacology Columbia Coll. Pharm. Sci., N.Y.C., 1965—. Fellow A.A.A.S.; mem. Am. Assn. Cancer Research, Soc. Gen. Physiologists, Am. Soc. Exptl. Pathology, Am. Pharm. Assn. Investigations, publs. on exptl. cancer chemotherapy, including studies of human tumor transplants growing in exptl. animals. Home: 29 Redfield St., Rye, N.Y. 10580.*

MERKULOV, Ivan Iosifovich, Russian ophthalmologist; b. Koshibeevo (now Ryazan Oblast), 1897; grad. Kharkov Med. Inst., 1921; postgrad. in Germany, 1927-28; D.Med. Sci. Asst., lectr. Ukrainian Research Inst. Ophthalmology, Kharkov, 1929-39, prof., 1940-45, dir., 1946—; head chair eye diseases Kharkov Postgrad. Med. Inst., 1944—. Mem. USSR Acad. Med. Scis. (corr.), Kharkov Oblast Soc. Ophthalmologists (chmn. 1949—). Author: Malignant Neoplasms of the Orbit, 1940; Stereoscopic Atlas of Neoplasms of the Eyelids and Eye; co-author, editor Problems of Neuroophthalmology, 1959. Co-editor Ophthalmology sect. Large Med. Ency., 2d edit; mem. editorial bd. Ophthalmology Jour.; mem. editorial council Herald of Ophthalmology. Address: Kharkov Postgrad. Med. Inst., ulitsa Artema 8, Kharkov, Ukrainian SSR, USSR.

MERLE (or MORLEY), William, Brit. meteorologist; s. William Merle; became rector of Driby, Eng., 1331; fellow Merton Coll., Oxford; author: Temperies aeris Oxoniae pro septennio scilicet a Januario MCCCXXXVII ad Januario MCCCXLIV, 1344; De pronosticationis aeris, 1340. Made 1st systematic and regular weather record. Died 1347.

MERLINI, Edoardo, Italian geophysicist; b. Novara, Italy, Jan. 15, 1924; s. Giacomo and Angelica (Realdi) M.; Elec. engring. degree, Politecnico Milano (Italy), 1948; postgrad. in electronics Scuola di Ingegneria, Pisa, Italy; m. Sandra Occhetta, Feb. 11, 1952; children—Elettra, Piera. Chief elec. engr. test room Scotti Brioschi, Novara, 1950-51; head lab. electronics AGIP-DIREZIONE Mineraria, Milano, Italy, 1951-57, sect. head in charge seismic rec. sect., 1958-59, head research and devel. dept. for geophys. exploration, elec. well logs, and new devices for geophys. exploration, 1959—; prof. electronics Indsl. Sch., Novara, 1950-64. Mem. Comitato Elettrotecnico Italiano (sec. 31th subcom.). Research, publs. on devel. stratigraph for detecting underground reflecting layers, also phase filter for filtering seismic signals by phase. Home: 3/B Via Bordolano, S. Donato. Office: AGIP-Direzione Mineraria, S. Donato Milanese, Milano, Italy.*

MERRELL, David John, Am. geneticist, educator; b. Bound Brook, N.J., Aug. 20, 1919; s. Edward Peterson and Elizabeth (Herder) M.; B.S., Rutgers U., 1941; M.A., Harvard, 1947, Ph.D., 1948; m. Jessie Clark, Mar. 19, 1945; children—Edward Clark, David Wilson, James Hart, Ann Elizabeth. Chemist, Am. Cyanamid Co., Bound Brook, 1941-42; teaching fellow Harvard, 1946-48; faculty U. Minn., Mpls., 1948—, prof. dept. zoology, 1964—. Mem. Genetics Soc. Am., Am. Genetics Assn., Soc. for Study Evolution, Am. Soc. Human Geneticists, Am. Soc. Naturalists, Am. Eugenics Soc., A.A.A.S., Am. Inst. Biol. Scis., Phi Beta Kappa, Sigma Xi. Author: Evolution and Genetics, 1962; also articles. Research in genetics of populations, ecol. genetics, behavior genetics.*

MERRET, Christopher (Merrett), naturalist, physician; b. Winchcomb, Eng., Feb. 16, 1614; s. Christopher Merret; B.A., Oriel Coll., Oxford (Eng.) U., 1634; M.B., Gloucester Hall, Oxford, Eng., 1636, M.D., 1643. Lived in London, Eng.; became research librarian Coll. Physicians, 1654; Gulstonian lectr., 1654. Fellow Royal Soc., 1663. Genus, Merrettia named in his honor. Author: Catalogus librorum, instrumentorum, . . . , 1660; Pinax rerum naturalium Britanicum (listed many Brit. plants and animals for 1st time), 1666, 67; also articles on vegetable physiology tin mining, wines; engaged in controversy with Stubbe about practice of apothecaries. Died Aug. 19, 1695.

MERRIAM, Alan Parkhurst, Am. anthropologist, educator; b. Missoula, Mont., Nov. 1, 1923; s. Harold Guy and Doris (Foote) M.; B.A., Mont. State U., 1947; Mus.M., Northwestern U., 1948, Ph.D. in Anthropology, 1951; m. Barbara Williams, Aug. 29, 1947; children—Virginia Claire, Paige Alison, Cynthia Williams. Instr. anthropology Northwestern U., 1953-54, asst. prof., 1956-58, asso. prof., 1958-62; asst. prof. U. Wis., Milw., 1954-56, U. Minn., summer 1957; prof. Ind. U., Bloomington, 1962—, chairman department of anthropology, 1966—. Mem. com. on human resources in Central Africa, NRC, 1958; spl. cons. on polit. situation in Republic of Congo, U. S. Govt., 1960; mem. Pres. Kennedy's Task Force for Africa, 1960; mem. adv. bd. Inter-Am. Inst. for Mus. Research, Tulane U., 1961—; U. S. del. 1st Internat. Congress Africanists, Accra, Ghana, 1962, 2d Internat. Congress, Dakar, 1967; mem. joint com. on Africa, Social Sci. Research Council and Am. Council Learned Socs., 1960-66, chmn., 1962-66. Fellow Am. Anthrop. Assn. (editorial council 1957-59), A.A.A.S., African Studies Assn. (founding, com. fine arts and humanities, 1960—, chmn. 1960-62, proposed jour. com. 1962, policy and planning com. 1963—, dir.); mem. Internat. African Inst., Am. Folklore Soc. (counselor 1956-58, rev. editor Jour. Am. Folklore 1957-58), Central States Anthrop. Soc. (pres. 1960-61, exec. com. 1961-64), African Music Soc., Soc. for Ethnomusicology (co-founder, counselor 1958—, v.p. 1960-62, pres. 1962-64), Internat. Folk Music Council. Author: (with assistance Robert J. Benford) A Bibliography of Jazz, 1954; Congo: Background of Conflict, 1961; The Anthropology of Music, 1964; Ethnomusicology of the Flathead Indians, 1967. Field research on Flathead Indians, Western Mont., 1950, 58, Belgian Congo, Ruanda-Urundi, 1951-52, 59-60. Home: R.R.3, Bloomington, Ind. 47401.*

MERRIAM, C(linton) Hart, Am. naturalist; b. N.Y.C., Dec. 5, 1885; s. Hon. Clinton L. and Caroline (Hart) M.; student Sheffield Sci. Sch. (Yale), 1874-77; M.D., Coll. Phys. and Surg. (Columbia), 1879; m. Virginia Elizabeth Gosnell, Oct. 15, 1886; children—Dorothy (Mrs. Henry Abbot), Zenaida (Mrs. M. W. Talbot). In med. practice, 1879-85; chief U. S. Biol. Survey, 1885-1910; resigned to conduct biol. and ethnol. investigations under a spl. trust fund established by Mrs. E. H. Harriman, 1910-39. Naturalist, Hayden's survey, 1872; asst. U. S. Fish Commn., 1875; visited Arctic seal fishery from Newfoundland, 1883, as surgeon S.S. Proteus; visited Alaska, 1891, as U. S. Bering Sea Commr., and investigated the fur seal on Pribilof Islands; conducted many biol. explorations in Far West. Fellow Am. Ornithologists' Union (pres. 1900-02), A.A.A.S.; a founder Nat. Geog. Soc.; mem. Nat. Acad. Scis., Am. Philos. Soc., Am. Soc. Naturalists (pres. 1924-25), Washington Acad. Scis., Biol. Soc. Washington (pres. 1891, 92), Anthropol. Soc. Washington (pres. 1920-21), Am. Soc. Mammalogists (pres. 1919-21), Zool. Soc. London (fgn.). Author: The Birds of Connecticut, 1877; Mammals of the Adirondacks, 1882-84; Results of Biol. Survey of San Francisco Mountain Region and Desert of Little Colorado in Arizona, 1890; Biological Reconnaissance of Idaho, 1891; Geographic Distribution of Life in North America, 1892; Trees, Shrubs, Cactuses and Yuccas of Death Valley Expedition, 1893; Laws of Temperature Control of Geographic Distribution of Terrestrial Animals and Plants, 1894; Monographic Revision of the Pocket Gophers (Geomyidae), 1895; Revision of the American Shrews, 1895; Synopsis of Weasels of North America, 1896; Biological Survey of Mount Shasta, Calif., 1899; Life Zones and Crop Zones of the United States, 1898; Indian Population of California, 1905; Distribution and Classification of the Mewan Indians of California, 1907; Totemism in California, 1908; The Dawn of the World, 1910; Review of the Grizzly and Big Brown Bears of America, 1917; G. K. Gilbert, Geologist, 1918; The Acorn, a Neglected Source of Food, 1918; A California Elk Drive, 1921; Earliest Crossing of the Deserts of Utah and Nevada to Southern California—Route of Jedediah H. Smith in 1826, 1923; First Crossing of the Sierra Nevada—Jedediah Smith's trip from California to Salt Lake in 1827, 1923; The Name of Mount Rainier, 1924; Baird, the Naturalist, 1924; Source of the Name Shasta, 1926; The Buffalo in Northern California, 1926; Classification and Distribution of the Pit River Indian Tribes of California, 1926; William Healey Dall, 1927; Annikadel History of the Universe as told by the Modesse Indians of Calif., 1928; also papers on zool., bot. and ethnol. subjects. Died Mar. 19, 1942.

MERRIAM, John Campbell, Am. palaeontologist; b. Hopkinton, Ia., Oct. 20, 1869; s. Charles Edward and

Margaret Campbell (Kirkwood) M.; B.S., Lenox Coll., 1887; Ph.D., U. Munich, 1893; Sc.D., Columbia, 1921, Princeton, 1922, Yale, 1922, U. Pa., 1936, U. State of N.Y., 1937; Ore. State Coll., 1939; LL.D., Wesleyan U., 1922, U. Cal., 1924, N.Y. U., 1926, U. Mich., 1933, Harvard, 1935, George Washington U., 1937, U. Ore., 1939; m. Ada Gertrude Little, Dec. 22, 1896 (dec. Apr. 1949); children—Lawrence Campbell, Charles Warren, Malcolm Landers; m. 2d, Margaret Louise Webb, Feb. 20, 1941. Instr. palaeontology and hist. geology U. Cal., 1894-99, asst. prof., 1899-1905, asso. prof., 1905-12, prof., 1912-20, dean faculties, 1920; chmn. NRC, 1919; pres. Carnegie Instn., Washington, 1920-38, emeritus since 1939. Fellow A.A. A.S., Geol. Soc. Am. (pres. 1910), Am. Palaeontol. Soc. (pres. 1917); mem. Nat. Acad. Scis., Am. Philos. Soc., Am. Acad. Arts and Scis., London Zoöl. Soc. (corr.), other sci. socs. Author: Primitive Characters of the Triassic Ichthysauria, 1904; The Thalattosauria, a Group of Marine Reptiles fromt he Triassic of California, 1905; Cave Exploration, 1906; Triassic Ichthysauria (with special reference to the American forms), 1908; The Occurrence of Human Remains in California Caves, 1909; The Occurrence of Twisted Horned Antelopes in the Tertiary of Northwestern Nevada, 1909; The Story of the Calaveras Skull, 1910; Synopsis of Lectures in Palaeontology, 1910; The Relation of Palaeontology to the History of Man (with particular reference to the Am. problem), 1910; The Fauna of Rancho La Brea, Part I, Occurrence, 1911; Part II, Canidae, 1912; The Horses of Rancho La Brea, 1913, Discovery of Human Remains in an Asphalt Deposit at Rancho La Brea, 1914; Extinct Faunas of the Mojave Desert (their significance in a study of the origin and evolution of life in America; 1915; Relationships of Pliocene Mammalian Faunas from the Pacific Coast and Great Basin Provinces of North America, 1917; Science in Mobilization, 1917; The Beginnings of Human History Read from the Geological Record; The Emergence of Man, 1919; The Function of Educational Institutions in Development of Research, 1920; Earth Sciences as the Background of History, 1920; The Research Spirit in the Everyday Life of the Average Man, 1920; Common Aims of Culture and Research in the University, 1922; The Place of Education in a Research Institution, 1925; The Responsibility of the Federal and State Governments for Recreation, 1926; International Coöperation in Historical Research, 1926; Medicine and the Evolution of Society, 1926; Inspiration and Education in National Parks, 1927; The Place of Geology Among the Sciences, 1929; Institutes for Research in the Natural Sciences, 1929; Significance of the Border Area between Natural and Social Sciences, 1929; The Living Past, 1930; The Unity of Nature as Illustrated by the Grand Canyon, 1931; The Felidae of Rancho La Brea (with Chester Stock), 1932; Spiritual Values and the Constructive Life, 1933; Responsibility of Science to Government, 1934; Ultimate Values of Science, 1935; Science and Human Values; Time and Change in History, 1936; The Most Important Methods of Promoting Research, as Seen by Research Foundations and Institutions; Geography and History Among the Sciences, as Influencing Research in the Americas, 1937; Application of Science in Human Affairs; Influence of Science upon Appreciation of Nature; Some Aspects of Cooperative Research in History, 1938; Contribution of Geology to Shaping of Ideas on the Meaning of History; Science and Belief; The Development of Cultural and Social Values through the Relation of Science to Other Major Fields of Activity, 1939; also numerous other papers. Died Oct. 30, 1945.

MERRILL, Edward Wilson, Am. chem. engr.; b. New Bedford, Mass., Aug. 31, 1923; s. Edward Clifton and Gertrude (Wilson) M.; A.B., Harvard, 1944; D.Sc., Mass. Inst. Tech., 1947; m. Genevieve de Bidart, Aug. 19, 1948; children—Anne, Francis. Faculty, Mass. Inst. Tech., 1950-—, prof. chem. engring., 1964-—; vis. lectr. chemistry Harvard, 1956-59. Cons. Mass. Gen. Hosp., Boston, 1964-—, to dir. Nat. Inst. Arthritis and Metabolic Diseases, 1965-—; asso. staff medicine, cons. Peter Bent Brigham Hosp., Boston, 1962-—. Mem. Am. Soc. Arts and Scis., Am. Chem. Soc., Am. Inst. Chem. Engrs., Am. Inst. Physics, Am. Soc. for Artificial Internal Organs. Patentee in field. Research, publs. on rheology of synthetic and biol. polymers; blood viscosity related to composition; design membrane oxygenators for blood, artificial kidney; synthesis of non-clotting surfaces for blood; inventor viscosimeters. Home: 50 Sparks St., Cambridge, Mass. 02139.*

MERRILL, Elmer Drew, Am. botanist, educator; b. East Auburn, Me., Oct. 15, 1876; s. Daniel C. and Mary A. (Noyes) M.; B.S., U. Me., 1898, M.S., 1904, Sc.D., 1925; student dept. medicine George Washington U., 1900-01; Sc.D., Harvard, 1936; LL.D., U. Cal., 1936; m. Mary Augusta Sperry, May 21, 1907; children—Lynne, Dudley Sperry, Wilmans Noyes, Ann. Asst. in natural sci. U. Me., 1898-99; asst. agrostologist U. S. Dept. Agr., Washington, 1899-1902; botanist Bur. Agr., Manila, P.I., 1902, Bur. Agr. and Bur. Forestry, 1902-03, Bur. Govt. Labs., 1903-05, Bur. Science, from 1906; asso. prof. botany and head of dept. U. Philippines, 1912-19, prof. 1916-19; dir. Bur. Science, Manila, 1919-23; dean Coll. Agr. and dir. Agrl. Expt. Sta., U. Cal., 1923-29; prof. botany Columbia, 1930-35; dir. N.Y. Bot. Garden, 1930-35; prof. botany, dir. Arnold Arboretum, adminstr. bot. collections Harvard, 1935-46, Arnold prof. botany, 1946-48, emeritus. Specialist in taxonomy and phytogeography of Philippine, Polynesian and Indo-Malayan plants. Mem. Am. Acad. Arts and Scis., Nat. Acad. Scis., Am. Philos. Soc., Royal Asiatic Soc. (Malayan br.); hon. mem. Deutsche Bot. Gesellschaft, Netherlands Bot. Soc., Royal Netherlands Geol. Soc., Acad. Sci., Inst. France, Inst. Genevoise, Swedish Acad. Sci.; fgn. mem. Linnean Soc., London (medalist 1939). Contbr. numerous papers on botany of N.Am., China, Philippines, Malaya and Polynesia. Died Feb. 25, 1956.

MERRILL, George Perkins, Am. geologist; b. Auburn, Me., May 31, 1854; s. Lucius and Anne Elizabeth (Jones) M.; B.S., U. Me., 1879, M.S., 1883, Ph.D., 1889; student Wesleyan U., 1879-80; Johns Hopkins U., 1886-87; Sc.D., George Washington U., 1917; m. Sarah P. Farrington, Nov. 1883 (dec. 1894); m. 2d, Katherine L. Yancey, 1900. Asst. chemist Wesleyan U., 1879-80; asst. in geol. dept. U. S. Nat. Mus., 1881, head curator dept. geology, 1897-—; prof. geology and mineralogy George Washington U., 1893-1915. Expert spl. agt. of 12th Census in stonequarry statistics. Hon. corr. mem. A.I.A. Recipient J. L. Smith Gold medal Nat. Acad. Scis., for researches in meteorites. Author: Stones for Building and Decoration, 1891, 1897, 1903; Rocks, Rockweathering and Soils, 1897, 1907; The Non-Metallic Minerals—Their Occurrence and Uses, 1904, 1910; also contributions to a History of American Geology, 1905, and History of American State Geological and Natural History Surveys, 1920; Catalogue and Handbook Meteorite Collection in the U. S. Nat. Museum, 1915; Handbook of Gems and Precious Stones (with others), 1922; The First 100 Years of American Geology, 1924. Noted for theory of explosive origin of meteors; made spl. study of Coon Butte or Meteor Crater, Ariz., 1906. Died Aug. 15, 1929.

MERRILL, John Ellsworth, Am. astronomer, educator; b. Parsonsfield, Me., May 10, 1902; s. John and Mabel (Buckpitt) M.; A.B. (Augustus Howe Buck fellow), Boston U., 1923; postgrad. (Augustus Howe Buck fellow) Harvard; M.S., Case Inst., 1927; M.A., Princeton, 1929, Ph.D. (Thaw fellow), 1931; m. Esther Mary Ives, Sept. 3, 1925; children—Laura-May Esther (Mrs. Richard A. Donnenwirth), John Raymond. Instr. math. Case Inst., 1924-28; curator physics and astronomy Buffalo Mus. Sci., 1932-36; asso. prof. astronomy Hunter Coll., 1937-50; prof. astronomy Ohio Wesleyan U., 1950-59, Howard-Perkins prof., 1955-59; prin. scientist in astronomy Franklin Inst. Labs., Phila., 1959-63; chmn. dept. math. and astronomy, dir. Morrison Obs., Central Meth. Coll., Fayette, Mo., 1963-—. Vis. asst. prof. Princeton, 1942-45, research asso., 1947-—; adj. prof. U. Pa., Phila., 1959-—; lunar photography Bosscha Obs., Indonesia, 1961; cons. higher edn. Meth. Ch., 1955-60. Am. Philos. Soc. research grantee, 1941-42. Fellow A.A.A.S. (councilman 1956-60); mem. N.Y. Acad. Scis., Am. Astron. Soc. (com. on edn. in astronomy 1951-61), Am. Assn. U. Profs., Internat. Astron. Union (U. S. del., Dublin 1955, Moscow 1958, Berkeley 1961, Hamburg 1964, pres. commn. 42, 61-—), Phi Beta Kappa, Sigma Xi. Author: Revision of astronomy vol. Collier's Popular Sci., 1938; Tables for Solution of Light Curves of Eclipsing Binaries, 1950; (with Henry Norris Russell) Determination of the Orbital Elements of Eclipsing Binaries, 1952; Monographs for the Solution of Light Curves of Eclipsing Binaries, 1953. also articles, chpts. in books. Address: Morrison Obs., Fayette, Mo. 65248.*

MERRILL, John Putnam, Am. physician; b. Hartford, Conn., Mar. 10, 1917; s. Arthur Hodges and Olive (Grinnell) M.; A.B., Dartmouth, 1938; M.D., Harvard, 1942; postgrad. Cambridge (Eng.) U., Institut Pasteur, Paris; m. Suzanne Strauss, Sept. 19, 1943; children—John Putnam, Stephen Grinnell, Ann Lawrence. Staff, Peter Bent Brigham Hosp., Boston, 1947-—, sr. asso. in medicine, 1957-—, physician, 1961-—; faculty Harvard, 1949-—, asso. clin. prof. medicine, 1962-—; established investigator Am. Heart Assn., 1950-57; investigator Howard Hughes Med. Inst., 1957-—. Mem. sci. adv. bd. Nat. Kidney Disease Found.; nat. cons. to surgeon gen. U.S.A.F.; research collaborator Brookhaven Nat. Labs., 1965-66. Recipient Alvarenga prize Phila. Coll. Physicians, 1960; Distinguished Achievement award Modern Medicine, 1965. Diplomate Am. Bd. Internal Medicine. Fellow A.C.P.; mem. A.M.A., Am. Soc. Clin. Investigation (pres. 1962-63), Am. Clin. and Climatol. Soc., Am. Physiol. Soc., Assn. Am. Physicians, Am. Heart Assn. (chmn. exec. com. kidney sect. since 19——), Am. Fedn. Clin. Research (pres. Boston chpt. 1952), N.Y. Acad. Scis., Internat. Soc. Nephrology (exec. com., pres. 1964-—), Am. Acad. Arts and Scis. (Amory prize 1962), Med. and Metaphys. Soc., Phi Beta Kappa, Alpha Omega Alpha, others. Publs. on renal physiology and role of pathology in hypertension; 1st successful use of artificial kidney in U. S.; 1st successful transplantation of kidney between identical twins, non-identical twins, and from cadaver to living persons; immunology of tissue transplantation. Home: 55 Loring Rd., Weston, Mass. Office: Peter Bent Brigham Hosp., Boston 02115.*

MERRILL, Malcolm H(endricks), Am. physician; b. Richmond, Utah, June 28, 1903; s. L. Edgar and Clara (Hendricks) M.; B.S., Utah State Agrl. Coll., 1925; M.S., St. Louis U., 1927, M.D., 1932; M.P.H., U. Cal., 1946; m. Thelma Holdaway, Aug. 11, 1926; children—M. Donald, L. Bruce, Jean. Asst., Rockefeller Inst. Med. Research, 1932-35; with Cal. Dept. Pub. Health, San Francisco, 1937-65, successively chief bur. venereal diseases, chief div. labs., dep. dir., acting dir. and dir. pub. health, 1954-65; dir. health services AID, Washington, 1965-67, dep. asst. adminstr. Office of War on Hunger, 1967-—. Mem. exptl. therapeutics study sect. USPHS, 1948-53, mem. nat. tng. adv. com., 1957-64; cons. pub. health for India, TCA, 1952; lectr. Sch. Pub. Health, U. Cal., 1947-—. Mem. cancer control adv. com. NIH, 1955-57; mem. nat. adv. health council USPHS, 1957-61; mem. mission to Russia, 1957; mem. tech. adv. com. INCAP-PASB, Guatemala, 1955, 56, 58, 59. Diplomate Am. Bd. Preventive Medicine, mem. bd., 1965-—. Fellow Am. Pub. Health Assn. (pres. 1959-60, chmn. tech. devel. bd. 1963-66, speaker of council 1967-—), Conf. State and Provincial Pub. Health Lab. Dirs. (chmn. 1950); mem. A.M.A., A.A.A.S., State and Territorial Pub. Health Lab. Dirs., U. S.-Mexico Border Pub. Health Assn. (pres. 1957-58), Assn. State and Territorial Health Officers (pres. 1961-62), Sigma Xi, Delta Omega (nat. pres. 1951-52), Alpha Omega Alpha, Phi Kappa Phi, Pi Delta Epsilon. Research, publs. in fields of bacteriology, immunology, virology, pub. health. Home: 1425 4th St. S.W., Washington 20024. Office: New State Building, AID, Washington.*

MERRILL, Paul Willard, Am. astronomer; b. Mpls., Aug. 15, 1887; s. Charles Wilbur and Kate Amelia (Kreis) M.; A.B., Stanford, 1908; Ph.D., U. Cal., 1913; m. Ruth L. Currier, Sept. 12, 1913. Fellow and asst. Lick Obs., Cal., 1908-13; instr. astronomy U. Mich., 1913-16; asst. and asso. physicist U. S. Bur. Standards, Washington, 1916-18; astronomer Mt. Wilson Obs., Carnegie Inst., 1919-52. Dir. City of Pasadena, 1927-31. Mem. Nat. Acad. Scis. (Draper Medal 1946), Am. Philos. Society, American (Russell lecturer 1955, president 1956-58), Royal (fgn. asso.) astron. socs., Astron. Society of the Pacific (Bruce Medal 1946), Am. Phys. Soc. A.A.A.S., Am. Assn. Variable Star Observers, Phi Beta Kappa, Sigma Xi. Author: Space Chemistry, 1963; contbr. articles. Extensive work on wave lengths of spectrum lines, red and infrared photography, stellar spectroscopy, and interstellar matter; authority on red stars; discovered presence of technetium in stars of spectral type S; discovered many stars. Died Los Angeles, July 16, 1961.

MERRILL, William Meredith, Am. geologist, educator; b. Detroit, Dec. 1, 1918; s. Frederick Ivan and Isobel (LaBombard) M.; B.S., Mich. State Coll., 1946; M.A., Ohio State U., 1948, Ph.D. (Bownocker fellow), 1950; m. Gypsie Virginia Smith, Feb. 23, 1943; children—Russell B., W. Wood, Douglas G. Asst. geologist to geologist Geol. Survey Ohio, 1946-50; faculty geology U. Ill., Urbana, 1950-58, asso. prof., 1955-58; prof., chmn. dept. geology Syracuse (N.Y.) U., 1958-63, U. Kan., Lawrence, 1963-—. Geologist, party chief Geol. Survey Nfld., 1954; research asso. Ohio State U. Research Found., Red Rock Lake, Greenland, 1955-57; geologist Research Council Alta. (Can.), summers 1958-62; vis. geoscientist Am. Geol. Inst., 1962, 64, capt. vis. team Geo-study, 1962-63, chmn. geo-study panel on preparation secondary sch. earth sci. tchrs., 1964-—, mem. steering com. Earth Study Sci. Curriculum Project, 1964-—. Fellow A.A.A.S., Geol. Soc. Am.; mem. Internat. Assn. Sedimentologists, Am. Assn. Petroleum Geologists, Soc. Econ. Paleontologists and Mineralogists, Soc. Econ. Geologists, Glaciological Soc., Am. Assn. U. Profs., Nat. Assn. Geology Tchrs., Sigma Xi. Research, publs. on Miss., Pa., Ohio, Colo. rocks, structure of glacier ice, non-marine Cretaceous and Tertiary rocks of West Central Alta. Home: 224 Dakota St., Lawrence, Kan. 66044.*

MERRIMAN, Daniel, Am. biologist; b. Cambridge, Mass., Sept. 17, 1908; s. Roger Bigelow and Dorothea (Foote) M.; grad. Groton Sch., 1927; student Harvard, 1927-30; B.S., U. Wash., 1933, M.S., 1934; Ph.D., Yale, 1939; m. Mary Wieland, Sept. 18, 1934. Aquatic biologist Conn. State Bd. Fish and Game, 1936-38; faculty biology Yale, 1938-—, asso. prof., 1946-—, dir. Bingham Oceanographic Lab., 1942-66, master Davenport Coll., 1946-66; trustee Bermuda Biol. Sta., 1944-—, Woods Hole Oceanographic Instn., 1944-64; dir. Sears Found. Marine Research, 1966-—. Fellow N.Y. Zool. Soc.; mem. A.A.A.S., Am. Soc. Anatomy, Am. Soc. Limnology and Oceanography, Am. Soc. Zoology, Conn. Acad. Arts and Scis., Hist. of Sci. Soc., Soc. Ichthyology and Herpetology, Sigma Xi. Research, numerous publs. in field marine biology, particularly ichthyology and history of oceanography. Home: 45 Lincoln St., New Haven 06511.*

MERRIMAN, Gaylord Maish, Am. mathematician; b. Elmwood Place, O., July 24, 1901; s. Roy Gaylord and Alice (Maish) M.; A.B., U. Cin., 1923, A.M., 1924, Ph.D., 1926; m. Emily Stowitts, Mar. 28, 1934; 1 dau., Lucey. Faculty Grinnell Coll., 1928-29; faculty U. Cin., 1929-—, prof. math., 1948-—, head dept., 1957-62. Vis. prof. U. N.C., Chapel Hill, 1960; NSF Inst. lectr. various univs.; mem. advanced placement com. CEEB, 1955-63. NRC fellow Harvard, 1926-28. Mem. Am. Math. Soc., Math. Assn. Am. (gov. 1959-62), Am. Assn. U. Profs., Sigma Xi, Phi Beta Kappa. Author: To Discover Mathematics, 1942; Calculus, 1954. Research on orthogonal polynomials, especially Fourier series; teaching of math., direction of research. Home: 536 Evanswood Pl., Cin. 45220.*

MERRIMAN, Mansfield, Am. engr., mathematician; b. Southington, Conn., Mar. 27, 1848; s. Mansfield and Lucy (Hall) M.; Ph.B., Sheffield Sci. Sch., Yale, 1871, C.E., 1872, Ph.D., 1876; Sc.D., U. Pa., 1906; LL.D., Lehigh U., 1913; m. Wanda Kubale, June 5, 1875 (dec. 1889), 2nd Anna Rosina Godshalk, May 24, 1891 (dec. 1907), 3rd Bazena Treat, June 6, 1910; children—Thaddeus, Lucy, Richard, Alice Pauline, Richard Mansfield, Norman Nathaniel. Asst. engr. U. S. Engr. Corps, 1872-73; instr. civil engring. Sheffield Sci. Sch., 1875-78; asst. U. S. coast and Geod. survey, 1880-85; prof. civil engring. Lehigh U., 1878-1907; cons. civil and hydraulic engr., from 1907. Author: Elements of the Method of Least Squares, 1877; Mechanics of Materials, 1885; (with H. S. Jacoby) Roofs and Bridges, 1890; (with R. S. Woodworth) Higher Mathematics, 1896; Strength of Materials, 1897; Precise Surveying and Geodesy, 1899; Elements of Sanitary Engineering, 1898; Elements of Hydraulics, 1912. Asso. editor, Appleton's Universal Cyclopedia, 1895. Research on hydraulics, strength of materials, pure mathematics; devel. theory of least squares (important to bridge triangulation). Died N.Y.C., June 7, 1925.

MERRISON, Alexander Walter, English physicist; b. London, Eng., Mar. 20, 1924; s. Henry and Violet (Mortimer) M.; B.Sc., Ph.D., King's Coll., London; m. B. G. Le Marquand, Mar. 6, 1948; children—Jonathan, Timothy. Research scientist Signals Research & Devel. Establishment, 1944-46; research with nuclear reactors and nuclear physics Harwell (Eng.) Atomic Energy Research Establishment, 1946-51; instr. U. Liverpool (Eng.), 1951-57, prof. physics, 1960—; physicist European Nuclear Research Council, Geneva, Switzerland, 1957-60; dir. Nuclear Physics Labs., Daresbury Nat. Inst. for Research in Nuclear Sci., 1962—. Leverhulme fellow. Research, publs. on accelerator constrn., elementary particle physics. Address: Nuclear Physics Research Lab., U. Liverpool, Liverpool 3, Eng.

MERRITT, Arthur Donald, Am. geneticist; b. Shawnee, Okla., June 11, 1925; s. Arthur B. and Margaret (Harris) M.; A.B., George Washington U., 1949, M.D., 1952; m. Doris Honig, May 5, 1953; children—Kenneth Arthur, Christopher Ralph. Fellow medicine Duke, 1955, chief resident, instr. medicine, 1956-57; clin. asso. Nat. Inst. Arthritis and Metabolic Diseases NIH, USPHS, Bethesda, Md., 1957-58, chief med. investigations sect., clin. investigations br. Nat. Inst. Dental Research, 1958-60; clin. instr. medicine George Washington U. Sch. Medicine 1959-60; asso. prof. medicine, biochemistry, chmn. med. genetics program Ind. U. Med. Center, Indpls., 1961—, prof. medicine, prof. and chmn. dept. med. genetics, 1966—. Diplomate Am. Bd. Internal Medicine. Mem. Genetics Soc. Am., Central Soc. Clin. Research, A.C.P., Am. Soc. Human Genetics, Am. Fedn. Clin. Research, N.Y. Acad. Scis., A.M.A., Pub. Health Service Clin. Soc., A.A.A.S. Research, publs. on biochem., genetic studies of enzymes, metabolic pathways, certain diseases such as cystic fibrosis, diabetes mellitus, aortic stenosis. Home: 6139 Autumn Lane, Indpls. 46220. Office: Ind. U. Med. Center, 1100 W. Michigan St., Indpls. 46207.*

MERRITT, Ernest George, Am. physicist; b. Indpls., Apr. 28, 1865; s. George and Paulina Tate (McClung) M.; student Purdue U., 1881-82; M.E., Cornell U., 1886; grad. student, Cornell, 1888-89, U. Berlin, 1893-94; m. Bertha A. Sutermeister, Apr. 10, 1901; children—Louise S. (Mrs. Ralph H. Brandt), Julia S. (Mrs. J. G. Hodge), Virginia S. (Mrs. J. T. Emlen, Jr.), Grace S. (Mrs. Jürg Waser), Howard S. Instr. in physics Cornell U., 1889-92, asst. prof., 1892-1903, prof., 1903-35, head dept., 1918-35, emeritus since 1935, dean grad. sch., 1909-14. Engaged in research on anti-submarine devices at U. S. Naval Exptl. Sta., New London, Conn., 1917-18. Fellow Am. Acad. Arts and Scis., Am. Phys. Soc. (pres. 1914-15), A.A.A.S.; mem. Nat. Acad. Scis., Sigma Xi. Asso. editor Phys. Rev., 1893-1913. Contbr. papers on investigations in luminescence and radio. Died June 5, 1948.

MERRITT, H(iram) Houston, Am. physician; b. Wilmington, N.C., Jan. 12, 1902; s. Hiram Houston and Dessie (Cline) M.; student U. N.C., 1919-20; A.B., Vanderbilt U., 1922; M.D., Johns Hopkins, 1926; A.M. (hon.), Harvard, 1942; m. Mabel Carmichael, Aug. 2, 1930. Practice medicine, specializing in neurology, N.Y.C., 1944—; prof. neurology Columbia Coll. Phys. and Surgs., 1948-63, Moses prof. neurology, 1963—, dean faculty medicine, v.p. charge med. affairs, 1959—; dir. service neurology Neurol. Inst., 1948—. Mem. Nat. Inst. Neurol. Diseases and Blindness Council NIH, 1950-58, 60-64; cons. surg. gen. Dept. Army, 1957—; dir. Nat. Fund For Med. Edn. Diplomate Am. Bd. Psychiatry and Neurology (past pres., dir.). Fellow Am. Acad. Arts and Scis., Royal Soc. Medicine London (hon.); mem. Am. Neurol. Assn. (past pres.), Assn. for Research in Nervous and Mental Disease (past pres.), Nat. Adv. Gen. Med. Scis. Council. Chief editor: Archives of Neurology, 1962—. Publs. on investigation of the biochemical relationship of blood and cerebrospinal fluid; chem. structure of anticonvulsants in treatment of epilepsy; studies of infections of the nervous system. Home: 7 Beechwood Rd., Bronxville, N.Y. 10708. Office: 630 W. 168th St., N.Y.C. 10032.*

MERRITT, Lynne Lionel, Jr., Am. chemist; b. Alba, Pa., Sept. 10, 1915; s. Lynne Lionel and Pauline (Brown) M.; B.S., Wayne U., 1936, M.S., 1937; Ph.D., U. Mich., 1940; m. Lucille Elizabeth Widman, Dec. 18, 1937; children—Margaret (Mrs. David H. Bowen), Lucille (Mrs. I. Clay Williams), Lynn Robert, Linda. Faculty, Wayne U., Detroit, 1936-37, 39-42, U. Mich., Ann Arbor, 1937-39; faculty Ind. U., Bloomington, 1942—, prof. chemistry, 1953—, asso. dean faculties, 1962-65, v.p., dean research and advanced studies, 1965—. Pres., dir. Ind. Instrument & Chem. Corp., Bloomington, 1959—; dir. Bur. Instl. Research, 1960-65. Guggenheim Found. fellow Cal. Inst. Tech., 1955-56; Fulbright research fellow Centre National de la Recherche Scientifique, Bellevue, France, 1963. Fellow A.A.A.S.; Mem. Coblenz Soc., Am. Chem. Soc., Am. Assn. U. Profs., Ind. Acad. Sci., Sigma Xi, Phi Beta Kappa, Phi Kappa Phi, Gamma Alpha, Phi Lambda Upsilon. Author: (with H. H. Willard, J. A. Dean) Instrumental Methods of Analysis, 1948, 4th edit., 1965. Home: 916 S. Highland Av., Bloomington, Ind. 47401.*

MERSENNE, Marin, French natural philosopher, mathematician, theologian; b. La Soultière/Sarthe, nr. Oizé, France, Sept. 8, 1588; ed. Collège du Mans, Jesuit Sch. of La Flèche (France), U. Paris. Entered novitiate Order of Minimes, Roman Catholic ch., Paris, 1611; tchr. philosophy Mimin convent, Nevers, France, 1614-19, convent L'Annonciade, Paris, from 1620; traveled to Rouen, France, 1625, Flanders and Holland, 1630, Champagne, France, 1639, Spain, 1644, Italy, 1645, So. France, 1646. Author: Questiones celeberrimae in Genesim, 1623; La verité des sciences contre les sceptiques on pyrrhoniens, 1625; Traité de l'harmonie universelle, 1627-36; Nouvelles pensées de Galilee, 1630; Questions théologiques, physiques, morales, et mathématiques, 1634; Questions inouyes, 1634; Cogitata phisico-mathematica, 1644; L'impiété des deistes, athées, et libertins de ce temps, 1674; many others. Important link between European scholars before sci. jours. were founded (met with many natural philosophers in Paris, correspondence disseminated sci. ideas throughout Europe); research in math., physics, astronomy; investigated cycloid curves, echoes, speed of sound, vibrating strings, theory of music, harmonics, mus. instruments; determined frequency of mus. note, showed that raising note an octave doubles its frequency; attacked atheists, deists; opposed astrology, alchemy, divination and magic; sought to construct a mechanistic physics. Died Paris, Sept. 1, 1648.

MERTE, Hanns-Jürgen, German ophthalmologist; b. Jena, Germany, Aug. 17, 1921; s. Willy and Antonie (Popp) M.; ed. Jena, Berlin (Germany) univs.; M.D.; m. Gertrud Richter, July 25, 1956; 1 dau., Birgit. Asst. Univ. Ophthal. Clinic, Munich, Germany, 1945-48, physician, from 1948; from instr. to prof. ophthalmology U. Munich, from 1953, also mem. senate; ophthalmologist Munich municipal hosps., from 1963. Mem. German, Australian, Munich ophthal. socs., German Soc. Labor Medicine, Union Bavarian Ophthalmologists, Munich Med. Assn. Co-author: Augenheilkunde in Klinik und Praxis; Pathologie der Laboratoriumstiere; Almanach fur Augenheilkunde. Research, publs. on exptl. and clin. work on anaphylaxie of cornea, glaucoma, therapeutic problems. Home: Föhrenweg 5, Munich 9. Office: Ismaningerstrasse 22, Munich 8, W. Germany.

MERTENS, Robert, zoologist; b. St. Petersburg (now Leningrad, USSR), Dec. 1, 1894; s. Robert and Eugenie (Brunst) M.; Ph.D., U. Leipzig (Germany); m. Karline Bergmann, June 6, 1921; 1 dau., Xenia. Asst. Senckenberg Mus., Frankfort, Germany, from 1919, curator, from 1925, dir., 1946-60, emeritus, from 1960; instr. U. Frankfort, 1932-39, prof., from 1939. Recipient Gold medal Senckenberg Natural Sci. Researchers. Mem. Leopoldina Soc., Indian Acad. Zoology (corr.), Am. Mus. Natural History, Am. Soc. Ichthyologists and Herpetologists. Research, publs. on herpetology; travels and research in zoology on all continents. Home: Georg-Speyerstrasse 31, Frankfort/Main. Office: Senckenberg-Museum, Senckenberg-Anlange 25, Frankfort/Main, W. Germany.

MERTON, Robert K, Am. sociologist; b. Phila., July 5, 1910; A.B., Temple U., 1931; M.A., Harvard, 1932, Ph.D. 1936; LL.D., Temple U., 1956; Dr. hon. causa U. Leyden, 1965; L.H.D., Emory U., 1965; LL.D., Western Res. U., 1966; Litt.D., Colgate U., 1967; m. Suzanne Mae Carhart, Sept. 8, 1934; children—Setphanie C. (Mrs. Richard L. Russell), Robert C., Vanessa H. Faculty, Harvard, 1934-39, Tulane U., 1939-41; faculty Columbia, 1941—, Giddings prof. sociology, 1963—, chmn. dept., 1962-65, asso. dir. Bur. Applied Social Research, 1942—. Mem. com. on selection Guggenheim Found., 1963—; NIH lectr., 1964, Gilman lectr. Johns Hopkins, 1962. Recipient prize for distinguished scholarship in humanities Am. Council Learned Socs., 1962. Guggenheim fellow, 1962. Fellow Am. Acad. Arts and Scis., Am. Philos. Soc., World Acad. Art and Sci., Nat. Acad. Edn.; mem. Nat. Acad. Scis., Century Assn. Author: Science Technology and Society in 17th Century England, 1938, Mass. Persuasion; 1946; Social Theory and Social Structure, 1949, 57; Reader in Bureaucracy, 1952; Continuities in Social Research, 1950; Freedom to Read, 1954; The Focused Interview, 1956; The Student-Physician, 1957; Sociology Today, 1959, Contemporary Social Problems, 1961, 66; On the Shoulders of Giants, 1965; On Theoretical Sociology, 1967. Adv. editor Sociology, Harcourt, Brace & World. Research in mass communications; sociol. theory; sociology of professions and science. Home: 111 Pinecrest Dr., Hastings-on-Hudson, N.Y. 10706. Office: Fayerweather Hall, Columbia U., N.Y.C. 10027.

MERTZ, Edwin Theodore, Am. biochemist; b. Missoula, Mont., Dec. 6, 1909; s. Gustav Henry and Louise (Sain) M.; B.A., U. Mont., 1931; M.S. in Biochemistry, U. Ill., 1933, Ph.D., 1935; m. Mary Ellen Ruskamp, Oct. 5, 1936; children—Martha Ellen, Edwin Theodore. Research biochemist Armour & Co., Chgo., 1935-37; instr. U. Ill., Urbana, 1937-38; research asso. pathology U. Ia. Med. Sch., 1938-40; instr. agrl. biochemistry U. Mo., Columbia, 1940-43; research chemist Hercules Powder Co., Wilmington, Del., 1943-46; prof. biochemistry Purdue U., Lafayette, Ind., 1946—. Cons. biochemistry Ind. State Hosps., 1956—; cooperator Western Fish Nutrition Lab., Cook, Wash., 1955—. Mem. Am. Soc. Biol. Chemists, Am. Inst. Nutrition, Am. Chem. Soc. Am. Assn. Cereal Chemists, Am. Soc. Animal Sci., A.A.A.S. Author: Elementary Biochemistry, 1966; (with Herbert E. Parker) Laboratory Experiments in Biochemistry, 1964; also numerous articles. Research on amino acid requirements weanling pigs, fingerling salmon, lysine requirements humans; determined activation mechanism highly purified native and altered forms plasminogen; co-discoverer of high lysine maize. Home: 1149 Hillcrest Rd., West Lafayette, Ind. 47906.*

MERTZ, Robert Theodore, Am. mathematician; b. Floral Park, N.Y., Jan. 18, 1928; s. Pierre and Eunice (Hanhart) M.; A.B., Harvard, 1948; A.M., Columbia, 1951, Ph.D. in Applied Math., 1961; m. Alice Ann Deardorff, June 6, 1953; children—Susan Elizabeth, John Pierre. Applied mathematician problem analyst, digital computer programmer IBM, N.Y.C., 1950—. Mem. Am. Math. Soc., Soc. for Indsl. and Applied Math., A.A.A.S., Assn. for Computing machinery. Research on formulation, solution and computer programming problems in missile trajectories, satellite orbits, Braille translation, portfolio selection, gen. linear, quadratic and non-linear optimization. Home: 51 Mansfield Av., Darien, Conn. 06820. Office: 590 Madison Av., N.Y.C. 10022.*

MÉRY, Jean, French surgeon; b. Vatan, France, Jan. 6, 1645; 1st surgeon Hôtel-Dieu, Paris, France; mem. French Acad. Scis. Author: New System of the Circulation of the Blood, 1700; also other med. works. Described 2 glands near bulb of corpus spongiosum which communicate with cavernous part of urethra (Méry's glands and Cowper's glands), 1684. Died Paris, Feb. 18, 1722.

MERYMAN, Harold Thayer, Am. biophysicist; b. Washington, Feb. 5, 1921; s. Richard S. and Charlotte (Bates) M.; grad. Groton Sch., 1939; student (Harvard Prize scholar) Harvard, 1939-43; M.D., L.I. Coll. Medicine, 1946; m. Mary-Lane Latimer, Apr. 10, 1948; children—Richard, Louise, Henry, Charlotte. Staff, Naval Med. Research Inst., Bethesda, Md., 1947-54; faculty dept. biophysics, Yale, 1954-57; head cryobiology br. Naval Med. Research Inst., Bethesda, 1957—. Spl. cons. USPHS, 1963—. Am. Cancer Soc. fellow, 1954-56. Mem. Cryobiological Soc. (founding mem.), Internat. Inst. Refrigeration (gen. sec. commn. 6C 1963), N.Y. Acad. Sci., Biophys. Soc., Am. Soc. Cell Biology, Electron Microscope Soc., A.A.-A.S. Author: Cryobiology, 1966. Editorial bd. Cryobiology, 1964—. Research, publs. on effects freezing on biol. materials, freeze-drying, therapy frostbite; helped develope method freezing whole blood. Home: Tucker Lane, Sandy Spring, Md. Office: Naval Med. Research Inst., Bethesda, Md. 20014.*

MERZBACHER, Eugen, physicist; b. Berlin, Germany, Apr. 9, 1921; s. Siegfried and Lilli (Wilmersdoerffer) M.; licentiate U. Istanbul, 1943; A.M., Harvard, 1948, Ph.D., 1950; m. Ann Townsend Reid, July 11, 1952; children—Celia Irene, Charles Reid, Matthew Allen, Mary Letitia. Came to U. S., 1947, naturalized, 1953. High sch. tchr., Ankara, Turkey, 1943-47; mem. Inst. for Advanced Study, Princeton, N.J., 1950-51; vis. asst. prof. Duke, 1951-52; faculty U. N.C., Chapel Hill, 1952—, prof., 1961—, acting chmn. physics dept., 1965—. Vis. scientist Am. Inst. Physics, 1962—; mem. subcom. of com. on nuclear sci. Nat. Acad. Scis.-NRC, 1964—; mem. adv. bd. U. S. Eastern Theoretical Physics Conf., 1962—. NSF Sci. Faculty fellow U. Copenhagen (Denmark), 1959-60. Fellow Am. Phys. Soc. (exec. com. Southeastern sect. 1963—); mem. Am. Assn. Physics Tchrs., A.A.A.S., Am. Assn. U. Profs., Sigma Xi. Author: Quantum Mechanics, 1961; also articles. Editorial bd. Am. Jour. Physics, 1965—. Research on applications of quantum mechanics to study atoms and nuclei especially beta decay, neutron spectroscopy, nuclear reactions, ionization and stopping power problems, radiation damage in solids. Home: 1396 Halifax Rd., Chapel Hill, N.C. 27514.*

MESCHAN, Isadore, Am. radiologist; b. Cleve., May 30, 1914; s. Julius and Anna (Gordon) M.; B.A., Western Res. U., 1935, M.A., 1937, M.D., 1939; m. Rachel Farrer, Sept. 3, 1943; children—David, Eleanor Jane, Rosalind, Joyce. Instr., Western Res. U.,

1946-47; prof., head dept. radiology U. Ark., Little Rock, 1947-55; prof., dir. dept. radiology Bowman Gray Sch. Medicine, Winston-Salem, N.C., 1955—. Fellow Am. Roentgen Ray Soc.; mem. Am. Coll. Radiology (chmn. com. on radiation biology 1964—), A.M.A., Radiology Soc. N.Am., N.C. Radiol. Soc., So. Med. Assn., Soc. Nuclear Medicine, Assn. U. Radiologists, Sigma Xi, Alpha Omega Alpha. Author: Atlas of Normal Radiographic Anatomy, 1951; Roentgen Signs in Clinical Diagnosis, 1956; (with R. Meschan) Synopsis of Roentgen Signs, 1962; Roentgen Signs in Clinical Practice, 1966; also numerous articles. Editor: The Radiologic Clinics of North America, 1965. Research on radioactive materials, radiation biology, diagnostic radiology. Home: 2716 Bartram Rd., Winston-Salem, N.C. 27106.*

MESCHIA, Giacomo, physiologist; b. Milan, Italy, Feb. 7, 1926; s. Carlo and Ada (Marca) M.; M.D., U. Milan, 1950; m. Irene Battaglia, June 17, 1961; 1 son, James F. Came to U. S., 1953, naturalized, 1964. Asst. prof. physiology Milan U., 1951-56; with Yale U., 1953-65, asst. prof. physiology, 1959-65; asso. prof. physiology U. Colo., Denver, 1965—. Mem. N.Y. Acad. Sci., A.A.A.S., Sigma Xi. Studies, publs. on respiratory function of placenta, maternal and fetal blood circulation and composition, differentiation between osmosis and simple diffusion of water. Home: 2401 E. 7th Av., Denver 80206.*

MESCHKOWSKI, Herbert, German mathematician; b. Berlin, Germany, Feb. 13, 1909; s. Eduard and Emma (Schäfer) M.; Ph.D., U. Berlin; m. Magdalena Meitz, Feb. 12, 1936; children—Inge, Christa, Helmut, Katrin. From agrégé, to full prof., to prof. at large U. Berlin, from 1954. Mem. German Math. Soc. Author: Wandlungen des mathematischen Denkens, 1956; Differenzengleichungen, 1959; Das Christentum im Jahrhundert der Naturwissenschaften, 1961; Denkweisen grosser Mathematiker, 1961; Hilbertsche Räume mit Kernfunktion, 1962. Home: Thielallee 66, Berlin-Dahlem. Office: 2. Math. Inst. FU, Thielallee 52, Berlin 33, W. Germany.

MESCON, Herbert, Am. physician; b. Toronto, Ont., Can., Apr. 4, 1919 (parents Am. citizens); s. Morris and Bella (Paleschuck) M.; B.S., Coll. City N.Y., 1938; M.D., Boston U., 1942; m. Barbara Jeanne McKenzie, Sept. 15, 1946; children—Susan Lee, Gary Lawrence, Robin Lenore, Stanley Richard. Fellow dermatology U. Pa. Hosp., 1948-51, Damon Runyon cnacer research fellow, 1949-51, asst. instr. dermatology, 1948-50, instr., 1950, asso., 1951-52, instr. Grad. Sch. Medicine, 1950-52; attending dermatologist VA Hosp., Wilmington, Del., 1951-52; prof., chmn. dept. dermatology Boston U. Med. Sch., 1952—; chief dermatology, dir. genito-infectious diseases Mass. Meml. Hosp., 1952—; cons. dermatopathology VA Hosp., Boston, 1955—; area cons. dermatology VA, 1959—; sr. cons. Lemuel Shattuck Hosp., Boston, 1957—. Mem. com. cutaneous medicine NRC, 1960—. Diplomate Am. Bd. Dermatology and Syphilology. Mem. Am. Assn. Pathologists and Bacteriologists, Soc. Investigative Dermatology (pres. 1963), New Eng., Boston dermatol. socs., Am. Acad. Dermatology and Syphilology (adv. com. to FDA 1961—), A.M.A., Histochem. Soc., Am. Dermatol. Assn. Research, publs. on skin histochemistry and dermatopathology, sebaceous glands in disease. Home: 155 Lake Av., Newton Center, Mass. 02159. Office: 203 Commonwealth Av., Boston 02116.*

MESELSON, Matthew Stanley, Am. biologist; b. Denver, May 24, 1930; s. Hymen Avram and Ann (Swedlow) M.; Ph.B., U. Chgo., 1951; Ph.D., Cal. Inst. Tech., 1957. From research fellow to sr. research fellow Cal. Inst. Tech., 1957-60; mem. faculty Harvard, 1960—, prof. biology, 1964—. Recipient prize for molecular biology Nat. Acad. Scis., 1963, Eli Lilly award microbiology and immunology, 1964. Research on molecular biology. Address: Biological Labs., Harvard Univ., Cambridge, Mass.

MESHCHERYAKOV, Mikhail Grigorevich, Russian physicist; b. Sept. 17, 1910; grad. Leningrad U., 1936. With Radium Inst., USSR Acad. Scis., 1937-47; prof. Moscow U., 1954—; asso. Joint Nuclear Research Inst., Dubna, 1956—. Decorated Order of Lenin (2). Mem. USSR Acad. Scis. (corr.). Author: On the Absorption of Fast Neutrons by Heavy Nuclei, 1945; co-author: Investigation of Interaction of Protons with Protons at High Energies, 1955; Polarization of Protons with Energy of 660MEV in Nuclear Scattering, 1956. Research and publs. on physics of high-energy particles, absorption of fast neutrons by heavy nuclei, high-energy nuclear processes in accelerators, proton-proton interaction, polarization of protons. Address: Joint Nuclear Research Inst., Dubna, Moskovskaya Oblast, USSR.

MESHITSUKA, Gisuke, Japanese chemist; b. Shimane, Japan, Dec. 17, 1919; s. Kamenosuke Iwata and Mumeno (Hara) M.; Bachelor, Osaka (Japan) U., 1943, D.Sc., 1953; m. Tokiko, Dec. 28, 1942; children—Gyosuke, Shunsuke, Keisuke. Asst., Osaka U., 1943-48, asst. prof., 1949-59; prof. Naniwa High Sch., Osaka, 1948-49; head Tokyo Met. Isotope Research Center, 1959-66, pres., 1966—. Mem. Radiation Research Soc., Chem. Soc. Japan, Soc. Polymer Sci. Japan. Research, numerous publs. on osmotic pressure, measurement and exptl. formula polymerization, living polymer by metallic sodium radiation chemistry, methanol radiolysis, yields and mechanisms. Home: 1-634 Marukodori, Kawasaki, Kanagawa, Japan. Office: 1-899 Fukazawa, Setagaya, Tokyo, Japan.*

MESHKE, Edna Dorothy, Am. textile scientist; b. Morristown, Minn., Apr. 22, 1906; d. Reinhold F. and Bertha (Migge) Meshke; B.S., U. Minn., 1927, Ph.D., 1942; M.A., Columbia, 1933. Tchr. home econ. pub. schs., Lamberton, Bemidji, Mpls., 1927-34; instr. N.D. Agrl. Coll., N.Y. State Tchrs. Coll., Buffalo, 1934-40; prof., head home economics dept. Butler U., Indpls., 1943-45; asso. prof., head home economics edn. dept., U. Md., 1946-48; prof., dept. chmn. U. Cal., Santa Barbara Coll., 1948-53, prof. and tchr. trainer, 1953-55, professor of textiles, 1955-61; prof., head dept. clothing and textiles Purdue U., West Lafayette, Ind., 1961-64; vis. prof. textiles and clothing Okla. State U., 1966-67. cons. fabrics and apparel design. Mem. Am. Assn. Textile Chemists and Colorists, Phi Upsilon Omicron, Omicron Nu, Pi Lambda Theta. Contbr. profl. publs. Research on fabrics as engineered materials, interrelationships of fabric geometry, stress, strain properties, anthropometric data; currently ind. study to extend scope and applications of research findings to garment designs in custom studios, indsl. prodn. plants. Address: 715 1st St. S.W., Faribault, Minn. 55021.*

MESMER, Franz (or Friedrich) Anton, physician; b. Iznang am Bodensee, nr. Rodolfzell, Baden, Germany, May 23, 1734; studied for priesthood, Dillingen and Ingolstadt, Germany; doctor's degree U. Vienna, 1766. Began practice using magnets to cure disease; later discarded this method when his theories aroused antagonism in Viennese med. profession, in 1778; went to Paris to practice with new method which he believed was based on animal magnetism; investigated and discredited by com. French Acad. Scis., including A. Lavoisier, J. Guillotin, and B. Franklin, 1784; forced to leave Paris, 1785; ret. to Versailles, France, then Switzerland. Author: De planetarum influxu in corpus humanum, 1766; Mémoire sur la découverte du magnétisme animal, 1779; Histoire abrégée du magnétisme animal, 1783; Mémoire de F. A. Mesmer sur les découvertes, 1799; Mesmerismus, 1814. Believed stars have magnetic influence on human beings; held that magnetic fluid exerts force, which he labelled animal magnetism, on man; claimed to possess control over this force; his theory of animal magnetism laid found. for modern hypnosis and suggestion therapy. Died Meersburg, Germany, Mar. 5, 1815.

MESNAGER, Augustine-Charles-Marie, French civil engr.; b. Paris, June 11, 1862; civil engr.; prof. Ecole des Ponts et Chausées; became prof. Conservatoire des Arts et Métiers, 1913. Mem. French Acad. Scis. (v.p. 1933). Author: Cours de béton armé, 1921; Table pour le calcul rationnel des planchers sans nervure, 1929. Studied effect of pressure on transparent blocks using polarized light; studied durability problems of reinforced concrete, 1914, theory of 3 dimensional elasticity. Died Paris, Feb. 6, 1933.

MESNARD, Guy, French physicist; b. Champagnac, France, Feb. 22, 1923; s. Raymond and Julia (Couillaud) M.; Agregation des Sciences Physiques, Ecole normale supérieure, 1947, Doctorat ès Sciences, 1952; m. Monique Doeuvre, Dec. 20, 1950; children—Eric-Paul, Isabelle, Alain, Yann, Catherine. Prof., Faculté des Sciences, Lyons, France, 1957—; cons. Decorated Officer des palmes académiques. Research and numerous publs. on solid state physics, electronics. Home: 16, Avenue de Grange-Blanche, 69-Tassin-la-Demi-Lune, France. Office: 43, Boulevard du 11 Novembre 1918, 69-Villeurbanne, France.*

MESNIL, Félix-Étienne-Pierre, French zoologist; b. Omonville, France, Dec. 12, 1868; prof. Pasteur Inst., Paris; mem. French Acad. Scis., French Acad. Medicine, Acad. Colonial Scis. Author: les Trypanosomes et les trypanosomiases, 1913. Studied flagellate protozoa especially trypanosomes and diseases they caused. Died Paris, Feb. 15, 1938.

MESNY, René Marie, French physicist; b. Brest, France, 1874; dir. Laboratoire National de Radioélectricité. Research on short waves and ultra high frequency waves, also radiogoniometry (wireless direction determination). Died 1949.

MESSEL, Harry, physicist; b. Levine Siding, Man., Can., Mar. 3, 1922; s. James Nikola and Katrina (Maslanka) M.; grad. Royal Mil. Coll. Can., 1942; B.A. with honors in Math., Queen's U., Kingston, Ont., Can., 1948, B.Sc. with honors in Engring. Physics, 1948; Ph.D. in Theoretical Physics, Nat. U. Ireland, 1951; m. Patricia Iona Pegram, Nov. 1, 1949; children—Naomi Lee, Wendy Jane, Iona Sue. Sr. lectr. theoretical physics U. Adelaide (Australia), 1951-52; apptd. chair physics, head sch. physics U. Sydney (Australia), 1952—, founder, dir. Sci. Found. for Physics (formerly Nuclear Research Found.), 1954—; joint dir. Cornell-Sydney U. Astronomy Centre, 1964—. Overseas exchange fellow in math. St. Andrew's U. (Scotland), 1948-49, research fellow Dublin (Ireland) Inst. Advanced Studies, 1949-51. Author: Selected Lectures in Modern Physics for School Science Teachers, 1958; (with S. T. Butler) From Nucleus to Universe, 1960; (with S. T. Butler) Space and the Atom, 1961; A Modern Introduction to Physics, 3 vols., 1960-61; (with S. T. Butler) A Journey through Space and the Atom, 1962; (with S. T. Butler) The Universe of Time and Space, 1963; Science for High School Students, 1964; (with S. T. Butler) An Introduction to Modern Physics, 1964; (with S. T. Butler) Space Physics and Radio Astronomy, 1964; (with S. T. Butler) The Universe and its Origin, 1964; Senior Science for High School Students, 1965; (with S. T. Butler) Atoms to Andromeda, 1965; (with S. T. Butler) Apollo and the Universe, 1967; (with S. T. Butler) Inner and Outer Space, 1968. Research, publs. theoretical nuclear physics, problems asso. with high-energy nuclear particles. Office: U. Sydney, Sydney, Australia.*

MESSEL, Rudolph, chemist; b. Darmstadt, Germany, Jan. 14, 1848; s. L. Messel; D.Sc. nat., U. Tübingen (Germany); Ph.D. Came to England, 1870. Asst. to J. C. Calvert, Manchester, Eng., then to Sir Henry Roscoe; entered chem. industry and introduced numerous new processes; mng. dir. Spencer, Chapman & Messel, Ltd. Mem. bd. studies in chemistry U. London (Eng.); bd. govs. Imperial Coll. Sci. and Tech., London; bd. mgrs. Royal Inst. Fellow Royal Soc., 1912; mem. Soc. Chem. Industry (pres., past fgn. sec.), Chem. Soc. London (past v.p.). Publns. in relation to indsl. chemistry. Was the first, prior to Winkler, to devise conjointly with Squire, a successful process for the manufacture of sulphuric anhydride by the catalytic process. Died Apr. 18, 1920.

MESSER, August, German psychologist, philosopher; b. Mainz, Hesse-Darmstadt, Feb. 11, 1867; s. Joseph and Margarethe (Kapp) M.; ed. univs. Giessen, Heidelberg (both Germany), Strasbourg (France); m. Paula Platz, 1909. Tchr. secondary sch.; docent philosophy U. Giessen, from 1899, prof. extraordinary, 1904-10, full prof., from 1910. Author: Experimentell-Psychologische Untersuchungen über das Denken, 1906; Empfindung und Desken, 1908; Psychologie, 1914. Mem. Würzburg Sch. psychology; developed system combining content psychology and act psychology. Died 1937.

MESSICK, Samuel James, Am. psychologist; b. Phila., Apr. 3, 1931; s. Samuel James and Carolyn (Funk) M.; A.B., U. Pa., 1951; M.A., Princeton, 1953, Ph.D., 1954; m. Bett Alice Kerr, July 12, 1952; children—Kathy Ellen, Samuel Christopher, Marjorie Alison, Jonathan Douglas. Ednl. Testing Service Psychometric fellow Princeton, 1951-54; Ford Found. fellow U. Ill., 1954-55; USPHS fellow Menninger Found., Topeka, 1955-56; fellow Center for Advanced Study in the Behavioral Scis., Stanford, Cal., 1962-63; chmn. personality research group Ednl. Testing Service, Princeton, N.J., 1956—. Mem. Sch. Math. Study Group Panel on Tests, 1963—. Fellow Am. Psychol. Assn.; mem. Am. Statis. Assn., Psychometric Soc., Soc. for Research in Child Devel., Psychonomic Soc., A.A.A.S., Phi Beta Kappa, Sigma Xi. Editor: (with H. Gulliksen) Psychological Scaling, 1960; Measurement in Personality and Cognition (with J. Ross), 1962; (with S. Tomkins) Computer Simulation of Personality, 1963; (with A. Brayfield) Decision and Choice, 1964; editorial bd. Ednl. and Psychol. Measurement, 1961—, Brit. Jour. Statis. Math. Psychology, 1963—, Psychol. Bull., 1964—, Jour. Exptl. Research in Personality, 1964—, Multivariate Behavioral Research, 1964—, Child-Devel., 1964—. Research, numerous publs. on personality measurement, response styles, cognition, cognitive styles, social perception, scaling, multi-dimensional scaling, affect and personality, probability learning. Home: 19 Pin Oak Dr., Trenton, N.J. 08630. Office: Ednl. Testing Service, Princeton, N.J.*

MESSIER, Charles-Joseph, French astronomer; b. Badonviller, France, June 26, 1730; visited Paris, 1751; employed in DeLisle's obs. Fellow Royal Soc., 1764; mem. Bur. Longitudes, French Acad. Scis., also acads. in St. Petersburg, Berlin. Compiled catalogue containing over 100 nebulae, 1781. Discovered 21 comets; measured length of pendulum beating seconds at 45° latitude, 1775; recorded various celestial objects many of which were star clusters. Died Paris, Apr. 12, 1817.

MESSINI, Mariano, Italian physician; b. Foligno, Italy, Sept. 12, 1901; s. Ruggiero and Luisa (Gregori) M.; grad. Padua U., 1924; m. Maria Sabina Ercole, Oct. 25, 1939; children—Pietro, Ruggiero, Francesco. Asst. to Luigi Sabbatani, inst. pharmacology Padua U.; asst., 1st asst., med. clinics, Padua, Rome U., 1928-39; chair, Med. Faculty Rome, 1939—; lectr. Rome U., 1938, 46-47, now dir. inst. systematic med. therapy and med. hydrology, postgrad. schs. for diseases of liver and metabolism and for med. hydro climatology, research center on liver and metabolic diseases; dir. Center for Prevention and Treatment of Rheumatic and Joint Diseases, Ministry of Health; chmn. Nat. Inst. Diabetes and Metabolic Diseases. Mem. Nat. Health Council; mem. permanent com., ofcl. pharmacopoeia; mem. research com. Constituent Assembly, from 1946. Recipient Guido Baccelli Gold medal for clin. merits, 1942; Gold medal in pub. health, 1963; Gold medal Ministry Edn., 1963; encomio solenne for treatise on therapy. Mem. Italian Soc. Internal Medicine, Rome Med. Acad., Accademia Lancisiana, Italian Soc. for Progress of Sci. (councillor), Italian Soc. Pharm. Scis. (councillor), Union Therapeutique (founder), Italian Med. Soc. for Hydroclimatology, Thalassology and Phys. Therapy (pres.), Internat. Soc. Med. Hydrology and Climatology (pres.). Author: Trattato di terapia clinica con note di diagnostica, 5 vols.; (with V. Meccoli) Terapia, 4th edit., 1952; (with others) Trattato di idroclimatologia clinica, 1950-51; (with G. Di Lollo) Acque minerali del mondo, 1957; (with M. Cairella) Chemioterapia e terapia antibiotica, 1964, Dietetica, 1967; also papers. Founder, editor La clinica terapeutica, Epatologia, La Clinica termale. Research on liver and atherosclerosis, liver cirrhosis, liver regeneration, liver and mineral waters, phospholipid synthesis in liver, hepatic microlithiasis, biliary acids, transformation of iron salts into ferrous phosphate and reticulo-histiocytic system, hemochromatosis, hemolytic effects as function of temperature, tropism of sulphur for joint tissues, uric acid arthrosis, muscle function and potassium, thymus and lymph glands. Home: 11 Rovereto, Rome, Italy.*

MESTER, László, carbohydrate chemist; b. Szászvaros, Hungary, May 26, 1918; s. Mihàly and Mária (de Sarkàny) de Mester; student Faculty Chem. Engrig., Tech. U. Budapest, 1936-40, D.Eng., 1943, D.Chem. Sci., 1956; m. Madeleine de Szadeczky-Kardoss, Sept. 20, 1947; children—Laszlo, Gabor, András. With Tech. U. Budapest, 1940-56, Tech. U. Berlin-Charlottenburg, 1957-58, Faculté de Pharmacie, U. Paris, 1958-60; sci. research dir. Centre National de la Recherche Scientifieque, Paris, Institut de Chimie des Substances Naturelles, Gif-sur-Yvette, France, 1961—. Vis. scientist NIH, Bethesda, Md., 1961, Instituto Oswaldo Cruz, Rio de Janeiro, Brazil, 1963. Mem. Chem. Soc. London, Am.., Austrian chem. socs., Chem. Soc. France, Biochem. Soc. France. Author: Dérivés hydraziniques des glucides, 1967; also numerous articles. Applications of Formazan reaction in carbohydrate chemistry; structure of sugar phenylhydrazones and osazones; synthesis of new nitrogenous sugar derivatives; structure and role of sugar fragments of fibrinogen in blood clotting. Home: 36, rue Dailly, Parc de Béarn, Saint-Cloud-92, France. Office: Institut de Chimie des Substances Naturelles, Gif-sur-Yvette-91, France.*

MESUE MAJOR, see Ibn Masawaih.

MESUE THE YOUNGER (Masawaih al-Mardini), Muslim physician; b. 925. From Mardin, Upper Mesopotamia; flourished in Baghdad then in Egypt at court of Fatimid caliph al-Hakim. Author: De medicinis laxativis; De consolatione medicinarum et correctione operationum earundem; De egritudinibus; Antidotarium sive Grabadin medicamentorum compositorum (12 parts). Wrote complete pharmacopoeia based on Muslim knowledge; a very popular work, it was standard pharmacy textbook in the West for centuries; also wrote on purgatives, emetics, remedies and distillation of empyreumatic oils. Died Egypt, 1015.

METCALF, Clell Lee, Am. entomologist; b. Lakeville, O., Mar. 26, 1888; s. Abel Crawford and Catherine (Fulmer) M.; A.B., Ohio State U., 1911, A.M., 1912; D.Sc., Harvard, 1919; m. Cleo Esther Fouch, Dec. 31, 1908; children—Robert Lee, James Richard. Asst., Ohio State U., 1911-12; asst. entomologist N.C. Dept. Agr., 1912-14; asst. prof. entomology Ohio State U., 1914-19, prof., 1920-21, prof. entomology, head dept. U. Ill., since 1921; chmn. div. biol. scis., 1936-38, sec., since 1938; cons. entomologist, Me. Expt. Sta., summers 1915-17; tchr. biology Cornell U., summers, 1918, 19; field entomologist N.Y. State Mus., summer 1929; v.p. Illini Pest Control and Service Co. Fellow Entomol. Soc. Am. (pres. 1934), A.A.A.S.; mem. Am. Assn. Econ. Entomologists (v.p. 1940), Ill. Acad. Sci., Sigma Xi. Author: Destructive and Useful Insects (with W. P. Flint), 2d edition, 1939; Key to the Principal Orders and Families of Insects (with Zeno Payne Metcalf), 1928; Fundamentals of Insect Life (with W. P. Flint), 1932; Insects—Man's Chief Competitors (with W. P. Flint), Century of Progress Series, 1932. Contbr. bulls. and articles on biology and entomology. Died Aug. 21, 1948.

METCALF, Donald, Australian med. scientist; b. Mittagong, Australia, Feb. 26, 1929; s. Donald Davidson and Enid (Thomas) M.; B.Sc. in Medicine, Sydney (Australia) U., 1951, M.B.B.S., 1953, M.D., 1961; m. Josephine Emily Lentaigne, Aug. 14, 1954; children—Katherine, Mary-Ann, Penelope, Johanna. Carden fellow in cancer research Walter and Eliza Hall Inst., Melbourne, Australia, 1954-56, head cancer research unit, 1958-66, asst. dir., 1966—; research asso. pathology Harvard Med. Sch., Boston, 1956-58; vis. prof. Roswell Park Meml. Inst., Buffalo, 1966. Mem. sci. council Internat. Agy. for Cancer Research, Lyon, France, 1965—. Recipient Ency. Brit. award for medicine Australia, 1966. Mem. Coll. Pathologists Australia, Australian Med. Assn., Am. Assn. for Cancer Research. Author: The Thymus, 1966; also numerous articles. Analysis of structure and function of thymus gland, influence of thymus on white cell formation in spleen and lymph nodes, influence of thymus on immune responses and leukemia devel.; discovered thymic hormone. Home: 268 Union Rd., Balwyn, Victoria, Australia. Office: Walter and Eliza Hall Inst., Royal Melbourne Hosp. P.O., Melbourne, Australia.*

METCALF, George Forrest, Am. engr.; b. Milw., Dec. 7, 1906; s. George D. and Minnie A. (de Young) M.; B.S. in Elec. Engring., Purdue U., 1928, D.Eng., 1956; m. May Carroll, Apr. 8, 1931; children—Thomas R., Carroll May (Mrs. Peter B. Hutchinson). With Gen. Elec. Co., 1928-62, with engring. services div., 1953-55, gen. mgr. missile and ordnance systems dept., 1955-58, regional v.p. Washington def. activities, 1958-62; v.p. research and engring. Martin Co., 1962—. Decorated Legion of Merit (U. S.); Order Brit. Empire. Fellow I.E.E.E., Am. Inst. Aeros. and Astronautics (asso.). Contbns. include devel. of low grid-current vacuum tube, 1930; pioneer in devel. of all-metal vacuum tubes used in radio receivers 1934-36; collaborated in theory and application of velocity-modulated tubes. Home: Stillwater Rd., Gibson Island, Md. 21056. Office: Martin Co., Friendship Internat. Airport, Md. 21240.*

METCALF, Joel Hastings, Am. astronomer; b. Meadville, Pa., Jan. 4, 1866; s. Lewis Herbert and Anna (Hicks) M.; grad. Meadville Theol. Sch., 1890; student Harvard Div. Sch., 1890; Ph.D., Allegheny Coll., 1892; Manchester Coll. (U. of Oxford), 1903; D.D., Meadville Theol. School, 1920; m. Elizabeth S. Lochman, Sept. 22, 1891. Ordained Unitarian ministry, 1890; minister, Burlington, Vt., 1893-1903, First Congl. Ch., Taunton, Mass., 1904-10, Unitarian Ch., Winchester, Mass., 1910-20, First Parish of Portland, Me., 1920—. Recipient 5 medals Astron. Soc. Pacific, Gold medal Astron. Soc. Mexico. Author: World Stories, 1909. Discovered about 41 minor planets, several variable stars, 6 comets (2 periodic); built 16 inch double telescope for Harvard Obs. Died Feb. 21, 1925.

METCALF, Maynard Mayo, Am. zoologist; b. Elyria, O., Mar. 12, 1868; s. Eliab Wight and Eliza Maria (Ely) M.; A.B., Oberlin Coll., 1889; Ph.D., Johns Hopkins, 1893; Sc.D., Oberlin, 1914; m. Ella M. Wilder, Sept. 10, 1890; children—Fern Wilder, Mildred Ella (Mrs. William P. Beetham). Asso. prof. and prof. biology Women's Coll., Balt., 1893-1906; prof. zoology Oberlin Coll., 1906-14 (leave of absence for zool. study in Germany and Naples, 1906-08); research asso., prof. zoology Johns Hopkins, 1925—; collaborator marine invertebrates U. S. Nat. Mus. Mem. numerous sci. socs. Contbr. zool. memoirs in Am. and German jours., chiefly on Protozoa, Tunicata and Mollusca, also numerous papers on geog. distbn. of animals since the Triassic, and of An Outline of the Theory of Organic Evolution, as well as articles on economic theory. Died Apr. 1940.

METCALF, Robert Lee, Am. entomologist; b. Columbus, O., Nov. 13, 1916; s. C. L. and Cleo (Fouch) M.; B.A., U. Ill., 1939, M.A., 1940; Ph.D., Cornell U., 1943; m. Esther Rutherford, June 22, 1940; children—Esther Lee, Robert Alan, Michael Rutherford. Faculty, U. Cal., Riverside, 1946-68, prof. entomology, entomologist, 1952-68, chmn. dept. entomology, 1951-63, faculty research lectr., 1959, vice chancellor, 1963-66; prof. entomology U. Ill., Urbana, 1968—. Chmn. panel on pesticides NSF U. S.-Japan Coop. Sci. Program; cons. WHO, AID, U. S. Dept. Agr., TVA. Recipient Order of Cherubini, U. Pisa, 1966. Mem. Nat. Acad. Scis., Am. Chem. Soc. (Charles F. Spencer award 1966), Entomol. Soc. Am., A.A.A.S., Am. Mosquito Control Assn., Phi Beta Kappa, Sigma Xi. Author: Destructive and Useful Insects, 1962; Organic Insecticides—Their Chemistry and Mode of Action, 1955; also numerous articles. Research in insect nervous system and its susceptibilities to action of insecticides, basic biochemistry of insect nerve transmission, devel. of lipoid soluble N-methylcarbamate anticholinesterase insecticides. Address: Dept. Entomology, U. Ill., Morrill Hall, Urbana, Ill. 61801.*

METCALF, Zeno Payne, Am. zoologist; b. Lakeville, O., May 1, 1885; s. Abel Crawford and Catherine (Fulmer) M.; A.B., Ohio State U., 1907; D.Sc., Harvard, 1925; m. Mary Luella Correll, Oct. 20, 1909; 1 dau., Katharine (Mrs. Micou F. Browne). Instr. entomology Mich. State Agrl. Coll., 1907-08; with N.C. Dept. Agr., 1908-12; prof. zoölogy N.C. State Coll. and entomologist expt. sta., 1912-50, also dir. instrn. sch. of agr., 1923-44, dir. grad. studies, 1940-50; asso. dean grad. sch. U. N.C., 1943-50, research prof. of zoology and entomology, 1950—; instr. biol. lab. Ohio State U., 1916-18, U. Mich., 1920; vis. prof. zoölogy Duke, 1935-36. Fellow A.A.A.S., Micros. Soc. (v.p. 1922; pres. 1927), N.C. Acad. Sci. (v.p. 1914; pres. 1921); mem. Entomol. Soc. Am., Assn. Econ. Entomologists (chmn. Cotton States br. 1940), Ornithol. Union, Ecol. Soc., Soc. Systematic Zoologists, Am. Soc. Limnology and Oceanology, Am. Mus. Natural History, Soc. Herpetologists, Wilson Ornithol. Club, Assn. S. Eastern Biologists, Am. Biol. Assn., Soc. for Study of Evolution, Sigma Xi, other sci. socs. Author: Insect Pests in Rural Efficiency Guide, 1918; Key to Insects, 1918; Key to the Family Fulgoridae, 1923; General Zoölogy, 1927; Economic Zoölogy 1927; Text Book of Economic Zoölogy, 1930; Introduction to Zoölogy, 1932; General Catalgue of Hemiptera Tettigometridae, 1932; Cixiidae, 1936; Araeopidae, 1943; Derbidae, 1945; Achilixidae, 1945; Meenoplidae, 1945; Kinnaridae, 1945; Achilixidae, 1945; Meenoplidae, 1945; Kinnaridae, 1945; Dictyopharidae, 1946; Fulgoridae, 1947; Achilidae, 1948; The Fulgorina of Barro Colorado, 1938; Bibliography of the Homoptera of the World, 1943; Cercopidae of Cuba, 1944; Homoptera of Kartabo, 1945; Homoptera of Guam, 1946, Center of Origin Theory, 1946; Cuban Flatidae (with S. C. Bruner), 1948; Catalog of the Hemiptera, Fulgoroidea Fascicle IV, 1957; Issidae, 1958. Editor Homoptera, Biol. Abstracts; mem. editorial bd. Catalog of Hemiptera of World, editorial bd. ecology, 1935-37, and editorial bd. Ecol. Monographs, 1940-42. Died Jan. 5, 1956.

METCHNIKOFF, Elie (Ilia Ilich Mechnikov), bacteriologist; b. Ivanovka, Kharkov Province, Russia, May 15, 1845; grad. Kharkov U.; student, Giessen and Munich; m. 1st (dec. 1873); m. 2d, Olga, 1875. Became docent U. Odessa, 1867, prof., 1870-82, resigned to do research, 1882; prof. U. St. Petersburg, 1868-70; dir. Bacteriological Inst., Odessa, 1886; asso. Pasteur Inst., Paris, 1888, dir., 1895-1916. Recipient (with Paul Ehrlich) Nobel prize for medicine and physiology, 1908; mem. French Acad. Scis., 1904. Author: Aetiologie und Prophylaxe der Infektionskrankheiten, 1897; Immunite, 1897; La Vieillesse, 1903; Die Lehr von den Phagozyten und deren esperimentelle Grundlagen, 1904; Quelques remarques sur le lait aigri, 1905; Genuine Lactobocilline, a Medicine and Food, 1906; Bactériotherapie, vaccination, sérothérapie, 1908. Pioneer research on infectious diseases; formulated theory of phagocytosis and discovered phagocytes (cells which devour infective organisms); made microscopic studies of diseases of blood; studied longevity in relation to bacteria of alimentary canal; believed inflammation to be fundamental defense reaction of organism; studied intracellular digestion. Died Paris, France, July 15, 1916.

METCOFF, Jack, Am. physician; b. Chgo., Feb. 2, 1917; s. Samuel and Ella (Rappaport) M.; B.S., Northwestern U., 1938, M.D., 1942, M.S., 1944; M.P.H., Harvard, 1944; m. Elinor Gluck, Apr. 16, 1943; children—Donald, Jill. With Harvard, Med. Sch., 1948-56, asso. prof., 1953-56; prof. Northwestern U. Med. Sch., 1956-63; prof., chmn. dept. pediatrics Chgo. Med. Sch., 1963—; with USPHS, 1944-45; chmn. div. pediatric Michael Reese Hosp., Chgo., 1956—; vis. prof. pediatrics, Mexico, 1963, S.Am., 1964, Greece, 1967. Cons. govt. agys, founds., assns. Mem. Am. Acad. Pediatrics (com. on fetus and newborn), Am. Soc. for Clin. Investigation, Am. Pediatric Soc., Soc. for Pediatric Research, Central Soc. for Clin. Research, Federated Soc. Biology and Medicine, Am. Inst. Nutrition, Midwestern Soc. for Pediatric Research, Am. Assn. for Maternal and Infant Health, A.A.A.S., Assn. Am. Med. Colls., Assn. for Ambulatory Pediatrics, Internat., Am. socs. nephrology, Infant Welfare Soc. Editor: Ann. Conf. on the Kidney, 1950-67. Research, publs. on kidney physiology and disease, fluid and electrolyte therapy, salt and water chem. and physiol. feature of treatment, dehydration and other disturbances of salt and water in body, metabolism in human malnutrition. Home: 860 Hibbard Rd., Winnetka, Ill. 60093. Office: Michael Reese Hosp. and Med. Center, Dept. Pediatrics, 29th St. and Ellis Av., Chgo. 60616.*

METFORD, William Ellis, Brit. inventor; b. Taunton, Eng., Oct. 4, 1824; s. William and M. E. (Anderdon) M.; apprenticed to W. M. Peniston, Bristol and Exeter Ry. With Wilts, Somerset and Weymouth R.R., 1846-50; joined T. E. Blackwell in traffic devel., Bristol, Eng., 1850; later engr. Wycombe R.R.; apptd. to E. India Ry., 1857; returned to Eng. Mem. Inst. Civil Engrs. (asso.). Suggested hollow-based exploding bullet for Enfield rifle 1852-53; pioneered substitution of shallow grooving in rifles and hard cylindrical bullet for deep grooving and soft bullets, 1865; began design of breech loading rifle, 1870; his ideas adopted by Brit. War Office (Lee-Metford rifle), 1888. Died 1879.

METILDI, Pasquale Frederic, physician; b. Celano, Italy, Feb. 12, 1899; s. Angelo and Felicia (Silvestri) M.; M.D., U. Rochester; m. Doris Howell, Dec. 3, 1931; children—Sandra June, Doris Catherine, Frederic Howell, Leonard Angelo. Comdt., M.C., USMCR. Mem. A.M.A., Asso. Mil. Surgeons U.S.A., Am. Coll. Physicians, World Med. Assn., Phi Beta Kappa, Alpha Omega Alpha. Author: Studies on Pulmonary Acoustics, Syphilitic Aortitsis. Home: 340 Beresford Rd. Office: 277 Alexander St., Rochester, N.Y.*

METIUS, Adriaen (Adriaanszoon), Dutch mathematician; b. Alkmaar, Holland, Dec. 9, 1571; s. Adriaan Anthonitz; M.D., 1625; studied under Tycho Brahe, Denmark; prof. math. and medicine, Franeker, 1598 until his death. Author: Primum mobile, 1630-31; also various other works in math. and astronomy. Developed (with father) approximation of value of pi. Died Sept. 1635.

METIUS, Jacobus, Dutch astronomer; b. Alkmaar, Netherlands, 1580; s. Adriaan Metius; invented refracting telescope, 1609; built mirrors, including burning mirror. Died Alkmaar, June 1628.

METOCHITES, Theodoros, Byzantine natural philosopher; b. circa 1260; s. Georgio Metochites; grand chancellor or treas. for Andronicos II, 1282-1328; retired to Monastery of St. Savior, Chora, 1330, restored and decorated monastery; astron. tutor Nicephoros Gregoras. Author: Commentary on Ptolemy or Astronomical Introduction; also writings on natural history, philosophy, astronomy, rhetorical pieces, letters, poetry. Died 1332.

METON, b. circa 440 B.C.; flourished Athens; originated cycle of nineteen years (Metonic cycle) embracing 235 lunar months on which Greek calendar was based until 46 B.C.; Metonic period or year was used in computing date of Easter.

METRODOROS OF CHIOS, philosopher, physician; flourished circa 330 B.C.; pupil of Democritos; studied and may have practiced medicine; tchr. of Hippocrates and Anaxarchos. Author of hist. works. Sought explanations of meteorol. and astron. phenomena, also sought to combine Atomism and Eleatic denial of reality of change.

METROPOLIS, Nicholas Constantine, Am. physicist, educator; b. Chgo., June 11, 1915; s. Constantine and Catharine (Ganas) Metropoulos; B.S., U. Chgo., 1936, Ph.D., 1941; m. Patricia Hendrix, Oct. 15, 1955; children—Katharine, Penelope, Christopher. Research instr. U. Chgo., 1941, physicist Metall. Project, 1942, asst. prof. Inst. Nuclear Studies, 1946-48, prof. physics Enrico Fermi Inst. Nuclear Studies, 1957-65, dir. Inst. Computer Research, 1958-65; physicist Manhattan Project, Columbia, 1942, Los Alamos Sci. Lab., 1943-46, 48-57, Mem. Greenhouse Operation, Eniwetok, 1951; mem. Edward Teller's group for study thermonuclear problems AEC, 1944-52, chmn. AEC computer adv. group, 1959—; mem. rev. com. Applied Math. div. Argonne Nat. Lab., 1960—; U. S. del. Internat. Conf. on Information Processing, 1959; UN tech. adviser Sci. Mission to India, 1961; vis. prof. U. Colo., Boulder, 1964-65. Fellow Am. Phys. Soc., Am. Math. Soc.; mem. Assn. for Computing Machinery (council 1960-62), Sigma Xi. Contbr. articles on math., physics, chemistry to profl. jours. Designer electronic computers Maniac I, II, III. Office: Los Alamos Sci. Lab., Los Alamos 87544.*

METTAUER, John Peter, Am. physician; b. Prince Edward County, Va., 1787; s. Francis Joseph and Jemimah (Gaulding) M.; M.D., U. Pa., 1809; m. Mary Woodard; m. 2d, Margaret Carter, Apr. 14, 1825; m. 3d, Louisa Mansfield, 1833; m. 4th, Mary E. Dyson; at least 10 children. Practiced medicine, Prince Edward County, 1809-34, 37-55; prof. surgery Washington Med. Coll., Balt., 1835-36; pioneer in genitourinary surgery; developed use of lead sutures for treating vesico-vaginal fistula; established Prince Edward Med. Inst. (became med. dept. Randolph-Macon Coll.), 1837. Author: Continued Fever in Middle Southern Virginia from 1816 to 1829, published 1843. Died Nov. 22, 1875.

METTLER, Frederick Albert, Am. anatomist; b. N.Y.C., June 13, 1907; s. Frederick and Anne (Mathews) M.; A.B., Clark U., 1929, Sc.D., 1951; A.M., Cornell U., 1931, Ph.D., 1933; M.D., U. Ga., 1937. Instr., Cornell Med. Coll., 1930-33, Med. Coll. St. Louis, 1933-34; faculty U. Ga. Sch. Medicine, 1934-42, prof. anatomy, 1938-42; faculty Columbia Coll. Phys. and Surg., 1941—, prof. anatomy, 1951—; Commonwealth Fund vis. prof. L.I. Med. Coll., 1943-44; coordinator research N.Y. Dept. Mental Hygiene, 1948-49, dir. research, 1949. Cons., N.J. State Hosp., Greystone Park, 1947—, VA, 1949—. Fellow N.Y. Acad. Medicine, N.Y. Zool. Soc.; mem. Assn. Anatomy, Physiology Soc., Assn. Phys. Anthropology, Soc. for Exptl. Biology and Medicine, Harvey Soc., Assn. Research in Nervous and Mental Disorders (editorial bd. 1943), Assn. Eastern Electroencephalographers, Ga. Med. Soc., N.Y. Acad. Scis. (asso.). Author: Neuroanatomy, 1941; Fundamentals of Neurology, 1947; The Medical Sourcebook, 1959; also numerous articles. Editorial bd. Jour. Comparative Neurology, 1946—. Research on exptl. prodn. various types neurologic disorders, psychosurgery, basic contbns. in anatomy, physiology, pathology nervous system. Home: Pippin Hill, Blairstown, N.J. Office: 630 W. 168th St., N.Y.C. 10032.*

METZ, Charles Baker, Am. biologist; b. N.Y.C., Dec. 27, 1916; s. Charles William and Blanche (Stafford) M.; B.A., Johns Hopkins, 1939; Ph.D., Cal. Inst. Tech., 1942; m. Mary Ethel Bucheimer, July 3, 1940; children—Rinda M., Richard S. Instr. biology Wesleyan U., Middletown, Conn., 1942-46; NRC fellow Ind. U., Bloomington, 1945-46; instr. zoology Yale, 1946-47, asst. prof., 1947-52; asso. prof. zoology U. N.C., Chapel Hill, 1952-53; asso. prof. zoology Fla. State U., 1953-57, prof., 1957, asso. dir. Oceanographic Inst., in charge Alligator Harbor Marine Lab., 1958-62, mem. Inst. for Space Bioscis., 1961-64; prof. zoology Inst. Molecular Evolution, U. Miami, Coral Gables, Fla., 1964—. Instr. embryology Marine Biol. Lab., Woods Hole, Mass., summers 1947-52, trustee, 1956-64; vis. asst. prof. U. Cal., Berkeley, 1952. Fellow A.A.A.S.; mem. Am. Soc. Zoologists (sec. 1961-63), Soc. for Exptl. Biology and Medicine, Am. Soc. Naturalists, Bermuda Biol. Assn. (corp. mem.), Soc. for Study Devel. and Growth, Soc. Gen. Physiologists, Soc. for Study Fertility, Internat. Inst. Embryology, Internat. Soc. for Cell Biology, Sigma Xi. Home: 7220 S.W. 124th St., Miami, Fla. 33134. Office: 519 Anastasia Av., Coral Gables, Fla. 33134.*

METZ, Charles Franklin, Am. chemist; b. Deerfield, Ind., Sept. 23, 1904; s. Walter Frank and Mary (Eltzroth) M.; B.S. in Chem. Engring., S.D. Sch. Mines, 1926; M.S., U. Colo., 1928, Ph.D., 1936; D.Sc. (hon.), S.D. Sch. Mines and Tech., 1963; m. Catherine Elizabeth Vowell, June 3, 1927; 1 son, Charles Vowell. Instr., S.D. Sch. Mines, 1926-27, U. Colo., 1927-28; asso. prof. Park Coll., Parkville, Mo., 1928-35; faculty Colo. State U., Ft. Collins, 1936-44; scientist Manhattan Engring. Dist., Los Alamos, 1944-45; prof. chemistry Colo. A. and M. Coll., Ft. Collins, 1945-46; group leader chem. and instrumental analysis Los Alamos Sci. Lab, U. Cal., 1946—. Mem. A.A.A.S., Am. Chem. Soc., Sigma Xi, Sigma Pi Sigma. Author: (with Glen R. Waterbury) Treatise on Analytical Chemistry, vol. 9, 1962; also articles. Developed methods of analysis for trace impurities in plutonium for analysis of various alloys of plutonium also uranium; developed dry boxes for handling radioactive materials, devel. methods analysis for actinide elements, alloys and compounds and materials of strategic importance in AEC programs. Home: 912 Hillcrest Dr., Sante Fe 87501. Office: Los Alamos Sci. Lab., Los Alamos.*

METZ, Charles William, Am. zoologist; b. Sundance, Wyo., Feb. 17, 1889; s. William Summerfield and Jennie (Gammon) M.; A.B., Pomona Coll., 1911, D.Sc., 1958; postgrad. Stanford U., 1911-12; Ph.D., Columbia, 1916; m. Blanche Elizabeth Stafford, Aug. 20, 1913; children—Charles Baker, William Stafford, Jane Gammon (Mrs. Emile Zuckerkandl), Alburn Stafford. Mem. staff dept. exptl. evolution Carnegie Instn. Washington, Cold Spring Harbor, N.Y., 1914-30, dept. embryology, Balt., 1930-40; vis. prof. Johns Hopkins, 1930-37; prof. zoology U. Pa., Phila., 1940-59, chmn. dept., 1940-55, prof. emeritus, 1959; research at Marine Biol. Lab., Woods Hole, Mass., 1957—. Mem. Nat. Acad. Scis., Am. Philos. Soc., Am. Acad. Arts and Scis., Internat. Soc. for Cell Biology, Am. Soc. Zoologists, Genetics Soc. Am., Soc. for Study Evolution. Research in cytogenetics with emphasis on nature of chromosomes and their role in heredity, devel., sex determination and evolution. Home: 28 Hyatt Rd. Office: Marine Biol. Lab., Woods Hole, Mass. 02543.*

METZENBAUM, Myron Firth, Am. surgeon; b. Cleve., O., Apr. 1, 1876; s. Joseph and Fannie (Firth) M.; B.S., Adelbert Coll., 1897; M.D., Western Res. U., 1900; grad. study, Vienna Med. U., 1900, 23, U. Paris, 1927, Berlin and London, 1927; m. Elsa Fuldheim, 1912; children—Louise, Jane. Intern St. Alexis Hosp., Cleve., 1900; lectr. anatomy Western Res. U., 1902-05; specialist in ear, nose and throat, oral surgery and reconstructive surgery of head and neck. Diplomate Am. Board Otolaryngology, Am. Board Plastic Surgery (founders group). Fellow A.C.S.; mem. A.M.A., Am. Acad. Otolaryngology, Cleve. Otolaryngol. Club, Acad. Medicine Cleve., European Congress of Reconstructive Surgery, Am. Soc. Plastic and Reconstructive Surgery, Am. Soc. Plastic and Reconstructive Surgery. Recipient medal U. S. Govt. for research in radium, St. Louis Expn., 1904. Established Cleveland's ambulance system under Police Dept., 1909 (adopted throughout U. S.); research papers on radium, anesthesia, nose, throat and larynx surgery and reconstructive nasal surgery; developed and introduced method of administering ether-air or drop ether anesthesia, 1900; pioneer in use of scopolamin (twilight sleep) in gen. surgery; author of surg. method for resetting dislocated cartilage of nose in young children and dry method for treatment of sinus infections in children. Died Jan. 25, 1944.

METZENBERG, Robert Lee, Am. biochemist; b. Chgo., June 11, 1930; s. Robert L. and Eleanor (Loeb) M.; A.B., Pomona Coll., 1951; Ph.D., Cal. Inst. Tech., 1956; m. Helene Worthington Fox, June 26, 1954; children—Howard, Stan. Postdoctoral fellow U. Wis., Madison, 1955-56, faculty, 1956—, asso. prof. biochemistry, 1963—; vis. fellow dept. zoology U. Zurich (Switzerland), 1959-60. Markle scholar, 1958-63. Recipient Career Devel. award USPHS, 1963—. Mem. Am. Chem. Soc., Am. Soc. Biol. Chemists, Am. Eugenics Soc., A.A.A.S., Phi Beta Kappa, Sigma Xi. Research on pathway of biosynthesis of aromatic amino acids, mechanism of action of enzyme of urea biosynthesis, carbamyl phosphate synthetase, purification, properties and control of synthesis of enzyme invertase in Neurospora, control enzymes of sulfur utilization by repression. Home: 2802 Colgate Rd., Madison, Wis. 53705.*

METZGER, Jacques, French chemist; b. Charmes, France, Nov. 19, 1921; s. Jean and Jeanne (Lemaire) M.; Ingénieur-chimiste, Ecole Nationale Supérieure des Industries Chimiques, Nancy, France, 1943, Doctorat ès-Sciences, 1948; m. Marie-José Condé, Aug. 23, 1946; children—Sylvie, Edith, Eric. Faculty, U. Saarbrücken, Saar, 1948-52; prof. U. Aix-Marseille, 1953-54; prof., 1954—, dir. Petrochemistry Inst, Faculté Scis., 1964—. Cons. to pvt. cos. Mem. Société Chimique de France, Société de Chimie-Physique de France, Sté de Chimie Industrielle de France, Faraday Soc. Author: Notions élémentaires de Chimie Générale Centre Documentation Pédagogique, 1962; also numerous articles. Research on heterocyclic chemistry, including 5-membered heteroaromatic series, including structure and reactivity, mechanism of radical-metallic derivatives; structure and reactivity of anhydrobases, intermediates in syntheses of cyanine dyes; photochromic derivatives of thiazol; hydride transfer reactions in heterocyclic series; petrochemistry including oxidation of parafines, mechanism of oxidation, oxidation of olefines in homegeneous catalysis, mechanism of oxidation. Home: 280 Bd. Michelet Marseilles 8°. (B. du Rh) 13 France. Office: Faculté de Sciences, Traverse de la Baraqse (13°), 13-France.*

METZLER, Lloyd Appleton, Am. economist; b. Los Springs, Kan., Apr. 13, 1913; s. Leroy Michael and Lulu (Appleton) M.; B.S., U. Kan., 1935, M.B.A., 1938; M.A., Harvard, 1941, Ph.D., 1942; m. Edith Bean, Jan. 18, 1944; children—Margaret Jane, Richard Appleton. Asst. instr. U. Kan., 1935-37; instr., tutor Harvard, 1937-42; economist OSS, Washington, 1944; economist Fed. Res. Bd., 1944-46; adviser on monetary reform Office of Mil. Govt., 1946; asst. prof. Yale, 1946-47; asso. prof. econs. U. Chgo., 1947-49, prof., 1949—. Lectr., USIS, Denmark, Sweden, Norway, 1951; mem. staff com. to study cost of living Ho. of Reps., 1945; cons., sec. treas., 1960-66. Editor: (with Howard Ellis) Readings in the Theory of International Trade, 1949. Research, publs. on application of Keynesian theory of employment to internat. trade and transfer; introduced theory of inventory cycles; research on relation between money, interest and prices, internat. trade and internat. monetary reform. Home: 18510 Highland Av., Homewood, Ill. 60430. Office: Dept. Econs., U. Chgo., 60637.*

MEULENBELD, Barend, Dutch mathematician; b. Almelo, Netherlands, Sept. 5, 1908; s. Frederik and Martha (Visser) M.; Doctor.Ex., State U., Groningen, Netherlands, 1930, Ph.D., 1936; m. Kornelia Fransina Braaksma, Sept. 26, 1950; children—Frederik, Essiena Martha (Mrs. Willem Borkent). Lector, Tech. Coll., Amsterdam, 1931-38, Royal Mil. Acad., Breda, Netherlands, 1938-45; vice prin. grammar sch., Breda, 1945-49; prof. math. U. Indonesia, Bandung, 1949-52; prof. Technol. U., Delft, Netherlands, 1952—; vis. prof. Wash. State U., Pullmann, 1962-63. Mem. Netherlands Math. Soc. Author several textbooks, articles. Research in number theory, spl. functions, integral transforms. Home: 32, Heemskerkstraat, Delft, Netherlands.*

MEURERS, Joseph, astronomer; b. Cologne, Germany, Feb. 13, 1909; s. Joseph and Bertha (Grohs) M.; ed. Fribourg, Göttingen, Bonn (all Germany) univs.; Ph.D. agrégé; m. Alice Jung, Aug. 17, 1943; children—Bernhard, Georg, Bruno. Asst. Bonn Obs., from 1936, meteorologist, from 1938; instr. astronomy and atmospheric physics U. Bonn, from 1946, prof. at large, from 1948; full prof. astronomy U. Vienna (Austria), dir. univ. obs., from 1962. Mem. Internat. Astron. Union. Research, publs. on sci. and tech., astronomy, astrometry, natural philosophy. Address: Türkenschanzstrasse 17, Sternwarte, Vienna 18, Austria.

MEURMAN, Olavi, Finnish botanist; b. Ilmajoki, Finland, July 19, 1893; s. Otto and Aina (Ignatius) M.; Ph.D., U. Helsinki (Finland); m. Xenia Erikson, June 10, 1922; children—Otto Erik, Lauri, Ann. Botanist, 1919-25; dir. exptl. bot. sta., 1925-35; dir., prof. Horticulture Inst. Piikkiö, Finland, 1935-60. Mem. Swedish Pomology Soc. (hon.), Finnish Acad. Sci., Swedish Royal Acad. Agr., Sigma Xi, others. Research, publs. on cytology and horticulture. Address: Piikkiö, Finland.

MEUSNIER DE LA PLACE, Jean-Baptiste-Marie-Claude, French engr.; b. Tours, France, June 19, 1754; participated in Lavoisier's expts., beginning 1784. Mem. French Acad. Scis. (reporter), 1776. Discovered theorem on centre of curvature of any plane sect. (named after him), 1777; improved theory for equilibrium of air balloons, 1783; designed dirigible, utilizing several small balloons and containing a protected hall and camp enclosure, 1785; invented engraving machine for paper money, 1792; studied (with Lavoisier) composition of water, method of preparing hydrogen. Died Mayence, France, June 17, 1793.

MEWALDT, Leonard Richard, Am. biologist; b. La-Crosse, Wis., May 31, 1917; s. N. H. and Helen (Canom) M.; student No. State Coll., Aberdeen, S.D., 1935-38; B.A., U. Ia., 1939; M.A., U. Mont., 1948; Ph.D., Wash. State U., 1953; m. Frances Lee Booth, Sept. 10, 1941; children—William Thomas, John Arthur. Research asst. U. Ia., 1939-40; asst. econ. analyst U. S. Dept. Commerce, Washington, 1940-42; teaching asst. U. Mont., 1946-48; teaching and research asst. Wash. State U., 1948-53; faculty San Jose (Cal.) State Coll. 1953—, prof. zoology, 1963—. Pres. bd. dirs. Point Reyes Bird Obs., Inverness, Cal. Mem. Brit., Am. (sec. 1964—) ornithologists unions, Cooper (pres. bd. govs. 1965—), Wilson ornithol. socs., A.A.A.S., Western Soc. Naturalists, Am. Soc. Zoologists, Nat. Audubon Soc., Northeastern, Eastern, Inland, Western bird banding assns., Sigma Xi. Research, publs. on biology Nucifraga and Zonotrichia, physiology reprodn. and migration in birds, orientation migration in birds, avian population dynamics. Address: Dept. Biol. Scis., San Jose State Coll., San Jose, Cal. 95114.*

MEYER, Adolf, psychiatrist, neurologist; b. Niederweningen, nr. Zurich, Switzerland, Sept. 13, 1866; s. Rudolf and Anna (Walder) M.; ed. Gymnasium, Zurich; Swiss Staatsexamen for practice of medicine, 1890; post-grad. studies at Paris, London, Edinburgh, Zurich, Vienna and Berlin, 1890-92; M.D., Zurich, 1892; LL.D., Glasgow Univ., 1901, Clark U., 1909; Sc.D., Yale, 1934; Harvard, 1942; m. Mary Potter Brooks, Sept. 15, 1902; 1 dau., Julia Lathrop. Came to U. S., 1892. Hon. fellow, then docent in neurology, U. of Chicago, 1892-95; pathologist to Ill. Eastern Hosp. for the Insane, Kankakee, 1893-95; pathologist and later dir. of clin. and lab. work, Worcester (Mass.) Insane Hosp. and docent in psychiatry, Clark U., 1895-1902; dir. Pathol. (psychiatric) Inst., N.Y. State Hosps., 1902-10; prof. psychiatry, Cornell U.

Med. Coll., 1904-09; prof. psychiatry, Johns Hopkins, and dir. Henry Phipps Psychiatric Clinic, Johns Hopkins Hosp., 1910-41, prof. emeritus since 1941; Salmon memorial lecturer, 1932; Maudsley lecturer, 1933; guest lecturer Acad. of Neurology and Psychiatry, Kharkow, U.S.S.R., 1933; Thomas Salmon medal for distinguished service in psychiatry, 1942. Hon. pres. Nat. Com. for Mental Hygiene and president Internat. Com. for Mental Hygiene since 1937; hon. vice-pres. Conf. on Method in Philosophy and the Sciences; hon. mem. Boston Soc. Neurology and Psychiatry, Royal Medico-Psychological Assn., New York Psychiatric Society (pres. 1905-07); New York and Washington psychoanalytic societies, mem. Assn. Am. Physicians, Am. Neurol. Assn. (pres. 1922), Am. Psychiatric Assn. (pres. 1927), Academie der Naturforscher zu Halle, Am. Inst. Criminal Law and Criminology, A.A.A.S., N.Y. Acad. Sciences, Assn. for Research in Nervous and Mental Diseases, Am. Orthopsychiatric Assn., Am. Psychopathol. Assn. (pres. 1912, 16) American Psychological Assn., Assn. of Anatomy, Harvey Society, New England Soc. Psychiatry. Author: Collected Papers, 4 vols., 1950-52; Psychobiology, 1957. Introduced idea of psychobiology, use of comprehensive case histories, also program for mental hygiene in U. S.; stressed totality of individual and necessity of considering all facets of individual in treatment. Died Balt., Mar. 17, 1950.

MEYER, Arthur William, Am. anatomist; b. Cedarburg, Wis., Aug. 18, 1873; s. Henry and Louise (Wiepking) M.; B.S., U. of Wis., 1898; M.D., Johns Hopkins, 1905; m. Esther Hartshorne Robinson, Dec. 28, 1907; children—Ruth Robinson, Robert Wiepking. Asst. and asso. in anatomy Johns Hopkins Med. Sch., 1905-07; asst. prof. anatomy, U. Minn., 1907-08; prof. anatomy Northwestern U. Med. Sch., 1908-09; prof. human anatomy Stanford U., 1909-38, prof. emeritus, 1938—. Research asso. Carnegie Instn., 1917-18. Fellow A.A.A.S.; mem. Assn. Am. Anatomists, Am. Soc. Zoologists. Author: An Analysis of the De Generatione Animalium of William Harvey, 1936; Rise of Embryology, 1939; Human Generation, 1956. Investigations on embryology, growth, lymphatics, anat. variations, attrition pathology, ednl. hist. and social problems. Died Jan. 18, 1966.

MEYER, Bernard Sandler, Am. plant physiologist, educator; b. Nantucket, Mass., July 20, 1901; s. J. Frederick and Florence (Hinkle) M.; B.A., Ohio State U., 1921, M.A., 1923, Ph.D., 1926; m. Grace Townsend, Sept. 8, 1948. Faculty botany Ohio State U., Columbus, 1923-27, 28—, prof., 1940—, chmn. dept. botany and plant pathology, 1946—, prof., chmn. dept. botany and plant pathology Ohio Agrl. Exptl. Sta., 1948—; asso. forest ecologist U. S. Forest Service, 1927-28. Mem. com. on biology and agr. NRC, 1945-51; cons. NSF, 1958—. Mem. Am. Soc. Plant Physiologists (past pres.), Am. Inst. Biol. Scis. (governing bd. 1958-64, v.p. 1961), A.A.A.S., Bot. Soc. Am. (certificate of merit 1959), Ohio Acad Sci., Ecol. Soc. Am., Phi Beta Kappa, Sigma Xi. Author: (with D. B. Anderson) Plant Physiology, 1952; (with Anderson, C. A. Swanson) Laboratory Plant Physiology, 1955; (with Anderson, R. H. Böhning) Introduction to Plant Physiology, 1960. Editor-in-chief Am. Jour. Botany, 1946-51. Research, publs. on water relations of plants. Home: 3782 Olentangy Blvd., Columbus, O. 43214.*

MEYER, Charles Ferdinand, Am. physicist; b. Balt., Jan. 15, 1887; s. Charles Ferdinand and Nanny (Gail) M.; A.B., Johns Hopkins, 1906, Ph.D. (fellow in physics), 1912; m. Marjorie Fleming, June 14, 1924. Asst. applied electricity Johns Hopkins, 1911-12, lectr. in physics, 1914-15; instr. physics Washington U., St. Louis, 1912-13, acting asst. prof., 1913-14; instr. physics U. Mich., Ann Arbor, 1915-17, 19-21, asst. prof., 1921-30, asso. prof., 1930-50; sci. adviser OSRD, Washington, 1945. Fellow Am. Phys. Soc.; mem. A.A.A.S. (life), Optical Soc. Am., Am. Assn. Physics Tchrs., Phi Gamma Delta. Author: The Diffraction of Light, X-rays and Material Particles, 1934. Research on spectroscopy, especially infrared absorption spectra gases. Address: Broadview Apts., 116 W. University Pkwy., Balt. 21210.*

MEYER, Erwin, German physicist; b. Königshute, Germany, July 21, 1899; s. Paul and Margarete (Schleiffer) M.; Ph.D. in Physics; dr. engring. honoris causa; m. Edith Bergau, May 10, 1932; 1 dau., Angelika. Prof. at large, full prof. Berlin-Charlottenburg Tech. Coll.; full prof. U. Göttingen, dir. Univ. Inst. Physics. Recipient Gauss-Weber medal. Mem. Acad. Sci. Göttingen, Am. Acoustics Soc., Inst. Radio Engrs. Research, publs. on physics of vibrations, microwave physics, acoustics. Home: Olfried Müller Weg 6, Göttingen. Office: Bürgerstrasse 42-44, Göttingen, W. Germany.

MEYER, Eugene, Am. physician; b. N.Y.C., June 7, 1915; s. Eugene and Agnes (Ernst) M.; A.B., Yale, 1937; M.D., Johns Hopkins, 1941; m. Mary A. Bradley, Dec. 27, 1940; children—Eugene Bradley, Ruth Emery, Anne, Elizabeth Ernst. Faculty, Johns Hopkins, Balt., 1949—, prof. psychiatry, 1966—, asso. medicine, 1955—; physician-in-charge psychiat. liaison service Johns Hopkins Hosp., 1951—. Pres. adv. bd. Md. Children's Center, Balt., 1965—. Chmn. bd. trustees William A. White Found., Washington. Mem. Am. Psychosomatic Soc. (past pres.), Md. Psychiat. Soc. (past pres.), Acad. Psychoanalysis, A.A.A.S., Washington Psychoanalytic Soc., Am., Balt. psychoanalytic assns., A.M.A., N.Y. Acad. Scis. Research, publs. on psychosomatic diseases and acute stress reaction on med. and surg. wards, plastic surgery, myasthenia gravis, disseminated lupus erythematosus, hysteria, analysis psychiat. consultation. Home: 809 W. Lake Av., Balt. 21210.*

MEYER, Frank, botanist, explorer; b. Amsterdam, Holland, Nov. 29, 1875; s. Jan Franciscus Meyer; staff Bot. Garden, Amsterdam; gardener, asst. to Hugo De Vries; studied two years, Eng.; came to U. S.; traveled and explored various areas of world, including Japan, Korea, Eastern and No. China, Manchuria, Mongolia, Siberia, Caucasus, Crimea, Russian and Chinese Turkestan, 1905-18. Author: Chinese Plant Names, 1911; Explorations in the Fruit and Nut Orchards of China. Introduced to U. S. wild Chinese peach, Chinese jujube, Tomopan seedless persimmon, Chinese dry-land elm, Chinese pistachio, Manchurian spinach, Tangri cherry; discovered source of chestnut bark disease, also a species of blight-resistant chestnut. Died China, 1918.

MEYER, Georg Wilhelm Biron Rudolf, German microbiologist; b. Gut Nienbuttel/Uelzen, Germany, Aug. 22, 1900; s. George and Marie (Krüger) M.; ed. Marburg, Göttingen (both Germany) univs.; Ph.D.; m. Käthe Pietschmann, Mar. 28, 1943. From asst. to prof. at large U. Göttingen. Mem. German Soc. Hygiene and Microbiology. Author: Zum Ertragsgesetz bei Aspergillus niger, 1930; Beiträge zur Kenntnis der Zellulosezersetzung unter niedriger Sauerstoffspannung, 1934; Methoden der Anaerobenzüchtung. Abderhaldens Handbuch der biologischen Arbeitmethoden. Home: Weenderstrasse 73 1, Göttingen. Office: Gosslenstrasse 16, Göttingen, W. Germany.

MEYER, Hans, German radiologist; b. Bremen, Germany, July 30, 1877; s. Engelbert and Beta (Klatte) M.; ed. univs. Munich, Kiel, Marburg (all Germany); M.D. honoris causa, U. Giessen (Germany); m. Rosemarie Nölke, 1936. Asst. clinics univs. Kiel, Basel, Bern (Switzerland), also asso. with Physiol.-Chem. Inst. Strasbourg, France, prior to 1920; lectr., prof. U. Kiel, 1911, 20; dir. Bremen Hosp., 1920-36; dir. Radiol. Inst., U. Marburg, 1945-50, prof., from 1942. Recipient Goethe Medal Art and Sci., Germany. Mem. German Röntgen Soc. (chmn. 1928, 33-34), Comité International de Photobiologie (hon. pres.), others. Author: Lehrbuch der Strahlentherapie, 5 vols., 1929-38. Pioneer med. radiology. Home: 1 Am Grassenburg, Marburg, West Germany.

MEYER, Hans Heinrich Joseph, German geographer; b. Hildburghausen, Germany, Mar. 22, 1858; s. Herrman Julius Meyer; ed. in polit. and natural scis.; traveled around world, 1881-82; became partner Bibliographisches Institut (publishers), 1884, dir., until 1914; made expdns. to E. Africa; discovered crater and glacier of Mt. Kilimanjaro, 1st (with L. Purtscheller) to ascend its Kibo summit, 1889; visited Canary Islands, 1894; studied volcanoes and glaciation in mountains of Ecuador, 1903; explored Ruanda-Urundi, then in German E. Africa, 1911; became mem. German colonial council, 1901; prof. colonial geography, Leipzig, Germany, 1915-28. Author: Ostafrikanische Gletscherfahrten, 1890, English edit., 1891; Der Kilimandscharo, 1900; Das deutsche Kolonialreich, 2 vols., 1909-10. Died Leipzig, July 5, 1929.

MEYER, Hans Horst, German pharmacologist; b. Insterburg, Germany, Mar. 17, 1853; prof. Dorpat, Estonia, Marburg, Germany, Vienna, Austria. Author: (with R. Gottlieb) Die experimentelle Pharmakologie als Grundlage der Arzneibehandlung, 1910. Research on pharmacological effects of iron, bismuth and aluminum, nature of acute phosphorus poisoning, heat regulation, diuretic and narcotic effects, theory of narcosis, sympathetic nervous system (especially as affected by drugs), migration of tetanus toxin in nerves. Died Vienna, Oct. 6, 1939.

MEYER, (Johannes) Horst (Max), physicist; b. Berlin, Germany, Mar. 1, 1926; s. Kurt H. and Gertrude (Hellwig) M.; B.Sc., U. Geneva, 1949; Ph.D., U. Zurich, 1953; m. Ruth Mary Hunter, Mar. 28, 1953; children—Richard, Christopher. Fellow, Swiss Fedn. for Research in Physics and Math., Oxford U., Eng., 1953-54, Nuffield fellow, 1955-57; lectr. Harvard, 1957-59; faculty Duke, 1959—, prof. physics, 1964—. Vis. prof. Tech. U., Munich, 1965. Alfred P. Sloan fellow, 1960-63. Fellow Am. Phys. Soc. Research, numerous publs. on low temperature physics, magnetism, solid and liquid helium, molecules trapped in enclosure compounds. Home: 1411 Anderson St., Durham, N.C. 27707.*

MEYER, James Wagner, Am. physicist; b. Rhineland, Mo., May 22, 1920; s. Julius C. and Laura (Wagner) M.; Ph.B. in Physics and Math., U. Wis., 1948, Ph.D., 1956; M.A. in Physics, Dartmouth, 1950; postgrad. U. Cal. at Berkeley; m. Carol Elizabeth Hamre, June 18, 1949; children—Sue Ann, Sara Elizabeth, Sandra Lee, Sharon Jane. Staff mem. Lincoln Lab., Mass. Inst. Tech., 1952-65, asso. head radar div., 1959-62, asso. head solid state div., 1962-63, head radio physics div., 1963-65; sr. scientist Ednl. Services, Inc., Watertown, Mass., 1965—. Mem. sci. adv. bd. to chief of staff USAF, 1959—; mem. study group IV, Internat. Radio Consultative Com., 1960—. Mem. Am. Phys. Soc., Am. Assn. Physics Tchrs., Am. Geophys. Union, A.A.A.S. Research, publs. on magnetic behavior solids at temperatures nr. absolute zero, change magnetic properties with time, developed (with A. L. McWhorter) 1st solid state maser to function as amplifier, developed spl. gas discharge tubes for high power def. radar. Home: Laurel Dr., Lincoln, Mass. 01751. Office: 47 Galen St., Watertown, Mass. 02172.*

MEYER, Joachim Ernst, German psychiatrist; b. Königsberg, Germany, July 2, 1917; s. Ernst and Käte (Schmieden) M.; m. Ruth Thwaites, July 11, 1953; children—Barbara, Marion. Staff dept. neuropathology Max Planck Inst. for Psychiatry, Munich, Germany, 1945-49, neuro-psychiat. and neurophysiol. dept. U. Freiburg (Germany), 1949-53; asst. prof. psychiatry and neurology U. Munich, 1953-63; prof. psychiatry U. Göttingen (Germany), 1963—, dir. Psychiat. Hosp., 1963—. Mem. Royal Medico-Psycho. Assn. (corr.). Author: Die Entfremdungserlebnisse, 1958; also articles. Editor: (with H. Feldmann) Anorexia Nervosa, 1965. Research in neuropathology, psychopathology, disorders of puberty; diagnostic schemes in psychiatry, war neuroses. Home: 3 Bonhoefferweg, Göttingen, West Germany.*

MEYER, Johann Friedrich, German chemist; b. Osnabrück, 1705. Apprenticed to apothecary, 1720, then went to Leipzig and Nordhausen, then Clausthel and St. Andreasberg in Harz; traveled to Frankfort-am-Main and Trier, then Halle, stayed 2 years before returning home to manage his grandmother's apothecary shop. Author: Chymische Versuche, zur näheren Erkenntnifs des ungelöshten Kalchs, der elastischen und electrischen Materie des allerreinsten Feuerwesens, und der ursprünglichen allgemeinen Säure, 1764; Alchymistische Briefe. Known for theory of acidum pingue and his view as to causticity of lime; experimented in attempt to accomplish transmutation. Died Nov. 1765.

MEYER, John Stirling, physician; b. London, Eng., Feb. 24, 1924; s. W. C. B. and Alice (Stirling) M.; B.S., Trinity Coll., Conn., 1944; M.D., C.M., McGill U., Can., 1948, M.Sc., 1949; m. Wendy Haskell, June 20, 1947; children—Jane, Anne, Elizabeth, Helen, Margaret. Instr., research asso. Harvard Med. Sch., 1955-57; prof., chmn. dept. neurology Wayne State U. Sch. Medicine, Detroit, 1957—; program dir. Wayne Neurol. Center for Cerebrovascular Research, 1963—; head neurol. dept. Detroit Gen. Hosp., 1957—, Harp Hosp., Detroit, 1963—; program dir. neurology Wayne U. Affiliated Hosps., 1957—. Cons. neurologist to hosps.; sec. nat. study extra-cranial causes strokes NIH, USPHS, 1964—, chmn. 1966—; mem. Pres. Lyndon Johnson's Commn. on Heart Disease, Cancer and Stroke, 1964, chmn. stroke subcom. Diplomate Am. Bd. Neurology and Psychiatry (asst. examiner 1961—). Mem. Am. Heart Assn. (mem. council on circulation, 1963—), fellow Council on Cardiovascular Disease), Am. Electroencephalograph Soc., Am. Neurol. Assn., A.M.A., Am. Assn. Neuropathologists, Am. Geriatrics Soc., Alpha Omega Alpha. Author: (with Loeb) Strokes due to Vertebro-basilar Arterial Disease, 1965; (with Gastaut) Cerebral Anoxia and the Electroencephalogram, 1961; also numerous articles. Research on causes and treatment of stroke; co-developer new methods for measuring cerebral blood flow and metabolism in living patient and test treatment. Home: 375 Lake Park Dr., Birmingham, Mich. 48009. Office: 3825 Brush St., Detroit 48201.*

MEYER, Karl, biochemist; b. Kerpen/Cologne, Germany, Sept. 4, 1899; s. Ludwig and Ida (Aaron) M.; M.D., U. Cologne, 1924; Ph.D. in Chemistry, U. Berlin (Germany), 1927; m. Marthe M. Ehrlich, Apr. 15, 1930; children—Robert F., Janet R. (Mrs. David Levy). Came to U. S., 1930, naturalized, 1937. Asst. to prof. Meyerhof, Kaiser Wilhelm Inst. for Physiology, Berlin-Dahlem, 1927-28; fellow Internat. Edn. Bd. dept. organic chemistry Eidgenossische Technische Hochschule, Zurich, Switzerland, 1928-29; Deutsche Forschungsgemeinschaft fellow Kaiser Wilhelm Institute für physikalische Chemie, Berlin-Dahlem, 1930; asst. prof. U. Cal., 1930-32; faculty Columbia, N.Y.C., 1933—, prof. biochemistry, 1952-67, emeritus, 1967—; prof. biochemistry Belfer Grad. Sch. Sci., Yeshiva U., N.Y.C., 1967—. Recipient Claude Bernard medal U. Montreal, 1952; Lasker award Albert and Mary Lasker Found., 1956; T. Duckett Jones Meml. award Helen Hay Whitney Found., 1959; Gairdner Found. award, 1960; N.Y. Med. Coll. award, 1961. Fellow Am. Acad. Arts and Sci.; mem. Am. Soc. Biol. Chemists, Am. Chem. Soc., Am. Rheumatism Soc., Chem. Soc. (London, Eng.), Harvey Soc. Contbr. numerous articles to tech. jours. Editorial bd. Jour. Biol. Chemistry, 1962—. Research on structure and chem. activity lysozyme, isolation and chem. structure hyaluronic acid, three types chondroitin sulfates, keratosulfates, mechanism three types hyaluronidases, patterns distbn. mucopolysaccharides in connective tissue and their changes with aging. Home: 642 Wyndham Rd., Teaneck, N.J. 07666. Office: Belfer Grad. Sch. Sci., Yeshiva U., 2469 Amsterdam Av., N.Y.C. 10033.*

MEYER, Karl Friedrich, exptl. pathologist, epidemiologist; b. Basel, Switzerland, May 19, 1884; s. Theodor and Sophie (Lichtenhahn) M.; student U.

Basel, 1902, Ph.D., 1924; student U. Zurich, 1902-05, U. Bern (Switzerland), 1906-08; D.V.M., U. Munich (Germany), 1905; hon. M.D. U. Basel, 1952, U. Zurich, 1937, Coll. Med. Evangelists, 1936; LL.D., U. So. Cal., 1946, U. Cal. at Berkeley, 1958; D.Sc., Ohio State U., 1958, U. Pa., 1959; m. Marion Lewis, Feb. 26, 1960; 1 dau., Charlotte Meyer (Mrs. Bartler P. Cardon). Came to U. S., 1910, naturalized, 1922. Faculty, U. Pa., 1910-13; dir. lab. and expt. farm Pa. Livestock San. Bd., Phila., 1912-13; prof. bacteriology and protozoology U. Cal. at Berkeley, 1914-15, asso. prof. tropical medicine Hooper Found. for Med. Research, 1915-21, prof. research medicine and acting dir., 1921-24, dir., 1924-54, prof. bacteriology, 1924-48, prof. exptl. pathology, 1948-54, prof., dir. emeritus, 1954—. Cons. Cal. State Dept. Pub. Health, 1920—; mem. nat. adv. health council NIH, USPHS, 1940-50. Decorated Officer l'Ordre de la Sante Publique; recipient plaque WHO, Sedgwick Meml. medal, 1946, 49; James D. Bruce medal, 1950; Nat. Acad. Scis., 1961; Lasker award, 1951; Walter Reed medal, 1956; Howard L. Ricketts award U. Chgo., 1960. Fellow Nat. (Jessie Stevenson Kovalenko medal 1961), N.Y., Cal. acads. scis., Acad. Pediatrics (asso.), A.A.A.S., Am. Acad. Arts and Scis., Soc. Am. Bacteriologists (past pres.), Assn. Immunology (past pres.), Soc. for Exptl. Biology and Medicine, Am. Soc. Tropical Medicine, Sigma Xi. Author: Practical Bacteriology, 1925; Disinfected Mail, 1962. Research, publs. on control botulism, isolation, epidemiology various infectious agts. Home: 260 San Leandro Way, San Francisco 94127.*

MEYER, (Julius) Lothar, German chemist; b. Varel, Germany, Aug. 19, 1830; M.D., Zurich, Switzerland; also studied at Würzburg, Heidelberg, Königsberg (all Germany); Ph.D., Breslau, Germany, 1858; became dir. chem. lab. Physiol. Inst. Breslau, 1859; named prof. natural history, Eberswalde, Germany, 1866; apptd. prof. chemistry, Karlsruhe, Germany, 1868-76; 1st prof. chemistry, Tübingen, Germany, from 1876. Recipient (with Mendeléev) Davy medal Royal Soc., 1882. Author: Die Gase des Blutes, 1857; Da sanguine oxydo carbonico infecto, 1858; Die modernen Theorien der Chemie, 1864; (with K. Seubert) Die Atomgewichte der Elemente . . . , 1883, Das natürliche System der Elemente, 1889; Grundzüge der theoretischen Chemie, 1890. Discovered (independently of Mendeléev) periodic law, 1869; evolved atomic volume curve which showed atomic volumes are functions of atomic weights, 1869; discovered chem. affinity of hemoglobin for oxygen; studied physiology of respiration; recognized that oxygen breathed is held in blood by some type of chem. union, 1857. Died Tübingen, Apr. 11, 1895.

MEYER, Lothar, chemist; b. Breslau, Germany (now Wroclaw), Poland, July 13, 1906; s. Gotthold and Selma (Heimann) M.; Dr.Engring., Inst. Tech., Breslau, 1930; m. Marion Meyer, Mar. 25, 1935. Came to U. S. 1946, naturalized, 1952. Asst., U. Göttingen, 1930-32; mgr. patent dept. Gesellschaft für Linde's Eismaschinen, Munich, Germany, 1932-39; research asso. U. Leiden (Holland), 1939-46; with U. Chgo., 1947—, prof. chemistry, 1953—; Gauss prof. U. Göttingen, 1961-62. Fellow Am. Phys. Soc.; mem. Faraday Soc. London, Sigma Xi. Research, numerous publs. on superfluid behavior of liquid helium, quantum hydrodynamics, chem. reactions at hot surfaces, crystal structures at low temperatures. Home: 5631 Dorchester Av., Chgo. 60637.

MEYER, Max Frederick, psychologist; b. Danzig, Germany, June 15, 1873; Ph.D., U. Berlin, 1896. Asst., U. Berlin, 1896-98; prof. exptl. psychology U. Mo., 1900-30, research prof. psychology, 1930-32. Exchange prof. U. Chile, 1929; in charge research Central Inst. for Deaf, St. Louis, 1929-30; vis. prof. psychology U. Miami, 1932-40. Fellow A.A.A.S.; mem. Am. Psychol. Assn., Acoustical Soc. Am. Author: Fundamental Laws of Human Behavior, 1911; Psychology of the Other One, 1921. Considered pre-Watsonian behaviorist; stressed public data not private consciousness as material for psychology; research on physiol. acoustics. Address: 3939 Loquat Av., Miami, Fla. 33133.

MEYER, Ovid Otto, Am. physician; b. Stevens Point, Wis., Dec. 17, 1900; s. Otto F. and Elmere E. (Belanger) M.; S.B., U. Wis., 1924; M.D., Columbia, 1926; m. Lyda Ann Henry, June 24, 1932; children—Nancy A., Mary Lynn, Ralph H.; asst. in medicine Harvard Med. Sch., 1929-32; asst. physician, Thorndike Memorial Lab., 1932; asst. prof. medicine, U. Wis. Med. Sch., 1932-37, asso. prof., 1937-44, prof., 1944—, chmn. dept. medicine 1945-64; asso. physician U. hosps., 1937-44, physician, 1944—. Fellow A.C.P.; mem. A.M.A., Internat. Soc. Hematology, Am. Soc. Clin. Investigation, Central Soc. Clin. Research, Soc. Exptl. Biology and Medicine, Am. Soc. Internal Medicine, Coll. Clin. Pharmacology and Chemotherapy, Assn. Am. Physicians, Sigma Xi, Phi Kappa, Alpha Kappa Kappa, Sigma Sigma. Contbr. articles on clin. medicine, hematology, clin. research to sci. publs.; spl. investigations on anticoagulants, leukemias, malignant diseases. Home: 1806 Summit Av. Office: 1300 University Av., Madison, Wis. 53706.*

MEYER, Peter, physicist; b. Berlin, Germany, Jan. 6, 1920; s. Franz and Frida (Lehmann) M.; dipl. ing. Tech. U., Berlin, 1942; Ph.D., U. Göttingen, 1948; m. Luise Schuetzmeister, July 20, 1946; children—Stephan S., Andreas S. Came to U. S., 1952, naturalized, 1962. Faculty, U. Göttingen, 1946-49; fellow U. Cambridge (Eng.), 1949-50; research asso. Max-Planck-Inst. Fúr Physik, Göttingen, 1950-52; research asso. Enrico Fermi Inst. Nuclear Studies, dept. physics U. Chgo., 1953-56, faculty, 1956—, prof., 1965—. Fellow Am. Phys. Soc.; mem. Am. Geophys. Union, Am. Astron. Soc., Sigma Xi. Research, numerous publs. on cosmic radiation, astro physics. Home: 2626 Park Dr., Flossmoor, Ill. 60422. Office: 933 E. 56th St., U. Chgo., Chgo. 60637.*

MEYER, Roland Kenneth, Am. zoologist; b. Fond du Lac, Wis., May 4, 1904; s. August John L. and Mabel (Case) M.; B.A., Milton Coll., 1926; M.A., U. Wis., 1928, Ph.D. in Zoology, 1930; m. Beulah L. Lanphere, June 15, 1924 (div. June 1946); children—Roland Kenneth, John Russell; m. 2d, Elva G. Shipley, June 2, 1947. NRC fellow in anatomy U. Rochester Med. Sch., 1931-32, spl. fellow in anatomy, 1933; mem. research staff Upjohn Co., 1934-35; asst. prof. zoology U. Wis., Madison, 1935-38, prof., 1938—, chmn. dept. zoology, 1962-64, researcher in endocrinology, 1930—; co-owner Endocrine Labs., Madison, 1948—. Endocrinology study sect. NIH, 1958-61, endocrinology-pharmacology panel, 1962-66, research career award com., 1966—, population research com., 1967—; mem. oral adv. group Internat. Planned Parenthood Fedn., 1963-67; cons. reproductive physiology Ford Found., India, 1964. Fellow A.A.A.S.; mem. Assn. Study Internal Secretions, Soc. Exptl. Biology and Medicine, Am. Physiol. Soc. (sect. editor endocrinology and metabolism 1964-66), Internat. Soc. for Research in Biology of Reprodn., Am. Assn. Anatomists, Am. Naturalists, Am. Soc. Zoologists, Am. Assn. U. Profs., Sigma Xi, Gamma Alpha, Phi Sigma, Phi Beta Pi (hon.). Corr. editor Jour. Reproduction and Fertility, 1964—. Contbr. articles to profl. jours. Co-patentee method for preparation gonadotrophic hormone, 1944. Home: 2532 Balden St., Madison, Wis. 53713.*

MEYER, Rudolf Hans Wilhelm, German meteorologist; b. Riga, Germany, Aug. 23, 1880; s. Alexander and Elisabet (Pohrt) M.; ed. Dorpat (now Tartu, Estonia), Berlin univs.; Ph.D.; m. Luise von Eltz, Oct. 12, 1909; children—Hildegard, Heinrich, Benita, Gerd. Asst. Tartu, Riga univs.; prof. agrégé; instr. in meteorology U. Riga; full prof. Riga, Poznan (Poland) univs. Mem. German, Riga (hon.) physics socs., German Soc. Meteorology. Author: Meteorologie für Landwirte, 1922; Haloerscheinungen, Abhandlungen der Herder-Instituts zu Riga, 1923; Handbuch der Geophysik, 1956-58; others. Address: Calsowstrasse 50, Göttingen, W. Germany.

MEYER, Viktor, German chemist; b. Berlin, Germany, Sept. 8, 1848; s. Jacques Meyer; Ph.D., U. Heidelberg (Germany), 1867; postgrad. U. Berlin, 1868-71; m. 1873. Asso. prof. U. Stuttgart (Germany), 1871-72; prof. chemistry Zurich (Switzerland) Poly. Inst., 1872-85, U. Göttingen (Germany), 1885-89, U. Heidelberg, 1889-97. Author: (with F. P. Treadwell) Tabellen zur qualitativen Analyse, 1884; Pyrochemische Untersuchungen, 1885; Thiophengruppe, 1888; Lehrbuch der organischen Chemie, 1893-96; Probleme der Atomistik, 1896. Contbr. sci. papers. Investigated nitric and nitrous compounds, organic iodine compounds; discovered aliphatic nitrocompounds, thiophene; determined formulas of nitrolic acids, pseudo-nitrols; discovered synthesis aromatic acids with sodium formiate; developed displacement method for determination vapor densities; showed dissociation into atoms of bromine and iodine molecules with rise of temperature; discovered oximes and explained their stereoisomerism; introduced name stereochemistry; found steric hindrance in chem. reactions. Died Heidelberg, Aug. 8, 1897.

MEYER, Walter Leslie, Am. chemist; b. Toledo, Feb. 28, 1931; s. Walter Dimsdale and Vera (Kilburn) M.; B.S. in Chemistry, U. Mich., 1953, M.S., 1955, Ph.D., 1957; m. Ann Margaret Weaver, June 26, 1954; children—Janet Elaine, Robert Leslie, Barbara Kay. Instr., research asso. U. Mich., 1957; NSF fellow U. Wis., 1957-58; faculty Ind. U., 1958-65, asst. prof. chemistry, 1960-65; asso. prof. chemistry U. Ark., Fayetteville, 1965—. Mem. Am. Chem. Soc., Chem. Soc. (London, Eng.), Phi Beta Kappa, Sigma Xi, Phi Lambda Upsilon. Research, publs. on chemistry and total synthesis of natural products, use of phys. methods in organic chem. research. Home: Fayetteville, Ark. 72701.*

MEYER, (Hans) Wilhelm, Danish otologist; b. Fredericia, Denmark, Oct. 25, 1824; ed. U. Copenhagen; monument erected in his honor in Copenhagen. Author: (paper) Über adenoide Vegetationen in der Nasenrachenhöhle (classic description of clin. features of adenoids, noting inhibitory influence on devel. of facial skeleton and giving methods of operative treatment), 1873-74. Died 1895.

MEYER, Willy, surgeon; b. Minden, Westphalia, Germany, July 24, 1858; M.D., U. Bonn, 1880; m. Lily O. Maass, Apr. 29, 1885. Came to U. S., 1884. Asst. to surg. clinic U. Bonn, 1881-84; prof. clin. surgery Woman's Med. Coll., N.Y., 1886-93; instr. and prof. surgery N.Y. Post-Grad. Med. Sch. and Hosp., N.Y.C., 1887—. Introduced into U. S. cystoscopy with modification of Nitze cystoscope, 1887; 1st to perform ureteral catheterization with modified Caspar instrument, 1896; 1st in U. S. to perform Bottini operation, 1897; introduced into U. S. new methods of gastrostomy; contbd. to surgery of esophagus and chest by improving cabinet for differential positive and negative pressure; suggested Meyer's solution for treatment of thromboangiitis obliterans. Died N.Y., Feb. 24 or 25, 1932.

MEYERAND, Russell Gilbert, Jr., Am. research physicist; b. St. Louis, Dec. 2, 1933; s. Russell Gilbert and Elsa Louise (Gebhardt) M.; B.S. in Elec. Engring., Mass. Inst. Tech., 1955, M.S. in Nuclear Engring., 1956, Sc.D. in Plasma Physics, 1959; m. Mary Grace Guillemin, June 16, 1956. Staff, Research Lab. Electronics, Mass. Inst. Tech., 1956-57, plasma physics research asst., 1957-58; prin. scientist United Aircraft Research Labs., East Hartford, Conn., 1958-64, chief research scientist, 1964-67, director of research, since 1967—. Cons. to atomic power equipment dept. Gen. Electric Co., 1955-56. Asso. fellow Am. Inst. Aeros. and Astronautics; mem. I.E.E.E. (sr.), Am. Phy. Soc., Sci. Research Soc. Am., N.Y. Acad. Scis., Sigma Xi. Research, publs. on generation of electrostatic potential gradients with applications as ion source, studies gas ionization by high-intensity laser beams. Home: 64 Littel Acres Rd., Glastonbury, Conn. 06033. Office: 400 Main St., East Hartford, Conn. 06108.*

MEYERDING, Henry William, Am. surgeon; b. St. Paul, Minn., Sept. 5, 1884; s. Henry John and Adelgunda (Rosenkranz) M.; B.Sc., U. of Minn., 1907, M.D., 1909, M.Sc. in orthopedic surgery, 1918; m. Lura Abbie Stinchfield, Feb. 12, 1912 (dec. Apr. 1960); children—Augustus (dec.), Edward Henry, Anne (dec.). House surgeon Mayo Clinic, 1911-12, attending physician, 1912-14, asst. orthopedist, 1914-15, asso. orthopedic surgeon, 1915, surgeon since 1915; orthopedic surgeon St. Mary's and Colonial hosps., 1915; instr. orthopedic surgery, Mayo Foundation, U. of Minn. Grad. Sch., 1918-20, asst. prof., 1920-22, associate professor, 1922-37, prof. 1937-49, now emeritus. Served in Minn. Nat. Guard, 1st lt. M.C., 1909, col., 1938. Recipient Gold medal, Am. Med. Assn., 1939; gold medal, Am. Cong. Phys. Therapy, 1939; First award, Chgo. Med. Soc., 1947; medal of honor, from the University of Bordeaux, 1952; Certificate of Merit, U. Minn., 1952. Diplomate Am. Bd. Orthopaedic Surgery. Fellow A.C.S. (gov. 1946-53), Internat. Coll. Surgeons (pres. U. S. sect. 1950-51, internat. president 1958), Acad. Surgery, Spain (hon.); mem. Am. Fracture Assn. (pres. 1952-56), Internat. Soc. Orthopaedic Surgery and Traumatology (nat. chmn. U. S. sect., pres. 6th congress 1948, chmn. U. S. delegations 1946-55), hon. mem., corr. mem. fgn., internat. and nat. profl. and scientific orgns. and assns. Italian, Brazilian, Argentine and including hon. memberships in: French Socs. Orthopedic Surgery and Traumatology, Internat. Surg. Soc., World Med. Assn., Netherlands Orthopaedic Soc., Belgian, Czechoslovak, Bordeaux, Madrid, Internat. surg. socs., Brazilian Acad. Medicine, Philippine Coll. Surgeons, Turkish Assn. Surgeons. Research on bone and joint surgery; bone tumors; radiology; sarcomas; fractures and dislocations; Dupuytren's contracture; Volkmann's ischemic contracture associated with supracondylar fracture; spondylotisthesis. Home: 1531 Sixth St. S.W. Office: Emeritus Room, Mayo Clinic, Rochester, Minn.

MEYERHOF, Otto, physician, biochemist; b. Hanover, Germany, Apr. 12, 1884; s. Felix and Bettina (May) M.; student, Freiburg, Strasbourg; M.D., Heidelberg, 1909; LL.D., U. Edinburgh, 1926; m. Hedwig Schallenberg, June 4, 1914; children—George Geoffrey, Bettina Ida (Mrs. Donald E. Emerson), Walter Ernst. Came to U. S., 1940, naturalized, 1946. Research worker, U. Heidelberg, Zool. Sta., Naples, Italy, 1909-11; lectr. U. Kiel (Germany), 1912-18, asso. prof., 1918-24; mem. Kaiser Wilhelm Inst. of Biologie, Berlin, 1924-29; dir. Kaiser Wilhelm Inst. of Physiology, Heidelberg, 1929-38; dir. research Centre Nationale, Paris, 1938-40; research prof., U. Pa., since 1940. Recipient (with A. V. Hill) Nobel prize in medicine and physiology, Fellow Royal Soc., 1937; mem. Harvey Soc. (hon.), Nat. Acad. Scis., Virchow Soc., other sci. socs., Sigma Xi. Author: Chemical Dynamics of Life Phenomena, 1924; Chemische Vorgänge im Muskel Springer, 1930; Respiration and Fermentation of Cells; Energy Transformation and Chemistry of Muscle; also papers. Research in chemistry of muscles; showed that lactic acid is formed by breakdown of glycogen in absence of oxygen in the working muscle, that some lactic acid is oxidized to reconvert the rest into glycogen in the resting muscle. Died Oct. 6, 1951.

MEYERHOF, Walter Ernst, physicist; b. Kiel, Germany, Apr. 29, 1922; s. Otto Fritz and Hedwig (Schallenberg) M.; Ph.D., U. Pa., 1946; m. Miriam G. Ruben, Aug. 21, 1947; children—Michael O., David L. Came to U. S., 1941, naturalized, 1946. Asst. prof. U. Ill., Urbana, 1946-49; faculty physics Stanford (Cal.), 1949—, prof., 1959—, acting exec. head, 1962-63. Fellow Am. Phys. Soc. Studies of nuclear energy levels and reaction mechanisms, especially threshold effects and multi-particle break-up reactions. Home: 213 Blackburn Av., Menlo Park, Cal. 94025. Office: Physics Dept., Stanford, Stanford, Cal. 94305.*

MEYERHOFF, Günther, German chemist; b. Hagen, Nov. 4, 1919; s. Friedrich and Elisabeth (Fromm) G.; ed. univs. of Rostock, Leipzig, Göttingen, Mainz (all Germany); Ph.D.; m. Claire Krug, Nov. 17, 1951; children—Martina, Michaela, Marie-Claire. Became instr., 1954; named prof. physics and chemistry U. Mainz, 1956. Mem. German Assn. Chemists, Deutsche Busengesellschaft, Faraday Soc. Research and publs. on phys. chemistry of polymers, polymerization kinetics. Home: Gabelsbergerstrasse 11, Mainz-Gonsenheim, Germany. Office: Institut physikalische Chemie-Universität Mainz, West Germany.

MEYERHOFF, Howard Augustus, Am. geologist; b. N.Y.C., May 27, 1899; s. Augustus Henry and Grace E. (Berry) M.; B.A., U. Ill., 1920; M.A., Columbia, 1922, Ph.D., 1935; LL.D., Drexel Inst. Tech., 1955; m. Sophie Theilen, Oct. 6, 1923; 1 son, Arthur Augustus. Curator paleontology Columbia, N.Y.C., 1921-24; prof. geology Smith Coll., 1924-49; adminstrv. sec. A.A.A.S. Washington, 1949-53; exec. dir. Sci. Manpower Commn., Washington, 1953-62; prof. geology, chmn. dept. U. Pa., Phila., 1963—. Geologist sci. survey, P.R., V.I., 1924-44; cons. P.R. Bur. Mines, 1935-42, Dominican Republic, 1938-49; dir. subsurface survey, Springfield, Mass., 1940-42; mem. exec. res. Office Emergency Planning; partner Geo-Surveys, Washington, 1955-61, owner, Phila., 1961—. Mem. A.A.A.S. (v.p. 1944, 59), Geol. Soc. Am. (chmn. Northeastern sect. 1965-67), other sci. and profl. orgns., Sigma Xi, Phi Beta Kappa. Author: Geology of Puerto Rico, 1932; (with G. W. Bain) Flow of Time in the Connecticut Valley, 1941, 2d edit., 1964; Natural Resources in Most of the World; also articles. Chmn. editorial bd. Science (Sci. Monthly), 1950-53. Specialist in Caribbean geology, mineral raw materials. Home: 225 Gulf Creek Rd., Radnor, Pa. 19088.*

MEYERS, Albert Irving, Am. chemist; b. N.Y.C., Nov. 22, 1932; s. Hyman and Sylvia (Greenberg) M.; B.A., N.Y. U., 1954, Ph.D., 1957; m. Joan Shepard, Aug. 10, 1957; children—Harold, Jill, Lisa. Research chemist Cities Service R and D, 1957-58; faculty La. State U., New Orleans, 1958—, prof. chemistry, 1964—; spl. fellow Harvard, 1965-66. Research Corp. grantee; NIH grantee; U. S. Army Research and Devel. Command grantee; Am. Cancer Soc. grantee. Recipient Distinguished Prof. award La. State U. Alumni, 1964. Mem. Am. Chem. Soc., Chem. Soc. (London), Sigma Xi, Phi Lambda Upsilon. Editorial bd. Internat Jour. Heterocyclic Chemistry. Research, publs. on methods for synthesizing new cyclic-nitrogen-containing molecules which may provide new medicinal agts.*

MEYERS, H(arold) Russell, Am. neurologist, neurosurgeon; b. Bklyn., Feb. 25, 1905; s. Harry Russell and Catherine (Ward) M.; B.A., Brown U., 1927, M.S., 1929; M.D., Cornell U., 1932; m. Mary Emma Conyers, Sept. 4, 1947; children—Harvey, Audrey, Mary, David, Russell, Jeffrey, Steven, Paul; m. 2d, Pauline M. Simmons, Apr. 12, 1960. Chmn. div. neurosurgery State U. Ia., 1946-63; chief neurology and neurosurgery The Highlands Clinic, Williamson W.Va., 1963—; chief, neurology and neurosurgery Appalachian Regional Hosps., 1963—. Instr. embryology and histology, exptl. psychology Brown U., 1926-28; instr. psychology N.Y.U., 1928-32; asst. physiology Cornell U. Med. Coll., 1931-32; instr. speech pathology Columbia, 1937-38; instr. neurology and neurophysiology L.I. Coll. Medicine, 1939-46; asst. prof. surgery, div. neurosurgery State U. Ia. Coll. Medicine, 1946-47, asso. prof., 1948-49, prof., 1949-63; vis. prof. neurology U. Cin., 1956, 62; vis. lectr. U. Denver; neurosurg. cons. VA Hosp., Iowa City. Examiner Am. Bd. Psychiatry and Neurology, 1952—; med. adv. bd. Nat. Soc. Multiple Sclerosis, 1953—; bd. dirs. Inst. Gen. Semantics, 1955—, pres. bd. trustees, 1957—. Recipient Lucien Howe prize in surgery N.Y. State Med. Soc., 1942. Diplomate Am. Bd. Psychiatry and Neurology, Am. Bd. Neur. Surgery. Fellow A.C.S. (chmn. adv. council neurosurgery 1958-60); mem. Internat. Soc. Gen. Semantics (pres. 1950-53, past bd. govs.), Am. Neurol. Assn. (2d v.p. 1953-54), A.M.A., Assn. Research Nervous and Mental Diseases, Central Neuropsychiat. Assn., Harvey Cushing Soc., Internat. League Against Epilepsy, Ia.-Midwest Neurosurg. Soc. (pres. 1958-59), Ia., Am. Fedn. Clin. Research, Soc. Exptl. Biology and Medicine, Nat. Soc. Study Communication, Speech Assn. Am., Internat. Neurol. Congress, Soc. Neurol. Surgeons, A.A.A.S., Am. Acad. Cerebral Palsy (pres. 1962-63), United Cerebral Palsy Assn. (med. research council 1958—). Soc. Advancement Gen. Systems Theory, The Horseshoe Club (Anglo-Am. Medicine), Phi Beta Kappa, Sigma Xi, Alpha Omega Alpha. Contbr. numerous articles med. jours. Demonstrated feasibility of operating on human basal ganglia; surg. treatment of Parkinsonism; pioneer in use of high frequency focused ultrasound in neurosurgery. Office: Highlands Clinic, Williamson, W.Va. 25661.*

MEYERSON, Ignace, psychologist; b. Warsaw, Poland, Feb. 27, 1888; ed. U. Paris, U. Heidelberg; Ph.D. Asst., later asst. dir. Ecole des hautes études Sorbonne, Paris; instr. Faculty Letters, U. Paris; titular prof. Faculty Letters, Toulouse, France; dir. studies Sorbonne; dir. Center for Research in Psychology. Dir. Jour. Gen. Psychology. Home: 9, rue Edouard-Detaille, Boulogne (Seine), France. Office: 54, rue de Varenne, Paris 7, France.

MEYNERT, Theodor Hermann, physician; b. Dresden, Saxony, June 15, 1833. Prosector mental hosp. Vienna, Austria, from 1866, prof. psychiatry, dir. sect. nervous diseases, from 1870. Author: Die Bau der Grosshirnrinde und Seine Örtliche Verschiedenheit, 1868. Advocated asylums with least possible restraint; contbr. studies on brain anatomy, functions; gave earliest description association neurones, 1868; described small bundle nerve fibers (Meynert's fasciculus), 1867, tract of nerve fibers (Meynert's commissure), 1867, solitary pyramid-shaped cells in cerebral cortext near calcarine fissure, 1868. Died Vienna, Austria, May 31, 1892.

MEYRATH, Joseph, microbiologist; b. Schieren, Luxembourg, Sept. 9, 1928; s. Nicolas and Marie (Schiffmann) M.; Dipl.Ing.Agr., Swiss Fed. Inst. Tech., 1953, Dr.Sc.techn., 1957; m. Anita Claudia Lareida, 1951; children—Nicolas, Marcel, Elizabeth, William. Staff, Swiss Fed. Inst. Water Supply and Sewage Treatment, Zurich, 1953-55, Swiss Fed. Inst. Tech., 1955-59, Swiss Hort. Exptl. Sta., Wädensvil, Switzerland, 1959; sr. lectr. U. Strathclyde, Glasgow, Scotland, 1959-67; prof. applied microbiology and biol. methods State U. Agr., Vienna, Austria, 1967—. Assessor in microbiology Higher Nat. Certificate in Biology, 1965—. Recipient Silver medal and prize for excellent doctorate thesis, 1957. Fellow Inst. Food Sci. and Tech.; mem. Inst. Biology, microbiol. socs. Britain, Holland, Switzerland, Austria, Biochem. Soc. Research, publs. on growth properties of micro-organisms especially effects of inoculum size, prodn. of microbial enzymes, citric acid fermentation, disinfection with ozone. Office: State U. Agr., Vienna, Austria.*

MEYTHALER, Friedrich Karl Wilhelm Georg, German physician; b. Offenburg, Mar. 18, 1898; s. Friedrich and Luise (Coblitz) M.; ed. univs. of Heidelberg, Munich, Wurzburg, Bonn, Rostock; M.D.; m. Hermine Baer, 1929; 5 children. Asso. prof., Erlangen, Germany; dir. Med. Clinic, Nuremburg (Germany), Erlangen hosps.; became instr., Rostock, Germany, 1933; dir. Polyclinic, 1937-38, asso. prof., 1939; dir. Erlangen Polyclinic, 1942-45; dir. Nuremburg Hosp. Mem. Com. Against Cancer. Recipient Ernst von Bergmann medal. Mem. Deutscher Städtetag, Kommunales Krankenhauswesen. Author: Kriesmalaria, 1942; Viruspneumonie, 1949; Balneologie und Balneotherapie, 1954; Wissenschaft für die Praxis, 1951-57; also articles. Editor: Beiträge zur praktischen Medizin. Founder, dir. Nürnberger Wissenschaftlichen Arztetagungen. Research on pathophysiology, intermediate liver metabolism, infectious diseases, diabetes, hepatitis, tropical diseases, malignant tumors, carbohydrate metabolism, Home: Ebrardstrasse 27. Office: Universität, Erlangen, West Germany.

MEZGER, Peter Georg, radio astronomer; b. Lindau, West Germany, Nov. 19, 1928; s. Viktor G. and Helene (Spaeth) M.; Diplom in physics, Technische Hochschule, Munich, Germany, 1955; Dr.Ing., Technische Hochschule, Darmstadt, Germany, 1963; m. Barbara Nothhelfer, May 21, 1953; children—Brigitte, Verena. Research asst. Ecole Normal Superieure, Paris, France, 1953-54; asst. prof. radio-astronomy U. Bonn (Germany), 1955-60; research scientist microwave physics research labs. Siemens & Halske AG, Munich, 1960-63; asso. scientist radio-astronomy Nat. Radio Astronomy Obs., Green Bank, W.Va., 1963—. Mem. I.E.E.E., A.A.A.S., Deutsche Phys. Gesellschaft, Deutsche Astron. Gesellschaft, Internat. Astron. Union, Internat. Sci. Radio Union. Research, publs. on theory and devel. of parametric amplifiers, evaluation of very large radio-telescopes, galactic sources, high-frequency spectra of extragalactic sources, radio radiation of moon and planets, detection of hydrogen radio recombination lines. Home: 135 Bollingwood Rd., Charlottesville 22903. Office: Edgemont Rd., Charlottesville, Va. 22901.*

MIALHE, Louis, French physician; b. Vabre, France, Nov. 5, 1807; Apothecary, Faculty Medicine, Paris, 1836, M.D., 1839; Aggregation in Medicine. Mem. French Acad. Medicine. Author: De l'albumine, 1852; La pepsine, 1860; De la destruction des acides organiques dans l'économie animale, 1866. First isolation of ptyalin (salivary enzyme), 1845; studied pepsin, albumin, cause of indigestion. Died Paris, 1886.

MIASNIKOV, Aleksandr Leonidovich, Russian physician; b. Sept. 18, 1899; grad. Moscow U., 1922. Prof., Inst. for Improvement Physicians, Novosibirsk, 1932-38, Army and Navy Med. Acad., 1940-48, 1st Moscow Med. Inst., 1948—; in charge Therapeutics Inst., USSR Acad. Med. Scis., 1948—. Mem. USSR Acad. Med. Scis. Research on liver diseases, cardiovascular system, vitamin metabolism, malaria, brucellosis.

MICHAEL, Ernest Arthur, mathematician; b. Zurich, Switzerland, Aug. 26, 1925; s. Jakob and Erna (Sondheimer) M.; came to U. S., 1939, naturalized, 1944; B.A., Cornell U., 1947; M.A., Harvard, 1948; Ph.D., U. Chgo., 1951; m. Colette Verger, Sept. 26, 1956 (div. Feb. 1966); children—Alan, David, Gerard; m. Erika Goodman Joseph, Dec. 4, 1966; 1 dau., Hillary. Mem. Inst. for Advanced Study, Princeton, N.J., 1951-52, 56-57, 60-61; postdoctoral fellow U. Chgo., 1952-53; faculty U. Wash., Seattle, 1953—, prof. math., 1960—. Mem. Am. Math. Soc. Research in gen. topology, especially topological spaces and existence theorems for continuous functions. Home: 16751 15th Av. N.W., Seattle 98177.*

MICHAEL, Gerhard Franz Adolf, German plant physiologist, agr. chemist; b. Magdeburg, Germany, Mar. 25, 1911; s. Franz and Anna (Ehrecke) M.; Ph.D., U. Berlin (Germany), 1935, Dr.phil.habil., 1941; m. Brigitte Blume, Dec. 28, 1960; 1 dau., Irmgard. Research fellow U. Leipzig (Germany), 1936, U. Königsberg (Germany), 1937; asst. U. Berlin, 1937-47; prof., dir. Inst. Agr. Chemistry, U. Jena (Germany), 1947-60; prof., dir. Inst. Plant Nutrition, Agr. U. Stuttgart-Hohenheim (Germany), 1960—; dir. div. plant physiology Inst. Plant Research, Gatersleben, 1959-60. Mem. Academie Nat. Sci. Leopoldina. Co-editor: Zeitschrift für Pflanzenernährung und Bodenkunde, 1951. Research, numerous publs. on plant physiology especially mineral nutrition of plants, effect of minerals on protein metabolism. Home: 12 Steckfeldstr., 7000 Stuttgart-Hohenheim, Germany.*

MICHAEL SCOTUS, mathematician; b. probably in Scotland, circa 1175; ed. Durham cathedral sch., Oxford, Paris; tchr. in Bologna, Italy; joined ct. of Frederick II, Palermo, Italy, 1200; undertook studies in astronomy, alchemy, Toledo, Spain, 1209; returned to Palermo, 1220; refused archbishoprics of Cashel, Ireland, 1223, Canterbury, 1227. Author: Liber physiognomiae, 1209; commentaries on the Sphere of Sacrobosco. Translator: Astronomy (Al-Bitruji), 1217; Zoology, circa 1220, De coelo et mundo, Ethics (all Aristotle); other Aristotelian works with Averroes' commentaries. Known as Father of Christian Averroism; depicted as magician in Inferno (Dante), Decameron (Boccaccio), Lay of the Last Minstrel (Scott). Died circa 1234.

MICHAELIDES, Nikolaus, Greek bacteriologist; b. Athens, 1880; s. Antonius and Kaliope Michaelides; M.D., U. Athens, 1901; postgrad. Inst. Pasteur, Paris, 1904-06. Asst., Univ. Inst., 1902-04; asst. to Profs. Koch and Behring, Berlin and Marburg, 1907-08; studied with Wassermann, 1908-09; physician Univ. Poliklinik, Athens, 1909-12; dir. Dermatology Clinic, U. Athens, 1912-33, lectr. hygiene and bacteriology, 1908-33, prof. bacteriology, 1933—. Mem. State Hygiene Council. Mem. Athens Med. Assn., Assn. Greek Bacteriologists and Hygienists (chmn.), Greek Dermatology and Venerologic Soc. Author: Über eine durch Zielfärbung nicht darstellbare Form des Tuberkelbazillus, 1907; Tuberkulin als diagnostisches Mittel der Tuberkulose, 1908-09; Kala-Azar in Greece, 1911; Die Wassermannsche und die Pirquetsche Reaktion bei Lepra, 1912; Über Amöben der Harnblase, 1918; Über die Spirochaeta ikterohaemorrhagica, 1919; Die Wirkung der Galle, 1927; Beitrag zur laboratorischen Diagnose der Echinokokkiase, 1929; Lehrbuch: Laboratorium Diagnostik, 1927; Der Einfluss der Opotherapie auf die Wassermannsche Reaktion, 1928; Die Lüsmaniose der Hunde in Griechenland, 1929; Das Cholestrin in den verschiedenen Perioden der Syphilis, 1933. Research on leprosy, Tb, particularly syphillis and tests for it.

MICHAELIS, Gustav Adolph, German obstetrician; b. Germany, 1798; with obstetric clinic, Kiel, Germany. Author: Das enge Becken (classic work), 1851. Described diamond-shaped area above intergluteal cleft (Michaelis's area or rhomboid), 1841; reported case of patient upon whom 4 cesarean operations were successfully performed; studied optic disc. Died 1848.

MICHAELIS, Leonor, physician, biochemist; b. Berlin, Germany, Jan. 16, 1875; s. Moriz and Hulda (Rosenbaum) M.; student U. Berlin, 1893-96 (M.D.), Freiburg, 1896-97; m. Hedwig Philipsthal, Apr. 12, 1905; children—Ilse, Eva M. Became asst. to Prof. Paul Ehrlich, then at Berlin, 1898-99; asst. Municipal Hosp., Berlin, 1899-1902; oberarzt Inst. for Cancer Research, Berlin, 1902-06; dir. lab., Berlin Municipal Hosp., 1906-22; became pvt. docent U. Berlin, 1905, prof., 1908; prof. biochemistry. Med. Sch., Nagoya, Japan, 1922-26; resident lectr. Johns Hopkins U., 1926-29; mem. Rockefeller Inst. Med. Research, 1929-40, emeritus. Fellow A.A.A.S., N.Y. Acad. Sci.; mem. Am. Soc. Biol. Chemists, Nat. Acad. Scis., Am. Chem. Soc. Author: Compendium der Entwicklungsgeschiente des Menschen mit Berücksichtigung der Wirbeltiere, 1898; Einführung in die Farbstoffchemie für Histologen, 1900; Dynamik der Oberflächen, 1909; Einführung in die Mathematik für Biologen u. Chemiker, 1912; Die Wasserstoff-Ionen-Concentration, 1914; Praktikum d. Physikalischen Chemie insb. der Kolloid-Chemie, 1920; Oxydations-Reductions-Potentiale, 1929. Developed Michaelis-Menten equation to explain rate of variation of enzyme-catalyzed reaction with concentration of reacting substance, 1913; discovered Janus green stain (for microscopic examination of certain cell structures); discovered that keratin (chief ingredient of hair) is soluble in thioglycolic acid, thus making possible devel. of home permanents. Died N.Y.C., Oct. 9, 1949.

MICHAELIS, Moritz, biochemist; b. Hamburg, Germany, Jan. 6, 1906; s. Carl and Flora (Wohlgemuth) M.; student U. Hamburg, 1923-25, U. Berlin,

1929-31; student U. Würzburg, 1925-29, Ph.D. in Chemistry, 1934; m. Miriam Kauffman, Oct. 9, 1945. Came to U. S., 1947, naturalized, 1956. Research asst. U. Stockholm (Sweden), 1934-35, Biochem. Research Lab., Helsinki, Finland, 1935-37, Biochem. Lab., Cardiff City Mental Hosp., Great Britain, 1937-40; research fellow McGill U., Montreal, Ont., Can., 1942-47; asst. prof. physiology U. Chgo., 1947-48; research biochemist Mt. Sinai Hosp., N.Y.C., 1948-50; faculty U. Md. Sch. Medicine, Balt., 1950—, asso. prof., 1962—. Fellow Chem. Inst. Can.; mem. Soc. Exptl. Biology and Medicine, Am. Soc. Biol. Chemists, Biochem. Soc. (Great Britain), Am. Chem. Soc., Am. Soc. Clin. Chemists, Sigma Xi. Research, publs. on components respiratory enzymes, mode action narcotics, nerve poiscns, biochem. changes in shock, biochemistry eye, mode action antibiotics, degradation cellulose by soil organisms. Office: 520 R, W. Lombard St., Balt. 21201.*

MICHAELIS, Peter, German biologist; b. Munich, Germany, May 28, 1900; s. Oscar and Thusnelde (Jäger) M.; ed. U. Munich; Ph.D.; m. Gertrud Aichele, 1932; children—Helga, Georg, Luitgard. Asst., U. Iena, U. Stuttgart (Germany); head sect. Kaiser Wilhelm Inst., also Exptl. Inst. for Cultures, Max Planck Inst. Mem. Max Planck Soc. Research and publs. on morphology, woody alpine plants, cytoplasmic inheritance, ecology, plant cytology. Home: Kolibriweg 5, 5 Köln-Vogelsang. Office: Max Planck Institut, Köln-Vogelsang, West Germany.

MICHAELS, Joseph Jules, Am. psychoanalyst; b. Boston, June 7, 1902; s. Abraham and Kate (Weinberger) M.; B.S., U. Mich., 1923, M.D., 1926; m. Anna Cahan, June 4, 1931; children—George Albert, John Edward. Instr. psychiatry U. Mich., also asst. physician State Psychopath. Hosp., Ann Arbor, 1931-32; Commonwealth Fund fellow in psychiatry Harvard, 1932-33, 34-35, Neuropsychiat. Clinic, U. Vienna, 1933-34; psychoanalytic tng. Psychoanalytic Inst., Vienna, 1935-36; asst. in psychiatry Harvard Med. Sch., Boston Psychopathic Hosp., Mass. Gen. Hosp., Beth Israel Hosp., 1936-37; instr. in psychiatry Harvard, 1938-58, clin. asso. in psychiatry, faculty medicine, 1958—; instr. in psychiatry Boston Psychopathic Hosp., 1938-42; pvt. practice neuropsychiatry, 1938—; faculty Boston Psychoanalytic Soc. & Inst., 1946—, pres. 1958-61; tng. analyst, 1948—; vis. psychiatrist Beth Israel Hosp., 1948—; cons. in psychiatry VA Hosp., Boston, 1948—. Decorated Legion of Merit. Diplomate Am. Bd. Neurology and Psychiatry. Mem. Am. Psychoanalytic Assn. (exec. councillor 1961-63, chmn. com. social problems 1962—, editorial bd. Jour. 1963-66), Am. Psychiat. Assn., Am. Orthopsychiat. Assn., Group for Advancement Psychiatry (mem. com. on social issues), A.M.A. Author: Disorders of Character—Persistent Enuresis, Juvenile Delinquency, Psychopathic Personality, 1955. Studies and publs. on relationship of enuresis and neuroticism; integration, regression in psychoanalysis; juvenile delinquency and its control; impulsive actions. Died Nov. 19, 1966.

MICHAL, Aristotle D(emetrius), mathematician; b. Smyrna, Asia Minor, May 1, 1899; s. Demetrius and Sophia (Chaousoglou) M.; came to U. S., 1911, naturalized, 1924; A.B., Clark U., 1920, A.M., 1921; Ph.D., Rice Inst., 1924; Nat. Research fellow in math., U. Chgo., Harvard, Princeton, 1925-27; m. Luddye Charlotte Kennerly, June 9, 1924; 1 dau., Thalia Charlotte. Instr. math., Rice Inst., 1923-25, U. Tex., summer, 1924; asst. prof. math. Ohio State U., 1927-29; asso. prof. math. Cal. Inst. Tech., 1929-38, prof. since 1938, dir. of research in math. analysis, geometry and applied math., also dir. Engring., Sci. Mgt. War Tng. program in advanced tng. in math. and mechanics, World War II. Fellow A.A.A.S.; mem. Am. Math. Soc. (council 1938-40; sec. Far West 1942-44), Math. Assn. Am. (corr.) Acad. Nacional de Ciencias Exactas, Fisicas y Naturales de Lima, Assn. of Symbolic Logic, Sigma Xi. Author: Matrix and Tensor Calculus with Applications to Mechanics, Elasticity and Aeronautics, 1947; Differential Equations in Abstract Spaces with Applications to Analysis; Geometry and Mechanics. Editor Mathematics Mag. since 1947. Contbr. numerous research papers to tech. jours. Died June 14, 1953.

MICHALOWSKI, Roman Edmund Hugo, Polish dermatologist, psychologist; b. Warsaw, Poland, Oct. 31, 1912; s. Aleksander and Stanislawa, (Raciborska) M.; student Faculty Medicine, U. Poznan (Poland), 1932-34; student U. Warsaw, 1934-38, physician, 1939, M.A., 1952; M.D., U. S.B., Wilno, 1939; m. Zofia Lachowicz, Sept. 6, 1939; 1 dau., Malgorzata. With U. Warsaw, 1934-39, asst. prof., 1949; staff Inst. Histology, U. Warsaw, 1934-38, Charles U., Prague, Czechoslovakia, 1937; asst. dept. dermatology U. Warsaw, 1941-49; cons. dermatologist 2 Municipal Hosp., Warsaw, 1958-61; prof. dermatology dir. dept. Med. Sch., Lublin, Poland, 1962—. Recipient Triple awards Med. Sch. Lublin, 1964-66. Mem. Lublin Med. Assn. (v.p. 1965—), Polish Dermatol. Soc. (directory bd. 1965—), Polish Psychol. Soc. Author: Diseases of Buccal Mucosa, 1956, 2d edit., 1960; also numerous articles. Editorial bd. Przeglad Dermatologiczny, 1950—. First description of lipids deposition in soft palate of man and its relation with aging; first descriptions of ultrastructure of blisters, in Duhring dermatitis herpetiformis and skeletal muscle fiber in diffuse scleroderma syndrome, labial heterotopia of salivary glands and cheilitis actinica, syndrome hypertrichcsis, seborrhea and calvities frontalis with hypoestrogenism in young women. Home: Warszawa, Podkowa Lesna, Wróbia 20, Poland. Office: Dermatol. Clinic, Dymitrow 11, Lublin, Poland.*

MICHAS, Panayiotis, Greek surgeon; b. Athens, Greece, Sept. 18, 1917; s. Athanse and Evel (Marketos) M.; ed. U. Athens; M.D.; m. Iris Psorulla, Sept. 30, 1943; children—Athanase, Alexandre. Prof. agrégé U. Athens; surgeon Athens Surg. Clinic. Mem. Internat. Surgery Assn., Internat. Coll. Surgeons, Internat. Assn. Cardiovascular Surgery, Hellenic Surgery Soc., Athens Med. Assn. Author: Sur le traitement des furoncles de la face; L'occlusion de l'artère pulmonaire comme un test fonctionnel de l'aspiration et circulation sur des maladies pulmonaires; Fibromes intrathoracaux; Sur le traitement des diverticules de l'oesophage; Expérience de 10 années sur le traitement chirurgical des rétrécissements mitraux. Home: 4, Skoufa, Athens. Office: Clinique chirurgicale, Athens, Greece.

MICHEEL, Fritz Karl Hermann, chemist; b. Strasbourg, France, July 3, 1900; s. Hermann and Marie (Stegemann) M.; Ph.D., U. Berlin; m. Hertha Falkenthal, Dec. 20, 1926; 2 children. Agrégé, U. Göttingen, 1931; became prof. at large, 1936, asso. prof., 1937, prof., 1946; joined Inst. Organic Chemistry, U. Munster, 1951. Author: Chemie der Zucker und Polysaccharide, 1956; also numerous articles. Research in chemistry of carbohydrates and proteins, various syntheses. Home: Auf dem Draun 17, Münster, West Germany.

MICHEJDA, Jan Victor, Polish biologist; b. Poznan, Poland, Feb. 28, 1927; s. Jan Stanislaw and Sophie (Jamrozik) M.; Ph.D., Poznan U., 1951, docent, 1962; m. Jadwiga Kurnatowska, 1953 (div. Apr. 1962); 1 son, Marek. Instr., adj. Lab. Pharmacodynamics Poznan Med. Acad., 1947-57; instr., adj. dept. animal physiology Poznan U., 1952-62, head dept., 1962—; research fellow, vis. prof. Carlsberg Lab., Copenhagen, Denmark, 1958. Rockefeller Found. Fellow Biol. Labs. Harvard, 1959-60; grad. dept. biochemistry Brandeis U., 1964-65; research asso. Polish-Am. Collaboration, Communicable Disease Center (Atlanta). Mem. Am. Soc. Zoologists, Polish Soc. Genetics, Polish Soc. Biochemists, Internat. Cell Research Orgn. Publs. on studies of hydro-ecology of springs and streams, biochem. basis of taxonomy of mollusks, bioenergetics of invertebrates, comparative biochemistry of respiratory enzymes, isozymes of dehydrogenases, trichinosis. Address: 10 Fredry, Poznan, Poland.*

MICHEL, Andrée Vieille, French sociologist; b. Golfe-Juan, France, Sept. 22, 1920; d. Charles Marie and Laure (Cardon) Vieille; Licence de droit, Faculty Law, Aix en Provence, France, 1941; Licence de philosophy, Faculty Philosophy, Grenoble, France, 1944; Ph.D. with honors, Faculty Human Scis., 1959; m. Andre Michel, Aug. 15, 1954. Head researches Nat. Center Sci. Research, Paris, 1951—, dir. group family sociology, 1963-66; mem. sociol. missions in Algeria, 1954, 63, 64; vis. prof., U. Algiers, 1963-64; vis. prof. Western Res. U., 1966-67; NSF research fellow Family Study Center, U. Minn., Mpls., 1967. Cons. expert Ministry Economy, Ministry Nat. Edn., 1958—, Algerian Govt., 1963—, French Family Planning movement, 1960-66, Internat. House Europe, 1965, Internat. Secretariat Vol. Service, 1966. Recipient several awards for participation in internat. congresses or seminars. Mem. French Sociol. Soc., Internat. Sociol. Assn. French Speaking Sociologists, Am. Sociol. Assn., Internat. Com. Family Research. Author: Les travailleurs algériens en France, 1957; Famille, Industrialisation, Logement, 1959; La condition dela Francaise d'Aujourd'hui (with G. Texier), 1964; also articles. Research on conditions of Algerian immigrants in France and their orgns., urban French families and status of women. Office: 15, Quai Anatole France, Paris 73, France.*

MICHEL, Raymond Paul Joseph, French endocrinologist; b. Marseille, France, Mar. 14, 1914; s. Jean Louis Eugene and Marie (Carvi) M.; Licence èn Science, Marseille, 1936, Doctorat Elat Pharmacié, 1944, Doctorat èn Sci., 1948; m. Odette Marie Pierrette Lila, July 24, 1943; 1 dau., Fancoise Andrée Jeanne. Asst., Faculty Médecine Pharmacy, Marseille, 1945-47, Coll. de France, Paris, 1947-53; asso. prof. Ecole de Médecine et Pharmacie, Dijon, France, 1953-56; prof. endocrinology Faculté de Pharmacie, Paris, France; dir. dept. cellular biochemistry Inst. de Biologie, Dijon. Recipient Palmes academiques. Mem. Société de Biologié, de Chimie biologique d'Endocrinologie, de Therapeutique, Am. Chem. Soc., Biochem. Soc. U.K., Soc. Endocrinology U.K. and U. S. A. Research, publs. on endocrinology, biochemistry and physiology of thyroid hormones. Home: 3 rue Charles Dickens, Paris 16e. Office: Endocrinologie Dept., Faculté de Pharmacie, 4 Avenue de l'observatoire, Paris 6e, France.*

MICHEL-LÉVY, Albert-Victor, French petrographer; b. Autun, France, July 3, 1877; s. Auguste Michel-Lévy; prof. petrography, Paris. Mem. French Acad. Scis. Author: Les Minéraux des roches, 1888. Research on composition of Vosges mountain range; produced artificial metamorphism in rocks using strong pressure under high temperature; micropyrotech. research (with Muraou) on gun powder. Died Paris, 1955.

MICHEL-LÉVY, Augustus, French geologist, mining engr.; b. Paris, Aug. 17, 1847; ed. U. Paris; became dir. French Geol. Survey, also nat. insp. mines, 1874; prof. Collège de France. Author: Micrographical Mineralogy, 1879; Structure and Classification of Eruptive Rocks, 1880; Synthesis of Minerals and Rocks, 1882. Mem. French Acad. Scis. First to introduce use of polarizing microscope to examine mineral structure; pioneered artificial prodn. of crystallized minerals; studied meteorites in France; introduced term, granulite. Died Paris, Sept. 24, 1911.

MICHELI DU CREST, Jacques Barthélemi, engr.; b. Geneva, Switzerland, 1690; fortifications engr. French army; correspondent of Réaumur, Maupertuis, other scholars; pub. 1st alpine panorama based on trigonometric measurements; studied temperature of interior of earth; improved Réaumur's ethyl alcohol thermometer. Died 1766.

MICHELIN, André, French inventor; b. Paris, 1853; engr. Arts et Manufactures; owner rubber factory, Clermont-Ferrand, France; improved and developed prodn. of tires; inventor removable air tire with balloon outer cover for bicycles, 1890, for automobiles, 1895; inventor double tire for heavy loads, 1908. Died 1931.

MICHELL, John, Brit. geologist; b. Nottinghamshire, Eng., 1724; M.A., Queen's Coll., Cambridge (Eng.) U., 1752, B.D., 1761; 1 dau. Fellow Queen's Coll., 1749-64, prof. geology, from 1762; rector of Thornhill, from 1767; lectr. Hebrew, Greek, arithmetic, geometry. Fellow Royal Soc., 1760. Author: Treatise on Artificial Magnets, 1750; Conjectures Concerning the Cause, and Observations Upon the Phenomena of Earthquakes, 1760; Recommendation of Hadley's Quadrant for Surveying, 1765; Proposal of a Method for Measuring Degrees of Longitude Upon Parallels of the Equator, 1767; Enquiry into the Probable Parallax and Magnitude of the Fixed Stars From the Quantity of Light Which They Afford Us, 1767. Invented torsion balance; studies in seismology, astronomy. Died April 21, 1793.

MICHELL, John Henry, mathematician; M.A. Asst. prof. math. U. Melbourne (Australia). Fellow Trinity Coll., Cambridge (Eng.) U. Fellow Royal Soc., 1902. Author: Theory of Free Stream Lines, 1890; The Highest Waves in Water Ship, 1898; The Small Deformation of Curves and Surfaces; also numerous papers on determination and distbns. of stress, stability in elastic solids. Research in math. physics. Died Feb. 3, 1940.

MICHELS, Antonius Mathias Johannes Friedrich, physicist; b. Amsterdam, Holland, Dec. 31, 1889; s. Jasper Stephanus and Jacoba (Vierling) M.; candidaat Physics and Math., candidaat Chemistry, Faculty of Scis., U. Amsterdam, 1913, doctoraal Physics and Math., 1919, doctoraal Chemistry, 1922, Dr. Physics and Math., 1924; m. C. A. M. Veraart, Nov. 27, 1930; children—Jacoba (Mrs. A. Nooy), Teun, Jans, Jan, Dorothy, Bob, Thys, Paul, Marlene. High sch. tchr., 1913-14; asst. exptl. physics part-time U. Amsterdam, 1918-28; high sch. tchr. part-time, 1918-28; mem. faculty U. Amsterdam, 1928—, prof. exptl. and applied physics, 1939-60, extraordinary prof. on behalf of van der Waals Fonds., 1960—; engaged in consulting. Hon. dir. Metrology Inst., 1947—. Served with Army, 1914-18, 39-40, 45-47. Decorated ridder Orde van de Nederlandsche Leeuw; recipient Gordon award for research in sci. Fellow Inst. Physics, A.A.A.S.; mem. Royal Acad. The Netherlands, Ned. Nat. Vereniging, Ned. Scheik. Vereniging, Kon. Inst. Ingenieurs, Phys. Soc., Farady Soc., Am. Phys. Soc., Am. Chem. Soc. Author numerous articles in field. Exptl. and theoretical research on effect of pressure on phys. properties and chem. reactions; devel. high pressure prodn. methods relating to ethylene polymerisation. Home: 92 Middenweg, Amsterdam-Oost. Office: 67 Valckenierstraat, Amsterdam, Holland.*

MICHELS, Nicholas Aloysius, Am. anatomist; b. St. Paul, Oct. 1, 1891; s. John Pierre and Catherine (Kraemer) M.; B.A., St. Thomas Coll., 1914; S.T.B., Cath. U. Am., 1918; M.A., U. Minn., 1920; Dr.Sc., Louvain (Belgium) U., 1922; postgrad. Sorbonne U., Paris, Siena (Italy) U., U. Chgo., 1925; m. Martha A. Tweeddale, June 18, 1929 (Dec. 1939); children—Adelle Virginia (Mrs. Sidney Parsons, Jr.), Horace Harvey; m. 2d, Hilda Datty, June 24, 1942. Prof., St. Thomas Coll., St. Paul, 1918-21; researcher N.Y. U., Bellevue Hosp., also Mt. Sinai Hosp., 1926-28; asst. prof. biology and histology St. Louis Med. Sch., 1926-27; asso. prof. anatomy Creighton U. Med. Sch., Omaha, 1927-29; faculty Jefferson Med. Coll., Phila., 1929—, prof. anatomy, 1948-62, emeritus prof., 1962—. Recipient Lindback Found. award for distinguished teaching, 1962. Mem. Am. Assn. Anatomists, Soc. Exptl. Biology and Medicine, N.Y. Acad. Sci., Internat., Am. socs. hematologists, Am. Coll. Cardiologists, Internat. Coll. Surgeons, A.A.A.S. Author: Blood Supply and Anatomy of Upper Abdominal Organs, 1955; contbg. author Handbook of Hematology, 1938; other books; also numerous articles. Research on mast cells from fish to man proving them living not degenerating cells, blood supply of kidney and suprarenal glands, segmentation of liver, variant blood supply small and large intestine, pathways collateral circulation liver and gastrointestinal tract. Home: R.D. 1, Hatfield, Pa. 19440. Office: 307 S. 11th St., Phila. 19107.*

MICHELS, Robert, economist, sociologist; b. Cologne, Germany, Jan. 9, 1876; ed. univs. of Paris, Munich, Leipzig, Halle. Dozent, U. Brussels, 1905-07; prof. econs., Turin, Italy, 1907-13; prof. econs. and statistics, Basel, Switzerland, 1913-28; prof. econ. sociology, Perugia, Author: Brautstandsmoral, 1903; Patriotismus und Ethik, 1906; Il proletario e la borghesia nel movimento socialista italiano, 1908; Zur Soziologie des Parteiwesens, 1910; La sociologia del partito politico nella democrazia moderna, 1912; I limiti della morale sussuale, 1912; Saggi economico-statistici sulle classi popolari, 1914; Problemi de sociologia applicata, 1919; Corso de sociologia politica, 1927; Der Patriotismus, 1929; Italien von heute, 1930; Il Boicottaggio, 1934; Political Parties. Formulated iron law of oligarchy; theorized in Democratic socialism. Died Rome, Italy, May 2, 1936.

MICHELS, Walter Christian, Am. physicist, educator; b. Utica, N.Y., June 14, 1906; s. Christian A. and Anna (Haigis) M.; E.E., Rensselaer Poly. Inst., 1927; Ph.D., Cal. Inst. Tech., 1930; m. Lorraine Elder, June 21, 1930 (dec. Mar. 1940); 1 dau., Leslyn Jane; m. 2d, Agnes K. Lake, June 4, 1941. Test engr. Utica Gas & Electric Co., 1926; teaching asst. Cal. Inst. Tech., 1927-29, teaching fellow, 1929-30; NRC fellow Princeton, 1930-32; asso. in physics Bryn Mawr (Pa.) Coll., 1932-34, asso. prof., 1934-46, Marion Reilly prof., 1946—, head dept. physics, 1936—. Chmn., Commn. on Coll. Physics, 1960—. Recipient Glover award Dickinson Coll., 1964. Mem. Am. Phys. Soc., Am. Assn. Physics Tchrs. (Oersted medal 1963, pres. 1956-57), Am. Assn. U. Profs., A.A.A.S., Franklin Inst., Am. Inst. Physics (governing bd. 1957-60), Fedn. Am. Scientists, Optical Soc. Am., Sigma Xi. Author: Advanced Electrical Measurements, 1941; (with A. L. Patterson) Elements of Modern Physics, 1951; Electrical Measurements and Their Applications, 1957; also articles. Sr. editor: Internat. Dictionary Physics and Electronics, 1956, editor-in-chief 2d edit., 1961; editor Am. Jour. Physics, 1959-66. Research on excitation of atoms by electrons, heat transfer between solids and gases, response of humans to phys. stimuli; inventor elec. measuring devices that discriminate against random disturbances (noise), elec. circuits that suppress unwanted variations; devel. new elementary approaches to presentation of phys. theory including relativity, wave behavior, planetary orbits. Home: 532 Red Fox Lane, Strafford, Wayne, Pa. 19087. Office: Bryn Mawr Coll., Bryn Mawr, Pa. 19010.*

MICHELSON, Adolf Michael, Brit. biochemist; b. North Shields, Eng., July 18, 1926; s. Adolf Michael and Jane Ellen (Johnson) M.; B.Sc. with 1st Class Honors, Kings Coll., Durham (Eng.) U., 1946; Ph.D., Pembroke Coll., Cambridge, Eng., 1949; D.Sc., King's Coll., 1960; postgrad. Cal. Inst. Tech.; m. Guillermina Gomez Hidalgo, Aug. 16, 1955; children—Paul Michael, Christina Maria, Helen Teresa, Catherine Françoise. Staff, U. Chem. Labs., Cambridge, Eng., 1946-56; with Arthur Guinness Son & Co. (Ltd.), Dublin, Ireland, 1956-62; staff Institute de Biologique Physico-Chimique, 1962—; dir. research CNRS, Paris, France, 1963—, chief Service de Biochemie Physique, 1966—. Author: Chemistry of Nucleotides, 1963; also numerous articles. Research on chemistry, biochemistry and biophysics of nucleotides, oligonucleotides and polynucleotides; devel. methods for synthesis of nucleosides, nucleotide anhydrides and coenzymes and polynucleotides; synthesis of nucleotide and polynucleotide analogues; phys. properties of modified polynucleotides and biol. effects. Home: 100 Av. Jean Jaurès, Chatenay-Malabry 92, France. Office: 13 rue Pierre Curie, Paris Ve, France.*

MICHELSON, Albert Abraham, physicist; b. Strelno, Germany, Dec. 19, 1852; s. Samuel and Rosalie (Przlubska) M.; grad. U. S. Naval Acad., 1873; postgrad. student at univs. of Berlin, 1880, Heidelberg, 1881, Collège de France and École Polytechnique, 1882; hon. Ph.D., Western Reserve, 1886, Stevens Inst. Tech., 1887; Sc.D., U. of Cambridge, 1899; LL.D., Yale, 1901, Franklin Bicentenary U. of Pa., 1906; Ph.D., Leipzig, 1909, Göttingen, 1911; LL.D., McGill U., 1921; m. 1st, Margaret McLean Hemingway, 1877 (div.); children: Albert Hemingway, Truman, Elsa; m. 2d, Edna Stanton, Dec. 23, 1899; children: Madeleine, Dorothy, Beatrice. Instr. U. S. Naval Acad., 1875-79; prof. of physics, Case Sch. Applied Science, Cleveland, 1883-89, Clark U., 1889-92; prof. and head of dept. of physics, U. of Chicago, 1892-1929, distinguished service prof., same, 1925-29; Lowell lecturer, 1899; exchange prof., U. of Göttingen, summer 1911, Université de Paris, 1920. Grand Prix, Paris Expn., 1900; Mattencci Medal, Soc. Italiana, Rome, 1904; awarded Nobel Prize in Physics, 1907; Elliott Cresson medal, 1912. Fellow Am. Acad. Arts and Sciences, A.A.A.S. (pres. 1910-11); Royal Society of London, 1902 (Rumford Medal, 1889, Copley Medal, 1907); mem. Am. Phys. Soc.; Nat. Acad. Scis. (Draper medal, 1916, pres. 1923-27); Am. Philos. Soc.; French Acad. Scis., 1900, others. Author: (brochure) Velocity of Light, 1902; Light Waves and Their Uses, 1903. Studies in Optics, 1927. Contbd. to found. for relativity; Michelson-Morley expt. helped destroy ether concept; designed echelon grating spectroscope, 1907; built an engine to rule diffraction gratings; contbd. important work on spectral lines; determined speed of light with extremely high degree of accuracy; measured a meter in terms of wave length of cadmium light; inventor interferometer for measuring distances by means of length of light waves; measured angular diameters of satellites of Jupiter, 1891; demonstrated that core of earth is molten, not rigid, 1916; made 1st measurement of a star's diameter (Alpha Orionis), 1920; showed that earth's viscosity is similar to that of steel. Died Pasadena, Cal. May 9, 1931.

MICHELSON, Truman, Am. ethnologist; b. New Rochelle, N.Y., Aug. 11, 1879; s. Albert Abraham and Margaret McLean (Heminway) M.; A.B., Harvard, 1902, A.M., 1903, Ph.D., 1904; postgrad. univs. of Leipzig and Bonn, 1904-05; studied with Prof. Boas of Columbia, 1909, 10; m. Katherine Harrison, July 18, 1903. Parker fellow Harvard, 1904-05; instr. Latin, U. Mo., 1905-06; pvt. research, 1906-09; clk. U. S. Immigration Commn., 1909; ethnologist Bur. Am. Ethnology, Washington, 1910—; prof. ethnology George Washington U., 1917-32; exec. officer dept. anthropology, 1927-32; tchr. anthropology Columbia, summer 1924. Expdns. to Algonquin tribes every season, 1910-32, to Algonquian Indians and Eskimos of James' and Hudson's Bays, 1935, 36, to Algonquin Indians of Northern shore of St. Lawrence River, 1937. Author: Kickapoo Tales (with William Jones), 1915; The Autobiography of a Fox Indian Woman, 1925; Notes on Fox Mortuary Customs and Beliefs, 1925; Contributions to Fox Ethnology, 1927; Buffalo Head Dance of Thunder Gens of Fox Indians, 1928; Thunder Dance of Bear Gens of Fox Indians, 1929; Notes on the Fox Wapanowiweni, 1932; Fox Miscellany, 1937. Made 1st sci. classification of Algonquin tribes on a linguistic basis. Deceased.

MICHENER, Charles Duncan, Am. entomologist, educator; b. Pasadena, Cal., Sept. 22, 1918; s. Harold and Josephine (Rigden) M.; B.S., U. Cal. at Berkeley, 1939, Ph.D., 1942; m. Mary Frances Hastings, Jan. 1, 1941; children—David H., Daniel R., Barbara J., Walter H. Tech. asst. entomology U. Cal., 1939-42; asst. curator Am. Mus. Natural History, N.Y.C., 1942-46, asso. curator, 1946-48; asso. prof. entomology U. Kan., Lawrence, 1948-50, prof., chmn. dept., 1950-61, Watkins Distinguished prof. entomology, 1961—. Entomologist, State of Kan., 1950-61; mem. faculty, bd. dirs. Orgn. for Tropical Studies, Costa Rica. Guggenheim fellow U. Parana (Brazil), 1955-56; Fulbright research grantee U. Queensland, 1957-58, Africa, 1966-67. Fellow Am. Acad. Arts and Scis., Am. Entomol. Soc. (hon.); mem. Entomol. Soc. Am., Kan. Entomol. Soc. (pres. 1950), Ecol. Soc. Am., Am. Soc. Naturalists, Linnaen Soc. London (hon. fgn.), Kan. Acad. Scis., Internat. Union for Study Social Insects, Soc. Systematic Zoologists (pres. elect 1968), National Acad. Sciences, Am. Inst. Biol. Scis., Phi Beta Kappa (pres. Kan. chpt. 1963-64), Sigma Xi, Alpha Zeta, Phi Sigma. Author: American Social Insects, 1951; Nest Architecture of Sweat Bees, 1962. Editor: Evolution, 1962-64; Am. editor Insectes Sociaux, 1955—. Research on social behavior of bees in U. S., Panama, Costa Rica, Brazil, Mexico, Australia, New Guinea, Fiji, also on bee systematics, chigger mites, saturniid moths. Home: Route 4, Lawrence, Kan. 66044.*

MICHIE, Donald, biologist, computer scientist; b. Rangoon, Burma, Nov. 11, 1923; s. James Kilgour and Marjorie Crain (Pfeiffer) M.; M.A. in Human Anatomy and Physiology (Open Classical scholar), Oxford U., 1949, D.Phil., 1952; children—Christopher, Susan, Jonathan, Caroline. Staff dept. zoology U. Coll., London, 1952-55, Royal Vet. Coll., London, 1955-58; sr. lectr. dept. surg. sci. U. Edinburgh (Scotland), 1958-62, dir. exptl. programming unit, 1965—, reader, 1962—; vis. asso. prof. elec. engring. Stanford, 1962. Spl. cons. USAF project for automatic lang. analysis Indiana U., 1962. Balliol Coll. War Meml. student, 1949. Mem. Brit. Computer Soc., Biometric Soc., Assn. Sci. Workers. Author: Computing Science, 1964; Science Research Council, 1965; (with A. Ortony, R. M. Burstall) Computer Programming for Schools—First Steps, 1967; also numerous articles. First (with Anne McLaren) to demonstrate hormonal induction of pregnancy and super-pregnancy; 1st case of determination of a morphological character via uterine environment; discovered and investigated (with J. G. Howard) immunological responsiveness of newborn to transplanted tissues; designer (with J. E. Doran) gen. problem-solving automaton; designer boxes (gen. trial-and-error learning automaton. Home: 15 Hope Park Terrace, Edinburgh, Gt. Britain.*

MICHON, Georges Charles, French biologist; b. Bourg, France, Apr. 19, 1926; s. Maurice and Aimee (Froment) M.; Bachelier Mathematiques, U. Paris (France), 1944, Dr. Veterinaire, 1952, Licencié ès Scis., 1953; Diplome, Ecole Nationale Veterinaire, 1950; m. G. Dassieu, July 20, 1953; children—Hubert, Elisabeth. Asst., Ecole Nationale Veterinaire Alfort, 1951-53; ingenieur Commissariat Energie Atomique, 1953-61, chief sect., 1962-65; head radiobiology Ecole Vétérinaire Alfort, 1955—. Permanent sec. Commn. Interministerielle des Radioéléments Artificiel, 1966—; expert Euratom, 1960—. Internat. Com. Radiol. Protection, 1963—. Named Laureat, Faculte de Médecine Paris, 1953, Laureat Academie Veterinaire de France, 1960, 64. Mem. Société des Experts Chimiste, Société Pathologie Comparée, Academie Veterinaire de France. Research and publs. on toxicology of radionuclides, inhibition of intestinal absorption of strontium, new techniques for detection and evaluation of radioactive pollution of foods and biol. materials, devel. new concepts in radioecology. Office: BPN 8, Gif sur Yvette, gl, France.*

MICHURIN, Ivan Vladimirovich, Russian horticulturist, geneticist; b. Dolgoje, Oct. 29 28, 1855; founder a nursery, Koslov (later named Michurinsk), taken over by state, after 1917, placed under his direction, later enlarged to include research lab. and tech. coll.; engaged in grafting expts. to produce improved fruit trees; formulated theory that hereditary changes can be induced by grafting; attempted to prove that acquired characters can be inherited; developed plants which thrive in arid regions and in intense cold. Died Michurinsk, Russia, June 7, 1935.

MICKELSEN, Olaf, Am. biochemist, educator; b. Perth Amboy, N.J., July 29, 1912; s. Fredrick and Marie (Nielsen) M.; B.S., Rutgers U., 1935; M.S., U. Wis., 1937, Ph.D., 1939; m. Edith L. Nielsen, Dec. 1, 1939 (dec. Feb. 1953); children—Elizabeth K. (Mrs. T. W. Kurczynski), Margaret L.; m. 2d, Clarice H. Lewerenz, Sept. 1, 1953. Chemist, U. Minn. Hosps., 1939-41, asso. scientist to asso. prof. Lab. Physiol. Hygiene, 1942-48; with Nutrition br. USPHS, 1948-51; chief lab. nutrition and endocrinology Nat. Inst. Arthritis and Metabolic Disease, NIH, Bethesda, Md., 1951-62; prof. dept. foods and nutrition Mich. State U., East Lansing, 1962—. Mem. Am. Inst. Nutrition (sec. 1963-66), Am. Chem. Soc., Brit. Nutrition Soc., Soc. for Exptl. Biology and Medicine, Am. Soc. Biol. Chemists, Phi Beta Kappa, Sigma Xi. Author: (with others) The Biology of Human Starvation, 2 vols., 1950; Nutrition Science and You, 1964. Editorial bd. Jour. Agrl. and Food Chemistry, 1955-58, Jour. Nutrition, 1959-62, Nutrition Rev., 1955—. Research in importance of calories in starvation rehab.; developed concept of chem. imprints in foods; showed direct rise in plasma cholesterol level with age. Home: 4412 Tacoma St., Okemos, Mich. 48864. Office: Home Econs. Bldg., Mich. State U., East Lansing, Mich. 48823.*

MICKEY, George Henry, Am. cytogeneticist; b. Claude, Tex., Jan. 26, 1910; s. Luke Ross and Clara (Pennington) M.; A.B., Baylor U., 1931; M.S., U. Okla., 1934; Ph.D. in Genetics, U. Tex., 1938; m. Alwilda Davis, Aug. 20, 1932; children—Wilda Rhea (Mrs. L. Arnold Wyse), Don Davis. Instr., La. State U., 1938-42, prof., chmn. zoology, 1956-59, dean Grad. Sch., 1959-60; faculty Cal. Inst. Tech., 1942-48, asso. prof., 1944-48, research fellow biology, 1948; asso. prof. Northwestern U., 1949-56; cytogeneticist New Eng. Inst. for Med. Research, Ridgefield, Conn., 1960—; prin. biologist Oak Ridge Nat. Lab., 1953. Mem. selection panel NSF, 1959. Guggenheim fellow, 1948. Fellow A.A.A.S.; mem. Am. Inst. Biol. Scis., Am. Soc. Naturalists, Genetic Soc. Am., Soc. for Study Evolution, Am. Soc. Zoologists, Am. Soc. Human Genetics, Am. Genetic Assn., Okla., Ill., N.Y., La. (past pres.) acads. scis., Beta Beta Beta (nat. pres. 1957-60), Phi Sigma. Author: Manual for Studies in General Zoology, part I and II, 1947, part I, rev., 1949, part II, rev., 1950; (with others) The Strawberry—History, Breeding and Physiology, 1966; also numerous articles. Described maturation male germ cells in insects and amphibia; discovered abnormal chromosome numbers in various tissues of insects; determined structure and div. cycle grasshopper chromosomes; research on parthenogenesis and sex determination in grasshoppers, abnormalities in reproductive system, mosaics and intersexuality in fruit flies, aspects of radiation genetics in Drosophila, radiofrequency effects in biol. systems in tissue culture and intact organisms, human disease syndromes related to chromosome aberrations. Home: 25 Sound View Rd. Office: Box 308, Ridgefield, Conn. 06877.*

MICKEY, Wendell Vaden, Am. geophysicist; b. Mickey, Tex., Sept. 4, 1920; s. Rhea Lewis and Vida (May) M.; student Wayland Bapt. Coll., 1938, Tex. U., 1947, Tulane U., 1959; B.A. in Math., Tex. Technol. Coll., 1959; postgrad. George Washington U., Mass. Inst. Tech.; m. Veda Jeanette Hutchison, May 13, 1953; children—Lynette, Wendell Vaden, Patrick, Randell. Geophysicist, Western Geophys. Co., 1947-59; chief spl. projects br. U. S. Coast and Geodetic Survey, Rockville, Md., 1959-65, chief vibrations and engring. projects br., 1965—. Cons. to govt. agys.; mem. seismic working group J. F. Kennedy Space Center, NASA, 1962—. Recipient U. S. Dept. Commerce Meritorious Service award, 1964. Mem. Washington Acad. Scis., A.A.A.S., European Assn. Exploration Geophysicists, Am. Geophys. Union, Soc. Exploration Geophysicists, Seismol. Soc. Am., Philos. Soc. Washington, Soc. Exptl. Stress Analysis. Research, numerous publs. on seismic effects peaceful application large nuclear explosives, acoustic energy transfer to seismic during launching advanced rocket systems, TNT explosive equivalence in detonations liquid and solid rocket propellants, structural response to vibratory motion, microearthquakes relative to earthquake occurrence prediction, seismic energy propagation characteristics destructive earth motion, regional accumulative strain phenomena from earthquakes, microseismic activity in vicinity space vehicle calibration facilities; devel. prediction techniques in interest pub. safety in excavation with explosives. Home: 7313 Mill Run Dr., Derwood, Md. 20855. Office: Environmental Sci. Services Adminstrn., Coast and Geodetic Survey, Washington Sci. Center, Rockville, Md. 20852.*

MICKLEY, Harold Somers, Am. chem. engr.; b. Seneca Falls, N.Y., Oct. 14, 1918; s. Harold Franklin and Marguerite Gladys (Somers) M.; m. Margaret Winifred Phillips, Dec. 20, 1946; children—Steven Phillips, Richard Somers. Chem. engr. Union Oil Co. Cal., 1941-42; with Mass. Inst. Tech., 1943—, prof., 1957—, Ford prof. engring, 1962—, chmn. faculty, 1962-64, dir. Center for Advanced Engring. Study, 1963—; cons. chem. engr., 1942—. Mem. com., project USN, 1946-55; mem. Lexington Project, AEC, 1948, Atlantis Project, Dept. Def., 1958. Recipient Naval Ordnance Devel. award, 1946. Mem. Am. Acad. Arts and Scis., Am. Chem. Soc., Am. Inst. Chem. Engrs. (mem. spl. lecture com. 1958—, edn. accreditation com. 1963—), Sigma Xi, Tau Beta Pi. Author: (with T. K. Sherwood, C. E. Reed) Applied Mathematics in Chemical Engineering, 1957. Editor: Recent Advances in Heat and Mass Transfer, 19—. Research, publs. on fluid mechanics, heat and mass transfer, chem. kinetics. Home: 48 Elizabeth Rd., Belmont, Mass. 02178. Office: Mass. Inst. Tech., 77 Massachusetts Av., Cambridge, Mass. 02139.*

MICKS, Don Wilfred, Am. biologist; b. Mt. Vernon, N.Y., Nov. 23, 1918; s. Wilfred Wallace and Bernice (Barbour) M.; B.S., N.Tex. State U., 1940, M.S., 1942; Sc.D., Johns Hopkins, 1949; m. Martha Millican, Feb. 15, 1944; children—Donald Frederick, Stephen Alan, Marjorie Ellen, Carol Jeanne. Faculty, Med. br. U. Tex., Galveston, 1949—, prof., chmn. preventive medicine, community health, 1966—. Fulbright sr. research scholar U. Pavia, 1953-54; scientist-biologist div. Environmental Health WHO, Geneva, Switzerland, 1958-59. Fellow Am. Pub. Health Assn., Royal Soc. Tropical Medicine and Hygiene, A.A.A.S.; mem. Assn. Tchrs. Preventive Medicine, Soc. Exptl. Biology and Medicine, Am. Soc. Tropical Medicine and Hygiene, A.A.A.S., Internat. Soc. Toxinology, Entomol. Soc. Am., Am. Mosquito Control Assn., Sigma Xi. Research, numerous publs. on factors influencing malaria transmission by mosquitoes, new agts. for control of mosquito vectors of human disease, biochemical methods for identifying and differentiating morphologically identical mosquito vectors of disease; pioneer method of producing wound-healing agt. derived from blood. Home: 1302 Ball St., Galveston, Tex. 77550.*

MIDDELDORPF, Albrecht Theodor, see von Middeldorpf, Albrecht Theodor.

MIDDLEBROOK, Gardner, Am. physician; b. Lowell, Mass., Dec. 6, 1915; s. Walter Clark and Ethel (Gardner) M.; grad. Phillips Acad., Andover, 1934; A.B., Harvard, 1938, M.D., 1944; m. Zulema Reggiardo, Oct. 20, 1964; children—Timothy, Peter, Anne, Alice. Asso., Rockefeller Inst., N.Y.C., 1945-52; dir. research and labs. Nat. Jewish Hosp., Denver, 1952-64; prof. microbiology U. Colo. Sch. Medicine, 1952-64; prof. internat. medicine U. Md. Sch. Medicine, 1965—. Cons. NIH, USPHS. Recipient Pasteur medal, 1954. Mem. Am. Soc. Clin. Investigation, Am. Bd. Pathology, Western Assn. Physicians, Harvey Soc., numerous others. Author: Chemotherapy of Tuberculosis, 1961. Publs. on cytochem. and biochem. analysis of tubercle bacilli; new serologic procedures and genetic analysis of drug resistance in Tb; exptl. areogenic infection and immunization. Home: 4000 N. Charles St., Balt. 21218. Office: 660 W. Redwood St., Balt. 21201.*

MIDDLEHURST, Barbara Mary, astronomer, astrophysicist; b. Penarth, Wales, U.K., Sept. 10, 1915; d. George Frederick and Gladys (Sadler) Middlehurst; B.A., Girton Coll., Cambridge, Eng., 1936, M.A., 1947. Observer, St. Andrews U. Obs., Scotland, 1951-54, lectr., 1954-59; vis. astronomer Hamburg (Germany) Obs., 1951; research asso. Ind. U., 1953-54, Yerkes Obs., Williams Bay, Wis., 1959-60, U. Ariz., Tucson, 1960-68; asso. editor Ency. Britannica, 1968—. Fellow Royal Astron. Soc. (London); mem. Am. Astron. Soc., Am. Geophys. Union, Internat. Astron. Union (sec. commn. 17 1967). Editor: (with G. P. Kuiper) series Stars and Stellar Systems, 1959; The Solar System, vol. 3, 1961, vol. 4, 1963. Research, publs. on photog. and photoelectric stellar photometry, stellar spectroscopy, lunar and planetary studies. Address: Ency. Britannica, 425 Michigan Av., Chgo. 60611.

MIDDLETON, Austin Ralph, Am. zoologist; b. Balt., Apr. 14, 1881; s. Christopher Byrne and George Anna (Belt) M.; grad. Balt. City Coll., 1901, Balt. Tchrs.' Tng. Sch., 1902; A.B., Johns Hopkins, 1910, Ph.D., 1915; m. Margaret Mary Lougbridge, July 3, 1917. Johns Hopkins scholar, 1912-14, fellow, 1914-15, fellow by courtesy, 1915-16; prof. zoölogy U. Louisville, since 1916, organized biol. labs., dir. depts. biology, 1916-28 prof. emeritus of biology since 1952. Member Johns Hopkins Sci. Expdn., Jamaica, B.W.I., 1910. Fellow A.A.A.S., Conf. on State Acads. of Sci., Am. Geog. Soc., Royal Soc. Arts; mem. Am. Soc. Zoölogists, Ecol. Soc. Am., Am. Soc. Mammalogists, Am. Soc. Parasitologists, Am. Acad. Polit. and Social Sci., Ky. Acad. Sci., Eugenics Soc. Am., Eugenics Research Assn. Author College Biology, 1925-29; also numerous research articles. Editor for Ky. of The Naturalist's Guide to the Americas, biographies in Am. Men of Science, Index Biologorum and Menchen und Menchenwerke. Organized tropical biol. expdn. to jungles of Honduras, C. Am., 1933, Author of plan for Univ. of Am., 1938. Died Apr. 11, 1956.

MIDDLETON, David, Am. physicist, applied mathematician; b. N.Y.C., Apr. 19, 1920; s. Charles Davies Scudder and Lucille (Davidson) M.; grad. Deerfield Acad., 1938; A.B. summa cum laude, Harvard, 1942; A.M., 1945, Ph.D. (NSF predoctoral fellow), 1947; m. Nadea Butler, May 26, 1945; children—Susan Terry, Leslie Butler, David Scudder Blakeslee, George Davidson Powell. Teaching fellow electronics Harvard, 1942, spl. research asso. Radio Research Lab., 1942-45, research fellow electronics, 1947-49, asst. prof. applied physics, 1949-54; cons. physicist, Concord, Mass., 1954—. Adj. prof. elec. engring. Columbia U., N.Y.C., 1960-61; adj. prof. applied physics, communication theory Rensselaer Poly. Inst. Conn., East Windsor Hill, 1961—; cons. Raytheon, Rand Corp., Office Naval Research, Inst. Def. Analyses, Bendix Corp., Litton Corp., Avco, Lincoln Lab., NASA-Electronic Research Center, Cambridge, Mass., Johns Hopkins, others. Nat. Electronics Conf. award (with W. H. Huggins), 1956. Fellow Am. Phys. Soc., I.E.E.E., A.A.A.S.; mem. Am. Statis. Soc., Am. Math. Soc, Am. Assn. U. Profs., Am. Acoustical Soc., optical Soc., N.Y. Acad. Scis., Phi Beta Kappa, Sigma Xi. Author: Introduction to Statistical Communication Theory, 1960, 2 vol. Russian edit., 1962; Topics in Communication Theory, 1965. Pioneer devel. statis. communication theory with applications to space, optics, seismology, system optimization, detection; 1st systematic application (with D. Van Meter) decision-theoretic methods in radar, radio, underwater sound. Address: 23 Park Lane, Concord, Mass. 01742; also 35 Concord Av., Cambridge, Mass. 02138.*

MIDDLETON, John Tylor, Am. botanist, educator; b. Chgo., Ill., Sept. 15, 1912; s. Walter Guy and Gertrude (Baldwin) M.; B.S., U. Cal. at Berkeley, 1935; Ph.D., U. Mo., 1940; m. Diana J. Clarkson, June 16, 1961; children—Peter Cornell, David Burke, Mary Russell, Sara Parke. Faculty, U. Cal., Los Angeles, Riverside, 1939—, plant pathologist, 1954—, prof., chmn., 1957-63, dir. Air Pollution Research Center, Riverside, 1962—. Cons. pathologist various pub. agys., pvt. cos.; chmn. Cal. Motor Vehicle Polution Control Bd., 1960-62, vice chmn., 1962-63, mem. 1964—; spl. cons. USPHS, 1963—; adviser WHO, 1963; mem. nat. adv. com. on air pollution U. S. Surgeon Gen., 1963—; del. U. S. State Dept., Council of Europe, 1964. Recipient Clean Air award Los Angeles Air Pollution Control dist. and Los Angeles C. of C., 1960. Mem. Air Pollution Control Assn., Am. Inst. Biol. Scis., Am. Phytopath. Soc., Internat. Assn. Plant Taxonomy, Mycol. Soc. Am., Torrey Bot. Club, Nederlandse Planteziektenkundige Vereniging, Sigma Xi, Alpha Zeta, Gamma Alpha. Research, publs. on air pollution and geofungi with spl. emphasis on phycomycetes; 1st to identify smog damage on plants in So. Cal. Home: 2040 Arroyo Dr., Riverside, Cal. 92506.*

MIDDLETON, William Shainline, Am. physician; b. Norristown, Pa., Jan. 7, 1890; s. Daniel Shepherd and Ann Sophia (Holstein) M.; M.D., U. Pa., 1911, Sc.D., 1946; Sc.D., U. Cambridge (Eng.), 1950; LL.D., Temple U., 1956; L.H.D., Franklin and Marshall Coll., 1957; Litt.D., Marquette U., 1958; m. Maude Hazel Webster, Sept. 30, 1921. Faculty, U. Wis. Med. Sch., Madison, 1912-60, prof. medicine, 1933-60, dean, 1935-55, dean emeritus, prof. emeritus, 1960—; chief med. dir. VA, 1955-63; vis. prof. medicine U. Okla., 1963-64; cons. research and edn. VA Hosp., Madison, 1964—. Recipient U. Pa. award of merit, 1943; Centennial award Northwestern U., 1951; Council award Wis. State Med. Soc., 1938. Fellow Coll. Physicians Phila., Royal Coll. Physicians (London, Eng.), Royal Soc. Medicine, (hon.); mem. A.C.P. (master, Alfred Stengel Meml. medal 1962), A.M.A., Assn. Am. Physicians, Am., Central socs. for clin. investigations, Am. Clin. and Climatol. Assn. Research, numerous publs. on diseases of blood, blood forming organs, cardio-vascular system, lung, med. history. Home: 2114 Adams St., Madison, Wis. 53711.

MIDGLEY, Thomas, Jr., Am. chemist; b. Beaver Falls, Pa., May 18, 1889; s. Thomas and Hattie Lena (Emerson) M.; M.E., Cornell U., 1911; D.Sc., Coll. Wooster, 1936; m. Carrie M. Reynolds, Aug. 3, 1911; children—Thomas 3d, Jane (Mrs. Edward Z. Lewis). With Nat. Cash Register Co., 1911; research work on automobile tires, 1912-14; supt. Midgley Tire & Rubber Co., Lancaster, O., 1914-16; with Charles F. Kettering, Dayton, O., later with Gen. Motors Research Corp., 1916-18; head fuel div. Gen. Motors Research Corp., 1918-23; gen. mgr. Gen. Motors Chem. Co., 1923; v.p. Ethyl Corp. since 1923 and Kinetics Chem., Inc., since 1930; dir. Ethyl-Dow Chem Co. since 1933. Vice pres. Ohio State U. Research Found., 1940-44; vice chmn. Nat. Inventors Council, 1940-44. Nichols medal Am. Chem. Soc., 1923, Perkins medal, 1937; Longstreth medal Franklin Inst., 1925; Priestly medal Am. Chem. Soc., 1941; Willard Gibbs Medal, 1942. Pres. Am. Chem. Soc. (chmn. bd. dirs.); mem. Nat. Acad. Scis., Discovered tetraethyl lead, used as gasoline anti-knock compound, also certain organic fluoride compounds for refrigerants which are nontoxic and noninflammable; holder many patents. Died Nov. 2, 1944.

MIDULLA, Mario, Italian physician; b. Catania, Italy, June 23, 1920; s. Carmelo and Adele (Sangiorgi) M.; M.D., magna cum laude, U. Rome (Italy), 1943, Specialization in pediatrics magna cum laude, 1945, Libera Docenza in Child Health, 1956, in Pediatrics, 1959; m. Velia Conversi, Feb. 18, 1952; children—Livia, Cecilia. Staff, Inst. Pediatrics, State U. Rome, 1943—, ordinary asst., 1964—, mem. faculty; practice medicine, specializing in pediatrics, Rome, 1945—; faculty Sch. Specialization in Infantle Neuropsychiatry, Sch. Profl. Nurses of Italian Red. Cross. Lederle fellow U. Cal. at Berkeley; NIH Internat. Postdoctoral fellow Baylor Col. Medicine, Houston, 1961-62. Mem. Italian Soc. Pediatrics. Author: (with G. Falchi) Il Gargoilismo, 1958; (with others) Manuale di Pediatria, 1965; also articles. Demonstrated very low level immunoglobulin M in Wiskott-Aldrich syndrome, 1963; isolated (with others) new herpes virus from marmoset monkeys, 1964; suggested (with others) new method for precipitation and concentration with salts of enteroviruses for use as antigens in animals, 1965. Home: 21 via R.R. Pereira, Rome, Italy.*

MIE, Gustav, German physicist; b. Rostock, Germany, 1868; ed. U. Rostock (Germany), also Heidelberg, Germany, Tech. Sch., Karlsruhe, Germany; student of Arnold; prof. physics U. Greifswald (Germany), 1902-17, U. Halle (Germany), 1917; prof., dir. phys. inst. U. Freiburg im Breisgau (Germany), 1924-35, emeritus, from 1935. Author: Molecules, Atoms, and Ether; Textbook of Electricity and Magnetism: Die Einsteinsche Gravitationstheorie; Naturwissenschaft und Theologie; Die Denkweise der Physik. Studied (with Arnold) commutation, 1899; research on electric waves between 2 parallel wires, 1900; developed kinetic theory of monoatomic bodies, 1903; set up curves of short circuit currents in direct current motors, 1907; research on Einstein's theories, then on relativity and quanta, after 1923. Died 1957.

MIEKK-OJA, Heikki M., Finnish physicist; b. Tampere, Finland, June 4, 1908; s. Wäinö Waldemar and Ida (Lehtonen) M.; Ph.D., U. Helsinki, 1940; m. Anu Mirjam Koivistoinen, May 15, 1937; children—Annikki, Ilkka, Susanna. With National Bureau of Standards, Finland, 1932-44; Outokumpu Company, 1945-50; Technical U., Helsinki-Otaniemi, 1950—. Consultant, Outokumpu Company, 1950-55; prof. Physical Metallurgy, 1950. Mem., Finnish Acad. Sci.; Acad. Technical Sci.; Inst. of Metals. Author: Metallioppi, 1960; about forty publ. articles in field. Specialist in physical metallurgy. Home: 46 Abrahaminkatu, Helsinki, Finland. Office: Dept. of Mining and Metallurgy, Technical University, Helsinki-Otaniemi, Finland.*

MIELE, Angelo, aerodynamicist; b. Formia, Italy, Aug. 21, 1922; s. Salvatore and Elena (Marino) M.; Dr. Civil Engring. U. Rome, 1944, Dr. Aero. Engring., 1946. Asst. prof. Poly. Inst. Bklyn., 1952-55; prof. Purdue U., 1955-59; dir. astrodynamics, flight mechanics Boeing Sci. Research Labs., Seattle, 1959-64; prof. astronautics Rice U., Houston, 1964—; cons. Allison div. Gen. Motors, Guided Missile div. Douglas Aircraft Co., 1955-58. Fellow Am. Inst. Aeros. and Astronautics; mem. Am. Astronautical Soc., Italian Rocket Assn., Italian Aerotech. Assn., Internat. Acad. Astronautics (corr.). Author: Flight Mechanics, vol. 1, Theory of Flight Paths, 1962, Theory of Optimum Aerodynamic Shapes, 1965; also numerous articles. Asso. editor Jour. Astronautical Scis., 1964—; editor-in-chief Jour. Optimization Theory and Applications, 1967—. Research in flight mechanics, astrodynamics, aerodynamics, optimization theory, optimum flight paths, optimum aerodynamic shapes. Home: 3333 Cummins Lane, Houston 77001.

MIELENZ, Klaus Dieter, physicist; b. Berlin, Germany, May 8, 1929; s. Walter Ernst and Ilse (Hassenkamp) M.; B.Sc., Humboldt U., Berlin, 1949; M.Sc., Free U., Berlin, 1952, Ph.D., 1955; m. Renate Anna Therese Zorn, Apr. 11, 1959; children—Frank Stephan, Monica Suzanne. Physicist, R. Fuess Sci. Instrument Co., Berlin, 1952-58, tech. mgr., 1960-63; physicist Nat. Bur. Standards, Washington, 1958-60, project leader optical masers, 1963—. Guest research worker Inst. for Applied Spectroscopy, Dortmund, Germany, 1957. Recipient Silver medal U. S. Dept. Commerce, 1966. Mem. Optical Soc. Am., German Soc. for Applied Optics. Designer vacuum gauges; research in thin-film; devel. spectroscopic instruments; research and devel. in application of continuous-wave gas lasers to length metrology. Home: 6 Waycross Ct., Kensington, Md. 20795. Office: Nat. Bur. Standards, Washington 20234.*

MIERS, Sir Henry (Alexander), mineralogist; b. Rio de Janeiro, Brazil, May 25, 1858; s. Francis C. Miers; ed. Eton Coll. (fellow); D.Sc., M.A., D.C.L. (hon.), Oxford (Eng.) U., also fellow Magdalen, Trinity colls.; Ph.D. (hon.), Christiana Coll.; D.Sc. (hon.), Manchester, Sheffield (both Eng.) univs.; LL.D., Liverpool (Eng.), Mich. univs. Asst., Brit. Mus., 1882-95; instr. crystallography Central Tech. Coll., S. Kensington, Eng., 1886-95; Waynflete prof. mineralogy Oxford U., 1895-1908; prin. U. London (Eng.), 1908-15; prof. crystallography, vice-chancellor U. Manchester, 1915-26. Fellow Royal Soc., 1896, Geol. Soc. (v.p. 1902-04, Wollaston medal 1934), Chem. Soc. (v.p. 1901-04); mem. Mineral. Soc. (pres. 1904-09), Brit. Assn. (pres. geol. sect. 1905). Author: (with R. Crosskey) The Soil in Relation to Health, 1893; A Visit to the Yukon Gold Fields, 1901; Mineralogy, 1902. Editor: Mineral. Mag., 1891-1900. Made authoritative survey on state of knowledge of mineralogy of his time. Died Dec. 10, 1942.

MIESCHER, Ernst, Swiss physicist; b. Basel, Switzerland, Oct. 6, 1905; s. Ernst and Helen (Gemuseus) M.; ed. U. Basel, U. Munich (Germany); Ph.D. Physicist, asso. prof. Research and publs. in molecular spectroscopy. Home: Schorenweg 18a, Basel. Office: Klingelbergstrasse 82, Basel, Switzerland.

MIESCHER, Johann Friedrich, Swiss physiologist; b. Basel, Switzerland, Aug. 13, 1844; prof., Basel; founder cellular chemistry and theory of intermediary metabolism; discovered nuclein, 1868, also protamin, various new tech. lab. aids; discovered relative amount of carbon dioxide in blood is decisive in regulation of breathing. Died Davos, Switzerland, Aug. 26, 1895.

MIESCHER, Peter Anton, physician; b. Zurich, Switzerland, Oct. 6, 1923; s. Guido and Lucie Helen (Mayer) M.; student U. Lausanne (Switzerland), 1942-43; M.D., U. Zurich, 1948; m. Annatina Lotscher, Aug. 26, 1950; children—Guido, Annatina. Came to U. S., 1959, naturalized, 1966. Research fellow U. Basle, 1948-49, chief sect. dept. medicine, 1954-59; vis. prof. N.Y. U. Sch. Medicine, N.Y.C., 1959, prof. medicine, head div. hematology, 1960——. Recipient Research award City of Basle, 1963. Mem. Am. Jour. Clin. Investigation, Am. Assn. Immunology, Am. Hematology Soc., Am. Soc. Rheumatism, Internat. Soc. Hematology. Editor: (with K. O. Vorlaender) Immunopathologie in Klinik and Forschung, 1957; (with H. Mueller Eberhard) Textbook in Immunopathology, 1967. Contbr. numerous articles to tech. jours. Characterized lupus erythematosus factor as antinuclear autoantibody; research on mechanisms cell damage by immune reactions, role antigen-antibody complexes, function reticuloendothelial system in eliminating aged red cells by phagocytosis, role analgesic abuse in chronic interstitial nephritis, treatment patients with auto-immune disorders, exptl. autoimmune disorders.*

MIETKIEWSKI, Kazimierz, Polish physician, histologist; b. Krasnica, Poland, July 2, 1906; s. Joseph and Stanislawa (Lesniewska) M.; physician diploma Faculty Medicine, Poznan (Poland) U., 1934, dr. medicine, 1937. Faculty, dept. histology Med. Faculty, Poznan, 1934——, prof. Med. Acad., 1954-59, prof. in chief, 1959——. Scholar dept. histology U. Lyon (France), 1937, Radium Inst., Paris, France, 1937. Decorated Polonia Restituta. Mem. Polish Endocrinological Soc., Sci. Soc. Poznan, Polish Anat. Soc., Histochem. Soc., Polish Biol. Soc., Assn. des Anatomistes Français, Société Française d'Histochimie. Research, publs. on exptl. endocrinology and neurosecretion, interrelationship between hormones and enzymes, histochem. studies liver and adrenal glands after transplantation exptl. cancer in lab. animals. Home: 18 ul. Kniewskiego, Poznan, Poland.*

MIETTINEN, Jorma Kalervo, Finnish biochemist; b. Helsinki, Finland, Aug. 11, 1921; s. Kalle Mikko and Hilda (Saarinen) M.; student U. Upsala, 1946, M.Sc., U. Helsinki, 1948, Ph.D., 1954; postgrad. Sorbonne, U. Cal. at Berkeley, U. Oxford; m. Irja Kaisa Marita Koivisto, June 24, 1943; children—Seppo Kalervo, Hannu Ilmari, Matti Juhani. Asst., Biochem. Research Inst., Helsinki, 1948-51, project leader, 1959-64, prof. radiochemistry, head dept., 1964——; research asst. to pres. Acad. Finland, 1951-58; Andre Mayer fellow FAO, 1958-59. Dir. Finnish AEC Research Project, 1959——; lectr. Tech. U. Helsinki, 1964——. Mem. Finnish Acad. Scis., Am. Chem. Soc. Editor: Biochemistry of Nitrogen, 1955. Research, numerous publs. on nitrogen fixation and amino acid metabolism in plants, nucleotide metabolism in yeast, paper chromatography and electrophoresis, devel. of an ambulatory whole body counter. Home: 8A6 Cygnaeuksenkatu, Helsinki 10. Office: 35 Unionkatu, Helsinki 17, Finland.*

MIETZSCH, Fritz, German chemist; b. Dresden, Germany, 1896; dir. pharm. research Bayer firm; developed thio-semicarbazone; closely connected with discovery of antebrin and sulfonamides. Died Leverkusen, Germany, Nov. 29, 1958.

MIGDAL, Arkadiy Baynusovich, Russian theoretical physicist; b. Mar. 11, 1911; grad. Leningrad U., 1936. Prof., Moscow Engring. and Physics Inst., 1944——; asso. USSR Acad. Scis., 1945——. Decorated Order of Lenin. Mem. USSR Acad. Scis. (corr.). Author: The Ionization of Atoms in Alpha and Beta Decay, 1941; Quadrupole and Dipole Gamma Radiation of Nuclei, 1945; Theory of Nuclear Reactions with Formation of Slow Particles, 1955; Quantum Kinetic Equation for Multiple Scattering, 1955; Bremsstrahlung and Pair Production in Condensed Media at High Energies, 1956. Research and publs. on nuclear physics and quantum mechanics; developer theory of dipole radiation of atomic nuclei, theory of ionization of atoms in nuclear reactions. Address: Moscow Engring. and Physics Inst., ulitsa Kirova 21, Moscow, USSR.

MIGDALSKA, Barbara Chojnacka, Polish physician; b. Warsaw, Poland, Nov. 18, 1928; d. Edmund and Stefani (Maciejowska) Chojnacki; High Living Certificate, 1947; grad. Warsaw Med. Sch., 1953; IId degree specialization in Lab. Diagnostic, 1957; M.D., 1962; m. Piotr Migdalski, Feb. 5, 1948; 1 son, Stefan. Asst. dept. biochemistry Inst. Tb, Warsaw, 1951-60; chief Lab. Adrenal Steroids, I Clinic Internal Diseases, Postgrad. Med. Sch., Warsaw, 1961——, faculty pathology adrenals. Mem. Polish Soc. Endocrinology, Polish Soc. Biochemistry. Contbg. author: Biochemical Diagnostic, 1967. Research, numerous publs. on lab. diagnosis of adrenal cortex pathology, including elaboration and evaluation of methods and functional tests; developed electrolie test, estimation of free corticoids without chromatography, math. method for evaluation of adrenal res. test. Home: 93/97 Marymoncka-Bielany, Warsaw, Poland.*

MIGICOVSKY, Bert B., Canadian research scientist; b. Winnipeg, Man., Can., Mar. 15, 1915; s. Samuel and Rebecca (Winestock) M.; B.S., U. Man., 1935, M.S., 1937; Ph.D., U. Minn., 1939; m. Geraldine Shnier, Mar. 15, 1943; children—John Franklin, Janet Francis. With Canadian Dept. Agr., 1940——, animal nutrition chemist, 1940, chemistry div., 1945-55, head animal chemistry unit biochemistry, 1955-64, asst. dir. gen. Research Insts. and Services, Ottawa, Ont., Can., 1964——. Fellow Chem. Inst. Can., Agrl. Inst. Can.; mem. Am. Assn. Biol. Chemists, Am. Chem. Soc., Can. Biochem. Soc., Can. Physiol. Soc., Nutrition Soc. Can. Contbr.; Radioisotopes and Environmental Circumstances, 1960; Use of Radioisotopes in Animal Biology, 1961; Hawk's Physiological Chemistry, 14th edit., 1965. Publs. on devel. of method for removal of radio elements from milk; studies of action of vitamin D; control of body cholesterol content. Home: 185 Patricia St. Office: Central Exptl. Farm, Ottawa, Ont., Can.*

MIHALYI, Elemer, biochemist; b. Deva, Rumania, Jan. 11, 1919; s. Elemer and Maria (Illes) M.; M.D., U. Kolozsvar (Hungary), 1943; Ph.D., U. Cambridge (Eng.), 1963; m. Elisabeth Sedony, Mar. 24, 1948; children—Christina, Julia. Came to U. S., 1949, naturalized, 1955. Instr., U. Kolozsvar, 1941-45, U. Budapest, Hungary, 1946-48; guest investigator Med. Nobel Inst., Stockholm, Sweden, 1948-49; research asso. Inst. Muscle Research, Woods Hole, Mass., 1949-51, Inst. Surg. Research, U. Pa., Phila., 1951-55; chemist Nat. Heart Inst., NIH, Bethesda, Md., 1955——. Mem. Am. Soc. Biol. Chemists, Am. Chem. Soc., Internat. Soc. Hematology. Contbr. articles to profl. jours. Developed technique of dissolution of fibrin gels in urea to compare physiochem. properties of fibrin and fibrinogen to show that thrombin does not alter structure of fibrinogen molecule; studies of proteolytic fragmentation of myosin to discover detailed submolecular structure. Home: 10210 Fleming Av. Office: NIH, Bethesda, Md. 20014.*

MIHELICH, John William, Am. physicist; b. Colorado Springs, Colo., Jan. 2, 1922; s. John and Amelia (Pirnat) M.; A.B., Colo. Coll., 1942; Ph.D., U. Ill., 1950; m. Jeanette Van Osdol, Dec. 22, 1946; children—John William, Kathryn Lee, Margaret Lucille. Asso. physicist Brookhaven Nat. Lab., N.Y.C., 1950-54; faculty physics U. Notre Dame, Ind., 1954——, prof., 1961——. With Lawrence Radiation Lab., Cal., summer 1955, Oak Ridge Nat. Lab., summers, 1956-61, Brookhaven Nat. Lab., 1962. Fellow Am. Phys. Soc.; mem. Sigma Xi. Contbr. numerous articles on low energy nuclear physics (internal conversion, beta and gamma-ray spectroscopy, nuclear spectroscopy, properties of nuclear levels). Home: 1122 Clermont Dr., South Bend, Ind. 46617.*

MIHOC, Gheorghe, Rumanian mathematician; b. 1906; rector Bucharest (Ruamania) U.; mem. Acad. Socialist Republic Rumania. Author: (with others) Dependence, 1937; Treatise on Actuarial Mathematics, 1942; Elements of Probability Calculation, 1952; numerous other works on theory of probabilities and math. statistics. Introduced (with O. Oniscescu) concept of chains with complete connections, 1935; studied Markov chains, generalization of Poisson equations.

MIKA, Leonard Aloysius, Am. microbiologist; b. Bay City, Mich., Apr. 17, 1917; s. Steve J. and Victoria (Kaczmarek) M.; Asso. Sci., Grand Rapids Jr. Coll., 1939; B.S., U. Mich., 1947, M.S., 1949; Ph.D., George Washington U., 1955. Staff, U. S. Army Biol. Labs., Ft. Detrick, Md., 1949——, sr. investigator, 1953-63, staff adminstr., 1963——. Sci. instr. Frederick (Md.) Acad. of Visitation, 1962——. Diplomate Am. Bd. Microbiology. Fellow Am. Acad. Microbiology; mem. Soc. Exptl. Biology and Medicine, Am. Soc. for Microbiology, N.Y. Acad. Scis., Research Soc. Am. (pres. Ft. Detrick br. 1966——), Am. Assn. Immunologists. Research, publs. on demonstration active role metabolites in bacterial mutations, mixed infections and interference phenomenon, propagation, differentiation certain viruses. Home: 922 Shawnee Dr., Frederick, Md. 21701. Office: U. S. Army Biol. Labs., Ft. Detrick, Md. 21701.*

MIKESELL, Raymond F., Am. economist; b. Eaton, O., Feb. 13, 1913; s. Otho F. and Josephine (Frech) M.; student Carnegie Inst. Tech., 1931-33; B.A., Ohio State U., 1933, M.A., 1935, Ph.D., 1939; m. Desyl DeLauder, July 6, 1937 (dec.); m. 2d, Irene Langdoc, Feb. 18, 1956; children—George D., Norman D. (twins). Asst. prof. U. Wash., Seattle, 1937-41; economist OPA, Washington, 1941-42, Treasury Dept., Washington, 1942-46; prof. U. Va., Charlottesville, 1946-57; W. E. Miner prof. econs. U. Ore., Eugene, 1957——. Active as staff mem. or cons. numerous internat. and nat. agys., including ICA, UN Econ. Commn. for Latin Am., Senate Fgn. Relations Com., Internat. C. of C., U. S. Dept. State, U. S. Dept. Commerce, UNESCO, OAS, 1946-63; cons. AID, 1964——; vis. prof. Grad. Inst. Internat. Studies, Geneva, Switzerland, 1964; adviser U. S. delegation UN Trade and Devel. Conf., Geneva, 1964; mem. panel advisers to sec. treasury U. S. Treasury Dept., 1965——. Mem. Am. Econ. Assn., Internat. Studies Assn., Soc. for Internat. Devel., Pi Kappa Alpha, Theta Tau, Alpha Kappa Psi. Author: (with Hollis Chenery) Arabian Oil, 1949; U. S. Economic Policy and International Relations, 1952; Foreign Exchange in the Postwar World, 1954; The Emerging Pattern of International Payments, 1954; (with Merlyn Trued) Postwar Bilateral Payments Agreements, 1955; Foreign Investments in Latin America, 1955; Promoting United States Private Investment Abroad, 1957; Agricultural Surpluses and Export Policy, 1958; Liberalization of Inter-Latin American Trade, 1958; (with Jack N. Behrman) Financing Free World Trade with the Sino-Soviet Bloc, 1958; Financing of Economic Development in Latin America, 1958; Intra-Regional Trade and Economic Development, 1958; (with R. L. Allen) Economic Policies Toward Less Developed Countries, 1961; Some Observations on the Operation of the Alliance for Progress: the First Six Months, 1962; (with Raymond Staepelaere) Common Market Competition in Manufactures, 1963; (with G. Kalmanoff) Public International Development Financing in Colombia, 1963; Mecanismos de Ayuda Economica Externa, 1964; Public International Lending in Development, 1966; (with R. W. Adler) Public External Financing of Development Banks, 1966. Editor, author several chpts. United States Private and Government Investment Abroad, 1962. Bd. editors Am. Econ. Rev., 1953-55; editorial bd. Middle East Jour., 1947-57. Contbr. articles to profl. jours. Home: 2290 Spring Blvd., Eugene, Ore. 97403.*

MIKESELL, William Henry, Am. psychologist; b. Westminster, Md.; s. William Augustus and Madalena (Harner) M.; A.B. cum laude, Western Md. Coll., 1909; B.D., Westminster Theol. Sem., 1912; M.A. with honors, Harvard, 1914; Ph.D., U. Ill., 1926; m. Patricia Rand Patterson, June 16, 1923; children—Ritchie, William Henry II. Instr. pub. speaking U. Tex., 1916-18; ednl. dir. YMCA, France, 1918-19; instr. pub. speaking, dramatics U. Ky., 1920-22; instr. pub. speaking U. Mo., 1922-23, U. Ill., 1923-26; dean liberal arts U. Wichita, 1926-29, chmn. psychology dept., 1926-47; chmn. psychology dept. Washburn U., 1947-53; dir. guidance Miss. Coll. for Women, 1953-58, Anderson Coll., 1958-64, Roswell (N.M.) Community Coll., 1964——. Mem. Kan. Mental Hygiene Soc., Am. Psychol. Assn. Author: Psychology and Life, 8 vols., 1933; Mental Hygiene, 1939; How To Study, 1940; Psychology of Adjustment, 1952; Techniques of Living, 1953; Power of High Purpose, 1961; Counseling for Ministers, 1961. Editor: Modern Abnormal Psychology, 1950. Research, publs. indicated suggestion lessens or augments energy outlay, depending whether positive or negative; auto-suggestion makes for more vigorous action than hetero suggestion; left-handers in a steadiness expt. make more mistakes than right-handers; hungry rats hoard food even when hungry while well fed rats do not. Home: 608 W. 1st St., Roswell, N.M. 88201.*

MIKHAYLOV, Aleksandr Aleksandrovich, Russian astronomer, gravimetrist; b. Apr. 26, 1888; grad. Moscow U., 1911. Prof., Moscow U., 1918-48; dir. Main Astron. Obs., USSR Acad. Scis., Pulkovo, 1947-64, chmn. astron. council, 1939-60. Decorated Order of Lenin (3). Mem. USSR Acad. Scis., Internat. Astron. Union (v.p. 1946-48), All-Union Astron. and Geodetic Soc. (past chmn. central council). Author: A Course on Gravimetry and the Theory of the Earth's Figure, 1939; Theory of Eclipses, 1954; Stellar Atlas of Stars up to 8.25 Magnitude, 2d ed., 1959. Mem. main editorial bd. Large Soviet Ency., 1949. Research and publs. on deflection of light rays in sun's gravitational field (Einstein's effect), theory of solar and lunar eclipses, cartographic projections, stellar atlases, developer method to determine shape of earth, 1942-45; compiled eclipse-prediction tables. Address: USSR Acad. Scis., Pulkovo Obs., Leningrad M140, USSR.*

MIKHEEV, Mikhail Aleksandrovich, Russian phys. power engr.; b. May 25, 1902; grad. Leningrad Poly. Inst., 1927. With Physico-Tech. Inst., USSR Acad. Scis., 1925-34; prof. Moscow Energy Inst., 1936; head heat exchange lab. Krzhizhanovsky Power Engring. Inst., 1933——; Decorated Order of Lenin; recipient Stalin prize, 1941, 51. Mem. USSR Acad. Scis. (corr., academician 1953——). Author: (with M. V. Kirpichev) Modelling Heat Equipment, 1936, Basses of Heat Transfer, 2d edit., 1949, numerous others. Research on heat transfer, theory of hydromechanics and heat analysis. Home: 1-aya Cheremushkinskaya 3. Office: Moscow Energy Institute, Moscow, USSR.

MIKHEEV, Vadim Vladimirovich, Russian neuropathologist; b. 1899; grad. Med. Faculty, 1st Moscow U., 1924; D.Med. Sci. Intern, asst., vis. lectr. dept. neuropathology 1st Moscow Med. Inst., 1924-36, head chair, dir. nervous diseases clinic, 1960—; head chair nervous diseases Arkhangelsk Med. Inst., 1936-45; head chair nervous diseases and psychiatry Moscow Med. Stomatological Inst., 1945-60. Decorated Order of Lenin. Mem. All-Russian (bd. mem.), Moscow (bd. mem.) socs. neuropathologists and psychiatrists. Author numerous works including: Neuropathology of Malignant Neoplasms, 1946; Cerebral Rheumatism, 1949; Neurorheumatism; Textbook of Nervous Diseases for Students of Medical Institutes, 1954; Stomatoneurology, 1958. Mem. editorial council Korsakov Jour. Neuropathology and Psychiatry. Address: 1st Moscow Med. Inst., ulitsa Karla Marksa 124, Krasnoyarsk, RSFSR, USSR.

MIKOYAN, Artyom Ivanovich, Russian aero. engr.; b. Sanain, Georgia, Aug. 5, 1905; grad. N.E. Zhukovskii Mil. Air Acad., 1936. Maj.-gen. in engring.-tech. service; dep. insp. aircraft plant, Moscow, 1937-38; with Exptl. Design Bur., 1941—, designer-in-chief aircraft engring., 1958—. Decorated Order of Lenin (4); recipient Stalin prize, 1946, 49, 52. Mem. USSR Acad. Scis. (corr.). Designer (with M. I. Guervich) fighter plane MIG-1, also later MIG's; pioneer of jet aviation in USSR; demonstrated his 1st turbojet plane, 1946; designer various supersonic planes, 1954—. Address: USSR Acad. Scis., Leninskii Prospect 14, Moscow, USSR.

MIKULASZEK, Edmund Julius, Polish microbiologist; b. Lwów, Poland, Sept. 21, 1895; s. Julius and Elisa (Gerber) M.; doctors degree in medicine Johannes Casimir U., Lwów, 1922. Staff, State Inst. Hygiene, Lwów, 1925-39, Vet. Inst. Lwów, 1939-44; prof. microbiology U. Warsaw Med. Faculty, 1945-50; prof. med. microbiology Med. High Sch., Warsaw, 1950-65; staff Akademia Medyczna, 1950—. Decorated Officer Cross of Order Polonia Restituta; recipient Grünwald-Gross, 1945; State Sci. award I.Cl., 1955. Mem. Polish Acad. Scis. (past chmn. microbiol. com.), Polish Microbiologist Soc. Author: Fundamentals of Immunochemistry, 1948; Immunologically Active Polysaccharides, 1952; Immunochemistry of Sugars and Polysaccharides, 1961; Bakterielle Polysaccharide, 1935; also numerous articles. Research on antigenic structure of pathogenic bacterial; correlation of chem. structure and immunological activity in bacterial polysaccharides. Home: Nr. 11, ul. Zloczowska, Warsaw. Office: Nr. 5, ul. Chalubinskiego, Warsaw, Poland.*

MIKULICZ-RADECKI, Johann von, see von Mikulicz-Radecki, Johann.

MIKULIN, Aleksandr Aleksandrovich, Russian aero. engr.; b. Feb. 2, 1895. Maj.-gen. in engr.-tech. service; designer Sci. Automotor Inst., 1923. Decorated Order of Lenin (3); recipient Stalin prize, 1941, 42, 43, 46. Mem. USSR Acad. Scis. (academician). Designer aircraft engines; introduced rotating blades to regulate superchargers, high pressure, feed and cool intake air; developed 1st turbo-compressor and variable pitch propeller; dir. jet engine devel., after 1945. Home: Pugoshvinikov p. 15. Office: USSR Acad. Scis., Leninski Prospect 14, Moscow, USSR.

MIKUNI, Masakichi, Japanese physician; b. Ryotsu City, Japan, Sept. 2, 1906; s. Taikichi and Kii (Goto) M.; M.D., Niigata (Japan) Med. Coll., 1933, D.Med. Sci., 1939; m. Masae Kajii, Dec. 8, 1936; children—Ikuo, Akio. Faculty, Niigata Med. Coll., 1935—, prof., 1945—; prof. Niigata U. Sch. Medicine, 1952—, dir. Niigata U. Hosp., 1966—, pres. Nursing Sch., 1965—. Mem. Japan Ophthal. Soc. (past pres.), Japan Chemotherapy Soc. (councilor). Author: Retinal Blood Pressure, 1958; Eye and Systemic Disease, 1965; Diary in Guadalcanal Island, 1966; numerous articles. Inventor device to measure caliber of retinal blood vessels, ophthalmodynamometer, suction cup apparatus for diagnosis of glaucoma. Home: 2-808 Suidocho. Office: Eye Dept., U. Hosp., 1 Asahimachi, Niigata City, Japan.*

MILAS, Nicholas Athanasius, chemist; b. Candia, Crete, Greece, Jan. 1, 1897; s. Athanasius E. and Mary Milas; came to U. S., 1912, naturalized, 1918; m. Georgia C. Despotes, Feb. 23, 1929; children—Beatrice Mary (Mrs. Frank Wyse), Helen Frances (Mrs. Karl Gundersen). Research asso. Mass. Inst. Tech., 1928-35, faculty, 1935—, asso. prof. chemistry, 1941-61, asso. prof. emeritus, lectr., 1962—; lectr. fgn. univs., 1952, 55. Cons. to industry on syntheses vitamins A and D, 1935-60, on organic peroxides and polymerizations, 1935—. Fellow Am. Acad. Arts and Scis., N.Y. Acad. Scis., A.A.A.S.; mem. Acad. Applied Sci., Am Assn. U. Profs. (past pres. local br.), Am., Croation chem. socs., Sci. History Soc., Sigma Xi, Gamma Alpha, Alpha Chi Sigma. Contbr. numerous articles to tech. jours, chpts. to books, encys. Patentee in field; pioneered devel. field organic peroxides and polyoxides; invented vitamin A and D syntheses; research on catalytic oxidations, hydroxylations and auto-oxidation in organic chemistry. Home: 34 Payson Terrace, Belmont, Mass. 02178. Office: Mass. Inst. Tech., Cambridge, Mass. 02139.*

MILAZZO, Giulio, Italian chemist; b. Palermo, Italy, Feb. 11, 1912; s. Nicola and Marie Anne (Macaluso) M.; Ph.D., U. Rome; m. Edmee Heinsius, July 4, 1938; children—Franca, Anna Maria, Cristina. Instr. electrochemistry Faculty Sci., U. Rome: chief phys. chem. sect. Inst. Health, Rome. Mem. Comité International de thermodynamique et cinétique électrochimique, Internat. Union Pure and Applied Chemistry. Research and numerous publs. on electrochemistry, thermodynamics, spectroscopy, chem. analysis. Home: piazza G. Verdi 9, 00198 Rome, Italy.*

MILBRATH, John A., Am. plant pathologist; b. Marlin, Wash., Sept. 23, 1909; s. August William and Mary Jane (Emery) M.; B.S., Wash. State U., 1934; Ph.D., Ore. State U., 1938; m. Margaret Lucille McCoy, July 29, 1938; children—Gene McCoy, Judy Jean. Plant pathologist Ore. State Exptl. Sta., Corvallis, 1938—, faculty, 1938—. Mem. Am. Plant Pathology Soc., Am. Inst. Biol. Sci. Research, numerous publs. on diseases of ornamental plants, virus diseases of potatoes, strain alfalfa mosaic, viruses of stone fruits and control methods, stony pit virus of pears. Home: 1740 Woodland Dr., Corvallis 97330. Office: Cordley Hall, 201 Central St., Corvallis, Ore. 97331.*

MILCH, Robert Austin, Am. research surgeon; b. N.Y.C., May 24, 1929; s. Henry and Pearl (Salzberg) M.; A.B., Columbia, 1949, M.D., 1953; m. Margot Wurtzburger, Aug. 14, 1960; children—Pamela Alexandra, Thomas Andrew. Research fellow Sloan-Kettering Inst., N.Y.C., 1953-54; surg. house officer Peter Bent Brigham Hosp., 1954-55, asst. resident surgery, 1957-58; clin. asso. surgery Nat. Cancer Inst. USPHS, Bethesda, Md., 1955-57; asst. resident orthopedic surgery Johns Hopkins Hosp., Balt., 1958-60, chief resident, 1960-61, orthopedic surgeon, 1961-67; faculty Sch. Medicine Johns Hopkins, 1961-67, asso. prof. orthopedic surgery, dir. orthopedic research lab., 1964-67; with Office Sci. and Tech., Exec. Office of Pres., Washington, 1967—. Recipient Francis F. Schwentker award Johns Hopkins Hosp., 1964. Diplomate Am. Bd. Orthopedic Surgery. Mem. Am. Acad. Orthopedic Surgeons, A.A.A.S., Am. Assn. Anatomists, Am. Assn. U. Profs., Am. Chem. Soc., A.C.S., Am. Fedn. Clin. Research, Am. Leather Chems. Assn., Am. Mgmt. Assn., numerous others. Author: (with H. Milch) Fracture Surgery, A Textbook of Common Fractures, 1959; also numerous articles. Editor: Surgery of Arthritis, 1964; (with V. A. McKusick) Symposium on Genetics of Congenital Deformity, 1964; Structural Organization of the Skeleton: A Symposium, 1966; editoral bd. Arthritis and Rheumatism, 1960-64; Monographs in the Surgical Sciences, 1963—; Clinical Orthopaedics and Related Research, 1963—. Research on pathogenesis of bone changes in metastatic cancer, pathogenesis of alcaptonuric ochronosis and arthritis, use of tetracyclines as markers for bone growth and devel., mechanism of action of aldehydes on fibrous proteins, plasticizing function of polysaccharides on collagen structures. Home: "Overlook", P.O. Box 5795, Pikesville, Md. 21208. Office: Exec. Office of Pres., Washington 20506.*

MILCU, Stephan Maius, Rumanian endocrinologist; b. Craiova, Rumania, Aug. 15, 1903; s. Athanasie Marin and Silvia (Sudiciu) M.; M.D., U. Bucharest, 1928; D.Med. Scis., Inst. Medicine and Pharmacology, Bucharest, 1954; m. Ioana Parhon, Jan. 3, 1933; children—Andrei, Ileana. Asst., Inst. Anatomy and Endocrinology, Bucharest, Rumania, 1927-35; asst. Clinic Endocrinology, 1935—, lectr., 1943-46, asst. prof., 1946-48, prof. endocrinology, 1948—; rector Med. Inst. Bucharest, 1953-55; dep. dir. Inst. Endocrinology, 1946-57, dir., 1957—. Recipient Rumanian Republic Star I class, 1963, Work Order I class, 1956. Mem. Rumanian Acad. (past gen. sec., past dep. chmn.), Rumanian Soc. of Endocrinology, Société de Biologie de Paris (corr.), Royal Soc. Medicine (affiliate), Purkynie Soc. Prague (hon.), Deutsch. Akad. Naturforscher Halle, Bulgarian Akad. Scis., Acad. Tiberna Roma. Author: (with Parhon, Goldstein) Handbook of Endocrinology, 1938; Endocrine Therapeutics, 1964; (with others) Clinical Endocrinology, 1966; (with A. Lungu) Hormones and Life, 1965; Endemic Goiter, 2 vols., 1955, 58; also numerous articles. Research on new genetic forms of gonadal dysgenesis, autoimmune goiter, ultrastructure and hormone-like substances of pineal body, Addison's disease. Home: 13 dr. Obedenaru. Office: 34, Bd. Aviatori, Bucharest, Rumania.*

MILES, (Arnold) Ashley, Brit. pathologist; b. York, Eng., Mar. 20, 1904; s. Harry and Kate (Hindley) M.; B.A., King's Coll., Cambridge, 1925; M.B., Ch.B., 1928, M.D., 1952; postgrad. St. Bartholomew's Hosp., London; m. Ellen Marguerite Dahl, Apr. 8, 1930. Demonstrator, London Sch. Hygiene and Tropical Medicine, 1929-31, dept. pathology U. Cambridge, 1931-35; reader bacteriology Brit. Postgrad. Med. Sch., U. London, 1935, prof. U. Coll. Hosp. Med. Sch., 1935-46; dir. dept. biol. standards Nat. Inst. for Med. Research, London, 1946-52, dep. dir. inst., 1947-52; dir. Lister Inst. Preventive Medicine, London, 1952—; prof. exptl. pathology U. London, 1952—; London sector pathologist Emergency Med. Service, 1940-46. Hon. dir. Med. Research Council unit on wound infections Accident Hosps., Birmingham, U.K., 1941-46; mem. Med. Research Council, 1957-61. Decorated comdr. Order Brit. Empire; created knight bachelor, 1966. Fellow Royal Coll. Physicians, Royal Soc., 1961 (biol. sec., v.p. 1963—); mem. Soc. for Gen. Microbiology (past pres.), Path. Soc. Gt. Britain and Ireland, Brit. Soc. for Immunology. Author: (with G. S. Wilson) Topley and Wilson's Principles of Bacteriology and Immunity, 3d edit., 1946, 4th edit., 1955, 5th edit., 1964; also numerous articles. Editor: Jour. Gen. Microbiology, 1946-51. Research on bacteriology of egg rot, endocarditis and wound infection in man, antigenic structure and toxicology of organisms of gas gangrene, brucellae and Proteus bacilli; studies on vascular reactions of inflammation and biochem. mediators concerned, mechanisms of non-specific immunity in early stages of microbial infection. Home: 7, Holly Pl., Hampstead, London, N.W.3. Office: Lister Inst. Preventive Medicine, Chelsea Bridge Rd., London S.W.1, Eng.*

MILES, E(rnest) P(ercy), Jr., Am. mathematician; b. Birmingham, Ala., Mar. 16, 1919; s. Ernest Percy and Ida (Duke) M.; B.S., Birmingham So. Coll., 1937; M.A., Duke, 1939, Ph.D., 1949; m. Audrey Adair Vicknair, June 13, 1945; children—Duke, Vick Adair. Faculty, Auburn U., 1949-58, U. Md., 1957-58; faculty Fla. State U., Tallahassee, 1958—, prof., dir. Computing Center, 1961—. Summer staff mem. NSF, 1958, cons., 1959—; task force chmn. Sci. Information Retrieval, Fla. Space Era Edn. Study, 1962-63; mem. Inter-U. Communications Council, 1965—; field reader U. S. Office Edn., 1966—. Fellow A.A.A.S.; mem. Am. Math. Soc., Math. Assn. Am., Soc. For Indsl. and Applied Math., Nat. Council Tchrs. Math., Assn. Computing Machinery, Assn. Ednl. Data Systems, Am. Assn. U. Profs., World Future Soc., Sigma Xi. Author: (with Dr. Hartford) A Study of Administrative Uses of Computers in College and Universities of the United States, 1962; (with others) Guidelines for Planning Computing Centers in Universities and Colleges, 1963. Research in math. largely in harmonic functions, partial differential equations, ill conditioned matrices, sci. edn., computer sci., curriculum revision needs. Home: 2804 St. Leonard Dr., Tallahassee, Fla. 32303.*

MILES, Grant Lewis, Australian chemist; b. Western Australia, Feb. 17, 1921; s. Henry Edgar and Linda (Gipping) M.; M.Sc., U. Western Australia, 1946, B.A., 1944; Ph.D., U. Cambridge (Eng.), 1948; m. Mary Joyce Hamilton, July 24, 1948; children—Christopher, Adrienne, Gabrielle; m. 2d, June Helen Hellens, Sept. 1, 1967. Research chemist Western Australia Dept. Indsl. Devel., 1942-46; Commonwealth Sci. and Indsl. Research Crgn. Sr. Research student Cambridge U., 1946-48; from research officer to prin. research officer Commonwealth Sci. and Indsl. Research Orgn., Harwell, Eng., 1948-56; head chemistry sect. Australian Atomic Energy Commn. Research Establishment, Lucas Heights, 1956-61, asst. dir., then asso. dir., 1962—. Fellow Royal Inst. Chemistry, Royal Australian Inst. Chemistry; mem. Australian Inst. Physics (asso.), Inst. Physics (U.K.). Author: (with F. S. Mortin) Chemical Processing, 1958; also articles. Research on treatment of W. Australian alonite deposits, chemistry of protactinium, processing of irradiated nuclear fuels including plutonian purification and non-aqueous processing. Home: 14 Deepwater, Woronora River, Sutherland. Office: Australian Atomic Energy Commn. Research Establishment, Lucas Heights, Sutherland, N.S.W. Australia.*

MILES, John Arthur Reginald, virologist; b. Sidcup, Eng., May 13, 1913; s. Albert Edward and Mary (Watson) M.; B.A., Cambridge U., 1934, M.A., M.B.B.Chir., Cambridge, St. Thomas's Hosp., London, Eng., 1938, M.D., 1951; m. Ruth Herbert French, Nov. 19, 1951; children—Deborah, Eleanor. Mem. staff St. Thomas's Hosp., London, 1939-42; Royal Army Med. Corps, 1942-45; Huddersfield lectr. U. Cambridge, 1946-50; med. research fellow Inst. Med. and Vet. Sci., Adelaide, S. Australia, 1951-55; prof. microbiology U. Otago, Duedin, New Zealand, 1955—; hon. dir. virus research unit New Zealand med. Research Council, 1955—, hon. dir. microbiology research, 1960—. Fellow Royal Australian Coll. Physicians, Royal Soc. New Zealand (pres. 1966—), Royal Soc. Medicine U.K.; mem. Soc. Gen. Microbiology U.K., N.Y. Acad. Sci., Wildlife Disease Assn. U.S.A., New Zealand Microbiology Soc., New Zealand Soc. Pathologists, Royal Australian Ornithologists Union. Research and publs. on ecology of some virus diseases transmitted by insects and with a reservoir in wild animals and similar studies on psittacosis; epidemiology of poliomyelitis; latent survival of certain insect transmitted viruses in birds and in animal cells in tissue culture. Home: 375 High St., Dunedin, New Zealand.*

MILES, Walter Richard, Am. psychologist; b. Silverleaf, N.D., Mar. 29, 1885; s. Thomas Elwood and Caroline (White) M.; A.B., Earlham Coll., 1908, D.Sc., Earlham, 1952; M.A., Ia. State U., 1910, Ph.D., 1913; M.A., Yale, 1952; m. Elizabeth Kirk, Sept. 1, 1908 (dec. July 1925); children—Thomas Kirk, Caretta E. (Mrs. F. L. Capers, Jr.), Marjorie H. (Mrs. R. D. McClelland); m. Catharine M. Cox, Sept. 9, 1927; 1 dau., Anna Mary (Mrs. E. O. Jones, III). Asso. prof. psychology Wesleyan U., Conn., 1913-14; research psychologist Carnegie Nutrition Lab., Boston, 1914-22; prof. exptl. psychology Stanford, 1922-32, prof. psychology Yale, 1931-53, fellow Jonathan Edwards Coll., 1935-53, emeritus fellow, 1954—; prof. psychology Istanbul U., Turkey, 1954-57; sci. dir. USN Submarine Med. Lab., Groton, Conn., 1957—. Psychology lectr. Tehran U., Iran, 1954; attending psychologist, New Haven Hosp. and Dispensary, 1932-54; U. S. del., 2d Internat. Congress Mental Hygiene, Paris, 1937; pres. Psychol. Corp., N.Y.C., 1939-44; chmn. bd. dirs. Am. Inst. Research, 1947-

54; vice chmn., chmn. commn. aviation medicine NRC, 1939-46; cons. RAF, 1945-46. Recipient Pres.' certificate of Merit, 1948; Warren Gold medal exptl. psychology, 1949; gold medal Am. Psychol. Found., 1962. Fellow Optical Soc. Am., Am. Psychol. Assn. (pres. 1932); mem. Nat. Acad. Sci., Am. Philos. Soc., Am. Physiol. Soc., Am. Assn. U. Profs. (pres. Yale chpt. 1934-35), Soc. Exptl. Psychology (pres. 1935), Sigma Xi, Phi Delta Kappa. Author: Effects of Alcohol on Man, 1924; (with Benedict, others) Human Efficiency Under Restricted Diet, 1918. Editor: Psychological Studies of Human Variability, 1936, sci. dir. in naval research, 1957——. Contbr. numerous articles to tech. jours.; inventor psychol. methods and apparatus. Home: Box 154 Harvard Terrace, Gales Ferry, Conn. 06335. Office: USN Submarine Med. Research Lab., Groton, Conn. 06342.*

MILFORD, Sidney Nevil, physicist; b. Melbourne, Australia, Feb. 11, 1925; s. Sidney James and Edith (Johnson) M.; B.S., Melbourne U., 1944, B.A., 1945; Ph.D., London U., 1949; m. Danuta Maria Einaugler, Sept. 28, 1950; children—Peter N., Richard Ian. Came to U. S., 1950, naturalized, 1963. Fellow, Institut D'Astrophysique, Paris, France, 1949-50, U. Chgo., 1950-51; chmn. physics dept. St. John's U., N.Y.C., 1952-62; head geoastrophysics Grumman Aircraft Engring. Corp., Bethpage, N.Y., 1962——; adj. prof. Adelphi U., Garden City, N.Y., 1964——. Cons. Republic Aviation Corp., Farmingdale, L.I., 1961-62, Allied Research Assos., Boston, 1960-62. Fellow Royal Astron. Soc.; mem. Internat. Astron. Union, Am. Inst. Aeros. and Astronautics, Am. Astron. Soc., Am. Phys. Soc., Am. Geophys. Union. Research, publs. on astrophysics, cosmic rays, and atomic scattering, calculated variations in lunar atmosphere; investigated collisions of high energy cosmic rays with interstellar gas. Home: 60 Nassau Dr., Great Neck, N.Y. 11021. Office: Grumman Aircraft Engring. Corp., Bethpage, N.Y. 11714.*

MILGROM, Felix, immunologist; b. Rohatyn, Poland, Oct. 12, 1919; s. Henryk and Ernestina (Cyryl) M.; student U. Lwow, 1937-41, U. Lublin, 1945; M.D., U. Wroclaw, 1947; m. Halina Miszel, Oct. 15, 1941; children—Henry, Martin Louis. Came to U. S., 1958, naturalized, 1963. Prof., head dept. microbiology Sch. Medicine Silesian U., Zabrze, Poland, 1954-57; research asso. Service de Chime Microbienne Pasteur Inst., Paris, France, 1957; research asso. prof. dept. bacteriology, immunology U. Buffalo Sch. Medicine, 1958-62; asso. prof., prof. and chmn. dept. microbiology State U. N.Y. Sch. Medicine, Buffalo, 1962——. Mem. Am. Assn. Immunologists, Am. Assn. U. Profs., Am. Acad. Microbiology, Sigma Xi. Author: Studies on the Structure of Antibodies, 1950; also numerous articles. Editor-in-chief Internat. Archives of Allergy and Applied Immunology, 1965——; contbg. editor: Vox Sanguinis, 1965——. Research, serology of syphilis, Tb and rheumatoid arthritis, organ and tissue specificity including blood groups, transplantation and autoimmunity. Home: 474 Getzville Rd., Buffalo 14226.*

MILHAUD, Gerard Marcel, physician, biochemist; b. Geneva, Switzerland, Oct. 2, 1922; s. Maurice E. and Rose (Engel) M.; Chem. Engr., Fed. Poly. Sch., Zurich, 1944; Ph.D., U. Zurich; M.D., U. Geneva; M.D., U. Paris, 1951, Ph.D., 1954; m. Vera Y. Pamm, July 16, 1951; children—Anne-Laurence, Sylvie-Beatrice. Research attache Centre National de la Recherche Scientifique, 1947-52; asst. Pasteur Inst., 1952-56, head lab., 1956-58; prof. Faculty Medicine, Paris, 1958——; biologist Hôpitaux, 1966——. Expert in pharmacology and toxicology to test new drugs; expert in pharmacology and toxicology Law Ct. of Paris. Decorated Chavalier Ordre National Merite; recipient Labbe award French Acad. Scis. Mem. Soc. Chimie Biol., Soc. Chimie Physique, Soc. Exptl. Biology and Medicine, Brazilian Acad. Sci. Fellow Royal Soc. Medicine. Research and numerous publs. in path of carbon dioxide in chemoautotrophy; biochem. mechanism of curarization; hereditary disorders of metabolism, including galactosemia, fructose intolerance, calcium losing enteropathy, thyrocalcitonin mechanism of hypocalcemia, effects on calcium metabolism, with therapeutic applications. Home: 5 rue des Saints-Peres, Paris 6e. Office: C.H.U. Saint-Antoine, 27 rue Chaligny, Paris 12e, France.*

MILHORAT, Ade Thomas, Am. physician; b. Hoboken, N.J., Jan. 12, 1899; s. Paul and Pauline (Becherer) M.; A.B., Columbia, 1924; M.D., Cornell U., 1928; m. Edith Caulkins Herrick, July 1, 1930; children—Thomas H., Edith H. (Mrs. John A. Boothby). Research fellow medicine U. Leipzig (Germany), 1930-32; faculty Cornell U. Med. Coll., 1932——, prof. clin. medicine, 1956-64, emeritus prof. clin. medicine, 1964——; mem. staff N.Y. Hosp., 1935——, attending physician, 1952-64, cons., 1964——; dir. Inst. for Muscle Disease, Inc., N.Y.C., 1958——. Chmn. med. adv. bd. Muscular Dystrophy Assns. Am., 1950——; mem. med. adv. bd. Muscular Dystrophy Assn. Can., 1954——. Recipient N.Y. Philanthropic League-Ann. Medicine award, 1961; Columbia Coll. award, 1964. Mem. Fedn. State Med. Bds. U. S. (hon.), L'Association Francaise Contre La Myopathie, (hon.), Am. Acad. Neurology, A.A.A.S., Am. Chem. Soc., Am. Congress Phys. Medicine and Rehab., Am. Inst. Nutrition, A.M.A., Am. Physiol. Soc., Am. Soc. for Clin. Investigation, Am. Soc. for Pharmacology and Exptl. Therapeutics, Harvey Soc., N.Y. County Med. Soc., Soc. for Exptl. Pathology and Medicine. Research, numerous publs. on voluntary muscle in health and disease, clin. and path. features of muscle diseases, especially muscular dystrophy, biochemistry of muscle in muscle disease, especially metabolism of creatine and creatinine. Home: 14 Terrace Pl., Pelham Manor, N.Y. 10803. Office: 515 E. 71st St., N.Y.C. 10021.*

MILINE, Radivoj, Yugoslavian physician; b. Lalic, Yugoslavia, Mar. 18, 1912; s. Josif and Evica (Zizakov) M.; M.D., Med. Sch., Belgrade, Yugoslavia, 1936; m. Milka Glavaski, Aug. 13, 1939; 1 son, Josif. Demonstrator histology and embryology Med. Sch., Belgrade, 1933-36; research fellow Med. Sch., Nancy, France, Paris, France, 1938; practice medicine, Becej, Yugoslavia, 1939-41, 47-49; faculty Med. Sch., Sarajevo, Yugoslavia, 1949-61, prof., 1952-61, chief dept. histology and embryology, 1949-61; prof., chief dept. histology and embrsyology Med. Sch., Novi Sad, Yugoslavia, 1961——, dean, 1966——; dir. Inst. for Univ. Teaching, U. Novi Sad, 1963——. Recipient Medal of Work, 1949, Ornament of Work II, 1959, Ornament of Merit for People with Silver Rays, 1965, October prize, 1964 (all Yugoslavia); Palms Academic (France), 1954. Mem. Internat. Soc. Bioclimatology and Meteorology, Société d'Endocrinolgie, Royal Med. Soc., Assn. des Anatomistses, Societa italiana di Anatomia, Anatomische Gesellschaft, Acad. Scis. and Art Bosnia and Hercegovina. Author: Ceruminogenic Apparatus of the External Ear, 1946; also numerous articles. Research on influence of sound and vibrations (audiogenic stress) and emotional stress on neuroendocrine system; effect of Night and darkness on hypothalamus, reactivity changes in histophysiology of epithalamopineal complex under stress conditions, epithalamopineal complex. Home: 2/1, Jovana Boskovica. Novi Sad, Yugoslavia.*

MILL, Hugh Robert, Brit. geographer, meteorologist; b. Thurso, Scotland, May 28, 1861; s. James and Harriet (Davidson) M.; ed. U. Edinburgh, D.Sc.; LL.D, St. Andrews, 1900; m. Frances MacDonald, 1889, m. 2d, Alfreda Dransfield, 1937. Chemist, physicist Scottish Marine Sta., 1884-87; univ. extension lectr., 1887-1900; dir. Brit. Rainfall Orgn., also editor Brit. Rainfall and Symon's Meteorological Magazine, 1901-19; rep. Internat. Council for Study of Sea, 1901-08; rainfall expert to Met. Water Bd., 1906-19; mem. bd. Trade Com. on Water Power of Brit. Isles, 1918-21. Librarian Royal Geog. Soc., 1892-1900, v.p., 1927-31; recorder sect. E, Brit. Assn., 1893-99, pres. 1901; hon. sec. Royal Meterol. Soc., 1902-07, pres., 1907-08; pres. Geog. Assn., 1932. Author: Realm of Nature, 1891; The Clyde Sea Area; The English Lakes, 1895; Hints on the Choice of Geographical Books, 1897; New Lands, 1900; The Siege of the South Pole, 1905; Historical Introduction to Sir Ernest Shackleton, 1923; Record of Royal Geographical Society, 1930; Hugh Robert Mill, an Autobiography, 1951. Collected records of rainfall in Brit. Isles for period 1677-1910; influenced reforms in teaching of geography; proposed internat. Ice Patrol of N. Atlantic; encouraged polar exploration. Died Dormans Park, East Grinstead, Sussex, Apr. 5, 1950.

MILL, John Stuart, economist, social philosopher; b. London, May 20, 1806; s. James and Harriet (Burrow) M.; ed. by his father; studied French, math., chemistry, botany in France, 1820, also jurisprudence, psychology in Eng., 1822; m. Harriet Hardy Taylor, 1851. Became clerk in examiner's office East India Service, 1823, then served as head, until 1858; elected to parliament, 1865-68; helped organize Utilitarian Soc., 1822. Author: System of Logic, 1843; Essays on Some Unsettled Questions in Political Economy, 1844; Principles of Political Economy, 2 vols., 1848; Essay on Liberty 1859; Thoughts on Parliamentary Reform, 1859; Representative Government, 1861; Utilitarianism, 1863; Auguste Comte and Positivismn 1865; Examination of Sir William Hamilton's Philosophy, 1865; The Subjection of Women, 1869; The Irish Land Question, 1870; Autobiography, 1873; also numerous articles to jours. Greatest theorist of Liberal Democracy; concerned with question of human liberty and opposed all forms of depostism, including that of the majority; influenced by his father and Jeremy Bentham; sought to merge utilitarian doctrines with humanitarianism; urged reform and meliorative democratic change; believed the first element of good govt. was virtue and intelligence of community; stressed the utility of freedon of thought and action; felt that ultimate goal of govt. must be the permanent good soc.; exerted great influence in economics, philosophy, and polit. sci. Died Avignon, France, May 8, 1873.

MILL, Theodore, Am. chemist; b. Hamilton, Ont., Can., Apr. 17, 1931 (parents Am. citizens); s. Joseph and Frances (Tick) M.; B.S. in Chemistry, Wayne State U., 1953; Ph.D., U. Wash., 1956; m. Aug. 10, 1957 (div.); children—Susan, Jeffrey. Post-doctoral fellow Hickrill Research Found., Katonah, N.Y., 1956-57; research chemist organic chems. dept. E. I. DuPont de Nemours & Co., Wilmington, Del., 1957-60; sr. organic chemist Stanford Research Inst., Menlo Park, Cal., 1960-63, chmn. dept. phys. organic chemistry, 1963——. Mem. Am. Chem. Soc., Chem. Soc. (London), Research Soc. Am., Sigma Xi. Research, publs. on reactions aldehydes with chromium ion, reactions fluorinated olefins, ketones and aldehydes with basic reagents, mechanisms reaction nitrogen fluorine compounds, free radical oxidation organic compounds. Home: 156 30th St., San Mateo, Cal. 94403. Office: Stanford Research Inst., Menlo Park, Cal.*

MILLAR, Ian Torrance, English chemist; b. Leicester, Eng., Aug. 4, 1927; s. William Hardwick and Eve (Turner) M.; Dip.Chem. Tech., Loughborough U. Tech., 1949; B.Sc. with 1st class honors, U. London, 1949; postgrad. Trinity Coll., Cambridge; Ph.D., 1953; m. Margaret Brenda Smith, July 4, 1952; children—David Hugh, John William Aiden. Faculty, U. Coll. N. Staffordshire, Eng., 1954-64, sr. lectr., 1961-64; faculty U. Keele (Eng.), 1964——, prof. organic chemistry, 1965——. Cons. in industry and medicine. Fellow Royal Inst. Chemistry, Royal Astron. Soc. Author: (with H. D. Springall) The Organic Chemistry of Nitrogen, 1967; also articles. Editor: (with D. Cohen) Titration in Non-Aqueous Media. Research in synthesis and properties of novel cyclic organic compounds, especially heterocyclic derivatives of phosphorus and arsenic, and polycyclic aromatic types, metal complexes. Home: Church Planatation, Univ., Keele, Staffs., Eng.; also Eccle Riggs Cottage, Broughton-in-Furness, Eng.*

MILLAR, John, Brit. physician; b. Scotland, 1733; M.D., U. Edinburgh (Scotland); m. Isabella Brisbane; 2 sons, including John; practiced medicine, Kelso, Scotland; apptd. physician Westminster Gen. Dispensary, 1774. Author: Observations on the Asthma and on the Whooping Cough, 1769; Observations on the Prevailing Diseases in Great Britain, 1770; Some Additional Observations on the Prevailing Diseases in Great Britain, 1777; Observations on Antimony, 1774; A Discourse on the Duty of Physicians, 1776; Observations on the Practice in the Medical Department of the Westminster Dispensary, 1777; Observations on the Management of the Diseases of the Army and Navy During the American War, 1783; Observations on the change of Public Opinion in Religion, Politics and Medicine, on the Conduct of the War, on the Prevailing diseases in Great Britain, and on Medical arrangements in the Army and Navy, 2 vols., 1804. Credited with earliest complete account of laryngismus stridulus (Millar's asthma), 1769. Died Feb. 25, 1805.

MILLAR, John, Scottish social philosopher; b. Shotts, Lanarkshire, Scotland, June 22, 1735; s. James Millar; student U. Glasgow, became disciple of Adam Smith; m. Margaret Craig; children include John, James, William. Became advocate, 1760; prof. law U. Glasgow, from 1761, achieved nat. recognition for his lectrs; proprietor Milheugh farm, from 1785. Mem. Soc. of Friends of People. Author: The Origin of the Distinction of Ranks, 1771; Historical View of the English Government, 1787. Sympathized with Am. and French Revolutions; deplored slave trade and criticized toryism; held that all social relations are determined by econ. orgn. of soc.; stressed social and econ. influences on English constl. history. Died Millheugh, May 30, 1801.

MILLARD, Naomi Adeline Helen (Mrs. Pierce Arthur Newton Millard), zoologist; b. Cape Town, S. Africa, July 16, 1914; d. Harold and Adeline Kate (Thompson) Bokenham; B.Sc., U. Cape Town, 1934, M.Sc., 1935, Ph.D., 1942; m. Pierce Arthur Newton Millard, Mar. 10, 1938; children—Mary (Mrs. A. B. Clark), Peter Harold. Faculty, U. Cape Town, 1936——, sr. lectr., 1959——. U. Cape Town fellow, 1952-61. Fellow Royal Soc. S. Africa; mem. Zool. Soc. So. Africa (hon. sec.). Author: (with Robinson) The Dissection of the Spiny Dogfish and the Platanna, 1945; also articles. Research on anatomy and devel. blood system Xenopus laevis, marine ecology, fouling ships' hulls, intertidal ecology, systematics S. African Hydrozoa. Home: 5 Pillans Rd., Rosebank, Cape. Office: Zoology Dept., U. Cape Town, Rondebosch, Cape, S. Africa.*

MILLARDET, Alexis, French botanist; b. Montmirey, France, Dec. 3 ,1830; prof. botany, universities of Nancy, Strasbourg, Bordeaux (all France). Mem. French Acad. Scis., 1888. Author: Le prothallium male des cryptogames vasculaires, 1869; Notes sur les vignes américaines, 1876; La Question des Vignes Américaines au Point du Vue Theoretique et Pratique, 1877. Planned hybridization of Am. and French grapevines. developed cupric treatment for mildew. Died Bordeaux, Dec. 14, 1902.

MILLER, Augustus Taylor, Am. physiologist; b. Arlington, Tex., Apr. 14, 1910; s. Augustus Taylor and Maude (Duckett) M.; B.S., Emory U., 1931, M.S., 1933; Ph.D., U. Mich., 1939; M.D., Duke, 1953; m. Adeline Helen Porombovics, Oct. 17, 1938; 1 son, Robert David. Research physiologist Maybury Sanatorium, Northville, Mich., 1936-39; faculty dept. physiology U. N.C., Chapel Hill, 1939——, prof., 1950——. Mem. adv. com. on environmental medicine Surgeon Gen. U. S. Army, 1958——; cons. USN 1950, USAF, 1965——. Mem. Am. Physiol. Soc., Elisha Mitchell Sci. Soc., Phi Beta Kappa, Alpha Omega Alpha. Author: (with L. E. Morehouse) Physiology of Exercise, 5th edit., 1967; Energy Metabolism, 1967. Publs. on contbns. to physiology of exercise, body composition and metabolism, regulation of cell metabolism, effects of hypoxia on cells, acclimatization to altitude. Home: 804 Old Mill Rd., Chapel Hill, N.C. 27514.*

MILLER, Benjamin LeRoy, Am. geologist; b. Sabetha, Kan., Apr. 13, 1874; s. Jacob J. and Mary (Moorhead) M.; student Morrill Coll., 1889-90, Washburn Coll., 1891-92; A.B., U. Kan., 1897; postgrad. U. Chgo., summer, 1898; Ph.D., Johns Hopkins, 1903; Sc.D., Moravian College, 1941; m. Mary A. Meredith, Sept. 15, 1904; children—Ruth Meredith (Mrs. Otto H. Spillman), Ralph LeRoy. Tchr., pub. schs. of Kan., 1894-95; asst. Kan. U. Geol. Survey, summer, 1896; prof. biology and chemistry Penn Coll., Oskaloosa, Ia., 1897-1900; spl. asst. Ia. Geol. Survey, summer, 1899; asso. in geology Bryn Mawr Coll., 1903-07; prof. geology Lehigh U., since 1907, Geologist, Md. Geol. Survey, 1900-11; asst. U. S. Geol. Survey, 1904-07, asst. geologist, 1907-13; asso. geologist, Pa. Geol. Survey, since 1919. Fellow A.A.A.S., Mineral. Soc. Am., Geol. Soc. Am., Ia. Acad. Scis., Geol. Soc. London; mem. Am. Inst. Mining and Metall. Engrs., Soc. Econ. Geologists, Seismol. Soc. Am., Am. Meteorol. Soc., Pa. Acad. Sci. (pres. 1925-26), Sigma Xi, Contbr. numerous reports on geol. survey results, pub. by U. S. Geol. Survey and state geol. surveys of Iowa, Md., Va., N.C. and Pa., also articles on econ. geology in tech. jours., especially on limestones, cement, graphite and other non-metallic products as well as articles on stratigraphic geology of Eastern Pa. in geol. periodicals; collaborator with Dr. George B. Shattuck in Geology and Geography of the Bahama Islands, in Bahama Islands, 1905. Author: Geology of Mining Districts of South America and Central America; Mineral Deposits of South America (with Dr. J. T. Singewald, Jr.), 1919. Spl. cons. editor Engring. and Mining Jour., 1920-22. Died Mar. 23, 1944.

MILLER, Bernard, Am. organic chemist; b. Monticello, N.Y., Sept. 1, 1930; s. Isidore and Sarah (Mandelbaum) M.; B.S., Coll. City N.Y., 1951; M.A., Columbia, 1953, Ph.D., 1955; m. Ruth Alice Kussner, Dec. 12, 1965. Postdoctoral fellow U. Wis., 1955-57; research chemist Am. Cyanamid Co., Stamford, Conn., 1957-60, sr. research scientist, Princeton, N.J., 1961-67; asso. prof. dept. chemistry U. Mass., Amherst, 1967—. Mem. Am. Chem. Soc., Phi Beta Kappa, Phi Lambda Upsilon. Research, publs. and patents on mechanisms of organo-phosphorus reactions, molecular rearrangements, reactions of blocked aromatic molecules. Home: 120 Columbia Dr., Amherst, Mass.*

MILLER, Carl Wallace, Am. physicist; b. Somerville, Mass., Mar. 28, 1893; s. Charles Nahum and Lula (Lombard) M.; A.B., Harvard, 1915, Ph.D., 1922; student U. Zurich (Switzerland), 1915-16; M.A. (hon.) Brown U., 1945; m. Edna Louise Savary, Jan. 1, 1920; children—Virginia (Mrs. Travis J. Covington), Carlton Stone. Instr. math. Harvard, 1917-18, asst. in physics, 1920-22; instr. physics N.Y. U., 1922-24; asst. prof. physics Brown U., Providence, 1924-29, asso. prof., 1929-45, prof., 1945-55, prof. emeritus, 1955—; research asso. Office Naval Research, Beavertale Project, Yale, 1952-54. Recipient Franklin L. Burr prize Nat. Geog. Soc., 1948. Fellow A.A.A.S., Am. Phys. Soc., Royal Photog. Soc. Gt. Britain; mem. Photog. Soc. Am. Author: Principles of Photographic Reproduction, 1942; A Scientist's Approach to Religion, 1947. Participant, Nat. Geog. Soc. eclipse expdn. to Siam, 1948 ; research on selnium rectifier and electrolytic condenser; early applications of radio circuits and electronic tubes, elec. instrument design; photog. densitometry, methods of photog. reprodn. in color, photogrammetry; physiol. optics, night vision, binoculars. Home: 32 Balton Rd., Providence 02906.*

MILLER, Charles Leslie, Am. civil engr.; b. Tampa, Fla., June 5, 1929; s. Charles Henry and Myrle (Walstrom) M.; B.S. in Civil Engring., Mass. Inst. Tech., 1951, M.S., 1958; m. Roberta Jean Pye, Sept. 10, 1949; children—Charles Henry, Stephen Leslie, Jonathan Lee, Matthew William. With Michael Baker, Jr., Inc., cons. engrs., Jackson, Miss., Rochester, Pa., 1951-55; faculty Mass. Inst. Tech., Cambridge, Mass., 1955—, prof. civil engring., head dept., founder Civil Engring. Systems Lab.; cons. engr., owner CLM/Research, C. L. Miller Cons. Engr., 1955—. Dir. Lockwood, Kessler & Bartlett, Inc., Spaulding and Slye, Inc. Cons., adviser govt. agys., indsl. cos.; mem. Latin Am. Sci. Bd. Recipient ENR Men Who Made Mark award; George Westinghouse award Am. Soc. Elec. Engrs. Mem. Am. Soc. C.E., Am. Soc. Elec. Engrs., Sigma Xi, others. Contbr. articles to tech. jours. Originated digital terrain model system, coordinate geometry system, integrated civil engring. system; advances in man-machine communications and approaches to application of computers to civil engring. practice; innovations in civil engring. edn. Home: 24 Colony Rd., Lexington, Mass. Office: Mass. Inst. Tech., Cambridge, Mass. 02139.*

MILLER, C(harles) Phillip, (Jr.), Am. physician; b. Oak Park, Ill., Aug. 29, 1894; s. Charles Phillip and Louise (Pebbles) M.; B.S., U. Chicago, 1916; M.D., Rush Med. Coll., 1919; M.S., U. Mich., 1920; m. Florence Lowden, Oct. 20, 1931; children—Phillip Lowden, Warren Pullman. Intern Presbyn. Hosp., Chicago, 1918-19; vol. research asst. pathology U. Mich., 1919-20, asst. prof., 1920; asst. resident physician Hosp. of Rockefeller Inst. for Med. Research, 1920-24, asst. pathology and bacteriology, 1924-25; vol. research asst. Institut für Infektionskranheiten Robert Koch, Berlin, 1926; asst. prof. medicine U. Chicago, 1925-30, asso. prof., 1930-41, prof., 1941-60, professor emeritus, 1960—; consultant to the Secretary of War, 1941-49; mem. Commission Meningitis and Commn. on Air Borne Infection, Army Epidemiological Bd., 1941-45; sr. sci. officer Office Sci. and Tech. Am. Embassy, London, 1948. Mem. NRC. (fellowship bd. 1947-53, exec. com. 1957-62); mem. Streptococcal Commn., Armed Forces Epidemiology Bd., 1949-54, now mem. Commission on Radiation and Infection; mem. divisional com. Biol. & Med. Scis., Nat. Sci. Found., 1955-60; adv. com. forestry U. Ill.; Neisseria sub-com. Nomenclature Com. Internat. Assn. Microbiologists. Member bd. of trustees Farm Found., 1943-64. Fellow American Coll. Physicians, Am. Acad. Microbiology; mem. Conf Bd. Asso. Research Councils, Com. Internat. Exchange of Persons 1950-53, Radiation Research Soc. (councilor 1957-60), Am. Forest Assn. Am. Clin. and Climathol. Assn., A.M.A., A.A.A.S., Am. Assn. Immunologists, Am. Soc. Exptl. Pathology (sec.-treas., 1930-34, councillor, 1934-36, v.p., 1936, pres., 1937), Am. Soc. Clin. Investn. (v.p. 1938), Assn. Am. Phys. (pres. 1956-57), Nat. Academy Science, American Society Experimental Biology (secretary 1933), Institute of Medicine, Am. Acad. Arts and Sciences, American Society for Microbiology, Society for Experimental Biology and Medicine (editorial board, 1953-58), Society of General Microbiology Great Britain, Sigma Xi. Studies of the biological properties of meningococcus and gonococcus; development of penicillin resistance by meningococcus in vitro and in vivo; research on antibiotics and effects of radiation on resistance to infection; bacteriology; immunology. Home: 5757 Kimbark Ave., Chgo. 60637.*

MILLER, Charles Walter, English physicist; b. Leeds, Eng., Mar. 25, 1915; s. Charles Walter and Margaret (Bailey) M.; B.Sc. with honors in Physics with Elec. Engring., Leeds U., 1936, M.Sc. in Physics, 1939, D.Sc., 1957; m. Evelyn Appleyard, Oct. 7, 1939; children—Margaret Evelyn, Carol Anne, Ronald Stephen Bradley. With Asso. Elec. Industries Ltd. (formerly Met.-Vickers Elec. Co. Ltd.), Manchester, Eng., 1936-68, group leader electronic group, 1956-64, group leader physics group, 1964-68; prof., head dept. physics The City U., London, 1968—. Recipient Inst. Elec. Engrs. premium, 1954; Brit. I.R.E., premium, 1954; Radio Industry Council award for tech. writing, 1957. Fellow Inst. Physics, Inst. Electronic and Radio Engrs. Contbg. author: Radiation Sources, 1964; Radiological Monitoring of the Environment, 1965. Research, publs. patents on high vacuum techniques; devel. vacuum pumps; war-time radar; microwave aerials and wave-guide components; particle accelerators including linear accelerators for X-ray therapy; high power radio valves, high energy radiography, magneto hydrodynamic generation, high voltage D.C. problems, voltage breakdown problems. Home: Lawnside, 15 Greenhill Rd., Timperley, Cheshire, Eng. Office: Dept. Physics, The City U., St. John St., London E.C. 1, Eng.*

MILLER, Daniel R., Am. psychologist; b. N.Y.C., Sept. 14, 1917; s. Herman and Gussie (Feinberg) M.; B.Social Service, Coll. City N.Y., 1938; M.A., George Washington U., 1942; Ph.D., Stanford, 1948; m. Jeanette Roman, Dec. 10, 1939; children—Jonathan, Celia, Emily. Faculty, U. Mich., Ann Arbor, 1948—. Mem. Am. Psychol. Assn., Am. Sociol. Assn., Soc. for Research in Child Devel. Author: (with G. E. Swanson) Changing American Parent, 1958, Inner Conflict and Defense, 1960; also numerous articles. Research on learning def. mechanisms, social origins patterns child rearing, basis for mut. attraction and efficiency enduring relationships. Home: 1829 Vinewood St., Ann Arbor, Mich.*

MILLER, Daniel Weber, Am. physicist, educator; b. Omaha, Jan. 24, 1926; s. Merritt Finley and Flora Grace (Ernst) M.; B.S. in Elec. Engring., U. Mo., 1947; Ph.D. in Physics (AEC predoctoral fellow), U. Wis., 1951; m. Carolyn Leigh Hunt, Dec. 26, 1947; children—Debra Leigh, Douglas Hunt. Research asso. Ind. U., Bloomington, 1951-52, faculty, 1952—, prof. physics, 1962—, asso. dean Coll. Arts and Scis., 1962-64, acting chmn. dept. physics, 1964-65. Cons. Los Alamos Sci. Lab., 1959-64, vis. staff mem., 1964—. NSF grantee, 1963—. Fellow Am. Phys. Soc.; mem. Midwestern Univs. Research Assn. (dir. 1964—), Sigma Xi. Research in exptl. nuclear physics including fast-neutron total cross sect., charged-particle investigations nuclear structure and nuclear reaction mechanisms, devel. separated-magnet isochronous cyclotron. Home: 1030 S. High St., Bloomington, Ind. 47401.*

MILLER, Dayton Clarence, Am. physicist; b. Strongsville, O., Mar. 13, 1866; s. Charles W. D. and Vienna (Pomeroy) M.; A.B., Baldwin U., 1886, A.M., 1889; D.Sc., Princeton, 1890, Miami, 1924, Dartmouth, 1927; LL.D., Western Reserve, 1927, Baldwin-Wallace Coll., 1933; D.Eng., Case, 1936; m. Edith C. Easton, June 28, 1893. Prof. natural sci. Baldwin U., 1888-89; asst. in math. and physics Case Sch. Applied Sci., Cleve., 1890-93, prof. physics, from 1893. Fellow Am. Phys. Soc. (pres. 1925-26), A.A.A.S. (v.p. 1908, gen. sec. 1910), Am. Acad. Arts and Scis., Ohio Acad. Scis.; mem. numerous sci. socs. Recipient socs. Awarded Longstreth medal, 1917; Elliott Cresson Gold Medal Franklin Inst., 1926; A.A.A.S. prize, 1925; Cleveland Distinguished Service medal, 1927. Author: Laboratory Physics, 1903; Boehm on The Flute and Flute-Playing, 1908; The Science of Musical Sounds, 1916; Bibliography of the Flute, 1935; Anecdotal History of Sound, 1935; Sound Waves, Shape and Speed, 1937; Sparks, Lightning, Cosmic Rays, 1939. Engaged in research (with Morley) to confirm Michelson-Morley exptl. results on speed of light, 1902-04; inventor photodeik for making sound waves visible, 1912, used it to study mus. sounds; developed musicology as a science; attempted to determine ether drift (for purposes of invalidating theory of relativity), 1921. Died Feb. 22, 1941.

MILLER, E. Willard, Am. geographer; b. Turkey City, Pa., May 17, 1915; s. Archie H. and Tessie (Master) M.; B.S., Clarion State Coll., 1937; A.M., U. Neb., 1939; Ph.D., Ohio State U., 1942; m. Ruby Marie Skinner, June 26, 1941. Faculty, Ohio State U., 1942-43, Western Res. U., 1943-44; with OSS, Washington, 1944-45; faculty Pa. State U., University Park, 1945—, prof., 1949—, head, dept. geography, 1945-63, asst. dean resident edn. Coll. Earth and Mineral Scis., 1964-67, asst. dean for resident instrn. and continuing edn., 1967—. Dir. NSF Acad. Year Inst., 1967—. NSF grantee, 1960-63. Fellow A.A.A.S., Am. Geog. Soc., Nat. Council Geog. Edn. (Ray Hughes Whitbeck award 1947), Internat. Inst. Arts and Letters; mem. Assn. Am. Geographers, Am. Soc. Profl. Geographers (pres. 1948-49), Am. Inst. for Mining, Metall. and Petroleum Engrs., Pa. Acad. Sci. (pres. 1966—), Sigma Xi, Beta Gamma Sigma, Pi Gamma Mu. Author: A Geography of Manufacturing, 1962, An Economic Atlas of Pennsylvania, 1964; (with Deasy, Griess, Case) The World's Nations, 1957; (with Langdon) Exploring Earth Environments, 1964; Energy Resources of the U. S., 1967; Mineral Resources of the United States, 1967. Editor: Global Geography, 1956. Geog. editor Thomas Y. Crowell Co., 1956—; contbg. editor Producers Monthly Mag., 1947—. Research in fields of minerals and mfg. geography with emphasis on indsl. localization and importance of minerals in economy of regions. Home: 845 Outer Dr., State College, Pa. 16801. Office: Mineral Scis., U. Pa., University Park, Pa. 16802.*

MILLER, Edward Calvin, Am. metall. engr.; b. Bonne Terre, Mo., Dec. 28, 1905; s. Adolph William and Emilie (Wolf) M.; B.S. in Metall. Engring., U. Mo. Sch. Mines, 1928; M.S., U. Ida., 1930; m. Jean Ainsworth, Feb. 2, 1947; children—Emily Jean, Edward Calvin, Kenneth A., Janet E. Metall. prodn. and devel. St. Joseph Lead Co., Mont., Pa., 1929-35; faculty Purdue U., 1935-41, 46-47, Wayne U., metallurgist materials test reactor, air-craft nuclear propulsion and homogeneous reactor projects Oak Ridge Nat. Lab., Union Carbide Corp., 1948-55, supt. inspection engring. dept., 1955—. Mem. exchange tour Soviet welding industry, 1962. Mem. Am. Soc. M.E., Am. Welding Soc. (nat. pres. 1967-68), Am. Soc. for Metals, Am. Nuclear Soc., Am. Soc. Testing and Materials, Am. Inst. Metall. Engrs., Sigma Xi, Tau Beta Pi, Phi Kappa Phi. Lectr., publs. on zirconium, liquid metal corrosion, codes, non-destructive testing, integrity of nuclear pressure vessels, welding in USSR, reactor vessel fabrication and inspn. Home: 8115 Chesterfield Dr., Knoxville, Tenn. 37919. Office: Oak Ridge Nat. Lab., Oak Ridge 37830.*

MILLER, Edward Furber, Am. mech. engr.; b. Somerville, Mass., Jan. 18, 1866; s. William Gibbs and Sarah (Furber) M.; S.B., Mass. Inst. Tech., 1886; D.Sc., R.I. State Coll., 1921; m. Mary Willard Reed, Sept 11, 1900. Tchr. mech. engring. Mass. Inst. Tech., 1886-92, prof. steam engring., 1892—, in charge dept. mech. engring., 1911—, dean of army officers, 1922—. Author: Steam Boilers (with Cecil H. Peabody), 1897; Problems in Thermodynamics and Heat-Engineering (with C. W. Berry and J. C. Riley), 1911; Notes on Power Plant Design (with James Holt), 3d edit.; Notes on Heat Engineering, 1931. Died June 12, 1933.

MILLER, Elizabeth Cavert, Am. biochemist; b. Mpls., May 2, 1920; d. William Lane and Mary (Mead) Cavert; B.S., U. Minn., 1941; M.S., U. Wis., 1943, Ph.D., 1945; m. James Alexander Miller, Aug. 30, 1942; children—Linda Ann, Helen Louise. With U. Wis., 1945—, asso. prof. dept. oncology, 1958—. Recipient Teplitz-Langer award, 1963; Lucy Wortham James award, 1965. Mem. Am. Assn. Cancer Research (dir., 1957-60), Am. Soc. Biol. Chemists, Sigma Xi. Asst., asso. editor Cancer Research, 1952-64. Studies, publs. on biochem. mechanisms involved in induction of tumors by chem. carciogens. Home: 5517 Hammersley Rd., Madison, Wis. 53711.*

MILLER, Erston Vinton, Am. plant physiologist; b. Hagerstown, Md., Aug. 8, 1898; s. Ernest Wilson and Clara (Nunamaker) M.; B.S., U. Md., 1919, M.S., 1921; Ph.D., Mich. State U., 1926; m. Elinor Case, Aug. 23, 1927; children—Rosemary, Alan, Carol Munger, Andrew. Instr. scis. Shanghai Am. Sch., Shanghai, China, 1921-24; plant physiologist U. S. Dept. Agr., 1928-47; prof. biology U. Pitts., 1947-67; cons. Universidad Central, Quito, Ecuador, 1966-67. Cons. to pvt. cos., assns. Mem. Am. Soc. Plant Physiologists, Am. Inst. Biol. Scis., A.A.A.S., Sigma Xi. Author: Within the Living Plant, 1953; Chemistry of Plants, 1957; (with James I. Munger) Good Fruits, 1967; also articles. Reported evolution of ethylene by decaying oranges and by decay-producing organisms. Home: 7616 Homestead Rd., Benzonia, Mich. 49616.*

MILLER, Foil Allan, Am. chemist; b. Aurora, Ill., Jan. 18, 1916; s. Fred Allen and Bertha (Milliren) M.; B.S., Hamline U., 1937; postgrad. U. Neb.; Ph.D., Johns Hopkins, 1942; m. Ruth Naomi Zeller, Sept. 4, 1941; children—Bruce Allan, Craig Foil. Fellow chemistry U. Minn., 1942-44; asso. and asst. prof. chemistry U. Ill., 1944-48; asst. prof. chemistry, head spectroscopy div. Mellon Inst., Pitts., 1948-58, sr. fellow in ind. research, 1958-67, adj. sr. fellow, 1967—; univ. prof., dir. Spectroscopy Lab., U. Pitts., 1967—; Lectr. in chemistry U. Pitts., 1952-64, adj. prof. chemistry, 1964-67. Co-recipient Pitts. Spectroscopy award, 1964. Mem. Am. Chem. Soc. (Pitts. award 1965), Optical Soc. Am., Coblentz Soc. Research, numerous publs. on infrared and Raman spectra, far infrared spectroscopy. Home: 4625 5th Av., Pitts. 15213.*

MILLER, Frank Ebenezer, Am. physician; b. Hartford, Conn., Apr. 12, 1859; s Ebenezer B. and Mayette (Deming) M.; A.B., Trinity Coll., Conn., 1881; M.D., Coll. Phys. and Surg. (Columbia), 1884; m. Emily Weston, Apr. 28, 1892. Intern N.Y., and Charity hosps., 6 mos., St. Francis Hosp., 2 yrs.; san. insp. Bd. Health, N.Y.C., 1886-89; asst. to various specialists in treatment of nose throat and ear; began practice, 1896; chief throat surgeon Bellevue Hosp., 1886, Vanderbilt Clinic, 1890-93; cons. physician St. Francis and St. Joseph's hosps. Author: Observations in Vocal Art Science, 1909; The Voice, Its Production, Care and Preservation, 1910; Vocal Art-Science, 1917; The Banner of Universal Harmony, 1919. Made sci. study of voice; originator vocal art-science method of voice prodn. Died Apr. 15, 1932.

MILLER, Frederic Howell, Am. mathematician; b. N.Y.C., June 17, 1903; s. Frederic William and Anna Margaret (Bergheim), M.; B.S., Cooper Union Inst. of Tech., 1926; M.S., Cornell, 1927; Ph.D., Columbia, 1932; m. Marie Glauser, July 30, 1927; 1 dau., Lois Ruth (Mrs. Reginald Bruce Collier). Instr. math., Cooper Union, 1927-29, Columbia, 1929-32; asst. prof. math., Cooper Union, 1932-42, asso. prof. of math., 1942-43, prof. and head dept. math., 1943—. Examiner math., Coll. Entrance Exam. Bd., 1949-51. Mem. Am. Soc. Engring. Edn. (math. div., chmn., 1948-49, dir., 1949-51, council, 1951-53), Math. Assn. Am. (gov., 1948-51), Am. Math. Soc. Author: Advanced Mathematics for Engineers, 1938; Calculus, 1939, Partial Differential Equations, 1941; College Algebra and Trigonometry, 1945; Analytic Geometry and Calculus, 1949. Studied electric circuits; linear differential equations. Died Jan. 11, 1964.

MILLER, Frederick Robert, Canadian physiologist; b. Toronto, Ont., Can., May 2, 1881; s. Allan Frederick and Elizabeth (Crean) M.; B.A., MA., M.B., M.D., U. Toronto, 1903-11, U. Munich (Germany), 1903-05, Cornell U., U. Liverpool (Eng.), Oxford (Eng.) U., 1911-12, U. Strasbourg (France) 1911; m. Lulor M. Porte, Sept. 1924 (dec.); 1 dau., Mary Elizabeth (dec.); m. 2d, Lillian E. Blong, June 15, 1938 (dec. Nov. 1950). Demonstrator physiology U. Toronto, 1907-10; prof. physiology U. Western Ont., London, Can., 1914-50. Fellow Royal Soc., 1932. Research, publs. on peripheral effects faradic stimulation various parts of brain, acetylcholine as chem. transmitter impulses at synapse. Died Toronto, Nov. 11, 1967.

MILLER, Freeman Devold, Am. astronomer; b. Somerville, Mass., Jan. 4, 1909; s. Rasmus Kjeldsberg and Ednah (Weeks) M.; S.B., Harvard, 1930, A.M., 1932, Ph.D., 1934; m. Marie Dresser, June 27, 1933. Dir. Swasey Obs., Denison U., 1934-40; faculty U. Mich., Ann Arbor, 1946—, prof. astronomy, 1956—, asso. dean Horace H. Rackham Sch. Grad. Studies, 1959-66. Mem. Am. Astron. Soc., Pacific, Internat. Astron. Union, Phi Beta Kappa, Sigma Xi. Research, publs. on structure and behavior comets. Home: 1614 Shadford Rd., Ann Arbor, Mich 48104.*

MILLER, Gail Lorenz, Am. biochemist; b. Pleasant Valley, Ia., Nov. 3, 1913; s. Rudolph L. and Sadie (Besse) M.; A.B., U. Ill., 1933, M.A., 1934; Ph.D., George Washington U., 1937; m. Mary Elizabeth Eshelman, Dec. 23, 1914. Instr., George Washington U., 1937-38, Cornell U. Med. Coll., 1938-39; Rockefeller Travelling fellow, Uppsala, Sweden, 1939-40; asst. Rockefeller Inst. Med. Research, Princeton, 1940-45; research chemist Biochem. Research Found., Newark, Del., 1945-46; asso. mem. Inst. Cancer Research, Phila., 1946-52; supervisory biol. chemist U. S. Army, Q.M. Research and Devel. Lab., Natick, Mass., 1952-61; research asso. U. Pitts., 1961-62; Merck Inst. for Therapeutic Research, 1962-64; biochemist VA Hosp., also asso. prof. microbiology U. Mich., Ann Arbor, 1964-65; research asso. Lankenau Hosp., Phila., 1965—. Mem. Am. Chem. Soc., Am. Soc. Biol. Chemists, A.A.A.S., Sigma Xi, Alpha Chi Sigma. Contbr. numerous articles to tech. jours. Synthesized gluathione; research on sulfur amino acids insulin, infectious chem. derivatives tobacco mosaic virus, phys. properties So. bean mosaic virus, electrophoretic properties influenza virus, biochem. methods, electron microscope counting virus particles, measurement cellulase activity, automatic rec. colorimetry; demonstrated electrophoretic isoenzyme systems in fungal carbohydrases. Home: 1001 City Av., Phila. 19151. Office: Lankenau Hosp., City and Lancaster Avs., Phila. 19151.*

MILLER, George Abram, Am. mathematician; b. Lynnville, Pa., July 31, 1863; s. Nathan and Mary Miller (Sittler) M.; A.B., Muhlenberg (Pa.) Coll., 1887, A.M., 1890; Ph.D., Cumberland U., 1893; student Univs. of Leipzig and Paris, 1895-97; m. Cassandra Boggs, Dec. 23, 1909, Prin. schs., Greeley, Kan., 1887-88; prof. mathematics, Eureka (Ill.) College, 1888-93; instr. mathematics, U. Mich., 1893-95, Cornell, 1897-1901; asst. prof. mathematics Stanford, 1901-02, asso. prof., 1902-06; asso. prof. mathematics U. Ill., 1906-07, prof., 1907-31 when retired; prof. mathematics U. Chicago summer 1912, U. Calif., summer 1913. Co-editor Am. Year Book School Science and Mathematics, and Ency. des Sciences Mathematiques. Winner internat., math. prize, 1900. Fellow Am. Acad. Arts and Sciences, A.A.A.S. (sec. Sect. A., 1907-12, chmn., 1921-22; chmn. math. sub-com. on com. of 100 on sci. research); mem. Nat. Acad. Sciences, Math. Assn. Am. (v.p., 1916, pres. 1921), Am. Math. Soc. (v.p. 1907-08), London Math. Soc., Deutsche Mathematiker Verein; corr. mem. Spanish Mathematic Soc.; hon. mem. Indian Mathematic Soc. Author: Determinants, 1892; Mathematical Monographs (co-author), 1911; Theory and Applications of Groups of Finite Order, 1916, rev. ed., 1938; Historical Introduction to the Mathematical Literature, 1916; College Teaching, 1919; Collected Works (Vol. 1), 1935, (Vol. II), 1938; also articles on the theory of groups and the history of mathematics in Am. and fgn. jours. Important in the development and use of determinants; research on abstract groups: showed a non-abelian group can have an abelian group of isomorphisms; proved the number of independent generators of every prime power group is an invariant of the group; completed the determination of the substitution groups of degrees 8 and 9; published his own list of 994 intransitive groups of degree 10. Home: 1203 W. Illinois St., Urbana, Ill.

MILLER, George Armitage, Am. psychologist, educator; b. Charleston, W.Va., Feb. 3, 1920; s. George E. and Florence (Armitage) M.; B.A., U. Ala., 1940, M.A., 1941; A.M., Harvard, 1944, Ph.D., 1946; m. Katherine James, Nov. 29, 1939; children—Nancy, Donnally James. Instr. psychology U. Ala., 1941-43; research fellow Psycho-Acoustic Lab., Harvard, Cambridge, Mass., 1944-48, asst. prof. psychology, 1948-51, asso. prof., 1955-58, prof., 1958—, chmn. dept., 1964—, dir. Center for Cognitive Studies, 1960—; vis. fellow Inst. Advanced Study, Princeton, 1950; asso. prof. psychology Mass. Inst. Tech., Cambridge, 1951-55. Fellow Center Advanced Study Behavioral Scis., Stanford, 1958-59; Fulbright research prof. Oxford (Eng.) U., 1963-64. Recipient Kenneth Craik award St. John's Coll., Cambridge, 1963. Mem. Am. Psychol. Assn. (Distinguished sci. Contbn. award 1963), Nat. Acad. Scis. Author: Psychology, the Science of Mental Life, 1962. Home: 14 Barberry Rd., Lexington, Mass. 02173. Office: 633 Kirkland St., Cambridge, Mass. 02138.*

MILLER, Gerrit Smith, Jr., Am. zoologist; b. Peterboro, N.Y., Dec. 6, 1869; s. Gerrit Smith and Susan (Dixwell) M.; A.B., Harvard U., 1894; m. Elizabeth Eleanor Page, 1897; m. 2d, Anne Chapin Gates, 1921. Asst. curator mammals U. S. Nat. Mus., 1898-1909, curator, 1909-40, asso. in biology, 1941—. Fellow A.A.A.S.; mem. Am. Acad. Arts and Scis., Am. Philos. Soc.; corr. mem. Acad. Natural Scis. (Phila.), Zool. Soc. London. Author: The Families and Genera of Bats; Catalogue of the Land Mammals of Western Europe in the British Museum; List of North American Land Mammals in the United States National Museum, 1911; List of North American Recent Mammals, 1923; also monographs and contbns. to sci. jours. Died Feb. 24, 1956.

MILLER, Glenn Harry, Am. phys. chemist; b. Pitts., Feb. 10, 1922; s. Harry M. and Margaretta (Elling) M.; B.S., Geneva Coll., 1943; Ph.D., Brown U., 1948; m. Mary A. McKay, Feb. 10, 1951; children—Laurie, Marilyn, Jeanne. Chemist, Oak Ridge Eastman Corp., 1944; sr. chemist butadiene div. Koppers, Inc., 1944-45; chemist Tex. Co., 1948-49; faculty U. Cal. at Santa Barbara, 1949—, chmn. dept., 1960-64, prof. phys. chemistry, 1963—. Mem. Am. Chem. Soc., Cal. Chemistry Tchrs. Assn., Sigma Xi. Asst. editor Jour. Phys. Chemistry, 1964-66. Research, publs. on polymerization, popcorn polymer systems, photochemistry, reactions in glow discharges. Home: 1400 Holiday Hill Rd., Goleta, Cal. 93017. Office: U. Cal., Santa Barbara, Cal. 93106.*

MILLER, Harold, English physicist; b. Derbyshire, Eng., Sept. 14, 1909; s. Ephraim Wheat and Alice (Winter) M.; M.A., St. Johns Coll., Cambridge U., Ph.D., 1934; m. Mary Bacon, July 21, 1937; children—John, Robert, Thomas Richard. Staff research dept. Electric and Mus. Industries Ltd., Hayes, Eng., 1934-42; physicist Sheffield Regional Hosp. Bd. (formerly Sheffield Radio Therapy Centre), 1942—, now chief physicist; Hon. lectr. radiol. physics U. Sheffield. Fellow Inst. Physics; mem. Brit. Inst. Radiology (v.p.), Hosp. Physicists Assn. Author: (with J. Walter) Text Book of Radiotherapy; also articles. Devel. TV transmission pick-up devices; application of physics to medicine, especially radiotherapy and radioactive isotopes. Home: 20 Blackbrook Rd., Office: 20 Claremont Crescent, Sheffield, 10, Eng.*

MILLER, Harry Milton, Jr., Am. parasitologist; b. Balt., June 3, 1895; s. Harry Milton and Clara Elizabeth (Mathes) M.; B.S., Ohio Wesleyan U., 1917; Ph.D., U. Ill., 1923; Sc.D. (hon.), U. Sao Paulo, 1951, U. Brazil, 1960, U. Rio Grande do Sul, 1960; m. Elva Anna Pumphrey, June 25, 1921; children—Harriet Ann (Mrs. Alan S. Correll), Robert Milton (dec.). Mem. faculty Washington U., St. Louis, 1923-32; with Rockefeller Found., N.Y.C., 1932—, asso. dir. med and natural scis., 1956—; hon. prof. univs., Brazil, Peru, 1959-60. Hon. tech. advisor Ministry Pub. Health, Uruguay, 1952—; coordinator Chile project Nat. Acad. Scis.-Rockefeller Found., 1960-61. Decorated Officer, Order So. Cross, Brazil; Chevalier, Legion d'Honneur; Merito Agronomico, Colombia; Commendador Al Merito Bernardo O'Higgins, Chile; Comdr., Order Daniel A. Carrion, Peru. Fellow A.A.A.S.; mem. Brazilian Soc. Genetics (hon.), Inst. Physiology, Brazil, Brazilian Soc. for Progress Sci. (hon.), Phi Beta Kappa, Sigma Xi, Gamma Alpha. Research and publs. on morphology and behavior of larval trematodes, immunity of mammals to larval tapeworms devel. of programs for research in basic scis. in Europe, Latin Am. Home: 1263 Old Nassau Rd., Jamesburg, N.J. 08831. Office: Rockefeller Found., 111 W. 50th St., N.Y.C. 10020.*

MILLER, Harvey Alfred, Am. botanist; b. Sturgis, Mich., Oct. 19, 1928; s. Harry Clifton and Carmen (Sager) M.; B.S., U. Mich., 1950; M.S., U. Hawaii, 1952; Ph.D., Stanford, 1957; m. Marjorie Rosemary Bunge, Sept. 12, 1953; children—Valerie Yvonne, Harry Alfred. Asst. botanist U. Mich. Expdn. to Aleutians, summers 1949-50; asst. U. Hawaii, 1950-53; asst. Stanford, 1953-55; instr. botany U. Mass., 1955-56; faculty Miami U., Oxford, O., 1956—, asso. prof., curator Herbarium botany dept., 1961—; vis. lectr. Coll. Guam, 1965. Prin. investigator Schooner Collegiate Rebel Expdn. to Mironesia and Polynesia, 1960, NSF Expdn. to Micronesia and Philippines, 1965; v.p. Marine Research Assos., Inc., 1962—; cons. botany AVCO Ordnance Div., 1963—; hon. staff mem. Hattori Bot. Lab., Nichinan, Japan. Mem. Am. (pres. 1964-65), Brit. bryological socs., Am. Soc. Plant Taxonomists, Sigma Xi, numerous others. Author: (with others) Bryoflora of the Atolls of Micronesia, 1963; also numerous articles. Research specialist in mosses and liverworts in Hawaiian Islands; only bryologist to work extensively in Micronesia, Caroline and Mariana Islands; described numerous new species of Hepaticae and mosses from Pacific areas. Home: 502 Sandra Dr., Oxford, O. 45056.*

MILLER, Helena Agnes, Am. biologist; b. Rudolph, O., Apr. 25, 1913; d. Royal James and Bertha (Hansen) Miller; B.A., B.S., Ohio State U., 1935, M.S., 1937; Ph.D., Harvard, 1945. Instr. botany, Wellesley Coll., 1945-48; faculty Duquesne U., Pitts., 1948—, prof. biology, 1959—, asst. to dean Arts and Sci., 1966; mgr. coop. house, dining hall Mt. Desert Island Biol. Labs., summer 1948. Participant, NSF Botany Conf., summer 1961. Recipient Woman of Year for Pitts. Area award Sigma Lambda Phi, 1959; First Lady of Day award Radio Sta. WRYT, 1962. Mem. Internat. Assn. Plant Taxonomy, Internat. Assn. Plant Morphologists, Soc. Study Evolution, Soc. Exptl. Biology, A.A.A.S., Am. Inst. Biol. Scis., Bot. Soc. Am., Nat. Assn. Biology Tchrs., Nat. Sci. Tchrs. Assn., Soc. Econ. Botany, Soc. Study Devel. and Growth, Taxonomic Soc. Am., Torrey Bot. Soc., Brit. Assn. Advancement Sci., Phi Beta Kappa, Sigma Xi, Sigma Delta Epsilon, Phi Epsilon Phi, Sigma Pi Sigma. Research, publs. on developmental anatomy of vascular plants; proposed plan for improving Am. ednl. system. Home: 532 Highview Rd., Pitts. 15234.*

MILLER, Herbert Chauncey, Am. physician, educator; b. East Orange, N.J., Nov. 2, 1907; s. Herbert C. and Mary (Alling) M.; A.B., Yale, 1930, M.D., 1934; m. Mary C. Thomas, June 16, 1934; children—Norman, Patricia (Mrs. Keith Libbey), Frances (Mrs. Burton Taylor, Jr.). Instr. pediatrics Yale, 1938-42, asso. prof., 1942-45; prof., chmn. dept. pediatrics U. Kan. Med. Sch., Kansas City, 1945—; dir. children's rehab. unit U. Kan. Med. Center, 1963—. Mem. Am. Pediatric Soc. Research on respiratory physiology in newborn; concepts of prediabetic state. Home: 6408 W. 66th St., Overland Park, Kan. Office: U. Kan. Med. Center, Kansas City, Kan. 66103.*

MILLER, Herman, Am. nuclear engr.; b. St. Paul, Dec. 16, 1919; s. S. Ruben and Fae (Powell) M.; B.S., Cal. Inst. Tech., 1943, M.S., 1945, A.E., 1948; m. Joane Soss, Dec. 5, 1943; children—Stacey, Loren, Kris. Chief aerodynamic devel. So. Cal. Coop. Wind Tunnel Cal. Inst. Tech., 1941-51; dep. chief propulsion wind tunnel Arnold Engring. Devel. Center, 1951-53; mgr. Ida. Engring. ANP dept. Gen. Electric Co., Palo Alto, Cal., 1953-55, mgr. design, projects sect., 1955-60, mgr. spl. purpose nuclear systems operation Atomic Products div., 1960-62, gen. mgr., 1962-66; owner Herman Miller Co., 1966; pres. Nat. Nuclear Corp., Palo Alto, Cal., 1968—. Mem. Am. Mgmt. Assn., Inst. Aero. Scis., Am. Nuclear Soc., Caltech Mgmt. Club (bd., founder), Sigma Xi. Publs. on aerodynamics, nuclear power, thermionics. Home: 175 Fawn Lane, Portola Valley, Cal. 94025. Office: 701 Welch Rd., Palo Alto, Cal. 94302.*

MILLER, Hugh, Scottish geologist; b. Cromarty, Scotland, 1802; s. Hugh Miller; self-educated; m.

Lydia Mackenzie Fraser, Jan. 7, 1837. Mason's apprentice, 1820-23; journeyman mason, 1823-34; later stone-cutter; became accountant Bank, Cromarty, 1834. Editor of anti-intrusionist paper, Witness, 1840-56. Instrumental in founding Free Church of Scotland. Mem. Royal Phys. Soc. Edinburgh (became pres. 1852). Author: Letters on the Herring Fishery, 1829; Old Red Sandstone, 1841; The Footprints of the Creator, 1847; My Schools and Schoolmasters, 1852; The Testimony of the Rocks; Rambles of a Geologist. Studied fossils of Old Red Sandstone; contributed to knowledge of Devonian geology and paleontology; popularized geology through writing. Died 1856.

MILLER, Jacques Francis Albert Pierre, physician; b. Nice, France, Apr. 2, 1931; s. Maurice Eugene and Fernande (Debarnot) M.; B.Sc. in Medicine with 1st class honors in Pathology, U. Sydney (Australia), 1953, M.B., B.S. with 2d class honors, 1955; Ph.D. in Exptl. Pathology, U. London, 1960, D.Sc., 1965; m. Margaret Denise Houen, Mar. 17, 1965; children—Jacqueline Suzanne, John Peter (dec.). Reginald Maney Lake and Amy Laura Bonamy scholar U. Sydney, 1957-58; Gaggin research fellow Chester Beatty Research Inst., Inst. Cancer Research, Royal Cancer Hosp., London, 1958-60, lectr., 1960-63; Eleanor Roosevelt Internat. fellow Nat. Cancer Inst., NIH, Bethesda, Md., 1963-65; reader exptl. pathology London U., 1965-66; head dept. exptl. pathology Walter and Eliza Hall Inst. Med. Research, Royal Melbourne (Australia) Hosp., 1966—. Recipient Langer-Teplitz Cancer Research award, 1965. Mem. Brit. Med. Assn., Brit. Soc. for Immunology, Transplantation Soc., Nat. Geog. Soc. Research, publs. on importance of thymus gland to devel. of immune system; relationhip of thymic deficiency to cancer susceptibility in mice. Home: 32 Burke Rd. N., East Ivanhoe, Melbourne. Office: Walter and Eliza Hall Inst. Med. Research, Royal Melbourne Hosp., Melbourne, Australia.*

MILLER, James Albert, Jr., Am. anatomist, physiologist; b. Peitaiho, China, June 21, 1907 (parents Am. citizens); s. James Albert and Mary (McGaw) M.; A.B., Coll. Wooster, 1928; Ph.D., U. Chgo., 1935; D.Sc., Coll. Wooster, 1961; m. Faith Stone, Mar. 23, 1935; children—David Albert, Janet Alice, (Mrs. Charles Armand Levie). Instr. biology Assiut Coll. (Egypt), 1928-31, Ohio U., Athens, 1935-37; instr. anatomy U. Mich., Ann Arbor, 1937-42; asst. prof. U. Tenn., Memphis, 1942-46; asso. prof. Emory U., Atlanta, 1946-48, prof., 1948-60; chmn. anatomy Tulane U., New Orleans, 1960—. NSF Sr. Postdoctoral fellow, 1958; Fulbright Research fellow, Finland, 1962. Recipient Research award Emory chpt. Sigma Xi, 1959; Research prize Assn. Southeastern Biologists, 1959. Mem. So. Soc. Anatomists (organizer, past 1st pres., exec. com. 1962—), Am. Assn. Anatomists, Am. Physiol. Soc., Am. Soc. Zoologists, Growth Soc., Soc. for Cryobiology. Research, publs. on hypothermia protecting newborn and adult mammals from brain damage or death during asphyxia; deleterious effect of intravenous transfusions during severe shock, beneficial effect of intra-arterial transfusions; reversible blocking of embryonic devel. by hyperbaric oxygen. Home: 1428 Bourbon St., New Orleans 70116.*

MILLER, James Edward, Am. meteorologist; b. McCune, Kan., Sept. 19, 1916; s. John A. and Winettie (Hooke) M.; A.B., Central Coll., Mo., 1937; postgrad. U. Tenn., 1938-40; M.S., N.Y. U., 1941; m. Margaret H. Kinsey, May 3, 1940 (div. Nov. 1957); children—Martha, Emily, Elinor; m. 2d., Marie Miller Necarsulmer, Sept. 2, 1960; 1 dau. Holly. Computer, Shell Petroleum Co., Mt. Vernon, Ill., 1937-38; observer U. S. Weather Bur., Knoxville, Tenn., Billings, Mont., 1938-40, forecaster, 1940; faculty N.Y. U., 1940—, prof., 1952—, chmn. dept. meteorology and oceanography, 1961—. Cons. U. S. Weather Bur., Washington, 1946—, Time, Inc., N.Y.C., 1964—. Mem. World Meteorol. Orgn. (exec. com. panel experts on meteorol. edn. and tng.), Am. (Meisinger award 1948), Royal (fgn.) meteorol. socs., Am. Geophys. Union. Author: (with A. F. Spilhaus) Workbook in Meteorology, 1942. Contbg. author: World Book Ency., Ency. Americana. Research, publs. on cyclogenesis, frontogenesis, vertical motions of the atmosphere, atmospheric energetics, severe weather phenomena. Home: 11 Robin Hill Rd., Scarsdale, N.Y.*

MILLER, James Roland, Am. soil chemist; b. Millington, Md., May 19, 1929; s. Roland and Edith (Judefind) M.; student Washington Coll., 1947-50; B.S., U. Md., 1951, M.S., 1953, Ph.D., 1956; m. Katherine Patricia Melvin, Dec. 18, 1954. With Agrl. Research Service, U. S. Dept. Agr., Beltsville, Md., 1956-58; faculty dept. agronomy U. Md., College Park, 1958-63, head dept., 1963—, prof. 1964—. Cons. NASA, 1965—; mem. soil test work group N.E. Soil Research Com., 1961—. Mem. Am. Soc. Agronomy (dir.), Soil Sci. Soc. Am., Am. Soc. Plant Physiologists, Am. Chem. Soc., Soil Conservation Soc. Am., Sigma Xi. Research, publs. on leaching of radiostrontium and radiocesium through soils; developer soil test method for determining available phosphorus in soils.*

MILLER, John Alfred, Brit. geophysicist; b. Davenham, Cheshire, U.K., Oct. 9, 1935; s. John and Lilly (Garner) M.; B.Sc. U. Hull (Eng.), 1957; M.Sc., U. Birmingham (Eng.), 1958; Ph.D., Cambridge (Eng.) U., 1961; m. H. Elizabeth Woodward, Oct. 11, 1958; 1 son, Marcus Hadyn Woodward. Jr. research fellow Churchill Coll., Cambridge, 1962-65, sr. research fellow, 1965—; asst. dir. research Dept. Geodesy and Geophysics, Madingley Rd., Cambridge, Eng., 1967—. Dir. Miller Bros. Farms. Smithson fellow Royal Soc., 1962-66. Fellow Geol. Soc. London, Chem. Soc. London. Research, numerous publs. on geochronology, advances in technique and application of potassium-argon age determination to rocks, ages of rocks from Brit. Isles, geochronological aspects of continental drift on rocks surrounding N. Atlantic. Home: 1 South Rd., Histon, Cambridge; also Thorn Tree Farm, Ainsworth Lane, Crowton, Cheshire, Eng. Office: Dept. Geodesy and Geophysics, Madingley Rise, Madingley Rd., Cambridge, Eng.*

MILLER, John Allen, Am. biologist; b. Ashland, O., Oct. 18, 1905; s. John Allen and Clara (Worst) M.; A.B., Ashland Coll., 1926, Sc.D., 1963; M.S., Ohio State U., 1927, Ph.D., 1932; m. Josephine Mary Sauder, June 27, 1951; children—John Allen III, Caryl Lee. Faculty Ohio State U., Columbus, 1929-63, prof. zoology, 1946-63, prof. emeritus, 1963; prof. biol. scis. U. Miami, Coral Gables, Fla., 1963—. Mem. A.A.A.S., Am. Soc. Zoologists, Ohio Acad. Sci., Inst. Animal Behavior, Sigma Xi, Gamma Alpha. Author: General Zoology Workbook, 7th edit., 1960; also articles. Research on animal behavior, anatomy and physiology of neuro-muscular mechanisms in animal reactions, devel. micro-techniques, electron microscope studies. Home: 12981 Nevada St., Coral Gables, Fla. 33156.*

MILLER, John G(eorge), Am. chemist; b. Phila., Oct. 18, 1908; s. Robert and Regina (Ramspacher) M.; A.B., U. Pa., 1929, Ph.D., 1932; m. Elizabeth Gregg Snyder, June 8, 1940; children—John Gregg, Margery Kampen. Asst. chemistry dept. U. Pa., Phila., 1930-31, faculty, prof., 1952—. Cons. Minerals & Chems. Philipp Corp. Menlo Park, N.J., 1944—, Smith Kline & French Labs., Phila., 1946-49, Pennsalt Chems. Corp., Phila., 1949—, Research and Devel. Lab., Franklin Inst. Phila., 1959-60; vis. examiner in chemistry Swarthmore (Pa.) Coll., 1950-52; spl. examiner personnel dept. City of Phila., 1955-56. Fellow, A.A.A.S.; mem. Am. Chem. Soc. (Service award Phila. sect. 1957, councilor 1965-67), Electrochem. Soc., Metachem. Club, Am. Phys. Soc., Phi Beta Kappa (pres. local chpt. 1950-52), Sigma Xi, Pi Mu Epsilon, Phi Lambda Upsilon, Sigma Tau Sigma (hon.), Alpha Chi Sigma. Patentee in phys.- and mechanisms, molecular structure, thermodynamics, intermolecular forces and surface chemistry. Home: 7801 Lincoln Dr., Phila. 19118.*

MILLER, Joseph Leggett, Am. physician; b. Kewanee, Ill., Nov. 24, 1867; s. James and Jane (Leggett) M.; B.S., U. Mich., 1893; M.D., Northwestern Med. Sch., 1895; m. 1901; 1 son. Began practice medicine, Chgo., 1895; attending physician Cook County Hosp., then St. Luke's Hosp.; prof. clin. medicine U. Chgo. Editor-in-chief, Archives Internal Medicine. Research and numerous publs. on arteriosclerosis, typhoid fever, arthritis, fgn. protein therapy, thyrotoxicosis. Died Chgo., Aug. 6, 1937.

MILLER, Julian Creighton, Am. horticulturist; b. Lexington, S.C., Nov. 29, 1895; s. Simeon Jeremiah and Plumie Elizabeth (Shull) M.; B.S., Clemson Coll. 1921, D.(hon.), 1961; M.S., Cornell U., 1926, Ph.D., 1928; m. Caroline Stone Leichliter, Dec. 26, 1923; children—Rodman B., Julian Creighton. Instr. horticulture, N.C. State Coll., 1921-23; county agrl. agt., S.C., 1923-25; asst., Cornell U., 1925-28; asso. prof. horticulture and research, Okla. A. and M. Coll., 1928-29; prof. horticulture and head dept., La. State U. 1929-66; developer techniques for breeding vegetable crops. Agrl. adviser to P.R. and Central Am.; plant exploration for Ipomea species, econ. medicinal plants in W.I. in cooperation with U. S. Dept. Agr., 1953; U. S. del. to Internat. Hort. Congress, 1955; developed process for dehydrating sweet potatoes for Army, World War II. Recipient Wilder Medal for breeding and intro. of Klonmore strawberry, also plaques for services rendered hort. field. Fellow A.A.-A.S.; mem. Am. Soc. Hort. Sci. (pres. 1942, chmn. So. Sect., 1948), Am. Genetic Assn., Potato Assn. Am. (pres. 1938), Assn. So. Agrl. Workers, So. Assn. Sci. and Industry, Am. Inst. Biol. Scis., La. Farm Bur. Fedn. (hon. life mem.), So. Seedsmen's Assn. (hon. life mem.), Sigma Xi, Phi Kappa Phi (provincal sec. So. sect. 1935-52; nat. regent), Omicron Delta Kappa, Alpha Gamma Rho, Alpha Gamma Delta, Alpha Zeta (Centennial hon. mem.). Contbr. numerous expt. sta. bulls., spl. feature articles on research, articles to sci. jours. Home: 338 Stanford Av., Baton Rouge 14, La.*

MILLER, Julian Malcolm, Am. chemist; b. Berkeley, Cal., Aug. 5, 1922; s. John Benjamin and Pauline (Goldman) M.; B.S., U. Cal. at Berkeley, 1944; Ph.D., Columbia, 1949; m. Prudence Loeb, July 26, 1960; children—Susan, Christopher. Faculty, Columbia, 1949—, prof., 1960—. Cons. Brookhaven Nat. Lab., 1950—. Mem. Am. Phys. Soc., Am. Chem. Soc. Author: (with G. Friedlander, J. Kennedy) Nuclear and Radiochemistry, 1964; also articles. Research on elucidation of processes involved in nuclear reactions. Home: 117 Glenwood Av., Leonia, N.J. 07605. Office: Dept. Chemistry, Columbia U., N.Y.C. 10027.*

MILLER, Kenneth Sielke, Am. mathematician; b. N.Y.C., June 4, 1922; s. Wilfred A. and Joan (Sielke) M.; B.S., Columbia, 1943, A.M., 1947, Ph.D., 1950; postgrad. Inst. for Advanced Study, Princeton; m. Elizabeth Barbara Autsch, May 17, 1953; children—Susan Barbara, Diane Elizabeth. With Manhattan Project, N.Y.C., 1943; prof. math. N.Y. U., 1950-64; sr. staff scientist Columbia, N.Y.C., 1964-67; sr. research asso. Riverside Research Inst., N.Y.C., 1967—; adj. prof. math. Fordham U., N.Y.C., 1964—. Cons. various bus. firms, U. S. Govt. Mem. Am. Math Soc., I.E.E.E. (sr.), Sigma Xi, Tau Beta Pi, Pi Mu Epsilon. Author: Linear Differential Equations in the Real Domain, 1963; Multidimensional Gaussian Distributions, 1964; also numerous articles. Research on differential equations, analog computers, random noise, signal detection and system studies; applications to indsl. and mil. uses via consulting work. Home: 25 Bonwit Rd., Port Chester, N.Y. 10573. Office: 632 W. 125th St., N.Y.C. 10027.*

MILLER, Lawrence Ingram, Am. plant pathologist; b. Jackson Center, O., May 12, 1914; s. Lawrence H. and Nellie J. (Ingram) M.; A.B., Oberlin Coll., 1936; M.S., Va. Poly. Inst., 1938; Ph.D., U. Minn., 1953; Mary Comfort McBryde, Dec. 18, 1939; children—Lawrence Ingram, Mary McBryde. Faculty, Va. Poly. Inst., Tidewater Research Sta., Holland, 1940—, prof. plant pathology, 1955—. Cooperator plant pest control div. U. S. Dept. Agr., 1955—. Freeport Sulphur Co. fellow, 1938-40; recipient Golden Peanut Research award Nat. Peanut Council, 1965. Mem. Va. Acad. Sci. (past chmn. agr. sci. sect., J. Shelton Horley Research award 1960), Am. Inst. Biol. Sci., Helminthological Soc. Washington, Soc. Nematologists, Am. Phytopath. Soc., Sigma Xi. Research, numerous publs. on plant pathology of peanut crop including fungus, virus, nematode and insect incitants of peanut diseases, ecology and morphological and physiol. variation and variability of plant parastic nematodes. Home: Box. 97, Holland, Va. 23391.*

MILLER, Leo Edward, Am. naturalist, explorer; b. Huntingburg, Ind., May 11, 1887; s. Bernhardt and Maria (Herrndorf) M.; ed. Bus. Coll., Indpls. Conservatory Music, Sch. Expression, Indpls.; m. Clarissa Amelia Kelsey, Apr. 2, 1918; children—Leo Edward, Spencer Kelsey. Made 6 expdns., including one with Theodore Roosevelt, to interior of Columbia, along Orinoco, Guianas, Bolivia, Argentina; staff Oakley Chem. Co., N.Y.; instr. Camp Dix, Dallas, also Camp Jackson, Columbia, S.C. Chmn., Bd. Edn., Stratford. Mem. Zool. Soc. Buenos Aires, New Haven County Hort. Soc. Author: The Wilds of South America, 1918; The Hidden People, 1920; In the Tiger's Lair, 1921; The Black Phantom, 1922; Adrift on the Amazon, 1923; The Jungle Pirates, 1925. Died Stratford, Conn., Oct. 6, 1952.

MILLER, Loye (Holmes), Am. biologist, paleontologist; b. Minden, La., Oct. 13, 1874; s. George and Cora (Holmes) M.; B.S., U. Cal. at Berkeley, 1898, M.S., 1904, Ph.D., 1912, LL.D., 1951; m. Anne Lucia Holmes, Aug. 1, 1901; children—Alden Holmes, Holmes Odell. Tchr., Boone's U. Sch., Berkeley, 1897-99, Punahou Sch. (formerly Oahu Coll.), Honolulu, 1900-03; faculty U. Cal. at Los Angeles (formerly Los Angeles State Normal Sch.), 1904—, prof. biology, 1919-43; research and extension lectr., cons. biology and paleontology U. Cal. at Davis, 1960—; coordinator U. Cal. at Los Angeles and Scripps Inst. Oceanography. Fellow Cal. Acad. Sci.; mem. Cooper Ornithol. Soc. (past pres., past chmn. bd.), Paleontol. Soc. (v.p. 1910—); hon. mem. Zool. Soc. San Diego, Audubon Soc. So. Cal., Western Soc. Naturalists, Sigma Xi, Phi Beta Kappa. Author: Life-long Boyhood, Recollections of a Naturalist Afield, 1950; also numerous publs. Initiated sch. avian paleontology in Western U. S.; initiated (with H. C. Bryant) park naturalist service in U. S. nat. parks; "brought to life" many species and genera of long extinct birds. Home: 821 Cherry Lane, Davis, Cal. 95616.*

MILLER, Max, Am. internist; b. New Haven, June 22, 1910; s. Morris and Bessie (Shulim) M.; B.S., Yale, 1931, M.D., 1935; m. Barbara Ann Foster, June 29, 1940; children—Claire L., Eric F. Asst. medicine Yale, 1936-37; faculty Western Res. U. Sch. Medicine, Cleve., 1937—, teaching fellow, 1937-40, prof. medicine, 1967—; staff U. Hosps., 1940—, asso. physician, 1948—, dir. clin. research center, 1962—; cons. Nat. Inst. Arthritis and Metabolic Diseases, 1965—, VA, hosps. Diplomate Nat. Bd. Med. Examiners, Am. Bd. Internal Medicine. Mem. A.A.A.S., Am. Diabetes Assn., Am. Fedn. Clin. Research, Am. Soc. Clin. Investigation, Central Soc. Clin. Research, Endocrine Soc., N.Y. Acad. Scis., Soc. Exptl. Biology and Medicine, Phi Beta Kappa, Alpha Omega Alpha, Sigma Xi, others. Editorial bd. Diabetes, 1955—. Research, publs. on carbohydrate metabolism and diabetes mellitus, use of oral hypoglycemic agts., intermediary metabolism of steroid hormones. Home: 2288 Chatfield Dr., Cleve. 44106. Office: 2065 Adelbert Rd., Cleve. 44106.*

MILLER, Milton Albert, Am. zoologist; b. Pittsburg, Mo., Dec. 2, 1907; s. William Albert and Effie (Houser) M.; student Crane Jr. Coll., 1925-26; A.B.

with honors, U. Ill., 1929; Ph.D., U. Cal. at Berkeley, 1934; m. Marion Pauline Christensen, June 22, 1941; children—Lauren Albert, Richard Noel. Instr., U. Hawaii, Honolulu, 1935-37, asst. prof., 1937-41; biologist Naval Biol. Lab., San Diego, Woods Hole (Mass.) Oceanographic Inst., 1941-45; faculty U. Cal., Davis, 1944—, prof., 1956—, chmn. dept. zoology, 1959-64, vice chmn. Coll. Letters and Scis., 1956-57. Fellow, A.A.A.S., Cal. Acad. Sci., Sigma Xi (pres. Davis chpt. 1950-51), Phi Delta Theta. Research, publs. on systematics, distbn. and ecology of isopod crustacea, marine fouling, biology and control of pocket gopher. Home: 639 A St., Davis, Cal. 95616.*

MILLER, Milton Howard, Am. psychiatrist; b. Indpls., Sept. 1, 1927; s. William and Helen (Lefkovits) M.; B.S., Ind. U., 1946, M.D., 1950; m. Harriet Sanders, June 27, 1948; children—Bruce, Jeffrey, Marcie. Faculty, U. Wis., Madison, 1955—, prof. psychiatry univ. hosps., 1963—, chmn. dept. psychiatry, 1962—; acting dir. Wis. Psychiat. Inst., 1961, dir., 1962—. Mem. Am., Wis. psychiat. assns., A.M.A., Wis., Dane County med. assns. Contbg. author Psychoanalysis and Current Biological Thought, 1965, Comprehensive Textbook of Psychiatry, 1967. Contbr. articles to med. jours. Home: 3682 Lake Mendota Dr., Madison, Wis. 53705.*

MILLER, Neal E(lgar), Am. psychologist; b. Milw., Aug. 3, 1909; s. Irving E. and Lily R. (Fuenfstueck) M.; B.S., U. Wash., 1931; M.S., Stanford, 1932; Ph.D., Yale, 1935; D.Sc., U. Mich., 1965; m. Marion E. Edwards, June 30, 1948; children—York, Sara. Social sci. research fellow Inst. Psychoanalysis, Vienna, Austria, 1935-36; asst. research psychologist Yale, 1933-35, instr., asst. prof., research asst. psychologist Inst. Human Relations, 1936-41, asso. prof., research asso., 1941-42, 46-50, prof. psychology, 1950-52, James Rowland Angell prof. psychology, 1952-66; prof. psychology Rockefeller U., N.Y.C., 1966—; fellow Berkeley College, 1955—. Expert cons. Am. Inst. Research, 1946-62; spl. cons. com. human resources, research and devel. bd. Office of Sec. Def., 1951-53; mem. tech. adv. panel Office Asst. Sec. Defense, 1954-57; expert cons. Operations Research Office and Human Resources Research Office, 1951-54. Chmn. bd. sci. overseers Roscoe B. Jackson Meml. Lab., Bar Harbor, Me. 1962—; bd. sci. counsellors Nat. Inst. Mental Health, 1957-61; fellowship com. Founds. Fund for Research in Psychiatry, 1956-61; mem. central council Internat. Brain Research Orgn., 1964; v.p. bd. dirs. Foote Sch., 1964—. Recipient Warren Medal for exptl. psychology, 1954, Newcomb Cleveland Prize, 1956; Nat. Medal of Sci., 1964. Fellow Am. Acad. Arts and Scis.; mem. Am. Psychol. Assn. (council reps., 1954-58, pres. exptl. div. 1952-53, pres. 1960-61), Eastern Psychol. Assn. (pres., 1952-53), NRC (div. anthropology and psychology 1950-53, chmn. 1958-60), Nat. Acad. Sci. (chmn. sect. psychology 1965—), Soc. Exptl. Psychologists, A.A.A.S., Sigma Xi, Phi Beta Kappa. Author: Frustration and Aggression (with J. Dollard et al), 1939; Social Learning and Imitation (with Dollard), 1941; Personality and Psychotherapy (with Dollard), 1950; Graphic Communication and the Crisis in Education, 1957. Contbr. chpts. in psychol. handbooks. Editor: Psychological Research on Pilot Tng., 1947. Research in elec. and chem. brain stimuli; analytical studies in neurotic symptoms, psychotherapy. Home: 227 Everit St., New Haven.*

MILLER, Norman F., Am. gynecologist; b. Iron Mountain, Mich., Aug. 14, 1894; s. R. T. and H. (Von Norman) M.; student Beloit Coll., 1913-15; B.S., U. Mich., 1920, M.D., 1920; m. Dorothy Kingsford, Dec. 31, 1921; children—Norman F. II, Edward K., Mary F., Groves. Asso. prof. obstetrics, gynecology, U. Iowa, 1926-28; prof., 1928-31; prof. chmn. dept. obstetrics, gynecology U. Mich., Ann Arbor, 1931-64, emeritus, 1964—; practice medicine specializing in gynecology. Recipient U. Mich. Distinguished Achievement award, 1963. Mem. Johnson County, Washtenaw County med. socs., A.M.A., Central Assn. Obstetrics and Gynecology, Mich. Soc. Obstetrics and Gynecology, Brit. Gynecol. Club, Edinburgh Obstet. Soc., A.C.S. (vice pres. 1965-66). Author: (with H. Avery) Gynecology for Nurses and Gynecologic Nursing, 5th edit., 1965; (with Haas, Evans) Human Parturition, 1958; also articles. Research on cancer and virology of female reproductive tract. Home: Rural Route 2, Box 390, Florence, Wis. 54121.

MILLER, Patrick, Scottish inventor; b. Glasgow, Scotland, 1731; s. William and Janet (Hamilton) M.; m. Miss Lindsay; children—Patrick, William, Janet (Mrs. John Thomas), Jean (Mrs. Leslie G. Jones), Thomas Hamilton. Banker, Edinburgh, Scotland; became dir. Bank of Scotland, 1767, dep. gov., 1790-1815. Author: (tract) The Elevation, Section, Plan and Views of a Triple Vessel, 1787; Treatise on Fiorin Grass, 1810. Studied ships with doubled and tripled hulls which were driven by paddle wheel operated by human power; fitted 2 ships with steam engines (of W. Symington) and successfully tested them at sea, 1788-89; improved naval cannon; introduced cultivation of fiorin grass into Scotland. Died Dec. 9, 1815.

MILLER, Paul R(only), Am. plant pathologist; b. Rockport, Ind., Apr. 30, 1905; s. Curtis C. and Eva Ann (Parr) M.; B.S., Ind. State U., 1929; M.S., Purdue U., 1931; Ph.D., George Washington U., 1938; m. Nancy Starkey Wright, Jan. 18, 1933; children—Richard Wright, Peter Parr, Merilyn. Tchr. biology and athletic coach Burritt Coll., Tenn. 1926-28; entered Bur. Plant Industry, U. S. Dept. Agr., Washington, 1931, investigator diseases of peanuts, tobacco and cotton, 1931-43, charge field work U. S. on Def. Biol. Warfare project, 1943-45, head, Plant Disease Survey, 1945; established Nat. Plant Disease Forecasting project, 1947, and since served as leader of project; apptd. bd. expert examiners by U. S. Civil Service Commn., 1949; head epidemiological investigations Agrl. Research Service, Dept. Agr., 1958—. Fellow Am. Phytopath. Soc. (pres.); mem. Washington Acad. Scis., Sigma Xi. Contbr. numerous sci. papers. Bd. editors Microbiol. Revs. Home: 1557 Farlow Av., Crofton, Md. Office: Plant Industry Station, Beltsville, Md.*

MILLER, Paul Theodore, Am. geologist, educator; b. Conway, Ia., Dec. 31, 1905; s. Porter C. and Grace (Dugan) M.; B.A., Simpson Coll., 1927; M.S., U. Ia., 1930, Ph.D., 1932; postgrad. U. Wis., Fla. State U.; m. Melba Dunkerton, June 7, 1932; 1 son, Kent D. Lab. asst. geology U. Ia., 1927-29, research asst., 1929-32, research asso., 1932-34; asst. to state geologist Ia., summers 1928-34; geologist Amerada Petroleum Corp., 1935; prof. geology and geography N.D. State Tchrs. Coll., Minot, 1935-36, Wis. State Coll., Superior, 1936-47; prof. geology, head dept. Ariz. State U., Tempe, 1947—. Fellow A.A.A.S.; mem. Am. Assn. Petroleum Geologists, Nat. Council Geography Tchrs. (past regional coordinator Ariz.), Nat. (past pres.), Am. assns. geology tchrs., Four Corners, N.M., Ariz., Central Ariz. (organizer 1959) geol. socs., Ariz. Acad. Sci. (past chmn. sect., editor), Ariz. Mineral. Soc. (hon.), Ariz. Coll. Assn. (past pres., exec. com.). Research, publs. Pleistocene gravel and loess deposits of Ia. Home: 33 E. 15th St., Tempe, Ariz. 85281.*

MILLER, Paul William, Am. plant pathologist; b. Mt. Vernon, Ind., May 2, 1901; s. Arthur J. and Florence (Seats) M.; B.S., U. Ky., 1923, M.S., 1924; Ph.D., U. Wis., 1929; m. Carrie Aleada Egge, Mar. 23, 1927. Instr. plant pathology, research asst. U. Wis., 1927-29; agt. U. S. Dept. Agr., Hood River, Ore., 1929-30, asso. plant pathologist Ore. State U., 1930-42, plant pathologist, 1942-63, research plant pathologist, 1963—. Mem. Phytopath. Soc., Am. Rose Soc., Sigma Xi. Research, numerous publs. on fireblight of apple, perennial canker of apples, walnut and filbert diseases, vegetable seed diseases, strawberry root rot, strawberry virus diseases. Home: 703 N. 30th St., Corvallis, Ore. 97330.*

MILLER, Philip, English botanist; b. Deptford or Greenwich, Eng., 1691; m. Mary Kennet; children—Philip, Charles. Florist St. George's Fields, Deptford, Eng.; gardener Chelsea (Eng.) Garden, 1722-1770. Fellow Royal Soc, 1729, mem. Bot. Acad., Florence. Author: The Gardener's and Florist's Dictionary, or a Complete System of Horticulture, 2 vols., 1724; Catalogus Plantarum, 1730; The Gardener's Dictionary, Vol. I, 1731, Vol. 2, 1739; The Method of Cultivating Madder, 1758. Described method of flowering bulbous plants in water-filled bottles, 1730. Died 1771.

MILLER, Richard Henry, Am. astrophysicist; b. Aurora, Ill., Aug. 31, 1926; s. Perl Hobart and Mary Ina (Seed) M.; S.B. in Elec. Engring., Ia. State Coll., 1946; Ph.D. in Physics, U. Chgo., 1957; m. Mary Alice Funk, June 17, 1952. With Cyclotron Project, U. Chgo., 1947-52; with Cal. Research & Devel. Corp., Livermore, 1952; with NRC of Brazil, 1953-54; faculty U. Chgo., 1957—, asso. prof. astronomy, dir. Inst. for Computer Research, 1963—, acting chmn. com. on information scis., 1965—. Mem. Internat. Astron. Union, Am. Astron. Soc., Am. Phys. Soc., Assn. for Computing Machinery. Research, publs. on accelerator design, pion physics, surface photometry of galaxies, stellar dynamics, computers. Home: 7337 South Shore Dr., Chgo. 60649.*

MILLER, Robert Charles, Am. physicist; b. State College, Pa., Feb. 2, 1925; s. Lawrence P. and Eva M. (Gross) M.; A.B., Columbia, 1948, M.A., 1952, Ph.D. (RCA fellow), 1956; m. Virginia Callaghan, Aug. 30, 1952; children—Robin, Jeffrey, Lauren. Chemist, Johns Mansville Research Center, Finderne, N.J., 1948-49; tech. staff Bell Telephone Labs., Murray Hill, N.J., 1954-63, dept. head, 1963-67; staff Inst. for Def. Analyses, Arlington, Va., 1967—. Lectr. physics Columbia, 1951-53. Fellow Am. Phys. Soc.; mem. N.Y. Acad. Sci., Sigma Psi. Research, publs. on verification Maxwellian velocity distbn. atoms and molecules, diffusion of impurities in semiconductors, domain walls in ferro-electrics, non linear optical studies dielectrics including co-discovery broad band tunable coherent light source; discovered technique for observing domain walls in barium titanate. Home: 1427 Woodacre Rd., McLean, Va. 22101. Office: 400 Army-Navy Dr., Arlington, Va. 22202.*

MILLER, Robert Clay, Am. chemist, educator; b. Wollaston, Mass., Feb. 26, 1923; s. Homer Blaine and Emma (Plummer) M.; B.S., Northeastern U., 1947; M.A., Columbia, 1948; Ph.D., Temple U., 1956; m. Caroline Delaney, Nov. 5, 1949; children—Barbara Howard, Anne Crane. Chemist, Socony Vacuum Oil Co., Paulsboro, N.J., 1948-56, E. I. du Pont de Nemours & Co., Inc., Wilmington, Del., 1956-58; asst. prof. chemistry St. Vincent Coll., Latrobe, Pa., 1958-59; faculty dept. chemistry DePaul U., Chgo., 1959—, asso. prof., 1963—, chmn. dept. chemistry, 1961-65. Mem. Am. Chem. Soc., Chem. Soc. (London), Sigma Xi. Research in organophosphorus chemistry. Home: 839 Park Av., Wilmette, Ill. Office: 1036 W. Belden St., Chgo. 60614.*

MILLER, Robert Cunningham, Am. marine biologist; b. Blairsville, Pa., July 3, 1899; s. Coursen Herbert and Alma (Gilmore) M.; A.B., Greenville Coll., 1920; A.M., U. Cal., 1921, Ph.D., 1923; m. Lea Van Puymbroeck, Sept. 15, 1937. Asst. prof. zoology Washington (Seattle), 1924-30, asso. prof., 30-36; prof., 1936-38; vis. prof. Lingnan (China) U., 1929-31; dir. Cal. Acad. Scis., San Francisco, 1938-63, sr. scientist, 1963—. Chmn. tuna industry com. Cal. Reconstrn. and Reemployment Commn., 1945-47; mem. Cal. Marine Research Commn., 1948-56, U. S. Com. on Oceanography of Pacific, 1948-55. Fellow A.A.A.S. (sec. Pacific div. 1944—), Cal. Acad. Scis.; mem. Cal. C. of C. (chmn. salmon resources commn. 1944-45, sardine industry com. 1945-46, central coast natural resources com. 1947-62, exec. com. 1962—), Am. Soc. Zoologists, Am. Geophys. Union, Western Soc. Naturalists (pres. 1936), Oceanographic Soc. Pacific (pres. 1941-47). Contbr. articles on marine biology and oceanography to tech. jours. Research on biology of marine wood-boring organisms, photobiology; and bird behavior and flight. Home: 3003 Dwight Way, Berkeley, Cal. 94704. Office: Cal. Acad. Scis., San Francisco 94118.

MILLER, Robert Demorest, Am. soil physicist; b. Omaha, Sept. 25, 1919; s. Merritt Finley and Grace (Ernst) M.; B.S., U. Mo., 1940; M.S., U. Neb., 1941; Ph.D., Cornell U., 1948; m. Beulah Wilson Cooper, Sept. 6, 1941; children—Leslie Grace, Anne Cosby, Melanie Randolph. Jr. asst. soil physicist U. Cal. at Berkeley, 1948-52; faculty Cornell U., Ithaca, N.Y., 1952—, prof. soil physics, 1956—, asst. to provost, 1964-65, dean univ. faculty, 1967—. Cons. div. applied scis. Harvard, 1955, Expt. Sta., Hawaiian Sugar Planter's Assn., 1958, pres.'s Sci. Adv. Com. Office Sci. and Tech., 1964-65. Fulbright Research scholar, postdoctoral fellow Royal Norwegian Soc. for Sci. and Indsl. Research, 1965-66. Fellow Am. Soc. Agronomy; mem. Soil Sci. Soc. Am., Internat. Soc. Soil Sci., Am. Geophys. Union. Research publs. on transient movements of water in soil, concurrent movements of electricity, ions and water in soils (electro-osmosis), role of exchangeable ions in clay swelling and in frost heaving in soils, conditions for equilibrium for water and ice in soils and porous media. Address: Cornell U., Ithaca, N.Y. 14850.*

MILLER, Robert E., Am. psychologist, educator; b. Warren, Pa., Apr. 29, 1926; s. Arch W. and Lorena (Merrill) M.; B.A., Allegheny Coll., 1949; M.S., U. Pitts., 1951, Ph.D., 1953; m. Eleanor Keir, Sept. 4, 1948; children—Barbara, Thomas. Faculty, U. Pitts., 1953—, prof., 1963—. Mem. Am. Psychol. Assn., A.A.A.S., Am. Assn. U. Profs., Phi Beta Kappa, Sigma Xi. Research, publs. on investigations on non-verbal communication processes both in monkeys and in man, effects of neuro-hormones on behavior of animals, psychophysiol. correlates of stress situations. Home: 540 Lucia Rd., Pitts. 15221.*

MILLER, Robert Lee, Am. marine geophysicist; b. Chgo., Apr. 6, 1920; s. Fred Lee and Gertrude (Byron) M.; A.B., U. Ill., 1942; Ph.D., U. Chgo., 1950; m. Dorothy L. Schick, June 12, 1942; children—Dolores (Mrs. Robert Pekrul), Doretta Doralee Doricia. Research asso. U. Chgo., 1950-52, faculty, 1953—, prof. geophys. scis., 1965—; vis. prof. Frankfort (Germany) U., 1958, Brown U., Providence, 1963; asso. Woods Hole Oceanographic Instn., Woods Hole, Mass., 1957—. Cons. to pvt. cos., govt. agy. Mem. Am. Geophys. Union, Marine Tech. Soc., N.Y. Acad Scis., Council for Wave Research, Sigma Xi. Author: (with E. C. Olson) Morphological Integration, 1957; (with J. S. Kahn) Statistical Analysis in the Geological Sciences, 1962; Papers in Marine Geology, 1964; also numerous articles. Application of statis. analysis to earth scis., treatment organisms by numerical analysis, devel. and application trend mapping and mosaic analysis; research on breaking waves and gravity shock waves, coastal dynamics. Home: 10526 S. Prospect St., Chgo. 60643.*

MILLER, Roland Drew, Am. physician; b. Chgo., Mar. 16, 1922; s. Roland B. and Fern (Drew) M.; A.B., DePauw U., 1943; M.D., Northwestern U., 1945; M.S., U. Minn., 1951; m. Elizabeth M. Lancaster, Dec. 18, 1943; children—Judith Lee (Mrs. Rome), Cheryl Sue, Randall Scott. Staff, Mayo Found., Rochester, Minn., 1948-61; asso. prof. medicine Mayo Found. U. Minn., 1961-65, prof., 1965—; asso. dir. Mayo Grad. Sch. (formerly Mayo Found.), 1961—. Cons. internal medicine, thoracic diseases, 1952—. Mem. Alpha Omega Alpha, Beta Theta Pi. Researcher in pulmonary diseases and function, med. edn. Home: 439 16th Av. S.W., Rochester, Minn. 55901.*

MILLER, Russell Cooper, Am. animal nutritionist, educator; b. Tower City, Pa., Feb. 2, 1901; s. Samuel Cooper and Mildred (Mandaville) M.; B.S., Pa. State Coll., 1922; Ph.D., Cornell U., 1925; m. Helen Elizabeth Weed, Aug. 25, 1925; children—David W., Carolyn Frances (Mrs. Robert C. Baldwin). Asst. dept. animal industry Ohio Agr. Expt. Sta., Wooster, 1922-

23, Cornell U., Ithaca, N.Y., 1923-26; faculty Pa. State U. Coll. Agr., University Park, 1926—, prof. agrl. and biol. chemistry, 1944-65, prof. emeritus dept. animal science, 1965—, head dept. animal industry and nutrition, 1959-65, chmn. div. animal sci. and industry, 1961-64. Vis. prof., cons. U. P.R. Coll. Agr., 1953, 59, 63. Mem. Am. Chem. Soc., Am. Soc. for Exptl. Biology Am. Soc. Animal Sci., Am. Inst. Nutrition, Sigma Xi, Alpha Zeta, Gamma Sigma Delta, Phi Lambda Upsilon. Contbr. numerous articles on nutrition to profl. jours. Home: 330 S. Patterson St., State College, Pa.*

MILLER, S. M., Am. sociologist; b. Phila., Nov. 21, 1922; s. Morris S. and Lena (Landau) M.; B.A., Bklyn. Coll., 1943; M.A., Columbia U., 1945; A.M. (Wyman fellow), Princeton, 1946, Ph.D., 1951; m. Jean T. Baker, Apr. 8, 1955; children—Jonathan, Edward. With Greene and Co., 1943, Bank of Manhattan Co., 1943-45, Nat. Bur. Econ. Research, 1945, U. S. Dept. Labor, 1945; asst. regional economist U. S. Fed. Pub. Housing Authority, 1946-47; instr. Princeton, 1946; lectr. Rutgers U., 1946-47, instr. econs., 1947-49; research asso. planning project for advanced tng. in social research Columbia, 1952-53; research sociologist Rockland County Mental Health Assn. and Community Mental Health Bd., Monsey, N.Y., 1957-58, cons., 1958-59; lectr. Rutgers U., 1958-61, Cornell U. extension, 1958-61; lectr. sociology and econs. Bklyn. Coll., 1949-52, chmn. social sci. group, 1953-57, asso. prof. sociology and anthropology, 1958-63; prof. sociology Maxwell Grad. Sch., also sr. research asso. Youth Devel. Center, Syracuse (N.Y.) U., 1961-65; prof. edn. and sociology N.Y. U., 1965—; program adviser in social devel., 1966—. Cons. to numerous colls., univs., govt. agys., welfare agys. Recipient internat. travel grants Internat. Sociol. Assn., Am. Sociol. Assn., Social Sci. Research Council; research grants NSF, Social Security Adminstrn., N.Y. State Div. for Youth, Rabinowitz Found., Stern Family Fund, Ford Found. Mem. Am. Sociol. Assn., Eastern Sociol. Soc. (mem. exec. com. 1963-66), Soc. for Study Social Problems (adv. editor Social Problems 1959-61, book rev. editor 1961-64), Internat. Group for Study Nat. Planning (mem. com. on tng., Warsaw, 1965), Soc. for Psychol. Study Social Issues, Internat. Sociol. Assn., Indsl. Relations Research Assn., Am. Econ. Assn., Sigma Xi, Alpha Kappa Delta. Author: Comparative Social Mobility: A Trend-Report, 1960. Co-author: Facts About Rockland County, 1958; The Dynamics of the American Economy, 1956; The School Dropout Problem: Syracuse, 1963; School Dropouts: A Commentary and Annotated Bibliography, 1964. Editor (with Alvin W. Gouldner) Applied Sociology: Opportunities and Problems, 1965. Asso. editor Am. Sociol. Rev., 1966—. Address: 510 E. 86th St., N.Y.C. 10028.*

MILLER, Samuel, Am. botanist; b. Lancaster, Pa., Oct. 4, 1820; m. Martha Isabel Evans, 1847; 9 children; justice of peace, Lebanon City, Pa.; plant breeder in Mo., from 1867; officer State (Mo.) Hort. Soc.; made extensive tests on various fruits and ornamental plants; developed Capt. Jack strawberry, useful as pollinizer. Died 1901.

MILLER, Seward Elmore, Am. physician; b. Perry, N.Y., Aug. 16, 1905; s. George W. and Luella M. (Bennett) M.; B.S., U. Mich., 1928, M.D., 1931; m. Helen E. Wood, Jan. 26, 1935; children—John Roger, Douglas Seward. Staff physician indsl. medicine French Hosp., Los Angeles, 1935-37; pathology resident U. S. Marine Hosp., Seattle, 1937-39, chief pathologist, Norfolk, Va., 1939-42, Balt., 1942-45; chief Communicable Disease Center Labs., Atlanta, 1945-50; regional dir. Dept. Health, Edn. and Welfare, Chgo., 1950-51; chief occupational health div. USPHS, Washington, 1951-56; dir. Inst. Indsl. Health, prof. internal medicine, U. Mich. Med. Sch., Ann Arbor, 1956-61, chmn. dept. indsl. health Sch. Pub. Health, 1958-61; prof. occupational medicine U. Cal., Los Angeles, 1961—. Cons. WHO, 1952—, Gen. Motors Research Labs., Santa Barbara, Cal., 1962—, AEC, Nev. Test Site, 1963—. Mem. A.M.A., Am. Indsl. Hygiene Assn., Am. Pub. Health Assn., Am. Soc. Clin. Pathologists, Acad. Occupational Medicine, Indsl. Med. Assn., Am. Coll. Preventive Medicine, Assn. Mil. Surgeons, Internat. Assn. on Occupational Health (permanent com. 1954—), Ramzzini Soc. Author: (with Rutherford T. Johnstone) Occupational Disease and Industrial Medicine, 1960. Editor: Textbook of Clinical Pathology, 7th edit., 1965. Home: 2816 Miradero Lane, Santa Barbara, Cal. 93105.*

MILLER, Shelby Alexander, Am. chem. engr.; b. Louisville, July 9, 1914; s. George Walter and Stella Katherine (Cralle) M.; B.S., U. Louisville, 1935; Ph.D., U. Minn., 1944; m. Jean Adele Danielson, Dec. 26, 1939; (div. May, 1948); 1 son, Shelby Carlton; m. 2d, Doreen Adare Kennedy, May 29, 1952. Asst. chemist Corhart Refractories Co., Louisville, 1935-36; teaching, research asst. chem. engring. U. Minn., 1935-39; devel., research chem. engr. E. I. duPont de Nemours & Co., Wilmington, Del., 1940-46; asso. prof. chem. engring. U. Kan., 1946-50, prof., 1950-55; Fulbright prof. chem. engring. King's Coll. Durham U., Newcastle upon Tyne, Eng., 1952-53; prof., chmn. chem. engring. U. Rochester, 1955—. Sec. Kan. Bd. Engring. Examiners, 1954-55. Treas. Lawrence (Kan.) League for Practice Democracy, 1950-52. Registered profl. engr., Del., Kan., N.Y. Fellow A.A.A.S.; mem. Am. Inst. Chem. Engrs. (past chmn. Kansas City sect.), Am. Chem. Soc. (chmn. Rochester sect.), Soc. Chem. Industry, Nat. Soc. Profl. Engrs., Am. Soc. Engring., Edn., Am. Assn. U. Profs., Sigma Xi, Sigma Tau, Phi Lambda Upsilon, Tau Beta Pi, Alpha Chi Sigma. Editor Chem. Engring. Edn., 1965—. Contbns. include analysis and design of clarifying filters. Contbr. articles to tech., profl. jours. Home: 347 Hillside Av., Rochester, N.Y. 14610.*

MILLER, Stanley Lloyd, Am. biochemist; b. Oakland, Cal., Mar. 7, 1930; B.S., U. Cal., 1951; Ph.D., U. Chgo., 1954; postgrad. (Jewett fellow) Cal. Inst. Tech., 1954-55. Instr. biochemistry Columbia Coll. Phys. and Surg., 1955-58, asst. prof., 1958-60; asst. prof. chemistry U. Cal. at San Diego, La Jolla, 1960-62, asso. prof., 1962—. Mem. A.A.A.S., Am. Chem. Soc. Research on anesthesia and enzyme mechanisms; expt. on origin of life produced simple amino acids from sterilized water exposed to hydrogen, ammonia, methane and elec. charge. Address: Dept. of Chemistry, University of Cal. at San Diego, La Jolla, Cal. 92037.*

MILLER, Walter E., Am. chemist; b. N.Y.C., Jan. 28, 1914; s. Louis and Hattie (Rose) M.; B.S., Coll. City N.Y., 1935, Ch.E., 1936; Ph.D., N.Y. U., 1941; m. Harriette Noschkes, Nov. 13, 1943; children—Larry, Diane, Michael. With N.Y. U., 1937-42, instr., 1941-42; faculty Coll. City N.Y., 1941-42, 46—, prof., 1966—. Cons. U. S. Army Chem. Corps., 1958—, pvt. industry. Mem. Am. Chem. Soc., A.A.A.S., N.Y. Acad. Scis., Am. Radio Relay League. Author: (with J. A. Babor) General Chemistry, 1965; also articles. Devel. ion exchange resin applications, demineralization of water, recovery of wastes. Home: 64 Morris Av., Haworth, N.J. 07641. Office: 139th St. and Convent Av., N.Y.C. 10031.*

MILLER, Warren E., Am. polit. scientist; b. Hawarden, Ia., Mar. 26, 1924; s. John Carroll and Mildred (Lien) M.; B.S., U. Ore., 1948, M.S., 1950; D.S.S., Syracuse U., 1954; m. Mildred Kiplinger, June 20, 1948; children—Jeffrey R., Jennifer L. Faculty, U. Mich., Ann Arbor, 1951-54, 56—, now prof. polit. sci.; faculty U. Cal., Berkeley, 1954-56. Exec. dir. Inter-U. Consortium for Polit. Research; mem. Behavioral Scis. div. NRC; mem. exec. com. Council Social Sci. Data Archives. Fellow Center For Advanced Study in Behavioral Scis., 1961-62. Mem. Am. Polit. Sci. Assn., A.A.A.S., Soc. for Psychol. Study Social Issues. Author: (with A. Campbell, G. Gurin) The Voter Decides, 1954; (with A. Campbell, P. Converse, D. Stokes) The American Voter, 1960, Elections and the Political Order, 1966. Research in analysis of electoral behavior, role of polit. instns. in representation, instl. devel. of resources for sci. research and tng. Home: 1511 Hillridge St., Ann Arbor 48103. Office: Inst. for Social Research, Ann Arbor, Mich. 48106.*

MILLER, Willet G., Canadian geologist; b. Ont., Can., M.A., U. Toronto; LL.D., Queen's U., Kingston, Ont., 1907, U. Toronto (Ont., Can.), 1913. Asst. in field geology Geol. Survey Can., 1891-93; lectr., later prof. geology Queen's U., 1893-1902; head field work in geology eastern Ont., Bur. Mines, 1897-1901; provincial geologist Ont., from 1902. Pres. Canadian Mining Inst., 1908-10; mem. Royal Ont. Nickel Commn., 1915-17; Canadian rep. Imperial Mineral Resources Bur., 1918-19, 22. Fellow Royal Soc. Can., Geol. Soc.; mem. geol. socs. Am., London, Brit. Assn. Advancement Sci., Am. Inst. Mining Engrs., Instn. Mining and Metallurgy (hon., Gold medal 1914), Soc. Econ. Geologists, A.A.A.S. Research, publs. on pre-Cambrian and econ. geology Ont., corundum-bearing rocks, iron ores, gold deposits, cobalt-silver ores, others. Died Feb. 1925.

MILLER, William Alfonso, Am. physicist; b. Portland, Ore., Mar. 16, 1912; s. William O. and Rose (Tunison) M.; B.S. in Elec. Engring., Ore. State U., 1932, M.S. in Physics, 1936; Ph.D., Purdue U., 1941; m. Florence N. Richards, Sept. 2, 1938; children—William, Jonathan, Diana, Sarah, David, Deborah. Research asso. Mass. Inst. Tech. Radiation Lab., Cambridge, 1941; tech. staff RCA Labs., Rocky Point, N.Y., 1942-61; sr. staff scientist Fairchild-Hiller Corp., Bay Shore, N.Y., 1961-65; sr. staff scientist radiometrics div. Polarad Electronics Corp., Long Island City, N.Y., 1965-67; cons., tech. asst. to gen. mgr. Geospace Electronics div. Sanders Assos., Plainview, N.Y., 1967—; staff OSRD, 1942-45. Mem. Am. Phys. Soc., Am. Geophys. Union, I.E.E.E. (sr.), N.Y. Acad. Sci., Soc. Photo-Optical Instruments Engrs., Sigma Xi (award commendation 1936), Sigma Pi Sigma, Pi Mu Epsilon. Contbr. articles to tech. jours. Patentee in fields radar, distance measuring equipment, microwaves, electronic circuitry, solid state devices and application. Research in systems design, communications systems, radio wave propagation, solar physics, optical and photo-optical systems engring. Home: Dogwood Lane, Miller Place, N.Y. 11764. Office: 1 Fairchild Av., Plainview, N.Y. 11803.*

MILLER, William Allen, English chemist; b. Ipswich, Eng., Dec. 17, 1817; s. William and Frances (Bowyer) M.; ed. Birmingham Gen. Hosp., King's Coll., London; M.D., London U., 1842; LL.D., Edinburgh, 1860; D.C.L., Oxford, 1868; LL.D., Cambridge, 1869; m. Eliza Forrest, 1842; 2 daus., 1 son. Worked in Liebig's lab., 1840; chem. demonstrator King's Coll., London, named prof. chemistry, 1845; assayer to mint and Bank of Eng. Recipient Gold medal Royal Astron. Soc., 1867. Fellow Royal Soc., 1845 (mem. council 1848-50, 1855-57, treas. 1861-70); founding mem. Chem. Soc., 1841. Author: Elements of Chemistry, 1855-57; also papers. Devised new methods for employing spectrum analysis to determine chem. identity of substances; demonstrated, with photographs, existence of characteristic differences in spectra of 25 metals; (with Sir William Huggins) made photospectroscopic analyses of heavenly bodies, also gathered 1st accurate data on solar and stellar chemistry, 1862. Died Liverpool, Sept. 30, 1870.

MILLER, William Hallowes, Brit. mineralogist; b. Velindre, Wales, Apr. 6, 1801; s. Capt. Miller; grad. St. John's Coll., Cambridge, 1826, fellow, 1829-44, 74, M.D., 1841; LL.D., Dublin, Ireland, 1865; D.C.L., Oxford U., 1876; m. Harriet Susan Minty, 1844; 2 sons, 4 daus. Prof. mineralogy Cambridge U., 1832-70; commr. for standard weights and measures; mem. internat. commn., 1870. Mem. Royal Soc., 1838 (fgn. sec. 1856-73, Royal medal 1870), French Acad. Scis. (corr.). Author work on crystallography, 1838, textbooks on hydrostatics and hydrodynamics, 1831, 35; also articles. Developed Millerian indices for defining crystal faces, 1838; helped reconstruct length and weight standards destroyed in 1834. Died Cambridge, Eng., May 20, 1880.

MILLER, William Lash, Canadian chemist; b. Galt, Ont., Can., 1866; grad. U. Toronto (Ont., Can.), 1887; postgrad. U. Berlin; Ph.D., univs. Munich and Leipzig, Germany; children—Adelaide, Frederica, W. R. Miller. Head dept. chemistry Toronto U., 1921-37, emeritus, 1937—, mem. staff 48 years. Mem. Am. Electrochem. Soc. (became pres. 1912), Canadian Inst. Chemistry (became pres. 1926), Royal Soc. Can. (became pres. 1935), A.A.A.S., Am. Chem. Soc. (hon.), Soc. Chem. Industry, Electro-Chem. Soc., Brit. Assn. for Advancement Sci. Asso. editor Jour. Phys. Chemistry, 1910-26. Research on chem. thermodynamics, chem. kinetics and electrochemistry. Died 1940.

MILLER, William Taylor, Jr., Am. chemist; b. Winston-Salem, N.C., Aug. 24, 1911; s. William Taylor and Katie Jane (McMahan) M.; A.B., Duke, 1932, Ph.D., 1935; m. Betty Stewart Robb, July 1951. Faculty, Cornell U., Ithaca, N.Y., 1936—, prof., 1947—; research group leader Manhattan Project at Columbia, Carbide & Carbon Chem. Corp., N.Y.C., 1943-46. Fellow A.A.A.S.; mem. Am. Chem. Soc., London Chem. Soc., Phi Beta Kappa, Sigma Xi. Author: Preparation Properties and Technology of Fluorine and Organic Fluorine Compounds, 1951. Contbr. chpt. Nat. Nuclear Energy Series, 1946. Research, publs. on devel. of carbon-fluorine chemistry, reactions of elementary fluorine with organic compounds and ionic and free-radical reactions of unsaturated carbon fluorine compounds. Home: 100 Sunset Park, Ithaca, N.Y. 14850.*

MILLER, William Wadd, III, Am. physiologist; b. Starkville, Miss., Oct. 4, 1932; s. William Wadd and Nettie (Sanders) M.; B.S., Miss. State U., 1954, M.S., 1957; Ph.D., 1962; m. Jane Pierce, July 21, 1957; children—William Wadd IV, Eva M. Faculty, Howard Coll., Birmingham, Ala., 1962-68, asso. prof. biology, 1965-68; faculty N.E. La. State Coll., Monroe, 1968—. Mem. A.A.A.S., Am. Soc. Zoology, Am. Inst. Biol. Scis, Sigma Xi. Publs. on research in physiology of reprodn., neurophysiology. Home: 43 Jana Dr., Monroe, La. 71201.*

MILLER, Willoughby Dayton, Am. bacteriologist, dentist; b. nr. Alexandria, O., Aug. 1, 1853; s. John H. and Nancy (Sommerville) M.; ed. U. Mich., U. Edinburgh (Scotland), U. Berlin (Germany); D.D.S., U. Pa., 1879; m. Caroline L. Abbott, Oct. 26, 1879; 2 sons, 1 dau. Became asst. chemistry to F. P. Abbott, Berlin, 1877; practiced in Berlin, 1879; became prof. operative dentistry U. Berlin, 1884, prof. extraordinary, 1894; state examiner for dentistry, Berlin; apptd. privy med. councillor by emperor, 1906; named dean Dental Coll., U. Mich., 1907. Mem. Nat. Dental Assn. Germany (pres.), Fédération Dentaire Internationale (pres.), Am. Dental Soc. Europe (pres.), Assn. Dental Faculties Germany (pres.); hon. mem. more than 40 dental socs. in U. S., Europe. Author: The Micro-organisms of the Human Mouth, 1890; Lehrbuch der Conservirenden Zahnheilkunde, 1896-98; also numerous articles. Contributed to knowledge of chem. and bacterial causes of dental and oral disease; introduced chemo-parasitic theory for origin of dental caries, 1889; studied use of antiseptics in dentistry, dental and oral diseases, etiology of dental abrasion and erosion. Died Newark, July 27, 1907.

MILLET, John Alfred Parsons, psychiatrist, psychoanalyst; b. Broadway, Eng., July 8, 1888; (parents Am. citizens); s. Francis David and Elizabeth (Greeley) M.; A.B., Harvard, 1910, M.D., 1914; m. Alice Murrell, May 21, 1913 (div. Aug. 1941); children—Jeanne Alice (Mrs. John Marshall), John Bradford, Elisabeth (Mrs. Henry Sanford); m. 2d, Carmen de Gonzalo Manice, Aug. 7, 1941. Practice medicine specializing in internal medicine, Buffalo, 1916-23, specializing in psychiatry and psychoanalysis, Stockbridge, Mass., 1923-30, N.Y.C., 1930—; faculty Buffalo U. Med. Sch., 1916-23, Cornell U. Med. Sch., 1933-36; faculty Columbia Coll. Phys. and Surg., 1931-32, asst. clin. prof. psychiatry, 1947-54, hon. cons. psychiat. clinic, for tng. and research; faculty N.Y. Sch. Psychiatry, N.Y.C., 1958-64, prof. psychiatry, 1960-64, prof. emeritus, 1965—, asst. dean, 1958-64; psychiatrist-in-chief Am. Rehab. Com., 1946—. Ofcl. cons. N.Y. State Dept. Mental Hygiene, 1956—, mem. council, 1951-56; Mem. A.M.A., Am. Psychiat. Assn., Acad. Medicine, Am. Acad. Psychoanalysis, N.Y. Soc. Clin. Psychiatry, N.Y. Acad. Sci., A.A.A.S., World, Pan Am. med. assns., World Fedn. for Mental Health (chmn. bd. U. S. com. 1965—), Group for Advancement Psychiatry, Assn. Psychiat. Medicine. Author: Insomnia: Its Causes and Treatment, 1938; also numerous articles, chpts. in books, book revs. Research on clin. endocrine disorders, allergies, peripheral vascular disorders. Address: 45 E. End Av., N.Y.C. 10028.*

MILLIAN, Stephen Jerry, Am. virologist; b. Okeechobee, Fla., Feb. 15, 1927; s. Edward Julius and Celia (Tepper) M.; B.S., Bklyn. Coll., 1949; M.S., Ohio State U., 1950, Ph.D., 1953; m. Lenore Zelda Fogelson, May 6, 1956; children—Betsy Lynn, Nancy Kim, Cynthia Jane, Melissa Dawn. Research virologist Armour Pharm. Corp., Chgo., 1953-57; staff virologist biologics research and devel. Chas. Pfizer & Co., Terre Haute, Ind., 1957-59; asso. cancer research scientist Roswell Park Meml. Inst., Buffalo, 1960-61; chief virus unit Bur. Labs., N.Y.C. Dept. Health, 1961—. Lectr., Hunter Coll.; virology lab. cons. First Army Med. Lab. Diplomate Am. Acad. Microbiology in Pub. Health and Lab. Virology. Mem. Harvey Soc., N.Y. Acad. Scis., Am. Pub. Health Assn., Assn. Mgmt. in Pub. Health, Sigma Xi. Research, publs. in diagnostic virology, serology, immunology; prodn. virus vaccines; isolation of infectious agts. from malignant tumors. Home: 61-25 98th St., Rego Park, N.Y. 11374. Office: 566 1st Av., N.Y.C.*

MILLICHAP, Joseph Gordon, physician; b. Wellington, Shropshire, Eng., Dec. 18, 1918; s. Joseph Profit and Alice (Flello) M.; M.B. (hon.), London U., 1946, M.D., 1951; m. Mary Irene Fortey, Feb. 26, 1946; children—Martin Gordon, Paul Anthony. Came to U. S., 1953, naturalized, 1965. Asso. prof. pharmacology U. Utah, 1954-55; vis. scientist pediatric neurologist NIH, Bethesda, Md., 1955-56; asso. prof. pediatrics and pharmacology, pediatric neurologist Albert Einstein Coll. Med., 1956-58; USPHS fellow in neurology and neuropathology Mass. Gen. Hosp., Boston, 1958-60; pediatric neurologist Mayo Clinic, Rochester, Minn., 1960-63; asso. prof. pediatric neurology and pharmacology U. Minn. Grad. Sch., 1960-63; head div. neurology Children's Meml. Hosp., Chgo., 1963—; prof. neurology and pediatrics Northwestern U. Med. Sch., 1963—. Cons. Surgeon Gen. USPHS, Nat. Inst. Neurol. Diseases and Blindness, Com. on Mental Retardation, A.M.A. Dept. Adverse Drug Reactions, Ill. Crippled Children's Sch., Chgo.; chmn. research com. Epilepsy Found., Washington. Mem. Am. Neurol. Assn., Acad. Neurology, Epilepsy Soc., Soc. Pharmacology and Therapeutics, Soc. for Pediatric Research, Soc. Exptl. Biology and Medicine, A.M.A., Am. Pediatric Soc., Chgo. Neurol. Soc. Author chpts. in Holt's Textbook of Pediatrics, 1964, Brennemann's Pediatrics, 1965, Physiological Pharmacology, 1965, Modern Therapy of Epilepsy, 1964. Research, publs. on causes and therapy of epilepsies, anticonvulsant drugs, chem. maturation of developing brain; discoveries of mechanism of action of carbonic anhydrase inhibitors, a new drug for febrile convulsions, abnormal aminoaciduria in children with petit mal epilepsy and treatment of infantile spasms. Home: 1350 N. State Pkwy., Chgo. 60610. Office: 707 W. Fullerton Av., Chgo. 60614.*

MILLIEZ, Paul Lucien, French physician; b. Mons en Baroeul, France, June 15, 1912; s. Lucien and Marie (Rivet) M.; student Faculty Scis. Paris, 1929; M.D., Faculty Medicine Paris, 1935; m. Jacqueline Alice Lemierre, June 28, 1939; children—Francoise, Jean, Jacques, Anne, Yvonne, Odile. Dir. gen. Cabinet Sec. Health, 1944; asst. dir. French Red Cross, 1944-45; head clinic Paris Faculty Medicine, 1945; aggregate prof., 1949-59, prof. med. pathology, 1959—; chief physician Beaujon Hosp., 1960—; head physician Broussais Hosp., 1962—. Mem. Internat. Assn. Med. Publs. (pres.), French Assn. Urban Medicine, Paris Coll. Medicine, Internat., French (v.p.) socs. nephrology. Author (with Bonenfant) Maladies Infectieuses. Research, numerous publs. on gravidic toxemiae, progressive degenerative nephritis, prognosis of isolated protein uriae, nephro-vascular arterial hypertension, pheochromocytomae, Conn's syndrome, treatment of nephrotic syndomes with nitrogen mustards. Home: 3 rue Rabelais, Paris 8. Office: 96, rue Didot, Paris 14, 75 France.*

MILLIGAN, Dolphus Edward, Am. chemist; b. Brighton, Ala., June 17, 1928; s. Reuben and Ruberta (Tucker) M.; B.S. in Chemistry, Morehouse Coll., 1949; M.S. in Chemistry and Math., Atlanta U., 1951; Ph.D., U. Cal. at Berkeley, 1958; m. Thedola Hayes, June 20, 1952; children—Deborah, Stephen, Charles. Faculty, Ft. Valley (Ga.) State Coll., 1951-52; fellow in fundamental research Mellon Inst., Pitts., 1957-63; phys. chemist Nat. Bur. Standards, Washington, 1963—. Recipient Prof. Arturo Miolati prize U. Padua, Italy, 1965. Mem. Am. Chem. Soc., Am. Phys. Soc. Application of matrix isolation technique to prodn. of free radicals for direct spectroscopic observation. Home: 1404 Primrose Rd. N.W., Washington 20012. Office: Nat. Bur. Standards, Washington 20234.*

MILLIGAN, W(infred) O(liver), Am. chemist; b. Coulterville, Ill., Nov. 5, 1908; s. John Winfred and Millie Mae (McMillan) M.; A.B., Ill. Coll., Jacksonville, 1930, Sc.D., 1946; M.A., Rice Inst., 1932, Ph.D., 1934; D.Sc., Tex. Christian U., 1960. Research chemist Harshaw Chem. Co., Cleve., 1934; cons. Houdry Process Corp., 1936-45, Humble Oil & Refining Co., Houston, 1945-62, Oak Ridge Nat. Lab., 1950—; prof. chemistry Rice U., 1934-63; dir. research Robert A. Welch Found., Houston, 1955—, editor Proc. Confs. on Chem. Research, also Research Bull., 1957—; pres. Tex. Christian U. Research Found., 1963-65. Chmn. Nat. Colloid Symposium, 1952-59; mem. Nat. Acad. Sci.-NRC coms. on application x-ray and electron diffraction, 1938-41, panel on permanent magnet materials, 1952—, clay mineralogy, 1953—, postdoctoral fellow chemistry, 1954-59; mem. Tex. Adv. Com. Atomic Energy, 1955—; investigator OSRD, 1943-45. Fellow Am. Inst. Chemists, Am. Phys. Soc.; mem. Am. Crystallographic Assn., Am. Chem. Soc. (dir. 1961—, chmn. several coms.), Faraday Soc., Phi Beta Kappa, Sigma Xi, Phi Lambda Upsilon, Alpha Chi Sigma. Asso. editor Jour. Phys. Chemistry, 1952, editorial bd. 1952-58. Contbr. papers tech. lit. Studies in catalysts; electron microscopy; adsorption of gases in solids; metallic oxides; x-ray and electron diffraction. Home: 2114 Rice Blvd., Houston 77005. Office: Robert A. Welch Found. Bank of the Southwest Building, Houston 77002.*

MILLIKAN, Clark Blanchard, Am. aero. engr., educator; b. Chgo., Aug. 23, 1903; s. Robert Andrews and Greta (Blanchard) M.; Ph.B., Yale, 1924; Ph.D., Cal. Inst. Tech., 1928; m. Helen Staats Battle, 1928 (div. 1958); children—Marcia (dec.), Robert (dec.), Michael; m. 2d, Edith Nussbaum Parry, Feb. 19, 1959; 1 dau., Virginia. Faculty, Cal. Inst. Tech., Pasadena, 1928-66, prof., 1940-66, dir. Grad. Aero. Labs., 1949-66; dir., mem. tech. adv. bd. Aerojet-Gen. Corp.; dir. Menasco Mfg. Corp., Burbank, Cal., Nat. Engring. Sci. Co., Nat. Engring. & Sci. Co., Pasadena, Cal. Chmn., USAF Sci. Adv. Bd., Space Systems Div. Adv. Group, subcom. on fluid mechanics NASA; mem. sci. adv. com. Ballistic Research Labs. U. S. Army, Naval Research Adv. Com.; cons. Inst. for Def. Analysis, Office Sec. Def. Sci. Bd. Recipient Exceptional Service award USAF, 1960. Fellow Am. Phys. Soc., Am. Acad. Arts and Scis., Am. Inst. Aeros. and Astronautics (hon., pres. 1937), Royal Aero. Soc. Gt. Britain; mem. Nat. Acad. Scis., Phi Beta Kappa, Sigma Xi, Tau Beta Pi. Author: Aerodynamics of the Airplane, 1941. Contbr. sci. and tech. articles to lit. Home: 690 Wendover Rd., Pasadena, Cal. 91103. Died Jan. 2, 1966.

MILLIKAN, Clark Harold, Am. neurologist; b. Freeport, Ill., Mar. 2, 1915; s. William Clarence and Louise (Chamberlain) M.; M.D., U. Kan., 1939; m. Gayle Margaret Gross, May 2, 1942; (div. 1966); children—Terri (Mrs. Gene Goodro), Clark William, Jeffrey Brent; m. 2d, Janet Tillotson Holmes, July, 1966. Resident in neurology State U. Ia., Iowa City, 1941-44, asst. in neurology, 1944-45, asso., 1945-46, asso. prof. neurology, 1946-49; cons. in neurology Mayo Clinic, faculty neurology U. Minn., Rochester, 1949—, head sect. neurology Mayo Clinic, 1954—, prof., 1958—. Spl. lectr. Radcliffe Infirmary, Oxford, Eng., 1963; chmn. Nat. Heart Inst.-Nat. Inst. Neurologic Diseases and Blindness Joint Council Subcom. on Cerebrovascular Diseases, 1964; adv. council div. regional med. programs USPHS; Conner lectr. Am. Heart Assn., 1961; Peter T. Bohan lectr. U. Kan., 1965. Mem. Am. Neurol. Assn. (asst. sec. 1959—), Assn. Research in Nervous and Mental Diseases (pres. 1961), A.M.A., Am. Acad. Neurologists, N.Y. Acad. Scis., A.C.P., Central Neuropsychol. Assn., Minn. Med. Assn., Royal Soc. Medicine, Am. Heart Assn. (chmn. council on cerebrovascular disease 1967—), Sigma Xi. Editor, chmn. Transactions Princeton Confs., 1961, 64. Research, publs. on causative mechanisms, diagnosis and treatment of occlusive cerebrovascular disease. Home: 1061 Plummer Lane, Rochester 55901. Office: Mayo Clinic, Rochester, Minn. 55901.*

MILLIKAN, Robert Andrews, Am. physicist; b. Morrison, Ill., Mar. 22, 1868; s. Silas Franklin and Mary Jane (Andrews) M.; A.B., Oberlin, 1891, A.M., 1893; Ph.D., Columbia, 1895; postgrad. univs. Berlin and Göttingen, 1895-96; numerous hon. degrees, including ones from Columbia, 1917, U. Dublin, 1924, Yale, 1925, Princeton, 1928, Harvard, 1932, U. Liege, 1930, U. Paris, 1939; m. Greta Irvin Blanchard, Apr. 10, 1902; children—Clark Blanchard, Glenn Allan (dec.), Max Franklin. Tutor physics, Oberlin, 1891-93; member physics staff U. Chgo., 1896-1921; dir. Norman Bridge Lab. Physics, chmn. exec. council Calif. Inst. Tech., Pasadena, 1921-45, prof. emeritus, v.p. bd. trustees, from 1945. Recipient Comstock prize Nat. Acad. Scis., 1913; Edison medal Am. Inst. E.E., 1922; Hughes medal Royal Soc. Gt. Britain, 1923; Nobel prize in physics, 1923; Faraday medal London Chem. Soc., 1924; Matteucci medal Societa Italiana della Scienze, 1925; Gold medal Am. Soc. Mech. Engrs., 1926; Messel medal Soc. Chem. Industry (British), 1928; Gold medal Holland Soc., 1928, Soc. Arts and Sciences, 1929, Radiol. Soc. N.Am., 1930; Gold medal Roosevelt Meml. Assn., 1932; Gold medal Franklin Inst., 1937; Joy Kissen Mookerjee Gold medal Indian Assn. for Cultivation Sci., 1939; Oersted medal Am. Assn. Physics Teachers, 1940; several decorations. Fellow Am. Acad. Arts and Scis., A.A.A.S. (pres. 1929); mem. Nat. Acad. Scis., Am. Philos. Soc., Am. Phys. Soc. (pres. 1916-18), many fgn. sci. socs., Sigma Xi, Phi Beta Kappa; asso. Royal Acad. Belgium. Author or co-author: A Course of College Experiments in Physics, 1898; Theory of Optics, 1900; Mechanics, Molecular Physics and Heat, 1901; A First Course in Physics, 1906; A Laboratory Course in Physics for Secondary Schools, 1906; Electricity, Sound and Light, 1908; The Electron, 1917, 25; Science and Life, 1923; Elements of Physics, 1917; Evolution of Science and Religion, 1927; A First Course in Physics for Colleges, 1928; Science and the New Civilization, 1930; Time, Matter, and Values, 1932; Electrons (+ and —), Protons, Photons, Neutrons, and Cosmic Rays, 1935, rev. edit., 1947; New Elementary Physics, 1936; Mechanics, Molecular Physics, Heat and Sound, 1937; Cosmic Rays, 1939; Autobiography, 1950. Contbr. to tech. jours. Carried out original research on x-rays and free expansion of gases; isolated electron and measured its electric charge, 1910; worked on verifying Einstein's photoelectric equations, also on determining Planks's constant, 1912-15, antisubmarine and meteorol. devices, World War I, hotspark spectroscopy, 1920-23, ionization chambers; his expts. in electricity led to determination of number of molecules in unit vol. of gas at a given pressure; studied cosmic rays, which he named. Died San Marino, Cal., Dec. 19, 1953.

MILLIKAN, Roger Conant, Am. chemist; b. Tiffin, O., Jan. 27, 1931; s. Robert F. and Laura (Grosvenor) M.; A.B., Oberlin Coll., 1953; Ph.D., U. Cal. at Berkeley, 1957; m. Mary Clark Stickell, Dec. 28, 1953; children—Ross E., Polly S., Sandra K., Jane G., Clark F. NSF fellow U. Cal. at Berkeley, 1955-56; phys. chemist Gen. Electric Co., Schenectady, 1956-67; prof. chemistry U. Cal. at Santa Barbara, Cal., 1967—. Fellow Am. Phys. Soc., Optical Soc. Am. Publs. on 1st correct interpretation of formic acid infrared spectrum; discovered (with W. E. Kaskan) molecule BO2 and showed it is responsible for green color of boron flames; 1st observation of pure vibrational fluorescence; determined rates of vibrational energy exchange for many gases. Home: 5475 Toltec Dr., Santa Barbara 93105.*

MILLINGTON, Sir Thomas, English physician; b. Newbury, Eng., 1628; s. Thomas Millington; M.A., Trinity Coll., Cambridge (Eng.) U.; M.D., Oxford (Eng.) U., 1659; fellow All Souls Coll., Oxford U.; named Sedleian prof. natural philosophy, Oxford U., 1675; ct. physician. Knighted, 1680. Fellow Coll. Physicians (censor 1678, Harveian orator 1679, treas. 1686-89, pres. from 1696); Fellow Royal Soc., 1663. Credited with discovery of sexuality in plants, 1676. Died Jan. 5, 1704.

MILLIONSHCHIKOV, Mikhail Dmitrievich, Russian applied physicist; b. Jan. 16, 1913; grad. Grozny Petroleum Inst., 1932; D.Tech. Sci. Instr., Grozny Petroleum Inst., 1932-34, Moscow Aviation Inst., 1934-43, Moscow Engring. and Physics Inst., 1943-49; with Inst. Mechanics, USSR Acad. Scis., 1944-49, acad. sec. dept. physicotech. problems of power engring., 1963-65; dep. dir. Kurchatov Inst. Atomic Energy, 1960—. Decorated Order of Lenin, 1963; recipient Stalin prize. Mem. USSR Acad. Scis. (v.p. 1962—). Author: The Theory of Uniform Isotropic Turbulence, 1941; co-author: Applied Gas Dynamics, 1948. Research and publs. on theory of turbulence, theory of filtration and applied gas dynamics. Address: USSR Acad. Scis., Leninsky prospect 14, Moscow, USSR.

MILLMAN, Jacob, elec. engr., educator; b. Russia, May 17, 1911; s. Philip and Gertrude (Nachschen) M.; came to U. S., 1913, naturalized, 1917; B.S., Mass. Inst. Tech., 1932, Ph.D., 1935; postgrad. U. Munich, Germany; m. Sally Dublin Millman, Oct. 11, 1936; children—Richard S., Jeffrey T. Faculty, Coll. City N.Y., 1936-42, 45-52; mem. staff radiation lab. Mass. Inst. Tech., Cambridge, 1942-45; prof. elec. engring. Columbia, N.Y.C., 1952—. Cons. on electronics to bus. firms. Recipient citation OSRD, 1945. Fulbright grantee U. Rome, Italy, 1959-60. Fellow I.E.E.E., Am. Phys. Soc. Author: (with S. Seely) Electronics, 1942, 2d edit., 1951; Vacuum-tube and Semiconductor Electronics, 1958; (with H. Taub) Pulse and Digital Circuits, 1956, Pulse, Digital and Switching Waveforms, 1964; (with C. C. Halkias) Electronic Devices and Circuits, 1967. Home: 7 Adrienne Pl., White Plains, N.Y. 10605. Office: Elec. Engring. Dept., Columbia U., N.Y.C. 10027.*

MILLMAN, Peter Mackenzie, Canadian astronomer; b. Toronto, Ont., Can., Aug. 10, 1906; s. Robert Malcolm and Edith (Middleton) M.; B.A., U. Toronto, 1929; A.M., Harvard, 1931; Ph.D., 1932; m. Margaret Bowness Gray, July 10, 1931; children—Barry Mackenzie, Cynthia Gray. Asst. astronomy, Harvard, 1929-31, asso. Obs., 1955—; Astronomer, David Dunlap Obs., lectr. U. Toronto, 1933-45; chief stellar physics div. Dominion Obs., Ottawa, Ont., 1946-55; prin. research officer, head upper atmosphere research Nat. Research Council Can., Ottawa, 1955—. Counsellor, Smithsonian Instn., Washington, 1966—. Recipient J. Lawrence Smith medal Nat. Acad. Scis., Washington, 1954. Fellow Royal Soc. Can., Canadian Aeros. and Space Inst., Meteoritical Soc. (past pres.), A.A.A.S.; mem. Royal Astron. Soc. Can. (past pres.), Internat. Astron. Union (past pres. commn. on meteors and meteorites), Am. Inst. Aeros. and Astronautics, Canadian Assn. Physicists, Am. Astron. Soc., Société Astron. de France, Am. Assn. Variable Star Observers.

Author: This Universe of Space, 1961; also numerous articles, newspaper columns. Pioneered photography of spectra of meteors and solar eclipses from aircraft; among 1st to use radar methods for astron. study meteors; radar patrol studies of influx of meteoritic material on earth; took charge Canadian centre for meteoritic and auroral studies during IGY. Home: 4 Windsor Av., Ottawa. Office: Nat. Research Council, Ottawa, 7, Ont., Can.*

MILLMAN, Sidney, physicist; b. Dawid-Gorodok, Russia, Mar. 15, 1908; s. Jacob and Nora (Berman) M.; came to U. S., 1922, naturalized, 1929; B.S., Coll. City N.Y., 1931; A.M., Columbia, 1932, Ph.D. in Physics, 1935; m. Dorothy Rosenfeld, Dec. 26, 1931; 1 son, Michael G. Tyndall fellow Columbia, N.Y.C., 1935-36, Barnard fellow, 1936-37, research asst., 1937-39, mem. sci. staff Radiation Lab., 1942-45; instr. physics Coll. City N.Y., 1939-41, Queens Coll., 1941 42; mem. tech. staff Bell Telephone Labs., Murray Hill, N.J., 1949-52. dir. phys. research, 1952-65, exec. dir. research, physics, univ. relations, 1965—. Fellow Am. Phys. Soc., I.E.E.E. Contbr. numerous articles on nuclear spins, hyperfine structure and magnetic moments to profl. publs. Inventor rising sun magnetron, contbr. to other devices in millimeter waves. Home: 17 Fairview Av., Summit, N.J. 07901. Office: Bell Telephone Labs., Murray Hill, N.J. 07974.*

MILLON, Eugene Auguste Nicolas, French chemist; b. Châlons-sur-Marne, France, 1812; M.D., Paris, 1836; with corps mil. pharmacy, 1836-37; instr. chemistry, Val-de-Grace, France 1837-40, prof. chemistry, 1841-47, pharmacist, 1847-48; asst. pharmacist, Gros-Caillou, 1840; 1st prof., mil. hosp., Lille, France, 1848-50; prin. pharmacist, Algiers, 1850. Author: (with F. Hoefer, J. Nicklès) Recherches sur l'acide nitrique, 1842; Eléments de chimie organique, 2 vols., 1845-48; Recherches chimiques sur le mer cure et les constitutions salines, 1846; La liberté du commerce de la boucherie, 1851. Editor Annuaire de chimie, 7 vols., 1845-51. Research on nitric acid and its action on metals; also on mercury salts and compounds resulting from their interaction with ammonia, chlorine dioxide; pub. his discovery of chlorites, 1841-43, iodine dioxide, 1844, Millon's reagent for detection of proteins, 1849. Died 1867.

MILLOTT, Norman, Brit. zoologist; b. Oct. 24, 1912; s. Reuben and Mary (Thislethwaite) M.; ed. U. Sheffield (Eng.), U. Manchester (Eng.), Cambridge U.; B.Sc., M.Sc., D.Sc., Ph.D.; m. Margaret Newns, Sept. 21, 1939; children—Susan, Judith, Stephanie. Asst., U. Manchester, 1935-36; Rouse-Ball scholar Trinity Coll., Cambridge, 1936-38; lectr. zoology U. Manchester, 1938-48; prof. zoology Faculty, U. India, 1948-55, Bedford Coll., U. London, 1955—. Mem. Marine Biol. Assn. U.K., Zool. Soc. London, Ray Soc. Research and publs. on physiology of invertebrates, especially echinoderms; histochemistry; animal pigmentation; morphology; ecology. Home: 2 Monahan Av., Purley, Eng. Office: Bedford Coll., Regents Park, London N.W.1, Gt. Britain.

MILLS, C. Wright, Am. sociologist; b. Waco, Tex., 1916; s. Charles G. and Frances (Wright) M.; B.A., U. Tex., 1939, M.A., 1939; Ph.D., U. Wis., 1941; m. Yaroslava Surmach, June 11, 1959. Asso. prof. sociology U. Md., 1941-45; fellow Guggenheim Found., 1945-46; dir. labor research div., bur. applied social research Columbia, 1945-48, asst. prof. sociology Columbia, 1946-50, asso. prof. sociology, 1950-56, prof. sociology, 1956—; vis. prof. Univ. Chgo., 1949; vis. prof. human relations Brandeis U., 1953; lectr. William Alanson White Inst. Psychiatry, 1954-56; Fulbright professor U. Copenhagen, Denmark, 1956-57. Special business consultant Smaller War Plants Corp., WPB, 1945. Mem. Am. Sociological Society, Industrial Relations Research Assn. (v.p. 1947-48). Author: The New Man of Power: America's Labor Leaders, 1948; The Puerto Rican Journey: New York's Newest Migrants (with Clarence Senior, Rose K. Goldsen), 1950; White Collar: The American Middle Classes 1951; Character and Social Structure: The Psychology of Social Institutions (with Hans Gerth), 1953; The Power Elite, 1956; The Causes of World War Three, 1958; The Sociological Imagination, 1959. Criticized previous sociol. approaches; called for scientific sociology of knowledge; studied soc. with synthesized analyses of power, politics and human complexities; leading commentator on depersonalization and alienation in contemporary mass soc. Died 1962.

MILLS, Charles Karsner, Am. neurologist; b. Phila., Dec. 4, 1845; s. James and Lavinia Ann (Fitzgerald) M.; ed. Central High Sch., Phila., 1860-64; M.D., U. Pa., 1869, Ph.D., 1871, LL.D., 1916; m. Clara Elizabeth Peale, Nov. 6, 1873; 1 dau., 3 sons. Prof. diseases of mind and nervous system Phila. Polyclinic, 1883-98, also a founder; clin. prof. nervous diseases Women's Med. Coll. of Pa., 1891-1902; prof. mental diseases and med. jurisprudence U. Pa., 1893-1901, clin. prof. nervous diseases, 1901-03, prof. neurology, 1903-15, emeritus prof. neurology, 1915; became chief of clinic for nervous diseases Univ. Hosp., 1874; apptd. neurologist Phila. Hosp., 1877. Med. witness in numerous medico-legal cases. Fellow Coll. Physicians of Phila.; mem. Royal Soc. Medicine of Eng. (corr.), Am. Neurol. Soc. (pres. 1877, 1924), Phila. Neurol. Soc. (founder 1883). Author: A Treatise on the Nervous System and Its Diseases, 1898; also papers. Described a case of unilateral ascending paralysis (Mill's disease), 1900, of unilateral descending paralysis, 1906; a founder Phila. sch. of neurology. Died May 28, 1931.

MILLS, Clarence Alonzo, Am. physician; b. nr. Miami, Ind., Dec. 9, 1891; s. Alonzo F. and Margaret (Wininger) M.; student Valparaiso U., 1913; B.A., U. S.D., 1917, LL.D., 1961; postgrad. U. Kan., U. Chgo.; Ph.D., U. Cin., 1920, M.D., 1922; m. Elith Clarissa Parrett, June 22, 1915; children—Russell Clarence, Marjorie Ruth (Mrs. Jack R. Porter), Don Harper. Instr., U. Kan., 1917-18, Marquette U. Med. Sch., 1918; instr. biochemistry U. Cin., 1919-22, fellow NRC, 1922-23, asst. prof. internal medicine, 1923-26, asso. prof., 1928-30, dir. lab. for exptl. medicine, 1925-62, James T. Heady prof. exptl. medicine, 1930-62, prof. emeritus, 1962—; asso. prof. medicine Peking (China) Union Med. Coll., 1926-28. Pres., Reflectotherm, Inc., Cin., 1951—; mem. health panel U. S. Tech. Conf. on Air Pollution, Washington, 1950; mem. panel on housings and bldgs. Bldg. Research and Adv. bd. NRC, 1953. Recipient Ohioana Library medal, 1942. Mem. A.M.A., Am. Physiol. Soc., Am. Soc. Biochemists, Am. Assn. Phys. Anthropology, A.A.A.S., Central Soc. Clin. Research, Research Club Cin., Sigma Xi, Alpha Omega Alpha, Alpha Kappa Kappa. Author: Living with the Weather, 1934; Medical Climatology, 1939; Climate Makes the Man, 1942, 43; Reflective Radiant Conditioning, 1950; Air Pollution and Community Health, 1954; This Air We Breathe, 1962; World Power Amid Shifting Climates, 1963; Collection of Published Works, 4 vols., 1962. Research, publs. on blood coagulation and hemorrhage, climate and effects on man, air pollution, radiant heating and cooling, food processes. Patentee fields radiant cooling and radiant heating by differential wavelengths. Address: 2311 Fairview Av., Cin. 45219.*

MILLS, Edmund James, Brit. chemist; b. London, Dec. 8, 1840; s. Charles F. and Mary Anne M.; ed. Royal Sch. Mines; D.Sc., London; LL.D., Glasgow, Scotland; m. Amelia Burnett. Asst. to John Stenhouse, 1861; chem. tutor Glasgow U., 1862-65; named prof. tech. chemistry, Glasgow, 1875; prof. emeritus tech. chemistry Glasgow Royal Tech. Coll. Fellow Royal Soc., 1874, Inst. Chemistry. Research and publs. on destructive distillation, fuel, Died Apr. 21, 1921.

MILLS, Harlan Duncan, Am. mathematician; b. Fort Dodge, Ia., May 14, 1919; s. Oral Harlan and Joy (Duncan) M.; B.Sc., Ia. State U., 1948, M.S., 1950, Ph.D., 1952; postgrad. U. Zurich; m. Luella Christine Sprecher, June 21, 1940; children—Kay Doreen (Mrs. Robert Van de Water), Wanda Jeanne (Mrs. George Ringold). Faculty, Ia. State U., 1950-52, Princeton, 1952-54, 57-59, N.Y. U., 1957-58; cons. Gen. Elec. Corp., 1954-57; pres. Mathematica, 1958-61; cons. RCA Labs., 1961-63; math. cons. IBM, 1964—. Mem. NSF Com. in Applied Math., 1956-58. Hon. fellow Wesleyan U. Mem. Am. Math. Soc., Math. Assn. Am., Inst. Math. Statistics, Am. Statis. Assn., Econometric Soc., Soc. for Indsl. and Applied Math., Operations Research Soc. Am., Assn. for Computing Machinery. Research in linear programming theory and game theory; applications of game theory to econ. and mgmt. problems, computer programming theory and round off error theory. Home: 5808 Brookside Dr., Chevy Chase, Md. 20015. Office: 18100 Frederick Pike, Gaithersburg, Md. 20760.*

MILLS, Harlow Burgess, Am. biologist; b. Le Grand, Ia., Aug. 20, 1906; m. Ernest M. and Anna (Burgess) M.; B.S., Ia. State Coll., 1929, M.S., 1931, Ph.D., 1934; m. Esther Winifred Brewer, Aug. 27, 1930; children—David Harlow, Gary Paul, Judith Anne. With U. S. Dept. Agr., 1929, 31; asst. prof. Tex. A. and M. U., 1930-31; naturalist, wildlife technician Yellowstone Park, 1934-35; asst. entomologist State of Mont., 1935-37; faculty zoology and entomology, head dept. Mont. State Coll., 1937-47; chief Ill. Natural History Survey, Urbana, 1947—. Chief scientist Latin Am. Office, NSF, 1962-63. Mem. Entomol. Soc. Am., Wildlife Soc., Ecol. Soc. Am., A.A.A.S., Sigma Xi, Phi Kappa Phi, Phi Mu Alpha. Contbr. numerous articles to sci., popular publs. Home: 1211 W. University St., Champaign, Ill. Office: Natural History Survey, Urbana, Ill.*

MILLS, Hiram Francis, Am. hydraulic engr.; b. Bangor, Me., Nov. 1, 1836; s. Preserved B. and Jane (Lunt) M.; C.E., Rensselaer Poly. Inst., 1856; A.M. (hon.), Harvard, 1899; m. Elizabeth Worcester, Oct. 8, 1873. Asst. engr. Bergen Tunnel, 1858, Bklyn. Water Works, 1859; water measurements, Cohoes, N.Y., 1859; with J. B. Francis, civil engr., Lowell, Mass., 1860-63; on important ry. and hydraulic work until 1867 (Hoosac tunnel, Deerfield dam, water power on Penobscot River at Bangor); hydraulic engr., Boston, 1867-9; chief engr. Essex Co., controlling water power of Merrimac River at Lawrence, 1869—; cons. engr. on hydraulic work in 10 states and Mexico, 1868—; cons. engr. Proprs. Locks and Canals on Merrimac River at Lowell, 1893, chief engr., 1894-1917; in charge of investigations of Mass. Bd. Health on purification of water supplies and sewage by filtration and otherwise, 1886-1914; designed and built Lawrence City filter, 1892-3; chmn. com. state bd. in originating and designing met. sewerage system and met. water supply; cons. engr. Wachusett dam and reservoir. Fellow Am. Acad. Arts and Scis. Died Oct. 4, 1921.

MILLS, Ivor Henry, English clinician, physiologist; b. London, Eng., June 13, 1921; s. John Henry and Priscilla (Young) M.; B.Sc., Queen Mary Coll., London (Eng.) U., 1942, Ph.D., 1946; B.A., Trinity Coll., Cambridge (Eng.) U., 1948, M.B., B.Chir., 1951, M.D., 1956, M.A., 1963; postgrad. St. Thomas's Hosp., London; m. Sydney Elizabeth Roberts, Jan. 4, 1947; children—Christopher, Diana. Research asst. Agrl. Research Council, U. N. Wales, 1943-46; house physician sr. med. casualty officer St. Thomas's Hosp., London, 1951-53, lectr., 1953-56, Med. Research Council (Eli Lilly) Research fellow, 1956-57, sr. lectr., 1959-61, reader in medicine, 1961-63; vis. scientist NIH, Bethesda, Md., 1958; prof. medicine U. Cambridge, 1963—, fellow Churchill Coll., 1963—. Mem. Soc. for Endocrinology (sec. 1964—), Renal Assn. Gt. Britain (exec. com. 1964—), Med. Research Soc., Internat. Nephrology Soc., Physiol. Soc. Gt. Britain, Royal Coll. Physicians (fellow). Author: Clinical Aspects of Adrenal Function, 1964; also numerous articles. Research on pre-natal mortality in wild rabbits, mechanisms controlling androgen prodn. in humans, mechanisms controlling renal excretion of sodium. Home: 6 Spinney Dr., Great Shelford, Cambridge. Office: Dept. Investigative Medicine, Addenbrooke's Hosp., Cambridge, Eng.*

MILLS, James Theodore, Sr., Am. plastic surgeon; b. Austin, Minn., Nov. 14, 1900; s. James Bishop and Lena (Erickson) M.; student St. Thomas Coll., St. Paul, 1914-19; B.S., U. Minn. 1921, M.D., 1925; m. Rosemary Zonne, Oct. 20, 1925; children—Hildegarde (Mrs. Chester Fullinwider, Jr.), James, Diana (Mrs. David T. Blackburn), Rosemary (Mrs. John Coats). Practice medicine specializing in plastic surgery, Dallas, 1932—; clin. prof., chmn. dept. plastic surgery U. Tex., Southwestern Med. Div., Dallas. Chmn. adv. bd. Crippled Children's Div. Tex. Vice chmn. Am. Bd. Plastic Surgery, 1953, chmn., 1954. Mem. Am. Assn. Plastic Surgeons (pres. 1948), Am. Soc. Plastic and Reconstructive Surgery (pres. 1953), Beta Theta Phi, Nu Sigma Nu. Home: 9307 Guernsey Lane Dallas 75220. Office: Doctors Bldg., 3707 Gaston Av., Dallas.*

MILLS, John, Am. elec. engr.; b. Morgan Park, Ill., Apr. 13, 1880; s. John and Sarah Elizabeth (Ten Broeke) M.; A.B., U. Chgo., 1901; A.M., U. Neb., 1904; B.S., Mass. Inst. Tech., 1909; m. Emma Gardner Moore, June 1, 1909; children—John, Marion, Theodora Ten Broeke. Fellow in physics U. Chgo., 1901-02, U. Neb., 1902-03; instr. physics Western Res. U., 1903-07, Mass. Inst. Tech., 1907-09; prof. physics Colo. Coll., 1909-11; with engring. dept. Am. Tel. & Tel. Co., 1911-15; with research dept. Western Electric Co., 1915-21, asst. personnel dir., 1921-23, personnel dir., 1923-24; dir. of publ. Bell Telephone Labs., Inc., 1925-45; adminstrv. asst. Cal. Inst. Tech. since 1946. Fellow Am. Phys. Soc., Am. Inst. Elec. Engrs., Inst. Radio Engrs.; mem. Phi Beta Kappa, Sigma Xi. Author: Electricity, Sound and Light, 1907; Introduction to Thermodynamics, 1909; Alternating Currents, 1911; Radio-Communication, 1917; Realities of Modern Science, 1919; Within the Atom, 1921; Letters of a Radio Engineer to His Son, 1922; Magic of Communication, 1923; Signals and Speech in Electrical Communication, 1934; A Fugue in Cycles and Bels, 1935; Electronics Today and Tomorrow, 1944; The Engineer in Society, 1946. Inventor of several methods for wire and radio-telephony; conceived and supervised design of Bell Telephone Exhibits at the world's fairs, Chgo., 1933, San Diego, 1935, Dallas, 1936, San Francisco, 1939-40, N.Y., 1939-40. Died June 14, 1948.

MILLS, Joseph William, geologist; b. Toronto, Ont., Can., June 9, 1917; s. Joseph William and Marguerite (Houle) M.; B.A. in Geology and Mineralogy, U. Toronto, 1939; Ph.D. in Geology, Mass. Inst. Tech., 1942; m. Yola L. Radochia, May 16, 1942; children—Andrea L., Joseph W. Came to U. S., 1950, naturalized, 1957. Mine and exploration geologist for Consol. Mining & Smelting Co., MicMac Mines Ltd., also O'Brien Gold Mines Ltd., 1942-50; mem. faculty Wash. State U., 1950—, prof. geology, 1962—, chmn. dept., 1961—; geologist div. mines and geology Wash. Dept. Conservation, summers 1959—. Mem. Geol. Assn. Can., Soc. Econ. Geologists, Nat. Assn. Geology Tchrs., N.W. Sci. Assn., Sigma Xi (pres. Wash. State U. chpt. 1964-65). Editor N.W. Sci., 1958—. Home: 2106 Orion Dr., Pullman, Wash.*

MILLS, Lewis Craig, Am. physician; b. Chgo., May 19, 1923; s. Lewis Craig and Jessie Alvina (Carlson) M.; student Tex. Technol. Coll., 1940-43; M.D., Baylor U., 1946; m. Jacqueline Hildebrand, Oct. 11, 1947; children—Karen, Maureen, Lew Ann, Lewis Craig. Faculty, Baylor U., Houston, 1953-57; faculty Hahnemann Med. Coll., Phila., 1957—, prof. medicine, 1964—, chief sect. endocrinology, 1957—. Recipient Lindback Found. award, 1964. Diplomate Am. Bd. Internal Medicine. Fellow A.A.A.S., A.C.P.; mem. Am. Diabetes Assn., Am. Fedn. Clin. Research, Am. Heart Assn., A.M.A., Am. Soc. for Pharm. and Exptl. Therapeutics, Physiol. Soc. Phila., Soc. Exptl. Biology and Medicine, So. Soc. Clin. Research, others, Alpha Omega Alpha. Research, publs. on connective tissue diseases, shock, influence of various drugs on vascular system. Home: 440 Levering Mill Rd., Merion, Pa. 19066. Office: 230 N. Broad St., Phila. 19102.*

MILLS, Robert Gail, Am. nuclear physicist; b. Effingham, Ill., Jan. 20, 1924; s. Gail A. and Helen (Taylor) M.; B.S. in Engring., Princeton, 1944; M.A., U. Mich., 1947; Ph.D., U. Cal. at Berkeley, 1952; m. Mary A. Steer, May 25, 1946; children—Susan Elizabeth, Robert William. Instr., Princeton, 1943-44, research staff, 1945-46, 54—, sr. staff mem. fusion research, 1954—; NRC fellow U. Zurich (Switzerland), 1953, tchr., 1954. Mem. adv. bd. Nat. Magnet Lab., Mass. Inst. Tech., 1963-67. Mem. Am. Phys. Soc., I.E.E.E. (chmn. nuclear sci. group 1966-67), Phi Beta Kappa, Sigma Xi. Editor: (with others) High Magnetic Fields, 1962. Research, publs. on controlled thermonuclear reactions. Home: 150 Prospect Av. Office: Box 451, Princeton, N.J. 08540.*

MILLS, Russell Clarence, Am. biochemist; b. Milw., Nov. 13, 1918; s. Clarence A. and Edith (Parrett) M.; B.S., U. Wis., 1940, M.S., 1942, Ph.D., 1944; m. Margaret A. Muth, June 19, 1940; children—Randolph, Elizabeth, Russell Clarence, Peter, Melissa. Asst. prof. biochemistry U. Kan., Lawrence, 1946-48, asso. prof., 1948-51, prof., 1951—, chmn. dept., 1948-62, asst. dean Grad. Sch., 1962, asso. dean Grad. and Med. Sch., 1963—. Mem. Am. Soc. Biol. Chemists, Am. Chem. Soc., Soc. Am. Microbiologists, A.A.A.S. Contbr. articles profl. jours. Home: 5407 Mission Dr., Shawnee Mission, Kan. 66208. Office: Univ. Kansas Med. Center, Kansas City, Kan. 66103.*

MILLS, T. Wesley, Canadian physiologist; M.A., Toronto (Ont., Can.) U.; M.D., McGill U., Montreal, Que., Can.; later student in Eng. and Germany; licentiate Royal Coll. Physicians. Prof. physiology McGill U., later emeritus prof. Fellow Royal Soc. Can. (pres. sect. IV twice), Soc. for Study Comparative Psychology Montreal (founder), Natural History Soc. (pres. Montreal), Soc. Am. Naturalists (v.p.). Author: Animal Physiology; Voice Production; Comparative Physiology; The Dog in Health and in Disease; The Nature and Development of Animal Intelligence; also research articles. Died Feb. 14, 1915.

MILLS, Sir William, Brit. engr.-inventor; b. Apr. 24, 1856; s. David Mills; apprenticed as marine engr., obtaining 1st class certificate; considerable experience at sea as engr. and repairing telegraph cables; m. Eliza Gandy Hodgson. Organizer, propr. Mill Munitions, Ltd., Birmingham, also William Mills (Sunderland), Ltd.; mng. dir. William Mills, Ltd., Birmingham. Chmn. James Wat Meml. Trust. Mem. Inst. Mech. Engrs., Inst. Metals, Birmingham C. of C. (council). Established 1st aluminium foundry in U.K.; large amount of research works in alloys, acknowledged pioneer of this industry; patentee in foundry connection and moulding; invented and introduced Mills hand grenade, 1915, used exclusively and successfully by Brit. and other allied troops throughout World War I, of which 75,000,000 were supplied; established Mills Munitions Co.; turned out large quantities of aluminum castings for airplanes; other inventions include Mills boat disengaging gear, several mfg. inventions; in field of sports invented telescopic seats, golf clubs, others. Died Jan. 7, 1932.

MILLS, William Harold, Am. mathematician; b. N.Y.C., Nov. 9, 1921; s. Frederick Cecil and Dorothy (Clarke) M.; A.B., Swarthmore Coll., 1943; M.A., Princeton, 1947, Ph.D., 1949; m. Joan Woolf Rounds, July 15, 1949; children—Charles Frederick, James Lawrence, Robert Clarke. Faculty, Yale, 1949-64, asso. prof. math., 1958-64; mathematician Inst. for Def. Analyses, Princeton, N.J., 1963—. Mem. Am. Math. Soc., Math. Assn. Am., Canadian Math. Congress. Research, publs. on number theory, algebra, theory of games; proved existence of function that represents only prime numbers. Home: 347 Prospect Av., Princeton 08540. Office: Inst. for Def. Analyses, Von Neumann Hall, Princeton, N.J. 08540.*

MILLS, William Hobson, English chemist; b. July 6, 1873; s. William Henry Mills; ed. Jesus Coll., Cambridge, Eng.; student U. Tübingen (Germany), 1899-1900, also D.Sc.; M.A., Cambridge; m. Mildred Gostling, 1903; 1 son, 3 daus. Head chem. dept. No. Poly. Inst., 1902-12; fellow, pres. Jesus Coll., Cambridge, 1940-48, emeritus reader stereochemistry U. Cambridge. Mem. adv. council Dept. Sci. and Indsl. Research, 1935-40. Fellow Royal Soc., 1923 (Davy medal 1933); mem. Brit. Assn. (became pres. sect. B 1932), Chem. Soc. (pres. 1941-44, Longstaff medal 1930). Contbr. articles to chem. jours. Died. Feb. 22, 1959.

MILLSAPS, Knox Taylor, Am. mathematician, educator; b. Birmingham, Ala., Sept. 10, 1921; s. Knox Taylor and Mae (Joyce) M.; B.S., Auburn U., 1940; Ph.D., Cal. Inst. Tech., 1943; m. Lorraine Marie Hartle, June 12, 1956; children—Melinda Marie, Mary Charmaine, Karla Marie, Knox Taylor. Asso. prof. aero. engring. Ohio State U., Columbus, 1946-48; physicist Wright Field, O., 1948-49, 50-51, chief mathematician, 1952-55; prof. physics Auburn (Ala.) U., 1949-52; prof. mech. engring. Mass. Inst. Tech., Cambridge, 1955-56; chief scientist Air Force Missile Devel. Center, Alamogordo, N.M., 1956-60; exec. dir. Air Force Office Sci. Research, Washington, 1960-63; chief scientist Office Aerospace Research, USAF, Washington, 1960-63; research prof. aero. engring. U. Fla., Gainesville, 1963—. Mem. Am. Inst. Aeros. and Astronautics, Am. Phys. Soc., Soc. for Indsl. and Applied Math., Am. Math. Soc., Math. Assn. Am., Am. Soc. for Engring. Edn., Soc. Am. Mil. Engrs., Air Force Assn., Am. Ordnance Assn., A.A.A.S., Sigma Xi, Kappa Sigma. Research on math. theories of fluid dynamics and heat transfer. Home: 3735 S.W. 5th Pl., Gainesville, Fla. 32603.*

MILLSPAUGH, Charles Frederick, Am. botanist; b. Ithaca, N.Y., June 20, 1854; s. John Hill and Marion E. (Cornell) M.; studied Cornell, 1872-73; M.D., New York Homoe. Med. Coll., 1881; m. Mary Louisa Spaulding, Sept. 19, 1877 (dec. 1907); m. 2d, Clara Isobel Mitchell, 1910. Practiced medicine, Binghamton, N.Y., 1881-90, Waverly, N.Y., 1890-91; botanist W.Va. U., 1891-93; curator dept. botany, Field Mus. Natural History, Chgo., 1894—; prof. med. botany Chgo. Homoe. Med. Coll., 1897—; professorial lectr. econ. botany U. Chgo., 1895—. Mem. Pan-Am. Com. on Med. Botany. Fellow Am. Acad. Arts and Scis.; hon. mem. N.Y. Homoe. Med. Soc., Faculty of Medicine Mexico, Faculty of Medicine Brazil, Binghamton Acad. Sci. Editor Homoeopathic Recorder, 1890-92. Author: American Medical Plants (illustrated), 1887; Weeds of West Virginia, 1892; Flora of West Virginia, 1892; Plantae Utowanae; Flora of St. Croix; Flora Sand Keys of Florida; Prenunciae Bahamenses; Plantae Yucatanae; Flora of West Virginia, 1891, 95, 1913; revised and enlarged MacIlvaine's 1,000 American Fungi, 1911. Explored in Mexico, also W.I., 1887, 94, 98, 1900, Brazil, 1888, uninhabited Bahamian islets, 1904, 05, 07, 11; authority on Antillean region. Died Sept. 16, 1923.

MILNE, (Edward) Arthur, Brit. mathematician, astrophysicist; b. Hull, Eng., Feb. 14, 1896; s. Sidney A. Milne; M.A., Trinity Coll., Cambridge, 1920, D.Sc., 1925; D.Sc. (hon.), Amsterdam; m. Margaret Campbell, 1928; 1 son, 2 daus.; m. Beatrice Renwick, 1945; 1 dau. Fellow Trinity Coll., 1919-25, Smith's Prizeman, 1922; asst. dir. Solar Physics Obs., Cambridge, 1920-24; lectr. Trinity Coll., Cambridge, 1924-25, univ. lectr. astrophysics, 1922-25; Beyer prof. applied math. Manchester (Eng.) U., 1924-28; prof. math. Wadham Coll., Oxford, 1928-50. Mem. ordnance bd. Ministry Supply, 1939-44. Recipient Johnson Meml. prize, 1931, Bruce medal Astron. Soc. Pacific, 1945, Hopkins prize Cambridge Philos. Soc., 1946. Became Fellow Royal Soc., 1926, recipient Royal medal, 1941; hon. mem. Am. Astron. Soc., Am. Acad. Arts and Scis. (fgn.), Calcutta Math. Soc.; mem. Royal Astron. Soc. (Gold medal 1935, pres. 1943-45), London Math. Soc. (pres. 1937-39). Author: The White Dwarf Stars, 1932; Relativity, Gravitation and World-Structure, 1935; Vectorial Mechanics, 1948; Kinematic Relativity, 1948; also articles. Research on surface of stars; proposed new systems of dynamics and electrodynamics, theory for devel. on universe based on inference; developed system of kinematic relativity (alternative to Einstein's gen. relativity theory). Died Dublin, Ireland, Sept. 21, 1950.

MILNE, John, English seismologist, geologist; b. Liverpool, Eng., Dec. 30, 1850; s. John and Emma (Twycross) M.; ed. King's Coll., London, Eng., Royal Sch. Mines, London, U. Freiberg (Germany); D.Sc. (hon.), Oxford (Eng.) U.; m. Tone Noritsune. Mining engr. Cornwall, Lancashire (both Eng.), central Europe; investigated mineral resources of Newfoundland, Labrador (both Can.), 1872-74; with C. J. Beke's expdn. to Mt. Sinai; prof. geology, mining Imperial Coll. Engring., Tokyo, Japan, 1875-95; 1st prof. seismology Imperial U., Tokyo; returned to Eng., 1895. Fellow Geol. Soc., Royal Soc., 1887; mem. Seismol. Soc. Japan (leading founder, sec. for 15 years), Brit. Assn. (sec. seismol. com. until 1913). Author: Earthquakes, 1883; Seismology, 1898; The Miner's Handbook; Crystallography. Contbr. papers to profl. publs. Invented seismograph, 1880; established seismic survey of Japan, also made seismic survey all Brit. ter.; built seismol. obs. on Isle of Wight; study of seismic waves led to new knowledge of composition of earth's interior, opened new realm of sci. inquiry. Died Shide, Isle of Wight, July 30, 1913.

MILNE, Joshua, Brit. actuary; b. 1776; actuary Sun Life Assurance Soc., 1810-43. Author: (with J. Heysham) A Treatise on the Valuation of Annuities and Assurances on Lives and Survivorships, on the Construction of Tables of Mortality, and on the Probabilities and Expectations of Life, 2 vols., 1815; also articles in Ency. Brit., 4th edit. Used statistics of John Heysham to construct Carlisle table of mortality, 1816, thus bringing about a revolution in actuarial sci.; 1st to compute accurately value of fines. Died 1851.

MILNE, Lorus Johnson, biologist; b. Toronto, Ont., Can., Sept. 12, 1912; s. Charles Stanley and Edna S. (Johnson) M.; B.A. with honours in Biology, U. Toronto, 1933; M.A., Harvard, 1934, Ph.D. in Biology, 1936; m. Margery (Joan) Greene, Sept. 10, 1939. Came to U. S., 1933, naturalized, 1942. Faculty Southwestern U., Tex., 1936-37, Randolph-Macon Woman's Coll., 1937-42, U. Vt., 1947-48; faculty U. N.H., Durham, 1948—, prof. zoology, 1951—. Cons., writer to bus. firms, pub. cos., 1950—; exchangee U. S.-S.Africa Leader Exchange Program, 1959; cons. to New Zealand Dept. Edn. under UNESCO, 1966. Recipient Eugene Saxton Meml. Found. Lit. award, 1954; Nash Motor Co. Conservation award, 1954. Ford Found. Fund for Advancement Edn. fellow, 1953-54. Fellow A.A.A.S. (Writing award 1947); mem. Am. Soc. Zoologists, Corp. Woods Hole Marine Biol. Lab., Explorers Club, Animal Behavior Soc., Nature Conservancy, Wilderness Soc., Sigma Xi. Author: (with Margery Milne) A Multitude of Living Things, 1947; Machine Shop Methods, 1950; Famous Naturalists, 1952; The Biotic World and Man, 1952, 58, 65; The Mating Instinct, 1954; The World of Night, 1956, 68; Paths Across the Earth, 1958; Animal Life, 1959; Plant Life, 1959; The Balance of Nature, 1960; (with Margery Milne, others) The Lower Animals: Living Invertebrates of the World, 1960; The Mountains, 1962; The Senses of Animals and Men, 1962; The Valley: Meadow, Grove and Stream, 1963; Because of a Tree, 1963; Water and Life, 1964; The Crab That Crawled Out of the Past, 1965; Gift from the Sky, 1967; Living Plants of the World, 1967; Patterns of Survival, 1967; The Ages of Life, 1968; also articles. Home: 1 Garden Lane, Durham, N.H. 03824.*

MILNE-EDWARDS, Alphonse, French zoologist; b. Paris, Oct. 13, 1835; s. Henri Milne-Edwards; M.D.; docteur ès scis., 1861; prof. zoology Paris Sch. Pharmacy; became prof. mammalogy and ornithology Paris Mus. Natural History, 1876, dir., 1891-1900; mem. Acad. Medicine, Soc. Agr., French Acad. Scis. (v.p. 1900). Author: Recherches anatomiques et paléontologiques pour servir à l'histoire des oiseaux fossiles de la France, 2 vols., 1867-72; Éléments de l'histoire naturelle des animaux, 2 vols., 1881-82. Research on fossil birds, deep-sea fauna, fauna of Madagascar, crustaceans. Died Paris, Apr. 21, 1900.

MILNE-EDWARDS, Henri, zoologist, physiologist; b. Bruges, Belgium, Oct. 23, 1800; s. William Edwards; ed. Paris, M.D., 1823; became prof. hygiene École centrale, Paris, 1832; prof. natural history Henry IV Coll; apptd. prof. Sorbonne, 1843; named prof. entomology Paris Mus. Natural History, 1841, prof. mammalogy, 1861, dir., 1864. Fellow Royal Soc, 1848 (Copley medal 1850). French Acad. Scis., 1838, Acad. Medicine, Soc. Agr. Author: Recherches anatomiques sur les crustacées, 1828; Histoire naturelle des crustacées, 3 vols., 1834-41; Leçons sur la physiologie et l'anatomie comparée de l'homme et des animaux, 14 vols., 1857-81; Histoire naturelle des corallinaires, 3 vols., 1858-60; Recherches pour servir à l'histoire naturelle des mammifères, 2 vols., 1868-74: (with J. Haime) Recherches anatomique, physiologique et zoologique sur les Polypiers. Editor, zool. dept., Annales des sciences naturelles, from 1837. Research on crustaceans, polyps, fossil birds; 1st to describe biol. principle of physiol. division of labor; held doctrine of special creation, not evolution, theory of his contemporary, C. Darwin. Died Paris, July 29, 1885.

MILNER, Christopher John, physicist; b. Sheffield, Eng., Apr. 3, 1912; s. Samuel Roslington and Winifred (Walker) M.; B.A., Cambridge (Eng.) U., 1933, M.A., Ph.D., 1937; m. Eirene Joyce Thorburn, June 3, 1937; children—Jocelyn (Mrs. H. Chey), Jessica, Hugh, Francis. Physicist, Research Lab., Brit. Thomson Houston Co. Ltd., Rugby, Eng., 1936-45, head physics sect., 1945-52; prof. applied physics, head Sch. Physics, U. New S. Wales, Kensington, Australia, 1952—, dean Faculty Sci., 1956-59, acting dean, 1960, 66. Fellow Inst. Physics, Phys. Soc. (London), Australian Inst. Physics. Research, publs. on electronics, vacuum devices and other applications of physics including klystron tubes, solar furnace design, vehicle-safety devices. Home: 16 Keston Av., Mosman, N.S.W. 2088. Office: U. New S. Wales, P.O. Box 1, Kensington, N.S.W., 2033 Australia.*

MILNER, Samuel Roslington, English physicist; b. Dodsworth, Eng., Aug. 22, 1875; s. Samuel Wilkinson and Ann (Roslington) M.; B.Sc., Univ. Coll., Bristol, Eng., 1895; postgrad. U. Göttingen (Germany); m. Winifred Walker; 1 son. Named 1851 Exhbn. scholar U. Manchester (Eng.), 1895-98, demonstrator in physics, 1898; became lectr. in physics U. Sheffield (Eng.), 1900, acting prof., 1917, prof. physics, 1921-40; asst. radiographer 3d No. Gen. Hosp., 1914-17. Fellow Royal Soc., 1922. Contbr. papers to sci. jours. Introduced cathode-ray tube and rotating mirror for study of spark discharges; explained mechanism of formation of durable liquid films from soap solutions; laid found. of modern theory of electrolytes; research in spectroscopy, electromagnetics, classical relativistic physics. Died 1958.

MILNE-THOMSON, Louis Melville, mathematician; b. London, Eng., May 1, 1891; s. Alexander and Eva (Milne) M.-T.; student Clifton Coll., 1904-09; M.A., Corpus Christi Coll., Cambridge U., 1913, Sc.D., 1934; M.A. (hon.) Brown U.; m. Gertrude Frommknecht, Sept. 12, 1914; children—Joan (Mrs. Peter Jarvis), Aileen (Mrs. John Laws), Margaret (Mrs. Cedric Motley). Came to U. S., 1956, naturalized, 1966. Faculty, Winchester Coll. (Eng.), 1915-21; prof., head math dept. Royal Naval Coll., Greenwich, Eng., 1921-56, Gresham prof. geometry, 1946-56; vis. prof. Brown U., 1956-58; with Math Research Center, U. S. Army, Madison, Wis., 1958-61;

vis. prof. math. U. Ariz., Tucson, 1961—. Named comdr. Order Brit. Empire. Mem. Royal Astron. Soc., London Math. Soc., Royal Soc. Edinburgh, Cambridge Philos. Soc., Inst. Aero. Scis. Author numerous books including Standard Table of Square Roots, 1928; Die elliptischen Funktionon von Jacobi, 1931; Calculus of Finite Differences, 1933; Theoretical Hydrodynamics, 1938, 5th edit., 1967; Theoretical Aerodynamics, 1948, 4th edit., 1966; Jacobian elliptic function tables, 1957; Plane Elastic Systems, 1960; Antiplane Elastic Systems, 1962; Russian-English Mathematical Dictionary, 1962. Research, publs. on theory of aircraft design, ship resistance, theory of fluid motion, elasticity and certain brs. of pure math. Home: 3727 E. Fifth St., Tucson 85716. Office: Istituto Matematico, U. Rome, 00185 Rome, Italy.*

MILNOR, John Willard, Am. mathematician; b. Orange, N.J., Feb. 20, 1931; s. J. Willard and Emily (Cox) M.; A.B., Princeton, 1951, Ph.D., 1954; Sc.D., Syracuse U., 1965, U. Chgo., 1967; m. Brigitte Weber, Jan. 5, 1954; children—Stefan, Daniel, Gabrielle. Mem. faculty Princeton, 1953—, prof. math., 1960—, Henry Putnam Univ. prof., 1962—, chmn. dept., 1963-65; vis. prof. U. Cal. at Berkeley, 1959-60, 67-68. Alfred P. Sloan fellow, 1955-59; recipient Field award Internat. Congress Math., 1962; Nat. Medal of Sci., 1967. Mem. Nat. Acad. Scis., Am. Acad. Arts and Scis., Am. Philos. Soc., Am. Math. Soc. Editor: Annals of Math., 1962—. Study of the topology of manifolds. Address: Department of Mathematics, Princeton University, Princeton, N.J. 08540.

MILONE, Carmelo, Italian physicist; b. Catania, Italy, Nov. 17, 1920; s. Pietro and Agata (Faro) M.; Physics D., Catania U., 1941; m. Sara Tamburino, Dec. 1, 1951; children—Agata, Pietro, Vincenzo, Maria. Faculty, Catania U., 1954—, prof. gen. physics, 1965—; asso. Istituto Nazionale di fisica Nucleare, 1960—, Centro Siciliano di Fisica Nucleare, 1965—. Mem. Società Italiana di Fisicia. Research, publs. on cosmic radiation, nuclear photoreactions, low energy nuclear physics. Home: 4 Europa, Catania, Italy.*

MILOVANOV, Viktor Konstantinovich, Russian biologist, zootechnician; b. 1904; grad. Moscow Zootech. Inst., 1928. Asso., All-Union Inst. Stock-Raising, 1931—; head lab. artificial insemination Ivanov All-Union Research Inst. Acclimatization and Hybridization of Animals, 1935-37. Recipient Stalin prize, 1951. Mem. All-Union Lenin Acad. Agrl. Sci. Author: Artificial Insemination of Farm Animals, 5th edit., 1940; Biology of Reproduction and Artificial Insemination in Farm Animals, 1962. Research and publs. on reprodn. biology and artificial insemination of farm animals. Address: All-Union Inst. Stock-Raising, P.O. Strelkovo, Podolska Dist., Moscow, USSR.*

MILSTONE, Jacob Haskell, Am. pathologist; b. St. Louis, June 30, 1912; s. Maurice and Sadie (Mackler) M.; A.B., Johns Hopkins, 1933, M.D., 1937; m. Vivian Kaufman, Dec. 9, 1942; children—Leonard Matthew, Alan David. Asst. bacteriology N.Y. U., N.Y.C., 1937-40; Commonwealth fellow Rockefeller Inst. for Med. Research, Princeton, N.J., 1940-42; Life Ins. Med. Research fellow Yale, 1946-48, research asst., 1948-49, faculty, 1949—, prof. pathology, 1967—; asso. and attending pathologist Yale-New Haven Hosp., 1949—. Mem. A.A.A.S., Soc. for Exptl. Biology and Medicine, A.M.A., Am. Assn. Pathologists and Bacteriologists, Am. Soc. Exptl. Pathologists, N.Y. Acad. Sci. Research, publs. on enzymes concerned with clotting blood and dissolution blood clots. Home: 88 Hall St., New Haven 06512.*

MILTON, John, English poet, polit. theorist; b. London, Dec. 9, 1608; s. John and Sarah (Jeffrey) M.; ed. St. Paul's Sch.; B.A., Christ's Coll., Cambridge, 1629, M.A., 1632; m. Mary Powell, 1643, m. 2d, Catherine Woodstock, 1656, m. 3d, Elizabeth Minshull, 1663. Retired to his father's estate at Horton after deciding not to enter the clergy, but rather to become a poet, 1632-38; went to Italy, met many notables including Galileo, 1638; returned to Eng., helped support Presbyns. to reform Eng. Ch.; became sec. for fgn. tongues in govt. of Oliver Cromwell, circa 1649; after the Restoration, forced into hiding, 1660; after amnesty, lived quietly and devoted reminder of life to writing. Author: Areopagitica, 1644; Of Education, 1644; Eikonoklastes, 1649; Tenure of Kings and Magistrates, 1649; Defense of the English People, 1651; Second Defense of the English People, 1654; Paradise Lost, 1667, 74; Paradise Regained, 1671; Samson Agonistes, 1671. Argued in favor of freedom of press; declared subjects could put to death an unworthy king; defended Commonwealth govt. and Oliver Cromwell; one of the great Eng. poets. Died Nov. 8, 1674.

MINAKAMI, Takeshi, Japanese geophysicist; b. Toyama Prefecture, Japan, 1909; grad. Tokyo U., 1934; D.Sc., 1944. Asst. prof., staff mem. Earthquake Research Inst.; prof. Tokyo U., 1948, with Asama Volcanic Mountain Obs., 1944—. Author: Volcanic Mountains and Earthquakes. Research and publs. on terrestrial gravitation of Fuji, preliminary phenomena of eruption, seismometric studies.

MINAKAWA, Osamu, Japanese physicist; b. Tokyo, Japan, 1908; grad. Tokyo U., 1933; D.Sc. Technician, Meteorol. Research Inst., Central Meteorol. Obs.; prof. Kobe U. Established Nat. Cosmic Ray Research Inst. on Mt. Norikura. Recipient Asahi Sci. Promotion Fund, 1949, 50. Research on longevity of meson and cosmic rays.

MINAMI, Shigeo, Japanese physicist; b. Osaka, Japan, Sept. 28, 1921; s. Tokutaro and Umeno (Nakao) M.; grad. Osaka U., 1944, D.Sc., 1955; m. Shoko Okada, Jan. 24, 1957; children—Fuyumi, Izumi. Research asso. physics Osaka City U., 1954-57, faculty, 1957—, prof. physics, 1964—; vis. prof. physics La. State U., Baton Rouge, 1962-64. Mem. Japanese, Am., Italian phys. socs. Contbg. author: Nuclear Physics, vol. 3, 1959. Research, numerous publs. on discovery of degeneracy of phase shifts in pion-nucleon scattering called Minami ambiguity, strong interactions of elementary particles.*

MINARD, David, Am. physician, educator; b. Fargo, N.D., May 23, 1913; s. Archibald Ellsworth and Gladys (Pease) M.; B.S., U. Chgo., 1935, Ph.D., 1937, M.D., 1943; M.P.H. cum laude, Harvard, 1954; m. Sarah Prince Zimmermann, Aug. 21, 1948; children—David IV, Michael D., Nicholas, Mary Rebecca. Instr. physiology U. Louisville Sch. Medicine, 1937-38; instr., asso. in physiology U. Ill. Coll. Medicine, 1938-40; commd. lt. (j.g.) M.C., USN, 1943, advanced through grades to capt., 1959, ret., 1963; med. officer amphibious forces, PTO, 1944-46; research Naval Med. Research Inst., Bethesda, Md., 1946-52, 52-63; head thermal stress control br. Div. Occupational Medicine, Bur. Medicine and Surgery, Washington, 1954-63; prof. occupational health, chmn. dept. U. Pitts. Grad. Sch. Pub. Health, 1963—, program dir. exec. health evaluation program, 1964—; cons. occupational health Western Pa. Hosp. Dept. Occupational Health, 1964—. Chmn. subcom. on thermal factors in environment Nat. Acad. Scis.-NRC, 1963—; mem. Commn. on Environmental Hygiene, Armed Forces Epidemiological Bd., 1965—. Recipient Gorgas medal Assn. Mil. Surgeons, 1960; certificate of merit USN Bur. Medicine and Surgery, 1963. Diplomate occupational medicine Am. Bd. Preventive Medicine. Mem. Am. Physiol. Soc., Am., Indsl. med. assns., Am. Acad. Occupational Medicine, A.A.A.S., N.Y., Washington acads. scis., Royal Soc. Medicine (affiliate). Contbr. chpts. to books, article to profl. jours. Research on human calorimetry, thermal regulation in humans exposed to heat stress, physiol. and clin. effects cold stress, physiology of histamine, intracranial circulation; described field method for assessing heat stress. Home: 719 Amberson Av., Pitts. 15232.*

MINC, Henryk, mathematician; b. Lódz, Poland, Nov. 12, 1919; s. Izrael and Haja (Zyngler) M.; M.A. with honors in Math., U. Edinburgh (Scotland), 1955, Ph.D. in Math., 1959; m. Catherine Taylor Duncan, Apr. 16, 1943; children—Robert Henry, Ralph Edward, Raymond. Tchr., Morgan Acad., Dundee, Scotland, lectr. Dundee Tech. Coll., 1956-58; faculty U. B.C., Vancouver, Can., 1958-60, asst. prof., 1959-60; asso. prof. U. Fla., Gainesville, 1960-63; prof. math. U. Cal. at Santa Barbara, 1963—. Mem. Math. Assn. Am., Am., Edinburgh math. socs., Polskie Tow. Mat., Société Math. de Belgique, Polish Inst. Arts and Scis. Author: (with Marvin Marcus) A Survey of Matrix Theory and Matrix Inequalities, 1964, Introduction to Linear Algebra, 1965; Modern University Algebra, 1966; Elementary Linear Algebra, 1968; New College Algebra, 1968; also articles. Research in theory nonassociative groupoids, combinatorial analysis, linear and multilinear algebra. Home: 4076 Naranjo Dr., Santa Barbara, Cal. 93105.*

MINC, Stefan, Polish chemist; b. Warsaw, Poland, Aug. 5, 1914; s. Jakub and Bronislawa (Galezowska) M.; M.Chemistry, Warsaw U., 1936, D.Chemistry 1947; m. Alicja Gajownik-Kopaczynska, Apr. 27, 1946; children—Jolanta, Joanna. Prof. phys. chemistry U. Gdansk (Poland), 1945-52; prof., head dept. phys. chemistry Warsaw U., 1952—; head dept., radiation chemistry Inst. Nuclear Research, Warsaw, 1957—, vice dir., 1956-62; dean Faculty Chemistry, 1947-52; Under sec. state Ministry of Higher Edn., 1962-66. Recipient State awards in electrochemistry and radiation chemistry. Mem. Polish Chem. Soc., Polish Acad. Scis. (past vice sec. dept.), Am. Electrochem. Soc., Soc. de Chimie Industrielle, Com. Internat. de Thermodyn. et de Cinetique Electrochim., Assn. F. et I. Joliot-Curie. Author: (with L. Stolarczyk) The Principles of Physical Chemistry of Colloids, 1956; also numerous articles. Research on influence of structure of electrolyte solutions on structure of double layers, structure of solutions and mechanism of their gamma-radiolysis.*

MINCK, Raymond Joseph, French physician; b. Strasbourg, France, Apr. 6, 1924; s. Lucien and Emma (Jundt) M.; certificate Gen. Biology and Biol. Chemistry, U. Strasbourg (France), M.D., 1948, Professeur Agrégé, 1955; m. Andrée Moyet, Mar. 20, 1945; children—Danielle, Jean-Louise, Francois, Catherine, Dominique, Marie-Christine. Asst., Faculty Medicine, Montpellier, 1945; asst. Faculty Medicine Strasbourg, 1946-52, chef de travaux, 1952-55; prof. agrégé medicine Faculty Strasbourg, 1955-65; prof. biology Biologiste des Hôpitaux, 1965—, chief service, 1965—. Mem. French Soc. Microbiology. Research, numerous publs. on bacterial cytology, L form bacteria, immunology, bacterial physiology. Home: 1 Chemin Doernelbruck, 67 Strasbourg, France.*

MINCZEWSKI, Jerzy, Polish chemist; b. Zamosc, Poland, Dec. 11, 1916; s. Henryk and Wanda (Klossowska) M.; M.Sc., Poly. Inst., Warsaw, 1944; D.Sc., Inst. Gen. Chemistry, Warsaw, 1956; m. Zofia Matuszewicz, Sept. 29, 1940; children—Malgorzata, Jan. Staff, Inst. Sugary Industry, Warsaw, 1946-47; research analyst Inst. Gen. Chemistry, 1947-55; chief dept. analytical chemistry Inst. Nuclear Research, Warsaw, 1956—; prof. analytical chemistry Poly. Inst. Warsaw, 1959—. Vice chmn. analytical chemistry commn. Polish Acad. Scis., 1955—. Mem. Polish Chem. Soc. Author: (with Z. Lada) Potentiometric Titration, 1957; (with Z. Marczenko) Analytical Chemistry, 1966; also numerous articles. Research on potentiometric titration in aqueous and nonaqueous media, analytical chemistry of trace-contaminations in high purity materials, especially chem. methods of separation and enrichment of traces, organic reagents in inorganic analysis. Home: 14 Wilcza, Warsaw, Poland.*

MINDER, Walter, Swiss physicist; b. Scheuren, Aug. 6, 1905; s. Jakob and Barbara (Geissbuhler) M.; student Fed. Tech. U. Zurich, 1925-27; Ph.D., U. Berne, 1931; m. Hedwig Muller, May 3, 1941; children—Markus, Christopher. Chief, Radium-Inst. Berne, Switzerland, 1931-64; lectr. U. Berne, extraordinary prof. med. radiation physics; chief radiation protection sect. Fed. Office of Health, Berne, 1964—. Mem. Swiss Phys. Soc., Swiss Soc. Radiology. Author: Radium Dosimetry, 1941; (with A. Liecht) X-Ray Physics, 1951; Dosimetry of Radiations of Radioactive Substances, 1961. Research, publs. on measurement and determination of dosages from radioactive substances and radiation-machines in radiation therapy; first to predict element 85 in desintegration products of radon; radiation chemistry in relation to radiation dosimetry. Home: 46 Sägeinaltstrasse, Liebefeld-BE, Switzerland. Office: 11 Falkenplaz, Berne, Switzerland.*

MINDERER, Raymond, German physician; b. Augsburg, Germany, circa 1570; M.D., Ingolstadt, Germany, 1597; army physician; practiced medicine, Augsburg; physician to emperors Mathias and Maximilian. Author: De pestilentia . . . , 1608; Aloedarium marocostinum . . . , 1616. Described green and blue vitriol, spirit and oil of vitriol; wrote on mineral waters, silver, vinegar; mentioned that blue paper can be reddened by sulphuric acid; credited with descriptions of lead acetate, ferric oxide; may have used solution of ammonium acetate made by dissolving ammonium carbonate in distilled vinegar. Died May 13, 1621.

MINDLIN, Raymond D(avid), Am. civil engr.; b. N.Y.C., Sept. 17, 1906; s. Henry and Beatrice (Levy) M.; student Ethical Culture Sch., N.Y.C., 1918-24; B.A., Columbia, 1928, B.S., 1931, C.E., 1932, Ph.D., (Bridgham fellow), 1936; m. Elizabeth Roth, Aug. 5, 1940 (dec. Nov. 1950), m. 2d, Patricia Kaveney, Jan. 13, 1953. Research asst. dept. civil engring., Columbia, 1932-38, instr. civil engring., 1938-40, asst. prof., 1940-45, asso. prof., 1945-47, prof. civil engring., 1947-67, James Kip Finch prof. applied sci., 1967—; cons. Bell Telephone Labs. Inc., N.Y.C., 1943-51; cons. physicist Dept. Terrestrial Magnetism, Carnegie Instn. Washington, 1940-42; cons. NDRC, 1941-42, Sect. T., OSRD, Applied Physics Lab. Johns Hopkins, 1942-45. Recipient Illig Medal, Columbia, 1932; U. S. Naval Ordnance Devel. award, 1945; Medal for Merit, 1946; Class of 1889 Sch. of Mines Medal, Columbia, 1947; Research prize, Am. Soc. C.E., 1958, von Karman medal, 1961; Timoshenko medal Am. Soc. M.E., 1964; Great Tchr. award Columbia, 1960; C. B. Sawyer award, 1967. Fellow Nat. Acad. Engring., Acoustical Soc. Am., Am. Soc. M.E., Am. Acad. Arts and Scis.; mem. soc. Exptl. Stress Analysis (exec. com. 1943-50; pres. 1947), Am. Soc. C.E. (sec. com. on applied mechanics 1940-42, chmn. 1942-45), Eastern Photoelasticity Conf. (exec. com. 1938-41), Am. Phys. Soc., Am. Math. Soc., A.A.A.S., U. S. Nat. Com. Theoretical and Applied Mechanics, mem. gen. assembly Internat. Union Theoretical and Applied Mechanics, Am. Soc. Advancement of Hebrew Inst. of Tech. of Haifa, Palestine (bd. dirs. 1946-50), Tau Beta Pi, Sigma Xi. Mem. adv. bd. Applied Mech. Reviews. Contbr. articles to tech. and sci. jours. Contbr. to advances in math. theory of the deformation and motion of solids. Home: R.F.D. 2, Box 4, Katonah, N.Y. 10536.*

MINER, Horace Mitchell, Am. social scientist, educator; b. St. Paul, May 26, 1912; s. James Burt and Jessie Leightner (Schulten) M.; A.B., U. Ky., 1933; A.M., U. Chgo., 1935, Ph.D., 1937; m. Agnes Genevieve Murphy, June 12, 1936; 1 dau., Denise Allison. Asso. curator Mus. Anthropology, U. Ky., 1932-33; dir. archeol. field party TVA, 1934; Social Sci. Research Council fellow, Que., Can., 1936-37, Timbuctoo, French West Africa, 1940; instr. anthropology and sociology Wayne State U., 1937-39; dir. archeol. field party U. Chgo., summer 1938; social sci. analyst, field research in Ia., U. S. Dept. Agr., 1939; Yale Inst. Human Relations fellow, Colombia, 1941-42; asst. prof. sociology U. Mich., 1946-47, asso. prof. sociology and anthropology, 1947-51, research asso. Mus. Anthropology, 1948—, prof., 1951—. Cons. Nat. Resources Planning Bd., 1941; mem. com. experts on indigenous labor ILO, 1949-58, 61—, conf. on indigenous labor, La Paz, Bolivia, 1951, Geneva, Switzerland, 1954, 62; mem. divisional com. for social scis. NSF, 1962; Fulbright lectr. Makerere Coll., Uganda, 1961-62. Decorated Legion of Merit, Bronze Star; recipient Social Sci. Research Council Demobilization award, 1945; Fulbright research award, Horace Rackham grant for field research, Algeria, 1950; Ford Found. grant-in-aid, 1956, Rockefeller grantee, 1957-58. Fellow Am. Sociol. Assn., Am. Anthrop. Assn., African Studies Assn.; mem. Internat. African Inst. (gov. body 1949-64), Soc. Applied Anthropology (pres. 1954-55), Phi Beta Kappa, Sigma Xi, Omicron Delta Kappa, Alpha Kappa Delta, Delta Tau Delta. Author: St. Denis, a French-Canadian Parish, 1939, 63; Culture and Agriculture, 1949; (with others) Principles of Sociology, 1952, 56; The Primitive City of Timbuctoo, 1953, 65; (with G. DeVos) Oasis and Casbah: Algerian Culture and Personality in Change, 1960. Home: 26 Harvard Pl., Ann Arbor, Mich. 48104.*

MINES, George Ralph, physiologist; b. May 13, 1886; s. H. R. Mines; ed. Sidney Sussex Coll., Cambridge, Eng.; M.A.; m. Marjory Rolfe, 1909; 1 son, 1 dau. Demonstrator physiology U. Cambridge, 1911-14, examiner in physiology for med. degrees, 1913-14, dir. physiology Balfour Lab., 1910-13; lectr. London U., 1912; spl. lectr. U. Toronto (Ont., Can.), 1914; prof. physiology McGill U., Montreal, Que., Can. Contbr. papers to jours. Died Nov. 7, 1914.

MINKH, Aleksey Alekseevich, Russian hygienist; b. 1904; grad. Med. Faculty, Saratov U., 1927; D.Med. Sci., 1937. Asst. chair hygiene Saratov Med. Inst., 1928-29, acting head, 1930-32; asst. dept. gen. hygiene 1st Leningrad Med. Inst., 1932-36, lectr., 1937-38; head sch. hygiene lab. Leningrad. Inst. Child and Juvenile Care, 1934-40; prof., 1938—; head chair hygiene Leningrad Stomatological Inst., 1938-41, Central Inst. Phys. Tng., 1946—; head chair food hygiene 2d Leningrad Med. Inst., 1941-42; head chair gen. hygiene Moscow Stomatological Med. Inst., 1946—. Mem. USSR Acad. Med. Sci. (corr.). Author: Guide to Practical Studies in Hygiene, 1950; Methods of Hygiene Research, 1954, 2d edit., 1961; Ionization of the Air and Its Hygienic Significance, 1958; co-author: Sports Medicine, 1957. Co-editor Hygiene sect. Large Med. Ency., 2d edit. Research and numerous publs. on gen. and exptl. hygiene, phys. tng., energetics of muscular activity; one of 1st in USSR to discover hygienic and physiol. significance of ionization of air. Address: Moscow Stomatological Med. Inst., Kalyaevskaya ulitsa 18, Moscow, USSR.

MINKIN, Semen Yudovich, Russian surgeon; b. 1898; grad. Kharkov Med. Inst., 1925; D.Med. Sci., 1939. Asst., head neurosurgery dept. Faculty Surgery Clinic, Kharkov Med. Inst., 1928-32, head exptl. surgery lab., 1932-39; asst. chair surgery and neurosurgery Ukrainian Psychoneurol. Acad., Kharkov, 1932-39; instr., dep. head Fedorov Faculty Surgery Clinic, Leningrad Mil. Med. Acad., 1940, acting head, 1941-45; prof., 1941—; head chair hosp. surgery Perm Med. Inst., 1948—. Research and numerous publs. on pathology and treatment of brain injuries, trophic ulcers, amputation pains, causalgia. Address: Perm Med. Inst., Kommunisticheskaya ulitsa 26, Perm, RSFSR, USSR.

MINKOWSKI, Hermann, mathematician; b. Alexota, Russia, June 22, 1864; student Berlin; Ph.D., Königsberg (now Kaliningrad, USSR), 1885. Asso. prof. Königsberg, 1894-96; prof. Zurich, Switzerland, Fed. Inst. Technology, 1896-1902; U. of Göttingen, Germany, 1902-09. Author: Raum und Zeit, 1907; Zwei Abhandlungen über die Grundgleichungen der Electrodynamik, 1909. Provided math. basis for general theory of relativity, particularly in his 4-dimensional math. form (known as Minkowski's World); devised geometrical method for number-theory study; introduced use of lattice as setting for algebraic theory. Died Göttingen, Jan. 12, 1909.

MINKOWSKI, Oscar, physician, pathologist; b. Alexota, Russia, Jan. 13, 1858; M.D., Königsberg (now Kaliningrad, USSR), 1881; became asso. prof., Strasbourg (now in France), 1891; named prof. medicine, Greifswald, Germany, 1905; prof., Breslau (now Wroclaw, Poland), 1909-26; Research on acidotic urine of diabetics, 1884, metabolism after excision of liver, 1885; demonstrated pituitary origin of acromegaly, 1887; artificially produced (with von Mehring) diabetes in dogs by removing pancreas, 1889; described (with Chauffard) familial hemolytic jaundice (Chauffard-Minkowski disease), 1900. Died Wiesbaden, Germany, June 18, 1931.

MINKOWSKI, Rudolph Leo B., astronomer; b. Strasbourg, France, May 28, 1895; s. Oscar and Marie (Siegel) M.; Ph.D. in Physics, U. Breslau, Germany, 1921; m. Luise Amalie David, Aug. 23, 1926; children—Eva Marie (Mrs. Thomas), and Herman Oscar. Came to U.S., 1935, naturalized, 1940. Asst. Physikalisches Staatsinstitut, Hamburg, Germany, 1922, privat-dozent, Hamburg, 1926-31, A. O. prof., Hamburg, 1931-35; staff Mt. Wilson Obs., Carnegie Inst., 1935-48, Mt. Wilson and Palomar Obs., Pasadena, Cal., 1948-60. U. Cal., at Berkeley, 1961—; vis. prof. U. Wis., 1960-61. Mem. Am., Royal astron. socs., Nat. Acad. Scis., Astron. Soc. Pacific. Research on supernovae, planetary nebulae; radio sources and their identification. Home: 1100 Siler Pl., Berkeley 94705. Office: U. Cal., Berkely, Cal. 94720.

MINNAERT, Marcel Gilles, astrophysicist; b. Brugge, Belgium, Feb. 12, 1893; s. Jozef and Jozefine (van Overberge) M.; Dr.biology, U. Gent (Belgium), 1914; Dr. physics and math., U. Utrecht (Netherlands), 1925; Hon. Dr., U. Heidelberg (Germany); m. Maria Bourgonje Coelingh, Dec. 20, 1929; children—Koenraad, Boudewyn. Lectr., U. Gent, 1916-18; with U. Utrecht, 1920—, prof. astronomy, dir. Obs., 1937-63. Decorated commandeur Ordre de la Découverte et de l'Invention; recipient Gold medal Astron. Soc. London; Bruce medal Astron. Soc. Pacific, Janssen medal U. Heidelberg. Fgn. mem. acads. of Boston, Brussels, N.Y., Rome, Uppsala, Flemish Acad.; mem. Acad. Sci. Amsterdam. Author: De Natuurkunde van Vrye Veld, 3 vols., 1938-43; also numerous articles, chpts. in books. Research on photometry of Fraunhofer lines; interpretation of Fraunhofer lines and quantitative analysis of solar atmosphere; photometry of other solar phenomena; photometry of moon. Home: 25 bis Zuifenstraat, Utrecht, Netherlands.*

MINOT, Ann Stone, Am. chemist, physiologist; b. Bath, N.H., Apr. 25, 1894; A.B., Smith Coll., 1915; Ph.D., Radcliffe Coll., 1923. Lab. asst. Mass. Gen. Hosp., 1915-20; researcher Harvard Med. Sch., 1923-26; instr. physiology Wellesley Coll., 1925-26; research asso. pharmacology Vanderbilt U. Sch. Medicine, Nashville, 1926-30, asst. prof. pediatric research, 1930-40, asso. prof., 1940-43, asso. prof. biochemistry in charge clin. chemistry, 1943-50, prof., 1950-60, emeritus prof. clin. chemistry, 1960—, research asso. endocrinology, 1960—. Mem. Am. Physiol. Soc., Soc. Pediatric Research (hon.). Research on hormone influence on bone growth, fluid balance of diarrhea of infants, muscular diseases, especially muscular dystrophy and myasthenia gravis; 1st to use guanidine in treatment of myasthenia gravis, 1938. Address: Dept. of Biochemistry, Vanderbilt University, Nashville 37203.*

MINOT, Charles Sedgwick, Am. anatomist; b. W. Roxbury, Boston, Mass., Dec. 23, 1852; s. William and Katherine (Sedgwick) M.; B.S., Mass. Inst. Tech., 1872; postgrad. univs. of Leipzig, Paris and Würzburg, 1873-76; studied physiology, conducted research on muscles and growth, under Ludwig at Leipzig; S.D., Harvard, 1878; LL.D., Yale, 1899, U. Toronto, 1904, St. Andrew's U., Scotland, 1911; Sc.D., Oxford U., 1902; m. Lucy Fosdick, June 1, 1889. Lectr. embryology, instr. oral pathology and surgery Harvard, 1880-83, instr. histology and embryology, 1883-87, asst. prof., 1887-92, prof., 1892-1905, named James Stillman prof. comparative anatomy, 1905, dir. anat. labs., 1912; Harvard exchange prof. univs. Berlin and Jena, 1912-13. Fellow A.A.A.S. (gen. sec. 1885, pres. 1900), Am. Acad. Arts and Scis.; pres. Boston Soc. Natural Scis., from 1897; mem. Nat. Acad. Scis., Brit. Assn. for Advancement of Sci. (corr.), Acad. Turin (corr.), Biol. Soc. Paris (corr.), other fgn. socs. Author: Human Embryology, 1892; Bibliography of Vertebrate Embryology, 1893; A Laboratory Text-Book of Embryology, 1903, 2d edit., 1910; Age, Growth and Death, 1908; Die Methode der Wissenschaft, 1913; Problème Moderne de Biologie, 1913 (also translated into English). Described structure of placenta; gave law of cytomorphosis; inventor 2 forms of automatic microtomes. Died Nov. 19, 1914.

MINOT, Francis, Am. physician; b. 1821; Hersey prof. theory and practice of physic Harvard Med. Sch., until 1892; described and analyzed 46 cases of hermorrhagic disease of newborn, 1852. Died 1899.

MINOT, George Richards, Am. physician; b. Boston, Dec. 2, 1885; s. James Jackson and Elizabeth (Whitney) M.; A.B., Harvard, 1908, M.D., 1912, S.D. (hon.), 1928; m. Marian Linzee Weld, June 29, 1915; children—Marian Linzee, Elizabeth Whitney, Charles Sedgwick. House officer Mass. Gen. Hosp., Boston, 1912-13, mem. staff, 1915-23, then bd. consultation; asst. resident physician Johns Hopkins Hosp., 1913-14; asst. in medicine and research fellow Physiol. Lab., Johns Hopkins Med. Sch., 1914-15; asso. in medicine Peter Bent Brigham Hosp., 1923-28, then cons.; chief of med. service Collis P. Huntington Meml. Hosp., 1923-28; prof. medicine Harvard, 1928-48; vis. physician Boston City Hosp., 1928; dir. Thorndike Meml. Lab. (Harvard) of Boston City Hosp. to 1948, cons., since 1948. Recipient Charles Mickle fellowship U. Toronto, 1928; Cameron prize U. Toronto, 1930; Gold medals Nat. Inst. Social Scis., 1930, Popular Sci. Monthly, 1930, Humane Soc. Mass., 1935; John Scott medal City of Phila., 1933; (with W. P. Murphy and George H. Whipple) Nobel prize in medicine and physiology, 1934, for work on liver treatment of anemias; scroll award Asso. Grocery Mfrs. of Am., 1936. Hon. fellow Royal Coll. Physicians, Edinburgh, Royal Coll. Physicians (London, Moxon medal 1933), N.Y. Acad. Medicine, Inst. of Medicine Chgo., Royal Soc. of Medicine, London; v.p. étranger Société Française d'Hematologie, 1938-39; fellow Am. Philos. Soc., Phila., A.C.P.; mem. A.M.A., Assn. Am. Physicians (Kober Gold medal, 1928, pres. 1937-38), Am. Soc. Clin. Investigation, Am. Acad. Arts and Sciences, Am. Clin. and Climatol. Assn. (pres. 1932-33, medal 1939), Nat. Acad. Sciences, Med. Library Assn. Am. (v.p. 1938-39), Phi Beta Kappa; hon. mem. Royal Acad. Med. (Belgium), Kaiserlich Leopold Caroline Deutsche Akademie der Naturforscher (Halle), Soc. Biol. Chemists (India), Finland Soc. Internal Medicine (Helsingfors). Author: Pathological Physiology and Clinical Description of the Anemias (with William B. Castle), 1936. Contbr. papers, chiefly on blood. Discovered curative effect of liver on pernicious anemia, 1926; helped develop thrombin; discovered (with L. W. Smith) changes in blood of persons exposed to certain chemicals; discovered (with S. Cobb and M. Strauss) relationship between vitamin B deficiency and alcoholic polyneuritis. Died Brookline, Mass., Feb. 25, 1950.

MINSKY, Hyman Philip, Am. economist; b. Chgo., Sept. 23, 1919; s. Sam and Dora (Zakon) M.; B.S., U. Chgo., 1941; M.P.A., Harvard, 1948, Ph.D., 1954; m. Esther Depardo, Nov. 9, 1955; children—Diana Hillary, Alan Depardo. Asst. prof. Brown U., 1949-55; asst. prof. U. Cal., Berkeley, 1956-57, asso. prof., 1957-65; prof. Washington U., St. Louis, 1965—. Mem. Am. Econ. Assn., Econometric Soc. Editor: California Banking in a Growing Society, 1966. Research, publs. on conditions of financial stability and instability, impact of monetary-financial factors in system behavior, evolution of banking system, central banking in a complex environment, growth and cycle models. Home: 541 Warren Av., St. Louis 63130.*

M'INTOSH, William Carmichael, zoologist; b. St. Andrews, Scotland, Oct. 10, 1838, ed. M.D., U. Edinburgh (Scotland); LL.D., univs. St. Andrews and Edinburgh; D.Sc., univs. Oxford and Durham. Past dir. Univ. Museum; dir. Gatty Marine Lab.; past physician, also cons. physician Perth Dist. Mental Hosp., subcommnr., sci. reporter Royal Com. on Trawling, 1884; prof. natural history U. St. Andrews, 1882-1918. Recipient Gold medal Edinburgh Fisheries Exhbn., 1882, London Fisheries Exhbn., 1883; Linnean Gold medal, 1924. Fellow Royal Soc. (Royal medal 1899), Royal Soc. Edinburgh (Neill medal 1869), Linnaean Soc.; licentiate Royal Coll. Surgeons Edinburgh; licentiate Midwifery; hon. mem. Royal Zool. Soc. Ireland, Natural History Soc. Glasgow, Scottish Natural History Soc., Psychol. Soc. Paris, France, Soc. Centrale d'Acquicult. de France, Soc. Royal Zoology de Belique; mem. Ray Soc. (pres.). Author: Observations and Experiments on Shore Crab, 1861; (monograph) British Annelids, 4 vols., 1874-1923; The Marine Invertebrates and Fishes of St. Andrews, 1875; Report to Royal Commission on Trawling, 1884; The Resources of the Sea, 1899; The Annelids of the Challenger, 1885; (with Dr. Masterman) Food Fishes of Britain, 1897; also numerous memoirs and papers on zoology (mainly illus. by the author) with hundreds of plates and woodcuts. Died Apr. 1, 1931.

MINTS, Aleksandr Lvovich, Russian radio engr.; b. Rostov-on-Don, Dec. 27, 1895; grad. Don U., Rostov-on-Don, 1918, Moscow Inst. Communications Engring., 1932. Chief engr. charge design and constrn. high and super high power broadcasting stas. USSR, 1924-43; prof. Leningrad. Communication Engring. Inst., 1930-38; dir. Radiotech. Inst.. USSR Acad. Scis., 1946—; charge research, design, devel. particle accelerators at Dubna, Moscow and Serpukhov, 1946-68. Decorated Order of Lenin (4); recipient Stalin prize, 1946, 51, Popov Gold medal, 1950, Lenin prize, 1959. Mem. USSR Acad. Scis. (pres. sci. council on particle accelerator problems 1967—). Co-author: Design Principles of Plate and Grid Modulation Systems, 1926,29; author High Power Broadcasting Stations, 1936, 45; Scientific and Design Problems of High Energy Particle Accelerators, 1956; Design Principles of Super High Energy Particle Accelerators, 1961,65, 68. Address: Radiotechnical Inst., USSR Acad. Scis., ulitsa 8, Marta 12, Moscow, USSR.*

MINTY, George James, Jr., Am. mathematician; b. Detroit, Sept. 16, 1929; s. George J. and Williamina (Goodall) M.; B.S., Wayne State U., 1949, M.A., 1951; Ph.D., U. Mich., 1959; m. Emiko Shibata, Dec. 27, 1959; 1 dau., Michiko. Instr., Duke, 1958-59, U. Wash., 1959-60; asst. prof. U. Mich., 1960-64; mem. Courant Inst., N.Y. U., 1964-65; prof. Ind. U., Bloomington, 1965—. Sloan Research fellow, 1966. Mem. Am. Math. Soc., Math. Assn. Am., Soc. for Indsl. and Applied Math. Research, publs. on graph-theory especially network-programming, nonlinear functional analysis. Office: Math. Dept., Ind. U., Bloomington, Ind. 47401.*

MINTZ, Sidney Wilfred, Am. anthropologist, educator; b. Dover, N.J., Nov. 16, 1922; s. Solomon and Fannie (Tulchin) M.; B.A. in Psychology, Bklyn. Coll., 1943; Ph.D. in Anthropology, Columbia, 1951; M.A. (hon.), Yale, 1963; m. Jacqueline Wei, June 6, 1964; children—Eric, Elizabeth (by previous marriage). Faculty, Yale, New Haven, 1951—, prof. anthropology, 1963—; vis. prof. anthropology Mass. Inst. Tech., Cambridge, 1964-65. Mem. Am. Anthrop. Assn., Am. Ethnol. Soc., Royal Anthrop. Inst. Gt. Britain and Ireland, Sigma Xi. Author: Worker in the Cane, 1960; also articles, revs. on Caribbean ethnology and culture history, econ. anthropology. Home: 100 York Sq., New Haven 06511.*

MIQUEL, (Antonin) Pierre, French bacteriologist; b. Montmiral, France, 1850; student pharmacy, Toulouse, France, Paris; M.D., 1883; studied under Schutzenberger; joined staff Montsouris Obs., Paris, 1876. Author: (with L. Cambier) Traité de bactériologie, pure et appliquée à la médecine et à l'hygiène,

1900. Author: Les Organismes vivants de l'atmosphère, 1883; also articles. Studied bacteria in air, water, soil. Died 1922.

MIRBEL, Charles François Brisseau de, see de Mirbel, Charles François Brisseau.

MIRCHINK, Mikhail Fedorovich, Russian oil geologist; b. June 15, 1901; grad. Moscow Mining Acad, 1930. Lectr. oil field geology Azerbaijan Indsl. Indsl. Inst., 1932; prof. Moscow Mining Acad., 1943; dir. Inst. Geology and Processing of Oil Deposits, 1961-—; bur. mem. geology dept. USSR Acad. Scis., 1960-—. Decorated Order of Lenin (2); recipient Stalin prize, 1949, 50. Mem. USSR Acad. Scis. (corr.). Author: Stratigraphic Deposits of Oil, 1943; Oil Field Geology, 1946; Scientific Basis for Development of Oil Deposits, 1948. Research on regional geology of Caucasus and Russian Platform oil-bearing areas, specialist in exploitation and discovery of oil fields. Address: Moscow Petroleum Institute, Leninsky prospect 6, Moscow, USSR.

MIROYIANNIS, Stanley Demetrius, anatomist; b. Metelin, Greece, Nov. 24, 1908· s. Demetrius George and Androniki John (Mandanis) M.; naturalized U. S. citizen; ed. Gymnasium of Metelin, Greece, 1917-21; S.B., N.W. Nazarene Coll., Nampa, Ida., 1927; postgrad. Mass. State Coll., 1927-28; A.M., Boston U., 1929, Ph.D., 1939; m. Grad. asst. biology, Boston U., 1929-31, teaching fellow, 1931-36; lecturer and also head dept. gross anatomy New Eng. Inst. Anatomy, Boston, 1935-49; cons. biology Northeastern U., Boston, 1935-36, asst. prof., 1936-38, asso. prof., 1938-40, prof. and head dept., 1940-48; prof. advanced biology, Grad. Sch. Mass. Coll. Pharmacy, Boston, 1948-50; writer med. div. Merck & Co., Inc., Rahway, N.J., 1950-52; prof., chmn. dept. anatomy Des Moines (Ia.) Still Coll., 1952-—, anatomy prof. chmn. div. basic scis., 1954-—. Fellow A.A.A.S., Ia. Acad. Sci., Am. Med. Writers Assn., Acad. Zool.; mem. N.Y. Acad. Scis., Teratology Soc., Assn. Tropical Biology, Mus. Natural History, Nat. Soc. for Med. Research, Royal Soc. Health (London), Am. Inst. Biol. Scis., Am. Soc. Exptl. Biology, Nat. Assn. Biology, Research Officers Assn., Am. Soc. Mammalogists, Am. Assn. Profl. Biologists, Assn. Mil. Surgeons U.S.A., Am. Assn. U. Profs., Am. Soc. Zoölogists, Soc. Systematic Zoology, Gen. Research System Soc., Mil. Govt. Assn., Res. Officers Assn., Med. Service Corps. Assn., Am. Assn. Developmental Biology, Am. Micros. Soc., Am. Soc. Morphologists, So. Soc. Anatomists, Japanese, Mexican assns. anatomists. Sigma Xi, Iota Tau Sigma, Kappa Phi Kappa (chmn. pre-med. and predental com. Northeastern U.). Author: A Laboratory Course in General Zoology; The Cell and Fundamental Tissues; Microscopic Anatomy; 501 Questions and Their Answers in Gross Anatomy, 1959; also sci. papers.*

MIRSKY, Alfred Ezra, biochemist; b. Flushing, L.I., Oct. 17, 1900; s. Michael David and Frieda (Ittelson) M.; grad. Ethical Culture Sch.; A.B., Harvard, 1922; Ph.D., U. Cambridge, 1926; M.D. (hon.), Gothenburg, 1954; m. Reba Paeff, May 25, 1926; children —Reba, II (Mrs. Robert Goodman), Jonathan. Prof. and librarian Rockefeller Univ.; Member Am. Philosophical Society, Nat. Acad. Sci., Am. Chem. Soc., Soc. Biol. Chemists, Soc. Gen. Physiologists, Genetics Soc., Am. Naturalists Soc., Soc. Exptl. Biology and Medicine, Harvey Soc., Am. Acad. Arts and Scis., Phi Beta Kappa. Editor: Cell Physiology. Research on proteins; nucleoproteins; hemoglobin; nucleus and cell chemistry; chromosomes. Home: 350 Central Park W., N.Y.C. 10025. Office: Rockefeller U., 66th St. and York Av., N.Y.C. 10021.

MISCH, Peter, geologist; b. Berlin, Germany, Aug. 30, 1909; s. Georg and Clara (Dilthey) M.; Ph.D., U. Göttingen, 1932; m. Nicoletta C. Rosenthal, Apr. 15, 1947; children—Hanna C. (Mrs. Stephen French), Felix G., Anthony A. Came to U. S., 1946, naturalized, 1952. Geologist, Himalayan expdn. to Nanga Parbat, 1934-35; prof. Nat. Sun Yatsen U., Canton, China, 1936-38, Yunnan, China, 1939-40; prof. Nat. Peking U., Kumming, Yunnan, 1940-46; faculty U. Wash., Seattle, 1947-—, prof. geology, 1950-—. Guggenheim fellow, 1954-55. Fellow Geol. Soc. Am. (vice chmn. Cordill sect. 1960, chmn. 1966); mem. Am. Assn. Petroleum Geologists (asso. editor 1965-—, distinguished lectr. 1953), Am. Geol. Inst. (lectr. 1963), Mineralogy Soc. Am., Geol. Soc. London, Geochem. Soc., Am. Geophys. Union, Geologische Verein, N.W. Sci. Assn., Sigma Xi. Research, numerous publs. on structure young mountain belts, rock metamorphism, origin granitic rocks; geol. expln. Pyrenees, Himalayas, mountains S.W. China, No. Cascade Mountains, Eastern Great Basin. Home: 5726 N.E. 60th St., Seattle 98115.*

MISES, Ludwig, economist; B. Lemberg, Austria, Sept. 29, 1881; s. Arthur Edler von and Adele (Landau) M.; D.Law, Akademisches Gymnasium and U., Vienna, Austria, 1905; hon. degrees N.Y. U., 1963, U. Freiburg (Germany), 1964, Grove City Coll., 1957; m. Margit Herzfeld, July 3, 1938. Came to U.S., 1940, naturalized, 1946. Econ. adviser Austrian C. of C., 1909-34; founder, v.p. Oesterreichisches Inst. für Konjunturforschung, 1926-38; faculty U. Vienna, 1913-38; prof. internat. econ. relations Grad. Inst. Internat. Studies, Geneva, Switzerland, 1934-40; vis. prof. economy N.Y. U., N.Y.C., 1946-—. Guest lectr. univs., orgns. in Eng., France, Germany, Italy, Netherlands, Mexico, Peru, Argentina, Guatemala, Costa Rica. Decorated Austrian Ehrenzeichen fuer Wissenschaft and Kunst, 1962. Author: Human Action, 1949; Planning for Freedom, 1952; The Theory of Money and Credit, 1953; Socialism, An Economic and Sociological Analysis, 1951; Omnipotent Government, 1944; The Anticapitalistic Mentality, 1956; Theory and History, 1957; The Ultimate Foundation of Economic Science, 1962. Developed Austrian theory of trade cycle; elaborated critical analysis of problem econ. calculation in frame of non-market economy; investigated operation govt. interference with market phenomena. Home: 777 West End Av., N.Y.C. 10025.*

MISHUSTIN, Yevgeniy Nikolaevich, Russian microbiologist; b. Feb. 22, 1901; grad. Moscow Timiryazev Agrl. Acad., 1924. Asso., Inst. Microbiology, USSR Acad. Scis., 1939-—. Decorated Order of Lenin; recipient Stalin prize, 1951. Mem. USSR Acad. Scis. (corr.). Author: The Scientific Principles of Siloing Fodder, 1933; Course on Agricultural Microbiology, 1934; The Ecological and Geographical Variability of Soil Bacteria, 1947; Thermophylic Microorganisms in Nature and Practice, 1950; Microorganisms and the Self-Purification of the Soil, 1954; Microorganisms and Soil Fertility, 1956; Biological Means of Increasing the Effective Fertility of Soils, 1960; Modern Problems in the Theory and Practice of Siloing Fodder, 1960; Soil Cultivation and Its Effective Fertility, 1960. Research and publs. on agrl. microbiology. Address: Inst. Microbiology, USSR Acad. Scis., Leninsky prospect 33, Moscow, USSR.

MISLOW, Kurt Martin, chemist; b. Berlin, Germany, June 5, 1923; s. Max and Ida (Bingen) M.: B.S., Tulane U., 1944; Ph.D., Cal. Inst. Tech., 1947; m. Wesley Metcalf, Oct. 22, 1948 (div. July, 1965); 1 son, Christopher. Came to U. S., 1940, naturalized, 1946. Faculty, N.Y. U., N.Y.C., 1947-65, prof. chemistry, 1960-64; Hugh Stott Taylor prof. Princeton, 1964-—; vis. prof. Stanford, 1962, U. N.H., 1963; lectr. Coll. City N.Y., 1953-55. Mem. adv. panel for chemistry NSF, 1963-66. Guggenheim fellow, 1956; Sloan fellow, 1959. Mem. Am. Chem. Soc. Author: Introduction to Stereochemistry, 1965; also numerous articles. Editorial bd. Jour. Organic Chemistry, 1965-—. Research in stereochemistry, chem. and phys. consequences different spatial distbn. atoms in a molecule. Home: 86 Western Way, Princeton, N.J. 08540.*

MISNER, Charles William, Am. physicist; b. Jackson, Mich., June 13, 1932; s. Francis de Sales and Madge (Mee) M.; B.S., U. Notre Dame, 1952; M.A., Princeton, 1954, Ph.D.., 1957; m. Susanne Elisabeth Kemp, June 13, 1959; children—Benedicte Elisabeth, Francis Frithjof, Timothy Charles, Christopher Kemp. Faculty, Princeton, 1956-63; faculty U. Md., College Park, 1963-—, prof. physics and astronomy, 1966-—; vis. faculty mem. Bohr Inst., Copenhagen, Denmark, 1959, 67, Brandeis U., 1960, Cambridge (Eng.) U., 1966-67. Cons. Lawrence Radiation Lab., U. Cal. at Livermore, 1964-66. Sloan fellow, 1958-62. Named Md.'s Outstanding Young Scientist of 1965, Md. Acad. Scis., 1965; recipient Sci. Centennial award U. Notre Dame, 1965. Mem. Am. Phys. Soc., Am. Math. Soc., Royal Astron. Soc., Am. Assn. U. Profs., A.A.-A.S., Sigma Xi. Research, publs. on gen. relativity and applications to relativistic astrophysics and cosmology. Home: 1009 Crest Park Dr., Silver Spring, Md. 20903. Office: Dept. Physics and Astronomy, U. Md., College Park 20742.*

MISRA, Ayodhya, Indian mycologist; b. Fyzabad, India, July 29, 1917; s. Mohan Lal and Gopi (Devi) M.; B.Sc., Agra (India) Coll., M.Sc. (Merit scholar), 1941; Ph.D., Inst. Agr., Minn. U., 1948; m. Bhawani Kumari, May 13, 1945; children—Ramona, Ashok Kumar, Krishna Kumar. Student demonstrator botany Agra Coll., 1939-41; asst. prof. botany B.R. Coll., Agra., 1941-42; research scholar Inst. Agr., U. Minn., 1946-48; asst. systematic mycologist Dept. Plant Protection, Quarantines and Storage, New Delhi, India, 1948-57; prof. mycology, head dept. Bihar Agr. Coll., P.O. Sabour, Bhagalupur, India., 1957-—; prin. investigator U.S.P.L. research project on Helminthosporium species, 1965-—. Mem. bd. studies and sci. and agr. faculties in several univs. Govt. India scholar. Fellow Indian Bot. Soc., Nat. Acad. Scis. India; Mem. Bihar Acad. Agrl. Scis. (sec., editor 1962-—), Indian Phys. Soc. (editor), Am., Indian phytopath. socs., Sigma Xi. Research, publs. on physiology, pathogenicity and nature of host resistance in several pathogenic fungi, variability in plant pathogens and disease devel.; 1st to report that a temperature of 92° F. and low humidity suppressed linseed rust. Home: Hathipur, Lakhimpur, U.P., India. Office: Bihar Agr. Coll., Sabour (Bhagalpur), Bihar, India.*

MISRACHI, Elia (Mizrahi, Elijah ben Abraham (Re'em), Turkish Jewish mathematician; b. probably in Constantinople, circa 1455; student math. and astronomy under Mordecai Comtino; chief rabbi over Turkish Jews, circa 1495-1526; previously taught Talmud, math. and astronomy. Author: Sefer ha-Mispar (Book of Numbers) (simple and practical arithmetic); Commentary to Ptolemy's Almagest (lost); also commentary on Euclid. First Hebrew writer to deal with sum of cubes of 1st natural numbers. Died Constantinople, 1526.

MITCHEL, Ormsby MacKnight, Am. astronomer; b. Morganfield, Ky., July 28, 1809; s. John and Elizabeth (MacAlister) M.; grad. U. S. Mil. Acad., 1829; LL.D., Harvard, 1851, Washington Coll., 1853, Hamilton Coll., 1856; m. Louisa (Clark) Trask, 1831. Asst. prof. math. U. S. Mil. Acad., 1829; chief engr. Rittle Miami R.R., 1836-37; prof. math., philosophy, astronomy Cincinnati Coll., 1836-46; published mag. Sidereal Messenger, 1846-1848; adj. gen. Ohio, 1848; inventor chronograph, 1848; chief engr. Ohio & Miss. R.R., 1848-53; dir., largely responsible for erection Cin. Obs.; largely responsible for erecting 2d largest telescope, and largest on Western continent under auspices of Cin. Astron. Soc., 1845; made approximately 50,000 observations of faint stars between 1854-59; discovered duplicity of stars (Antares); dir. Dudley Obs., Albany, N.Y., 1859; apptd. brig. gen. U. S. Vols., 1861; placed in command Dept. of South and X Army Corps, 1862. Author: Planetary and Stellar Worlds, 1848; Popular Astronomy, 1860. Died Beaufort, S.C., Oct. 30, 1862.

MITCHELL, Allan Charles Gray, Am. physicist; b. Houston, Oct. 1, 1902; s. Samuel Alfred and Milly Gray (Dumble) M.; B.S., U. Va., 1923, M.S., 1924; Ph.D., Cal. Inst. Tech., 1927; postgrad. univs. Munich and Göttingen, 1927-28; m. Georgianna Peck Fales, Sept. 8, 1926; children—Georgianna (Mrs. L. A. Rivlin), Priscilla. Asst., U. Va., 1920-24; teaching fellow, Cal. Inst. Tech., 1924-27; fellow, Bartol Research Found., 1928-31; asst. prof. physics, N.Y. U., 1931-34, asso. prof. and chmn. dept., 1934-38; prof. physics and head dept., Ind. U., 1938-63. Research assoc., Mass. Inst. Tech., 1940, U. Chgo., 1942; ofcl. investigator OSRD, Indiana U., 1942-44, physicist Applied Physics Lab., Johns Hopkins, 1944-46; mem. Project Vista, Cal. Inst. Tech., 1951; bd. govs. Argonne Nat. Lab., 1949-52; Past pres., dir. Midwestern U. Research Assn. Fellow Am. Phys. Soc. (council 1943-47), A.A.A.S., Ind. Acad. Scis.; mem. Phi Beta Kappa, Sigma Xi, Alpha Chi Sigma, Beta Theta Pi. Author: Resonance Radiation and Excited Atoms (with M. W. Zemansky), 1934; chpts. in Beta and Gamma Ray Spectroscopy, 1955. Asso. editor Jour. Chem. Physics, 1932-34, Phys. Rev., 1941-44. Contbr. sci. articles to jours. Research in nuclear physics; chem. physics. Died Nov. 1963.*

MITCHELL, Charles Leslie, orthopedic surgeon; b. Victoria, B.C., Can., July 1, 1901; s. Albert Hugh and Mary Elizabeth (Bunting) M.; student Victoria affiliate, McGill U., 1917-19; M.D., U. Toronto (Ont., Can.), 1925; m. Irene Tennant, Oct. 5, 1928; children—David C., Joyce (Mrs. Francis A. Mulligan). Came to U.S., 1925, naturalized, 1939. Asso. surgeon div. orthopaedics Henry Ford Hosp., Detroit, 1928-32, surgeon-in-charge div. orthopaedic surgery, 1932-—, chmn. dept. orthopaedic surgery, 1958-—. Diplomate Am. Bd. Orthopaedic Surgery. Fellow A.C.S.; mem. Am. Orthopaedic Assn. (pres. 1958), Am., Detroit acads. orthopaedic surgeons, Am. Rheumatism Assn., Clin. (past pres.), Mich. orthopaedic socs., Central, Detroit surg. assns., Canadian Physiol. Assn., Am., Mich., Wayne County med. assns., Detroit Acad. Medicine, Internat. Soc. Orthopaedic Surgery and Traumatology, Nat. Acad. Sci. (chmn. com. prosthetics edn. and information), Kappa Sigma. Contbr. articles to profl. jours. Home: 34 Hendrie Lane, Grosse Pointe Farms, Mich. 48236. Office: Henry Ford Hosp., Detroit 48202.*

MITCHELL, Clifford, Am. physician; b. Nantucket, Mass., Jan. 28, 1854; s. Francis Macy and Ellen M.; A.B., cum laude, Harvard, 1875; postgrad. Chgo., Med. Coll., 1876-77; M.D., Chgo. Homoe. Med. Coll., 1878; m. Susan P. Lillie, May 1878 (dec.); m. 2d, Anna L. Proctor, Sept. 2, 1908 (dec.); m. 3d, Sarah Celeste Dorland, June 24, 1912. Practice limited to urinology and diseases of kidneys; prof. clin. urinology and renal diseases Chgo. Homoe. Med. Coll., Hahnemann Med. Coll., Gen. Med. Coll. Author: Manual of Urinary Analysis, 1897-1902; Renal Therapeutics, 1898; Diseases of the Urinary Organs, 1903; Modern Urinology, 1911. Discovered (with Frederick G. Germuth) reagent for detection of uranium. Died Oct. 19, 1939.

MITCHELL, Dana Paul, Am. physicist, educator; b. Fowler, Ind., May 2, 1899; s. Charles R. and Anna (Sniff) M.; B.S., Tri State Coll., 1919, Sc.D., 1959; Ph.D., Columbia U., 1936; m. Marjorie Vandervort, Sept. 27, 1924; 1 son, Dana D. Faculty, Columbia, 1921-—, physicist submarine warfare New London (Conn.) Lab., 1941-42, asso. prof. physics, 1945-—, exec. dir. Columbia Radiation Lab., 1946-50, sr. research sci. Hudson Labs., Dobbs Ferry, N.Y., 1962-—; Physicist proximity fuse research Carnegie Instn., Washington, 1940-41; asst. dir. atomic bonb work U. Cal., Los Alamos Lab., 1943-45. Recipient Exceptional Service Devel. award U. S. Naval Ordnance Dept., 1945. Fellow Am. Phys. Soc.; mem. Sigma Xi. First exptl. proof of wave like nature of neutrons. Home: 679 Warburton Av., Yonkers, N.Y. 10701. Office: 145 Palisade St., Dobbs Ferry, N.Y. 10522.*

MITCHELL, David Farrar, Am. dentist; b. Arkansas City, Kan., Dec. 15, 1918; s. Lester David and Lucille (Farrar) M.; student U. Tex., 1937-38; B.S., U. Ill., 1940, D.D.S., 1942; Ph.D., U. Rochester, 1948; m. Trone Hawkes, Feb. 6, 1943; children—Dana S., Lindsay T., David Farrar. NIH fellow, USPHS Sr. research fellow U. Rochester (N.Y.), 1946-48; asso.

prof., chmn. div. oral histopathology U. Minn., 1948-55; prof., chmn. dept. oral diagnosis Ind. U., Indpls., 1955—. Cons. to state and fed. govt. agys. Recipient Ann. award for meritorious teaching Ind. U. Sch. Dentistry, 1959. Fellow A.A.A.S.; mem. Am. Dental Assn., Internat. Assn. Dental Research, Am. Acad. Oral Pathologists, Soc. for Exptl. Biology and Medicine, Sigma Xi, Omicron Kappa Upsilon. Research, publs. on exptl. oral pathology, dental caries, periodontal disease, biol. reactions to dental materials, vital dye staining of developing bones and teeth; clin. research in oral diagnosis, roentgentology, prevention and control of periodontal diseases and pain. Home: 4451 Broadway, Indpls. 46205. Office: 1121 W. Michigan St., Indpls. 46202.*

MITCHELL, Edgar William John, Brit. physicist; b. Kingsbridge, Devon, Eng., Sept. 25, 1925; B.Sc., U. Sheffield, 1945; Ph.D., U. Bristol, 1950. Physicist, Met. Vickers Elec. Co. Ltd., 1946-48, 50-51; faculty U. Reading (Eng.), 1951—, prof., 1961—. Fellow Inst. Physics, Phys. Soc. U.K. Research, publs. on semiconductor physics, defects in solids. Office: J. J. Thomson Phys. Lab., U. Reading, Reading, Eng.*

MITCHELL, Elisha, Am. geologist, botanist; b. Washington, Conn., Aug. 19, 1793; s. Abner and Phoebe (Eliot) M.; grad. Yale, 1813; attended Andover Theol. Sem., 1817-18; D.D. (hon.), U. Ala., 1838; m. Maria Sybil North, Nov. 19, 1819; 7 children. Tutor, Yale, 1816-17; prof. math. and natural philosophy U. N.C., 1818-25, prof. chemistry, mineralogy and geology, 1825-57; licensed to preach, 1817; ordained to ministry Presbyn. Ch., 1821; made geol. and bot. excursions throughout N.C.; contbd. articles to Am. Jour. of Sci., other publs.; 1st to measure height of highest mountain in U. S. east of Rockies, Black Mountain (now called Mitchell's Peak), N.C. Author: Elements of Geology, 1842. Killed by fall during storm while exploring Black Mountain, June 27, 1857.

MITCHELL, F. L., Brit. clin. chemist; b. Hesketh Bank, Eng., Dec. 21, 1921; s. Frederick Dearden and Annie (Howarth) M.; B.Sc. with honors, U. Liverpool (Eng.), 1949; Ph.D., U. Sheffield (Eng.), 1954; m. Andrey Kathleen Bradshaw, July 29, 1948; children—Hazel Kathleen, Anne Christine. Biochem. research asst. Jessop Hosp., Sheffield, 1949-51; sr. biochemist Sheffield and Region Endocrine Investigation Centre, 1951-57; head biochemistry Maryfield Hosp., U. St. Andrews, Dundee, Scotland, 1957-64, lectr. biochemistry, 1957-64; faculty U. Edinburgh (Scotland), 1964—, reader clin. chemistry, 1966—. Lectr. chem. pathology Sheffield Tech. Coll., 1952-57. Mem. Whitley council Nat. Health Service, 1962-64; mem. adv. group Ministry Health, Chemistry and Pathology, 1966—. Fellow Royal Inst. Chemistry; mem. Coll. Pathology, Assn. Clin. Biochemistry (past sec. profl. com.), Biochem. Soc., Endocrine Soc., Royal Inst. Chemistry. Publs. on isolation and identification of tissue estrogens; devel. methods for separation and measurement of steroids in human infant urine; steroid metabolism in fetus and neonate. Home: 15 Carnethy Av., Edinburgh 13, Scotland.*

MITCHELL, George Washington, Jr., Am. physician; b. Balt., Apr. 30, 1917; s. George Washington and Katharyne Eugenia (Diggs) M.; A.B., Johns Hopkins, 1938, M.D., 1942; m. Anne Jenkins Shriver, Dec. 19, 1942 (div. 1954); children—Beverly Shriver, George Washington III, Anne Jenkins, Edward Diggs; m. 2d, Mary Elizabeth McKay, Sept. 14, 1957; children—Bruce McKay, Katharyne Wilcox. Gynecologist in chief New Eng. Center Hosp., Boston, 1950—; prof. obstetrics and gynecology Tufts U. Med. Sch., 1954—, chmn. dept., 1956—; surgeon in chief gynecology Boston Dispensary, 1952—; cons. Boston City, St. Margaret's, Carney, Chelsea Naval, Boston Floating hosps. Diplomate Am. Bd. Obstetrics and Gynecology. Mem. Am. Gynecol. Soc., A.C.S., Am. Coll. Obstetricians and Gynecologists, Am. Assn. Obstetricians and Gynecologists, Am. Soc. Study Sterility, Am. Soc. Cytology, Soc. Pelvic Surgeons, Sigma Xi. Study of female urology; cytology. Home: 99 Stanton Av., Auburndale, Mass. Office: 171 Harrison Av., Boston.*

MITCHELL, Helen S., Am. nutritionist; b. Bridgeport, Conn., Sept. 21, 1895; d. Walter L. and Minnie (Swift) M.; B.A., Mt. Holyoke Coll., 1917; Ph.D., Yale, 1921. Research, prof. physiology, nutrition Battle Creek Coll., 1921-35; research prof. nutrition U. Mass., 1935-41; chief nutritionist Office Def. Health Welfare Services, Washington, 1941-43; chief nutritionist Office Fgn. Relief Rehab., Washington, 1943-44; dean. U. Mass. Sch. Home Econs., 1946-60; exchange prof. Hokkaido U., Sapporo, Japan, 1960-62; research cons. Harvard Sch. Pub. Health, 1964-67. Nutrition cons. Head Start, Office Econ. Opportunity. Mem. Mass. Heart Assn. (chmn. nutrition com. 1963-66), Am. Inst. Nutrition, Am. Pub. Health Assn., Am. Dietetic Assn., Am. Home Econs. Assn., Sr. author: (with others) Nutrition in Health and Disease, 15th edit., 1968. Research, numerous publs. on quantitative relations of iron and copper in nutritional anemia; prodn. and control of galactose cataract in rats, relation of protein to human growth at adolescence, publ. of recommended and non-recommended nutrition books for lay readers. Home: 48 Park St., Rural Route 1, Box 33A, Mattapoisett, Mass. 02739.*

MITCHELL, Henry, Am. engr.; b. Nantucket, Mass., Sept. 16, 1830; s. William and Lydia (Coleman) M.; ed. pvt. schs.; A.M. (hon.), Harvard, 1867; m. Mary Dawes; m. 2d, Margaret Hayward, 1873; 1 dau.; m. 3d, Mary Hayward, 1877. Asst. to commrs. on harbor encroachments of N.Y., 1859; cons. engr. U. S. commn. on Boston harbor; later mem. of commn., mem. adv. council, to bd. harbor commrs. of Boston; prof. Am. Inst. Tech.; prof. phys. hydrography Mass. Inst. Tech., 1869, Agassiz Sch. of Sci., 1873; mem. U. S. adv. councils on harbors of Portland, Me., Providence, R.I., Norfolk and Portsmouth, Va., Phila.; apptd. by Pres. Grant to represent Coast and Geodetic Survey in bd. of engrs. for improvement of mouth of Mississippi, 1874; later mem. Mississippi River commn. Fellow Am. Acad. of Scis. of Boston; mem. Nat. Acad. Scis. Contbr. many reports U. S. Coast and Geodetic Survey. Inventor tide gage usable in strong currents, apparatus for study of subsurface current; studied Gulf Stream current between Fla. and Cuba; elucidated tidal movement of N.Y. harbor; disproved theory of emergence of northeastern shores of Am. continent. Died 1902.

MITCHELL, Howard Hawks, Am. mathematician; b. Marietta, O., Jan. 14, 1885; s. Oscar Howard and Mary Hoadley (Hawks) M.; Ph.B., Marietta Coll., 1906, Sc.D., 1935; Ph.D., Princeton, 1910; m. Emma Vestine White, Sept. 18, 1912. Instr. math. Yale, 1910-11; instr. math. U. Pa., 1911-14, asst. prof., 1914-21, prof. since 1921. Mem. Am. Math. Soc. (v.p. 1932-33, editor Trans. 1925-30), A.A.A.S. (v.p. 1932), Am. Philos. Soc., Phi Beta Kappa, Sigma Xi. Died Mar. 13, 1943.

MITCHELL, James Alfred, geologist; s. M. and Jane (Petrie) M.; grad. Royal Sch. Mines, Eng.; studied astronomy and meteorology at Lord Ross's Obs., Birr Castle, Ireland; carried on further researches in geology and chemistry at Harvard, and in paleontology at Johns Hopkins; A.M. (hon.), Mt. St. Mary's Coll., 1888; Ph.D., Niagara U., 1894; m. Margaret J. Willson, 1889. Worked on reports for State Geol. Survey and U. S. Weather Bur.; prof. geology, mineralogy and physics Mt. St. Mary's Coll., Emmitsburg, Md.; lectr. St. Josephs Acad., Emmitsburg. 1st to discover fossiliferous footprints in the Newark system of the Jura Trias in Md.; mem. Washington Acad. Scis. Died 1902.

MITCHELL, John, physician, map maker; probably born Brit. Isles; ed. U. Edinburgh (Scotland); came to Va., 1725; justice of peace Middlesex County (Va.), 1738; practiced medicine in Va., circa 1725-46; went to Eng., 1746. Fellow Royal Soc., 1748. Made Map of the British & French Dominions in North America with the Roads, Distances, Limits, and Extent of the Settlements, published, London, Eng., 1755 (most important map in Am. history, used in various treaties, border adjustments up to 1932, basis for Webster Ashburton Treaty, 1842, Wis.-Mich. boundary dispute, 1926, others); described many plants; introduced new flora to Britain; friend of Linnaeus and of Franklin; developed method of treating yellow fever thought to have saved more than 6,000 lives in Phila. during epidemic of 1793. Died 1768.

MITCHELL, John Kearsley, Am. physician, chemist; b. Shepherdstown, Jefferson County, Va. (now W.Va.), May 12, 1793; s. Alexander and Elizabeth (Kearsly) M.; grad. U. Edinbrugh (Scotland); M.A., U. Pa., 1819; m. Sarah Matilda Henry, 1822; 9 children, including Silas Weir. Ship's surgeon on voyages to China and East Indies, 1819-21; prof. medicine and physiology Phila. Med. Inst., 1824; lectr. chemistry Franklin Inst., 1833-38; prof. medicine Jefferson Med. Coll., Phila., 1841-58. Author: Indecision, a Tale of the Far West and Other Poems, 1839; On the Cryptogamous Origin of Malarious and Epidemical Fevers (important contbn. to parasitic concept of infectious diseases), 1849. Early protagonist of germ theory in U. S.; credited with 1st description of neurotic spinal arthropathy, 1831. Died Phila., Apr. 4, 1858.

MITCHELL, Joseph Stanley, English physician, radiotherapeutist; b. Birmingham, Eng., July 22, 1909; s. Joseph Brown and Ethel (Arnold) M.; student U. Birmingham Med. Sch., 1926-28, 31-33, D.Sc. (hon.); M.B., B.Chir. (Cantab), St. John's Coll., Cambridge U., 1934, Ph., Cambridge U., 1937, M.D. (Cantab), 1957; m. Lilian Mary Buxton, Sept. 4, 1934; children—Christopher Buxton, Janet Buxton. Beit Meml. fellow Colloid Sci. Lab., Cambridge, Eng., 1934-37, resident radiol. officer Christie Hosp., Manchester, Eng., 1937-38; asst. in research dept. medicine Cambridge U., 1938; radiotherapist Emergency Med. Service, Cambridge, 1939; chief med. investigations Montreal lab. NRC Can., 1944-45; prof. radiotherapeutics U. Cambridge (Eng.), 1946-57, Regius prof. physics, 1957—; hon. dir. radiotherapeutic centre Addenbrooke's Hosp., Cambridge, 1946—. Fellow St. John's Coll. Cambridge. Named comdr. Order Brit. Empire. Fellow Royal Soc., 1952, Royal Coll. Physicians, Faculty of Radiologists; mem. Anglo-German Med. Soc. (pres. English sect.). Author: Studies in Radiotherapeutics, 1960; The Treatment of Cancer, 1965. Research, publs. on uses of radio-cobalt 60 in radiotherapy, disturbances of nucleic acid metabolism produced by therapeutic doses of x- and y-radiation, clin. and lab. studies of chem. radiosensitizers, attempts to develop a radioactive drug for use in cancer treatment. Home: Thorndyke, Huntingdon Rd., Cambridge, Eng.*

MITCHELL, Maria, Am. astronomer, educator; b. Nantucket Island, Mass., Aug. 1, 1818; d. William and Lydia (Coleman) Mitchell; LL.D., Hanover Coll., 1882, Columbia, 1887. Assisted father in his chronometer ratings during her youth; apptd. librarian Town of Atheneum, Nantucket Island, Mass., 1836; discovered new comet, Oct. 1847; recipient gold medal from King of Denmark; 1st woman elected to membership Am. Acad. Arts and Scis., hon. mem., 1848, later fellow; apptd. computer Am. Ephemeric and Nautical Almanac; 1st prof. astronomy Vassar Coll., 1865-88; mem. Am. Philos. Soc.; elected to Am. Hall of Fame, 1922. Died Lynn, Mass., June 28, 1889.

MITCHELL, Sir Peter Chalmers, zoologist; b. Dunfermline, Scotland, Nov. 23, 1864; s. Rev. Alex Mitchell; M.A., Aberdeen (Scotland) U.; exhibitioner, then hon. student Christ Ch., Oxford (Eng.) U.; student univs. Berlin and Leipzig (Germany); D.Sc., LL.D.; m. Lilian Pritchard. Univ. demonstrator comparative anatomy, also asst. to Linacre prof., Oxford U., 1888-91; organizing sec. tech. instrn. Oxfordshire County Council, 1891-93; lectr. biology Charing Cross Hosp., 1892-94, London Hosp., 1894; examiner biology Royal Coll. Physicians, 1892-96, 1901-02; sec. Zool. Soc. London, 1903-35; examiner zoology U. London, 1903. Mem. Com. Fishery Investigations, also Com. Sleeping Sickness. Decorated comdr. Brit. Empire. Fellow Royal Soc., 1906, Zool. Soc. (hon. v.p.); mem. Cremation Soc. (v.p.). Author: Outlines of Biology, 1894; Thomas Henry Huxley, 1900; The Childhood of Animals, 1912; Evolution and the War, 1915; Report on the Propaganda Library in Intelligence Division of War Office, 1917; Centenary History of the Zoological Society of London, 1929; Materialism and Vitalism in Biology, 1930; also numerous articles in profl. jours. Directed numerous improvements in London Zool. Gardens. Died July 2, 1945.

MITCHELL, Philip Henry, Am. physiol. chemist; b. Southbury, Conn., Dec. 13, 1883; s. Henry Painter and Phoebe (Stoddard) M.; Ph.B., Sheffield Sci. Sch. (Yale), 1904; Ph.D., Yale, 1907; m. Alice Hinman Friend, May 30, 1910; children—Margery Fuller, Edith Stoddard. Instr. physiology Brown, 1907-11, asst. prof., 1911-20, asso. prof., 1920-26, prof., 1926-49; dir. Woods Hole Biol. Sta., U. S. Bur. Fisheries, summers, 1914-20; cons. chemist U. S. Dept. Agr., 1913-17, investigator Conn. Bd. Fish and Game, 1923-24. Mem. Woods Hole Marine Biol. Lab. Corp. Mem. A.A.A.S., Am. Soc. Biol. Chemists, Sigma Xi. Author: Text book of General Physiology for Colleges, 1923, 4th edit., 1948; Text Book of Biochemistry, 1946; also many articles on purine metabolism, physiology of shellfish, especially oysters, permeability of cells and tissues, chemistry of sea water. Died Feb. 2, 1955.

MITCHELL, Richard Scott, Am. mineralogist; b. Longmont, Colo., Jan. 28, 1929; s. Clarence Floyd and Margaret May (Hartman) M.; student Scottsbluff Jr. Coll., 1946-47, U. Neb., 1947-48; B.S., U. Mich., 1950, M.S. in Mineralogy, 1951, Ph.D., 1956. Faculty, U. Va., Charlottesville, 1953—, prof. geology, 1963—, acting chmn. dept. geology, 1964—. Fellow Geol. Soc. Am., Mineral. Soc. Am.; mem. Va. Acad. Scis. (past sec. geology sect.), Am. Crystallographic Assn., Geochem. Soc., Nat. Assn. Geology Tchrs., Va. Assn. Profl. Geologists, A.A.A.S., Mineral. Assn. Can., Sigma Xi (sec. Va. chpt. 1959-62), Sigma Gamma Epsilon. Contbr. numerous articles to tech. jours. Discovered numerous structural polytypes silicon carbide, cadmium iodide and others determined crystal structures for some largest known polytypes. Home: 510 14th St., Charlottesville, Va. 22903.*

MITCHELL, Rodger D., Am. biologist; b. Wheaton, Ill., July 22, 1926; s. Chester F. and Marie (Yates) M.; B.S., Kan. State Coll., 1950; Ph.D., U. Mich., 1954; m. Adelle Virtue, Sept. 5, 1955; children—Irene, Annette, Scott. Asst. prof. U. Vt., 1954-57; asso. prof. U. Fla., Gainesville, 1957—. NSF faculty fellow U. Cal., Berkeley, 1959-60; Fulbright research fellow Ibaraki U., Mito, Japan, 1965-66. Fellow A.A.A.S.; mem. Soc. for Study Evolution, Soc. Systematic Zoology, Am. Micros. Soc. Research, publs. on morphology of chigger mites and related water mites, devel. of analytical methods for study of mites parasitic on aquatic insects. Home: 919 N.W. 10th Av., Gainesville, Fla. 32601.*

MITCHELL, Roland Burnell, Am. biomed. scientist; b. Denton, Tex., Mar. 24, 1910; s. Robert Marion and Robbie (Hawkins) M.; B.S., N. Tex. State U., 1932; M.A., U. Tex., 1937, Ph.D., 1939; m. Julianne Still, Sept. 6, 1938; children—Bonnie Ann, Susan. Sci. tchr. Tex. pub. schs., 1928-34; supr. sci. tchr. S.W. Tex. Tchrs. Coll., 1935-37; bacteriologist Bur. Plant Industry, Dept. Agr., 1937-42, Tex. Pub. Health Labs., 1942; asst. dir. bur. labs. Fla. Bd. Health, 1946-48; prof. pub. health bacteriology U. Fla., 1947-48; chief dept. aerobiology Sch. Aviation Medicine, USAF, 1948-51, chief dept. microbiology, 1951-57, acting dir. med. sci. div., 1954-57, dir., 1957-59; chief med. scis. dept. Sch. Aerospace Medicine, Aerospace Med. Div., Brooks AFB, Tex., 1959-61, chief biosystems research div. USAF Sch. Aerospace Medicine, Aerospace Med. Div., 1961-63; dir. biol. scis. Office Dep. for Research and Development, Hdqrs. Aerospace Med. Div., 1963—. Exec. dir. 3d Internat. Symposium on bioastronautics and exploration of space, 1964. Diplomate Am. Bd. Microbiology. Fellow Am. Acad. Microbiology, Am. Pub. Health Assn.; mem. Am. Soc. Microbiology, Soc. Exptl. Biology and Medicine, Aeromed. Assn., Assn. Mil. Surgeons, Sigma Xi. Research on aerobiology of upper atmosphere, astromicrobiology, devel. and evaluation of rapid lab. methodology for med. diagnostic procedures. Home: 234 Larkwood Dr., San Antonio 78209. Office: Biol. Scis. Directorate, Office Headquarters, Aerospace Medical Division, Brooks AFB, Tex. 78235.*

MITCHELL, Ross Galbraith, Brit. pediatrician; b. Wallington, Eng., Nov. 18, 1920; s. Richard Galbraith and Ishobel (Ross) M.; M.B., Ch.B., U. Edinburgh, 1944, M.D., 1953; m. June Phylis Butcher, Sept. 16, 1950; children—Andrew Ross, Lindsay Margaret, Alison Wendy, Christine Ann. Rockefeller travelling fellow in physiology Mayo Clinic, Rochester, Minn., 1952-53; lectr. child health U. St. Andrews, 1953-55; pediatrician Dundee Royal Infirmary, 1955-62, Royal Aberdeen (Scotland) Hosp. for Sick Children, 1963—; prof. child health U. Aberdeen, 1963—. Chmn., Scottish Adv. Council for Child Care, 1965—. Fellow Royal Coll. Physicians Edinburgh; mem. Assn. Physicians, Brit. Paediatric Assn., Paediatric Research Soc., Neonatal Soc., Am Acad. Cerebral Palsy (fgn. corr.). Author: (with R. W. B. Ellis) Disease in Infancy and Childhood, 6th edit., 1968; also numerous articles. Research on children with cerebral palsy, child nutrition, metabolism of histamine. Home: Harecraig, Culter House Rd., Milltimber, Aberdeen. Office: Dept. Child Health, Med. Sch., Foresterhill, Aberdeen, Scotland.*

MITCHELL, Samuel Alfred, astronomer; b. Kingston, Can., Apr. 29, 1874; s. John C. and Sarah (Chown) M.; M.A., Queen's U., Can., 1894, LL.D., 1924; Ph.D., Johns Hopkins, 1898; LL.D., U. Western Ont., 1940; m. Milly Gray Dumble, Dec. 28, 1899; 1 son, Allan Charles Gray. Instr. and adj. prof. Columbia, 1899-1913; prof. astronomy and dir. McCormick Obs., U. Va., 1913-45, dir. emeritus 1945-60. Astronomer, eclipse expdns., Georgia 1900, Sumatra, 1901, Spain, 1905, Ore., 1918, Cal., 1923, Conn., 1925, Norway, 1927, Niuafoou Island, 1930, Quebec, 1932, Canton Island, 1937; research asso. Yerkes Obs. (U. Chgo.), summers, 1907, 09, 10, 11, asst. prof. astrophysics, 1912-13. Recipient Watson medal, 1948. Mem. Nat. Acad. Scis., Am. Philos. Soc., Am. Acad. Arts and Scis., A.A.A.S. (v.p. 1921); fellow and foreign asso. Royal Astron. Soc.; hon. mem. Am. Assn. Variable Star Observers; mem. Am. Astron. Soc. (v.p. 1925-27); research asso. Carnegie Instn., 1934-45, Internat. Astron. Union, Astronomisches Gesellschaft, Société Astronomique de France, Royal Astron. Soc. Can., Phi Beta Kappa, Sigma Xi. Author: Parallaxes of 260 Stars, 1920; Eclipses of the Sun, 1923, 5th edit., 1950; Parallaxes of 440 Stars, 1927; Fundamentals of Astronomy (with C. G. Abbot), 1927; Solar Eclipses, 1929, 36; Variable Stars, 1935; Parallaxes of 650 Stars (with D. Reuvl), 1940. Used high-speed photography for astronomical observations; calculated distances of over 1,000 stars; studied solar eclipses. Died Feb. 22, 1960.

MITCHELL, Samuel Augustus, Am. geographer, publisher; b. Bristol, Conn., Mar. 20, 1752; s. William and Mary (Alton) M.; m. Rhoda Ann Fuller, Aug. 1815. Prepared textbooks, maps, geog. manuals, including Mitchell's Geographic Reader 1840; Map of the United States and Territories, 1861; published A New American Atlas, 1831; Mitchell's Traveller's Guide Through the United States, 1832; published successful series of sch. geography books; began series of Tourist's Pocket Maps of different states, 1834; published A New Universal Atlas, 1847; an outstanding figure in devel. of Am. geography. Died Phila., Dec. 18, 1868.

MITCHELL, S(ilas) Weir, Am. neurologist; b. Phila., Feb. 15, 1829; s. John Kearsley and Matilda (Henry) M.; ed. U. Pa.; M.D., Jefferson Med. Coll., 1850, LL.D., 1910; M.D. (hon.), Bologna, 1888; LL.D., Harvard, 1886, Edinburgh, 1895, Princeton, 1896, Toronto, 1906; m. Mary Middleton Elwyn; m. 2d, Mary Cadwalader, 1875; children—John Kearsley, Langdon Elwyn M. Asst. to father (a physician), 1851-53; practiced medicine, Phila.; prominent as physiologist and neurologist; connected with Phila. Polyclinic and Coll. for Grads. in Medicine, Phila. Orthopaedic Hosp., Infirmary for Nervous Diseases. Fellow Am. Acad. Arts and Scis., Coll. Physicians, Phila. (pres.), Royal Soc., 1908, Royal Med. Chirurg. Soc.; mem. Brit. Med. Soc. (hon.), French Acad. Medicine (hon. fgn. asso.), Acad. Natural Scis., Phila. Author: (with G. R. Moorehouse, W. W. Keen) Gunshot Wounds and Other Injuries of Nerves, 1864; Reflex Paralysis, 1864; Wear and Tear, 1871; Fat and Blood, 1877; Lectures on the Diseases of the Nervous System, Especially in Women, 1881; Clinical Lessons on Nervous Diseases, 1897; Nurses and Their Education, 1902; The Evolution of the Rest Treatment, 1904; Rest Treatment and Psychic Medicine, 1908; Collected Poems, 1896; Hugh Wynne, 1898; Adventures of François, 1899; Autobiography of a Quack, 1900; Youth of Washington, 1904; The Red City, 1907; also papers. Founder of neurology in Am.; advocated treatment of functional neuroses by prolonged rest and physiotherapy (Weir Mitchell rest cure), 1877; gave 1st accurate description of erythromelalgia (Mitchell's disease), 1872, 1st description of ascending neuritis, 1872, of post-paralytic chorea, 1874; 1st to relate astigmatism to headache, 1874, also of pain in traumatic neuralgia to weather, 1877; showed (with E. T. Reichert) that snake venom is protein in nature, 1886. Died Jan. 4, 1914.

MITCHILL, Samuel Latham, Am. naturalist, physician; b. North Hempstead, L.I., Aug. 20, 1764; s. Robert and Mary (Latham) M.; M.D., U. Edinburgh (Scotland), 1786; m. Catherine Akerly, June 23, 1799. Mem. N.Y. Legislature, 1791, 98, 1810; prof. natural history, chemistry, agr. Columbia, 1792, asso. prof. botany, 1793-95; gave mineral collection to Columbia mus.; a founder Soc. for Promotion of Agr., Arts and Manufactures; made 1st geol. and mineral. survey in U. S., on banks of Hudson River, 1796; a founder Med. Repository, 1797, editor, 23 years; mem. U. S. Ho. of Reps. from N.Y. (Democrat), 7th, 9th, 12th congresses; mem. U. S. Senate (Democrat) from N.Y., 1804-09; commd. to supervise constrn. of a steam war-vessel during War of 1812; prof. chemistry Coll. Physicians and Surgeons, N.Y.C., 1807, prof. natural history, 1808-20, prof. botany and materia medica, 1820-26; an organizer Rutgers Med. Coll., v.p., 1826-30; a founder N.Y. Lit. and Philos. Soc., 1814; prin. founder Lyceum of Natural History, 1817; surgeon gen. N.Y. State Militia, 1818. Author: Explanation of the Synopsis of Chemical Nomenclature and Arrangement, 1801; The Present State of Medical Learning in the City of New York, 1797; A Sketch of the Mineralogical History of N.Y., 1797, 1800, 02; most notable contbns. include papers on fishes of N.Y., the origin of Indians, Indian poetry, Indian antiquities. Described and classified 166 fish species, chiefly from fresh and salt waters nr. N.Y.C.; added to knowledge of Am. vegetable materia medica, soap manufacture, disinfectants; 1st Am. tchr. of chemistry to use Lavoisier's nomenclature; Mitchill's theory of doctrine of septon and septic acid led to many papers and gave impulse to Davy's discoveries. Died Bklyn., Sept. 7, 1831.

MITCHISON, Denis Anthony, Brit. bacteriologist; b. Oxford, Eng., Sept. 6, 1919; s. Gilbert Richard and Naomi (Haldane) M.; student Trinity Coll., Cambridge, Eng., 1936-39; M.R.C.S., L.R.C.P., U. Coll. London, 1943, M.B., B.Sc., 1944, M.R.C.P., 1965, M.C. Path., 1966; m. Ruth Sylvia Gill, Sept. 9, 1940; children—Susan, Graeme, Terence, Clare. House physician Addenbrooke's Hosp., Cambridge, 1943; house physician Royal Berkshire Hosp., Reading, Eng., 1944; trainee pathologist sector IV, Watford, Eng., 1944-47; asst. to pathologist Hosp. for Diseases of Chest, Brompton, Eng., 1947; asst. lectr. bacteriology Postgrad. Med. Sch. London, 1948-50, lectr., 1950-64, reader in bacteriology, 1964—; hon. cons. Hammersmith Hosp. Mem. NRC; hon. dir. Med. Research Council unit for Research on Drug Sensitivity in Tb; mem. panel on Tb, WHO. Mem. Internat. Union Against Tb (past mem. com. of treatment). Contbr. numerous articles to sci. jours. Devel. and application of treatment of Tb with antibacterial drugs. Home: 14 Marlborough Rd., Richmond Surrey, Eng. Office: Dept. Bacteriology, Royal Postgrad. Med. Sch., Ducane Rd., London W.12, Eng.*

MITEREV, Georgiy Andreevich, Russian hygienist; b. 1900; grad. Med. Faculty, Saratov U., 1925; D.Med. Sci., 1945. Head, Melaksa Rayon Health Dept., Ulyanovsk Oblast, 1925-32; state med. insp. Central Volga Kray, chief physician Samara Central and Gen. Hosp., 1932-39; USSR peoples commissar, then minister health, 1939-47; dir. Erisman Health-Hygiene Research Inst., Moscow, 1947-54; prof., head chair gen. hygiene Moscow Pharm. Inst., 1950-58, Sechenov 1st Moscow Med. Inst., 1960—. Decorated Order of Lenin (3). Author: Twenty Five Years of Soviet Public Health, 1952. Research and numerous publs. on hygiene, sanitation, pub. health orgn. Address: Sechenov 1st Moscow Med. Inst., B. Pirogovskaya ulitsa 2-6, Moscow, USSR.

MITOLO, Michele, Italian surgeon; b. Foggia, Italy, Mar. 22, 1903; s. Vincenzo and Maddalena (Magrone) M.; M.D., U. Bari (Italy); m. Ermelinda Ricciardi, Jan. 16, 1937; 1 son, Vincenzo. Prof. human physiology U. Bari. Recipient Marzotto prize for medicine and surgery, 1959. Research on physiology and biochemistry of nutrition, vitamins, hormones, muscle and of central nervous system. Home: Corso Sonnino 140/B, Bari, Italy.

MITRA, Asoke Nath, Indian physicist; b. Rajshahi, Bengal, India, Apr. 15, 1929; s. Jatindra Nath and Ramarani (Bose) M.; B.A. with honors in Math., U. Delhi, 1947, M.A. in Math., 1949, Ph.D. in Physics, 1952; Ph.D. in Theoretical Physics, Cornell U., 1955; m. Anjali Ghosh, Nov. 27, 1956; children—Bani, Buli. Lectr. physics Delhi (India) U., 1949-52, reader, 1960-62, prof., 1962—; reader physics Aligarh U., 1955-60. Vis. prof. physics Ind. U., 1962-63; asso. Internat. Center For Theoretical Physics, Trieste. Fellow Nat. Inst. Scis. (India); mem. Am. Phys. Soc. Research, publs. on solution to three-body problem using factorable but realistic two-body forces, and application to observable properties of three-nucleon systems confirming belief that two-nucleon forces are mainly responsible for nuclear structure, study of elementary particle structures through postulated substructures called quarks using para statistics. Home: care Prof. J. N. Mitra, Govind-Bhawan, 4-Daryaganj, Delhi 6, India.*

MITRA, Girija Bhushan, Indian physicist; b. Calcutta, India, May 1, 1923; s. Krishna Pada and Snehalata (Bose) M.; B.Sc., Ripon Coll., Calcutta, 1942; M.Sc., Sci. Coll. Calcutta; D.Phil. (Calcutta), Indian Assn. for Cultivation Sci., 1952; D.Sc.. (Calcutta), Indian Inst. Tech., Kharagpur, India, 1967; m. Sadhana Basu, Feb. 6, 1961; 1 dau., Paramita. Phys. asst. Govt. Test House, Calcutta, 1945-47; research asst. Indian Assn. for Cultivation Sci., 1947-53; lectr. physics Indian Inst. Tech., Kharagpur, 1953-60, asst. prof. physics, 1960-65, asso. prof., 1965—. Fellow Indian Phys. Soc., Indian Inst. Metals, Indian Assn. for Cultivation Sci. Contbr. numerous articles to sci. jours. Developed theory of 4th moment of line profiles for study of strain distbn., showed particle size dependence of effective absorption co-efficient of random media, developed theory of diffraction by bent crystals, studies microwave analogue of X-ray diffraction, studied defects in deformed metals and minerals. Home: 284/B Rash Behari Av., Calcutta 19, India. Office: Dept. Physics, Indian Inst. Tech., Kharagpur, India.*

MITSCHERLICH, Alexander, German chemist; b. Berlin, May 28, 1836; s. Eilhard Mitscherlich; prof., Hanover-Münden, Germany; developed process for extracting cellulose from wood. Died Obersdorf, Germany, May 31, 1918.

MITSCHERLICH, Eilhard, German chemist; b. Neuende, Germany, Jan. 7, 1794; ed. Jenver, Heidelberg (both Germany), Paris, Göttingen, Germany; worked with H.F. Link in Berlin, 1818; studied with Berzelius in Stockholm, 1820-21; lectr. in chemistry, Berlin; prof. chemistry Friedrich-Wilhelm-Institut, Berlin. Fellow Royal Soc., 1828 (Royal medal 1828); mem. Berlin Acad. Scis., French Acad. Scis., 1852. Author: Mirchondi historia Thaheridarum historicis nostris huiusque incognitorum Persiae principum, 1814; Lehrbuch der Chemie, 2 vols., 1829-30; also articles in Annalen (Poggendorf), other jours. Investigated crystalline structure; discovered isomorphism, 1819, dimorphism, 1826, relation between chem. compositions of crystalline form and bearing of isomorphism on determination of atomic weights; experimented on variation in inclination of optic axes in biaxial crystals with a temperature change; improved accuracy of organic and inorganic analysis; discovered permanganic acid, double crystalline form of sulphur (one of 1st observed cases of dimorphism), selenic acid and found its salts isomorphous with sulfates; studied manganic acid, decomposition products of benzaldehyde and benzoin; discovered monoclinic sulfur, 1823, nitrobenzene, 1833; experimented on artificial minerals; one of 1st chemists to recognize contact action (now called catalysis). Died Schönberg, Germany, Aug. 28, 1863.

MITSCHERLICH, Karl Gustav, German pharmacologist; b. Jever, Germany, Nov. 9, 1805; brother of Eilhard Mitscherlich; prof., Berlin. Author: Verhalten des Kupfersulfats auf Tierköper, 1837; Lehrbuch der Arzneimittellehre, 2 vols., 1840-43. Conducted 1st systematic expts. on animals for evaluation of pharmacological studies. Died Berlin, Mar. 19, 1871.

MITSUDA, Hisateru, Japanese chemist; b. Osaka, Japan, May 27, 1914; s. Kyuji and Ai (Omura) M.; grad. Faculty Agr., Kyoto (Japan) U., 1934, B.Agr., 1937, Dr.Agr., 1949; m. Kiku Minamioji, Apr. 23, 1939; children—Teruko (Mrs. Kenzo Takeda), Hisayoshi. Faculty, Kyoto U., 1937—, prof. Inst. for Chem. Research, 1952-55, prof. dept. agrl. chemistry, 1955—. Mem. Com. Essential Amino acids, 1958—, Japan Sci. Council, 1965—; vice chmn. Com. Vitamins, 1956—. Recipient prize Dir.-Gen. Sci. and Tech. Agy., 1959, Japanese Soc. Food and Nutrition, 1962, Japanese Soc. Vitaminology, 1963, Cultural prize Kyoto newspaper, 1966. Mem. Japanese Soc. Agr. Chemistry (counsilor), Biochem. Soc. (dir.), Vitaminological Soc. (counsilor), Food and Nutrition Soc. (dir.), Am. Chem. Soc., Inst. Food Technology. Author: Essential of Nutrition Soc., 1959; Methods in Nutritional Chemistry, 1961; also numerous articles. Editor: (with Y. Sakurai, K. Shibazaki) Food Preservation, 1966. Research in proteins, enzymes, vitamins and foods; invented enriched rice, various methods to prevent appearance of cloudiness in beer, to utilize Chlorella for human food, and to prevent oxidative deterioration in processed food. Home: 8 Kitazonocho, Shimogamo, Sakyoku, Kyoto, Japan.*

MITSUI, Shingo, Japanese agrl. chemist; b. Tokyo, Japan, 1910; grad. Tokyo U., 1933; Dr.Agr. Technician, Agrl. Lab., Agr. and Forestry Ministry; prof. Tokyo U. Ministry Edn. grantee, 1951. Research and publs. on burned soil as fertilizer, use of isotopes in study of soil fertilization.

MITSUMA, Kenzo, Japanese physician; b. Yamagata Prefecture, 1852; entered Tokyo Med. Sch., 1870. After grad. joined Tokyo Hosp.; became dir. Met. Police Hosp., 1880; later founder hosp. Discovered diphtheria virus, 1875. Died 1894.

MITSUTANI, Akio, Japanese chemist; b. Okayama, Japan, Oct. 25, 1929; s. Mitsuo and Mayumi (Susukida) M.; grad. Kyoto (Japan) U., 1953, Doctor degree, 1960; m. Ikuko Yamantani, Apr. 8, 1957; children—Atsushi, Minako. Staff, Kurashiki Rayon Co., Osaka-Shi, Japan, 1953—, sr. staff planning div., 1963-65, sr. staff co-ordinating chamber, 1965—.

Mem. Japan Chem. Soc., Catalysis Soc. Japan. Contbg. author: Catalytic Engineering, vol. IX, X, 1966. Research, publs., patents on reaction kinetics of catalytic vinylation of acetic acid and ethylene oxidation, effect of irradiation on solid catalysts; discovered new catalyst for isoprene synthesis; pioneered industrialization of fluidized bed synthesis of vinyl acetate. Home: 11-10 3-Chome, Nogami, Takarazuka-Shi, Japan. Office: 8 Umeda, Osaka-Shi, Japan.*

MITSUYOSHI, see Koyu, Yoshida Schichibei.

MITTAG-LEFFLER, (Mangus) Gosta, Swedish mathematician; b. Stockholm, Sweden, Mar. 16, 1846; s. J. O. Leffler and Gustava Mittag; Dr. Phil., U. Uppsala (Sweden), 1872; hon. doctorates from univs. Bologna, Oxford, Cambridge, Christiania, Aberdeen, St. Andrews; m. Signe Lindfors, 1882. Prof., U. Helsinki (Finland), 1877-81; prof. math. U. Stockholm, 1881-1911, rector, 1886-92-93; founder Acta Mathematica, 1883, chief editor for 45 years; founder Mittag-Leffler Inst., Djursholm, for devel. math. teaching in Scandinavian countries. Fellow Royal Soc., 1896; hon. mem. Philos. Soc. Cambridge, London Math. Soc., Royal Instn., Manchester Lit. and Philos. Soc., Royal Irish Acad.; mem. French Acad. Sci., 1900, also acad. scis. in Göttingen, Belgium, Stockholm, Christiana, Copenhagen, Helsinki. Author: Sur la représentation analytique des fonctions monogenes uniformes d'une variable indépendante, 1884; Sur la représentation analytique d'une branche uniforme fonction monogène, 1900-20. Research in analysis, also gen. theory functions and analytic representation of a one-valued function (Mittag-Leffler theorem); gave proof of Cauchy's theorem. Died July 1927.

MITTLER, Sidney, Am. biologist; b. Detroit, Aug. 2, 1917; s. Max and Ida (Shulman) M.; B.S., Wayne U., 1938, M.S., 1938; Ph.D., U. Mich., 1944; m. Leonore Broder, Aug. 15, 1942; children—Jeanne Ellen, Judith Gail, Michele Barbara. Instr., Bowling Green State U., 1945-46; asst. prof. Ill. Inst. Tech., 1946-52, research biologist Armour Research Found., 1952-60; prof. biology No. Ill. U., DeKalb, 1960——. Recipient award sci. merit Chemistry Dept., Armour Research Found., Ill. Inst. Tech., 1955. Fellow A.A.A.S.; mem. Genetics Soc. Am., Am. Soc. Zoologists, Ill. Acad. Sci., Am. Inst. Biologists. Research, numerous pubs. on modification sex ratio in white fly, nutrition and influence on tumor appearance in Drosophila, toxicity ozone, prodn. tumors from gasoline engine exhaust, breakage chromosomes by X-ray radiation. Home: 309 Fairmont Dr., DeKalb, Ill. 60115.*

MITTWOCH, Ursula, geneticist; b. Berlin, Germany; d. Eugen and Hermine (Lipman) Mittwoch; B.Sc., U. London, 1947, Ph.D., 1950; m. Bernard Springer, Dec. 21, 1954; 1 dau., Caroline. Researcher, Galton Lab., U. Coll., London, Eng., 1947——. Mem. Royal Soc. Medicine, Linnean Soc., Genetical Soc., Soc. for Exptl. Biology, Soc. for Developmental Biology. Author: Sex Chromosomes, 1967; also numerous articles. Research on changes in blood cells in Mongolism and other forms of mental retardation, sex differences in cells and chromosomal basis leading to differentiation of sexes in embryos. Home: 14 Cyprus Av., London N.3. Office: Galton Lab., U. College, London, Eng.*

MIURA, Momshige, Japanese psychiatrist; b. Shizuoka Prefecture, Japan, 1891; grad. Kyoto U., 1918, also Kyushu Med. Coll.; M.D., 1929. Became asst. prof. Kyoto U., 1925, prof., 1936-54, emeritus prof., 1954——; later instr. Osaka Med. Coll. Research in psychopathology and symptomatology.

MIURA, Taiei, Japanese psychopathologist; b. Nagaoka, Japan, 1901; grad. Keio U., 1925; doctorate, 1929; M.D. Became instr. neurology sect. med. dept. Keio U., 1937, prof., 1953; named dir. Sakuramachi Hosp., 1943. Recipient French Medal Legion d'Honour, 1954. Japanese-French Med. Assn. (chmn.). Author: Diagnosis and Remedy of Neurology; Mental Physiology. Translator: Introduction to Experimental Medical Science (Claude Bernard).

MIURA, Yoshiaki, Japanese biochemist; b. Tokyo, Japan, Apr. 9, 1915; s. Kinnosuke and Oshie (Miyake) M.; M.D., Tokyo U., 1941; Ph.D., 1950; m. Reiko Hirata, June 27, 1944; children—Etsuko, Michiko. Research fellow Tokyo U. Faculty Medicine, 1950-52; Rockefeller Found. fellow U. Pa. Sch. Medicine, 1952-53; asso. prof. biochemistry Tokyo U. Faculty Medicine, 1953-60; prof., chmn. dept. biochemistry Chiba (Japan) U. Sch. Medicine, 1960——. Sec., VIIth Internat. Congress of Biochemistry, Tokyo, 1967. Mem. Japanese, French biochem. socs., N.Y. Acad. Scis. Author: (with others) Textbook of Biochemistry, 1965; also articles. Research on protein synthesis, especially silk synthesis, function of cell nuclei. Home: 1-8-17 Nishikata Bunkyoku, Tokyo, Japan. Office: 33 Inohanacho, Chiba, Japan.*

MIVART, St. George Jackson, English biologist; b. London, Nov. 30, 1827; s. James Edward Mivart; ed. King's Coll., London, St. Mary's Coll., Oscott, Lincoln's Inn; doctorate in philosophy from Pius IX, 1876; Ph.D., Louvain, Belgium, 1884; called to bar at Lincoln's Inn, 1851, but did not practice; became lectr. comparative anatomy St. Mary's Hosp. Med. Sch., 1862; prof. biology Roman Cath. Univ. Coll., Kensington, Eng., 1874-77; prof. natural history, U. of Louvain, 1890-93. Fellow Linnean Soc. (v.p. 1892), Zool. Soc., Royal Soc., 1869; mem. Royal Instn., Metaphys. Soc. Converted to Roman Catholicism, 1844, excommunicated, 1900. Author: On the Appendicular Skeleton of the Primates, 1867; On the Genesis of Species, 1871; Lessons in Elementary Anatomy, 1873; Man and Apes, 1873; The Common Frog, 1874; Contemporary Evolution, 1876; Lessons from Nature, 1876; The Cat: an Introduction to the Study of Backboned Animals, 1881; Nature and Thought, 1882; The Origin of Human Reason, 1889; American Types of Animal Life, 1894. Leading critic of Darwin's and Huxley's theory of natural selection; explained appearance of new species as due to innate plastic power of "individuation"; tried to reconcile evolution and Catholicism by emphasizing distinction between organic and inorganic matter; his attempts led to his excommunication, 1900; attributed devel. of human mind to action of divine power; made important studies of carnivores and insectivores. Died London, Apr. 1, 1900.

MIWA, Hirohide, Japanese nuclear engr.; b. Akashi, Hyogo, Japan, Jan. 12, 1924; s. Yuzo and Kiyo (Abe) M.; grad. Tokyo (Japan) Imperial U., 1946; Philos. D. Engring., Tokyo U., 1959; m. Miyoko Koike, Apr. 24, 1949; children—Hiroaki, Keiko, Kenichiro. Chief nuclear instrument sect. Kobe Industries Corp. (Japan), 1957-64, vice mgr., 1964-66, vice directing mgr. research div., 1966——; prof. Kobe U., 1961-67, Nagoya U., 1967——. Recipient Grand prize Japan Invention Assn., 1952. Mem. Phys. Soc. Japan, Soc. Applied Physics Japan, Atomic Energy Soc. Japan, Health Physics Soc. Author: Oyo Hoshasen Keisoku, 1961; Hoshasei Tsuisekishi, 1953; also numerous articles. Optimum design procedures for applied radiation gauging; electronics for nuclear instrumentation considering radiation detectors as current generator; improvements and devels. nuclear instruments. Home: 20-4 1 chome, Uenomaru, Akashi. Office: Kobe Indsl. Corp., Okubocho, Akashi, Hyogoken, Japan.*

MIWA, Mitsuo, Japanese physicist; b. Shizuokaken, Japan, Nov. 3, 1904; s. Gyujiro Oguri and Jun Miwa; M.Sc., Tokyo U., 1928, Ph.D., 1942; m. Toshiko Suzuki, Oct. 13, 1931; children—Hiroshi, Satoshi, Takashi, Tomoo, Nobuo. Research asso. Inst. Metals, Tohoku U., Sendai, 1931-35; research asso. Japan Found. for Cancer Research, 1935-41, head dept., 1941; lectr. physics Tokyo U., 1935-42; prof. physics Tokyo Bunrika U., 1942-52; prof. physics Tokyo U. Edn., 1953——, dean faculty of sci., 1962-64. Mem. Phys. Soc. Japan (v.p. 1956-57). Research on electron diffraction, radiation dosimetry, photoreaction with betatron. Home: Miyasaka 3-17-5, Setagaya, Tokyo, Japan.*

MIXNER, John Paulding, Am. physiologist; b. Bridgeton, N.J., Nov. 1, 1915; s. Melvin B. and Frances (Smith) M.; B.Sc., Rutgers U., 1936, M.Sc., 1938; Ph.D., U. Mo., 1943; m. Exa L. O'Neal, June 18, 1942; children—Jean Frances, Jack Bradford. Instr. zoology La. State U., 1943-47; faculty Rutgers U., New Brunswick, N.J., 1947——, prof. dairy sci., 1956——, chmn. dept. dairy sci., 1961-63, chmn. dept. animal scis., 1963——. Recipient N.Y. Farmers Award Bronze medal, 1959. Mem. Am. Physiol. Soc., Endocrine Soc., Soc. for Exptl. Biology and Medicine, Am. Dairy Sci. Assn., Am. Soc. Animal Sci., N.Y. Acad. Scis., Sigma Xi. Contbg. author: Reproduction in Domestic Animals, 1959; also numerous articles. Research on mammary growth and lactation, thyroid activity dairy cattle, frozen semen in dairy bulls, hepatic and renal functions dairy cattle. Home: 34 Whitehall Rd., East Brunswick, N.J. 08816. Office: Dept. Animal Scis., Rutgers U., New Brunswick, N.J. 08903.*

MIXTER, Samuel Jason, Am. surgeon; b. Hardwick, Mass., 1855; s. William and Mary (Ruggles) M.; S.B., Mass. Inst. Tech., 1875; M.D., Harvard, 1879; m. Wilhelmina Galloupe, 1879; children—William Jason, Charles Galloupe, George, Samuel. Began practice at Boston, 1879; cons. surgeon Mass. Gen. Hosp., Mass. Charitable Eye and Ear Infirmary. Inventor tube similar to glass drainage tube for use in intestinal operations (Paul's, or Paul-Mixter's, tube), circa 1895. Died Jan. 19, 1926.

MIYA, Tom Saburo, Am. pharamacologist; b. Hanford, Cal., Apr. 6, 1923; s. Katsunasuke and Harue (Takeachi) M.; B.A., U. Neb., 1947, M.S., 1948; Ph.D., Purdue U., 1952; m. Midori Sakamoto, Aug. 14, 1948; 1 dau., Pamela Anne. Faculty, Purdue U., Lafayette, Ind., 1948-56, prof. pharmacology, 1958——, head pharmacology, 1961——; head dept. pharmacology U. Neb., 1956-57. Cons. Internat. Research & Devel. Co., Mattawan, Mich., 1965——. Mem. Am. Assn. Coll. Pharmacy (chmn. sect. tchrs. biol. scis 1965-66), Am. Pharm. Assn. (Research achievement award in pharmacodynamics 1964), Am. Soc. Pharmacology and Exptl. Therapeutics, Am. Chem. Soc., Soc. for Exptl. Biology and Medicine, N.Y. Acad. Sci., Acad. Pharm. Scis., A.A.A.S., Am. Assn. U. Profs., Sigma Xi, Rho Chi, Phi Lambda Upsilon, Kappa Psi. Author: (with Hoick, Yim, Myers) Laboratory Guide in Pharmacology, 2d edit., 1964; also numerous articles. Research on mechanisms action tranquilizers, tolerance devel. and detoxication mechanisms, effect endocrines on drug action. Home: 1836 Happy Hollow Rd., West Lafayette, Ind. 47906. Office: Dept. Pharmacology, Purdue U., Lafayette, Ind. 47907.*

MIYABE, Naomi, Japanese geophysicist; b. Tokyo, Japan, Oct. 2, 1901; s. Naoya and Eiko (Watanabe) M.; student Faculty Sci., Tokyo U., 1923-27, D.Sc., 1936; m. Yasuko Ishizuka, Dec. 29, 1929. Mem. Earthquake Research Inst., Tokyo U., 1927-41, asst. prof., 1936-41; prof. physics Nagoya U., 1941-50; mem. geog. survey inst. Ministry of Constrn., 1950-60, chief divs. topography, geography, geodesy, 1956-60; tech. assistance expert UN, Somalia, 1961-62; non-regular mem. Tokyo Inst. Civil Engring., 1962——; prof. physics Komazawa U., 1964——. Profl. mem. resources research com. Nat. Agy. for Sci. and Tech., 1960——; mem. com. for study preventing measure for landsubsidence Tokyo Met. Govt., 1959——. Decorated Fourth Order of Sacred Treasure. Mem. Geodetic Soc. Japan, Seismol. Soc. Japan, Volcanological Soc. Japan, Phys. Soc. Japan, Am. Geophys. Union. Author: Chikaku-no-Hendo, 1938; also numerous articles. Research on acute and chronic crustal movements with or without destructive earthquakes, landsubsidence caused by ground water. Home: 1-6 Kita 4-Chome, Yachiyodai, Chiba-ken, Japan.*

MIYAGAWA, Yoneji, Japanese physician; b. Aichi Prefecture, Japan; grad. Tokyo U., 1910; M.D. Became mem. Infectious Disease Research Inst., prof. Tokyo U., 1927, dir. Infectious Disease Research Inst., 1934, ret., 1945, now emeritus prof.; dir. Toshiba Tosei Hosp. Recipient Imperial Acad. Prize, 1955. Author: The Parasite in the Human Body; New Internal Medical Science; Cultural Zone of Mongolia. Differentiated seasonal chancre from syphilis and discovered it is caused by an intracellular filterable germ in granular form (Miyagawa Germ). Died Dec. 26, 1959.

MIYAHARA, Shohei, Japanese physicist; b. Tokyo, 1914; D.Sc. Became asst. prof. Nagoya U., 1943; named prof. Hokkaido U., 1949. Recipient Japan Metal Sci. Soc. prize. Mem. Japan Sci. Council. Author: Magnetism and Magnet; On Strong Magnetic Substances. Research on magnetism of semiconductors.

MIYAJI, Masashi, Japanese astronomer; b. Hiroshima Prefecture, Oct. 7, 1902; grad. Tokyo U., 1925. Asst. engr. Tokyo Astron. Obs., now tech. ofcl.; former dir. astron. obs. in Java; now prof. Tokyo U. Recipient Asahi Cultural prize for thesis on study of change of longitude, 1949. Research and numerous publs. on measurement of reception error during time of radio reporting, change in time of transmission of short-wave radio signals.

MIYAKE, Shizuo, Japanese physicist; b. Okayama Prefecture, 1911; grad. Tokyo U., 1933; D.Sc. Asst., Tokyo U.; researcher Phys. and Chem. Research Inst.; prof. Tokyo U. Tech.; lectr. Tokyo U. Mem. Japan Phys. Soc. Author: X-Rays; Rochelle Salt; The Diffraction of X-Rays. Research on x-ray crystal structure by x-ray and electron diffraction, cathode ray reflection from crystal surface, inner electric potential of sphalerite, iron corrosion by electron diffraction.

MIYAKE, Yasuo, Japanese geochemist; b. Okayama, Japan, Apr. 17, 1908; s. Ishiro and Kazu (Miyake) M.; Dr.Sci., Sch. Chemistry, Tokyo (Japan) Imperial U., 1931; m. Suzu Nakano, Nov. 5, 1931; 1 dau., Haruna. Research asst. Hokkaido U., 1931-35; chief chemist Meteorol. Obs., 1935-41; dir. geochem. lab. Meteorol. Research Inst., Tokyo, Japan, 1941——; prof. Tokyo Kyoiku U., 1957——; vis. prof. U. Cal. Scripps Inst. Oceanography, La Jolla, 1956, Cal. Inst. Tech., 1956. Mem. Geochem. Soc. Japan (pres. 1966——). Author: Elements of Geochemistry, 1965; also numerous articles. Chem. research Western N. Pacific Ocean, meteoric precipitation especially on radioactive elements. Home: 73 Shimmeicho, Suginami, Tokyo. Office: Meteorol. Research Inst., 34 Kita-4, Suginami, Tokyo, Japan.*

MIYAMOTO, Shotaro, Japanese astronomer; b. Hiroshima Prefecture, 1912; grad. Kyoto U., 1936; Dr. Sci.; Asst. prof. Kyoto U., later apptd. prof. Author: On Equation and Method of Calculation; Observation and Study of the Stars. Authority on theory of corona of sun and ionosphere; developed Miyamoto theory which has world wide recognition.

MIYAMURA, Setumi, Japanese seismologist; b. Nagasaki, Japan, May 6, 1915; s. Rekizou and Sadako (Ohta) M.; D.Sc., U. Tokyo (Japan), 1960; m. Isoko Kondo, Oct. 17, 1944. Asst., Earthquake Research Inst., U. Tokyo, 1939-47, asst. prof., 1947-63, prof., 1963——. Mem. Am. Geophys. Union, Seismol. Soc. Am., Soc. Exploration Physicists (U. S. A.), Geodetic Soc., Seismol. Soc., Internat. Union Geodesy and Geophysics, Internat. Assn. Geodesy, Internat. Assn. Seismology and Physiology Earth's Interior, Volcanological Soc. (Japan). Author: Seismicity and Geotectonics, 1962; also numerous articles. Research on seismicity of Japan and of the world by instrumental observations and geotectonic study, devel. seismometry including radio telemetering and array sta. processing, recent crustal movement. Home: 2-51 Simo-uma, Setagaya, Tokyo, Japan.*

MIYATA, Akira, Japanese elec. engr.; b. Aichi, Japan, Oct. 22, 1900; s. Shigema and Tei (Miyata) M.; D.Engring., Tokyo (Japan) Imperial U., 1940; m. Etsuko Yamada, Feb. 11, 1929 (dec. 1949); children—Hajime, Fujiko (Mrs. Yoshimaru Suda), Ryozo, Yoshio Yukiko, Kimiko; m. 2d, Kiyo Obata, May 21, 1950. Research fellow Inst. Phys. and Chem. Research, Tokyo, 1924-60; dir. tech. research inst. Japan Steel & Tube Corp., 1960-63; prof. elec. engring. Kogakuin Tech. Coll., 1965—. Mem. Govt. Com. Atomic Fusion, 1958; cons. tech. adviser various corps., 1960—. Recipient Hattori Hokokai award, 1943; Inventors prize spl. grade, 1933, 56, Govt. Shiju Hosho, 1956; named Tokyo Met. Person of Merit in Invention, 1956. Mem. Japanese Assn. Surface Treatment of Metal (trustee). Author: Anodic Oxidation, 1954; also articles. Invented new process of anodic oxidation of aluminum, new type electrolytic capacitor, energy stored type spot welder. Home: 15-23 Wakabayashi 3 chome, Setagaya, Tokyo, Japan.*

MIZELL, Merle, Am. biologist; b. Chgo., Apr. 25, 1927; s. Harry and Eleanor (Ellman) M.; B.S., U. Ill., 1950, M.S., 1952, Ph.D., 1957; m. Lorraine Busse, Oct. 26, 1958; children—L. Tracy, Michael Jon. Asst., U. Ill., 1955-57; faculty dept. biology Tulane U., New Orleans, 1957—, asso. prof., 1964—. Mem. A.A.A.S., Am. Soc. Zoologists, Am. Inst. Biol. Scis., Am. Assn. for Cancer Research, Am. Assn. U. Profs., Sigma Xi. Research, publs. on exptl. embryology, cellular and tumor differentiation, effects of regenerating appendages on Lucke tumor, induction of regeneration in vertebrates, viral theory of carcinogenesis, infectious nucleic acids as agts. of normal and abnormal differentiation. Home: 644 Fielding Av., Gretna, La. 70053. Office: Biology Dept., Tulane University, New Orleans 70118.*

MIZRAHI, Elijah ben Abraham (Re'em), see Misrachi, Elia.

MIZUHARA, Shunzi, Japanese biochemist; b. Okayama, Japan, Aug. 24, 1915; s. Jinjiro and Isono (Matsushima) M.; student Okayama U. Med. Sch., 1937-41, Dr.Med. Sci., 1949; m. Shigeko Negi, Nov. 12, 1941; children—Mari Mizuhara, Kazuyuki. Prof. biochemistry Okayama U. Med. Sch., 1954—. Mem. Japanese Biochem. Soc., Am. Chem. Soc. Author: Seikagaku, 1965; Shin Seikagaku, 1966; also numerous articles. Research on mechanism of thiamine-catalyzed reactions; discovered several new cystein conjugates.*

MIZUSHIMA, San-ichiro, Japanese chemist; b. Tokyo, Japan, Mar. 21, 1899; grad. Tokyo U., 1923; D.Sc., 1930. Asst. prof., Tokyo U., 1923, prof., 1938, former dir. Radiation Research Inst.; chief Steel Research Inst., Yawata Iron & Steel Co. Mem. Japan Sci. Council, 1940. Recipient award for research on bi-polar element and molecular structure Imperial Acad., 1938, Cultural medal Imperial Palace, Tokyo, 1961. Mem. Japan Acad., Supreme Sci. Acad. (Spain) (hon.), Royal Physics and Chemistry Acad. (hon.). Author: Various Problems of Molecular Structure; Spectroscopic Chemistry. Research on abnormal dispersion and absorption of electric waves, molecular structure and their application.

MIZUSHIMA, Usaburo, Japanese cytogeneticist; b. Tokyo, Japan, Feb. 15, 1903; s. Sanemon and Kiku (Yamamoto) M.; Sc.D., U. Tokyo, 1950; m. Yoshiko Amino, Dec. 15, 1937; children—Keiko (Mrs. Roger Keyes), Yasuhiko. Subengr., Nat. Agrl. Expt. Sta., Konosu, Japan, 1931-35, engr., 1935-41; asst. prof. Inst. Agrl. Research Tohoku U., Sendai, Japan, 1941-50, prof. genetics, plant breeding, 1950-66, prof. emeritus, 1966—, councilor, 1962—. Recipient Japanese Soc. Agr. Ando prize. Mem. Breeding Soc. Japan (v.p. 1962-66), Genetics Soc. Japan. Author: Kariogenetic Study on Brassiceae, 1950; Introduction to Statistical Analysis of Experimental Data, 1952; also numerous articles. Reported karyogenetical relationships among species of Brassica and allied genera in tribe Brassiceae of Cruciferae and gave suggestions for their practical breeding; classified world's cultivated varieties into several groups on basis of hybrid sterility; gave guiding principles for practical rice breeding by means of hybridization between varieties of remote origin. Home: 30-237 Nagamachi-Otoya, Sendai. Office: Faculty Agr., Tohoku U., 210 Kita-6-Bancho, Sendai, Japan.*

MIZUSHINA, Tokuro, Japanese chem. engr.; b. Hokkaido, Japan, Jan. 2, 1920; s. Kichiro and Ikuyo (Suzuki) M.; student Kyoto (Japan) U., 1940-42, D.Engring., 1953; m. Miwako Kamei, Apr. 21, 1945; children—Masako, Motofumi. Faculty, Kyoto U., 1944—, prof. dept. chem. engring., 1956—. Mem. Soc. Chem. Engrs. Japan. Research, publs. on heat transfer to high viscous liquids, non-Newtonian fluids and liquid metals, simultaneous heat and mass transfer. Home: 24, Kowakicho, Matsugasaki, Sakyoku, Kyoto, Japan.*

M'KENDRICK, John Gray, Brit. physiologist; b. Aberdeen, Scotland, 1841; s. James M'K.; M.D., Ch.M., Aberdeen U.; studied Edinburgh; hon. LL.D. Aberdeen and Glasgow; m. Mary Souttar, 1867; 2 sons, 2 daus. Asst. prof., lectr., physiology, Edinburgh; Fullerian Prof., physiology, Royal Institute; prof., physiology, Glasgow U., 1876-1906; emeritus prof.; examiner physiology, U. London, Victoria U., U. Birmingham, U. Oxford, Cambridge U., Aberdeen U., Durham U., 1916. Fellow Royal Soc., 1884 (councillor); Fellow Royal Soc., Edinburgh (councillor); Fellow Royal College Physicians, Edinburgh; pres., Physiology Section, Brit. Assn. Author: Animal Physiology, 1876; Lectures on the History of Physiology, 1879; A Text-Book of Physiology, 1888; Life in Motion, or Muscle Nerve, 1892; Physiology, 1896; Life of Helmholtz, 1899; Boyle Lecture on Hearing, 1899; Science and Faith, 1899; Christianity and the Sick, 1901; various papers. Studies in physiological acoustics and experimental phonetics. Died Jan. 2, 1926.

MKRTCHYAN, Sergey Sedrakovich, Russian geologist; b. 1911; grad. Azerbaijan Oil Inst., 1932. Mem. geol. survey parties, 1932-47; instr. Yerevan U., 1935-55, prof., 1955—; chief engr. Armenian Geol. Bd., 1943-47; asso. Inst. Geol. Sci., Armenian Acad. Scis., 1947-50, dir., 1950—, acad. sec., 1956—. Recipient Stalin prize, 1950. Mem. Armenian Acad. Scis. Research and publs. on pattern of distbn. of copper molybdenum and copper pyrite deposits in Caucasus Minor. Address: Yerevan University, ulitsa Abovyana 104, Yerevan, Armenian SSR, USSR.

M'LACHLAN, Robert, English entomologist; b. London, Apr. 10, 1837; s. Hugh M'Lachlan; ed. Ilford, Eng. Editor Entomologist's Monthly Mag., from 1864. Fellow Royal Soc. (mem. council), Linnean Soc., Entomol. Soc.; mem. Zool. Soc., Royal Hort. Soc.; pres., sec., treas. Entomol. Soc. London. Author: Catalogue of British Neuroptera, 1870; A Monographic Revision and Synopsis of the Trichoptera of the European Fauna, and Supplement, 1874-84; article on insects in 9th edit. Ency. Brit.; also more than 150 papers on Neuroptera and other entomol. subjects. Formed important collection of Neuroptera; named many Am. insects. Died May 23, 1904.

M'LENNAN, Sir John Cunningham, Canadian physicist; b. Ont., Can., Apr. 14, 1967; s. David and Barbara (Cunningham) M'L.; ed. Toronto and Cambridge; B.A., Ph.D., D.Sc., univs. Manchester and Liverpool; LL.D., univs. Toronto and McGill; m. Elsie Monro Ramsay. Mem. faculty U. Toronto, from 1892, prof., 1907-32, dean Sch. Grad. Studies and Research, 1930-32, prof. emeritus and vis. prof. physics from 1932. Lectr. to Dominion Govt. Can. on metric system, 1906; sci. adviser Brit. Admiralty, 1919; mem. Nat. Research Council Can., 1916-31; mem. internat. com. Bur. Weights and Measures, Sevres, from 1929, mem. administrv. com., from 1933. Fellow Royal Soc., 1915 (Royal medallist, Bakerian lectr. 1928, v.p., mem. council 1933-34), Royal Soc. Can. (pres. 1924, Flavelle medallist 1926), A.A.A.S.; mem. Royal Cana-Inst. (pres. 1916-17), Brit. Assn. Advancement Sci. (pres. sect. A 1923). Research and numerous papers on radioactivity, elec. conduction in gasas, prodn. and liquefaction of helium, electric conductivity of metals at lowest temperatures, also spectroscopy. Died Jan. 9, 1935.

M'MURRICH, James Playfair, anatomist; b. Oct. 16, 1859; s. John and Janet (Dickson) M'M.; B.A., U. Toronto, 1879, M.A., 1881; Ph.D., Johns Hopkins, 1885; m. Katie Moodie (Vickers), 1882; 1 dau., 1 son. Prof. biology Haverford (Pa.) Coll., 1886-89; asst. prof. animal morphology Clark U., Worcester, Mass., 1889-92; prof. biology U. Cin., 1892-94; prof. anatomy U. Mich., 1894-1907; prof. anatomy U. Toronto (Ont., Can.), 1907-30, dean sch. grad. studies, 1922-30. Chmn. Biol. Bd. Can., 1926-34. Fellow Royal Micros. Soc., Royal Soc. Can. (pres. 1922); corr. mem. Zool. Soc.; pres. A.A.A.S., 1922. Author: Text-book of Invertebrate Morphology, 1894; The Development of the Human Body, 1902; Leonardo da Vinci, Anatomist, 1930; also papers on morphology of Actinaria, vertebrate morphology, embryology of Crustacea and mollusks. Editor: Atlas and Text-book of Human Anatomy (Sobotta), 1906-07. Co-editor: Human Anatomy (Morris), 4th edit., 1906. Died Feb. 9, 1939.

MNDZHOYAN, Armenak, chemist; b. Sarkamish, Turkey, Nov. 23, 1904; s. Levon and Arigah (Bajatjan) M.; student Inst. Chemistry and Pharmaceutics, Moscow, USSR, 1928, Yerevan Med. Inst., 1933; D.Chemistry, Ministry High Edn., Armenia, 1942; m. Varduhi Afrikjan, Sept. 20, 1935; children—Elize (Mrs. Hovhannes Agababian). Head, Chem.-Pharm. Research Lab., Ministry Health of Armenian S.S.R., 1924-38; head chair of organic chemistry Yerevan Med. Inst., 1931-46; head chair organic chemistry Yerevan State U., 1938-50; dir. Inst. Fine Organic Chemistry Armenian Acad. Scis., 1945—, academician sec. chemistry dept., 1963—. Decorated Order Lenin, Order Red Banner from Presidum Supreme Soviet, USSR, 1958. Mem. Armenian Acad. Scis. (v.p. 1953-60), Mendeleyev Chem. Soc., socs. of pharmacologists, biochemists and physiologists. Author: (with others) Syntheses of Heterocyclic Compounds, vol. I, 1956, vol. II, 1957, vol. III, 1958, vol. IV, 1958, vol. V, 1960, vol. VI, 1964, vol. VII, 1966; also numerous articles. Patentee on methods of preparation chem. substances and new drugs including gangleron, quateronum, arpenalum, ditilium, subecholinum, fubromeganum, mesphenalum. Home: 22a Barekamutjan St. Office: 26 Tbilissi Av., Yerevan, Armenian SSR, USSR.*

MöBIUS, August Ferdinand, German mathematician, astronomer; b. Schulpforta, Germany, Nov. 17, 1790; ed. univs. Leipzig, Göttingen (student of K. F. Gauss), Halle (all Germany). Faculty U. Leipzig, 1815-68, became extraordinary prof., 1816, dir. obs., ordinary prof. higher math., astronomy, 1844-68. Author: De computandis occulation ibus fixarum per planetas, 1815; Der Barycentrische Calcul (new treatment of analytical geometry by introduction of homogeneous co-ordinates and principle of duality), 1827; Die Hauptsätze der Astronomie, 1836; Lehrbuch der Statistik (2 vols.), 1837; Die Elemente der Mechanik des Himmels, 1843; also numerous articles in Crelle's Jour., 1828-58. A founder of topology (developed paradoxical math. figure known as Möbius Strip); leader in introducing modern projective geometry methods. Died Leipzig, Sept. 26, 1868.

MöBIUS, Karl August, German zoologist; b. Eilenburg, Germany, Feb. 7, 1825; prof., Kiel, Germany; dir. Mus. Natural History, Berlin; discovered symbiosis in marine animals; promoted artificial cultivation of pearls. Died Berlin, Apr. 26, 1908.

MöBIUS, Paul Julius, German psychiatrist; b. Leipzig, Germany, Jan. 24, 1854; student theology and philosophy; Ph.D.; M.D., 1877. Practiced psychiatry in Leipzig; also lectr. Author: Uberdenphysiologischen Schwachsinn des Weibes, 1900. Known for his pathographies of gt. men as Rousseau, Nietzsche, Schopenhauer, Goethe; also attracted attention with his theory that woman is by nature a feeble minded creature; described a form of periodic migraine associated with paralysis of extraocular muscle of eye (known as Möbius' disease or periodic ophthalmoplegic migraine), 1884. Died Jan. 8, 1907.

MOBSY, Hakon, Norwegian oceanographer; b. Kristiandsand S, Norway, July 10, 1903; s. Salve and Mette (Nodeland) M.; cand.real., U. Oslo, 1930, Dr.philos., 1934; m. Alfhild Mowinckel, Mar. 1, 1930; children—Mette Marie (Mrs. Peter Haugan), Kaya (Mrs. Jacob Irgens) M.; Asst. to Fridtjof Nansen, 1923-27; asst. prof. Geophys. Inst., Bergen, Norway, 1927—, prof., 1947—; dean Faculty Scis., U. Bergen, 1954-59, rector, 1966—. Pres., subcom. on oceanographical research NATO, 1960-65, mem. sci. com., NATO, 1965—. Mem. Internat. Assn. Phys. Oceanography (past pres.), acads. scis. Oslo, Bergen, Gothenburg, Helsingfors. Research, numerous publs. on atmospheric radiation, evaporation from oceans, heat and salt balance in sea, mixing of water masses in sea, phys. processes in sea, formation of bottom water, ocean currents and their determination, oceanographic instrumentation. Home: 59 Kalfarveien, Bergen, Norway. Office: Geofysisk Inst., 70 Allegahen, 5000 Bergen, Norway.*

MOCK, Harry Edgar, Am. surgeon; b. Muncie, Ind., Oct. 27, 1880; s. John D. and Mary Minerva (Jackson) M.; student Franklin (Ind.) Coll., 1900-02, D.Sc., 1926; B.S., U. Chgo., 1904; M.D., Rush Med. Coll., 1906; m. Golda M. Taylor, Dec. 25, 1908; children—Harry Edgar, William Byford Taylor, Marjorie (Mrs. Wayne L. Gregory), Charles Jackson, John Edward. Intern Cook County Hosp., 1908; practiced medicine, Chgo., 1906—; attending surgeon St. Luke's Hosp.; asso. prof. surgery Northwestern Med. Sch.; cons. surgeon B.&O. R.R., Rock Island R.R., I.C. R.R. Fellow A.C.S., Chgo. Inst. Medicine, Chgo. Surg. Soc.; mem. Am. Assn. Indsl. Physicians and Surgeons (hon.), Am. Bd. Surgery, Am. Assn. Surgery of Trauma, Am. R.R. Surgeons Assn., A.M.A., Ill. State, Chgo. med. socs., Ind. Soc., Am. Acad. Orthopedic Surgery. Author: Industrial Medicine and Surgery, 1919; Skull Fractures and Brain Injuries; also many articles on surgery. Editor: Principles and Practices of Physical Therapy, 1932. Died 1959.

MOCK, John Edwin, Am. nuclear engr.; b. Altoona, Pa., Sept. 29, 1925; s. Daniel Raymond and Sarah (Lorenz) M.; student Pa. State U., 1942-44; B.S., U. S. Mil. Acad., 1947; B.S. in Chem. Engring., M.S., Purdue U., 1950, Ph.D. in Nuclear Engring., 1960; M.S. in Chemistry, Ohio State U., 1953; M.S. in Internat. Affairs, George Washington U., 1966; m. Jeannette Daly, Oct. 25, 1947; children—Donna Jean, Susan Jean. Commd. 2d lt. USAF, 1947, advanced through grades to lt. col., 1964; project engr. Air Research and Devel. Command, Wright Field, 1950-54; sci. adviser Hdqrs., USAF Europe, Wiesbaden, Germany, 1955-58; chmn. ionization panel Def. Atomic Support Agy., Washington, 1960-64; asso. prof. math. USAF Acad., Colorado Springs, Colo., 1964-65; sci. adminstr. Advanced Projects Research Agy., Washington, 1966—; asso. prof. math. George Washington U., Washington, 1961-64. Sci. adviser to govt. agys., pvt. cos. Mem. Am. Nuclear Soc. (Mark Mills award 1960), Am. Phys. Soc., Am. Geophys. Union, A.A.A.S., N.Y. Acad. Scis., Sigma Xi. Research, publs. on propulsion systems, combustion processes, control systems, corrosion control, high altitude nuclear effects, atmospheric re-entry physics, atomic and molecular reactions rates, space physics. Home: 1121 20th Av., Altoona, Pa. 16601. Office: Advanced Research Projects Agy., The Pentagon, Washington 20301.*

MOCKLE, J(erry) Auguste, Canadian pharmacist, univ. ofcl.; b. Doucet, Que., Can., Sept. 16, 1926; s. Joseph George Louis and Alice (Delisle) M.; Pharm. B., U. Montreal, 1952; Pharm. D., U. Paris (France), 1955; m. Marthe Trudeau, Aug. 18, 1951; 1 son, Daniel. Faculty, U. Montreal (Que.), 1955—, prof., 1966—, vice dean faculty pharmacy, 1965—. Recipient Charles LeRoy award Academie de Pharmacie de Paris, 1955; Gold medal awards College des Pharmaciens de la Province de Quebec, 1952, Horner Pharm. Labs., 1952. Mem. Am. Soc. Pharmacognosy, Soc. Econ. Botany, Am. Chem. Soc., A.A.A.S., Internat. Soc. Chemotherapy, Federation Internationale Pharmaceutique, Société de Pharmacologie du Canada. Author: Contribution a l'etude des plantes medicinales du Canada, 1955; also articles. Home: 3975 St. Zotique. Office: 2900 Mt.-Royal, Montreal, Que., Can.*

MODEST, Edward Julian, Am. chemist; b. Boston, Sept. 9, 1923; s. Herman J. and Leah (Rosenwald) M.; A.B. cum laude, Harvard, 1943, A.M. in Chemistry, 1947, Ph.D., 1949; m. Bernice N. Goldsmith, June 19, 1947; children—Geoffrey Alan, Vicki Ellen, Andrew Paul. Research asst. in pathology Children's Hosp. Med. Center, Boston, 1949-52; research asso., 1952-—; head, labs. organic chemistry Children's Cancer Research Found., Boston, 1952-—; research asso. in pathology Harvard Med. Sch., Boston, 1956-—; adj. prof. pharm. chemistry U.R.I., Kingston, 1968-—. Cons. Nat. Cancer Inst., NIH, 1959-62, 65-—. Mem. Am. Chem. Soc., Am. Assn. Cancer Research, A.A.A.S., Assn. Harvard Chemists. Editorial bd. Year Book of Cancer, 1958-—, Jour. Medicinal Chemistry, 1966-—. Research, numerous publs. on cancer chemotherapy, medicinal chemistry, synthesis of nitrogen heterocyclic compounds, structural organic chemistry, biochem. pharmacology. Home: 122 Andrew St., Newton Highlands, Mass. 02161. Office: 35 Binney St., Boston 02115.*

MOEBIUS, Wilfried Friedrich Hugo Max, German physician; b. Lohmen, Germany, Oct. 30, 1914; s. Max Friedrich and Elisabeth (Andrae) M.; ed. U. Leipzig, U. Vienna; M.D., 1940; m. Gertraude Antonie Maehler, Oct. 13, 1940; 1 son, Jorg Steffen. Faculty, U. Leipzig, 1940-56; prof. Med. Acad., Magdeburg, 1956-58, Med. Acad., Erfurt, 1958-59; prof. Friedrich Schiller U., Jena, also dir. Women's Hosp., 1959-—. Recipient Hufeland medal, 1964. Mem. Deutsche Gesellschaft für Klinische Medizin, Deutsche Gesellschaft für Gynakologie, Deutsche Rontgengesellschaft, Internat. Fertility Assn. Author: Beitrag zur Radiumbehandlung in der Gynakologie, 1951; Geburtshilfliche Rontgendiagnostik 1957; (with Harald Hilmerth) Der hohe Gradstand, 1957; Die Geburt, 1961; Geburtshilfliche Strahlen-und Roentgen-Diagnostik, 1967; also articles. Research on therapy of women's diseases, obstetrics and gynecology; application of X-ray diagnostics in gynecology. Home: 18 Bachstrasse, Jena, Germany.*

MOELLER, Dade William, Am. environmental engr.; b. Grant, Fla., Feb. 27, 1927; s. Robert A. and Victoria (Bolton) M.; B.C.E., Ga. Inst. Tech., 1947, M.S. in Civil Engrng., 1948; Ph.D., N.C. State U., 1957; m. Betty Jean Radford, Oct. 7, 1949; children—Garland Radford, Mark Bolton, William Kehne, Matthew Palmer, Elisabeth Anne. Commd. jr. asst. san. engr. USPHS, 1948, advanced through grades to san. engr. dir., 1961; research engr. Los Alamos Sci. Lab., 1949-52; staff asst. Radiol. Health Program, Washington, 1952-54; research asso. Oak Ridge Nat. Lab., 1956-57; chief radiol. health tng. Taft San. Engring. Center, Cin., 1957-61; officer in charge Northeastern Radiol. Health Lab., Winchester, Mass., 1961-66; ret., 1966; asso. dir. Kresge Center for Environmental Health, Harvard Sch. Pub. Health, Boston, 1966-—. Cons. on radiol. health Prof. Examination Service, Am. Pub. Health Assn., 1960-62, WHO, 1965-—; mem. Nat. Council Radiation Protection and Measurements. Diplomate Am. Bd. Health Physics (chmn. 1967-—), Am. Environmental Engring. Intersoc. Bd. Fellow Am. Pub. Health Assn.; mem. Health Physics Soc., Am. Nuclear Soc., A.A.A.S. Research, publs. on treatment and handling of radioactive wastes with particular emphasis on reactor cooling water systems; establishment of lab. systems for large scale analyses of radionuclides in all types of environmental media. Home: 27 Wildwood Dr., Bedford, Mass. 01730. Office: 665 Huntington Av., Boston 02115.*

MOELLER, Hugo Charles, Am. physician; b. Schleswig, Ia., July 9, 1923; s. John D. and Freida (Jensen) M.; S.B., U. Chgo., 1945, M.D., 1948, Ph.D. (USPHS fellow), 1951; B.A., Drake U., 1946; m. Tia Lobell, July 4, 1962. Faculty, U. Colo. Sch. Medicine, 1953-55, Northwestern U. Med. Sch., 1955-56; faculty U. Cal. Med. Center, San Francisco, 1956-—, asst. clin. prof. medicine, 1963-—. Diplomate Am. Bd. Internal Medicine, Am. Bd. Gastroenterology; Fellow A.C.P.; mem. A.M.A., A.A.A.S., Am. Chem. Soc., Am. Gastroscopic Soc., Am. Fedn. Clin. Research, Am. Gastroent. Assn., Am. Geriatric Soc., Am. Therapeutic Soc., Am. Soc. Pharmacology and Exptl. Therapeutics, Soc. Exptl. Biology and Medicine, N.Y. Acad. Scis., Western Pharmacology Soc., Western Gastroenterology Research Club, Western Soc. Clin. Research, Instituto National de Gastroenterologiz (hon. mem. 5 Argentine congress 1959), Sociedad Argentian de Gastroenterologiz, Sociedad Argentin de Hidrologia y Climatologia Medica, Sigma Xi, Phi Beta Kappa, others. Research, publs. on pathogenesis of ulcerative colitis; enzyme studies, particularly lysozyme and cholinesterase. Home: 1795 14th Av., San Francisco 94122.*

MOELLER, Therald, Am. chemist, educator; b. North Bend, Ore., Apr. 3, 1913; s. Edward and Kate (Wickham) M.; B.S., Ore. State Coll., 1934; Ph.D., U. Wis., 1938; m. Ellyn Clara Elizabeth Stephenson, May 17, 1935; children—Karlyn Ann (Mrs. Charles E. Coverdale), Kathryn Jean, Stephen Therald. Faculty, Mich. State Coll., East Lansing, 1938-40; faculty U. Ill., Urbana, 1940-—, prof. inorganic chemistry, 1953-—. Cons. Argonne Nat. Lab., various indsl. concerns; lectr. Mem. Am. Chem. Soc. (chmn. div. inorganic chemistry 1961), Chem. Soc. Author: Inorganic Chemistry, 1952; Qualitative Analysis, 1958; The Chemistry of the Lanthanides, 1963; (with J. C. Bailar, J. Kleinberg) University Chemistry, 1965; (with D. F. Martin) Laboratory Chemistry, 1965. Editor: vol. 5 Inorganic Syntheses, 1957. Research in syntheses and evaluation of compounds, particularly lanthanides and gallium family elements and of nonmetals; structure and bonding in lanthanide complexes; isomer and substitution patterns in cyclic phosphorus and sulfur-nitrogen compounds; inorganic polymers. Home: 303 Flora Dr., Champaign, Ill. 61820. Office: Noyes Lab., U. Ill., Urbana, Ill. 61801.*

MOERSCH, H(erman) J(ohn), Am. physician; b. St. Paul, Mar. 14, 1895; s. Julius and Cecilia (Kurffier) M.; B.S., U. Minn., 1918, M.D., 1921, M.S., 1925; m. Charline Buck, Sept. 26, 1925; children—Richard, Donald. 1st asst. in medicine, Mayo Clinic, 1924-26, mem. staff dept. medicine, 1926-—, head sect. med. and personal endoscopy 1938-—. Instr. medicine Mayo Found., U. Minn., 1925-28, asst. prof., 1928-33, asso. prof., 1933-47, prof. 1947-—; chmn. sect. medicine Mayo Clinic 1948-—. Recipient Great Cross Merit (W. German Republic), Rudolph Schindler Award, Chevalier Jackson award; gold medal, Am. Coll. Chest Physicians 1965. Mem. A.M.A., Am. Assn. Thoracic Surgeons, Am. Broncho-Esophagol. Soc., Am. Gastroscopic Soc., A.C.P., Am. Coll. Chest Physicians (dir. edn. and research), Central Soc. Clin. Research, Am. Gastroent. Assn., Sigma Xi, Alpha Kappa Kappa, Alpha Omega Alpha. Contbr. articles to med. jours. on diseases of esophagus and chest. Home: 1064 Plummer Lane, Rochester, Minn.

MOERTEL, Charles G., Am. physician; b. Milw., Oct. 17, 1927; s. Charles H. and Alma (Soffel) M.; student U. Ill., 1945-46, Northwestern U., 1948-49; B.S., M.S., U. Ill.; M.S., U. Minn., 1957; m. Virginia Claire Sheridan, Mar. 20, 1953; children—Charles Stephen, Christopher, Henther, David. Fellow Mayo Grad Sch. Medicine, 1954-57; asst. to staff Mayo Clinic, 1957-58, cons. in internal medicine, 1958-—; faculty U. Minn. Grad. Sch., Rochester, Minn., 1959-—. Diplomate Am. Bd. Internal Medicine; Mem. A.C.P., Am. Assn. for Cancer Research, Am. Soc. for Clin. Oncology, A.M.A. Author: Multiple Primary Malignant Neoplasms, 1967; also articles. Research in clin. gastroenterology especially gastrointestinal cancer, controlled observations of treatment of advanced gastrointestinal cancer. Home: 48 Skyline Dr. Office: Mayo Clinic, Rochester, Minn.*

MOESSBAUER, Rudolf Ludwig, see Mössbauer, Rudolf Ludwig.

MOFFETT, Robert Bruce, Am. chemist; b. Madison, Ind., June 8, 1914; s. Robert Bowman and Ellen (Elliott) M.; A.B., Hanover Coll., 1937, Sc.D., 1959; M.A., U. Ill., 1939, Ph.D., 1941; m. Martha Jane Powell, Aug. 30, 1941; children—Ellen Katherine (Mrs. Brian William Hamoton), Robert Bruce. Grad. asst. and fellow U. Ill., Urbana, 1937-41; research asso. Northwestern U., Evanston, Ill., 1941-43; research chemist, asst. dir. research George A. Breon and Co., Kansas City, Mo., 1943-47; research chemist Upjohn Co., Kalamazoo, 1947-—. Mem. Am. Chem. Soc., A.A.A.S., Sigma Xi, Phi Lambda Upsilon, Phi Kappa Phi. Research, numerous publs. on organic and medicinal chemistry, structure of Benzopyrylium salt, steroids, anticholinergics, antiviral agents, central nervous system depressants and stimulants, antiulcer agents, heterocyclic compounds. Home: 2895 Bronson Blvd. Office: Upjohn Co., Unit 7150-25-6, Kalamazoo 49001.*

MOGABGAB, William Joseph, Am. physician; b. Durant, Okla., Nov. 2, 1921; s. Anees and Maude (Jopes) M.; B.S., Tulane U., 1942, M.D. 1944; m. Joy Roddy, Dec. 24, 1948; children—Robert W., Ann, Kay, Edward R., Jean. With Tulane U., 1948-49, asso. prof. medicine, 1956-62, prof., 1962-—; vis. physician Charity Hosp. of La., New Orleans, 1949-51, sr. vis. physician, 1961; vis. investigator, asst. physician Hosp. Rockefeller Inst. Med. Research, N.Y.C., 1951-52; fellow Nat. Found. for Infantile Parlaysis, 1951-52; chief infectious disease VA Hosp., Houston, 1952-53; asst. prof. medicine Baylor U. Coll. Medicine, 1952-53; head virology div. U. S. N., Great Lakes, Ill. 1953-55; cons. infectious disease VA Hosp., New Orleans, 1956. Asso. mem. Commn. on Influenza, Armed Forces Epidemiol. Bd. Diplomate Am. Bd. Internal Medicine, Am. Bd. Microbiology. Fellow A.C.P., Am. Acad. Microbiology; mem. Soc. Exptl. Biology and Medicine, So., Central socs. clin. research, Am. Fedn. Clin Research, Am. Soc. Cell Biology, Am. Soc. for Microbiology, Am. Soc. Clin. Investigation, Infectious Disease Soc. Am. Research, publs. on agts. of and vaccines for respiratory infections, various viruses. Home: 39 Lark St., New Orleans 70124. Office: 1430 Tulane Av., New Orleans 70112.*

MOGILEVCHIK, Zakhar Kuzmich, Russian hygienist; b. 1895; grad. Med. Faculty, 1st Moscow U., 1923; D.Med. Sci. Asst., lectr. dept. hygiene Minsk Med. Inst., 1924-30, head chair gen. hygiene, 1931-—. Mem. USSR Acad. Med. Scis. (corr.), Belorussian Soc. Hygienists (chmn.). Editor: Pub. Health in Belorussia; co-editor Hygiene sect. Large Med. Ency., 2d edit.; mem. editorial council Hygiene and Sanitation. Research and publs. on hygiene of housing, water supply, soil, pub. utilities, housing and communal hygiene, contamination processes in soils. Address: Minsk Med. Inst., Leninsky prospect 7, Minsk, Belorussian SSR, USSR.

MOHL, Hugo von, German botanist; b. Stuttgart, Württemberg, Apr. 8, 1805; s. Benjamin Ferdinand M.; ed. U. Tübingen, U. Munich. Prof., physiology, Berne, Switzerland; prof. botany, U. Tübingen, 1832-72. Founded weekly, Botanische Zeitung, 1843. Fellow Royal Soc., 1868; corr. mem., French Acad. Scis., 1838. Author: Beiträge zur Anatomie und Physiologie der Gewächse, 1834; Mikrographie oder Anleitung zur Kenntnis und zum Gebrauch des Mikroskops, 1846; Grundzüge der Anatomie und Physiologie der vegetabilischen Zelle, 1851; Vermischte Schriften botanischen Inhalts, 1845. Research on anatomy and physiology of plant cells; discovered and named protoplasm, 1846; founder of plant cell theory. Died Tübingen, Germany, Apr. 1, 1872.

MOHLENBROCK, Robert H., Am. botanist; b. Murphysboro, Ill., Sept. 26, 1931; s. Robert Herman and Elsie (Treece) M.; B.A., So. Ill. U., 1953, M.S., 1954; Ph.D., Washington U., St. Louis, 1957; m. Beverly Ann Kling, Oct. 19, 1957; children—Mark William, Wendy Ann, Trent Alan. Asst. prof. botany So. Ill. U., Carbondale, 1957-60, asso. prof., 1960-66, prof., chmn. dept., 1966-—. Mem. Am. Soc. Plant Taxonomists, Internat. Assn. Plant Taxonomists, Am. Fern Soc., Assn. for Tropical Biology, New Eng. Bot. Club, Ill. Acad. Sci., Sigma Xi. Author: (with J. W. Voigt) A Flora of Southern Illinois, 1959, Plant Communities of Southern Illinois, 1963; Ferns of Illinois, 1966. Taxonomic research in Ill. plants and tropical legumes. Home: 629 Surrey Lane, Carbondale, Ill. 62901.*

MOHLER, Fred Loomis, Am. physicist; b. Wilbraham, Mass., Aug. 23, 1893; s. John Fred and Sarah (Loomis) M.; B.A., Dickinson Coll., 1914, Sc.D., 1947; Ph.D., Johns Hopkins, 1917; m. Pearl Worthington, Aug. 17, 1920 (dec. 1954); children—Wilmer Worthington, Aug. 17, 1920 (dec. 1954); children—Wilmer Worthington, Emily (Mrs. Dewey M. Stowers). With Nat. Bur. Standards, Washington, 1917-43, 45-60, researcher atomic physics, chief, atomic physics sect., 1917-43, chief, Mass. spectrometry sect., 1946-60; operations analyst 9th Bomber Command, Europe, 1943; cons. Manhattan dist., Oak Ridge, Tenn., 1945. Recipient Medal of Freedom, U. S. Army, 1945. Mem. Phi Beta Kappa, Sigma Xi, Kappa Sigma, Gamma Alpha. Research in atomic physics; electric discharges in gases; mass spectrometry. Home: 2853 Brandywine St., Washington 20008.*

MOHLER, Hermann, chemist; b. Liestal, Mar. 2, 1900; s. Hermann and Mina (Schaub) M.; ed. U. Basel (Switzerland), U. Berne (Switzerland); Ph.D.; m. Trude Acker. Food controller for Basel region; asst., also chemist; dep. of canton; head Zurich Chem. Labs.; instr. phys. chemistry U. Basel; prof.; tech. asst. UNESCO; prof. Coll. Arts and Scis. Baghdad; prof. U. Basel; dir. research Knorr Food Products, 1954-—. Mem. Swiss Assn. Chemistry (pres.), German Chem. Assn. (corr.). Author: Lösungsspektren, Beziehungen der Chemie zum neuen Weltbild der Physik; Das Absorptionsspektrum der chemischen Bindung; Elektrentheorie der Chemie; Chemische Optik; also numerous articles. Research in physical and food chemistry, food tech., food irradiation, tech. of water and sewage, chem. engring., famine. Address: Oskar Biderstrasse 10, Zurich, Switzerland.

MOHLER, Orren Cuthbert Am. astronomer; b. Indpls., July 28, 1908; s. Charles Mikesell and Mary (Culp) M.; A.B., Mich. State Normal Coll., 1929; A.M., U. Mich., 1930, Ph.D., 1933; Sc.D., Eastern Mich. U., 1957; m. Helen Jenkins Beal, June 10, 1935; children—Alice Beal (Mrs. William George Delana), Jane Radcliffe (Mrs. Jeffery Wessels Barry). Observer, McMath-Hulbert Obs., U. Mich., Lake Angelus, 1933; astronomer, instr. Cook Obs., Swarthmore Coll., 1933-40; faculty U. Mich., Ann Arbor, 1940-—, prof. astronomy, 1962-—, chmn. dept. astronomy, dir. observatories. Fulbright Research scholar Institut d'Astrophysique, Liége, Belgique, 1960-61. Mem. Am. Astron. Soc., Royal Astron. Soc. Can., Phi Beta Kappa, Sigma Xi. Author: (with A. K. Pierce, R. R. McMath., L. Goldberg) Photometric Atlas of the Near Infrared Solar Spectrum, lambda-8465 to lambda-25242, 1950; A Table of Solar Spectrum Wave Lengths, 11984A to 25578A, 1955; also numerous articles. Observed fine detail in atmosphere of sun, photometry and measurement of wave lengths of lines in solar spectrum, prodn. motion pictures illustrating and rec. solar activity, motion picture records of solar, lunar, and planetary changes, emmission lines of ionized calcium and neutral helium in solar spectrum and their variation with position on solar disc. Home: 405 Awixa Rd., Ann Arbor, Mich. 48104.*

MOHLING, Franz, Am. physicist; b. Jersey City, July 22, 1930; s. E. Gunther and Florence (Robinson) M.; B.S., Rensselaer Poly. Inst., 1951; Ph.D., U. Wash., 1958; m. Judith Ann Holland, Sept. 28, 1962; children—Shanti Indira, Hans Torleif. Research asso. Columbia, 1958-60, Cornell U., 1960-61; with U. Colo., 1961—, asso. prof., 1965—. U. S. Govt. Fulbright Research scholar in India, 1963-64. Mem. Am. Inst. Physics, Sigma Xi. Research, publs. on devel. micros. theories of liquid heliums (both isotopes); quantum statistics. Home: 432 Marine St., Boulder, Colo. 80302.

MOHN, Henrik, Norwegian meteorologist; b. Bergen, Norway, May 15, 1835; ed. Christiana (now Oslo) U. Became observer at Christiana Astronomical Obs.; prof. there, from 1866; dir. Meteorological Inst. there (which he helped found), 1866-1913; head expdn. to northern coast of Norway, 1876-78; set up Bossekop meteorological station in Lapland, 1882-83. Author: Grundzüge der meteorologie, 1875. Editor, Yearbook of Norwegian Meteorological Inst., from 1867. Research on oceanography and meteorology of N. Atlantic regions, dynamics of atmosphere; developed hypothesis of arctic currents and drift. Died 1916.

MOHNIKE, Gerhard Ludwig Jakob, German physician; b. Wiesbaden, Germany, Jan. 6, 1918; s. Ludwig and Emmi (Maurer) M.; student univs. Marburg/Lahn, Jena; M.D. U. Greifswald, 1942; m. Annemarie Hoffmann, May 21, 1942; children—Dagmar, Wolfgang, Klaus. Faculty, U. Greifswald, 1951-66, prof., 1955-66, prof. with chair diabetology, endocrinology, metabolic diseases, 1964-66; dir. Inst. Diabetes, Karlsburg/Greifswald, 1961-66. Recipient Nat. prize, E. Germany. Mem. N.Y. Acad. Scis., European Soc. Diabetology, Problem Commn. Endocrinology and Metabolic Diseases, (chmn.), others. Author: (with Katsch, John) Aceton bis Zucker; (with Schmidt, Friedrich, Schicht) Kost für Zuckerkranke; also numerous articles in handbooks, med. dictionaries, other profl. lit. Research on diabetes, other metabolic diseases. Died Berlin, Mar. 8, 1966.

MOHOROVICIC, Andrija, Croatian geologist; b. Volosko, Istria, Jan. 23, 1857; in studies of wave patterns of Balkan earthquake of 1909, deduced that earth's structure is layered, that earthquake waves travel more quickly through more rigid outer layer, and that separation between 2 layers is sharp (Mohorovicic discontinuity). Died Zagreb, Yugolslavia, Dec. 18, 1936.

MOHR, Charles (Carl) Theodor, botanist; b. in Esslingen, Würtemberg, Germany, Dec. 28, 1824; s. Louis M.; ed. Paedagogium Esslingen, Volksschule Denkendorf; studied chemistry and natural scis. Poly. Sch., Stuttgart, 1842-43; Ph.D., U. Ala., 1890; m. Sophia Roemer, Mar. 12, 1852. Accompanied A. Kappler on exploring expdn. of Dutch Guiana, 1845; one of pioneers Cal. gold fields, 1849; pharmacist, Louisville, 1853-57, Mobile, Ala., 1857-92; explored forests of Gulf states for 10th census, 1880-81; ret. from business to engage in forestry and bot. work exclusively, 1892; botanist Geol. Survey Ala., 1884—; agt. Div. Forestry, U. S. Dept. Agr., 1889—. Author: The Timber Pines of the Southern United States, 1896, 1897; Plant Life of Alabama, 1901 (Contributions U. S. Nat. Herbarium, Vol. VI.) Contbr. The Forests of Alabama and Their Products, the Grasses and Other Forage Plants of Alabama in Handbook of Alabama. (Saffold Berney). Died Asheville, N.C., July 17, 1901.

MOHR, Courtney Balthazar Oppenheim, Australian physicist; b. Melbourne, Australia, Nov. 5, 1906; s. Walter Balthazar and Alma (West) M.; B.A., M.Sc., U. Melbourne, 1930; Ph.D., Cambridge (Eng.) U., 1933; m. Christina Mary Fear, Sept. 24, 1938; children—Walter Edwin, Frederick Courtney, Geoffrey Arnold. Lectr., U. Cape Town (S. Africa), 1936-46; faculty U. Melbourne, 1947—, prof. physics, 1961—. Fellow Inst. Physics, Phys. Soc., Australian Inst. Physics. Research, publs. on structure of atoms and nuclei, especially as revealed in collision processes. Home: 32 Walmer St., Melbourne E.4., Victoria, Australia.

MOHR, Georg, Danish mathematician; b. Copenhagen, Denmark, 1640. Author: Euclides danicus, 1672. Gave systematic treatment of geometric constructions using only compasses. Died 1697.

MOHR, Hans, German botanist; b. Altburg, Germany, May 11, 1930; s. Friedrich and Rosine (Strinz) M.; Ph.D., U. Tubingen (Germany); m. Elisabeth (Kraut, July 18, 1957. Postdoctoral fellow, U. S. A.; asst., later instr. U. Tubingen; prof. botany U. Freiburg in Brisgau, W. Germany. Mem. German Bot. Soc., Am. Soc. Plant Physiology, Bot. Soc. Am. Research and publs. on developmental physiology, especially control of devel. by light, especially related to control of gene function; molecular biology of devel. Home: Mattenstrasse 6, Kappel (Freiburg in Brisgau), W. Germany. Office: Schänzlestrasse 9, Freiburg in Brisgau, W. Germany.

MOHR, Jan Gunnar Faye, geneticist; b. Paris, France, Jan. 10, 1921; s. Hugo Lous and Elna Faye M.; grad., Med. School, U. Oslo, 1948; studied genetics, U. Stockholm. Dir., U. Institute of Medical Genetics, Oslo U., 1959; prof., Medical Genetics, Dir. Institute of Medical Genetics, Copenhagen U., 1964. Specialist in study of medical genetics. Office: 14 Tagensvej, Copenhagen, Denmark.*

MOHR, John Luther, Am. biologist, educator; b. Reading, Pa., Dec. 1, 1911; s. Luther Seth and Anna (Davis) M.; A.B., Bucknell, U., 1933; postgrad. Oberlin Coll.; Ph.D. in Zoology, U. Cal. at Berkeley, 1939; m. Frances Edith Christensen, Nov. 23, 1939; children—Jeremy John, Christopher Charles. Teaching asst. U. Cal. at Berkeley, 1934-38, technician, 1938-42; research asso. Pacific Islands Research Project, Stanford, 1942-44; research asso. Allan Hancock Found., U. So. Cal., Los Angeles, 1944-47, faculty, 1947—, prof. biology, 1957—, head dept., 1959-62; vis. asst. prof. invertebrate zoology Friday Harbor Labs., U. Wash., summer 1956, prof., summer 1957. Chief zool. party on research vessel Eltanin, Drake Passage, summer 1962; research asso. invertebrate zoology Los Angeles County Mus., 1964—; mem. panel on invertebrates Smithsonian Oceanographic Sorting Center, Washington, 1963—. Guggenheim fellow, 1957-58. Fellow A.A.A.S., Am. Geog. Soc.; mem. Am. Assn. U. Profs., Am. Microscopic Soc., Am. Soc. Mammalogists, Am. Soc. Parasitologists, Am. Soc. Tropical Medicine and Hygiene, Am. Soc. Zoologists, Arctic Inst. N.Am. (asso.), Assn. Tropical Biology (charter), Biol. Stain Commn., Ecol. Soc. Am., Marine Biol. Assn. (life), Soc. Protozoologists (charter, exec. com. 1948-50), Soc. Systematic Zoology (charter), Western Soc. Naturalists (pres. 1960-61), Ore., So. Cal. (bd. advisers, chmn. invertebrates sect.) acads. sci., Sigma Xi (exec. com. 1963—), Phi Sigma. Research on protozoology, parasitology, biol. consequences of marine pollution, biology of polar seas. Home: 3819 Chanson Dr., Los Angeles 90043.*

MOHR, Karl Friedrich, German chemist; b. Koblenz, Germany, Apr. 11, 1806; s. Karl Mohr; D.Sc., Heidelberg, Germany, 1831; studied in Berlin and Bonn, Germany; m. Jacobine Derichs, May 30, 1833; 3 sons, 2 daus. Profl. pharmacist, Koblenz, 1832-57; became private docent U. Berlin, 1864; prof. pharmacy, Bonn, 1867-79. Author: Lehrbuch der pharmaceutischen Technik, 1847; Lehrbuch der chemischanalytischen Titriermethode, 2 parts, 1855-56; Geschichte der Erde, 1866; Chemische Toxicologie, 1874; also articles, including Ueber die Natur der Wärme, 1837. Editor: Pharmacopoea Universalis, 1845. Noted for work in volumetric analysis, analytical and pharmaceutical chemistry; introduced use of oxalic acid in alkalimetry, also chromic acid as indicator in determination of halogens; developed concept of different forms of energy and their mut. connection similar to Joule's concept of conservation of energy, 1837; inventor cork-borer, pinch-clamp, specific-gravity balance, Mohr's salt. Died Bonn, Sept. 29, 1879.

MÖHRES, Franz Peter, German zoologist; b. Cologne, Germany, Apr. 24, 1912; s. Paul and Gertrud (Bosbach) M.; Ph.D., U. Cologne; m. Hildegard Krahmer, Oct. 16, 1944; children—Ursula, Angela, Georg, Elisabeth, Michael, Heribert. Successively asst., instr., prof. zoology, U. Tübingen (Germany), also dir. Inst. Zool. Physiology. Mem. Görres Soc., Soc. for Study Mammals, German Zool. Soc., Leopoldina Soc. Research and publs. on animal physiology, animal orientation, vertebrate zoology. Home: Gartenstrasse 79, Tübingen, W. Germany.

MOHS, Frederic Edward, Am. surgeon, educator; b. Burlington, Wis., Mar. 1, 1910; s. Frederic Carl and Grace (Tilton) M.; B.Sc., U. Wis., 1931, M.D., 1934; m. Mary Ellen Reynolds, June 18, 1934; children—Frederic Edward, Thomas James, Jane Ann (Mrs. Jack J. Lincicum). Bowman cancer research fellow U. Wis. Med. Sch., Madison, 1935-38, asst. prof. chemosurgery, 1942-48, asso. prof., 1948—; chief chemosurgery clinic Univ. Hosp. Mem. A.M.A., Am. Assn. Cancer Research, A.A.A.S., Pacific Dermatologic Assn. (hon.), Dane County Med. Soc. (pres. 1959-60), Nu Sigma Nu, Delta Chi. Author: Chemosurgery in Cancer, Gangrene and Infection, 1956; also articles. Discovered, developed clin. use of chemosurgery for removal skin cancers utilizing frozen sects. Home: 3616 Lake Mendota Dr., Madison 53705. Office: 1300 University Av., Madison, Wis. 53706.*

MOHS, Friedrich, German mineralogist; b. Gernrode, Germany, Jan. 29, 1773; ed. Halle, Germany, mining acad., Freiburg, Germany, also in Austria; became prof. mineralogy, Graz, Austria, 1811, Freiburg, 1816; named prof. mineralogy, supt. Imperial Cabinet, Vienna, 1826. Author: Grundriss der Mineralogie, 2 vols., 1822-24. Devised scale of mineral hardness named after him, 1822. Died Agardo, nr. Belluno, Italy, Sept. 29, 1839.

MOIGNO, François Napoléon Marie, French physicist, mathematician; b. Guémené, France, Apr. 15, 1804; student Ste. Anne d'Aurray; student theology Montrouge. Mem. Soc. Jesus, 1822-43; fled to Switzerland when revolution began, 1830; tchr. math. Ste. Geneviève, Paris, from 1836; toured Europe, 1843; chaplain Lycée Louis-le-Grand, 1848-51. Translated English and Italian sci. works into French. Became sci. editor Presse, 1850, Pays, 1851; founder Cosmos, 1852, Les Mondes, 1862. Research and publs. on analytical mechanics, saccharimetry, molecular physics, integral calculus, also faith and sci. Died St. Denis, France, July 14, 1884.

MOIR, R(obert) Y(oung), Canadian chemist; b. Estevan, Sask., Can., Oct. 30, 1920; s. Robert Clare and Hester (Young) M.; B.A., Queen's U., 1941, M.A., 1942; Ph.D., McGill U., 1946; m. Marie A. Anderson, Sept. 7, 1946; children—James Robert, Lynn Marie, Michael Edward. Research chemist Dominion Rubber Co., 1943-44, 46-49; faculty Queen's U., Kingston, Ont., Can., 1949—, prof. chemistry, 1964—. Fellow Chem. Inst. Can. (examining bd. for profl. qualification 1959-64); mem. Am. Chem. Soc. Research, publs. on geometry non-rigid molecules, steric effects in cyclohexandes, diphenyl ethers, phthalideisoquinoline alkaloids. Home: 223 Victoria St., Kingston, Ont., Can.*

MOISE, Edwin Evariste, Am. mathematician, educator; b. New Orleans, Dec. 22, 1918; s. Edwin Evariste and Josephine (Boatner) M.; student La. State U., 1935-37; B.A., Tulane U., 1940; Ph.D. in Pure Math., U. Tex., 1947; M.A. (hon.), Harvard, 1960; m. Mary Lorena Leake, May 28, 1942; children—Edwin Evariste, Claire Mary. Instr. math. U. Mich., Ann Arbor, 1947-49, from asst. prof. to prof., 1951-60; James Bryant Conant prof. edn. and math. Harvard, Cambridge, Mass., 1960—; temporary mem. Inst. Advanced Study, Princeton (N.J.), 1949-51, 56-57. Recipient Henry Russel award U. Mich., 1954. Fellow Am. Acad. Arts and Scis.; mem. Am. Math. Soc. (mng. editor bull. 1958-63), Math. Assn. Am., A.A.A.S., Am. Assn. U. Profs., Phi Beta Kappa. Author: Elementary Geometry from an Advanced Standpoint, 1963; (with F. L. Downs, Jr.) Geometry, 1964; The Number Systems of Elementary Mathematics, 1966; Calculus, 1966. Home: 25 Pine Ridge Rd., Waban, Mass. 02168. Office: Harvard, Cambridge, Mass. 02138.*

MOISIL, Grigore C., Rumanian mathematician; b. 1906; prof. Bucharest (Rumania) U.; chmn. Group of Mathematicians of Latin Expression; mem. Acad. Socialist Republic Rumania. Author: La mécanique analytique des systèmes continus, 1929; Logique modale, 1942; Algebraic Theory of Automatic Mechanisms, 1959. Research in math. logics, math. analysis, algebra, geometry, mechanics, automatic mechanisms, circuits with relays and transistors.

MOISSAN, Ferdinand Frédéric Henri, French chemist; b. Paris, Sept. 28, 1852; ed. Mus. d'Histoire Naturelle; Sch. Pharmacy, Paris; Faculty Sci., Paris; student of Berthelot, Frémy. Qualified as profl. pharmacist, 1879; taught at Sch. Pharmacy, Paris, 1879-1883; prof. toxicology and inorganic chemistry there, 1886-99; prof. inorganic chemistry U. Paris, from 1900. Decorated Legion of Honor; recipient Nobel prize in chemistry, 1906. Mem. French Acad. Scis., 1891 (Lacaze Prize, 1887). Author: Reproduction du diamant, 1893; Etude complète des carbones amorphes et des graphites, 1898; L'isolement du fluor; Carbure de calcium; Le flour, 1900; Traité de chimie minerale, 5 vols., 1904-06. First to isolate flourine, 1886; inventor electric arc furnace, 1892; simplified manufacture of acetylene; synthesized ruby, diamonds, 1893; reduced oxides of uranium, chromium, tungsten; prepared metal carbides, silicon carbide, nitrides, borides, hydrides; founder chemistry of high temperature. Died Paris, Feb. 20, 1907.

MOJSISOVICS VON MOJSVAR, Edmund, Austrian geologist, paleontologist; b. Vienna, Austria, Oct. 18, 1839; became chief geologist Imperial Geol. Inst., 1870, asst. dir., 1892-1900. Editor (with Melchior Neumayr) Beiträge zur Paläontologie Österreich-Ungarns und des Orients, from 1880. Died 1907.

MOLARD, François Emmanuel, French engr.; b. Jura, France, 1774; ed. École polytechnique, Paris; became asst. dir. Paris Conservatory Arts and Crafts, 1817; dir. Sch. Arts and Crafts, Compiegne, France; transferred to Chalons-sur-Maine, France, 1805; organized sch., Angers, 1811. Introduced to France cable cars and turning cranes for use in mines; studied applied mechanics; numerous inventions, including machines for cutting faces of wheels. Died Paris, 1829.

MOLARD, (Claude) Pierre, French engr.; b. Jura, France, 1759; a founder Conservatory of Arts and Crafts, became dir., 1801. Mem. French Acad. Scis., 1816. Several inventions, including loom for damask, machine for drilling numerous rifle barrels simultaneously, machine for making parallel designs. Died Paris, 1837.

MOLCAN, Jan, Czechoslovakian psychiatrist; b. Trencín, Czechoslovakia, Sept. 11, 1925; s. Ladislav and Alzbeta (Matejcík) M.; student Comenius U., Bratislava, Czechoslovakia, 1943-44; M.D., Masaryk U., Brno, Czechoslovakia, 1950, Ph.D., 1965; m. Irene Sobansky, Jan. 15, 1953; children—Michaela, Peter. Practice medicine, 1950-51; staff Inst. Brain Diseases, Pezinok, Czechoslovakia, 1951-54; staff psychiat. dept. Comenius-U. Hosp., Bratislava, 1955—, chief biochem. research lab., 1957—; lectr. psychiatry Med. Sch., Law Coll., Arts Faculty, 1964—. Mem. Purkynje Soc. Czechoslovakia, Biochem. Soc., Czechoslovakian Acad. Scis., Czechoslovakian Sociol. Soc. Author: Musico-therapeutics in Psychiatry, 1954; General Psychiatry, and Med. Psychology (with others),

1963; also articles. Biochem. research in psychoses especially proteins and animo-acids, action psychopharmacological drugs. Home: 25 Záhradnícka. Office: 13 Mickiewiczova, Bratislava, Czechoslovakia.*

MOLDAUER, Peter Arnold, physicist; b. Vienna, Austria, June 18, 1923; s. Carl and Else (Kellner) M.; came to U. S., 1938, naturalized, 1944; B.S., Northeastern U., 1944; M.A., Harvard, 1947; Ph.D., U. Mich., 1956; m. Anne Bradlee Childs, Aug. 27, 1949; children—Wendy Sprague, Thomas Weld, Peter Shaw. Instr., U. Conn., 1955-57; asso. physicist Argonne (Ill.) Nat. Lab., 1957-67, senior physicist, since 1967—. Fellow of the Am. Phys. Soc. Translated: (with O. Laporte) Optics, 1954. Author: (with S. Yiftah, D. Okrent) Fast Reactor Cross Sections, 1960. Research on theory of nuclear reactions, neutron cross sections, optical models; theory of relativistic wave equations, higher spin particles; interpretation of quantum theory. Office: Argonne Nat. Lab., Argonne, Ill. 60440.*

MOLDAVE, Kivie, biochemist; b. Kiev, Russia, Oct. 22, 1923; s. James and Lucille (Shapiro) M.; came to U. S., 1937, naturalized, 1946; A.B., U. Cal. at Berkeley, 1947; M.S., U. So. Cal., 1950, Ph.D., 1952; m. Rose H. Spiro, June 6, 1949; children—Peter, Anne. USPHS Research fellow U. of Wis., 1952-53, Faculte des Scis., U. Paris, (France), 1953-54; faculty Tufts U., Boston, 1954-66, prof. biochemistry, 1964-66; prof., chmn. dept. biochemistry Pitts. Sch. Medicine, 1966—. Mem. Am. Soc. Biol. Chemists, Am. Chem. Soc., Research Soc., Sigma Xi, Phi Lambda Epsilon. Research, publs. on enzymes involved in protein and nucleic acid biosynthesis. Home: 5856 Aylesboro Av., Pitts. 15217.*

MOLDENHAWER, Jakob Paul, German botanist; b. Hamburg, Germany, Feb. 11, 1766; prof. philosophy, Kiel, Germany, also supt. Fruit Tree Sch. Author: Lentamen in historian plantarum Theophrasti, 1791; Beiträge zur Anatomie der Pflanzen, 1812. Studied plant anatomy; 1st description of structure of lactiferous plants. Died Kiel, Aug. 22, 1827.

MOLDENKE, Harold Norman, Am. naturalist, botanist; b. Watchung, N.J., Mar. 11, 1909; s. Charles Edward and Sophia Meta (Heins) M.; B.S. with 1st honors, Susquehanna U., 1929; M.A., Columbia, 1932, Ph.D., 1934; m. Alma Lance Ericson, Sept. 2, 1942; 1 son, Andrew Ralph. With N.Y. Bot. Garden, 1932-52, instr., lectr., 1958—, curator, adminstr. Herbarium, 1949-52; faculty Columbia, 1936-42, 46-52, Hunter Coll., evenings, 1947-50; faculty Westfield (N.J.) Adult Sch., 1952—; dir. Trailside Nature and Sci. Center, Union County Park Commn., Mountainside, N.J., 1952—, exec. officer Trailside Mus. Assn., 1956—. NRC fellow, 1935-36. Recipient Alumni Achievement award Susquehanna U., 1960. Fellow A.A.A.S.; mem. John Burroughs Meml. Assn., N.J. Audubon Soc. (Conservation award 1958, mem. natural sci. adv. bd. 1959—), Torrey Bot. Club (past treas., past mem. council), Bot. Soc. Am. Sigma Xi, Pi Gamma Mu. Author: Plants of the Bible, 1952; American Wildflowers, 1949; A Resume of the Verbenaceae, Avicenniaceae, Stilbaceae, Symphoremaceae, and Eriocaulaceae of the World, 1959; also numerous articles. Research on plant families Verbenaceae, Avicenniaceae, Stilbaceae, Symphoremaceae, Chloanthaceae, Eriocaulaceae, arrow poison plants, plants strategic to war effort, state floras, identity of shamrock, plants of Bible. Home: 303 Parkside Rd., Plainfield, N.J., 07060. Office: Trailside Nature and Sci. Center, Mountainside, N.J. 07092.*

MOLDOLA, Raphael, English chemist; b. London, Eng., July 19, 1849; s. Samuel Moldola; certificate of proficiency Royal Coll. Chemistry, 1868, postgrad., 1872; D.Sc., Oxford U.; LL.D., St. Andrews; m. Ella Frederica Davis, 1886. Jr. asst. lab. of J. Stenhouse; technician for color mfg. firm, Brentford; demonstrator, lectr., from 1874; asst. to N. Lockyer, from 1874; in charge Nicobar Island br. of expdn. to observe total eclipse of Apr. 1875; sci. chemist in coal tar dye factory, 1877; prof. chemistry Finsbury Tech. Coll., from 1885; prof. organic chemistry U. London, from 1912; H. Spencer lectr. Oxford, 1910. Fellow Royal Soc. (Davy medal 1913, council 1896-98, v.p. 1914-15), 1886, Inst. Chemists (pres. 1912-15), Phys. Soc.; mem. Inst. Brewing (hon.), Entomol. Soc. (pres. 1896-97), Chem. Soc. (pres. 1905-07), Brit. Assn. (council), Dyers and Colorists, Soc. Chem. Industry, Chem. Soc. Spain (hon.), Chem. Soc. France (hon.). Translator, editor: Studies in the Theory of Descent (Weismann), 1882-83; Report on the East Anglian Earthquake of 1884; The Chemistry of Photography, 1891; Coal and what we get from it, 1891; The Chemical Synthesis of Vital Products, 1904; Chemistry, also over 300 articles. Discovered veridine, 1878, samine blue; stated laws of substitution in naphthalene series; research on Azo compounds; studied constitution of diazoanido compounds; also amidoamidines, derivatives of phenol, mobile nitro group, imidazol bases; invented method to improve ways of obtaining complex Azo compounds, thereby allowing him to produce whole range of dyes; applied Darwinian theory to problems of animal coloration; research in spectrum analysis (with N. Lockyer). Died Nov. 16, 1915.

MOLE, Robert Hereward, med. scientist; b. Moukden, Manchuria, Jan. 27, 1914; s. Richard Howard and Grace (Moscrop) M.; B.A., U. Oxford (Eng.), 1935, B.M., B.Ch., 1938; m. Norah Grace Evers, Sept. 22, 1939; children—Antonia (Mrs. Geoffrey Cannon), Richard Hugh, Robert Peter. Clin. pathologist Royal Infirmary, Liverpool, Eng., 1947; sci. staff radiobiol. research unit Med. Research Council, Harwell, Eng., 1948—. Fellow Royal Coll. Physicians; mem. Coll. Pathologists, Path. Soc. Great Britain, Assn. Radiol. Research, Brit. Inst. Radiology, Radiation Research Soc. Author: (with G. J. Neary, R. J. Munson) Chronic Radiation Hazards, 1957; also numerous articles. Research on damaging effects of high environmental temperature and ionizing radiation on man and other mammals. Home: Heath Barrows, Bayworth Lane, Boar's Hill, Oxford, Eng. Office: Med. Research Council, Radiol. Research Unit, Harwell, Didcot, Eng.*

MOLENGRAAFF, Gustaaf Adolph Frederick, Dutch geologist; b. Nymegen, Holland, Feb. 27, 1866; student math., physics at Leyden, also geology, natural history at Utrecht; Ph.D. in botany and mineralogy, 1886. Instr., U. Amsterdam, 1888, prof. geology, from 1891; geologist Borneo expdn., 1893-94; state geologist South African Republic of Transvaal, 1897; cons. geologist, Johannesburg, Africa, from 1905; prof. geology Tech. U. Delft, Netherlands, 1910-12; head Netherlands expdn. Recipient medal Borneo, 1897, Wollaston, 1936, comdr. of Draper, 1938. Mem. Royal Acad. Scis. Amsterdam, Deutsche Akademie Der Naturforscher, Royal Acad. Belgium, A.A.A.S. Research, publs. on Bushveld complex (Bushveld Plutonic Series, pub. in Engring. Trans. 1904). Died 1942.

MOLESCHOTT, Jacob, physiologist; b. Bois-le-duc, Netherlands, Aug. 9, 1822; prof., Zurich, Switzerland, Turin, Italy, then in Rome, from 1879; elected Italian senator, 1876. Author: Physiologie der Nahrungsmittel, 1850; Lehre der Nahrungsmittel, 1850; Physiologie des Stoffwechsels, 1851; Der Kreislauf des Lebens, 1852; Physiologie Skizzenbuch, 1861; Kleine Schriften, 2 vols., 1880-87. Leading rep. of sci. materialism; studied blood, respiration, innervation of heart; discovered significance of phosphorus in metabolism of nerve cells. Died Rome, May 20, 1893.

MOLIERE, Paul Friedrich Gaspard Gert, German physicist; b. Butzbach, Germany, Apr. 7, 1909; s. Curt and Gertrud (Hoepfner) M.; Ph.D., U. Berlin; m. Marianne Dünkel, Aug. 10, 1938; 1 son, Rainer. Sci. collaborator K.W.I fur physikalischen Chemie, 1936; K.W.I. fur Physik, 1941; became prof. at large Rio de Janeiro (Brazil) C.B.P.F., 1952; named dir. Instituto de Fisica teorica, Sao Paulo, Brazil, 1954; became prof. theoretical physics U. Sao Paulo, 1956; named mem. European Council for Nuclear Research, Geneva, Switzerland, 1957; apptd. prof. theoretical physics U. Tübingen, 1959. Mem. Deutsche physikalische Gesellschaft Württ-Baden-Pfalz. Research and publs. on theory of cosmic ray dispersion, theory of multiples diffusion. Home: Wilhelmstrasse 163, Tübingen, W. Germany.

MOLIÈRES, see de Molières, Abbé Joseph Privat.

MOLINA, Juan Ignacio, naturalist; b. Guaraculen, Chile, July 20, 1740; ed. Santiago, Chile; became mem. Soc. Jesus, circa 1755; sent to Italy, 1767; tchr., Bologna, Italy; founder library, Talca, Chile. Author flora of Chile, 1782, also hist. and geog. works on Chile. Studied natural resources of Chile; eponym of several genera of plants, also town in Chile. Died Bologna or Imola, Italy, 1829.

MOLINA-PASQUEL, Claudio, Mexican physician; b. Veracruz, México, July 14, 1905; s. Mario and Sara (Pasquel) Molina; degree Med. Sch., U. Nacional Autónoma de México, 1930; postgrad. La. State U.; m. Betriz Torres-Izábal, Sept. 7, 1934; children—Beatriz, Claudio, Laura, Alberto, Jorge, Lucía, María, Gen. practitioner silver and gold mines El Oro Mining Co., State of Zacatecas, Mexico, 1931-32, oil field in coast of Veracruz, 1932-38, oil field Cía Mexicana de Petróleo el Aguila subsidiary Shell Oil Co., 1931-37; practice internal medicine, specializing in parasitic diseases, 1938—; adj. instr. Clínica Médica de la Escuela Nacional Medicina de Mexico, 1943-45; head clin. sect. Inst. Health and Tropical Diseases, 1944—; prof. infectious and parasitic diseases Sch. Medicine, U. Mexico, 1965—. WHO fellow to study epidemiology of onchocerciasis in Central Africa, 1959. Mem. Asociación de Médicos del Hospital de Enfermedades de la Nutrición, Sociedad de Investigación Pediátrica, Asociaiación Médica del Am. Brit. Cowdraw Hosp., Asociación de Médicos del Instituto Nacional de la Nutrición, Asociación de Investigación Pediátrica. Research, publs. on diagnosis and treatment of diseases caused by intestinal parasites, especially amebiasis, also tropical eosinophylia; discovered 1st cases of larva migrans visceralis in Mexico.*

MOLINARI, Ettore, Italian chemist; b. Milan, Italy, Oct. 13, 1926; s. Alessandro and Antonia (Rozza) M.; D.degree in Chemistry, U. Rome (Italy), 1949; m. Marta Baldin, June 30, 1956; children—Paola, Alessandra. Postdoctoral fellow Princeton, 1950-52; faculty U. Rome, 1953-65, asst. prof., 1959-65; prof. gen. and inorganic chemistry U. Bari (Italy), 1966—. Mem. Consiglio Superiore della Pubblica Istruzione, 1962-65. Mem. Italian Chem. Soc. Research, publs. on chem. kinetics, heterogeneous catalysis, plasma chemistry. Home: 239 Re David. Office: Istituto Chimico dell'Universita, Bari, Italy.*

MOLINES, Allan (Mullen), Irish anatomist; b. North of Ireland; B.A., M.B., Dublin (Ireland) U., 1676, M.D., 1684. Went to Barbados, 1690. Fellow Royal Coll. Physicians Ireland, Royal Soc., 1683; mem. Dublin Philos. Soc. Author: An Anatomical Account of the Elephant accidentally burnt in Dublin on 17 June 1681 together with a Relation of New Anatomical Observations on the Eyes of Animals, 1682; On the Injection of Mercury into the Blood; On a Black shining Sand brought from Virginia; Anatomical Observations on the Heads of Fowls. Described vascularity of lens of eye; discovered several structures in tunics of eye; studied human and comparative anatomy. Died 1690.

MOLISCH, Hans, botanist; b. Brünn (now Brno, Czechoslovakia), Dec. 6, 1856; prof., Graz, Austria, Prague, Czechoslovakia, Vienna, Austria, Sendai, Japan. Author: Leuchtende Pflanzen, 1904; Mikrochemie der Pflanzen, 1913; Pflanzenphysiologie, 1917; Anatomie der Pflanze, 1920; Im Lande der aufgehenden Sonne, 1927; Als Naturforscher in Indien, 1930; Die Lebensdauer der Pflanze, 1930; Abhandlungen, 1940. Studied comparative plant anatomy, gen. botany, morphology and physiology of plants; discovered test for detecting grape sugar (named after him). Died Vienna, Dec. 8, 1937.

MOLL, Albert, psychiatrist; b. Lissa, Prussia, May 4, 1862. Physician Berlin, Germany. Author: Rapport in der Hypnotismus, 1889; Untersuchungen über den Libido Sexualis, 1897; Die Nervöse Weib, 1898; Ärztliche Ethnik, 1902; Gesundbeten Medicin und Okkultismus, 1902; Einfluss der Gross-stadtischen Lebens und der Verkehrs auf den Nervensystem, 1902; Geschlectsleben der Kinds, 1908. Editor Handbuch der Sexualwissen, 1912. Introduced hypnotic psychotherapy into Germany; made studies on sexual problems; advocated sci. investigations of occultism.

MOLLAND, Jacob, Norwegian pharmacologist; b. Oslo, Norway, May 17, 1909; s. Jacob and Laura (Heyerdahl) M.; Cand.real., U. Oslo, 1935, D.Ph., 1940, M.D., 1948; postgrad. Yale; m. Ruth Andersen, Mar. 24, 1945; children—Ragnar, Ove. Prof. pharmacology, dir. Inst. Pharmacology, U. Oslo, 1950—. Mem. Norwegian and Nordic Pharmacopoea Commns., 1950—. Mem. Am. Soc. Pharmacology and Exptl. Therapeutics, Biometric Soc. Author: Kjemi for sykepleiersker, 1944, 51; (with E. Botolfsen) Kjemi for medisinere, odontologer og veterinaerer, 1947; also articles. Norwegian editor Acta pharmacologica et toxicologica, 1950—. Research on bacterial metabolism. Home: 7 Husebyveien, Oslo 3, Norway. Office: Blindern, Oslo 3, Norway.*

MOLLER, Christian, Danish physicist; b. Notmark, Denmark, Dec. 22, 1904; s. Jorgen Hansen and Maria (Terkelsen) M.; Student U. Copenhagen (Denmark), 1923-29, D.Ph., 1932; m. Kirsten Pedersen, June 22, 1931; children—Ole, Mette (Mrs. Klaus Flebo-Hansen). Asst. prof. Niels Bohr Inst., U. Copenhagen, 1932-40; faculty U. Copenhagen, 1940—, prof. physics, 1943—; vis. prof. Purdue U., Lafayette, Ind., 1948-49, Carnegie Inst. Tech., Pitts., 1957-58, U. Tex., Austin, 1964. Dir. theoretical study group CERN, 1954-57; dir. Nordic Inst. for Theoretical Atomic Physics, 1957. Recipient Order Dannebrog, 1953; Ole Romer medal, 1966. Mem. various Scandinavian acads. Author: The Theory of Relativity, 1952; also numerous articles. Research on theory of relativity and quantum theory. Home: 42A Frolishsvej, Copenhagen, Denmark.*

MÖLLER, Fritz, German meteorologist; b. Rudolstadt, Thur., Germany, May 16, 1906; s. Fritz and Hermine (Bergmann) M.; student U. Jena (Germany), 1924-26, U. Göttingen (Germany), 1926-27; D.phil. nat., U. Frankfurt (Germany), 1929, D.habil., 1935; m. Irmgard Kühne, July 13, 1935; children—Hans, Oskar, Gerhard. Meteorologist, Reichsamt f. Wetterdienst Berlin, Germany, 1935-38; observator, lectr. U. Frankfurt, 1938-49; prof. meteorology, dir. Inst. Meteorology, U. Mainz (Germany), 1949-60; prof. meteorology, dir. Inst. Meteorology U. München, (Germany), 1960—. Mem. Internat. Assn. Meteorol. Physics of Atmosphere (pres. radiation commn. 1957—), Leopoldina Acad. Natural Scis., Am. Meteorol. Soc., Am. Geophys. Union, Verband Deutscher Meteorol. Gesellschaften, Deutsche Geophys. Gesellschaft, Deutsche Union Geodesy Geophysics (pres. 1963—). Editor: Handbuch Geophysik, vol. 9, 1942-61. Research, numerous articles on atmospheric dynamics and thermodynamics, radiation, satellite meteorology. Home: 21 Friedrichst., 8 München 13, Germany.

MöLLER, Göran Lars, Swedish immunologist; b. Vittangi, Sweden, Aug. 30, 1936; s. Sven Gottne and Edla (Falck) M.; M.D., Karolinska Institutet, Stockholm, Sweden, 1963, Sci. degree Med., 1963; m. Erna Birgitta Irmgard Lindell, June 10, 1960; children—Gunnar, Elisabeth. Research fellow dept. tumor biology Karolinska Inst., 1958-63, asst. prof. transplantation immunology, 1963-—, asst. prof. dept. bacteriology, 1965-—; head lab. transplantation immunology dept. surgery Serfimerlasarettet, Stockholm, 1965-—. Mem. Brit. Soc. for Immunology, Internat. Soc. for Blood Transfusion and Transplantation Soc. Research, publs. on differentiation of transplantation antigens in mice; biol. effects of isoantibodies; elucidation of mechanism of immunological enhancement; homeostatic function of humoral antibodies in immune response; mechanism of allogeneic inhibition and cell-mediated immunity. Home: 14 Morabergsvägen, Sahsjöbaden, Sweden.*

MOLLER, Poul Flemming, Danish radiologist; b. Elsinore, Jan. 22, 1885; s. Julius and Anna (Nielsen) M.; M.D., U. Copenhagen (Denmark); m. Rigmor Rördam Holm; children—Edith, Bent, Jnga. Surgeon, radiologist, 1916-20; head radiology dept. Frederiksberg Hosp., 1920-30; prof. radiology, head radiation dept. U. Hosp. Copenhagen. Hon. mem. Faculty Radiology London, Brit. Radiol. Soc., Am. X-Ray Soc.; mem. Mexican, Peruvian socs. radiology. Author: Silicosis in Porcelain Works; Massive-flurosis of Bones and Ligaments; Congenital Thoracic Cyst and Deformities in the Roentgen Picture; Roentgen Picture of the Tabetic Arthropaties. Home: Kochsvej 18, Copenhagen. Office: H. C. Orstedvej 70, Copenhagen, Denmark.

MOLLGAARD, Holger Christian, Danish surgeon; b. May 20, 1885; s. Eske and Regittze (Skeel) M.; M.D., U. Copenhagen (Denmark); m. Else Bruun, Feb. 25, 1914. Physician, prof., sec. gen., later pres. Danish Food Orgn., World War I. Mem. Swedish Agr. Soc., Acad. Tech. Sci., Danish Acad. Tech. Sci. (co-founder), European Zootech. Soc. Author: Physiological Foundation of Lungs Surgery; Methodic and Mathematical Theory of Quantitative Measuring of Metabolism; Chemotherapy of Tuberculosis; Theory of Growth. Address: Espérance Allé 15, Charlottenlund, Denmark.

MOLLIARD, Marin, French biologist; b. Chatillon, France, June 8, 1866; became titulary prof. plant physiology Sorbonne, 1913, dean faculty scis., 1920; mem. French Acad. Scis., 1923. Author: Recherches sur les cécides florales, 1895. Research on relationships between nutrition and form of plants; discovered conditions for growing morels. Died Paris, July 24, 1944.

MOLLO-CHRISTENSEN, Erik Leonard, aero. engr.; b. Bergen, Norway; s. Axel and Helga (Holmboe) M.-C.; S.M., Mass. Inst. Tech., 1949, Sc.D., 1953; m. Johanna Waller, Nov. 20, 1948; children—Jan E., Peter, Anne. With Norwegian Def. Research Establishment, 1949-51; with Mass. Inst. Tech., Cambridge, 1953-—, prof. meteorology, 1964-—; sr. research fellow Cal. Inst. Tech., 1957-58. Mem. Nat. Com. for Fluid Mechanics Films; mem. air-sea interaction panel Nat. Acad. Scis.-NRC. Mem. Am. Phys. Soc., Am. Meteorol. Soc., Am. Inst. Aeros. and Astronautics, Am. Geophys. Union. Research, publs. on aero-elasticity and unsteady fluid mechanics, jet noise, wind generated waves, flow instability. Home: 10 Barberry Rd., Lexington, Mass. 02173. Office: Mass. Inst. Tech., Cambridge, Mass.*

MOLLWEIDE, Karl Brandan, German mathematician; b. Wolfenbuttel, Germany, Feb. 2, 1774. Prof., Leipzig, Germany. Author: Prufung der Farbenlehre des Herrn von Goethe, 1810; Sarstellung der optischen Irrtumer in Goethes Farbenlehre, 1821; Commentaiones mathematico-philologicae, 1813; De quadratis magicis, 1816. Introduced into math. trigonometric equations named after him. Died Leipzig, Mar. 10, 1825.

MOLODENSKY, Mikhail Sergeevich, Russian geophysicist, gravimetrist, geodesist; b. June 16, 1909; grad. Moscow U., 1932. Sr. asso. Central Research Inst. Geodesy, Aerial Surveying and Cartography, 1932-46; with Geophys. Inst., USSR Acad. Scis., 1946-56, dir. Inst. Geophysics, 1956-—, now sr. asso. Recipient Stalin prize, 1946, 51, Lenin prize, 1963. Mem. USSR. Acad. Scis. (corr.). Author: Fundamental Problems of Geodetical Gravimetry, 1945; Elastic Movement, Free Mutation, and Some Questions on the Structure of the Earth, 1953. Research and publs. on earth's structure, elastic tides, free nutation; evolved theory for practical geodetic application of measurements of earth's gravitational field; designer gravimeter for relative determination of gravitational acceleration forces. Address: Inst. Geophysics, B. Gruzinskaya 10, Moscow, USSR.

MOLONEY, P(eter) J(oseph), Canadian chemist; b. Penetang, Ont., Can., June 29, 1891; s. Henry and Elizabeth (Byrnes) M.; B.A., U. Toronto, 1912, M.A., 1915, Ph.D., 1924; m. Angelina Chapman, July 6, 1916; children—Mary, Henry, Oliver, John, Peter. Chemist, Canadian Dept. Agr., Ottawa, Can., 1917-19; faculty Connaught Med. Research Labs. U. Toronto (Ont.), 1919-—, cons., 1961-—. Recipient medal Am. Diabetic Assn., 1964; Mickel award U. Toronto Faculty Medicine, 1964. Named to Order Brit. Empire. Mem. Am. Chem. Soc., Royal Soc. Can., Biochem. Soc. Eng., Canadian Physiol. Soc. Research, publs. on toxins, toxoids, and prophylaxis, antibodies in insulin and antibody-induced diabetes mellitus, sulfated insulin for treatment of insulin-antibody-resistant diabetes. Home: 56 St. Mary St., Toronto, 5, Ont., Can.*

MOLONEY, William Curry, Am. physician; b. Boston, Dec. 19, 1907; s. Francis and Elizabeth (Curry) M.; student Tufts U., M.D., 1932; D.Sc. (hon.), Coll. Holy Cross, 1961; m. Josephine O'Brien, Feb. 22, 1934; children—Patricia (Mrs. Thomas McGrath), William Curry, Elizabeth (Mrs. Barry Kane), Thomas. Faculty, Tufts U., Sch. Medicine, Boston, prof. medicine, dir. 1st, also 3d med. services, Boston City Hosp., dir. hematology lab.; clin. prof. medicine Harvard Med. Sch., 1967-—; physician Peter Bent Brigham Hosp., Boston; dir. research hematology Childrens Cancer Research Found., Boston. Recipient Bronze medal Mass. div. Am. Cancer Soc., 1964. Mem. A.M.A., Mass. Med. Soc., Soc. for Exptl. Medicine and Biology, Am. Soc. for Cancer Research. Research, numerous publs. on radiation effects on blood, leukemogenesis in rat. Home: 14 Roanoke Av., Jamaica Plain, Mass. 02130. Office: Peter Bent Brigham Hosp., 721 Huntington Av., Boston 02115.*

MOLOTKOV, Vladimir Gerasimovich, Russian pathoanatomist; b. Vitebsk, 1903; grad. Med. Faculty, Smolensk U., 1928; post-grad. Smolensk Med. Inst., 1928-30; D.Med. Sci. Asst., lectr. dept. path. anatomy Smolensk Med. Inst., 1930-41, prof., head chair path. anatomy, 1945-—, dean, dep. dir. for sci. work, 1945-55. Mem. All-Union (bd. mem.), Smolensk Oblast (chmn.) socs. pathoanatomists. Author over 50 works. Address: Smolensk Med. Inst., ulitsa Glinki 3, Smolensk, RSFSR, USSR.

MOLOY, Howard Carman, gynecologist; b. Thedford, Ont., Can., 1903; M.Sc., London, Ont., 1928; m.; 3 children. Asst. clin. prof. obstetrics and gynecology Columbia U. Coll. Phys. and Surg.; instr. path. U. Western Ont., 1927-29; intern, asst. resident Sloane Hosp., 1929-35, asst. attending obstetrician and gynecologist, 1935-—; asso. attending obstetrician and gynecologist Vanderbilt Clinic, 1935. Diplomate Am. Bd. Obstetrics and Gynecology. Fellow Am. Gynecol. Soc.; mem. N.Y. Obstet. Soc. Contbg. author: New Loose-leaf Surgery (Nelson); author: (with William E. Caldwell) Gynecol. Surgical Obstetrics; Clinical and Roentgenologic Evaluation of the Pelvis in Obstetrics, 1951. Developed (with Caldwell) precision stethescope; classified (with Caldwell) varieties of female pelvis using form and measurements (Caldwell-Moloy classification), 1933. Died 1953.

MOLTESEN, Peter, Danish Sylviculturist; b. Horning, Apr. 9, 1914; s. Niels Peter and Bedil M.; ed. Royal Vet. and Agrl. Coll.; license in sylviculture; m. Marie, Oct. 18, 1945; children—Jens, Dodil. Asst., Coll. Sylviculture, 1941-42; asst. insp., later insp. State forests, 1944-47; prof. sylviculture, 1948-—; pres. Danish Forestry Council, 1949-—. Mem. Assn. Tech. Sci., Danish Forestry Council. Author: Forests and Forest Industries Statistics, 1952; Acer pseudoplatanus, 1958; Length of Filling Season, 1957; From Large to Small Stands in Forestry, 1957; Chemical Debarking, 1962. Research in forest products, especially influence of growing conditions, inheritance of wood quality. Home: 3, Thorsvej, Birkerod, Denmark. Office: 23, Rolighedsvej, Copenhagen 5, Denmark.

MOLYNEUX, Samuel, Brit. astronomer, politician; b. Chester, Eng., July 18, 1689; s. William Molyneux; M.A., Trinity Coll., Dublin, 1710; m. Elizabeth Capel, 1717. Visited Eng. and Holland; sent to Hanover, Germany; sec. to George, Prince of Wales; mem. Parliament from Bossiney, Eng., 1715, St. Mawes, Eng., 1726, Exeter, Eng., 1727; mem. Irish Parliament, 1727; named lord of admiralty, 1727; privy councillor of Eng. and Ireland. Fellow Royal Soc., 1812. Contbg. author: Optics (Dr. Robert Smith), 1738; his description of zenith sector and jour. of Kew obs. printed by Rigaud, 1832. Built reflecting telescopes, 1724; attempted to determine stellar annular parallax. Died Apr. 13, 1728.

MOLYNEUX, William, Irish mathematician, astronomer; b. Dublin, Ireland, 1656; ed. Trinity Coll., Dublin, Middle Temple, London, 1675; LL.D., U. Dublin; m. dau. of Sir. William Domvilla, 1678; 1 son, Samuel. Surveyor-gen. crown bldgs., chief engr.; began astron. observations and corr. with Mr. Flamsteed, king's astronomer, 1681; apptd. to inspect fortresses in Flanders, 1685; rep. of Dublin; chosen mem. U. Dublin, 1695; promoted linen and woolen manufacture. Fellow Royal Soc., 1685. Author: Sciothericum telescopium, 1686; Theorem for Finding the Foci of Optic Glasses, 1692; Dioptrica nova (1st English work on optics), 1692. Died Oct. 11, 1698.

MOMENT, Gairdner Bostwick, Am. zoologist; b. N.Y.C., May 4, 1905; s. Alfred G. and Laura (Bostwick) M.; A.B., Princeton, 1928; Ph.D., Yale, 1932; m. Ann Reed Faben, June 26, 1937; children—Charles G., Sarah G., James F., Ann R., Jane O. Faculty, Goucher Coll., Balt., 1932-—, prof., 1945-—; vis. asso. prof. Johns Hopkins, 1944-45. Asso. program dir. NSF, Washington, 1960-61; sec.-gen. XVI Internat. Congress Zoology, Washington, 1963; mem. Commn. for Undergrad. Edn. in BioScis., 1963-—; mem. corp. Bermuda Biosta., 1958-—. Research grantee Am. Philos. Soc., 1946-47, Am. Cancer Soc., 1948, NIH, 1957-65. Fellow A.A.A.S.; mem. Am. Soc. Zoologists (sec. 1957-60), Am. Inst. Biol. Sci. (film tchr., organizer film series on evolution and animal function 1961, chmn. edn. com. 1960-63), Md. Acad. Sci. (chmn. sci. council 1963), Acad. Zoology India (exec. council 1964-—), Phi Beta Kappa, Sigma Xi. Author: General Zoology, 1958, 2d edit., 1967; Frontiers Modern Biology, 1962. Co-ordinator, Voice of Am. Biol. Series, 1961; editor, Harper's Ency. Sci. Contbr. research articles on factors limiting growth and regeneration in annelids and urodeles, use of polyvinyl alcohol to quiet protozoa, protective variation and reflexive selection in animals. Home: 363 Evesham Av., Balt. 21212.*

MOMMAERTS, Wilfried F. H. M., physiologist; b. Broechem, Belgium, Mar. 4, 1917; s. Hendrik D. and Maria (Van Damme) M.; B.A., U. Leiden (The Netherlands), 1937, M.A., 1939; Ph.D., U. Kolozsvar (Hungary), 1943; m. Elizabeth Batyka, July 29, 1944; children—Robert, Edina, Quentin. Came to U. S., 1948, naturalized, 1956. Faculty, Am. U., Beirut (Lebanon) Sch. Medicine, 1945-48, Duke, 1948-53, Western Res. U., 1953-56; Rockefeller Found. fellow U. Coll., London, Eng., 1956; prof. medicine and physiology, U. Cal. at Los Angeles, 1956-—, chmn. dept. physiology, 1966-—, dir. Los Angeles County Heart Assn. Research Lab., 1956-—; cons. U. S. Army Med. Service. Mem. Am. Soc. Biol. Chemists, Am. Physiol. Soc., Biochem. Soc. (Great Britain), Am. Heart Assn., Biophys. Soc., Pacific Slope Biochemistry Conf., Am. Soc. Cell Biology, Cardiac Muscle Soc., Nat. Inst. Gen. Med. Sci. (chmn. physiol. tng. com.), Alpha Omega Alpha. Author: Muscular Contraction, 1950. Editorial bd. Circulation, 1950-53, 60-65, Jour. Applied Physiology, 1957-60. Research, publs. on chem. changes in contracting muscle and their correlation with energetics; cardiac contraction from fundamental viewpoints; also light-scattering and circular dichroism in investigation of protein structure. Home: 215 Denslow, Los Angeles 90049.*

MONACELLI, Mario, Italian dermatologist; b. Fabriano, Italy, May 1, 1900; s. Curzio and Beatrice (Barriera) M.; ed. U. Rome; M.D.; m. Adelheid Dolar, Aug. 10, 1939; children—Erika, Manfredo, Piero. Prof., U. Rome; dir. Dermatology Clinic, U. Rome. Research and publs. on psoriasis, eczema. Home: via Nomentana 222, Rome, Italy.

MONACHESI, Elio D(avid), sociologist, educator; b. Macerata, Italy, July 19, 1905; s. Armando and Nera (Vitaloni) M.; came to U. S., 1912, naturalized, 1931; A.B., U. Mo., 1927, A.M., 1928; Ph.D., U. Minn., 1931; m. Marjorie F. Diddy, Jan. 31, 1936; children—Livia (Mrs. James Coming), Aleeta (Mrs. John Lager). Undergrad. asst. in polit. sci. U. Mo., 1926-27, grad. asst. sociology, 1927-28; faculty sociology U. Minn., Mpls., 1928-—, prof., 1945-—, chmn. dept. sociology, 1951-—; post-doctoral fellow Social Sci. Research Council, Boston and Italy, 1932-34. Vis. prof. U. Wash., summer 1956, U. Ore., summer 1958. Fellow A.A.A.S., Am. Sociol. Assn.; mem. Sociol. Research Assn., Nat. Assn. Social Workers, Am. Assn. U. Profs., Midwest Sociol. Soc. (pres. 1958-59), Phi Beta Kappa, Alpha Pi Zeta, Alpha Kappa Delta. Author: Prediction Factors in Probation, 1932; (with Edith H. Baylor) The Rehabilitation of Children, 1939; (with Don Martindale) Elements of Sociology, 1951; (with Starke R. Hathaway) Analyzing and Predicting Juvenile Delinquency with the MMPI, 1953; (with Starke R. Hathaway) An Atlas of Juvenile MMPI Profiles, 1961; (with Starke R. Hathaway) Adolescent Personality and Behavior, 1963. Home: 2249 Folwell St., St. Paul 55108. Office: Dept. Sociology, U. Minn., Mpls. 55455.*

MONACO, Lawrence Henry, Am. biologist; b. Phila., Mar. 31, 1925; s. Daniel and Jennie (Leggieri) M.; A.B., LaSalle Coll., 1949; M.S., U. Notre Dame, 1952, Ph.D., 1954; m. Rita Marie Hein, Aug. 4, 1954; children—Vincent J., Mary R., Loretta A., Theresa A. Instr., Del Mar Coll., Corpus Christi, Tex., 1954-57, Villa Madonna, Covington, Ky., 1957-58; faculty Dutchess Community Coll., Poughkeepsie, N.Y., 1958-—, prof., chmn. dept. biology, 1962-—. Cons. Jr. coll. biology depts., 1959, 60, 63, 65. Mem. Am. Soc. Zoologists, Am. Soc. Parasitologists, Albertus Magnus Guild, Sigma Xi. Author: (with Aaron P. Seamster) Laboratory Manual in Zoology, 1955; also articles. Discovered numerous new species gill parasites fish; responsible for creation new subfamily gill parasites. Home: 86 S. Grand Av., Poughkeepsie, N.Y. 12603.*

MONAELESSER, Adolph, physician, surgeon; b. Laxey, Isle of Man, June 22, 1855; s. Maurice and Emilie (Schyar) M.; student 1869-73, Greifswald (Germany) U., 1873-75, U. Berlin, 1875-76, Breslau and Bonn, Germany, 1876-77, U. Paris, 1877-79; M.D., Eclectic Med. Coll., 1882; M.D., Coll. City N.Y., 1886; Sc.D., Lincoln Meml. U., 1930; m. Bettina Hofker, Aug. 6, 1887; 1 son, Mozart, Came to U. S., 1879, naturalized, 1922. Asso. curator to Dr. N. M. Miller, City Hosp., N.Y.C., 1884-87; attending surgeon St. Elizabeth's Hosp., N.Y.C., 1887-—; surgeon in chief A.R.C. (served in Cuba, Spanish-Am. War), 1893-1903; pathologist to Commn. for Investigation of Crime, Am. Bar Assn., 1923. Fellow A.M.A., N.Y.

Acad. Medicine, N.Y. Acad. Scis., N.Y. Micros. Soc., Institut Pasteur (Paris). Author: Medical Service During the Cuban Insurrection and Spanish American War, 1899; Effets du venim de cobra modifie sur les tumeurs cancereuses, 1930. Research in therapeutic value of snake venom in nerve affections and malignant growths. Died Mar. 27, 1936.

MONAKOW, Constantin, see von Monakow, Constantin.

MONARDES, Nicolas, Spanish botanist, physician; b. Seville, Spain, 1507; ed. Alcala U.; Author: De secanda vena in pluritide inter Graecos et Asalees concordia, 1539; De rosa et partibus ejus, 1565; Dos libros de las cosas que se traen de las Indias Occidentales, que sirven al uso de medicina . . . ; Historia medicinal, 1565-74. Recorded early bot. and zool. descriptions of W. and E. Indies; acclimatized various Am. plants in Seville; eponym of bot. genus Monarda in Linnaeus' class of Diandria. Died circa, 1578.

MONASTERIO, Gabriele, Italian physician; b. Reggio Calabria, Italy, Dec. 23, 1903; s. Luigi and Margherita (Rubino) M.; M.D., U. Genova (Italy), 1927; m. Clara Gentili, Sept. 25, 1935. Research fellow Ludwig Spiegler Inst. Biol. Chemistry, Vienna, Austria, 1928-29; research fellow microbiology Robert Koch Inst., Berlin, Germany, 1931; asst. prof. Med. Clinic, U. Pisa (Italy), 1933—, chair med. pathophysiology, 1948-55, chair med. Clinic, 1955—. Recipient Marzotto prize, 1954; Gold medal for pub. edn., 1965. Mem. Italian Soc. Nuclear Biol. Medicine (sec. 1956—), Italian Soc. Nephrology (pres. 1960—). Author: Le nefropatie mediche, 1954; Radioisotopi nell'indagine medica, 1960; also numerous articles. Research on hemoglobin metabolism in jaundice, techniques for determination blood cholesterol, glycerides and fecal fat, identification renal diabetes as a tubular dysplasia; identification of indirect reacting bilirubin as bilirubinic acid; devel. of methods for early detection of tumors using radioisotopes, low-nitrogen diet in treatment of chronic uremia. Home: Via Cardinale Maffi, Pisa, Italia.*

MONASTERO, Salvatore, Italian agronomist; b. Palermo, Italy, Jan. 1, 1900; s. Rosario and Vita (Saso) M.; Ph.D.; m. Elena Pirrone, Dec. 10, 1940; children—Vita, Riccardo, Ettora, Roberto. Prof. agrarian entomology U. Palermo. Mem. Palermo Acad. Sci. Research and publs. on plant parasites; biol. control of olive fly. Home: via Villa Franca 99, Palermo. Office: Instituto Entomologia, Agraria, Palermo, Italy.

MÖNCH, Ernst, German physicist; b. Grünwald, Germany, Nov. 2, 1909; s. Ernst and Rosa (Fink) M.; Ph.D., Munich Tech. Coll.; m. Isabella Ittlinger, Nov. 16, 1948; children—Ernst, Ludwig. Engr., Gesellschaft für Lindes Eismaschinen Munich, 1935-37; asst. Munich Tech. Coll., 1937-46; prof. agrégé, 1946-48; prof., 1948—; prof. Tucuman U., Argentina, 1949-52. Mem. Gesellschaft för Angewandt Mathematik und Mechanik, Sociedad Argentina de Ensayo de Materiales, Soc. for Exptl. Stress Analysis, Deutscher Alpenverein. Author: Praktische Spannungsoptik, 1959; also articles, films. Research on photoelasticity, engrng. mechanics. Home: Reinekestrasse 25, Munich, West Germany.

MONCRIEF, John A., Am. surgeon; b. Manila, P.I., July 22, 1924; s. William Henry and Utah Lee (Ensley) M.; student the Citadel, 1942-43, Cornell U., 1943-44; M.D., Emory U., 1948; m. Constance Jane Knudsen, June 17, 1949; children—John A., Christian Lee, Constance Helen. Bn. surgeon 17th F.A., Korea, 1950; chief gen. surgery 8054 Evacuation Hosp., chief gen. surgery 10th Sta. Hosp., Korea, 1950-51; chief gen. surgery U. S. Army Hosp., Ft. Sill, Okla., 1954-55; chief clin. div. U. S. Army Research Unit, 1955-57; chief surg. research br. U. S. Army Med. Research and Devel. Command, 1960-61; now col. M.C.; comdr., dir. U. S. Army Surg. Research Unit, Brooke Army Med. Center, Ft. Sam Houston, Tex., 1961—. Instr. surgery Washington U. Med. Sch., St. Louis, 1954-57, Emory U. Med. Sch., Atlanta, 1957-60. Cons. for thermal trauma to surgeon gen. Diplomate Am. Bd. Surgery. Mem. Am. Assn. for Surgery of Trauma, A.C.S. (trauma com.), Am. Fedn. for Clin. Research, A.M.A., Am., Western surg. assns., Assn. Mil. Surgeons, N.Y. Acad. Scis., Soc. Univ. Surgeons. Author: (with S. E. Order) The Burn Wound, 1965. Research, numerous publs., particularly regarding treatment of burns, operative blood loss, skin transplants. Office: U. S. Army Surgical Research Unit, Brooke Army Medical Center, Ft. Sam Houston, Tex.*

MOND, Ludwig, chemist; b. Kassel, Germany, Mar. 7, 1839; s. Moritz B. Mond; ed. Polytech. Sch., Cassel, also Marburg, Heidelberg univs.; hon. Ph.D., Padua, 1892, Heidelberg, 1896; hon. D.Sc., Oxford; m. Frida Lowenthal, 1866; children—Robert Ludwig, Alfred Moritz. Came to Eng., 1862, naturalized, 1880; engaged with Leblanc Soda industry; introduced process for recovering sulphur from alkali waste; partner (with J. T. Brunner) in mfg. ammonia soda by Solvay Process (perfected process and alkali works became largest in world), at Winnington, Northwich, Cheshire, from 1873; founded, endowed Davy Faraday Research Lab. of Royal Instn., v.p. Fellow Royal Soc., 1891, Royal Inst. Chemistry (v.p.), Chem. Soc.; mem. Accademia dei Lincei, Rome; pres. Soc. Chem. Industry, Chem. Section, Brit. Assn., 1896. Contbr. numerous papers to jours. Made and patented many inventions of great sci. and comml. importance, including process for mfr. of chlorine in conjunction with ammonia soda process, method of producing gas for power and heating purposes, new form of gas battery, Mond process for mfg. pure nickel; investigated, discovered (with Langer and Quincke) new chem. compounds. Died Dec. 11, 1909.

MOND, Sir Robert Ludwig, Brit. chemist; archeologist; b. Widnes, Lancshire, Eng., Sept. 9, 1867; s. Ludwig Mond; ed. Cheltenham Coll., St. Peter's Coll., Cambridge (Eng.) U., Zurich (Switzerland) Poly., univs. Edinburgh and Glasgow; M.A.; LL.D. (hon.) univs. Liverpool (Eng.) and Toronto (Can.); m. Helen Edith Levis, 1898 (dec. 1905); two daus.; m. 2d, Marie Louise Le Manach. Dir. S. Staffordshire Mond Gas Co., Internat. Nickel Co. Can., Ltd, Mond Staffordshire. Hon. sec. Davy-Faraday Research Lab. of Royal Instn.; treas. Palestine Exploration Fund; hon. treas. Brit. Sch. Archaeology at Jerusalem; v.p. Archaeol. Inst., U. London, also Archaeol. Inst., Liverpool; Spiers meml. lectr. on Faraday. Recipient Messei medal Soc. Chem. Industry, 1936. Fellow Royal Soc., 1938, Soc. Engrs., Philos. Soc. Engr., Royal Soc. Edn., Geol. Soc., Chem. Soc., Zool. Soc.; mem. Iron and Steel Inst., Egyptian Exploration Soc. (pres.), French Soc. Chem. Industry (pres.), Inst. Physics, Faraday Soc. (past pres.), de l'Inst. Ass. Etr. Acad Inscriptions Belles Lettres. Author (with Oliver H. Myers) The Bucheum, 1934; also papers on electrolytic, phys. and chem. problems; reports on excavations; assisted in publn. of Aramaic Papyri discovered at Assuan. Research on metallic carbonyls and action of nitric oxide on metallic carbonyls; pure and applied chemistry, electro-chemistry, color photography; assisted father in discovery of new carbonyls; carried out series of excavations in cemetery at Thebes, Palestine and Britanny. Died Oct. 22, 1938.

MONDEVILLE, Henri de, see de Mondeville, Henri.

MONETTE, John Wesley, Am. physician, historian; b. Shenandoah Valley, Va., Apr. 5, 1803; s. Samuel and Mary (Wayland) Monett; M.D., Transylvania U., 1825; m. Cornelia Newman, Dec. 10, 1828; 10 children. Practiced medicine; 1st to suggest quarantine as means of preventing spread of yellow fever; mayor, councilman, Washington, Miss. Author: An Account of the Epidemic of Yellow Fever that Occurred in Washington, Mississippi, in the Autumn of 1825, 1827; Observations on the Epidemic of Yellow Fever of Natchez and the Southwest, 1842; Oil of Turpentine as an External Irritant, 1827; History of the Discovery and Settlement of the Valley of the Mississippi by the Three Great European Powers, Spain, France, and Great Britain, and the Subsequent Occupation, Settlement, and Extension of Civil Government by the United States until the Year 1816, 2 vols., 1846. Died Mar. 1, 1851.

MONEY, John William, med. psychologist; b. Morrinsville, New Zealand, Aug. 7, 1921; s. Frank and Ruth Mary (Read) M.; M.A., U. New Zealand, 1943, diploma with honors 1st class, 1944; postgrad. U. Pitts.; Ph.D., Harvard, 1952. Came to U S., 1947, naturalized, 1962. Jr. lectr. U. Otago, New Zealand, 1945-47; faculty Johns Hopkins, Balt., 1951—, asso. prof. med. psychology and pediatrics, 1959—; psychologist Johns Hopkins Hosp., 1955—. Cons. Hiltman Assos., Balt., 1965—, Sexology, N.Y.C., 1961—. Co-recipient Hofheimer prize Am. Psychiat. Assn., 1956; recipient Gold medal Children's Hosp. Phila., 1966. Mem. Soc. for Sci. Study Sex, Royal Soc. Medicine, Am. Psychopath. Assn., Sigma Xi. Author: The Psychologic Study of Man, 1957; (with D. Alexander, H. T. Walker, Jr.) A Standardized Road-Map Test of Direction Sense, 1965. Editor: Reading Disability: Progress and Research Needs in Dyslexia, 1962; Sex Research: New Developments, 1965; The Disabled Reader: Education of the Dyslexic Child, 1966; also numerous articles. Research on psychology hermaphroditism and related sexual disorders, glandular disorders childhood, reading disability, behavioral genetics and chromosome abnormalities; developed theory psychosexual differentiation. Home: 2104 E. Madison St., Balt. 21205.*

MONEY, William Lang, Am. biologist; b. Centerville, R.I., Mar. 15, 1914; s. Joseph William and Edith (Dewell) M.; A.B., Brown U., 1941; Ph.D., Harvard, 1947; m. Lillian Lee Pieper, Aug. 28, 1946; children—William Lang, Peter Albert. Research fellow Harvard, also Mass. Gen. Hosp., 1947-48; staff Sloan-Kettering Inst., N.Y.C., 1949-50, asso., 50-60, asso. mem., 1960-66; staff Cornell U. Med. Sch., N.Y.C., 1949-50, 51-66, asso. prof. biology, 1957-66; prof. zoology U. Ark., Fayetteville, 1966—. Fellow Am. Cancer Soc., 1947-49; Spl. fellow NIH, 1951, Nat. Cancer Inst., 1952. Mem. A.A.A.S., Am. Thyroid Assn., Endocrine Soc., Harvey Soc., Sigma Xi. Research, numerous publs. on endocrine, normal and abnormal thyroid, reprodn. physiology. Home: Mt. Comfort Rd. Office: Dept. Zoology, U. Arkansas, Fayetteville, Ark. 72701.*

MONEYMAKER, Berlen Clifford, Am. geologist; b. Knoxville, Tenn., Sept. 15, 1904; s. Thomas Jefferson and Esther (Varner) M.; B.S., U. Tenn., 1928, M.S. (fellow in geology), 1929; postgrad. Yale, 1930-33; m. Hilda Fuller, July 15, 1939; children—Joan Fuller (Mrs. Selcuk Kutuk), Marcia Hilda. Asst. lab. instr. U. Tenn., 1927-28, instr. geology, 1929-30; asst. instr. geology Yale, 1930-33; field asst. in geology TVA, Knoxville, 1933-34, asst. geologist, 1934-35, asso. geologist, 1935-36, geologist, asst. chief geologist, 1936-41, chief geologist, 1941—. Cons. geologist P.R. Reconstrn. Adminstrn. and Water Resources Authority, 1936—. Mem. Geol. Soc. Am., Mineral. Soc. Am., Seismol. Soc. Am., Brit. Assn. for Advancement Sci., Am. Soc. C.E., Am. Geophys. Union, Am. Assn. Petroleum Geologists, A.A.A.S., Nat. Speleological Soc., Mineral. Assn. Can., Am. Ordnance Assn., Assn. Engrng. Geologists, Am. Inst. Profl. Geologist, Soc. Am. Mil. Engrs., Soc. Econ. Geologists, Tenn., N.C. acads. sci., Tech. Soc. Knoxville. Research and numerous publs. on engrng. geology, earthquakes, pre-Cambrian rocks of So. Appalachians; origin and devel. solution channels and caves in limestones indicating solution may take place in phreatic zone; pioneered application detailed geologic information to engrng. projects. Home: 4037 Stillwood Dr. S.W., Knoxville 37919. Office: Arnstein Bldg., TVA, Knoxville, Tenn. 37902.*

MONGE, Carlos, Peruvian physician; b. Lima, Peru, Dec. 13, 1884; s. José and Eleodora (Medrano) M.; student Colegio Nacional de Guadalupe, 1897-1901, London Sch. Tropical Medicine, 1911, U. Paris, 1912-13; M.S., U. San Marcos, 1914; postgrad. U. Chgo., 1941; m. Cristina Cassinelli; children—Luis, Carlos, Cristina. Prof., Faculty Medicine, U. San Marcos, Lima, Peru, 1919—; dir. Inst. Andean Biology, 1932, 40. Author: Algunos apuntes sobre la hematología de la enfermedad de Carrión, 1911; Les érythrémies de l'altitude, leurs rapports avec la maladie de Váquez, étude physiologique et pathologique, 1929, also numerous articles. First to describe Andes disease (now called Monge's disease) common on high Andean plateau; research on nervous system, content of blood and gen. physiology peculiar to dwellers on Andean plateau. Home: Calle Emilio Fernández 611. Office: Calle de Trinidad 373, Lima, Peru.

MONGE, Gaspard, French mathematician; b. Beaune, France, May 10, 1746; s. Jacques Monge; ed. Oratorian schs. in Beaune and Lyons; student practical sch. attached to Mézières Sch. Mil. Engrng., 1762. Tchr. physics, Lyons, 1762, math., Mézières, 1768, physics, 1771-82; apptd. prof. hydraulics, Lycée, Paris, 1780; examiner naval candidates École de Marine, 1783; minister of marine, 1792-93; accompanied Napoleon on Egyptian expdn.; a founder École Poly., 1794-95, prof. math.; prof. math. École Normale; apptd. mem. Senate under Napoleon, 1799. title Comte de Péluse; stripped of all honors in Bourbon restoration. Mem. French Acad. Scis., 1780. Author: Traité élementaire de la statique, 1788; Feuilles d'analyse appliquée à la geometrie, 1795; Essais sur les Géométrie descriptive, 1799; Application de l'analyse à la géometrie des surfaces du 1er and 2e degré 1807; Géométrie sur les plans et les surfaces courbes, 1812. Inventor of descriptive geometry; began study of perspective and polarity; undertook important researches in differential geometry, especially in theory of curvature; also provided solutions to partial differential equations by means of his theory of surface, Died Paris, France, July 28, 1818.

MONIE, Ian Whitelaw, anatomist; b. Paisley, Scotland, May 24, 1918; s. John Whitelaw and Isabella Kinnear (Miller) M.; M.B., Ch.B., U. Glasgow (Scotland), 1940; m. Barbara Jane Harris, July 30, 1956; children—Margaret, Ian, Peter, Gail. Came to U. S., 1952, naturalized, 1957. Faculty, U. Glasgow, 1942-47, lectr. anatomy, 1942-47; faculty U. Man. (Can.), 1947-52, asso. prof., 1950-52; faculty U. Cal. Med. Center, San Francisco 1952—, prof. anatomy, 1960—, chmn. dept. anatomy, 1963—. State curator for unclaimed dead No. Cal., 1960-66. Mem. Teratology Soc. (past pres.), Am. Assn. Anatomists, Anat. Soc. Gt. Britain and Ireland, Assn. Am. Med. Colls., Sigma Xi. Research, publs. on congenital abnormalities of cardiovascular, urogenital and nervous systems. Home: 1494 Plymouth St., San Francisco 94112.*

MONIZ, Antônio Caetano de Abreau Freire de Egas, see de Egas Moniz, Antônio Caetano de Abreau Freire.

MONK, George Spencer, physicist; b. London, Eng., June 18, 1884; s. John William and Jane Elizabeth (Spencer) M.; came to U. S., 1897, naturalized, 1920; S.B., U. Chgo., 1913, Ph.D., 1923; m. Ardis Ethelyn Thomas, Dec. 17, 1913; 1 son, George Davies.

Staff, Mt. Wilson Obs., 1913-16, Yerkes Obs., 1916-17, Ore. Agrl. Coll., 1918-19, U. Ore., 1919-20; with U. Chgo., 1920-49, asso. prof., 1942-49; sect. chief Manhattan Project, Chgo., 1943-45; sr. physicist Argonne (Ill.) Nat. Lab., 1949-52. Fellow Am. Phys. Soc., Optical Soc. Am.; mem. Sigma Xi. Author: Light-Principles and Experiments, 1937. Patentee in field. Contbr. articles to tech. jours. on spectroscopy and optics. Home: 828 10th St., Boulder, Colo. 80302.*

MONNET, Paul Auguste, French physician; b. Montreal, France, June 23, 1914; s. Paul Clement and Mathilde (Deguerry) M.; student Faculty Medicine, Lyons, France, 1932-40; m. Renée Banette, Apr. 12, 1939; children—Marc, Guy, Colette, Chantal, Alain. Chief lab. Institut Pasteur Lyons, 1953-61, prof. bacteriology, 1960-65, prof. pediatrics, 1965—. Mem. Conseil sup. Hygiene publique de France, 1961—, Commn. sanitaire pour la protection de l'enfance, 1963—. Decorated officer Palmes Academiques. Mem. Société Med. des Hôp. de Lyon, Société Francaise de Pediatrie, Société Francaise de Microbiologie. Author: Dietelique du nourrisson normal et pathologique, 1955; Les Antibiotiques en Pediatrie, 1966; also numerous articles. Research on therapy of infantile gastroenteritis especially polymyxin in therapy of cases due to Escherichia coli, surg. treatment of Tb cervical adenitis, craniostenosis. Home: 106 Pt Herriot, Lyons, Rhone, France.*

MONNIER, Denys, Swiss chemist; b. Geneva, Switzerland, Oct. 23, 1903; s. Alfred and Genevieve M.; Diploma Ing.Chemist, U. Geneva, 1927, Doct.-es Sciences, 1932; m. Eberle Beltg, July 16, 1930; children—Claude, Edmond, Jacques. Faculty, U. Geneva, 1945—, prof. mineral and analytical chemistry, 1959—. Hon. mem. Spanish Soc. Legal Medicine. mem. Editorial com. Analytica-Chemica ACTA, also Chimi Minerale. Author 4 books on analytical chemistry, including Chimie et éléments de chimie analytique. Research in spectrophotometry, high speed. separations using mercury, electro-analysis, fluorometry, neutron activation; trace and ultra trace estimation. Office: U. of Geneva, Rue de Candolle 3, Geneva, Switzerland.*

MONNIER, Louis Marie Marcel Maurice Alexandre, French physiologist; b. Jura, Aug. 25, 1904; s. Marcel and Madeleine (Jarrosson) M.; ed. Sorbonne, Paris, Washington U., St. Louis; dr.honoris causa U. Recife, Brazil; m. Andree Dumont, 1927; children—Francois, Jean-Pierre. Staff, Nat. Center for Sci. Research; instr. Sorbonne, later prof. functional physiology, prof. gen. physiology; vis. prof. Rockefeller Inst., N.Y.C. Recipient Acad. Palms, Nat. Legion of Honor, Order So. Cross. Mem. Paris Acad. Medicine, Besancon Acad. Sci., Royal Soc. Sci. Liège, Italian Soc. Biology, Soc. Biology, Med. Electronics and Biophysics, Biology Assn. Author: L'excitation électrique des tissus, 1934; Actualités neurophysiologiques, 1959; La machine humaine, 1962; also articles. Home: 2, square Montsouris, Paris 14, France.

MONNIER, Marcel, Swiss physiologist, neurophysiologist; b. Ch. de Fas, Switzerland, May 28, 1907; s. Henri and Anna (Kindlimann) M.; student U. Geneva (Switzerland), U. Zurich (Switzerland); postgrad. Northwestern U.; m. Gilberte Monod, Aug. 1, 1937; children—Philippe, François, Bernard, Vincent. Asst. prof. physiology U. Zurich, 1942-49; head dept. applied neurophysiology U. Geneva, 1949-56; head dept. physiology U. Basel (Switzerland), prof. physiology Inst. Physiology, Basel, 1956—. Recipient Prix Dejerine, Sociéte francais de Neurologie, 1940; Prix Bizot, U. Geneva, 1941. Affiliate Royal Soc. Medicine. Author: Physiologie du tronc cérébral, 1944; Der Retinale Blutdruck im gesunden und kranken Organismus, 1946; Topographische Tafein des Hirnstamms der Katze und des Affen für exp.-physiol. Physiologie Untersuchungen, 1949; Atlas for Stereotaxic Brain Research, 1961; Physiologie und Pathophysiologie des vegetativen Nervensystems, 1963; Functions of the Nervous System, 1968; also articles. Research on reticular apparatus and thalamus in animals and man, humoral transmission of sleep and wakefulness, electro-psychopharmacology. Home: Riehen b/Basel, Aeussere Baselstrasse 91, Basel, Switzerland.*

MONOD, Hugues J.V.G.A., Jr., French physiologist; b. Paris, Apr. 19, 1929; s. Guy H.F. and F. (Westercamp) M.; M.D., Facultee de Medecine, Paris, 1956, Prof. agrégé, 1961; m. Janine Doucet, Feb. 24, 1953; children—Alain G., Isabelle L., Xavier A. Research staff Laboratoire de Physiologie du Travail, Centre National de la Recherche Scientifique, 1954-61, now under dir.; prof agrégé physiology Ecole Nationale de Medecine, Anviers, France, 1961-65; prof. agrégé Faculté de Medecine, Paris, 1965—; biologist Hopitaux de Paris. Mem. Soc. Biotypologie et Biometria Humaine (sec.), Societé d'Ergonomie de Langue francoise, Association des Physiologistes de Langue francaise, Ergonomics Research Soc., Internat. Ergonomics Assn., A.A.A.S. Research and numerous publs. on muscular physiology including static contraction, work capacity of local muscular groups, oxygen consumption of local muscular works, oxygen content of venous muscular blood; work physiology and ergonomics, including fatigue, sitting posture. Home: 22 Av. de La Bourdonnais, Paris (7). Office; 91 Bd de l'Hopital, Paris (13) France.

MONOD, Jacques Lucien, French biochemist; b. Paris, Feb. 9, 1910; B.S., 1931; D.Sc., Paris, 1941; hon. Sc.D., U. Chgo., 1965. Asst. prof. zoology U. Paris, 1934-45; Rockefeller fellow Cal. Inst. Tech., 1936; mem. staff Pasteur Inst., from 1945, head dept. microbial physiology, apptd. dir. cellular biochemistry, 1954; apptd. prof. Faculty Scis. U. Paris, 1957. Recipient (with F. Jacob and A. Lwoff) Nobel prize in medicine and physiology, 1965. Mem. Nat. Acad. Scis. Studied induced synthesis of bacterial enzyme, from 1946; introduced (with Jacob) theories of ribonucleic acid and operon, which added to understanding of molecular genetic mechanisms, 1961. Address: Pasteur Institute, Paris, France.

MONOD, Théodore, French zoologist; b. Rouen, France, Apr. 9, 1902; s. Wilfred and Dorina (Monod) M.; Doct.Sc.Nat., Sorbonne, Paris, France, 1926; diploma Ecole Nat. Langues Orientales, Paris, 1938; Doct. (hon.), U. Köln (Germany), 1965, U. Neuchatel (Switzerland), 1967; m. Olga Pickova, Mar. 22, 1930; children—Béatrice (Mrs. Jean-Claude Morlot), Cyrille, Ambroise. Asst., Mus. Nat. D'Histoire Naturelle, 1922-42, prof., 1942; dir. Inst. Francais d'Afrique Noire, 1938-64; prof. faculty sci. U. Dakar, Senegal, 1957-59, dean faculty sci., 1957-58. Named officier Légion d'Honneur (France), 1958; comdr. Ordre du Christ (Portugal); comdr. Mérite Saharien; comdr. Ordre Nat. (Sénégal); officier Ordre Mérite Nat. (Mauritanie); recipient Gold medal Soc. Geography, 1952, Royal Geog. Soc., 1960, Am. Geog. Soc., 1961. Mem. Institut de France, Académie de Marine, Académie des Scis. d'Outre-Mer. Author: Méharées, 1937; L'Hippopotame et le Philosophe, 1943; Bathyfolages, 1954; also numerous articles. Research on fishes, crustaceans, Sahara. Home: 14 quai d'Orleans, Paris, IV, Office: 57 Rue Cuvier, Paris Ve, France.*

MONOSZON, Naum Abramovich, Russian engr., physicist; b. 1913; cand. Tech. Scis. grad. Leningrad Poly., 1938; With Elektrosila, Leningrad, 1938-41, 44-46; div. head Research Inst. Electrophys. Apparatus, 1946—; head lab. high energy dept. Joint Nuclear Research Inst., Dubna. Recipient Stalin prize, 1949, 52, Lenin prize, 1959. Author: The Electromagnetic Feed System of the Synchrophasatron at the Joint Nuclear Research Institute, 1960.

MONRAD, Carl Corydon, Am. chem. engr.; b. Buffalo, Jan. 15, 1905; s. Charles Olaf and Wilhelmina (Lagergren) M.; B.S. in Engring., U. Mich., 1927, M.S., 1928, Ph.D., 1930; m. Christine Alberta Clark, Sept. 26, 1930; children—Margaret Eleanor, Donald Richard. Chem. engr. Standard Oil Co. (Ind.), Whiting, 1930-37; faculty Carnegie-Mellon U., Pitts., 1937—, prof. chem. engring., 1941—, head dept. chem. engring., 1946-65, asso. dean, 1965—. Group exec. Nat. Def. Adv. Commn., Washington, 1940-41; chief butadiene and styrene br. Office of Rubber Dir., Washington, 1942-44; cons. to various petroleum cos.; regional chmn. Engrs. Council for Profl. Devel., 1956-60. Recipient certificate of appreciation U. S. Army, 1951. Mem. Am. Inst. Chem. Engrs., Am. Chem. Soc., Am. Soc. for Engring. Edn., Sigma Xi, Tau Beta Pi, Alpha Chi Sigma, Iota Alpha, Phi Kappa Phi. Research, publs. on heat transfer in furnaces, evaporation of caustic soda, condensing vapor, heat transfer in annuli, transfer processes in falling films, petroleum refining processes, monomers for synthetic rubber. Home: 339 Inglewood Dr., Pitts. 15228.*

MONRAD-KROHN, Georg Herman, Norwegian physician; b. Bergen, Norway, Mar. 14, 1884; s. Hjalmar and Alette Wilhelmine (Dahl) M.-K.; studied in Germany, France; M.D., 1919; m. Maria Gronquist, 1906; m. 2d, Celia Middleton, 1917; m. 3d, Elisabeth Bergljot Hücke Nobel Coucheron Nielsen, 1925; m. 4, Signe Bommen. House physician Riks Hosp., Bergen Municipal Hosp., 1911-12; resident med. officer London County Mental Hosp., Bexley, 1914-16; asst. physician Bethlem Royal Hosp., London, 1916-17; physician Riks Hosp., 1917, prof., chief physician Nerve Clinic, 1922-54; supervising physician State Foster-Home for Epileptics, 1930-54; mem. Med. Faculty, 1941-45. Fellow Royal Coll. Physicians (London); mem. Norwegian Neurol. Soc. (pres.), Oslo Sci. Acad., Royal Coll. Surgeons (Eng.); hon. mem. Med. Soc. Copenhagen, Swedish Med. Soc., neurol. socs. of Paris, Eng., Copenhagen, N.Y. Author: The Clinical Examination of the Nervous System, 1921; On the Dissociation between Voluntary and Emotional Innervation in Facial Paresis of the Central Origin in Brain, 1924; The Neurological Aspect of Leprosy, 1923; Homo Sapiens and the Peace Problem, 1946; Dysprosody, 1947; The Third Element of Speech, 1957. Research on neurology, speech and nervous system. Address: Oslo, Norway.

MONRO, Alexander (Primus), Brit. physician; b. London, Sept. 8, 1697; s. John Monro; M.D., U. Edinburgh; studied under Hawksbee and Whiston in London, dissection under Cheselden, under Boerhaave in Leiden, Netherlands; m. Isabella MacDonald, 1725; children—Alexander secundus, Donald. Became prof. anatomy and surgery Surgeons Col., 1719; apptd. 1st prof. anatomy U. Edinburgh, 1720-64 (formally inducted 1725), lectr., after 1764; physician on field after battle of Prestonpons, 1745. Mem. Medico-Chirurg. Soc. Edinburgh (editor 1st vol. Trans. 1732, 1st sec.). Author: Osteology, A Treatise on the Anatomy of the Human Bones, 6 edits., to 1758; An account of the Inoculation of Small-pox in Scotland, 1764; Collected Works, 1781. Showed that jaundice is caused by obstruction of bile duct. Died July 10, 1767.

MONRO, Alexander (Secundus), Scottish anatomist; b. Edinburgh, Scotland, May 20, 1733; s. Alexander (Primus) and Isabella (MacDonald) M.; entered U. Edinburgh, 1752, M.D., 1755; student London, Paris, Leyden, Berlin; 1 son, Alexander tertius. Coadjutor to his father as prof. anatomy and surgery; lectr., Edinburgh, 1759-1808. Licentiate Coll. Physicians Edinburgh. Mem. Philos. Soc. Edinburgh (sec.). Author: Observations on the Structure and Functions of the Nervous System, 1783; The Structure and Physiology of Fishes explained and compared with those of Man and Other Animals; Description of the Bursae Mucosae of the Human Body, their Structure, Accidents and Diseases and Operations for their Cure, 1788; Experiments on the Nervous System with Opium and Metalline Substances to determine the Nature and Effects of Animal Electricity, 1793; Three Treatises on the Brain, the Eye and the Ear, 1797. Described connection between lateral ventricles of brain (foramen of Monro), 1783. Died Oct. 2, 1817.

MONRO, Alexander (Tertius), Scottish anatomist; b. Edinburgh, Nov. 5, 1773; s. Alexander Monro (Secundus); M.D., U. Edinburgh, 1797; postgrad. in anatomy under Wilson, London, also Paris; m. dau. of Carmichael Smyth, 1800; m. 2d, dau. of David Hunter, 1836; 12 children. Asst. to his father, 1798, named joint prof. (with father) anat. medicine, surgery U. Edinburgh, 1800, taught alone, from 1808, prof., 1817-46. Author: Observations on Crural Hernia, 1803; Morbid Anatomy of the Human Gullet, Stomach and Intestines, 1811; Outlines of the Anatomy of the Human Body, 1813; Engravings of the Thoracic and Abdominal Viscera, 1814; Observations of the different kinds of Small-pox, 1818; Morbid Anatomy of the Brain, 1827; Anatomy of the Urinary Bladder and Perinaeum in the Male, 1842. Died Mar. 10, 1859.

MONRO, Thomas Kirkpatrick, Scottish physician; b. Arbroath, Scotland, Sept. 6, 1865; s. William and Jane (Kirkpatrick) M.; M.A., M.D., LL.D., Glasgow (Scotland) U.; m. Jane Christian, 1905 (dec. 1951); 2 sons, 1 dau. Pathologist Victoria Infirmary; examiner in pathology Glasgow U., regius prof. medicine, 1913-36, later emeritus; examiner in medicine Dublin (Ireland), Durham (Eng.), St. Andrews (Scotland) univs.; physician, clin. lectr., later mgr. Glasgow Royal Infirmary; prof. medicine, dean med. faculty St. Mungo's Coll.; sr. physician Western Infirmary, Glasgow. Dir. Glasgow Royal Maternity and Women's Hosp.; mem. exec. com. orphan homes Scotland, colony for epileptics; gov. Royal Tech. Coll., Glasgow; pres. Scottish Western Asylums Research Inst., Med. and Dental Defence Union Scotland Ltd. Fellow Royal Faculty Physicians and Surgeons Glasgow (hon., examiner, pres.); mem. Assn. Physicians Gt. Britain and Ireland (hon., pres.), Royal Medico-Chirurg. Soc. Glasgow (hon., pres.), Brit. Med. Assn. Author: History of the Chronic Degenerative Diseases of the Central Nervous System, 1895; Raynaud's Disease, 1899; Manual of Medicine, 5th edit., 1925; The Physician as Man of Letters, Science and Action, 1933. Sr. editor: Glasgow Med. Jour. Died Jan. 10, 1958.

MONROE, Russell Ronald, Am. psychiatrist; b. Des Moines, June 7, 1920; s. Ronald Russell and Mildred (Schmidt) M.; B.S., Yale, 1942, M.D., 1944; m. Lillian Brooks, June 23, 1945; children—Constance Ellen, Nancy Brooks, Russell Ronald. Faculty, Tulane U., 1950-60; asso. psychoanalyst Columbia, 1950-53; prof., dir. grad. tng. Psychiat. Inst., U. Md. Sch. Medicine, Balt., 1960—. Mem. Acad. Psychoanalysis, Am. Psychiat. Assn., Am. Psychosomatic Soc., Sigma Xi, Alpha Omega Alpha. Research, numerous publs. on correlation between brain mechanisms and human behavior, psychotherapeutic techniques and psychiat. edn. Home: 201 Bolton Pl., Balt. 21201.*

MONROY, Alberto, Italian biologist; b. Palermo, Italy, July 26, 1913; s. Antonio and Emma (Jaforte) M.; M.D., U. Palermo, 1937; m. Anna Oddo, June 19, 1943; children—Gabriella, Valentina, Beatrice. Asst. prof. anatomy Med. Sch., Palermo, 1938-44; head physiology lab. Stazione Zoologica, Naples, Italy, 1945-51; prof. comparative anatomy faculty scis. U.

Palermo, 1952—. Chmn. biology com. EUROTOM, 1963—. Mem. Internat. Inst. Embryology (sec. 1961—), N.Y. Acad. Scis. Author: Chemistry and Physiology of Fertilization, 1965; also numerous articles. Research in chem., molecular aspects of embryonic devel. Home: 91 Via Villafranca, 90140 Palermo. Office: Istituto di Anatomia Comparata, via Archirafi 20, Palermo, Italy.*

MONSELISE, Shaul Paul, plant physiologist; b. Milano, Italy, Apr. 23, 1920; s. Maurizio and Pierina (Ottolenghi) M.; M.Sc., Hebrew U., Jerusalem, Israel, 1944, Ph.D., 1950; m. Rachel Moshejov, Feb. 19, 1948; children—Nira-Beracha, Edna, Dan-Yoel. Faculty, Faculty Agr., Hebrew U., 1945—, asso. prof. horticulture, 1963—, head dept. citricultures, 1959—; research asso. U. Cal. at Riverside, 1955 E. Malling Research Sta., Kent, Eng., 1964; research fellow Cal. Inst. Tech., Pasadena, 1956. Leader research projects sponsored by Ford Found., U. S. Dept. Agr., IAEA; mem. steering bd. citrus industry Ministry Agr., Israel, 1965—. Mem. Israeli (pres. 1962—, Am. bot. socs., Am. Soc. Hort. Sci. (Alex Laurie award in floriculture and ornamental horticulture 1964), A.A.A.S., Am., Scandinavian socs. plant physiologists, Internat., Fla. State hort. socs., Am. Weed Soc., Internat. Orgn. Citrus Virologists. Research, publs. on gladiolus and iris, growth of citrus shoots and roots, illumination and carbon dioxide in citrus groves, growth analysis of seedlings, nucleic acid and chlorophyll in citrus leaves, seed biology, flower induction, physiology and enzymology of citrus fruits, natural growth regulators, effects ionizing radiation on citrus fruits. Home: 64 Jacob, Rehovot, Israel.*

MONSEN, Frederick Imman, explorer; b. Bergen, Norway, July 8, 1865; s. Hans and Anna Sophia (von Branneberg) M.; ed. under pvt. tutors; Ph.D., U. Christiania, 1910; studied art; came to U. S., 1880; m. Harriet Van Anden, Jan. 15, 1892; children—Frederick Courtenay, Hans Shavenau. Engaged in western exploration, geol. survey, 1887; with Salton Sea expdn., 1891; explored Lower Cal., 1892, Death Valley, other Cal. deserts, 1893; made investigations among N.M. Indians, 1894-95; artist and topographer Yosemite Nat. Park Boundary Survey, 1896, Alaska Coast Survey, 1897, Yukon Exploration, 1898; investigations among Indian tribes of Ariz., N.M., Cal., Mexico, 1896—; Mexican, Central and S. Am. travels and exploration, 1906-09; travels and observations in W.I. and the Orinoco River, Venezuela, 1909-11, Senora and Chihuahua Deserts, Mexico, 1914. Fellow Royal Geog. Soc., Eng., Am. Nat., and Phila. geog. socs.; mem. A.A.A.S., Am. Anthrop. Assn. Professional lectr. on Am. history and exploration, on ethnol. and history of Indians of Southwestern U. S., geog. subjects; author and illustrator of many reports and mag. articles on exploration and travel among primitive people and in little known lands. Died Nov. 10, 1929.

MONTAGNA, William, biologist; b. Roccacasale, Italy, July 6, 1913; s. Charles and Adele (Giannangelo) M.; came to U. S., 1927, naturalized, 1927; A.B., Bethany Coll., 1936, D.Sc., 1960; Ph.D., Cornell U., 1944; D. Biol. Sci., Universita degli Studi of Sassari (Sardinia), 1964; m. Martha Helen Fife, Sept. 1, 1939; children—Eleanor, Margaret, John and James (twins). Faculty, Brown U., 1948-63, prof., 1952-63, L. Herbert Ballou prof. biology, 1960-63; prof., head div. exptl. biology U. Ore. Med. Sch., dir. Ore. Regional Primate Research Center, Beaverton, 1963—; vis. prof. U. Cin., 1958, Wayne State U., 1959; lectr. dermatology Mass. Gen. Hosp., Med. Sch. Harvard, 1961-62. Mem. various coms. Nat. Acad. Scis., NIH; cons. gen. med. sci. NIH; 1st Carl Herzog lectr. Am. Dermatol. Assn., 1963. Recipient Spl. award Soc. Cosmetic Chemists, 1957; Gold award Am. Acad. Dermatology, 1958; Order Merit Italian Republic (Cavaliere), 1963. Fellow A.A.A.S., N.Y. Acad. Scis.; mem. Acad. Dermatology and Syphilology, Am. Assn. Anatomists, Am. Assn. Zoology, Histochem. Soc., Internat. Coll. Exptl. Dermatology, Soc. Gerontology, Soc. Investigative Dermatology, Brit. Assn. Dermatology, Sigma Xi, Phi Kappa Phi. Research, numerous publs. on structure, function of human skin, description of several cutaneous systems in detail, comparative primate skins, phenomenon of baldness. Home: 5323 S.W. Hewett Blvd., Portland, Ore. 97221. Office: 505 N.W. 185th Av., Beaverton, Ore. 97005.*

MONTAGU, (Montague Francis) Ashley, anthropologist, social biologist; b. London, Eng., June 28, 1905; s. Charles Ashley and Mary (Plot) M.; student U. of London, 1922-25, U. Florence, 1928-29; Ph.D., Columbia, 1937; m. Marjorie Helen Peakes, Sept. 18, 1931; children—Audrey, Barbara, Geoffrey. Came to U. S., 1930, naturalized, 1940. Research asso. British Mus. Natural History, London, 1926-27, curator phys. anthropology Wellcome Hist. Med. Mus., 1929-30; asst. prof. anatomy N.Y. U., 1931-38; asso. prof. anatomy Hahnemann Med. Coll. and Hosp., Phila., 1938-49; chmn. dept. anthropology Rutgers U., 1949-55; vis. lectr. dept. social sci., Harvard, 1945; sr. lectr. Vets. Adminstrn. Postgrad. Training program since 1946; lectr. New Sch. Social Research, 1931, 1948-59; vis. prof. U. Del., 1955; Regents prof., U. Cal. at Santa Barbara, 1962; dir. research N.J. com. phys. development and health, 1953-57; family affairs editor, anthrop. adv. NBC-TV, 1954. Chairman of Anisfield-Wolf Award Committee. Guardsman with the Welsh Guards in 1919. Recipient of the first prize, and also Morris Chaim prize, Centennary 2d Dist. Dental Soc., N.Y., 1936; Chicago Forum Literary contest, 1943. Expert witness on legal, sci. problems relating to race since 1930. Produced, financed, wrote and directed film, One World or None, 1946. Rapporteur responsible for drafting statement on race for UNESCO, 1949-50; cons. UNESCO, 1949. Hon. corr. mem. anthrop. societies Paris and Florence. Fellow Am. Assn. for the Advancement of Sci., Royal Soc. Medicine London; mem. PEN, Am. Assn. Phys. Anthropologists, International Society for the study of Race Relations, Am. Assn. Anatomists, Hist. Sci. Soc., Am. Soc. Study Child Growth and Development, Sigma Xi. Author: Coming into Being Among the Australian Aborigines, 1937; Man's Most Dangerous Myth: The Fallacy of Race, 1942; Edward Tyson, M.D., F.R.S., (1650-1708); and the rise of human and comparative anatomy in England, 1943; Introduction to Physical Anthropology, 1945; Adolescent Sterility, 1946; On Being Human, 1950; On Being Intelligent, 1951; Statement on Race, 1952; Darwin, Competition and Cooperation, 1952; The Natural Superiority of Women, 1953; The Direction of Human Development, 1955; Immortality, 1955, Biosocial Nature of Man, 1956; Anthropology—And Human Nature, 1957; Man: His First Million Years, 1957; The Reproductive Development of the Female, 1957; Education and Human Relations, 1958; The Cultured Man, 1958; Human Heredity, 1959; Handbook of Anthropometry, 1960; Man in Process, 1961; Prenatal Influences, 1962; The Humanization of Man, 1962; Race, Science, and Humanity, 1963; (with J. Lilly) The Dolphin in History, 1963; Life Before Birth, 1964; The Science of Man, 1964; (with E. Steer) Anatomy and Physiology, 1959; (with C. L. Brace) Man's Evolution, 1965; The Idea of Race, 1965; The Human Revolution, 1965; Up the Ivy, 1966; The American Way of Life, 1967. Adv. editor Acta Genet, Medic and Gemell. Developed theory of adolescent sterility. Home: Cherry Hill Rd., Princeton, N.J. 08540.*

MONTAGUE, Joel Benjamin, Jr., Am. sociologist; b. Edgerton, Mo., July 17, 1912; s. Joe B. and Susan (Sodeman) M.; B.S. in Edn., Central Mo. State Coll., 1935; M.A., Colo. State Coll., 1938; Ph.D., Mich. State U., 1947; postgrad. London Sch. Econs., 1950-51; m. Evelyn Catherine Perry, June 20, 1941; children—Ann Marie, Charles Rama V. High sch. prin., tchr., Holt, Mo., 1935-37; ednl. adviser Civilian Conservation Corps, Colo.-Wyo. Dist., 1938-40; instr. U. Tenn., 1946; faculty Wash. State U., Pullman, 1946—, prof. sociology, 1959—; vis. prof. U. New South Wales, Sydney, Australia, 1964. Mem. Am., Pacific sociol. assns., Am. Assn. U. Profs., Alpha Kappa Delta. Author: Class and Nationality: English and American Studies, 1963; also articles. Research on comparative (U. S., Eng., Australia) studies of social stratification, edn., professionalism and orgn. of med. services. Home: 1603 Charlotte St., Pullman, Wash. 99163.*

MONTAGUE, Joseph Franklin, Am. physician, surgeon; b. N.Y.C., Aug. 6, 1895; s. Joseph and Mary Frances (LeDieu) M.; M.D., N.Y. U., 1917. Began practice, 1917, specialist in intestinal diseases. Formerly mem., lectr. Rectal Clinic, Bellevue Med. Coll.; surgeon N.Y. State Nautical Sch., 1919, N.Y. State Indsl. Commn., 1920; founder Med. Writers' Inst.; cons. Instrument Research Inst., 1958—; prof. affiliate in communications Colo. State U. Fellow Soc. Nuclear Medicine, Internat. Coll. Surgeons, Am. Geog. Soc., Am. Pub. Health Assn., Am. Geriatric Soc., Am. Med. Writers Assn. (pres. N.Y. chpt.), Am. Coll. Nutrition, Am. Coll. Gastroenterology (asso.); mem. A.M.A., Am. Med. Authors (pres.), Soc. Tech. Writers and Editors, World Med. Assn., N.Y. Acad. Sci., A.A.A.S., Order Mil. Surgeons, Health Guild Am., Am. Anthrop. Assn., Am. Soc. Tropical Medicine, Am. Assn. History Medicine, Am. Chem. Soc., Instrument Soc. Am., Am. Astron. Soc., Authors Guild Am., Assn. Profl. Translators, Soc. Tech. Writers, Gourmet Soc. (pres), Am. Polar Soc., Royal Soc. Health (London), Nat. Assn. Sci. Writers, Nu Sigma Nu. Author: Pruritus of the Perineum, 1923; The Modern Treatment of Hemorrhoids, 1926; Troubles We Don't Talk About, 1927; Enfermedades que no se habla, 1927; Maux dont On Ne Parle Pas, 1927; Taking the Doctor's Pulse, 1928; Psyllium Seed—The Latest Laxative, 1932; I Know Just the Thing for That, 1934; Why Bring That Up, 1935; How to Conquer Constipation, 1938, Broadway Stomach, 1940; Nervous Stomach Trouble, 1941; How to Overcome Nervous Stomach Trouble, 1943; Desordenes Nerviosos del Estomago, 1948; Mali Chi Si Taccioni, 1956; Leiden Von Denen Man Nicht Soricht, 1956; Disturbi Nervosi dello Stomaco, 1957; How to Overcome Colitis, 1957; How to Conquer Nervous Stomach Trouble, 1964; The Why of Albert Schweitzer. Editor in chief Health Digest mag.; cons. editor Clin. Medicine, Med. Digest. Contbr. Jour. A.M.A., Internat. Clinics, Internat. Record of Medicine. Pioneer in writing popular health books and in application of motion pictures to surg. edn.; inventor sigmoidoscopes and proctologic instruments. Address: 104 E. 40th St., N.Y.C. 10016.*

MONTALEMBERT, Marc René, Marquis de, see de Montalembert, Marc René, Marquis.

MONTALENTI, Giuseppe, Italian biologist; b. Asti, Italy, Dec. 13, 1904; s. Paolo and Ida (Bertola) M.; Sc.D., U. Rome, 1926; m. Luciana Fratini, Mar. 19, 1964. Asst., Inst. Zoology, U. Rome, 1926-37, Inst. Zoology, U. Bologna, 1937-39; head dept. zoology, Zool. Sta., Naples, 1939-44; prof. genetics U. Naples, 1944-60; prof. genetics U. Rome, dir. Inst. Genetics, 1960—. Mem. Accademia Nazionale dei Lei, Nat. Soc. Sci. and Letters of Naples, Accademia Nazionale dei XL, Royal Acad. Scis. Sweden (fgn. mem.), Internat. Union Biol. Scis. (sec. gen. 1953-58, pres. 1958-61). Author: Il Prodromo di Nicola Stenone, 1928; Lazzaro Spallanzani, 1928; Elementi di Genetica, 1939; Problemi di Biologia della Riproduzione, 1945; Compendio di Embriologia, 1945, 54; Storia della Biologia e della Medicina, 1962; L'Evoluzione, 1965. Publs. on exptl. parthenogenesis, physiology of fertilization in lampreys; genetic studies of animals, man; history of biology. Home: 25 via Asmara, Rome, Italy.*

MONTANARI, Geminiano, Italian astronomer; b. Modena, Italy, 1633; ed. Florence; mathematician to Alphonso IV, duke of Modena; prof. math., Bologna; named prof. astronomy, Padua, 1674. Author: Pensieri phisicimatematici, 1665; Physikalischen Spekulationem, 1674. Research on capillary tubes, 1667, also on power of winds, hearing horns; made astron. discoveries (using reticle telescope and hydrometer); determined mountain heights by means of barometer (credited as discoverer of method). Died 1687.

MONTE, see dal Monte, Guidubaldo.

MONTEL, Paul Antoine, French mathematician; b. Nice, France, Apr. 29, 1876; s. Aristide and Anais (Magiolo) M.; Agrégé, Ecole normale supérieure, 1897; D.Sc., univs. Warsaw, Cluj, Cernauti, Sofia, Lima, Liège; m. Berthe Perrinel, Feb. 9, 1953. Prof., Lyceum of Poitiers, 1898-1901, Lyceums at Nantes, Paris, 1904-11; asst. lectr., lectr., prof. U. Paris, dean Faculté des Sciences, 1911-46; prof. École normale supérieure des beaux-arts, 1913-33, also dir. Annales scientifiques; asst. prof. analysis École polytechnique, 1913-18. Mem. Bd. Inventions, 1915-18, Inst. Applied Sci. Research, Nat. Def.; titular mem. Bur. des Longitudes. Mem. French Acad. Scis., Royal Belgian Acad., Rumanian Acad., Acad. Scis., Rumania, Polish Acad. Scis. and Letters, acads. scis. Peru, Buenos Aires, Inst. Coimbra, Royal Soc. Liege, Warsaw Soc. Scis. and Letters, Argentine Sci. Soc.; hon. mem. French, Rumanian Belgian, Polish math. socs. Research and publs. on analysis, geometry, mechanics. Address: 79, rue du Faubourg-Saint-Jacques, Paris 14e, France.

MONTESQUIEU, Charles Louis de Secondat, Baron de La Brède et de, French polit. philosopher; b. La Brède, nr. Bordeaux, France, Jan. 1689; s. Jacques de Secondat and Marie Françoise de Pénel; ed. Oratorian Sch. Juilly (France), 1700-05, U. Bordeaux; m. Jeanne Lartigue, 1715; children—Charles, Denise. Counsellor parliament, from 1714; inherited uncle's title and position, president à mortier, 1716-26; went to Paris, 1722; toured Europe to observe people and ways of life, from 1728; returned to La Brède, devoted himself to writing, circa 1731. Fellow Royal Soc., 1730; mem. French Acad., 1727. Author: Lettres persanes (one of early books of Philosophe movement), 1721; Considérations sur les causes de la grandeur des Romains et de leur décadence (one of 1st important essays on philosophy of history), 1734; De l'Esprit des Lois (comparative study of republics, monarchies and despotism), 1748. Believed that laws are concrete social facts which arise from nature of world, that one must seek spirit of laws to understand human socs. and that entire universe is regulated by laws; felt purpose of polit. sci. is to find truth and rationality in situations and from this evolve principles to guide legislators (concept distinct from earlier theories); other theories: climate and circumstances determine forms of govt.; powers of govt. should be separated and balanced, liberty is right of doing what laws permit; his theories used by moderate reforming party in France for many years. Died Paris, Feb. 10, 1755.

MONTESSUS DE BALLORE, Fernand de, see de Montessus de Ballore, Fernand.

MONTGOLFIER, Jacques Étienne, French inventor; b. Vidalon-lès-Annonay, France, Jan. 7, 1745; s. Pierre and Marguerite M.; studied architecture under Soufflot. Practiced architecture; later took over father's paper bus. Mem. French Acad. Scis. Recipient (with bro. Joseph) prize for their work; Order Saint-Michel from Louis XVI. Author: (with brother Joseph) Mémoires sur la machine aérostatique, 1784, Ballons aérostatique, 1784, les Voyageurs aériens, 1784. Built (with bro.) 1st successful man-carrying balloon using heated air, 1783. Died Serrières, France, Aug. 2, 1799.

MONTGOLFIER, Joseph Michel, French inventor; b. Vidalon-lès-Annonay, France, Aug. 26, 1740; s. Pierre and Marguerite M.; m. Therese Filhol; 5 children. Worked in family paper factory; founder textile plant, Voiron, which went bankrupt during revolution; joined Bur. Arts and Manufactures, Paris, after revolution; adminstr. Conservatory Facts and Trades. Recipient (with brother) prize for work; named Chevalier de la legion d'honneur by Napoleon. Mem. French Acad. Scis. Author: Discours sur l'aérostat, 1784; (with his brother Jacques) Mémoires sur la machine aérostatique, 1784, Ballons aérostatique, 1784, les Voyageurs aériens, 1784; Note sur le belier hydraulique, 1803. Built (with bro. Jacques) 1st successful man-carrying balloon using heated air, 1783; invented parachute, also device which raised water to 60 ft. Died Balarue-les-Bains, France, June 26, 1810.

MONTGOMERY, Deane, Am. mathematician, educator; b. Weaver, Minn., Sept. 2, 1909; s. Richard and Florence (Hitchcock) M.; B.S., Hamline U., 1929, hon. dr., 1954; M.S., U. Ia., 1930, Ph.D., 1933; hon. degree Yeshiva U., 1961; m. Katherine Fulton, July 14, 1933; children—Mary R. (Mrs. Henry D'Arcy Heck), Richard D. NRC fellow Harvard, 1933-34; NRC fellow Inst. for Advanced Study, Princeton, N.J., 1934-35, mem., 1948-51, prof. math., 1951—; asst. prof. math. Smith Coll., 1935-38, asso. prof., 1938-42, prof., 1942-46; asso. prof. math. Yale, 1946-48. Vis. asso. prof. Princeton, 1943-45; mem. project NDRC, 1945-46. Guggenheim fellow, 1941-42. Mem. Am. Math. Soc. (pres. 1960-62), Nat. Acad. Scis., Math. Assn. Am., Am. Philos. Soc., Am. Acad. Arts and Scis. Study of topology and topological groups. Home: 55 Rollingmeade St., Princeton, N.J. 08540.

MONTGOMERY, Donald Joseph, Am. physicist, educator; b. Cin., June 11, 1917; s. Robert John and Stella (Steffen) M.; Chem.E., U. Cin., 1939, Ph.D., 1945; postgrad. Cornell U.; m. Mary Miller, July 27, 1951; children—Denis Broyles, Malcolm David, Steven Michael, Laurence Matthew. Instr. elec. engring. U. Cin., 1942-44; physicist Flight Propulsion Lab., Nat. Adv. Com. for Aeros., Cleve., 1944-45; research asso. physics, asst. prof. Princeton, 1945-46; sci. liaison officer Office Naval Research, London, Eng., 1947-48; research fellow physics U. Manchester (Eng.), 1947-48; chief spl. problems br. Interior Ballistics Lab., Ballistic Research Labs., Aberdeen Proving Ground, Md., 1948-50; head gen. physics sect. Textile Research Inst., Princeton, N.J., 1950-53; asso. prof. physics Mich. State U., East Lansing, 1953-56, prof., 1956-61, research prof. physics, prof. engring. research, 1961—. Cons. Chemstrand Research Center, Durham, N.C., 1956—; spl. asst. to dir. Office Grants and Research Contracts, NASA, 1964-65; Recipient Distinguished Faculty award Mich. State U., 1961. Fulbright lectr. in physics, Guggenheim fellow U. Grenoble (France), 1959-60. Fellow Am. Phys. Soc.; mem. Am. Nucleonic Soc. (charter), Am. Assn. Physics Tchrs., A.A.A.S., Biophys. Soc., Textile Research Inst., Am. Soc. for Engring. Edn., N.Y. Acad. Scis., Sigma Xi. Author: Cosmic Ray Physics, 1949; also chpts. in books, articles, revs. Research on cosmic physics, electrostatics and solid-state physics. Home: 2391 Shawnee Trail, Okemos, Mich. 48864. Office: Physics-Math. Bldg., Mich. State U., East Lansing, Mich. 48823.*

MONTGOMERY, Hamilton, Am. physician; b. Chgo., May 21, 1898; s. Frank Hugh and Caroline (Williamson) M.; A.B., Harvard, 1918, M.D., 1922; M.S. in Dermatology and Syphilology, U. Minn., 1927; m. Beatrice A. Jennison, June 19, 1924; children—Beatrice (Mrs. Wolfe Frankl), Hamilton Montgomery. Practice medicine specializing in dermatology, Chgo., 1922-25; faculty Mayo Found., U. Minn. Grad. Sch., Rochester, 1929—, prof. dermatology and syphilology, 1949-59, emeritus prof., 1959—. Diplomate Am. Bd. Dermatology and Syphilogy, Mem. Minn., Chgo. dermatol. socs., Soc. for Investigative Dermatology, Am. Dermatol. Assn., Am. Acad. Dermatology and Syphilogy (mem. com. on nomenclature 1950—), Sigma Xi, Nu Sigma Nu. Author: (with Oliver S. Ormsby) Diseases of the Skin, 1943, 8th edit., 1954; Dermatopathology, 2 vols., 1967; also numerous articles. Research on pathology, clin. states skin diseases, description new diseases. Home: 1510 Damon Ct. Office: Mayo Clinic, Rochester, Minn. 55901.*

MONTGOMERY, Hugh, Am. physician; b. Austin, Tex., Apr. 17, 1904; s. Thomas H. and Priscilla (Braislin) M.; B.S., Haverford Coll., 1925; M.D., Harvard, 1930; m. Esther Howland, June 28, 1930; children—Charles H., Priscilla (Mrs. Elmer Makay), Susan White (Mrs. David G. Howell). Practice medicine, specializing in vascular disease, Phila., 1933—; chief vascular sect. Hosp. U. Pa., 1937—; prof. medicine U. Pa., 1960—. Mem. Assn. Am. Physicians, Am. Physiol. Soc., Am. Clin. and Climatol. Assn., Am. Soc. Clin. Investigation. Contbr. numerous articles to profl. jours. Established pH of glomerular and tubular fluid; pioneer in amperimetric studies of oxygen tension in intact, living human tissue and effect of drugs on peripheral circulation. Home: 119 Glenn Rd., Ardmore, Pa. 19003. Office: Gates Bldg., U. Hosp., Phila. 19104.*

MONTGOMERY, John Atterbury, Am. chemist; b. Greenville, Miss., Mar. 29, 1924; s. D. Cameron and Ruth (Atterbury) M.; A.B. cum laude, Vanderbilt U., 1946, M.S., 1947; Ph.D., U. N.C., 1951; m. Jean Kirkman, July 19, 1947; children—John Atterbury, Elaine Porter, Kirkman, Ruth Adrianne. Faculty, U. N.C., 1947-52; chmn. So. Research Inst., Birmingham, Ala., 1952-55, head organic synthesis sect., 1955-56, head organic chemistry div., 1956-64, dir. organic chemistry research, 1964—. Mem. Am. Chem. Soc., Am. Assn. Cancer Research, Sigma Xi. Contbr. chpt. to Advances in Carbohydrate Chemistry, 1963; numerous articles to profl. jours. Research in synthesis of heterocyclic compounds for evaluation as potential medicinals, mechanism of action of anticancer agts. Home: 3596 Springhill Rd., Birmingham 35223. Office: 2000 9th Av. S., Birmingham, Ala. 35205.*

MONTGOMERY, John Dickey, Am. polit. scientist; b. Evanston, Ill., Feb. 15, 1920; s. Charles William and Lora (Dickey) M.; B.A., Kalamazoo Coll., 1941, M.A., 1942, LL.D., 1962; M.A., Harvard, 1948, Ph.D., 1951; m. Jane Ireland, Dec. 19, 1954; children—Faith, Patience, John. Chmn. govt. and law Babson Inst., Wellesley, Mass., 1947-57, dean faculty, 1953-57; head acad. instrn. sect. Viet Nam project Mich. State U., Saigon, 1957-59; research fellow Council on Fgn. Relations, N.Y.C., 1959-60, asso. dir. African studies program, 1962-63; dir. Devel. Research Center, African Studies Program, Boston U., 1960-63; prof. pub. adminstrn. Harvard, 1963—. Mem. exec. bd. Study Fellowships in Internat. Devel. 1963—. Guggenheim fellow, 1955. Mem. Soc. for Internat. Devel. (chmn. com. on internat. devel. research 1965—), Am. Soc. for Pub. Adminstrn. (mem. comparative adminstrn. group 1961—, exec. com. 1963—), Am. Polit. Sci. Assn., Am. Soc. for Polit. and Legal Research. Author: Forced to be Free, 1957; Cases in Vietnamese Administration, 1958; Politics of Foreign Aid, 1962; (with William Siffin) Approaches to Development, 1966; Foreign Aid in International Politics, 1967; also articles. Comparative studies mil. govt.; research on polit. democratization Germany and Japan, relationships between fgn. aid and politico-econ. devel., adminstrn. U. S. fgn. policy. Home: 33 Whitney Rd., Newtonville, Mass. 02160. Office: Harvard, Littauer Center, Cambridge, Mass. 02138.*

MONTGOMERY, Philip O'Bryan, Jr., Am. pathologist; b. Dallas, Aug. 16, 1921; s. Philip O'Bryan and Francis (Hench) M.; B.S., So. Meth. U., 1942; M.D., Columbia, 1945; m. Ruth Ann Rogers, June 20, 1953; children—Philip O'Bryan III, Carter Rogers, Will Stuart, Harold Hench. Faculty, Southwestern Med. Sch., Dallas, 1950-51, 52—, prof. pathology, 1961—; research asst. in pathology and cancer research Cancer Research Inst., New Eng. Deaconess Hosp., Boston, 1951-52; pathologist Parkland Meml. Hosp., Dallas, 1952—; cons. pathologist various hosps., 1952—. Mem. pathology B study sect. NIH. Recipient Career Devel. award NIH, 1962—; Sci. adv. com. Damon Runyan Fund, 1966—. Diplomate Am. Bd. Pathology. Fellow Am. Soc. Clin. Pathologists, Coll. Am. Pathologists, Royal Microscop. Soc., N.Y. Acad. Scis.; mem. Am. Assn. Pathologists and Bacteriologists, Am. Assn. Cancer Research, Internat. Acad. Pathology Am. Acad. Forensic Scis., Soc. Exptl. Biology and Medicine, Law-Sci. Acad. Am. (co-founder), Law-Sci. Found. (co-founder), Am. Soc. Exptl. Pathology, A.A.A.S., Optical Soc. Am., Am. Assn. U. Profs. Am. Soc. Cell Biology, Tissue Culture Assn., Pan Am., Am., Aerospace, So. med. assns., Internat. Fedn. Med. Electronics, Biophys. Soc., Internat. Soc. Cell Biology, others. Research, publs. on path. aspects of medicolegal cases; ultraviolet irradiation and microscopy; time-lapse photography; cell ultra-structure; carcinogenesis and nucleolar structure and function. Home: 6343 Kalani Pl., Dallas 75240.*

MONTGOMERY, Raymond Braislin, Am. oceanographer, b. Phila., May 5, 1910; s. Thomas Harrison and Priscilla (Braislin) M.; A.B., Harvard, 1932; S.M., Mass. Inst. Tech., 1934, Sc.D., 1938; m. Mary Eleanor Perkins, Apr. 22, 1944; children—Kate, Mary T., Eleanor. Phys. oceanographer Woods Hole Oceanographic Instn., 1940-49; vis. prof. Brown U., 1949-54; faculty Johns Hopkins, 1954—, prof. oceanography, 1961—. Mem. Am. Soc. Limnology and Oceanography, Oceanographical Soc. Japan. Research on ocean currents and distbns. of water characteristics. With T. Cromwell and E. D. Stroup discovered Pacific Equatorial Undercurrent. Home: 5722 Cross Country Blvd., Balt. 21209.*

MONTGOMERY, Rex, biochemist; b. Halesowen, Eng., Sept. 4, 1923; s. Fred and Jennie (Holloway) M.; B.Sc. with 1st class honors, U. Birmingham (Eng.), 1943, Ph.D., 1946, D.Sc., 1963; m. Barbara W. Price, Aug. 9, 1948; children—Ian, David, Jennifer Win, Christopher. Came to U. S., 1948, naturalized, 1963. Research chemist Dunlop Rubber Co., 1946-47; sci. officer Brit. Govt., 1947-48; postdoctoral fellow Ohio State U., 1948-49; Sugar Research Found. fellow U. S. Dept. Agr., 1949-51; research asso. U. Minn., 1951-55; faculty U. Ia., Iowa City, 1955, prof. biochemistry, 1964—. Mem. Biochem. Soc. (London), Am. Soc. Biol. Chemists, Am. Chem. Soc., Chem. Soc. (London), Sigma Xi. Author: (with Fred Smith) The Chemistry of Plant Gums and Mucilages, 1959; also numerous articles. Research on structures polysaccharides and glycoproteins. Home: 5 Princeton Ct., Iowa City 52240.*

MONTGOMERY, William Fetherston, Irish obstetrician; b. Ireland, 1797; practiced medicine, also prof. obstetrics, Dublin; contbr. papers to med. publs.; described epithelial depressions in uterine mucosa (Montgomery's cups), 1837. Died 1859.

MONTIAS, John Michael, economist; b. Paris, France, Oct. 3, 1928; s. Santiago and Gisele (Robin) M.; B.A., Columbia, 1947, M.A., 1949, certificate Russian Inst., 1950; m. Mania Montias; 1 son, Jean-Luc. With UN, Geneva, Switzerland, 1950, Beirut, 1950-52, N.Y.C., 1952-53; research supr. Mid-European Studies Center, 1953-54; faculty Yale, 1958—, prof. econs., 1964—; vis. asso. prof. Ind. U., 1960-61. Mem. Council on Fgn. Affairs, Am. Econ. Assn., Econometric Soc., A.A.A.S. (mem. directing bd. 1967—). Author: Central Planning in Poland, 1962; Economic Development in Rumania, 1967; also articles. Constructed math. models for representation of actual planning procedures in Soviet-type economies; studied relation between consistency and econ. efficiency of plans from a set of material balances; simulated indsl. process of Soviet-type economy with gen. characteristics of Rumania using computer to shed light on factors underlying trends in imports of capital goods; research on econ. devel. in Czechoslovakia, Rumania. Home: 19 Howard Av., New Haven 06519. Office: Econ. Growth Center, 52 Hillhouse Av., New Haven.*

MONTMORT, Pierre Rémond de, see de Montmort, Pierre Rémond.

MONTREUIL, Jean, French biochemist; b. Lille, France, Oct. 11, 1920; s. Maurice and Jeanne (Liagre) M.; pharmacien U. Lille, 1945, D.Sc., 1952; m. Janine Dupagny, July 4, 1945; children—Michele, Francoise, Jacques. Faculty, U. Lille, 1948—, prof. biochemistry Faculty of Sci., 1962—; staff Lille Inst. Cancer Research, 1948—, chief cellular biochemistry dept., 1966—. Decorated officier des Palmes Académiques. Mem. Société de Chimie Biologique, Société Chimique de France, Société de Biologie France. Author: Biochimie Generale, 1958; also numerous articles. Discovered new polysaccharides, lactotransferin and A-globulins from human milk; studies on metabolism and structure of nucleic acids and glycoproteins of normal and cancerous tissues. Home: 145 Jules Boucly, Flers-Lez-Lille. Office: Faculté des Sciences, B.P. 36, 59 Lille Distribution, France.*

MONTUCLA, Jean Étienne, French mathematician; b. Lyons, France, Sept. 5, 1725. Began law practice in Toulouse, France; editor and royal censor for sci. works, Paris; asso. with Gazette de France; French (1796), Berlin acads. scis.; 1st historian of math. in France. Author: Histoire des recherches sur la quadrature du cercle, 1754; Histoire des mathématiques, 2 vols., 1758 (completed by J. J. de Lalande, publ. in 4 vols., 1799-1802). Reedited Récréations mathématiques (Jacques Ozanam), 1778. Died Versailles, France, Dec. 18, 1799.

MONTY, Kenneth James, Am. biochemist; b. Sanford, Me., Sept. 11, 1930; s. Leo James and Evelyn (Delahunt) M.; B.A., Bowdoin Coll., 1951; Ph.D., U. Rochester, 1956; m. Barbara Helen Oles, June 15, 1952; children—Melissa Lee, Stuart James. Faculty, Johns Hopkins, 1955-63; prof., head dept. biochemistry U. Tenn., Knoxville, 1963—. Mem. A.A.A.S., Am. Chem. Soc., Am. Soc. Biol. Chemists, Am. Assn. U. Profs., N.Y. Acad. Scis., Sigma Xi. Research, publs. on chemistry of cell nucleus and chromosomes, effects of radiation on biol. systems, metabolism of trace metals, molecular genetics, control of cellular processes. Home: 920 Marlboro Rd., Knoxville, Tenn. 37919.*

MOODY, Herbert Raymond, Am. chemist; b. Chelsea, Mass., Nov. 19, 1869; s. Luther Richmond and Mary Emily (Sherman) M.; S.B., Mass. Inst. Tech., 1892; A.M., Columbia, 1900, Ph.D., 1901; m. Edna Wadsworth, Aug. 20, 1895. Asst. labs. Chelsea High Sch., 1887-88; asst. gen. chemistry Mass. Inst. Tech., 1892-94, instr. analyt. chemistry, 1894-95; instr. science Gilbert Sch., Winsted, Conn., 1895-99; prof. chemistry Hobart Coll., 1901-05; prof. chemistry Coll. City of N.Y., 1905-20, prof. chem. engring., 1921, prof. chemistry and dir. of dept. 1922-38, prof. emeritus. Mem. Div. Chemistry and Chem. Tech. of NRC, 1936-41. Fellow Am. Inst. of Chemistry; mem. Am. Chem Soc., London Soc. Chem. Industry, London Chem. Soc., Societe de Chimie Industrielle, Phi Beta Kappa. Author: Reactions at the Temperature of the Electric Arc, 1901; College Text-book of Quantitative Analysis, 1914; Chemistry of the Metals, 1923. Died Oct. 20, 1947.

MOODY, Lewis Ferry, Am. hydraulic engr.; b. Phila., Jan. 5, 1880; s. Carlton Montague and Elizabeth Eddy (Lewis) M.; B.S., Towne Sci. Sch. (U. Pa.), 1901, M.S., 1902; m. Eleanor Carman Greene, June 22, 1909; children—Mary Elizabeth, Lewis Ferry, Arthur Maurice Greene, Eleanor Lowry (Mrs. Edw. M. Broadhurst). Instr. mech. engring. U. Pa., 1902-04; engring. staff hydraulic dept. I. P. Morris Co., Phila., 1904-08; asst. prof. mech. engring., later prof. hydraulic engring. Rensselaer Poly. Inst., 1908-16, also ind. practice; cons. engr. I. P. Morris Co. (now Baldwin Lima Hamilton), 1911-46, Worthington Pump & Mchy. Corp. 1938-49; prof. hydraulic engring. Princeton. Fellow A.A.A.S.; mem. Am. Soc. M.E., Franklin Inst. (Elliott Cresson award 1945). Am. Soc. for Engring. Edn., Sigma Xi. Author: Lectures on Machine Design, 1942; sect. in Handbook of Applied Hydraulics, 1942; also papers. Inventor numerous improvements in hydraulic turbines, pumps and accessories; holder many patents for inventions, including spiral draft tube, Moody spreading draft tube, Moody spiral pump, new high speed turbine. Died Apr. 18, 1953.

MOODY, Paul, Am. inventor; b. Byfield Parish, Newbury, Mass., May 21, 1779; s. Paul and Mary Moody; m. Susannah Morill, July 13, 1800; 3 children. Established (with Francis C. Lowell) cotton mill and other machinery plant, Waltham, Mass.; secured patent for mechanism to wind yarn from bobbins or spools, 1816; perfected soapstone rollers, doubled efficiency of Horrock's dressing machine, 1818; granted patents for machines to make cotton roping, also to rope and spin cotton, 1821; supt. cotton mills, East Chelmsford (now Lowell), Mass., 1823; under his direction the manufacture of cotton machinery was continued and improved designs of machinery perfected at Lowell Machine Works, 1825. Died Lowell, July 8, 1831.

MOOG, Florence, Am. biologist; b. Bklyn., Jan. 24, 1916; d. George Alfred and Freda (Ott) Moog; A.B. summa cum laude, N.Y. U., 1936; A.M., Columbia, 1938, Ph.D., 1944. Med. records clk. U. S. Dept. Labor, 1937-38; instr. biology U. Del., 1940; faculty Washington U., St. Louis, 1942—, prof. biology, 1958—. Mem. study sect. on human embryology and devel. NIH, 1966—. Recipient A.A.A.S.-Westinghouse award for distinguished sci. writing, 1948; Merck sr. postdoctoral fellow Cambridge, U., 1954-55. Mem. Am. Soc. Zoologists, Am. Soc. for Cell Biology, Soc. for Developmental Biology, Histochem. Soc., Soc. Gen. Physiologists. Author: Structure and Development of Vertebrates, 1949; (with T. S. Hall) Life Science, 1955; also articles. Functional differentiation of small intestine in vertebrates; discoveries in pattern of accumulation of enzymes necessary for intestinal absorption, localization of such enzymes and identification of hormonal and other factors controling devel. of enzymes. Home: 4466 W. Pine St., St. Louis 63108.*

MOON, Henry Dukso, Am. pathologist; b. San Francisco, Sept. 28, 1914; s. Yang M. and Chan S. (Lee) M.; A.B., U. Cal. at Berkeley, 1935, M.A., 1937, M.D., 1940; m. Lona A. Lowe, Nov. 22, 1941; children—Nancy L., Henry B., Thomas L. Instr. pathology U. Cal. at San Francisco, 1943-44, asst. clin. prof. pathology, 1947-52, lectr. pathology, oncology, 1952-53, acting chmn. dept. legal medicine, 1953, asso. prof. pathology, 1953-58, prof., 1958—, chmn. dept. pathology U. Cal. Med. Sch., 1956—; pathologist U. Cal. Hosp., 1953—; vis. pathologist San Francisco Gen. Hosp., 1952—; Laguna Honda Home, San Francisco, 1952—; chief pathology service VA Hosp., San Francisco, 1947-51, now cons. pathologist; cons. Surg. Gen. USPHS, NIH, 1954-56. Diplomate Am. Bd. Pathology. Mem. A.A.A.S., A.M.A., Am. Coll. Pathologists, Am. Assn. Pathologists and Bacteriologists, Am. Soc. Exptl. Pathology, Am. Fedn. Clin. Research, Internat. Acad. Pathology, Endocrine Soc., Soc. Exptl. Biology and Medicine, N.Y. Acad. Sci. Research, publs. on biologic effects of ACTH, role of pituitary hormones in exptl. neoplasia, histogenetic mechanisms in arteriosclerosis; immunologic studies of cytolysis by sensitized lymphoid cells. Office: Department of Pathology, U. Cal. Medical Center, San Francisco 94122.*

MOONEY, James, Am. ethnologist; b. Richmond, Ind., Feb. 10, 1861; s. James and Ellen (Devlin) M.; ed. pub. schs.; m. Ione Lee Gaut, 1897; entered office daily newspaper, 1879; had already begun Indian studies which became life work; m. Ione Lee Gaut, 1897; moved to Washington, 1885, became mem. Bur. Am. Ethnology; conducted extended field investigations among So. and Western Indian tribes, particularly Cherokee and tribes of the Great Plains; prepared govt. Indian exhibits for Chgo., Nashville, Omaha and St. Louis expns. Author: (monographs) Funeral Customs of Ireland; Holiday Customs of Ireland; Sacred Formulas of the Cherokee; Siouan Tribes of the East; The Messiah Religion and the Ghost Dance; Calendar History of the Kiowa Indians; Myths of the Cherokee; also contbr. to New International and Catholic encys. Co-author: Hand book of American Indians, 2 vols., 1907-10. Died Dec. 22, 1921.

MOORA, Kharri Albertovich, Russian archeologist; b. Ekhavere (now Tartu Rayon), 1900; grad. Philosophy Faculty, Tartu U., 1925; D.Hist. Sci. Instr., Tartu U., 1927-30, acting prof. archeology of Estonia and neighboring countries, 1930-37, dean Philosophy Faculty, Tartu U., 1940, prof., head dept., 1944-50, dean History and Philology Faculty, 1948-49; prof., 1938—; Estonian dep. minister edn., 1940-41; acting dir. Talinn Hist. Mus., 1942-44; head archeology sect. Inst. History, Estonian Acad. Scis., 1947—, asso., 1950—, also bur. mem. dept. social sci. Lectr., Sweden, 1947. Mem. Estonian Acad. Sci., Internat. Union Prehistory and Early Hist. Knowledge (mem. permanent council). Author: The Iron Age in Latvia until Approximately 500 A.D., 1929, 38 (in German); Estonian National Costumes of the 19th and Early 20th Century, 1957; co-author: History of the Estonian SSR, vol. 1, 1961. Editor: The Ethnic History of the Estonian People. Died 1968.

MOORE, Arthur Dearth, Am. elec. engr.; b. Fairchance, Pa., Jan. 7, 1895; s. William Edgar and Ella (Hall) M.; B.S. in Elec. Engring., Carnegie Inst. Tech., 1915; M.S., U. Mich., 1923; m. Mary Josephine Shaffer, Aug. 19, 1920; children—Jeanne (Mrs. John D. Goodman), Arthur (Mrs. Wilma Morris), Josephine. With Westinghouse Electric Co., 1915-16 from instr. to prof. elec. engring., U. Mich., Ann Arbor, 1916-63, prof. emeritus, 1963—; vis. lectr. Northwestern U., Duke, Queen's U., Can. Fellow I.E.-E.E.; mem. Am. Soc. Elec. Engrs., Sigma Xi, Tau Beta Pi (past nat. pres.), Eta Kappa Nu (eminent mem.). Author: Fundamentals of Electrical Design, 1927; Heat Transfer Notes for Electrical Engineering, 1942; Fluid Mapper Patterns, 1956; Fluid Mapper Manual, 1961; Electrostatics, 1968; also articles. Invented hydrocal, hydrodynamic device for solving heat transients; developed numerous fluid mapper techniques for solving potential field problems; discovered electropherics and magnetospherics; invented new forms of induction-type electrostatic generators and accessories. Home: 718 Onondaga, Ann Arbor, Mich. 48104.*

MOORE, Arthur Ulric, Am. animal behaviorist; b. Gt. Barrington, Mass., Mar. 23, 1903; s. Arthur Leon and Gertrude Tyler (Platt) M.; grad. Ethical Culture Schs., N.Y.C., 1921; A.B., Cornell U., 1927, M.A., 1928, Ph.D., 1936; m. Frances Miriam Goodnough, Feb. 13, 1926; children—Royall Tyler, Kent Trowbridge. Faculty, Washington and Lee U., 1929-32, Ia. State Tchrs. Coll., 1936-38; instr. Cornell U., Ithaca, N.Y., 1932-36, research asso., mgr. Liddell Lab. Comparative and Physiol. Psychology, 1936-63, sr. research asso. psychobiology Grad. Sch. Nutrition, 1963—. Mem. Pavlovian Soc., Animal Behavior Soc., Am. Zool. Soc., Am. Psychol. Assn., N.Y. Acad. Scis., Am. Assn. U. Profs, Sigma Xi. Research, publs. on long-term effects of early protein malnutrition on central nervous system, measured by responses to classical (Pavlovian) and operant conditioning, effects of continued Pavlovian conditioning on central nervous system particularly immature system, effects of early disruption of mother-young bond in sheep and goats parallel to study of human deprivation. Home: 36 Freese Rd., R.D. 2, Ithaca, N.Y. 14850.*

MOORE, Benjamin LaBree, Am. physicist; b. Louisville, Feb. 10, 1915; s. Paul Homer and Amy (LaBree) M.; A.B., Davidson Coll., 1934; M.A. (Teaching fellow), Vanderbilt U., 1935; Ph.D., Cornell U., 1940; m. Bessie Evelyn Cupp, Sept. 12, 1943; children—Christopher Warren, Marilyn Hopkins. Physicist, Bur. Ordnance, Navy Dept., 1940-42; instr. Naval Mine Warfare Sch., 1942-44; sr. physicist Tenn. Eastman Corp., 1944-46; research fellow Harvard, 1946-49, asst. dir. Computation Lab., Harvard, 1949-50; staff mem. Los Alamos Sci. Lab., 1950-51, asst. div. leader, 1951-60, asso. div. leader, 1960—; extra mural prof. elec. engring. Washington U., St. Louis, 1957-59. Fellow A.A.A.S.; mem. Am. Phys. Soc., Am. Assn. Physics Tchrs., Sigma Xi. Research on mine warfare, separation of uranium isotopes by electromagnetic means, atomic weapons; design and devel. large scale digital computing machines. Home: 3504 Arizona Av. Office: Los Alamos Sci. Lab., Box 1663, Los Alamos 87544.*

MOORE, Burton Evans, Am. physicist; b. Westerville, O., Apr. 8, 1866; s. Royal and Rachel (Evans) M.; A.B., Otterbein U., 1888; A.M., Cornell, 1890; student, univs. of Strassburg and Berlin, 1893-94; Ph.D., U. Göttingen, 1907; m. Harriette Clemens, Sept. 1, 1897 (dec. 1909); m. 2d, Hanna Eberle, Dec. 16, 1911. Instr. physics Lehigh U., Pa., 1891-92. U. Ill., 1895-96; instr. physics U. Neb., 1896-1902, asst. prof., 1902-06, prof., 1906—. Research in excitation stages in Open Arc Spectra. Died July 15, 1925.

MOORE, Carl Richard, Am. zoologist; b. Green County, Mo., Dec. 5, 1892; s. Johnathan Newton and Sarah Francis (Harris) M.; A.B., Drury Coll., 1913, M.S., 1914; Sc.D., 1948; Ph.D., U. Chgo. 1916; postgrad. Marine Biol. Lab., Woods Hole, Mass., summers 1914, 15, 16; m. Edith Naomi Abernethy, July 2, 1920; children—Howard Frederick, Harris Mason, Ellen Abernethy. Lab. instr. Drury Coll., Springfield, Mo., 1911-14; asst., later asso. in zoology U. Chgo., 1914-18, instr. in zoology, 1918-22, asst. prof., 1922-25, asso. prof., 1925-28, prof. since 1928. Recipient 1st Francis Amory award, 1941, medal from Endocrine Soc., 1955. Mem. Endocrine Soc. (pres. 1944-46), Am. Soc. Zoologists (v.p. 1926), Am. Soc. Naturalists, A.A.A.S. (v.p. sect. F, 1943), Inst. Medicine, Marine Biol. Lab. Corp., Sigma Xi, Assn. for Study Internal Secretions (pres. 1944-46), Soc. Exptl. Biology and Medicine, Nat. Acad. Scis. Mem. editorial bd. Biol. Bull., 1926—, mng. editor, 1926-29; editorial bd. Physiol. Zoology. Isolated (with T. F. Gallagher and F. C. Koch) 1st testicular extract containing male sex hormone (testosterone, androsterone), circa 1929. Died Oct. 16, 1955.

MOORE, Carl Vernon, Am. physician, educator; b. St. Louis, Aug. 21, 1908; s. Charles and Mary (Kamp) M.; student Elmhurst Coll., 1924-27, LL.D., 1955; A.B., Washington U., St. Louis, 1938, M.D., 1932; m. Dorothy Adams, May 25, 1934; 1 dau., Judith (Mrs. James H. Frisbie). Asst. prof. medicine Ohio State U., 1935-38; faculty Washington U., St. Louis, 1938—, prof. medicine, 1946—, dean Sch. Medicine, 1953-55, chmn. dept. medicine, 1955—, vice chancellor for med. affairs, 1964-65. Recipient Goldberger award in clinical nutrition A.M.A., 1958; Stratton medal Internat. Soc. Hematology, 1964. Mem. Am. Soc. Clin. Investigators (pres. 1953-54), Assn. Am. Physicians (pres. 1963-64), Am. Soc. Hematology (pres. 1959-60). Internat. Soc. Hematology (pres. 1966-68). Editor: Jour. Lab. and Clin. Medicine, 1944-49; co-editor Progress in Hematology, 1964. Research on iron metabolism and nutrition, pathogenesis of anemias. Home: 6944 Pershing St., University City, Mo. 63130. Office: 660 S. Euclid, St. Louis 63110.*

MOORE, Charles J(ames), Am. chemist; b. Flint Hill, Va., Aug. 9, 1875; s. John Randolph and Elizabeth Jane (Green) M.; B.S., Va. Mil. Inst., 1895; Ph.D., U. Va., 1901; A.M., Harvard, 1909; m. Sophie Schwartz, June 25, 1919; children—Elizabeth Jane, Charles James. Instr. math. Horner Sch., N.C., 1896-98; prof. chemistry and geology Western Md. Coll., 1901-02; instr. U. Ga., 1902-04, adj. prof., 1904-07; Austin teaching fellow Harvard, 1908-12; instr. chem. N.Y. U., 1912-14; asst. prof. chem. Hunter Coll., 1914-17, asso. prof., 1917-20, prof., 1921-45, emeritus since 1945; chief. chem. Bur. Soils, U. S. Dept. Agr., 1920-21. Recipient Jackson-Hope medal. Fellow A.A.A.S.; mem. Am. Chem. Soc. Author: Logarithmic reduction tables for analytical chemists, 1913; Exercises in organic chemistry; Lecture table demonstrations of common gases. Research in aliphatic metal amines. colloidal materials in soils, purification of mercury, atomic weight of phosphorus. Died Jan. 25, 1950.

MOORE, C(harles) Ulysses, Am. pediatrist; b. Alden, Ia., Jan. 1, 1877; s. Henry Vrooman and Sarah Eugenia (Balcom) M.; M.Di., Ia. State Tchrs.' Coll., 1901; B.A., U. Tex., 1906; postgrad. U. Colo. Sch. Medicine, 1906-07; M.D., U. Minn., 1910, M.S. in Pediatrics, 1916; postgrad. Harvard, 1915; m. Nettie B. Rosenberry, Sept. 11, 1907; children—Mary Katherine (adopted), Charles Balcom. Practiced medicine, Carthage, S.D., 1910-14; located at Portland, Ore., 1916; instr. pediatrics U. Ore. Med. Sch., 1919-23, also dir. nutritional research lab.; pediatrist Multnomah County Hosp., Crittenden-Wemme Home; lectr. nutrition Utah Agrl. Coll., summers 1923, 24. Organizer Ore. Infant Welfare Soc., started its clinics, 1920. Fellow A.C.P., A.M.A. Author: Nutrition of Mother and Child, 1923, 4th edit., 1935; also numerous articles on rickets, breast feeding, beriberi, pyloric obstruction and vitamin research. Deceased.

MOORE, Daniel Charles, Am. physician; b. Cin., Sept. 9, 1918; s. Daniel Clark and May (Strebel) M.; A.B., Amherst Coll., 1940; M.D., Northwestern U., 1944; m. Betty Maxine Tobias, Aug. 5, 1945; children—Barbara, Nancy, Daniel, Susan. Dir. dept. anesthesiology Mason Clinic, Seattle, 1947—; chief anesthesia Virginia Mason Hosp., Seattle, 1947—; clin. asso. prof. anesthesia U. Wash. Sch. Medicine, 1956-64, clin. prof., 1965—. Diplomate Am. Bd. Anesthesiology. Mem. Am. Soc. Anesthesiology (pres. 1958-59), A.M.A. (sec. sect. anesthesiology 1961-62, vice chmn. 1961-62), Wash. State Soc. Anesthesiologists (pres. 1949-50), King County Med. Soc., Acad. Anesthesiology, Beta Theta Pi, Nu Sigma Nu. Author: Regional Block, 1953, 57, 61, 65; Stellate Ganglion Block, 1954; Complications of Regional Anesthesia, 1955; Anesthetic Techniques for Obstetrical Anesthesia and Analgesia, 1964. Clin., exptl. research on all phases of regional block anesthesia to further its safe usage. Home: 611 Evergreen Point Rd., Medina, Wash. 98039. Office: 1118 9th Av., Seattle 98101.

MOORE, David Gillis, Am. geologist; b. Long Beach, Cal., July 11, 1925; s. Return F. and Fay (Boeme) M.; A.B., U. So. Cal., 1950, M.S., 1952; Ph.D., U. Groningen, Holland, 1966; m. Claire M. Harding, Apr. 7, 1945; children—Kathleen Ann, Laurie Claire, Jennifer Francis, Patricia Helen, Theresa Fay. Jr. research geologist Scripps Instn. Oceanography, La Jolla, Cal., 1952-55; geol. oceanographer USN Electronics Lab., San Diego, Cal., 1955—. Cons. Gen. Oceanographics, Inc., 1953—. Fellow Geol. Soc. Am.; mem. Soc. Econ. Paleontologists and Mineralogists, Am. Assn. Petroleum Geologists, Soc. Exploration Geophysicists. Research, publs. on properties of recent marine sediments indicative of environment of deposition, strength and bearing capacity of shallow and deep sea sediments, structure and sediment distbrn. of continental shelf of So. Cal., origin and structure of continental margins of N.Am., Western Europe, other regions by continuous seismic reflection profiling, structure of Cal. Continental Borderland, mechanisms and rates of implacement of sediments in submarine basins marginal to continents. Home: 9440 La Jolla Shores Dr., La Jolla, Cal. 92037. Office: USN Electronics Lab., San Diego 92152.*

MOORE, Dwight Munson, Am. botanist, educator; b. Zanesville, O., Dec. 10, 1891; s. Newton Hoffman and Mary A. (Munson) M.; B.S., Denison U., 1914, M.S., 1921; postgrad. U. Montpellier (France); Ph.D., Ohio State U., 1924; m. Elizabeth Alice French, Sept. 5, 1922; 1 son, Dwight French. Tchr., prin. high schs., 1914-17, 20; instr. biology Denison U., 1920-23; instr. botany Ohio State U., 1923-24; prof. botany U. Ark., Fayetteville, 1924-57; prof. forestry Ark. A. and M. Coll., College Heights, 1957-61; prof. botany, head biology dept. Ark. Poly. Coll., Russellville, 1961—. Coop. agt. U. S. Forest Service, 1945-62; vis. prof. botany U. Saigon, 1958-59; vis. prof. biology U. Alaska, summer 1964. Mem. A.A.A.S., Am. Inst. Biol. Scis., Bot. Soc. Am., Am. Fern Soc. (past v.p.), Am. Soc. Plant Taxonomists, Internat. Assn. for Plant Taxonomy, Ark. Acad. Sci. (pres. 1932-34, 63-64, editor Proc. 1940-50), Sigma Xi, Lambda Chi Alpha. Author: Trees of Arkansas, 1950, 60; (with others) Arkansas Natural Resources, 1942; (with others) Browse Plants of Southern Forests, 1961. Contbr. numerous articles on Ark. ferns and other flora to profl. jours. Home: 506 Vandeventer Av., Fayetteville, Ark. 72701. Office: Ark. Poly. College, Box 529, Russellville, Ark. 72802.*

MOORE, Edward James, Am. physicist; b. Chili, N.Y., June 13, 1873; s. Thomas and Margaret (Hill) M.; A.B., Oberlin, 1903, A.M., 1906; Ph.D., U. Chgo., 1913; m. Amelia May Eade, July 12, 1905; children—Margaret Carolyn, Edward James. Began as tutor math., Oberlin, 1903, and advanced to asso. prof. physics, 1910; prof. physics, U. Buffalo, since 1919, dean Grad. Sch. Arts and Scis., 1939-46, dean emeritus since 1946. Fellow A.A.A.S.; mem. Am. Phys. Soc., Sigma Xi. Contbr. on tech. topics. Specialized in molecular physics and the electron theory; perfected (with J. A. Demuth) an autographic system for recording employees' time, known as Symbol System. Died Mar. 11, 1948.

MOORE, Eliakim Hastings, Am. mathematician; b. Marietta, O., Jan. 26, 1862; s. David Hastings and Julia Sophia (Carpenter) M.; A.B., Yale, 1883, Ph.D., 1885, Sc.D., 1909; U. Berlin, 1885-86; Ph.D. (hon.), U. Göttingen, 1899; LL.D., U. Wis., 1904; Math.D., Clark U., 1909; Sc.D., U. Toronto, 1921, Northwestern U., 1927; m. Martha Morris Young, June 21, 1892; children—David Hastings (dec.), Eliakim Hastings. Tutor math. Yale, 1887-89; asst. prof. math. Northwestern U., 1889-91, asso. prof., 1891-92; prof. math. U. Chgo., 1892-1931, head dept., 1896-1931. Fellow A.A.A.S. (pres. 1921), Am. Acad. Arts and Scis. (asso.). Author: Introduction to a Form of General Analysis, 1910. Editor: Trans. Am. Math. Soc., 1899-1907; asso. editor Proc. Nat. Acad. Scis., 1915. Discovered group of automorphisms of any finite group, 1893; one of largest contbrs. to study of groups of linear homogeneous substitutions on n variables; proved that every finite communative field contains a finite number of distinct elements, 1893; called for system of general analysis to be produced, 1906. Died Dec. 30, 1932.

MOORE, Forris Jewett, Am. chemist; b. Pittsfield, Mass., June 9, 1867; s. Forris Jewett and Ellen S. (Wightman) M.; B.A., Amherst, 1889; Ph.D., U. Heidelberg (Germany), 1893; m. Emma B. Tod, Aug. 9, 1892. Lab. asst. Amherst Coll., 1889-90; instr. in chemistry Cornell U., 1893-94; asst. in chemistry Mass. Inst. Tech., 1894-95, instr., 1895-1902, asst. prof., 1902-10, asso. prof., 1910-12, prof. organic chemistry, 1912—. Lectr. Lecturer organic chemistry Harvard, 1910-11, 17-18, 18-19. Fellow Am. Acad. Arts and Scis. Author: Outlines of Organic Chemistry, 1910; Experiments in Organic Chemistry, 1911; A History of Chemistry, 1918. Died Nov. 20, 1926.

MOORE, Francis Daniels, Am. surgeon; b. Evanston, Ill., Aug. 17, 1913; s. Philip Wyatt and Caroline (Daniels) M.; grad. Harvard, 1935, M.D., 1939; hon. degree Nat. U. Ireland, 1959, U. Glasgow (Scotland), 1965; m. Laura Bartlett, June 24, 1935; children—Nancy Hill, Peter Moore, Sarah Warren, Caroline Moore, Francis D. Surgeon, investigator, tchr.; Moseley prof. surgery Harvard Med. Sch., 1948—. Chmn. surgery study sect. NIH, 1957-60; chmn. com. on metabolism in trauma U. S. Army. Recipient Blakeslee award for med. writing, 1965. Mem. Am. Surg. Assn., Am. Soc. for Clin. Investigation. Author: Metabolic Care of the Surgical Patient, 1959; The Biology of Tissue Transplantation, 1963; Give and Take, 1964; also articles. Measurement total body liquids and solids with isotopes; vagotomy and duodenal ulcer, metabolic response to surgery, tissue transplantation. Home: 371 Walnut St., Brookline, Mass. 02146. Office: 721 Huntington Av., Boston 02115.*

MOORE, George Eugene, Am. physician; b. Mpls., Feb. 22, 1920; s. Jesse and Elizabeth (MacRae) M.; B.A., U. Minn., 1942, M.A., 1943, B.S., 1944, M.B., 1946, M.D., 1947, Ph.D., 1950; m. Lorraine Phyllis Hammell, Feb. 22, 1945; children—Allan Bruce, Laurie Ann, Linda, Donald, Cathy Lee. Dir. tumor clinic, asso. prof. surgeon U. Minn. Hosps., 1951-52; dir., chief surgeon Roswell Park Meml. Inst., Buffalo, 1952-67; dir. pub. health research N.Y. State Dept. Health, 1967—; clin. prof. surgery State U. N.Y. at Buffalo, 1962—. Cons. Nat. Cancer Chemotherapy Center, 1960—, M.D. Anderson Hosp. and Tumor Inst., Houston, 1962—; mem. Nat. Council on Regional Med. Programs, 1966—. Recipient Buffalo Evening News award for outstanding citizen, 1958; Buffalo Club award for outstanding sci. achievement, 1959; Modern Medicine Distinguished Achievement award, 1962; Chancellor's medal State U. N.Y. at Buffalo, 1963; Charles Evans Hughes award in pub. adminstrn. N.Y. State, 1963; Bronfman prize award Am. Pub. Health Assn., 1964. Diplomate Am. Bd. Surgery. Mem. A.C.S., Am. Assn. Cancer Research, Halsted Soc., N.Y. Acad. Scis., Am. Surg. Assn. Author: Diagnosis and Localization of Brain Tumors, 1953; also numerous articles. Devised method for diagnosis and localization of brain tumors, exptl. and clin. advances in causes, nature, and treatment of malignant diseases, growth of human cells outside of body. Home: 3 Lehn Springs Dr., Williamsville, N.Y. 14221. Office: 666 Elm St., Buffalo 14203.*

MOORE, George Thomas, Am. botanist; b. Indpls., Feb. 23, 1871; s. George T. and Margaret (Marshall) M.; B.S., Wabash Coll., 1894; A.B., Harvard, 1895, A.M., 1896, Ph.D., 1900; m. Emma L. Hall, Dec. 30, 1896 (dec. Jan. 1934); children—Harriet Hall, Thomas Gaunt; m. 2d, Katherine H. Leigh, Feb. 20, 1937. Asst. in cryptogamic botany Harvard; tchr. Radcliffe Coll.; in charge botany Dartmouth Coll., 1899-1901; became physiologist and algologist bur. plant industry, Dept. Agr., 1901, in charge Lab. Plant Physiology, 1903-05; in charge of botany Marine Biol. Lab., Mass., 1909-19; prof. applied botany and plant physiology Shaw Sch. Botany (Washington U.) and physiologist to Mo. Bot. Garden, 1909-1912; dir. Mo. Bot. Garden since 1912. Fellow A.A.A.S.; mem. Am. Philos. Soc., Washington Acad. Scis., St. Louis Acad. Sci., bot. socs. Am., Washington, Soc. Am. Bacteriologists, Sigma Xi, Phi Beta Kappa. Reviser for algae Century Dictionary. Contbr. to sci. jours. and govt. bulls. Discoverer of a method for preventing pollution of water supplies by algae and certain pathogenic bacteria; perfected method for inoculating the soil with bacteria which enable certain crops to use atmospheric nitrogen. Died Nov. 27, 1956.

MOORE, Harold Emery, Am. botanist; b. Winthrop, Mass., July 7, 1917; s. Harold Emery and Helen (Margeson) M.; B.S., U. Mass., 1939; M.A., Harvard, 1940, Ph.D., 1942. Tech. asst. Gray Herbarium, Harvard, 1947-48; faculty L. H. Bailey Hortorium, Cornell U., Ithaca, N.Y., 1948—, prof., dir., 1960—. Recipient Founder's medal Fairchild Tropical Garden, 1954. John Simon Guggenheim Meml. fellow, 1946-47, 55-56. Mem. Palm Soc. (dir., editor Principes 1957—), Internat. Assn. for Plant Taxonomy (mem. standing com. on stblzn. 1965—), Am. Inst. Biol. Sci., A.A.A.S., Am. Soc. Plant Taxonomists, Assn. for Tropical Biology, Bot. Soc. Am., New Eng. Bot. Club, Soc. for Econ. Botany, Soc. for Study Evolution. Author: African Violets, Gloxinias, and their Relatives, 1957; also numerous articles. Research on taxonomy cultivated plants. Home: 31 Genung Rd., R.D. 2, Ithaca, N.Y. 14850.*

MOORE, Harvey Cleaver, Am. anthropologist; b. Port Penn, Del., Mar. 13, 1918; s. Harvey Enos and Lina (Bendler) M.; B.A., U. Del., 1938; Ph.D., U. N.M., 1950; m. Sarah Morehead, Aug. 16, 1948. Faculty, Am. U., Washington, 1951—, prof. anthropology, 1958—, chmn. dept. anthropology, 1965—; research asso. Bur. Social Sci. Research, Washington, 1951-54. Fellow Am. Anthrop. Assn., Royal Anthrop. Inst., A.A.A.S., Washington Acad. Scis., African Studies Assn., Am. Ethnol. Soc., Assn. Current Anthropology; mem. Anthrop. Soc. Washington (past pres.), Author articles on Navaho; contbg. author: American Historical Anthropology, 1967. Field studies on Sub-Saharan Africa, Am. S.W.; research on culture change, including religious, polit. and econ. aspects. Home: 1508 44th St. N.W., Washington 20007.

MOORE, Henry Frank, Am. biologist; b. Phila., June 4, 1867; s. John P. and Emma C. (Frank) M.; A.B., Central High Sch., Phila., 1885; Ph.D., U. Pa., 1895; m. Annie Florence Dennis, Apr. 13, 1903. Naturalist, Internat. Fishery Commn., 1893-95; chief naturalist Steamer Albatross, 1896-1903; sci. asst. U. S. Bur. Fisheries, 1903-11, in charge of sci. inquiry, 1911-15, and dep. commr., 1915-23. Mem. Inter-departmental Bd. Internat. Ice Patrol, 1915-23, NRC, 1917-25, Internat., Com. Marine Fishery Investigations, 1919-23. Recipient of various awards for investigations. Fellow A.A.A.S.; hon. mem. Société Internationale Protectrice des Pecheurs d'Éponges; mem. Am. Geophys. Union, N.C. Forestry Assn., Acad. Natural Sci., Phila. Writer on zoology and fisheries. Died Jan. 8, 1948.

MOORE, Hugh Kelsea, Am. chem. engr.; b. Andover, Mass., Jan. 3, 1872; s. Albert Weston and Sarah Frances (Norton) M.; student Mass. Inst. Tech., 1893-96; D.Sc (hon.), U. Me., 1924; m. Mary Esther Tebbetts, Jan. 1, 1902; children—Mrs. Katherine Burgess Durell, Hugh Kelsea, Jr., Dorothy Esther. Began with Electro-Chem. Co, Rumford Falls, Me., 1897; with Moore Electro-Chem. Co., also Am. Electro-Chem. Co. until 1903; with Burgess Sulphite Fibre Co. (later Brown Co.) as chief chemist and chem. engr., 1903-34. Mem. div. chemistry NRC. Pres. Am. Inst. Chem. Engrs., 1925-26 Gold medalist, 1925. Author: Incomplete Hydrogenation of Cotton Seed Oil, 1917; Testing of Lubricating Oils, 1917; Chemical Engineering Aspect of Renovating a Sulphite Mill, 1918; Analysis of the Explosion Process of Recovering Soda Salts from Black Liquor, 1919; Accident Prevention in the Mill, 1919; Fundamentals of Electrolytic Diaphragm Cells, 1920; The Use and Value of Physical and Chemical Constants, 1920; Scientific Facts about Pure and Impure Milk, 1921; The Production of Hydrochloride Acid by Direct Union of Hydrogen and Chlorine, 1922; Development of Taxation, 1923; Fundamental Principles of Multiple Effect Evaporative Separation, 1923. Conducted studies in evaporation and separation, and in 1929, designed, built and operated a ten-effect multiple effect evaporator; research in refrigeration; conducted investigations in electrolysis, inventing and patenting the unsubmerged diaphragm cell, 1897, which revolutionized that industry; invented and patented new method of making calcium arsenate, 1925-27, stationary furnace for recovery of soda content from black liquor, 1913-15, a new acid resisting hydraulic cement 1926-27, a new process of converting sodium sulphate into caustic soda and other chemicals 1932-34, new metal filter cloth and method manufacture, 1934; took out about 45 other patents relating to pulp making, pulp bleaching, evaporation, continuous process of hydrogenating oil, mfg. of sodium sulphide, refrigeration. Died Dec. 18, 1939.

MOORE, J. George, Am. gynecologist; b. Berkeley, Cal., Sept. 17, 1917; s. John George and Mercedes (Sullivan) M.; A.B., U. Cal. at Berkeley, 1939; M.D., U. Cal. at San Francisco, 1942; m. Mary Louise Laffer, Feb. 8, 1946; children—Barbara, Terence, Bruce, Martha. Faculty, U., Ia., 1950-51, U. Cal. at Los Angeles, 1951-65; prof., chmn. dept. obstetrics and gynecology Columbia Coll. Phys. and Surg., N.Y.C., 1965—: cons. St. Luke's Hosp. Center, N.Y.C., 1966—. Diplomate Am. Bd. Obstetrics and Gynecology. Mem. Am. Fedn. Clin. Research, Am. Fertility Soc., Soc. Gynecol. Investigation (pres. 1966-67), Am. Coll. Obstetricians and Gynecologists, Am. Assn. Obstetricians and Gynecologists (exec. council 1965-68), A.C.S., Am. Gynecol. Soc. (exec. council 1966—), Soc. Pelvic Surgeons, Sigma Xi, Alpha Omega Alpha, Nu Sigma Nu, others. Research, publs. on cancer of cervix and its mgmt., complications of pregnancy, therapeutic abortion. Home: 1 Apawamis Av., Rye, N.Y. 10580. Office: 622 W. 168th St., N.Y.C. 10032.*

MOORE, J. Percy, zoologist; b. Williamsport, Pa., May 17, 1869; s. John P. and Emma (Frank) M.; B.S., U. Pa., 1892, Ph.D., 1896; m. Kathleen Carter, May 16, 1892, children—Percy Warren, Kathleen, Elinor (Mrs. J. Logan Irvin), Caroline. Scientific asst. U. S. Bur. Fisheries, 1890-19; with U. Pa., 1890-65, prof., 1912-39, prof. emeritus, 1939—; asst. curator, corresponding sec. Acad. Natural Scis., Phila., 1902-39, trustee, 1938-57, hon. life trustee, 1957-65; instr. biology Hahnemann Med. Coll., 1896-98, Marine Biol. Lab., Wood's Hole, Mass., 1901-02; with Ludwick Inst., 1902-56, pres., 1942-56; hon. research asso. Smithsonian Instn. Fellow A.A.A.S.; mem. Soc. Systematic Zoology, Soc. Study Evolution, Am. Soc. Naturalists, Am. Soc. Zoologists, Ecol. Soc. Am., Soc. Zool. de France, Phila. Acad. Scis., Am. Philosoph. Soc., Sigma Xi. Author: (with W. A. Harding) Fauna of British India, Hirudinea, 1927, (with H. F. Nachtrieb, E. E. Hemingway) The Leeches of Minnesota, 1912; asso. editor Jour. of Morphology, 1937-40. Research, publs. on taxonomy, morphology, ecology of leeches, other worms; mosquito control; fish, other marine organisms. Died Mar. 1, 1965.

MOORE, James Edward, Am. surgeon; b. Clarksville, Pa., Mar. 2, 1852; s. George W. and Margaret (Ziegler) M.; ed. Poland (Ohio) Union Sem., U. Mich.; M.D., Bellevue Hosp. Med. Coll., N.Y. U., 1873; postgrad., N.Y., 3 years; studied in Germany, France, Eng., 1885-87; m. Louie C. Irving, Feb. 1887. Practice medicine, Ft. Wayne, Ind., 1873-74, Emelenton, Pa., 1876-82, Mpls., 1882—; specialized in surgery, 1888—; surgeon-in-chief Northwestern Hosp., 1897—; prof. surgery U. Minn., 1904—, chief dept. surgery, 1908—. Fellow Am. Surg. Assn. (became v.p. 1905), A.C.S. (gov.). Author: Moore's Orthopedic Surgery, 1898; General Principles of Surgical Treatment (American Practice of Surgery), 1906. First specialist in surgery exclusively, west of N.Y. Died 1918.

MOORE, James Gregory, Am. geologist; b. Palo Alto, Cal., Apr. 30, 1930; s. George Raymond and Grace (Hauch) M.; B.S., Stanford, 1951; M.S., U. Wash., 1952; Ph.D., Johns Hopkins, 1954; m. Florence M. Gooch, Aug. 30, 1952; children—Dugan Lee, Nikki Ann. Geologist, U. S. Geol. Survey, Menlo Park, Cal., 1956—; scientist-in-charge Hawaiian Volcano Obs., 1962-64; presdl. appointee for investigation eruption Taal Volcano, Philippines, 1965. Fellow Geol. Soc. Am.; mem. Geochemistry Soc., Internat. Assn. Volcanology. Research in geologic investigation in Sierra Nev., Cal., Western Nev., delimited regional variation of granitic rocks in western U. S., mechanism of recent volcanic eruption in Hawaii and Philippines, processes of ocean-bottom volcanic activity. Home: 4045 Ben Lomond St., Palo Alto, Cal. 94306. Office: 345 Middlefield St., Menlo Park, Cal. 94025.*

MOORE, Sir John (William), Irish physician; b. Dublin, Ireland, Oct. 23, 1845; s. William Daniel Moore; M.A., M.D., M.Ch., Dublin U.; D.Sc. (hon.), Oxford (Eng.) U.; m. Ellie Ridley, 1876 (dec. 1878); m. 2d, Louisa E. Armstrong, 1881; 1 son, 3 daus. Cons. physician Meath Hosp., Cork St. Fever Hosp., Drumcondra Hosp., Coombe Hosp. (all Dublin), Dental Hosp. Ireland; prof. practice medicine Royal Coll. Surgeons Ireland; scholar, diplomate in state medicine Trinity Coll., Dublin. Fellow Royal Meteorol. Soc., Swedish Soc. Physicians (hon.), Royal Coll. Physicians Ireland (pres. 1898-1900); mem. Royal Acad. Medicine Ireland (pres. 1918-21). Author: Text Book of the Eruptive and Continued Fevers, 1892; Meterology: Practical and Applied, 1894, 2d rev. edit., 1910; Smallpox (Twentieth Century Practice of Medicine), 1898; (with others) Manual of Public Health for Ireland, 1875. Editor: Dublin Jour. Med. Sci., 1873-1920. Died Oct. 12, 1937.

MOORE, John Alexander, Am. biologist; b. Charles Town, W.Va., June 27, 1915; s. George Douglas and Louise (Blume) M.; A.B., Columbia, 1936, M.A., 1939, Ph.D., 1940; m. Betty Clark, June 4, 1938; children—Sally, Nancy (dec.). Tutor, Bklyn. Coll., 1939-41; instr. Queens Coll., Flushing, N.Y., 1941-43; faculty Barnard Coll., also Columbia, N.Y.C., 1943—, prof. biology, 1950—; research asso. herpetology Am. Mus. Natural History, N.Y.C., 1942—. Fulbright fellow, 1952-53; Guggenheim fellow, 1959-60. Mem. Nat. Acad. Scis., Am. Acad. Arts and Scis., Soc. for Study Evolution (past pres.). Author: Principles of Zoology, 1957; Heredity and Development, 1963; A Guide Book to Washington, 1963; also numerous articles. Editor: Physiology of the Amphibia, 1964; Ideas in Modern Biology, 1965. Research on evolution in frogs and other animals, genetic control of early devel. Home: 80 LaSalle St., N.Y.C. 10027.*

MOORE, John Coleman, Am. mathematician; b. Staten Island, N.Y., May 27, 1923; s. Henry Coleman and Ruth (Fritz) M.; S.B., Mass. Inst. Tech., 1948; Ph.D., Brown U., 1952. Faculty Princeton, 1952-—, prof. math., 1961-—, co-chmn. dept., 1962-63; NSF fellow Princeton, 1953-54, U. Paris (France), 1954-55. Chmn. com. math. instrn. NRC, 1960-62. Mem. A.A.A.S., Am. Assn. U. Profs., Am. Math. Soc. (asso. editor trans. 1958-—, asso. editor proc. 1958-61), Math. Assn. Am. (chmn. panel pregrad. tng. com. undergrad. program 1959-63), Société Mathematique de France, Sociedad Mathemática Mexicana. Contbr. articles algebraic topology and homological algebra. Editor Annals Math. Studies, 1956-—. Home: 65 College Rd. W., Princeton, N.J. 08540.*

MOORE, Jonas, English mathematician; b. Whittle, Eng., Feb. 8, 1617; began study math., 1640; 1 son, Jonas. Clk. to Dr. Burghill, Durham, Eng.; named math. tutor to Duke of York, 1647; became surveyor for draining of fens, 1649; surveyed English coast, also mapped Cambridgeshire; named insp. fortifications of Tangier, 1663; later apptd. surveyor-gen. ordnance; founder math. sch. Christ's Hosp. Fellow Royal Soc., 1674. Author: Arithmetick, 1650; Modern Fortification, or Elements of Military Architecture, 1673; A Mathematical Compendium, 1674; A New System of the Mathematiks, pub. 1681; History of the Great Level of the Fennes . . . , 8 vols., pub. 1685; Translator: Treatise of Artillery (Moretti), 1683. Died Aug. 25, 1679.

MOORE, Joseph Curtis, Am. mammalogist; b. Washington, D.C., June 5, 1914; s. William Lovejoy and Esther (Hardgrove) M.; B.S., U. Ky., 1939; M.S., U. Fla., 1942, Ph.D., (Grad. Sch. scholar, Dudley Beaumont Meml. fellow), 1953; m. Evelyn Ray Lannert, June 7, 1940; children—Rosalind Devon, Diana Karen, Melliny Shannon. Park biologist Everglades Nat. Park, Homestead, Fla., 1949-55; research fellow Am. Mus. Natural History, N.Y.C., 1955-61; curator mammals Field Mus. Natural History, Chgo., 1962-—. Pres. Fla. Acad. Scis., 1955. Recipient Phi Sigma medallion biol. research U. Fla., 1949. Fellow Zool. Soc. London; mem. Am. Soc. Mammalogists (life; bd. dirs. 1951, 61-62, 63-64), Soc. Systematic Zoologists, Soc. Study Evolution. Author: (with George H. H. Tate) A Study of the Diurnal Squirrels, Sciurinae, of the Indian and Indochinese Subregions, 1965. Contbr. Fieldiana: Zoology. Editor quar. mag., Everglades Natural History, 1953-55; editorial bd. Systematic Zoology, 1960-61, 63-66. Spl. research squirrel family Sciurinae, manatees and beaked whales. Home: 318 Shabbona Dr., Park Forest, Ill. 60466. Office: Field Museum of Natural History, Roosevelt Rd. and Lake Shore Dr., Chgo. 60605.*

MOORE, Joseph Haines, Am. astronomer; b. Wilmington, O., Sept. 7, 1878; s. John Haines and Mary A. (Haines), M.; A.B., Wilmington Coll., 1897, Ph.D., Johns Hopkins, 1903; m. Fredrica Chase, June 12, 1907; children—Mary Kathryn, Margaret Elizabeth. Asst., Lick Obs., 1903-06, asst. astronomer, 1906-09, asst. astronomer, 1913, asso. astronomer, 1918-23, astronomer since 1923, asst. dir., 1936-42, dir., 1942-46, mem. 5 eclipse expdns., 1918-32; acting astronomer in charge D. O. Mills expdn. to Chile, 1909-13. Fellow A.A.A.S. (v.p.), Royal Astron. Soc., Cal. Acad. Sci.; mem. Nat. Acad. Sci., Am. Astron. Soc. (v.p. 1942), Astron. Soc. Pacific (pres. 1920 and 28), Internat. Astron. Union. Contbr. astron. papers. Died Mar. 15, 1949.

MOORE, Keith Leon, Canadian anatomist; b. Brantford, Ont., Can., Oct. 5, 1925; s. James Henry and Gertrude (McCombe) M.; B.A. in Sci., U. Western Ont., London, 1949, M.Sc., in Anatomy, 1951, Ph.D. in Microscopic Anatomy, 1954; m. Marion Edith McDermid, Aug. 20, 1949; children—Warren, Pamela, Karen, Laurel, Joyce. Lectr. anatomy U. Western Ont., 1954-56; faculty U. Man., Winnipeg, Can., 1956-—, prof., head dept. anatomy, 1965-—. Cons. Children's Hosp., Winnipeg, 1959-—, Internat. Acad. Cytology, 1962-—. Fellow Internat. Acad. Cytology; mem. Anat. Soc. Gt. Britain and Ireland, Canadian (pres. 1967-—), Am. assns. antomists, Canadian Cytology Council, Editor: The Sex Chromatin; editorial adv. bd. Acta Cytologica, 1962-—. Research, publs. on devel. buccal smear sex chromatin test used to diagnose sex in doubtful cases and chromosome abnormalities, teratology. Home: 54 Wilton St., Winnipeg 9, Man., Can.*

MOORE, Matthew Thibaud, Am. physician; b. Phila., Sept. 12, 1901; s. Joseph B. and Sarah (Gottlieb) M.; student U. Pa., 1920-21, Harvard, 1921-23; M.D., Temple U., 1927; m. Stella Chalfin, Feb. 2, 1927. Faculty, U. Pa. Sch. Medicine Grad. Div., Phila., 1936-—, prof. neuropathology, 1960-—, neuropathologist dept. neuropathology, 1936-—; attending chief psychiatrist Phila. Psychiat. Hosp., 1937-—; neuropsychiatrist-in-chief Drs. Hosp., Phila., 1940-—; psychosurgeon Del. State Hosp., Farnhurst, 1951-—; vis. neuropathologist, 1952-—; sr. attending neurologist Albert Einstein Med. Center, 1954-—; cons. neuropsychiatrist Home for Jewish Aged, 1958-—. Recipient 25 year service award Phila. Psychiat. Center, 1962; Distinguished Alumnus award Temple U., 1964. Diplomate Pan Am. Med. Assn. Fellow A.M.A., Am. Psychiat. Assn., A.C.P., Am. Assn. Neuropathologists (v.p. 1957-58, pres. 1958-59), Am. Acad. Neurology, A.A.A.S., Am. Geriatrics Soc.; mem. No. Med. Assn. (v.p. 1949, pres. 1950), Assn. Research in Nervous and Mental Diseases, Nat. Com. Mental Hygiene, Nat. Multiple Sclerosis Soc., Nat. Council on Family Relations, Am. Epilepsy Soc., Am. Soc. Med. Psychiatry (pres. elect 1966-67, pres. 1967-68), Soc. of Biological Psychiatry, Med.-Surg. Acad. Athens (Greece), N.Y. Acad. Scis., Sigma Xi, others. Research, publs. on treatment, encephalographic studies of mental disorders, effects of brain infections and lesions and their treatment with vitamins, abdominal epilepsy, cerebral tumors, aneurysms. Address: 1813 Delancey Pl., Phila. 19103.*

MOORE, Maurice L(ee), Am. medicinal chemist; b. Laurel Hill, Fla., Sept. 11, 1909; s. Daniel Richmond and Maggie Alice (Steele) M.; B.S., U. Fla., 1930, M.S., 1931; Ph.D., Northwestern U., 1934; m. Charlotte M. Holg, Dec. 2, 1933; children—Daniel, Margaret, Amanda Lee. Asst. instr., Northwestern U., 1931-34; research fellow in organic chemistry, Yale, 1934-36; organic research chemist Sharp and Dohme, Phila., 1936-43; dir. organic chemistry Frederick Stearns & Co., 1943-47, asst. dir. research, 1945-47; dir. research labs. Smith, Kline & French, Phila., 1947-51; v.p. Vick Chem. Co., N.Y., 1951-59; dir. new product devel. Sterling Drug, Inc., N.Y., 1959-64; v.p. Winthrop Labs., N.Y., 1959-61, exec. v.p.) 1961-—; v.p. Glenbrook Labs., N.Y., 1959-64. Asso. trustee U. Pa. Fellow A.A.A.S., Am. Inst. Chemists (vice chmn. Pa. chpt., 1938-40, sec. 1940-42); mem. Am. Chem. Soc. (sec. 1943-47; chmn. 1947; mem. exec. com., 1942-48), Am. Pharm. Assn., Soc. Chem. Industry, Cancer Chemotherapy Nat. Com. (chmn. Industry com.), N.Y. Acad. Sci., Assn. Coll. Honor Socs., Sigma Xi; Delta Chi, Phi Kappa Phi, Alpha Epsilon Delta (nat. pres. 1930-32, dir. expansion and nat. historian, 1932-34, nat. sec. and historian since 1934), Alpha Chi Sigma, Gamma Sigma Epsilon, Phi Sigma. Contbr. research articles on chemistry to profl. jours. Patentee many chem. compounds and their derivatives. Home: 7 Brookside Circle, Bronxville, N.Y. 10708. Office: 90 Park Av., N.Y.C., 10016.*

MOORE, Merrill, Am. psychiatrist; b. Columbia, Tenn., Sept. 11, 1903; s. John Trotwood and Mary Brown (Daniel) M.; B.A., Vanderbilt U., 1924, M.D., 1928; m. Ann Leslie Nichol, Aug. 14, 1930; children—Adam G. N. Moore, John Trotwood, Leslie and Hester. Intern. St. Thomas Hosp., Nashville, 1928-29; teaching fellow neurology Harvard Med. Sch., 1930-31, asst. in neuropathology, 1931-32, research fellow psychiatry, 1936-42; neurol. house officer, Boston City Hosp., 1930-31, neurol. physician, 1930-31; asst. physician Boston Psychopathic Hosp., 1932-35; grad. asst. psychiat. clinic Mass. Gen. Hosp., 1933-34; vis. psychiatrist Boston City Hosp.; clin. asso. psychiatry Harvard Med. Sch.; research asso., Boston Psychopathic Hosp. Fellow Am. Psychiat. Assn., Am. Neurol. Assn., Mass. Med. Soc., Am. Psychopathol. Assn., A.A.A.S.; mem. A.M.A., Phi Beta Kappa. Author: The Noise That Time Makes, 1929; Six Sides to a Man, 1935; M: one thousand autobiog. sonnets; Clinical Sonnets, 1949; Illegitimate Sonnets, 1950; Case Record From a Sonnetorium, 1952; More Clinical Sonnets, 1952; A Doctor's Book of Hours, 1954; Dance of Death, 1957; The Phoenix & The Bees (poems), 1958; also other vols. of poetry, prose essays. Contbr. articles on alcoholism, syphilis, suicide, psychiatry and conchology. Died Sept. 20, 1957.

MOORE, Norman Slawson, Am. physician, educator; b. Ithaca, N.Y., Apr. 17, 1901; s. Veranus A. and Mary Louise (Slawson) M.; A.B., Cornell U., 1923, M.D., 1926; m. Bernice A. Barkee, June 28, 1932. Practice medicine specializing in internal medicine, Ithaca, 1929-40; instr. anatomy Cornell U. Med. Coll., Ithaca, 1929-30, physician-in-chief infirmary and clinic, 1940-67, emeritus, 1967-—, chmn. dept. clin. and preventive medicine, 1943-67, prof. clin. medicine Sch. Nutrition, 1945-67, attending physician Tompkins County Meml. Hosp., 1930-40, cons. physician, 1940-67. Chmn., Tompkins County Bd. Health, 1947-67; mem. pub. health council State N.Y., 1955-—, sec. temporary health ins. bd. Dept. Civil Service, 1957-59, hosp. rev. and planning council, 1962-—, health resources commn., 1967-—; mem. Gov. N.Y. Com. to End Polio by Vaccination, 1957-59; pres., Empire State Med., Ednl. and Sci. Found., 1961-—; mem. panel on orgn. health care services Nat. Adv. Commn. Health Manpower, 1966-—. Diplomate Am. Bd. Internal Medicine. Fellow A.C.P., A.M.A. (house dels. 1950-—), chmn. council on vol. agys. 1967-—), N.Y. Acad. Medicine; mem. Harvey Soc., Am. Coll. Health Assn. (sec.-treas. 1959-—), Tompkins County Med. Soc. (past pres.), Med. Soc. State N.Y. (council 1955-, pres. 1960-61, dir. sci. activities 1967-—), Sigma Xi, Alpha Omega Alpha. Asst. editor N.Y. State Jour. Medicine, 1953-58, cons. editor, 1958-63; cons. editor Student Medicine, 1952-—. Contbr. articles to med. jours. on internal medicine. Home: 128 Pleasant Grove Rd. Office: 505 E. Seneca St., Ithaca, N.Y. 14850.*

MOORE, Raymond Cecil, Am. paleontologist, geologist; b. Roslyn, Wash., Feb. 20, 1892; s. Bernard Harding and Winnifred (Denney) M.; A.B., Denison University, 1913, Sc.D., 1935; Ph.D., U. of Chicago, 1916; m. Georgine Watters, 1917; 1 dau., Marjorie Ann; m. 2d, Lilian Botts, 1936. Instructor in geology, Denison Univ., 1912-13; member United States Geological Survey, 1913-49; instr. geology, U. of Chicago, 1916; asst. prof., 1917; asst. prof. geology, 1916-18, asso. prof., 1918-19, prof. since 1919, U. of Kan., head. dept., 1920-39, 40-41, 52-54; state geologist of Kan., 1916-54; principal geologist Kan. Geol. Survey, 1954-—. Geologist govt. expdn. which made trip by boat through Grand Canyon of the Colorado, Ariz., 1923. Research asso. in petroleum geology, auspices Am. Petroleum Inst., 1926-27. Geologist Gen. MacArthur's staff, Tokyo, Japan, 1949. Fulbright profs. Rijks-Universiteit te Utrecht, 1951-52; Walker-Ames prof. U. Wash., 1957; Solon E. Summerfield distinguished prof. geology, 1958-—; consultant Humble Oil and Refining Company also to the Ariz. Power Authority on Colo. River Devel. Maj. U. S. Army, Fuels and Lubricants Div. Office Quartermaster General, Washington. Hayden Meml. Geol. Medal, Phila. Acad. Natural Science, 1956; medal Paleontological Society, 1963. Fellow Geological Soc. Am. (v.p. 1956-57, pres. 1957-58), Paleontological Soc. (pres. 1947), mem. Am. Assn. Petroleum Geologists (Powers medal 1959), Soc. Systematic Zoology (pres.), Kan. Geol. Soc., Soc. Econ. Paleontology and Mineralogy (honorary member, president 1928), Am. Acad. Arts and Scis., geol. socs. of London (Wollaston medal 1968), Belgium, France, Holland, Switzerland, Germany, Argentina, Japan, Internat. Geological Congress (president commission on stratigraphy 1952-60), American Commission on Stratigraphic Nomenclature (chmn. 1947-49), Paläontologische Gesellschaft, Germany (hon.), Accademia Nazionale di Scienze Lettere ed Artidi Modena, Italy (hon.), Det Norske Videnskaps-Akademi, Académie Royale de Belgique (Prix Paul Fourmarier 1966), Sigma Xi. Author: Historical Geology, 1949; Invertebrate Fossils, 1952. Organizer-editor internat. treatise on invertebrate paleontology; various other books and articles on petroleum, regional and structural geology, physiography, stratigraphy, and invertebrate paleontology (especially Paleozoic crinoids, bryozoans, and corals). Editor: Bull. of Am. Assn. Petroleum Geologists, 1920-26, Jour. Paleontology, 1930-39, Jour. Sedimentary Petrology, 1931-39; Univ. Kansas Paleontological Contributions, since 1946. Devised classification system of layered sedimentary deposits and their organic contents. Home: 1640 Stratford Rd., Lawrence, Kan. 66040.*

MOORE, Raymond Hugh, Am. chemist; b. Spokane, Wash., May 29, 1918, s. Edwin James and Pearl (Baker) M.; B.S. magna cum laude, Gonzaga U., 1940; postgrad. U. Pitts.; m. Margaret Helen Diehl, Sept. 5, 1941; children—Carol (Mrs. Stanley Cass), James Ray. Faculty, U. Pitts., 1941-42; jr. chemist U. S. Bur. Mines, Bruceton, Pa., 1942, asst. chemist, College Park, Md., 1942-44, Albany, Ore., 1944-47; sr. scientist Gen. Electric Co., Richland, Wash., 1947-65; sr. specialist Battelle N.W., Richland, 1965-—. Mem. Am. Chem. Soc. Publs. on studies of chem. equilibria in salt systems, metal diffusion, solid-state electrodiffusion, patentee, metallurgy, extractive metallurgy and chem. separations processing. Home: 827 W. 23d Pl., Kennewick, Wash. 99336. Office: P.O. Box 999, Richland, Wash. 99352.*

MOORE, Raymond John, Canadian botanist; b. Hamilton, Ont., Can., Oct. 26, 1918; s. Joseph and May (Rothwell) M.; B.A., McMaster U., 1941; M.A., U. Va., 1943, Ph.D., 1946. With Central Exptl. Farm, Can. Dept. Agr., Ottawa, Ont., 1944-—, research scientist 3, 1965-—. Mem. Bot. Soc. Am., Genetics Soc. Can., Canadian Bot. Assn. Research, publs. in cytotaxonomy, evolutionary relationships of species using number and morphology of chromosomes together with gen. morphology of plants, reviser classification of groups concerned; studied family Loganiaceae (Buddleia), Caragana, N.Am. Cirsium (thistles). Home: 606 O'Connor St., Ottawa 1. Office: Plant Research Inst., Central Exptl. Farm, Ottawa 3, Ont., Can.*

MOORE, Richard Bishop, Am. chemist; b. Cin., May 6, 1871; s. William Thomas and Mary A. (Bishop) M.; went to Eng. with parents, 1878; student Argyle Coll., London, Eng., 1881-83, St. Edmund's Coll., London, 1883-85, Institut Keller, Paris, 1885-86, Univ. Coll., London, 1886-90, U. Chgo., 1896-97, B.S., 1896; D.Sc., U. Colo., 1916; m. Callie Pemberton, June 11, 1902; m. 2d, Georgie Elizabeth Dowell, June 18, 1924. Lived in Southport and London, Eng., 1878-95; instr. chemistry Oswestry High Sch., Eng., 1890-91, Birkbeck Inst., London, 1891-93; asst. in chemistry U. Chgo., 1896; instr. chemistry U. Mo., 1897-1905; prof. chemistry Butler Coll., Indpls. 1905-11; soil scientist Lab. Phys. and Chem. Investigations, Bur. Soils, Washington, 1911-1912; phys. chemist in charge of chemistry and metallurgy of rare metals U. S. Bur. Mines, 1912-19, chief chemist and chief div. mineral tech., 1919-23; gen. mgr. Door Co., engrs., N.Y., 1923-26; dean of science, head chem. dept. Purdue, 1926-—. Made survey for U. S. Geol. Survey, of thermal waters of Yellowstone Nat. Park for radio-active properties, 1906; with Sir William Ramsey, London, Eng., 1907-08; in charge all helium work for U. S. Bur. Mines, 1918-23; mem. U. S. Helium Bd., 1920-23. Author: A Laboratory Chemistry, 1904; also papers on radio-activity, inorganic phys.

chemistry, and rare gases. Pioneer in advocating helium for use in dirigibles and balloons; supervised prodn. of 1st radium salts in U. S.; influenced policy of helium gas conservation. Died Jan. 20, 1931.

MOORE, Robert Allan, Am. pathologist; born Chicago, Ill., July 12, 1901; s. Ellis Philip and Nelly (Clymer) M.; A.B., Ohio State U., 1921, M.D., 1928, M.Sc., 1927, D.Sc., 1956; Ph.D., Western Res. U., 1930; D.Sc. (hon.), Ohio State U., 1954, Union Coll., 1954, Waynesburg, 1957; L.H.D., U. Miami, 1956; LL.D. (honorary), Long Island University, 1959; m. Ruth Miller, June 15, 1922; children—Richard Allan, Calvin Cooper. Instr. pathology, O. State U., 1924-28; research fellow, pathology, Western Res. U., 1928-30, instr., 1930-33; asst. prof. pathology, Cornell U., 1933-37, asso. prof., 1937-39; prof. of pathology, Washington U., St. Louis, Mo., 1939-54; dean, Washington U., Sch. Medicine, 1946-54; vice chancellor schs. of health professions, Univ. Pitts., 1954-57, prof. pathology 1954-57; pres. Downstate Med. Center, dean Coll. Medicine U. of N.Y., 1957-66; Guiteras lectr., Am. Urol. Assn., 1950; Poynter lectr. U. Neb., 1951; Mellon lectr., U. Pitts., 1951; Luis Guerrero lect., U. Santo Tomas, 1952; Macgregor lectr. U. Western Ontario, 1952; Ballenger lectr., Southeastern Sec., Am. Urol. Assn., 1956; sr. cons. path. surgeon gen., AUS; mem. com. pathol., 1942-58, Nat. Research Council; civilian adviser on epidemic diseases to secretary of war, 1942-46; spl. consultant to surgeon gen., U. S. Army; scientific adv. bd. Army Inst. Pathol. (chmn. 1953); mem. adv. com. VA; adv. com. med. pub. health Rockefeller Found.; mem. Am. Bd. Pathology (pres. 1951-53); advisory committee Cancer Control USPHS (chmn. 1952-55); hon. cons. surgeon gen. USN, 1956-59; Nat. adv. council Health Research Facilities, USPHS, 1956-60. Coordinating dir. U. Pitts. Health Center, 1954-57; mem. bd. trustees Nat. bd. Medical Examiners, 1948-60, pres. 1954-57, China Med. bd. N.Y., 1955—; adv. council Med. Edn., 1950-56; adv. com. Nat. Com. Resettlement Fgn. Physicians, 1956-63; adv. bd. Med. Specialties, pres. 1953-57; Mem. Am. Assn. Pathol. and Bacteriol. (president 1952), Federation Biol. Societies, Club for Rsrch. on Aging, Soc. Exptl. Pathol., Am. Soc. Clin. Pathol., Am. Soc. Cancer Research. Gerontological Soc. (pres. 1951), Internat. Soc. Geographic Pathology pres. 1952-54), Coll. Am. Pathol., Am. Soc. Clin. Pathologists, Mexican Association of Pathologists (hon.), Alpha Omega Alpha, Sigma Xi. Author: Textbook of Pathology, 1944, 1951. Contbr. sci. jours. Christian Fenger lectr. Chgo. Inst. Medicine, 1947. Research, publs. on pathology of urinary and male genital organs, med. edn. Home: 3500 Snyder Av., Bklyn. 3.*

MOORE, Robert Lee, Am. mathematician; b. Dallas, Nov. 14, 1882; s. Charles Jonathan and Louisa Ann (Moore) M.; B.S., U. Tex., 1901, M.A., 1901; Ph.D., U. Chgo., 1905; m. Margaret MacLellan Key, Aug. 19, 1910. Teaching fellow U. Tex., 1901-1902; tchr. high sch., Marshall, Tex., 1902-03; asst. prof. math. U. Tenn., 1905-06; instr. Princeton, 1906-08, Northwestern U., 1908-11; instr. U. Pa., 1911-16, asst. prof., 1916-20; asso. prof. pure math. U. Tex., Austin, 1920-23, prof., 1923-37, distinguished prof., 1937-53, prof. math and astonomy, 1953-59, prof. math., 1959—. Colloquium lectr., 1929; vis. lectr. Am. Math. Soc., 1931-32. Fellow A.A.A.S. (v.p. 1947); mem. Am. Math. Soc. (asso. editor Transactions 1913-26, pres. 1936-38), Math. Assn. Am., Nat. Acad. Scis. Author: Foundations of Point Set Theory, vol. XIII, 1932, rev., 1962. Editor: Colloquium Publs., 1929-36. Research on foundations of mathematics; functions of real variables; point sets. Home: 904 W. 23d St., Austin, Tex.

MOORE, Robert Thomas, Am. zoologist; b. Haddonfield, N.J., June 24, 1882; s. Henry Dyer and Mary J. (Smith) M.; A.B., U. Pa., 1903; A.M., Harvard, 1904; postgrad. U. Munich; D.S. (hon.), Occidental Coll., Los Angeles, 1949; m. Selma Helena Muller, Dec. 22, 1903; children—Terris, Karlene; m. 2d, Margaret Forbes Cleaves, June 17, 1922; 1 dau., Marilynn; step-children—Waddell Austin, Paul Austin. Editor Cassinia, ofcl. publ. Del. Valley Ornithol. Club, 1911-16; breeder of silver black foxes; owner Borestone Mountain Fox Ranch, Onawa, Me., 1915-30, Western bus. inc., 1923, as Big Bear Fox Ranch of Cal.; asso. dept. vertebrate zoology Cal. Inst. Tech., 1929-50; asso. in vertebrate zoology and dir. zool. lab. Occidental Coll., Los Angeles 1950-55; v.p. Moore Securities Co. (Phila.); pres. Big Bear Fox Ranching Co. until 1928; dir. Guanajuato Reduction & Mines Co., Empire Lumber Co., Cowichan Lake Lumber Co. Founder of World's 1st Nat. Silver Fox Show, Boston, 1919; leader of ornithol. expdn. to Ecuador, 1927, zool. expdn. to S. Ecuador, 1929 (made 1st successful ascent of Mt. Sangai, active volcano, large zool. collection from hitherto unexplored regions; zool. species new to science), to Mexico for Cal. Inst. Tech., 1933, 34, 36, 37, 38, 42, 43, 45 (secured many birds new to science); lectured in Cultural Relations Mexico, 1942-45; chmn. Galapagos Com., 1934-38; instrumental in having a large part of Galapagos Archipelago set aside by Ecuador as sanctuary for zool. life. Fellow Royal (London), Am. geog. socs., Am. Ornithol. Union (mem. council); mem. Am. Com. for Internat. Wild Life Protection, Acad. Natural Scis., Am. Nat. Fox Breeders Assn. (bd. govs., 1st hon. pres.), Am. Fox Breeders Assn. (bd. govs.), Phi Beta Kappa. Author: Eileen, a Sonnet Seguence, 1946; chairman of authors: Check List of Mexican Birds, 1950; Biotic Provinces of Mexico (with A. E. Goldman). Editor in chief Poetry Awards. Contbr. more than 60 articles on zoology, breeding and exploration. Died Oct. 30, 1958.

MOORE, Samuel Wilson, Am. surgeon; b. Mooresville, N.C., Apr. 19, 1906; s. Nicholas Gibbon and Margaret (White) M.; B.S., Davidson Coll., 1926; M.D., Harvard, 1930; m. Mary Jameson Posey, Aug. 22, 1952; children—Samuel Posey, Nicholas, David. Practice medicine specializing in surgery, N.Y.C., 1939-42; staff N.Y. Hosp., 1939-42, 45—, attending surgeon, 1949—; with Cornell U. Med. Coll., 1933—, clin. prof. surgery, 1956—. Cons. surgery staffs Mt. Vernon Hosp., 1953—, Hosp. for Spl. Surgery, 1955—; cons. staffs Sharon (Conn.) Hosp., Lawrence Hosp., Mooresville, N.C.; courtesy staff Drs. Hosp. Diplomate Nat., Am. bds. surgery. Fellow A.M.A., A.C.S., N.Y. Acad. Medicine; mem. Med. Soc. County N.Y., Med. Soc. State N.Y., Soc. U. Surgeons (past pres.), N.Y. Soc. for Cardiovascular Surgery, N.Y. Surg. Soc., Internat. Soc. Surgery, So. Surg. Assn., Soc. for Surgery Alimentary Tract, Pan Am. Med. Assn., N.Y. Acad. Scis., others. Research, publs. on arterial disease, abdominal surgery, ulcers, malignancies. Home: 920 Fifth Av., N.Y.C. 10021. Office: 525 E. 68th St., N.Y.C. 10021.*

MOORE, Stanford, Am. biochemist; b. Chgo., Sept. 4, 1913; s. John Howard and Ruth (Fowler) M.; B.A., Vanderbilt U., 1935; Ph.D., U. Wis., 1938; M.D. (hon.), U. Brussels, 1954; Dr. hon. causa, U. Paris, 1964. U. Wis. Alumni Research Found. fellow, 1935-39; asst. Rockefeller Inst. Med. Research, N.Y.C., 1939-42, asso., 1942-49, asso. mem., 1949-52, mem., prof., 1952—. Tech. aide OSRD, NDRC, 1942-45; cons. Chem. Corps., U. S. Army, 1945-58; chmn. panel on proteins, com. on growth NRC, 1947-49; vis. prof. (Franqui chair) U. Brussels, Belgium, 1950-51; vis. investigator U. Cambridge (Eng.), 1951; sec. commn. on proteins Internat. Union Pure and Applied Chemistry, 1953-57. Mem. Am. Chem. Soc. (recipient chromatography award, 1963), Am. Soc. Biol. Chemists (treas. 1956-59, pres. 1966-67), Brit., Belgian biochem. socs., Harvey Soc., A.A.A.S., Nat. Acad. Scis., Am. Acad. Arts and Scis., Belgian Royal Acad. Medicine, Phi Beta Kappa, Sigma Xi. Editorial bd. Jour. Biol. Chemistry, 1950-60. Research on chem. structure of proteins, ribonuclease, chromatographic methods; designer automatic equipment for chromatographic determination of amino acids in proteins, physiol. fluids, foods. Home: 200 E. 66th St., N.Y.C. 10021.*

MOORE, Walter John, Am. chemist; b. N.Y.C., Mar. 25, 1918; s. Walter John and Ruth (Hart) M.; student Edinburgh (Scotland) U., 1935-36; B.S., N.Y. U., 1937; Ph.D., Princeton, 1940; m. Patricia Bacon, Nov. 4, 1943; children—Anthony, Julia, Catherine. NRC fellow Cal. Inst. Tech., 1940-41; civilian staff Manhattan Project, 1942-46; faculty Cath. U., Washington, 1946-51; Guggenheim and Fulbright fellow U. Bristol (Eng.), 1952-53; prof. chemistry Ind. U., Bloomington, 1953-63, research prof. chemistry, 1964—; NSF fellow U. Paris (France), 1958-59; Fulbright fellow U. Queensland (Australia), 1966; vis. prof. Harvard, 1960, U. Brasil, Rio de Janeiro, 1962-63. Chmn. com. on phys. chemistry NRC, 1964—; mem. solid state scis. panel NRC, 1965—. Mem. Am. Chem. Soc. (James Flack Norris award in chem. edn. 1965), Biophys. Soc., Faraday Soc., Soc. Chimie Physique, A.A.A.S., Phi Beta Kappa. Author: Physical Chemistry, 1952; Seven Solid States, 1967; also numerous articles. Editorial bd. Jour. Phys. Chemistry, 1964—. Research in phys. chemistry of solid state, conductivity, diffusion in oxides at high temperatures, biol. electrochemistry, biophysics of learning and memory. Home: Bender Rd., Rural Route 2, Bloomington, Ind. 47401.*

MOORE, Walter Leon, Am. civil engr.; b. Estrella, Cal., Mar. 12, 1916; s. Leon Wallace and Nellie (Munson) M.; B.S. in Engring., Cal. Inst. Tech., M.S. in Civil Engring., 1938; Ph.D. in Mechanics and Hydraulics, State U. Ia., 1951; m. Reta Mae Nunn, Nov. 28, 1942; children—Claire Louise, Catherine Adele, Geneva Elaine, James Walter. Jr. engr., U. S. Corps Engrs., Los Angeles, 1938-39, Soil Conservation Service, Pasadena, Cal., 1939-40; research analyst, research engr. Lockheed Aircraft Corp., Burbank, Cal., 1940-47; asso. prof. civil engring. U. Tex., Austin, 1947-53, prof., 1953—, chmn. dept., 1959—. Mem. organizing group Univs. Council on Water Resources; mem. Nat. Com. for Fluid Mechanics Films; cons. hydraulic engring. Mem. Am. Soc. C.E. (Collingwood prize 1944, dir. Tex. sect., 1954-56, pres. Tex. 1956-57, Am. Soc. for Engring. Edn. (dir. civil engring. div. 1953-54), Tex. Soc. Profl. Engrs., Sigma Xi, Tau Beta Pi, Xi Epsilon. Research, publs. on hydraulic energy dissipation, scour of cohesive sediments, flow around objects in a velocity gradient field, hydraulics of drilling with jet bits, rainfall runoff problems and mechanics of water storage reservoirs. Home: 4508 Crewsway, Austin, Tex. 78731.*

MOORE, Wilbert Ellis, Am. sociologist; b. Elma, Wash., Oct. 26, 1914; s. Lavergne W. and Bertha (Maffit) M.; B.A., Linfield Coll., 1935; M.A., U. Ore., 1937; A.M., Harvard, 1939, Ph.D., 1940; m. Dorothy Mary Hewitt, June 15, 1936 (div. Jan. 1957); children—Dorothy Marjory, Flo Heawitt; m. 2d, Jeanne Brindle Bailey, May 1957 (dec. Oct. 1959); m 3d, Jeanne Ellen Yates, Mar. 12, 1960. Grad. asst. U. Ore., 1937-39; tutor, asst. sociologist Harvard, 1938-40; faculty Pa. State U., 1940-43; faculty Princeton, 1943-64, prof., 1951-64, vis. prof., 1964—; sociologist Russell Sage Found., N.Y.C., 1964—. Mem. Am. Philos. Soc., Eastern Sociol. Soc. (pres. 1953), Am. Sociol Assn. (pres. 1965-66, MacIver award 1963), Sigma Xi. Author: Economic Demography of Eastern and Southern Europe, 1946; Industrial Relations and the Social Order, 1951; Industrialization and Labor, 1951; Economy and Society, 1955; The Conduct of the Corporation, 1962; Man, Time, and Society, 1963; Social Change, 1963; The Impact of Industry, 1965. Home: 396 Riverside Dr., Princeton, N.J. 08540. Office: 230 Park Av., N.Y.C. 10017.*

MOORE, William Robert, Am. chemist; b. Mpls., July 18, 1928; s. Cecil Robert and Lena (Segal) M.; student U. Minn., 1946-47; B.S., U. Cal. at Los Angeles, 1950; Ph.D., U. Minn., 1954; m. Phylis Grace Tocco, Feb. 2, 1956; children—Karen Elizabeth, Steven Robert, Patricia Ann. With Mass. Inst. Tech., Cambridge, 1954—, asso. prof. chemistry, 1964—. Cons., Union Carbide, 1962—. Mem. Am. Chem. Soc., A.A.A.S. Contbr. articles to tech. jours. Discovered and developed methods for separating nuclear spin isomers and isotopes of hydrogen by gas chromatography; synthesized and studied mechanisms of formation and chemistry of various allenes and highly-strained small-ring compounds. Home: 7 Page Rd., Lexington, Mass. 02173. Office: Dept. Chemistry, Mass. Inst. Tech., 77 Massachusetts Av., Cambridge, Mass. 02139.*

MOOREHEAD, Warren King, Am. archeologist; b. Siena, Italy, Mar. 10, 1866 (parents Am. citizens); ed. Denison U., D.Sc., spent 3 yrs. in study under Dr. Thomas Wilson, Smithsonian Instn.; 4 yrs. in investigation of Ohio mounds at own expense; M.A. (hon.), Dartmouth, 1901; D.Sc., Oglethorpe U., 1927; m. Evelyn Ludwig, Nov. 10, 1892; children—Ludwig King, Singleton Peabody. In charge work in Ohio Valley, Utah, Colo. and N.Mex. for Chgo., Expn.; made valuable finds in altar mounds of Scioto Valley, explorations New Eng., Tex., etc.; formerly curator mus. Ohio State U. and Hist. Soc.; now dir. dept. archeology, Phillips Acad. Former mem. U. S. Bd. Indian Commrs. investigating Indian reservations for Dept. Interior until bd. abolished, 1933; exploration of Cahokia Mounds for U. Ill., 1920-23, 27, Etowah Mounds investigation, 1925-27. Author: Hopewell Explorations, Fort Ancient; The Stone Age; The American Indian; Stone Ornaments of American Indians; Archaeology of Maine; Cahokia Mounds; Archaeology of the Arkansas Valley. Died Jan. 5, 1939.

MOORJANI, Madhov Naraindas, food technologist; b. Sind, India, Mar. 28, 1923; s. Naraindas M. and Vasibai Moorjani; B.Sc. with honors, D.J. Sind Coll., Karachi, India, 1944, M.Sc., 1947; Ph.D., Indian Inst. Sci., Bangalore, India, 1949; m. Mohini Moorjani, June 18, 1951; children—Archana, Ravi. Demonstrator for chemistry practicals D. J. Sind Coll., 1944-47; research scholar Indian Inst. Sci., Bangalore, 1947-48; research staff food tech. Central Food Technol. Research Inst., Mysore, India, 1949—, lectr. on fish tech., 1952—, sr. sci. officer, 1962—. Sr. Research fellow Colombo Plan, Commonwealth Sci. and Indsl Research Orgn., Homebush, Australia, 1955-56. Mem. Assn. Food Technologists, Nutrition Research Soc. India. Research, publs. on processed protein rich foods based on milk substitutes of vegetable origin, fish flour, Fricola, fish macaroni, tech. fish protein concentrates, changes in nitrogenous constituents of fish during chilling, improvement of freeze-dried shrimp; devel. new methods and techniques in fish processing preservation. Home: 2945 4th Main Rd., Vontikoppal, Mysore-2. Office: Central Food Technol. Research Inst., Mysore-2, S. India.*

MOORREES, Coenraad Frans August, dentist; b. The Hague, Holland, Oct. 23, 1916; s. Coenraad F. A. and Theodora (Mensinga) M.; Tandarts, U. Utrecht, Holland, 1939; D.D.S., U. Pa., 1941; A.M. (hon.), Harvard, 1959; m. Louise H. van der Mey, Nov. 21, 1939; children—Louise Frances Theodora, Alexander Mensinga. Came to U. S. 1947, naturalized, 1956. Faculty, Forsyth Dental Center, Harvard, Boston, 1947—, prof. orthodontics, 1964—. Mem. Nat. Inst. Dental Research, Am. Acad. Dental Sci., Am. Assn. Orthodontists, Am. Assn. Phys. Anthropologists, Dutch Soc. Study Orthodontics, Fedn. Dentaire Internationale, Groupement International pour la Recherche Scietifique en Stomatologie, Internat. Assn. Dental Research, Internat. Soc. Cranio-Facial Biology, Internat. Coll. Dentists, Finnish, Netherlands dental socs., Deutsche Gesellschaft für Kieferorthopädie, Nederlands Vereniging and Tandartsen, Sigma Xi, Omicron Kappa Upsilon, Delta Sigma Delta, others. Mem. adv. editorial bd. Jour. Dental Research, 1962-64. Research, publs. racial characteristics of dentition and its growth and devel. from longitudinal observations; orofacial forces. Home: 4 Peacock Farm Rd., Lexington, Mass. 02173. Office: 140 The Fenway, Boston 02115.*

MOQUIN-TANDOU, (Christian-Horace-Bénédict) Alfred, French botanist; b. Montpellier, France, May 7, 1804: Docteur ès scis.; prof. Toulouse (France) Faculty Scis., Paris Faculty Medicine; mem. Acad. Medicine, French Acad. Scis., 1851. Author: Eléments de tératologie végétale, 1841; Histoire naturelle des mollusques terrestres et fluviatiles de France, 1855. Died Paris, Apr. 15, 1863.

MORA, Peter Tibor, biochemist; b. Szolnok, Hungary, July 18, 1924; s. Lajos and Ilona (Medgyessy) M.; Ph.D., U. Budapest, 1948; m. Leora May Knapp, Sept. 29, 1951. Came to U.S., 1948, naturalized, 1952. Postdoctoral fellow chemistry Princeton, 1948-50; chem. research E. I. duPont de Nemours & Co., Wilmington, Del., 1950-54; cancer research Lankenau Inst. for Cancer Research, Fox Chase, Phila., 1955; biochemistry research Oxford U., 1955-56; cancer research, head macromolecular biology sect. Nat. Cancer Inst., NIH, Bethesda, Md., 1956—, instr. NIH Grad. Program, 1958—. Mem. Am. Chem. Soc. Contbr. articles to sci. jours. Discovered chem. polymerization of carbohydrates; use of synthetic polysaccharides and derivatives in controlling interactions of biologic macromolecules such as enzyme-substrate, virus-receptor cell and antigen-antibody interactions; studies of macromolecular control of cell differentiation and of cancer formation. Home: 4006 W. Underwood St., Chevy Chase, Md. 20015. Office: Nat. Cancer Inst., NIH, Bethesda, Md. 20014.*

MORAND, Max, French physicist; b. Romorantin, France, Nov. 28, 1900; s. J. E. and Yvonne (Tribot) M.; D.Sc., U. Paris, 1927; m. Geneviève Sutra, Dec. 26, 1962. Prof., Faculty Scis., U. Liège (Belgium), 1930-45, U. Paris, 1945—. Author: Introduction mathematique aux théories physiques modernes; Les rayons cosmiques. Research, publs. on nuclear physics, emulsion plates, cosmic rays, high energy physics, elementary particle physics. Home: 8 Rue Octave Feuillet, Paris XVI. Office: 9 Quai St. Bernard, Paris V, France.*

MORAND, Sauveur François, French surgeon; b. Paris, Apr. 2, 1697; s. Jean Morand; ed. Coll. Mazarin; a son, Jean-François-Clément. Prin. surgeon Invalides; visited London, 1729; surgeon Hosp. La Charite, became royal censor, surgeon-in-chief, 1730; pensioner, prof. anatomy French Acad. Scis., 1722. Recipient Order of St. Michael from king, 1715. Fellow Royal Soc., 1728, mem. acads. of Petersburg, Stockholm, Bologna, Florence, Rouen. Author: Traité de la taille au haut appareil, 1728; Opuscules de chirurgie, 1768-72; also articles. Studied Chaselden's lateral method of lithotomy and introduced it to France; 1st successful operation for temporosphenoidal abscess following ear infection, 1752; 1st description of cleidocranial dysostosis, 1766. Died Paris, July 21, 1773.

MORANI, Valentino, Italian chemist; b. Rome, Dec. 2, 1899; s. Alessandro and Elisabetta (Helbig) M.; Ph.D. in Pure and Applied Chemistry; m. Wanda Segni, Nov. 28, 1936; 1 dau., Livia. Past prof. U. Sassari; instr. agrl. chemistry, dir. exptl. sta. for agrl. chemistry U. Rome. Mem. Potash Inst. Berne, Florence Acad. Agriculturists. Research and publs. on agrl. chemistry, agrl. industry, gen. agronomy. Home: via Antonio Gallonio 9, Rome.

MORAX, Victor, ophthalmologist; b. Morges, Switzerland, 1866; M.D., 1894; certified physician, 1900; became mem. staff Pasteur Inst., 1894, head labs., 1935; founder Internat. League against Trachoma. Author: Précis of Ophthalmology; also articles on conjunctiva and cornea. Isolated (independently of Karl Axenfeld) diplobacillus which causes a form of chronic conjunctivitis (Morax Axenfeld haemophilus), 1896. Died 1935.

MORAY, Sir Robert, mathematician, naturalist; b. Highlands of Scotland, circa 1608; s. Mungo and dau. of George Halkett of Pifinan Moray; ed. U. St. Andrews (Scotland); also studied in France; m. sister of Earl of Balcarries, circa 1652-53. Mil. service under Louis XIII; privy council under King Charles II. Mem. Royal Soc. (a founder 1661; pres. several times, v.p. at time of his death). Publs. on chemistry, medicine, horticulture, fuel, whale fishing, coal mining, water wheels, tide mills, math. and surveying instruments, various mech. devices. Died Whitehall, Eng., July 4, 1673.

MORCILLO RUBIO, Jesús, Spanish physicist; b. Tarancon, Spain, Nov. 16, 1921; s. Gabriel and Alejandra (Rubio) Morcillo; Ph.D., U. Madrid (Spain); m. Gloria Ortego, June 20, 1951; children—Gloria, Marla Jose, Jesus, Iberto, Juan Gabriel, Riansares. Asst. phys. chemistry U. Madrid; agrégé Inst. Phys. Chemistry; research chief Inst. Chemistry and Physics; prof. atomic molecular structure and spectroscopy U. Madrid. Mem. Royal Spanish Soc. Physics and Chemistry, Nat. Assn. Chemists, Nat. Assn. Chemists. Author: Teoria de grupos y fissica molecular, 1950; Aspectos modernos de la espectroscopia molecular, 1951; Applicaciones praticas de la espectroscopia infraroja, 1963; also articles. Mem. Royal Spanish Soc. Physics and Chemistry, Nat. Assn. Chemists, Coll. Chemists. Home: avda de America 49, Madrid, Spain.

MORCOS, Selim A., Egyptian oceanographer; b. Cairo, Egypt, Sept. 2, 1928; student U. Cairo, U. Alexandria; m. Mary B. Ghobrial, Feb. 4, 1965; 1 son, Hany. Research asso. (Alexander von Humboldt award), Oceanographic Inst., U. Kiel (Germany), 1956-59; lectr. oceanography dept., sci. faculty U. Alexandria (Egypt), 1959—; UNESCO fellow Internat. Indian Ocean Expdn., 1962; prof. Naval Inst., Suez Canal Authority. Brit. Council bursar, 1965. Mem. Cairo Geog. Soc., Alexandria Archeol. Soc. Author: (in Arabic) Sunken Civilization, An Account on Submarine Archeology, 1965; also articles in field. Research on hydrography of Suez Canal. Address: 49 Semouha, Alexandria, Egypt.*

MORDELL, Louis Joel, mathematician; b. Phila., Jan. 28, 1888; s. Phineas and Annie (Feller) M.; M.A., St. John's Coll., Cambridge, U., 1945; LL.D., U. Glasgow, 1956; D.Sc., Mt. Allison U., 1959; m. Mabel Elizabeth Cambridge, May 25; 1916; children—Frances Kathleen, Nicholson Smith, Donald Louis. Lectr., Birkbeck Coll., London, 1913-20, Manchester Coll. Tech., 1920-22; reader Manchester U., 1922-23, prof. math., 1923-45; prof. math. Cambridge U., 1945-53; vis. prof. numerous univs. in U. S., Can., Africa; fellow St. Johns Coll., Cambridge. Fellow Royal Soc., 1924. (Sylvester medal 1949); mem. London Math. Soc. (pres. 1943-45, Demorgan medal 1941, Berwick prize 1946), Norwegian (fgn. mem.), Uppsala, Bologna acads. sci., Am., Canadian math. socs., Am. Math. Assn., Cambridge Philos. Soc. Author: Three Lectures on Fermats Last Theorem, 1921; A Chapter in the Theory of Numbers, 1947; Reflections of a Mathematician, 1959; also numerous articles. Research in theory of numbers, allied subjects. Home: 1 Bulstrode Gardens. Office: St. John's Coll., Cambridge, Eng.*

MORE, Robert Hall, Canadian pathologist; b. Kitchener, Ont., Can., Dec. 16, 1912; s. Robert Hall and Nellie (Lackner) M.; M.D., U. Toronto, 1939; M.Sc., McGill U., 1942; m. Dorothy Charlotte McOrmond, 1943; children—David, Patricia, Christopher. Miranda Fraser prof. comparative pathology McGill U., Montreal, 1950-51, Strathcona prof. pathology, chmn. dept., 1967—; prof., head dept. pathology Queen's U., 1951-66; pathologist-in-chief Kingston (Ont.) Gen. Hosp.; cons. pathology Ont., Hotel Dieu, St. Mary's of the Lake, Dept. VA, Kingston Mil. hosps. chmn. sci. subcom. Canadian Heart Found. Mem. Royal Coll. Physicians and Surgeons, Ont. Assn. Pathology (pres. 1959), Internat. Acad. Pathology (pres. 1967), Canadian Arthritis and Rheumatism Soc. (mem. council on research, profl. edn.), Canadian Assn. Pathologists (chmn. program com. 1963-65), Am. Assn. Pathologists and Bacteriologists, Am. Soc. Exptl. Pathologists, Pathol. Soc. Great Britain and Ireland, Am. Heart Soc. (mem. council on arteriosclerosis). Editorial bd. Am. Jour. Pathology Studies, numerous publs. of mechanisms by which fluids of blood are deposited in tissues in rheumatic diseases and cause deformities of joints and heart valves, mechanisms by which blood constituents accumulate inside arteries to lead to arteriosclerosis. Home: 1212 Pine Av. W., Montreal 25, Office: 3775 University St., Montreal 2, Que., Can.*

MORE, Sir Thomas, English statesman, polit. theorist; b. Cheapside, Feb. 7, 1478; s. John and Agnes (Graunger) M.; ed. St. Anthony's Sch.; resided in home of Thomas Morton, archbishop of Canterbury and lord chancellor, circa 1790-91; entered Canterbury Hall, Oxford, circa 1492; law student New Inn, London, 1496; m. Jane Colt, 1505; children—Margaret, Elizabeth, Cecily, John; m. 2d, Alice Middleton; 1 stepdau., Alice. Apptd. lectr. law Furnival's Inn; maintained lit. interest, met many scholars including Erasmus; lived nr. Charterhouse, lectr. in ch. of St. Lawrence, 1499-1503; after comtemplating becoming priest, instead entered politics, circa 1503; elected mem. parliament, 1504; traveled to Louvain and Paris, 1505; elected bencher Lincoln's Inn, 1509, reader, 1511; apptd. under-sheriff, London, 1510, then king's ambassador, Flanders, 1514; mem. Hampshire Commn. of Peace, from 1510; became officer of crown by Henry VIII; nominated master of requests, 1517; sub.-treas. to king, from 1521; speaker House of Commons, 1523; high steward Oxford U., 1524, Cambridge U., 1525; chancellor of ducy, Lancaster, 1525; became lord chancellor of Eng. (1st layman not a noble to hold position), 1529, refused to approve divorce of King Henry VIII from Queen Catherine, then resigned, 1532. Author: Life of John Picus, Earl of Mirandula, 1510; Utopia (commentary on contemporary affairs through discussion of ideal state or utopia), 1516; History of Richard III, 1528-33; Dialogue of Comfort against Tribulation, 1534; Treatise on the Passion, 1534. Known for his polit. career, also as polit. theorist; introduced term Utopia (nowhere); progenitor of utopian lit. Refused to recognize Henry VIII as head of Ch. of Eng., beheaded by king's orders, 1535.

MOREAU, Jean Jacques, French mathematician; b. Blaye, France, July 31, 1923; s. Joseph and Henriette (Bonneau) M.; Licence ès Sciences, U. Poitiers (France), 1944, Agrégation de Mathematiques, 1945; Doct. ès Sci. Math., U. Paris, 1949; m. Louise Raynal, July 7, 1956. In charge research Centre Nat. Recherche Scientifique, 1945-51; head math. works U. Montpellier (France), 1951-53, lectr., prof., 1956-60, titular prof., 1960—; lectr. math. methods physics U. Poitiers, 1953-56. Recipient Palmes Acad. Mem. Am. Math. Soc., Math. Soc. France. Research, publs. on dynamics of incompressible media, vortex theory, unilateral constraints in mechanics of continua, functional analysis and convexity theory. Home: 4 B rue Nozeran, Montpellier (34), France.*

MOREAU, Mireille, French plant pathologist; b. Chambry, France, June 15, 1925; d. Pierre and Lise (Vasseur) Froment; Baccalauréat in Elementary Math. and Philosophy, 1943; Licence ès Sciences Naturelles, U. Caen (France), 1946; Doctorat ès Sciences Naturelles, U. Paris, 1956; m. Claude Moreau, July 16, 1946; children—Francis, Joel. Tech. asst. mycology Nat. Center Sci. Research, Caen, 1945-46; certified asst. Lab. Mycology and Tropical Phytopathology, Ecole Pratique des Hautes Etudes, Paris, 1946-56, Lab. Crytogamy, Nat. Mus. Natural History, Paris, 1956-63; lectr., dir. Lab. Plant Biology, Brest (France) Faculty Scis., 1963—, founder dept. research mycology and phytopathology, 1963. Recipient Xavier Bernard award Acad. Agr., 1958; laureate Soc. Encouragement Nat. Industry. Mem. Bot. Soc. France, Mycol. Soc. France, Soc. Biology (Rennes chpt.), Soc. Mediterranean Phytopathology. Author: The Withering of Carnations, 1957. Systematic studies of mushrooms, especially tropical varieties; studies vascular diseases of plants caused by parasitic mushrooms. Home: 13 rue Van Gogh, Brest. Office: Faculte des Sciences, Brest, France 29.*

MOREAUX, Arsène Jean Joseph Marie René, French entomologist; b. Nancy, France, 1889; s. Adrien and Marie-Anna (Bagnard) M.; ed. Faculty Sci. and Medicine; M.D.; m. Marie-Hélène Colin, Sept. 2, 1925; children—Geneviève, Annette, José, Charles, Jean, Hélène, Colette, Bernadette. Practice medicine, specializing in otorhinolaryngology; dir. otorhinolaryngology services 4th Army; biologist, instr. Nancy Sch. Agronomy. Decorated Palms Acad., Order Merit in Agr., Enamel Epidemics medal. Mem. French Otorhinolaryngol. Soc., Stanislas Acad., Soc. for Comparative Pathology. Research and numerous publs. in medicine, biology, agronomy. Address: 20, rue Verlaine, Nancy, France.

MOREHEAD, Robert Page, Am. physician; b. Lasker, N.C., Sept. 4, 1910; s. Robert Page and Dorcas Ann (Vernon) M.; B.S., Wake Forest Coll., 1931, M.A., 1932, B.S. in Medicine, 1934; M.D., Jefferson Med. Coll., Phila., 1936; m. Dorothy Ann Myers, May 18, 1946; children—Robert Page III, John Myers, Dorothy Ann. Practice medicine, specializing in pathology, Winston-Salem, N.C., 1936—; faculty Bowman Gray Sch. Medicine, 1941—, prof., chmn. dept. pathology, 1946—, mem. faculty adv. council, 1943—, mem. exec. council, 1959—; chief pathology N.C. Bapt. Hosp., 1941—, pathologist Med. Center, 1963—. Recipient Gaston County award for best sci. exhibit at Med. Soc. N.C. meeting, 1961; Bronze award Am. Soc. Clin. Pathologists, Coll. Am. Pathologists for exhibit, 1960. Mem. A.M.A., So. Med. Assn., A.C.P., Coll. Am. Pathologists, Am. Soc. Clin. Pathologists, Am. Assn. Pathologists and Bacteriologists, Am. Assn. Cancer Research, Internat. Acad. Pathology, A.A.A.S., Am. Cancer Soc. Author: Human Pathology: An Introduction to Medicine, 1965; also numerous articles. Distinguished contbr. articles on third biol. type of tumor, denoted as intermediate as opposed to benign and malignant. Home: 1051 Arbor Rd., Winston-Salem 27104. Office: Bowman Gray Sch. Medicine, Winston-Salem, N.C. 27103.*

MOREHOUSE, Daniel Walter, Am. astronomer; b. Mankato, Minn., Feb. 22, 1876; s. Aaron and Sabra Ann (Burleson) M.; N.W. Christian Coll., Excelsior, 1895-97; S.B., Drake U., 1900, S.M., 1902; S.B., U. Chgo., 1902; Ph.D., U. Cal., 1914; LL.D., Butler U., 1932; m. Myrtle Slayton, June 9, 1903; children—Charles Aaron, Vega Lorraine, Frances Roberta. Prof. physics and astronomy Drake U., 1900-41; pres., 1923-30, 30-41, dean Coll. Liberal Arts, 1922-30; instr. astronomy, U. Cal., 1911-12. Recipient Donahue Comet medal, 1908. Fellow Ia. State Acad. Sci. (pres. 1921-22). Discovered Comet (c), 1908 (Morehouse), Sept. 1, 1908. Died Jan. 21, 1941.

MOREIRA-FILHO, Hermes, Brazilian botanist; b. Curitiba, Brazil, Mar. 26, 1929; s. Hermes Augusto and Maria Sibylla Bürmann (Gama) Moreira; Farmaceutico, U. Paraná, 1949, Doutor em Farmácia e Biovimica, 1957; m. Ita Moema Valente, Janeiro 4, 1955. children—Cleide, Deise Maria, Marcia. Faculty, U. Paraná (Brazil), 1952-65, prof. Sch. Pharmacy, 1962-65; prof. Sch. Pharmacy and Biochemistry U. Fed. Paraná, 1965—. Mem. Instituto de Historia Natural do Paraná, Conselho Nacional de Pesquisas, Sociedad Botanica do Brasil, Sociedad Brasileira para o Prgresso da Ciencia, Sociedad Latino Americana de Ficologia. Research, publs. on bacillariophyceae (diatoms), marine and fresh water, diatoms of Atlantic Coast S. of Brazil, larval lagoons of Anopheles mosquito, in digestive tracts of mollusks, epiphyte diatoms in marine algae. Home: 619 Apto 6 Largo A. Parodi Curitiba, Paraná, Brazil.*

MOREL, Augustin Benoit, psychiatrist; b. Vienna, Nov. 22, 1809; degree, 1839; resident psychiatrist, Paris. Author: Traité des maladies mentales, 1852-53; Influence de la constitution géologique du sol sur la production du cretinisme, 1855; Traité des dégénérescences physiques intellectuelles et morales de l'espece humaine, 1857; Mélanges d'anthropologie pathologique, 1859; Du goitre et du crétinisme, 1864. Developed clin. methods for treating mental disorders and detecting mental degeneration; described degenerative ear deformed by partial obliteration of folds, marked by thin edge, gen. prominence (Morel ear), 1857. Died Rouen, France, Mar. 30, 1873.

MOREL, Georges Michel, French physiologist; b. Magnat, France, Apr. 16, 1916; s. Georges Victor and Alice (Maziére) M.; student Ecole Nationale Superieure de Chimie, Paris, 1935-37; student Sorbonne, Paris, 1937-42, Licencic es Sciences, 1948, Dr. ès Sciences, 1948; m. Josée Treillard, Oct. 14, 1955; children—Francois, Jeanne. Research fellow Harvard, 1950-51; research dir. Nat. Inst. Agronomical Research, 1953-62; chargé de cours Sorbonne, 1962—. Decorated chevalier de La Legion d'Honneur, Mem. Am. Chem. Soc., Am. Bot. Soc., Soc. Française de Physiologie Vegetive, Soc. de Biologie. Contbg. author: Ency. Biologie La Pleiade, 1964; author: La Multiplication Vegetative Masson, 1967; also articles. Discovered factor underlying cell division and differentiation in growing point of higher plants (making possible their cultivation in vitro on synthetic medicum); produced virus free plants, clonal propagation of orchids. Home: 22 rue du plateau, Saint Antoine, Versailles, Yvelines 78, France. Office: C.N.R.A. route de Saint Cyr, Versailles 78, France.*

MORELAND, Donald Edwin, Am. plant physiologist; b. Enfield, Conn., Oct. 12, 1919; s. Albert Sinclair and Ruth (Cowan) M.; B.S., N.C. State Coll., 1949, M.S., 1950, Ph.D., 1953; m. Verdie Stallings, Nov. 6, 1954; 1 dau., Donna Faye. Teaching asst. N.C. State Coll., 1949-50, AEC fellow, 1950-52; plant physiologist State U. N.Y. Coll. Forestry, Syracuse, 1952-53; plant physiologist Agrl. Research Service U. S. Dept. Agr., asst. prof. N.C. State U., Raleigh, 1953-61, asso. prof. botany, sr. plant physiologist, 1961-65, prof. botany, crop. sci., prin. plant physiologist, 1965—. Mem. toxicology study sect. NIH, Pub. Health Service, 1963-67. Fellow A.A.A.S.; mem. Am. Soc. Plant Physiologists, Bot. Soc. Am., Weed Soc. Am., N.C. Acad. Sci., Sigma Xi, Phi Kappa Phi, Alpha Gamma Rho, Gamma Sigma Delta. Studies, numerous publs. on translocation of mineral elements in woody plants, mechanisms of action of herbicides and growth regulators at cellular and molecular level and biochem. mechanisms of selective action of herbicides; metabolism of herbicides in plants and effects of herbicides on photosynthesis, behavior of mitochondria, nucleic acid biochemistry. Home: 1508 Pineview Dr., Raleigh, N.C. 27606.*

MORELLI, Carlo, Italian geophysicist; b. Trieste, Italy, Oct. 10, 1917; s. Carlo and Anna (Nachtigall) M.; Ph.D. in Math.; m. Antonietta Carlevaris, May 22, 1948; 1 dau., Maria Luisa. Dir. exptl. Geophysics Obs., Trieste; prof. applied geophysics U. Trieste. Contbr. numerous articles to sci. jours. Address: Osservatorio Geofissico, Trieste, Italy.

MORENG, Robert Edward, Am. poultry scientist; b. N.Y.C., Jan. 29, 1922; s. Joseph and Martha (Schlosser) M.; B.S., U. Md., 1944, M.S., 1948, Ph.D., 1950; m. Miriam Tittmann, Aug. 12, 1950; children—George R., Nathan T., Jon C., Diane M., Michael Q., Charles C., Joseph P. Asst. prof. poultry husbandry, N.D. Agr. Coll., Fargo, 1950-55; prof., head poultry sci. dept. Colo. State U., Ft. Collins, 1955—. Named Hon. Farmer, Future Farmers Am., 1962. Mem. Poultry Sci. Assn. Am., A.A.A.S., Soc. for Exptl. Biology and Medicine, Genetics Soc. Am., Radiol. Research Soc., Sigma Xi. Research, numerous publs. on avian embryo, effects of embryonic X-irradiation and low temperature exposure, reproductive performance and growth of turkeys. Home: Route 2, Box 79, Ft. Collins, Colo. 80521.*

MORESTEAD, Thomas, Brit. surgeon; b. circa 1375; became body surgeon to King Henry V, 1415, also to 2 other kings; chief mil. surgeon to English Army, Agincourt; organized alliance of mil. surgeons and physicians into Acad. Medicine, London, 1432. Attempted to reform medicine and surgery of the period. Died 1450.

MORESTIN, Hippolyte, surgeon; b. Martinique, 1869; ed. Paris. Author: Treatise on Diseases of the Joints. Contbd. to wound and reconstrn. theory, especially that of facial area, during World War I; noted for methods in sliding hernia, 1900, operation for reconstrn. of female breast, 1902, resection of wrist, 1902, inter-ilio-abdominal amputation, 1903, spino facial anastamosis. Died Paris, 1919.

MORET, Léon Marie Louis, French geologist; b. Annecy, France, July 4, 1890; s. Joseph and Marie (Magdelain) M.; lic.Sc., M.D., U. Lyons, 1919; Sc.D., U. Strasbourg (France), 1926; m. Elisabeth Denarie, Sept. 21, 1921; 4 children. Asst. geology U. Strasbourg, 1919-21; chief geol. studies Faculty Scis., 1921-23; lectr. geology and mineralogy Faculty Scis., U. Grenoble (France), 1923-31, prof., 1931—; prin. collaborator Service Geol. Charts. Became corr. mem. French Acad. Scis., 1949, Belgian Geol. Soc.; mem. French Geol. Soc. (became v.p. 1939, prize 1930), Sci. Soc. Dauphiné (pres. 1939-41). Recipient prize Institut de France, 1926. Author: Géologie du Massif des Bornes, 1934; (with Gignoux) Description géologique du bassin supérieur de la Durance, 1938, Géologie dauphinoise, 1944; Manuel de Paléontologie animale, 1940; Manuel de Paléontologie végétale, 1943. Research in alpine geology, paleontology of invertebrates. Office: Geol. Lab., Place Notre-Dame, Grenoble (Isère), France.

MORET, Pierre Raymond, Swiss physician; b. Fribourg, Switzerland, June 6, 1923; s. Léon and Angela (de Agostini) M.; student medicine U. Fribourg, 1944-48, U. Paris (France), 1948; grad. U. Lausanne (Switzerland), 1951, M.D., 1955; postgrad. U. Fribourg, French Hosp., Bellevue Hosp., N.Y.C., Center Cardiology, Geneva, Switzerland; m. Anne Marguerite Digier, Sept. 6, 1952; children—Jacques, Françoise, Philippe. Asst. dir. Center Cardiology, Hopital Cantonal, Geneva, 1958—, chief pulmonary function lab, chief research unit, 1958—; chargé de cours U. Geneva, 1962—. Mem. adv. bd. chemotherapy Research Bull., 1962—. Recipient Prix Bizo Geneva, 1960; Prix mondial Nessim Habif, 1965. Mem. Am. Coll. Cardiology, Swiss Cardiac Soc., Swiss Soc. Angiology (sec.), Swiss Soc. for Lung Diseases (asst. gen.). Author: Circulation Pulmonaire, 1963; also articles. Editor: (with L. Widmer) The Arterial Wall: Biochemistry and Elasticity, 1964; (with L. Mahaim) Electro-Vecto-Phonocardiography, 1966. Research, publs. on pulmonary circulation, influence of age on arterial wall elasticity, coronary circulation. Home: 10 Av. Peschier. Office: Center Cardiology, Hopital Cantonal, Geneva, Switzerland.*

MORETON, Robert Dulaney, Am. radiologist; b. Brookhaven, Miss., Sept. 24, 1913; s. Robert D. and Lena (Durfey) M.; student Tulane U., 1931-32; B.S., Millsaps Coll., 1934; postgrad. U. Miss. Med. Sch., 1934-36; M.D., U. Tenn., 1938; fellow radiology Mayo Found., 1940-42; m. Alma Williamson, Sept. 21, 1945. Instr. physiology U. Miss. Med. Sch., 1939-40; staff radiologist Scott and White Clinic and Hosp., also Santa Fe Hosp., Temple, Tex., 1942-50; sr. partner Bond Radiol. Group, Ft. Worth, 1950-65; chmn., dir. dept. radiology Harris Hosp., 1960-65; Ft. Worth Childrens Hosp., 1961-65; cons. St. Joseph, USPHS, John Peter Smith hosps.; prof. clin. radiology Southwestern Med. Sch., Dallas, 1960-65; asst. to dir., prof. radiology U. Tex., M.D. Anderson Hosp. and Tumor Inst.; v.p. U. Cancer Found., Houston, 1965—. Mem. Tex. Bd. Health, 1961—, vice chmn., 1963; mem. Tex. Adv. Com. Atomic Energy, 1955-61; founding mem. Carter Blood Bank, Ft. Worth, 1959; founding mem. Radiation and Research Found. S.W., 1957, mem. exec. com. Radiation Center, 1958—. Recipient gold medal Assn. Mil. Surgeons, 1949. Fellow Am. Coll. Radiology (exec. com. 1957-59); mem. Am. (chmn. radiology sect. 1964-65; gold medal for exhibit 1949, So. (chmn. bd. councilors and exec. com. 1959-60, chmn. ins. com. 1956-63, 65—; pres. 1964), Tex. (chmn. sect. radiology 1950, council med. jurisprudence 1958-62), Indsl. med. assns., Am. Roentgen Ray Soc. (certificate of merit of exhibit 1949), Radiol. Soc. N. Am. (chmn. bd. 1963, pres. 1965; certificate of merit of exhibit 1949). Contbr. profl. jours. Home: 1600 Holcombe Blvd., Houston 77025. Office: 6723 Bertner Av., Houston 77025.*

MORETTI, Jean, French biochemist; b. Avignon, France, June 15, 1914; s. Paul and Amélie (Odidier) M.; Docteur ès Scis. physiques, 1948; Faculté des Scis. de Marseilles, 1948. Joined Soc. Jesus, 1940; ordained priest Roman Catholic Ch., 1950; research staff Centre Nat. de la Recherche scientifique, 1955—, head research, 1961—; chief lab. proteins Faculté Médecine, Paris, France, 1955—. Named Chevalier du mérite militaire, 1960; named Lauréat de l'Académie des Scis., 1965. Mem. société de chimie biologique de France. Research, numerous publs. on isolation of haptoglobin from human, rabbit and rat serum, and determination of phys. and chem. properties, their half-life, anabolic and catabolic pathways; proteins involved in response to injury of connective tissue. Home: 42 Rue de Grenelle, Paris 7°, Office: 45 Rue des Saints-Pères, Paris 6°, France.*

MORETZ, William Henry, Am. physician; b. Hickory, N.C., Oct. 23, 1914; s. Joseph Alfred and Elizabeth (Leonard) M.; B.S., Lenoir Rhyne Coll., 1934, D.Sc. (hon.), 1960; postgrad. U. N.C. Sch. Medicine; M.D., Harvard, 1939; m. Laura Schlums, Dec. 7, 1947; children—William Henry, John D., Robert L., Richard E., Elizabeth L., David L. Instr. surgery U. Rochester (N.Y.), 1944-47; asst. prof. U. Utah, 1947-49, asso. prof., 1949-55; prof. surgery, chmn. dept. Med. Coll. Ga., Augusta, 1955—. Cons. to VA Hosp., U. S. Army Hosp. Fellow A.C.S. (gov. 1965); mem. Ga. Surg. Soc., Soc. U. Surgeons, Western, So., Am. surg. assns., Soc. for Surgery Alimentary Tract. Contbr. articles to profl. jours. Popularized use of peritoneal aspiration for diagnosis of acute surg. conditions in abdomen; developed concept of prevention of pulmonary emboli by partially occluding inferior vena cava. Home: 2345 McDowell St., Augusta, Ga. 30904.*

MOREUX, Théophile (Abbe Théophile), French meteorologist, astronomer; b. Argent, France, 1867; founder, dir. obs., Bourges, France. Author: les Tremblements de Terre, 1909; A l'assaut du pôle Sud, 1911; Les Merveilles des mondes, 1911; Les autres mondes sont-ils habités? 1912; les Eclipses, 1912; les Enigmes de la création, 1912; Comment prévoir le temps, 1919; Cosmographie; Atlas du ciel; Astronomie moderne. Popularized astronomy through his many works; research on sun; developed theory of moon's origin in cluster of meteors, 1921. Died Bourges, 1954.

MOREY, Samuel, Am. inventor; b. Hebron, Conn., Oct. 23, 1762; s. Israel and Martha (Palmer) M.; m. Hannah Avery, 1 child. Participated in constrn. Conn. River locks between Windsor (Conn.) and Okott Falls, engr. in charge, Bellows Falls, Vt.; obtained 1st patent for steam-operated spit, 1793; patented rotary steam engine, 1795; patented windmill, water wheel, steam pump; built stern wheel steamboat, ran from Hartford, Conn. to N.Y.C., 1794; attempted to persuade Robert Fulton to adopt his steamboat model, claimed his ideas were stolen by Fulton; patented internal combustion engine, 1826; propelled boat Aunt Sally by vapor engine on Fairlee (Vt.) Pond (now known as Lake Morey), 1820. Died Fairlee, Apr. 17, 1843.

MORFIT, Campbell, chemist; b. Herculaneum, Mo., Nov. 19, 1820; s. Henry Mason and Catherine (Campbell) M.; attended Columbian Coll. (now George Washington U.); m. Maria Clapier Chancellor, Apr. 13, 1854 (dec. Apr. 1855); 1 dau. Left sch. to go to pvt. chemistry lab. of James Curtis Booth, Phila.; became indsl. chemist, owner of bus. in Phila.; prof. applied chemistry U. Md., 1854-58, offered to set up chemistry dept. in conjunction with med. sch. (offer rejected); pub. his research in various sci. jours.; prepared (with Booth) Ency. of Chemistry, 1850; went to Eng., 1861. Author: A Treatise on Chemistry Applied, 1856; Chemical and Pharmaceutical Manipulations, 1857. Research on guano, salt, sugar, coal, gums, glycerine; devoted himself to improving tech. processes of condensed food rations, paper mfg., soap, candles, oil refining. Died South Hampstead, Eng., Dec. 8, 1897.

MORGAGNI, Giovanni Battista (Giambattista), Italian anatomist, pathologist; b. Forli, Italy, Feb. 25, 1682; M.D., U. Bologna (Italy), 1701; research at univs. Pisa, Padua (Italy). Apptd. lectr. anatomy U. Bologna, 1706, lectr., demonstrator theoretical medicine, 1710, prof. anatomy, 1715. Fellow Royal Soc., 1722; mem. French Acad. Scis., 1731, Russian Imperial Acad., Berlin Acad. Author: Adversaria anatomica, 1706-19; De sedibus et causis morborum per anatomen indigatis, 1761. Studied, by postmortem exams., disease and its effect on body, making him one of greatest path. anatomists; correlated autopsy findings with chem. case histories; research in pulmonary diseases, meningitis, palpitation of heart; discovered that apoplexy is due to break in or destruction of a blood vessel in brain; first to describe causes of sclerosis of cerebral arteries; described cystic remnant of Müllerian duct, heart block, superior nasal concha (Morgagni's concha), expanded portion of olfactory nerve on anterior lobe of cerebrum, cuneiform cartilage of larynx (tubercle of Morgagni), openings in diaphragm due to congenital defects (foramina Morgagni). Died Padua, Italy, Dec. 6, 1771.

MORGAN, Agnes Fay, Am. chemist; b. Peoria, Ill., May 4, 1884; B.S., U. Chgo., 1904, M.S., 1905, Ph.D., 1914; postgrad. (fellow) U. Mont., 1907-08; LL.D., U. Cal., 1959; m. 1908; 1 child. Instr. chemistry Hardin Coll., 1905-07, U. Wash., 1910-13; asst. prof. nutrition U. Cal. at Berkeley, 1915-19, asso. prof. household sci., 1919-23, prof., 1923-28, prof. home econs., biochemist, 1938-54, emeritus prof. nutrition, emeritus biochemist Agrl. Expt. Sta., 1954—. With OSRD, 1942-45, Bur. Human Nutrition and Home Econs., U. S. Dept. Agr., 1944. Recipient Garvan medal Am. Chem. Soc., 1949. Fellow Inst. Nutrition (Borden award 1954); mem. A.A.A.S., Soc. Biol. Chemistry. Author: (with I. S. Hall) Experimental Food Study, 1938. Research on heat effect on protein, action of vitamins especially B and D. Address: 1620 Spruce St., Berkeley, Cal. 94709.

MORGAN, Arthur Ivason, Jr., Am. chem. engr.; b. Berkeley, Cal., May 21, 1923; s. Arthur J. and Agnes Fay Morgan; B.S., U. Cal. at Berkeley, 1943, M.S., 1948; Ph.D., Eidgenossische Technische Hochschule, Zurich, Switzerland, 1952; m. Lillian Kozak, Feb. 14, 1948; children—Arthur Ivason III, Fay. Chem. engr. Western Regional Research Lab., U. S. Dept. Agr., Albany, Cal., 1952-62, chief Engring. Lab., 1962—. Lectr., U. Cal. at Berkeley, 1967—; sci. lectr. Inst. Food Tech. (Babcock-Hart award 1968), Chgo. Mem. Am. Inst. Chem. Engrs., Am. Chem. Soc., Inst. Food Technologists. Research, publs., patents in factors affecting evaporator fouling, low moisture equilibra; invented foam-mat drying, peeled wheat, evaporator, instant applesauce flakes. continuous freeze drying; developed reverse osmosis concentration of foods, san. dejuicing, evaporation control, microwave blanching, puffing. Home: 1202 Keith Av., Berkeley, Cal. 94708. Office: 800 Buchanan St., Albany, Cal. 94710.*

MORGAN, Clifford Thomas, Am. psychologist; b. Minotola, N.J., July 21, 1915; s. Samuel and Ethel (Dixon) M.; A.B., Maryville Coll., 1936; M.A., U. Rochester, 1937, Ph.D., 1939; m. Jean Chase Snow, Nov. 28, 1946; children—Peter, Patricia, Michael. Instr., Harvard, 1939-43; faculty Johns Hopkins, 1943-59, prof., 1948-59, chmn. dept. psychology, 1946-54; lectr. U. Wis., Madison, 1959-62, U. Cal., Santa Barbara, 1962-65. Fellow Am. Psychol. Assn.; mem. Am. Physiol. Soc., Soc. Exptl. Psychologists, A.A.A.S. Author: Physiological Psychology, 3d edit., 1965; Introduction to Psychology, 1961. Editor: Psychonomic Sci., 1964. Discovered that epileptogenic seizures in rats are due to auditory stimulation and established acoustic frequencies involved; established that hoarding behavior in rats is controlled by hunger and environmental temperature; measured effects of intensity on human pitch perception. Home: 1047 La Vista Rd., Santa Barbara, Cal. 93105.*

MORGAN, Conway Lloyd, English biologist, psychologist; b. London, Eng., Feb. 6, 1852; s. James Arthur and Mary (Anderson) M.; ed. Sch. Mines, London, Royal Coll. Sci.; D.Sc., Bristol, 1910; LL.D.; m. Emily Charlotte Maddock, 1878; 2 sons. Pvt. tutor in N.Am. and S.Am.; lectr. English and phys. sci. Diocesan Coll., Rondebosch, South Africa, 1878-83; prof. zoology and geology Univ. Coll., Bristol, 1884, prin., 1887-1909; 1st vice chancellor Bristol U., 1909-10, prof. psychology and ethics, 1910-19, prof. emeritus, 1919-36. Lectr. Clark U., Harvard; Gifford lectr., 1922-23. Fellow Royal Soc., 1899. Author: Animal Biology, 1887; Animal Life and Intelligence, 1890; Introduction to Comparative Psychology, 1895; Psychology for Teachers, 1895; Habit and Instinct, 1896; Animal Behavior, 1900; The Interpretation of Nature, 1905; Instinct and Experience, 1912; Emergent Evolution, 1923; Life, Mind and Spirit, 1926; Mind at the Crossways, 1929; The Animal Mind, 1930; The Emergence of Novelty, 1933. A founder of comparative psychology; among 1st to study animal psychology and to apply systematically methods of sci. expt. to subject; enunciated principle of parisimony, stated one should seek explanations for psychol. activity in most primitive psychol. faculties rather than interpreting them as consequence of most highly developed intellectual abilities; divided mental activity into 3 developmental stages (vague, effective and self consciousness); wrote on philos. meaning of evolution, especially as applied to human nature and psychic devel.; concerned with philosophic and sci. concept of emergency of novelty. Died Hastings, Mar. 6, 1936.

MORGAN, Councilman, Am. virologist; b. Boston, Sept. 6, 1920; s. William O. and Christiana (Councilman) M.; B.S., Harvard, 1943; postgrad. Columbia, 1943-46; m. Hallee Perkins, June 17, 1945; children—Hallee P., Christiana C., Hilary E., Councilman S. Instr. pathology Columbia, 1947-48, staff, 1953-—, prof. microbiology, 1963-—; sr. asst. surgeon Nat. Inst. Arthritic and Metabolic Diseases, NIH, USPHS, Bethesda, Md., 1948-51; asst. attending physician Presbyn. Hosp., N.Y.C., 1958-—. Mem. Am. Assn. Immunologists, Am. Assn. Pathologists and Bacteriologists, Am. Soc. for Microbiology, Electron Microscope Soc. Am. Asso. editor Virology, 1959-61; adv. editor Jour. Exptl. Medicine, 1963-—; editorial bd. Am. Jour. Pathology, 1967-—, Jour. Virology, 1968-—. Pioneered electron microscopic studies of viral structure and devel.; among 1st to describe internal structure of viruses; studied pathology of infected cells as revealed by electron microscopic examination of thin sections. Home: 198 Elm St., Tenafly, N.J. 07670. Office: 630 W. 168th St., N.Y.C. 10032.*

MORGAN, Sir Gilbert Thomas, Brit. chemist; b. Essenden, Eng., Sept. 22, 1872; s. Thomas and Malie Louise (Corday) M.; grad. in sci. univs. of London, Birmingham, Dublin; D.Sc., London; LL.D., Birmingham, Eng., St. Andrews, Scotland; Sc.D., Dublin, Ireland; m. Kathleen Nembhard Desborough, 1912. Asst. chemist Read Holliday and Co., dye mfrs., Huddersfield, Eng., 1889-94; became prof. chemistry Royal Coll. Sci., Dublin, 1912, Finsbury Tech. Coll., 1916; named Mason prof. chemistry Birmingham (Eng.) U., 1919; became 1st dir. Chem. Research Lab., Teddington, Eng., 1925; ret., 1937; hon. asso. Manchester (Eng.) Coll. Tech.; asso. Royal Coll. Sci. Recipient Gold medal Worshipful Co. Dyers, 1921, medal Soc. Chem. Industry, 1939. Fellow Royal Soc., 1915, City and Guilds of London Inst., Royal Inst. Chemistry; mem. Soc. Chem. Industry (became pres. 1931), Chem. Soc. (became pres. 1933), Royal Dublin Soc. (corr.). Author: British Chemical Industry: its Rise and Development, 1938; Achievements of British Chemical Industry in the Last 25 Years, 1939; Organic Compounds of Arsenic and Antimony, 1918; Inorganic Chemistry: a Survey of Modern Developments, 1936; also numerous articles. Research on co-ordination compounds, cerium, thorium, thiazine dyes; prepared series of new derivatives of benzenazobeta-napthol; assisted in extraction of ceria from cerita; studied chem. reactions under high pressure, synthetic resins. Editor, Jour. Chem. Soc., 1903-06; became editor chemistry sect. Ency. Britannica, 1929. Died Feb. 1, 1940.

MORGAN, Herbert Rollo, Am. astronomer; b. Medford, Minn., Mar. 21, 1875; s. Henry D. and Olive Sabre (Smith) M.; B.A., U. Va., 1899, Ph.D., 1901; m. Fannie Evelyn Wallis, May 25, 1904; 1 dau., Amy Eleanor (Mrs. George Hoffman). Fellow in astronomy Leander McCormick Obs., 1896-1901; prof. math. Pantops Acad., Charlottesville, Va., 1900-01; computer Naval Obs., Washington, 1901-05; prof. astronomy and math. Pritchett Coll., also dir. Morrison Obs., 1905-07; asst. astronomer U. S. Naval Obs., 1907-24, astronomer, 1925-28, sr. astronomer, 1928-29, prin. astronomer, 1929-44; research asso., Yale, 1947-57. Mem. Am. Astron. Soc. (v.p. 1940-42), Am. Geophys. Union, Internat. Astron. Union, A.A.A.S. (v.p. 1935-36). Author publ. U. S. Naval Obs., Vol. XIII, and co-author of Vols. IX, XIV, XV. Contbr. to sci. jours. Died June 11, 1957.

MORGAN, Herbert Roy, Am. physician, educator; b. St. Paul, June 19, 1914; s. Frank Wesley and Grace (Peake) M.; A.B., U. Cal. at Los Angeles, 1936, M.A., 1938; M.D., Harvard, 1942; m. Cynthia A. Walser, June 12, 1948; children—Stephanie Grace, Margaret Ann, Susan Aline. NRC sr. fellow in med. scis. Thorndike Meml. Lab., Boston City Hosp., dept. internal medicine Harvard Med. Sch., Boston, 1946-48; asso. prof. epidemiology Sch. Pub. Health, asst. prof. internal medicine Sch. Medicine, U. Mich., 1948-50; prof. microbiology, chmn. dept. microbiology, asso. prof. medicine U. Rochester (N.Y.) Sch. Medicine and Dentistry, 1950-—. Cons. to surgeon gen. USPHS, 1959-—; spl. cons. Nat. Cancer Inst., 1962-64; mem. microbiology tng. com. NIH, USPHS, 1964-—; mem. virology and rickettsiology study sect. NIH, 1959-63. Served with AUS, 1943-46. Diplomate Am. Bd. Preventive Medicine and Pub. Health, Am. Bd. Microbiology. Mem. Am. Soc. Microbiology, Am. Assn. Cancer Research, Am. Assn. Immunologists, Am. Soc. Clin. Investigation, Am. Pub. Health Assn. Research on antigens of typhoid bacteria, latent infection, tumor viruses. Home: 94 Branford Rd., Rochester, N.Y. 14618.*

MORGAN, Hugh Jackson, Am. physician; b. Nashville, Jan. 25, 1893; B.S., Vanderbilt U., 1915; M.D., Johns Hopkins, 1918; D.Sc., U. N.C., 1946, U. So. Cal., 1953; 4 children. Asst., resident physician Rockefeller Inst., 1922-24, traveling fellow, 1924-25; asso. prof. medicine Vanderbilt U. Sch. Medicine, Nashville, 1924-28, prof. clin. medicine, 1928-35, prof., 1935-58, prof. emeritus, 1958-—, physician-in-chief Univ. Hosp., 1935-58; chief cons. Thayer VA Hosp., 1946-58. Chief med. cons. to U. S. surgeon gen.; mem. nat. adv. health council USPHS, 1948-—. Diplomate Am. Bd. Internal Medicine. Fellow A.C.P. (pres. 1947); mem. A.M.A., A.A.A.S., Soc. Clin. Investigation, Am. Clin. and Climatol. Assn. (pres. 1953), Assn. Am. Physicians (pres. 1951). Research on infectious diseases, internal medicine; removed thymus in treatment of type of muscular debility. Address: 15 White Bridge Rd., Nashville 37205.

MORGAN, Ira Lon, Am. physicist; b. Ft. Worth, Aug. 3, 1926; s. I. M. and Winnie (Osborne) M.; B.A., Tex. Christian U., 1949, M.A., 1951; Ph.D., U. Tex., 1954; m. Mary Esther Massey, Aug. 7, 1948; children—Marilon, David Stanton, Carol Ann. Instr., Tex. Christian U., Ft. Worth, 1948-51; research scientist U. Tex., Austin, 1951-56; exec. v.p., dir. research Tex. Nuclear Corp., Austin, 1956-67, pres., dir., 1967-—. AEC fellow, 1954-56. Mem. Am. Phys. Soc., Am. Nuclear Soc. (exec. com. isotopes and radiation div. 1965-—), Sigma Xi, Sigma Pi Sigma. Research, publs. on nuclear physics, radiobiology, applied physics and asso. instrumentation especially on neutron inelastic scattering, gamma ray spectroscopy, particle accelerators, high energy proton biol. effects; invented neutron proportional counters and beam pulsing techniques. Home: 4005 Northills Dr., Austin 78731. Office: 9101 Hwy. 183, Austin, Tex. 78756.*

MORGAN, John, Am. physician; b. Phila., June 10, 1735; s. Evan and Joanna (Biles) M.; grad. Coll. of Phila. (now U. Pa.), 1757; M.D., U. Edinburgh (Scotland), 1763; m. Mary Hopkinson, Sept. 4, 1765. Admitted to Academie Royal de Chirurgie de Paris (France), 1764, Fellow Royal Soc., 1765; mem. Belles-Lettres Soc. of Rome (Italy); licentiate Royal Coll. Physicians, London and Edinburgh; established med. sch. in connection with U. Pa., 1765, apptd. prof. theory and practice of physic; author oration A Discourse upon the Institution of Medical Schools in America, 1765; pub. Four Dissertations on The Reciprocal Advantages of a Perpetual Union between Great Britain and her American Colonies (won a gold medal), 1766; dir. gen. hosps. Continental Army, 1775, physician in chief, 1775, dir. hosps. East of Hudson River, 1776-77; physician Pa. Hosp.; mem. Am. Philos. Soc.; Phila. Coll. Physicians was an outgrowth of his suggestion (organized 1787). Author: A Recommendation of Inoculation, According to Baron Pimsdale's Method, 1776. Foremost med. educator of pre-revolutionary period; proposed separation medicine from surgery and drug selling and requiring liberal edn. for med. students; did more than any other man of his time to bring advanced sci. med. knowledge of Europe to colonies. Died Phila., Oct. 15, 1789.

MORGAN, John Jacob Brooke, Am. psychologist; b. Norristown, Pa., Aug. 23, 1888; s. George Custer and Inez (Brooke) M.; A.B., Taylor U., 1911; A.M., Columbia, 1913, Ph.D., 1916; B.D., Drew Theol. Sem., 1914; m. Rose Davis, Mar. 23, 1913; children—Burton Davis, James Newton; m. 2d, Sarah Smith, June 22, 1934; 1 dau., Nancy Wynn. Instr. psychology Princeton, 1916-17; asst. prof. psychology U. Minn., 1919-20; grading and testing specialist ednl. dept., U. S. A., 1920-21; dir. psycholog. clinic State U. Ia., 1921-24; asso. prof. psychology Northwestern U., 1925-33, prof., 1933-45. Fellow A.A.A.S., Am. Assn. Applied Psychology; mem. Am., Mid-Western (pres. 1934) psychol. assns., Sigma Xi, Alpha Pi Zeta. Author: The Psychology of the Unadjusted School Child, 1924, rev. edit., 1936; (with A. R. Gilliland) An Introduction to Psychology, 1927; The Psychology of Abnormal People, 1928, rev. edit., 1936; General Psychology for Professional Students (with Gilliland and Stevens), 1930; Strategy in Handling People (with Ewing T. Webb), 1930; Child Psychology, 1931, rev. edit., 1942; Workbook in Abnormal Psychology, 1931; Making the Most of Your Life (with E. T. Webb), 1932; Keeping a Sound Mind, 1934; Psychology: A General Textbook, 1941; Workbook in General Psychology, 1941; also (monograph) The Overcoming of Distractions and Other Resistances (Archives of Psychology), 1916; also articles. Died Aug. 17, 1945.

MORGAN, Karl Ziegler, Am. health physicist; b. Enochsville, N.C., Sept. 27, 1907; s. Jacob L. and Virginia (Shoup) M.; B.S., U. N.C., 1929, M.S., 1931; Ph.D., Duke, 1934; student Lenoir Rhyne Coll., 1925-27, D.Sc., 1967; m. Helen Lee McCoy, Aug. 2, 1937; children—Karl Ziegler, Eric Lee, Joan Elen, Diana. With Westinghouse Electric, 1930-31; chmn. physics dept. also cooperative research program Duke, Lenoir Rhyne Coll., 1934-43; mem. health physics div. U. Chgo., 1943; dir. health physics div. Oak Ridge Nat. Lab., 1943-—. Recipient 1st Gold medal Royal Acad. Sci. Sweden, 1962; Distinguished Alumni award Lenoir Rhyne Coll., 1967. Mem. A.A.A.S., Am. Assn. for Physicists in Biology and Medicine, Am. Bd. Health Physicists, Am. Chem. Soc., Am. Coll. Radiology, Am. Inst. E.E., Am. Indsl. Hygiene Assn., Am. Inst. Biol. Scis., A.M.A., Am. Nuclear Soc., Am. Phys. Soc., Am. Pub. Health Assn., Am. Standards Assn., Am. Soc. for Testing Materials, Health Physics Soc., Internat. Soc. Radiology, Med. Research Council, Nat. Acad. Scis., Soc. Nuclear Medicine, Nat. Safety Council, Radiation Research Soc., Radol. Soc. N.Am., Research Soc. Am., Royal Soc. Medicine, Research, numerous publs. on cosmic ray showers, health physics problems of instrumentation, air contamination and internal dose; an organizer of profession of health physics; dir. research programs in physics of dosimetry, radiation physics, internal dose, health physics tech., radioactive waste disposal and radiation ecology. Home: 110 Plymouth Circle. Office: P.O. Box X, Oak Ridge 37831.*

MORGAN, Leon Owen, Am. chemist; b. Oklahoma City, Oct. 25, 1919; s. Leon Oscar and Catherine Joan (Reitermann); B.S., Oklahoma City U., 1941; M.A., U. Tex., 1943; Ph.D., U. Cal. at Berkeley, 1948; m. Betty Boyd, Dec. 27, 1942; children—Joseph, Boyd, Robert, Mary Katherine. Chemist, metall. lab. U. Chgo., 1944-45; research asso. Radiation Lab., U. Cal. at Berkeley, 1945-47; faculty U. Tex., Austin, 1947-—, prof. chemistry, 1962-—. Asso. editor Jour. Phys. Chemistry, 1964-65. Research, publs. in field of nuclear and electron paramagnetic resonance and relaxation, fast reaction rate processes. Home: 3307 River Rd., Austin 78703.*

MORGAN, Lewis Henry, Am. anthropologist; b. Aurora, N.Y., Nov. 21, 1818; s. Jedediah and Harriet (Steele) M.; grad. Union Coll., 1840, LL.D., 1873; m. Mary Elizabeth Steele, Aug. 13, 1851. Admitted to N.Y. bar; legal adviser of r.r. under constrn. between Marquette, Mich. and Lake Superior iron region, 1855; mem. N.Y. State Assembly, 1861-68, N.Y. State Senate, 1868-69; known as father of Am. anthropology; leading mem. Grand Order of the Iroquois (chief purposes were to study and perpetuate Indian lore, to educate Indians, reconcile them to conditions imposed on them by civilization); succeeded in defeating ratification of fraudulent treaty by which Seneca would have given up their lands to Ogden Land Co.; adopted by Hawk clan of Seneca, 1847, given name Tayadawahkugh; entrusted by U. State N.Y. with executing enlargement of its Indian collection for which an appropriation had been made, 1849. Author: League of the Ho-de-no-sau-nee, or Iroquois (1st sci. account of an Indian tribe), 1851; pub. Laws of Consanguinity and Descent of the Iroquois, 1859; Systems of Consanguinity and Affinity of the Human Family; Ancient Society or Researches in the Lines of Human Progress, 1877; studied various ruins, visited some of existing pueblos, 1878; wrote On the Ruins of a Stone Pueblo on the Animas River in New Mexico, 1880; Houses and House-Life of the American Aborigines, 1881; The American Beaver and His Works, 1868; instrumental in organizing anthropology sect. A.A.A.S., 1875, 1st chmn.; mem. Nat. acad. Scis.; pres. A.A.A.S., 1879. Studied Iroquois and other tribal kinship systems; divided culture devel. into savage, barbarian, and civilized stages. Died Rochester, N.Y., Dec. 17, 1881.

MORGAN, Meredith Walter, Am. optometrist; b. Kingman, Ariz., Mar. 22, 1912; s. Meredith Walter and Florence (Forsyth) M.; A.B., U. Cal. at Berkeley, 1941, M.A., 1939, Ph.D., 1941; m. Ida Marcia Engelking, Mar. 7, 1937; children—Linda Meredith. Practice optometry, Richmond, Cal., 1934-60; faculty U. Cal. at Berkeley, 1942-—, prof. optometry, 1951-—, dean Sch. Optometry, 1960-—. Named Alumnus of Year, U. Cal. Sch. Optometry, 1953; Beverly Meyer Nelson Achievement award Edn. Found. in Ophthal. Optics, 1959. Fellow Am. Acad. Optometry (past pres., editor Synopsis Series 1960-—), Am. Optometry Assn. (com. on standards 1950-60), A.A.A.S., Optical Soc. Am., Sigma Xi. Author: (with H. B. Peters), Optics of Ophthalmic Lenses, 1948; also numerous articles. Chmn. editorial council Am. Jour Optometry, 1956-57. Research on nervous control accommodation, relationship accomodation and convergence. Home: 11 Silver Leaf Ct., Lafayette, Cal. 94749. Office: Sch. Optometry, U. Cal., Berkeley, Cal. 94720.*

MORGAN, Millett Granger, Am. physicist, engr.; b. Hanover, N.H., Jan. 25, 1915; s. Frank Millett and Mary (Granger) M.; B.A. with honors in Physics, Cornell U. (Geo. W. Lefevre Scholar), 1937, M.Sc. in Engring., 1938; E.E., Stanford, 1939, Ph.D., 1946; m. Eleanor Walbridge, June 29, 1937; children—Millett Granger II, Deborah J., Janet Eleanor, Beverly Sue. Instr. elec. engring. Thayer Sch. Engring., Dartmouth Coll., 1941, asst. prof., 1947-53, prof., 1953-—, asst. dean. 1947-49, dir. research, 1949-64, dir. Radiophysics Lab., 1964-—; research specialty ionospheric radio; engaged as instr. elec. engring. Mass. Inst. Tech., 1942; research and devel. engr. Submarine Signal Co., Boston, 1942-44, Cal. Inst. Tech., Pasadena, Cal., 1944-45; staff engr., lectr. elec. engring. U. Cal., 1946-47. Nat. Acad. Scis. chmn. Ionospheric Physics Panel adv. to U. S. A. Nat. Com. for Internat. Geophys. Year 1957-58; vis. scientist Ellsworth Sta., Antarctica, 1957-58; mem. internat. relations com. Space Sci. Bd., 1964-—; cons. NASA, 1963-—. Mem. I.E.E.E. (chmn. wave propagation com. 1955-57), Internat. Sci. Radio Union sec. U. S. A. nat. com. 1961-64, chmn. 1964-—), Sigma Xi. Research, publs. on polarization of ionospherically propagated radio waves; collaborator in confirming theory of whistlers. Home: Box 92, Hanover, N.H. 03755. Office: Radiophysics Lab., Dartmouth College, Hanover, N.H. 03755.*

MORGAN, Russell Hedley, physician; b. London, Ont., Can., Oct. 9, 1911; s. Alfred Hedley and Edith (Rowe) M.; B.A., U. Western Ont., 1934, M.D., 1937, D.Sc., 1963; m. Mary Stella McManus, Jan. 25, 1938; children—Monica May (Mrs. Harvey Walters), Mary Margaret. Faculty, U. Chgo., 1942-44, 46; sr. surgeon USPHS, 1944-46; prof. radiology Johns Hopkins, 1946-—; radiologist in chief Johns Hopkins Hosp., Balt., 1946-—, prof. radiol. sci., 1960-—. Sr. cons. to surgeon gen. on radiation, 1958-—; chmn. Nat. Adv. Com. on Radiation, 1958-—. Mem. A.M.A., Am. Coll. Radiologists, Am. Phys. Soc., Radiation Research Soc., Health Physics Soc., Am. Pub. Health Assn. Author: (with H. E. Hilleboe) Mass Radiation of Chest, 1944; (with K. E. Corrigan) Handbook of Radiology, 1955; (with others) Physical Foundations of Radiology, 1963; also numerous articles. Developed autonomatic exposure control equipment for X-rays, automatic X-ray fluoroscopic brightness control, image amplifier-tv X-ray intensifier, radiant energy measuring devices. Home: 4000 N. Charles St., Balt. 21218. Office: Johns Hopkins Hosp., Balt. 21205.*

MORGAN, Samuel Pope, Am. mathematician; b. San Diego, July 14, 1923; s. Samuel Pope and Beatrice (Summers) M.; B.S., Cal. Inst. Tech., 1943, M.S., 1944, Ph.D., 1947; postgrad. U. Cal., Berkeley; m. Caroline Annin, Jan. 23, 1948; children—Carol, Lesley, Alison, Diane. Faculty, U. Cal., Berkeley, 1943-44, Cal. Inst. Tech., 1944-47; tech. staff Bell Telephone Labs., Murray Hill, N.J., 1947-—, head math. physics dept., 1959-67, dir. computing sci. research, 1967-—. Fellow I.E.E.E.; mem. Am. Phys. Soc., Am. Math. Soc., A.A.A.S., Soc. for Indsl. and Applied Math., Assn. for Computing Machinery. Patentee in field. Research, publs. in electromagnetic theory, especially wave guides for microwaves, radar antennas, heat flow, analytical mechanics, numerical methods, spl. math. functions. Home: 9 Raleigh Ct., Morristown, N.J. 07960. Office: Bell Telephone Labs., Murray Hill, N.J. 07974.*

MORGAN, Thomas Hunt, Am. zoologist; b. Lexington, Ky., Sept. 25, 1866; s. Charlton H. and Ellen Key (Howard) M.; B.S., State Coll. Ky., 1886, M.S., 1888; Ph.D., Johns Hopkins, 1890; LL.D., 1915; LL.D., U. Ky., 1916, McGill U., 1921, U. Edinburgh, 1922, U. Cal., 1930; Sc.D., U. Mich., 1924; Ph.D., Heidelberg U., 1931; M.D. (hon.), U. Zurich, 1933; D.honoris causa, U. Paris, 1935; m. Lilian V. Sampson, 1904; children—Howard Key, Edith Sampson, Lilian Vaughn, Isabel Merrick. Prof. biology Bryn Mawr Coll., 1891-1904; prof. exptl. zoology Columbia, 1904-28; dir. William G. Kerckhoff Labs. Biol. Scis., Cal. Inst. Tech., 1928-45. Recipient Nobel prize for physiology and medicine, 1933. Fellow Royal Soc., 1919, A.A.A.S. (pres. 1929-30); mem. Nat. Acad. Science (pres. 1927-31), Am. Soc. Naturalists, Am. Soc. Zoologists, Soc. Exptl. Biology and Medicine, N.Y. Acad. Sciences; corr. mem. or fgn. asso. numerous European socs. Author: Regeneration, 1901; Evolution and Adaptation, 1903; Experimental Zoology, 1907; Heredity and Sex, 1913; Mechanism of Mendelian Heredity, 1915; Critique of the Theory of Evolution, 1916; The Physical Bases of Heredity, 1919; The Theory of the Gene, 1926; Experimental Embryology, 1927; The Scientific Basis of Evolution, 1932; Embryology and Genetics, 1933; also monographs and papers on biol. and embryol. subjects. One of founders of modern genetics; using Drosophila, demonstrated phys. basis of heredity and importance of gene; described phenomena of linkage and crossing-over; proved linear arrangement of genes along chromosomes; other research and publs. on devel., exptl. embryology, regeneration, evolution, and adaptation. Died Dec. 4, 1945.

MORGAN, Walter Clifford, Am. biologist; b. Ledyard, Conn., Dec. 22, 1921; s. Walter Clifford and Margaret (Allyn) M.; B.Sc., U. Conn., 1946, Ph.D., 1953; M.Sc., George Washington U., 1949; m. Helen Jean Naden, Dec. 28, 1948; children—Nancy Ruth, Margaret Louise, Elizabeth Jane. Animal husbandman Nat. Cancer Inst., Bethesda, Md. 1946-49; research asso. Columbia, 1951-53; asst. prof. U. Tenn., 1953-54; asso. prof. poultry genetics, physiology S.D. State U., Brookings, 1954-58, prof., 1958-—. Fellow A.A.A.S.; mem. Am. Inst. Biol. Sci., Am. Genetic Assn., N.Y., Tenn., S.D. acads. sci., Poultry Sci. Assn., World's Poultry Sci. Assn., Am. Poultry Hist. Soc., Sigma Xi. Studies, publs. on effect of inbreeding poultry, mating systems, discovered and first reported transmission of persistent right oviducts in chickens; discovered and reported two mutations in chickens and three tail mutations in mice; described effect of gamma irradiation on developing chick embryo. Home: 1610 1st St., Brookings, S.D. 57006.*

MORGAN, William Wilson, Am. astronomer; b. Bethesda, Tenn., Jan. 3, 1906; s. William Thomas and Mary McCorkle (Wilson) M.; student Washington and Lee U., 1923-26; B.S., U. Chgo., 1927, Ph.D., 1931; m. Helen Montgomery Barrett, June 2, 1928 (dec. 1963); children—Emily Wilson, William Barrett. Instr. Yerkes Obs., U. Chgo., Williams Bay, Wis., 1932-36, asst. prof., 1936-43, asso. prof., 1943-47, prof., 1947-—; dir. Yerkes and McDonald Observatories, 1960-63; mng. editor Astrophysical Jour., 1947-52. Mem. Am. Astron. Soc., Nat. Acad. Scis., Am. Acad. Arts and Scis., Pontifical Acad. Scis., Royal Danish Acad. Scis. and Letters. Author: (with P. C. Keenan, Edith Kellman) An Atlas of Stellar Spectra, 1943. Contbr. research articles profl. publs. Devised (with Philip C. Keenan) stellar classification which permits more accurate determination of distance of remote stars; (with Stewart Sharpless and Donald Osterbrock) discovered spiral structure of Milky Way Galaxy; developed method for determining stellar constn. of galaxies from study of their forms; (with Janet Lesh) demonstrated existence of super-giant galaxies.*

MORGENROTH, Julius, German bacteriologist; b. Bamberg, Germany, Oct. 19, 1871. Discovered (with E. Bumke) use of eucupine as local anesthetic, 1918; pioneered (with Paul Ehrlich) chemotherapy, discovered new information about immune serum, including complement (immune factor produced by reaction to injected material); produced optachin, a therapeutic agt., 1911. Died Berlin, Dec. 20, 1924.

MORGENSTERN, Dietrich, German statistician; b. Ratzeburg, Germany, Sept. 26, 1924; s. Kurt and Dora (Garbe) M.; ed. Berlin Tech. Coll.; engring. diploma, math. diploma, Ph.D.; m. Elisabeth Schlüter, Apr. 3, 1959. Successively instr., cons. scientist, asso. prof., Munster, Germany; prof., Freiburg, West Germany. Mem. Gesellschaft für angewandte Math. und Mech. Author: (with I. Szabo) Einführung in Warscheinlichkeitsrechnung und Statistik. Research on mechanic theory. Home: Am Hagmättle 5, 78 Freiburg, West Germany.

MORGENTHALER, Frederic Richard, Am. physicist; b. Cleve., Mar. 12, 1933; s. Frederick H. and Anna (Welke) M.; S.B., S.M., Mass. Inst. Tech., 1956, Ph.D., 1960; m. Barbara Pullen, May 17, 1958; children—Ann Welke, Janet Nason. Faculty, Mass. Inst. Tech., Cambridge, 1960-—, prof. elec. engring., 1968-—; research fellow applied physics Harvard, 1960-61, vis. lectr. applied physics, 1961. Recipient Air Force Cambridge Research Labs. award, 1960. Mem. I.E.E.E., Am. Phys. Soc., Sigma Xi, Tau Beta Pi, Eta Kappa Nu. Contbr. articles in field to sci. jours. Research, publs. on microwave solid state physics, magnetism, electromagnetic theory with emphasis on magnetic resonance and linear and nonlinear interactions among photons, magnons and phonons. Patentee microwave pulsed resonance oscillator. Office: Mass. Inst. Tech., Cambridge, Mass. 02139.

MORGENTHAU, Hans Joachim, political scientist; educator; b. Coburg, Germany, Feb. 17, 1904; s. Ludwig and Frieda (Bachmann) M.; student U. Berlin, U. Frankfort, U. Munich, 1923-27; magna cum laude, U. Munich, 1927; summa cum laude, U. Frankfort, 1929; grad. work, Grad. Inst. for Internat. Studies, Geneva, 1932; LL.D., Clark University, 1962, Ripon Coll., 1962; m. Irma Thormann, June 3, 1935; children—Matthew, Susanna. Came to U. S., 1937, naturalized, 1943. Admitted to bar, 1927; practiced law, 1927-30; asst. to law faculty U. Frankfort, 1931; acting pres. Labor Law Court, Frankfort, 1931-33; instr. polit. sci. U. Geneva, 1932-35; prof. internat. law, Inst. Internat. and Econ. Studies, Madrid, Spain, 1935-36; instr. govt. Brooklyn Coll., 1937-39; asst. prof. law history and polit. sci. U. Kan., Kansas City (Mo.), 1939-43; admitted to Mo. bar; vis. asso. prof. polit. sci. U. Chgo., 1943-45, asso. prof., 1945-49, prof., 1949-61, prof. polit. sci. and modern history, 1961-—, Albert A. Michelson Distinguished Service prof., 1963-—, dir. Center for Study of Am. Fgn. Policy, 1950-—; vis. prof. U. Cal. at Berkeley, 1949, Harvard, 1951, 59, 60, 61, Northwestern U., 1954, Columbia, Yale, 1956, Princeton Inst. for Advanced Study, 1958, Washington Center Fgn. Policy Research, 1958-60. Cons. Dept. of State, 1949-51, 61-—, Department of Defense, 1961-—. Member American Philosophical Society, American, International political sci. assns., Am. Society Internat. Law, Am. Acad. Arts and Scis., Am. Assn. U. Profs. Author, coauthor numerous books relating to field, 1929-—, among latest: Scientific Man vs. Power Politics, 1946; Politics Among Nations, 1948, 3d edit., 1960; Principles and Problems of International Politics (with Keneth W. Thompson), 1950; In Defense of the National Interest, 1951; Dilemmas of Politics, 1958; The Purpose of American Politics, pub. in 1960; Politics in the 20th Century, 3 vols., 1962. Contbr. articles to philos., law and polit. sci. jours. nat. newspapers, and Ency. Britannica. Study of international relations; U. S. foreign policy; the U. S. position in Vietnam; power politics; 20th century politics. Home: 5542 S. Dorchester Av., Chgo. 37.

MORHOF, Daniel George, German historian; b. Wismar, Germany, Feb. 6, 1639; studied under John Micraelius, Joachim Fabricius, John Sithman, at Stetin, Germany, 1655; m., 1671; became prof. poetry Rostock, 1660; prof. poetry and eloquence, Kiel, Germany, 1666, prof. history, 1673; named librarian, 1680. Author: Stentor hyaloclastes, 1672; Polyhistor, sive de notitia auctorum et rerum commentarii (survey of all learning from antiquity to his own time), 3 vols., 1688-92; De Metallorum Transmutatione, 1683. Described phenomena of resonance; broke a glass shell with aid of human voice. Died Lübeck, Germany, July 30, 1691.

MORI, Alberto, Italian geographer; b. Como, July 5, 1909; s. Assunto and Alinda (Arrighetti) M.; Ph.D.; m. Aida Perris, Sept. 12, 1932; children—Vittoria, Ettore, Silvia, Luisa, Sandro. Prof. geography Faculty Letters, U. Pisa. Mem. Italian Geog. Soc. (counselor), Soc. for Geog. Studies (counselor), Italian Anthropology Soc. Author: Il mare nei suoi aspetti economici, La Dalmazia—Nuovo Atlante geografico, La Sardegna. Home: via Risorgimento 16. Office: Istituto di Geografia dell Università, Pisa, Italy.

MORI, Shigeki, Japanese physician; b. Kobe City, Japan, 1893; grad. Med. Dept., Kyoto U., 1919. Asst. prof. Kyoto U., prof. Med. Dept., 1940; prof. Kumamoto Med. U., from 1926; founder Research Inst. Constn. Medicine. Mem. Japan Cancer Soc. (chmn.), Japan Constn. Research Soc. (chmn.), Japan Secretion Sci. Soc. (councilor), Japan Research Soc. Constn. Medicine (founder 1932). Author: Experimental Oncology; General Principles of Pathology; Principles on Each Branch of Pathology.

MORI, Shigeyoshi, Japanese mathematician; b. Settsu, Osaka Prefecture, Japan; flourished circa 1600; studied math., China; retainer to Terumasa Ikeda; later in service of Hideyoshi Toyotomi; believed to have taught math., Kyoto, Japan. Author: Kijo-Ransho (1st math. book pub. in Japan during Genna Era 1615-24, covers principles of abacus and its use in land surveying). Credited with introducing book Sampo-Toso (Origin of Mathematics) into Japan; founder abacus calculating method in Japan.

MORICONI, Emil John, Am. chemist; b. N.Y.C., Feb. 12, 1922; s. Emilio and Virginia (Armaniaco) M.; B.S., Manhattan Coll., 1943; M.S., N.Y. U., 1948; Ph.D., Fordham U., 1952; postgrad. Cornell U.; m. Antonietta C. Yon, June 14, 1952; children—E. Steven, Carolyne Anne, Maria Antonietta, Isabel Susan, Katherine May, David Yon. Research chemist Koppers Co., 1943-46; asso. prof., chmn. sci. dept. Marymount Coll., N.Y.C., 1949-55; faculty Fordham U., N.Y.C., 1955-—, chmn. dept. chemistry, 1960-—, prof., 1966-—. Fellow N.Y. Acad. Sci.; mem. Chem. Soc., Chem. Soc. (Gt. Britain), Am. Chem. Soc., Sigma Xi, Phi Lambda Upsilon. Research, numerous publs. on organic reactions and mechanisms, ozonolysis polycyclic aromatics. Home: 348 W. 260th St., N.Y.C. 10471.*

MöRIKOFER, Walter, Swiss meteorologist; b. Basle, May 24, 1892; s. Peter-Paul and Mathilde (Widmer) M.; Ph.D., U. Basle; m. Elsbeth Kreis, Nov. 21, 1928; children—Walter-Andreas, Suzanne-Elisabeth. Asst., Meteorol. Inst., U. Basle, 1911-18; physico-meteorologist Davos Obs., dir. research on meteorol. radiation, medi. climatology, 1929-—; hon. prof. med. meteorology U. Basle. Sec., later pres. Internat. Commn. on Radiation; treas., sec.-gen., later pres. Internat. Com. Photobiology; hon. pres. various socs.; pres. Naturforschende Gesellschaft Davos. Author: Das Hochgebirgsklima, 1932; 4 chpts. in Traite de climatologie biologique et medicale, 1934; Meteorolog. Strahlungsmessmethoden, 1939; Meteorologie und Meteobiologie des Alpenfohns, 1950. Home: Davos, Rusticana, Davos-Platz. Office: Observatoire, Davos-Platz, Switzerland.

MORIN, Arthur Jules, French physicist; b. Paris, Oct. 17, 1795; ed. Ecole polytechnique, Ecole d'application, Metz, France; commd. lt. Engr. Battalion, 1819, served to div. comdr. (gen.), 1855; became dir. Conservatoire des Arts et Métiers, Paris, 1952; mem. Soc. Civil Engrs. (pres.), French Acad. Scis., 1843 (pres. 1864). Author: Hydraulique, 1837; Aide-mémoire de mécanique practique, 1838; Mémoir sur la pénétration des projectiles, 1838; Mémoir sur les pendules, balistiques, 1839; Fundamental Ideas of Mechanics and Experimental Data . . . , 1850; Etudes sur la ventilation, 1863. Constructed device for recording free fall; experimented on friction, 1833, dynamometric brakes, 1836; turbines, 1838, penetration force of projectiles, application of ballistic pendulum, 1839, air resistance, 1842, ventilation, 1863. Died Paris, Feb. 7, 1880.

MORIN, Jean, French physicist; b. Meung-sur-Loire, France, Apr. 26, 1705; prof. philosophy Chartres; became canon of Chartres, 1732; mem. French Acad. Scis., 1736. Author: Sur l'électricité des corps dans laquelle on développe le vrai mécanism des phénomèns, 1748. Discovered a liquid phosphorus, 1726; research on Morin apparatus, static electricity. Died Chartres, Mar. 28, 1764.

MORIN, Jean Baptiste, French astrologer, physician; b. Villefranche, France, 1583; M.D., Avignon, France, 1613. Sent to investigate Hungarian mines by Claude Dormy, 1615; physician to a bishop, an abbot, and in 1621 to duke of Luxemburg; became Royal prof. math., Paris, 1630; apptd. to draw up horoscope of new born Louis XIV. Author: Réfutation des thèses . . . d'A. Fillon dit le soldat philosophe et E. de Claves . . . contre la doctrine d'Aristote, 1624; Nova mundi sublunaris anatomia, 1619; Astrologicarum domorum cabala detecta, 1623; Ad australes et boreales astrologos pro astrologia restituenda epistolae, 1628; Longitudinum terrestrium necnon coelestium nova et hactenus optata scientia, 1634; Astronomia iam a fundamentis . . . restituta, 1640; Dissertatio de atomis et vacuo contra Petri Gassendi philosophiam Epicuream, 1650; Astrologia Gallica (posthumous), 1661. Developed method of determining longitudes at sea using distance of moon from a star; divided universe into elemental, ethereal and celestial matter; proposed 3 layers of earth which corresponded in reverse order to 3 regions of air; criticized Ptolemy; rejected Cardan and Kepler; denied existence of vacuum and Toricelli's proof; advocated medicine of Paracelsus; opposed astronomy of Copernicus. Died Paris, 1656.

MORIN, Ugo, Italian mathematician; b. Trieste, Feb. 7, 1901; became prof., 1942; past prof. U. Trieste; prof. analytical and descriptive geometry U. Padua; dir. Sch. Design, U. Padua. Address: via San Pietro 2/a, Padua, Italy.

MORINAGA, Toshitaro, Japanese agriculturist; b. Toyama Prefecture, 1895; grad. Tokyo U., 1919; postgrad. in agrl. physiology in U. S. Asst. prof. Kyushu U., later prof.; dir. Agrl. Lab., Agr. and Forestry Ministry, now chief physiol. heredity dept. Agrl. Technique Research Inst. Mem. Japan Sci. Council, 1950. Author: Agricultural Science. Research on breeding sci., physiology and heredity of rice plants.

MORINIGO, Fernand Bernardino, physicist; b. Parana, Argentina, June 1, 1936; s. Marcos Augusto and Maria (Vazquez) M.; B.S., U. So. Cal., 1957; Ph.D., Cal. Inst. Tech. 1963; m. Hildegard Helen Weeren, Nov. 28, 1963. Research fellow Cal. Inst. Tech., 1962-—; faculty Cal. State Coll., Los Angeles, 1963-—, asso. prof. physics, 1966-—; vis. prof. U. Freiburg (Germany), 1964-65. Mem. Am. Phys. Soc., Phi Beta Kappa. Research, publs. on nuclear physics, beta-decay, nuclear reaction theory. Home: 1715 Los Robles Av., San Marino, Cal. 91108. Office: 5151 State College Dr., Los Angeles 90032.*

MORINO, Yonezo, Japanese chemist; b. Osaka, Japan, Aug. 31, 1908; s. Matakichi Tamaki and Masa (Kasano) M.; M.S., Imperial U. Tokyo, 1931, D.Sc., 1937; m. Yoshi Torii, Nov. 21, 1936. Asst., Faculty Sci., Imperial U. Tokyo, 1937-40, asst. prof., 1940-43; prof. Imperial U. Nagoya, 1943-45; prof. U. Tokyo, 1945-—. Recipient Chem. Soc. Japan Sakurai prize, 1942; Japan Acad. prize, 1964. Mem. Chem. Soc. Japan (v.p. 1961-62), Phys. Soc. Japan, Crystallographic Soc. Japan. Author chpt. Raman Effect and Infrared Spectra monograph series Experimental Chemistry, part I, vol. 6, 1943. Research, numerous publs. on rotational isomerism in dihalogenoethanes by Raman effect; force constants from means sq. amplitudes of vibrations by gas electron diffraction, anharmonic potential functions of polyatomic molecules by microwave and infrared spectra. Home: No. 807, Akamon Abitashion, Hongo-5, 29-13, Bunkyo-ku, Tokyo. Office: Hongo-7, Bunkyo-ku, Tokyo, Japan.*

MORISON, John Miller Woodburn, Brit. radiologist; b. Beith, Scotland, Mar. 13, 1875; s. John M. and Elizabeth (Woodburn) M.; M.D., Glasgow U.; D.M.R.E., Cambridge U., 1924; m. Elizabeth Radcliffe, 1911; 2 sons. Prof. radiology, examiner in radiology U. London; dir. radiol. dept. Royal Cancer Hosp., London. Fellow Royal Coll. Physicians, Edinburgh, Faculty Radiologists; pres. Brit. Assn. Radiologists, radiol. sects. Royal Soc. Medicine, Brit. Med. Assn.; hon. mem. various fgn. radiol. socs. Contbr. sect. on radiology to Recent Progress in Medicine and Surgery, 1919-33, edit. by Sir John Collie, 1933. Research and publs. on X-ray diagnosis of cancer, diaphragmatic hernia, elevation of diaphragm, massive lung collapse, bone tumors. Died Sept. 3, 1951.

MORISON, Robert, Brit. physician, botanist; b. Aberdeen, Scotland, 1620; s. John and Anna (Gray) M.; M.A., Ph.D., U. Aberdeen, 1638; M.D., Angers, France, 1648; apptd. dir. Royal Garden, Blois, France, 1650; toured France searching for new plants; returned to Eng. with Charles II, 1660; nominated physician to Charles II; became botanist and superintendent of all royal gardens; became Regius prof. botany Oxford U., 1669. Author: Praeludium botanicum, 1669; Plantarum unbelliferarum distributio nova, 1672; Plantarum historia universalis oxoniensis, 2 vols., 1680-99 (completed by Jacob Bobart). Revived study of systematic botany; made 1st plant classification according to form and structure of fruit alone; eponym of genus Morisonia. Died London, Eng., Nov. 10, 1683.

MORITA, Susumu, Japanese physicist; b. Mie-Prefecture, Japan, Sept. 9, 1918; s. Osamu and Kimiko (Minota) M.; student Osaka (Japan) U., 1938-41; m. Fumiko Kihara, Oct. 5, 1946; children—Junichi, Tomoji. Faculty, Kyushu U., Fukuoka, Japan, 1941-61, prof. physics, 1960-61; prof. Tohoku U., Sendai, Japan, 1961-—. Mem. Phys. Soc. Japan, Am. Phys. Soc. Author: Neutron Spectroscopy, 1960; Nuclear Reaction by Neutrons, 1961; also articles. Research on reaction mechanism in low energy nuclear physics. Home: No. 302 Kawauchi-komuin-jutaku, Sendai. Office: Dept. Physics, Faculty of Sci., Tohoku U., Sendai, Japan.

MORITANI, Ichiro, Japanese chemist; b. Okayama, Japan, June 4, 1916; s. Minbunosuke and Kaya (Murakam) M.; B.Eng., Osaka (Japan) U., 1940, D.Eng., 1952; m. Kiyoko Funakubo, June 1, 1945; Faculty, Osaka U., 1946-—, prof. faculty engring. sci., 1962-—, trustee, 1965-—. Mem. Chem. Soc. Japan, Am. Chem. Soc., Chem. Soc. London. Author: New Synthetic Organic Reactions, 1962; Solvolytic Reactions, 1966; also articles. Research on rearrangement of aromatic compounds, steric effect in elimination reactions, theory and application of absorption chromatography, steric and electronic effect in cyclohexyl derivatives, reaction carbene, mechanism and application to synthetic chemistry, reaction cyclopropene derivatives. Home: 4-6-17 Hagormo, Takaishi, Osaka, Japan.*

MORITZI, Alexander, Swiss naturalist; b. Chur, Switzerland, 1806; tchr., canton sch.; Author: Réflexions sur l'espèce en histoire naturelle, 1842; Die Pflanzen der Schwiez, 1844. Set forth difficulties of differentiating plant and animal species; hinted possibility of evolution. Died 1850.

MORIWAKI, Daigoro, Japanese geneticist; b. Osaka, Japan, Oct. 12, 1906; s. Shigeru and Nami (Katsura) M.; student Tokyo (Japan) Imperial U., 1926-29, Sc.D., 1938; m. Miwa Yamasaki, Nov. 8, 1929; children—Kazuo, Ryoji, Tsuguto. Prof. biology Tokyo Met. Higher Sch., 1931-46, prin., 1946-50; prof. biology Faculty Sci., Tokyo Met. U., 1949-—, dean faculty, 1959-65. Rep., Japan, Permanent Internat. Com. on Genetics Congresses, 1958-63; mem. Japan Sci. Council, 1959-62, chmn. nat. coms. genetics, 1957-—. Mem. Genetics Soc. Japan (past pres., prize 1958), Zool. Soc. Japan, Soc. for Study Evolution, Japan Radiation Research Soc. Research, numerous publs. on genetics of various species of Drosophila, gene analysis of mutants, genetic effects of radiation, population genetics. Home: 2-79 Shimouma-chô, Setagaya-Ku. Office: 2-1, Fukazawa-chô, Setagaya-Ku, Tokyo, Japan.*

MORIYAMA, Hideo, Japanese biochemist; b. Kanagawa-ken, Japan, June 28, 1902; s. Kashitaro and Fude (Ohkubo) M.; student Tokyo (Japan) Imperial U., 1925-29, M.D., 1944; m. Koto Fukuda, Dec. 25, 1933; children—Yoichi, Rubiko (Mrs. Shozo Nakazawa), Matsuyo. Research staff Inst. for Infectious Diseases, Tokyo U., 1929-36, Shanghai Sci. Inst., 1936-42, Aero. Research Inst., Yokosuka, Japan, 1942-45; founder Shonan Hygiene Inst., Kamakura, 1946. Author: The Nature of Viruses and the Origin of Life, 1955; Immunity, 1955; Life, its Nature and Origin, 1958; Immunity and Immune Reactions, 1958; also numerous articles. Research on immunochemistry, viruses, aviation medicine, origin of life, unknown phys. factor (X-agt.) whch effects both living and non-living objects. Home: 1439 Kokufu-Hongo, Oisomachi, Kanagawa-ken. Office: 1913, Omachi, Kamakura, Kanagawa-ken, Japan.*

MORLAND, Sir Samuel, English mathematician, inventor; b. Berkshire, Eng. 1625; s. Thomas Morland; ed. Winchester Sch., also Magdalene Coll., Cambridge (Eng.) U. (fellow and tutor), 1649; m. Suzanne de Milleville, 1657; m. 2d, Carola Harnett, Oct. 26, 1670 (dec. 1676); m. 3d, Anne Feilding, Nov. 16, 1676 (dec. 1680); m. 4th, Mary Aylip, Feb. 1, 1687. Sent (with Whitelock) on embassy to Queen of Sweden, 1653; master of mechanics to Charles II; apptd. clerk of signet by Cromwell, 1665; visit to France regarding King's waterworks, 1682. Recipient medal for loyalty from Charles II. Fellow Royal Soc., 1704. Author: New Method of Criptography, 1666; Four Diagrams of Fortifications, 1670; New and Useful Instrument for Addition and Subtraction, 1672; Description and Use of Two Arithmetick Instruments, 1673. Invented 2 math. machines, 1666, speaking tube, 1671, plunger-pump used to raise water to the top of Windsor castle, 1675; endeavoured to use high-pressure steam as power and suggested it be used in propulsion of vessels. Died Dec. 30, 1695.

MORLEY, Edward Williams, Am. chemist; b. Newark, 1838; s. Sardis Brewster and Anna C. (Treat) M.; A.B., Williams Coll., 1860, A.M., 1863, LL.D., 1901; Ph.D. (hon.), U. Wooster, 1878; LL.D., Adelbert Coll., 1891, Lafayette Coll., 1907, U. Pitts., 1915; ScD., Yale, 1909; m. Isabella E. Birdsall, Dec. 24, 1868. Prof. chemistry Western Res. Coll., Hudson, O. (afterward removed to Cleve. and named Adelbert Coll.), 1869-1906; prof. chemistry, Cleve. Med. Coll., 1873-88. Recipient Davy medal Royal Soc., 1907. Fellow A.A.A.S. (pres. 1895-96), Am. Acad. Arts and Scis. (asso.); mem. Royal Inst. (London) (hon.). Author: On the Densities of Oxygen and Hydrogen, and on the Ratio of Their Atomic Weights, 1895. Has devised improved apparatus for gas analysis; asso. with Michelson in expt. on ether drift and velocity of light (negative result led to theory of relativity); invented a measuring instrument, interferometer; made accurate determination of ratio of combining weights of oxygen and hydrogen; density of oxygen and hydrogen. Died Feb. 24, 1923.

MORLEY, Frank, mathematician; b. Woodbridge, Suffolk, Eng., Sept. 9, 1860; s. Joseph R. and Elizabeth (Muskett) M.; A.B., King's Coll., Cambridge, 1883, A.M., 1886, ScD., 1898; m. Lilian Janet Bird, of Hayward's Heath, Sussex, Eng., July 11, 1889; children—Christopher D., Felix M., Frank V. Master, Bath Coll., Eng., 1884-87; instr., 1887-88, prof. pure math., 1888-1900, Haverford Coll., Pa.; prof. math. Johns Hopkins, 1900-28. Author: (with James Harkness) Elementary Treatise on the Theory of Functions, 1893; Introduction to the Theory of Analytic Functions, 1898; (with F. V. Morley) Inversive Geometry, 1933. Wrote on spl. waves of fifth order. Died Oct. 17, 1937.

MORLEY, Frederick H. W., Australian agronomist; b. Sydney, Australia, July 14, 1918; s. Harold B. and M. H. (King) M.; H.D.A., Hawkesbury Agrl. Coll., 1937; B.V.Sc., U. Sydney, 1942; Ph.D., Ia. State U., 1950; m. Helen R. Browne, Dec. 5, 1947. Vet. officer New S. Wales Dept. Agr., Sydney, Trangie, Australia, 1942-53; geneticist, agronomist Commonwealth Sci. and Indsl. Research Orgn., Canberra, Australia, 1954-—. Mem. Australian Vet. Assn., Inst. Agrl. Sci. and Animal Prodn. Research, publs. on genetic analysis of Australian merino sheep leading to recommendations for breed improvement, taxonomy and evolution of Trifolium especially pasture ecology and agronomy, effects of phytoestrogens on mammalian fertility. Home: 31 Colvin, Canberra ACT. Office: Commonwealth Sci. and Indsl. Research Orgn., Clunies Ross, Canberra, ACT, Australia.*

MORLEY, Sylvanus Griswold, Am. archeologist; b. Chester, Pa., June 7, 1883; s. Benjamin Franklin and Sarah Eleanor Constance (de Lannoy) M.; C.E., Pa. Mil. Coll., 1904, Ph.D. (hon.), 1921; A.B. Harvard, 1907; research fellow in Central Am. Archeology, Harvard, 1907-08, A.M., 1908; m. Alice Gallinger Williams, Dec. 30, 1908 (div.); 1 dau., Alice Virginia; m. 2d, Frances Louella Rhoads, July 14, 1927. Engaged in field work in C.Am. and Mexico for Sch. Am. Archeology, 1909-14; research asso. Carnegie Instn. of Washington, 1915-17, asso., 1918; in charge Carnegie Instn. archeol. expdns. to C.Am.; dir. Chichen Itza Project, 1924-40. Fellow Royal Anthrop. Inst. (Eng.) (hon.); mem. Soc. Am. Archaeology, Am. Philos. Soc. Author: Introduction to Study of Maya Hieroglyphs, 1915; Inscriptions at Copan, 1920; Guide Book to the Ruins of Quirigua, 1935; The Inscriptions of Petén, 5 vols., 1937, 38; The Ancient Maya; also articles. Specialist in Maya hieroglyphic writing and gen. problems in Middle Am. archeology. Died Sept. 2, 1948.

MORLEY, William, see Merle, William.

MORO, Antonio Lazzaro, Italian naturalist; b. San Vito, Italy, Mar. 16, 1687; Author: De' crostacei e degli altri marini corpi che si trovano su montio, 1740. Held that fossil shells were thrown up from sea by volcanic activity like that which had caused a new island in Greek archipelago in 1707. Died Apr. 13, 1764.

MORO, Ernst, Austrian pediatrician; b. Laibach, Austria, Dec. 8, 1874; credited with isolating Lactobacillus acidophilus, 1900; described skin reaction from tuberculin (now a diagnostic test), 1908. Died 1951.

MOROGUES, Sebastien de, see de Morogues, Sebastien.

MOROTWITZ, Harold Joseph, Am. biophysicist; b. Poughkeepsie, N.Y., Dec. 4, 1927; s. Philip Frank and Anna (Levine) M.; B.S., Yale, 1947, M.S., 1950, Ph.D., 1951; m. Lucille Rita Stein, Jan. 30, 1949; children—Joanna Lynn, Eli David, Joshua Alan, Zachary Adam, Noah Daniel. Physicist, Nat. Bur. Standards, 1951-53, Nat. Heart Inst., Bethesda, Md., 1953-55; faculty Yale, New Haven, 1955-—, asso. prof. biophysics, 1960-—. Lectr., U. Md., 1952-53. Mem. A.A.A.S., Biophys. Soc. (exec. com. 1965), N.Y. Acad. Scis. Author: Life and the Physical Sciences, 1964; (with Waterman) Theoretical and Mathematical Biology, 1965; Energy Flow in Biology, 1968; also numerous articles. Research on effect ultraviolet light on bacterial mutations, application laws thermo-dynamics to biology, molecular biology pleuropneumonia-like organisms, structure of plasma membrane. Home: Ox Bow Lane, Woodbridge, Conn. 06525. Office: Yale, New Haven 06520.*

MOROZKIN, Nikolay Ivanovich, Russian therapist, infectionist; b. 1893; grad. Med. Faculty, Moscow U., 1916; D.Med. Sci., 1940. Instr. infectious diseases Smolensk Med. Inst., asso. health epidemiology lab. All-Union State Pub. Health Inspectorate,

Moscow, 1923-39; head chair infectious diseases Gorky Med. Inst., 1939-51; prof., 1940—; dep. dir. for sci. work, head acute respiratory infections clinic Inst. Infectious Diseases, USSR Acad. Med. Scis., 1952—. Del., 2d Internat. Congress on Infectious Pathology, Milan, 1959, 3d, Bucharest, 1962. Mem. USSR Acad. Med. Scis. (corr.). Author numerous works including: Fifteen Years of Experience in the Use of Hemotherapy in the Clinical Treatment of Infections, 1952; Influenza, 1958. Address: Inst. Infectious Diseases, USSR Acad. Med. Scis., Tsitadel 11, Kiev, Ukrainian SSR, USSR.

MOROZOV, Mikhail Akimovich, Russian virologist; b. Jan. 23, 1879; grad. Med. Faculty, Moscow U., 1904. Founder, Voronezh Pasteur Sta., 1914, dir., 1918-23; dir. Central Smallpox Inst., 1923-30; prof., head smallpox dept. Gamalyea Inst. Epidemiology and Microbiology, USSR Acad. Med. Scis., 1930—. Recipient Stalin prize, 1952. Mem. USSR Acad. Med. Sci. Author: New Data on the Etiology of Paravaccines, 1940; Die Farbung der Paschenschen Korperchen durch Versilberung, 1946; Smallpox, 1948; The Morozov Method of Viroscopy, 1956; co-author: Specific Agglutination and Lysis of Viruses, 1953; Atlas of Virus Morphology. Mem. editorial council Problems of Virology; co-editor-Microbiology sect. Large Med. Ency., 2d edit. Research and numerous publs. on poxlike diseases of animals, etiology and pathogenesis of virus diseases: influenza, scarlet fever, measles, poliomyelitis, porphyria; diagnosis and etiology of smallpox and smallpox vaccine; developer technique to produce a stable combined dry smallpox vaccine, 1943, method to stain bacteria with tannin-chrysoidin, 1944; discovered virus-like corpuscles in spinal fluid of schizophrenia patients, 1954; pub. hypothesis on virus nature of schizophrenia, 1955; inventor method to stain virus pathogens with silver. Address: Gamaleya Inst. Epidemiology and Microbiology, USSR Acad. Med. Scis., Malaya Shukinskaya ulitsa 13, Moscow, USSR.

MORPURGO, Giacomo, Italian physicist; b. Florence, Italy, Dec. 14, 1927; s. Augusto and Maria (Castelnuovo) M.; Laurea in Physics, U. Rome (Italy), 1948; m. Gemma Mignone, June 12, 1956; children—Giulio, Paola, Luisa. Lectr., U. Rome, 1949-52, lectr. Postgrad. Sch. Nuclear Physics, 1953-57; research asso. U. Chgo., 1952-53; mem. Inst. Advanced Studies, Princeton, N.J., 1957-58; prof. theoretical physics U. Parma (Italy), 1958-59, U. Florence (Italy), 1960-62; prof. chair structure of matter U. Genoa (Italy), 1962—. Bd. dirs. Nat. Inst. Nuclear Physics. Mem. Am., Italian phys. socs. Author: Lezioni sulle forze nucleari, 1955; (with C. Franzinetti) An Introduction to the Physics of Elementary Particles, 1957; also articles. Research on theoretical low energy nuclear physics, elementary particle and high energy physics, simmetry properties in field theory. Address: U. Genoa, Genoa, Italy.*

MORQUIO, Luis, physician; b. Montevideo, Uruguay, 1867; prof. faculty medicine Montevideo, Uruguay; dir. Children's Clinic, Inst. Pediatrics and Child Care; pres. Soc. Pediatrics. Described an ossification with multiple discrete centers, causing moderate dwarfishness and bodily deformities (Morquio-Brailsford disease, or Morquio's disease), 1929. Died 1935.

MORRELL, Clarence Allison, biochemist; b. Rochester, N.Y., Mar. 15, 1899; s. Richard and Jessie (Howes) M.; B.A. with honors, U. Toronto, 1924, M.A., 1925; Ph.D., Harvard, 1930; m. Jessie Brown Gibson, June 28, 1930; 1 dau., Sheila Mary (Mrs. Kenneth M. Croft). Mem. West African Yellow Fever Commn., Rockefeller Found., 1928-29; travelling fellow Royal Soc. Can., U. Coll., London, Eng., 1932-33; chemist, pharmacologist Dept. Nat. Health, Ottawa, Ont., Can., 1930-32, 33-45, dir. FDA, 1946-65; cons., Ottawa, 1965—; dir. CIBA Co. Ltd. Gold medallist Profl. Inst. Pub. Service Can., 1962. Fellow Royal Soc. Can., Chem. Inst. Can.; mem. Consumers Assn. Can. (v.p. 1965), Am. Soc. for Pharmacology and Exptl. Therapeutics, Can. Physiol. Soc., Can. Pharm. Soc. Research, numerous publs. in application of statis. methods to biol. assays, chem. and phys. methods for testing of drugs. Address: 376 Hamilton Av., Ottawa, Ont., Can.*

MORREY, Charles Bradfield, Am. bacteriologist; b. Chesterhill, O., Nov. 5, 1869; s. John Cheetham and Mary Jenkinson (Wright) M.; B.A., Ohio State U., 1890; M.D., Starling Med. Coll., Columbus, O., 1896; studied in Europe; m. Grace Hamilton Jones, 1898; children—Marion (Mrs. O. C. Richter), Jessie (Mrs. Michael Condoide), Charles Bradfield. With Ohio State U., 1899, founded dept. bacteriology, 1903, prof. bacteriology, head dept., 1904, emeritus prof., 1935-54; cons. on medicine, agr., industry; pioneered in use of vaccine therapy. Mem. Soc. Am. Bacteriologists, Am. Chem. Soc., A.A.A.S., Sigma Xi. Author: Laboratory Exercises in General Bacteriology, 1906, 10th edit. 1929; Fundamentals of Bacteriology, 4th edit. 1929; also various brochures. Died Apr. 21, 1954.

MORREY, Charles Bradfield, Jr., Am. mathematician; b. Columbus, O., July 23, 1907; s. Charles Bradfield and Grace (Jones) M.; A.B., Ohio State U., 1927, M.A., 1928; Ph.D., Harvard, 1931; m. Frances E(leanor) Moss, June 28, 1937; children—Robert A(rthur), Carolyn G(race), Walter T(homas). Faculty, U. Cal., Berkeley, 1933—, prof. math., 1945—, chmn. dept., 1949-54, 57-58. Mathematician, Aberdeen Proving Ground, Md., 1942-45. NRC fellow, 1931-33. Mem. Nat. Acad. Scis., Am. Math. Soc. (pres. 1967-68), Am. Acad. Arts and Scis. Author: University Calculus, 1962; (with Murray H. Protter) First Course in Calcuus, 1963, College Calculus, 1964, Modern Mathematical Analysis, 1964; Multiple Integrals in the Calculus of Variations, 1966; Analytic Geometry, 1966; Calculus for College Students, 1967; also research articles. Study of calculus of variations, area of surfaces; elliptic partial differential equations; multiple Denjoy integrals. Home: 210 Yale Av., Berkeley, Cal. 94708.*

MORRIS, Earl Halstead, Am. archeologist; b. Chama, N.M., Oct. 24, 1889; s. Scott Neering and Juliette Amanda (Halstead) M.; A.B., U. Colo., 1914, A.M., 1916, D.Sc., 1942; postgrad. Columbia, 1916-17; m. Ann McCheane Axtell, Sept. 8, 1923; children—Elizabeth Ann, Sarah Lane; m. 2d, Lucile Bowman, June 4, 1946. Archaeologist, Am. Mus. Natural History 1917-24, Carnegie Instn., 1924-56, leader archeol. expdns. to Guatemala, 1914, 34; leader expdns. to N.M. and Ariz. for U. Colo., 1913-16, 22, 24-28; explored Aztec Ruin in N.M. for Am. Mus. Natural History, 1916-24; dir. excavations in Yucatan for Carnegie Instn., 1924-29. Recipient Norlin medal U. Colo., 1931. Mem. Am. Ethnol. Soc., Am. Anthrop. Assn., N.Y. Acad. Scis., Phi Beta Kappa, Sigma Xi. Author: The Temple of the Warriors, 1931. Research and articles on southwestern archaeology. Died June 24, 1956.

MORRIS, Fred John, Am. physicist; b. Chgo., Dec. 6, 1919; s. Harry and Lillium (Richardson) M.; B.S., Tex. Coll. Arts and Industries, 1942; M.A., U. Tex., 1944, Ph.D., 1951; m. Vera Walsh, Sept. 5, 1942; children—Nansi, Gary Kim. Tech. observer U. S. Naval Air Force, 1944-46; instr. physics U. Tex., Austin, 1946-51; pres., dir. research Electro-Mechanics Co., Austin, 1951—. Sci. adviser Joint Chiefs of Staff Joint Spectrum Evaluation Group, Washington, 1959; cons. research and engring. program Dept. Def., 1962-64, Colgate U., 1959. Mem. Am. Phys. Soc., Soc. Am. Mil. Engrs., Am. Geophys. Union, A.A.A.S., I.E.E.E., Optical Soc. Tex., Soc. Exploration Geophysicists. Research, publs. in radio-frequency interference analysis and control, magnetics and magnetic field measurement instrumentation; developed magnetic field intensity meter. Home: 1200 Westlake Dr., Austin 78746. Office: P.O. Box 1546, Austin, Tex. 78767.*

MORRIS, George Cooper, Am. physician; b. Evanston, Ill., Feb. 15, 1924; s. George C. and Mary (Terrell) M.; student Washington and Lee Coll., 1942-43, Duke, 1943-44; M.D., U. Pa., 1948; m. Jean Carson, June 29, 1946; children—Anne Elisabeth, Penelope Carson, Susan Terrell, George Cooper III, Marian Shopf. Asst. instr. surgery U. Pa., 1949-50; faculty Baylor U. Coll. Medicine, Houston, 1950—, asso. prof. surgery, 1963—, dir. Surg. Research Lab., 1957-62. Markel scholar, 1957. Recipient Bronze Bucranium award for film U. Padova (Italy), 1961; Golden Eagle award for film Com. on Internat. Non-Theatrical Events, 1962, 63. Mem. Am. Coll. Chest Physicians (prize certificate for film 1964, Coll. Film award 1960), Am. Coll. Cardiology (Cummings Humanitarian award 1966), Halsted Soc., Internat. Cardiovascular Soc., Am. Surg. Assn., Am. Heart Assn., A.C.S. Research, numerous publs. on heart and artery surgery. Home: 2529 Reba Dr., Houston 77025.*

MORRIS, Harold Paul, Am. biochemist; b. Salem, Ind., May 8, 1900; s. Ora Alvin and Mary (Hollowell) M.; B.S. in Biochemistry, U. Minn., 1925; M.S. in Genetics, Kan. State U., 1926; Ph.D. in Biochemistry, U. Minn., 1930; m. Mary Sisson Dey, May 31, 1928; children—Joseph Alvin, Harold Elbert, John Emory. Research asst. genetics Kan. State U., 1925-26; asst. animal nutrition U. Ill., 1926-28; fellow biochemistry U. Minn., 1928-30, NRC fellow, 1930-31; research asso. bur. fisheries U. S. Dept. Commerce, 1931-33; jr. bacteriologist U. S. Dept. Agr. Bur. Home Econs., 1933-34; asso. biochemist FDA, 1934-38; chemist Nat. Cancer Inst., 1938-41, sr. nutrition chemist Lab. Biochemistry, 1941-49, biochemist, head nutrition and carcinogenesis sect., 1949—. Mem. NIH Incentive awards Bd., 1957-61, chmn., 1960-61, bd. dirs. Credit Union Bd., 1959—, pres. Recreation and Welfare Assos., Bethesda, Md., 1961. Recipient Superior Service award Dept. Health Edn., and Welfare, 1956. Fellow A.A.A.S.; mem. Am. Chem. Soc., Am. Inst. Nutrition, Soc. Biol. Chemists, Assn. Cancer Research, Soc. for Exptl. Biology, Societa Italiana di Can. (corr. fgn.), Japan Cancer Assn. (fgn. corr.), Sigma Xi, Alpha Zeta, Gamma Alpha, Gamma Sigma Delta, Phi Kappa Phi, Phi Lambda Upsilon. Research, numerous publs. on nutritional requirement mice, mechanisms chem. carcinogenesis, devel., biology and biochemistry chemically induced rat hepatomas of different growth rates, minimal deviation hepatomas. Home: 1112 Noyes Dr., Silver Spring, Md., 20910. Office: NIH, Bethesda, Md. 20014.*

MORRIS, Henry Madison, Jr., Am. hydraulic engr.; b. Dallas, Oct. 6, 1918; s. Henry Madison and Ida (Hunter) M.; B.S. in Civil Engring. with distinction, Rice U., 1939; M.S., U. Minn., 1948, Ph.D., 1950; m. Mary Louise Beach, Jan. 24, 1940; children—Henry Madison III, Kathleen Louise, John David, Andrew Hunter, Mary Ruth, Rebecca Jean. Asst. hydraulic engr. Internat. Boundary and Water Commn., 1939-42; instr. civil engring. Rice U., Houston, 1942-46; instr. civil engring. U. Minn., Mpls., 1946-50, asst. prof., 1950-51, research project leader St. Anthony Falls Hydraulic Lab., 1947-51; prof., head dept. civil engring. U. S.W. La., Lafayette, 1951-56, Va. Poly. Inst., Blacksburg, 1957—. Fellow Am. Soc. C.E., A.A.A.S.; mem. Am. Soc. Engring. Edn. (mem. applied hydraulics com. 1958-61, 64—, chmn. 1960-61), Nat. Soc. Profl. Engrs., Am. Geophys. Union, Va. Acad. Sci., Am. Meteorol. Soc., Creation Research Soc. (charter, dir.), Phi Beta Kappa, Sigma Xi, Tau Beta Pi, Chi Epsilon. Author: Rio Grande Water Conservation Investigation, 1942; Hydraulics of Flow in Culverts, 1948; The Genesis Flood, 1961; Applied Hydraulics in Engineering, 1963; The Twilight of Evolution, 1964; Studies in the Bible and Sci., 1965; also articles, bulls. Research on methodologies for hydraulics of flow of fluids over rough boundaries, applications in design of pipelines, river basin planning, correlation of geologic, other data with framework of sci. creation and catastrophism. Home: 106 Eastview Terrace, Blacksburg, Va.*

MORRIS, John Gottlieb, Am. entomologist; b. York, Pa., Nov. 14, 1803; s. John and Barbara (Myers) M.; student Coll. of N.J., 1820-22; B.A., Dickinson Coll., 1823; attended Princeton Theol. Sem., 1825-26, Gettysburg Theol. Sem., 1826-27; m. Eliza Hay, Nov. 21, 1827; several daus. Ordained to ministry Lutheran Ch., 1827; pastor 1st English Luth. Ch., Balt., 1827-60; librarian Peabody Inst., 1860-65; pastor 3d Ch., 1864-73; founder Lutheran Observer, 1831; often pres. Md. Synod; pres. Gen. Synod, 1843, 83; dir. Gettysburg Sem., a founder Lutherville, Md. (a Balt. suburb), 1851, Luth. Hist. Soc.; pres. entomol. sect. A.A.A.S., many years; Author: Catalogue of the Described Lepidoptera of North America, 1860; Synopsis of the Described Lepidoptera of the United States, 1862; Bibliotheca Lutherana, 1876. Known for entomology works. Died Lutherville, Oct. 10, 1895.

MORRIS, John McLean, Am. surgeon; b. Kuling, China, Sept. 1, 1914 (parents Am. citizens); s. DuBois and Alice (Buell) M.; A.B., Princeton, 1936; M.D., Harvard, 1940; M.A., Yale, 1961; m. Marjorie Stout Austin, Feb. 14, 1951; children—Marjorie, Christina, Constance, Robert, Virginia. Am. Cancer Soc. fellow Radiumhemmet, Stockholm, Sweden, 1951-52; attending gynecologist Yale-New Haven Hosp., 1952—; faculty Yale, 1952—, prof. gynecology, 1961—; cons. gynecologist Meriden Hosp., 1952—, VA Hosp., West Haven, Conn., 1953—, Middlesex Meml., New Britain Gen. hosps., 1954—, Uncason-Thames Hosp., 1959—. Diplomate Nat. Bd. Med. Examiners, Am. Bd. Surgery, Am. Bd. Obstetrics and Gynecology. Fellow A.C.S., Am. Coll. Obstetricians and Gynecologists; mem. Am. Gynecol. Soc., Am. Fertility Soc., Soc. Pelvic Surgeons, New Eng., St. Paul surg. socs., New Haven Obstet. Soc., Conn., New Haven County med. socs., Sigma Xi, Alpha Omega Alpha. Author: (with Robert E. Scully) Endocrine Pathology of the Ovary, 1958. Research, publs. on field surgery and radiation therapy in gynecol. cancer, dosimetry, radiation sensitivity, ovarian tumors, endocrinology, inter-sexuality, urinary diversion procedures, agts. affecting devel. of fertilized ovum. Home: Cedar Rd., Woodbridge, Conn. 06525. Office: 333 Cedar St., New Haven 06510.*

MORRIS, J(oseph) Anthony, Am. virologist; b. nr. Marboro, Md., Sept. 6, 1918; s. Charles L. and Essie (Stokes) M.; B.Sc., Cath U. Am., 1940, M.Sc., 1942, Ph.D., 1947; m. Ruth Savoy, Nov. 1, 1942; children—Carol Ann, Marilyn, Joseph, Larry. With Josiah Macy Jr. Found., N.Y.C., 1943-44; civilian OSRD, 1944; with U. S. Dept. Interior and U. S. Dept. Agr., Beltsville, Md., 1944-47; virologist Walter Reed Army Inst. Research, Washington, 1947-56; chief viral hepatitis research U. S. Army Command, Japan, 1956-59; chief respiratory virology, div. biol. standards NIH, Bethesda, Md., 1959—; instr. Am. U., Washington, 1943-46; Asso. mem. Commn. on Influenza, Armed Forces Epidemiological Bd., 1961—. Mem. Am. Soc. Microbiologists, Soc. Tropical Medicine and Hygiene, Soc. for Exptl. Biology and Medicine, Assn. Am. Immunologists, N.Y. Acad. Scis. Research, numerous publs. on etiology of virus and rickettsial diseases; discovered respiratory synctium virus a maj. cause of human respiratory disease. Home: 23-E Ridge Rd., Greenbelt, Md. 20770. Office: NIH, Bethesda, Md. 20014.*

MORRIS, Joseph Chandler, Am. physicist; b. New Orleans, May 29, 1902; s. Joseph Chandler and Margaret Moore (West) M.; B.S., Tulane U., 1921, M.S. 1923; M.A., Princeton, 1926, Ph.D., 1928; m. Grace Elwood Oldfather, June 9, 1934; children—Grace Elwood (Mrs. Donald M. Williamson), Joseph Chandler. Instr. physics Princeton, 1923-25, 1926-28, fellow, 1925-26, asst. prof., 1928-38; instr. physics Tulane U., New Orleans, 1921-23, prof., 1939—, head dept. physics 1945-60, v.p., 1949—. Dir., v.p., Central Gulf S.S. Corp. Dir. office sci. personnel, NRC, 1941-43; asst., later asso. dir. tng. program San Diego Labs. Div. war research U. Cal., 1943-45; personnel procurement officer Applied Physics Lab., Johns Hopkins, 1945; cons. Nat. Roster Sci. and Specialized Personnel, 1941-45, Office of Edn., 1949—; mem. bd. NSF, 1950-66; pres., Greater New Orleans Ednl. Television Found. Bd. dirs. Council on Library Resources, Inc., Internat. City Bank and Trust Co., New Orleans, Grad. Research Center S.W. Mem. A.A.A.S., Am. Assn. Physics Tchrs., Am. Phys. Soc., I.E.E.E., Sigma Xi. Research on mass spectroscopy; ionization potentials of hydrocarbon gases; ionization of gases. Home: 1654 State St., New Orleans 70118.

MORRIS, Kelso Bronson, Am. phys. inorganic chemist; b. Beaumont, Tex.; s. Isaiah H. and Frances E. W. (Kelso) M.; M.S., Cornell U., 1937, Ph.D., 1940; 1 son, Kenneth Bruce; m. 2d, Marlene I. Cook, 1961; children—Gregory Alfred, Karen Denise, Lisa Frances. Asso. prof. chemistry Howard U., Washington, 1946-61, prof., 1961—, head dept., 1965—; prof. chemistry Air Force Inst. Tech., 1960-61. Fellow Tex. Acad. Sci., A.A.A.S., Washington Acad. Scis.; mem. Am. Chem. Soc., Am. Assn. U. Profs., Nat. Sci. Tchrs. Assn., Nat. Assn. for Research in Sci. Teaching, N.Y. Acad. Scis., Sigma Xi, Beta Kappa Chi. Research, and publs. on properties of molten materials and chemistry of salts in water solution. Home: 1448 Leegate Rd. N.W., Washington 20012.*

MORRIS, Lewis Coleman, Am. surgeon; b. Claremont, Va., Jan. 23, 1872; s. Edward W. and Matilda (Coleman) M.; ed., Randolph-Macon Coll.; M.D., U. Va., 1892. Became demonstrator anatomy, Va., 1893; began pvt. practice, Birmingham, Ala., 1893; founder (with bro. Edward) Morris Sanatorium; prof. surgery Birmingham Sch. Medicine; dean med. sch. U. Ala.; helped in founding state hosp. for Tb., Birmingham. Urged a study of pellagra be formed. Died Mar. 23, 1923.

MORRIS, Lucien Ellis, Am. anesthesiologist; b. Mattoon, Ill., Nov. 30, 1914; s. James Lucien and Pearl (Ellis) M.; A.B., Oberlin Coll., 1936; M.D., Western Res. U., 1943; m. Jean Pinder, June 27, 1942; children—James, Robert, Sara, Donald, Laura. Instr., Wis. Gen. Hosp., 1948-49; faculty State U. Ia., 1949-54, asso. prof., 1951-54; prof. anesthesiology U. Wash., Seattle, 1954-60, clin. prof., 1961-68; practice medicine specializing in anesthesiology, Seattle, 1960-68; dir. anesthesia research labs. Providence Hosp., Seattle, 1960-68, dir. med. edn. and research, 1965-68; prof. anaesthesia Faculty of Medicine, U. Toronto (Ont., Can.), 1968—. Traveling med. faculty to Israel and Iran, WHO and Unitarian Service Com., 1951; mem. com. on anesthesia NRC, 1956-61. Fellow Am. Coll. Anesthesiologists; mem. Am. Soc. Pharmacology and Exptl. Therapeutics, Soc. for Exptl. Biology and Medicine, Assn. U. Anesthetists, Am. Soc. Anesthesiologists, Am. (past dir.), Wash. State (past pres.) socs. anesthesiologists, World Fedn. Socs. Anesthesiologists (del. 1960-64). Diplomate Am. Bd. Anesthesiology. Author: Code of Operations, State University of Iowa Hospitals, 1952; also numerous articles, chpts. in books. Invented copper kettle anesthesia vaporizer; research in anesthesiology. Home: 99 Dinnick Crescent, Toronto 12. Office: St. Michael's Hosp., 30 Bond St., Toronto 1, Ont., Can.*

MORRIS, Robert Franklin, Canadian population ecologist; b. Woodstock, N.B., Can., Dec. 10, 1916; s. Ernest H. and Janet (Beveridge) M.; B.Sc., U. N.B., Can., 1938; M.Sc., Syracuse U., 1940; Ph.D., U. Mich., 1947; m. Cecilia Livingstone, Mar. 29, 1942. With Canadian Dept. Agr., Fredericton, N.B., 1941-60; with Canadian Dept. Forestry, Fredericton, 1961—, prin. scientist, 1964—. Recipient Gold medal Entomol. Soc. Can., 1962. Author: (with Clark, Geier, Hughes) Insect Populations in Theory and Practice, 1966. Devel. sampling techniques, life table methods, and key-factor population models in forest entomology; contbns. to interpretation of population and mortality data. Home: 425 Waterloo Row. Office: Forest Research Lab., Box 4000, Fredericton, N.B., Can.*

MORRIS, Robert James, Am. chemist; b. New Auburn, Wis., Mar. 10, 1915; s. Arthur and Madie (Pasco) M.; B.S., U. Ida., 1936, M.S., 1938; Ph.D., Ohio State U., 1947; m. Venetia Hazel Smith, Mar. 9, 1940; children—Janice Carol (Mrs. Stephen Neal Parker), William James. Engr., Ida. Planning Commn., 1936; faculty U. Ida., 1936-40, Ohio State U., 1941-44; research asst. OSRD, 1944-45, research asso., 1945-47; faculty U. Nev., Reno, 1947—, prof., 1961—; research asso. Desert Research Inst., 1961—. Cons. AEC, Fleishmann Coll. Agr.; cons., research dir. cosmetic div. Sea & Ski Corp., 1965—. Recipient Distinguished Service award OSRD, 1946. Mem. Am. Chem. Soc., Sigma Xi, Phi Kappa Phi, Phi Lambda Upsilon, Sigma Pi Sigma, Sigma Tau. Research, and publs. in devel. of infrared filters for night communications, spectral investigations of dye structures, absorption studies on desert plants, ultra-violet analysis of vegetable oils in adulteration, structural studies of organo-uranium complexes with concurrent developmental studies of flotation techniques and colorimetric analysis procedures, aminoacid dihydroxyacetone complexes, natural product investigations, cosmetic formulations. Home: 1445 Hillside Dr., Reno 89507.*

MORRIS, Rosemary Shull, Am. biochemist; b. Los Angeles, Aug. 11, 1929; d. Charles Phillip and Bernice (Bower) Shull; B.A., U. Cal. at Berkeley, 1950; B.S., U. So. Cal., 1953, M.S., 1956, Ph.D., 1959. Research asso. in biochemistry and nutrition U. So. Cal., Los Angeles, 1959; jr. research biochemist U. Cal. at Los Angeles, 1959-61; research chemist Eastern Utilization Research and Devel. Div., U. S. Dept. Agr., Washington, 1961-66; research chemist div. nutrition FDA, Washington, 1966-67, human nutrition research div., Beltsville, Md., 1967—. Mem. Am. Inst. Nutrition, Am. Chem. Soc., Am. Oil Chemists Soc. Research on lipid metabolism, vitamin E, allergens, trace elements.*

MORRIS, Thomas Martin, Am. engr.; b. West New York, N.J., May 26, 1916; s. Thomas and Meta (Jacobsmuhlen) M.; B.S., Columbia, 1938, M.S., 1940; Ph.D., U. Mo., 1950; m. Patricia Flynn, Aug. 11, 1948; children—Mary Pamela, Meta Jo, Thomas Flynn, Paul Martin. Research engr. Anaconda Co. (Mont.), 1940-44; faculty Mo. Sch. Mines, Rolla, 1947-59, prof. engring., 1956-59; head dept. metall. engring. U. Ariz., Tucson, 1959—. Cons. to various metal cos. Mem. Am. Inst. Mining Engrs. Basic research in unit processes and unit operations in metallurgical engineering. Home: 7131 E. 32d St., Tucson 85721.*

MORRISH, Allan Henry, Canadian physicist; b. Winnipeg, Man., Can., Apr. 18, 1921; s. Stanley and Agnes (Payne) M.; B.Sc. with honors, U. Man., 1943; M.A., U. Toronto, 1946; Ph.D., U. Chgo., 1949; m. Hilda Gertrude Fiske, Sept. 16, 1952; children—John Stanley, Allan Richard. Faculty, U. B.C., Vancouver, Can., 1949-52; research asst. McGill U. Radiation Lab., Montreal, Que., Can., 1952-53; with dept. elec. engring. U. Minn., Mpls., 1953-64, prof. dept. elec. engring., 1959-64; prof. U. Man., Winnipeg, 1964—, head dept. physics, 1966—. Cons. Honeywell, Inc., Hopkins, Minn., 1956-57, 59-63. NRC Can. postdoctoral fellow U. Bristol (Eng.), 1950-51; Guggenheim fellow U. Oxford (Eng.), 1957-58. Mem. Am. Phys. Soc., Canadian Assn. Physicists, Phys. Soc., Sigma Xi. Author: The Physical Principles of Magnetism, 1965; also articles. Editor: (with R. J. Prosen, S. W. Rubens) Magnetic Materials Digest, 1964. Precision measurement of nuclear reaction energy, single photon counting technique; discoveries in soft cosmic-ray radiation employing nuclear emulsions, including trident cross section, minimum ionization; interaction of 222 pi-mesons with nuclei; devel. supersensitive quartz balance to demonstrate existence of single-domain magnetic particles; growth and study of highly perfect single crystals of pure and doped hematite; use of Mossbauer effect to investigate ferrimagnets. Home: 71 Agassiz Dr., Winnipeg 19, Man., Can.*

MORRISON, James Douglas, phys. chemist; b. Glasgow, Scotland, Nov. 9, 1924; s. James Kinloch and Rose (Wheeler) M.; B.Sc. with honors, Glasgow U., 1945, Ph.D., 1948, D.Sc., 1958; m. Christine Barbara Maria Mayer, Sept. 5, 1947; children—Richard J. S., Gordon T. G., Alan A. R. Asst. lectr. Glasgow U., 1946-48; staff Commonwealth Sci. and Indsl. Research Orgn., Clayton, Australia, 1949—, chief research scientist, 1965—; Found. prof. chemistry La Trobe U. Melbourne, 1967—; vis. prof. Princeton, 1964. Harkness fellow Commonwealth Fund, U. Chgo., 1956-57. Fellow Royal Australian Chem. Inst. (Rennie and H. G. Smith Meml. medal 1961), Australian Acad. Sci.; mem. Royal Soc. Victoria (mem. council 1960—), Chem. Soc. London, Sigma Xi. Research, publs. on precise measurement of chem. bond lengths in molecules using X-ray crystallography, molecular energetics by collision methods, application mass spectrometer to various chem. problems. Home: 40 Central Av., Mooroolbark, Victoria. Office: Commonwealth Sci. and Indsl. Research Orgn., Div. Chem. Physics, P.O. Box 160, Clayton, Victoria, Australia.*

MORRISON, John Allan, research mathematician; b. Beckenham, Eng., June 10, 1927; s. Reginald David and Edith (North) M.; B.Sc. with 1st honors in Math., U. London, 1952; Sc.M., Brown U., 1954, Ph.D., 1956; m. Barbara Ann Cotter, Sept. 10, 1955. Research asst. applied math. Brown U., 1952-56; research mathematician Bell Telephone Labs., Murray Hill, N.J., 1956—. Mem. Am. Math. Soc., Soc. for Indsl. and Applied Math., Sigma Xi. Research, publs. in math. physics and theory of nonlinear oscillations. Home: 28 Ashwood Rd., New Providence, N.J. 07974. Office: Bell Telephone Labs., Murray Hill, N.J. 07971.*

MORRISON, Lester Marvin, Am. physician; b. London, Eng., Sept. 18, 1907 (parents Am. citizens); s. Harold and Sophia (Beck) M.; student McGill U., 1926-29; M.D., Temple U., 1933; m. Rita J. Rosenthal, May 5, 1938. Practice medicine, specializing in internal medicine, Los Angeles, 1945—; lectr. medicine Sch. Medicine Loma Linda U., 1945—; sr. attending physician Los Angeles County Gen. Hosp., 1945—; cons. medicine St. John's Hosp., 1945-53, Hollywood Presbyn. Hosp., 1948-52, Fairview State Hosp., 1963—, U. Cal. Med. Center, 1965—; pres. Crenshaw Research Found., 1959—; pres., dir. Inst. Arteriosclerosis Research, 1965—; pres., owner Crenshaw Hosp., Los Angeles, 1948-52, Whittier (Cal.) Hosp., 1953-61, Doctors Hosp., Santa Ana, Cal., 1954—, Sun Valley (Cal.) Hosp., 1959-62, Pasadena (Cal.) Community Hosp., 1963—. Diplomate Am. Bd. Internal Medicine, Am. Bd. Gastroenterology. Mem. Am. Coll. Gastroenterology. Author: Low Fat Way to Health and Longer Life, 1958; (with R. G. Hubler) Trial and Triumph: Maimonides, 1965. Contbr. numerous articles to profl. jours. Innovator spl. dietary, medicinal and nutritional treatment of liver, arteriosclerosis, intestinal parasites (giardiasis) and gastritis. Home: 7012 LaPresa Dr., Hollywood, Cal. 90028. Office: 10921 Wilshire Blvd., Los Angeles 90024.*

MORRISON, Roger Barron, Am. geologist; b. Madison, Wis., Mar. 26, 1914; s. Frank Barron and Elsie B. (Bullard) M.; B.S., Cornell U., 1933, M.A., 1934; postgrad. U. Cal. at Berkeley, Stanford, Ph.D., U. Nev., 1964; m. Harriet Louise Williams, Apr. 7, 1941; children—John Christopher, Peter Hallock, Craig Brewster. Geologist ground water br. water resources div. U. S. Geol. Survey, 1939-42, mil. geology br. geologic div., 1942-47, gen. geology br. geologic div., 1947-60, geologic div. Southwestern states br., Denver, 1960—. Mem. Geol. Soc. Am., Am. Geog. Soc., Friends of Pleistocene, Internat. Assn. for Quaternary Research, Sigma Xi. Study, interpretation and dating of late Cenozoic (particularly Pleistocene and recent) deposits and landforms; chronology of pluvial lakes Western U. S. and correlation with glacial history; trancontinental and intercontinental correlation of Quaternary deposits; principles of soil stratigraphy. Home: 13150 W. 9th Av., Golden, Colo. 80401. Office: U. S. Geol. Survey, Fed. Center, Denver 80225.*

MORRISON, Samuel, Am. physician; b. Phila., Jan. 4, 1904; s. Morris and Annie (Lipsitz) M.; student Balt. Poly. Inst., 1922; A.B., Johns Hopkins, 1925; M.D., 1929; m. Mary Jane Selser, July 18, 1941. Faculty, U. Md. Med. Sch., Balt., 1930—, asso. prof. medicine and gastroenterology, 1939—; practice medicine, Balt., 1930—; chief medicine St. Joseph's Hosp., Balt., 1956-58; v.p. Med.-Chirurgical Faculty, State of Md., 1960. Cons. to surgeon gen., VA, USPHS. Mem. Balt. Med. Soc. (pres. 1965), A.M.A., A.C.P., Am. Gastroent. Assn., Am. Soc. Internal Medicine, Soc. Med. Cons. to Armed Forces. Author: (with J. Friedenwald, T. Morrison) Secondary Gastro-intestinal Disorders, 1938; History of the Enema and History of Gastric Intubation, 1940. Research, numerous publs. on dissolution of gallstones, mechanism of prodn. of ulcers in human beings, sugar metabolism, leukemia, diseases of liver, vital dyes, oxyntic cells and pernicious anemia, diet after surgery, Pavlov pouch studies, hydrogen ion concentration studies, gastric mucosal function, various psychophysiol. mechanisms, vagotomy, thyroid-pituitary functions. Home: 3799 Juniper Rd., Balt. 21218. Office: 11 E. Chase St., Balt. 21202.*

MORROW, Andrew Gleen, Am. surgeon; b. Indpls., Nov. 3, 1922; s. Henry B. and Doris (Gordon) M.; A.B., Wabash Coll., 1943; M.D., Johns Hopkins, 1946; m. Phyllis Ruth Perry, June 2, 1945; children—Elizabeth Ruth, Andrew Glenn, Katherine Gordon. Faculty, Johns Hopkins, 1947-52, 54—, asso. prof. surgery, 1960—; clin. asst. thoracic surgery U. Leeds, Eng., 1952-53; chief, clinic surgery Nat. Heart Inst., Bethesda, Md., 1953—. Diplomate Am. Bd. Surgery, Am. Bd. Thoracic Surgery. Fellow A.C.S. (bd. govs. 1963-66), Am. Coll. Cardiology; mem. A.M.A., Soc. Vascular Surgery, Am. Assn. Thoracic Surgery, Halsted Soc., Am. Heart Assn. (exec. com. council on cardiovascular surgery), Am. Fedn. Clin. Research, Soc. U. Surgeons, Soc. Thoracic Surgeons Gt. Britain and Ireland, Am. Surg. Assn., Internat. Cardiovascular Soc., Soc. Clin. Surgery. Editorial bd. Jour. Surg. Research, 1960—, Circulation, 1960—, Am. Heart Jour., 1963—. Research, publs. on devel. surgery of heart and study of allied problems in cardiovascular physiology. Home: 227 W. Montgomery Av., Rockville, Md. 20850. Office: Nat. Heart Inst., Bethesda, Md. 20014.*

MORROW, James Edwin, Jr., Am. zoologist; b. Bklyn.; Nov. 7, 1918; s. James Edwin and Elizabeth (Hodenpyl) M.; A.B., Middlebury Coll., 1940, M.S., 1942; M.S., Yale, 1944, Ph.D., 1949; m. Catherine Joan O'Brien, July 7, 1950; 1 son, Matthew James. Research asst. Bingham Oceanographic Lab., Yale, 1949-54, research asso., 1954-60, ichthyologist New Zealand Expdn., 1948, E. African Expdn., 1949-50, chief sicentist S. Am. Expdn., 1953, chief scientist, dir. Seychelles Expdn., 1957-58; faculty U. Alaska, College, 1960—, prof. fisheries biology, 1963-67, prof. zoology, head dept. biology, 1967—. Fellow A.A.A.S.; mem. Am. Soc. Ichthylogists and Herpetologists, Am. Soc. Zoologists, Am. Fisheries Soc., Sigma Xi. Contbr. articles to tech. jours. Research, publs. on ecology and classification fishes. Home: Box 105, College, Alaska, 99735.*

MORROW, JoDean, Am. engr.; b. Woodbine, Ia., Oct. 16, 1929; s. Wade Rankin and Lovina (Swackhammer) M.; B.S. in Civil Engring., Rose Poly. Inst., 1950; Ph.D. in Theoretical and Applied Mechanics, U. Ill., 1957; m. Sally Jean Coonrod, Dec. 23, 1950; children—JoDean, Daniel L., Theodore A., Linda S. Faculty, U. Ill., Urbana, 1957—, prof. theoretical and applied mechanics, 1964—. Cons. to pvt. cos. Am. Soc. for Testing Materials, Am. Soc. Metals, Soc. Automotive Engrs., Sigma Xi. Research, publs. on fatigue resistance metals; developed techniques for exptl. determination cyclic stress-strain and fracture properties metals. Home: 408 N. McKinley St., Champaign, Ill. 61821. Office: Talbot Lab., U. Ill., Urbana, Ill. 61803.*

MORROW, Prince Albert, Am. dermatologist; b. Mt. Vernon, Ky., Dec. 19, 1846; s. William and Mary (Cox) M.; A.B., Princeton Coll., Ky., 1864; M.D., Univ. Med. Coll. (N.Y. U.), 1874, A.M. (hon.), 1883; m. Lucy B. Slaughter, Apr. 23, 1874. Attending surgeon City Hosp., N.Y., 1884-1904; clin. prof. venereal diseases Univ. and Bellevue Hosp. Med. Coll., 1884-1913, lectr. dermatology, 1882, 83, clin. prof. genitourinary diseases, 1886-90, emeritus prof., 1890; attending physician skin and venereal dept., N.Y. Hosp., 1890-1904; cons. dermatologist St. Vincent's, City hosps. Mem. Am. Soc. for San. and Moral Prophylaxis (pres.). Author: Venereal Memoranda, 1885; Drug Eruptions, 1887; Atlas of Skin and Venereal Diseases, 1888-89; System of Genito-Urinary Diseases, Syphilology and Dermatology (3 vols.), 1892-94; Leprosy, 1899; Social Diseases and Marriage, 1904. A founder, editor Jour. Cutaneous and Genito-Urinary Diseases. Specialized in dermatology, venereal and genito-urinary diseases; pioneer in educating pub. to need of disseminating knowledge of sex hygiene. Died Mar. 17, 1913.

MORROW, Thomas Vaughan, Am. eclectic physician; b. Fairview, Ky., Apr. 14, 1804; s. Thomas and Elizabeth (Vaughan) M.; attended Transylvania U., Ref. Med. Coll. of N.Y.; m. Isabel Greer. Pres., dean, prof. materia medica, obstetrics, theory and practice of medicine Ref. Med. Coll. of Ohio, 1830-39; organized Ref. Med. Sch. of Cin., 1842, became Cin. Eclectic Med. Inst., 1845, dean, treas., prof. physiology, pathology and theory and practice, 1845-50; advocated eclectic system of medicine, founder of 1st schs. of that cult in West; lectr. on eclectic system of medicine; pres. Nat. Eclectic Med. Assn., 1848; wrote articles, editorials for Western Med. Reformer and Eclectic Med. Jour. Died Cin., July 16, 1850.

MORSE, Albert Pitts, Am. entomologist; b. Sherborn, Mass., Feb. 10, 1863; s. Leonard Townsend and Phoebe Adeline (Knapp) M.; studied entomology privately; student Cornell U., summer 1890, Marine Biol. Lab., Woods Hole, Mass., also Harvard, 1906; m. Annie McGill, July 29, 1893; children—Catherine (Mrs. Starr Truscott), Leonard Henshaw. Became asst. zoologist Wellesley Coll., 1888, curator Zool. Mus., lectr. econ. entomology, 1890-1933; instr. zoology Tchr.'s Sch. Sci., Boston, 1901-10; curator natural history Peabody Mus., Salem, Mass., 1911—; research asst. Carnegie Instn., Washington, 1903, 05. Fellow Entomol. Soc. Am., A.A.A.S.; mem. Ornithologists' Union, Am. Assn. Econ. Entomologists, Boston Soc. Natural History, Am. Fern Soc., New Eng. Bot. Club, Cambridge, N.Y. entomol. clubs, Nuttall Ornithol. Club, Morse Sci. Club Boston, Hort. Soc. New Eng., Birdbanding Assn. Author: Annotated List of Birds of Wellesley and Vicinity; Pocket List of Birds of Eastern Massachusetts; Notes on New England Acridiidae; The North America Odonata; Researches on North American Acridiidae; Manual of the Orthoptera of New England; Orthoptera of Maine; also articles. Discovered many new species. Died Wellesley, Mass., Apr. 29, 1936.

MORSE, Anthony Perry, mathematician; b. Ithaca, N.Y., Aug. 21, 1911; s. Frank Lincoln and Cora Martha (Perry) M.; A.B., Cornell U., 1933; Ph.D., Brown U., 1937; m. Mary Rojier Evans, July 24, 1934 (div.); children—Frank L., Mary L., Peter E., Frederick A., Martha R.; m. 2d, Barbara Elise Vernon, May 5, 1956; 1 dau., Suzanne R. Instr. math. Brown U., 1936-37, Princeton, 1937-38; mem. Inst. for Advanced Study, 1937-39; faculty U. Cal. at Berkeley, 1939—, prof. math., 1949—; mathematician theory sect. Aberdeen Proving Ground, 1944-45. Miller fellow, 1951-52. Mem. Am. Math. Soc., N.Y. Acad. Sci., Sigma Xi. Author: A Theory of Sets, 1965; also articles. Research on real function theory, measure theory, covering and differentiation, set theory. Home: 28 La Campana, Orinda, Cal. 94563. Office: Math. Dept., U. Cal., Berkeley, Cal. 94720.*

MORSE, Chandler, Am. economist; b. Bklyn., Mar. 29, 1906; s. Ernest Chandler and Clementine (Ayer) M.; B.A., Amherst Coll., 1927; M.A., Harvard, 1928; m. Katrina Rosina Pease, Oct. 17, 1931. With Fed. Res. System, OSS, 1927-47; faculty Williams Coll., 1947-50; asso. prof. econs. Cornell U., Ithaca, N.Y., 1950-58, prof., 1958—, dir. Social Sci. Research Center, 1954-56, dir. Modernization Workshop, 1962-66. Chmn., Econ. Survey Mission to Basutoland, Bechuanaland Protectorate and Swaziland, 1959; cons. govt. agys. Mem. Am. Econ. Assn., African Studies Assn., Assn. For Comparative Econs. Author: (with others) Basutoland, Bechuanaland Protectorate and Swaziland: Report of an Economic Survey Mission, 1960; (with H. J. Barneff) Scarcity and Growth, 1963. Editor: Fact and Theory in Economics: The Essays of Morris A. Copeland, 1958. Research on natural resources and econ. growth, understanding of role of instns. and instl. change in econ. devel., econ. interpretation of African Socialism, role of direct fgn. investment in econ. devel. Home: 115 Ellis Hollow Creek Rd., Ithaca, N.Y. 14850.*

MORSE, Edward Sylvester, Am. zoologist; b. Portland, Me., June 18, 1838; s. Jonathan K. and Jane Seymour (Beckett) M.; ed. Lawrence Sci. Sch. (Harvard), A.M., 1892; Ph.D. (hon.), Bowdoin Coll., 1871; D.Sc., Yale, 1918; L.H.D., Lufts Coll., 1922. Prof. comparative anatomy and zoology Bowdoin Coll., 1871-74; prof. zoology, Imperial U., Tokyo, Japan, 1877-80; dir. Peabody Museum, Salem, Mass., 1880-1925. Lectr. Harvard, 1872-73; keeper Japanese pottery Mus. Fine Arts, Boston, 1892-1925; authority on Japanese ceramics. Fellow Am. Acad. Arts and Scis.; mem. numerous sci. socs., A.A.A.S. (pres. 1886), Nat. Acad. Scis. Author: First Book of Zoology, 1875; Japanese Homes and Their Surroundings, 1886; Catalogue of the Morse Collection of Japanese Pottery (Mus. of Fine Arts, Boston), 1901; Glimpses of China and Chinese Homes, 1902; Mars and Its Mystery, 1906; Shell Mounds of Omori; Japan Day by Day, 1917. Proved brachiopods were worms and not mollusks; inventor mus. shelf bracket widely used, apparatus for using sun's rays to heat and ventilate a worm with fresh warm air in winter; 1st to expound Darwin's theory in Japan; excavated mounds in Kinki Dist. and Hokkaido; studied Japanese manners and customs; 1st scientist to refute theory that Ainus were original inhabitants of Japan. Died Dec. 20, 1925.

MORSE, Ellen Hastings, Am. nutritionist, educator; b. Durham, N.H., Oct. 10, 1908; d. Fred Winslow and Lelia (White) Morse; B.A., Wellesley Coll., 1930; M.A., Smith Coll., 1938; M.S., U. Mass., 1949; Ph.D., U. Conn., 1960. Asst. curator Smith Coll., 1931-46; research fellow U. Mass., 1946-49; nutritionist Colo. Expt. Sta, 1949-51, N.Y. State Health Dept., 1951-52; research asst. U. Conn., 1952-60; asso. prof. nutrition, dir. nutrition research, U. Vt., Burlington, 1960—. Mem. Am. Inst. Nutrition, Am., Vt. dietetic assns., Am., Vt. home econs. assns., Sigma Xi. Research, publs. on ascorbic acid needs of women, dietary fats in controlled diet expts., amino acid utilization, lipid metabolism. Home: 38-A University Heights, Burlighton, Vt. 05401.*

MORSE, Harld Marston, Am. mathematician; b. Waterville, Me., Mar. 24, 1892; s. Howard Calvin and Ella (Marston) M.; A.B., Colby Coll., 1914, Dr. honoris causa, 1935; M.A., Harvard, 1915, Ph.D., 1917, H.D.Sc., 1965; Dr. honoris causa, Sorbonne, 1946, Kenyon Coll., 1948, U. Pisa, 1948, Technische Hochshule Vienna, 1952, U. Rennes, 1953, U Md., 1955, Bklyn. Poly. Inst., 1955, U. Notre Dame, 1956, Boston Coll., 1958, Fordham U., 1958, Williams Coll., 1959, LaSalle Coll., 1960, St. Peter's Coll., 1961; LL.D., Xavier Coll., 1961; Litt.D., Yeshiva U., 1962; m. Celeste Phelps, June 20, 1922; children—Meroe, Dryden; m. 2d, Louise Jefferys, Jan, 13, 1940; children—Julia, William, Elizabeth, Peter, Louise. Benjamin Peirce instr. Harvard, 1919-20, asst. prof., 1926-28, asso. prof., 1928-29, prof., 1929-35; asso. prof. Cornell U., 1920-25, Brown U., 1925-26; prof. math. Inst. for Advanced Study, Princeton, N.J., 1935-62, research contract USAF Office Sci. Research, 1962-66. Chmn. 1st div. math. NRC, 1950; vis. prof. City U. N.Y., 1965-66. 1st bd. dirs. NSF, 1950-54. Recipient medal of merit Pres. Truman, 1945; Nat. medal Sci., Pres. Johnson, 1965. Fellow Am. Acad. Arts and Scis., Nat. Acad. Scis., A.A.A.S.; mem. Am. Philos. Soc., Am. Math. Soc. (pres. 1940-42, Bocher prize 1932), Circolo Math. di Palermo, Academia delle Scienze Bologna (fgn. corr.), Internat. Math. Union (v.p. 1958-62), Heidelberg (asso., corr.), Rumanian (corr.) acads. sci., Istituto Lombardo Milan (corr.), Acad. Scis. Paris (corr.), Academia Nazionale dei Lincei (corr.), Phi Beta Kappa, Sigma Xi. Author: Functional Topology and Abstract Variational Theory, 1938; Topological Methods in the Theory of Functions of a Complex Variable, 1947; Lectures on Analysis in the Large, 1947; also articles. Devel. variational theory in the large with application to equilibrium problems in math., physics. Home: 40 Battle Rd., Princeton, N.J. 08540.*

MORSE, Harmon Northrop, Am. chemist; b. Cambridge, Vt., Oct. 15, 1848; s. Harmon and Elizabeth Murray (Buck) M.; A.B., Amherst Coll., 1873, LL.D., 1915; Ph.D., U. Göttingen, 1875; m. Caroline Augusta Brooks, Dec. 13, 1876 (dec. 1887); m. 2d, Elizabeth Dennis Clarke, Dec. 24, 1890. Asst. in chemistry Amherst Coll., 1875-76; asso. prof. chemistry, Johns Hopkins, 1876-91, prof. analytical chemistry and adj. dir. chem. lab., 1891, prof. inorganic and analytical chemistry, dir. chem. lab. Research asso. Carnegie Instn. of Washington. Recipient Avogadro medal. Fellow Am. Acad. and Scis. Author: Exercises in Quantitative Chemistry, 1905; The Osmotic Pressure of Aqueous Solutions, 1914, also numerous articles. Research on osmotic pressure; made accurate determinations of atomic weight, including cadmium and zinc. Died Sept. 8, 1920.

MORSE, Jedidiah, Am. geographer; b. Woodstock, Conn., Aug. 23, 1761; s. Jedidiah and Sarah (Child) M.; grad. Yale, 1783; studied theology New Haven, 1783-85; S.T.D. (hon.), U. Edinburgh (Scotland), 1794; m. Elizabeth Breese, May 14, 1789, 11 children including Samuel Finley Breese, Sidney Edwards, Richard. Licensed to preach Congl. Ch., 1785, ordained to ministry, 1786; pastor in Charlestown, Mass., 1789-1819; commd. by sec. war to study conditions of the Indian nations, 1820, mem. Phi Beta Kappa; known as father of Am. Geography. Author: Geography Made Easy, 1784 (1st pub. Am. geography); The American Geography, 1789; The American Universal Geography; Elements of Geography, 1795; The American Gazeteer, 1797; A New Gazeteer for the Eastern Continent, 1802; Annals of the American Revolution, 1824; co-author: A Compendious History of New England, 1804; A Report to the Secretary of War—On Indian Affairs, Comprising A Narrative of a Tour Performed in the Summer of 1820, 1822. Died New Haven, June 9, 1826.

MORSE, Jerome Gilbert, Am. nuclear engr.; b. N.Y.C., Oct. 22, 1921; s. Edward and Etta (Burke) M.; B.S., Coll. City N.Y., 1942; M.S., U. Pa., 1947; Ph.D., Ill. Inst. Tech., 1951; m. Elizabeth Tarlowsky, June 19, 1949; children—David Lyle, Evan Elliott, Lauren. Research chemist Gen. Electric Co., Schenectady, 1951-52; asst. prof. chemistry U. Miami (Fla.), 1952-55; with Martin Co., 1955—, tech. dir. Hispan Martin subsidiary, Paris, France, 1964-67, prin. research scientist aerospace div., Denver, 1967—. Cons. Oak Ridge Inst. Nuclear Studies, 1960-64. Mem. Am. Nuclear Soc. (chmn. isotopes and radiation div. 1963-64), Am. Chem. Soc., Sigma Xi, Phi Lambda Upsilon. Editor: (with Harwood, Hausner, Rauch) Effects of Radiation on Materials, 1958. Research, publs. on devel. radioisotope-fueled thermoelectric generators leading to world's first nuclear power plants used on satellites in outer space, arctic and antarctic nuclear powered automatic weather stas., undersea navigation beacon. Home: 5066 Tule Lake Dr., Littleton, Colo. 80120. Office: Dept. 1609, Martin Co., Denver 80201.*

MORSE, John Lovett, Am. physician; b. Taunton, Mass., Apr. 21, 1865; s. Erastus and Sarah Seabury (Basset) M.; A.B., Harvard, 1887, A.M., M.D., 1891; Adelaide M. Fairbrother, Sept. 3, 1906; 1 son, Lovett. Practice medicine, Boston, 1892-1937; asst. clin. medicine Harvard, 1896-1900, instr. pediatrics, 1903-06, asst. prof., 1906-11, asso. prof., 1911-15, prof., 1915-21, prof. emeritus, 1921—; cons. physician Children's Hosp., Infant's Hosp. Mem. Am. Pediatric Soc. (twice head). Author numerous med. texts. Research on children's diseases. Died Apr. 3, 1940.

MORSE, Philip McCord, Am. physicist; b. Shreveport, La., Aug. 6, 1903; s. Allen Crafts and Edith (McCord) M.; B.S., Case Inst. Tech., 1926, Sc.D. (hon.), 1940; M.A., Princeton, 1927, Ph.D., 1929; m. Annabelle Hopkins, Apr. 26, 1929; children—Conrad Philip, Annabella (Mrs. Hugh Fowler). Instr. physics Princeton, 1929-30, Internat. research fellow, 1930-31; faculty physics Mass. Inst. Tech., Cambridge, 1931—, prof., 1939—, asso. editor Tech. Rev., 1936-46, dir. Computation Center, 1955—, dir. Operations Research Center, 1958—, chmn. faculty, 1958-60. Dir. Brookhaven Nat. Lab., 1946-48; mem. div. 6, NDRC, asst. dir. Office Field Service, OSRD, 1943-45; dir. operations research group USN, 1942-46; dir. research weapons systems evaluation group Office Sec. Def., 1949-50; mem. Ordnance Research Adv. Com., 1950-56; chmn. NATO Adv. Panel Operations Research, 1960-64; chmn. adv. panel operations research OECD, 1962—. Recipient medal for merit Pres. Truman, 1946. Fellow Operations Research Soc. Am. (1st pres. 1952-53), Am. Acad. Arts and Scis., Am. Inst. Physics (trustee 1948-50, 53-57), Am. Phys. Soc. (council 1935—), Acoustical Soc. Am. (pres. 1950-51); mem. Nat. Acad. Scis., Internat. Fedn. Operations Research Socs. (internat. sec. 1961-64), Nat. Research Soc. Am. (trustee 1950-58), Sigma Xi, Tau Beta Pi. Author: (with E. U. Condon) Quantum Mechanics, 1929; Vibration and Sound, 1936, rev. edit. 1946; (with G. E. Kimball) Methods of Operations Research, 1950; (with H. Feshbach) Methods of Theoretical Physics, 1953; Queus, Inventories and Maintenance, 1957; Thermal Physics, 1962, 64. Editor: Annals of Physics, 1957—, Notes on Operations Research, 1959; editorial bd. Physics Today, 1947-52, Sci., 1960-64, Bull. Atomic Scientists, 1948-64. Research, publs. in quantum mechanics, theoretical physics, acoustics, operations research; originated Morse Potential for diatomic molecules; directed 1st operations research group in U. S. Home: 126 Wildwood St., Winchester, Mass. 01890. Office: Mass. Inst. Tech., Cambridge, Mass. 02139.*

MORSE, Robert Warren, Am. physicist, govt. ofcl.; b. Boston, May 25, 1921; s. Walter L. and Ethel (Prince) M.; B.S., Bowdoin Coll., 1943; Sc.M., Brown U., 1947, Ph.D., 1949; m. Alice Muriel Cooper, Jan. 25, 1943; children—Robert W., Pamela Dean, James Prince. Faculty, Brown U. Providence, 1949-64, prof. physics, 1958-64, chmn. dept. physics, 1960-62, dean coll., 1962-64; Howard Found. fellow, Royal Soc. Mond Lab., Cambridge U., 1954-55; asst. sec. for research and devel. USN, 1964—. Mem. com. on undersea warfare Nat. Acad. Scis., Washington, 1957-64, chmn., 1962-64; chmn. interagy. com. on oceanography Fed. Council for Sci. and Tech., Washington, 1964—. Fellow Am. Phys. Soc. (chmn. div. solid state physics 1963-64); mem. Acoustical Soc. (pres. 1965—), Am. Acad. Arts and Scis., Sigma Xi. Contbr. articles to tech. jours. on ultrasonics, superconductivity, properties of metals, underwater acoustics. Home: 2646 S. June St., Arlington, Va. 22202. Office: Dept. Navy, Pentagon, Washington 20350.*

MORSE, Samuel Finley Breese, Am. inventor; b. Charlestown, Mass., Apr. 27, 1791; s. Jedidiah and Elizabeth (Breese) M.; grad. Yale, 1810, LL.D.; attended Royal Acad., London, Eng., 1811-15; m. Lucretia Walker, Sept. 29, 1818; m. 2d, Sarah Griswold, Aug. 9, 1848; 8 children. Painter in Eng.; engaged in portrait painting in U. S., 1815-29; entered politics as member of Native American party (also sympathized with South during Civil War);

a founder N.A.D.., 1st pres., 1826-42, also pres., 1861; made study trip to Europe, 1829-31; prof. painting, sculpture N.Y.U., 1832; recipient Silver medal French Acad. Scis. Invented electro-magnetic recording telegraph; invented sending and receiving apparatus, code (Morse Code), by 1832; worked out system of electro-magnetic relays to be placed in telegraph line weak points, 1836; Congress voted $30,000 for an exptl. line from Washington to Balt., 1843, line completed, May 24, 1844; Morse's rights to profits from his invention were upheld in courts; electrician Cyrus W. Field's Co., engaged in laying Trans-Atlantic Cable, 1857-58; inventor pressure pump for fire engines, electric telegram. Died N.Y.C., Apr. 2, 1872.

MORSE, Sidney Edwards, Am. inventor; b. Charlestown, Mass., Feb. 7, 1794; s. Jedidiah and Elizabeth (Breese) M.; A.B., Yale, 1811; attended Litchfield (Conn.) Law Sch., Andover Theol. Sem., 1817-20; m. Catharine Livingston, Apr. 1, 1841, 2 children. A founder, Recorder (1st religious newspaper in Boston), 1816; a founder N.Y. Observer, 1823, sr. editor, propr., 1823-58; editor (with father) A New System of Modern Geography . . . Accompanied by an Atlas, 1822; patentee (with brother Samuel) flexible piston pump; inventor process of cerography (map of Conn. was 1st example), 1839; patentee (with son) bathometer, 1866. Author: The New States, or a Comparison of the . . . Northern and Southern States: With a View to Expose the Injustice of Erecting New States at the South (collection of reprinted articles), 1813; An Atlas of the United States, 1823; Cerographic Atlas of the United States, 1842-45; A System of Geography for the Use of Schools, 1844. Died Dec. 23, 1871.

MORTARA, Giorgio, Italian statistician; b. Mantua, Apr. 4, 1885; s. Lodovico and Clelia (Vivanti) M.; D.Law; m. Laura Ottolenghi, Apr. 4, 1919; children—Alberto, Guido, Valerio, Marcella. Prof. statistics U. Messina, Milan and Rome; cons. Brazilian Inst. Geography and Statistics; prof. hon. causa U. Brazil. Mem. dei Lincei Acad., Internat. Inst. Statistics and Polit. Economy, Internat. Union for Study Population (hon. pres.). Contbr. writings on demography, statistics, econs. Address: via Oglio 9, Rome, Italy.

MORTENSEN, Otto Axel, Am. med. educator; b. Milw., June 4, 1902; s. Morton Hansen and Anna (Petersen) M.; B.S., U. Wis., 1927, M.S., 1928, M.D., 1929; m. Lila Alvilda Embretson, Jan. 1, 1931; children—Margaret (Mrs. William H. Wolberg), Charles Otto, Peter Andrew. Faculty anatomy U. Wis., Madison, 1930—, prof., 1946-55, prof., chmn. dept. anatomy, 1955—, asst. dean Med. Sch., 1949-52, asso. dean, 1952—. Mem. Gov.'s Edn. Adv. Com., 1955—; cons. med. edn. San Marcos U., Lima, Peru, 1957-58; dir. Madison Gen. Hosp. Cons. career research awards com. NIH, 1962—. Mem. State Med. Soc. Wis. (hon.), Phi Beta Kappa, Sigma Xi, Alpha Omega Alpha, Phi Kappa Phi, Phi Chi. Research on relations of cerebrospinal fluid to peripheral nerves and lymphatics, electromyography of motor unit. Home: 3201 Lake Menodta Dr., Madison, Wis. 53705.*

MORTENSEN, Raymond A(rchie), Am. chemist; b. Marquam, Ore., Sept. 17, 1896; s. William and Margaret (Jensen) M.; A.B., Pacific Union Coll., 1919; M.S., U. So. Cal., 1925; Ph.D., Stanford, 1933; postgrad. U. Chgo., 1938; m. Marion Paap, May 31, 1923; 1 son, John William. Instr. physics, chemistry Pacific Union Coll., 1919-24, prof. chemistry, 1925-37; prof. biochemistry Loma Linda U. (Cal.), 1938—, head dept., 1939—. Fellow A.A.A.S.; mem. Am. Chem. Soc., Sigma Xi, Phi Lambda Upsilon, Phi Kappa Phi. Asso. editor Med. Arts and Scis., 1947—. Research on photochem. reactions, behavior of lead in animal body, metabolism of glutathione, cholesterol, and pyruvate. Contbr. sci. jours. Home: 25095 Prospect Av., Loma Linda, Cal. 92354.*

MORTENSEN, Theodor, Danish zoologist; b. Harlose, Denmark, Feb. 22, 1868; s. John G. and Petra (Jensen) M.; ed. Frederiksborg, Denmark, U. Copenhagen (Denmark); m. Valborg Blomberg, Apr. 10, 1901; 1 son, 1 dau. Asst., Danish Biol. Sta., 1895-96, later Zool. Mus. Copenhagen; mem. Marine Zool. Expdns. to Siam, 1899-1900, W.I., 1905-06, Pacific, 1914-16, Kei Islands, 1922, Java, Mauritius, S. Africa, St. Helena, 1929-30, Red Sea, 1936, 37; dir. dept. invertebrates Zool. Mus., Copenhagen, 1917-35. Recipient Prize Czar Nicolai II, Internat. Zool. Congress Monaco, 1913, Linnean gold medal Linnean Soc. London, 1951. Author: Echinoidea of the Danish Ingolf Expedition, I-II, 1903-07; Swedish South Polar, 1910; German, 1909; Echinodermenlarven of Plankton Expedition, 1898; Ctenophora of the Danish Ingolf Expedition, 1912; Studies of the Development and Larval Forms of Echinoderms, 1921; Handbook of the Echinoderms of the British Isles, 1927; Monograph of the Echinoidea, I-V, 1928-52; Discovery Reports, Echinoidea and Ophiuroidea, 1936; also numerous articles on echinoderms. Died Apr. 3, 1952.

MORTIMER, Clifford Hiley, limnologist, oceanographer; b. Whitchurch, Somerset, Eng., Feb. 27, 1911; s. Walter Herbert and Bessie (Russell) M.; B.S., U. Manchester, 1932, D.Sc., 1946; Ph.D., U. Berlin, 1935; m. Ingeborg Margarete Closs, Mar. 5, 1936; children—Christine Ann, Alison Margaret. Mem. staff Freshwater Biol. Assn., 1935-41, 46-56, v.p., 1966—; with Royal Naval Sci. Service, Oceanographic Group at Admiralty Research Lab., Teddington, 1941-46; sec. Scottish Marine Biol. Assn., dir. Millport Marine Sta., 1956-66; distinguished prof. zoology, dir. Center For Gt. Lakes Studies U. Wis., Milw., 1966—. Brittingham vis. prof. U. Wis., Madison, 1962-63. Recipient Naumann medal Internat. Assn. Limnology and Oceanography. Fellow Royal Soc., 1949. Research, publs. in environmental factors which control life in natural waters, lakes and oceans, oscillatory water movements asso. with internal boundary waves on density discontinuities within the water body. Office: 3230 E. Kenwood Blvd., Milw. 53201.*

MORTON, Avery Adrian, Am. chemist; b. St. Lawrence, S.D., Nov. 27, 1892; s. Levi P. and Amanda (McLane) M.; A.B., Cotner Coll., 1913; postgrad. U. Chgo., Harvard; Ph.D., Mass. Inst. Tech., 1924; m. Zelma Owen, Sept. 5, 1925; children—Mary (Mrs. Joachim A. Weissfeld), Elizabeth (Mrs. John A. Germer). Faculty, Mass. Inst. Tech., Cambridge, 1919—, prof. chemistry, 1940-57, prof. emeritus, 1957—. Staff, Govt. Rubber Program for Research on Polymerization, 1943-54. Mem. Am. Chem. Soc., Am. Acad. Arts and Sci., Am. Inst. Chemists, A.A.A.S., N.Y. Acad. Scis. Author: Laboratory Techniques in Organic Chemistry, 1938; Chemistry of Heterocyclic Compounds, 1946; Solid Organoalkali Metal Reagents, 1964; also numerous articles. Research on preparation and reactions solid organosodium compounds. Home: 182 Standish Rd., Watertown, Mass. 02172. Office: Mass. Inst. Tech., Cambridge, Mass. 02139.*

MORTON, Conrad Vernon, Am. botanist; b. Fresno, Cal., Oct. 24, 1905; s. Walter Crow and Noma (Bartholomew) M.; B.A., U. Cal. at Berkeley, 1928. Aid, div. plants, U. S. Nat. Mus., Smithsonian Instn., Washington, 1928-38, asst. curator, asso. curator, 1938-48, curator div. ferns, 1948—; asso. editor Am. Fern Jour., 1940-47, editor-in-chief, 1948-61, asso. editor, 1962—; editor for pteridophyta Biol. Abstracts, 1946—. Guggenheim fellow, 1954. Mem. Phi Beta Kappa, Sigma Xi, Phi Sigma. Contbr. numerous articles in field. Research in systematics of ferns of U. S. and tropical Am.; studies in gesneriaceae and solanaceae. Home: 2480 16th St. N.W., Washington 20009. Office: Smithsonian Instn., Washington 20560.*

MORTON, Donald John, Am. plant pathologist; b. Bklyn., Jan. 11, 1931; s. Elwood S. and Gladys (Hassler) M.; B.S. with honors and distinction, U. Del., 1952; M.S., La. State U., 1954; Ph.D., U. Cal. at Berkeley, 1957; m. Ann Mayo Tilden, Aug. 16, 1953; children—Saundra Kay, Donald John, Mary Ann. Asst. prof. N.M. State U., University Park, 1957-58; asst. prof. N.D. State U., Fargo, 1959-61; plant pathologist U. S. Dept. Agr., Tifton, Ga., 1961-65; asso. prof. U. Del., Newark, 1965—. Mem. Am. Phytopath. Soc., Mycol. Soc. Am., Air Pollution Control Assn. Research and numerous articles on devel. method for adding nematode killing chems. to fields by irrigation water; discovered particles causing a virus disease of artichode; worked out technique for rapidly examining barley embryos for smut infection; studied serological methods for distinguishing between certain plant pathogenic microorganisms. Home: 13 Hillvale Circle, Wilmington, Del. 19808. Office: Dept. Plan Pathology, U. Del., Newark 19711.*

MORTON, George Ashmun, Am. electronic engr.; b. New Hartford, N.Y., Mar. 24, 1903; s. Walter H. and Laura (Johnson) M.; B.S. in Elec. Engring., Mass. Inst. Tech., 1926, M.S. in Physics, 1928, Ph.D. in Physics, 1932; m. Lucy M. Groat, Sept. 15, 1934; children—Walter G., George Ashmun, Grace M., Lewis H. Research asso. Mass. Inst. Tech., 1927-33; research engr. RCA Mfg. Co., Camden, N.J., 1933-41, research sect. head lab. div., Princeton, N.J., asso. dir. physics-chemistry lab., 1954-60, dir. Conversion Devices Lab., Electronic Components and Devices, 1961—. Recipient Zworykin award I.R.E., 1962; David Richardson award Optical Soc. Am., 1967. Fellow I.E.E.E. (mem. adminstv. com. Profl. Group on Nuclear Sci. 1962—, Overseas Premium award, 1937-38), Am. Phys. Soc.; mem. A.A.A.S., Am. Standards, Assn. (mem. nuclear standards bd. 1956—), Sigma Xi. Author: (with V. K. Zworykin) Television, 1940; (with others) Electron Optics and the Electron Microscope, 1945. Contbr. articles on tv, electronics, electron optics, infrared imaging, computers, nucleonics to tech. jours. Patentee in above fields. Home: 200 Library Pl. Office: RCA Labs., Princeton, N.J. 08540.*

MORTON, Harry Edward, Am. microbiologist; b. Wayne, Mich., Aug. 4, 1906; s. Edward and Stella (Hasselback) M.; student Eastern Mich. U., 1923-26; B.S., U. Mich., 1930, M.S., 1931, Sc.D., 1936; m. Leah Spencer, Aug. 25, 1934; children—David Harry, Roger Spencer. Bacteriologist, Parke, Davis & Co., 1926-27; faculty U. Pa. Sch. Medicine, Phila., 1931—, prof. bacteriology, 1950—, sr. research microbiologist Inst. for Coop. Research, 1952-61; chief microbiology div. Pepper Lab. Clin. Medicine, Hosp. of Univ. of Pa., 1967—. Mem. Am. Acad. Microbiology, Am. Assn. Pathologists and Bacteriologists, Am. Chem. Soc., Am. Pub. Health Assn., Am. Soc. Microbiologists, Biol. Photog. Assn., Electron Microscope Soc. Am., Sigma Xi. Contbg. author: Bacterial and Mycotic Infections of Man, 1958—; also numerous articles. Devised safer method for cultivation bacteria under anaerobic conditions, invented stainless steel closures for replacement cotton plugs in culture vessels for bacteriological use; described developmental cycle of diphtheria bacillus; research on pleuro-pneumonia-like organisms. Home: 4114 School Lane, Drexel Hill, Pa. 19026. Office: Hosp. of U. Pa., Phila. 19104.*

MORTON, Henry, Am. chemist; b. N.Y.C., Dec. 11, 1837; s. Henry Jackson and Helen (McFarlan) M.; A.B., U. Pa., A.M., 1857; Sc.D.; Ph.D., LL.D., Princeton; m. Clara Whiting Dodge, 1863. Pub. translation hieroglyphic text of Rosetta Stone, 1859; condr. expdn. to observe and make photographs of total solar eclipse, Ia.; 1868; prof. chemistry Phila. Dental Coll., 1863, U. Pa., 1868-70; resident sec. Franklin Inst. and editor of its journal, 1864-70; pres. Stevens Inst. Tech., 1870-1902. Mem. Nat. Acad. Scis. Research and numerous articles on flourescence, galvanic batteries, pneumatic pyrometer, X-rays, photometry, liquid air; 1st to show certain feature noted in photographs in eclipses was artifact of photog. process, not optical phenomenon; developed 1st mech. engring. curriculum in U. S., new plastic materials for filling teeth; foremost patent witness for sci. topics of his day. Died May 9, 1902.

MORTON, Henry H., Am. surgeon; b. Hoboken, N.J., 1861; s. Edmond Ludlow and Josephine (Holdich) M.; M.D., L.I. Coll. Hosp., 1882; postgrad. in Prague, Munich, Berlin, Vienna and Paris. Emeritus prof. genito-urinary diseases L.I. Coll. Hosp.; attending genitourinary surgeon St. Peters Hosp., Bklyn. Mem. Com. on Venereal Diseases, Surgeon Gen.'s Office, World War I. Author: Genito-Urinary Diseases and Syphilis, 1902 (6 edits.). Authority on venereal diseases. Died May 3, 1940.

MORTON, Howard McIlvain, Am. oculist, aurist; b. Chester, Pa., May 23, 1868; s. Charles J. and Annie E. (Coates) M.; B.S., Lafayette Coll., 1888, M.S., 1891; M.D., U. Pa., 1891; postgrad. Royal Ophthalmic Hosp., London, Charite Hosp., Berlin; m. Lucretia Jarvis, Dec. 9, 1891. Interne, St. Luke's Hosp., Bethlehem, Pa., 1891; practiced medicine, Phila., from 1891; prof. diseases of eye and ear Hamline U., 1893-95; formerly oculist and aurist, Mpls. City and Swedish hosps.; chief dept. eye surgery, Mpls. Gen. Hosp.; chief eye and ear surgeon Wells Meml. Clinic, St. Barnabas, Fairview hosps.. Fellow A.C.S.; mem. Internat. Congress Ophthalmologists, A.M.A., Am. Acad. Ophthalmology and Oto-Laryngology, Assn. Mil. Surgeons U. S., Internat. Soc. for Prevention Blindness. Author: Visual Neurology; (monograph) also over 100 articles. Inventor of Morton perimeter and other instruments. Died July 19, 1939.

MORTON, J. E., New Zealand zoologist; b. Morrinsville, New Zealand, July 18, 1923; s. Ronald Bampton and Ethel (Sceats) M.; B.Sc., Auckland U. Coll., 1945, M.Sc., 1947; Ph.D., London U., 1952, D.Sc., 1959; m. Patricia Helen Lees, Mar. 1, 1956; children—Clare Patricia, Robert Paul. Jr. lectr. U. Auckland, 1948-50; lectr. zoology U. London, 1952-59; prof., head dept. zoology U. Auckland, 1960—. Author: Molluscs, 1958; also numerous articles. Research on morphology adaptations and evolution of Molluscs, ecology of Molluscs and other sea shore invertebrates, ecology and zonation patterns of intertidal shores, biology of coral shores in S.W. Pacific. Home: 120 Aberdeen Rd., Castor Bay, Auckland. Office: U. Auckland, Auckland, New Zealand.*

MORTON, Jack A., Am. elec. engr.; b. St. Louis, Sept. 14, 1913; s. Mack Ray and Minette (Hirshfeld) M.; B.Sc., Wayne U., 1935, D.Sc. (hon.), 1956; M.Sc., U. Mich., 1936; postgrad. Columbia, 1937-41; Ph.D. (hon.), Ohio State U., 1954; m. Helen Read, May 27, 1938; children—Kim, Mack. Asst. dir. electronic apparatus devel. Bell Telephone Labs., Inc., Murray Hill, N.J., 1952-53, dir. transistor devel., 1953-54, dir. devel. solid state devices, 1954-55, dir. device devel., 1955-58, v.p., 1958—. Mem. planning bd. Hillsborough (N.J.) Twp. Recipient Univ. Alumni award Wayne U., 1951, Certificate of Merit, 1958; Distinguished Alumnus citation U. Mich., 1953. Fellow I.E.E.E. (David Sarnoff award 1965); mem. MacKenzie Honor Soc., Sigma Xi, Eta Kappa Nu, Alpha Delta Psi, Phi Kappa Phi, Tau Beta Pi. Advances in communications tech., including research on coaxial cable and microwave amplifiers; devel. improved radar receivers; inventor microwave vacuum tube. Home: Box 1-A, South Branch, N.J. 08881. Office: Bell Telephone Labs., Murray Hill, N.J. 07971.*

MORTON, John Jamieson, Jr., Am. physician; b. Holyoke, Mass., May 19, 1886; s. John Jamieson and Nellie (Taylor) M.; B.A., Amherst, 1907, Sc.D., 1947; M.D., Johns Hopkins U., 1913; m. Nancy Barnard, Oct. 7, 1919; 1 son, John Henderson. Practiced orthopedic surgery, Boston, 1919-21; asst. prof. surgery Yale, 1921-23; prof. surgery U. Rochester (N.Y.) Sch. Medicine and Dentistry, 1923-53, surgeon in chief, 1923-53. Dir. Atomic Bomb Casualty Commn., Japan, 1953-54; bd. sci. advisers Jane Coffin Childs Cancer Research, 1939-65. Recipient Albert Kaiser medal, 1954. Mem. Am. Soc. Control Cancer (pres. 1938-41), Soc. Clin. Surgeons (pres. 1939-41), Am. Surg. Soc. (v.p. 1950), Am. Bd. Surgery Founders Group, Phi Beta Kappa. Research, numerous publs. on intestinal obstruction, sympathetic nervous system, diseases of blood vessels, diseases of bone, radiation, cancer. Home: 1915 Westfall Rd., Rochester, N.Y. 14618.*

MORTON, Maurice, chemist; b. Latvia, June 3, 1913; s. Mendel and Fanny (Chisling) M.; B.S., McGill U., Montreal, Que., Can., 1934, Ph.D., 1945; m. Lilian Rosenbloom, Dec. 21, 1933; children—Jay Dennis, Ruth, John Alex. Came to U. S., 1948, naturalized, 1954. Chief chemist Canadian Johns-Manville Co., Asbestos, Que., 1936-41; chief chemist Congoleum Can., Ltd., Montreal, 1942; prof., head chemistry dept. Sir George Williams Coll., Montreal, 1945-48; asst. dir. rubber research U. Akron (O.), 1948-53, dir. Inst. Rubber Research, prof. polymer chemistry, 1953-65, dir. Inst. Polymer Sci., prof. polymer chemistry, 1965—. Cons. to industry, govt. agys. Recipient Naturalized Am. award Akron Bar Assn., 1960. Mem. Am. Chem. Soc., A.A.A.S., Sigma Xi, Alpha Chi Sigma. Author: Introduction to Rubber Technology, 1959. Research, numerous publs. on processes by which important synthetic rubbers are formed from simple compounds by chem. reaction polymerization. Home: 500 Delaware Av., Akron, O. 44303.*

MORTON, Mortimer Erving, Am. physician; b. Columbus, O., July 24, 1919; s. Ernest and Beatrice (Becher) M.; A.B. in Chemistry, U. Cal. at Los Angeles, 1940; M.A., U. Cal. at Berkeley, 1941; Ph.D., 1942; M.D., U. So. Cal., 1945; m. Helene Steinberg, June 12, 1965; 1 dau., Julia Diane. Staff, U. S. Naval Med. Research Inst., 1946, Crocker Radiation Lab., U. Cal. at Berkeley, 1946-47; clin. radiochemist U. S. Naval Radiation Lab., San Francisco, 1946-48; dir. radioisotope units VA Hosp., Van Nuys, Cal., 1948-50, Long Beach, Cal., 1950-55; asso. prof. biophysics U. Cal. at Los Angeles, 1950-55, clin. instr. medicine, 1952-55; practice nuclear medicine, Long Beach, Garden Grove, Cal., Beverly Hills, Cal., 1955—. Mem. Soc. for Exptl. Biology and Medicine, Sigma Xi. Contbr. articles to tech. jours. Research, publs. on application radioisotopes to clin. medicine, diagnosis, therapy, fission product and plutonium radio-chemistry, thyroid metabolism as studied with radioactive iodine and bromine, pyrimidine metabolism. Office: 12511 Brookhurst St., Garden Grove, Cal. 92540.*

MORTON, Newton Ennis, Am. geneticist; b. Camden, N.J., Dec. 21, 1929; s. Newton and Laura (Jones) M.; B.A., U. Hawaii, 1951; M.S., U. Wis., 1952, Ph.D., 1955; m. Nancy T. Okazaki, Feb. 12, 1949; children—Teru, Peter, Amy, John, Robert. Geneticist, Atomic Bomb Casualty Commn., Hiroshima, Japan, 1952-53; faculty U. Wis., Madison, 1956-62, asso. prof., 1960-62; prof., chmn. dept. genetics U. Hawaii, 1962-65, prof., 1962—; dir. genetics research project Hawaii Dept. Health, 1958-59; dir. Hospedaria de Imigrantes, Brazil, 1962-63. Nat. Cancer Inst. fellow, 1955-56. Recipient Lederle Med. Faculty award, 1958. Mem. Genetics Soc. Am., Am. Soc. Human Genetics, (1st William Allan Meml. award 1962), Am. Soc. Naturalists, A.A.A.S., Sigma Xi. Author: (with C. S. Chung, M. P. Mi) Genetics of Interracial Crosses in Hawaii, 1966; also numerous articles. Devel. and application methods population genetics to human material including segregation analysis and linkage, effects inbreeding, polymorphisms, population structure, formal genetics diseases and blood groups. Home: 2828 Kahawai St., Honolulu 96814. Office: 2538 The Mall, Honolulu 96822.*

MORTON, Richard, English physician; b. Ribbesford, Eng.; baptized July 30, 1637; s. Robert Morton; B.A., Oxford, 1657, M.A., 1659, M.D., 1670; children—Richard, Sarah, Marcia. Vicar of Kinver, Eng., 1659-62; physician-in-ordinary to king; fellow Coll. Physicians; censor, 1690, 91, 97. Author: Phthisiologia: seu Exercitationes de phthisi (clin. observations on diseases of lung, effects of prolonged jaundice, fevers, gout), 1689. Specialist on chest disorders; pioneer in use of quinine. Died Surrey, Eng., Aug. 30, 1698.

MORTON, Richard Alan, Brit. biochemist; b. Liverpool, Eng., Sept. 22, 1899; s. John and Ann (Humphreys) M.; B.Sc. with honors, U. Liverpool, 1921, Ph.D., 1923, D.Sc., 1928; hon. D.Sc., U. Coimbra, U. Wales, U. Trinity (Dublin); m. Myfanwy Heulwen Roberts, July 21, 1926; 1 dau., Ruth Gillian (Mrs. Peter S. Lewis). Faculty, U. Liverpool, 1924—, Johnston prof. biochemistry, 1944-66, hon. research asso. in food sci., 1967—; vis. prof. Ohio State U., 1930; Loeb lectr. St. Louis U., 1960; Babcock lectr. U. Wis., 1966. Chmn., Brit. Com. on Chem. Edn., 1961-65, Food Additives and Contaminants Com., 1963—. Fellow Royal Soc., 1950 (mem. council 1949-61); mem. Biochem. Soc. (chmn. 1959-61, hon.), Nutrition Soc., Chem. Soc., Soc. Chem. Industry, Faraday Soc., Soc. Analytical Chemists. Author: Radiation in Chemistry, 1928; Absorption Spectra of Vitamines and Hormones, 1942; Biochemistry of Quinones, 1965; Protein Utilization by Poultry, 1967; also numerous articles. Research on ultraviolet absorption spectroscopy of organic compounds; vitamins A, D, E; quinones, polyprenal alcohols. Home: 39 Greenhill Rd., Liverpool, Eng.*

MORTON, Samuel George, Am. physician, naturalist; b. Phila., Jan. 26, 1799; s. George and Jane (Cummings) M.; M.D., U. Pa., 1820; M.D., U. Edinburgh, 1823; m. Rebecca Pearsall, Oct. 23, 1827; 7 children, including James St. Clair. Became mem. Acad. Natural Scis. Phila., circa 1820, recording sec., 1825-29, corr. sec., 1831, v.p., 1840, pres., 1849-51; prof. anatomy Pa. Med. Coll., 1839-43. Author: Synopsis of the Organic Remains of the Cretaceous Group of the U. S., 1834; Illustrations of Pulmonary Consumption, 1834; Crania Americana, 1839; Crania Egyptiaca, 1839; Human Anatomy, Special, General and Microscopic, 1849. Collected large number human skulls for comparative study, concluded that races of man were of diverse origin; credited with describing new species of hippopotamus; advised fresh-air treatment for Tb patient. Died Phila., May 15, 1851.

MORTON, Thomas George, Am. surgeon; b. Phila., 1835; s. Samuel George and Rebecca Grellet (Pearsall) M.; M.D., U. Pa., 1856; practiced medicine, specializing in gen. surgery, Phila., 1856-60; established mil. hosps., 1861-65; apptd. commr. to erect state insane asylum, 1876; became chmn. com. on lunacy, 1886; named commr. state pub. charities, 1883. Mem. Pa. Anti-vivisection Soc. (pres. 1880), Pa. Soc. for Prevention of Cruelty to Children (v.p.). Research and publs. on blood transfusion, other med. topics; gave 1st satisfactory description of metatarsalgia (Morton's disease), 1876. Died 1903.

MORTON, William James, Am. neurologist; b. Boston, Mass., July 3, 1845; s. William Thomas Green Morton; grad. Harvard, 1867; M.D., Harvard Med. Sch., 1872; postgrad., Vienna, Austria, 1873-74. Physician in Capetown, S. Africa, 1874-76; practiced medicine, specializing in neurology, N.Y.C., from 1878; prof. diseases of mind and nervous system and of electrotherapeutics, N.Y. Post Grad. Med. Sch., 1890-1909. Convicted of using mails to defraud in connection with Canadian mining stock, 1913; pardoned by Pres. Wilson. Editor Jour. Nervous and Mental Diseasee, 1882-86. Pioneer in use of X rays for treatment of skin disorders and cancerous growths in U. S.; developed high frequency "static-induced" or "Morton current," which by producing x-rays was valuable to medicine; contributed to modern practice of iontophoresis; research in electrotherapeutics. Died Miami, Fla., Mar. 26, 1920.

MORTON, William Thomas Green, dentist; b. Charlton, Mass., Aug. 9, 1819; s. James and Rebecca (Needham) M.; studied dentistry Coll. Dental Surgery, Balt., 1840-42; also studied dentistry Harvard Med. Sch.; M.D. (hon.), Washington U. of Medicine, Balt.; m. Elizabeth Whitman, May 1844; 4 children, including William James. While experimenting with mesmerism and nostrums became involved with sulfur ether which he later connected and linked to use in dental anaesthesia; used ether in drops as local anaesthetic during filling of a tooth, 1844, extracted a tooth with this method, 1846; etherized a patient from whom Dr. John Warren removed vascular tumor from left side of neck, 1846; applied for patent to protect his rights, received patent for 14 years, 1846; issued weekly circular Morton's Letheon, 5 editions, under his direction, 1846; recipient Montyon prize French Acad. Medicine, 1847; although others also discovered anaesthesia, he convinced surg. world of value of discovery of a surg. anaesthetic; during Civil War worked in various hosps.; made many improvements on crude methods of attaching false teeth. Elected to Hall of Fame for Gt. Americans, 1920. Author: Remarks on the Proper Mode of Administering Sulphuric Ether by Inhalation, 1847; On the Loss of the Teeth and the Modern Way of Restoring Them, 1848; On the Physiological Effects of Sulphuric Ether, and Its Superiority to Chloroform, 1850. Died NY.C., July 15, 1868.

MORUZZI, Giuseppe, Italian physiologist; b. Campagnola, July 30, 1910; s. Giovanni and Bianca (Carbonieri) M.; M.D., Faculty of Medicine; m. Vittoria Venturini, Oct. 11, 1941; children—Giovanni, Paolo. Asst. in physiology, Rockefeller Inst. scholar Parma and Bologna, Brussels and Cambridge, 1937-39; vis. prof. Northwestern U., Chgo., 1948-49; instr. Sienna Inst. Physiology, 1942, Parma, 1945, prof. physiology Ferrara, 1947, Pisa, 1948—. Mem. Am. Physiol. Soc. (hon.), Am. Philos. Soc., Argentine Biology Soc., Norske Vidanskaps-Akademi, Oslo, Nat. dei Lincei Acad. Research in reticular formation, sleep, neurophysiology, cerebellum. Address: via S. Zeno 31, Pisa, Italy.

MORVEAU, Baron Louis Bernard Guyton de, see Guyton de Morveau, Baron Louis Bernard.

MOSANDER, Carl Gustav, Swedish chemist; b. Kalmar, Sweden, Sept. 10, 1797; army surgeon; prof. chemistry and mineralogy, Caroline Med. Inst., Stockholm; asst. to Berzelius. Research on so-called rare earth elements; discovered cerium and lanthanum, 1839, didymium, 1842, erbium, terbium, 1843. Died Angsholm, Sweden, Oct. 15, 1858.

MOSANSKY, Aristid, Czechoslovakian zoologist, ornithologist; b. Bardejov, Czechoslovakia, Oct. 12, 1928; s. Severin and Edita (Uhlig) M.; Ingeneer, High Sch. Forest and Wood Engring., Zvolen, 1953; Candidate Biol. Scis. (hon.); m. Noémi Szekerák, Dec. 27, 1952; 1 son, Ladislav. Head, zool. dept. East-Slovak Mus., Kosice, Czechoslovakia, 1955—. Mem. mammalogical sect., staff Nat. Mus., Prague. Mem. coms. Slovak Zool. Soc. at Slovak Acad. Sci., Czechoslovak Biol. Soc. at Czechoslovak Acad. Sci, Czechoslovak Ornith. Soc. in Prague. Author: (with Ferianc O. and Feriancová-Hanák) Illustration to the Monography of Vertebrates of Slovakia, parts II, IV; coauthor: Fauna of Czechoslovakia, Birds I and III; also articles. Contbd. to knowledge of hist. devel. and present situation of geog. and ecol. extension of continental vertebrates and knowledge of their biometricy and taxonomy in Western Karpats; knowledge of anthropogenical influences on avifaunes and fauna in Western Karpats. Home: 42/Iv.tr. Slov. Národ. Povstainia, Kosice, -Nové Mesto. Office: 5. Hviezdoslavova, Kosice, Czechoslovakia.

MOSBY, Hakon, Norwegian geophysicist; b. Kristiansand, July 10, 1903; s. Salve and Mette C. (Nodeland) M.; Ph.D., U. Oslo; m. Alfhild Mowinckel, Mar. 1, 1930; children—Mette Marie, Kaya. Asst., Kridtjob Nansen, 1923; asst. prof. Bergen Geophys. Inst., 1927; prof. phys. oceanography Bergen Mus., 1947, U. Bergen, 1948; dir. Bergen Geophys. Inst., 1948-58, 63—; pres. Norway Geophysics Commn., 1949-50, 53-54, 56-57, 63—; dean faculty sci. U. Bergen, 1954-59. Pres. oceanographic sub-com. NATO, 1960—. Pres., Internat. Assn. Phys. Oceanography, 1954-60. Research on Arctic and Antarctic waters, phys. problems of sea including heat balance, turbulent diffusion, vertical convection, bottom water formation, evaporation, sea-air interaction. Home: Kalfarveien 59. Office: Allégaten 70, Bergen, Norway.

MOSCA, Gaetano, Italian polit. scientist; b. Palermo, Italy, Apr. 1, 1858; student law U. Palermo. Became prof. constl. law Palermo U., 1885, Rome, 1888, Turin, Italy, 1896, Rome, 1922; ret., 1933; mem. Italian Chamber of Deps., 1908; sec. of state for colonies Salandara Cabinet, 1914-16; royal appointment as life senator, 1919. Mem. Royal Acad. Turin. Author: Sulla teorica dei governi e sul governo parlamentare, 1884; Elementi di scienza politica, 2 vols., 1896-1923; Die herrschende Klasse, 1950; Saggi di storia delle dottrine politiche, 1927. A leader of Elitists who believed state is always ruled by small governing class, regardless of polit. system, especially in highly developed socs.; postulated cyclical theory of govt. Died Rome, Nov. 8, 1941.

MOSCHOPULOS, Manuel, Byzantine mathematician, humanist; b. Constantinople; flourished 1282-1328; disciple of Maximos Planudes; outstanding philologist of his time, thus indirectly a founder of Western humanism. Author scholia on 1st two books of Iliad, on Hesiod, Pindar, Euripides, Theocritus, other authors, also a grammatical catechism popular in West during early Renaissance, a treatise on invention of magic squares (earliest Greek treatise which explained math. properties of sqs. and method of bldg. them without reference to Arabic math., written at request of Nicolar Rhabdas).

MOSCICKI, Ignace (Ignacy), chemist; b. Mierzanów near Plock, Poland, Dec. 1, 1867; student Poly. Sch., Riga, Latvia; returned to Warsaw, Poland after grad.; fled to Eng., 1892; became asst. prof. electrochemistry and electrophysics U. Freiburg (Switzerland), 1897; named prof. electrochemistry Poly. Sch. Lwów, Poland, 1912; founder chem. research insts. and factories to manufacture synthetic fertilizers, Poland; 3d pres. Republic of Poland, 1926-39; exiled and became Swiss citizen, 1939. Research and over 600 patents on method of manufacturing synthetic nitric acid in highly charged electric furnace, manufacture chem. fertilizers, apparatus to provide mountain air for lowland apartment dwellers (helpful for patients with pulmonary complaints). Died Versoix, Switzerland, Oct. 2, 1946.

MOSCONA, Aron Arthur, biologist; b. Haifa, Israel, July 4, 1922; s. David and Lola (Krochmaal) M.; M.Sc., U. Jerusalem (Israel), 1947, Ph.D., 1950; m. Malka H. Kampinsky; 1 dau., Anne. Came to U. S., 1958, naturalized, 1964. Research fellow Strangeways Research Lab., Cambridge, Eng., 1950-52; asso. prof. physiology Hebrew U., Jerusalem, 1953-54; vis. investigator Rockefeller Inst., N.Y.C., 1955-57; faculty U. Chgo., 1958—, prof. dept. zoology, 1960—; vis. prof. Stanford, 1959, U. Montreal (Que., Can.), 1960, U. Palermo (Italy), 1966. Lillie fellow Marine Biol. Lab., Woods Hole, Mass., 1960. Fellow Internat. Soc. Cell Biologists, N.Y. Acad. Scis.; mem. Soc. Zoologists, Growth Soc., Tissue Culture Assn., Internat. Inst. Embryology, Am. Soc. Anatomists. Editor: Experimental Cell Research, Current Topics in Developmental Biology. Research, publs. on synthesis of tissues from cells outside body and on genetic regulatory processes in cells, as means of understanding mechanisms of devel. and differentiatism in tissues, organs and organisms. Office: Dept. Zoology, U. Chgo., Chgo. 60637.*

MÖSE, Josef Richard, Austrian physician; b. Vienna, Austria, Oct. 26, 1920; s. Josef and Anna (Weikert) M.; ed. univs. Göttingen, Germany, Berlin, Germany, Prague, Czechoslovakia; Dr.med., 1944; m. Gisela Korbe, Dec. 9, 1944; 1 dau., Angelika. With Hygiene Inst., U. Graz (Austria), 1945—, prof., dir., 1961—, rector, 1966-67. Research and publs.

on microbiology, cancer, antibiotics. Home: 6 Kaltenbrunngasse, Graz, Austria.*

MOSEI, Muramatsu Kudayu, Japanese mathematician; flourished 1663; studied under Hirada Yasuhide; retainer of Hsano, Lord of Akō. Author: Sanso, 1663-84; Sampō Chokkai. Studied polygons mensuration of circle; estimated pi to be 3.14.

MOSELEY, Henry, English mathematician; b. Newcastle-under-Lyme, Eng., July 9, 1801; s. William W. and Margaret (Jackson) M.; ed. Naval Sch., Portsmouth, Eng.; B.A., St. John's Coll., Cambridge, Eng., 1826, 7th Wrangler, M.A., 1836; LL.D., 1870; m. Harriet Nottidge, Apr. 20, 1835; a son, Henry Nottidge. Ordained deacon, 1827, priest, 1828; became curate at W. Monkton, 1828; chaplain King's Coll., London, 1831-33, prof. natural philosophy, 1831-44; one of 1st royal insps. of normal schs., 1844; resident canon at Bristol Cathedral, 1853; named vicar of Olveston, Eng., 1854; named chaplain to Queen Victoria, 1855. Fellow Royal Soc., 1839; mem. Council Mil. Edn., Instn. Naval Architects (v.p.), French Acad. Scis. (corr.). Author: A Treatise on Hydrostatics, 1830; A Treatise on Mechanccs, Applied to the Arts . . . , 1834; Lectures on Astronomy, 1839; Mechanical Principles of Engineering and Architecture, 1843; Astro-Theology . . . , 1847; also numerous articles. Developed formula for calculating dynamic stabilities of war ships; studies on motion of lead on roof of Bristol Cathedral under changing temperatures (led to his theory that movement of glaciers could be explained in same way). Died Olveston, Jan. 20, 1872.

MOSELEY, Henry Gwyn Jeffrey, English physicist; b. Weymouth, Eng., Nov. 23, 1887; s. Henry Nottidge Moseley; ed. Trinity Coll., Oxford (Eng.) U., Millard scholar, honors in natural scis., 1910. Lectr., demonstrator in physics U. Manchester (Eng.), also studied under E. Rutherford, 1910-12, resigned to devote himself to research; lieutenant in Royal Engineers, 1914. Research in field of radioactivity and on x-ray spectra; discovered relationship between x-ray spectra and atomic numers (Law of Atomic Numbers), 1914; measured wavelengths and showed units to be equal to atomic number, varying from 1 in hydrogen to 92 in uranium; showed that only 3 elements were missing between aluminum and gold, also predicted their spectra; his results (completed by Siegbahn and Barka) clearly indicated that main properties of an element are not determined by atomic weight, but by a whole number defining its nuclear charge (supplied the structure of electron rings of atom); this discovery landmark in spectrum analysis and devel. of periodic law of elements. Killed in action during Brit. campaign, Gallipoli, Turkey, Aug. 10, 1915.

MOSELEY, Henry Nottidge, Brit. naturalist; b. Wandsworth, Eng., Nov. 14, 1844; s. Henry Moseley; ed. Harrow; B.A., Oxford (Eng.) U., 1868; M.A., 1872; student medicine Univ. Coll., London, Eng.; LL.D., McGill U., Montreal, Can., 1844; m. Miss Jeffreys, 1881. With govt. expdn. to Ceylon, 1871, Challenger expdn., 1872-76; traveled to Cal., Ore., 1877; asst. registrar U. London, from 1879; Linacre prof. human and comparative anatomy Oxford U., from 1881. Fellow Royal Soc., 1877 (Royal medal, 1887, twice mem. council), 1879, Zool. Soc., Anthop. Inst., Linnean Soc., Royal Geog. Soc., Brit. Assn. Montreal (pres. sect. D 1884); mem. Marine Biol. Assn. Author: On Oregon, 1878; On the Structure of the Sylasteridae, 1878; Notes by a Naturalist on the Challenger, 1879. Contbr. papers to sci. publs. Research on invertebrates; discovered system of tracheal vessels in Peripatus that furnished new clue tp origin of tracheae; memoir of Peripatus constituted important contbn. towards knowledge of phylogeny of anthropods; investigations on living corals led to establishment of group of hydrocorallin; discovered eyes on shells of several species of chiton. Died Nov. 10, 1891.

MOSELEY, Herbert Frederick, Canadian surgeon; b. Cape Breton Island, N.S., Can., Jan. 28, 1906; s. Herbert Charles and Gertrude (Reid) M.; B.A., McGill U., 1926; postgrad. Oxford U., M.A., M.Ch., 1934, D.M., 1938; m. Pauline Stilwell, Aug. 17, 1934, (dec. Aug. 1936); 1 dau., Suzanne; m. 2d, Josephine Deems, May 26, 1945 (div. Dec. 1963); m. 3d, Helen MacArthur, Jan. 17, 1964. Rhodes scholar, Que., 1927; Harmsworth Sr. scholar Merton Coll., Oxford, 1930; Province Que. Travelling scholar, 1932; Hunt Travelling scholar Cambridge, 1936; asso. prof. surgery McGill U., Montreal, 19——; sr. surgeon Royal Victoria Hosp., dir. accident service. Mem. Am. Assn. Surgery of Trauma, Canadian, Que. orthopaedic assns., Soc. Internat. Chirurg de Orthopedie et Traumatologie, Royal Coll. Surgeons Eng. Author: Shoulder Lesions, 1945, 3d edit., 1968; Recurrent Dislocation of the Shoulder, 1961; An Atlas of Musculoskeletal Exposures, 1955. Editor: Textbook of Surgery, 3d edit., 1959; Accident Surgery, vols. 1-3, 1962, 64, 66. Clin., lab. investigations on shoulder, other joints. Home: 1460 McGregor St., Montreal 25. Office: 10 Surgical, Royal Victoria Hosp., Montreal, Que., Can.*

MOSELEY, Robert David, Jr., Am. radiologist; b. Minden, La., Feb. 29, 1924; s. Robert David and Lettie E. (Looney) M.; M.D., La. State U., 1947; m. Janet C. Watson, Mar. 15, 1947; children—Robert David III, Richard Havard, Marianne Lee. Staff mem. Los Alamos Sci. Lab., 1951-52, radiologist, asso. chief staff med. center, 1951-52; mem. staff dept. radiology U. Chgo., 1954——, prof., chmn. dept., 1958——. Tech. adviser U. S. delegation 2d Internat. Conf. Peaceful Uses of Atomic Energy, Geneva, 1958. Diplomate Am. Bd. Radiology. Fellow Chgo. Roentgen Soc. (sec.-treas. bd. trustees), Am. Coll. Radiology (chmn. com. pub. health, chancellor); mem. A.M.A., Assn. U. Radiologists (past pres.), Am. Roentgen Ray Soc., Radiation Research Soc., Radiol. Soc. N.Am., Inter Am. Coll. Radiology, A.A.A.S., Royal Soc. Medicine Eng. (affiliate), Sigma Xi, Sigma Nu, Phi Chi. Research in diagnostic radiologic instrumentation. Home: 4901 S. Greenwood Av., Chgo. 60615. Office: 950 E. 59th St., Chgo. 60637.*

MOSELEY, Vince, Am. physician; b. Orangeburg, S.C., Oct. 29, 1912; s. William Lawrence and Jessie (Vince) M.; student Clemson U., 1929-31; A.B., Duke, 1933, M.D., 1936; m. Matilda Holleman, Oct. 11, 1938; children—Robert Dwight, Julia C. (Mrs. Allan Strickland), Kelsey E., William Vince, Matilda Raine, Esther Jane, Lawrence, Selma. Fellow dermatology-syphilology Duke Sch. Medicine, 1939-40, asst. bacteriology, instr. phys. diagnosis and physiology, 1940; asso. in medicine, fellow Gastrointestinal Clinic, U. Pa. Sch. Medicine, 1940-41; chief med. service gen. hosp., Panama, 1942-45; asst. chief med. sect. Letterman Gen. Hosp., San Francisco, 1945-47; asso. Med. Coll. S.C., Charleston, 1947, faculty 1947-—, prof. medicine, 1949-—, dean clin. medicine, 1960-—, chmn. out-patient clinic, 1948-—, physician-in-charge Gastrointestinal Clinic, 1947-—; physician-in-chief Roper Hosp., Charleston, 1949-—. Med. cons. Naval Hosp., Charleston, 1948-—, Ft. Jackson Army Hosp., Columbia, S.C., 1949-—. Fellow A.C.P.; mem. Am. Clin. and Climatol. Assn., A.M.A., So. Med. Assn., Am. Fedn. for Clin. Research, Soc. Exptl. Biology and Medicine. Research, publs. on gastrointestinal diseases and antibiotic studies. Home: 51 E. Battery St., Charleston, S.C. 29401.*

MOSEMAN, John Gustav, Am. plant pathologist; b. Oakland, Neb., Dec. 7, 1921; s. John Gerhart and Bertha (Hopp) M.; B.S., U. Neb., 1943; M.S., Wash. State U., 1948; Ph.D., Ia. State U., 1950; m. Marjorie Jean Bell, May 31, 1948; children—David R., Barbara J., Thomas B. Plant pathologist N.C. State Coll., U. S. Dept. Agr., Raleigh, 1950-54, research plant pathologist Plant Industry Sta., Beltsville, Md., 1954-—. Mem. Am. Phytopath. Soc., Am. Soc. Agronomy, Am. Inst. Biol. Scis. Research, publs. on genes conditioning reaction of plants to pathogens and pathogenicity of plant pathogen to plant host; devel. host varieties resistant to plant pathogens. Home: 1918 Blackbriar St., Silver Spring, Md. 20903. Office: Plant Industry Sta., Beltsville, Md. 20705.*

MOSER, Helmut, German physicist; b. Heidelberg, Aug. 17, 1903; s. Alfred and Emilie (Hepting) M.; Ph.D., U. Heidelberg; m. Maria Westermann, Aug. 14, 1930; children—Volker, Gisela. Physicist, Fed. Inst. Physics and Tech., 1928-—. Mem. Internat. Orgn. Legal Metrology, Cons. Com. on Thermometry. Author: Forschung und Prüfung, 1962. Research and publs. on surface tension, phosphorescence, thermometry, calorimetry. Home: Jasperallee 15. Office: Bundesallee 100, Braunschweig, West Germany.

MOSER, Jurgen Kurt, mathematician; b. Konigsberg, Germany, July 4, 1928; s. Kurt E. and Ilse (Strehlke) M.; Dr. rer. nat., Universitat Gottingen, 1952; m. Gertrude Courant, Sept. 10, 1955; children—Nina, Lucy. Came to U. S., 1955, naturalized, 1958. Asst., Gottingen U., 1954-55, N.Y. U., 1956-57; faculty Mass. Inst. Tech., 1957-—, prof., 1960-—. Cons. Lincoln Lab., IBM; dir. Inst. Math. Scis., 1967. Sloan fellow, 1961. Fellow Am. Acad. Scis., Am. Math. Soc., Soc. Indsl. and Applied Math. Research in ordinary and partial differential equations and application, in particular to spectral theory, celestial mechanics, stability theory. Office: 251 Mercer St., N.Y.C. 10012.*

MOSER, Leo, mathematician; b. Vienna, Austria, Apr. 11, 1921; s. Robert and Laura (Feurstein) M.; B.Sc., U. Man., Can., 1944; M.A., U. Toronto, 1945; Ph.D., U. N.C., 1950; m. Eva Moser, Sept. 10, 1946; children—Barbara, Melanie, David, Sheryl. Faculty, U. Man., U. Toronto, U. N.C., Tex. Tech. Coll.; faculty U. Alta., Edmonton, 1951-—. Fellow Royal Soc. Can. Research, numerous publs. in number theory, combinational analysis. Home: 9015 138 St., Edmonton, Alta., Can.*

MOSES, Campbell, Jr., Am. physician; b. Pitts., Feb. 12, 1917; s. Campbell and Kirsten (Jensen) M.; B.S., U. Pitts., 1939, M.D., 1941; m. Lois Haseltine, Aug. 28, 1940; children—Campbell Warren, James Robison, Robert Haseltine, Ann Louise. Faculty, U. Pitts., 1942-—, asso. prof. medicine, 1949-—, dir. Addison H. Gibson Lab., 1948-65, dir. postgrad. medicine, 1960-68; med. dir. Am. Heart Assn., N.Y.C., 1968-—. Mem. Am. Soc. for Study Ateriosclerosis (past pres.), Am. Heart Assn. (chmn. com. on med. edn. 1962-66), Am. Physiol. Soc., Soc. for Exptl. Biology and Medicine, A.C.P., Council on Arteriosclerosis, Am. Heart Assn., Am. Diabetes Soc. Author: Atherosclerosis, Mechanisms as a Guide to Prevention, 1963; also numerous articles. Research on role dietary cholesterol and polyunsaturated fat in exptl. atherosclerosis, acute hypertension in acceleration exptl. atherosclerosis, contbn. changes in thyroid and adrenal function to hypercholesterolemia of aging, thyroid and thyroid analogues in preventing serum cholesterol increases with aging. Office: 44 E. 23d St., N.Y.C. 10010.*

MOSES, Lincoln Ellsworth, Am. statistician; b. Kansas City, Mo., Dec. 21, 1921; s. Edward Walter and Virginia (Holmes) M.; student San Bernardino Valley Jr. Coll., 1937-39; A.B., Stanford, 1941, Ph.D., 1950; m. Jean Runnels, Dec. 26, 1942; children—Katherine, James, William, Margaret, Elizabeth. Asst. prof. edn. Columbia Tchrs. Coll., 1950-52; faculty Stanford, 1952-—, prof. statistics, 1959-—, exec. head dept. statistics, 1964-—. Mem. halothane com. Nat. Acad. Scis.-NRC, 1963-—. Author: (with H. Chernoff) Elementary Decision Theory, 1959; (with R. V. Oakford) Tables of Random Permutations, 1963. Home: 120 Carmel Way, Portola Valley, Cal.*

MOSES, Montrose James, Am. cytologist; b. N.Y.C., June 26, 1919; s. Montrose Jonas and Dorothy (Herne) M.; B.S., Bates Coll., 1941; A.M. in Zoology, Columbia, 1942, Ph.D. in Zoology, 1949; m. Constance Roy, July 1949; children—Mollie Constance, Catherine Corcoran. Asso. cytochemist, dept. biology Brookhaven Nat. Lab., Upton, N.Y., 1948-52, cytochemist, 1952-55, guest biologist, 1955-64; vis. investigator Rockefeller Inst. Med. Research, 1954, asst., 1955, asso., 1956, asst. prof., 1957-59; asso. prof., dept. anatomy Duke U. Sch. Medicine, Durham, N.C., 1959-66, prof., 1966-—, dir. research tng. program, 1963-66. Vice chmn. Gordon Conf. on Cell Structure and Metabolism, 1957, chmn., 1958; participant 9th Internat. Congress for Cell Biology, Scotland, 1957, 10th Internat. Genetics Congress, Montreal, 1958, 1st Internat. Congress Electron Microscopy, Berlin, 1958, 9th Internat. Bot. Congress, Montreal, 1958, 1st Internat. Congress Electron Microscopy, Berlin, 1958, 9th Internat. Bot. Congress, Montreal, 1959, Conf. on Biochem. and Biophys. Mechanisms in Prodn. of Radiation-Induced Chromosome Aberrations, P.R., 1961, Internat. Conf. Genes and Chromosomes, Buenos Aires, 1964, Internat. Symposium on Nuclear Physiology and Differentiation; mem. NRC, 1962-64; cons. microbiol. tng. com., div. gen. med. scis. NIH, 1962-63. Fellow A.A.A.S.; mem. Am. Soc. for Cell Biology (sec.), Am. Soc. Naturalists, Am. Soc. Zoologists, Genetics Soc. Am., Histochem. Soc. (mem. council), Am. Assn. Anatomists, Soc. Developmental Biology (pres. 1968-69), Biophys. Soc., Internat. Soc. for Cell Biology, Sigma Xi. Editorial bd. Jour. Cell Biology, 1968-—. Research on electron microscopy and cytochemistry of nucleus and chromosomes, also spermatogenesis, particularly in aflagellate forms. Home: 152 Pinecrest Rd., Durham, N.C. 27705.*

MOSES BEN MAIMON, see Maimonides.

MOSETTI, Ferruccio, Italian geophysicist; b. Trieste, Italy, Mar. 28, 1929; s. Francesco and Melany (Stanta) M.; Ph.D., U. Trieste, 1951, Libera docenza, 1962; m. Giorgia d'Henry, June 23, 1952; 5 children. Geophysicist, Osservatorio Geofisico of Trieste, 1950-—, dir., 1957; prof. oceanography Bari U., 1960-64; prof. terrestial physics U. Trieste, 1965-—. Mem. Internat. Soc. Geonomy, Internat. Soc. Climatology, Assn. Ital. Meteorologia e Geofinica. Author: Oceanografia, 1964; also numerous articles. Research on natural cyclic changes in geophysics, climatology, ecology by statis. and math. methods, applied geophysics, hydrology, tides and thermic fluctuations; described new law for natural fluctuation periods. Home: via del Pesce 3. Office: viale Romolo Gessi 4, Trieste, Italy.*

MOSHER, Harry Stone, Am. chemist; b. Salem, Ore., Aug. 31, 1915; s. Daniel Harrison and Maude (Stone) M.; B.A., Willamette U., 1937; M.S., Ore. State Coll., 1939; Ph.D., Pa. State Coll., 1942; m. Carol Beth Walker, June 23, 1944; children—Janet Lee, Stephen Eric, Leslie Jean. Instr., Willamette U., 1939-40; asst. prof. Pa. State Coll., 1942-46; faculty Stanford, 1947-—, prof. chemistry, 1955-—. Cons. Stanford Research Inst., 1960-—. Recipient Distinguished Alumni award Willamette U., 1966. Mem. Am. Chem. Soc. (mem. adv. bd. Advances in Chemistry 1965-—), Chem. Soc. (London, Eng.), A.A.A.S., Sigma Xi, Phi Lambda Upsilon. Contbg. author: Heterocyclic Compounds, vol. I, 1950; also numerous articles. Research on synthesis nitrogen containing heterocyclic organic compounds, fundamental studies on stereochem. forces controlling pathway organic chem. reactions, elucidation chem. structure neurotoxin Tarichatoxin-tetrodotoxin isolated from embryos Cal. newt, Taricha torosa. Home: 713 Mayfield Av., Stanford, Cal. 94305.*

MOSHKIN, Panteleymon Afanasevich, Russian chem. technologist; b. Feb. 13, 1891; grad. Moscow Higher Tech. Sch., 1918. Instr., Moscow Higher Tech. Sch., 1919-30; prof. Moscow Chem. Tech. Inst., 1928-31; lab. head Moscow Plastics Research Inst., 1943-—. Recipient Stalin prize, 1948. Mem. USSR Acad. Scis. (corr.). Author: Sulfurous Petroleum Paraffins as Raw Material for Synthetic Fatty Acid Production, 1957; Higher Alcohols of the Fatty Series from the Products of Paraffin Wax Oxidation, 1959; Synthesis of Alkyllactates of Phosphinous Acids, 1963. Research and publs. on indsl. methods to chem. synthesis; developer methods for extraction and analysis of phenol in primary humus coal tar; devised indsl. methods to synthesize semiproducts and plasticizers for plastics. Address: Moscow Plastics Research Inst., Perovsky pr. 41, Moscow, USSR.

MOSHKOV, Valentin Nikolaevich, Russian specialist in kinesiatrics; b. 1903; grad. Moscow Inst. Phys. Culture, 1927, 2d Moscow Med. Inst., 1931; D. Med. Sci. Asst., head dept. kinesiatrics Inst. Physiotherapy, 1931-41, prof., 1943——; head dept. kinesiatrics Central Inst. Balneology, 1943——; head chair kinesiatrics Central Postgrad. Med. Inst., Moscow, 1943-—. Chmn., Problems Commn. for Med. Problems of Phys. Culture and Sport. Mem. USSR Acad. Med. Scis. (corr.), All-Union (chmn. 1961——), Moscow (chmn. 1961——) socs. for med. control and kinesiatrics. Author: Kinesiatrics in Hypertony, 1950; Kinesiatrics in the Clinical Treatment of Internal Diseases, 1952; General Principles of Kinesiatrics, 1954; Kinesiatrics in Health Resorts and Sanatoria, 1955; Kinesiatrics in the Clinical Treatment of Nervous Diseases, 1959. Mem. editorial bd. Problems of Physiotherapy and Phys. Culture. Research and numerous publs. on theory and methods of kinesiatrics and its use in complex treatment of health resort patients; developer use of kinesiatrics in treatment of cardiovascular disorders, hypertony and endarteritis, kinesiatrics as means of functional and pathogenic therapy in clin. treatment of internal diseases. Address: Central Postgrad. Med. Inst., pl. Vosstaniya 1-2, Moscow, USSR.

MOSHKOVSKY, Shabsay Davidovich, Russian epidemiologist, parasitologist, chemotherapist; b. 1895; grad. Med. Faculty, 1st Moscow U., 1919. Asso., dept. protozoology Inst. Med. Parasitology and Tropical Medicine, USSR Peoples Commissariat Health, 1921-35, head, 1935-46; prof., head chair med. parasitology Central Postgrad. Med. Inst., Moscow, 1935——; head dept. protozoology Martsinovsky Inst. Parasitology and Tropical Medicine, USSR Ministry Health, 1946-—; head protozoology course Moscow U., 1949——. Mem. USSR Acad. Med. Scis. (corr.), All-Russian Soc. Epidemiologists, Microbiologists and Infectionists (dep. chmn. 1957——). Author: Functional Parasitology, 1946; Allergy and Immunity, 1947; Basic Laws Governing Epidemiology, 1950; Protozoa: The Nature of Protozoa and the Limits of Protozoology, 1957; A Regular Feature of Immunity in Infectious Diseases: The Law of Reinoculation, 1957. Mem. editorial bd. Med. Parasitology and Parasitic Diseases; mem. editorial council Antibiotics; co-editor Epidemiology and Infectious Diseases sects. Large Med. Ency., 2d edit. Research and publs. on correlation between immunity and allergy; introducer concept of immunological states, leukocyte profile methods, math. analysis methods into epidemiology; formulated rule of reinoculation, prin. of physiol. imitation; introducer concept of functional parasitology; founder semiotics of filterable viruses; devised new methods to stain blood and blood parasites, laws governing hemogram changes in infectious diseases, new method to investigate biol. testing and efficient use of chemotherapeutic agts. Address: Central Postgrad. Med. Inst., pl. Vosstaniya 1-2, Moscow, USSR.

MOSHMAN, Jack, Am. statistician; b. Richmond Hill, N.Y., Aug. 12, 1924; s. Morris and Sadye (Posner) M.; B.A., N.Y. U., 1946; M.A., Columbia U., 1947; Ph.D., U. Tenn., 1953; m. Annette Gordon, Aug. 10, 1947; children—Gordon S., Marc L., Sheri, Ira H. Instr. math. Queens Coll., 1946-47, U. Tenn. 1947-52; sr. statistician U. S. AEC, 1948-50, Oak Ridge Nat. Lab., 1950-54; mem. tech. staff Bell Telephone Labs., Inc., N.Y.C., 1954-57; v.p. C-E-I-R, Inc., Washington, 1957-66; mng. dir., mgmt. scis. EBS Mgmt. Consultants, Washington, 1966-67; v.p. Leasco Systems and Research, Bethesda, Md., 1967-—. Professorial lectr. operations research George Washington U., Washington, 1959-61. Fellow Brit. Interplanetary Soc.; mem. Assn. for Computing Machinery (past v.p.), NRC-Nat. Acad. Scis., Inst. Math. Statistics, Am. Statis. Assn. (past sec., sect. phys. and engring. scis., past mem. council), Operations Research Soc. Am., Inst. Mgmt. Scis., A.A.A.S., Biometric Soc., Phi Beta Kappa, Pi Mu Epsilon, Kappa Pi Sigma. Contbg. author: Advances in Computers, Vol. V, 1964; Mathematical Models for Digital Computers, Vol. 2, 1967. Research and publs. on Monte Carlo methods for generation of pseudo-random numbers; developed projection and interpretation techniques for extrapolating early election returns to complete totals. Gen. editor Internat. Series on Computing Scis., 1961-65. Editor, Faith, Hope and Parity. Home: 7008 Carmichael, Bethesda 20034. Office: 4833 Rugby Av., Bethesda, Md. 20014.

MOSIER, H(enry) David, Jr., Am. physician; b. Topeka, May 22, 1925; s. Harry David and Josephine (Johnson) M.; student Ga. Inst. Tech., 1944, U. Mich., 1946, U. Colo., 1947; B.S., U. Notre Dame, 1948; M.D., Johns Hopkins, 1952; m. Nadine Oclea Merilatt, Aug. 24, 1949 (div. Sept. 1963); children—Carolyn Josephine, William David, Daniel Thomas, Christine Elizabeth; m. 2d, Marjorie Knight Armstrong, Sept. 26, 1963. Asst. pathology U. So. Cal., 1954-55; fellow pediatric endocrinology Johns Hopkins Hosp., Balt., 1955-57; faculty U. Cal. at Los Angeles, 1957-63, asso. prof., 1961-63; asso. prof. pediatrics U. Ill., Chgo., 1963——; dir. research Ill.

state Pediatric Inst., Chgo., 1963——. Cons. Pacific State Hosp., Pomona, Cal., 1957——. Mem. Endocrine Soc., Soc. for Pediatric Research, Midwest, Western socs., for pediatric research, Soc. for Exptl. Biology and Medicine, Soc. for Research in child Devel., Sigma Xi. Research, numerous publs. in endocrinology and metabolism in animals and humans particularly phys. growth and mechanisms regulating it, role central nervous system in phys. growth. Home: 690 Irving Park Rd., Chgo. 60613. Office: 2021 Santa Monica Blvd., Santa Monica, Cal.*

MOSIER, Jeremiah George, Am. soil scientist; b. Pike County, O., Jan. 8, 1862; s. David and Amanda Rachel (Brill) M.; student Nat. Normal U., Lebanon, O., 1883-85; B.S., U. of Ill., 1893; m. Lydia C. Miller, June 22, 1892. Tchr. in rural schs., Champaign, County, Ill., 1885-88, and village sch., Sandorus, Ill., 1893-94; asst. in geology U. Ill., 1894-97; instr. soil physics, 1902-05, asst. prof., 1905-11, prof., 1911—, also chief of soil physics Agrl. Expt. Sta.; engaged in farming, 1897-98; instr. in high sch., Urbana, 1899-1900, Champaign, 1900-02; in charge of detailed soil survey of State of Ill. Author: Laboratory Manual of Soil Physics, 1911; Soil Physics and Management; Climate of Illinois; Soils and Crops. Died 1922.

MOSS, Arthur J., Am. physician; b. St. Paul, May 12, 1914; s. David and Anna Moss; B.S., U. Minn., 1935, M.B., 1937, M.D., 1938, M.S., 1942; m. Alice Sylvia Litman, Oct. 19, 1941; children—Stephanie, Patricia, Tom. Health service physician U. Minn. Health Service, Mpls., 1946; practice medicine, specializing in pediatrics, Inglewood, Cal., 1946-60; chmn. dept. pediatrics Los Angeles Harbor Gen. Hosp., Torrance, Cal., 1948-51; head dept. pediatrics Methodist Hosp., Los Angeles, 1951; faculty U. Cal. at Los Angeles, 1952——, prof. pediatrics, 1964——, dir. fibrocystic center Med. Center, 1963——, acting chmn. dept. pediatrics, 1966——. Diplomate Am. Bd. Pediatrics, Am. Bd. Pediatric Cardiology. Fellow Am. Coll. Cardiology; mem. Am. Acad. Pediatrics, Am., Cal., Los Angeles heart assns., Am., Cal., Los Angeles County med. assns., Soc. for Exptl. Biology and Medicine, Sigma Xi. Author: (with Forrest H. Adams) Problems of Blood Pressure in Childhood, 1962, Heart Disease in Infants and Children, 1967; also articles. Editorial bd. Pediatrics Digest, 1962——. Home: 2701 Forrester Dr., Los Angeles.*

MOSS, James Mercer, Am. physician; b. Bradley, Ga., Dec. 15, 1917; s. Fred A. and Rosa M. (Mercer) M.; M.D., U. Va, 1941; m. Rachel S. Bybee, Sept. 6, 1941; children—James Marion, Fred Aubrey (dec.), William W., Robert F. Fellow endocrinology Duke, 1946, instr., 1946-47; faculty Georgetown U., Washington, 1949——, prof. medicine, 1962——; chief diabetic clinic Georgetown U. Hosp., 1949——, chief diabetic clinic, vis. physician D.C. Gen. Hosp., 1950-55; chief medicine, Circle Terrace Hosp., Alexandria, Va., 1962——; practice medicine specializing in internal medicine, Alexandria, 1949——. Recipient 2d award 1st Interstate Sci. Assembly, 1954, 3d award So. Med. Assn., 1960; Gold medal Am. Podiatry Assn., 1962; 2d award Va. Acad. Gen. Practice, 1964. Mem. No. Va. Heart Assn., (pres. 1964-65), Med. Alumni Assn. U. Va. (pres. 1965-66), Va. Soc. Internal Medicine (past pres.), Alexandria Med. Soc. (past pres.), Med. Council Washington Met. Area (past pres.), Diabetes Assn. D.C. (past pres.), Med. Soc. Va. (past v.p.), A.C.P. (past Va. chmn.), Am. Diabetes Assn., A.M.A., Endocrine Soc., Am. Coll. Cardiology. Author: Fundamentals of Diabetic Management, 1962; also numerous articles. Research on diabetes mellitus with evaluation oral hypoglycemic drugs. Home: 319 Mansion Dr., Alexandria 22302. Office: 3805 Florence Dr., Alexandria, Va. 22305.*

MOSS, Melvin Lionel, Am. anatomist; b. N.Y.C., Jan. 3, 1923; s. Maurice and Ethel (Lander) M.; A.B., N.Y. U., 1942; D.D.S., Columbia, 1946, Ph.D., 1946; m. Elaine Schweidel, Sept. 2, 1945; children—Noel Morrow, James Andrew. Faculty, Columbia, N.Y.C., 1946——, professor, 1967——. Recipient Lederle Med. Faculty award, 1954-56. Mem. Am. Assn. Anatomists, Am. Assn. Phys. Anthropologists, Internat. Assn. Dental Research, Am. Soc. Zoologists, Sigma Xi, Omicron Kappa Upsilon. Research, numerous publs. on skeletal growth and comparative biology calcification, emphasizing processes common to all such mineralizations in vertebrates, invertebrates and plants.*

MÖSSBAUER (or MOESSBAUER), Rudolf Ludwig, physicist; b. Munich, Germany, Jan. 31, 1929; s. Ludwig and Erna (Ernst) M.; Hauptdiplom, Technische Hochschule, Munich, 1955, Ph.D., 1958; m. Elizabeth Pritz, 1956; children—Peter, Regine. Came to U. S., 1960. Asst., Rodenstock Optics Factory, Munich, 1949; research asst. Max-Planck Inst., Heidelberg, Germany, 1955-57; research asst. Technische Hochschule, Munich, 1959-60; research fellow Cal. Inst. Tech., 1960-61, sr. research fellow, 1961, prof. physics, 1961——. Participant 2d All-Union Conf. Nuclear Reactions, Moscow, 1960, Conf. Moessbauer Effect, Urbana, Ill., 1960, Paris, France, 1961. Recipient award Research Corp., 1960, (with Hofstadter) Nobel prize for physics, 1961, Rontgenpreis U. Giessen (Germany), 1961, Elliott Cresson medal Franklin Inst., 1961. Mem. Am. Phys. Soc. Author numerous articles. Discovered action of gamma-ray emission and absorption, now called Mössbauer

effect; devised method for producing gamma rays of very precise wave length; his work has provided means of testing some predictions of gen. relativity and solid-state physics. Home: 1041 E. Beverly Way, Altadena, Cal. Office: California Inst. Tech., Pasadena 4, Cal.

MOSSMAN, Harland Winfield, Am. anatomist; b. Portland, N.Y., May 7, 1898; s. Herdmen X. and Lucy (Fuller) M.; B.S., Allegheny Coll., 1920; M.S., U. Wis., 1922, Ph.D., 1924; m. Ruth Hannah Jackson, June 24, 1924; children—Archie S., Malcolm H., Ardith R. (dec.). Faculty, U. Wis., Madison, 1924——, prof. anatomy, 1953——. Fellow A.A.A.S.; mem. Am. Assn. Anatomists, Am. Soc. Zoologists, Am. Soc. Mammalogists, Soc. for Study Evolution, Soc. for Exptl. Biology and Medicine, Internat. Inst. Embryology, Soc. for Study Fertility. Author: (with W. J. Hamilton, J. D. Boyd) Human Embryology, 1945; also numerous articles. Asso. editor Am. Jour. Anatomy, 1958——. Research on comparative embryology, anatomy, and physiology of genital systems of mammals especially placenta and ovaries. Home: 2902 Columbia Rd., Madison, Wis. 53705.*

MOSSO, Angelo, Italian physiologist; b. Turin, Piedmont, May 31, 1846; student U. Turin, 1870. Worked with J. Moleschott, Turin, M. Schiff, Florence, C. Ludwig, Vienna, C. Bernard, Ranvier and E. J. Marcy, Paris; prof. U. Turin, 1877; head physiology dept., Turin; founder Internat. Inst. Physiology, Turin (under his direction became one of most important centers for exptl. studies); founder Italian Archives of Biology. Mem. French Acad. Scis. (corr.). Author: La Plura; La Fatica; L'educazione fisica della gioventi; La temperatura del cervello; Le origine della civiltà mediterranea; Vita moderna dei italiani. Research on circulation of blood and on muscular fatigue; proved fatigue results from toxic products of muscular contraction, 1890; invented erograph and sphygmomanometer; established mountain lab. to study physiol. effects of high altitude. Died Turin, Italy, Nov. 24, 1910.

MOSTELLER, Frederick, Am. statistician, educator; b. Clarksburg, W.Va., Dec. 24, 1916; s. W. Roy and Helen (Kelley) M.; B.S., Carnegie Inst. Tech., 1938, M.S., 1939; A.M., Princeton, 1941, Ph.D., 1946; m. Virginia Gilroy, May 17, 1941; children—William Samuel, Gale Robin. Instr. math. Princeton, 1942-44, research mathematician applied math. panel Statis. Research Group, 1944-45; faculty Harvard, Cambridge, 1946——, prof. math. statistics, 1951——, chmn. dept. statistics, 1957——. Nat. tchr. probability and statistics Continental Classroom, NBC-TV, 1961. Fund for Advancement Edn. fellow U. Chgo., 1954-55; fellow Center for Advanced Study in Behavioral Scis., 1962-63. Fellow Am. Acad. Arts and Scis., Am. Philos. Soc., Inst. Math. Statistics, Am. Statis. Assn. (v.p. 1962-64), Am. Soc. for Quality Control, A.A.A.S.; mem. Internat. Statis. Inst., Am. Math. Soc., Math. Assn. Am., Psychometric Soc. (pres. 1957-58), Biometric Soc., Royal Statis. Soc., Am. Anthrop. Assn., Am. Sociol. Assn., Sociol. Research Assn. Author: (with others) Sampling Inspection, 1948, The Preelection Polls of 1938, 1949, Stochastic Models for Learning, 1955, Probability with Statistical Applications, 1961, Inference and Disputed Authorship: The Federalist, 1964. Research, publs. on math. statistics and application to indsl., social sci. and med. problems. Home: 28 Pierce Rd., Belmont, Mass. 02178. Office: 2 Divinity Av., Cambridge, Mass. 02138.*

MOSTERT, Paul Stallings, Am. mathematician; b. Morrilton, Ark., Nov. 27, 1927; s. J.F.T. and Lucy (Stallings) M.; A.B., Southwestern U., 1950; M.S., U. Chgo., 1951; Ph.D., Purdue U., 1953; m. Kathleen Gray, Dec. 20, 1947; children—Paul Theodore, Richard Stallings, Kathleen, Kristina. Faculty, Tulane U., New Orleans, 1953——, prof. math., 1962; vis. prof. U. Tuebingen, (German), 1962-63. Mem. Am. Math. Soc. Author: Analytic Trigonometry, 1960; (with K. H. Hofmann) Splitting in Topological Groups, 1963, Elements of Compact Semigroups, 1966; also articles. Research in compact topological semigroups, topological groups, topological transformation groups, category theory, theory sheaves. Home: 2227 Calhoun St., New Orleans 70118.*

MOSTOW, George Daniel, Am. mathematician; b. Boston, July 4, 1923; s. Isaac J. and Ida (Rotman) M.; B.A., Harvard, 1943, M.A., 1946, Ph.D. 1948; m. Evelyn Davidoff, Sept. 1, 1947; children—Mark Alan, David, Carol, Jonathan. Instr. math. Princeton, 1947-48, mem. Inst. Advanced Study, 1947-49, 56-57; asst. prof. Syracuse U., 1949-52; asst. prof. math. Johns Hopkins, 1952-53, asso. prof., 1954-56, prof. math., 1957-62; prof. math. Yale, 1962——; vis. prof. Conselho National des Pesquisas, Instituto de Matematica, Rio de Janeiro, Brazil, 1953-54, U. Paris, 1967, Hebrew U. Jerusalem, 1968. Fulbright research scholar, Utrecht U., Netherlands, also John Simon Guggenheim fellow, 1957-58. Mem. Am. Math. Soc., Phi Beta Kappa, Sigma Xi. Editor Am. Jour. Math, asso. editor Annals of Math., 1957-63, Trans. Am. Math. Soc., 1958——. Contbr. research articles. Contbns. to theory of lie groups and lie algebras. Home: Beechwood Rd., Woodbridge, Conn. Office: Yale U., New Haven, Conn.*

MOSZKOWSKI, Steven Alexander, physicist, educator; b. Berlin, Germany, Mar. 13, 1927; s. Richard and Ruth (Bamberger) M.; B.S., U. Chgo., 1946,

Ph.D., 1952; m. Lena Iggers, Aug. 29, 1952; children—Benjamin, Richard, Ronald. Came to U. S., 1940, naturalized, 1945. Research asst. Columbia, N.Y.C., 1952-53; faculty physics U. Cal. at Los Angeles, 1953—, prof., 1963—. Cons. Rand Corp., 1953—, Oak Ridge Nat. Lab., 1962—. Guggenheim fellow, 1961-62. Fellow Am. Phys. Soc.; mem. Phi Beta Kappa. Author: (with C.S. Wu) Beta Decay. Research on beta decay, nuclear models, many body problem. Home: 3283 Inglewood Blvd., Los Angeles 90066.*

MOTCHANE, Leon, mathematician; b. St. Petersburg (now Leningrad), USSR, June 19, 1900; s. Edmond and Henriette (Morgueev) M.; B.A., U. St. Petersburg, 1917; postgrad. U. Lausanne (Switzerland), 1921; Docteur ès sciences, U. Paris; children—Didier, Jean-Loup. Asst. physics U. Lausanne, 1921-22; in industry, also sociol. studies, 1922-39; dir. Institut des Hautes Études Scientifiques, Bures, France, 1958—. Mem. Math. Soc. France, Statis. Soc., Am. Math. Soc. Research, publs. on gen. topology, set theory, others. Home: 19 Residence gratien, Bures. Office: Institut des Hautes Études Scientifiques, Bures sur Yvette 91, France.*

MOTHES, Kurt, German biochemist, physiologist; b. Plauen, Germany, Nov. 3, 1900; s. Albin and Anna (Gemeinhardt) M.; Dr.phil., U. Leipzig, 1925; Dozent, U. Halle (Germany), 1928, Dr.med.h.c., 1960, Dr.rer.-nat.hc., 1965; Dr.agr.h.c., U. Kiel, 1960; Dr.phil.h.c., U. Vienna, 1965; m. Hilda Eilts, July 13, 1929; children—Uta, Georg, Winrich and Heinrich (twins). Faculty, dir. Bot. Inst., U. Königsberg, 1935-45; prof. botany U. Halle, 1935-45, prof. biochemistry, 1963-65; head dept. chem. physiology German Acad. Scis., Gatersleben, 1949-57; dir. Bot. Inst. U. Halle, 1958-63; dir. Inst. Biochemistry of plants, German Acad. Sci., Halle, 1958—. Recipient Cothenius medal Deutsche Akademie der Naturforscher Leopoldina, 1960; Hoest-Madsen medal Fédération Internationale Pharmaceutique, 1962. Gergor-Mendel medal CSSS Acad. Scis., 1965; Otto-Warburg medal Gesellschaft für Physiologische Chemie, 1965; Carl-Mannich medal Deutsche Pharmazeutische Gesellschaft, 1965, Döbereiner medal, 1965. Mem. Berlin, Halle, Leipzig, Heidelberg, Munich, Budapest, Bucharest, Pilani, Vienna acads. scis., Deutsche Akad. der Naturforscher Halle (pres.). Research, numerous publs. on biosynthesis of alkaloids, metabolism of proteins, regulation of proteinsynthesis, growth by kinins and roots. Home: 23 Hoher Weg. Office: Weinbergweg, Halle, Germany.

MOTLEY, Hurley Lee, Am. physician; b. Silex, Mo., July 23, 1904; s. Jasper and Della (Jamieson) M.; A.B., U. Mo., 1930, A.M., B.S. in Medicine, with distinction, 1932, Ph.D., 1934; M.D., Harvard, 1936; m. Cornealia Grace Ellis, Aug. 10, 1941; children—Dale Ellis, Susan Ann, James Hurley. Faculty, U. Mo., 1936-47, asso. prof. medicine, dir. cardio-respiratory lab. Jefferson Med. Coll., 1947-52; prof. medicine, dir. cardio-respiratory lab. U. So. Cal., Los Angeles, 1952—; dir. research Hosp. Good Samaritan, Los Angeles, 1962—; mem. staff Hollywood Presbyn., Los Angeles County, Children's, Orthopedic hosps., Barlow Sanatarium (all Los Angeles); practice medicine specializing in chest diseases, Phila., 1947-52, Los Angeles, 1952—. Mem. nat. med. and research adv. council City of Hope Med. Center, Duarte, Cal., 1958—; cons. Surgeon Gen. Army, 1961—; mem. sci. com. Los Angeles Air Pollution Control Dist., 1955—. Recipient Citation Merit in medicine U. Mo. Alumni Assn., 1960; Clean Air award Los Angeles County Bd. Supervisors, 1960. Fellow A.C.P., Am. Coll. Chest Physicians; mem. Am. Physiol. Soc., Am. Trudeau Soc., A.M.A., Western Soc. Clin. Research, Am. Fedn. Clin. Research, Coll. Physicians Phila., Am. Indsl. Hygiene Assn., Los Angeles Acad. Medicine, Pi Beta Kappa, Sigma Xi. Contbr. numerous articles on cardio-respiratory physiology, pneumoconiosis, emphysema, artificial respiration, pulmonary circulation, pressure breathing, air pollution, oxygen therapy. Pioneered use intermittent positive pressure breathing as therapy in chronic pulmonary disease. Home: 2003 N. Serrano Av., Los Angeles 90027. Office: 1212 Shatto St., Los Angeles 90017.*

MOTODA, Sigeru, Japanese marine biologist; b. Tokyo, Japan, Jan. 10, 1908; s. Joseph Sakunoshin and Tei (Koide) M.; M.Agr., Hokkaido (Japan) Imperial U., 1933, D.Agr., 1948; m. Michiko Nii, Dec. 31, 1944; 1 son, Andrew Susumu. Research asso. Hokkaido Imperial U., 1937-40, faculty, 1940—, prof. marine biology, 1950—. Mem. consultative com. Indian Ocean Biol. Centre, South India, 1963-65. Mem. Internat. Council Sci. Unions (mem. sci. com. oceanic research 1964—), Am. Soc. Limnology and Oceanography, Marine Biol. Assn. U.K., Marine Biol. Assn. India, Oceanographical Soc. Japan, Plankton Soc. Japan. Author: Sea and Plankton, 1944; also numerous articles. Research on marine zooplankton taxonomy and ecology. Home: 22-5 Aoyagicho, Hakodate, Hokkaido, Japan.*

MOTOKAWA, Koichi, Japanese physiologist; b. Ishikawa Prefecture, 1903; grad. Tokyo U., 1929; M.D. Became instr. Tokyo U., 1939; prof. Tohoku U., 1940. Recipient Japan Acad. prize for Study of Electrical Diagram of Brain, 1954, Asahi Cultural prize for Research on Color Sensation, 1953. Author: Medicine—Methods of Biological Electrical Experiments; also others. Made analytical studies of brain pulse which showed new results; developed original theory of color sense.

MOTT, Sir Frederick Walker, Brit. neuro-pathologist; b. Brighton, Eng., Oct. 23, 1853; s. Henry and Caroline (Fuller) M.; M.B., U. Coll., London, Eng., B.S., 1881, M.D., 1886, also fellow; LL.D., Edinburgh (Scotland) U., 1919; m. Georgina Soley, 1885; 4 daus. Asst. prof. physiology Liverpool (Eng.) U., 1883; lectr. physiology Charing Cross Hosp. Med. Sch., 1884, later lectr. pathology, asst. physician, physician, lectr. medicine; pathologist in charge Claybury Lab. (transferred to Maudsley Hosp., Dehmarh Hill, Eng.), 1895-1923, resigned 1923; tchr. Maudsley Hosp., lectr. morbid psychology U. Birmingham (Eng.), from 1923. Pathologist London County asylums; Fullerian prof. physiology Royal Inst. and Am. Psychiat. Soc. Recipient Stewart prize Med. Assn., 1903; Fothergill Gold medal and prize, 1910, Moxon Gold medal, 1918. Fellow Royal Soc., 1896, Royal Coll. Physicians; mem. neurol. and psychiat. socs. France, Belgium, Netherlands (corr.), Royal Medico-Psychol. Assn. Author: Nature and Nurture in Mentlai Development; Brain and the Voice; War Neuroses and Shell Shock. Editor: Archives of Neurology and Psychiatry. Research, publs. on nervous system, paths of conduction in spinal cord, localization of cerebral cortex (especially relating to eye movements), effect of acute anemia on brain; established connection of disease of nervous system, called gen. paralysis of insane, with syphilis, asso. with presence of specific spirochaete, also asso. syphilis with other mental disorders; demonstrated close relationship between nervous system and sexual organs manifested in dementia praecox, also asso. deficient mental condition with degeneration of thyroid and other endocrine glands. Died Birmingham, Eng., June 8, 1926.

MOTT, Sir Nevill Francis, Brit. physicist; b. Leeds, U.K., Sept. 30, 1905; s. Charles Frances and Lilian (Reynolds) M.; student Clifton Coll., 1918-23; B.A., Cambridge U., 1927, M.A., 1930; D.Sc. (hon.), Louvain U., 1947, Grenoble U., 1950, Poitiers U., 1953, Paris U., Bristol U., Ottawa U., 1955, Liverpool U., Reading U., 1960, Sheffield U., 1961, London U., 1963; m. Ruth Horder, Mar. 22, 1930; children—Elizabeth, Alice. Lectr. in physics Manchester U., 1929-30; lectr. math. Cambridge U., 1930-33; prof. physics Bristol (Eng.) U., 1933-54; master Gonville and Caius Coll., Cambridge, 1959-66. Chmn. adv. com. on sch. edn. Nuffield Found., London, Eng, 1962—. Knighted, 1962. Fellow Royal Soc., 1936; mem. Athenaeum. Author: (with H.S. W. Massey) The Theory of Atomic Collisions, 1933; (with H. Jones) The Theory of the Properties of Metals and Alloys, 1936; (with R. W. Gurney) Electronic Processes in Ionic Crystals, 1940. Home: 31 Sedley Taylor Rd., Cambridge, U.K.*

MOTT, Valentine, Am. surgeon; b. Glen Cove, L.I., N.Y., Aug. 20, 1785; s. Henry and Jane (Way) M.; M.D., Columbia, 1806; M.D. (hon.), U. Edinburgh (Scotland); LL.D., Univ. State N.Y., 1851; m. Louisa Mums, 1819; 9 children. Prof. surgery Columbia, 1811-13; prof. surgery Coll. Physicians and Surgeons, 1813-26, 30-35, mem. staff during 1850's; a founder Rutgers Med. Coll.; a founder med. dept. Univ. State N.Y., 1840, prof. surgery and anatomy until 1850, emeritus, from 1852. Hon. fellow Imperial Acad. Medicine of Paris (France); mem. Paris Clin. Soc. Author: Motts Cliniques, 1860; Pain and Anaesthetics, 1862. Co-editor Med. Mag., 1814-15, Med. and Surg. Reporter, 1818-20. 1st to tie innominate artery with aim of preventing death from subclavian aneurysm, 1818; successfully tied common iliac artery for an aneurysm of external iliac, to perform successful amputation of hip joint, 1824; studied treatment of jaw and surgical use of anesthesia. Died N.Y.C., Apr. 26, 1865.

MOTT, William E., Am. physicist; b. Pawling, N.Y., Feb. 28, 1926; s. Raymond A. and Myrtle (Gregory) M.; student Vassar Coll., 1946-47; A.B., Coll. of Wooster, 1949; M.S., Carnegie Inst. Tech., 1951, Ph.D., 1953; m. Mary E. Steinhilper, Feb. 21, 1953; children—Katherine Elise, William Gregory. Research physicist Gulf Research & Devel. Co., Pitts., 1953-55, supr. nuclear applications sect., 1955-66; asst. dir. for tech. programs, div. isotopes devel. U. S. AEC, Washington, 1966—. Recipient Arthur H. Compton award in physics Coll. Wooster, 1949. Mem. Am. Nuclear Soc. (exec. com. isotopes and radiation div.), Am. Phys. Soc., A.A.A.S., Health Physics Soc., Phi Beta Kappa, Sigma Xi. Contbr. articles to profl. jours., chpts. to books; patentee in field. Research in nuclear well logging, natural radioactivity of rocks, radioisotope and radiation applications. Home: 6707 Tildenwood Lane, Rockville, Md. 20852. Office: U. S. AEC, Washington 20545.*

MOTULSKY, Arno Gunther, physician, geneticist; b. Fischhausen, Germany, July 5, 1923; s. Herman and Rena (Sass) Molton; student Central YMCA Coll., Chgo., 1941-43, Yale, 1943-44; B.S., M.D., U. Ill., 1944-47; m. Gretel C. Stern, Mar. 22, 1945; children—Judy, Harvey, Arlene. Staff mem. in charge clin. investigation, dept. hematology Army Med. Service Grad. Sch., Walter Reed Army Med. Center, Washington, 1952-53, also research asso. internal medicine George Washington Sch. Medicine, 1952-53; instr., asst., asso. prof., dept. medicine U. Wash. Sch. Medicine, Seattle, 1953-61, prof. medicine, prof. genetics, head div. med. genetics, dir. Genetics Clinic, Univ. Hosp. and Children's Med. Center, dir. med. genetics tng. program, 1961—; vis. prof. U. Cal., San Francisco, 1960; attending physician King County, VA, Univ. hosps., Seattle; cons. physician Children's Orthopedic Hosp. and Med. Center, Seattle; cons. Madigan Army Hosp., Tacoma, Wash., 1955—, subcom. on blood transfusion, com. on blood NRC, 1958-63, study sect. human ecology NIH, 1961-65; ad hoc cons. Nat. Found., WHO, Children's Bur. Spl. Commonwealth Fund fellow in human genetics Galton Lab., Univ. Coll., London, Eng., 1957-58; John and Mary Markle scholar in med. sci., 1957-62. Diplomate Am. Bd. Internal Medicine. Fellow A.C.P.; mem. Internat. Soc. Hematology, Am. Fedn. for Clin. Research, A.M.A., A.A.A.S., Genetics Soc. Am., N.Y. Acad. Scis., Western Soc. Clin. Research, Am. Soc. Human Genetics, Am. Soc. for Clin. Investigation, Western, Am. assns. physicians. Mem. editorial bds. Blood, Humangenetik, Annals Internal Medicine. Research and numerous publs. on elucidation of mechanisms in various hereditary anemias, devel. methods for studies of biochem. traits in human blood, role of genetic variation in drug reactions, role of malaria in distbn. of enzyme deficiency in man, role of genetics in various diseases, genetic-environmental interactions in disease. Home: 10618 Durland Av., Seattle 98125.*

MOTYKA, Josef, Polish botanist; b. Kaclowa, Poland, Mar. 23, 1900; s. Wojciech and Anna (Radziak) M.; phil.doctor, Jagiellonian U., Kraków, Poland, 1925; m. Zofia Lecyk, Nov. 21, 1939; children—Zbigniew, Maria. Asst., Jagiellonian U., 1924-29; adj. J.K. U., Lwów, USSR, 1929-39; docent Lwów U., 1939-41; prof. Maria Curie-sklodowska U., Lublin, Poland, 1945—. Decorated Golden Cross of Distinction, Officers Polonia Restituta-Cross. Mem. Polish Bot. Soc. Author: Monograph of Usnea, 2 vols., 1936-46; Plant Ecology, 1962; Lichen Flora of Poland, 1-4 vols., 1960-65; also popular sci. books, articles. Described numerous new species; studied genera Alectoria, Evernia, Ramalina, Thamnolia, also physiographical autoecology of many species of Phanerogames, ecology of meadows; developed new method of causal analysis in geobotany. Home: 17 Glowackiego, Lublin, Poland.*

MOTZ, Lloyd, Am. astrophysicist, educator; b. Susquehanna, Pa., June 5, 1910; s. Solomon and Minnie (Seltzer) M.; B.S., Coll. City N.Y., 1930; Ph.D., Columbia, 1936; m. Minne R. Motz, June 14, 1934; children—Robin Owen, Julie Ann. Instr. physics Coll. City N.Y., 1931-41; faculty astronomy Columbia, N.Y.C., 1935—, prof., 1961—. Organized, supervised optical firms, 1941-47; cons. various optical firms. Recipient 1st prize for new theory of structure of fundamental particles Gravity Research Found., 1960. Fellow Am. Phys. Soc., Royal Astron. Soc.; mem. Am. Astron. Soc., A.A.A.S., Phi Beta Kappa, Sigma Xi. Author: This is Astronomy, 1958; This Is Outer Space, 1960; The Essentials of Astronomy, 1964; The World of the Atom. Editor: Molecules, Crystals and Statistical Mechanics. Research on internal structure of stars, optical systems, elementary particles and quantum mechanics. Home: 815 W. 181st St., N.Y.C. 10033.*

MOTZKIN, Theodore S., mathematician; b. Berlin, Germany, 1908; s. Leo Pauline (Rosenblum) M.; student U. Berlin, 1924-27, U. Goettingen, 1928, U. Paris, 1930; Ph.D., U. Basel, 1934; m. Naomi Orenstein; children—Aryeh Leo, Joseph J. E. Ihanan, Gabriel G. H. Faculty, U. Jerusalem, 1936-48, Boston Coll., 1950; prof. math. U. Cal., Los Angeles, 1950—. Vis. prof. U. Jerusalem, 1962; Rockefeller U., 1966; cons. U. Chgo., 1953. Harvard research fellow, 1948; NSF sr. postdoctoral fellow U. Copenhagen, 1963. Mem. Am., Denmark, France, Israel (pres. 1936-48), London, Switzerland math. socs. Contbr. to abstract structures, polynomial algebra and geometry, convexity and approximation theory.

MOUAT, Frederic John, physician; b. Maidstone, Eng., 1816; s. James Mouat; studied in London, Edinburgh, Scotland, Paris; M.C.S., 1838; M.D., Edinburgh, 1839, LL.D., 1886. Became asst. surgeon Bengal Army, 1840, surgeon, 1853; prof. chemistry, materia medica Bengal Med. Coll., 1841-45, med. jurisprudence 1845-49, medicine and clin. medicine, 1849-53, medicine, 1853; became 1st physician Med. Coll. Hosp., 1853; named insp. gen. jails in Lower Provinces, 1855; became surgeon maj., 1860; ret., 1870. Fellow Royal Coll. Surgeons; mem. Bethune Soc. Calcutta (founder 1851), Royal Statis. Soc. (pres. 1890-92). Author: Rough Notes of a Trip to Reunion, The Mauritius and Ceylon, 1852; Andaman Island, 1859-63. Introduced chaulmoogra oil for treatment of leprosy to Western medicine, 1854. Died Jan. 12, 1897.

MOUCHEZ, Amédée Ernest Barthélémy, French astronomer; b. Madrid, Spain, Aug. 24, 1821 (of French parents); ed. French Naval Acad.; served in French navy to 1778; dir. Paris obs., from 1878; founder Montsouris Obs., Paris; apptd. rear adm. in French navy, 1880. Mem. French Acad. Scis., 1875, Bur. Longitudes. Author: La photographie astronomique et la carte du ciel; Rio de la Plata; Description et instruction nautique . . . Observed transit of Venus, 1874; planned internat. photog. map of sky which made possible more precise studies of star positions; made hydrographic investigations along Atlantic coast of Am.; made coastal surveys of Brazil and Algeria. Died Wissous, France, June 25, 1892.

MOUCHOT, Augustin-Bernard, French physicist; b. Semur, France, circa 1821; prof. Tours (France) lycée. Author: La chaleur solaire, 1869. Used solar heat to power steam engine, Tours, 1875; exhibited furnaces and distillation equipment which used his sun receiver, Paris, 1878. Died Paris, 1911.

MOUFET, Thomas (or Moffat, Moffett, Muffet), English naturalist, chemist, physician; b. London, 1553; s. Thomas and Alice (Ashley) Moffett; B.A., Caius Coll.; M.A., Trinity Coll., Cambridge (Eng.) U., 1576; M.D, Basel, 1578, Cambridge, 1582; m. Jane Wheeler Dec. 23, 1580 (dec. 1600); m. 2d, Catherine Brown. Practiced medicine, London and Ipswich; elected M.P. for Wilton, 1597. Mem. Royal Coll. Physicians, London. Author: De I et praestantia chemicorum medicamentorum dialogus apologeticus, 1584; Nosomantica Hippocratea, 1588; Insectorum sive minimorum animalium theatrum . . . (compiled from writings of Wotton, Gesner, Penny), 1634. Studied and defended use of chem. medicines; defended Paracelsus against Galenists. Died Wilton, June 25, 1605.

MOULDER, James William, Am. microbiologist, educator; b. Burgin, Ky., Mar. 28, 1921; s. Webb and Mayme (Downey) M.; S.B., U. Chgo., 1941, Ph.D., 1944; m. Dolly Petterson, Jan. 9, 1942; children—Linda J. (Mrs. Stephen P. Hubbell), John E., Carol A., Susan K. Faculty microbiology, biochemistry U. Chgo., 1944—, 57, prof. microbiology, 1957—, chmn. dept., 1960—. Cons. to U. S. Army Chem. Corps, 1956—. Recipient Eli Lilly award in microbiology and immunology, 1954. Mem. N.Y. Acad. Sci., Am. Soc. Microbiology, Am. Acad. Microbiology, Am. Soc. Biol. Chemists, A.A.A.S. Author: The Biochemistry of Intracellular Parasitism, 1962; The Psittacosis Group as Bacteria, 1964. Editor: Jour. Infectious Diseases, 1957—. Research on biochemistry of psittacosis-lymphogranuloma group of microorganisms; intermediary metabolism. Home: 313 Douglas St., Park Forest, Ill. 60466. Office: 5724 Ellis Av., Chgo. 60637.

MOULE, George Russell, Australian veterinarian; b. Mt. Morgan, Queensland, Australia, Apr. 29, 1914; s. John William and Catharine (Heale) M.; student Queensland Agrl. Coll., 1934-35; B.V.Sc., U. Queensland, 1941, D.V.Sc., 1956; m. Shirley Isabel Davidson, Apr. 18, 1945; children—Shirley Elizabeth, Wendy Margaret, Dorothy Merrilyn. Vet. officer Queensland Dept. Agr., 1942-50, dir. sheep husbandry, 1950-57; gen. mgr. Esperance Plains, W.A., 1957-58; sr. prin. research scientist Ian Clunies Ross Animal Research Lab., Prospect, New South Wales, Australia, 1958-63; dir. prodn. research Australian Wool Bd., Sydney, Australia, 1964—. Mem. expert panel on infertility of livestock FAO, 1962—; lectr. U. Queensland, 1950—, U. Sydney, 1958—. Mem. Australian Vet. Assn., Assn. Soc. Animal Prodn., New Zealand Soc. Animal Prodn. Editor: Field Experiments with Sheep, A Manual of Techniques, 1965. Research, numerous publs. on ecology of pastoral prodn. by sheep in Australia, reprodn., neonatal survival, reprodn. by ram, and pasture animal relationships of breeding flocks with spl. reference to quality of available forage, studies on control of testes temperature. Home: 68 Chapman Av., Beecroft, New South Wales. Office: 261 George St., Sydney, New South Wales, Australia.*

MOULTON, Charles Robert, Am. chemist; b. Clifton, Pa., Sept. 16, 1884; s. Charles Lewis and Maria Ross (Harper) M.; academic certificate Lewis Inst., 1903; B.S. in Chem. Engring., U. Ill., 1907; M.S. in Agr., U. Mo., 1909, Ph.D., 1911; m. Edith Ione Lehnen, June 24, 1911; children—Ruth Elizabeth, Marjorie. Asst. agrl. chemistry U. Mo., 1907-10, instr. in agrl. chemistry, 1910-11, asst. prof., 1912-18, asst. in animal nutrition, Inst. of Animal Nutrition, 1917-18, prof., 1918-22; dir. Dept. of Nutrition, Inst. Am. Meat Packers, 1923-33; lecturer Inst. Meat Packing, U. Chgo., 1926-32; lectr. Schs. of Speech and Edn., Northwestern U., 1933-37; cons. chemist since 1935; curator dept. of chemistry Mus. of Sci. and Industry, Chgo., 1942-43, research asso., 1942-43; tech. adviser Chicago O.S.R.D. Patent Group, 1943-46; asst. dir. Chgo. Patent Group, Argonne Natl. Lab., since 1946. Fellow A.A.A.S., Chem. Soc., Am. Inst. Nutrition, Inst. Food Technologists, Am. Soc. Animal Prodn., Research Council, Sigma Xi. Author: Meat Through the Microscope; also sect. on meat, meat products, poultry, eggs, fish in Allen's Commercial Organic Analysis. Editor and joint author: The Service of Science in the Packing Industry; co-author (with H. P. Armsby): The Animal as a Converter of Matter and Energy. Tech. editor Meat Mag., 1934-35, mng. editor, 1935-37, editor, 1937-40; cons. editor Nat. Provisioner, 1941-42. Died Dec. 4, 1949.

MOULTON, Forest Ray, Am. astronomer; b. Le Roy, Mich., Apr. 29, 1872; s. Belah G. and Mary C. (Smith) M.; A.B., Albion Coll., 1894, Sc.D., 1922; Ph.D. summa cum laude, U. Chgo., 1899; LL.D., Drake U., 1939; Sc.D., Case Sch. Applied Sci., 1940; m. Estelle Gillette, Mar. 25, 1897; 2 sons, 2 daus. Faculty in astronomy U. Chgo., 1898-1926, prof., 1912-26; adminstrv. sec. A.A.A.S., 1937-48. Research asso. Carnegie Instn., 1908-23. Fellow Royal Astron. Soc., A.A.A.S. (editor 25 symposium vols.), Am. Philos. Soc., Am. Acad. Arts and Sciences; mem. Nat. Acad. Scis., Am. Math. Soc. (asso. editor Trans. 1907-12); Am. Astron. Soc.; hon. fgn. asso. Brit. Assn. Adv. Science. Author: Celestial Mechanics, 1902, 14; Introduction to Astronomy, 1905, 16; Descriptive Astronomy, 1911; Periodic Orbits, 1920; New Methods in Exterior Ballistics, 1926; Differential Equations, 1929; Astronomy, 1931; Consider the Heavens, 1935; Autobiography of Science (with J. J. Schifferes), 1945. Contbr. and editor The World and Man, 1937. Contbr. to math. and astron. jours. Research (with Chamberlin) on planetesimal or spiral-nebula hypothesis of origin of solar system; found formula indicating southerly deviation in bodies falling from rest nr. earth's surface, 1914. Died Dec. 8, 1952.

MOULTON, Jack E., Am. pathologist, veterinarian; b. Seattle; Mar. 4, 1922; s. E.C. and M.V. (Klement) M. student U. Wash., 1942-43; B.S., Wash. State U., 1947, D.V.M., 1949; Ph.D., U. Minn., 1953; m. Idell F. Dudley, June 26, 1949; children—William Scott, Sally Rae, Colette Marie. Instr., U. Minn., 1949-52; faculty U. Cal. at Davis, 1952—, prof. pathology, 1964—. Mem. Am. Vet. Med. Assn., Research workers Animal Diseases, Phi Zeta, Alpha Psi. Author: Tumors in Domestic Animals, 1961; articles also. Research on cytopathogenesis canine hepatitis virus infection cell. Home: 18 Parkside Dr., Davis, Cal. 95616.*

MOULTON, James Malcolm, Am. biologist; b. West Haven, Conn., July 25, 1921; s. James William and Anna Kimball (Young) M.; B.S., U. Mass., 1947; M.A., Harvard, 1950, Ph.D., 1952; m. Hope Isabel Kibbe, June 25, 1949; children—Nancy Martha, John Newell, James Roby. Instr. biology Brown U., 1951; instr. anatomy Johns Hopkins Med. Sch., 1951-52; faculty Bowdoin Coll., Brunswick, Me., 1952—, now prof. biology, acting chmn. biology, 1959-60, 66. Vis. scientist U. Edinburgh (Scotland), 1967; asso. in marine biology Woods Hole Oceanographic Instn., 1955—. Memhard scholar Woods Hole Marine Biol. Lab., 1949; Guggenheim fellow, Fulbright scholar U. Queensland (Australia), 1960-61. Mem. A.A.A.S., Am. Soc. Zoologists, Am. Fisheries Soc., Soc. Ichthyologists and Herpetologists, Am. Micros. Soc., Animal Behaviour Soc. Research, publs. in acoustical biology of marine organisms, including sound prodn. of schooling clues, behaviour in relation to sound, hearing; morphology in exptl. embryology and cytology. Home: 11 McKeen St., Brunswick, Me. 04011.*

MOUNT, Donald Irvin, Am. biologist; b. Miamisburg, O., Sept. 20, 1931; s. Ralph Edwin and Mary (Prether) M.; B.S., Ohio State U., 1953, M.S., 1957, Ph.D., 1960; m. Miriam Ann Pospesel, Dec. 21, 1953; children—Martha Jane, David Ralph. Fisheries research biologist USPHS, Cin., 1960-67; dir. Nat. Water Quality Lab., Duluth, Minn., 1967—. Recipient Superior Service award Dept. Health, Edn. and Welfare, 1965. Mem. Am. Fisheries Soc., Am. Soc. Zoologists, A.A.A.S., Ohio Acad. Sci., Sigma Xi. Contbr. articles to profl. jours. Developer apparatus to control dissolved oxygen in water, apparatus to mix toxicants in water at various concentration, autopsy methods to detect mortality of fish due to zinc and endrin. Home: 2013 Lawn St., Duluth 55812. Office: 6201 Congdon Blvd., Duluth, Minn. 55804.*

MOUNT, Lester Adran, Am. neurosurgeon; b. Lebanon, O., Mar. 23, 1910; s. Jesse Adran and Dorothea (Franz) M.; B.S., U. Cin., 1932, B.M., 1934, M.D., 1935; m. Ruth Maxwell Baker, June 16, 1934; children—Philip Maxwell, Melinda, David Jefferson. Practice medicine, specializing in neurol. surgery, N.Y.C., 1941—; asso. attending neurosurgeon Vanderbilt Clinic Presbyn. Hosp., 1947-60, attending neurosurgeon, 1960—; faculty Columbia Coll. Phys. and Surgs., 1941—, asso. profl. clin. neurol. surgery, 1955—; chief neurosurgery Project Hope. Diplomate Am. Bd. Neurol. Surgery. Mem. Neurosurg. Soc. Am. (past pres.), Soc. Neurol. Surgeons, Harvey Cushing Soc., A.C.S., A.M.A., Assn. Research Nervous and Mental Disease, Sociedad Chilena de Neurologia, Psiquiatria y Neurocirugia, Benemerita Sociedad Medico, Quirurgua del Guayas (Ecuador), others. Contbr. chpts. to books. Research, publs. on diagnosis, treatment, operative procedures, instruments in neurosurgery; aneurysms, collateral cerebral circulation. Home: 48 Greenfield Av., Bronxville, N.Y. 10708. Office: 710 W. 168th St., N.Y.C. 10032.

MOUNT, Wayne Delano, Am. atmospheric physicist; b. West Allis, Wis., Dec. 15, 1927; s. Jess Raymond and Grace (Delano) M.; B.S., Mass. Inst. Tech., 1952, M.S., 1953, Ph.D., 1958; student Weber Coll., 1948-50; m. Mary Claire Lindgren, Feb. 19, 1949; children—Robin, Bruce Delano, Lisa. Staff, Geophysics Research Directorate, Air Force Cambridge Research Center, Bedford, Mass., 1953-62, chief technique devel. sect. atmospheric research, 1958-60, chief application and devel. br., 1960-62; head atmospheric physics dept. Sperry Rand Research Center div. Sperry Rand Corp., Sudbury, Mass., 1962—. Mem. Am. (past program chmn.), Royal meteorol. socs., Am. Geophys. Union, A.A.A.S. Contbg. author: Proceedings of the International Symposium on Numerical Weather Prediction in Tokyo, 1962. Research, publs. on phys. and dynamical processes of atmosphere, millimeter wave radiation from constitutents in atmosphere; developed objective techniques for modeling environment; established relation between spatial and temporal predictability of atmosphere as function of density and type of data; co-invented system for probing atmosphere remotely to measure its thermal structure. Home: Tower Rd., Lincoln Mass. 01773. Office: Sperry Rand Research Center, North Rd., Sudbury, Mass. 01776.*

MOUNTCASTLE, Vernon Benjamin, Am. neurophysiologist; b. Shelbyville, Ky., July 15, 1918; s. Vernon Benjamin and Marguerite (Waugh) M.; B.S. in Chemistry, Roanoke Coll., 1938; M.D., Johns Hopkins, 1942; m. Nancy Clayton Pierpont, Sept. 6, 1945; children—Vernon B. III, Anne Clayton, George Earle Pierpont. House physician Johns Hopkins Hosp., 1942-43, fellow in physiology Johns Hopkins Sch. Medicine, 1946-48, faculty, 1948—, prof., 1959—, dir. dept. physiology, 1964—; lectr. U. Coll., London, 1959; vis. lectr. Coll. de France, Paris, 1959. Chmn. physiology study sect. NIH, 1958-61. Mem. Nat. Acad. Sci., Am. Physiol. Soc., Sigma Xi, Sigma Chi, Alpha Omega Alpha, Phi Chi. Research on neural mechanisms in sensation and perception, physiology of central nervous system. Home: 4 Midvale Rd., Balt. 21210.*

MOURANT, Arthur Ernest, Brit. geneticist; b. Jersey, Channel Islands, Apr. 11, 1904; s. Ernest Charles and (Emily) Gertrude (Bray) M.; B.A., Exeter Coll., Oxford, Eng., 1925, postgrad.; M.A., D.Phil., Oxford U., 1931, B.M., B.Ch., 1943, D.M., 1948. Staff, Geol. Survey Gt. Britain, 1929-31, Jersey Path. Lab., 1935-38; med. officer Nat. Blood Transfusion Service, 1944-45, Galton Lab. Serum Unit, Cambridge, Eng., 1945-46; dir. Blood Group Reference Lab., London, 1946-65; dir. Serological Population Genetics Lab., London, 1965—; hon. adviser Nuffield Blood Group Centre, Royal Anthrop. Inst., 1951-65; hon. mem. staff Lister Inst., London, 1950-65; hon. sr. lectr. hematology St. Bartholomews Hosp. Med. Sch., 1965—. Recipient Huxley Meml. medal Royal Anthropol. Inst., 1961. Fellow Royal Soc., 1966, Royal Coll. Physicians, Coll. Pathology. Author: The Distribution of Human Blood Groups, 1945; (with A. Kopec, K. D. Sobczak), The ABO Blood Groups 1958; also numerous articles. Research on distbn. blood groups of all systems in human populations throughout world and their interpretation in terms of racial origin, natural selection and hybridization, genetics newly discovered blood groups, devel. blood grouping techniques. Home: 5 Mercier Rd., London S.W.15. Office: Serological Population Genetics Lab., St. Bartholomew's Hosp., West Smithfield, London E.C.1, Eng.*

MOUREAU, Paul Léon Joseph Benoit Gérard, Belgian toxicologist; b. Liège, Jan. 9, 1904; s. Léon and Guillemine (Pirnay) M.; M.D. agrégé, U. Liege; Dr. honoris causa U. Toulouse and Bordeaux; m. Marie-Thérèse De Boosere, Apr. 25, 1929. Asst., agrégé, instr., prof. legal medicine, toxicology, immuno-hematology, human genetics; instr. Pasteur Inst., Paris; med. expert for courts and tribunals. Decorated Order Leopold, Legion of Honor; recipient Count Launoit prize, Pasteur Inst. medal of Paris. Mem. Internat. Acad. Legal Medicine (sec. gen., treas.), Royal Acad. Medicine Belgium, Acad. Paris (laureate), various sci. socs. Author: Contribution à l'étude des facteurs d'individualisation du sang humain et leurs applications en med. légale; La transfusion sanguinaire, 1952; Determination of Blood Groups in Blood Stains, 1963; others. Home: 11 quai Churchill. Office: 1 rue des Bonnes Villes, Liège, Belgium.

MOUREU, Charles, French chemist; b. Mourenx, France, Apr. 19, 1803; prof. École supérieure de pharmacie, then Coll. de France; mem. Acad. Medicine, French Acad. Scis. Author: Notions fondamentales de chimie organique, 1902. Research on unsaturated hydrocarbons, acetylene derivatives, presence of rare gases in mines and wells; developed theory of antioxidation, leading to current use of antioxidants in vegetable oil and rubber industries; studied condensation products of acrolein; discovered rubene. Died Biarritz, France, June 13, 1929.

MOURSUND, Andrew Fleming Jr., Am. mathematician; b. Fredericksburg, Tex., Dec. 4, 1901; s. Andrew Fleming and Therese (Wahrmund) M.; B.A., U. Tex., 1923, M.A., 1927; Ph.D., Brown U., 1932; m. Lulu Amelia Vorleck, June 29, 1931; children—Robert Andrew, David Garvin, Anne Loreen, Peter Douglas. Tchr. high schs., Tex., 1924-27; instr. math. Tex. Tech. Coll., 1927-28; teaching research fellow Brown U., 1928-31; faculty U. Ore., Eugene, 1931—, prof., 1942—, dept. head, 1939—. Mem. Am. Math. Soc., Math. Assn. Am., Inst. Math. Statistics, Sigma Xi. Research, publs. in Fourier series and summability. Home: 1953 Moss St., Eugene, Ore. 97403.*

MOUSSU, Auguste Leopold (called Gustav), French pathologist; b. St.-Laurent-en-Gatines, France, Jan. 1, 1864; prof. pathology Vet. Sch. Alfort, also Inst. Agronomy; mem. French Acad. Scis., 1934, Acad. Agr. Research on function of salivary glands, tubercular diseases; showed functional independence of thyroid and parathyroid glands, 1897. Died Oct. 16, 1945.

MOUTON, Gabriel, French mathematician; b. Lyons, France, 1618; D. Theology; became perpetual vicar St.-Paul's Ch., Lyons, 1654. Author: Observationes diametrorum solis et nunae apparentium, 1670. Determined diameter of sun at apogee, 1661; built precise astron. pendulum, preserved at Lyons; developed 1st, though incomplete, idea of natural longitude measurement; proposed system of measurements similar to metric system, 1670; found method for summation of series of numbers by differences (early step in devel. of Newton's and Leibniz's calculus); calculated logarithms, sines, tangents. Died Sept. 28, 1694.

MOUZON, James Carlisle, Am. physicist, elec. engr., educator; b. San Antonio, Jan. 8, 1907; s. Edwin DuBose and Mary (Mike) M.; A.B., So. Meth. U., 1927, recipient of honorary Doctor of Science, 1966; Ph.D. cum laude (fellow), Cal. Inst. Tech., 1932; m. Elizabeth Nancy Walker, Sept. 10, 1932; children—Elizabeth (Mrs. James R. Butterfield, Margaret (Mrs. Wayne T. VanWagoner). Head sci. dept. Wesley Jr. Coll., 1927-28; grad. asst., fellow Cal. Inst. Tech., 1929-32; faculty Duke, 1932-44; sr. physicist Brown Instrument div. Mpls. Honeywell, Phila., 1944-45, head elec. sect., 1945-47, dir. research, 1947-49; planning div. Research and Devel. Bd., Washington, 1949-50; chief atomic warfare div. Office Asst. for Operations Analysis, USAF, Washington, 1950-53; chief intelligence div. Operations Research Office, Johns Hopkins, 1953-57; mem. research planning staff Engring. Research Inst., U. Mich., Ann Arbor, 1957, prof. elec. engring., 1957-—, cons. Willow Run Labs., 1957-60, dir. Ford Found. Engring. Faculty Devel. Program, 1959-—, asso. dean Coll. Engring., 1960-66, dir. engring. Brazilian Program, 1961-67. Fellow Am. Phys. Soc., A.A.A.S.; mem. I.E.E.E., Am. Soc. Physics Tchrs., Am. Soc. for Engring. Edn. (chmn. Mich. sect. 1964-65), Sigma Xi, Tau Beta Pi, Eta Kappa Nu, Sigma Alpha Epsilon. Contbr. articles on ionization of gases, electronics, nuclear physics, instrumentation and control, mil. operational problems to profl. publs. Patentee instrumentation, submarine detection and control. Home: 2687 Apple Way, Ann Arbor, Mich. 48104.*

MOVAT, Henry Zoltán, pathologist; b. Timisoara, Rumania, Aug. 11, 1923; s. Erwin and Piroska (Kubitschka) M.; student U. Vienna, 1942-45; M.D., U. Innsbruck, 1948; M.S., Queen's U., Kingston, Ont., 1954, Ph.D., 1956, F.R.C.P., 1967; m. Ilse Hirselandt, Dec. 29, 1956; children—Ronald, Kenneth, Douglas. Faculty, U. Toronto, Ont., 1957-—, prof. pathology, 1965-—, chmn. div. gen. and exptl. pathology, 1967-—. Mem. Am. Soc. Exptl. Pathology, Am. Assn. Immunologists, Am. Assn. Pathology, Soc. Exptl. Biology and Medicine. Research, numerous publs. in ultrastructure of renal diseases, mechanism of inflammation, mechanism of anaphylaxis. Home: 17 Truxford Rd., Don Mills, Ont. Office: 100 College St., Toronto 5, Ont., Can.*

MOVIUS, Hallam Leonard, Jr., Am. anthropologist; b. Newton, Mass., Nov. 28, 1907; s. Hallam Leonard and Alice Lee (West) M.; A.B., Harvard, 1930, M.A., 1932, Ph.D., 1937; Docteur Honoris Causa, U. Bordeaux (France), 1961; m. Nancy Ch. de Crespigny, Sept. 25, 1936; children—Geoffrey Hallam, Alice Vierville. Mem. faculty Harvard, 1930-—, asso. prof., curator palaeolithic archaeology Peabody Museum, 1950-57, prof. anthropology, 1957-—; field expdns. to Czechoslovakia, 1930, Central Europe (Am. Sch. Prehistoric Research summer fellow), 1931, Palestine, 1932, Ireland, 1932-36, Burma and Java, 1937-38, Eastern France, 1948, Western Europe, 1949, 53, 54-55; field dir. Abri Pataud project, Southwestern France, 1958, 59, 60-61, 62, 63-64. Viking Fund medalist in archaeology, 1949. Fellow Am. Acad. Arts and Scis., Soc. Antiquaries London, Am. Anthrop. Assn., Geol. Soc. Am.; mem. Nat. Acad. Scis.; hon. mem. Schweizerischen Gesellschaft für Urgeschichte, Soc. Prehistorique de L'Ariege, Inst. Italiano di Prehistoria o Protostoria; corr. mem. Deutsches Archäologisches Inst., Prehistoric Soc. Eng., Dept. Archaeology Govt., India, Mus. Nat. d'Histoire Naturelle, Paris. Author numerous articles in field. Research on Stone age archaeology in the Old World. Home: 17 Larchwood Dr., Cambridge, Mass. 02138.

MOWRER, Ernest Russell, Am. sociologist; b. Lost Springs, Kans., Aug. 18, 1895; s. James Theodore and Anna (Smith) M.; A.B., U. Kans., 1918; M.A., U. Chgo., 1921, Ph.D. (Research fellow), 1924; m. Harriet C. Rosenthal, Oct. 12, 1924. Faculty, Coe Coll., 1922-23, Ohio Wesleyan U., 1924-25; research sociologist Wieboldt Found., Chgo., 1926-28; faculty Northwestern U., Evanston, Ill., 1928-—, prof. sociology, 1943-—; vis. prof. Mich. State U., 1940, U. Chgo., 1948, U. Ariz., 1964. Social Sci. Research Council fellow, 1925-26, grantee, 1933. Fellow Am. Sociol. Assn. (sec.-treas. 1947-49); mem. Sociol. Research Assn., Am. Statis. Assn., Am. Assn. U. Profs., Alpha Kappa Delta. Author: Family Disorganization, 1927; (with Harriet R. Mowrer) Domestic Discord, 1929; The Family, 1932; Disorganization, Personal and Social, 1942; (with William Dobriner, et al) The Suburban Community, 1958; (with Ernest W. Burgess, Donald J. Bogue, et al) to Urban Sociology, 1964; Life Styles in Suburbia, 1966. Contbg. editor Social Sci. Abstracts, 1928-33; mng. editor Am. Sociol. Rev., 1947-49; editor Crofts Sociol. Series, 1940-50. Research, publs. on family interaction, marital disorgn., unification of maj. forms deviant behavior into a common pattern. Home: 4037 Fairway Dr., Wilmette, Ill. 60091. Office: Dept. Sociology, Northwestern U., Evanston, Ill. 60201.*

MOWRER, Orval Hobart, Am. psychologist; b. Unionville, Mo., Jan. 23, 1907; s. John A. and Sallie (Todd) M.; A.B., U. Mo., 1929; Ph.D., Johns Hopkins, 1932; NRC fellow Northwestern U., 1932-33, Princeton, 1933-34; Sterling fellow, Yale, 1934-36; m. Willie Mae Cook, Sept. 9, 1931; children—Linda, Kathryn, Todd. Instr. psychology, mem. research staff Inst. Human Relations, Yale, 1936-40; asst. prof. edn. Harvard, 1940-43, asso. prof., 1943-48; research prof. psychology U. Ill. 1948-—; spl. cons. USPHS. Served as clin. psychologist OSS, 1944-45. Recipient Certificate of Merit, U. Mo., 1956. Diplomate Am. Bd. Examiners Profl. Psychology. Fellow Am. Psychol. Assn. (pres. 1954; dir.; pres. personality, social psychology, clin. and abnormal psychology divs. 1953); mem. Am. Acad. Psychotherapists, Am. Assn. U. Profs., Am. Psychol. Found. (pres. 1959-60), Sigma Xi. Author: Frustration and Aggression (with Dollard, et al,) 1939; Learning Theory and Personality Dynamics, 1950; Psychotherapy—Theory and Research, 1953; Learning Theory and Behavior, 1960; Learning Theory and the Symbolic Processes, 1960; The Crisis in Psychiatry and Religion, 1961; The New Group Therapy, 1964. Editor: Patterns of Modern Living (3 Vols.), 1950; Morality and Mental Health, 1967; Harvard Editorial Rev., 1945-48. Home: 610 W. Vermont St., Urbana, Ill. 61802.*

MOYA, Frank, Am. physician; b. N.Y.C., Jan. 20, 1929; s. Frank O. and Candita (Soto) M.; B.A., N.Y. U., 1949; M.D., State U. N.Y., 1953; m. Elizabeth Strelecki, Sept. 27, 1963; children—Richard, Jonathan, Casey, Maria, Elizabeth. Asst. prof. Columbia Coll. Phys. and Surg., 1960-62; prof., chmn. dept. anesthesiology U. Miami (Fla.) Sch. Medicine, 1962-—. Mem. Am. Soc. Anesthesiologists (gen. chmn. ann. session arrangements 1966), Assn. U. Anesthetists (sec. 1966-—), Am. Soc. Pharmacology and Exptl. Therapeutics. Research, numerous publs. on maternal physiology and methods anesthesia for obstetrics and resuscitation newborn. Home: 1430 S. Bayshore Dr., Miami, Fla. 33132.*

MOYER, Andrew Jackson, Am. microbiologist; b. Star City, Ind., Nov. 30, 1899; s. Edward R. and Minnie (McCloud) M.; A.B., Wabash Coll., 1922; postgrad. U. Wis., 1922-23; M.S., N.D. Agrl. Coll., 1925; Ph.D., U. Md., 1929; m. Dorothy Randall Phillips, Apr. 4, 1931. With Dept. Agr. 1929-—, last assignment No. Utilization Research, and Devel. Div., Fermentation Sec., U. S. Dept. Agr., Peoria, Ill. Recipient Lasker group award, 1946. Mem. Bot. Soc. Am., Ill. Acad. Sci., Sigma Xi. Contbr. to sci. jours. Research on physiology of fungi, mold fermentations, gluconic, lactic, kojic, glauconic, citric and itaconic acids from glucose, penicillin; devised methods for producing substantial increase in yields of penicillin, making large scale prodn. possible; discovered methanol process for submerged mold fermentation of crude carbohydrates to citric acid; improved fermentaion of glucose and Molasses to fumanic acid; patentee in field. Died Feb. 17, 1959.

MOYER, Burton Jones, Am. physicist; b. Greenville, Ill., Feb. 24, 1912; s. Jacob and Mabel (Jones) M.; A.B., Seattle Pacific Coll., 1933, D.Sc., 1954; Ph.D., U. Wash., 1939; m. Lela I. Brushwood, June 22, 1937; children—Burton Jones, John H., Robert P., Lela Virginia. Faculty physics and math. Greenville Coll., 1939-42; with Manhattan Dist. Project, U. Cal. Radiation Lab., also at Oak Ridge, 1942-45; physicist Lawrence Radiation Lab., U. Cal. at Berkeley, 1945-—, faculty, 1947-—, prof., 1954-—, chmn. dept., 1962-—. Fellow Am. Phys. Soc. Research, publs. in nuclear structure and reactions, meson physics, elementary particle properties. Home: 878 Arlington Av., Berkeley, Cal. 94707.*

MOYER, John Henry, III, Am. physician; b. Hershey, Pa., Apr. 1, 1917; s. John Henry II and Anna Mae (Gruber) M.; B.S., Lebanon Valley Coll., 1939; M.D., U. Pa., 1943; m. Mary Hughes, Sept. 3, 1948; children—John Henry, IV, Michael, Carl, Anna Mary, Nancy Elizabeth, Mary Louise. Pharmacology and medicine fellow U. Pa., 1948-49; faculty Baylor U. Coll. Medicine, 1950-57; prof., chmn. dept. medicine Hahnemann Med. Coll. and Hosp., Phila., 1957-—; cons. in internal medicine various hosps.; med. adviser Social Security Adminstrn., 1964-—. Recipient Presdl. citation, 1964. Diplomate Am. Bd. Internal Medicine. Fellow Am. Coll. Cardiology (Susan and Theodore R. Cummings Humanitarian award 1962, 65, 66, trustee), Am. Coll. Chest Physicians, Am. Coll. Clin. Pharmacology and Chemotherapy (pres. 1964-66), A.C.P., Coll. Physicians Phila., N.Y. Acad. Scis.; mem. A.A.A.S., Am. Soc. Pharmacology and Exptl. Therapeutics, Am. Therapeutic Soc. (Oscar B. Hunter award 1959, pres. 1965-66), Am. Heart Assn., Assn. Am. Med. Colls., Assn. Profs. Medicine, Gerontological Soc., Nat. Assn. Standard Med. Vocabulary, Soc. Exptl. Biology and Medicine, So. Soc. Clin. Research, Am. Med. Authors, Hellenic Soc. Cardiology (hon.), Sigma Xi, others. Editor: Hypertension, 1959; (with others) The Theory and Practice of Auscultation, 1964; (with Leon Cander) Aging of the Lung: Perspectives, 1964; (with Albert Brest) Cardiovascular Drug Therapy, 1965; (with L. C. Mills) Shock and Hypertension, 1965; editorial cons. Am. Jour. Cardiology, 1960; editor in chief Cyclopedia of Medicine, Surgery and Specialties, 1963-65. Research, publs. on renal dynamics and pulmonary physiology, clin. pharmacology of antihypertensives, tranquilizers, cholesterol control, cerebral metabolism and hemodynamics. Home: 305 Llandrillo Rd., Bala Cynwyd, Pa. 19004. Office: 230 N. Broad St., Phila. 19102.*

MOYER, Kenneth Evan, Am. psychologist; b. Chippewa Falls, Wis., Nov. 19, 1919; s. John Evan and Margaret (Lashway) M.; A.B. with honors, Park Coll., 1943; M.A., Washington U., St. Louis, 1948, Ph.D., 1951; m. Doris Virginia Johonson, May 29, 1943; children—Robert Stephen, Cathy Lita. Faculty, Carnegie-Mellon U., Pitts., 1949-—, prof. psychology, 1961-—; cons. on higher edn. Govt. of Norway, 1954. Mem. research adv. com. Pa. Commonwealth Mental Health Found., 1956-—. Recipient Carnegie Found. award for excellence in teaching, 1954. Mem. Am., Pitts. (past dir.) psychol. assns., Psychonomic Soc., So. Soc. Philosophy and Psychology, Sigma Xi, Theta Kappa Theta. Research, publs. on endocrinology emotion, startle response and avoidance behavior; demonstrated young children have a capacity for prolonged attention spans if proper toys are used, that adrenal glands are not essential for effects electroconvulsive shock on behavior. Home: 252 Gates Dr., Munhall, Pa. 15121. Office: Carnegie-Mellon U., Pitts. 15213.*

MOYERS, Robert Edison, Am. physiologist, dentist; b. Sidney, Ia., Nov. 12, 1919; s. Albert Edison and Ruth (Johns) M.; B.S., U. Ia., 1942, M.S., 1947, D.D.S., 1942, Ph.D., 1949; m. Barbara Renee Quick, July 15, 1956; children—Mary Katherine, Martha Jane. Practice dentistry, Lake City, Ia., 1942-43; asso. in orthodontics U. Ia., 1945-49; prof., head dept. orthodontics U. Toronto (Ont., Can.), 1949-53; prof. dentistry, chmn. dept. orthodontics U. Mich., Ann Arbor, 1953-65, dir. Center for Human Growth and Devel., 1965-—. Cons. WHO, 1958-—, U. S. Vets. Hosp., Ann Arbor, Mich., 1956-—. Fellow A.A.A.S.; mem. Am. Dental Assn., Am. Assn. Orthodontists, European Orthodontics Soc., N.Y. Acad. Sci., Sigma Xi. Author: Handbook of Orthodontics, 1963; also articles. Editor: (with P. Jay) Orthodontics in Mid-Century, 1958. Pioneered research in electromyography facial and jaw muscles relating changes in elec. activity to functions muscles and devel. various usages facial and jaw muscles; research devel. teeth and bones, face and head. Home: 1035 Country Club Rd., Ann Arbor, Mich. 48105.*

MOYLE, Walter, English naturalist; b. Bake, Eng., Nov. 3, 1672; s. Sir Walter Moyle; ed. Oxford, Temple; studied birds of Belgium; worked to form complete collection of birds for Royal Soc., rectifying Ray's errors; author various works edited by T. Serjeant, 2 vols., 1726, by A. Hammond, 1727. Died June 9, 1721.

MOYLS, Benjamin Nelson, Canadian mathematician; b. Vancouver, B.C., Can., May 1, 1919; s. Benjamin James and Jessie (Walker) M.; B.A., U. B.C., 1940, M.A., 1941; A.M., Harvard, 1942, Ph.D., 1947; m. Ina Elizabeth Barbour, Nov. 7, 1942; children—Gregory Nelson, Peter William. Faculty dept. math. U. B.C., Vancouver, 1947-—, prof., 1959-—, asst. dean grad. studies, 1967-—. Mem. Am. Math. Soc., Math. Assn. Am., Soc. Indsl. and Applied Math., A.A.A.S., Can. Math. Congress, Edinburgh Math. Soc. Research in linear transformations on algebras of matrices, maximum and minimum values of functions of hermitian forms. Home: 2016 Western Pkwy., Vancouver 8, B.C., Can.*

MOYNIHAN, Sir Berkeley George Andrew, surgeon; b. Malta, Oct. 2, 1865; s. Capt. Moynihan; student Royal Naval Sch.; M.D., univs. Ghent and Buenos Aires; M.Ch., univs. Dublin and Cairo; D.Sc., U. Belfast; D.C.L. (hon.), univs. Oxford and Durham; LL.D. (hon.), univs. Leeds, Edinburgh, Bristol, St. Andrews, McGill, Toronto and Winnipeg; m. Isabella Wellesley Jessop, 1895 (dec. 1936); one son, two daus. Arris and Gale lectr., 1898-1900; Hunterian prof., 1919-20; Hunterian orator, 1927; Romanes lectr. Oxford U., 1932; Walker lectr. U. St. Andrews (Scotland), 1933; Linacre lectr. Cambridge U., 1936; emeritus prof. surgery U. Leeds (London); cons. surgeon Leeds Gen. Infirmary. Chmn. Army Med. Adv. Bd.; pres. d'honneir XIX Congrès français de Chirurgie. Served in World War I; hon. maj. gen. Army Med. Service. Decorated knight comdr. Order St. Michael and St. George; Companion of the Bath; knight justice Order St. John Jerusalem; Order of Merit (Chila); grand cordon Order Nile. Fellow Royal Coll. Surgeons (pres. 1926-32), Royal Coll. Surgeons Edinburgh, Royal Coll. Surgeons Ireland, A.C.S.; hon. fellow Am. Surg. Assn., French Acad. Medicine, med. socs. Bologna, Lombardy, Norway, Poland, Serbia, acads. medicine N.Y., Cleve.; corr. mem. Soc. Chir. de Paris. Author: Abdominal Operations, 4th edit., 1925; Gallstones and their Surgical Treatment, 2d edit., 1906; Duodenal Ulcer, 2d edit., 1912; The Pathology of the Living and other Essays; The Spleen and its Diseases, 1921. Credited with introducing the expression hunger pain, 1901, to describe the pain felt by patients with duodenal ulcer several hours after a meal; specialized in techniques of abdominal, pancreatic, gastric operations. Died Sept. 7, 1936.

MOYSE, Alexis, French biologist; b. Arcueil, Oct. 2, 1912; s. Auguste and Marie (Hery) M.; Ph.D. agrégé, Faculty of Sci., Paris. Dir. research Nat. Center for Sci. Research; dir. photosynthesis labs.; prof. Faculty of Sci., Paris. Author: Respiration et metabolisme azoté; Etude de physiologie folliaire; Biologie et physicochimie; Les produits de la fixation de CO2 par les végétaux; Relations entre la photosynthèse et la respiration. Research on carbon metabolism, relations between structure and function; metabolism of algae; photosynthesis; chloroplasts activities; photophosphorylation. Home: 24 rue Ferdinand-Jamin, Bourg-la-Reine, Seine. Office: Faculté des Sciences, Laboratoire de Physiologie Végétable, Orsay, Seine-et-Oise, France.

MOZLEY, Robert F., Am. physicist; b. Boston, Apr. 18, 1917; s. Fred and Mabel (Snow) M.; A.B., Harvard, 1938; M.S., U. Cal. at Berkeley, Ph.D., 1950; m. Anita Ventura, June 23, 1967; children—(by previous marriage) Peter Snow, John Cochran. Elec. engr. Radar Sperry Gyroscope Co., N.Y.C., 1941-45; asst. Radiation Lab., U. Cal. at Berkeley, 1945-50; faculty Princeton, 1950-53, asst. prof., 1952-53; faculty Stanfrod, 1953—, prof. physics, 1963—. Mem. Am. Phys. Soc., Am. Assn. U. Profs., Research, publs. in elementary particle physics. Home: 607 Laurel Av., Menlo Park, Cal. Office: S.L.A.C., Stanford, Cal.*

MRAZEC, Ludovic, Rumanian geologist; b. 1867; founder Rumania's Geol. Inst.; mem., chmn. Rumanian Acad. Author: On the Formaticn of Oil Fields in Rumania (explained by theory of organic origin), 1907; more than 200 other works on mineralogy, petrography, tectonics. Died 1944.

MROZOWSKI, Stanislaw, physicist, educator; b. Warsaw, Poland, Feb. 9, 1902; s. Josef and Amelia (Zlotnicka) M.; Ph.D., U. Warsaw, 1931; Docteur honoris causa, U. Bordeaux (France), 1964; m. Irena M. Tarwacka, June 26, 1926. Came to U. S., 1939, naturalized, 1948. Research fellow U. Warsaw, 1929-33, faculty, 1933-39; research fellow U. Cal. at Berkeley, 1939-40; research asso. U. Chgo., 1940-45; asso. prof., chmn. physics dept. George Williams Coll., 1942-45; head physics dept. research and devel. div. Gt. Lakes Carbon Corp., 1945-49; prof. physics, dir. carbon research lab. State U. N.Y. at Buffalo, 1949—, chmn. physics dept., 1959-64. Chmn., Am. Carbon Com., 1957-63, exec. com., 1963—; Fulbright prof. Keio and Nagoya univs., Japan, 1963-64; vis. prof. U. Karlsruhe (Germany), summer 1968. Recipient M. Kernbaum prize in physics Warsaw Sci. Soc., 1932. Fellow Am. Phys. Soc., Optical Soc. Am.; mem. Polish Phys. Soc. (chmn. Warsaw sect. 1938-39; hon. mem.), Am. Chem. Soc., Am. Assn. Physics Tchrs., Polish Inst. Arts and Scis. in Am. (v.p. council 1964-65, pres. 1965—). Editor: Proc. Confrs. on Carbon, 1955-62; editor-in-chief Carbon, 1962—. Contbr. numerous articles to profl. jours. Discovered nuclear isotope effect in atomic and band spectra, 1931, (with Jenkins) interference effect, 1940; research on atomic and molecular spectroscopy, electronic processes in chars, carbons and graphites, bend model, 1950, mech. and thermal properties carbons, paramagnetic resonance in carbons. Home: 109 Westfield St., Buffalo 14226.*

MUAN, Arnulf, chemist; b. Meldal, Norway, Apr. 19, 1923; s. Anders O. and Ingeborg (Engen) M.; Dipl.Chem., Tech. U. Norway, 1948; Ph.D., Pa. State U., 1955; m. Hildegard Hoss, Jan. 29, 1960; children—Michael, Ingrid. Came to U. S., 1949, naturalized, 1962. Instr., Tech. U. Norway, 1948-49; Norwegian Govt. fellow Pa. State U., 1949-50, research asso., 1950-51, 52-55, faculty, 1955—, prof. metallurgy, 1962—, prof. mineral sci., head dept. geochemistry and mineralogy, 1966—; research fellow U. Oslo (Norway), 1951-52. Mem. Am. Chem. Soc., Am. Soc. for Metals, Am. Inst. Metall. Engrs., Am. Ceramic Soc. (Ross Coffin Purdy award 1958), Mineral. Soc. Am., Norwegian Chem. Soc., Norwegian Metall. Soc., Sigma Xi. Author: (with E. F. Osborn) Phase Equilibria among Oxides in Steelmaking, 1965; also numerous articles. Research on delineation phase equilibria at high temperatures in multicomponent oxide and silicate systems involving components of more than one oxidation state; determination of thermodynamic properties of compounds and solutions at high temperatures. Home: 1121 Walnut Circle, State College, Pa. 16801. Office: M.S. Bldg., University Park, Pa. 16802.*

MUCH, Hans Christian R., German physician; b. Zechlin, Germany, 1880; prof., dir. Serology Inst., Hamburg, Germany. Author: Die pathologische Biologie, 1911; Über die unspezifische Immunität, 1921; Homöopathie, 1926. Described non-acid-fast gram-positive granules (Much's granules) in sputum of tubercular patients, now considered modified tubercle bacilli, 1907. Died 1932.

MUCKENHAUSEN, Eduard, German geologist; b. Enzen, Feb. 17, 1907; s. Jakob and Annan Maria (Esser) M.; Ph.D., U. Bonn; m. Elise Ludwig, Apr. 12, 1938; 1 dau., Gabriele. Prospector for Geol. Inst., Prussia; head edn. div. Geol. Office N. Rhine and Westphalia; prof., dir. Inst. at Bonn. Author: Die wichtigsten Böden der Bundesrepublik Deutschland, 1959; Entstenhung Eigenschaften und Systematik der Böden der Budesrepublik Deutschland, 1961. Research in soil genetics, systematics, geography, mineralogy of clay. Home: Lotharstrasse 113. Office: Nussallee 13, Bonn-Rhein, West Germany.

MUCKENHOUPT, Carl Frederick, Am. physicist, educator; b. Poughkeepsie, N.Y., Nov. 7, 1901; s. Frederick William and Sophrona (Geller) M.; A.B., Williams Coll., 1922; B.S. in Elec. Engring., Mass. Inst. Tech., 1924, Ph.D., 1929; m. Sarah Joanna Boell, Jan. 12, 1929; children—Frederick William (dec.), Benjamin, Joanna Margaret (Mrs. Robert Duncan Enzmann). With power transmission dept. Westinghouse Electric and Mfg. Co., 1925-26; research mech. div. Gen. Electric Research Lab., 1929-31; asst. prof. elec. engring. Northeastern U., Boston, 1929-35, prof. physics, chmn. dept., 1935-46, dir. research, 1959-61, dean research adminstrn., 1959-61, prof. elec. engring, 1961-66. Cons. engring. electronics, radio and elec. circuits, 1929—; chief scientist Office Naval Research, Boston, 1946-59. Fellow A.A.A.S.; mem. Am. Soc. Engring. Edn., Am. Phys. Soc., Phi Beta Kappa, Sigma Xi, Eta Kappa Nu, Tau Beta Pi (nat. council 1954-58). Author book on modern physics. Research, publs. on turbine found. vibration, elec. heating of large cylindrical bodies, high frequency resonances in capacitors, minesweeping devices. Home: 332 Winchester St., Newton Highlands, Mass. 02161. Office: Northeastern U., Boston 02115.*

MUCKERMANN, Hermann, German anthropologist, eugenicist; b. Bückenburg, Germany, Aug. 30, 1877. Jesuit priest, until 1926; head eugenics dept. Kaiser Wilhelm Inst. for Anthropolcgy, 1927-33; prof. anthropology and social ethics Berlin-Charlottenburg Tech. Sch., from 1948; prof. anthropology Free U. Berlin; dir. Inst. for Natural and Cultural anthropology, Berlin-Dahlem. Mem. Max Planck Soc. for Advancement Sci., Internat. Population Union, others. Author: Kind und Volk, 16th edit., 1933-34; Vererbung und Entwicklung, 2d edit., 1947; Die Familie im Lichte der Lebensgesetze, 1952; Der Sinn der Ehe, 3d edit., 1952; Von Sein und Sollen des Menschen, 1954; Grundriss der Biologie; Der Biologische West der Mütterlichen Stillpflicht; Eugenishe Ehebenatung; Um das Leben der Ungeborenen; Eugenik. Editor: Studien aus dem Institut für natur-und geisteswissenschaftliche Anthropologie; Humanismus und Technik. Research on basic biology, eugenics, cultural anthropolcgy. Home: 9 Kammgasse, Berlin-Frohnau. Office: Institute for Natural and Cultural Anthropology, 24 Ihnestrasse, Berlin-Dahlem, Germany.

MUDD, Stuart, Am. microbiologist, educator; b. St. Louis, Sept. 23, 1893; s. Harvey Gilmer and Margaret (Clark) M.; B.S., Princeton, 1916; A.M., Washington U., St. Louis, 1918; M.D., Harvard, 1920; m. Emily Borie Hartshorne, Sept. 12, 1922; children—Emily Borie (Mrs. James Mitchell), Stuart Harvey, Margaret Clark, John Hodgen. Research fellow Harvard, 1920-23; asso. Rockefeller Inst., 1923-25; asso. pathology Henry Phipps Inst., U. Pa., Phila., 1925-31, asst. prof. exptl. pathology, 1925-31, asso. prof. bacteriology, 1931-34, prof., 1934-51, prof. microbiology, 1951-59, prof. emeritus, 1959—; chief microbiologic research program VA Hops., Phila., 1959—. Mem. council Biochem. Research Found. Franklin Inst., 1946—. Recipient Guggenheim Honor Cup award, 1944; Consejero de Honor, Consejo Superior de Investigaciones Scientificas, Madrid, Spain, 1964. Fellow A.A.A.S., Am. Pub. Health Assn., Coll. Physicians Phila., N.Y. Acad. Sci.; mem. World Acad. Art and Sci. (charter, v.p. 1962—), Am. Human Serum Assn. (pres. 1940-41), Soc. Am. Bacteriologists (pres. 1943), Histochem. Soc. (pres. 1952), Internat. Assn. Microbiol. Socs. (pres. 1958-62), Am. Assn. Pathologists and Bacteriologists, Am. Assn. Immunologists, Am. Soc. Exptl. Pathology, Soc. Exptl. Biology and Medicine, Harvey Soc., Am. Physiol. Soc., Phi Beta Kappa, Sigma Xi, Alpha Omega Alpha. Editor: Blood Substitutes and Blood Transfusion, 1942; The Population Crisis and the Use of World Resources, vol. II, 1964. Contbr. articles on exptl. pathology, immunology, microbiology to profl. jours. Home: 734 Millbrook Lane, Haverford, Pa. 19041. Office: VA Hosp., Phila. 19104.*

MUDGE, Gilbert Horton, Am. physician; b. N.Y.C., Apr. 9, 1915; s. Alfred E. and Alice (Horton) M.; B.A., Amherst Coll., 1936; M.D., Columbia, 1941, D.M.Sc., 1948; m. Eleanor Mackenzie, July 12, 1941; children—George A., Gilbert H., John T., Eleanor W. Faculty, Columbia, 1948-55, asso. prof. medicine, 1954-55; prof. pharmacology and exptl. therapeutics Johns Hopkins, 1955-62; dean, prof. exptl. therapeutics, Dartmouth Med. Sch., Hanover, N.H., 1962-65, prof. medicine, 1965—. With pharmacology and exptl. therapeutics study sect. USPHS, 1957-60, regulatory biology panel NSF, 1960-65. Mem. Soc. Exptl. Biology, Am Physiol. Soc., Am. Soc. Clin. Investigation (v.p. 1959-60), Life Ins. Med. Research Fund, Assn. Am. Physicians, Am. Soc. Pharmacology and Exptl. Therapeutics. Editorial bd.: Am. Jour. Pharmacology and Experimental Therapeutics, 1958-65. Contbr. articles to tech. jours. on exptl. medicine. Study of electrolyte physiology; renal function. Home: River Rd., Lyme, N.H. Office: Dartmouth Med. Sch., Hanover, N.H.

MUDGE, Thomas, inventor; b. Exeter, Eng., 1717; apprenticed to watchmaker George Graham, 1731; clock maker, London; apptd. watchmaker to King, 1776; inventor clock escapement, circa 1755, free anchor escapement (now in gen. use), 1765; built elaborate, accurate chronometers, from 1771, marine chronometers, from 1776. Died 1794.

MUDGE, William, English engr.; b. Plymouth, Eng., Dec. 1, 1762; s. John Mudge; ed. Royal Mil. Acad., Woolwich, Eng., 1777; LL.D., U. Edinburgh, 1818; m. Margaret Jane Williamson; 1 dau., 4 sons. Commd. 2d lt. Royal Arty., 1779; named lt. gov. Woolwich Royal Mil. Acad., 1809; apptd. to ordnance Trigonometrica Survey, 1791, dir., 1798; commd. brevet maj. Brit. Army, 1801, served to maj. gen., 1819. Fellow Royal Soc., 1798; mem. French Acad. Sci., 1816, Copenhagen Acad. Sci. Author: General Survey of England And Wales, 1805; An Account of the Trigonometrical Survey . . . , 1809; also numerous articles. Participated in drawing of trigonometric map of Gt. Britain; prepared maps of Wales and Scotland; supervised extension of meridian line to Scotland. Died Apr. 17, 1820, London.

MUECKE, Dietrich Karl August, German physiol. chemist; b. Rybnik, Oct. 6, 1920; s. Albert and Alice (Wycisk); student U. Göttingen; Dr.med., U. Leipzig, 1948; m. Erika Wagner, May 27, 1955. Faculty U. Leipzig, 1956-59; prof., dir. Inst. Physiol. Chemistry, U. Rostock, 1959—. Affiliate Royal Soc. Medicine (London); mem. Internat. Soc. Bioclimatology, E. German Biochem. Soc. Author: Einführung in mikrobiologische Bestimmungsverfahren, 1955; also numerous articles. Research on ultra-micromethod for B12 determination; micromethod for determination and isolation of humic acids; metabolism of neurospora crassa; biochemistry and biology of humic acids and their metabolism in mammals. Address: 32 Klement-Gottwald Strasse, Rostock, East Germany.*

MUEHLBERGER, Clarence Weinert, Am. toxicologist; b. Chgo., July 16, 1896; s. Otto and Rosa (Weinert) M.; B.S. in Chem. Engring., Armour Inst. Tech. (now Ill. Inst. Tech.), 1920; M.S. in Chemistry, U. Wis., 1922, Ph.D. in Chemistry and Toxicology, 1923; m. Mary Ellen Finn, Sept. 15, 1923; 1 son, Robert Mortelle. State toxicologist for Wis., 1923-30; toxicologist, dir. Cook County (Ill.) Coroner's Lab., Chgo., 1930-41; asst. dir. Northwestern U. Law Sch. Sci. Crime Detection Lab., Chgo., 1930-41; chief Mich. Crime Detection Lab., State Health Labs., Lansing, 1941—; asso. prof. toxicology U. Wis., Madison, 1923-30, Northwestern U. Med. Sch., Chgo., 1930-35; professorial lectr. in toxicology U. Chgo., Med. Sch. 1935-41; extramural lectr. in toxicology U. Mich. Med. Sch., Ann Arbor, 1941—. Fellow Am. Acad. Forensic Scis.; mem. Am. Chem. Soc., Pharmacology Soc., Sigma Xi, Tau Beta Pi. Research on selenium chemistry and toxicology, 1920-23; measurement of alcohol intoxication in human body, 1935-56; toxicity of tellurium. Home: 463 Rosewood Av., East Lansing, Mich. Office: Mich. Dept. Health Labs., 3500 N. Logan St., Lansing, Mich.

MUEHLEMANN, Hans Rudolph, Swiss dentist; b. St. Moritz, Switzerland, Aug. 26, 1917; s. Hans and Elsa (Müller) M.; student U. Geneva (Switzerland), 1938-40, U. Berne (Switzerland), 1940; D.D.S., U. Zurich, 1942, M.D., 1946; postgrad. U. Ill., U. Minn.; m. Maria Stoller, Sept. 25, 1948; 1 dau., Marietta. Asso. prof. periodontology U. Zurich, 1951, prof. operative dentistry and periodontology, 1953—, dir. Dental Inst., 1963—. Editor: Helv. Odont. Acta, 1956—. Research, numerous publs. on devel. of aminefluorides for caries prevention, tooth mobility, quantitative methods in dental calculus formation, short term animal caries experimentation. Home: 8 Beustweg Ch 8032, Zurich, Switzerland.*

MUEHLHAUSE, Carl O., Am. physicist; b. Balt., Sept. 17, 1918; s. William M. and Virginia (Greer) M.; B.S., U. Va., 1940; M.S., U. Ill., 1941, Ph.D., 1943; m. Carolyn M. Lyon, June 17, 1947; children—Pamela Ann, Carl Gregory. With Argonne (Ill.) Nat. Lab., 1946-52, Brookhaven Nat. Lab., Upton, N.Y., 1952-56, Republic Aviation, 1957-58, Nat. Bur. Standards, Gaithersburg, Md., 1958—. Fellow Am. Phys. Soc., N.Y. Acad. Scis.; mem. Am. Nuclear Soc. Research, publs. in nuclear physics, reactor mgmt., radiation detection. Home: 9105 Seven Locks Rd., Bethesda, Md. 20034. Office: Nat. Bur. Standards, Gaithersburg, Md. 20234.*

MUELLER, Achim, German chemist; b. Detmold, Germany, Feb. 14, 1938; s. Karl and Berta (Heierhoff) M.; Vordiplom., U. Goettingen, (Germany), 1961, Diplom in Chemistry, 1962, Dr.rer.nat., 1965; m. Helga Fischer, Dec. 15, 1961; children—Harald, Henning, Foern. Dozent, dept. inorganic chemistry U. Goettingen, 1962—. Mem. Gesellschaft Deutscher Chemiker, Deutsche Bunsen-Gesellschaft. Research, numerous publs. on thermochemistry, X-ray diffraction, quantum chemistry, molecular spectra, molecular vibration, mean amplitudes of vibration, vibration spectra of inorganic solids. Home: 28 Schlesierring, Goettingen, Germany.*

MUELLER, Alfred Don, Am. clin. psychologist; b. Port Washington, Wis., Nov. 7, 1893; s. Michael H. and Anna (Majerus) M.; Ph.B., U. Wis., 1918, Ph.M., 1919; postgrad. Carnegie Inst. Tech.; A.M., Yale, 1926, Ph.D., 1927; m. Marie S. Struve, July 19, 1919; children—Alfred Jerome, Charles Wendell. Prof. edn. and psychology State Tchrs. Coll., Worcester, Mass., 1923-29; asso. prof. ednl. psychology U. Tenn., 1929-38; dir., psychologist Knoxville (Tenn.) Child Guidance and Adult Adjustment Clinic, 1938-47; clin. psychologist med. teaching group Kennedy VA Hosp., Memphis, 1947-50, chief clin. psychologist, 1950-58, chief psychology service, 1958-63; pvt. psychol. practice, Memphis, 1963—. Prof. psychology Ft. Sanders Hosp., Nurses Tng. Sch., 1945-47; staff asso., clin. cons. Carrol Turner Sanitorium, Memphis, 1950-63; mem. Tenn. Bd. Examiners in Psychology, 1954-59, chmn., 1958-59. Fellow A.A.A.S., Am. Psychol. Assn.; mem. Am. Assn. U. Profs., Acad. Polit. and Social Sci., Nat. Soc. Coll. Tchrs. Edn., Nat. Soc. for Study Edn., Phi Kappa Phi, Phi Delta Kappa, Pi Gamma Mu. Author: Progressive Trends in Rural Education, 1926; Teaching in Secondary Schools, 1927; A Vocational and Socio-Educational Survey of New England High Schools, 1929; Principles and Methods of Adult Education, 1937; Psychology and You, 1963; contbr. to American Psychology: In Persons and Anecdotes, 1965; also articles. Organized 1st child guidance and adult adjustment clinic in Tenn., 1938; psychol. research on paraplegic, arthritis patients, pain studies. Address: 4035 Tutwiler Rd., Memphis, Tenn. 38122.*

MUELLER, Charles Richard, Am. phys. chemist, educator; b. St. Louis, June 22, 1925; s. Charles Adolph and Mamie (Bruno) M.; A.B., Washington U., St. Louis, 1948; Ph.D., U. Utah, 1951; m. Amelia Ann Mae Atee, Feb. 10, 1947; children—Karen Mary, Joan Amelia. Faculty chemistry dept. Purdue, Lafayette, Ind., 1951-—, prof., 1964-—. Mem. Am. Chem. Soc., Am. Assn. U. Profs., Sigma Xi. Contbr. articles to profl. jours. Developer semi-localized orbital theory of chem. bonding, application of scattering theory and molecular beam technique to problem of forces between molecules. Home: 471 Robinson St., West Lafayette, Ind. 47906. Office: Chemistry Dept., Purdue U., Lafayette, Ind. 47901.*

MUELLER, Donald Weier, Am. physicist; b. Cin., June 30, 1907; s. William Henry and Amelia (Weier) M.; A.B., Cornell U., 1929, postgrad.; postgrad. Princeton; m. Frances Wentworth Clough, June 9, 1944; children—Frances Wentworth, Elizabeth French, Henry Weier, Carolin Graf. Physicist, Hartford-Empire Co., 1937-43, 45-46; instr. USAAF program U. N.M., Albuquerque, 1943; sci. staff mem. Manhattan Project, Los Alamos, 1943-45; supervising staff mem. Los Alamos Sci. Lab., 1946-63, meson facility devel. staff, 1963-—. Mem. Am. Phys. Soc. Research, publs. on Van de Graaff electrostatic generator, spherical implosions; group demonstration small thermonuclear device; design staff linear accelerator for meson physics.*

MUELLER, Edward, Am. chemist; b. South Bend, Ind., Oct. 16, 1883; s. Frederick William and Anna Margaret (Sack) M.; B.S., Purdue, 1902; A.M., Harvard, 1905, Ph.D., 1907; postgrad. Heidelberg (Germany) U., 1908; m. Georgiana Crane, Aug. 21, 1913. Chemist Norfolk & Western Ry. Co., Roanoke, Va., 1902-04; asst. in chemistry Harvard, 1905-07; instr. chemistry Washington U. Med. Sch., 1907-10, Tufts Coll., 1910-11; instr. chemistry Mass. Inst. Tech., 1911-13, asst. prof., 1913-20, asso. prof., 1920-29; chem. cons., since 1929; in charge of chemistry Franklin Tech. Inst., 1942-46. Fellow A.A.A.S., Am. Acad. Arts and Scis., mem. Am. Chem. Soc. Translator: Holde's Hydro-carbon Oils and Saponifiable Fats and Waxes, 1915, 2d edit., 1922. Contbr. to sci. publs. Redetermined atomic weights of potassium and chromium. Died Aug. 9, 1954.

MUELLER, Emil, Swiss mycologist; b. Winterthur, Switzerland, May 3, 1920; s. Emil Jakob and Helena (Wanger) M.; ing.agr., Swiss Fed. Inst. Tech., 1944, Dr.sc.tech., 1950; m. Elisabeth Amans, 1953; children—Martin, Marianne, Andreas, Verena. Agrl. service and agrl. tchr. Graubünden, 1944-48; agrl. service and agrl. tchr. Agrl. Research Stas. of Waedenswil and Oerlikon, Swiss Fed. Inst. Tech., 1948-53, curator bot. collections, 1953-—, private dozent, 1966-—, dozent mycology, 1961-—. Author: (with A. von Arx) Die Gattungen der amerosporen Pyrenomyceten, 1954; Die Gattungen der didymosporen Pyrenomyceten, 1962; also numerous articles. Research on systematic mycology; description of new species, especially ascomycetes. Home: Breitackerstrasse 15, 8702 Zollikon, Switzerland. Office: Inst. for Special Botany, Swiss Fed. Inst. Tech., Universitätsstrasse 2, 8006, Zurich, Switzerland.*

MUELLER, Erwin Wilhelm, see Müller, Erwin Wilhelm.

MUELLER, George Edwin, Am. physicist; b. St. Louis, July 16, 1918; s. Edwin and Ella Flora (Bosch) M.; B.S., Mo. Sch. Mines, 1939; M.S., Purdue U., 1940; postgrad. Princeton; Ph.D., Ohio State U., 1951; m. Maude Rosenbaum, Dec. 27, 1941; children—Karen Ann, Jean Elizabeth. Tech. staff Bell Telephone Labs., 1940-46; prof. elec. engring. Ohio State U., Columbus, 1946-56; cons. electronics Ramo-Wooldridge, Inc., 1955-57; with Space Tech. Labs., Los Angeles, 1958-63, v.p. for research and devel., 1962-—, asso. adminstr. for manned space flight NASA, Washington. Fellow Am. Inst. Aeros. and Astronautics, I.E.E.E.; mem. Am. Phys. Soc., N.Y. Acad. Sci., Nat. Space Club, Sigma Xi, Tau Beta Pi, Sigma Pi Sigma, Eta Kappa Nu. Author: (with E. R. Spangler) Communication Satellites, 1963; also tech. papers. Patentee, research on electron tubes and antennas; adminstr. devel. and testing ballistic missiles, manned space flights. Home: 3043 West Lane Keys N.W., Washington 20007. Office: NASA, Washington 20546.*

MUELLER, Gerhard O. W., criminologist; b. Eigenrieden, Germany, Mar. 15, 1926; s. Alfred A. and Ella (Lother) M.; came to U. S., 1950, naturalized, 1953; student U. Kiel (Germany), 1947-49; J.D., U. Chgo., 1953; LL.M., Columbia, 1955; m. Ruth H. L. Pieper, Dec. 15, 1949; children—Mark H., Marla L., Monica R., Matthew A. Instr., U. Wash. Sch. Law, 1953-54; faculty W.Va. U. Coll. Law, 1955-57, asso. prof. law, 1956-57; sr. fellow Yale Law Sch., 1957-58; faculty N.Y. U. Sch. Law, 1958-—, prof. law, 1960-—, dir. comparative criminal law project, 1958-—, chmn. dept. criminal law, 1963-—. Fulbright lectr. U. Freiburg (Germany), 1958; spl. cons. U. S. Senate Com. on Judiciary, 1965-—; dir. criminal law program Universite Internationale de Scis. Comparees, Luxembourg, 1961-64; v.p. Centro-Italo-Statunitense di Studio Judiciaria, Milan, Italy, 1961-—. Rockefeller grantee, 1957-58. Mem. Assn. Internationale de Droit Penal (past pres., Am. nat. sec., v.p.), Assn. for Psychiat. Treatment Offenders (sec. 1959-—), Am. Fgn. Law Assn. (dir. 1959-65), Assn. Am. Law Schs. (chmn. com. on comparative law 1964), Am. Soc. Criminology (v.p. 1963-67, pres. 1967-68). Author numerous books including: Essays in Criminal Science, 1960; Legal Regulation of Sexual Conduct, 1961; El Derecho Penal, 1963; (with E. Wise) International Criminal Law, 1965; (with J. Hall) Criminal Law and Procedure, 1965; also numerous articles. Co-editor: Jour. Offender Therapy, 1961-—; asso. editor Jour. Criminal Law, Criminology and Police Sci., 1957-—; editor Am. Series Fgn. Penal Codes and Publs. of Comparative Criminal Law Project, 1958-—. Comparative method in law and criminology; original criminological research. Home: 161 Sagamore Rd., Millburn, N.J. 07041. Office: 40 Washington Sq. S., N.Y.C. 10003.*

MUELLER, Ivan I., geodesist; b. Budapest, Hungary, Jan. 9, 1930; s. Laszlo and Irene (Fischer) M.; Dipl. Engring., Tech. U. Budapest, 1952; Ph.D., Ohio State U., 1960; m. Marianne Gelei, Sept. 9, 1950; children—Julianne, Elizabeth. Came to U. S.; naturalized, 1963. Instr., dept. geodesy Coll. Mil. Engring., Tech. U. Budapest, 1952-53, asst. prof., 1954-56; design engr. C. H. Sells, Inc., cons. engrs., N.Y.C., 1957-58; instr. dept. geodetic sci. Ohio State U., Columbus, 1959, asst. prof., 1960-63, asso. prof., 1963-66, prof., 1966-—, researcher Research Found., 1960-—; vis. scientist Am. Geophys. Union, Washington, 1964; vis. prof. Tech. U. Berlin (Germany), 1965. Mem. Am. Geophys. Union (chmn. com. on sci. papers), Am. Soc. Photogrammetry, Sigma Xi. Author: Physical Geodesy I: The Variation of Gravity in Space (in Hungarian), 1954; The Gradients of Gravity and their Application in Geodesy, 1960; Introduction to Satellite Geodesy, 1964; also papers. Research on satellite geodesy. Home: 4290 Cambourne Rd., Columbus 43221. Office: 164 W. 19th Av., Columbus, O. 43210.*

MUELLER, Richard Gustav, East German chemist; b. Hartha, Sachsen, East Germany, July 17, 1903; s. Bruno and Luise (Sawall) M.; Dr.phil., U. Leipzig (Germany), 1931; m. Lotte Reichardt, Aug. 18, 1934; 1 dau., Hilde. Staff, Chemische Fabrik von Heyden, A.G., Radebeul, Sachsen, 1931-45, VEB Chemische Fabrik von Heyden, Radebeul, 1945-54; staff, dir. Institut für Silikon- und Fluorkarbonchemie, Radebeul, 1954-—; prof. Technische U. Dresden (Germany), 1954-—. Recipient Nationalpreis, 1951. Mem. Deutsche Bunsengesellschaft, Chemische Gesellschaft in der Deutschen Demokratischen Republik (Clemens-Winkler medal 1962), Am. Chem. Soc. Research, numerous publs. on preparation of compounds with 4 silicon atoms bonded on one carbon atom, new compounds with penta- and hexacovalent silicon atoms; discovered direct synthesis of organo-halogenosilanes. Home: 4a Hoflossnitzstr., Office: 35 Wilh.-Pieck-Str., Radebeul, Sachsen, East Germany.*

MUENCH, Hugo, Am. physician; b. St. Louis, Oct. 17, 1894; s. Hugo and Eugenia (Thamer) M.; A.B., Cornell U., 1915; M.D., Washington U., St. Louis, 1918; Dr.P.H., Johns Hopkins, 1932; A.M. (hon.), Harvard, 1947; m. Helen Ruth Harrison, Dec. 28, 1920; 1 son, James Frederick. Field staff Internat. Health div. Rockefeller Found., 1921-46; prof. biostatistics Harvard Sch. Pub. Health, 1946-61, emeritus, 1961-—; sr. cons. biostatistics Lemuel Shattuck Hosp., Boston, 1961-—. Cons. NIH, 1947-—. Recipient Lemuel Shattuck award Mass. Pub. Health Assn., 1964. Mem. Am. Pub. Health Assn., Am. Statis. Assn., Inst. Math. Statistics, A.A.A.S., Am. Epidemiological Soc. Author: Catalytic Models in Epidemiology, 1959; also articles. Research on statis. methods as applied to epidemiologic and clin. problems. Home: 100 Memorial Dr., Cambridge, Mass. 02142. Office: 170 Morton St., Boston 02130.*

MUETTERTIES, Earl Leonard, Am. chemist; b. Elgin, Ill., June 23, 1927; s. Earl Conrad and Muriel (Carpenter) M.; B.S., Northwestern U., 1949; A.M., Harvard, 1951, Ph.D., 1952; m. JoAnn Mary Wood, Mar. 3, 1956; children—Eric J., Mark C., Gretchen A., Maria C., Martha A., Kurt A. Staff central research dept. E. I. du Pont de Nemours & Co., Wilmington, Del., 1952-—, research supr., 1957-65, asso. dir. basic scis., 1965-—. Mem. Am. Chem. Soc. (mem. adv. bd. Monographs 1964-—, exec. com. div. inorganic chemistry 1965-—, award in inorganic chemistry 1965), Chem. Soc. London. Author: Chemistry of Boron and its Compounds, 1967; also numerous articles. Editor: Inorganic Syntheses, vol. X, 1967; editorial bd. Inorganic Chemistry, 1962-—, Inorganic Syntheses, 1960-—. Patentee in field. Research in fields of boron hydrides, organometallics, fluorine chemistry, molecular structure. Home: 1137 S. Concord Rd., West Chester, Pa. 19380. Office: Central Research Dept. Exptl. Sta., E. I. du Pont de Nemours & Co., Wilmington, Del. 19898.*

MUFTIC, Mahmud Kamal, Yugoslavian physician; b. Sarajevo, Yugoslavia, Jan. 14, 1919; s. Saleh Firuz and Umihane (Vranic) M.; M.D., Zagreb Med. Sch., 1944; m. Isaad M. Atia, June 22, 1952; 1 son, Nejad. With Hygiene Inst., Zagreb, Yugoslavia, 1945-47, AMA Labs., Cairo, Egypt, 1948-54, Hygiene Inst., Baghdad, Iraq, 1955-59, Inst. Exptl. Medicine, Borstel, Germany, 1960-62, Hygiene Inst., Lausanne, Switzerland; asst. prof. U. Lausanne, 1962-64; head med. dept. Schering AG, Berlin, Germany, 1964-—. Mem. Am. Soc. Clin. Hypnosis, Brit. Med. Hypnosis Soc., Am. Inst. Med. Hypnosis (award), Palestinian, Iraqui med. assns. Research, numerous publs. in biochemistry: role of enzymes, mechanism of hypnosis, immunopath. reactions, hypnogenetic diseases. Home: P.O. Box 333, Geneva-Cornavine, Switzerland. Office: Mueller Strasse 172, Berlin 65, Germany.*

MUGELI, Henri, Swiss physicist; b. Fuet, July 22, 1894; s. Richard and Henriette (Küpfer) M.; Ph.D., U. Neuchatel; Dr. hon. causa U. Besancon, 1960; m. Anny Jakob, July 18, 1923. Asst., later head works Physics Inst. U. Neuchatel, from 1916, agrégé in physics, 1928, prof. metallography and solid physics, 1940, sec. Faculty of Sci., 1945-47, dean, 1947-49, asst. dean, 1949-51; creator Swiss Labs. for Horological Research, 1921, asst. dir., 1941, dir., 1950, hon. dir., 1962-—. Recipient Léon du Pasquier prize Acad. of U. Neuchatel. Mem. Swiss Soc. Natural Sci., Swiss Physics Soc., Swiss Chronometry Soc., Inst. Metals. Research and publs. on dielectric anomalies of silicon glass, properties of metals and chronometry, study of friction in lubrication, corrosion and processes of protection of metals. Home: 51 rue de Bel-Air. Office: 2 rue A. L. Bréguet, Neuchatel, Switzerland.

MUHAMMAD AL-ZUHRT (Muhammad ibn Abu Bakr al-Zuhri), Arabian geographer; b. al-Zuhra; flourished circa 1140; lived at Granada, Spain, circa 1140. In his gen. treatise on geography he divides world into 7 areas.

MUHAMMAD BEN ABD AL-BAGI AL BAGDADI, Arabic mathematician; flourished circa 1100; author commentary on 10th book of Euclid's Elements which was translated by Gerard of Cremona as Liber judei.

MUHAMMAD BEN 'ALI IBN FARAH AL SAFRA (al-Safra), Arabic botanist; flourished circa 1199; collected and studied plants in meridional Spain; founder bot. garden, Guadix, Spain.

MUHAMMAD BEN IBRAHIM AL-FAZARI, translator; s. Ibrahim (Abu Ishaq) b. Habib b. Sulayman b. Samura ibn Gundab al-Fazari; knew some of Indian sci., including sect. of Siddharta (astron. works of 8th century); which he was ordered to translate from Sanskirt to Arabic by Caliph al-Mansur. Died circa 800.

MUHLENBERG, Gotthilf Henry Ernst, Am. botanist; b. Trappe, Pa., Nov. 17, 1753; s. Henry Melchior and Anna Maria (Weiser) M.; student U. Halle (Germany); M.A. (hon.), U. Pa., 1780; D.D., Princeton, 1787; m. Mary Catharine Hall, July 26, 1774. Ordained to ministry Lutheran Ch., 1770; pastor, Phila., 1774-79, Lancaster, Pa., 1779-1815; sec., Pres. Ministerium of Pa.; became 1st pres. Franklin Coll., 1787. Author: Catalogue of the Plants of North America; Index flora Lancastriense. Co-author: English-German and German-English Dictionary, 12 vols., 1812. Listed over 1100 plants growing nr. Lancaster, 1778-91; pioneer in N.Am. botany, classified native plants according to accepted sci. principles; studied econ. and medicinal value of plants; recognized need of cooperation among botanists in compiling flora of N.Am. Died Lancaster, May 23, 1815.

MUHLER, Joseph Charles, Am. dental educator; b. Ft. Wayne, Ind., Dec. 22, 1923; s. Howard J. and Lauretta (Zurbuck) M.; B.S., Ind. U., 1945, D.D.S. with distinction, 1948, Ph.D. in Chemistry, 1951; m. Majetta Jean Stewart, Feb. 2, 1949; children—Joseph Charles II, James Patrick. Asst. prof. Ind. U., Indpls., 1951-55, asso. prof., 1955-59, prof., 1959-61, research prof. basic scis., 1961-—, co-chmn. dental grad. program Sch. Dentistry, 1963-—. Cons., Ind. Bd. Health, 1951-—, Procter and Gamble Co., 1949-—, Mead Johnson Co., 1960-—, Gen. Foods Corp., 1964-—; cons. biochemistry Sch. Aviation Medicine, Brooks AFB, Tex., 1959-61; cons. preventive dentistry Office Surgeon Gen., U. S. Army, 1961-—, Ft. Knox, U. S. Army, 1962-—, Bur. Medicine and Surgery, USN, 1964-—. Fellow Am. Coll. Dentists, A.A.A.S.; mem. European Assn. for Fluorine Research (hon.), Am. Chem. Soc., Am. Dental Assn. (mem. council Nat. Bd. Dental Examiners 1955-58), Internat. Assn. for Dental Research (sec.-treas. 1961-64), Soc. for Exptl. Biology and Medicine, Ind. Acad. Sci., Am. Assn. Univ. Profs., Am. Soc. Dentistry for Children, Am. Pub. Health Assn., Pierre Fauchard Soc., Sigma Xi, Omicron Kappa Upsilon, Phi Lambda Upsilon. Author: (with M. K. Hine and H. G. Day) Preventive

Dentistry, 1954; (with M. K. Hine) A Symposium on Preventive Dentistry, 1956, Fluorine and Dental Health: The Pharmacology and Toxicology of Fluorine, 1959; (with C. S. Rohrer, E. E. Campaigne) An Introduction to Chemistry, 1957, Introduction to Experimental Chemistry, 1958; A Textbook of Biochemistry for Students of Dentistry, 1959; (with W. C. Hess) Laboratory Manual of Biological Chemistry for Students of Dentistry, 1959; Control of Dental Caries in the Adult Dentition, 1964; Fifty-Two Pearls and Their Environment, 1966; also many articles. Editor Jour. Ind. Dental Assn., 1958—. Home: 420 Buckingham Dr., Indpls. 46208.

MÜHLMANN, Wilhelm Emil, German ethnologist, sociologist; b. Düsseldorf, Germany, Oct. 1, 1904; prof., Mainz, Germany. Author: Rassen- und Völkerkunde, 1938; Staatsbildung und Amphiktyonien in Polynesien, 1938; Methodik der Völkerkunde, 1938; Geschichte der Anthropologie, 1948. Research on ethnic assimilation, culture sociology of Polynesia, gen. methods found. of anthropology.

MUHLY, Harry Townsend, Am. mathematician; b. Balt., Oct. 10, 1916; s. Harry Edward and Anna (Townsend) M.; Ph.D., Johns Hopkins, 1940; m. Martha Isabel Fields, May 26, 1941; children—Paul Scott, Alan Eric, Catherine Ruark. NRC fellow Princeton, 1940-41; faculty U. S. Naval Acad., 1941-50; faculty U. Ia., Iowa City, 1950—, prof., 1955-64, prof., chmn. dept. math., 1964—; lectr. NSF Insts., Western Mich. U., summers 1958, 59, 61, 63, 65. Naval Research fellow Harvard, 1947-48. Mem. Math. Assn. Am. (past bd. govs.), Am., London math. socs., Soc. Indsl. and Applied Math., Societe Math. de France. Author: (with S. Saslaw) Plane and Spherical Trigonometry, 1946; Fundamental Concepts of Mathematics, 1961. Research, publs. in commutative algebra. Home: 1107 Kirkwood Ct., Iowa City, Ia. 52240. Died Dec. 22, 1966.*

MUIR, John, geologist, explorer, naturalist; b. Dunbar, Scotland, Apr. 21, 1838; s. Daniel and Anne (Gilrye) M.; ed. in Scotland and at U. Wis.; hon. degrees Harvard, U. Wis., U. Cal., Yale; m. Louise Strentzel, 1880. Discoverer Muir Glacier, Alaska; visited Arctic regions on U. S. steamer Corwin in search of DeLong expdn.; labored many yrs. in cause of forest preservation and establishment of nat. reservations and parks. Fellow A.A.A.S.; mem. Washington Acad. Sci., Am. Acad. Arts and Letters; pres. Sierra Club. Author: The Mountains of California, 1894; Our National Parks, 1901; Stickeen, the Story of a Dog, 1909; My First Summer in the Sierra, 1911; The Yosemite, 1912; Story of My Boyhood and Youth, 1913; pub. posthumously: Travels in Alaska, 1915; The Cruise of the Corwin, 1917; Steep Trails, 1918; also articles. Editor: Picturesque California. Traveled and studied in Russia, Siberia, Manchuria, India, Australia and New Zealand, 1903-04, in S. America, 1911, in Africa, 1911-12, engaged in bot. and geol. studies. Died Los Angeles, Dec. 24, 1914.

MUIR, Matthew Moncrieff Pattison, Scottish chemist; b. Glasgow, Nov. 1, 1848; ed. univs. Glasgow, Tübingen (Germany); M.A. (hon.), Cambridge U.; m. Florence Haslam, 1873; 2 sons. Fellow Gonville and Caius Coll., Cambridge. Author: (with T. E. Thorpe) Qualitative Analysis and Laboratory Practice, 1874; Chemistry for Medical Students, 1878; Chemistry, 1883; Elements of Thermal Chemistry, 1885; (with D. J. Carnegie) Practical Chemistry, 1887; (with Charles Slater) Elementary Chemistry, 1887; A Treatise on the Principles of Chemistry, 2d edit., 1889; Chemistry of Fire, 1893; The Alchemical Essence and Chemical Element, 1894; The Story of the Chemical Elements, 1896; A Course of Practical Chemistry, 1897; The Story of the Wanderings of Atoms, 1898; The Story of Alchemy, 1902; The Elements of Chemistry, 1904; A History of Chemical Theories and Laws, 1906; Roger Bacon, his Relations to Alchemy and Chemistry, 1914; Men and Women of Letters in Norwich a hundred years ago, 1924. Co-editor: Watts' Dictionary of Chemistry. Died Sept. 2, 1931.

MUIR, Sir Robert, Scottish pathologist; b. Balfron, Scotland, July 5, 1864; s. R. and Susan Cameron (Duncan) M.; M.A., Edinburgh (Scotland) U., 1884, M.B., C.M., 1888, M.D., 1890, Sc.D. (hon.), LL.D., D.C.L. Became sr. asst. to prof. pathology Edinburgh U., 1892, lectr. path. bacteriology, 1894; prof. pathology St. Andrews, 1898-99, U. Glasgow, 1899-1936; became pathologist Edinburgh Royal Infirmary, 1892. Recipient Lister medal, 1936. Fellow Royal Soc., 1911 (Royal medal 1929), Royal Coll. Physicians, Royal Coll. Physicians Edinburgh. Author: (with J. Ritchie) Manual of Bacteriology; Studies on Immunity; Text-book of Pathology, also papers. Died Mar. 30, 1959.

MUIR, Sir Thomas, Scottish mathematician; b. Stonebyres, nr. Lanark, Scotland, Aug. 25, 1844; M.A., LL.D., D.Sc., U. Glasgow (Scotland); m. Margaret Bell, 1876 (died 1919); two sons, two daus. Tutor math. College Hall, St. Andrews, 1868-71; asst. prof. math. U. Glasgow, 1871-74; headmaster math. and sci. depts. Glasgow High Sch.; supt. gen. edn. Cape of Good Hope, S. Africa, 1892-1915; vice chancellor U. Cape Good Hope to 1901. Pres. S. African Assn. Advancement Sci., 1910. Recipient Keith medal for math. research Royal Soc. Edinburgh, 1884, 99, Gunning-Victoria prize for sci. work, 1916. Fellow Royal Soc., 1900; hon. fellow Royal Scottish Geog. Soc. Author: Text Book of Determinants, 1882; History of Determinants, 1890; Theory of Determinants in the Historical Order of Development, vol. I, 1906, vol. II, 1911, vol. III, 1920, vol. IV, 1923; others. Contbr. History of Determinants, 1930, also papers to societies or sci. jours. Died Mar. 21, 1934.

MUIRHEAD, John Charles, Canadian physicist; b. Brandon, Man., Can., Dec. 31, 1932; s. Samuel Andrew and Dora (Brandt) M.; B.Sc., Brandon Coll., 1952; m. Fern Partington Gorrie, Sept. 11, 1952; children—Deborah Lynn, Murray John, Eric Andrew. Physicist, Def. Research Bd. Can., Ottawa, Ont., 1953-57, Ralston, Alta., Can., 1957—. Research, publs. on low, high speed aerodynamics, including devel. specialized test equipment; originator smoke-stream technique for shock tube air movements; inventor high-speed electronic shutter, surface tension gauges for transient pressures, calipers for storage oscilloscope, improved blast pressure gauge, devel. shock wave valves; experimenter in ferroelectric phenomena. Home: 898 13th St. N.E., Medicine Hat, Alta. Office: Def. Research Establishment, Suffield, Ralston, Alta., Can.*

MUKAI, Shigeru, Japanese metallurgist; b. Uwamizawa, Hokkaido, Japan, Nov. 25, 1913; s. Takeshi and Tomi (Toda) M.; B.Engring., Kyoto (Japan) U., 1937, D.Engring., 1955; m. Chieko Okabe, Apr. 28, 1940; children—Tomoko, Yoshiko, Takeki. Mining engr. Furukawa Mining Co. Ltd., 1937-40; prof. Kurume (Japan) Tech. Coll., 1940-50; asst. prof. Kyushu U., Fukuoka, Japan, 1950-56; prof. metallurgy Kyoto (Japan) U., 1956—, dir. mineral processing, 1956—. Mem. Flotation Research Assn. Japan (pres. 1957—), Mining and Metall. Inst. Japan (councillor 1962-65, 66—), Am. Inst. Mining, Metall. and Petroleum Engrs., Mining and Metall. Inst. Japan. Research, numerous publs. on flotation of sulphide and non-sulphide minerals, electric separation of minerals; devel. process recovering high quality low ash coal in normal atmospheric humidity by electrostatic concentration. Home: 16-5 Imakaidocho Matsugasaki, Sakyoku, Kyoto, Japan.*

MUKAIBO, Takashi, Japanese chemist; b. Dairen City, Japan, Mar. 24, 1917; s. Seiichiro and Tsumako (Morino) M.; B.S., U. Tokyo (Japan), 1939, Dr.Engring., 1954; m. Nobuko Mizuta, Jan. 19, 1944; children—Ryuichi, Hiroshi, Atsushi. Faculty engring. U. Tokyo, 1943—, asso. prof. chemistry, 1947-54, prof., 1959—, chmn. dept. nuclear engring., 1959-63, chmn. dept. indsl. chemistry, 1964—; sci. attache Japanese Embassy, Washington, 1954-58. Mem. adv. com. Japan AEC, 1939—; mem. U. S.-Japan Intergovtl. Com. on Cooperation in Sci., 1940—. Mem. Japan Chem. Soc., Japan Electrochem. Soc., Japan Atomic Energy Soc., Am. Chem. Soc., Electrochem. Soc. (U. S.). Author: Materials Science, 1952; also articles. Research in kinetics of formation of calcium carbide, reactions of carbon with halogens and halides, kinetics of oxidation of metals at high temperatures, electrochem. reactions. Home: 2-37-30 Daizawa, Setagaya, Tokyo, Japan.*

MUKERJEE, Sachimohan, Indian microbiologist; b. Calcutta, India, Feb. 8, 1909; s. Chandra Bhusan and Basanta K. (Banerjee) M.; M.B., U. Calcutta, 1935, D.Phil., 1960, D.Sc., 1966; m. Bina Bhaumik, Apr. 11, 1942; 1 son, Monish. Research staff Indian Inst. for Med. Research, Calcutta, 1935-56; sr. sci. officer Indian Inst. for Biochemistry and Exptl. Medicine, Calcutta, 1956-63, dept. dir., 1963—. Mem. sci. group on cholera research WHO, 1962—, dir. Internat. Reference Centre for Vibrio Phagetyping, 1963—; mem. cholera expert group Indian Council Med. Research, 1963—. Recipient G. J. Watumull Meml. award, 1962; Basanti Devi Amir Chand Sr. award Indian Council Med. Research, 1963. Fellow Egyptian Pub. Health Assn. (hon.); mem. Assn. Microbiologists India (mem. central council 1958—), Am. Soc. for Microbiology. Research, numerous publs. on devel. phage-typing of cholera vibrios, live oral vaccine for more effective prophylaxis against cholera. Home: 29 Ponchanontola Lane, Calcutta 34. Office: 4 Raja Subodh Mullick Rd., Calcutta 32, India.*

MUKHERJEA, Amiya Kumar, Indian med. protozoologist; b. Ranchi, India, Mar. 1, 1916; s. Kali Pada and Sarajubala (Maulick) M.; M.B., U. Calcutta (India) U., 1941; D.T.M., Faculty Tropical Medicine and Hygiene, West Bengal, 1949; D.Phil. in Parasitology, Calcutta, U., 1956; m. Maitrayee Banerjee, Aug. 12, 1946; children—Kasturi, Ashoke, Kanai. Asst. research officer Indian Council Med. Research, 1950-51; research fellow Nat. Inst. Scis., India, 1951-54; head dept. protozoology Indian Inst. Biochemistry and Exptl. Medicine, 1954; asst. prof. protozoology Sch. Tropical Medicine, Calcutta, 1955-62, prof., mem. bd. examiners, 1962—; vis. prof. All India Inst. Hygiene and Pub. Health, Calcutta, 1962-65. Mem. Assn. Microbiologists India, Assn. Parasitologists India, Indian Sci. Congress Assn., Indian Council Med. Research (mem. sci. adv. com. 1966—). Research, publs. on cytochemistry and immunity of protozoa; protozoa living in alimentary and genital tracts, blood of man and animals; described 2 new species of Coccidia; developed new methods of staining parasitic protozoa, fungi and spirochaetes; demonstrated occurrence of polar body in eggs of Ascaris lumbricoides. Home: 5/3 Cornfield Rd., Calcutta 19, West Bengal, India.*

MUKHERJEE, Shyamapada, Indian chemist; b. Ranchi, India, Nov. 4, 1919; s. K. C. and Prafulla K. (Banerjee) M.; B.Sc. with honors, Patna (India) U., 1937, M.Sc., 1939; A.I.I.S.C., Indian Inst. Sci., Bangalore, India, 1942; Ph.D., U. Cambridge (Eng.), 1948; m. Gouri Banerjee, Mar. 1, 1949; children—Pracheta, Chandreyee, Atreyee. Asst. devel. officer Ministry Industry and Supply, Govt. of India, Delhi, 1948-49; faculty Nat. Sugar Inst., Kanpur, India, 1949—, prof. sugar chemistry, 1958—. Fellow Sugar Technologists Assn. India medals and prizes for best papers, editor 1950—). Research, numerous publs. on polysaccharide chemistry; improvements in mfr. sugar from sugar cane and utilization of by-products of cane sugar industry. Home: 113/40 Swaroop Nangar, Kanpur. Office: Nat. Sugar Inst., Kalyanpur, Kanpur, India.*

MULAY, Laxman Nikakantha, chemist; b. Rhuri, India, Mar. 5, 1923; s. N.R. and Anasuyabai (Joshi) M.; B.Sc. with honors, U. Bombay, 1943, M.Sc., 1946, Ph.D., 1950; m. Indumati Karandikar, Dec. 28, 1945. Came to U. S., 1952, naturalized, 1967. Daxina Merit Research fellow Karnatak Coll., India, 1943-45, instr., lectr., 1946-53; research asso. Northwestern U., 1953-55; research fellow Harvard, 1955-57; asst. prof. U. Cin., 1958-63; asso. prof. chemistry Pa. State U., University Park, 1963—. Mem. N.Y. Acad. Scis., Am. Chem. Soc. (chmn. Central Pa. sect. 1966-67), Am. Phys. Soc., Chem. Soc. (London), Sigma Xi, Phi Lambda Upsilon. Contbg. author: Treatise on Analytical Chemistry, 1963; also numerous articles. Research on magnetic ions in solution, of species free radical concentration in normal and cancer tissues, mobile electron density in organometallic compounds. Home: 255 Corl St., State College, Pa. 16801. Office: Pa. State U., University Park, Pa. 16802.*

MULDAWER, Leonard, Am. physicist; b. Phila., Aug. 6, 1920; s. Isaac Jacob and Sadie (Kaufman) M.; A.B., Temple U., 1942, A.M., 1944; Ph.D., Mass. Inst. Tech., 1948; m. Marcea Rosen, Dec. 17, 1950; children—Julia L., Richard W., Elizabeth A. Faculty, Temple U., Phila., 1948—, prof. physics, 1961—, dir., originator NSF In-Service Inst., 1958-64, research project dir. Office Ordnance Research, 1951-54, Air Force Office Sci. Research, 1957-59, AEC, 1961—. Mem. Am. Phys. Soc., Am. Crystallographic Assn., Am. Soc. for Metals, Am. Assn. Physics Tchrs., A.A.A.S., Franklin Inst., Am. Assn. U. Profs., Sigma Xi. Contbr. articles to tech. jours. Pioneered research on effect thermal vibration atoms on diffuse scattering X-rays from crystals; discovered strong temperature effect on spectral reflectivity beta-brass type alloys; research effect 3d element on phas transformations in alloys, X-ray diffraction, elec. resistivity, optical properties metals, phase transformations, physics metals. Home: 5 Radcliff Rd., Bala-Cynwyd, Pa. 19004. Office: Physics Dept., Temple U., Phila. 19122.*

MULDER, Derk, Dutch phytopathologist; b. Haarlem, Netherlands, Apr. 19, 1913; s. G. H. and J. C. (van der Laan) M.; Dr., U. Amsterdam, 1943; m. Vibeke Vogel Hansen, Aug. 11, 1951; children—Joke M. Torben J. D., Wijnand A., Clara S., Lubkea M., Axel G. E. Phytopathologist, Fruit Exptl. Sta., 1943-51, Inst. Phytopath. Research, Wageningen, Netherlands, 1951-56; FAO expert, Damascus, Syria, 1956-58; phytopathologist Tea Research Inst., Talawakele, Ceylon, 1958-62, Fruit Exptl. Garden, Wilhelminadorp Fruit Research Sta., Kesteren, Netherlands, 1963—. Mem. Netherlands bot., phytopath. socs. Author: Nutritional Disorders of Fruit Trees, 1953. Research on zinc deficiency of fruit trees in Holland, Syria, virus diseases of cherry, oilspot disease of tea, bacterial and virus diseases of Guatemala Grass. Home: 11 Hertenlaan, Wageningen. Office: 95 Tielsestraat, Kesteren, Netherlands.*

MULDER, Donald William, Am. physician; b. Rehoboth, N.M., June 30, 1917; s. Jacob D. and Gertrude (Hofstra) M.; A.B., Calvin Coll., Grand Rapids, Mich., 1940; M.D., Marquette U., 1943; M.S., U. Mich., 1946; m. Cortrude Ellens, Feb. 22, 1943. Practice medicine, specializing in neurology, Rochester, Minn.; cons. neurology Mayo Clinic, 1950—, bd. govs. 1963—, chmn. research, lab. com., 1963—; chmn. sects. neurology, 1966—, faculty Mayo Grad. Sch. U. Minn., 1951—, prof. neurology, 1964—; civilian cons. Lackland AFB (San Antonio). Mem. A.M.A., Am. Acad. Neurology, Am. Assn. EMG and Electrodiagnosis, Am. Neurol. Assn., Am., Central psychiat. assns., A.C.P., Sigma Xi. Research, numerous publs. on neuromuscular diseases, organic brain syndromes. Home: Route 1. Office: 200 1st St. S.W., Rochester, Minn. 55902.*

MULDER, Gerardus Johannes, Dutch chemist; b. Utrecht, Netherlands, Dec. 27, 1802; prof., Utrecht, 1841-68. Author: Proeve eener algemeene physiologische Scheikunde, 1843-50. Studied proteins (which he named 1838), also agrl. chemistry; discovered chlorophyll; identified caffein and fibrin. Died Apr. 18, 1880.

MULERIUS, Nicolaus, physician, astronomer; b. Bruges, Belgium, 1564; s. Pierre des Muliers; ed. Leiden, Netherlands; M.D., 1589; at least 1 son, Pierre; prof. medicine and math., univs. Groningen, Leeuwarden (both Netherlands). Author: Instututionum astronomicarum libri duo . . . , 1616; Notae breves (for an edit. Copernicus' De revolutionibus), 1617. Criticized Copernican system for removing earth from center of universe but agreed that earth rotates on its axis; related tides to moon, phys. spirits to sun. Died Groningen, Sept. 5, 1630.

MULHOLLAND, Henry Bearden, Am. physician; b. Knoxville, Tenn., Jan. 9, 1892; s. John Henry and Martha (Bearden) M.; high sch. edn. by tutors; student U. Toronto, 1917; M.D., U. Va., 1920; m. Elizabeth Caldwell Brown, Oct. 19, 1925; children—Elizabeth Brown, John Henry. Mem. faculty U. Va. Med. Sch., 1922, became prof. internal medicine, 1937, now prof. emeritus, acting head dept. internal medicine, 1942-49, asst. dean, 1942-58, cons. medicine U. Va. Hosp.; active practice internal medicine; research in diseases of metabolism, Wurzburg, Germany, and Copenhagen, Denmark, also visited various clinics in Europe, 1927-28. Mem. U. S. del. to WHO, Geneva, Switzerland, 1951; mem. sub-com. Hoover Task Force, 1954. Recipient Algeron Sullivan award, 1962; Thomas Jefferson award U. Va., 1962. Master A.C.P.; mem. A.M.A. Am. Clin. and Climatol. Assn., Am. Diabetes Assn. (past pres.), Phi Beta Kappa. Research, publs. on metabolic diseases; diabetes mellitus, internal medicine, clin. problems. Home: 1817 Fendall Av. Office: 1400 Jefferson Park Av., Charlottesville, Va. 22903.

MULINOS, Michael George, physician; b. Kefalonia, Greece, Nov. 24, 1897; s. George G. and Helen (Couros) M.; came to U. S., 1908, naturalized, 1914; A.B., Columbia, 1921, A.M., 1922, Ph.D., 1927, M.D. 1924; m. Joyce Leora Stevens, June 30, 1927; children—Elenie M., Steven M. Research asso., instr. pediatrics U. Minn., also Mayo Found. Grad. Sch., 1925-27; research asso. Hull Physiol. Labs., U. Chgo., 1926; faculty Columbia Coll. Phys. and Surg., 1927-44, asso. prof., 1932-45; asso. prof. physiology and pharmacology N.Y. Med. Coll., N.Y.C., 1944-55; med. research dir. Interchem. Corp., Union, N.J., 1945-47, Comml. Solvents Corp., N.Y.C., 1953-63; cons. on med. research to chem. and pharm. industry, N.Y.C., 1947-53, 63—. Fellow N.Y. Acad. Medicine, N.Y. Acad. Scis., Am. Coll. Angiology; mem. A.M.A., Soc. for Exptl. Biology and Medicine, Am. Soc. Pharmacology and Exptl. Therapeutics, Harvey Soc., Am. Med. Writers Assn., Am. Geriatric Soc., Am. Coll. Clin. Pharmacology and Chemotherapy, Sigma Xi, Phi Chi. Author: Outline of Pharmacology, 1945; also numerous articles. Research on physiology of digestive and autonomic nervous systems, toxicology, clin. pharmacology for testing of new drugs, drugs in inflammation. Home 549 Birch Av., Westfield, N.J. 07090. Office: 342 Madison Av., N.Y.C. 10017.*

MULLAN, John Francis, neurosurgeon; b. Derry, Ireland, May 17, 1925; s. John and Catherine (Gilmartin) M.; M.B., B.Ch., B.A.O., Queens U. (Ireland), 1947; m. Vivian Dunn, June 2, 1959; children—Joan, John C., Brian. Came to U. S., 1956, naturalized, 1962. Practice medicine, specializing in neurosurgery, Chgo., 1956—; faculty U. Chgo., 1956—, prof. neurol. surgery, 1963—; cons. neurosurgeon Cook County Hosp., Manteno (Ill.) State Hosp. Recipient McClintock Teaching award U. Chgo., 1961. Diplomate Am. Bd. Neurosurgery. Fellow Royal Coll. Surgeons (Eng.), A.C.S.; mem. Harvey Cushing Soc. Author: Essentials of Neurosurgery, 1961. Pub. lowest mortality figure yet recorded in surg. treatment of brain tumors; introduced operation of percutaneous cordotomy for relief of intractable pain, stereotactic thrombosis of intracranial aneurysm, trans oral surgery of basal brain tumors. Home: 6911 Bennett Av., Chgo. 60649. Office: 950 E. 59th St., Chgo. 60637.*

MULLENDERS, William, Belgian botanist; b. Antwerp, Belgium, Nov. 20, 1913; s. Henri and Julia (Schmoele) M.; B.philosphy, U. Louvain, 1937, Licence science, 1945, Dr.science, 1953; m. Simone Elisabeth Meeus, Oct. 10, 1946; 1 dau., Geneviève-Anne. Asst. bot. div. Institut National Etude agronomique du Congo Belge, 1946-49; sr. asst. Institut Recherche Scientifique pour Industrie et Agr., Brussels, Belgium, 1950-53; prof. botany U. Louvain (Belgium), 1953—; vis. scientist Inst. Arctic and Alpine Research, U. Colo., Boulder, 1965, 68-69. Sci. adviser to prehist. and hist. socs. Mem. A.A.A.S., Internat. Assn. for Quaternary Research, Deutsch Quartärvereinigung. Author: La végétation de Kaniama, 1954; (with A. Focan) Carte des sols et de la végétation de Kaniama, 1955; (with others) Flore de la Belgique, 1967; also articles. Research on plant communities and vegetation mapping in Central Africa and Belgium, E. France; palynology, study of pollen and spores with respect to hay-fever, vegetation history of Quaternary, systematics, prehist. and hist. monuments, neolithic agr. Home: 175, rue des Deux Petits Chemins, Héverlé, Belgium. Office: U. Louvain, Louvain, Belgium.*

MÜLLER, Adam Heinrich, German polit. economist; b. Berlin, June 30, 1779; ed. in theology and law U. Gottingen. Statesman in Berlin; founder journal Phöbus, 1808; consul for Austria, Leipzig, 1816-27; journalist, Vienna, 1827-29. Author: Elemente der Staatskunst, 3 vols., 1810; Die Theorie der Staatshaushaltung, 2 vols., 1812; Vermischte Schriften, 1812; Versuche einer neuen Theorie des Geldes, 1816; Von der Notwendigkeit einer theologischen Grundlage der geschichten Staatswissenschaft, 1820. Criticized rationalistic approach to polit. economy and reflected romantic-sch. theories of organic bases of state. Influenced devel. of Nat. Socialist polit. theory; stressed subordination of individual to state. Died Vienna, Jan. 17, 1829.

MULLER, Christian Georg, Swiss physician; b. Berne, Switzerland, Aug. 11, 1921; s. Max and Gertrude (Adrian) M.; M.D., Berne U., 1947; m. Madeleine Schaetti, Oct. 18, 1947; children—Henriette, Regine, Jacques. Asst. in clinics of Zurich, Berne, 1947-53; head physician Psychiat. Clinic, Lausanne U., 1953-57, dir., chmn., 1961—; head physician Psychiat. Clinic, Zurich U., 1957-61. Mem. Swiss Acad. Med. Sci., Swiss Soc. Psychiatry, Swiss Soc. Psychoanalysis. Author: (with others) Alterspsychiatrie, 1967. Co-editor, Swiss Archives of Neurology, Neurosurgery and Psychiatry, Social Psychiatry, Jour. for Psychotherapy and Med. Psychology; adv. bd. Der Nervenarzt. Research, publs. in psychotherapy of psychoses, gerontopsychiatry. Address: Hopital de Cery, 1008 Prilly, Lausanne, Switzerland.*

MULLER, Claus, German mathematician; b. Solingen, Germany, Feb. 20, 1920; s. Michael F. and Grete (Porten) M.; student univs. Bonn, Munich, Berlin; Dr.rer.nat., 1944; m. Irmgard Döring, Apr. 17, 1947; children—Christine, Michael, Axel. With univs. Göttingen, Bonn, Germany, 1945-47, Cambridge (Eng.), 1948, Hull (Eng.), 1949, U. Bonn, 1950-55; prof., dir. inst. pure and applied math. Aachen (Germany) Inst. Tech., 1955—, dean, 1957-59. Chmn., Nuclear Research Center, Jülich, Germany, 1964-66. Mem. German Math. Assn., Am. Math. Soc. Author: Mathematische Theorie Elektromagnetischer Schwingungen, 1957; also numerous articles. Research on math. analysis of theory of electromagnetic waves, especially problems of diffraction; generalization of spherical harmonies and Bessel functions to higher dimensions. Home: 7b Hauptstrasse, Richterich bei Aachen. Office: Rhenish-Westphalian Inst. Tech., Aachen, West Germany.*

MULLER, Cornelius Herman, Am. botanist; b. Collinsville, Ill., July 22, 1909; s. Cornelius Bernhard and Olga (Dietze) M.; B.A., U. Tex., 1932, M.A., 1933; Ph.D., U. Ill., 1938; m. Katherine Kinsel, July 15, 1939; 1 son, Robert Neil. Ecologist, Ill. State Natural History Survey, 1938; asst. botanist div. plant exploration and introduction Bur. Plant Industry, U. S. Dept. Agr., 1938-42, asso. botanist spl. guayule research project, 1943-45; faculty U. Cal at Santa Barbara, 1945—, prof. botany, 1956—, acting dean grad. div., 1961-62. Cons. dry lands agr. in Mexico, Ford Found., 1965. Mem. Ecol. Soc. Am., Bot. Soc. Am., Am. Soc. Plant Taxonomists, Internat. Assn. for Plant Taxonomy, Linnean Soc. London. Author: The Central American Species of Quercus, 1942; The Oaks of Texas, 1951; also numerous articles. Research on classification Am. oaks, basic nature biotic community; reconstrn. evolutionary history certain species and groups oaks, exploration and description vegetation no. Mexico; biochem. interactions between plants. Home: 2770 Las Encinas Rd., Santa Barbara, Cal. 93105.*

MÜLLER, D(etlev), plant physiologist; b. Hamburg, Germany, Apr. 20, 1899; s. Frederik Immanuel and Lonny (Scholz) M.; M.Sc., U. Copenhagen (Denmark), 1923, dr.phil., 1928; m. Clara-Marie Christensen Elmvad, Dec. 22, 1927; children—Mogens, Hanne (Mrs. Niels Jacob Brandt), Kirsten (Mrs. Claud Müller). Faculty, Royal Agr. Coll., Copenhagen, 1923-49, prof., 1935-49; faculty U. Copenhagen, 1923-35, prof., head dept. plant physiology, 1949—. Decorated Knight of Dannebrog (Denmark). Mem. Kungl. Vetenskapssoc. Uppsala, Deutsche Akad. Naturforscher Leopoldina; hon. mem. Societas Biochem. Biophys. et Microbiol. Fenniae, Zoolog.-Bot. Gesellschaft Wien. Discovered enzymes glucose oxidase, 1925, alcohol dehydrogenase in yeast, 1933, mannitol dehydrogenase in yeast, 1937; discovered dependence of photosynthesis on light and temperature, 1928; analysis of prodn. of dry matter in Danish Beech forest, 1954, virgin tropical rain forest, 1965; 1st film of spore-shooting in basidiomycetes, 1954. Home: 140 Gothersgade, Copenhagen 1123, Denmark.*

MÜLLER, Erwin Wilhelm, physicist, educator; b. Berlin, Germany, June 13, 1911; s. Wilhelm and Käthe (Teipelke) M.; Dipl. Ing., Tech. U. Berlin, 1935, Dr.-Ing. in Physics, 1936, Dr.-Ing. habil., 1950; Dr. rer. nat. h.c., Free U. Berlin, 1968; m. Klara E. Thüssing, Feb. 13, 1939; 1 dau., Jutta B. Came to U. S., 1951, naturalized, 1962. Research physicist Siemens Research Lab., Berlin, 1935-37; physicist research and devel. Stabilovolt Co., 1937-45; faculty Tech. Inst., Altenburg, Germany, 1945-47, Free U., Berlin, 1951-52; div. chief Kaiser-Wilhelm Inst., Max-Planck Inst., Berlin, 1947-55, research prof. physics, 1955-68, Evan-Pugh research prof., 1968—. Recipient C. F. Gauss medal Braunschweigische Wissenschaftliche Gesellschaft, 1952; award Instrument Soc. Am., 1960; H. B. Potts gold medal Franklin Inst., 1964. Fellow Am. Phys. Soc.; mem. German Phys. Soc., N.Y. Acad. Sci., Am. Vac. Soc., Sigma Xi. Contbr. to Handbuch der Physik, vol. XXI, 1956, Advances in Electronics, vol. 13, 1960; contbr. articles to profl. jours. Inventor field electron microscope, field ion microscope; discoverer phys. effects of field desorption, field ionization, field evaporation. Home: 659 Glenn Rd., State College, Pa. 16801. Office: Osmond Lab., University Park, Pa. 16802.*

MÜLLER, Baron Ferdinand von, see von Müller, Baron Ferdinand.

MÜLLER, Franz Joseph (Baron von Reichenstein), mineralogist; b. Nagyszeben, Transylvania (now Sibiu, Rumania), July 1, 1740; ed. Vienna, Austria; chief insp. mines, Transylvania; discovered a new metal, 1782, sent specimens to Bergman, then to Klaproth, who named it tellurium. Died Vienna, Oct. 12, 1825.

MÜLLER, Fritz, zoologist; b. Windischholzhausen, Mar. 31, 1821; businessman, prof., Desterro, Brazil. Author: Für Darwin, 1864. An early Darwinian; pioneer in theory that individual devel. recapitulates phylogenic devel. (recapitulation theory); research on devel. of crustaceans, also olfactory organs of butterflies, mimicry. Died Blumenau, Brazil, May 21, 1897.

MÜLLER, Fritz Paul, German entomologist; b. Meerane, Germany, May 25, 1913; s. Paul Otto and Maria (Müller) M.; doctorate U. Rostock, 1938; m. Elfriede Buchholtz, Jan. 15, 1940; children—Erika (Mrs. Lemke), Wolfgang. Entomologist, Biol. Inst. for Land and Forestry, Berlin-Dahlem, 1945-48; entomologist Inst. Phytopathology, Naumburg, 1948-55; faculty agr. U. Rostock, 1955—, prof. entomology and agrl. zoology, 1958—. Mem. German Entomol. Soc., German Soc. for Applied Entomology. Research, numerous publs. on taxonomy and biology of aphids, particularly aphid species with econ. importance, relations between aphids and host plants; description of new aphid species of Europe, S. Africa. Home: 3 Georg Büchner-Strasse, 25 Rostock, Germany.*

MÜLLER, Georg Elias, German psychologist; b. Grimma, Germany, July 20, 1850; prof., Czernowitz, now USSR, also Göttingen, Germany. Author: Zur Theorie der sinnlichen Aufmerksamkeit, 1873; Zur Analyse der Gedachtnistätigkeit und des Vorstellungsverlautes, 3 vols., 1911-17; Komplextheorie und Gesalttheorie, 1923; Abriss der Psychologie, 1924; Beiträge zur Psychophysik der Farbenhempfindungen, 1934. A founder of exptl. psychology; studied physiology of senses and memory; rejected Gestalt psychology. Died Göttingen, Dec. 23, 1934.

MÜLLER, Gerhard Friedrich, geographer, historiographer; b. Herford, Westphalia, Nov. 11, 1705. Historiographer, 1746, also keeper of archives, 1766, Russia. Fellow Royal Soc., 1730, corr. mem. French Acad. Scis., 1761; perpetual sec. Acad. Scis., St. Petersburg. Pub. history of Siberia, also collection of Russian histories; 1st to use ethnographic data (1st step toward a sci. of man). Died Moscow, Russia, 1783.

MÜLLER, German Karl, German petrologist; b. Schramberg, Germany, Feb. 9, 1930; s. German and Helene (Wagner) M.; student U. Cologne (Germany), 1948-51; Dr.rer.nat., U. Bonn (Germany), 1952; m. Renate Hoffmann, Sept. 14, 1951; children—Almuth, Gerrit, Helgard, Folke, Thorsten, Sören, Sönke, Arne, Lars. Petrologist, Turkish Govt., Ankara, 1953; chief sedimentary lab. Mobil Oil in Germany, 1954-57; petrologist Tex. Africa Exploration Co., Ethiopia, 1958-59; asst. prof. Tübingen (Germany), U., 1959-64; prof. petrology U. Heidelberg (Germany), 1964—. Mem. Deutsche Mineralogische Gesellschaft (gen.-sec. 1965—). Author: Methoden der Sediment-Untersuchung, 1964; also numerous articles. Research on petrology and geochemistry of sediments. Home: 4 Silcherweg, 6901 Bammental, Heidelberg, Germany.*

MÜLLER, Hans Gerhard, German meteorologist; b. Bad Gandersheim, Oct. 27, 1905; s. Heinrich and Marie (Böhlke) M.; ed. univs. Gottingen, Munich, Hamburg; Ph.D.; m. Marlen, Sept. 22, 1934; children—Inge, Hartmut, Gisela, Gudrun, Hans-Gerhard, Ilse. Physicist and meteorologist; prof. U. Munich, head Atmospheric Physics Inst.; adminstr. for German Research Service for Aerial Navigation. Mem. Wiss. Gesellschaft fur Luftfahrt, German Meteorol. Soc., German Geophys. Soc. Research and publs. on cloud physics, atmospheric optics, high atmosphere, phys. meteorology, thermal and solar radiation. Home: Haselsberger Strasse 219 D, Schleissheim. Office: Flughafen, Munich, West Germany.

MÜLLER, Heinrich, German anatomist; b. 1820; credited with discovery of visual purple, 1851; showed (with Rudolph Albert von Kölliker) that each contraction of frog's heart produces electric current, 1856; described superior and inferior tarsal muscles, orbitalis muscle which bridges inferior orbital fissure, 1858; eponym of fibers of Müller (in retina), other anat. structures. Died 1864.

MÜLLER, Herbert F., physician; b. Cologne, Germany, July 4, 1924; s. Reiner and Elisabeth (Kühner) M.; M.D., U. Cologne, 1951; m. A. Karola Kümpel, June 18, 1953; children—Marcus, Bruno. Sr. psychiatrist Douglas Hosp., Montreal, Que., Can., 1959—, also dir. electroencephalograph labs.; asst. prof. psychiatry McGill U., 1966—. Mem. Canadian, Am. electroencephalographic socs., Can. Psychiat. Assn., Am. Psychiat. Assn. Contbr. articles to profl. jours. Research regarding electroencephalography in psychiatry, in particular concerning geriatrics, adolescents, psychotropic drugs. Home: 118 Arlington Av. Office: Douglas Hospital, Montreal 19, Que., Can.*

MULLER, Hermann Joseph, Am. geneticist; b. N.Y.C., Dec. 21, 1890; s. Hermann Joseph and Frances (Lyons) M.; A.B., Columbia, 1910, A.M., 1911, Ph.D., 1916, D.Sc., 1949; postgrad. Cornell Med. Coll.; D.Sc., Edinburgh U., 1940, U. Chgo., 1959, Swarthmore Coll., 1964; M.D. (hon.), Jefferson Med. Coll., 1963; m. Jessie M. Jacobs, June 11, 1923 (div. 1934); 1 son, David Eugene; m. 2d, Dorothea J. Kantorowicz, May 20, 1939; 1 dau., Helen Juliette. Instr. biology Rice Inst., 1915-18, in charge dept., 1916-18; instr. zoology Columbia, 1918-20; asso. prof. zoology U. Tex., 1920-25, prof., 1925-36; sr. geneticist Inst. Genetics, Leningrad, 1933-34, Moscow, 1934-37; research asso., lectr. Inst. Animal Genetics U. Edinburgh, 1937-40; research asso. biology Amherst Coll., 1940-42, vis. prof., 1942-45; prof. zoology Ind. U., 1945-64, Distinguished Service prof., 1953—, emeritus, 1964—; mem. Inst. for Advanced Learning in Med. Scis., City of Hope Med. Center, Duarte, Cal., 1964-65; vis. prof. depts. zoology, genetics U. Wis., 1965-66. Recipient Kimber Genetics award, 1955; Darwin medal, 1959; Alexander Hamilton award, 1960; Nobel laureate in physiology, medicine, 1946. Mem. Nat., Royal Swedish, Royal Danish acads. sci., Royal Soc. London, Japan Acad., Genetical Soc. Japan, Nat. Inst. Sci. India, A.A.A.S., Am. Humanist Assn. (pres. 1955-59), Am. Soc. Naturalists, Soc. Study of Evolution, Am. Soc. Zoologists, Soc. Exptl. Biology and Medicine, Am. Genetic Assn., Genetics Soc. Am., Am. Soc. Human Genetics, Am. Assn. U. Profs., Am. Philos. Soc., Am. Acad. Arts and Sci., World Acad. (v.p.), Phi Beta Kappa, Sigma Xi. Author: (with others) The Mechanism of Mendelian Heredity, 1915, 1922; Out of the Night, 1935; (with others) Genetics, Medicine and Man, 1947; Studies in Genetics, 1962. Research, publs. on breeding expts. on fruit fly, Drosophila; analysis of arrangement and method of recombination of hereditary units; explanation of so-called mutations in the evening primrose, 1917; studies on mutation and evolution; theory of gene; artificial transmutation of the gene by x-rays; prodn. of chromosome changes; heredity in man. Died Apr. 5, 1967.

MÜLLER, Horst Friedrich, German chemist; b. Leipzig, Apr. 1, 1907; s. Paul and Anny (Wünscher) M.; Ph.D., U. Leipzig; m. Ilse Heybey; 1 son, Frank-Michael. Asst. at univ.; collaborator Siemens und Haltke, Berlin; asst. and instr. chemistry Physics and Chemistry Inst., U. Marburg; dir. Polymers Inst., U. Marburg. Mem. various sci. assns. Research and publs. on physics and chemistry of polymers. Home: Marbach Schülstrasse 26. Office: Marbacherweg 15, Marburg, West Germany.

MULLER, Jean Herman, Swiss physician; b. Lausanne, Switzerland, Nov. 5, 1908; s. Herman Adolf and Maria Betty (Christen-Herter) M.; student univs. Lausanne, Bern, London; M.D., 1934; postgrad. Paris, Hamburg, Stockholm, London, N.Y.; m. Beate Schmidt, May 18, 1940. Head radiol., histopath. depts. Women's Hosp., U. Zurich, Switzerland, 1939—, prof., 1949—; with Zurich cyclotron group Fed. Inst. Tech., 1943-58. Cons. hosps. Chur, Glarus, Switzerland, 1940-66; Swiss del. UN confs. on peaceful uses atomic energy, Geneva, UNESCO conf. radioisotopes, Paris, 1958; mem. Swiss Bd. Radioprotection. Mem. Royal Soc. Medicine London, Swiss Soc. Nuclear Tech., other Swiss sci. socs.; hon. mem. Swiss Soc. Gynecology, French Clin. Cytology, Anat. socs. Author: Radioactive Isotope Therapy, with particular Reference to the Uses of Radiocolloids, 1962; also numerous articles. Research in nuclear medicine especially therapeutic uses of artificial radioactive isotopes, radiodiagnostics—first description of aseptic osteonecrosis of styloid process of cubitus, conventional radiotherapy, gynecol. pathology. Address: 34 Rigistrasse, 8006 Zurich, Switzerland.*

MÜLLER, Johannes Peter, German physiologist, anatomist; b. Koblenz, Germany, July 14, 1801; student medicine U. Bonn (Germany). Became privatdocent U. Bonn., 1824, asso. prof. physiology, 1826, prof., 1830; prof. anatomy and physiology U. Berlin, 1833-58, became rector, 1847. Founder (with Magendie) modern physiology; research in anatomy, mechanisms of speech, voice, hearing; chem. and phys. properties of lymph, chyle and blood; discovered Müllerian duct in females; chondrin of cartilage; studied hermaphroditism, embryology, metamorphoses of echinoderms; stated principle that sensation following stimulation of sensory nerve depends on nature of sense organ rather than stimulation (law of specific nerve energies or law of specific irritability). Author: Handbuch der Physiologie des Menschen, 1833-40; Zur vergleichenden Physiologie des Gesichtssins des Menschen, 1826; Bildungsgeschichte der Genitalien, 1830; De glandularum secernentium structura, 1830; Vergleichende Anatomie der Myxinoden, 1834-43. Died Berlin, Apr. 28, 1858.

MÜLLER, Karl, German bryologist; b. Allstedt, Germany, Dec. 16, 1818; wrote a standard work based on moss collection of 10,000 types. Author: Synopsis muscorum frondosorum, 2 vols., 1849-51; Genera muscorum frondosorum, 1901; Antäus oder die Natur im Spiegel der Menschheit, 1902. Died Halle, Germany, Feb. 9, 1899.

MÜLLER, Karl Ottfied, archaeologist; b. Brieg, Silesia, Aug. 28, 1797; ed. Breslau (now Wroclaw, Poland), Berlin (under Böckh); named prof. langs. Magdalenum, Breslau (now Wroclaw, Poland), 1817; became prof. ancient literature, U. Göttingen, Germany, 1819; visited Italy, 1839; went to Greece to study remains of ancient Athens, 1840; began excavating Delphi. Author: Aegineticorum liber, 1817; Geschichten hellenischer Stämme und Städte: Orchomenos und die Minyer, 1820; Die Dorier, 1824; Prolegomena zu einer wissenschaftlichen Mythologie, 1825; Die Etrusker, 1828; Handbuch der Archäologie der Kunst, 1830; also history of ancient Greek lit., 1840. Helped develop new concept of Hellenism by viewing Greek life as a whole; introduced new standards of accuracy in cartography of ancient Greece; laid found. for sci. study of myths. Died 1840.

MÜLLER, Klaus Jürgen, German geologist; b. Berlin, Feb. 6, 1923; s. Arthur and Traute (Schwandt) M.; Ph.D., U. Berlin; m. Eva-Maria Globig, July 2, 1952. Prof. agrégé U. Berlin, 1954; research chief Nat. Acad. Sci., U. S., 1954-56; prof. U. Tokyo; 1962, Berlin Tech. Coll., 1960, U. Bonn, 1964—. Research and publs. on geology and paleontology, especially paleozoics. Home: Eichenweg 2, 5301 Röttgen Kr. Office: Friedrich-Wilhelms-Universität, Nusallee 8, 53 Bonn, West Germany.

MÜLLER, Ludwig Robert, German neurologist; b. Augsburg, Germany, Apr. 26, 1870; prof., Erlangen, Germany. Author: Die Lebensnerven und die Lebenstriebe, 1930; Über den Schlaf, 1940. Studied vegetative nervous system, mid-brain hypophysis system, basis antagonism between vagus and sympathetic nerves.

MÜLLER, Otto Friedrich, Danish biologist; b. Copenhagen, Denmark, Mar. 11, 1730; became tutor to young nobleman, 1753; commd. to continue publ. of Oeder's Flora Danica by Frederick V; prof. botany, Copenhagen. Author: Fauna insectorum friedrichsdaliana, 1764; Flora friedrichsdaliana, 1767; Zoologiae danicae prodromus, 1776; Zoologia danica, 1777-1806; also other works on earth worms, infusoria. First description of diatoms; 1st classification of animalcules into genera and species; invented naturalist's dredge; introduced terms, bacillus, spirillum; advocated theory of spontaneous generation; discovered many new species of infusoria. Died Copenhagen, Dec. 26, 1784.

MÜLLER, Paul Hermann, Swiss chemist; b. Olten, Switzerland, Jan. 12, 1899; s. Gottlieb and Fanny (Leypoldt) M.; Ph.D. in Phys. Chemistry and Botany, U. Basel (Switzerland), 1925; m. Friedel Ruegsegger, Oct. 6, 1927. Asst. chemist Lonza Power Plant, 1916-17; with J. R. Geiger, exptl. lab., Basel, 1925—. Recipient Nobel prize in physiology and medicine, 1948. Mem. Swiss Soc. Chemistry; hon. mem. Schweizerische Naturforschende Gesellschaft, Société de Chimie Industrielle Paris. Publs. on devel. of DDT and other insecticides. Rediscovered DDT and its use as an insecticide. Home: Glaserbergstr. 78, Basel, Switzerland. Died Basel, Oct. 12, 1965.

MÜLLER, Rolf, German astronomer; b. Potsdam, Jan. 26, 1898; s. Gustav and Johanna (Schultress) M.; Ph.D., U. Berlin; m. Eleonore Droescher, Jan. 26, 1932; 1 dau., Renate. Astrophysicist, Potsdam Obs., 1924; mem. solar observation expdn., Sumatra, 1926; head Obs. of Andes, La Paz, Bolivia, 1928-30; mem. expdn., Iceland, 1939; dir. Wendestein Solar Obs., 1946-63. Mem. German Radiolocalization Soc., German Astron. Soc., Royal Astron. Soc. Research and publs. on astrophysics, variable stars, solar corona, history of astronomy, solar physics. Address: Postfack 8, 8204 Degerndorf-Inn, West Germany.

MULLER, Siemon William, geologist; b. Russia, May 9, 1900; s. Wilhelm Peter and Evdokiya (D'yachkova) M.; Candidate of Commerce, Comml. Coll., Vladivostok, Russia, 1917; B.A. in Geology, U. Ore., 1927; M.A. in Geology, Stanford, 1929, Ph.D., in Geology, 1930; m. Vera Vilamovsky, June 23, 1928; 1 son, Eric. Faculty, Stanford, 1928—, prof., 1941—. Cons. Alaskan div. USAAF, 1944-45. Fulbright Research fellow, 1956-57; Guggenheim fellow, 1957. Mem. Paleontol. Soc. Am. (pres. 1963), Cal. Acad. Scis. (past sec. trustee), Geol. Soc. Am., Geol. Soc. London, Am. Assn. Petroleum Geologists. Research on stratigraphy and paleontology of early Mesozoic invertebrates, permafrost and engring. problems. Home: 2255 Webster St., Palo Alto, Cal. Office: Box 5846, Stanford U., Stanford, Cal.

MULLER-EBERHARD, Hans Joachim, immunologist; b. Magdeburg, Germany, May 5, 1927; s. Adolf and Emma (Jenrich) M-E.; M.D., U. Goettingen, Germany, 1953; D.M.Sc., U. Uppsala, Sweden, 1961; m. Ursula Fleck, Dec. 30, 1953; children—Monika, Kristina. Faculty, Rockefeller Inst., staff Rockefeller Inst. Hosp., N.Y.C., 1954-56, 59-63; Swedish Med. Research Council fellow, 1957-59; permanent vis. lectr. (docent) immunochemistry U. Uppsala, 1961; mem. dept. exptl. pathology Scripps Clinic and Research Found., La Jolla, Cal., 1963—. Mem. allergy and immunology study sect. NIH, 1965—; mem. immunology and microbiology research study com. Am. Heart Assn., 1966—; mem. adv. bd. Henry M. and Lillian Stratton Found., 1964—; vis. prof. Med. Faculties Sweden, 1965. Recipient Parke Davis Meritorious award Am. Soc. Exptl. Pathology, 1966. Affiliate fellow Am. Acad. Allergy, Royal Soc. Medicine London; mem. Am. Soc. Clin. Investigation, Am. Assn. Immunologists, Am. Soc. Exptl. Pathology, Harvey Soc., Sigma Xi. Research, publs. on serum proteins, factors in resistance to infections, antibodies, complement and other subjects in fields of immunology and immunochemistry. Home: 1237 Muirlands Vista Way. Office: 476 Prospect St., La Jolla, Cal. 92037.*

MÜLLER-FREIENFELS, Richard, psychologist; b. Bad Ems, Germany, Aug. 7, 1882; prof., Stettin, Poland, 1930-33, Berlin, 1933-38. Author: Psychologie der Kunst, 3 vols., 1912-33; Grundzüge einer Lebenspsychologie, 2 vols., 1924; Hauptrichtungen der gegenwärtigen Psychologie, 1929; Allgemeine Sozial- und Kultur-Psychologie, 1930; Menschenkenntnis und Menschenbehandlung, 1940. Founder of "Lebenspsychologie." Died Weilburg, Germany, Dec. 12, 1948.

MULLER-HEGEMANN, Dietfried, physician; b. Ljubliana, Yugoslavia, May 5, 1910; s. Otmar and Louisa (Carl) H.; ed. U. Munich, U. Vienna, U. Königsberg; M.D., U. Berlin, 1937; m. Marianne Bornschein, Aug. 15, 1964; children—(from previous marriage) Ellinor, Corinna, Kai. Dir. neurol.-psychiat. clinic U. Leipzig, 1957—; med. dir. Wuhlgarten Psychiat. and Neurol. Hosp., Berlin, 1964—; prof. psychiatry and neurology German Acad. for Continuing Edn. of Physicians, Berlin, 1964—. Decorated Verdienter Arzt des Volkes, other state awards. Author: Psychotherapie, 1961; Neurologie und Psychiatrie, 1966; also articles. Editor-in-chief Jour. Psychiatrie, Neurologie und medizinische Psychologie. Address: Krankenhaus Wuhlgarten, 114 Berlin, East Germany.*

MULLIGAN, Joseph Francis, Am. physicist; b. N.Y.C., Dec. 12, 1920; s. Joseph Lawrence and Mary (Collins) M.; student Fordham Coll., 1938-39, 41-43; A.B., Boston Coll., 1945, M.A., 1946; Ph.L., Weston Coll., 1947; Ph.D. in Physics, Cath. U. Am., 1951; S.T.L., Woodstock (Md.) Coll., 1954. Joined Soc. Jesus, 1939. Ordained priest Roman Catholic Ch., 1953; instr. physics St. Peter's Coll., Jersey City, 1946-47; theology student, Münster, Germany, 1954-55; mem. faculty Fordham U., 1955—, asso. prof. physics, 1963—, chmn. dept., 1956-64; dean Grad. Sch. Arts and Scis., also dean liberal arts faculty, 1964—. Mem. adv. com. grad. fellowship program Nat. Def. Edn. Act, 1960-63. NSF fellow, U. Cal. at La Jolla, 1961-62. Mem. Am. Phys. Soc., Am. Assn. Physics Tchrs., A.A.A.S., Am. Assn. U. Profs., Am. Assn. Jesuit Scientists (past pres.), N.Y. Acad. Sci., Sigma Xi. Contbns. to knowledge of the theoretical structure of molecules on the basis of quantum mechanics; contbr. articles on the fundamental atomic constants, particularly the velocity of light. Address: Fordham Univ., E. Fordham Rd., Bronx, N.Y. 10458.*

MULLIGAN, Richard Michael, Am. physician; b. Sherburne, N.Y., July 1, 1912; s. Frank J. and Grace (Farley) M.; A.B., Cornell U., 1933, M.D., U. Rochester, 1937; m. Martha M. Moser, Feb. 26, 1938; 1 son, Richard Martin. Faculty, U. Colo. Sch. Medicine, Denver, 1939—, prof. pathology, 1948—. Cons. to hosps., govt. agys. Mem. Soc. for Exptl. Biology and Medicine, Am. Soc. for Exptl. Pathology, Am. Assn. for Cancer Research, Tissue Culture Assn., Colo. Soc. Clin. Pathologists, Sigma Xi, Alpha Omega Alpha. Author: (with Williams, Wilkins) Neoplasms of the Dog, 1949; (with Lea, Febiger) Syllabus of Human Neoplasms, 1951; also numerous articles, chpt. in book. Research on neoplasia in man and animals. Home: 756 Fairfax St., Denver 80220.*

MULLIKEN, Robert Sanderson, Am. chem. physicist, educator; b. Newburyport, Mass., June 7, 1865; s. Samuel Parsons and Katherine W. (Mulliken) M.; B.S., Mass. Inst. Tech., 1917; Ph.D., U. Chgo., 1921; Sc.D., Columbia, 1939; Ph.D., U. Stockholm, Sweden, 1960; Sc.D. (hon.), Marquette, U., Cambridge U., 1967; m. Mary Helen von Noe, Dec. 29, 1929; children—Lucia Maria (Mrs. William W. McGrew), Valerie Noe. Jr. chem. engr. Bur. Mines, U. S. Dept. Interior, Washington, 1917-18; asst. in rubber research N.J. Zinc Co., Palmerton, Pa., 1919; asst. prof. physics Washington Sq. Coll., N.Y. U., 1926-28; asso. prof. physics, U. Chgo., 1928-31, prof. physics, 1931-61, prof. chemistry, 1961, Ernest deWitt Burton Distinguished Service prof., 1956-61, distinguished service prof. physics, chemistry, 1961—; distinguished research prof. chem. physics Fla. State U., 1965—; John van Geuns vis. prof. Amsterdam (Netherlands) U., 1965. Dir. editorial work, information Plutonium Project, U. Chgo., 1942-45; sci. attache U. S. Embassy, London, Eng., 1955; Baker lectr. Cornell U., 1960; Silliman lectr. Yale, 1965; sci., editorial cons. Recipient Bronze medal award Liège, 1948; Coll. City N.Y. Alumni Assn. award, 1965. NRC fellow, U. Chgo.,

Harvard, 1921-25; Guggenheim fellow, Germany, Europe, 1930, 1932-33; Fulbright scholar Oxford U., Eng., 1952-53. Nobel prize in chemistry, 1966. Fellow Am. Phys. Soc., A.A.A.S.; mem. Am. Chem. Soc. (Gilbert N. Lewis Gold medal award Cal. sect. 1960, Theodore W. Richards Gold medal award Northeastern sect. 1960, Peter Debye award 1963, J. G. Kirkwood award New Haven sect., Willard Gibbs award Chgo. sect. 1965), Nat. Acad. Scis., Am. Philos. Soc., Royal Soc. Gt. Britian (fgn.), Am. Acad. Arts and Scis., Soc. de Chimie Physique (Paris) (hon.). Fellow Chem. Soc. Gt. Britain (hon.). Research, publs. on isotope separation, activity and spectroscopy of electrons in molecular formation and stblzn. Home: 6825 S. Dorchester Av., Chgo. 60637.*

MULLIKEN, Samuel Parsons, Am. chemist; b. Newburyport, Mass., Dec. 19, 1864; s. Moses J. and Sarah D. (Gibbs) M.; S.B., Mass. Inst. Tech., 1887; Ph.D., U. Leipzig, 1890; postgrad. Clark U., 1891; m. Katherine W. Mulliken, June 27, 1893; children—Robert S., Samuel G. P., Katherine F. Asst. in chemistry Bryn Mawr Coll., 1892; instr. and acting head of chem. dept. Clark U., 1892-94; instr. organic chemistry and organic analysis Mass. Inst. Tech., 1895-1904, asst. prof., 1905-13, asso. prof. organic chem. research, 1913-26, prof. organic chemistry, 1926-—. Author: Laboratory Experiments on the Class Reactions and Identification of Organic Substances, 1896; The Compounds of Carbon with Hydrogen and Oxygen, 1904; The Commercial Dye-stuffs, 1909; The Compounds of Carbon with Nitrogen, Hydrogen, and Oxygen, 1916; The Compounds of the Higher Orders, 1922. Died Oct. 24, 1934.

MULLIN, Charles James, Am. physicist; b. St. Louis, Feb. 26, 1916; s. Charles Francis and Marguerite (Olsen) M.; B.S., St. Louis U., 1938; Ph.D., U. Notre Dame, 1942; m. Mary Joan Drake, Aug. 4, 1940; children—Kathleen Ann (Mrs. Daniel S. Tankersley), Charles James, Susan Patricia. Instr. physics St. Louis U., 1941-43; research asso. Harvard, 1943-45; faculty physics U. Notre Dame, Notre Dame, Ind., 1945-—, prof., 1955-—, chmn. dept., 1963. Fellow Am. Phys. Soc.; mem. Am. Assn. Physics Tchrs., Sigma Xi. Contbr. articles to physics jours. Research in nuclear and elementary particle physics; electromagnetic interactions with nuclei. Home: 52755 Highlands Dr., South Bend, Ind. 46635.

MULLIN, Francis Joseph, Am. physiologist; b. El Paso, Tex., Dec. 16, 1906; s. Joseph Peter and Charlotte (Norville) M.; A.B., U. Mo., 1929; S.M., U. Chgo., 1933, Ph.D., 1936; Sc.D., Blackburn Coll., 1965; m. Alma Gray Hill, May 29, 1935; children—Michael Mahlon, Mark Hill. Faculty dept. physiology U. Tex. Sch. Medicine, 1935-38; U. Chgo., 1938-51, Chgo. Med. Sch., 1951-54; pres. Shimer Coll., Mt. Carroll, Ill., 1954-—. Mem. Am. Physiol. Soc., Soc. for Exptl. Biology and Medicine, A.A.A.S., Phi Beta Kappa, Sigma Xi. Author: (with others) Sleep Characteristics, 1937. Contbr. articles to sci. jours. Research on physiology of sleep, depth, effect of drugs and motility during sleep, neuro-muscular and circulatory changes asso. with ionic content of cerebral spinal fluid, effects of drugs on skin sensitivity and body temperature, ionic changes in blood caused by asphyxia. Home: Shimer Coll., Mt. Carroll, Ill. 61053.*

MULLIN, R. E., Canadian forester; b. Toronto, Ont., Can., Mar. 25, 1920; s. Alexander and Gertrude R. (Brown) M.; B.Sc. in Forestry, U. Toronto, 1942; M.F., U. Mich., 1946; Ph.D., Yale, 1961; m. Golda Joyce Robinson, Aug. 28, 1948; children—David A., Robert F., Catherine E., Stephen J. With Ont. Dept. Lands and Forests, Maple, Ont., 1946-—, research on nurseries and planting, 1953-—; instr. Kemptville Agrl. Sch., 1946-52. Mem. Canadian Inst. Forestry, Soc. Am. Foresters, Soil Sci. Soc. Am. Research, publs. on growth and handling of forest tree nursery stock, packaging and storage inventory, root pruning, methods of planting as related to species and sites in Ont. Home: 223 St. Leonards Av., Toronto 12, Ont. Office: Dept. Lands and Forests, Maple, Ont., Can.*

MULLINS, William Wilson, Am. phys. metallurgist; b. Boonville, Ind., Mar. 5, 1927; s. Thomas Clinton and Ruth (Wilson) M.; Ph.B., U. Chgo., 1949, M.S. in Physics, 1951, Ph.D., 1955; m. June Bonner, June 26, 1948; children—William Wilson, Oliver Clinton, Timothy Bonner, Garrick Russell. Reesearch physicist, then adv. physicist Westinghouse Research Labs., 1955-60; asso. prof. metall. engring. Carnegie Inst. Tech., 1960-63, prof., head dept., 1963-—; spl. research surfaces and phase transformations. Chmn. phys. metallurgy Gordon Conf., 1966. Fulbright and Guggenheim fellows U. Paris (France), 1961-62. Mem. Am. Inst. Metall. Engrs. (Mathewson gold medal 1963), Am. Inst. Mining, Metall. and Petroleum Engrs., Am. Phys. Soc., Am. Soc. Engring. Edn., N.Y. Acad. Scis., Am. Civil Liberties Union, Sigma Xi, Alpha Sigma Nu. Research on Brownian motion and biased diffusion on solid surfaces and shape changes thereby produced; surface energy and equilibrium shapes of crystals; mechanisms of crystal growth; periodic precipitation in solids. Contbr. articles in field. Home: 1904 Brushcliff Rd., Pitts. 15221.*

MULNARD, Jacques, Belgian anatomist, embryologist; b. Wavre, Belgium, Feb. 15, 1922; s. Alfred and Victorine (Germain) M.; M.D., U. Libre de Bruxelles (Belgium), 1946, Agrégé de l'enseignement Superieur, 1956; m. Felicie Alice Alhadeff, Nov. 16, 1955; children—Marianne, Martine. Asst., Faculty Medicine and Pharmacy, Lab. Anatomy and Human Embryology, U. Libre de Bruxelles, 1946-56, chef de travaux, 1956-64, charge de cours, 1958-61, prof. anatomy and embryology, 1961-—. Decorated Officer de l'Ordre Leopold; recipient Medaille Civique 1940-45; Rockefeller Found. fellow, 1957. Mem. Internat. Soc. for Cell Biology, Internat. Inst. Embryology, Royal Soc. Medicine London, Assn. des Anatomists de langue Francaise, Société Royale Zoologique de Belgique, Société de Biologie Paris. Research, publs. on cytochemistry of insect devel., histo- and cytochemistry of embryonic devel. mammals, in vitro cultivation of mammalian ovum and embryo; microcinematography of mouse devel., exptl. embryology of mammals, exptl. analysis of organogenesis in mammals. Home: 180 Av. Montjoie, Brussels, Belgium.*

MULOCK, Houwer Adriaan Willem, Dutch ophthalmologist; b. Goes, May 2, 1884; s. Johannes E. and Francoise J. (Ochtman) M.; M.D., U. Amsterdam; m. Brigitta C. Schottee de Vries; children—Jan, Francoise, Johannes. Asst. to Prof. Staub; ophthalmologist, 1910-12; prof. ophthalmology, Batavia, 1932-47; pathologist-anatomist Royal Dutch Ophthalmol. Hosp., Utrecht, 1947-58. Decorated Order Dutch Lion; recipient Snellen medal. Author: Keratitis filamentosa und chronische arthritis; the pathology and therapy of fistula cornese; Zur Frage der klinischen Diagnose des Aderhautangions. Address: Midland, Flat 98, Zeist, Netherlands.

MULSANT, Martial-Etienne, French naturalist; b. Mornant, France, Mar. 2, 1797. Prof. natural history lycée at Lyon, France; asst. librarian, Lyon. Corr. mem. French Acad. Scis., 1873. Author: Histoire naturelle des Coléoptères de France, 1839-74; (with Verreaux) Histoire naturelle des oiseaux-mouches, 1874-76. Distinguished ornithologist and entomologist; research on life cycle of humming birds. Died Lyon, Nov. 4, 1880.

MULSOW, Frederick William, Am. physician; b. Baldwin, Kan., Oct. 6, 1882; s. Frederick William and Mary (Keel) M.; A.B., Baker U., 1911; A.M., Kan. U., 1915; Ph.D., Chgo. U., 1918; M.D., Rush Med. Sh., 1920; m. Maude A. Emery, June 20, 1922. Prin., Wathena High Sch., 1911-13; fellow Kan. U., 1913-14; food bacteriologist Kan. U., 1914-15; prof. bacteriology Wash. State Coll., Pullman, 1915-16; fellow, U. Chgo., 1916-17, instr., 1917-18; research fellow McCormick Inst., 1919-20; prof. pathology and bacteriology U. Ia., 1921-22, head dept. pathology and bacteriology, 1922-26; pathologist in charge pathology lab. St. Luke's Hosp., Cedar Rapids, Ia., 1926-47; practice internal medicine, Cedar Rapids, 1947-—; dir. pvt. clin. lab., Cedar Rapids, 1947-—. Mem. Ia. Soc. Pathologists (past pres.), A.A.A.S., A.M.A., Am. Coll. Pathologists, A.C.P., Ia., Linn County med. socs., N.Y. Acad. Scis., Ia. Clin. Med. Soc., Ia. Soc. Pathologists (past pres.). Research, publs. on metastatic cancer of skeletal muscles, low incidence of cancer of stomach in Ia., peptic ulcers of aged, death from radium therapy of cancer of uterus, cancer and sarcoma in physicians doing X-ray diagnosis, food poisoning outbreaks, culture media for gonococcal diverticula of appendix operation. Home: 359 Garden Dr. S.E., Cedar Rapids, 32403. Office: Highley Bldg., Cedar Rapids, Ia. 52401.*

MUMA, Martin Hammond, Am. entomologist; b. Topeka, July 24, 1916; s. Harold Hammond and Elsie Virginia (Martin) M.; student Western Md. Coll., 1935, Frostburg State Tchrs. Coll., 1936; B.S., U. Md., 1939, M.S., 1940, Ph.D., 1943; m. Katharine Elizabeth Short, Sept. 14, 1940; children—Bonnie Louise, Leslie Martin, Merrie Lynn, Sallie Anne, Cherie Kay, Elsie Virginia. Asso. entomologist U. Fla. Citrus Exptl. Sta., Lake Alfred, 1951-58, entomologist, prof., 1958-—. Mem. Ecol. Soc. Am., Soc. Systematic Zoologists, Animal Behavior Soc. Author: Common Spiders of Maryland, 1943. Research, numerous publs. on devel. Japanese beetle baits, bait substitutes, automatic traps in Md.; biol. controls for citrus insects and mites in Fla.; devel. control measures for insects and mites infesting livestock; revised, modernized taxonomy of arachnids in U. S., N. Am. Home: 401 W. Lake Summit Dr., Winter Haven, Fla. 33880. Office: Citrus Expt. Sta., Lake Alfred, Fla. 33850.*

MUMENTHALER, Marco, Swiss neurologist; b. Berne, July 23, 1925; s. Johann Jakob and Lydia (Piccoli) M.; ed. univs. Zurich, Paris, Amsterdam, Basle; M.D.; m. Livia Morandini, Nov. 19, 1949; children—Maia, Manuela, Isabel. Asst. in various Swiss hosps.; fgn. asst. Paris Hosp.; clinic chief Univ. Neurol. Clinic, Zurich; head labs. for research Zurich Neurol. Clinic; vis. asso. Inst. Neurol. Diseases and Blindness, NIH, Bethesda, Md.; dir. neurol. services Berne U. Med. Clinic. Author: Die Ulnarisparesen, 1961. Research in neuromuscular diseases. Home: Neufeldstrasse 17. Office: Hopital de l'Ile, Berne, Switzerland.

MUMFORD, George Saltonstall, Jr., Am. astronomer; b. Milton, Mass., Nov. 13, 1928; s. George S. and Alice (Herrick) M.; A.B., Harvard, 1950; M.A., U. Ind., 1952; Ph.D., U. Va., 1955; m. Nancy Carey, Dec. 22, 1949; children—Barbara, Elizabeth, Robert, George. Instr., Randolph-Macon Woman's Coll., 1952-53; faculty Tufts U., Medford, Mass., 1955-—, asso. prof. astronomy, 1962-—; vis. astronomer Kitt Peak Nat. Obs., Tucson, 1962-—. Mem. Am. Astron. Soc., Astron. Soc. Pacific, A.A.A.S., Am. Assn. Variable Star Observers, Am. Phys. Soc., Sigma Xi. Contbr. articles to tech. jours. Co-discoverer eclipsing nature dwarf novae U Geminorum, EX Hydrae, Z Camelopardalis, old nova EM Cygni; research with photelectric observation on nature short-period light variations, changes in duration eclipses. Home: Pegan Lane, Dover, Mass. 02030. Office: Bromfield-Pearson Bldg., Tufts U., Medford, Mass. 02155.*

MUMMERY, Peter William, Brit. physicist; b. Parkgate, Eng., Mar. 27, 1926; s. John and Alice (Bailey) M.; B.A., Queens Coll., Cambridge (Eng.) U., 1947, M.A., 1950; m. Anne Foster, Mar. 4, 1950; Staff, Atomic Energy Research Establishment, Harwell, Eng., 1947-59; head indsl. power div. A.E.E., Winfrith, Eng., 1959-62, dep. dir., 1962-66; reactor group chief physicist reactor group hdqrs. U.K. Atomic Energy Authority, Risley, Eng., 1966-67, dep. dir. tech. policy, 1967-—. Decorated officer Order Brit. Empire. Fellow Inst. Physics; mem. Brit. Nuclear Energy Soc. (mem. bd.), Am. Nuclear Soc. Joint editor Progress in Nuclear Energy, 1958-—. Research, publs. on neutron physics processes in nuclear reactors, especially design. Home: Buckland Ripers Manor, Buckland Ripers, nr. Weymouth, Dorset, Eng. Office: U.K. Atomic Energy Authority, Risley nr. Warrington, Lancashire, Eng.*

MÜNCH, Guido, astronomer, educator; b. Las Casas, Mexico, June 9, 1921; s. August and Maria (Paniagua) M.; M.S. in Math., U. Mexico, 1944; Ph.D. in Astronomy, U. Chgo., 1946; children—Fred, Charles, Amelia, Christopher. Came to U. S., 1947, naturalized, 1957. Astronomer, U. Mexico Obs., 1946-47; faculty U. Chgo., 1947-51; faculty Cal. Inst. Tech., Pasadena, 1951-—, prof. astronomy, 1959-—; staff mem. Mt. Wilson and Palomar Obs., 1951-—. Mem. Am., Royal astron. socs., Astron. Soc. Pacific, Am. Acad. Arts and Scis., Nat. Acad. Scis. Contbr. chpts. to Compendium of Stellar Astronomy, 1960, numerous articles to profl. jours. Research on theory of radiative transfer and stellar atmospheres, stellar statistics, interstellar absorption lines, structure of galaxies, planetary atmospheres. Address: 1201 E. California St., Pasadena, Cal. 91109.*

MUNCH, James Clyde, pharmacologist; b. Farmer City, Ill., Feb. 20, 1896; s. Henry and Nellie (Jackson) M.; B.S., Ill. Wesleyan U., 1915, M.S., 1916; Ph.D., George Washington U., 1924; m. Soula Clanton Robinson, Dec., 1928; children—Margaret Clyde (Mrs. H. J. McWhinnie), James Clanton. Toxicologist, Ill. Wesleyan U., 1915-16, U. Louisville Med. Sch., 1916-17; pharmacologist, Bur. Chemistry, U. S. Dept. Agr., 1917-28, Bur. Biol. Survey, 1928-44; prof. physiology and pharmacology, Temple U., Phila., 1934-44, dir. research Coll. Pharmacy, 1931-54, dir. research Dental Sch., 1944-51; med. dir. Strong-Cobb, 1951-53, Kay Pharms., 1958-—. Hahnemann Med. Coll. and Hosp., Phila.; dir. Pharmacology and research Sharp & Dohme, 1928-36, Wyeth, 1936-37; cons., 1938-52; Pa. Sec. Health and Welfare, various racetracks for detecting doped horses; cons., med. dir. Vaponefrin Co., 1938-—; mem. adv. com. U. S. Bur. Narcotics; sec. pharmacology 4th Pan Am. Congress, 1957; mem. 1st Internat. Pharmacology Congress, Stockholm, Sweden, 1961. Served as lt., U. S. Army, 1917-19. Decorated Lys-Schieldt; recipient medal of honor Kiwanis Club, 1932. Fellow Am. Coll. Cardiology; mem. Pan Am. Med. Assn. (sec., N.Am. pres.), Internat. Assn. Forensic Toxicologists, Interam. Congress Toxicology and Occupational Medicine, Soc. Toxicologists (charter), A.A.A.S. (life), Am. Pharm. Assn. (life, v.p. 1932-33), Assn. Mil. Surgeons (life), Nat. Pest Control Assn. (hon.), Am. Statis. Assn. (life), Brit. Assn. Advancement Sci., Brit. Soc. Study Addiction, Fedn. Internat. Pharmacists, Internat. Physiol. Soc., A.M.A. (asso.), Coll. Physicians Phila., European Soc. Study Drug Toxicity, Sigma Xi. Developed antidote for thallium poisoning, 1931, several methods bioassay, 1931; studies stability various drug preparations and preparation sustained-action materials; research in commercial applications of pharmacology as in detection of doping race horses. Home: 1270 N.E. 94th St., Miami Shores, Fla. 33138. Office: 300 N.E. 59th St., Miami, Fla. 33138.*

MUNDINGER, Fritz, German neurosurgeon; b. Freiburg/Br., Germany, June 13, 1924; s. Karl Friedrich and Luise (Isele) M.; Dr.med., Freiburg Med. Sch., 1951; m. Liselotte Dreyer, May 16, 1923; children—Friedrich Alexander, Peter Matthias. Faculty, Neurosurg. U. Hosp., Freiburg, 1950-—, prof. neurosurgery, 1964-—, head radioisotope dept., 1958-—. Mem. Soc. Neurology, Psychiatry and Neurosurgery Argentina (hon.), Med. Soc. Pontifice U. Rio de Janeiro (hon.), Med. Soc. Instituto dos Bancarios Rio de Janeiro (hon.), Soc. of Neurosurgery Finland (fgn. mem.), Soc. Neurology, Psychiatry and Neurosurgery Chile (corr.) Soc. Endocrinology and Metabolism Chile (corr.), German Soc. Neurosurgery, German Med. Assn. for Irradiation Protection, German Soc. Nuclear Medicine. Author: (with T. Riechert) Hypophysen-tumoren, Hypophysektomie, 1967. Publs. on stereotaxic procedures and apparatus in brain surgery, path. brain conditions and use of radioisotopes for diagnosis and treatment brain diseases. Home: 54 Jacobi Strasse. Office: 55 Hugstetter Strasse, Freiburg, Germany.*

MUNGALL, Allan George, Canadian physicist; b. Vancouver, B.C., Can., Mar. 12, 1928; s. Robert and Annie May (Herbert) M.; B.A.Sc., U. B.C., 1949, M.A.Sc., 1950; Ph.D., McGill U., 1954; m. Shirley Nowell Rennie, Apr. 29, 1950; children—David Robert, Carolyn Jean, James Edward. Geophysicist, Cal. Standard Co., Calgary, Alta., Can., 1950; physicist NRC, Ottawa, Ont., Can., 1950-52, 54—; research asst. McGill U., 1954, Mem. Canadian Assn. Physicists, Assn. Profl. Engrs. Ont., I.E.E.E., Profl. Inst. Pub. Service Can. Research, publs. on photometry, colorimetry, theory reflections from randomly reflecting surfaces, noise in travelling wave amplifiers, velocity of light, radio distance measurement, group and phase velocity propagation surface E-M waves, dielectric properties materials, applications dielectric measurements in med. physics, atomic beam frequency and time standards. Home: 1918 Haig Dr., Ottawa 8. Office: NRC, Ottawa, Ont., Can.*

MUNGER, Bryce Leon, Am. anatomist; b. Everett, Wash., May 20, 1933; s. Leon C. and Lina (Eaton) M.; student U. Wash., 1951-54; M.D., Wash. U., 1958; m. Donna Grace Bingham, July 20, 1957; children—Ailene Annette, Darcy Leon, B. Kirtley. Investigator exptl. pathology USAF M.C., Washington, 1959-61; asst. prof. anatomy Wash. U. Sch. Medicine, 1961-65; asso. prof. anatomy U. Chgo., 1965-66; prof., chmn. dept. anatomy Milton S. Hershey Med. Center, Pa. State U. Coll. Medicine, 1966—. Mem. Am. Assn. Anatomists, Am. Soc. Cell Biology, A.A.A.S., Phi Beta Kappa, Sigma Xi, Alpha Omega Alpha. Research, publs. dealing with structure and function of cells of pancreatic islets, by light and electron microscopy, electron microscopy of other secretory cells such as sweat glands, recent studies have included structure and cytochemistry of peripheral sensory receptors in mammalian skin. Home: 78 Laurel Ridge Rd., Hershey, Pa. 17033.*

MUNGER, Edwin Stanton, Am. geographer; b. Chgo., Nov. 19, 1921; s. Royal F. and Mia (Stanton) M.; B.S., M.S., U. Chgo., Ph.D., 1951; 1 dau. from previous marriage, Elizabeth Stanton; m. Elisabeth Meyer, Sept. 15, 1958. Field asso. Am. Univs. field staff, 1954-60; faculty U. Chgo., 1956-60; prof. geography Cal. Inst. Tech., Pasadena, 1960—. Mem. Am. delegation to Pugwash Conf. Internat. Affairs, Addis Ababa, 1966, Malmo, 1967; cons. Peace Corps, Uganda, 1966, Botswana, 1967; mem. Africa com. Nat. Acad. Sci. Fellow African Studies Assn. U. S.; mem. African and Am. Assn. (dir.), Am. Friends Africa (dir.), Inst. Current World Affairs (dir., gov.), African-Am. Inst. Author several books, numerous articles. Research on politics, econ. devel. in Africa. Office: 1201 E. California St., Pasadena, Cal. 91106.*

MUNIER-CHALMAS, (Charles-Philippe-) Ernest, French geologist; b. Tournus, France, Apr. 7, 1843; License ès sci., Caen. Asst. dir. Lab. Geol. Research, Faculté des Scis., Paris, 1879, prof. geology, 1891; lectr. geology École normale, 1882; dir. studies for geology École des Hautes Études. Mem. French Acad. Scis., 1903. Author: Excursions dans le bassin parisien, 1900; Nouvelle observations sur le dimorphisme des Foraminifères. Authority of geology of Paris basin; wrote on paleontology and stratigraphy. Died Saint-Simon, nr. Aix-les-Bains, France, May 25, 1903.

MUNK, Walter Heinrich, geophysicist; b. Vienna, Austria, Oct. 19, 1917; s. Hans and Rega (Brunner) M.; B.S., Cal. Inst. Tech., 1939, M.S., 1940; Ph.D., U. Cal. 1947; m. Edith Horton, June 20, 1953; children—Edith, Kendall. Asst. prof. geophysics U. Cal., 1947-49, asso. prof., 1949-54, prof. Inst. Geophysics and Scripps Instn., La Jolla, Cal., 1954—, asso. dir. Inst. Geophysics and Planetary Physics (statewide) and dir. La Jolla labs., 1959—. Guggenheim fellow Oslo U., 1948, Cambridge, 1955, 62. Mem. Nat. Acad. Scis, Am. Acad. Arts and Scis., Am. Geophys. Union (v.p. oceanography sect. 1954-56, pres. 1960-61), Am. Philos. Soc. (Arthur Day award). Study of rotation of earth; tides; ocean waves; ocean currents and wind stress. Home: 9530 La Jolla Shores Dr., La Jolla, Cal. 92037.

MUNNECKE, Donald Edwin, Am. plant pathologist; b. St. Paul, May 30, 1920; s. Harry M. and Maywood (Schreiber) M.; B.S., U. Minn., 1942, M.S., 1949, Ph.D., 1950; m. Elaine A. Miller, Dec. 12, 1942; children—Dianne, Douglas, Thomas, Janet. Research asso. U. Minn., St. Paul, 1949-51; faculty U. Cal. at Los Angeles, 1951-61, asso. prof., 1958-61; faculty U. Cal. at Riverside, 1961—, prof. plant pathology, 1965—. Guggenheim fellow, 1965; Fulbright fellow, Germany, 1965. Mem. Am. Phytopath. Soc., Am. Inst. Biol. Scis. Research, publs. on fate of fungicides after application to soil, including chem. form, influence of other organisms. Home: 1960 Wetherly, Riverside, Cal. 92506.*

MUNOZ, John Joaquin, microbiologist, govt. ofcl.; b. Guatemala City, Guatemala, Dec. 23, 1918; s. Juan and Carmen (Valdes) M.; came to U. S., 1938, naturalized, 1954; B.S. in Chemistry, La. State U., 1942; M.S. in Bacteriology, U. Ky., 1945; Ph.D. in Bacteriology, U. Wis., 1947; m. Margaret Allen, June 21, 1947; children—William A., Maureen C., John R., Michael R. Asst. prof. U. Ill. Med. Sch., Chgo., 1947-51; research asso. Merck-Sharp & Dohme, West Point, Pa., 1951-57; prof. microbiology Mont. State U., Missoula, 1957-61, chmn. dept., 1957-61, dir. Stella Duncan Lab., 1957-61; research microbiologist NIH, Nat. Inst. Allergy and Infectious Diseases, Rocky Mountain Lab., Hamilton, Mont., 1961—. Diplomate Am. Acad. Microbiology. Fellow Am. Acad. Microbiology; mem. Am. Assn. Immunologists, soc. for Exptl. Biology and Medicine, N.Y. Acad. Sci., Am. Soc. for Microbiology, Sigma Xi, Phi Sigma. Research, publs. on mechanism of anaphylaxis, detection, mixtures of antigens and antibodies, mechanisms of hypersensitivity. Home: 1003 S. 2d St. Office: Rocky Mountain Lab., Hamilton, Mont. 59840.*

MUNRO, Robert, Scottish physician, archaeologist; b. Ross-shire, Scotland, July 21, 1835; M.A., M.D., LL.D., Edinburgh (Scotland) U.; m. Anna Taylor, 1875 (dec. 1907). Practiced medicine, Kilmarnock, Scotland, until 1886; Rhind lectr. archaeology, 1888; Dalrymple lectr. archaeology U. Glasgow, 1910; Munro lectr. anthropology, prehistoric archaeology U. Edinburgh, 1911. Fellow Royal Soc. Edinburgh; mem. Brit. Assn. (pres. anthrop. sect. 1893), Royal Irish Acad. (hon.), Royal Soc. Antiquaires Ireland (hon.), Société Royale des Antiquaires du Nord (hon.), Société d'Arch de Bruxelles (hon.), Friesch Genootschap (hon.), Société d'Anthropologie de Paris (asso.), anthrop. socs. Berlin, Vienna (corr.). Author: Ancient Scottish Lake-Dwellings, 1882; The Lake-Dwellings of Europe, 1890; Rambles and Studies in Bosnia-Herzegovina and Dalmatia, 1895; Prehistoric Problems, 1897; Historic Scotland and its Place in European Civilisation, 1899; Man as Artist and Sportsman in the Palaelithic Period, 1904; Archaeology and False Antiquities, 1905; Prehistoric Britain, 1914; Darwinism and Human Civilization, with special reference to the origin of German Military Kultur, 1917; From Darwinism to Kaiserism, 1919; numerous contbns. sci., med. jours. Died July 18, 1920.

MUNRO, William Delmar, Am. mathematician; b. Cedaredge, Colo., Nov. 22, 1916; s. Fred Osburn and Elia (Little) M.; B.A., U. Colo., 1938; M.A., U. Minn., 1940, Ph.D., 1947; m. Anne Gillette, June 13, 1951; children—Douglas, Victoria, Margaret. Faculty, U. Minn., 1938-43, 45—, now prof. math.; project engr. Mpls.-Honeywell, 1943-45; vis. research mathematician U. Cal. at Los Angeles, 1957-58; vis. prof. Johns Hopkins, 1959-60. Cons. to pvt. cos. Mem. Am. Math. Soc., Math. Assn. Am. Author: (with M. L. Stein) Computer Programming, A Mixed Language Approach, 1964; A Fortran Introduction to Programming and Computers, 1966; also articles. Application computers to math. and phys. problems., methods solving equations, solutions of problems in heat transfers, fluid flow, applied probability, microwave theory. Home: 41 Clarence Av., S.E., Mpls. 55414.*

MUNROE, Charles Edward, Am. chemist; b. Cambridge, Mass., May 24, 1849; s. Enoch and Emeline Elizabeth (Russell) M.; S.B., summa cum laude, Harvard, 1871; Ph.D., George Washington U., 1894, LL.D., 1912; m. Mary Louise Barker, June 20, 1883; children—Mrs. Winifred M. Mathews, Russell Barker, (George) Treadway Barker, Mrs. Dorothy Rouzer, Mrs. Charlotte Dolph. Asst. in chemistry Harvard, 1871-74; prof. chemistry U. S. Naval Acad., 1874-86; chemist to torpedo corps U. S. Naval Torpedo Sta. and War Coll., 1886-92; head prof. chemistry George Washington U., 1892-1918, dean Corcoran Sci. Sch., 1892-98, and dean faculty grad. studies, 1893-1918. Mem. U. S. Assay Commn. 1885, 90, 93; visitor U. S. Naval Acad., 1898; organized and directed on Anacostan Island a vol. torpedo corps, 1898; expert spl. agt. in charge chem. industries of the U. S. for censuses of 1900, 05, 10; cons. expert U. S. Geol. Survey, U. S. Bur. of Mines and Civil Service Commn.; chmn. advisory com. Am. Ry. Assn. for drafting of regulations governing transportation of explosives, 1905; supt. denatured alcohol exhibit Jamestown Expn., 1907, and mem. jury on chemicals; apptd. by Swedish Acad. Sciences, 1900, to nominate candidate for Nobel prizes in chemistry; chmn. com. on explosives investigations NRC, 1918-28; chief explosives chemist U. S. Bur. Mines, 1919-33; cons. specialist on explosives, U. S. Forest Service, 1934-38. Fellow Am. Inst. Chemists (hon.), Am. Inst. Chemistry, Chem. Soc., London, Am. Acad. Arts and Sciences, Soc. Chem. Industry, Eng., A.A.A.S.; pres. Am. Chem. Soc., 1898-99, Washington Chem. Soc., 1895-96; chmn. com. on explosives Am. Soc. Testing Materials. Research and publs. on chemistry and explosives; inventor smokeless powder; discovered Monroe effect, leading to new understanding of detonation wave of nitroglycerine. Died Dec. 7, 1938.

MUNSON, James Eugene, Am. inventor; b. Paris, N.Y., May 12, 1835; ed. Amherst; studied shorthand and became an expert stenographer. Settled in N.Y., 1857, ct. stenographer over 30 yrs.; devised simplified shorthand in Munson System; inventor process of setting and justifying type automatically, and machines for doing same; also assisted in inventing a machine for operating typewriting machines by telegraph. Author: The Complete Phonographer, 1866; Dictionary of Practical Phonography, 1875, 1906; Phrase Book of Practical Phonography; The Art of Phonography, 1898; A Shorter Course in Munson Phonography, 1900; First Phonographic Reader, 1904; Phonographic Dictation Book, 1904; Munson's Pocket Dictionary of Phonography, 1906. Died 1906.

MUNSON, John P., biologist; b. Jolster Sunfjord, Norway, Feb. 21, 1860; s. Peter and Elizabeth (Dvergsdal) M.; came to U. S., 1864; B.S., U. Wis., 1887, M.S., 1892; Ph.B., Yale, 1891; Ph.D. (fellow), U. Chgo., 1897; m. Sophie Josephine Mikkelsen, Dec. 30, 1897; 1 dau., Esther Ingeborg. Master in English, Augustana Coll., Sioux Falls, S.D., 1889-91; investigator in biology, Woods Hole, Mass., 1894; hon. fellow in biology Clark U., 1897; head Dept. of Biology Wash. State Tchrs.' Coll., 1899—; dir. zoölogy Seaside Lab., Port Renfrew, B.C., 1903. Lectr. 7th Internat. Zoöl. Congress, Boston, 1907, 8th Internat. Zoöl. Congress, Graz, Austria, 1910; research Christiana, Berlin, Naples, 1910. Recipient Walker 1st prize Boston Soc. Natural History, 1911. Fellow A.A.A.S., Western Soc. Naturalists, Royal Soc. (London). Author: Education through Nature, 1903; Spermatogenesis of the Butterfly, 1906. Collaborator on Am. Jour. Anatomy. Spent many yrs. in comparative cell studies, 25 plates completed. Died Feb. 27, 1928.

MUNSON, Paul Lewis, Am. pharmacologist; b. Washta, Ia., Aug. 21, 1910; s. Lewis Sylvester and Alice (Orser) M.; B.A., Antioch Coll., 1933; M.A., U. Wis., 1937; Ph.D., U. Chgo., 1942; M.A. (hon.), Harvard, 1955; m. Aileen Geisinger, Mar. 7, 1931 (div. 1948); 1 dau., Abigail; m. 2d, Mary Ellen Jones, Aug. 15, 1948; children—Ethan Vincent, Catherine Laura. Research fellow U. Chgo., 1939-42; research biochemist Wm. S. Merrell Co., Cin., 1942-43; research biochemist Armour Labs., Chgo., 1943-47, head endocrinology research, 1947-48; research asso. Yale Sch. Medicine, 1948-50; faculty Harvard Sch. Dental Medicine, 1950-66, prof. pharmacology, 1963-65, lectr., 1965-66; prof., chmn. pharmacology U. N.C. Sch. Medicine, Chapel Hill, 1965—. Mem. Nat. Bd. Med. Examiners, 1967—; mem. gen. med. study sect. USPHS, 1966—. Mem. Am. Soc. Pharmacology and Exptl. Therapeutics, Am. Soc. Biol. Chemists, Endocrine Soc. (council, 1963-65), Am. Chem. Soc., Internat. Assn. Dental Research (councillor 1957-59), Biometrics Soc., Biochem. Soc., (Great Britain), Am. Dental Assn., Am. Assn. U. Profs., Sigma Xi. Editorial bd. Biochem. Medicine, 1967—. Research, numerous publs. on hormones, hormone antagonists, biol. and chem. assay methods, isolation of natural products; discovered thyrocalcitonin, dehydroepiandrosterone sulfate in normal human urine, inhibitory effect of morphine on anterior pituitary gland; prepared 1st ACTH used in medicine and 1st internat. standard for ACTH. Home: 407 Clayton Rd., Chapel Hill, N.C. 27514.*

MÜNSTER, Arnold, German physicist; b. Oberursel, Jan. 5, 1912; s. Rudolf and Auguste (Schlüter) M.; ed. Münster, Berlin, Heidelberg; Ph.D.; m. Lilly Curtius; children—Johannes, Thomas, Nikolaus. Instr., head metall. labs., Frankfurt; asso. prof. U. Paris; prof. U. Frankfurt; dir. Institut fur theoret. physikal. chemie. Recipient medal Free U. Brussels. Mem. Bunsen Soc., German Soc. Chemists, Physics Soc. Author: Risenmoleküle, 1952; Statistiche Termodynamik kondensierter Phasen, 1962; Les liquides simples, 1964; Münter und Dupuis, Traité de mécanique statique. Research on light scattering and diffuse scattering of X-rays; critical phenomena; thermodynamics; statis. mechanics; theory of irreversible processes. Home: Lessingstrasse 5. Office: Institut fur theoret. physik. Chemie der Univ., Jügelstrasse 11, Frankfurt, West Germany.

MÜNSTERBERG, Hugo, psychologist; b. Danzig, Germany, June 1, 1863; s. Moritz Münsterberg; Ph.D., Leipzig, 1885, M.D., Heidelberg, 1887; A.M., Harvard, 1901; LL.D., Washington U., 1904; Litt.D., Lafayette Coll., 1907; m. Selma Oppler, 1887. Instr. and asst. prof. U. Freiburg, 1887-91; prof. psychology Harvard, 1892-1916; also dir. Psychol. Lab.; exchange prof. at U. Berlin, 1910-11. Organizer and 1st dir. of Amerika-Institut of the German Govt., 1910-11; v.p. Internat. Congress Arts and Sciences, St. Louis, 1904, Internat. Psychol. Congress, Paris, Internat. Philos. Congress, Heidelberg, 1908. Fellow Am. Acad. Arts and Sciences; mem. Washington Acad. Sciences, Psychol. Assn. (pres. 1898), Philos. Assn. (pres. 1908). Author: Psychology and Life, 1899, Grundzüge der Psychologie, 1900; American Traits, 1902; The Americans, 1904; Eternal Life, 1905; Science and Idealism, 1906; Philosophie der Werte, 1908; The Eternal Values, 1909; Psychology and the Teacher, 1909; American Problems, 1910; Psychology and Industrial Efficiency, 1912; American Patriotism, 1913; Grundzüge der Psychotechnik, 1914; Psychology and Social Sanity, 1914; The War and America, 1914; The Peace and America, 1915; Psychology: General and Applied. Editor of Harvard Psychological Studies, 1903—. Pioneer in applied psychology. Died Dec. 16, 1916.

MUNTING, Abraham, Dutch physician, botanist; b. Groningen, Netherlands, June 19, 1626; ed. univs. Franeker, Leiden, Utrecht; M.D., Angers, France; rector Groningen U., became prof. botany and chemistry, 1658. Author: Hortus et universae materiae medicae gazophylacium, 1664; De vera antiquorum herba Brittannica, 1681; Naauwkeurige Beschryving der Aardgewassen, 1696. Rediscovered ancient herb britannica mentioned by Pliny as cure for scurvy used by Roman soldiers. Died Groningen, Jan. 31, 1783.

MÜNTZING, Arne, Swedish geneticist; b. Göteborg, Sweden, Mar. 2, 1903; s. Natanael and Elin (Esselius) M.; Ph.D., U. Lund (Sweden), 1930; Ph.D., U. Brno (Czechoslovakia), 1965; m. Gudrun Lewis-Jonsson, July 22, 1933; children—Lars, Jonas, Eva (Mrs. Johan Akesson), Karin, Hans Erik. Asst., Inst. Genetics, Lund, 1926-29, prof. genetics, dir., 1938-—; 1st asst. Hilleshög Sugar Beet Breeding Inst., Landskrona, Sweden, 1929-31; head cytogenetic dept. Svalöf Plant Breeding Inst., 1931-38. Mem. Swedish Research Council Sci., 1955-62, Swedish Research Council Atomic Research, 1964-—. Recipient Darwin plaque, Halle, Germany, 1959; Mendel plaque, Brno, 1965. Hon. mem. genetic socs. of India, Japan, Egypt, Am. Bot. Soc., Bot. Soc. Edinburgh; mem. acads. sci. in Sweden, Denmark, Finland, Germany, Scandinavian Assn. Geneticist (pres. 1960-—). Author: Ärflighetsforskning, 1953, 64; Genetics: Basic and Applied, 1967; also numerous articles. Research on genetics in wild and cultivated plant material especially species formation, effects of deviating chromosome constns., possibilities of creating new types. Home: Nicolovius Väg 10, Lund, Sweden.*

MUNZ, Philip Alexander, Am. botanist; b. Saratoga, Wyo., Apr. 1, 1892; s. Alexander and Caroline (Wolf) M.; A.B., U. Denver, 1913, A.M., 1914; Ph.D., Cornell U., 1917; Sc.D., Pomona Coll., 1960; m. Alice Virginia McCully, June 10, 1925; children—Robert A., Frederick W. Faculty, Pomona Coll., Claremont, Cal., 1917-44, prof. botany, 1925-44, dean faculty, 1941-42; prof. botany and horticulture Cornell U., Ithaca, N.Y., 1944-46; dir. Rancho Santa Ana Bot. Garden, Claremont, 1947-61, dir. emeritus, 1961-—; prof. botany Claremont Grad. Sch., 1951-61. Guggenheim fellow, 1936-37. Recipient Henry A. Gleason award N.Y. Bot. Garden, 1966. Mem. A.A.A.S., Bot. Soc. Am. (certificate merit 1958), Am. Soc. Plant Taxonomists, Cal. Acad. Sci., New Eng., Torrey bot. clubs. Author several books including: (with Keck) A California Flora, 1959; Onagraceae of North America, 1965; also articles. Research on identification Cal. plants, evening primrose family of new world, larkspurs and delphiniums of Asia and Africa. Home: 4141 Via Padova, Claremont, Cal. 91711.*

MÜNZER, Hans Georg, German mathematician; b. Munich, Apr. 23, 1906; s. Richard and Leopoldine (Furtner) M.; Ph.D., U. Gottingen; m. Elfriede Rotherbach, Apr. 9, 1934. Asst., statistics inst. U. Gottingen, 1931, dir., 1933, agrégé, 1937, instr., 1939, prof., 1944, dir. inst. math. statistics, 1945-46; chair statistics in social scis. Faculty of Berlin Free Univ., also dir. Tng. Sch. in Statistics, 1956, dir. statistics and math. inst., 1958-—. Mem. Internat. Inst. Statistics, German Soc. for Applied Math., German Soc. for Research on Enterprise. Research and publs. on probability, statis. method, use and application of statistics, calculus of probabilities and applications to problems of econs., biology and medicine. Home: Berner Strasse 46, Berlin 45. Office: Garystrasse 21, Berlin 33, West Germany.

MURAKAMI, Ujihiro, Japanese pathologist; b. Nagoya City, Japan, May 15, 1910; s. Tomojiro and Take (Murakami) Furukawa; M.D., Nagoya Med. Sch., 1935; D.Med. Sci., Nagoya Imperial U., 1939; m. Kazuko Kato, Oct. 9, 1938; children—Kyoko (Mrs. Kazuro Nakazawa), Junko Kato, Nobuyuki, Seiko. Asso., Nagoya Med. Sch., 1935-40; faculty Nagoya Imperial U. Sch. Medicine, 1940-—, prof., 1956-—, mem. Research Inst. Environmental Medicine, 1950-—, dir., 1967-—. Research fellow Harvard Sch. Pub. Health, 1957-58; mem. Central Pharm. Affairs Council, 1963-—. Recipient 12th Chunichi Bunka Sho prize, 1959; 3d prize for studies on mechanisms manifesting congenital anomalies Japan Assn. Human Genetics, 1962. Mem. Am. Teratology Soc. Author: Clinico-Genetic Study of Disorders of the Nervous System, 1957; numerous articles. Research on disorders of nervous system, their phenogenesis, exptl. teratology, principles governing formation of congenital malformations. Home: 4, 2-Chome, Higashisugi-Cho, Kita-ku, Nagoya City, Japan.*

MURALT, Johannes von, see von Muralt, Johannes.

MURATA, Kiku, Japanese food chemist; b. Niigata Prefecture, Japan, Sept. 18, 1912; d. Hideo and Iso Murata; diploma, High sch. tchr.'s certificate, Japan Women's Coll., 1933; D.Sc., Osaka U., 1945. Asst. food analysis Keio U., Tokyo, 1933-35; research asst. organic chemistry Osaka (Japan) U., 1935-42; research staff nutrition chemistry Osaka Municipal Inst. for Hygiene, 1942-54; prof. Japan Women's U., 1948-54; prof. Osaka City U., 1949-—. Mem. Vitamin B Research Com. Japan, 1945-—, Essential Amino Acid Com. Japan, 1957-—; councilor Tanabe Amino Acid Research Fund, 1961-—. Research fellow Phila. Gen. Hosp., 1959-60. Mem. Japan Vitamin Soc. (Yamanouchi award for studies on thiaminase 1957), Japanese Soc. Food and Nutrition, Agrl. Chem. Soc. Japan, Japanese Biol. Soc., Japan Home Econs. Assn. Author: Textbook of Nutrition, 1957; Advances in Vitaminology I, 1959; Japanese Literature on Beriberi and Thiamine, 1965; also articles. Determined chem. structure of diazonium compound of thiamin, pigment used for determination of thiamine; purified enzyme, thiaminase; discovered decomposition mechanisms of thiamine using thiaminase; isolated some antioxidants from fermented soybeans, including an isoflavone which had previously not been isolated from natural sources. Home: 608 Honyakushi, Nara City, Nara Prefecture, Japan. Office: Oikedori, Nishi-ku, Osaka City, Osaka Prefecture, Japan.*

MURAWSKI, Krzysztof Wojciech, Polish biochemist; b. Warsaw, Poland, Jan. 11, 1930; s. Feliks and Kazimiera (Stangreciak) M.; M.D., Warsaw Med. Sch., 1953; m. Sylvia Anna Winawer, Aug. 30, 1958; children—Jan Grzegorz, Antoni Wladyslaw. Staff dept. biochemistry Inst. Hematology, Warsaw, 1951-—, chief dept., 1961-—, sec. sci., 1964-—; asst. prof. biochemistry Inst. Hematology, 1966-—; vis. biochemist dept. animal biochemistry Moscow State U. (USSR), 1955-56, Inst. Biol. Chemistry, U. Rome (Italy), 1963-64; cons. biochemist Psychoneurol. Inst., Pruszków, Poland, 1957-66. Mem. Polish Biochem. Soc., (sec. gen. 1965-—), Polish Hematological Soc. (com. auditors 1966-—). Author: (with B. Bogdanikowa, W. Dzierzkowa-Borodej) Klinika bialek krwi, 1960; also articles. Research on structure of dextran and its use as substitute of blood plasma, genetics and biochemistry of plasma proteins, normal and abnormal human hemoglobins. Home: ul. Czarnieckiego 38. Office: ul. Chocimska 5, Warsaw, Poland.*

MURCHISON, Sir Roderick Impey, Brit. geologist; b. Tarradale, Scotland, Feb. 19, 1792; s. Kenneth M.; ed. Great Marlow; hon. D.C.L., Oxford; hon. LL.D. Cambridge, Dublin; m. Charlotte Hugonin, Aug. 29, 1814. Served Brit. army, 1807-14; traveled on continent, 1816-18; moved to London; attended lectrs., Royal Institution, 1824; geological field-work around Nursted, Kent, 1825; traveled in highlands, 1827, in Auvergne and northern Italy, 1828, Germany and eastern Alps, 1829, southern Wales, 1831; Germany and the Boulonnais, 1839, Russia, Italy and eastern Alps, 1847-8; dir.-general, geological survey, 1855; traveled in Scottish highlands, 1858, 59 and 60; dir., Royal School of Mines, 1852. Awarded Order of St. Anne and of Stanislaus; Wollaston medal; prix Cuvier; Brisbane medal; knighted 1846; Commander of Bath, 1863; baronet, 1866. Fellow Royal Soc., 1826 (Copley Medal); Fellow Geological Soc. (sec., pres., 1831) corr. mem. French Acad. Scis., 1844. Author: The Silurian System (2 vol.), 1839; (with von Keyserling and de Verneuil) The Geology of Russia and the Ural Mountains, 1845. Research on lower fossiliferous strata of England and Wales; differentiated the Silurian and Devonian systems (with Sedgwick), which he named; introducted name Permian for system of rocks, 1841; predicted discovery of gold in Australia, 1844. Died London, Oct. 22, 1871.

MURDOCH, Joseph, Am. mineralogist; b. Washington, Feb. 19, 1890; s. John and Abby (Stuart) M.; A.B., Harvard, 1911, S.M., 1912, Ph.D., 1915; m. Maude E. Russell, Oct. 3, 1914; 1 dau., Barbara (Mrs. Horace Pease Phillips). With Russell & Co., Cambridge, Mass., 1915-28; faculty U. Cal. at Los Angeles, 1928-—, prof. mineralogy, 1939-67, prof. emeritus, 1967-—. Fellow Mineral. Soc. Am. (pres. 1960); m. Geol. Soc. Am., Am. Crystallographic Soc., A.A.A.S. Author: Microscopic Determination of Ore Minerals, 1915; also articles. Developed microscopic identification of opaque minerals by reflected light from polished surface; identification and description of new minerals. Home: 1130 Georgina Av., Santa Monica, Cal. 90402.*

MURDOCK, Carleton Chase, Am. physicist; b. Cooperstown, N.Y., July 29, 1884; s. Benjamin Franklin and Myrtle Emma (Chase) M., Jr.; B.S., Colgate U., 1907; A.M., Cornell U., 1910, Ph.D., 1919; m. Dorothy Lee Waugh, Aug. 28, 1923; children—Franklin Waugh, Edmund Lee. Tutor physics U. Me., 1907-08; faculty physics Cornell U., Ithaca, N.Y., 1909-52, prof., 1930-52, dean univ. faculty, 1945-52, prof. physics emeritus, 1952. Mem. symbols, units and nomenclature com. Nat. Acad. Sci.-NRC, 1955-63. Mem. Sigma Xi (exec. com. 1943-48). Research, publs. on shape of Debye-Scherrer lines, interpretation of Laue photographs, multiple Laue spots. Home: 319 Wait Av., Ithaca, N.Y. 14850.*

MURDOCK, George John, Am. inventor; b. New Berlin, N.Y., Apr. 17, 1858; s. Chester and Elizabeth (Armstrong) M.; acad. and engrng. edn.; m. Jeanette P., Waterman, Apr. 23, 1883; at least 2 children. Discovered in 1879 that electric lamp carbons when isolated from atmospheric air were of much longer life, he took out in 1883 first patent in U. S. on enclosed form of arc lamp, now commonly used; prior to 1885 developed complete system of electric lighting, including dynamo, regulator for arc lamps, arc lamps, and other accessories; other patented inventions include bolt machines, files, and holder button, and button fastener (with A. L. Lesher), an exhaust turbine, an electric surface gage, magnetic drill holder, electric ry. signal indicator, and many other tools, and instruments that have come into common use; constructed 1st gasoline tank with a rubber composite cover, 1903; inventor self-sealing fuel tanks for war airplanes of the type used by U. S. and fgn. govts. in World War I. Elected to membership in many Am. and fgn. socs. Contbr. to tech. press on subjects relating to electricity and mechanics. Died July 25, 1942.

MURDOCK, George Peter, Am. anthropologist, educator; b. Meriden, Conn., May 11, 1897; s. George Bronson and Harriet (Graves) M.; grad. Phillips Acad., 1915; A.B., Yale 1919, Ph.D., 1925; postgrad. Harvard Law Sch.; m. Carmen Swanson, Sept. 4, 1925; 1 son, Robert Douglas. Instr. sociology U. Md., 1925-27; asst. prof. sociology Yale, 1928-34, asso. prof. ethnology, 1934-39, prof. anthropology, 1939-60, chmn. dept. anthropology, 1937-43, 54-57; Andrew Mellon prof. social anthropology U. Pitts., 1960-—. Chmn., Pacific Sci. bd. NRC, 1953-57, Div. Behavioral Scis., 1964-66. Recipient Viking Fund medal and award in gen. anthropology, 1949. Fellow Am. Acad. Arts and Scis., Am. Sociol. Soc., Am. Anthrop. Assn. (pres. 1955), Am. Geog. Soc.; mem. Am. Ethnol. Soc. (pres. 1952-53), Soc. for Applied Anthropology (pres. 1947), Societe' des Oceanistes, Nat. Acad. Scis., Sigma Xi, Beta Theta Pi. Author: Our Primitive Contemporaries, 1934; Outline of Cultural Materials, 1938; Ethnographic Bibliography of North America, 1941; Social Structure, 1949; Outline of South American Cultures, 1951; Africa, 1959. Editor: The Evolution of Culture (Julius Lippert, 1931; Studies in the Science of Society, 1937; Ethnology: An Internat. Jour. Cultural and Social Anthropology, 1962-—. Work on classifying and indexing known cultures of world. Home: 4150 Bigelow Blvd., Pitts. 15213.*

MURDOCK, John Carey, Am. economist; b. Blackwell, Okla., Dec. 10, 1922; s. Frank Elbert and Nannine (Watt) M.; B.S., U. Okla., 1947; M.S., U. Wis., 1951, Ph.D., 1955; m. Jean Boardman, Oct. 15, 1949; children—John B., Robert C. Faculty, U. Mo., Columbia, 1951-—, chmn. dept., 1962-64, prof. econs., dean Grad. Sch., 1967-—. Project dir. NASA study on location of research, 1964-67; cons. industry, 1962-—. Gulf Refining Co. fellow, 1957; U. Mo. fellow, 1958; Community Studies fellow, 1960-61. Mem. Am. Econs. Assn., Regional Sci. Assn. Author: (with J. Graves) Regions and Research, 1966; also articles. Analysis of changing mix of inputs with mine depletion; theory of structural change in spatially concentrated economies. Home: 110 E. Parkway, Columbia, Mo. 65201.*

MURDOCK, William, English engr., inventor; b. Lugar, nr. Auchinleck, Scotland, Aug. 21, 1754; s. John Murdock; ed. as gunner; m. Miss Paynter (dec. 1790); children—William, John. Began work under Boulton and Watt, Soho, Eng., 1777; became manager of Boulton and Watt works, 1800; later became partner; built exptl. gas apparatus, Soho, 1800; lighted foundry with gas, 1803. Recipient Rumford gold medal 1808. Patentee stone pipes, 1810. Inventor: coal-gas lighting, 1792, sun and planet motion, bell-crank engine, iron cement; built 3 steam engines. Died Handsworth, Eng., Nov. 15, 1839.

MURGOCI, Munteanu Gheorghe, Rumanian geologist, mineralogist; an organizer Internat. Assn. Soil Sci.; mem. Rumanian Acad. Author: Uber die Einschlüsse von Grant—Vesuvianfels in dem Serpentin des Paringu-Massivis (Rumänien), 1901; La grande nappe de charriage des Carpathes Méridionales, 1905; Selected Works, 1917. Founder soil sci. in Rumania; studied tectonics of So. Carpathians; discovered Getic ground-layer water. Died 1925.

MURGULESCU, Ilie, Rumanian chemist; b. 1902; prof. Bucharest (Rumania) U.; chmn. Acad. Socialist Republic Rumania, 1963-66. Author: Text Book of Analytical Chemistry, 2 vols., 1937; Sur la dissociation du carbonate de calcium, 1946. Research on physico-chem. composition of ionic liquids, molecular structure of coordinative combinations, synetics of processes of thermic and photo-chem. decomposition of complex oxalates.

MURIE, Adolph, Am. ecologist; b. Moorhead, Minn., Sept. 6, 1899; s. Adolph Nelson and Marie (Frimanslund) M.; student Fargo (N.D.) Coll., 1922, B.Sc., Concordia Coll., Moorhead, 1925; M.S., Ph.D., U. Mich.; m. Louise Gillette, Sept. 8, 1932; children—Gail, Jan. Asst. field biologist Alaska studies U. S. Biol. Survey, 1922-23; asst. curator mammals U. Mich., 1929-34; field ecologist U. S. Nat. Park Service, 1934-65. Fellow Arctic Inst. N.Am.; mem. Am. Soc. Mammalogists, Wildlife Soc., Wilderness Soc., Nat. Parks Assn., Defenders of Wildlife, Nat. Audubon Soc., Cooper Ornithol. Soc., Wilson Ornithol. Soc., Alaska Conservation Soc., Sierra Club. Author: Moose of Isle Royale, 1934; Following Fox Trails, 1936; Ecology of the Coyote in the Yellowstone (Best Wildlife Publ. award Wildlife Soc. 1941); The Wolves of Mount McKinley, 1944; A Naturalist in Alaska (John Burroughs award 1963), 1961. Study of predators; distribution and life histories of mammals and birds; range and big game. Address: Moose, Wyo. 83012.

MURIS, Jean de, see de Muris, Jean.

MURNAGHAN, Francis D., mathematician; b. Omagh, County Tyrone, Ireland, August 4, 1893; s. George and Angela (Mooney) M.; B.A., Nat. U. Ireland, 1913, M.A., 1914; D.Sc. (honoris causa), 1940; Ph.D., Johns Hopkins, 1916; m. Ada May Kimbell, June 23, 1919; children—Francis D., Mary Patricia. Came to U. S., 1914, naturalized, 1928. Instr. math. Rice Inst., Houston, 1916-18; faculty Johns Hopkins, 1918-48; prof. math. Instituto Técnico de Aeronáutica, Sao José dos Campos, 1949-59; cons. to U. S.

Navy, 1955-63; dir. Math. Inst., Rutgers, 1926; vis. prof., U. Chgo., 1928, 30, U. Pa., 1929, Duke, 1941, Brown U., 1943-44, Dublin Inst. Advanced Studies, Acad., Am. Phys. Soc., A.A.A.S. (v.p., chmn. Sect. A, 1943); mem. Nat. Acad-Scis; Am. Philos. Soc., Am. Math. Soc., Math. Assn. Am., London Math. Soc., Edinburgh Math. Soc., Acad. Bras das Cienciais, Acad. Nac de Cien. de Lima, Sigma Xi, Phi Beta Kappa, Gamma Alpha. Author: Vector Analysis and the Theory of Relativity, 1922; (with Joseph S. Ames) Theoretical Mechanics, 1929; (with H. Bateman and H. L. Dryden) Hydrodynamics, 1932; Theory of Group Representations, 1938; Analytic Geometry, 1946; Differential and Integral Calculus, 1947; Applied Mathematics, 1948; Finite Deformation of an Elastic Solid, 1951; Algebra elementar e Trigonometria, 1955; The Laplace Transformation, 1962; The Calculus of Variations, 1962; The Unitary and Rotation Groups, 1962. Research on elastic behavior of solids under large stress; equation of state of an isotropic solid; theory of representations of symmetric and unitary groups. Address: 6202 Sycamore Rd., Balt. 12.*

MURNAGHAN, Maurice Francis, Irish physiologist; b. Dublin, Ireland, Sept. 1, 1917; s. Daniel F. and Helena (Whelan) M.; student U. Coll., Dublin, 1936-44; M.B., B.Ch., B.A.O., Nat. U. Ireland, 1942, B.Sc., 1943, M.Sc., 1944, M.D., 1961; m. Ellen Eleonore Elsen, July 5, 1952; children—Siobhan Maria Martha, Christa Isolde Maria, Renata Anna Maria Lieselotte. Research fellow dept. pharmacology Oxford (Eng.) U., 1945-46; sr. lectr. dept. pharmacology U. Coll. Dublin, 1947-48, prof., head dept. physiology, 1964—, prof., head dept. pharmacology U. Ottawa (Ont., Can.), 1948-64. Dep. registrar Med. Council Can., 1960-64. Recipient Birmingham Gold medal U. Coll. Dublin, 1938; Irish Red Cross fellow, 1944. Licentiate Med. Council Can. Fellow Royal Acad. Medicine Ireland; mem. Physiol. Soc., Am. Pharmacological Soc., Canadian Physiol. Soc., Pharmacological Soc. Can. (past pres.). Author: Materia Medica, 1949; also articles. Research on elucidation of mechanism of paralysis produced by wood tick and effect various drugs on heart and blood vessels. Home: 7 Crannagh Rd., Rathfarnham, Dublin 14, Ireland.*

MURPHEY, Bradford, Am. physician; b. Kansas City, Mo., May 23, 1891; s. James Jay and Mary Alice (Hayes) M.; A.B., U. Neb., 1918; M.D., Neb. Coll. Medicine, 1920; D.Sc., Colo. Coll., 1939; m. Margaret Griffin, Nov. 9, 1921; children—Bradford Griffin, Murray Griffin. Asst. physician Elgin (Ill.) State Hosp., 1922; physician Chgo. State Hosp., 1923; sr. physician N.J. State Hosp., Morris Plains, N.J., 1924-25; Rockefeller Found. fellow in psychiatry Boston Psychopathic Hosp., Boston, in child psychiatry Judge Baker Found., Boston, in neurology Neurol. Inst. N.Y., in pediatrics Children's Hosp., Boston, 1925-26; chief cons. psychiatrist New Eng. Home for Little Wanderers, Boston, 1926; dir. Bemis Taylor Found. Child Guidance Clinic, Colorado Springs, Colo., 1927-38; dir. Children's Service Center, Child Guidance Clinic, Wilkes-Barre, Pa., 1938-40; pvt. practice psychiatry, Denver, 1940—; asso. clin. prof. psychiatry U. Colo., 1940-64, prof. emeritus, 1964—. Diplomate Am. Bd. Psychiatry and Neurology. Fellow Am. Psychiat. Assn., Am. Orthopsychiat. Assn., A.M.A., A.C.P.; mem. Denver, Colo. med. socs., Colo. Neuropsychiat. Soc., Colo. Conf. Social Work, Rocky Mountain Pediatric Soc., Denver Clin. Path. Soc., Colo. Mental Hygiene Soc., Alpha Omega Alpha. Research, publs. on clarification of dynamics of personality devel. in children in terms of family relationships and effects of relationship between early childhood experiences and illness, delinquency and crime in later life. Home: 345 Vine St., Denver 80206. Office: Republic Bldg., Denver, 80202.*

MURPHREE, Henry Bernard Scott, Am. physician; b. Decatur, Ala., Aug. 11, 1927; s. Henry Bernard and Nancy (Burrus) M.; student Mass. Inst. Tech., 1944-45; B.A., Yale, 1950; M.D., Emory U., 1959; m. Dorothy Elaine Simmons, Nov. 14, 1953; children—Julie Elizabeth, Susan Louise, Jefferson Van. Research asst. Phipps Clinic Johns Hopkins, 1950; staff psychologist U. S. Naval Hosp., Bethesda, Md., 1951; div. psychologist 2d Marine Div., Fleet Marine Forces, 1952-53, psychologist U. S. Submarine Base, Med. Research Lab., New London, Conn., 1953-55; instr. psychopharmacology Emory U. Sch. Medicine, 1959-61; asst. chief neuropharmacology N.J. Bur. Research in Neurology and Psychiatry, Princeton, 1961—; psychologist Koff Psychiat. Clinic, Atlanta, 1955-61; mem. staff N.J. Neuropsychiat. Inst., 1962—; mem. cons. staff Princeton Hosp., 1964—; lectr. pharmacology Hahnemann Med. Coll., 1965—. Mem. Am. Soc. Pharmacology and Exptl. Therapeutics, N.Y. Acad. Sci., A.A.A.S., Am. Chem. Soc., Soc. Biol. Psychiatry, Am. Coll. Neuropsychopharmacology, Sigma Xi. Cons. editor Effects of Smoking and Nicotine on the Central Nervous System, 1967. Research, numerous publs. on psychometrics, validation of personnel selection techniques, psychol. characteristics of selected groups, reliability and validity of scholastic examinations, pharmacology, activities of drugs in CNS, antagonism and antidoting, exptl. psychotherapeutics, differential effects of drugs in different populations, neurophysiology, multidimensional and time-series analysis of EEG, digital computer methods, medicinal chemistry, novel class of vasodilators. Home: R.D., Skillman, N.J. 08558. Office: Neuropharmacology, Bur. Research in Neurology and Psychiatry, Box 1000, Princeton, N.J. 08540.*

MURPHY, Cornelius Bernard, Am. chemist; b. Worcester, Mass., Dec. 10, 1918; s. Francis B. and Mary (Lane) M.; B.S., Holy Cross Coll., 1941, M.S., 1942; Ph.D., Clark U., 1952; m. Gertrude C. Tracy, Feb. 18, 1943; children—Cornelius Bernard, Francis M., Kathleen M., Joanne M. Asst. prof. Holy Cross Coll., 1945-52; phys. chemist Research Lab. Am. Cyanamid Co., Stamford, Conn., 1952-55; with Gen. Elec. Co. Advanced Tech. Labs., Schenectady, N.Y., 1955-65, project chemist, 1963-65; mgr. materials analyses Xerox Corp., Webster, N.Y., 1965—; mem. founding com. Internat. Conf. on Thermal Analysis. Cons. Limited Warfare Lab., U. S. Army. Mem. Am. Chem. Soc., Electrochem. Soc., N.Y. Acad. Scis. Abstractor, Chem. Abstracts. Research, numerous publs. on chelation of glycine and glycine peptides, thermal methods of analysis, conversion of gaseous materials to condensation nuclei and utilization in chem. analysis, analytical applications of nuclear chemistry, devel. liquid oxygen-compatible potting compounds, invention of biplastic thermostat actuator. Home: 42 Clarke's Crossing, Fairport, N.Y. 14450. Office: 800 Phillips Rd., Webster, N.Y. 14580.*

MURPHY, Edwin Daniel, Am. pathologist; b. Bklyn., July 30, 1917; s. Daniel Joseph and Rose (Brockhaven) M.; B.S. summa cum laude, St. John's U., N.Y.C., 1939; M.D., Yale, 1943; m. Mary E. Nissley, Apr. 2, 1942; children—Donal Brian, Kathleen Rosemary, Sheila Patrice, Stephen Robert, Janette Christine, Suzanne Elisabeth, Brian David. Jane Coffin Childs fellow Yale, 1944-46; instr. U. Tenn. Med. Sch., Memphis, 1946-48; head unit on gynecologic pathology Nat. Cancer Inst., NIH, Bethesda, Md., 1948-53; research asso. Jackson Lab., Bar Harbor, Me., 1953-57, staff scientist, 1957—, staff sci. dir., 1956-57, asst. dir. research, 1957-58; guest prof. U. Frankfurt (Germany), 1963-64. Fulbright Sr. Research scholar, 1963-64. Mem. Am. Assn. for Cancer Research, Sigma Xi, Alpha Omega Alpha. Contbg. Author: Biology of the Laboratory Mouse, 1966; also articles. Devised practical methods for exptl. induction carcinoma uterine cervix in mice by chem. carcinogens; described promoting effect estrogenic hormone and inhibiting effect castration; a spontaneously occurring model for human Hodgkin's disease in mice. Home: 11 High St. Office: Jackson Lab., Bar Harbor, Me. 04609.*

MURPHY, Francis Daniel, Am. physician; b. New Diggings, Wis., Nov. 7, 1895; s. Michael J. and Mary (Driscoll) M.; B.S., Marquette U., 1918, M.D., 1920, LL.D., 1961; M.D. in Medicine, U. Pa., 1924; m. Madaline McNamara, June 27, 1925; children—Joan Ellen (Mrs. Thomas Foley), Francis Daniel. Prof., chmn. dept. medicine Marquette U. Sch. Medicine, Milw., 1928-58; med. dir. dept. medicine Milwaukee County Gen. Hosp., 1924-58; chief staff St. Joseph's Hosp., 1936-42, 47-54. Named Marquette U. Alumnus of Year, 1956; recipient Marquette Med. Alumni award, 1952; Edward R. Wehr Research grantee, 1955; Francis D. Murphy Library, Milw. County Gen. Hosp. named in his honor, 1959; Chair medicine established in his honor Marquette U. Fellow A.C.P., Am. Coll. Cardiology; mem. A.M.A. (certificate of honor 1933), Central Soc. Clin. Research, Chgo. Soc. Internal Medicine, Am. Therapeutic Soc., Am. Diabetic Assn., Am. Heart Assn., A.A.A.S., Am. Geriatrics Soc., Am. Soc. for Study Arteriosclerosis, Am. Found. High Blood Pressure, Alpha Omega Alpha, Alpha Sigma Nu. Author: The Diagnosis and Treatment of Acute Medical Disorders, 1944, 4th edit. titled Medical Emergencies, 1952; also numerous articles. Research in nephritis and hypertension, use of antihypertensive drugs and artificial kidney hypertension and heart disease. Home: 610 Honey Creek Pkwy., Wauwatosa, Wis. 53213.*

MURPHY, G. W., Am. chemist; b. Hot Springs, Ark., Jan. 2, 1919; s. Andrew Jackson and Mildred (Jones) M.; A.B., U. Ark., 1940; Ph.D., U. N.C., 1946; m. Priscilla Eaton, Apr. 5, 1945 (div. Feb. 1961); children—Priscilla Mary, Janet Eaton; m. 2d, June Richey, Sept. 2, 1967; adopted children—Cindy, Sally. Faculty, U. Okla., Norman, 1946-51, 56—, prof. chemistry, 1959—, chmn. dept., 1960—; vis. scientist Argonne (Ill.) Nat. Lab., 1951-53; prof., chmn. chemistry dept. State U. N.Y., Albany, 1953-56. Fulbright scholar, Deutscheforschungs-gemeinschaft scholar, vis. prof. Heidelberg (Germany) U., 1963. Mem. Am. Assn. U. Profs. (past pres. local chpt., past state pres.), Electrochem. Soc. (Young Author's prize 1950), Am. Chem. Soc., Sigma Xi. Research, publs. theory of solutions, electrochemistry, water demineralization. Home: 739 S. Flood St., Norman, Okla. 73069.*

MURPHY, Gardner, Am. psychologist; b. Chillicothe, O., July 8, 1895; s. Edgar Gardner and Maud (King) M.; B.A., Yale, 1916; A.M., Harvard, 1917; Ph.D., Columbia, 1923; m. Lois Barclay, Nov. 27, 1926; children—Alpen, Margaret (Mrs. Fred A. Small). Asst. prof. psychology Columbia, 1929-40; prof. psychology Coll. City N.Y., 1940-52; dir. research Menninger Found., Topeka, 1952—, Henry March Pfieffer research-tng. chair in psychiatry, 1963—; editor coll. dept. (psychology) Harper & Row, N.Y.C., 1931-65. Cons. Dept. Agr., 1942-45; UNESCO cons. Ministry Edn., New Delhi, India, 1950. Fellow A.A.A.S., N.Y. Acad. Scis.; mem. Am. (pres. 1943-44, pres.-elect div. 24), Southwestern (pres. 1954), Kan. psychol. assns., Am. Soc. for Psychical Research (pres. 1962—), Soc. for Psychical Research London (past pres.), Soc. for Psychol. Study Social Issues (chmn. 1937-39), Sigma Xi. Author: Personality: A Biosocial Approach to Origins and Structure, 1947; In the Minds of Men, 1953; Human Potentialities, 1958; (with Charles Solley) Development of the Perceptual World, 1960; (with Robert Ballou) William James on Psychical Research, 1960; (with L. Dale) Challenge of Psychical Research: A Primer of Parapsychology, 1961; Freeing Intelligence through Teaching: A Dialectic of the Rational and the Personal, 1962; (with Herbert Spohn) Encounter with Reality: New Forms for an Old Quest, 1968. Well known for his researches in psychology of personality and social psychology; also parapsychology. Home: 915 25th St. N.W., Washington 20037. Office: Dept. Psychology, George Washington U., Washington 20006.*

MURPHY, Glenn, Am. nuclear engr.; b. Boulder, Colo., Jan. 17, 1908; s. Peter F. and A. Myrtle (Eggleston) M.; B.S. in Civil Engring., U. Colo., 1929, M.S., 1930; C.E., 1937; M.S., U. Ill., 1932; Ph.D., Ia. State U., 1935; m. Frances Pearce, Aug. 18, 1934. Instr. civil engring. U. Colo., 1929-30; spl. research asst. U. Ill., 1930-32; faculty Ia. State U., Ames, 1932—, prof., head aerospace engring., 1952-55, head theoretical and applied mechanics, 1955-60, Anson Marston Distinguished prof. engring., 1956—, head nuclear engring., 1959—. Mem. Am. Soc. for Engring. Edn. (pres. 1962-63), Sigma Xi, Phi Kappa Phi, Tau Beta Pi, Chi Epsilon, Pi Tau Sigma. Author several textbooks including Similitude in Engineering, 1950; Properties of Engineering Materials, 3d edit., 1957; Elements of Nuclear Engineering, 1961; Elementos de Ingenieria Nuclear, 1962. Developed theory of similitude; study of structural analysis; radiation effects. Office: Dept. Nuclear Engring., Ia. State U., Ames, Ia. 50010.

MURPHY, James B(umgardner), Am. pathologist; b. Morganton, N.C., Aug. 4, 1884; s. Patrick Livingston and Bettie (Bumgardner) M.; B.S., U. N.C., 1905, D.Sc., 1927; M.D., Johns Hopkins, 1909; hon. Dr. U. Louvain (Belgium), 1927; D.Sc., Oglethorpe U., 1938; m. Ray Slater, 1919; children—James Slater, Ray Livingston. Med. intern Pathol. Inst., N.Y.C., 1909-10; asst. in pathology and bacteriology Rockefeller Inst., 1910-13, asso., 1913-15, asso. mem., 1915-23, life mem., 1923—, in charge Lab. Cancer Research; mem. Nat. Adv. Cancer Council 1938-44; mem. bd. visitors N.Y. State Institute for Study of Malignant Diseases; bd. Sloan-Kettering Inst. for Cancer Research; com. on growth NRC. Lectr. various Am. univs. Mem. bd. mgrs. Meml. Hosp, N.Y., Mt. Desert Hosp. (pres. 1928, v.p. 1929—). Mem. Fedn. Am. Socs. for Exptl. Biology, Am. Soc. for Exptl. Pathology, N.Y. Acad. Medicine, A.A.A.S., Am. Soc. for Clinic Investigation, Assn. Am. Physicians, Am. Assn. for Cancer Research (council, v.p. 1921, pres. 1922), Assn. Française pur L'Étude du Cancer, Leewenhoek-Vereeniging, Am. Cancer Soc., Rosce Jackson Meml. Lab. of Bar Harbor (mem. bd.), Nat. Acad. Scis., Sigma Xi. Contbr. numerous articles to med. and sci. jours., dealing with tissue grafting, cancer immunity, also role of lymphocyte in tuberculosis, and studies in X-ray effects, nature of malignant tumors of fowls, cancer inhibitor from normal tissues. Mem. adv. bd., chmn. editorial com. of Cancer Research. Died Aug. 24, 1950.

MURPHY, John Benjamin, Am. surgeon; b. Appleton, Wis., Dec. 21, 1857; s. Michael and Ann (Grimes) M.; M.D., Rush Med. Coll., 1879; postgrad., Germany, 1882-84; LL.D., U. Ill., 1905, Catholic U. Am., 1915; M.Sc., U. Sheffield (Eng.), 1908; m. Jeanette C. Plamondon, Nov. 25, 1885. Practiced medicine, Chgo., 1879-82, 84—; with Rush Med. Coll., Chgo., 1884—, prof. surgery, 1905-08; with Northwestern U., 1884—, prof. surgery, 1901-05, 08-16, also head dept. surgery and clin. surgery Med. Sch.; chief surgeon Mercy Hosp. Recipient Laetare medal Notre Dame U., 1902. Fellow A.C.S., Am. Surg. Assn., Royal Coll. Surgeons Eng.; pres. A.M.A. Pub. notes on clin. consultations; pioneer in work on gall bladder; developed method of repairing injured blood vessels; revolutionized intestinal surgery with device for linking severed ends of intestine; adapted Italian method of relaxing tubercular lungs; studied joint diseases. Died Mackinack Island, Mich., Aug. 11, 1916.

MURPHY, John Joseph, Am. physician; b. Scranton, Pa., Oct. 2, 1920; s. John Joseph and Ida (Neher) M.; B.S., U. Scranton, 1942; M.D., U. Pa., 1945; m. Alice Joan McHale, Sept. 18, 1944; children—Madeline, John, Peter, Alice Marie, Patricia, Genevieve. Practice medicine specializing in urology, Phila., 1953—; faculty U. Pa., Phila., 1956—, prof. urology Sch. Medicine, dir. div. urology U. Hosp., also U. Grad. Hosp., 1964—. Clin. fellow Am. Cancer Soc., 1951-52, 52-53; I. S. Ravdin Traveling fellow, 1952-53. Mem. Pa. Med. Soc., A.M.A., Am. Urol. Assn., A.C.S., Am. Soc. for Exptl. Pathology, Soc. U. Surgeons, Am. Assn. for Surgery Trauma, Am. Assn. Genitourinary Surgeons, Société Internationale d'Urologie, Soc. U. Urologists (charter). Sigma Xi. Research, numerous publs. on urinary tract infections, methods of urinary diversion, bladder substitution, dysfunction, and dynamics. Home: 239 Winding Way, Merion Station, Pa. 19066. Office: 3400 Spruce St., Phila. 19104.*

MURPHY, Lois Barclay, Am. psychologist; b. Lisbon, Ia., Mar. 23, 1902; d. Wade Crawford and May (Hartley) Barclay; A.B., Vassar Coll., 1923; B.D., Union Theol. Sem., 1928; Ph.D., Columbia, 1937; m. Gardner Murphy, Nov. 27, 1926; children—Alpen G., Margaret (Mrs. Fred A. Small). Faculty, Sarah Lawrence Coll., Bronxville, N.Y., 1928-52; research psychologist Menninger Found., Topeka, 1952—; lectr. City Coll. Grad. Clin. Program, 1947-51, William Alanson White Inst. Psychiatry, 1949-56, Bank St. Coll. for Tchrs., 1937-52, Kan. State U., 1964—, Head Start Program, 1965—. Cons. BM Inst., Ahmedabad, India, 1950, 55; mem. Gov.'s Com. on Mental Retardation, 1964, U. S. com. World Orgn. for early childhood Edn., com. on early childhood care reconsidered Nat. Inst. Mental Health, 1964—; cons. Nat. Inst. for child Health and Human Devel., 1966-67, Children's Hosp. Infancy Project, Washington, 1965—. Recipient Children's Service award Toy Mfrs. U. S., 1962, Child Study Devel. Assn. award, 1962. Mem. Topeka Psychoanalytic Assn. (asso.), N.Y. Acad. Scis., Am. Orthopsychiat. Assn., Nat. Assn. for Edn. Young Children, Soc. Projective Techniques, Soc. for Psychol. Study Social Issues, Am. Pub. Health Assn., Phi Beta Kappa. Author: Social Behavior and Child Personality, 1937; (with G. Murphy, T. Newcomb) Experimental Social Psychology, 1937; (with B. Biber, L. P. Woodcock, I. S. Black) Life and Ways of the Seven-to-Eight-Year-Old, 1952; (with Henry Ladd) Emotional Factors in Learning, 1944; Personality in Young Children, vols. 1, 2, 1956; The Widening World of Childhood, 1962; also articles. Research on healthy aspects of personality and social devel. in childhood; introduced use play techniques for study normal child, integration of qualitative and quantitative studies, inventories of vulnerability factors, coping resources. Home: 2104 Carnahan, Topeka 66605. Office: Menninger Found., Topeka, 66601.*

MURPHY, Robert Cushman, Am. zoologist; b. Bklyn., Apr. 29, 1887; s. Thomas D. and Augusta (Cushman) M.; Ph.B., Brown U., 1911, D.Sc., 1941; M.A., Columbia, 1917; D.honoris causa, San Marcos U., Lima, Peru, 1925; D.Sc., L.I. U., 1964; m. Grace E. Barstow, Feb. 17, 1912; children—Alison M. (Mrs. Mathews), Robert Cushman, Amos Chafee Barstow. With Am. Mus. Natural History, N.Y.C., 1921-55, Lamont curator of birds, 1947-55, emeritus, 1955—; research and field work in Atlantic and Pacific Oceans, Antarctic. Pres., Biol. Lab. Cold Spring Harbor, L.I., N.Y., 1940-52. Recipient John B. Burroughs Meml. Assn. medal 1938; Hutchinson medal Garden Club Am., 1941, Elliot medal Nat. Acad. Sci., 1943; Raimondi medal Geog. Soc. Lima, 1953. Fellow N.Y. Acad. Sci. (life mem., past v.p.), Am. Geog. Soc. (life mem., recipient Cullum medal, 1940), A.A.A.S., Linnaean Soc. N.Y., Am. Ornithologists' Union (past pres., Brewster medal, 1937), Zool. Soc. London (fgn.); hon. fellow Cal. Acad. Scis., Royal Soc. New Zealand, Royal Soc. Hungary; mem. Am. Philos. Soc., N.Y. Hist. Soc., Nat. Audubon Soc. (Past pres.). Author: Bird Islands of Peru, 1925; (with others) Problems of Polar Research, 1928; Oceanic Birds of South America (2 vols.), 1936; Logbook for Grace, 1947-65; (with Dean Amadon) Land Birds of America, 1953; Fish-Shape Paumanok, 1964; Rare and Exotic Birds, 1964; also numerous articles. Pioneered in study of ecology and biogeographic zonation of oceanic birds; research in Peru improved Guano prodn. there. Home: Briarlea, Old Field, Setauket, L.I., N.Y. 11785. Office: Am. Mus. Natural History, N.Y.C., 10024.*

MURPHY, Robert Francis, Am. anthropologist; b. Rockaway Beach, N.Y., Mar. 3, 1924; s. John E. and Marion (Nolan) M.; B.A., Columbia U., 1949, Ph.D. in Anthropology, 1954; m. Yolanda Bukowska, Apr. 1, 1950; children—Pamela Ann, Robert Steven. Asso., U. Ill., Urbana, 1953-55; asso. prof. dept. anthropology U. Cal. at Berkeley, 1955-63; prof. Columbia, 1963—. Mem. anthropology adv. com. NSF, 1966—. Mem. Am. Anthrop. Assn., Am. Ethnol. Soc., Kroeber Anthrop. Soc. Author: The Trumai Indians of Central Brazil, 1956; Mundurcu Religion, 1958; Headhunter's Heritage, 1960; also articles. Research on ethnology of Mundurucu Indians of Brazil, Tuareg of N.Africa, Shoshone Indians of Western U. S., social structural and social change theory. Home: 328 Allaire Av., Leonia, N.J. Office: Schermerhorn Hall, Columbia, N.Y.C.*

MURPHY, William Parry, Am. physician; b. Stoughton, Wis., Feb. 6, 1892; s. Rev. Thomas Francis and Rose Anna (Parry) M.; A.B., U. of Ore., 1914; M.D., Harvard, 1920; D.Sc. (hon.), Gustavus Adolphus Coll., 1963; m. Harriett Adams, Sept. 10, 1919; children—Priscilla Adams (dec.), William Parry. Interne, R.I. Hosp., Providence, 1920-22; asst. resident phys., Peter Bent Brigham Hosp., Boston, 1922-23, jr. asso., 1923-28, asso. in medicine, 1928-35, sr. asso. in medicine, 1935-58, consultant in hematology, 1958—; assistant in medicine, Harvard Univ., 1923-28, instr. in medicine, 1928-35, associate in medicine, 1935-48, lecturer on medicine, 1948-58, lecturer on medicine emeritus, 1958—; engaged private practice medicine, Brookline, 1923—; consultant in internal med., Melrose, Quincy hosps., Emerson Hosp., Concord (all Mass.), Del. State Hosp. Farnhurst. Served with Med. Corps, U. S. Army, 1917-18. Awarded Cameron prize, 1930; Nobel prize in Medicine, 1934; bronze medal for scientific exhibit, Am. Med. Assn., 1934; Comdr. Order of the White Rose, 1st rank (Finland), 1934; gold medal of Humane Soc. of State of Mass., 1935; Distinguished Achievement award City of Boston, 1965; National Order of Merit, Carlos J. Finlay (Cuba). Diplomate in Internal Medicine, 1937. Mem. U. Ore. Med. Alumni Assn. (hon. mem.), Am. Med. Assn., Am. Soc. Clinical Investigation, Assn. Am. Physicians, A.A.A.S., N.Y. Acad. Sciences, Soc. Finnish Physicians for Internal Diseases (hon.), Nat. Inst. Social Sciences, Kaiserlich Leopold Caroline Deutsche Acad. der Naturforscher, Alpha Kappa Kappa, Sigma Xi; hon. mem. Robert K. Duncan Club of Mellon Inst. Author: Anemia in Practice—Pernicious Anemia, 1939. Contbr. scientific articles to med. jours., also author numerous papers. Research on diabetes and diseases of blood, especially pernicious anemia. Codiscoverer (with Dr. George R. Minot) of liver treatment for pernicious anemia. Home: 97 Sewall Avenue. Office: 1101 Beacon St., Brookline 46, Mass.

MURRAY, Daniel H., chemist; b. Ft. Frances, Ont., Can., Aug. 10, 1919; s. Alexander John and Alice Mary (Sara) M.; Phm B., U. Toronto, 1946, B.S.P., 1947, B.A., 1948, M.A., 1949, Ph.D., 1958. Lectr., Ont. Coll. Pharmacy, Toronto, 1949-50, asst. prof. pharm. chemistry, 1950-53; prof. pharm. chemistry, asst. dean State U. N.Y. at Buffalo, 1953-54, dean, prof. pharm. chemistry, 1954-60, dean and prof. medicinal chemistry, 1960—. Sec.-treas. Canadian Conf. Pharm. Faculties, 1950-53. Mem. Brit. Inst. Radio Engrs., Am., N.Y. pharm. assns., Am. Assn. Colls. Pharmacy, Rho Chi. Cons. editor Drug Merchandising, 1952-53. Research on tracer syntheses; carbohydrates; nucleosides. Home: 4748 Smiley Terrace, Clarence, N.Y. Office: Sch. Pharmacy, State U. N.Y. at Buffalo, Buffalo.

MURRAY, George Redmayne, English physician; b. Newcastle-upon-Tyne, June 20, 1865; s. William and Frances Mary (Redmayne) M.; M.B., Trinity Coll., Cambridge, 1889, M.D., 1896; student Univ. Coll. Hosp., London, 1887-88; D.C.L., Durham; M.D. (hon.), Dublin; m. Annie Bickersteth, 1892; 3 sons, 1 dau. Pathologist, Hosp. for Sick Children, Newcastle-upon-Tyne, 1891-96; lectr. bacteriology Durham U. Coll. Medicine, 1891-93, Heath prof. comparative pathology, 1893-1908, mem. statutory commn., 1935-37; physician Royal Victoria Infirmary, Newcastle-upon-Tyne, 1896-1908, later cons. physician; physician Manchester Royal Infirmary, 1908-25, later cons. physician; prof. systematic medicine Victoria U., Manchester, 1908-25, later emeritus prof. medicine; mem. Dept. Com. of Home Office on Dust in Cardrooms, 1927-31, Dept. Com on Compensation for Cardroom Workers, 1937, Com. on Sickness among Cardroom Operatives, 1930; cons. physician Barrowmore Tb Colony. Fellow Royal Coll. Physicians (Goulstonian lectr. 1899, Bradshaw lectr. 1905, council 1914-17); mem. Assn. Physicians Gt. Britain (pres. 1936-37), Harveian Soc. (hon.) (London). Author: Diseases of the Thyroid Gland, 1905, also numerous articles. First to treat myxoedema by hypodermic injection of thyroid gland extract; showed genetic influence of incidence of mammary cancer in rats, 1933; important work on dust diseases in cardroom workers. Died Mobberley, Cheshire, Eng., Sept. 21, 1939.

MURRAY, George Robert Milne, Brit. botanist; b. Arbroath, Scotland, Nov. 11, 1858; s. George and Margaret (Sayles) M.; student of Anton de Bary, Strasbourg (now in France) U., 1875; m. Helen Welsh, 1884; 1 son, 1 dau. Became asst. keeper bot. dept. Brit. Mus. Natural History, 1876, keeper, 1895-1905; lectr. botany St. George's Hosp. Med. Sch., 1882-86, Royal Vet. Coll., 1890-95; naturalist Solar Eclipse Expdn. to W.I., 1886; sci. dir. Nat. Antarctic Expdn., 1901. Fellow Royal Soc., Linnean Soc. Author: (with Alfred W. Bennett) Hand-book of Cryptogamic Botany, 1889; Introduction to the Study of Seaweeds, 1895; also articles on fungi and vegetable parasitism in Ency. Brit., 9th edit. Editor Phycological Memoirs, 1892-95, The Antarctic Manual, 1901. Died Dec. 16, 1911.

MURRAY, Grover Elmer, Am. geologist; b. Maiden, N.C., Oct. 26, 1916; s. Grover Elmer and Lucy (Lore) M.; B.S., U. N.C., 1937; M.S., La. State U., 1939, Ph.D., 1942; m. Nancy Beatrice Setzer, June 21, 1941; children—Martha, Barbara Elizabeth. Research geologist La. Geol. Survey, 1939-41; geologist Magnolia Petroleum Co., Jackson, Miss., 1941-48; prof. dept. geology La. State U., 1948-55, chmn. dept., 1950-53. Boyd prof. geology, 1955-66, v.p., dean of academic affairs, 1963-65, vice pres. academic affairs La. State U. System, 1965-66; pres. Tex. Technol. Coll., Lubbock, 1966—; prof. geology camp, Colo., 1949, 51, La., 1961; dir. U. Tex. geology camp, East Tex., 1949, 51, vis. lectr. U. Tex., 1958. Mem. Internat. Commn. on Stratigraphy; part-time cons. geologist, 1948—; dir. Nat. Sci. Found. project for basic geologic studies in Northeastern Mexico, 1958-61. Mem. Nat. Sci. Bd.; chmn. U. S. Nat. Com. on Geology, 1965—. Fellow Geol. Soc. Am. (chmn. symposium on sedimentary vols. in Coastal Plain, U. S. and Mexico, 1951, commr. Am. Stratigraphic Commn., 1951-54, program chmn. New Orleans Meeting 1955; councillor 1961-64); mem. Am. Assn. Petroleum Geologists (chmn. com. geol. names and nomenclature 1952-54, distinguished lectr. 1954, editor 1959-63; mem. Am. Stratigraphic Commission, 1957-63, president 1964-65; Society of Economic Paleontologists and Mineralogists (editor Jour. Paleontology 1951-54, chmn. research com. 1958-59, president 1963-64), American Geol. Institute (visiting geoscience lecturer 1959-60), Paleontol. Society, Orgn. Tropical Studies (dir.), Gulf Univs. Research Corp. (pres., chmn. bd.), Orgn. Trop. Studies (dir.), Soc. Exploration Geophysicists, American Institute Profl. Geologists, Paleontol. Research Institute, Geol. and Mining Soc., Norsk Geologisk Forening. Asociación Mexicana de Geólogos Petroleros, Sociedad Geólogica Mexicana, Sigma Xi, Sigma Gamma Epsilon, Omicron Delta Kappa. Author: Geology of Atlantic and Gulf Coastal Province of North America, 1961. Contbr. articles to ednl., sci. jours. Contbd. devel. data and concepts about tertiary stratigraphy So. U. S., geology Atlantic, Gulf coastal province N.Am., salt domes and petroleum geology Gulf Mexico Basin, geology northeastern Mexico, indigenous Precambrian oil and gas, gen. concepts regarding rock and timerock geol. units. Home: 2909 19th St., Lubbock, Tex. 79410.*

MURRAY, Henry Alexander, Am. psychologist; b. N.Y.C., May 13, 1893; s. Henry A. and Fannie M. (Babcock) M.; grad. Groton Sch., 1911; A.B., Harvard, 1915; M.D., Columbia, 1919, M.A., 1920; Ph.D., Cambridge (Eng) U., 1928; L.H.D., Lawrence U., 1964; D.Psychology, Louvain, Belgium, 1966; m. Josephine L. Rantoul, May 30, 1916; 1 dau., Josephine L. Faculty, Harvard, Cambridge, Mass., 1926—, prof. psychology, 1950-62, prof. emeritus, 1962—. Mem. bd. Francis G. Wickes Found., 1963—; patron C. G. Jung Inst., Zurich, 1964—. Mem. Am. Acad. Arts and Scis., Am. Psychol. Assn. (award for distinguished sci. contbn., 1961), Am. Psychoanalytic Assn., Am. Psychosomatic Soc., Soc. for Projective Techniques, Societe Francaise de Psychologie (Paris), Royal Medico-Psychol. Assn., Melville Soc. (past pres.), Phi Beta Kappa. Editor, co-author: Explorations in Personality, 1938; Assessment of Men, 1948; Personality in Nature, Society and Culture, 1953; Myth and Mythmaking, 1960. Research, numerous publs. in blood chemistry, especially alkalosis after parathyroidectomy and pyloric occlusion, physiol. ontogeny, devel. of methods for assessing personality, especially the matic apperception test. Home: 22 Francis Av., Cambridge, Mass. 02138.*

MURRAY, Sir James, Irish physician; b. County Londonderry, Ireland, 1788; s. Edward Murray; student, Dublin; M.D., Edinburgh, 1829; M.D. (hon.), Dublin U., 1832; at least 1 son, John Fisher. Resident physician to lord-lt. of Ireland; insp. anatomy, Dublin; founder manufactory for fluid magnesia. Licentiate Coll. Surgeons Edinburgh. Mem. Dublin Coll. Author: (paper) On the Danger of using Solid Magnesia, and on its great value in a Fluid State for internal use, 1817; Heat and Humidity, 1929; Observations on Fluid Magnesia, 1840. Discovered magnesia; suggested electricity as therapeutic agt. Died Dec. 8, 1871.

MURRAY, Johannes Andreas, botanist, physician; b. Stockholm, Sweden, Jan. 27, 1740; early edn. in Stockholm and Upsala; grad. Güttingen, 1763. Extraordinary prof., 1764; magister philosophiae, 1768; ordinary prof. medicine, insp. Botanic Garden, 1769. Author: Apparatus Medicaminum tam Simplicium quam Praeparatorum et Compositorum in praxeos adjumentum consideratus, 1795. Works on medicine, materia medica, botany, others. Died May 22, 1791.

MURRAY, Sir John, oceanographer, marine naturalist; b. Coburg, Ont., Can., Mar. 3, 1841; s. Robert and Elizabeth (Macfarlane) M.; LL.D., D.Sc., Ph.D., U. Edinburgh (Scotland); m. Isabel Henderson, 1889; two sons, three daus. Naturalist on board whaling ship to Spitzbergen and Arctic regions, 1868; one of naturalists on board H.M.S. Challenger during exploration of phys. and biol. conditions of great oceans basins, 1872-76; 1st asst. staff apptd. to undertake publn. sci. results Challenger expdn., 1876-82, apptd. editor, 1882; took part in Triton and Knight Errant explorations, Faröe Channel, also other deep sea and marine expdns.; numerous explorations in tropical oceanic islands; took part, paid expenses Michael Sara N. Atlantic Expdn., 1910. Brit. del. Internat. Hydrographic Conf., Stockholm, Sweden, 1899. Recipient Cuvier prize Inst. de France; Humboldt medal Gesellschaft für Erdkunde, Berlin, Germany; Royal medal Royal Soc.; founder's medal Royal Geog. Soc.; Neill, also Makdougall-Brisbane medals Royal Soc. Edinburgh; Cullum medal Am. Geog. Soc.; Clarke medal Royal Soc. New S. Wales; Lutke medal Imperial Russian Soc. Geography; Livingstone medal Royal Scottish Geog. Soc.; Vega medal Swedish Anthrop. and Geog. Soc. Fellow Royal Soc., 1896 (Royal medal); hon. mem. numerous Brit. and fgn. sci. socs. Author: The Ocean, a General Account of the Science of the Sea, 1913, also numerous papers. Research on deep sea deposits with observations of surface organisms especially the Foraminifera and Radiolaria; study of deposits from ocean floor, also the part played by certain surface organisms in forming these deposits; devised apparatus for sounding and registering the temperature of great depths; bathymetrical survey of fresh water lochs in Scotland. Died Mar. 16, 1914.

MURRAY, John Wolcott, Am. chemist; b. Flushing, N.Y., Jan 9, 1909; s. Harris King and Arabella (Prime) M.; A.B., Colgate U., 1930; Ph.D., Johns Hopkins, 1933; m. Ruth Evelyn Terborgh, Dec. 25, 1938; children—John Harris, Beulah Mae (Mrs. Maury Willis Fincham). Instr., Johns Hopkins, 1933-34; asst. Rockefeller Inst. for Med. Research, 1934-39; chief chemist Thomasville Stone & Lime Co. (Pa.), 1939-42; faculty Va. Poly. Inst, Blacksburg, 1942—, prof., 1953—. Mem. Am. Chem. Soc., Nat. Speleological Soc., Sigma Xi. Author: (with P. C. Scherer) Laboratory Manual for Physical Chemistry, 1954. Research, publs. on Raman Spectra and mech. models as related to molecular structure, behavior of electrolytes and water in models of living cells, factors governing mineral composition of cave deposits. Home: 101 York Dr., Blacksburg, Va. 24060.*

MURRAY, Margaret Ransone (Mrs. Burton LeDoux), Am. cell biologist; b. nr. Yorktown, Va., Nov. 16, 1901; d. Archibald Campbell and Harriet (Ransone) Murray; A.B, Goucher Coll., 1922; M.S., Washington U., St. Louis, 1924; Ph.D., U. Chgo., 1926-28; m. Burton LeDoux, Jan. 19, 1941. Asso. prof. biology and physiology Fla. State Coll. for Women, 1928-29; faculty Columbia Coll. Physicians and Surgeons, 1929—, prof. anatomy, 1959—. Sec., Tissue Culture Commn., 1946-50; mem. fellowship rev. panel NIH, 1960-63. Recipient NIH Research Career award, 1962; Sci. citation Goucher Coll., 1954; Sci. medal Free U. Brussels, 1964; Golden Hope Chest award Nat. Multiple Sclerosis Soc., 1964. Commonwealth Fund Travelling fellow, 1963-64. Fellow N.Y. Acad. Scis.; mem. Tissue Culture Assn. (past pres.), Nat. Multiple Sclerosis Soc. (mem. research evaluation and programming com. 1963—), Japanese Tissue Culture Assn. (hon.), Am. Assn. Anatomists, Am. Assn. Cancer Research, Internat. Soc. Cell Biology, Am. Soc. Cell Biology, Internat. Inst. Embryology, N.Y. Acad. Medicine, Growth Soc., British Royal Soc. Medicine (affiliate). Author: (with Kopech) A Bibliography of the Research in Tissue Culture, 1884-1950, 1953; also numerous articles. Developed tissue culture method to allow organized portions of nervous system to be maintained for periods of months. Home: 23 Haven Av., N.Y.C. 10032.*

MURRAY, Raymond G., anatomist; b. Tokyo, Japan, May 12, 1916; s. David A. and Annie L. (Foster) M.; B.S., Monmouth Coll., 1937; Ph.D., U. Chgo., 1942; m. Lorraine E. Laxson, Aug. 16, 1938 (dec. 1955); children—Joel Elizabeth, Janeth A. (Mrs. Peter Schwandt), Andrea M., April E., Elaine T., Stephanie; m. 2d, Assia S. Tzankoff, Nov. 30, 1956. Research asst. Office Naval Research, 1942-43; research asso. Manhattan Dist. Project, 1943-46; instr. Tufts U. Med. Sch., 1946-48; asst. prof. Northwestern U. Dental Sch., 1948-49; asso. prof. Ind. U., Bloomington, 1949-65, prof, 1965—. Mem. Am. Assn. Anatomists, A.A.A.S., Am. Soc. for Cell Biology, Electron Microscope Soc. Am., Ind. Acad. Sci., Am. Assn. U. Profs., Sigma Xi. Research, publs. on lymphocytes, thymus gland, ultrastructure of sensory receptors. Home: 1910 E. 1st St., Bloomington, Ind. 47401.*

MURRAY, Raymond LeRoy, Am. physicist; b. Lincoln, Neb., Feb. 14, 1920; s. Ray Annis and Bertha (Mann) M.; B.S., U. Neb., 1940, M.S., 1941; Ph.D., U. Tenn., 1950; postgrad. U. Cal. at Berkeley, 1941-43; m. Ilah Mae Rengler, June 16, 1941; children—Stephen, Maureen, Marshall. Physicist, U. Cal. Radiation Lab., Berkeley, 1942-43; asst. dept. supt. Tenn. Eastman Corp., Oak Ridge, 1943-47; research physicist Carbide & Carbon Chem. Co., Oak Ridge, 1946-50; prof. physics N.C. State Coll., Raleigh, 1950-57, Burlington prof. physics, 1957—, head dept. physics, 1960-63, head dept. nuclear engring., 1963—. Acting dir. Nuclear Reactor Project, 1956-57; cons. Oak Ridge Nat. Lab., 1950—, Alco Products, Inc., 1954-62, AMF Atomics, 1955—, Westinghouse, 1958-61, Internat. Atomic Energy Agy., 1963—. Adv. com. on radiation N.C. Bd. Health, 1958-59; mem. Gov.'s Sci. Adv. Com. N.C., 1961—. Recipient Oliver Max Gardner award, 1965. Fellow Am. Nuclear Soc. (dir. 1958-61, chmn. edn. com. 1961-62), Am. Phys. Soc.; mem. Am. Assn. Physics Tchrs., Am. Soc. Engring. Edn., Phi Beta Kappa, Sigma Xi, Pi Mu Epsilon, Phi Kappa Phi. Author: Introduction to Nuclear Engineering, 1954, 61; Nuclear Reactor Physics, 1957; A New Approach to Physics, 1964. Mem. editorial adv. bd., U. S. exec. editor Jour. Nuclear Energy, 1963—. Contbns. to devel. of the electromagnetic method for separating uranium isotopes; design and use of nuclear chain reactors for research, power, and tng.; research in theory and analysis of nuclear reactor phenomena. Contbr. numerous articles profl. jours. Home: Box 5596 State College Sta., Creedmoor Rd., Raleigh, N.C.

MURRELL, William, physician; b. Nov. 26, 1853; s. William Kenrick Murrell; ed. Univ. Coll., London, Eng.; M.D.; Sharpey physiol. scholar and demonstrator physiology, Univ. Coll., 1875-78; laureat de l'Acad. Méd., Paris, France, 1881. Examiner material medica U. Edinburgh (Scotland), 1882-87; examiner Royal Coll. Physicians, London, 1886-90, U. Glasgow (Scotland), 1899-1902; hon. life gov., physician' lectr. clin. medicine and principles and practice of medicine Westminster Hosp.; mem. Faculty of Medicine, U. London. Vice pres. Internat. Med. Congress, 1887. Fellow Royal Coll. Physicians. Author: Manual of Pharmacology and Therapeutics, 1896; What to do in Cases of Poisoning, 10th edit., 1907; Massotherapeutics, 5th edit., 1980; Prevention of Consumption, 1895; Bronchitis, 1890; Angina Pectoris, 1880; Forensic Medicine and Toxicology, 6th edit., 1903; Materia Medica, 3 vols., 1900. Editor: (Fothergill) Handbook of Treatment, 4th edit., 1897.With Sir Lauder Brunton credited with introducing amyl nitrate and nitro-glycerine as a remedy for relief of agina pectoris, 1879. Died June 28, 1912.

MURRILL, William Alphonso, Am. botanist; b. Campbell County, Va., Oct. 13, 1869; s. Samuel Leroy and Virginia Daniel (Woodroof) M.; B.S., Va. Poly. Inst., 1887; B.S., Randolph-Macon Coll., Va., 1889, A.B., 1890, A.M., 1891; Ph.D., Cornell U., 1900; research student N.Y. Bot. Garden, 1900-04; m. Edna Lee Luttrell, Sept. 1, 1897. Prof. natural sci. Bowling Green Sem., Va., 1891-93, Wesleyan Female Inst., Va., 1893-97; scholar in botany Cornell U., 1897-98, asst. in botany, 1898-99, asst. cryptogamic botanist, 1899-1900; tchr. biology DeWitt Clinton High Sch., N.Y.C., 1900-04; asst. curator N.Y. Bot. Garden, 1904-05, 1st asst., 1906-07, asst. dir., 1908-22. Recipient Gold medal Holland Soc. of N.Y., for mycological work, 1923. Mem. Torrey Bot. Club, Bot. Soc. AM., N.Y. Acad. Scis., Am. Phytopathol. Soc., Sigma Xi. Author: North American Flora, Vol. IX, parts 1-7, 1907, 08, 10, 15, 16, Vol. X, parts 1-3, 1914, 17; Northern Polypores, 1914; American Boletes, 1914; Southern Polypores, 1915; Western Polypores, 1915; Tropical Polypores, 1915; Edible and Poisonous Mushrooms, 1916; Billy, the Boy Naturalist, 1918; Three Young Crusces, 1918; The Naturalist in a Boarding-school, 1919; The Natural History of Staunton, Virginia, 1919; also (pocket guides), Stars, Rocks, Trees, Reptiles, Autobiography, Florida Plants, Florida Animals, Historic Foundations of Botany in Florida (and America) and (illustrated guides), Ferns, Flowers, Pore Fungi, Familiar Trees; also many botany pamphlets and articles in scientific journals. Editor Mycologia, 1909-24; asso. editor North American Flora. Named and described 1,700 species of fungi new to science. Has made extensive bot. explorations in Europe, tropical Am., S.A.; on Pacific Coast, secured over 70,000 specimens; completed important studies of Florida fungi, Florida hawthorns and a botanic survey of Alachua County, Fla. Died Dec. 25, 1957.

MURTY, Gollakota Suryanarayana, agrl. scientist; b. Kakinada, Andhra Pradesh, India, Aug. 13, 1913; s. Gollakota Jagannadham and Gollakota (Sreeharamma) M.; B.Sc., Andhra U., 1934; M.Sc. in Botany, Lucknow U., 1936; Ph.D. in Genetics, U. Minn., 1948; m. Mahalakshmi Chavali, Feb. 14, 1930; children—Parvathi (Mrs. Ramarao), Srihari (Mrs. Ramanaiah), Sumangali (Mrs. Somayajulu), Bhavant, Jagannadham, Surya Prakash. With Indian Agrl. Research Inst., New Delhi, 1940-58; dir. tobacco research Ministry Agr., 1958-63; plant breeding expert FAO, Damascus, Syria, 1963—. Research, publs. on Lysenko's theory of vernalization and phasic devel. in crop plants, work on hybrid corn, developer improved wheat varieties, tobacco cultivation and curing. Home: Gollakotavari House, Ramaraopeta, Kakinada, Andhra Pradesh, India. Office: FAO Mission, P.O. Box 256, Damascus, Syria.*

MUSA, Antonius, physician; flourished Rome, 23 B.C.; originally slave of Emperor Augustus, whom he cured of gout through cold baths; physician to Virgil, Horace, Maecenas, Agrippa; author works on materia medica frequently quoted by Galen, extant in fragments. Probably 1st to recommend cold baths for therapeutic purposes.

MUSACCHIA, Xavier Joseph, Am. physiologist; b. Bklyn, Feb. 11, 1923; s. Castrense and Orsolina (Mazzola) M.; B.S., St. Francis Coll., 1943; M.S., Fordham U., 1947, Ph.D., 1949; m. Betty Louise Cook, Nov. 22, 1950; children—Joseph, Mary, Thomas, Laura. Grad. fellow Fordham U., 1947-49; instr. Marymount Coll., 1948-49; faculty St. Louis U., 1949-65; prof. physiology, Space Scis. Research Center, sr. investigator U. Mo., Columbia, 1965—. Fellow A.A.A.S.; mem. Am. Physiol. Soc., Soc. Exptl. Biology and Medicine, Am. Soc. Zoology, Am. Soc. Mammalogy, Am. Soc. Ichthyology and Herpetology, Am. Micros. Soc., Am. Assn. U. Profs., Sigma Xi. Research, publs. on physiology of fat metabolism, hibernation, intestinal absorption, role of ionizing radiation in hibernation of animals. Home: 1805 Ridgemont St., Columbia, Mo. 65201.*

MUSAJO, Luigi, Italian chemist; b. Bari, Nov. 8, 1904; s. Francesco and Teresa (Mongardi) M.; Ph.D., U. Bologna; m. Carla Giacobazzi, Mar. 3, 1957; 1 son, Francesco-Giovanni. Prof., U. Modena; prof. U. Padua, 1950; prof., dir. pharm. chem. inst. U. Padua, also dean faculty pharmacy. Decorated Order Merit Italian Republic; recipient Gold medal of Merit for Teaching. Mem. Inst. Pub. Instrn., dei Lincei Acad., Rome, Venetian Inst. Sci., Letters and Arts, Acad. Padua, Acad. Bari, Società Chimica Italiana (pres. Venetian sect.). Research and publs. on organic and biol. chemistry, pharmaceutics. Home: Prato della Valle 17. Office: via Marzolo 5, Padua, Italy.

MUSCHEL, Louis Henry, Am. immunologist; b. N.Y.C., July 4, 1916; s. Maurice and Betty (Tobey) M.; B.S., N.Y. U., 1936; M.S., Yale, 1951, Ph.D., 1953; m. Anne Orzel, Oct. 22, 1946; 1 dau., Ruth Josephine. Joined U. S. Army, 1941, advanced through grades to lt. col., 1961; chief, dept. serology Walter Reed Army Inst. Research, Washington, 1958-62; faculty U. Minn., Mpls., 1962—, prof. microbiology, 1964—. Mem. Am. Assn. Immunologists, Brit. Soc. Immunology, N.Y. Acad. Scis., A.A.A.S., Am. Soc. Microbiology, Tissue Culture Assn., Soc. Exptl. Biology and Medicine, Soc. for Gen. Microbiology, Sigma Xi, Phi Beta Kappa, Phi Lambda Upsilon. Research, publs. on bactericidal action of serum and its role in host defenses, natural bactericidal and viricidal antibodies, applications of complement-fixation technique. Home: 1648 Eleanor Av., St. Paul 55116.*

MUSCHLITZ, Earle Eugene, Jr., Am. chemist; b. Palmerton, Pa., Apr. 23, 1921; s. Earle Eugene and Ferne (Altemose) M.; B.S., Pa. State U., 1941, M.S., 1942, Ph.D., 1947; m. Barbara Pfahler, Sept. 17, 1953; children—Robert Earle, Karl William. Instr., Cornell U., 1947-51; faculty U. Fla., Gainesville, 1951—, prof. chemistry, 1958—, head div. phys. chemistry, 1960—. Cons. NIH, 1962-66. Mem. Am. Chem. Soc., Am. Phys. Soc., Faraday Soc., Sigma Xi, Alpha Chi Sigma, Sigma Pi Sigma. Research, publs. on collisions of highly excited atoms and molecules in gases, ion-molecule reactions in gases, determination of intermolecular forces by scattering measurements, formation of negative ions in gases by electron impact, identification of ionic species by mass spectrometry. Address: U. Fla., Gainesville, Fla.*

MUSCHOLL, Erich Otto Rudolf, German pharmacologist; b. Biskupitz, Germany, July 3, 1926; s. Erich Georg Günther and Hanne (Bartsch) M.; M.D., U. Mainz, 1952; Brit. Council scholar, U. Edinburgh (Scotland), 1956-57; m. Hilde Elisabeth Osburg, May 28, 1960; children—Walter, Ilse. Faculty dept. pharmacology U. Mainz, Germany, 1959—, prof. pharmacology, toxicology, 1965—. Research, numerous publs. on mode of action of drugs upon sympathetic nervous system. Address: 5 Hafenstrasse, Mainz, West Germany.*

MUSEN, Peter, physicist-astronomer; b. Nikolaiev, Russia, Jan. 29, 1912; s Vladimir and Nadezda (Volkov) M.; Diploma Math., U. Belgrade, 1955, Ph.D. in Math., 1937; m. Ludmila Sawitski, July 5, 1946, Came to U. S., 1949, naturalized, 1955. Sci. collaborator Obs. Belgrade (Yugoslavia), 1939-42, Astronomisches Recheninstitut, Berlin, Heidleberg, Germany, 1942-49; faculty U. Cin., 1949-59, asso. prof., 1958-59; physicist-astronomer NASA, Silver Spring, Md., Greenbelt, Md., 1959—; prof. astronomy U. Md., College Park, 1962—. Recipient NASA award, 1963. Mem. Am. Astron. Soc., Am. Geophys. Union, Am. Astronautical Soc. Contbr. articles to tech. jours. Developed theory artificial satellites used for Vanguard I, modified form Hansen's lunar theory, theory planetary perturbations 1st and higher orders in rectangular coordinates; investigated lunar and solar effects on artificial satellites. Home: 9304 Orbit Lane, Lanham, Md. 20801. Office: Goddard Space Flight Center, Greenbelt, Md.*

MÜSER, Helmut August, German physicist; b. Bochum, Dec. 18, 1915; s. August and Toni (Thiel) M.; Ph.D., U. Frankfurt; m. Margaret Hohmann, Jan. 27, 1945. Research, U. Prague; acad. career U. Frankfurt, prof., 1957—. Mem. Assn. German Physics Socs. Author: Einführung in die Halbleiterphysik, 1960. Research and publs. on solid state physics. Home: Am. Dornbusch 18, Vockenhausen. Office: Robert-Mayer Strasse 2, Frankfurt, West Germany.

MUSGRAVE, William, English physician; b. Charlton, Eng., circa 1655; M.D., Oxford U., 1689; practiced medicine, Exeter, Eng.; fellow New Coll. Fellow Royal Soc., 1683 (sec. 1685), Royal Coll. Physicians. Author: Antiquitates Britanno-Belgicae, 1-4 vols., 1719-20; also med. tracts on arthritis. Died Exeter, 1721.

MUSHIN, William Woolf, Brit. physician; b. London, Eng., Sept., 1910; s. Moses and Jessie (Kalmenson) M.; M.B., B.S., U. London, 1933; M.A., U. Oxford (Eng.), 1946; m. Betty H. Goldberg, Dec. 1939; children—Susan, Jeremy, Dilys, Elizabeth. Faculty, London Hosp. Med. Coll., 1929-33; 1st asst. Nuffield dept. anesthetics U. Oxford, 1943-47; prof. anesthetics U. Wales, Cardiff, 1947—. Hon. fellow Faculty Anesthetics Royal Australian Coll. Surgeons, Royal Coll. Surgeons Ireland, Royal Coll. Surgeons S. Africa; fellow Royal Coll. Surgeons Eng. Author: (with R. R. Macintosh, G. Epstein) Physics for the Anesthetist, 1943, 63; Thoracic Anesthesia, 1965; Automatic Ventilation of Lungs (with P. Thompson), 1968; also numerous articles. Research on pharmacology of anesthetic drugs, phys. prins. of anesthesia, methods of artificial automatic ventilation. Home: 30 Bettws-Y-Coed Rd., Cardiff. Office: Dept. Anaesthetics, Welsh Nat. Sch. Medicine, Cardiff, U.K.*

MUSKAT, Irving Elkin, chemist; b. Dec. 25, 1905; s. Samuel and Celia (Glass) M.; A.B., Marietta Coll., 1924, D.Sc., 1963; M.S., U. Chgo., 1925, Ph.D., 1927; m. Dorothy Ruth Gaston, Sept. 19, 1929; children—Leslie Eloise (Mrs. Clark Perkins), Lindsay Ann.

Faculty, U. Chgo., 1928-34; dir. chem. exhibits Century of Progress, 1931-33; NRC fellow Rockefeller Inst., 1933-34; Petroleum fellow Mellon Inst., 1934-35; dir. chem. research Pitts Plate Glass Co., 1935-42; dir. research Vulcan Co., 1942-51; pres. Marco Chemicals Company Linden, N.J., 1942-53; cons. Celanese Corp., New York, N.Y., 1953-55; pres. Elkin Chem. Co., Miami, Fla., 1954—; v.p. research U. Miami, 1960-63, pres. Indsl. Research Center, 1960-63. Chmn. adv. bd. Dade County Port Authority, 1958-61. Chmn. Inter-Am. Cultural and Trade Ctr., Miami, 1961—. Mem., Am. Chem. Soc.; AAAS; A.I.C.; London Chem. Soc. Contbr. articles to sci. jours. Patentee in field. Research in synthetic resins and plastics; major patents in allyl resins; polyester resins and glass reinforced resins; styrene-maleic anhydride resins; cross-linking thermoplastic resins and production of glass reinforced products. Home: 4975 N. Kendall Dr., Miami 33156. Office: Interama-Internat. Airport, Miami, 33159.*

MUSKAT, Morris, physicist-engr.; b. Riga, Latvia, Apr. 21, 1906; s. Sam and Celia (Glass) M.; came to U. S., 1911, naturalized, 1914; B.A., M.A., Ohio State U., 1926; Ph.D. in Physics, Cal. Inst. Tech., 1929; m. Fern Bonita Metzger, July 3, 1936; children—Phyllis Ann (Mrs. David F. Goddard), David Allen, Robert Elliott, Rosalyn Jean. Instr. physics and chemistry Bowling Green (O.) U., 1926-27; dir. physics div. Gulf Research & Devel. Co., Pitts., 1929-50; tech. coordinator prodn. dept. Gulf Oil Corp., Pitts., 1950-61, tech. adviser, exec., Coral Gables, Fla., 1961—. Named Distinguished Lectr., Soc. Petroleum Engrs., 1962-63. Fellow Am. Phys. Soc., A.A.A.S., N.Y. Acad. Scis.; mem. Am. Geophys. Union, Am. Petroleum Inst. (certificate appreciation 1965), Am. Inst. Mining and Metall. Engrs. (Lucas medal 1953), Phi Beta Kappa, Sigma Xi, Pi Mu Epsilon. Author: Flow of Homogeneous Fluids Through Porous Media, 1937; Physical Principles of Oil Prodn., 1949. Research, numerous publs. on oil reservoir mechanics, lubrication, geophysics, modern physics. Home: 8190 S.W. 108th St., Miami, Fla. 33156. Office: 95 Merrick Way, Coral Gables, Fla. 33134.*

MUSKETT, Arthur Edmund, Brit. mycologist, plant pathologist; b. Norwich, Eng., Apr. 15, 1900; s. Arthur and Alice (James) M.; D.Sc., Imperial Coll. Sci., London, 1938; m. Hilda Elizabeth Smith, Apr. 7, 1926; children—Arthur Henry, John Edmund, Doreen Elizabeth Ann, Brian David. Staff, Ministry Agr., N. Ireland, 1923-65, head plant pathology div., 1938-65; staff Queens U., Belfast, Ireland, 1924-65, prof. mycology and plant pathology, 1945-65, dean Faculty Agr., 1950-57. Decorated Order Brit. Empire. Fellow Inst. Biologists (found.); mem. Royal Coll. Sci. (asso.), Royal Irish Acad., Central Garden Assn. No. Ireland (chmn.). Author: (with J. Colhoun) Flax Diseases, 1947; (with J. P. Malone) Seed Borne Fungi, 1964; also numerous articles. Research on control of diseases of potato, apple, strawberry; investigation and control of seed borne diseases of cereals and flax; study Irish fungal flora, micro-fungi contaminants, in seeds of cereals and flax. Home: 254 Saintfield Rd., Belfast 8, No. Ireland.*

MUSKHELISHVILI, Nikolay Ivanovich, Russian mathematician, mech. engr.; b. Tbilisi, Feb. 16, 1891; grad. St. Petersburg U., 1914; D.Physico-Math. Sci., 1934. Asst., instr. Petrograd U., 1917; prof., Tbilisi U. and Polit. Inst., 1922—; a founder Tbilisi Math. Inst., 1935. Head Soviet delegation 9th Internat. Congress on Applied Mechanics, Brussels, 1956. Decorated Order of Lenin (4); recipient Stalin prize, 1941, 47. Mem. USSR (chmn. Georgian br. 1938-41, Presidium mem. 1957—), Georgian (pres. 1941—) acads. scis., Georgian Znanie Soc. (chmn. 1956). Author: Periodic Orbits and Closed Geodetic Lines, 1927; Singular Integral Equations, 1946; Course in Analytic Geometry, 1947; Some Fundamental Problems of the Mathematical Theory of Elasticity, 1954. Research and publs. on theory of elasticity, integral equations, boundary value problems of theory of functions; one of 1st to apply theory of functions of complex variable to problems of theory of elasticity; applied potential theory to problems involving behavior of elastic materials and torsion and bending of steel beams. Address: Tbilisi University, prospect Chavchavadze 45, Tbilisi, Georgian SSR, USSR.

MUSPRATT, James, Brit. chemist; b. Dublin, Aug. 12, 1793; s. Evan and Sarah (Mainwarings) M.; ed. comml. sch., Dublin; wholesale chemist and druggist's apprentice, 1807; m. Julia Connor, Oct. 6, 1819; 10 children, including Richard, Frederick, James Sheridan, Edmund Knowles. Manufactured acids and other products, Dublin, then in Liverpool, Eng.; founder (with Josias Gamble) alkali industry, St. Helens, 1828. Improved methods of chem. manufacture; introduced Leblanc process of alkali manufacture into Eng. Died Seaforth Hall, nr. Liverpool, May 4, 1886.

MUSPRATT, James Sheridan, Brit. chemist; b. Dublin, Ireland, Mar. 8, 1821; s. James and Julia (Connor) M.; ed. Andersonian U., Glasgow (Scotland), U. Coll., London, Eng.; m. Susan Cushman, 1848 (dec. 1859). Placed in charge chem. dept. Thompson's Manufactory, Manchester, Eng., 1838; went into bus. in Am. and failed; returned to Europe; joined Liebig's Lab., Giessen, Germany; traveled in Germany; returned to Eng., 1848; founder Liverpool (Eng.) Coll. Chemistry; received share in father's bus., 1857. Editor: Dictionary of Chemistry, 1854-60. Translator Plattner's treatise on blowpipe, 1845. Author: Outlines of Analysis, 1849; also articles. Research on sulphites, nitraniline, toluidine, chemistry of vegetation. Died Liverpool, Apr. 3, 1871.

MUSSELMAN, Luther Kyner, Am. physician; b. Gettysburg, Pa., June 18, 1895; s. John Elmer and Euphemia (Rogers) M.; B.S., Pa. Coll., 1915; M.D., Johns Hopkins, 1919; Ph.D., Yale, 1923; m. Amelia Helen Brodeur, June 18, 1925; 1 son, Luther Kyner. Faculty, Yale Sch. Medicine, 1921—, clin. prof., 1946-63, clin. prof. emeritus, 1963—; attending obstetrician, gynecologists Yale-New Haven Hosp., 1925-63, cons. obstetrician, gynecologist, 1963—. Cons. Fairfield Hills Hosp., Newtown, Conn., 1935-67. Diplomate Am. Bd. Obstetricians and Gynecologists. Fellow A.C.S., Am. Coll. Obstetricians and Gynecologists, Acad. Internat. Medicine; mem. New Haven Med. Assn., Conn. State, New Haven County med. socs., A.M.A., New Haven Obstet. Soc., Phi Beta Kappa, Sigma Xi, Phi Delta Theta, Phi Beta Pi, Gamma Alpha, Beta Beta Beta. Research, publs. on obstetrics and gynecology; studies on bioelectrical phenomena in menstruation and ovulation; malformations of embryo; natural immunity in newborn. Home: 192 Livingston St., New Haven 06511. Office: 39 Trumbull St., New Haven 06510.*

MUSSELMAN, Merle McNeil, Am. surgeon, educator; b. Topeka, Sept. 19, 1915; s. Charles Albert and Jimmie (Drake) M.; student U. Omaha, 1933-35; B.S., U. Neb., 1937, M.D., 1939; M.S., U. Mich., 1949; m. Dorothy Gregg, Oct. 5, 1940; children—Charles G., Ann, Jane, Mary. Instr. surgery, U. Mich., Ann Arbor, 1949-54; dir. surgery Wayne County Gen. Hosp., Eloise, Mich., 1950-54; prof. surgery U. Neb., Omaha, 1954—, chmn. dept. surgery, 1956—. Diplomate Am. Bd. Surgery. Fellow A.C.S. (bd. govs. 1958-61); mem. Am. Fedn. for Clin. Research, Am. Assn. for Surgery Trauma, A.M.A., Am. Surg. Assn., Frederick A. Coller Surg. Soc. (pres. elect 1964-65), Internat. Soc. Surgery, Soc. U. Surgeons, A.A.-A.S., Sigma Xi, Alpha Omega Alpha, many others. Author: Terramycin, 1965. Editorial bd. Am. Jour. Surgery, 1957—, Jour. Trauma, 1960—. Research, publs. on nutritional deficiencies, fat embolism, wound healing, antibiotic therapy. Home: 119 S. 51st St., Omaha 68132.*

MUSSEN, Paul Henry, Am. psychologist; b. Paterson, N.J., Mar. 21, 1922; s. Harry and Taube (Meyers) M.; A.B., Stanford, 1942, M.A., 1943; Ph.D., Yale, 1949; m. Ethel Foladare, Oct. 30, 1953; children—Michele, James. Asst. prof. psychology U. Wis., 1949-51; faculty Ohio State U., Columbus, 1951-55; Fund for Advancement Edn. fellow U. Cal. at Berkeley, 1955-56, faculty, 1956—, now prof., research psychologist, 1956—. Mem. Am. Psychol. Assn. (mem. council reps. 1958—), Assn. for Research and Child Devel. (governing council 1962—). Author: (with J. Conger, J. Kagan) Child Development and Personality, 1963; also articles, chpt. in book. Research on socialization and identification. Home: 120 Hill Rd., Berkeley, Cal. 94708.*

MUSSER, Marc James, Am. physician; b. Terre Haute, Ind., July 3, 1910; s. Marc James and Margaret (Gallagher) M.; A.B., U. Wis., 1932, M.D., 1934; m. Alice Balcuns, Sept. 23, 1957; children—Marc D., Barbara (Mrs. Frank Bilek), Stephen J., William M. Instr. neuropsychiatry U. Wis. Med. Sch., Madison, 1938-40, asst. prof. neuropsychiatry and internal medicine, 1940-53, prof. internal medicine, 1953-58; prof. internal medicine Baylor U. Coll. Medicine, Houston, 1958-59; chief staff VA Hosp., 1957-59; dir. research service, asst. chief med. dir. for research and edn. in medicine VA, Washington, 1959-64, dep. chief med. dir., 1964-66, exec. dir. N.C. Regional Med. Program, Durham, 1966—; prof. medicine Duke Sch. Medicine, 1966—; adj. prof. Bowman Gray, U. N.C. schs. medicine. Decorated Legion of Merit. Diplomate Am. Bd. Internal Medicine. Fellow A.C.P.; mem. A.M.A., Central Soc. for Clin. Research, N.C. Soc. Internal Medicine, Sigma Xi. Research in psychosomatic medicine. Home: 2756 McDowell St., Durham 27705. Office: 4019 N. Roxboro Rd., Durham, N.C. 27704.*

MUSSEY, Reuben Dimond, Am. surgeon, educator; b. Pelham, N.H., June 23, 1780; s. John and Beulah (Butler) M.; A.B., Dartmouth, 1803, M.B., Med. Dept., 1805, LL.D., 1854; M.D., U. Pa., 1809; m. Mary Sewell, before 1807; m. 2d, Mehitable Osgood, 1813; 9 children, including William Hand Francis. Tchr. theory and practice of medicine, materia medica, obstetrics Dartmouth, 1814-20, prof. anatomy and surgery, 1822-38; proved union was possible in cases of intra-capsular fracture, 1830; 1st to tie both carotid arteries successfully; lectr. on anatomy and surgery Coll. Physicians and Surgeons, Fairfield, N.Y., 1836-38; prof. surgery Med. Coll. Ohio, 1838; founder Miami Med. Coll., prof. surgery, 1852-57; fellow Med. Coll., Phila.; mem. A.M.A. (pres. 1850), N.H. Med. Soc. (pres.), Mass. Med. Soc., Am. Acad. Arts and Scis. Died Boston, June 21, 1866.

MUSSO, Hans, German chemist; b. Camby, Aug. 17, 1925; s. Alexander and Ellinor (Gernhardt) M.; Ph.D., U Gottingen; m. Eve-Maria Thouison, Nov. 14, 1953; children—Andraeas, Dieter, Peter. Chemist, asst. and prof. agrégé U. Gottingen, 1961—; prof. organic chemistry U. Marburg Lahn. Recipient prize in chemistry Acad. Sci. Gottingen, 1961. Mem. German Chem. Soc., Chem. Soc. London. Research and publs. in field. Home: Grosseelheimerstrasse 19a. Office: Bahnhofstrasse 7, Marburg, West Germany.

MUSTAKALLIO, Eero, Finnish bacteriologist; b. Helsinki, July 11, 1907; s. Henrik and Hanna (Gummerus) M.; M.D., U. Helsinki; m. Kaisa Pelkonen, Apr. 21, 1935; children—Arto, Sole. Asst. in serobacteriology U. Helsinki, 1936-45; lectr. U. Helsinki, 1944-46; prof. U. Turku, 1945—. Decorated Liberty Cross, Order Lion. Mem. Finnish Biochem. Soc., Finnish Soc. Biophysics and Microbiology, Scandanavian Pathology and Bacteriology Soc. Author: Untersuchungen über die MN, AA und OAB Blutgruppen in Finnland, 1937; Untersuchungen über die Blutgruppen bei den Nordfinnen und Lappen, 1938; Om blodgruppen hos nötkreatur, 1949; Urinary excretion of Vitamin B1 and B2, nicotinic and pantothenic acid and biotine carriers of fish tapeworm, 1960; Complementa fixation tests for antisera prepared against epithelial and fibroblast cell strains, 1961. Research in med. microbiology and immunity. Home: Läntinen, Rantakatu 21 A 6. Office: Kiinamyllynkatu 8-10, Turku, Finland.

MUSTAKALLIO, Kimmo Kalvervo, Finnish physician; b. Helsinki, Finland, July 7, 1931; s. Martti Joeli and Heddi (Sjoberg) M.; licenciate in medicine, U. Helsinki, 1956, 2d hon. degree, 1966; m. Marita Nordberg, Dec. 21, 1956; 1 son, Sami. Faculty, U. Central Hosp., Helsinki, 1961—, asso. prof. dermatology, 1966—. Recipient Duodecim award, 1954. Mem. Finnish, No. dermatol. socs., Finnish Pharmacological Soc., Finnish Histochem. Soc., Histochem. Soc., Soc. for Investigative Dermatology, CP Soc. Finland (pres. 1965—). Research, numerous publs. on enzymes and lipids of epidermis, histochemistry of tumor genesis, localization od drug-, hormone-, and toxin-targets, cutaneous side-effects of drugs, barrier function of skin against bacteria, infectious diseases and psoriasis, psoriatic arthritis, sarcoidosis, Kveim reaction, connective tissue diseases, investigative applications of epidermal abrasion, dermo-epidermal separation with suction, monochromatic ultra-violet photography of skin. Home: Valikatu 2 B 19, Helsinki 17, Finland.*

MUSTAKALLIO, Sakari, Finnish physician; b. Kuopio, Finland, Jan. 9, 1899; s. Johannes and Haana (Rahm) Schwartzberg; Dr.Med., U. Helsinki, 1934; m. Lea Lyyli Santaholma, Jan. 5, 1927; children—Hilkka Lea (Mrs. Pentti Parvio), Helvi Riitta (Mrs. Matti Valanki), Päivi Annikki, Irma Tellervo, Marja Sirkku. Pathologist, Finnish Red Cross Hosp., 1932-37; dir. radiotherapy dept. Gen. Hosp., 1935-58, U. Center Hosp., 1958— (both Helsinki); faculty U. Helsinki, 1938—, prof. radiology, 1950—, dean med. faculty, 1960-63. Mem. fedn. bd. U. Helsinki Center Hosp., 1957-62; mem. State Com. Radiol. Protection, 1955, chmn., 1960—. Recipient several profl., pub. service awards. Mem. Finnish (chmn. 1950-62, hon. chmn. 1962—), Scandinavian (bd. mem. 1933-62) radiol. socs., Finnish Cancer Soc. (chmn. 1952—), Internat. Club Radiotherapists; hon. corr. mem. German Radiol. Soc. Author: Nurses' textbook IV: Cancer, 1940; Cutaneous Cancer in Finland, 1946; also numerous articles. Research on local fibroid ostitis, nasopharyngeal fibrom, malignant giant cell tumors of bones, radiodiagnosis and radiotherapy especially larynx and mammary cancer, periarthritis humeroscapularis. Home: 10 Ilmarinkatu. Office: 4 Haartmaninkatu, Helsinki, Finland.*

MUSTAKAS, Gus Carl, Am. chem. engr.; b. Gary, Ind., July 29, 1919; s. Carl and Despina (Delardos) M.; student U.W.Va., 1936; B.Ch.E., Ill. Inst. Tech., 1941; m. Calliope Dikos, June 29, 1952; children—Stephanie, Michael. Metallurgist, Carnegie-Ill. Steel Corp., Gary, 1941-43; research chem. engr. No. Region Research Lab., Agrl. Research Service, U. S. Dept. Agr., Peoria, Ill., 1945-62, prin. chem. engr., 1962—. Mem. Am. Inst. Chem. Engrs., Am. Chem. Soc., Am. Oil Chemists Soc., Am. Soybean Assn. Contbr. articles to sci. jours. Pioneer in chem. engring. research developing new products, processes from oilseeds, cereal grains and new spl. crops; devel. econ. and practical methods of treating soybeans making them more widely usable as foods including fgn. markets and developing countries; engaged in coop. programs with A.I.D., UN (UNICEF). Patentee in field. Home: 9169 N. Picture Ridge Rd. Office: 1815 N. University St., Peoria, Ill. 61614.*

MUSTARD, William Thornton, Canadian surgeon; b. Clinton, Ont., Can., Aug. 8, 1914; s. James Thornton and Pearl (MacDonald) M.; M.D., U. Toronto, 1937, M.S., 1947; m. Elise Dunbar Howe, Oct. 25, 1941; children—Neil, Susan, Shirley, Sharon, Sandra, Charlotte, Charles. Practice medicine, specializing in surgery, Toronto, Ont., 1947—; chief cardiovascular surgery Hosp. For Sick Children; asso. prof. surgery U. Toronto. Mem. Order Brit. Empire. Fellow A.C.S., Royal Coll. Surgeons, Am. Heart Assn., (council clin. cardiology); mem. Canadian Med. Assn., Am. Assn. Thoracic Surgery, Am. Acad. Pediatrics, Canadian Cardiovascular Soc., Am. Soc. Artificial Internal Organs. Author: Technics in Treatment of

Congenital Heart Disease-Advances in Surgery, 1965. Editor: (with C. E. Welch) Textbook of Pediatric Surgery, 1962. Research, numerous publs. on glass tubes in arteries; pioneer artificial heart-lung machine, congenital heart surgery; devel. new hip operation for poliomyelitis, operation to correct blue babies. Home: 482 Russell Hill Rd., Toronto 7. Office: 123 Edward St., Toronto 2, Ont., Can.*

MUSTEL, Evald Rudolfovich, Russian astrophysicist; b. June 3, 1911; grad. Moscow U., 1935. Instr., Moscow U., 1933-44, prof., 1944-51; asso. Crimean Astrophys. Obs., USSR Acad. Scis., 1946—, mem. astronom. council, 1957—. Del., Internat. Geophysics Com., Paris, 1962. Mem. USSR Acad. Scis. (corr.). Co-author: Theoretical Astrophysics, 1952; author: The Magnetic Fields of Nova Stars, 1956. Chief editor: Astronom. Jour., 1964—. Research and publs. on theory of radiant equilibrium of stellar atmospheres regarding their absorption factor; described physical picture of processes occurring during flare of nova stars. Address: Astronomical Council, USSR Acad. Scis., Pyzhevsky p. 3, Moscow, USSR.

MUTH, Herman Karl Wilhelm, German biophysicist; b. Bad Vilbel, Feb. 3, 1915; s. Wilhelm Konrad and Elise Marie (Kohl) M.; Ph.D., U. Frankfurt/Main; m. Margret Hack, July 21, 1945; children—Barbara, Petra, Evelyn, Corinna, Tanja Daniela. With Max Planck Inst. Biophysics, Frankfurt, 1942-59, 1st asst., rep. for dir. inst., 1947-59; prof. U. Frankfurt, 1955, asso. prof., 1958; prof. biophysics and elementary physics U. Sarre, also dir. Univ. Inst. Biophysics, Hamburg, 1959. Mem. Verband Deutscher Physikalischer Gesellschaften, Deutsche Gesellschaft für Biophysik, Deutsche Röntgengesellschaft, Deutsche Gesellschaft für Lichtforschung. Research and publs. in biophysics, biology, radiation. Home: Eichelscheiderstrasse 45, 6651 Jägersburg (Saar). Office: Universitätskliniken, Institut für Biophysik, 665 Hamburg (Saar), West Germany.

MUTHIAH, Palaniappan P. L., Indian leather technologist; b. Kulipirai, Madras, India, May 28, 1936; s. Palaniappan and Annapurani Muthiah; B.Sc. Algappa Arts Coll., Karaikudi, South India, 1958; B.Sc. in Tech., A.C. Coll. Tech., Madras, South India, 1960, C.L.R.I., 1963, M.Sc., 1964; m. M. Valliammai, Sept. 9, 1965. Govt. of India Sr. Stipendiary Central Leather Research Inst., Adyar, Madras, 1960-61, Jr. Research fellow Council of Sci. and Inds. Research, 1961-63, Sr. Research fellow, 1963-65, scientist, 1965—. Fellow Indian, Royal micros. socs., Indian Chem. Soc.; mem. Am. Leather Chemists Assn. Contbr. chpt. Lecture Notes on Leather, 1965. Research, publs. on waterproofing; developed processes for improving water and wear resistances of leather, processes for rapid tanning of leather, noncracky upper leather, tacky leather, selection of samples from skin, leather for research processes, method for better utilization of vegetable tanned bellies and shoulders, locational differences in skins, hides and leather at molecular, ultra structural, optical and macro levels of organization, replica technique for study topography of skins from living animals and human beings. Home: 29 Thiruvenkada Mudali St., Periamet, Madras-3. Office: CLRI, Adyar, Madras-20, Madras, South India.*

MUTING, Dieter, German physician; b. Breslau, Germany, Nov. 11, 1921; s. Joseph and Elisabeth Müting; student U. Breslau Med. Sch., 1939-45; Dr.med., U. Leipzig (Germany), 1945; m. Elisabeth Reuter, Apr. 23, 1949; children—Angelika, Christiane, Michael, Reinhard, Beate. Staff med. dept. U. Hosp., Greifswald, 1948-54, asso. prof., 1954; asso. prof. medicine med. dept. U. Hosp. Homburg/Saar, Germany, 1955-57, asso. prof. med. dept., 1958-64, prof. internal medicine, 1964—. Mem. Royal Soc. Medicine London, N.Y. Acad. Scis., Am. Diabetes Assn., Assn. des Diabétologues de Langue Francaise. Author: Aminosäurenhaushalt des Menschen, 1958; Eiweissstoffwechsel bei Leberkrankheiten, 1963; also numerous articles. Research on protein metabolism in normal and diseased men, diabetes mellitus, liver diseases, metabolic disorders; devel. new biochem. analytical methods. Home: Am. Gedünner, 665 Homburg/Saar, Germany.*

MUTIS, José Celestino, Spanish botanist; b. Cadiz, Spain, Apr. 6, 1731; studied medicine, Cadiz, Seville; sent to New Granada, 1760, collected and studied plants, specimens of which he sent to Linnaeus; prof. anatomy, Madrid; prof. botany, supt. royal garden, Madrid. Early disciple of Linnaeus in Spain; 1st to distinguish various species of cinchona. Died Sept. 2, 1808.

MUUS, L. T., Danish phys. chemist; b. Odder, Denmark, June 10, 1919; s. Laurits Tage and Mary Jane (Crafoord) M.; M.Sc., Tech. U. Denmark, 1942, Dr.-techn., 1955; m. Gudrun Jensen, Sept. 20, 1952; children—Lars, Niels. Chem. engr. F. L. Smidth & Co., 1942-44; research chemist Danish Inst. Textile Research, 1944-47; asso. prof. Tech. U. Denmark, 1947-53; research chemist polychem. dept. E. I. du Pont de Nemours & Co., 1953-57, research asso., 1957-58, research supr., 1958-60; prof. chemistry, Chem. Inst., Aarhus, Denmark, 1960—. Danish del. sci. com. NATO, 1963—. Mem. Chem. Soc., Phys. Soc. Research, publs. on kinetics, thermodynamics, magnetic resonance, solid state physics, polymer chemistry and physics. Home: Jagtvej 31, 8270 Hojbjerg, Denmark. Office: Chem. Inst., Langelandsgade, 8000, Aarhus, Denmark.*

MUYBRIDGE, Eadweard, photographer; b. Kingston-on-Thames, Eng., Apr. 9, 1830; s. John and Susannah Muggeridge. Photographer on U. S. Coast and Geodetic Survey, Pacific coast, 1872; engaged by Leland Stanford (Governor of Cal.) to ascertain whether at any point a running horse has all 4 feet off ground, May 1872, used camera operated by string stretched across horse's path, definitely proved that all 4 feet are off ground at certain times; performed series of expts. designed to make more detailed study of moving horse, 1872-78; continued expts., using men, dogs and birds, 1878-81; developed zoopraxiscope (machine which reproduced moving figures on screen), 1879; worked on animal motion studies with Dr. E. J. Marey, Paris, France, 1881-82; did series of electro-photographic expts. in animal movement under sponsorship of U. Pa., 1884-86; lectured at World's Columbian Expn., Chgo., 1893. Author: The Horse in Motion, 1878; Animal Locomotion; An Electro-Photographic Investigation of Consecutive Phases of Animal Movements, 1872-85, 11 vols., published 1887; Descriptive Zoopraxography, 1893; The Human Figure in Motion, 1901. Died Kingston, Eng., May 8, 1904.

MYANT, Nicolas Bruce, Brit. physiologist; b. Cardiff, Wales, U.K., Oct. 26, 1917; s. Joseph Nicolas and Margaret (Riches) M.; B.M., B.Ch., Balliol Coll., Oxford U., 1943, B.Sc. in Physiology, 1944, D.M., 1950; m. Audrey Palmer, Dec. 3, 1943; children—Christopher, Martin, Ruth. Demonstrator physiology U. Oxford, 1941-42; mem. sci. staff Brit. Med. Research Council, London, 1947—; faculty U. London, 1964—; hon. lectr. Royal Postgrad. Med. Sch., London, 1962—. Mem. Royal Coll. Physicians, Med. Research Soc., Biochem. Soc. Research, numerous publs. on hormone secreted by thyroid gland including transport in blood, passage from mother to fetus, its action on biochem. processes in adult and fetal organs and its excretion in bile; disorders of cholesterol metabolism. Home: Sheepcote Cottage, Kingsway, Chalfont St. Peter, Bucks, Eng. Office: Med. Research Council, Hammersmith Hosp., Ducane Rd., London, W.12, Eng.*

MYASNIKOV, Aleksandr Leonidovich, Russian therapeutist; b. Krasny Kholm (now Kalinin Oblast, 1899; grad. Med. Faculty, Moscow U., 1922; D.Med. Sci. Asst. dept. faculty therapy Petrograd U., 1925-32, prof., 1932—; prof. chair internal diseases Novosibirsk Med. Inst., Novosibirsk Postgrad. Med. Inst., 1932-38; head chair faculty therapy 3d Leningrad Med. Inst., 1938, Naval Med. Acad., 1940-48; prof. Naval Faculty, Pavlov 1st Leningrad Med. Inst., 1932-38; head Ostroumov chair hosp. therapy Sechenov 1st Moscow Med. Inst., 1948—; dir. Inst. Therapy, USSR Acad. Med. Scis., 1948—, chmn. learned council dept. clin. medicine, 1956—, former Presidium mem. and acad. sec. Del., Internat. Congress Therapeutists, Stockholm, 1954, Madrid, 1956, Phila., Brussels, 1958, London, 1959, Internat. Congress Cardiologists, Mexico, 1962, 6th Internat. Congress on Nutrition, Eng., 1963. Decorated Order of Lenin. Mem. USSR Acad. Med. Scis., All-Russian Soc. Therapeutists (chmn. 1956—), Rumanian Acad. Sci. (corr.), Czech Purkinje Med. Soc., Internat. Soc. for Internal Medicine (com. mem.). Author (monographs): Liver and Bile Duct Diseases, 1934; Visceral Malaria, 1936; Liver Diseases, 1940, 49; The Clinical Aspects of Brucellosis, 1944; The Clinical Aspects of Alimentary Dystrophy, 1945; Epidemic Hepatitis, 1946; Hypertonia, 1954; (textbook) The Propedeutics of Internal Diseases, 1944, 4th edit., 1957. Exec. editor: Problems of Pathology of Cardiovascular System; coeditor Internal Diseases sect. Large Med. Ency., 2d edit. Research and numerous publs. on pathology of liver, hypertonia, arteriosclerosis; developer new theory of epithelial and mesenchymal forms of hepatitis; new methods to diagnose arteriosclerosis; devised clin. classification for arteriosclerosis based on its devel. by stages. Home: Noroslobodskaya 57-65. Office: Inst. Therapy, USSR Acad. Med. Scis., Petroverigsky p. 10, Moscow, USSR.

MYCIELSKI, Jan, Polish mathematician; b. Wisniowa, Poland, Feb. 7, 1932; s. Jan and Helena (Bal) M.; M.S. in Math., U. Wroclaw, 1955, Ph.D., 1957; m. Emilia Przezdziecka, Apr. 25, 1959. With Inst. Math., Polish Acad. Scis., Wroclaw, 1956—, docent, 1963—; vis. prof. Case-Western Res. U., Cleve., U. Colo., Boulder, 1967. Attache de recherche Centre National de la Recherche Scientifique, Paris, France, 1957-58; asst. prof. U. Cal., Berkeley, 1961-62. Recipient Stefan Banach prize, 1965. Mem. Polish, Am. math. socs. Research, numerous publs. on powers of natural numbers, free subalgebras in topological algebras, various notions of compactness in topology and theory of models, applications of infinite positional games in founds. of math. Home: 6, m.5 Plac PKWN, Wroclaw, Poland.

MYERS, Charles Roger, Canadian psychologist; b. Calgary, Alta., Can., Feb. 12, 1906; s. Charles Allan and Sophie (Rogers) M.; B.A., U. Toronto (Ont., Can.), 1927, M.A., 1929, Ph.D., 1937; m. Helen K. Cooper, July 30, 1932; children—Douglas, Kathy (Mrs. Thomas E. Krogh). Faculty, U. Toronto, 1931—, prof. psychology, 1948—, chmn. dept., 1956—, asso. dean Sch. Grad. Studies, 1956-62; cons. psychologist Ont. Dept. Health, 1931-63. Diplomate Am. Bd. Examiners in Profl. Psychology. Mem. Canadian (pres. 1950), Ont. (pres. 1947) psychol. assns., Am. Assn. State Psychology Bds. (pres. 1966) Am. Psychol. Assn. Author: Toward Mental Health in School, 1939; also articles. Research on causation of mental retardation, community mental health services, personnel selection, history of psychology. Home: 145 Sheldrake Blvd., Toronto 12. Office: 100 St. George St., Toronto 5, Ont., Can.*

MYERS, Charles Samuel, English psychologist; b. London, Eng., Mar. 13, 1873; s. W. Myers; ed. Gonville and Caius Coll., Cambridge U.; St. Bartholomew's Hosp., London, Eng.; M.A., M.D., Sc.D., U. Cambridge; D.Sc. (hon.) U. Manchester (Eng.), also U. Pa.; LL.D. (hon.), U. Calcutta; m. Edith Seligman; two sons, three daus. Mem. Cambridge Anthrop. Expdn. to Torres Straits and Sarawak, 1898-99; house physician St. Bartholomew's Hosp., 1899-1900; prof. psychology King's Coll., London, 1906-09; past dir. psychol. lab., reader exptl. psychology Cambridge U.; cons. psychologist British Armies, France, 1914-19; hon. sci. adviser (past prin.) Nat. Inst. Indsl. Psychology. Hon. fellow, formerly fellow Gonville and Caius Coll.; decorated comdr. Order British Empire. Fellow Royal Soc., 1915; mem. Brit. Psychol. Soc. (1st pres.), Brit. Assn. (pres. sect. J, 1922, 31), Internat. Congress Psychology (pres. 1923). Author: An Introduction to Experimental Psychology; A Text Book of Experimental Psychology; Present-day Applications of Psychology; Mind and Work; Industrial Psychology in Great Britain; Business Rationalisation, 1932; A Psychologist's Point of View, 1933; In the Realm of the Mind, 1937; Shell Shock in France 1914-18, 1940; (with Henry J. Welch) Ten Years of Industrial Psychology; also numerous articles. Editor Brit. Jour. Psychology, 1911-24. With Kendall, Mason and Allers isolated nine related steroid hormones from adrenal cortical extracts, 1936, one of which was compound E, later named cortisone, 1939; studies in shock treatment. Died Oct. 12, 1946.

MYERS, Dean Wentworth, Am. surgeon; b. Ionia County, Mich., Apr. 27, 1874; s. David Wallace and Rebecca Jane (Macomber) M.; M.D., U. Mich. Homeo. Med. Sch., 1899, postgrad., 1899-1903; m. Cora Louise Owen, Aug. 29, 1900 (dec. May 1904); 1 dau., Dorothy Louise; m. 2d, Eleanor Sheldon, Aug. 19, 1922. Asst. dept. ophthalmology and oto-laryngology U. Mich. Homeo. Med. Sch., 1899-1903, prof. otolaryngology, 1907-08, prof. ophthalmology and otolaryngology, 1908-22; practiced, Grand Rapids, Mich., 1903-07. Fellow A.C.S. (gov. 1920-26); mem. A.M.A., Pan-American Assn. Ophthalmology, Am. Inst. Homeopathy, Am. Homeo. Ophthal., Otol. and Laryngol. Soc. (sec. 1910-14; pres. 1914-15), Mich. State Homeo. Med. Soc. (pres. 1910-11), Michigan, Washtenaw County med. socs. Contbr. numerous articles to med. jours. Widely recognized for successful surgery of the eye, and one of first surgeons in America to remove cataractous lens in its closed capsule; 1st to establish the exact center of rotation of eye by a series of X-Ray photographs of a needle passed directly through eyeball. Died July 2, 1955, Ann Arbor.

MYERS, Frank Evans, Am. physicist; b. Portland, Ore., Dec. 31, 1906; s. Henry Lewis and Alice (Evans) M.; B.A. in Physics, Reed Coll., 1927; M.S., N.Y. U., 1930, Ph.D., D.Sc. (hon.), Beloit Coll., 1965; m. Ionemary Williams, July 25, 1931. Teaching fellow physics N.Y. U., N.Y.C., 1928-30, faculty 1930-37, asst. prof., 1937-45, asso. prof., 1946; research asso. Mass. Inst. Tech., Cambridge, 1940-41; physicist Ordnance Lab., Frankford Arsenal, Phila., 1941-45, cons., 1950-58; prof. Lehigh U., Bethlehem, Pa., 1946-58, head dept., dir. curriculum in engring. physics, 1947-58, dean Grad. Sch., 1956-58; asso. lab. dir. Argonne (Ill.) Nat. Lab., 1958—, cons., 1955-58. Fellow Am. Phys. Soc., A.A.A.S.; mem. Am. Soc. Engring. Edn., Am. Assn. Physics Tchrs., Am. Nuclear Soc., Phi Beta Kappa, Sigma Xi, Omicron Delta Kappa. Editor: Jour. Applied Physics and Physics letters, 1965—. Research and publs. in exptl. nuclear physics, especially electron scattering and polarization, and ballistics. Home: 242 W. Washington St., Lombard, Ill. 60148. Office: 9700 S. Cass Av., Argonne, Ill. 60440.*

MYERS, Garry Cleveland, Am. psychologist; b. Sylvan, Pa., July 15, 1884; s. John Abner and Sarah (Besore) M.; A.B., Ursinus Coll., 1909; postgrad., U. Pa., 1909-10; Ph.D., Columbia, 1913; m. Caroline Elizabeth Clark, June 26, 1912; children—Jack, Elizabeth (Mrs. Kent L. Brown), Garry Cleveland (dec.). Supervising prin. pub. schs., Mercersburg, Pa., 1905-06; prof. sociology, edn. Juaniata Coll., 1912-14; prof. edn. Bklyn. Tng. Sch. for Tchrs., 1914-18; prof. edn., psychology Cleve. Sch. Edn., 1920-27; head, dept. parent edn. Cleve. Coll. Western Res. U., 1927-41; editor-in-chief Weedon's Modern Ency. (now Jr. Britannica), Chgo., 1931, Children Activities, Chgo., 1934-46; editor, co-founder Highlights for Children, v.p. Highlights for Children, Inc., Honesdale, Pa., 1946—; daily columnist King Features Syndicate, N.Y.C., 1932—. Fellow A.A.A.S., Am. Ednl. Research Assn., Am. Psychol. Assn., Soc. Research in Child Devel.; mem. Am. Soc. Mag. Editors. Author: Modern Parent, 1930, Modern Family, 1934, Building Personality in Children, 1931; (with Caroline Elizabeth Clark) Homes Build Persons, 1950, Myers Mental

Measure, 1920, Series Work Books in Arithmetic, 1927, For Beginning the School Day, 1966; Headwork for Preschoolers, 1968; Headwork for Elementary School Children, 1968. Asso. editor Education mag. since 1960——. Home: Boyds Mills, Pa. 18406. Office: 803 Church St., Honesdale, Pa. 18431.*

MYERS, George Sprague, Am. zoogeographer, ichthyologist; b. Jersey City, Feb. 2, 1905; s. Harvey Derwood and Lily Vale (Sprague) M.; student St. John's Mil. Sch., Ossining, N.Y.; A.B., Stanford U., 1930, A.M., 1931; Ph.D., 1933; m. Martha Ruth Frisinger, Sept. 25, 1926 (div. 1954); children—Thomas Sprague, John William; m. 2d, Irma Ann Block Zimmerman, Feb. 14, 1957 (div. 1965). Vol. asst. in herpetology and ichthyology depts., Am. Mus. Natural History, 1922-24; asst. in natural history mus. and zoölogy dept., Stanford U., 1926-32; asst. curator in charge Div. Fishes, U. S. Nat. Mus. (Smithsonian Instn.), 1933-36; asso. prof. biology and curator zoöl. collections, Stanford U., 1936-38, prof. biology and head curator zoöl. collections, 1938——; went with expdn. to Mexican Border and West Tex., 1929, Death Valley, 1930; ichthyologist Hancock Pacific Expdn. to Panama, Colombia, Peru, Ecuador and Galapagos Islands, 1938; leader Crocker-Stanford Deep-Sea Expdn., 1938; del. and mem. fisheries orgn. com. 6th Pacific Sci. Congress, 1939; cons., lectr. ichthyology and fisheries biology, Museu Nacional do Brasil, Brazilian Div., Fish and Game, Rio de Janeiro, 1942-44. Dir. survey Brazilian marine fishes of comml. importance, 1943-44. Mem Pacific Sci. Conf., Washington, 1946; West Coast Adv. Commn., Pacific Sci. Bd. (NRC), 1946-50; mem. Bikini Sci. Resurvey (USN) 1947; del. 7th Pacific Sci. Congress, New Zealand, 1949; fish work in Brazil, 1950, Colombia, 1958, 60, Europe, 1960, Nicaragua, 1963. mem. organizing com. 1st Internat. Congress of Oceanography, 1959; permanent adv. com. on fisheries FAO, UN, 1962——. Medalist of Société Nationale d'Acclimatation de France, 1936. Fellow Cal. Acad. Scis. (2d v.p. 1945-52), Am. Geog. Soc. N.Y., A.A.A.S., Herpetologists League, Zool. Soc. India; mem. Ecol. Society Am., Am. Soc. Ichthyologists and Herpetologists (pres. 1950-52), Am. Soc. Oceanography and Limnology, Soc. for Study Evolution, Soc. Systematic Zoologists, Zool. Soc. London (corr. mem.), Sigma Xi. Editor: Stanford Ichthyological Bull. 1938——; asso. editor: The Aquarium, Phila., 1932-56; mng. editor Aquariam Jour., San Francisco, 1952-54. Writer of papers and monographs on ichthyology, herpetology. Contbns. on evolution and classification of fishes, leading to better understanding of hist. dispersal patterns of fish groups through geological time, establishing fact that dispersal patterns of modern groups of fresh-water fishes are older and more conservative than those of mammals; also contbns. on fish habits and behavior, amphibian taxonomy and zoogeography. Office: Div. Systematic Biology, Stanford University, Cal.*

MYERS, Howard Milton, Am. oral biologist; b. N.Y.C., Dec. 12, 1923; s. Charles and Rose (Nassberg) M.; D.D.S., Western Res. U., 1949; M.S., U. Cal. at San Francisco, 1953, Ph.D. in Pharmacology (NIH fellow), U Rochester, 1957; M.A. in Edn., San Francisco State Coll., 1964; m. Muriel Anne Epstein, Sept. 2, 1945; children—Clifford Raymond, Nancy Rose, Stephen Andrew. Instr. U. Cal. Sch. Dentistry, San Francisco, 1951-54, faculty, 1957——, prof. oral biology, 1965——, dean, 1962-65; spl. research fellow Kavolinska Inst., Stockholm, Sweden, 1964-65; cons. dental research Stanford Research Inst., 1963——; mem. dental tng. com. Nat. Inst. Dental Research, Bethesda, Md., 1965——. Fellow A.A.A.S.; mem. Internat. Assn. Dental Research, Western Pharmacology Soc., N.Y. Acad. Scis., Sigma Xi. Research and publs. on bones and teeth with tracers showing selective uptake fluoride by chalky defect areas teeth; showed ligand structure of alizarin is responsible for its binding to bone, relationship alizarin to bone is similar to tetracycline and bone. Home: 220 San Anselmo St., San Francisco 94127.*

MYERS, Jack Duane, Am. physician; b. New Brighton, Pa., May 24, 1913; s. Louis Albert and Esther Fern (McCabe) M.; A.B., Stanford 1933, M.D., 1937; m. Jessica Helen Lewis, Aug. 31, 1946; children—Judith, John, Jessica, Elizabeth, Margaret. Asst. prof. medicine Emory U., 1946-47; instr. to asso. prof. medicine Duke, 1947-55; prof. medicine, chmn. dept. U. Pitts., 1955——. Diplomate Am. Bd. Internal Medicine (chairman 1967——). Mem. Assn. Am. Physicians, Am. Physiol. Soc., Am. Clin. and Climatol. Assn., Am. Soc. Clin. Investigation (sec. 1954-57), A.C.P., A.M.A. (chmn. sect. exptl. medicine and therapeutics 1960). Research and publs. on splanchnic and hepatic blood flow and metabolism of man in health and disease. Home: Oak Hill Farms, Allison Park, Pa. 15101. Office: University of Pittsburgh School of Medicine, Pitts. 15213.*

MYERS, Jack Edgar, Am. biologist; b. Boyds Mills, Pa., July 10, 1913; s. Garry Cleveland and Caroline (Clark) M.; B.S., Juniata Coll., 1934; M.S., Mont. State Coll., 1935; Ph.D., U. Minn., 1939; m. Evelyn DeTurck, June 19, 1937; children—Shirley Ann, Jacquelyn, Linda Caroline, Kathleen. NRC fellow Smithsonian Instn., 1940-41; faculty zoology U. Tex., Austin, 1941——; prof. zoology, 1948——, prof. botany, 1955——. Guggenheim fellow, 1959. Mem. NRC, Soc. Gen. Physiologists, Am. Soc. Plant Physiologists, Bot. Soc. Am., Phycological Soc., A.A.A.S., Tex. Acad. Sci., Sigma Xi. Author: (with F. A. Matsen and N. H. Hackerman) Premedical Physical Chemistry, 1947; (with others) Algal Culture: From Laboratory to Pilot Plant, 1953. Sci. editor Highlights for Children, 1960——. Research and publs. on photosynthesis; physiology, biochemistry of algae and its possible uses in space travel. Home: 2408 Vista Lane, Austin, Tex. 78703.*

MYERS, Jay Arthur, Am. physician; b. Croton. O., Nov. 25, 1888; s. Charlie and Clara (Baker) M.; B.S., Ohio U., 1912, M.S., 1913, LL.D., 1954; Ph.D., Cornell U., 1914; M.D., U. Minn., 1920; m. Faithe Lavonne McCracken, June 23, 1909; children—Wathena Fay (Mrs. Cranford W. Ingham), Marguerite LaVonne (Mrs. A. Thomas Whaley), Carlton Jay, Jack Arthur. Mem. faculty University of Minnesota, 1914-57, professor, 1931-57, professor emeritus 1957——, chief of chest clinic, 1924-57, attending specialist in diseases of chest Students Health Service, 1920-57 in practice specializing in tuberculosis and diseases of chest, 1920——; chief of tuberculosis service Mpls. Gen. Hosp., 1924-57, cons. in tuberculosis VA, Mpls., 1946-57, Nat. Insts. of Health Tb Therapy Study sect., 1946——, United States Public Health Service, 1949——. Recipient of Dearholt medal Miss. Valley Conf. on Tb, 1950; Varrier-Jones Memorial award, 1956; Certificate of award, Minn. State Med. Assn., 1957, U. Minn., 1957; William A. Howe Honor award, 1957; Am. Sch. Health Assn. and Harrington award, Minnesota Junior Association of Commerce, 1958, Will Ross medal Nat. Tuberculosis Assn., 1964. Member of Selective Service Med. Adv. Bd. No. 1, Mpls., 1940-45. Mem. Nat. Tuberculosis Assn., A.M.A., A.C.P., Am. Assn. Anatomists, Am. Pub. Health Assn., A.A.A.S., Am. Coll. Chest Physicians (gold medalist 1947), Am. Sch. Health Assn. (chmn. Tb. com.), Am. Assn. Thoracic Surg., Am. Thoracic Soc., World Med. Assn. (a founder), Am. Coll. Health Assn., Minn. Pub. Health Assn., Minn. State Med. Assn., Phi Beta Kappa, Sigma Xi. Author of numerous books relating to field, 1924——. Mem. editorial bd. Geriatrics, 1945——, Sect. XV (tuberculosis) Excerpta Medica, 1949——, Jour. of Sch. Health; editor in chief Diseases of the Chest, 1946——, Journal-Lancet, 1949——. Recorded natural history of Tb in the human body from birth to old age. Home: 48 Clarence Av. S.E., Mpls. Office: Mayo Meml. Bldg., U. Minn., Mpls. 55455.*

MYERS, Jerome Keeley, Jr., Am. sociologist; b. Lancaster, Pa., Oct. 6, 1921; s. Keeley and Nellie (Negley) M.; B.A., Franklin and Marshall Coll., 1942; M.A., Yale, 1947, Ph.D., 1950; m. Betsy Virginia Nolan, June 17, 1944; children—Joseph N., Keeley J., Susan L. From instr. to prof. sociology Yale, 1949——. Cons., USPHS, Am. Nurses' Found. Mem. Am. Sociol. Assn., Eastern Sociol. Soc., Phi Beta Kappa. Author: (with B. H. Roberts) Family and Class Dynamics in Mental Illness, 1959; also articles. Research on relationships between social class and devel., manifestation, extent, distbn. and treatment of mental illness. Home: 10 Esterly Rd., Hamden, Conn. 06518. Office: Linsly-Chittenden Hall, Yale, New Haven 06520.*

MYERS, Lawrence Stanley, Jr., Am. chemist, biophysicist; b. Memphis, Apr. 29, 1919; s. Lawrence Stanley and Jane (May) M.; B.S., U. Chgo., 1941, Ph.D., 1949; m. Janet Vanderwalker, June 13, 1942; children—David L., Frederick L., Lee S. Jr. chemist, supr. metall. lab. Manhattan Engring. Dist. at U. Chgo., 1942-44, asst. chemist Clinton labs., Oak Ridge, 1944-46; chemist Inst. for Nuclear Studies, 1947-48; asso. chemist to group leader Argonne Nat. Lab. (Ill.), 1948-52; faculty U. Cal., Los Angeles, 1952——, asst. prof. radiology, 1953——, asst. prof. biophysics and nuclear medicine, 1960——, chief, radiobiology div. Lab. Nuclear Medicine and Radiation Biology, 1959——. Recipient Certificates of Appreciation OSRD, 1945, Manhattan Engring. Dist., 1945; Army-Navy Prodn. award, 1945. Fellow A.A.A.S.; mem. Am. Chem. Soc., Radiation Research Soc., Biophys. Soc., N.Y. Acad. Sci., Sigma Xi. Editor (with others): Radiobiology at the Intra-Cellular Level: Proceedings of First UCLA Conf. on Radiobiology, 1959. Studies, publs. on neutron absorption methods of analysis: trace element analyses; analysis of biol. materials for radioactivity; radiation biophysics; dissociation constants of complex ions; electron resonance studies; pulse radiolysis. Office: Lab. Nuclear Medicine and Radiation Biology, 900 Veteran Av., Los Angeles 90024.*

MYERS, Mabel Adelaide, Am. biologist; b. Indpls., Oct. 8, 1900; d. William Park and Bertha Helen (Klusman) Myers; A.A., Fullerton Jr. Coll., 1919; B.A., Pomona Coll., 1921, M.A., 1922; Ph.D., Cornell U., 1926. Faculty Palo Verde Valley U. High Sch., 1922; 24, Fullerton Union High Sch., 1926-28; faculty biol. scis. and geology Fullerton Jr. Coll., 1928-46; faculty Life Sci. div. San Diego State Coll., 1946——, prof., 1952——, chmn. dept. bacteriology, 1955-60; coordinator grad. studies in biol. scis., 1960——. Bd. dirs. San Diego Bio-Med. Research Inst. Pres. Orange Co. Health and Welfare Council, 1945-46. Fellow San Diego Soc. Natural History (past pres.), mem. Am. Assn. U. Profs., Western Soc. Naturalists, Am. Pub. Health Association; member Western Society Naturalists, Audubon Soc., A.A.A.S., Am. Inst. Biol. Sci., Am. Assn. Profl. Biologists, N.Y. Acad. Sci., Delta Kappa Gamma, Sigma Delta Epsilon (past chpt. pres.; nat. bd. dirs.), Phi Sigma. Research and publs. on tonsillar structures of Anurans, life history of the moth; studies on the control of the pear slug and the leafhopper; slime, bacteria. Home: 6234 Mary Lane Dr., San Diego 92115.*

MYERS, Raymond Reever, Am. chemist; b. New Oxford, Pa., Jan. 23, 1920; s. John Clayton and Neta (Reever) M.; A.B., Lehigh U., 1941, Ph.D., 1952; M.S., U. Tenn., 1942; m. Hilma Louise Stirk, May 19, 1943; children—John C. II, Hilma Louise, Kristen J. Research chemist Central Research Lab., Monsanto Co., 1942-46; staff Jefferson Chem. Co., 1946-50; research asso. Lehigh U., 1952-53, faculty, 1953-65, research prof., 1960-65; chmn. chemistry dept., prof. Kent (O.) State U., 1965——; research dir. Paint Research Inst., Kent, 1964——. Cons. to govt. agys., pvt. cos. Recipient Morrison award N.Y. Acad. Scis., 1958. Mem. Am. Chem. Soc. (past chmn. organic coatings and plastics chem. div., past chmn. Lehigh Valley sect.), Soc. Rheology (editor 1963——). Contbg. author: Rheology, Theory and Applications, vol. 3; Handbook of Adhesives; Methods in Carbohydrate Chemistry; Starch Chemistry and Industry. Developed concept of catalysis by synchronization, new silicon polymer, viscometer, coordination catalyst for drying; related stability and interfacial concepts of dispersions to their mech. properties; guided research in paint industry and starch industry. Home: 1244 Woodhill Dr., Kent, O. 44240.*

MYERS, Ronald Elwood, Am. neurologist; b. Chicago Heights, Ill., Sept. 24, 1929; s. Elwood Jacob and Ella (Krabbe) M.; A.B., U. Chgo., 1950, Ph.D., 1955, M.D., 1956; m. Jane Elisabeth Adams, June 20, 1957; children—Susan Elisabeth, Ardith Ann, Ronald Neal, Peter Owen. Research officer Walter Reed Army Inst. Research, Washington, 1957-60; fellow in neurology and physiology Johns Hopkins U. Sch. Medicine, Balt., 1960-63; research Spring Grove State Hosp., Balt., 1963-64; chief Lab. of Perinatal Physiology, San Juan, P.R., 1964——; hon. asso. prof. neurology U. P.R., San Juan, 1964. Mem. Am. Assn. Anatomists, Am. Physiol. Soc., Am. Acad. Neurology, Am. Assn. Neuropathologists. Research and publs. in definition of function of cerebral commissures, studies in cerebral localization of vision, touch, audition, neurology of sensory-motor function, exptl. neuropathology, connectionism of brain. Home: 30 Emajagua St., Santurce, P.R. Office: P.O. Box 5095, Pta. de Tierra Sta., San Juan 00906.*

MYERS, Victor Caryl, Am. biochemist; b. Buskirk Bridge, N.Y., Apr. 13, 1883; s. Adam Young and Mary Evelyn (Defandorf) M.; B.A., Wesleyan U., 1905, M.A., 1907, D.Sc., 1930; Ph.D., Yale, 1909; m. Marion Christine Smith, Sept. 7, 1910. Adj. prof. physiol. chemistry and exptl. physiology, dir. labs. Albany Med. Coll. (Union U.), 1909-11; with N.Y. Post-Grad. Med. Sch. and Hosp. as lectr. on chem. pathology, 1911-12, prof. pathol. chemistry, 1912-22, acting dir. labs., 1917-19, prof. biochemistry and dir. dept., 1922-24; prof. and head biochemistry State U. of Ia., pathol. chemist to Univ. hosps., 1924-27; prof. biochemistry, dir. dept. Sch. Med. Western Res. U., since 1927; visiting biochemist Cleve. City Hosp., since 1927; sec. Med. Faculty, 1929-44; Mem. council on dental therapeutics Am. Dental Assn. Fellow A.A.A.S., A.M.A., (asso.), N.Y. Acad. Medicine (asso.); mem. Am. Soc. Biology Chemists (sec. 1919-23; councilor 1924) Fedn. Am. Socs. Exptl. Biology (exec. sec. 1922), Soc. Exptl. Biol. Medicine (councilor, 1921-23; sec.-treas., mng. editor, 1923-24, chmn. Ia. br. and v.p., 1927), Am. Physiol. Soc., Am. Inst. Nutrition, Am. Gastroenterol. Assn., Harvey Soc., Am. Chem. Soc., Internat. Assn. Dental Research, other sci. socs., Phi Beta Kappa, Sigma Xi. Author: Essentials of Pathological Chemistry, 1913; Practical Chemical Analysis of Blood, 1921, 24; Laboratory Directions in Biochemistry, 1942. Asso. editor Jour. Lab. and Clin. Medicine and Gastroenterology; asso. editor Cyclo. of Medicine; sectional editor Biology Abstracts. Died Oct. 7, 1948.

MYERS, William Graydon, Am. biophysicist; b. Toledo, Aug. 7, 1908; s. Leo J. and Anna (Johnson) M.; B.A., Ohio State U., 1933, M.Sc., 1937, Ph.D. in Phys. Chemistry, 1939, M.D., 1941; m. Florence Lenahan, Dec. 24, 1940. Grad. asst. chemistry Ohio State U., Columbus, 1933-37, Comly asst. in med. research, 1939-42, intern U. Hosp., 1941-42, research asso. U. Research Found., 1945-49, Julius F. Stone fellow in medicine, 1945-49, research asso. prof., 1949-53, research prof. med. biophysics depts. medicine, physiology, radiology, 1953——. Cons. nuclear medicine Oak Ridge Nat. Lab., 1948——, Lawrence Radiation Lab., 1961——, U. S. Naval Med. Sch., 1952——. Recipient Alumni Achievement award Ohio State U. Coll. Medicine, 1961; Lucy Wortham James award for research James Ewing Soc., 1966. Mem. Am. Chem. Soc., Am. Phys. Soc., Am. Inst. Physics, Am. Radium Soc., Am. Physiol. Soc., A.C.P. (life), Am. Fedn. for clin. Research, Am. Assn. for Cancer Research, Am. Nuclear Soc. (charter), A.A.A.S., Sigma Xi (pres. Ohio State chpt. 1959-60), Phi Eta Sigma, Phi Lambda Upsilon, numerous others. Developed radiocobalt-60, radiogold-198, chromium-51 for radiation therapy, 1946-58, radiostrontium-87m for studies bone metabolism, 1956-66, radioiodine-125 (1959) for med. applications, radioiodine-125 x-rays, 1960, radioiodine-

123 (1962); research, publs. on synthesis radioactive compounds, 1945-—, radiation-induced mutations, 1945-50. Contbr. chpts., sects. to books. Home: 2724 Wexford Rd., Columbus, O. 43221.*

MYERS, William Shields, Am. chemist; b. Albany, N.Y., Dec. 15, 1866; s. Benjamin F. and Elizabeth (Shields) M.; B.Sc., Rutgers U., 1889, M.Sc., 1894, D.Sc., 1908; studied at Munich, Berlin, under von Hofmann, and at London, under Sir William Ramsay, 1890-92; m. Annie Tayler Lambert, Sept. 11, 1889; 1 son, W. Lambert. Asst. chemist N.J. Expt. Sta., 1888-89; chemist, Lister Chem. Works, 1892-93; instr. and later asso. prof. chemistry Rutgers Coll., 1893-1901; dir. Chilean Nitrate Com. for U. S. and Colonies, 1901-26 (agr. consumption of Chilean nitrate in U. S. increased nearly forty fold under his adminstrn., 1st to use motion pictures in presenting qualities and character of Chilean nitrate to North Am. farmers). Chmn. Com. on Survey of Coll. of Agr., Rutgers U.; pres. Rutgers League of N.J.; mayor New Brunswick, N.J., 1904-06; mem. jury award Jamestown Expn., 1907; mem. mng. com. N.J. State Coll. Agr. since 1920; Life fellow Chem. Soc., London, 1891; life mem. Soc. Chem. Industry of Gt. Britain; mem. Am. Chem. Soc., A.A.A.S. Contbr. papers to jours. on water, soils and clays of N.J. Editor and pub. of monographs on sci. fertilization, water transportation and freights; joint author, with Prof. E. B. Voorhees, of plan for systematic devel. of agrl. edn. in Mexico, accepted by govt.; spent 6 months in Europe in study of crop prodn. and soils, Britain, France, Germany, Italy and Denmark, 1926; asso. of Myron T. Herrick in promoting agrl. co-operative socs. Author: Some Causes of the Depression and Some Aids to Convalescence for Sunday Times, New Brunswick, N.J.; Bureaucrats Song In Washington and Out; The Cult of Incompetence; Philosophies of Governments, Christian and Barbarian, 1939. Founder (with L. F. Loree) Jour. of Soil Science. Died Jan. 10, 1945.

MYHRE, Eivind Kristian, Norwegian physician; b. Oslo, Norway, Aug. 4, 1920; s. Haakon Paul and Ida (Bjerke) M.; M.D., U. Oslo, 1949, Doctor Medicin, 1962; m. Siri Nergaard, Sept. 22, 1945; children—Jan Eivind, Anne Mette. Practice gen. medicine, 1949-50; gen. hosp. service, 1950-51; staff dept. pathology Det Norske Radium-hospital, 1951-54; faculty U. Oslo, 1954-—, staff Inst. for Gen. and Exptl. Pathology, 1957-—, sr. pathologist, 1964-—, prof. medicine, 1964-—. Mem. Path. Soc. Norway (chmn.). Research, publs. on clin. and path. relationship of diseases, hormones in relation to cancer, histochem. studies of uterine mucosa in patients with menstrual disorders and infertility. Home: 18A Konventveien, Oslo, Norway.*

MYLLER, Alexandru, Rumanian mathematician; b. 1879; prof. Iasi (Rumania) U.; mem. Acad. Socialist Republic Rumania. Author: Sur la théorie des équations intégrales, 1905; Géometrie différentielle aréolaire, 1929. Creator math. sch. of geometry at Iasi U.; research on differential and integral equations, differential geometry. Died 1965.

MYLON, Ernest, physician; b. Schneidemuehl, Germany, Jan. 20, 1895; s. Emil and Reisa (Ausbach) M.; M.D., Berlin U.; m. Hildegard Dorothea Joseph; children—Inge (Mrs. Marvin Jaffe), Peter. Staff, U. Berlin, 1922-35, asso. prof. 1931-35; guest prof. U. Belgrade, Yugoslavia, 1933-38; research fellow Yale, 1938-39, faculty, 1939-—, asso. prof. biochemistry, 1951-63, asso. prof. pathology emeritus, 1963-—; prof. emeritus U. Bonn (Germany), 1957-—; chief pathologist Lawrence Meml. Hosp., New London, Conn., 1951-65. Mem. Am., Conn. med. assns., Soc. Clin. Pathology, Am. Soc. Clin. Genetics, N.Y. Acad. Socs., Sigma Xi. Author: The Determination of the Hydrogen Ion Concentration, 1928; also numerous articles. Research on traumatic shock, renal hypertension, biochem. methods, phys.-chem. methods. Home: 358 Central Av., New Haven 06575.*

MYLONAS, George Emmanuel, archaeologist; b. Smyrna, Asia Minor, Dec. 9, 1898; s. Emmanuel Basil and Maria (Tenekides) M.; Diploma, U. Athens, 1922, D.Phil., 1927; Ph.D., Johns Hopkins, 1929; m. Lella Papazoghlou, May 2, 1925; children—Nike Maria, Ione Doris, Daphne Eirene. Came to U. S., 1931, naturalized, 1937. Sec. Am. Sch. Classical Studies, Athens, Greece, 1923-28; faculty U. Chgo., 1929, U. Ill., 1931-33, 39-40; faculty Washington U., St. Louis, 1933-—, prof. art and archaeology, 1963-65, Rora Muy prof. humanities, 1965-—; ann. prof. Am. Sch. Classical Studies, Athens, 1951-52, 63-64; Fulbright prof. archaeology U. Athens, 1954; lectr. archaeology Inst. Am., 1932-—, Brown U., 1961. Named comdr. Order of King George A Greece, 1955; Gold cross grand comdr. Order Phoenix King Constantine Greece, 1965. Fulbright sr. fellow, Inst. for Advanced Study Princeton, 1951-52, 54; Guggenheim fellow, 1955-56. Fellow Am. Philos. Soc., Am. Council Learned Socs., Am. Acad. Arts and Scis.; mem. Am. Archaeol. Inst. (pres. 1956-60), Archtl. Soc. Am. (asso.), Acad. Greece, Scarab, Phi Beta Kappa, Eta Sigma Phi. Excavated at Olynthos, Aghios Kosmas, Eleusis, Mycenae. Author: Mycenae; The Capital City of Agamemnon, 1957, The Protoattic Amphora of Eleusis, 1958, Aghios Kosmas, 1959, Eleusis and the Eleusinian Mysteries, 1962, The Walls and Gates of Mycenae, 1965, The Homeric Hymn to Demeter and Her Sanctuary at Eleusis, 1939; Mycenae and the Mycenaean Age, 1966. Home: 5314 Waterman St., St. Louis 63112.*

MYNSICHT, Hadrian (real name Sümenicht), physician; b. Ottenstein, Brunswich, 1603; M.D. (under name Tribudenius), Helmstedt, Germany; physician to Duke of Mecklenburg; named imperial poet-laureate, 1631. Author: Theusaurus et armanentarium medicochymicium . . . , 1631. First to crystallize tartar emetic by boiling cream of tartar with roasted antimony sulphide; aware of potassium sulphate, silver nitrate, calomel; follower of Paracelsus. Died Oct. 1638.

MYRBÄCK, Karl David Reinhold, Swedish biochemist; b. Vendel, Sweden, Sept. 7, 1900; s. Herman C. A. and Helena (Lundgren) M.; Filosofie kandidate, Stockholm (Sweden) U., 1923, Filosofie licentiate, 1923, Dr.Phil., 1927; m. Signe Maria Karlsson, June 1, 1927; children—Barbara (Mrs. Peter Pierce Manley), Karl Erik. Faculty, U. Stockholm, 1926-67, prof. fermentation chemistry, 1932-47, prof. organic chemistry and biochemistry, 1947-63, prof. biochemistry, 1963-67. Recipient Scheele medal Stockholm Chem. Soc., 1930. Mem. Royal Swedish Acad. Sci. (vice sec. 1952-—), Royal Swedish Acad. Engring. Sci., Heidelberger Akad. Wiss., Deutsche Akad. Naturforsch. Leopoldina. Author: Homogene Katalyse, II, 1939; (with E. Bamann) Methoden der Fermentforschung, 4 vols., 1940-41; (with G. Angel, O. Stelling, R. Steenhoff, och R. Winbladh) Handbok i kemisk teknologi, 4 vols., 1947-49; (with J. B. Sumner) The Enzymes, 4 vols., 1950-52; Enzymatische Katalyse, 1953; (with H. Willstaedt, E. Virgin) Kemi med. biokemi, 1953; (with P. D. Boyer, H. Lardy) The Enzymes, 8 vols., 1959-63; Biokemi, 1966. Editor-in-chief Acta Chemica Scandinavica, 1947-—. Research, numerous publs. on isolation of coenzyme of alcoholic fermentation (now adenine-nicotinamide-dinucleotide), chemistry and biochemistry of polysaccharides especially starch, gen. enzymology including dependence of enzyme activity on hydrogen ion concentration, substrate concentration, chem. background of action of poisons (inhibitors) on enzymes. Home: 45, Tegnérgatan, 111 61 Stockholm, Sweden.*

MYRDAL, Gunnar Karl, Swedish economist; b. Gustafs, Sweden, Dec. 6, 1898; s. Carl Adolf and Anna Sofia (Carlson) Petterson; grad. Law Sch., Stockholm U., 1923; Juris Dr. in Economics, 1927; hon. LL.D., Harvard University, 1938; Dr. Lit., Fisk U., 1947; J.D., Nancy U., 1950, L.H.D., 1950; L. Columbia, 1954, New Sch. Social Research, N.Y., 1956; LL.D., Leeds U., 1957, Yale 1959, Brandeis Coll., 1962, Howard University, 1962, Edinburgh, 1964, Swarthmore, 1964, Sir George Williams U., 1967, Mich. U., 1967; Dr. Social Sci., Univ. Birmingham, 1961; Doctor Humanities, Wayne State University, 1963; D.D., Lincoln University, 1964; Fil. Dr., Stockholm Univ., 1966; m. Alva Reimer, Oct. 8, 1924; children—Jan, Sissela, Kaj. Practiced law, Sweden; appointed docent in polit. econs. Stockholm U., 1927; spent much of the time, 1925-29, in England, Germany, France; travel as Rockefeller fellow in U. S., 1929-30; associate prof., Post-Grad. Inst. Internat. Studies, Geneva, 1930-31; apptd. acting prof., Stockholm U., 1931; apptd. to Lars Hierta chair of polit. econ. and public finance, Stockholm U., 1933; gave Godkin lecture, Harvard, spring 1938; dir. Study of Am. Negro Problem for Carnegie Corp. of America, 1938-42. Professor of internat. economy University of Stockholm. Active in public affairs in Sweden; advisor to govt. on economic, social and fiscal policy, 1933-38; mem. Swedish Senate (for Social Dem. Party); mem. board of dirs. Swedish Bank; mem. Population, Housing and Agrl. commns., Sweden; minister commerce Sweden, 1945-47; exec. sec. U.N. Economic Commn. for Europe, 1947-57. Mem. Royal Acad. of Science in Sweden; fellow Econometric Soc.; hon. mem. American Economic Assn.; member American Academy of Arts and Sciences. Author: Cost of Living in Sweden, 1830-1930, 1929; Monetary Equilibrium, 1939; Population: A Problem for Democracy, 1940; An American Dilemma: The Negro Problem and Modern Democracy, 1944; The Political Element in the Development of Economic Theory, 1953; Development and Under-development, The Mechanism of National and International Inequality, 1956; An International Economy, Problems and Prospects, 1956; Economic Theory and Under-developed Regions, 1957; Beyond the Welfare State, 1960; Challenge to Affluence, 1963; Asian Drama, 1968; also several other books on econ. theory published in Sweden, and govt. reports. Home: 31 Vasferlanggaton. Office: Stockholm U., Stockholm, Sweden.

MYREN, Johannes, Norwegian physician; b. Bremanger, Norway, July 7, 1915; s. Lars and Jorgine (Myren) M.; grad. in medicine U. Oslo (Norway), 1946, dr.med., 1957; m. Randi Bjornstad, Jan. 11, 1947; children—Frode, Torunn. Physician specializing in internal medicine, county dist., 1946-50; asso. prof. Rikshospitalet, 1961; fellow medicine U. Ala. Med. Center, Birmingham, 1963-64; asso. prof. medicine dept. IX, Ullevaal Hosp., Oslo, Norway, 1964-—. Mem. Norwegian Med. Assn., Soc. Gastroenterology Norway. Research and numerous articles on preventive medicine, health and social problems in country dist., exptl. carbon tetrachloride injuries of liver in mice, transplantation of liver tissue, exptl. liver injuries, functional activity enzymes in response to these injuries, mucosal alterations in stomach diseases, gastric secretion, small intestinal mucosal alterations. Home: 119 D, Nadderudveien. Office: U. Oslo, Ullevaal Hosp., Oslo, Norway.*

MYREPOS, Nicholas, Byzantine physician; b. Alexandria, flourished 13th century. Physician to Joannes III, Ducas Batatzes (emperor 1222-54). Author: De Compositione Medicamentorum (collection of pharmaceutical recipes, mainly derived from Antidotarium of Nicholas of Salerno, but with elaborations, contains 2656 recipes classified by med. properties into 48 groups), circa 1280.

MYRES, Sir John (Linton), Brit. anthropologist; b. Preston, Eng., July 3, 1869; s. W. M. and Jane (Linton) M.; ed. New Coll., Oxford (Eng.) U.; D.Litt. (hon.), 1929; D.Sc. (hon.), U. Wales, 1920, U. Manchester (Eng.), 1933; Ph.D. (hon.), U. Athens (Greece), 1937; m. Sophia Florence Ballance; 2 sons, 1 dau. Lectr. classical archaeology Oxford U., 1903-07, sec. com. anthropology, 1905-07; Gladstone prof. Greek, lectr. ancient geography U. Liverpool (Eng.), 1907-10; Wykeham prof. ancient history, 1910-39. Sather prof. classical lit. U. Cal., 1914, 27; Frazer lectr. Cambridge (Eng.) U., 1947. Recipient Huxley Meml. medal, 1933, Victoria medal Royal Geog. Soc., 1953. Fellow Brit. Acad., Soc. Antiquaries; mem. Royal Anthrop. Inst. Gt. Britain and Ireland (hon. sec. 1900-03), Brit. Assn. Advancement Sci. (gen. sec. 1919-32), Internat. Congress Anthrop. and Ethnol. Scis. (gen. sec. 1934-47). Author: A Catalogue of the Cyprus Museum, 1899; A History of Rome, 1902; The Dawn of History, 1911; Handbook of Cesnola Collection of Antiquities from Cyprus, 1914; Geographical History, 1952; Herodotus, Father of History, 1953; Homer and his Critics, 1959. Archaeol. expdns. Greece, Asia Minor, 1893, Crete, 1893, 95, 98, 1903; excavations Cyprus, 1894, 1913; reorganized govt. mus., 1894; research on Mediterranean geography, archaeology, anthropology. Died Mar. 6, 1954.

MYSELS, Karl Joseph, chemist; b. Krakow, Poland, Apr. 14, 1914; s. Adolph and Janina H. (Rosenberg) Meisels; Licencié-és-Sciences, Ingenieur Chimiste, U. Lyon (France), 1937; Ph.D., Harvard, 1941; m. Estella R. Katzenellenbogen, Mar. 28, 1953. Came to U. S., 1938, naturalized, 1942. With patent and engring. depts. Shell Devel. Co., San Francisco, 1940-42; research asst., asso. Stanford U., 1941-45; instr. N.Y. U., 1945-47; asst. prof. chemistry U. So. Cal., Los Angeles, 1947-50, asso. prof., 1950-54, prof., 1954-66; asso. dir. research R. J. Reynolds Tobacco Co., Winston-Salem, N.C., 1966-—. NSF faculty fellow, 1957-58, sr. postdoctoral fellow, 1962-63. Cons., Continental Oil Co., 1956-65, Monsanto Co., 1957-65. Rennebohm lectr. U. Wis., 1964. Recipient Kendall Co. award in colloid chemistry Am. Chem. Soc., 1964. Guggenheim fellow for study biol. membranes, 1965-66. Mem. Am. Chem. Soc., Am. Inst. Chemists (chmn. Western chpt. 1961-62), A.A.A.S., Faraday Soc. Author: (with C. S. Copeland) Introduction to the Science of Chemistry, 1952; (with K. Shinoda, S. Frankel) Soap Films, 1959; Introduction to Colloid Chemistry, 1959. Contbd. to clarification of behaviour of aluminum soaps and their gels, of structure of micelles of assn. colloids, of thinning, equilibrium and bursting of soap films, to theories of dielectric constant, transport phenomena and light scattering, to improving methods of electrophoresis, diffusion conductivity and dialysis; discovered reduction of hydrodynamic drag by macromolecules. Home: 3141 Bonhurst Dr., Winston-Salem 27106. Office: Research Dept., R. J. Reynolds Tobacco Co., Winston-Salem, N.C. 27102.*

N

NABARRO, David Nunes, English physician; b. London, Eng., Feb. 27, 1874; s. Jacob Nunes and Hannah (Ricardo) N.; B.Sc., M.D., D.P.H., U. Coll. and Hosp., London; m. Florence Nora Webster, 1914; 1 son, 1 dau. Asst. prof. pathology, bacteriology U. Coll., London, 1899-1910; pathologist Evelina Hosp. for Children, West Riding Asylum, Wakefield, Eng.; sci. asst. in pathology U. London, 1904-24; dir. path. dept., bacteriologist Gt. Ormond St. Hosp., 1912-39, later cons. Mem. Royal Soc. Sleeping Sickness Commn., Uganda, 1903; pathologist Emergency Med. Service, 1940-46. Fellow Royal Coll. Physicians, Royal Soc. Medicine; mem. Brit. Pediatric Assn. (hon.), Assn. Clin. Pathologists (pres.), London Jewish Hosp. Med. Soc., Med. Soc. Study Venereal Disease, Path. Soc. Gt. Britain and Ireland (sr.), Soc. Tropical Pathologists Paris (France). Author: The Laws of Health, 1905; trans. and amplifications of Trypanosomes et Trypanosomiases (Laveran and Mesnil), 1907. Contbr. articles to sci. publs. (With David Bruce, A. Castellani) discovered cause and method of transmission of sleeping sickness. Died Sept. 30, 1958.

NABARRO, Frank Reginald Nunes, physicist; b. London, Eng., Mar. 7, 1916; s. Stanley and Leah (Cohen) N.; B.A., Oxford (Eng.) U., 1937, B.A. in Maths. with 1st class distinction, 1938, B.Sc., 1940, M.A., 1946; D.Sc., U. Birmingham (Eng.), 1952; m. Margaret Constance Dalziel, June 25, 1948; children—David, Ruth, Jonathan, Mairi, Andrew. Sr. exptl. officer Army Operational Research Group, 1940-45; Royal Soc. Warren Research fellow, lectr. physics U. Bristol (Eng.), 1945-49; lectr. grade 1 metallurgy U. Birmingham, 1949-53; prof., head dept. physics U. Witwatersrand, Johannesburg, S. Africa, 1953-—; vis. prof. metallurgy Case Inst., Cleve., 1964-65. Decorated Order Brit. Empire; recipient Beilby Meml. award, 1950; Overseas fellow Churchill Coll., Cambridge U., 1966-67. Author: Theory of Crystal Dislocations, 1967; also articles. Research on destructive effects of weapons, influence of crystal dislocations, mech. and other properties of solids. Home: 32 Cookham Rd., Auckland Park, Johannesburg, S. Africa.*

NACE, Harold Russ, Am. chemist; b. Collingswood, N.J., July 5, 1921; s. Harold Harris and Ruth (Russ) N.; B.S. in Chemistry, Lehigh U., 1943; postgrad. Inst. Paper Chemistry, 1943-44; Ph.D. in Organic Chemistry, Mass. Inst. Tech., 1948; M.A., Brown U., 1957; m. Mary Alice Griffith, June 24, 1944. Research chemist Merck and Co., Inc., Rahway, N.J., 1944-45, Mayo Clinic, Rochester, Minn., 1949, Jackson Lab., E.I. duPont de Nemours and Co., Deepwater, N.J., 1956-57; faculty Brown U., 1948—, prof. chemistry, 1959—. Cons., William S. Merrell Co., Cin., 1960—. Fellow N.Y. Acad. Scis.; mem. Am. Chem. Soc., A.A.A.S., Phi Beta Kappa, Tau Beta Pi, Sigma Xi. Research and publs. on application known organic chem. reactions to steroid hormones resulting in new steroids, fundamental mechanisms elimination reaction by which unsaturated compounds are produced. Home: 28 Chapin Rd., Barrington, R.I. 02806. Office: Dept. Chemistry, Brown U., Providence 02912.*

NACE, Raymond Lee, Am. geologist, hydrologist; b. Los Angeles, Oct. 13, 1907; s. Percy F. and Alice (Carson) N.; B.S., U. Wyo., 1935, M.A., 1936; postgrad. Columbia, 1936-37, Ph.D., 1960; postgrad. Yale, 1939-41; m. Edna Helen Lane, July 31, 1935; children—Samuel L., Robert L. Instr. geology U. Wyo., 1937-39; asst. instr. Yale, 1939-41; with U.S. Geol. Survey, 1941—, asso. chief water resources div., Washington, 1956-62, research hydrologist, 1963—. Mem. adv. com. arid zone research UNESCO, 1963-65; head del., delegation mem. to congresses and tech. meetings of UN agys., 1962—; chmn. U.S. Nat. Com. for Internat. Hydrological Decade, 1964-67. Recipient Distinguished Service award U.S. Dept. Interior, 1959. Fellow Geol. Soc. Am., A.A.A.S.; mem. Sigma Xi. Research, publs. on occurrence and behavior of ground water in heterogeneous basalt, selection of sites for atomic-energy installations, effects of radioactive wastes in water phase of environment; 1st discovery of fossils of ichthyosaurs in Late Cretaceous rocks of N.Am. Home: 9226 E. Parkhill Dr., Bethesda, Md. 20014. Office: U.S. Geol. Survey, Washington 20242.*

NACHBIN, Leopoldo, Brazilian mathematician; b. Recife, Pernambuco, Brazil, Jan. 7, 1922; s. Jacob and Lotty (Drechsler) N.; Ph.D., U. Brazil, Rio De Janeiro, 1947; Dr. degree (hon.), U. Pernambuco, 1966; m. Maria da Graça M. Mousinho, July 28, 1956; children—André, Léa, Luis. Prof., Inst. for Pure and Applied Math., Rio de Janeiro, 1952—; vis. prof. Brandeis U., Waltham, Mass., 1960-61; U. Paris (France), 1961-63; prof. U. Rochester (N.Y.), 1963—; vis. prof. U. Chgo., 1966. Head div. math. research NRC Brazil, 1955-56. Recipient Moinho Santista prize in math., Brazil, 1962. Mem. Brazilian Acad. Scis. Author: The Haar Intergral, 1965; Topology and order, 1965; Elements of Approximation Theory, 1966; also numerous articles. Editor, Summa Brasiliensis Mathematicae, 1950—, Notas de Mathematica, 1950—. Research on functional analysis, approximation theory, topological vector spaces, differential and convolution operators, topological ordered spaces. Home: Rua Prudentade Morais 1420, Rio de Janeiro 37· (Ipanema), GB, Brazil. Office: Dept. Math., U. Rochester, Rochester, N.Y. 14627.

NACHMANSOHN, David, biochemist; b. Jekaterinoslaw, Russia, Mar. 17, 1899; s. Moses and Regina (Klinkowstein) N.; M.D., U. Berlin (Germany), 1926; M.D. (hon.), Free U. Berlin, 1964; m. Edith Berger, July 29, 1909; 1 dau., Ruth (Mrs. Joseph Rothschild). Came to U. S., 1939, naturalized, 1945. Research fellow biochemistry U. Berlin, 1924-26; fellow Kaiser Wilhelm Inst. Biology, 1926-30; research at Sorbonne, Paris, France, 1933-39; mem. faculty Yale Sch. Medicine, 1939-42; mem. faculty Columbia Coll. Phys. and Surg., 1942—, prof. biochemistry, 1954—. Mem. med. adv. bd. Hadassah, 1950-55; mem. nat. bd. Friends Hebrew U., 1966—. Recipient Pasteur medal Soc. Biol. Chemistry, Paris, 1952; Neuberg medal Am. Soc. European Chemists, 1953; Nicloux medal Soc. Biol. Chemistry, 1962. Med. Nat. Acad. Scis., Am. Soc. European Chemists (pres. 1959-61), Harvey Soc. (bd. dirs. 1966—), Am. Acad. Scis., German Acad. Scis., N.Y. Acad. Medicine. Author numerous articles, monograph in field. Spl. research chem. and molecular basis nerve activity, molecular biology excitable membranes, proteins and enzymes associated with bioelectrogenesis, explanation mechanism nerve gases and insecticides, devel. effective antidotes. Home: 350 Central Park W., N.Y.C. 10025.

NACHOD, Frederick Constantine, chemist; b. Leipzig, Germany, Oct. 4, 1913; s. Hans and Lili (Von Hoesslin) N.; student U. Paris (Germany), 1931-32, U. Freiburg (Germany), 1932-33, U. Leipzig (Germany), 1933-37; D.Sc., U. Utrecht (Germany), 1938; m. Emy Henning, Sept. 28, 1940; children—Thomas C., James H. Came to U.S., 1939, naturalized, 1944. Research chemist Baker & Co., Inc., Newark, 1939; instr. Coll. City N.Y., 1939-40; research chemist Permutit Co., Birmingham, N.J., 1940-44; cons. Smith, Kline & French Labs., 1944; sr. chemist Atlantic Refining Co., Phila., 1944-46; staff Sterling-Winthrop Research Inst., Rensselaer, N.Y., 1946—, chem. liaison staff dir., 1964—; adj. prof. chemistry Rensselaer Poly. Inst., 1952—. Decorated Pour-Le-Merite (France). Mem. Am. Chem. Soc., A.A.A.S., Sigma Xi, Phi Lambda Upsilon. Author: Ion Exchange, Theory and Application, 1949; (with Braude) Structure of Organic Compound by Physical Methods, I, 1955; (with Schubert) Ion Exchange Technology, 1955; (with Phillips) Structure of Organic Compounds by Physical Methods, II, 1962, Organic Electronic Spectral Data, Vol. IV, 1963; also numerous articles. Research on structure complicated organic molecules with application modern physio-chem. methods. Home: P.O. Box 345, Kinderhook, N.Y. 12106. Office: Sterling-Winthrop Research Inst., Rensselaer, N.Y. 12144.*

NACHTIGALL, Eduard F., Austrian metallurgist; b. Vienna, Austria, Jan. 9, 1908; s. Franz and Julie Nachtigalli; Ph.D., U. Vienna; m. Margarete Walter von Waltersheim, Oct. 29, 1939. Asst. Vienna Tech. Coll., 1931-36, Aciéries Schoeller Bleckmann, Ternitz, 1937-38; head dept. I. G. Farbenindustrie, Bitterfeld, 1938-47; dir. exptl. sta. Metallwerk Plansee, Reutte/Tyrol, 1947-51; dir., mgr. Braunau Metal Works, 1951—; instr. Leoben Sch. of Mines, also Vienna Tech. Coll. Mem. Internat. Center Devel. Aluminum (Paris, France), European Wrought Aluminum Assn. (Zurich, Switzerland), Austrian Standarization Commn. Author: Theorie des Passivität von Magnesium; Eigenschaften von Molybdän und Wolfram bei niedrigen Temperaturen; Die elekrische Leitfähigkeit von Aluminum; Interkristalliner Zerfall der Metalle im Atomreaktor; Schwingfestigkeit plattierter Metalle. Address: Partstrasse, Braunau, Austria.

NACHTRIEB, Norman Harry, Am. chemist; b. Chgo., Mar. 4, 1916; s. Norman D. and Minnie (Barnard) N.; S.B., U. Chgo., 1936, Ph.D., 1941; m. Marcia Binford, Aug. 22, 1953; I dau., Marianna C. Chemist, Pitts. Plate Glass Co., Barberton, O., 1941-43; alternate group leader Los Alamos Sci. Lab., 1944-46; faculty U. Chgo., 1946—, prof. chemistry, 1952—, chmn. dept., 1962—. Mem. sci. personnel and edn. divisional com. NSF, 1964-66. NSF fellow, 1959-60. Fellow Am. Phys. Soc.; Mem. Am. Chem. Soc., A.A.A.S. Author: Principles and Practice of Spectro-chemical Analysis, 1950. Asso. editor Jour. Chem. Physics, 1956-58; adv. editor chemistry Ency. Brit., 1955—. Research on rates of atom movements in liquids and crystalline solids (diffusion), effects of high pressures on rate processes, spectrochem. analysis of metals, phys. chemistry of molten salts, electrochem. studies in aqueous and non-aqueous media, solvent extraction; devel. trace spectroscopic methods. Home: 8400 W. 131st St., Palos Park, Ill. 60464. Office: 5747 Ellis Av., Chgo. 60637.*

NACHTSHEIM, Hans, German geneticist; b. Koblenz/Rhine, Germany, June 13, 1890; s. Friedrich and Anna (Mallmann) N.; Ph.D., M.D. (hon.), univs. Bonn, Munich; m. Dorothea Freiwald, 1923. With zool. inst., Freiburg, Munich univs. 1914-21; dept. head Berlin genetic research, agrl. inst. of Univ., 1921-40; prof., 1923; head. dept. exptl. hereditary pathology Kaiser Wilhelm Inst. Anthropology, Derlin-Dahlem, 1941-45; prof., dir. inst. for genetics Berlin U., 1946-49; dir. inst. comparative hereditary biology hereditary pathology German Research Inst., prof. gen. biology and genetics, dir. Inst. for genetics Free U. Berlin, 1949—; dir. inst. comparative biology heredity pathology Max Planck Inst., 1953-60. Mem. German Acad. for Population Research, Hamburg U. Author: Vom Wildtier zum Haustier, 2d edit., 1949; Für und Wider die Sterilisierung aus engenischer Indikation, 1952; also papers on cell research, eugenics, heredity. Authority on comparative genetic pathology; conducted genetic expts. with animals. Address: 54 Hüningerstrasse, Berlin-Dahlem, Germany.

NADACHOWSKI, Franciszek, Polish ceramist; b. Warsaw, Poland, Sept. 9, 1923; s. Adam and Zofia (Ilaszewicz) N.; student Poly. Sch., Lvov, USSR, 1944-45; M.Sc.Chem.Engring., Poly Sch., Gliwice, Poland, 1948; Dr.Eng., Acad. Mining and Metallurgy, Cracow, Poland, 1960; m. Ewa Fulinska, June 15, 1948; children—Adam, Dorota. Research asst. Metall. Inst., Gliwice, Poland, 1949-53; dept. head Refractories Inst., Gliwice, 1953-63; asso. prof. Ceramic Faculty, Acad. Mining and Metallurgy, Cracow, 1963—, head dept. refractories tech., 1965—. Tech. expert, Minex, also Cekop, export cos., 1958-60; mem. ceramic sect. mineral. commn. Polish Acad. Scis., 1964—, mem. commn. for tech. scis., 1965—. Author: Magnesite Refractories, 1955; also articles. Research on relation between microstructure and technol. properties of basic refractories; detection 4 barium minerals in phase system of barium, calcium and magnesium orthosilicates; devel. barium aluminate cement and determination of its novel features; devel. refractory bricks from sintered pure calcites, equilibrium phase composition of burnt aluminosilicate materials. Home: 2, Zawiszy Czarnego, Gliwice, Poland. Office: 30, Al. Mickiewicza, Cracow, Poland.*

NADAI, Arpad Ludwig, mech. engr.; b. Budapest, Hungary, Apr. 3, 1883; student Fed. Swiss Tech. Inst., Zurich, 1902-06; D.Eng., Tech. U., Berlin, 1912; m. Elisabeth Vally Witte, Sept. 20, 1913 (dec.). Came to U. S., 1929, naturalized, 1936. Prof. applied mechanics U. Gottingen, Germany, 1923-29; cons. mech. engr. Westinghouse Research Labs., Pitts., from 1929; cons. David Taylor Model Basin, Navy Dept., Washington, 1942-46. Recipient Eugene Bingham medal Soc. Rheology, 1952, Stephen Timoshenko medal, 1958; Elliott Cresson medal Franklin Inst., 1960. Fellow Am. Soc. M.E. (Worcester Reed Warner gold medalist 1947). Author: Die Elastischen Platten, 1925; Plasticity (in German, English, Russian), 1931; Theory of Flow and Fracture of Solids, 1950; also papers on applied mechanics, elasticity, plasticity, strength, creep of metals, geomechanics.

NADAR, Félix (born Félix Tournachon), French photographer; b. Paris, Apr. 5, 1820; designer, writer, photographer; 1st to photograph from balloon, 1858; 1st to photograph under artificial light, in sewers and catacombs of Paris, circa 1860; turned to flying, after 1863; organizer 1st captive balloon obs. service and balloon post, during siege of Paris. Died 1910.

NADAS, Alexander Sandor, physician; b. Budapest, Hungary, Nov. 12, 1913; s. Sandor and Margit (Roth) N.; M.D., U. Budapest, 1937; M.D., Wayne U., 1945; m. Elizabeth McClearen, Nov. 21, 1941; children—Trudi A., Elizabeth P., John A.M. Came to U.S., 1939, naturalized, 1945. Pvt. practice medicine, Greenfield, Mass., 1945-49; cardiologist, physician The Sharon Cardiovascular Unit, also sr. physician in medicine Children's Hosp., Boston, 1955—; clin. prof. pediatrics Harvard Med. Sch., 1964—; cons. pediatric cardiology Boston Lying-In-Hosp., Mass. Gen. Hosp., Boston; Fulbright vis. prof. of pediatrics U. Groningen, The Netherlands, 1956-57. Mem. Mass. Med. Soc., Am. Heart Assn., Soc. Pediatric Research, Am. Pediatric Soc. Author: Pediatric Cardiology, 1957, rev. edit., 1963. Home: 11 Benton St., Wellesley, Mass. Office: 300 Longwood Av., Boston.*

NADCHATRAM, M., Malaysian entomologist; b. Malaysia, Dec. 27, 1966; s. Muthuvelu and Sinnamah Nadchatram; student Inst. Acarology, U. Md., 1961; m. Maheswari, Jan. 2, 1963; 1 son, Santha Kumar. Lab. asst. Inst. for Med. Research, Kuala Lumpur, Malaysia, 1948-61, sr. asst., 1961-66, acarologist, 1966—. Hon. warden wildlife nat. park Templer Park, 1963—; hon. curator entomology Nat. Mus. Malaysia, 1964—. Fellow Zool. Soc. London (Eng.); mem. Malayan Soc. Parasitologists. Research, publs. on life histories and natural behaviour of parasitic mites, host-parasite relationship of ticks and mites; described over 70 species of Trombiculid mites, genera and subgenera of scrub typhus and scrub-itch mites. Home: 38 Jalan Chermai, Kuala Lumpur, Malaysia.*

NADEIN, Aleksandr Pavlovich, Russian surgeon; b. Kharkov, 1893; grad. Med. Faculty, Kharkov U., 1917; D.Med. Sci. Asst., lectr. dept. topographical anatomy and operative surgery Azerbaijan U., 1924-29; head surg. dept. Krasin Hosp., Leningrad, 1931-33; asst. dept. emergency surgery, lectr. 3d Surg. Clinic, 2d Leningrad Med. Inst., 1934-37; head chair gen. surgery Leningrad Stomatological Inst., 1937-42; prof., 1937—; head chair operative surgery and clin. anatomy Leningrad Postgrad. Med. Inst., 1943—. Author: Essays on Purulent Surgery of the Male Pelvis; co-author: Manual of Operative Surgery for Rural District Surgeons. Research and numerous publs. on tissue preservation and transplantation. Address: Leningrad Postgrad. Med. Inst., ulitsa Saltykova-Shchedrina 41, Leningrad S-15, USSR.

NADEL, Eli Maurice, Am. research physician; b. N.Y.C., Oct. 9, 1918; s. Joseph and Rae (Gross) N.; B.S., U. City N.Y., 1937, M.S. in Edn., 1939; M.D., L.I. Coll. Medicine, 1945; fellow pathology Washington U. Sch. Medicine, 1948-50; fellow in biochemistry, Jefferson Med. Coll., 1950-51; m. Ruthe Galler, June 3, 1943; children—Robert Eric, James Oliver, Amy Elizabeth. From asst. surgeon to med. dir. USPHS, 1946-65, ret.med.dir., 1965; mem. staff NIH, 1946-58, asst. to asso. dir., 1959-60; staff Nat. Cancer Inst., 1960-65; asst. dir., 1960-61, chief diagnostic research br., 1961-65; chief research in pathology and lab. medicine VA Central Office, 1965-68, cons. research and edn., 1968—; clin. research prof. Georgetown U. Med. Center, 1965-68; prof. pathology, asso. dean St. Louis U. Sch. Medicine, 1968—; vis. scientist numerous instns. Mem. Am. Assn. Pathology and Bacteriology, A.M.A., Am. Assn. Cancer Research, Endocrine Soc., Internat. Acad. Pathology, Am. Soc. Clin. Pathology, Fedn. Clin. Research, Am. Cancer Soc., A.A.A.S., Assn. Am. Med. Colls., Coll. Am. Pathologists, Am. Humanist Assn., Sigma Xi. Research and publs. on exptl. malaria and leukemia, metastases and circulating tumor cells, correlative enzymology and diagnostic research, biology of guinea pig leukemia, steroids in man and guinea pig, med. edn. Home: 18 Lynnbrook, Frontenac, Mo. 63131. Office: 1402 S. Grand Blvd., St. Louis 63104.*

NADEL, Jay Alan, Am. physician; b. Phila., Jan. 21, 1929; s. Julius and Lillian (Lazarus) N.; A.B.,

Temple U., 1949; M.D., Jefferson Med. Coll., 1953; m. Judy Weisman, Mar. 20, 1960; children—Deborah Beth, William Benjamin. Practice medicine, specializing in internal medicine, chest diseases, San Francisco, 1958—; staff mem. Cardiovascular Research Inst., also faculty medicine Sch. Medicine U. Cal., 1958—; asso. prof., 1966—; vis. scientist U. Lab. Physiology, Oxford, Eng., 1961-62. Mem. A.A.A.S., A.C.P. (asso.), Western Soc. Clin. Research, Am. Physiol. Soc., Alpha Omega Alpha. Studies, publs. chpts. in books on pulmonary physiology, particularly on airway size; pulmonary circulation and disease; lung microcirculation. Home: 2371 Pacific Av., San Francisco 94115.*

NADZHMIDDINOV, Tursun Khodzhaevich, Russian physician; born Tashkent, 1901; grad. Med Faculty, Central Asian U., Tashkent, 1928; D.Med. Sci., 1944. Tchr. secondary schs., Tashkent, 1919-28; asst. hosp. therapy clinic Tashkent Med. Inst., 1930-39, head chair tropical diseases, 1944-52, head chair infectious diseases Med. Faculty, 1952—; lectr., 1936-44; prof., 1944—. Author numerous works on parasitology, tropical and infectious diseases. Address: Tashkent Med. Inst., ulitsa Karla Marksa 103, Tashkent, Uzbekistan SSR, USSR.

NAEGELE, Franz Karl, German obstetrician; b. Dusseldorf, Germany, 1778; M.D., l'École de Médecine de Dusseldorf, 1800; children—Herman, Francois, Joseph, Maximilien. Practiced in Barmen, Germany for several years; became prof., Heidelberg, Germany, 1807. Author: Erfahrungen aus dem Gebiete der Frankheiten des Weiblichen Geschlechts, 1812; Über den Mechanisms der Geburt, 1822; Der Weibliche Becken, 1825, 50; Lehrbuch der Geburtshülfe, 1830; Das schräg verenckte Becken uebst einen Anhange über die wichtigsten Fehler des Weiblichen Beckens, 1839; Methodologie der Geburtshülfe, 1848; also numerous articles. Dir., Heidelberger Klinische Annalen, 1845. Described biparietal obliquity of fetal head with regard to superior pelvic strait, 1819, pelvis contracted in one of oblique diameters with ankylosis of one sacroiliac synchondrosis and imperfect devel. of sacrum, 1839. Died 1851.

NAEGELI, Charles-Guillaume, Swiss botanist; b. Kilchberg, 1817. Author: Recherches de physiologie botanique, 1855-58; Le microscope, 1865-67; Les Hieraciums. Discoverer fern antherozoid. Died Munich, Germany, 1891.

NAEGELI, Otto, Swiss hematologist; b. Ermatingen, Switzerland, July 9, 1871; prof., Tübingen, Germany, also Zurich, Switzerland. Author: Blutkrankheiten und Blutdiagnostik, 1908; Research in hematology; discovered leukopenia in typhus, also myeloblasts; studied blood formation and pernicious anemia; described method of stopping nosebleed by pulling head up with physician's hands under occiput and jaw of patient (Naegeli's maneuver), 1907; made bot. studies. Died Zurich, Mar. 11, 1938.

NAESER, Charles Rudolph, Am. chemist; b. Mineral Point, Wis., Nov. 13, 1910; s. William Adolph and Clara (Smith) N.; B.S., U. Wis., 1931; M.S., U. Ill., 1933, Ph.D., 1935; m. Elma Meyer, Dec. 26, 1936; children—Charles Wilbur, Margaret Ann. From instr. to prof. George Washington U., Washington, 1935—, prof., 1947, chmn. dept. chemistry 1947—; chief chemistry group, geochemistry and petrology br. U.S. Geol. Survey, 1952-55; sci. adviser U.S. Army, Heidelberg, Germany, 1950-51; cons. Office Saline Water, U.S. Dept. Interior. Recipient, award Am. Inst. Chemists, Washington sect., 1961. Fellow A.A.A.S., Washington Acad. Sci. (sci. teaching award 1961); mem. Am. Chem. Soc., Geochem. Soc., Chem. Soc. London, Geol. Soc. Washington, Sigma Xi, Delta Chi, Alpha Chi Sigma, Phi Lambda Upsilon. Research in rare earths geochemistry, inorganic chemistry. Home: 6654 Van Winkle Dr., Falls Church, Va. 22044.*

NAFE, John Elliott, Am. geophysicist; b. Seattle, July 2, 1914; s. Arthur Edward and Muriel (Elliott) N., B.S., U. Mich., 1938; M.S., Wash. U. 1940; Ph.D., Columbia, 1948; m. Sarah Gilpin Underhill, Sept. 6, 1941; children—Mary Malcolm, Katharine Elliott. Instr. physics Columbia, 1946-49; asst. prof. physics U. Minn., 1949-51; dir. research Hudson Labs. of Columbia, 1951-53, research scientist Lamont Geol. Obs., 1953-55, adj. asso. prof. geology at univ., 1953-58, prof. geology, 1958—, chmn. dept., 1962-65; chief scientist oceanographic cruises, 1951—. Fellow Geol. Soc. Am., A.A.A.S. (council), Am. Phys. Soc., Am. Geophys. Union; mem. Seismol. Soc. Am. (editorial com. Bull.), Royal Astron. Soc., Soc. Exploration Geophysicists, Am. Assn. Physics Tchrs., Sigma Xi. Contbr. numerous papers in field. Research in seismic refraction and other geophys. studies at sea; dispersion of long period Rayleigh waves and polar phase shift; measured hyperfine structures of hydrogen, deuterium, and tritium. Home: Ludlow Lane, Palisades, N.Y. 10964. Office: Lamont Geological Observatory, Palisades, N.Y. 10027.*

NAFIS, Ibn al-, see Al-Quarashi.

NAG, Biswa Ranjan, physicist; b. Comilla, East Pakistan, Oct. 1, 1932; s. Satya Ranjan and Sailabala (Mujaumder) N.; B.Sc., Presidency Coll., Calcutta, India, 1951; M.Sc., U. Coll. Tech., Calcutta U., 1954, D.Phil., 1961; M.S., U. Wis., 1960; m. Mridula Roy Choudhury, June 5, 1964. Electronic engr. Indian Statis. Inst., Calcutta, 1954-55; faculty U. Coll. Tech., 1956—, reader, 1964—. Fellow Indian Phys. Soc.; mem. I.E.R.E. (U.K.). Research, publs. on application of analogue computers for study of electronic oscillators, transport characteristics of semiconductors and ionosphere under hot electron conditions; design of multistate devices using oscillators; devel. microwave techniques for study of semiconductors. Home: 27 B Convent Rd., Calcutta, West Bengal, India.*

NAGAI, Nagayoshi, Japanese pharmacologist; b. 1845; retainer Tokushima Clan, Shikoku;; sent by govt. to study in Germany, 1871; D.Sc., D. Pharmacology; m.; apptd. prof. Tokyo U. Sch. Medicine, 1883; mem. Imperial Acad.; a founder Japan Pharmacological Soc. Pioneer in organic chemistry and pharmacology in Japan; isolated ephedrine from a Chinese herb; co-founder Nihon Women's Coll. Died 1929.

NAGAI, Yoshio, Japanese chemist; b. Tokyo, Japan, Nov. 18, 1905; s. Kikuji and Sadako (Nagasaki) N.; grad. U. Tokyo, 1930, D. Tech., 1945; m. Fumiko Kuratsuka, Nov. 2, 1937; children—Kemi, Yuki. Faculty, U. Tokyo, 1936—, prof. chemistry, 1944-66, prof. emeritus, 1966—; prof. synthetic organic chemistry Salitama (Japan) U., 1966—; prof. Tokyo Inst. Sci., 1937—. Mem. council Tokyo Met. City for Atomic Energy Application for Peace, 1956—; vice chmn. Japan Com. on Chemistry of Aromatics, 1951—. Mem. Soc. Organic Synthetic Chemistry Japan (past pres.), Chem. Soc. Japan (past v.p., Sci. medal 1966). Author: Textbook of Organic Chemistry (with Takeshi Yoshino), 1955; also numerous articles. Research on syntheses of polycyclic compounds, preparation of indathrone with world highest yield of 66%, preparations of new violongthrone and violongthrene and their isomers, new synthesis of quinacridone, new azo-dyes for straight polypropylene fiber. Home: 1-4-15 Setagaya, Setagaya-ku, Tokyo, Japan. Office: Saitama U., Shimo-okubo, Saitama-ken, Japan.*

NAGAKUBO, Sekisui, (named Genju, called Gengobe), Japanese geographer; b. Hitachi Province; ed. Mito, Japan; produced 1st complete maps of Japan, over 20 year period, also maps of world; contbr. topography to Dai Nihon-shi (Japanese history compiled by order of Mitsukuni Tokugawa, Lord of Mito). Died 1801.

NAGAOKA, Hantaro, Japanese physicist; b. Omura, Japan, 1865; grad. physics sect. Tokyo (Japan) U., 1887; studied in Germany, 1893-96. Became asst. prof. Tokyo U., 1887, prof., 1896; named 1st pres. Osaka U., 1931; Recipient Cultural medal, 1941. Mem. Imperial Acad., Phys. Research Inst. A founder Japanese physics; research in math., spectroscopy, electromagenetism, atomic mechanisms; studied magnetization of nickel steels, 1890-98, influence of temperature on its magnetization, 1903, vibrations in seismic disturbances, 1906, Zeeman effect in weak magnetic fields, induction coefficients of solenoid, 1911, spectrum of mercury and its isotopes, 1923, math. theory of electric induction, atomic models, atomic spectra. Died 1950.

NAGAYO, Mataro, Japanese physician; b. Tokyo, 1878; s. Sensai Nagayo; grad. med. dept. Tokyo U., 1904; postgrad., Germany; M.D.; became prof. Tokyo U., after 1910, later dean med. dept., apptd., univ. pres., 1934; dir. Infectious Disease Research Inst.; mem. Imperial Acad.; v.p. Far E. Tropical Medicine Assn., 1921. Research on cancer, also tsutsugamushi disease (rickettsial fever). Died 1941.

NÄGELI, Karl Wilhelm von, botanist; b. Kilchberg, nr. Zurich, Switzerland, Mar. 27, 1817; studied under A.P. Candolle at Geneva; grad. with bot. thesis Zurich, 1840. Became privatdozent, later prof. extraordinary U. Zurich; prof. botany U. Freiburg-im-Breisgau, from 1852, U. Munich, 1858-91. Fellow Royal Soc., 1881. Author: series of papers in Zeitschrift für wissenschaftliche Botanik, 1844-46; Die neuern Algensysteme, 1847; Gattungen einzelliger Algen, 1849; Pflanzenphysiologische Untersuchungen, 1855-58; Mechanischphysiologische Theorie der Abstammungslehre, 1884. Early writer on evolution; postulated that evolution may occur in jumps; believed environment was not cause of variation, that idioplasm was material basis of heredity, also that evolutionary changes were not random, but in particular direction, such as increased size; made significant studies on cell growth; and nuclei of cells; demonstrated by chem. analysis the presence of nitrogenous matter in protoplasm; postulated hypothetical micella as found. of structure of starch grains, cell walls; investigated mode of growth in numerous plants; discovered antheridia and spermatozoids of ferns; theorized that new cells, at least in vegetative organs, are formed by division; gave first accurate account of apical cell; wrote on vascular plants and introduced theory of meristem; corresponded with Mendel, but appears not to have realized importance of his work for heredity. Died Munich, Germany, May 10, 1891.

NÄGELIN, see Carion, Johannes.

NAGELL, Trygve, mathematician; b. Oslo, Norway, July 13, 1895; s. Otto and Marie (Olsen) N.; M.S., U. Oslo, 1920, dr.philos., 1926; m. Bianca Guiduzzi, July 10, 1930 (div. 1943); children—Björn Henrik, Sigrid Margareta. Faculty, U. Oslo, 1921-31; prof. math. U. Uppsala (Sweden), 1931-62. Decorated knight comdr. Nordstjärneorden (Sweden), Sankt Olavorden (Norway); Kong Haakon VII's Frihetskors. Mem. acads. sci. in Oslo, Stockholm, Uppsala, Trondheim. Author: L'analyse indéterminée de degré supérieur, 1929; Lärobok i algebra, 1949; Elementär talteori, 1950; Introduction to Number Theory, 1951; also numerous articles. Editorial bd. Norsk matematisk tidsskrift, Archiv der Mathematik, Acta mathematica, Acta Arithmetica. Research on theory of algebraic numbers, including rational points on curves of genus one, arithmetical properties of binary cubic and biquadratic forms, algebraic numbers of given discriminant, properties of units, diophantine equations. Home: 60 Dragarbrunnsgatan, Uppsala, Sweden.*

NAGELSCHMIDT, Karl Franz, physician; b. Germany, 1875; credited with introducing method of diathermy treatment, also term diathermy, 1909; believed to have been 1st to introduce high frequency electric currents and radiation in coagulation and diathermy, 1909. Died 1952.

NAGIEV, Murtuza Fatullaogly, Russian chem. technologist; b. Sarab, Azerbaijan, 1908; grad. Azerbaijan Indsl. Inst., 1935; D.Chem Sci., 1944. With Azerbaijan Petroleum Indsl. Research Inst., 1938-40; instr. Azerbaijan Indsl. Inst., 1940-45, prof., 1945—; with Petroleum Inst., Azerbaijan Acad. Scis., 1945-59, dir., 1946-59, acad. sec. dept. physicotech. sci., 1956, chief learned sec., 1962—. Mem. Azerbaijan Acad. Scis. (v.p. 1957-62). Author: Modern Engine Fuels, 1954; The Theory of Recycle Processes in Chemical Engineering, 1965. Research and numerous publs. on petroleum chemistry and technology; developer theory of recirculation processes for determining most profitable ways of comprehensive refining of chem. raw materials. Address: Azerbaijan Acad. Scis., Baku, Azerbaijan SSR, USSR.*

NAGLE, Darragh Edmund, Am. physicist; b. N.Y.C., Feb. 25, 1919; s. Percival Edmund Darragh and Mabel Mary (Russel) N.; B.S., (scholar 1937, 38) Cal. Inst. Tech., 1940; A.M., Columbia, 1942; Ph.D., Mass. Inst. Tech., 1947; m. Avery Leeming, Jan. 29, 1950; children—Carol Joy, Darragh Joseph, Patricia Gay. Lectr., Columbia, 1940-43; instr. Mass. Inst. Tech., Cambridge, 1945-49; asst. prof. physics U. Chgo., 1949-56; staff mem. Los Alamos Sci. Lab., 1956—, asso. div. leader intermediate energy physics, tech. dir. meson factory project, 1963—; prof. physics U. N.M.; tech. rep. 2d Internat. Atoms for Peace Conf., Geneva, 1958. Fulbright fellow, 1949; Guggenheim fellow, 1952. Fellow Am. Phys. Soc. Research and publs. on hyperfine structure of hydrogen. Home: 698 46th St. Office: P.O. Box 1663, Group P-11, Los Alamos 87544.*

NAGLER, Benedict, neuropsychiatrist; b. Czermowitz, Austria, Mar. 14, 1900; s. Samuel Oswald and Charlotte Josephine (Schorr) N.; M.D., U. Hamburg, 1925; m. Hilde Laub, Oct. 20, 1927; children—Ralph Lewis, Eva (wife of Dr. Barron M. Hirsch). Came to U.S. 1935, naturalized, 1941. Intern, resident neurology, psychiatry, internal medicine, Hamburg, Berlin, 1924-31; practice medicine, specializing in psychiatry and neurology, Berlin, 1931-33, Tunis, N.Africa, 1943, Newark, 1935-43; cihef neurology, psychiatry service VA Hosp., Richmond, Va., 1946-53; chief neurology div. Psychiatry and Neurology Service VA, Washington, 1953-57; supt. Lynchburg Tng. Sch. and Hosp., Colony, Va., 1957—; asst. prof. psychiatry and neurology Med. Coll. Va., 1946—; asso. prof. clin. neurology Georgetown U., 1953-57, professorial lectr., 1957—; lectr. dept. neurology and psychiatry U. Va., 1957—; cons. Nat. Inst. Neurol. Diseases and Blindness, VA, NRC. Bd. dirs. Partridge Schs. and Rehab. Center, Devel. Council Sweet Briar Coll., Florence Crittenton Home; trustee Woodrow Wilson Rehab. Found. Diplomate Am. Bd. Psychiatry and Neurology (asso. examiner). Fellow Am. Psychiat. Assn., Am. Acad. Neurology (past councillor, past chmn. com. on problems mental retardation), Am. Assn. Mental Deficiency (councillor, chmn. com. internat. activities); mem. A.M.A., So. Psychiat. Assn., So. Electroencepholographic Soc. (past pres.), N.Y. Acad. Scis., Fedn. Am. Scientists, Am. Epilepsy Soc. (past councillor), Assn. For Research Nervous and Mental Diseases, Am. Assn. Med. Supts. Pub. Mental Hosps., Am. Acad. Mental Retardation (pres. elect), Am. Med. EEG Soc., Internat. Assn. Sci. Study Mental Deficiency (councillor, 1964—). Translator: (from German to English) Cerebral Function in Infancy and Childhood, 1963. Cons. editor: Am. Jour. Mental Deficiency, 1961—; mem. editorial bd. Staff Am. Psychiat. Assn. Publ., 1964—. Psychiat. studies in art of the mentally ill and emotional impact of organic illnesses; co-op. treatment trials in multiple sclerosis and cerebral arteriosckrosis; epidemiology studies in multiple sclerosis, cerebral malfunction and brain damage in infants and children. Home: P.O. Box 61. Office: Lynchburg Tng. Sch. and Hosp., Colony, Va. 24537.*

NAGUIB, Mohammed Ibrahim, Egyptian biologist; b. Cairo, Egypt, Jan. 24, 1926; s. Ibrahim and Zeinab (Said) N.; B.Sc., Cairo U., 1946, M.Sc., 1949, Ph. D.,

1952, D.Sc., 1968; m. Nawal Abulnaga, July 18, 1957; children—Manal, Ihab. Faculty, Faculty Sci., Cairo U., 1946—, asst. prof., 1962—. Establisher, dir. botany dept. U. Libya, 1957-61; exec. dir. dept. sci. relation Egypt Sci. Council, 1962-63; staff Middle Eastern Regional Radioisotope Centre, Cairo, 1963-64. Recipient Egyptian State Prize for Sci. Research in biology, also Republic Star 3d class, 1965. Mem. Bot. Soc. Egypt, Atomic Energy Soc. Egypt. Sec. editorial bd. Jour. Botany U.A.R. Research, publs. on effects of some organic compounds on metabolic pathways of higher plants and fungi particularly carbohydrate, nitrogen and phosphorus metabolism with reference to role of various nutritive conditions on respiration in relation to sugar absorption and keto acid prodn. particularly in fungi. Home: 58 Abbasiah St., Cairo, Egypt.*

NAGY, Esther Maria Kovacs, Hungarian paleobotanist, palynologist; b. Déva, Hungary, July 5, 1914; d. Ernst and Esther (Kolumban) Kovacs; Dr.Phil., U. Szeged (Hungary), 1940; Kandidate Biol. Scis., Hungarian Acad. Scis., 1957, D.Biol. Scis., 1965; m. Laszló Nagy, Sept. 16, 1943; children—Laszló, Gabor. Asst., Pedagogical High Sch., 1937-38; tchr. Tchr. Tng. Colls., Papa, Hungary, Debrecen, Budapest, 1939-51; scientist Hungarian Geol. Inst., Budapest, 1951—, head paleontol. dept., 1958—; lectr. U. Budapest, 1963—. Named Excellent worker geol. research, 1960, 65. Mem. Hungarian Biol. Soc., Hungarian Geol. Soc., Internat. Orgn. Paleobotany, Hungarian Acad. Scis. (mem. bot. com. 1957—, sec. com. 1957-62, editorial commn. Acta Botanica 1958—, sec. geol. com. 1965—). Author: Papa, 1940; Palynologische Untersuchung der am Fusse des Matra-Gebirges gelagerten oberpannonischen Braunkohle, 1958; also articles. Reconstructed vegetation of Quartenary and Neogen using spores, pollens and remains of microplankton organisms found in Hungarian Quarternary and Tertiary layers; discovered a swamp or bog forest of Pliocene in Hungary was similar to present Taxodieceae forest nr. estuary of Mississippi; determined stratigraphical identification of paleo-vegetation. Home: 104 Ménesi, Budapest XI. Office: 14 Népstadion, Budapest, XIV, Hungary.*

NAHAS, Gabriel Georges, physiologist; b. Alexandria, Egypt, Mar. 4, 1920; s. Bishara and Gabrielle (Wolf) N.; B.A., U. Toulouse (France), 1937, M.D., 1944; M.S., U. Rochester, 1949; Ph.D., U. Minn., 1953; m. Marilyn Cashman, Feb. 13, 1954; children—Michele, Anthony, Christiane. Came to U. S., 1946, naturalized, 1961. Rockefeller Found. fellow U. Rochester, 1947-48; Mayo Found. fellow Mayo Clinic, 1949-50; research fellow U. Minn., 1950-53; chief lab. exptl. surgery Marie Lannelongue Hosp., Paris, 1953-55; asst. prof. physiology U. Minn., 1955-57; chief respiratory sect. Walter Reed Army Inst. Research, 1957-59; lectr. George Washington U., 1957-59; asso. prof. anesthesiology Columbia Coll. Phys. and Surg., 1959-62, prof., 1962—. Mem. com. on trauma NRC, 1964—; med. adv. bd. Council on Circulation, Am. Heart Assn., 1961—; Recipient Presdl. medal of Freedom with gold palm, 1945; Legion of Honor (France); Order Brit. Empire; Order of Orange Nassau (The Netherlands). Mem. Am. Physiol. Soc., Am. Soc. for Pharmacology and Exptl. Therapeutics, Soc. for Artificial Internal Organs, Sigma Xi. Editor: In Vivo and In Vitro Effects of Amine Buffers, 1961; Regulation of Respiration, 1963; (with D.V. Bates) Respiratory Failure, 1964; Current Concepts in Acid-Base Measurements, 1966. Research, numerous publs. on measurement of oxyhemoglobin concentration, control of acidity of body fluids with organic buffer, effect of disturbances of acid-base equilibrium on secretion and metabolic activity of catecholamines. Home: 114 Chestnut St., Englewood, N.J. 07631. Office: 630 W. 168th St., N.Y.C. 10032.*

NAHUM, Louis Herman, physiologist; b. Vilna, Lithuania, Aug. 16, 1892; s. Samuel and Tyba (Koscher) N.; Ph.B., Sheffield Sci. Sch. Yale, 1912, M.D., 1916; m. Stella Koskoff, Jan. 10, 1918; children—Harriet, Jeremy. Staff, Rockefeller Inst., 1917; faculty Yale Med. Sch., 1919—, prof. physiology, 1946—. Recipient Warren award Harvard, 1940. Fellow, Sci. Council, Am. Heart Assn., 1963; Editor, Conn. Medicine, 1958—. Mem. Am. Physiol. Soc. Research and numerous publs. on nature and exptl. basis of electrocardiogram. Home: 85 Loomis Pl, New Haven 06511.*

NAIMARK, Mark Aronovich, Russian mathematician; b. Odessa, USSR, Dec. 5, 1909; s. Aron Jakovlevich and Esther (Faivush) N.; grad. Odessa U., 1933, postgrad., 1933-36, candid. Phys. Math. Sci., 1936; Doctorant, Steklov Math. Inst., 1938-41, D. Phys. Math. Sci., 1941, Prof., 1942; m. Larisa P. Scheerbakova, June 1, 1932; children—Boris Markovich, Leonid Markovich. Faculty, Odessa U., 1933-38, reader, 1936-38; research fellow Seismol. Inst., also other insts. Acad. Sci. USSR, 1941-62, prof., 1941-62; prof. Steklov Math. Inst., Acad. Scis. USSR, 1962—; reader German Pedagog. Inst., Odessa, 1935-36, chmn. maths., 1936-38; reader Moscow Pedagog. Inst., 1939-41; prof. math. Moscow Stee Inst., and other higher sch. insts., 1942-54; prof math. Moscow Physico-Tech. Inst., 1954—. Mem. Moscow Math. Soc. Author: (with I.M. Gelfand) Unitary Representations of Class Groups, 1950; Lin. Differential Operatiors, 1954; Normed Rings, 1956; Linear Representations of the Lorentz Group, 1958; also numerous articles. Research on generalized extensions and generalized spectral functions of symmetric operators in Hilbert space, spectral representations of singular non-self-adjoint differential operations, algebras with involutions and their representations in Hilbert space, infinite dimensional representations of complex semi-simple Lie groups, Home: Moscow W-333, USSR, Ulitsa Gubfina 4, KU 95.*

NAIR, N. Chandrasekharan, Indian botanist; b. Perunna, Kerala, India, Aug. 19, 1927; s. A. Narayana Pillai and Ammukutty (Amma) N.; B.Sc. St.Berchmans College, Changanacherry, 1949; M.Sc., Birla Coll., Pilani, 1951, Ph.D., 1956; m. T. Suseela Bai, May 28, 1954; children—Harikumar, Jaya Shri, Srikumar. Research asst. Birla Coll., Pilani, 1951-52, lectr. botany, 1952-53, asst. prof. botany, 1953-61; systematic botanist Botan. Survey of India, Dehra Dun, 1961-66, regional botanist, Calcutta, India, 1966—. Mem. Indian Botan. Soc., Internat. Soc. Plant Morphologists, Rajasthan Acad. Scis. Contbr. numerous articles in field to sci. jours. Research, publs. field floral anatomy, angiosperm embryology and floristic studies; described new species of flowering plants. Home: Elanjimuttam 1, Perunna, Changana Cherry, Kerala, India. Office: 76 Lower Circular Rd., Calcutta 14, India.*

NAIR, Pandarathil Madhavan, Indian chemist; b. Kondayoor, Kerala, India, Aug. 15, 1925; s. K. P. Unnikannan and Pandarathil Karthiayani N.; B.Sc., Madras Christian Coll. U. Madras, 1945, M.Sc., 1949; student Indian Inst. Sci., 1950-52; Ph.D., U. Ark., 1956; m. Tekkekalathil Kamala Feb. 10, 1951; children—Chandrasekhar, Unnikrishnan. Demonstrator, resident tutor Madras Christian Coll., Tambaram, India, 1945-47; asst. prof. Pachaivappa's Coll., Madras, India, 1949-50; A. A. Noyes fellow Cal. Inst. Tech., 1956-57; research fellow U. Cal., Berkeley, 1957-58; jr. scientific officer Nat. Chem. Lab., Poona, Maharashtra, India, 1959-60, sr. scientific officer, 1960-64, scientist, 1964—. Mem. Indian, Am. chem. socs. Contbr. articles in field to sci. jours. Research field physical, structural organic chemistry; studies on molecular properties, mechanisms of some organic reactions, solvent effects on reaction rates and organic structure determination with the aid of physical methods, particularly nuclear magnetic resonance spectroscopy. Home: D-II/29 NCL Colony, Poona 8, Maharashtra, India. Office: Nat. Chem. Lab., Poona 8, Maharashtra, India.*

NAIRNE, Edward, English physicist; b. Sandwich, Eng., 1726; instrument maker, London; became a propr. Royal Instn., 1800. Fellow Royal Soc., 1776. Author: Electrical Experiments Made to Show Advantages of Pointed Conductors, 1783; Description and Use of the Electrical Machine, 1787; also numerous articles. Invented 1st electric machine capable of receiving 2 electricities at same time (Nairne's elec. machine), 1782; invented process of artificial desiccation using sulphuric acid under receiver of air pump; improved astron. apparatus at Greenwich, Eng. Died London, Sept. 1, 1806.

NAJARIAN, Haig Hagop, Am. biologist; b. Nashua, N.H., Jan. 5, 1925; s. Hagop M. and Antaram (Shamlyian) N.; B.S., U. Mass., 1948; M.A., Boston U., 1949; Ph.D., U. Mich., 1953; m. Mary A. Been, May 26, 1957; children—Andrea Cheryl, John Varant, Steven Douglas. Instr., Wayne State U., Detroit, 1956-57; asst. prof. biology Northeastern U., Boston, 1953-55; asso. research parasitologist Parke, Davis & Co., Detroit, 1955-57; scientist, WHO, Baghdad, Iraq, 1958-59; asst. prof. microbiology U. Tex. Med. Br., Galveston, 1960-66; asso. prof. biology U. Me. in Portland, 1966-68, prof. biology, chmn. div. sci. and math., 1968—. USPHS fellow U. Tex., 1959-60. Mem. A.A.A.S., Am. Soc. Parasitologists, Royal, Am. socs. tropical medicine and hygiene, Am. Micros. Soc., Sigma Xi. Author: Textbook of Medical Parasitology, 1967; also numerous articles. Research on parasitic organisms. Home: 173 Pleasant Av., Portland, Me. 04103.*

NAJJAR, Victor Assad, microbiologist; b. Zalka, Lebanon, Apr. 15, 1914; s. Assad M. and Hala (Ashkar) N.; M.D., Am. U., Beirut, 1935; m. Mathilde Montgomery, July 8, 1948; children—Jennifer Lee, Julie Abigail, Victor Montgomery. Came to U.S., 1938, naturalized, 1945. Mem. NRC microbiology unit U. Cambridge, 1948-49; pediatrician Johns Hopkins, Balt., 1949-57; prof., head, dept. microbiology Vanderbilt U., Nashville, 1957—. Ervine McQuarrie lectr. U. Minn., 1957; mem. sci. bd. Babies' Hosp. Research Center, Wilmington, N.C., 1960-65, St. Jude Hosp., Memphis, 1962—. Recipient Mead Johnson award, 1951, Distinguished Service medal Republic of Lebanon, 1956. NRC fellow Washington U., 1946-48. Mem. Soc. Pediatric Research, Soc. Clin. Investigation, Am. Soc. Biol. Chemists, Am. Pediatric Soc., Sigma Xi. Author, editor: Carbohydrate Metabolism, 1953, Fat Metabolism, 1954; Immunity & Virus Infection, 1959, Researcher in enzymechemistry and immunochemistry. Home: 517 32d Av. S., Nashville, 37212.*

NAKADA, Minoru P., Am. physicist; b. Los Angeles, Jan. 15, 1921; s. Ginzo and Kagi (Ikehara) N.; student U. Cal. at Berkeley, 1939-41, U. Cal. at Los Angeles, 1941-42, U. Utah, 1943; B.A., U. Cal. at Berkeley, 1947, Ph.D., 1952; m. Rose Kazuko Enomoto, Mar. 21, 1953; children—Patricia, Scott, Mari Rebecca. Physicist, Lawrence Radiation Lab., U. Cal. at Livermore, 1952-61, Jet Propulsion Lab., Pasadena, Cal., 1961-62; physicist Goddard Space Flight Center, NASA, Greenbelt, Md., 1962—. Mem. Am. Phys. Soc. Am. Geophys. Union. Research, and publs. on V-particles by study decay products and measuring lifetimes; neutron interactions with nuclei by measuring scattering 14 Mev neutrons; calculations synchrotron radiation from artificial radiation belt; diffusion theory formation outer radiation belt. Home: 12404 Lima Dr., Silver Spring, Md. 20904. Office: Code 641, NASA, Goddard Space Flight Center, Greenbelt, Md. 20771.*

NAKAGAKI, Masayuki, Japanese chemist; b. Japan, Apr. 19, 1923; s. Sengoro and Shizue (Yamakawa) N.; B.Sc., Imperial U. Tokyo (Japan), 1945; D.Sc., U. Tokyo, 1950; m. Hisako Yoshitake, May 3, 1952. Lectr., U. Tokyo, 1951-54; prof. Osaka (Japan) City U., 1954-60; vis. prof. Wayne State U., Detroit, 1955-57; prof. Kyoto (Japan) U., 1960—. Mem. Chem. Soc. Japan, Pharm. Soc. Japan. Author: Theories of Liquid and Solid for Chemists, 1955; Colloid and its Applications, 1955; Physical Chemistry of Pharmaceutics, 1965; also numerous articles. Proposed theory on stability of foam, theories of light scattering; discovered that Catalin, a drug, can retard aging of eye lens capsule. Home: 43 Fukakusa-Jutaku, Nishidate, Fushimi, Kyoto, Japan.*

NAKAGAWA, Senkichi, Japanese mathematician; b. Kanazawa, Japan, 1876; grad. math. sect. Tokyo U., 1898; postgrad. in geometry, Germany, 1891; D.Sc.; apptd. asst. prof. Tokyo U., after 1891, prof., 1914. Specialist in non-Euclidean geometry. Died 1942.

NAKAGAWA, Shigeo, Japanese physicist; b. Tokyo, Japan, Apr. 24, 1906; grad. Tokyo U., 1930; D.Sc.; m. Kyoko Sasaki; with central research inst. then phys. and chem. research inst. S. Manchurian Ry.; prof. Rikkyo U. Recipient Asahi subsidy for sci. research for Study of Cosmic Rays, 1949—. Author: Physical Experiments on Atomic Nucleus; Radioactivity, Vol. 10, Science of Physical Experiments. Research on measurement of energy of gamma rays; specialist in cosmic rays and atomic nucleus.

NAKAGAWA, Yoshinari, Am. physicist; b. Tacoma, Wash., Dec. 13, 1923; s. Shigetaro and Kuni (Toya) N.; B.S., Japanese Naval Acad., 1941; B.S., U. Tokyo, 1949, Ph.D. in Physics, 1957; m. Namiko Kishimoto, Nov. 18, 1955. Research fellow Geophys. Inst., U. Tokyo, 1949-51; research asst. hydrodynamics lab., U. Chgo., 1952-55; research asso. Enrico Fermi Inst. Nuclear Studies, 1955-60, Stevens Inst. Tech., 1960-61; chargé de recherche Inst. d'Astrophysique, U. Paris, 1961-62; vis. fellow Joint Inst. Lab. Astrophysic, U. Colo., 1962-63; program scientist Lab. Atmospheric Scis., Nat. Center Atmospheric Research, Boulder, Colo., 1963-65, sr. scientist High Altitude Obs., 1965—. Mem. Am. Phys. Soc., Am. Geophys. Union, Am. Meteorol. Soc., A.A.A.S. Research and publs. on geophys. phenomena by hydrodynamical models, hydromagnetic research, geophys. and astrophys. fluid, magneto fluid dynamics; pioneered hydromagnetic expts. Home: 1755 Table Mesa Dr. Office: Nat. Center for Atmospheric Research, Boulder, Colo. 80302.*

NAKAI, Takenoshin, Japanese botanist; b. Yamaguchi Prefecture, Japan, 1882; grad. sci. dept. Tokyo U.; doctorate, 1930; instr., asst. prof., Tokyo U., named prof., 1927, also dir. bot. gardens; dir. Bogor Bot. Gardens, Java, during World War II; then dir. Nat. Mus. Sci., Tokyo. Recipient Katsura Meml. prize (for research on Korean forests) Imperial Acad. Research, numerous publs. on flora of E. Asia, plant classification. Died Dec. 6, 1952.

NAKAIZUMI, Masonori, Japanese physician; b. Nagoya, Japan, 1895; s. Yukinori Nakaizumi; grad. Tokyo U., 1919; doctorate Sci. Research Inst., 1926; postgrad. study of ultraviolet ray treatment, Europe, U.S. Became prof. Tokyo U., 1932, dean Faculty Medicine, 1952-54. Became mem. council on radioactivity Ministry Welfare, 1954. Recipient Hattori prize for research on Roentgen ray treatment, 1938. Author: The Science of Ultraviolet Ray Treatment.

NAKAMURA, Kiyo-o, Japanese meteorologist; b. Yamaguchi Prefecture, Japan, 1855; grad. physics sect. Tokyo U., 1879; D.Sc.; with geol. bur. Home Ministry; named dir. Central Meteorol. Obs., 1895; dir., co-founder Tokyo Sch. Physics; mem. Imperial Acad. Author: Climate of Japan. Helped establish Japanese meteorol. survey system. Died 1931.

NAKAMURA, Masahiro, Japanese bacteriologist; b. Kurume, Japan, Dec. 8, 1919; s. Heizaburo and Yaeno (Mitsuyasu) N.; grad. Kurume Med. Coll., 1941, D.M.S. in Bacteriology, Kyushu U., 1949; m. Toshiko, Oct. 20, 1945; children—Hiroko, Jun, Akiko. Faculty, Kurume U. Sch. Medicine, Fukuoka, Japan, 1948—, prof. bacteriology, 1960—. Rockefeller Found. fellow in virology U. Mich., 1958-59. Mem. Japan Leprosy Soc. (Sakurane prize 1953), Japan Bacteriological Soc., Soc. Japanese Virologist. Author (with others) Toda, 1967. Research, publs. on extraction of infectious viruses of Japanese encephalitis and dengue fever and related problems, relationship between mycoplasma and leukemia, animal expts. on transmission of human leprosy. Home: 1432 Noda, Kurume, Fukuoka, Japan.*

NAKAMURA, Seiji, Japanese physicist; b. Fukui Prefecture, Japan, 1869; grad. sci. dept. Tokyo U., 1892; postgrad. in crystallography, Germany, 1903; D.Sc.; prof. First Higher Sch.; prof. Tokyo U., later dean sci. dept., ret., 1930; mem. Japan Acad. Author many books. Specialist in radiation and electricity. Died July 18, 1960.

NAKAMURA, Yoji, Japanese physicist; b. Aichi-Prefecture, Japan, Aug. 10, 1924; s. Kenkichi and Haru (Kawai) N.; B.Sc. in Physics, Tohoku U., 1946; D.Sc. in Chemistry, Kyoto U., 1961; m. Matsuko Kameda, May 5, 1955; children—Rika, So. Faculty, Kyoto (Japan) U., 1956-—, prof. dept. metal sci. and tech., 1963-—. Cons., IBM Research Center, N.Y.C., 1964-65, dept. head, 1966. Mem. Phys. Soc. Japan, Am. Phys. Soc., Japan Inst. Metals. Research and publs. in magnetism of metal and alloys, origin of invar-type alloys, 1st observation of relaxation Mössbauer spectra in very fine magnetic particles. Home: 32, Iseyado-cho, Yamashina-Otowa, Kyoto, Japan.*

NALBANDOV, Andrew Vladimir, endocrinologist; b. Simferopol, Russia, July 17, 1912; s. Vladimir Sergei and Alexandra (Rosher) N.; came to U.S., 1934, naturalized, 1941; Dipl. Ing., U. Munich (Germany), 1932; M.S., Okla. State U., 1935; Ph.D., U. Wis., 1940; m. Olga Oliver, Nov. 27, 1935. Research asst. U. Wis., 1937-40; mem. faculty U. Ill., 1940-—, prof. physiology and zoology, 1951-—, chmn. div. genetics, 1958-—. Cons. NIH, NSF, Nat. Acad. Sci. Recipient Borden award Poultry Sci. Assn., 1959. Fellow A.A.A.S.; mem. Endocrine Soc., Soc. Reprodn. and Fertility, Am. Soc. Anatomists, Soc. Animal Prodn., Am. Soc. Zoologist, Soc. Exptl. Biology and Medicine, Sigma Xi. Author: Reproductive Physiology, 2d edit., 1964. Editor: Advances in Neuroendocrinology, 1963. Edittorial bd. Endocrinology, Jour. Reprodn. and Fertility. Studies in the comparative aspects of the mechanisms controlling sexual reprodn., especially the hypothalmo-hypophys. ovarian or testicular axis. Home: 1113 W. William St., Champaign, Ill. Office: Animal Genetics Labs., Urbana, Ill.*

NALIVKIN, Dmitriy Vasilevich, Russian geologist, paleontologist; b. Aug. 25, 1889; grad. Petrograd Mining Inst., 1915; D.Geol. Sci Asso., Geol. Com. (now All-Union Geol. Research Inst.), 1917-49; prof. chair hist. geology Leningrad Mining Inst., 1920-—, head, 1930-—; dir. lab. limnology USSR Acad. Scis., 1946-53. Decorated Order of Lenin (3); recipient Stalin prize, 1946, Lenin prize, 1957. Mem. USSR (chmn. Presidium Turkmenia br. 1946-51, Karpinsky gold medal), Turkmenia (hon.) acads. scis. Author: Outline Geology of Turkestan, 1926; Devonian Deposits of the USSR, 1947; Theory of Facies, parts 1-2, 1955-56. Chief editor geol. survey maps of USSR; editor geol. map of world. Research and publs. on hist. geology, stratigraphy and paleontology, facies, stratigraphy and paleography of Paleozoic in Urals, Central Asia, Russian Platform. Address: Leningrad Mining Inst., 21-ya Liniya 2, Leningrad, USSR.

NAM, Charles Benjamin, Am. sociologist; b. Lynbrook, N.Y., Mar. 25, 1926; s. Samuel and Yetta (Huff) N.; B.A., N.Y. U., 1950; M.A., U. N.C., 1957, Ph.D., 1959; m. Marjorie Lee Tallant, Jan. 1, 1956; children—David Wallace, Rebecca Jane. Statistician U.S. Bur. Census, Washington, 1950-53, chief, edn. and social stratification br., 1957-64; statistician USAF, Montgomery, Ala., 1953-54; research asst. U. N.C., Chapel Hill, 1954-57; prof. sociology Fla. State U., Tallahassee, 1964-—, dir. Population and Manpower Research Center, 1967-—. Mem. Am. Sociol. Assn., Population Assn. Am. (dir. 1966-—), Internat. Union for Sci. Study Population, Am. Statis. Assn., So. Sociol. Soc. Author: (with John K. Folger) Education of the American Population, 1967. Editor, contbr.: Population and Society, 1967. asso. editor Demography, 1965-—. Prin. contbr. to devel. demographic statistics on ednl. status of population; contbns. to understanding of links between population change and changes in socs. Home: 2309 Armistead Rd., Tallahassee, Fla. 32303.*

NAMBA, Susumu, Japanese applied physicist; b. Okayama, Japan, Feb. 7, 1928; s. Taiji and Teruno (Namba) N.; grad. Osaka U., 1950; Dr.Engring., Tokyo U., 1959; Dr.Sc., Kyoto (Japan) U., 1962; m. Sumiko Ihara, Apr. 15, 1950; children—Kimiko, Hiroki. Chief semiconductor lab. Inst. Phys. and Chem. Research, Tokyo, 1950-—. Vis. lectr. Tokyo (Japan) U., 1960-—; prof. Osaka (Japan) U., 1967-—. Mem. Japan Soc. Applied Physics, Inst. Elec. Communication Engrs. Japan. Research and publs. on interferometer type gas analyzer, electro-optic effect in crystals, thin film tech., electron, ion and laser beams, photo-sensitization in solids. Home: 4-8-5, Naka-Ochiai, Shinjuku-ku, Tokyo. Office: 2028-8 Honkomagome, Bunkyo-ku, Tokyo, Japan.*

NAMIAS, Jerome, Am. meteorologist; b. Bridgeport, Conn., Mar. 19, 1910; s. Joseph and Sadie (Jacobs) N.; student Mass. Inst. Tech., 1932-33, U. Mich., 1934-35; M.S., Mass. Inst. Tech., 1941; m. Edith Paipert, Sept. 15, 1938; 1 dau., Judith Ellen. Asso. dir. Nat. Meteorol. Center; U.S. Weather Bur., Washington, 1941-—, chief, extended forecast div., 1941-—. Asso. in meteorology Woods Hole (Mass.) Oceanographic Instn., 1954-—; internat. lectr., cons. Recipient citation Sec. Navy, 1943, Meritorious Service award U.S. Dept. Commerce, 1950, Rockefeller Pub. Service award, 1955, Gold medal Dept. Commerce, 1965. Fellow Am. Meteorol. Soc. (Meisinger award 1938, Extraordinary Sci. Accomplishment award 1955), Washington, N.Y. acads. sci., Am. Geophys. Union; mem. Royal Meteorol. Soc. Gt. Britain, Mexican Geophys. Union, Sigma Xi. Author: An Introduction to the study of Air Mass and Isentropic Analysis, 1940. Research, publs., monographs on devel. techniques for analysis of hemispheric weather maps at surface and aloft; discovery of geog. and seasonal variations of jet stream and of its effects on weather; devel. techniques for extending range of weather forecasts to periods from a week to a month; discovery of how ocean interacts with atmosphere to produce climatic fluctuations of the order of a month to a few years. Home: 800 4th St. S.W., Washington 20024. Office: care U.S. Weather Bur., ESSA, Extended Forecast Div., Silver Spring, Md. 20910.*

NAMIKAWA, Shigesuke, Japanese agrl. botanist; b. Kyoto Prefecture, Japan, 1897; grad. Tokyo U.; apptd. tech. expert, agrl. exptl. sta., Nagaoka, Niigata Prefecture, 1925; named dir. Chugoku Dist. Wheat Exptl. Sta., Himeji, 1932. Evolved Norin Number 1 rice strain for regions too cold for other rices. Died Kyoto Prefecture, 1937.

NANCELIUS, Nicolas, French physician; b. 1539; from Noyon, France; physician to Leonore Bourbon, abbess of Fontevrault. Author: Analogia microcosmi ad macrocosmum; Life of Peter Ramus, 1600; other books. Died 1610.

NANDA, Krishna Kumar, Indian botanist; b. Pind Dadan Kahan, (now Pakistan), Jan. 1, 1918; s. Ram Nath and Mohan (Devi) N.; B.Sc., Govt. Coll., Lahore, India, 1939; M.Sc., B.H. U., Varansi, India, 1947; Ph.D., Bombay (India) U., 1947; m. Santosh Kumari, Nov. 21, 1941; children—Poornima (Mrs. Ravi K. Gupta), Mridula, Ashok Kumar, Vinod Kumar. Hon. researcher Indian Agr. Research Inst., New Delhi, 1941-45; civilian gazetted officer Q.M.G.'s Br., New Delhi, 1945-46; hort. asst. Dept. Agr., Govt. India, New Delhi, 1946-47; asso. prof. crop husbandry B.R. Coll., Agra, India, 1947; lectr. botany Delhi U., 1947-60; sr. research officer F.R. Inst., Dehra Dun, India, 1960-63; plant physiologist Indian Agr. Research Inst., 1963; prof. plant physiology Panjab U., Chandigarh, India, 1963-—. Mem. panel experts for seawater agrl. research projects Central Salt and Marine Research Inst., Bhavnagar, India, 1965-—, I.C.A.R. panel for plant physiology and biochemistry, 1966-—. Fulbright-Smith-Mundt scholar, jr. research botanist U. Cal. at Los Angeles, 1955-56; Tata Trust Fund scholar, 1947-49. Mem. Indian Soc. for Plant Physiology (editor-in-chief), Indian Sci. Congress. Research, publs. on developmental process controlling rate and magnitude of growth, uptake of mineral nutrients and distbn. of metabolites, branching pattern of plants, concept of photothermic quantum in plants, existence of endogenous rhythm in flowering of Biloxi soybean, possible role of ascorbic acid in growth and reprodn., physiology of gibberellin action, growth and devel. of Imatiene balsamina. Home: F/9, Sector 14, Punjab U., Chandigahr-14, Panjab, India.*

NANNEY, David Ledbetter, Am. biologist; b. Abingdon, Va., Oct. 10, 1925; s. Thomas Grady and Pearl (Ledbetter) N.; A.B., Okla. Baptist U., 1946; Ph.D., Ind. U., 1951; m. Jean Ruth Kelly, June 15, 1951; children—Douglas Paul, Ruth Elizabeth. Faculty, U. Mich., 1951-58; NSF Postdoctoral fellow Cal. Inst. Tech., 1958-59; prof. zoology U. Ill., Urbana, 1959-—. Fellow A.A.A.S.; mem. Genetics Soc. Am., Soc. Am. Zoologists, Soc. Protozoologists, Am. Inst. Biol. Scis. (past bd. govs.), Soc. Naturalists, Soc. Developmental Biology. Author: (with Herbert Stern) The Biology of Cells, 1965. Studies, publs. on ciliated protozoan Tetrahymena pyriformis with spl. reference to mating types, antigenic types and cortical variations. Home: 703 W. Indiana St., Urbana, Ill. 61801.*

NANNFELDT, John Axel, Swedish botanist; b. Tralleborg, Sweden, Jan. 18, 1904; s. Johan Frithiof and Antonia (Andersson) N.; Fil.dr., U. Uppsala (Sweden), 1932; Dr.is.sci. (hon.), U. Caen, 1957; m. Ragnhild Birgitta Grundell, July 13, 1935. Asst. prof. U. Uppsala, 1932--39, prof. botany, 1939-—, prorector (v.p.), 1963-—. Named knight Royal Swedish Order Polar Star, 1946, comdr., 1957; recipient Linnous medal Royal Physiographical Soc. Lund, 1966. Mem. Swedish Acad. Sci., Royal Soc. Sci. Uppsala, Norwegian Acad. Sci., Royal Danish Soc. Sci., Bot. Soc. Edinburgh (hon.), Brit. Mycology Soc., S.W. (pres. 1965), Danish, Polish (corr.) bot. socs., others. Research, publs. on taxonomy of fungi especially discomycetes, taxonomy and phytogeography of Scandinavian vascular plants. Home: 17B Sibyllegatan, Uppsala, Sweden.*

NANNONI, Angelo, Italian surgeon; b. Jussa, nr. Florence, Tuscany, 1715. Prof., chief surgeon Hosp. Florence. Author: Trattato Chirurgio Sopra la Semplicità del Medicare, 1770. Distinguished as surgeon and lectr.; perfected method of lateral cutting for stone. Died Florence, 1790.

NANSEN, Fridtjof, Norwegian explorer, anatomist; b. Fröen, Norway, Oct. 10, 1861; Ph.D., U. Christiana (now Oslo), 1887; m. Eva Sars, Sept. 1889. First trip to Arctic, 1882; curator Museum Natural History, Bergen, Norway, 1882; expdn. across Greenland, 1888; curator Mus. Comparative Anatomy, U. Christiana, 1892; expdn. to Arctic on Fram, 1893-96; prof. zoology U. Christiana, 1897; expdn. to N. Atlantic, 1900; dir. Internat. Commn. for Study Sea, 1901; Norwegian minister to Britain, 1906-08; prof. oceanography U. Christiana, 1908; dir. repatriation of prisoners after World War I, famine relief for Russia, 1921-23. Norwegian del. to League of Nations Rector, St. Andrew's (Scotland) U., 1925. Recipient Nobel Peace prize, 1922. Author: First Crossing of Greenland, 1890; Eskimo Life, 1891; Farthest North, 1897; The Norwegian North Polar Expedition, 6 vols., 1900-06; Norway and the Union with Sweden, 1905; Through Siberia, 1914; Armenia and the Near East, 1928. Known for his 1888 trek across Greenland showing it to be an ice-sheet and experiment with ice-drift on the Fram, which showed Polar Basin to be packed with ice; sci. data collected facilitated future Arctic exploration; doctoral dissertation first showed that root fibers of spinal cord are in both ascending and descending branches. Died Oslo, Norway, May 13, 1930.

NANSON, Eric Musard, surgeon; b. Geraldine, New Zealand, Jan. 4, 1915; s. Gerald Boucher and Florence (Bell) N.; M.B., Ch.B., Otago U., New Zealand, 1939; m. Vera Morrow, Nov. 8, 1945; children—Jennifer Anne, John Boucher, Judith Catherine. Practice medicine, specializing in surgery, Saskatoon, Sask., 1954-—; prof. surgery U. Sask., 1954-—; surgeon-in-chief U. Hosp., 1954-—; cons. surgeon City, St. Paul's hosps., Saskatoon Sanatorium. Fellow Royal Colls. Surgeons (Eng., Can.), A.C.S.; mem. Canadian Assn. Clin. Surgeons, Internat. Surg. Group, Canadian, Brit. med. assns., Am., Pan Pacific surg. assns., Canadian Cardiovascular Soc., Am. Coll. Chest Physicians, Canadian Soc. Clin. Investigation, Canadian Gastroent. Assn. Research, numerous publs. on function of stomach, study of heart disease, clotting of blood, clin. disease of chest. Home: 802 Idylwyld Crescent. Office: Dept. Surgery, U. Hosp., Saskatoon, Sask., Can.*

NANZAN, see Ajima, Naonobu

NAPIER, John, Scottish mathematician; b. Merchiston Castle, nr. Edinburgh, Scotland, 1550; s. Archibald and Janet (Bothwell) N.; ed. St. Andrew's U., circa 1563-66; also probably in Low Countries, France and Italy; m. Elizabeth Stirling, 1572; 1 son, 1 dau.; m. 2d, Agnes Chisholm, circa 1579; 5 sons, 5 daus. Named landlord of Merchiston, Edentellie, Gartnes, 1572; lived at Gartnes, 1573-1608; became 8th Laird of Merchiston, inherited Merchiston Castle, 1608; pursued math. in his leisure. Author: A Plaine Discovery of the Whole Revelation of St. John, 1593; Mirifici logarithmorum canonis descripto, 1614; Rabdologiae seu numerationis per virgulas libri duo, 1617; Mirifici logarithmorum canonis constructio, 1619. Best known as inventor of logarithms; found exponential expressions for various numbers, especially trigonometric functions; invented Napier's rule of circular parts for solution of spherical triangles, also mech. devices, Napier's bones, for computing logarithms, and decimal point to express decimal fractions; introduced decimal notation. Died Merchiston, Apr. 4, 1617.

NAQVI, Ali Mehdi, astrophysicist, physicist; b. Delhi, India, Dec. 13, 1926; s. Mehdi Hasan and Nazar (Zaidi) N.; B.Sc., Meerut Coll., Agra U., India, 1944; M.Sc., U. Dehli, 1946; M.A. 1949; Ph.D., Harvard, 1951; m. Lenore Elaine Brooks, Oct. 14, 1949; children—Sarah, Ali Javed. Came to U.S., 1947, naturalized, 1961. Lectr., Delhi Poly., 1944-46; lectr. U. Delhi, 1946-47, 53-55; prof., head dept. physics and astronomy, U. Sind, Hyderabad, Pakistan, 1955-58; prof. physics Thiel Coll., Greenville, Pa., 1958-60; research asso. Harvard Obs., Cambridge, Mass., 1959-65; prin. scientist, head atomic and molecular physics dept. Geophysics Corp. Am., Bedford, Mass., 1960-64; sr. staff scientist Aerospace Corp., San Bernardino, Cal., 1964-—. Nat. Inst. Sci. India Sr. Research fellow, 1951-53. Fellow Royal Astron. Soc. (London); mem. Internat. Astron. Union, Am. Astron. Soc., Am. Phys. Soc., Inst. Physics and Phys. Soc. (London), Am. Geophys. Union, Pakistan Assn. for Advancement Sci. (sec. physics math. astronomy sect. 1956-57). Contbg. author: Atmospheric Processes, 1963. Research, publs. on effect of mut. spin-orbit and spin-spin interactions on energy levels and forbidden transition probabilities for a number ions belonging to p-electron configurations, applied results to identification spectral lines solar corona, artificial satellites navigation, nuclear explosions, detection, atomic and molecular processes in upper atmosphere and astron. objects; calculated wave functions of ions in lithium through cesium isoelectronic sequences, transition probabilities importance to astrophysics and upper atmospheric physics. Home: 109 Masters Av., Riverside, Cal. 92507. Office: Aerospace Corp., San Bernardino, Cal. 92402.*

NARABAYASHI, Chinzan (named Tokitoshi, called Shingobe), Japanese physician; b. Nagasaki, 1643; studied medicine under Dutch physician Hoffmann, Nagasaki, 1688; interpreter in Dutch, Nagasiaki Magistrate's Office; founder Narabayashi Sch. Surgery; declined Shogunate's invitation to become physician to Shogun Tsunayoshi. Author: Geka Soden (based on Ambroise Paré's book on surgery). Died 1711.

NARABAYASHI, H(irotaro), Japanese physician; b. Kobe, Japan, Sept. 4, 1922; s. Shigeo and Kogiku (Yamaguchi) N.; grad. Tokyo Imperial U., 1946; m. Atsuko Jyogo, Oct. 10, 1965; children—Riichiro, Reiko, Yosuke. Faculty, Juntendo Med. Sch., Tokyo, 1956—, prof. neurology, 1964—; dir. Neurol. Clinic Naka-Meguro, Tokyo, 1956—. Mem. Japanese Soc. Neurology (officer), Japanese Soc. Neurosurgery, Japanese Soc. for Neurology and Psychiatry (prize 1955), Japanese Soc. Stereoencephalotomy, Japanese Soc. EEG, Internat. Stereoencephalotomy (dir. bd.), French Soc. Neurology (hon.). Research, publs. on clin. neurophysiol. study on function and analysis of extrapyramidal motor system or disorders, technique and application of stereoencephalotomy on human brain, clin. physiol. research and neurosurg. treatment of cerebral palsy, physiol. study on system of muscle tone control. Home: 5-12 Naka-Meguro, Meguro-Ku, Tokyo, Japan.*

NARABAYASHI, Soken (named Takafusa, pseudonym Wazan), Japanese physician; b. Hizen, Japan, 1802; s. Eitetsu; trained in Dutch learning; in service Lord Naomasa (Kanso) Nabeshima, at whose command he introduced vaccination for smallpox in Japan at Hizen; friend of Philipp Franz von Siebold. Died 1852.

NARAHARI, Kashmirian physician, grammarian; composed dictionary of materia medica known variously as Rajanighantu, Nighanturaja, or Abhidhanacudamani, 1235-50.

NARANJO, Plutarco, Ecuadorian pharmacologist, allergist; b. Ambato, Ecuador, June 18, 1921; s. Enrique and Antonia (Vargas) N.; M.D., Faculty Medicine, U. Quito (Ecuador), 1949; postgrad. U. Utah, Harvard, Yale, U. Kan.; m. Enriqueta Banda, July 24, 1947; children—Alexis, Anita, Plutarco. Faculty, U. Quito, 1941—, prof. pharmacology, 1949-62, head dept., 1962—, dir. Inst. Natural Scis., 1957-66; head dept. pharmacology U. Cali (Colombia), 1953-56; dir. exptl. medicine Life Labs., 1956—; practiced medicine specializing in allergies, Quito, 1951—. Mem. human health adv. bd. Dow Internat., 1965—. Recipient Universidad Central Ann. awards, 1949, 52, 59. Mem. Latin Am. Soc. Pharmacology, Am. Acad. Allergy, Am. Coll. Allergy. Author several books including: Manual of Pharmacosology, 1965; Montalvo: a Bibliographic Study, 2 vols., 1966; also numerous articles. Bd. dirs., editor Revista Ecuatoriana de Medicina y Ciencias Biológicas, 1963—, Ciencia y Naturaleza, 1956—; coeditor Chemotherapía, 1961—, Alergía, 1959—. Research on allergy producing molds of Equador, causative factors of respiratory allergy in tropical countries of S.Am., influence of climate in asthma, pharmacological effects of some antibiotics, plants of Ecuador containing psychotomimetic substances. Home: 851 San Ignacio, Quito, Ecuador.*

NARASIMHA, Rao, Indian chemist, microbiologist; b. M'puram, Cuddapah, India, June 12, 1913; s. Rao Virabhadra Polur and Pitchamma Palaparti; Inter, Presidency Coll., Madras, 1931; B.Sc., Annamaldi U., 1933; M.Sc., Ann. U., 1937; Ph.D., Bombay (India), 1943; Biochem. Engr., U. Wis., 1955; m. Saraswathi Gudalar, May 4, 1934; children—Uma, Nalini. Lady Tata scholar Indian Inst. Sci., Bangalore, India, 1939-40, research asst., 1940-44, faculty, 1944—, asst. prof. biochemistry, 1963-—; indsl. planning officer drugs and dressings India Govt., 1944-45; exchange visitor U. Wis., 1954-55, Inst. Microbiology, Rutgers U., 1955-56; cons. chemist and chem. engr., 1943—. Mem. Govt. Sandal Spike Disease Research Com., 1959—. Recipient Prof. Sudborough medal, 1941, Sir. M.O. Forster medal, 1943, Sir Ratanji Ranchotji Desai's Gold medal, 1943. Mem. Biochem. Soc. Bangalore (past v.p.), Sigma Xi. Research and numerous publs. on chemotherapy of tropical diseases and chemistry of plant and mold products, total synthesis of pinene; discovered antimicrobial properties of benzyl isothiocyanates and allied compounds. Home: 29, E. Park Rd., Malleswaram, Bangalore 3, India.*

NARAYANA, Hindu mathematician; flourished 1356; s. Nrisimha (or Narasimha); lived under rule of Mohammad Ibn Taghlaq (ruled 1325-51), and rule of Firuz Shāh (ruled 1351-88). Author: Ganita-Pāti-Kaumudi (or Ganita-Kaumudi) (Elucidation of Arithmetic); Bija-Ganitavatamsa (Ornament of Algebras). Used letters of alphabet or names of colors to represent unknown quantities;; proposed that initial letters of positive numbers be written to indicate them, negative numbers should be marked with superposed dots.

NARDELLI, Mario, Italian chemist; b. Parma, Italy, an. 6, 1922; s. Liberato and Irma (Roffi) N.; Laurea in Chemistry, U. Parma, 1946; m. Olga Zanichelli, June 4, 1951; 1 son, Paolo. Faculty, U. Parma, 1948—, prof. structural chemistry, 1963—, dir. Structural Chemistry Inst., 1963—. Mem. Italian Chem. Soc., Italian Physics Soc., Chem. Soc. (London, Eng.). Research and numerous publs. on X-ray crystal structure analysis of inorganic, organic and metal-coordination compounds. Home: 21 Viale Martiri della Libertà, Parma, Italy.*

NARDO, Giandomenico, Italian naturalist; b. Venice, Italy, Mar. 4, 1802; s. Francesco and Angela (Favetta) N.; sem. student, Chioggia, Italy; med. degree U. Padua, 1827; m. Chiara Rizzi, 1849; 1 dau., Angela. Named dir. hosp. for cholera victims, Venice, 1826; became physician Central Orphanage, Chioggia, 1838, dir., 1849. Research on marine life, classification of sponges. Died Apr. 7, 1877.

NARIANI, Tulsidas Khimandas, Indian virologist; b. Karachi, India, July 14, 1921; s. K.L. and Rochibai; B.Sc. Agr. with honors, Bombay (India) U., 1943; postgrad. Asso. Indian Agrl. Research Inst., New Delhi, India, 1947, (Colombo Plan fellow), Glasshouse Crops Research Inst., Littlehampton, U.K., 1963-64; m. Asha Vachani, May 29, 1949; children—Meena, Shobha, Neelam. Jr. lectr. Agr. Coll., Sakrand, Sind, India, 1943-45; staff Indian Agrl. Research Inst., New Delhi, 1948—, virus pathologist, 1962—. Fellow Indian Phytopathological Soc.; mem. Pusa Agrl. Research Soc. Research and publs. on virus diseases of fruits, vegetables, legumes and cucurbits, transmission, virus-vector relationships, inhibition of plant viruses, purification and serology of viruses, triesteza and greening viruses of citrus; discovered several new virus diseases from India and reported others for first time. Home: D-38 Pusa Inst., New Delhi, Delhi-12, India.*

NARITA, Kiichi, Japanese chemist; b. Kyoto, Japan, Mar. 25, 1927; s. Toyoji, and Minori (Hirai) N.; B.S., Kyoto U., 1951, D.Engring., 1958; m. Hisako Noda, Apr. 22, 1952; children—Izumi, Takayasu, Masaki. With Kobe Steel Ltd., (Japan), 1951—, chief engr. sect. chem. metallurgy, 1960-65, mgr. sect. fundamental research on steel making, 1965—. Mem. Japan Soc. Promotion of Sci., Chem. Soc. Japan, Japan Inst. Metals, Iron and Steel Inst. Japan (Watanabe prize 1962). Author: Chemical Analysis of Zirconium in Iron and Steel, 1963; Vanadium in Iron and Steel, 1965; Nickel in Iron and Steel, 1965; also articles. Research on determination of silicon, aluminum, zirconium and their compounds in iron and steel, morphology of silicic acids in aqueous solution, origin and identification of nonmetallic inclusion in steel, solidification, vacuum degassing and melting of steel, thermochem. study on chem. reactions in steel-making process; Patentee vanadium alloying method in steel by using its compound. Home: 3-71-2 Inano-cho, Itami, Hyogo, Japan. Office: 1-36-1, Wakinohama-cho, Kobe, Hyogo, Japan.*

NAROLL, Raoul, anthropologist; b. Toronto, Ont., Can., Sept. 10, 1920; s. Albert Abe and Tessa (Soskin) N.; A.B., U. Cal. at Los Angeles, 1950, M.A., 1952, Ph.D., 1953; m. Frada Kaufman, Aug. 27, 1941; 1 dau., Maud Margaret. Fellow, Center for Advanced Study in Behavioral Sci., 1954-55; with Human Relations Area Files, Washington, 1955-56; faculty San Fernando Valley State Coll., Northridge, Cal., 1957-62, Northwestern U., Evanston, Ill., 1962-67; prof. anthropology State U. N.Y., Buffalo, 1967—. Mem. Am. Anthrop. Assn., Am. Ethnol. Soc., Am. Hist. Assn., Am. Statis. Assn. Author: (with Ronald Cohen) Handbook of Method in Cultural Anthropology; Data Quality Control, 1962; also articles. Developed a measure of social devel., measures of data inaccuracy (bias) and of trait diffusion through borrowing or migration, probability sampling methods for world-wide samples of primitive tribes; studied effect of warfare and culture stress on cultural evolution, characteristics of cycles of rise and fall of civilization. Home: 4695 Main St., Snyder, N.Y. 14226. Office: Dept. Anthropology, State U., N.Y., Buffalo 14214.*

NARSIMHAN, Ganapathi, Indian chem. engr.; b. Nagpur, India, Sept. 15, 1924; s. P.S. Ganapathi and Kamalam (Iyer) Aiyar; B.Sc., Madras (India) U., 1944; B.Sc., Nagpur U., 1946, M.Sc. Tech., 1950, Ph.D., 1957; m. Parvathi, May 5, 1944; children—Shekar, Sridhar, Jayaram. Lectr., Nagpur U., 1946-57; chem. engr. Nat. Chem. Lab., Poona, India, 1957-65; prof., head dept. chem. engring. Indian Inst. Tech., Kanpur, India, 1965—. Mem. Indian Soc. Theoretical and Applied Mechanics (founding), Indian Inst. Chem. Engrs. Research and publs. on mass transfer, applied thermodynamics and kinetics, theory non-catalytic reactions, generalized methods of property estimation, chem. reactor analysis for hydrocarbon oxidation. Home: 622, Indian Inst. Tech., IIT P.O., Kanpur, India.*

NASANEN, Reino Olavi, Finnish chemist; b. Turku, Finland, June 10, 1908; s. Oskar Aleksander and Johanna (Ahlgren) N.; cand.phil. U. Turku, 1933; Ph.D., U. Helsinki (Finland), 1939; m. Ilona Leino, June 20, 1939; children—Perttu, Outi, Timo, Jyrki, Risto, Johanna. With Inst. Tech., Helsinki, 1936-40, prof. phys. chemistry, 1951-55; with U. Helsinki, 1940-50, prof. inorganic chemistry, 1956—, dir. chem. lab., 1967—. Mem. Finnish Acad. Scis., Finnish, Am. chem. socs., Chem. Soc. (London), Internat. Com. for Electrochem. Thermodynamics and Kinetics. Research, numerous publs. on coordination compounds, particularly in solution. Home: 3 Päivärinnankatu, Helsinki 25, Finland. Office: 5 Hallituskatu, Helsinki 17, Finland.*

NASH, Leonard Kollender, Am. chemist; b. N.Y.C., Oct. 27, 1918; s. Adolph and Carol (Kollender) N.; B.S. summa cum laude, Harvard, 1939, M.A., 1941, Ph.D., 1944; m. Ava Cline Byer, Mar. 3, 1945; children—Vivian Carol, David Byer. Research asso. Columbia U., Manhattan Project, 1944-45; instr. U. Ill., 1945-46; faculty Harvard, 1946—, prof. chemistry, 1958—. Mem. Adv. Council on Coll. Chemistry, 1964-67. Mem. Am. Acad. Arts and Scis. Author: The Nature of the Natural Sciences, 1963; The Atomic-Molecular Theory, 1950; Plants and the Atmosphere, 1952; Elements of Chemical Thermodynamics, 1962; Stoichiometry, 1966; also articles. Innovator in chem. edn.; developer of microanalytical techniques. Home: 11 Field Rd., Lexington, Mass. 02173. Office: Dept. Chemistry, Harvard U., Cambridge, Mass.*

NASH, Manning, Am. anthropologist; b. Phila., May 4, 1924; s. Abraham and Molly (Sukonik) N.; B.S. with honors, Temple U., 1949; M.A., U. Chgo., 1952, Ph.D., 1955; m. June C. Bousley, Sept. 20, 1952; children—Eric, Laura. Instr., U. Cal. at Los Angeles, 1955-56; asst. prof. U. Wash., Seattle, 1956-57, vis. prof., 1962-63; faculty U. Chgo., 1957—, prof. anthropology, 1963—. Mem. Am. Anthropol. Assn., Am. Ethnol. Soc., Royal Anthropol. Inst. Gt. Britain and Ireland. Author: Machine Age Maya, 1958; The Golden Road to Modernity, 1965; Peasant and Primitive Economic Systems, 1966; also articles. Research on theory structure and dynamics of small non-western econ. systems, analysis process modernization in Latin Am., and S.E. Asia. Home: 5805 S. Dorchester St., Chgo., 60637.*

NASINI, Raffaello, Italian chemist; b. Sienna, Italy Aug. 20, 1854; student, Pisa, Italy, Rome; asst. to Stanislao Cannizzaro; became prof. chemistry U. with Giacomo Ciamician. Mem. French Acad. Scis., 1930. Research on rotary power refraction indexes, dispersion power of organic compositions, function of their constn., rare gases of boric acid suffioni, or hot springs, of Tuscany. Died Rome, Mar. 30, 1931.

NASIR AL-DIN (or Nasir Eddin), Mohammed-ibn-Hassan, Persian astronomer, mathematician; b. Toos, Khorasan, 1201. Supt. obs., Maragha, Azerbaijan (built for him by Mongol ruler, Hulagu). Author: Ilchanic Tables; treatises on algebra, geometry, arithmetic. Translator: Elements (Euclid). Wrote 1st Moslem text on trigonometry as sci. ind. of astronomy; pub. proof of parallel postulate. Died 1274.

NASMYTH, James, British engr.; b. Edinburgh, Scotland, Aug. 19, 1808; s. Alexander Nasmyth; ed. Edinburgh, in engring., London. Began in father's workshop; personal asst. to Maudslay, 1829-30; opened workshop, Edinburgh, 1834; became head foundry, Patricroft, Eng., 1853. Named chmn. Royal Small Arms Commn. Author: (with James Carpenter) The Moon Considered as a Planet, a World and Satellite, 1874. Built 6 inch diameter reflecting telescope, also steam engine able to carry 6 people, 1827, self-acting nut-milling machine, 1829; invented reversible roll, 1833, shaper, 1836, also planing machine, hydraulic punching machine; one of originators of steam hammer with adjustable height of drop, designed 1839, patented 1842; 1st to observe mottled appearance of sun's surface, 1860. Died London, May 7, 1890.

NASON, Alvin, Am. biochemist; b. Coatsville, Pennsylvania, June 10, 1919; s. Samuel and Rose (Vilensky) N.; B.S., Cornell U., 1940; A.M., Columbia, 1948, Ph.D. (Lalor fellow), 1949; m. Thelma Stein, Oct. 18, 1944; children—Deborah Rose, Steffi Ruth, Jean, Gerson, Benjamin. Research asst. genetics Carnegie Inst., 1940; chemist Chromium Corp. Am., 1941-42; agt. field lab. tung investigation Dept. Agr., 1942-43; teaching asst. Columbia, 1946-48; asst. prof. biology, McCollum-Pratt Inst., Johns Hopkins, 1949-54, asso. prof., 1954-58, prof., 1958—, asso. dir. Inst., 1959—. Mem. panel NSF; cons. Nat. Institutes of Health. Mem. vis. com. biology Brookhaven Nat. Laboratory. Recipient travel award NRC to 4th Internat. Biochem. Congress, Vienna, Austria, 1958. Served to 1st lt. USAAF, 1943-46. Mem. Am. Soc. Biol. Chemistry, Am. Chem. Soc., Soc. Am. Bacteriology, Am. Soc. Plant Physiology, Soc. Gen. Physiologists, A.A.A.S., Am. Assn. U. Profs. Author: Textbook of Modern Biology. Editor-in-chief Analytical Biochemistry, 1959—. Contbr. numerous articles. Research on vitamin metabolism; metalloflavoproteins; enzymology and mechanism of terminal electron transport and inorganic nitrogen assimilation. Home: 2002 Sulgrave Av., Balt. 21209.

NASSAU, Jason John, astronomer; b. Smyrna, Asia Minor, Mar. 29, 1892; s. John and Maria (Christie) N.; C.E. and M.S., Syracuse U., 1915, Ph.D., 1920, D.Sc. (hon.), 1940; D.Sc. (hon.), Case Inst. Tech., 1965; LL.D., Lake Erie Coll., 1956; m. Laura Alice Johnson, June 27, 1920; children—James, Sherwood.

Came to U.S., 1910, naturalized, 1917. Instr., Syracuse U., 1919-21; asst. prof. astronomy Case Inst. Tech., Cleve., 1921-24, asso. prof., 1924-29, prof., 1929-62, prof. emeritus, 1962—; dir. Warner and Swasey Obs., East Cleveland, O., 1924-59, head grad. div. Case Inst. Tech., 1936-40, acting head physics dept., 1941-45. Mem. Internat. Astron. Union (chmn. U.S. nat. com. 1948-56), Cleve. Astron. Soc. (pres. 1922-25, 32-62); corrr. mem. Nat. Acad. Athens. Author: Practical Astronomy, 1932, 36, 38. Research on stellar spectra and galectic structure. Home: 2019 Taylor Rd., East Cleveland 44112. Died May 11, 1965.*

NASSAU, Kurt, chemist; b. Stockerau, Austria, Aug. 25, 1927; s. Julius and Frieda (Hauser) N.; B.Sc. with honors U. Bristol, Eng., 1948; m. Julia Wechsler, June 21, 1949. Came to U.S., 1948, naturalized, 1954. Research and devel. chemist, fat and oil Glyco Products Co., Inc., Williamsport, Pa., 1948-54; research in human metabolism Walter Reed Army Research Inst., Washington, 1954-56; research solid state chemistry and crystal growth Bell Tel. Labs., Murray Hill, N.J., 1959—; instr. U. Pitts., 1956. Mem. N.J. Mineral. Soc. (pres. 1966), Am. Chem. Soc., Am. Crystallographic Assn., Sigma Xi. Research and publs. in solid state chemistry related to crystal growth; co-inventor of first continuous solid state laser; several patents on lasers; discovery of a new type of ferroelectricity in lithium niobate. Home: Round Top Rd., Bernardsville, N.J. 07924. Office: Bell Tel. Labs., Murray Hill, N.J. 07971.*

NASSE, Christian Friedrich, German physician; b. Bielefeld, Germany, 1778; grad. medicine, Halle, Germany, 1800; practiced in Gottingen, Leipzig, Dresden, Weimer (all Germany); became prof. medicine, Halle, 1815; became prof., dir. med. clinic, Bonn, Germany, 1819-51. Author: Handbuch der speciellen Therapie, 1830-38; Handbuch der Allgemeinen Therapie, 1840; Die Behandlung der Gemuthskrkranken und Irrendurch Nichtarzte; also numerous articles. First German to use percussion and auscultation in his clinic; 1st to use microscope in clin. demonstrations; pioneered treatment of the insane; formulated law of Nasse (hemophilia occurs only in males and is passed on by unaffected females), 1820; active in movement stressing physiol. aspects of clin. medicine. Died 1851.

NASSET, Edmund Sigurd, Am. physiologist; b. Willmar, Minn., Feb. 24, 1900; s. Aslak O. and Martha (Alvig) N.; student Dickinson State Coll., 1921-23; A.B., with honors, St. Olaf Coll., 1925; M.S., Pa. State U., 1927; postgrad. U. Cal. at Berkeley, 1928; Ph.D., U. Rochester, 1931; D.Med.-Sci. (hon.), Woo Sok U. Coll. Medicine, Seoul, Korea, 1965; L.H.D., Dickinson State Coll., 1966; m. Elizabeth Hope Custer, June 8, 1928; children—Laura Louise, (Mrs. Joal Cronenwett), Susan Elizabeth (Mrs. Thad C. Stanford). Faculty, U. Rochester, (N.Y.), 1931-65, prof. physiology, 1948-65, prof. emeritus, 1965—; vis. prof. George Washington U., 1966, FAO nutrition adviser Govt. of India, 1961-63; research physiologist U. Cal. at Berkeley, 1966; mem. gen. medicine study sect. NIH, 1964-68; cons. health physics div. Oak Ridge Nat. Lab., 1964—. Fulbright lectr., India, 1954-55. Mem. Am. Physiol. Soc., Inst. Nutrition, Soc. for Exptl. Biology and Medicine, A.A.A.S. Author: Food and You, 1951; Your Diet Contbr. sect. on digestive system to Med. Physiology, 10th, 11th, 12th edits., 1956, 61, 68. Discovered enterocrinin; demonstrated that feeding dried thyroid gland reduces drastically secretion of acid gastric juice; research on regulation of amino acid mixture for intestinal absorption; essential amino acid requirement for adult animal. Home: 1254 Grizzly Peak Blvd., Berkeley, Cal. 94708.*

NASTUK, William Leo, Am. physiologist; b. Passaic, N.J., June 17, 1917; s. William and Mary (Tulchak) N.; B.Sc., Rutgers U., 1939, Ph.D., 1945; m. Ruth Alden Lester, Apr. 1, 1950; children—William Lester, John Andrew, Mary Alden. Faculty, Columbia U., N.Y.C., 1945—, prof. physiology, 1960—. Cons., USPHS, 1960-64; mem. physiol. tng. com. NIH, 1964-66; mem. sci. adv. bd. Myasthenia Gravis Found., 1963—. Recipient Am. Physiol. Soc. Travel award, 1950. Mem. Am. Physiol. Soc., N.Y. Acad. Scis., A.A.A.S., Soc. Exptl. Biology and Medicine, Assn. Research in Nervous and Mental Diseases, Harvey Soc., Sigma Xi, Phi Lambda Upsilon. Editor: Physical Techniques in Biological Research, Vol. 4, 1962, Vol. 5, 1964, Vol. 6, 1963. Editorial bd. Physiol. Rev., 1960-66. Developed spl. instruments and techniques used in neurophysiol. research; research and publs. in basic processes in neuromuscular transmission including chem. agts., and drugs active at this site, immunological changes appearing in myasthenia gravis; co-author automimmune hypothesis concerning cause of myasthenia gravis. Home: 70 Haven Av., N.Y.C. 10032.*

NASU, Noriyuki, Japanese marine geologist; b. Fukuoka, Japan, Feb. 27, 1924; s. Senkichi and Hisako (Yamamoto) N.; B.Eng., U. Tokyo (Japan), 1946, B.S., 1950; M.S., Scripps Instn. Oceanography, U. Cal. at Los Angeles, 1953, Ph.D., 1955; D.Sc., U. Tokyo, 1961; m. Kyoko Nozaka, Apr. 21, 1957; 1 child, Keiko. Staff, U. Tokyo, 1951—, prof. Ocean Research Inst., 1963—. Mem. Geol. Soc. Japan, Oceanographical Soc. Japan. Author: (with K. Watanabe, S. Obi) Earth Science; also articles. Asso. editor Jour. Geology, U. Chgo. Research on marine geology; discovered sediment fill ancient river channel formed during glacial sea-level lowering (Paleo Kuji River), Japanese Pacific Sea Shelf on trench slopes along Pacific coast of Japan. Home: 1-2-A17, Taishido, Setagaya, Tokyo, Japan.*

NASZ, István, Hungarian virologist, microbiologist; b. Turkeve, Hungary, May 3, 1927; s. Istvan and Julianna (Szabó) N.; M.D., U. Medicine, Budapest, Hungary, 1951, Med. Specialist Lab. Examinations, 1955, C.Sc., 1956, D.Sc., 1964; m. Sarolta Albrecht, Aug. 2, 1954; 1 dau., Ildikó. Faculty, Inst. Microbiology U. Medicine, Budapest, 1948—, asso. prof. virology and microbiology, 1963—, Sec. microbiol. com. Hungarian Acad. Sci., since 1959—; mem. microbiol. com. Ministry of Health, 1960—; mem. Internat. Com. Nomenclature of Viruses, 1966—. Mem. Hungarian Microbiol. Assn. (pres. 1960—). Author: Adenoviruses and their Pathogenicity, 1967; also numerous articles. Editorial bd. Acta Microbiologica, 1965—. Research in different biol. properties and path. role of adenoviruses; first isolation of adenoviruses from excised tonsils in Hungary; description of heterotypic hemagglutination-inhibition reaction, immuno-osmophoresis in adenovirus group, hemagglutination spectra, cytopathic effects of adenoviruses. Home: 8 Erömü, Budapest XI, Hungary.*

NATADZE, Georgiy Mikhaylovich, Russian hygienist; b. 1892; grad. Med. Faculty, Kiev U., 1921; D.Med. Sci. Med. officer, founder Central Health-Hygiene Lab. for Transcaucasian R.R., asso. various hygiene insts., asst., lectr. chair gen. hygiene Tbilisi Med. Inst., 1922-38; prof., head chair gen. hygiene Tbilisi Med. Inst., 1938—; founder, former dep. dir. Inst. Nutrition, Tbilisi; founder Nutrition Research Inst., Georgian Ministry Health, 1958, head nutrition research lab., 1958—. Mem. USSR Acad. Med. Scis. (corr.), Georgian Soc. Hygienists (chmn.). Author: The Principles of Hygiene, 1951 (translated into Korean). Co-editor Hygiene sect. Large Med. Ency., 2d edit.; mem. editorial council Hygiene and Sanitation, Problems of Nutrition. Research and numerous publs. on gen. and food hygiene, vitaminology, regional pathology, energy consumption of agrl. and indsl. workers. Address: Tbilisi Med. Inst., ulitsa Melikashuili 16, Tbilisi, Georgian SSR, USSR.

NATADZE, Revaz Grigorevich, Russian psychologist; b. 1903; grad. psychology dept. Philosophy Faculty, Tiflis U., 1924; D. Pedagogical Sci., 1939. Head chair psychology Tbilisi U., 1950—. Mem. Georgian Acad. Scis. (corr.), RSFSR Acad. Pedagogical Sci. (corr.). Author numerous works in Georgian, Russian, English and German, including: Psychology: Textbook for Secondary School, 8th edit., 1956; General Psychology: For Students in Special Psychology Departments of Universities, 1956; The Adapting Effect of Imagination, 1958; The Founder of Scientific Psychology in Georgia, 1962. Address: Tbilisi University, prospect Chavchavdze 45, Tbilisi, Georgian SSR, USSR.

NATARAJAN, Adayapalam Tyagarajan, geneticist; b. Mannargudi, India, June 15, 1928; s. A. and Savitri (Tyagarajan) N.; M.A., Annamalai U., 1953, Indian Agr. Research Inst., 1955; Ph.D., Delhi U., 1958, Stockholm (Sweden) U., 1966; m. Satya Nirula, Feb. 11, 1966. Staff, Indian Agr. Research Inst., New Delhi, India, 1958-65, cytogeneticist, 1962-65; docent radiobiology Stockholm U., 1966—. Regional expert I.A.E.A., Bangkok, Thailand, 1966. Rockefeller Found. postdoctoral fellow Brookhaven Nat. Lab., 1958-59; Swedish Inst. fellow genetics dept. Forest Inst., Stockholm, 1960. Mem. Indian Soc. Genetics, Genetics Soc. Am. Research, numerous publs. on mechanism of induction of chromosome breakage and mutation in higher organisms using both chem. mutagens and radiations. Home: Radjursvägan 5, Ekerö, Sweden.*

NATH, Madhab Chandra, Indian biochemist; b. Hashara (now E. Pakistan), Oct. 23, 1905; s. Nibaran Chandra and Sonatara Nath; B.Sc. with honors in Chemistry, 1929, M.Sc. in Chemistry, 1930, D.Sc. in Biochemistry, 1937; m. Indu Nath, Feb. 15, 1932; children—Indira (Mrs. N.K. Nath Talukdar), Nirmalendu, Kamalendu, Manika, Mukti. Tata Meml. Research scholar, Dacca U., 1934-37, lectr. physiol. chemistry, 1937-46; Chitnavis prof. biochemistry Nagpur (India) U., 1946—, head dept. biochemistry, 1946—. Recipient Elliott prize Royal Asiatiac Soc. Bengal, 1941; G.J. Watumull award in biochemistry, 1964. Fellow Royal Inst. Chemistry London, Nat. Inst. Scis. India, Instn. Chemists India; mem. Soc. Exptl. Biol. Medicine (N.Y.), Indian Sci. Congress Assn., Soc. Biol. Chemists India, Physiol. Soc. India. Research, numerous publs. on metabolism of carbohydrates, lipids and vitamins; demonstrated that continued injection of ketone bodies can induce diabetes in normal animals and that such diabetes can be prevented by prior injection of GCA; showed hydrogenated fat diet gives rise to atherosclerosis with increased cholesterol/lipid phosphorus ratio and decreased blood coagulation time and that atherogenesis can be prevented with GCA; showed GCA is a precursor of vitamin C and that it contributes to biosynthesis of thiamine. Home: The Terraces, Amravati Rd., Nagpur-1, India.*

NATH, Rajendra Lal, Indian biochemist; b. Silchar, India, Jan. 1, 1913; s. Nabin Chandra and Sarabala Nath; B.Sc., Calcutta U., 1935, M.Sc. (Gold medal), 1937; Ph.D., London (Eng.) U., 1951; m. Bibha Nath, Jan. 17, 1939; children—Ransan Kumar, Anjan Kanti. Faculty, B.W. Med. Sch., Dibrugarh, 1939-51, Assam Med. Coll., Dibrugarh, 1951-57; prof. biochemistry, head dept. Sch. Tropical Medicine, Calcutta, India, 1957—; vis. prof. biochemistry Post Grad. Med. Inst., Calcutta, 1962—. Mem. Indian Chem. Soc., Indian Sci. Congress Assn., Soc. Biol. Chemists India, Instn. Chemists India. Collaborator, Internat. Jour. Enzymolegia, 1960—. Research, publs. on study of mechanism of action of enzymes, methodology in several clin. biochem. estimations has been improved and biochem. studies on blood and other body fluids have carried out in various tropical diseases. Home: 3 Priyanathchatterjee St., Calcutta 56. Office: Sch. Tropical Medicine, Calcutta 12, India.*

NATHAN, Helmuth Max, physician; b. Hamburg, Germany, Oct. 26, 1901; s. Neumann and Regina (Seligmann) N.; B.S., Freiburg U. (Germany), 1922; M.D., Hamburg U., 1925; m. Irene Nelson, Jan. 17, 1927; 1 dau., Ruth (Mrs. Peter V. Norden). Came to U.S., 1936, naturalized, 1943. Asso., Jewish Hosp., Hamburg, 1933-36; Cancer Research fellow Beth Israel Hosp., N.Y.C., 1936-37; attending surgeon Sydenham Hosp., Bronx Municipal Hosp. Centre, 1941-53; faculty Albert Einstein Coll., N.Y.C., 1954—, prof. surgery, 1960—; prof. surgery Hamburg U., 1950; practice surgery, N.Y.C., 1937—. Recipient Deneke medal U. Hamburg, 1932; Salonmon Heine medal, 1936. Mem. A.M.A., N.Y. County, Virchow (pres. 1966—) med. socs., Internat. Coll. Surgeons, Am. Coll. Cardiology, Am. Coll. Gastroenterology, N.Y. Acad. Medicine, Acad. Sci., Acad. Gasteroenterology, Acad. Facial Plastic Surgery, Am. Soc. Facial Plastic Surgery, Am. Soc. Abdominal Surgeons, Am. Assn. Mil. Surgeons, Soc. Med. Geriatrics. Co-author: Should the Patient Know the Truth? 1954; also articles. Developed theory of septic focus of circumscribed circulatory systems; 1st case of encephalitis in bacterial endocarditis, infections, internal hernias, injuries of heart and big vessels, medicine and art, others. Home: 327 Central Park W., N.Y.C. 10025. Office: 667 Madison Av., N.Y.C. 10021.*

NATHANSON, Neal, Am. epidemiologist, virologist; b. Boston, Sept. 1, 1927; s. Robert B. and Leah (Rabinowitch) N.; B.A., Harvard, 1949, M.D., 1953; m. Constance Allen, June 8, 1954; 1 dau., Katherine. 1 son, John A. Chief polio surveillance unit USPHS, 1955-57; research asso., asst. prof. anatomy Johns Hopkins U., Balt., 1957-63, asso. prof. epidemiology, 1963-1968, prof., 1968—. Fellow Am. Pub. Health Assn.; mem. A.A.A.S., Am. Soc. Tropical Medicine, Am. Soc. Microbiology, Sigma Xi. Mng. editor Am. Jour. Epidemiology, 1964—. Research and publs. poliomyelitis, pathogenesis of arbovirus encepholitis, and other neurotropic viruses. Home: 220 Cable St., Balt. 21210.*

NATHORST, Alfred Gabriel, Swedish explorer, geologist, paleobotanist; b. 1850; mem. Nordenskjöld's expdn. to Greenland, 1883; dir. div. Riks-Mus. Natural History, Stockholm, Sweden, 1885-1918; head expdn. to Bear Island, Spitsbergen, King Charles Island, 1898; traveled to Jan Mayen, Greenland, 1899, also to Franz Josef Fiord; discovered King Oscar Fiord. Died 1921.

NATTA, Giulio, Italian chemist; b. Imperia, Italy, Feb. 26, 1903; s. Francesco and Elena Crespi; Dr. Chem. Engring., Milan Poly. Inst. Tchr. gen. and inorganic chemistry Pavia U.; prof. phys. chemistry Rome U., indsl. chemistry Turin U.; prof. indsl. chemistry Milan Polytech. devel. thermo-plastic resins; numerous macromolecule discoveries. Recipient (with Ziegler) Nobel prize for chemistry, 1963, also various gold medals, hon. degrees. Mem. Lincei Acad., Lombard Inst. Sci. and Letters; also hon. memberships in fgn. chem. socs. Contbr. articles on phys. and indsl. chemistry tech. and sci. periodicals. Discovered stereospecific polymerization, 1954, theory prompting much present work in stereospecific catalysis and stereoregular polymers; devised techniques for synthesis of methylic alcohol under pressure, also new techniques for prodn. of aldehydes and synthetic rubber. Home: 54 Via Mario Pagano, Milan. Office: Polytech. Inst., Piazza Leonardo Da Vinci, Milan, Italy.

NATTERER, Jean, naturalist; b. Laxenburg, Austria, 1787; conservator Geol. Mus., Vienna, Austria. Author: Faune des mammifères du Bresil, 1840. Collected zool. and geol. specimens from S. Am. Died Vienna, 1843.

NATTERER, Johann August, physicist; b. Vienna, Austria, 1821; condensed nitrogen, carbon dioxide, other gases under pressure; determined compressibility of gases from 1300 to 2700 atmospheres, 1850-54; inventor practical method of determining critical temperature of carbon dioxide and other gases. Died 1901.

NATVIG, Haakon Valde, Norwegian physician; b. Stavanger, Norway, Oct. 26, 1905; s. Jacob Birger and Rachel (Soma) N.; M.D., U. Oslo (Norway), 1931, D. degree, 1941; m. Dorothea Sendstad, Dec. 30, 1931; children—Harald, Ketil. Practice gen. medicine, 1932-34; research asst. U. Oslo, 1933-39, prof. hygiene, dir. Inst. Hygiene, 1952—; dir. food control div. Oslo City Bd. Health, 1940-51. Pres., Norwegian Indsl. Health Service, 1946—, Norwegian Nutrition Found., 1955—, Nordic Com. for Analyses of Foods, 1962—. Travel grantee, Germany, 1936, Gt. Britain, 1946, U.S., 1948. Recipient U. Golden medal for sci. research, 1935, 45; Royal Order of St. Olav, Knight of 1st Class, 1965. Mem. Norwegian Med. Sci. Soc., Norwegian, Finland (hon.) indsl. med. socs., Nutrition Soc. (Gt. Britain). Author: Sinkfeber i messingindustrien, 1936; Die Bedeutung der Vitamine für die Resistenz der Vitamine, 1941; Textbook in Hygiene, 1958 Kjokkenhygiene, 1967; also numerous articles. Co-editor, Health Mag., 1934-44, editor, 1944—. Research on preventive medicine especially occupational diseases, vitamins, nutrition, food infections and intoxications, food analyses, infectious diseases, hemoglobin levels of Norwegian population. Home: Osthornvei Oslo, 8. Office: 8 Gydas vei, Oslo 3, Norway.*

NAUDAIN, Glenn Garnet, Am. chemist; b. Algona, Ia., Oct. 31, 1894; s. William Eliason and Ella (Foster) N.; B.S., Ia. State U., 1917, M.S., 1922, Ph.D., 1923; postgrad. (fellow) Mellon Inst.; B.E., Kan. State Coll., 1927; M.A. in Coll. Administrn., N.Y. U., 1939; D.Sc., Academica de Sciencias And Arts, Rio de Janeiro, 1937; m. Lillian Belle Gamrath, Sept. 15, 1921. Prof., Kan. State Coll., 1925-26; prof. chemistry, head dept. Winthrop Coll., Rock Hill, S.C., 1926-50; cons. chemist, Rock Hill, 1950—. Fellow A.A.A.S.; mem. Royal Soc. Arts (London), Am. Chem. Soc. (past state pres.), S.C. Acad. Sci. (past mem. exec. council, past pres.), Am. Assn. U. Profs., Sigma Xi, Phi Kappa Phi, Phi Lambda Upsilon, Gamma Sigma Delta, Lambda Sigma Kappa, Alpha Gamma Tau, Zeta Alpha. Research, publs. on poisonous gases for chem. warfare, mil. food packaging and vitamin study. Home: 919 Charlotte Av., Rock Hill, S.C. 29730.*

NAUDÉ, Stefan Meiring, S. African physicist; b. De Doorns, S. Africa, Dec. 31, 1904; s. Charl François and Annie (Lötter) N.; B.Sc., U. Stellenbosch, 1924, M.Sc., 1925; Ph.D., U. Berlin, 1928; D.Sc.h.c., U. Potchefstroom, 1957, U. Witwatersrand, 1959, U. Cape Town, 1959, U. Stellenbosch, 1962, U. O.F.S., 1967 (all S. Africa); m. Josephine Lodewica Ziervogel, June 22, 1929; children—Charl Francois, Cecilia Johanna (Mrs. Emil Loubser), Annette Isebella (Mrs. Siegfried Maré), Frederik Ziervogel, Josephine Stefanie, Bernadine Rinda. Instr., research fellow U. Chgo., 1929-31; sr. lectr. U. Cape Town, 1931-33; prof. physics U. Stellenbosch, 1934-45; dir. Nat. Phys. Research Lab., Pretoria, S.Africa, 1946-50; v.p. Council for Sci. and Indsl. Research, Pretoria, 1950-52, pres., 1952—. Recipient Havenga prize for sci. research S.African Akademie, 1951; S. African Medal for Research, 1960. Mem. Physics Soc. Am., Deutsche Physikalische Gesellschaft, Royal Soc. S. Africa (past pres.), Asso. Sci. and Tech. Socs. S. Africa (past pres.), S. African Akademie vir Wetenskap en Kuns (v.p.). Research, publs. in phys. chemistry in field of heat, atomic and molecular spectra, discovered nitrogen 15 isotope, 1929; analyzed several molecular spectra, including sulphur and nitrogen. Home: 420 Friesland Av., Lynnwood, Pretoria. Office: P.O. Box 395, Pretoria, Transvaal, S. Africa.*

NAUDIN, Charles, French botanist; b. Autun, France, Aug. 14, 1815; D.Sc. Dir. Thuret villa and lab., Antibes, France, from 1878; also with Jardin des Plantes, from 1852, Revue Horticole. Mem. French Acad. Scis., 1863, French Agrl. Soc. Author: Les espèces affines et la théorie de l'évolution, 1874. Work on hereditary traits in plants (1852) allowed him to state more clearly than previously uniformity of 1st generation crosses, identity of reciprocal crosses, reappearance of parental type in 2d generation; work establishes him as forerunner of Gregor Mendel; anticipated Darwin by writing that nature has brought forth diverse species from small number of initial types (believed however, that this process was preordained and followed definite and correlative courses). Died Antibes, Mar. 19, 1899.

NAUMANN, Johann Friedrich, German ornithologist; b. Ziebigk, nr. Köthen, Germany, Feb. 14, 1780; s. Johann Andreas Naumann; dir. mus., Köthen. Author: Taxidermie, 1815; Die Eier der Vögel Deutschlands und der benachbarten Länder, 1819-28; Naturgeschichte der Vögel Deutschlands, 13 vols., 1820-60. Noted for book on birds which he illus. with 500 copper engravings. Died Ziebigk, Aug. 15, 1857.

NAUMANN, Karl Friedrich, German mineralogist, geologist; b. Dresden, Germany, May 30, 1797; s. Johann Gottlieb Naumann; prof. Freiburg Bergakademie; prof. mineralogy and geography U. Leipzig (Germany). Mem. French Acad. Scis., 1870. Author: Beiträge zur Kenntnis Norwegens, 2 vols., 1824; Lehrbuch der Kristallographie, 2 vols., 1830; (with Cotta) Geognostiche Spezialkante, 1834-43; Elemente der Mineralogie, 1846; Elemente der theoretischen Kristallographie, 1856. Died Nov. 26, 1873.

NAUMOV, Nikolai Pavloic, Russian biologist, zoologist; b. Moscow, USSR, Nov. 25, 1902; s. Paul Alexandrovic and Anna (Malinina) N.; ed. Timirjasev Agrl. Acad., Moscow, candidate sci. 1934, Sc.D., 1941, prof. zoology, 1949; m. Vera Potechina, Apr. 1926; 1 dau., Natalja Pojarkova. With lab. in Govt. Cultural Base of Com. of Nord., 1928-31, Zool. Inst. of Moscow U., 1931--40, Inst. Evolutionary Morophology of USSR Acad. Scis., 1939-45, Inst. Epidemiology and Microbiology, 1946-56, biol. faculty U. Moscow, 1951—. Head lab. med. zoology Inst. Epidemiology and Microbiology; chief vertebrate zoology dept., dean biol. faculty Moscow U. Mem. Moscow Soc. Investigators of Nature, Zool. Soc. USSR (hon.). Author: Wild Reindeer, 1933; (with A. Formosov, Kiris) Ecology of Squirrels, 1934; (with Kurilowicz) Soviet Tungussia, 1934; Mammals of Tunguska District, 1934; Essays of Comparative Ecology of Mouse-like Rodents, 1948; Ecology of Animals, 1955, 2d edit., 1963. Investigated and described new forms of East Siberian mammals; studied comparative ecology and population dynamics of certain small rodents; investigated natural nidi of plague in Kazachstan and elaborated method of sanitation; studies of means of communication among animals, other problems of population biology. Office: Biological Faculty, Univ. of Moscow, Moscow W-234, USSR.*

NAUNYN, Bernard, German physician; b. Berlin, Sept. 2, 1839; ed. U. Bonn (Germany); M.D., U. Berlin, 1863. Prof. internal medicine, Dorpat, (now Tartu, Estonia), 1869-71; faculty Bern, Switzerland, 1871; prof., Konigsberg, Germany, 1872-88; succeeded Kussmaul at Strasbourg, France, 1888. Author: Klinik der Cholelithiasis, 1892; (monograph) Der Diabetes mellitus, 1898. Founder (with Schmiedelberg) Archives für experimentelle Pathologie und Pharmakologie. Research on diseases of metabolism, diseases of liver and pancreas, gall stones, diabetes; introduced term, acidosis, to describe metabolic condition of acid formation in diabetes; demonstrated true erythremia may be present in chronic cardiac patient; introduced concept of cholangitis as inflammation of lining membrane of smallest bile-ducts causing obliteration of their lumina, explaining catarrhal jaundice and syphilitic hepatitis as primary and secondary forms of infectious cholangitis; treated this by drainage of bile tracts. Died Baden-Baden, Germany, July 30, 1925.

NAUSIPHANES OF TEOS, philosopher; b. circa 360 B.C.; student of Pyrrhon of Elis; tchr., Teos; follower of Democritos; tchr. of Epicuros.

NAUTA, Walle Jetze Harinx, anatomist; b. Medan, Indonesia, June 8, 1916; s. Haring Jelles and Janneke (Bos) N.; M.D., U. Utrecht (Netherlands), 1942, Ph.D., 1945; m. Ellie Plaat, July 1, 1942; children—Tjalda (Mrs. John Auer), Janneke, Haring J. W. Came to U. S., 1951, naturalized 1958. Instr., U. Utrecht, 1942-46; asst. prof. U. Leiden (Netherlands,) 1946-47; asso. prof. U. Zurich (Switzerland), 1947-51; Neuro-physiologist Walter Reed Army Inst. Research, 1951-64; prof. anatomy U. Md., 1955-64, Mass. Inst. Tech., 1964—. Bd. dirs. Founds. Fund Research Psychiatry, 1959-62. Recipient Karl Spencer Lashley award Am. Philos. Soc., 1964. Mem. Am. Assn. Anatomists, Biol. Stain Commn., A.A.A.S., Neuro scis. Research Found. (asso.). Spl. research devel. histopathol. techniques for tracing neural connections in brain and spinal cord, also exptl. research mammalian brain orgn. Home: 196 Kent Rd., Newton, Mass. 02168. Office: Dept. Psychology, Mass. Inst. Tech., Cambridge, Mass. 02139.

NAVES, Yves René, chemist; b. Auch, France, Nov. 10, 1902; s. Julien Joseph and Marie Eva (Dossat) N.; Chem.Engr., U. Toulouse (France), 1924, D.Sc., 1925; m. Georgette Suran, July 6, 1926 (dec. Jan. 1942); m. 2d., Lise-Monique Yersin, July 7, 1945; 1 dau., Renée Georgette. Research chemist Etablissements Antoine Chiris, Grasse, France, 1927-30, head research and analysis lab., 1930-35; tech. dir. Soc. Biotix, Pantin, France, 1936-37; research chemist Givaudan S.A., Vernier-Geneva, Switzerland, 1938-54, sci. dir., 1954—; lectr. chem. tech. U. Geneva, 1944-52; prof. plant chemistry U. Neuchatel (Switzerland), 1965—. Mem. French Chem. Soc. (past mem. council), Swiss Chem. Soc., Soc. technique parfumeurs de France, Institut Nat. Genevois, Assn. suisse Prof. Univ. Decorated Chevalier Légion d'Honneur (France), 1951; recipient Fritzsche award Am. Chem. Soc., 1952. Author: (with G. Mazuyer) Natural Perfume Materials, 1947; also numerous articles, chpt. in books. Patentee in field. Research on essential oils and natural perfumes, terpene and sesquiterpene chemistry. Home: 14 Chemin des Erables, 1213 Petit-Lancy, Geneva. Office: Givaudan Societé Anonyme, 1214 Vernier, Geneva, Switzerland.*

NAVIER, Claude-Louis-Marie, French engr.; b. Dijon, France, Feb. 15, 1785; student Poly. Sch. Chief engr. bridges and rds.; prof. analysis and mechanics Poly. Sch., also Sch. Bridges and Rds.; Mem. French Acad. Scis., 1824. Author: Mémoire sur laflexion des verges élastiques courbes, 1819; Mémoire sur les lois de l'équilibre et du mouvement des corps solides élastiques, 1821. A founder bldg. mechanics; studied water lifting gear, 1818, tensile strength, 1826, flow of elastic liquids in tubes, 1830; outlined (with Cauchy, Poisson) theory of elasticity, 1821; built bridge over Seine; studied tensor analysis. Died Paris, Aug. 21, 1836.

NAVIER, Pierre-Toussaint, French physician, chemist; b. St.-Dizier, France, Nov. 1, 1712; M.D., U. Rheims (France), 1741; practiced medicine, Châlons-sur-Marne, France; mem. French Acad. Scis., 1742; founder Soc. Scis. Author: Mémoire sur la dévouverte de l'ether nitreux, 1741; Observations sur le ramollissement des os en général, 1757. Discovered nitrous acid, nitrous ether; produced some compounds of iron and mercury previously considered impossible; research on bone diseases, smallpox, plagues. Died July 16, 1779.

NAVROTSKY, Vasiliy Korneevich, Russian hygienist; b. 1897; grad. Khrakov Med. Inst., 1921; D.Med. Sci., 1939. Sr. asso., dep. head Ukrainian Central Inst. Labor Hygiene and Occupational Diseases, 1927-41, dir., 1944-53; lectr. chair labor hygiene Kharkov Med. Inst., 1930-42; head chair labor hygiene Novosibirsk Postgrad. Med. Inst., 1942-43; prof., 1942—; head chair hygiene Kharkov Postgrad. Med. Inst., 1944—. Mem. USSR Acad. Med. Scis., Kharkov Soc. Hygienists (chmn.). Author: Labor Hygiene in the Coal-Mining Industry, 1933. Co-editor Hygiene sect. Large Med. Ency., 2d edit.; mem. editorial council Labor Hygiene and Occupational Diseases, Hygiene and Sanitation. Research and numerous publs. on exptl. gen. and indsl. toxicology, indsl. environment, silicosis, statistics of occupational diseases, labor hygiene in soda, coke-chem., aniline dye, coal and ore mining industry, labor physiology in ore industry and metallurgy. Address: Kharkov Postgrad. Med. Inst., ulitsa Artema 8, Kharkov, Ukrainian SSR, USSR.

NAYLOR, Aubrey Willard, Am. plant physiologist; b. Union City, Tenn., Feb. 5, 1915; s. Harry Joseph and Clara (Isbell) N.; B.S., U. Chgo., 1937, M.S., 1938, Ph.D., 1940; m. Frances Valentine Lloyd, Dec. 26, 1940; children—Virginia Dawson, Edith-Margaret Lloyd. Staff, Bur. Plant Industry, U.S. Dept. Agr., Chgo., 1938-40; instr. U. Chgo., 1940-44, Northwestern U., Evanston, Ill., 1944-45, Boyce Thompson Inst. for Plant Research, Yonkers, N.Y., 1945-56; asst. prof. U. Wash., Seattle, 1946-47, Yale, 1947-52; faculty Duke, Durham, N.C., 1952—, prof. plant physiology, 1958—; vis. prof. U. Bristol (Eng.), 1958-59. Program dir. for metabolic biology NSF, Washington, 1961-62; mem. Commn. on Undergrad. Edn. in Biol. Scis., 1964—. NRC fellow in biology, 1945-46; NSF sr. postdoctoral fellow, 1958-59. Mem. Am. Soc. Plant Physiologists (pres. 1961), Bot. Soc. Am., Scandinavian, Japanese, Am. (pres. 1961) Socs. plant physiologists, Soc. for Exptl. Biology, Biochem. Soc., A.A.A.S. Editor: South Atlantic Quarterly, 1966-68; editorial bd. Ann. Rev. Plant Physiology, 1964—. Research and publs. on control plant growth and flowering, action of growth regulating compounds, enzyme control of differentiation, use maleic hydrazide for tobacco, fluorescent lamps for plant growth. Home: 2430 Wrightwood Av., Durham, N.C. 27705.*

NAYLOR, Derek, mathematician; b. Eng., Nov. 9, 1929; s. John Samuel and Miriam (Armstrong) N.; B.Sc., U. London, 1951, Ph.D., 1953; m. Kathleen Hilda Pemberton, Aug. 15, 1960; children—Marvin Brian, Steven Terence, Andrew John. Aerodynamicist, A.V. Roe & Co., Manchester, Eng., 1954-56; asst. prof. U. Toronto (Ont., Can.), 1956-62; sr. lectr. U. Strathclyde, Glasgow, Scotland, 1962-63; prof. U. Western Ont., London, 1963—. Fellow Inst. Math. and Applications; mem. Canadian Math. Congress (past mem. council), London Math. Soc. Author: (with G.F.D. Duff) Differential Equations of Applied Mathematics, 1966; also articles. Research in aerodynamics, fluid dynamics, applied math., boundary problems of math. physics, eigenfunction expansions. Home: 594 Middlewoods Dr., London, Ont., Can.*

NEAL, Josephine Bicknell, Am. physician; b. Belmont, Me., Oct. 10, 1880; d. Alton J. and Mary (Alexander) Neal; A.B., Bates Coll., 1901, hon. D.Sc., 1926; M.D., Cornell U., 1910; D.Sc. (hon.), Russell Sage Coll., 1937. Licensed to practice medicine, N.Y. State, 1913, practice limited to consultation in neurology, 1918—; asst. in meningitis div., research lab., Dept. Health, N.Y.C., 1910-14, in charge of div., 1914—; instr. in medicine, Cornell U. Med. Coll., 1914-20; instr. in medicine, Coll. Phys. and Surg., Columbia, 1922-27, clin. prof. neurology, 1929-44; attending physician Children's Tb Clinic and Vanderbilt Clinic, 1922-27; cons. N.Y. Infirmary for Women and Children, from 1925, Neurol. Inst. of N.Y., 1936-44, Vanderbilt Clinic, 1936-44; dir. dept. infectious diseases Neurol. Inst., 1937-39; vis. pjhysician on neurol. service, Willard Parker Hosp., from 1937; asst. attending neurologist, Neurol. Inst. and Vanderbilt Clinic, from 1939; cons. St. Vincent's Hosp.; asso. dir. research lab., N.Y. Dept. Health, 1937-44; dir. William J. Matheson Survey of Epidemic Encephalitis, 1927-29; exec. sec. Matheson Commn. for Encephalitis Research, 1929—; sec. Internat. Com. for Study of Infantile Paralysis, 1929-32. Diplomate Am. Bd. Psychiatry and Neurology. Fellow. A.C.P.; mem. N.Y. State, N.Y. County med. socs., A.M.A., Am. Pub. Health Assn., Assn. for Research in Nervous and Mental Diseases, N.Y. Acad. Medicine, Phi Beta Kappa. Author chpts. in Abt's System of Pediatrics; Tice's Practice of Medicine;

Barr's Modern Medical Therapy in General Practice; 1st, 2d, and 3d Reports on Epidemic Encephalitis (Matheson commn.); Poliomyelitis (Internat. Commn. for Study of Infantile Paralysis); The Human Cerebral Spinal Fluid; Infections of the Central Nervous System; chpt. Viral Diseases of the Central Nervous System in Cyclo. of Medicine, 1940; Encephalitis, A Clinical Study, 1942; also about 75 articles on acute infections of central nervous system. Died Mar. 19, 1955.

NEAL, Marcus Pinson, Am. pathologist; b. Heflin, Ala., Sept. 23, 1887; s. William Alexander and Ella Jane (Pinson) N.; student U. Ala., 1904-07; M.D., U. Coll. Medicine, Richmond, Va., 1912; m. Mathilde Frances Evers, Apr. 25, 1917 (dec. Feb. 1962); 1 son, Marcus Pinson. Asst. and clin. pathologist Von Ruck Research Lab. for Tb and Winyah Sanitorium, Asheville, N.C., 1913-16; mem. faculty Northwestern U. Med. Sch., 1916-20, Schs. of Medicine and Dentistry, State U. Ia., 1920-22; faculty U. Mo., 1922-63, prof., chmn. dept. pathology, 1922-58, prof. emeritus, 1958—. Diplomate Am. Bd. Pathology. Fellow A.M.A.; mem. So. Med. Assn., Am. Soc. Clin. Pathologists, Am. Soc. Control of Cancer, Am. Med. Writers Assn., Internat. Acad. Pathology, Mo. Acad. Sci., Fifty Year Club Am. Medicine, Omega Upsilon Phi, Sigma Xi, Alpha Omega Alpha, others. Mem. editorial bd. Miss. Valley Med. Jour., 1935-60. Research and publs. on fat necrosis, diseases of the male breast, leukocytes in disease; leukemia and pernicious anemia; tumor studies. Home: 812 Maupin Rd., Columbia, Mo. 65201.*

NEANDER, Michael, German mathematician; b. Joachimsthal, Bohemia, Apr. 3, 1529; student, Wittenberg, Germany; doctorate, Jena, Germany, 1558; became med. prof., Jena, 1560. Author: Synopsis of Weights and Measures, according to the Romans, Athenians . . . (on metrology), 1555; Methodorum in omni genere Artium brevis et succinta; Physicesen sylloge physicia rerum eraditarum ad omnem vitam . . . sphaerica elementa, cum computo ecclesiastico. Died Jena, Oct. 23, 1581.

NEATBY, Edwin Awdas, English physician; b. Barnsley, Eng., 1858; M.D.; m. 2 daus. Dean Missionary Sch. Medicine; cons. physician diseases of women London Homeo. Hosp., Buchanan Hosp., St. Leonards, Leaf Hosp., Eastbourne, Eng.; mem. council Africa Inland Mission. Fellow Royal Soc. Medicine, Royal Soc. Tropical Medicine, Brit. Homeo. Soc. (pres.), Royal Hort. Soc. Author: The Place of Operation in the Treatment of Uterine Fibroids, 1911; (with T.M. Neatby) A Manual of Tropical Diseases and Hygiene for Missionaries; (with T.G. Stonham) A Manual of Homoeotherapeutics; also papers. Died Dec. 1, 1933.

NECKAM, Alexander (Necham or Nequam), English philosopher; b. St. Albans, Eng., Sept. 1157; ed. St. Albans; student Sch. Petit Pons, Paris; became prof., Paris, 1180; schoolmaster, Dunstable, Eng.; became abbot of Cirencester, 1213. Author: De naturis rerum (includes popular ency. sci. knowledge); De laudibus divinae sapientiae; De utensilibus (vocabulary arranged as reading book); De contemptu mundi. First European to describe use of magnet as compass, circa 1180. Died Kempsey, Eng., 1217.

NECKER, Louis, Swiss physicist, mathematician; b. Geneva, Switzerland, Aug. 31, 1730; studied math. under d'Alembert; prof. math. and exptl. physics, Geneva; mem. French Acad. Scis., 1756. Author: De electricitate, 1747; also articles on electricity in Diderot's Ency. Died Geneva, July 31, 1804.

NEDELEC, Nerve, French natural philosopher; bachelor in theology; licentiate, 1307; Dominican monk; apptd. prior, a French province, 1302; attended Dominican deliberations, Paris, 1303; lectr. on sentences, Paris, 1303; elected master gen. of order, 1318. Author: De materia coeli (on influence of heavens on earth, concluding that if motions of heavens stop, so would those on earth; also, no generation, only decomposition would occur). Died Aug. 7, 1323.

NEEDHAM, Dorothy Mary Moyle, English biochemist; b. London, Eng., Sept. 22, 1896; d. John and Ellen (Daves) Moyle; M.A., Girton Coll., Cambridge U., 1923, Ph.D., 1930, Sc.D., 1945; m. Joseph Needham, Sept. 13, 1924. Research worker Food Investigation Bd., 1920-24, Dept. Sci. and Indsl. Research, 1924-25, Med. Research Council, 1946-52, Agrl. Research Council, 1955-62; chem. and biochem. mem. Brit. Sci. Mission in China, 1944-45. Broodbank fellow Cambridge U., 1952-55; Beit Meml. Research fellow Beit Trust, 1925-28; fellow Lucy Cavendish Coll., Cambridge U., 1964. Recipient Foulerton Research award, 1961-62, Leverhulme award 1948. Fellow Royal Soc., 1948; mem. Biochem. Soc. Author: Biochemistry of Muscle, 1934. Publs. on biochemistry of muscle contraction; elucidation of some essential steps in phosphorylation cycles for energy prodn. Address: Master's Lodge, Gonville and Caius Coll., Cambridge U., Cambridge, Eng.*

NEEDHAM, James George, Am. naturalist, entomologist; b. Virginia, Ill., Mar. 18, 1869; s. John and Barbara (Cooper) N.; B.S., Knox Coll., 1891, M.S., 1893, Litt.B., 1921; Ph.D., Cornell U., 1893; Sc.D., Lake Forest Coll., 1930; m. Anna Belle Taylor, July 26, 1894. Asst., Johns Hopkins Marine Biol. Lab., 1894; instr. biology Knox Coll., 1895-96; prof. biology Lake Forest Coll., 1898-1907; prof. entomology and limnology Cornell U., 1907-36, emeritus, from 1936. Recipient Thurston medal Caius Coll., 1934. Mem. Ecol. Soc. Am. (v.p.), Limnological Soc. Am. (pres.). Author: General Biology, 1909; The Life of Inland Waters, 1915; A Guide to the Study of Fresh-water Biology, 1928; (with H. B. Heywood) Handbook of the Dragonflies of North America, 1929; Manual of the Dragon Flies of China, 1929; The Animal World, 1931; (with J. R. Traver and Y. C. Hsu) The Biology of Mayflies, 1935. Studied dragon flies at Archbold Biol. Sta., Fla. Died Ithaca, N.Y., July 25, 1957.

NEEDHAM, John Tuberville, naturalist; b. London, Eng., Sept. 10, 1713; s. John and Margaret (Lucas); ed. Douai (formerly Doucey, France). Ordained priest Roman Catholic Ch., Douai, 1732; founder, dir. Acad. Scis., Brussels, Belgium; collaborated with Buffon in sci. work. Fellow Royal Soc., 1746; mem. French Acad. Scis., 1768. Author: Recherches physiques et métaphysique sur la nature et la religion, 1769. Discovered (with Spallanzani) apparently dead vinegar eels, rotifers, tardigrades revived when placed in water; believed in spontaneous generation, erroneously believed he produced living organisms in sealed tube of beef broth, 1740, refuted, 1770; discovered pollen grains in water expand, extrude papillae at pores and break open to release their contents, 1740; studied micrographics. Died Brussels, Belgium, Dec. 30, 1781.

NEEDHAM, Joseph, English biochemist, orientalist; b. London, Eng., Dec. 9, 1900; s. Joseph and Alicia (Montgomery) N.; M.A., Gonville and Caius Coll., Cambridge U., 1925, Ph.D., 1926, Sc.D., 1932; D.Sc. (hon.), U. Brussels, 1954; m. Dorothy M. Moyle, Sept. 13, 1924. Demonstrator in biochemistry Cambridge U., 1928-33, Sir William Dunn Reader in biochemistry, 1933-66; fellow Caius Coll., 1924-66, librarian, 1959-60, pres., 1959-65, master, 1966—; dir. Brit. Sci. Mission in China, counsellor Brit. Embassy Chungking, 1942-46, dir. div. natural sci. UNESCO, 1946-48. Decorated Order of Brilliant Star, China. Fellow Royal Soc., 1941; mem. Internat. Acads. History Sci. and Medicine, Internat. Inst. Embryology, Academia Sinica (fgn.). Author: Chemical Embryology, 3 vols., 1931; Order and Life, 1936; Biochemistry and Morphogenesis, 1942; Science Outpost, 1946; Science and Civilization in China, 7 vols., 1954—; Heavenly Clockwork, 1960; Clerks and Craftsmen in China and the West, 1968; The Grand Titration: Science and Society in China and the West, 1968; Within the Four Seas: the Dialogue of East and West, 1968. Study, publs. on induction phenomena in embryolog. biochemistry; discovery of indirect induction; pioneering studies of science in Chinese civilization. Address: Master's Lodge, Gonville and Caius Coll., Cambridge U., Cambridge, Eng.*

NEEDHAM, J(oseph) Garton, Am. psychologist; b. Brandon, Vt., Nov. 22, 1908; s. Charles VanNess and Florence (Garton) N.; A.B. with distinction Boston U., 1930; A.M., Harvard U., 1932, Ph.D., 1933; m. Eleanor Nina Beverstock, Sept. 9, 1933; children—Deborah Garton (Mrs. John Henry Bratzler), Bruce Herbert. NRC fellow Princeton U., 1933-34; resident psychologist N.J. State Colony, New Lisbon, 1934-36; with Simmons Coll., 1936—, prof. psychology, 1946—, v.p., 1946—, dean instrn., 1957-65. Mem. Am., Eastern, psychol. assns., Soc. Psychol. Study of Social Issues, Sigma Xi. Author: (with W.F. Vaughan) Student's Manual for General Psychology, 1936, rev., 1939. Research and publs. on early exptl. studies of factors influencing psychophys. judgmental process; research upon socio-econ. factors in disease and defect; studies of student attitudes and institutional behavior. Home: 39 Elmwood Rd., Wellesley, Mass. 02081.*

NEEL, James Van Gundia, Am. geneticist; b. Hamilton, O., Mar. 22, 1915; s. Hiram A. and Elizabeth (Van Gundia) N.; A.B., Coll. Wooster, 1935, D.Sc., 1959; Ph.D., U. Rochester, 1939, M.D., 1944; m. Priscilla Baxter, May 6, 1943; children—Frances, James Van Gundia, Alexander. Instr. zoology Dartmouth Coll., 1939-41; fellow in zoology NRC, 1941-42; faculty U. Mich., Ann Arbor, 1948—, prof. human genetics, chmn. dept., 1956—, prof. internal medicine, 1957—, Russel lectr., 1966. Cons. USPHS, 1956—, NRC, 1949—, WHO, 1957—. Recipient Albert Lasker award Am. Pub. Health Assn., 1960. Mem. Nat. Acad. Sci., Am. Soc. Human Genetics (pres. 1953-54 Allen award, 1965), Assn. Am. Physicians, Am. Philos. Soc., Genetics Soc. Am., Japanese Soc. Human Genetics, Brazilian Soc. Genetics, Phi Beta Kappa, Sigma Xi, Alpha Omega Alpha. Human genetic studies: mutation rates, blood abnormalities, radiation and inbreeding effects, and population genetics of primitive man. Home: 2235 Belmont St., Ann Arbor, Mich. 48104.*

NÉEL, Jean Maurice, French chemist; b. Laval, France, Mar. 4, 1925; s. Jean Maurice and Gabrielle (Groupil) N.; Piplôme Ingénieur Chimiste, École Supérieure de Physique et Chimie, Paris, 1948; Doctorat ès Scis. Physiques, Faculty Scis., Paris, 1957; m. Andrée Bouvier, July 27, 1954; children—Laurent, Francois. Scholar, macromolecular chemistry lab. of G. Champetier, Nat. Center Sci. Research, Paris, 1948-50, instr., asst. to Champetier, 1950-60, research master at the center, 1960-62; prof. Faculty Scis., Nancy, France also prof. organic indsl. chemistry Nat. Superior Sch. Indsl. Chemistry, Nancy, 1962—. Mem. Chem. Soc. France, Soc. Phys. Chemistry, Am. Chem. Soc. (polymer, rubber chemistry divs.). Author: Structure chimique des polyosides, 1966; also papers. Research on physico-chem. aspects of macromaolecules, systhesis of macromolecules, conformational analysis, molecular interactions. Home: 24, ave. du Chateau, 54 Villers les Nancy, France.*

NEEL, Joe Kendall, Am. limnologist; b. Tacoma, Va., June 12, 1915; s. George Frank and Evelyn (Herold) N.; B.S., U. Ky., M.S., 1938; Ph.D. (Rackham fellow), U. Mich., 1947; m. Erma Bernice Baugh, Aug. 20, 1942; 1 son, Joe Kendall. Faculty U. Ky., 1938-39, 1947-50; staff USPHS, 1942-46, 1950-63; dir. potamological Inst., U. Louisville, 1963-66, prof. biology, U.N.D., Grand Forks, 1966—. Mem. Am. Soc. Limnology and Oceanography, Am. Micros. Soc., Ecol. Soc. Am., Am. Inst. Biol. Scis., A.A.A.S. Author: (with others) Limnology in North America, 1963. Research and publs. on phys. and chem. interrelations in streams, sewage treatment by lagoon or oxidation pond method, reservoirs and their effects on water quality in rivers, effects varied pollutants in natural waters, algal influences on water quality. Home: 2221 Chestnut St., Grand Forks, N.D. 58201.*

NÉEL, Louis Eugène Félix, French physicist; b. Lyons, France, Nov. 22, 1904; s. Louis Antoine and Marie Antoinette (Hartmayer) N.; Agrégé de l'Université, Ecole Normale Supérieure, 1928; Docteur ès-Sciences, Strasbourg, France, 1932; m. Hélène Hourticq, Sept. 14, 1931; children—Marie-Francoise, Marguerite Guély, Pierre. With Faculté des Sciences, Strasbourg, 1928-45, prof., 1937-45; prof. Faculté des Sciences Grenoble (France), 1945—; dir. Lab. Electrostatics and Physics of Metal, 1940—; dir. Institut Polytechnique, Grenoble, 1954—; dir. Centre d'Etudes Nucléaires, Grenoble, 1957—. French rep. sci. council NATO; Recipient Legion of Honor, Gold medal Nat. Center Sci. Research. Mem. French Acad. Scis., acads. of sci. of Moscow, Halle, Royal Soc. London, Rumanian Acad., Royal Netherlands Acad. Scis., Am. Acad. Arts and Scis., French Soc. Physics (hon. pres.), Internat. Union Pure and Applied Physics. Research and numerous publs. on magnetic properties of solids; introduced sci. ideas of ferromagnetism and antiferromagnetism; discoveries of certain magnetic properties of fine grains and crystals, directional order of magnetism, magnetic hauling. Home: 41, avenue Maréchal Randon, Grenoble (38) Office: Centre d'Etudes Nucléaires, B.P. 269, rue des Martyrs, Grenoble, France.

NEELY, (Robert) Dan, Am. plant pathologist; b. Senath, Mo., Oct. 6, 1928; s. Robert Dewey and Annie (Robinson) N.; student Westminster Coll., Fulton, Mo., 1946-47; B.S. in Agr., U. Mo., 1950, Ph.D. in Botany, 1957; m. Betty Ann Slaughter, Dec. 27, 1953; children—Claire, Jean. Asst. plant pathologist Ill. Natural History Survey, Urbana, 1957-60, asso. plant pathologist, 1960-65, plant pathologist, 1965—. Mem. Am. Phytopath. Soc., Mycol. Soc. Am., Internat. Shade Tree Conf. (dir. 1962-66), Am. Forestry Assn., Ill. Acad. Sci. Research on etiology and control of Dutch elm disease, sycamore anthracnose, the juniper rusts, and leaf bloth of horsechestnut; devel. techniques used to measure fungistatic and fungicidal properties of chems. and a bioassay to detect movement of fungitoxic chems. through vascular system of plants. Home: 1306 W. Springfield Av., Champaign, Ill. 61820. Office: Natural Resources Bldg., Urbana, Ill. 61801.*

NE'EMAN, Yuval, Israeli physicist; b. Tel-Aviv, Israel, May 14, 1925; s. Gedalia and Zipora (Ben Ya'acov) N.; B.Sc. in Engring., Israel Inst. Tech., Haifa, 1945; Dip.Mech.Ing., 1946; Dip.E.M., Ecole de Guerre, Paris, 1952; Ph.D. in Physics, U. London, 1961; Diploma, Imperial Coll., London, 1961; m. Dvora Rubinstein, June 28, 1951; children—Anath, Tid'al Avdath. Hydrodynamical design engr. pump factory, 1945-46; with Hagana (Jewish Underground), Israel, 1946-48; commd. maj. inf. Israel Def. Forces, 1948, advanced through grades to col., 1955; dep. dir. intelligence div., 1955-57; mil. attache London Embassy, 1958-60; resigned, 1960; with Tel-Aviv U., 1955—, head physics dept., 1963—, prof. physics, v.p., 1965—; sci. dir. Soreq Research Establishment, Israel Atomic Energy Commn., 1962-63; vis. prof. Cal. Inst. Tech., 1964-65; asso. Internat. Center for Theoretical Physics, Trieste, Italy. Mem. Israeli, Am. phys. socs., Inst. Physics and Phys. Soc. (London), Inst. Strategic Studies (London). Author: (with M. Gell-Mann) The Eightfold Way, 1964; also articles. Research in nuclear and particle physics; orginated theory of unitary symmetry for classifying elementary particles; research in astrophysics. Home: 6 Ehud, Tel Aviv, Israel.*

NEES VON ESENBECK, Christian Gottfried, German botanist; b. Erbach, Germany, Feb. 14, 1776; M.D.; became prof. botany, Erlangen, Germany, 1818, Bonn, 1819, Berlin, 1848; dir. Erlangen Bot. Gardens. Author: Handbuch der Botanik, 2 vols., 1820-22; Bryologia germanica, 2 vols., 1823-31; Naturgeschichte der europäischen Lebermoose, 4 vols., 1833-38; Hymenopterorum icheumonibus affinium monographia, 1834. Research on fresh water algae, mushroom classification. Died Breslau, Germany (now Wroclaw, Poland), Mar. 16, 1858.

NEES VON ESENBECK, Theodor Friedrich Ludwig, German botanist, pharmacognosist; b. Erbach, Germany, July 26, 1787; prof., Bonn, Germany. Author: Handbuch der mediz.-pharmaz. Botanik, 3 vols., 1830-32; Genera plantarum florae germanicae, 1833-60. Worked in descriptive botany. Died Hyères, France, Dec. 12, 1837.

NEF, John Ulric, chemist; b. Herisau, Switzerland, June 14, 1862; s. John Ulric and Anna Katharine (Mock) N.; came to U.S., 1868; A.B., Harvard, 1884; Kirkland fellow, Harvard, 1884-87; Ph.D., U. Munich, 1886; m. Louise Bates Comstock, May 17, 1898. Prof. chemistry and dir. chem. lab. Purdue, 1887-89; asst. prof. chemistry and acting head chem. lab. Clark U., 1889-92; prof. chemistry U. Chgo., 1892-96, head dept., from 1896. Fellow Am. Acad. Arts and Scis. Pioneer in research on bivalent carbon, fulminates, mechanism of organic reactions; demonstrated that Couper's method of writing structural formula of organic compounds was correct, thus contbd. to usefulness of Kekulé system. Died Carmel, Cal., Aug. 13, 1915.

NEFF, William Duwayne, Am. physiol. psychologist; b. Lomax, Ill., Oct. 27, 1912; s. Lyman M. and Emma (Jacobson) N.; A.B., U. Ill., 1936; Ph.D., U. Rochester, 1940; m. (Ernestine Anderson), Aug. 14, 1937 (div. Dec. 1960); children—Carol Jean (Mrs. William Fritsch), Peter Lyman; m. 2d, Frances Palmer Anderson, Aug. 23, 1961. Research asso. Swarthmore Coll., 1940-42; civilian scientist NDRC, 1942-46; faculty U. Chgo., 1946-61; dir. psychophysiology lab. Bolt, Beranek and Newman, Cambridge, Mass., 1961-63; prof. psychology Ind. U., Bloomington, 1963-64, research prof., 1964—, dir. Center for Neural Scis., 1965—; cons. to govt. agys.; sci. liason officer Office Naval Research, London, 1953-54, Am. Standards Assn., 1959—, Deafness Research Fedn. 1959-64. Mem. Nat. Acad. Scis., Acoustical Soc. Am., Am. Physiol. Soc., A.A.A.S., Soc. Exptl. Psychology, Psychonomic Soc., Am. Acad. Neurology, Royal Soc. Medicine, Internat. Soc. Audiology, Internat. Brain Research Orgn. Editor: Contribution to Sensory Physiology, 1965—. Research and publs., chpts. in books on central neural mechanisms of hearing and learning. Home: 3505 Bradley St., Bloomington, Ind. 47401.*

NEGOVSKY, Vladimir Aleksandrovich, Russian pathophysiologist; b. 1909; grad. 2d Moscow Med. Inst., 1933; D.Med. Sci. Head, Lab. Exptl. Physiology of Resuscitation, USSR Acad. Med. Scis., 1936—. Recipient State prize. Author numerous works including: Reviving the Vital Functions of a Body in a State of Agony or Clinical Death, 1943; Pathophysiology and Therapy of Agony and Clinical Death, 1954; Resuscitation and Artificial Hypothermia, 1960; Principles of Reanimatology, 1966; External Cardiac Massage and Expirational Respiration, 1966. Mem. (corr.) Toulouse Acad. Scis.; hon. mem. Polish Surg. Soc., German Soc. Clinical Med. Address: Lab. Exptl. Physiology of Resuscitation, USSR Acad. Med. Scis., ulitsa 25 Oktyabrya 9, Moscow, USSR.*

NEGRO, Andalo di, see di Negro, Andalo.

NEHARI, Zeev, mathematician; b. Berlin, Germany, Feb. 2, 1915; s. Saul and Gusta (Rath) Weissbach; M.Sc., Hebrew U., Jerusalem, 1939, Ph.D., 1941; m. Varda Fishzon, Apr. 8, 1938. Came to U.S., 1947, naturalized, 1954. Instr., Hebrew U., 1946-47; mem. research project Harvard, 1947-48; faculty Washington U., St. Louis, 1948-54; prof. math. Carnegie Inst. Tech., Pitts., 1954—. Mem. Am. Math. Soc., Math. Assn. Am. Author: Conformal Mapping, 1952; Introduction to Complex Analysis, 1961. Home: 1540 Beechwood Blvd., Pitts. 15217. Office: Carnegie Inst. Tech., Pitts. 15213.*

NEHER, Henry Victor, Am. physicist; b. Page, Kan., Nov. 4, 1904; s. John Allen and Lena (Kuns) N.; B.A., Pomona Coll., 1926; Ph.D., Cal. Inst. Tech., 1931; m. Sara Elizabeth Yoder, June 29, 1929; children—Philip, Andrew, Merilyn (Mrs. David Harkrider), Eleanor, Stephen. Mem. faculty Cal. Inst. Tech., Pasadena, 1933—, prof. physics, 1944—; staff mem., research on radar Mass. Inst. Tech., 1940-45; mem. vis. staff Phys. Research Lab., Ahmedabad, India, 1955-56. Mem. Am. Phys. Soc., Am. Assn. Physics Tchrs., Am. Geophys. Union, Phi Beta Kappa, Sigma Xi, Tau Beta Pi. Author: (with John D. Strong) Procedures in Experimental Physics, 1936; Progress in Cosmic-Ray Physics, 1953. Research on cosmic rays; use of balloons in atmosphere, and space probes. Home: 885 N. Holliston St., Pasadena, Cal. 91104.*

NEHNEVAJSA, Jiri, behavioral scientist; b. Dyjakovice, Czechoslovakia, Aug. 9, 1925; s. Stepan Jan and Josefa (Hrubosoval) N.; student Masaryk U., Brno, Czechoslovakia, 1945-48, U. Lausanne (Switzerland), 1948-49; Ph.D., U. Zurich (Switzerland), 1953; m. Vera Marie Jelinkova, Jan. 13, 1949; children—Peter Bruce, David, Michael George. Came to U.S., 1951, naturalized, 1956. Editor Svobodne noviny, Brno, Czechoslovakia, 1945-48; corr. Radio Brno, 1946-48; mem. research staff conservation human resources project Columbia Grad. Sch. Bus., 1951, asst. prof. sociology, 1956-61; instr., then asst. prof. sociology U. Colo., 1951-56; chmn. products control com. System Devel. Corp., Paramus, N.J., 1960-61; prof. sociology U. Pitts., 1961—, chmn. dept., 1962—; cons. in field, 1956—; pres., dir. Overseas Systems Corp., N.Y.C., 1960-61; cons. social affairs dept. Pan Am. Union, 1962-63. Mem. sci. adv. bd. Consad Research Corp., 1965—; chmn. internat. bd. Nat. Research Colombia, 1966—. Mem. Am. Sociol. Assn., Am. Assn. Pub. Opinion Research, Am. Inst. Aero. and Astronautics, Air Force Assn., Am. Ordnance Assn., United World Federalists, Czechoslovak Acad. Arts and Scis. in Am. Co-author: Sociometry Reader, 1961; also articles, chpts. in books. Co-editor, co-author: Message Diffusion, 1956. Research on issues of peace and war; devel. research methods for analysis of future contingencies and problems of socio-polit. devel. Home: 1520 W. Ingomar Rd., Pitts. 15237.*

NEI, Tokio, Japanese microbiologist; b. Hokkaido, Japan, May 24, 1913; s. Seisaku and Fusano (Inazuka) N.; grad. Med. Sch., Hokkaido U., M.D., 1943; m. Yukiko Tamura, Oct. 17, 1942; children—Kimiko, Akiko, Keiko. Prof., Inst. Low Temperature Sci., Hokkaido U., Sapporo, 1948—, dir., 1954—. Mem. Soc. Cryobiology, Japan Soc. Microbiology, Japan Soc. Electron Microscopy. Research, numerous publs. on mechanism of freezing injury of microorganisms and mammalian cells, freeze-drying of biol. materials. Home: 1883 Fushimi-cho, Sapporo, Japan.*

NEI, Wada, Japanese mathematician; b. 1787; perfected yenri, or circle-principle, by developing an integral calculus useful for ordinary mensuration; worked on maxima, minima, roulettes. Died 1840.

NEIBURGER, Morris, Am. meteorologist; b. Hazleton, Pa., Dec. 5, 1910; s. Samuel and Rose (Krakusin) N.; student Crane Jr. Coll., 1928-29, Central YMCA Coll., Chgo., 1934-35; S.B., U. Chgo., 1936, Ph.D., 1945; fellow Mass. Inst. Tech., 1938-41; m. Beatrice Francis Rosenzweig, June 18, 1932; 1 son, Carl David. With U.S. Weather Bur., 1930-40; instr. Mass. Inst. Tech., 1940-41; faculty U. Cal. at Los Angeles, 1941—, prof. meteorology, 1954—, chmn. dept., 1956-62. Sr. meteorologist Air Pollution Found., Los Angeles, 1954-56; cons. Los Angeles Air Pollution Control Dist.; mem. Nat. Acad. Scis. com. on internat. programs in meteorology and hydrology, 1962-64; mem. U.S. delegation 4th Internat. Congress World Meteorol. Orgns. Fellow A.A.A.S. (mem. council 1962-66), Am. Geophys. Union (pres. sect. meteorology 1961-64); mem. Am. Meteorol. Soc. (councilor 1954-56, 64—, pres. 1962-64, Meisinger award 1946), Royal Meteorol. Soc. London (fgn.), Geochem. Soc., Sigma Xi. Contbr. chpts. to books, articles to profl. jours. Research on Cal. stratus clouds, air pollution, cloud physics, air trajectories and pressure systems. Home: 1220 Kenter Av., Los Angeles 90049.*

NEIDHARDT, Frederick Carl, Am. biologist; b. Phila., May 12, 1931; s. Adam Fred and Carrie (Fry) N.; B.A., Kenyon Coll., 1952; Ph.D., Harvard, 1956; m. Elizabeth Robinson, June 9, 1956; children—Richard Frederick, Jane Elizabeth. Research fellow Pasteur Inst., Paris, France, 1956-57; H.C. Ernst Research fellow Harvard Med. Sch. 1957-58, instr., 1958-59, asso., 1959-61; faculty Purdue U., West Lafayette, Ind., 1961—, prof., asso. head dept. biol. scis., 1965—. Cons., U.S. Dept. Agr., 1964-65; mem. grant study panel NIH, 1965—. Mem. Am. Soc. Microbiology, Am. Soc. Biol. Chemists, Am. Inst. Biol. Scis., N.Y. Acad. Sci., Soc. Gen. Physiology, Sigma Xi, Phi Beta Kappa. Research and publs. on mechanisms by which bacterial cells regulate synthesis enzymes and nucleic acids during growth. Home: 1628 Western Dr., West Lafayette, Ind. 47906.*

NEIDLEIN, Richard, German pharm. chemist; b. Schwäb. Hall, Württenberg, Germany, Oct. 25, 1930; s. Konrad and Emmy (Frenz) N.; student U. Tübingen (Germany), 1953-56; Ph.D., 1958; Diplomchemist, U. Marburg (Germany), 1960; m. Edda Kuhlmann, Aug. 30, 1962; 1 son, Axel. Asst., Inst. Pharm. Chemistry, U. Tubingen, 1956-58; asst. Inst. Pharm. Chemistry U. Marburg, 1959-63; asst. prof., 1964—. Mem. Deutsche Pharmazeutische Gesellschaft, Gesellschaft Deutscher Chemiker. Research and publs. on solvolysis of cyclic sulfonates, additions to isonitriles, chemistry of various nitrogen compounds. Home: 52 Under dem Gedankenspiel, 3550 Marburg-Wehrda, Germany. Office: 6 Marbacher Weg, 3550 Marburg/1., Hessen, Germany.*

NEILE, William, English mathematician; b. Dec. 7, 1637; s. Sir Paul Neil; became gentleman commoner Wadham Coll., Oxford, 1652; Fellow Royal Soc. 1663. Credited with being 1st to rectify a curve line (a semi-cubical parabola) absolutely, 1657; also rectified cubical parabola; developed a theory of motion. Died Aug. 24, 1670.

NEILSON, James Beaumont, Scottish inventor; b. Shetleston, Scotland, June 22, 1792; s. Walter and Marion (Smith) N.; student phys. and chem. sci. Andersonian U., Glasgow, Scotland; m. Barbara Montgomerie, circa 1815. Began as enginewright of colliery, Irvine, Scotland, 1814; became foreman Glasgow Gasworks, 1817. Fellow Royal Soc., 1846. Discovered substitution of hot blast for a refrigerated one in a furnace produced 3 times as much as iron with the same amount of fuel, tested at Clyde Ironworks and patented, 1828; improved manufacture of gas. Died Queen's Hill, Scotland, Jan. 18, 1865.

NEISSER, Albert Ludwig Siegmund, German dermatologist; b. Schweidnitz, Germany, Jan. 22, 1855; ed. at Erlangen, Germany and Breslau, Germany (now Wroclaw, Poland); M.D., U. Breslau, 1877; prof. dermatology, Breslau, from 1882. Discovered bacillus causing gonorrhea, 1879; named it gonococcus, 1882, and proved its specific nature; devel. of diagnosis, therapy and prophylaxis of gonnorrhea; etiology and path. anatomy leprosy; confirmed and expanded Hansen's work on leprosy bacillus; discovered (simultaneously with F. Schaudinn, P.E. Hoffmann) that Treponema pallidum is cause of syphilis; developed (with Carl Bruck, August von Wassermann) Wassermann test for syphilis, 1906; dir. research in biology, histology, protozoology, bacteriology, X-rays; genus Neisseria and a syringe named after him. Died Breslau, July 23, 1916.

NEISSER, Hans Philipp, economist; b. Breslau, Germany (now Wroclaw, Poland) Sept. 3, 1895; s. Gustav and Else (Silberstein) N.; student U. Freiburg, 1913, U. Munich, 1915; Dr. jur., U. Breslau, 1919, Dr. rer. pol., 1921; m. Charlotte Schroeter, Dec. 22, 1923; children--Marianne (Mrs. Frank Selph), Ulric. Came to U.S., 1933, naturalized, 1939. Economist. govt. commns. Germany, 1922-27; lectr., dep. dir. research U. Kiel, 1927-33; prof. monetary theory U. Pa., 1933-43; prin. economist OPA, Washington, 1941-43; prof. econs. grad. faculty New Sch. Social Research, N.Y.C., 1943-65. Fellow Econometric Soc., Royal Econ. Soc.; mem. Am. Econ. Assn., Internat. Phenomenological Soc. Author: The Exchange Value of Money, 1927; Some International Aspects of the Business Cycle, 1936; (with F. Modigliani) National Incomes and International Trade, 1953; An Essay on the Sociology of Knowledge, 1965. Publs. on monetary theory, bus. cycle theory, internat. econs. Home: 3135 Johnson Av., Riverdale, N.Y. 10463.*

NEISSER, Max, German bacteriologist, hygienist; b. Legnice, Silesia, June 19, 1869; introduced stain for differential diagnosis of diptheria bacillus, 1897, also of staphylococci; developed test for living and dead cells, 1900; described (with R. Lubowski but independently of Friedrich Wechsberg) complement deviation, 1901. Died Frankfort/Main, Germany, Feb. 25, 1938.

NEKRASOV, Boris Vladimirovich, Russian chemist; b. Sept. 18, 1899; grad. Plekhanov Inst. Popular Economy, 1924; with Moscow Textile Inst.; prof. Moscow Kalinin Inst. Non-ferrous Metals and Gold. Corr. mem. Acad. Scis. USSR. Author: Course of General Chemistry, 2 vols., 1935, 12th edit., 1955; Theory of the Structure of Borances, 1940; Electro Affinity of Chemical Elements, 1946; Unusual Valency of Some Metals, 1956. Research on relationship of structure and properties of chem. compounds; proposed theory of structure of borances, 1948, also explanation of trans influence in complex compounds. Home: Zubovskii bulv. 16/20, Moscow. Office: Kalinin Moscow Inst. of Non-Ferrous Metals and Gold, Moscow, USSR.

NEL, Louis Taylor, South African geologist; b. Wolmaransstad, South Africa, Feb. 24, 1895; s. Paul and Mabel (Taylor) N.; M.Sc. cum laude, U. Stellenbosch, South Africa, 1920, D.Sc., 1927; postgrad. U. Freiburg, U. Bonn, 1930; m. Muriel Isabelle Malherbe, Mar. 15, 1932; children—Paul Malherbe, Muriel Mabel (Mrs. J.G.M. Antelme), Louis Hubert Alvin. With Geol. Survey of South Africa, 1920-55, dir., 1948-55, ret., 1955; geol. adviser Atomic Energy Bd. South Africa, 1955—. Bd. curators Transvaal Mus., 1944-63; mem. mgmt. com. Govt. Metall. Lab., 1948-60; mem. Nat. Com. for Advancement Sci., 1955-62. Recipient Prix Spendiaroff, Internat. Geol. Congress, 1929. Fellow Geol. Soc. London; mem. Geol. Soc. S. Africa (Draper Meml. medal 1943, council 1937-64, pres. 1942), S. African Assn. Advancement Sci. (life), S. African Acad. Sci. and Arts (Havenga prize 1955). Author: The Geology of the Country Around Vredefort, 1927; Geology of the Postmasburg Manganese Deposits, 1929; Geology of the Klerksdorp' and Ventersdorp Districts, 1935; also articles, ofcl. publs. Research on Witwatersrand System, Dredefort Dome, genesis of gold and uranium deposits, resources of nuclear raw materials in S. Africa. Home: 49 Barnstable Rd., Lynnwood Manor, Pretoria. Office: Nat. Nuclear Research Center, Pelindaba, Pretoria, S. Africa.*

NÉLATON, Auguste, French physician, surgeon; b. Paris, June 17, 1807; prof. clin. surgery Med. Faculty, U. Paris, 1851-67; surgeon to Napoleon III, 1866; became senator 1868. Mem. French Acad. Medicine, French Surg. Soc. Author: Élements de pathologie chirugicale, 5 vols., 1844-60; De l'influence de la position dans les maladies chirurgicales, 1851. Described line from anterior superior iliac spin to tuberosity of ischium, the tip of greater trochanter is above this line in femur dislocation (Nélaton's line), 1847; attempted cauterization of tumors using Heider's galvanocautery, 1850; desribed pelvic hematocele, 1851; invented probe tipped with porcelain which became stained when it touched an embedded bullet (Nélaton's probe), 1837; developed method for removing calculus without lithotrity. Died Paris, Sept. 21, 1873.

NELSENS, Louis Henri Frédéric, Belgian chemist; b. Louvain, Belgium, July 11, 1814; pupil of Liebig. Asst. to Dumas, at Sorbonne, then in Eng.; prof. physics and chemistry École de medecine vétérinaire, Brussels, Belgium. Author: Mémoire sur l' emploi de l'iodure de potassium, pour combattre les affections saturnines, mercurièlles et les accidents consecutifs de la syphilis, 1865; Des operatonnerres à pointes, à conducteurs et à raicordements terrestres multipter, 1871; also a work on Van Helmont. Contbr. articles to bull. of Brussels Acad. Proved trichloracetic acid and acetic acid have similar formulas; inventor multiple ground connection lightning rod. Died Brussels, Apr. 20, 1886.

NELSON, Bardin Hubert, Am. sociologist; b. Strong, Ark., Aug. 25, 1921; s. William Hubert and Eron (Bardin) N.; B.S., La. State U., 1942, M.A., 1943, Ph.D., 1950; m. Barbara June Campbell, Aug. 12, 1940; children—Bardin Hubert, Howard Campbell. Dir. Bur. Testing and Guidance, La. State U., 1948-50; faculty Tex. A. and M. U., College Station, 1950—, prof. sociology, 1955—; vis. prof. numerous colls., univs. Cons. AID, Dominican Republic, Thailand; cons. Ford Found., U. Aleppo. Editorial rev. bd. Rural Sociology, 1965—. Mem. Am. Sociol. Assn., Rural Sociol. Soc.. Author: Rural Sociology, 1960; Consumer Behavior and the Behavioral Sciences, 1966; Readings in Marketing, 1967; also articles. Research on factors influencing image formation, analysis of social systems with emphasis on motivational qualities of social structure, attitudinal studies, cross cultural studies. Home: 705 Dexter St., College Station, Tex. 77840.*

NELSON, Benjamin (Nathaniel), Am. historian, sociologist; b. N.Y.C., Feb. 11, 1911; s. Marks and Mary (Finesmith) N.; B.A., Coll. City N.Y., 1931; M.A., Columbia, 1933, Ph.D., 1944; m. Marie Alma Louise Poole, Nov. 30, 1959. Permanent fellow univ. seminars Columbia, 1945—, vis. asso. prof. history, contemporary civilization, 1952-53, vis. lectr. sociology, 1963-65; Faculty, U. Chgo., 1945-48; cochmn. Social Sci. Program, chmn. European heritage sci. Humanities Program, U. Minn., 1948-56; prof. history, social sci. Hofstra Coll., 1956-59; prof. history, sociology; chmn. dept. sociology, anthropology State U. N.Y., Stony Brook, Oyster Bay, 1959-65; prof. sociology, history New Sch. For Social Research Grad. Faculty, N.Y.C., 1966—; mem. exec. bd. Conf. Methods in Sci. and Philosophy. Mem. Soc. for Sci. Study Religion (council), A.A.A.S., Am. Anthrop. Assn., Am. Sociol. Assn., Am. Hist. Assn., Mediaeval Acad. Am., Am. Philos. Assn., Am. Soc. Study Religion. Author: The Idea of Usury: From Tribal Brotherhood to Universal Otherhood, 1949; (with others) An Introduction to Social Science: Personality-Work-Community, 3d edit., 1961. Editor: (with others) Freud and the 20th Century, 1957; Psychoanalysis and the Future, 1957; Social Issues of Psychoanalytic Review, 1962, 63, 65; Sigmund Freud: On Creativity and the Unconscious, 1958; sr. cons. editor Harper Torchbooks, 1957—; bd. editors Social Research, 1966—; contbg. editor Psychoanalysis in America: Historical Perspectives, 1966. Research on social-cultural-behavioral scis., social and cultural psychiatry, theory and techniques of paradigmatic psychotherapy, cultural and polit. strains underlying major historic crises over philos. founds. of sci. Home: 29 Woodbine Av., Stony Brook, N.Y. 11790. Office: 66 W. 12th St., N.Y.C. 11790.*

NELSON, Carl Truman, Am. physician; b. Providence, June 27, 1908; s. Edward and Bessie Nelson; A.B., Harvard, 1935, D.M.D., 1932, A.M., 1937, M.D., 1941; m. Evelyn E. Lewis, June 26, 1937. With Harvard, 1932-41, instr., 1935-41; faculty Columbia, 1946—, prof. medicine, chmn. dept., 1951—; dir. dermatology Presbyn. Hosp., N.Y.C., 1951—. Mem. Am. Acad. Dermatology (past pres.), Am. Fedn. Clin. Research, Am. Assn. Immunologists, N.Y. Acad. Medicine, Am. Dermatol. Assn., Alpha Omega Alpha. Research, publs. on tissue electrolyte changes in anaphylaxis, sarcoidosis. Home: 211 Central Park W., N.Y.C. 10024.*

NELSON, Eino, Am. pharmacokineticist; b. Seattle, Mar. 14, 1919; s. Karl Verner and Sanni (Quivaheu) N.; B.S., U. Wash., 1947; Ph.D., U. Wis., 1954; m. Susannah Sarah Finigan, Nov. 1, 1947; children—Karl Thomas Arthur, Susannah Sarah, Timothy John Edward, Jennifer Ruth. Instr. U. Wis., 1954-55; faculty U. Cal., San Francisco, 1953-62; prof., dept. chmn. State U. N.Y. at Buffalo, 1962—. Fellow A.A.A.S., Am. Soc. Clin. Pharmacology and Chemotherapy; mem. Am. Pharm. Assn., Am. Chem. Soc., Am. Soc. Pharmacology and Exptl. Therapeutics, N.Y. Acad. Sci., Sigma Xi, Phi Lambda Upsilon, Rho Chi. Publs. on application of biology, math., and phys. chemistry to aspects of pharmacology. Home: 42 E. Spring St., Williamsville, N.Y. 14221.*

NELSON, John Brockway, Am. animal pathologist; b. Newburyport, Mass., Dec. 9, 1894; s. William Thomas and Sarah (Piper) N.; B.S., Mass. Agrl. Coll., 1917; A.M., Harvard, 1923; Ph.D., U. Mo., 1924; m. Mary Graves, Oct. 5, 1924; children—John Brockway III, Marshall G., Sarah P., Elizabeth T. Asst. bacteriologist Pub. Service Labs., Lexington, Ky., 1919-22; instr. microbiology Mass. Agrl. Coll., 1924; faculty Rockefeller Inst. (now Rockefeller U.), N.Y.C., 1925—, asso. mem. 1938-65, emeritus, 1965. Mem. dysentery commn. Army Epidemiol. Bd., India, Burma, 1944. Recipient Griffin award Animal Care Panel, 1958; citation Am. Coll. Lab. Animal Medicine, 1964. Mem. A.A.A.S., Animal Care Panel, Soc. Exptl. Biology and Medicine, Am. Soc. Microbiology, Am. Soc. Exptl. Pathology, Sigma Xi. Numerous publs. on research in microbiol. and pathol. studies of naturally acquired disease in lab. animals. Home: 498 Stockton Rd., Princeton, N.J. 08540. Office: Rockefeller U., N.Y.C. 10021.*

NELSON, John White, Am. pharmacologist; b. Lafayette, Ind., Sept. 1, 1916; s. John Thomas and Ada (Frazer) N.; B.S., Purdue U., 1938, Ph.D., 1945; M.S., U. Fla., 1939; m. Addie Crosby, Aug. 4, 1939; children—John Crosby, Thomas Richard. Research asst. Eli Lilly Co., Indpls., 1939-40; asst. prof. U. Ga., Athens, 1940-44; research asso. Purdue U., Lafayette, Ind., 1944-45; asso. prof. Ore. State U., Corvallis, 1945-47; faculty Ohio State U., Columbus, 1947—, prof. pharmacology, 1953—; lectr. pharmacology Doctor's Hosp., Columbus, 1955—. Mem. Am. Pharm. Assn., Acad. Pharm. Scis., Am. Soc. Pharmacology, Sigma Xi, Rho Chi. Research and publs. in pharmacology. Home: 176 Deland Av., Columbus, O. 43214.*

NELSON, Paul Andrew, Am. physician; b. Mpls., Nov. 25, 1920; s. Andrew A. and A. Lydia (Anderson) N.; A.B., Gustavus Adolphus Coll., 1941; M.B., U. Minn. Sch. Medicine, 1945, M.D., 1946, M.S. in Phys. Medicine, Mayo Found. U. Minn. Grad. Sch., 1951. With Mayo Clinic, Rochester, Minn., 1948-51, first asst. neurology and phys. medicine, rehab. sects., 1950-51; head dept. phys. med. and rehab. Cleve. Clinic, 1951—. bd. dirs. Council and League for Nursing Cuyahoga County; chmn. profl. services adv. bd. Cuyahoga chpt. United Cerebral Palsy Assn. Diplomate Am. Bd. Phys. Medicine and Rehab. Fellow A.C.P.; mem. Cleve. Acad. Medicine, Ohio, Am. med. assns., Am. Acad., Am. Congress of Phys. Medicine and Rehab., A.A.A.S., Sigma Xi. Mem. editorial bd. Archives of Phys. Medicine and Rehab., 1955—, chmn., 1958—, editor, 1962—; editorial bd. Cleve. Clinic Quar., 1957—; mem. med. adv. bd. Cleve.-Marshall Law Rev., 1957—; editor: Proc. Third Internat. Congress Phys. Medicine, 1962. Research, publs. on applications of ultrasound, treatment of lymphedema and cervicothoracic outlet syndromes, cerebral palsy within a community. Home: 3547 Ingleside Rd., Shaker Heights, O. 44122. Office: 2020 E. 93d St., Cleve. 44106.*

NELSON, Richard Carl, Am. physicist; b. Stillwater, Minn., May 1, 1915; s. Carl Elis and Minnie (Martinson) N.; A.B., U. Minn., 1935, Ph.D., 1938; m. Margaret Starr, Sept. 4, 1943; children—Daniel R.S., Sarah L. Agt., U.S. Dept. Agr., Gainesville, Fla., 1939; research fellow U. Minn., Mpls., 1940-41; spectroscopist Armour and Co., Chgo., 1942; chief chemist, mgr. pectin product div. Citrus Concentrates, Inc., Dunedin, Fla., 1943-46; research asso. Northwestern U., 1946-49; faculty Ohio State U., Columbus, 1949—, prof. physics, 1962—. Mem. com. Q.M. Research and Devel. Adv. Bd., 1959-62; cons. 3M Co., St. Paul, 1960—. Recipient NIH Research Career award, 1964—. Fellow Am. Phys. Soc.; mem. N.Y. Acad. Sci. Research and publs. on properties dyes and organic pigments in relation to their function as energy converters in photosynthesis and other dye-sensitized processes, organic photo- and semiconductors. Home: 117 E. Riverglen Dr., Worthington, O. 43085. Office: 174 W. 18th St., Columbus, O. 43210.*

NELSON, Richard Robert, Am. plant pathologist; b. Austin, Minn., May 23, 1926; s. Victor Robert and Marie (Chapman) N.; B.A., Augsburg Coll., 1950; student St. Olaf Coll., 1945-47; Ph.D., U. Minn., 1954, postdoctoral fellow, 1953-55; m. Sally Ann Hicks, Dec. 5, 1947; children—Richard R., Scot Charles, Shelly Elizabeth, Mark William. Plant pathologist U.S. Dept. Agr., Raleigh, N.C., also faculty N.C. State U., Raleigh, 1955-66; prof. Pa. State U., University Park, 1966—. Recipient Research award Rockefeller Found., 1958- NSF, 1962, 64. Mem. Am. Phytopath. Soc., Mycol. Soc. Am., Genetics Soc., Elisha Mitchell Sci. Soc., N.C. Acad. Sci., Sigma Xi, Gamma Alpha. Research and numerous publs. on genetics and evolution of pathogenic and sexual mechanisms in fungi, taxonomy of fungi, biol. species concept, nature of gene action, predicting genetic potentials in fungi, role of toxins in pathogenic fungi, factors affecting origin and evolution of species, genetic pools for virulence in fungi. Home: 733 N. Mckee St., State College, Pa. 16802. Office: Buckhout Lab., Pa. State U., University Park, Pa. 16801.*

NELSON, Richard Stuart, English physicist; b. London, Eng., May 1, 1937; s. Richard and W.E. (Harrison) N.; B.S. with 1st. class honors in physics, U. Reading, 1958; m. Veronica Mary Beck, May 29, 1965. Sci. officer, metallurgy div. Atomic Energy Research Establishment, Harwell, Berkshire, U.K., 1958-64, sr. sci. officer, 1964—. Fellow Inst. Physics. Research, publs. on atomic collisions in solids and radiation damage; discoverer phenomena of channeling of protons in crystals; studies on implantation of fgn. impurity atoms into solids. Home: 14, Valley Close, Goring-on-Thames, Oxfordshire, U.K. Office: Metallurgy Div., AERE, Harwell, Berkshire, U.K.*

NELSON, Robert Stuart, Am. physician; b. Atlantic City, Apr. 7, 1911; s. Kent and Edith (Wills) N.; B.S., B.M., U. Minn., 1934, M.D., 1935; m. Mary Agnes Groves, July 11, 1936; children—Mary Sheila (Mrs. Jack Pearson), Patricia Wills (Mrs. F. Scott Catlett), Roberta Jean. Commd. 1t. M.C., U.S. Army, 1935, advanced through grades to col., 1945; chief, med. service 98 Gen. Hosp., Munich, Germany, 1949-53, European Hepatitis Research Center, 1950-53; Europe cons. gastroenterology, 1951-53; chief, gastroenterology service Brooke Gen. Hosp., San Antonio, 1954-55; ret., 1955; faculty U. Tex., Houston, 1956—, prof. medicine, 1964—; clin. asso. prof. medicine Baylor U. Coll. Medicine, Houston, 1961—. Fellow A.C.P.; mem. Am. Gastroent. Assn., Am. Soc. for Gastrointestinal Endoscopy (Schindler award 1960, v.p. 1965), A.M.A., Am. Assn. Cancer Research. Author: Gastroscopic Photography, 1966. Devel., publs. on diagnostic and treatment methods for liver, stomach, and gastrointestinal diseases. Home 5213 Memorial Dr., Houston 77007.

NELSON, Samuel James, Canadian geologist; b. Vancouver, B.C., Can., June 2, 1925; s. Robert Thomas and Jenny (Leonard) N.; B.A.Sc., U. B.C., 1948, M.A.Sc., 1950; Ph.D., McGill U., 1952; m. Marjorie Hart, Aug. 22, 1953; children—Marilyn Jean, Eric Robert. Asst. prof. U. N.B., 1952-54; faculty U. Alta., 1954-65, prof. geology, 1963—; prof. geology U. Calgary, 1963—. With Geol. Survey Can., 1946-53, Nfld. Geol. Survey, 1954; cons. to oil cos. Mem. Alta. Soc. Petroleum Geologists, Paleontol. Soc., Profl. Engrs. Alta., Geol. Assn. Can., Sigma Xi. Author: Ordovician Paleontology of Northern Hudson Bay Lowland, 1963; Mississippian Faunas, Western Canada, 1962; (with others) Atlas of Geological History of Western Canada, 1965; also articles. Research and publs. on exploration of geology and fossils of Hudson Bay; description of fossils from Permocarboniferous rocks of Western Can. Home: 1620 Cayuga Dr., Calgary, Alta., Can.*

NELSON, Thomas Morgan, psychologist; b. Naguanee, Mich., Aug. 13, 1924; s. Axel Julius and Myrtle (Thomas) N.; B.A., Mich. State U., 1949, M.A., 1953, Ph.D., 1958; m. Joyce Marjorie Armitage, Dec. 1957; children—Anna Louisa, John Audun, Aurnir Bartley. With Mich State U., 1957-64, research asso. psychology, 1961-64; with U. Alta., Can., 1964—, prof., head dept. psychology, 1967—; acting head dept. psychology, 1965-66; mem. vision com. Nat. Research Council Acad. Sci. Mem. Psychonomic Soc., A.A.A.S., Sigma Xi. Research and publs. on field sensation and perception, especially visual processes. Address: Dept. Psychology, U. Alta., Edmonton, Alta., Can.*

NELSON, Werner Lind, Am. agronomist; b. Manlius, Ill., Oct. 17, 1914; s. Carl Herbert and Ida (Carlson) N.; B.S., U. Ill., 1937, M.S., 1938; Ph.D., Ohio State U., 1940; m. C. Jeanette Wilcox Nov. 16, 1940; children—John, Jean. Asst. agronomist U. Ida., Moscow, 1940-41; faculty N.C. State Coll., Raleigh, 1942-54, prof., 1947-54; dir. N.C. Soil Testing Lab., 1949-52, in charge soil fertility research, 1952-54; Midwest dir. Am. Potash Inst., Lafayette, Ind., 1955-67; sr. v.p., 1967—; regional dir. Found. for Internat. Potash Research, Washington, 1962-67; vis. prof. agronomy U. Ill., 1961; mem. Freedom from Hunger Fertilizer adv. panel FAO, 1961—. Recipient bronze tablet, Gamma Sigma Delta award U. Ill., 1937. Fellow Am. Soc. Agronomy (pres.; Agronomic Service award 1964), A.A.A.S.; mem. Soil Sci. Soc. Am. (past pres., past asso. editor, past chmn. div Internat. Soil Sci. Soc., Farm House (asso.), Sigma Xi, Gamma Sigma Delta, Phi Kappa Phi, Alpha Zeta, Phi Eta Sigma. Author: (with S. L. Tisdale) Soil Fertility and Fertilizers, 1956, rev., 1966; also chpts. in books, numerous sci. papers. Home: 1800 Happy Hollow Rd. Office: 402 Northwestern Av., West Lafayette, Ind. 47906.*

NELSON, Wilbur C(lifton), Am. aero. engr.; b. Flint, Mich., May 9, 1913; s. Gustaf F. and Clara Louise (Linden) N.; B.S.E., U. Mich., 1935, M.S.E., 1937; grad. student Cal. Inst. Tech., 1935; m. Pauline Elizabeth Woodward, Nov. 26, 1937; children—Patricia Anne, Bruce Arthur, John Allan, Gregory Karl, Douglas Patrick, David Erik. Engr., Lockheed Aircraft Corp., Burbank, Cal., 1936; CAA, Washington, 1937-39; project engr., Engring. Projects, Inc., Dayton, O., 1939-40; asst. prof., Ia. State Coll., 1940-42, prof. aero. engring. and head dept., 1942-46; project supr. Johns Hopkins Applied Physics Lab., 1945; prof. aero. engring., 1946—, chmn., 1953—, U. Mich. U.S. Navy Tech. Liaison England, 1948; cons. Swedish Govt. Stockholm, 1949; USAF cons. Europe, 1950-51; NATO-AGARD, 1953—; mem. sci. adv. bd. AEPG, 1958-61; chmn. adv. com. on launch vehicles NASA, 1963-65. Recipient Naval Ordnance Devel. Award, 1946. Registered profl. engr. Licensed pvt. pilot. Mem. Inst. Aero. Scis., Am. Soc. Engring. Edn., Nat. Soc. Profl. Engrs., Sigma Xi, Tau Beta Pi, Phi Kappa Phi, Delta Chi. Author: Airplane Propeller Principles, 1944. Contbr. articles to tech. jours. and revs. Applied research in aircraft propulsion, aerodynamics, aircraft design and space systems. Home: 1540 Cedar Bend Dr., Ann Arbor, Mich.*

NELSON, Wilbur Lundine, Am. petroleum engr.; b. Chanute, Kan., Oct. 1, 1904; s. Charles Oscar and Mary (Lundine) N.; student U. Kan., 1923-26; B.S. in Chem. Engring., U. Mich., 1927, M.S., 1928; hon. degree U. Tulsa, 1954; m. Marian G. Roe, Feb. 7, 1930; 1 son, Robert Andrew Nelson. Design engr., cost estimator Smith Engring. Co., Oil City, Pa., Kansas City, Mo., 1928-30; prof. petroleum refining U. Tulsa, 1930—; expert Gyro Process Corp., also Chem. Research Corp., Detroit, 1946-59; Owner, Nelson Refinery Constrn., Cost Index, 1949—, Nelson Refinery Operating Cost Index, 1959—. Cons. to pvt. cos. Recipient award for best tech. series pub., 1950. Mem. Am. Assn. Cost Engrs. (1st award of merit), Am. Chem. Soc., Am. Inst. Chem. Engrs. Author: Petroleum Refinery Engineering, 1936; Venezuelan and Other World Crude Oils, 1952; also numerous articles, other books. Tech. and metal editor Oil and Gas Jour., Tulsa, 1936—. Developed true boiling point method of evaluating crude oils, 1928; invented vacuum decomposition process, 1928; studies in productivity in design, constrn. and operation of process units, yields, design and operation of hydro-cracking plants. Home and office: 1916 E. 38th St., Tulsa 74105.*

NEME, Bussamara, Brazilian physician; b. Sao Paulo, Brazil, Sept. 9, 1915; s. Sad and Rosa (Samara) N.; M.D., U. Sao Paulo, 1941; m. Ruth Lange de Morretes, May 30, 1948; children— Eduardo, Paulo. Faculty, U. Sao Paulo Med. Sch., 1947-60; pvt. docent obstetrics and gynecology Med. U. Brazil, 1960-64; prof. obstetrics Cath. U. Sorocaba Med. Sch., Sao Paulo, 1964-66; prof. obstetrics and gynecology U. Campinas Med. Sch., Sao Paulo, 1966—. Mem. numerous Brazilian, fgn. med. assns. Author: Spinal Anesthesia in Obstetrics, 1950; Basal Temperature, 1953; Effects of Spinal Anesthesia upon Contracility of Pregnant Human Uterus, 1960; also numerous articles. Research on effects of nervous system reactions to uterine automatism. Home: 441 Alameda os Guaiases, Sao Paulo, Brazil.*

NEMEC, Pavel, Czechoslovakian microbiologist, biochemist; b. Prague, Czechoslovakia, May 11, 1912; s. Bohumil and Bozena (Ullrich) N.; RNDr., Charles U., Prague, 1938; postgrad. Komensky U., Bratislava, Czechoslavakia, 1947; m. Maria Placatkova, Aug. 10, 1939; 1 son, Pavel. Demonstrator microbiology Charles U., Prague, 1937-38; head dept. indsl. research Ministry of Industry Bratislava, 1945-48; dir. Research Inst., Dynamite-Nobel Chem. Works, Bratislava; faculty Slovak Poly. U., Bratislava, 1950—, prof. microbiology and biochemistry, 1951—; head dept. microbiology Slovak Acad. Sci., Bratislava, 1956—, dir. Biol. Inst., 1963-65, corr. mem., 1955—. Recipient Nat. award for sci., 1949. Mem. Czechoslovak Microbiol. Soc., Czechoslovak Biochem. Soc., Czechoslovak Cytological Soc. Author: Introduction Into General Microbiology, 1953; Principles of Microbiological Techniques, 1954; also numerous articles. Codiscoverer of specifically active antiprotozoal antibiotics. Home: 29/a Frana Krala, Bratislava, Czechoslovakia.*

NEMIR, Paul, Jr., Am. surgeon; b. Tex., Aug. 30, 1920; s. Paul and Ida (Courry) N.; A.B., U. Tex., 1940; M.D., 1944; m. Helen Powell Pratt, Sept. 10, 1949; children—Virginia Pratt, Paula Hammill, Helen Powell. Asso. surgery Grad. Sch. Medicine, U. Pa., 1951-53, faculty, 1953—, asso. prof. surgery, 1955—, dir. Bockus Research Inst., Grad. Hosp., 1959-61, dean Grad. Sch. Medicine, 1959-64, dir. div. grad. medicine Sch. Medicine, 1964—. Recipient Honors Achievement award Angiology Research Found., 1964-65. Mem. Soc. U. Surgeons, Soc. for Vascular Surgery, Am. Assn. Thoracic Surgery, Am., Surg. Assn., Internat. Surg. Soc., Internat. Cardiovascular Soc. Research on role hemoglobin and hemoglobin derivatives in intestinal obstruction, pancreatitis, shock. Home: 152 Lakeview Dr., Media, Pa. 19063. Office: Med. Labs., 36th St. and Hamilton Walk, Phila. 19104.*

NEMYTSKY, Viktor Vladimirovich, Russian mathematician; b. Smolensk, 1900; grad. Moscow U., 1925, postgrad., until 1939; D.Physico-Math. Sci., 1935. Instr., Moscow U., 1925-35, prof., 1935—. Co-author: The Qualitative Theory of Differential Equations, 1949. Research and publs. on theory of functions of real variable and topology, qualitative theory of differential equations, non-linear equation theory. Address: Moscow University, Leninskie gory, Moscow, USSR.

NENADKEVICH, Konstantin Avtonomovich, Russian chemist, mineralogist; b. June 2, 1880; grad. Moscow U., 1902; with Acad. Scis. USSR, 1906—, in various geol. depts., including geol. and mineral. mus., geol. inst., inst. of mineralogy and geochemistry of rare metals; also corr. mem. Recipient Stalin prize, 1948. Author: Question of the USSR Soda Industry, 1924; Electrolytic Methods of Separating Nickel and Cobalt, 1945. Developed methods of obtaining rare metals from ores; established age of uranite, 1926; devised tech. of prodn. of metallic bismuth from domestic raw materials.

NENITESCU, D. Costin, Rumanian chemist; b. 1902; prof. Bucharest (Rumania) Polytechnic Inst.; mem. Acad. Socialist Republic Rumania, numerous fgn. sci. socs. and acads. Author textbook on gen. chemistry, 1963, textbook on organic chemistry, 6th edit., 1965. Isolated some hydrocarbons in petroleum and determined their constants.

NERLOVE, Marc L., Am. economist; b. Chgo., Oct. 12, 1933; s. Samuel Henry and Evelyn (Andelman) N.; B.A., U. Chgo., 1952; M.A., Johns Hopkins U., 1955, Ph.D., 1956; m. Mary Ellen Lieberman, Feb. 5, 1956; children—Susan, Miriam. Analytical statistician U.S. Dept. Agr., Washington, 1956-57; asso. prof. U. Minn., Mpls., 1959-60; prof. Stanford, 1960-65, Yale, 1965—; F.W. Taussig Research prof., Harvard, Cambridge, Mass., 1967-68. Mem. adv. coms. Bur. Census, CAB. Recipient awards Am. Farm Econ. Assn., 1956, 58, 61. Fellow Am. Statis. Assn., Econometric Soc. (mem. council); mem. Am. Econ. Assn., Phi Beta Kappa, others. Author: Dynamics of Supply, 1958; Dist. Lags and Demand Analysis, 1958; also articles. Formulation and estimation of dynamic econ. models, particularly those involving adjustment processes or expectation formation. Office: P.O. Box 2125, Yale Sta., New Haven 06520.*

NERNST, Walther Hermann, German phys. chemist; b. Briesen, West Prussia, June 25, 1864; ed. Zurich, Berlin, Graz, Würzburg univs.; Ph.D. summa cum laude, Würzburg U., 1887. Became asst. to Ostwald, Graz, 1887, and later in Leipzig; prof. phys. chemistry U. Göttingen, from 1891; prof. U. Berlin, from 1905; dir. Inst. for Exptl. Physics, Berlin, 1924-33. Recipient Nobel prize in chemistry, 1920. Fellow Royal Soc., 1932. Author: Theoretische Chemie von Standpunkte der Avogadroschen Regel und der Thermodynamik, 1893; Die Theoretischen und experimentallen Grundlagen des neuen Wärmesatzes, 1918; Das Weltgebäude im Lichte der neueren Forschung, 1921. Devised theory of electric potential and conduction on electrolytic solutions, 1889; developed third law of thermodynamics (at absolute zero temperature, the entropy of every substance in perfect equilibrium is zero, thus pressure, volume, and surface tension all become independent of temperature), 1906; explained how hydrogen and chlorine explode on exposure to light, 1918; invented an elect. piano, also an elec. incandescent lamp, Nernst's lamp, which used little current and required no vacuum. Died Muskau, nr. Berlin, Germany, Nov. 18, 1941.

NERODE, Anil, Am. mathematician; b. Los Angeles, June 4, 1932; s. Nirad Ranjan and Agnes (Spencer) N.; B.A., U. Chgo., 1949, B.S., 1952, M.S., 1953, Ph.D., 1956; m. Sondra Raines, Feb. 12, 1955; children—Christopher Curtis, Gregory Daniel. Group leader automata and weapon systems Lab. Applied Sci., U. Chgo., 1954-57; mem. Inst. for Advanced Study, Princeton, 1957-58, 62-63; vis. asst. prof. math. U. Cal. at Berkeley, 1958-59; faculty Cornell U., 1959—, prof. math., 1965—, acting dir. Center for Applied Math., 1965-66. Cons., Inst. for Def. Analyses, 1962—, Center for Naval Analyses, 1962—. Mem. Am. Math. Soc. (asso. editor Proceedings 1962-65), Soc. Indsl. and Applied Math., Am. Math. Soc., Math. Assn. Am., Assn. Symbolic Logic. Research in math. logic, theory computability, automata theory. Home: 406 Cayuga Heights Rd., Ithaca, N.Y. 14850.*

NERVI, Pier Luigi, Italian civil engr.; b. Sondrio, June 21, 1891; s. Antonio and Maria Luisa (Bartoli) N.; Dr. Civil Engring., U. Bologne, 1913; Dr. Arch., U. Buenos Aires, 1950; Doctor of Laws (hon.), U. of Edinburgh, 1960; honorary degrees, Technischen Hochschule, Munich, 1960, Warsaw University, 1961, Harvard, 1962, Dartmouth College, 1962; married Irene Calosi, April 27, 1924; children—Antonio, Mario, Carlo, Vittorio. Designer technical department Society Construzioni Cementizie, Bologne, 1913-15, designer and works supt., Florence, Italy, 1918-22; chief designer, works supt. Soc. Ing. Nervi & Nebbiosi, Rome, Italy, 1922-32; chief designer, works supt. Soc. Ing. Nervi & Bartoli, Rome, 1932—, also pres., adminstr.; chief prof. charge dept. tech. and technique of constrn. Faculty Architecture, Rome U., 1946-61. Mem. superior council pub. works Ministry Pub. Works, Rome. Served as engring. officer, 1915-18. Recipient Bronw medal Franklin Inst. of Phila., 1957, Exner medal Austrian Technological Soc. of Vienna, 1958; Gold medal, Royal Institute of British Architects, London, 1960; Gold medal American Inst. Architects, 1964; Cavaliere Ordine Civile di Savoia, 1964. Member Am. Acad. Arts and Scis. (hon. fgn.), Internat. Union Architects (pres. Italian sect.), mem. fgn. sect. Royal Acad. Fine Arts of Stockholm; hon. mem. A.I.A., Am. Acad. Arts, Sci. and Letters. Designed Palermo Ry. Sta., 1946, Vienna Sports Palace, 1956, Olympic Stadium, Rome, 1960, also UNESCO bldgs., Paris; developed techniques for use of metal and reinforced concrete. Home: Lungotevere Arnaldo da Brescia n. 9, Rome, Italy.

NES, William Robert, Am. biochemist; born Oxford, Eng., May 16, 1926 (parents Am. citizens); s. William H. and Mary (Lineback) N.; student U. South, 1943-45; B.A., U. Okla., 1946; Ph.D., U. Va., 1950; m. Estelle Jeanne Shirley, May 16, 1946; children—Shirley Anne, William David. Postdoctoral fellow Mayo Found., Rochester, Minn., 1950-51; vis. scientist NIH, Bethesda, Md., 1951; staff sci. Nat. Inst. Arthritis and Metabolic Diseases, Bethesda, 1951-58; postdoctoral fellow James Forrestal Research Center, Princeton, 1954, Inst. for Exptl. Cancer Research, U. Heidelberg, Germany, 1955; U. Wales, Gt. Britain, 1956; asso. prof. biochemistry Clark U., also sr. sci. Worcester Foudn. for Exptl. Biology, Shrewsbury, Mass., dir. tng. program for steroid biochemistry, 1958-64; prof. chemistry and pharm. chemistry U. Miss., University, 1964—. Cons. NSF, metabolic biology, 1966—; Danforth asso. Mem. Am. Soc. Biol. Chemists, Am. Chem. Soc., Endocrine Soc., A.A.A.A.S., N.Y., Miss. acads. sci., Sigma Xi, Alpha Chi Sigma, Phi Delta Chi. Numerous publs. on research in chemistry and biochemistry of steroids, terpenoids, related naturally occuring molecules. Home: Box 397, University, Miss. 38677.*

NESBITT, Robert Edward Lee, Jr., Am. obstetrician, gynecologist; b. Albany, Ga., Aug. 21, 1924; s. Robert E.L. and Anne Louise (Hill) N.; B.A., Vanderbilt U., 1944; M.D., Vanderbilt Med. Sch., 1947; m. Verne A. Feeback, June 14, 1947; children—Marcia, Mary Beth. Asst. prof. Johns Hopkins, 1954-56, chief Obstetric Pathology Lab., 1955-65, faculty Albany (N.Y.) Med. Coll., Union U., 1956-61; prof., chmn. dept. obstetrics-gynecology State U. N.Y., Upstate Med. Center, Syracuse, 1961—; obstetrician, gynecologist-in-chief Crouse-Irving Hosp., 1963—; chief obstetrics and gynecology State U. Hosp., 1964—, chmn. med. staff and med. bd., 1964-66. Diplomate Am. Bd. Obstetrics and Gynecology (asso. examiner). Fellow Am. Assn. Maternal and Child Health, Am. Coll. Obstetricians and Gynecologists, A.C.S., Venezuelan Obs.. - Gyn. Soc. (hon.); mem. A.M.A., Soc. for Gynecol. Investigation, Pan Am. Med. Assn. (med. ambassador goodwill), N.Y. Acad. Scis., Med. Soc. N.Y. State, Onondaga County Med. Soc., Pub. Health Council N.Y. State; hon. mem. S.W., Fla. obstet. and gynecol. socs., others. Named one of ten Outstanding Young Men U. S. C. of C., 1957. Author: Perinatal Loss in Modern Obstetrics, 1957. Research and publs. on cytologic, cytochem. and histochem. study of early cervical cancer, perinatal and placental pathology, cytologic and hormonal studies in normal and high-risk obstet. patients, exptl. prodn. of abruptio placentae, reproductive endocrinology, animal experimentation, induced endocrine insults upon pregnant and nonpregnant ewes, and hormonal influence on placentation, fetal growth and devel. Home: 112 Circle Rd., Syracuse, N.Y. 13210.*

NESHEIM, Robert Olaf, Am. animal scientist; b. Monroe Center, Ill., Sept. 13, 1921; s. Olaf and Sena Marie (Willms) N.; B.S. with highest honors, U. Ill., 1943, M.S., 1950, Ph.D. in Animal Nutrition, 1951; m. Emogene Pearl Sullivan, July 13, 1946; children—Barbara Jean, Susan Kay and Sandra Lynne (twins). Fieldman, Halderman Farm Mgmt. Service, Wabash, Ind., 1946-48; asst. animal sci. U. Ill., 1948-51; swine research specialist Gen. Mills, Inc., 1951-52; mgr. swine feed research Quaker Oats Co., 1952-59, mgr. livestock feed research, 1959-64; prof. animal sci., head dept. U. Ill., 1964—. Mem. Am. Soc. Animal Sci., Am. Inst. Biol. Scientists, Animal Nutrition Research Council (vice chmn. 1964), Am. Dehydrators Research Council, Alpha Zeta, Phi Kappa Phi (Bronze tablet 1943). Research in amino acid, energy and animal nutrition. Home: 5 Monterey Ct., Champaign, Ill. 61820. Office: Dept. Animal Sci., Univ. Ill., Urbana, Ill. 61803.*

NESMEYANOV, Alexandr Nikolaevitch, Russian organic chemist; b. Moscow, Russia, Sept. 9, 1899; s. Nikolai Vasilievich and Ludmila Danilovna (Rudnitskaia) N.; grad. Moscow State U., 1922, grad. study, 1922-24; Hon. Dr., Calcutta, Jena universities, University of Paris (France), 1964; m. Nina Vladimirovna Koperina, June 6, 1926; children—Olga, Alexandrovna, Nikolai Alexandrovich; m. 2d, Vinogradora Marina Anatolievra, 1962. Asst. prof. Moscow State U., 1924-30, lectures reader, 1930-34, prof., 1934—, head organic chemistry chair, chemistry dept., 1944—, rector, 1948-51; founder organic chemistry lab. Inst. Fertilizers and Insectofungicides, 1930, head, 1930-34; co-founder Lab. Metallo-Organic Chemicstry, 1934; prof. head organic chemistry chair Inst. Fine Tech. of Chemistry, 1938-41; dir. N.D. Zelinski Organic Inst., 1939-54; dir. Inst. Organic Elements' Compounds, 1954—; dep. Supreme Soviet of USSR, 1950-62. Recipient Stalin prize, 1943, M.V. Lomonoso Gold medal, 1962. Fellow Nat. Inst. Scis. of India (hon.); mem. USSR Acad. Scis. (pres. 1951-61); hon. mem. academies of science of Germany, N.Y., Bulgaria, Roumania, Hungary, Czechoslovakia, Poland, Armenia, Tajik, Turkmen, N.Y. (life), Soc. Chem. Industry (hon.), Am. Acad. Arts and Scis. (fgn. hon.), Royal Soc., 1961, Edinburgh Royal Soc. One of most influential scientists in USSR; research, publs. in chemistry of some 27 metallo-organic compounds (lithium, titanium, cadmium, others); devised diazo method of synthesis of metallo-organic compounds, 1929; discovered new classes of compounds; demonstrated relationship between structure and reactivity of metallic derivatives of tautomeric systems. Home: Leninski Gory, University of Moscow, K Flat 105, Moscow. Office: Leninski Prospekt 14, Moscow, USSR.

NESS, Norman Frederick, Am. geophysicist; b. Springfield, Mass., Apr. 15, 1933; s. Herman Hugo and Eva (Carlson) N.; B.S. in Geophysics, Mass.

Inst. Tech., 1955, Ph.D., 1959; m. Amelia Mercaldi, Aug. 25, 1956; children—Elizabeth Ann, Stephen Andrew. Faculty U. Cal., Los Angeles, 1959-61; Nat. Acad. Sci. -NRC postdoctoral asso. NASA-Goddard Space Flight Center, Greenbelt, Md., 1960-61, geophysicist, 1961—. Lectr. math. U. Md., 1962-64, vis. asso. prof., 1965. Recipient John Adam Fleming award Am. Geophys. Union, 1965. Fellow Am. Geophys. Union; mem. A.A.A.S., Sigma Xi. Measurement interplanetary and planetary magnetic fields by satellites; research on earth's magnetic field. Home: 10222 Greenforest Dr., Silver Spring, Md. 20903. Office: NASA, Goddard Space Flight Center, Greenbelt, Md. 20701.*

NESSLER, Julius, German agrl. chemist; b. 1827; contbd. to analytical methods; eponym of Nessler jar, Nessler reagent; authority on winegrowing. Died 1905.

NESTEROV, Anatoly Innoketjevich, Russian therapeutist; b. 1895; grad. Med. Faculty, Tomsk U., 1920; D.Med. Sci., 1929. Intern, asst., lectr. therapeutics clinic Tomsk U., 1920-36; prof. chair hosp. therapy Tomsk Med. Inst., 1931-36, dir. Sochi Clin. Research Inst., head chair rheumatology Central Postgrad. Med. Inst., 1936-39; prof., 1931—; head clinic Moscow Central Inst. Health Resort Therapy, cons. on sci. health resort problems USSR Peoples Commissariat Health, 1939-41; head chair hosp. therapy Novosibirsk Med. Inst., 1941-43; dir Moscow Inst. Physiotherapy, head therapeutic clinic, 1943-50; head chair propedeutics internal diseases Pirogov 2d Moscow Med. Inst., 1947-52, prof. dept. faculty therapy. Chmn. learned council RSFSR Peoples Commissariat Health, 1943-50, dir. Rheumatism Research Inst., 1958—. Chmn., All-Union Com. for Study Rheumatism and Joint Diseases and their Treatment; chmn. All-Union Rheumatism Socs., 1966—; founder Rheumatism Research Inst., 1958. Decorated Order of Lenin (2); Order Labour Red Banner. Named hero of socialist labour, 1966. Mem. USSR Acad. Med. Scis. (acad. sec. 1950-53, v.p. 1953-57), All-Union Soc. Therapeutists (chmn.), Dutch Soc. Rheumatologists (hon.), Czechoslovak Purkinje Med. Soc. (hon.), Am. Rheumatism Assn., Canadian, Swedish, Turkish, Polish, Italian rheumatism socs. Author: The Classification of Rheumatic Diseases and Joint Diseases of Various Origins, 1935; Blood Capillaries and Capillaroscopy, 1939; An Outline Study of Rheumatism and Joint Diseases, 1951; Theory of the Pathogenesis of Rheumatism, 1952; co-author: Principles of Combined Treatment in Hospitals, 1946; Clinical Aspects of Collagenous Diseases, 1961, 2d edit., 1966. Mem. editorial bd. Therapeutic Archives, Large Med. Ency., 2d edit. Research and numerous publs. on rheumatic diseases, use of hormones in prophylaxis and treatment of internal diseases, theory of pathogenesis of rheumatism, clin. aspects of internal diseases, physiotherapy; developer infectious-allergy theory of rheumatism, capillaroscopic method of examination, methods to determine strength of capillaries, capillaromesenchymal tests, method to register venous pulse; compiler classification of diseases. Address: Inst. for Rheumatism, Petrovka 25, Moscow, USSR.*

NESTEROVICH, Nikolay Dmitrievich, Russian botanist, dendrologist; b. 1903; grad. Belorussian Forestry Inst., 1931; D.Biol. Sci. With Belorussian Forestry Inst., 1931-35, lectr., 1948-50; learned sec., head dept. dendrology Bot. Gardens, Belorussian Acad. Scis., 1937-40, with Inst. Biology, 1948-56, now head dep. arboraceous plants, acad. sec. dept. biol. sci., 1956—. Mem. Belorussian Acad. Scis. (Presidium mem.) 1956—). Author: The Mineral Nutrition and Fructification of Arboraceous Plants, 1957. Research and publs. on introduction and acclimatization of arboraceous plants, theoretical aspects of introduction and acclimatization. Address: Belorussian Acad. Scis., Minsk, Belorussian SSR, USSR.

NETER, Erwin, microbiologist; b. Mannheim, Germany, May 26, 1909; s. Julius and Marie (Oppenheimer) N.; M.D., U. Heidelberg, 1934; m. Dena Louise Wilson, May 26, 1945; 1 son, Stephen. Came to U.S., 1934, naturalized, 1939. Dir. bacteriology Children's Hosp., Buffalo, 1936—; prof. clin. microbiology State U. N.Y. at Buffalo, 1965—; cons. bacteriologist Roswell Park Meml. Inst., Buffalo, 1951—. Mem. A.M.A., Am. Assn. Immunologists, Soc. Am. Bacteriologists, Am. Pub. Health Assn. Author: Medical Microbiology, 1966; also numerous articles. Research on intestinal infections, infections of urinary tract, toxins of bacteria. Home: 32 Granger Pl., Buffalo 14222. Office: Children's Hosp., 219 Bryant St., Buffalo 14222.*

NETSKY, Martin George, Am. neurologist; b. Phila., May 15, 1917; s. George Nathan and Clara (Sherman) N.; A.B., U. Pa., 1938, M.S., 1940, M.D., 1943; m. Margaret Pease, Sept. 30, 1946. Practice medicine, specializing in neurology, N.Y.C., 1946-55, Winston-Salem, N.C., 1955-61, Charlottesville, Va., 1962—; asso. neuropathologist Montefiore Hosp., 1950-55, asso. attending physician div. neuropsychiatry, 1952-55; cons. pathologist U.S. Naval Hosp., St. Albans, N.Y., 1950-55; asso. pathologist N.C. Baptist Hosp., 1955-61, chief neurol. service, 1957-61; faculty Bowman Gray Sch. Medicine, 1955-61, prof. neurology and neuropathology, chmn. dept. neurology, 1960-61; cons. U.S. Army Hosp., Ft. Bragg, N.C., 1960-61; vis. prof. pathology Faculty Med. Sci. Chiengmai (Thailand) Hosp., 1961-62; prof. neuropathology Sch. Medicine U. Va., 1962—; pathologist U. Va. Hosp., 1962—; mem. neurol. sci. research tng. com. NIH, 1959-63. Diplomate Am. Bd. Psychiatry and Neurology. Mem. Assn. Research in Nervous and Mental Diseases, A.M.A., Am. Acad. Neurology, Am. Assn. Neuropathologists (past pres), Am. Neurol. Assn., Assn. Am. Med. Colls., Nat. Soc. Study Edn., Am. Assn. U. Profs., Am. Assn. Pathologists and Bacteriologists, Am. Soc. Exptl. Pathology, Société Française de Neurologie, Sigma Xi, Alpha Omega Alpha. Author: (with others) Atlas of Tumors of Nervous System, 1956. Mem. editorial bd. Neurology, 1963. Studies, publs. on care and treatment of patients with neurol. disorders; neoplastic, degenerative, congenital disorders of the nervous system. Home: "Franklin", R.F.D. 7, Charlottesville, Va. 22901.*

NETTER, Arnold-Juste, French physician; b. Strasbourg (now France), 1855; became aggregate prof., 1904; hon. physician, hosps. of Paris; prof. Faculty Paris; mem. Acad. Medicine. Research on pneumococcus, epidemic encephalitis, meningitis; stressed importance of Kernig's sign in diagnosis of meningitis, 1898; popularized use of collargol (or argentum credé) to treat blood-borne infections, 1902. Died Paris, 1936.

NETTLER, Gwynn, sociologist, psychologist; b. N.Y.C., July 7, 1913; s. Harry Lester and Dorothy (Wald) N.; A.B., U. Cal. at Los Angeles, 1934; M.A., Claremont Coll., 1936; Ph.D., Stanford, 1946; m. Audrey Cutler Michaels; children—Bruce Lovelace, Shauna Lovelace. Practice clin. psychology, San Francisco, Beverly Hills, Cal., 1955-57; dir. child welfare study Community Council Houston, 1957-59; asso. indsl. psychologist, Dando, S. Am., Mexico City, Mexico, 1959-61; sr. clin. psychologist Nev. Mental Health Center, Reno, 1961-63; prof. sociology U. Alta., Edmonton, Can., 1963—. Mem. Am., Pacific, Western psychol. assns., Am., Pacific sociol. assns., Brit. Psychol. Soc., Mind Soc., A.A.A.S. Research and publs. on socio-psycho. correlates of response to crime, dimensions of anti-social sentiment, sources and consequences of feelings of estrangement from one's soc., relationship between information about an ethnic group and attitude toward it. Home: 11665 Saskatchewan Dr., Edmonton, Alta., Can.*

NETTLESHIP, Anderson, Am. physician; b. Fayetteville, Ark., Oct. 19, 1910; s. William Lloyd and Elizabeth (Davis); B.S.c., U. Ark., 1932; M.D., Johns Hopkins, 1935; m. Mae Banwell, June 7, 1952; children—Martin Anderson, William Allan, Ann Melissa. NRC research fellow Duke, 1936-37, Vanderbilt U., 1937-38; practiced medicine specializing in pathology, Springfield, Mo., 1938-40; instr. Nat. Cancer Inst., USPHS, 1940-44; asso. prof. pathology U. Okla., 1944-45; prof., chief dept. U. Ark., Little Rock, 1946-58; chief lab. service VA Hosp., Fayetteville, Ark., 1958—; sr. asso. Antaeus Lineal Research Group, Fayetteville, 1959—.Franklin P. Mall scholar, 1931-32; Marle fellow tropical medicine, 1944. Mem. Phi Beta Kappa, Alpha Omega Alpha, Alpha Epsilon Delta. Author: Principles of Cancer Practice, 1952; Words for Joy but of Sorrow; also articles. Research in prodn. exptl. sarcomas in tissue cultures, discovery of urethane as a carcinogen. Address: Old Johnson Rd., Fayetteville, Ark. 72702.*

NETTLESHIP, Edward, English ophthalmic surgeon, dermatologist; b. Kettering, Eng., Mar. 3, 1845; s. Henry John and Isabella (Hogg) N.; ed. Moorsfield Eye Hosp., Royal Agrl. Coll., King's Coll., London Hosp. Med. Sch.; m. Elizabeth Endacott, 1869. Became prof. vet. surgery Agrl. Coll., Cirencester, 1867; named curator Moorsfield Hosp. mus. and library, 1871; ophthalmic surgeon St. Thomas' Hosp.; surgeon Royal London Ophthalmic Hosp.; insp. met. poor-schs. for ophthalmia (led to reforms), 1873. ret. from practice, 1902, engaged in research thereafter. Nettleship medal established in his honor, 1902, recipient, 1909. Fellow Royal Soc., 1912; mem. Royal Coll. Vet. Surgeons, Royal Coll. Surgeons, Ophthal. Soc. U.K. (founder 1880, pres. 1895). Author: Treasury of Human Inheritance, 1922. Research in heredity of eye and disorders of vision; described a urticaria which occurs in infants and leaves permanent pigmented stains (Nettleship's disease), 1869. Died Hindhead, Eng., Oct. 30, 1913.

NETTLETON, Richard Ellsworth, Am. physicist; b. New Haven, June 1, 1930; s. William Ellsworth and Henrietta (Bradley) N.; B.A., Amherst Coll., 1951; Sc.M., Brown U., 1953, Ph.D., 1955. Physicist, Nat. Bur. Standards, Washington, 1959-64, Nat. Acad. Sci.-NRC asso., 1955-57; mem. tech. staff Sandia Lab., Albuquerque, 1965; vis. fellow Cornell U., 1963. Welch Found. fellow, Rice U., 1957-59. Mem. Am. Phys. Soc., N.Y. Acad. Scis., Sigma Xi, others. Theoretical investigations of heat conduction, diffusion, and viscosity in liquids and solids and cross effects resulting from mut. interference of these processes; devised new techniques in stats. mechanics and in thermodynamics of irreversible processes to further these ends in the case of liquids in solids, lattice dynamics were used to bring order from fragmented picture presented by existing expts. on ferroelecs. Home: 1826 Mesa Vista Rd. N.E., Albuquerque 87106. Office: Org. 5155, Sandia Corp., Sandia Base, Albequerque 87115.*

NETTO, Eugen, German mathematician; b. Halle, Germany, June 30, 1846; student of Kronecker; prof., Strasbourg (now in France), Berlin, Giessen (both Germany). Author: Substitutionstheorie und ihre Anwendungen auf die Algebra, 1882; Vorlesungen über Algebra, 2 vols., 1896-1900; Lehrbuch der Kombinatorik, 1901; Elementaire Algebra, 1904. Died Giessen, May 13, 1919.

NETZ, Heinrich Josef, German physicist; b. Oberhausen, Germany, Dec. 28, 1896; s. Heinrich and A. (Jung) N.; ed. Hanover, Munich (both Germany) Tech. Colls.; engring. degree; m. Charlotte Hasler, May 26, 1923; 1 son, Paul-heinz. Engr., pvt. industry; prof. Author: Messungen und Untersuchungen an wärmetechischen Anlagen und Maschinen; Dampfkessel; Wärmewirtschaft; Formein der Technik; others. Home: Adelheidstrasse 28, Munich. Office: Arcisstrasse 21, Munich, W. Germany.

NEUBAUER, Hans Johann Franz, plant physiologist; b. Vienna, Austria, Jan. 4, 1911; ed. Klosterneuberg (Austria) Profl. Sch., U. Vienna; Ph.D.; m. Karoline Muhlbacher; children—Gernot Otfried, Juta Karoline, Hans Hartung. Scientist, reporter Agrl. Exptl. Sta. Vienna, 1938-40; asst. Pharmacology Inst., U. Vienna, 1940-48; bot. expert, Afghanistan, prof. botany U. Kabul (Afghanistan), 1948-52, in conjunction with Bonn (Germany) U., 1963—; prof. pharmacology, plant physiology, Indonesia, dir. bot. dept. U. Bandung (Indonesia), 1952-63. UNESCO expert, Pakistan, 1963. Mem. German Botany Soc. (Berlin), Botany Assn. Berlin, London Soc. Exptl. Biology, Internat. Soc. Morphologists (Delhi, India), Vienna Soc. Zoology and Botany. Contbr. articles bot. jours. Address: P.O. Box 233, Kabul, Afghanistan.

NEUBER, Gustav Adolf, German surgeon; b. Tondern, Germany, June 24, 1850. Author: Die aseptische Wundbehandlung in meinen chirurgischen Privat-Hospitälern, 1886; Kurze Beschreibung der aseptischen Wundbehandlung, 1892. Promoted asepsis in Germany. Died Kiel, Germany, Apr. 14, 1932.

NEUBERT, Theodore J(ohn), Am. phys. chemist; b. Rochester, N.Y., Jan. 10, 1917; s. John H. and Josephine B. (Eich) N.; B.S., U. Rochester, 1939; Ph.D., Brown U., 1942. Prof. dept. chemistry Ill. Inst. Tech., Chgo., 1949—; cons. solid state sci. div. Argonne Nat. Lab., 1947—. Recipient award for excellence in undergraduate teaching Ill. Inst. Tech., 1966. Mem. Am. Phys. Soc., Am. Chem. Soc., Am. Assn. U. Profs. Investigated effect of fast neutron bombardment on phys. properties of graphite used in nuclear reactors; studies of color centers which can be introduced into alkali halide type crystals by ionizing radiation or other means.*

NEUFELD, Fred, bacteriologist; b. Germany, 1861; devised method of typing pneumococci on basis of capsular swelling, circa 1902; described and named bacteriotropins, 1904. Died 1945.

NEUFELD, Jacob, physicist; b. Lodz, Poland, Apr. 15, 1906; s. Icek Majer and Tauba Maria (Lubinski) N.; M.S., U. S.C., 1930; Sc.D., U. Pa., 1935; m. Myra Gene Stallard, Sept. 23, 1946; 1 son, John Langdon. Research engr. Well Surveys, Inc., Tulsa, 1947-48; physicist Oak Ridge Nat. Labs., 1948—. Mem. Am. Phys. Soc. Research, publs. on fundamentals of Maxwell's theory; passage of charged particles through dispersive media; plasma-beam instability; dosimetry of ionizing radiations. Home: 113 Cedar Lane. Office: P. O. Box X, Oak Ridge Nat. Lab., Oak Ridge, Tenn. 37830.*

NEUGEBAUER, Theobald Ludwig (Lewis), Hungarian theoretical physicist; b. Budapest, Hungary, May 30, 1904; s. Viktor Emil and Maria (Chimani) N.; ed. U. Budapest, 1922-27; Ph.D., 1931. Asst., U. Budapest, 1930-41, docent, 1935-41, prof. theoretical physics, 1949—; librarian Technical Hochschule, Budapest, 1942-49. Recipient Kossuth prize, 1950, 1st prize Hungarian Acad. Scis., 1964. Author: (with K.F. Novobatzky) Theoretische Elektrizitätslehre und Wellenoptik, 1957; textbooks in Hungarian, sci. works in German, numerous works for laymen in Hungarian. Research on wave mech. theory of electro-optical Kerr effect and other optical effects, theory of crystal lattices and of molecules, also theory of ball lighting, light scattering of solutions and in cosmological problems; work with superconductivity, biophysics (autocatalysis of protein molecules and viruses), light scattering with double frequency, nuclear forces. Home: 6 Szász Karoly u. Budapest II. Office: 5-7 Puskin u. Budapest VIII, Hungary.*

NEUKOM, Hans, Swiss chemist; b. Winterthur, Switzerland, Aug. 19, 1921; s. Salomon and Margarit (Spalinger) N.; Diploma in Chemistry, Swiss Fed. Inst. Tech., Zurich, 1945, Ph.D., in Chemistry, 1949; m. Ruth Schanz, Jan. 17, 1953; children—Hans Peter, Gabriela. Research asso. Corn Products Refining Co., Argo., Ill., 1949-51; chemist Meypro Ltd., Weinfelden, Switzerland, 1952; sr. research chemist Internat. Minerals & Chem. Corp., Skokie, Ill., 1953-55; faculty Swiss Fed. Inst. Tech., Zurich,

1956—, prof., head dept. agrl. chemistry, 1965—. Research and publs. on chemistry and composition of foods, utilization of agrl. products. Home: 7 Karrenstr., Kusnacht, Zurich, Switzerland.*

NEUMAN, Leo Handel, physician; b. 1868; tchr. Albany Med. Coll.; contbr. papers on gastrointestinal and cardiac disorders to med. publs. Credited with theory of portable electro-cardiograph developed by Steinmetz and others in 1917. Died 1941.

NEUMAN, Vojtech, Czechoslovakian veterinarian; b. Humenne, Czechoslovakia, May 10, 1921; s. Herman and Gizela (Neuman) N.; Grad., Vet. Faculty, Brno, Czechoslovakia, 1950, C.Sc., 1960; m. Zlata Treitelova, Feb. 9, 1946; children—Jiri, Zuzana. Asst. lectr. biochem. dept. Vet. Faculty, Brno, 1949-57, asst. prof., docent, 1957-65, prof. toxicology, 1965—, head toxicological sect., 1951—. Chmn. cons. subcom. Ministry of Agr.; forensic expert for toxicology, 1950-58. Mem. Czechoslovak Biochem. Soc., Czechoslovak Biol. Soc. of Jan Evangelista Purkyne, Czechoslovak Soc. Vet. Surgeons. Author: (with V. Jelinek, O. Svoboda) Forensic Veterinary Surgery, 1958. Research and publs. on function, pharmacodynamic effect and toxicity of various kinds of drugs, role of glycoproteins and mucoproteins in relation to immunological processes in impaired liver, enzymology and its relation to toxicity of various drugs, prevention of mass toxicoses, Brdicka polarographic protein reaction with spl. regard to diagnosis of liver impairment following toxicosis. Died Sept. 19, 1965.

NEUMAN, William Frederick, Am. biophysicist; b. Potoskey, Mich., June 2, 1919; s. Reinhart F. and Louise (Grauel N.; B.S., Mich. State U., 1940; Ph.D., U. Rochester, 1943; m. Margaret Wrightington, June 10, 1943; children—Russell, Carol, Peter. With U. Rochester, 1944—, prof. radiation biology and biophysics, 1958—, chmn. dept., 1965—. Recipient Eli Lilly award, 1955; Claude Bernard medal U. Montreal, 1962; Lancet lectr. U. Minn., 1958; Rainbow lectr. Western Res. U., 1959; Kappa Delta award, 1964; Biol. Mineralization award, 1965. Mem. Am. Coll. Orthopedic Surgeons, Internat. Assn. Dental Research, Am. Chem. Soc., Soc. Am. Biol. Chemistry, Soc. Pharmacology and Exptl. Therapeutics, Radiation Research Soc., Biophysics Soc. Author: Chemical Dynamics of Bone Mineral, 1958. Research, publs. on phys. chemistry of bone mineral; bone metabolism; action of parathyroid hormone; metabolism of radioactive fission products; origins of life. Home: 40 Alpine St., Rochester, N.Y. 14620.*

NEUMANN, Bernhard Hermann, mathematician; b. Berlin, Charlottenburg, Germany, Oct. 15, 1909; s. Richard and Elsie (Aronstein) N.; student U. Freiburg (Germany), 1928-29, Dr. phil., U. Berlin (Germany), 1932; Ph.D., Cambridge (Eng.) U., 1935; D.Sc., U. Manchester (Eng.), 1954; m. Hanna von Caemmerer, Dec. 22, 1938; children—Irene Dorothy (Mrs. Daram Pal Dhall), Peter Michael, Barbara Edith, Walter David, Daniel Richard. Asst. lectr. U. Coll., Cardiff, Wales, 1937-40; lectr. U. Coll., Hull, Eng., 1946-48; faculty U. Manchester, 1948-61, reader, 1954-61; vis. prof. Courant Inst., N.Y. U., 1961-62; prof., head dept. math. Inst. Advanced Studies, Australian Nat. U., Canberra, 1962—; vis. lectr. Australian univs., Tata Inst. Fundamental Research, Bombay, India, 1959; vis. prof. U. Wis., Madison, 1966-67. Recipient Prize Wiskundig Genootschap te Amsterdam, 1949; Adams prize U. Cambridge, 1953. Fellow Royal Soc., 1959, Australian Acad. Sci.; mem. London (hon. editor Proc. 1959-61, past mem. council, past v.p.), Australian (mem. council 1963—, pres. 1964-66, v.p. 1966—), Am. math. socs., Wiskundig Genootschap te Amsterdam (prize 1949), Math. Assn., Math. Assn. Am., Canadian Math. Congress, Glasgow Math. Assn. Asso. editor Pacific Jour. Math., 1964-66. Research, numerous publs. in theory of groups and other areas in algebra especially infinite groups. Home: 20 Talbot St., Forrest, Canberra, A.C.T., Australia.*

NEUMANN, Caspar, German chemist; b. Züllichau, Silesia, July 11, 1683; student tech., mining, pharmacy, chemistry in Germany, Holland, Eng., 1711-13. Worked for several apothecaries; became lab. asst. to Dr. Cyprian, 1713; later worked with Geoffroys, France; became ct. apothecary to King Frederick William I, 1719; named prof. practical chemistry Collegium Medico-Chirurgicum, Berlin, 1723; became mem. Higher Coll. Medicine and supreme surveyor of apothecaries in Prussia, 1724, Hofrat, 1733. Fellow Royal Soc., 1725; mem. Berlin Acad.; mem. Inst. of Bologna, Acad. Naturae Curiosorum. Author: Praelectiones chemicae seu chemia . . . , 1740; Allgemeine Grundsätze der theoretisch-practischen Chemie, . . . , 1755; The Chemical Works of Casper Neumann . . . , 1759; Lectiones chymicae von salibus alkalino-fixis und vom camphora . . . , 1727; also numerous articles. Prominent phlogistonist; studied nature of osteocolla and proved it was petrified root of a tree; 1st to mention that calomel blackens when exposed to sunlight; described properties of zinc; studied qualities of fixed alkalis of camphor, amber, opium, alcohol. Died Berlin, Oct. 20, 1737.

NEUMANN, Franz Ernst, German mineralogist, physicist, mathematician; b. Joachimstal, Brandenburg, Germany, Sept. 11, 1798; ed. U. Berlin, U. Jena, Germany; Ph.D., 1826. Became prof. mineralogy and physics U. Königsberg (now Kaliningrad, USSR), 1828. Fought against Napoleon, 1815. Fellow Royal Soc., 1862; mem. French Acad. Scis. Author: Beiträge zur Kristallonomie (included devel. improved math. methods for analyzing crystal structures), 1823; Beiträge zur Theorie der Kugelfunktionen, 1878; Vorlesungen über Theorie des Magnetismus, 1881; Vorlesungen über mathematische Physik, 7 vols., 1881-94; Einleitung in die theoretische Physik, 1883; Theoretische Optik, 1885; Theorie das Elasticät, 1885. Formulated law on molecular heat of a compound (Neuman's Law), 1831, law of eletctromagnetic induction using results of Faraday and Henry, theory of reflection and refraction of light, of double refraction and of crystal colors in polarized lights; measurements of heat flow in metal bars; research on variation of hardness of crystals at different faces; discovered Neumann bands in crystal structure of iron and steel, 1848; studied method for determination specific heat. Died Königsberg, May 23, 1895.

NEUMANN, (Louis) Georges, French veterinarian, parasitologist; b. Paris, Oct. 22, 1846; prof. Toulouse (France) Vet. Sch.; mem. French Acad. Scis., 1918, Acad. Medicine. Author: Traité des maladies parasitaires des animaux domestiques. Fundamental research on acarines; studied parasitic diseases of domestic animals, also history of vet. medicine. Died St.-Jeande-Luz, France, June 28, 1930.

NEUMANN, Hanna, mathematician; b. Berlin, Germany, Feb. 12, 1914; b. Hermann and Katharine (Jordan) von Caemmerer; Staats Examen, U. Berlin, 1937; postgrad. U. Gottingen; D.Phil., Oxford U., 1943, D.Sc., 1955; m. Bernhard Hermann Neumann, Dec. 1938; children—Irene (Mrs. D. Pal Dhall), Peter M., Barbara E., (Mrs. A. R. Cullingworth), Walter D., Daniel R. Lectr., sr. lectr. pure math. U. Hull, 1946-57; sr. lectr. Manchester Coll. Sci. and Tech., 1957-63; vis. lectr. Courant Inst. Math. Scis., N.Y. U., 1961-62; professional fellow Inst. Advanced Studies, Australian Nat. U., Canberra, 1963-64, prof., head dept. pure math., 1964—. Mem. Am., Edinburgh, London, Australian math. socs., Math. Assn. Great Britain. Author: Varieties of Groups, 1967. Contbr. articles to profl. jours. Research, publs. on theory of groups. Home: 20 Talbot St., Forrest, A.C.T. Australia. Office: Australian Nat. University, Canberra, A.C.T., Australia.*

NEUMANN, Karl Gottfried, German mathematician; b. Königsberg, Germany (now Kaliningrad, Russia), May 7, 1832; s. Franz Ernst Gottfried Neumann; ed. Königsberg; became instr. math., Halle, Germany, 1858; named prof., Basel, Switzerland, 1863, Tübingen, Germany, 1865, Leipzig, Germany, 1868-1911. Author: Vorlesungen über Riemanns Theorie der Abelschen Integrale, 1865; Das Dirichletsche Prinzip, 1865; Die Haupt und Brennpunkte eines Linsensystems, 1866; Theorie der Besselschen Funktionen, 1867; Die Prinzipien der Elektrodynamik, 1868; Über die Prinzipien der Galilei-Newtonschen Theorie, 1870; Die elektrischen Kräfte, 2 vols., 1873-98; Vorlesungen über die mechanische Theorie der Wärme, 1875; Untersuchungen über das logarithmische und Newtonsche Potential, 1879; Hydrodynamische Untersuchungen, 1881; Über die Methode des arithmetischen Mittels, 2 vols., 1887-88; Allgemeine Untercuchungen über das Newtonsche Prinzip der Fernwirkungen, 1896. Co-founder Mathematischen Annalen, 1868. Developed potential theory; said to be founder logarithmic potentials; studied math. physics, higher analysis. Died Leipzig, Mar. 27, 1925.

NEUMANN, Wilhelm Paul, German chemist; b. Würzburg, Oct. 29, 1926; s. Wilhelm and Margarete (Bertram) N.; Dipl. Chem., U. Wurzburg, 1950, Dr.rer.nat., 1952; m. Mechtild Maier, Feb. 7, 1957; children—Brigitte, Albrecht, Doris. Research chemist U. Wurzburg, 1952-55, Max Planck Inst. Coal Research, Mülheim-Ruhr; faculty U. Giessen, 1959—, unscheduled prof. chemistry, 1965—. Mem. Soc. German Chemists (chmn. Giessen sect.). Author: Organic Chemistry of Tin, 1967; also numerous articles. Research on biochemistry of snake and insect venoms, organoaluminum chemistry, organic chemistry of germanium, tin, lead, chemistry of organic free radicals. Address: 2 Ostanlage, 63 Giessen, West Germany.*

NEUMAYER, Georg von, see von Neumayer.

NEUMEYER, Martin Henry, Am. sociologist; b. Jackson, Mo., Oct. 8, 1892; s. Henry and Martha (Bohnsack) N.; A.B., DePauw U., 1919; B.D., Garrett Theol. Sem., 1921; A.M., Northwestern U., 1922; Ph.D., U. Chgo., 1929; m. Esther S. Sternberg, July 15, 1919. Ordained to ministry Methodist Ch., 1917; instr. Chgo. Tng. Sch., 1923-27; faculty U. So. Cal., Los Angeles, 1927—, prof. sociology, 1937-62, head dept., 1954-58, prof. emeritus, 1962—. Mem. adv. bd. Nat. Child Guidance Found., 1955—; life mem. Los Angeles County Fedn. Community Coordinating Councils, 1951—; Fellow Internat. Inst. Arts and Letters; mem. Am. Sociol. Assn.; mem. Pacific Sociol. Assn., Nat. Council on Crime and Delinquency, Alpha Kappa Delta (nat. v.p. 1948-54, nat. pres. 1954-58, exec. com. 1958-64), Pi Gamma Mu, Theta Phi. Author: (with L.D. Osborn) Community and Society, 1933; (with Esther S. Neumeyer) Leisure and Recreation, 1936, 49, 58; Juvenile Delinquency in Modern Society,1949, 55, 61; Social Problems in a Changing Society, 1953; also articles. Contbg. editor Dictionary of Sociology, 1944; Mng. editor Sociology and Social Research, 1934-60, asst. editor, 1960-61, editor 1961—; editor publs. Cal. Community Councils, 1963—. Research on leisure, recreation and play, juvenile delinquency. Home: 4707 S. Keniston Av., Los Angeles 90043.*

NEURATH, Hans, biochemist; b. Vienna, Austria, Oct. 29, 1909; s. Rudolf and Hedda (Samek) N.; Ph.D., U. Vienna, 1933; m. Susi R. Spitzer, Nov. 10, 1960; 1 son, Peter F. (by previous marriage). With U. Coll., U. London, 1934-35; faculty div. agrl. biochemistry U. Minn., 1935-36, dept. chemistry Cornell U., 1936-38; faculty Duke, Durham, N.C., 1938-50, chmn., prof. biochemistry U. Wash., Seattle, 1950—. Cons. NIH, USPHS. Recipient certificate of appreciation Austrian Ministry Edn., 1958. Fellow Am. Acad. Arts and Scis.; mem. Nat. Acad. Scis., Am. Chem. Soc., Am. Soc. Biol. Chemists, Sigma Xi. Editor: (1st edit. with K. Bailey) The Proteins, 2d edit.; Biochemistry. Research, numerous publs. on elucidation of structure and function of proteins and enzymes, mode of action of proteolytic (protein-digesting) enzymes. Home: 5752 60th St. N.E., Seattle 98105.*

NEUSE, Eberhard Wilhelm, chemist; b. Berlin, Germany, Mar. 7, 1925; s. Eberhard O. and Felicitas (Winde) N.; B.S., Tech U. of Hannover, 1948, M.S., 1950, Ph.D., 1953; m. Hildegard G. Berger, Oct. 25, 1963. Came to U.S., 1957. Research asst. Tech. U. of Hannover, Germany, 1951-53; head application lab. Neynaber AG, Bremerhaven, Germany, 1954-57; research asso. plastics lab. Princeton U., 1957-59; with Douglas Aircraft Co., Inc., Santa Monica, Cal., 1960—, head polymer lab. missile and space systems div., 1960-66, chief plastics and elastomers sect., 1966—; cons. chem. industry. Mem. Am. Chem. Soc., N.Y. Acad. Scis., Sigma Xi. Research, publs. on synthesis of heterocylic pharmaceuticals; chemistry of organometallic and lipoid compounds; thermoplastic resin systems; discovery of new heterocyclic, ironorganic and other organometallic polymers. Home: 11904 Tennessee Pl., Los Angeles 90064. Office: Douglas Aircraft Corp., Santa Monica, Cal. 90406.*

NEUSSER, Edmund von, see von Neusser.

NEUSTADT, Richard Elliott, Am. polit. scientist; b. Phila., June 27, 1919; s. Richard M. and Elizabeth (Neufeld) N.; A.B., U. Cal. at Berkeley, 1939; M.A., Harvard, 1941, Ph.D., 1951; m. Bertha Cummings, Dec. 21, 1945; children—Richard M., Elizabeth. With OPA, 1942; mem. starf Budget Bur., Washington, 1946-50; mem. White House Staff, Washington, 1950-53; prof. govt. Columbia, N.Y.C., 1954-65; prof. govt., dir. Inst. Politics, John F. Kennedy Sch. Govt., Harvard, Cambridge, Mass., 1966—; asso. mem. Nuffield Coll., Oxford (Eng.) U., 1965—. Cons. to pvt. cos., govt. agys., Pres. Kennedy, 1961-63, Pres. Johnson, 1964-67. Recipient Woodrow Wilson award Am. Polit. Sci. Assn., 1961. Fellow Am. Acad. Arts and Scis.; mem. Council on Fgn. Relations, Inst. for Strategic Studies, Am. Philos. Soc. Author: Presidential Power, 1960; also articles. Developed conceptual framework for analysis of govtl. operations at chief-of-govt. level, especially Am. presidency and Brit. premiership. Home: 10 Traill St., Cambridge, Mass. 02138.*

NEUTS, Marcel F., mathematician; b. Ostend, Belgium, Feb. 21, 1935; s. Achille J. and Marceline (Neuts) N.; Lic. Math. U. Louvain, 1956; M.S., Stanford U., 1959, Ph.D., 1961; m. Olga Alida Topff, June 27, 1959; children—Christopher J., Myriam A., Catherine J., Deborah A. Came to U.S., 1962. With Lovanium U., Leopoldville, 1956-57, instr., 1956-57; research asst. U. Louvain, Belgium, 1957-58, 1960-62; with Purdue U., 1962—, prof., 1968—; cons. Gen. Motors Research Labs., 1965—; asso. editor Operations Research, 1965—. Commn. Relief in Belgium fellow, 1958-60. Mem. Inst. Math. Statistics, Soc. Belge Math., A.A.U.P., Am. Soc. Operations Research, Math. Assn. Am., Internat. Assn. Statistics in Phys. Scis., Sigma Xi. Research, publs. on theory of Queues and Semi-Markov processes; order statistics and numerical methods in stochastic processes. Home: 2001 Carlisle Rd., West Lafayette, Ind. 47906.*

NEVA, Franklin Allen, Am. physician; b. Cloquet, Minn., June 8, 1922; s. Lauri A. and Anna (Lahti) N.; S.B., U. Minn., 1944, M.D., 1947; M.A. (hon.), (Research fellow 1950-53) Harvard, 1964; m. Alice Leona Hanson, July 5, 1947; children—Karen, Kristin, Erik. Faculty U. Pitts., 1953-55; faculty Harvard Sch. Pub. Health, Boston, 1955—, John LaPorte Given prof. tropical pub. health, 1964—. Recipient Bailey K. Ashford award Am. Soc. Tropical Medicine and Hygiene, 1965. Diplomate Am. Bd. Internal Medicine. Mem. Soc. Exptl. Biology and Medicine, Infectious Disease Soc. Am. Research, publs. on action rickettsial and viral toxins, typhoid fever, growth of viruses and parasites in-vitro. Home: 232 Grove St., Wellesley, Mass. 02181. Office: 25 Shattuck St., Boston 02115.*

NEVEU, Jacques Jean-Pierre, mathematician; b. Brussels (Belgium), Nov. 14, 1932; s. Henri and Madeleine (Bastin) N.; student U. Brussels; D.Sc., U. Paris, 1955; m. Monique Rene, Nov. 7, 1959; 2 children. Research asso. Nat. Center Sci. Research, Paris, 1953-59; Columbia U., N.Y.C, 1956, U. Cal. at Berkeley, 1957, 65; prof. U. Paris, 1959——; with Syracuse (N.Y.) U., 1961, U. Göttingen (Germany), 1963; prof. Inst. Electricity, Paris. Mem. French, Am. math. socs., Inst. Math. Statistics. Author: Mathematical foundations of the calculus of probability, 1965. Study of probability theory. Address: 11 Rue Pierre Curie, Paris 5, France.*

NEVILLE, William Evans, Am. surgeon; b. Fairbury, Neb., Apr. 13, 1919; s. William Dennis and Ruth (Evans) N.; student Creighton U., 1936-39; B.S., U. Neb., 1949, M.D., 1943; m. Artis Jane Kaiser, Feb. 14, 1958; children—Laura, Maureen, Sheila, William, Brian. Faculty, Western Res. U., 1951-62; mem. thoracic surgery staff Cleve. City Hosp., 1951-62; asso. prof. surgery U. Ill. Coll. Medicine, Chgo., 1962——; chief, cardiopulmonary surgery, dir. surg. research Hines (Ill.) VA Hosp., 1962——. Mem. A.C.S., Central, Western surg. assns., Am. Assn. Thoracic Surgery, Soc. Thoracic Surgeons, Soc. Vascular Surgery, Internat. Vascular Soc., Am. Heart Assn., others. Research and devel. heart-lung machine; investigations in physiology of patients undergoing open heart surgery or other thoracic operations. Home: Box 237, Hines VA Hosp., Hines, Ill. 60141.*

NEWALL, Hugh Frank, English astrophysicist; b. Gesthead-on-Tyne, Eng., June 21, 1857; s. Robert and Mary (Pattinson) Newal; ed. Trinity Coll., Cambridge; M.A.; Hon. D.Sc., Durham, Eng.; m. Margaret Arnold, 1881 (dec. 1930); m. 2d, Bertha Surtees Phillpotts, 1931 (dec. 1932). Asst., Cambridge; demonstrator exptl. physics Cavendish Lab. 1886-90; dir. Solar Physics Obs., 1913-28, prof. astrophysics, 1909-28; fellow Trinity Coll., Cambridge; emeritus prof. astrophysics, Cambridge; astronomer Obs., Cambridge. Fellow Royal Soc., 1902; mem. Royal Astron. Soc. (pres. 1907-09), Cambridge Philos. Soc. (pres. 1914-16), Spectroscopisti Italiani (fgn.), Internat. Astron. Union (became v.p. 1925). Research and publs. on physics and astrophysics. Died Feb. 22, 1944.

NEWALL, Robert Stirling, Brit. engr., astronomer; b. Dundee, Scotland, May 27, 1812; D.C.L., U. Durham (Eng.), 1887; m. Mary Pattinson, Feb. 14, 1849; 4 sons, 1 dau. Research on rapid generation of steam under employ of Robert McCalmont. Fellow Royal Astron. Soc., Royal Soc., 1875; mem. Inst. Mech. Engrs. Author: Observations on the Present Condition of Telegraphs, 1860; Facts relating to the Submarine Cable, 1882. Invented and patented wire ropes, 1840; laid many sumarine telegraph cables including 1st successful cable between Dover and Calais, 1851; invented brakedrum and cone for laying cables in deep seas, 1853; supplied armor for 1st armored under sea cable, 1857; prepared drawings of sun, 1848-52. Died Apr. 21, 1889.

NEWBERRY, John Strong, Am. geologist; b. Windsor, Conn., Dec. 22, 1822; s. Henry and Elizabeth (Strong) N.; grad. Western Res. Coll., 1846, LL.D. (hon.), 1867; grad. Cleve. Med. Sch., 1848; m. Sarah Brownell Gaylord, Oct. 22, 1848; 6 children. Asst. surgeon, geologist on expdn. from San Francisco Bay to Columbia River, May 1855; prof. geology Columbian U. (now George Washington U.), 1856-57; physician, naturalist on mil. exploration expdn. of Colorado River, 1857-58; an organizer, prof. geology and paleontology Sch. of Mines, Columbia, 1866-92; geologist State of Ohio, 1869-74. Recipient Murchison medal Geol. Soc. London, 1888. Incorporator, mem. Nat. Acad. Scis., pres. A.A.A.S., 1862; v.p. Internat. Geol. Congress, 1891. Author: Report on the Colorado River of the West, Explored (1857-58), 1861; Report of the Exploring Expedition from Santa Fe to the Junction of the Grand and Green Rivers, 1876; Fossil Fishes and Fossil Plants of the Triassic Rocks of New Jersey, 1888; The Paleozoic Fishes of North America, 1889. Died New Haven, Dec. 7, 1892.

NEWBIGIN, Marion I., Brit. geographer; ed. Univ. Coll., Aberystwyth, Wales, Univ. and Sch. Medicine, Edinburgh; B.Sc., London, 1893, D.Sc., 1898; prof. biology and zoology Sch. Medicine for Women, Edinburgh. Author: Colour in Nature, 1899; Life by the Sea-Shore, 1901; Tillers of the Ground, 1910; Modern Geography, 1911; Man and his Conquest of Nature, 1912; Introduction to Physical Geography, 1912; Animal Geography, 1913; The British Empire beyond the Seas, 1914; Geographical Aspects of Balkan Problems, 1915; A New Geography of Scotland, 1920; Aftermath: A Geographical Study of the Peace Terms, 1920; Frequented Ways, 1922; Commercial Geography, 1923; Mediterranean Lands, 1924; Canada, 1927; Regional Geography of the World, 1929; Southern Europe, 1932. Editor Scottish Geog. Mag. Died July 20, 1934.

NEWBURGH, Robert Warren, Am. biochemist; b. Sioux City, Ia., Mar. 22, 1922; s. Oscar and Ida (Kreutz) N.; B.S., State U. Ia., 1949; Ph.D., U. Wis., 1953; m. Marilyn Jean Gould, Aug. 30, 1947; children—Kirk, Elizabeth. Research asso. Ore. State

U., Corvallis, 1952-54, mem. faculty, 1954——, prof. chemistry, 1961——, asst. dir. Sci. Research Inst., 1962——; vis. asso. prof. U. Conn. Storrs, 1960-61. Scholar, Am. Cancer Soc., 1958-61. Mem. Am. Soc. Biol. Chemistry, Am. Chem. Soc., Soc. Developmental Biology, A.A.A.S., Sigma Xi. Author: (with V.H. Cheldelin) The Chemistry of Some Life Processes, 1964; also numerous articles. Research on phospholipid and carbohydrate metabolism of insects; biochemistry of devel. Home: 630 Merrie Dr., Corvallis, Ore. 97330.*

NEWBY, Hayes Augustus, Am. audiologist; b. Marion, O., Apr. 2, 1914; s. Alva Wilbur and Mary Elsie (Hayes) N.; A.B., Ohio Wesleyan U., 1935; M.A., U. Ia., 1939, Ph.D., 1947; m. Jean Louise Herbert, June 20, 1936; children—Jeffrey Hayes, Scott Herbert, Nancy Jean. Asst. prof. speech La. Poly. Inst., 1940-41; instr. U. Ia., 1945-47; faculty Stanford (Cal.), 1947-67, prof. speech and drama, 1956-57, prof. speech pathology and audiology Sch. Medicine, 1957-67, dir. div., 1963-67; prof. communication arts and scis., dir. Speech and Hearing Center, Queens Coll. of City U. N.Y., Flushing, N.Y., 1967——. dir. San Francisco Hearing and Speech Center, 1948-56; also cons. to govt. and state agys. Past dir., past pres. Am. Bd. Examiners in Speech Pathology and Audiology. Fellow Am. Speech and Hearing Assn. (past pres.), A.A.A.S.; mem. Acoustical Soc. Am., Speech Assn. Am., Am. Pub. Health Assn., Am. Assn. U. Profs., Am. Fedn. TV and Radio Artists, Phi Beta Kappa, Sigma Xi, Omicron Delta Kappa, Theta Alpha Phi. Author: Audiology, 1958, rev., 1964; also articles. Pioneered devel. clin. and acad. field audiology; research in audiology. Home: 300 E. 40th St., N.Y.C. 10016. Office: Speech and Hearing Center, Queens Coll. of City U. N.Y., Flushing, N.Y. 11367.*

NEWCOMB, Simon, astronomer; b. Wallace, N.S., Can., Mar. 12, 1835; s. John Burton and Emily (Prince) N.; came to U.S., 1853; B.S., Lawrence Sci. Sch., Harvard, 1858; hon. degrees Yale, Harvard, Columbia, Edinburgh, Glasgow, Princeton, Cracow, Johns Hopkins, Heidelberg, Cambridge, Oxford, other univs. m. Mary Caroline Hassler, Aug. 4, 1863. Tchr. in Md., 1854-56; computer Nautical Almanac, 1857; apptd. prof. math. USN, 1861; assigned to U.S. Naval Obs.; negotiated contract for and supervised construction of 26-inch equatorial telescope; sec. U.S. Transit of Venus Commn., 1871-74; observed transit of Venus at Cape of Good Hope, 1882; dir. Nautical Almanac Office, 1877-97; ret., 1897; prof. mathematics and astronomy, Johns Hopkins, 1884-94. Recipient Royal Astron. Soc. Gold medal, 1874, Huygens Gold medal, Dutch Soc. Scis., 1878; Bruce medal Astron. Soc. of Pacific, 1898, Schubert prize (Russia), Sylvester prize Johns Hopkins. Fellow or mem. Royal Soc., 1877 (gold medal 1890), A.A.A.S. (pres.), Am. Math. Soc. (pres.), Astron. and Astrophys. Soc. Am. (pres.), Nat. Acad. Scis. (v.p., fgn. sec.), Inst. of France (1st native Am. fgn. asso. since Franklin), other sci. socs. Author: Elements of the Four Inner Planets and the Fundamental Constants of Astronomy, 1895; The Stars, 1901; Astronomy for Everybody, 1903; Reminiscences of an Astronomer, 1903; Spherical Astronomy, 1906; Side Lights on Astronomy, 1906; also various other books on astron. and econ. topics, mag. articles. Editor Am. Jour. Math., from 1874. Published new and extremely accurate tables of motions of stars, planets and moon; determinations resulted in new set of astron. constants, revision of world's nautical almanacs; improved upon work of Leverrier and all other tabulations, completed, 1899; attacked Olbers' hypothesis that bodies of planetoid zone formed through breakup of planet that once orbited between Mars and Jupiter, 1860; worked (with Michelson) to determine velocity of light. Died Washington, D.C., July 11, 1909.

NEWCOMBE, Curtis Lakeman, zoologist; b. Port Williams, N.S., Can., Mar. 23, 1905; s. Arthur Fowlis and Alma (Beattie) N.; B.A., Acadia U., 1926, M.A., 1927; M.S., W.Va. U., 1929; Ph.D., U. Toronto, 1933; m. Dorothy Mae Middleton, Aug. 31, 1940. Came to U.S., 1927, naturalized, 1939. Faculty U. Md., 1933-40, William and Mary Coll., 1940-46, U. Cal., Berkeley, 1951-54, U. Utah, 1954-55; dir. Va. Fisheries Lab., Yorktown, 1940-46; zoologist Cranbrook Inst. Sci., Bloomfield Hills, Mich., 1946-49; biologist USPHS, San Francisco, 1949-53; sr. scientist U.S. Naval Radiol. Def. Lab., San Francisco, 1955-63; prof. biology, dir. Frederic Burk Found. Research Center, San Francisco State Coll., 1963——. Recipient Merit award Cal. Conservation Council, 1952, Navy Superior Accomplishment award, 1962. Fellow A.A.A.S., Cranbrook Inst. Sci., Cal. Acad. Sci.; mem. Cal. Conservation Council (pres.), Ecol. Soc. Am. (past chmn. Western div.). Conducted investigations of fresh and salt water environments, growth of invertebrates, hazards from exposures to ionizing radiation and effects of shock energy on terrestrial organisms. Home: 8 Middle Rd., Lafayette, Cal. 94549. Office: 1400 Holloway St., San Francisco 94132.*

NEWCOMBE, Howard B., Canadian biologist; b. Kentville, N.S., Can., Sept. 19, 1914; s. Edward Borden and Mable (Outebridge) N.; B.Sc., Acadia U., 1935; Ph.D., Imperial Coll. Tropical Agr., McGill U., 1939, D.Sc., 1966; m. Beryl Honor

Callaway, Feb. 14, 1942; children—Kenneth Donald, Charles Philip, Richard William. 1851 Sci. Research scholar John Innes Hort. Instn., London, Eng., 1939-40; sci. officer Brit. Ministry Supply, 1940-41; research asso. Carnegie Instn. of Washington, 1946-47; with Atomic Energy of Can. Ltd., Chalk River, Ont., 1947——, head biology br. Vis. prof. Ind. U., 1963. Mem. Genetics Soc. Am. (sec. 1956-58), Genetics Soc. Can. (pres. 1965), Am. Soc. Human Genetics (pres. 1964-65), Royal Soc. Can. Mem. editorial bds. Radiation Biology, Human Genetics Abstracts, Mutation Research. Research, numerous publs. on chromosome breakage in plants, mutations in microorganisms, radiation effects in mammals and fish, human population genetics, use of linked med. and vital records to derive family data on statis. scale for genetic research. Home: 67 Hillcrest Av., Deep River, Ont. Office: Biology Br., Atomic Energy of Can. Ltd., Chalk River, Ont., Can.*

NEWCOMEN, Thomas, English inventor; b. Darthmouth, Eng., Feb. 28, 1663; s. Elias Newcomen; m. Hannah Waymouth, July 13, 1705; partner with Thomas Savery. Invented atmospheric steam engine with John Calley (or Cawley), 1705; improved steam engine for pumping water from mines (invented by Thomas Savery), 1725. Died London, Aug. 5, 1729.

NEWCOMER, Harry Sidney, Am. physician, biophysicist; b. Willets Point, N.Y., Oct. 16, 1887; A.B., U. Wis., 1907, A.M., 1910, postgrad., 1910-12; postgrad., Freiburg, 1912-13; M.D., Johns Hopkins, 1915; postgrad., Tübingen, 1914, Paris, 1923-24, Uppsala, 1924; m., 1922; 4 children; research pathologist Pa. Hosp., Phila., 1915-16, sci. dir. dept. mental and nervous diseases, 1919-21, chief syphilis clinic, 1920-22; biophysicist Henry Phipps Inst., Pa., from 1922. Fellow N.Y. Acad.; mem. Optical Soc., A.M.A. Research in biol. physics, physiol. optics, x-ray fluorescence; made pharm. and med. studies of relaxant and hypotensive drugs; devised small colorimeter with colored glass standards for determination of hemoglobin (Newcomer's hemoglobinometer), circa 1923.

NEWCOMER, Victor Delbert, Am. dermatologist; b. Kalamazoo, Apr. 5, 1916; s. Delbert E. and Anna (Leghorn) N.; B.S., Western State Tchrs. Coll., Kalamazoo, 1940; M.D., U. Mich., 1943; m. Kathryn Jordan, Aug. 17, 1951; children—Dianne, Eileen, Donald. Fellow, asst. instr. dermatology U. Pa. Hosp., 1947-49; chief dermatology sect. VA Center, Wadsworth Hosp., Los Angeles, 1949-55; vis. cons. in dermatology student health service U. Cal. at Los Angeles, 1955——, VA Med. Center, Sawtelle Hosp., 1955——; asst. clin. prof. medicine U. Cal. at Los Angeles Sch. Medicine, 1951-54, asso. prof., 1955-61, prof., 1961——. Diplomate Am. Bd. Dermatology and Syphilology. Mem. A.M.A., Am. Acad. Dermatology and Syphilology, A.A.A.S., Am. Dermatol. Soc., N.Y. Acad. Scis., Pacific Dermatologic Assn., Soc. for Investigative Dermatology, Internat. Soc. for Tropical Dermatology, Internat. Soc. for Human and Animal Mycology. Co-editor: Therapy of Fungus Diseases, 1955; Modern Dermatologic Therapy, 1959; The Evaluation of Therapeutic Agents and Cosemetics, 1964. Research and numerous publs. on atopic dermatitis and treatment of fungal infections; systemic mycoses and immunological response. Home: 3314 Serra Rd., Malibu, Cal. 90265. Office: 10833 Le Conte Av., Los Angeles 90024.*

NEWELL, Frank William, Am. ophthalmologist; b. St. Paul, Jan. 14, 1916; s. Frank J. and Hilda (Turnquist) N.; B.S., Loyola U., Chgo., 1938, M.D., 1940; M.Sc. in Ophthalmology, U. Minn., 1942; m. Marian Glennon, Sept. 12, 1942; children—Frank W., Mary Susan, Elizabeth Ann, David Andrew. Faculty, Northwestern U., Chgo., 1946-53; faculty U. Chgo., 1953——, chmn. sect. ophthalmology, 1953——, prof. surgery, 1955——; sr. attending, chmn. div. ophthalmology U. Chgo. Hosps. and Clinics, 1953——. Mem. Council Research Glaucoma and Allied Diseases Alfred Sloan Found. 1958——, sci. adv. bd. Nat. Council Combat Blindness, 1959——. Bd. dirs. Nat. Soc. for Prevention Blindness, Ophthalmic Pub. Co. Diplomate Am. Bd. Ophthalmology. (also mem..). Mem. A.M.A. (chmn. sect. ophthalmology 1964-65), A.S.C. (chmn. adv. council ophthalmic surgery 1963-66), Assn. Research Ophthalmology (trustee 1963——), Am. Acad. Ophthalmology and Otolaryngology (1st v.p. 1966, Honor Key 1953), Am. Assn. U. Profs. (chmn. 1967-68), Phi Beta Pi. Author: (with Daniel Syndacker) Refraction, 1952, 2d edit., 1964; Ophthalmology, Principles and Concepts, 1965; contbg. author: Practice of Surgery (Lewis-Walters), 1961. Editor translations Glaucoma Symposium, vols. I-V, 1955-61; editor-in-chief Am. Jour. Ophthalmology, 1965——. Home: 4500 N. Mozart St., Chgo. 60625.*

NEWELL, Gordon Ewart, English zoologist; b. Whitstable, Eng., Apr. 16, 1908; s. Joseph and Gertrude (Rowden) N.; B.Sc., King's Coll., London, 1932, Ph.D., 1939, D.Sc., 1957; m. Winifred Young, July 31, 1938; children—Richard Charles, Peter, Anthony, Andrew, Janetta, Roger. Mem. faculty Queen Mary Coll., London, 1933——, prof., head dept. zoology, 1965——. Recipient Terr. Decoration. Mem. Zool. Soc. London, Marine Biol. Assn., Soc. Exptl. Biology, Soc. Study Animal Behavior. Author: (with

A. J. Grove) Animal Biology, 1942; (with R. C. Newell), Marine Plankton, 1966; also articles. Research on ecology, behavior, physiology of marine animals.*

NEWELL, Gordon Frank, Am. mathematician; b. Dayton, O., Jan. 26, 1925; s. Floyd B. and Edith (Field) N.; B.S., Union Coll., Schenectady, 1945; Ph.D., U. Ill., 1950; m. Barbara L. Pacyga, Jan. 29, 1949; children—Amy Bell, Jeffrey Edward. Faculty, Brown U., Providence, 1953-66; prof. U. Cal., Berkeley, 1966—. Alfred P. Sloan research fellow, 1956-59, Fulbright scholar to Australia, 1963. Mem. Am. Math. Soc., Soc. Indsl. and Applied Math., Operations Research Soc. Am. Research, publs. on devel. of math. theories to describe the behavior of hwy. traffic. Home: 217 Willamette Av., Kensington, Cal. 94708. Office: Dept. Civil Engring., U. Cal., Berkeley, Cal. 94720.*

NEWELL, Homer Edward, Am. physicist, mathematician; b. Holyoke, Mass., Mar. 11, 1915; s. Homer Edward and Annie Abigail (Davis) N.; A.B., Harvard, 1936, grad. study, 1936-37; Ph.D., U. Wis., 1940; D.Sc., Central Methodist College, 1963; m. Janice May Hurd, February 12, 1938; children—Judith Deborah, Sue Ellen, Jennifer Dianne, Andrew David. Grad. asst. mathematics dept. U. Wis., 1937-40; instr. to asst. prof. dept. mathematics U. Md., 1940-44; theoretical physicist, mathematician Naval Research Lab., head Rocket-Sonde Research Br., mem. Upper Atmosphere Rocket Research Panel, mathematics dept. rep. U. Md. Extension, 1952-55; acting supt., Atmosphere and Astrophysics Div., Naval Research Lab., 1955-58, science program coordinator for Project Vanguard (earth satellite project); asst. dir. space scis. Nat. Aeros. and Space Adminstrn., 1958-60, deputy director space flight programs, 1960-61, dir. Office Space Scis., 1961-63, asso. adminstr. for space sci. and applications, 1963—; lectr. U. Md., 1951-58, Nat. Park Coll., 1962. Recipient Pendray award Am. Rocket Soc., 1958; Career Service award Nat. Civil Service League, 1965; President's award for distinguished federal civilian service, 1965. Fellow American Assn. Advancement Service, member of the Am. Inst. Aeronautics and Astronautics, Am. Geophysical Union, Phi Beta Kappa, Sigma Xi. Author of books, including: High Altitude Rocket Research, 1953; Vector Analysis, 1955; Space Book For Young People, 1958; Guide to Rockets, Missiles, and Satellites, 1958; Sounding Rockets, 1959; Window in the Sky, 1959; Express to the Stars, 1961; also articles for trade mags., reviews. Studies of differential equations; work on exploration of upper atmosphere with rockets, artificial earth satellites and space probes. Home: 3704 33d Pl. N.W., Washington 20008. Office: NASA, 4th and Maryland Avs. S.W., Washington 20546.*

NEWELL, Martin Joseph, Irish mathematician; b. Galway, Ireland, June 8, 1910; s. Martin J. and Annie (Corr) N.; B.Sc., U. Coll., Galway, 1929, M.S., 1930, D.Sc., 1952; student St. John's Coll., Cambridge U., 1930-32; m. Nora Winston, July 4, 1937; children—Michael, Martin, Eamon, John. Lectr. math. U. Coll. Galway, 1935-55, prof. math., 1955-60, pres., 1960—. Mem. governing bd. Sch. Theoretical Physics, Dublin Inst. Advanced Studies, 1950-65. Mem. Royal Irish Acad. Author: University Algebra; (with M. Power) Notes on Calculus, 1947. Research on group representations. Address: University College, Galway, Ireland.*

NEWHOUSE, Albert, mathematician; born Cambrai, France, May 31, 1914; s. Moritz and Paula (Grünewald) N.; student U. Hamburg (Germany), 1934-37; Ph.D., U. Chgo., 1940; m. Alice Jean Miles, Sept. 3, 1943; children—Anthony Albert, Andrea Jean. Came to U.S., 1937, naturalized, 1943. Instr., U. Ala., Tuscaloosa, 1941-42, U. Neb., Lincoln, 1942-44, Rice U., Houston, 1944-46; faculty U. Houston, 1946—, prof., 1951—. Cons. to indsl. cos.; reviewer Computing Revs., 1960—. Fellow A.A.A.S.; mem. Am. Math. Soc., Math. Assn. Am., Soc. for Indsl. and Applied Math., Assn. for Computing Machinery, Sigma Xi, Pi Mu Epsilon. Author: (with Benner, Yates, Rader) Topics in Modern Algebra, 1962; also articles. Research in modern algebra and computing scis. Home: 7907 Braesview Lane, Houston 77071.*

NEWHOUSE, Walter Harry, Am. geologist; b. Fisher, Pa., Dec. 13, 1897; s. Edward Winfield and Hattie (Elder) N.; B.S., Pa. State Coll., 1921; M.S., Mass. Inst. Tech., 1923, Ph.D., 1926; m. Grace Edna Brown, June 30, 1923; 1 son, Jan. Faculty, Mass. Inst. Tech., 1923-46, prof. geology, 1944-46; prof. geology U. Chgo., 1946-63, chmn. dept. geology, 1946-57; pvt. researcher, Tucson, 1963—. Mem. Ill. State Bd. Natural Resources and Conservation, 1947-63; chmn. panel on geology and geophysics Joint Research and Devel. Bd., 1947-50; chmn. com. on geophysics adv. to Office Naval Research, 1947-50. Mem. Geol. Soc. Am., Soc. Econ. Geologists, Am. Acad. Arts and Scis., Am. Mineral. Soc., Geochem. Soc. Editor: Ore Deposits as Related to Structural Features, 1942; co-editor Jour. Geology, 1956-62. Research, numerous publs. on structural relations and mineral paragenesis ore deposits, three dimensional relations structural features in layered metamorphic rocks, three dimensional map. Address: 4233 N. Flowing Wells Rd. 55, Tucson 85705.*

NEWKIRK, Gordon Allen, Jr., Am. astrophysicist; b. Orange, N.J., June 12, 1928; s. Gordon Allen and Mildred (Fleming) N.; B.A., Harvard, 1950; M.A., U. Mich., 1952, Ph.D., 1953; m. Nancy Buck, Apr. 11, 1956; children—Sally, Linda, Jennifer. Research asst. Obs., U. Mich., Ann Arbor, 1950-53; astrophysicist Upper Air Research Obs., Sacramento Peak, N.M., 1953; sr. research staff High Altitude Obs., Boulder, Colo., 1955—; prof. adj. dept. astrogeophysics U. Colo., Boulder, 1961-65, dept. physics and astrophysics, 1965—. Mem. Internat. Astron. Union, Am. Astron. Soc., A.A.A.S., Research Soc. Am., Union Radio Scientifique Internationale, Am. Geophys. Union, Sigma Xi. Research and publs. on phys. state solar corona, scattering light by dust in upper atmosphere earth; devel. techniques observing solar corona outside eclipse from balloons and satellites; developed new type coronograph. Home: Sunshine Canyon. Office: High Altitude Obs., Boulder 80302.*

NEWLANDS, John Alexander Reina, English chemist; b. London, Eng., 1838; student of Hofmann, Newlands; joined Garibaldi's army in Italy, 1860; became chemist, sugar refinery, Victoria Docks. Recipient Davy medal Royal Soc., 1887. Author: Discovery of the Periodic Law. First to arrange elements according to atomic number, also noticed connection between every eighth, 1864. Died London, July 29, 1898.

NEWMAN, Ezra Theodore, Am. physicist; b. N.Y.C., Oct. 17, 1929; s. David and Fannie (Slutsky) N.; B.A., N.Y. U., 1951; M.A., Syracuse U., 1955, Ph.D., 1956; m. Sally Faskow, Apr. 20, 1958; children—David, Dara. Faculty, U. Pitts., 1956—, prof. physics, 1966—; vis. prof. Syracuse U., 1960-61, Kings Coll., U. London (Eng.), 1964-65. Cons. Wright-Patterson AFB 1957. Mem. Am. Phys. Soc. Research and publs. on gen. theory of relativity, theory of gravitational radiation. Home: 1308 Denniston Av., Pitts. 15217.*

NEWMAN, Sir George, English bacteriologist, physician; b. Oct. 23, 1870; M.D., Edinburgh (Scotland) U.; also ed. King's Coll., London, Eng.; several hon. degrees; m. Adelaide Constance Thorp, 1898. Sr. demonstrator bacteriology, infective diseases King's Coll., London, 1896-1900; emeritus lectr. pub. health St. Bartholomew's Hosp., U. London; med. officer Firsburg, Bedfordshire County, Eng.; chief med. officer Bd. Edn. 1907-35, Ministry Health, 1919-35; mem. numerous health coms. Yale lectr., 1927, Linacre lectr., 1928, Gresham lectr., 1929, Halley Stewart lectr., 1930; treas. gen. Med. Council. Recipient Fothergill Gold medal, 1935. Fellow Royal Coll. Surgeons (hon.), Royal Coll. Physicians (Bisset-Hawkins Gold medal 1935), Royal Soc. Edinburgh, N.Y. Acad. Medicine (hon.); mem. Soc. Apothecaries (hon. freeman), Pub. Health Assn. Australasia (hon.). Author: Bacteriology of Milk, 1903; Bacteriology and the Public Health, 1904; Infant Mortality, 1906; Hygiene and Public Health, 1917; An Outline of the Practice of Preventive Medicine, 1919; Recent Advances in Medical Education in England, 1923; Interpreters of Nature, 1927; Rise of Preventive Medicine, 1932; The Building of a Nations's Health, 1939; others. Died May 26, 1948.

NEWMAN, Horatio Hackett, Am. zoologist; b. nr. Seale, Ala., Mar. 19, 1875; s. Albert Henry N. and Mary Augusta (Ware) N.; B.A., McMaster U., Toronto, 1896, D.Sc., 1933; spl. student U. Toronto, 1896-97; Ph.D., U. Chgo., 1905; m. Isobel Currie Marshall, 1907; children—Elizabeth Ware, Marshall Thornton; m. 2d, Marie E. Heald, June 5, 1954. Instr. biology and Latin, Des Moines (Ia.) Coll., 1897-98; instr. biology and chemistry Culver (Ind.) Mil. Acad., 1900-04; instr. zoology U. Mich., 1905-08), asst. prof. elect, 1908; prof. and head dept. zoology U. Tex., 1908-11; asso. prof. zoology and embryology U. Chgo., 1911-17, prof. zoology, 1917-40, emeritus from 1940, dean Colls. Sci., 1915-22. Head of instrn. force in physiology Marine Biol. Lab., Woods Hole, Mass., 1909-12. Fellow A.A.A.S.; mem. Am. Soc. Zoologists, Am. Soc. Naturalists, Sigma Xi. Author: The Biology of Twins, 1917; Vertebrat Zoology, 1919; Readings in Evolution, Genetics, and Eugenics, 1921; The Physiology of Twinning, 1923; Outlines of General Zoology, 1924; The Gist of Evolution, 1926. Editor, contbr. to The Nature of the World and of Man, 1926; Evolution Yesterday and Today, 1932; Twins—A Study of Heredity and Environment (with F. N. Freeman and K. J. Holzinger), 1937; The Phylum Chordata, 1939; Multiple Human Births, 1940. Died Aug. 29, 1957.

NEWMAN, Louis Benjamin, Am. physician; b. N.Y. C., Apr. 5, 1900; s. Morris A. and Mollie (Benzuly) N.; M.E., Ill. Inst. Tech., 1921; M.D., Rush Med. Coll., 1932; m. Rose Manilow, Jan. 21, 1951. Staff, Cook County Hosp., Chgo., 1933-42, U.S. Naval Hosp., Oakland, Cal., and Seattle, 1942-46, VA Hosp., Hines, Ill., 1946-53; chief phys. medicine and rehab. service VA Research Hosp., Chgo., 1953-67; cons. hosps. and VA hosps. Chgo. area, 1967—; prof. phys. medicine and rehab. Northwestern U. Med. Sch., Chgo., 1946—. Mem. med. adv. bd. Am. Rehab. Found., Armour Research Found., Nat. Found., Inc., Assn. Phys. and Mental Rehab., Nat. Multiple Sclerosis Soc., Nat. Parkinson Found., others. Recipient B'nai B'rith award, 1953, U.S. Presdl. Commendation, 1956; John E. Davis award Assn. Phys. and Mental Rehab., 1956, Distinguished Achievement award, 1966; Distinguished Service award, Ill. Inst. Tech., 1957; Meritorious Service award VA, 1958; others. Diplomate Am. Bd. Phys. Medicine and Rehab. Mem. Am. Acad. Phys. Medicine and Rehab. (pres. 1959, ofcl. rep. Nat. Stroke Congress, 1964—); Am. Congress Phys. Medicine and Rehab. (v.p. 1961, Distinguished Service key, Gold key 1963), A.M.A., Am. Soc. M.E., Assn. Mil. Surgeons, Internat. Soc. Rehab. of Disabled, Assn. Am. Mil. Engrs., World Med. Assn., I.E.E.E., A.A.A.S., N.Y. Acad. Scis., Am. Geriatrics Soc., Am. Assn. Electrodiagnosis and Electromyography, Am. Inst. Ultrasonics in Medicine, others. Research and publs. on the application of high frequency currents in medicine; rehab. in neuromuscular and musculoskeletal disorders; devel. of techniques in rehab. the severely disabled, including assistive devices to aid in performing self-care, also developed myometer for evaluating muscle strength, hydrotherapy tank, chest expansometer for respiration, others. Home: 400 E. Randolph St., Chgo. 60601.*

NEWMAN, Louis Leon, chem. engr.; b. Dec. 16, 1898; B.S., Carnegie Inst. Tech., 1921, M.E., 1941; m. Lucie Leonie Seyler, 1941. Research on gasoline recovery from natural gas H.A. Fisher Co., Wooten-Hughes Co., 1922-24; research and devel. manufactured gas processes Semet-Solvay Engring. div. Allied Chem. & Dye Corp., 1925-38; asst. prof. fuel tech Pa. State U., 1939-42; chief maintenance and repair sect., facilities div., chems. br. WPB, 1942-43; with Bur. Mines, Washington, 1943—, asst. chief coal technologist, 1955-59, chief coal technologist, 1959—. Mem. synthetic fuels mission to Germany, Tech Indsl. Intelligence Com., 1945; cons. synthetic ammonia prodn. ICA, 1950—; cons. com. on pub works peat mission to USSR, U.S. Ho. Reps., 1957 Bur. Mines liaison rep. div. chemistry and chem tech. NRC-Nat. Acad. Scis., 1959-61; chmn. Gordon Research Conf. on Coal Sci., 1963; head U.S. delegation Internat. Peat Congress, Russia, 1963. Mem Am. Gas Assn., Inst. Gas Engrs. (Gt. Britain), Inst Fuel (Gt. Britain), Chem. Engrs. Club Washington Sigma Xi. Research, publs. on underground gasification of coal, use of continuous gasification of coa to generate synthesis gas for prodn. of synthetic ammonia, methanol, liquid fuels, high-caloric-value gas reducing gases for smelting ores. Patentee in field Home: 4523 Que Lane N.W., Washington 20007 Office: Bur. of Mines, Washington 20240.*

NEWMAN, Marshall T(hornton), Am. anthropologist b. New Bedford, Mass., July 15, 1911; s. Horatio Hackett and Isobel Curry (Mashall)) N.; student Dartmouth, 1929-31; Ph.B., U. Chgo., (sr. honor schola in anthropology, 1933, 1933, A.M., 1935; Bolto fellow in anatomy, Western Res., 1935-36; Ph.D. Harvard, 1941; m. Avolena Mae Barton, June 16 1945; children—Douglas Ayer, Gregory, Bruce. Archaeol. work in Ill., 1932, 33, 34, field work in Fla. 1933, 34, in Ariz., 1935; teaching fellow, Harvard 1938-41; supr. Project 8, Inst. of Andean Research in Peru, S. Am., 1941-42; asso. curator phys. anthropology, U.S. Nat. Mus., Smithsonian Instn., Washington, 1942-62; prof. anthropology Portland Stat Coll., Ore., 1962—, chairman, department of anthropology, 1964-66; prof. anthropology U. Wash. Seattle, 1966—; integrated biol. nutritional studie in Peru, S. Am., 1956; vis. lectr. U. Mich., 1958-59 Fellow Am. Anthropol. Assn.; mem. Am. Assn. Phys Anthropologists (v.p.), Anthrop. Soc. Washington Sigma Xi. Asso. editor: Am. Jour. Phys. Anthropology, 1959-62. Contbr. articles on dermatoglyphics Indian skeletal material from N. and S. Am., phys and physiol. adaptation of man to his environment race formation, growth and nutrition. Studied biological adaptation of man to his environment (heat cold, high altitude, nutrition, disease); human growt and maturation, human body. Address: Savery Hal U. Wash., Seattle.*

NEWMAN, Maxwell Herman Alexander, Englis mathematician; b. London, Eng., Feb. 7, 1897; Herman and Sarah (Pike) N.; M.A., Cambridge (Eng U., 1924; m. Lyn Irvine, Dec. 28, 1934; children—Edward Irvine, William Maxwell. Fellow St. John Coll., 1923-45, lectr. Cambridge U., 1927-45; pro U. Manchester (Eng.), 1945-64. Fellow Royal Soc 1939 (Sylvester medal 1959); mem. London Mat Soc. (past pres., De Morgan medal 1962), Math. Ass (past pres.). Author: Topology of Plane Sets Points, 1939; also numerous articles. Research geometrical topology. Home: Cross Farm, Comberto Cambridge, Eng.*

NEWMAN, Melvin Spencer, Am. organic chemis b. N.Y.C., Mar. 10, 1908; s. Jacob K. and M (Polack) N.; B.S., Yale, 1929, Ph.D., 1932; m. Be trice N. Crystal, June 30, 1933; children—Kiefe Susan, Beth, Robert. Instr. Ohio State U., 1936-3 Elizabeth Clay Howald scholar, 1939-40, asst. pro 1940-44, prof., 1944—; Fulbright lecturer Universi Glasgow, 1957-67. Guggenheim fellow, 1951. Recip ent award American Chemical Society 1961. Mem. An British chem. socs., A.A.A.S., Nat. Acad. Sci., Sign Xi. Asso. editor Jour. Am. Chem. Soc., Jour. Orgar Chemistry and Organic Syntheses. Research on the retical and synthetic organic chemistry. Home: 2239 Onandaga Dr., Columbus, O. 43221.

NEWMAN, Monroe, Am. economist; b. Bklyn., Jan. 31, 1929; s. David A. and Ida Mary (Leight) N.; B.A., Antioch Coll., 1950; M.A., U. Ill., 1953, Ph.D., 1954; m. Ruth Zielinski Feb. 6, 1951. Mem. research staff AFL, 1947-48; examiner NLRB, 1949-50; asst. U. Ill., 1950-54; research analyst Assn. Casualty and Surety Companies, 1954-55; mem. faculty Pa. State U., 1955-—, prof. econs., 1961-—, head dept., 1958-62; vis. research professor of economics University Pitts., 1964-65. Economist, Pres.'s Appalachian Regional Commn., 1964-65, research dir., 1965-—. Mem. American Economic Association, Industrial Relations Research Assn. Co-author: Insurance and Risk, 1964. Author, or co-author articles, monographs. Home: 625 W. Nittany Av., State College, Pa. Office: Pennsylvania State U., University Park, Pa.*

NEWMAN, Morris, Am. mathematician; b. N.Y.C., Feb. 25, 1924; s. Isaac and Sarah (Cohen) N.; A.B., N.Y. U., 1945; A.M., Columbia, 1946; Ph.D., U. Pa., 1952; m. Mary Aileen Lenk, Sept. 18, 1948; children—Sally Ann, Carl Lenk. Instr. U. Del., 1948-51; mathematician Nat. Bur. Standards, Washington, 1951-—, chief numerical analysis sect., 1965-—, editor math. sect. Jour. Research, 1966-—, chmn. Grad. Sch. dept. math., 1960-—. Professorial lectr. Catholic U., 1960-62, U. Md., 1956-—. Recipient Gold Medal award U.S. Dept. Commerce, 1966. Mem. Am. Math. Soc., Math. Assn. Am., London Math. Soc., Washington Acad. Scis., Pi Mu Epsilon, Sigma Xi. Research and numerous publs. in group theory, number theory and matrix theory, classification of groups of matrices with coefficients from rings. Home: 10703 Clermont Av., Garrett Park, Md. 20766. Office: Nat. Bur. Standards, Washington 20234.*

NEWMAN, Robert, physician; b. Königsberg, Prussia; s. Gustav Lebrecht and Rosalie Jacobine (Molkentin) N.; grad. Long Island Coll. Hosp., 1863; Bellevue Hosp. Med. Coll., 1869; m. Ada B. K. Blackwell, Oct. 1877. Served in European War, 1849; commd. State's vol. surgeon, 1863; held many dispensary and hosp. appointments; pres. faculty and prof. N.Y. Sch. Phys. Therapeutics; a founder Medico-Legal Soc., 1878, 1st v.p. Mem. Am. Electro-Therapeutic Assn. (pres. 1896, chmn. exec. council). Author: Electricity in Genito-Urinary Diseases; also monographs and papers on electrolysis and electro-therapeutics. Inventor of electrodes for treatment of stricture by electrolysis, also of devices for use in electrotherapy. Deceased.

NEWMARK, Nathan M(ortimore), Am. civil engineer; b. Plainfield, N.J., Sept. 22, 1910; s. Abraham S. and Mollie (Nathanson) N.; B.C.E., Rutgers U., 1930, D.Sc. (hon.), 1955; M.S., U. Ill., 1932, Ph.D., 1934; m. Anne Mae Cohen, Aug. 6, 1932; children—Richard Alan, Linda Beth, Susan Adele. Faculty U. Ill., 1930-—, successively research asst. civil engring., research asso., research asst. prof., research prof. structural engring., 1943-56, prof. and head dept. civil engineering, 1956-—, chmn. digital computer lab. grad. coll., 1951-57; cons. structural, machine analysis and design govt. agencies, indsl. orgns.; editor series of texts, civil engring. and engring. mechanics Prentice Hall, Inc., 1949-—; cons. antiseismic design Latio Americana tower, Mexico City, Mex. Cons. NDRC, office Field service OSRD, World War II; sci. adv. bd. USAF, 1945-49; mem. U. S. Nat. Com. Theoretical and Applied Mechanics, 1953-58; tech. adv. bd. Dept. Commerce, 1963-64; mem. com. to form Nat. Acad. of Engring., 1964-—. Recipient Presdl. Certificate of Merit; award Concrete Reinforced Steel Inst., 1956; Vincent Bendix award, Am. Soc. Engring. Edn., 1961; Order of Lincoln, Lincoln Acad. of Ill., 1965. Fellow Am. Acad. Arts and Scis., Am. Geophys. Union, A.A.-A.S.; mem. Nat. Acad. Scis., National Academy of Engring. (charter), American Soc. C. E. (J. James R. Croes medalist 1945, Moisseiff award, 1950, Norman medal 1958, E. E. Howard award 1958, Theodore von kármán award 1962; chmn. engring. mechanics div. 1953), A.S.M.E. (chairman of the applied mechanics division 1954; honorary member), American Railway Engineering Association, also Column Research Council, Welding Research Council, Research Council Bolted Structural Joints, Reinforced Concrete Research Council, Am. Concrete Inst. (bd. dir. 1949-52, Wason medalist 1950), American Physical Soc., Internat. Assn. Bridge & Structural Engrs., Earthquake Engring. Research Inst., Am. Soc. Engring. Edn., Seismol. Soc. Am., Phi Beta Kappa, Sigma Xi. Author articles engring. subjects. Research on applied mechanics; dynamics and design; structural analysis; arch bridges; highway bridge floors; buckling; theory of plates and elasticity; fatigue of metals; soil mechanics; effects of earthquake and nuclear blast forces; numerical methods of analysis; high speed digital computers. Home: 1705 S. Pleasant St., Urbana, Ill.

NEWPORT, George, English naturalist, anatomist; b. Canterbury, Eng., July 4, 1803; s. of a wheelwright; student Mus. Natural History; entered London (Eng.) U., 1832. House surgeon Chichester Infirmary, 1835-37; practice surgery, London. Fellow Royal Soc., 1846 (Royal medal 1851), Entomol. Soc. (pres. 1844-45), Coll. Physician (hon.), Linnean Soc. Author: Observations on the Anatomy, Habits, and Economy of Athalia Centifoliae, the Saw-fly of the Turnip and of the means adopted for the Prevention of its Ravages, 1838; List of Specimens of Myriapoda in the British Museum, 1844; On the Impregnation of the Ovum in Amphibia, 1851. Catalogue of the Myriapoda in the British Museum, 1856; also articles. Studied insect structure and behavior; discovered impregnation of ovum by spermatozoa in higher animals results from penetration. Died Apr. 7, 1854.

NEWSOM, Carroll Vincent, Am. mathematician; b. Buckley, Ill., Feb. 23, 1904; s. Curtis Bishop and Mattie F. (Fisher) N.; A.B., Coll. Emporia, 1924, L.H.D., 1957; M.A., U. Mich., 1927, Ph.D., 1931; recipient hon. degrees, some 20 colls. and univs., 1951-—; m. Frances J. Higley, Aug. 15, 1928; children—Jeanne Carolyn (Mrs. W. A. Challener, III), Walter Burton, Gerald Higley. Mem. faculties U. Mich., U. N.M., Oberlin Coll., 1927-48; asst. commr. for higher edn. State N.Y. 1948-50, asso. commr. for higher and profl. edn., 1950-55; exec. v.p. New York University, 1955-56, president, 1956-62; sr. v.p. Prentice-Hall, Inc., Englewood Cliffs, N.J., 1962-64, vice chmn. bd., 1962-64, pres., 1964-65; chairman board Hawthorn Books, 1964-—; dir. RCA, Nat. Broadcasting Co., Lowenstein & Sons, Incorporated. Member profl. coms. on ednl. standards several assns. and orgns., also coms. on accrediting. Former mem. bds. several orgns. in field of edn. Currently: pres. Council Higher Ednl. Instns. in N.Y.C., 1958-—; chmn. AAU Com. Urban Renewal, 1958-—; mem., dir. N.Y. World's Fair-1964 Corp., 1959-—; dir. Guggenheim Found.; v.p. Edison Found.; trustee Nat. Fund Graduate Nursing Edn., 1960; several other coms. and bds. Briarcliff Coll. Fellow A.A.A.S. (pres. S.W. div., 1940-41, mem. nat. council, 1945-46); mem. American Geographical Soc. (member of council), many, sometime officer several, nat. profl. assns. and orgns. Author or co-author several books in field of mathematics and gen. edn., latest being: (with Howard Eves) Foundations and Fundamental Concepts of Mathematics, 1958; Mathematical Discourses, 1964; also articles to professional jours. Contbr. to yearbook Nat. Soc. for the Study of Edn., 1952. Editor: A Television Policy for Education, 1952. Studies of asymptotic behavior of function defined by power series. Home: Box 51, New Fairfield, Conn.

NEWSON, Henry Winston, Am. physicist; b. Lawrence, Kan., Nov. 26, 1909; s. Henry Byron and Mary (Winston) N.; B.S., U. Ill., 1931; Ph.D. in Chemistry, U. Chgo., 1934; m. Meta Thode, Aug. 31, 1934; children—Meta Mary, Caroline. Fellow, Lawrence Radiation Lab., U. Cal., 1934-36; instr. chemistry U. Chgo., 1936-39, instr. physics, 1939-41; sr. physicist Metal Lab., 1941-43; staff Clinton Labs., Oak Ridge, 1943-44, Hanford (Wash) Engr. Works, 1944-45, Los Alamos Sci. Lab., 1945-46; chief physicist Oak Ridge Nat. Lab., 1946-48; prof. Duke, Durham, N.C., 1948-—, dir. Nuclear Structure Lab., 1961-—; cons. in atomic energy, 1948-—. Mem. adv. com. on reactor safeguards AEC, 1958-—, vice chmn., 1962, mem. nuclear cross sect. adv. group, 1954-—; mem. N.C. Gov.'s Sci. Adv. Com., 1961-—. Recipient Service award War Dept., Manhattan Dist. C.E., 1945; Am. Nuclear Soc. medal for participating in first nuclear chain reaction expt. at Stagg Field, 1962. Fellow Am. Phys. Soc.; mem. Sigma Xi. Author: (with J.H. Gibbons) Fast Neutron Physics part I, 1960, part II, 1963. Research on precision nuclear spectroscopy, neutron spectroscopy, statis. nuclear shell theory; invented (with colleagues) overall control system still followed in modern nuclear reactors; first observed neutron disintegration of fluorine and predicted discovery of induced radioactivity; participated in devel. cyclotron at Berkeley, atomic bombs at Los Alamos, slow chain reaction at Stagg Field. Home: 1111 N. Gregson St., Durham, N.C. 27701.*

NEWSTEAD, Robert, English entomologist; b. Sept. 11, 1859; m. M.Sc.; asst. to Miss Ormerod; tutor rural economy The Coll., Chester, Eng.; prof. entomology Liverpool (Eng.) U.; mem. Imperial Bur. Entomology; mem. Big Game and Sleeping Sickness Com.; mem. Com. for Econ. Preservation of Birds. Fellow Royal Soc., 1912, Entomol. Soc.; aso. Linnaean Soc.; mem. Royal Hort. Soc. (hon.), Chester and N. Wales Archaeol. Soc. (hon. curator, chmn.), Assn. Econ. Biologists (pres. 1912-13), N. Western Fedn. Museums and Art Galleries (pres. 1938), Chester Soc. Natural Sci., Lit. and Art, (pres. 1912-14, 29-31). Author: Monograph of the Coccidae of the British Isles; also guide to study of tsetse flies, various papers on econ. entomology, pathogenic animals, archaeology. Died Feb. 16, 1947.

NEWTON, Alfred, ornithologist, zoologist; b. Geneva, Switzerland, June 11, 1829; s. William and Elizabeth (Milnes) N.; B.A., Magdalene Coll., Cambridge (Eng.) U., 1853. Drury traveling fellow, 1854-63, 77; voyages to Lapland, 1855, Iceland, 1858, West Indies, 1857, N. Am., Spitzbergen, 1864, Brit. Isles; 1st prof. zoology, comparative anatomy Cambridge, 1866-1907. Recipient Royal medal, 1900, Gold medal Linnaen Soc., 1900. Fellow Royal Soc., 1870 (v.p.); mem. Zool. Soc. (v.p.), Brit. Assn. (chmn. com. studying bird migration, close-time com.), Cambridge Philos. Soc. (pres.), Brit. Ornithologists Union (a founder). Author: The Zoology of Ancient Europe, 1862; Ootheca Wolleyana, 1864-1905; Aves in the Record of Zoological Literature, vols. 1-4; Zoology, 1874, 2d edit., 1894; Birds of Greenland (Arctic Manual, 1875); A Dictionary of Birds, 1893-96; appendix Ornithology of Iceland (Baring-Gould), 1863. Editor: British Birds (Yarrell), 4th ed. vols 1, 2, 1871-72; The Ibis (new series), 1865-70; Zool. Record, 1870-72. Research on breeding places of sea bids; sought out last nesting place of Great Auk (with J. Wolley), Iceland; foremost authority of time on ornithology; through Brit. Assn. influenced parliamentary legislation for protection of birds, especially mascarene of Sandwich Is. (now Hawaii). Died June 7, 1907.

NEWTON, Amos Sylvester, Am. chemist; b. Shingletown, Cal., July 26, 1916; s. James Albert and Mabel (Fox) N.; B.S. with honors U. Cal. at Berkeley, 1938; M.S., U. Mich., 1939, Ph.D., 1941; m. Elizabeth Powers, Apr. 25, 1942; children—Mabel Anne (Mrs. James C. Webb), Margaret. Tchg. asst. U. Mich., 1938-40; chemist Eastman Kodak Co., Rochester, N.Y., 1941-42, Berkeley, Cal., 1946-62; chemist Ia. State Coll. Manhattan Project, 1942-46; cons. Lawrence Radiation Lab. U. Cal., Berkeley, 1946-62; chemist, 1962-—; lectr. U. Cal. Ext. Div., 1964, Western Res. U., 1960. Mem. Am. Chem. Soc., A.A.A.S., Radiation Research Soc. (councillor 1960-63), Phi Beta Kappa, Sigma Xi, Pi Mu Epsilon, Phi Lambda Upsilon. Studies, publs., also chpts. in book on fission products of uranium and thorium; phys. chemistry of uranium and thorium compounds, hydrides, nitrides, halides; effects of ionizing radiation on alcohols, ethers, esters, organic halides; the states of excitation of molecules and ions resulting from low energy electron bombardment of various substrate molecules. Address: U. Cal., Berkeley, Cal. 94720.*

NEWTON, Sir Charles Thomas, English archeologist; b. Bredwardine, Eng., Sept. 13, 1816; s. Newton Dickinson Hand Newton; B.A., Christ Church, Oxford (Eng.) U., 1837, M.A., 1840, D.C.L., Worcester Coll., 1875; LL.D., Cambridge (Eng.) U., 1875; m. Mary Severn, Apr. 27, 1861. Apptd. curator antiquities Brit. Mus., 1840; vice consul Mytilene, Asia Minor, 1852-53; acting consul Rhodes, 1854; Brit. consul, Rome, Italy, 1859-61; keeper Greek and Roman antiquities Brit. Mus., 1861-85; became prof. archeology U. Coll., London, 1880; made excavations in Calymnos, 1854-55. Mem. Accademia dei Lincei, Rome (hon.), French Acad. Scis. (corr.), Archeol. Inst. Berlin (hon. dir.). Author: Method of the Study of Ancient Art, 1850; History of Discoveries at Halicarnassus, Cnidus and Branchidae, 1862-63; Travels and Discoveries in the Levant, 2 vols., 1865; Essays on Art and Archaeology, 1886; also numerous articles. Discovered tomb of Mausolus at Halicarnassus (one of 7 wonders of ancient world), 1856; bronze serpent of Delphi at Constantinople; excavated at Rhodes. Died Margate, Eng., Nov. 28, 1894.

NEWTON, Chester Whittier, Am. meteorologist; b. Los Angeles, Aug. 17, 1920; s. Chester Whittier and Atta (Stewart) N.; student Phoenix Jr. Coll., 1937-39; B.S., U. Chgo., 1946, S.M., 1947, Ph.D., 1951; m. Harriet Rose Rodebush, May 8, 1948; children—Lesley, David, Matthew, Alice. Weather observer U.S. Weather Bur., Phoenix, 1939-42; weather officer USAAF, CBI, 1943-46; meteorologist U.S. Weather Bur. Thunderstorm project, Chgo., 1948; synoptic analyst U. Chgo., 1948-51, U. Stockholm (Sweden), 1951-53, Woods Hole (Mass.) Oceanographic Instn., 1953; faculty U. Chgo., 1953-61; chief sci. U.S. Weather Bur. Nat. Severe Storms Project, Kansas City, Mo., 1961-63; program sci. Nat. Center Atmospheric Research, Boulder, Colo., 1963-—; affiliate prof. Pa. State U., 1966. Mem. steering com. Earth Sci. Curriculum Project, 1965. Mem. Am. Meteorol. Soc., Am. Geophys. Union, Royal Meteorol. Soc. (fgn. mem.), Sigma Xi. Contbr. articles to encys., profl. jours. Asso. editor Meteorol.. Monographs, 1964, Jour. Applied Meteorology, 1962. Research on structure of jet stream and Gulf stream, upper-level waves and cyclones, role of disturbances in atmospheric gen. circulation, process of formation of fronts between air masses, structure and mechanics of severe thunderstorms and squall lines particularly phys. process governing devel. and movements of thunderstorms. Home: 1380 Bluebell Av. Office: Nat. Center for Atmospheric Research, Boulder, Colo. 80302.*

NEWTON, Henry Jotham, Am. inventor; b. Hartleton, Pa., Feb. 9, 1823; s. Sr. Jotham and Harriet (Wood) N.; m. Mary Gates, 1850, 3 children. Partner piano firm Lights, Newton & Bradbury, 1853; pres. Henry Bonnard Bronze Col., 1884. Treas. photog. sect. Am. Inst. City N.Y., chmn., after 1873; pres. 1st Soc. Spiritualists N.Y.; founder, treas. Theosophical Soc., 1875. Research in photography; improved the dryplate process, pioneered in preparation of ready-sensitized paper, credited with paraffin paper process; effected 1st sci. cremation of human body in Am. Died Dec. 23, 1895.

NEWTON, Hubert Anson, Am. mathematician; b. Sherburne, N.Y., Mar. 19, 1830; s. William and Lois (Butler) N.; grad. Yale, 1850; LL.D., U. Mich., 1868; m. Anna C. Stiles, Apr. 14, 1859. Tutor, Yale, July 1852, became head math. dept., 1853, prof. math., 1855-90. Fellow Royal Soc., 1892; mem. Nat. Acad. Scis., Am. Philos. Soc. (Lawrence-Smith medal for meteoric studies), Royal Astron. Soc., London, A.A.A.S. (v.p. 1875, pres. 1885), Conn. Acad. Arts and Scis. (pres.); a founder Am. Metrol. Soc. Author: Investigations on the Construction of Certain Curves by Points; Certain Transcendental Curves . . . ; The Metric System of Weights and Measures (advocating adoption of metric system), 1868; also papers asso. editor. Am. Jour. Science; editor math. and astron. definitions Webster's Internat. Dictionary, 1890. Authority on theory transcendental curves; known for astron. research on comets, meteors; verified Olmsted's hypothesis that meteors were the more visible part of stream of bodies in fixed solar orbit. Died New Haven, Aug. 12, 1896.

NEWTON, Sir Isaac, English mathematician, physicist, astronomer; b. manor house at Woolsthorpe, near Grantham, in parish of Colsterworth, Lincolnshire, Eng., Dec. 25, 1642 (old style; Jan. 4, 1643 new style); baptized, Colsterworth, Jan. 1, 1642/3; he was posthumous, premature, and only child of Isaac (dec. Oct. 1642) and Hannah (Ayscough) N. (dec. 1679), who were married Apr. 1642; mother remarried, Jan. 27, 1645/6 to Rev. Barnabas Smith (dec. 1653), rector of North Witham, about one mile from Woolsthorpe, where she went to live, leaving her 3 year old son in the care of his grandmother Ayscough and under the guardianship of his uncle James Ayscough; children by mother's 2d marriage were Benjamin, Mary (Mrs. Pilkington), Hannah (Mrs. Barton); Newton never married; ed. local day schs. Skillington and Stoke; student Old King's (grammar) Sch., Grantham, circa 1654-58, 60-61; admitted as subsizar Trinity Coll., Cambridge (Eng.) U., June 5, 1661, matriculated July 8, 1661; elected scholar, Apr. 28, 1664; his tutor was Benjamin Pulleyn(e); B.A. (in a class of 26 men), Jan. 1664/5; elected minor fellow Trinity Coll., Oct. 1, 1667, major fellow, Mar. 16, 1667/8; M.A., July 7, 1668. Elected 2d holder (succeeding Isaac Barrow) Lucasian chair math., Cambridge U., Oct. 29, 1669; retained fellowship without taking holy orders by patent from crown, Apr. 1675 (resigned chair and fellowship, Dec. 10, 1701); proposed by Seth Ward as Fellow Royal Soc., Dec. 21, 1671, elected Jan. 11, 1671/2, formally admitted Feb. 8, 1674/5, mem. council, 1699, pres., 1703-27; apptd. deputy (1 of 8) by Cambridge U. Senate to present case against James II for university's refusal to confer M.A. degree on Alban Francis, a Benedictine monk, without his taking the usual oaths of allegiance and supremacy, Apr. 11, 1687; argued case before High Commn. Ct., Westminster, Lord Jeffreys presiding, Apr. 21-May 7, 1687; elected Whig mem. Parliament for Cambridge U., Jan. 1688/9 to Feb. 1689/90, Nov. 26, 1700 to July 2, 1702, defeated by Tory candidate in univ. election, May 17, 1705; created knight by Queen Anne, Cambridge, Apr. 16, 1705; among Bishop Moore's assessors at Bentley's trial, 1714. Author: Philosophiae naturalis principia mathematica (pub. summer 1687 at Halley's urging and expense, under auspices of Royal Soc.), 2d edit., 1713, 3rd edit., 1726, 1st English edit., 1729; Opticks, 1704, 1st Latin edit., 1706, 2d edit., 1718, 3rd edit., 1721, 4th edit., 1730; Tractatus de quadratura curvarum, 1704; Enumeratio linearum tertii ordinis et methodus differentialis, 1704; Arithmetica universalis, sive de compositione et resolutione arithmetica liber, 1707; De analysi per equationes numero terminorum infinitas, 1711; Lectiones opticae (English edit., 1728; Latin edit., 1729); The Chronology of Ancient Kingdoms Amended, to which is prefixed a Short Chronicle from the First Memory of Kings in Europe to the Conquest of Persia by Alexander the Great, 1728; Observations upon the Prophecies of Daniel and the Apocalypse of Saint John, 1733; Lexicon propheticum, 1737; Historical Account of Two Notable Corruptions of the Scripture (sent to John Locke, Nov. 1690), published in an imperfect form under the title: Two Letters from Isaac Newton to M. le Clerc, 1754. At the time Newton went up to Cambridge U., he had already shown an aptitude for mech. invention, but no evidence of math. precocity; at best, he was probably acquainted with arithmetic, reduction of fractions and use of proportions (no algebra, trigonometry, or geometry); at Cambridge, where the curriculum was severely scholastic, studied logic, ethics, rhetoric and Aristotelian philosophy; then in last undergraduate year, 1664, studied the new philosophy by reading Descartes, Galileo, Gassendi, Boyle, Henry More, Charleton, Hobbes, Glanville, Streete, Digby, Kepler, Hooke and others; the period late 1665-66, the so-called annus mirabilis, was crucial for development of Newton's sci. career, for it was then that he developed founds. of his work in light, gravity, and math.; probably to understand a book on astrology, studied trigonometry, and then went on to read math. works of Euclid, Oughtred, Descartes, Wallis, Viète, Schooten, and others; as an autodidact deriving his knowledge from reading with little or no outside help, assimilated existing math. tradition and began to move beyond it in creative math. endeavors of his own, 1664-66; found methods of binomial expansion and approximation by infinite series; elaborated fluxional (differential and integral) calculus; reduced his math. discoveries to a treatise by 1669. Also in 1664-66, began to develop his theory of colors and performed prismatic expts. establishing refrangibility (refraction) of colored light; lectured on his theory of colors in his initial Lucasian lectures, spring 1670; analyzed white light into constituent parts using prism; showed it to be heterogeneous mixture of rays, each having its own index of refraction; argued that colors are not qualifications of light but original and connate properties, particular rays being disposed to exhibit particular colors; thought (wrongly) that problem of chromatic aberration insuperable in refraction telescopes, since all rays coming from one point of an object cannot be collected by lens at a common focus; developed theory of reflecting telescopes and constructed 1st reflecting telescope, 1668; sent one to Royal Soc., 1670; also sent an account of his theory of colors, 1672; as a result, became involved in controversies with Hooke, Huygens, Pardies, Lucas, Linus; vexed, expressed intention of withdrawing from Royal Soc., Mar. 8, 1673-4; but later went on to explain colors of thin and thick plates, inflection of light, double refraction, polarization, and binocular vision; rejected undulatary theories of light and suggested a corpuscular theory (although to explain his fits of easy reflection and refraction he assumed there were waves associated with his particles); though basically ready in 1672, he did not allow his Opticks, which presented results of optical researches, to be published until 1704, the year after death of Hooke. Stimulated by reading Galileo and Descartes, Newton began to develop his own views on dynamics, 1665-66; determined quantitively the law of centrifugal force for a body in uniform circular motion; declared that centrifugal forces of planets, assuming their orbits to be circular, must vary inversely as the squares of their distances from the sun; began to think of gravity extending to orb of moon; computed, by expts. with conical and vertical pendulums, the acceleration due to gravity at surface of earth; on assumption that force of gravity would vary inversely with the square of the distance, compared force of gravity at surface of earth with moon's centrifugal force, and found them to agree "pretty nearly," thus laying founds. for what was later to become concept of universal gravitation; apparently did little to develop his dynamical ideas from 1667-79; then, pressed by letter from Hooke of Jan. 17, 1679-80 about orbit of a body continually deviating from its inertial path under influence of central force varying inversely as the square of the distance, Newton demonstrated, by applying the principles of mechanics he had already developed, that it would be an ellipse, or precisely the path Kepler had assigned to planets; in response to a question communicated this result to Halley in Aug. 1684; by 1686, Newton had obtained accurate values for radius of earth and for acceleration due to gravity; he had come to think in terms of centripetal instead of centrifugal forces; and had proved that a sphere in attracting external bodies acts as though its mass were concentrated at its center; Newton's Principia is founded on his 3 laws or axioms of motion: 1. every body continues in its state of rest or of uniform motion in a straight line, unless it is compelled to change that state by forces impressed upon it, 2. change of motion is proportional to, and in the direction of, motive force impressed, 3. to every action there is always an equal and opposite reaction; by applying these axioms to Kepler's laws, Newton demonstrated that the force holding the planets in elliptical orbits varies inversely as the square of the distance between planet and sun; also demonstrated that there is a force between any two bodies (universal gravitation), which varies directly as their masses and inversely as the square of the distance between them; applied these results to explain comets, tides, the motion of satellites, and many other phenomena; distinguished between mass and weight; the Principia completed Copernican revolution by supplying the necessary new physics for the new astronomy, and by providing a synthesis of celestial and terrestrial mechanics; for the future it stood as symbol of what human reason, properly applied, could accomplish. Newton also conducted further investigations of lunar theory, 1690's; suffered from insomnia and a long nervous illness, 1694-95; involved in acrimonious controversy with astronomer John Flamsteed, 1705-12; and with Leibniz over priority in discovery of the calculus and over several other philosophical and theological issues, 1705-16. Died Kensington, England, Mar. 20, 1726 o.s. (Mar. 31, 1727 n.s.); buried Westminster Abbey.

NEWTON, John, English astronomer, mathematician; b. Great Oundle, Eng., 1622; s. Humphrey Newton; B.A., St. Edmund Hall, Oxford (Eng.) U., 1641, M.A., 1642; D.D.; 2 children—Thomas, John. Tchr. pure and applied math. before Restoration; King's chaplain and rector of Ross, Eng., 1661-78; rector of Upminster, Eng., 1662-78; named canon of Hereford, 1663; rector Ross Sch. Author: Institutio mathematica, 1654; Tabulae mathematicae, 1654; Astronomia britannica, 1657; Trigonometria britannica, or the doctrine of triangles in two books, 1658; Geometrical Trigonometry, 1659; Mathematical Elements in three Parts, 1660; Practical Geometry or the Art of Surveying, 1667; Art of Practical Gauging, 1669; Introduction to the Art of Logic, 1671; Cosmographia, 1679; Introduction to Astronomy. Used decimals in his math. calculations. Died Ross, Eng., Dec. 25, 1678.

NEWTON, Michael, physician; born Malvern, Eng., June 4, 1920; s. Frank Leslie and Alice (Henderson) N.; B.A., Cambridge (Eng.) U., 1942, M.A., 1946, M.B., B.Ch., 1944; M.D., U. Pa., 1943; m. Niles Polk Rumely, Mar. 27, 1943; children—Elizabeth Willoughby, Frances Lees, Edward Robson, Warren Polk. Came to U.S., 1946, naturalized, 1949. Practiced medicine specializing in obstetrics and gynecology U. Miss.. Sch. Medicine, Jackson, 1955-66; clin. prof. obstetrics and gynecology U. Chgo. Sch. Medicine, 1966—. Fellow A.C.S., Am. Coll. Obstetricians and Gynecologists (dir.); mem. A.M.A., Am. Fertility Soc., Central Assn. Obstetricians and Gynecologists. Contbr. numerous articles to tech. jours. Research on physiology and practical mgmt. lactation, physiology and psychology labor and delivery, maternal deaths and postpartum blood loss, gynecol. disorders and operative complications, cancer cervix and corpus uteri. Home: 330 W. Diversey Pkwy., Chgo. 60657.*

NEWTON, Niles Rumely, Am. behavioral scientist; b. N.Y.C., Jan. 19, 1923; d. Edward Aloysius and Fanny (Scott) Rumely; B.A., Bryn Mawr Coll., 1945; M.A., Columbia, 1948, Ph.D., 1952; m. Michael Newton, Mar. 27, 1943; children—Elizabeth Willoughby, Frances Lees, Edward Robson, Warren Polk. Research asso. Sch. Medicine, U. Pa., 1952-55; asst. prof. U. Miss. Sch. Medicine, Jackson, 1955-66; asst. prof., div. psychology, dept. neurology and psychiatry, Northwestern U. Med. Sch., Chgo., 1966—. Mem. sci. group on physiology lactation WHO, Geneva, Switzerland, 1963. Mem. Am. Psychosomatic Soc., Am. Psychol. Assn., Acad. Psycosomatic Medicine, Soc. Research in Child Devel. Author: Family Book of Child Care, 1957; Maternal Emotions, 1955; also articles. Editorial adv. bd. Child and Family, 1962—, Your New Baby, 1964—; contbg. editor Baby Talk, 1966—. Research on behavioral aspects reprodn. especially lactation and labor. Home: 330 W. Diversey Pkwy., Chgo. 60657.*

NEWTON, Robert Russell, Am. physicist; b. Chattanooga, July 7, 1918; s. Harry Russell and Myrtle (Wardrep) N.; B.S., U. Tenn., 1940; M.S., Ohio State U., 1942, Ph.D. in Physics, 1946; m. Doris Lanelle Cole, June 10, 1944; children—Robert Cole, Frederick Page. Instr., U. Tenn., 1942-44; research asso. Allegany Ballistics Lab., 1944-45; mem. tech. staff Bell Telephone Labs., 1946-48; asso. prof. U. Tenn., 1948-55; prof. physics Tulane U., 1955-57; staff Applied Physics Lab., Silver Spring, Md., 1957—, mem. prin. profl. staff, 1959—, supr. Space Research and Analysis br., 1964—. Mem. Am. Phys. Soc., Am. Inst. Aeros. and Astronautics, Am. Geophys. Union, Sigma Xi, Tau Beta Pi. Author: (with J.B. Rosser, G.L. Gross), The Mathematical Theory of Rocket Flight, 1947; also numerous articles. Research in molecular structure by analysis of spectra, electron processes in solids, elec. discharge through gases, exterior ballistics unguided rockets, dynamics orbital motion and orientation satellites, accurate tracking methods for satellites, stablzn. system for satellite orientation, applications satellites to navigation, gravity field and figures earth. Home: 701 Quaint Acres Dr., Silver Spring 20904. Office: 8621 Georgia Av., Silver Spring, Md. 20910.*

NEWTON, Roger Gerhard, physicist; b. Landsberg/Warthe, Germany, Nov. 30, 1924; s. Arthur and Margaret (Blume) Neuweg; student U. Berlin (Germany), 1945-46; A.B. summa cum laude, Harvard, 1949, M.A., 1950, Ph.D., 1953; m. Ruth Gordon, June 18, 1953; children—Julie, Rachel, Paul. Came to U.S., 1946, naturalized, 1949. Mem. Inst. for Advanced Study, Princeton, N.J., 1953-55; asst. prof. physics Ind. U., Bloomington, 1955-58, asso. prof., 1958-60, prof., 1960—. NSF sr. postdoctoral fellow, 1962-63, U. Rome, 1963-64. Fellow Am. Phys. Soc. Author: Complex j-Plane, 1964; Scattering Theory of Waves and Particles, 1966. Co-editor: Proc. High Energy Physics Conf., 1956. Contbr. numerous articles on field theory, scatter theory. Home: 1023 S. Ballantine St., Bloomington, Ind. 47401.*

NEY, Edward Purdy, Am. physicist; b. Mpls., Oct. 28, 1920; s. Otto Fred and Jessamine (Purdey) N.; B.S., U. Minn., 1942; Ph.D., U. Va., 1946; m. June Virginia Felsing, June 20, 1942; children—Judith, John, Arthur, William. Asst. prof. U. Va., 1946, asso. prof., 1947; faculty U. Minn., Mpls., 1947—, prof., 1955—. Recipient, Distinguished Teaching award, U. Minn., 1963. Fellow, A.A.A.S., Am. Phys. Soc.; mem. Am. Astron. Soc., Am. Geophys. Union. Author: Electromagnetism and Relativity, 1962. Research in cosmic rays and atmospheric physics, astrophysics. Home: 1925 Penn Av. South, Mpls.*

NEYMAN, Jerzy, statistician; b. Bendery, Rumania, Apr. 16, 1894; s. Czeslaw and Kazimiera (Lutoslawska) N.; ed. U. Kharkov, Russia, 1912-16; Ph.D., U. Warsaw, Poland, 1923; student U. London, 1925-26, U. Paris, 1926-27; D.Sc., U. Chgo., 1959; LL.D., U. Cal.; Ph.D., U. Stockholm; m. Olga Solodovnikova, May 4, 1920; 1 son, Michael John. Came to U. S., 1938. Lectr., Inst. Technology, Kharkov, 1917-21; statistician Inst. Agriculture, Bydgoszcz, Poland, 1921-23; lectr. Coll. Agr., Warsaw, 1923-24, U. Warsaw, 1928-34; spl. lectr. Univ. Coll., London, 1934-35, reader in statistics, 1935-38; spl. lectr. U. Paris, 1936, U. London, Paris, Universite Libre de Bruxelles, U. Warsaw, U. Amsterdam, U. Cambridge, U. Coll. London, U. Stockholm, U. Uppsala, 1950; prof. math., charge Statis. Lab., U. Cal. at Berkeley, 1938-41, dir. Statis. Lab., 1941—; statis. expert U. S. Mission to observe Greek elections, 1946; del. Internat. Statis. Conf., 1947, Internat. Congress Philos. of Sci., Paris, 1949. Fellow Royal Statis. Soc.; mem. Nat. Acad. Scis., Royal Swedish Acad. Scis., Inst. Math. Statistics (pres. 1949), Am., French, Polish math. socs., Am. Statis. Assn. (v.p. 1947), Inst. Statistics, Econometric Soc., Internat. Statis. Inst. Author several math. and statis. books published in Poland and France. Co-editor: Statis. Research Memoirs, Annals of Math. Statistics. Study of mathematical statistics; theories of testing hypotheses and of estimation; application to genetics; medical diagnosis; astronomy; weather modification experiments; statistical treatment of agricultural experiments. Home: 954 Euclid Av., Berkeley, Cal.

NEYMAN, Leonid Robertovich, Russian elec. engr.; b. St. Petersburg, 1902; grad. Elec. Engring. Faculty, Leningrad Poly. Inst., 1930; D.Tech. Sci., 1940. Inst., Elec. Engring. Faculty, Leningrad Poly. Inst., 1930-40, prof., 1940-46, dean, 1946-50, head chair theoretical principles of elec. engring., 1951—; dir. strong currents group Leningrad Electrophys. Inst., 1931-35; prof., 1940—; with lab Power Engring. Inst., USSR Acad. Scis., 1946. Mem. USSR Acad. Scis. (corr.). Author: The Electric Parameters of Ferroaluminum Conductors, 1935; The Physical Principles of Electrical Engineering, 1941; The Theoretical Principles of Electrical Engineering, 1948, 50, 51; The Electromagnetic Field Laboratory Handbook, 1950, 62. Mem. editorial bd. Electricity. Address: Leningrad Poly. Inst., Politekhnicheskaya ulitsa 3, Leningrad, USSR.

NEYMARK, Jouri Isanovich, Russian applied mathematician; b. Ukraine, USSR, Nov. 24, 1920; s. I.G. and E. (Perfiljeva) N.; candidate phys. and math. scis. Gorky State U. (USSR), 1947, doctor tech. scis., 1957; m. Valentina V. Ushachova, 1946; children—Tatjana, Aleksander. Faculty, Gorky State U., 1946—, prof., head chair, 1961—. Author: Linear Systems Stability, 1949; Dynamics of Nonholonomic Systems, 1967; also numerous articles. New methods of stability research of descrete and distributed systems; devel. point transformations method with applications to theory of nonlinear oscillations; devel. some aspects on nonholonomic systems dynamics. Address: Gorky State University, Computing Dept., Control Theory Subdepartment, Gorky, USSR.*

NEYNABER, Roy Harold, Am. physicist; b. Highland Park, Mich., July 4, 1926; s. Clifford and Ruth (Hanshew) N.; student Wayne State U., 1943-44, Ohio State U., 1945; B.S. in Elec. Engring., U. Wis., 1949, M.S. in Physics, 1951, Ph.D. in Physics, 1955; m. Claire Conant, Feb. 2, 1951; children—Mark Steven, Scott Andrew, Karen Beth. Sr. staff scientist Gen. Dynamics/Convair, San Diego, 1955—. Mem. Am. Phys. Soc., Sigma Xi, Tau Beta Pi. Condr. 1st expts. of x-ray scattering from cold-worked metals to be explained by double Bragg scattering; molecular beam techniques to measure atomic collision cross sections, co-developer techniques to superimpose two molecular beams for measuring reaction rates at energies from few hundredths to several hundred electron volts. Home: 4471 Braeburn Rd., San Diego 92116. Office: Space Science Lab., M.Z. 596-51, Gen. Dynamics/Convair, San Diego 92112.*

NGAI, Shih Hsun, physician; born Wuchang, China, Sept. 15, 1920; s. Chih F. and Shen (Shih) N.; M.B., Nat. Central U. Sch. Medicine, China, 1944; m. Hsueh-hwa Wang, Nov. 6, 1948; children—Mae, Janet, John. Came to U.S., 1946, naturalized, 1953. Mem. staff Presbyn. Hosp., N.Y.C., 1949—, attending anesthesiologist, 1965—; faculty Columbia Coll. Phys. and Surg. 1949—, prof anesthesiology, 1965—. Mem. com. on anesthesia NRC-Nat. Acad. Scis., 1961—; cons. NIH, 1963-67. Mem. Am. Physiol. Soc., Am. Soc. for Pharmacology and Exptl. Therapeutics, Assn. U. Anesthetists, Soc. for Exptl. Biology and Medicine, Am. Soc. Anesthesiologists, A.M.A., Sigma Xi. Author: Manual of Anesthesiology, 1959; Metabolic Effects of Anesthesia, 1962; Contbr. to Physiol. Pharmacology, 1963, Handbook of Physiology, 1964. Modern Trends in Anesthesia, 1966; Advances in Anesthesiology — Muscle Relaxants, 1967. Home: 281 Edgewood Av., Teaneck, N.J. 07666. Office: 622 W. 168th St., N.Y.C. 10032.*

NIAUSSAT, Pierre Marie Jules, French physician; b. Jonzac, France, Nov. 21, 1921; s. Jules and Germaine (Ansault) N.; M.D., U. Bordeaux, 1948, Ph.D., 1966; m. Marie-Madeleine Lalanne, Aug. 1, 1946; children—Marie-Christine, Jacques-Noel. Surgeon, Med. Corps, French Navy, 1949-55, anesthesiologist Naval hosps., Bordeaux, Paris, 1946-58; head div. biology and ecology Research Center, Armed Forces Health Service, Paris, 1959—; guest sci. Naval Med. Research Inst., Bethesda, Md., 1963-64. Sec.-gen. Assn. for Study of Water Pollution, 1964. Decorated Chevalier, Legion of Honor; Tropical Medicine prize, 1947. Mem. French Soc. Anesthesiology, Zool. Soc. France, Soc. Plant Physiology, Soc. Comparative Pathology. Research and numerous publs. on animal hypometabolism, action of serotonin in vegetal growth, radiosensibility of venomous anthropods. Home: 23 rue de l'Orangerie, Versailles 78, France. Office: 57 rue Cuvier, Paris V, France.*

NICAISE, Jules Edouard, French physician, surgeon; b. Port-en-Bessin, France; practiced surgery and medicine, Paris; ret. 1894. Author: Les Ecoles des medicine et le fondation des universities ad Moyen Age, 1891; Premieres status des chirurgiens de Paris, 1892; le Pharmacie et la mathierre medicale aux XIV° Siecle, 1892; also numerous biographies. Studied history of medicine and surgery, especially med. edn. in Middle Ages. Died 1896.

NICANDER (or Nicandros of Colophon), Greek pharmacist, toxicologist; b. Colophon, Ionia; flourished 275 B.C.; contemporary of Aratos and Theocritos; priest of Apollo in Claros, nr. Colophon; didactic, bucolic poet. Author (in Greek): Collection of Cures; Prognostica; Treatise on Agriculture; Treatise on Apiculture; Treatise on Snakes; Theriaca (on poisonous animal, effects of venoms, antidotes); Alexipharmica (on poisons and remedies). First to refer to therapeutic use of leeches.

NICE, Charles Monroe, Jr., Am. physician; b. Parsons, Kan., Dec. 21, 1919; s. Charles Monroe and Margaret (McClenahan) N.; A.B., U. Kan., 1939, M.D., 1943; M.S., U. Colo., 1946; Ph.D., U. Minn., 1956; m. Mary Ellen Cranmer, Dec. 21, 1940; children—Norma Jane (Mrs. Dennis E. Murphy), Pamela, Deborah, Julianne, Charles Monroe III, Thomas, Mary Ellen, Rebecca. Faculty, U. Minn., Mpls., 1951-58; asso. prof. radiology, 1951-58; prof., chmn. dept. radiology Tulane U., New Orleans, 1958—; dir. dept. diagnostic radiology Charity Hosp., New Orleans, 1958—. Mem. A.A.A.S., Am. Assn. U. Profs., Assn. U. Radiologists, Royal Soc. Medicine, Am. Coll. Radiologists, A.M.A. Research, publs. on radioresistance in tumors, diagnostic and therapeutic radiology. Home: 508 Millaudon St., New Orleans 70118.*

NICHOL, Charles Adam, pharmacologist; b. Fergus, Ont., Can., May 3, 1922; s. Walter Laidlaw and Marianne (Wingate) N.; B.S., U. Toronto, 1944; M.S., McGill U., 1946; Ph.D., U. Wis., 1949; m. Marion Aldred, Dec. 27, 1947; children—Paul, Catherine, Charles. Asst. prof., dept. pharmacology Western Res. U., Cleve., 1949-53, Yale, 1953-56; research prof. Sch. Pharmacy, State U. N.Y., Buffalo, also dir. dept. exptl. therapeutics Roswell Park Meml. Inst., Buffalo, 1956—. Cons. Cancer Chemotherapy Nat. Service Center, 1957-61, Nat. Cancer Inst., 1967-68. Mem. Am. Soc. Biol. Chemistry, Am. Assn. for Cancer Research, A.A.A.S., Am. Chem. Soc., Am. Soc. for Pharmacology and Exptl. Therapeutics N.Y. Acad. Sci., Soc. Exptl. Biology and Medicine, Sigma Xi. Contbr. articles to jours. Research on cancer chemotherapy, mechanism of action of folic acid antagonists, drug resistance in bacterial and mammalian cells, nutrition and tumor growth, corticosteroids and transaminase enzymes, drug-metabolizing enzymes. Home: 163 Lehn Springs Dr., Williamsville, N.Y. 14221. Office: 666 Elm St., Buffalo 14221.*

NICHOL, John Pringle, Scottish astronomer; b. Huntley Hill, Scotland, Jan. 13, 1804; s. John N. and Jane (Forbes) N.; ed. King's Coll., Aberdeen, Scotland; LL.D., U. Glasgow, 1837; m. Miss Tullis, 1831; children—John, a dau.; m. 2d, Elizabeth Pease, July 6, 1853. Became parish schoolmaster, Dun, Scotland, 1821; licensed as preacher and practiced as minister for short period; headmaster Hawick Grammar Sch.; editor Fife Herald; headmaster Cupar Acad.; rector Montrose Acad., 1827-34; named Regius prof. astronomy, U. Glasgow (Scotland), 1836; lectr. in U.S., 1848-49. Mem. Royal Astron. Soc., Royal Soc. Edinburgh. Author: Views on the Architecture of the Heavens, 1838; Pheonomena of the Solar System, 1838; The Stellar Universe, 1847; The Planetary System, 1848; The Planet Neptune, 1855. Editor: A Cyclopaedia of the Physical Sciences, 1857. Refitted Glasgow obs. with better equipment, 1840; lectured on Donati's comet, 1858; believed nebula to be gaseous masses until resolution of nebula in Orion by Lord Russe's telescope; earliest to suggest using photography to study sun spots; studied phys. features of moon. Died nr. Rothesay, Scotland, Sept. 19, 1859.

NICHOLAS, (Brother) Gerardus, Am. speleologist; b. Phila., Dec. 20, 1927; s. Edward J. and Florence (Delaney) N.; B.S., Cath. U. Am., 1950; M.Sc., U. Pitts., 1954; Ph.D., U. Notre Dame, 1960; D.Sc. (hon.), U. San Carlos, 1962, U. Kyoto, 1962. Joined Bros. of Christian Schs., 1945; faculty La Salle High Sch., Cumberland, Md., 1950-57, La Salle Coll., Phila., 1957-58, 63—; faculty U. Notre Dame (Ind.), 1959-62, U. New South Wales, Sydney, Australia, 1963. Fellow A.A.A.S.; mem. Nat. Speleological Soc. (pres. 1957-61), Nat. Assn. Biology Tchrs. (v.p. 1961), Soc. Systematic Zoology, Soc. Mammalogy, Ecol. Soc. Am., Am. Soc. Zoologists, Cave Research Found., Am. Assn. U. Profs., Pa., N.Y. acads. sci. Author (with George Moore) Speleology - The Study of Caves, 1964. Editor: Internat. Jour. Speleology, 1964—; asst. editor Am. Biology Tchr., 1953-56. Explored numerous caves throughout world, studying ecology and systematics of fauna. Address: LaSalle Coll., Phila. 19141.*

NICHOLAS, John Spangler, Am. biologist; b. Pitts., Mar. 10, 1895; s. Samuel Trauger and Elisabeth Ellen (Spangler) N.; B.S., Pa. Coll., 1916, M.S., 1917; Ph.D., Yale, 1921; m. Helen Benton Brown, Dec. 17, 1921. Instr. biology, Pa. Coll., 1915-17; master Gettysburg Acad., 1916-17; instr. anatomy, U. of Pitts., 1921-22, asst. prof., 1922-26; asst. prof. biology, Yale, 1926-32, asso. prof. comparative anatomy, 1932-35, Bronson prof. comparative anatomy, 1935-39, Sterling prof. biology, 1939-63, Sterling prof. biology emeritus, 1963, chmn. dept. zoology, 1946-56; fellow Trumbull College, 1933-45, master, 1945-63; hon. fellow St. Catherine's Coll., Cambridge, Eng.; dir. Osborn Zool. Lab., 1946-56; Mellon lectr. U. Pitts., 1948; Woods Hall lectr., 1953; Rockefeller Inst. lectr., 1956; vis. prof., 1956-63; speaker Humanities Seminar, U. Mass., 1958; mem. planning commn. Internat. Congress Developmental Biology; mem. Brookhaven Biol. vis. commn. Chmn. Div. Biology and Agr., NRC, 1948, vice chmn., 1949; sec. 7th Internat. Congress for Cell Biology; adviser Nat. Selective Service, 1948-53. Bd. trustees, Sheffield Sci. Sch. of Yale U., sec., 1941-56; mem. sci. adv. bd. Wistar Inst. of Anatomy and Biology, N.Y. Zoöl. Soc.; NRC rep. on Nat. Roster of Sci. and Specialized Personnel, 1940-47; cons. Nat. Resources Planning Bd., 1940-43, War Manpower Commn., 1943-48, Sci. Manpower Commn., v.p., 1953-54 mem. exec. com., 1954-58, pres., 1955-58; mem. vis. com. Brookhaven Nat. Labs., 1955-58, chmn. 1958. Seminar moderator Gettysburg Coll., 1961. Fellow A.A.A.S., Conn. Acad. Arts and Scis., Am. Inst. Biol. Scis. (governing bd.); member NSF (div. biol. sci., 1951-53), Am. Acad. Arts and Scis., Nat. Acad. Scis. (chmn. Anatomy and Zoology sect. 1955-58), Soc. for Cell Biology, Growth Soc., l'Institute International d'Embryologie, Am. Soc. Physiology, Am. Soc. Zoology, Am. Biol. Soc. (pres. 1945), Am. Philos. Soc. (councilor 1954-57), Am. Soc. Anatomy, Am. Cancer Assn., Soc. for Exptl. Biology and Medicine, Phi Beta Kappa, Sigma Xi (nat. exec. com. 1954-57). Assn. editor Jour. of Morphology, 1933-36; editor Biol. Abstracts, Vertebrate Embryology 1938-63; co-editor Anat. Memoirs 1940-63; mem. editorial bd. Jour. Exptl. Zoology, 1946, mng. editor, 1947; pres. Yale Jour. Biology and Medicine, 1954-62; adv. editor Exerpta Medica, Yale Sci. Mag.; numerous papers on exptl. embryology. Effect of surrounding tissues on establishment of symmetry in developing amphibian limb; ability of embryonic rat tissue to develop after grafting to various abnormal sites in rat hosts, on chick chorioallantois, and in tissue culture; regulation in early stages of devel. of rat (separation and fusion of blastomeres). Died Sept. 11, 1963.

NICHOLAS OF CUSA (Nicolaus Cusanus), German clergyman, natural philosopher; b. Kues, Reinland, 1401; studied law at Padua; became cardinal, 1448, Bishop of Brixen, also papal legate, circa 1450, travelled through most of Holy Roman empire, preaching and instituting monastic reform, 1451-52. Author: De docta ignorantia (rejection of classical cosmos, introduced concept of universe with center everywhere and circumference nowhere), 1440; Idiotae libri quatuor (book 4 contains dialog on static expts.), 1450; also math. treatises on quadrature of circle, calendar reform, improvement of Alphonsine Tables, theory of numbers. Primarily a philosopher; suggested Earth might rotate on its axis and that observer on Sun would see the Earth revolving about him; suggested weighing earth and seeds, the resulting plants, then the plant ashes and earth in which they had grown (thought ashes would show weight of element earth in them, expt. later discussed by Helmont and Boyle); emphasized significance of weight measurements and timing; suggested pulse count be used as diagnostic technique; among first to work on exposure of false Decretals. Died Lodi, Italy, Aug. 11, 1464.

NICHOLAS OF POLAND, physician; studied medicine, Montpellier, France, probably lived there 20 years; mem. Dominican order; at ct. of Leszek the Black, king of Poland, after 1270. Author: Antipocras (med. poem with prose prologue); possibly wrote Experiments (a similar work). Versed in Greek and Arabic med. knowledge; attacked physicians steeped in Galen and Hippocrates; mentioned occult force of magnet, other magical lore. Died Circa 1278.

NICHOLAS OF SALERNO (or Nicolaus Salernitanus; also Nicolaus Praepositus) physician, flourished 12th century; wrote Antidotarium (often called Antidotarium parvum); it is in 2 versions (1 for physicians, 1 for apothecaries); contains about 150 recipes for preparation of drugs, their application and action; it is a collection of recipes transmitted by Greek and Latin writers with additional recipes from Arabic sources; a very popular work, often translated and adapted, it became the basis of all later pharmacopoeias.

NICHOLAS II OF SALERNO, see Richard of Salerno.

NICHOLLS, Cyril Minchin, chem. and nuclear engr.; b. Krugersdorp, S. Africa, June 28, 1914; s. Charles Beresford and Anne (Jennings) N.; B.Sc. in Engring., U. Witwatersrand, Johannesburg, S. Africa, 1936, D.Sc., 1958; m. Hermione Joan Martin, Sept. 21, 1961; children—Richard Michael, Alexandra Louise, Charles Martin. Chem. engr. Angio-Alpha Cement Ltd., S. Africa, 1936-41; chem. warfare adviser U.K. Royal Air Force, 1941-46; with U.K. Atomic Energy Authority at Atomic Energy Research Establishment,

Harwell, 1946—, dep. head chem. engring. div., 1955—. Decorated Order Brit. Empire. Fellow Royal Inst. Chemistry; mem. Instn. Chem. Engrs. (mem. council 1966—). Editor: Progress in Nuclear Energy, vol. 3, 1960, vol. 4, 1961, vol. 5, 1963; (with R.B. Scott, W.H. Denton) Technology and Uses of Liquid Hydrogen,, 1964. Research, publs. on processing of nuclear fuels, chem. and nuclear engring. Home: Larchwood, Lincombe Lane, Boar's Hill, Oxford, Eng. Office: Atomic Energy Research Establishment, Harwell, Didcot, Berks., Eng.*

NICHOLLS, Doris Margaret McEwen (Mrs. Ralph William Nicholls), Canadian biochemist; b. Bayfield, Ont., Can., Jan. 24, 1927; d. Frederick William and Ellen (Peck) N.; B.Sc., U. Western Ont., 1949, M.Sc., 1951, Ph.D. in Biochemistry, 1956, M.D. in Medicine, 1959; m. Ralph William Nicholls, June 28, 1952. NRC Research fellow U. Western Ont., London, Can., 1953-55, asso. research, 1955-62; teaching asso. dept. physiology George Washington U., Washington, 1959-60; dir. clin. investigation unit Dept. Vets. Affairs, Westminster Hosp., London, 1960-65; asso. prof. biochemistry dept. biology York U., Toronto, Ont., 1965—. Cons. Nat. Heart Inst., Bethesda, Md., 1959-60. Mem. Biochem. Soc. (London, Eng.), Canadian Biochem. Soc., Canadian Physiol. Soc., Canadian Soc. Cell Biology, Chem. Inst. Can., Canadian Med. Assn., A.A.A.S., N.Y. Acad. Sci., Mt. Desert Island Biol. Soc., Alpha Omega Alpha. Research, publs. on adrenal phosphorus metabolism in cold stress, Quabain on phospholipid metabolism, cholesterol in regenerating sciatic nerve, antihypertensive drugs preventing kidney deterioration, nephrotic kidney synthesizing protein differently from normal, protein biosynthesis. Office: Dept. Biology, York U., Toronto, 12, Ont., Can.*

NICHOLLS, Frank, English physician; b. London, Eng., 1699; s. John Nicholls; B.A., Exeter Coll., Oxford (Eng.) U., 1718; M.A., 1721, M.B., 1724, M.D., 1729; m. Elizabeth Mead; 5 children, including John. Lectr. anatomy, Oxford U.; practice medicine, Cornwall, Eng.; lectr. anatomy, London; became Gulstonian lectr., 1734, 36, Harveian orator, 1739, Lumleian lectr., 1748-49; named physician to George II, 1753. Fellow Royal Soc., 1728, Coll. Physicians. Pub. compendium of lectures, 1732. Demonstrated small structure of blood vessels; showed arteries were supplied with nerves and proposed they regulated blood pressure; 1st to use corroded preparations in which part of an organ is left prominent after injection; proved inner and middle coat of artery could rupture with outer coat intact, demonstrating formation of chronic aneurysm. Died Jan. 7, 1778.

NICHOLLS, John Van Vliet, Canadian physician; b. Montreal, Que., Can., Dec. 5, 1909; s. Albert George and Lucia (Van Vliet) N.; B.A., McGill U., 1930, M.Sc., 1935, M.D., 1934; m. Flora Adeline Fowke, June 30, 1937; children—Elizabeth Caroline (Mrs. Ronald Toop), John Frederick, Peter John. Pvt. practice medicine specializing in ophthalmology, Montreal, 1938-41, 45—; with McGill U., 1938—, asso. prof., 1958—; ophthalmologist Royal Victoria Hosp., Montreal. Bd. dirs. Can. Nat. Inst. Blind Fellow A.C.S., Royal Coll. Physicians and Surgeons Can.; mem. Montreal (pres. 1952-54), Canadian (pres. 1955-56) ophthalmology socs., Assn. Ophthalmology Province Que. (mem. council), Am. Ophthalmol. Soc., Canadian Med. Assn., Am. Acad. Ophthalmology and Otolaryngology, N.Y. Acad. Scis., Canadian Phsiol. Soc., Sigma Xi, Alpha Omega Alpha. Research, publs. on vascular and morphological pathology of eye, reading disabilities in children, vision and flying performance. Home: 5690 Queen Mary Rd., Montreal 29. Office: 1414 Drummond St., Montreal 25, Que., Can.*

NICHOLS, Alexander Vladimir, Am. biophysicist; b. San Francisco, Oct. 9, 1924; s. Vladimir Ivan and Mary (Beshenovan) N.; B.A., U. Cal., Berkeley, 1949, Ph.D. (San Joaquin County Heart Assn. fellow), 1955; m. Sally Frances Streets, Oct. 9, 1955; children—Robert, Kyra, Alexander. Faculty, U. Cal., Berkeley, 1961—, asso. prof. med. physics and biophysics, 1964—. Mem. Sigma Xi. Author: (with J.W. Gofman, V. Dobbin) Dietary Prevention and Treatment of Heart Disease, 1958. Co-discoverer lipolytic action of heparin-induced lipemia "clearing factor"; evaluated influence of dietary carbohydrates and fats on serum levels of lipoproteins. Home: 266 Grizzly Peak Blvd., Kensington, Cal. 94708. Office: Donner Lab., U. Cal., Berkeley, Cal. 94720.*

NICHOLS, Donald Richardson, Am. physician; b. Mpls., Feb. 22, 1911; s. Arthur R. and Augusta (Fisher) N.; A.B., Amherst Coll., 1933; M.D., U. Minn., 1938, M.S. in Medicine, 1942; m. Margery Spicer, Mar. 7, 1942; children—Virginia, John; m. 2d, Mary Jean Scholberg, Mar. 2, 1957; children—Arthur, Edwin, Mary Jean, Barbara. Practice medicine specializing in internal medicine, Rochester, Minn., 1940—; cons. Mayo Clinic, 1943—, head sect. infectious diseases, 1961—; faculty Mayo Grad. Sch. Medicine, U. Minn., 1948—, prof. clin. medicine, 1965—. Fellow A.C.P.; mem. A.M.A., Am. Pub. Health Assn., Infectious Diseases Soc. Am., Am. Health Assn., Infectious Diseases Soc. Am., Am. Thoracic Soc., Central Soc. for Clin. Research, N.Y. Acad. Scis., Minn. Soc. Internal Medicine, Sigma Xi, Nu Sigma Nu. Research, numerous publs. on absorption, diffusion, and excretion antibiotic agts. in human body; developed new programs for treatment human infections. Home: 1068 Plummer Lane, Rochester 55902. Oéice: 200 1st St. S.W., Rochester, Minn. 55902.*

NICHOLS, Edward Leamington, Am. physicist; b. Leamington, Eng., Sept. 14, 1854 (parents Am. citizens); s. Edward Willard and Maria (Watkinson) N.; B.S., Cornell, 1875; postgrad. univs. Leipzig, Berlin, 1875-78; Ph.D., U. Göttingen, 1879; fellow Johns Hopkins U., 1879-80; LL.D., U. Pa., 1906; D.Sc., Dartmouth, 1910; m. Ida Preston, May 25, 1881; children—Elizabeth (Mrs. Montgomery Hunt Throop), Robert Preston. With Edison at Menlo Park, N.J., 1880-81; prof. physics and chemistry, Central U. of Ky., 1881-83; prof. physics and astronomy U. Kan., 1883-87; prof. physics Cornell, 1887-1919, emeritus, from 1919. Fellow Am. Acad. Arts and Scis., pres. A.A.A.S., 1907, Am. Phys. Soc., 1908-09, Kan. Acad. Sci., 1884-86. Author: The Galvanometer, 1894; A Laboratory Manual of Physics and Applied Electricity (with Merritt, Bedell and others), 2 vols., 1895; The Elements of Physics, 3 vols. (with W. S. Franklin), 1896; Outlines of Physics, 1897; Studies in Luminescence (with Merritt), 1910; Fluorescence of the Uranyl Salts (with Howes), 1919; Cathodo-Luminescence and the Luminescence of Incandescent Solids (with Howes and Wilber), 1928; also numerous papers on exptl. physics. Founder Phys. Rev., editor, 1893-1912. Died West Palm Beach, Fla., Nov. 10, 1937.

NICHOLS, Ernest Fox, Am. physicist; b. Leavenworth, Kan., June 1, 1869; s. Alonzo Curtis and Sophronia (Fox) N.; B.Sc., Kan. Agrl. Coll., 1888; M.Sc., Cornell U., 1893, D.Sc., 1897; postgrad. U. Berlin, 1894-96, Cambridge U., 1904-05; D.Sc., (hon.), Dartmouth, 1903; LL.D., Colgate, Clark, Wesleyan, 1909, Vt., 1911, Pitts., 1912, Dennison, 1914, Dartmouth, 1916; m. Katharine Williams West, June 16, 1894. Prof. physics, Colgate U., 1892-98; prof. Dartmouth, 1898-1903, pres., 1909-16; prof. exptl. physics Columbia, 1903-09; prof. physics Yale, 1916-20; dir. pure science Nela Research Labs., Nat. Lamp Works, Cleve., 1920-21; pres. Mass. Inst. of Tech., 1921; dir. Nela Research Lab., from 1929. Research asso. Carnegie Instn., Washington, 1907-09, Bur. of Ordnance, Navy Dept., 1917-19. Recipient Rumford medal Am. Acad. Arts and Scis., 1905. Fellow Am. Acad. Arts and Scis., A.A.A.S. (v.p. 1903); mem. Nat. Acad. Scis. Collaborator Astrophys. Jour. Contbr. many papers on radiation and other phys. subjects. Conducted (with J.F. Hull) 1st measurement of radiation pressure, 1901; devised sensitive radiometer for measurement of planetary heat and determination of light pressure; succeeded (with James D. Tear) in bridging gap between radio electric waves and infrared radiation, 1923; produced Hertzian waves. Died Apr. 29, 1924.

NICHOLS, George Jr., Am. physician; b. N.Y.C., May 15, 1922; s. George and Jane (Morgan) N.; student Harvard, 1939-42; M.D., Columbia, 1945; m. Nancy T. Pinks, July 1, 1944; children—George, III, Susan L. Dir. health and med. care program for students, Harvard Med. Sch., Boston, 1953-55, faculty medicine, 1953—, Markle scholar in med. scis., 1957-62, sec. faculty, 1959-61, clin. prof. medicine, 1965—, asso. dean for acad. affairs, 1962-65; investigator Howard Hughes Med. Inst., 1955-57; chief medicine Cambridge (Mass.) City Hosp., 1965—. Mem. A.A.A.S., A.C.P., Am. Physiol. Soc., Am. Soc. for Clin. Investigation, Endocrine Soc., Assn. Am. Physicians, N.Y. Acad. Scis., Sigma Xi, Alpha Omicron Alpha. Contbr. numerous articles to tech. jours. Pioneered systematic examination cell metabolism in bone; described aerobic prodn. lactic acid by bone; demonstrated enzyme collagenase is a constituent of bone cells. Home: Procter St., Manchester, Mass. 01944. Office: Cambridge City Hosp., Cambridge, Mass. 02139.*

NICHOLS, Robert Cosby, Am. psychologist; b. Hopkinsville, Ky., Jan. 24, 1927; s. Robert T. and Lulye (Coxby) N.; B.A., U. Louisville, 1949, M.A., 1950; Ph.D., U. Ky., 1953; m. Virginia Kimbel, Sept. 3, 1949; children—David, James. Clin. psychologist VA Hosp., Houston, 1953-55, Downey, Ill., 1955-56; asst. prof. psychology Purdue U., 1956-61; research psychologist Nat. Merit Scholarship Corp., Evanston, Ill., 1961-64, dir. research, 1964-67, v.p. research, Mem. Am. Psychol. Assn., Am. Sociol. Assn., Am. Edn. Research Assn., Am. Personnel and Guidance Assn. Research, publs. on methods of personality assessment, resemblance of twins in intelligence and personality, effects of ednl. experiences. Home: 8 Salem Circle, Evanston 60203. Office: 990 Grove St., Evanston, Ill. 60201.*

NICHOLS, Roy Woodward, English metallurgist; b. Rawdon, Eng., June 9, 1923; s. Ernest and Sarah (Woodward) N.; Certificate Air Engring., Loughborough Coll., 1943; Mech. Engr., Leeds (Eng.) Coll. Tech., 1942; B. Metallurgy with 1st class honors, U. Sheffield (Eng.), 1949; m. Jacqueline Woods, Aug. 30, 1946; children—Mark A., Paul S., Claire A. Mech. engring. draughtsman Kirkstall Forge Ltd., Leeds, 1939-42; research metallurgist Ministry Supply, Woolwich, Eng., 1949-56; research metallurgist, dep. head labs. U.K. Atomic Energy Authority Reactor Materials Lab., Culcheth, Eng., 1956—. Chmn. coordinated research group Navy Dept., 1964—, joint nuclear power steels study group, 1961—, welding tech. group Ministry Tech., 1966—. Fellow Instn. Metallurgists; mem. Internat. Inst. Welding (mem. commn. X 1962—), Brit. Welding Research Assn. (mem. research bd. 1962—), Inst. Metals, West Scotland Iron and Steel Inst. Contbg. author: Design of Gas-cooled Graphite Moderated Reactors, 1962; Materials for Nuclear Engineers, 1960. Research, publs. on devel. steels for nuclear pressure vessels and fuel elements, zirconium alloy pressure tubs; assessed failure conditions for engring. materials in creep, fatigue and fast fracture; developed an analysis of prevention of such failures; studies on effects of irradiation by neutrons, of welding and thermal treatment on engring. properties. Home: 13 Winwick Lane, Lowton Warrington, Lancashire. Office: Reactor Materials Lab., Culcheth Warrington, Lancashire, Eng.*

NICHOLSON, George Edward, Jr., Am. statistician; b. Bklyn., June 21, 1918; s. George Edward and Ada (White) N.; A.B., U. N.C., 1940, M.A., 1941, Ph.D., 1948; m. Virginia Williams, 1942; children—George Edward III, Molly, Britain, Patrick. Teaching fellow U. N.C., 1940-41; instr. Ga. Inst. Tech., 1941-43; research math. Columbia, 1944-46; with U. N.C., Chapel Hill, 1947—, prof. statistics, 1956—, chmn. dept., 1952—. Cons. Hq. USAF, dir. U. N.C. USAF Operations Analysis Unit; cons. Esso Research & Devel. Co.; chmn. SHAPE LWSR Operations Com. Recipient Medal of Freedom, 1947. Fellow Inst. Math. Statistics (exec. sec. 1954—); mem. Am. Statis. Assn., A.A.A.S., Psychometric Soc., Operations Research Soc. Am. Contbr. articles to sci. jours. Devel. and application of statis. methods in various fields of natural and social sci. Home: 234 McCauley St., Chapel Hill, N.C. 27515.*

NICHOLSON, Hayden Coler, Am. physiologist; b. Redwood Falls, Minn., Feb. 4, 1904; s. Ernest Crawford and Alma (Bordeaux) N.; A.B., U. Mich., 1925, M.S., 1927, M.D., 1939; m. Marian Louise Lawless, Nov. 1, 1929; 1 dau., Barbara Louise (Mrs. Jerry Titel). Faculty physiology U. Mich., Ann Arbor, 1929-46, asso. prof., 1940-46; exec. sec. com. on growth NRC, Washington, 1946-50; dean U. Ark. Sch. Medicine, Little Rock, 1950-55; exec. dir. Hosp. Council of Greater N.Y., N.Y.C., 1955-62; dean U. Miami (Fla.) Sch. Medicine, 1962—. Mem. A.A.A.S., Soc. for Exptl. Biology and Medicine, Am. Physiol. Assn., Aerospace Med. Assn., Am. Pub. Health Assn., A.M.A., So., Fla. med. assns., N.Y. Acad. Scis., Sigma Xi, Alpha Omega Alpha, Phi Sigma, Alpha Kappa Kappa. Research on chem. regulation of respiration, localization and mechanism of action of respiratory center. Home: 610 Gondoliere Av., Coral Gables, Fla. 33143. Office: 1600 N.W. 10th Av., Miami, Fla. 33136.*

NICHOLSON, Henry Alleyne, Brit. zoologist, paleontologist; b. Penrith, Eng., Sept. 11, 1844; s. John Nicholson; ed. U. Göttingen (Germany), Edinburgh (Scotland) U.; M.D., D.Sc., Ph.D. Lectr. natural history Extra-Mural Sch. Medicine, Edinburgh, from 1869; prof. natural history U. Toronto (Ont., Can.), from 1871, U. Durham (Eng.), 1874, St. Andrews (Scotland) U., 1875-82; regius prof. natural history Aberdeen (Scotland) U., from 1882. Swiney lectr. geology Brit. Mus., 1877-82, 90-94. Recipient Lyell medal Geol. Soc., 1888. Fellow Royal Soc., 1897. Author: Essay on the Geology of Cumberland and Westmoreland, 1868; Monograph of the British Graptolitidae, Introduction, 1872; Reports on the Paleontology of Ontario, 1874-75; Ancient life-History of the Earth, 1877; Tabulate Corals of the Palaeozoic Period, 1879; Structure and Affinities of Monticuliporoids, 1881; Manual of Zoology, 7th edit., 1887; Monograph of the British Stromatoporoids; Manual of Palaeontology, 3d edit., 1889. Died Aberdeen, Scotland, Jan. 19, 1899.

NICHOLSON, Jesse Thompson, Am. orthopaedic surgeon; b. Camden, N.J., Apr. 28, 1903; s. Joseph Lippincott and Elizabeth (Thompson) N.; B.S., U. Pa., 1925, M.D., 1928; m. Edith Rose, Apr. 8, 1942; children—Elizabeth Thompson, Edith Davis, Joseph Lippincott, Virginia. Orthopaedic surgeon Pa. Hosp., Phila., 1938—, Grad. Hosp., 1934—, Phila. Children's Hosp., 1935—, Lankenau Hosp., 1953—; faculty orthopaedic surgery U. Pa. Grad. Sch. Medicine, 1940—, prof., chmn. grad. div., 1965—; cons. Phila. Gen. Hosp., Naval Hosp., (all Phila.), Atlantic City Seashore House. Research dir. N.Y. Cancer Research Inst., Inc., 1952-57; mem. orthopaedic sect. NRC, 1956-59; med. adviser Phila. chpt. Nat. Found. Poliomyelitis, 1935-55. Diplomate Am. Bd. Orthopaedic Surgery. Fellow A.C.S.; mem. Am. Rheumatism Assn., Am. Acad. Orthopaedic Surgeons (treas. 1954-59), Am. Orthopaedic Assn. (pres. 1964-65), Am. Assn. for Surgery of Trauma (pres., 1963), Internat. Soc. Orthopaedic Surgery and Traumatology, A.M.A. (chmn. ortho. sect. 1955). Co-editor: Bone as a Tissue, 1959; asso. editor Jour. Bone and Joint Surgery, 1947-49, 57-60, 61-64. Contbr. med., surg. jours. Home: 516 Oakley Rd., Haverford, Pa. Office: 419 S. 19th St., Phila. 19146.*

NICHOLSON, John William, English mathematician; b. 1881; s. John William and Alice Emily (Kirton) N.; M.Sc., U. Manchester (Eng.); M.A., Trinity

Coll., Cambridge (Eng.) U., Oxford (Eng.) U.; D.Sc., U. London (Eng.); m. Dorothy Wrinch, 1922; 1 dau. Fellow, tutor, dir. studies in math., physics Balliol. Coll., Oxford U., prof. math. U. London, 1912-21; lectr. Cavendish Lab., Cambridge U., also Queen's U., Belfast, Ireland. Gov. tech. optics dept. Imperial Coll. Sci.; mem. com. on radio-telegraphic investigations. Fellow Royal Soc. (council), 1917, Royal Astron. Soc. (council); mem. Röntgen Soc. (pres.), Phys. Soc. London (v.p.), London Math. Soc., Société Française de Physique, Brit. Assn. (com. on math. tables), Brit. Sci. Instruments Research Assn. (council). Author: (with Dendy and others) Problems of Modern Science, 1921; (with Arthur Schuster) The Theory of Optics, 1924; (with Joseph Larmor and others) Scientific Papers of S. B. McLaren, 1925. Contbr. papers to sci. publs. Died Oct. 10, 1955.

NICHOLSON, Peter, English mathematician, architect; b. Prestonkirk, Eng., July 20, 1765; ed. East Lothian village sch; m.; 1 son, Michael Angelo; m. 2d; 1 son, 1 dau. Apprentice to cabinet maker, Linton, 4 years; journeyman, Edinburgh; went to London, 1785, opened sch. for mechanics, Soho; architect County Cumberland, 1805; architect, Glasgow, 1800-08; returned to London, 1810, gave pvt. lessons in math., surveying, geography, nav., mech. drawing, fortification; moved to Morpeth, 1829; went to Newcastle-on-Tyne, Eng., 1832, opened a sch. Recipient Gold Medal Soc. Arts, 1814. Mem. Newcastle Soc. for Promotion of Fine Arts (pres.). Author: The Carpenter's New Guide, 1792; Principles of Architecture, 1795-98; Essays on Combinatorial Analysis, 1818; The Rudiments of Algebra, 1819; Popular Course of Pure and Mixed Mathematics, 1822; Carpentry, Joining, and Building, 1851. Introduced improved mech. processes of bldg.; inventor centrolinead, circa 1814; 1st to apply orthographical projection to solids in gen.; devised new method of extracting cube root; 1st to notice Grecian moldings are conic sections. Died Carlisle, Eng., June 18, 1844.

NICHOLSON, Richard Benjamin, Am. physicist; b. Tacoma, Sept. 8, 1928; s. Benjamin Lester and Flora (Faver) N.; B.S., U. Puget Sound, 1950; postgrad. U. Hawaii; M.S., Cornell U., 1960; Ph.D., U. Mich., 1963; m. Suzanne Marie Berven, Aug. 22, 1952; children—Ellen Marie, Lisa Kathryn, Paul Richard, Ingrid Maija. Physicist, Puget Sound Naval Shipyard, 1952-53, Lawrence Radiation Lab., 1965; physicist Atomic Power Devel. Assos., Detroit, 1954-61, cons., 1961-66; asso. prof. nuclear engring. U. Wis., 1963-66; physicist Argonne Nat. Lab., Idaho Falls, Ida., 1966—. Mem. Am. Nuclear Soc., Am. Phys. Soc., Sigma Xi, Phi Kappa Phi. Research, publs. on theory of fast neutron reactors and plasma physics, theory of Doppler temperature co-efficient of reactivity in fast reactors and theory of nuclear accidents in fast reactors, new solutions of Vlasov equations for hot collisionless plasmas. Home: Route 4, Box 323B. Office: Argonne Nat. Lab., Idaho Falls, Ida.*

NICHOLSON, Seth Barnes, Am. astronomer; b. Springfield, Ill., Nov. 12, 1891; s. William Franklin and Martha (Ames) N.; B.S., Drake U., LL.D., 1949; Ph.D., U. Cal., 1915; m. Alma M. Stotts, May 29, 1913; children—Margaret Ruth, Donald Seth, Jean Cary. Instr. physics, Drake U., 1911-12; fellow in astronomy U. Cal., 1912-13; instr., 1913-15; astronomer Mt. Wilson Obs. (Carnegie Instn., Washington), Pasadena, 1915-57, ret.; lectr. astronomy, U. Cal., summers, 1915, 23, 28, 44. Mem. Nat. Acad. Scis., A.A.A.S., Am. Astron. Soc., Astron. Soc. Pacific (pres. 1935-36, editor publs.). Internat. Astron. Union, Sigma Xi, Phi Beta Kappa. Discovered 9th, 10th, 11th, 12th satellites of Jupiter (1914, 38, 51); specialized in solar astronomy and stellar radiation; studied (with Pettit) surface temperature of planets with thermocouple which they devised. Died Los Angeles, July 2, 1963.

NICHOLSON, William, English physicist; b. London, 1753; ed. in N. Yorkshire; entered service of E. India Co. at age of 16; made 2 or 3 voyages to E. Indies, before 1773; settled in London; traveller for Wedgwood; schoolmaster in Soho Sq., 1775; patent agt.; water works engr. for Portsmouth and Gosport. Author: Dictionary of Practical and Theoretical Chemistry, 1808; First Principles of Chemistry, 1790; Introduction to Natural Philosophy, 1781; Navigator's Assistant, 1784; also numerous other books, articles. Editor, Journal Natural Philosophy, 1797-1815. Invented hydrometer (Nicholson's hydrometer), machine for printing on linen, 1790, aerometer, circa 1789, printing press; improved slide rule, doubler (an elec. amplifier); used antiphlogistic theory; built (with Carlisle) 1st voltaic pile in Eng.; discovered water could be dissociated by electricity. Died Bloomsbury, Eng., May 21, 1815.

NICHOLSON JEFFERSON, Carlos Walter, Peruvian geographer and military scientist; b. Arequipa, Peru, July 21, 1892; s. Walter and Sara (Jefferson) N.; ed. U. San Agustín, Arequipa; military colleges in Peru and France; B.S., Ecole Normale Supérieure, Paris, France; m. Lía Jahnsen, Mar. 19, 1951. Attached to Peruvian Military Mission in France, 1909-12; in Germany, 1928-32; prof. U. Arequipa, 1932-40; Guggenheim grant studies in U.S.A., 1940-41; prof. U. San Agustín, 1941-46; Ambassador to China, 1946-68; prof. U. San Agustín, 1948-54; prof. U. San Marcos, Lima, 1954-61; prof. emeritus, San Marcos, 1961—. Mem. Am. Geographical Soc.; Geographic Soc. of Lima; Geological Soc. of Peru; Pan-Am. Congress of Mines and Geology; UNESCO consultant on oceanography. Author: Organización de Formaciones Andinas, 1917; Ensayos de geografia politica del Perú, 1935; many articles and studies. Major contribution to study of geography of Peru and its ocean. Address: 163 San Antonio, Miraflores, Lima, Perú.*

NICKEL, Erwin J. K., mineralogist; b. Frankenstein, Germany, May 11, 1921; s. Julius and Martha (Hauke) N.; student Breslau, Germany (now Wroclaw, Poland), Vienna, Posen; Dr.rer.nat., 1944; habilitation U. Heidelberg (Germany), 1951; m. Herta Knappe, Sept. 28, 1947; children—Andreas, Barbara, Ekkehart. Main asst. dept. mineralogy and petrography U. Heidelberg, 1947-51; lectr. U. Munster (Germany), 1951-56; prof. mineralogy, dir. Inst. Mineralogy and Petrography, U. Fribourg (Switzerland), 1966—. Research and publs. on petrography, philosophy of nature. Home: 18, Av. Moléson, Fribourg, Switzerland.*

NICKERSON, John Lester, physiologist; b. Halifax, N.S., Can., Oct. 8, 1903; s. Obed Frederick and Olivia Agnes (Mills) N.; B.A., Dalhousie U., Halifax, 1925, M.A. (MacGregor fellow), 1928; Ph.D., Princeton, 1935; m. Annie McLeod Cook, Oct. 5, 1929. Came to U.S., 1939, naturalized, 1945. Bursar, Canadian NRC, 1927-28; instr. physics Princeton, 1928-30; asst. prof. physics Mt. Allison U., Sacksville, N.B., Can., 1931-38, prof., 1938-39, dean men, 1936-39; faculty physiology Columbia, 1939-56, prof., 1950-56; prof., chmn. dept. physiology and pharmacology Chgo. Med. Sch., 1956-62, chmn. dept. physiology, dir. div. biophysics, 1962—. Liason officer, biophysicist U.S. Office Naval Research, London, Eng., 1952-53; responsible investigator USPHS, 1947—; project dir. contracts USAF, 1960—; mem. Internat. Physiol. Congress, Denmark, 1950, Internat. Cardiological Congress, Paris, France, 1950, 3d, Brussels, Belgium, 1958, 1st World Congress on Ballistocardiography and Cardiovascular Dynamics, Amsterdam, 1965. Fellow Royal Soc. Medicine; mem. N.Y. Acad. Sci., Assn. Am. Med. Colls., Research Def. Soc. Gt. Britain, Am. Phys. Soc., Am. Inst. Physics, Am. Physiol. Soc., Harvey Soc., Soc. for Exptl. Biology and Medicine, Am. Heart Assn., Chgo. Heart Assn., Ballistocardiograph Research Soc. (pres. 1962-65), Sigma Xi. Research, publs. on radioactivity, spectrophotometry, ballistocardiology, body fluid status during trauma, x-ray kymography. Patentee decade photometer. Home: 101 Sterling Lane, Wilmette, Ill. 60091. Office: Inst. for Med. Research, Chgo. Med. Sch., 2020 W. Ogden Av., Chgo. 60612.*

NICKERSON, John T. R., food technologist; m. Helen V. Lefasz, Oct. 27, 1945; children—George T., Claire C. Research asso. Mass. Inst. Tech., Cambridge, 1948-50, asst. prof., 1950-54, asso. prof., 1954-66, prof. food tech., 1966—. Adviser, Swedish Inst. Food Processing Research, Gothenburg, 1958. Fellow Am. Pub. Health Assn.; mem. Am. Chem. Soc., Soc. Am. Microbiologists, Inst. Food Tech. Contbr. chpts. to Food Processing Operations; Am. Soc. Heating, Refrigerating and Air Conditioning Engrs. Data Book. Research, numerous publs. on irradiation preservation of foods, microbiology of foods, control of food-borne disease-causing bacteria by application of heat or irradiation treatment. Home: 111 Highland Rd., Somerville, Mass. Office: 77 Massachusetts Av., Cambridge, Mass.*

NICKERSON, Walter John, Am. microbiologist, biochemist, educator; b. Plainfield, N.J., Aug. 6, 1915; s. walter John and Elizabeth (Smyth) N.; B.S., West Chester State Coll., 1937; M.A., Harvard, 1940, Ph.D., 1942; m. Helen Diggles Kelsall, Sept. 6, 1941 (div.); 1 son, Kenneth Warwick; m. 2d, Thelma Cancro, Oct. 5, 1962. Faculty, Wheaton Coll., Norton, Mass., 1924-50; asso. prof. microbiology Rutgers U., New Brunswick, N.J., 1950-54, prof. microbial biochemistry, 1954—. Microbiologist, Blue Dolphin No. Labrador Expdn., 1954; cons. USPHS, 1955-59, 62—; research collaborator Brookhaven Nat. Lab., AEC, 1956-57; del. numerous internat. sci. congresses. Fellow A.A.A.S., Am. Acad. Microbiology; mem. Soc. Am. Bacteriologists (pres. N.J. br. 1953-54), Soc. Gen. Physiologists, Biochem. Soc. (Gt. Britain), Soc. Gen. Microbiologists (Gt. Britain), Am. Soc. Biol. Chemists, Am. Chem. Soc., Chemists Club (N.Y.C.). Editor: Biology of Pathogenic Fungi, 1947; Biochemistry of Morphogenesis, 1959; editorial bd. Chronica Botanica, 1943-56, Sci., 1952-54, Sci. Monthly, 1952-54. Research, publs. on biochem. bases of morphogenesis, particularly in microorganisms. Home: 213 Hazelwood Av., Middlesex, N.J. 08846. Office: Inst. Microbiology; Rutgers U., New Brunswick, N.J. 08903.*

NICKLES, Maurice, French geologist; b. Nancy, France, Aug. 3, 1907; s. René and Jeanne (Guyot) N.; M.S. in Geol. Engring., U. Nancy; m. Marie-Louise Berne, Oct. 5, 1935; children—Francoise, Marie-Claude, Béatrice, Marie-Hélène, Elisabeth, Solange, Bernadette, Louis-Charles. Geologist in mine service French W. Africa; head geologist French Equatorial Africa, cons. geologist Cameroons; dir. Geol. Center (France Overseas), Chatenay-Malabry. Head geologist, geol. map France project; attache Nat. Mus. Natural History. Publs. on geology, malacology of intertropical Africa. Home: 18, bis, rue Henri-Barbusse, Paris 5. Office: B.R.G.M. 74 rue Fédération, Paris 15, France.*

NICKON, Alex, chemist; born Choljow, Poland, Oct. 6, 1927; s. Steve and Maria Nickon; came to U.S., 1955, naturalized, 1961; B.Sc. in Chemistry, U. Alta., 1949; M.A., Harvard, 1951, Ph.D., 1953; m. Beulah Monica Godby, Aug. 22, 1950; children—Dale Beverly, Linda Cheryl, Leanne Marie. Vis. lectr. Bryn Mawr Coll., 1953; NRC Can. fellow Birkbeck Coll., U. London (Eng.), 1953-54, Ottawa, Ont., Can., 1954-55; faculty Johns Hopkins, Balt., 1955—, prof. chemistry, 1964—. Alfred P. Sloan Found. fellow, 1957-61; NSF Sr. fellow Imperial Coll., London, 1963-64. Mem. Am. Chem. Soc., Brit. Chem. Soc., Phi Lambda Upsilon. Contbr. articles to tech. jours. Editor: Jour. Organic Chemistry, 1965—. Research on stereochemistry and structures natural products, mechanism photosensitized oxygenations, chemistry carbanions and carbonium in bridged-ring systems, mechanisms terpene cyclizations, photochemistry.*

NICKSON, James Joseph, Am. radiation therapist; b. Portland, Ore., Dec. 31, 1915; s. Delbert Harry and Elvira W. (Wallin) N.; B.S., U. Wash., 1936; postgrad. Harvard Med. Sch., 1936-38; M.D., Johns Hopkins, 1940; m. Margaret Jane Hofrichter, June 12, 1939; children—Robert Frazier, James Bridges. Mem. metall. lab. U. Chgo., 1942-47; sr. fellow NRC, 1947-49, now cons.; dir. dept. radiation therapy Meml. Hosp. and mem. Sloan-Kettering Inst., N.Y.C., 1950-65; prof. Sloan-Kettering div. Cornell Med. Sch., 1952-55; prof. Cornell U. Med. Coll., 1955-65; chmn. dept. radiation therapy Michael Reese Hosp. and Med. Center, Chgo., 1965—, mem. Michael Reese Med. Research Inst., 1965—; cons. USPHS. Fellow Am. Coll. Radiology; mem. Radiation Research Soc., A.A.A.S., A.M.A., Radiol. Soc. N.A., N.Y. Acad. Scis., Sigma Xi, Phi Beta Kappa. Editor: Symposium on Radiobiology, 1952. Contbr. articles profl. jours. Research on distbn. plutonium in man; devel. and application techniques of controlled clin. trials in man with reference to evaluation of efficacy of treatment in cancer. Home: 2901 South Pkwy., Chgo. 60616. Office: 2929 Ellis Av., Chgo. 60616.*

NICKUL, Karl Voldemar, Finnish ethnologist; b. Oulu, Finland, Dec. 31, 1900; s. Johan Wilhelm and Friederike (Niemann) N.; candidate in philosophy U. Helsinki (Finland); m. Annikki Outinen, Mar. 24, 1936; children—Kirsti, Erkki. With Land Registry Office Helsinki, 1924-63. Decorated Order of White Rose. Mem. Finnish League for Peace (sec.), Federalist Union European Ethic Communities (mem. central com.), Soc. for Promotion of Lapp Culture (sec.), Mus. Tromso (corr.), Lapp Internordic Council (pres. Finnish sect.), Finnish Geography Soc., Author: The Skolt Lapp Community Suenjelsijd in the Year 1938, 1948; Report on Lapp Affairs, 1952; Research and publs. on ethnography of Lapps. Address: Kimmeltie 11 C 31, Tapiola, Finland.

NICLAUSE, Michel Charles, French phys. chemist; b. Nancy, France, Mar. 13, 1923; s. Charles and Yvonne (Martin) N.; B.Sc., phys. chemist engr., U. Nancy, 1945, D. phys. chemistry, 1953; m. Colette Masson, Aug. 6, 1957; children—Benoit, Mireille, Jean-Charles. With U. Nancy, 1945—, prof., 1957—. Mem. Comité de la Recherche Scientifique, 1967—. Recipient Grand Prix Scientifique, l'Academie de Stanislas, 1955; Officer des Palmes Académiques, 1967. Mem. Am. Chem. Soc., Société de chimie physique, Combustion Inst. Research, numerous publs. on kinetics of thermal or photochem. decomposition or oxidation in organic chemistry, acceleration or inhibition mechanism of free radicals reactions. Home: 42, rue de Beauregard, 54-Nancy, France.*

NICOL, James, Brit. geologist, mineralologist; b. nr. Innerleithen, Scotland, Aug. 12, 1810; s. James and Agnes (Walker) N.; ed. univs. Edinburgh, Bonn, Berlin; m. Alexandrina Anne Macleay Downie, 1849. Became asso. with Queen's Coll., Cork, Ireland, 1841, prof. geology, 1849; prof. natural history U. Aberdeen, 1853-78. Fellow Geol. Soc. London (asst. sec. 1847), Royal Soc. Edinburgh. Author: A Guide to the Geology of Scotland, 1844; Manual of Mineralogy, 1858; Elements of Mineralogy, 1858; The Geology and Scenery of the North of Scotland, 1866; also a geol. map of Scotland, 1858. Editor: Select Writings of Charles Maclaren, 1869. Contbr. papers to jours. Studied geology of valley of Tweed and Scottish uplands, especially stratigraphy of 2 great masses of gneisses and schists in north-west highlands, both of which he assigned to a single group of pre-Cambrian rocks. Died London, Apr. 8, 1879.

NICOL, William, Scottish physicist; b. 1768; prof. physics U. Edinburgh (Scotland); popular lectr. natural philosophy. Improved technique of preparing wood for micros. exam.; invented prism of Iceland spar (Nicol prism) for studying polarization of light, 1828; made lenses of garnet and other precious stones; studied fluid filled cavities in crystals, microscopic structure of fossil wood. Died Edinburgh, 1851.

NICOLAIER, Arthur, German physician, bacteriologist; b. Cosel, Prussian Silesia, Feb. 4, 1862; ed.

Heidelberg, Berlin, Göttingen (all Germany) univs., 1880-85; M.D., 1885. Privat dozent U. Berlin, prof., from 1900; asst. Ebstein Clinic, from 1890; prof., from 1894; head dept. medicine U. Göttingen, from 1897. Author: (with Ebstein) Über die experimentelle Erzeugung von Harnsteinen, 1891; Handbuch der Praktischen Medizin, 1900. Credited with discovery of tetanus bacillus (though not able to obtain it as pure culture), 1884; (with Ebstein) made important contbn. to understanding of kidney stones.

NICOLAS, (Léon-Marie-Joseph) Gustave, French botanist; b. Nans, France, June 28, 1879; ed. Besancon, France; Licence ès scis., Algeria; Docteur ès scis., 1909. Became prof. École supérieure, Algiers, Algeria, 1906; prof. botany Nancy (France) Faculty Scis., 1919; named prof. agrl. botany Toulouse Faculty Scis., 1920, titulary prof., 1927. Mem. French Acad. Scis., 1944 (prize in plant physiology 1923). Research on respiration of higher plants, influence of parasitic fungi on respiration of plant tissues, improvement of cultivated plants; developed hybrid wheats which brought increased yield in southwestern France. Died Apr. 24, 1944.

NICOLAS, Pierre François, French chemist; b. Barrois, France, Dec. 26, 1743; insp. mines; prof. chemistry U. Nancy (France); also prof. univs. Strasbourg (now in France), Caen (France), Mem. French Acad. Scis., 1796. Author: Historie des maladies épidémiques, 1781; Recherches et expériences sur le diabète sucré, 1803. Editor: Nouveau dictionaire universel de médecine, 1772. Research on sugar diabetes, epidemic diseases. Died Caen, Apr. 18, 1816.

NICOLAU, S. Stefan, Rumanian physician; b. 1896; prof. Bucharest (Rumania) U.; mem. Acad. Socialist Republic Rumania, numerous fgn. sci. acads. and socs. Author: Poliomyelitis, 1961; (with others) Rabies, 1962; Elements of Special Inframicrobiology, 1962. Founder Rumanian sch. inframicrobiology; research on some viral diseases, biology of viruses, epidemic hepatitis.

NICOLAU, Stefan George, Rumanian physician, dermatologist; b. Ploiesti; Rumania, June 10, 1874; s. Gheorghe and Zamfira Nicolau; M.D., U. Bucharest, 1899; m. Claire Dulermez, 1909; 1 dau., Yvonne. Chief clinic U. Bucharest (Rumania), 1907-11, prof. dermatology, 1919-51; prof. dermatology U. Cluj, 1918; dir. Inst. Dermato-Venerology, Bucharest, 1951-—. Recipient Rumanian Socialist Republic State prize. Fellow Rumanian Acad. Scis.; mem. Rumanian Dermatol. Soc. (pres. 1950), sci. socs. of dermatology France, Italy, Denmark, U.S., Czechoslovakia, Turkey, Hungary, Spain, Uurguay, Germany, U.S.S.R., Internat Soc. Tropical Dermatology. Author: (with others) Dermatology and Venereology, 1955; (with M. Blumenthal, P. Vulcan) Investigations of the Epidemiology of Leprosy in Rumania, 1955; (with A. Avram) The Cutaneous Mycetoma, 1960; Pre-cancer of the Skin, 1963; (with L. Balus) The Precancer of the Skin, 1967; also numerous articles. Research in clin. dermatology, syphilography, cancer of skin, immunoallergic disorders of skin, histopathology of skin diseases, treatment of syphilis; mycetoma, dermatomycosis; laid found. for internat. research with his description of follicular and peri-follicular eruption in scurvy; developed theory of oscillating immunity in syphilis. Home: 6, Str. Boteanu. Office: Central Dermato-Venerologie, str. dir. Grozovici nr.1, Bucharest, Rumania.

NICOLAUS, Rodolfo Alessandro, Italian chemist; b. Naples, Italy, Aug. 1, 1920; s. Oscar and Rosa (Bippert) N.; Doctor, Faculty Sci., U. Naples, 1946; Ph.D., Faculty Sci., U. Rome, 1955; m. Adriana Siniscalco, Dec. 27, 1947; children—Oscar, Clara, Barbara. With dept. organic chemistry U. Naples, 1948-53, prof., 1958-—, dir. dept., 1959-—; asst. dept. organic chemistry U. Rome, 1953-58. Fellow Acad. Scis. of Nat. Soc. Research, numerous publs. on structure and biogenesis of natural black pigments (melanins); isolation and structure of betacianines and betaxantins from Centrospaerme, structure of porphyrin a and bile pigments; synthesis of pyrrolic acids. Home: 56/A Posillipo, Naples, Italy.*

NICOLE, François, French mathematician; b. Paris, Dec. 23, 1683; mem. French Acad. Scis., 1707 (became dir. 1730, 34, 41). Author: Essai sur la théorie des roulettes, 1706; Traité du calcul des différences finies, 1717-27; Traites des Lignes de troisieme ardre, 1729. Studied trisection of angle, rectification of cissoid curve, infinitesimal calculus and calculus of probabilities, theory of spheric epicycloids; demonstrated Newton's theorems; 1st to show 3d degree curves can be considered 3 different projections. Died Paris, 1758.

NICOLESCU, Miron, Rumanian mathematician; b. 1903; prof. Bucharest (Rumania) U.; mem. Acad. Socialist Republic Rumania (chmn. 1966-—). Author: Polyharmonic Functions, 1936; Mathematical Analysis, 3 vols., 1958-60; (with others) Text-book of Mathematical Analysis, 2 vols., 1963-64. Research on structure of some classes of functions; creator theory of polyharmonic functions; laid founds. of theory of polycaloric functions.

NICOLET, Jean, explorer; b. Cherbourg, France, 1598; s. Thomas and Marguerite (de la Mer) N.; m. Marguerite Couillard, Oct. 7, 1637. Came to New France (now Can.) with Samuel de Champlain, 1618; lived among Indians on Allumette Island, upper Ottowa River; ofcl. interpreter among Nipissing Indians, 1624; returned to Can., 1633, became ofcl. interpreter for colony at Three Rivers; 1st known white man to discover Lake Michigan and Wis., 1634, hoped to find Chinese, instead found Winnebago Indian tribe; he was called Manitou-iriniou (wonderful man) by Indians. Drowned in storm on St. Lawrence River, Nov. 1, 1642.

NICOLET, Marcel, Belgian astrophysicist; b. Basse-Bodeux, Belgium, Feb. 26, 1912; grad. with distinction, U. Liege, 1934, doctorate summa cum laude, 1937; postgrad. Lichtklimatisches Observatorium, Switzerland, 1937, 38-39. Asst. Belgian Institut Royal meteorologique, 1935-46, meteorologist, dir. radiation dept., 1946-—; sci. faculty U. Brussels, 1946-—; dir. Institut d'Aeronomie spatiale de Belgique; sec. gen. Spl. Com. for Internat. Geophys. Year, 1953-—; sometime research prof. Pa. State U., 1951-56, vis. prof., 1957; lectr. U.S. and Europe; internat. mem. mixed commn. on ionosphere Internat. Council Sci. Unions; pres. Astronomy and Geophysics Commn. of Belgium. Recipient honors Agathon de Potter Found., Royal Acad. Scis., Guggenheim award Internat. Acad. Astronautics. Member of Belgian Com. Geodesy and Geophysics, Belgian Com. Radiosciences, Belgian Com. Astronomy; fgn. mem. Internat. Union Geodesy and Geophysics, Internat. Assn. Meteorology (past sec. radiation commn.), Internat. Astronomical Union, Internat. Radioscientific Union, Internat. Assn. Geomagnetism and Aeronomy (pres.). Asso. editor: Jour. Geophys. Research. Editorial adv. bd. Jour. Atmospheric and Terrestrial Physics. Contbr. numerous articles to sci., tech. jours. Authority on phys. state of outer atmosphere, auroral spectrum, relationship of solar and terrestrial phenomena with atomic processes. Address: 30, Avenue Den Doorn, Ukkel, Belgium.*

NICOLLE, Charles Jules Henri, French physician, bacteriologist; b. Rouen, France, Sept. 21, 1866; ed., Rouen, France; med. degree, 1893; studied under Pasteur. Became dir. Pasteur Inst., Tunis, Tunisia, 1903; named prof. Collège de France, 1932. Recipient Nobel prize for medicine and physiology, 1928. Mem. French Acad. Scis. Author: General Elements of Microbiology; Techniques of Microbiology; (with Remlinger) Toxins and Antitoxins; Naissance, vie et mort des maladies infectieuses; Destin des maladies infectieuses; also numerous articles. Isolated virus causing infantile kala azar; introduced (with Felix Mesnil) trypan for treatment of typanosomiasis, 1906; discovered typhus is transmitted by body louse, 1909, also Nicollia, a genus of protozoan blood parasites, named after him; proved (with L. Blaizot, E. Conseil) relapsing fever could be transmitted by body louse, 1910-11; produced typhus in monkey and guinea pigs by injecting blood from affected subjects, 1910-11; reported (with L. Blaisot, A. Cuenod) filtration of trichoma virus, 1912. Died Tunis, Feb. 28, 1936.

NICOLLET, Joseph Nicolas, explorer, mathematician; b. Cluses, Savoy, France, July 24, 1786; ed. Coll. of Cluses. Apptd. astron. asst. French Bur. Longitude, 1822; prof. math. Coll. Louis-le-Grand; came to U.S., 1832; made 1st expdn. (ascent up Mississippi), 1843. Discovered comet in constellation expdn. for survey of upper Missouri River, 1838, head 2d expdn., 1839. Author: Report Intended to Illustrate A Map of the Hydrographical Basin of the Upper Mississippi (a map of region N.W. of the Mississippi), 1843. Discovered comet in constellation Pegasus, 1821. Died Washington Sept. 11, 1843.

NICOLOFF, Demetre Matthew, Am. surgeon; b. Lorain, O., Aug. 31, 1933; s. Matthew Eftin and Blaga (Korora) N.; B.A., Ohio State U., 1954, M.D., 1957; Ph.D. in Surgery, U. Minn., 1965; m. Ardelle F. Risch, Sept. 16, 1962; children— Stephanie Kaye, Alexander Demetre. Intern, U. Minn. Hosp., 1957-58, surg. resident, 1958-65; instr. surgery U. Minn. Hosp., VA Hosp., Mpls. 1965-—. Mem. A.M.A., Phi Beta Kappa, Alpha Omega Alpha. Clin. and exptl. research on factors involved in prodn. duodenal ulcers; exptl. investigation of factors involved in origin of pulmonary edema. Home: 5628 14th Av. S., Mpls. 55417. Office: VA Hosp. Mpls. 55417.*

NICOMACHOS OF GERASA; mathematician; b. Gerasa, Arabia Petraea; flourished circa 100. Probably traveled to Alexandria to study doctrines of Neopythagoreans. Author: Introductio arithmetica (earliest work which considers arithmetic an autonomous discipline; gave Pythagorean theory of numbers); Enchiridion harmonices. Studied mystical properties of numbers, perfect numbers, music; mentions sieve of Eratosthenes, Pythagorean doctrines; his work contains one of earliest examples of Greek multiplication tables; probably introduced proposition that cubical numbers are always equal to sum of successive odd numbers.

NICOMEDES, mathematician; flourished 2d century B.C.; probably lived at Pergamum; studied Hippias' quadratix and used it to square the circle; criticized Eratosthenes' doubling of cube; discovered cochloidal or conchoidal curves, which he used to solve problems of doubling cube and trisecting angle.

NICOT, Jean, French biologist, philologist; b. Nîmes, France, 1530; ed. Nîmes; ambassador to Portugal, 1559-61. Author Thrésor de la langue française, pub., 1606. Nicotine, plant genus Nicotina named in his honor. Introduced tobacco culture in France; intended the plant for therapeutic purposes rather than to be smoked. Died Paris, May 5, 1600.

NIEDERLAND, Teofil Rudolf, clin. chemist; b. Maria Radna, Rumania, Jan. 14, 1915; s. Rudolf and Eugenia (Popovici) N.; M.D., Charles U., Prague, 1939; D.Sc., Komensky U., Bratislava, 1960; m. Zdenka Klusacek, Sept. 24, 1942. With II. med. clinic Charles U., Prague, 1939-40, I. med. clinic, 1945-47; Rockefeller fellow Washington U. Med. Sch., St. Louis, 1947-48; with II. med. clinic Komensky U., Bratislava, 1940-45, head dept. biochemistry, 1950-57, prof., 1952-—, head III. med. clinic, dir. research lab. pharma-biochemistry, 1957-—. Mem. Czechoslovakian Med. Soc. J.E. Purkyne (pres. Slovak council); corr. mem. Slovak Acad. Scis. Author: (with E. Brixova) The Biochemical Symdromology in Clinical Hepatology, 1962; The Effect of Salicylates on the Organism, 1963; also numerous articles in field. Research on metabolic effect of various drugs and chem. agts. on biochem. processes especially in liver, causes of fatty liver. Home: 5. Nam. 1.maja. Office: 11. Hlboka, Bratislava, Czechoslovakia.*

NIEDERLE, Lubor, Czechoslovakian archeologist; b. Klatovy, Czechoslovakia, Sept. 20, 1865; prof., Prague, Czechoslovakia. Author: Slovanske starozitnosti, 4 vols., 1901-25 (partially trans. into French as Manuel de l'antiquité Slave, 2 vols., 1923-26); Zivot starych Slovanu, from 1911; Rukovet' slovanske, archeologie, 1913. Authority on origin, early civilization of Slavs. Died Prague, June 14, 1944.

NIEDNER, Franz Friedrich, German scientist; b. Frankfort, Germany, Sept. 17, 1905; s. Franz Friedrich and Gisela (Friedrich) N.; M.D.; m. Gisela Schöss, May 27, 1937; children—Helga, Niels, Franz, Ingrid, Karin, Astrid. State exams. at Kiel; specializing in surgery at Iena; instr. surgery U. Vienna; prof. surgery U. Tübingen (Germany); doctor in chief Municipal Clinic. Mem. German Surgery Soc., Internat. Soc. Cardiovascular Surgery, German Blood Circulation Soc., Internat. Soc. Gastro-enterology; corr. mem. Swiss Cardiology Soc. Author: Die Chirurgie des Herzens und der gross Gefässe; also articles on heart surgery, biliary ducts and papilla vateri. Research on cardiovascular surgery, gastroenterology. Home: Memelstrasse 48, 791-Neu-Ulm. Office: Steinhövelstrasse 9, 79-Ulm-Donau, West Germany.

NIELSEN, Alvin Herborg, Am. physicist, educator; b. Menominee, Mich., May 30, 1910; s. Knud Herborg and Maren (Nielsen) N.; B.A., U. Mich., 1931, M.S., 1932, Ph.D., 1934; m. Jane Ann Evans, Dec. 29, 1942; 1 dau., Margaret Arthur. Research fellow Ohio State U., 1934-35, research asso. Research Found., OSRD, 1944-46; faculty U. Tenn., Knoxvile, 1935-—, prof., 1946-—, head dept. physics, 1956-—, dean Coll. Liberal Arts, 1963-—. Cons. Oak Ridge Gaseous Diffusion Plant, Oak Ridge Nat. Lab., 1947-—; speaker Internat. Molecular Spectroscopy Symposia, 1954, 63; vis. lectr. NSF Insts., Birmingham-So. Coll., 1959-61; mem. com. on basic research adv. U. S. Army Research Office, 1964-—. Research grantee Am. Acad. Arts and Scis., 1942, Research Corp. Am., 1946-47, NSF, 1962; Fulbright research scholar Astrophys. Inst., U. Liege, 1951-52. Fellow Am. Phys. Soc. (pres. Southeastern sect. 1953-54, exec. com. 1962-65), A.A.A.S., Optical Soc. Am.; mem. Am. Assn. Physics Tchrs., Am. Assn. U. Profs., Tenn. Acad. Sci. (exec. com. 1959-61), Knoxville Sci. Club, Coblentz Soc., Sigma Xi, Phi Kappa Phi, Sigma Pi Sigma, Pi Delta Phi. Asso. editor Jour. Chem. Physics, 1955-57. Research, publs. on molecular spectroscopy to profl. jours. Home: 3605 Timberlake Rd., Knoxville, Tenn. 37920.*

NIELSEN, Carl Eby, Am. physicist; b. Los Angeles, Jan. 22, 1915; s. Charles H. and Josie (Musselman) N.; A.B., U. Cal. at Berkeley, 1934, M.A., 1940, Ph.D., 1941; m. Imogene Herron, June 26, 1938; children—Paul Herron, Sylvia Jean (Mrs. Cornelius Johnson), Robert Kent. Instr. physics U. Cal. at Berkeley, 1941-45, lectr., 1945-46; asst. prof. U. Denver, 1946-47; faculty Ohio State U., Columbus, 1946-—, prof. physics, 1964-—. Cons. thermonuclear div. Oak Ridge Nat. Lab., 1961-—. Mem. Am. Assn. Physics Tchrs., Am. Phys. Soc., Phi Beta Kappa, Sigma Xi. Contbr. articles to profl. jours. Developed simplified diffusion cloud chamber, 1st one-component cloud chamber; discovered negative mass instability as a mode of collective behavior in ion beams and plasmas. Home: 8030 Sawmill Rd., Dublin, O. 43017.*

NIELSEN, Donald Rodney, Am. soil physicist; b. Phoenix, Oct. 10, 1931; s. Irven R. and Irma (Chase) N.; B.S., U. Ariz., 1953, M.S., 1954; Ph.D., Ia. State U., 1958; m. Joanne Joyce Locke, Sept. 26, 1953; children—Cynthia Lynn, Pamela Kay, Barbara Anne, Wayne Edward. Research asso. Ia. State U., Ames, 1956-58; asst. prof. water sci. U.

Cal. at Davis, 1958-63, aso. prof., 1963—. Collaborator, U.S. Dept. Agr. Agrl. Research Service. Sr. postdoctoral fellow NSF, 1965-66. Mem. Am. Geophys. Union, Am. Soc. Agronomy, Western Soil Sci. Soc. Research, numerous publs. on explanation of soil water use, saturated and unsaturated flow of water through soil in relation to plant growth, methodology of evaluating soil phys. conditions for plant growth, movement of solutes in soil. Home: 1004 Pine Lane, Davis, Cal. 95616.*

NIELSEN, E(iner) Steemann, Danish plant physiologist, marine biologist; b. Copenhagen, Denmark, June 13, 1907; s. Peder and Karoline (Steemann) N.; M.Sc., U. Copenhagen, 1931, Ph.D., 1934; m. Elise Andersen, Aug. 28, 1935; children—Birgitte Steemann (Mrs. Knud Nielsen), Per, Mette. Mem. Danish Dana Expdn., 1928-30; asst. Danish Fishery Inst., 1932-44; prof. botany Royal Danish Sch. Pharmacy, 1944—; lectr. biol. oceanography U. Copenhagen, 1962—. Recipient Ridder Dannebrogsordenen, 1953. Mem. Royal Danish, Norwegian acads. sci., Scandinavian Soc. for Plant Physiology (past sec.), Internat. Council for Exploration Sea (chmn. plankton com. 1955-60), Sci. Com. Oceanographic Research. Research, numerous publs. on geog. distbn. of plankton algae, photosynthesis of plankton algae and other aquatic plants, organic prodn. in sea; first described carbon-14 technique for measuring primary prodn. due to plankton algae. Home: 6 Engbakken, Virum, Denmark. Office: Royal Danish Sch. Pharmacy, Bot. Lab., 2 Universitetsparken, Copenhagen O, Denmark.*

NIELSEN, Erik Tetens, Danish biologist; b. Frederiksberg, Denmark, Feb. 12, 1903; s. Valdemar P.T. and Kirstine (Olsen) N.; cand.mag., U. Copenhagen, 1929, Dr.phil., 1936; m. Astrid Sehested, Sept. 6, 1929; children—Hedvig, Kirsten. Chief, Privat Lab., Pilehuset, Denmark, 1927-48, Archbold Sta., Fla., 1948, Fla. Bd. Health, 1949-55, 56-61, Dept. Agr., Iraq, 1955-56; amanuensis Molslaboratory, Femmöller, Denmark, 1961—; staff Zool. Inst., U. Arhus, Denmark, 1964—. Mem. Entomologisk Forening, Copenhagen, Dansk Naturhistorisk Forening, Royal Entomol. Soc. London, Arhus Entomologklub. Author: Livets vrimmel, 1948; Moerus de Bembex, 1948; Lige ud ad landevejen til Sahara, 1949; Insektvandringer, 1964. Research, publs. on ethology, ecology and physiology of insects, behavior of solitary bees and wasps, ecol. methods, influence of phys. factors on devel. and behavior of insects, motivation of migration of butterflies, swarming of mosquitoes and stridulation of grasshoppers and crickets. Office: Molslaboratoriet, Femmöller, Denmark.*

NIELSEN, Etlar Lester, Am. geneticist; b. Tyler, Minn., June 30, 1905; s. Niels Christian and Karen (Pedersen) N.; B.S., U. Minn., 1928, Ph.D., 1936; m. Sarah Creecie Dyal, Dec. 22, 1940; children—Niels Christian, Karen Sarah, Anna Christine (Mrs. Jim Pharo). Faculty, U. Ark., 1936-41; faculty U. Wis., Madison, 1953—, prof agronomy, 1962—; jr. botanist Soil Conservation Service, U.S. Dept. Agr., Ames, Ia., 1935, agronomist Agrl. Research Service, Madison, Wis., 1946-60, geneticist, Wis. Agrl. Expt. Sta., 1960—; agt. Bur. Plant Industry, Madison, 1941-42, asst. agronomist, 1942-43, asso. agronomist, 1943-46. Cons. forage crops Ministries Agr., Norway, Denmark, 1953. Rockefeller grantee, 1960. Fellow Am. Soc. Agronomy (vis. scientist panel 1965-67); mem. Bot. Soc. Am., Am. Soc. Agronomy, Crop Sci. Soc., Sigma Xi. Correlated progeny behavior of apomictic and sexual forms of Ky. bluegrass with devel. female gametophyte; correlated gross morphology and polyploidy in switchgrass; cytogenetics and progeny analysis of smooth bromegrass and timothy; cytogenetics of wide crosses of grasses; devel., with co-workers, improved agrl. varieties of smooth bromegrass and timothy; taxonomy of Amelanchier. Home: 2732 Regent St., Madison, Wis. 53705.*

NIELSEN, Harald Herborg, Am. physicist; b. Menominee, Mich., Jan. 25, 1903; s. Knud and Maren (Nielsen) N.; student St. Olaf Coll., Northfield, Minn., 1923-25; B.S., U. of Mich., 1926, A.M., 1927, Ph.D., 1929; student U. of Copenhagen (Universitetets Institut for Teoretisk Fysik), 1929-30; Dr. Honoris Cause, U. Dijon, France, 1965; m. Martha Ann Evans, Sept. 18, 1943; 1 son, Peter Herborg. Am. Scandinavian fellow, 1929-30; mem. faculty Ohio State U., since 1930, prof. physics since 1943, chairman department of physics, 1946-67; tech. rep. N.D.R.C., 1942-44; Guggenheim Memorial fellow, 1949-50; Fulbright lectr. U. Paris, 1958-59; Dept. of State Sci. attaché, Stockholm, Sweden, 1952-53; pres. commn. on symbols, units and nomenclature Internat. Union of Pure and Applied Physics, 1952-60, v.p., 1960—, sec. triple commn. spectroscopy, 1960-63; president, 1963-64, vice pres. U.S.A. nat. com., 1958-61, secretary U.S.A. committee, 1961—; chmn. NRC com. symbols, units and nomenclature, 1956-61, mem., 1962—, mem. div. chem. and chem. tech. com. phys. chem.; cons. to sci. adviser to Sec. State, 1958—. Award of Cross of Leopold, Belgium; medal of U. Liege (Belgium), 1949; Cross of Order of Knight of Dannebrog (Denmark), 1957. Fellow Am. Phys. Soc., A.A.A.S.; mem. American Association Physics Teachers American Scandinavian Found. (associate), Optical Soc. Am., Am. Nat. Rebild Com., Societé Royale des Sciences de Liege, Belgium, Societe Francaise de Physique. Royal Danish Academy of Sciences and Letters, Phi Beta Kappa, Sigma Xi. Editor: Jour. Molecular Spectrocopy. Asso. editor Phys. Rev., 1945-48, Jour. of Chem. Physics, 1952-55. Research in molecular spectroscopy; infrared absorption spectra of polyatomic molecules; vibration-rotation energies of polyatomic molecules; hindered rotation in molecules. Home: 199 Rustic Pl., Columbus 14, O.

NIELSEN, J(ens) Rud, physicist; born Copenhagen, Denmark, Sept. 22, 1894; s. Niels F. and Marie (Johansen) N.; M.S., U. Copenhagen, 1919; Ph.D. (Am.-Scandinavian Found. fellow), Cal. Inst. Tech., 1924; m. Gertrude Siegmund, Oct. 19, 1923; children—John Rud, Thomas Rud (dec.), Mary Ruth (Mrs. Lejeune Wilson). Came to U.S., 1922, naturalized, 1930. Instr., Royal Tech. Coll., Copenhagen, 1919-22; prof. physics Rumboldt State Coll., Arcata, Cal., 1923-24; faculty physics U. Okla., Norman, 1924—, prof., 1930-44, research prof., 1944—, dir. U. Okla. Research Inst., 1941—. Cons. Phillips Petroleum Co., 1945-63. Recipient Distinguished Service citation U. Okla., 1962. Guggenheim Meml. fellow, 1931-32; Rask-Oersted fellow, 1932-33. Fellow Am. Phys. Soc., Optical Soc. Am., A.A.A.S.; mem. Am. Assn. Physics Tchrs., Am. Assn. U. Profs. Asso. editor Am. Jour. Physics, 1940-43, Jour. Chem. Physics, 1950-53, Jour. Optical Soc. Am., 1950-61. Contbr. numerous articles on spectroscopy and molecular structure to U. S., fgn. sci. jours. Home: 310 S. University Blvd., Norman, Okla. 73069.*

NIELSEN, Johannes, Danish psychiatrist; b. Sjorring, Denmark, June 19, 1924; s. Niels Pedersen and Marthine (Thomsen) N.; M.D., U. Arhus, Denmark, 1958; m. Paulette Louise Jaubert, May 21, 1960. Registrar, Demographic-Genetic Research unit, Psychiat. Inst., U. Arhus, 1961-63, Arhus State Hosp., 1963—, chief cytogenetic lab., 1963—; chief cytogenetic lab. U. Arhus, 1963—. Co-ordinator cross nat. study delirium tremens in Nordic Capitals, Nordic Com. for Alcohol Research, 1962-66. Mem. Danish Psychiat. Assn. Contbg. author: Cross-National Investigation of Delirium Tremens in Nordic Countries, 1965. Publs. on discovery of increased serum-concentration of magnesium in patients with delirium tremens and manic-depressive patients treated with lithiumcarbonate; studies on delirum tremens in Copenhagen, prevalence of mental disorders in old age, mental disorders in married couples; community psychiat. and epidemiological studies in island of Samso, Denmark; prevalence and incidence and study of Klinefelter's syndrome; found correlation between twin frequency and frequency of Klinefelter's syndrome, between diabetes mellitus and Klinefelter's syndrome; discovered higher fall out of chromosomes in patients with senile dementia compared with control persons of same age. Home: 13, II Tjelevje, Arhus, Denmark.*

NIELSEN, Karl Ove, Danish physicist; b. Hundborg, Denmark, July 27, 1920; s. Hans Peter and Edele (Kirstine) N.; cand.polyt., Tech. U. Denmark, Copenhagen, 1947; Dr.phil., U. Copenhagen (Denmark), 1958; m. Gunvor Nikoline Rudjord, July 7, 1949; children—Hans Peter, Tone Rudjord, Espen. Research asst. Niels Bohr Inst., Copenhagen, 1949-57; head Danish Reactor 2, Danish AEC, Research Establishment Risö, 1957-61; prof. physics U. Aarhus (Denmark), 1961—, dean Faculty Math. and Natural Scis., 1965-66, chmn. bd. dirs. Inst. Physics, 1962-64, 67—. Mem. Utilized Nuclear Physics Soc., Danish Acad. Tech. Scis., Danish Soc. Natural Scis. Author: Electromagnetic Isotope Separation and Some Uses of Separated Isotopes in Nuclear Investigations, 1958; also articles. Devel. electromagnetic isotope separators and accelerators for heavy ions, including ion sources for radioactive materials; research on penetration of heavy particles in matter and use of separated isotopes in nuclear and atomic investigations; problems with isotope separators coupled to accelerators for investigation of very short-lived isotopes. Home: 47 Solbakken, Risskov, Denmark. Office: Inst. Physics, U. Aarhus, Aarhus C, Denmark.*

NIELSEN, Svend Woge, vet. pathologist; b. Herning, Denmark, Apr. 4, 1926; s. Peter Woge and Marie (Nielsen) N.; D.V.M., Royal Vet. Coll., Copenhagen, 1951; M.S., Ohio State U., 1957, Ph.D., 1959; m. Lorraine C. Beaman, July 18, 1952; children—Karen Beth, Peter Kevin, Ingrid Woge. Came to U.S., 1951, naturalized, 1958. Faculty, Ont. Vet. Coll., Guelph, 1952-55; faculty Ohio State U., 1955-60; prof. U. Conn., Storrs, 1960—; vis. lectr. Wash. State U., 1964, Royal Vet. Coll., 1965. Sect. chmn. 3d Quadrennial Conf. on Cancer U. Perugia, Italy, 1965. Nat. Cancer Inst. USPHS Splty. fellow U. Cambridge, Eng., 1967-68. Recipient Krabbe award Royal Vet. Coll., 1951. Diplomate Am. Coll. Vet. Pathologists. Mem. Am. Vet. Med. Assn., A.A.A.S., Conf. Research Workers Animal Disease, Internat. Acad. Pathology, Phi Zeta, Alpha Psi, Omega Tau Sigma. Research, numerous publs. on comparative oncology and nutritional pathology. Home: Davis Rd., Storrs, Conn. 06268.*

NIEMANN, Albert, German physician; b. Germany, 1880; isolated cocaine from Peruvian coca leaves, circa 1860; described a form of xanthomatosis (Niemann-Pick disease), 1914. Died 1921.

NIEPCE, Joseph Nicéphore, French physicist, photographer; b. Chalon-sur-Saone, France, Mar. 7, 1765; studied under Peres de l'Oratoire; m. Agnes Romero, Aug. 4, 1794; 1 son, Isidor. Officer in 42d regt., inf., 1792-94; mechanic; landed propr., Gras, France; became partner with L. Daguerre to perfect photog. inventions. Recipient prize Société d'Encouragement pour l'industrie nationale, 1817. Author: Notice sur l'héliographie, 1829. Used glass plates coated with bitumen to make permanent heliotypes, 1824; made 1st photoengravings using heliographic process, also 1st permanent photograph, 1826; added iris diaphragm to camera; built (with his brother Claude) Pyreolophore motor, developed method for extraction of sugar from pumpkins and beet roots; devel. (with Daguerre) heliographic processes, 1829-33; experimented with ferrochloride, manganese oxide, phosphorous for fixing agt. in photography, paper, glass, metal and stone for copying pictures. Died Saint-Loup-de-Varennes, July 3, 1833.

NIEPER, Hans Alfred, German internist; b. Hanover, Germany, May 23, 1928; s. Ferdinand and Margarete (Krauss) N.; M.S., U. Hamburg, 1953. Mem. med. dept. Hosp. Friederikenstift and Hamelin City, Germany, also mem. cancer research team Freiburg U. Clinic, 1955-56; staff Paul Ehrlich State Inst. Exptl. Therapy, Frankfort, Germany, 1958-61, chem. labs. Aschaffenburg City Hosp., 1961-63; med. staff Silbersee Clinic, Hanover, 1964—; also with pvt. labs. Mem. German Soc. Circulation Research, Soc. Agressologie, Paris, France. Author: (with Blumberger) Electrolyte Transport Therapy of Cardiovascular Diseases, 1965; (with Royston) Heart Infarction Syndromes, 1967. Research, publs. on med. cancer therapy. Home: 21 Sedanstrasse, Hanover, Germany. Office: Silbersee Clinic, Hanover, FRG, Germany.*

NIER, Alfred O(tto) C(arl), Am. physicist; b. S. Paul, May 28, 1911; s. August Carl and Anna J. (Stoll) N.; B.S., U. of Minn., 1931, M.S., 1933, Ph.D., 1936; m. Ruth E. Andersen, June 19, 1937; children—Janet, Keith. Nat. Research fellow, Harvard, 1936-38; asst. prof. physics, U. of Minn., 1938-40, asso. prof., 1940-43, prof. since 1946; physicist The Kellex Corp., N.Y. City, 1943-45. Mem. A.A.A.S., Am. Physical Soc., Minn., Nat. acads. scis., Geochem. Soc., Am. Geophys. Union, Am. Philos. Soc., American Assn. Physics Teachers, Sigma Xi. Research activities include development of the mass spectrometer and its application to problems in physics, chem., geol. and med.; first to separate rare isotope of uranium, U-235, 1940; with J. R. Dunning, E. T. Booth and A. V. Grosse of Columbia, demonstrated it was source of atomic energy when uranium is bombarded with slow neutrons; other work in electronic circuit design and thermal diffusion of gases. Home: 2279 Hoyt Av. W., St. Paul 8.*

NIERENBERG, William Aaron, Am. physicist; b. N.Y.C., Feb. 13, 1919; s. Joseph and Minnie (Drucker) N.; B.S., Coll. City N.Y., 1939; postgrad. U. Paris (France); M.A., Columbia, 1942, Ph.D., 1947; m. Edith Meyerson, Nov. 21, 1941; children—Victoria Jean, Nicholas Clarke Eugene. Physicist, Manhattan Project, N.Y.C., 1941-45; instr., Columbia, 1948-50, project dir. Hudson Labs., Dobbs Ferry, N.Y., 1953-54; asst. prof. U. Mich., 1948-50; prof. U. Cal. at Berkeley, 1953-54, prof., 1954-65; dir. Scripps Instn. Oceanography, La. Jolla, Cal., 1965—; prof. associe U. Paris, 1960-62. Asst. sec. gen. for sci. NATO, Paris, France, 1960; cons. pvt. cos. and govt. agys. Fellow Am. Phys. Soc.; mem. Am. Geophys. Union, Am. Acad. Arts and Scis. Author: (with L.N. Ridenour) Modern Physics for the Engineer, 1961; also numerous articles. Research in nuclear physics, gaseous diffusion, cascade theory, propagation sound in ocean and related geophysics. Home: 9581 La Jolla Farms Rd., La Jolla 92037. Office: Scripps Instn. Oceanography, La Jolla, Cal. 92037.*

NIERING, William Albert, Am. ecologist; b. Scotrun, Pa., Aug. 28, 1924; s. George William and Emma (Everitt) N.; B.S., Pa. State U., 1948, M.S., 1950; Ph.D., Rutgers U., 1952; m. Catherine Mildred Sullivan, Dec. 10, 1928; children—William Stephen, Hugh Leslie. Faculty, Conn. Coll., New London, 1952—, prof. botany, 1964—, dir. Conn. Arboretum, 1966—, mem. Kapingamarangi Atoll, Caroline Islands Expdn., 1954; asso program dir. environmental biology NSF, 1967-68. Fellow A.A.A.S; mem. Ecol. Soc. Am., Bot. Soc. Am., Sigma Xi. Author: Nature in the Metropolis, 1960; The Life of the Marsh - The North American Wetlands, 1966. Vegetation studies, publs. on No. N.J., Kapingamarangi Atoll, No. Minn. spruce-fir forests, Conn. Arboretum, palynology of No. N.J., S.E. Conn., herbicide studies on rights of way, vegetation of S.W., especially saguaro. Home: Woodridge Circle, Gales Ferry, Conn. 06335. Office: Dept. Botany, Conn. Coll., New London, Conn. 06320.*

NIESET, Robert Thomas, Am. biophysicist, photog. film co. exec.; b. Gibsonburg, O., Aug. 6, 1912; s. Henry and Anna (Reineck) N.; A.B., Cath. U., 1934, M.A. in Philosophy, 1935; Ph.D. in Biophysics (Rackham fellow), U. Mich., 1943; m. Mary Elizabeth Young, June 21, 1939; children—James R., Anne, Jane, Marjorie, Richard. Research asso. engring. U. Mich., 1941-43, sr. scientist engring. research, 1944-

46, sr. fellow Am. Cancer Soc., 1946-47; dir. research Erwood Corp., Chgo., 1943-44; asso. prof. physics Tulane U., New Orleans, 1947-51, prof., 1951-62, dir. biophysics program, 1947-62, chmn. physics dept., 1960-62; v.p. research and devel., dir. Kalvar Corp., New Orleans, 1956-—; dir. Delta Capital Corp., New Orleans. Cons. Alton Ochsner Med. Found. Hosp., New Orleans, 1945-62, Woodrow Wilson Fellowship Found., 1960-62, also various indsl. firms, 1949-60. Recipient Naval Ordinance Devel. award, Manhattan Dist. citation, OSRD citation (all 1945). Fellow Am. Phys. Soc.; mem. I.E.E.E., A.A.A.S., Sigma Xi. Contbr. articles in biophysics, mass spectrometry, photography to profl. jours.; developed freeze-dry technique for preparing tissues for microscopic study, 1942; patentee photog. equipment, materials, information storage technique. Home: 1776 State St., New Orleans 70118. Office: 909 S. Broad St., New Orleans 70125.*

NIETHAMMER, Günther Theodor, German ornithologist; b. Waldheim/Saxe, Germany, Sept. 28, 1908; s. Konrad and Katharina Boehmer; ed. univs. Tübingen and Leipzig; Ph.D.; m. Ruth Filtzer, Sept. 1, 1933; children—Jochen, Gerd, Michael, Rolf. Ornithologist, Zool. Soc. of Museum Berlin, Bonn, Vienna; div. head, instr. U. Bonn; at Koenig (Germany) Mus., 1950-—. Mem. Soc. German Ornithologists (v.p.). Author: Handbuch der deutschen Vogelkunde, 3d edit., 1942. Address: Koblenzerstrasse 162, Bonn, West Germany.

NIETZKI, Rudolf, chemist; b. Heilsberg, E. Prussia, Mar. 9, 1847; indsl. chemist, prof., Basel, Switzerland; research on chinone and dyes; gave structural explanation in chemistry of dyes. Died Nekkargemünd, Sept. 28, 1917.

NIEUWENTYT, Bernhard, mathematician, biologist; b. West Graafdyk, Holland, 1654; burgomaster Purmerend, Holland. Author: Analysis infinitorum, 1695; Réfutation de Spinoza. Follower of Descartes. Died 1781.

NIEUWLAND, Julius Arthur, chemist, botanist; b. Hansbeke, Belgium, Feb. 14, 1878; grad. Notre Dame U., 1899; doctorate philosophy, Catholic U., 1904. Joined Congregation of Holy Cross ordained priest, 1903; Prof. Botany U. Notre Dame, 1904-18, dean Coll. Sci., 1918-22, prof. organic chemistry, 1918-1936, curator bot. herbaria. Recipient medal and awards for research in synthetic rubber, 1906-31. Author: Some Reactions on Acetylene (led to discovery of the poison gas Lewisite). Founder Am. Midland Naturalist, 1918, also editor. Developed Duprene, a synthetic rubber; synthesized organic compounds; research on taxonomy and anatomy of flowering plants and ferns. Died Washington, D.C., June 11, 1936.

NIEWIADOMSKI, Henryk, Polish chemist; b. Brzezany, Poland, June 22, 1905; s. Marian and Maria (Moraczewska) N.; M.Sc., Lwow Tech. U. (Poland), 1927, Dr.Ing., 1936; m. Eufemia Hempel, Feb. 21, 1935. Chemist, Potassium Mines, Kalusz, 1927-30; head lab. Potatoes Starch Factory, Torun, Poland, 1931; tech. dir. Rice Starch Factory, Tczew, Poland, 1932-39, Edible Fat Factory, Gdansk, Poland, 1946-49; prof. Tech. U., Gdansk, 1949-—, head dept. fat chemistry and tech., 1950-—. Pres. sci. council Indsl. Fat Inst., Warszawa, Poland, 1960-—. Decorated Polonia Restituta Cross (twice). Mem. Internat. Soc. for Fat Research (past pres.), Internat. Soc. for Research on Nutrition and Vital Substances Sci. Council, Polish Acad. Sci. (mem. com. food chemistry and tech. 1957—). Research, numerous publs. in chemistry and tech. of minor substances in vegetable fats; new methods of instrumental analysis. Home: 2 Morska, Sopot, Poland. Office: 11 Majakowskiego, Gdansk 6, Poland.*

NIGAM, Satgur Saran, Indian chemist; b. Allahabad, India, Feb. 19, 1921; s. Nanak Saran and Satyavati Nigam; B.Sc., U. Allahabad, 1940, M.Sc., 1942; Ph.D., U. Saugar, 1953; Ph.D., U. London, 1955; D.I.C., Imperial Coll. London, 1955; m. Lakshmi Nigam, May 9, 1944; children—Sahab Saran, Soami Saran. Lectr., Meerut (India) Coll., 1946-47; sr. lectr. U. Saugar, Sagar, India, 1947-57, reader organic chemistry, 1957-—; research asso. Imperial Coll. Sci. and Tech., London, 1953-55. Mem. Nat. Natural and Synthetic Perfumery Materials Com., Indian Standards Instn., 1963-—; Indian del. Congress Internat. Orgn. Standardization, 1964. Fellow Royal Inst. Chemistry (London), Brit. Chem. Soc., Indian Nat. Acad. Scis. Contbr. numerous articles to profl. lit. Achieved first total synthesis of linolenic acid by acetylenic routes (with B.C.L. Weedon); chem. analysis of essential Indian oils. Home: CT/12-2 Gournager, Sagar, Madhya Pradesh. Office: Chemistry Dept., U. Saugar, Sagar, India.*

NIGGLI, Paul, Swiss mineralogist; b. Zofingen, Switzerland, June 26, 1888; prof. mineralogy and petrography U. Zurich (Switzerland); mem. French Acad Scis., 1946. Author: Lehrbuch der Mineralogie und Kristallchemie, 1941-42, 2 vols.; Grundlagen der Stereochemie, 1945; Probleme der Naturwissenschaften, 1949. Studied structure and chem. composition of minerals, stereochemistry. Died Zurich, Jan. 13, 1953.

NIGHTINGALE, Florence, English nurse, hosp. reformer; b. Florence, Italy, May 12, 1820; d. William Edward and Frances (Smith) Nightingale; ed. in nursing, Inst. Protestant Deaconesses, Kaiserswerth, 1851; visited and admired nursing instn. of Sisters of St. Vincent de Paul. Spent childhood in Eng.; after completing edn. spent much time examining the working of hosps., reformatories, other charitable instns.; became supt. Hosp. for Invalid Gentlewomen, 1853; during Crimean War vol. services and organized nursing dept., Scutari, Turkey, 1854; devoted herself to eliminating sanitation problem in the wards; returned to Eng., 1856; established an instn. for tng. nurses at St. Thomas Hosp., also King's Coll. Hosp. from funds received in recognition of services during war; following war devoted her attention to question of army sanitation reform and army hosps., also to sanitation in India and among English poorer classes. Recipient Order of Merit from King Edward VIII; Cross of Merit (Germany); Secours aux blessés Militaires (France). Author; Notes on Matters Affecting the Health, Efficiency and Hospital Administration of the British Army, 1858; Notes on Hospitals, 1859; Notes on the Sanitary State of the Army in India, 1871; Life or Death in India, 1874. Publns. gave enormous stimulus to the study of sanitation problems in Eng.; raised nursing standards and status of profession. Died London, Eng., Aug. 13, 1910.

NIGON, V., French biologist; b. Metz, France, Dec. 10, 1920; s. V. and E. (Deutscher) N.; student U. Strasbourg (France), 1935; B.Sc., U. Paris, 1936, postgrad., 1936-39, 44-48, Dr.S., 1948; Lic.Sc., U. Lyon (France), 1940, postgrad., 1940-41; postgrad., U. Toulouse (France), 1941-44, m. M. T. Dubreuil, Nov. 30, 1940; 1 dau., Marie-Françoise. Became research asso., U. Paris, 1944, research dir., 1947; faculty U. Lyon (France), 1949-—, prof. Faculty Scis., 1953-—. Author: Reproductive Biology of Nematodes; Traité de Zoologie. Research publs. on sex determination of free-living nematodes, exptl. embryology of nematodes, metabolism of nucleic acids and proteins in nematodes, insects and sea-urchins. Home: 6 rue de la Pouponnière. Office: 43 Blvd. du 11 Novembre 1918, 69 Villeurbanne, France.*

NIGRELLI, Ross Franco, Am. comparative pathologist; b. Pittston, Pa., Dec. 12, 1903; s. Castrenza and Emanuela (Franco) N.; B.S., Pa. State U., 1927; M.S., N.Y. U., 1929, Ph.D., 1936; m. Margaret R. Carrozza, Dec. 31, 1927; 1 dau., Emmanuela Irma (Mrs. Philip Boon Dobrin). Teaching fellow N.Y. U., 1927-31, vis. instr. dept. biology, 1943-45, vis. asst. prof., 1945-49, adj. asso. prof., 1949-58, adj. prof., 1958-—, research fellow N.Y. Zool. Soc., 1931-32; parasitologist N.Y. Aquarium, 1932-34; pathologist N.Y. Aquarium, N.Y. Zool. Soc., 1934-—, chmn. dept. marine biochemistry and ecology, 1958-63, dir. research Osborn Labs. Marine Scis., 1964-—; instr. biology Coll. City N.Y., 1936-43; Mem. tech. adv. commn. Atlantic States Fishery Commn., 1946-—; cons. Fish and Wildlife Service, 1943-48, Bingham Oceanographic Lab., Yale, 1943-60, U. S. Pure Food and Drug Adminstrn., 1945-—; U.S. Naval Applied Sci. Lab., Bklyn, 1962-—, spl. panel ocean resources, com. on oceanography NRC, 1958-59, Inst. Animal Resources, Nat. Acad. Sci.-NRC, 1954-56; Mem. Pres. Sci. Adv. Com., Panel on Biol. Oceanography. Fellow A.A.A.S., N.Y. Acad. Scis., N.Y. Zool. Soc, Soc. d'Encouragement pour la Recherche et l'Invention (Order of Merit 1963); mem. Soc. Protozoologists (pres. 1947-49), Am. Soc. Zoologists, Am. Soc. Parasitologists, Am. Soc. Micros. Soc., Am. Soc. Microbiologists, Acad. Microbiology, Pathologists and Bacteriologists, Cancer Assn., Internat. Acad. Pathologists, Internat. Soc. Toxinology, Systematic Zoologists, Internat. Soc. Icthyology and Herpetology, Atlantic Fishery Biologists, Bermuda Biol. Station, Royal Soc. Medicine (affiliate), Sigma Xi, others. Research in causes of death of aquatic organisms, marine biochemistry in search for products with potential drug action. Home: 111 W. 183d St., N.Y.C. 10453. Office: N.Y. Aquarium W. 8th St., Bklyn. 11224.*

NIINO, Hiroshi, Japanese geologist; b. Tochigi Prefecture, 1905; grad. Tohoku U., 1930; D.Sc.; prof. Fisheries Inst., Ministry Agr. and Forestry; apptd. prof. Tokyo Fisheries U., 1949; instr. Scripps Marine Inst., U. Cal., 1953; studied coal mines and oil deposits under sea, Mexico, Europe; returned to Japan, 1954; organizer Gt. Japan Submarine Resources Exploitation Assn. (for exploitation of natural resources in Japan's continental ledge), 1954; instr. Tohoku U., also Tokyo Edn. U. Author: The Sea and Its Resources; The Geology of the Sea. Recipient Asahi Sci. Scholarship fund, 1952. Research on continental shelf; participant investigations of eruption of Myojin Reef aboard Shimpo Maru, Sept. 1952.

NIJENUIS, Albert, mathematician; b. Eindhoven, Netherlands, Nov. 21, 1926; s. Hendrik and Lijdia (Koornneef) N.; Candidaat, U. Amsterdam (Netherlands), 1947, Doctorandus, 1950, Doctor cum laude, 1952; m. Marianne Dannhauser, Aug. 14, 1955; children—Erika, Karin, Sabien. Came to U.S., 1952, naturalized, 1959. Asso., Math. Center Amsterdam, 1951-52; asst. Inst. for Advanced Study, 1955, mem. 1961-62; instr., research asso. U. Chgo., 1955-56; faculty U. Wash., Seattle, 1956-63, prof., 1961-63; prof. math. U. Pa., Phila., 1963-—. Mem. Inst. for Advanced Study, 1961-62; Fulbright lectr. U. Amsterdam, 1963-64. Fulbright grantee, 1952-53, 63-64; Guggenheim fellow, 1961-62. Mem. Am. Math. Soc., Math. Assn. Am., Royal Netherlands Acad. Scis. (corr.). Research, publs. in differential geometry and deformation theory in algebra, especially tensors, holonomy groups, graded lie algebras.*

NIJLAND, Albertus Antonie, Dutch astronomer; b. Oct. 30, 1868; D.Sc. in math. Utrecht U., 1896, Dr. astronomy, 1897; LL.D., St. Andrews, 1911; m. M.M.J. Moll, 1904; 1 dau. Became observator Utrecht Obs, 1895, dir, from 1898; prof. astronomy Utrecht U., then rector magnificus, 1911-12; mem. Royal Geodetic Commn., 1915; with eclipse expds., U. S., 1900, Sumatra, 1901, Spain, 1905, Limburg, 1912. Fellow Dutch Sci. Soc., Haarlem, 1909, Royal Acad. Amsterdam. Author: On Certain Functions; On the Cluster G.C. 4410; Results of Observations of Jupiter; Observations of Cepheids; (with D. J. Korteweg) Recueil des observations astronomique de Chr. Huygens; De Bouw ran het Heelal. Research, publs. on variable stars, solar eclipses, comets, meteors. Died Aug. 18, 1936.

NIKITIN, Evgenii Evgenievich, Russian physicist; b. Saratov, USSR, May 9, 1933; s. Evgenii K. and Serafima (Spiridonova) N.; prof., Dr. Sci., Saratov U., 1955; postgrad. Inst. Chem. Physics; 1 son. Jr. research worker Inst. Chem. Physics, Acad. Sci., Moscow, USSR, 1958-61, sr. research worker, 1961-—; lectr. Moscow State U., 1960-61. Mem. internat. adv. com. Confs. on Physics of Electronic and Atomic Collisions, 1965-—. Author: Teoriya termicheskogo raspada molekul, 1964 (English translation Theory of Thermally Induced Gas Phase Reactions, 1966); also numerous articles. Editorial bd. Chem. Phys. Letters, Theoret. chim. acta, 1965-—. Research on non-equilibrium chem. kinetics in gases, theory of non-adiabatic transitions. Office: Inst. Chem. Phys., 2-b Vorobyevskoye chaussee, Moscow, V-334, USSR.*

NIKKILA, Esko A., Finnish physician; b. Helsinki, Finland, Aug. 3, 1926; M.D., U. Helsinki, 1951, dissert., 1953. Faculty, U. Helsinki, 1955-—, prof. medicine, head 3d dept. medicine, 1962-—. Research, numerous publs. on plasma lipoproteins and lipids in relation to coronary heart disease, red cell enzymes in different disorders, plasma, plasma insulin regulation; discovered enzymatic defect in hereditary fructose intolerance. Office: Univ. Hosp., IIId Dept. Medicine, Helsinki, Finland.*

NIKOLAEV, Anatoliy Petrovich, Russian obstetrician, gynecologist; b. Tarasha (now Kiev Oblast), 1896; grad. Med. Faculty, Kiev U., 1917; D.Med. Sci., 1939. Prof., 1933-—; prof. Poltava Med. Inst., 1933-36; dep. dir. Stalino Research Inst. Mother and Child Care, 1936-41; dep. dir., dir. Inst. Obstetrics and Gynecology, USSR Acad. Med. Scis., Moscow, Leningrad, 1944-54; dep. dir. for sci. work Kiev Inst. Mother and Child Care, 1954-—. Decorated Order of Lenin; recipient Stalin prize, 1952. Author: Theory and Practice of Hypnosis from the Physiological Standpoint, 1927; Manual of Examination Technique, Diagnosis and Therapy in Gynecology, 1937; Pavlov's Teaching and the Major Problems of Obstetrics and Gynecology, 1951; Prophylaxis and Therapy of Intrauterine Fetal Asphyxia, 1952; Outline Theory and Practice of Painless Birth, 1953, 59, 60; Weak Labor Activity and Its Treatment, 1956. Editor: Practical Obstetrics, 1958; co-editor Obstetrics and Gynecology sects. Large Med. Ency., 2d edit.; mem. editorial council Obstetrics and Gynecology. Research and numerous publs. on obstetrics, measures against fetal asphyxia, painless birth, toxicosis of pregnancy, psychoprophylaxis and psychotherapy of labor pains, neurohumoral regulation of labor activity. Address: Kiev Inst. Mother and Child Care, Vozdukhoflotskoe sh. 24, Kiev, Ukrainian SSR, USSR.

NIKOLAEV, Anatoliy Vasilievich, Russian chemist; b. Orenburg, Russia, Nov. 27, 1902; s. Vasilie and Marie (Neshoroschkova) N.; grad. Leningrad U., 1924; D.Sc., U. Moscow, 1940; m. Sophie Lvin, 1924; m. 2d, Vera Maximadgie, 1944; 1 son, Sergey. Chief salt lake expdn. in Kulunda Steppes, USSR Acad. Scis., Siberia, 1927-33; chief lab. Inst. Gen. and Inorganic Chemistry, Moscow, 1934-58, dir. Inst. Inorganic Chemistry, Siberian Div., Novosibirsk, 1957-—; asst. prof. U. Leningrad, 1928; prof. Inst. Non-Ferrous Metals and Gold, Moscow, 1944-58, Novosibirsk U., 1958-—. Recipient Vernadskie award, 1947. Mem. USSR Acad. Scis., Mendeleev Soc. Novosibirsk Dist. (pres. 1958). Author: The Salt Lakes of Irtiesh District, 1931; The Salt Lakes of Kulunda, 1934; Thermography, 1944; Physiocochemical Study of Natural Borates, 1947; Protective Films on Salts and Their Use, 1944. Editor: Radiation Shielding, 1961. Research in physico-chem. analysis of salt systems, their formation in nature, and their industrial processing; studies of radio chemistry, thermography, and chemistry of rare-earth elements; developed thermal analysis of complex compounds of platinum. Home: Academist St. 24, Novosibirsk 72. Office: Inst. Inorganic Chemistry, Novosibirsk 90, USSR.*

NIKOLAEV, Oleg Vladimirovich, Russian surgeon, endocrinologist; b. Kazan, 1903; grad. Med. Faculty, 1st Moscow U., 1924; D. Med. Sci., 1938. Head exptl. surgery lab. Inst. Exptl. Biology, 1927-31; head surg. dept. clinic All-Union Inst. Exptl. Endocrinology, 1931-41, 42—, also mem. learned council; 2d prof. chair gen. surgery and chair operative surgery and topographical anatomy Kiev Med. Inst., 1941-42; prof. Central Postgrad. Med. Inst., 1942—; sr. asso. Central Neurosurg. Inst., 1942-45. Mem. Georgian Goiter Out-Patients Clinic; mem. Endocrinology and Endemic Goiter Problems Commns. Decorated Order of Lenin. Mem. All-Union (Presidium mem.), Moscow (Presidium mem.), Estonian (hon.) socs. endocrinologists. Author numerous works including: Endemic Goiter, 2d edit., 1949; Surgery of the Endocrine System, 1952; co-author: Diseases of the Thyroid Gland, 1961; Hormone Active Tumors of the Suprarenal Glands, 1963. Editor: Problems of Endocrinology and Hormone Therapy. Address: Central Postgrad. Med. Inst., pl. Vosstaniya 1-2, Moscow, USSR.

NIKOLAEV, Viktor Arsenievich, Russian petrographer; b. Dec. 6, 1893; with Mining Inst., Petrograd, 1918; mem. staff Geologic Com. (now All-Union Geology Inst.), 1920-47; prof. Central Asian Indsl. Inst., 1933-45; prof. Leningrad Mining Inst., 1947—. Recipient Lenin prize, 1958; named honored scientist Georgian SSR. Corr. mem. Acad. Scis. USSR (researcher Lab. Pre-Cambrian Regions 1951—); pres. All-Union Mineral. Soc., 1955—. Research on alkaline formations in Talass area of Alatau Range, other regions of Central Asia; advanced theory of vulcanism for Tian-Shan area.

NIKOLSKII, Georgii Vasilievich, Russian ichthyologist; b. May 6, 1910; grad. Moscow U., 1930; with Aral Sea Fish Economy Sta.; mem. staff Moscow U., 1932—, prof., 1940—. Recipient Stalin prize, 1950. Corr. mem. Acad. Scis. USSR. Author: Fish of the Aral Sea, 1940; Biology of Fish, 1944; Specialized Ichthyology, 1954. Authority on classification, zoogeography, biology of fish; contbd. to devel. of system for increasing fish productivity in Amur River basin; research on dynamics of quantities and feeding of fish; dir. expdns. for study of ichthyofauna and pisciculture in Aral Sea, 1937-47, Pechora River, 1941-44, Amur River, 1945-49.

NIKOLSKY, Boris Petrovich, Russian phys. chemist; b. 1900; grad. Leningrad U., 1924. Instr., Leningrad U., 1924-39, prof., 1939—. Mem. USSR Acad. Scis. (corr.). Author: Vitreous Electrode Theory, 1935; Laws of Ion Exchange between the Solid State and Solutions, 1939. Research and publs. on ion exchange processes between aqueous solutions and various solid systems, soils, ionites (developer theory of the processes); devised ion exchange theory of vitreous electrodes. Address: Leningrad University, Universitetskaya n. 7-9, Leningrad, USSR.

NILSON, Lars Fredrik, Swedish chemist; b. Östergötland, Sweden, May 27, 1840; ed. Uppsala; prof. analytical chemistry, Uppsala, 1878, Stockholm, Sweden, 1883; agrl. chemist, Stockholm. Determined atomic weight and specific heat of beryllium (results supported Mendeléeff and Meyer's classification of elements); discovered scandium, 1879; prepared pure thoria, 1883, also (with Petterson) titanium in an impure state by reducing chloride with metallic sodium in steel bomb, 1887; isolated ytterbium; proved indium had 3 chlorides; studied chem. fertilizers. Died Stockholm, May 14, 1899.

NILSON, Sven, Swedish naturalist; b. nr. Landskrona, Sweden, 1787; dir. zool. mus., Lund, Sweden; became dir. zool. mus., Stockholm, Sweden, 1819. Author: Ornithologica suecica, 2 vols., 1817-22; Historia molluscorum Sueciae, 1822. Research on vertebrates, birds of Sweden, fossilized crustacea. Died Lund, 1883.

NILSSON, Bernt Ove, Swedish anatomist; b. Falun, Sweden, Jan. 8, 1929; s. Johan Bernhard and Gurli (Högman). N.; M.D., Karolinska Instituet, Stockholm, Sweden, 1959, certificate, 1960; m. Irene Brann, Dec. 16, 1956; children—Sven Ludvig, Karin Elisabet. Asso. prof. Karolinska Institutet, 1961-62; asso. prof. U. Uppsala (Sweden), 1963—. Mem. Scandinavian Assn. Electron Microscopy. Author: (with Wirsén) Techniques in Light Microscopy, 1965; also numerous articles. Research on early stages of pregnancy, mechanism of blastocyst attachment onto uterine epithelium, process of implantation. Home: 12A Götgatan, Uppsala, Sweden.*

NIMS, Leslie Frederick, Am. physiologist; b. Salt Lake City, June 15, 1906; s. Leslie White and Lillian (Lane) N.; B.S., U. Utah, 1927, M.A., 1928; Ph.D., Yale, 1931; m. Elsea Winabeth Stephenson, June 15, 1929; children—Winafred Sisson (Mrs. Dwight Schoeffler), Nancy Lou (Mrs. James Tasi), Leslie Irving. Faculty Yale, New Haven, 1931-47, asso. prof., 1945-50, dir. Yale Aeromed. Research Unit, 1941-47; acting head biology dept. Brookhaven Nat. Lab., Upton, L.I., N.Y., 1947-50, sr. scientist, 1950—. Alexander Forbes lectr. Marine Biol. Labs., Woods Hole, Mass., 1965. Mem. A.A.A.S., Am. Chem. Soc., Am. Physiol. Soc., Am. Inst. Biol. Scis., Radiation Research Soc., Biophysics Soc., Am. Soc. Naturalists, N.Y. Acad. Sci., Sigma Xi, Phi Kappa Phi, Alpha Xi Sigma. Contbr. chpts. to books, numerous articles to profl. jours. Research in thermodynamics of NaCl solutions, ionization constants of weak acids, pH measurements in vivo of blood and brain, phys. theory of decompression sickness, carbohydrate metabolism after body X-irradiation, material transfers across biol. barriers, meaning of fluxes in biol. systems when estimated with tracers. Home: Library Lane, Brookhaven, L.I., N.Y. 11719. Office: Brookhaven Nat. Lab., Upton, L.I., N.Y. 11973.*

NININGER, Harvey Harlow, Am. meteoriticist; b. Conway Springs, Kan., Jan. 17, 1887; s. James Buchanan and Mary Ann (Firestone) N.; student Northwestern State Normal Coll., Okla., 1907-09; A.B., McPherson Coll. 1914, D.Sc., 1937; M.S., Pomona Coll., 1916; postgrad. U. Cal. at Berkeley; LL.D., Ariz. State U., 1963; m. Nancy Adeline Delp, June 5, 1914; children—Robert D., Doris Elaine (Mrs. John C. Banks), Margaret Ann (Mrs. Glenn I. Huss). Prof. biology LaVerne Coll., 1914-18, Southwestern Coll., 1919-20, McPherson Coll., 1920-30; instr. Pomona Coll., summer 1916; spl. field agt. U.S. Bur. Entomology, 1918-19; field agt. entomology Kan. State Agrl. Coll., summer 1919; dir. meteorite expdn. to Mexico, 1929; founder, dir. Rocky Mountain Summer Sch., McPherson (Kan.) Coll., 1922-30; curator meteorites Colo. Mus. Natural History, Denver, 1930-46; established Nininger Meteorite Lab., Denver, 1930; established Am. Meteorite Mus., Ariz., dir., 1946-60; made extensive explorations meteorite crater, Ariz., 1946-53; meteorite and tektite explorations, Australia, P.I., S.E. Asia, 1958-59; survey meteorite collections, Europe, 1960; tektite survey South Viet Nam, 1958-64. Recipient Distinguished Citizens' award City of Denver; Leonard Meml. medal, 1967. Mem. Kan. Acad. Sci. (pres. 1924-25), Geol. Soc. Am., Meteoritical Soc. (co-founder, pres. 1937, 41), Ariz. Acad. Sci. (charter). Author: Oour Stone-Pelted Planet, 1933; A Comet Strikes the Earth, 1942; Chips From the Moon, 1943; The Nininger Collection of Meteorites, 1950; Out of the Sky, 1952; Ask a Question About Meteorites, 1961; numerous papers on meteorites; contbr. to Ency. Brit. Discovered many new meteorites, new varieties of meteorites, explosion products; assembled largest collection of meteorites and tektites in world; organized theory of lunar origin of tektites. Address: P.O. Box 420, Sedona, Ariz. 86336.*

NININGER, Robert D., Am. geologist; b. Brookings, S.D., Mar. 28, 1919; s. Harvey Harlow and Addie (Delp) N.; A.B. magna cum laude with high honors in Geology (John Woodruff Simpson fellow), Amherst Coll., 1941; A.M. in Econ. Geology, Harvard, 1942; m. Eleanor Frances Jones, June 28, 1943; children—Sally E., Susan M., Robert H., Catherine F. Geologist, U.S. Geol. Survey, Washington, 1942-43; dep. asst. dir. for exploration div. raw materials U.S. AEC, N.Y.C., 1947-54, asst. dir. for exploration, 1955-61, asst. dir. for procurement operations, Washington, 1962—. Tech. adviser Internat. Conf. on Peaceful Uses Atomic Energy, Geneva, Switzerland, 1955, 58. Mem. Am. Nuclear Soc., Am. Inst. Mining, Metall. and Petroleum Engrs., Soc. Econ. Geologists, A.A.A.S., Geol. Soc. Am. Author: Minerals for Atomic Energy, 1954; Exploration for Nuclear Raw Materials, 1956; also articles. Assisted in devel., orgn. and direction uranium geology and exploration program to provide supply fissionable material. Home: 14267 Crossway Rd., Rockville, Md. 20853. Office: U.S. AEC, Washington 20545.*

NINO, Pedro Alonso (El Negro), Spanish navigator; b. Moguer, Spain, 1468; pilot in Columbus' expdns. to New World, 1492, to Trinidad and coast of Paria, 1498; sailed (with royal permission, accompanied by Cristobal Guerra) in search of treasure in New World, 1499-1500; accused of witholding part of royal share, but died before conclusion of trial, circa 1505.

NIPHER, Francis Eugene, Am. physicist; b. Port Byron, N.Y., Dec. 10, 1847; s. Peter and Roxalana P. (Tilden) N.; Ph.B., State U. Ia., 1870, A.M., 1873; LL.D., Washington U., 1905; m. Matilda Aikins, July 1, 1873; children—Mary E. (Mrs. E. N. Birge), Edith C. (Mrs. H. M. Pollord), Elma F. (Mrs. J. C. Dawson), Clara Ellen, Edwin Tilden. Instr. in phys. lab. State U. Ia., 1870-74; prof. physics Washington U., St. Louis, 1874-1914, emeritus, from 1914. Author: Theory of Magnetic Measurements, 1886; Electricity and Magnetism, 1895; Introduction to Graphical Algebra, 1898; Experimental Studies in Electricity and Magnetism, 1914. Research asso. Carnegie Instn., Washington. Fellow A.A.A.S.; mem. Am. Phys. Soc., Am. Philos. Soc. Phila., French Soc. Physics, Royal Soc. Arts, Acad. Sci. St. Louis (pres. 1884-90). Author: Theory of Magnetic Measurements, 1886; Electricity and Magnetism: A Mathematical Treatise for Advanced Undergraduate Students, 1895; Experimental Studies in Electricity and Magnetism, 1914. Showed that positive or reversed photog. pictures can be best produced by devel. in light, 1889; developed perfect pictures on most sensitive plates with developing bath fully exposed to direct sunlight; made extensive study of elec. discharge; demonstrated that elec. potential of earth is constantly subject to considerable local variation; discovered causes of magnetic disturbances in earth's field; produced similar disturbances in field of bar magnet surrounded by electrified air; showed that vaporized water can be decomposed by elec. current; proved experimentally that gravitational attraction can be reduced by electrification of masses of matter; converted gravitational attraction between masses of matter into repulsion when masses were completely shielded from each other by metal. Died Kirkwood, Mo., Oct. 6, 1927.

NIPKOW, Paul Gottlieb, German inventor; b. Liauenburg, Germany, Aug. 22, 1860; named hon. pres. German TV Soc., 1934; pioneer in devel. of TV; inventor rotating disc with spiral of small holes which partitioned picture (Nipkow disc), 1884, basic to TV prodn. until invention of electronic scanning device, 1923. Died Berlin, Germany, Aug. 24, 1940.

NIRENBERG, Marshall Warren, Am. biochemist; b N.Y.C., Apr. 10, 1927; s. Harry Edward and Minerva (Bykowsky) N.; B.S. in Zoology, U. Fla., 1948, M.S., 1952; Ph.D. in Biochemistry, U. Mich., 1957; m. Perola Zaltzman, July 14, 1961. Postdoctoral fellow Am. Cancer Soc. at NIH, 1957-59; postdoctoral fellow USPHS at NIH, 1959-60; mem. staff NIH, 1960—, research biochemist, 1961-62, research chemist, head sect. biochem. genetics Nat. Heart Inst., 1962—. Recipient Molecular Biology award Nat. Acad. Scis., 1962, medal Dept. Health, Edn. and Welfare, 1963, Modern Medicine award, 1964, Nat. Med. Sci., President Johnson, 1965. Mem. Am. Soc. Biol. Chemists, Am. Chem. Soc. (Paul Lewis award enzyme chemistry 1964, Harrison Howe award 1964), Biophys. Soc., Washington Acad. Scis. (Biol. Sci. award 1963), Sigma Xi. Editor Biopolymers, 1963—. Spl. research mechanism protein synthesis, genetic code, nucleic acids, regulatory mechanisms in synthesis macromolecules. Home: 7723 Old Chester Rd., Bethesda, Md. 20034. Office: Nat. Heart Inst., Nat. Insts. Health, Bethesda, Md. 20014.

NISBET, Robert Alexander, Am. sociologist; b. Los Angeles, Sept. 30, 1913; s. Henry S. and Cynthia (Jenifer) N.; A.B., U. Cal. at Berkeley, 1936, M.A., 1937, Ph.D., 1939; m. Emily P. Heron, July 15, 1936; children—Martha Swift (Mrs. Robert F. Rehrman), Constance Emily. Faculty, U. Cal., Berkeley, 1939-53; prof. sociology U. Cal., Riverside, 1953—, dean Coll. Letters and Sci., 1953-63, vice chancellor acad. affairs, 1959-63; vis. prof. U. Bologna, Italy, 1956-57; Guggenheim vis. fellow Princeton, 1963-64. Cons. socio-behavioral scis. Pacific State Hosp., Pomona, Cal., 1962—. Recipient award of merit Republic of Italy, 1957. Mem. Am., Pacific (pres. 1962-63) sociol. assns., A.A.A.S., Authors Guild, Phi Beta Kappa. Author: Community and Power, 1935, 62; Human Relations and Administration, 1957; Emile Durkheim, 1964; (with Robert K. Merton) Contemporary Social Problems, 1961; 66; The Sociological Tradition, 1966. cons. editor Random House, Inc., 1964-67. Publs. research on relationship between social community and polit. power, origins of modern sociol. ideas in 19th century ideology.

NISBET, William, Brit. physician, med. writer; b. 1759; practised in Edinburgh, Scotland; lectr. on venereal diseases Edinburgh, winter 1786. Settled in London, by 1801. Fellow Royal Coll. Surgeons Edinburgh. Author: First Lines of the Theory and Practice in Venereal Diseases, 1787; The Clinical Guide, 1793; An Inquiry into the History, Nature, Causes and Different Modes of Treatment hitherto pursued in the Cure of Scrophula and Cancer, 1795; A practical Treatise on Diet, 1801; The Edinburgh School of Medicine, 4 vols., 1802; A General Dictionary of Chemistry, 1805 (rev. and completed by another writer). Described ulcerative nodules in penis following lymphangitis caused from chancroid, 1787. Died 1822.

NISHIJIMA, Kazuhiko, Japanese physicist; b. Japan, Oct. 4, 1926; s. Ayao and Sachiko (Hirose) N.; B.S., U. Tokyo, 1948; Ph.D., U. Osaka (Japan), 1954; m. Hideko Yamura, Nov. 1, 1954; children—Fuyuhiko, Tokihiko. With Osaka City U., 1950-59; prof. U. Ill., Urbana, 1959-67; prof. U. Tokyo (Japan), 1967—. Staff Max-Planck-Institut für Physik, Göttingen, Germany, 1956-57, Inst. for Advanced Study, Princeton, N.J., 1957-58. Recipient Nishina prize, 1956; Japan Acad. prize, 1964. Mem. Phys. Soc. Japan, Am. Phys. Soc. Author: Fundamental Particles, 1963; also articles. Discovered strangeness, 1953; predicted 2 kinds of neutrinos, 1957; studies in field theory of bound states, weak interactions. Home: 5-1-4 Kyonan-Cho Musashino, Tokyo, Japan.*

NISHIKAWA, Joken (named Tadahide, called Jirozaemon), Japanese astronomer, geographer; b. Nagasaki Prefecture, 1648; studied Confucian lit. under Soju Nambu, also astronomy, calendar sci.; pursued knowledge of Western sci.; astron. and calendar sci. expert for Shogunate govt., Edo, Japan, from 1718; Author: numerous books including, Kai-tsusho-ko (reference book on overseas affairs). Introduced Western learning into Japan. Died 1724.

NISHIMURA, Edwin Takayasu, Am. physician; b. Sacramento, Sept. 12, 1918; s. Manzo and Kajiu (Nishimura) N; A.B., U. Cal., Berkeley, 1940; M.D., Wayne State U., 1945; m. Mayme Semba, Apr. 10, 1943; children—Janet Marie, Thomas Gerard, Mark Richard. Faculty, Northwestern U. Med. Sch., 1953-64, U. Cal., Los Angeles, 1964-65; faculty U. Hawaii Sch. Medicine, 1965—, prof., chmn. dept., 1966—. Recipient Career Devel. award USPHS,

1959-—. Mem. Am. Soc. Clin. Pathology, Am. Assn. Pathologists and Bacteriologists, Am. Soc. Exptl. Pathology, Soc. Cell Biology. Research, numerous publs. on cell physiology of exptl. cancer cells, enzyme abnormalities, especially catalase in patients and in exptl. animals, enzymes and immunity. Home: 3770 Lurline Dr., Honolulu 96816.*

NISHINA, Yoshio, Japanese physicist; b. Okayama Prefecture, Japan, Dec. 6, 1890; elec. degree, Tokyo U.; postgrad in Europe, 1921; student of Niels Bohr; mem. Physics and Chemistry Research Inst., Tokyo, apptd. sect. chief., 1928, dir. constrn. of cyclotron, 1937; chief Research Inst., also Japan Sci. Council. Mem. Acad. Author: Quantum Dynamics. Pioneer in study of cosmic rays, also in devel. of sci. studies in Japan; an originator Klein-Nishina formula relating to Compton effect, 1928; produced 1st artificial uranium 237 in Japan; tchr. of Hideki Yukawa. Died Tokyo, Jan. 10, 1951.

NISHIWAKI, Masaharu, Japanese biologist, marine mammalogist; b. Tokyo, Japan, Jan. 23, 1915; s. Kenji and Yoshiko (Sakai) N; B.S., U. Tokyo, 1939, D.Sc., 1951; m. Haruko Taku, June 27, 1943; children—Yoko, Yoshiharu. Asso. dept. fisheries U. Tokyo, 1939-48; research mem. Whales Research Inst., Tokyo, 1948-65; prof. Ocean Research Inst., U. Tokyo, 1965-—, coll. lectr., 1953-65; research insp. Fisheries Agy. for Antarctic Whaling Expdn., 1947-48, 48-49, 55-56; cons. to labs., RyuKyu Island govt., U.S. govt. Mem. Mammalogical Soc. Japan (dir. 1953-—), Japan Soc. for Prevention of Cruelty to Animals (councilor 1962-—), Japanese Assn. Anatomists, Japanese Soc. Sci. Fisheries, Oceanographical Soc. Japan. Author: Physical Explorations of Whale Body, 1954; Whaling in Antarctic, 1957; Tales of Whales and Dolphins, 1963; Whales and Pinnipeds, 1965; Whales, Dolphins and Porpoises, 1966; also numerous articles. Research on assessment marine mammals and their classification, methods age determination of whales, protection of animals in Japan. Home: 17 Kawada-cho, Shinjuku, Tokyo, Japan.*

NISSEN, Rudolph, surgeon; b. Neisse, Schlesien, Germany, Sept. 9, 1896; s. Franz and Margarethe (Borchert) N.; student univs. Munich, Marburg, Breslau; M.D., U. Breslau, 1920; m. Ruth Becherer, May 29, 1933; children—Renate Marina (Mrs. Bernard Jaffe), Tim Oliver. Asst., U. Breslau, 1920, U. Freiburg, 1920-21; asst. surg. dept., instr. surgery, asso. surgeon U. Munich, 1921-33; asso. surgeon U. Berlin, 1927-33; instr., asso. prof. surgery Med. Faculty Berlin, 1927-33; prof. surgery, head dept., chief surgeon 1st surg. clinic U. Istanbul (Turkey), 1933-39; research fellow surgery Mass. Gen. Hosp., Boston, 1939-40; attending surgeon, chief div. Jewish Hosp. Bklyn., 1941-52; chief surgeon Maimonides Hosp., Bklyn., 1944-50; prof. surgery, head dept. U. Basel (Switzerland), 1952-67; chief surgeon Burgerspital Basel, 1952-67. Recipient Rudolph Virchow medal Rudolph Virchow Med. Soc., 1960, Paracelsus medal, 1967. Hon. mem. numerous med. and surg. socs. Author over 30 books, including (with G. Brandt, H. Kunz) Intra- and Post-operative Complications in Surgery, 1965; (with H. Hellner, K. Vossschulte) Textbook of Surgery, 1967; also chpts. to books, over 475 articles on thoracic and abdominal surgery to profl. jours. Did 1st total pneumonectomy, 1931. Home: 45 Höhenstrasse, 4125 Riehen, Switzerland. Office: Nonnenweg 31, Basel, Switzerland.*

NISSL, Franz, German neurologist; b. Frankenthal, Germany, Sept. 9, 1860; M.D., U. Munich (Germany), 1895. Docent, U. Munich, 1896-1903, became prof., 1903; dir. psychiat. clinic, Heidelberg, Germany, 1904-1918; a founder, asso. Deutschen Forschungsintituts für Psychiatrie, Munich. A founder histopathology; described histopathology of gen. paralysis, 1904, degeneration of ganglion cells, (Nissl's degeneration), 1892, chromophilic bodies in nerve cells (Nissl bodies). Died Munich, Aug. 11, 1919.

NITSCH, Jean Paul, French plant physiologist; b. Mulhouse, France, Sept. 15, 1921; s. Edouard Maurice and Suzanne (Lesage) N.; Ingénieur Agronome, Inst. Nat. Agronomy, Paris, France, 1947; Ph.D., Cal. Inst. Tech., 1951; m. Colette Kirchner, July 4, 1953; children—Pierre-Marcel (dec.), Agnès, Nicolas. Research fellow Harvard, 1950, 53-55; asst. prof. Cornell U., 1955-58; asso. dir. Phytotron, Gif-sur-Yvette, France, 1958-66, sci. dir., 1966-—. Vis. prof. U. Cal. at Davis, 1964-65; lectr. Sorbonne, 1966, Inst. Nat. Agronomy, 1967-—. Mem. Société Française de Physiologie Végétale, Am. Scandinavian socs. plant physiologists, Bot. Soc. Am., Bot. Soc. France. Research, numerous publs. on hormonal interactions between cells in plants, especially in processes of bud formation, fruit setting, growth and dormancy of trees, extraction and identification of natural growth factors, biochem. investigation of growth-regulatory effects produced by climatic factors, constrn. of French Phytotron, induction of flowering in pieces of tissues and fruit devel. in excised flowers grown in test tubes. Home: 48 rue G. Vatonne. Office: Phytotron, 91 Gif-sur-Yvette, France.*

NITTA, Yoshihiro, Japanese chemist; b. Tsurugi-Machi, Ishikawa Prefecture, Japan, Nov. 8, 1922; s. Yoshiaki and Fuji (Nishiguchi) N.; Ph.D., U. Kyoto (Japan), 1957; m. Shigeko Tada, Dec. 15, 1948; 2 children. With Chugai Pharm. Co., Tokyo, 1946-—, dir. chemistry dept., 1963-66, dir. products devel. dept., 1966-—. Mem. Chem. Soc. Japan, Pharm. Soc. Japan. Research, numerous publs., numerous patents on new drugs, contbg. to devel. pharm. industry in Japan, glucuronic acid, heterocyclic compounds. Home: 1300 Izumi, Komae-machi, Kitatama-gun, Tokyo. Office: 41-8 Takada-3-chome, Toshima, Tokyo, Japan.*

NITZE, Max, German surgeon, urologist; b. Berlin, Sept. 18, 1848; ed. Heidelberg, Würzburg, Leipzig (all Germany); with State Hosp., Dresden, Germany, 1874-78; then worked in Vienna, later Berlin. Author: Über eine neue Beobachtungs- und Untersuchungsmethode für Harnblase und Rectum, 1879; Lehrbuch der Kystoskopie, 1889; articles on cystoscopy and photocystoscopy. Pioneer in urology; provided basis for modern surgery of genito-urinary tract by inventing (with Viennese instrument maker Lietner) electrically lighted cystoscope, 1877; improved cystoscope by introducing system of prisms, 1894, also adapted it for excision of bladder tumors and other operations, 1896. Died Berlin, Feb. 23, 1906.

NIVEN, Charles, Brit. mathematician, natural philosopher; ed. Aberdeen, Cambridge; M.A., D.Sc.; m. Mary Stewart; 1 son, 2 daus.; fellow Trinity Coll., Cambridge; prof. math. Queen's Coll., Cork, Ireland; prof. natural philosophy Aberdeen U., 1880-1922; fellow Royal Soc., 1880. Contbr. papers to sci. jours. Died May 11, 1923.

NIVEN, Charles Franklin, Jr., Am. microbiologist; b. Clemson, S.C., July 22, 1915; s. Charles F. and Ida (Wiley) N.; B.S., U. Ark., 1935; Ph.D., Cornell U., 1939; m. Treva Landers, Sept. 10, 1939; children—Katherine (Mrs. Lary Howell), William, Robert, Margaret. Faculty, Cornell U., Ithaca, N.Y., 1937-46, asso. prof., 1945-46; bacteriologist Hiram Walker & Sons, Inc., Peoria, Ill., 1946; staff Am. Meat Inst. Found., Chgo., 1946-64, asso. dir., 1957-61, sci. dir., 1961-64; faculty U. Chgo., 1946-64, prof. microbiology, 1961-64; dir. research Del Monte Corp., San Francisco, 1964-—. Recipient Indsl. Achievement award Inst. Food Technologists, 1960. Mem. Am. Soc. for Microbiology (past mem. council), A.A.A.S., Am. Pub. Health Assn., Am. Chem. Soc., Soc. Gen. Microbiology. Editorial bd. Jour. Bacteriology, 1951-60, Applied Microbiology, 1961-—, Jour. Infectious Diseases, 1962-67, Jour. Food Sci., 1964-—. Research, publs. on bacterial nutrition, physiology and metabolism, bacterial taxonomy, food microbiology, food sci. Home: 7424 Potrero Av., El Cerrito, Cal. 94530. Office: 215 Fremont St., San Francisco 94119.*

NIVEN, Ivan Morton, mathematician; b. Vancouver, B.C., Can., Oct. 25, 1915; s. James and Agnes (Morton) N.; came to U. S., 1936, naturalized, 1942; B.A., U. B.C., 1934, M.A., 1936; Ph.D., U. Chgo., 1938; m. Betty Mitchell, Sept. 30, 1939; 1 son, Scott. Harrison Research fellow U. Pa., 1938-39; instr. U. Ill., 1939-42; faculty Purdue U., 1942-47, asso. prof., 1943-47; faculty U. Ore., Eugene, 1947-—, prof. math., 1950-—; vis. prof. U. Cal. at Berkeley, 1964-65. Mem. math. adv. panel Office Naval Research, 1955-58, NSF, 1958-61. Fellow A.A.A.S.; mem. Am. Math. Soc., Math. Assn. Am. (Earle Raymond Hedrick lectr. 1960), Sigma Xi. Author: Irrational Numbers, 1956; (with Herbert S. Zuckerman) Theory of Numbers, 1960; Calculus, 1960; Numbers: Rational and Irrational, 1961; Diophantine Approximations, 1963; Mathematics of Choice, 1965; also articles. Research on theory irrational numbers, algebraic numbers, Diophantine approximation; simplification several theories. Home: 3940 Hilyard St., Eugene, Ore. 97405.*

NIVEN, William, mineralogist; b. Bellshill, Lanarkshire, Scotland, Oct. 6, 1850; s. William and Sarah (Brown) N.; ed. common schs.; came to U.S. 1879; m. Nellie Blanch Purcell, Jan. 26, 1886; children—William Albert, David Sumner, Norman Sumner, Kingsley Burns, Harold Andrew, Mrs. Elna Blanche Harrison, Francis Joseph, Malcolm, Robert Nelson. Asst. commr. Ariz. to New Orleans Expn. Fellow Am. Geog. Soc., Royal Soc. Arts, London; hon. life mem. Am. Mus. Natural History; titled mem. Sci. Soc. Antonio Alzate, Mexico. Engaged in mineral investigations; discovered 3 new minerals, yttrialite, thorogummite and nivenite, in Llano County, Tex., 1889, the new mineral aguilarite, Guanajuato, Mex., 1891; discovered remains of prehistoric city or nation, State of Guerrero, Mex., 1891; discovered buried prehistoric cities beneath Valley of Mexico, 1911. Died Austin, Tex., June 3, 1937.

NIWELINSKI, Józef Marian, Polish biologist; b. Zakopane, Poland, Nov. 25, 1920; s. Aleksander and Emilia (Lojewska) N.; grad. in pharmacy Jagellonian U., Krakow, Poland, 1947, grad. in biology, 1952, d. pharmacy, 1949; m. Janina Danuta Romanowska, Aug. 17, 1950; children—Agnieszka, Aleksander, Maria, Barbara. Vol. sci. worker Jagellonian U., Med. Faculty, 1947-48, elder asst., 1948-49; adj. Polish Acad. Scis., 1953-54; asst. prof., 1954; fellow Rockefeller Found. dept. histochemistry U. London (Eng.), 1958-59; asso. editor sci. quar. Folia Biologica, 1953-58; editor-in-chief Folia Histochemica et Cytochemica, Polish Histochem. and Cytochem. Soc., 1963-—; head dept. histochemistry Med. Acad., Krakow, 1966-—; Adminstr. Krakow City Council pharm. dept., 1949-51. Mem. Copernicus Soc. Polish Naturalists, Polish Histochem. and Cytochem. Soc., Internat. Fertility Assn. Co-author: Textbook of Histochemical Methods, 1963; also articles. Research on synthesis of haemopoietic thiouracil derivatives: demonstration of stimulating effect of crystalline cholesterol on gametogenesis and hormone secretion in mammalian gonads, forelimb regeneration in newts including localization of oxidative enzymes, hereditary anomalies of metabolism, gen. and comparative endocrinology. Home: Sw. Sebastiana 17, m. 4, Krakow, Poland.*

NIXON, Dennis Andrew, English physiologist; b. London, Eng., Feb. 16, 1924; s. Andrew and Ethel (Dix) N.; B.Sc., U. Coll., U. London, 1950, M.Sc., 1952; Ph.D., St. Mary's Hosp. Med. Sch., 1959; m. Marion Putnam, Aug. 6, 1955; 1 dau., Ann Helen Sarah. Staff physiology dept. St. Mary's Hosp. Med. Sch., London, 1950-—, lectr., 1956-—; tchr. physiology U. London, 1960-—. Mem. Physiol. Soc., Royal Soc. Medicine, Zool. Soc., Neonatal Soc., Blair-Bell Research Soc., Soc. for Development Biology. Research, publs. on reproductive physiology especially placental and foetal physiology; developed techniques for perfusing pregnant uterus and maintaining foetus detached from placental and maternal connections. Home: 60 Norfolk Rd., New Barnet, Hertfordshire, Eng. Office: Physiology Dept., St. Mary's Hosp. Med. Sch., Paddington, London, W.2., Eng.*

NIXON, Eugene Ray, Am. chemist; b. Mt. Pleasant, Mich., Apr. 14, 1919; s. William S. and Grace (Brookens) N.; Sc.B. summa cum laude, Alma Coll., 1941; Ph.D., Brown U., 1947; m. Phyllis R. Jones, June 10, 1945; children—Cynthia L., Emily E. Research chemist Manhattan Project, 1942-46; instr. chemistry Brown U., 1947-49; faculty U. Pa., Phila., 1949-—, prof. chemistry, 1965-—, vice dean grad. sch., 1958-62, acting chmn. dept. chemistry, 1965-66; vis. prof. U. London (Eng.), 1963-64. Vis. lectr. Bryn Mawr Coll., 1957-58; cons. U.S. Army, 1965-—. Mem. Am. Chem. Soc., Am. Phys. Soc., Soc. Applied Spectroscopy (Jour. award 1965), Sigma Xi. Research, publs. on phys. chemistry, molecular structure and molecular spectroscopy, properties of crystals, intermolecular interactions, chemistry of phosphides, free radicals, fluorination reactions. Home: 36 Springhouse Lane, Media, Pa. 19063. Office: Dept. Chemistry, U. Pa., Phila. 19104.*

NIXON, John, English mining engr.; b. Barlow, Eng., May 10, 1815; employed in colliery, S. Wales; demonstrated advantages of Welsh coal to manufactures in Nantes, France, thus creating French, also later English, markets for Welsh coal; launched Welsh steam coal trade with mining operations in S. Wales; produced (with assos.) about 1,250,000 tons of coal a year by 1897; inventor Billy Fairplay machine for measuring proportion between large and small coal; introduced longwall mining system as replacement for wasteful pillar, or pillar and stall system. Died London, June 3, 1899.

NIZET, Alphonse Henri, Belgian physician; b. Liège, Belgium, Oct. 25, 1919; s. Joseph and Charlotte (Radoux) N.; M.D., U. Liège; m. Yvonne Laval, Aug. 24, 1945; children—Cécile, Jean, Michel, François. Agt., Fonds nat. de la recherche sci., 1945-60; agrégé, 1951; lectr. U. Liège, 1950-60; prof. U. London (Eng.), 1958; prof. clin. medicine U. Liège, 1961-—. Author numerous publns. in internal and exptl. medicine, especially hematology, nephrology and endocrinology. Home: 93B, blvd. Emile de Laveleye. Office: Inst. de médicine, blvd. de la Constitution, Liège, Belgium.

NOACK, Hugdieter Johann Georg, German physician; b. Gera, Germany, Oct. 23, 1919; s. Bruno Ernst and Catherina (Fischer) N.; student U. Freiburg (Germany); Dr.med., U. Leipzig (Germany), 1945, Dr.med.habil., 1953; m. Sigrid Dembowski, Oct. 1, 1943; children—Ragna, Ernst Joachim Hugdieter, Sigrid Kirsti. Staff women's clinic U. Leipzig, 1949-53, faculty, 1953-58, prof., 1958; dir. womens clinic Med. Sch., Magdeburg, Germany, 1958; dir. evang. Johannes Hosp., Bielefeld, Germany, 1959-—; prof. U. Münster (Germany), 1960-—. Author: Hormonelle Regulationen in der Frühschwongerschaft, 1958; (with K.H. Sommer) Eine kleine Frauengymnastik, 1958; (with E.J. Klaus) Frau und Sport, 1961; also numerous articles. Research on reduction of maternal and fetal mortality, phys. condition of women, menstrual cycle, endocrinology in early pregnancy. Home: 29 Droegestr. Office: 103 Schildescher str., D4800 Bielefeld, Germany.*

NOACK, Walter, engr.; b. Nuremberg, Germany, 1881; ed. Fed. Poly. Sch., Zurich, Switzerland; pioneer in technique of high compression vapors at high temperatures; conceived and installed 1st power house equipped with gas turbines, Neuchâtel, Switzerland; 1st to propose turbocompressors to supply airplane motors, during World War I. Died Baden, Switzerland, 1945.

NOBACK, Charles Robert, Am. anatomist; b. N.Y.C., Feb. 15, 1916; s. Charles Victor and

Beatrice (Cerny) N.; B.S., Cornell U., 1936; M.S., N.Y. U., 1938; postgrad. Columbia; Ph.D., U. Minn., 1941; m. Eleanor Louise Loomis, Nov. 23, 1938; children—Charles Victor, Margaret Beatrice, Ralph Theodore, Elizabeth Louise. Asst. prof. anatomy U. Ga., 1941-44; faculty L.I. Coll. Medicine, 1944-49, asso. prof., 1948-49; faculty Columbia Coll. Phys. and Surg., 1949—, asso. prof. anatomy, 1953—. Fellow N.Y. Acad. Scis. (rec. sec. 1962-63); mem. A.A.A.S., Am. Assn. Anatomists, Histochem. Soc., Am. Assn. Phys. Anthropologists, Harvey Soc., Am. Acad. Neurology, Sigma Xi. Author: The Human Nervous System, 1967; also numerous articles. Research on nervous system, regeneration and degeneration, evolution, devel. and comparative anatomy brain, developmental and reproductive studies primates including man, gorilla and baboon. Home: 116 7th St., Cresskill, N.J. 07626. Office: 630 W. 168th St., N.Y.C. 10032.*

NOBACK, Gustave J(oseph), Am. anatomist; b. N.Y.C., May 29, 1890; s. Alfred and Marie (Mirejovsky) N.; B.S., Cornell U., 1916; A.M., U. Minn., 1920; Ph.D.,1923; m. Hazel Ogden Kilborn, June 17, 1917; 1 son, Richardson Kilborn. Asst. in histology and embryology Cornell U., 1914-16; mgr. med. dept. Macmillan Co., Chgo., 1917-18; instr. anatomy U. Minn., 1918-21; asso. prof. anatomy, Med. Coll. Va., 1921-22, prof., 1922-24; with N.Y. U., 1924-45, prof. anatomy, 1930-45; chmn. biol. and geol. scis. Grad. Sch.; dean, Essex Coll. Medicine, Newark, 1945; resigned 1945; asso. prof. Cornell Univ. Coll. 1946-50; prof. anatomy and head dept. U.P.R. Sch. Medicine, 1950-53, also chmn.. bd. anatomy; ret., 1953. Fellow A.A.A.S., Am. Geo. Soc., N.Y. Acad. Medicine, (asso.) mem; Assn. Anatomists, Am. Assn. Phys. Anthropologists, Harvey Soc., Soc. Med. Jurisprudence (pres.; Sec. 1936-41), Gerontological Soc. Am. Artists Profl. League (pres.), Biol. Photographers Assn., Soc. for Study of Arterio-Sclerosis, Sigma Xi. Contbr. to jours. Research work in human growth and morphological and physiol. changes incident to birth, especially respiratory and vascular systems, age changes in tissues and organs. Died Sept. 8, 1955.

NOBEL, Alfred Bernhard, Swedish inventor; b. Stockholm, Sweden, Oct. 21, 1833; s. Emanuel Nobel; studied under engr. John Ericsson, U.S., 1850-54; student engring., St. Petersburg, Russia. Worked for his father, St. Petersburg, 1854-59; family returned to Sweden and operated chem. works at Helenborg, 1859-64 (1st commercial prodn. of nitroglycerin); plant destroyed by explosion, 1864; built new works on barge in Lake Mälaren; also in Winterviken, Sweden, Krummel, Germany, 1865, Ardeer, Scotland (became largest dynamite works in world), 1871; formed (with bros. Ludwig and Robert) Nobel Bros. firm operating petroleum wells, Baku, Russia; purchased ordnance works, Bofors, Sweden. With father, assisted in devel. torpedoes and mines; invented dynamite, 1866-67, blasting gelatine (colloidal solution of nitrocellulose in nitroglycerin), 1875; produced ballistite, 1888; inventions in ordnance; studied metallurgy; numerous patents including rubber, silk and leather substitutes. Left fund for establishment of annual awards to men and women who worked for benefit of mankind in fields of chemistry, physics, physiology or medicine, literature, and peace. Died San Remo, Italy, Dec 10, 1896.

NOBEL, Jacobus de, Dutch physicist; b. Velsen, Netherlands, July 11, 1909; s. J. and S.E.J. (Tyssen) de N.; Cand. in Sci., Leiden (Netherlands), 1930, Doct. Sci., 1936, Doctor's Degree, 1954. Asst., Kamerlingh Onnes Lab., Leiden U., 1936-38, researcher Found. Fundamental Research Materials, 1951—, faculty, 1963—, extraordinary lectr., 1965—; tchr. Ned. Lyceum, The Hague, Netherlands, 1938-39, 40-43. Mem. Dutch Phys. Soc. Research, publs. on heat conductivities of steels and other alloys at low temperatures; heat conductivity, elec. resistivity at low temperature in strong magnetic fields of single crystals of tungsten; heat conduction of steels, alloys, transition metals in noble metals and specific heats of such alloys, soft solders and a stainless steel. Home: 111 Mozartstr., Leiden, Netherlands.*

NOBILE, Umberto, Italian engr., explorer; b. Lauro, Italy, Jan. 12, 1885; s. Vincenzo and Maria (La Torraca) N.; Ph.D. in Engring. and Nautical Sci.; m. Carlotta Ferraiolo. Prof. aero. constrns. Naples (Italy) U., prof. emeritus aeronautics, 1960—; mgr. aircraft works; supr. constrn. of airships in USSR, 1932-37. Mem. Pontifical Acad. Sci., Venice Royal Inst. Sci., Italian Aerotech. Assn. Author: Da Roma all' Alaska attraverso il Polo Nord, 1927; L'Italia al Polo Nord, 1930; La preparazione e i risultati scientifici della spedizione polare dell'Italia, 1929; Posso dire la verità, 1945; Quello che ho visto nella Russia societica, 1946. Made (with Amundsen) 1st polar expdn. with a dirigible, 1926; designed and built airships, including Norge and Italia; reached N. Pole twice. Address: 45 via Sabotino, Rome, Italy.

NOBILI, Leopoldo, Italian physicist; b. Reggio nell'Emilia, Italy, 1784; capt. Italian army, 1814; prof. physics Ducal Mus., Florence, Italy, from 1831; mem. French Acad. Scis., 1833. Inventor high sensitivity galvanometer in equilibrium, 1825, thermopile, 1830, also astatic galvnometer not affected by earth's magnetic field; described prismatically-colored metal films deposited electrolytically from lead and other salt solutions (Nobilli's rings), 1826; discovered generating color rings on basis of electrolytic deposits, 1830; studied (with Melloni) heat waves, 1831, also developed thermomultiplier. Died Florence, Aug. 17, 1835.

NOBLE, Sir Andrew, Scottish physicist, b. Greenock, Scotland, Sept. 13, 1831; s. George and Geils Moore (Donald) N.; ed. Edinburgh, (Scotland) Acad., Royal Mil. Acad., Woolwich, Eng.; m. Margary Durham Campbell, 1854; 4 sons, 2 daus. Joined Elswich Ordnance Co., 1861; with Sir W.G. Armstrong, Whitworth & Co., Ltd. (merger Elswich Ordnance Co. and Hydraulic Engring. Works), beginning 1863, became vice chmn., 1882, chmn., 1900. Recipient Royal medal Royal Soc., 1880, Albert medal Royal Soc. Arts, 1909. Author: Artillery and Explosions, 1906. Invented chronoscope device for measuring small time intervals, 1862, chronometer (revolutionized prodn. of gun powder), new types of arty. for army, 1858-60; laid groundwork for modern sci. of ballistics; determined pressures created and analyzed gas and residues left when arty. piece fired; studied pressures needed to produce rotation of projectile in weapons with rifled barrels, 1873; originated (with Abel) pyrostatic law of burning pressure of powder considering covolume. Died Aryllshire, Scotland, Oct. 22, 1915.

NOBLE, Elmer Ray, Am. parasitologist; b. Pyong Yang, Korea, Jan. 16, 1909 (parents Am. citizens); s. William Arthur and Martha (Wilcox) N.; A.B., U. Cal. at Berkeley, 1931, M.A., 1933, Ph.D., 1936; m. Mary Nancy Burnell, May 28, 1932; children—Carolyn (Mrs. Dennis R. Cogan), Elaine (Mrs. Michael Dvortcsak), Ellen (Mrs. Paul McKaskle), David Ray. Instr., Santa Barbara (Cal.) State Coll., 1936-38, asst. prof., 1938-43; faculty U. Cal. at Santa Barbara, 1943—, prof. zoology, 1948—, chmn. dept. biol. scis., 1947-51, dean letters and sci., 1951-58, vice chancellor, 1958-61; vis. research investigator Marine Biol. Lab., Plymouth, Eng., 1955-56, U. Hong Kong, 1962, Bur. Comml. Fisheries, Honolulu, 1962-63. Cons. in higher edn. to govt. Indonesia., 1960. Mem. Am. Soc. Parasitologists, Soc. Protozoologists, Am. Micros. Soc. Author: Parasitology, The Biology of Animal Parasites, 1961, 2d edit., 1964; (with Glenn A. Noble) Laboratory Manual in Parasitology, 1962; also articles. Research in life cycles and morphology of sporozoan parasites of fishes, amoebic parasites of domestic animals, life cycles and nutrition of trypanosome parasites, ecol. and population studies of fish parasites, animal parasitism of deep-sea fishes. Home: 1250 Dover Lane, Santa Barbara, Cal. 93103.*

NOBLE, G. Kingsley, Am. explorer; b. Yonkers, N.Y., Sept. 20, 1894; s. G. Clifford and Elizabeth (Adams) N.; A.B., Harvard, 1917, A.M., 1918; Ph.D., Columbia, 1922; m. Ruth Crosby, Aug. 13, 1921; children—G. Kingsley, Alan Crosby. Leader Harvard expdn. to Guadeloupe, 1914, to Newfoundland, 1915; zoölogist Harvard expdn. to Peru, 1916; leader Am. Mus. expdn. to Santo Domingo, 1922; lectr. vertebrate paleontology Columbia; curator of herpetology Am. Mus. Natural History, from 1919, curator exptl. biology, from 1928. Vis. prof. U. Chgo., 1931, N.Y. U., 1939. Fellow N.Y. Zool. Soc., N.Y. Acad. (v.p. 1929); mem. Ecology Soc., Philos. Soc., Assn. for Study Internal Secretions, other sci. socs. Author: The Biology of the Amphibia, 1931. Asso. editor Jour. Morphology. Research in herpetology, ornithology, neurology, endocrinology, exptl. morphology; studied anatomy and physiology of vertebrates, amphibian life histories. Died Dec. 9, 1940.

NOBLE, Merrill Emmett, Am. psychologist; b. Las Vegas, N.M., July 25, 1923; s. Merrill Emmett and Martha (Van Petten) N.; B.A., N.M. Highlands U., 1947; M.A., Ohio State U., 1949, Ph.D., 1951; m. Joy Lind, July 18, 1953; children—Margaret Lind, Eric Severin. Faculty, Kan. State U., Manhattan, 1954—, prof. psychology, 1961—, chmn. dept., 1962—. Cons. research dept. Menninger Found., 1962—, Greater Kansas City (Mo.) Mental Health Found., 1962—. Mem. bd. Environmental Research Found. Mem. Psychonomic Soc., Am. Psychol. Assn., A.A.A.S., Sigma Xi. Author: Skilled Performance, 1959; contbg. author: Experimental Methods and Instrumentation in Psychology, 1966; also numerous articles. Research on learning and skilled performance in animals and man. Home: 2832 Illinois Lane, Manhattan, Kan. 66502.*

NOBLE, William, physicist; flourished 17th century; chaplain Christ Ch. Coll., Oxford, Eng.; pioneer in study of harmonics of vibrating string; produced (with Pigott) harmonic overtones by matching oscillating note formation with oscillating strings. Died 1681.

NOCARD, Edmond-Isidore-Etienne, French veterinarian, biologist; b. Provins, France, Jan. 22, 1850; prof. pathology Sch. d'Alfort, became dir., 1887; prof. cantagious diseases; a collaborator of Pasteur. Developed method for early diagnosis of glanders in horses by injection of mallein; discovered Bacillus psittacosis which he thought caused psittacosis (later found to be Bacillus aortrycke), bacillus causing bovine glanders; showed that bacillus of avian Tb. is identical to that in mammals, that meat and milk from tubercular cattle could cause infection in man; early user of tuberculin for a diagnosis in cattle; described organism causing pseudotuberculosis in sheep, cattle and horses (Corynebacterium pseudotuberculosis, also Preisz-Nocard bacillus), 1885; studies (with Pierre Roux) on bovine pneumonia which led to discovery of pneumococcus. Died Saint-Maurice, France, Aug. 2, 1903.

NOCE, Robert Henry, Am. neuropsychiatrist; b. Phila., Feb. 19, 1914; s. Sisto Julius and Madeline (Saulino) N.; A.B., Kenyon Coll., 1935; M.D., U. Louisville, 1939; postgrad. U. Pa. Grad. Sch. Medicine, 1947, Langley-Porter Psychiat. Inst., 1949, 52; m. Jane Wolstenholme, Nov. 4, 1957; children—Lee, Lynn. Practice medicine specializing in psychiatry, Modesto, Cal., 1952-65, Watsonville, Aptos, Cal., 1965—, Salinas, Cal., 1966—; mem. staffs Warren (Pa.) State Hosp., 1946-48, Met. State Hosp., Norwalk, Cal., 1948-50; dir. clin. services Pacific State Hosp., Spadra, Cal., 1950-52; asst. supt. psychiat. services Modesto State Hosp., 1952-64; chief psychiatry Monterey County Hosp. Decorated Presdl. Citation; recipient Albert Lasker award, Am. Pub. Health Assn., 1957. Fellow Am. Psychiat. Assn.; mem. A.M.A., Western Soc. Electro-encephalography, Phi Beta Kappa, Delta Phi, others. Research, publs. on reserpine and its use in the mentally ill and retarded. Home: P.O. Box 178, Aromas, Cal. 95004. Office: 30 Brennan St., Watsonville, Cal. 95076.*

NOCKOLDS, Stephen Robert, English geochemist; b. St. Columb Major, Eng., May 10, 1909; s. Stephen and Hilda (Tomlinson) N.; student U. Manchester (Eng.), 1926-30; Ph.D., U. Cambridge (Eng.), 1932; m. Hilda Jackson, July 25, 1932. Asst. lectr. Manchester (Eng.) U., 1932-37; faculty U. Manchester, 1932-37; faculty Cambridge U., 1937—, reader, 1957—, fellow Trinity Coll., 1950—. Fellow Royal Soc., 1959, Geol. Soc. London, Mineral. Soc. Am.; mem. Mineral. Soc. Gt. Britain, Geochem. Soc. Joint editor Geol. Mag., 1950—, Jour. Petrology, 1960—; asso. editor Geochimica et Cosmochimica Acta, 1950—. Publs. on petrological and mineral. investigations of calc-alkali igneous rocks, geochemistry of trace elements in igneous rocks and minerals, petrogenesis of igneous rock series. Home: Old Manor House, Linton, Cambridgeshire, Eng. Office: Dept. Mineralogy and Petrology, Downing Pl., Cambridge, Eng.*

NODDACK, Ida Eva Tacke, German chemist; b. Lackhausen/Wesel, Germany, Feb. 25, 1896; d. Adelbert and Hedwig (Danner) Tacke; Dipl.-Ing., Tech. U. Berlin-Charlottenburg (Germany), 1919, D.Engring., 1921; m. Dr. Walter Noddack, May 20, 1926. Chemist, Allgemeine Electrische Gesellschaft, 1921-23, Siemens & Halske, Berlin, 1924-25, Physikalisch-Technische Reichsanstalt, 1925-35; collaborator Inst. Phys. Chemistry, U. Freiburg (Germany), 1935-41, U. Strasbourg (France), 1947-55; ofcl. Inst. Geochem. Research, Bamberg, Germany, 1956—. Recipient 1st prize dept. chemistry and metallurgy Tech. U. Berlin, 1919, Scheele medal Swedish Chem. Soc., 1934. Mem. Assn. German Chemists (Justus-Liebig medal, 1931); hon. mem. Spanish Soc. Physics and Chemistry, German Acad. Nat. Sci. (Leopoldina Halle), Inst. Soc. Food Research, Fraunhofer Soc. Author: Haüfigkeiten der Elemente; Algegenwart der Elemente; Spurenelemente; also publns. on Rhenium, geochemistry. Discoverer (with husband) of elements Rhenium, and Masurium, 1925. Home: Stegaurach uber Bamberg. Office: 5 Urbanstrasse, Bamberg, West Germany.

NODDACK, Walter Karl Friedrich, German chemist; b. Berlin, Germany, Aug. 17, 1893; s. Oskar and Anna (Seiler) Noddack; ed. U. Berlin, also Bergakademie, 1920; m. Ida Tacke, 1926. Became asst. Inst. for Phys. Chemistry, U. Berlin, 1922; govt. councillor Physikalisch Technische Reichsanstalt, Berlin, 1922-35; prof., dir. Inst. for Phys. Chemistry, U. Freiberg, 1935-41, U. Strasbourg (France), 1941-46; prof. chemistry Philosophische Theologische Hochschule, Bamberg, Germany, 1946-57; dir. Ofcl. Research Inst. for Geochemistry, Bamberg, 1957—. Recipient Gold medal Philosophy Faculty, U. Berlin, 1920, Justus Liebig Commemorative medal Assn. German Chemists; Scheele medal Swedish Chem. Soc., 1934; Bundesverdientskreuz 1st class, 1959. Mem. German Acad. Natural Scientists Leopoldina Halle, Soc. German Chemists. Contbr. numerous articles to tech. jours. Discovered (with wife) elements masurium and rhenium. Died Bamberg, Dec. 7, 1960.

NOÉ, Adolf Carl, paleobotanist; b. Graz, Austria, Oct. 28, 1873; s. Adolf Gustav and Marie (Krauss) von Noé; student U. Graz, 1894-94 Ph.D. (hon.), 1923; student U. Göttingen, 1897-99; A.B., U. Chgo., 1900, Ph.D., 1905; hon. mem. Innsbruck, 1922; m. Mary Evelyn Cullaton, July 3, 1901; children—Mary Helen, Valerie. Demonstrator in paleobotany, U. Graz, 1895-97; came to U.S., 1899, naturalized, 1904; instr. science and modern langs. Burlington Inst., 1901-02; instr. German, Stanford, 1901-03; instr., asst. prof. German lit., U. Chgo., 1903-23, asst. prof. paleobotany, 1923-24, asso. prof., from 1924. Geologist Ill. State Geol. Survey, from 1921, Ia. Geol. Survey, 1923-25, Ky. Geol. Survey, 1922; mem. Allen & Garcia Coal Commn. to Soviet Russia, 1927. Recipient Golden medal U. Vienna, 1923. Author: Fossil Flora of Northern

Illinois, 1926; Golden Days of Soviet Russia, 1931; Ferns, Fossils and Fuels, 1931. Research on geology of coal, stratigraphy of coal measures, geology of Ill., Ia., Tex., Mexico, Donetz Basin, Russia, morphology of fossil plants, carboniferous flora of Am. Died Apr. 10, 1939.

NOEGGERATH, Emil Oscar Jacob Bruno, German gynecologist, obstetrician; b. Bonn, Germany, Oct. 5, 1927; s. Jacob John and (Primavesi) N.; M.D., U. Bonn., 1852; passed spl. state exams., Berlin, 1853; m. Rolonda; 1 dau., 3 sons. Became 1st asst. obstet. clinic, Bonn, 1854; began pvt. practice Neuwied-on-the-Rhine, 1856; came to St. Louis, 1857; gen. practice, N.Y.C.; staff German Hosp., Mt. Sinai Hosp.; prof. obstetrics and gynecology N.Y. Med. Coll.; returned to Germany, 1885. Mem. Am. Gynecol. Soc. (a founder), N.Y. Obstet. Soc. (a founder). Author: Die Latente Gonorrhoe im weiblichen Geschlect, 1872; (with Jacobi) Contributions to Midwifery, 1859. Founder (with Jacobi), Jour. Obstetrics, co-editor, 1868-71; One of 1st to use pathology to examine uterine tissues removed by curettage; developed surg. aseptic technique for gynecol. exams; originated new method in ovariotomy; demonstrated importance of latent gonorrhea in fertility of women, 1876. Died Wiesbaden, Germany, May 3, 1895.

NOEL, Nicholas, physician; b. Reims, France, Mar. 27, 1746; ed. U. Paris; blacksmith's apprentice; came to U.S., 1776; surgeon Gen. Nathan Greene's div.; surgeon to frigate Boston; organized mil. hosps., Phila.; returned to France, 1784; house surgeon, hosp. in Reims; insp. mil. hosps., Belgium, 1793-5; (founder acad. medicine, botanic garden, Reims. Author: Journal d'un chirurgien pendant la guerre pour l'independence des colonies Anglaises de l'Amerique du Nord, 1787; Traité historique et pratique de l'inoculation, 1789; Dissertation sur la nécessité de réunir les connaissances chirurgicales et médicales, 1804. Died Reims, May 11, 1832.

NOE-NYGAARD, Arne, Danish geologist; b. Ribe, Denmark, July 30, 1908; s. Soren and Ingrid (Iversen) Noe-Nygaard; M.Sc., Copenhagen U., 1933, Dr. phil., 1937; m. Ellinor Bro Larsen, Nov. 3, 1938; children—Nanna (Mrs. Henrik Jeppesen), Karsten, Bodil. Chief boringarchive Geol. Survey Denmark, 1938-40, dept. geologist, 1940-42; prof. mineralogy U. Copenhagen, dir. U. Mineral. Mus., 1942—. Bd. mem. Rask-Orsted Found., Carlsberg Found., Artic Inst. Denmark. Mem. royal socs. in Copenhagen, Oslo, Stockholm. Author 2 textbooks. Research, publs. on regional geol. work in East Greenland, 1929-34; petrological work in basaltic volcanics belonging to N. Atlantic Plateau Basalt region. Home: No. 26, Nor. Stranding, Helsingor, Denmark. Office: Ostervoldgade 5-7, Copenhagen, Denmark.*

NOER, Rudolf J., Am. surgeon; b. Menominee, Mich., Apr. 25, 1904; s. Peter Juul and Ellen Marion (Peabody) N.; A.B., U. Wis., 1924; M.D., U. Pa., 1927; m. Anita May Showerman, June 22, 1933; children—Richard Juul, Philip Douglas. Practice gen. medicine, Wabeno, Wis., 1929-32; asst. in anatomy U. Wis., 1932-34, resident in surgery, 1934-37; from fellow to prof. surgery Wayne State U. Coll. Medicine, 1937-52, asst. dean, 1949-52; prof. surgery, chmn. dept. U. Louisville Sch. Medicine, 1952—, chief dept. surgery Louisville Gen. Hosp., 1952—; surgeon-in-chief Childrens Hosp., Louisville, 1961—; sr. cons. surgery Louisville VA Hosp.; area cons. surgery Columbus area VA. Diplomate Am. Bd. Surgery. Fellow A.C.S. (1st v.p. 1964-65), Am. (v.p. 1963-64), So. (1st v.p. 1965-66) surg. assns.; mem. Western, Central (pres. 1955) surg. assns., Am. Assn. Surgery Trauma (pres. 1963-64), Am. Assn. Anatomists, Internat. Surg. Soc. Author articles on intestinal obstruction, pancreatitis, diverticular disease colon, cancer of breast. Editor: Jour. Trauma. Home: 224 Travois Rd., Louisville 40207.*

NOETHER, Emmy, mathematician; b. Erlangen, Germany, Mar. 22, 1882; m. Max Noether; prof., Göttingen, Germany, Bryn Mawr, Pa. Research in theory of commutative fields, ideal theory; applied arithmetical concepts to some linear algebras. Died Bryn Mawr, Pa., Apr. 14, 1935.

NOETHER, Gottfried Emanuel, math. statistician; b. Germany, Jan. 7, 1915; s. Fritz A. and Regina (Würth) N.; came to U.S., 1939, naturalized, 1942; B.A., Ohio State U., 1940; M.A., U. Ill., 1940; Ph.D., Columbia, 1949; m. Emiliana Pasca, Aug. 1, 1942; 1 dau., Monica Gail. Instr., N.Y. U., 1949-51; faculty Boston U., 1951-68, prof., 1957-68; head dept. statistics U. Conn., Storrs, 1968—. Recipient Fulbright award, Tübingen, Germany, 1957-58, Vienna, Austria, 1965-66. Fellow A.A.A.S., mem. Am. Statis. Assn., Inst. Math. Statistics; mem. Math. Assn. Am. Author: Guide to Probability and Statistics, 1961; Elements of Nonparametric Statistics, 1967; also articles. Research in theory and methodology of nonparametric statistics. Address: Dept. Statistics, U. Conn., Storrs, Conn. 06268.*

NOETHER, Max, German mathematician; b. Mannheim, Germany, Sept. 24, 1844; ed. Heidleberg, Giessen, Göttingen (both Germany); became prof. math. U. Erlangen (Germany), 1875; dir. Mathematische Annalen. Mem. French Acad. Scis., 1903. Author: Zur Grundlegung der Theorie der Algebraischen Raumkurven, 1882; Bericht über der Entwicklung der algebraischen Funktionen, 1894; also biographies of Sylvester, Birdsch, Sophuslie, C.H. Nermite. Studies on devel. theory of algebraic functions. Died Erlangen, 1921.

NÖGGERATH, Johann Jakob, German geologist, mineralogist, seismologist; b. Bonn, Germany, Oct. 10, 1788; commr. mines, various Rhine provinces, 1814-15; named prof. mineralogy U. Bonn, later prof. geology, dir. mus. natural history, chief, mining dept. Carboniferous plant Noeggerathia named in his honor. Author: Über aufrecht im Gebirgsgestein eingeschlossene fossile Baumstämme und andere Vegetabilien, 1819-21; Das Gebirge in Rheinland-Westphalen, nach mineralogischem chemischem Bezuge, 4 vols., 1822-26; Die Enstehung der Erde, 1843; Der Laacher See und seine vulkanischen Umgebungen, 1870. Died Bonn, Sept. 13, 1877.

NOGGLE, Glenn Ray, Am. botanist; b. New Madison, O., July 25, 1914; s. Glenn J. and Jennie (Ray) N.; A.B., Miami U., Oxford, O., 1935; M.A., U. Ill., 1942, Ph.D., 1945; m. Ruth Chambers, June 30, 1945; 1 dau., Jean. Research asso. agronomy U. Ill., 1945-46; asst. prof. plant physiology U. Va., 1946-48; sr. scientist biology div. Oak Ridge Nat. Lab., 1948-52; biochemist charge photosynthesis project So. Research Inst., Birmingham, Ala., 1952-54; biochemist C. F. Kettering Found., also asso. prof. Antioch Coll., 1954-57; prof. botany, Univ. Fla., also botanist, head dept. Agr. Exptl. Sta., 1957-64; prof., head dept. botany and bacteriology N.C. State U., Raleigh, 1964—. Mem. Am. Soc. Plant Physiologists (exec. sec.-treas. 1956-60), Assn. Southeastern Biologists (rep. on A.A.A.S. council 1958-61), Assn. Tropical Biologists, A.A.A.S., Bot. Soc. Am., Sigma Xi. Research, publs. on role of inorganic nutrients in growth, devel. plants, also synthesis and separation of carbon-14 labeled carbohydrates. Home: 2346 Churchill Rd., Raleigh, N.C. 27608.*

NOGUCHI, Hideyo, bacteriologist; b. Inawashiro, Japan, Nov. 24, 1876; apprentice to Dr. Kanae Watanbe, physician in Wakamatsu, Japan; went to Tokyo, 1896; passed govt. exam. for med. practice 1897; studied under Dr. Morinosuke Chiwaki, also Dr. Susumu Sato, Juntendo Hosp.; M.D., Sc.D. Went to Newchuang, China as quarantine doctor when an epidemic of bubonic plague broke out, 1899; came to U.S., 1900; joined lab. pathology U. Pa., 1900; mem. staff Rockefeller Inst., 1904 until his death. Mem. Imperial Acad. Author: The Action of Snake Venom upon Cold-blooded Animals, 1904. Confirmed (with Simon Flexner) that Treponema pallidum causes syphilis, 1910; discovered spirochete in brain of dementia paralytica patient, 1913; began studying yellow fever in S. America, 1918; studied Oroya fever and verruga peruana of Peru and Ecuador, showing them to be caused by the same organism, a species of Bartonella; also studied viruses, snake venom, toxins. Died Brit. W. Africa, May 21, 1928.

NOLAN, Patrick Joseph, Irish physicist; b. Tyrone, Ireland, Aug. 11, 1894; s. Martin and Bridget (Owens) N.; B.Sc., U. Coll., Dublin, Ireland, 1914, M.Sc., 1915, postgrad.; Ph.D., Cavendish Lab., Cambridge U. (Eng.), 1920; D.Sc., Nat. U. Ireland, 1930; m. Una Hurley, Apr. 18, 1922. Faculty, U. Coll. Dublin, 1920-64, prof. geophysics, 1953-64. Mem. governing bd. Sch. Cosmic Physics, Dublin Inst. for Advanced Studies, 1950—; mem. Nat. Com. for Geodesy and Geophysics, 1950—. Mem. Royal Irish Acad. Research, numerous publs. on atmospheric electricity and condensation nuclei, equilibrium atmospheric ionization, measurement diurnal variation ion and nucleus concentration in atmosphere, air-earth current measurements; discovered extremely small nuclei in corona discharge; co-inventor photo-electric counter for measuring concentration condensation nuclei. Home: 35 Eglinton Rd., Donnybrook, Dublin 4, Ireland.*

NOLAN, Thomas Brennan, Am. geologist; b. Greenfield, Mass., Mar. 21, 1901; s. Frank Wesley and Anna (Brennan) N.; Ph.B. in Metallurgy, Yale, 1921, Ph.D. in Geology, 1924; LL.D. (hon.), U. St. Andrews, Fife, Scotland, 1962; m. Mabelle Orleman, Dec. 3, 1927; 1 son, Thomas Brennan. Geologist, U.S. Geol. Survey, 1925-44, asst. dir., Washington, 1944-56, dir. 1956-65, research geologist, 1965—. Recipient Spendiaroff prize Internat. Geol. Congress, 1933; K.C. Li medal and prize Columbia, 1954; Rockefeller Pub. Service award Princeton, 1961; Silver medal Tokyo Geog. Soc., 1965. Fellow Geol. Soc. Am. (v.p. 1960, pres. 1961), Soc. Econ. Geologists (v.p. 1945, pres. 1950), Mineral. Soc. Am., Am. Geophys. Union, Royal Soc. Edinburgh (hon.); mem. Geol. Soc. London (fgn. mem.), Internat. Union Geol. Scis. (v.p. 1965), Nat. Acad. Scis., Am. Philos. Soc., Am. Acad. Arts and Scis., Am. Inst. Mining, Metall. and Petroleum Engrs., Geochem. Soc., Mineral. Assn. Can., Geol. Soc. Washington, Mining and Metall. Soc. Am., Sigma Xi. Contbr. articles and reports to profl. and govtl. jours. Home: 2219 California St. N.W., Washington 20008. Office: U.S. Geol. Survey, Washington 20242.*

NOLAND, Edward William, Am. sociologist, educator; b. Romney, W.Va., Dec. 30, 1910; s. Edward Wilson and Eliza Elizabeth (Yost) N.; diploma Potomac State Coll., 1928; A.B., W.Va. U., 1930, M.A., 1935; postgrad. U. Chgo.; Ph.D., Cornell U., 1944; m. Minnie Merle Martin, Aug. 10, 1954. Chmn. dept. math. Alleghany High Sch., Cumberland, Md., 1932-40; instr. sociology and statistics Cornell U., 1941-44, asst. prof., 1945; asso. dir. research Yale Labor and Mgmt. Center, 1945-46; v.p. R.S. Dickson & Co., Charlotte, N.C., 1946-49; coordinator indsl. relations Am. Yarn & Processing Co., Mt. Holly, N.C., 1946-49; prof. sociology and statistics, research prof. Inst. Research Econs. and Bus., U. Ia., 1949; prof., chmn. dept. sociology and anthropology, research prof. Inst. Research in Social Sci., lectr. exec. program Sch. Bus. Adminstrn., U. N.C., Chapel Hill, 1949-63, chmn. div. social scis., 1956-62, Kenan prof. sociology, 1961-63; prof., head dept. sociology Sch. Humanities, Social Sci. and Edn., prof. administrv. scis. Krannert Grad. Sch. Indsl. Adminstrn., Purdue U., West Lafayette, Ind., 1963—. Lectr., vis. prof. Am., fgn. univs.; mem. com. on math. in social scis. Social Sci. Research Council, Dartmouth Coll., summer 1952. Mem. Am., So. (pres. 1959-60) sociol. socs., Am. Statis. Assn., Regional Sci. Assn., Sociol. Research Assn., Theta Chi, Omega Delta Gamma, Kappa Delta Pi, Alpha Kappa Delta, Sigma Phi Omega. Author: (with E.W. Bakke) Workers Wanted, 1949; (with R.P. Calhoon, Whitehill) Human Relations in Management, 1958; also articles. Asso. editor Social Forces, 1949-63. Home: 2114 S. 7th St., Lafayette, Ind.*

NOLAND, Lowell E(van), Am. zoologist; b. Lee, Ind., July 15, 1896; s. David Evan and Mary Emma (Matthews) N.; B.A., DePauw U., 1917; student U. of Montpellier, France, 1919, Marine Biol. Lab., Woods Hole, Mass., 1922; M.A., U. Wis., 1921, Ph.D., 1924; m. Ruth Wayland Chase, Sept. 6, 1923; children—Wayland Evan, Ruth Mary. Editor and pub. Rolling Prairie (Ind.) Record, 1912-13; high sch. tchr., Mitchell, Ind., 1917-18, 1919-20; asst. in zoology, U. Wis., 1920-21, instr., 1921-25, asst. prof., 1925-31, asso. prof., 1931-35, prof. 1935-66, prof. emeritus, 1966—, chmn. dept. zoology, 1945-48, prof. integrated liberal studies 1948-66; research, U. S. Bur. Fisheries Sta., Beaufort, N.C., and Bass Biol. Lab., Englewood, Fla., 1934; vis. prof., U. Hawaii, Honolulu, 1944. Fellow A.A.A.S.; mem. Am. Soc. Zoologists, Micros. Soc. (pres. 1939), Am. Soc. Naturalists, Soc. Protozoologists (pres. 1954-55), Soc. Systematic Zoology, Société des Amis de la Maison Française, Phi Beta Kappa, Sigma Xi, Gamma Alpha, Phi Sigma, Tau Kappa Epsilon. Author lab. manual on fetal pig; also publs. on classification and environmental relations of one-celled animals, especially ciliates. Home: 1723 Summit Av., Madison, Wis. 53705.*

NOLAND, Wayland Evan, Am. chemist; b. Madison, Wis., Dec. 8, 1926; s. Lowell Evan and Ruth (Chase) N.; B.A., U. Wis., 1948; M.A., Harvard, 1950, Ph.D., 1952. DuPont fellow U. Minn., Mpls., 1951-52, faculty, 1952—, prof. chemistry, 1962—. Cons. petroleum tech. Sun Oil Co., Marcus Hook, Pa., 1958—. Recipient Distinguished Teaching award U. Minn. Inst. Tech., 1964. Mem. A.A.A.S., N.Y., Wis. acads. sci., Arts and Letters, Am. Chem. Soc., Am. Assn. U. Profs., Phi Beta Kappa, Sigma Xi, Phi Kappa Phi, Phi Lambda Upsilon, Phi Eta Sigma, Alpha Chi Sigma, Gamma Alpha. Research, numerous publs. on chemistry organic nitrogen compounds, synthesis serotonin, synthetic methods, new rearrangements, reaction mechanisms, structure determination. Home: 820 Washington Av., S.E., Mpls., 55414.*

NOLL, Hans, biologist; b. Basel, Switzerland, June 14, 1924; s. Johannes Ernst and Hedwig (Staehelin) N.; student Denmark Poly. Inst., U. Copenhagen, 1948-49; Ph.D., U. Basel, 1950; m. Johanne Bindesboll, Jan. 20, 1949; children—Kirsten, Lucas, Barbara, Elizabeth. Came to U.S., 1951, naturalized, 1957. Asso. div. Tb, Pub. Health Research Inst., N.Y.C., 1954-56; faculty U. Pitts., 1956-64; prof. biol. scis. Northwestern U., Evanston, Ill., 1964—. Mem. study sect. NIH; cons. Ill. div. Am. Cancer Soc. Recipient Lifetime Career award NIH, USPHS, 1964. Sr. Research fellow NIH, USPHS, 1959-63, Career prof. Am. Cancer Soc., 1966. Mem. Am. Chem. Soc., Am. Soc. Biol. Chemists, A.A.A.S., N.Y. Acad. Scis., Biochem. Soc. (London), Am. Soc. Microbiology. Research, numerous publs. on chem. structure of cord factor, toxic principle of tubercle bacilli; isolated and identified vitamin K2(45); discovered polyribosomes, multiple ribosomal structures responsible for protein biosynthesis; established tape mechanism of protein synthesis. Home: 2665 Orrington Av., Evanston, Ill. 60201.*

NOLL, Walter, mathematician; b. Berlin, Germany, Jan. 7, 1925; s. Franz and Martha (Janssen) N.; Diplom-Ingenieur, Technische Universität, Berlin, 1951; Licencié ès sciences, U. Paris (France), 1950; Ph.D., Ind. U., 1954; m. Helga I. Schoenberg, Apr. 1, 1955; children—Virginia, Peter. Came to U.S., 1955, naturalized, 1961. Wissenschaftlicher asst. Technische Universität Berlin, 1951-55; instr. U. So. Cal., 1955-56; faculty Carnegie Mellon U., Pitts., 1956—, prof. math., 1960—; vis. prof. Johns Hopkins, 1962-63. Mem. Soc. for Natural Philosophy (founding), Am. Math. Soc., Math. Assn. Am., Sigma Xi. Author: (with C. Truesdell) The Non-Linear Field Theories of Mechanics, 1965; (with B.D. Coleman,

H. Markovitz) Viscometric Flows of Non-Newtonian Fluids, 1966; also articles. Developed gen. math. theory for describing mech. and thermodynamical behavior deformable bodies; research math. structure space and time in classical and relativistic physics. Home: 308 Field Club Ridge Rd., Pitts. 15238.*

NOLL, Walter Friedrich Heinrich, German chemist; b. Iena, Germany, Mar. 9, 1907; s. Alfred and Hedwig (Ritthausen) N.; ed. univs. Iena and Heidelberg (Germany); Ph.D.; m. Eva Maria Rohde; children—Klaus-Dieter, Monika-Renate. Research dir. U. Göttingen (Germany); asst., instr. Hanover (Germany) Tech. Coll.; collaborator, div. head Farbenfabriken Bayer Leverkusen (Germany), 1937—. Author: Chemie und Technologie der Silicone, 1960; also publns. on chemistry of crystals, geochemistry, physicochem. mineralogy, mineral synthesis, silicates and silicones. Home: Westkott-Strasse 8. Office: Farbenfabriken Bayer, Anorganische Abteilung, Leverkusen, West Germany.

NOLLA, Jose Antonio Bernabe, Puerto Rican phytopathologist; b. Camuy, P.R., May 24, 1902; s. Jose and Rosa (Cabrera) N.; B.S.A., U. P.R., 1923; M.S., Cornell U., 1926, Ph.D., 1932; m. Olga R. de Arellano, Aug. 21, 1937; children—Olga, Carlos Conde III, Jose A. M. Agrl. agt. P.R. Dept. Agr., 1923-24; asst. plant pathologist P.R. Agrl. Exptl. Sta., 1924-26; plant breeder, 1927-30; asst. commr. agr. P.R., 1933-38; dir. Tobacco Inst., P.R., 1936-38; dir. U. Agr. Exptl. Sta., 1938-43; field supt. Ramirez de Arellano Sugar Cane Estates, Mayaguez, P.R., 1943-54; dir. Central Igualdad Sugar Co., Mayaguez, 1946—; prof. agronomy Coll. Agr., U. P.R., Mayaguez, 1958—. Mem. agrl. mission Cauca Valley, Colombia, S. Am., 1928; spl. adviser Ministry Agr., Venezuela, 1936; sugar cane cons. AID, Ivory Coast, Africa, 1963. Mem. Sigma Xi, Gamma Sigma Delta, Alpha Theta, Beta Beta Beta. Contbr. articles to tech. jours. on tobacco diseases, disease resistance, inheritance of color in eggplant, inheritance resistance to tobacco mosaic virus. Home: P.O. Box 820, Mayaguez, P.R., 00708.*

NOLLER, Carl Robert, Am. chemist; b. St. Louis, Nov. 10, 1900; s. Charles and Mary (Laessig) N.; B.S., Washington U., St. Louis, 1922, M.S., 1923; Ph.D., U. Ill., 1926; m. Edna Amelia Rasmussen, Apr. 16, 1927; children—Robert Marcus, Eric Charles, Martha Louise. Instr., U. Ill., 1926, Northwestern U., 1926-27; chemist Eastman Kodak Co., 1927-29; prof. chemistry, Stanford, 1929—; vis. lectr. Harvard, 1938-39; acting head dept. chemistry Stanford, 1959-60. John Simon Guggenheim Meml. fellow, 1930. Mem. Am. Chem. Soc., A.A.A.S., Am. Assn. U. Profs. Author: Chemistry of Organic Compounds, 1951, 3d edit., 1965; Textbook of Organic Chemistry, 1951, 3d edit., 1966; Structure and Properties of Organic Compounds, 1962; also numerous articles. Research on structure, chem. properties organic compounds including natural products, Grignard reagents, medicinals. Home: 13060 La Paloma Rd., Los Altos Hills, Cal. 94022. Office: Dept. Chemistry, Stanford, Stanford, Cal. 94305.*

NOLLET, Jean Antoine, French physicist; b. Pimpré, France, Nov. 19, 1700; student, Clermont, Beauvais, Paris (all France); studied under Dufay. Began teaching exptl. physics, 1735; prof. physics, Turin, Italy; lectr., Bordeaux, Versailles (both France); named prof. Royal Coll., Navarre, France, 1753; became prof. physics, Paris, 1738; became tchr. arty. sch., Mézières, 1761. Fellow Royal Soc., 1734; mem. French Acad. Scis. Author: Leçons de physique expérimentale, 1743; L'Art des experiences, 3 vols., 1770. Discovered osmosis, 1748; introduced Leyden jar into France and improved it; improved electric machine and invented 1st electroscope, 1747; described lens-grinding machine; observed and described double exchange in solutions, 1748; observed appearance of condensation when vessel is evacuated; studied steam and showed sound's pitch is not changed in fluids; 1st to notice significance of sharp points on elec. conductors. Died Paris, Apr. 12, 1770.

NOMOTO, Otohiko, Japanase physicist; b. Tokyo, Japan, Jan. 20, 1912; s. Otoki and Turu (Wakimizu) N.; B.Sc., Tokyo U., 1934, D.Sci., 1944. Asst., Mil. Sci. Research Lab., Tokyo, 1934-37; lectr. Army Gunnery Coll., Tokyo, 1937-38; scientist Army Weather Bur., Tokyo, 1938-39; research mem. Kobayasi Inst. Phys. Research, Tokyo, 1939—; prof. physics Tokyo U. Agr. and Tech., Tokyo, 1949-65; prof. applied physics Def. Acad., Yokosuka, Japan, 1965—. Mem. Phys. Soc. Japan, Acoustical Soc. Japan, Am. Phys. Soc., Acoustical Soc. Am., I.E.E.E. Editor: (with J. Saneyoshi, Y. Kikuchi) Ultrasonics Handbook, 1966. Research, numerous publs. on ultrasonic physics including diffraction of light by ultrasonic waves, visualization of ultrasonic waves, velocity, dispersion and absorption of sound in liquids and gases, molecular mechanisms of sound propagation in fluids, mechanisms of phys. effects of ultrasonics. Home: 2-37-24 Izniu, Suginamiku, Tokyo, Japan. Office: 1-10-20 Hasirimizu, Yokosuka, Kanagawaken, Japan.*

NOMURA, Danji, Japanese biochemist; b. Japan, Dec. 6, 1920; s. Shoichi and Turuyo Nomura; grad. Kyoto U., 1943, Ph.D., 1957; m. Ikue Nomura, June 6, 1947; children—Miyako, Manabu, Osamu, Hiroshi. Asst., Kyoto, U., 1947-50; asst. prof. Yamaguchi U., 1950-62; fellow Stockholm (Sweden) U., 1959-60; prof. Yamaguchi U. (Japan), 1962—, head prof. dept. agr. and food chemistry, 1964—; lectr. Shimonsek Woman Coll., 1962—, Ube Coll. Food Sci., 1961—. Mem. Agr. Chem. Soc. Japan, Japanese, Am. chem. socs. Author: (with S. Adachi, K. Yamfuji, Hans von Euler) Grundlagen der Redukton-Chemie und Biochemische Ergebnisse an Ascorbin saure; also articles. Research on citrus juice which established industry citrus juice in Japan; introduced reductone chemistry to Japan; discovered citrus debittering enzyme, naringinase. Home: 6-29 Yudaonsan, Yamaguchi, Yamaguchi-Ker, Japan.

NONNE, Max, German neurologist; b. Hamburg, Germany, Jan. 13, 1861; prof., Hamburg. Author: Syphilis und Nervensystem, 1902. Introduced Nonne-Apelt reaction into diagnostics; described elephantiasis congenita hereditaria (Nonne-Milroy-Meige disease, or hereditary lyphedema), 1891; credited with 1st description of hereditary edema of legs (Milroy's, or Meige's, disease), 1891. Died Aug. 12, 1939.

NOOJIN, Ray Oscar, Am. physician; b. Birmingham, Ala., Nov. 29, 1912; s. Ray Oscar and Lillian (Dunnam) N.; A.B., U. Ala., 1933; M.D. (Acad. scholar), U. Chgo., 1937; m. Martha Elizabeth Gunning, May 12, 1938; children—Barbara Dade (Mrs. Richard Lee Walthall), Ray Oscar, Martha Elizabeth. Practice medicine, specializing in dermatology, Birmingham, 1945—; prof., chmn. dept. dermatology U. Ala. Med. Center, 1945—. Recipient W.B. Saunders prize in pathology and bacteriology, 1935; Williams and Wilkins prize in clin. pathology, 1936. Diplomate Am. Bd. Dermatology (mem. bd.). Fellow Gorgas Soc.; mem. A.M.A., So. Med. Assn., Southeastern Dermatol. Soc. (past pres.), Am. Acad. Dermatology and Syphilology (past dir.), Soc. for Investigative Dermatology, Am. Dermatol. Assn. (past dir.), Phi Beta Kappa, Sigma Xi, Alpha Omega Alpha, Alpha Epsilon Delta, Gamma Sigma Epsilon. Contbr. chpt. to Pediatric Therapy, 1964. Editor: Dermatology For Students, 1962. Research, publs. in clinical dermatology and basic sci. as related to dermatology. Home: 3425 Argyle Rd., Birmingham 35213. Office: 2661 10th Av. S., Birmingham, Ala. 35205.*

NOORDEN, Carl von, see von Noorden.

NORBECK, Edward, anthropologist; b. Prince Albert, Sask., Can., Mar. 18, 1915; s. Gabriel and Hannah (Norman) N.; came to U.S. 1923, naturalized 1941; student State Coll. Wash., 1936-37; B.A. Oriental Lang. and Lit., U. Mich., 1948, M.A. Oriental Civilizations (U. fellow), 1949, Ph.D. Anthropology, 1952; m. Margaret Roberta Field, Feb. 18, 1950; children—Hannah Field, Edward Crosby, Seth Peter. From instr. to asst. prof. U. Utah, 1952-54; asst. prof. U. Cal., Berkeley, 1954-60; vis. prof. Tokyo (Japan) U., 1959; asso. prof. anthropology Rice U., Houston, 1960-62, prof., 1962—, chmn. dept. anthropology, sociology, 1960—, dean humanities and social scis., 1966-67, dir. Center for Research in Social Change and Econ. Devel., 1966—, dir. New World Symposium on Archaeology, 1962, lectr. anthropology NSF, 1961--63, 64-68. Cons. to various welfareinsts, Houston, 1960—. NSF sr. postdoctoral fellow, 1958-59; numerous research grants from founds.; univs. Fellow Am. Anthropol. Assn.; mem. Am. Ethnol. Soc., Soc. for Asian Studies, Japanese Soc. for Ethnology, others, Phi Beta Kappa, Phi Kappa Phi, Sigma Xi. Author: Takashima, A Japanese Fishing Community, 1954; Pineapple Town—Hawaii, 1959; Religion in Primitive Society, 1961; Changing Japan, 1965. Co-editor, contbr. Prehistoric Man in the New World, 1964; The Study of Personality: An Interdisciplinary Approach, 1968; editor Rice Univ. Studies, 1963—. Sociocultural research on Japan and Hawaii; also studies on cultural aspects of religion. Home: 2420 Locke Lane, Houston, Tex. 77019.*

NORBERG, Richard Edwin, Am. physicist; b. Newark, Dec. 28, 1922; s. Arthur Edwin and Melita (Roefer) N.; B.A., DePauw U., 1943; M.A., U. Ill., 1947, Ph.D., 1951; m. Patricia Ann Leach, Dec. 27, 1947; children—Karen Elizabeth, Craig Allan, Peter Douglas. Research asso., control systems lab. U. Ill., 1951-53, asst. prof., 1953; vis. lectr. physics Washington U., St. Louis, 1954—, mem. faculty, 1955—, prof. physics, 1958—; chmn. dept. 1963—. Fellow Am. Phys. Soc. Research, publs. on application of nuclear magnetic resonance to study of metals and other simple solids, and liquid helium. Home: 7134 Princeton St., St. Louis 63130.*

NORCIA, Leonard Nicholas, Am. biochemist; b. Mountain Iron, Minn., Jan. 1, 1916; s. Frank and Raffaela (Gliatta) N.; B. Chemistry, U. Minn., 1946, Ph.D. in Physiol. Chemistry, 1952; m. Violet A. Rohrer, June 2, 1950; 1 dau., Hope Irene. Research fellow Hormel Inst., U. Minn., Austin, 1952-55; biochemist Okla. Med. Research Found., Oklahoma City, 1955-60; asst. prof. research biochemistry U. Okla., Sch. Medicine, Oklahoma City, 1956-58, asso. prof., 1958-60; asst. prof. biochemistry Temple U. Sch. Medicine, Phila., 1960—. Mem. Am. Chem. Soc., Am. Oil Chemists Soc., Am. Heart Assn., Sigma Xi, Alpha Chi Sigma. Research, publs. in intestinal absorption and endogenous excretion of lipids, transport of lipids in body fluids, tissue lipid composition as influenced by diet or altered homeostasis, cholesterol metabolism and auto-oxidation, biol. antioxidants. Home: 413 Roberts Av., Glenside, Pa. 19038. Office: Dept. Biochemistry, Temple U., Phila. 19140.*

NORDAL, Arnold Sofus, Norwegian chemist; b. Syvde, Norway, May 19, 1909; s. Jakob Kornelius and Anne (Eline) N.; grad. in pharmacy U. Oslo (Norway), 1936, Dr.Uhilos., 1947; m. Alma Margrete Ringstad, Dec. 31, 1939; children—Per Erik, Helge Jostein, Anne Elisabeth. Faculty, U. Oslo, 1940—, prof. pharmacology, 1948—; research biologist Radiation Lab., U. Cal. at Berkeley, 1953-54, research biologist Scripps Inst. Oceanography, La Jolla, 1966-67. UN adviser to Burma Pharm. Industry Bd., Rangoon, 1957-59, 61, Govt. of Afghanistan, Kabul, 1959; mem. Scandinavian Pharmacopoea Commn., Norwegian Pharm. Commn.; participant UN Internat. Program Opium Research. Mem. Norwegian Acad. Scis. Author: A Pharmacognostical Study of Sedum acre, 1946. Research and publs. in carbohydrates (heptuloses), organic acids and their relation to photosynthesis, biosynthesis of organic compounds and metabolism of succulent plants; popular medicine, cultivation of medicinal plants. Home: 15 Bjerkasen, Blommenholm, Norway. Office: U. Oslo, Blindern, Oslo 3, Norway.*

NORDÉN, Ake, Swedish physician; b. Lund, Sweden, Aug. 4, 1915; s. Johan Emil and Karine (Montén) N.; med. lic., Med. Sch., U. Lund, 1943, M.D., 1951; m. Kerstin Maria Rhedin, Dec. 21, 1941; children—Jan Henrik, Knut Fredrik. Rockefeller Found. fellow Sch. Medicine, Duke, 1948-50; asso. prof. medicine Lund U. Sch. Medicine, 1955-65, investigator clin. hematology Swedish Med. Research Council, 1965—. Author: Sporotrichosis, 1951; Mycobacterium balnei, 1954; (with T. Karlefos) Studies on Iron Dextran Complex, 1958; also numerous articles. Research on chronic infections, blood diseases, population studies on diabetes; discovered cause of swimming pool disease. Home: 34 Neptunusgatan, Lund, Sweden.*

NORDEN, John, English topographer; b. 1548; apptd. by Queen Elizabeth to travel through and map Eng. and Wales; surveyor crown woods and forests, including those of Berkshire, Devonshire, Surrey, 1600-18; worked in Ireland, circa 1605; surveyed Windsor and surrounding area, 1607; surveyor, duchy of Cornwall, 1616-17. Author: Speculum Britanniae, first parte . . . Middlesex, 1593; A Plott of the Six Escheated Counties of Ulster, 1605; A Description of the Honor of Winsor . . ., 1607: England, an Intended Guyde for English Travellers, 1625. Made 1st plan for complete series of histories of English counties. Died circa 1625.

NORDENFELT, Thorsten Wilhelm, Swedish engr., inventor; b. Sweden, 1842; promoted sale of Swedish iron in London, 1862-66; founder factories for prodn. of ammunition, machine guns, torpedoes, submarine torpedo boats, in Sweden, Britain, France, early 1880's; founder Maxim-Nordenfeldt Co.; named royal chamberlain by Oscar II of Sweden, 1885. Built several submarines, after 1885; inventor machine gun for def. against torpedo boats. Died 1920.

NORDENSKJÖLD, Baron Nils Adolf Erik, geologist, explorer; b. Helsingfors, Finland, Nov. 18, 1832; s. Nils Gustav Nordenskjöld; Ph.D., U. Helsingfors, 1857; married; children—Baron Nils Erland Herbert, Gustav Erik Adolf. Dismissed from U. Helsingfors for polit. views, 1855; 57; mem. expdn. Spitsbergen, 1958, 61; mem. staff Swedish State Museum, 1858; leader expdns. to Spitsbergen, 1864, 68, to Greenland, 1870; preliminary North-East Passage expdns., 1872, 75-76; successful North-East Passage from Karlskrona through Bering Strait to Yokahama and Suez Canal, 1878-79; leader expdn. to Greenland, 1883. Mem. Swedish Parliament, 1870-72; created baron, 1880. Author: Letters, 1880; Voyage of the Vega, 5 and 2 vol. editions, 1881; Facsimile-Atlas, 1889; Bidragtil Nordens äldsta historia, 1892; Periplus, an Essay on the Early History of Charts and Sailing Directions, 1897. A leader in Arctic exploration; 1st to sail through North-East Passage; contbr. study early history of cartography. Died Stockholm, Sweden, Aug. 12, 1901.

NORDENSKJÖLD, Baron Nils Erland Herbert, Swedish ethnologist; b. Stockholm, Sweden, July 19, 1877; s. Nils Adolf Erik and Anna (Mannerheim) N.; ed. Uppsala, Sweden; with Mus. Natural History, Stockholm, 1906-08; named dir. ethnographic div. Göteborg (Sweden) Mus., 1913. Recipient Loubat prize, 1912, Wahlberg Gold medal, 1912. Author: Fran högfjäll och urskogar, 1902; Indianlif i el gran chaco, 1910; Indianerna pa Panamanäset 1928; also articles, comparative ethnographical studies. Visited Patagonia, 1899, Argentina and Bolivia, 1904-05, Bolivia, 1908-09, S.Am. interior, 1913. Died June 3, 1928.

NORDENSKJÖLD, (Nils) Otto (Gustaf), Swedish explorer, geologist; b. Sjogle, Sweden, Dec. 6, 1869; ed. U. Uppsala (Sweden); Ph.D.; m. Karen Berg; 1 son, 3 daus. Became prof. geography U. Gothenburg (Sweden), 1905; explorer, Arctic and Antarctic; leader sci. expdns., Terra del Fuego, 1895-97, Alaska, Yukon, 1898, Antarctic Regions, 1901-04, West Greenland, 1909, S.Am., Patagonia, 1920, also Spitzbergen, Iceland. Hon. corr. mem. Royal Geog. Soc. London; hon. mem. geog. socs. of Paris, Berlin,

Vienna, Rome, Madrid, Budapest, Brussels, Amsterdam. Author: Antarctica: The Swedish Antarctic Expedition, 1901-04; Die Polarwelt; La Terre de Feu; Sudamerika. Research on remains of Createaceous fossils, tertiary vertebrates and plants in Antarctica, former extension of land-ice in S.Am., extinct animal neomylodon brought to Europe, Antarctic ice conditions, Swedish petrology. Died June 2, 1928.

NORDHEIM, Lothar Wolfgang, physicist; b. Munich, Germany, Nov. 7, 1899; s. Moritz and Anna (Tandler) N.; Ph.D., U. Goettingen (Germany), 1923; D.Sc., Karlsruhe (Germany) Inst. Tech., 1951, Purdue U., 1962; m. Gertrud Poeschl, Jan. 12, 1935 (dec. June 1949); 1 son, Erik. Asst., U. Goettingen, 1922-27, lectr., 1928-33; research fellow U. Cambridge, 1927-28, U. Copenhagen, 1928, Inst. H. Poincare, Paris, 1933, Teylers Stichting Found., Harlem, Holland, 1934; vis. prof. Ohio State U., 1930, U. Moscow, U.S.S.R., 1932; Purdue U., 1935-37, U. Heidelberg (Germany), 1949; prof. physics Duke, 1937-56; prin. physicist, dir. physics div. Oak Ridge Nat. Lab., 1943-47; resident cons. Los Alamos Sci. Lab., 1950-52; chmn. theoretical physics dept. Gulf Gen. Atomic, Inc., San Diego, 1956-65, sr. research adviser, 1960—. Mem. adv. com. on reactor physics, AEC, 1962—; Recipient certificate of merit Dept. War, 1947. Fellow Am. Phys. Soc., Am. Nuclear Soc. (dir. 1961-64); mem. A.A.A.S., Fedn. Am. Scientists (charter mem., vice chmn. 1955). Editorial bd. Annals of Physics, 1957—, Nuclear Sci. and Engring, 1959—. Research in quantum mechanics, solid state, cosmic radiation, nuclear physics, and reactor physics. Home: 804 Muirlands Dr., La oJlla, Cal. 92037. Office: Gulf Gen. Atomic, Inc., P.O. Box 608, San Diego 92112.*

NORDIN, Borje Edgar Christopher, English physician; b. London, Eng., Mar. 14, 1920; s. Helge Edgar and Katharine (Wright) N.; M.B., B.S., U. Coll. Hosp., London, 1950, M.D., 1952, M.R.C.P., 1954, Ph.D., 1960, F.R.C.P., 1966; m. Wencke Backer, June 3, 1944; children—Antony, Veronica, Christine; m. 2d, Margaret Cecil, Feb. 2, 1958; children—Katharine, Richard. Med. registrar Postgrad. Med. Sch., London, 1952-55; vis. fellow Presbyn. Hosp., N.Y.C., 1956; lectr. in medicine, sr. lectr. U. Glasgow, 1957-64, Who Traveling fellow, 1964; cons. physician Gen. Infirmary, Leeds, 1964—, dir. MRC Mineral Metabolism unit, 1964—; reader U. Leeds, 1965—. Cons. U.K. Atomic Energy Authority; editorial sec. Calcified Tissue Research. Mem. Med. Research Soc. London, Royal Soc. Medicine, Royal Coll. Physicians. Author: (with D. A. Smith) Diagnostic Procedures in Disorders of Calcium Metabolism, 1965; also numerous articles. Developed new diagnostic methods for disorders of bone disease and calcium metabolism which permit more precise diagnosis and better treatment; elucidated importance of calcium in human nutrition and its possible relation to fractures in elderly; developing med. methods for treatment of kidney stones. Home: Farnley Old Hall, Otley, Yorkshire, Eng. Office: Med. Research Council Mineral Metabolism Unit, Gen. Infirmary, Leeds, Eng.*

NORDIO, Sergio, Italian pediatrician; b. Trieste, Italy, Mar. 7, 1924; s. Luigi and Gabriella (Zollia) N.; student U. Padua (Italy), 1943-49; m. Felicita Chiarelli, Dec. 21, 1952. Staff, Pediatric Clinic, U. Genoa (Italy), since 1949—, asst. prof. pediatrics, since 1958—. Mem. Italian, Swiss, Argentine, Spanish (hon.) pediatric socs., European Club for Pediatric Research, A.A.A.S. Author: (with Antener) Rickets, 1964; (with DePrà) Metabolism of Water and Electrolytes, 1967; (with Antener) Metabolic and Renal Dwarfisms, 1966; also numerous articles. Research on metabolic diseases, calcium-phosphorus metabolism, intestinal digestion and absorption, carbohydrates and amino acids metabolism, liver function. Home: 6/8 L. DaVinci Sq., Genoa, Italy.

NORDLIE, Robert Conrad, Am. biochemist; b. Willmar, Minn., June 11, 1930; s. Peter C. and Myrtle (Spindler) N.; B.S.Ed., St. Cloud State Coll., 1952; M.S., U. N.D., 1957, Ph.D., 1960; m. Sally Ann Christianson, Aug. 23, 1959; children—Margaret Ann, Melissa Jane, John Conrad. Research fellow Inst. for Enzyme Research U. Wis., 1960-62; Hill research prof. biochemistry U. N.D. Med. Sch. Grand Forks, 1962—. Nat. Heart Inst. fellow, 1958-60; Nat. Cancer Inst. fellow, 1960-62. Mem. Am. Chem. Soc. (sec. Red River Valley sect.), Am. Biol. Chemists, A.A.A.S., N.D. Acad. Sci., Sigma Xi. Studies, publs. on enzyme function, notably that of glucose-6-phosphatase and its response to diabetes, insulin, corticosteroid levels; carbohydrate metabolism. Home: 162 Columbia Ct., Grand Forks, N.D. 58201.*

NORDLING, Carl Leif Arne, physicist; b. Edmonton, Alta., Can., Feb. 6, 1931; s. Jarl and Karin (Thoren) N.; Ph.D., U. Uppsala, Sweden, 1959; m. Gunhild Soderstrom, Jan. 1, 1954; children—Marianne, Dag, Fredrik. Research asst. U. Uppsala, Inst. Physics, 1954-59, asst. prof. physics, 1959—; asst. editor Nuclear Instruments and Methods, 1960—. Mem. Swedish Phys. Soc., Royal Acad. Arts and Scis. Uppsala. Author: Problemsamling i Fysik, 1965; also numerous articles. Exptl. studies on electronic structure of atoms, molecules, solids by X-ray produced photo and Auger electrons; new approach to elementary analysis and chem. structure research based on electron spectroscopy. Home: 45 B Malma Ringvag. Office: 7 Thunbergsvagen, Uppsala, Sweden.*

NORDMANN, Charles, French astronomer; b. 1881; dir. Paris Obs.; sci. writer for Le Revue Parisienne, Le Matin, other French publs.; inventor heterochromatic photometer; made spectral studies confirming theories of Secchi; expert in star photography; espoused theory that excess electrification caused mental illness among residents of N.Y.C., 1923. Died 1940.

NORGAARD, Gunnar, Danish geodeist; b. Nexö, Nov. 23, 1903; s. Jens Peter and Amalia (Kornbeck) N.; Ph.D, U. Copenhagen (Denmark); m. Else Fournais, Apr. 28, 1930; children—Evalis, Kirsten, Birgit. Geodesist, 1929—; head sect. Copenhagen Geodesic Inst., 1938-46, state geodesist, 1946; prof. geodesy U. Lund (Sweden), 1957; head geophys. mission UNESCO to Pakistan, 1951; expts. in pvt. labs., 1954. Publns. on measure of gravity and constrn. of gravimeters. Address: Slotsbjergby Slagelse, Denmark.

NORICUS, Bernardus, historiographer; b. probably Austria; ordained priest, Passau, 1300; in Bavaria, 1314, Avignon, 1319; later writer, corrector, dir. manuscript sch., Kremsmünster. First important historiographer of monastery at Kremsmünster; ofcl. chronicler, from 1312. Influenced historiography and lit. of his time. Died after 1326.

NORIN, Erik Stephen, Swedish geologist; b. Karlshamn, Sweden, Mar. 15, 1895; s. Emil Fredrik and Wendla (Eriksson) N.; Fil.kand., Stockholms (Sweden) Högskola, Stockholm U., 1916, Fil.lic., 1923; Fil. dr. U. Lund (Sweden), 1937; m. Anna Göthilda Ambolt, Mar. 24, 1936; children—Elsa (Mrs. Karl Erik Rybäck), Anders. Geol. research Shansi Province, China, 1920-22, Kashmir-Ladakh, N.W. India, 1924-25; chief geologist Sino-Swedish Expdn., 1927-35; prof. mineralogy and petrology, dir. Mineral.-Geol. Inst., U. Uppsala (Sweden), 1945-61. Decorated comdr. Royal Order N. Star. Mem. Royal Physiographic Soc. Lund, Swedish Royal Acad. Sci., Royal Soc. Sci. Uppsala. Author: Rep. Sino-Swedish Expedition, Ser. III, vol. 1, 1937, vol. 6, 1941, vol. 7, 1946, Ser. I. vol. 1 (with N. Ambolt), 1968; The Sediments of the Central Tyrrhenian Sea, 1958; also articles. Topographical and geol. surveys in Shansi Province, China, T'ien Shan ranges, Lop Nuur region, Tarim Basin of Chinese Turkistan, Northwestern part of Tibetan drainageless plateau-land; research on Quaternary stratigraphy and volcanism as recorded in sediments of Tyrrhenian Sea and its border regions. Home: 24B Sturegatan, Upsala, Sweden.*

NORIN, Torbjorn, Swedish chemist; b. Ornskoldsvik, Sweden, Sept. 16, 1933; s. Einar and Lisa (Wik) N.; Civ.ing., Royal Inst. Tech., Stockholm, Sweden, 1957, tekn.lic., 1962, tekn.dr. and docent, 1964, prof., 1967; m. Ingegerd Andersson, July 18, 1959; children—Martin, Elisabeth, Magdalena. Research asst. Royal Inst. Tech., Stockholm, 1957-61, 63-64, docent, 1964-65, head, wood chemistry div., 1966—; vis. sci. Dyson Perrins Lab., Oxford, Eng., 1961-62; dir. research, head wood chemistry dept. Swedish Forest Products Research Lab., Stockholm, 1966—. Mem. Swedish Chem. Soc. (Nordblad-Ekstrand medal 1966). Research, publs. on chemistry of terpenes and other natural products and to chemistry of 3 membered ring compounds. Home: Gronviksvagen 12, Stockholm-Bromma. Office: Drottning Kristinas vag 61, Stockholm, Sweden.*

NÖRING, Friedrich Karl, German geologist; b. Niederrodenbach, Germany, Aug. 6, 1912; s. Otto Heinrich and Johanna (Horst) N.; student Johann-Wolfgang-Goethe U., Frankfurt, Germany, 1930-33; Dr. rer.nat., Friedrich-Wilhelm U., Berlin, Germany, 1937, Dipl.-Geol., 1938; m. Elsa Greta Maria Jacobi, Aug. 21, 1948; 1 dau., Beate Carola. Asst., Inst. Geol.-paleont., U. Berlin, 1937, Geol. Staatsinst., Hamburg, Germany, 1938; staff Deutscher Verein von Gas-uno Wasserfachmännern, Berlin, 1938-39, Bezirksgeologe Reichsamt für Bodenforschung, Berlin, 1939-45; staff Hessisches Landesamt für Bodenforschung, Wiesbaden, Germany, 1947—, 1st dir., 1965—; faculty hydrogeology Tech. U., Darmstadt, Germany, 1962—. Mem. Deutscher Verein von Gas- und Wasserfachmännern Frankfurt (recipient DVGW-Ehrenring 1960), Deutsche Geologische Gesellschaft Hanover. Research and numerous publs. in hydrogeol. field using measurements of distbn. of run-off in dry periods. Home: 9 Lessingstr. Office: 9 Leberberg, Wiesbaden, Hesse 62, Germany.*

NÖRLUND, Niels Erik, Danish mathematician; b. Slagelse, Denmark; Oct. 26, 1885; s. Alfred Christian and Sophie (Holm) N.; Ph.D., U. Copenhagen, 1910; Hon.Dr.ing., Darmstadt, 1936; Hon.Dr.Sci., U. London, 1937, U. Dijon, 1950; Hon. Ph.D., U. Lund, 1941, U. Oslo, 1951; m. Agnete Weaver, Dec. 28, 1912; children—Hildur (Mrs. Eivind Ouwren), Elisabeth (Mrs. Armas Aavatsmark). Prof. math. U. Lund, 1912-22; prof. math. Copenhagen U., 1922-56, rector, 1933-34; dir. Danish Geodetic Inst., 1928-55. Pres., Rask-Orsted Found., 1928-64. Decorated Grand Cross Order Denmark (Norway, Iceland); comdr. Legion of Honor (France); Grand Cross Royal Norwegian Order St. Olav. Fellow Royal Soc., 1938; mem. Royal Danish Acad. Sci. (past pres.), Internat. Council Sci. Unions (past pres.); fgn. mem. acads. sci. of Paris, Rome, Napoli, Strassbourg, Stockholm, Oslo, Helsinki, Lund. Author: Vorlesungen über Differenzenrechnung, 1924; Leçons sur les séries d'interpolation, 1926; La somme d'une fonction, 1927; Leçons sur les équations aux différences finies, 1929; The Map of Denmark, 1942; The Map of Iceland, 1944; Hypergeometric Functions, 1963. Contbd. to study of factorial analysis and homogeneous linear difference equations, calculus of finite differences, class of new functions having properties similar to those of gamma function. Home: 6 Malmögade, Copenhagen, Denmark.*

NORMAN, Alfred Merle, English zoologist; b. 1831; s. John Norman; ed. Christ Ch., Oxford; M.A., LL.D., D.C.L.; rector Houghton-le-Spring, Eng., also rural dean. Recipient medal French Inst. Fellow Royal Soc., 1890, Linnean Soc. (Gold medal 1906). Research and publs. on marine zoology; collected invertebrate fauna of N. Atlantic and Arctic oceans (now in Brit. Mus.). Died Oct. 26, 1918.

NORMAN, Arthur Geoffrey, biochemist, univ. ofcl.; b. Birmingham, Eng., Nov. 26, 1905; s. Arthur and Charlotte (Mant) N.; B.Sc., U. Birmingham, 1925, Ph.D., 1928; M.S. (Rockefeller fellow), U. Wis., 1932; D.Sc., U. London (Eng.), 1933; m. Marian Esther Foote, Sept. 5, 1933; children—Anthony Westcott, Stephen Trevor. Came to U.S., 1930, naturalized, 1946. Biochemist, Rothamsted Expt. Sta., 1928-30, charge biochemistry sect., 1933-37; prof. soils Ia. State U., 1937-46; div. chief U.S. Army Biol. Labs., Ft. Detrick, Md., 1946-52; prof. botany U. Mich., Ann Arbor, 1952—, dir. bot. gardens, 1955-65, v.p. for research, 1964—. Fellow Royal Inst. Chemistry (Gt. Britain); mem. Sigma Xi. Author: Biochemistry of Cellulose, 1938. Editor: Agronomy Monographs, I-VI; Advances in Agronomy, vols. 1-20, 1948-68. Home: 3475 Woodland Rd., Ann Arbor, Mich. 48104.*

NORMAN, Charles, Am. biologist; b. N.Y.C., Jan. 16, 1916; s. Samuel and Yetta (Saft) N.; B.S., U. Miami (Fla.), 1948, M.S., 1949; Ph.D., State U. Ia., 1954; m. Beverley Salzhauer, Feb. 24, 1946; children—Ellen Sandra, Rebecca Robin, Samuel Jason. Faculty, W. Va. U., 1953—, prof. biology, 1961—; vis. prof. Mass. Inst. Tech., 1965-66. Cons. AID, U.S. Dept. Agr., 1963, 65. Research Fellow Am. Phys. Soc., 1959, Oak Ridge Inst. Nuclear Studies, 1964; Spl. Research fellow NIH, 1965-66. Mem. Am. Soc. Zoologists, Am. Soc. for Cell Biology, A.A.A.S., Am. Inst. Biol. Scis., Soc. for Study Fertility, Sigma Xi. Contbr. articles to tech. jours. Formulated maintenance media for conservation mammalian sperm at room temperatures; discovered that visible light is spermicidal; research in physiology and biochemistry mammalian germ cells, physiology reprodn., molecular biology differentiation and devel. Home: 344 Laurel St., Morgantown, W. Va. 26501.*

NORMAN, Nicolai, Norwegian physicist; b. Trondheim, Norway, Apr. 17, 1919; s. Christian August and Anna (Lie) N.; B.Sc., U. Oslo (Norway), 1942, M.S., 1947, Dr.Phil., 1956; m. Marit Orrung, Dec. 31, 1947; children—Britt Elisabeth, Hanne Marit. Research asst. U. Oslo, 1948, research asso., 1950-55, prof. physics, 1964—; research asst. Norwegian Inst. Tech., Trondheim, 1949; research physicist Central Inst. for Indsl. Research, Oslo, 1956-64. Mem. Norwegian Acad. Sci., Norwegian Phys. Soc. Research, publs. on X-ray diffraction techniques, structure of crystals, and amorphous materials, fibre structures, order-disorder phenomena in alloys. Home: 17 Rektorhaugen, Oslo, Norway.*

NORMAN, Philip Sidney, Am. physician; b. Pittsburg, Kan., Aug. 4, 1924; s. P. Sidney and Mildred, (Lawyer) N.; B.A., Kan. State Coll., 1947; M.D. cum laude, Washington U., St. Louis, 1951; m. Marion Birmingham, Apr. 15, 1955; children—Margaret Reynolds, Meredith Andrew, Helen Elizabeth. USPHS fellow Rockefeller Inst., N.Y.C., 1954-56; faculty Johns Hopkins, Balt., 1956—, asso. prof. medicine, 1964—; physician Johns Hopkins Hosp., 1959—. Mem. N.Y. Acad. Scis., Am. Fedn. Clin. Research Soc. for Exptl. Biology and Medicine, Am. Acad. Allergy, Am. Assn. Inmmunologists, Soc. for Clin. Investigation. Research on chemistry plasmin system blood, antibiotic treatment chronic bronchitis, active prin. ragweed pollen, treatment hay fever. Home: Manor Rd., Baldwin, Md. 21013. Office: Johns Hopkins Hosp., Balt. 21205.*

NORMAN, Robert, English natural philosopher; flourished circa 1590; Author: The Newe Attractive, 1581; Safegarde of Saylers, 1590. Built 1st inclination compass; observed that magnetization of steel did not change its weight and that steel needle can swim against surface tension on water; measured magnetic inclination and declination at different points on the earth.

NORMAND, Augustin, French shipbuilder; b. probably Le Havre, 1792; engaged in shipbuilding, Le Havre; with Barnes established postal service to Corsica; built schooner Napoleon (1st propellor driven steamship, achieving speed of 11 knots), 1843. Died 1871.

NORMAND, Sir Charles William Blyth, Brit. meteorologist; b. Edinbourgh (Scotland), Sept. 10, 1889; s. John and Mary (Baxter) N.; M.A., D.Sc., U. Edinbourgh; m. Alison MacLennan, Nov. 27, 1920; children—Alan, Colin. Imperial meteorologist at Simla, 1913-26; dir. observatories in the Indies, 1927-44; spl. govt. work in Indies, 1944-45; sec. gen. Internat. Ozone Commn., 1948-59; mem. com. for meteorol. research sponsored by Air Ministry, 1946-58. Decorated companion Order Brit. Empire. Fellow Nat. Inst. India; mem. Royal Meteorol. Soc. (pres. 1951-53). Address: 56 Holywell St., Oxford, Eng.

NORMANDY, Alphonse René Le Mire, chemist; b. Rouen, France, Oct. 23, 1809; studied medicine; studied chemistry under Gmelin, Germany; went to Eng., circa 1843; practiced as cons. and analytical chemist. Fellow Chem. Soc. Author: Guide to the Alkali-metrical Chest, 1849; Introduction to Rose's Chemical Analysis, 1849; Handbook of Chemical Analysis, 1850; The Chemical Atlas, 1855; The Dictionaries of the Chemical Atlas, 1857. Patentee indelible inks and dyes, 1839, method for hardening soap, 1841, apparatus for distilling sea water for drinking, 1851. Died May 10, 1864.

NORMANT, Henri Marie, French chemist; b. Plozevet, France, June 25, 1907; s. Jean and Anne (Gentric) N.; Agrégé des Sciences Physiques, Docteur ès Sciences, Faculté des sciences, Caen; m. Madeleine Sosson, Sept. 6, 1932; children—Annick, Jean, Francoise Claude, Yves. Prof. lycées of Roanne and Angers; lectr. Faculté des Sciences, Caen; prof. Faculté des Sciences de Lille, then at Faculté des Sciences, Paris (Lab. de Synthèse Organique, Sorbonne, Paris). Mem. French Acad. Scis., Rhone-Poulenc Soc. (sci. counselor). Author: Cours de Licence; also numerous articles.; contbg. author: Advances in Organic Chemistry, vol. II, 1960. Research on heterocyclic oxygen compounds, organometallic compounds; discovered vinyl magnesium, 1954; studies on HMPT solvents. Home: 40 bis, rue Violet, Paris 15, France.*

NORMANUS, see de Muris, Jean.

NORO, Genjo, Japanese physician, botanist; b. Mie Prefecture, Ise Province, Japan, 1693; studied medicine under Genshu Yamawaki, Kyoto, Confucian classics under Temmin Namikawa, botany under Jakusui Ino; began to learn Dutch at request of Shogun, 1720; collected bot. specimens throughout Japan. Author: Oranda Honzo Wakai (1st Japanese book on Western botany), 1750. Died 1760.

NÖRRENBERG, Johann Gottlieb Christian, German physicist; b. Pustenbach/Gummersbach, Germany, Aug. 11, 1787; prof., Darmstadt, Tübingen (both Germany); developed polarization apparatus named after him. Died Stuttgart, Germany, July 20, 1862.

NORRIS, Albert Stanley, physician; b. Sudbury, Ont., Can., July 14, 1926; s. William and Mary (Zell) N.; student U. Western Ont., 1945-47, M.D., 1951; m. Dorothy Alwynne James, Sept. 2, 1950; children—Barbara, Robert, Kimberely. Came to U.S., 1957, naturalized, 1963. Chief resident Boston City Hosp., teaching fellow Harvard Med. Sch., 1955-56; pvt. practice medicine specializing in psychiatry, 1956-57; instr. Queens U., Kingston, Ont., 1956-57; faculty U. Ia., Iowa City, 1957-64, prof., 1965—; asso. prof. U. Ore. Med. Sch., Portland, 1964-65; dir. out-patient clinic, 1965. Fellow Am. Psychiat. Assn.; A.M.A., Soc. Med. Psychiatry, Assn. Am. Med. Colls. Research, publs. on constl. factors in psychiat. illness, particularly capillary morphology and its relations to prognosis, clin. research on depression and psychosomatic obstetrics and gynecology, in chronic renal dialysis. Home: 914 Talwrn Ct., Iowa City 52240.*

NORRIS, Charles, Am. physician; b. Hoboken, N.J., Dec. 4, 1867; s. Joseph Parker and Frances Ann (Stevens) N.; Ph.B., Sheffield Sci. Sch. (Yale), 1888; M.D., Coll. Phys. and Surg. Columbia, 1892; postgrad., Kiel, Gottingen, Berlin and Vienna, 1894-96; m. Eugenie Gebhart, Sept. 3, 1908. Instr. pathology Coll. Phys. and Surg., 1896-1904; dir. labs. Bellevue Hosp., 1904-18; chief med. examiner City of N.Y., from 1918; prof. forensic medicine, N.Y. U., 1933. Authority in forensic medicine. in med. jurisprudence; discovered (with T. Flournoy and A. M. Pappenheimer) a spirochete in a case of Am. relapsing fever, pub. 1906. Died Sept. 11, 1935.

NORRIS, Charles H(ead), Am. civil engr.; b. Pendleton, Ore., Mar. 30, 1910; s. Charles Head and Louise May (Hansen) N.; B.S., U. Wash., 1931; S.M., Mass. Inst. Tech., 1932, Sc.D., 1942; m. Martha Marie White, Aug. 18, 1931; children—Charles Head, John Matthew. Mech. engr. Am. Steel Foundries, Chgo., 1936-38; asst. civil engring. Mass. Inst. Tech., 1934-36, instr., 1938-41, asst. prof. structural engring., 1941-46, asso. prof., 1946-51, prof. structural engring., dep. head structural div. civil engring. dept., 1951-55, prof., head structural engring. div., exec. officer dept. civil and san. engring., 1957-62; chmn., prof. civil engring. U. Wash., Seattle, 1962-65, dean Coll. Engring., 1965—. Expert, Office Chief Engrs., Dept. Army, 1947-62; cons. Armed Forces Spl. Weapons Project, 1950-60, Dept. Air Force, 1952-60. Miscellaneous engring. projects. Recipient Structural Sect. prize Boston Soc. Civil Engrs., 1950. Fellow Am. Soc. C.E.; mem. Soc. Exptl. Stress Analysis, Phi Beta Kappa, Sigma Xi, Tau Beta Pi, Chi Epsilon (hon.). Author: Elementary Structural Analysis (with John B. Wilbur), 2d edit., 1960; chpt. Handbook of Experimental Stress Analysis (with John B. Wilbur), 1950; sr. author Structural Design for Dynamic Loads, 1959. Contbr. tech. jours. Home: 4001 N.E. Belvoir Pl., Seattle 98105.*

NORRIS, Dale Melvin, Jr., Am. biologist; b. nr. Essex, Ia., Aug. 19, 1930; s. Dale M. and Opal (Klepinger) N.; student Tarkio Coll., 1948-50; B.S., Ia. State U., 1952, M.S., 1953, Ph.D., 1956; m. Eleanor Ann Brown, Sept. 9, 1951 (dec. May 1963); children—Kathleen, Elizabeth; m. 2d, Mary Price Wilmar, June 12, 1965; children—Diana, Mark. Research asst. Ia. State U., Ames, 1952-56; asst. prof. U. Fla., Gainesville, 1956-57; faculty U. Wis., Madison, 1958—, prof., 1966—, acting chmn. dept. entomology, 1964, chmn. research adv. com. Coll. Agr., 1965—. Mem. Entomol. Soc. Am., Am. Phytopath. Soc., A.A.A.S. Cited for thesis 1953), Am. Assn. U. Profs. Research, publs. on chemotherapy of trees against diseases and insects, chemistry of host plant selection by insects, symbiontic inter-relationships between insects and microbes in wood, insect transmission of fungal pathogens of trees. Home: 101 S. Rock Rd., Madison, Wis. 53705.*

NORRIS, Forbes Holten, Jr., Am. physician; b. Richmond, Va., May 1, 1928; s. Forbes Holten and Veda (Vose) N.; S.B., Harvard, 1949, M.D., 1955; m. Louise Eldred Kelly, Feb. 12, 1955; children—Heather, Holten, Brian. Nat. Found. fellow, 1953; Kirkland fellow Nat. Paraplegia Found., 1953-55; clin. asso. NIH, 1956-61; faculty U. Rochester, 1961-66, asso. prof. neurology, acting chmn. neurology, 1963-66; asso. dir. Inst. Neurol. Sci., Pacific Med. Center, San Francisco, 1966—; vis. research fellow U. London (Eng.), 1966. USPHS Spl. fellow, 1961-63. Mem. San Francisco Med. Soc., A.M.A., Cal. Med. Assn., Am. Assn. for Electromyography, Am. Acad. Neurology. Author: The EMG, 1963; also articles. Co-editor (with Lord Brain) The Remote Effects of Cancer on the Nervous System, 1965; (with L. T. Kurland) Amyotrophic Lateral Sclerosis, 1968. Research on disordered nervous system especially diseases of spinal cord, nerves and muscles. Home: 85 Seafirth Rd., Tiburon, Cal. 94920. Office: Pacific Med. Center, San Francisco 94115.*

NORRIS, James Flack, Am. chemist; b. Balt., Jan. 20, 1871; s. Richard and Sarah Amanda (Baker) N.; A.B., Johns Hopkins, 1892, Ph.D., 1895; Sc.D., Bowdoin Coll., 1929; m. Anne Bent Chamberlin, Feb. 4, 1902. Asst., Mass. Inst. Tech., 1895-96, instr., 1896-1900, asst. prof. organic chemistry, 1900-04, prof., dir. research lab. organic chemistry, from 1916; prof. chemistry, Simmons Coll., Boston, 1944-15, Vanderbilt U., 1915-16. Lectr. organic chemistry, Harvard, 1912-14, Clark U., 1913-14, Bowdoin, 1929. Medalist Am. Inst. Chemists, 1937. Sec. Soc. of Arts of Boston, Mass., 1902-04; pres. Am. Chem. Soc., 1925-26 (pres. Northeastern sect., 1905-06); v.p. Internat. Union of Pure and Applied Chemistry, 1925-28; chmn. div. chemistry and chem. tech. NRC, 1924-25, mem. exec. bd., 1925-33; fellow Am. Acad. Arts and Scis. (v.p. from 1934), A.A.A.S. Author: The Principles of Organic Chemistry; Experimental Organic Chemistry; Text-book of Inorganic Chemistry for Colleges (with R. C. Young); Laboratory Exercises in Inorganic Chemistry (with K. L. Mark). Research on reactivity of atoms and groups in organic compounds, pyrolysis temperatures, thermal decomposition of triphenylmenthyl alkyl ethers, prodn. of alcohols from butenes and pentenes through interaction with sulfuric acid, polymerization of amylenes, rearrangement of isopropylethylene to trimethylethylene, pyrogenic decomposition of 2-pentene and trimethylethylene. Died Aug. 4, 1940.

NORRIS, Kenneth S., Am. biologist; b. Los Angeles, Aug. 11, 1924; s. Robert D. and Jessie (Matheson) N.; A.B., U. Cal. at Los Angeles, 1945, M.A., 1948, Ph.D., 1959; m. Phyllis Strout, Feb. 27, 1953; children—Susan, Nancy, Barbara, Richard. Curator, Marineland of the Pacific, Cal., 1953-59; faculty U. Cal. at Los Angeles, 1960—, asso. prof. zoology, 1963—. Vice pres. Oceanic Found., 1960—; research bd. San Diego Zoo; cons. Brain Research Inst., Biol. Instrumentation Adv. Council, U.S. Naval Ordnance Test Sta., Interuniv. Council, Santa Catalina Island Marine Sta.; chmn. U. Cal. Natural Land and Water Res. System. Recipient award Brain Research Inst., 1966. Mem. Ecol. Soc. Am. (Mercer award 1964), A.A.A.S., Herpetologists League, Soc. for Study Evolution, Wilderness Soc., Nat. Parks Assn., Sigma Xi. Author: Whales, Dolphins and Porpoises, 1966; also articles. Research on biology, gen. ecology, thermal relations of desert reptiles, Pleistocene history of western N.Am., thermal biology of intertidal fishes, fish transp., aquarium mgmt., relationships between anemones and anemone fish; tooth growth and replacement in herbivorous fish; zoogeography of Baja, Cal.; color matching in reptiles; distbn. and gen. biology of Pacific cetaceans. Home: 5 Fruittree Rd., Portuguese Bend, Cal. Office: Dept. Zoology, U. Cal., Los Angeles.*

NORRIS, Robert Matheson, Am. geologist; b. Los Angeles, Apr. 24, 1921; s. Robert DeWitt and Jessie Ellen (Matheson) N.; A.B., U. Cal. at Los Angeles, 1943, M.A., 1949, Ph.D., 1951; m. Virginia Grace Oakley, Jan. 5, 1952; children—Donald Oakley, James Matheson, Elizabeth Ann. Asso. marine geology Scripps Inst. Oceanography, U. Cal. at La Jolla, 1951-52; faculty U. Cal. at Santa Barbara, 1952—, asso. prof. geology, 1960—, chmn. dept. geology, 1960-63; geologist U.S. Geol. Survey, 1956-58; shipbd. geologist downwind expdn. to S.E. Pacific, Internat. Geophys. Year, 1957-58. Fulbright Research scholar New Zealand Oceanographic Instn., 1961-62. Mem. Geol. Soc. Am., Am. Assn. Petroleum Geologists, Soc. Econ. Paleontologists and Mineralogists, Nat. Assn. Geology Tchrs., Meteoritical Soc., Geol. Soc. New Zealand. Research, publs. on origin and evolution certain desert sand dunes, nature and origin sediments on some undersea ridges, rate shoreline changes. Home: 4424 Nueces Dr., Santa Barbara, Cal. 93105.*

NORRISH, Ronald George Wreyford, English phys. chemist; b. Cambridge, Eng., Nov. 9, 1897; s. Herbert Norrish; ed. Emmanuel Coll., Cambridge (Eng.) U.; hon. D.de l'U Sorbonne, Paris, France, 1958; D.Sc. (hon.) Leeds (Eng.) U., 1965, Sheffield (Eng.) U., 1965; m. Annie Smith, 1926; two daus. Research fellow Emmanuel Coll., 1925-31, now fellow of coll.; prof. phys. chemistry, dir. dept. Cambridge U., 1937-65, emeritus, 1965—, past Humphrey Owens Jones lectr. phys. chemistry. Mem. sci. adv. council Ministry of Supply, 1942-45. Recipient Meldola medal Inst. Chemistry, 1926; Leverhulme research fellow, 1935; Lewis medal Combustion Inst., 1964; liveryman Worshipful Company Gunmakers, 1961; Recipient (with M. Eigen and G. Porter) Nobel prize in chemistry, 1967. Served with British Army, 1916-19. Mem. Chem. Soc. (council 1933-36; Liversidge lectr. 1958; Faraday Meml. lectr. and medal 1954), Faraday Soc. (council 1933-36, pres. 1953-55), Royal Inst. Chemistry (v.p. 1957-59), Fellow Royal Soc., 1936 (Davy medal 1958, Bakerian lectr. 1966), mem. British Assn. (pres. sect. chemistry 1960-61); hon. mem. Polish Chem. Soc., Royal Soc. Scis., Uppsala, Sweden; corr. mem. Acad. Scis., Göttingen, Germany, Royal Soc. Scis., Liege; fgn. hon. mem. Polish Acad. Scis. Contbr. profl. jours. Research in extremely fast chem. reactions; by disturbing chem. equilibrium with short energy pulses, made analysis of reactions of one ten billionth of a second possible. Address: Emmanuel Coll., Cambridge, Univ., Cambridge, Eng.

NORTH, Elisha, Am. physician; b. Goshen, Conn., Jan. 8, 1771; s. Joseph and Lucy (Cowles) N.; studied medicine, Hartford, Conn., under Lemuel Hopkins, also under Benjamin Rush, U. Pa., 1793-95; M.D., Conn. Med. Soc., 1813; m. Hannah Beach, 1797; 8 children. Practiced medicine, Goshen, until 1812, then New London, Conn. Established 1st eye dispensary, New London, 1812. Author: A Treatise on a Malignant Epidemic Commonly Called Spotted Fever, 1811; Outlines of the Science of Life, 1829; Uncle Toby's Pilgrim's Progress in Phrenology, 1836. Pioneer in practice of vaccination in U.S.; instigated 1st use of Kine-pox for vaccination purposes in N.Y.C.; coped successfully with spotted fever epidemic in New Eng., 1806-10. Died Dec. 29, 1843.

NORTH, Harper Qua, Am. physicist; b. Los Angeles, Jan. 24, 1917; s. Richard L. and Mary Elizabeth (Qua) N.; B.S., Cal. Inst. Tech., 1938; M.A., U. Cal. at Los Angeles, 1940, Ph.D., 1947; m. Mary Ellen Foster, July 15, 1955; children—Anita M., James S. Research asso. Gen. Electric Co., 1940-49; dir. semiconductor div. Hughes Aircraft Co., Culver City, Cal., 1949-54; pres. Pacific Semiconductors, Inc., Lawndale, Cal., 1954-62, chmn. bd., 1961-62; chmn. bd. TRW Electronics, Lawndale, 1962; v.p. research and devel. TRW Inc., Los Angeles, 1962—. Mem. Nat. Export Expansion Council, 1962-64. Bd. dirs. Cal. Inst. Cancer Research, 1964—, pres., 1965—; bd. dirs. U. S. A. Standards Inst., 1968—. Fellow Am. Phys. Soc., I.E.E.E.; mem. Electronic Industries Assn. (pres. 1964-66), Sigma Xi. Pioneer research in germanium for semiconductor devices; discovered prin. varactor amplifier; invented semiconductor diode glass package. Home: 1132 Granvia Altamira, Palos Verdes Estates, Cal. 90274. Office: 1 Space Park, Redondo Beach, Cal. 90278.*

NORTH, Robert Carver, Am. polit. scientist; b. Walton, N.Y., Nov. 17, 1914; s. Arthur W. and Irene (Davenport) N.; A.B., Union Coll., 1936; M.A., Stanford, 1948, Ph.D., 1957; m. Woesha Roe-Cloud, Aug. 14, 1943; children—Woesha Kristina (Mrs. Thomas J. Hampson), Mary Davenport, Elizabeth Katrynka, Robert Cloud, Renya Catarina. Tchr. English, history Milford (Conn.) Prep. Sch., 1939-42; research asst. Hoover Instn., Stanford, 1947-49, research asso., 1950-57, mng. editor Pacific Spectator, 1951-57, asso. prof. polit. sci. Stanford, 1957-62, prof., 1962—, dir. Studies in Internat. Conflict and Integration, 1958—. Mem. Am. Polit. Sci. Assn., Am. Hist. Assn., Phi Beta Kappa. Author: Kuomintang and Chinese Communist Elites, 1952; Moscow and Chinese Communists, 1953; (with X. J. Eudin) Soviet Russia and the East, 1920-27, 1957; (with Eudin) M. N. Roy's Mission to China, 1963; (with others(Content Analysis, 1963; Chinese Communism, 1966; also numerous articles. Devel. methods of quantitative analysis in study of internat. relations, measurement of varying differences between nation states along crucial dimensions of capability; automated content analysis of diplomatic documents via computer. Home: 12610 Page Mill Rd., Los Altos Hills, Cal. 94022. Office: Polit. Sci. Dept., Stanford U., Stanford, Cal. 94305.*

NORTH, Wheeler James, Am. marine biologist; b. San Francisco, Jan. 2, 1922; s. Wheeler Orrin and Florence (Ross) N.; B.S., Cal. Inst. Tech., 1940, B.Sc., 1950; M.S., U. Cal. Scripps Inst. Oceanography, Ph.D., 1953; m. Barbara Alice Best, Apr. 25, 1964; children—Hannah Catherine, Wheeler Orrin. Electronics engr. Navy Electronics Lab., San Diego, 1946-48; asst. research biologist U. Cal., La Jolla, 1956-63; sr. research scientist Lockheed Cal. Co., San Diego, 1963; asso. prof. environmental health engring. Cal. Inst. Tech., 1963—. Cons. govtl., acad., indsl. orgns. Mem. A.A.A.S., Am. Soc. Zoology, N.Y. Acad. Sci., Soc. Gen. Physiology, Am. Geophys. Union, Internat. Oceanographic Found., Am. Littoral Soc., Am. Malac Soc., Western Soc. Naturalists, San Diego Soc. Nat. Hist., Am. Mus. Nat. Hist. Research, publs. on metabolism, ecology, biochemistry of marine organisms, especially kelp; marine pollution. Home: Box 204, La Jolla, Cal. 92037. Office: Cal. Inst. Tech., Pasadena, Cal. 91109.*

NORTHEN, Henry T., Am. botanist; b. Butte, Mont., Apr. 12, 1908; s. John Alfred and Sigrid (Anderson) N.; B.S., Wash. State Coll., 1932, M.S., 1934; Ph.D., University of California, 1936; married Rebecca J. Tyson, Aug. 9, 1937; children—Elizabeth S. (Mrs. Richard B. Lyons). Philip T., Thomas H. Faculty U. Wyo. since 1936; prof. botany since 1947, head dept. since 1951. Fellow A.A.A.S.; mem. Am. Soc. Plant Physiol., Bot. Soc. Am., Colo.-Wyo. Acad. Sci., Phi Beta Kappa, Sigma Xi. Author: Plant Science, 3d edit., 1967; (with wife) The Secret of the Green Thumb, 1954; The Complete Book of Greenhouse Gardening, 1956. Research on the molecular basis of stimulation and anesthesia. Home: 1722 Rainbow, Laramie, Wyo.*

NORTHOVER, Francis Henry, mathematician; b. Gosport, Eng., Feb. 26, 1921; s. Henry James and Noel Mary (Eden) N.; M.A., Cambridge U., 1949; Ph.D., London U., 1952; m. Eunice Mary Needham, Jan. 15, 1949; children—Janet, Sally, Hazel, Philip, Stephen. With Royal Navy Sci. Service, 1941-45; lectr. math. Brit. Tech. Colls., 1947-52; asso. prof. math. Meml. U., Nfld., 1953-55; sr. lectr. Liverpool (Eng.) U., 1955-57; asso. prof. Carleton U., Ottawa, Can., 1957-63, prof. math., 1964—. Cons. Canadian Def. Research Bd., 1955-60, GM Def. Research Labs., Santa Barbara, Cal., 1961—. Mem. Canadian Math. Congress. Author: Applied Diffraction Theory; also articles. Research in applications of math. analysis and techniques to fields of wave propagation, diffraction theory, electromagnetic theory. Home: 2339 Ridgecrest Pl., Ottawa 8, Can.*

NORTHROP, John Howard, Am. chemist; b. Yonkers, N.Y., July 5, 1891; s. John I. and Alice B. (Rich) N.; B.S., Columbia, 1912, M.A., 1913, Ph.D., 1915, Sc.D., 1937; Sc.D., Harvard, 1936, Yale, 1937, Princeton, 1940, Rutgers U., 1941; LL.D., U. Cal., 1939; m. Louise Walker, June 18, 1917; children—John, Alice Havemeyer (Mrs. Frederick C. Robbins). Mem. Rockefeller Inst. for Med. Research, 1925-62, prof. emeritus Inst., 1962—; prof. bacteriology, biophysics U. Cal., Berkeley, 1949-59, prof. emeritus, 1959—; lectr. Johns Hopkins, 1937, 40, Columbia, 1938, U. Cal., 1939. Recipient Stevens prize, 1931, Chandler medal, 1937, Lion award, 1944, Alexander Hamilton medal, 1961 (all Columbia); (with W. M. Stanley and J. B. Sumner) recipient Nobel Prize in chemistry, 1964. Fellow chem. Soc. London; mem. Nat. Acad. Sci. (Elliott medal 1944), Am. Philos. Soc., A.A.A.S., Soc. Philomathique (Paris). Author: Crystalline Enzymes, 1939. Research on isolation and crystallization of enzymes; isolated 1st bacterial virus; research established nucleoprotein nature of enzymes and their adherence to laws of chem. reaction; 1st to produce enzyme trypsin in lab.; also crystalline form of diphtheria antitoxin, 1941. Home: 838 San Luis Rd., Berkeley, Cal. 94707.

NORTHROP, John Knudsen, Am. aero. engr.; b. Newark, Nov. 10, 1895; s. Charles Wheeler and Helen C. (Knudsen) N.; ed. pub. schs. Santa Barbara, Cal.; m. Inez M. Harmer, anuary 30, 1918 (div. 1948); children—Bette (Mrs. Paul G. Johansing), John H., Ynez S. (Mrs. Robert W. Koch); m. 2d, Margaret Bateman, Dec. 23, 1950. Designer, Loughead Aircraft Co., Santa Barbara, Cal., 1916-17, 1919-20; designer, project engr., Douglas Aircraft, Santa Monica, Cal., 1923-26; co-founder Lockheed Aircraft, Los Angeles, 1927, served as chief engr., 1927-28; v.p., chief engr. Northrop Aircraft Co. (div. United Aircraft), Burbank, Cal., 1929-31, The Northrop Corp. (now El Segundo div., Douglas Aircraft), 1932-37; co-founder, pres. and dir. Northrop Aircraft, Inc., Hawthorne, Cal., 1939-52; engineering cons., 1953—. Trustee Occidental Coll., Northrop Inst. Tech. Fellow Am. Inst. Aeros. and Astronautics (pres. 1948); mem. Soc. Automotive Engrs. Designer: Lockheed Vega, 1927; co-designer: Northrop Alpha, 1929; Gamma, Delta, 1932; BT-1, 1935; A-17, A-17A, 1935; N3PB, 1939; P-61, 1941; Northrop Flying-Wing, Airplanes, 1940-45; F89 Scorpion, Snark, 1948. Died 1967.

NORTHROP, Paul Allen, Am. physicist; b. Mehoopany, Pa., Sept. 9, 1896; s. George Henry and Florence Esther (Kibbe) N.; B.S., John B. Stetson U., 1916; Ph.D., Cornell, 1926; m. Grace Vivian Michael, June 28, 1922 (dec. Apr. 1958); children—Lucille Vivian (Mrs. John L. Carter), Theodore George, John Allen, Elizabeth Ann (Mrs. Carl G. Jockusch); m. 2d, Lillian Frances Wright, Feb. 21, 1959; stepchildren—Lila Kathryn Towle, Martha Susan Towle. High sch. tchr., St. Cloud, Fla., 1916-17; prin. high sch., Wauchula, Fla., 1917-18; prof. physics Vassar Coll., Poughkeepsie, N.Y., 1924-52; staff mem. Lincoln Lab., Mass. Inst. Tech., 1952-62; prof. physics U. Miami, Coral Gables, Fla., 1962—. Staff Manhattan Project, N.Y.C., 1944-45, Sandia Corp., Albuquerque, N.M., 1949-50. Home: 552 Alminar Av., Coral Gables, Fla. 33146.*

NORTHROP, Stuart Alvord, Am. geologist; b. Danbury, Conn., Mar. 14, 1904; s. Henry Eugene and Lydia (Alvord) N.; student Robert Coll., Constantinople, 1921-23; B.S. cum laude, Yale, 1925, Ph.D. (Binney fellow), 1929; m. Ivah M. Lewis, Mar. 11, 1930; 1 dau., June Sandra (Mrs. Earl E. Weaver). Faculty, U. N.M., Albuquerque, 1928—, chmn. dept. geology, 1929-61, prof. geology, 1930-62, acting dean Grad. Sch., 1961-62, research prof. geology, 1962—, curator Geology Mus., 1940—, editor U. N.M. Publ. in geology, 1945—; state collaborator seismology U.S. Coast and Geodetic Survey, 1941—; geologist U.S. Geol. Survey, 1943—. Fellow Geol. Soc. Am.; mem. Paleontol. Soc., Seismol. Soc. Am., Am. Assn. Petroleum Geologists, Sigma Xi. Author: Paleontology and Stratigraphy Silurian Gaspe, 1939; Minerals of New Mexico, 1942; New Mexico's Fossil Record, 1962; U. N.M. Contributions in Geology, 1966; also numerous articles. Research in stratigraphy and paleontology, chiefly Paleozoic rocks and fossils, history of mining and minerals in N.M., history of fossil discovery in N.M., N.M. earthquakes. Home: 1804 Las Lomas Rd. N.E., Albuquerque Home: 1404 Valencia Dr. N.E., Albuquerque 87110.*

NORTHRUP, Edwin Fitch, Am. electrothermic engr.; b. Syracuse, N.Y., Feb. 23, 1866; s. Ansel Judd and Eliza Sophia (Fitch) N.; A.B., Amherst, 1891; postgrad. Cornell U., 1891; Ph.D., ohns Hopkins, 1895; D.Sc., Lehigh U., 1932; m. Margaret Jane Stewart, Oct. 9, 1900. In practical elec. work in the West, 1895-96; prof. physics U. Tex., 1896-97; became asst. to Prof. H. A. Rowland, Balt., 1898, in devel. of his multiplex printing telegraph system, later chief constructing engr. Rowland Printing Telegraph Co., until 1902; sec. Leeds & Northrop Co., mfrs. of elec. instruments, Phila., 1903-10; mem. physics faculty Princeton, 1910-20; v.p. and tech. adviser Ajax Electrothermic Corp., Trenton, N.J. Recipient Bronze medal Paris Expn., 1900; Edward Longstreth medal, 1912; Elliott Cresson medal, 1916; Edward Goodrich Acheson Gold medal, 1931. Author: Methods of Measuring Electrical Resistance, 1912; Laws of Physical Science, 1917; Zero to Eighty, 1937. Extended research upon elec. conductivity and properties of matter at elevated temperatures; inventor of Ajax-Northrup high frequency induction furnace; his furnace made possible high temperatures without influencing composition of smelting, 1916. Patentee of methods and numerous devices for inductive heating used throughout the world; developed means for producing high speed linear motions with polyphase currents. Died Apr. 29, 1940.

NORTON, Dorita A., Am. biophysicist; b. Bridgeport, Conn., Mar. 15, 1931; d. Arthur Clinton and Dorita (Besette) Atkins; B.S., Columbia, 1954, M.A., 1956; Ph.D., Bryn Mawr Coll., 1958. With Nat. Cash Register Co., Dayton, O., 1958-59; sr. cancer research scientist Roswell Park Meml. Inst., Buffalo, 1959-62, asso. cancer research scientist, 1962-67, asst. to inst. dir., 1963-67; Head, Dept. of Biophysics, Medical Foundation of Buffalo, Buffalo, 1967—; faculty State U. N.Y. at Buffalo, 1963—, asso. prof. biophysics, 1967—. Mem. Am. Crystallographic Assn., Biophys. Soc., N.Y. Acad. Scis., Buffalo Endocrine Club, Sigma Xi (exec. com. Buffalo chpt.). Research, numerous publs. on crystal and molecular structure determination of steroid compounds and complexes by means of x-ray crystallographic analysis, relationships between steroid molecular and physiol. function, steroid complex formation, formation and properties of artificial lipoprotein membranes. Home: 10 Saybrook Pl., Buffalo 14209. Office: Medical Foundation of Buffalo, Buffalo, 14203.*

NORTON, Horace Wakeman, Am. statistician; b. Lansing, Mich., Jan. 17, 1914; s. Horace Wakeman and Mabel (Reeves) N.; B.Sc., U. Wis., 1935; M.S., Ia. State U., 1937; Ph.D., U. London (Eng.), 1940; m. Anne Dorsey Wallace, Dec. 20, 1937 (div. June 1961); children—Karl, David, Philip, Paul; m. 2d, Winifred McCue, Apr. 6, 1963. Research asso. Inst. Meteorology, U. Chgo., also agt. agrl. marketing service U.S. Dept. Agr., 1940-42; statistician U.S. Weather Bur., Washington, 1942-47, U.S. AEC, Oak Ridge, 1947-50; prof. statis. design and analysis U. Ill., Urbana, 1950—. Adv. com. on nuclear materials safeguards AEC. Mem. Am. Statis. Assn., Am. Genetic Assn., Inst. Math. Statistics, Biometric Soc., Am. Soc. Human Genetics, Inst. Nuclear Materials, Mgmt., Am. Inst. Biol. Scis., A.A.A.S., Am. Soc. Animal Sci., Ill. State Acad. Sci., Sigma Xi. Research, numerous publs. on statis. design expts. and analysis of data. Home: 1502 S. Orchard St., Urbana, Ill. 61801.*

NORTON, John Pitkin, Am. agrl. chemist; b. Albany, N.Y., July 19, 1822; s. John Treadwell and Mary (Pitkin) N.; studied chemistry; m. Elizabeth Marvin, Dec. 15, 1847; 2 sons. Apptd. prof. agrl. chemistry Yale, 1846, helped found dept. sci. edn. which later became Sheffield Sci. Sch. Author: Elements of Scientific Agriculture, 1850; also numerous papers, particularly in field of chemistry of crops. Died Farmington, Conn., Sept. 5, 1852.

NORTON, John T(orrey), Am. metallurgist; b. Medford, Mass., Nov. 13, 1898; s. Charles Ladd and Frances (Torrey) N.; B.S., Mass. Inst. Tech., 1918, Sc.D., 1933; m. Rose Eleanor Demmon, Aug. 17, 1929. Instr. physics Mass. Inst. Tech., 1920-26, asst. prof., 1926-29, asso. prof., 1930-41, prof. metallurgy and physics, 1941—. Fellow Am.-Scandinavian Found., Am. Acad. Arts and Scis.; mem. Am. Inst. Mining Metall. and Petroleum Engrs., Am. Phys. Soc., Am. Soc. for Metals, Brit. Inst. Metals, Am. Soc. Testing Materials, Am. Ordnance Assn., Sigma Xi. Office: Mass. Institute of Technology, Cambridge, Mass.

NORTON, Thomas, English alchemist; identification of this author contested, but most recent research indicates that he was mem. of prominent Bristol family b. circa 1437 and died 1514. Author: The Ordinall of Alchimy (written in 1477 and published in Ashmole's Theatrum Chemium Britannicum, 1652); De transmutatione Metallarum; De Lapide Philosophorum. Norton's Ordinall one of classics of alchemy; recognized importance of color, odor, taste as guides in chemical analysis.

NORTON, Thomas Herbert, Am. chemist; b. Rushford, N.Y., June 30, 1851; s. Robert and Julia Ann Grañger (Horsford) N.; A.B., Hamilton, 1873, Sc.D., 1895; Ph.D., U. Heidelberg, 1875, Dr. Natural Sci., 1936; postgrad. univs. Berlin and Paris, 1876-78; m. Edith Eliza Ames, Dec. 27, 1883; 1 son, Robert Ames. Mgr. large chem. works, Paris, France, 1878-83; prof. chemistry U. Cin., 1883-1900; traveled 12,000 miles on foot through Europe and Asia (1st to traverse Greece and Syria by foot); apptd. by President McKinley to establish U.S. consulate at Harput, Turkey, 1900; sent by U.S. Govt. to investigate conditions in Armenia, 1904, to Persia, 1904-05; Am. consul at Smyrna, Turkey, 1905-06; at Chemnitz, Saxony, 1906-14; detailed under Dept. of Commerce to report on chem. industries of Europe, 1911-12; to further devel. of Am. chem. industries, especially dyestuffs, 1915-17; editor Chem. Engr., 1917-18; chemist E. I. du Pont de Nemours & Co., 1917-20; editor Chemicals, 1920-29; research chemist Am. Cyanamid Co. from 1930. Recipient Lavoisier medal La Société Chimique de France, 1937. Fellow A.A.A.S. (sec. council, 1892; gen. sec., 1893; v.p., 1894), Chem. Soc. (London); mem. Washington, N.Y. acads. scis., 1897-99), Nat. Inst. Social Scis., Internat. Inst. of China (sec. of council), Soc. Chem. Industry, Soc. Chimique de France, Deutsche Chem. Gesellschaft, Am. chem. socs., Phi Beta Kappa, Sigma Xi. Author: Report on Chemical Industry, Paris Exposition, 1878; Utilization of Atmospheric Nitrogen, 1912; Chemical Industries of Belgium, Holland, Norway and Sweden, 1913; Dyestuffs for American Textile Industry, 1915; Cottonseed Industry in Foreign Countries, 1915; The Dyestuff Census, 1916; Training Materials of Latin America, 1917; Reflections at the 70th Milestone, 1921; also many papers on chem., tech. and econ. subjects in jours. Prepared (with W.F. Hillebrand) cerium in coherent form by electrolyzing fused cerous chloride, 1875; issued dyestuff census after cutoff of German dyestuffs by World War I which formed found. of Am. dye industry. Died White Plains, N.Y., Dec. 2, 1941.

NORTON, William A., Am. astronomer; b. East Bloomfield, N.Y., 1810; grad. U.S. Mil. Acad., 1831; m. Elizabeth Emery Stevens, 1839. Asst. prof. natural and exptl. philosophy U.S. Mil. Acad., West Point, 1831-33; prof. natural philosophy and astronomy Coll. City N.Y., 1833-39; prof. natural philosophy and civil engring. Brown U., Providence, 1850-52; prof. civil engring. Sheffield Sci. Sch., Yale, 1852-53. Mem. Nat. Acad. Scis. Author: An Elementary Treatise on Astronomy, 1839; First Book of Natural Philosophy and Astronomy, 1858; also articles. Research on hot-air engine, cause of comet tails, deflection of beams exposed to transverse strains, molecular physics. Died 1833.

NORTON, William Thompson, Am. chemist; b. Damariscotta, Me., Jan. 27, 1929; s. Carroll P. and Josephine (Eales) N.; A.B., Bowdoin Coll., 1950; M.A., Princeton, 1952, Ph.D., 1954; m. Lila Mazur, Sept. 13, 1957; children—William M., Adam E. Research chemist E.I. duPont de Nemours & Co., Wilmington, Del., 1953-57; faculty Albert Einstein Coll. Medicine, Bronx, N.Y., 1957—, asso. prof. neurology, 1964—. Mem. research adv. com. United Cerebral Palsy Found. Mem. Am. Chem. Soc., A.A.A.S., N.Y. Acad. Scis., Am. Soc. Biol. Chemists, Sigma Xi. Research in neurochemistry, lipids and the myelin sheath in normal and diseased brains; devel. new histochem. methods for lipids; determined lipid structures by infra-red spectra; isolated three new galactolipids of brain. Home: 19-A Barstow Rd., Great Neck, N.Y. 11020. Office: Albert Einstein Coll. Medicine, Bronx, N.Y. 10461.*

NORWOOD, Richard, mathematician, astronomer; b. circa 1590; seaman until 1627; tchr. math., London, Eng., circa 1627-40; tchr., surveyor, Bermuda, 1640; made 1st survey, Bermudas, 1616, 2d survey, 1622. Author: Trigonometry, 1631; The Seaman's Practice, 1637; Fortification or Military Architecture, 1639; Norwood's Epitome, 1667. Measured circumference of earth; measured arc of meridian, 1635. Died 1675.

NOSE, Yukihiko, surgeon; b. Sorachi, Hokkaido, Japan, May 7, 1932; s. Minoru and Haru (Murakami) N.; M.D., Hokkaido U., 1957, Ph.D. in Medicine, 1962; m. Marian J. MacDonald, Mar. 15, 1965; 1 dau., Kimi W. Physician charge div. artificial organs and lab. dept. surgery Hokkaido U. Sch. Medicine, 1961-62; research asso. surg. research lab. Maimonides Hosp., Bklyn., 1962-64; 1st postgrad. fellow dept. artificial organs Cleve. Clin. Found., 1964-66, asso. staff, 1966—, charge div. research dept. artificial organs, 1964—. Mem. Am. Soc. Artificial Internal Organs, Japanese Soc. for Artificial Organs, A.A.A.S., Japanese Surg. Soc., Japanese Soc. for Thoracic Surgeons, N.Y. Acad. Scis. Editorial bd. Jour. Biomed. Materials Sci. Contbr. numerous articles to profl. jours. Developer artificial heart to be implanted inside body; investigator for implantable aux. ventricle, developer artificial kidney washing machine for home dialysis, artificial liver utilizing liver homogenate for patients, physiologic access plug. Home: 17500 Euclid St., Cleve. 44112. Office: 2020 E. 93d St., Cleve. 44106.*

NOSTRADAMUS (or Michel de Notredame or Nostredame), French astrologer, physician; b. St. Remy, France, Dec. 14, 1503; ed. Avignon, France; M.D., Med. Sch. Montpellier (France), 1529. Prof. Montpellier; practiced medicine Agez, France, later Salon; began to make predictions, circa 1547; invited to ct. by Catherine de' Medici (drew horoscopes of her sons); apptd. physician-in-ordinary to Charles IX (because of accuracy of predictions). Author: almanac containing weather predictions, 1550; Prophéties (astrological predictions written in rhymed quatrains, read for long time after Nostradamus' death; many have been believed to foretell distant future), 1555. Noted for cures he effected with medicine invented by him during plague at Aix and Lyons (both France), 1546-47; conducted alchem. expts. dealing with benzoic acid, 1556. Died Salon, July 2, 1566.

NOTESTEIN, Frank Wallace, Am. demographer; b. Alma, Mich., Aug. 16, 1902; s. Frank Newton and Mary (Wallace) N.; student Alma Coll., 1919-20; B.S., Coll. of Wooster, 1923, LL.D., 1946; Ph.D., Cornell U., 1927; D.Sc., Northwestern U., 1953, Princeton, 1963; LL.D., Alma Coll., 1964; D.Sc., U. Mich., 1967.m. Daphne Limbach, Oct. 8, 1927. Instr. econs. Cornell U., 1926-27; fellow Social Sci. Research Council, 1927-28; research asso. Milbank Meml. Fund, 1928-29, mem. tech. staff, 1929-36; lectr. Office Population Research, Princeton, 1936-41, dir., 1941-59, prof. demography, 1945-59, vis. research asso., 1959—. Pres., Population Council, N.Y.C., 1959-68; cons., dir. population div. Dept. Social Affairs, UN, 1946-48; adviser to minister of health Govt. India, 1955-56; mem. tech. adv. com. on population U.S. Censuses, 1940, 50, 60, chmn., 1950. Fellow Am. Acad. Arts and Scis.; mem. Council on Fgn. Relations, Am. Statis. Assn., Population Assn. Am., Internat. Statis. Inst., Union Internationale pour l'Etude Scientifique de la Population, Am. Philos. Soc., Am. Social. Soc., A.A.A.S., Phi Beta Kappa, Sigma Xi. Author: (with Regine Stix) Controlled Fertility, 1940; (with Taeuber, Kirk, Coale, Kiser) The Future Population of Europe and the Soviet Union: Population Projections 1940-1970, 1944; also numerous articles. Research in internat. comparative demography, trends and differences in components of population change, particularly problems in regulation of human fertility. Home: 24 Roper Rd., Princeton, N.J. 08540. Office: 230 Park Av., N.Y.C. 10017.*

NOTEVARP, Olav Tollefsen, Norwegian chemist; b. Gvarv, Norway, Feb. 13, 1900; s. Tollef K. and Gunhild (Sannes) N.; ed. Tech. High Sch. of Norway; study tours in Europe, U.S., Can.; m. Thyra Sofie Swenogard, 1924. Pvt. asst. to Prof. C.N. Riiber, 1921-23; sci. asst. Inst. for Organic Chemistry, 1923; research chemist Gollaug Factory, Lier, Belgium, 1923-30; mgr. State Ichthyology Sta., from 1930; dir. Directorate of Fisheries Chem.-Tech. Research Inst., 1948-50; named prof. indsl. chemistry Tech. High Sch. of Norway, 1949. Mem. Tech. Sci. Research Soc., Norwegian Chem. Soc. (pres. from 1954). Research and publs. on chem. and tech. studies of fish and marine oils; patentee prodn. methods for fish meal, herring oil.

NOTHNAGEL, Carl Wilhelm Hermann, physician; b. Alt-Leitzegöricke, Sept. 28, 1841; prof., Vienna, Austria. Editor: Handbuch der speziellen Pathologie und Therapie, 2 vols., from 1894. Research on Addison's disease, heart and stomach diseases, physiology and pathology of nervous system. Died Vienna, July 7, 1905.

NOTT, Josiah Clark, Am. physician; b. Columbia, S.C., Mar. 31, 1804; s. Abraham and Angelica (Mitchell) N.; grad. S.C. Coll. (now U. S.C.), 1824; studied medicine under Dr. James Davis, Columbia; attended Coll. Physicians and Surgeons, N.Y.C., 1825; M.D., U. Pa., 1827; m. Sarah Deas, 1832; 8 children. Demonstrator anatomy U. Pa., 1827-29; practiced medicine, Columbia, 1829-35; an organizer Mobile (Ala.) Med. Soc., 1841; prof. anatomy U. La., 1857-58; a founder Med. Coll. of Ala., 1858, prof. surgery, 1858-61. Mem. N.Y. Obstet. Soc., 1868. Author: (with George R. Gliddon) Types of Mankind, 1854, Indigenous Races of the Earth, 1857; Contributions to Bone and Nerve Surgery, 1866; contbr. paper on anatomy to New Orleans Med. Jour., 1844. Best known for his conclusions on yellow fever (believed it was caused by living organism and had nothing to do with atmospheric conditions), credited with being 1st to suggest its transmission by mosquito, 1848. Died Mobile, Mar. 31, 1873.

NOTT, Otis Fessenden, Am. physician; b. Ballston Springs, N.Y., Mar. 6, 1825; s. Oran Gray and Lucy (Kingman) O.; A.M. (hon.), Union Coll., 1851; grad. N.Y. Med. Coll., 1852; m. Frances Helen Cooke, 1867. Took up landscape painting, 1843, also taught landscaping; ship surgeon in Panama and Pacific service; practiced medicine, N.Y.C., 1860; police surgeon for N.Y.C., 1861-71; pres. med. bd. N.Y. Police Dept., 1869-71; lectr. Coll. Physicians and Surgeons, N.Y.C., 1871-90; 1st to cure stricture in male urethra. Author: Essays Lessons in Landscape, 1856; Stricture of the Male Urethra: Its Radical Cure, 1878; Classroom Lectures on Syphilis and the Genito-Urinary Disease, 1878; Clinical Lectures on the Physiological Pathology of Syphilis and Treatment of Syphilis, 1881. Died New Orleans, May 24, 1900.

NOUET, Nicolas-Antoine, French astronomer; b. Pompey, France, 1740; named prof. Obs. Paris, 1796; calculated 1st elliptic of Uranus; determined latitude and longitude of French cities, 1789. Died Chambery, France, 1811.

NOUVEL, Jacques Charles Georges, French zoologist; b. Nancy, France, Jan. 31, 1909; s. Adrien and Emilenne (Borchard) N.; M.D., veterinarian, M.A., U. Paris (France); 1 dau., Geneviève. Asst., 1935, then under dir. Museum Labs., 1947; then prof. France Overseas Operation and Inst. Vet. Medicine in Tropics; prof. animal ethnological study Nat. Mus. of Natural History, 1956—; dir. Zool. Park, Paris, also menagerie at Garden of Plants. Decorated Legion of Honor, Palmes Acad., Order of Merit (agr.). Mem. French Vet. Acad. Publns. on psycho-physicology and comparative pathology. Address: parc zoologique de Paris, 53 av. de saint-Maurice, Paris 12, France.

NOVACHENKO, Nikolay Petrovich, Russian orthopedist, traumatologist; b. Burin (now Sumy Oblast), 1898; grad. Kharkov Med. Inst., 1922; Cand. Med. Sci., 1937; D.Med. Sci., 1940. Intern, head x-ray sect., clinic head, dep. dir for sci. and med. work Ukrainian Research Inst. Orthopedics and Traumatology, dir., 1940—; asst. dept. orthopedics and traumatology Kharkov Postgrad. Med. Inst., 1925-34, lectr., 1934-40; prof., 1940—; head chair orthopedics and traumatology Kharkov postgrad. Med. Inst., 1941—; founder Kharkov Prosthesis Research Inst., 1945. Chmn. Orthopedic Commn. of learned med. council Ukrainian Ministry Health. Decorated Order of Lenin (2). Mem. USSR Acad. Med. Scis. (corr.), Kharkov Med. Soc. (chmn. 1945—), Oblast (chmn. 1945—), Ukrainian (chmn.), All-Union (dep. chmn.) socs. traumatologists and orthopedists, All-Union Soc. Surgeons (bd. mem.). Author numerous works including: Constant Traction, 1940; Vascularization of a Bone Graft, 1946; Sanazine Treatment of Osteoarticular Tuberculosis, 1951; Principles of Orthopedics and Traumatology, 1961. Chief editor Orthopedics, Traumatology and Prosthetics. Address: Ukrainian Research Inst. of Orthopedics and Traumatology, Pushkinskaya ulitsa 80, Kharkov, Ukrainian SSR, USSR.

NOVAK, Ervin Károly (Charles), Hungarian microbiologist, mycologist; b. Budapest, Hungary, May 31, 1932; s. Károly (Charles) and Elisabeth (Máday) N.; Diplom. Microbiologist, Faculty Natural Sci., Loánd Eötvös U. Sci., 1955, Dr.rer.nat., 1962; m. Margit Eigner, 1954; 1 son, Béla; m. 2d, Magdolna Lengyel, Feb. 8, 1961; 1 dau., Ágota. Asst., Károly Than Tech. Sch. Light Indsl. Chemistry, Lab. Organic Chemistry, Budapest, 1955-57; head dept. mycology State Inst. Hygiene, Budapest, 1957—. Lectr. med. mycology József Fodor Tng. Coll. Hygiene Sch. Med. Lab. Assts., Budapest, 1958-63, lectr. gen., inorganic and organic chemistry, 1959-66. Mem. Hungarian Microbiol. Soc. (sec. mycol. sect.), Assn. Hungarian Chemists (pres. biochem. sect.). Author: (with J. Zsolt, B. Pazonyi, A. Pelc) The Yeasts, 1961; contbg. author: Manual of Laboratory Assistants, vol. I, 1961, vol. II, 1962. Research on carbohydrate metabolism of yeasts and action of inhibitors, simple metabolic inhibitors, antibiotics, chemotherapeutics, etiology and pathomechanism of mycoses. Home: 30 Zichy J. str., Budapest VI. Office: 2-6 Gyáli str., Budapest IX, Hungary.*

NOVAK, Franc, Yugoslavian physician; b. Kranj, Yugoslavia, June 2, 1908; s. Alojz and Franja (Smid) N.; M.D., 1931; specialist obstetrics and gynecology, 1938; Sc.D., 1966; doctor honoris causa U. Torino (Italy), 1965; m. Vida Bernot, Aug. 30, 1946; children—Branko, Ziva. Vol. physician Gynecol. Hosp., Ljubljana, Yugoslavia, 1932-38; asst. Oncological Inst., Ljubljana, 1938-42; partisan's physician, 1942-45; lectr. U. Clin. for Gynecology and Obstetrics, U. Ljubljana, 1945-47, U. Belgrade, 1948-51; prof. U. Clinic for Gynecology and Obstetrics, Ljubljana, 1952—, dir., 1955—. Mem. Assn. Italian Gynecologists (hon.), Slovenian Med. Assn. (hon.), Med. Assn. Czecho-Slovakia (hon.), Assn. Chilean Gynecologists (hon.), Slovene Acad. Scis. and Art (corr.), Slovenian Med. Soc. (pres. gynecol. sect.), Assn. Gynecologists and Obstetricians Yugoslavia (past pres.). Author: Fiziologija zanositve, 1948; co-author: Porodilystuo Medicinska knjiga, 1962; also numerous articles. Editor: Ginekologija, 1961. Developed abdominal operation for cervical cancer of uterus; also other gynecol. operations. Home: 7, Valvazorjeva, Ljubljana, Yugoslavia, 22406.*

NOVAK, Milan Vaclav, Am. microbiologist, educator; b. Cobb, Wis., Dec. 24, 1907; s. Philip and Barbara (Vavrina) N.; B.A., Macalester Coll., 1929, D.Sc., 1947; M.S., U. Minn., 1930, Ph.D., 1932, B.S., 1936, M.B., 1938, M.D., 1938; m. Dorothy Flint, July 28, 1934; children—Mary Dayle (Mrs. William Sand), John Lotus, Raymond William. Asst. prof. bacteriology U. Tenn., 1932-33; instr. U. Minn., 1934-40, dir. Blood Bank, 1938-40; asso. prof. bacteriology and pub. health U. Ill., Chgo., 1940-43, prof. bacteriology, 1943—, head dept. microbiology Coll. Medicine, 1943-64, asso. dean Grad. Coll., Med. Center, 1950—, dir. Blood Bank, 1942-54. Cons. VA Hosp., Hines, Ill., 1946—; pres. Tb Inst. Chgo., 1960-62. Cited for blood filter discovery Chgo. Sun-Times, 1944; recipient testimonial and plaque for work in Tb prevention Tb Inst. Chgo., 1962. Mem. Midwest Grad. Deans (chmn. 1962), Council Grad. Deans and Land Grant Colls. Nat. Assn. State Univs. (pres. 1964), Sigma Xi, Phi Beta Pi. Research on Tb prevention, storage and adminstrn. blood and plasma, antibiotics. Home: 51 N. Lincoln St., Lombard, Ill. 60148.*

NOVAL, Joseph James, Am. biochemist; b. Bklyn., June 14, 1930; s. William F. and Ann (McGettigan) N.; B.S., Manhattan Coll., 1952; M.S., St. John's U., 1954; Ph.D., Rutgers U., 1957; m. Martha C. Mastrandrea, May 30, 1955; children—William Vito, Theresa Ann, Elizabeth Jane, Susan Valarie. Postdoctoral fellow Rutgers U., 1957-58; chief, neurochemistry sect. Bur. Research in Neurology and Psychiatry, Princeton, N.J., 1958—. Mem. A.A.A.S., Am. Chem. Soc., Am. Acad. Neurology, Assn. Research Nervous and Mental Disorders. Research, publs. on synthesis ACTH, biochem. effects phenothiazine tranquilizers; discovered keratinase. Home: 36 Hart Av., Hopewell, N.J. 08525. Office: P.O. Box 1000, NJNPI, Princeton, N.J. 08540.*

NOVARO, Giacomo Filippo, Italian surgeon; b. Sarreta, Italy, 1843; asst., surg. clinic, Turin, 1873-85; became prof. clin. surgery, Sienna, 1885, Bologna, 1890, Genoa, 1898. Author numerous works on surgery. Made important larynx and intestinal operations. Died 1934.

NOVELLI, Guerino David, Am. biochemist; b. Agawam, Mass., Nov. 6, 1918; s. David and Adele (Baiardi) N.; B.S., U. Mass., 1940; M.S., Rutgers U., 1942; Ph.D., Harvard, 1949; m. Ruth Bunker, Oct. 22, 1943; children—Gail Ruth, Dean David. Asso. prof. microbiology Western Res. U. Sch. Medicine, 1953-56; prin. biochemist, group leader in enzymology biology div. Oak Ridge Nat. Lab., 1956—. Mem. physiol. chemistry study sect., radiation study sect. USPHS; mem. AEC Com. on DNA, RNA and Proteins; lectr. in botany, faculty mem. biomed. mem. com. on etiology Am. Cancer Soc.; research prof. biochemistry U. Ga. scis. U. Tenn.; Fellow A.A.A.S.; mem. Am. Soc. Biol. Chemists, Am. Chem. Soc., Am. Soc. Microbiology, Am. Soc. Biol. Chemists, Am. Chem. Soc., Am. Soc. Microbiology. Research, numerous publs. on biosynthesis of Coenzyme A; contbd. to gen. field of mechanism of protein biosynthesis in bacteria, plants and animals. Home: 398 East Dr. Oak Ridge 37831. Office: Biology Div., Oak Ridge Nat. Lab., Oak Ridge 37830.*

NOVICK, Alvin, Am. biologist; b. Flushing, N.Y., June 27, 1925; s. Isidore and Lena (Janowitz) N.; A.B. cum laude in Biochem. Scis., Harvard, 1948, M.D., 1951; Staff, Harvard, 1952-57; faculty Yale, 1957—, asso. prof. biology, 1965—. Mem. Am. Soc. Zoologists, Animal Behavior Soc., Am. Soc. Mammalogists, Ecol. Soc. Am., Soc. Gen. Physiologists, Conn. Acad. Arts and Scis., Sigma Xi. Asso. editor Jour. Morphology, 1961-63, Ecol. Monographs, 1963—, Ecology, 1964—. Research, publs. on acoustic orientation in animals; measured input echoes from electrophysiol. ear and brain records, tng. expts. Home: 769 Yale Sta., New Haven 06520.*

NOVICK, Robert, Am. physicist; b. N.Y.C., May 3, 1923; s. Abraham and Carolyn (Weisberg) N.; M.E.,

Stevens Inst. Tech., 1944, M.S., 1949; Ph.D., Columbia, 1955; m. Bernice Lehrman, July 2, 1947; children—Beth, Amy, Peter. Microwave engr. Wheeler Lab., Inc., 1946-47; faculty Columbia, 1952-54, 57, 60—, prof. physics, 1962—, research asso., 1954-57, dir. Radiation Lab., 1960—; asst. prof. U. Ill., Urbana, 1957-59, asso. prof., 1959-60. Cons. to labs., research insts; chmn. subpanel on atomic and molecular physics Nat. Acad. Sci., 1964-65. A.P. Sloan fellow, 1958—. Fellow Am. Phys. Soc., I.E.E.E.; mem Am. Astron. Soc. Research, publs. on atomic physics, atomic collisons, quantrum electronics, atomic frequency standards, nuclear spins and moments, X-ray astronomy. Home: 366 W. 245th St., Riverside, N.Y.C. 10471.*

NOVICOMENSIS, see Villanovanus, Arnaldus.

NOVIKOFF, Alex Benjamin, biochemist; b. Chernigov, Russia, Feb. 28, 1913; s. Jack and Anna (Trgtyakoff) N.; came to U.S., 1913, naturalized, 1918; B.S., Columbia, 1931, M.A., 1933, Ph.D., 1938; m. Rosalind Shaftel, June 1, 1939; children—Kenneth Reed, Laurence Eric. Asst. prof. biology dept. Bklyn. Coll., 1931-48; prof. exptl. pathology, asso. prof. biochemistry U. Vt. Coll. Med., 1948-53; sr. investigator Waldemar Med. Research Found., 1954-55; research prof. Albert Einstein Coll. Medicine, Yeshiva U., 1955—; Recipient Columbia Distinguished Service award, 1960; USPHS, Nat. Cancer Inst. Career award, 1962. Fellow A.A.A.S.; mem. Am. (pres. 1962-63), Japan socs. cell biology, Histochem. Soc. (pres. 1958-59), N.Y. Soc. Electron Microscopists (pres. 1961-62), Societe Francaise de Microscopic Electronique, Japanese Histochem. Soc., Am. Soc. Biol. Chemists, Am. Assn. Cancer Research, Am. Soc. Exptl. Pathology, Am. Soc. Zoologists, Electron Microscopic Soc. Am., Soc. for Developmental Biology, Harvey Soc., Am. Assn. Pathologists and Bacteriologists, Sigma Xi. Editorial bds. Jour. Histochemistry and Cytochemistry, 1955—, Cancer Research, 1959-64, Jour. Cell Biology, 1965—, Jour. Exptl. Zoology, 1964-66, Jour. Morphology, 1957-62. Research, publs. on biochem. cytology of normal and malignant cells. Home: 51 White Oak St., New Rochelle, N.Y. 10801. Office: Albert Einstein Coll. Medicine, Bronx, N.Y. 10461.

NOVIKOFF, Michel, zoologist; b. Moscow, Russia, Mar. 27, 1876; s. Michel and Marie (Ogorodnikoff) N.; ed. U. Heidelberg (Germany); m. Valentine Volkoff, 1897. Became asst. prof. U. Moscow, 1906; named prof. Polytech. High Sch., Moscow, 1911; named prof., chmn. zoology Lab., U. Moscow, 1917, dean faculty sci., 1918, rector U. Moscow, 1919-20; with dept. Russian Parliament (Imperial Duma), 1912-17; rector free Russian U., Prague, Czechoslovakia, 1923—; prof. zoology and comparative anatomy Charles U., Prague, 1927—. Mem. Royal Soc. Sci. Prague, Zoologic Soc. Prague, Assn. Anatomists, Deutsche Zoologische Gesellschaft. Research and publs. on various problems of zoology and anatomy especially ocular organs, methods of comparative anatomy.

NOVIKOV, Igor Dmitrievich, Russian astrophysicist; b. Moscow, USSR, Nov. 10, 1935; s. Dmitriy Vasilievich and Ksenia (Sheremetsinskaya) N.; astronomer, Moscow U., 1959, candidate phys. and math. scis., 1963; m. Eleonora Viktorovna Kotok, Oct. 17, 1959; 1 dau., Elena Igorevna. Sci. sec. Stenberg Astron. Inst., U. Moscow, 1962-63; sr. sci. worker astron. council USSR Acad. Scis., Moscow, 1963-66; sr. sci. worker Inst. Pricladnoy Mathem., Moscow, 1966—; lectr. relativistic astrophysics Moscow U., 1965—. Mem. Soviet Astron. and Geodesical Soc., Soviet Soc. Knowledge. Author: (with Ya.B. Zeldovich) Relativistic Astrophysics 1967; also articles. Editorial bd. Jour. Earth and Universe, 1965—. Research on theory relativistic stages of slow and catastrophic evolution of ordinary and supermassive stars, theory asymetrical relativistic collapse, evolution of inhomogeneities in expanding universe and formation of galaxies, spectrum of gen. radiation in metagalaxy, research on gravitational relativistic theory. Home Kahovka Str. 29, Bldg. 2, Moscow, USSR.*

NOVIKOV, Ivan Ivanovich, Russian physicist; b. Jan. 16, 1916; ed. Moscow U.; staff various sci. divs. USSR Navy, 1940-48; prof. Moscow Engring. and Physics Inst., 1950—, dir., 1956—. Recipient Stalin prize. Corr. mem. Acad. Scis. USSR (dep. chief sci. sec. Praesidium 1954-57, dir. thermophysics inst. Siberian br. 1957—). Chief editor Atomic Energy, 1956—. Research on thermodynamics of gases and of atomic energy.

NOVIKOV, Peter Sergeevich, Russian mathematician; b. 1901; grad. Moscow U., 1925; D.Physico-Math. Sci., 1935. With Moscow Inst. Chem. Tech., 1929-34; asso. Math. Inst., USSR Acad. Scis., 1934—; prof., 1939—. Decorated Order of Lenin; recipient Lenin prize, 1957. Mem. USSR Acad. Scis. Author: On Algorithmic Unsolved Problems of Word Identity in the Theory of Groups (Lenin prize 1957), 1957. Editor: (symposium) Higher Algebra and Mathematical Analysis, 1963. Research and publs. on set theory, math. physics, logic, theory of great numbers. Address: Math. Inst., USSR Acad. Scis., 1-y Akademichesky prospect 28, Moscow, USSR.

NOVITSKI, Edward, Am. biologist; b. Wilkes-Barre, Pa., July 24, 1918; s. Edward Joseph and Mary (Kalita) N.; B.S., Purdue U., 1938; Ph.D., Cal. Inst. Tech., 1942; m. Esther Ellen Rudkin, Apr. 23, 1943; children—Charles Edward, Barbara Jo, Ellen Mary, Paul David. Asso. prof. zoology U. Mo., 1950-56; head biologist Oak Ridge Nat. Lab., 1956-58; prof. biology U. Ore., Eugene, 1958—; editor Drosophila Information Service, 1960—. Mem. genetics com. NIH, 1964-67; mem. vis. com. biology Brookhaven Nat. Lab., 1967—. Guggenheim fellow U. Rochester, 1947; NSF Sr. Postdoctoral fellow U. Zurich, 1962, CSIRO, Canberra, Australia, 1967. Mem. Am. Inst. Biol. Scis., Genetics Soc. Am., Biometrics Soc., Am. Assn. U. Profs., Am. Soc. Naturalists, Sigma Xi. Research, numerous publs. in behavior and structure of chromosomes at cell div. in Drosophila melanogaster (fruit fly), effects of abnormal behavior on gene frequencies in populations, biochem. genetics of Drosophila. Home: 1690 E. 26th St., Eugene, Ore. 97403.*

NOVOSELOVA, Aleksanrdra Vasilevna, Russian chemist; b. Mar. 11, 1900; grad. Moscow U., 1924. With Moscow U., 1930—, docent, 1935, prof., 1946—. Decorated Order of Lenin; recipient State prize, 1948. Mem. USSR Acad. Scis. (corr.). Co-author: The Structure and Conversion of Beryllium Fluoride Compounds, 1955: The Polymorphism of Beryllium Fluoride, 1956; The Polymorphism of Beryllium Chloride, 1960. Research and publs. on rare elements. Address: Moscow University, Leninskie gory, Moscow B-234, USSR.*

NOVOZHILOV, Dmitriy Antonovich, Russian surgeon, orthopedist; b. 1895; grad. Mil. Med. Acad., Leningrad, 1924, postgrad., 1928-30, 31-39; D.Med. Sci. Head surg. dept. Saki Mil. Sanatorium, 1931-39; lectr. dept. field surgery Kuybyshev Mil. Med. Acad., 1939-40; dep. head chair field surgery Leningrad Mil. Med Acad., 1942-44; dep. head chair naval surgery and traumatology Naval Med. Acad., 1945-52; sci. dir. Turner Child Orthopedics Research Inst., Leningrad, 1953—. Mem. problems commn. learned med. council RSFSR Ministry Health. Decorated Order of Lenin. Mem. All-Union (bd. mem.), Leningrad (chmn.) socs. orthopedists and traumatologists. Author: Problems of Pain in Traumatology and Orthopedics; Bone and Joint Injuries. Mem. editorial bd. Orthopedics, Traumatology and prosthetics. Research and numerous publs. on field surgery, traumatology, orthopedics. Address: Turner Child Orthopedics Research Inst., Leningrad, USSR.

NOVOZHILOV, Valentin Valentinovich, Russian mechanics specialist; b. 1910; grad. Leningrad Physicomech. Inst., 1931. With various research instns., Leningrad, 1931-46; instr. Leningrad U., 1946-49, prof., 1949—; asso. lab. math. and econ. problems Engring. Econs. Inst., 1962. Mem. USSR Acad. Scis. (corr.). Author: The Principles of Non-Linear Elasticity Theory, 1948; The Theory of Thin Coatings, 1951. Research and publs. on theories of elasticity, plasticity and coatings, also their application in shipbldg. Address: Leningrad University, Universitetskaya n. 7-9, Leningrad, USSR.

NOVY, Frederick George, Am. bacteriologist; b. Chgo., Dec. 9, 1864; B.S., U. Mich., 1886, M.S., 1887, Sc.D., 1890, M.D., 1891, LL.D., 1936; studied in Koch's lab., Berlin, 1888, Pasteur Inst., Paris, 1897, at Prague, 1894; LL.D. U. Cin., 1920; m.; children—Robert L., Frank O., Marguerite F., Frederick G., Frances L. Asst., chemistry, 1886; instr. hygiene and physiol. chemistry U. Mich., 1887-91, asst. prof., 1891-93, jr. prof., 1893-1902, prof. bacteriology, 1902-35, also dir. hygienic lab., dean Med. Sch. Mem. U.S. Commn. to investigate plague, 1901; mem. State Bd. Health, 1897-99. Recipient Bausch and Lomb's 250,000th microscope A.A.A.S., 1936. Hon. fellow N.Y. Acad. Medicine, Internat. Coll. Surgeons; mem. Nat. Acad. Scis., Am. Philos. Soc., A.M.A. (mem. council on pharmacy and chemistry, 1905-30, Gold medal 1930); hon. mem. Assn. Am. Immunologists, Harvey Soc., Pathol. Soc. (Phila.), Am. Soc. Tropical Med., Soc. Am. Bacteriol., Am. Trudeau Soc., Assn. Am. Physicians, Am. Acad. Tropical Medicine, Am. Soc. Biological Chemists, Société de pathologie exotique, Paris, Sociéte royale des scis. médicales at naturelles, Bruxelles (asso. mem.); corr. mem. Société de Biologie Paris. Author: Cocaine and Its Derivatives, 1887; Laboratory Work in Bacteriology, 1899; Laboratory Work in Physiological Chemistry, 1898; Cellular Toxins (with Dr. V. C. Vaughan), 1902. Pioneer in bacteriology in U.S.; gave 1st coll. course in bacteriology; discovered, isolated anaerobic bacillus responsible for fatal septicemia in animals, 1894; inventor Novy jar, other lab. equipment; studied spirochetes, proved that in different geog. areas different strains cause relapsing fevers; investigated trypanosomes, microbic respiration. Died Aug. 8, 1957.

NOWACKI, Edmund Kazimierz, Polish biochemist; b. Ostrzeszów, Poland, Dec. 24, 1930; s. Stefan and Jadwiga (Jankowska) N.; student U. Wroclaw (Poland), 1951-53; M.S. in Biochemistry, Jagiellonian U., Cracov, Poland, 1956; Ph.D., Coll. Agr., Poznán, Poland, 1960; Docent, Polish Acad. Scis., 1965; m. Danuta Christine Graboñ, Apr. 24, 1961; children—Margarette, Joanna. Staff, Polish Acad. Scis. Inst. Plant Genetics, Poznañ, 1956-66, head lab. plant physiology, 1966; research asso. German Acad. Scis. Inst. Plant Biochemistry, 1959-60; Rockefeller Found. fellow Mich. State U., 1961-63; head lab. phytochemistry Inst. Soil Sci. and Cultivation of Plants, Przebedowo, Poland, 1966—. Cons., Plant Breeding Inst., Szelejewo, Poland, 1964—. Mem. Polish Bot. Soc., Polish Genetical Soc., Soc. Naturalists. Author: Szyfr Zycia, 1965; also articles. Research on biosynthesis of lupine alkaloids, transformation of lysine to lupanine, interconversion of alkaloids in plants, genetics of chem. characters in plants. Home: Przebedowo 1, Murowana Goslina/nr. Poznan, Poland. Office: 17. Dabrowskiego, Poznan, Poland.*

NOWACKI, Pawei Jan, nuclear engr.; b. Berlin, Germany, June 25, 1905; s. Jan and Katarzyna (Misiek) N.; grad. with distinction, Tech. U. Lwow, 1929, D.S. with distinction, 1936; m. Irena Natalia Sztark, Jan. 5, 1963; children—Marck, Irena (Mrs. Nowacka-Jaskierska). Elec. engr. Siemens Co. Berlin, 1930-33; tech. dir. Cable Factory, 1937-39; with Royal Aircraft Establishment, Farnborough, Eng., 1942-46; prof. chair electric machines Tech. U. Wroclaw, 1947-53; prof. Tech. U. Warsaw, Poland, 1951—, prof., chair nuclear engring., 1958—; dir. gen. Inst. Nuclear Research, Warsaw, 1958—. Head planning Polish State Power Bd., 1947-52. Recipient Polish state award I. degree in sci., 1956, others. Mem. Polish Acad. Scis., Royal Swedish Acad. Tech. Scis. (corr., Merit medal 1958); Inst. Elec. Engrs. (asso.), Internat. Fedn. Atomic Control (pres. 1966—). Author numerous books including: H.T. Transmission Lines, 1951; Symmetrical Components, 1952; Physique des Plasmas À Basses Températures, 1965. Research, publs. on theory of long transmission lines in power engring., nonlinear control systems, multichannel pulse systems in electronic, theory and design of plasma lamps. Home: 34 m.4., Walowa, Warsaw. Office: IBJ, Swierk nr. Otwock, Poland.*

NOWACKI, Werner, Swiss crystallographer; b. Zürich, Switzerland, Mar. 14, 1909; s. Karl and Anna Nowacki; Dr.sc.math., Eidg. Techn. Hochschule, Zürich, 1935; m. Trudy Kaeser; children—Anneliese, Rainer. Faculty, U. Goettingen, 1932-36; faculty U. Bern (Switzerland), 1936—, prof. crystallography and structural scis. dept. crystallography, 1956—; research fellow Cal. Inst. Tech., 1947. Fellow Am. Mineral. Soc. Author: Moderne Allgemeine Mineralogie, 1950; Fouriersynthese von Kristallen, 1952; Fourier Synthesis of Crystals, 1962; Crystal Data Systematic Tables, 2d edit., 1967; also numerous articles. Research on regular partition of space into convex polyhedra, packings of ellipses and ellipsoids, classification of all crystallized substances according to symmetry, crystal structural principles of organic compounds, crystallochem. principles of structures of sulfosalts. Office: 6, Sahlistrasse, Bern, CH-3012, Switzerland.*

NOWELL, Peter Carey, Am. pathologist; b. Phila., Feb. 8, 1928; s. Foster and Margaret (Matlack) N.; B.A., Wesleyan U., 1948; M.D., U. Pa., 1952; m. Helen Walker Worst, Sept. 9, 1950; children—Sharon, Timothy, Karen, Kristin, Michael. Pathologist, U.S. Naval Radiol. Def. Lab., San Francisco, 1954-56; faculty pathology dept. Sch. Medicine, U. Pa., Phila., 1956—, prof., 1964—. Mem. rev. com. bio-med. div. Argonne (Ill.) Nat. Lab., 1964—. Recipient USPHS Career Research award, 1964. Mem. Am. Soc. Exptl. Pathology (Parke-Davis award 1964), Am. Assn. for Cancer Research, Am. Soc. Exptl. Pathology, Radiation Research Soc., Transplantation Soc. Research, numerous publs. on lymphocytes and transplantation immunity, carcinogenesis and chromosomes of cancer. Home: 345 Mt. Alverno Rd, Media, Pa. 19063. Office: Pathology Dept., Sch. Medicine, U. Pa., Phila. 19104.*

NOWICK, Arthur Stanley, Am. physicist; b. N.Y.C., Aug. 29, 1923; s. Hyman and Clara (Sperling) N.; A.B., Bklyn. Coll., 1943; postgrad. Johns Hopkins; A.M., Columbia, 1948, Ph.D., 1950; m. Joan Franzblau, Oct. 30, 1949; children—Jonathan, Steven, Alan, James. Instr., Inst. for Study Metals, U. Chgo., 1949-51; faculty Yale, 1951-57, asso. prof. metallurgy, 1955-57; mgr. metallurgy research group IBM, Yorktown Heights, N.Y., 1957-66; adj. prof. metallurgy Columbia, 1957-66, prof. metallurgy, 1966—. Recipient Achievement award in metallurgy Am. Soc. for Metals, 1963. Fellow Am. Phys. Soc.; mem. Am. Inst. Mining, Metall. and Petroleum Engrs., Sigma Xi. Research, publs. on solid-state physics, particularly on defects in crystals, in internal friction, thin films. Home: 5 Putnam Rd., Scarsdale, N.Y. 10583. Office: Krumb Sch. Mines, Columbia U., N.Y.C.*

NOWINSKI, Wiktor Waclaw, biochemist; b. Lodz, Poland, Nov. 26, 1903; s. Leon and Regina (Bornsztajn) N.; Dr.phil.nat., U. Berne, Switzerland, 1933; Ph.D., U. Cambridge, Eng., 1938; m. Christine Gajewska, May 23, 1933. Chief div. chem. embryology dept. gen. anatomy U. Buenos Aires, Argentina; faculty U. Tex. Med. Sch., Galveston, 1948—, research prof. biochemistry, 1961—, dir. div. biochem. research, 1965—; vis. prof. Coll. of France, Paris, Nat. U. Buenos Aires. Cons. Oak Ridge Inst. Nuclear Studies, 1949-59. Fellow N.Y. Acad. Scis., A.A.A.S.; mem. Argentine Biol. Soc. (corr.), Biochem. Soc., Internat. Soc. Cell Biology, Am. Physiol. Soc., others. Author: Fundamental Aspects of Normal and Malignant Growth, 1960; also numerous articles. Research on biochemistry of early embryonic devel. and growth, active substances regulating growth, mitotic stimulating factor in compensatory hypertrophy of kidney, detoxifying effect of vitamin C in treatment of alcoholics with Antabus, relations between compensatory hypertrophy of kidney and increase in enzymatic activity. Home: 3816 Av. Q, Galveston, Tex. 77550.*

NOWOTNY, Alois Henry, chemist; b. Gyongyos, Hungary, July 30, 1922; s. Alois and Roza (Rusznyak) N.; Ph.D., Pazmany P. U., Budapest, Hungary, 1947; m. Anna Maria Urban, Mar. 2, 1960; children—Sue Zinner, Andrea. Came to U.S., 1960, naturalized, 1965. Clin. chemist Hosp. Gyöngyös, 1947-48; asst. prof. U. Budapest Sch. Medicine, 1948-51; research asso. Hungarian Blood Center, Budapest, 1951-54, vice chmn. research dept., 1954-56; research asso. Wander Research Inst., Freiburg, West Germany, 1956-60; sr. research asso. City of Hope Med. Center, Duarte, Cal., 1960-62; prof. immunochemistry Temple U. Sch. Medicine, Phila., 1962—. Mem. Am. Assn. Immunologists, Hungarian Physiol. Soc., Hungarian Chem. Soc., Am. Soc. Microbiology, Am. Acad Microbiology. Research, numerous publs. on chem. structure of some natural products, analysis of relationship between structure and biol. functions.*

NOWOTNY, Hans, Austrian metallurgist; b. Linz, Austria, Sept. 27, 1911; s. Leopold and Anna (Priesner) N.; Sc.D., Vienna (Austria) Poly.; m. Hildegard Nowotny, Feb. 11, 1944; 1 son, Axel. Asst., Poly. Vienna, also Karlsruhe Poly., 1934-40; with Max Planck Inst. Metallurgy, Stuttgart (Germany), 1941-45; prof. U. Vienna, then hon. prof., 1952-58; titular prof. Vienna Poly., 1947-52; titular prof., dir. Physico-Chem. Inst., Vienna, 1958—. Recipient Lavoisier medal French Chem. Soc. Mem. Austrian Göttingen acad. scis. Publs. on structural, mineral and metall. chemistry. Research organic structural chemistry and metallurgy. Home: Tulpengasse 2, Vienna 8. Office: Währingerstrasse 42, Vienna 9, Austria.

NOWY, Herbert, physician; b. Znaim, Czechoslovakia, Apr. 27, 1916; s. Franz and Rosa (Lang) N.; M.D., German U. at Prague, 1941; m. Elisabeth Ritt, May 31, 1941; 1 dau., Elisabeth Sibylle. Asst., Inst. Exptl. Pathology, German U., 1943-45; med. faculty U. Munich, 1945—, prof. internal medicine, 1961—; research fellow Columbia, N.Y.C., 1954-55. Mem. German Soc. Circulatory Research, German Soc. Internal Medicine. Research, numerous publs. on chem. changes in heart muscle in cardiac hypertrophy and insufficiency, proteins, nucleic acids, analyses of cardiac proteins, application of dye dilution technique. Home: 16 Kornblumenweg, 8 Munich 90, Germany.*

NOYES, Arthur Amos, Am. chemist; b. Newburyport, Mass., Sept. 13, 1866; s. Amos and Anna Page (Andrews) N.; S.B., Mass. Inst. Tech., 1886, S.M., 1887; Ph.D., U. Leipzig, 1890; LL.D., U. Me., 1908, Clark U., 1909, U. Pitts., 1915; Sc.D., Harvard, 1909, Yale, 1913. Asst. analytical chemistry Mass. Inst. Tech., 1887-88, instr., 1890-92, instr. organic chemistry, 1892-94, asst. and asso. prof., 1894-99, prof. theoretical chemistry, 1899-1919, dir. Research Lab. Phys. Chemistry, 1903-07, 09-19, acting pres., 1907-09, dir. Gates Chem. Lab., Cal. Inst. Tech., from 1915. Recipient Davy medal Royal Soc. of London, 1927. Fellow A.A.A.S. (pres. 1927), Am. Acad. Arts and Scis.; mem. Am. Chem. Soc. (Willard Gibbs medal 1915, Richards medal 1932), NRC (chmn. 1918), Nat. Acad. Scis. (editor Proc. 1915-16). Author: Qualitative Chemical Analysis of Inorganic Substances, 1895; Laboratory Experiments on the Class Reactions and Identification of Organic Substances (with S. P. Mulliken), 1899; The General Principles of Physical Science, 1902; Electrical Conductivity of Aqueous Solutions, 1907; Chemical Principles (with M. S. Sherrill), 1921; Qualitative Analysis for the Rare Elements (with W. C. Bray), 1927; Chemistry of Solutions, 1932. Editor Rev. of Am. Chem. Research, 1895-1901. Research (with others) on chem. properties of rare elements, developed complete system of chem. analysis which included them; studied properties of electrolytic solutions; developed (with W.R. Whitney) process for recovery of alcohol and ether vapors which had previously been lost in manufacture of photog. film; founder research lab. of phys. chemistry Mass. Inst. Tech.; noted for work in thermodynamic chemistry, ionic theory, qualitative analysis. Died June 3, 1936.

NOYES, Henry Drury, Am. physician; b. N.Y.C., 1832; grad. N.Y. U., 1851, A.M., 1854; grad. Coll. Phys. and Surg., 1859; m. Isabella Beveridge, 1859; m. 2d, Anna M. Grant, 1870. On surg. staff N.Y. Hosp. nearly 3 years; specialist eye and ear diseases; prof. ophthalmology and otology Bellevue Hosp. Med. Coll., from 1866, also prof. ophthalmology Univ. Bellevue Hospital Med. Coll. Member Am. Med. Assn.; a founder (sec. 10 years, pres. 5 years) Am. Ophthalmol. Soc. Author: Text Book on Diseases of the Eye. First to make known in U.S. use of cocaine as local anesthetic in ophthalmic surgery; 1st to study and record retinitis connected with diabetes mellitus, 1867-68. Died 1900.

NOYES, Richard Macy, Am. phys. chemist; b. Champaign, Ill., Apr. 6, 1919; s. William Albert and Katharine (Macy) N.; A.B. summa cum laude, Harvard, 1939; Ph.D., Cal. Inst. Tech., 1942; m. Winninette Arnold, July 12, 1946. Instr., Cal. Inst. Tech., 1942-44, research asso., 1942-46; faculty Columbia, 1946-58, asso. prof., 1954-58; prof. U. Ore., Eugene, 1958—, head dept. chemistry, 1960-61, 63-64, 66—; vis. prof. U. Leeds (Eng.), 1955-56, Victoria U., Wellington, New Zealand, 1964, Max Planck Institut für physikalische Chemie, Göttingen, Germany, 1965. Mem. at large div. chemistry and chem. tech. NRC, 1963—. Guggenheim fellow, 1955-56; Fulbright fellow, New Zealand, 1964; NSF Sr. fellow, 1965. Mem. Am. Chem. Soc. (chmn. div. phys. chemistry 1962, mem. com. on nominations and elections council 1962—), Am. Phys. Soc., Chem. Soc. (London, Eng.), Phi Beta Kappa, Sigma Xi. Research, numerous publs. on mechanisms reactions organic compounds especially isotopic exchange with elementary halogens, diffusion controlled reactions, reactions of diatomic molecules, thermodynamics of ion hydration. Home: 2014 Elk. Av., Eugene, Ore. 97403*

NOYES, Robert Wallace, Am. physician; b. Berkeley, Cal., June 3, 1919; s. Chester Bromley and Mary (Schmidt) N.; A.B., U. Cal. at Berkeley, 1941, M.D., 1944; postgrad. med. tchr. tng. program U. Ill. Sch. Medicine, 1965; m. Mary Holley, May 30, 1941; children—Martha Holley, Paul David. Research fellow Free Hosp. for Women, Harvard Med. Sch., Boston, 1948-49; asst. resident Stanford U. Hosp., 1949-50, faculty, Med. Sch., 1951-61, asso. prof. Obstetrics and Gynecology, 1960-61; research prof. Agr. Research Council, U. Cambridge (Eng.), 1957-58, Inst. Research Agronomique Station de Recherches de Physsiologie Animale Jouy-en-Josas, France, 1961; prof. obstetrics and gynecology, chmn. dept. Vanderbilt U. Sch. Medicine, 1961-65; prof. anatomy, chmn. dept. U. Hawaii Sch. Medicine, Honolulu, 1965—. John and Mary R. Markle scholar med. sci. mem. Am. Fertility Soc. (dir. 1961-64; Ortho Found. award 1955, Squibb prize 1959, Rubin award 1959, 61, 62. Ortho Medal, 1965), Soc. Study Fertility (Eng.). Am. Coll. Obstetricians and Gynecologists, Soc. Gynecol. Investigation, A.M.A., Am. Assn. U. Profs., Soc. Profs. Obstetrics and Gynecology, Internat. Fertility Assn., Sigma Xi. Home: 2342 Ferdinand Av., Honolulu 96822.*

NOYES, William Albert, Am. chemist; b. nr. Independence, Ia., Nov. 6, 1857; s. Spencer W. and Mary (Packard) N.; A.B. and S.B., Ia. (now Grinnell) Coll., 1879, D. Sc., 1929; Ph.D., Johns Hopkins, 1882; LL.D., Clark U., 1909; Chem.D., U. Pitts., 1920; m. Flora E. Collier, Dec. 24, 1884 (dec.); children—Helen Mary, Ethel, William Albert, m. 2d, Mattie L. Elwell, June 18, 1902 (dec); 1 son, Charles Edmund; m. 3d, Katharine Haworth Macy, Nov. 25, 1915; children—Richard Macy, Henry Pierre. Prof. chemistry U. Tenn., 1883-86; prof. Rose Poly. Inst., 1886-1903; chief chemist Nat. Bur. Standards, 1903-07; prof. chemistry and dir. chem. lab. U. of Ill., 1907-26 (emeritus). Recipient Nichols medal, 1908, Willard Gibbs medal, 1919, Priestly medal, 1935. Fellow Am. Acad. of Arts and Scis.; mem. Nat. Acad. Scis., Am. Philos. Soc., Am. Chem. Soc. (sec. 1903-07, pres. 1920, editor Jour. 1902-17, Sci. Monographs, from 1919); chmn. chem. sect. A.A.A.S., 1896, 1918. Author: Organic Chemistry for the Laboratory, 1897; Elements of Qualitative Analysis, 1888; Organic Chemistry, 1903; Kurzes Lehrbuch der organischen Chemie (translation), 1907; Text-book of Chemistry, 1913; Laboratory Exercises in Chemistry, 1917; College Textbook of Chemistry, 1919; Building for Peace, 1923; Building for Peace, II; Pour la Paix (translation), 1924; Organic Chemistry, 1926; Modern Alchemy (with W. Albert Noyes, Jr.), 1932; also many papers. Editor Chem. Abstracts, 1907-09, Chem. Revs., 194-26. Developed standard methods of analysis and standard specifications for chemicals which formed basis of U.S. Bur. Standards procedures; research in organic synthesis, structure of camphor, electronic theory of valence, atomic weights of chlorine and hydrogen, oxidation of benzene derivatives; noted for analytical methods for determination of phosphorus, titanium, manganese in steel. Died Oct. 24, 1941.

NOYES, William Albert, Jr., Am. chemist, educator; b. Terre Haute, Ind., Apr. 18, 1898; s. William Albert and Flora (Collier) N.; student U. Ill., 1916-17, D.Sc., 1964; A.B., Grinnell Coll., 1919, D. Sc., 1946; D.-ès-Sc., U. Paris (France), 1920, D.Sc., 1957; postgrad. U. Geneva (Switzerland), U. Cal.; D.Sc., U. R.I., 1953, Ind. U., 1959, U. Ottawa, 1960, U. Montreal, 1961, Carleton U. (Ont. Can.), U. Rochester, 1965; m. Sabine Onnilon, June 10, 1921; 1 son, Claude Charles. Faculty, U. Cal., 1921-22, U. Chgo., 1922-29, Brown U., 1929-38; prof. phys. chemistry U. Rochester (N.Y.), 1938-40, chmn. dept. chemistry, 1939-55, Charles Frederick Houghton prof., 1940-60, dean Grad. Sch., 1952-56, acting dean Coll. Arts and Sci., 1956-58, distinguished univ. prof. chemistry, 1961-63, prof. emeritus, 1963—; Ashbel Smith prof. chemistry U. Tex., Austin, 1963—. Chmn. div. chemistry and chem. tech. NRC, 1947-53, chmn. various coms., 1948-50, 62—. mem. spl. mission to implement vis. prof. program UAR, 1959; mem. U.S. Nat. Commn. for UNESCO, 1951-62, v.p., 1956-58; mem. panel com. on sci. and astronautics U.S. Ho. Reps., 1961; sr. chemist emeritus Argonne Nat. Lab. Recipient King's medal Brit. Empire, 1948; medal for merit U.S. Govt., 1948; Austin M. Patterson award for service to chem. lit., 1963; officer Legion of Honor, France, 1953. Mem. Nat. (council 1947-50), Ill. acads. scis., Am. Chem. Soc. (pres. 1947, chmn. com. on internat. activities 1962-63, Priestley medal 1954, Willard Gibbs medal Chgo. sect. 1957), Am. Acad. Arts and Scis., Am. Phys. Soc., Am. Philos. Soc. (council 1963-66), Internat. Union Pure and Applied Chemistry (pres. 1959-63), Phi Beta Kappa, Sigma Xi, Sigma Alpha Epsilon, Alpha Chi Sigma, Phi Lambda Upsilon (hon.), others. Author: (with W.A. Noyes) Modern Alchemy, 1932; (with P.A. Leighton) Photochemistry of Gases, 1941; (with S. Weiss) Traité de Chimie Physique, 1925; Spectroscopie et les Réactions Initiées par la Lumieré, 1938. Editor: Chem. Revs., 1939-49, Jour. Am. Chem. Soc., 1950-62, Jour. Phys. Chemistry, 1952-64. Research, publs. in electrochemistry, discharge through gases, vapor pressures, thermal reactions, effects of radiant energy on various chem. processes, photochemistry and reaction kinetics, especially reaction of free radicals produced by absorption of light. Home: 5109 Lucas Lane, Austin, Tex. 78731.

NOZAKI, Hitosi, Japanese chemist; b. Okayama Pref., Japan, Jan. 1, 1922; s. Tutazi Terao and Matu Nozaki; B.Sc., Kyôto U., 1943, Ph.D., 1949; m. Yôko Udo, Mar. 29, 1960; children—Ikuko, Miwako, Kyôko, Yûko. Research fellow Kyôto U., 1943-46, faculty, 1946—, prof. dept. indsl. chemistry Faculty Engring., 1963—; research asso. Cornell U., Ithaca, N.Y., 1957. Mem. Chem. Soc. Japan, Am. Chem. Soc. Research, numerous publs. on organic reactions and synthetic applications. Home: 11-27 Tennô-Tyô, Takatuki-Si, Osaka-Pref., Japan. Office: Dept. Indsl. Chemistry, Faculty Engring., Kyôto U., Yosida, Sakyô-Ku, Kyôto, Japan.*

NOZOE, Tetsuo, Japanese chemist; b. Sendai, Japan, May 15, 1902; s. Jyuichi Kinoshita and Toyoko Nozoe; B.Sc. in Chemistry, Tohoku Imperial U., 1926; D.Sc., Osaka Imperial U., 1936; m. Kyoko Horiuchi, Nov., 1927; children—Takako (Mrs. S. Masamune), Shigeo, Yoko (Mrs. Ishikura), Yuriko. Research asst. Central Research Inst., Formosan Govt., Taihoku, 1927-29; asst. prof. Taihoku Imperial U., 1929-37, prof. organic chemistry, 1937-45; prof. Nat. Taiwan U., 1945-48; prof. dept. chemistry Tohoku U., Sendai, Japan, 1948-66, emeritus, 1966—. Recipient Majima award Chem. Soc. Japan, 1944; Asahi Newspaper award, 1952; Japan Acad. award, 1953, Order of Cultural Merit, Japanese Govt., 1959. Author: Non-benzenoid Aromatic Compound, 1959; also numerous articles. Discovered novel aromatic system; research on chemistry of troponoids, azulenoids and related compounds. Home: 83 Kita-8-bancho, Sendai, Japan.*

NUCK, Anton, Dutch anatomist, surgeon; b. Harderwyk, Holland, 1650; practiced medicine, the Hague; became prof. anatomy and surgery, Leiden, Netherlands, 1687. Author: Adrenographia, 1692; Sialographia et ductum aquosorum anatome nova, 1695. Suggested that dental instruments for extraction be shaped to conform to individual teeth; noted for anat. descriptions of various glands; described diverticulum or sac of peritoneum accompanying round ligament of uterus through inguinal canal (canal of Nuck, or Nuck's diverticulum), 1691. Died 1692.

NUERNBERGK, Erich Ludolf, botanist; b. Cairo, Egypt, Mar. 30, 1900; s. Ludwig and Helene (Schilling) N.; ed. in botany; m. Agnes Helene, Aug. 1, 1934; children—Dorothea Beate, Detley. Asst., U. Munich (Germany); instr. U. Fribourg (Germany); now. prof. botany U. Hamburg (Germany). Rockefeller fellow. Mem. German Bot. Soc., Union Applied Botany, German Orchid Soc. Author: Physiologie des Tagesschlaf der Pflanzen, 1925; Phototropisms und Wachstrum der Pflanzen, 1932-35; Kunstlicht und Pfanzenkultur, 1961. Research on physiology and technique of protected cultivation of plants; phys. and ecol. plant physiology; physiology of orchids. Home: Reye 6. Office: Jungiusstrasse 6, Hamburg 36, West Germany.

NUMA POMPILIUS, 2d king of Rome; b. Rome, Italy, 715 B.C.; ordered that an abdominal sect. to deliver child be performed on women dying in late pregnancy or childbirth (lex regia; under Caesars called lex Caesaria). Died 673 B.C.

NUMATA, Makoto, Japanese plant ecologist; b Ibaraki, Japan, Nov. 27, 1917; s. Kamenosuke and Tami Hayashi N.; grad. Tokyo U. Lit. and Sci., 1942; D.Sc., Kyoto U., 1953; m. Sadako Sagara, Apr. 8, 1949; children—Michiko, Takako. Lectr., Tokyo Higher Normal Sch., 1947-49, Tokyo U. Lit. and Sci., 1948-50; asso. prof. Chiba U., 1949-63, prof., 1964—, leader Himalayan expdn. Eastern Napal, 1963; prof. Tohoku U., 1964—. Mem. nature conservation com. Japan Sci. Council, 1962—; councilor Natural Park Council, Ministry Pub. Welfare, 1962—; Mem. Missâo Cientifica Japonesa, 1966-67. Recipient research prize Tokyo U. Lit. and Sci.,

1950; Edn. Service award Chiba Prefecture, 1953. Mem. Ecol. Soc. Am., Am. Inst. Biol. Sci., Ecol. Soc. Japan, Bot. Soc. Japan, Japan Soc. Grassland Sci., Japan Forestry Soc., History Sci. Soc. Japan. Author: Methodology of Ecology, 1953; Standpoint of Ecology, 1958; Ecological Study and Mountaineering of Mt. Numbur in Eastern Napal, 1965; also numerous articles. Editor: Plant Ecology, 1959; History of Modern Biology, 1962; Applied Ecology, 1963. Statis. analyses of structure of plant communities; structure and succession of coastal dune communities, grasslands, bamboo forests, pine and broadleaved forests; ecology of temperate bamboo forests; quantitative measurement of plant succession by degree of succession; weed ecology; biota and ecology of Nepal Himalaya. Home: 2/74Bentencho, Chiba-shi. Office: 33/1 Yayoicho, Chiba-shi, Japan.*

NUMEROF, Paul, Am. chemist; b. Phila., May 7, 1922; s. Jacob and Sophie (Saale) N.; B.S., Temple U., 1944; M.S., Carnegie Inst. Tech., 1948, D.Sc., 1949; m. Claire S. Slachowitz, Feb. 25, 1945; children—Rita Ellen, Norman David. With spl. engr. detachment Manhattan project, Los Alamos, 1944-45; research asso. Squibb Inst. for Med. Research, New Brunswick, N.J., 1949-50, head radioisotope dept., 1950-57; mgr. radiopharm. services E.R. Squibb & Sons, N.Y.C., 1957—; adj. asst. prof. Drexel Inst. Tech., Phila., 1950-57. Mem. com. on standards radioisotopes in medicine Internat. Pharmacopaeia, WHO, 1959—; cons. on nuclear medicine N.J. Commn. on Radiation Protection, 1967—; mem. adv. com. N.J. Commn. on Radiation Protection, N.J. Dept. Health, 1967—. Mem. Soc. Nuclear Medicine (pres. Greater N.Y. chpt. 1966-67), Am. Chem. Soc., Health Physics Soc., A.A.A.S., Am. Nuclear Soc., Sigma Xi. Research, publs. on devel. radioactive pharms. for clin. diagnosis and treatment of disease, pharm. agts. for physiol. studies. Patentee radioactive pharms. Home: 10 Ryan Rd., Edison, N.J. 08817. Office. 25 Kennedy Blvd., East Brunswick, N.J. 08816.*

NUNEZ, Pedro, Portuguese mathematician; b. Alcacer-do-Sol, Portugal; student Lisbon, Portugal, Salamanca, Spain; apptd. insp. gen., Goa, Portuguese India, 1519; became royal cosmographer, Portugal, 1529; became prof. moral philosophy U. Lisbon, 1530, prof. philosophy, 1541-44; prof. math. U. Coimbra (Portugal), 1544-62. Author: De crepusculis, 1542; De arte atque ratione navigandi, 1546; Livro de algebra en arithmetica y geometria, 1567. Translated parts of Ptolemy. Invented nonius, an instrument for accurate measurement of small angles which was replaced by vernier, also loxodromic line; studied astronomy, nav. geometry, map projection. Died Coimbra, Aug. 11, 1578.

NUNEZ-MONTIEL, Otto Luis, Venezuelan virologist; b. Maracaibo, Venezuela, May 24, 1932; s. José Luis and Ana (Montiel) Núñez; B.S., Venezuela Central U., M.D., 1956; m. Josephyne Aranda, Nov. 25, 1957; children—Jody Lisa, Mónica, Luis Vocem Otto Luis. Jr. investigator, research fellow Yale, 1957-58; staff IVIC, Caracas, Venezuela, 1958—, head gastroenteritis study dept., 1960-63, chief microbiology dept., 1963—; hon. prof. virology Colls. Sci. and Medicine, Venezuelan Central U. Adviser, Commn. on Respiratory Diseases, Ministry Pub. Health, Venezuela. Decorated Coat of Arms, City of San Juan, P.R. Mem. Internat. Acad. Pathology, Am. Soc. for Microbiologists, Soc. Pediatric Investigation, Venezuelan Soc. Pathology, Venezuelan Soc. Microbiology, Latin Am. Assn. Microbiology, Am. Soc. Tropical Medicine and Hygiene, N.Y. Acad. Scis. Research, publs. on ECHO virus in cultivated cells and their different states of formation using electron-microscope, enteroviruses in Venezuela. Home: 957 Calle La Cantera, La Trinidad, Caracas. Office: IVIC, Caracas, Venezuela.*

NUNGESTER, Walter James, Am. immunologist, educator; b. Lima, O., Feb. 22, 1901; s. William A. and Margery (Woodworth) N.; B.S., U. Mich., 1923, M.S., 1924, Sc.D., 1928; M.D., Northwestern U., 1934; m. Lucile E. Roush, Sept. 21, 1924; children—Margery Patricia (Mrs. Peter S. Wright), Nancy Ann (Mrs. Robert E. Meader). Instr. bacteriology Northwestern U. Med. Sch., Chgo., 1928-33, asst. prof., 1933-36; asso. prof. immunology U. Mich., Ann Arbor, 1936-47, prof., 1947—, chmn. dept. microbiology, 1952—. Cons. Parke-Davis Co., 1946—, sec. army, 1957—. Recipient Sci. award Am. Acad. Tb Physicians, 1952; Centennial Distinguished Merit award Northwestern U., 1959. Mem. Am. Soc. for Microbiology (pres. 1951). Author articles on pathogenesis of exptl. pneumococcus pneumonia, action of anti-tumor serum in vivo, infrared spectrophotometry in differentiating tubercle bacilli, host def. mechanisms in rat, guinea pig and man to profl. publs.; discoverer role of mucin in lowering host resistance. Home: 2109 Tuomy Rd., Ann Arbor, Mich. 48104.*

NUNN, John Francis, English anesthetist; b. Leeds, Eng., Nov. 7, 1925; s. Francis and Lillian (Davies) N.; student Wrekin Coll., 1939-43; M.B., Ch.B., Birmingham U., 1948, Ph.D., 1959; m. Sheila Ernestine Doubleday, Sept. 25, 1949; children—Geoffrey, Carolyn, Shelley. Research fellow U. Birmingham, 1955-56; Leverhulme Research fellow Royal Coll. Surgeons, Eng., 1957-63; lectr. U. London. Postgrad. Med. Sch., 1960-64; cons. anesthetist Hammersmith Hosp., London, 1960-64; external sci. staff Med. Research Council Gt. Britain, 1964; prof. anesthesia U. Leeds, 1964—; hon. cons. anesthetist United Leeds Hosps., 1964—. Mem. bd. faculty of anesthetists Royal Coll. Surgeons of Eng., 1966—. Fellow Royal Soc. Medicine, Assn. Anesthetist mem. Med. Research Soc. Author: (with G.R. Kelman) Computer Written Physiological Tables, numerous articles. Cons. editor Brit. Jour. Anaesthesia, 1960—. Investigation of respiratory and circulatory function in anesthetized patient, together with devel. work on analytical methods which are required for this work. Home: East Garth, School Lane, Collingham, Wetherby, Yorkshire, Eng. Office: U. Leeds, Leeds, Eng.*

NUNNEMACHER, Rudolph Fink, Am. zoologist; b. Milw., Mar. 21, 1912; s. Henry Jacob and Gertrude Anita (Fink) N.; B.Sc., Kenyon Coll., 1934; M.A., Harvard, 1936, Ph.D., 1938; m. Sylvia Acken Hendricks, Dec. 29, 1938; children—Robert, Sallie, Gretl, Dorothea. Instr. histology Okla. U. Med. Sch., 1938-39; mem. faculty Clark U., 1939—, prof. zoology, 1956—, chmn. dept. biology, 1958—, dir. Evening Coll., 1953-54, dir. Summer Sch., 1954-56. Mem. Am. Soc. Zoology, N.Y. Acad. Scis., Electron Microscopy Soc. Am., Bermuda Biol. Sta., Sigma Xi. Research on comparative study electron microscopy nerves; analysis by electron microscopy of composition, in numbers and sizes of individual nerve fibers, of major portions of nervous system of crabs and related forms with spl. emphasis on optic nerve. Home: Putnam Hill Rd., Sutton, Mass. 01527. Office: Clark Univ., Worcester, Mass. 01610.*

NURMIA, Matti Juhani, physicist; b. Rauma, Finland, Aug. 26, 1930; s. Veli Manne and Elli (Söderman) N.; M.A., U. Helsinki, (Finland), 1952, Ph.D., 1957; m. Toini Inkeri Tyrväinen, Apr. 6, 1952; children—Erkki Juhani, Ilkka Antero. Physicist, Atomic Energy Co., Finland, 1955-60; asst. prof. physics U. Ark., 1957-58; asso. prof. physics U. Helsinki, 1959-67; physicist Seismol. Lab., 1958-62; vis. research prof. Okla. State U., 1962-63; vis. physicist Lawrence Radiation Lab., Berkeley, Cal., 1965, physicist, 1966—. Mem. Am. Phys. Soc., Phys. Soc. Finland, Geophys. Soc. Finland, Astron. Soc. Finland. Research, publs. on nuclear physics, geophysics, solid state physics, mixed crystals; satellite observations; discovered new isotopes and alpha emitters; developed exptl. techniques and instrumentation for seismology. Office: Lawrence Radiation Lab., Berkeley, Cal. 94720.*

NURMIKKO, Veikko Toivo, Finnish biochemist; b. Uusikaupunki, Finland, Dec. 17, 1918; s. Emil H. and Martta (Friman) N.; Ph.D., U. Helsinki (Finland), 1955; m. Maija-Liisa Liljeblad, July 24, 1941; children—Arto, Turo, Reima, Mika. Research asst. Biochem. Inst., Helsinki, 1947-56; docent U. Helsinki, 1956-61; faculty U. Turku (Finland), 1956—, prof. biochemistry 1957—, head dept. biochemistry, 1959—; research biochemist U. Cal. at Berkeley, 1961-62, 64-66. Recipient Gust. Komppa award, 1955. Mem. Finnish, Am. chem. socs., Biochem. Soc. (London). Author: Growth Chemistry of Lactic Acid Bacteria, 1964; also articles. Research, publs. on metabolism of symbiotic growth of bacteria, formation and control of enzymes as function of bacterial growth cycle, search of new enzymes (pantothenate hydrolase). Home: 1 C 31 Multavierunk., Turku, Finland.*

NURNBERGER, John Ignatius, Am. physician; b. Chgo., Apr. 9, 1916; s. L. L. and Anna M. (Burke) N.; B.S., Loyola U., 1938; M.S. (Frederick Robert Zeit fellow), Northwestern U., 1942, B.M., 1942, M.D., 1943; research fellow cytochemistry of nervous system Med. Nobel Inst. Cell Research and Genetics, Stockholm, Sweden, 1949-50; m. Mary Cornelia McNulty, Aug. 23, 1943; children—John, Joanna, David Cornelia, Raoul. Dir. labs., chief cyto-chemistry lab. Inst. Living, Hartford, Conn., later ednl. dir.; former asst. clin. prof. medicine (neurology) and psychiatry Yale Sch. Medicine; prof., chmn. dept. psychiatry, acting dean Ind. U. Med. Sch., also dir. Inst. Psychiat. Research, Ind. U. Med. Center. Mem. tng. com. basic biol. scis. Nat. Inst. Mental Health. Fellow Am. Psychiat. Assn.; mem. Central Neuropsychiat. Assn., Soc. for Biol. Psychiatry, Assn. Research Nervous and Mental Diseases, A.M.A., Am. Neurol. Assn., Histo-chem. Soc., Nat. Multiple Sclerosis Soc. (research rev. panel), Alpha Kappa Kappa, Alpha Omega Alpha. Author: An Introduction to the Science of Human Behavior, 1963. Mem. editorial bd. Jour. Nervous and Mental Diseases. Work in analysis of cellular chem. responses to low temperature and metabolic forms of stress in brain with spl. emphasis on nucleic acids, cellular protein and amino acids, chromosomal changes in mental and emotional disorders. Home: 5215 Washington Blvd., Indpls. 46220. Office: 1100 W. Michigan St., Indpls. 46207*

NURSALL, John Ralph, Canadian zoologist; b. Regina, Sask., Can., Dec. 25, 1925; s. John and Helen (Little) N.; B.A., U. Sask., 1947, M.A., 1949; Ph.D., U. Wis., 1953; m. Mary Stewart, Dec. 29, 1953; children—John Colin, Alan Stewart, Catherine Mary. Instr., U. Sask., 1948-49; faculty U. Alta., Edmonton, 1953—, prof., head dept. zoology, 1964—. Nuffield Found. traveling fellow, 1962-63. Mem. A.A.A.S., Am. Soc. Zoologists, Canadian Soc. Zoologists, Soc. for Study Evolution, Zool. Soc. London, Sigma Xi. Research, publs. on productivity of lakes and reservoirs, means by which fish-like animals swim, evolution and distbn. of fossil fishes, consequences of evolution. Home: 12912 65 Av., Edmonton, Alta., Can.*

NUSSBAUM, Mirko M., physicist; b. Belgrade, Yugoslavia, July 24, 1930; s. Munio and Sara (Alkalai) N.; came to U.S., 1944, naturalized, 1952; A.B., Rutgers U., 1954; M.S., U. Chgo., 1956; Ph.D., Johns Hopkins, 1962; m. Audrey Powell, Dec. 27, 1959. Instr., Johns Hopkins, 1961-62; research asso. Columbia, 1962-64; asst. prof. U. Pa., Phila., 1964—; guest research asso. Brookhaven Nat. Lab., 1961—. Mem. Am. Phys. Soc., Phi Beta Kappa, Sigma Xi. Co-discoverer eta meson S^0 meson; determination of omega meson lifetime, sigma lambda relative parity. Home: 4431 Osage Av., Phila. 19104.*

NUSSBAUM, Rudi Hans, physicist; b. Furth, Germany, Mar. 21, 1922; s. Hermann and Eleonore (Mager) N.; Ph.D., U. Amsterdam (Netherlands), 1954; m. Laureen Klein, Oct. 15, 1947; children—Ralph E., Fred D., Doreen Ellen. Came to U.S., 1957, naturalized, 1963. Scientist, Inst. Nuclear Research, Amsterdam, 1952-54; research asso. physics dept. Ind. U., 1955-56; sr. fellow European Center for Nuclear Studies, Geneva, Switzerland, 1956-57; asst. prof. U. Cal. at Davis, 1957-59; asso. prof. physics Portland (Ore.) State Coll., 1959-65, prof., 1965—; vis. prof. U. Wash., Seattle, 1965-66. Cons. Tektronix, Inc., 1961-63. Mem. Am. Assn. Physics Tchrs. (past pres. Ore. sect.), Am., Netherlands Phys. Soc., Am. Assn. U. Profs. Author: (monograph with J.B. Gerhart, R.H. Nussbaum), Motion, 1965; also articles. Research on energy levels in atomic nuclei using radioactive emission of beta and gamma rays (nuclear spectroscopy), detailed properties of solid state matter using Mössbauer effect. Home: 344 N.W. Macleay Blvd., Portland, Ore. 97210.*

NUTTALL, George Henry Falkiner, biologist; b. San Francisco, July 5, 1862; s. Robert Kennedy and Magdalena (Parrott) N.; ed. U.S., Germany, Eng., France, Switzerland; M.D., U. Cal., 1884; Ph.D., U. Göttingen, 1890; hon. degrees Cambridge, other univs.; m. Paula von Oertzen, Apr. 26, 1894. Asst. and later asso. in hygiene, Johns Hopkins, 1891-94; hon. asst. Hygiene Inst., Berlin, 1894-99; univ. lectr. bacteriology and preventive medicine, later reader in hygiene U. Cambridge (Eng.), 1900-06, Quick prof. biology, from 1906, also dir. Molteno Inst. Fellow Christ's Coll., Magdalent Coll., Cambridge; examiner in pub. health and tropical medicine and hygiene Cambridge, tropical medicine U. Liverpool, also Royal Army Med. Corps. Fellow Royal Soc., 1904; fgn., hon. or corr. mem. Am. Acad. Arts and Scis., Royal Acad. Medicine, Turin, Royal Soc. Medicine, Budapest, Soc. Biology, Paris, Microbiology Soc., Leningrad, Harvey Soc., N.Y., other sci. socs. Author: Hygienic Measures in Relation to Infectious Diseases (published in both English and German), 1893; Blood Immunity and Blood Relationship, 1904; The Bacteriology of Diphtheria (with Graham Smith and others), 1908, 13; Ticks (with C. Warburton and others), from 1808; also papers. Founder, chief editor Jour. Hygiene, 1901-37, Jour. Parasitology, 1908-33. Discovered defibrinated blood possesses strong bactericidal property against anthrax bacilli which disappears when blood is heated to 55 degrees Centigrade, 1888, thus founded study of humoral immunity; discovered (with W.H. Welch) causative agt. of gas gangrene, 1892; demonstrated role of arthropods in spread of bacterial and parasitic disease, 1899; investigated precipitin reaction, 1901; showed that bacteria in alimentary canal are required for perfect digestion, 1905; authority on tropical diseases; worked on cures for piroplasmis; investigated (with C. Warburton and L. E. Robinson) anatomy, biology, life history, systematics of ticks. Died London, Dec. 16, 1937.

NUTTALL, Thomas, English naturalist; b. Settle, Eng., Jan. 5, 1786; s. Jonas Nuttall. Studied printing with uncle, Liverpool, Eng.; came to U.S. (Phila.), 1808; inspired by Benjamin Smith Barton to pursue botany; investigated flora up Missouri River, 1809-11; went along Arkansas and Red rivers in Ark., La., Indian territories, 1818-20; with Wyeth Expdn. to mouth of Columbia River, 1834-35; curator Harvard Bot. Gardens, 1822-33; delivered paper to Phila. Acad. Natural Scis. entitled Observations on the Geological Structure of the Valley of the Mississippi (1st attempt in Am. to correlate by means of fossils geog. formations widely separated geographically). Fellow Linnaean Soc., London; mem. Am. Philos. Soc., Phila. Acad. Natural Scis. Author: The Genera of North American Plants, and a Catalogue of the Species, to the Year 1817, 1818; A Journal of Travels into the Arkansas Territory during the Year 1819, 1821; A Manual of the Ornithology of the United States and Canada, 1832; An Introduction to Systematic and Physiological Botany; wrote supplement to Michaux's North American Sylva, 3 vols., 1846; also many papers. Explored vast areas of U. S. never before visited by botanists. Died Liverpool, Eng., Sept. 10, 1859.

NUTTIN, Joseph Remi, Belgian psychologist; b. nr. Bruges, Belgium, Nov. 7, 1909; s. Remi C. and

Adelaide (Toeloose) N.; Dr. in Philosophy and Letters, U. Louvain (Belgium), 1934, Dr. philosophiae in Psychology, 1941. Faculty, U. Louvain, 1941—, prof., 1946—; vis. prof. Kan. U., 1956; pres. Psychol. Inst., U. Louvain, 1965—, dean Faculty Psychology, 1968—, dir. lab. gen. exptl. psychology; dir. Internat. Center Research in Human Motivation. Pres. psychology commn. Nat. Found. Research in Belgium. Decorated Comdr. Order Leopold. Mem. Internat. Union Psychol. Sci. (chmn. com. on publ. and communication 1966—), Belgian Psychol. Soc. (pres.), Association de psychologie de langue francaise (past pres.), Internat. Union Psychology (exec. com.). Fellow Brit. Psychol. Soc. Author: Psychoanalysis and Personality, 1953; Tâche, réussite et échec: une theorie du comportement humain, 1953; Structure de la Personnalité, 1965; Reward and Punishment in Human Behavior, 1968; also numerous articles. Research on human learning with new interpretation of law of effect; developed interaction theory of motivation in which fundamental needs are conceptualized in terms of patterns of behavioral relationships with environment; study of perception of outcomes of a person's actions. Home: 112 Tiense straat, Louvain, Belgium.*

NUTTING, Charles Cleveland, Am. zoologist; b. Jacksonville, Ill., May 25, 1858; s. Rufus and Margaretta Leib (Hunt) N.; A.B., Blackburn U., 1880, A.M., 1882; m. Lizzie B. Hersman, Aug. 10, 1886; 1 dau., Elizabeth H.; m. 2d, M. Eloise Willis, June 16, 1897; children—Willis D., Carl B. Engaged in explorations for Smithsonian Instn. in C. Am., 1881-82; prof. zoology, curator Mus. Natural History, State U. Ia., from 1886, prof. and head dept. of zoology, from 1890. Engaged in spl. scientific researches in Costa Rica, 1882; Nicaragua, 1883; Fla., 1885; Saskatchewan River, 1891, W.I., 1888, 93, Plymouth, Eng., Naples, Italy, 1895; Cal., 1905, 09; Barbados, 1917-18; member of civilian sci. staff U.S.S. Albatross during Hawaiian cruise, 1902; collaborator, Reports of Siboga Expdn., 1909-11; dir. Barbados-Antigua Expdn. from State U. Ia., 1918, Fiji-New Zealand Expdn., 1922. Author: Narrative of Bahama Expedition from University of Iowa, 1893; American Hydroids, Parts 1, 2, 3, 1900, 1904; Report on Gorgoniacea of the Siboga Expedition, 1910; Narrative Barbados-Antigua Expedition, 1918; Narrative of the Fiji-New Zealand Expedition, 1924. Collected birds and antiquities of Nicaragua and Costa Rica. Died Jan. 23, 1927.

NUTTLI, Otto William, Am. geophysicist; b. St. Louis, Dec. 11, 1926; s. Otto Peter and Marie Wehinger) N.; student U. Wyo., 1944-45; B.S., St. Louis U., 1948, M.S., 1950, Ph.D., 1953. Faculty, St. Louis U., 1952—, prof., 1962—; vis. research seismologist U. Cal.., Berkeley, 1964. Mem. Am. Geophys. Union, Seismol. Soc. Am. (chmn. Eastern sect. 1960-61), Soc. Exploration Geophysicists, Royal Astron. Soc., A.A.A.S., Sigma Xi. Research, numerous publs. on structure of earth's interior utilizing travel-time, amplitude, related properties of seismic waves. Home: 5422 Finkman St., St. Louis 63109.*

NUYTS, Jean René, physicist; b. Brussels, Belgium, Sept. 13, 1936; s. Emile Nicolas and Georgette (François) N.; Licence en sci. physiques, Université Libre de Bruxelles, 1958, Ph.D., 1962; m. Laure Isabelle Christoffel, Aug. 25, 1962; children—Arielle, Iolaine. Aspirant, Fonds National de la Recherche Scientifique, Brussels, Paris, 1958-62; physicist Centre Européen pour la Recherche Nucléaire, Geneva, 1962-65, Inst. Advanced Study, Princeton, N.J., 1965-67, Faculté des Sciences, 91 Orsay, France.* Research in theory of high energy physics, emphasis on symmetry groups and weak interactions.

NUZHDIN, Nikolay Ivanovich, Russian biologist; b. 1904; grad. Yaroslavl Pedagogical Inst., 1929. Asso., Inst. Genetics, USSR Acad. Scis., 1935—; prof. Moscow Timiryazev Agrl. Acad., 1948—. Mem. USSR Acad. Scis. (corr.). Author: Hereditary Changes and Ontogenesis, 1945; A Critique of Idealistic Gene Theory, 1950; Darwin and Michurinian Biology, 1952; The Role of Hybridization and Variability, 1946; Editor, co-author: Symposium of Works on Radiobiology, 1955. Research and publs. on genetics, evolutionary theory, radiobiology. Address: Inst. Genetics, USSR Acad. Scis., Leninsky prospect 33, Moscow, USSR.

NYBORG, Wesley LeMars, Am. physicist; b. Ruthven, Ia., May 15, 1917; s. Isaac and Leva (Larson) N.; A.B., Luther Coll., 1941; M.S., Pa. State U., 1944, Ph.D., 1947; m. Beth Woolsey, Sept. 8, 1945; 1 dau., Elsa Beth. Faculty, Brown U., 1950-60; prof. U. Vt., Burlington, 1960—; vis. scientist Oxford U., 1960-61. USPHS fellow Sch. Advanced Study Mass. Inst. Tech., 1956-57. Fellow Acoustical Soc. Am.; mem. I.E.E.E., Am. Phys. Soc., Am. Assn. Physics Tchrs., Biophys. Soc., Sigma Xi. Contbr. chpt. to Physical Acoustics, 1965. Research, numerous publs. on acoustics, ultrasonics and biophysics of cells and macromolecules, properties of high amplitude sound, changes brought about by sound in structures and in rates of reactions. Home: 270 Shore Rd., Burlington, Vt. 05401.*

NYE, Robert Eugene, Jr., Am. physiologist; b. Cin., Feb. 6, 1922; s. Robert Eugene and Mary (Sutherland) N.; A.B., Ohio U., 1943; M.D., U. Rochester, 1947; m. Frances Elizabeth Thomsen, Aug. 7, 1948; children—David Andrew, Christopher John, Peter Benjamin. Fellow in medicine U. Rochester, Strong Meml. Hosp., 1947-50; house physician Hammersmith Hosp., London, Eng., 1951; fellow U. Rochester, 1951-54, instr., 1954-56; faculty Dartmouth Med. Sch., Hanover, N.H., 1956—, asso. prof. physiology, 1963—; cons. in medicine, asso. staff Mary Hitchcock Meml. Hosp., 1961—; cons. in cardio-pulmonary research VA Hosp., White River Junction, Vt., 1957—. Mem. Am. Physiol. Soc., Am. Fedn. Clin. Research, N.Y. Acad. Scis., A.A.A.S., Grafton County Med. Soc., Sigma Xi. Research, publs. on alterations in circulation of blood as a result of various types of heart disease, factors which influence distbrn. of air flow within lung, effect of unequal distbrn. of various characteristics within lung upon effectiveness of gas transfer between lung air and lung blood. Home: Church St., Norwich, Vt. 05055. Office: Dept. Physiology, Dartmouth Med. Sch., Hanover, N.H. 03755.*

NYGAARD, Agnar Petter, Norwegian biochemist; b. Trondheim, Norway, Nov. 22, 1919; s. P.O. and G. (Nordbotten) N.; Chem. Engring. degree Trondheim (Norway) U. Tech., 1942; Ph.D., research asst., Ore. State U., 1951; Dr. Phil., U. Oslo (Norway), 1962; m. Gerd Sivertsen, July 29, 1944; children—Randi, Toril, Inger Anne. Chem. engr. O. Mustad & Son, Oslo, 1945-47; research asst. Cornell U., 1951-53; Research fellow Med. Nobel Inst., Stockholm, Sweden, 1953-55; asst. prof. U. Oslo, 1955-62; vis. prof. U. Ill., 1963; prof., chmn. dept. biochemistry, U. Bergen (Norway), 1964—. Research and publs. on structured lipid factor in the respiratory chain of some respiratory enzymes, and involved in the characterization of factors in protein synthesis. Home: 27 Vestre Solheia, Söreidgrend, Norway.*

NYHAN, William L., Am. physician; b. Boston, Am. Mar. 13, 1926; s. William L. and Mary (Cleary) N.; M.D., Columbia U., 1949; M.S., U. Illinois, 1956; Ph.D., U. Ill., 1958; m. Christine Murphy, Nov. 20, 1948; children—Christopher, Victoria, Abigail. Fellow, Nat. Foundation, 1955-58; asst. prof., pediatrics, 1958-61, assoc. prof., 1961-63, Johns Hopkins; prof. pediatrics and biochemistry, chmn. dept. pediatrics, U. Miami, Fla., 1963—. Mem., Am. Chem. Soc., A.A.-A.S., Assn. Cancer Res.; Pediatrics Soc.; Soc. Pharmacol; Soc. Pediatrics Res. Author numerous publications in field. Described two new inborn errors of metabolism: hyperglycinemia and a hyperuricemic disease; contributed to intermediary metabolism of cancer cell and to metabolism in general in children. Home: 1008 Alhambra Circle, Coral Gables, Florida. Office: University of Miami, School of Medicine, 1600 NW 10th Ave., Miami, Fla.

NYHOLM, Karl-Georg, Swedish biologist; b. Norrtälje, Sweden, Jan. 25, 1912; s. K. J. and G. (Eriksson) N.; Dr. Phil., U. Uppsala; m. Gurli Nyholm, June 28, 1956; one son, Per Georg. Assoc. prof. Stockholm U.; Prof. zoology, U. Gothenburg; prof. zoology, U. Uppsala, 1964. Mem. Swedish Intergovernmental oceanographic commission. Author about 25 published articles. Research on embryology of Collenterates; taxonomy of Echinoderms; taxonomy and ecology of Forminifera. Home: 17 Tegnérgatan, Uppsala, Sweden. Office: Zoological Inst., Uppsala, Sweden.*

NYHOLM, Ronald Sydney, chemist; b. Broken Hill, Australia, Jan. 29, 1917; s. Eric Edward and Gertrude (Woods) N.; D.Sc. with 1st Class Honors, Sydney (Australia) U., 1938, M.Sc., 1942; Ph.D., U. Coll., London, 1950, D.Sc., 1952; m. Maureen Richardson, Aug. 6, 1948; children—Peter, Elizabeth, Rosemary. Research chemist Eveready, Ltd., (Australia), 1938-39; faculty Sydney Tech. Coll., 1940-47; Imperial Chem. Industries Research fellow U. Coll., London, 1947-50, prof. chemistry, 1955—, head dept., 1963—; asso. prof. U. New S. Wales, Sydney, 1952-55. Mem. U. Sci. and Tech. Bd.; mem. Sci. Research Council, 1967—. Named Knight batchelor, 1967. Recipient H.G. Smith medal Royal Australian Chem. Soc., 1955; Royal medal Royal Soc. New S. Wales, 1959. Fellow Royal Soc., 1958, Royal Inst. Chemistry; mem. Chem. Soc. (Corday-Morgan medal 1952, Tilden lectr. 1961, pres. elect 1968), Faraday Soc., Brit. Assn. for Advancement Sci. (sect. pres. 1966). Contbg. author to books. Research, numerous publs. on chemistry of transition metals, structure in broadest sense magnetism and spectra. Home: 21 Manor Rd., S., Esther, Surrey, U.K. Office: Chemistry Dept., U. Coll., Gouver St., London WC1, U.K.*

NYHUS, Lloyd Milton, Am. surgeon; b. Mt. Vernon, Wash., June 24, 1923; s. Lewis G. and Mary (Shervem) N.; student Mt. Vernon Jr. Coll., 1941-42, Willamette U., 1943-44; B.A., Pacific Lutheran Coll., 1945; M.D., Med. Coll. Ala., 1947; m. Margaret Sheldon, Nov. 25, 1949; children—Sheila Margaret, Leif Torger. Instr., dept. surgery U. Wash. Sch. Medicine, 1954-56, asst. prof., 1956-59, asso. prof., 1959-64, prof., 1964-67; Warren H. Cole prof., head dept. surgery U. Ill. Coll. Medicine, 1967—; surgeon-in-chief U. Ill. Hosps.; cons. surgeon Presbyn.-St. Lukes Hosp., West Side VA Hosp., Chgo., Hines (Ill.) VA Hosp. Guggenheim fellow, Lund, Sweden, Glasgow, Scotland, 1955-56; 1st vis. prof., Honolulu, 1961; lectr. 65th Japanese Surg. Soc., 1965; cons. surgery study sect. A. to surgeon gen. NIH, 1965—. Diplomate Am. Bd. Surgery. Fellow A.C.S. (counselor state chpt. 1960-66), Royal Soc. Medicine; mem. A.M.A., Internat. Soc. Surgery (titular mem.), Soc. Univ. Surgeons (pres. 1967-68), Am. Gastroenterol. Assn., Soc. for Surgery Alimentary Tract, Am. Physiol. Soc., Sigma Xi, Phi Beta Pi. Editor: (with H. N. Harkins) Surgery of the Stomach and Duodenum, 1962, Hernia, 1964; Rev. of Surgery, 1964—; asso. editor Quar. Rev. Surgery, 1958-61; Rev. editorial bd. Am. Jour. Digestive Diseases, 1961-61, Northwest Medicine, 1964-67, Scandinavian Gastroenterology, 1966—, Am. Surgeon, 1967—. Research, publs. in profl. jours. and books on gastric physiology, anatomy and surg. repair of groin hernias. Home: 310 Maple Row, Northbrook, Ill. Office: 840 S. Wood St., Chgo. 60612.*

NYLANDER, Friedrich, botanist; b. 1822; prof., Helsinki, Finland, Paris, France. Author: Synopsis methodica lichenum, 1850-60. Investigated and classified lichen. Died Paris, 1889.

NYLANDER, Gunnar Per, Swedish surgeon; b. Huskvarna, Sweden, Aug. 2, 1919; s. Martin and Emma (Forsberg) N.; License as doctor U. Uppsala (Sweden), 1947, M.D., 1953; m. Karin Barre, Oct. 7, 1944; children—Gisela, Per, Olle. Faculty, U. Uppsala, 1953—, asso. prof. surgery, 1957-66; research prof. Swedish Med. Research Council, 1966—. Author: Placental Transfer of Iron, 1953; Thyroid Hormone Activity and Gastrointestinal Function, 1963; also numerous articles. Research on placental physiology, prostatic physiology and pathophysiology, vascular pathophysiology, gastrointestinal motility and secretion, thyroid morphology and vascular anatomy. Home: Arosgot. 7, Uppsala, Sweden.*

NYMAN, Carl John, Jr., Am. chemist; b. New Orleans, Oct. 21, 1924; s. Carl John and Dorothy (Kraft) N.; B.S., Tulane, 1944, M.S., 1945; Ph.D., U. Ill., 1948; m. Betty Marjorie Spiegelberg, July 15, 1950; children—Gail Katherine, John Victor, Nancy Kraft. Faculty, Wash. State U., Pullman, 1948—, prof., 1961—; vis. asst. prof. Tulane U., summer 1950, vis. asso. prof., summer 1959. Vis. fellow Cornell U., 1959-60; summer prof. Hanford Labs. Operation, Gen. Electric Co., Richland, Wash., 1963; Am. Chem Soc. Petroleum Research Fund fellof Imperial Coll., London, Eng., 1966-67. Mem. Am. Chem. Soc., Am. Assn. U. Profs., Sigma Xi. Author: (with G.B. King) Problems for General Chemistry and Qualative Analysis, 1966 (with R.E. Hamm) Chemical Equilibrium, 1968. Studies, publs. on various equilibria which occur in aqueous and non aqueous solutions; properties of oxygen carrying inorganic complexes and polynuclear coordination complexes have been investigated. Home: 2100 Orion Dr., Pullman, Wash. 99163.*

NYMAN, Gustaf Arthur, Finnish chemist; b. Notsjö, Finland, July 4, 1906; s. Karl Hugo and Alvine Luise (Jörgensen) N.; B.Sc., Tech. U., Helsinki, Finland, 1932, Dr., 1936, Docent organic chemistry, 1937; m. Linnéa Granlund, Dec. 27, 1963; children—(by previous marriage) Inger Birgitta (Mrs. Nils Gabriel Lybeck), Birgitta Margareta (Mrs. Lars Gustav Palén). With Ahlström Oy, Warkaus, Finland, 1942-54; prof. organic chemistry Tech. U. Finland, Otaniemi, 1954—, dean chemistry dept., 1956-62. Cons. Neste Oy, Finland, 1962—. Decorated RSL 1, KSL, MM. Mem. Tech. Sci. Acad. Finland, Sv. Tech. Sci., Acad. Finland, Finnish Chem. Assn., Pulp and Timber Assn. Research, publs. on organic chemistry, terpen chemistry, furan chemistry and prodn. Home: Källgard, Domsby, Finland. Office: Tech. U., Otanieni, Finland.*

NYREN, Magnus, Swedish astronomer; b. Västra Furlen, Sweden, Feb. 21, 1837; ed. Uppsala, Sweden; became instr. Pulkovo Obs., 1868, vice dir., 1890-1908; apptd. councilor of state, 1888. Recipient Valz prize French Acad. Scis., 1899. Author: Détermination du coefficient condtant de la précession aux moyens d'étoiles de faible éclat, 1870; Bestimmung der Nutation der Erdachse, 1872; Die Polhöhe von Pulkowa, daselbst 1873. Improved determination of constants for aberration, precision and nutation as basis for fundamental catalogue of Pulkova Obs.; indicated that latitude is a variable for purposes of cataloguing positions of celestial bodies. Died Stockholm, Sweden, Jan. 16, 1921.

NYSTEN, Pierre Humbert, physician; b. Liège, Belgium, 1771; ed. Paris; became aid in anatomy, Paris, 1798; sent on mission to Spain; upon return became physician l'Hôpital des Enfants. Author: Nouvelles expériences faites sur les organes musculaires de l'homme et des animaux à sang rouge, 1803; Dictionary of Medicine, 1806; Recherches sur les maladies des vers à soie, 1808. Described course of rigor mortis beginning in muscles of mastication, then in face and neck, arms and upper trunk, lower trunk, thighs, legs and feet, 1812. Died Paris, 1818.

NYYSSÖNEN, Aarne Olavi, Finnish forester; b. Karttula, Finland, Aug. 9, 1921; s. Herman Vilhelm and Helmi (Vilpponen) N.; B.Forestry, U. Helsinki (Finland), 1948, M. Forestry, 1950, licentiate of For., D. Forestry, 1954; m. Aino Katri Huttunen, Apr. 6, 1952; 1 son, Mikko. Research asst. Acad. Finland, 1948-56; prof. Forest Research Inst. Finland, 1957-59; prof. forest mensuration and mgmt.

U. Helsinki, 1959—, vice dean Faculty Agr. and Forestry, 1963—. Mem. Soc. Forestry Finland, Soc. Geography Finland. Research, publs. on growth and yield of Finnish pine stands, rotation and its determination, accuracy of stand vol. estimation through different methods, efficiency of forest inventory methods. Home: 107 Lepolantie, Helsinki 66. Office: 40B Unioninkatu, Helsinki 17, Finland.*

O

OAE, Shigeru, Japanese chemist; b. Okayama, Japan, Jan. 18, 1920; s. Isokichi and Shina (Takahara) O.; B.Sc., Waseda U., Tokyo, Japan, 1943; D.Sc., Osaka (Japen) U., 1954; m. Akiko Isoe, Sept. 22, 1946; children—Keiko, Margaret Naomi. Research chemist Hodogaya Chem. Works Co., Tokyo, 1943-46; asst. Osaka U., 1946-50, asst., asst. prof., 1953-55; research fellow U. Kan., 1950-53; asso. U. Pa., 1955-57; asst. prof. Drexel Inst. Tech., 1957-58; vis. chemist Brookhaven Nat. Lab., 1958-60; chmn. dept. Chem. Radiation Center, Osaka, 1960-62; prof. Osaka City U., 1962—, chmn. dept. applied chemistry, 1963-64; vis. prof. Stevens Inst. Tech., Ore. State U., 1966. Recipient Valedictorian award Waseda U., 1943. Mem. Am., Japanese chem. socs., Japanese Pharm. Soc., Soc. Social Responsibilities in Sci., Sigma Xi. Author: (with M. Murakami, Y. Yukawa) Name Reactions in Organic Chemistry, vol. I., 1954, vol. II, 1955; (with C.C. Price) Sulfur Bonding, 1962; Elimination, 1965; Physical Organic Chemistry, 1966; also numerous articles. Research on organic sulfur compound reactions, mechanism of alkaline fusion of benzenesulfonic acid to form phenol was clarified; postulated hypothesis of metabolic process of methionine. Home: 2509 Hirao Mihara-cho, Minamikawachigun, Osaka-fu. Office: 459 Sugimoto-cho Sumiyoshi-ku, Osaka, Japan.*

OAKES, William Francis, Am. psychologist; b. Edmond, Okla., May 11, 1926; s. Francis Gibson and Wilma (Snyder) O.; B.A. cum laude, U. Wichita, 1951, M.A., 1952; Ph.D., U. Minn., 1956; m. Harriet Juckem, May 2, 1956; children—William Steven, Gretchen Ann. Residence counselor, research asst. U. Minn., 1952-53, head counselor, asst. dir. Pionner and Centennial Halls, 1953-55, instr., Ford Found. teaching intern, 1955-56, research asso. Bur. Instnl. Research, 1956-57; faculty U. Wichita, 1957-61, asso. prof., 1960-61; faculty U. Hawaii, Honolulu, 1961-67, prof. psychology, 1944-67, chmn. dept., 1963-67; prof. psychology, dean instrn. freshman program Grad. Center, City U. N.Y., 1967—. Mem. Am., Midwestern, Western, Hawaii (pres. 1966—) psychol. assns., Psychonomic Soc., Am. Assn. U. Profs., Hawaii Acad. Sci. Research, publs. on role of awareness in verbal learning and influence of social rewards and punishments on individual behavior in group interaction. Home: 20 Chestnut Ridge Way, Dobbs Ferry, N.Y. 10522. Office: 33 W. 42d St., N.Y.C. 10036.*

OAKLEY, Kenneth Page, English anthropologist, geologist; b. Amersham, Eng., Apr. 7, 1911; s. Tom Page and Dorothy (Thomas) O.; B.Sc. in Geology and Anthropology, 1933, D.Sc., 1955. Geologist, Geol. Survey, Gt. Britain, 1934-35; became asst. keeper dept. geology (paleontology) Brit. Mus., 1935, prin. sci. office, 1947-55, sr. prin., 1955—, in charge anthropology sub-dept., 1959—. Vis. prof. U. Chgo., 1956; expdns. to E. Africa, 1947, S. Africa, 1953. Recipient Rosa Morison Meml. medal Univ. Coll., London, 1933; Wollaston fund, 1941; Prestwich medal Geol. Soc. London, 1963; Henry Stopes Meml. medal Geol. Assn., 1952. Fellow or mem. Geol. Soc. London (sec. 1946-49), Royal Anthrop. Inst. (sec. Swanscombe Skull com. 1936-38), Soc. for Promotion Nature Reserves, Brit. Assn. (pres. anthrop. sect. 1961), Italian Inst. Human Paleontology (corr. mem.). Author: Man the Tool-maker, 1949; Frameworks for Dating Fossil Man, 1964; The Problem of Man's Antiquity, 1964; also articles. Contbg. author: The Fluorine-dating Method (in Year-book of Physician Anthropology for 1949), 1951. Co-author: (paper) The Solution of the Piltdown Problem, 1953. Research in human paleontology, archeology, dating of fossil man; proposed determination of relative age of bones by their fluorine content, used to confirm age of Swanscombe skull and Galley Hill skeleton; demonstrated (with others) fraudulence of famous Piltdown man (Eoanthropus), showing it to be jaw of modern ape altered to resemble human cranial bones with which it had been found. Address: 64 Queen's Gate, London S.W. 7, Eng.

OBAL, Francis George, Hungarian neurophysiologist; b. Budapest, Hungary, Sept. 28, 1916; s. Francis and Anne (Bárczy) O.; M.D., U. Budapest, 1940, med.habil., 1948; postgrad. U. Heidelberg (Germany); Cand. of Sci., Hungarian Acad. Sci., 1952, D.Sc., 1967; m. Magdalene Nagy, July 28, 1947; children—Francis, Georgette. Staff pathophysiol. dept. U. Budapest, 1936, 39-40, dept. physiology U. Heidelberg, 1937--39, U. Kolozsvár, Cluj, Rumania, 1940-45; prof. gen. and exptl. pathology Med. Faculty, U. Kolozsvár, Marosvásárhely, Tirgu- Mures, Rumania, 1945-54, head physiology, 1947-54, head pharmacology dept., 1949-54; head physiology dept. State Inst. Neurosurgery, Budapest, 1954-58; prof. physiology Med. U. Szeged (Hungary), 1958—, head Inst., 1958—. EEG cons. State Rys. Health Service, 1956—. Named Eminent Physician, Govt. of Hungary, 1957. Mem. Hungarian Physiol. Soc. (leading mem.), Hungarian EEG Soc. (founding, leading mem.), Pavlovian Soc. N.Am. (corr.), Hungarian Biol. Soc. Editorial com. Kisérletes Orvostudonmány, 1962—, Acta Physiol. Acad. Hung. Sci., 1966—. Research, numerous publs. on normal and path. EEG's, ECG's, electrophysiol. features of epilepsy, physico-chem. characteristics of stored blood in spleen, villi in intestinal reabsorptions, effects of tissue and bacterial toxins in infectious diseases, conditioned reflexes with respect to drug action, neurocybernetic mechanisms in vegetative homeostasis. Home: 12 Kazinczy, Szeged, Hungary.*

OBATA, Yataro, Japanese chemist; b. Osaka, Japan, Dec. 8, 1907; s. Kametoshi and Sei Obata; grad. Hokkaido (Japan) U., 1930; dr.degree Tokyo (Japan) U., 1944; m. Nori-ko Kimura, Feb. 5, 1937; children—Teruo, Masa-ko, Kiyo-ko. Asst., Inst. Indsl. and Sci. Research, Osaka U., 1942-46; prof. Hokkaido U., Sapporo, 1946—. Recipient prize Japan Agrl. Soc. Mem. Chem. Soc. Japan (editorial bd. 1953-64, councilor), Agr. Chem. Soc. Japan (Hokkaido regional sect. chmn.), Soc. for Analytical Chemistry (v.p.), Japan Vitamin Soc. (councilor). Author: Aromas in Human Life, 1959; Foods of Japanese People, 1962; Outline of Biochemistry, 1957; Coloring Matters, Flavors and Tastes of Foods, 1961; also numerous articles. Research on foods flavoring additives sulfur containing compounds, prevention mechanism of occurring of off-flavor in beer. Home: 2-8 Hiragishi, Sapporo, Japan.*

OBAYASHI, Tatsuzo, Japanese physicist; b. Wakayama, Japan, Apr. 10, 1926; s. Goro and Miyoko Obayshi; M.Sc., U. Tokyo (Japan), D.Sc., 1959; Ph.D. in Physics, U. Toronto (Ont., Can.), 1957; m. Tomi Obayashi, Nov. 3, 1961; children—Akihiko, Chiemi, Masami. With Radio Research Lab., Japan, 1948-61, head sci. program Hiraiso Radio Obs., 1951-61; prof., dir. Ionosphere Research Lab., Kyoto (Japan) U., 1961—. Mem. ionosphere research com. NRC; mem. Japanese URSI. Mem. Rocket Soc., Astron. Soc., Geophysics Soc., Am. Geophys. Union, Internat. Acad. Astronautics. Author: (with others) Space Science, 1961; also numerous articles. Research on world radio communication, ionosphere, geomagnetism, solar cosmic rays, space physics.*

OBER, Frederick Albion, Am. ornithologist; b. Beverly, Mass., Feb. 13, 1849; s. Andrew Kimbal and Sarah (Hadlock) O.; ed. pub. schs.; explored Lake Okeechobee region, Fla., 1872-74, Lesser Antilles, W.I., 1876-78, 80, Mexico, 1881, 83, 85, later Spain, N. Africa, S.Am.; commr. Chgo. Expn., 1891-92. Author: Travels in Mexico, 1884; A Guide to the West Indies, 1907; Heroes of American History, 1907; Smithsonian Instn. Discovered 22 new bird species, 1872-76. Died Hackensack, N.J., May 31, 1913.

OBERDORFER, Erich Otto, German botanist; b. Fribourg, Germany, Mar. 26, 1915; s. Otto and Lydia (Sigmund) O.; ed. U. Tübingen; Ph.D., U. Fribourg, also named hon. prof.; m. Kalre Bärth, Aug. 10, 1931; children—Martin, Ursel, Hanne, Gertrud, Ilse, Renate. Prof. at gymnasium until 1937; curator Baden Inst. Protection Nature until 1958; dir. Natural Sci. collections at Karlsruhe; instr. plant sociology U. Fribourg, 1951—. Mem. German Botany Soc.; corr. mem. Pollichia. Author: Pflanzensoziologische Exkursionsflora fur Süddeutschland, 1962; Süddeneustche Pflanzengesellsschaften, 1957; Pflanzensoziologische Studien in Chile, 1960. Home: Karl-Schrempp Strasse 4, Karlsruhe; Office: Erbprinzenstrasse 13, Karlsruhe, West Germany.

OBERHÄUSER, Georg, instrument maker; b. Anspach, Bavaria, 1798; lived in Paris, from 1815; built topog. instruments, instruments and machines for making and producing ordnance maps; built microscope with magnification of 500 power, 1837. Died 1868.

OBERHUMMER, Eugen, hist. geographer; b. Munich, Bavaria, Mar. 29, 1859; m. Hermine Drächsler, 1888; children—Fritz, Helene, Ernst, Wilfrid. Prof., U. Munich, 1903, later at U. Vienna. Author: Akarnanien, Ambrakia, Amphilochien, Leukas im Altertum, 1887; Konstantinopel unter Sultan Suleiman dem Grossen, 1902; Die Insel Cypern, eine Landskunde auf historischer Grundlage, 1903; Die Stellung der Geographie zu der historischen Wissenschaft, 1904; Wolfgang Lazius, 1906; Hellas als Wiege der wissenschaftlichen Erdkunde, 1913; Die Türken und das Osmanische Reich, 1917; Die Brixener Globen von 1522, 1926. Leading researcher on ancient and oriental geography. Died Vienna, Austria, May 4, 1944.

OBERMAYER, Maximilian Ernst, physician; b. Leoben, Austria-Hungary, Jan. 6, 1896; s. Herman and Augusta (Pichler) O.; M.D., Carl Franzen's U., Graz, Austria, 1924. Asst. dermatology U. Clinic, Graz, 1925-29; Douglas Smith Found. fellow U. Chgo., 1930, asst. prof. dermatology and syphilology, 1930-41; practice dermatology, Los Angeles, 1941—; asso. clin. prof. U. So. Cal., Los Angeles, 1941-45, clin. prof., 1945-61, prof. emeritus, 1961—; sr. staff mem. Los Angeles County Gen. Hosp., Los Angeles, 1942-61. Sr. cons. USN Hosp., San Diego, 1954-61; cons. in leprosy Cal. Dept. Health, 1954-62. Recipient Civis Academicus, Carl Franzen's U., 1961. Mem. all local, regional, nat. dermatol. socs., Swedish (corr.), Mexican (hon.), Austrian (hon.), Venezuelean (hon.), dermatol. socs. Author: (with S.W. Becker) Modern Dermatology and Syphilology, 1947, Dermatology Sifilologia Modernas, 1945; Psychocutaneous Medicine, 1955; Medicina Psicocutanea, 1956; also numerous articles. Research on dermatology and syphilology, psychosomatic aspect diseases of skin. Home: 515 S. Genesee Av., Los Angeles, 90036. Office: 3875 Wilshire Blvd., Los Angeles 90005.*

OBERMEIER, Otto Hugo Franz, German bacteriologist; b. Spandau, Germany, 1843; doctorate U. Berlin, 1866; asst. Charité Krankenhaus; practiced in Berlin; discovered Borrelia recurrentis (Spirillum recurrentis), the causative agt. of European louse-borne form of relapsing fever, 1872. Died 1873.

OBERT, Edward Fredric, Am. mech. engr., educator; b. Detroit, Jan. 18, 1910; s. Edward and Jessie (Funderburg) O.; B.S., Northwestern U., 1933, M.E., 1934; M.S., U. Mich., 1940; m. Helen Florence Boyce, Sept. 14, 1935. Engr. mfg. Western Electric Co., Chgo., 1929-30, insp. engring. material, naval material, Chgo., 1934-37; prof. mech. engring. Northwestern U., Evanston, Ill., 1937-58; prof., chmn. dept. mech. engring. U. Wis., Madison, 1858—. Editor mech. engring. series Internat. Textbook Co., 1954—. Mem. Am. Soc. M.E. (past chmn. Chgo. sect.), Am. Soc. Engring. Edn. (8th Westinghouse award 1953), Sigma Xi, Tau Beta Pi, Pi Tau Sigma. Author: Internal Combustion Engines, 1949; Concepts of Thermodynamics, 1960; Elements of Thermodynamics and Heat Transfer, 1962; Thermodynamics, 1963; also articles; Research on thermodynamics especially equations of state, availability, virial coefficients and compressibility factor; theory of combustion engines. Home: 3534 Lake Mendota Dr., Madison, Wis. 53705.*

OBERTH, Hermann Julius, physicist; b. Sibiu, Transylvania, June 25, 1894; s. Julius Gotthold and Valerie Emma (Krasson) G.; ed. univs. Munich, Göttingen, Heidelberg, Klausenburg; m. Mathilde Hummel, 1918; children—Julius, Erna, Ilse, Adolph Prof., S.L. Roth Coll., Mediasch, 1925-38; rocket research for army and air force Vienna Tech. U., 1938-40, Dresden, Germany, 1940-41; adv. engr. Peenemünde rocket base, 1941-43, Westfälisch-Anhaltische Sprengstoff A.G., 1943-45; researcher on Brienzersee (Switzerland), 1949, for Italian Navy La Spezia, 1950-53, for U.S. Army Redstone Arsenal, Huntsville, Ala., 1955-58. Recipient R.E.P.-Hirsch prize French Astronautical Soc., 1929, medal German Soc. for Space Research, 1950, Diesel medal Assn. German Inventors, 1954, Am. Astronautical Soc. Award, 1955, Edward Pendray award, 1956, Bundesverdienstkreuz (I Klasse), German Fed. Republic, 1961. Author: Die Rakete zu den Planetenraumen, 1923; Wege zur Raumschiffahrt, 1929; Forschung und Jenseits, 1932; Menschen im Weltraum, 1955; Das Mondauto, 1957, 1958; Stoff und Leben, 1959. Conducted rocket research which led to V-2. Address: Untere Kellerstrasse 758, Feucht bei Nürnberg, W. Germany.

OBICIUS, Hyppolitus, Italian physician, natural philosopher; b. Ferrara, Italy; prof. medicine U. Ferrara, then U. Belluno (Italy); 1st physician City of Belluno. Author: Dialogus tripartitus (on arts and scis., distinguished evil, superstition and the natural as 3 varieties of magic, maintains div. into depts. of knowledge and superstition of the 13th century), 1605; Atroastronomicon (held that incorrect understanding of Galen had led Guilio Cesare Claudini to question astrological explanation of critical days in 1612), 1618.

OBREIMOV, Ivan Vasilievich, Russian physicist; b. Mar. 8, 1894; grad. Petrograd U., 1915; with State Optical Inst., 1919-24; became asso. with Leningrad Physicotech. Inst., 1924; dir. Ukranian Physicotech. Inst. 1929-37; with inst. elemento-organic compounds Acad. Scis. USSR, 1954—, also mem. Recipient Stalin prize, 1946; Vavilov Gold medal, 1959. Author (in Russian): Application of Gresnel Diffraction in Physical and Technical Measurements, 1945; Identification of Hydrocarbons by Dispersion Curves, 1955; Formation of Ultramicroscopic Heterogeneity During Plastic Deformation of Rock Salt, 1956; Double Refraction in Organic Crystals, 1957. Specialist in physics of molecular and crystal spectroscopy. Address: AN SSSR, Leninsky prosp. 14, Moscow, USSR.

O'BRIEN, Brian John, physicist, space scientist; b. Sydney, Australia, Feb. 27, 1934; s. Richard Ignatius and Thelma (Hoban) O'B.; B.Sc. with honours, U. Sydney, 1954, Ph.D., 1957; m. Avril Searle, Apr. 10, 1959; children—Richard John, Rosalind Anne, John (dec.), Caroline Maria. Came to U. S., 1959. Dep. chief physicist Australian Nat. Antarctic Research Expdn., Melbourne, 1958-59; asst. prof., then asso. prof. State U. Ia., 1959-63; prof. space sci. Rice U., 1963—; vis. prof. physics U. Sydney, 1964. Mem. ad hoc com. rocket-satellite research Nat. Acad. Scis., 1965; cons. manned space sci. joint working group space radiation, Goddard Space Flight Center, 1964—; cons. Apollo program

NASA Manned Space Sci., 1965—, European Space Research Orgn., 1963—. Mem. Australian Speleological Fedn. (pres. 1956-59), Am. Geophys. Union, A.A.A.S., Com. Space Research, Internat. Sci. Radio Union, Internat. Union Geodesy and Geophysics. Author numerous articles, monographs in field; contbr. chpts. to books. Co-discoverer artificial radiation belt produced by Starfish nuclear explosion, 1962. Study of solar-terrestrial relationships, including auroral causes and effects; heavy cosmic radiation; Van Allen and artificial magnetically trapped radiation; thermoelectric and thermomagnetic refrigeration. Home: 8014 Rowan Lane, Houston 77036.

O'BRIEN, Donough, physician; b. Edinburgh, Scotland, May 9, 1923; s. Arthur John Rushton and Catherine (Aikman) O'B.; B.A., Cambridge U., 1943, M.A., 1944, M.D., 1952; M.B., London U., 1946, B.Chir., 1947; m. Madeline Mary Walker, Apr. 1952; children—Turlough, Rushton, Quentin. Came to U.S., 1957, naturalized, 1962. Practice medicine specializing in pediatrics, Denver, 1957—; faculty U. Colo. Med. Center, 1957—, prof. dept. pediatrics, 1964—. Markle scholar med. scis., 1958. Mem. Am. Pediatric Soc., Soc. Pediatric Research, Am. Acad. Pediatrics, Am. Diabetes Assn. Author: (with F.A. Ibbott, D. Rogerson) Laboratory Manual of Pediatric Micro and Ultramicro Techniques, 1962. Research, numerous publs. primarily in field of inborn errors of metabolism. Home 1 S. Cherry St., Denver 80222. Office: 4200 E. 9th Av., Denver 80220.*

O'BRIEN, Morrough Parker, Am. cons. engr.; b. Hammond, Ind., Sept. 21, 1902; s. Morrough and Lulu (Parker) O'B.; B.S., Mass. Inst. Tech., 1925; D.Sc., Northwestern U., 1959; D.Eng., Purdue U., 1961; m. Roberta Libbey, May 16, 1931 (div.); children—Sheila, Morrough; m. 2d, Mary Wallner Kremers, 1963. Engr., Hudson River Regulating Dist., 1925-27; research asst. Purdue U., 1925-27; Freeman scholar, Am. Soc. C.E., 1927-28; asst., Royal Coll., Engring., Stockholm, Sweden, 1927-28; asst. prof., asso. prof., U. of Cal., 1928-59, prof. emeritus, 1959—, chmn. dept. mech. engring., 1937-43, dean coll. engring., 1943-59, dean emeritus, 1959—; dir. research and engring. Air Reduction, Inc., 1947-49, cons. engr. Aerospace and Defense group Gen. Elec. Co., 1950—. Dir. McGraw-Hill Pub. Co. Mem. U.S. Coastal Engring. Research Bd.; chmn. council wave research Engring. Found.; mem. Army Sci. Adv. Panel. Mem. Inst. Aeros. and Astronautics, Am. Soc. C.E., Am. Soc. M.E., Nat. Soc. Profl. Engrs., Inst. Mech. Engrs. (Gt. Britain), Am. Soc. Engring. Edn., Newcomen Soc. (Gt. Britain), Soc. Hist. Tech., Hist. Sci. Soc., Sigma Xi, Tau Beta Pi, Delta Tau Delta. Author: Applied Fluid Mechanics, 1937; contbr. to World of Engineering, 1964. Research on turbojet engines; hydraulic machinery; shoreline phenomena; management of research and development. Home: P.O. Box 423, Berkeley, Cal.*

O'BRIEN, Richmond Desmond, neurobiologist; b. Sydenham, U.K., May 29, 1929; s. Joseph Andrew and Louise (Stevens) O'B.; B.Sc., Reading U. (U.K.), 1950; Ph.D. in Chemistry, U. Western Ont., 1954, B.A. in Gen. Arts, 1956; m. Ann Margaret Thom, Mar. 16, 1961; 1 son, Ian Richard. Came to U.S., 1960, naturalized, 1966. Soil specialist Ont. Agrl. Coll., Guelph, Can., 1950-51; chemist Pesticide Research Inst., London, Ont., 1954-60; vis. asso. prof. U. Wis., 1958-59; faculty Cornell U., Ithaca, N.Y., 1960—, prof. entomology, 1964—, chmn. dept. biochemistry, 1964-65, chmn. sect. neurobiology, 1965—. Cons. Am. Cyanamid, 1960—; mem. toxicology study sect. USPHS, 1964—. NRC fellow Inst. Animal Physiology, Cambridge, U.K., 1956-57; Guggenheim fellow, 1967-68. Fellow A.A.A.S.; mem. Biochem. Soc. (U.K.), Am. Chem. Soc., Entomol. Soc. Am., Sigma Xi. Author: Toxic Phosphorus Esters, 1960; (with L.S. Wolfe) Radiation, Radioactivity and Insects, 1964; Action and Metabolism of Insecticides, 1967; also articles. Elucidation biochem. and physiol. mechanisms selective toxicity of chem. substances, penetration solutes into vertebrates and insects. Home: R.D. 3, Trumansburg, N.Y. 14886. Office: Langmuir Lab., Cornell U., Ithaca, N.Y. 14850.*

O'BRIEN, Thomas Doran, Am. chemist; b. Washington, Mar. 31, 1910; s. Thomas A. and Madge (Goggen) O'B.; B.S., George Washington U., 1935, M.S., 1938; Ph.D., U. Ill., 1941; m. Bernice Thomas, Nov. 24, 1935; children—Thomas D., Michael John. Research chemist U.S. Naval Research Lab., 1940-42, Allied Chem. & Dye, 1942-43; prof. chemistry Tulane U., 1943-45; faculty U. Minn., 1945-55; prof., head chemistry dept., dir. acad. research Kan. State U., 1955-60; prof. chemistry, dean Grad. Sch., U. Nev., Reno, 1960—. Mem. Am. Chem. Soc., Am. Assn. U. Profs., Sigma Xi. Research, publs. on mechanisms and structures of inorganic coordination compounds. Home: 2240 Arcane St., Reno 89507.*

OBROSOV, Aleksandr Nikolaevich, Russian physiotherapist; b. 1895; grad. Med. Faculty, Moscow U., 1925; D.Med. Sci., 1948. Asst. dept. physics State Inst. Physiotherapy RSFSR, 1926-30, mgr. lab. physic tech., 1930-51; asst., lectr. dept. physiotherapy Central Postgrad. Med. Inst., Moscow, 1930-41; dir., sci. mgr., 1951-58; sci. dir. Central Inst. Balneology and Physiotherapy, USSR Ministry Health, 1958—; prof. physiotherapy, 1949—. Presidium mem. learned council, also chmn. Physiotherapy and Balneology Commn, RSFSR Ministry Health. Mem. USSR Acad. Med. Scis. (corr.), All-Union Sci. Soc. Physiotherapists and Balneologists (chmn.), Internat. Assn. Med. Hydrol. (v.p.); hon. mem. Czechoslovak Purkinje Med. Soc., Bulgar Soc. Physiotherapists, Polish Soc. Klimatobalneology, Am. Assn. Med. Hydrol. Research and numerous publs. on theory of neurochumoral mechanism of action of physical factors based on Pavlov's nervism principles; developer new methods of physiotherapy; inventor apparatus for treating wounded in field hosps. Address: Central Inst. Balneology and Physiotherapy, USSR Ministry Health, Kalinin prospect 50, Moscow G-314, USSR.*

OBRUCHEV, Sergey Vladimirovich, Russian geologist; b. Feb. 3, 1891; s. V.A. Obruchev; grad. Moscow U., 1915. Geol. researcher in Yenisey Basin, 1917-24, northeastern Siberia, 1926-35, mountain ranges of Eastern Sayan, Khamar-Daban, Eastern Tuva, also other geol. and geomorphology areas of USSR Acad. Scis. (corr.). Author: The Kolyman-Indigirka Region: A Geographical and Geological Outline, 1931; New Orographical Study of Northeastern Asia: An Outline of the Tectonics of Northeastern Asia, 1938 Fundamental Features of the Tectonics and Stratigraphy of the Eastern Sayan, 1942; The Orography and Geomorphology of the Eastern Half of the Eastern Sayan, 1946; New Materials on the Orography of Northeastern Tuva, 1955. Research and publs. on orography, geomorphology, geol. structure of northeastern Asia. Address: USSR Acad. Scis., Leninsky prospect 14, Moscow, USSR.

OBRUCHEV, Vladimir Afanasevich, Russian geologist; b. Tver, Russia, Sept. 28, 1863; children—Vladimir Vladimirovich, Dimitri Vladimirovich, Sergei Vladimirovich. Prof. geology Tomsk Tech. Inst., 1901-12, Mining Acad., Moscow, 1922-28; organizer, dir. inst. on permafrost All-Union Acad. Scis., 23 years. Author: Istorija geologicheskogo issledovanija Sibiri, 5 vols., 1931-49; Geologija Sibiri, 3 vols., 1935-38; also articles. Research on geography and geology of deserts and ranges of east-central Asia and adjacent Siberia; his discovery of asphaltum lakes in western approach to Gobi desert led to 1st comml. petroleum in central Asia east of Fergana valley. Died 1956.

OBST, (Karl Gustav) Erich, German geographer; b. Berlin-Steglitz, Germany, Sept. 13, 1886; Ph.D., Breslau, 1909; student U. Jena; Asst. in geography Colonial Inst., Hamburg, 1908, lectr., from 1909; leader East Africa expdn. Geog. Soc., Hamburg, 1910-12; lectr. U. Marburg, 1912; prof. U. Constantinople, also dir. meteorol. inst., 1915; asso. prof. U. Breslau, 1919; asso. prof. Tech. U. Hanover, 1921, prof., 1922, 46, founder, dir. Geog. Inst., 1923; dir. Leibniz Acad., Hanover, from 1926; prof. U. Graz, 1928, Univ. and Tech. U. of Breslau, 1938-45; U. Frankfort-am-Main, 1950, emeritus, 1953. Mem. expdn. to Russia, 1924, expdns. to South Africa, 1932-33, 35-36. Mem. Geog. Soc. Hanover (chmn. 1925), Acad. Beneficial Scis. (Erfurt), Leopoldina; hon. mem. numerous geog. soc., including Frankfort, Munich, Finland, Inst. for Cultural Sci. (and knowledge of fgn. countries) Munich-Starnberg). Author: Die Wirtschaftsreiche in Vergangenheit und Zukunft, 1922; Die Deutschetum in Südafrika, 1924; England, Europa, und die Welt, 1927; Grundzüge der physikalischen Erdkunde, 1950.

OBUKHOV, Aleksandr Mikhaylovich, Russian geophysicist; b. May, 5, 1918; grad. Moscow U., 1940, doctorate in physical-math. scis.; With Geophys. Inst., USSR Acad. Scis., 1940-56, dir. Inst. Atmospheric Physics, 1956—. Mem. USSR Acad. Scis. (corr.). Author: Energy Distribution in the Turbulent Flow Spectrum, 1941; Turbulence in an Atmosphere of Non-Uniform Temperature, 1946; Temperature Field Structure in a Turbulent Flow, 1949; co-author: The Basic Laws of Turbulent Interosculation in the Atmospheric Surface Layer, 1954. Editor: Oceanic and Atmospheric Hydrodynamics, 1963. Research and publs. on statis. theory of turbulence and its use in meteorology. Address: Inst. Atmospheric Physics, USSR Acad. Scis., Pyhzevksy 3, Zh-17, USSR.*

O'CEALLAIGH, Cormac, Irish physicist; b. Dublin, Ireland, July 29, 1912; s. Seamus and Maire (Ferran) O'C.; B.Sc., U. Coll., Dublin, 1933, M.Sc., 1934, Ph.D., 1943; m. Millie Carr, Aug. 14, 1939; children—Niamh, Dairine, Nuala. Lectr. exptl. physics U. Coll., Cork, 1938-49, 1951-53; research U. Bristol, 1949-51; sr. prof. in charge cosmic ray sect. Sch. CosmicPhysics, Dublin Inst. for Advanced Studies, 1953—; vis. prof. Tata Inst. Fundamental Research, Bombay, 1956-66. Mem. emulsion experiments com. CERN. Mem. Royal Irish Acad. Publs. on research in cloud chamber studies of low energy nuclear physics and electron scattering, elementary particles and their decay modes using nuclear emulsion technique; application of statistics to measurements and investigation of ionization in nuclear emulsion; discovered various decay modes of K-mesons. Home: 46 Killiney Rd., Killiney, County Dublin. Office: 5 Merrion Sq., Dublin 2, Republic of Ireland.

OCHOA, Severo, biochemist; b. Luarca, Spain, Sept. 24, 1905; s. Severo and Carmen (deAlbornoz) O.; B.A., Malaga (Spain) Coll., 1921; M.D. with honors, U. Madrid, 1929; D.Sc., Washington U., St. Louis, 1957, Wesleyan U., Middletown, Conn., U. Oxford (Eng.), U. Salamanca (Spain), 1961, Gustavus Adolphus Coll., 1963, U. Pa., 1964, Brandeis U., 1965, Yeshiva U., 1966, Granada (Spain) U., 1967, Oviedo (Spain) U., U. Mich., 1967; Dr. Honoria Causa, U. Brazil, 1957; LL.D., U. Glasgow (Scotland)), 1959; D.Med. Sci., U. Santo Tomas (P.I.), 1963; m. Carmen Garcia Cobian, July 8, 1931. Came to U.S., 1941, naturalized, 1956. Faculty, U. Madrid, 1931-36, Kaiser-Wilhelm Inst., Heidelberg, 1936-37; investigator Marine Biol. Lab., Plymouth, Eng., 1937; demonstrator, Nuffield research asst. U. Oxford Med. Sch., 1938-40; faculty Washington U. Sch. Medicine, St. Louis, 1941-42; faculty N.Y. U. Sch. Medicine, 1942—, prof. biochemistry, chmn. dept., 1954—. Mem. nat. com. for biochemistry NRC, 1955-59, pres., 1958; mem. bd. sci. advisers Jane Coffin Childs Fund for Med. Research, 1961—; pres. Internat. Union Biochemistry, 1961-67; mem. various govt. adv. coms., panels. Recipient Neuberg medal award in biochemistry Soc. European Chemists, 1951; Charles Meyer Price award Société de Chimie Biologique, 1955; Borden award in med. scis. Assn. Am. Med. Colls., 1958; Nobel prize in physiology and medicine (with A. Kornberg), 1959. Fellow A.A.A.S., N.Y. Acad. Scis., N.Y. Acad. Medicine, Am. Acad. Arts and Scis; mem. Am. Soc. Biol. Chemists (pres. 1958-59, editorial bd. jour. 1951-61), Am. Chem. Soc., Harvey Soc. (pres. 1953-54), Soc. Exptl. Biology and Medicine, U.S. Nat. Acad. Scis., Am. Philos. Soc., Alpha Omega Alpha, also numerous fgn. socs. Research on intermediary metabolism, especially enzymatic reactions of carbohydrates and fatty acid metabolism; photosynthesis, particularly description of light-catalyzed reduction of coenzymes, their utilization for carbohydrate synthesis, key reactions of carbon dioxide fixation; (with A. Kornberg) 1st to synthesize nucleic acid in a test tube. Home: 530 E. 72d St., N.Y.C. 10021.

OCHSNER, Albert John, Am. surgeon; b. Baraboo, Wis., Apr. 3, 1858; s. Henry and Judith (Hottinger) O.; B.Sc., U. Wis., 1884; M.D., Rush Med. Coll., Chgo., 1886; student medicine U. Vienna, also U. Berlin, 1887-88; LL.D., U. Wis., 1909; m. Marion H. Mitchell, Apr. 3, 1888. Began practice medicine, Chgo., 1889; became chief surgeon Augustana, St. Mary's hosps., 1896; prof. clin. surgery med. dept. U. Ill., Chgo., 1900-25; instr. histology, then asst. chief of clinics Rush Med. Coll. Fellow Am. Surg. Assn., A.C.S., Royal Coll. Surgeons (Ireland), Royal Micros. Soc. (London). Author: Handbook on Appendicitis, 2d edit., 1906; Clinical Surgery for the Instruction of Practitioners and Students, 6th edit., 1905; Organization, Management and Construction of Hospitals, 1907; Surgery of Thyroid and Parathyroid Glands, 1910; Yearbook of Surgery, 1917-23; Treatise on Surgical Diagnosis and Treatment, 1918. Research on femoral hernia and hernia of young children; studied causes and prevention of diffuse peritonitis complicating appendicitis, 1901. Died July 25, 1925.

OCHSNER, Alton, Am. surgeon; b. Kimball, S.D., May 4, 1896; s. Edward Phillip and Clara (Shontz) O.; A.B., U. of S.D., 1918, hon. D.Sc., 1936; M.D., Washington U., 1920; LL.D. (hon.), Tulane University, 1966; m. Isabel Kathryn Lockwood, Sept. 13, 1923; children—Alton, John Lockwood, Mims Gage, Isabel. Interne Barnes Hosp., St. Louis, 1920-21, Augustana Hosp., Chicago, 1921-22; exchange surg. asst., Kantons Hosp., Zurich, Switzerland, 1922 23, staedtisches Krankenhaus, Frankfurt am Main, 1923-24; visited European and Am. clinics, 1924-25; instr. in surgery, Northwestern U. Med. School, 1925-26; asst. prof. surgery, U. of Wis., 1926-27; prof. surgery and chmn. dept. of surgery, Tulane U., 1927-56, William Henderson Prof. Surgery, 1938-56, prof. clin. surgery, 1956-61, prof. emeritus, 1961—, dir. sect. on gen. surgery Ochsner Clinic, Found. Hosp.; cons. surgeon Touro Infirmary; senior vis. surgeon, Charity Hosp.; cons. in thoracic surgery, USPHS, Hosp., Vets. Hosp., New Orleans, cons. surgeon I.C. Hosp., New Orleans, Dir. Eversharp, Inc., Nat. Airlines, Inc. Decorated Order Vasco Nunez de Balboa (Panama); recipient Times-Picayune Loving Cup as New Orleans' outstanding citizen; Distinguished Service award by A.M.A.; named Distinguished Salesman-at-Large of New Orleans, Fellow A.C.S. (regent, past president). Am., So. (past pres.) surg. assns., Royal College Surgeons in Ireland (honorary), Royal College of Surgeons of England (hon.); mem. Internat. Society of Surgery (past president), A.M.A., Am. Assn. Thoracic Surgery (past president), Soc. for Vascular Surgery (past pres.), Internat. Surg. Soc. Clin. Surg., Southeastern Surg. Assn. (past president), So. Surg. Assn. (past pres.), Orleans Parish Medical Soc. Exptl. Biol. and Med., So. Med. Assn., Am. Cancer Soc. (dir.), Pan-Pacific Surgical Association (pres. 1963), La. State Med. Assn. Am. Acad. Orthopedic Surgeons (hon), numerous fgn. sci. assns. (honorary), Sigma Xi, Phi Beta Kappa. Author: Smoking and Health. Writer of the section on the intestines, Nelson's Loose Leaf Surgery, 1928, section on diseases of veins, Lewis' System of Surgery, sect. on Thoracic Surgery and Mediastinum, Brennemann's Pediatrics. Monograph on Varicose Veins. Chief editor Internat. Surgical Digest; editor surg. sect. The Cyclopedia of Medicine, Surgery and Specialties; asst. editor Surgery of the Emergency; co-editor of Surgery; asso. editor Lewis

Practice of Surgery. Research on vascular and thoracic surgery. Home: 1347 Exposition Blvd. Office: Ochsner Clinic, New Orleans.

OCKENFUSS, see Oken, Lorenz.

OCKER, William C., Am. inventor; b. Phila., 1876; s. John and Margaret (Roschman) O.; ed. pub. schs., Phila.; m. Doris McLeod, June 1926; children— William C., Doris Ann. Served in Spanish Am. War, Philippine Insurrection; assigned to aviation sect. Signal Corps., 1912; became capt. aviation sect. Res., 1917; commanding officer Chandler Field; in charge flying instrn. Gerstner Field, 1918, then Wilbur Wright Field; with office of dir. air. service, Washington, 1920; commanding officer 56th Service Squadron, Langley Field, to 1924; comdr. 46th Sch. Squadron, 1930-37; with Barksdale Field, 1937. Mem. Nat. Aero. Assn. Author: (with C.J. Crane) Blind Flight in Theory and Practice. Advanced flying by instrument; called father of blind flying; developed flight integrator; inventor (with G.R. Smith) new airplane propeller, 1938; inventor (with C.J. Crane) pre-flight reflex trainer, 1941. Died 1942.

OCKHAM, William, see William of Occam.

O'CONNELL, Daniel Joseph Kelly, astronomer; b. Rugby, Eng., July 25, 1896; s. Daniel and Rosa (Kelly) O'C.; B.Sc., U. Coll., Dublin, 1919, M.Sc., 1920; postgrad. Harvard; D.Sc., Nat. U. Ireland, 1948. Faculty, Riverview Coll. Obs., New South Wales, Australia, 1933-52, dir., 1938-52; dir. Vatican Obs., Italy, 1952—. Pres. commn. on photometric double stars Internat. Astron. Union, 1955-61. Fellow Royal Astron. Soc.; mem. Royal Irish Acad., Am. Astron. Soc., Royal Soc. New South Wales (hon. mem.), Pontifical Acad. Scis. (pres.). Author: The Green Flash and Other Low Sun Phenomena, 1958; also numerous articles. Editor: Stellar Populations, 1958. Research, publs. on seismology, variable stars, eclipsing binaries. Address: Vatican Obs., Castel Gandolfo, Rome, Italy.*

O'CONNELL, Walter Edward, Am. psychologist; b. Reading, Mass., Aug. 2, 1925; s. Walter Edward and Margaret (Turner) O'C.; B.A., U. Mass., 1950; M.A., U. Tex., 1952, Ph.D., 1958; m. Gloria June Kane, Aug. 5, 1960; stepchildren—Beverly, Sherry (Mrs. Ray Housley); 1 dau., Vickie. Clin. psychologist VA Hosp., Waco, Tex., 1958-66, research asso., Houston, 1966—; lectr. Baylor U., 1959-66, U. St. Thomas, Houston, 1966—. Del., Internat. Assn. Individual Psychology. Diplomate Am. Bd. Examiners in Profl. Psychology. Fellow Am. Psychol. Assn.; mem. Am. Acad. Psychotherapists (editorial bd. Voices), Am. Cath. Psychol. Assn., Am. Soc. Adlerian Psychology (dir.), Acad. Religion and Mental Health, Sigma Xi, Psi Chi, Alpha Kappa Delta, others. Research, publs. in personality change; invention and testing of psychotherapeutic methods and theories. Home: 6405 Westward, Houston 77036. Office: VA Hosp., Houston 77031.*

O'CONNOR, Donald Joseph, Am. civil engr.; b. N.Y.C., Nov. 7, 1922; s. William and Helen (Ryan) O'C.; B.C.E., Manhattan Coll., 1944; M.C.E., Poly. Inst. Bklyn., 1947; Eng.Sc.D., N.Y. U., 1956; m. Anita Lordi, Oct. 30, 1948; children—Dennis, Arlene Jeannette. Instr., Manhattan Coll., 1946-47, Poly. Inst. Bklyn., 1948-50; faculty civil engring. Manhattan Coll., 1952—, prof. civil engring., 1964—. Cons. on stream pollution and assimilation capacity and diffusion studies to govt. agys. and engring. firms. Recipient Founders Day award N.Y. U., 1956. Mem. Am. Soc. C.E. (Rudolph Hering award 1958, 66), Am. Geophys. Union, N.Y. State Water Pollution Control Assn., Am. Soc. Engring. Edn., Am. Assn. U. Profs., Am. Soc. Limnology and Oceanography, Am. Assn. Profs. San. Engrs. (dir.). Author: (with W.W. Eckenfelder) Biological Waste Treatment, 1961; also articles. Research on pollution and purification of natural bodies of water, stream, estuary and ocean disposal of waste waters, estuarine water quality and pollution, distbn. of non-conservative contaminants in estuaries. Home: 307 Dunham Pl., Glen Rock, N.J. 07452. Office: Civil Engring. Dept., Manhattan Coll., N.Y.C. 10471.*

O'CONNOR, James Malachy, Irish physiologist; b. Limerick, Ireland, Feb. 9, 1886; s. Michael and Kathleen (Bowler) O'C.; B.A., Royal U., 1907, M.B., 1909; M.D., Nat. U., 1911, D.Sc. (hon.), 1956; m. Genevieve McGilligan, July 24, 1919; children—Malachy, Michael, John, Peter. Asst., Pharmacol. Inst., Heidelberg, 1910-21; asst. in physiology U. Coll., Dublin, 1911-19, prof. physiology, 1920-36. Mem. Royal Irish Acad. (pres. 1956-59), Royal Acad. Medicine Dublin, Physiol. Soc. London. Research, publs. on adrenalin content of blood, regulation of body temperature, action of insulin. Home: 7 Palmerston Villas, Dublin 6, Ireland.*

O'CONNOR, Johnson, Am. psychometrician; b. Chgo., Jan. 22, 1891; s. John and Nellie (Johnson) O'C.; A.B., Harvard, 1913, A.M., 1914; m. Ruth Davis, Dec. 17, 1913 (dec. Feb. 8, 1920); 1 son, Chadwell; m. 2d, Eleanor Manning, June 3, 1931. Astron. math. research with Percival Lowell, 1911-18; metall. research, Am. Steel & Wire Co., Worcester, Mass., 1918-20; elec. engring. with Gen. Elec. Co., West Lynn, Mass., 1920-22, organized Human Engring. Lab. for Gen. Elec. Co. to study applicants and new employees, 1922; lectr. on psychology, Stevens Inst. Tech., 1928-31, asso. prof. and dir. of psychol. studies, 1931-46; lectr. psychology, Mass. Inst. Tech., 1928-31, asst. prof. psychology, 1931-34, organized, 1930, the Human Engring. Lab. at Stevens Inst. Tech., which became Johnson O'Connor Research Found., 1942; pres. dir. Human Engring. Lab., Boston, Chgo., Phila., Ft. Worth, Tulsa, Los Angeles, Johnson O'Connor Research Found., N.Y.C., Los Angeles, Baytown, Tex., Detroit, Fundación de Investigaciones Johnson O'Connor, A.C., Johnson O'Connor Research Found. Wertheim fellow, 1927. Fellow Am. Acad. Arts and Scis. Author: Born That Way, 1928; Psychometrics, 1934; Johnson O'Connor English Vocabulary Builder, 1937 (Vol. 1, new edit., 1949, vol. 2, new edit. pub. 1951); Unsolved Business Problems, 1940; The Too Many Aptitude Woman, 1941; Structural Visualization, 1943; Aptitudes and the Languages, 1944; Ideaphoria, 1945; The Unique Individual, 1947; Johnson O'Connor Science Vocabulary Builder, 1955; also mag. articles: Vocabulary and Success; Redirecting Americans etc. Work on separation gen. intelligence into inherent and acquired factors; discovered and isolated 19 inherent aptitudes, probably unit genetic Mendelian traits; followed descent through 3 generations. Home: 381 Beacon St., Boston 02116. Address: Johnson O'Connor Research Foundation, Inc., 11 E. 62d St., N.Y. City 10021.*

O'CONNOR, Neil, psychologist; b. Geraldton, Australia, Mar. 23, 1917; B.A. with honors, U. W. Australia, Perth, 1937; M.A., U. Oxford (Eng.), 1941; Ph.D., U. London (Eng.), 1951; m. Margaret Edge, July 30, 1949; 2 children. With Med. Research Council, London, 1949—, devel. psychol. research unit Maudsley Hosp., London. Fellow Brit. Psychol. Soc.; mem. Exptl. Psychology Soc. Author: (with J. Tizard) Social Problems of Mental Deficiency, 1956; (with B. Hermelin) Speech and Thought in Severe Subnormality, 1963; also numerous articles, book chpts. Editor: Recent Soviet Psychology, 1961. Introduced concepts of social rehab. of educationally subnormal adolescents; research on use exptl. psychology in study severely subnormal adolescents; introduced physiol. psychology in study subnormal. Home: 66 Alton Rd., London S.W. 15. Office: Inst Psychiatry, Maudsley Hosp., Denmark Hill, London S.E. 5, Eng.*

O'CONNOR, Rod, Am. chemist; b. Cape Girardeau, Mo., July 4, 1934; s. Jay H. and Flora (Winters) O'C.; B.S., S.E. Mo. State Coll., 1955; Ph.D., U. Cal. at Berkeley, 1958; m. Shirley Ann Sander, Aug. 7, 1955; children—Mark Alan, Kara Ann, Shanna Suzanne. Asst. prof chemistry U. Omaha, 1958-60, Mont. State Coll., 1960-63; asso. prof. chemistry Mont. State U., 1963-66; asso. prof., coordinator gen. chemistry Kent O. State U., 1966—; on leave as staff asso. Adv. Council on Coll. Chemistry, Stanford, 1967-68. cons. insect venoms Hollister-Stier Labs., Spokane, 1963—. Fellow A.A.A.S.; mem. Internat. Soc. Toxinology, Am. Chem. Soc., Sigma Xi. Author: Laboratory Study of Natural Products, 1964; Laboratory Introduction to Analysis of Organic Compounds, 1966. Publs. on method for obtaining pure venom from wasps, hornets and bees, determination of chem. composition. Home: 903 El Cajon Way, Palo Alto, Cal. 94303. Office: Dept. Chemistry, Stanford U., Stanford, Cal. 94305.*

O'CONOR, Cornelio Marcelo, Argentinian plastic surgeon; b. Rosario de Santa Fe, Argentina, Nov. 9, 1915; s. Cornelio P. and Marcela (Rooney) O'C.; M.D., Universidad Nacional del Litoral, 1947; m. Chapelle Evelyn Power, May 16, 1951; children—Daniel, Federico, Carolina, Santiago. Chief dept. pastic surgery Nat. Rehab. Center, Buenos Aires; chief plastic and reconstructive surgery service Rivadavia Hosp., Buenos Aires; asst. resident in plastic surgery Royal Victoria Hosp., Montreal, 1949; resident in plastic surgery Franklin Hosp., San Francisco. Ofcl. del. Internat. Congress Plastic Surgery, Washington; sec., pres. various congresses, Argentina. Recipient Geigy Surgery prize. Mem. Argentine (pres. 1963), Latin Am. socs. plastic surgery, Buenos Aires Soc. Surgery Argentina Assn. Surgery (Gold medal, diploma of honor 1962). Author: Ulceras por Decúbito, 1963; also numerous articles. Home: 542 Peru Acassuso, Provincia de Buenos Aires. Office: 858 Arroyo, Buenos Aires, Argentina.*

ODDIE, Thomas Harold, physicist; b. Ballarat, Australia, July 19, 1911; s. Thomas Alfred and Marion (James) O.; B.Sc., U. Melbourne, 1932, M.Sc., 1933, D.Sc., 1944; m. Marion Taylor, July 25, 1935 (dec. 1961); children—Susan Annabelle (Mrs. John Sypkens), Catherine Jane, Pamela Louise (Mrs. John Larritt); m. 2d, Phyllis Anne Isom, Feb. 12, 1966. Physicist, Commonwealth X-Ray and Radium Lab., Melbourne, Australia, 1932-40; components engr. Philips Elec. Industries, Sydney, Australia, 1940-47; office-in-charge tracer elements sect. Commonwealth Sci. and Indsl. Research Orgn., Melbourne, 1947-52; faculty U. Ark. Sch. Medicine, Bowman Gray Sch. Medicine, Winston-Salem, N.C., 1952-56; research physicist unit clin. investigation Royal North Shore Hosp., Sydney, 1956-60; prof. radiology U. Ark. Med. Center, Little Rock, 1960—. Cons. Oak Ridge Inst. Nuclear Studies, 1952-56, 60—. Fellow Inst. Physics (London), Australian Inst. Physics; mem. Soc. Nuclear Medicine, Am. Thyroid Assn. Research, publs. on handling and measurement of radium in clin. use, clin. applications of radioactive isotopes, kinetics of iodine metabolism, diagnosis and treatment of thyroid diseases. Home: Route 1, Box 67A, Little Rock 72204.*

ODDY, Harold Grant, chemist, educator; b. St. Marys, Ont., Can., Aug. 15, 1899; s. Joseph Henry and Ellen (Aiken) O.; B.A., McMaster U., 1920, M.A., 1921; Ph.D., U. Toronto, 1923; m. Helyn Alexander, June 25, 1932. Came to U.S., 1925, naturalized, 1935. Lectr. chemistry U. Alta., 1923-25; asst. prof. chemistry U. Fla., 1925-26, Ore. State Coll., 1926-28; faculty chemistry U. Toledo, 1928—, prof., 1942—, head dept., 1944-65. Mem. Am. Chem. Soc., Ohio Acad. Sci. Contbr. articles on Friedel-Crafts reactions, thiodiacetic acid and analogs, betaine to profl. jours. Home: 3410 Valleston Pkwy., Toledo 43607.*

O'DELL, Boyd Lee, Am. biochemist; b. Hale, Mo.; Oct. 14, 1916; s. Orvis I. and Flossie (Hoover) O'D.; A.B., U. Mo., 1940, A.M., 1940, Ph.D., 1943; m. Vera L. Stone, Dec. 2, 1944; children—Ann Louise, David Lee. With Parke, Davis & Co., Detroit, 1943-46; faculty U. Mo., Columbia, 1946—, prof. agrl. chemistry, 1954—. Cons. USPHS Nutrition Study Sect., 1963—. Guggenheim fellow, 1964. Mem. Am. Soc. Biol. Chemists, Am. Inst. Nutrition, Am. Chem. Soc., Soc. Exptl. Biology and Medicine, Phi Beta Kappa. Editorial bd. Jour. Nutrition, 1965—. Research, publs. in, contbr. to isolation and characterization of folic acid and its role in prevention of anemia and abnormalities in newborn animals; produced zinc deficiency in chickens and demonstrated effect of phytic acid on zinc availability; elucidated role of copper in connective tissue metabolism. Home: 709 Morningside St., Columbia, Mo. 65201.*

ODENBACH, Frederick Louis, Am. meteorologist; b. Rochester, N.Y., Oct. 21, 1857; s. John and Elizabeth (Minges) O.; A.B., Canisius Coll., 1881; studied in Europe. Joined Soc. of Jesus, 1881; ordained priest, 1892; prof. physics and chemistry St. Ignatius Coll., Cleve., 1893-1903, later prof. astronomy and meteorology; founder, dir. Meteorol. Obs., 1895; prof. astronomy John Carroll U. (formerly St. Ignatius Coll.), Cleve.; founder Jesuit Seismol. Service, 1909. Invented ceraunograph, electric seismography, 1899; 6th observer of Helvetian halo, 1901. Died Cleve., Mar. 15, 1933.

ODENING, Klaus Heinz Friedrich, German parasitologist; b. Leipzig, Germany, Dec. 29, 1932; s. Will and Hanna (Ulbricht) O.; ed. U. Rostock, 1950-52, U. Halle, 1952-54, Dr.rer.nat., 1958; Dr.rer.nat.habil., East Berlin Humboldt U., 1965; m. Waltraud Wiegand, Aug. 18, 1956; children—Karen, Sven, Frank. Asst., U. Jena, 1954-59; head asst. Zool. Research Sta., German Acad. Scis., East Berlin Zool. Gardens, 1959—; lectr. Zool. Inst., East Berlin Humboldt U., 1966. Mem. German Zool. Soc., Biol. Soc. East Germany, Parasitology Soc. East Germany. Research, numerous publs. on helminths, mainly on digenetic trematodes. Office: 41, Am Tierpark, 1136 Berlin, East Germany.

ODIER, Giovanni Battista, see Hodierna, Giovanni Battista.

ODIER, Louis, Swiss physician; b. Geneva, Switzerland, Mar. 17, 1748; M.D., Edinburgh U., 1770. Prof. medicine Acad. of Geneva, 1799. Mem. French Acad. Scis. (corr.). Author: Epistola physiologica, 1770; Réflexions sur l'inoculation de la vaccine, 1880. Introduced vaccination into France and Switzerland; worked to purify city air, to fight prison fever, to prevent spread of poisonous fumes from nitric gas; wrote manual of practical medicine circa 1798. Died Geneva, Apr. 13, 1817.

ODING, Ivan Augustovich, Russian metallurgist; b. Riga, Latvia, July 6, 1896; grad. Technol. Inst. Petrograd, 1921. Sr. asst. chair metallography and metal working Petrograd Tech. Inst., 1921-23; staff Petrograd Elektrosila Plant, 1923-30; prof. chair metallography Leningrad Poly. Inst., 1930-42; dir. Central Research Inst. Tech. and Machine Bldg., 1942-47; dep. dir. Inst. Mechanics, 1947-53; asso. Inst. Metallurgy, USSR Acad. Sci., 1947-53; asso. Inst. Metallurgy, USSR Acad. Scis., 1953-56; head chair metallography and metal tech. Moscow Power Engring. Inst., 1956—. Recipient Stalin prize, 1946; named Honored Scientist of R.S.F.S.R., 1956; Order of Lenin, twice. Corr. mem. USSR Acad. Scis. Author: Metal Strength, 1937; Modern Methods of Metal Testing, 1944; Permissible Stresses in Machine Building and Cyclic Metal Strength, 1947. Research on strength of metals; devel. methods for testing mech. properties of metals, including cyclical viscosity and relaxation. Founder lab. for metal research, 1923. Home: B. Ordynka 34/38, Moscow, & USSR. Office: A. A. Baykev Inst. Metallurgy, USSR Acad. Scis., Leninskii Prospekt, 49, Moscow, USSR.

ODINGTON, Walter (or Walter of Evesham), English mathematician, astronomer; b. North Oxfordshire, Eng.; as monk in Benedictine abbey of Evesham, compiled its calendar, 1301; made astron. observations, Oxford, Eng., 1316; asso. with Merton Coll., by 1330. Author: Declaratio de motu octave sper, 1316; De speculatione musices (earliest Latin treatise on mensural music); Icocedron (attack on contemporary alchemists, includes explanations of various chem. processes, discussions of separation of 4 elements, attempts at quantitative measurement in degrees of virtues such as heat and dryness which failed through lack of exptl. precision); Tractatus de multiplicatione specierum in visu secundum omnem modum; Liber quintus geometrie per numeros loco quantitatum; numerous other works.

ODLING, William, English chemist; b. London, Sept. 5, 1829; originally trained in medicine Guy's Hosp.; M.B., U. London, 1851; M.A.; fellow Worcester Coll. Fullerian prof. chemistry Royal Inst., 1868-72; Waynflete prof. chemistry Oxford U., 1872-1912. Fellow Royal Soc., 1859. Author: Manual of Chemistry, Descriptive and Theoretical, 1861; Lectures on Animal Chemistry, 1866; Outlines of Chemistry, 1869; A Course of Six Lectures on the Chemical Changes of Carbon, 1869; Chemistry, 1884; Laurent's Chemical Method. Systematically studied and classified silicates; work on atomic weights led him to suggest that atomic weight of oxygen should be 16, not 8, as previously held; did early research on problems of valence and bonding and table of elements. Died Feb. 17, 1921.

ODOBESCU, Alexandru, Rumanian archeologist; b. Bucharest, Rumania, June 23, 1834; prof., Bucharest. Author: Scieri literare si istorice, 1887; Opere complete, 3 vols., 1906. Pioneer in Rumania in classical philology and archeology. Died Nov. 10, 1895.

O'DOHERTY, Desmond Sylvester, physician; b. Dublin, Ireland, July 27, 1920; s. Sean and Mary O'Doherty; came to U.S., 1922, naturalized, 1934; A.B., LaSalle Coll., 1942; M.D., Jefferson Med. Coll., 1945; m. Marcella J. duBoce, Aug. 15, 1951; children—Marianne, Desmond Patrick. Faculty, Georgetown U. Med. Center, Washington, 1952—, prof., 1961—, chmn. dept. neurology, 1959—, dir. Muscular Dystrophy Clinic, 1954—, med. dir. hosp., 1966. Cons. to gov. agys. Fellow Am. Acad. Neurology; mem. Assn. for Research Nervous and Mental Disease, Am. Neurol. Assn., Am. Epilepsy Assn., A.M.A. Author: (with Estelle Ramey) Electrical Studies of Unanesthetized Brain, 1961; also articles. Research on new treatments for Parkinsonism and epilepsy, mechanism brain changes asso. with vascular spasm, changes in spinal fluid enzymes with certain neurologic diseases; demonstrated suppressor area in temporal lobe brain. Home: 3540 N. Valley St., Arlington, Va. Office: 3800 Reservoir Rd. N.W., Washington 22207.*

ODQVIST, Folke Karl Gustaf, Swedish mech. engr.; b. Stockholm, Sweden, July 29, 1899; s. Karl Albert and Inez Mathilda Chr (Wahlén) O.; Mech.Engr., Royal Inst. Tech., 1922; Ph.D., U. Stockholm, 1928; Dr.Tech. h.c., U. Helsinki, 1966; m. Vera Traugott, Oct. 26, 1926; children—Ann-Marie (Mrs. Frithiof I. Niordson), Kerstin (Mrs. Anders E.G. Björkman), Eva (Mrs. Carl S. G. L. Bark), Sven O. F. Employed in industry, 1924-36; with Royal Inst. Tech., Stockholm, 1922-66, prof. solid mechanics, 1936-66, prorector, 1943-66, dean student, 1942-50. Cons. Swedish indsl. and mil. insts.; mem. govtl. coms.; mem. bd. Nat. Inst. for Materials Testing, 1947-67, pres., 1956-67. Decorated comdr. Order No. Star. Mem. Swedish Assn. Engrs. and Architects, Swedish Royal Acad. Engring. Scis., Swedish Royal Acad. Scis. (v.p. 1957—), Polish Academia Nauk, (fgn.), Internat. Union Theoretical and Applied Mechanics (mem. bur. 1952-64, pres., 1956-60). Author: Strength of Materials, 1948; (with J. Hult) Creep Strength, 1962; Mathematical Theory of Creep and Creep Rupture, 1966; also numerous articles. First complete solution of boundary value problems of hydrodynamics of viscous fluids, 1928, 30; stress distbn. in neighborhood of contact of solids, plasticity with isotropic strain hardening and ductile creep rupture, creep at combined stresses, effective width of reinforced elastic plates, creep in plates and membranes, influence of primary creep on creep rupture; developed primary creep theory, 1952. Home: 7 D Torstensonsvägen, Djursholm, Sweden. Office: Royal Inst. Tech., Stockholm 70, Sweden.*

O'DRISCOLL, Kenneth Francis, Am. chemist; b. N.Y.C., July 22, 1931; s. Frederick T. and Mary (Antonelli) O'D.; B.Chem. Engring., Pratt Inst., 1952; M.A., Princeton, 1957, Ph.D., 1958; m. Patricia Burns, June 26, 1954; children—Teresa, Kenneth Daniel, Carolyn, Michael. Devel. engr. DuPont Co., Arlington, N.J., 1952-53; faculty Villanova (Pa.) U., 1958—, prof. chemistry, 1965-66; asso. prof. chem. engring. State U. N.Y. at Buffalo, 1966—; vis. prof. dept. polymer chemistry Kyoto (Japan) U., 1964-65. Recipient Petroleum Research Fund Internat. award, 1964. Mem. Am. Chem. Soc., Sigma Xi. ionic mechanisms. Office: Chem. Engring. Dept., State U. N.Y., Buffalo 14214.*

ODUM, Howard Thomas, Am. ecologist; b. Durham, N.C., Sept. 1, 1924; s. Howard Washington and Anna Louise (Kranz) O.; A.B., U. N.C., 1947; Ph.D., Yale, 1951; m. Virginia Millie Wood, Sept. 6, 1947; children—Frances Ann, Mary Louise. Faculty, U. Fla., 1950-54, Duke, 1955-56; dir. Inst. Marine Sci. div. U. Tex., Port Aransas, 1956-63; chief scientist P.R. Nuclear Center, San Juan, 1963-66; prof. zoology, botany, environmental sci. depts. U. N.C., 1966—. Mem. coms., adviser USPHS; also lectr. Grantee NSF, NIH, AEC, Office Naval Research, various Tex. State Agys. Fellow A.A.A.S., Tex. Acad. Sci. (v.p.); mem. Ecol. Soc. Am. (George Mercer award 1957), Am. Soc. Limnology and Oceanography, Am. Meteorol. Soc., Geochem. Soc., Am. Inst. Biol. Sci., Soc. Gen. Systems Theory, Internat. Soc. Limnology, Sigma Xi. Research, publs. on measurement of energy flows in marine and forest environmental systems; concepts of biol. circuits; engring. ecol. systems. Home: Ridgefield Park, Chapel Hill, N.C. 27514.*

O'DWYER, Joseph, Am. physician; b. Cleve., Oct. 12, 1841, ed. McGill U., grad. Coll. Phys. and Surg., Columbia U., 1866; m. Catherine Begg; 8 children. Resident physician City Hospital, N.Y., 1866; joined staff N.Y. Foundling Asylum, 1872. First to use intubation for asphyxia in diphtheria (O'Dwyer tube), 1885, also to use diphtheria serum; treated stenotic diseases of larynx. Died N.Y.C., Jan. 7, 1898.

OEDING, Per, Norwegian microbiologist; b. Oslo, Norway, Feb. 5, 1916; s. Harald and Esther (Friedrichsen) O.; M.D., U. Oslo; m. Berit Otters, May 10, 1941; children—Per Harald, Kristin. Mem. Norwegian expdn. to Tristan da Cunha, then doctor Bacteriological Clinic, U. Oslo; instr. microbiology U. Bergen (Norway), 1949-57, prof., 1967—. Author numerous papers cn microbiology. Studies in serology and immunochemistry regarding straphylococcal and micrococcal antigens. Home: Professorvie 4, Minde, Bergen. Office: Haukeland Sykehus, Bergen, Norway.

OEHLERT, Wolfgang, German pathologist; b. Leipzig, Germany, Sept. 4, 1922; s. Otto Kurt and Rosel (Schwarze) O.; student U. Würzburg, U. Giessen (Germany), U. Hamburg, U. Köln; Med. Staatsexamen, U. Köln, 1952, Habilitation, 1960; m. Gisela Knoop, Dec. 22, 1953; 2 children. Staff, Pathologisches Inst., U. Köln (Germany), 1952-57, Inst. fur. Med. Isotopenforsch., 1957-59, faculty U. Freiburg (Germany), 1960—, prof. pathology, 1966—, head physician Path. Inst., 1960—. Author: Handbuch Allg. Pathologie; (with W. Maurer, B. Schultze) Autoradiography; Biochem. Taschenbuch. Rauen; Autoradiography; also articles. Research in pathology of tumors, proliferation kinetics in physiology and pathology, autoradiographic technique. Home: 14 Lindenau Str., Kirchzarten/Br., W. Germany. Office: 19 Albertstr., Freiburg/Br., West Germany.*

OEHME, Reinhard Friedrich Arthur, physicist; b. Wiesbaden, Germany, Jan. 26, 1928; s. Reinhold and Katharina (Kraus) O.; Diplom Physiker, Universität Frankfurt (Germany), 1948; Dr.rer.nat., Universität Göttingen (Germany), 1951; m. Mafalda Pisani, Nov. 5, 1952. Asst., Max Planck Inst. für Physik, Göttingen, 1949-53; research asso. Enrico Fermi Inst. Nuclear Studies, U. Chgo., 1954-56, faculty dept. physics and inst., 1958—, prof. physics, 1964—; mem. Inst. Advanced Study, Princeton, N.J., 1956-58; vis. prof. Instituto de Fisica Teorica, Sao Paulo, Brasil, 1952-53, U. Md., 1957, U. Vienna, 1961, Imperial Coll., London, Eng., 1963-64, Internat. Centre for Theoretical physics, Trieste, Italy, 1966, 68. Vis. scientist Brookhaven Nat. Lab., 1957, 60, 62, 65, 67, Lawrence Radiation Lab. U. Cal., Berkeley, 1959, CERN, Geneva, Switzerland, 1961, 63, 64. J.S. Guggenheim fellow, 1963-64. Fellow Am. Phys. Soc. Author: (with others) Kosmische Strahlung, 1953; Werner Heisenberg und die Physik unreser Zeit, 1961; Strong Interactions and High Energy Physics, 1964; Dispersion Relations and Causality, 1964; Preludes in Theoretical Physics, 1965; High Energy Physics and Elementary Particles, 1965; also numerous articles. Research, publs. on formulation and proof of dispersion relations for elementary particle scattering, formulation of charge conjugation and time reversal non-invariance of weak interactions, theory of elementary particle reactions, high energy scattering, higher dynamical symmetries. Office: Enrico Fermi Inst., U. Chgo., Chgo. 60637.*

OEMLER, Arminus, Am. agriculturist; b. Savannah, Ga., Sept. 12, 1827; s. Augustus Gottlieb and Mary Ann (Shad) O.; grad. with honors Dresden Technische Bildungsanstalt, 1848; M.D., U. City N.Y., 1856; m. Elizabeth P. Heyward, Apr. 10, 1856, 6 children. Joined Confederate Army, commd. capt. 2d Company, deKalb Riflemen; made 1st map of Chatham County (Ga.); founder 1st comml. oyster packing plant in South, Wilmington Island, Ga.; discoverer presence of nitrogen-fixing bacteria in nodules of leguminous plants, 1886, discouraged from further research by U.S. Dept. of agr. (actual discovery made in Germany 2 years later). Died Savannah, Aug. 8, 1897.

DENOPIDES OF CHIOS, mathematician, astronomer; flourished 450 B.C.; introduced improvements in elementary geometry; discovered what were to be incorporated as the 12th and 23rd propositions of 1st book of Euclid; invented (according to Eudemus) cincture of zodiac circle and long year; determined either position or div. of ecliptic; studied periods of revolution or calendar periods; fixed length of year at a little more than 365 1/4 days; followed Pythagorean practice in assigning planets to fixed orbits in zodiac.

OERIU, Simion, Rumanian chemist; b. Jassy, Rumania, Dec. 19, 1902; s. Lupu and Perla (Lebovici) O.; student U. Vienna (Austria), 1925-30; D.Chem. Scis., U. Bucharest (Rumania), 1930; m. Serafina Oeriu, Nov. 23, 1923; 1 son, Ion. Lab. chief Tb. Center, 1930-39; chief chemotherapy dept. Inst. Dr.I. Cantacuzino, Bucharest, 1946—; head chemotherapy dept. Acad. Socialist Republic Rumania, 1948—; prof. biochemistry Gen. Faculty Medicine, Bucharest, 1947—. Decorated Apararea Patriei, 1949, Steaua Republicei Populare, 1955, 23 August, 1961, Ordinul Muncii, 1963. Mem. Acad. Socialist Republic Romania (corr.), Société de Biochimie-France, Internat. Assn. Biochemistry, Société Européenne de Biologie Basel, Soc. Gerontology. Author: Chimia biologica, 2 vols., 1952; Proteins in Development and Senscence, vol. I. of Advances in Gerontological Researches, 1964; also numerous articles. Research on Oeriu concepts on the inhibition of microbial resistance and the role of disulfide bonds as a factor in aging process; sci. basis for inhibition of early aging. Home: 15, Herastrau, Raion 30, Decembrie, Bucharest, Rumania.*

ÖERSTED, Anders Sandoë, Danish naturalist; b. Radkjobing, Denmark, 1816; s. Hans Christian Ørsted; Ph.D., 1844; prof. botany; traveled to C.Am., 1846-48. Author: De regionibus marinis (classic work in marine ecology), 1844; Observations on a hitherto Unknown General Extension of Microscopic Plants in the Ocean (described phytoplankton and its role in sea metabolism, 40 years before Hensen); Research on the Flora and Geography of Central America, 1864; Les Chênes de l'Amérique tropicale, 1868; also papers on botany and mycology. Died 1872.

OERSTED, Hans Christian, Danish physicist, chemist; b. Rudkoebing, Island of Langeland, Denmark; Aug. 14, 1777; s. Soeren-Christian O.; assisted his father, an apothacary, when he was a boy; studied medicine, physics, astronomy, Ellersen College, U. Copenhagen, from 1793; received pharmacy degree, 1797; Ph.D., 1799. Asst. med. faculty, Ellersen College, U. Copenhagen, 1800; traveled and lectured in various European countries on scholarship, 1801-03; prof. physics, U. Copenhagen, 1806—; 1st dir., Polyteckniske Laeranstalt, Copenhagen; privy councilor, 1850. Founded Danish Soc. for Propagation of Natural Scis., 1824 (this Soc. has awarded Oersted Medal since 1908). Created knight of order of Danebrog. Fellow Royal Soc., 1821 (Copley Medal, 1820); mem. many academies and sci. socs.; perpetual sec., Royal Soc. of Sciences, Denmark, from 1815 (published revue of work of Soc.); French Acad. Sci., 1823; Am. Assoc. of Physics Tchrs. established Oersted Medal, 1937. Author: Experimenta circa effectum conflictus electrici, 1820. Founder of sci. of electromagnetism; discovered that magnetic needle was deflected at right angles to wire carrying an electric current, 1819; continued studies of this phenomenon, thereby demonstrating connection between electricity and magnetism; constructed (with Fournier) thermo-electric pile, 1823; discovered piperidine, 1820; 1st to prepare metallic aluminum, 1825; studied compressibility of liquids and gases; showed that sulphur dioxide is more compressible than air, hence perfect gas law of Boyle does not hold, 1806; suggested that acidity and alkalinity could be related to opposing electrical forces of the voltaic pile, 1804; developed hypothesis that element "heat" was the basis of light (when combined with air), of electric fluid (when combined with water) and of magnetic fluid (when combined with earth); unit of electromagnetic force, the oersted, is named after him; postulated that a magnet should exert a force on an element of a circuit (later confirmed by Davy). Died Copenhagen, Denmark, Mar. 9, 1851.

OERTEL, see Ortelius, Abraham.

OERTEL, Gerhard Friedrich Martin, geologist; b. Leipzig, Germany, Apr. 22, 1920; s. Friedrich and Hildegard (Szyllwahsszy) O.; student Bonn U., 1939-40, Graz U., 1940, Bonn. U., 1941-45; Dr.rer.nat., Bonn U., 1945, Privatdozent, 1950; m. Irmgard Maria Elisabeth Blüher, Oct. 15, 1946; children—Donata, Ulrich, Martin. Came to U.S., 1956. Asst., Bonn U., 1946-50; geologist Companhia dos Petróleos de Portugal, Lisbon, 1953-56; geologist in Portuguese India, Portuguese Overseas Ministry, 1953-56; asso. prof. geology Pomona Coll., Claremont, Cal., 1956-60; asso. prof. geology U. Cal. at Los Angeles, 1960-66; prof., 1966—. Brit. Council fellow to U. Edinburgh, Scotland, 1948-49; John Simon Guggenheim fellow, 1966-67. Mem. Geol. Soc. Am., A.A.A.S., Geologische Vereinigung, Deutsche Geologische Gesellschaft, Sigma Xi, Research, publs. in origin of granites, structures of volcanoes, theory of deformation of geol. bodies, origin of laterite soils, geology of Goa, math. model of fossil populations, origin of slaty clavage. Home: 11408 Cashmere St., Los Angeles 90049.*

OESTER, Yvo Thomas, physician; b. Ill., Dec. 21, 1908; s. Frank L. and Caroline (Hildmann) O.; B.S., U. Notre Dame, 1931, M.S., 1933; Ph.D., U. Chgo., 1938, M.D., 1943; m. Elizabeth F. Elkin,

Mar. 19, 1939; children—Michael, Miriam (Mrs. Vincent Hale), Yvonne. Chemist, Ill. Racing Bd., Fla. Racing Commn., Mich. Racing Commn., 1934-64; faculty Stritch Sch. Medicine, Loyola U., Chgo., 1947—, prof. pharmacology, 1956—; mem. staff phys. medicine and rehab. Cook County Hosp., Chgo., 1951—; asso. chief staff gen. med. and surg. service VA Hosp., Hines, Ill., 1963—. Loyola U. rep. Council Participating Instns. (now Asso. Midwest Univs.), Argonne Nat. Lab., 1948—. Mem. A.M.A. Soc. Nuclear Medicine, Am. Soc. Pharmacology and Exptl. Therapeutics, Am. Chem. Soc., Soc. Toxicology, Am. Coll. Clin. Pharmacology. Author: (with J.H. Mayer) Motor Examination of Peripheral Nerve Injuries, 1960; contbg. author Electrodiagnosis and Electromyography, 1956, Peripheral Nerve Regeneration, 1956. Research, publs. on neutron activation, applied to human specimens; whole body counting; neuropharmacology, neuromyal junctions; sensory deprivation, relation to motor patterning, electromyography devel. in clin. use. Home: 5009 Willow Spring Rd., LaGrange, Ill. 60525. Office: Box 98, Hines, Ill. 60141.*

OETJEN, Robert Adrian, Am. physicist; b. Detroit, Mar. 31, 1912; s. Frederick Tobias and Vida Pearl (Au) O.; B.A., Asbury Coll., Wilmore, Ky., 1936; M.S., U. Mich., 1938, Ph.D., 1942; m. Dorothy Mae Myers, June 3, 1936; children—Barbara Jane (Mrs. Lawrence Miller), Charlene Mae, Margaret Mary (Mrs. William K. Williams III), Charles William, Marilyn Ruth. Chemist, Mellon Inst. Indsl. Research, 1936-37; physicist Texaco Research Labs., Beacon, N.Y., 1941-46; mem. faculty Ohio State U., 1946—, prof. physics, 1957—, asso. dean Coll. Arts and Scis., 1960-62, 64-68, asst. v.p. for learning resources, 1968—, asst. to dir. Ohio State U. Research Found., 1953-55. Chief scientist Tokyo (Japan) office NSF, 1962-64; mem. U. S. delegation Internat. Commn. Optics, 1965—. Fulbright grantee, 1955-57. Fellow Optical Soc. Am.; mem. Am. Phys. Soc., Am. Assn. Physics Tchrs., Phys. Soc. Japan, Applied Physics Soc. Japan, Korean Phys. Soc., Japan Soc., Asia Soc. Contbns. to design and techniques for calibrating infrared spectographs for radiation of wave lengths between 1 and 1000 microns; absorption measurements and their relation to the structure of molecules. Home: 516 E. Schreyer Pl., Columbus, O. 43214.*

OETTINGER, Anthony Gervin, mathematician; b. Nuremberg, Germany, Mar. 29, 1929; s. Albert and Marguerite (Bing) O.; came to U.S., 1941, naturalized, 1947; A.B., Harvard, 1951, Ph.D., 1954; Henry fellow, U. Cambridge (Eng.), 1951-52; m. Marilyn Tanner, June 20, 1954; children—Douglas, Marjorie. Mem. faculty Harvard, 1955—, asso. prof. applied math., 1960-63, prof. linguistics, Gordon McKay prof. applied math., 1963—; cons. Arthur D. Little, Inc., 1956—, President's Sci. Adv. Com., 1961—, Bellcomm, Inc., 1963—. Mem. Assn. Computing Machinery (mem. council 1961—, chmn. com. U.S. Govt. relations, 1964-66, editor computational linguistics sect. Communications 1964-66, pres. 1966-68), Soc. Indsl. and Applied Math. (mem. council 1963-66), Am. Acad. Arts and Scis., I.E.E.E., Am. Math. Soc., Phi Beta Kappa, Sigma Xi. Author: A Study for the Design of an Automatic Dictionary, 1954; Automatic Language Translation; Lexical and Technical Aspects, 1960. Editor Proc. of a Symposium on Digital Computers and Their Applications, 1962. Research on switching theory, artificial intelligence, automatic lang. translation, computational linguistics, and tech. aids to creative thought. Home: 65 Elizabeth Rd., Belmont, Mass. 02178. Office: 33 Oxford St., Cambridge, Mass. 02138.*

OETTLÉ, Alfred George, South African med. researcher; b. Port Elizabeth, South Africa, June 22, 1919; s. Carl Maximilian and Ruth (Pudney) O.; B.Sc., U. Witwatersrand, 1938, B.Sc. (Hons.), 1940, M.B., B.Ch., 1942; m. Elsa Madeleine Oettlé, Dec. 20, 1947; children—George Julien, Alison Hilda, Elizabeth Marguerite, Charl Andrew, Edmund Eric, Caroline Helena. Lectr. histology Med. Sch., U. Witwatersrand, Johannesburg, South Africa, 1944-48, 51; Nuffield Dominion Travelling Med. fellow U. Oxford, 1949; grad. asst. in haematology U. Oxford, 1950; Cancer Research fellow S. African Inst. for Med. Research, Johannesburg, 1951-59; head cancer research unit Nat. Cancer Assn. of S. Africa, Johannesburg, 1959—. Recipient Bronze medal So. Transvaal br. Brit. Med. Assn., 1942. Mem. Internat. Union Against Cancer. Numerous publs. on studies in biol. microscopy; epidemiology and geog. pathology of cancer in various African races. Home: 44 Park St., Oaklands, Johannesburg. Office: S. African Inst. Med. Research P.O. Box 1038, Johannesburg, South Africa.

OFER, Shimon, Israeli physicist; b. Jerusalem, Oct. 18, 1927; s. Jacob and Sara (Yaffe) Friedman; M.S. in Physics, Hebrew U., 1954, Ph.D., 1957; m. Mira Zvi-Tov, Aug. 12, 1952; children—Dov, Shai. Research fellow Brookhaven Nat. Lab., N.Y., 1957-59; faculty Hebrew U., Jerusalem, 1959—, asso. prof., 1964—, chmn. dept. physics, 1966-67; vis. scientist Bell Telephone Labs., Murray Hill, N.J., 1965-66. Mem. Am. Phys. Soc. Research, numerous publs. on use of techniques of Mossbauer effect and angular correlation of Y-rays for investigation of magnetic properties of materials and of spin relaxation phenomena. Address: 36 Hapalmach St., Jerusalem, Israel.*

OFFICER, Julius Earle, Am. virologist; b. Rupert, Ida., Mar. 20, 1927; s. J. E. and Ellen (Dahlberg) O.; B.S., U. Ida., 1951; postgrad. Ind. U.; Ph.D., George Washington U., 1958; children—Jana Lynn, Jay E., Lori Jean; m. 2d, Sue Wilson, Aug. 13, 1966. Research asst. Ind. U. Med. Center, 1951-52; prin. investigator U.S. Army Biol. Labs., Frederick, Md., 1954-66; asso. prof. microbiology U. So. Cal., Los Angeles, 1966—; lectr. anatomy, physiology, microbiology Frederick Community Coll., 1964-66. Mem. Am. Soc. Microbiology, N.Y. Acad. Scis., Research Soc. Am. (Ft. Dietrick award 1958), Am. Soc. Immunologists, Sigma Xi. Contbr. articles to tech. jours. Research in molecule biology of viruses, mode of action of interferon, genetic contbrn. and replication RNA viruses. Home: 35 San Miguel Av., Pasadena, Cal. Office: Dept. Microbiology, U. So. Cal. Sch. Medicine, Los Angeles 90330.*

OFFNER, Franklin Faller, Am. biophysicist; b. Chgo., Apr. 8, 1911; s. I. H. and Jennie (Faller) O.; B.Chemistry, Cornell U., 1933; M.S., Cal. Inst. Tech., 1934; Ph.D., U. Chgo., 1938; m. Janine Y. Zurcher, Sept. 22, 1956; children—Laurens, Alexandra, Sylvia, Robin. Pres. Offner Electronics Inc., Chgo., 1939-63; prof. biophysics Northwestern U., 1963—. Laureate in tech. Lincoln Acad. Ill. Fellow I.E.E.E.; mem. Am. Electroencephalographic Soc., Soc. Electroencephalographic Langue Francais, Am. Phys. Soc., Biophys. Soc., Sigma Xi. Contbr. research papers biophysics, electronics. Research and patents in electronic controls for aircraft, electronic instruments for biomed. research, guided missiles; math. theory of nerve conduction. Home: 1890 Telegraph Rd., Bannockburn, Deerfield, Ill. 60015. Office: Technological Inst., Northwestern Univ., Evanston, Ill.*

OFFNER, Walter Worthington, cons. engr.; b. Munich, Germany, Feb. 15, 1906; s. Maximilian F. and Maria (Bentey) O.; grad. Tech. U., Augsburg, Germany, 1927; postgrad. U. Chgo., 1930-31, U. Wash., 1939-40; m. Estelle Kerr Hadden, June 26, 1948. Engr. tube and test dept. Gen. Elec. X-Ray Corp., Chgo., 1930-37; founder Indsl. X-Ray Labs., Seattle, 1938, pres., 1940—, partner, 1943—; pres. Standard X-Ray Sales Co., Seattle, 1940, v.p., 1941; founder, pres., tech. dir. X-Ray Engring. Co., Mill Valley, Cal., also San Francisco, 1946-56; founder, pres. X-Ray Engring. Internat., 1956—; mng. dir. X-Ray Engring. Australia Pty., Ltd.; v.p. Radiografias Industrales, S.A., Mexico; v.p. X-Ray Engring. Hawaii, Ltd.; v.p., dir. X-Ray Engring. India, Ltd., Bombay; officer, dir. affiliated cos. radiographic weld control first high-pressure oil pipeline in U.S., 1946, 30-inch gas pipeline, 1947; supervising, cons. engr. radiographic weld inspection Trans-Arabian Oil Pipe Line, 1947-50, Iraq Petroleum 30 inch pipeline, 1950, Tacagua Oil Pipe Line, Venezuela, 1950, Arabian Am. Oil Co. projects, Saudi Arabia, 1960—. U.S. del. Internat. Inst. Welding. Chmn., Ultrasonic Com. to survey and develop ultrasonic methods for ship hull constrn., 1961; chmn. weld flaw evaluation com. Nat. Acad. Sci., 1964. Fellow Am. Soc. M.E.; mem. Nat. Soc. Profl. Engrs., Am. Gas Assn., Am. Ordnance Assn., Am. Petroleum Inst., Am. Soc. Metals, Am. Welding Soc., Am. Soc. Testing Materials. Contbr. to Ency. Americana, 1965-66, also articles to profl. lit. Home: 650 N. San Pedro Rd., San Rafael, Cal. 94901. Office: 420 Market St., San Francisco 94111.*

OFFORD, Albert Cyril, mathematician; b. London, Eng., June 9, 1906; s. Albert Edwin and Hester (Sexton) O.; Ph.D., St. John's Coll., Cambridge U., 1936; D.Sc., U. Coll., London, 1937; m. Marguerite Yvonne Pickard, Aug. 4, 1945; 1 dau., Margaret Alison. Fellow, St. John's Coll., 1937-40; lectr. U. Coll., Banger, Wales, 1940-41; lectr. King's Coll., Newcastle on Tyne, 1941-45, prof. math., 1945-48; prof. math. U. London, 1948—. Fellow Royal Soc., 1952; mem. London Math. Soc., Royal Soc. Edinburgh, Cambridge Philos. Soc. Research in math. analysis, application of probability theory in analysis. Office: London Sch. Econs. and Polit. Sci., Houghton St., London W.C. 2, Eng.*

OFTEDAHL, Christoffer Gedde, Norwegian geologist; b. Oslo, Norway, June 28, 1917; s. Knut T. and Astrid (Kvepild) O.; Cand.real., U. Oslo, 1942, Dr.Philos., 1949; m. Nanna Liv Hansen, June 27, 1942; children—Knut Reidar, Lars Christoffer. Research asst. Mineralogisk Inst., U. Oslo, 1942-48, research asso., 1948-52; state geologist Geol. Survey Norway, 1952-59; prof. gen. geology Tech. U. Norway, Trondheim, 1959—. Mem. Norwegian Research Council for Sci. and Liberal Arts, 1962—, chmn. continental shelf com., 1965-67; chmn. Norwegian Group Volcanology and Seismology, 1962—, Norwegian Nat. Com. for Geology, 1965—, Norwegian Upper Mantle Com., 1965—. Recipient Reusch medal Norsk Geologisk Forening. Mem. Internat. Union Geodesy and Geophysics, Geol. Soc. Norway, Geologiska Föreningen i Stockholm, Kongelige Norske Videnskapsselskab i Trondhjem, Teknisk Videnskapsakademi i Norge. Author: (monographs) The Feldspars, 1949, The Lavas, 1952, The Cauldons, 1953; (with P. Holmsen) Ytre Rendal og Storelvdal (geol. maps with description). Research and publs. on igneous rocks and structures of Permian Oslo Rift Valley, including its feldspars, volcanics, magna formation, calderas; Eocambrian arkoses in Central Norway; regional studies of various phases of Caledonian orogenic zone, including its massive sulphide ores. Home: Solhögdveien 22, Trondheim, Norway.*

OFTEDAL, Per, Norwegian radiation geneticist; b. Stavanger, Norway, Oct. 18, 1919; s. Lars and Alice (Stephansen) O.; student U. Oslo, 1941-43, cand.real., 1952, D.philos., 1959; student U. Uppsala (Sweden), 1944; m. Gudrun Julie Ödegaard, Dec. 7, 1957; children—Terje, Hanne Bergljot, Marte Alice. Chmn. genetics lab. Norsk Hydro's Inst. for Cancer Research, Oslo, 1961—; fellow Norwegian Cancer Soc., 1952-61; vis. scientist Inst. Advanced Learning, City of Hope Med. Center, Duarte, Cal., 1964-65. Mem. Royal Radiation Council, 1959—, Royal Reactor Safety Com., 1960—. Research, publs. on fruit fly tumors, genetics and morphology; metabolism and genetic effects of radioactive phosophorus and yttrium, low dose radiation genetics, and radiation hygiene. Home: Gamle Ringeriksvei 50F, Bekkestua, Norway. Office: Norsk Hydro's Inst. for Cancer Research, Norwegian Radium Hosp., Montebello, Oslo, Norway.*

OGANESYAN, Leon Andreevich, Russian internist, med. historian; b. 1885; grad. Med. Faculty, Kharkov U., 1909; D.Med. Sci. Prof., head chair propedeutics of internal diseases Yerevan Med. Inst., 1923—; head history of medicine and biology sect. Armenian Acad. Scis., 1944—, head cardiology sect., 1955—. Mem. USSR Acad. Med. Scis., Armenian Acad. Scis. Author numerous works including: History of Medicine in Armenia from Ancient Times To the End of the 13th Century, 1927; Mineral Springs of Armenia, 1936; History of Medicine in Armenia from Ancient Times to the Present, 5 vols., 1946-47; co-author: Manual on Diagnosis of Internal Diseases, 1943; Latin-Russian-Armenian Medical Dictionary, 1951; Specific Pathology of Internal Diseases, 1956. Co-editor: History of Medicine sect. Large Med. Ency., 2d edit.; mem. editorial council Clin. Medicine, Therapeutic Archives. Yerevan Med. Inst., ulitsa Kirova 2, Yerevan, Armenian SSR, USSR.

OGATA, Koan (originally Akira), Japanese physician; b. Bitchu Province, Okayama Prefecture, Japan, 1810; studied Dutch medicine under Nobumichi Tsuboi and Shinsai Udagawa, Edo, also under Johannes Niemann, Nagasaki, 1836; began practice medicine, Osaka, 1938; founder Tekitekisai-juku (sch. of Dutch learning); named physician in ordinary to Shogun, also head Shogunate med. Sch., Edo., 1862; author books on Western medicine, also transl. of a Dutch work on internal medicine; popularized vaccination. Died 1863.

OGATA, Koreichi, Japanese physicist; b. Kumamotoshi, Japan, Jan. 5, 1912; s. Jyuemon and Yaeko (Sakaki) O.; B.Sc., Osaka (Japan) U., 1936, D.Sc., 1944; m. Kumiko Shimizu, Oct. 25, 1941; children—Koreaki, Koreharu. Faculty, Osaka U., Toyonaka-shi, Osaka-fu, Japan, 1936—, prof. physics, 1956—. Recipient Nishina Meml. award, 1955. Mem. Am., Japan phys. socs., Mass Spectroscopy Soc. Research, publs. on precise atomic mass determination with mass spectroscopical methods; mass spectrometrical studies on long-life nuclear decay, fission, spallation. Home: 1-78 Nigawa-cho, Nishinomiya-shi, Hyogo-ken, Japan. Office: 1-1 Machikaneyama-cho, Toyonaka-shi, Osaka-fu, Japan.*

OGATA, Masanori, Japanese physician; b. Kumamoto Prefecture, apan, 1854; grad. Kumamoto Med. Sch.; postgrad., Germany; doctorate, 1888; lectr. Kumamoto Med. Sch.; dean med. dept. Tokyo U., 1890-1901; mem. Imperial Acad. Studied bubonic plague in Formosa, 1896; 1st to show that the flea, in particular Xenopsylla cheopis, is main vector of bubonic plague, 1897; studied tsutsugamushi rickettsial fever, endemic in Japan. Died 1919.

OGATA, Tomio, Japanese serologist, med. historian, b. Osaka, Japan, Nov. 3, 1901; s. Keijiro and Tomoka (Miura) O.; M.D., U. Tokyo (Japan), 1926, D.Med.Sci., 1929; m. Haruko Nishiyama, Feb. 10, 1938; children—Natsuko (Mrs. Yuzuru J. Homma), Hiroaki. Rockefeller fellow med. research, 1934-35; serology research U. Chgo., also Mt. Sinai Hosp., N.Y.C., 1934-35; faculty U. Tokyo, 1936—, prof. serology, 1949-62, prof. emeritus, 1962—, chmn. dept. serology, 1936-62, councilor, 1958-62, dir. New Central Med. Library, 1961-62; founder, pres. Ogata Inst. for Med. and Chem. Research, Tokyo, 1962—; exec. dir. Research Inst. for Sch. Edn., Tokyo, 1963—. Participant study group on internationally acceptable minimum standards med. edn. WHO, Geneva, 1961. Recipient Prof. Minami prize for serological study syphilis, 1946; Empress Teimei prize for serological study leprosy, 1955; Mainichi prize for outstanding publs., 1947. Mem. Internat. (dep. pres. 8th congress Tokyo 1960), Japaneses socs. blood transfusion, A.M.A. (hon.), N.Y. Acad. Scis.; Am. Assn. Immunologists, Am. Soc. for Microbiology, Internat. Soc. History Medicine. Author: Gempaku Sugitas Rangaku Kotohajimie, 1941, rev., 1959; Biography of Koan Ogata, 1942, revised edit., 1963; Theoretical Serology, 1965; also numerous articles. Research in antigen-antibody reactions in vitro and vivo, serological tests for syphilis and leprosy; discovered therapeutic activity bacterial toxins in humans. Home: 3-435 Komagome, Toshima, Tokyo. Office: Ogata Inst., 1-11-4 Higashi-Kanda, Chiyoda, Tokyo, Japan.*

OGAWA, Joseph Minoru, Am. plant pathologist; b. Sanger, Cal., Apr. 24, 1925; s. Shosaku and Naomi (Yamaka) O.; B.S., U. Cal., Davis, 1950, Ph.D., 1954; m. Margie H. Kawasaki, Nov. 7, 1954; children—Julie M., Martin K., Jo Ann. Faculty, U. Cal., Davis, 1953—, asso. prof., 1963—. Recipient plaque Cal. Freezers Assn., 1966. Mem. Am. Phytopath. Soc., Am. Inst. Biol. Sci., Sigma Xi. Studies relating to devel. and control of plant pathogens on stone fruits, hops and tomatoes, tropical fruits. Home: 806 Linden Lane, Davis, Cal. 95616.*

OGAWA, Kazuo, Japanese anatomist, cytochemist, cell biologist; b. Shiga Ken, Japan, May 10, 1928; s. Hideichi and Masako (Yuki) O.; M.D., Kyoto U., 1954; Ph.D. in Anatomy, Kobe (Japan) U. Sch. Medicine, 1959; m. Yoko Hirasawa, Jan. 13, 1958; children—Kazuoki, Kazuto, Kazuteru. Research asso. U. Tex., Galveston, 1956-57, Rockefeller fellow, 1957-58; fellow neuropathology Montefiore Hosp., N.Y.C., 1958-59; instr. Kobe U. Sch. Medicine, 1959-60; instr. Kyoto U. Sch. Medicine, 1959-63, asso. prof., 1963-64; prof. anatomy Kansai Med. Sch., Osaka, Japan, 1964—. Roosevelt Internat. Cancer fellow Yale Sch. Medicine, 1963-64. Mem. Internat. Brain Research Orgn., Am. Assn. Anatomists, Am. Soc. for Cell Biology, Japanese Soc. for Electron Microscopy, Am., Japanese histochem. socs., Japanese Assn. Anatomists, Japanese Tissue Culture Soc. Author: Modern Cytology, 1964; also articles. Editor: Tissue Culture, Theory and Applied, 1963. Pioneered studies of biochem. functions of biol. orgn. at ultrastructural level using electron microscopy in field of morphology. Home: 21-2 Konmachi, Katsura, Ukyoku, Kyoto, Japan. Office: 1 Fumizonocho, Moriguchi, Osaka, Japan.*

OGAWA, Toru, Japanese physicist; b. Tokushima, Japan, Mar. 9, 1924; s. Masuji and Machie Ogawa; Rigakushi, Kyoto U., 1949, Rigakuhakase, 1961; m. Toshiko Suzue, Apr. 1, 1954; children—Mikiko, Atsushi, Keiko. Faculty electronics Doshisha U., Kyoto, 1953—, prof., 1961—; prof. Kyoto U., 1967—. Mem. com. radio tech. Ministry of Post Office, Tokyo, 1960—; fellow Sch. for Advanced Study, Mass. Inst. Tech., 1960-61; vice chief rocket observation br. Inst. Space and Aero. Sci., Tokyo U., 1963—; mem. com. quantum electronics Inst. Elec. Communication Engrs. Japan, 1966—. Research, publs. on time and frequency standard (atomic clock), optical pumping and Laser, observation of disturbances in ionosphere, magnetic waves in upper atmosphere. Home: 8 Nagaredacho, Kamitakano, Sakyoku, Kyoto. Office: Yoshidahonmachi, Sakyoku, Kyoto, Japan.*

OGAWA, Yoshiki, Japanese metallurgist; b. Tokyo, 1902; s. Takuji Ogawa; grad. Tokyo U., 1925; doctorate, 1934; m. Hide Ogawa. Became instr., then asst. prof. Tohoku U., 1927; named prof. Kyushu U., 1937; prof. engring. Tokyo U., from 1942; sent as mem. Atomic Commn. to inspect atomic reactors in Europe and Am., 1954. Author: The Foundation of Nonferrous Refinement; Treatment of Surface of Metals; also papers of refinement of nonferrous metals and its application. Pioneer in metall. exploitation of uranium in Japan. Died Tokyo, Mar. 27, 1959.

OGDEN, Eric, physiologist; b. London, Eng., Jan. 26, 1903; s. Michael Guy and Fanny (Bridge) O.; B.Sc., U. Coll., London, 1925, M.R.C.S., L.R.C.P., 1928; m. Olga Ivanoff, Sept. 18, 1928; children—Michael, Eleanor (Mrs. Harry C. Poehlmann) Kathryn (Mrs. Robert Kingwell). Came to U.S., 1929, naturalized, 1947. Faculty, U. Coll., London, 1928-29, Coll. Medicine, Dentistry and Pharmacy U. Cal. at Berkeley and San Francisco, 1929-43; practice medicine, Berkeley, 1940-43; prof. physiology, cons. John Sealy Hosp. U. Tex., Galveston, 1943-49; prof. physiology and medicine, chmn. dept. Ohio State U., Columbus, 1949-63; chief, div. environmental biology NASA, Ames Research Center, Moffet Field, Cal., 1963—. Lectr. aero. and astronautical engring. Stanford U., 1963—. Mem. Am. Heart Assn. (Research Service award 1966, dir.), Aerospace Med. Assn. Contbr. articles to nat. sci. jours. Home: 1795 Stanford Av., Menlo Park, Cal. 94025. Office: Div. Environmental Biology, Ames Research Center, Moffet Field, Cal.*

OGDEN, Francis Barber, engr., b. Boonton, N.J., Mar. 3, 1783; s. Matthias and Hannah (Dayton) O.; m. Louisa Pawnall, 1837. Aide-de-camp to Gen. Andrew Jackson, Battle of New Orleans, 1815; apptd. U.S. consul, Liverpool, Eng., 1829-40; consul, Bristol, Eng., 1840-57. 1st to apply principles of expansive power of steam and use of right angular cranks in marine engines; built 1st low pressure condensing engine with 2 cylinders, Leeds, Eng., 1817; assisted John Ericson (inventor of Moniter) in devel. and promotion screw propellor for steam powered craft. Died Bristol, Eng., July 4, 1857.

OGILVIE, Alan Grant, Brit. geographer; b. Edinburgh, Scotland, 1887; s. Sir Francis Ogilvie; B.A., M.A., B.Sc., Magdalen Coll., Oxford (Eng.) U.; student Imperial Coll., London, Eng., also univs. Berlin (Germany) and Paris (France); m. Evelyn Decima Willes, 1919; 1 dau. Demonstrator geography Oxford U., 1912-14; reader geography U. Manchester (Eng.), 1919-20; chief Hispanic Am. div. Am. Geog. Soc., 1920-23; lectr. U. Edinburgh, 1923-23, reader, 1924-31, prof., 1931—. Hon. sec. Royal Scottish Geog. Soc., 1925-41, pres., 1946-50; pres. Inst. Brit. Geographers, 1951-52, geog. sect. Brit. Assn., 1934, phys. geography sect. Internat. Geog. Congress, Lisbon, Portugal, 1949. Mem. Peace Delegation, Paris, 1919. Served with Royal F.A., 1911-19. Albert Kahn Travelling fellow, 1914; decorated Order British Empire, 1919. Author: Some Aspects of the Boundary Settlement at the Peace Conference, 1922; Geography of the Central Andes, 1922; Europe and its Borderlines, 1957. Editor, part author: Great Britain, 1928, contbg. editor Geog. Rev. Devel. list of geog. core areas from which each European state 1st achieved identity. Died Feb. 10, 1954.

OGLE, Kenneth Neil, Am. biophysicist; b. Lake City, Colo., Nov. 27, 1902; s. Wesley Harlan and Luella (Moore) O.; A.B., Colo. Coll., 1925, D.Sc. honoris causa, 1963; A.M., Dartmouth, 1927, Ph.D., 1930; M.D. honoris causa, U. Uppsala, Sweden, 1962; m. Elizabeth Bartlett, Sept. 18, 1934; children—Betsy Jane (Mrs. Donald W. Jordan, Jr.), Nancy Moore (Mrs. Richard F. Brubaker). Research fellow dept. physiol. optics Dartmouth, 1930-34, faculty Dartmouth Eye Inst. and Med. Sch., 1934-47; research cons. sect ophthalmology Mayo Clinic, Rochester, Minn., 1947-67, cons. biophysics sect., 1947-67, head, biophysics sect., 1958-67; faculty Mayo Grad. Sch. Medicine, U. Minn., 1947-67, prof. physiol. optics, 1953-67, prof. ophthalmology Med. Sch., 1968—. Cons., mem. com. NIH; mem. ednl. adv. com. Ednl. Found. in Ophthalmic Optics, Am. Bd. Opticianry, 1961—; mem. optical aids research adv. com. Am. Found. for Blind, 1954—; mem. com. on vision Armed Forces-NRC, 1947—, exec. com., 1953—. Recipient Beverly Myers Nelson Achievement award Am. Bd. Opticianry, 1957. Fellow Am. Psychol. Assn.; asso. fellow Am. Acad. Ophthalmology and Otolaryngology (instr. 1952—, certificate award 1960); mem. Nat Soc. for Prevention Blindness (com. on basic and clinic. research 1950—), Internat. Acad. Opticianry (council 1963—), A.A.A.S., Optical Soc. Am. (asso. editor Jour. 1954—, Tilyer medal 1967), Assn. for Research in Opthalmology (Proctor medal 1962), Am. Physiol. Soc., Biophys. Soc., Psychonomic Soc., Minn. Acad. Sci., Minn. Psychol. Assn., Sigma Xi, Pi Kappa Alpha, Gamma Alpha. Author: Researches in Binocular Vision, 1950 (recipient certificate of award Book Clinic Phila. 1951); Optics: An Introduction for Ophthalmologists, 1961; sect. in The Eye, 1962; Oculomotor Imbalance in Binocular Vision and Fixation Disparity, 1967. Asso. editor Investigative Ophthalmology, 1962—; hon. editor Vision Research, 1961—. Research, publs. on binocular vision, various disorders of vision and eye movements, fixation disparity, pupillography. Home: 625 Memorial Pkwy., Rochester. Office: care Mayo Clinic, Rochester, Minn. 55901. Died Rochester, Feb. 22, 1968.

OGNEV, Boris Vladimirovich, Russian surgeon, anatomist; b. 1901; grad. Med. Faculty, Kazan U., 1924; D.Med. Sci., 1935. Asst. dep. topographical anatomy and operative surgery Kazan U., 1924-31; asst., lectr., 2d prof. chair operative surgery and topographical anatomy Central Postgrad. Med. Inst., Moscow, 1931-38, head chair, 1939—, prof., 1935—; head clin. dept. Inst. Exptl. Cancer Pathology and Therapy, USSR Acad. Med. Scis. Recipient Burdenko prize. Mem. USSR Acad. Med. Scis. (corr.), Internat. Soc. Surgeons (hon.), Internat. Soc. Oncologists and Microbiologists. Author: The Blood Vessels of the Central and Peripheral Nervous System in Man; The Blood Vessels of the Heart in Normal and Pathological States; co-author: Topographical and Clinical Anatomy, 1960. Mem. editorial bd. Surgery, Herald of USSR Acad. Med. Scis.; co-editor Surgery sect. Large Med. Ency., 2d edit. Research and numerous publs. on anatomy and surgery, cardiovascular system in normal and path. states, symmetry in nature. Address: Central Postgrad. Med. Inst., pl. Vosstaniya 1-2, Moscow, USSR.

OGSTON, Sir Alexander, Scottish surgeon, bacteriologist; b. Apr. 19, 1844; M.D., C.M., LL.D., Aberdeen U.; LL.D., Glasgow U.; m. Mary Jane Hargrave, 1867; 1 son, 2 daus.; m. 2d, Isabella Margaret Matthews; 1 son, 3 daus. Served in Sudan, 1885, South Africa, 1899-1900, Serbia, 1915, Italy, 1916-17; extra surgeon to king in Scotland; emeritus prof. surgery Aberdeen U.; hon. surgeon to Queen Victoria and King Edward VII in Scotland. Mem. Brit. Med. Assn. (pres. 1914-15). Credited with discovery of Staphylococcus pyogens aureus, circa 1881. Died Feb. 1, 1929.

OGUR, Maurice, Am. microbiologist; b. N.Y.C., Nov. 29, 1914; s. Ellis M. and Edith (Levine) O.; B.A., Bklyn. Coll., 1934; M.A., Columbia, 1937, Ph.D., 1948; m. Slyvia Bregman, May 23, 1937; children—Jonathan David, Barbara Ruth. Faculty, Bklyn. Coll., 1935-52, asst. prof. chemistry, 1949-52; research asso., cons. Bot. Labs., U. Pa., 1948-50; asso. prof. Microbiology and Biol. Research Lab., So. Ill. U., Carbondale, 1953-61, prof., 1961—, chmn. dept., dir. lab., 1964—. Fulbright Research scholar, France, 1960-61; OEEC fellow, 1961. Mem. A.A.A.S., Am. Chem. Soc., Am. Soc. Microbiology, Am. Soc. Biol. Chemists, Sigma Xi (chpt. pres. 1965—). Contbr. articles to tech. jours. Developed methods nucleic acid assay; discovered polyploid yeasts; related nucleic acid synthesis to cell div. in lily pollen and in yeast; developed methods for assaying cytochrome deficient mutants, mutation rate and phenomic lag; discovered tricarboxylic acid cycle mutants. Home: 1219 W. Hill St., Carbondale, Ill.*

OGURA, Joseph H., Am. surgeon; b. San Francisco, May 25, 1915; s. Kikuji and Hannah (Akita) O.; A.B., U. Cal. at Berkeley, 1937, M.D., 1941; postgrad. U. Washington; m. Ruth Miyoko Miyamoto, July 15, 1942; children—Susan Anne, John William, Peter Joseph. Faculty, Washington U., St. Louis, 1948—, prof., head, dept. otolaryngology, 1966—; otolaryngologist in chief Barnes, St. Louis Children's hosps.; cons. NIH, VA, USAF, St. Louis City, Jewish, St. Johns hosps. others; nat. cons USAF. Diplomate Am. Bd. Otolaryngology. Fellow A.C.S.; mem. Am. Acad. Ophthalmology and Otolaryngology, Am. Soc. Ophthalmologic and Otolaryngologic Allergy, A.M.A., Am. Laryngol., Rhinol. Otol. Soc., Am. Soc. for Head and Neck Surgery (pres. 1963-65), Am. Acad. Facial Plastic and Reconstructive Surgery Inc., Am. Coll. Chest Physicians, Pan-Am. Med. Assn., Sigma Xi, Alpha Omega Alpha, others. Research, numerous publs. on disorders of larynx, nose and surg. treatment; laryngoscopy, mechanisms of swallowing, larynx transplantation, nasopulmonary mechanics. Home: 1038 Winwood Dr., St. Louis 63124.*

OHARA, Itaru, physician; b. Oakland, Cal., Nov. 20, 1922; s. Toru and Rin Ohara; M.D., Tohoku (Japan) U., 1948, Ph.D., 1955; m. Yoko Ohara, May 5, 1950. Fellow cardiovascular surgery Columbia Coll. Phys. and Surg., 1951-52, U. Cal. at Los Angeles, 1958-60, Baylor U. Coll. Medicine, 1960-61; instr. Tohoku U. Sch. Medicine, 1961—, Postgrad. Sch. Medicine, 1962—, Nurse Sch., 1964—; chief surgeon 2d dept. surgery Tohoku U. Hosp., 1964—; vis. surgeon Ishinomaki (Japan) Red Cross Hosp., 1963—, Ogasawara Hosp., Shiogama, Japan, 1965—. Mem. Japan Med. Soc., Japan Surg. Soc., Japan Thoracic Surg. Soc., Soc. Angiology. Contbg. author: Textbook of Surgical Technique, 1962; also numerous articles. Research in vascular surgery, first successful transplantation of a synthetic prosthesis for abdominal aortic aneurysm in Japan; physiol. studies and treatment of occlusion of blood vessels; studies on whale heparin. Home: Kawauchi 30 ninmachi 49-93, Sendai, Japan. Office: Kita 4 bancho, Sendai, Tohoku U. Hosp., 2d Surg. Dept.*

OHLE, Waldemar, German biologist; b. Hamburg, Germany, Feb. 10, 1908; s. Waldemar and Alma (Schultz) O.; Ph.D., U. Kiel; m. Anni Becker, May 8, 1937; 1 son, Wulf. Prof. limnology U. Kiel; dir. research group gen. limnology, hydro-biology div. Max-Planck Soc. Mem. Internat. Limnology Assn., Assn. Limnology and Oceanography, Assn. German Chemists. Author: (theses on metabolism) Stoffhaushalt; Produktionsbiologie und Wasserchemie der Binnengewasser einschliessl. ihrer Sedimente. Research on prodn. biology of phytoplankton; water chemistry; physiology of lakes. Address: Friedrichstrasse 12, Plön, West Germany.

OHLIN, Lloyd Edgar, Am. sociologist, educator; b. Belmont, Mass., Aug. 27, 1918; s. Emil and Elise (Nelson) O.; B.A., Brown U., 1940; M.A., Ind. U., 1942; Ph.D., U. Chgo., 1954; m. Helen Barbara Hunter, Jan. 27, 1946; children—Janet A., George H., Robert N., Nancy E. Sociologist, Ill. Parole and Parson Bd., Joliet, Chgo., 1947-53; dir. Research Center on Correctional Practice, U. Chgo., 1953-56; prof. sociology Columbia Sch. Social Work, N.Y.C., 1962-67, dir. Research Center, 1962-64; prof. criminology Harvard Law Sch., 1967—. Spl. asst. to sec. health, edn. and welfare, Washington, 1961-62; rep. on Pres.'s Com. on Juvenile Delinquency and Youth Crime, mem., 1962-64; asso. dir. Pres.'s Commn. on Law Enforcement and Adminstrn. of Justice, 1965-67; cons. Am. Bar Found. Survey on Criminal Justice in U.S., 1955-58; cons. com. on juvenile delinquency Nat. Inst. Mental Health, 1959-64. Ford Found. Travel grantee, Europe, 1964-65. Mem. Am. Sociol. Soc., Nat. Council on Crime and Delinquency (profl. council 1960-66, research council 1967—), Am. Correctional Assn., Phi Beta Kappa, Phi Kappa Psi. Author: Selection for Parole, 1951; Sociology and the Field of Corrections, 1956; (with Richard Cloward) Delinquency and Opportunity, 1961. Editor: Research in Crime and Delinquency, 1963—. Publs. on prediction of outcome of parole, using tech. devices; theories on effects of social orgn. on delinquent subcultures in communities and inmate subcultures in corrective instns. Home: 9 Red Coat Lane, Lexington, Mass. Office: Harvard Law Sch., Cambridge, Mass. 02138.*

OHLMACHER, Albert Philip, Am. physician; b. Sandusky, O., Aug. 19, 1865; s. Christian John and Anna (Scherer) O.; M.D., Northwestern U., 1890; m. Grace M. Peck, June 14, 1890. Prof. comparative anatomy and embryology Coll. Physicians and Surgeons, Chgo., 1891-94; prof. pathology Chgo. Polyclinic, 1892-94; prof. pathology and bacteriology Ohio Wesleyan U., Cleve., 1894-97; dir. path. lab.

Ohio Hosp. for Epileptics, Gallipolis, 1897-1901, supt., 1902-05; prof. pathology Northwestern U., 1901-02; dir. biol. lab. Frederick Sterns & Co., Detroit, 1905-07; practiced medicine, from 1907. Research in epilepsy, treatment of infections by bacterial or vaccine therapy. Died Nov. 10, 1916.

OHLSON, Margaret Alexander, Am. nutritionist; b. Chgo., May 5, 1901; s. Anton and Esther J. (Anderson) O.; student Coll. Puget Sound, 1919-22; B.A., Wash. State U., 1923; M.S., U. Ia., 1930, Ph.D., 1934. Food supr. YWCA, Salt Lake City, 1923-24; dietitian Wash. Vets. Hosp., Retsil, 1924-27; dietitian Sacred Heart Hosp., Spokane, 1927-29; instr., dietitian U. Ind. Hosps., 1930-32; asst. prof. Mich. State Coll., 1934-35; asso. prof. Ia. State Coll., 1935-44; prof., head dept. food and medicine Mich. State U., 1944-56; prof. internal medicine, dir. nutrition services U. Ia. Coll. Medicine and Hosps., Iowa City, 1956-65, ret., 1965; research asso. (post tenure) div. child health, dept. pediatrics U. Wash. Sch. Medicine, 1966—. Mem. food and nutrition bd. NRC, 1952-58; cons. in dietetics USAF Med. Corps, 1952-56; com. on food tech. to Q.M.C., 1954-56; cons. VA, 1955-58, adv. com. dietetic service, 1958-65, chmn., 1964-65. Recipient Borden award in nutrition, 1942-50; Am. Dairy Assn. Service award, 1953; Cooper award in dietetics, 1965. Fellow Am. Assn. Pub. Health; mem. Am. Inst. Nutrition, Am. Dietetic Assn. (pres. 1951-52, rep. to Internat. Congress Dietitians, 1952, pres. 2d Congress, Rome, 19—), Gerontol. Soc., A.A.A.S., Sigma Xi, Omicron Nu (editor 1942-44, pres. 1955-57). Author: Handbook of Experimental and Therapeutic Dietetics, 1962; also numerous articles. Research on food requirements of healthy people and adaptations to therapeutic dietetics. Home: 2919 27th Av. W., Seattle, Wash. 98199.

OHM, Georg Simon, German physicist; b. Erlangen, Bavaria, Mar. 16, 1787; ed. Erlangen; taught math. and physics, Jesuit school in Cologne, from 1817; prof. polytechnic school, Nurnberg, 1833; prof. physics, U. Munich, from 1849. Fellow Royal Soc., 1842 (Copley Medal, 1841). Author: Die Bestimmung des Gesetzes, 1826; Die galvanische Kette, mathematisch beiarbeitet, 1827. Found Ohm's law relating resistance to voltage and current strength, 1827 (the flow of a current through a conductor is directly proportional to the potential difference and inversely proportional to the resistance); his work little appreciated for many years; practical unit of electrical resistance named after him; also formulated theory of sirens, 1844; and investigated interference phenomena of polarized rays in crystals, 1852. Died Munich, Bavaria, July 7, 1854.

OHMORI, Seiichi, Japanese plastic surgeon; b. Tokyo, Japan, Nov. 8, 1906; s. Otogoro Oki and Rui Omori; M.D., U. Tokyo, 1931, Med.Ph.D., 1937; m. Shigeko Tanikawa, Dec. 5, 1936; children—Mrs. Mitio Inokuti, Kitaro, Yuki, Taki, Fuki. Chief plastic surgery unit U. Tokyo, 1960—, clin. asso. prof., 1966—, prof. plastic surgery, 1967—; chief in charge plastic surgery service Tokyo Met. Police Hosp., 1927—. Recipient Certificate of Appreciation, Saigon Minister of Health, 1965. Mem. Am. (corr.), Japan socs. plastic and reconstructive surgery, Deutsche Gesellschaft für die Aesthetische Medizin und ihre Grenzgebiete (corr.). Author: Tumor of Skin, 1963; also numerous articles. Promoted devel. plastic and reconstructive surgery in Japan. Home: 4-20-4, Otuka, Bunkyo-ku, Tokyo, Japan.*

OHOYAMA, Tetuo, Japanese physicist; b. Osaka, Japan, Feb. 26, 1924; s. Chiyo-o and Toshi (Maeda) O.; Rigakusi, U. Tokyo, 1946; Rigaku Hakusi, Kyoto U., 1962; m. Kazuko Tsuchiya, Apr. 1, 1948; children—Chiyoko, Kôzô, Kenzi. Asst. prof. Toritsu Women's Coll., Tokyo, 1946-49; with Tokyo Met. U., 1949—, asso. prof., 1962—; lectr. U. Electro-Communications, Tokyo, 1963-64. Mem. Phys. Soc. Japan (mem. com.), Japan Inst. Metals. Research, publs. on magnetic properties of intermetallic compounds of iron group transition elements with IVB group elements, discovered new phases in manganese-germanium and iron-germanium systems with proposed phase diagrams. Home: 1-35 Kokuryo 8, Chofu-Shi, Tokyo, Japan.*

OISHI, Kyukei, Japanese economist; b. 1721; believed to have been retainer Takasaki Clan, Gumma Prefecture. Author: Jikata Hanreiroku (record of econ. practices, farmland and tax systems of Edo Period); also revised and added to Den-en Ruisetsu (Masayo Komiyama). Died 1794.

OJEMANN, Ralph Henry, Am. psychologist. educator; b. Peoria, Ill., Aug. 30, 1901; s. Rolf Gerdes and Johanna C. (Menninga) O.; B.S., U. Ill., 1923, M.S., 1924; Ph.D. (Edn. fellow), U. Chgo., 1929; m. Freda Elizabeth Metzger, Aug. 27, 1925; children—Robert Gerdes, George Alvin, Kathryn Elizabeth (Mrs. Paul A. McKean). Faculty, U. Ia., Iowa City, 1929-65, prof. psychology, 1951-65; dir. child psychology, preventive psychiatry div. Ednl. Research Council Greater Cleve., 1965—; vis. prof. Cleve. State U., 1965—. Chmn. parent edn. com. Nat. Congress Parents and Tchrs., 1942-52; mem. U.S. Joint Commn. on Mental Illness and Health, 1957-60. Recipient Leadership Merit citation U. Chgo. Alumni Assn., 1955. Mem. Am. Ednl. Research Assn. (chmn. com. on mental health in edn. 1956-65), Ia. Commn. on Children and Youth (chmn. 1960-64), Ia. Assn. for Mental Health (Merit award 1952, pres. 1960-62), Sigma Xi, Phi Delta Kappa. Author: Developing a Program for Education in Human Behavior, 1959; Personality Adjustment of Individual Children, 1954. Research, publs. in field ednl. psychology and preventive psychiatry. Home: 19101 Van Aken Blvd., Shaker Heights, O. 44122.*

OKA, Hiko-Ichi, Japanese geneticist; b. Wakayama City, Japan, Dec. 22, 1916; s. Shigeo and Yaeko (Yamada) O.; D.Ag., Hokkaido U., 1954; m. Fumiko Takamatsu, Nov. 1942; 1 dau., Mary. Asst. researcher Ohara Inst. for Agr. Research, Kurashiki, Japan, 1940-42, Nat. Taiwan U., Taipei, 1942-43; prof. Taichung Agrl. Coll., 1943-45, Taiwan Provincial Coll. Agr., Taichung, 1945-54; researcher Nat. Inst. Genetics, Misima, Japan, 1954—. Recipient Japan Agrl. Sci. prize, 1963; Indian Jour. Genetics prize, 1964. Mem. Japanese Soc. Breeding, Japanese Soc. Genetics, Academic Sinica Taiwan (corr.). Author: (with others) Ecology and Evolution (in Japanese), 1966. Research, numerous publs. on evolutionary dynamics in wild and cultivated rice populations, analysis of genetic basis of hybrid sterility in rice, survey of character variations and isolating barriers in rice species, responses to environmental conditions of rice. Home: 182 Sekido-Takamatsu, Wakayama-City, Japan. Office: Nat. Inst. Genetics, 1111 Yata, Misima, Sizuoka-ken, Japan.*

OKA, Kiyoshi, Japanese mathematician; b. Wakayama Prefecture, Japan, 1901; grad. Kyoto U., 1925; postgrad., France; D.Sc. Became asst. prof. Kyoto U., 1929, Hiroshima Bunrika U., 1932; prof. Nara Women's Coll., 1929—. Recipient prize Acad. Japan for complete solution to problem of regular pherer, 1951; recipient Asahi Cultural prize for study of theory of multivariant functions, 1954. Solved problem of approximation of functions, also problem of Cousin.

OKA, Martti Juhani, Finnish physician; b. Käkisalmi, Finland, June 8, 1925; s. Walter and Irene (Förnäs) O.; M.D., Turku U., 1955; m. Solveig Charpentier, Sept. 15, 1950; children—Marita, Irma, Anna. Asst. physician Rheumatism Found. Hosp., Heinola, 1951-53, Turku (Finland) U. Hosp., 1953-54, Kivelä Hosp., Helsinki, Finland, 1955-58; ward physician Central Hosp., Kuopio, Finland, 1959-62; asst. head Turku U. Hosp., 1963-65; head dept. medicine Middle Finland Central Hosp., Jyväskylä, 1966—; guest scientist Nat. Inst. Arthritis and Metabolic Diseases, NIH, Bethesda, Md., 1955—; fellow in medicine study group on rheumatic diseases N.Y.U.-Bellevue Med. Center, 1954-55. Mem. Finnish Assn. for Internal Medicine, Finnish Assn. for Rheumatology, Finnish Assn. for Endocrinology, Finnish Assn. for Clin. Chemistry and Physiology, Scandinavian Assn. for Internal Medicine. Research, publs. on rheumatology, corticosteroids in plasma, tryptophan metabolism in diseases, electrocardiography. Address: Middle Finland Central Hosp., Jyväskylä, Finland.*

OKABE, Kinjiro, Japanese elec. engr.; b. Nagoya, Japan, 1896; grad. textile sect. Nagoya Higher Engring. Sch.; grad. elec. engring. sect. Tokyo U., 1923; D.Eng., 1927. Became asst. prof. Tokyo U., 1925; prof. Osaka U., 1935—. Recipient prize Asahi Shimbun for ultra super short-wave research, 1937; Cultural medal, 1944. Developed partitioned anode magnetron method for generating ultra super short-wave generation, 1927; inventor electron beam magnetron.

OKAJIMA, Saburo, Japanese chemist; b. Tokyo, Japan, Apr. 23, 1908; s. Takesaburo Goto and Chizu (Goto) O.; student Tokyo Imperial U., Tokyo, 1928-31; Dr.Engring. Tokyo U., 1948; m. Keiko Yokota, Apr. 13, 1938; children—Yuko, Asako, Itoko, Keita. Researcher, Teijin, Ltd., 1931-34; prof. Ryojun Coll. Engring., Port Arthur, 1936-48, Yamagata U., 1948-53; prof. chemistry Tokyo Met. U., 1953—. Mem. Chem. Soc. Japan (chief editor 1965-67), Soc. Polymer Sci. Japan, Soc. Textile and Cellulose Ind. Japan (pres. 1967-68). Research, publs. on fine structure of chem. fibers, dichroic orientation of fibers, application of continuous high temperature steaming to finishing and dyeing of rayon and synthetic fibers. Patentee in field. Home: 3-26, Nishihara, Shibuya-ku, Tokyo, Japan.*

OKAMOTO, Michio, Japanese neuroanatomist; b. Kyoto, Japan, Nov. 25, 1913; s. Suekichi and Kana (Sakane) O.; M.D., Kyoto U., 1941, D.M.S., 1945; m. Akiko Okido, Nov. 7, 1942; children—Hiroko, Takako. Prof. anatomy Mie Prefectural U., 1951-58, Kobe Med. Coll., 1956-60; prof. anatomy, chmn. dept. anatomy, Kyoto U. Faculty Medicine, 1960—, dir. Brain Research Inst., 1967—. Mem. Japan Anatomist Assn. (dir.), Japan Neuro-surgeon Assn., (council), Japan Clin. Neurology Assn., Japan Histochemistry Assn., Internat. Anat. Nomenclature Com., Internat. Brain Orgn. Author: (with others) Anatomy, 1963; (with J. Nakai) Morphology of Neurogolia, 1961, Tissue Culture, 1965. Research, publs. on autonomic nervous system and lymph, comparative anatomy of spinal cord, tissue culture, cytochemistry and electron-microscopical study of nervous system. Home: 38 Higashi-hinokuchicho, Sakyo-Ku, Kyoto, Japan.*

OKAMURA, Kintaro, Japanese botanist; b. Tokyo, 1867; grad. bot. sect. Tokyo U., 1890; Dr. Sci.; rector govt. Fisheries Tng. Inst., 1924-31; compiled 1st botany textbooks in Japan; pioneer in seaweed research in Japan; studied nori (laver). Died 1935.

OKAMURA, Seizo, Japanese chemist; b. Tokyo, Japan, Jan. 1, 1914; s. Katsutaro and Sumako Saeki; student Kyoto U., 1934-37, D.Engring., 1943; m. Taeko Ito, May 29, 1944; children—Chieko, Saeko. Prof. polymer chemistry Kyoto U., 1944—; sci. supr. Ministry Edn., Tokyo, 1961-62; lectr. Okayama U., Kyoto Tech. U., Ritsumeikan U., 1956—. Mem. Internat. Union Pure and Applied Chemistry (mem. macromolecular commn. 1963—), Royal Swedish Acad. Engring. Scis. (corr.), Chem. Soc. Japan (Progressive Research award 1945, Sci. award 1963), Soc. Polymer Sci. Japan, Am. Chem. Soc., Faraday Soc., U.S. Am.-Japan Soc. Author: Emulsion-polymerization and its Application, 1941; Radiation Research, 1955; Window in Trip, 1955; Polyvinylacetate, 1960; Radiation Chemistry, 1966; also numerous articles. Adv. editorial bd. Jour. Polymer Sci., 1965—, (series) Comprehensive Chem. Kinetics, 1965—, Jour. Revs. in Polymer Sci., 1965—. Research on mechanism of emulsion polymerization of water-soluble monomer, free ionic mechanism of radiation-induced polymerization; discovered solid state polymerization from crystalline monomer, process for making paper from synthetic fiber, new processes for formation of fiber or film directly from emulsion. Home: 24, Minamigosho-machi, Okazaki, Sakyoku, Kyoto, Japan.*

OKAYA, Yoshiharu, physicist; b. Osaka, Japan, Feb. 11, 1927; s. Tokiharu and Fumiko (Nagaoka) O.; B.S., Osaka U., 1947, Ph.D. (Yukawa fellow), 1953; m. Ikuko Sawabe, Nov. 17, 1950; children—Michiko, Marie, David A., Catherine Y. Faculty, Pa. State U., 1953-61, asso. prof. physics, 1958-61; research physicist research div. IBM, Yorktown Heights, N.Y., 1961-67; prof. chemistry State U. N.Y., Stony Brook, 1967—. guest scientist Brookhaven Nat. Lab., 1958. Cons. Allied Chem. Corp., 1957-59. Mem. Am. Crystallographic Assn., Crystallographic Soc. Japan. Contbg. author: Computing Methods and Phase Problems in Crystallography, 1960; also numerous articles. Crystal and molecular structure determination with X-ray and neutron diffraction techniques; established steric configuration aureomycin; research on biol. implication molecular configurations. Home: 55 Tompkins Av., Hastings-on-Hudson, N.Y. 10706. Office: Chemistry Dept., State U. N.Y., Stony Brook, N.Y. 11790.*

O'KEEFE, John Aloysius, Am. astronomer; b. Lynn, Mass., Oct. 13, 1916; A.B., Harvard, 1937; Ph.D., U. Chgo., 1941; m., 1941; 9 children. Prof. math. and physics Brenau Coll., 1941-42; mathematician Office Chief Engr., Washington, 1942-43 1942; geodetic engr. U. S. Lake Survey, N.Y., 1943-44; with map service U. S. Army, 1945-50; mathematician, 1950-58; astronomer NASA, 1958—. Mem. Astronomy Inst., Internat. Union Geodesy and Geophysics, Astronomy Soc., Geophys. Union. Research on moon, tekites, variable stars; proposed theories for interpretation of findings of Vanguard I, artificial earth satellite, 1958. Office: Goddard Space Flight Center, Greenbelt, Md. 20771.

O'KELLEY, G. Davis, Am. chemist; b. Birmingham, Ala., Nov. 23, 1928; s. Grover C. and Ruth (Davis) O'K.; A.B., Howard Coll., 1948; Ph.D. in Chemistry, U. Cal. at Berkeley, 1951; m. Genie Rae Slocum, June 23, 1950; children—Joanne Ruth, Kevin Davis. Chemist, U. Cal. Radiation Lab., 1949-51; lead chemist Cal. Research & Devel. Co., 1951-54; sr. chemist chemistry div. Oak Ridge Nat. Lab., 1954-59, group leader, 1959—; prof. chemistry dept. U. Tenn., 1964—. Mem. photomultiplier tube steering com. U.S. AEC, 1963-67; mem. subcom. on radiochemistry Nat. Acad. Scis.-NRC. Mem. A.A.A.S., Research Soc. Am., Am. Phys. Soc., Am. Chem. Soc. Author: (with N.R. Johnson, E. Eichler) Nuclear Chemistry, 1963; also numerous articles. Studied energy levels of nuclei by determination of radiation emitted in radioactive decay, radioactive content of meteorites, lunar materials and samples of geochem. interest and implications of such data in understanding origin and history of these materials; contbd. to devel. of new techniques of nuclear spectroscopy. Home: 8228 Corteland Dr., Knoxville, Tenn. 37919. Office: P.O. Box X, Oak Ridge 37830.*

OKEN (Ockenfuss), Lorenz, biologist, physiologist; b. Bolsbach, Baden, Aug. 1, 1779; ed. Würzburg, Göttingen univs. Became privant-dozent, Göttingen; extraordinary pof. med. scis., U. Jena, 1807-28; prof.

U. Munich, 1828-32; prof. U. Zürich, 1833-51. Founded biol. jour, Isis, 1816. Author: Gundriss der Naturphilosophie, der Theorie der Sinne, und der darauf gegundeten Classification der Thiere, 1802; Die Zeugung, 1805; Abriss der Biologie, 1805; (with D. G. Keiser) Beitrage zur vegleichenden-Zoologie, Anatomie und Physiologie, 1806-07; Über die Bedeutung der Schadelknochen, 1807; Über das Universum als Fortsetzung des Sinnensystems, 1808; Erste Ideen zur Theorie des Lichts, der Finsterniss, der Farben und der Warme, 1808; Grandzeichnung des naturlichen Systems der Erze, 1809; Uber den Werth der Naturgeschichte, 1809; Lehrbuch der Naturphilosophie, 1809-11; Esquisse d'un Systeme d'Anatomie, de Physiologie, et d'Histoire Naturelle, 1812; Lehrbuch der Naturgeschichte, 1813; Handbuch der Naturgeschichte zum Gebrauch bei Vorlesungen, 1816-20; Naturgeschichte fur Schulen, 1821; Allgemeine Naturgeschichte, 14 vols., 1833-42. One of the most important "nature philosophers" of 19th century; tried to unify the natural scis.; his work foreshadowed theories of cellular structure of organisms and of protoplasmic basis of life; believed skull to be result of several vertebrae fused together; influenced growth of study of anatomy; advocated meetings of scientists to share and publicize their views. Died Zurich, Switzerland, Aug. 11, 1851.

OKHOTSIMSKII, Dmitrii Evgenevich, Russian mech. engr.; b. 1921; grad. Moscow State U., 1946; became prof. theoretical mechanics dept. Moscow State U., 1959; named dept. chief Math. Inst., USSR Acad. Scis., 1953. Recipient Lenin prize, 1957. Corr. mem. USSR Acad. Scis. Research and publs. on devel. earth satellites, mechanics of precision explosions, motion of fluid filled bodies. Office: Theoretical Mechanics Dept., Moscow State U., Moscow, USSR.

OKI, Michinori, Japanese chemist; b. Hyogo, Japan, Mar. 30, 1928; s. Tetsunosuke and Fusa (Murakami) O.; B.Sc., U. Tokyo, 1950, D.Sc., 1953; postgrad. U. Ill.; m. Fusae Yanagi, Apr. 12, 1957; children—Yukari, Yoriko. Faculty, Tokyo (Japan) Met. U., 1956; faculty U. Tokyo, 1956—, prof. chemistry, 1962—. Mem. teaching of chemistry com. IUPAC. Mem. Chem. Soc. Japan (Young Chemist award), Am. Chem. Soc. Author: Stereochemistry, 1961. Research, publs. on relationship between conformation and estrogenic action, conformation and reactivity of organic compounds. Home: 1324-29, 2-Chome, Hasune, Itabshi, Tokyo. Office: 1-3, 7-Chome, Hongo, Bunkyo, Tokyo, Japan.*

OKITE, George Torao, Am. pharmacologist; b. Seattle, Jan. 18, 1922; s. Kazue and Fusao (Muguruma) O.; student U. Cin., 1943-44; B.A., Ohio State U., 1948; Ph.D., U. Chgo., 1951; m. Fujiko Shimizu, Dec. 20, 1934; children—Ronald Hajime, Sharon Mariko, Glenn Torao. Research asso. U. Chgo., 1953-58, faculty, 1958-63, asst. prof. pharmacology, 1958-63; asso. prof. Northwestern U. Med. Sch., Chgo., 1963-66, prof., 1966—. USPHS Post-doctoral research fellow, 1952. Mem. Am. Soc. Pharmacology and Exptl. Therapeutics, A.A.A.S., Am. Assn. U. Profs. Asst. editor Jour. Pharmacology and Exptl. Therapeutics, 1965—. Research and publs. on mechanism of action of digitalis heart drugs, metabolism of digitalis in patients and animals, biochemistry of carcinogenesis; biochem. studies on metabolic disorders. Home: 128 Heather Lane, Wilmette, Ill. 60091. Office: 303 E. Chicago Av., Chgo. 60611.*

OKLADNIKOV, Alexei, Russian archeologist; b. 1908; head archeol. expdn. Marr Inst. History and Material Culture, Kolyma, Siberia; in charge excavations in Yakutsk area, discovered settlement of Stone Age; also discovered flutelike mus. instrument made of swan bone. Address: Marr Inst. of History and Material Culture, Moscow, USSR.

OKSALA, Arvo Alfred, Finnish physician; b. Sortavala, Finland, July 2, 1920; s. Arvi Adiel and Xenia (Feldmann) O.; M.D., U. Helsinki (Finland), 1947, med.sci.d., 1951; m. Kaija Kuusamo, Jan. 6, 1947; children—Illkka, Lauri, Meri. Asst. surgeon ophthalmic dept. U. Hosp., Turku, Finland, 1948-50; head ophthalmic dept. Deaconesses' Hosp., Pori, Finland, 1951-54; head ophthalmic dept. Central-Finland Regional Hosp., Jyväskylä, 1954-61; head, clin. prof. ophthalmology U. Hosp., Turku, Finland, 1961—; faculty U. Turku, 1954—, clin. prof., 1961—. Mem.. Finnish, German ophthalmol. socs., Oxford Ophthal. Congress, Internat. Coll. Surgeons, Jules Gonin Club. Author: Studies on the Interstitial Keratitis Associated with Congenital Syphilis Occurring in Finland, 1951; also numerous articles. Research in ultrasound diagnosis in ophthalmology for use when optic exploration of eye is impossible. Home: 6-7 Linnankatu. Office: Ophthalmic Dept., U. Hosp., Turku, Finland.*

OKSALA, Tarvo Anopoika, Finnish geneticist; b. Jyväskylä, Finland, Feb. 24, 1915; s. Kaarle Johannes and Aina (Sirén) O.; M.A., Helsinki (Finland) U., 1939, Ph.D., 1943. Faculty, Helsinki U., 1949—, docent, 1945—; prof. genetics Turku (Finland) U., 1960—, dean Sch. Math. and Natural Scis., 1963-66. Mem. Scandinavian Soc. Genetics (mem. bd. 1960—). Publs. on chromosomes studies in insects; problems of mitosis and meiosis, tumor cytology; genetics studies in mammals. Home: 6B Porthaninkatu, Turku 2, Finland.*

OKSNER, Alfred Nikolaevich, Russian lichenologist; b.Yelisavetgrad (now Kirovograd, Ukraine), 1898; grad. biol. dept. Kiev Inst. Pub. Edn., 1924; Cand. Biol. Sci., 1935; D.Biol. Sci., 1942. Former mem. faculty Azerbaijan U., also other Ukrainian higher ednl. instns.; prof., 1943—; head dept. sporophytes Inst. Botany, Ukrainian Acad. Scis. Mem. Ukrainian Bot. Soc. (bd. mem., Presidium mem. 1960—). Author: Material on the Lichen Flora of Belorussia, 1924; Lichen Flora of the Ukraine, 2 vols., 1956. Research and numerous publs. on floristics, systematics, lichen ecology, plant phytogeography; developer lichen classification key, 1936. Address: Inst. Botany, Ukrainian Acad. Scis., Kiev, Ukrainian SSR, USSR.

OKUBO, Akira, oceanographer; b. Tokyo, Japan, Feb. 5, 1925; s. Masayoshi and Setsuko (Tohriyama) O.; B.Engring., Tokyo Inst. Tech., 1947, M.Sci., 1949; Ph.D., Johns Hopkins, 1963; m. Keiko Fujisaki, Mar. 25, 1954. Chief chem. sect. Japan Metoerol. Agy., Tokyo, 1950-58; research asso. Chesapeake Bay Inst., Johns Hopkins, Balt., 1963—. Cons., USPHS, U.S. Dept. Health, Edn. and Welfare, 1962—; panel mem. com. on oceanography Nat. Acad. Scis.-NRC, 1966—. Mem. Am. Geophys. Union, Japan Oceanographical Soc. Research, publs. on oceanic diffusion any contaminant due to oceanic turbulence, mixing processes in estuaries and lakes. Home: 500 W. University Pkwy., Balt. 21210.*

OKUBO, Susumu, physicist; b. Tokyo, Japan, Mar. 2, 1930; s. Yoshitaro and Yoshi (Sekiguchi) O.; M.S., U. Tokyo, 1952; Ph.D., U. Rochester, 1957; m. Mary C. Catlin, June 30, 1965. Research asso. U. Rochester, (N.Y.), 1957-59, sr. research asso., 1962-64, prof. physics, 1964—; vis. physicist Cern, Geneva, Switzerland, 1960-61; asst. prof. U. Naples (Italy), 1959-60. NSF fellow, 1960-61. Fellow Am. Phys. Soc. Research, numerous publs. on symmetry properties in weak interaction and also in strong interactions elementary particles, especially investigations based upon unitary symmetry group, general unitary symmetry mass formula. Home: 1051 Highland Av., Rochester, N.Y. 14620.*

OKUDA, Kunio, Japanese physician, biochemist; b. Tokyo, Japan, May 21, 1921; s. Kinmatsu X. and Hatsue (Hashida) O.; M.D., Manchuria Med. Coll., 1944; D.Med.Sc., Chiba U. Sch. Medicine, 1951; m. Hinae Katsumata, May 25, 1946; children—Hiroaki, Keiko. Med. staff Chiba (Japan) Nat. Hosp., 1945-51; post-doctoral fellow Chiba U. Sch. Medicine, 1947-51; faculty Yamaguchi Med. Coll., Ube, Japan, 1951-63, asst. prof. medicine, 1953-63; fellow biochemistry Johns Hopkins Sch. Hygiene and Pub. Health, 1953-56, asst. prof. biochemistry, 1958-60; prof. medicine, Kurume U. Sch. Medicine (Japan), 1963—. Mem. Am. Inst. Nutrition, Soc. for Exptl. Biology and Medicine, Japan Vitamin Soc. (award 1963), Japanese Soc. Internal Medicine, Internat. Soc. Hematology. Author: (with M. Takahara, A. Nishimoto, W.C. Caccamise) Handbook of Medical English, 1960; (record with M. Takahara, R.H. Ingram, Jr.) Spoken Medical English, 1963; also numerous articles. Research on metabolism of vitamin B12, mechanism of intestinal absorption of vitamin B12; physiology of liver; discovered etiologic agt. of an epidemic type myelitis. Home: Shojima, Kurume City, Fukuoka-Ken, Japan.*

OKUI, Seiichi, Japanese biochemist; b. Ibaraki, Japan, Mar. 11, 1922; s. Yuichiro (Murata) and Shizu Okui; Kimiko Hara, Dec. 23, 1946; children—Eiichi, Tomoko. Asst. U. Tokyo, 1946-50, asst. prof., 1950-59; prof. hygienic chemistry Tohoku U. Sch. Medicine, 1959—; lectr. Tokyo Coll. Pharmacy, 1950-58; head pharm. office Tohoku U. Hosp., Sendai, 1963—. Mem. Pharm. Soc. Japan (award for encouragement of research 1955, past mem. editorial bd.), Biochem. Soc. (past councilor). Author: (with H. Tsukamoto) Detection of Poisons, 1958; Hygienic Chemistry, 1963 (with T. Tsukamoto, C. Ukita, T. Murata, O. Hoshino); also articles. Research on chem. characterization of carbohydrates, toxicology of organophosphorous compounds, lipid factors in mitochondria, metabolism and enzymes. Home: 15-6 Asahigaoka 2-chome, Sendai, Japan.*

OKULITCH, Vladimir Joseph, geologist; b. St. Petersburg, Russia, June 18, 1906; s. Joseph Constantine and Alexandra (Drozhilov) O.; B.A.Sc. in Geology, U. B.C., Vancouver, Can., 1931, M.A.Sc., 1932; Ph.D., McGill. U., Que., Can., 1934; m. Susanne Kouhar, Jan. 19, 1934; children—Andrw, Peter. Research fellow Harvard, 1934-36; asst. prof. U. Toronto (Ont., Can.), 1936-44; faculty U. B.C., Vancouver, 1944—, prof. paleontology, 1949—, head dept. geology, 1953-64, dean faculty sci., 1964—; cons. geologist, Vancouver, 1944—. Fellow Royal Soc. Can.; mem. Geol. Soc. Am., Paleontology Soc., Royal Astron. Soc. Can. Author: North American Pleospongia, 1943; also numerous articles; contbg. author: Treatise on Invertebrate Paleontology, 1955. Research on early invertebrates, stratigraphy Ordovician rocks Ont. and Que., Cambrian rocks Canadian Rocky Mountains. Home: 1843 Knox Rd., Vancouver 8, B.C., Can.*

OLAH, George Andrew, chemist; b. Budapest, Hungary, May 22, 1927; s. Julius and Magda (Krasznai) O.; Ph.D., Tech. U. Budapest, 1949; m. Judith Agnes Lengyel, July 9, 1949; children—George John, Ronald Peter. Came to U.S., 1964. Mem. faculty Tech. U. Budapest, 1949-54; asso. dir. Central Chem. Research Inst., Hungarian Acad. Scis., 1954-56; research chemist Dow Chem. Can. Ltd., 1959, asso. scientist, 1959-63, research scientist, 1963-64; research scientist Dow Chem. Co., Framingham, Mass., 1963-65; vis. prof. chemistry Ohio State U., 1963, Heidelberg U., 1965; prof. chemistry, chmn. dept. Western Res. U., Cleve., 1965—. Fellow Chem. Inst. Can.; mem. Am. (award petroleum chemistry 1964), Brit. chem. socs. Author books, papers in field; patentee in field. Home: 21876 Shelburne Rd., Shaker Heights, O. 44122. Office: Dept. Chemistry, Western Reserve Univ., Cleve.*

OLAH, Leslie Von, cytologist; b. Kassa, Hungary, Feb. 25, 1904; s. Stephen and Etelka (Demeter) O.; Agrl. Engr., Debrecen, Hungary, 1925; Ph.D., Stephan Tisza U., 1934; m. Elena Bakody, Feb. 20, 1938; children—Elena (Mrs. George Geistauts), Amanda (Mrs. Ben van Rooy). Head, Inst. Plant Genetics, Budapest, 1938-45; cens. Chamber Agr., Carinthina, Austria, 1946-48; research fellow La Plata U., Buenos Aires, 1948-53; head Treub Lab., Buitenzorg, Java, prof. genetics U. Indonesia, 1953-58; prof. Duquesne U., 1958-59; prof. botany dept. So. Ill. U., Carbondale, 1959—. Recipient medal Hungarian Acad. Sci., 1942. Mem. A.A.A.S., Am. Inst. Biol. Sci., Genetics Soc. Can., Genetics Soc. Am., Torrey Bot. Club. Research, numerous publs. on meiosis and behavior of nucleoli and satellite chromosomes during meiosis of Corypha palms, Dodecatheon species, comparison of meiosis of Amorphophallus species, new mitotic agt, digitonin, found and applied for study of orgn., functional and ultrastructure of phragmoplast and cell plate formation. Home: 610 Thompson St., Carbondale, Ill. 62903.*

OLBERS, Heinrich Wilhelm Matthäus, German astronomer, physician; b. Arbergen, nr. Bremen, Oct. 11, 1758; studied medicine U. Göttingen. Practiced medicine at Bremen, constructed pvt. obs. there. Fellow Royal Soc., 1804; mem. French Acad. Sci., 1810. Some of his work described in publs. of Baron Von Lach, other papers appeared in Annals of Bode, 1782-1829 and Encke, 1832. Devised improved method to determine orbits of comets, 1779; rediscovered Uranus, 1781, confirmed identity of Ceres, 1802; discovered planetoids Pallas, 1802 and Vesta, 1807, also 5 comets (all but one of which had already been observed at Paris); presented hypothesis that asteroid belt between Mars and Jupiter was in fact fragmented remains of previously existing planet; 1st to maintain that either light is dimmed by passage through space or universe if finite since night sky would not otherwise be dark; proposed theory about comet tails that anticipated concept of radiation pressure. Died Bremen, Mar. 2, 1840.

OLCOTT, Harold Saft, Am. biochemist; b. Denver, July 24, 1909; s. Jacob and Florence (Saft) Olcovich; B.S. in Chem. Engring., U. Denver, 1928, M.S., 1929; Ph.D., U. Ia., 1931; m. Bernice Rosenthal, Mar. 15, 1934; children—Harold S., Dana (Mrs. Robert E. Quittner), James L., Margaret L. NRC fellow Yale, 1931-32; research asso. U. Ia., 1932-36; sr. research fellow Mellon Inst., Pitts., 1936-41; chemist Wester Utility Research Devel. Lab., U.S. Dept. Agr., 1941-55; prof. marine food sci. U. Cal. at Berkeley, 1955—. Recipient Superior award U.S. Dept. Agr., 1952. Mem. Am. Chem. Soc. (Eli Lilly award in biochemistry 1936), Am. Soc. Biol. Chemists, Am. Inst. Nutrition, Biochem. Soc. (London, Eng.), Japan Soc. Sci. Fisheries, A.A.A.S., Inst. Food Technologists. Research, numerous publs. on autoxidation unsaturated fats, antioxidants, vitamin E, muscle proteins, food sci. Home: 969 Hilldale Av., Berkeley, Cal. 94708.*

OLDEKOP, Werner Ewald, physicist; b. Tallinn, Estonia, Jan. 1, 1927; s. Arnold Michael and Apollonia (Badendieck) O.; Diplom, U. Göttingen (Germany), 1951, Dr.rer.nat., 1952; m. Hannelise Heinrich, Apr. 17, 1954; children—Gerhard Reinhold, Vera Brigette. With Siemens Co., Erlangen (Germany), 1953-65, head nuclear reactor physics research, 1961-65; prof. Space Tech. and Reactor Theory, Brunswick (Germany) Tech. U., 1966—; sci. dir. elec. propulsion, space power systems, German Aeronautics and Space Research Inst., Brunswick, 1966—. Mem. German Phys. Soc., German Soc. Space and Missile Tech. Publs. in field. Research on nuclear reactor theory: neutron spectra in heterogeneous media, higher order perturbation theory; design, development work on D2O-moderated thermal thorium breeder reactors and nuclear reactors for space power systems with in-core thermionic energy conversion. Home: 2 Bergiusstrasse, Brunswick. Office: Tech. U., 14 Pockelstrasse, Brunswick, West Germany.*

OLDENBURG, Henry, natural philosopher; b. Bremen, Germany, probably between 1617-20; s. Heinrich O.; ed. Paedagogium, Bremen, Gymansium Illustre, Album Studiorum Gymnasii, from 1633; Master of Theology, 1639; studied at Utrecht, 1641; first wife died, 1666; m. 2d Dora Katherina Durie, Aug. 11, 1668; children—Rupert, Sophia. Acquired facility in several European languages; sent to Eng. as envoy of senate of Bremen, 1653-54; became acquainted with R. Boyle, S. Hartlib, others, and his interests changed from theology and classics to science; tutor, Oxford U., 1656; participated in meetings of Oxford Scientific Club which met at John Wilkins lodgings; traveled on continent 1657-60, 1661, and became acquainted with those interested in scientific, philosophical, and medical topics, esp. in Paris; original fellow, Royal Soc. (mem. council; sec. till 1677); as secretary, kept the records of Royal Soc., managed its correspondence, and edited 12 vols. of its Philosophical Transactions, 1664-77; became focus of European scientific intelligence; invented the scientific article as new means of disseminating knowledge (original articles were letters less salutation and closing); has properly been called a servant of science; became literary agent for Robert Boyle, editing his works and translating for him. Died, Kent, Eng., Sept. 1677.

OLDENBURGER, Rufus, Am. mech. engr.; b. Grand Rapids, Mich., July 6, 1908; s. Teunis and Jennie (Feenstra) O.; A.B., U. Chgo., 1928, M.S., 1930, Ph.D., 1934; m. Eleanor Sonia Wolf, May 11, 1913; 1 son, Derek. Instr. math. U. Mich., 1930-31, Case Inst. Tech., 1931-33; instr. to prof. math. Ill. Inst. Tech., 1934-48; chmn. dept. math. DePaul U., 1948-50; chief mathematician, devel. engr. Woodward Governor Co., 1942-54, dir. research, 1954-56; prof. elec. and mech. engring. Purdue U., Lafayette, Ind., 1956-57, prof. mech. engring., 1957—, dir. Automatic Control Center, 1960—. U.S. rep. Internat. Math. Congress, Oslo, Norway, 1936, Mexican Nat. Math. Congress, Guadalajara, 1945, German Automatic Control Congress, Heidelberg, 1956; pres. Am. Automatic Control Council, 1957-60; mem. provisional com. Internat. Fedn. Automatic Control, 1956-57, exec. council, 1959; vis. prof. Nat. U. Mexico, summers 1940, 58, U. Paris, 1952, U. Tokyo (Japan), summer 1957, U. Cal. at Los Angeles, summer 1959, Tokyo Inst. Tech., 1963; chmn. U.S. delegation Internat. Automatic Control Congress, Moscow, USSR, 1960; chmn. U.S. Automatic Control Mission to Japan, 1961; U.S. del. Internat. Automatic Control Congress, Basel, Switzerland, 1963; vis. prof. elec. engring. U. Witwatersrand, Johannesburg, S. Africa, summer 1963; spl. adviser aero. systems div. U.S. Air Force, 1963—. Recipient award Internat. Fedn. Automatic Control, 1959. Fellow A.A.A.S. (council 1956-57); mem. Am. Soc. M.E. (exec. com. instrument and regulators div. 1953-58, chmn. div. 1957, award 1959, Distinguished Service award Automatic Control div. 1961, Distinguished Service Machine Design award 1964), Chgo. Geog. Soc. (dir. 1943-50), Am. Inst. E.E., Am. Math. Soc., Math. Assn. Am., Soc. Engring. Edn., Japan Soc. Mech. Engrs. (hon.), Swiss Soc. Automatic Control (corr.), Phi Beta Kappa, Sigma Xi (pres. Rockford 1954), Tau Beta Pi (hon.), Eta Sigma Phi. Author: (with L.E. Grinter, others) Engineering Preview, 1945; Mathematical Engineering Analysis, 1950; (with John G. Truxal and others) McGraw-Hill Control Handbook, 1957; Optimal Control, 1966; also numerous articles. Author, editor: Frequency Response, 1956. Asso. editor Applied Mechanics Revs., 1962—; mem. hon. editorial adv. bd. Automatica, 1962—. Solved convergence problem for infinite powers of matrices, 1940; inventor non-linear governors, linear and nonlinear automatic controllers, adjustable hydraulic governor; discovered rapid algebraic solution theory, 1942, optimal control, 1944, signal stblzn., 1954, infinite product distributed systems, 1960. Home: 1516 Marilyn Av., West Lafayette, Ind. 47906. Office: Sch. Mech. Engring., Purdue U., Lafayette, Ind. 47907.*

OLDFIELD, Richard Charles, Brit. psychologist; b. London, Eng., Sept. 26, 1909; s. Francis Dupré and Frances Sophia Henrietta (Cayley) O.; B.A., Cambridge U., 1931, M.A., 1935; m. Kathleen Constance Blanche Balfour, Aug. 23, 1933; children—Frances Elizabeth (Mrs. Roderick Whitfield), Margaret Cayley. Lectr. psychology Oxford (Eng.) U., 1946-50, prof. psychology, dir. Inst. Exptl. Psychology, fellow Magdalen Coll., 1956-66; prof. psychology Reading U., 1950-56; dir. speech and communication research unit Med. Research Council, Edinburgh (Scotland) U., 1966—. Mem. Exptl. Psychology Soc., Royal Soc. Medicine. Author: Psychology of the Interview, 1941; also numerous articles. Research in perception, language and its disorders. Home: Woodhall Cottage, Pencaitland, East Lothian, Scotland. Office: 31 Buccleuch Pl., Edinburgh 8, Scotland.*

OLDHAM, Thomas, geologist; b. Dublin, Ireland, May 4, 1816; s. Thomas and Margaret (Bagot) O.; B.A., Trinity Coll., Dublin, 1836; student engring., also attended geol. lectures of prof. Jamieson, Edinburgh, 1836-38; m. Miss Dixon, 1850; 5 sons, 1 dau. Apptd. to geol. dept. of ordinance survey, Ireland, 1839; asst. prof. engring. Trinity Coll., 1844; prof. geology, from 1845; local dir. for Ireland geol. survey of United Kingdom, 1846; supt. geol. survey of India, from 1850; retired, returned to Eng., 1875. Recipient Royal medal from emperor, Austria, 1875. Fellow Geol. Soc. (pres. Dublin 1846), Royal Soc., 1848 (Royal medal); mem. Royal Irish Acad., Royal Asciatic Soc. Bengal (pres. 4 times). Contrb. articles to jours. Discovered in Cambrian rocks of Bay Head, Ireland, the fossils named after him (Oldhamia), 1849. Died Rugby, Eng., July 17, 1878.

O'LEARY, Austin John, physicist; b. Lindsay, Ont., Can., Dec. 28, 1906; s. John and Sarah (McGinley) O'L.; B.Sc., Queen's U., 1928, M.Sc., 1929; Ph.D., Columbia, 1932; m. Martha Herold, May 28, 1942. Came to U.S., 1929, naturalized, 1934. Faculty, Coll. City N.Y., 1931—, prof. physics, 1957—. Mem. Am. Phys. Soc., Am. Assn. Physics Tchrs., A.A.A.S., Sigma Xi. Research, publs. on radioactivity and x-rays. Home: 431 E. 20th St., N.Y.C. 10010.*

O'LEARY, James Lee, Am. neurologist; b. Tomahawk, Wis., Dec. 8, 1904; s. James and Mary E. B. (Whalen) O'L.; S.B., U. Chgo., 1925, Ph.D., 1928, M.D., 1931; m. Nancy Lucas Blair, June 5, 1939; children—Mary, Nancy Lee. Instr., U. Chgo., 1926-28; faculty Washington U., Sch. Medicine, St. Louis, 1928—, prof., 1946—. Mem. Am. Acad. Neurology, Am. Neurol. Assn., Am. Epilepsy Soc., Am. EEG Soc. Home: 415 N. Hanley Rd., St. Louis 30, Mo. 63130.*

O'LEARY, Paul Arthur, Am. dermatologist; b. Bklyn., Nov. 11, 1891; s. Jeremiah J. and Anna Belle (Coy) O'L.; student Dartmouth, 1910-11; M.D., L.I. Coll., 1915; m. Ruth Youmans, June 18, 1921; children—Paul Arthur, Patricia. Intern L.I. Coll. Hosp., 1915-16; asso., sect. dermatology Mayo Clinic, Rochester, Minn., 1917-24, head sect. dermatology and syphilology, from 1924; prof. dermatology Mayo Found. Grad. Sch., U. Minn., from 1924. Diplomate Am. Bd. Dermatology and Syphilology. Fellow A.C.P.; mem. So. Minn. Med. Assn., Dermatol. Assn. (Pres. 1946), A.M.A. (chmn. sect. dermatology 1936-37), Pan-Am. Med. Assn., A.A.A.S., Minn. Acad. Medicine, Am. Acad. Dermatology and Syphilology (pres. 1938-39). Editor in chief Archives of Dermatology, from 1947. Died Rochester, June 20, 1955.

OLECH, Eli Morton, oral surgeon; b. Tokmok, Russia, Feb. 12, 1902; s. Morris Moses and Anna (Terkov) O.; D.D.S., U. Ill., 1923, M.S. in Oral Surgery, 1930; m. Lilyan Carmen Anaya, May 2, 1947; 1 dau., Lillian Ann. Faculty, U. Ill. Coll. Dentistry, Chgo., 1923—, prof. oral and maxillofacial surgery, dir. minor oral surgery, clin. prof. surgery Coll. Medicine, 1948-65, prof. emeritus, 1965; chief dental sect. and oral surgery Highland Park Hosp. Recipient Loyalty award Alumni Assn. of U. Ill., 1963. Fellow Am. Coll. Dentists, A.A.A.S.; mem. Chgo., Ill., Am. socs. oral surgeons, Internat. Assn. for Dental Research, Chgo., Ill., Lake County dental socs., Am. Dental Assn., A.M.A. (affiliate mem.), Sigma Xi, Omicron Kappa Upsilon. Research, publs. on cyst of jaws, fractures, bone diseases, fgn. bodies, use of sulfonimides and antibiotics, replantation of avulsed teeth. Home: 334 Roger Williams Av. Office: 1893 Sheridan Rd., Highland Park, Ill. 60035.*

OLÉRON, Pierre, French psychologist; b. Belfort, France, Oct. 4, 1915; s. Victor and Adrienne (Millot) O.; diploma, U. Paris, 1946, dr., 1956; m. Geneviève Clairouin, Apr. 11, 1949. Research fellow Nat. Center Sci. Research, Paris, 1947-56; asso. prof., prof. Psychol. Inst., U. Paris, 1948—; asst. prof. Faculty Humanities, Paris, 1956-59, asso. prof., 1959-60, prof., dir. lab. psychol. genetics, 1960—. Recipient Bronze medal Nat. Center Sci. Research, 1957; Dagnan-Bouveret prize French Inst., 1958. Mem. French Soc. Psychology, Assn. Sci. Psychology French Lang. Author: Les Composantes de l'Intelligence, 1957; Recherches sur le Développement Mental des Sourds-Muets, 1957; Les activités intellectuelles, 1964; also numerous articles. Editor: Monographies Française de Psychologie, 1957—. Research in psycholinguistics, intellectual devel. in relation to acquisition of lang., psychology of handicapped children, French non-verbal intelligence test (Borelli-Oléron scale). Home: 8 allée Bernadotte, Sceaux 92, France. Office: 17 rue de la Sorbonne, Paris 5, France.*

OLESEN, Poul, Danish physicist; b. Aalborg, Denmark, Apr. 28, 1939; s. Viktor and Herdis Olseen; Ph.D., Niels Bohr Inst., U. Copenhagen, 1966. Faculty, Niels Bohr Inst., U. Copenhagen, 1964—, adjunkt, 1966—. Author: (with B. Lautrup) Elementary Quantum Electrodynamics, 1965; also articles. Reviewer: Zentralblatt für Mathematik. Research in theoretical high energy physics; investigations of whether local field theory (causal relativistic description of fundamental particles) is valid; calculation of life time of eta and pi mesons at high energies. Home: 2 Juliusvej, Gentofte, Denmark. Office: 17 Blegdamsvej, Copenhagen, Denmark.*

OLESON, Norman Lee, Am. physicist; b. Detroit, Aug. 19, 1912; s. Christian Gad and Mathilde (Halvorsen) O.; student Wayne State U., 1931-34; B.S., U. Mich., 1935, M.S., 1937, Ph.D., 1940; m. Gabrielle Dorothy Sauve, June 18, 1939; children—Karen O. (Mrs. Lorin Milton Castleman), Norman Lee, Richard Paul. Research physicist Gen. Electric Co., Nela Park, East Cleveland, O., 1946-48; prof. physics U.S. Naval Postgrad. Sch., Monterey, Cal., 1948—; research fellow Johns Hopkins, 1948-51; vis. prof. physics Queen's U., Belfast, Ireland, 1955-56; vis. prof. nuclear engring. Mass. Inst. Tech., 1967-68 Cons. plasma physics Lawrence Radiation Lab., U. Cal., Livermore, 1958—. Fellow Am. Phys. Soc.; mem. A.A.A.S., Am. Assn. U. Profs, Sigma Xi. Research, publs. on scattering of fast electrons through solids, phenomena, of moving striations in glow discharges, rare gas plasmas in magnetic fields, plasma oscillations and resonance phenomena in low pressure elec. discharge, plasmas from plasma guns. Home: Route 3, Box 388, Carmel, Cal. 93921. Office: Dept. Physics, U.S. Naval Postgrad. Sch., Monterey, Cal. 93940.*

OLINER, Arthur Aaron, Am. physicist; b. Shanghai, China, Mar. 5, 1921 (parents Am. citizens); s. Saul and Sarah (Schulsohn) O.; B.A., Bklyn. Coll., 1941; Ph.D., Cornell U., 1945; m. Frieda Ginsberg, June 16, 1946; children—Marian A., Eric J. Research asso. Microwave Research Inst., Poly. Inst. Bklyn., 1946-53, faculty, 1953—, prof. electrophysics, 1957—, dept. head, 1967—; Walker-Ames vis. prof. U. Wash., Seattle, summer 1964. Dir. Merrimac Research & Devel., Inc.,Irvington, N.J. Chmn. adv. panel Nat. Bur. Standards, 1960-64; tech. cons. to pvt. cos. Guggenheim fellow Ecole Normale Superieure, Paris, 1965-66. Recipient Instn. premium Instn. Elec. Engrs., London, Eng., 1963. Fellow I.R.E. (past chmn. profl. group on microwave theory and techniques); mem. Internat. Sci. Radio Union (past chmn. U.S.A. commn. I), I.E.E.E. (Microwave prize 1967), Am. Phys. Soc. Contbg. author: Microwave Scanning Antenna, 1966; also numerous articles. Research on equivalent circuits for microwave structures, precise measurement waveguide quantities, antennas, surface wave, leaky waves and properties periodic structures, electromagnetic radiation and diffraction. Home: 545 Westminster Rd., Bklyn. 11230.*

OLIPHANT, Marcus Laurence Elwin, Australian physicist; b. Adelaide, Australia, Oct. 8, 1901; s. Harold George and Beatrice (Tucker) O.; B.Sc. with honors, U. Adelaide, 1927; Ph.D., U. Cambridge, Eng., 1929; hon. D.Sc., U. Toronto, U. Belfast, U. Melbourne, U. Birmingham, U. New South Wales; LL.D. (hon.), U. St. Andrews; m. Rosa Louise Wilbraham, May 23, 1924; children—Michael John, Vivian Margaret. Asst. dir. research Cavendish Lab., Cambridge, 1935-37, fellow, lectr. St. John's Coll., 1935-37; Poynting prof. physics, dir. phys. labs. U. Birmingham (Eng.), 1937-40; dir. Research Sch. Phys. Scis., Australian Nat. U., 1950-63, head research unit on physics of ionized gases, 1964—. Mem. Brit. AEC, 1939-46; mem. Australian Com. on Atomic Energy, 1952-63. Decorated knight Brit. Empire. Fellow Royal Soc., 1937 (Hughes medal 1943, Bakerian and Rutherford meml. lectr. 1955); mem. Australian Acad. Sci. (pres. 1954-56), N.Y. Acad. Scis. (hon. mem.). Led team in Eng. which developed magnetron and microwave radar, 1939-43; discovered deuterium reactions basic to thermonuclear physics, independently discovered synchrotron process for acceleration of charged particles; research on surface physics, electricity in gases, nuclear and high energy physics; discovered two new light isotopes of helium and hydrogen; verified laws of conservation of mass energy. Home: 37 Colvin St., Hughes, Canberra. Office: Australian Nat. U., Box 4, Canberra, Australia.

OLIVA, H., Spanish physician; b. Madrid, Spain, Nov. 22, 1933; s. Horacio and Maria (Aldamiz) O.; B.Sc., Seville U., 1950; m. Maria del Milagro Argota, Aug. 2, 1960; M.D., Madrid U., 1956, Ph.D., 1961. Research fellow Inst. Pathology, Dusseldorf, Germany, 1962; asso. chief pathology Jimenez Diaz Found., Madrid; pvt. practice medicine, Madrid. Mem. Spanish Soc. Pathol. Anatomy, Internat. Soc. Nephrology. Author: Ultraestructura celular, 1964. Research, publs. on fibrinoid degeneration in spleen; foam cells in glomerulonephritis. Home: 45 Victor Pradera. Office: 2 Reyes Catolicos, Madrid, Spain.*

OLIVE, Lindsay Shepherd, Am. biologist; b. Florence, S.C., Apr. 30, 1917; s. Lindsay Shepherd and Sarah (Williamson) O.; A.B., U. N.C., 1938, M.A., 1940, Ph.D., 1942; m. Anna Jean Grant, Aug. 28, 1942. Instr. biology U. N.C., Chapel Hill, 1942-44; pathologist U.S. Dept. Agr., Beltsville, Md., 1944-45; asst. prof. biology U. Ga., Athens, 1945-46; asso. prof. biology La. State U., Baton Rouge, 1946-49; asso. prof. biology Columbia, 1949-57, prof., 1957-67; prof botany U. N.C., Chapel Hill, 1967—. Guggenheim fellow, 1956. Fellow A.A.A.S.; mem. Mycol. Soc. Am. (pres. 1966), Bot. Soc. Am., Soc. Protozoologists, Torrey Bot. Club (pres. 1962). Contbr. to Ency. Americana, chpts. to books on microbiology and genetics, articles to sci. jours. Classification, evolution, cytology, genetics, sexual processes of fungi; discovered new order (Protostelida) of Mycetozoa. Address: Dept. Botany, U. N.C., Chapel Hill, N.C. 25714.*

OLIVECRONA, Herbert, Swedish neurosurgeon; b. Visby, Sweden, July 11, 1891; s. Axel and Ebba (Morner) O.; ed. U. Uppsala, U. Stockholm, 1909-18; M.D. (hon.), U. Athens; m. Ulla Dagmar Lofberg, Feb. 21, 1942; 4 children (by previous marriage). Asst., Surg. Clinic, U. Leipzig (Germany), 1918; Swedish Am. Found. fellow Johns Hopkins Hosp., Balt., 1919-20; asst. and resident surgeon Serafimerlasarttet, Stockholm, 1920-30, surgeon-in-chief neurosurg. dept., 1930-35, dir. Neurosurg. Clinic, 1935-60; prof. neurosurgery Royal Caroline Inst., 1935-60. Mem. Société de Neurologie Warszawa, Societas Neurologiae Estoniensis, Harvey Cushing Soc., Hungarian Soc. Neurology, Deutsche Gesellschaft für Neurochirurgie, Royal Coll. Surgeons, Royal Coll. Physicians, Brit. Soc. Neurol. Surgeons, Société de Neurologie (Paris), Societa radio-neuro-chirurgica Italiana. Author: Die chirurgische Behandlung der Genirntumoren, 1927; Die Parasagittalen Meiingeome, 1934; Gefässgeschwülste und Gefässmissbildungen des Gehirns, 1936; Chirurgie der Hirntumoren, 1940; Surgical Treatment of Angina Pectoris, 1947. Founded process of hypophysectomy for treatment of cancer; improved method of neurosurgery. Home: Orevägen 9, Alsten, Sweden. Office: Karolinska Sjukhuset, Stockholm 60, Sweden.

OLIVER, Andrew, Am. natural scientist; b. Boston, Nov. 13, 1731; s. Andrew and Mary (Fitch) O.; grad. Harvard, 1749; m. Mary Lynde, May 28, 1852; judge Inferior Ct. Common Pleas, Essex County, Mass., 1761-65; mem. Mass. Gen. Ct., 1762-67; mem. Am. Philos. Soc.; founder Am. Acad. Arts and Sci. Author: An Essay on Comets, in Two Parts, 1772. Died Dec. 6, 1799.

OLIVER, Charles Augustus, Am. ophthalmologist; b. Cin., Dec. 14, 1853; s. Geoge Powell and Maria Louisa (Oliver) O.; A.B., U. Pa., 1873, M.D., 1876, A.M., 1878; A.M. (hon.), Lafayette, 1900; m. Mary Schermerhorn Henry, June 6, 1888; surgeon Wills Eye Hosp., Phila. Hosp. Author: Correlation Theory of Color Perception; Ophthalmic Methods in Recognition of Nerve Diseases, 1895. Co-author: Text Book of Ophthalmology. Died 1911.

OLIVER, Clarence Paul, Am. geneticist; b. nr. Union City, Tenn., Nov. 8, 1898; s. Benjamin Wesley and Mary E. (Massey) P.; B.A., U. Tex., 1925, M.A., 1930, Ph.D., 1931; m. Cecile Wharton Worley, July 29, 1931; children—Peter Lee, George Benjamin. Asst. prof. Washington U., St. Louis, 1930-32; faculty U. Minn., Mpls., 1932-46, prof. zoology, 1945-46, dir. Dight Inst. Human Genetics, 1941-46; prof. zoology U. Tex., Austin, 1946—, chmn. dept., 1947-59, dir. Genetics Found., 1952—, Ashbel Smith professor in zoology, 1963—. Cons. sect. morphology and genetics NIH, 1949-52, chmn., 1955-58, mem. com. genetics tng., 1959-63, mem. com. career devel. awards, 1964—; pres. Genetics Inc., Austin, 1958—; panelist on sci. and tech., com. on sci. and astronautics U.S. Ho. of Reps., 1960—. Mem. Genetics Soc. Am. (pres. 1958), Am. Soc. Human Genetics (pres. 1953), Phi Beta Kappa, Sigma Xi, Phi Kappa Phi. Editor: (with W.S. Stone) Genetics, 1957-62. Research, publs. on relationship between amount of X-ray treatment and rate of induced mutations, crossing over between alleles of lozenge in Drosophila, interaction effects in compounds of lozenge alleles. Home: 305 E. 32nd. St., Austin, Tex. 78705.*

OLIVER, Douglas Llewellyn, Am. anthropologist; b. Ruston, La., Feb. 10, 1913; s. Robert Campbell and Henrietta (Whitehurst) O.; A.B., Harvard, 1934; Ph.D., U. Vienna, 1935; children—Susan (Mrs. Edward Rollins), Andrew, Amelia, Jonathan. Research asso. Peadbody Mus., Harvard, Cambridge, Mass., 1936-42, faculty dept. anthropology, 1948—. Author: The Pacific Islands; Studies in Anthropology of Bougainville; A Solomon Island Society. Home: 10 Channing St., Cambridge, Mass.*

OLIVER, Francis Wall, botanist; b. Richmond, Eng., May 10, 1864; s. D. Oliver; ed. Univ. Coll., London; M.A., Trinity Coll., Cambridge; D.Sc.; LL.D., Aberdeen; m. Mildred Alice Thompson, 1896; 2 sons. Prof. botany Egyptian U., Cairo; Quain prof. botany Univ. Coll., London, 1888-1929. Recipient Linnean medal, 1925. Fellow Royal Soc., 1905; mem. Brit. Assn. (pres. bot. sect. 1906), Brit. Ecol. Soc. (pres. 1915-16), Egyptian Bot. Soc. (pres. 1929-35). Co-author: Tidal Lands. Editor: Kerner's Natural History of Plants, annals of Botany. Research and publs. on plant ecology of deserts and seashore, morphological and fossil botany. Died Sept. 14, 1951.

OLIVER, Jack Ertle, Am. geophysicist; b. Massillon, O., Sept. 26, 1923; s. Chester L. and Marie (Ertle) O.; A.B., Columbia, 1947, M.A., 1950, Ph.D., 1953; m. Gertrude van der Hoeven, Apr. 16, 1964; 1 dau., Cornelia. Research asst., then research asso. Columbia, 1947-55, mem. faculty, 1955—, prof. geology, 1961—. Terrestrial physicist USAF Cambridge (Mass.) Research Labs., 1951; cons. Advanced Research Projects Agy., 1960—, U.S. Arms Control and Disarmament Agy., 1962—, USAF Tech. Applications Center, 1959—; mem. polar research com., also com. on seismography, mem. nat. commn. upper mantle program, panel solid earth problems, Nat. Acad. Sci.; earth scis. panel NSF, 1962-65; mem. USAF Sci. Adv. Bd.; mem. Air Force Office Sci. Research Contractor's Research and Evaluation Panel; U.S. delegate Test Ban Conf., Geneva, Switzerland, 1958-59, inter-govtl. meeting seismology and earthquake engring. UNESCO, Paris, France, 1964, U.S.-Japan Earthquake Prediction Conf., Tokyo, 1964, Palisades, N.Y., 1966; mem. consultative com. seismology and earthquake engring. UNESCO, 1964—. Carnegie Instn. fellow in seismology, 1964. Fellow Am. Geophys. Union (pres. seismology sect. 1964—); mem. Seismol. Soc. Am. (pres. 1964-65, bd. dirs. 1961—), Acoustical Soc. Am., Am. Phys. Soc., Soc. Exploration Geophysicists, Geol. Soc. Am., A.A.A.S., Sigma Xi. Interpretation of certain portions of seismic wave train in terms of modes of propagation in surface wave guide; exploration of earth structure using seismic surface waves; explanation of ultra-long period microseisms; invention of 2-dimensional technique of model seismology; study of elastic properties of Arctic pack ice; study of microearthquakes, seismic waves generated by nuclear explosions. Home: Iroquois Av., Palisades 10964. Office: Lamont Geol. Obs., Palisades, N.Y. 10964.

OLIVER, James, inventor; b. Whitehaugh, Scotland, Aug. 28, 1823; s. George and Elizabeth (Irving) O.; m. Susan Doty, May 30, 1840; 2 children. Came to U.S., 1835; settled nr. Geneva, N.Y.; at Mishawaka, Ind., 1836-55; became iron-molder; began manufacture plows, 1855. Devised process of chill hardening iron castings of plows (which made plows more durable and efficient); granted patent for chill-hardening process, 1869. Died 1908.

OLIVER, James A(rthur), Am. zoologist; b. Caruthersville, Mo., Jan. 1, 1914; s. Arthur L. and Mary E. (Roberts) O.; student U. Tex., 1932-34; A.B., U. Mich., 1936, A.M., 1937, Ph.D. (Univ. fellow 1938-40, Hinsdale scholar 1940-41), 1941; m. Elizabeth Kimball, May 3, 1941; children—Patricia A., Dexter K. Instr. No. Mich. Coll. Edn., 1941-42; asst. curator Am. Mus. Natural History, N.Y.C., 1942-47, asso. curator, 1947-48; asst. prof. zoology U. Fla., 1948-51; curator reptiles N.Y. Zool. Soc., 1951-59, asst. dir. Zool. Park, 1958, dir., 1958-59; research asso. Am. Mus. Natural History, 1948-59, dir., 1959—. Adv. com. Mianus River Gorge Conservation Commn.; bd. dirs. Caribbean Conservation Corp.; steering com. Biol. Sci. Curriculum Study, 1962—, exec. com., 1964—; com. mus. resources N.Y. Commr. Edn., 1961-64; organizing com. 16th Internat. Congress Zoologists, 1963; mem. Am. Commn. Internat. Wildlife Protection. Recipient Outstanding Achievement award U. Mich., 1963. Fellow N.Y. Zool. Soc., Rochester Mus. Assn.; mem. Am. Soc. Ichthyologists and Herpetologists (bd. govs., editorial bd.), Herpetologists League, A.A.A.S., Am. Assn. Museums, N.Y. State Assn. Museums (council, v.p. 1965-66), Internat. Council Museums (exec. com.), Assn. Dirs. Systematic Collections, Dirs. Sci. Museums. Author: The Natural History of North American Amphibians and Reptiles, 1955; Snakes in Fact and Fiction, 1958; Prevention and Treatment of Snake Bite, 1952; also sci. papers. Home: Sherman, Conn. Office: Am. Mus. Natural History, N.Y.C. 10024.*

OLIVER, James Edward, Am. mathematician; b. Portland, Me., July 27, 1829; s. James and Olivia (Cobb) O.; grad. Harvard, 1849; m. Sarah T. Van Petten, June 28, 1888; asst. on American Nautical Almanac; prof. math. Cornell U.; Mem. A.A.A.S., Am. Philosophical Soc., Nat. Acad. Scis. Co-author advanced textbook on algebra and trigonometry; contbr. papers to math. Died Ithaca, N.Y., Mar. 27, 1895.

OLIVER, Leslie Claremont, English neurosurgeon; b. London, Eng., Feb. 5, 1909; s. Joseph Oliver and Gertrude (Lewis) O.; M.B., B.S., Guy's Hosp. Med. Sch., 1933; m. Regine Dequidt, May 13, 1949; children—Georgina Jane, Lawrence Mitchell. Surg. registrar Bristol Gen. Hosp., tchr. surgery Bristol U., 1936-37; resident, surgeon, tutor West London Hosp., 1936-37, 44-45; 1st asst. in neurosurgery London Hosp., 1939; neurosurgeon Emergency Med. Service, 1939-41; surgeon Mil. Hosp. for Head Injuries, 1941-42; cons. neurosurgeon Charing Cross Hosp., 1959—, U. London, West London Hosp., 1956—, Royal No. Hosp., 1947—, West End Hosp., 1947—; founder, sr. neurosurgeon N.East Regional Neurosurg. Center. WHO fellow, U.S., 1950. Fellow A.C.S., Royal Coll. Surgeons; mem. Soc. Brit. Neurol. Surgeons, Soc. Neurosurgery France. Author: Essentials of Neurosurgery, 1952; Parkinson's Disease and Its Surgical Treatment, 1953; Basic Surgery, 1958; Parkinson's Disease, 1967. Research, publs. on blood transfusion into aorta, sect. of lateral part of spinal cord for Parkinsonian tremor, treatment of brain abscess by aspiration only, arrest of bleeding by absorbable alginates, skull repair with acrylic resin; discovery of supranuclear pathway for corneal reflex; operations for Parkinson's disease. Home: 105 Baron's Keep, London, W.14. Office: 94 Harley St., London W.1, Eng.*

OLIVEREAU, Madeleine, French endocrinologist; b. Paris, France, Nov. 6, 1924; s. Andre and Raymonde (Olivereau) O.; License es Sc., U. Paris, 1945, Doctorat Sci., 1953. Mem. Nat. Center Sci. Research, Paris, 1945—, chargée de recherches à l'Institut Oceanographique, 1955-60, maitre de recherche, Laboratoire de Physiologie, 1960—. Author: Hypophyse et glande thyroide chez les Poissons, 1954; also numerous articles. Research on endocrine modifications eel and salmon during migration and reprodn., histological and histochem. studies on various endocrine glands; identification and physiology of cell types of anterior pituitary in fish. Office: 195 rue Saint Jacques 75, Paris 5, France.*

OLIVI, Peter (Petrus Joannis Olivi), French physicist, theologian; b. Sérignan, France, 1248 or 1249; s. Jean Olivi; probably student, Paris; joined Franciscans, Béziers, France, 1260 or 61. Author: Quaestiones; Postilla super Genesim; Commentary on the Sentences of Peter the Lombard; Tractatus de quantitae. Explained theory of impetus (anticipated concept of inertia); studied theory of adhesion. Died Narbonne, France, 1298.

OLIVIER, Charles Pollard, Am. astronomer; b. Charlottesville, Va., Apr. 10, 1884; s. George Wythe and Katharine Roy (Pollard) O.; B.A., U. Va., 1905, M.A., 1908, Ph.D., 1911; m. Mary Frances Pender, Oct. 18, 1919 (dec. July 1934); children—Alice Dorsey (Mrs. Arnold Edmund Hayes), Elise Pender (Mrs. Harry E. Ferris); m. 2d, Ninuzza Seymour, Oct. 23, 1936; m. 3d, Margaret F. Austin, July 24, 1950. Fellow, McCormick Obs., U. Va., 1905-09; fellow, then asst. Lick Obs., U. Cal., 1909-1910; prof. astronomy and physics Agnes Scott Coll., Decatur, Ga., 1911-14; asst. prof. U. Va., 1914-23, asso. prof., 1923-28; prof. astronomy, dir. Flower Obs., U. Pa., Phila., 1928-54, dir. Cook Obs., 1940-54, prof. emeritus, 1954—; sci. staff Aberdeen Proving Ground, Md., 1918-19, cons. 1943-44. Vol. Naval Eclipse Expdn. to Spain, 1905. Recipient Research prize U. Va., 1926. Fellow A.A.A.S.; mem. Meteor Soc. (founder, dir. 1911—), Rittenhouse (pres. 1930-31, 1942-43), Am. astron. socs., Am. Philos. Soc., Internat. Astron. Union (pres. meteor sect. 1924-34), Société Astronomique de France, Va. Acad. Sci., Brotherhood of St. Andrew, Raven Soc., Phi Beta Kappa, Sigma Xi, Sigma Alpha Epsilon. Author: Meteors, 1925; Comets, 1931; also articles to tech. jours., Grolier Ency., Ency. Brit., Ency. Americana. Home: 521 N. Wynnewood Av., Narberth, Pa. 19072.*

OLIVIER, Georges Henri, French physician, anthropologist; b. Paris, France, June 9, 1912; s. Maurice and Therèse (Delprat) O.; Doctor in Medicine, Faculty Medicine, U. Paris, 1938, D.Sc., Sorbonne, 1955; m. Marie-Louise Dalle, June 8, 1945; children—Charles, Maurice, Eric, Jean-Baptiste (dec.). Mil. physician, E. Africa, 1939-41, Cameroons, 1943-45, Indochina, 1947-50; became prof. agrégé Faculty Medicine, Lille, France, 1952, Paris, 1955; prof. anthropology Faculté des Sciences, Paris, 1960—. Author: Les populations du Cambodge, 1956; Pratique anthropologique, 1960; Anthropologie des Tamouls, 1961; Anatomie anthropologique, 1965; l'Evolution et l'homme, 1965. Anthropol. interpretation of anat. characteristics in phylogenesis and ontogenesis; studies in different Asiatic populations. Home: 39 avenue des Gobelins, Paris 13, France. Office: 45 rue des Saints-Pères, Paris 6, France.*

OLIVIER, Guillaume-Antoine, French naturalist, entomologist; b. Arcs, France, Jan. 19, 1756; M.D., Montpellier, 1773. With Bruguière on sci. expdn. to Persia and Turkey, 1792; prof. zoology Alfort Vet. Sch., 1801. Mem. French Acad. Scis., 1799; Société d'Agriculture. Author: Entomologie ou Histoire naturelle des insectes coléoptères, 1789-1809; Dictionnaire d'histoire naturelle des insectes, papillons, crustacés, etc., 1789-1825; Voyage dans l'Empire ottoman, l'Egypte et la Perse, 1801-07. An editor Encyclopédie méthodique. Made valuable collection of natural history objects from Persia and Turkey; did work on life cycles of coleopteron beetles. Died Lyon, France, Oct. 1, 1814.

OLIVIER, Louis John, Am. parasitologist; b. Grand Rapids, Mich., Nov. 3, 1913; s. Louis J. and Mary (DeJong) O.; B.S., U. Mich., 1936; M.S., N.Y. U., 1938, Ph.D., 1940; m. Margaret Cort, June 5, 1937; children—Donald, Robert, James. Sci. dir. USPHS, 1944-66; mem. staff Pan Am. Health Orgn., Washington, 1966—. Adviser on tropical diseases, del. WHO, 1962-64. Mem. Am. Soc. Parasitologists (pres. 1967), Am. Soc. Tropical Medicine and Hygiene, Phi Beta Kappa, Sigma Xi. Research, numerous publs. on life cycles of parasites, epidemiology and biology of schistosomiasis. Home: 9312 Elmhirst Dr., Bethesda, Md. 20014. Office: Pan Am. Health Orgn., 525 23d St., Washington 20035.*

OLIVIER, Santiago Raul, Argentinian biologist; b. La Plata, Argentina, June 5, 1923; s. Ricieri and Eugenia (Palermo) O.; B.S., Nat. Coll. La Plata, 1942; Dr. Natural Scis., U. La Plata, 1949; m. Maria Luisa Scorpati, Jan. 17, 1951; children—Mirela Marisa, Hilda Ines. Faculty Nat. Scis. U. La Plata, 1948-49, prof. limnology, 1957-58; with Ministry of Agrarian Affairs, Province of Buenos Aires, 1946-60, head limnology and fish culture dept., office of research and conservation, 1956-60; dir. Inst. Marine Biology, Mar del Plata, 1964—. Author: (with J. C. Garcia Palou) Habitantes de Nuestro Mar, 1966; also articles. Research on gen. limnology, marine ecology, zooplankton, rotifera, cladocera, chaetognatha, benthonic and shore communities. Home: Alvear 3306. Office: Inst. Marine Biology, Casilla 175, Mar del Plata, Argentina.*

OLIVO, Olivero Marion, Italian anatomist; b. Trieste, Italy, May 24, 1896; s. Antonio and Maria (Kuk) O.; ed. in medicine and surgery; m. Eletta Porta, July 23, 1931; children—Franco, Chiara, Paola. Asst. in human anatomy; prof. gen. biology and histology; instr. gen. biology U. Turin (Italy); prof. histology and embryology, later anatomy U. Bologna (Italy). Decorated Silver medal for mil. valor; laureate of Bressa prize Turin Acad. Sci.; laureate of Nat. prize dei Lincei Acad.; laureate of Sacchetti prize Bologna; recipient Gold medal of merit. Mem. Acad. dei Lincei, Bologna, Turin acad. scis., Turin Accademia Medica. Author papers on histology, embryology, tissue cultures in vitro. Research in exptl. embryology; electron microscopy; tissue culture; growth. Home: via Vallescura 24. Office: via Irnerio 48, Bologna, Italy.

OLKIN, Ingram, Am. statistician; b. Waterbury, Conn., July 23, 1924; s. Julius and Caroline (Bander) O.; B.S., Coll. City N.Y., 1947; M.A., Columbia, 1949; Ph.D., U. N.C., 1951; m. Anita Mankin, May 19, 1945; children—Vivian L., Rhoda J., Julia A. Faculty, Mich State U., 1951-58, asso. prof., 1956-58; vis. asso. prof. U. Chgo., 1955-56; vis. asso. prof. Stanford, 1958-59, prof. statistics, 1961—, asso. prof., chmn. dept. statistics U. Minn., 1960-61. Fellow Inst. Math. Statistics, Am. Statis. Assn.; mem. Am. Math. Soc., Math. Assn. Am., Psychometric Soc. Editor, contbg. author: Probability and Statistics, Essays in Honor of H. Hotelling, 1960. Research, publs. on multivariate analysis and applications to social scis., research on matrix theory. Home: 950 Lathrop Pl., Stanford, Cal. 94305.*

OLLIER, (Louis Xavier Edouard) Léopold, French surgeon; b. Vans, France, Dec. 2, 1830; M.D., U. Montpellier (France), 1856; became surgeon Hotel-Dieu, Lyons, France, 1860; named prof. Faculty Medicine, Lyons, 1877; mem. French Acad. Scis., 1874, also Acad. Medicine. Author: Traité expérimental et clinique de la régénération des os et de la production artificielle du tissu osseux, 1867; Traité des résections et des opérations conservatrices qu'on peut pratiquer sur le système osseux, 3 vols., 1885-91. Discovered dyschondroplasia (Ollier's disease); formulated Ollier's law that arrest of growth in a bone jointed at its extremities by ligaments to a parallel bone involves growth disturbance in the other bone; discovered osteogenetic layer of periosteum (Ollier's layer), 1858; gave 1st description of intermediate thickness skin graefts (Ollier's grafts), 1872. Died Lyons, Nov. 2, 1900.

OLMO, Harold Paul, Am. horticulturist; b. San Francisco, July 31, 1909; s. Frank and Bertha (Hashagen) O.; B.S., U. Cal. at Berkeley, 1931, Ph.D., 1934; m. Helen Frances Miller, Sept. 21, 1939; children—Paul, Jeanne (Mrs. John E. Briscoe), Daniel. Instr. viticulture, jr. viticulturist U. Cal., Davis, 1938-39, asst. prof., 1945-46, asso. prof., 1946-52, prof. viticulture, viticulturist, 1952—. Mem. Am. Soc. Hort. Sci., Am. Soc. Enologists, Genetics Soc. Am., A.A.A.S. Author: Survey of the Grape Industry of Western Australia, 1955; (with R. M. Brooks) Register of New Fruit and Nut Varieties, pub. yearly; also numerous articles. Prodn., testing of new grape varieties, cytogenetics of fruit plants, particularly grapes and blackberries. Home: Box 102, Davis, Cal. 95616.*

OLMSTED, Charles Edward, Am. ecologist, educator; b. Holyoke, Colo., Apr. 23, 1908; s. Charles Edward and Alice (Johnson) O.; A.B., U. Neb., 1929; M.Sc., U. Okla., 1931; postgrad. U. Chgo.; Ph.D., Yale, 1936; m. Hazel A. M. Wiggers, June 27, 1937; children—Charles Edward III, Joanna Belle. Faculty botany U. Chgo., 1934—, prof., 1948—, chmn. dept. botany, 1953—. Vis. prof. U. So. Cal., 1951, U. Wyo., 1959; mem. adv. panel on environmental biology NSF, 1960-63; mem. Ill. Bd. Natural Resources and Conservation, 1962—. Mem. Ecol. Soc. Am. (pres. 1959-60), A.A.A.S. (commn. on sci. edn.), Phi Beta Kappa, Sigma Xi. Bot. editor Ecology, 1942-46; editor Bot. Gazette, 1946-59. Research on ecology of photoperiodism, physiol. ecology of grasses and deciduous trees, ecology of Lake Michigan sand dunes. Home: 5641 Drexel Av., Chgo. 60637.*

OLMSTED, Denison, Am. natural philosopher; b. nr. East Hartford, Conn., June 18, 1791; s. Nathaniel and Eunice (Kingsbury) O.; A.B., Yale, 1813, M.A., 1816; m. Eliza Allyn, 1818; m. 2d, Julia Mason, 1831; 7 children. Tchr. Union Sch., New London, Conn., 1813-15; apptd. tutor Yale, 1815; prof. chemistry U. N.C., 1817; apptd. state geologist and mineralogist N.C., 1822, made 1st survey of and reports on state's natural resources; prof. math., natural philosophy Yale, 1825, prof. natural philosophy and astronomy, 1836-59. Author: Introduction to Natural Philosophy, 2 vols., 1831-32; Compendium of Natural Philosophy, 1833 (100 edit.); Introduction to Astronomy, 1839; A Compendium of Astronomy (for schs.), 1839; Letters on Astronomy Addressed to A Lady, 1840; Rudiments of Natural Philosophy and Astronomy, 1844 (also pub. in raised letters for the blind); also papers dealing with meteoric showers of Nov. 13, 1833, study of hailstorms. Inventor a gas process called gas light from cotton seed, patented, 1827, also a lubricant for machinery made from lard and rosin. Died New Haven, Conn., May 13, 1859.

OLMSTED, John M. H., Am. mathematician; b. Ithaca, N.Y., June 28, 1911; s. Everett Ward and Bula (Hubbell) O.; B.A. summa cum laude, U. Minn., 1934, M.A., 1936; A.M. (jr. fellow math.) Princeton, 1938, Ph.D., 1940; m. Victoria Louise Mercer, Dec. 21, 1936 (dec. Aug. 1951); children—Hugh Mercer, Emily Mercer, David Mercer, Gordon Mercer; m. 2d, Elizabeth Hiatt, Dec. 17, 1951 (div. 1955); 1 dau., Jane Isabel; m. 3d, Cynthia Sundal Dickman, Dec. 16, 1955; stepchildren—Cynthia Louise Dickman (Mrs. Richard John Haas), Rolf Dickman Olmsted. Instr., Princeton, part-time 1937-40; faculty U. Minn., Mpls., 1940-60, prof., 1955-60, asso. chmn., 1957-60; vis prof. San Jose State Coll., San Jose, Cal., summers 1959-60; prof., chmn. dept. math. So. Ill. U., Carbondale, 1960—. Mem. Math Assn. Am. (gov. 1965—, mem. Putnam prize competition com. 1961-63, vis. lectr. 1962-63), Am. Math. Soc., Am. Assn. U. Profs., Am. Civil Liberties Union, Phi Beta Kappa, Sigma Xi. Author: Solid Analytic Geometry, 1947; Intermediate Analysis, 1956; Real Variables, 1959; Advanced Calculus, 1961; The Real Number System, 1962; (with B. R. Gelbaum) Counterexamples in Analysis, 1964; Calculus with Analytic Geometry, 2 vols., 1966; also articles to tech. jours, Ency. Americana. Asso. editor, Am. Math. Monthly, 1956-64; Classroom Notes, 1960-63. Home: Route 2, Carbondale, Ill. 62901.*

OLOMUCKI, Martin, chemist; b. Warsaw, Poland, Nov. 5, 1923; s. David and Maria (Kac) O.; ed. Polytechnic, Lodz, Poland; licence-ès-sci., 1950, dr. sci., 1957, U. Paris; m. Anna Gawartin, May 16, 1947; 1 dau., Hélène Eugénie. Research fellow, Institut National de Recherche Chimique Appliquée, Le Bouchet, France, 1948-63; vice-dir., Lab. of biochem., Collège de France, Paris, 1963—. Mem., Société Chimique de France. Author numerous published articles. Research on synthesis and properties of acetylenic animes, animoacids and bataines; pharmacological activities of unsaturated ammonium salts; conversion of unsaturated amines into heterocyclic compounds; guanidine derivatives; protein reagents and bifunctional enzyme inhibitors. Home: 31, ave. Parmentier, Paris 11, France. Office: Collège de France, Place Marcelin Berthelot, Paris 5, France.*

O'LOUGHLIN, Bernard James, Am. radiologist; b. Baudette, Minn., Oct. 31, 1914; s. Bernard Benjamin and Mary Elizabeth (Early) O'L.; B.S., Coll. St. Thomas, 1937; M.S., Creighton U., 1941, M.D., 1942; Ph.D., U. Minn., 1950; m. Margaret Mary Delaney, June 15, 1942; children—Susan Margaret, Sara Mary, Bernard Benjamin, John Delaney, Kevin James, Rory Michael, Brian Joseph, Brigid Eileen. Faculty, U. Minn., 1948-52, asso. prof., 1952; faculty U. Cal. at Los Angeles Sch. Medicine, 1952-64, prof., 1960-64; vis. prof. faculty medicine Oxford U., 1961-62; prof., chmn. dept. radiol. scis. Cal. Coll. Medicine, Los Angeles, 1964—. Mem. A.M.A., Minn., Pacific Northwest radiol. socs., Am. Coll. Chest Physicians, Am. Coll. Radiologists, Am. Coll. Angiology, Los Angeles Guild Cath. Physicians (pres. 1959-60), others. Research, publs. on shock-induced leukopenia, cardiovascular diseases, lung cancer, tuberculin reaction reversal. Home: 216 Crest, Huntington Beach, Cal. 92646.*

OLPIN, A. Ray, Am. physicist, educator; b. Pleasant Grove, Utah, June 1, 1898; s. Albert Ray and Alvira (Smith) O.; A.B., Brigham Young U., 1923; Ph.D., Columbia, 1930; D.Sc., Cal. Coll. Medicine, 1960; LL.D., U. Utah, 1964; m. Elva Chipman, Apr. 12, 1922; children—Helen Rae (Mrs. Ivan D. Callahan), Barbara Ann (Mrs. William R. Hooks), Virginia (Mrs. Neal H. Adams), Howard Ray. Instr. physics Brigham Young U., 1922-24; asst. in physics Columbia, N.Y.C., 1924-25; mem. tech. staff Bell Telephone Labs., N.Y.C., 1925-33; lectr. Bklyn. Poly. Inst., 1931-33; dir. research Kendall Mills, Charlotte, N.C., 1933-39; prof. physics, dir. indsl. research, exec. dir. Research Found. field dir. engring. expt. sta. Ohio State U., Columbus, 1939-46; pres. U. Utah, Salt Lake City, 1946-64, pres. emeritus, 1964. Cons. Anthony Wayne Research Found., 1944-45, research and engring. div. Army Chem. Corp., 1949-50. Mem. Nat. Assn. State Univs., 1946-64, pres., 1961-62; charter mem. Asso. Rocky Mountain Univs.; mem. adv. com. Rocky Mountain Region Inst. Internat. Edn., 1952—; mem. nat. council Nat. Planning Assn.; mem. Utah Assn. for UN. Recipient Distinguished Service award Salt Lake City C. of C., 1956; Distinguished Alumni awards, Brigham Young U., U. Utah, citation Western Polit. Sci. Assn. Fellow Am. Physics Soc.; A.A.A.S., N.E.A., Utah Edn. Assn., Phi Beta Kappa, Sigma Xi, Phi Kappa Phi, Phi Alpha, Omega, Sigma Pi Sigma, Tau Kappa Alpha. Asso. editor Jour. Applied Physics, 1937-38; editor new materials sect. Rev. Sci. Instruments, 1941-42. Author: (with others) Physics In Industry, 1937; also numerous articles. Inventor device for enhancing sensitivity of photo-electric cells; participated in 1st demonstration of tv, 1927. Home: 2431 Beacon Dr., Salt Lake City 84108.*

OLSEN, Arthur Martin, Am. physician; b. Chgo., Aug. 27, 1909; s. Martin I. and Aagot (Rovelstad) O.; A.B., Dartmouth, 1930; M.D., U. Chgo., 1935; M.S. in Medicine, U. Minn., 1938; m. Yelena Pavlinova, Sept. 16, 1936; children—Margaret Ann, David Martin, Karen Yelena, Mary Elizabeth. Practice medicine specializing in diseases of chest, broncho esophagology. Fellow in medicine Mayo Found., U. Minn., 1936-40, trustee, 1962—; cons. in medicine Mayo Clinic, Rochester, Minn., 1940—, head internal medicine sect., 1949—; prof. medicine grad. sch. medicine U. Minn., 1957—. Dirs. Central Life Assurance Co., Des Moines Recipient Billings Gold medal A.M.A., 1955. Mem. Am. Coll. Chest Physicians (pres. elect, chmn. bd. regents), Internat. Congress on Diseases of Chest (pres. 1968), Minn. Tb and Health Assn. (pres.). Research, publs. on diseases of lungs and esophagus, particularly on manometric studies of esophageal motility. Home: Box 87 Oronoco, Minn. Office: Mayo Clinic, Rochester, Minn.*

OLSEN, Clayton Edward, physicist; b. Volksrust, Transvaal, S. Africa, Sept. 19, 1920; s. Oscar Peter and Mary Emily (Buckingham) O.; B.Sc., Trinity Coll., Hartford, Conn., 1938-42; postgrad. Bklyn. Poly. Inst., 1947-49; Ph.D., Ohio State U., 1955; m. Julia Alice Davis, Mar. 17, 1951; children—Hugh Edward, David Evan, Joyce Ann. Profl. chemist Am. Cyanamid Co., Stamford, Conn., 1946-49; staff Los Alamos Sci. Lab., U. Cal. at Los Alamos, 1955—. Fellow Am. Inst. Chemists; mem. Am. Chem. Soc., Am. Inst. Chemists, Am. Phys. Soc. Contbg. author: Plutonium, 1960; Digest, 1965; Magnetism and Magnetic Materials, 1965. Research, publs. on elec. and magnetic properties of transuranic elements and rare earths, magnetic and crystallographic structures of intermetallic compounds. Home: 3227 Woodland Rd. Office: Los Alamos Sci. Lab., U. Cal., Box 1663, Los Alamos 87544.*

OLSEN, Jorgen Lykke, physicist; b. Copenhagen, Denmark, May 10, 1923; s. Bernhard Kristian and Paula (Lykke) O.; student Trinity Coll., Dublin, Ireland, 1941-42; M.A., D.Phil., Oxford (Eng.) U., 1951; m. Marianne Bär, Jan. 4, 1951; children—Annette, Richard. Staff, Admiralty Research Lab., 1942-43; staff Swiss Fed. Inst. Tech., Zürich, 1952—, asso. prof. exptl. and low temperature physics, 1961—. Fellow Inst. Physics (U.K.); mem. Am., Swiss phys. socs., Glaciol. Soc., Internat. Inst. Refrigeration, Internat. Union Pure and Applied Physics (corr. mem. com. on very low temperature physics 1967—). Author: Electron Transport in Metals, 1962; also articles. Research on low temperature properties of normal and superconducting metals. Home: 48 Eierbrechstrasse, Zürich, Switzerland.*

OLSEN, Malling Erik, Danish bacteriologist; b. Copenhagen, Denmark, May 20, 1923; s. Valdemar and Edel (Malling) O.; ed. Royal Vet. and Agrl. Coll., Copenhagen; M.D. in Vet. Medicine; m. Birthe Liisberg, Mar. 20, 1948. Asst. hygiene and bacteriology dept. Royal Coll., 1948-52; head Enigheden Labs., Copenhagen, 1952-56; head vet. dept.; head Danish Vet. Service Dept., 1956—. Mem. Danish Vet. Assn. Author: On Coliform Bacteria in Milk with Special Reference to the Detection, 1952, also bacteriology and vet. hygiene manuals. Work on hygiene, bacteriology, milk control; civil defense; radioactive contamination. Home: Lyngbyvej, 387 C, Gentofte, Copenhagen. Office: 37, Nyropsgade, Copenhagen 5, Denmark.

OLSEN, O(liver) Wilford, Am. parasitologist, educator; b. Brigham City, Utah, Aug. 4, 1901; s. Oliver Woodruff and Pauline (Korth) O.; B.A., Brigham Young U., 1929; M.A., U. Minn., 1931, Ph.D., 1936; postgrad. Harvard; m. Ione Palfreyman, Sept. 11, 1929; children—Barbara Ann (Mrs. Raymond Slanker), John Leslie. Instr., U. Hawaii, 1931-32, U. Minn., 1936-39; parasitologist U.S. Dept. Agr., Angleton, Tex., 1939-48; prof., head dept. zoology Colo. State U., Ft. Collins, 1948—. Parasitologist, U.S. Fish and Wildlife Service, Pribilof Islands, Alaska, summers 1951-62. Recipient Meritorious citation for work on liver flukes in cattle U.S. Dept. Agr., 1945. Mem. Am. Soc. Parasitologists, Am. Micros. Soc., Helminthology Soc. Wash. Author: Animal Parasites: Their Biology and Life Cycles, 1943. Developed medication for control of liver flukes in cattle; discovered life cycle of hookworms in fur seals. Home: 1608 Remington St., Ft. Collins, Colo. 80521.*

OLSEN, Orjan Mikael, Norwegian physician, explorer; b. Hattfjelldal, Norway, Sept. 5, 1885; s. Ole Tobias and Christine Bernhardine (Dahl) O.; B.A., U. Innsbruck (Austria), Ph.D., 1922. Traveled in Europe, China, Japan, Siberia, Central Asia, S. and W. Asia, Palestine, N. and S. Africa, Australia, N.Am., Polynesia; asst. univ. zool. lab., 1914-20; with Sultanim Hydrobiology, Alexandria, 1923; Author: Nordlige Traekfugle i Syd-Afrika, 1913; Hvaler og Hvalfangst i Syd-Afrika, 1913; Et primitivt folk, 1915; Til Jeniseis Kilder, 1915; De store Opdagelser, 6 vols., 1929-30; La conquête de la terre, 6 vols., 1934-37; I Sydhavsparadiset, 1930, Swedish edit., 1931; Eventvurlandet, 1931. Discovered Balaenoptera Brydei (whale).

OLSEN, Stanley J., Am. vertebrate paleontologist; b. Akron, O., June 24, 1919; s. John Mons and Louise (Marquardt) O.; grad. high sch.; m. Eleanor Louise Vinez, June 20, 1942; 1 son, John Wilfred. Staff dept. vertebrate paleontology Mus. Comparative Zoology, Harvard, 1945-56; dir. geology Fla. Bd. Conservation, Tallahassee, 1956-68; faculty Fla. State U., 1957—, asso. prof., dept. anthropology, 1968—; research asso. Mus. No. Ariz. Mem. Soc. Vertebrate Paleontology (past pres.), Soc. Study Evolution, Soc. Mammalogists, Soc. for Am. Archaeology, Am. Soc. Systematic Zoologists, Soc. Icthyologists and Herpetologists. Contbg. author: Captain Cousteau's Underwater Treasury, 1959. Research, publs. on interpretation of vertebrate remains as applied to stratigraphy, osteological analysis of animal bones from archaeol. sites, preparation of field manuals on animal osteology. Home: 2301 Don Andres Av., Tallahassee 32304.*

OLSEN, Sterling Robertson, Am. chemist; b. Spanish Fork, Utah, Sept. 24, 1916; s. Joseph Michael and Rhoda Myrl (Robertson) O.; B.S., Brigham Young U., 1938; Ph.D., Ohio State U., 1942; m. Grace Holley, Dec. 17, 1941; children—Jerilynn (Mrs. Stephen Bruce Oldroyd), Karen, John Sterling, Marsha Rae. Chemist, Agrl. Research Service, Prosser, Wash., 1945-48, Ft. Collins, Colo., 1948—; vis. prof. Ia. State U., 1955; vis. fellow U. Melbourne, Victoria, Australia, 1966. Recipient Nat. Hoblitzelle award for agr., 1955; Superior Service award U.S. Dept. Agr., 1962. Fellow Am. Soc. Agronomy; mem. Western Soc. Soil Sci. (past pres.), Sigma Xi, Phi Lambda Upsilon, Phi Kappa Phi, Gamma Sigma Delta. Research, numerous publs. on methods for measuring available phosphorus soil and reactions of fertilizer phosphorus with soil, diffusion of phosphorus in soils and devel. of prins. governing soil-root environment. Home: 1909 Crestmore Pl., Ft. Collins, Colo. 80521.*

OLSHAUSEN, Robert von, see von Olshausen, Robert.

OLSON, Carl, Am. vet. pathologist; b. Sac City, Ia., Sept. 15, 1910; s. Carl and Olga (Larson) O.; D.V.M., Ia. State U., 1931; M.S., U. Minn., 1934, Ph.D., 1935; m. Elizabeth Armstrong, June 18, 1934; children—Mary Olga (Mrs. Richard Norman), Carl Edmund, Evelyn Joyce (Mrs. Stephen Britten), Robert Ossian. Fellow comparative pathology Mayo Found., Rochester, Minn., 1931-35; asst. prof. N.Y. State Vet. Coll., Cornell U., Ithaca, N.Y., 1935-37; research prof. vet. sci. U. Mass., Amherst, 1937-45; prof., chmn. dept. animal pathology and hygiene U. Neb., Lincoln, 1945-56; prof. vet. sci. U. Wis., Madison, 1956—, chmn. dept., 1957-64. Cons. to govt. agys. Recipient centennial citation Ia. State U., 1958. Mem. Am. Assn. Pathologists and Bacteriologists, Soc. Exptl. Pathologists, Internat. Acad. Pathologists, Am. Coll. Vet. Pathologists, Am. Assn. Avian Pathology, Am. Vet. Med. Assn. Contbg. author: Diseases of Poultry; Veterinary Necropry Procedures, 1954; Equine Medicine and Surgery, 1963. Research, numerous publs. on infectious disease and transmissible tumors of domestic animals; established 1st transplantable lymphoid tumor of chicken; relation of bovine wart virus to urinary bladder tumors and neoplasia in horse; contbd. to solution etiology boving hyperkeratosis and highly chlorinated naphthalene poisoning. Home: 921 University Bay Dr., Madison, Wis. 53705.*

OLSON, Everett Claire, Am. vertebrate paleontologist, educator; b. Waupaca, Wis., Nov. 6, 1910; s. Claire Myron and Aimee (Hicks) O.; S.B., U. Chgo., 1932, M.S., 1933, Ph.D., 1935; m. Lila Richardson Baker, July 15, 1939; children—Claire (Mrs. Claire McAleer), George Everett, Mary Ellen. Faculty vertebrate paleontology U. Chgo., 1935—, prof., 1953—, asso. dean phys. scis., 1950-60, chmn. dept. geology, 1956-60. Mem. Soc. Vertebrate Paleontology (pres. 1948), Soc. for Study Evolution (pres. 1965), Phi Kappa Psi. Author: (with R. L. Miller) Morphological Integration, 1958; The Evolution of Life, 1965. Research on evolutionary Permian vertebrates in U.S. and USSR, physiol. basis for explanation of evolution. Home: 17936 Gottschalk St., Homewood, Ill. Office: Rosenwald Hall, University of Chgo., Chgo. 60430.*

OLSON, F(ranklin) C(arl) W(ester), Am. oceanographer; b. Waukegan, Ill., Mar. 15, 1910; s. C. Gottfrid and Ingeborg (Wester) O.; B.S., U. Chgo., 1933; Ph.D., Ohio State U., 1950; m. Mary Ann Joyner, Apr. 13, 1957; children—Storrs, Susan, Carl Robert, Terri. Physicist, Am. Can Co., Maywood, Ill., 1934-42; research asso. Northwestern Technol. Inst., Evanston, 1943-46, Franz Theodore Stone Inst. Hydrobiology, Put-in-Bay, Ohio, 1947-50; asso. prof. oceanography Fla. State U., 1950-57; head physics br. U.S. Naval Mine Def. Lab., Panama City, Fla., 1957-60, mem. tech. analysis staff, 1963-66, head Environmental Scis. br., 1966—; mem. tech. staff advanced mil. systems RCA, Princeton, N.J., 1960-63. Cons. lake currents, estuarine diffusion. Mem. Sigma Xi. Author: (with C. Olin Ball) Sterilization in Food Technology, 1957. Studies, publs. on heat transmission in canned food, effects on microorganisms, epidemiologic studies on control of airborne infections; turbulence, shallow water oceanography. Home: Rural Route 3, Box 359 A. Office: U.S. Naval Mine Def. Lab., Panama City, Fla. 32401.*

OLSON, Gerald Walter, Am. soil scientist; b. Gothenburg, Neb., Mar. 22, 1932; s. Walter George and Mabel (Bergquist) O.; student Hastings (Neb.) Coll., 1950-52, Cotner Coll., Lincoln, Neb., 1953-54; B.S., U. Neb., 1954, M.S., 1959; Ph.D., U. Wis., 1962; m. Mary Lee Gruber, Jan. 21, 1961; children—Bradford Gerald, David Gerald. Research asst. soil sci. U. Neb., 1956-57; with Community Devel. Project P.R., 1957; fgn. service officer trainee ICA, Washington, India, 1957-58; research asst. soil sci. U. Wis., 1959-62; party chief soil survey field party Florence and Menominee Counties, 1959-62; soil technologist, asst. prof. soil sci. Cornell U., Ithaca, N.Y., 1962—. Fellow A.A.A.S.; mem. Soc. Agronomy, Soil, Sci. Soc. Am., Brit. Soc. Soil Sci., Soil Conservation Soc. Am., Assn. Am. Geographers, Sigma Xi, Gamma Sigma Delta. Research, publs. on microbial respiration in soils; morphology and genesis of fragipans in Wis., U.S.A.; econ. uses of soil, sewage disposal, watershed planning, food prodn., relationships between man and soils. Home: 1818 Slaterville Rd., Ithaca, N.Y. 14850.*

OLSON, Harry F(erdinand), Am. acoustical engineer; b. Mt. Pleasant, Ia., Dec. 28, 1902; s. Frans O. and Nelly (Benson) O.; B.E., U. Ia., 1924, M.S., 1925, Ph.D., 1928, E.E., 1932; D.Sc. (honorary) Iowa Wesleyan University, 1959; m. Lorene E. Johnson, June 11, 1935. Acoustical research Radio Corp. Am., Princeton, since 1928, dir. Acoustical Lab., RCA Labs., 1945—, also staff v.p. acoustical and electromech. research pioneered in research and development of various types of directional microphones; lectr. acoustical engring. Columbia, 1939-42. Recipient John H. Potts medal of Audio Engring. Soc., 1949, Warner award, Soc. Motion Picture and TV Engrs., Scott award, Engineers Club, Phila. 1956; John Ericsson medal American Society of Swedish Engrs. Fellow Inst. Radio Engrs. (chairman professional group audio 1957-58), Acoustical Society America (pres. 1952), Am. Phys. Soc., Soc. Motion Picture and TV Engrs., Audio Engring. Soc. (pres. 1960); mem. of National Academy of Sciences; Tau Beta Pi; Sigma Xi. Author: Dynamical Anaologies, 1943; Elements of Acoustical Engring., 1947; Musical Engineering, 1952; Acoustical Engineering, 1958; Music, Physics and Engineering, 1966. Contbr. articles sci. jours. Patentee in field. Research in electronics; general acoustics. Home: 71 Palmer Sq. W. Office: RCA Laboratories, Princeton, N.J.

OLSON, Jerry S., Am. ecologist; b. Chgo., Mar. 22, 1928; s. Howard R. and Anne (Westrom) O.; Ph.B., U. Chgo., 1947, B.S., 1948, M.S., 1949, Ph.D., 1951; m. Margaret Houston Ford, July 22, 1950; children—Karen, Martha. Forest ecologist Conn. Agrl. Exptl. Sta., New Haven, 1952-58; geobotanist Oak Ridge Nat. Lab., 1958—; lectr. biology U. Tenn., 1962-64, prof. biology, 1964—; lectr. U. London, Eng., 1963. Adviser, Nat. Acad. Sci., 1961, 63-64, Internat. Biol. Program, 1946—. Recipient Rockefeller award in statistics, 1951-52; Guggenheim fellow, 1962-64. Fellow A.A.A.S.; mem. Sci. Research Soc. Am., Biometric Soc., Ecol. Soc. Am. (George Mercer award 1958), Am. Geophys. Union, Internat. Soc. Soil Sci., Am. Meteorol. Soc., Am. Water Resources Assn., Phi Beta Kappa. Studies, publs. on geology, ecology, and soil formation, especially of sand dunes, dormancy and growth analysis, forest devel., nutrients, isotopes in landscapes, systems ecology, radiation and nuclear safety. Home: 110 W. Farragut Rd. Office: Bldg. 2001, Oak Ridge 37830.*

OLSON, John M(elvin), Am. biophysicist; b. Niagara Falls, N.Y., Sept. 18, 1929; s. Carl Melvin and Ann (Perry) O.; B.A., Wesleyan U., Conn., 1951; Ph.D. in Biophysics, U. Pa., 1957; m. Caroline Nell Claypool, June 20, 1953; children—James Carol, Jon Charles, Ann Elizabeth. USPHS fellow U. Utrecht (Netherlands), 1957-58; faculty Brandeis U., Waltham, Mass., 1958-61, asst. prof., 1959-61; staff Brookhaven Nat. Lab., Upton, N.Y., 1961—, asso. biophysicist, 1964-65. biophysicist, 1965—. Mem. Biophys. Soc., Am. Chem. Soc., Fedn. Am. Scientists, Am. Soc. Biol. Chemists, A.A.A.S. Contbg. author: The Chlorophylls, 1966. Research, publs. on quantum efficiency cytochrome oxidation in photosynthetic bacteria, path of energy transfer in green photosynthetic bacteria, characterization bacteriochlorophyll-protein complex isolated from green bacteria. Office: Biology Dept., Brookhaven Nat. Lab., Upton, N.Y. 11973.*

OLSON, Kenneth Barrie, Am. physician; b. Seattle, Jan. 21, 1908; s. Donald B. and Hattie (Palmer) O.; B.S., U. Wash., 1929; M.D., Harvard, 1933; m. Emma H. Tallman, Apr. 4, 1937; children—Karen B., Kenneth Barrie. Practice medicine, specializing in oncology, Albany, N.Y.; faculty Albany Med. Coll. Union U., 1950—, prof. medicine, head subdept. oncology, 1960—. Mem. A.M.A., A.C.P., Am. Soc. Clin. Oncology, Am. Assn. Cancer Edn., Am. Assn. Med. Colls., Whipple Soc., Ewing Soc., Am. Soc. Clin. Oncology (dir.), Am. Assn. Cancer Research. Research, numerous publs. on effect of protamine on heparin, use of alkaline phosphatase in patients with jaundice, chemotherapy of cancer. Home: 14 London Pkwy., Loudonville, N.Y. 12211. Office: Albany Med. Coll., Union U., New Scotland Av., Albany, N.Y. 12208.*

OLSON, Raymond V(erlin), Am. agronomist, educator; b. Cavalier, N.D., Oct. 4, 1919; s. Henry L. and Lena B., (Bobert) O.; student N.D. Sch. Forestry, 1937-39; B.S., N.D. State Agrl. Coll., 1941; M.S., U. Wis., 1942, Ph.D., 1947; m. Jean Schumacher, May 25, 1943; children—Nancy, Peter, Susan. Chemist Hercules Powder Co., Wilmington, Del., 1942-45; asso. prof. soils Kan. State Univ., 1947-50, prof., 1950—, head dept. agronomy, 1952—; chief Kan. State U. project, dean faculty agriculture Ahmadu Bello U., Zaria, No. Nigeria, 1964-66. Chmn. Nat. Soil Research Com., 1957-58. Fellow Am. Soc. Agronomy; mem. Soil Sci. Soc. Am. (chmn. com. chem. analysis; asso. editor proc. 1954-56), Soil Conservation Soc. Am., Sigma Xi. Contbr. articles to profl. jours. Research on the nature of deficiency of boron and iron in agricultural soils. Home: 2007 Anderson Av., Manhattan, Kan.*

OLSON, Robert Eugene, Am. biochemist; b. Mpls., Jan. 23, 1919; s. Ralph William and Minnie (Holtin) O.; A.B., Gustavus Adolphus Coll., 1938; Ph.D., St. Louis U., 1944; M.D., Harvard, 1951; m. Catherine Silvoso, Oct. 21, 1944; children—Barbara Lynn, Robert E., Mark Alan, Mary, Carol. Grad. student research asst. biochemistry St. Louis U. Sch. Medicine, 1938-43, asst. biochemistry, 1943-44; instr. biochemistry and nutrition Harvard Sch. Pub. Health, 1946-47, research fellow Nutrition Found., 1947-49, research fellow Am. Heart Assn., 1949-51, established investigator, 1951-52; house officer Peter Bent Brigham Hosp., Boston, 1951-52; prof., head dept. biochemistry and nutrition Grad. Sch. Pub. Health, lectr. medicine Sch. Medicine U. Pitts., 1952-65, dir. Nutrition Clinic, Falk Clinic, 1953-65; mem. sr. staff Presbyn. Hosp.; dir. metabolic unit, 1960-65; Doisy prof. biochemistry, dir. dept. St. Louis U. Sch. medicine, 1965—; cons. Mercy Hosp., U. Pitts. Med. Center; asso. medicine St. Margaret's Meml. Hosp., Pitts., dir. metabolic unit, 1954-60; cons. div. research grants USPHS, 1954—. Vis. scholar dept. biochemistry Oxford U., 1961-62; mem. sci. adv. com. Nat. Vitamin Found., 1955-58, 62-65. Fulbright award 1961-62; Guggenheim Found. award, 1961-62. Diplomate Nat. Bd. Med. Exam., Am. Bd. Nutrition (president 1962-63). Fellow American Public Health Association (chairman of the food, nutrition sect. 1960-61); mem. Am. Assn. Cancer Research, Am. Heart Assn., A.M.A. (mem. council food and nutrition 1959-66, vice chmn. 1962-66), Royal Society of Health (London), N.Y. Acad. Scis., Am. Fedn. Clin. Research, American Soc. Clinical Investigation, Boyleston Medical Society, American Chemical Soc. (president biochemistry group Pitts. section 1960-61), American Society Biol. Chemists, Am. Inst. Nutrition, A.A.A.S. (sec. med. sci. sect. 1965-67), Soc. Exptl. Biology and Medicine, Am. Soc. Clin. Nutrition (pres. 1961-62), Am. Society Study Liver Diseases, Sigma Xi, Phi Lambda Upsilon, Alpha Omega Alpha, Alpha Sigma Nu. Associate editor Nutrition Reviews, 1954-56, Am. Jour. Medicine, 1956-64, Circulation Research, 1956-67, Am. Heart Jour., 1958-65, American Journal of Clinical Nutrition, 1960-68, Biochem. Medicine, 1967—. Publs. on discoveries of actions of adrenal cortical hormones, fat-soluble vitamins; role of coenzyme Q, amino acid nutrition. Home: 9060 Clayton Rd., St. Louis 63117. Office: 1402 S. Grand Blvd., St. Louis 63104.

OLSON, Sigurd Ferdinand, Am. conservationist, educator, author; b. Chgo., Apr. 4, 1899; s. Lawrence J. and Ida (Mae) O.; B.S., U. Wis., 1920; M.S., U. Ill., 1932; D.Sc. (hon.), Northland Coll., 1961, Macalester Coll., 1963; L.H.D. (hon.), Hamline Coll, 1962; D.Sc., Carlton Coll., 1963; m. Elizabeth Dorothy Uhrenheldt, Aug. 8, 1921; children—Sigurd Thorn, Robert Thorne. Head biology dept. Ely Jr. Coll., Minn., 1922-35, dean, 1936-45; chief, zool. dept. Shrivenham Army U., Eng., 1945; cons. Pres.'s Quetico-Superior Com., Washington, 1947—; cons. wilderness preservation to dir. Nat. Park Service, sec. interior, Washington, 1962—; mem. com. to supervice sci. research in nat. parks Dept. Interior, Washington, 1964—. Cons. Izaak Walton League Am., 1947—; mem. adv. com. nat. parks, monuments and hist. sites, Citizens Conservation Com. Recipient citations Am. Camping Assn., 1958, hall of fame Izaak Walton League Am., 1959. Mem. Nat. Parks Assn. (citation 196), pres. 1951-56), Wilderness Soc. (vp. 1962), Ecol. Soc. Am., N.Y. Acad. Scis. Nature Conservancy, Western Fed. Outdoors. Author: Singing Wilderness, 1956; Listening Point, 1958; The Lonely Land, 1960; Runes of the North, 1962; also numerous articles. Studies in ecology of wilderness of N. America, interpreted in light of expanding population and impact of technology. Home 106 Wilson E., Ely, Minn. 55731. Office: Dept. of Interior, Washington.*

OLSON, Stanley William, Am. physician, coll. dean; b. Chgo., Feb. 10, 1914; s. David W. and Agnes (Nelson) O.; B.S. cum laude, Wheaton Coll., 1934, LL.D., 1953; M.D. magna cum laude, U. Ill., 1938; M.S. in Internal Medicine, U. Minn., 1943, postgrad. (Mayo Found. fellow), 1946-47; m. Lorraine C. Lofdahl, June 26, 1936; children—Patricia Ann, Richard David, Robert Dean. Fellow in medicine Mayo Found., 1940-43, 1st asst. to dir., 1946-47, asst. dir., 1947-50; cons. medicine Mayo Clinic, 1947-50; prof. medicine, dean Coll. Medicine, dir. research and ednl. hosps. U. Ill., Chgo., 1950-53; prof. medicine, dean Coll. Medicine, Baylor U., Houston, 1953-66; chmn. dean's com. VA Hosp., Houston, 1953-66; chmn. med. bd. Jefferson Davis and Ben Taub Gen. hosps., Houston, 1953-66; mem. exec. com. Meth. Hosp., Houston, 1953-66, St. Luke's Episcopal Hosp., Houston, 1961-66. dir. Tenn. Mid-S. Regional Med. Program, 1967—; prof. medicine Vanderbilt U. Sch. Medicine, 1967—; clin. prof. medicine Meharry Med. Coll., 1967—. Cons., mem. numerous med. adv. panels; sec.-treas. Baylor Med. Found., 1962-66; v.p. DeBakey Med. Found., 1962-66; Recipient Distinguished Service award Wheaton Coll. Alumni Assn., 1956; Outstanding Achievement award U. Minn., 1964. Diplomate Am. Bd. Internal Medicine. Fellow A.C.P.; mem. Assn. Am. Med. Colls. (v.p. 1960-61), Sigma Xi, Alpha Omega Alpha. Research, publs. on pernicious anemia, trichinosis, and med. edn. for nat. def. Mem. adv. editorial bd. Heart Bull., 1960-66. Home: 678 Timber Lane, 37215. Office: 110 21st Av. S., Nashville 37203.*

OLSSON, Olle, Swedish radiologist; b. Landskrona, Sweden, Jan. 9, 1911; s. Tage and Inez Paulina (Andersson) O.; Cand.med., Lund (Sweden) U., 1932, Lic.med., 1937, Dr.med., 1943; m. Brita Wally Person, Apr. 16, 1938; children—Sven Christer Owman, Per Torben Owman, Gorel Ann-Christine (Mrs. Torbjorn Hellstrom), Py Bodil Ann-Charlotte Owman. diagnostic radiology, 1949—, med. dir. hosp., Faculty, U. Hosp., Lund, 1943—, prof., chmn. dept. 1956—; vice chmn. bd. Acta Radiologica, 1951—, co-editor, 1952—. Chmn. planning bd. Internat. Commn. on Radiation Units and Measurements, 1962—; mem. expert commns. WHO, 1953—. Mem. Swedish Assn. Med. Radiology (past chmn., past pres.), Internat. Assn. Radiology (mem. bd. 1959—). Author: (with others) Rontgendiagnostik, 1963; also numerous articles, chpts. in books; contbg. author: Roentgen Examination of the Kidney and the Ureter, 1962. Co-editor: Ency. of Radiology, 1959; Handbuch der medizinischen Radiologie, 1959; Scandinavian Textbook of Roentgen Diagnosis, 1963; Der Radiologe, 1960—. Research on diagnostic radiology in field urology. Home: Linnegatan 14B. Office: Roentgendiagnostic Dept., U. Hosp., Lund, Sweden.*

OLSSON, Ragnar, Swedish zoologist, endocrinologist; b. Stockholm, Sweden, Sept. 10, 1923; s. Georg Isidor and Charlotta (Schmidt) O.; fil.kand., Stockholm U., 1948, fil.dr., 1958; m. Karin Widtskiöld, July 11, 1953; children—Monika, Gunilla. Faculty, Stockholm U., 1947—, asso. prof. zoology, 1963—, dir. div. structure research dept. zoology, 1964—; research fellow comparative endocrinology Swedish Natural Sci. Research Council, 1962. Recipient Florman prize Royal Swedish Acad. Sci., 1962. Mem. Internat. Organizing Com. for Comparative Endocrinology, 1964—. Author: Chordaternas morfologi och systematik, 1968; also articles. Research on secretory phenomena in brain of primitive chordates, origin of vertebrate endocrine organs and hormonal functions in primitive chordates. Home. 2 Järpstigen, Lidingö 1, Sweden. Office: Dept. Zool., 70A, Radmansgatan Stockholm Va, Sweden.*

OLSUFEV, Nikolay Grigorevich, Russian parasitologist, microbiologist; b. Feb. 9, 1905; grad. Entomology Faculty, Inst. Applied Zoology and Phytopathology Leningrad, 1930; D.Biol. Sci. Head lab. vectors of especially dangerous infections Inst. Epidemiology and Microbiology, USSR Acad. Med. Scis., 1946-48, head tularemia lab., 1948—. Mem. USSR Acad. Med. Scis. (corr.). Author: Gadflies, Fauna of the USSR, 1937; The Parasitology of Tularemia, 1943; Tularemia, 1959; Listeriosis, 1960; Erisipeloid, 1960; co-author, editor: The Effectiveness of Vaccination against Tularemia, 1953; co-author: Tularemia, 1960. Research and numerous publs. on role of blood-sucking Diptera in transmission of anthrax and tularemia; determined role of species of mammals highly susceptible to tularemia in distbn. of causative organism in nature, demonstrated importance of Ixodidae as carriers of causative agt. Address: Gamaleya Inst. Epidemiology and Microbiology, USSR Acad. Med. Scis., Gamaleya St. 2, Moscow, D-98, USSR.*

OLSZEWSKA, Maria Joanna, Polish cytologist; b. Dabrówka, Poland, Apr. 21, 1929; d. Tadeusz and Stefania (Strzelbika) Skwarczynski; Engr., Sch. Agr., Lodz, Poland, 1949; Magister Botany Faculty Scis., U. Lódz, 1950, D. Biology, 1956, Docent Plant Cytology and Anatomy, 1960; m. Waclaw Olszewski, Oct. 27, 1948; 1 dau., Joanna; m. 2d, Kazimierz Opaek, June 13, 1965. Asst., Sch. Agr., Lódz, 1949-50, Pedagogozical Coll., Lodz, 1950-53; staff U. Lódz, 1953—, asst. prof. plant anatomy and cytology, 1961—, head lab. cytochemistry, 1961—; head chair plant anatomy and cytology U. M. Curie-Skodowska, Lublin, Poland, 1961-65. Polish fellow Lab. Animal Morphology, U. Brussels (Belgium), 1957; Rockefeller Found. fellow, 1959-60. Recipient award for research on kinetin Poish Acad. Scis., 1958, for research on synthesis of nucleic acids and protein in interphasic nucleus Polish Histo- and Cytochem. Soc., 1965. Mem. Societas Scientiarum Lodzenisis (sec. dept. III 1965—), Polish Histo- and Cytochem. Soc. (pres. Lódz br. 1964—), Societas Scientiarum Lublinensis (corr.), Polish Histo- a. Cytochem. Soc., Copernicus Soc. Research, publs. on synthesis of nuclear and nucleolar RNA; discovered proteins responsible for morphogenesis are rich in methionine in unicellular agae Acetabularia; interpretation of cytokinesis in meristems as local lyse of cytoplasm by hydrolytic enzymes; localization of hydrolytic enzymes in Golgi-apparatus in plant cells, presence of DNA in chloroplasts. Home: 38/43 Wierzbowa, Lódz, Poland.*

OLSZEWSKI, Karol Stanislov (or Olszevski) Polish chemist; b. Broniszow, Galicia (now Poland), 1846; prof. chemistry U. Cracow (Poland); succeed (with Wroblevski) in condensing oxygen at 25 atmospheres pressure by means of cold obtained by evaporization of liquid methylene in vacuum, 1882; under same conditions condensed nitrogen, 1883, also carbon monoxide, hydrogen by continuous expansion of previously condensed and cooled gas. Died 1915.

OLSZEWSKI, Stanislaw Marian, Polish physicist; b. Warsaw, Poland, Dec. 8, 1932; s. Pawel and Teodora (Zuchniewska) O.; M.S. in Theoretical Physics, U. Warsaw, 1954; chemist-engr. Poly. Inst., Warsaw, 1954; dr. degree, U. Paris (France), 1962; m. Halina Kowalska, Jan. 10, 1957; 1 son, Mateusz. Research worker Inst. Phys. Chemistry, Polish Acad. Scis., Warsaw, 1955—, privatdocent, 1964, head div., 1965—; lectr. physics Acad. Cath. Theology, Warsaw, 1964—. Mem. Polish Chem. Soc. (award 1957). Research, publs. on proposition concerning devel. of fundamental equations of quantum mechanics, theory of electron interaction in organic unsaturated molecules, gen. quantum method for calculating cohesive energy of monatomic solids. Home: 24/69 Niemcewicza. Office: 44/52 Kasprzaka, Warsaw, Poland.*

OLUM, Paul, Am. mathematician; b. Binghamton, N.Y., Aug. 16, 1918; s. Jacob and Rose (Citlen) O.; A.B. summa cum laude, Harvard, 1940, Ph.D., 1947; M.A., Princeton, 1942; m. Vivian Goldstein, June 8, 1942; children—Judith Ann, Joyce Margaret, Kenneth Daniel. Theoretical physicist Manhattan project Princeton, 1941-42, Los Alamos Sci. Lab., 1943-45; Frank B. Jewett fellow Harvard, 1947-48, Inst. for Advanced Study, Princeton, N.J., 1948-49; faculty Cornell U., Ithaca, N.Y., 1949—, prof. math., 1957—, chmn. dept., 1963-66; mem. Inst. for Advanced Study, 1955-56; vis. prof. U. Paris (France), 1962-63, Hebrew U., Jerusalem, Israel, 1963. Mem. adv. com. Office Ordnance Research, NRC, 1958-61. NSF Sr. fellow Stanford, 1966-67. Mem. Am. Math. Soc., Math. Assn. Am., Am. Assn. U. Profs., Phi Beta Kappa. Contbr. articles to tech. jours., monograph. Research on obstruction theory in algebraic topology, homotopy classification, continuous mappings, homotopy type, degrees mappings and self-equivalence spaces, van Kampen's theorem, psychol. math. problem aircraft landing. Home: 424 Hanshaw Rd., Ithaca, N.Y. 14850.*

OLVER, Frank William John, mathematician; b. Croydon, Eng., Dec. 15, 1924; s. John A. and Susan (Barnes) O.; B.Sc., U. London, 1945, M.Sc., 1948, D.Sc., 1961; m. Grace E. Smith, Sept. 25, 1948; children—Peter J., Linda M. (dec.), Sally E. Came to U.S., 1961, naturalized, 1967. Exptl. officer Brit. Nautical Almanac Office, 1945; with Nat. Phys. Lab., Eng., 1946-61, head numerical methods group, 1959-61; mathematician Nat. Bur. Standards, Washington, 1961—, asso. editor Jour. Research, 1966—; instr. numerical analysis U. Md., 1963-65. Mem. Am. Math. Soc., Soc. Inds. and Applied Math., Sigma Xi. Author: (with others) Modern Computing Methods, 1961; also articles. Editor: SIAM Jour. on Numerical Analysis, 1964—. Asymptotic analysis, especially devel. theory of error bounds; numerical analysis, especially zeros of polynominals and approximation theory; spl. functions, especially asymptotic expansions. Home: 10803 Margate Rd., Silver Spring, Md. 20901. Office: Nat. Bur. Standards, Washington 20234.*

OMALIUS D'HALLOY, Jean-Baptiste-Julien, Belgian geologist; b. Liège, Feb. 16, 1783; studied in Paris. Supt. of Dinant; sec. of Liège; gov. of Namur, 1814. Mem. Acad. Brussels (pres. 1850), French Acad. Scis. (corr.), Royal Belgian Acad. Author: Description géologique des pays situés entre le pas de Calais et le Rhin, 1808; Mémoires pour servir à la description géologique des Pays-Bas, de la France et de quelques contrées voisines, 1828; Eléments de géologie, 1831; Introduction à la gèologie ou première partie des éléments d'histoire naturelle inorganique, 1833; Des roches considérées minéralogiquement, 1941; Coup d'oeil sur la géologie de la Belgique, 1842; Des races humaines ou eléments d'ethnographie, 1845. Prepared geol. map of French empire for Napoleon, 1813; pioneer in modern geology; authority on geology of Holland and Belgium; work on ethnography and metamorphism; proposed systematic subdiv. of geol. formation in earth's crust, 1830. Died Brussels, Belgium, Jan. 15, 1875.

OMAR KHAYYAM (Ghiyathuddin Afblfath 'Omár ibn Ibrahim al-Khayyami), Persian mathematician, astronomer, poet; b. Nishapur, Khurasan, circa 1050; studied under Imam Muaffak. Recieved ann. salary from his friend Nizam al-Mulk (vizier of Seljuk sultan Alp Arslan) for pursuit of math. studies; Arslan's successor (Malik Shah) called him (with others) to new obs. at Ray to reform Persian calendar, Malik Shah provided him with means of living so that he could continue his studies and writings, also named him royal astronomer; lived most of his life in Nishapur; said to have been authority on philosophy, jurisprudence and history, only few brief tracts remain; also poet and agnostic. Author: L'Algèbre d'Omar Alkhayyami (Arabic and French texts by F. Weepcke), 1851. His work contains geometric and algebraic solutions of equations of 2d degree, also admirable classification of equations including cubic, coupled with systematic apptempt to solve them all; provided partial geometric solutions to most of them, faltering on cubics with negative roots; new Persian calendar for which he was largely responsible was extremely accurate, having error of one day each 5000 years; may have discovered binomial theorem; prepared astron. tables. Died Nishapur, circa 1123.

O'MARA, Joseph George, Am. botanist; b. Boston, Sept. 7, 1911; s. Joseph James and Ethel (Morton) O'M.; B.S., U. Mass., 1933; M.S., Harvard, 1934, Ph.D., 1936; m. Frances Richardson Rust, Aug. 15, 1938. Geneticist, U. Mo., 1936-50; prof. genetics Ia. State U., 1950-59, chmn. dept. genetics, 1959-65; head dept. botany Pa. State U., University Park, 1965—, head dept. biology, 1967. Mem. Genetics Soc. Am., Soc. for Study Evolution, Am. Soc. Human Genetics. Research, publs. on methods of adding chromosomes from one species to another individually, substitution individual chromosomes from one species for individual chromosomes in a different species. Home: 2210 N. Oak Lane, State Coll., Pa. 16801. Office: Life Sciences I, Pa. State U., University Park, Pa 16802.*

O'MEARA, Barry Edward, surgeon; b. Ireland, 1786; s. Jeremiah O'Meara; probably ed. London; m. Theodosia Boughton Leigh, 1823 (dec. 1830); m. 2d. Became asst. surgeon 62d army regiment, 1804; apptd. sr. med. officer Fraser's expdn. to Egypt, 1807; asst. surgeon H.M.S. Victorious; surgeon on ships Espiègle, Goliath, Bellerophon, 1815; named physician to Napoleon on St. Helena, 1815, diagnosed his disorder as liver disease caused by climate in St. Helena, dismissed by govt. for partisan attitude toward Napoleon, 1818; espoused cause of Queen Caroline. Author: Napoleon in Exile, 1822; Observations upon the Authenticity of Bourrienne's Memoirs, 1831; also numerous pamphlets against Sir Hudson Lowe, gov. of St. Helena. Died June 3, 1836.

OMER, Guy Clifton, Jr., Am. physicist; b. Mankato, Kan., Mar. 20, 1912; s. Guy Clifton and Margaret (Callahan) O.; B.S. in Elec. Engring., U. Kan., 1936, M.S., 1937; Ph.D. in Physics cum laude, Cal. Inst. Tech., 1947; m. Martha Grace Steer, Sept. 13, 1942; children—Guy Clifton III, Richard William. Instr. physics U. Hawaii, 1941-43; asst. prof. physics and astronomy Occidental Coll., 1947-48, U. Ore., 1948-49; asst. prof. physics and phys. scis. U. Chgo., 1949-55; prof. phys. scis., physics and astronomy U. Fla., Gainesville, 1955—; vis. prof. astronomy U. Cal. at Los Angeles, 1964-65. Mem. Internat., Am., astron. socs., Am. Phys. Soc., Am. Assn. Physics Tchrs., Seismol. Soc. Am., History Sci. Soc., Soc. for History Tech., Sigma Xi, Phi Mu Epsilon, Tau Beta Pi. Author: (with H.L. Knowles, B.W. Mundy, W.H. Yoho) Physical Science: Men and Concepts, 1962; also articles. Research on non-homogeneous cosmology, Coma Cluster galaxies. Home: 1080 S.W. 11th Terrace, Gainesville, Fla. 32601.*

OMHOLT, Anders Kristian, Norwegian physicist; b. Oslo, Norway, Nov. 27, 1926; s. Ingvald and Margot (Anderson) O.; Cand.Real., U. Oslo, 1953, Dr.Philos., 1959; m. Ragnhild Gram, June 29, 1953; children—Lise Margrethe, Knut Erik, Jan Egil. Research asso. Norwegian Inst. Cosmic Physics, Tromsö, Oslo, 1953-58, Yerkes Obs., U. Chgo., 1956-57; lectr. U. Oslo, 1959-60, prof. physics, 1963—, chmn. Inst. Physics, 1964-65; head Norwegian Inst. Cosmic Physics, Oslo, 1960-62; dir. Auroral Obs., Tromsö, 1966—. Mem. Norwegian Geophys. Assn. Research and publs. on aurora borealis and its effects on ionosphere especially on excitation and ionization mechanisms in aurora. Home: Mortensnes 26. Office: Auroral Obs., Tromsö Norway.*

OMMAYA, Ayub Khan, neurosurgeon; b. Mian Channun, Pakistan, Apr. 14, 1930; s. Sultan Nadir Khan and Ida (Counil) O.; M.B., B.S., King Edward Med. Coll., Lahore, 1953; B.A., Balliol Coll., Oxford, 1956, M.A., 1960; F.R.C.S., U. London, 1958; m. Wendy Preece, Aug. 19, 1961; children—David Shamyl Khan, Alexander Emyr Khan. Sr. house surgeon Radcliffe Infirmary, Oxford, 1956-57, sr. resident house surgeon neurosurgery, 1959-61; house physician Nat. Hosp. For Nervous Diseases, London, 1958-59; vis. scientist Nat. Inst. Neurol. Diseases and Blindness NIH, 1961-64, asso. neurosurgeon br. Surg. Neurology, 1964—, mem. head injury planning com., 1965—; asst. prof. surgery Med. Sch. Georgetown U., 1964—. Recipient Harper Nelson Gold medal K.E. Med. Coll., Lahore, 1953, Jane Willis Kirkaldy prize U. Oxford, 1956. Rhodes scholar Pakistan, 1954; Hunterian prof. Royal Coll. Surgeons Eng., 1965. Mem. Soc. Brit. Neurol. Surgeons, Congress Neurol. Surgeons U.S.A. Developed techniques for topical cooling of brain; demonstrated exptl. limitations of freezing in animal brain as a surg. tool; invented device and technique for chronic implantation in human head and spine; discovered importance of Whiplash type of mechanism in prodn. of brain injury and protective effect of a neck collar in preventing exptl. concussion in the monkey. Home: 5808 Greenlawn Dr. Office: 9000 Wisconsin Av., Bethesda, Md. 20014.*

OMODEO, Pietro, Italian zoologist; b. Celfalu, Palermo, Sept. 27, 1919; s. Adolfo and Eva (Zona) O.; Dr. Nat'l. Sci. degree; m. Miriam Donadoni, Jan. 25, 1947; children—Adolfo, Clara, Eugenio, Giovani, Maria. Titular prof. biology and zoology Faculty of Medicine, U. Sienna (Italy). Mem. Italian Zool. Union, Biogeography Soc., World Fedn. Sci. Researchers. Author: Cardiologia dei Lumbricidi; Citologia e Sistematica degli oligocheti; Storia della biologia. Research on regulatory mechanisms in animals; systematics of the oligochaeta; caryology. Home: via Casato di Sopra 57. Office: Inst. di Biologia, piazza S. Agostino 4, Sienna, Italy.

OMORI, Fusakichi, Japanese seismologist; b. Fukui Prefecture; s. Kanesuke (retainer of Fukui clan); grad. Tokyo U., 1890; postgrad. abroad in seismology; named prof. sci. dept. Tokyo U., 1897; mem. Imperial Acad.; made expdns. to study earthquakes in Italy, Mexico, Hong Kong; developed Omori formula for computing seismic tremors, also Omori-type seismograph; noted for study in quake zones; attempted to found internat. seismol. soc. Died en route from Hawaii to Japan, 1923.

OMURA, Hideo, Japanese biologist; b. Tokyo, Japan, Feb. 3, 1906; s. Genichi and Chiyeko (Kageyama) O.; grad. Tokyo Imperial U., 1929; doctor degree Kusyu U., 1951; m. Kimiko Asakura, Feb. 1, 1933; 1 dau. Michiko (Mrs. Tadashi Kobayashi). With Bur. Fisheries, Japanese Govt., 1929—, chief 1st research sect. div. researchers Fisheries Agy., 1950-54, dir. Whales Research Inst., Tokyo, 1954—. Mem. sci. com. representing Japan, Internat. Whaling Commn., 1951—; lectr. Nihon U., Tokyo, 1952-62, Tokyo U., 1962—. Mem. Japanese Soc. Sci. Fisheries, Oceanographical Soc. Japan, Japan Whaling Assn. (dir.). Author: (with Yoshio Matsuura, Ichiro Miyazaki) Whales—Their Science and Practice of Whaling, 1942; also articles. Research on N. Pacific Right whale, also Bryde's whale discovering that it is distributed widely in warm waters of world between 40°N and 40°S. Home: 3-44, Denenchofu, Ota-ku, Tokyo. Office: 1-3 Fukagawa, Etchujima, Koto-ku, Tokyo, Japan.*

ONCLEY, John Lawrence, Am. chemist; b. Wheaton, Ill., Feb. 14, 1910; s. Lawrence and Emma Arena (Hunsche) O.; A.B., Southwestern Coll., 1929, Sc.D., 1954; Ph.D., U. Wis., 1932; A.M. (hon.), Harvard Univ., 1946; m. Genevieve Reese, June 14, 1933; children—Genevieve Louise, Nancy Anne. Asst. U. of Wis., 1929-31, Coffin Fellow, 1931-32; Nat. Research Fellow, Mass. Inst. Tech., 1932-34; instr. U. of Wis., 1934-35; with research lab., Gen. Electric Co., 1935; instr. Mass. Inst. Tech., 1935-43; research asso., Harvard Med. School, 1939-41, associate, 1941-43, assistant professor, 1943-46, associate professor, 1946-50, professor biol. chemistry, 1950-62; prof. chemistry and biol. chemistry, University Michigan, Ann Arbor 1962—, director biophysics research div. Inst. Science and Technology, 1962—. Received award in pure chemistry, Am. Chem. Society, 1942. Pres. commn. on proteins, section of biochem., Internat. Union Pure and Applied Chem., 1951-55. Chmn. commn. plasma fractionation, Protein Found., 1961—. Fulbright and Guggenheim fellow. Mem. Am. Chem. Soc., Am. Acad. Arts and Scis. (sec. 1959-62), Biophys. Soc. (pres. 1962-63), Michigan Acad. Sci., Arts and Letters, Nat. Acad. Scis., N.Y. Acad. Sciences, A.A.A.S., Sigma Xi, Alpha Chi Sigma, Gamma Alpha, Phi Lambda Upsilon. Editor-in-chief: Biophys. Sci.-A Study Program, 1959; editor Biophys. Jour., 1963-66. Contbr. articles to provl. jours. Research on biophysical chemistry of protein systems; liquids and proteins; dielectric properties of gases; fractionation and interactions of proteins and liproproteins. Home: 9 Heatheridge, Ann Arbor, Mich.

ONDRACKOVA, Jana, Czechoslovakian phonetist; b. Prague, Czechoslovakia, May 18, 1924; d. Jan and Frantiska (Sourková) Sláma; Ph.D., Karlova U., Prague, 1950, sci. worker, 1960; m. Frantisek Ondrácvek, Feb. 20, 1944. Dept. head lab. phonetics Inst. Czech Lang., Czechoslovakian Acad. Scis., Prague, 1953—; lectr. phonetics Karlova U. Mem. Internat. Soc. Phonetic Scis., Phonetic Soc. Japan, Czechoslovakian Soc. Sci. Cinematography. Author: Rentgenologicky vyzkum artikulace ceskych vokálu (X-ray research of articulation of Czech vowels), 1964. Research, publs. on physiol. activity of speech organs, rhythmical structure of continuous speech. Home: 15 Nad Primaskou, Prague 10, Czechoslovakia.*

O'NEAL, Robert Munger, Am. physician, educator; b. Wiggins, Miss., Oct. 7, 1922; s. Charles Ernest and Myra Mae (Munger) O'N.; B.S., U. Miss., 1943; M.D., U. Tenn., 1945; m. Mildred Estell Morris, June 14, 1947; children—Julia Ann, Clarence Ernest, Margaret Alice, John Munger. Am. Cancer Soc. fellow in pathology Mass. Gen. Hosp., Boston, 1952-54; faculty pathology Washington U. Sch. Medicine, St. Louis, 1954-61, asso. prof., 1959-61; pathologist Greenwood (Miss.) Hosp., 1956-57; prof. pathology, chmn. dept. Baylor U. Coll. Medicine, Houston, 1961—. Pathology test com. Nat. Bd. Med. Examiners, 1963-68; heart tng. com. Nat. Heart Inst., 1963-68. Recipient Lederle Med. Faculty award 1958-Trudeau Soc., Sigma Xi, Alpha Omega Alpha. Research, publs. on vascular and pulmonary diseases. Home: 4086 Breakwood St., Houston 77025.*

O'NEAL, Russell Dewitt, Am. physicist; b. Columbus, Ind., Feb. 15, 1914; s. Clyde and Tessie (McClintic) O'N.; B.A., DePauw U., 1936; M.S., U. Ill., 1938, Ph.D., 1941; m. Monique Tinlot, June 5, 1948; children—William John, Lucy Jean. Faculty, U. Ill., 1936-42; mem. staff radiation lab., Mass. Inst. Tech., 1942-43, sect. head, 1943-45, group leader, 1945; project physicist Eastman Kodak Co., Rochester, N.Y., 1945-48; head aerophysics group Willow Run Research Center, Mich., 1949, asst. dir., 1950, dir., 1951-52; asst. div. mgr. Consol. Vultee Aircraft Corp., Fort Worth, 1952-53, program dir., 1953-55; with Bendix Corp., Detroit, 1955-66, v.p. engring. and research, 1960-63, v.p. aero-space systems, 1963-66; asst. sec. of army for research and devel., 1966—; dir. Atomic Power Devel. Assos., Detroit, Atomic Indsl. Forum, N.Y.C. Fellow Am. Phys. Soc., I.E.E.E. Contbr. articles to profl. jours. Important work includes initial half life and energy measurements on tritium, pioneer neutron cross section measurements, devel. first radar corner reflectors, directional couplers, FM test equipment. Home: 1805 Windmill Lane, Alexandria, Va. 22307. Office: The Pentagon, Washington 20310.*

O'NEILL, Gerard Kitchen, Am. physicist; b. Bklyn., Feb. 6, 1927; s. Edward Gerard and Dorothy (Kitchen) O'N.; B.A., Swarthmore Coll., 1950; Ph.D., Cornell U., 1954; m. Sylvia Turlington, June 17, 1950 (div. 1966); children—Janet, Roger, Eleanor. Faculty, Princeton (N.J.), 1954—, prof. physics, 1965—. Fellow Am. Phys. Soc. Bd. editors Rev. Sci. Instr., 1963-65. Research in fast-neutron time-of-flight spectrometry, accelerator design, K-meson physics; devel. storage-rings for high energy physics; expts. on quantum-electrodynamic limits. Home: 127 McCosh Circle, Princeton, N.J. 08540.*

ONIMUS, Ernest-Nicolas-Joseph, French physician; b. Mulhouse, France, 1840. Author: Théorie dynamique de la chaleur dans les sciences biologiques, 1866; Emploi de l'électricité dans les maladies nerveuses, 1868. Photographed phases of heartbeat on collodion (with Martin); worked toward popularization of sci. Died Cap d'Ail, 1915.

ONNES, Heike Kamerlingh, see Kamerlingh Onnes, Heike.

ONODERA, Konoshin, Japanese chemist; b. Tsu, Mie-ken, Japan, Oct. 5, 1910; s. Sakuemon and Chiyo (Yokoo) O.; B. in Agrl. Biochemistry, Kyoto (Japan) Imperial U., 1935, Ph.D., 1949; m. Yukari Yamada, Oct. 24, 1937; children—Akifumi, Mizuyo, Koji, Fumi. Research asst. dept. agr. chemistry Kyoto Imperial U., 1937-40, research asso., 1940-44, faculty, 1944—, prof. biol. chemistry, 1960—; research asso. Ohio State U., 1954-56. Mem. Agrl. Chem. Soc. Japan (dir. 1965—), Japanese Biochem. Soc. (dir. 1965—), Am. Chem. Soc., Sigma Xi. Research, numerous publs. on mucopolysaccharides, preparation of amino sugars, isolation of intact DNA from nuclear-polyhedrosis virus of silkworm, synthesis of nucleosides. Home: 17-12 Ebisudani-cho, Yamashina, Kyoto, Japan.*

ONSAGER, Lars, chemist; b. Oslo, Norway, Nov. 27, 1903; s. Erling and Ingrid (Kirkeby) O.; Chem.E., Norges Tekniske Hogskole, 1925, Dr.Technicae, 1960; Ph.D. (Sterling fellow), Yale, 1935; Sc.D., Harvard, 1954, Rensselaer Poly. Inst., 1962, Brown U., 1962; Dr. der Naturwissenschaften, Rheinisch-Westfaehlisch Technische Hochschule, Aachen, Germany, 1962; m. Margarete Arledter, Sept. 7, 1933; children—Erling Frederick, Inger Marie (Mrs. Kenneth Roy Oldham), Hans Tanberg, Christian Carl. Came to U.S., 1928, naturalized, 1945. Asso. in chemistry Johns Hopkins, 1928; instr. chemistry Brown U., 1928-33; faculty Yale, New Haven, 1930—, J.W. Gibbs prof. theoretical chemistry, 1945—. Mem. Neurosci. Assos., 1962—. Recipient Rumford medal, 1953; Lorentz medal, 1958; G.N. Lewis medal, 1962; John G. Kirkwood medal, 1962; J.W. Gibbs medal, 1962; T.W. Richards medal, 1964; Peter Debye award Am. Chem. Soc., 1965. Fulbright scholar Cambridge (Eng.) U., 1951-52. Fellow A.A.A.S., Am. Phys. Soc., N.Y. Acad. Scis.; mem. Am. Philos. Soc., Nat. Acad. Scis., Royal Norwegian, Norwegian (Oslo), Royal Swedish, Conn. acads. sci., Norwegian Acad. Tech. Sci., Am., Norwegian (hon.) chem. socs., Am. Acad. Arts and Scis., Acad. Scis., Sigma Xi, Alpha Chi Sigma. Research, publs. on theory electrolytic conduction, theory dielectrics, protonic semiconductors, metals, superfluids, reciprocal relations in irreversible processes, crystal statistics. Home: 841 Whitney Av., New Haven 06511.*

ONUCHIC, Nelson, Brazilian mathematician; b. Brodosqui, S.P., Brazil, Mar. 12, 1926; s. Francisco and Maria (Doles) O.; grad. in physics, U. Mackenzie, Sao Paulo, Brazil, 1951,; D.Sc., U. Sao Paulo, 1957; m. Lourdes de la Rosa, Jan. 15, 1955; children—Maria Inês, José Nelson, Luiz Fernando, Paulo Eduardo. Faculty, Aeros. Inst. Tech., 1951-58, asst. pro., 1956-58; prof. math. analysis, Faculty Philosophy, Sci. and Letters, Rio Claro, Brazil, 1959--66; prof. applied maths. U. Sao Paulo, Sao Carlos, Brazil, 1966—. John Simon Guggenheim Found. fellow Research Inst. for Advanced Studies, Balt., 1961-62. Mem. Brazilian Acad. Scis. (asso.), Am. Math. Soc., Soc. Math. Sao Paulo. Reviewer on math Zentralblatt für Math., 1966—. Research, publs. on gen. topology and uniform spaces, theory concerning behavior of ordinary differential equations including asymptotic behavior and stability. Address: Departamento de Matematicia, Escola de Engenharia, Sao Carlos, S.P., Brazil.*

ONUIGBO, Wilson Ikechuku Beniah, Nigerian physician; b. nr. Onitsha, Nigeria, Apr. 28, 1928; s. Joseph and Margaret (Imonugo) O.; student U. Ibadan (Nigeria), 1951-53; M.B., Ch.B., U. Glasgow (Scotland), 1957; Ph.D., U. London (Eng.), 1961; m. Edith Odukwe, Sept. 30, 1960; children—Chinye, Nwamaka, Unoma, Kenechi. Med. officer Ministry of Health, Eastern Nigeria, 1958-61; research fellow U. Glasgow, 1962; lectr. pathology U. Lagos (Nigeria), 1963; specialist in pathology Ministry of Health, Eastern Nigeria, Enugu, 1964—. Fellow Royal Soc. Medicine, Royal Soc. Tropical Medicine and Hygiene; mem. Coll. Pathologists (London), Nigerian Med. Assn. (treas. Eastern br. 1966—), N.Y. Acad. Scis., Path. Soc. Gt. Britain and Ireland, Brit. Acad. Forensic Scis. Research, publs. on anat., clin., and epidemiological aspects of cancer, emphasizing role of lymph rather than blood in its dissemination. Home: 8 Nsukka Lane. Office: Gen. Hosp., Enugu, Nigeria.*

ONWUMECHILLI, Cyril Agodi, Nigerian physicist; b. Inyi, Nigeria, Jan. 20, 1932; s. Nwaime and Aku (Orji) O.; B.Sc., U. Coll., Ibadan, Nigeria, 1953, B.Sc. in Physics, 1954, Ph.D., 1958; diploma nuclear sci. and engring. Argonne (Ill.) Nat. Lab., 1960; m. Cecilia Bedeaka Anyadibe, July 7, 1958; children—Chibugo, Chukwuka, Anayo. Lectr., U. Coll., Ibadan, 1958-62; prof. physics U. Ibadan, 1962-66, dean faculty sci., 1965-66; prof., head physics U. Nigeria, Nsukka, 1966—. Mem. Sci. Council for Africa, 1964—, Commonwealth Consultative Com. on Space Research, 1961—; rep. Africa, Internat. Com. for Internat. Years of Quiet Sun, 1963—. Mem. Internat. Union Geodsey and Geophysics (chmn. Nigerian nat. com. 1963—), Sci. Assn. Nigeria (past auditor, W. African Sci. Assn. (past treas.), Internat. Assn. Geomagnetism and Aeronomy, Am. Geophys. Union. Contbg. author: Physics of Geomagnetic Phenomena, 1966. Research, publs. on variations in earth's magnetic field especially near Equator, equatorial electrojet. Home: Enugu Inyi, Inyi Post Office, Awgu Div. Office: U. Nigeria, Nsukka, Eastern Nigeria.*

OORT, Jan Hendrik, Dutch astronomer; b. Franeker, Netherlands, Apr. 28, 1900; s. Abraham H. and Ruth Hannah (Faber) O.; D.Sc., U. Groningen; hon. degrees U. Copenhagen, 1946, Glasgow U., 1950, Oxford U., 1951, U. Louvain, 1955, Harvard, 1957, U. Brussels, 1959, U. Cambridge, 1960, U. Bordeaux, 1961, Australian Nat. U., 1963; m. Johanna Maria Graadt van Roggen, May 24, 1927; children—Coenraad Jan, Marijke (Mrs. J. T. de Smidt), Abraham Hans. Asst., U. Groningen, 1921-22, Yale Obs., 1922-24; conservator Leiden (Netherlands) Obs., 1924-30, asst. dir., 1935-45, dir., 1945—; faculty U. Leiden, 1926—, prof., 1945—. Named comdr. Order Leopold II (Belgium), knight Order of the Nederlandse Leeuw. Recipient Vetlesen prize, Columbia U., 1966. Fellow Royal Soc., 1959; mem. Acad. Scis. Amsterdam, Nat. Acad. Scis., Internat. Astron. Union (pres. 1958-61), Pontifical Acad. Sci., French Acad. Scis. Research, publs. on structure and dynamics of stellar systems especially Galactic System; discovery rotation of Galatic System, large scale structure and motions of interstellar gas discovered by radio-astron. observations, origin of comets. Home: Sterrewacht 5, Leiden, Netherlands.*

OOSTENBRINK, Michiel, Dutch nematologist; b. Dwingeloo, Netherlands, July 20, 1921; s. Dirk Jan and Femmechien (Joosten) O.; Jr. in Agr., U. Wageningen (Netherlands), 1948, Dr. in Nematology, 1950; m. Albertje Meyers, 1945; children—Gien, Dick, Geeske, Tiemo, Erik, Rudi. Staff, Netherlands Plant Protection Service, Wageningen, 1945—, head diagnostic dept., 1952—; reader nematology, head dept. Wageningen Agrl. U., 1956—, dir. internat. postgrad. nematology course, 1965—; vis. prof. various countries. Mem. Comité de Biologie de Sol (mem. bd.), Soc. European Hematologists (mem. bd.), Nematologica (bd.). Co-author: Nematology, 1960. Research, numerous publs. on plant nematology; devel. techniques including soil extraction apparatus; promoted nematology as sci. in several countries. Home: 52 Prof. Ritzema Bosweg, Wageningen. Office: 15 Binnenhaven, Wageningen, Netherlands.*

OOSTERKAMP, Wybe Johannes, Dutch physicist; b. The Hague, Netherlands, June 17, 1912; s. Johannes and Wytske (van der Veen) O.; M.A., Tech. U. Delft (Netherlands), 1934, Ph.D., 1939; m. Elisabeth H. van der Sluys, July 13, 1939; children—Hans, Willem Jan, Rinze, Imma, Wietske. Physicist, Philips Research Lab., Eindhoven, Netherlands, 1933-51, prin. physicist, 1951—; asso. prof. applied nuclear physics Tech. U. Eindhoven, 1961-67. Fellow Inst. Physics London. Research, numerous publs. on X-ray generation, detection and rec. especially for med. applications. Home: Van Oldenbarneveltlaan 14, Eindhoven. Office: Philips Research Labs. N.V., Philips' Gloeilampenfabrieken, Eindhoven, Netherlands.*

OOSTING, Henry John, Am. botanist; b. Holland, Mich., Mar. 12, 1903; s. John Henry and Minnie (Bouwman) O.; A.B. Hope Coll., 1925; M.A., Mich. State U., 1927; Ph.D., U. Minn., 1931; m. Cornelia Ossewaarde, Aug. 17, 1927; children—Jan Kurt, Marta Joy (Mrs. Antonio Aliffi). Teaching asst. Mich. State U., 1925-27; instr. U. Minn., 1927-32; faculty Duke 1932—, prof., 1949—, chmn. dept. botany, 1953-63. Mem., A.A.A.S., Am. Inst. Biol. Sci., Bot. Soc. Am., Ecol. Soc. Am. (bus. mgr. 1950—, pres. 1956). Author: The Study of Plant Communities, 1948, rev. edit., 1956. Editor: Ecological Monographs, 1950—. Research, publs. on nature, causes of plant succession, vegetation N.C., various montane virgin forests. Home: 2642 University Dr., Durham, N.C. 27707.*

OOTA, Kunio, Japanese pathologist; b. Kobe, Japan, Feb. 23, 1913; s. Heishiro and Oshine Oota; grad. Med. Faculty, U. Tokyo (Japan), 1933, Medicinae Doctor, 1937; m. Chizu Yanagita, Oct. 12, 1940; children—Hidehiko, Masako. Faculty, U. Tokyo, 1937-41; prof. pathology, 1963—; pathologist Cancer Inst., Tokyo, 1946-48; prof. Tokyo Med. and Dental Coll., 1948-63. Mem. Japanese Path. Soc., Japanese Assn. Cancer Research, Internat. Acad. Pathology, Union Internationale Contre le Cancer. Author: (with T. Miyaji) Clinical Histopathology, 1956; also numerous articles. Research on pathology of carcinoma in situ, mastopathy, gastric carcinoma, cancer of lung, Friend's disease, electron microscopy of human tumors. Home: 2-31-2 Hakusan, Bunkyo-ku, Tokyo, Japan.*

OPARIN, Aleksandr Ivanovich, Russian biochemist; b. Mar. 3, 1894; grad. natural sci. dept. Physico-Math. Faculty, Moscow U., 1917; D.Biol. Sci. Specialized in biol. chemistry under A.N. Bakh, 1917-29; prof., head chair plant biochemistry Moscow U., 1929—; an organizer, asso. Bakh Inst. Biochemistry, USSR Acad. Scis., 1935-46, dir., 1946—, v.p., 1961-67, sec. dept. biol. sci., 1949-56. Del., Internat. Congress Biochemists, Brussels, 1955. Decorated Order of Lenin; recipient Bakh prize, Mechnikov prize, 1950. Mem. USSR Acad. Scis. (2d Mechinkov gold medal 1960), World Fedn. Scientists (v.p. 1952-66). Author: Enzyme Action in the Living Cell, 1934; Biological Theory of Tea Production, 1935; The Origin of Life on Earth, 1936, 41, 57; Life, Its Nature, Origin and Development, 1961; The Chemical Origin of Life, 1964. Co-editor Chemistry sect. Large Med. Ency. Research and numerous publs. on theory of enzyme activity in plant cells, theory of origin of life on earth, enzymology and indsl. biochemistry; devised rational biochem. basis for indsl. prodn. of tea, sugar, bread, tobacco, wine. Address: Leninsky prospect 33, Moscow, USSR.*

OPDYKE, David Franklin, Am physiologist; b. Montpelier, O., Sept. 11, 1915; s. Harry W. and Alta L. (Schmell) O.; B.S., Heidelberg Coll, O., 1937; Ph.D., U. Ind., 1942; m. Florence E. Rivenburgh, June 24, 1944; children—David W., Nancy E. Head dept. physiology Merck Inst. Therapeutic Research, Rahway, N.J., 1951-56; prof., head dept. physiology Seton Hall Coll. Medicine and Dentistry (now N.J. Coll. Medicine and Dentistry), Jersey City, 1956—, asst. dean, 1963-67; cons. physiologist St. Francis Hosp. and Sanatorium, Roslyn, L.I., N.Y., 1958-60. Vis. prof. physiology Airlangga U. Med. Sch., Surabaja, Indonesia (U.S. AID-U. Cal. contract), 1961-63. Mem. Am. Physiol. Soc., Am. Heart Assn., Soc. Exptl. Biology and Medicine, Sigma Xi. Editorial bd. Jour. Circulation, 1951-67. Research, numerous publs. on mode of insulin action, hemodynamics of congenital heart disease, genesis of irreversible shock, physiology of spleen, control of cardiac output. Home: 18 Fair Hill Rd., Westfield, N.J. 07090. Office: 24 Baldwin Av., Jersey City 07304.*

OPHEIM, Odd, Norwegian otologist; b. Norway, Dec. 11, 1904; s. Alf and Kirsten (Sandberg) O.; M.D., U. Oslo (Norway); m. Astrid Opheim, Dec. 27, 1932; children—Kari, Inger-Lise. Doctor on board whaling boat in The Antarctic, then engaged in pvt. practice; intern oto-rhino-laryng. service; work in surgery and ophthalmology; head of clinic, prof. Faculty of Medicine, U. Oslo. Mem. Norwegian Med. Assn., Scottish Otol. and Laryng. Soc. Author: The Pneumatic Conditions of the Human Temporal Bone; Corrosive Injuries of the Oesophagus; The Aural Harmonies in Normal and Pathological Hearing Morbus Meuniere; The Treatment of Otoscleroses. Home: Hammerstads gt 23 c III, Oslo. Office: Rikhospitalet, Oslo, Norway.

OPHÜLS, William, Am. pathologist; b. Bklyn., Oct. 23, 1871; s. Carl Julius and Clara (Wilhelms) O.; student U. Würzburg (Germany), 1890-93, U. Berlin, 1894; M.D., U. Göttingen (Germany), 1895; m. Emmy Feldmann, May 6, 1903; children—Clara Louise, Ernst Carl, Elinor Helen, Gertrud. Asst., Path. Inst., Göttingen, 1896-97; prof. path. and bacteriology U. Mo., 1897-98, Cooper Med. Coll., San Francisco, 1898-1912; became prof. pathology Stanford, 1919; named pathologist Lane Hosp., San Francisco, 1898, pres. San Francisco Bd. Health, 1907-10. Research and publs. on Tb., coccidioidal granuloma, arteriosclerosis, nephritis; described (with P.K. Brown) a case of severe leukopenia (reduction in number of leukocytes in blood), 1901. Died Apr. 27, 1933.

OPICINUS DE CANISTRIS, see de Canistris, Opicinus.

OPIE, Eugene Lindsay, Am. pathologist, microbiologist; b. Staunton, Va., July 5, 1873; s. Thomas and Sallie (Harman) O.; A.B., Johns Hopkins, 1893, M.D., 1897, LL.D., 1940; Sc.D., Yale, 1931; LL.D., Washington U., 1940; m. Gertrude Lovat Simpson, Aug. 6, 1902; m. 2d, Margaret Lovat Simpson, Sept. 16, 1916; children—Thomas Lindsay, Ann Lovat (Mrs. Joseph Hughes), Helen Lovat (Mrs. Charles C. Colby). Gertrude Eugenie (Mrs. Herald H. Dinkens). Mem. Rockefeller Inst. for Med. Research, N.Y.C., 1904-10, bd. sci. dirs., 1928-32, researcher, 1941—; faculty Washington U., St. Louis, 1910-23, U. Pa., 1923-32, Cornell Med. Coll., 1932-41; bd. sci. dirs. Internat. Health div. Rockefeller Found., 1935-38. Mem. research council Pub. Health Research Inst., N.Y.C. Recipient Gerhard, Trudeau medals, 1929; medal Soc. Puertorriqueña de Tisólogos, 1938; Weber-Parkes medal and award Coll. Physicians, 1945; Banting medal, 1946; medal N.Y. Acad. Medicine, 1960. Mem. Nat. Tb Assn. (pres. 1929), Assn. Am. Pathologists (pres. 1917), Am. Assn. Immunologists (pres. 1929), Harvey Soc., (pres. 1936-38), Nat. Acad. Scis. (Jessie Stevenson Kovalenko medal 1959). Author: Diseases of the Pancreas, 1902; Epidemic Respiratory Disease, 1921. Co-editor Jour. Exptl. Medicine, 1904-10. Research on pathogenesis and epidemiology of Tb, pancreatic diseases, inflammation and immunity, pharmacology of inflammation. Home: 404 E. 66th St., N.Y.C. 10021. Office: 66th St. and York Av., N.Y.C. 10021.*

OPITZ, Guenter, German chemist; b. Berlin, Mar. 17, 1926; s. Emil and Marie (Karmrodt) O.; Diplom, U. Tübingen, 1955, Dr.rer. nat., 1956; m. Isolde Löschmann, 1962; children—Ingeborg, Kirsten. Faculty, U. Tübingen, 1960—, unscheduled prof., 1966—. Mem. Soc. German Chemists. Author: (with H. Hellmann) a-Aminoalkylierung, 1960; also numerous articles. Research on scope and mechanism of Mannich reaction; chemistry of enamines, dienamines, ketene acetals, sulfenes; cycloadditions. Address: 34 Käsenbachstrasse, 74 Tübingen, West Germany.*

OPLER, Marvin Kaufmann, Am. anthropologist, social psychiat. educator; b. Buffalo, June 13, 1914; s. Arthur and Fanny (Coleman-Haas) O.; student U. Buffalo, 1931-33; A.B., U. Mich., 1935; Ph.D., Columbia, 1938; m. Charlotte Fox, Dec. 30, 1935; children—Ruth Ellen (Mrs. Lewis Curtis Perry), Lewis Alan. Chmn. dept. sociology, anthropology Reed Coll., 1938-43; chief community analysis War Relocation Authority, U.S. Dept. Interior, 1943-46; pub. panel chmn. NLRB, 1942-43; lectr. Inst. Psychoanalysis, Los Angeles, 1945-47; asso. prof. dept. sociology, anthropology Stanford, 1948-50, dept. Social Relations, Harvard, 1950-52; prof. anthropology, social psychiatry Cornell U. Med Coll., 1952-58; vis. prof. sociology, anthropology New Sch. for Social Research, William Alanson White Inst., N.Y.C., 1956-58; prof. depts. anthropology psychiatry, sociology State U. N.Y., Buffalo, 1958—. Prin. investigator Midtown Manhattan Mental Health Study, 1952-62; exec. com. mem. Northeastern Anthrop. Conf., 1960—. Social Sci. Research Council fellow, 1949. Mem. Internat. Assn. Social Psychiatry (exec. com. 1964—), Am. Anthropol. Assn., Northeastern Anthropol. Assn. (pres. 1965-66), Academy for chiatry and Human Values, 1956; Culture and Social Psychiatry, 1967; (with others) Impounded People, 1946; (with others) Mental Health in the Metropolis, 1962; (with others) The Image of Man in Medicine and Anthropology, 1963; also other books, numerous articles in encys., profl. jours. Editor: Culture and Mental Health, 1959; Internat. Jour. Social Psychiatry, 1955—; asso. editor Am. Anthropologist, 1963—; mem. editorial bds. numerous profl. jours. Research on Ute Indians and modern ethnic groups, evolution of culture, social psychiatry; contbns. to theory of personality. Home: 22 Summerwood Ct., Buffalo 14223.*

OPLER, Morris Edward, Am. anthropologist; b. Buffalo, May 16, 1907; s. Arthur and Fanny (Haas) O.; A.B., U. Buffalo, 1929, M.A., 1930; Ph.D., U. Chgo., 1933; m. Lucille Ritter, July 29, 1957. Vis. lectr. Reed Coll., 1937-38; asst. prof. Claremont Colls., 1938-42; asst. prof. Harvard, 1946-48; prof. Cornell U., Ithaca, N.Y., 1948—, dir. South Asia program, 1948-66. Guggenheim fellow, 1942; Nat. Endowment for the Humanities sr. fellow, 1968-69. Mem. Am. Anthrop. Assn. (pres. 1962-63), Phi Beta Kappa, Sigma Xi, numerous others. Author books, numerous articles. Home: 124 Snyder Hill Rd., Ithaca, N.Y. 14850.*

OPPENHEIM, Antoni Kazimierz, mech. engr.; b. Warsaw, Poland, Aug. 11, 1915; s. Tadeusz and Zuzanna (Zuckerwar) O.; student Warsaw (Poland) Inst. Tech., 1933-39; diploma Ing., City and Guilds, Coll.; D.I.C., Imperial Coll. Sci. and Tech., U. London (Eng.); Ph.D., 1945; m. Lavinia Stephens, July 18, 1945; 1 dau., Terry Ann. Research asst. City and Guilds Coll., 1942-46, lectr., 1946-48; asst. prof. mech. engring. Stanford, 1948-50; faculty U. Cal., Berkeley, 1950—, prof. aeronautical scis., 1958—, research prof. Miller Inst. for Basic Research in Sci., 1961-62; lectr. Polish U. Coll., London, 1943-48; staff cons. Shell Devel. Co., 1952-60; vis. prof. Sorbonne, Paris, 1960-61. Mem. adv. com. on fluid mechanics NASA, 1963—. NSF Sr. Postdoctoral fellow, 1960-61. Recipient medal U. Brussels, 1961. Fellow Am. Rocket Soc. (pres. No. Cal. sect. 1957); mem. Internat. Acad. Astronautics (corr.), Am. Soc. M.E., Am. Inst. Aeros. and Astronautics, Am. Phys. Soc., Am. Soc. Engring. Edn., Inst. Mech. Engrs. (Great Britain, Water Arbitration prize 1948), Heat Transfer and Fluid Mechanics Inst. (chmn. 1958), Sigma Xi, Psi Tau Sigma, Tau Beta Psi. Research, numerous publs. on generalized entropy chart for compressible fluid flow calculation, analysis of free piston engines, network method radiation analysis, devel. gaseous detonation, vector polar method in gas wave dynamics, generation pressure waves in reacting gaseous mixtures. Home: 54 Norwood Av., Berkeley, Cal. 94707.*

OPPENHEIM, Hermann, German neurologist; b. Berlin, Jan. 1, 1858; prof., Berlin. Author: Lehrbuch der Nervenkrnakheiten, 1894. Described dystonia musculorum deformans, or torsion spasm (Zeihen-Oppenheim disease), 1911. Died May 22, 1919.

OPPENHEIM, Irwin, Am. chem. physicist, educator; b. Boston, June 30, 1929; s. James L. and Rose (Rosenberg) O.; A.B. summa cum laude, Harvard, 1949; postgrad. Cal. Inst. Tech.; Ph.D., Yale, 1956. Physicist, Nat. Bur. Standards, Washington, 1953-60; chief theoretical physics Gen. Dynamics/Convair, San Diego, 1960-61; asso. prof. chemistry Mass. Inst. Tech., Cambridge, 1961-65, prof., 1965—. Lectr. physics U. Md., 1953-60; vis. asso. prof. physics U. Leiden, 1955-56; vis. prof. Weizmann Inst. Sch., 1958-59, U. Cal., San Diego, 1966-67; Van der Waals prof. U. Amsterdam, 1966-67; cons. Inst. for Def. Analyses, Nat. Bur. Standards, Lincoln Lab., Air Force Cambridge Research Center, Gen. Dynamics, N.Am. Aviation, Bell Telephone Labs., 1961—. Fellow Am. Phys. Soc., Washington Acad. Sci.; mem. Phi Beta Kappa, Sigma Xi. Author: (with J.G. Kirkwood) Chemical Thermodynamics, 1961. Research in quantum statis. mechanics, statis. mechanics of transport processes, thermodynamics. Home: 60 Brattle St., Cambridge, Mass.*

OPPENHEIM, Maurice, dermatologist; b. Vienna, Austria, Jan. 1, 1876; s. Karl and Regine (Steiner) O.; M.D., U. Vienna, 1899; m. Karoline Schaefer, May 11, 1914; 2 daus. Intern Allgemeines Krankenhaus, Vienna, 1899; asst., dermatology clinic U. Vienna, 1902-07, named prof., 1927; chief of ambulatorium Gen. Ins. Co. for Workmen, 1907-14; chief skin and venereal dept. Wilhelmin Hosp., Vienna, 1918-38; lectr. U. Paris, also postgrad. sch. Columbia U.; prof. dermatology Chgo. Med. Sch. Recipient Gold medal Exhbn., Düsseldorf, Germany, 1930, Budapest Internat. Dermatol. Congress, 1935, distinctions from hygiene exhbns. at Vienna, 1912, 27, 37, Graz, Austria, 1932, As Ember, Budapest, 1928, 32. Mem. Am. Dermatol. Assn., dermatol. socs. Argentina, Berlin, Denmark, Poland, Italy, Hungary, France, Am. Acad. Dermatology and Syphilology Internat. Author: Short Textbook of Pathology and Treatment of Venereal Diseases, 1902; Atlas of Venereal Diseases of Portio Vaginalis Uteri and Vagina, 1908; (with Finger) The Atrophies of Skin, 1910; Practicum der Haut-und Geschlechtskrankheiten, 6th edit., 1937; Vademecum of Care of Venereal Diseases and Diseases of Skin for Nurses, 1922; (with Rille and Ullman) Berufskrankheiten der Haut, 3 vols., 1923-26; Affections of Skin by Occupation, Sport, Climate, and Cosmetic Factors, 1937; also articles. Developed (with R. Müller) complement-fixation test for diagnosis of gonorrhea (Müller-Oppenheim reaction), circa 1906; described necrobiosis lipoidica diabeticorum (Oppenheim-Urbach disease), 1932. Died 1949.

OPPENHEIM, U(ri) P(hilip), Israeli physicist; b. Tel-Aviv, Israel, Sept. 25, 1926; s. Jacob D. and Dora (deJong) Ophen; M.Sc., Hebrew U. Jerusalem (Israel), 1951, Ph.D., 1955; m. Els Levenbach, Feb. 16, 1949; children—Isaac, Eytan, David. Research physicist Weizmann Inst. Sci., 1951-59, cons., 1959—; faculty Technion, Haifa, Israel, 1959—, asso. prof. physics, 1966—; research fellow Cal. Inst. Tech., Pasadena, 1957-59. Mem. Israel Phys. Soc., Optical Soc. Am., Internat. Combustion Inst. Patentee in field. Research, publs. on devel. thin films for use in infrared interferometers; measurement absolute intensity of molecular lines and bands in infrared; determinated emissivity of hot gases important in combustion and atmospheric transmission; developed new methods to interpret observed spectral emissivities in terms of spectral band model theories. Home: 74 Horev St., Haifa, Israel.*

OPPENHEIMER, Carl, German biochemist; b. Berlin, Feb. 21, 1874; prof., Berlin; later indsl. chemist. Author: Die Fermente, 4 vols., 1900; Grundriss der organischen Chemie, 1895; Grundriss der anorganischen Chemie, 1898; Technologie der Fermente, 1929. Editor: Handbuch der Biochemie. Authority on enzymes. Died Dec. 24, 1941.

OPPENHEIMER, Carl Henry, Jr., Am. oceanographer; b. Los Angeles, Nov. 13, 1921; s. Carl Henry and Marion Vivian (Hess) O.; B.A., U. So. Cal., 1947, M.A., 1948; Ph.D., U. Cal. at Los Angeles, 1951. Asst. marine biologist Scripps Instn. Oceanography, U. Cal. at La Jolla, 1951-55; sr. research scientist Pan Am. Petroleum Corp. Research Center, 1955-57; research scientist, lectr. marine microbiology U. Tex., 1957-61; asso. prof. Inst. Marine Sci., U. Miami (Fla.), 1961-64; faculty Fla. State U., Tallahassee, 1964—, chmn. dept. oceanography, 1965—, prof., 1964—. Fulbright fellow U. Oslo, (Norway), 1952-53. Mem. Am. Soc. Microbiologists, Soc. Limnology and Oceanography, Soc. Gen. Microbiology, Geochem. Soc., Sigma Xi. Research, numerous publs. on marine microbial ecology, geomicrobiology, marine bacteriology, petroleum microbiology, problems of sewage disposal in Los Angeles and San Diego, pollution of Oslofjord, Norway, Texas bays, origin and composition of petroleum, origin of oil. Home: 1501 Chocksacka Nene, Tallahassee 32301.*

OPPENHEIMER, Franz, German sociologist; b. Berlin, Germany, Mar. 30, 1864; M.D., U. Berlin, 1885; Ph.D., U. Kiel, 1908. Engaged as gen. practitioner, 1885-96; Univ. lectr., Berlin, 1909, prof., 1917; prof. Frankfort U., 1919, emeritus, 1929—; guest prof. many schs. including Paris and

Palestine, Adviser to Ministry of War, later War Office, World War I. Author: System der Soziologie, 4 vols., 1922-29; The State (authorized trans. 1914); Die Siedlungsgenossenschaft, 3d edit., 1922; Wege zur Gemeinschaft, 1924. He thought of society as an organism in a state of normality when justice predominates; believed value judgments could be made about social structure through comparison with norms established by sociology, psychology and history. Died Los Angeles, Aug. 9, 1943.

OPPENHEIMER, Hillel (Heinz) Reinhard, plant physiologist; b. Berlin, Germany, Apr. 4, 1899; s. Franz O. and Martha Oppenheimer; student U. Berlin, 1917-19, U. Freiburg, 1919, U. Frankfurt, 1920-21; Dr.phil., U. Vienna, 1922; m. Carmela Lewité, Dec. 29, 1932; children—Ruth (Mrs. Amitai Katz), Yehudith (Mrs. Paul Mogle), Michal. Head plant protection service L. Spaeth Nurseries, Berlin, 1924-25; keeper Herbarium A. Aaronsohn, Zikhron-Yaaqov, Israel, 1926-31; head sect. plant physiology later citriculture Agr. Research Sta., Rehovot, Israel, 1933-53; James de Rothschild prof. horticulture and plant physiology Faculty Agr., Hebrew U., Rehovot, 1953-—, dean Faculty Agr., 1952-54. Mem. Israel Forest Research Com., 1960-—. Recipient Israel State Prize in Agr., 1959. Mem. Palestine Bot. Soc. (past pres.), Am. Soc. for Hort. Sci., Société botanique de France, Internat. Dendrological Union. Author: Florula Transiordanica, 1930; Citrus Growing, 1957; also numerous articles. Founder, Palestine Jour. Botany, 1935, editor, 1935-53. Research on germination inhibitors in fruits, osmotic and elastic properties of plant cells, drought tolerance of plant cells; tree physiology, including water relations in semi-arid surroundings, root structure and growth, cambial activity; citricultural physiology, including timing of irrigation, foliar analysis, rootstock selection, response to pruning, fruit prodn.; plants of Nr. E. Home: 4 Bustnai St., Rehovot, Israel.*

OPPENHEIMER, J. Robert, Am. physicist; b. New York, N.Y., Apr. 22, 1904; s. Julius and Ella (Freedman) O.; A.B., Harvard U., 1925, student Cambridge U., Eng., 1925-26; Ph.D., Göttingen U., Germany, 1927; m. Katherine Harrison, Nov., 1940; children—Peter, Katherine. National Research fellow, 1927-28; Internat. Edn. Bd. fellow, U. of Leyden and Zurich, 1928-29; asst. prof. physics, U. Cal. and Cal. Inst. Tech., 1929-31, asso. prof., 1931-36, prof., 1936-47. Dir. Los Alamos Sci. Lab., Los Alamos, N.M., 1943-45; director and professor of physics, Institute for Advanced Study, Princeton, 1947-—. Chmn. gen. adv. com. AEC, 1946-52. Recipient Fermi award AEC, 1963. Fellow Phys. Soc. (pres. 1948), Philos. Soc., Am. Acad.; mem. Royal Danish (fgn.), Brazilian (fgn.), Japanese (hon.) acads. Important in devel. of quantum theory, understanding of cosmic rays, fundamental particles and relativity; introduced use of symmetrical and antisymmetrical functions in scattering problems; worked out theory of neutron stars; instrumental in devel. of atomic bomb. Died Princeton, N.J., Feb. 20, 1967.

OPPENHEIMER, Jane Marion, Am. developmental biologist; b. Phila., Sept. 19, 1911; d. James H. and Sylvia (Stern) Oppenheimer; B.A., Bryn Mawr Coll., 1932; Ph.D., Yale, 1935. Sterling fellow Yale, 1935-36, Am. Assn. U. Women Sarah Berliner research fellow., 1936-37; research fellow U. Rochester, 1937-38; faculty Bryn Mawr (Pa.) Coll., 1938-—, prof. biology, 1953-65, Class of 1897 prof., 1965-—, acting dean Grad. Sch., 1946-47. Mem. study sect. History of Life Scis., NIH, 1966-—. Rockefeller Found. fellow, 1950-51; John Simon Guggenheim Meml. fellow, 1942-43, 52-53; NSF Sr. Postdoctoral fellow, 1959-60. Mem. A.A.A.S. (past sect. sec.) Am. Soc. Zoologists (past treas., past mem. exec. com., chmn. div. developmental biology 1967), Am. Assn. U. Women (past mem. com. on award internat. study grants), Internat. Fedn. U. Women (mem. panel experts comm. for award internat. fellowships 1955-—), Am. Assn. Anatomists, Am. Soc. Naturalists, Am. Soc. Cell Biologists., Internat. Inst. Embryology, Soc. for Developmental Biology, Am. Assn. for History Medicine, History Sci. Soc. Author: New Aspects of John and William Hunter, 1946; Essays in the History of Embryology and Biology, 1967; also numerous articles, book revs. Editor: (with B.H. Willier) Foundations of Experimental Embryology, 1964. Translator: Biology (K. von Frisch), 1965. Asso. editor Jour. Morphology, 1956-60, Quar. Rev. Biology, 1963-64; editorial bd. Am. Zoologists, 1965-—, Excerpta Medica, Jour. History of Biology, 1967-—, Quar. Rev. Biology, 1968-—. Research on factors controlling differentiation in embryos of bony fishes. Office: Dept. Biology, Bryn Mawr Coll., Bryn Mawr, Pa. 19010.*

OPPERT, Julius (Jules), Assyriologist; b. Hamburg, Germany, July 19, 1825; ed. Heidelberg, Bonn, Berlin univs.; grad. Kiel, 1847. Taught German in Laval and Reims, France, 1848-52; on archeol. expdn. to Mesopotamia, 1852-54; prof. sanscrit Sch. Languages, Nat. Library, Paris, 1857; prof. Assyrian philology and archeology Coll. France, 1869. Author: Déchiffrement des inscriptions cunéiformes, 1861; Eléments de la grammaire assyrienne, 1868; Le peuple et la langue des Mèdes, 1870; L'Immortalité de l'ame chez les Chaldéens, 1875; Salomon et ses successeurs, 1877; (with J. Ménant) Doctrines juridiques de l'Assyrie et de la Chaldée, 1877. Important contbr. to decipherment of Persian cuneiform inscriptions; recognized Turanian character of original language of Assyria; studied Assyrian history and mythology. Died Paris, France, Aug. 21, 1905.

OPPOLZER, Theodor Egon, see von Oppolzer, Theodor Egon.

O'RAIFEARTAIGH, Lochlainn Seamus, Irish physicist; b. Dublin, Ireland, Mar. 11, 1933; s. Tarlach and Aine (Morrissey) O'R.; B.Sc., Nat. U. Ireland, Dublin, 1953, M.Sc., 1956; Ph.D., Zurich (Switzerland) U., 1960; m. Treasa Donnelly, Aug. 25, 1958; children—Conor, Finbar, Cormac, Una. Scholar, Dublin Inst. for Advanced Studies, 1956-57, travelling studentship, 1957-58, faculty, 1961-—; research asso. Zurich U., 1958-60; visitor Madras (India) Inst. for Math. Sci., 1964; vis. asso. prof. Syracuse U., 1964-66, prof., 1966-67; mem. Princeton Inst. for Advanced Study, 1967-68. Mem. Royal Irish Acad. Contbg. author: Symmetry Groups in Nuclear and Particle Physics, Group Theory and Its Applications. Research, publs. on gen. relativity, local and non-local quantum field theory, parastatistics, Lie groups, symmetry in elementary particle physics. Home: 7 Iris Grove, Mt. Merrion, Dublin. Office: Inst. for Advanced Studies, Burlington Rd., Dublin 4, Ireland.*

ORBACH, Raymond L., Am. physicist; b. Los Angeles, July 12, 1934; s. Morris Albert and Mary Ruth (Miller) O.; B.S., Cal. Inst. Tech., 1956; Ph.D., U. Cal. at Berkeley, 1960; m. Eva Hannah Spiegler, Aug. 26, 1956; children—David Miller, Deborah Hedwig, Thomas Randolph. NSF postdoctoral fellow Oxford U., 1960-61; asst. prof. applied physics Harvard, 1961-63; prof. physics U. Cal. at Los Angeles, 1963-—. Cons. Aerospace Corp., 1965-—. Alfred P. Sloan Found. fellow, 1963-67. Mem. Am. Phys. Soc., Phys. Soc. (London), Phi Beta Kappa, Sigma Xi, Tau Beta Pi. Author: (with A.A. Manenkov) Spin-Lattice Relaxation in Ionic Solids, 1966; also articles. Research on magnetic properties of solids, interaction between paramagnetic ions and lattice vibrations, influence of lattice vibrations on hyperfine coupling, anharmonic forces in solids, magnetic resonance of localized momento in metals; discovered new spin-lattice relaxation process, predicted anomalously long lifetimes for high energy lattice vibrations. Home: 339 Arno Way, Pacific Palisades, Cal. 90272. Office: Dept. of Physics, U. Cal., Los Angeles 90024.*

ORBELI, Leon Abgarovich, physiologist; b. Yerevan, Armenia, 1883; studied under Ivan Pavlov; head Pavlov Inst. Physiology, Pavlovo, nr. Leningrad, USSR; head Soviet Delegation to 17th Internat. Physiol. Congress, London, Eng., 1947. Recipient Stalin prize; Metchnikov Gold medal, 1948. Mem. Acad. Scis. USSR (1st v.p.). Author: Lectures on the Physiology of the Nervous System; Conditioned Reflexes in Dogs. Founder sch. of evolutionary physiology of nervous system; conducted research with results which were applied to army medicine. Office: Pavlov Inst. Physiology, Pavlovo, USSR.

ORBISON, James Lowell, Am. pathologist, educator; b. Bronson, Kans., Mar. 8, 1915; s. Franklin Lloyd and Mary (Ireland) O.; A.B., Ottawa U., Kans., 1937, D.Sc. (hon.), 1966; M.S., Mich. State U., 1939; M.D., Northwestern U., 1944; m. Olga Dianich, June 21, 1941; children—Margaret Chase, James Graham. Faculty, Western Res. U., Cleve., 1947-55; George H. Whipple prof. pathology, chmn. dept. U. Rochester (N.Y.), 1955-—. Mem. sci. adv. bd. Armed Forces Inst. Pathology, 1957-62; mem. pathology tng. com. NIH, 1960-66, chmn., 1963-66; mem. intersoc. com. research potential in pathology, 1959-—, pres., 1961-—; Nat. Acad. Sci., 1966-—. Recipient Centennial Merit award Northwestern U., 1959. Alumni citation to faculty U. Rochester, 1966. Diplomate Nat. Bd. Med. Examiners. Mem. Am. Assn. Pathologists and Bacteriologists (council 1964-—), Internat. Acad. Pathology (pres. 1963-64), Am. Soc. Exptl Pathology, Sigma Xi, Phi Chi, Alpha Omicron Alpha. Editor: (with David Smith) Pathologic Physiology and Anatomy of Peripheral Blood Vessels, 1962. Research on peripheral vascular disease and connective tissue. Home: 105 Stonybrook Dr., Rochester, N.Y. 14618.*

ORCHIN, Milton, Am. chemist; b. Barnesboro, Pa., June 4, 1914; s. Morris and Mary (Rifkin) O.; B.A., Ohio State U., 1936, M.A., 1937, Ph.D., 1939; m. Ruth Wilner, June 4, 1941; children—Morton Lewis, Michael David. Chemist, FDA, Cin. and Chgo., 1939-42; chief organic chemistry sect. U.S. Bur. Mines, 1943-53; asso. prof. chemistry U. Cin., 1953-56, prof., head dept., 1956-62, prof. chemistry, dir. basic sci. lab., 1962-—. Guggenheim fellow, 1947-48. Mem. A.A.A.S. (chmn. chemistry sect. 1963), Am. Chem. Soc. (chmn. Cin. sect. 1962; Eminent Chemist award Cin. sect. 1957), Chem. Soc. London, Phi Beta Kappa, Sigma Xi. Contbr. numerous research publs.; co-author (with H. H. Joffe) 3 books on spectroscopy and symmetry. Patentee in field. Research on nature of binding between transition metals and organic ligands. Home: 1858 Northcut Av., Cin. 45237.*

ORCUTT, Frederic Scott, Am. biologist; b. Sioux City, Ia., Nov. 27, 1907; s. Robert William and Florence (Waters) O.; student Antioch Coll., 1926-27; B.S. in Chemistry, U. Wis., 1931, M.S. in Biochemistry, 1933, Ph.D. in Bacteriology and Biochemistry, 1935; m. Katharine Luvinia Krauskopf, Sept. 28, 1935; children—Barbara Sue (Mrs. William T. Keeton), Frederic Scott. Research asso., Squibb fellow in anesthesia and pharmacology U. Wis., 1935-36; faculty biology Va. Poly. Inst., Blacksburg, 1936-—, prof., 1945-69, emeritus, 1968-—, head dept., 1959-66. Cons. Oak Ridge Inst. Nuclear Studies, 1949-50, participant, 1950-55; vis. prof. bacteriology U. Va. Biology Sta., 1951, 55, U. Hawaii, 1953-54; vis. prof. Am. Physiol. Soc. Inst. Coll. Tchrs., summer 1957. Mem. Am. Soc. Microbiology (co-organizer Va. br. 1942, pres. 1946-49, nat. councilor 1949-52), Am. Acad. Microbiology (charter), Am. Inst. Biol. Scis. (governing bd. 1958-68), A.A.A.S., Assn. Southeastern Biologists, Va. Acad. Sci., Sigma Xi, Phi Sigma (nat. exec. sec.-treas. 1952-—), Phi Lambda Upsilon, Gamma Alpha, Alpha Kappa Lambda. Editorial bd. Jour. Applied Microbiology. Contbr. articles to profl. jours. Home: 1305 Hillcrest Dr., Blacksburg, Va. 24060.*

ORD, George, Am. naturalist, philologist; b. Phila., Mar. 4, 1781; s. George and Rebecca (Lindemeyer) O.; m. 1815; 2 children, including Joseph Benjamin. Employed in father's ship chandler and rope making business, 1806-29; attempted (with Charles Waterton) to discredit Audubon, circa 1824; accompanied Thomas Say, Titian Peale, William Maclure on extensive field trip to Ga. and Fla., 1818; mem. Am. Philos. Soc.; pres. Phila. Acad. Scis., 1851-58. Completed book by friend Alexander Wilson, American Ornithology or, the Natural History of Birds of the United States, 9 vols., 1808-14, edited vol. 8, wrote entire text vol. 9, published another edit. of work with much additional material, 1824-25; prepared memoirs of Say and C.A. Lesueur; prepared anonymous account of zoology of N. Am. for 2d Am. edit. New Geographical and Commercial Grammar (William Guthrie), 1815; disposed of manuscripts on philology to Latham of London who used them in compilation new edit. Johnson's Dictionary, circa 1860. Contbd. personal library to Coll. Physicians Phila., $16,000 to Pa. Hosp. Died Phila., Jan. 24, 1866.

ORD, William Miller, English physician; b. Brixton Hill, Eng., Sept. 23, 1834; s. George and Harriet (Clark) O.; student of John Simon at St. Thomas's Hosp. Med. Sch.; M.B., London U., 1857; M.D., 1877; m. Julia Rainbow, 1859 (dec. 1864); 1 son; m. 2d, Jane Toul; 2 daus. House surgeon, surg. registrar, demonstrator anatomy St. Thomas's Hosp., became lectr. zoology, asst. physician, joint lectr. physiology, 1860, dean med. sch., 1876-87, physician, 1877-98, cons. physician, from 1898. Fellow Royal Coll Physicians, Royal Med. and Chirug. Soc.; pres. Med. Soc. London, 1885; chmn. com. to investigate myxoedema Clin. Soc. London, 1883; treas. Clin. Soc. Author: Influence of Colloids upon Crystalline Forms and Cohesion, 1879; On some Disorders of Nutrition related with Affections of the Nervous System, 1885; also papers. Introduced term myxoedema, 1878, showed its cause was mainly atrophy or fibrosis of thyroid gland; studied neurotic dystrophies, neurotic origin of gout, arthritis. Died Salisbury, Eng., May 14, 1902.

ORDAL, Z(akarias) John, Am. microbiologist; b. Sioux Falls, S.D., Mar. 23, 1913; s. Zakarias J. and Sina (Wulfsberg) O.; student River Falls State Tchrs. Coll., 1933-34; B.A., Luther Coll., 1935; Ph.D., U. Minn., 1940; m. Dorothy Lenore Munson, June 12, 1942; children—John, Thomas, Carolyn, Mary. Muellhaupt fellow Ohio State U., 1941; Hormel Research Found. fellow U. Minn., 1942; asso. in bacteriology U. Ill. Coll. Medicine, 1942-44, asst. prof. bacteriology, pub. health, 1944-46; supr. bacteriology group Process and Product Research div. Owens-Ill. Glass Co., Toledo, 1946-47; chief bacteriology sect., research and devel. div. Econs. Labs., Inc., St. Paul, 1947-49; asso. prof. food microbiology, dept. food sci. U. Ill., 1949-57, prof. food microbiology, 1957-—, asst. dir. Ill. Agrl. Expt. Sta., 1966-—. Recipient F. C. Vibrans Sr. Scientist award Am. Meat Inst. Research Found., U. Chgo. Fellow Am. Acad. for Microbiology; mem. Am. Soc. for Microbiology, Am. Pub. Health Assn., Inst. Food Tech., Sigma Xi, Phi Kappa Phi, Gamma Sigma Delta. Author: (with L. W. Slanetz, C. O. Chichester, A. R. Gauffin) Microbiological Quality of Foods, 1963.. Research, numerous publs. on sporulation, germination requirements and properties of bacterial endospores; prodn. microorganisms using food processing wastes; characterization of bacterial cell injury and requirements for recovery from injury; isolation and characterization of lipolytic enzymes in lactobacilli and in propionic bacteria. Home: 206 E. Mumford Dr., Urbana, Ill. 61801.*

ORDONEZ, Castor, biologist; b. Palencia, Spain, Mar. 28, 1880; s. Nicanor and Rufina (Cabiedes) O.; A.B., Spain, 1901; A.M., 1902; Ph.D., St. Mary's Sem., Perryville, Mo., 1915; Sc.D., DePaul U., 1917. Naturalized Am. citizen, 1926. Mem. faculty St. Vincent's Sem., P.I., 1904-10; editor weekly mag. Eco de Leyte y Samar, 1911-12; became tchr. De Paul U., 1914, later dir. dept. biology, trustee. Author: Educational Biology; Genetics and Eugenics, 1932; also lab. manuals. Compiled Abridged Calculus, 1908. Research on rat cancer, cross between turkey and chicken; discoveries in uses of electricity. Died June 28, 1938.

ORDONEZ, Ezequiel, Mexican geologist; b. Lerma, Mexico, Apr. 10, 1867 s. Margarito and Eduwigis (Aguilar) O.; ed. Escuela Nacional de Ingenieros, Mexico City; m. Margarita Jullian, Apr. 16, 1904; 3 children. Became curator Engring. Sch. Mineral. Collection, 1886; named asst. geologist Geologic Commn., 1889; geologist Instituto Geológico de México, 1893-97, under-dir., 1898-1906, dir., 1916-17, from 1945; geologist Cia. Real del Monte y Pachuca, 1909-14; chief geologist Huasteca Petroleum Co., 1915-32; prof. petroleum geology and phys. geology of Mexico, Escuela Nacional de Ingenieros; geologist Comisión Coordinadora e Impulsore de la Investigación científica de México, from 1943; cons. geologist Mexican Petroleum Co. of Cal., from 1901 Mex. del. 4 internat. geol. congresses. Hon. mem. Am. Inst. Mining and Metallurg. Engrs., Am. Assn. Petroleum Geologists, Geol. Soc. Am., other sci. socs. Died 1950.

ORDWAY, Frederick Ira III, Am. astronautical researcher; b. N.Y.C., Apr. 4, 1927; s. Frederick Ira and Frances Antoinette (Wright) O.; S.B., Harvard, 1949; postgrad. Universite d'Alger (Algeria), 1950, U. Paris (France), 1950-51, 53-54, Universidad de Barcelona (Spain), 1953, Universitat Innsbruck Austria), 1954, Air U., 1952-63, Alexander Hamilton Bus. Inst., 1952-58, Indsl. Coll. Armed Forces, 1953, 63; m. Maria Victoria Arenas, Apr. 13, 1950; children—Frederick Ira IV, Albert James Aliette Marisol. Various geol. and engring. positions Mene Grande Oil Co., San Tome, Venezuela, 1949-50, Orinoco Mining Co., Venezuela, 1950, Reaction Motors, Inc., Lake Denmark, N.J., 1951-53, guided missile div. Republic Aviation Corp., 1954-55; pres. Gen. Astronautics Research Corp. (formerly Gen. Astronautics Corp.), Washington and Huntsville, Ala., 1955-59, 63—; pres. Astronautical Devel. Co., Huntsville, 1964—; v.p. Nat. Research & Devel. Corp., Atlanta, 1955-59; asst. to dir. Saturn systems office Army Ballistic Missile Agy., Huntsville, 1959-60; chief space systems information br. George C. Marshall Space Flight Center NASA, 1960-63; v.p. Space Sci. & Tech. Information Center, Huntsville, 1963-64; cons. to firms, including Ten Eyck Assos., Washington Assoc. (both Washington), Gen. Elec. Co. (Phila.) Ency. Brit., Am. Coll. Dictionary. MGM Brit. Studios, Ltd., Hawk Films. Ltd., 1965-67, Smithsonian Instn., Nat. Air & Space Mus., 1966. Fellow Royal Astron. Soc., A.A.A.S., Brit., So. African interplanetary socs., Canadian Inst. Aeros. and Astronautics; asso. fellow Am. Inst. Aeros. and Astronautics (dir. Ala. sect.); mem. Aerospace Med. Assn., Agrupación Astronáutica Espanola, Am. Astron. Soc., Am. Geophys. Union, Asociaciön Argentina Interplanetaria, Aviation Space Writers Assn., Internat. Soc. Aviation Writers, Sociedad Astronómica de Espana y América (titular), Société Française d'Astronautique, Alliance Française (v.p. Huntsville chpt.). Author: (with C. C. Adams) Space Flight, 1958; (with Ronald C. Wakeford) International Missile and Spacecraft Guide, 1960; Annotated Bibliography of Space Science and Technology, 1962; (with J. P. Gardner, M. R. Sharpe, Jr.) Basic Astronautics: An Introduction to Space Science, Engineering and Medicine, 1962; (with Adams, Wernher von Braun) Careers in Astronautics and Rocketry, 1962; (with Gardner, Sharpe, R. C. Wakeford) Applied Astronautics: An Introduction to Space Flight, 1963; (with Wakeford) Conquering the Sun's Empire, 1963; Life in Other Solar Systems, 1965; (with Roger A. MacGowan) Extra-Solar Intelligence, 1965; Intelligence in the Universe, 1966; (with Wernher von Braun) History of Rocketry and Space Travel, 1967. Editor: Advances in Space Science and Technology, vols. I-VII, 1959-65; Introduction to Astrodynamics (R. M. L. Baker, M. W. Makemson), 1960 (with others) From Peenemünde to Outer Space, 1962; (with others) Astronautical Engineering and Science, 1963. Mem. editorial bd. IX Internat. Astronautical Congress Proc., 2 vols., 1959, Xth Congress Proc., 2 vols., 1960. Contbr. numerous articles profl. jours, also chpts. in books, sects. in Ency. Brit., Ency. Americana. Devel. liquid propellant rocket powerplants, guided missiles, space carrier vehicles; studies in unmanned artificial satellite missions. Home: 4118 Shelby Av. S.E., Huntsville 35801. Office: Research Park, Huntsville, Ala. 35801.*

ORDWAY, Thomas, Am. physician; b. Dorchester, Mass., May 7, 1877; s. George Francis and Julia Maria (Gilbert) O.; A.B., Harvard, 1900, M.A., 1901, M.D., 1905; studied under Pierre and Marie Curie, 1913; m. Mary Olive, Apr. 6, 1906; 1 son, Thomas. Asst. in pathology Harvard Med. Sch., Boston City Hosp., dir. Bender Hygienic Lab., Albany, N.Y., 1909-11; prof. bacteriology and pathology Albany Med. Coll., 1909-11, apptd. dean, asst. prof. medicine, 1915, prof. 1931-37; physician in charge Huntington Hosp., also physician in charge, instr. Harvard Med., physician Boston City Hosp., 1911-15; physician Albany Hosp., 1915-37. Del. Internat. Cancer Congress, Brussels; chmn. N.Y. State Med. Retirement Bd., 1921-52. Diplomate Am. Bd. Internal Medicine. Fellow A.M.A.; mem. N.Y., Mass., Albany County med. assns., Am. Assn. for Cancer Research, Am. Assn. Pathologists and Bacterologists, Am. Soc. Clin. Investigation, A.A.A.S., Assn. Am. Physicians, Boston Soc. Med. Scis. Author: Diseases of the Blood, 1930. Research on diseases of blood, pancreatitis of cat, rat metabolism, blood in cancer, tumors of fowls, Wasserman reaction, early diagnosis of pneumonia; gave early description of occupations injuries caused by radium, 1916; described remission in leukemia produced by radium treatment, 1917. Died Albany, May 12, 1952.

ORE, Aadne, Norwegian physicist; b. Oslo, Norway, Nov. 10, 1916; s. Mikal Östensen and Christiane (Samuelsen) O.; cand.real.cum.laude, U. Oslo, 1941; M.S., Yale, 1946, Ph.D., 1947; m. Anne-Wenche Smith, June 24, 1952; children—Espen, Martin, Christian-Emil. Vis. fellow Princeton, 1947-48; fellow, instr. U. Chgo., 1948-49; asso. prof. U. Bergen (Norway), 1949-53; faculty U. Oslo, 1953—, prof. physics, 1960—; research asso. Yale, 1955-56. Mem. Norwegian Acad. Sci. and Letters (past officer), Norwegian (past officer), Am. phys. socs. Author: (with O. A. Höeg) University of Oslo 1911-1961, 1961; also articles. Editorial bd. Physica Norvegia, 1961—. Research on atom- and molecule-like compounds containing one or two positrons, entropy of radiation, ionization produced by radiations, nonradiative energy transfer between molecules. Home: 14 Jerpefaret, Oslo 3, Norway. Office: U. Oslo, Blindern, Norway.*

ORE, Oystein, mathematician; b. Oslo, Norway, Oct. 7, 1899; s. Mikal Beer and Christiane (Samuelsen) O.; B.A., Oslo Cathedral Sch., 1922; Ph.D., Oslo U., 1924; M.A. (hon.), Yale, 1929; m. Gudrun Lundevall, Aug. 24, 1930; children—Elisabet (Mrs. Philip Moore Orville), Berit (Mrs. William Richard Lytle). With Mittag Leffler Math. Inst., Djursholm, Sweden, 1923-24; Internat. Edn. Bd. fellow, 1924-26; asst. prof. U. Oslo, 1926-28; faculty Yale, 1927—, prof. math., 1929—, chmn. dept., 1936-45. Mem. NRC, 1939-42. Decorated Knight Order St. Olav (Norway). Mem. Norwegian Acad. Sci., Am. Acad. Arts and Scis. Author: Les corps algébriques et le thèorie des ideaux, 1934; L'Algèbre abstraite, 1936; Number theory and its History, 1948; Cardano, the Gambling Scholar, 1953; Niels Henrik Abel, Mathematician Extraordinary, 1957; Theory of Graphs, 1962; Graphs and Their Uses, 1963; The Four-Color Problem, 1967; also numerous articles. Research in theory of numbers, abstract algebra, theory of graphs and networks, history of sci. Home: 26 Hall St., Hamden, Conn. 06517. Office: Hall Grad. Studies, Yale, New Haven 06520.*

OREKHOVICH, Vasiliy Nikolaevich, Russian biochemist; b. 1905; grad. North Caucasian U., Rostov-on-Don, 1927. Asst., lectr. Sverdlov Communist U., Moscow, 1931-35; asso. All-Union Inst. Exptl. Medicine (now Inst. Biol. and Med. Chemistry), USSR Acad. Med. Scis., 1936-41, lab. head, 1941-49, dir., 1949—, acad. sec. dept. med. and biol. sci., 1957-60. Mem. USSR Acad. Med. Scis. (Presidium mem. 1953—, v.p. 1960—). Author: Procollagens: Their Chemical Composition, Properties and Biological Function, 1952; Modern Concepts of Protein and Its Significance in Biology and Medicine; The Role of the Biological Sciences in Medicine, 1963. Editor Chemistry sect. Large Med. Ency., 2d edit.; mem. editorial bd. Problems of Med. Chemistry. Research and numerous publs. on biochemistry of protein, protein metabolism of animals, protein enzyme changes in regenerative and embryo tissues, theory of tumor growth, susceptibility of body to tumors, chemistry of connective tissue; discoverer group of tissue proteins (procollagens). Address: USSR Acad. Med. Scis., Solyanka 14, Moscow, USSR.

ORESME, Nicole, French mathematician; b. probably at Allemagne, nr. Caen, France, circa 1323; became theol. student Coll. Navarre, U. Paris, 1348; mastership in theology, circa 1356. Grand master Coll. Navarre, 1356; instr. to Dauphin Charles; named archdeacon of Bayeux, 1361; canon Cathedral of Rouen, France, 1362, became dean, 1364; named canon La Sainte Chapelle, 1363; named Bishop of Lisieux, France, 1377. Author: De proportionibus proportionum; Ad pauca respicientes; De configurationibus qualitatum et motuum; De commensurabilitate sive incommensurabilite motuum cell; Algorismus proportionum; Tractatus contra iudiciarios astronomus; Le livre de divinacions; Tractatus de origine, naturi, jure et mutationibus monetarum; also French translations and commentaries on Aristotle, including Ethics, Politics, Economics, Livre du Ciel et du Monde; translation of Quadripartitum (Ptolemy). Developed a calculus of proportions and applied to terrestrial and celestial kinematics and dynamics; developed technique of graphing qualities of forms and applied it to problems in kinematics; discussed possible existence of plurality of worlds and possible rotation of earth; although he ultimately rejected these notions, he held them possible because of omniptotence of God; contbd. to monetary theory; attacked astrologer's ability to predict future; used metaphor of heavens as mechanical clock. Died Lisieux, July 11, 1382.

ORFILA, Mathieu-Joseph-Bonaventure, physician, chemist; b. H. Mahon, Majorca, Apr. 24, 1787; studied medicine, Valencia, Spain, 1805, Barcelona, 1806, also Paris; M.D., 1811; m. Mlle. Lesueur; prof. chemistry, Barcelona; named prof. medicine Faculty Medicine, Paris, 1819, prof. chemistry, 1923, dean, 1831-34; apptd. physician to Louis XVIII, 1816; named mem. Royal Council for Pub. Instrn., 1834; founder Orfila Mus. Comparative Anatomy. Recipient 1st prize in chemistry and physics U. Valencia. Mem. French Acad. Scis. (corr.), Acad. Medicine (pres.). Author: Traité de toxicologie générale; Traité de médecine légale; Traité de chimie; Traité des exhumations juridiques; Traité des secours à donner aux personnes empoissonnésou asphyxiées. A founder of medico-legal toxicology; showed that toxins can be found in parenchymas and organic liquids. Died Paris, Mar. 12, 1853.

ORIANI, Barnaba, Italian astronomer; b. Garegnano, Italy, July 17, 1752; studied astronomy under Joseph Louis Lagrange; student Coll. San Alessandro. Ordained priest, 1776; joined staff Brera Obs., Milan, Italy, 1776, became asst. astronomer, 1778, dir., 1802; apptd. dir. Milan Obs. by Napoleon; apptd. (with De Cesans, Reggio) to measure arc of meridian between zeniths of Rimini and Rome. Pres., Napoleon's Commn. to Regulate New System of Weights and Measures. Became corr. mem. French Acad. Sci., 1804, Fellow Royal Soc., 1795. Mem. Berlin Acad. Author: Tables of Uranus, 1785; Trigonometria sphaerica, 2 vols., 1806-10; Istruzione suelle misure e sui pesi, 1831. Asso. with Theatine monk Piazzi in astron. studies, for over 37 years; used his calculations to prove Uranus is a planet rather than a comet. Died Milan, Nov. 12, 1832.

ORIBASIOS OF PERGAMON, physician; b. Pergamon, flourished 325-circa 395; studied medicine under Zeno of Cyprus (according to Eunapios); m.; 4 children, probably including Eustathios; friend of emperor Julian whom he accompanied to Gaul, 355; apptd. quaestor of Constantinople; mem. Julian's expdn. against Persia; banished by emperors after Julian's death in 363, but returned before 369. Author: Collecta medicinalia, also an abridged version; Euporista (or De facile parabilibus); an epitome of Galen's writings made at Julian's request, with works of other med. writers, 70 or 72 vols., completed after 361, partly contained in Collecta medicinalia. Discovered salivary gland.

ORKIN, Louis R., Am. anesthesiologist; b. N.Y.C., Dec. 23, 1915; s. Samuel David and Rebecca (Rish) O.; B.A. (with honors), U. Wis., 1937; M.D., N.Y. U., 1941; m. Florence Fine, Mar. 5, 1938; 1 dau., Rita Louise. Mem. faculty N.Y. U., 1950-55; prof., chmn. dept. anesthesiology Albert Einstein Coll. Medicine, Yeshiva U., N.Y.C., 1955—; dir. anesthesia W.W. Backus Hosp., Norwich, Conn., 1948-49; anesthesiologist Bellevue Hosp., 1950-55; dir. dept. anesthesiology Bronx Municipal Hosp. Center, N.Y.C., 1955—. Cons. govtl. brs., and hosps.; mem. com. on anesthesia NRC, 1965-67. Decorated Bronze Star. Diplomate Am. Bd. Anesthesiology. Fellow Am. Coll. Anesthesiologists, N.Y. Acad. Medicine, Am. Coll. Chest Physicians, N.Y. Acad. Scis.; mem. A.M.A., Assn. U. Anesthetists, Soc. Pharm. and Exptl. Therapeutics, Alpha Omega Alpha, Sigma Xi., others. Editor: Management of Patient in Shock, Clinical Anesthesia (F.A. Davis), 1965; asso. editor Survey of Anesthesiology, 1958—; editor Clin. Anesthesia, 1962. Research, publs. on care of anesthetized patient and pharmacology of drugs related to anesthesia, relationship of anesthesia to shock and microcirculation. Home: 11 Stuyvesant Oval, N.Y.C. 10009. Office: Bronx Municipal Hosp. Center, N.Y.C. 10461.*

ORLAND, Frank J., Am. oral microbiologist, educator; b. Little Falls, N.Y., Jan. 23, 1917; s. Michael and Rose Orland; A.A., U. Chgo., 1937; B.S., U. Ill., 1939, D.D.S., 1941; S.M., U. Chgo., 1945, Ph.D., 1949; m. Phyllis Mrazek, May 8, 1943; children—Frank, Carl, June, Ralph. Faculty, U. Chgo., 1942—, instr. to asso. prof. microbiology, 1950-58, prof. dental surgery, 1958—, research asso. microbiology, 1958-64, dir. Zoller Meml. Dental Clinic, 1954-66. Mem. study sect., program planning com. Nat. Inst. Dental Research, NIH, 1955-59; chmn. com. on advanced edn. Am. Assn. Dental Schs., 1966-67. Recipient research essay award Chgo. Dental Soc., 1955. Diplomate Am. Bd. Microbiology. Mem. Am. Dental Assn. (chmn. council dental therapy 1961-62), Internat. Assn. Dental Research, Am. Soc. Microbiology, Am. Assn. Dental Editors, Inst. Medicine Chgo., A.A.A.S. Editor, Jour. Dental Research, 1958—. Research, publs. on dental caries; demonstration that tooth decay cannot occur in germfree animals, but does with lactic acid producing bacteria, antigenic analysis of oral lactobacilli, electron microscopy study of Entamoeba gingivalis ultrastructure. Home: 519 Jackson Blvd., Forest Park, Ill. 60130.*

ORLANDI, Francesco, Italian physician; b. Assisi, Italy, Mar. 26, 1927; s. Paolo and Natalia (Mariani) O.; M.D., U. Perugia, 1951, L.D., 1958. Asst. dept. internal medicine U. Perugia, 1951-57, prof. Postgrad. Sch. Gastroenterology, 1957—; dir. dept. internal medicine S. Maria della Pieta Hosp., 1966—; prof. nutrition sci. U. Camerino, Marche, Italy, 1966—, sec. Liver and Drugs Center, 1967; dir. Nuclear Medicine Center Hosp. of Camerino, 1966—. Mem. Soc. Italiana Med. Interna, Soc. Italiana Microscopia Elettronica, Soc. Francaise Microscopie Electronique, European Assn. for Study of Liver. Electron microscopic research, publs. on human liver in diabetes, cholestasis and drug injuries; discovered new type of needle for biopsies of human liver.

Home: Villa Napoleoni. Office: 54, Via Lili, Camerino, Marche, Italy.*

ORLOV, Iurii Aleksandrovich, Russian paleontologist, histologist; b. Simbursk Guberniya, USSR, 1893; grad. natural sci. dept., physico-math. faculty Petrograd U., 1917; instr. histology Perm U., 1917-24, Leningrad Mil. Med. Acad., 1925-35; prof. paleontology Leningrad U., 1929-45; sr. asso. inst. paleontology USSR Acad. Scis., dir., 1945—, also mem. acad. Author: Perunilnae, 1947; In the World of Ancient Animals. Chief editor Paleontologichesky zhurnal, 1959—. Research on fauna of western Siberia, comparative morphology of invertebrate nervous systems, paleontology of vertebrates and invertebrates. Office: Paleontol. Inst., USSR Acad. Scis., Leninsky prospekt 33, Moscow, USSR.

ORLOV, Yurij Feodorovich, Russian physicist; b. Moscow, Russia, Aug. 13, 1924; s. Feodor Pavlovich and Klavdia (Lebedeva) O.; grad. Moscow State U., 1957, Sc.D., 1963; m. Irina Alexandrovna Lagunova, May 17, 1962; children—Dimitrij, Alexander, Lev. Scientist, Inst. Exptl. and Theoretical Physics, Moscow, 1953-56; asst. Moscow Phys. Tech. Inst., 1953-56; chief theoretical lab. Phys. Inst., Yerevan, 1956—, lectr. Yerevan State U., 1956—; scientist Inst. Nuclear Physics, Novosibirsk, 1963-64. Research, publs. on theory of accelerators, theory of nonlinear oscillations, nonlinear focusing systems. Home: 5 Markarian. Office: 18 Barekamutian, Yerevan, Armenia, USSR.*

ORLOWSKI, Witold Jan, Polish physician; b. Warsaw, Poland, Apr. 30, 1918; s. Mieczystaw and Eliza (Abaxdujew) O.; student Copernicus Coll., Lodz, Poland, 1936, U. Warsaw, 1936-39; B.A., U. Lodz, 1947; M.B., Med. Acad. Lodz, 1950, M.D., 1952, habilitation, 1961; m. Maria Bartkowska, Dec. 6, 1966. Clin. asst. ophthalmology U. Lodz, 1947-50; clin. asst. ophthalmology Mil. Hosp., Warsaw, 1951-54, dir. div. ophthalmology, 1955-64; prof. ophthalmology, chmn. dept., Med. Acad. Poznan (Poland), 1965—. Mem. Polish, French, Belgian ophthal. socs. Author: Z nowszych zagadnien anestezji w chirugii oka, 1959; Lekksyon zespolow i objawow chorobowych, 1959; also numerous articles. Editor Wiadomosci Lekarskie, 1952-53; asso. editor Klinika Oczna, 1956—; editor-in-chief Zyjmy Dluze, 1958-59; cons. editor Wiadomosci Lekarskie, 1965—. Introduced modern ophthalmic anesthesiology and ophthalmic electromyography in Poland; invented cataract operation, expulsion of cataract. Home: 91c m. 1, ul Grochowska, Poznan 8, Poland.*

ORMEROD, Eleanor Anne, English entomologist; b. Salbury Park, Eng., May 11, 1828; d. George and Sarah (Latham) O.; LL.D., Edinburgh, 1900; asst. in forming collection to illustrate econ. entomology Royal Hort. Soc., 1868, cons. entomologist, 1882-92; participant Internat. Polytechnic Exhbn., Moscow, 1872; spl. lectr. econ. entomology Royal Agr. Coll., 1881-84; mem. (with Huxley) com. econ. entomology Royal Agr. Soc., 1882-86. Recipient Silver Flora medal Royal Hort. Soc., 1870, Gold Medal of Honor, Moscow U. Fellow Meteorol. Soc. (1st woman). Author: Notes for Observations of Injurious Insects, 1877; Manual of Injurious Insects, 1881; Guide to the Methods of Insect Life, 1884; Flies Injurious to Stock, 1900; A Textbook of Agricultural Entomology, 1892. Contbd. to devel. of econ. entomology as a science. Died St. Albans, Eng., July 19, 1901.

ORMONDROYD, Jesse, Am. mech. engr. b. Phila., Pa., Feb. 7, 1897; s. Herbert and Jeanette Ellson (Wrighton) O.; A.B., U. of Pa., 1920; m. Kathleen Felton, June 2, 1921; children—Edward, Ruth. Research and design engr. Westinghouse Electric & Mfg. Co., 1920-30, mgr. of exptl. div. of turbine works, 1930-37; prof. engring., U. of Mich., 1937-67, works, 1930-37; prof. engring., U. of Mich., 1937-67, professor engineering emeritus, 1967—, chairman dept. engring. mechanics, 1963-64; engr. on vibration problems. Mem. Jr. and Community College Study Commission for Michigan, 1956-57, In A.S., U. S. Army, 1917-19. Life fellow Am. Society Mech. Engineers (gen. lecturer 1953 and 1955); member Am. Society for Engring. Edn., Franklin Inst., Newcomen Soc. N.A., Phi Beta Kappa, Sigma Xi. Author: (with Karelitz and Garrelts) Problems in Mechanics, 1939. Author of several papers on theory of vibration in various technical publications. In charge of mfg. design of 200-inch telescope mounting. Designed radar foundation; missile tracking ships in Pacific Ocean; partition tubes. Captain U.S.N.R. active duty Feb. 17, 1941 to Apr. 16, 1946. Address: 604 Robin Rd., Ann Arbor, Mich. 48103.*

ORMSBEE, Richard Armstrong, Am. biochemist; b. Walla Walla, Wash., Jan. 6, 1915; s. Eugene Richard and Franc (Armstrong) O.; B.A., Mont. State U., 1936; postgrad. U. Cal. at Berkeley; M.S., Wash. State U., 1938; Ph.D., Brown U., 1941; m. Sue Yancey Green, May 6, 1944; children—Richard Berryman, Stuart Colquhoun. Spl. research asso. biol. chemistry dept. Harvard Med. Sch., 1941-43; tech. aide OSRD, Washington, 1943-45; exec. sec. Chem-Biol Coordination Center, Nat. Acad. Sci., 1945; asso. mem. Sloan-Kettering Inst., 1945-47; biochemist Rocky Mountain Lab., NIH, 1948-57, sr. scientist, 1957-60, scientist dir., 1960—. Lectr., Mont. State U., 1955—. Fellow A.A.A.S.;mem. Am. Chem. Soc., Tissue Culture Assn., N.Y. Acad. Scis., Am. Assn. Immunologists. Contbg. author: Viral and Rickettsial Infections of Man, 1965; also numerous articles. Research on growth and metabolism protozoa, purification and mechanism enzyme succinic dehydrogenase, toxicology and therapy mustard gas, cancer chemotherapy, growth and meabolism pathogenic rickettsia, rickettsial disease in mammals, action antibiotics on rickettsia, use ion exchange chromatography for purification viruses and rickettsia, ECHO viruses, rickettsial disease in wild animals, rikettsial vaccines. Office: Rocky Mountain Lab., NIH, Hamilton, Mont. 59840.*

ORO, Juan, biochemist; b. Lerida, Spain, Oct. 26, 1923; s. Juan and Maria Florensa (Rue) Oro-Vallverdu; licenciate in Chem. Scis., U. Barcelona, 1947; Ph.D., Baylor U., 1956; m. Francisca Forteza-Gasol, May 19, 1948; children—Maria Elena, Juan, Jaime, David. Faculty, U. Houston, 1955—, prof., 1963—. Research vis. chemist Lawrence Radiation Lab. U. Cal., Berkeley, summer 1962; Am. Chem. Soc. tour lectr., 1963-64; mem. exobiology study group space sci. bd. Nat. Acad. Scis., 1964-65. Mem. Am. Chem. Soc., Am. Soc. Biol. Chemists, Nat. Assn. Chemists (Spain), Royal Acad. Scis. and Arts Barcelona, Spanish Soc. Biochemistry, Geochem. Soc., Meteoritical Soc., Am. Soc. Oceanography, A.A.A.S., Sigma Xi. Research, numerous publs. on origin of life, abiol. synthesis of adenine, guanine, deoxyribose, amino acids and polypeptides; analysis of organic compounds in meteorites and Precambrian sediments. Home: 11306 Endicott St., Houston 77035.*

ORR, Clyde, Jr., Am. chem. engr.; b. Lewisburg, Tenn., Oct. 1, 1921; s. Clyde and Brownie (Lusby) O.; B.S., U. Tenn., 1944, M.S., 1948; Ph.D., Ga. Inst. Tech., 1953; m. Mary Ruth Gardner, Dec. 24, 1944; children—Donald Steven, Douglas Wayne, Dorothy Jeanne, Barbara Lynne. With TVA, 1946-47; research engr. Ga. Inst. Tech., Atlanta, 1948-52, faculty, 1952—, prof. chem. engring. 1962-66, Regents prof., 1966—, head micromeritics br. Engring. Expt. Sta., 1953—. Mem. Am. Inst. Chem. Engrs., Am. Chem. Soc., N.Y. Acad. Scis. Author: (with J.M. Dalla Valle) Fine Particle Measurements, 1959; Between Earth and Space, 1959; Particulate Technology, 1966; also numerous articles. Research on phys. properties (including specific surface area and (porosity) of powders and composite materials prepared therefrom; also devel. instruments for evaluation of such properties; elucidation of behavior of small airborn particles and condensation nuclei. Home: 3281 Pl., N.E., Atlanta 30324. Office: 778 Atlantic Dr., N.W., Atlanta 30332.*

ORR, H(iram) Winett, Am. surgeon; b. West Newton, Pa., Mar. 17, 1877; s. Andrew Wilson and Frances J. (Winnett) O.; student U. Neb., 1892-95; M.D., U. Mich., 1899; m. Grace Douglass, Sept. 7, 1904; children—Douglass, Willard, Josephine, Dorothy, Gwenith. In practice, Lincoln, Neb., from 1899; lectr. history of medicine Coll. Medicine, U. Neb., from 1903; chief med. insp. Lincoln pub. schs., 1908; supt. Neb. Orthopedic Hosp., 1911-17, chief surgeon, 1919-47, cons. surgeon since 1947; cons. surgeon dept. orthopedic surgery Lincoln Gen. Hosp. Recipient Distinguished Service award (posthumously) A.C.S. Mem. A.M.A. (chmn. orthopedic sect., 1921-22), Neb. Med. Assn. (pres.), Am. Orthopedic Assn. (editor, pres.), Soc. Internat. de Chirug Orthopedique (U.S.), Am. Med. Library Assn. (hon.), Assn. Bone and Joint Surgeons (hon.), other med. socs., Sigma Xi. Author: The Treatment of Osteomyelitis and Other Infected Wounds by Drainage and Rest, 1927; Osteomyelitis and Compound Fractures, 1928; Osteomyelitis and Compound Fractures and other Infected Wounds, 1929; A New Era in the Treatment of Osteomyelitis and Other Infections, 1930; Selected Pages from the History of Medicine in Nebraska, 1952. Contbr. to Sajous Cyclo., 1931. Editor Western Med. Rev., 1899-1906, Jour. Orthopedic Surgery, 1919-21. Introduced plaster cast technique for broken bones into France during World War I; insured use of limb in simple fractures by use of traction pins to hold fracture set. Died Oct. 11, 1956.

ORR, Hugh, inventor; b. Lochwinnoch, Scotland, Jan. 2, 1715; s. Robert Orr; m. Mary Bass, Aug. 4, 1742; 10 children, including Robert. Came to U.S., 1737. Began work for scythemaker, Bridgewater, Mass., 1741; became owner of shop, 1745; became firearms mfr., 1748; built foundry for casting cannons, 1775; became mem. Mass. Senate, 1786. Invented trip hammer (1st in colonies), machine to clean flaxseed, 1753, machines for carding and roping wool, 1785-87; introduced European mechanics to U.S., new types of machinery. Died Bridgewater, Mass., Dec. 6, 1798.

ORR, Jack Edward, Am. pharmacist, educator; b. Delphi, Ind., Dec. 11, 1918; s. Forrest Howard Sr. and Dorothy Marie (Smith) O.; B.S., Purdue U., 1940; Ph.D., U. Wis., 1943; m. Maxine Kennard, Sept. 6, 1942; 1 dau., Judith Elaine. Pharmacist, 1940-43; instr. pharmacy Ohio State U., 1943-44, asst. prof., 1946-47; prof. pharm. chemistry U. Utah, 1947-52; dean. prof. pharmacy Mont. State U., 1952-56, U. Wash., 1956—; state chemist, Wash., 1956—. Served from ensign to lt. USNR, 1944-46. Mem. Am. Assn. Colls. Pharmacy (exec. com. 1955-57, 61-63, chmn. com. constn. and by-laws 1956-57, com. future enrollment plans, 1958-59, faculty tchrs. seminars 1952-54, chairman sect. on tchrs. chemistry 1957, pres. 1964-65), Am. Pharm. Assn., Wash. State Pharm. Assn., Sigma Xi. Contbr. articles profl. jours. Research on chemistry of natural products. Home: 4245 N.E. 74th St., Seattle 98115.

ORSONI, Luciano, Italian elec. engr.; b. Milan, Italy, May 22, 1910; s. Carlo and Angiolina (Baderna) O.; E.E., Engring. U. Milan, 1934; m. Maria Calvi, June 24, 1948; children—Mariangela, Carla, Paola. Tech. officer Italian Air Force, 1938-46; elec. engr. for research and devel. in nuclear energy Montecatini Edison Ltd., Milan, 1947-56, exec. mgr. nuclear dept., 1956—; gen. mgr. Sorin nuclear research and engring. co., Milan, 1956—; prof. reactor physics Politecnico di Torino, 1963—. Decorated Cavaliere al merito della Repubblica. Mem. Forum Italiano per l'Energia Nucleare, Associazione Elettrotecnica e Elettronica Italiana, Am. Nuclear Soc. Author: Reattori Nucleari, 1958; (with C. Lo Surdo) Analisi del Reattore Nucleare, 1964; also articles. Research on nuclear reactors static and dynamic theory, heavy water prodn. by electrolysis. Home: 56 Alberto Mario. Office: 1/2 Largo Donegani, Milan, Italy.*

ORTEGA Y GASSET, José, Spanish philosopher; b. Madrid, Spain, May 9, 1883; s. José Ortega y Munilla; ed. pvt. tutors, also Jesuit sch., Mira del Palo; Ph.D., U. Madrid, 1904; influenced by neo-Kantian thought as student philosophy univs. Leipzig, Berlin, Marburg (all Germany), 1904-08. Tchr. at Escuela Superior de Magisterio; prof. metaphysics U. Madrid, 1910-36; became nat. celebrity as result of speech denouncing Restoration, Regency, and Alfonso XIII (led to found. League Polit. Edn. and its publ., Espana), at Teatro de la Comedia, 1914; on lectr. tour, Argentina, 1917, again to South Am., 1928; active in overthrow Spanish monarchy, worked with Assn. Service to Republic; elected dep. province Leon, 1931; refused Franco's offer to become Spain's ofcl. philosopher; became avowed opponent of Franco regime; in voluntary exile in France, Holland, Argentina, Peru (prof. philosophy U. San Marcos, Lima, 1941), Portugal (lectr. U. Lisbon, from 1943), 1936-45; returned to Spain, 1945; founder Instituto de Humanidades, Madrid, 1948. Author: Adán en el paraiso, 1910; Meditaciones del Quijote, 1914; España invertebrada, 1921; El tema de nuestro tiempo, 1923; La rebelion de las masas, 1929; Historia came sistema; others. Founder jours.: Faro, Europa, Revista de Occidente, El Sol, España. Examined problems of decline and decadence, and place of traditional European culture in rapidly changing world; argued that predominance of masses and the vulgar is characteristic of our times; also if chaos is to be avoided, men must be governed in liberal spirit by intellectual elite; made important contbns. to philosophy of history. Died Madrid, Oct. 18, 1955.

ORTELIUS, Abraham (Oertel or Wortels), geographer; b. Antwerp, Apr. 14, 1527. Entered Antwerp Engravers' Guild, 1547; seller of antiquities and maps, ca. 1554; traveled in Europe extensively, met Mercator, became interested in cartography, ca. 1560; apptd. geographer to King Philip II of Spain, 1575. Author: Theatrum orbis terrarum, 1570; Additamentum theatri orbis terrarum, 1573; Deorum Dearumque capita ex vetustis numismatibus . . . ex museo A. Ortelli, 1573; Synonymia geographica, 1578; Parergon, 1579; Itinerarium per nonnullas Galliae Beligicae partes, 1584; Nomenclator Ptolemaicus, 1584; C.J. Caesaris omnia quae extant, 1595; Aurei saeculi imago, sive Germanorum veterum vita, 1595. Ortelius and Mercator were the leading geographers of their age; Ortelius' Theatrum was the standard atlas in Europe for generations; also collected antiquities. Died Antwerp, June 28, 1598.

ORTEN, James M., Am. biochemist; b. Farmington, Mo., Nov. 29, 1905; s. Alvin Luther and Ida Judith (Jackson) O.; student U. Mo., 1923-25; B.S., U. Denver, 1928, M.S., 1929; Ph.D., U. Colo. Sch. Medicine, 1932; postgrad. Yale; m. F. Aline Underhill, Sept. 10, 1932. Research asst. U. Denver, 1928-29; teaching fellow U. Colo., 1929-32; research fellow Yale, 1932-37; faculty Wayne State U. Sch. Medicine, 1937—, prof. biochemistry, 1956—. Fellow A.A.A.S.; mem. Am. Inst. Nutrition (past sec., past asso. editor), Soc. for Exptl. Biology and Medicine (past pres. Mich. sect.), Mich. Acad. Arts, Sci. and Letters (past chmn. med. sect.), Detroit Physiol. Soc. (past pres.), Detroit Bio-Organic Club (past pres.), Am. Soc. Biol. Chemists, N.Y. Acad. Scis., Canadian Biochem. Soc., Am. Chem. Soc., Sigma Xi, Phi Beta Kappa. Author: (with I.S. Kleiner) Human Biochemistry, 1958; (with others) Laboratory Manual of Biochemistry, 1945; Textbook of Biochemistry, 1962, 66; also numerous articles. Research on factors producing exptl. anemias and polycythemia, metabolism citric acid and related organic acids; discovered that cobalt produces a polycythemia; metabolic inter-relations between succinic, fumaric and malic acids with citric acid contbg. to citric acid cycle concept metabolism; research on exptl. diabetes and alcohol metabolism; biosynthesis of porphyrins and hemoglobin. Home: 412 Maison Rd., Grosse Pointe Farms, Mich. 48236. Office: Wayne State U. Sch. Medicine, 1401 Rivard St., Detroit 48207.*

ORTHNER, Hans, neurologist; b. Ried, Austria, Aug. 7, 1914; s. Franz and Maria (Gober) O.; M.D., U. Innsbruck; m. Annette Hanse, July 4, 1963; children—Sigrid, Franz-Helmuth, Hans-Jörg, Gertraud, Christoph-Martin, Maria-Renate. Specialist in neurology and psychiatry; dir. neuro-pathology sect. U. Göttingen Neuro-Psychiat. Clinics. Author papers on neuropathology, pathology, study of the brain, stereo-toxic operations, neurology, psychiatry and legal medicine. Research on quantitative brain research, stereotaxic operations; multiple sclerosis, slow-virus infections; neuropathology, especially neuroendocrinology; basic brain mechanisms; disturbances of consciousness and memory. Home: Robert Koch strasse 24. Office: von Seiboldstrasse 5, Göttingen, West, Germany.

ORTHULAIN, see Hortulain.

ORTMANN, Arnold Edward, naturalist; b. Magdeburg, Prussia, Apr. 8, 1863; s. Edward Fran and Bertha (Lorenz) O.; ed. univs. Jena, Kiel (both Germany), Strasbourg (France); Ph.D., Jena, 1885; Sc.D., U. Pitts., 1911; m. Anna Zaiss, Dec. 5, 1894. Came to U.S., 1894. Served in German Army, 1882-83; zoologist and paleontologist; on collecting expdn. to Zanzibar, Africa, 1890-91; curator invertebrate paleontology Princeton, 1894-1903; curator invertebrate zoology Carnegie Mus., Pitts., from 1903; instr. geol. geography U. Pitts., 1909-10, prof. phys. geography, 1910-25, prof. zoology, from 1925. Mem. Princeton Arctic (Peary relief) expdn., 1899. Author: Flora Hennebergica, 1887; Grundzüge der Marinen Tiergeographie, 1896; Continuation of Die Decapodon in Bronn's Klassen und Ordnungen des Tierreiches, 1898-1900; Tertiary Invertebrates of the Princeton Expdn. to Patagonia, 1902. Died Jan. 3, 1927.

ORTON, Edward Francis Baxter, Am. geologist; b. Deposit, N.Y., Mar. 9, 1829; A.M., Hamilton Coll., 1848; studied at Andover and Harvard; Ph.D., LL.D., Ohio State U.; m. Mary M. Jennings, 1855; m. 2d, Anna D. Torrey, 1875. Entered ministry; held chair natural science, Albany State Normal Sch., 1856; principal, preparatory acad. Chester, N.Y., 1859-65; prof. natural science, Antioch Coll., 1865-1899; became state geologist of Ohio, 1869; pres. Antioch Coll., 1872-73; pres. Ohio State U., 1873-81, prof. geology, 1873-99. Mem. Geol. Soc. Am. (became pres. 1897), A.A.A.S. (pres. 1898-99). Author: vols. 5, 6, 7, Geology of Ohio, (with others)) vols. 1, 2, 3; also numerous geol. articles. Died Columbus, Ohio, Oct. 16, 1899.

ORTON, James, Am. naturalist; b. Seneca Falls, N.Y., Apr. 21, 1830; s. Azariah Giles and Minerva (Squire) O.; grad. Williams Coll., 1855, Andover Theol. Sem., 1858; m. Ellen Foote, 1859. Ordained to ministry Presbyn. Ch., 1860; instr. natural history U. Rochester (N.Y.), 1866; prof. natural history Vassar Coll.; made expdns. to S.Am., 1867, 73, 76. Author: The Miner's Guide and Metallurgist's Directory, 1849; Comparative Zoology, Structural and Systematic, 1876. Collected specimens in equatorial Andes and Amazon River region. Died Lake Titacaca, Peru, Sept. 25, 1877.

ORVIS, Alan LeRoy, Am. physicist; b. Cleve., May 2, 1921; s. Harvey Willard and Helen (Gerlach) O.; B.S., Westminister Coll., 1944; M.S., Case Inst. Tech., 1949; Ph.D., U. Tex., 1952; m. Jennie Morgan, Aug. 30, 1947; children—James Alan, Joan Morgan. Instr., Case Inst. Tech., 1947-49, U. Tex., 1950-52; cons. in biophysics Mayo Clinic, Rochester, Minn., 1952—; asso. prof. biophysics Mayo Grad. Sch. Medicine, U. Minn., 1966—. Mayo Found. rep. to Asso. Midwest Univs., Argonne (Ill.) Nat. Lab., 1955-65; chmn. adv. com. on radiation safety Minn. Bd. Health, 1963—. Mem. A.A.A.S., Am. Phys. Soc., Am. Geophys. Union, Am. Meteorol. Soc., Health Physics Soc., Am. Assn. Physicists in Medicine, Minn. Acad. Sci., Minn. Radiol. Soc., Am. Assn. Physics Tchrs., Sigma Xi, Sigma Pi Sigma. Contbg. author: Dynamic Clinical Studies with Radioisotopes, 1964; Instrumentation in Nuclear Medicine, 1966; also articles. Devel. counting systems for biomed. radioisotopes tracer research and math. analysis data, measurement endogenous radioactivity in human body. Home: 2002 Crest Lane, Merrihils, Rochester 55901. Office: Biophysics sect. Mayo Clinic, Rochester, Minn. 55902.*

ORZALESI, Nicola, Italian ophthalmologist; b. Florence, Italy, Sept. 6, 1938; d. Francesco and Franca (Morghen) O.; M.D., U. Bari, 1961. Asst. in ophthalmology U. Genova, 1961-65, 1st asst. in ophthalmology, 1965—, dir. dept. ophthalmology Lab. Electron Microscopy, 1961-66; research fellow Nat. Council Research of Italy, 1965-66; adj. tchr. ophthalmology U. Sassari. Mem. Italian Soc. Anatomy and Histology, Italian Soc. Electron Microscopy, Italian Soc. Ophthalmology. Research, numerous publs. on biochemistry of eye tissues, lens proteins, lysozyme, soluble macromolecules of cornea, adenovirus and eye, ultrastructure of eye tissues with particular reference to retina; discovery of filamentous structures within photoreceptors; discovery and illustration of cycle of renewal of outer segments of photoreceptors; correlated biochem. and ultrastructural studies on effects of retinotoxic substances; demonstration of functional role of retinal pigment epithelium. Home and Office: Via Mascheroni 3, Milano, Italy.*

ORZECHOWSKI, Gerhard Richard Theodor, German pharmacologist, physician; b. Rosenberg, Nov. 14, 1902; s. Waldemar and Elisabeth (Ziegler) O.; M.D., U. Breslau; m. Marianne Hundrieser, Oct. 5, 1935; children—Rolf, Marianne. Agrégé, 1934; prof. at large in pharmacology, 1939; specialist internal medicine, 1947; head internal medicine div. Nat. Hosp. in Schleswig, 1950; head sci. and clin. div. Dr. Madaus & Co., Cologne. Decorated crosses of merit both World Wars. Mem. Deutschen Pharmakologischen Gesellschaft, Gesellschaft Deutscher Naturforscher und Arzte. Collaborator with editor: Medicinia Experimentalis, also Pharmazeutischen Rundschau. Author papers on sympathicomimetica, antihistamines, antibiotics, steroids, saponine, senföle. Home: Wiehler Strasse 7. Office: Osterheimer Strasse 198, Köln-Bruck, West Germany.

OSANN, Emile, German physician; b. Weimar, Germany, 1787. Author: Description physiomedicale des sources minérales les plus connues des pays les plus favorises de l'Europe, 1839-41. First to use mineral baths scientifically. Died 1842.

OSARA, Nils Arthur, forest economist b. Ikaalinen, Finland, Nov. 29, 1903; s. Arthur G. and Emilia (von Bonsdorff) O.; B. Forestry, Helsinki U., 1924, M.F., 1926, Dr. Forestry, 1936; m. Anna Liisa Harjanne, Apr. 16, 1950; 1 dau., Anna-Kaisa Emilia. Asst., U. Helsinki, 1924-26, asso. prof., 1937-53; asst. Forest Research Inst., 1928-36, prof. forest economy, 1938-48; officer Central Forestry Ass. Tapio, 1936-38, mng. dir., 1947-52; dir. gen. Finnish State Bd. Forestry, Helsinki, 1952-60; with FAO, Rome, 1962—, dir. forestry and forest products div., 1963—. Head dept. for fuel, timber, transport Ministry of Supply, 1940-47, minister supply and agr., 1943-44. Recipient decoration from several countries. Hon. mem. Norwegian, Royal Scottish forestry socs., Soc. Am. Foresters; corr. mem. Swedish Forestry Soc.; mem. Royal Swedish Acad. Agr. and Forestry. Research on timber measurement, and trade, wood and fuel consumption, status of small-sized forest holdings, forestry adminstrn. and legislation. Home: Viale Astronomia 19, Rome. Office: FAO, Caracalla, Rome, Italy.*

OSBORN, Elburt F., Am. geologist; b. Winnebago County, Ill., Aug. 13, 1911; s. William Franklin and Anna (Sherman) O.; B.A., DePauw U., 1932; M.S., Northwestern U., 1934; Ph.D., Cal. Inst. Tech., 1937; D.Sc., Alfred U.; m. Jean Thomson, Aug. 12, 1939; children—James F., Ian C. Teaching fellow geology Northwestern U., 1932-34, instr., 1937; teaching fellow Cal. Inst. Tech., 1934-37; petrologist Geophys. Lab. Carnegie Inst. Washington, 1938-42; phys. chemist NDRC, 1942-45; research chemist Eastman Kodak Co., Rochester, N.Y., 1945-46; with Pa. State U., 1946—, prof. geochemistry, chmn. div. earth scis. Coll. Mineral Industries, 1946-52, asso. dean, 1952-53, dean, 1953-59, v.p. for research, 1959—. NSF fellow Cambridge (Eng.) U., 1958. Fellow A.A.A.S., Ceramic Soc. (pres. 1965), Geophys. Union, Geol. Soc., Mineral Soc. (pres. 1961); mem. Geol. Inst., Geochem. Soc., Pa. Research Corp., Soc. Econ. Geology (v.p. 1965), Ceramic Soc., Am. Chem. Soc., Geol. Soc. Washington, Internat. Assn. Volcanology, numerous others. Author: (with Arnulf Maun) Phase Equilibria Among Oxides in Steelmaking, 1965. Lab. research at high temperatures applicable to geol. problems and to processes and reactions in steel and ceramics industry. Home: 330 E. Irvin Av., State College, Pa. 16801. Office: Old Main Bldg., University Park, Pa. 16802.*

OSBORN, Gerald, Am. chemist; b. Tangier, Ind., Apr. 11, 1903; s. Luther and Mary Elizabeth (Hadley) O.; A.B., Eastern Michigan University, 1927. Doctor of Laws, 1965; Master of Science, U. Mich., 1929, Ph.D., 1939; m. Dorothy D. Dunlap, Aug. 25, 1927; children—James Maxwell, Arthur Hadley. High sch. tchr., prin., Homer, Mich., 1924-27; tchr. chemistry Eastern Mich. U., 1927-39; head chemistry dept. Western Mich. U., Kalamazoo, 1939-58, dean Sch. Liberal Arts, 1956—, acting pres., 1960-61; Fulbright lectr. U. S. Ednl. Found., Manila, Philippines, 1962-63; cons. Nat. Com. Sci. Edn., 1940-42. Pres. Westminister Found. Mich. Mem. Am. Chem. Soc., A.A.A.S., Central Assn. Sci. and Mathematics Tchrs., Nat. Sci. Tchrs. Assn., Mich. Schoolmasters Club, Mich. Coll. Assn. (past Pres.), Sigma Xi. Chemistry editor: School Science and Mathematics, 1951—. Research on free radicals and energy of activation. Home: 629 Campbell Av., Kalamazoo, Mich.*

OSBORN, H(enry) Fairfield, Am. paleontologist; b. Fairfield, Conn., Aug. 8, 1857; s. William Henry and Virginia Reed (Sturges) O.; A.B., Princeton, 1877, Sc.D., 1880; LL.D., Trinity, 1901, Princeton, 1902, Columbia, 1907, Union U., Schenectady, 1928; D.Sc., Cambridge U., 1904, Yale, 1923, Oxford, 1926, N.Y., 1927, Ph.D., Christiana, 1911; hon. doctorate U. Paris, 1931; m. Lucretia Thatcher Perry; children—Virginia Sturges (Mrs. Robt. Gordon McKay), Alexander Perry, Henry Fairfield, Josephine Adams (Mrs. Jay Coogan), Gurdon Saltonstall. Asst. prof. natural sci. Columbia, 1881-83, prof. comparative anatomy, 1883-90, Princeton Da Costa prof. biology, 1891-96, zoölogy, 1896-1910, research prof. zoölogy, from 1910, dean faculty pure sci., 1892-95. Curator dept. vertebrate paleontology Am. Mus. Natural History, 1891-1910, hon. curator, from 1910; recipient medals Nat. Inst. Soc. Scis., 1913; 1901-08, pres. trustees, 1908-33, hon. pres., from 1933; vertebrate paleontologist U.S. Geol. Survey, 1900-24, sr. geologist, from 1924; with Canadian Geol. Survey, 1900-04; chmn. exec. com. N.Y. Zoöl. Soc., 1896-1903, active in founding N.Y. Zoöl. Park; chmn. zoöl. and palentol. adv. coms. Carnegie Instn., Washington, 1902; elected sec. Smithsonian Instn., 1906, but declined. Elector N.Y. U. Hall of Fame, 1910; recipient medals Nat. Inst. Soc. Scis. 1913; Hayden Meml. Geol. award, 1914; Gaudry (Geol. Soc. of France), 1918; Cullom (Am. Geog. Soc.), 1919; Pasteur Inst., 1921; Société Nationale d'Acclimatation de France, 1923; Roosevelt Meml. Assn., 1923; Holland Soc., 1925; Wollaston (Geol. Soc. of London), 1926; Daniel Giraud Elliot medal, 1929. Fellow Royal Soc., 1926 (Darwin medal 1918), N.Y. Acad. Scis., Am. Geog. Soc., Am. Acad. Arts and Scis.; pres. Am. Soc. Naturalists, Am. Soc. Paleontologists, A.A.-A.S.; v.p. Am. Philos. Soc.; councilor Nat. Acad. Scis.; mem. French Acad. Scis., 1927; mem. officer many other sci. socs. Author: From the Greeks to Darwin, 1894: Evolution of Mammalian Molar Teeth, 1907; The Age of Mammals, 1910; Huxley and Education, 1910; Men of the Old Stone Age, 1915; Origin and Evolution of Life, 1917; Impressions of Great Naturalists, 1924; The Earth Speaks to Bryan, 1925; Evolution and Religion in Education, 1926; Creative Education, 1927; Man Rises to Parnassus, 1927; Fifty-two Years of Research, 1930; Cope, Master Naturalist, 1931; also 8 memoirs; over 860 scientific and educational papers. Editor: A Naturalist in the Bahamas (John I. Northrup Memorial), 1910; Fifty Years of Princeton '77, 1927. Died Nov. 6, 1935.

OSBORN, Herbert, Am. biologist; b. Lafayette, Wis., Mar. 19, 1856; s. Charles Paine and Harriet Newell (Marsh) O.; B.S., Ia. State Coll., 1879, M.S., 1880, D.Sc., 1916; LL.D., U. Pitts., 1930; Ohio State U., 1936; m. Alice Isadore Sayles, Jan. 19, 1883; children—Morse Foster, Herbert Tirrill, Evelyn, Dorothy, Margaret Stanton. Asst., Ia. State Coll., 1879-83, asst. prof., 1883-85, prof. zoology and entomology, 1885-98; entomologist, of expt. sta., 1890-98; state entomologist of Ia., 1898; prof. zoölogy and entomology Ohio State U., 1898-1916, research prof., 1916-33, emeritus, from 1933, also dir. Lake Lab., 1898-1918. Dir. Ohio Biol. Survey, from 1912; spl. agt. div. entomology U.S. Dept. Agr., 1885-94; cons. entomologist Me. Expt. Sta., from 1913; cons. entomologist Tropical Plant Research Found., 1925; collaborator U.S. Bur. Entomology, from 1930. Fellow A.A.A.S. (v.p. Sect. F, 1917); pres. Ia. Acad. Scis., 1887 (sec. and editor Proc., 1890-98), Assn. Econ. Entomologists, 1898, Ohio Acad. Scis., 1904-05, Am. Micros. Soc., 1907-09, Entomol. Soc. Am., 1911 (mng. editor Annals, 1908-28), Soc. Promotion Agrl. Science, 1917-18; fellow Cal. Acad. Sci.; mem. Internat. Entomol. Congress, Am. Soc. Naturalists, Am. Entomol. Soc., Am. Soc. Zoölogists, Biol. Soc. Washington, Société Entomologique de France, Entomol. Soc. Washington, Sigma Xi, Phi Kappa Phi, Alpha Zeta, Gamma Alpha. Author: Pediculi and Mallophaga of Man and Lower Animals, 1891; Insects Affecting Domestic Animals, 1896; The Hessian Fly in the United States, 1898; The Genus Scaphoideus, 1900; Economic Zoölogy, 1908; Agricultural Entomology, 1916; Leafhopper of Ohio, 1928; Fragments of Entomol. History, 1937, Part II, 1946; Meadow and Pasture Insects, 1939. Contbr. particles Insects, Parasitic, and Insects, Poisonous, in Wood's Reference Handbook of the Med. Sciences (new edit.), 1903, 15; Neotropic Homoptera in the Carnegie Museum, 7 parts, 1923-39; also many papers in jours. Died Sept. 20, 1954.

OSBORN, Richard Kent, Am. physicist; b. Ft. Wayne, Ind., Mar. 12, 1919; s. George Burr and Florence (Stowell) O.; B.S., Mich. State U., 1948, M.S., 1949; Ph.D., Case Inst. Tech., 1952; m. Elizabeth Rich, June 16, 1945; children—Mary R., D. Richard, David C., Ann S. Physicist, Oak Ridge Nat. Lab., 1951-57; faculty U. Mich., Ann Arbor, 1957—, prof. nuclear engring, 1959—; lectr. physics U. Tenn., Knoxville, 1951-57. Mem. Am. Phys. Soc., Am. Nuclear Soc., Am. Soc. Engring. Edn., N.Y. Acad. Scis., Phi Kappa Phi, Sigma Xi. Author: (with S. Yip) Foundations of Neutron Transport Theory; also articles. Research on meson field theory, theory beta-decay, models nuclear structure, kinetic theory radiation and particle distbns. in nuclear reactors and plasmas. Home: 1330 Ardmoor St., Ann Arbor, Mich. 48103.*

OSBORNE, Freleigh Fitz, geologist; b. Nogales, Ariz., Nov. 7, 1903; s. Walter and R.M. (Freleigh) O.; B.A.Sc., U. B.C., 1924, M.A.Sc., 1925; Ph.D., Yale, 1928; m. Agnes A. Jardine, Oct. 8, 1928; 1 son, Freleigh. Jardine Fitz. Engr., geologist Sally Mine, Beaverdell, B.C., Can., 1926; instr. U. Ia., 1928-30; prof. McGill U., 1930-47; prof. U. Laval, Cité Universitaire, Que., Can., 1947; geologist Que. Dept. Mines, Quebec, 1932—. Mem. Geo. Soc. Am. Soc. Econ. Geologists, Geol. Assn., Can., Royal Soc. Can. Editor: Geochronology in Canada, 1964; also numerous articles. Research on geology Grenville region, geology nr. Quebec City. Hime: 1700 Parc Chesnaye, Sillery, Quebec 6, Que.*

OSBORNE, Harold Smith, Am. elec. engr.; b. Fayetteville, N.Y., Aug. 1, 1887; s. Cyrus Pearl and

Ella Sophia (Smith) O.; B.S., Mass. Inst. Tech., 1908, Eng.D. (Austin Research fellow), 1910; m. Mary Agnes Wilson, Aug. 14, 1918 (dec. Jan. 1932); children—Margaret Ellen (Mrs. Hugh J. McLane), Mary Agnes Wilson (Mrs. Perry Dunlap Smith, Jr.); m. 2d, Dorothy Brockway, Mar. 24, 1938. With Am. Tel. & Tel. Co., N.Y.C., 1910-52, plant engr., 1940-42, asst. chief engr., 1942-43, chief engr., 1943-52; cons. in telecommunications, S. Am., Central Am., 1952——. Mem. various coms. govt. agys. Recipient numerous awards, including Howard Coonley medal Am. Standards Assn., 1956, 75th Anniversary medal Am. Soc. M.E., 1956, medal Acad. Achievement, 1961, Silver Medal award Am. Soc. Planning Ofcls., 1962. Fellow I.E.E.E., Acoustical Soc. Am., Am. Phys. Soc., A.A.A.S., Standards Engrs. Soc. (hon.); mem. Am. Soc. for Engring. Edn., Montclair Soc. Engrs. (hon.), Inst. Elec. Communications Engrs. Japan (hon.), Eta Kappa Nu, Tau Beta Pi (hon.). Active part in maj. devels. of telephone service, including transcontinental telephone service, new systems and facilities for long distance telephone and telegraph service, application of radio to Bell system services, overseas radio telephony, radio broadcast and tv networks, direct distance dialing throughout U.S. Address: 375 Highland Av., Upper Montclair, N.J. 07043.*

OSBORNE, Louis Shreve, physicist; b. Rome, Italy, Sept. 8, 1923; s. Algeron Ashburner and Marjorie (Adams) O.; S.B., Cal. Inst. Tech., 1944; Ph.D., Mass. Inst. Tech., 1950; m. Helen I. Innes, June 6, 1950; children—Marc A., Brian I., Duncan T. With Mass. Inst. Tech., Cambridge, 1948——, prof. physics, 1963——. Guggenheim fellow, Fulbright fellow, 1959-60. Mem. Am. Phys. Soc., Am. Acad. Arts and Scis. Research, publs. on high energy particle reactions in cosmic rays using nuclear emulsions, photoprodn. mesons at energies below 300 million electron volts and up to 6 billion electron volts. Home: 52 Peacock Farm Rd., Lexington, Mass. 02173. Office: Mass. Inst. Tech., Cambridge, Mass. 02139.*

OSBORNE, Thomas Burr, Am. biochemist; b. New Haven, Aug. 5, 1859; s. Arthur D. and Frances (Blake) O.; A.B., Yale, 1881; Ph.D., 1885; hon. Sc.D., 1910; m. Elizabeth Annah Johnson, 1886; children—Arthur Dimon, Francis Blake. Research chemist Conn. Agr. Expt. Sta., 1886; research asso. Carnegie Instn., Washington, Yale; dir. 2d Nat. Bank New Haven. Recipient Gold medal Paris, 1900, John Scott medal, 1922. Fellow Am. Acad. Arts and Scis.; mem. Am. Soc. Biol. Chemists (became pres. 1910). Author: Proteins and the Wheat Kernel, 1907; The Vegetable Proteins, 1909; also numerous articles on chemistry, nutritive value of vegetable proteins. Asso. editor Jour. Biol. Chemistry. Discovered vitamins A and D. in cod liver oil; authority on nutrition. Died Jan. 29, 1929.

OSBOURN, Raymond Allen, Am. physician; b. Hyattsville, Md., Sept. 6, 1911; s. Herbert Allen and Emma Irene (Boswell) O.; B.S., Georgetown U., 1933, M.D., 1940; postgrad. Duke, U. Pa.; m. Dorothea E. Schumann, Apr. 22, 1942; children—Raymond Voll, Dorothea Elise, Ruth Anne, Mary Theresa. Fellow in dermatology Georgetown U. Med. Center, Washington, 1945-47; practice medicine specializing in dermatology, Washington, 1947——; cons. in dermatology Walter Reed Army Hosp., Glenn Dale Tb Sanitarium, Mt. Alto VA Hosp.; asso. clin. prof. dermatology Georgetown U., 1954——. Recipient Brennan medal Georgetown U., 1933; Alumni Achievement award Georgetown Alumni D.C., 1960. Diplomate Am. Bd. Dermatology. Fellow A.M.A., Am. Acad. Dermatology; mem. Georgetown Clin. Soc. (pres. 1955-56), Georgetown Med. Alumni Assn. (pres. 1953-55). Author: Syllabus of Medical Mycology, 1954; also articles. Research on embryology of skin; co-inventor, patentee insulated test patch. Home: 8204 Kerry Rd., Chevy Chase, Md. 20015. Office: 1835 Eye St., N.W., Washington 20006.*

OSEBOLD, John William, Am. immunologist; b. Great Falls, Mont., Jan. 9, 1921; s. Anthony J. and Rosella (Le Vert) O.; B.A., Wash. State U., 1943, D.V.M., 1944; M.S. Ore. State U. 1951; Ph.D., U. Cal. at Davis, 1953; m. Ginevra Huffman, Jan. 29, 1944; children—William R., Michael J. Fed. veterinarian in charge brucellosis diagnostic lab. U.S. Dept. Agr., 1944-46; practice vet. medicine, Corvallis, Ore., 1946-49; instr. vet. medicine and vet. microbiology Ore. State U., 1949-50; faculty U. Cal. at Davis, 1950——, prof. immunology, 1962——, chmn. dept. vet. microbiology, 1965——. NIH grantee, 1956——. Mem. Am. Vet. Med. Assn, Am. Assn. Immunologists, Am. Soc. for Microbiology, Reticulo-Endothelial Soc., Sigma Xi, Phi Kappa Phi. Contbg. author: Reproduction in Domestic Animals, 1968; Introduction to Livestock Production, 1966; also numerous articles. Research on dynamics infection and resistance in diseases affecting man and animals, antibody response, cellular immunity, serology, serum prophylaxis, immunization.

OSELLADORE, Guido, Italian surgeon; b. Chioggia, Italy, July 2, 1894; s. Domenico and Elviro (Pagan) O.; M.D. Degree; m. Luisa Panajotti; children—Domenico, Elvira. Dir. Surg. Clinic, U. Milan (Italy). Mem. Internat. Coll. Surgeons (pres. Italian div.), Italian Surgery Soc., Lombardy Soc. Surgery. Author

numerous papers and reports to congress. Home: via Revere 2. Office: via F. Sforza 35, Milan, Italy.

OSGOOD, Edwin Eugene, Am. biochemist, med. educator; b. Fall River, Mass., Jan. 25, 1899; s. William Pleasants andLydia Lee (Smith) O.; student McMinville (now Linfield) Coll., 1916; B.A., U. Ore., 1923, M.A., M.D., 1924; grad. study, Mayo Clinic, Rochester, Minn., 1923, 26 U. Vienna, 1927-28, Basel, Freiburg, London, 1928; m. Mable Maru Wilhelm, May 30, 1934; children—Barbara Delight, Beverly Maru, Edwin Boyd, Brenda Gay, Beatrice Joy. Asst. in biochemistry, U. Ore., 1919-21, instr. in biochemistry, 1921-25, asso. in same, 1925-28, asso. in medicine, 1925-29, asst. prof. biochemistry, 1928-33, asst. prof. medicine, 1929-39, asso. prof., 1939-47, prof. 1947——, dir. Med. Sch. Labs., 1928-36; staff mem. Multinomah Co. and Doernbecher hosps. 1928——, head div. exptl. medicine, 1936-64, asso. head, 1964——. Recipient bronze medal sci. exhibit A.M.A., 1929, hon. mention, 1934, certificate merit, 1938. Distinguished Achievement award, Modern Medicine mag., 1957; U. Ore. Med. School Alumni Assn. meritorious achievement award, 1962; Gov.'s N.W. Scientist award for research in leukemia and Osgood growth prediction charts, 1962; N.W. Sci. award for unraveling human chromosome Ore. Mus. Sci. and Industry, 1963; Robert Roesler de Villiers award for research in leukemia, 1963. Master A.C.P., Fellow Internat. Soc. Hematology (councilor U. S. 1950-52), N.Y. Acad. Sci.; mem. A.M.A., Am. Heart Assn., Pacific Interurban Clin. Club, Soc. Exptl. Biology and Medicine, N. Pacific Soc. Internists (pres. 1950-51), Soc. Clin. Investigation, Western Assn. Physicians (v.p., 1958-59), Am. Soc. Hematology (v.p., 1958-59), Sigma Xi, Alpha Kappa Kappa, Alpha Omega Alpha. Author: Textbook of Laboratory Diagnosis, 1931 (3d edit. 1940). Co-author of Atlas of Hematology, 1937. Contbr. to Jour. A.M.A., Jour. Lab. and Clin. Medicine, Archives Internal Medicine, etc. Originator of method of culture of human marrow; developed method to keep human blood cells living over 10 yrs. in culture; Alpha-N concept cell div., cancer and aging. Home: 2157 N.E. 28th Av., Portland, Ore. 97212.*

OSGOOD, Robert Bayley, Am. surgeon; b. Salem, Mass., July 6, 1873; s. John C. and Martha E. (Whipple) O.; A.B., Amherst Coll., 1895, Sc.D., 1935; M.D., Harvard, 1899; m. Margaret Chapin, Apr. 29, 1902; 1 dau., Ellen. Practiced medicine, Boston from 1899; chief of staff of orthopedic dept. Mass. Gen. Hosp. and Children's Hosp.; John B. and Buckminster Brown prof. orthopedic surgery Harvard Med. Sch. Mem. adv. com., services for crippled children Children's Bur., Department Labor, Washington (chmn. 1938-40); mem. med. adv. bd. Armour Found. and Alfred I. Du Pont Inst., 1937-46. Fellow A.C.S., Royal Coll. Surgeons (hon., Eng.); mem. A.M.A., Mass. Med. Soc., Am. Orthopedic Assn. (pres. 1920-21), Boston, N.E. (pres. 1928-29) surg. socs.; Nat. Inst. Social Sci., Am. Acad. Arts and Scis.; corr. mem. Belgian Orthopedic Assn., Internat. Soc. Orthopedic Surgery, Lique Internationale Contre le Rhumatisme; hon. mem. Brit., Italian, Scandanavian, Australian orthopedic assns., Royal Soc. Medicine of Eng. Author: Diseases of the Bones and Joints (with J. E. Goldthwait and C. F. Painter), 1909; Fundamentals of Orthopedic Surgery in General Medicine and Surgery (with Nathaniel Allison), 1931; The Medical and Orthopedic Management of Chronic Arthritis (with Ralph Pemberton), 1934). Described painful lesions of tibial tuberosity in children and adolescents (osteochondrosis of tuberosity of tibia or Osgood-Schlatter disease), 1903. Died Oct. 2, 1956.

OSGOOD, Thomas Harris, physicist; b. Louth, Eng., Apr. 6, 1900; s. Frederick W. and Annie (Harris) O.; M.A., St. Andrews U. (Scotland), 1921, B.Sc., 1923; M.S. (Commonwealth Fund fellow), U. Chgo., 1927; Ph.D., U. Cambridge (Eng.), 1927; m. Dorothy Ewart Marston, Aug. 4, 1928; children—Elizabeth Harris (Mrs. Lewis Eric Russell), Thomas Marston. Came to U.S., 1928, naturalized, 1940. Asst. lectr. U. Manchester (Eng.), 1927-28; asst. prof. U. Pitts., 1928-33; prof. U. Toledo, 1934-41; war research Columbia, 1942-43; prof., head dept. physics Mich. State U., East Lansing, 1941-50, dir. div. math. and phys. scis., 1946-59, dean Grad. Sch., 1950-59, prof. physics, 1961——, dir. Abrams Planetarium, 1964——. Cons. Gen. Motors Corp., 1956-58; sci. attache U.S. Embassy, London, Eng., 1959-61. Fellow Am. Phys. Soc.; mem. Am. Inst. Physics, Am. Assn. Physics Tchrs., Sigma Xi, Sigma Chi. Author: (with others) An Outline of Atomic Physics, 1933, 3d edit., 1955; Atoms, Radiation and Nuclei, 1964. Editor Am. Jour. Physics, 1948-59. Research, publs. on nuclear physics, x-rays, double-stars. Home: 720 N. Harrison Rd., East Lansing, Mich. 48823.*

OSGOOD, Wilfred Hudson, Am. naturalist; b. Rochester, N.H., Dec. 8, 1875; s. Marion Hudson and Harriet Amanda O.; A.B., Stanford, 1899; Ph.D., U. Chgo., 1918. Biologist, U.S. Dept. Agr., 1897-1909; in charge U.S. biol. investigation in Alaska, 1899-1909; asst. curator mammalogy and ornithology Field Mus. Natural History, Chgo., 1909-21, curator zoölogy, 1921-40. Conducted biol. explorations in Alaska, Can., many parts U.S., Venezuela, Peru, Chile, Argentine, Brazil, Ethiopia, Indo-China; studied in European museums, 1906, 10, 30; spl. U.S. investigator fur-seal question, 1914; leader Field Museum

Abyssinian Expdn., 1926-27, Magellanic Expdn., 1939-40. Fellow A.A.A.S., Am. Ornithologists Union; founder and 1st pres. Cooper Ornithol. Club Cal.; sec. Biol. Soc. Washington, 1900-09; corr. mem. London Zool. Soc., Brit. Ornithol. Union; mem. Am. Soc. Mammalogists (pres. 1924-26), Chgo. Zool. Soc., Geog. Soc. Chgo., NRC (biology div.). Author: Revision of Pocket Mice, 1900; Revision of Mice of Genus Peromyscus, 1909; Biological Investigations Alaska and Yukon, 1909; Fur Seals of Pribilof Islands, 1915 (joint author); Monographic Study of Caenolestes, 1921; Mammals of Asiatic Expeditions, 1932; Artist and Naturalist in Ethiopia, 1936 (joint author); Mammals of Chile, 1943; and about 180 shorter papers on classification, anatomy, and habits of mammals and birds. Contbr. zool. definitions to Webster's New Internat. Dictionary. Died June 20, 1947.

OSGOOD, William Fogg, Am. mathematician; b. Boston, Mar. 10, 1864; s. William and Mary Rogers (Gannett) O.; A.B., Harvard, 1886, A.M., 1887; postgrad. Göttingen (Germany); Ph.D., U. Erlangen (Germany), 1890; LL.D., Clark U., 1909; m. Therese Ruprecht, July 17, 1890; children—William Ruprecht, Frieda Bertha (Mrs. Walter Silz), Rudolf Ruprecht; m. 2d, Mrs. Céleste Phelps Morse, Aug. 19, 1932. Instr. Harvard, 1890-93, asst. prof. math., 1893-1903, prof., 1903-33, emeritus; prof. math. Nat. U. Peking (China), 1934-36. Mem. Internat. Commn. on Teaching of Math. Mem. Nat. Acad. Scis., Am. Philos. Soc., Am. Math. Soc. (pres. 1904-05), Deutsche Mathematiker - Vereinigung, Leopoldinisch-Carolinisch Deutsche Akademie der Naturforscher, Circolo Matematico de Palermo; corr. mem. Math. Soc. of Charkow, Göttinger Gesellschaft der Wissenschaften; hon. mem. Calcutta Math. Soc.; mem. Phi Beta Kappa. Author: Introduction to Infinite Series, 3d edit., 1906; Lehrbuch der Funktionentheorie, 1905-07; First Course in Differential and Integral Calculus, 1907; Madison Colloquium Lectures, 1914; Analytic Geometry (with W. C. Graustein), 1921; Advanced Calculus, 1925; Functions of Real Variables, 1936; Functions of a Complex Variable, 1936; Mechanics, 1937; also monographs in math. jours. Editor Annals of Math., 1899-1902, Trans. Am. Math. Soc., 1900-10. Died July 22, 1943.

OSIANDER, Andreas (original name, Hosemann), German philosopher, theologian, amateur of science; b. Gunzenhausen, Bavaria, Dec. 19, 1948; studied at Leipzig, Altenburg, Ingolstadt; m. 1525. Ordained priest, 1520; apptd. Hebrew tutor in Augustinian convent at Nuremberg; publicly joined Lutheran party, circa 1522; apptd. preacher, St. Lorenz Kirche; attended Marburg Conference, 1529, Augsburg diet, 1530, signing of Schmalkald articles, 1547; left Nuremberg, 1548, with introduction of Augsburg Interim; went to Breslau, then Königsberg, where he became preacher and 1st prof. of theology at Königsberg U. Author: A Harmony of the Gospels, 1537; De lege et evangelio, 1550; De justificatione, 1550. Osiander, who was entrusted with supervising the printing of Copernicus' De revolutionibus, suggested to Copernicus that he present his heliostatic system as merely a more convenient calculating device (not as true picture of universe). Copernicus refused, but Osiander suppressed Copernicus' own introduction and added an unsigned preface containing his own positivistic notions, thereby causing for a long time a great deal of confusion among Copernicus' readers as to what his real views were. Died Königsberg, East Prussia, Oct. 15, 1552.

OSIANDER, Friedrich Benjamin, German physician; b. Germany, 1759; student, Tübingen, Germany, 1775-79, Strasbourg (now in France), 1 semester, Cassel, Germany, under G.W. Stein, 1781; at least 1 son, Johan Friedrich; practiced medicine; became prof. medicine and obstetrics, Göttingen, 1792, also dir. clinics; mem. Sociatät der Wissenschaften, Hanoverian Hofrath. Author: Handbuch der Entbindungskunst, 1818-21. Introduced lower-segment method of caesarean section, circa 1812. Died 1822.

OSINSKI, Joseph Herman, Polish plant physiologist; b. Dobrzykowo, Poland, Mar. 4, 1738; tchr. math., physics, philosophy Warsaw schs.; organized 1st chem. lab. in Warsaw. Author: Physics, Experimentally Confirmed . . ., 1777; Varieties of Air, 1783. Used green plants to study properties of gases; showed that animal respiration causes increased concentration of carbon dioxide in enclosed areas, also that carbon dioxide cannot support a flame or life; developed methods for purifying air from carbon dioxide. Died Warsaw, Mar. 13, 1802.

OSLER, William, physician; b. Bondhead, Ont., Canada, July 12, 1849; s. Featherstone Lake and Ellen Frere (Pickton) O.; ed. Trinity Coll., Toronto, 1868; M.D., McGill U., 1872; postgrad. Univ. Coll., London, univs. Berlin and Vienna, 1872-74; LL.D., McGill, 1895, Aberdeen, 1898, Toronto, 1899, Yale, 1901, Harvard, 1904, Johns Hopkins, 1905, Edinburgh, 1898; D.C.L., Trinity U., Toronto, 1902, Durham, 1913; Sc.D., Oxford, 1904, Liverpool, 1910, Dublin, 1912; m. Grace Linzee (Revere) Gross, May 1892; one son, Edward. Prof. medicine McGill U., 1874-84; prof. clin. medicine U. Pa., 1884-89; prof. principles and practice of medicine Johns Hopkins, 1889-1905; hon. prof. medicine, from 1905; physician-in-chief Johns Hopkins Hosp., 1889-1905; Regius prof. medicine Oxford U., from 1905. Student (fellow) of Christ Church Coll., Oxford. Fellow Royal Soc., 1898, also Royal Coll. Physicians. Author: The Cerebral Palsies of Children, 1889; Chorea and Choreiform Affections, 1894, Lectures on Abnormal Tumors, 1895; Angina Pectoris and Allied States, 1897; The Principles and Practice of Medicine, 1892; Cancer of the Stomach, 1900; Science and Immortality (Ingersoll lecture Harvard U.), 1904; Aequanimitas and Other Addresses, 1900; Counsels and Ideals, 1906; An Alabama Student and Other Essays, 1908; The Evolution of Modern Medicine, 1921; Incunabula Medica: a Study of the Earliest Printed Medical Books, 1467-80, 1923. Editor of System of Medicine. Renowned as physician and medical historian; numerous medical observations include those on blood platelets and abnormally high red blood cell count in polycythemia; made special study of angina pectoris; authority on spleen and heart. Died, Oxford, Eng., Dec. 29, 1919.

OSMOND, Floris, French metallurgist; b. Paris, 1849. Worked in Creusot Labs. Author: Théorie cellulaire des propriétés de l'acier, 1894. Studied tempering of steel, establishing existence of points of transformation and detailing constituents of iron products; can be considered founder of metallography; devised 2 main procedures of investigation, microscopic test method for alloys (with Sorby) and method of micrographic analysis of steel micro-sections. Died Saint-Lieu, France, 1912.

OSMOND, Humphry Fortescue, psychiatrist; b. Milford, Surrey, Eng., July 1, 1917; s. George William and Dorothy (Gray) O.; student Guys Hosp. Med. Sch., London, Eng., 1936-42; m. Amy Edith Roffey, Nov. 12, 1947; children—Helen Lavinia, Euphemia Janet, Julian Fortescue. Practice medicine, specializing in psychiatry, Weyburn, Sask., Can., 1951-61, Princeton, N.J., 1963—; clin. dir. Sask. Hosp., 1951-53, dir., 1953-61; dir. research Bur. Research Neurology and Psychiatry N.J., 1963—; cons. mental hosp. design Am. Psychiat. Assn., Nat. Inst. Mental Health, States of Pa., N.J., Fla., N.Y., Provinces of Ont., Sask. Mem. Royal Medico Psychol. Assn., Am. Psychiat. Assn., Group For Advancement Psychiatry, Collegium Internationale Neuropsychopharmacologtum. Author: (with A. Hoffer) Chemical Basis Clinical Psychiatry, 1960; How to Live With Schizophrenia, 1966; also numerous articles. Publs. on devel. of biochem. approach to schizophrenia; studies of psychotomimetic properties of some derivatives of adrenalin; extensive research on mescaline in which Aldous Huxley participated. Home: Box 1000, Princeton, N.J., 08540.*

OSOL, Arthur, chemist; b. Riga, Latvia, Dec. 1, 1905; s. Peter and Caroline (Irbit) O.; came to U.S., 1906, naturalized, 1915; Ph.G., Phila. Coll. Pharmacy and Sci., 1925, B.S. in Chemistry, 1928; M.Sc. in Chemistry, U. Pa., 1931, Ph.D., 1933; LL.D., Eastern Bapt. Coll., 1964; m. Amelia Virginia Lebo, Dec. 28, 1928. Faculty Phila. Coll. Pharmacy and Sci., 1928—. prof. chemistry, 1937—, pres., 1963—. Vice pres., dir. W. Phila. (University City) Corp. 1963—; dir. University City Sci. Center, 1964—, Phila. Sci. Council, 1950—; chief, chem. br. Phila. Tech. Def. Div., 1951-54. Bd. govs. Chem. Ednl. Projects. Recipient Honor scroll Pa. chpt. Am. Inst. Chemists, 1954; Ann. award Alumni Assn. Phila. Coll. Pharmacy and Sci., 1962; Pharmacy Achievement award Am. Pharm. Assn. Phila., 1964. Mem. Sigma Xi, Rho Chi, Kappa Psi. Author, editor: U.S. Dispensatory, 1937—; Blakiston Med. Dictionaries, 1947—; Remington's Pharm. Scis., 1936—. Contbr. numerous articles to sci. publs. As chmn. subcom. U.S. Pharmacopeia Revision Com. introduced infrared and other spectrophotometric and nonaqueous analytical methods for drugs, 1950—. Home: 128 Colwyn Lane, Bala-Cynwyd, Pa. 19004. Office: 43d St. and Kingsessing Av., Phila. 19104.*

OSOVETS, Samuil Markovich, Russian physicist; b. 1911; grad. Kharkov Electro-tech. Inst., 1935; candidate's degree in tech. scis., 1946; engr., then dir. research lab., electro-tech. factory, Kharkov, 1935-41; with All-Union Electro-tech. Inst., 1944-48, Inst. Atomic Energy, Acad. Scis. USSR, 1948—. Recipient Lenin prize, 1958, for research on powerful impulse discharges in a gas to obtain high-temperature plasma. Address: AN SSSR, Leninsky prosp. 14, Moscow, USSR.

OSSERMAN, Elliott Frederick, Am. physician; b. N.Y.C., Aug. 1, 1925; s. Hyman A. and Elizabeth (Bauman) O.; B.A., Columbia U., 1945, M.D., 1947; m. Nancy Smith Kringel, Oct. 16, 1949; children—Pamela Beth, Douglas Kringel, Deborah Ray. With Columbia U. Coll. Phys. and Surg., N.Y.C., 1953—, asso. prof. medicine, 1962—; staff Francis Delafield Hosp., 1953—, vis. physician, 1959—; staff Presbyn. Hosp., 1956—, asso. attending physician, 1963—; vis. investigator Service de Chimie Microbienne, Institut Pasteur, Paris, 1958; faculty research asso. Am. Cancer Soc., 1963—. Mem. cancer chemotherapy study sect. NIH, USPHS, 1963—. Recipient Joseph Mathers Smith prize, 1957. Diplomate Am. Bd. Internal Medicine, Nat. Bd. Med. Examiners. Mem. A.A.-A.S., Harvey Soc., Am. Assn. for Cancer Research, Am. Assn. Immunologists, Am. Soc. for Clin. Investigation, Internat. Soc. Hematology, Am. Soc. Hematology. Research and publs. on structure of serum gamma globulins; immunochem., biochem. and clin. studies of plasma cell dyscrasias; discovered lysozymuria in moncytic leukemia. Home: 150 Columbus Av., Closter, N.J. 07624. Office: Francis Delafield Hosp., 99 Ft. Washington Av., N.Y.C. 10032.

OSSERMAN, Robert, Am. mathematician; b. N.Y.C., Dec. 19, 1926; s. Herman Aaron and Charlotte (Adler) O.; B.A., N.Y. U., 1946; postgrad. U. Zurich, U. Paris; M.A., Harvard, 1948, Ph.D., 1955; m. Maria Anderson, June 15, 1952; 1 son, Paul. Teaching fellow Harvard, 1949-52, vis. lectr., research asso., 1961-62; instr. U. Colo., 1952-53; faculty Stanford, 1955—, prof., 1966—; mem. N.Y. U. Inst. Math. Scis., 1957-58; head math. br. Office Naval Research, 1960-61; Fulbright lectr. U. Paris, 1965-66. Mem. Am. Math. Soc. Author: Two-Dimensional Calculus, 1968. Research, publs. on differential geometry, complex variables, differential equations, especially border-line areas; discovered number of new properties of mineral surfaces, surface formed physically by soap films. Home: 680 Salvatierra St., Stanford, Cal. 94305.*

OSSIANNILSSON, Frej, Swedish entomologist; b. Halsingborg, Sweden, Dec. 8, 1908; s. Karl and Naemi (Arnman) O.; F.K., U. Lund (Sweden), 1935, F.M., 1936, F.L., 1943, F.D., 1949; m. Astrid Ingegärd Hildur Elisabeth Malmström, May 1, 1938; 1 dau., Maj. Vindeland. Amanuensis, Zool. Inst., Lund, 1938-40; with Swedish State Inst. Plant Protection, 1940-49, chief demonstrator, 1948-49; faculty Agrl. Coll. Sweden, Uppsala, 1949—, prof., 1963—. Decorated knight Order of North Star. Mem. entomol. socs. Stockholm, Lund and Helsingfors, Nordiska Jordbruksforskares Förening, Ecol. Soc. Am. (asso.). Author: Svensk insektfauna; Hemiptera Homoptera Auchenoorhyuncha, 1946-47; also numerous articles. Established that sound-producing apparatus of cicadas is also present in smaller relatives such as leafhoppers and studied calls of latter; taxonomy and distbn. of Swedish hemiptera and role of some as plant virus vectors. Home: 9 Källparksgatan, Uppsala, Sweden.*

O'STEEN, Wendall Keith, Am. neuroanatomist; b. Meigs, Ga., July 3, 1928; s. Wellna Hubert and Lillian (Powell) O'S.; B.A., Emory U., 1948, M.S., 1950; Ph.D. (So. Fellowship scholar), Duke, 1958; m. Mildred Virginia Reichert, Nov. 22, 1951; children—Lisa Diane, Kerry Keith. Asst. prof. biology Emory Jr. Coll., Valdosta, Ga., 1948-49; instr. biology Emory U., Atlanta, 1950-51; asst. prof. biology Wofford Coll., Spartanburg, S.C., 1951-55; chief biologist, pathology br. U.S. Army Chem. Labs., 1953-55; faculty Med. Br., U. Tex., Galveston, 1958-67, asso. prof., 1964-67; faculty Sch. Medicine, Emory U., Atlanta, 1967—, vis. lectr. U. Miami Sch. Medicine, 1963. Asst. dir. Nat. Inst. Mental Health Med. Student Research Tng. program U. Tex., 1958-67. Mem. Am. Assn. Anatomists, Soc. Am. Zoologists, A.A.A.S., A.M.A., Tex. Acad. Sci., Am. Assn. U. Profs., Sigma Xi, Phi Sigma. Research, publs. on histological and radioautographic study of amphibian regeneration with spl. emphasis on intestinal regeneration, growth of tissue and organs in diffusion chambers, spl. emphasis on normal and dystrophic muscle, central nervous system histochemistry, neuroendocrine effects of biogenic amines. Home: 1745 Ange lique Dr. N.E., Decatur, Ga. 30033. Office: Dept. Anatomy, Emory U., Atlanta 30322.*

OSTER, Ludwig Friedrich, astrophysicist; b. Konstanz, West Germany, Mar. 8, 1931; s. Ludwig F. and Emma (Schwarz) O.; Dipl.phys., U. Freiburg (West Germany), 1954; Dr.rer.nat., U. Kiel (West Germany), 1956; m. Rose-Marie Gunhild Hagetorn, Mar. 29, 1956; children—Ulrika L., Mattias T. Came to U.S., 1958, naturalized, 1964. Research fellow U. Kiel, 1956-58; research asst. physics dept. Yale, 1958-60, faculty, 1960—, asso. prof. astronomy dept., 1965—; vis prof. Bonn (Germany) U., 1966; vis. fellow Joint Inst. for Lab. Astrophysics U. Colo., 1966-67. Cons. Republic Aviation Corp., 1960-61, Boeing Co., 1962-65. Mem. Am. Phys. Soc., Am., German astron. socs., Internat. Astron. Union, Sigma Xi. Sci. editor Scripta Technica, 1960—. Research, publs. in astrophysics especially solar physics and theoretical radioastronomy, plasma physics especially radiation theory. Home: 415 Sunnyside Lane, Boulder, Colo. 80302.*

OSTERBERG, Charles Lamar, Am. oceanographer; b. Miami, Ariz., June 15, 1920; s. Arthur Edward and Grace (Johnson) O.; B.S., Ariz. State Coll., 1948, M.A., 1949; postgrad. Purdue U., U. Wash.; M.S., Ore. State U., 1960, Ph.D., 1962; m. Betty Avonne Peltier, Nov. 15, 1945; children—Cheryl Ann, David Arthur, John Charles. Research asst. Lowell Obs., Flagstaff, Ariz., 1949-53; research asst. Atmospheric Research Obs., Flagstaff, 1953-56; tchr. Flagstaff High Sch., 1956-59; NSF fellow Ore. State U., Corvallis, 1959-60, USPHS fellow, 1960-62, faculty, 1962—, prof. oceanography, 1966-67; marine biologist U.S. AEC, Washington, 1967—. Vis. lectr. Am. Geophys. Union, 1964. Mem. Am. Soc. Limnology and Oceanography (pub. relations com. 1965—), A.A.A.S., Sigma Xi, Phi Sigma. Research, publs. on role zooplankton in transport of fallout into depths of ocean; used radionuclide content to follow Columbia River water far out to sea and to determine rate of transport. Home: 9312 Gue Rd., Damascus, Md. 20750. Office: Div. Biology and Medicine, U.S. AEC, Washington.*

OSTERGAARD, Erling, Danish gynecologist; b. Copenhapen, Denmark, Nov. 9, 1907; s. Hans Christian and Camilla (Christeensen) O.; M.D., U. Copenhagen; m. Lilith Nicole Scheel, Aug. 19, 1938; children—Ole, Anne Marie, Claus. Asst. gynecol. and obstet. service, 1933-49; asst. dept. hormones Serum Inst., 1936-40; head State Gynecol. Dept., also head obstetrics Frederiksberg Hosp., Copenhagen, 1949—. Mem. Danish Obstet. and Gynecol. Soc. (pres.); hon. mem. French Gynecol. Soc., Royal Soc. Medicine (London). Author: Les Substances antigonadotropes. Research on gynecology, especially cancer of uterus and endocrinology. Home: Mathilde Fibigersvej 20. Office: Frederiksberg Hosp., Copenhagen, Denmark.

OSTERHOUT, Winthrop John Vanleuven, Am. physiologist; b. Bklyn., Aug. 2, 1871; s. John Vanleuven and Annie Loranthe (Beman) O.; A.B., Brown U., 1893, A.M., 1894, Ph.D., 1926; student U. Bonn, 1895-96; Ph.D., U. Cal., 1899; Sc.D. (hon.), Harvard, 1925; m. Anna Marie Landstrom, June 17, 1899; children—Anna Maria (Mrs. Theodore M. Edison), Olga (Mrs. Harold B. Sears); m. 2d, Marian Irwin, Feb. 27, 1933. Instr. botany Brown U., 1893-95, Woods Hole, Mass., 1894-95; instr. botany U. Cal., 1896-1901, asst. prof., 1901-08, asso. prof., 1908-09; asst. prof. botany Harvard, 1909-13, prof., 1913-25; mem. Rockefeller Inst. Med. Research, 1925-39, mem. emeritus, 1939—. Trustee Marine Biol. Lab., Woods Hole. Fellow A.A.A.S.; mem. Nat. Acad. Sci., Am. Philos. Soc., Bot. Soc. Am., Am. Physiol. Soc, Soc. Exptl. Biology and Medicine, Am. Chem. Soc., Am. Soc. Naturalists, Am. Acad. Arts and Scis., Washington, N.Y. acads. scis., Bot. Soc. Edinburgh, Kungliga Fysiog. Sallskapet Lund, Leopold-Carolin Deutsche Akad. d. Naturforscher (Halle). Author: Experiments with Plants, 1905; Nature of Life, 1924; others. Co-editor Jour. Gen. Physiology, 1919—. Research in plant physiology; photosynthesis, fertilization, permeability, spindle formation, antagonism and electrical conductivity. Died Apr. 1964.

OSTERTAG, Berthold, German psychiatrist; b. Berlin, Germany, Feb. 28, 1895; s. Robert and Margarete (Hertwig) O.; ed. univs. Tübingen and Berlin; M.D.; m. Ilse Kobel, 1924; children—Christel, Lisa, Bernhart. Asst. Tübingen Inst. Pathology, Berlin Inst. Pathology; asst., head div. neuropathology Univ. Clinic of Neurology of Berlin; 1st asst. Inst. Neuropathology, Buch Psychiat. Sanatarium; dir. Pathology Inst., Rudolf Virchow Hosp., 1934; lectr., pof. pathology U. Berlin; instr. U. Tübingen until 1945, asso. prof., 1959—. Mem. Soc. Pathology, Neurology, Psychiatry and Neuro-pathology. Author: Einteilung und Charakteristik der Hirngewäsche, 1936; Path. der raumfordernden Prozesse des Schädelbinnenraums, 1941; Die Sektion des Gehirns und des Rückenmarks und ihrer Hülen, 1944; Körpeliche Erbkrankeiten, Ihre Pathlogische und Differentialddiagnose, 1940; also articles. Research on the brain. Home: Hornschstrasse 4. Office: Calwerstrasse 3, Tübingen, West Germany.

OSTERWALD, Frank William, Am. geologist; b. Casper, Wyo., Feb. 11, 1922; s. George R. and Blanche (Luckey) O.; B.A., U. Wyo., 1944, M.A., 1947; Ph.D., U. Chgo., 1951; m. Doris Ray Beck, Sept. 3, 1947; children—Ray William, Frank Edward, Carl Robert, Becky Ann. Asst. geologist Geol. Survey Wyo., Laramie, 1947-52; asst. prof. geology U. Wyo., 1948-52; geologist Mineral Deposits br. U.S. Geol. Survey, Denver, 1952-58, Engring. Geology br., 1958—. Fellow Geol. Soc. Am.; mem. Am. Assn. Petroleum Geologists, Soc. Econ. Geologists, Assn. Engring. Geologists, Am. Inst. Profl. Geologists. Author: (with Doris B. Osterwald) Wyoming Mineral Resources, 1952; (with G.W. Walker) Geology of Uranium Bearing Veins in the United States, 1963; also articles. Analysis of structural control of uranium dists. to enable discovery of new dists.; devel. of principles to use in applying geology to make mining safer and more economical; discovery of principles and techniques to forecast some coal-mine bumps in Utah coal fields. Home: 40 S. Dover St., Lakewood, Colo. 80226. Office: U.S. Geol. Survey, Bldg. 25 Fed. Center, Denver 80225.*

OSTFELD, Adrian Michael, Am. physician; b. St. Louis, Sept. 2, 1926; s. Simon and Margaret (Fisman) O.; student Washington U., 1943-44, 46-47, M.D., cum laude, 1951; m. Ruth Vogel, Dec. 31, 1950; children—Barbara, Richard, Robin. Instr., Cornell U. Med. Coll., 1955-56; faculty U. Ill. Coll. Medicine, Chgo., 1956—, prof. preventive medicine, 1963—, head dept., 1966—, staff Research and Edn. Hosps., 1956—. Spl. cons. Surgeon Gen., USPHS, 1964—; cons. council on drugs A.M.A., 1964—. Mem. Soc. for Clin. Investigation, N.Y. Acad. Scis., Royal Soc. Medicine, Am. Soc. for Pharmacology and Exptl. Therapeutics, Am. Pub. Health Assn., Am. Assn. U. Profs., Sigma Xi, Alpha Omega Alpha. Author: The Common Headache Syndromes, 1962; also numerous articles. Research on relevance social and psychol. factors in cardiovascular disease, headache, effect drugs on higher mental processes, risk factors for stroke. Home: 203 Fairfield, Elmhurst, Ill. 60126. Office: 1853 W. Polk St., Chgo. 60612.*

OSTROM, John H., Am. paleontologist; b. N.Y.C., Feb. 18, 1928; s. William C. and Norma (Beebe) O.; B.S. in Geology, Union Coll., 1951; Ph.D., Columbia, 1960; m. Nancy Grace Hartman, June 14, 1952; children—Karen Ann, Alicia Jane. Lectr., Bklyn. Coll., 1955-56; instr. Beloit Coll., 1956-58, asst. prof. geology, 1958-61; asst. prof. geology, asst. curator vertebrate paleontology Yale, 1961-66, asso.

prof., asso. curator, 1966—; research asst. vertebrate paleontology Am. Mus. Natural History, N.Y.C., 1951-56, research asso., 1965—. John Simon Guggenheim fellow, 1966-67. Mem. Soc. Vertebrate Paleontology (editor Bull. 1962—), Soc. for Study Evolution, A.A.A.S., Soc. Systematic Zoology, Nat. Assn. Geology Tchrs., Sigma Xi. Author: The Strange World of Dinosaurs, 1964; (with John McIntosh) Marsh's Dinosaurs - The Collections from Como Bluff, 1966; also articles. Editor Am. Jour. Sci. 1967—. Research, discovery of new fossil vertebrates, particularly Mesozoic reptiles. Home: 15 Rentell Rd., Hamden, Conn. 06514. Office: Dept. Geology, Yale, New Haven 06520.*

OSTROM, Theodore Gleason, Am. mathematician; b. Nicollet, Minn., Jan. 4, 1916; s. Lloyd W. and Blanch M. (Gleason) O.; B.A., U. Minn., 1937, B.S., 1939, M.A., 1939, Ph.D., 1947; m. Charlotte Elizabeth Williams, June 19, 1949; children—Katherine Ann, David W., Nancy C., Susan E. Faculty, Mont. State U., Missoula, 1947-60, chmn. dept. math., 1954-60; prof. math. Wash. State U., Pullman, 1960—. Mem. Am. Math. Soc., Math. Assn. Am., Am. Assn. U. Profs., Sigma Xi. Research, publs. on theory of finite projective planes, especially semi-translation planes; co-discoverer procedure for constructing one plane from another; with A. Wagner, proof that all doubly transitive finite projective planes are Desarguesian. Home: 1402 Lower Dr., Pullman, Wash. 99163.*

OSTROWSKI, Alexander Marcus, mathematician; b. Kiev, Ukraine, Russia, Sept. 25, 1893; s. Marcus and Vera (Rashevski) O.; student U. Marburg (Germany); Ph.D., U. Göttingen, 1921; hon. dr. Zürich Polytechnicum, U. Besancon (France); m. Margret Sachs, Feb. 12, 1949. Asst. to Felix Klein, Güttingen, Germany, 1919-20, lectr., 1923-27; asst. to E. Hecke, Hamburg, Germany, 1921-23; lectr., 1922-23; Rockefeller fellow, Oxford U., Cambridge U., 1924; prof., Basel, Switzerland, 1927-58; staff Bur. Standards, Washington. Author: Textbook of Analysis, 3 vols.; Monograph on Solution of Equation, 2d edit., 1967; also numerous articles. Home: Certenago, Montagnola, Ti. 6926, Switzerland. Office: Math. Inst., U. Basel, Rheinsprung 21, Switzerland.*

OSTROWSKI, Kazimierz Ludwik, Polish histologist; b. Lwow, Poland, Oct. 24, 1921; s. Franciszek and Katarzyna (Borodajko) O.; grad. Med. Faculty Warsaw (Poland), 1950; m. Leontyna Calewiez, July 25, 1945. Faculty histology dept. Med. Sch. Warsaw, 1945—, head dept., 1960—, prof., 1965—; sci. vice sec. med. sect. Polish Acad. Sci., 1965—. Mem. Internat. Soc. for Cell Biology, Transplantation Soc., Cryobiol. Soc. Editor: Preservation and Transplantation of Tissues in Clinic, 1964. Research, publs. on induction of osteogenesis under influence of urinary bladder mucosa, histochemistry of enzymes, tissue transplantation. Home: Marszatkouska 28 m.80, Warsaw, Poland.*

OSTWALD, Wilhelm, German chemist; b. Riga, Latvia, Sept. 2, 1853; Ph.D., U. Dorpat, 1878. Named prof. Poly. Inst. Riga, 1882; with U. Leipzig (Germany), 1887-1906, dir. Electrochem. Inst., 1898; lectr. Harvard, 1905. Recipient Nobel prize for chemistry, 1909. Author: Lehrbuch der allgemeine Chemie, 1885-87; Elektrochemie, 1896; Vorlesungen über Naturphilosophie, 1902; Grosse Männer, 1909; Die Forderung des Tages, 1910; Der Farbenatlas, 1918; Die Farbenlehre, 1918; Die Pyramide der Wissenschaften, 1929. Founder, Zeitschrift für physikalische Chemie, 1887. A founder modern phys. chemistry; research in catalysis, rates of chem. reactions, chem. equilibirum, elec. conductivity of organic acids; using previous work of Gibbs and Arrhenius showed catalysts hasten chem. reactions by lowering energy of activation without altering energy relationships of substances; obtained nitric acid from ammonia (important for explosives prodn. in Germany); developed a new theory of color; demonstrated that ions could act as catalysts; an admirer of Mach, he refused for a long time to accept atomic theory. Died Grossbothen, near Leipzig, Saxony, Germany, Apr. 4, 1932.

OSTWALD, Wolfgang, chemist; b. Riga, Latvia, May 17, 1883; s. Wilhelm Ostwald; prof., Leipzig, Germany. Author: Grundriss der Kolloidchemie, 1909; Die Welt der vernachlässigten Dimensionen, 1915. Editor: Handbuch der Kolloidwissenschaft, 5 vols., 1924-32. A founder of colloid chemistry. Died Leipzig, Nov. 22, 1943.

OSUGI, Jiro, Japanese chemist; b. Okayama Prefecture, Japan, Oct. 9, 1919; s. Shigeru and Yoshie (Kumada) O.; B.S., Kyoto (Japan) U., 1943, postgrad.; D.Sc., 1953; m. Emiko Kikuchi, Jan. 16, 1949; children—Takeshi, Yumiko. Faculty sci. Kyoto U., 1948—, prof. chemistry, 1962—; vis. prof. Ohio State U., 1960; prof. chem. engring. Tokushima U., 1965-67. Recipient 16th Mainichi Press prize, 1964. Mem. Internat. Conf. on High Pressure (orgn. com. 1965—), Japan Soc. Material Sci. (chmn. high pressure div. 1966—). Author: Measurements of Physical Properties under High Pressure, 1966; also articles. Editor in chief Rev. Phys. Chemistry Japan, 1962—. Research on chem. reactions under high pressure, kinetic of fast chem. reaction, physio-chemistry of phase transition under high pressure, synthesis of new compounds and measurements of its properties; prodn. high pressures. Home: 36 Nishida-cho, Jodoji, Kyoto, Japan.*

O'SULLIVAN, Desmond Gerard, English biol. and med. chemist; b. Essex, Eng., Jan. 3, 1925; s. William Edward and Dorothy (Ford) O'S.; B.Sc. with 1st class honors, U. London, 1943, Ph.D., 1947; diploma Imperial Coll. Sci. and Tech., 1947; m. Dorothy Mary Wilson, Apr. 7, 1958; children—Richard Dominic, Christopher Jerome, Mary Colette, Rosalind Clare. Process devel. chemist May and Baker Ltd., Dagenham, Eng., 1943-44; lectr. biochemistry Middlesex Hosp. Med. Sch., London, 1946-57; head research unit Dept. Sci. and Indsl. Research, 1957-62; head chemotherapy research unit Courtauld Inst. Biochemistry, London, 1962—. Fellow Royal Inst. Chemistry (mem. council), Inst. Chemistry Ireland, Soc. Dyers and Colorists, Inst. Biology, Chem. Soc.; mem. Biochem. Soc. Author: Viruses and the Chemotherapy of Viral Diseases, 1965; also articles. Collaborated in discovery and study of potent chem. inhibitors of small RNA- and pox viruses, antidotes to organic phosphate poisons, vitamin A acid; studied spectra of biologically important compounds; developed theory of cytochem. staining and of certain enzymic processes. Home: 14, Brimstone Close, Chelsfield, Kent, Eng. Office: Courtauld Inst. Biochemistry, Middlesex Hosp., Med. Sch., London W.1., Eng.*

OTA, Zensuke, Japanese physician; b. Okayama City, Japan, Apr. 25, 1930; s. Sadao and Masako (Hatodani) O.; M.D., Okayama Med. Coll., 1954; m. Hiroko Namba, Mar. 9, 1962; children—Yusuke, Kosuke. Asso. dept. internal medicine Okayama U. Med. Sch., 1958-59, asst., 1959-60, lectr., 1960-67, asst. prof. internal medicine, 1967—; cons. physician Okayama Meml. Hosp. Recipient Yuki prize Okayama U. Med. Sch., 1966. Mem. Internat., Japanese socs. hematology, Japanese Soc. Virology, Japanese Soc. Cancer, Japanese Soc. Reticuloendothelial System, others. Research, numerous publs. on hematology; discovered crystallizations in human neutrophils and leukemic cells; established cytological criteria for classification of human leukemia type, Japanese encephalitis virus, electron microscopically clarified its morphology and devel. process, oncogenic viruses; demonstrated viruses in radiation or chem.-induced murine leukemia. Home: 7-18 Chuocho, Okayama, Japan.*

OTERO, Aenlle Enrique, Spanish chemist; b. Ribadeo, Lugo, Oct. 20, 1913; s. Antonio and Rita Aenlle) O.; ed. univs. Saint-Jacques-de-Compostelle and Madrid; m. Maria Rosa Pastor; children—Enrique, Maria Rosa, Maria Dolores, Maria del Pilar, Maria Isabel. Chemist, Spnish Oceanography Inst.; head div. Sci. Research Council. Decorated Order Civil Merit, Cross Merit World War I, Oder Cisneros, Legion of Honor (France); recipient plaque Alphonse X the Wise, Gold medal Paris, France. Author: Analisis de Grases y Ceras; Introduccion a la Fisico-Quimica; Los estudios regionales en el Desarollo; Sanidad y Desarrolo. Research on humectation processes; emulsions; molecular associations in surfaces; phys. chemistry of surfaces; monomolecular layers; detergency. Home: Gran Via 7, Salamanca, Spain. Office: Univ. Barcelona, Barcelona, Spain.

OTHMER, Donald Frederick, Am. chem. engr.; b. Omaha, May 11, 1904; s. Frederick George and Fredericka Darling (Snyder) O.; student Armour Inst., Chicago, 1921-23; B.S., U. of Neb., 1924; M.S., U. of Mich., 1925, Ph.D., 1927; Dr. Engring. (hon.), Univ. of Nebraska, 1962; married Mildred Jane Topp, November 18, 1950. Development engr. Eastman Kodak Co. and Tenn. Eastman Corp., 1927-31; instr. Poly. Inst., Brooklyn, 1932-33, prof., 1933—, distinguished prof., 1961—, head department of chemical engring., 1937-61; honorary professor University Conception, Chile; licensed profl. engr. in N.Y., N.J., O., Pa.; cons. chemical engr., and licensor of process patents, 1931—, to numerous companies, also to governmental depts. of U.S., Canada, Mexico, Cuba, Puerto Rico, Central and S.A., Norway, Sweden, Finland, Denmark, Germany, France, Eng., Belgium, Switzerland, Italy, Spain, S. Africa, India, Burma, Yugoslavia, Korea, Japan, Taiwan, Dominica, P.I., in field of chem. engring.; cons. UN, Office Saline Water U. S. Dept. Interior, Chem. Corps and Ordnance Dept., U. S. Army, Spl. Devices Div. USN, WHO, U. S. Dept. Health, Edn. and Welfare, WPB, Dept. of State, sci. adv. bd. U.S. Army Munitions Command, and other depts. of U.S. and fgn. govts. Recipient Tyler award Am. Inst. Chem. Engrs., 1958; Barber Coleman award N.Y. Acad. Sci., 1958. Fellow A.A.A.S., Am. Inst. Cons. Engrs., Am. Inst. Chemists, N.Y. Acad. Scis., mem. Am. Inst. Chem. Engrs. (chmn. N.Y. sect. 1944, past dir.), Am. Chem. Soc. (council 1945-47, chairman of the chemical processes division), Society Chemical Industry, Am. Soc. M.E., Am. Soc. Engring. Edn. (dir. engrs. joint council 1957-60), Societe de Chimie Industrielle, Chemurgic Council (dir. 1963), Japan Soc. Chem. Engrs., Newcomen Soc., Am. Arbitra. Assn. (panel mem.), Deutsche Gesellschaft für Chem. Appar. (hon.), Chemists Club (life), Sigma Xi, Tau Beta Pi, Phi Lambda Upsilon, Iota Alpha, Alpha Chi Sigma, Lambda Chi Alpha. Contbr. numerous articles to tech. jours. Co-editor: Kirk-Othmer Ency. Chem. Technology, 17 vols., vol. 17, 1960; editor Fluidization, 1956. Coauthor: Fluidization and Fluid Particle Systems, 1960. Mem. adv. bd. Perry's Chem. Engrs.'s Handbook. Lectr. sci. and engring. fields; lectr. Swiss univs. for Am. Swiss Found. Sci. Relations, 1950; lecture tour Chem. Inst. Can., 1944-52, Am. Chem. Soc. Designer of plants and processes for numerous corps. in U.S. and fgn. countries. Holder of about 80 U.S. and fgn. patents on methods, processes and engring. equipment in mfg. of rayon, plastics, wood distillation, refrigeration, wallboard, chemicals from pulping liquors, water desalination, petrochemicals, sugar refining, etc. Research and devel. on processes for prodn. and treating of acetic acid. Home: Coudersport, Pa.; also 140 Columbia Heights, Bklyn. 11201. Office: 333 Jay St., Bklyn. 11201.

OTIS, Arthur Brooks, Am. physiologist; b. Grafton, Me., Sept. 11, 1913; s. Will Howe and Carrie (Brooks) O.; A.B., U. Me., 1935; M.Ed., Springfield Coll., 1937; Sc.M., Brown U., 1939, Ph.D., 1941; m. Eileen Macomber, Aug. 24, 1942; 1 son, Chandler Brooks. Research asso. cellular physiology State U. Ia., 1941-42; faculty U. Rochester Sch. Medicine and Dentistry, 1942-51, asst. prof. physiology, 1947-51; Fulbright Research scholar Cambridge (Eng.) U., 1950-51; asso. prof. physiology and surgery Johns Hopkins Sch. Medicine, 1952-56; prof., chmn. physiology dept. U. Fla. Coll. Medicine, Gainesville, 1956—; Fulbright Research prof. U. Nijmegan, Holland, 1964-65. Mem. Am. Physiol. Soc., Soc. Gen. Physiologists. Research, numerous publs. on physiology, gas exchange and mechanics breathing. Home: 2123 N.W. 4th Pl., Gainesville, Fla. 32601.*

OTIS, Elisha Graves, Am. inventor; b. Halifax, Vt., Aug. 3, 1811; s. Stephen and Phoebe (Glynn) O.; m. Susan Houghton, June 2, 1834; m. 2d, Mrs. Elizabeth Boyd; at least 2 children, including Charles R., Norton P. Mfr. wagons and carriages in Vt., 1838-45; constructed and invented a turbine waterwheel, Albany, N.Y.; with bedstead factory, Bergen, N.J., master mechanic, 1851; in charge of erection and installation of machinery in new factory, Yonkers, N.Y.; devised, incorporated unique features into the elevator (1st one with safety features); established own shop, Yonkers (thought to be the beginning of elevator bus.); demonstrated his safety elevator at Am. Inst. Fair, N.Y., 1854; patented railroad car trucks and brakes, 1852; patented steam plow, 1857, bake oven, 1858; established Otis Elevator Co., after invention, patenting of steam elevator, 1861. Died Yonkers, Apr. 8, 1861.

OTIS, Fessenden Nott, Am. physician; b. 1825; prof. genitourinary and venereal diseases Coll. Phys. and Surg., Columbia U., N.Y.C., 1871-90. First to cure stricture; first to use local anesthesia in urology, 1884. Died 1900.

OTIS, Leon Spelly, Am. psychologist; b. Chgo., July 2, 1924; s. Nicholas and Mary (Kitcheos) O.; B.A., U. Chgo., 1951, Ph.D., 1956; m. Mary C. Contos, July 25, 1948; children—N. Gregory, Stephanie. USPHS fellow clin. psychology U. Ill., 1954-55, research psychologist Med. Sch., 1955-56; Career Tchr. fellow USPHS State U. Ia., 1956-57; asst. prof. Johns Hopkins, 1957-60; program dir. psychopharmacology sect. Stanford Research Inst., Menlo Park, Cal., 1960-61, chmn. biobehavioral scis. dept., 1961—; spl. cons. Psychopharmacology Service Center, Nat. Inst. Mental Health, 1958-62, Cal. Dept. Mental Hygiene, 1962-63. Mem. A.A.A.S., Am., Western psychol. assns., Sigma Xi. Sr. editor: Psychopharmacology Handbook Series, 1953-63. Research on etiology of emotional behavior, early experience and behavioral devel., brain lesions and learning, equipment design and devel., memory, psychopharmacology. Home: 748 Mayview St., Palo Alto, Cal. 94303. Office: 333 Ravenswood Av., Menlo Park, Cal. 94025.*

OTSUKI, Bansui (or Gentaku), Japanese physician; b. Sendai, Japan, 1757; studied Dutch medicine under Gempaku Sugita and Ryotaku Maeno, also Dutch lang., in Nagasaki; at least 1 son, Banri. Author: Rangaku Kaitei (1st systematic Dutch grammar by a Japanese), 2 vols.; other books, also revised Gempaku Sugita's book on anatomy in 13 vols. Died 1827.

OTT, Isaac, Am. physician; b. Northampton County, Pa., Nov. 30, 1847; s. Jacob and Sarah Ann (La Barre) O.; student Lafayette Coll., 1864-67, A.M. (hon.), 1876; M.D., U. Pa., 1869; postgrad. univs. Leipzig and Berlin, 1869; fellow biology Johns Hopkins, 1879; m. Katherine K. Wykoff, Oct. 14, 1886. In practice of medicine, Easton, Pa., from 1869; resident physician St. Mary's Hosp., Phila., 1871; lectr. physiology U. Pa., 1878; prof. physiology, Medico-Chirurg. Coll., Phila., from 1894, dean, 1895-96; cons. neurologist Norristown (Pa.) Asylum. Author: Action of Medicines, 1878; Modern Antipyretics, 1892; Contributions to Physiology and Pathology of Nervous System; Cocaine, Veratria and Gelsemium, 1872; Text-Book of Physiology, 1904; Internal Secretions, 1901. Described body's thermoinhibitory mechanism located in brain, 1887. Died Jan. 1, 1916.

OTT, Walther Henry, Am. biologist; b. Hermiston, Ore., Sept. 30, 1911; s. Henry John and Minnie (Walther) O.; B.S., Ore. State U., 1934, M.S., 1936; Ph.D., Pa. State U., 1942; m. Maxine Peterson, Aug. 27, 1936; children—Ruth Edna (Mrs. Richard G. Lewman), Arline Ann. Instr., Okla. State U., Stillwater, 1936-37; head poultry nutrition dept. Merck Inst. for Therapeutic Research, Rahway, N.J., 1942-54, head biol. control dept., 1945-52, asso. farm mgr., 1954-56, dir. animal husbandry, 1956—. Fellow N.Y. Acad. Scis., A.A.A.S.; mem. Poultry

Sci. Assn. (chmn. industry com. 1962-67, dir. 1967—). Am. Inst. Nutrition, Am. Chem. Soc., Am. Nutrition Research Council, Biometric Soc., N.Y. Acad. Scis. Research, numerous publs. on nutritive requirements of chickens, tolerance of poultry and farm animals to feed additives, techniques of biol. assays and exptl. designs; patentee penicillin and bacitracin in animal feeds. Home: 1874 Quimby Lane, Westfield, N.J. 07090. Office: Merck Inst. for Therapeutic Research, Rahway, N.J. 07065.

OTTAVIANI, Gaetano, Italian physician; b. Sommacampagna, Italy, Feb. 26, 1902; M.D.; m. Elisabetta Grigolini, Sept. 29, 1939; children—Valeria, Livia. Prof. anatomy; instr. histology and embryology. Mem. Italian Anatomy Soc., Deutsche anatomische Gesellschaft, Verona (Italy) Acad. Sci. Author papers on anatomy, especially on lymphatic systems and nerves. Home: viale Basetti 6. Office: Inst. anatomia Ospedale Maggiore, Parma, Italy.

OTTAVIANI, Pierfrancesco, Italian physician; b. Parma, Italy, Apr. 8, 1918; s. Menotti and Anna (Vernizzi) O.; Med. Degree, U. Parma (Italy), 1945; m. Edda Canossa, Sept. 28, 1944; children—Raimonda, Michela, Giuseppe, Anna Elisabetta, Gianluca. Libera docenza path. medicine U. Rome, 1956-59, libera docenza clin. medicine, 1959—, dep. dir. Istituto di Semeiotica Medica, since 19—. Mem. Italian Med. Assn., Italian Soc. for Hematology. Author: (with A.G. Dettori, G. Manai) La malattia trombocitopenica, 1955; (with G. Mani, F. Mandelli) L'Angioemofilia, 1961; Il trattamento delle malattie ermorragiche, 1963; also numerous articles. Research on hematology, especially coagulation; 1st to demonstrate existence of thromboplastin in red blood cells. Home: 35 via Priscilla, Rome, Italy.*

OTTE, Henry Martin, metallurgist; b. Madras, India, Apr. 6, 1929; s. Hans and Greta (Behavy) O.; B.Sc., U. Birmingham, 1949, Ph.D., 1952, D.Sc., 1966; m. Pamela R. Brettell, Dec. 21, 1956; children—Julian Martin Lewis, Steven Gregory Severin. Came to U.S., 1952, naturalized, 1966. Post-doctoral fellow Case Inst. Tech., Cleve., 1952-53; research asso. Columbia, 1953-54, U. Ill., 1954-55; U. and Imperial Chem. Industries fellow Birmingham U., Eng., 1955-57; prin. scientist RIAS (Martin Co.), Balt., 1957-64; mgr. materials research lab. Martin Co., Orlando, Fla., 1964—. Mem. Am. Inst. Mining and Metall. Engrs., Am. Soc. for Metals, Inst. Metals London, Am. Crystallographic Assn., Electron Microscope Soc. Am., Internat. Soc. Stereology. Author: (with M.P. Drazin) Tables for Determining Cubic Crystal Orientations, 1964. Editor: (with S.R. Locke) Materials Science Research, vol. 2, 1965. Research on transformation and deformation of metals and alloys using primarily x-ray diffraction techniques and electron microscopy; theoretical and exptl. contbns. to crystallographic study of martensitic transformations, especially in steels and to investigation of diffraction of stacking faults. Home: 1662 Summerland Av., Winter Park, Fla. 32789. Office: Mail Point 105, Orlando, Fla. 32805.*

OTTENSOOER, Fritz, immunohematologist; b. Nuremberg, Germany, July 19, 1891; s. Eugen and Ida (Salmonsen) O.; M.D., U. Heidelberg (Germany), 1915; Ph.D., U. Munich, 1924; m. Berta Lehmann, June 9, 1921. Asst., Inst. für exp. Ther., Frankfurt, Germany, 1926-30; dept. head Inst für Hygiene, Bern, Switzerland, 1930-40; dept. head Lab. Paulista de Biologia, Sao Paulo, Brazil, 1941—; chief researcher Conselho Nacional de Pesquisas, 1959-63; chief researcher Fundaçao de Amparo à Pesquisa do Est. de S.P., 1963—. Lectr. on blood groups; genetic counselor, 1959—. Mem. Internat. Soc. Hematology, Am. Acad. Forensic Sci., N.Y. Acad. Sci. Research, numerous publs. on blood groups especially in vaccines; established formula to estimate race mixture from gene frequencies; discovered anti-N in seeds of Vicia graminea. Home: 2494 Av. Paulista, Sao Paulo, S.P., Brazil.*

OTTEWILL, Ronald Harry, English chemist; b. London, Eng., Feb. 8, 1927; s. Harry Archibald and Violet (Bucklee) O.; B.Sc., Queen Mary Coll., U. London, 1948, Ph.D., 1951; Ph.D., M.A., U. Cambridge (Eng.), 1955; m. Ingrid Geraldine Roe, Aug. 31, 1952; children—Geraldine Astrid, Adrian Christopher. Staff dept. colloid sci. U. Cambridge, 1955-63, asst. dir. research, 1958-63; faculty U. Bristol (Eng.), 1964—, reader colloid sci. 1966—. Fellow Royal Inst. Chemistry, Cambridge Philos. Soc.; mem. Faraday Soc., Chem. Soc. Research, publs. on surface and colloid chemistry, adsorption of vapors on liquid surfaces, optical properties of colloids, formation of colloidal particles, especially factors influencing stability of colloidal dispersions. Office: Sch. Chemistry, U. Bristol, Bristol 8, Eng.

OTTO, Gerhard, German chemist; b. Ludwigshafen, Rhine, Germany, Jan. 28, 1901; s. Richard and Maria (Kaefer) O.; Dr.-Ing., Techn. High Sch. Karlsruhe, Germany, 1928; m. Lydia Strassner, Oct. 19, 1929; children—Soren, Peter. Chemist leather tech. dept. Badische Anilin & SodaFabrik A.G. Ludwigshafen, Mannheim, Germany, 1929-66, in charge devel. and research, 1951-64. Guest lectr. Darmstadt-Inst. Tanning Chemistry, 1949—. Recipient award for outstanding sci. work German Soc. Leather Chemists and Technicians, 1961. Mem. Spanish (hon.), German (hon.) socs. leather trades chemists, German Soc. Leather Chemists and Technicians (past pres.). Contbg. author: Study of Interactions Hide Tannin Dyestuff, 1954; Färben des Leders, 1962. Research, numerous publs. on interactions in system hide tannin dyestuff, charge phenomena on surface of leather, chemistry of tanning metal compounds; introduced non swelling acids in leather trade; invented quick tannage with compounds of sequestering agts. and chrome. Home: 13 Paul-Martin-Ufer, 68 Mannheim-Neuostheim, Germany.*

OTTO, Herbert, German pathologist; b. Bremen, Germany, July 31, 1922; s. Johannes and Meta (Bumann) O.; Dr.med., 1952; m. Gerda Horand, Dec. 19, 1957; children—Andreas, Joachim. Faculty, Erlangen (Germany) Pathologisches Institut, 1947—, prof. pathology, 1963—. Research, publs. on problems in pathology, especially lung and occupational pathology. Home: 12 Halerstrasse, Erlangen-Buckenhof-Germany.*

OTTO, Herbert Arthur, psychologist, social worker; b. Berlin, Germany, May 19, 1922 (parents Am. citizens); s. Arthur Curt and Frieda W. (Franck) O.; student U. Ark., 1940-42; A.B., U. Mich., 1946; M.S.W., Tulane U., 1950; M.S., Harvard, 1955; Ph.D., Fla. State U., 1956; m. Sarah Maude Thorpe, May 17, 1952; children—Frieda, Herbert Arthur Curt, Sarah Maude. Asst. prof., dir. Mental Health Edn. Program, U. Ga., 1952-60; asso. prof., dir. Sch. Social Work Tng. Program, dir. Human Potentialities Research Program, U. Utah, Salt Lake City, 1960; asso. prof. Miami U., Oxford, O., summers 1961-66. Cons. Ga. Dept. Pub. Health, Utah Council Aging. Mem. Am. Soc. For Group Psychotherapy and Psychodrama, Am. Assn. Humanistic Psychology (dir.), Nat. Assn. Social Workers, Acad. Certified Social Workers, Am. Assn. Marriage Counselors, Acad. Religion and Mental Health. Author: (with Nina Garton) The Development of Theory and Practice in Social Casework, 1964; Your Potentialities: The Undiscovered Self, 1966. Editor: Exploration in Human Potentialities, 1966. Research, publs. primarily in field of human potentialities; pre-marital counseling, developed Otto Inventory of Personal Resources, Otto Family Strength Survey, Life Goals Inquiry. Home: 74 Virginia St. Office: Grad. Sch. Social Work, U. Utah, Salt Lake City 84112.*

OTTO, John Conrad, Am. physician; b. nr. Woodbury, N.J., Mar. 15, 1774; s. Bodo and Catherine (Schweighauser) O.; grad. Coll. of N.J. (now Princeton), 1792, U. Pa., 1796; m. Eliza Tod, 1802; 9 children, including William Tod. Physician, Phila. Dispensary, 1798-1803; physician Orphan Asylum, Magdalen Asylum, many years; physician Pa. Hosp., 1813-34; mem. com. of 12 leading physicians appt. to deal with cholera epidemic in Phila., 1832; mem. Phila. Coll. of Physicians, 1819, censor, many years, v.p., 1840-44. Described hemophilia, 1803, showed that it is hereditary, that females are not affected but can transmit the disease to male offspring; advocated internal use of sodium sulfate to stop hemophiliac bleeding; studied epilepsy. Died Phila., June 26, 1844.

OTTO, Ulf, Swedish child psychiatrist; b. Stockholm, Sweden, Oct. 8, 1931; s. Eyvind and Lony (Fiby) O.; Med.dr., Karolinska, Inst., 1958, specialist in child psychiatry, 1962; m. Yvonne Louise Smith, Sept. 18, 1956; children—Gisela, Christina, Camilla. Head physician child psychiatric dept. Central Hosp., Kristianstad, Sweden, 1962—; also mil. psychiatric expert. Mem. Svenska Lakarsallskapet (sec. child-psychiatric sect.). Research and publs. on suicidal attempts among children and youth; mil. psychiatric investigations; brain damage syndrome; bed wetting. Address: Central Hosp., Kristianstad, Sweden.*

OTTOMAN, Richard Edward, Am. radiologist; b. Guthrie, Okla., Aug. 3, 1910; s. Adolph and Fern (Christian) O.; student Jackson Jr. Coll., 1932-34; M.D., U. Mich., 1941; m. Mary Elizabeth Merrill, Nov. 27, 1943; 1 dau., Bonnie Ann. Staff physician Birmingham VA Hosp., Van Nuys, Cal., 1946-47, resident radiology, 1947-50; asst. radiologist Birmingham Hosp., 1950—; asst. radiologist to chief therapeutic radiology Long Beach VA Hosp., 1950-52; affiliate cons. St. John's Hosp., 1960—; cons. radiology and radioisotopes St. Joseph's Hosp., Burbank, Cal., 1950—; cons. radiology Valley Hosp., Van Nuys, 1956—; instr. radiology U. So. Cal., 1950-52; asst. prof. radiology, asso. prof. radiology and anatomy U. Cal. at Los Angeles, 1954-61, prof. radiology and anatomy, vice chmn. dept. radiology, 1961—. Trustee, sec.-treas. James T. Case Radiologic Found. Diplomate Am. Bd. Radiology. Fellow Am. Coll. Radiology, Am. Coll. Clin. Pharmacology and Chemotherapy; mem. Radiol. Soc. N.Am., Soc. Nuclear Medicine, A.M.A., Am. Fedn. for Clin. Research, A.A.A.S., Pacific Fertility Soc., Am. Nuclear Soc., Am. Club Therapeutic Radiologists, Am. Roentgen Ray Soc. Research and publs. on gonadal radiation and infertility, lung cancer in females, treatment malignant tumors, other radiol. subjects. Home: 2260 Westridge Rd., Los Angeles 90024.*

OUCHI, Atsuyoshi, Japanese elec. engr.; b. Tokyo, Oct. 10, 1919; s. Kyusaburo and Tamako (Yasukawa) O.; B.E., Tokyo U., 1942, D.Eng., 1962; m. Kazu Ikemaga, Apr. 24, 1949; children—Mari, Yoshihiro. With Nippon Electric Co. Ltd., Tokyo, 1942—, mgr. med. electronics dept., electronics application div., 1965—. Profl. mem. Council of Electronics Industry, Ministry of Internat. Trade and Industry of Japan, 1965—, Council Electronics Engring., Ministry of Sci. and Engring., 1966—. Mem. Japan Invention Assn. (Kantodist. spl. award 1962, All Japan award 1963), Machinery Promotion Assn. (1st award 1966), I.E.E.E. (sr.), Japan Soc. Med. Electronics and Biol. Engring. (dir.), Inst. Elec. Communication Engrs. Japan (editorial adviser), Inst. Elec. Engring. Japan, Acoustical Soc. Japan, Information Processing Soc. Japan, Japan Soc. Ultrasonic in Medicine. Contbr. numerous articles to profl. jours. Devel. measuring equipment, including sinusoidal wave oscillator, pulse modulated communication equipment, battery-less endoradiosonde, oscillator circuit, other med. electronics devices. Home: 64-9 Kamiodamaka, Kawasaki-city, Kanagawa-ken, Japan. Office: 10, 1-chome, Nisshincho, Fuchu-city, Tokyo, Japan.*

OUCHI, Hyoe, Japanese economist; b. Awaji Island, Hyogo Prefecture, Japan, Aug. 29, 1888; grad. econs. dept. Tokyo U., 1913; D. Econs.; oldest son, Tsutomu. Entered Finance Ministry, 1913; ofcl. tour of U.S., 1916; apptd. instr. Tokyo U., 1918, asst. prof., 1919, prof., 1923, resigned (after arrest in Professors' Group Case 1938), 1944, returned, 1945, emeritus prof., 1949—; became pres. Hosei U., resigned, 1959. Mem. Acad. Japan. Author: General Features of Science of Finance. Translator: The Wealth of Nations (Adam Smith). Under his leadership, Japanese finances were systemized by Marxian methods.

OUDEMANS, Jean-Abraham-Chrétien, Dutch astronomer; b. Amsterdam, Netherlands, Dec. 16, 1827; ed. U. Utrecht (Netherlands); head Geog. Service to Java; prof. astronomy U. Utrecht; mem. Acad. Scis. Amsterdam, French Acad. Scis. (corr.); research on orbits of periodic comets, theory of telescopes. Died Utrecht, Dec. 14, 1906.

OUGHTRED, William, English mathematician; b. Eton, Eng., Mar. 5, 1575; s. Benjamin Oughtred; entered King's Coll., Cambridge, 1592; eldest son, Benjamin. Became fellow King's Coll., 1595; ordained priest, circa 1603; given living of Shalford, Eng., 1605; tchr. of mathematician John Wallis, astronomer Seth Ward, also Christopher Wren. Author: Clavis Mathematicae (introduced modern multiplication and proportion signs), 1631; Circles of Proportion, 1632; Trigonometria, 1657; Opuscula Mathematica, 1676. Invented abbreviations of trigonometric functions; originated horizontal instrument for delineation of dials in any plane, circa 1600; adapted Edmund Gunter's logarithmic scale to it (1st rectilienar logarithmic slide rule), circa 1620. Died Albury, Eng., June 30, 1660.

OUNSTED, Christopher, English physician; b. London, Eng., Aug. 8, 1921; s. Laurence John and Vera (Hopkins) O.; B.A., Oxford (Eng.) U., 1943, B.M., B.Ch., 1945; diploma in Child Health, 1946, D.Medicine, 1951, diploma Psychol. Medicine, 1956; M.R.C.P., 1966; m. Margaret Wilder, July 14, 1945; children—Mary, Teresa, Martin, Madeleine. Research asst. pediatric neurology Oxford Regional Hosp. Bd., 1948-55; sr. registrar Warneford Hosp., Oxford, 1955-57; physician in charge Park Hosp. for Children, Oxford, 1957—; clin. lectr. U. Oxford, 1950—. Mem. Royal Medico Psychol. Assn. Author: (with R. Norman, J. Lindsey) Biological Factors in Temporal Lobe Epilepsy, 1966; also articles. Research on biol. aspects of normal and abnormal human devel., genetics of convulsive disorders. Home: Church Farm House, Upper Wolvercote, Oxford. Office: Park Hosp., Old Rd., Oxford, Eng.*

OUTERBRIDGE, Alexander Ewing, Jr., Am. metallurgist; b. Phila., July 31, 1850; s. Alexander Ewing and Laura C. (Harvey) O.; ed. Episcopal Acad. Phila.; attended lectures on physics and chemistry as asst. to Prof. Henry Morton, U. Pa.; m. Mary Ely Whitney, 1880 (dec. 1881); 1 son, George Whitney; m. 2d, Margaret Hall Dunn, Jan. 29, 1905. Apptd. asst. in assay lab. U.S. Mint., Phila., 1868; sent to New Orleans to establish assay dept. of br. mint, 1879-80; metallurgist A. Whitney & Sons Car Wheel Works, Phila., 1880-88; metallurgist William Sellers & Co., Inc., Phila., from 1888. Lectr. indsl. econs. Wharton Sch. Finance, U. Pa.; apptd. prof. metallurgy Franklin Inst., 1901. Mem. Franklin Inst. Phila. Recipient Elliott Cresson Gold medal and John Scott legacy medal and premium from Franklin Inst. and City of Phila. for original discoveries in molecular physics of iron. Extensive contbr. to newspapers and tech. mags. Died Jan. 15, 1928.

OUTHIER, Renauld, French astronomer; b. La Marre-Jousserans, France, Aug. 16, 1694. (With Maupertius, others) made sci. expdn. to Lapland, 1736. Mem. French Acad. Scis. (corr.) 1731, Royal Soc. Berlin. Helped verify flattening of earth at poles. Died Bayeux, France, Apr. 12, 1774.

OUTHWAITE, Leonard, Am. anthropologist; b. Sierra Madre, Cal., July 12, 1892; s. Joseph Husband and Annette (Boyce) O.; student Yale, 1915; A.B., U. Cal. at Berkeley, 1916; m. Lucille Conrad, Mar. 1, 1936; children—Joan (Mrs. Lelan F. Sillin) (by previous marriage), Ann, Lynn (Mrs. Richard S. Pulsifer). Tchr. anthropology U. Cal. at Berkeley, 1916-17; expdn. Hawaii, Santa Cruz, 1916; mem. com. on classification of personnel U. S. Army, 1917-19; with Bur. Indsl. Research, 1919-21; mem. staff Laura Spelman Rockefeller Meml., 1922-28; tchr. personnel adminstrn. Columbia U., Sch. Bus. Tchrs., N.Y.C.,

1921-24; anthropologist Rockefeller Found., 1933-35; cons. mus. orgn. and design Franklin Inst., Phila., 1935, Mus. Natural History, Phila., 1935; studies for N.Y. Zool. Soc., N.Y.C., 1938-45; research on Am. cultural instns. for Inst. Pub. Adminstrn., 1952—. Sec., Pres.'s Commn. on Demobilization, 1943. Fellow Internat. Inst. Arts and Letters; mem. Phi Beta Kappa, Psi Upsilon. Author: Atlantic Circle, 1929; Unrolling the Map, the History of Geographic Exploration, 1935; The Atlantic, A History of the Ocean, 1957; Museums for Life; also numerous tech. reports. Home: 176 E. 77th St., N.Y.C. 10021. Office: 55 W. 44th St., N.Y.C. 10036.*

OUTTEN, L(ora) M(ilton), Am. biologist; b. Pocomoke City, Md., Aug. 17, 1913; s. L.P. and D. Elizabeth (Blades) O.; A.B., Western Md. Coll., 1934, M.A., 1937; M.S., Cornell U., 1950, Ph.D., 1956; postgrad. Harvard. Tchr. high schs., Worcester County, Md., 1934-36, Ridgeway, Va., 1940-41, Buckingham, Md., 1943-44, Chincoteague, Va., 1944-46; faculty Mars Hill (N.C.) Coll., 1946—, prof. biology, 1956—. Fellow A.A.A.S.; mem. Am. Inst. Biol. Scis., Am. Soc. Naturalists, Am. Soc. Zoologists, Am. Soc. Ichthyologists and Herpetologists, Ecol. Soc. Am., Soc. Systematic Zoology, Genetics Soc. Am., Am. Soc. Limnology and Oceanography, Internat. Assn. Limnology, N.C. Acad. Sci., N.E.A., Am. Nature Study Soc., Freshwater Biol. Assn., History of Sci. Soc., Assn. for History of Medicine, Am. Micros. Soc., Sigma Xi. Research, publs. on life history, ecology, behavior, devel., morphology and distbrn. of fishes, effects of insecticides on honeybees, devel. of mollusks, distbn. of organisms in slow and swift streams, distbn. of fishes in southeastern U.S., Caribbean, Bahamas, Bermuda, Panama, Britain, Hawaii, Australia, New Zealand. Home: Box 722-C, Mars Hill, N.C. 28754.*

OVARY, Zoltan, immunologist; b. Kolosvar, Hungary, Apr. 13, 1907; s. Elemér and Olga (Purjesz) O.; student Reformed Coll. Kolozsvár, 1918-25; M.D., Faculty Medicine, Paris, 1935; Licencié ès Sciences, Faculty Scis., Paris, 1939. Came to U.S., 1954, naturalized, 1967. Fellow Pasteur Inst., Paris, 1936-39, Clinica Medica, U. Rome (Italy), 1947-54; research fellow Inst. Biologio, Sao Paulo, Brazil, 1952; Fulbright fellow Johns Hopkins, Balt., 1954, research asso., 1955-59; faculty N.Y. U. Sch. Medicine, 1959—, prof. dept. pathology, 1964—, N.Y. Health Research Council Career scientist, 1961—. Mem. Internat. Coll. Allergy (founding), Soc. for Exptl. Biology and Medicine, N.Y. Acad. Scis., A.A.A.S., Am. Assn. Immunologists, Brit. Assn. for Immunologists, Harvey Soc. Research, numerous publs. on nature of antigenicity, passive cutaneous anaphylaxis which permits detection of very small amounts of antibodies or antigens. Home: 343 E. 30th St., N.Y.C. 10016.*

OVCHINNIKOV, Nikolay Mikhaylovich, Russian dermatologist, venereologist, microbiologist; b. 1899; grad. Med. Faculty, 1st Moscow U., 1923; D.Med. Sci., 1938. Asso., Moscow City Bacteriological Inst., 1929-38; sr. asso. Central Skin and Venereal Diseases Inst., 1935-39, head exptl. dept., 1939-42, prof., 1939—; head exptl. dept. Inst. Skin Tb, 1942-51; head microbiology dept. Central Skin and Venereal Diseases Inst., USSR Ministry Health, 1947—, also chmn. Serological Commn. Learned sec. Dermatology Problems Commn., USSR Acad. Med. Scis. Decorated Order of Lenin. Mem. Moscow Soc. Dermatologists and Venereologists (chmn. serological sect.). Author: Gonococcus and Laboratory Diagnosis of Gonorrhea; Morphology and Biology of Gonococcus, 1938; Serological Studies of Syphilis and Gonorrhea; Experimental Syphilis, 1956. Mem. editorial bd. Herald of Dermatology and Venereology Address: Central Skin and Venereal Diseases Inst., ulitsa Korolenko 3, Moscow, USSR.

OVCHINNIKOV, Pavel Nikolaevich, Russian botanist; b. 1903; grad. Leningrad U., 1925; Asso., Bot. Inst., USSR Acad. Scis., 1931-41; dir. Inst. Botany, Tadzhikistan Acad. Scis., acad. sec. dept. natural sci., 1957—. Mem. Tadzhikistan Acad. Scis. Author: History of the Vegetation of Southern Central Asia, 1940; The Main Arboreal Varieties of Tadzhikistan, 1948; a compiler of Flora of the USSR, Flora of Transbaykal, Flora of Uzbekistan. Research and publs. on taxonomy, floristics, bot. geography. Address: Tadzhikistan Acad. Scis., Dushanbe, Tadzhikistan SSR, USSR.

OVERBECK, Fritz Theodor, German botanist; b. Worpswede, Aug. 2, 1898; s. Fritz and Hermine (Rohte) O.; ed. univs. Kiel and Heidelberg (Germany); Ph.D.; m. Hertha Quincke, Sept. 22, 1927; children—Gertrud, Reinhard, Gesine. Prof. U. Frankfort (Germany), 1934-35; dir. Botanical Instn., Hanover (Germany) Tech. Coll., 1935-39; prof. agrégé, dir. Bonn (Germany) Inst. Botany, 1939-51; dir. U. Kiel Bot. Inst., 1951—. Author: Die Moore Niedersachsens, 1950; Das Groose Moor b. Grifhorn, 1951; Mittelgeirgsflora. 1935, also articles on mechanism of diaphragm of plants. Research on analysis pollen and swamp study; history of vegetation, marsh sci.; physiol. anatomy of plants; pollen analysis. Address: Bartelsallee 16, Kiel, West Germany.

OVERBEEK, Gerhard Anthony, Dutch pharmacologist, endocrinologist; b. Amsterdam, Oct. 27, 1911; s. Hendrik Jan and Jacoba (Gravestein) O.; student Municipal U. Amsterdam (Netherlands), 1928-35; Ph.D., U. Leyden (Netherlands), 1939; m. Maria Angenita de Wild, Sept. 3, 1935; children—Hendrik Jan, Angenita (Mrs. M.E. Whitward), Catherine M., Margaretha H. Asst. dept. pharmacology U. Leyden, 1936-42; with N.V. Organon, Oss, Netherlands, 1942—, head dept. pharmacological research, 1946-66, asst. research dir., 1966—. Recipient Dr. Saal van Zwanenberg prize, 1966. Mem. Dutch Soc. Endocrinology (pres. 1960—), European Soc. for Study Drug Toxicity (treas. 1963—), Dutch Soc. Physiology and Pharmacology, Dutch Soc. Radiology., Brit. Biochem. Soc. Author: Anabole Steroide, 1966; also numerous articles. Research on hypophysis, adrenals, and fatigue, anabolic steroids, progestational steroids. Home: 80 van Slichtenhorststraat, Nijmegen, Netherlands. Office: N.V. Organon, Kloosterstaat, Oss, Netherlands.*

OVERBEEK, Jan Theodoor Gerard, Dutch phys. chemist; b. Groningen, Netherlands, Jan. 5, 1911; s. Adam Adolf and Johanna Cornelia (van Rijssel) O.; Cand.Chem., U. Utrecht (Netherlands), 1930, Chem. Drs., 1933, Ph.D., 1941; D.Sc. (hon.), Clarkson Coll. Tech., 1967; m. Johanna Clasina Edie, Aug. 18, 1936; children—Reina E. (Mrs. H.A.C. Le Poole), Antoinetta W. (Mrs. N. De Both), Marijke, Titia E. Asst., U. Ghent (Belgium), 1935-36; with U. Utrecht, 1936-41, prof. phys. chemistry, 1946—; sci. officer N.V. Philips, Eindhoven, Netherlands, 1941-46. Sec., chmn. Faculty Sci., Utrecht; cons. to indsl. cos.; vis. prof. Mass. Inst. Tech., 1952-53, 66-67, U. So. Cal., 1959-60. Mem. Kon. Nederl. Chem. Vereniging, Nederland. Natuurk Ver., Am. Chem. Soc., Faraday Soc. Author: (with E.J.W. Verwey) Theory of Stability of Lyophobic Colloids, 1948; (with H.R. Kruyt) Introduction to Physical Chemistry, 1965; (with A.L. Loeb, P.H. Wiersema) The Electrical Double Layer, 1960; also numerous articles. Research in colloid and surface chemistry, stability of suspensions and emulsions, elec. forces, Van der Waals (dispersion) forces, soap solutions, aggregates (micelles) of soap molecules, soap films especially ultrathin soap films; membranes including biol. cell membranes; polymers especially polyelectrolytes; biosynthesis of proteins. Home: 35 Zweerslaan, Bilthoven, Netherlands. Office: 19 Sterrenbos, Vanthofflaboratory, Utrecht, Netherlands.*

OVERBERGER, Charles Gilbert, Am. chemist; b. Barnesboro, Pa., Oct. 12, 1920; s. Charles E. and Beatrice (McAnulty) O.; B.S., Pa. State U., 1941; Ph.D., U. Ill., 1944; m. Mildred Elizabeth Chase, 1966; children—Erica M., Carla L., Charles T., Ellen A. Faculty chemistry Poly. Inst. Bklyn., 1947—, prof., 1952—, head dept. chemistry, 1955-64, dean sci., dir. Polymer Research Inst., 1964-67. Mem. materials adv. bd. Nat. Acad. Scis.-NRC, 1961-66, mem. at large, 1962—; chmn. bd. trustees Gordon Research Confs., 1965—. Mem. Am. Chem. Soc. (dir.), Chem. Soc., Am. Inst. Chemists, A.A.A.S., Soc. Chem. Industry, N.Y. Acad. Scis., Sigma Xi, Phi Kappa Pi, Alpha Chi Sigma, Phi Eta Sigma, Sigma Pi, Phi Lambda Epsilon. Research in macromolecules and synthesis of these behaving as natural molecules, catalysis of polymerization, new types of ionic polymerization. Home: 2222 Fuller Rd., Ann Arbor, Mich. 48105.*

OVEREND, George William, Brit. chemist; b. Shropshire, Eng., Nov. 16, 1921; s. Harold George and Hilda (Parry) O; D.Sc., Ph.D., U. Birmingham (Eng.); m. Gina Olava Caldman, 1949; children—Edmund, Sheila, Desmond. Lectr., U. Nottingham (Eng.), 1946-47; chemist Dunlop Rubber Co., also British Rubber Producers' Research Assn., 1947-49; lectr. chemistry U. Birmingham, 1949-55; instr. organic chemistry Birkbeck Coll., 1955-57; prof. chemistry U. London (Eng.), also dir. chemistry dept. Birkbeck Coll., 1957—. Jubilee Meml. lectr. Fellow Royal Inst. Chemistry; mem. Chem. Soc. London, Soc. Chem. Industry, British Assn. Advancement Sci. Author articles on organic chemistry. Research organic chemistry, particularly the chemistry of carbohydrates. Home: 37, The Ridgeway, Kenton, Middlesex, Eng. Office: Birkbeck Coll., Malet St., London W.C.1, Eng.

OVERHAGE, Carl F. J., physicist; b. London, Eng., Apr. 2, 1910; s. Carl and Augusta (Lichteiker) O.; came to U.S., 1928, naturalized, 1940; B.S., Cal. Inst. Tech., 1931, M.S., 1934, Ph.D., 1937; m. Katya Grusina, Aug. 16, 1940. Physicist, acting dir. research Technicolor Motion Picture Corp., Hollywood, Cal., 1937-42; staff mem., asst. group leader radiation lab. Mass. Inst. Tech., Cambridge, 1942-45, div. head Lincoln Lab., 1955-56, dir., 1957-64, prof. engring., 1961—; research supr., asst. dir. color tech. div. Eastman Kodak Co., Rochester, N.Y., 1946-54. Cons. to govt. agys., Air Force. Recipient Presdl. Certificate of Merit, 1948; Exceptional Service award Dept. Air Force, 1958. Research in electronics, photography. Home: 17 Gray Gardens E., Cambridge, Mass. 02138.*

OVERMAN, John Reagan, Am. physician; b. Marion, Ind., Dec. 29, 1921; s. Ivan J. and Bessie (Seegar) O.; student George Washington U., 1945-47; B.S., Duke, 1950, M.D., 1950; m. Natalie Ann McChrystal, Dec. 26, 1945; children—Ann Louise, Paul Reagan, Nancy Jean, John Robert. Staff, NIH, 1952-54; asst. Rockefeller Inst. Med. Research, 1954-56, asst. physician Hosp., 1954-56; research in viruses and internal medicine, Durham, N.C., 1956-64, Bethesda, Md., 1964—; faculty Sch. Medicine Duke, 1956-64, prof. microbiology, asst. prof. medicine, 1964; staff Duke Pvt. Diagnostic Clinic, 1958-64; asso. dir. collaborative research Nat. Inst. Allergy and Infectious Diseases, Bethesda, Md., 1964—; asso. clin. prof. medicine George Washington U. Med. Sch., 1964—. Recipient Borden Undergrad. Research award, 1950; Lederle Med. Faculty award, 1956. Mem. Soc. For Exptl. Biology and Medicine, Am. Assn. Immunologists, Harvey Soc., Sigma Xi. Author: (with others) Zinsser's Microbiology, 1964. Research, publs. on first demonstration of mumps-induced encephalitis in animals; contbns. to techniques of counting virus particles by electron microscopy. Died Nov. 7, 1967.

OVERMAN, Ralph Theodore, Am. nuclear chemist; b. Clifton, Ariz., Aug. 9, 1919; s. Cecil Vaughn and Elva (Cole) O.; A.B., Kan. State Coll., 1939, M.S., 1940; Ph.D., La. State U., 1943; D.Sc., Phila. Coll. Pharmacy and Sci., 1959; m. Frances Henson, June 30, 1945; children—Ralph Theodore, Ann Frances. Instr., La. State U., 1942-43; head sci. dept. La. Coll., 1943-44; plant supt. Fercleve Corp., Oak Ridge, 1944-45; sr. research chemist Oak Ridge Nat. Lab., 1945-48; chmn. spl. tng. div. Oak Ridge Inst. Nuclear Studies, 1948-65; owner Ralph T. Overman Cons. Services, Oak Ridge, 1965—. Mem. adv. bd. Advances in Chemistry series Chem. Monograph series Am. Chem. Soc., 1963—, The Sci. Tchrs., 1964—. Named Hon. prof. Nat. U. San Marcos, Lima, Peru, 1962. Fellow Am. Nuclear Soc., A.A.A.S.; mem. Am. Nuclear Soc., Am. Chem. Soc., A.A.A.S., Radiation Research Soc., Health Physics Soc, Internat. Platform Assn. Author: (with H.M. Clark) Radioisotope Techniques, 1960; Basic Concepts of Nuclear Chemistry, 1963; also articles. Editorial adv. com. Jour. Nuclear Sci. and Engring., 1961—. Developed methods and techniques of prodn. and use of radioisotopes; developed applications for radiation and radioisotopes in medicine, industry, chemistry. Home: 109 Pelham Rd., Oak Ridge 37830. Office: Box 367, Oak Ridge 37830.*

OVERMAN, Richard Roll, Am. physiologist; b. Richmond, Ind., Nov. 10, 1916; s. Oliver Sumner and Camilla (Black) O.; A.B., DePauw U., 1939; postgrad. Harvard, 1939-40; A.M., Princeton, 1942, Ph.D., 1943; m. Sarah Frances Pope, June 20, 1942; children—Sally Brooks, Karen Wendel. Instr. physiology Columbia Coll. Phys. and Surg., 1943-45; faculty U. Tenn. Coll. Medicine, Memphis, 1945—, prof. clin. physiology, 1953—, prof. physiology and biophysics, 1958—, dir. clin. physiology labs., 1952-64, dir. research maternal spl. care unit, 1954-60, prof., chmn. div. radiation biology, 1963—, asst. dean research and grant affairs, 1964-66, asso. dean Coll. Medicine, 1966—. Cons. to labs. and hosps. Recipient NSF travel award, 1953; Rockefeller Found. travel award, 1960-67. Mem. Am. Physiol. Soc. (Travel award 1947), A.A.A.S., A.M.A., Soc. for Exptl. Biology and Medicine, So. Soc. for Clin. Research, Radiation Research Soc., Sigma Xi. Research, publs. on afferent nervous factor in traumatic shock, leading to present approach to shock treatment; discovered factors responsible for altered capillary and cellular permeability in disease; pioneered in gen. physiol. and biochem. studies in primates, in radio-protective drug utilization in large animals. Home: 3725 Spottswood Av., Memphis 38111.*

OVERZIER, Claus, German neurologist; b. Cologne, Germany, July 19, 1918; s. Carl and Christine (Goettert) O.; ed. univs. Fribourg, Bonn, Jena, Vienna, Cologne; M.D.; m. Kriemhild Stieve, Sept. 12, 1945; 1 dau., Brigitte. Asst., Pathol. Inst., U. Berlin (Germany), Berlin Neurol. Clinic doctor Clinic and Polyclinic, Berlin; agrégé; prof.; head doctor; now dir. clinic and polyclinic U. Mainz (Berlin). Referendary in Congress for Endocrinologists, Anatomists and Gynecologists (Die Intersexualität). Author papers on endocrinology. Home: Neumannstrasse 4. Office: Langenbeckstrasse 1, Mainz, West Germany.

OWEN, Charles Archibald, Jr., Am. physician; b. Assiut, Egypt, Dec. 3, 1915 (parents Am. citizens); s. Charles Archibald and Margaret Jane (Corette) O.; A.B., Monmouth Coll., 1936, D.Sc., 1959; M.D., State U. Ia., 1941; Ph.D., U. Minn., 1950; m Edna Maude Stonier, June 8, 1939; children—Nancy Anne, Martha Elizabeth, Mary Margaret. Research asst. State U. Ia., 1938-39; fellow internal medicine Mayo Found., U. Minn., 1946-49, faculty, 1950—, prof. med. research, 1960—; cons. clin. pathology Mayo Clinic, Rochester, Minn., 1950—. Mem. Am. (Gold medal 1939), Minn. (pres. 1963-64) socs. clin. pathologists, A.A.A.S., Am. Assn. Clin. Chemists, Am. Chem. Soc., A.M.A., Am. Gastroent. Soc., Am. Physiol. Soc., Am. Thyroid Assn., Internat., Am. socs. hematology, Coll. Am. Pathologists, N.Y. Acad. Sci., Soc. Exptl. Biology and Medicine, Soc. Exptl. Pathology, Soc. Nuclear Medicine, Sigma Xi, Alpha Omega Alpha, Phi Rho Sigma. Author: Diagnostic Radioisotopes, 1959; also numerous articles on blood coagulation and radiobiochemistry. Home: 215 15th Av. S.W., Rochester 55901. Office: Mayo Clinic, Rochester, Minn. 55901.*

OWEN, David Dale, geologist; b. New Lanark, Scotland, June 24, 1807; s. Robert and Ann (Dale) O.; attended ednl. inst. of Philip Emanuel von Fellenberg, Berne, Switzerland, 1824-27; M.D., Ohio Med. Coll., Cin., 1836; m. Caroline Neef, Mar. 23, 1837; 4 children. Came to U.S., 1827; state geologist Ind., 1837-38; made survey of Dubuque

(Ia.) and Mineral Point (Wis.) dists. under U.S. appointment, 1838 (report published as House Document 239, 1840); U.S. geologist to survey Chippewa Land Dist., 1847-52; published Report of a Geological Exploration of a Part of Iowa, Wisconsin and Minnesota and, Incidentally, a Portion of Nebraska Territory, 1852; state geologist Ky., 1854-59, Ark., 1857-60; apptd. state geologist Ind., 1860, died before taking office. Author: Geological Survey of Kentucky, 4 vols., 1856-61; The Report of a Geological Reconnaissance of Indiana Made During the Years 1859 and 1860 Under the Direction of the Late D. D. Owen (published by Richard Owen), 1862. 1st to point out rich mineral nature of Ia., Wis. lands, also that lead and zinc ores were limited to the magnesium limestone; 1st to give name subcarboniferous to beds underlying Ind. coal. Died New Harmony, Ind., Nov. 13, 1860.

OWEN, Edwin Augustine, Brit. physicist; b. Blaenau Ffestiniog, Aug. 5, 1887; s. John and Ellen (Loyd) O.; ed. Bangor Coll., Trinity Coll., Cambridge U., London Univ. Coll.; M.A., Sc.D., D.Sc., M.Sc.; 1 son, John. Asst. meteorology dept. Nat. Physics Lab., 1912-15; with Munitions Ministry, 1915-18; head radiology Nat. Physics Lab., 1919-26; prof. physics Univ. Coll. N. Wales, Bangor, 1926-54, prof. emeritus physics, 1954—. Recipient Röntgen award, 1927, Silvanus Thompson medal, 1945. Mem. Physics Soc., Inst. Physics, Cambridge Philos. Soc., Hosp. Physicists Assn., Inst. Quarrying, Brit. Inst. Radiology, Inst. Metals. Author articles radiology, X-rays, structure of crystalls in metals and alloys. Home: Penbre, College Rd., Bangor. Office: Univ. Coll., Bangor, North Wales, Eng.

OWEN, George, Brit. geologist; b. Henllys, Wales, 1552; s. William and Elizabeth (Herbert) O.; m. Elizabeth Philipps, 1573; 10 children; m. 2d, Ann Gwillim; 7 children. Admitted to Barnard's Inn, 1573; became sheriff of Pembroke County, 1587, 1602; apptd. by crown to survey property of Sir John Perrott, 1592. Author of numerous books including: Description of Pembrokeshire (contains description of geology of S. Wales), 1603; also numerous articles. Often called father of English geology. Died 1613.

OWEN, George E., Am. physicist; b. St. Louis, Jan. 7, 1922; s. George Ernest and Ruth (Spradling) O.; B.S., Washington U., St. Louis, 1943, M.S., 1948, Ph.D., 1950; student U.S. Naval Postgrad. Sch.; m. Deha Gursey, Oct. 3, 1959; 1 son, Stephen. Asst. prof. physics U. Pitts., 1950-51; faculty physics Johns Hopkins, Balt., 1951—, prof., 1960—. Mem. Am. Phys. Soc. Author: Fundamentals of Scientific Mathematics, 1961; Introduction to Electromagnetic Theory, 1963; Fundamentals of Electronics, 3 vols., 1967. Confirmed Fermi theory of Beta decay, 1949; research in nuclear reactions, stripping theory, 1955. Home: 4000 N. Charles St., Balt. 21218.

OWEN, Ray David, Am. biologist; b. Genesee, Wis., Oct. 30, 1915; s. Dave and Ida (Hoeft) O.; B.S., Carroll Coll. (Wis.), 1937, Sc.D., 1962; Ph.D., U. Wis., 1941; Sc.D., U. Pacific, 1966; m. June J. Weissenberg, June 24, 1939; children—David G., Griffith H. Asst. prof. genetics and zoology U. Wis., 1944-47; Gosney fellow Cal. Inst. Tech., Pasadena, 1946-47, asso. prof. div. biology, 1947-53, prof. biology, 1953—, also chmn.; research participant Oak Ridge Nat. Lab., 1957-58; chmn. genetics study sect. NIH; cons. Oak Ridge Inst. Nuclear Studies, U. S. AEC. Mem. Nat. Acad. Scis., Genetics Soc. Am. (pres.), Am. Assn. Immunologists, Am. Soc. Human Genetics, Western Soc. Naturalists, Am. Soc. Zoologists, Am. Genetics Assn., Nat. Inst. Allergy and Infectious Diseases (bd. sci. counselors), Sigma Xi. Author: (with A. M. Srb) General Genetics, 1952, 2d edit., 1965. Contbr. articles to sci. jours. Genetic contbns. include research in blood groups, tissue transplantation, immunology, especially devel. immunological tolerance to transplants. Home: 1583 Rose Villa St. Office: 1201 E. California St., Pasadena, Cal. 91106.*

OWEN, Sir Richard, English comparative anatomist, paleontologist, zoologist; b. Lancaster, Eng., July 20, 1804; s. Richard and Catherine Owen; M.D., Edinburgh, Scotland; hon. degrees, Oxford, Cambridge, Dublin; m. Caroline Clift, 1835. Became prosector to Obernethy, St. Bartholomew's Hosp., 1825, lectr. anatomy, 1829; named asst. conservator Hunterian Mus., 1827, joint conservator, then soil conservator, 1842-56; 1st Hunterian prof. comparative anatomy and physiology Royal Coll. Surgeons, 1836-56. Fellow Royal Soc., 1834, Royal medal, 1846, Copley medal, 1851; mem. French Acad. Scis., 1839, Microscop. Soc. (became 1st pres. 1840). Recipient Prix Cuvier, 1857. Author: Odontography, 1840-45; Lectures on Comparative Anatomy and Physiology of Invertebates, 1843; History of British Fossil Mammals and Birds, 1846; On the Anatomy of Invertebrates, 1866-68; Researches on Fossil Remains of Extinct Mammals of Australia, 1877-78; Memoirs on Extinct Wingless Birds of New Zealand, 1879. Made important contributions to every dept. of comparative anatomy, zoology, including teeth of mammals, paleontology; 1st to describe of sponge, Euplectella (Venus's flower basket); discovered parasite in muscles of man in trichnosis; gave many important technical descriptions of both vertebrate and invertebrate animals; described both living and extinct forms; 1st to distinguish analogy and homology in vertebrate skeleton; his essay on parthenogenesis or "virgin reproduction" was a pioneer work; developed models of extinct animals for Crystal Palace Expn.; as last of the "naturephilosophers," opposed Darwin's theory of evolution by natural selection. Died Richmond Park, Eng., Dec. 18, 1892.

OWEN, Robert, Brit. economist, management scientist, social theorist; b. Newton, Montgomeryshire, Wales, May 14, 1771; s. Robert and Anne (Williams) O.; self-educated; m. Anne Dale, Sept. 30, 1771; children—Robert Dale, William, David Dale, Richard. Became asst. in sadlery at age 11, mgr. cotton mill, Manchester, at age 19; formed Chorlton Twist Co., 1794-95; moved to New Lanark, Scotland, 1799; became part-owner of father-in-law's cotton mills, and set up modern indsl. community; inspired passage of ill-fated Factory Act of 1819; campaigned unsuccessfully for widespread establishment of communities on New Lanark model; founder exptl. Community of Equality, New Harmony, Ind., 1825, but admitted its failure, 1827; upon return to Brit., promoted various assns., including Equitable Labor Exchange; again resident in Am., 1844-47; convened Congress of Advanced Minds of World, 1857. Author: New View of Society, or Essays on the Principle of the Formation of the Human Character, 1813-16; Report to County of Lanark, 1821; Book of the New Moral World, 1826-44; Revolution in Mind and Practice, 1849; Letters to the Human Race, 1850; Autobiography, 1857-58. A leading utopian socialist of 19th century; believed man's character is molded by environmental conditions; saw idealistically planned community living as solution to insecurity and demoralization in indsl. age; pioneer in management, especially in stressing importance of personnel and the human factor in industry. Died Newtown, Nov. 17, 1858.

OWEN, Walter Shepherd, Brit. metallurgist; b. Liverpool, Eng., Mar. 13, 1920; Ph.D., U. Liverpool; m. Carol Wood, 1953; 1 dau., Ruth. With English Electric Co., 1941-46; lectr. U. Liverpool, 1946-51, 53-54; mem. research team Mass. Inst. Tech., 1951-57; holder title chair metallurgy U. Liverpool, 1957—. Author articles in Brit. and Am. jours. Home: 6 Lansdowne Rd., Wallasey, Eng. Office: Univ. Liverpool, Liverpool 3, Eng.

OWENS, Arthur Neal, Am. plastic surgeon; b. Heflin, Ala., Aug. 5, 1899; s. James Arthur and Laura (Neal) O.; B.S., U. Ala., 1924; M.D., Emory U., 1926; m. Georgia May Little, Sept. 30, 1931; children—Laura (Mrs. George Lewis), Alice (Mrs. John C. Pryor), Alice C. (Mrs. Lee Johnson), Arthur Neal. Gen. surg. tng. with Dr. Joseph Bloodgood, Balt., 1928-33, also tng. plastic surgery; study plastic surgery with Dr. Harold Gillies, London, 1933; prof. plastic surgery Tulane U. Sch. Medicine; former head dept. plastic surgery Eye, Ear, Nose and Throat Hosp., New Orleans; cons. staff Touro Infirmary, Ill. Central R.R. Hosp., USPHS Hosp., Crippled Children's Hosp.; surg. staff Hotel Dieu, Mercy Hosp., Sara Mayo Hosp.; vis. surgeon Charity Hosp. Pres., bd. dirs. Audubon Park Natatorium. Recipient Honor award Am. Med. Writers. Diplomate Am. Bd. Plastic Surgery. Fellow A.C.S., Internat. Coll. Surgeons (exec. council, v.p. U.S. sect.; chmn. plastic surgery sect.); mem. Am. Assn. Plastic Surgeons (past pres., trustee), Am. Soc. Plastic and Reconstructive Surgeons (past pres.), A.M.A., So. Med. Assn., Am. Cancer Soc. (bd. dirs.), Sigma Xi, Sigma Chi, Phi Chi, Alpha Phi Omega. Developed use of celluloid in correction defects of contour, surg. tub for treatment of burns and infected wounds, protective pressure dressing for wound treatment, surg. fabric, use of periosteal grafts. Home: 6 Audubon Pl., New Orleans. Office: American Bank Bldg., New Orleans 70130.*

OWENS, Guy, Am. physician; b. Amarillo, Tex., Jan. 25, 1926; s. Guy Fistzhugh and Mary Helen (Virgin) O.; B.S. magna cum laude, Tufts U., 1946; M.D., Harvard, 1950; m. Janet Parkinson, June 11, 1949. With Vanderbilt U., Nashville, 1951-60, NRC, Rockefeller fellow, 1957-58, Markle scholar, 1958-60, asst. prof. anatomy and surgery, 1958-60; chief neurosurg. Vanderbilt U. Hosp. Out Patient Dept., 1958-60; chief dept. neurosurgery Roswell Park Meml. Inst., Buffalo, 1960—; staff VA Hosp., Buffalo, 1963—; asst. research prof. surgery State U. N.Y. at Buffalo; practice medicine specializing in neurosurgery, Nashville, 1958-60, Buffalo, 1960—. Mem. Am. Physiol. Soc., Am. Assn. for Cancer Research, Soc. U. Surgeons, A.C.S., Harvey Cushing Soc. Contbr. numerous articles to tech. jours. Defined cause convulsions during anesthesia; developed technique for therapy strokes, effective chemotherapeutic regime for primary brain tumors; pioneered isolation brain perfusion in mammals. Home: 87 Oakland Pl., Buffalo 14222. Office: 666 Elm St., Buffalo 14203.*

OWENS, Robert Bowie, Am. elec. engr.; b. Anne Arundel County, Md., Oct. 29, 1870; s. James and Maria Louise (Bowie) O.; grad. Charlotte Hall Mil. Sch., Md., 1886; Johns Hopkins, 1887-89; E.E., Columbia, 1891, A.M., 1899; B.Sc., ad eundem, McGill U., 1900, M.Sc., 1900, D.Sc., 1903; research student Cambridge (Eng.) U., 1899. Supt. Greenwich Gas & Elec. Co., 1889-91; prof. elec. and steam engring. U. Neb., 1891-98; Tyndal fellow in physics Columbia, 1898-1901; Macdonald prof. elec. engring. McGill U., 1898-1909; elec. engr. So. Power Co., 1909-10; sec. Franklin Inst., Phila., 1910-24, also editor Jour. Franklin Inst., dir. Bartol Research Found., 1921-24; pres. Fox Hall Farm, Harwood, Md., from 1927. In charge of and operated all telephone and telegraph communications between A.E.F., France and Eng., and all Am. owned cables (Western Union and Comml.) between Eng. and U.S., June-Dec. 1918. Mem. Internat. Elec. Congress and Internat. Jury of Awards, World's Fair, Chicago, 1893; dir. Electricity and Machinery Bldg., Trans-Miss. Expn., 1898 (Gold medal); mem. Internat. Elec. Congress and Internat. Jury Awards, La. Purchase Expn., 1904 (commemorative medal). Fellow Royal Soc. Can.; mem. Md. Acad. Scis. (editor jour.), Am. Inst. E.E. (v.p.), Canadian Soc. Civil Engrs. (pres. elec. sect.). Discoverer of Alpha ray; inventor radio direction finding, electromagnetic system for guiding ships and aeroplanes, differentiating machine, electric accelerometer; 1st to detect thorium emanation. Died Nov. 1, 1940.

OWENS, William Abbott Jr., Am. psychologist; b. Duluth, Minn., June 13, 1914; s. William Abbott and Sarah Jane (Hines) O.; B.A., Winona State Coll., 1935; Ph.D., U. Minn., 1940; m. Barbara Louise Ramsey, July 26, 1941; children—Scott Ramsey. Instr. psychology Ia. State U., Ames, 1940-42, from asso. prof. to prof., head psychology dept., 1946-59; prof. psychology Purdue U., Lafayette, Ind., 1959—. Adviser employee relations research Standard Oil Co., Chgo., 1956—. Recipient Centennial commendation Winona State Coll., 1960. Fellow Am. Psychol. Assn.; mem. Sigma Xi, Kappa Delta Pi, Psi Chi, Sigma Alpha Epsilon. Studies of age changes in mental abilities, creativity in sci., integration in behavioral scis.; aptitude tests. Home: 615 Carrolton Blvd., West Lafayette, Ind. 47906.*

OWMAN, Christer, Swedish anatomist; b. Lund, Sweden, Feb. 9, 1939; s. Olle Olsson and Wally (Person) O.; M.D., U. Lund, 1964; m. Ulla Britt Ericsson, June 10, 1962; children—Fredrik, Caroline. Faculty, U. Lund, 1959—, research asso., asst. prof. anatomy, 1965—, periodically dep. asst. surgeon, dept. neurol. surgery U. Hosp., 1962—. Research, publs. on embryology of nervous system and endocrine structures, histochem. and biochem. investigations on cellular localization, pharmacology and functional significance of histamine and various adrenergic neurohumors, basic investigations on induced deep hypothermia, especially with regard to its practical clin. application in neurosurgery. Home: Storskolevägen 15. Office: Biskopsgatan 5-7, Lund, Sweden.*

OWREN, Paul A., Norwegian physician; b. Faaberg, Norway, Aug. 27, 1905; s. Peder A. and Anna (Nermo) O.; M.D., U. Oslo (Norway), 1947; m. Marit Rodland, June 14, 1935 (dec. 1964); 1 dau., Inger Anna. Faculty, U. Oslo, 1943—, prof. medicine, 1949—, dir. Inst. for Thrombosis Research, 1955-64. Chmn. med. sect. Norwegian Research Council for Sci. and Humanities, 1958-61. Decorated comdr. Royal Order St. Olav, 1953. Hon. fellow A.C.P.; mem. Norwegian Soc. Internal Medicine (past pres.), Internat. Soc. Hematology (past v.p.), Royal Norwegian Soc. Scis., Acad. Sci. and Letters, Royal Soc. Scis. (Uppsala). Author: The Coagulation of Blood, 1947; also numerous articles. Research on blood diseases, blood coagulation, thrombosis, coronary heart disease, anticoagulan therapy; discovered blood clotting factor; devel. methods for controlling anticoagulant therapy. Home: Bjerkasen 44, Blommenholm pr., Oslo. Office: Rikshospitalet, Oslo, Norway.*

OZANAM, Jacques, French mathematician; b. Bouligneux, 1640. Prof. math. at Lyon, then Paris. Mem. French Acad. Scis., 1707. Author: Table des sinus, tangentes et sécantes, 1670; Dictionnaire mathématique ou l'idée general des mathématiques, 1690; Cours de mathématiques qui comprend toutes les parties de cette science les plus utiles et les plus nécessaires, 5 vols., 1693; Récréations mathématiques et physiques, 2 vols., 1694; Méthode pour tacer les cadrans; Géométrie pratique; Nouveaux éléments d'algèbra, ou Principes généraux pour résoudre toutes sortes de problèmes des mathématiques, 8 vols., 1702. Contbd. proof of theorem that neither sum nor difference of two 4th powers can be 4th power proof of theorem relative to imaginary roots, method to determine cubic and sur solid roots of binomial; one of 1st French mathematicians to develop method of repeated formulation of geometric means for computing common logarithms. Died Paris, Apr. 3, 1717.

OZAWA, Mitsuru, Japanese mathematician; b. Aichi-Pref., Japan, Oct. 30, 1923; s. Takuzo and Yaa (Huma) O.; B. in Math., Tokyo (Japan) Bunrika Daigaku, 1947; D.Sc., Osaka (Japan) U., 1956; m. Kazuko-Abe, Jan. 28, 1951; children—Shin, Kaku. Asst., Tokyo Gakugei Coll., 1947-50; faculty Tokyo Inst. Tech., 1950—, prof. math. 1965—; vis. prof., NSF Sr. Fgn. scientist fellow Washington U., St. Louis, 1965-66. Mem. Math. Soc. Japan, Am. Math. Soc. Research, numerous publs. on complex function theory, theory of open Riemann surfaces, conformal mappings, analytic mappings including value distbn. theory, univalent function theory, partial differential equations. Home: 922 Hiromachi, Hiromachi, Nakano-Ku, Tokyo, Japan.*

OZAWA, Yasutomo, Japanese applied physicist; b. Hokkaido, Japan, Aug. 24, 1919; s. Torao and Nobu (Watanabe) O.; M.Sc. in Engring., Hokkaido U., 1942; D.Sc. (Brit. Council scholar) Imperial Coll. and Sci. and Tech. U. London (Eng.), 1953; m. Yoko Shima, Nov. 2, 1946; 1 son, Tomohisa. Faculty, Research Inst. Applied Electricity, Hokkaido U., 1945-—, prof. applied nuclear physics lab. Sch. Engring., 1958-—, chmn. nuclear engring. lab., also accelerator and direct energy conversion lab., 1962-—. Councillor steering com. joint utilization Japanese Atomic Research Inst. for Govtl. U., 1964-—; commr. Inst. Reactor Research, Kyoto U., 1965-—; cons. research inst. plasma physics Nagoya U., also Electrotech. lab. Ministry of Trade and Commerce, 1964-—. Mem. Inst. Elec. Engrs. Japan, Atomic Energy Soc. Japan, Japan Soc. Applied Physics, Inst. Elec. Communication Engrs. Japan. Author: Electromagnetism, 1967; also numerous articles. Research on theoretical and exptl. verification of thermal instability of dielectrics and frequency-selective effects for inhomogenous medium under high frequency fields; new measurements of pulsed nuclear radiation; minimum entropy increase formalism for magnetohydrodynamical elec. power generation; stability problem of magnetoplasma in velocity space. Home: W.27, N.1, Sapporo, Japan.*

OZOL, Alfred Martinovich, Russian agronomist, phytoculturist; b. 1898; grad. Don Inst. Agr. and Reclamation, Novocherkassk, 1927; D. Biol. Scis. With Bot. Gardens, Forestry Inst., Council for Coordinating Sci. Activities of Union Republics Acad. Scis., USSR Acad. Scis., 1937-52; dir. Inst. Biology, Latvian Acad. Scis., 1952-—; prof. botany. Mem. Latvian Acad. Scis. Author: Darwin's Teachings and the Evolution of Biological Science, 1959. Research and publs. on physiol. and ecol. studies of adaptive variability during introduction and acclimatization of arboreal and bushy plants of southern origin. Address: Inst. Biology, Latvian Acad. Scis., Meistarn St. 10, Riga, Latvian SSR, USSR.

P

PABST, Adolf, Am. mineralogist, crystallographer; b. Chgo., Nov. 30, 1899; s. Fridolin and Emma (Hoffmann) P.; grad. Francis W. Parker Sch., Chgo., 1918; A.B., U. Ill., 1925; Ph.D., U. Cal., Berkeley, 1928; m. Gudrun-Lisabeth Berg, June 5, 1929. Faculty mineralogy U. Cal., Berkeley, 1927-—, prof., 1944-67, prof. of mineralology emeritus, since 1967. Am.-Scandinavian Found. fellow, Oslo, 1928-29; Guggenheim fellow, 1938-39; Fulbright scholar, Vienna, 1955-56. Mem. Mineral. Soc. Am. (pres. 1951, Roebling medal 1965), Geol. Soc. Am. (v.p. 1951), Crystallographic Soc. Am. (pres. 1948-49); Am. Crystallographic Assn., Mineral. Soc. London, Mineral. Assn. Can., Sigma Xi (v.p. Cal. chpt. 1947), Phi Beta Kappa, Theta Tau (hon. mem.). Author: Minerals of California, 1938; Mineral Tables, 1938; also articles. Research in crystallography; crystal structure. Home: 36 Highgate Rd., Berkeley, Cal. 94707.

PABST, Charles Frederick, Am. physician, dermatologist; b. N.Y.C, Dec. 3, 1887; s. Charles and Margaret (Connorton) P.; M.D., L.I. Coll. Hosp., 1909; intern Brooklyn Hosp., 1910-12; unmarried. Student skin diseases in Puerto Rico and Venezuela; conducted clinic for skin diseases at Brooklyn and Greenpoint hosps., 1914-28; attending dermatologist and chief of clinic for skin diseases at Greenpoint Hosp., 1915-57, consultant dermatologist, 1957-—. Recipient award from Med. Soc. of State N.Y. Fellow A.M.A., Am. Acad. of Dermatology and Syphilology; mem. N.Y. State Med. Soc., Kings County Med. Soc., Alumnus Club L.I. Hosp., Brooklyn Hosp. Contbr. numerous articles on skin diseases and regarded as an authority on the subject. An expert swimmer, and saved several persons from drowning, at different times, on L.I. beaches. Gave U. S. Govt., 1934, nonpatented inexpensive formula for fireproofing ships, clothing and other fabrics; called attention to widespread prevalence of ringworm infection of feet, and started health campaign against bare feet; originated term "athlete's foot"; secured almost universal adoption of distinctive shape and color for bichloride of mercury tablets; pointed out dangers of overexposure to summer sun and gave the term "heliophobe" to individual whose skin will not tan. Address: 15 Clark St., Bklyn. 11201.

PACAULT, Adolphe, French chemist; b. Paris, Oct. 30, 1918; s. Auguste and Marthe (Marne) Pacault; Docteur ès sciences, Faculté des Sciences de Paris, 1946; m. Nicole Mury, July 31, 1943; children—Joëlle, Gillees, Hervé. Asst., Faculté des Sciences, Paris, 1949; master research Centre National de la Recherche Scientifique, 1949; became prof. Faculté des Sciences, Bordeaux, France; dir. Institut de Magnéto-Chimie, Centre Natinal de la Recherche Scientifique, 1962-—. Recipient Girard prize, Cahours prize French Acad. Scis.; Silver medal Service Naval Health. Author: Eléments de thermodynamique Statistique, 1963; Les Carbones, 1965; also numerous articles. Research in magneto-chemistry, geometric and electronic properties of carbons, evolution of systems. Home: Brivazac, Pessac (33), France. Office: Institut de Magnéto-Chimie, Brivazac (33), France.*

PACCHIONI, Antonio, Italian anatomist; b. Reggio, Italy, June 13, 1665; student philosophy, math., medicine, Reggrio; physician of Tivoli, 1692; moved to Rome, circa, 1700; became asso. with Lancisi, prof. Rome, Tivoli, Italy. Author: Dissertatio epistolaris ad Lucam Schroeckium de glandules conglobatis durae meningis humanae, 1705; Opera omina, pub. 1741. Described small arachnoid elevations (called Pacchionian bodies) forming clusters on surface of dura mater which produced depressions (called Pacchionian depressions) on inner surface of cranium (he believed they were glands for prodn. of lymph), 1705; Pacchionian foramen (incisura tentorii) also named for him. Died Rome, Italy, Nov. 5, 1726.

PACCHIONI, Giuseppe, Italian zoologist; b. Capri, Italy, July 14, 1893; s. Amilcar and Elisabeth (Ganolf) P.; Dr. Zoology degree; m. Giuseppina Redeschini, Nov. 24, 1927; children—Mariarenza, Ann Elisabeth. Prof. zoology U. Turin (Italy). Mem. Accademia Peloritana. Publns. on prevention infectious diseases of animals, causes of disease, pathogeny and diagnosis. Home: via Malmusi 2, Modena, Italy. Office: via Nizza 52, Turin, Italy.

PACCIOLI, see De Borgo, Luca.

PACE, Donald Metcalf, Am. physiologist; b. Wilkes-Barre, Pa., Nov. 8, 1906; s. James Weir and Gertrude (Metcalf) P.; student Temple U., 1924-26; B.S., Susquehanna U., 1928, D.Sc. (hon.), 1962; M.S., Duke, 1929, Ph.D., 1931; m. Norma Roberta Holland, Apr. 27, 1936; children—Donald Metcalf, James Robert, Norma Jean. Faculty, Johns Hopkins, 1931-42; faculty U. Neb., Lincoln, 1942-—, prof. zoology and physiology, 1963-—, dir. Inst. for Cellular Research, 1951-67; prof. pharmacology, physiology, dir. cellular research U. Pacific, Stockton, Cal., 1967-—. Recipient Achievement award Susquehanna U., 1960. Fellow N.Y. Acad. Sci., A.A.A.S.; mem. Am. Inst. Biol. Scis., Am. Physiol. Soc., Am. Soc. Zoologists, Gerontological Soc., Internat. Soc. for Cell Biology, Neb. Acad. Sci. (pres. 1947-48), Soc. Exptl. Biology and Medicine, Soc. Gen. Physiologists, Soc. for Cryobiology, Soc. Study Growth and Devel., Soc. Protozoologists, Tissue Culture Assn., Sigma Xi (pres. Neb. 1958-59). Author: (with B.W. McCashland, C.C. Riedesel) Laboratory Manual for Vertebrate Physiology, 1964; (with B.W. McCashland, P.A. Landolt) Physiology and Anatomy, 1965; (with B.W. McCashland) College Physiology, 1964. Studies, publs. on cell nutrition, growth, metabolism; standardization of tissue cultures; toxicity of air pollution; ozone toxicity and tolerance; cell effects of carcinogens. Home: 8643 Robin Lane, Stockton, Cal. 95205.*

PACHECO, Carlos R., Mexican physician; b. Mexico City, Mexico, Oct. 2, 1921; s. Rogerio R. and Carmen (Escobedo) P.; B., Sch. Medicine, Morelos Sch., 1939; m. Ana Maria Hinojosa, June 27, 1947; children—Ana, Pilar, Carlos, Fernando, Manuel, Isabel, Martha, Federico. Thoracic surgeon Mexico City Gen. Hosp., 1946-54, chief neumology service, 1954-61; chief teaching dept. Neumology Hosp., N.M.C.I.M.S.S., 1961-66, chief exptl. surgery, 1966-—; practiced medicine specializing in chest diseases, Mexico City, 1966-—; staff Eugenio Sué Clinic, 1946-—; prof. respiratory diseases Faculty Medicine, Nat. U. Mexico, 1946-56, prof. clin. scis., 1956-—. Felow A.C.S., Am. Coll. Chest Physicians, Academia Nacional de Medicina, Academia Mexicana de Cirugía, Soc. Mexican Neumologists and Thoracic Surgeons (pres. 1963-—); mem. Am. Trudeau Soc. Author: (with Raul Cicero) Temas de Patología) del Torax, 1959; also numerous articles. Research on lung diseases and exptl. surgery in dogs. Home: 338 Temistocles, México. Office: 355 Eugenio Sue, México, D.F. México.*

PACINI, Filippo, Italian anatomist, histologist; b. Pistoia, Italy, May 25, 1812; prof., Pisa, Italy; prof. medicine Faculty of Florence, Italy, 1849-—. Author: Novos organos des robertos no corpo humano, 1840. Rediscovered peripheral nerve terminations described earlier by Abraham Vaten (Pacinian corpuscles or bodies), 1835; studied retina; discovered cholera vibrio in intestine. Died Florence, Jan. 9, 1883.

PACINOTTI, Antonio, Italian physicist, inventor; b. Pisa, Italy, June 17, 1841; ed. U. Pisa; studied under Ricardo Felici; became astronomer, Florence, Italy, 1862; named prof. U. Bologna (Italy), 1864, Cagliari, Italy, 1873, Pisa, 1882. Mem. London Soc. Elec. and Telegraphical Engrs. Built model of electromotor, 1859; invented dynamo with ring winding (independently discovered and used by Z. T. Gramme), 1860; introduced 1st generator, Bologna, 1869; 1st to build induction direct current; built magneto-elec. machines; used small metal rolls instead of brushes in his machines. Died Pisa, Mar. 25, 1912.

PACIOLI, Luca (or Paciolus, Luca de Borgo), Italian mathematician; b. Borgo San Sepolcro, Tuscany, Italy, circa 1445; took voyage to Europe; with Ct. of Louis the Moor, Milan, Italy where he worked with Leonardo da Vinci until arrival of French; moved to Venice, 1464; became Franciscan monk; became tchr. math., Perugia, Italy, 1475, later at Florence, Naples, Pisa, Bologna, Venice, Rome (all Italy). Author: Summa de arithmetica, geometria, proportioni et proportionalita (introduced double-entry bookkeeping, theory of probability), 1494; Libellus in tres partiales tractatus, divisus quorumcumque corporum regularium et dependentium activae percutationis, 1508; Divina proportione, (with plates by Leonardo da Vinci; used letters for numerical quantities), 1509. Translator: Elements (Euclid), 1508. Studied fundamental operations of arithmetic; simple and quadratic equations in algebra although he recognized positive roots only; understood rules of signed numbers; his math. methods differed from the Greeks in that he recognized constant union between algebra and geometry; in his math. writings algebraic symbols began to appear. Died Rome, circa 1517.

PACK, George Thomas, Am. surgeon; b. Antrim, O., May 14, 1898; s. Charles Kenerly and Tacy (Dugan) P.; B.S., Ohio State U., 1920; M.D. cum laude, Yale, 1922; diploma in reconstructive surgery Hopitale de Paris, 1927; LL.D., U. Ala., 1942; M.D. honoris causa, Coll. Physicians and Surgeons, Costa Rica, 1942, U. Buenos Aires (Argentina), 1951; hon. degrees other fgn. univs.; m. Helen Weigelt, 1945; children—Kathleen (Mrs. George Mechir), George Thomas, Christopher Charles, Tacy, Helen Clytie, Johnathan. Asst. physiology Ohio State U., 1920-21; instr. pharmacology, toxicology Yale, 1921-23, asst. prof. clin. surgery Sch. Medicine, 1932-39; asso. prof. pathology U. Ala., 1923-24, lectr. minor surgery, 1923-28, prof. pathology, asso. dean Sch. Medicine, 1924-28; asst. prof. clin. surgery Cornell U. Med. Coll., N.Y.C., 1936-50, asso. prof. clin. surgery, dept. surgery, 1950-64; clin. prof. surgery N.Y. Med. Coll., N.Y.C., 1944-60. Cons. numerous N.Y., N.J. hosps.; hon. prof. S.Am. univs.; vis. lectr. numerous U.S., fgn. univs. Recipient award for distinguished achievement Modern Medicine, 1963; numerous fgn. decorations. Curie Found. fellow U. Paris, 1927, Rockefeller fellow Meml. Cancer Center, 1928-31. Diplomate Nat. Bd. Med. Examiners, Am. Bd. Radiation Therapy. Fellow A.A.A.S., Am. Coll. Surgeons, A.M.A., N.Y. Acad. Medicine; mem. Am. Assn. Cancer Research, Am. Assn. Pathologists, Bacteriologists, Am. Physiol. Soc., Am. Radium Soc., Soc. Exptl. Biology and Medicine, Soc. Nuclear Medicine, Phi Beta Kappa, Sigma Xi, others. Author: A Syllabus of Pathology, 4th edit., 1928; (with A. Hobson Davis) Burns. Types, Pathology and Management, 1930; (with Edward M. Livingston) End Results in the Treatment of Gastric Cancer, 1939; (with Irving M. Ariel) Tumors of the Soft Somatic Tissues, 1958, Tumors of the Stomach, 1967; others. Editor: (with Irving M. Ariel) The Treatment of Cancer and Allied Diseases, 2d edit., 10 vols., 1958-64; (with Gray H. Twombly) Endocrinology of Neoplastic Diseases, 1947; Cancer and Allied Diseases of Infancy and Childhood, 1959; mem. editorial bd. various publs. Research on incidence, also surg. treatment gastric cancer, tumors of extremities. Home: 131 S. Woodland St., Englewood, N.J. Office: 139 E. 36th St., N.Y.C.*

PACKARD, Alpheus Spring, Am. zoologist, entomologist, geologist; b. Brunswick, Me., Feb. 19, 1839; s. Alpheus Spring and Frances Elizabeth (Appleton) P.; grad. Bowdoin, 1861; LL.D.; 1901; Me. Med. Sch., 1864; studied under Agassiz, Lawrence Sci. Sch., 3 yrs.; S.B., Harvard (out of course), 1864; m. Elizabeth Derby Walcott, Oct. 1867. Asst. surgeon Me. Vet. Vols., 1864-65; librarian and custodian Boston Soc. Natural History, 1865-66; curator Essex Inst., 1866; curator, afterward dir. Peabody Acad. Sci., 1867-78; State entomologist Mass., 1871-73; mem. U.S. Entom. Commn., 1877-82; an hon. pres. Zool. Congress, Paris, 1889; a founder, editor-in-chief, Am. Naturalist, 20 years. Author: Guide to the Study of Insects, 1869; Observations on the Glacial Phenomena of Labrador and Maine, 1891; A Text Book of Entomology, 1898; Lamarck, the Founder of Evolution, His Life and Work, 1901, French edit., 1903; A Naturalist on the Labrador Coast, 1891. Described 50 genera and about 580 species. Died 1905.

PACKARD, Edward Newman, Am. physician; b. Dorchester, Mass., May 14, 1883; s. Edward Newman and Elizabeth (Ford) P.; Ph.B., Syracuse U., 1906, M.D., 1909; m. Mary Bissell Betts, Nov. 28, 1919 (dec.); children—John Mallory, Ann (Mrs. Anthony Thornton Ladd), Mary Betts (Mrs. John Benjamin Reely). Practice medicine specializing in diseases of chest, Saranac Lake, N.Y., 1913-17, 19-42, 50-—; med. dir. Trudeau (N.Y.) Sanatorium, 1946-50; lectr. medicine U. Rochester Sch. Medicine, part-time 1947-55; med. supt. Saranac Lake Rehab. Guild, 1954-65; attending physician Will Rogers Meml. Hosp., 1957-—. Cons. sanatoriums in Saranac Lake area, 1919-42, VA., 1946; instr. Trudeau Sch. Tb, 1925-—. Fellow A.C.P.; mem. Saranac Lake (past pres.), Franklin City med. socs., Nat. Tb Assn., Am. Trudeau Soc., Am. Clin. and Climatol. Assn., Am. Assn. for Thoracic Surgery. Discovered embryo of hookworm in human blood, bronchial obstruction caused lung tissue collapse in Tb.*

PACKCHANIAN, Ardzroony Arthur, microbiologist, immunologist; b. Van, Armenia, Nov. 17, 1900; s. Arthur and Sophia (Tumanyan) P.; naturalized, Am. citizen, 1928. Research bacteriologist, protozoologist Sch. Medicine, U. Washington, St. Louis, 1933-34; protozoologist NIH, USPHS, 1936-41; faculty U. Tex. Sch. Medicine, Galveston, 1941-—, prof. microbiology, 1957-—, dir. Microbiology Lab., 1946-65. Fellow A.M.A., Royal Soc. Tropical Medicine and Hygiene; mem. A.A.A.S., Soc. Bacteriologists, Soc.

Parasitologists, Soc. Mammalogists, Soc. for Exptl. Biology and Medicine, Soc. Protozoologists, Am. Assn. U. Profs. Research, publs. on leishmaniasis, African human sleeping sickness, chemotherapy, cultivation trypanosomes, spirochetes, tissue culture, relapsing fever, rat-bite fever, toxicity tests, Chagas' and Wil's diseases. Home: Foundation Apts., Galveston, Tex. 77551.*

PACKE, Christopher, English chemist; b. circa 1635-45; children include son, Edmund. Set up lab. in London, 1670; called himself prof. of chemical medicine; practiced as quack under patronage of Robert Boyle and Dickinson. Author: Mineralogia; or an Account of the Preparation, Manifold Vertues, and Uses of a Mineral Salt, both in Physick and Chyrurgery . . . to which is added a short Discourse of the Nature and Uses of the Sulphurs of Minerals and Metals in Curing Diseases, 1693; Medela Chymica; or an Account of the Vertues and Uses of a Select number of Chymical Medicines . . . as also an Essay upon the Acetum Acerrimum Philosophorum, or Vinegar of Antimony, 1708. Translator: De Succo Pancreatico of R. de Graaf, 1674; Praxis Catholica by R. Couch, 1680; Eremita Suburbanus, Penotus, Glauber. Known as English translator of works of J. R. Glauber, 1689; believed in transmutation and thought his vinegar of antimony was means of transmutation.

PACKE, Christopher, English physician, geologist; b. St. Albans, Eng., Mar. 6, 1686; s. Christopher Packe; M.D., Cambridge (Eng.) U., 1717; m. Mary Randolph, July 30, 1826; 1 son, Christopher. Practice medicine, Canterbury, Eng., 1726-49. Fellow Royal Soc., 1726; mem. Coll. Physicians. Author: A Dissertation upon the Surface of the Earth, 1737; A New Philosophico-chorographical Chart of East Kent, 1743. Application of hachuring by parallel lines drawn in direction of slope to show pattern of valleys and their use in draining waters. Died Nov. 15, 1749.

PACKER, Donald MacGregor, Am. physicist; b. Wellsville, O., Jan. 27, 1908; s. Frank Jesse and Carrie (Henderson) P.; A.B., Miami U., Oxford, O., 1933; M.A., U. Rochester, 1935, Ph.D., 1938; m. Francis Ann Coffee, Mar. 3, 1939; children—Donald Macgregor, Frank C., Nancy A. Rockefeller Found. research asso. Washington U. Sch. Medicine, St. Louis, 1938-39; with U.S. Naval Gun Factory, Washington, 1939-46; with rocket spectroscopy br. atmosphere and astrophysics div. U.S. Naval Research Lab., Washington, 1946—, supervisory physicist, 1955—. Recipient U.S. Meritorious Civilian Service award, 1946; Progress medal Photog. Soc. Am., 1959. Fellow Optical Soc. Am.; mem. Am. Geophys. Soc., Sci. Research Soc. Am. Research, publs. on measurement of altitudes of night airglow emissions in upper atmosphere, ultraviolet spectroscopy and spectrophotometry of sun and terrestrial sources, optical instrumentation for space research. Home: 6420 15th St., Alexandria, Va. 22307. Office: U.S. Naval Research Lab., Washington 20390.*

PADGETT, Earl C., Am. plastic surgeon; b. Greenleaf, Kan., July 8, 1893; s. John Manson and Martha (McGinnis) P.; B.S., U. Kan., 1916; M.D., Washington U., 1918; m. Winona Youmans, June 1, 1922; children—Joyce, Patricia, Earl, Calvin. Intern, then asst. resident and resident in surgery Washington U., St. Louis, 1919-22; asst. to Dr. V. P. Blair, St. Louis, 1922-24; in practice of plastic surgery, Kansas City, Mo., since 1926; successively instr., asst. prof., asso. and prof. clin. surg. U. Kan. Sch. Medicine, 1925-43; clin. prof. oral surgery U. Kan. City Sch. Dentistry, since 1926; mem. staff Mercy Hosp., Kansas City Gen. Hosp., Providence Hosp.; exec. staff, St. Luke's Hosp. Mem. founders' group Am. Bd. Surgery, Am. Bd. Plastic Surgery; fellow A.C.S.; mem. Am., Western surg. assns., A.M.A., Am. Assn. for Surgery of Trauma, Sigma Xi. Contbr. articles, especially on skin graftings to med. jours. Inventor dermatone for cutting skin grafts, 1938, developed with mech. aid of George J. Hood. Died Dec. 2, 1946.

PADOA, Emanuele, Italian biologist; b. Livorno, Italy, Sept. 21, 1905; s. Corrado and Argia (Favata) P.; Ph.D., U. Pisa (Italy); m. Lina Galeotti, Dec. 28, 1935. Prof. comparative anatomy Faculty of Sci., U. Florence; prof. gen. biology U. Sienna. Publns. on gen. biology, comparative anatomy of vertebrates. Research on sex determination and differentiation of vertebrates. Home: via s. Ammirato 105. Office: via Romana 17, Florence, Italy.

PADOVER, Saul K., Am. polit. scientist; b. Austria, Apr. 13, 1905 (parents Am. citizens); s. Keva and Frumet (Goldmann) P.; B.A., Wayne U., 1928; M.A., U. Chgo., 1930, Ph.D., 1932; m. Margaret Fenwick, Apr. 13, 1957. Research asso. U. Cal., Berkeley, 1932-36; asst. to U. S. Sec. Interior, Washington, 1938-43; intelligence officer FCC, Washington, 1943-44, OSS, 1944-45; editorial writer Newspaper PM, N.Y.C., 1946-48; research asso. Stanford, 1948; vis. prof. Sorbonne, Paris, France, 1949; prof. polit. sci. New Sch. for Social Research Grad. Faculty, 1949—, dean, 1950-55; vis. prof. U. Tokyo, U. Malaya, 1960, U. Kyoto, Japan, 1965. Decorated Bronze Star medal; French Legion of Honor. Mem. Am. Polit. Sci. Assn., Authors Guild, Phi Beta Kappa. Author numerous books including: Democracy by Jefferson, 1939; Jefferson (biography), 1940; The Complete Jefferson, 1941; A Jefferson Profile, 1955; The Washington Papers, 1956; The Genius of America, 1960; The Meaning of Democracy, 1963. One of developers of study of meaning and growth of democracy in U.S.; revived study of Thomas Jefferson with books pub. on subject. Home: 129 Amity St., Bklyn. 11201. Office: 66 W. 12th St., N.Y.C. 10011.*

PAETZOLD, Roland, German chemist; b. Grosshartmannsdorf, July 13, 1931; s. Erhard and Else (Voggenreiter) P.; ed. Tech. U. Dresden, Dipl.-Chem., 1955, Dr.rer.nat., 1958, Dr.rer.nat. habil., 1963; m. Ruth Haberecht, 1 dau., Kerstin. Asst., Inst. Inorganic Chemistry, Tech. U. Dresden, 1955-58, chief asst., 1958-63; dozent in inorganic and analytical chemistry Inst. Inorganic Chemistry, U. Jena, 1963-65, prof. inorganic and analytical chemistry, 1965—. Contbr. articles tech. jours. Work in synthesis of selenuium and tellurium compounds and their structural investigation by means of Ramanand infrared-spectroscopy. Home: 33 Friedrich-Engels Strasse, 69, Jena, East Germany.*

PAFFENGOLTS, Konstantin Nikolaevich, Russian geologist; b. Mar. 17, 1893; grad. Petrograd Mining Inst., 1919. Mem. staff geol. com. All-Union Geol. Inst., 1919—. Recipient Stalin prize; Order of Lenin. Mem. Armenian SSR Acad. Scis., 1943. Author: Geology of Armenia, 1948; Dashkesan and Zaglik, Magnetite and Alunite Deposits in the Gandsha Uezd, Armenian SSR, 1928; co-author A Geological Outline of the Basins of the Fedchenko Glacier and Tanymas River, 1935; The Seismotectionics of Armenia and the Adjacent Parts of Caucasia Minor, 1946; New Data on the Age of the Effusive Rocks of the Central Caucasus, 1956. Student of Transcaucasia and Pamir; work in hydrology and geol. engring. Address: An Arm. SSR, Yerevan, Armenia SSR, USSR.

PAGE, Charles Grafton, Am. inventor; b. Salem, Mass., Jan. 25, 1812; s. Jeremiah Lee and Lucy (Lang) P.; grad. Harvard, 1832; m. Priscilla Webster, Sept. 23, 1844; at least 5 children. Examiner U.S. Patent Office, 1841-52, 61-68; prof. chemistry, med. dept. Columbian Coll. (now George Washington U.), 1844-49. Devised self-acting circuit breaker (probably 1st to apply it to produce extreme alterations necessary in induction machines), circa 1837; demonstrated tones produced by rapid magnetization and demagnetization of iron bar, 1837; his inventions incorporated in a coil machine by Daniel Davis, Jr., 1838; completed small reciprocating electro-magnetic engine by 1846; developed induction apparatus (which in principle is modern induction coil); granted spl. Congressional appropriation to continue work on larger scale, 1849; built several large stationary reciprocating electro-magnetic engines of both vertical and horizontal type; established (with J.J. Greenough and Charles L. Fleischmann) Am. Poly. Jour. Science, Washington, 1852; patented design of reciprocating electro-magnetic engine, 1854; examiner of patents U.S. Patent Office, 1861-68. Author: Psychomancy, Spirit-Rappings and Table-Tippings Exposed, 1853; History of Induction: The American Claim to the Induction Coil and its Electrostatic Developments, 1867. Died Washington, May 5, 1868.

PAGE, Charles Hunt, Am. sociologist, educator; b. Tonawanda, N.Y., Apr. 12, 1909; s. Ralph and Laura (Hunt) P.; A.B., U. Ill., 1931; Ph.D., Columbia, 1940; m. Leonora McClure, June 15, 1936. Staff, Birch-Wathen Sch., N.Y.C., 1931-33; instr. sociology Coll. City N.Y., 1933-40, prof., chmn. dept. sociology and anthropology, 1952-53; field sec. Nat. Refugee Service, 1940-41; lectr. sociology Columbia, 1941-42; faculty sociology Smith Coll., Northampton, Mass., 1946-60, prof., 1950-52, 53-60, chmn. dept., 1951-52, 53-60; prof. sociology, chmn. dept. sociology and anthropology Princeton (N.J.), 1960-65; provost, prof. sociology Adlai E. Stevenson Coll., U. Cal., Santa Cruz, 1965-67; prof. sociology U. Mass., Amherst, 1968—. Cons. editor sociology Doubleday, Inc., 1952-55, Random House, Inc., 1955—. Mem. Am. Sociol. Assn. (council 1958-64), Eastern Sociol. Soc. (pres. 1965-66). Author: Class and American Sociology, 1940; (with R.H. MacIver) Society, 1949; (with M. Berger and T. Abel) Freedom and Control in Modern Society, 1954; Sociology and Contemporary Education, 1964. Editor: Am. Sociol. Rev., 1958-60. Publs. on an analysis of dual stratification of social system in U.S., and settings of Am. sociology; devel. of a system of sociolog. theory (with MacIver). Home: 7 Hampton Terrace, Northampton, Mass. 01060. Office: Dept. Sociology, U. Mass., Amherst, Mass. 01003.*

PAGE, Ernest Winslow, Am. obstetrician, gynecologist; b. Berkeley, Cal., Oct. 9, 1909; s. Clarence W. and Julia (Phillips) P.; A.B., Stanford, 1930, M.D., 1934; m. Dorothy Murdock, June 16, 1933; children—Marilyn (Mrs. Jerry South), Nancy (Mrs. C.R. Ostrom), Kathy (Mrs. Dennis Elliott), Martha (Mrs. David Blunt). Faculty, U. So. Cal., 1936-38; faculty U. Cal. Med. Center, San Francisco, 1938—, prof., 1954—, chmn. dept. obstetrics and gynecology, 1956—. Program cons. Nat. Inst. Child Health and Human Devel., 1962—. Mem. Am. Physiol. Soc., Am. Gynecol. Soc., Am. Assn. Obstetricians and Gynecologists, Soc. Gynecologic Investigation (pres. 1955), Pacific Coast Obstet. and Gynecol. Soc., Soc. Exptl. Biology and Medicine, Western Soc. Clin. Research, Am. Fedn. Clin. Research, A.M.A., Am. Coll. Obstetricians and Gynecologists, Am. Assn. Maternal and Child Health (pres. 1962), Phi Beta Kappa, Sigma Xi, Alpha Omega Alpha. Research, numerous publs. in physiology and biochemistry of human reprodn. Home: 312 Filbert St., San Francisco 94111.*

PAGE, Irvine Heinly, Am. physician; b. Indianapolis, Ind., Jan. 7, 1901; s. Lafayette and Marian (Heinly) P.; B.A. in Chemistry, Cornell U., 1921, M.D., 1926; Doctor of Laws, John Carroll University, 1956; D.Sc. (hon.), Union University, Boston University; D. Sc., Ohio State U., 1960, U. of Brazil, 1961; m. Beatrice Allen, Oct. 28, 1930; children—Christopher, Nicholas. Mem. interne staff, Presbyn. Hosp., New York, 1926-28; head chem. div., Kaiser Wilhelm Inst., Munich, Germany, 1928-31; asso. mem. Hosp. of Rockefeller Inst. for Med. Research, 1931-37; dir. Lilly Lab. for Clin. Research and Lilly Clinic 1937-44. Dir. Research Div., Cleve. Clinic Found., 1945-66, sr. cons., 1966—. Past mem. nat. adv. heart council, U.S.P.H.S.; chmn. governing bd. Methods in Medical Research. Recipient Lasker award American Heart Assn., 1958; alumni award of distinction Cornell U. Med. Coll., 1961; John Phillips Memorial award A.C.P., 1962; Gairdner award, 1963; distinguished service award A.M.A., 1964; Am. Acad. Achievement award, 1966; Oscar B. Hunter Meml. award, 1966; Passano Found. award, 1966. Mem. Central Society for Clinical Research, American Heart Assn. (past pres.), A.M.A. (past chmn. sect. on exptl. medicine), Am. Society Biol. Chemists, Am. Physiol. Soc., Am. Chem. Soc., A.A.A.S. (vice pres.), Nat. Research Council (subcom. on shock), Am. Acad. Arts and Sciences, American Society for Study of Arteriosclerosis, Sigma Xi. Author: Chemistry of the Brain, 1937; Hypertension, 1943; Arterial Hypertension—Its Diagnosis and Treatment, 1945; Experimental Renal Hypertension, 1948; Neurochemistry, 1955; Connective Tissue, Thrombosis and Atherosclerosis, 1959; Strokes, 1961; also articles profl. jours. Editorial bd. various profl. jours.; editor-in-chief Modern Medicine. Research in areas of brain chemistry, hypertension and arteriosclerosis; synthesized angiotension with F. M. Bumpus and H. Schwartz, 1956; developed drugs for treatment of hypertension, also concept that showed similar causative factors of hypertension and arteriosclerosis (called them diseases of regulation). Home: 2258 Coventry Rd., Cleve. 44118. Office: 2020 E. 93d St., Cleve. 44106.*

PAGE, James D., Am. psychologist; b. Rome, N.Y., May 6, 1910; s. Anthony and Josephine (Pace) P.; A.B., Columbia U., 1932, M.A., 1934, Ph.D., 1935; m. Dorothy S. Skene, Sept. 3, 1938; children—Margaret, Bruce D. Instr., U. Rochester, 1937-41; faculty Temple U., Phila., 1941-44, 46—, prof., dir. grad. tng. in clin. psychology, 1948—, dir. Psychol. Clinic, 1946-47. Cons., VA, 1955—. Mem. Am., Eastern psychol. assns., Sigma Xi. Author: (with C. Landis) Modern Society and Mental Diseases, 1938; Abnormal Psychology, 1947; Psychopathology, 1966; also articles. Research on cross-cultural, biosocial constants in mental disorders, efficacy of social group work in prevention of delinquency. Home: 649 Lindley Rd., Glenside, Pa. 19038. Office: Temple U., Phila. 19122.*

PAGE, Leigh, Am. physicist; b. South Orange, N.J., Oct. 13, 1884; S. Edward D. and Cornelia (Lee) P.; Ph.B., Sheffield Sci. Sch. (Yale), 1904; Ph.D., Yale, 1913; m. Mary Cholmondeley Thornton, June 27, 1910; children—Thornton Leigh, Barbara Helen, Marjory. Instr. physics Yale, 1912-16, asst. prof., 1916-22, prof. math. physics, from 1922. Apptd. judge Einstein contest by Sci. Am., 1921. Fellow Am. Acad. Arts and Scis., Am. Phys. Soc., A.A.A.S.; mem. Am. Math. Soc., NRC, Sigma Xi, Phi Gamma Delta, Gamma Alpha. Author: An Introduction to Electrodynamics, 1922; An Introduction to Theoretical Physics, 1928; (with N.I. Adams, Jr.) Principles of Electricity, 1931; Electrodynamics, 1940; also numerous professional papers. Developed new theory of electromagnetism. Died Sept. 14, 1952.

PAGE, Lorne Albert, Am. physicist; b. Buffalo, July 28, 1921; s. John Otway and Laura (Stewart) P.; B.Sc., Queen's U., Kingston, Can., 1944; Ph.D., Cornell U., 1950; m. Muriel Emily Jamieson, Sept. 7, 1946; children—J. Douglas, Kenneth L., James F., Donald S., David K. Faculty, U. Pitts., 1950—, prof. physics, 1958—. Guggenheim fellow Upsala U., Sweden, 1957-58; Alfred P. Sloan research fellow, 1961-63. Fellow Am. Phys. Soc. Research, publs. on measurements in photon and electron physics, law of electron scattering; developed method for analyzing circular polarization of high energy X-rays; measured first inherent polarization of a positive beta-particle; positronium atom in material; calculations on polarization prodn. and measurement. Home: 157 Lloyd Av., Pitts. 15218.*

PAGE, Robert Griffith, Am. physician, educator; b. Bryn Mawr, Pa., Mar. 25, 1921; s. Edward Crozer and Elizabeth (Griffith) P.; grad. St. Paul's Sch., Concord, N.H., 1939; A.B., Princeton, 1943; M.D., U. Pa., 1945; m. Mary Elizabeth Kent, Sept. 20, 1947; children—Robert Griffith, Elizabeth Wilmer, Mary Kent. Practice medicine, specializing in internal medicine, Chgo., 1953—; asst. instr. dept. pharmacology, U. Pa., 1949-50, instr., 1950-51, asso. pharmacology, 1951-53; vis. prof. pharmacology U. Rangoon Med. Coll., Burma, 1951-53; asst. prof. medicine U. Chgo., 1953-58, asso. prof., 1958—, asst. dean med. edn., 1957-63, asso. dean div. biol. sci., 1963—. Bd. dirs. Schweppe Found. Fellow A.A.A.S., A.C.P., Inst. Medicine Chgo.; mem. A.M.A, Am. Fedn. Clin. Research, Am. Heart Assn., Am. Soc. Pharmacology and Exptl. Therapeutics, Assn. Am. Med. Colls., Central Clin. Research Club, Central Soc. Clin. Research. Research cardiac physiology and pharmacology. Home: 5637 Woodlawn Av., Chgo. 60637.*

PAGE, Thornton Leigh, Am. astrophysicist; b. New Haven, Aug. 13, 1913; s. Leigh and Mary (Thornton) P.; B.S., Yale, 1934; Ph.D. (Rhodes scholar), Oxford (Eng.) U., 1938; m. Lou Williams, Aug. 28, 1948; children—Tanya, Mary Anne, Leigh II. Chief asst. Oxford U. Obs., 1937-38; instr. to asst. prof. astronomy U. Chgo., 1938-50; dep. dir. operations research office Johns Hopkins, Balt., 1950-58; prof. astronomy Wesleyan U., Middletown, Conn., 1958—, dir. Van Vleck Obs., 1959—; physicist Naval Ordnance Lab., Washington, 1941-43. Nat. Acad. Scis. research fellow Harvard-Smithsonian Obs., 1965-66. Decorated Bronze Star, Legion of Merit. Fellow Royal Astron. Soc.; mem. Am. Astron. Soc. (council), A.A.A.S., Astron. Soc. Pacific, Am. Assn. Physics Tchrs., Am. Geophys. Inst., Operations Research Soc. Am., Sigma Xi. Author: Physical Sciences, 1949; Stars and Galaxies, 1962; Wanderers in Space, 1964. Research on spectra of nebulae, masses and evolution of galaxies. Home: 319 Washington Terrace, Middletown, Conn. 06457.*

PAGEAU, Gérard, Canadian biologist; b. Montreal, Que., Can., Sept. 3, 1933; s. Honorat and Laurenza (Fournier) P.; B.A., U. Montreal, 1955, B.Sc. in Biology, 1958, M.Sc., 1959, Ph.D., 1967; m. Irène Couture, Nov. 20, 1963; 1 dau., Marie. Asst. biologist Dept. Pêcheries de Québec, 1957; faculty Inst. Botanique, U. Montreal, 1957-59; gen. biology tchr. Externat Classique Sainte-Croix, Montreal, 1959; staff Que. Dept. Tourism, Fish and Game, Wildlife Service Research Lab., Montreal, 1963—. Research, publs. on ichthyology, including warm-water fish ecology, smallmouth bass; limnology, including physico-chem. analysis, aquatic plants, mapping, biometrics. Home: 6463 Chouinard St. Office: 5075 Fullum St., Montreal, Que. 2, Can.*

PAGEL, Bernard Ephraim Julius, astronomer; b. Berlin, Germany, Jan. 4, 1930; s. Walter T. U. and Magdalene (Koll) P.; B.A., Sidney Sussex Coll., Cambridge U., 1950, M.A., 1954, Ph.D., 1955; postgrad. U. Mich.; m. Annabel Ruth Toby, July 12, 1958; children—Celia Ann, David Benjamin, Jonathan Francis. Research fellow Sidney Sussex Coll., 1953-56; prin. sci. officer Royal Greenwich Obs., 1956-61, sr. prin. sci. officer, 1961—. Astrophysicist, Sacramento Peak Obs., Sunspot, N.M., 1960; vis. reader astronomy U. Sussex, 1966—. Mem. Royal Astron. Soc. (council), Internat. Astron. Union. Research and publs. on examining temperature distbn. in outer layers of sun, theory of formation of spectral lines in outer layers of stars, determination of chem. composition of stars, deductions from chem. composition on theory of nuclear synthesis of elements. Home: Red Dwarf, Western Rd., Hailsham, Sussex. Office: Royal Greenwich Observatory, Herstmonceux, Sussex, Eng.*

PAGEL, Walter, pathologist, sci., medicine historian; b. Berlin, Germany, Nov. 12, 1898; s. Julius Leopold and Marie (Labaschin) P.; M.D., U. Berlin, 1922, U. Basel (Switzerland), 1961, U. Heidelberg (Germany), 1966; m. Magda Koll, Aug. 8, 1923; 1 son, Bernard. Lectr. pathology, history medicine U. Heidelberg, 1930-33, emeritus prof., 1956—; pathologist Papworth Village Settlement, Cambridge, Eng., also hon. sec., history of sci. lectures com. Cambridge U., 1933-39; cons. pathologist Central Middlesex Hosp., London, 1939-56, Clare Hall Hosp., Hertfordshire, Eng., 1956-65, hon., 1965—. Fellow Coll. Pathology, Soc. Apothecaries (hon. faculty medicine); mem. Am. Assn. History Medicine (hon.), Académie International pour l'Histoire des Sciences, Internat. Acad. History Medicine, Royal Soc. Medicine (hon., sect. history medicine). Author: Jo. Bapt. Van Helmont, Einführung in die Philosophische Medizin des Barock, 1930; Virchow und die Grundlagen der Medizin des XIX Jahrhunderts, 1931; Religious Motives in the Medical Biology of the 17th Century, 1935; The Religious and Philosophical Aspects of van Helmont's Science and Medicine, 1944; Paracelsus, An Introduction to Philosophical Medicine in the Era of the Renaissance, 1958; Das Medizinische Weltbild des Paracelsus seine Zusammenhänge mit Neuplatonismus und Gnosis, 1962; William Harvey's Biological Ideas, 1967; (with F. A. H. Simmonds, N. Macdonald, E. Nassau) Pulmonary Tuberculosis, 4th edit., 1966. Research, publs. on tissue changes attributable to hypersensitivity and immunity, philos., religious background to sci., med. discovery in 16th, 17th centuries, speculative basis modern disease, cellular pathology, history iatrochemistry, alchemy. Address: 58 Millway, London, N.W. 7, Eng.*

PAGENSTECHER, Alexander, German ophthalmologist; b. Wallau, Germany, Apr. 21, 1828; s. Frederic Pagenstecher. Introduced yellow mercuric oxide ointment for eye (Pagenstecher's ointment), 1862; improved cataract operation by extracting lens in closed capsule. Died Wiesbaden, Germany, Dec. 31, 1879.

PAGENSTECHER, Hermann, German physician; b. Langenschwalbach, Germany, Sept. 16, 1844; s. Frederic Pagenstecher; m. Bertha Wilhelmi, 1873; 2 sons, 1 dau. Passed exams., Wiesbaden, Germany, also Würzburg, Germany, 1867. Became med. asst. U. Hosp., Greifswald, Germany, 1868; asst. Wiesbaden Ophthalmic Hosp.; stayed in London, 1869-70; asst. surgeon during war, 1870-71; began 2d stay in London, 1872; staff Eye Hosp. until 1880; dir. Ophthalmic Hosp., Wiesbaden; also dir. pvt. clinic. Author: Atlas of the Pathological Anatomy of the Eyeball, 1874; Research and publs. on cataract operations, eye massage, eye injuries, treatment of asthenopia, ptosis operation, operations on cataract, diseases of vitreous body, optic nerves, glaucoma. Died Wiesbaden, Apr. 11, 1932.

PAGET, George Edward, English physician; b. Gt. Yarmouth, Eng., Dec. 22, 1809; s. Samuel and Sarah Elizabeth (Tolver) P.; student Gonvill and Caius Coll., Cambridge, Eng., 1827-31, M.B., Cambridge, 1833, M.L., 1936, M.D., 1838; student medicine, Paris, 1832; m. Clara Fardell, Dec. 11, 1851; 10 children. Physician, Addenbrooke's Hosp., 1839-84; bursar Caius Coll., Cambridge; became Linacre lectr. medicine St. John's Coll., 1851; prof. medicine Cambridge, 1872-92. Fellow Royal Soc., 1873, Royal Coll. Physicians; mem. Cambridge Philos. Soc. (pres. 1855-56; mem. council of senate). Research and numerous articles on aphasia, gastric epilepsy, variations of body temperature. Died Jan. 16, 1892.

PAGET, Sir James, English surgeon, pathologist; b. Gt. Yarmouth, Eng., Jan. 11, 1814; s. Samuel and Sarah Elizabeth (Tolver) P.; ed. St. Bartholomew's Hosp., London; D.C.L. Oxford (Eng.) U.; LL.D., Cambridge (Eng.) U.; M.D., Dublin (Ireland), Bonn, Würzburg (both Germany) univs.; m. Lydia North, 1844; 4 sons, 2 daus. Sub-editor Med. Gazette, 1837-42; demonstrator morbid anatomy St. Bartholomew's Hosp., from 1839, lectr. gen. anatomy, physiology, from 1843, asst. surgeon, from 1847, lectr. physiology Med. Sch., from 1859, surgeon, from 1861, lectr. surgery, from 1865, cons. surgeon, from 1871; fellow Royal Coll. Surgeons, 1843, Arris and Gale prof. anatomy, 1847-52; surgeon extraordinary to Queen Victoria, 1858; sergeant-surgeon extraordinary, 1867-77, sergeant-surgeon, 1877; vice-chancellor London (Eng.) U., 1883-95. Fellow Royal Soc., 1851; mem. Clin. Soc. (pres. 1869), Royal Med. and Chirurg. Soc. (pres. 1875), Path. Soc. London (pres. 1887), French Acad. Scis., 1885. Author: A Descriptive Catalogue of the Pathological Specimens contained in the Museum of the Royal College of Surgeons of England, 1847; A Descriptive Catalogue of the Anatomical Museum of St. Bartolomew's Hosp., 1847; Lectures on Tumours, 1851; Lectures on Surgical Pathology, 1853; Clinical Lectures and Essays, 1875; Studies of Old Case Books, 1891. Showed how pathology might be applied to elucidate clin. problems (no sci. bacteriology then); link between Hunterian surgery, modern devels. (due to recognition of part by micro-organisms in prodn. disease); described cancerous disease of nipple and areola of breast (Paget's disease), 1874; described osteitis deformans (also known as Paget's disease); 1st to advocate enucleation of tumors. Died Dec. 30, 1899.

PAGLIARO, Luigi, Italian physician; b. Vita, Italy, Apr. 4, 1931; s. Antonino and Concetta (Orlando) P.; Laureato in Medicina e Chirugia, U. Palermo (Italy), 1954; m. Vincenza Nicosia, Apr. 23, 1966. Faculty, U. Palermo, 1960—, asst. prof. medicine Clinica Medica Generale, 1961—, dir. labs. clin. chemistry, named chief liver disease and metabolism sect. 1959—, dir. in charge Postgrad. Sch. Gastroenterology, 1964, also cons. internal medicine Clinica Neuropsichiatrica, Clinica Chirurgica Generale. Mem. Clinica Medica di Palermo (editor Diagnosi e Terapia), Italian Soc. Gastroenterology (asso. editor Archivio Italiano di Malattie dell'Apparato Digerente). Author: L'epatite dà virus come problema di Santià Pubblica (with G. D. 'Alessandro, L. Bevere); also numerous articles. Described for 1st time clin. behaviour and significance of serum Alpha-hydroxy-butyric dehydrogenase (enzyme useful in diagnosis of myocardial infarction), familial occurrence of congenital spleno-portal venous anomalies; clin. and biochem. studies on liver and kidney diseases, diabetes; described occurrence of viral acute hepatitis in some relatives of cases with persistent hepatitis (suggesting prolonged persistence of virus). Home: 34/B Aquileja, Palermo, Italy.*

PAIC, Mladen, Yugoslavian physicist; b. Zagreb, Yugoslavia, Dec. 12, 1905; s. Filip and Paulina (Simon) P.; student Tech. Faculty, Zagreb, 1925-29, D.Tech.Sci., 1932; D.Sc., Sorbonne, Paris, France, 1933; m. Valerie Deutsch, Dec. 14, 1936; 1 son, Guy. Head physics Lab., Inst. Alfred Fournier, Paris, 1932-39; head X-ray lab. Cie Alais, Froges et Camargues, Paris, 1932-46; prof. physics Faculty Scis., Zagreb, 1946—, dir. Inst. Physics, 1960—; head nuclear physics lab. I, Inst. Ruder Boskovic, Zagreb, 1950-65. Recipient Nikola Tesla award, 1956; also 3 Yugoslav decorations for profl. activities. Mem. Yugoslav Soc. Math. and Physics, Soc. Francaise de Phys., Am. Phys. Soc., Faraday Soc. Author: Physical Measurements, 1948; General Physics: Mechanics, Heat and Thermodynamics, Electricity, Optics, Atomic Physics, 1966; also articles. Research on new thiomercuric sulphates, 1932, action of supersonic waves on microorganisms, 1935, semimicroradiography, 1941, glass electrode for differential titrimetrie, 1938, neutron generator, 1957, neutron yield of D-D reaction, 1963, discrimination of alpha-particles from protons in nuclear emulsions, 1964; 1st approximate determination by ultracentrifugation of molecular weight of hemolysin, 1938, of diptheria antitoxin, 1939; quantitative measurements of segragations in alloys using X-ray absorption, 1942. Home: 30 Krajiska, Zagreb, Croatia, Yugoslavia.*

PAIGE, Sidney, Am. geologist; b. Washington, Nov. 2, 1880; s. Nathaniel and Rosa Elizabeth (Goldsmith) P.; student U. Mich., 1901-03; postgrad. geology, Yale, 1908; m. Hildegard Brooks, Mar. 20, 1909; 1 son, Potter Brooks; m. 2d, Frances Hall, Mar. 1, 1924; children—Sidney Hall, Henry Hall. With engrs. Nicaragua Canal Commn., 1898-1900; U.S. Geol. Survey, 1903-07; geologist, Panama Canal Commn., 1907; with U.S. Geol. Survey, 1909-26; cons. geologist, 1926-33; adviser Bur. Mines, Dept. Economy, Republic of Turkey, 1933-35; prin. geologist N. Atlantic Div., U.S. Army Engrs., 1935-46. Vis. prof. engring. geology Columbia, 1946-58; dir. and exec. sec. Com. Geog. Exploration, Joint Research and Devel. Bd., Washington, 1947-49; now cons. geologist. Mem. Soc. Econ. Geologists, N.Y. Acad. Sci., Am. Soc. C.E., Geol. Soc. Am., Geol. Soc. Washington, Am. Inst. Mining and Metall. Engrs., Am. Assn. Petroleum Geologists, Sigma Xi, Chi Psi. Contbr. papers on stratigraphy structure, ore deposits, and petroleum in Alaska and U.S. Home: Forest St., Alpine, N.J.*

PAILER, Matthias, Austrian chemist; b. Leobersdorf, Austria, Apr. 4, 1911; s. Matthias and Magdalena (Schicker) P.; Doctor degree, U. Vienna (Austria), 1936; m. Hildegard Eisenhut, Oct. 19, 1940; 1 dau., Renate. Asst., Technische Hochschule, Vienna, 1936-38; faculty U. Vienna, 1938—, prof. chemistry, 1959—. Recipient R. Wegscheider-Preis, Austrian Acad. Sci., 1953. Mem. Verein osterreichischer Chemiker, Gesellschaft deutscher Chemiker. Research, numerous publs. on natural organic products especially alkaloids, naturally occurring substances containing nitro groups, furocoumarins and other lactones, synthetic work in heterocyclic chemistry, shale oils and tobacco smoke. Home: 23/10 Pfeilgasse, Vienna VIII, Austria.*

PAINE, John Alsop, Am. archeologist; b. Newark, Jan. 14, 1840; s. John Alsop and Amanda (Kellogg) P.; A.B., Hamilton, 1859, A.M., 1862, Ph.D., 1874; grad. Andover Theol. Sem., 1862; postgrad. Sheffield Sci. Sch. (Yale), Sch. Mines, Columbia, 1866-67, univs. Leipzig, Halle (both Germany), 1869, 70. Ordained, 1867; engaged by bd. regents to enlarge flora of State of N.Y., 1862-67; prof. natural sci. Robert Coll., Constantinople, 1867-69; prof. natural history and German, Lake Forest (Ill.) U., 1870-71; Coll. Phys. and Surgeons, and asso. editor The Independent, 1871-72; archeologist for 1st expdn. of Palestine Exploration Soc. east of the Jordan and Dead Sea, 1872-74; edited and published Jour. Christian Philosophy, 1882-84; staff Century Dictionary, 1887, 88; curator Met. Mus. Art, N.Y.C., 1889-1906. Author: Catalogue of Plants Found in Oneida County and Vicinity, 1865; Fifth Statement, Containing "Identification of Mount Pisgah," 1875 (Palestine Exploration Soc.); Handbook of Sculptural Plaster Casts and Bronze Reproductions in the Metropolitan Museum of Art. Researches in the archeaology of Chaldea, Asia Minor, Syria, Palestine, and Egypt, in the history and antiquities of unsuccessful attempts made in the Sixteenth Century to colonize the eastern coast of N. America, in the chemistry and radio-activity of rare elements. Died July 24, 1912.

PAINE, John Randolph, Am. surgeon; b. Dallas, Nov. 18, 1906; s. Randolph and Maude (Smith) P.; A.B., Harvard, 1927, M.D., 1931; M.S., U. Minn., 1936, Ph.D., 1938; m. Dorothy Turner, Sept. 13, 1932; children—Jonathan Turner, Judith. Instr. to prof. surgery U. Minn., 1936-47; prof. surgery U. Buffalo, State U. N.Y., Buffalo, 1947—; head, dept. surgery Buffalo Gen. Hosp., 1947—. Bd. visitors Roswell Park Meml. Inst. for Malignant Diseases, 1960—. Mem. Am. Surg. Assn., Am. Assn. for Thoracic Surgery, A.C.S. Researcher in naso-gastric intubation and suction, 1931-35, elasticity of lungs, 1935-38. Home: 177 Beard Av., Buffalo 14214. Office: 100 High St., Buffalo 14203.*

PAINE, Martyn, Am. physician; b. Williamstown, Vt., July 8, 1794; s. Elijah and Sarah (Porter) P.; A.B., Harvard, 1813, M.D., 1816; LL.D., U. Vt., 1854; m. Mary Ann Weeks, 1825, 1 dau., 2 sons. A promoter of med. coll. U. City N.Y., 1841-67, prof. insts. of medicine, 1841-50, prof. therapeutics and materia medica, after 1850; leading prof. of therapeutics in nation sent by faculty colleagues to Albany to influence passage of legislation permitting dissections in N.Y. (act passed 1854). Mem. Royal Soc. Prussia, med. socs. Leipzig, Sweden, Soc. Naturalists and Physicians of Dresden. Author: On the Cholera Asphyxia as It Appeared in the City of N.Y. in 1832; Medical and Physiological Commentaries,

1840-44; Essays on the Philosophy of Vitality and on the Modus Operandi of Remedial Agents, 1842; Institutes of Medicine, 1847; Materia Medica and Therapeutics, 1848; Review of Theoretical Geology, 1856. Died N.Y.C., Nov. 10, 1877.

PAINE, Richmond Shepard, Am. physician, educator; b. Washington, Aug. 3, 1920; s. Howard Shepard and Kate (Richmond) P.; A.B., Swarthmore Coll., 1941; M.D., Harvard, 1944; m. Mary Louise Collyer, Sept. 4, 1945; children—Howard S., Emily C., Martha R., Diana B. Asst. neurologist, asso. neurologist Children's Hosp., Boston, 1954-62; neurologist Children's Hosp. of D.C., Washington, 1962-—; prof. pediatric neurology George Washington U. Sch. Medicine, Washington, 1962-—; cons. neurology NIH, Walter Reed Army Med. Center, Convalescent Hosp. for Children. Chmn. med. adv. bd. United Cerebral Palsy of Washington, 1964-—. Member American Pediatric Society, American Academy of Pediatrics, Am. Acad. Neurology, Am. Acad. for Cerebral Palsy, Am. Soc. Human Genetics. Author: (with B. Crothers) The Natural History of Cerebral Palsy, 1959; (with T.E. Oppé) Neurological Examination of Children, 1966. Research, numerous pubs. in developmental neurology of infants and children, cerebral palsies, mental retardation and other chronic neurol. handicaps and their biochem. and cytogenetical correlations, learning disabilities. Home: 1416 33d St. N.W., Washington 20007. Office: 2125 13th St. N.W., Washington 20009.*

PAINE, Thomas, polit. writer; b. Thetford, Norfolk, Eng., Jan. 29, 1737; s. Joseph and Frances (Cocke) P.; m. Mary Lambert, Sept. 27, 1759; m. 2d, Elizabeth Ollive, Mar. 26, 1771. Excise officer, Eng., chosen as agt. for excisemen to agitate for higher pay, 1772; left for Phila. after meeting Benjamin Franklin in Eng., 1774; edited, contributed to Robert Aitken's Pa. Mag., 1775; pioneer in movement for abolition of Negro slavery; published pamphlet Common Sense, Phila., Jan. 10, 1776, urged immediate declaration of independence; expanded ideas in Common Sense in Public Good, 1780; edited Crisis, 12 issues, 1776-83, supported colonial cause; sec. to com. on fgn. affairs Continental Congress, 1777-79; published Dissertations on Government, the Affairs of the Bank and Paper Money, 1786; went to Eng., 1787; defended measures taken in revolutionary France and urged Eng. to overthrow monarchy and establish Republic; published 1st part of Rights of Man, 1791, 2d part, 1792; popular with English radicals; suppressed by William Pitt, outlawed and tried for treason; went to France as French citizen, 1792; elected to Nat. Conv. of France; deprived of citizenship when Robespierre came to power, 1793, arrested and imprisoned, 1793-94; wrote 1st part of Age of Reason, being an Investigation of True and Fabulous Theology, 1794, 2d part, 1796; released from prison at request of Am. minister James Monroe; seat in Nat. Conv. restored, 1795; wrote Agrarian Justice, 1795; The Decline and Fall of the System of Finance in England, 1795; Dissertation on the First Principles of Government, 1795; Letter to George Washington, 1796; returned to U. S., 1802. Polemicist of 18th century reform and revolutionary movements; writings present anti-monarchical ideas, polit. thought based on natural law, and deistic religious thought; pamphlets especially influential in Am. Revolution. Died N.Y.C., June 8, 1809.

PAINLEVÉ, Paul, French mathematician; b. Paris, Dec. 5, 1863; Ph.D., Ecole Normale Supérieure, 1887. Tchr. rational mechanics Faculty Scis., Lille, France, 1887-92; lectr. math., Paris, 1892, asso. prof., 1895; became lectr. Ecole Normal Supérieure, 1897, prof. gen. math., 1903; prof. math., Sorbonne, Paris; minister pub. instrn. and inventions Briand's cabinet, 1915-16; named minister of war to Ribot, 1917; became head of govt. after Ribot's resignation, 1917; resigned; formed (with Herriot) Cartel des gauches, 1924; became premier again, 1925; minster of war, 1926-29; minister of air, 1930-33. Recipient Grand Prize in Math., French Acad. Scis., 1890. Mem. French Acad. Scis., mem. acads. of sci of France, Bologna, Stockholm, Uppsala, Reale Accademia dei Lincei, Rome, French Bur. Longitudes. Author: Leçcons sur le frottement 1895; Leçons sur la theorie analytique des equations differentielles, 1897; les Axiomes de la mécanique, 1922; Cours de mécanique professé à l'Ecole polytechnique, 1930. Research in curvatures and algebraic surfaces; application of theory of continuous groups to functions; contributed to devel. of Cauch-Lipschitz theory of differential equations. Died Paris, Oct. 29, 1933.

PAINTER, Edgar Page, Am. chemist; b. Schuyler, Neb., Oct. 2, 1909; s. James B. and Ida (Page) P.; B.S., S.D. State U., 1932; Ph.D., U. Minn., 1939; m. Ruth Coburn Robbins, July 4, 1940; children—Jane C., Page R. Faculty, U. Cal., Davis, 1947-—, prof. chemistry, 1953-—. Mem. Am. Chem. Soc., A.A.A.S., Sigma Xi, Phi Lambda Upsilon, Gamma Alpha. Research, numerous publs. on poisonous nature of selenium compounds in plants, factors influencing compositions of oilseeds, substitution of carbohydrates, asymmetric synthesis; prepared amino acids with selenium in place of sulfur. Home: 815 Miller Dr., Davis, Cal. 95616.*

PAINTER, Terence John, English chemist; b. Dartford, Eng., Aug. 10, 1933; s. Arthur Harold and Florence (Jones) P.; B.Sc. with honors, Bristol (Eng.) U., 1954; M.A., Queen's U., Can., 1956, Ph.D., 1957. Harold Hibbert Meml. fellow McGill U., Montreal, Que., Can., 1957-58; asst. research officer NRC, Halifax, N.S., Can., 1958-60; sr. research fellow Lister Inst. Preventive Medicine, London, Eng., 1960-64; vis. scientist Tech. U. Norway, Trondheim, 1964-65; lectr. biochemistry Royal Free Hosp., London U. Med. Sch., 1965-—; Vis. scientist Swiss Fed. Inst. Tech., Zurich, 1966-—; cons. UN FAO, 1962-—. Mem. Chem. Soc., Biochem. Soc., Sigma Xi. Research, publs. on chemistry of wood and cellulose, structure and properties of seaweed mucilages, structure of human blood-group substances, artificial enzyme systems, theory of structure and hydrolysis of heteropolysacharides, chemistry of glycoproteins. Home: 32 Shenley Rd., Dartford, Kent., Eng. Office: Royal Free Hosp. Sch. Medicine, 8 Hunter St., London W.C.I., Eng.*

PAINTER, Theophilus Shickel, Am. zoölogist; b. Salem, Va., Aug. 22, 1889; s. Franklin Verzelius Newton and Laura Trimble (Shickel) P.; A.B., Roanoke Coll., Salem, Va., 1908; A.M., Yale, 1909, Ph.D., 1913, hon. Sc.D, 1936; studied U. of Würtzburg, 1913-14; m. Anna Mary Thomas, Dec. 29, 1917; children—Elizabeth Tyler, Anne Trimble, Theophilus S., Joseph Thomas. Instr. in zoölogy, Yale, 1914-16; adj. prof. zoölogy, U. of Texas, 1916-21, prof., 1922-—, acting pres., 1944-46, pres., 1946-52. Mem. 10th F.A., Conn. N.G., 1916; 1st lt. S.C., U. S. Army, later capt. A.S., till 1919. Mem. Am. Soc. Zoölcgists, Nat. Acad. Sci., Sigma Xi. Am. editor 10th edit. Vade-Mecum. Contbr. articles on cytology and chromosomes of man. Research on cytogenetics; cytpchemistry; experimental zoology. Home: 105 W. 32d St., Austin, Tex.

PAIS, Abraham, physicist; b. Amsterdam, Holland, May 19, 1918; s. Jesaja and Kaatje (van Kleeff) P.; B.Sc., U. Amsterdam, 1938; M.Sc., U. Utrecht, 1940, Ph.D., 1941; m. Lila Atwill, Dec. 15, 1956 (div. 1962); 1 son, Joshua. Came to U.S., 1946, naturalized, 1954. Research fellow Inst. Theoretical Physics, Kopenhagen, Denmark, 1946; prof. Inst. for Advanced Study, Princeton, N.J., 1950-63; prof. Rockefeller Inst., 1963-—; staff mem. Lawrence Radiation Lab., Berkeley, Cal., 1961-—; vis. prof. physics Columbia, 1955, adj. prof., 1963-—; vis. prof. CERN, Geneva, Switzerland, Berkeley, Cal., 1960. Guggenheim fellow, 1960. Fellow Am. Phys. Soc.; mem. Nat., Royal acads. scis. (Netherlands). Research, publs. on physics of fundamental particles, high energy physics, field theory. Address: Rockefeller Inst., N.Y.C. 10021.*

PAIXHANS, Henri Joseph, French inventor; b. Metz, France, Jan. 22, 1783; student Poly. Sch., Spl. Sch. Arty. Arty. comdr. with Napoleon in Austria, Prussia, Poland, Russia; held provincial polit. positions after 1815; later with mil. depts. of French govt. Decorated Croix d'Hanneur. Author: Considérations sur l'état actuel de l'artillerie des places et sur les améliorations dont elle parait susceptible, 1815; Observations sur la loi de recruitment et de l'avancement de l'armée de terre, 1822; Nouvelles forces maritimes (stressed advantages of longer shells), 1822; Expériences faites par la marine francaise sur une arme nouvelle; changements qui paraissent devoir en résultes sur le système naval (brought out need for armoring ships), 1825; Force et faiblesse militaires de la France, 1830; Fortifications de Paris, ou Paris doit-il être fortifié et quels seront les moyens de defense?, 1834. Died Jouy aux Arches, France, Aug. 19, 1854.

PAKE, George Edward, Am. physicist; b. Jeffersonville, O., Apr. 1, 1924; s. Edward Howe and Mabel (Fry) P.; B.S., Carnegie Inst. Tech., 1945, M.S., 1945; Ph.D., Harvard, 1948; m. Marjorie Elizabeth Semon, May 3, 1947; children—Warren E., Catherine E., Stephen G., Bruce E. Faculty physics Washington U., St. Louis, 1948-56, prof., 1953-56; prof. physics Stanford, 1956-62; provost, prof. physics Washington U., St. Louis, 1962-—. Chmn., St. Louis County Bus. and Indsl. Devel. Commn., 1964-—. Fellow Am. Phys Soc.; mem. A.A.A.S., Sigma Xi. Author: (with E. Feenberg) Quantum Theory of Angular Momentum, 1953; Paramagnetic Resonance, 1962; also articles on magnetic resonance. Research on application of nuclear magnetic resonance to structure and properties of solids; application of electron spin resonance to structure of molecules and to magnetic properties of crystalline free radicals. Home: 40 Picardy Lane, St. Louis 63124.

PAKISER, Louis Charles, Jr., Am. geophysicist; b. Denver, Feb. 8, 1919; s. Louis C. and Lila (Hanson) P.; Geol. Engr., Colo. Sch. Mines, 1942; postgrad. Nancy U., France, 1945, Stanford, U. Colo.; m. Helen L. Meineke, Oct. 9, 1939. Jr. geophysicist Carter Oil Co. (now Humble), Tulsa, 1942-43, geophys. computer, 1946-49; nat. exec. dir. Am. Vets. Com., Washington, 1950-52; geophysicist U.S. Geol. Survey, Lakewood, Colo., 1952-65, chief br. crustal studies, 1960-65, acting chief scientist Nat. Center for Earthquake Research, Menlo Park, Cal., 1965-67, chief Office Earthquake Research and Crustal Studies, chmn. executive committee, 1967-—. Distinguished lectr .Soc. Exploration Geophysicists, 1964. Mem. Ad Hoc Panel on Earthquake Prediction Office Sci. and Tech., 1964-—. Recipient Outstanding Performance award U.S. Geol. Survey, 1961, 63. Organizer, dir. major seismic-refraction study structure earth's crust and upper mantle, 1960-—. Home: 655 Riviera Dr., Los Altos, Cal. 94022. Office: 345 Middlefield Rd., Menlo Park, Cal. 94025.*

PAL, Benjamin Peary, Indian agrl. botanist; b. Mukandpur, Punjab, India, May 26, 1906; s. Rala Ram and Indra (Devi) P.; M.Sc. with honors, U. Rangoon, 1929; Ph.D., Cambridge (Eng.) U., 1932. Second econ. botanist Imperial Inst. Agr., 1933-37, imperial econ. botanist, 1937-44; head div. botany Indian Agrl. Research Inst., 1937-50, dir., 1950-65; dir.-gen. Indian Council Agrl. Research, New Delhi, India, 1965-—. Mem. edn. commn., head task force on agrl. edn. Govt. of India, 1964-67; chmn. spl. adv. com. on food and agr. Indian Dept. Atomic Energy, 1963-—; mem. Sci. Adv. Com. to Cabinet, 1956-—. Recipient Padma Shri, Pres. of India, 1958; Rafi Ahmed Kidwai Meml. prize in agrl. botany Indian Council Agrl. Research, 1962; Fellow Linnean Soc. London, Royal Hort. Soc., Nat. Inst. Scis. India (Srinivasa Ramanujan medal 1964, v.p. 1960-61); mem. Indian Sci. Congress (past pres. botany and agr. sects.), Indian Soc. Genetics and Plant Breeding (pres. 1947-48), Indian Bot. Soc. (pres. 1950, Birbal Sahni medal 1962), Hort. Soc. India (pres. 1962-64), Delhi Agr. Hort. Soc. (pres. 1963-65), Genetics Soc. Japan (hon.). Author: Beautiful Climbers of India, 1960; Charophyta, 1962; The Rose in India, 1966; Wheat, 1966; also numerous articles, monographs. Research on wheat breeding and genetics; developed high yielding and high quality, rust resistant wheats. Home: B-8 I.A.R.I., New Delhi. Office: Indian Council Agrl. Research, Krishi Bhavan, New Delhi, India.

PAL, L., Hungarian physicist; b. Gyoma, Hungary, Nov. 7, 1925; s. Imre and Erzsébet (Varga) P.; Qual. summa cum laude for M.Sc., Eötvös Lóránd U., Budapest, Hungary, 1949; D.S., Lomonosov U., Moscow, USSR, 1953; m. Angela Danóci, Jan. 5, 1963; 1 dau., Catherine. Head solid state physics dept. Central Research Inst. for Physics, Budapest, 1953-56, 1st dept. dir., 1956-—; prof. atomic physics Eötvös Lóránd U. Recipient Kossuth prize, 1965. Mem. Hungarian Atomic Energy Commn.; mem. sci. council Joint Inst. for Nuclear Research in USSR. Mem. Hungarian Acad. Scis. (corr.), Internat. Union PAP (pres. nat. com.). Editorial bd. Phys. Stat. Solidi, Kernenergie. Research and publs. on magnetism of solids, magnetic anistropy and magnetic phase-transformation, theory of stochastic processes in nuclear reactors. Home: 21 Széher, Budapest, Hungary. Office: Budapest 114. P.O.B. 49.*

PALACIOS, Julio, Spanish physicist; b. Paniza, Spain, Apr. 12, 1891; s. Miguel and Eusebia (Martinez) P.; Ph.D., U. Madrid, 1914; Sc.D. (hon.), U. Toulouse (Frace), 1943; m. Elena Calleya, Dec. 15, 1927; children—Elena Maristany, Carmen Hornung, Pilar Garcia De La Barga, Ana Arche, Rosario. Prof. physics U. Madrid, 1916-—; guest prof. U. Lisbon, 1947-61. Recipient Juan March award for physics, 1958. Mem. Royal Acad. Scis. (pres. 1966), Inst. Phys. Sci. (dir. 1960), Spanish Nat. Assn. Physics, Geog. Soc. Author: Termodinámica, 1958; Electricidad y Magnetismo, 1959; Relatividad, 1960; Termodinámica Aplicada, 1961; Mecánica, 1962; Dimensional Analysis, 1964; also numerous articles. Home: 3 Isaac Pearl, Madrid. Office: Instituto de Ciencias Físicas, Ciudad Univeristaria, Madrid 3, Spain.*

PALADE, George Emil, cytologist; b. Jassy, Rumania, Nov. 19, 1912; s. Emil George and Constana (Cantemir) P.; M.D., U. Bucharest (Rumania), 1940; m. Irina Malaxa, June 12, 1941; children—Georgia Teodora, Philip Theodore. Came to U.S., 1946, naturalized, 1952. With Rockefeller Inst., N.Y.C., 1946-—, vis. investigator, 1946-48, asst., 1948-51, asso., 1951-53, asso. mem., 1953-56, mem., prof., 1956-—, head cytology dept., 1961-—. Recipient Warren prize Mass. Gen. Hosp.; Passano award A.M.A., 1964. Fellow Am. Acad. Arts Scis.; mem. Nat. Acad. Scis., Electron Microscope Soc. Am., Am. Soc. Cell Biology, Am. Assn. Anatomy, Histochem. Soc. Important works include electron micros. studies on fine structure of animal cells which led to discovery of ribosomes, mitochondrial structure, and current description of intracellular membranous systems (with K.R. Porter); isolation and biochem. characterization of various cell fractions (microsomes, ribosomes, with P. Siekevitz); work on secretory cells on blood capillaries. Home: 164 E. 74th St. Office: Rockefeller Inst., York Av. and 66th St., N.Y.C., 10021.*

PALAIS, Richard Sheldon, Am. mathematician; b. Lynn, Mass., May 22, 1931; s. Saul and Elizabeth (Robinson) P.; B.A., Harvard, 1952, M.A., 1954, Ph.D., 1956; m. Eleanor Wiener Galland, May 9, 1954; children—Julie Michelle, Robert Andrew, David Galland. Instr., U. Chgo., 1956-58; NSF postdoctoral fellow, mem. Inst. for Advanced Study, Princeton, N.J., 1958-60, sr. NSF fellow, 1963-64; faculty Brandeis U., Waltham, Mass., 1960-—, prof. math., 1965-—, chmn. dept., 1966-—. Mem. Am. Math. Soc. (Editor Transactions 1966-—). Author: Seminar on the Atiyah-Singer Index Theorem, 1965;

also articles. Research in topological transformation groups, differential topology, calculus of variations, nonlinear analysis. Home: 70 Temple St., West Newton, Mass. 02165. Office: Dept. Math., Brandeis U., Waltham, Mass. 02154.*

PALASSA, Pierre-Bernard, French geologist; b. Oloron, France, June 5, 1745; worked with Lavoisier; clergyman. Mem. French Acad. Scis. Author: Essai sur la minéralogie des Monts-Pyrénées, 1781. First to recognize gen. parallelism of stroke of strata with chain of Pyrenees; believed river erosion caused formation of valleys. Died Ogenne, France, Apr. 9, 1830.

PALAY, Sanford Louis, Am. anatomist, educator; b. Cleve., Sept. 23, 1918; s. Harry and Lena (Sugarman) P.; A.B., Oberlin Coll., 1940; M.D. (Hoover prize scholar), Western Res. U., 1943; A.M. (hon.), Harvard, 1961. Teaching fellow medicine, research fellow anatomy Western Res. U., 1945-46; NRC fellow med. scis. Rockefeller Inst., 1948, vis. investigator, 1953; faculty anatomy Yale, 1949-56, asso. prof., 1955-56; chief sect. on neurocytology Lab. Neuroanat. Scis., NIH, Bethesda, Md., 1956-61, chief lab., 1960-61, mem. cell biology study sect., 1959-61, 62-65, fellowship bd., 1958-61; Bullard prof. neuroanatomy Harvard Med. Sch., 1961—. Chmn., Gordon Research Conf. on Cell Structure and Metabolism, 1960; vis. investigator Middlesex Hosp., Bland Sutton Inst., London, Eng., 1961; Ramsay Henderson Trust lectr. U. Edinburgh, 1962. Fellow Am. Acad. Arts and Scis.; mem. Am. Assn. Anatomists, Histochem. Soc., Electron Microscope Soc. Am., A.A.A.S., Am., Internat. socs. for cell biology, Societe Francaise de Microscopie Electronique (hon.), Phi Beta Kappa, Sigma Xi, Alpha Omega Alpha. Editor: Frontiers in Cytology, 1958; Editorial bd. Exptl. Neurology, Jour. Cell Biology, 1962—; Brain Research, 1965—; bd. co-editors Exptl. Brain Research, 1965—; sci. council Jour. Neuropharmacology, Progress in Neuropharmacology, 1961-66; adv. bd. Jour. Neuropathology and Exptl. Neurology, 1963—. Research, publs. on fine structure nervous system, neurosecretion, electron microscopy of fat absorption. Home: 8 Temple Rd., Concord, Mass. 01742. Office: Dept. Anatomy, Harvard Med. Sch., 25 Shattuck St., Boston 02115.*

PALERM, Angel, anthropologist; b. Ibiza, Spain, Sept. 11, 1919; s. Antonio and Maria Josefa (Vich) P.; B.A., U. Barcelona (Spain), 1936; M.History, U. Mexico, 1949; Anthropologist, Escuela Nacional de Antropología, Mexico, 1951; m. Carmen Viqueira, Sept. 18, 1941; children—Armando, Juan Vicente, Jacinta, Angel. Program specialist Pan Am. Union, 1952-58, chief social sci. div., 1957-58; exec. officer to sec. gen. OAS, 1958-61; dir. Dept. Social Affairs, 1961-66; dir. Esculela Graduados Ciencias Sociales U. Iberoamericana, Mexico City, Mexico, 1966—. Professional lectr. Am. U., 1960-65; vis. prof. Cath. U., 1961-63; prof. ethnology Escuela Nacional de Antropología, 1966—. Mem. com. Latin Am., Nat. Acad. Sci.-NRC, 1956-66; mem. com. on anthropology Pan Am. Inst. Geography and History, 1956-58. Mem. Am. Anthrop. Assn., Sociedad Mexicana de Antropología, Sociedad Brasileña de Antropología. Author: (with I. Kelly) The Tajín Totonac, 1952; The Agricultural Basis of Urban Civilization, 1958; Teoría etnológica. 1967; also articles. Research on ancient and modern ethnography of Totonac tribe, pre-colombian agrl. techniques, process of socio-cultural change in Latin Am., human ecology of Acolhuacan. Home: 83-202 Av., México, México D.F. 11, México. Office: Universidad Ibero-americana, México.*

PALETTA, Francis X., Am. surgeon; b. New Kensington, Pa., Nov. 4, 1915; s. Frank and Mary (Fillipelli) P.; student Duquesne U., 1932-34; M.D., Marquette U., 1938; m. Gertrude Hendricks, Apr. 1, 1946; children—Geraldine, Mary, Francis, Christian, David, Richard, Kathlee, Stephen. Practice medicine, specializing in plastic surgery, St. Louis; faculty St. Louis U. Sch. Medicine, prof. clin. surgery, dir. plastic surgery. Recipient Alumni award Marquette Sch. Medicine, 1942. Diplomate Am. Bd. Plastic Surgery. Mem. Am. Soc. Plastic and Reconstructive Surgery (pres.), Am. Assn. Plastic Surgeons, Soc. Head and Neck Surgeons, Am. Soc. Surgery of Hand. Contbg. author Reconstructive Plastic Surgery (J. Converse). Research, publs. in cancer, hypothermia, physiology circulation of hand, clin. plastic surg. problems. Home: 822 Hawbrook St., Glendale, Mo. 63122. Office: 634 N. Brand St., St. Louis 63101.*

PALFI, Gábor, Hungarian biologist; b. Gyula, Hungary, May 2, 1918; s. Ferenc and Julianna (Csösz) P.; Ph.D., Eötvös Lóráđ U., Budapest, Hungary, 1959; Candidate Biol. Scis., Hungarian Acad. Scis., 1961; m. Margit Báyer, Sept. 12, 1950; children—Zsófia, Ferenc, János. Documentator, Documentation Centre Med. Scis., Budapest, 1954-55; research officer Research Inst. for Agronomy, Szeged, Hungary, 1955-60; research officer Inst. for Plant Physiology, U. Szeged, 1960—. Research, publs. on nitrogen, amino acid and protein metabolism of wheat, maize and rice; induced poliploidy of Capsicum annum and its amino acid metabolism. Home: 45 Fósika, Szeged, Hungary.*

PALFYN, Jean, Belgian anatomist, surgeon; b. Courtrai, Belgium, Nov. 28, 1650; student at sch. surgery, Gand; later moved to Paris; became reader surgery and anatomy Sch. Surgery, Ghent, Belgium, 1708. Author: Description anatomique des parties de la femme que servent à la génération; avec un traité des montres de Fortunio Liceti, et une description anatomiques de deux enfants monstruex, nés à Gand, en 1703, 1708; Traité des malades des yeux Leyden, 1714; Nouvelle ostéogie, ou description exacte et curieuse des os du corps humain, avec des planches fort exactes que les représentent, . . ., 1732; Anatomie chirurgicale description exacte des parties du corps humain, avec des remarques utiles aux chirurgiens, 1734. Developed new forceps, 1721. Died Ghent, Apr. 21, 1730.

PALGOV, Nikolay Nikitich, Russian geographer; b. 1889; grad. geodetic and hydrotech. dept. Moscow Bldg. Sch., 1913, Moscow regional studies courses, geol. and mineral. course All-Union Inst. Tech. Edn.; postgrad. Moscow Geol. Survey Inst.; Cand. Geog. Sci., 1942; D.Geog. Sci., 1951. Sr. cartographer, head topog. and geodetic groups Kazakhstan Geol. Bd., Alma-Ata, Semipalatinsk, 1928-38; asso. geog. sect. Kazakhstan br. USSR Acad. Scis., 1939, sr. asso., 1940—; head geog. sect. Kazakhstan Acad. Scis., 1945—; instr. Kazakhstan Women's Pedagogical Inst., 1953-54, prof., 1954-56; prof. Kazakhstan U., Alma-Ata, 1957—. Mem. Kazakhstan Acad. Scis., All-Union Geog. Soc. (chmn. Kazakhstan br.), Kazakhstan Znanie Soc. (bur. mem. geog. sect.). Author: Kazakhstan, 1953: Glacial Rivers of the Trans-Ili Alatau; The Nature of Kazakhstan: Outline Physical Geography of the Kazakh SSR; The Present Glaciation of the Trans-Ili Alatau, 1958. Exec. editor (monograph) Problems of Geography in Kazakhstan, 1960. Research and publs. on past and present glaciation, Kazakhstan glacial water resources, glacial areas of Trans-Ili, Alatau, Dzhungar Alatau, Kokshaal Alatau mountain ranges. Address: Kazakhstan Acad. Scis., Alma-Ata, Kazakhstan SSR, USSR.

PALISA, Johann, astronomer; b. Troppau, Czechoslovakia, Dec. 6, 1848; dir. Vienna Obs. Author: Sternlexikon von $-1°$ bis $+19°$ Deklination, 1902. Made studies of double stars and comets; discovered numerous asteroids. Died Vienna, May 2, 1925.

PALISSOT DE BEAUVOIS, Ambrois-Marie-François-Joseph, French botanist, explorer; b. Arras, France, July 27, 1752; studied under Jussieu; Mem. French Acad. Scis., 1783, Soc. Agr. Author: La flore d'Oware et Bénin, 1802, 09; Essai d'une nouvelle agrostographie, 1812. Studied African flora, 1802-07, also entomology and herpetology. Died Paris, Jan. 21, 1820.

PALISSY, Bernard, French natural philosopher, potter; b. La Capella Biron, Lot et Garonne, 1499 or 1510; worked as glass-maker, portrait painter, and surveyor; apptd. ouvrier de terre et inventeur des rustiques figulines du roi, ca. 1562; created pieces for Catherine de'Medici and the French court. Author: Discours admirables de la nature des eaux et des fontaines . . . , 1580. After 16 years of experiments with enamels, produced a pure white enamel which provided excellent ground for decorative art; produced Palissy ware, widely admired and imitated; noted for pieces reproducing scriptural and mythological subjects in low relief and for pieces decorated with insects, reptiles, and plants copied from nature; developed production of faience in France; lectured in Paris on scientific subjects; emphasized fact and demonstration; developed a theory of crystallization; classified salts; claimed that plants extract salts from soil; investigated formation of underground springs and recognized identity between fossil forms and living forms; denied existence of void; explained volcanoes and earthquakes by hypothesis of subterranean fire; proposed an explanation of rainbows; held that water played fundamental role in nature. Sent to Paris Bastille as Huguenot, and died there hanged, strangled and burned for heresy, July 4, 1589.

PALITZSCH, Johann Georg, German astronomer; discovered reappearance of Halley's comet, 1758, periodic variability of star Algol, 1782. Died 1788.

PALKOVITS, Miklós, Hungarian physician; b. Budapest, Hungary, Dec. 5, 1933; s. Ferenc and Maria (Mészáros) P.; M.D., U. Med. Sch., Budapest, 1958, candidate Med. Scis., 1966; m. Maria Gábor, Sept. 6, 1963. Prof.'s asst. anat. dept. U. Med. Sch., Budapest, 1960-62, lectr., 1966—; sci. co-worker dept. pathophysiology Inst. Exptl. Med. Research, Hungarian Acad. Scis., Budapest, 1962-64. Mem. Internat. Soc. for Stereology (Eastern Europe rep. 1963—), Anatomische Gesellschaft, European Soc. for Comparative Endocrinology. Author: Morphology and function of the Subcommisural Organ, 1965; also numerous articles. Research on central nervous regulation of the salt- and water-balance, examination of activity of circumventricular organs, quantitative histology, stereological problems in histology, quantitative cytology, karyometry. Home: 5 Szilágyi Erzsébet fasor, Budapest II. Office: 58 Tüzoltó utca, Budapest IX, Hungary.*

PALLADIN, Aleksandr Vladimirovich, Russian biochemist; b. Moscow, Sept. 10, 1885; grad. natural sci. dept. Physico-Math. Faculty, St. Petersburg U., 1908; D.Biol. Sci. Prof. physiology Stebutov agrl. courses, Petrograd, 1914-16, Kharkov Inst. Agr. and Forestry, 1916-21; head chair biochemistry Kharkov Med. Inst., 1921-31, Kiev U., 1933-54; organizer, dir. Kiev Inst. Biochemistry, Ukrainian Acad. Scis. (formerly Ukrainian Biochem. Inst., Kharkov), 1931—. Del., Internat. Conf. on Peaceful Uses Atomic Energy, Geneva, 1955, 20th Internat. Congress Physiologists, Brussels, 1956. Decorated Order of Lenin (4); recipient Lenin prize, 1929. Mem. USSR (Presidum mem. 1960-65), Ukrainian (pres. 1946-62) acads. scis., USSR Acad. Med. Scis., Soc. for Cooperation among Scientists (co-founder), All-Union Soc. Physiologists, Biochemists and Pharmacologists (chmn. central council 1955-59); hon. mem. Belorussian, Polish, Hungarian, Bulgarian, Rumanian acads. scis. Author: The Principles of Nutrition, 3d edit., 1927; The Chemical Nature of Vitamins, 3d edit., 1941; A Textbook of Biological Chemistry, 1924, 12th edit., 1946; Research on the Biochemistry of Muscles and the Nervous System under Various Conditions, Part 2; 1946; Cerebral Metabolism under Various Functional Conditions, 1952; Problems Biochemistry Nervous System, 1964. Co-editor: Large Med. Ency., 1st and 2d edits.; editor Ukrainian Biochem. Jour., 1926—. Research and numerous publs. on biochemistry of muscular and nervous systems, creatine, biochemistry of muscular fatigue, metabolic disorders, cerebral biochemistry, avitaminosis, hypovitaminosis, vitamin conversion in tissues of animal organisms. Address: Kiev Inst. Biochemistry, Ukrainian Acad. Scis., Leontovitsch Str. 9, Kiev, Ukrainian SSR, USSR.

PALLADINO, Nunzio Joseph, Am. nuclear engr.; b. Allentown, Pa., Nov. 10, 1916; s. Joseph and Angelina (Trentalange) P.; B.S., Lehigh U., 1938, M.S., 1939, certificate bus. and mgmt., 1955, D.Eng., 1964; m. Virginia Marchetto, June 16, 1945; children—Linda, Lisa, Cynthia. Steam turbine design engr. Westinghouse Electric Corp., 1939-42, 45-46, mgr. reactor design subdiv., 1950-59; sr. engr. Oak Ridge Nat. Lab., 1946-48; staff asst. Argonne Nat. Lab., 1948-50; prof., head dept. nuclear engring. Pa. State U., University Park, 1959-66, prof., dean Coll. Engring., 1966—. Mem. AEC Com. on Reactor Safeguards; mem. Pa. Com. on Atomic Energy Devel. and Radiation Control. Fellow Am. Nuclear Soc.; mem. Am. Soc. M.E. (Prime Movers award 1956), Am. Soc. E.E., Am. Inst. Chem. Engrs., Sigma Xi, Tau Beta Pi, Pi Mu Epsilon, Phi Kappa Phi. Author: (with others) Shippingpor Pressurized Water Reactor, 1958; (with others) Technology of Nuclear Reactor Safety; also articles. Devel. of reactor designs for nuclear submarines, nuclear power plant. Home: 333 West Park Av., State College, Pa. 16801.*

PALLAS, Peter Simon, German geologist, naturalist, zoologist; b. Berlin, Germany, Sept. 22, 1741; s. Simon Pallas; ed. U. Berlin, U. Halle, U. Göttingen, U. Leiden; m. twice. Joined Geol. Coast Study Eng., 1761; apptd. naturalist to expdn. patronized by Catherine II, through Russia, Siberia, 1768-74; traveled through So. Russia, Crimea, 1793-94; became prof. natural history Imperial Acad. Scis., St. Petersburg, Russia, 1768. Fellow Royal Soc., 1764; mem. French Acad. Scis. Author: Elenchus zoophytorum, 1766; Miscellanea zoologica, 1866; Spicelegia zoologica, 1767-1804; Reisen durch verschiedene Provinzen; Des Russischen Reichs, 3 vols., 1771-76; Sammlungen Historischer Nachrichten uber die Mongolischen Volkerschafter, 2 vols., 1776-1802; Novae species quadrupedum, 1778-79; Icones insectorum praesertim rossiae siberiaeque pecularium, 1781-1806; Zoographia rossoasiatica, posthumous, 3 vols., 1831; Bemerkungen auf einer Reisen durch die sudlichen Statthalterschaften des russischen Reichs, 1799-1801; Vocabularium linguarum totius orbis, 1787-89; also articles. Discovered deposits of extinct mammoths, rhinoceroses, 1768-74, amphioxus (which he believed to be a slug and named Limax lenceolatus), 1778; pioneered study of Siberian topography, fauna, flora. Died Berlin, Sept. 8, 1811.

PALLAS, Simon, German surgeon; b. 1694; prof. surgery, Halle, also Berlin, Germany; recognized echinococcosis of liver as parasitic disease. Died 1770.

PALLOTTA, Arthur Joseph, Am. pharmacologist; b. Medford, Mass., June 3, 1927; s. Arthur and Alma (Lucia) P.; B.S., Georgetown U., 1950, M.S., 1953, Ph.D., 1956; m. Mary I. Jameson, Dec. 17, 1949; children—Deborah, Mary J., Karen, Arthur, Pamela, Thomas, Margaret. Research pharmacologist Hazelton Labs., Inc., Falls Church, Va., 1955-58, chief pharmacology-biochemistry dept., 1958-59, staff cons., 1959-60, sr. pharmacologist, 1960-61; dir. research, v.p. Bionetics Research Labs., Inc., Falls Church, 1961—. Research asso. Georgetown U., 1955—; lectr. Cath. U., 1957—. Mem. A.A.A.S., Am. Assn. Clin. Chemists, Am. Chem. Soc., Am. Fedn. Clin. Research, Am. Soc. Artificial Internal Organs, Am. Soc. Zoologists, Animal Care Panel, N.Y. Acad. Scis., Soc. Toxicology, Va. Acad. Sci., Soc. Cryobiology. Research, publs. on blood poisoning, evaluation drugs, poisons, chems.; mass culture techniques for microorganisms. Home: 601 Abbott Lane. Office: 101 W. Jefferson St., Falls Church, Va. 22046.*

PALMA, Antonio, Italian surgeon, embryologist; b. Rome, Italy, Dec. 21, 1931; s. Italo and Giuseppina

(Tondi) P.; M.D., U. Rome, 1959; m. Calisti Verbena, June 13, 1960; children—Piera Laura, Paolo Silverio. Faculty, faculty medicine, U. Rome, 1959—, prof. histology and embryology, 1965-66, prof. surg. anatomy, 1966—, specialist gen. surgery, 1967—; asst. dept. surgery Hosp. Policlinico Umberto I, Rome, 1964—. Mem. Soc. Gt. hystochemistry, Soc. Italian Anatomy, Soc. Italian Biol. Specialists. Author: (with V. Virno, G. Patriarca) Il Mediartruo uel feto, 1965; also numerous articles. Embryological surg. studies of clin. value of vagotomy; exptl. and histochem. studies of liver regeneration; microphys. studies of bone tissue; histochem. studies of placental barrier. Home: 331 via Aequa Bullicante, Rome, Italy.*

PALMÉN, Erik Herbert, Finnish meteorologist, oceanographer; b. Vasa, Finland, Aug. 31, 1898; s. Eskil and Sally (Skog) P.; M.Sc., U. Helsinki (Finland), 1921, Ph.D., 1927; m. Synnöve Maria vön Hellens, July 28, 1923; children—Anne Marie (Mrs. Walter Victor Hackman), Lars Johan. Asst., Finnish Inst. Marine Research, 1922-39, dir., 1939-46; faculty U. Helsinki, 1930-48, prof. meteorology, 1947-48; prof. Acad. Finland, Helsinki, 1948—; vis. prof. U. Chgo., 1946-48, 49-50, U. Cal. at Los Angeles, 1952, 56-62, U. Stockholm (Sweden), 1948, 54. Recipient Symons medal Royal Meteorol. Soc., 1957; Buys Ballot medal Royal Netherlands Acad. Scis., 1964; Rossby prize Swedish Geophys. Soc., 1966. Mem. Finnish Soc. Scis., Finnish Geog. Soc. (hon.), Am. Meteorol. Soc. (Rossby award 1960, hon.). Research, numerous publs. on interaction between atmosphere and sea and in ocean currents, aerology of extratropical cyclones, tropopause structure, dynamics of jet streams, gen. atmospheric circulation, structure of tropical cyclones and energy conversion in extratropical and tropical cyclones. Home: 1B Hogbergsgatan, Helsinki, Finland.*

PALMÉN, Ernst, Finnish zoologist; b. Helsinki, Finland, Apr. 22, 1916; s. L. Ilmari and Anni Emilia (Pelkonen) P.; M.Sc., U. Helsinki, 1943, D.Sc., 1944; m. Leena Katri Malkki, Dec. 12, 1943; children—Meri Helena, Ernst Harri Juhani, Anna Katri. Asst., dept. zoology U. Helsinki, 1940-49, prof. zoology, 1955—, dean Faculty Math. and Natural Scis., 1960-68; curator Tvärminne Zool. Sta., 1949-55. Mem. Finnish Zool. and Bot. Soc. (staff 1947—, pres. 1965—). Editor: Annales Zoologici Fennici. Research, publs. on taxonomy and distbn. of Fenno-Scandian Coleoptera, Chilopoda, Diplopoda and terrestrial Isopoda, also on Chilopoda, Diplopoda and terrestrial Isopoda of Newfoundland; papers on taxonomy and ecology of Chironomids (Diptera). Home: Mannerheimintie 16 A 12, Helsinki 10, Finland.*

PALMER, Alice Eugenia (Mrs. Lawrence Arthur Pratt), Am. dermatologist; b. Chgo., Sept. 17, 1910; d. Charles Grover and Eugenia Marie (Sundquist) Palmer; B.Sc. (Julius Stieglitz fellow chemistry 1928-32), U. Chgo., 1932, M.S. in Physiology, 1937; M.D., Wayne State U., 1938; m. Lawrence Arthur Pratt, Sept. 2, 1935; children—Lawrene Alice (Mrs. Dwight H. Porter, Jr.), Dorothy Jane. Spl. instr. pathology Wayne State U. Coll. Medicine, 1938; Attending physician dermatology Detroit Receiving Hosp., 1942-63; individual practice dermatology, Detroit, 1942-63; asst. clin. prof. dermatology, then asso. prof. dermatology Wayne State U. Sch. Medicine, 1946-53; chmn. dept. dermatology Grace Hosp., Detroit, 1956-63; sr. cons. dermatology Jennings Hosp., Detroit, 1954-63; cons. Detroit Rehab. Inst., 1960-63; med. educator AID, 1963—, chief med. edn. project, Vietnam, 1963—, med. cons. Far East; vis. prof. dermatology U. Saigon Faculty Medicine, 1963—. Diplomate Am. Bd. Dermatology (preceptor 1956—). Mem. A.M.A., Am. Acad. Dermatology, Am. Assn. U. Profs., Internat. Soc. Tropical Dermatology, Detroit Derm. Soc., Sigma Xi, Alpha Omega Alpha. Office: David Whitney Building, Detroit 48226.*

PALMER, Alonzo Benjamin, Am. physician; b. Richfield, N.Y., Oct. 6, 1815; s. Benjamin and Anna (Layton) P.; grad. Coll. Physicians and Surgeons, Fairfield, N.Y., 1839; LL.D. (hon.), U. Mich., 1881; m. Caroline Wright, July 19, 1843; m. 2d, Love Root, 1867. City physician Chgo., 1852-55, became ofcl. med. adviser to city health officer; prof. materia medica, therapeutics and diseases of women and children U. Mich., 1854, prof. pathology and practice of medicine, 1860-87, dean med. dept., 1875-87; prof. pathology and practice medicine Berkshire Med. Instn., Pittsfield, Mass., 1864-67; prof. practice medicine Bowdoin Coll., 1869-79; served as surgeon 2d Mich. Inf., 1861; pres. Mich. State Med. Soc., 1872-73. Author: Observations on the Cause, Nature, and Treatment of Epidemic Cholera, 1854; Treatise on the Science and Practice of Medicine, or the Pathology and Treatment of Internal Diseases, 2 vols., 1882; A Treatise on Epidemic Cholera, 1885; The Temperance Teachings of Science, 1886. Editor Peninsular Jour. Medicine and Collateral Scis., also Peninsular and Independent Med. Jour., 1853-60. Died Ann Arbor, Mich., Dec. 23, 1887.

PALMER, Carroll Edwards, Am. epidemiologist; b. Fairmont, Minn., Nov. 3, 1903; s. Roy Wentworth and Grace (Edwards) P.; B.S., Hamline U., 1925, D.Sc. (hon.), 1959; M.A. (Teaching fellow anatomy), U. Minn., 1927, M.D. (Teaching fellow Pediatrics), 1928, Ph.D., 1929; M.D. (hon.), U. Oslo (Norway), 1956; m. Margaret Ann Michaelson, June 30, 1928; children—Gaela Wentworth, and Richard Roy Palmer. Instr., asso. dept. biostatistics Johns Hopkins Sch. Hygiene, 1930-36; statistician Johns Hopkins Hosp., 1935-36; commd. asst. surgeon USPHS, 1936, advanced through grades to med. dir., 1950; cons., 1932-36; dir. research Child Hygiene, 1936-42; dir. research Tb, 1942-67; prof. bio-statistics School of Public Health, University of Cal. at Berkeley, 1967—. Dir. Tb Research Office, WHO, Copenhagen, Denmark, 1949-55. Recipient Weber-Parkes prize Royal Coll. Physicians, London, Eng., 1957; Trudeau medal Nat. Tb Assn., 1964. Diplomate Am. Bd. Preventive Medicine and Pub. Health (founders group 1949). Fellow Soc. for Growth and Devel.; mem. Am. Assn. Anatomists, Am. Trudeau Soc., Soc. for Research in Child Devel. (sec.-treas. 1936-48), Am. Pub. Health Assn., Am. Acad. Pediatrics, Am. Epidemiological Soc., Visindafelag Islendinga (Societas scientiarum Islandica corrs. mem.), Internat. Epidemiol. Assn., Sigma Xi, Alpha Omega Alpha, Delta Omega. Research on fungus infections, mycobacterial infections, immunity in Tb, especially evaluation of BCG vaccination of human populations. Hime: 2591 Hilgard Av., Berkeley, Cal. 94709.*

PALMER, Charles Skeele, Am. chemist; b. Danville, Ill., Aug. 4, 1858; s. William Randall and Clara E. (Skeele) P.; A.B., Amherst, 1879, A.M., 1882; Ph.D., Johns Hopkins, 1886; postgrad. U. Leipzig (Germany), 1892-93; m. Harriet B. Warner, Sept. 30, 1886 (dec. 1932); children—Mrs. Helen W. P. Bissell, Leigh W., Mrs. Winifred W. P. Bennett. Prof. chemistry U. Colo., 1887-1902; pres. Colo. Sch. Mines, 1902-03; chief chemist Washoe Smelter, Anaconda, Mont., 1903-04; asso. editor Engineering and Mining Journal, 1904-06; chem. engr. for large textile mills; fellow, Mellon Inst. of Industrial Research, U. Pitts., 1915-17. Cons. chemist United Fuel Gas Co., Charleston, W.Va., 1920. Author: Chemical Oxidation Tables, 1897. Defined chemical terms in Webster's Internat. Dictionary, 1890 edition. Translated 1st edition Nernst's "Theoretical Chemistry," 1895. Invented basic process for cracking oils to gasoline, 1900, patented, 1907, sold to Standard Oil Co. of Ind., 1916. Died Nov. 30, 1939.

PALMER, Claude Irwin, Am. mathematician; b. Barry County, Mich., May 31, 1871; s. Clark Hubbard and Martha Thompson (Kellogg) P.; A.B., U. Mich., 1903; postgrad. U. Chgo.; m. May Belle Hill, Aug. 31, 1897; children—Ethel May, Clark Francis. Tchr. in dist. and high schs., Mich., 8 yrs.; mem. faculty Armour Inst. Tech. (now Ill. Inst. Tech.), Chgo., 1903—, becoming prof. math., dean students. Author: Arithmetic with Applications, 1912; Geometry with Applications, 1912; Algebra with Applications, 1912; Trigonometry and Logarithms, 1913; Plane Trigonometry with Tables, 1914; Plane Geometry, 1915; Solid Geometry, 1918; Analytic Geometry, 1921; Practical Calculus, 1923; College Algebra, 1928. Died Apr. 8, 1931.

PALMER, Eddy Davis, Am. physician; b. Montclair, N.J., Nov. 7, 1917; s. Lubin and Marjorie (Maxfield) P.; A.B., Dartmouth, 1939; M.S., Tulane U., 1940; M.D., U. Rochester, 1943; m. Jeanne Walther, June 16, 1943; children—Hannah, Thomas F., Jonathan E. Commnd. 1st Lt. M.C., U.S. Army, 1945, advanced through grades to col., 1960; chief, gastroenterology services several Army gen. hosps. including Walter Reed, Brooke Gen, also faculty Georgetown U. Sch. Medicine, Baylor U. Sch. Grad. Studies, 1945-65; ret., 1965; chief gastroenterology VA Hosp., East Orange, N.J., 1965—; acting dir. gastroenterology div. N.J. Coll. Medicine, East Orange, 1966—. Decorated Legion of Merit; recipient Schindler award Am. Gastroscopic Soc., 1955. Mem. Am. Gastroent. Assn., Am. Soc. Gastrointestinal Endoscopy, Am. Soc. Tropical Medicine, William Beaumont Soc. Author: Stomach Disease as Diagnosed by Gastroscopy, 1949; The Esophagus and Its Diseases, 1952; Pearls for the Student of Gastroenterology, 1955; Clinical Gastroenterology, 1957, 58, 63, 65; Diagnosis of Upper Gastrointestinal Hemorrhage, 1961; (with H.W. Boyce) Manual of Gastrointestinal Endoscopy, 1964. Research, publs. on aggressive diagnostic approach to gastrointestinal hemorrhage; histopathology gastritis, feasibility psychotherapy for duodenal ulcer. Home: 143 Eagle Rock Way, Montclair, N.J. 07042. Office: VA Hosp., East Orange, N.J. 07019.*

PALMER, Frederic, Am. physicist; b. Brookline, Mass., Oct. 17, 1878; s. Frederic and Mary (Towle) P.; A.B., Harvard, 1900, A.M., 1904, Ph.D., 1913; m. Helen Wallace, June 19, 1907; children—Frederic III Helen (Mrs. Wm. H. Avery); m. 2d, Mary Clark Fox, July 28, 1950. Tchr., Asheville (N.C.) Sch., 1900-01, Worcester (Mass.) Acad., 1901-03; faculty Haverford (Pa.) Coll., 1904—, prof. physics, 1916—, ret., 1945. Home: 1 College Lane, Haverford, Pa.*

PALMER, George David, Jr., Am. chemist; b. Helena, Ark., Dec. 7, 1897; s. George David and Sophia (Ladd) P.; grad. Ark. State Agrl. and Mech. Coll., 1916; B.S., Clemson Coll., 1919; A.M., Johns Hopkins, 1921, Ph.D., 1924; m. Maude Roberts Collins, Dec. 28, 1928; 1 son, George David III. Technician, Bur. Standards, Washington, 1919; head dept. chemistry Guilford Coll., 1921-22; asst. prof. Kan. State U., 1924-27; faculty U. Ala., University, 1927—, prof. organic chemistry, 1946—. Honored for initiation of establishment So. Research Inst., 1966. Fellow A.A.A.S., Am. Inst. Chemists; mem. Am. Chem. Soc. (sec. chmn. 1942—), Ala. Acad. Sci. (past pres.), So. Assn. Sci. and Industry (sec. 1942—, founder, pres., award 1947), Gamma Sigma Epsilon. Author: Inside of the Atom, 1936; Introduction to Formula System of Organic Chemistry, 1940; also articles, chpts. in books. Patentee process for producing sulfur organic compounds. Research on sulfur organic compounds and lignin. Home: 704 11th St., Tuscaloosa, Ala. 35401. Office: Chem. Bldg., University, Ala., 35486.*

PALMER, Henry Procter, English radio astronomer; b. Monmouth, Gt. Britain, Sept. 16, 1926; s. William Nathaniel and Dorothy (Procter) P.; B.A., Queen's Coll., Oxford, Eng., 1947, D.Phil., 1952; m. Esmé Kemp, Aug. 1, 1951; children—Alice, Judith, Christopher. Research asst. meteorology Clarendon Lab., Oxford, 1947-52; faculty U. Manchester (Eng.), 1952—, reader physics and radio astronomy, 1967—; research staff radioastronomy Nuffield Radio Astronomy Labs., Jodrell Bank, Eng., since 19——. Fellow Royal Astron. Soc. London. Author: (with R.D. Davies) Radio Studies of the Universe, 1959; (with F.D. Kahn) Quasars, 1966. Editor: (with R.D. Davies, M.I. Large) Radio Astronomy To-Day, 1963; Research, publs. on physics of ice clouds; devel. frostpoint hygrometer, radio link interferometers; observations of angular sizes and structures of discrete radio sources and quasars. Home: Bredon, Sta. Rd, Goostrey, Crewe, Cheshire. Office: Nuffield Radio Astronomy Labs., Jodrell Bank, Macclesfield, Cheshire, Eng.*

PALMER, Howard, Am. explorer; b. Norwich, Conn., Nov. 28, 1883; s. George S. and Ida Amelia (Cooke) P.; B.A., Yale, 1905; LL.B., Harvard, 1908; unmarried. Admitted to Mass. bar, 1908; sec. dir. Palmer Bros. Co., mfrs. bed comfortables, New London, 1918-28. Fellow Royal Geog. Soc.; mem. New London County Hist. Soc. (v.p.), Am. Inst. of Mining Engrs.; corr. mem. Geographic Soc. Phila. Author: Mountaineering and Exploration in the Selkirks, 1914; A Pioneer of the Canadian Alps, 1931. Joint Author: A Climber's Guide to the Rocky Mountains of Canada, 1921, 40. Editor: Life on a Whaler, for New London County Hist. Soc., 1929; also editor of American Alpine Journal, 1930-33, Contbr. to Harvard Handbook of Travel, 1917, 35, Ency. Britannica 1929, and periodicals on history and exploration. A pioneer explorer, Selkirk Mts., B.C., 1907-15, ascending 50 new or little-visited peaks; made 1st conquest of Mt. Sir Sandford, 1912; Canadian Govt., confirmed his names for fifty new mountains, glaciers, etc., and named in his honor a peak, glacier and river; visited remote sections Canadian Rockies, 1916-27, ascending a score of new peaks; assisted in organizing Mt. Logan expdn. to Alaska, 1925; lecturer in mountaineering; made studies and measurements of movements of glaciers pub. by Royal Geog. Soc. (London) and Smithsonian Instn. Died Oct. 24, 1944.

PALMER, Howard Benedict, Am. chemist; b. Indpls., July 10, 1925; s. Claude Earl and Katharine (Benedict) P.; B.S. in Chemistry, Carnegie Inst. Tech., 1948; Ph.D., U. Wis., 1952; m. Katharine Douglas Watson, June 30, 1951; children—Andrew Stuart, Jeffrey Howard, David James. Research asso. Brown U., Providence, 1952-53, instr., 1953-55; faculty Pa. State U., University Park, Pa., 1955—, prof. fuel tech., 1960-65, prof. fuel sci., 1966—, head dept., 1959-65; vis. scientist Imperial Coll., London, Eng., 1963. Mem. Am. Chem. Soc., Am. Phys. Soc., Combustion Inst., Inst. Fuel (London), Chem. Soc. (London), N.Y. Acad. Scis., Sigma Xi, Phi Lambda Upsilon, Phi Kappa Phi. Research, publs. on thermal decomposition kinetics of bromine, chlorine, nitrosyl chloride, sulfur dioxide, carbon disulfide and hydrazine in shock waves; radiative recombination of atoms, thermal decomposition of gaseous hydrocarbons, kinetics of carbon formation, spectroscopy and energy distbns. in reactions in atomic diffusion flames. Home: 311 Adams Av., State Coll., Pa. 16801. Office: M.I. Bldg., Pa. State U., University Park, Pa. 16802.*

PALMER, Jeffress Gary, Am. physician; b. Bklyn., Oct. 7, 1921; s. William Ware and Margaret (Boswell) P.; B.S., Emory U., 1942, M.D., 1944; m. Jane Ann Cartwright, Feb. 2, 1951; children—Kristin Cartwright, Julie Mitchell. Am. Cancer Soc. fellow in hematology U. Utah, 1949-52; faculty dept. medicine U. N.C., Chapel Hill, 1952—, prof. internal medicine, 1964—, head div. hematology, 1952—. Mem. med. adv. com. Leukemia Soc., N.Y.C., 1964—, also bd. dirs. Mem. A.M.A., N.C., Durham-Orange County med. socs., So. Soc. for Clin. Investigation, Am. Soc. Hematology, N.Y. Acad. Scis., A.A.A.S., Reticuloendothelial Soc. Research, publs. on relation between spleen and number circulating blood cells, chemotherapeutic agts. in leukemia and other blood diseases, relation between leukocytes, bacterial products and fever, mechanisms white cell prodn. and destruction. Home: Morgan Creek Rd., Chapel Hill, N.C. 27514.*

PALMER, Katherine Evangeline Hilton Van Winkle, Am. paleontologist; b. Oakville, Wash., Feb. 4, 1895; d. Jacob Outwater and M. Edith (Hilton) Van Winkle; B.S., U. Wash., 1918; Ph.D., Cornell U., 1925; m.

Ephraim Laurence Palmer, Dec. 24, 1921; children—Laurence V.W. (dec.), R. Robin. Asst. geologist U. Ore., 1918; asst. prof. paleontology and historic geology U. Wash., 1922; curator paleontology Oberlin Coll., 1928-29; faculty geology dept. Cornell U., Ithaca, N.Y., 1918-35, lectr., 1942-46; technician zoology N.Y. State Mus., 1945-46; specialist zoology Redpath Mus., McGill U., 1950-51, Provence Mus., Quebec, 1951; dir. Paleontol. Research Inst., Ithaca, 1952-—sec.-treas. Cushman Found. Foraminferal Research, Ithaca, 1955-—. Fellow Paleontol. Soc. Am., Geol. Soc. Am., A.A.A.S.; mem. Am. Assn. Petroleum Geologists, Geol. Soc. France, Soc. Econ. Paleontologists and Mineralogists (hon.), Am. Malacol. Union (hon. life mem.), Sigma Xi, Phi Kappa Phi, Alpha Delta Pi, Sigma Delta Epsilon. Research, publs. on paleontology, spl. work on fossil and living mollusks. Home: 206 Oakhill Rd. Office: 109 Dearborn Pl., Ithaca, N.Y. 14850.*

PALMER, Theodore Sherman, Am. naturalist; b. Oakland, Cal., Jan. 26, 1868; s. Henry A. and Jane (Day) P.; A.B., U. Cal., 1888, M.D., Georgetown U., 1895; m. Bertha M. Ellis, Nov. 21, 1911. First asst. ornithologist U.S. Biol. Survey, 1890-96, in charge Death Valley Expdn., 5 months in 1891, asst. chief, 1896-1902, 10-14, asst. in charge game reservation, 1902-10, 14-16, expert in game conservation, 1916-24, biologist, 1924-28, sr. biologist, 1928-33; asso. in zoölogy U.S. Nat. Mus., from 1933. Fellow A.A.A.S., Am. Ornithologists Union (sec. 1917-37), Cal. Acad. Sci.; mem. Am. Soc. Naturalists, Am. Bison Soc., Am. Fisheries Soc., Am. Forestry Assn., Am. Genetic Assn., Baird Ornithol. Club, Nat. Parks Assn., Save Redwoods League, Am. Game Protective Assn., Internat. Com. Bird Protection, Washington Acad. Scis., Wilson Ornithol. Club, Sigma Xi; asso. mem. Soc. Am. Foresters, Am. Mus. Natural History, Boone and Crockett Soc.; corr. mem. Ornithol. Gesellschaft in Bayern, Royal Hungarian Inst. Ornithol.; hon. mem. Cooper Ornithol. Club, Internat. Assn. Game Commrs., Soc. Preservation Fauna Empire (London); pres. Biol. Soc. Washington, 1909-10, Audubon Soc. D.C., 1924-41; v.p. Nat. Assn. Audubon Socs., 1905-35, Am. Soc. Mammalogists, 1928-34. Author: Jack Rabbits of United States (2d edit.), 1897; List of Generic and Family Names of Rodents, 1897; Legislation for the Protection of Birds Other than Game Birds (2d edit.), 1902; Review of Economic Ornithology in the United States, 1900; Index Generum Mammalium, 1904; Hunting Licenses, 1904; Chronology and Index American Game Protection, 1912; Game as a National Resource, 1922; Place Names of the Death Valley Region, 1948; Chronology of the Death Valley Region, 1951; Biographies of members of the American Ornithologists Union, 1884-1954, 1954; also numerous papers on game protection; (with Henry Oldys and others) Laws Regulating the Transportation and Sale of Game, 1900; Digest of Game Laws for 1901; Game Birds and Eggs for Propagation, 1904. Chmn. com. which prepared regulations under federal migratory bird law, 1913; prepared preliminary draft of treaty for protection of migratory birds in U.S. and Can., 1916, whaling treaty act, 1936. Died July 23, 1955.

PALMER, Walter Lincoln, Am. physician, educator; b. Evanston, Ill., June 29, 1896; s. Walter Aaron and Alice (Bonney) P.; student Colo. Coll., 1914-17; B.S., U. Chg., 1918, M.D., 1921, M.S., 1919, Ph.D., 1926; m. Elisabeth Ricketts, May 15, 1926; children—Robert Howard, Donald Walter, Elizabeth Bonney, Henry Ricketts. Asso. prof. medicine U. Chgo., 1930-41, prof. medicine, 1941-54, Richard T. Crane prof. medicine, 1954-61, prof. emeritus, 1961-—. Mem. adv. bd. Med. Specialties, v.p., 1954-55; Chester M. Jones lectr., vis. prof. pro tem Mass. Gen. Hosp., Boston, 1959. Decorated comdr. Order Hipólito Unanue (Chile). Diplomate Am. Bd. Internal Medicine (Mem. 1947-55, chmn. 1951-55). Mem. American Clinical and Climatological Association, A.C.P. (pres. 1956-57, master 1961, Alfred Stengel Meml. award 1963), Am. Gastroent. Assn. (pres. 1946-47, Julius Friedenwald award 1963), Inst. Medicine Chgo. (pres. 1960), Sigma Xi, Alpha Omega Alpha, Nu Sigma Nu, Beta Theta Pi; Hon. mem. Colo., Miss. Valley med. socs., Soc. Grad. Surgeons Los Angeles County Hosp., La Fundación Lucas Sierra, Sociedad Chilena de Gastroenterología, Sociedad Médica de Valparaiso, Sociedad Médica de Santiago, La Sociedad de Gastroenterología del Peru. Chmn. editorial bd. Gastroenterology, 1951-56. Contbr. numerous articles on research in intestinal diseases, especially peptic ulcer, inflammation, neoplasms. Home: 1320 E. 58th St., Chgo. 60637.*

PALMER, Walter Walker, Am. physician, educator; b. Southfield, Mass., Feb. 27, 1882; s. Henry Wellington and Alma Roxana (Walker) P.; B.S., Amherst, 1905, Sc.D., 1922; M.D., Harvard, 1910; Sc.D., Columbia, 1929; m. Francesca Gilder, Oct. 12, 1922; children—Helena Francesca Gilder, Gilder, Walter de Kay. H. P. Walcott fellow in medicine, instr. in physiol. chemistry Harvard, 1913, asst. in medicine, also resident physician Mass. Gen. Hosp., 1913-15; asst. in medicine Rockefeller Inst., N.Y. C., 1915-17; asso. prof. medicine Columbia, 1917-19, also acting dir. med. service Presbyn. Hosp.; asso. prof. medicine Johns Hopkins Med. Sch., 1919-21, also asso. vis. physician Johns Hopkins Hosp.; Bard prof. medicine Columbia, 1921-47; cons. Presbyn. Hosp., 1947-—; dir. Pub. Health Research Inst. of N.Y., 1947-—. Mem. Nat. Bd. of Med. Examiners, 1921-43. Mem. A.M.A. (council pharmacy and chemistry), Soc. for Clin. Investigation, Assn. Am. Physicians, N.Y. Acad. Medicine, Harvey Society of N.Y.C. (pres. 1926-27), Phi Beta Kappa. Contbr. to profl. jours. Research on diabetes, nephritis, other metabolic disorders. Died Oct. 28, 1950.

PALMIERI, Luigi, Italian physicist; b. Faicchio, Italy, Apr. 22, 1807; student, Caiazzo; degree in architecture, Naples, Italy. Tchr. secondary schs.; named prof. physics Royal Naval Sch., Naples, 1845, U. Naples, 1847; joined staff Mt. Vesuvius Obs., 1848, became dir., 1854; dir. Phys. Obs., Naples; prof. meteorol. and terrestrial physics, U. Naples. Author: Les Lois et critiques de l'électricité atmosphériques, 1885. Invented rain gauge, seismograph, electrometer, seismometer, pluviometer; observed eruption of Mt. Vesuvius, 1872. Died Naples, Sept. 9, 1896.

PALSER, Barbara (Frances), Am. botanist; b. Worcester, Mass., June 2, 1916; d. George Norman and Cora (Munson) P.; A.B. magna cum laude, Mt. Holyoke Coll., 1938, A.M., 1940; Ph.D., U. Chgo., 1942. Faculty, U. Chgo., 1942-65, prof. botany, 1960-65; asso. prof. botany Rutgers U., New Brunswick, N.J., 1965-66, prof., 1966-—; vis. prof. Duke, 1962. Bot. adviser Ency. Brit., 1957-59. Mem. Bot. Soc. Am., Torrey Bot. Club, A.A.A.S., Am. Inst. Biol. Sci., N.J. Acad. Sci., Internat. Soc. Plant Morphology, Council Biol. Editors, Am. Forestry Assn. Asso. editor Bot. Gazette, 1952-59, editor, 1959-65. Research, publs. on anatomy and morphology vascular plants. Home: 18 Charlotte Dr., Somerville, N.J. 08876. Office: Dept. Botany, Rutgers U., New Brunswick, N.J. 08903.*

PALTRINIERI, Sebastiano, Italian veterinarian; b. San Felice sul Panaro, June 11, 1901; s. Mauro and Aurelia (Morselli) P.; M.D. in Vet. Medicine; m. Noemi Rossi, Apr. 22, 1934. Prof. of bacteriology, 1932, of spl pathology and clin. vet. medicine, 1934; instr. infectious diseases univs. Bologna and Messina, 1932-35; dir. U. Camerino Clinic Vet. Medicine, 1936, asso. prof. 1937; dir. U. Pisa Vet. Clinic, 1939-—, prof., 1941-—, past dean Faculty of Vet. Medicine. Pres. Order Veterinarians Province Pisa. Publns. on infectious diseases and vet. medicine. Home: Iungarno Mediceo 73. Office: via Savi 2, Pisa, Italy.

PALUCCI, Natale Giuseppe, surgeon; b. Florence, Italy, 1719; surgeon to Emperor of Austria, Vienna. Mem. French Acad. Scis. Author: Remarques sur la lithotomie, 1750; Methodus curandae fistulae lacrymalis, 1762. Studied lithotomy, eye infection; invented operating instruments for removing cataracts. Died Vienna, July 28, 1797.

PALUMBO, Louis Thomas, Am. physician, surgeon; b. N.Y.C., Jan. 25, 1909; s. Joseph and Christina (Marchesino) P.; A.A., Crane Jr. Coll., 1929; B.S. cum laude, Loyola U., Chgo., 1931, M.S. cum laude in Gross Anatomy, 1932, M.D. cum laude, 1934; m. Helen I. Thompson, June 21, 1934. Spl. trainee gen. surgery and allied spltys., Balt., 1937-42; chief, surg. service VA Hosp., Des Moines, Ia., 1946-—, dir. animal surg. research, 1961-—, dir. surg. residency tng. program, 1946-—. Clin prof. surgery State U. Ia. Coll. Medicine, 1956; sr. surg. cons. VA Hosp., Knoxville, Ia., 1946-—. Recipient Superior Accomplishment award VA, 1955, Commendation award for Otustanding Contbn. to VA Medicine and Original Contbn. to Surgery chief med. dir. VA, 1960; spl. citation and commendation certificate Italian Community Appeal, Chgo. and Loyola Stritch Sch. Medicine, 1962. Diplomate Am. Bd. Surgery. Fellow A.C.S.; mem. Internat. Soc. Surgery N. Am. chpt., Societe Internationale de Chirurgie (titular mem.), Soc. for Surgery Alimentary Tract, Ia. Acad. Surgery (founder mem. 1956, pres. 1959), Central, Western surg. assns. Author: Low Back Pain and Sciatica, 1954; Management Disorders of the Autonomic Nervous System, 1955; Surgical Service Guide, 1959. Editor: VA surg. staff bulls. 1946-—. Publs. on discovery of certain sympathetic pathways to eye, surg. treatment of duodenal ulcer, treatment for relief of angina pectoris and vascular insufficiency of upper extremities. Home: 3330 Douglas Av., Des Moines 50310. Office: VA Hosp., Des Moines, Ia. 50308.*

PALVA, Tauno Kalevi, Finnish physician; b. Hyvinkaa, Finland, Oct. 17, 1925; s. Akseli Emil and Olga (Palander) P.; M.D., Med. Sch., Turku (Finland) U., 1950, D.Med. Scis., 1952; m. Vuokko Annikki Vaisala, Aug. 17, 1951; children—Kirsti Ilona, Kimmo Kaleui. Asst. surgeon otolarynology Turku U., 1950-57, fellow, 1955-63, asso. surgeon ENT-Clinic, 1952-63; fellow otolaryngology Washington U., St. Louis, 1952-53; prof. otolaryngology Oulu (Finland), 1963-—, dean med. faculty, 1964-65, substitute rector, 1965-—. Mem. Collegium Oto-rhino-Laryngologicum. Author: Finnish Speech Audiometry, 1952; Mediastinoscopy, 1964; also numerous articles. Research in audiology and ear pathology, tech. improvements in chronic ear surgery, furtherance of mediastinoscopy. Home: 3 Nyttystie, Oulu, Finland.*

PAN, Kuan, Chinese chemist; b. Taiwan, China, May 29, 1907; s. Tzu-len and (Chien) P.; B.S., Taihoku Imperial U., 1933; D.Sc., Tohoku (Japan) Imperial U., 1944; m. Sueh-chih, Nov. 24, 1934; children—Yung-Hsin, Su-Huei, Wen-Huei, Yung-Chou. Prof., Tainan (Taiwan) Engring. Coll., 1939-45; prof. dept. chemistry Nat. Taiwan U., Taipei, 1945-—. Mem. Chinese, Am. chem. socs., Chem. Soc. Japan, Phys. Chemistry in Chinese. Research, publs. on thermodynamics of silver sulfate from E.M.F. measurements, physio-chem. study on hydrolysis of thorium nitrate, kinetics of reactions between titanium III and Mercury II; kinetics of oxidation of hypophosphorous acid by chromate. Home: 8 Lane 60, Kuling St. Office: Dept. Chemistry, Nat. Taiwan U., Taipei, Taiwan, China.*

PANAYOTIS, Pantazopoulos Epaminondas, Greek physician; b. Vytina, Greece, Feb. 5, 1919; s. Pantazopoulos E. and Helen (Crysantopulu) P.; grad. Sch. medicine Athens (Greece) U., 1942, Doctorate Medicine, 1952; m. Popy Hermogenus, Dec. 3, 1959; children—Epaminondas, Demetrius, Mary-Helen. With Athens U. Hippocrateion Hosp., 1943-—, asst. prof. otorhinolaryngology, 1965-—; staff Johns Hopkins, Balt., 1952, Pa. U., 1953, U. Vienna, 1964, Royal Ear Nose and Throat Hosp., London, Ear Nose and Throat Clinic, U. Lausanne (Switzerland), 1965; practice medicine, specializing in otorhinolaryngology, Athens. Research, publs. on otorhinolaryngology. Home: 59 Guizi, Psychikon, Athens, Greece.*

PANCOAST, Joseph, Am. anatomist, surgeon; b. nr. Burlington, N.J., Nov. 23, 1805; s. John and Ann (Abbott) P.; M.D., U. Pa., 1828; m. Rebecca Abbott, June 2, 1829; 1 son, William Henry. Conducted Phila. Sch. Anatomy, 1831-38; elected physician Phila. Hosp., 1835; vis. surgeon, 1838-45; prof. surgery Jefferson Med. Coll., 1838, prof. anatomy, 1841-74; mem. staff Pa. Hosp., 1854-64. Mem. Am. Philos. Soc., Med. Soc. Pa., Phila. County Med. Soc. Author: Treatise on Operative Surgery, 1844, 3d edit, 1852. Performed operation for remediation of exstrophy of bladder by plastic abdominal flaps, 1859, also for soft and mixed cataracts, for correction of occlusion of nasal duct; originated an abdominal tourniquet. Died Phila., Mar. 7, 1882.

PANDER, Christian Heinrich, German anatomist, embryologist; b. Germany, July 12, 1794; Dr.Med., Würzburg, Germany, 1817. Author: Beiträge zur Entwickelungsgeschichte des Hünchens in Eie, 1817. A founder of embryology; studied (with Karl Ernst von Baer) evolution of chick; observed germinal membrane of egg (blastoderm) develops three layers, serous, vascular, and mucous. Died 1865.

PANETH, Friedrich Adolf, chemist; b. Vienna, Austria, Aug. 31, 1887; s. Joseph and Sophie (Schwab) P.; student U. Munich, U. Glasgow; Ph.D., U. Vienna, 1910; m. Else Hartmann, Dec. 6, 1913; children—Eva, Heinrich Rudolph. Asst., Inst. for Radium Research, Vienna, 1912-18; prof. Prague Inst. Tech., 1918, Hamburg U., 1919-22, U. Berlin, 1922-29; George Fisher Baker lectr. Cornell U., 1926-27; prof., dir. chem. labs. U. Koenigsberg, 1929-33; guest Imperial Coll. of Sci. and Tech., London, 1933-38; reader in atomic chemistry U. London, 1938; prof. chemistry Durham U., 1939-53; head chemistry div. Joint Brit.-Canadian Atomic Energy Team, Montreal, 1943-45; dir. Max Planck Inst. for Chemistry, Mainz, Germany, 1953-—. Pres. Joint Commn. on Radioactivity, Internat. Council of Sci. Unions, 1949-55. Recipient Lieben prize Vienna Acad. scis., 1916, Lavoisier medal Chem. Soc. of France, 1952, Stas medal Chem. Soc. of Belgium, 1953, Liebig medal Gesellschaft Deutscher Chemiker, 1957, Auer von Welsbach medal Verein Osterreichischer Chemiker, 1957. Fellow Royal Soc. London; mem. Am. Acad. Arts and Scis. (hon.), Soc. Austrian Chemists (hon.), Austrian Acad. Scis. (corr.). Author: Radioelements as Indicators and Other Selected Topics in Inorganic Chemistry, 1928; Manual of Radioactivity (with G. Hevesy), 1938; The Origin of Meteorites, 1940; also articles in field. Improved technique for isolating and measuring minute traces of helium; discoverer, eopnym of histological cells; studied age of rocks by measuring helium formed by break-up of radium; also studied helium ratios of meteors. Died Sept. 17, 1958.

PANGBORN, Rose Marie Valdes, Am. food technologist; b. Las Cruces, N.M., Aug. 19, 1932; d. Leo J. and Rosalie (Page) Valdes; B.S., N.M. State U., 1953; M.S., Ia. State U., 1955; m. Jack Pangborn, Dec. 1, 1956. Research asst. food tech. Ia State U., Ames, 1953-55; food technologist sensory evaluation foods dept. food sci. and technology U. Cal. at Davis, 1955-—, lectr., 1960-67, assistant professor, since 1967-—. Cons. USPHS, Chile, 1966. Internat. Travel grantee, 1962, 65, 66, 68. Mem. Inst. Food Technologists, A.A.A.S., Am. Soc. for Testing and Materials, Sigma Xi, Iota Sigma Pi, Sigma Delta Epsilon. Author: (with M.A. Amerine, E.B. Roessler) Principles of Sensory Evaluation of Food, 1965; also numerous articles. Research on environmental, nutritional, physiol. and psychol. factors influencing human responses to taste and odor stimuli, sensory properties of fluid milk, factors affecting taste of drinking water, influence of cigarette smoking on taste, odor, oral tactile and oral thermal responses. Home: 337 Mills Dr., Davis, Cal. 95616.*

PANHARD, René, French engr., inventor; b. 1841; took over (with Levassor) machine tool factory Perrin-Panhart, 1886; founder Panhard Co. Built one of 1st automobiles, 1891; built valveless motor invented by Knight-Kilbourne, 1907; 1st to build chassis, 1891; devised (with Levassor) transmission gear. Died 1907.

PANIGEL, Maurice, biologist; b. Cairo, Egypt, Nov. 11, 1926; s. Jacques and Louise (Aboulafia) P.; Docteur en Médecine, U. Paris, 1953, Docteur ès Sciences, 1955; m. Jeanne Silly, Aug. 2, 1950; children—Marianne, Catherine, Valérie, Jean-Louis. Research worker Centre National de la Recherche Scientifique, 1949-54; lectr. Faculté des Sciences, Poitiers, France, 1956-61, Paris, 1961-62; prof. U. Paris, 1962—; sci. dir. Institut National de la Santé et Recherche Médicale, Paris, 1962—. Mem. French Soc. Anatomy, French Soc. Med. Biology, French Soc. Physiology, French Soc. Endocrinology. Research and publs. on endocrine and other aspects of viviparity in reptiles; physiol. studies on fetal circulation in human and mammalian placentae, placental perfusion and hemodynamics; electron microscopy of placental membrane in normal and path. human placenta. Home: 88, Boulevard de Courcelles, Par 17, France. Office: Hopital Saint Antoine, Paris (12), France.*

PANIKKAR, Nedumangattu Kesava, Indian oceanographer; b. Kottayam, South India, May 17, 1913; s. Sankunni Menon and Janaki (Amma) P.; B.A., Madras U., 1933, M.A., 1935, M.Sc., 1935, D.Sc., 1938; postgrad. U. London, U. Cambridge; m. Parukutty Amma, Oct. 1945; children—Mohan, Ranjini. Prof. zoology U. Travancore, 1943-44; dir. U. Zool. Lab., U. Madras, 1944-46; spl. officer for organizing fisheries research Govt. India, 1946-50; dir. Central Marine Fisheries Research Inst., Mandapam, 1950-57; fisheries devel. adviser Govt. India, 1957-62; dir. Indian Ocean Expdn., 1962-65; dir. Nat. Inst. Oceanography, Council Sci. and Indsl. Research, India, 1965—. Mem. gov. bd. Indian Council Agrl. Research; faculty sci., various Indian univs.; Indian rep. at various internat. congresses; pres. Intergovtl. Oceanographic Commn., 1963-65; pres. Indo-Pacific Fisheries Council, 1955-57. Fellow Indian, Nat. acads. scis., Nat. Inst. Scis. (sec., pres.), Zool. Soc. India. Research on zoology, comparative physiology, fisheries, oceanography, especially regarding India. Office: Nat. Inst. Oceanography, Rafi Marg, New Delhi 1, India.*

PANIZZA, Bartolomeo, Italian anatomist; b. Vicenza, Italy, Aug. 15, 1785; prof. U. Pavia (Italy); mem. French Acad. Scis., 1835; discovered cortical center of occipital lobe. Died Apr. 17, 1867.

PANJE, Rama Rao, Indian agronomist, botanist; b. South Kanara, India, Jan. 5, 1909; s. Mangesth Rao and Girija Bai (Benegal) P.; B.A. with honors, Presidency Coll., Madras, India, 1931, M.A., M.Sc., 1934; m. Varada Molahalli, May 17, 1938; children—Uma A. (Mrs. Kagal), Usha. Bot. asst. Sugarcane Research Sta., Patna, Bihar, India, 1934-40, agronomist, U.P. Muzaffarnager, Shahjahanpur, India, 1940-49; botanist Sugarcane Breeding Inst., Coimbatore, India, 1949-59; dir. Indian Inst. Sugarcane Research, Lucknow, India, 1959—. Mem. bd. agr. Indian Council Agr. Research, 1960—; mem. Devel. Council for Sugar Industry, 1965—, High Power Com. for Sugar Industry, 1963—. Mem. Internat. Soc. Sugar Cane Technologists. Research, numerous publs. on biosystematics and evolution of Saccharum spontaneum, germination, growth, ripening, flowering phenomena and cultural techniques for sugarcane; worked out new methods of cane culture for Bihar and U.P.; collected and assembled wild canes for breeding and assessing their characters; set up World Collection and Germ Plasm Bank for Sugarcane in India. Office: Indian Inst. Sugarcane Research, Lucknow, India.*

PANNARIA, Francesco, Italian physicist; b. Rome, Italy, May 7, 1898; s. Pietro and Nicolina (Biagini) P.; Ph.D., U. Rome; m. Angiola Intonti, July 13, 1940. Past head Center Study Chemistry and Physics of Exptl. Inst.; now dir. Labs. Ferrous Metals and Physics of Solid Matter, Rome. Mem. Internat. Center Comparison and Synthesis, Internat. Com. Thermodynamics and Electrochemistry. Decorated Order Merit Italian Republic. Publns. on subnuclear physics and structure of matter; collaborator in math. with F. Severi; pub. by Academia dei Lincei. Address: via Francesco Giambullari 8, Rome, Italy.

PANNEKOEK, Antonie, Dutch astronomer; b. Gelderland, Netherlands, 1873; prof. astronomy U. Leyden (Netherlands). Recipient gold medal, Royal Astron. Soc., 1950. Author: Anthropogenese, 1945; De Groei van ons Wereldbeeld, 1951. Research on astrophysics, structure of our galaxy, extra-galactic systems and cosmogeny.

PANNETIER, Guy, French chemist; b. Trans, France, Aug. 26, 1917; s. Georges and Rose (Michel) P.; Docteur ès Sciences; Agrégé de Sciences Physiques; Laureât de l'Academie des Sciences; m. Janine Christophe, Apr. 11, 1950; children—Sylvie, Bruno, Laurence. Prof., Faculté des Sciences de Paris, École Polytechnique. Mem. Chem. Soc. France (sec. gen.). Author: Atomistique, liaison chimiques, 3d edit.; Cenetique Chimique. Research, publs. on hydrozine chemistry, solid state physics, theory of combustion, molecular spectroscopy. Home: 26 Charles Peguy-Sceaux (92), France. Office: 1 Guy de la Brosse, Paris (5), France.*

PANOFF, Robert, Am. nuclear engr.; b. N.Y.C., Aug. 16, 1921; s. Emanuel and Lillie Panoff; B.S. in Elec. Engring., Union Coll., Schnectady, 1942; D.Sc. (hon.), Allegheny Coll., 1959; m. Kathleen D. Beck, Sept. 19, 1953; children—Kathleen Dorothy, Robert Michael, James Brian, Thomas Andrew, Stephen Edward, Timothy John. Engaged in design ships power systems, submarine propulsion systems Bur. Ships, Navy Dept., 1942-46, new design submarine propulsion systems, 1947-49; mem. staff comdr. Joint Task Force I for Atomic Weapons Test, Bikini, 1946-47; with Joint Bur. Ships/AEC Naval Reactor Group, 1949-54, head submarine nuclear propulsion plant, design group, 1954-64, also sr. submarine nuclear propulsion plant project officer; asst. mgr. naval reactors for submarine projects AEC, 1962-64. Prin. officer, dir. MPR Assos., Inc., Engrs., Washington, 1964—. Recipient Distinguished Civilian Service award USN, 1959. Development of nuclear power plants for generation of electricity; development of nuclear propulsion plants for naval vessels, oceanographic vessels, merchant ships. Address: 1140 Connecticut Av. N.W., Washington 20036.

PANOFSKY, Hans Arnold, meteorologist; b. Kassel, Germany, Sept. 18, 1917; s. Erwin and Dora (Mosse) P.; A.B., Princeton, 1938; Ph.D., U. Cal. at Berkeley, 1941; m. Margaret Ann Riker, July 24, 1943; children—Ruth Alice, Anne Davison. Instr., Wilson Coll., Chambersburg, Pa., 1941-42; asso. prof. N.Y. U., 1942-51; faculty Pa. State U., University Park, Pa., 1951—, Evan Pugh Research prof., 1966—; vis. prof. U. Minn., Mpls., 1961. Cons. govt., pvt. orgns. Guggenheim fellow, 1960. Fellow Am. Geophys. Union, Royal, Am. (Meisinger award 1965) meteorol. socs. Author: Introduction to Dynamic Meteorology, 1955; (with Glenn Brier) Some Applications of Statistics to Meteorology, 1958; (with J. Lumley) Structure of Atmospheric Turbulence, 1964; also articles. Research on atmospheric effects on air pollution, atmospheric turbulence applied to stability of structures, weather forecasting, relations between stratosphere and troposphere. Home: 1179 Oneida St., State College, Pa. 16801. Office: Deike Bldg., University Park, Pa. 16802.*

PANOFSKY, Wolfgang Kurt Hermann, physicist, educator; b. Berlin, Germany, Apr. 24, 1919; s. Erwin and Dorothea (Mosse) P.; came to U.S., 1934, naturalized, 1942; A.B., Princeton, 1938; Ph.D., Cal. Inst. Tech., 1942; D.Sc., Case Inst. Tech., 1963, U. Sask., 1964; m. Adele Irene DuMond, July 21, 1942; children—Richard J. and Margaret A. (twins), Edward F., Carol E., Steven T. Dir. project OSRD, Cal. Inst. Tech., 1942-43; cons. Manhattan Dist., Los Alamos, 1943-45; faculty U. Cal. at Berkeley, 1945-51; prof. physics Stanford, 1951-63, dir. high energy physics lab. 1953-61, dir., prof. Stanford Linear Accelerator Center, 1961—, prof. physics, 1963—. Cons. div. mil. application AEC, 1945-60, physics panel NSF, 1954-58, USAF Sci. Adv. Bd., 1955-57, Office Sci. and Tech., Ednl. Office Pres.; mem. Pres., Sci. Adv. Com., 1958-64, Advanced Research Projects Agy., 1958-62; fgn. service officer Dept. State, 1959, chmn. tech. working group on high-altitude detection, Geneva, Switzerland, 1959; mem. rev. com. Particle Accelerator div. High-Energy Physics div. Argonne Nat. Lab., 1958-60, 63—. Recipient Ernest Orlando Lawrence Meml. award AEC, 1961. Guggenheim fellow, 1959. Fellow Am. Phys. Soc. (council 1956-60, pres. div. of particles and fields 1968—); mem. Am., Nat. acads. scis., Phi Beta Kappa, Sigma Xi. Author: (with M. Phillips) Classical Electricity and Magnetism, 1955, 2d edit., 1962; also numerous articles. Research on x-rays and natural constants, accelerator design, nuclear physics, high-energy particle physics. Home: 25671 Chapin Av., Los Altos Hills, Cal. 94022. Office: Stanford Linear Accelerator Center, Stanford U., Stanford, Cal. 94305.*

PANSKY, Ben, Am. anatomist; b. Milw., Feb. 18, 1928; s. Abraham and Leah (Namerofsky) P.; B.A. in Zoology, U. Wis., 1948, M.S., 1950, Ph.D. in Anatomy, 1954; m. Julia Beverly Gossin, May 5, 1953; 1 son, Jonathan Hugh. Instr. anatomy U. Wis., 1950-53; faculty N.Y. Med. Coll., N.Y.C., 1953—, asso. prof., 1960—, dir. basic scis. Sch. Nursing, 1957-63. Mem. A.A.A.S., Am. Assn. Anatomists, N.Y. Acad. Sci., Am. Assn. U. Profs., A.M.A. (affiliate), Sigma Xi, Phi Chi. Author: (with E.L. House) A Functional Approach to Neuroanatomy, 1960; A Review of Gross Anatomy, 1964; Anatomy Study Wheels, 1964; also articles. Discovered insulin in thymus gland; research on carbohydrate metabolism as related to leukemia, anatomy of chimpanzee ear as related to space travel. Home: 144-60 Sanford Av., Flushing, N.Y. 11355. Office: N.Y. Med. Coll., 106th St. and Fifth Av., N.Y.C. 10029.*

PANT, Divya Darshan, Indian botanist; b. Ranikhet, India, Oct. 18, 1919; s. Ambika Datt and Tikkodevi (Joshi) P.; B.Sc., Lucknow U., M.Sc., 1945; D.Phil., Allahabad U., 1950; m. K.S. Radaha Aiyar, Dec. 4, 1947; children—Vijaya, Priya Darshan, Kusum. Mem. faculty Allahabad (India) U., 1945—, prof. botany, head dept., 1966—. Travel grantee Brit. Council U. Interchange Programme. Fellow Indian Bot. Soc., Linnean Soc. London (Eng.); mem. Internat. Soc. Plant Morphologists, Internat. Assn. Plant Taxonomists, Palaeobot. Soc., Vigyan Parishad India. Author: Studies in Gymnospermous Plants, 1962. Research, publs. on stomatal devel., root structure, foliar anatomy and nodal structure in diverse living plants, structure of Palaeozoic fossil plants; classification of fossil pollen and spores and gymnosperms. Home: 106, Tagore Town, Allahabad (U.P.*) India. Office: dept. botanay, The University, Allahabad-2, India.*

PANT, Narendra Chandra, Indian entomologist; b. Dehradun, India, Jan. 28, 1924; s. Ramdutt and Chandrawati (Joshi) P.; B.Sc., Agr. Coll., U. Agra (India), 1944, M.Sc., 1946; diploma Imperial Coll., Royal Coll. Sci., London (Eng.) U., 1950, Ph.D., 1950; m. Madhuri Joshi, Feb. 5, 1952; children—Vibha, Vinita, Sandeep. Quarantine entomologist Govt. of India, Bombay, 1950-52; lectr. zoology U. Delhi (India), 1952-56; insect physiologist Indian Agrl. Research Inst., New Delhi, 1956-63; prof. entomology Postgrad. Sch., Indian Agrl. Research Inst., 1963—; vis. prof. labs. in U.K., U.S., Can., Japan, 1964-65. Fellow Entomol. Soc. India (life). Editor: Entomology in India, 1964; chief editor Indian Jour. Entomology, 1962-64, 67—. Research, publs. on nutrition and microbiology of several insects, control of land snails, insect-plant relations with reference to plant varieties resistant to insects, behaviour Apis indica. Home: C-42, I.A.R.I., Delhi 12, India.*

PANTAZIS, George, Greek zoologist; b. Mytilini, Greece, July 8, 1906; s. Panayotis and Maria (Lidorikioti) P.; ed. univs. Athens, Leipzig, Munich, Paris and Naples; M.D., Ph.D.; m. Lela Katoga, July 4, 1934. Formerly prof. parasitology Athens Sch. Hygiene, then prof. zoology, 1933-58; prof. gen. biology U. Athens, 1958—. Vice pres. Atomic Energy Commn. Decorated Order Phoenix; Order Merit Italian Republic; Legion of Honor (France). Mem. Hellenic Biol. Soc. (pres.). Author: Zoologie, 1939; Anatomie comparee, 1952; Radiobiology, 1952. Home: Odos Aristotelous 189. Office: Odos Sina 3, Athens, Greece.

PANTELL, Richard Harris, Am. elec. engr., educator; b. N.Y.C., Dec. 25, 1927; s. Abraham Louis and Rebecca (Leader) P.; B.S., M.S., Mass. Inst. Tech., 1950; Ph.D., Stanford, 1954; m. Leona Harriet Siff, Dec. 19, 1951; children—Susan Ellen, Laurie Judith. Test engr. Gen. Elec. Co., Schenectady, 1947-49; faculty Poly. Inst. Bklyn., 1950-51; faculty Stanford, 1954-56, 57-62, prof. elec. engring., 1963—; vis. asst. prof. U. Ill., 1956-57. Cons. Internat. Tel.& Tel., Harlow, U.K., 1962-63; various industries, 1959—. Fulbright grantee, 1962. Mem. I.E.E.E., Am. Assn. U. Profs., Sigma Xi. Developer new types of microwave oscillators; research, publs. on ferroelectrics, lasers, and nonlinear optics. Home: 170 N. Balsamina Way, Menlo Park, Cal. 94025. Office: Microwave Lab., Stanford U., Stanford, Cal.*

PANTHEO, Giovanni Agostino, Italian alchemist; flourished 16th century. Priest in Venice. Author: Voarchadumia contra Alchímiam, 1530; Ars et Theoria Transmutationis metallicae cum Voarchadúmia, proportionibus, numeris, et iconibus rei accommodis illustrata, 1550; Lunario perpetuo, 1535. Wrote against spurious alchemy; described assay of gold; chemical preparation of various substances used in the arts.

PANTIC, Vladimir Radivoja, Yugoslavian cytologist, physician; b. Medjuluzje, S.R. Serbia, Apr. 5, 1921; s. Radivoje Radovana and Stanica (Urosevic) P.; Student, Vet. Faculty, 1946-50; Doctor's thesis, 1954; m. Olivera Jablan, July 11, 1949; children—Zoran, Sanja. Vet. faculty, Belgrade (Yugoslavia) U., 1950—, prof., 1962—; head dept. cytology and embryology Inst. for Biol. Research, Belgrade, 1963—; chief lab. for Electron Microscopy, Inst. for Application of Nuclear Energy in Agr., Vet. Medicine and Forestry, Zemun, Yugoslavia, 1964—. Recipient Republic prize for sci., 1960. Mem. Biol. Soc., Physiol. Soc., Anat. Soc., Vet. Soc., Serbian Biol. Soc. (pres. 1966—). Author: Embryology, 1962; The Cell, 1965; also articles; contbg. author, editor: Foundation of Electron Microscopy, 1962. Research on enzootic goitre, exptl. goitre and iodine prophylaxis, radiation effects, cytological and biochem. examination after internal and external irradiation; recovery effect of deoxiribonucleic acid, effects of hormones on nucleic acids and protein synthesis in endocrine glands, effects of hormones on differentiation and function of endocrine glands, cyclic changes in deer and roe-bucks thyroid, pituitary and parathyroid and their role in antler devel., transplantation of thyroid and pituitary, relation between hypothalamus, pituitary, thyroid, adrenals and gonads. Home: 13 Oblicev venac, Belgrade, S.F.R., Yugoslavia.*

PANTIN, Carl Frederick Abel, English zoologist; b. Blackheath, London, Eng., Mar. 30. 1899; s. Herbert and Emilie Pantin; M.A., Sc.D., Christ's Coll., Cambridge; hon. doctorates U. Sao Paulo (Brazil), also Durham; m. Amy Smith, 1917; 2 sons. Physiologist, Marine Biol. Lab., Plymouth, 1922-29; prof. zoology Cambridge U., fellow Trinity College Coll. Trustee Brit. Mus. Fellow Royal Soc., 1937, recipient Royal medal, 1950, Croonian lectr., 1952;

mem. Royal Soc. New Zealand, Brit. Assn. (pres. sect. D Edinburgh 1951), Brazilian Acad. Scis. (fgn.), Linnean Soc. London (pres. 1958-60, Trail award 1937, Gold medal 1964), Marine Biol. Assn. (pres.). Author: Notes on Microscopical Technique for Zoologists, 1946. Demonstrated that anemones respond like vertebrates on level of nervous conduction; compared hydrostatics of muscular contraction in these and other animals; studied relationship of the organism and its environment. Home: 25 Bentley Rd. Office: Trinity Coll., Cambridge, Eng.

PANUM, Peter Ludwig, Danish physiologist; b. Ronne/Bornholm, Denmark, Feb. 19, 1820; prof., Copenhagen, Denmark. Author: Experimentelle Untersuchungen zur Physiologie und Pathologie der Embolie, Transfusion und Blutmenge, 1864; Haandoi menneskets physiologi, 2 vols., 1865-72. Studied measles epidemic, Faroe Islands, 1846; precipitated protein by neutral salts, circa 1852; credited with 1st study of chem. products of putrefaction, thus influencing concept of putrid intoxication, 1856. Died Copenhagen, May 2, 1885.

PANZARINI, Rodolfo Niel Martino, Argentinian oceanographer; b. Buenos Aires, Argentina, July 22, 1910; s. Angel and Elisa (Mac Lennan) P.; student Naval Acad., 1925-31; M.S. in Oceanography, Scripps Instn. Oceanography, U. Cal., 1949; D. Oceanography, Argentina, 1952; postgrad. Naval War Coll., Buenos Aires, 1950; m. Haydée Josefina Chaves, Dec. 22, 1938; children—Silvia (Mrs. Alejandro Guillermo Varas), Nydia, Rodolfo, Ronaldo. Commd. in Argentine Navy, 1931, advanced through grades to rear adm.; ret., 1957; lectr. meteorology Naval Acad., 1944, Postgrad. Naval Sch., 1946; comdr. Argentine Antarctic expdns., 1950-51, 52-53; prof. phys. oceanography U. Buenos Aires, 1953—; dir. Argentine Antarctic Inst., 1955—. Del., Sci. Com. on Antarctic Research of Internat. Council Sci. Unions, 1958—, v.p., 1962—; v.p. Argentine Nat. Com. for IGY, 1957-58, Argentine Nat. Com. for Internat. Years of Quiet Sun, 1964-65. Fellow Nat. Acad. Geography (v.p. 1964—), Nat. Acad. Scis. Buenos Aires (v.p. 1963-65), Nat. Acad. Scis. Cordoba; mem. Am. Geophys. Union, Argentine Sci. Soc., Polar Soc. (U.S.). Author: La Naturaleza del Antártico, 1958; La Antártida, 1962; Introducción a la Oceanografia General, 1963; Nomenclatura del Hielo en el Mar, 1963; Compendio de Oceanografia Fisica, 1966; also articles. Organized Argentine Antarctic Inst. for sci. research in South Polar region, established sci. stas. in Antarctica to carry out research work, participated in coordination of internat. sci. cooperations. Home: General Paunero 1760, Martínez, Buenos Aires. Office: Instituto Antártico Argentino, Cerrito 1248, Buenos Aires, Argentina.*

PANZER, Wolfgang, German geographer; b. Munich, Germany, June 16, 1896; s. Friedrich and Helena (Klaubert) P.; ed. univs. Frankfurt, Heidelberg and Fribourg; Ph.D., dr. es sci. nat.; m. Martha Drener, June 20, 1925; children—Johannes, Christoph, Martin. Prof. agrege U. Giessen (Germany), U. Berlin (Germany), 1928; lectr. U. Cal. at Berkeley, 1929-30; asso. prof. at large U. Berlin, 1931; prof. Sun Yat Sen U., Canton, China, 1931-34; prof. U. Heidelberg, 1935-52; prof. U. Mainz (Germany), 1952—, now also dir. Geog. Inst. Hon. mem. Honover Geog. Soc., Dresden Geog. Assn. Author: (with N. Krebs) Länderkund Forschung; also works on geomorphology, glacial areas and geography Germany, Europe, The Mediterranean, N. Am., E. Asia and Pacific regions. Research on regional geography of N. Am., Mediterranean Sea, China, Western Pacific, geomorphology. Home: Ob. Laubenheimer Weg 13. Office: Geog. Inst. der Universität, Mainz 65, West Germany.

PANZONE, Rafael, Argentinian mathematician; b. Buenos Aires, Argentina, Apr. 4, 1932; s. Pélix Antonio and Maria Luisa (Musante) P.; student U. Buenos Aires, 1953-58, Dr. in Math.; m. Anges Ilona Benedek, Jan. 2, 1959; children—Susanne Helen, Pablo Andrés. Instr., U. Buenos Aires, 1958-61, asso. prof., 1962-66; asso. in research U. Ill., Champaigne-Urbana, 1967; prof. Nat. U. of South, Bahia Blanca, Argentina, 1967—. Mem. Nat. Council Research, Argentina, 1962—; cons. editor Revista UMA, 1964—. Recipient ODOL award Argentina, 1966. CNICT fellow U. Chgo., 1960-61. Mem. Union Matematica Argentina, Am. Math. Soc. Research, publs. on probability theory, measure theory, functional analysis. Office: 80 Colon, Bahía Blanca, Argentina.*

PAO, Richard Hsien Feng, civil engr.; b. Chekiang, China, Apr. 22, 1926; s. Chung-Tao and Yuching (Chang) P.; B.S. in Civil Engring., St. John's U., 1949; M.S., U. Ill., 1951, Ph.D., 1953; m. Miranda Lee, June 10, 1961. Came to U.S., 1950, naturalized, 1963. Design engr. J.G. White Engring. Corp., N.Y.C., 1953-54; faculty Rose Poly. Inst., Terre Haute, Ind., 1954—, prof., 1960—, chmn. civil engring. dept., 1961—; post-docotral fellow in fluid mechanics Harvard, 1964-65. Mem. Am. Soc. C.E., Am. Soc. for Engring. Edn., Am. Geophys. Union, A.A.A.S., Internat. Assn. for Hydraulic Research, Soc. for History Tech., Sigma Xi, Pi Mu Epsilon, Tau Beta Pi. Author: Fluid Mechanics, 1961; Fluid Dynamics, 1966; also articles. Research on incompressible fluid mechanics, statis. analysis hydrological data. Home: 4951 Dixie Bee Rd., Terre Haute, Ind. 47802.*

PAOLETTI, Antonio, Italian physicist; b. Rome, Italy, Sept. 9, 1930; s. Ciro and Maria (Paoletti) P.; Laurea in Fisica, U. Rome, 1953, Libera docenza in Fisica, 1961; m. Livia Ademollo, Sept. 14, 1961; children—Ciro, Francesco, Aldo. Asst. prof. Physics Inst., U. Rome, 1953-57; research physics Comitato Nazionale Energia Nucleare, Rome, 1957-64, dir. nuclear applied physics lab., 1964—. Cons. mem. physics com. NRC Italy, 1964—, cons. technol. com., 1964—. Mem. Italian Phys. Soc. Research, publs. on diffusion in liquid metals, scattering of thermal neutrons from solid systems with particular emphasis on magnetism, distbn. of magnetic electrons in some metals and alloys using polarized neutron technique. Home: I Monti, Anguillara, Rome, Italy. Office: Lab. F.N.A., C.S.N. Casaccia S. Maria di Galeria, Rome, Italy.*

PAOLETTI, Pietro, Italian physician; b. Milan, Italy, Jan. 31, 1934; s. Valente and Antonietta (Giorgetti) P.; M.D., U. Milano, 1957; m. Enrica Grossi, Feb. 22, 1962. Asst. prof. pharmacology U. Milano, 1957-59, asst. prof. neurosurgery, 1959—, prof. radio biology radioisotopes techniques Postgrad. Sch. Indsl. Pharmacology, 1960-61; research fellow in neurosurgery Harvard Med. Sch., 1962. Recipient Victor Emanuel II prize U. Milano, 1960; Bertarelli prize U. Milano, 1962. Mem. Italian Socs. Neurology, Neurosurgery, and Nuclear Medicine, N.Y. Acad. Scis., Deutsche Gesellschaft für Neurochirurgie, Congress Neurol. Surgeons, Soc. for Research into Hydrocephalus and Spina Bifida. Author: (with R. Villani) Lezioni di Neurochirurgia, 1965; also numerous articles. Research on mechanism of action of hypocholesteremic drugs, effects of radiation on central nervous system, brain tumor metabolism, dynamics of cerebral spinal fluid in hydrocephalus, brain scanning. Home: 12 Corso Indipendenza, Milan, Italy.*

PAOLETTI, Rodolfo, Italian pharmacologist; b. Milan, Italy, Aug. 23, 1931; s. Valente and Antonietta (Giorgetti) P.; M.D., U. Milan, 1955, Docente Radiobiology, 1960, Docente Pharmacology, 1962. Faculty, U. Milan, 1955—, asso. prof. pharmacology, 1960-65, dep. dir. Inst. Pharmacology, 1965—; vis. prof. dept. pharmacology George Washington U., Washington, 1960-61. Guest scientist NIH, Bethesda, Md., 1960-61; vis. scientist Australian Acad. Scis., 1964, Acad. Medicine Soviet Union, 1966. Mem. European Soc. for Biochem. Pharmacology (hon. sec. 1962), Italian (hon. sec. 1963), French (hon. mem.), socs. for atherosclerosis research. Author: (with D. Kritchevsky) Advance in Lipid Research, vol. 1, 1954; Gli isotopi radioattivi nella ricerca biologica, 1959; Lipid Pharmacology, 1964. also articles. Research on pharmacological activities of biphenyl acetates and cholesterol synthesis in mammals, role of catecholamines in adipose tissue; discovered desmosterol in growing nervous tissue and in exptl. and human brain tumors. Home: 43 Viale Regina Margherita, Milan, Italy. Office: Institute of Pharmacology, Via Andrea del Sarto, 21, Milan, Italy.*

PAPANICOLAOU, George Nicholas, physician, anatomist; b. Coumi, Greece, May 13, 1883; M.D., U. Athens (Greece); Ph.D., U. Munich (Germany); m., 1910. Asst., Oceanography Mus., Monaco, also physiologist Prince of Monaco oceanography expdn., 1911; with dept. pathology N.Y. Hosp., 1913-14; asst. in anatomy, med. coll. Cornell U., 1914-16, faculty, 1916—, prof. clin. anatomy, 1947-50, emeritus, 1950—, also dir. Papanicolaou Research Lab. Recipient Borden award Assn. Med. Colls., 1948, Amory award A.A.A.S., 1948; Wien award, 1953; Modern Med. award, 1954; Bertner award, 1955; Passano award, 1956; Alvarenga award, 1958 1957. Mem. Cancer Soc. (Honor medal 1952), Soc. Zoology, Soc. Exptl. Biology, Harvey Soc., Pub. Health Assn. (Lasker award 1950), Assn. Cancer Research, Assn. Anatomy, Acad. Athens, Inter-Soc. Cytology Council (past pres.). Developed (with C.R. Stockard) vaginal smear test which indicates histologic changes in vagina during menstrual cycle, 1917; extended test for detection of uterine cancer, 1940. Office: Cornell U. Med. Coll., N.Y.C.

PAPILIAN, Victor Victor, Rumanian anatomist, pathologist; b. Cluj, Rumania, Aug. 2, 1920; s. Victor and Katherine (Iorga) P.; M.D., U. Cluj, 1946. Faculty, U. Cluj, 1948-52, lectr., 1949-52; chief Central Prosecctral Service, Cluj, 1952—; researcher U. Paris, (France), 1958. Fellow Sci. Med. Soc. Rumania, Internat. Neurol. Soc. Author of several books in field normal and path. anatomy; also numerous articles. Research on histological and histochem. studies of demyelinating diseases of nervous system, exptl. allergical encephalomyelitis, theory of pathogenesis of allergies. Address: 3 Clinicilor, Cluj/Rumania.*

PAPIN, Denis, physicist, inventor; b. Blois, France, Aug. 22, 1647; s. Denis and Magdaleine (Pineau) P.; studied under Huygens; M.D., U. Angers, 1669. Asst. to Huygens, Lab. of Acad., Paris; left Paris and went to London, 1675; assisted Robert Boyle in experiments with air pump, 1676-79; prof. Marburg, Germany, 1688-95. Fellow Royal Soc., 1682; mem. French Acad. Scis. Author: Nouvelles Expériences du vuide, avec la description des machines qui servent à les faire, 1674; A New Digester or Engine for Softening Bones, 1680; A Continuation of the New Digester of Bones, 1687; Recueil de divers pièces touchant quelques machines, 1707 Ars nova ad aquam ignis adminiculo efficacissime elevandam, 1707. Improved air pump; invented steam digester for bones, 1679, condensing pump, safety valve; 1st to use steam to raise piston, 1690; built boat with paddle wheels driven by waterwheel, 1690; discovered principle of siphon; demonstrated that as pressure is raised or lowered below atmospheric pressure, boiling point is raised or lowered; (with Boyle and Hauksbee) demonstrated that air transmits sound vibrations. Died London, circa 1712.

PAPOULAR, Renaud, physicist; b. Cairo, Egypt, May 2, 1929; s. Jacques and Rachel (Maman) P.; Ingénieur, Ecole supérieure D'Electricité, 1952; Docteur es Sciences, Faculté des Sciences, Paris, 1955; m. Huguette Dowek, Oct. 15, 1950; children—Robert, André. Physicist, Centre National de Recherche Scientifique, Paris, 1953-56, Cie Francaise Thomas-Houston, Paris, 1956-59, Commissariat a l'Energie Atomique, 1959—; reader Ecole Supérieure D'Electricité, Paris, 1957-60; asso. prof. Faculté des Sciences, Orsay, France, 1961-66. Mem. Société Francaise de Physique. Author: Phénomènes Electriques dans les Gaz, 1963; (with J. Balazard), Application des Ondes Hyperfréquences et Infrarouges a l'Etude des Plasmas, 1965; also articles. Research on ionization of gases, plasma radiation, plasma diagnostics; invented microwave quasi-optical components. Home: 1 rue Yvonne, Bourg-La-Reine-92-France. Office 1 rue du Panorama-Fontenay-aux-Roses,92-France.*

PAPOUSEK, Dusan, Czechoslovakian chemist; b. Brno, Czechoslovakia, May 8, 1930; s. Antonin and Anna (Cechova) P.; Ph.D., U. Brno, 1957; m. Zlata Prochazkova, June 21, 1958; children—Roman, Ales. Research asst., sr. tchr. U. Brno, 1953-61; research scientist Czechoslovakian Acad. Sci., Prague, 1961—. Author: (with M. Horak) Characteristic Molecular Vibrations, 1968; also articles. Editorial bd. Collection Czechoslovakian Chem. Communications, 1965—. Research on propagation of ultrasonic waves in liquids, anharmonicity of molecular vibrations, applications of computers in molecular spectroscopy. Home: 61 Na Piskach, Prague. Office: 2 Fleming Sq., Prague, Czechoslovakia.*

PAPPALARDO, Romano Giuseppe, physicist; b. Genova, Italy, Nov. 23, 1932; s. Francésco V. and Maria (Canepa) P.; Dottore in Frisica, U. Pavia (Italy), 1956; m. Sheena MacColl, Mar. 15, 1962; children—Anna, Elena. Docente incaricato U. Pavia, 1955-56; Brit. Council scholar Bristol (Eng.) U., 1956-57; research asso., Fulbright scholar Pitts. U., 1957-59; vis. scholar Bell Telephone Labs., Murray Hill, N.J., 1959-61; research scientist Cyanamid European Research Inst., Geneva, Switzerland, 1961-66; resident research asso. chemistry div. Argonne (Ill.) Nat. Lab., 1966—. Mem. Optical Soc. Am. Research, publs. on electronic properties of transition metal ions of the iron series, rare-earths and actinides series, very air-sensitive organometallic-compounds. Home: 329 Lincoln Av., Downers Grove, Ill. 60515. Office: Argonne, Nat. Lab., Argonne, Ill. 60439.

PAPPAS, Alexis Constantin, chemist; b. London, Eng., Oct. 16, 1915; s. Miltiades C. and Polyxene (Acatos) P.; student U. Oslo (Norway), 1934-40, cand. real., Dr.philos., 1954; postgrad. Inst. du Radium, Paris, France, Mass. Inst. Tech., Cambridge; m. Ela Steen Mjoset, May 22, 1942; children—Jan Alexis, Tone Eleni. Head research N.A. Cassaccumulator-A/S Unitor, Oslo, 1941-46; research fellow U. Oslo, 1947-52, prof. radiochemistry, 1957-62, prof. nuclear chemistry, 1962—; research asso. Mass. Inst. Tech., 1949-51; CERN fellow, Uppsala, Copenhagen, Geneva, 1953-57, CERN vis. prof., 1959, 64; IAEA tech. asst. expert, Athens, Greece, 1963. Cons. nuclear chemistry group CERN, 1957-67; cons. Norwegian Center Inst. Indsl. Research, 1963—. Mem. Acad. Scis. Oslo, Norwegian Chem. Soc. (pres. 1966—). Research, numerous publs. on nuclear fission of low and high energies; discovery and characterization of short-lived radioactive isotopes; studies of delayed neutron emission; trace element distbn. in shebiosphere; assay of strontium-90 in human bones and diet in Norway. Home: 68B Blindernveien, Oslo-3, Norway.*

PAPPAS, Costas Ernest, Am. aero. engr.; b. Providence, Oct. 14, 1910; s. Ernest and Soffie (Rose) P.; B.S. in Mech. Engring, N.Y. U., 1933, Sc.M., 1934; m. Thetis Hero, June 9, 1940; children—Alceste Conrad. Asst. dir. sci. research Republic Aviation Corp., Farmingdale, N.Y., 1957-59, dir. sci. research, 1962-63, asst. to v.p., research and devel., 1959-64; pres. C.T. Engring. Assos., Huntington, N.Y., 1964—. Recipient, Wright Bros. award, 1943; Outstanding Achievement award Republic Aviation Corp.; certificate of distinction N.Y.U., 1955. Fellow Am. Inst. Aeros. and Astronautics asso., mem. membership com. 1951—), Brit. Interplanetary Soc; mem. A.A.A.S Contbr. articles to tech. jours. on compressibility effects on aircraft; originator of aerospace vehicle; patentee aircraft design.*

PAPPAS, George Demetrios, Am. cytologist, biologist; b. Portland, Me., Nov. 26, 1926; s. James and Anna (Dracopoulos) P.; B.A., Bowdoin Coll., 1947; M.Sc., Ohio State U., 1948, Ph.D., 1952; m. Bernice Levine, Jan. 23, 1952; children—Zoe, Clio. Vis. investigator Rockefeller Inst., N.Y.C., 1952-54; fellow in anatomy N.Y. U., N.Y.C., 1954-56; faculty Columbia Coll. Phys. and Surg., N.Y.C., 1956—, asso. prof. anatomy, 1963—. Recipient Career Devel. award NIH, 1965. Fellow A.A.A.S., N.Y. Acad. Sci.; mem. Am. Assn. Anatomists, Am. Soc. Cell Biology. Electron micros. studies of cells, specifically fluid transport mechanisms, cellular organelles, cell membranes, fine structure neurons. Home: 315 Central Park W., N.Y.C. 10025. Office: 630 W. 168th St., N.Y.C. 10032.*

PAPPENHEIM, Artur, German internist; b. Berlin, Dec. 13, 1870. Author: Grundriss der hämatologischen Diagnostik, 1911; Morphologische Hämatologie, 1919. Founder Folia Haematologica. Improved hematological dye technique. Died Dec. 31, 1916.

PAPPENHEIMER, John Richard, Am. physiologist; b. N.Y.C., Oct. 25, 1915; B.S., Harvard, 1936; Ph.D., Cambridge U., Eng., 1940; m. Helena F. Palmer, Sept. 2, 1949; children—Glenn Alwin, William de Kay, Rosamond Gilder, Frank Richard. Demonstrator in pharmacology U. Coll., London, 1939-40; instr. physiology Columbia, N.Y.C., 1941-42; fellow biophysics Johnson Found., U. Pa., 1942-45; asso. in physiology Harvard, 1946-49, asst. prof., 1949-52, vis. prof., 1953—, career investigator Am. Heart Assn., 1953—. Mem. Am. Acad. Arts and Scis., Nat. Acad. Scis., Am. Physiol. Soc. (pres. 1964-65), Soc. Gen. Physiology, Phi Beta Kappa (hon.). Editorial bd. Rev. Sci. Inst., 1949-51, Physiol. Rev., 1960-63; asso. editor Circulation Research, 1962—, Proc. Soc. Exptl. Biology and Medicine, 1955-61. Contbr. articles on circulatory, respiratory and renal physiology to profl. jours. Home: 15 Fayerweather St., Cambridge, Mass. 02138.*

PAPPER, Emanuel Martin, Am. anesthesiologist; b. N.Y.C., July 12, 1915; s. Max and Lillian (Weitzner) P.; A.B., Columbia, 1935; M.D., N.Y.U., 1938; Dr. Med. (hon.), Univ. Uppsala, Sweden, 1964; m. Julia Fisher, Dec. 21, 1939; children—Barbara Ellen, Richard Nelson. Fellow in medicine N.Y.U., 1938-39, fellow in physiology, 1940, instr. anesthesiology N.Y.U., 1942-46, asst. prof., 1946-49, asso. prof., 1949; professor of anesthesiology, chair Columbia and director anesthesiology service Presbyterian Hospital, 1949—; director of anesthesiology and visiting anesthesiologist of Francis Delafield Hosp., 1953—; cons. div. med. scis. Nat. Research Council, 1954—, Huntington (N.Y.) Hosp., 1949—; nat. cons. surgeon gen. U.S. Air Force, 1963—; mem. surgery study sect. Nat. Insts. Health, 1958-62; civilian cons. First Army, USN; mem. nat. heart council NIH, 1962—; mem. President's Commn. on Heart Disease, Cancer and Stroke, 1964. Diplomate American Board of Anesthesiology (dir. 1956—, chmn. 1964), Pan Am. Med. Assn. Fellow Faculty Anesthesiologists Royal Coll. Surgeons, Royal Soc. Medicine; mem. N.Y. Acad. Medicine (1st pres. sect. anesthesiology), Am. Soc. Anesthesiologists (2d v.p. 1961), N.Y. State Soc. Anesthesiologists (past president), Nat. Research Council (chmn. com. anesthesia 1962—), American College Anesthesiologists, Am. Soc. Pharmacology and Exptl. Therapeutics, A.M.A., N.Y. Acad. Scis., Am., N.Y. socs. anesthesiologists, A.A.A.S., Am. Assn. Thoracic Surgery, Harvey Soc., Am. Soc. Clin. Investigation, Am. Trudeau Soc., Assn. U. Anesthetists (co-founder; 1st pres.), Assn. Anaesthetists of Gt. Britain and Ireland (corr.), Swedish Society Anesthesiologists (honorary), Halsted Society. Phi Beta Kappa, Sigma Xi, Alpha Omega Alpha. Author 3 textbooks. Research and publs. on devel. concept of light gen. anesthesia; studies on metabolism and effects of anesthetics. Home: 2709 Arlington Av., Riverdale, N.Y. 10463. Office: 622 W. 168th St., N.Y.C. 10032.

PAPPER, Solomon, Am. physician; b. Bklyn., May 28, 1922; s. Max and Lillian (Weitzner) P; A.B., Columbia, 1942; M.D., N.Y.U., 1944; m. Renee Wolfson, Oct. 2, 1943; children—Robert Allen, Margaret Anne, Ellen Martha. Research fellow Harvard and Thorndike Meml. Lab., 1950-52, May Inst. at Jewish Hosp., Cin., 1952-53; from chief, metabolism sect. to asst. chief medicine Boston VA Hosp., 1953-60; from clin. instr. to asso. prof. medicine Tufts U. Med. Sch., 1953-60; instr. medicine Harvard Med. Sch., 1953-60; prof. medicine, prof. preventive medicine, also chmn. dept. Med. Coll. Va., 1960-62; prof. medicine, chmn. dept. U. N.M. Med. Sch., 1962-68; prof., co-chmn. dept. medicine U. Miami, 1968—, chief medicine Bernalillo Co. Indian Hosp., 1962-68, Miami VA Hosp., 1968—; cons. medicine Albuquerque VA Hosp., 1962-63. Mem. study sect. gen. medicine NIH, 1963—; mem. sci. adv. bd. Nat. Kidney Disease Found. Diplomate Am. Bd. Internal Medicine. fellow A.C.P.; mem. Am. Fedn. Clin. Research, Am., Western socs. clin. investigation, Assn. Study Liver Diseases, Am. Soc. Clin. Investigation, Endorine Soc., A.M.A., Western Assn. Physicians, Internat., Argentina (hon.) socs. nephrology, Phi Beta Kappa, Alpha Omega Alpha. Editor: The Kidney. Research in kidney failure, blood pressure. Address: VA Hosp.; Miami, Fla.*

PAPPOS OF ALEXANDRIA, Greek geometer; fl. ca. 300 at Alexandria. Author: A Mathematical Collection (only a mutilated portion of original 8 books is extant; it is a great source of knowledge of ancient Greek mathematics); also commentaries on Ptolemy's Almagest, Euclid's Elements, and the Analemma of Diodorus, all of which are now lost. His Mathematical Collection is a combination handbook and commentary on existing geometrical works, with historical annotations, improvements, and alterations of existing theorems; it is valuable for its discussions of: previous Greek achievements in higher geometry (esp. Euclid, Aristaeos, Appolonios); works by Autolycos, Theodosios, Menelaos, Aristatchos; various solutions of problem of two mean proportionals; a method of inscribing the five regular solids in a sphere; Archimedes spiral; Nicomedes cochloids; the quadratrix; Archimedes' semi-regular solids; isoperimetric figures; works on theoretical and practical mechanics by Archimedes, Philon of Byzantium, Heron, and Carpos; methods of squaring the circle and trisecting any angle. Pappos extended Euclid (Book I, Prop. 47) to any triangle; defined conic sections by means of directrix (and discussed focus-directrix property of conic sections); proved constancy of anharmonic ratios; determined measurement of surface area bounded by a spiral on a sphere; and presented Pappos' problem.

PAPY, Georges, Belgian mathematician; b. Brussels, Belgium, Nov. 4, 1920; s. Gaston and Jeanne (D'Hellt) P.; Dr. Math. Sci., agrégé, Free U. Brussels; m. Frédérique Lenger; children—Isabelle, René. Successively asst., asst. in research, researcher Nat. Found. Sci. Research, 1945-49; mem. faculty Free U. Brussels, 1949—, prof. math., 1961—. Pres. Belgian Algebra and Topology Center, 1958, Belgian Center Pedagogy and Math., 1961, also pres. Internat. Commn. Study and Betterment Teaching Math. Author: Premiers éléments de mathématique moderné; Groupes; Géométrie affine plane et nombres réels; Initiation aux espaces vectoriels; Mathématique moderne. Address: 57, av. de l'Université, Brussels 5, Belgium.

PAQUELIN, Claude-André, French physician; b. Avignon, France, Dec. 30, 1826; inventor thermocauterizer (pyrographer), used it in 1875 in treatment of chronic arthritis and neuritis. Died Paris, 1905.

PAQUOT, Charles, French chemist; b. Etrepagny, France, Oct. 30, 1912; s. René and Suzanne (Longfier) P.; Licencié ès Sciences, Ecole Normale Superieure, 1934, Agrégé, 1937, Docteur ès Sciences, 1944; m. Marie-Thérese Afchain, Dec. 21, 1942; children—Pierre, Gérard. Faculty, L'Ecole Normale Superieure, 1940-45; dir. lab. lipochemistry Centre National de la Recherche Scientifique, 1945—. Decorated Chevalier de la Legion d'Honneur, Grand Officier de l'Ordre Alphonse X le Sale, Medaille Chereul. Mem. Am. Chem. Soc., Am. Oil Chemists' Soc. Author: (with others) Les methodes analytiques des lipides simples, 1962; also numerous articles. Research in chemistry, physico-chemistry, analytical chemistry of fats, oils and detergents. Home: 77 BD Auguste Blanqui, Paris XIII, 75, France. Office: 2 Rue Henry Donant Thiais, 94, France.*

PARACELSUS, Theophrastus Bombastus von Hohenheim, physician, chemist, natural philosopher; b. Einsiedeln, Switzerland, probably May 1, 1493; s. Wilhelm von Hohenheim; ed. in adepta philosophia by father, also Johannes Trithemius, abbott of Spanheim; journeyman-scholar, from 1507; studied under Johannes Manardus, Nicolaus Leonicenus, Ferrara, Italy, 1513-16 (possible M.D. there). Apprentice miner; army surgeon wars, Low Countries, Italy, Scandinavia, Prussia, Tartary, countries under Venetian influence, possibly Nr. East, 1517-24; visited Salzburg, Austria, 1524-25, Strasbourg, France, 1526; apptd. municipal physician, Basel, Switzerland, 1527; lectr. U. Basel, 1527-28 (antagonized faculty and students by burning Canon of Avicenna, disputing authority of ancient and medieval authorities, lecturing in Swiss-German rather than Latin); left Basel after unsuccessful lawsuit, 1528; traveled through Europe, 1528-41; called to Salzburg by suffragen bishop Ernest of Wittelsbach, 1541. Author: works on syphilis including Von der Frantzosischen kranckheit (attacked use of guaiac wood as cure), 1529-30; Paragranum in quo columnae quatuor (insisted that medicine must be based on natural philosophy, astronomy, alchemy, virtue), 1530; Paramirum de Quinque Entibus (discussed tartaric diseases, also doctrine of imagination), 1531; Von der Bergsucht (book on miner's disease; 1st treatise in med. lit. systematically dealing with an occupational disease), 1533; Der Grossen Wundartzney (maj. surg. work. one of his few books pub. during lifetime), 1536; Labryinthus medicorum, circa 1539; Philosophia Sagax (theosophical work), circa 1537-38; many works on medicine, chemistry, natural philosophy, religion which form 3 folio vols. in 17th edit. of Huser, 13 vols. in modern edit. of med., philos. texts by Sudhoff, Matthiessen. Writings continually attacked then contemporary overreliance on authority of ancients; rejected teachings of univs.; insisted on importance of personal observations, expts. (largely meaning observations in nature, tests by fire, distillation analysis in lab.), at same time insisted on truthfulness of traditional macrocosm-microcosm universe and powers of stars; gave chem. oriented interpretation of Creation (in Philosophiae ad Atheniensis which may be spurious but is Paracelsian in tone); added salt to traditional sulphur, mercury of metals to form basic principles (tria prima used as 2d set elementary substances along with Aristotelian elements earth, air, fire, water); postulated internal archei which acted as alchemists within body separating pure from impure, regulating normal physiol. processes (imperfect functioning of archei resulted in disease, illness), concept resulted in view of local centers of disease in contrast to humoral theory which postulated disease as imbalance of humors throughout body; felt disease also possible from influence of stars; emphasized chem. prepared medicines in contrast to herbal remedies of past; rejected traditional uroscopy (called instead for chem. exam. of urine); noted sleep inducing properties of products of action of sulphuric acid or alcohol (ethers) on chickens. Paracelsus emerges as man calling for new exptl., observational medicine, sci. in opposition to sterile teachings of univs. but who accepted new-Platonic natural magic, alchemy of day as proper guide to truth; influence on medicine, chemistry enormous for next century and half. Died Salzburg, Sept. 24, 1541.

PARADE, Gustav-Wilhelm, German physician; b. Spiegelberg, Germany, May 2, 1901; s. Gustav and Anna (Wegener) P.; M.D.; m. Christine Gahrmann, 1929; children—Annelie, Christoph, Dietrich. Successively agrégé, asso. prof., full prof., prof. emeritus U. Sarre. Mem. German Soc. Internal Medicine, German Soc. Circulation and Cardiological Research, European Coll. Medicine; corr. mem. German Odontology Soc.; hon. mem. Deutsche Arbeitsgemeinschaft für Herdforschung. Publns. and research in cardiology and circulatory systems, hormones, also gen. internal medicine. Home: Neustadt (Wstr.). Office: Städtisches Krankenhaus Hetzelstift, Neustadt (Wstr.), West Germany.

PARASKEVAS, Michael, Greek surgeon; b. Piraeus, Greece, Nov. 4, 1904; s. John and Mary (Santou) P.; M.D., U. Berlin, 1927; m. Felicitas Lykissa, Feb. 3, 1945. Dir. surgery Piraeus Gen. Hosp., 1942-63; dir. 1st surg. clinic Evangelismos Med. Center, Athens, Greece, 1963—. Cons. surgeon Dromokaition Mental Hosp., 1948—. Mem. German (hon.), Greek (past pres.) surg. socs. Research, numerous publs. on distbn. and devel. of technique of reconstrn. of injured bile- and pancreatic ducts. Address: 34 Mithymnis, Athens, Greece 803.*

PARDEE, Arthur Beck, Am. biochemist, educator; b. Chgo., July 13, 1921; s. Charles A. and Elizabeth B. (Beck) P.; B.S., U. Cal. at Berkeley, 1942; M.S., Cal. Inst. Tech., 1943, Ph.D., 1947; m. Marjorie A. Maxstadt, June 30, 1947; children—Michael, Richard, Thomas, Elizabeth. Merck postdoctoral fellow U. Wis., 1947-49; faculty U. Cal. at Berkeley, 1949-61, asso. prof., 1947-61; NSF fellow Pasteur Inst., Paris, France, 1957-58; prof. biology, chmn. dept. biochem. scis. Princeton, 1961-67. Recipient Young Biochemists travel award NSF, 1952. Mem. Am. Chem. Soc. Lewis award 1960), Soc. Biol. Chemists (treas.), Am. Soc. Microbiologists, Japanese Bio-Chemists, Am. Soc. Microbiologists, Japanese Biochem. Soc. (hon.), Phi Beta Kappa, Sigma Xi. Author: (with R.W. Cowgill) Experiments in Biochemical Research Techniques, 1957. Editorial bd. Biochimica et Biophysica Acta, 1962—. Research, publs. on enzyme formation, inhibition, cell permeability, others concerning cell regulation of growth, function. Home: 291 Russell Rd., Princeton, N.J., 08540.*

PARDIES, Ignace Gaston, French mathematician, physicist; b. Pau, France, 1636; became mem. Soc. Jesus, circa 1652; lectr. Coll. Louis LeGrand, Paris. Author: Elements of Geometry, 1671; Statics, or the Science of Moving Forces, 1673; also correspondence with Newton (whose optics he opposed). Tried to develop Grimaldi's optics into complete theory of aether; attempted to effect a compromise between the ideas of Descartes and Aristotle. Died 1673.

PARDO, Efraín Guillermo, Mexican pharmacologist; b. Tingüindin, Michoacán, Mexico, Apr. 29, 1922; s. Efren and Josefina (Codina) P.; student Loyola U., 1938-39, U. So. Cal., 1940-42; M.D., Universidad Nacional Autónoma de México, 1948; m. Bertha Ortiz, May 22, 1948; children—Rebeca, Guillermo, Efrain, Raul, Carlos, Bertha, Jaime. Pub. Health fellow Instituto de Enfermedades Tropicales, 1948; asso. research pharmacologist U. Mich., Ann Arbor, 1949; prof. gen. physiology U. Nacional Autonoma de Mexico, Mexico City, Mexico, 1950-54, prof. pharmacology, 1955—, head dept. pharmacology, 1953-62; dir. Instituto Miles de Terapéutica Exptl., Mexico City, 1961—. Mem. Am. Soc. for Pharmacology and Exptl. Therapeutics, Sociedad Mexicana de Ciencias Fisológicas, Académia Nacional de Medicina. Author: Manual de Farmacología Terapéutica, 1960; also numerous articles. Research, numerous publs. on cardiac glycosides, ganglionic blocking agt. activity, devel. exptl. model of hypertension. Home: Mercaderes, 174, México 19, D.F., México.*

PARÉ, Ambroise, French surgeon; b. Bourg Hersent, near Laval, Mayenne, France, 1510 or 1517; a barber's apprentice; studied surgery at Hôtel-Dieu, Paris; in army as regimental barber-surgeon, from

1836; prosector at U. Paris; maitre-chirugien, Collège Saint Lôme; surgeon to Henry II, 1552; also to Francis II, Charles IX, and Henry III. Author: La méthode de traiter les playes faites par haquebuzades, 1545. Introduced improvements in treatment of gunshot wounds (used soothing ointments instead of boiling oil or cauterization), in operation of trepanning, and in amputation; reintroduced practice of tieing up divided arteries with ligatures and operated on articular concretions; improved obstetrical methods; invented several surgical instruments; suggested some forms of aneurysm caused by syphilis, 1575. Died, Paris, France, Dec. 20, 1590.

PARE, Charles Michael, English psychiatrist; b. Lancashire, Eng., Oct. 29, 1925; s. Frank William and Florence (Lee) P.; M.B., B.Ch., U. Cambridge (Eng.), 1948, M.A., 1950; M.R.C.P., U. London, 1950, D.P.M., 1957; M.D., U. Cambridge, 1956; m. Barbara Cowell, Jan. 28, 1950; children—Jane Helena, Caroline Louise, Christopher Frank Edward. Registrar, Maudsley Hosp., London, 1954-57, sr. registrar, 1957-59; USPHS travelling fellow Lab. Clin. Scis., NIH, USPHS, Bethesda, Md., 1959-60; cons. psychiatrist St. Bartholomew's Hosp., London, 1960—; hon. cons. psychiatrist Long Grove Hosp., Epsom, Surrey, Eng., 1960—. Fellow Royal Soc. Medicine; mem. Collegium Internationale Neuro-psychopharmacologicum, Royal Medico-psychol. Assn. Author: (with John Marks) The Scientific Basis of Drug Therapy in Psychiatry, 1964; A Practical Introduction to Psychiatry, 1964; also numerous articles. Research on genetics of homosexuality, brain amines in phenylketonuria, eticology and treatment of depressive illness. Home: Clay Hill Lodge, Meadway, West Hill, Epsom, Surrey, Eng. Office: 38 Devonshire St., London, W.1, Eng.*

PARENT, Antoine, French mathematician; b. Paris, Sept. 16, 1666; tchr. math., Paris; mem. French Acad. Scis., 1699. Author: Essais et recherches (includes 1st work on 3-dimensional analytic geometry), 1705; Research into Physics and Mathematics, 2 vols., 1705. Died Paris, Sept. 26, 1716.

PARENT-DUCHATFLET, Alexis-Jean-Baptiste, French physician; b. Paris, Sept. 29, 1790; ed. Paris Sch. Medicine, 1806; physician Hosp. Pitié. Became mem. Council for Pub. Sanitation, 1825. Author: Hygiène publique ou memoires sur les questions importantes de l'hygiène, 1836; Sur l'inflammation de l'arachnoide spinale. Research on inflamation of spinal chord, prostitution, pub. hygiene. Died Paris, 1836.

PARENTY, Henri, French engr.; b. Arras, France, Apr. 18, 1851; ed. Ecole polytechnique; engr. in state works; mem. French Acad. Scis., 1914. Studied emission of gases and steam under pressure; built steam quantity meter; proved existence of quantity limit as result of formation of nodes and loops, 1886. Died Dec. 16, 1921.

PARETO, Vilfredo, economist, sociologist; b. Paris, France, July 15, 1848; s. Raffaele Pareto; grad Turin (Italy) Poly. Inst., 1870. Engr., Rome, Italy, 1870-74; dir. Ferrovie Italiane; prof. polit. economy U. Lausanne (Switzerland), 1893-1923. Author: Cours d'économie politique, 2 vols., 1896-97; Les systems socialistes, 2 vols., 1902-03; Manuale di economia politica, 1906; Trattato di Soziologia generale, 2 vols., 1916; Compendio di Soziologia Generale, 1920; Transformazione della democrazia, 1921. Developed math. analysis of social, econ. problems, math. formula of econ. equilibrium using curves of indifference, concepts of logical action and non-logical action; proposed law of income distbn.; advanced cyclical theory social change involving polit., econ. and ideological shifts of power and circulation of elites. Died Celigny, Switzerland, Aug. 19, 1923.

PARETSKY, David, Am. microbiologist; b. Bklyn., Nov. 15, 1918; s. Joe and Ethel (Krupnik) P.; B.S. Coll. City N.Y., 1939; postgrad. Ia. State Coll.; Ph.D., 1948; m. Mary Ellen Edwards, Dec. 22, 1942; children—Jeremy, Sara, Daniel, Jonathan, Nicholas. Instr., Ia. State Coll., Ames, 1947-48; asst. prof. Rensselaer Poly. Inst., Troy, N.Y., 1948-51; faculty U. Kan., Lawrence, 1951—, prof., 1959—, chmn. dept. microbiology, 1957—; vis. prof. oncology U. Wis., 1964-65. Mem. Kan. Lab. Adv. Commn., 1958—; mem. rev. panel NIH, 1963-68; mem. panels on undergrad. tng. NSF, 1964—. Mem. A.A.A.S., Am. Acad. Microbiology, Am. Soc. for Microbiology (pres. Missouri Valley br. 1966-67, Am. Soc. for Biol. Chemists, Am. Chem. Soc., Sigma Xi (pres. Kan. chpt.), Phi Sigma. Research, publs. on rickettsial biochemistry, intermediary metabolism bacteria, mechanism of pathogensis of Q fever. Home: Route 1, Eudora, Kan. 66025. Office: Dept. Microbiology, U. Kan., Lawrence, Kan. 66044.*

PARHAM, Frederick William, Am. physician; b. New Orleans, Mar. 20, 1856; s. John Greenway and Mary E. (Blunt) P.; student Randolph-Macon Coll., Va., 1873-75; M.D., U. La., 1879; m. Mary K. Duncan, Dec. 15, 1892; children—Frederick D., Duncan, Mildred, Mary L., Lister, Landfried. Practiced in New Orleans, from 1879. Fellow Internat. Soc. Surgery, Am. Surg. Assn. (v.p. 1917), A.C.S. (bd. regents, 1920-26). Introduced (with E.D. Martin) steel or aluminum band to hold together ends of broken bone until union is established (Parham-Martin band), circa 1913. Died May 7, 1927.

PARHON, Constantin I., Rumanian physician; b. 1874; mem. Acad. Socialist Republic Rumania, numerous fgn. sci. socs. and acads. Author: Internal Secretions (1st treatise of its kind), 1909; Treatise of Endocrinologist, 3 vols., 1945-49; numerous other works on neurology, psychiatry, endrocinology. A founder of endocrinology; chem. and exptl. research on internal secretions, physiology of hormones, etiology of age.

PARIN, Vasiliy Vasilevich, Russian physiologist; b. Kazan, 1903; grad. Med. Faculty, Perm U., 1925; D.Med. Sci., 1939. Head chair physiology Perm Pedagogical Inst., 1930-33; prof., 1933—; head chair physiology 1st Moscow Med. Inst., 1933-41, dir., 1941-42; head chair physiology Moscow Med. Inst., RSFSR Peoples Commissariat Health, 1943-47; head physiology lab. Inst. Therapy, USSR Acad. Med. Scis., 1954-56, dir. Research Inst. Normal and Path. Physiology, 1960—, acad. sec., 1944-47, 57-60; Mem. Cybernetics head chair physiology Central Postgrad. Med. Inst., Moscow, 1956—, head chair clin. and exptl. physiology, 1956-60. Council, Med. Electronics Com. Del., Internat. Conf. on Med. Electronics, Paris, 1959; head Soviet delegation 20th Internat. Congress Physiologists, Holland, 1962 Decorated Order of Lenin. Mem. USSR Acad. Med. Scis. (v.p.), All-Union (dep. chmn.), Moscow (chmn.) socs. physiologists, All-Union Soc. Pathophysiologists (bd. mem.), All-Union Radio and Electronics Soc. (bd. mem.), Purkinje Czech Med. Soc. (hon.). Author: The Role of Pulmonary vessels in Regulating Circulation, 1946; Essays on the Clinical Physiology of Blood Circulation, 1960. Exec. editor Bull. Exptl. Biology and Medicine; mem. editorial council Path. Physiology and Exptl. Therapy; mem. editorial bd. Herald of USSR Acad. Med. Scis.; editor Biol. and Med. Electronics. Address: Central Postgrad. Med. Inst., pl. Vosstaniya 1-2, Moscow, USSR.

PARIS, François Edmond, French hydrographer, naval engr.; b. Paris, Mar. 2, 1806; dir. Nat. Map Depository; with d'Urville's voyage of 1826; named vice adm., 1864 mem. French Acad. Scis. (pres. 1876), Bur. Longitudes. Author: Essai sur la construction navale, 1848. Studied non-European naval constrn. methods. Died Apr. 18, 1893.

PARIS, René Raymond, French biochemist, botanist; b. Calais, France, May 12, 1907; s. Raymond and (Farrands) P.; student Reims Sch. Medicine and Pharmacy, 1927-28; Bachelier-Pharmacien, U. Paris (France), 1931, licencié ès sciences naturelles, 1932, Docteur es Scis. naturelles, 1935, Docteur en Pharmacie, 1938; m. Besnard, July 6, 1933; 1 son, Michel. Head lab. Paris Faculty Medicine, 1931-39; head projects Faculty Pharmacy, 1938-42, lectr., 1942-44; prof. Faculty Pharmacy, U. Paris, 1944—; dir. research Nat. Lab. Dept. Health, 1958—; emeritus prof. U. Montréal, 1961—. Recipient Gold medal Hosps. Paris, 1935; named chevalier Legion d'Honneur, 1953, officier l'ordre Nationale du Mérite, 1966, officier des palmes Academiques, 1954. Mem. Rio de Janero Acad. Pharmacy (hon.), Acad. Pharmacy, Bot. Soc. France (past pres.), Soc. Vegetale Physiology, Soc. Biol. Chemistry, Internat. Fedn. Pharmaceutics, Am. Soc. Pharmacognosy. Author: (with Moyse) Abrégé de Matière Medicale, 1953; Précis de Matierè Mèdicale, vol. I, 1964, vol. II, 1966; also numerous articles. Research on biochemistry and vegetal physiology especially polythenols, selections and improvement of medicinal plants, applications of chromatographie to analysis of medicinal plants, testing of medicine of vegetal origin; biochem. studies of numerous African plants used in popular medicine. Home: 104 Boulevard, Arago, Paris, France.*

PARISET, Étienne, French physician; b. Grand, France, Aug. 5, 1770; became physician Bicêtre Hosp., Paris, 1814, at Salpêtrière, France, 1814-40. Mem. French Acad. Scis., 1842, French Acad. Medicine (perpetual sec.). Author: Rapport sur la fièvre jaune de Cadix 1819. Translated Hippocrates. Research on yellow fever. Died Paris, July 3, 1847.

PARK, Charles Frederick, Jr., Am. geologist; b. Wilmington, Del., Dec. 18, 1903; s. Charles Frederick and Ida (Morey) P.; B.S. in E.M., N.M. Sch. Mines, 1926; M.S., Geology, U. Ariz., 1929; Ph.D. Geology, U. Minn., 1931; m. Eula Blair, Oct. 1, 1931; children—Frederick Blair, Allan Morey, Martha. Jr. geologist to geologist in charge metals sect. U.S. Geol. Survey, Washington, 1931-46; prof. Stanford, 1946-50, dean, 1950—, Donald Steel Prof., 1963—. Dir. Homestake Mining Co., San Francisco; cons. Bethlehem Steel Co., 1952—, Utah Constrn. & Mining Co., 1956-64. Mem. Soc. Econ. Geology (pres. 1963). Author: (with Roy A. MacDiarmid) Ore Deposits, 1964; Extensive mineral exploration, especially iron ore and magnesium ore; research on causes of localization of ore deposits, zoning phenomena. Home: 1431 Arcadia Pl., Palo Alto, Cal. 94303. Office: Sch. Earth Scis., Stanford U., Stanford, Cal. 94305.*

PARK, Charles Rawlinson, Am. physiologist; b. Balt., Mar. 2, 1916; s. Edwards Albert and Agnes (Bevan) P.; A.B., Harvard, 1937; M.D., Johns Hopkins, 1941; m. Jane Harting, Aug. 26, 1953; 1 son, Edwards A. USPHS fellow, Welch fellow internal medicine, dept. biochemistry Washington U., 1947-52; prof. physiology, chmn. dept. Vanderbilt U. Med. Sch., 1952—. Mem. Soc. Clin. Investigation, Am. Soc. Biol. Chemists, Am. Physiol. Soc., Assn. Am. Physiologists, Phi Beta Kappa, Sigma Xi, Alpha Omega Alpha. Research in physiology. Home: 2818 22d Av. C., Nashville.*

PARK, D., Brit. biologist; b. Leeds, Eng., June 28, 1929; s. Alfred and Lilian (Richardson) P.; B.Sc., U. Manchester, 1952, M.Sc., 1953, Ph.D., 1955, D.Sc., 1966; m. Bessie Maurine Prior, Aug. 2, 1958 (dec. 1965); children—Stephen, John, Nicolas, Michael Andrew. Lectr. botany dept. U. Manchester (Eng.), 1955-63; reader Queen's U., Belfast, N. Ireland, 1963—. Mem. Brit. Mycol. Soc., Soc. for Gen. Microbiology. Research and publs. on classification of ecol. groups of soil fungi, behavior in soil of plant pathogenic fungi with view to biol. control of plant disease, physiology and morphogenesis of fungi, aging phenomena in fungi; fine structure (electron microscopy) of fungi. Office: Queen's U., Belfast, N. Ireland.*

PARK, David Allen, Am. physicist; b. N.Y.C., Oct. 13, 1919; s. Edwin Avery and Frances (Paine) P.; A.B., Harvard, 1941; Ph.D., U. Mich., 1950; m. Clara Justine Claiborne, Aug. 18, 1945; children—Katharine, Rachel, Paul, Jessica. Faculty, Williams Coll., Williamstown, Mass., 1942-44, 51— prof. physics, 1959—; research asso. Radio Research Lab., Inst. for Advanced Study, 1950-51. Fulbright lectr. U. Ceylon, 1955-56; sr. visitor Cambridge (Eng.) U., 1962-63. Mem. Am. Phys. Soc., Am. Assn. Physics Tchrs., Am Assn. U. Profs. Author: Introduction to the Quantum Theory, 1964; Contemporary Physics, 1964; Introduction to Strong Interactions, 1966; also articles. Research on scattering elementary particles, solid state theory and devices, microwave techniques. Home: 4 Chapin Ct., Williamstown, Mass. 01267.*

PARK, Edwards Albert, Am. physician; b. Gloversville, N.Y., Dec. 30, 1877; s. William Edwards and Sara Billings (Edwards) P.; A.B., Yale, 1900; M.D., Columbia, 1905, M.A. (hon.), 1921; D.Sc., U. Rochester, 1936, Washington U., St. Louis, 1950, Laval U., 1955; LL.D., Johns Hopkins, 1951; m. Agnes Bevan, Aug. 2, 1913; children—Sara Bevan (Mrs. Henry Scattergood), Charles Rawlinson, David Chapman. Sterling Prof. pediatrics Yale Med. Sch., 1921-27; prof. pediatrics Johns Hopkins, Balt., 1927-47, prof. emeritus, 1947—. Recipient Order of Leopold, 1929; Howland medal, 1952; Bronze medal N.Y. Acad. Medicine, 1953; Borden award, 1945; Nutrition award A.M.A., 1962. Mem. Assn. Am. Physicians (Kober award 1950), Am. Pediatric Soc. Research on rickets and scurvy; bone growth in health and disease. Home: 1903 Ruxton Rd., Balt. 21204. Office: John Hopkins Hosp., Balt. 21205.*

PARK, Henry, English surgeon; b. Liverpool, Eng., Mar. 2, 1744; apprenticed to Liverpool Infirmary, also to Percival Pott, London; student Paris, Rouen, France; m. Miss Ranicar, 1776; 8 daus., 1 son; surgeon Liverpool infirmary, 1767-98. Author: Account of a New Method of Treating Diseases of the Joints of the Knee and Elbow, 8 vols., 1783. Pioneer in devel. of surgery for excision of carious joints. Died Liverpool, Jan. 28, 1831.

PARK, James Theodore, Am. microbiologist; b. Palo Alto, Cal., Aug. 3, 1922; s. Charles V. and Frances (Odenheimer) P.; A.B., Central Mich. U., 1943, D.Sc., 1963; M.S., U. Wis., 1944, Ph.D., 1949; m. Helen Sternberg, Dec. 13, 1952; children—Jane Frances, David Franklin, Elizabeth Ann. Biochemist, Ft. Detrick, Md., 1949-53, germ-free animal research unit Walter Reed Army Med. Center, 1953-57; asso. prof., then prof. microbiology Vanderbilt U. Sch. Medicine, 1958-62; prof. microbiology, chmn. dept. Tufts U. Sch. Medicine, 1962—. Mem. study sect. biochem. microbiology Nat. Inst. Allergy and Infectious Diseases, 1964—. NSF sr. postdoctoral research fellow Cambridge (Eng.) U., 1957-58. Mem. Am. Acad. Arts and Scis., Am. Soc. Biol. Chemists, Am. Soc. Microbiology, Biochem. Soc., Soc. Gen. Microbiology, Am. Chem. Soc., A.A.A.S., Sigma Xi. Mem. editorial bd. Jour. Bacteriology, 1964—. Contributed to understanding of the biosynthesis of bacterial cell walls and the mode of action of penicillin. Home: 11 Bradford Rd., Weston, Mass. 02193. Office: 136 Harrison Av., Boston 02111.*

PARK, Kilho, oceanographer; b. Kobe, Japan, Feb. 4, 1931; s. Bongjoo and Dooee (Chun) P.; B.S. in Fisheries, Pusan (Korea) Fisheries Coll., 1953; M.S., Tex. A. and M. U., 1957, Ph.D., 1961; m. Sookja Park, June 3, 1960; children—Arvin, Boyle. Faculty, Ore. State U., Corvallis, 1961—, asso. prof. oceanography, 1964—. Mem. Am. Geophys. Union, Am. Soc. Limnology and Oceanography, Geochem. Soc., A.A.A.S., Ore. Marine Biol. Soc., Am. Chem. Soc., Oceanological Soc. Korea, Oceanographical Soc. Japan, Geochem. Soc. Japan, Marine Biol. Assn. U.K., La Société Franco-Japonaise d'Oceanographie, Sigma Xi, Phi Lambda Upsilon. Research, publs. on chemistry of Northeastern Pacific Ocean, molecular identification of dissolved organic matter in seawater and electrolytic conductance of seawater.*

PARK, Mungo, Scottish explorer, physician; b. Foulshields on the Yarrow, Scotland, Sept. 10, 1771; s. Mungo Park; apprenticed to surgeon Thomas Anderson, Selkirk; student U. Edinburgh (Scotland), 1789-91, surg. diploma; m. Miss Anderson, Aug. 1799; several children, including Thomas. Asst. med. officer Worcestor E. Indiaman; sailed to Sumatra, 1792; leader expdn. to Niger, 1795-96; leader govt. expdn. (only survivor), 1805. Author: Travels in the Interior of Africa, 1799. Carried out bot. studies, discovered 8 new fish species in Sumatra; pioneer in exploration of Niger River, traced it for 2000 miles, its source for 300; incorrectly believed Niger and Congo rivers identical. Drowned under native attack in Niger rapids, 1806.

PARK, Orlando, Am. biologist, educator; b. Elizabethtown, Ky., Oct. 13, 1901; s. Samuel Thomas and Sophronia (Stealey) P.; B.S., U. Chgo., 1925, Ph.D., 1929; m. Alberta Fritsche, June 11, 1930; 1 dau., Patricia (Mrs. Manfred Engelmann). Asst. prof. zoology Kent State U., 1929-30; instr. zoology U. Ill., 1930-34; faculty zoology Northwestern U., Evanston, Ill., 1934—, prof. biology, 1942—, chmn. dept. biology, 1961-63. Cons. Oak Ridge Nat. Lab., 1953—; research asso. Chgo. Nat. History Mus., 1955—. Fellow A.A.A.S., Entomol. Soc. Am., Ecol. Soc. Am. (pres. 1943); mem. Chgo. Acad. Scis. (hon. curator), Nat. Speleological Soc. (spl. research award 1959), Sigma Xi, Hounds of Baskerville. Author: Manual of Animal Ecology and Taxonomy, 1939; Study in Neotropical Pselaphidae, 1942; (with others) Principles of Animal Ecology, 1949; Sherlock Holmes and John H. Watson, M.D., an encyclopedia of their affairs, 1962. Sectional editor Biol. Abstracts; Am. editor Pedobiologia. Research on devel., composition of natural animal and plant communities, ecology, life of pselaphid beetles. Home: 2201 Sherman Av., Evanston, Ill. 60201.*

PARK, Roswell, Am. surgeon; b. Pomfret, Conn., May 4, 1852; s. Roswell and Mary (Baldwin) P.; A.B., Racine (Wis.) Coll., 1872, A.M., 1875; M.D., Northwestern, 1876; M.D. (hon.), Lake Forest, 1892; A.M., Harvard, 1895; LL.D., Yale, 1902; m. Martha P. Durkee, 1880. Demonstrator anatomy Woman's Med. Coll., Chgo., 1877-79; adj. prof. anatomy Northwestern U., 1879-82; lectr. surgery, Rush Med. Coll., Chgo., 1882; prof. surgery U. Buffalo, also surgeon to Buffalo Gen. Hosp., from 1883. Attended President McKinley after he was shot, 1901. Author: Lectures on Surgical Pathology, 1891; History of Medicine, 1897; Text-Book of Surgery, 2 vols., 1896; The Principles and Practice of Modern Surgery, 1907. Disseminated new discoveries in pathology and bacteriology; promoted Lister's antiseptic techniques. Died Feb. 15, 1914.

PARK, Thomas, Am. zoologist, ecologist; b. Danville, Ill., Nov. 17, 1908; s. Samuel Thomas and Fronie (Stealey) P.; S.B., U. Chgo., 1930, Ph.D., 1932; m. Martha Alden Whitehead, July 31, 1928 (dec. June 1963); children—Sherley (Mrs. William Richard Hohmann), Judith (Mrs. Edgar Allan Barnett). Instr., asso. in biology Johns Hopkins, 1935-37, NRC fellow, 1933-35; faculty zoology U. Chgo., 1937—, prof., 1947—. Sci. attache Am. Embassy, London, Eng., 1949. Rockefeller Found. fellow Oxford (Eng.) U., 1948. Mem. A.A.A.S. (pres. 1961), Ecol. Soc. Am. (pres. 1959), Sigma Xi, Phi Delta Theta. Author: (with others) Principles of Animal Ecology, 1949. Has shown feasibility of exptl. lab population studies and analyses of statis. data, in some cases modeled in theoretical terms. Home: 5715 S. Blackstone Av., Chgo. 60637.*

PARK, William Hallock, Am. physician, bacteriologist; b. N.Y.C., Dec. 30, 1863; s. Rufus and Harriet (Hallock) P.; A.B., Coll. City N.Y., 1883; M.D., Coll. Phys. and Surg. (Columbia), 1886; postgrad. U. Vienna; 1889-90; LL.D., Queen's U., 1910; D.Sc., N.Y. U., 1926, Yale, 1929, Columbia, 1929. Prof. bacteriology and hygiene Univ. and Bellevue Hosp. Med. Coll. (N.Y. U.), 1897-1937; dir. N.Y. Health Dept. Bur. Labs. 1894-1937; cons. bacteriologist State Dept. Health, from 1914, med. examiner in bacteriology, from 1917; cons. bacteriologist, U. S. Quarantine Service, from 1921. Pres. Am. Pub. Health Assn., 1923. Author: Pathogenic Microörganisms, 10th edit., 1933; Public Health and Hygiene, 2d edit., 1927; Who's Who Among the Microbes, 1929; co-author: Bacteriology in Medicine and Surgery, 1899; Developed a diphtheria antitoxin, 1890; improved milk purification; studied child health; organized 1st municipal health lab. in N.Y.C., 1893-94; authority on pub. health aspects of Tb, pneumonia, diphtheria, poliomyelitis. Died N.Y.C., Apr. 6, 1939.

PARKE, Dennis Vernon, English biochemist; b. London, Eng., Nov. 15, 1922; s. William Thomas and Florence (Wyles) P.; B.Sc. with 1st class honors in Chemistry, London U., 1948, Ph.D., in Biochemistry, 1951; m. Doreen Joan Dunn, Feb. 28, 1943; children—Ann Lesley, Gerard Andrew Roger, Robert Jeremy Greville. Research officer microbiol, chemistry Glaxo Labs., 1948-49; faculty biochemistry dept. St. Mary's Hosp. Med. Sch., U. London, 1949-67; professor biochemistry department U. Surrey (England), Since 1967—. Fellow Royal Inst. Chemistry; mem. Chem. Soc., Biochem. Soc., Royal Soc. Medicine, Internat. Soc. for Biochem. Pharmacology. Author: Biochemistry of Foreign Compounds, 1967; also numerous articles. Research on metabolism of drugs, food additives, pesticides and other toxic chems., use of radioactively labelled compounds, effects of pregnancy on metabolism and activity of drugs. Home: Montrose 183A Cannon Lane, Pinner, Middlesex, Eng. Office: Biochemistry Dept.., U. Surrey, Guildford, Surrey, Eng.*

PARKER, Allan Elwood, Am. physicist; b. Bklyn., Feb. 12, 1909; s. S. Ridley and Vernie (Elwood) P.; A.B., Amherst Coll., 1929; Ph.D., Yale, 1933; m. Alice Frances Heywood, Mar. 19, 1932; children—Robert Allan Ridley, Peter Donald MacDougall, Allan Elwood II. Research asso. Columbia, N.Y.C., 1935-36; instr. Hunter Coll., N.Y.C., 1936-37; physicist Elec. Testing Labs., N.Y.C., 1937-42; prof. physics Worcester (Mass.) Poly. Inst., 1942—, head dept. physics, 1949—. Mem., Am. Physical Soc., Optical Soc. Am., Am. Assn. Physics Tchrs., Sigma Xi. Research on diatomic molecular spectra; developed methods and instruments for spectrophotometry. Home: 194 North St., Shrewsbury, Mass. 01545. Office: Worcester Polytechnic Institute, Worcester, Mass. 01609.*

PARKER, Earl Randall, Am. metallurgist; b. Denver, Colo., Nov. 22, 1912; s. Sam and Rebecca Rose (Presley) R.; Met.E., Colo. Sch. Mines, 1935; m. Mary Mildred Larkin, June 2, 1935; children—Robert Earl, Margaret Mary, William John. Research metallurgist Gen. Electric Research Lab., 1935-44; research metallurgist U. Cal. at Berkeley, 1944-45, asso. prof. metallurgy, 1945-49, prof., 1949—, chmn. div. mineral tech., Coll. Engring., 1953-57, dir. Inst. Engring. Research, 1957-64. Cons. various USN, USAF projs.; Campbell Meml. lectr., 1957, Robert S. Williams lectr., 1957. Recipient Mathewson Gold medal Am. Inst. Mining, Metall. and Petroleum Engrs., 1956 Mem. Am. Inst. Mining, Metall. and Petroleum Engrs., Am. Soc. Metals (trustee), Am. Ceramic Soc., Am. Phys. Soc. Author book, tech. papers in field. Research on brittle behavior of engineering structures; materials for spacecraft and missiles; mechanical behavior of metals. Home: 3375 Robinson Dr., Oakland 2, Cal.

PARKER, Eugene Newman, Am. physicist; b. Houghton, Mich., June 10, 1927; s. Glenn Hugh and Helen (McNair) P.; B.S. in Physics, Mich. State U., 1948; Ph.D., Cal. Inst. Tech., 1951; m. Niesje Meuter, Nov. 24, 1954; children—Joyce Marie, Eric Glenn. Faculty, U. Utah, 1951-55; faculty U. Chgo., 1955—, prof. physics, Fermi Inst., 1962—. Recipient Space Sci. award Am. Inst. Aeros. and Astronautics 1965. Mem. Am. Phys. Soc., Am. Astron. Soc., Am. Geophys. Union, Nat. Acad. Scis. Author: Interplanetary Dynamical Process, 1963, also articles. Research in theoretical behavior of gases, magnetic fields, and high energy particles in astrophys. universe; predicted existence of solar wind of atoms which move radially away from sun and might account for Van Allen particles and shape of geomagnetic field in space, 1959 (verified by Mariner 2 Venus probe). Home: 1323 Evergreen Rd., Homewood, Ill. 60430. Office: 933 E. 56th St., Chgo. 60637.*

PARKER, Frank Leon, Am. research engr.; b. Somerville, Mass., Mar. 23, 1926; s. Benjamin James and Bertha (Cohen) P.; student Norwich U., 1944-45, Pa. State Coll., 1945-46; B.S., Mass. Inst. Tech., 1948; Ph.D., Harvard, 1955; m. Elaine Marilyn Goldman, Aug. 22, 1954; children—Nina, Aaron, Stephan, David. Irrigation engr. U.S. Bur. Reclamation, Rivertown, Wyo., 1948; field engr. transmission lines Rockland Light & Power Co., Nyack, N.Y., 1949; research engr. Harvard, 1950-54; cons. hydraulic engr. Howard M. Turner, Boston, 1955; research engr. Oak Ridge Nat. Lab., 1956-60, head, radioactive waste disposal research, 1962-66; head, waste disposal research Internat. Atomic Energy Agy., Vienna, Austria, 1960-61; asso. prof. U. Tenn., 1966; asso. prof. water resources Vanderbilt U., Nashville, 1967—. Vis. expert, Israel, 1961, Pakistan, 1965; mem. U.S. Exchange Team to USSR, 1964. Mem. A.A.A.S., Health Physics Soc., Am. Soc. C.E., Harvard Soc. Engrs. and Scientists, Sigma Xi. Pioneer in establishment of total integrated radioactive waste mgmt. systems by disposing of wastes permanently in geologic formations; pioneer in use of radioactive tracer for hydrologic studies; research on water resources and waste disposal tech.*

PARKER, George, English astronomer; b. Eng., 1697; s. Thomas Parker; pupil, Abraham De Moivre, William Jones; D.C.L., Oxford (Eng.) U., 1759; m. Mary Lane, Sept. 18, 1722; children—Thomas, George Lane; m. 2d, Dorothy Nesbit, Nov. 1757. Mem. Parliament, Wallingford, 1722-27; built obs., Sherburn Castle, Oxfordshire, (Eng.) 1739; patron of James Bradley, Thomas Phelps. Fellow Royal Soc., 1722 (pres. 1752). Author: Remarks upon the Solar and Lunar Years, 1750; also other papers. Instrumental in changing computation of current chronology, 1762. Died Mar. 17, 1764.

PARKER, George Howard, Am. zoologist; b. Phila., Dec. 23, 1864; s. George Washington and Martha (Taylor) P.; S.B., Harvard, 1887, S.D., 1891, Colby Coll., 1935; spl. student at univs. Leipzig, Berlin, Freiburg, 1891-93; m. Louise Merritt Stabler, June 15, 1894. Asst. and instr. zoology Harvard, 1888-91, Parker fellow in Europe, 1891-93, instr., 1893-99, asst. prof., 1899-1906, prof. zoology, 1906-35, prof. emeritus since 1935. William B. Clark lectr. Amherst Coll., 1914; sent by U.S. Govt. to investigate Pribilof seal herd, 1914; exchange prof. to western colleges, 1921. Awarded Elliott medal Nat. Acad. Scis., 1937, Lewis prize Am. Philos. Soc., 1941. Fellow A.A.A.S. (v.p. 1916), Am. Acad. Arts and Scis. (pres. 1933); mem. Nat., Washington acads. scis., Am. Philos. Soc., Boston Soc. Natural History, Am. Zool. oc. (pres. 1903), Mass. Med. Soc., Am. Soc. Naturalists (pres. 1929), Am. Physiol. Soc., Assn. Am. Anatomists, Soc. Exptl. Biology and Medicine, Ecol. Soc. America, Marine Biol. Lab., Soc. Vert. Paleontology, Phi Beta Kappa, Sigma Xi (nat. pres. 1934-35); hon. mem. Am. Otol. Soc., Cal. Acad. Sciences, Cambridge Philos. Soc., Buffalo Soc. Natural History; corr. mem. London Zoöl. Soc., Acad. Natural Sciences Phila., N.Y. Acad. Sciences, Soc. Biol. (Paris), Peking Soc. Nat. History; asso. mem. Soc. Belge de Biol.; foreign mem. Linnaean Soc., London. Author: Biology and Social Problems, 1914; The Elementary Nervous System, 1919; Smell, Taste and Allied Senses in the Vertebrates, 1922; The Evolution of Man (co-author), 1922; What Evolution Is, 1925; Creation by Evolution (co-author), 1928; Human Biology and Racial Welfare (co-author), 1930; Humoral Agents in Nervous Activity, 1932; The Problem of Mental Disorder (co-author), 1934; Color Changes in Animals in Relation to Nervous Activity, 1936; The World Expands, 1946; Animal Colour Changes and Their Neurohumours, 1948. Contbr. articles to zool. journals, dealing chiefly with anatomy and physiology of nervous organs and animal reactions. Died Cambridge, Mass., Mar. 26, 1955.

PARKER, Herbert Edmund, Am. biochemist, educator; b. Springville, Tenn., Oct. 21, 1919; s. Charles Edmund and Bertha (Provo) P.; B.S.A., U. Tenn., 1942; M.S., Purdue U., 1948, Ph.D., 1950; m. Dortha Ellen Tudor, May 16, 1945; children—Richard E., Steven T., Paul Daniel. Faculty dept. biochemistry Purdue U., Lafayette, Ind., 1950—, prof., 1965—. Mem. Am. Chem. Soc., Am. Inst. Nutrition, Am. Soc. for Animal Scis., Sigma Xi. Author: (with Edwin Theodore Mertz) Laboratory Experiments in Biochemistry, 1948. Research, publs. on mineral nutrition of plants and animals, biochem. function of mineral elements, factors influencing utilization of mineral elements in living tissues. Home: Rural Route 9, Box 300, Lafayette, Ind. 47901.*

PARKER, Herbert M(yers), physicist; b. Eng., Apr. 13, 1910; s. William Henry and Elizabeth (Lancaster) P.; B.Sc., Manchester U., 1930, M.Sc., 1931, F.Inst.P., 1937; m. Margaret R. Fawthrop, Apr. 4, 1936; children—Elizabeth Anne, John, Linda Mary, Henry. Research physicist Manchester Com. on Cancer, 1932-34; physicist Holt Radium Inst., 1934-38, Tumor Inst. of Swedish Hosp., Seattle, 1938-42; research asso. Metall. Lab., U. Chgo., 1942-43; sect. chief health physics Clinton Labs., Oak Ridge, 1943-44; chief supr. health instruments Hanford (Wash.) works Gen. Elec. Co., 1944-48, mgr. health instruments, dir. radiol. scis., 1952-56, mgr. Hanford Labs., 1956-65; cons. Battelle-Northwest, Richland, Wash., 1965—. Mem. Wash. State Adv. Board Radiology Control. Recipient Moseley prize also Janeway medal Am. Radium Soc., 1955. Fellow Inst. Physics Gt. Britain, Am. Phys. Soc., A.A.A.S.; mem. Am. Radium Soc., Radiol. Soc. N.Am., Am. Coll. Radiology, Radiation Research Soc., Am. Nuclear Soc., Soc. Nuclear Medicine. Contbr. profl. jours. Published classic papers on radium therapy; developed radiation protection and radiological sciences research for the Plutonium Project branch of the atomic bomb project, including formulating the 'Rep' and 'Rem' dose units; led research on plutonium recycle concept for nuclear reactors. Home: 1936 Harris Av. Office: P.O. Box 999, Richland, Wash. 99352.*

PARKER, Jesse Elmer, Am. poultry scientist; b. Wetumpka, Ala., Nov. 23, 1911; s. Elmer Baird and Jessie (Cleveland) P.; B.S., U. Tenn., 1934; A.M., U. Mo., 1936, Ph.D., 1940; m. Charlotte Cornelia Krusekopf, July 9, 1938; children—Charlotte Louise, Jesse William, Mary Frances, Sarah Elizabeth, Jenny Lee, John Cleveland. Instr. poultry husbandry U. Mo., Columbia, 1936-39; poultry husbandman U. Tenn., Knoxville, 1939-44; prof., head poultry dept. N.D. State U., Fargo, 1944-46; prof., head poultry sci. dept. Ore. State U., Corvallis, 1946—. Mem. Poultry Sci. Assn. (past dir.), Am. Feed Mfrs. Assn. (feed survey com.), Sigma Xi, Alpha Zeta, Gamma Sigma Delta. Contbg. author: Fertility and Hatchability of Chickens and Turkeys, 1949; Reproduction of Farm Animals, 1962. Research, publs. on semen prodn. in domestic fowls, relation of semen characteristics to its fertilizing capacity, effect of photoperiod on reprodn. in male chickens, effects of nutrition on volume and fertility of avian semen; pioneered artificial insemination in comml. turkey prodn. Home: 642 N. 9th St., Corvallis, Ore. 97330.*

PARKER, Moses Greeley, Am. physician, physicist; b. Dracut, Mass., Oct. 12, 1842; s. Theodore and Hannah (Greeley) P.; student Phillips Acad., Andover, Mass., L.I. Coll. Hosp. Med. Sch.; M.D. Harvard Med. Sch., 1864; postgrad. U. Vienna, 1873-74, Paris, 1874-75. Asst. surgeon 2d U.S.C. Cav., 1864-65; located in Lowell, Mass., 1866; specialized as oculist and aurist; phys. St. John's Hosp., Lowell, 30 yrs.,

Lowell Gen. Hosp. (trustee 1898—) Dispensary and Ministry at Large, 10 yrs. Dir. and mem. exec. com. N.E. Telephone and Telegraph Co., from 1883, also interested in Bell Telephone Co. since its orgn. Mem. various med. societies. Made spl. study of electricity; 1st to photograph elec. current and show that it takes the form of spirals. Died Oct. 1, 1917.

PARKER, Neville, Australian psychiatrist; b. Brisbane, Australia, Jan. 2, 1928; s. Herbert and Barbara (Cherry) P.; student U. Queensland, 1945-52, U. Melbourne, 1953-54, U. London (Eng.), 1962-63; M.D., D.P.M., M.A.N.Z.C.P., A.B. Ps.C.; m. Joyce Helen Brenwood, Feb. 6, 1952; children—Ann, Sue, Janet, Bruce, David. Fellow psychiatry Royal Melbourne Hosp., 1954; psychiatrist Psychiat. Clinic Brisbane (Australia), 1955-58; cons., Brisbane, 1958—; vis. lectr. psychiatry and genetics U. Queensland, Brisbane, 1958—. Nuffield Dominion Trowlly fellow, 1962. Mem. Australian and New Zealand Coll. Psychiatrists, others. Author: An Introduction to Psychology; Elementary Psychiatry. Research, publs. on relationships which exist between twins, inheritance of neurosis and homosexuality. Home: 10 Lousdale St. Office: 97 Wickchan Terrace, Brisbane, Queensland, Australia.*

PARKER, Ralph Robinson, Am. entomologist; b. Malden, Mass., Feb. 23, 1888; s. Frank Howard and Marion Ellen (King) P.; B.S., Mass. State Coll., 1912, M.Sc., 1914, Ph.D., 1915, LL.D., 1943; D.Sc., Mont. State U., 1937; m. Adah L. Nicolet, 1916 (dec. 1931); children—Jane Louise, Robert Adams; m. 2d, Vivian Kaa, June 22, 1932. Asst. entomologist Mont. Bd. Entomology, 1915-21; spl. expert USPHS, 1921-47, dir., from 1947, dir. Rocky Mountain Lab., Hamilton, Mont., from 1928. Pres. Internat. Northwestern Conf. on Diseases of Nature Communicable to Man, 1947-48. Contbr. papers on Rocky Mountain spotted fever and other diseases. Developed (with R.R. Spencer) prophylactic vaccine of phenolized and emulsified organs of infected ticks for Rocky Mountain spotted fever, circa 1917. Died Sep. 4, 1949.

PARKER, Raymond Crandall, Canadian biologist; b. Newport, N.S., Can., Oct. 18, 1903; s. Albert Otto and Sabra (Parker) P.; B.A., Acadia U., 1924, D.Sc., 1955; Ph.D. (Sterling Research fellow), Yale, 1927. NRC and Rockefeller Found. fellow Kaiser-Wilhelm-Institut für Biologie, Berlin-Dahlem, Germany, 1927-29; research asst. Cancer Research Lab., U. Pa., Grad. Sch. Medicine, Phila., 1929-30; staff Rockefeller Inst. for Med. Research, N.Y.C., 1930-39, asso., 1931-39; head virus research lab. E.R. Squibb & Sons, New Brunswick, N.J., 1939-41; with Connaught Med. Research Labs., U. Toronto (Ont., Can.), 1941—, research mem., 1956—, faculty Sch. Hygiene, 1942—, prof. exptl. cytology, 1951—. Fellow A.A.A.S., N.Y. Acad. Sci., Royal Soc. Can.; mem. Harvey Soc. Am. Assn. Anatomists, Soc. for Exptl. Biology and Medicine, Tissue Culture Assn., Internat. Soc. for Cell Biology. Author: Methods of Tissue Culture, 1938, 2d edit.; 1950, 3d edit., 1961; also numerous articles. Research on tissue culture, cell nutrition, karyotype evolution of cells in cultures, carcinogenesis. Home: 58 Park Av., Oakville, Ont. Office: Connaught Med. Research Labs., U. Toronto, Toronto 5, Ont., Can.*

PARKER, Richard Anthony, Am. Egyptologist, educator; b. Chgo., Dec. 10, 1905; s. Thomas Frank and Emma Ursula (Heldman) P.; A.B., Dartmouth, 1930; Ph.D., U. Chgo., 1938; m. Gladys Anne Burns, Feb. 10, 1934; children—Michael (dec.), Beatrice Ann. Research asst. Oriental Inst., U. Chgo., 1938-42, research asso., 1942-46, asst. prof. Egyptology, 1946-48; Wilbour prof. Egyptology, Brown U., Providence, 1948—. Epigrapher, epigraphic and archtl. survey, Luxor, Egypt, 1938-40, asst. field dir., 1946-47, field dir., 1947-49; mem. vis. com. dept. Middle Eastern civilizations Harvard, 1950-61, dept. Egyptian art Boston Mus. Fine Arts, 1951—. Trustee Am. Research Center in Egypt. Mem. Oriental Soc., Egypt Exploration Soc., Société Francaise d'Egyptologie, Deutsches Archaologisches Institut, Phi Beta Kappa, Theta Chi. Author: (with Harold H. Nelson, others) Medinet Habu IV. Festival Scenes of Rameses III, 1940; (with Waldo H. Dubberstein) Babylonian Chronology 626 B.C. - A.D. 45, 1942, rev. edit. 626 B.C. - A.D. 75, 1956; The Calendars of Ancient Egypt, 1950; (with G.R. Hughes, others) Reliefs and Inscriptions at Karnak III: The Bubastite Portal, 1953; (with Hughes, others) Medinet Habu V: The Temple Proper, part I, 1957, Medinet Habu VI, part II, 1963; A Vienna Demotic Papyrus on Eclipse and Lunar-Omina, 1959; (with O. Neugebauer) Egyptian Astronomical Texts The Early Decans, I: 1960; A Saite Oracle Papyrus from Thebes, 1962; (with Neugebauer) Egyptian Astronomical Texts II: The Ramesside Star Clocks, 1964. Research establishing 2 lunar calendars in addition to civil calendar in ancient Egypt. Home: 38 Olive St., Providence 02906.*

PARKER, Robert Frederic, Am. physician; b. St. Louis, Oct. 29, 1907; s. Charles T. and Lydia (Gronemeyer) P.; B.S., Washington U., St. Louis, 1927, M.D., 1929; m. Mary Louella Warner, June 20, 1934; children—David Frederic, Jane Elanor (Mrs. Howard Hush). Asst. radiology Washington U., 1929-30, instr., 1932-33; asst. Rockefeller Inst., 1933-36; faculty Western Res. U., Cleve., 1936—, prof. microbiology, 1954—, asst. dean, 1964—. Mem. Nat. Bd. Med. Examiners, 1953-57. Mem. Cleve. Acad. Medicine (past sec., dir.) Am. Acad. Microbiology, Am. Soc. for Clin. Investigation, Central Soc. for Clin. Research. Research, numerous publs. on mechanisms immunity and quantitative aspects infection in viral disease, cultivation mammalian cells in vitro, viral infections in cultured cells. Home: 2819 Coleridge Rd., Cleve. 44118. Office: 2109 Adelhert Rd., Cleve. 44106.*

PARKER, Robert Hallett, Am. marine ecologist; b. Springfield, Mass., Feb. 14, 1922; s. Ralph C. and Mildred (Hallett) P.; student Duke, 1941-43, 49-50, Tex. Christian U., 1945-46; B.S., U. N.M., 1948, M.S., 1949; Magister Scientiarum, U. Copenhagen (Denmark), 1962, postgrad.; m. Harriet Elizabeth Logan, Dec. 23, 1945; foster children—Joycelene Bryan, Alfred Bryan, Dolly Bryan. Marine biologist Marine Lab., Tex. Game and Fish Commn., Rockport, 1950-51; jr. research ecologist Scripps Inst. Oceanography, U. Cal. at La Jolla, 1951-63; resident ecologist systematics-ecology program Marine Biol. Lab., Woods Hole, Mass., 1963-66; asso. prof. biology dept. Tex. Christian U., Fort Worth, 1966—. Cons. to pvt. cos. OEEC-NSF Sr. Sci. fellow, 1960. Mem. Am. Assn. Petroleum Geologists (Presd. award 1956), Am. Soc. Limnology and Oceanography, Geol. Soc. Am. Ecol. Soc. Am., Soc. Systematic Zoology, Marine Biologists Assn. U.K., Sigma Xi. Author: Zoogeography of the Gulf of California, 1964; also articles. Editor: (with van Andel, Shephard) Recent Sediments Northwest Gulf of Mexico, 1960; (with van Andel, Shor) Memoir 3, 1964. Devised means of recognizing environmental facies in ancient sediments through detailed study of marine environments in present day sediments; organic and inorganic biogeochemistry of marine invertebrates; research on basic structure marine benthic communities using computer techniques. Home 3601 Wren Av., Fort Worth 76133.*

PARKER, Vincent Eveland, Am. physicist; b. Kuala Lumpur, Malaya, Sept. 18, 1914 (parents Am. citizens); s. Walter Guy and Alam (Shell) P.; A.B., Evansville Coll., 1936; Ph.D., Ind. U., 1940; m. (Emma) Laverne Martin, June 18, 1938; (dec.); children—Don M., Rogers E., Paul J., David B.; m. 2d, Ruth Barker Boyer, Feb. 3, 1967. Faculty, Central Normal Coll., Danville, Ind., 1940-41; faculty U. Del., Newark, 1941-50; asst. microchemist Biochem. Research Found., Newark, Del., part time 1941-43; prof., head, dept. physics and astronomy La. State U., Baton Rouge, 1950-62; dep. dir. Oak Ridge Asso. Univs., 1962-67; dean Sch. Sci., Cal. State Polytechnic Coll., Pomona, since 1967—. Director of radiological defense for Delaware, 1950; exec. dir. La. Sci. Fair, 1955-58. Fellow A.A.A.S. (exec. council), Am. Phys. Soc. (chmn. Southeastern sect. 1958-59); mem. Am. Assn. Physics Tchrs. (pres. 1963), Am. Inst. Physics (governing bd. 1963—), Am. Chem. Soc., Am. Assn. U. Profs., Sigma Xi, Sigma Pi Sigma (nat. pres. 1954-59, nat. chancellor 1962—), Phi Lambda Upsilon, Alpha Chi Sigma, Omicron Delta Kappa, Phi Beta Chi, Pi Epsilon Phi. Asso. editor Am. Jour. Physics, 1958-61; fgn. editor Nuclear Energy Engring., since 1958—; editor Phys. Scis. sect. La. State U. Studies, 1951-62. Research on neutron physics; radioactivity; high energy particle accelerators. Home: 111 Elliott Circle, Oak Ridge 37830.*

PARKER, Willard, Am. surgeon; b. Lyndeborough, N.H., Sept. 2, 1800; s. Jonathan and Hannah (Clark) P.; A.B., Harvard, 1826, M.D., 1830; M.D., Berkshire Med. Instn.; LL.D., Princeton, 1870; m. Caroline Stirling, June 21, 1831; m. 2d, Mary Bissell, May 25, 1844; 5 children. Prof. anatomy and surgery Clin. Sch. Medicine, Woodstock, Vt., 1830-33; prof. surgery Berkshire Med. Instn., 1833-36; prof. anatomy, Geneva, N.Y., 1834-36; prof. surgery, Cin., 1836-37; prof. principles and practice of surgery Coll. Physicians and Surgeons, N.Y.C., 1839-70, established 1st coll. clinic in U.S., 1840, emeritus prof. surgery, 1870. Pres. N.Y. Acad. Medicine, 1856; Willard Parker Hosp. (N.Y.C.) named in his honor. Author: (med. monographs) Cystotomy, 1850; Spontaneous Fractures, 1852; Concussion of Nerves, 1856; Ligature of the Subclavian Artery, 1864; Cancer, 1873. Performed cystotomy for irritable bladder, 1850; tied subclavian artery for aneurysm on 5 occasions, 1864; 1st Am. to operate successfully on abscessed appendix, advocated opening of abscess at early stage. Died N.Y.C., Apr. 25, 1884.

PARKER, William Vann, Am. mathematician; b. Monroe, N.C., Dec. 22, 1901; s. Benjamin Franklin and Sara Della (Cox) P.; A.B., U. N.C., 1923, M.A., 1924; postgrad. Princeton; Ph.D., Brown U, 1931; m. Carolyn Adele Edwards, July 3, 1926; children—Edward Vann, Emily Anne (Mrs. Richard Leonard Shoemaker), Lola Jean (Mrs. Thomas Wayne Bishop). Asst. prof. U. South, 1924-25; asst. prof. U. N.C., 1925-31; instr. Princeton, 1926-27; prof. Miss. Woman's Coll., Hattiesburg, 1931-34; asst. prof. Ga. Inst. Tech., 1934-37; asso. prof. La. State U., 1937-39, prof., 1939-42, dept. head, 1942-47; prof. U. Ga., 1947-50; head prof. math. Auburn (Ala.) U., 1950—, dean Grad. Sch., 1953—. Mem. council Oak Ridge Inst. Nuclear Studies, 1953—. Mem. Am. Math. Soc., Math. Assn. Am., A.A.A.S., Acacia, Sigma Xi, Pi Mu Epsilon. Author: (with J.C. Eaves) Matrices, 1960. Contbr. articles on determinants and matrices, number theory and diophantine analysis to math. jours. Home: 304 Gardner St., Auburn, Ala. 36830.*

PARKES, Alan Sterling, English biologist; b. Rochdale, Eng., Sept. 10, 1900; s. Ebenezer Thomas and Helena Louisa (Banks) P.; B.A., Christ's Coll., Cambridge U., 1921, M.A., 1925, Sc.D., 1931; Ph.D., Manchester U., 1923; D.Sc., U. London, 1927; m. Ruth Deanesly, Sept. 27, 1933; children—John Francis Anthony, Katharine, Hilary. Beit Meml. research fellow, 1924-30; hon. asst. U. Coll., London, 1924-28, hon. lectr., 1929-31; mem. sci. staff Med. Research Council, London, 1932-61; Marshall prof. physiology of reprodn., Cambridge U., 1961-67. Decorated knight. Order Brit. Empire. Fellow Royal Soc., 1933; mem. Royal Soc. Medicine (pres. sect. endocrinology 1949-50), Brit. Assn. (sect. pres. 1958), Inst. Biology (pres. 1959-61), Soc. for Endocrinology (chmn. 1944-51), Soc. for Study Fertility (chmn. 1950-52, 64-66). Editor: Marshall's Physiology of Reproduction, 3d edit., 1952. Research, numerous publs. on physiology of reprodn. especially endocrinology of reprodn. and study of biol. effects of low temperatures. Home: 11 Adams Rd., Cambridge, Eng.*

PARKES, Alexander, English chemist, inventor; b. Birmingham, Eng., Dec. 29, 1813; apprentice to Messenger and Sons, brassfounders, Birmingham; in charge casting dept. Elkington works; recipient medal for celluloid invention, 1862, also medal Paris exhbn., 1867. Inventor, patentee numerous devices and methods connected with electrometallurgy; held 1st patent for electrodeposition of works of art, 1841; discovered cold vulcanization process, 1841, Parkes process of desilverizing lead with zincs, 1850; inventor celluloid (patented, 1855). Died London, June 29, 1890.

PARKES, Edmund Alexander, English physician; b. Bloxham, Eng., Mar. 29, 1819; s. William and Frances (Bresley) P.; ed. Christ's Hosp., London, Univ. Coll. and Hosp.; worked in lab. of uncle Dr. A. Todd; M.D., U. London, 1846; mil. surgeon, India, 1842-45; pvt. practice, London, 1855; founder, supt. civil hosp., Dardanelles, during Crimean War, 1855-56; prof. hygiene Army Med. Sch., Chatham, Eng., from 1860; Parkes Mus. founded in his honor, 1876. Fellow Royal Soc., 1861. Author: (essay) Remarks on the Dysentery and Hepatitis of India; Intestinal Discharges in Cholera; Early Cases of Cholera in London. Editor: Diseases of the Skin (Thompson), 1851; The Compositions of the Urine in Health and Disease. Founder modern hygiene as a science; contbd. to mil. hygiene. Died Bittern, nr. Southhampton, Eng., Mar. 15, 1876.

PARKES, Kenneth Carroll, Am. ornithologist; b. Hackensack, N.J., Aug. 8, 1922; s. Walter Carroll and Lillian (Capelle) P.; B.S., Cornell U., 1943, M.S., 1948, Ph.D., 1952; m. Ellen Stone, Sept. 6, 1953. Curator birds Cornell U., Ithaca, N.Y., part-time 1947-52, adminstv. bd. Lab. Ornithology, 1963-68; staff Carnegie Mus., Pitts., 1933—, asso. curator birds, 1955-61, curator birds, 1962—; adj. lectr. biology U. Pitts., 1963—. Fellow Am. Ornithologists Union (mem. council 1965-67), Acad. Zoology (Agra, India); mem. Wilson (council 1962-64, 66-68), Cooper ornithol. socs., Brit., Australasian, Netherlands ornithol. unions, Assn. Ornitologica del Plata (Argentina, corr.), Soc. Systematic Zoology, Soc for Study Evolution, other state and local ornithol. socs. Contbr. numerous articles, revs. to tech. jours., popular mags. Co-instituter new approach and terminology for study molts and plumage sequences birds; research on geog. variation birds especially in P.I. and New World tropics. Home: 5920 Walnut St., Pitts. 15232. Office: 4400 Forbes Av., Pitts. 15213.*

PARKIN, John Hamilton, Canadian mech, engr.; b. Toronto, Can., Sept. 27, 1891; s. Frederick and Lily (Hamilton) P.; B.A.Sc., University of Toronto, 1912, M.E., 1919, LL.D., 1961; m. Margaret Gertrude Locke, May 19, 1 dau., Margaret Lillian. Initiated aerodynamics research, designed, installed 4-foot wind tunnel U. Toronto, 1917, initiated, planned 1st undergrad. course aero. engring. in Can., giving all aero. lectures and instrn., 1928, asso. prof. mech. engring., 1926-29; made wind tests for aircraft designers and builders in devel. Canadian aircraft, 1917-29; asst. dir. div. physics N.R.C., 1929-37; dir. div. mech. engring. N.R.C. Can., 1937-57; 1st dir. Nat. Aero. Establishment, 1951-57. Apptd. mem. asso. com. aero. research N.R.C., 1920—, sr. cons. div. mech. engring., 1957-62, tech. adv. com. aero. Dept. of Nat. Def., 1924; apptd. one of 3 mems. representing Can., Commonwealth Adv. Aero. Research Council, 1947. Incorporator, gov.at-large Aviation League Can., 1928. Awarded Czowski Medal of Engring. Inst. of Can., 1937; U. Toronto Engring. Alumni Medal, 1951; 1st prize in Montreal Witness Model Aeroplane Meet, 1910; McCurdy award, Canadian Aero. Institute, 1956. Decorated Comdr. Order of Brit. Empire. Registered profl. engr., Providence Ont. Fellow Royal Aero. Soc., Inst. Aero Scis. (hon.), Am. Soc. M.E., Royal Soc. Can.; mem. Soc. Automotive Engrs. Author: Bell and Baldwin, Their Development of Aerodromes and Hydrodromes at Baddeck, N.S., 1964, also articles. Home: 290 Park Rd., Rockcliffe Park, Ottawa, Ont., Can.*

PARKINS, William Edward, Am. physicist; b. Bozeman, Mont., Mar. 1, 1916; s. William Edward and Emma (Rehsteiner) P.; B.S. in Elec. Engring., Mont. State Univ., 1937; Ph.D. in Physics, Cornell U., 1942; m. Alice Valentine Baird, Dec. 25, 1948; children—Janet Carol, Cheryl Leigh, April Elaine. McMullen research scholar in civil engring. Cornell U., 1937-39; research asso. Radiation Lab., U. Cal. at Berkeley, 1942-46; asst. prof. physics U. So. Cal., Los Angeles, 1946-48; with Atomics Internat. div. N.Am. Rockwell, Canoga Park, Cal., since 1948—, dir. organic reactors, 1959-60, asso. tech. dir., 1960-61, director of research and technology, 1961—. Spl. adviser to U.S. delegation Internat. Conf. on Peaceful Uses of Atomic Eenergy, 1955. Fellow Am. Nuclear Soc., Am. Phys. Soc.; mem. A.A.A.S., Am. Inst. Chem. Engrs., Am. Inst. Aeros. and Astronautics, Sigma Xi, Phi Kappa Phi, Tau Beta Pi. Phi Eta Sigma. Contbr. articles on physics and engring. to sci. publs.; patentee isotope separating apparatus. Home: 20120 Wells Dr., Woodland Hills, Cal. 91364. Office: 8900 De Soto St., Canoga Park, Cal. 91304.*

PARKINSON, James, English surgeon, paleontologist; b. Hoxton, Eng., Apr. 11, 1755; attended lectures of John Hunter; practiced medicine, Hoxton, from 1785; original mem. Geol. Soc. Author: Observations on Dr. Hugh Smith's Philosophy of Physic, 1780; Chemical Posketbook, 1801; Organic Remains of a Former World (1st attempt at familiar and sci. account of fossils with illustrations), 1804; Essay on Shaking Palsy, 1817; Elements of Oryctology, 1822; also 1st articles on appendicitis, 1812. First to recognize perforation as cause of death in appendicitis; described Parkinson's disease (paralysis agitans), 1817. Died London, Dec. 21, 1824.

PARKINSON, John, English botanist; b. Nottinghamshire, Eng., 1567; apothecary, Long Acre, 1616; named Botanicus Regius Primarius by Charles I; name given to Central Am. genus of leguminous trees by Plumier. Author: Paradisi in sole Paradisus Terrestris (descriptions nearly 1000 plants, 780 illustrated), 1629; Theatrum botanicum (descriptions of nearly 3800 plants, most complete English treatise on plants until time of J. Ray, contains new plant classification, includes most of L'Obel's work), 1640; Hortus Kewensis, 1810-13. Died London, Aug. 6, 1650.

PARKS, Harold Francis, Am. anatomist; b. Anna, Ill., Sept. 28, 1920; s. Guy Clay and Margaret (McCuminskey) P.; B.Ed., So. Ill. U., 1942; Ph.D., Cornell U., 1950; m. Margaret Bryner, Sept. 11, 1948; children—Edwin Thomas, Margaret Caroline. Tchr. music Martinsville (Ill.) High Sch., 1942; asst. band dir. Cornell U., 1942-43, dir. bands, 1943-46, teaching asst. comparative anatomy and histology and embryology, 1945-50; instr. then asst. prof. anatomy U. N.C. Med. Sch., 1950-54; asst. prof., then asso. prof. anatomy U. Rochester Sch. Medicine and Dentistry, 1954-61; prof. zoology Cornell U., 1961-64; prof. anatomy, chmn. dept. U. Ky. Med. Center, 1964—; vis. scientist Karolinska Inst., 1956. Mem. A.A.A.S., Am. Inst. Biol. Scis., Am. Soc. Zoologists, Am., So. socs. anatomists, Am. Soc. Cell Biology, Electron Microscopy Soc. Am., Sigma Xi. Contribution to understanding at the ultrastructural or submicroscopic level of biological mechanisms concerned with 1) phagocytosis, 2) glandular secretion, 3) the activities of the liver in fat metabolism, and 4) the decaying of teeth. Home: 286 Malabu Rd., Lexington, Ky. 40502.*

PARKS, John Louis, Am. physician; b. Muskogee, Okla., Jan. 4, 1908; s. John S. and Della N. (Northcutt) P.; B.A., U. Wis., 1930, M.S., 1932, M.D., 1934, tng. obstetrics, gynecology, 1935-37; m. Mary Dean Scott, Aug. 31, 1930; 1 son, John Scott. Instr. pathology U. Wis., 1937-38; prof. obstetrics and gynecology George Washington U., 1944—, dean Sch. Medicine, 1957, med. dir. hosp., 1957-65; cons. D.C. Gen. Hosp., Walter Reed Med. Center, N.I.H. Exec. com. Gorgas Meml. Inst. Decorated Eloy Alfaro Fundacion Internacional (Panama). Diplomatic Am. Bd. Obstetrics, Gynecology (past dir.). Fellow A.C.S. (bd. govs. 1956-59); hon. fellow Bklyn. Gynecol. Soc., Central, S. Atlantic assns. obstetricians and gynecologists, S.W., Fla., Wash., Miami, Panama obstet. and gynecol. socs., Soc. Obstetricians and Gynaecologists Can., La Societa Triveneta di Ostetrica e Ginecologia; mem. Am. Coll. Obstetrics and Gynecology (v.p. 1957), Assn. Am. Med. Colls., (mem. exec. council), Nat. Bd. Med. Examiners (exec. com., pres. 1965), Interstate Post-grad. Med. Assn. N. Am. (pres. 1966), Smith-Reed-Russell Soc., Am. Gynecology Soc. (treas. 1955-59), Am. Assn. Obstetrics and Gynecology (pres. 1961), So. Med. Assn., A.M.A., Sigma Xi, Osler Soc., Alpha Omega Alpha, Nu Sigma Nu, Alpha Delta Phi. Contbr. med. jours. Home: 4410 Dexter St., N.W. Office: George Washington U. Sch. Medicine, 1331 H St. N.W., Washington 20005.*

PARKS, Lloyd McClain, Am. pharmacist, educator; b. Scottsburg, Ind., Mar. 21, 1912; s. Othello C. and Mary (Wilson) P.; B.S., Purdue University, 1933, M.S., 1936, D.Sc. (hon.), 1962; Ph.D., U. Wis., 1938; m. Irene Comiskey, Aug. 24, 1940; 1 dau., Regina. Faculty U. Wis., Madison, 1938-56, prof., 1946-56; dean Ohio State U. Coll. Pharmacy, Columbus, 1956—. Mem. U.S. Pharmacopoeia Com. Revision N.Y.C., 1950-60. Mem. Am. Assn. Colls. Pharmacy (pres. 1961-62), Am. Council on Pharmaceutical Edn., Am. Pharm. Assn. (councilor 1964—), Am. Chem. Soc., A.A.A.S., Sigma Xi, Rho Chi (pres. 1960-62), Phi Lambda Upsilon, Kappa Psi. Author: (with P.J. Jannke, L.E. Harris) Inorganic Chemistry in Pharmacy, 1949. Conbr. articles on pharm. chemistry and phytochemistry to tech. jours. Home: 2846 Wellesley Dr., Columbus, O. 43221.*

PARKS, W(ilbur) George, Am. chemist, educator; b. Rockwood, Pa., Dec. 20, 1902; s. George and Ruby Pearl (Gardner) P.; A.B., U. Pa., 1926; M.A., Columbia, 1928, Ph.D., 1931; m. Margaret Mather Merriman, Nov. 24, 1928; children—Ann Mather, George Meriman. Faculty dept. chemistry, U. R.I., 1931—, pof. chemistry, 1939—, director of scientific criminal investigation laboratories; executive dir. adv. board Quartermaster Research and Development, Nat. Acad. Scis.-NRC, 1943—; dir. Gordon Research Confs.; mem. sci. adv. panel to sec. army. Trustee Colby Jr. Coll., New London, N.H. Fellow N.Y. Acad. Scis.; mem. Am. Inst. Chemists, Am. Inst. Chem. Engrs., A.A.A.S., Sigma Xi. Contbr. sci. papers Phys. Chemistry. Research on organic syntheses; solvent extractions; flameproofing; catalytic vapor phase oxidation; textile research. Home: 30 Lower College Rd., Kingston, R.I.

PARKS, William Arthur, Canadian geologist, paleontologist; b. Hamilton, Ont., Can., Dec. 11, 1868; ed. U. Toronto, fellow, 1893-97, instr., 1897-1902, lectr., 1902-05, asso. prof., 1905-15, prof. geology, head dept., 1922-36; prof. paleontology, dir. Royal Ont. Mus. Paleontology, 1915-22. Fellow Royal Soc., 1934, Royal Soc. Can. (pres. 1926); mem. Geol. Soc. Am., Paleontol. Soc. Am. (pres. 1927). Author: Building and Ornamental Stones of Canada, 2 vols., 1912-14; (with A. P. Coleman) Elementary Geology with Special Reference to Canada, 1922; American Stromatoporoids. Authority on dinosaurs of Alberta. Died Toronto, Oct. 3, 1936.

PARLATORE, Filippo, botanist; b. Palermo, Sicily, Aug. 8, 1816; ed. Palermo; asst. prof. anatomy; traveled in Europe; proposed establishment of gen. herbarium, 1841; dir. bot. garden, prof. botany Mus. Natural Scis. Author: Flora Palermitana, 1845; Flora Italiana, 1848-74, vols. 6-10 completed by T. Caruel. Studied flora of Mont Blanc, 1849, Scandinavia, 1851. Died Florence, Italy, Sept. 9, 1877.

PARLEE, Norman Allen Devine, metallurgist; b. South Farmington, N.S., Can., Mar. 23, 1915; s. Allen Chipman and Margaret (Foster) P.; B.S., Dalhousie U., 1935, M.S., 1937; Ph.D., McGill U., 1939; m. Eileen Vivian Elliott, Sept. 22, 1938; children—Cherie Eunice, Alan Lloyd. Came to U.S., 1952, naturalized, 1957. Research metallurgist Dominion Steel & Coal Corp., Sydney, N.S., 1939-41; acting chief metallurgist Trenton Steel Works, New Glasgow, N.S., 1941-42; asst. chief metallurgist Dominion Steel & Coal Corp., Can., 1942-46, dir. research and devel., 1946-51, asst. dir. metallurgy and research, 1951-52; prof. metall. engring. Purdue U., 1952-62; prof. extractive metallurgy Stanford, 1962—. Fellow Chem. Inst. Can.; mem. Am. Inst. Metall. Engring., Am. Soc. Metals, Can. Inst. Mining and Metallurgy, Engring. Inst. Can., Am. Assn. U. Profs., Profl. Engrs. N.S., N.S. Inst. Sci., Sigma Xi. Research, publs. on kinetics of gas reactions, thermodynamics and kinetics of gas-metal reactions. Home: 12145 Edgecliff Pl., Los Altos Hills, Cal. 94022. Office: Dept. Mineral Engring., Stanford U., Stanford, Cal.*

PARMENIDES OF ELEA, natural philosopher; b. Elea, Italy, circa 539 B.C.; s. Pyrrhes; influenced by teachings of Xenophanes and Pythagoras; leading member of Eleatic sch. of philosophy (founded by Xenophanes), lived in Athens, circa 550 B.C. Author: Nature (didactic poem) Part I, Preom, II, Truth, III, Opinion. Encouraged sci. research and skepticism; emphasized unreliability of senses as means to knowledge; recognized spherical form of earth, placed it at center of universe. Advocated philosophy of materialistic monism; opposed Heraclitos' doctrine of change (dialectic) with his principle of contradiction (change and motion are illusions of senses while reality is one, perfect, eternal, unchanging, indivisible, knowable only through intellect); opposed idea of void.

PARMENTER, Robert Haley, Am. physicist; b. Portland, Me., Sept. 19, 1925; s. L.F. and Esther (Haley) P.; B.S., U. Me., 1947; Ph.D., Mass. Inst. Tech., 1952; m. Elizabeth Kinnecom, Oct. 27, 1951; children—David Alan, Douglas Ian. Staff Mass. Inst. Tech., Cambridge, 1950-54; guest scientist Brookhaven Nat. Lab., N.Y., 1951-52; staff Lincoln Lab., Cambridge, Mass., 1952-54; mem. staff RCA Labs., Princeton, N.J., 1954-66, acting head gen. solid state research group, 1962-65; prof. physics U. Ariz., Tucson, 1966—. Mem. NASA Research Adv. Com. on Electrophysics, 1964—, chmn., 1967—. Fellow Am. Phys. Soc. (chmn. div. solid state physics 1967-68); mem. Sigma Xi, Tau Beta Pi. Research in theoretical solid-state physics; study of superconductivity; space-charge-limited current in insulators; group theory of crystals, electronic energy bands in ordered and disordered solids; acoustoelectric effect. Home: 1440 E. Ina Rd., Tucson. 85718.

PARMENTIER, Andrew, horticulturist, landscape gardener; b. Enghien, Belgium, July 3, 1780; s. Andre Joseph Parmentier; m. Sylvia Marie Parmentier, before 1814; 5 children, including Adele Bayer. Came to Am., 1824; purchased 24 acre triangular tract of land to establish bot. gardens and nursery, Bklyn., 1825; collected fgn. and domestic plants; introduced black beech tree, several species vegetables, shrubs and vines to U.S.; contbr. to New Eng. Farmer and N.Y. Farmer; earliest profl. landscape gardener in U.S.; laid out grounds and gardens for clients from Can. to Carolinas. Died Nov. 26, 1830.

PARMENTIER, Antoine Augustin, French agriculturist; b. Montdidier, France, Aug. 17, 1737; pharmacist's apprentice; practiced as apothecary, Paris; became insp.-gen. health, circa 1803; mem. French Acad. Scis., 1795. Author: Rural and Domestic Economy, 8 vols., 1790; also treatises on potato, chestnut, maize. Improved French agr. through sci. research; promoted potato cultivation and consumption in France; studied nutrition; supported Lavoisier's combustion theory. Died Paris, Dec. 13, 1813.

PARNAS, Josef, Polish physician; b. Lwow, Poland, June 14, 1909; s. Leon and Etylda Parnas; Dr., U. Lwow; Dr. honoris causa, U. Brno (Czechoslovakia); m. Zofia Mijal, Apr. 7, 1937; children—Witold, Josef. App. prof. U. Lwow, 1939-44; prof. Acad. Medicine, Lublin, Poland, 1944—; rector U. Lublin, 1948—; vice minister agr., Warsaw, Poland, 1946; vis. prof. U. Jerusalem (Israel), 1955, Instituto Superiore di Sanita, Rome, 1960, Svenska Institutet, Stockholm, Sweden, 1962. Expert, WHO, ILO, 1950—; mem. Internat. Com. Taxonomy of Brucella. Recipient Gold medal U. Lublin; Sci. Laureat of Voiewode, Adminstr. Council Poznan, 1961; Medal of 1000 Years of Przemysl; medal Charles Darvin, Acad. Scis. USSR, 1960; Rockefeller fellow, 1947. Fellow Royal Soc. Medicine London, Brit. Council; mem. Polish Acad. Sci., Polish Med. Assn., Polish Soc. Biochemistry, Polish Soc. Physiology, Polish Med. Assn. Poznan and Przemysl (hon.), Polish (past pres. Lublin), Am. socs. microbiology, Poud Assn. de Médicine Rurale, l'Association Internationale de la Medicine Rurale (past pres.), l'Office Internationale des Epizooties (asso.), Assn. Int. Med. Rura Bad-Kreuznach (past pres.), Purkine-Med. Soc. (hon.), Royal Soc. Tropical Medicine Brussels, Sociedad Argentina de Medicina. Research, numerous publs. on microbiology and epidemiology of anthropozoonoses; theory of foci in nature and med. ecology of tularemia, leptospirosis, ornithosis. Home: 1716 Chopin, Lublin, Poland.*

PARODI, Armando Santiago, Argentinian microbiologist, parasitologist; b. Argentina, May 6, 1909; s. Santiago and Adela (Trabucco) P.; M.D., Facultad de Medicina, 1932; m. Alicia Lalalame, Dec. 27, 1937; children—Gabriel, Alicia, Armando. Asst. prof. physiology, 1929-32; chief research Inst. Physiology, 1933; asst. Inst. Microbiology, Malbrán, Argentina, 1937-41, dir. virus dept., 1942-57; prof. microbiology and parasitology, Buenos Aires, Argentina, 1961—. Chief Influenza Centre, WHO, 1946-58, expert virus disease 1962—; dept. com. applied microbiology UNESCO, 1966. Rockefeller Found. fellow, 1939-40; recipient Mitre Gold medal. Fellow Am. Acad. Microbiology; mem. Soc. Argentine Biolog., Soc. Argentine Microbiologists, Soc. Enferm. Infec., Soc. Cient. Arg. Author: Hipófisis y Sangre, 1937; also numerous articles. Research on hematological alteration connected with hypophysis epidemiological and pathogenic research in influenza and psittacosis; isolated Junin virus (etiological agt. of Argentinian hemorrhagic disease) in human being, wild rodents and arthropods, and studied its characteristics; exptl. pathology of diseases in guinea pigs and mice; human relationships between Junin and Tacaribe virus; influence of Junin virus in immune mechanism of guinea pigs and mice. Home: 3603 Carbajal Capital. Office: 2155 Paraguay Capital, Buenos Aires, Argentina.*

PARODI, Hippolyte, French engr.; b. Bois-Colombes, France, Aug. 14, 1874; s. Alexandre Parodi; ed. Poly. Sch.; m. Aimée Lombard; children—Maurice, André, Marcel. Became asso. with Thomson-Houston Co., 1899 named dir. elec. service French ry. service between Paris and Orleans, 1906; prof. Nat. Conservatory Arts and Crafts, also Nat. Superior Sch. Aeronautics, Nat. Superior Sch. Telecommunications; cons. engr. French rys. (S.N.C.F.). Recipient prize in mechanics French Acad. Mem. French Acad. Scis., 1949, French Soc. Electricians (past pres.) French Soc. Radio-Electricians (past pres.), French Soc. Meteorology (past pres.). Constructed 1st high tension circuit of 220,000 volts in Europe, at hydroelectric works in central France; studied disturbances produced by vibrations of transmission devices in locomotives. Home: 80, rue Spontini, Paris 16, France.

PARR, Albert Eide, zoologist, oceanographer; b. Bergen, Norway, Aug. 15, 1900; s. Thomas and Helga (Eide) P.; student Royal U., Oslo, Norway, 1921-25; D.Sc. Yale, 1946, Colby Coll., 1956; m.

Ella Hage, December 31, 1925; children—Hedvig, Gifford Nils, Thomas, Victoria-Johanne. Came to United States, 1926, naturalized 1932. Assistant in Zoology, Bergen Museum, 1918-19; asst. Bur. of Fisheries, Norway, 1924-26; asst. New York Aquarium, 1926; curator, Bingham oceanographic collection, Yale, 1927-42; asst. prof. zoology, Yale, 1931-37, asso. prof., 1937-38, prof. of oceanography, 1938-42, sci. dir. oceanographic expdns., 1931-37, dir. marine research 1937-42, dir. Peabody Mus., 1938-42; dir. Am. Mus. of Natural History, 1942-59, sr. scientist, 1959—, also trustee. Trustee Scandanavian Seminar, Inc. Woods Hole (Mass.) Oceanographic Inst. Contbr. articles profl. jours. Specialist in evolution and adaptation, deep sea fishes, anatomy and classification of fishes, W. Indian fishes, instincts and biology of fishes, phys. oceanography, Sargasso weeds, environmental psychology. Home: Codfish Hill, Bethel, Conn.; also 176 Linden St., New Haven. Address: Am. Mus. Natural History, N.Y.C.

PARR, Leland Wilbur, Am. microbiologist; b. Cooksville, Ill., Nov. 2, 1892; s. Marion Elmer and Edna (Brigham) P.; student Drake U., 1913-15; S.B., U. Chgo., 1916, Ph.D., 1923; m. Grace Belle Ghormley, June 16, 1915; children—Patricia (Mrs. John Kibler Bash), Robert. Mem. faculties Assiut (Egypt) Coll., 1916-19, U. Chgo., 1922-23, 30, Am. U. Beirut (Lebanon), 1923-30; with Rockefeller Found. Field Research Lab., Andalusia, Ala., 1930-32; faculty George Washington U., Washington, 1932—, emeritus prof., 1958—, Cons. to Army, VA, NIH, USPHS, 1942-64; chmn. Conf. of Profs. Preventive Medicine, 1949—. Mem. Soc. Am. Bacteriologists (sec.-treas. 1945-49), Washington Acad. Scis. (pres. 1943-44), Am. Pub. Health Assn., A.A.A.S., Washington Acad. Medicine, Sigma Xi, Alpha Omega Alpha. Author: (with C.V. Kappers) An Introduction to the Anthropology of the Near East, 1934. Research, publs. in application of biol. methods to racial study; epidemiology of scarlet fever, diphtheria, Tb in Nr. East; intestinal microbiology. Address: 302 Scientists Cliffs, Port Republic, Md. 20676.*

PARR, Robert Ghormley, Am. chemist; b. Chgo., Sept. 22, 1921; s. Leland Wilbur and Grace (Ghormley) P.; A.B., Brown U., 1942; Ph.D., U. Minn., 1947; m. Jane Esther Bolstad, May 28, 1944; children—Steven Robert, Jeanne Karen, Carol Jane. Asst. prof. chemistry U. Minn., 1947-48; fellow U. Chgo., 1949, research asso., 1957; from asst. prof. to asso. prof. Carnegie Inst. Tech., Pitts., 1948-57, prof., 1957-62, chmn. gen faculty, 1960-61; prof. chemistry Johns Hopkins, Balt., 1962—; vis. prof. chemistry U. Ill., Urbana, 1962. Mem. chemistry adv. panel Air Force Office Sci. Research, 1960—; chmn. com. on postdoctoral fellowships in chemistry Nat. Acad. Sci.-NRC, 1961-63. Guggenheim fellow, Fulbright research fellow U. Cambridge, 1953-54;Alfred P. Sloan fellow, 1956-60. Fellow Am. Phys. Soc. (chmn. div. chem. physics 1963-64); mem. Am. Chem. Soc. (recipient Petroleum Research Fund award 1964), A.A.A.S., Phi Beta Kappa, Sigma Xi, Phi Lambda Upsilon, Pi Mu Epsilon. Author: Theory of Molecular Electronic Structure, 1963. Asso. editor Jour. Chem. Physics, 1956-58, Chem. Revs., 1961-63, Jour. Phys. Chemistry, 1963—; cons. editor Acad. Press Inc., 1966—. Research, publs. in theoretical chemistry especially in quantum theory of molecular structure. Home: 1009 Cowpens Av., Balt. 21204.*

PARR, Samuel Wilson, Am. chemist; b. Granville, Ill., Jan. 21, 1857; s. James and Elizabeth Fidelia (Moore) P.; B.S., U. Ill., 1884; M.S., Cornell U., 1885; postgrad. U. Berlin, 1900, Polytechnikum, Zürich, 1901; hon. Sc.D., Lehigh U., 1925, Ill. Coll., Jacksonville, 1929; m. Lucie A. Hall, Dec. 27, 1887; children—Elisabeth, Harold Lucien. Instr. gen. sci., Ill. Coll., 1885-86, prof., 1886-91; prof. applied chemistry U. Ill., 1891—; dir., Ill. State Water Survey, 1904-05; chemist on coal investigation Ill. Geol. Survey, 1905—. Recipient Chandler medal, 1926. Author: The Chemical Examination of Water, Fuel, Fluegases and Lubricants; also reports on composition and analysis of Ill. coals, bulls. Devised Parr calorimeter for determining heat value of coal and other hydrocarbons, widely used in America and Europe, also new type of calorimeter for determining and recording heat value of combustible gases; alloys with acid resisting properties, new calorimeter bomb with effective platinum substitution in its constrn. Died May 16, 1931.

PARRATT, Lyman George, Am. physicist; b. Salt Lake City, May 17, 1908; s. Delbert William and Mary (Wardrop) P.; A.B. with honors, U. Utah, 1928; Ph.D., U. Chgo., 1932; m. Rhea Gibson, Feb. 24, 1946; children—Carolyn, Portia Virginia. Asst. in physics U. Utah, 1928-29, U. Chgo., 1930-33; NRC fellow Cornell U., Ithaca, N.Y., 1933-35, faculty, 1935—, prof., 1948—, chmn. dept. physics, 1959—, mem. Lab. Nuclear Studies, 1946—, Lab. Atomic and Solid State Physics, 1959—, Materials Sci. Center, 1960—. Physicist, head engring. div. Naval Ordnance Lab., Washington, 1941-43; civilian with USAAF, overseas, 1942; group leader Los Alamos Sci. Lab., 1943-46; cons. govt. and industry. Fellow Am. Phys. Soc.; mem. Am. Inst. Physics, Am. Assn. U. Profs., Am. Assn. Physics Tchrs., Sigma Xi, Gamma Alpha, Sigma Pi Sigma. Research, publs. on x-ray spectroscopy, surface physics and chemistry, solid state physics, electronics, underwater ordnance, design torpedoes and submarine detection. Author: Probability and Experimental Errors, 1961; contbr. to Methods of Experimental Physics, vol. 6, 1959. Home: 513 Wyckoff Rd., Ithaca, N.Y. 14850.*

PARRENIN, Domenique, French missionary, physician; b. Pontarlier, France, 1665; joined Soc. Jesus; went to China, 1698; translated French works into Chinese; published series of plates, Manchu Anatomy, inspired by engravings in Anatomie de l'homme suivant la circulation du sang by Dionis. Died Pelsing, 1741.

PARRINGTON, Francis Rex, English zoologist; b. Neston, Cheshire, Eng., Feb. 20, 1905; s. Frank Harding and Bessie May (Harding) P.; ed. Sidney Sussex Coll., Cambridge (Eng.) U., 1924-27; m. Margaret Aileen Johnson, Aug. 7, 1946; children—Francis Glen, Rayne Mary Knox. Asst. dir. Mus. Zoology, Cambridge U., 1927-35, Strickland curator, 1928-38, dir., 1938—, faculty, 1935—, reader vertebrate zoology, 1962—; mem. Brit. Mus. E. African Expdn., 1930, Expdn. E. Africa, 1933. Balfour student, 1933-35. Fellow Royal Soc., 1962, Inst. Biology, Linnean Soc. Research, publs. on vertebrate paleontology, especially mammal-like reptiles, their dentitions, ears, origins, identification of early skull bone patterns. Home: 5, Sedley Taylor Rd., Cambridge, Eng.*

PARRISH, Henry Mack, Am. physician; b. Ocala, Fla., June 21, 1927; s. Joseph Gid and Winnie (Hunt) P.; B.S., Wake Forest Coll., 1949; M.D., U. Pa., 1953; M.P.H., Yale, 1956, Dr. P.H., 1959; m. Carole Anne Carr, Dec. 20, 1957; children—Cynthia Marie, Catherine Anne, Cheryl Lynn. Faculty, Bowman Gray Sch. Medicine, Winston-Salem, N.C., 1954-55, U. Vt., 1958-59, Ind. U., 1959-61; fellow pub. health, asst. physician U. Health, Yale, 1955-56, Am. Cancer Soc. fellow in epidemiology, 1957; research fellow in epidemiology U. Pitts., 1957-58; dir. med. edn. Marion County Gen. Hosp., Indpls., 1959-61; individual practice medicine, specializing in preventive medicine, Ocala, Fla., 1961-62; faculty U. Mo., Columbia, 1962—, prof., chmn. dept. community health, 1965—. Diplomate Am. Bd. Preventive Medicine. Fellow Am. Pub. Health Assn., Am. Coll. Preventive Medicine, A.A.A.S., Am. Geriatrics Soc.; mem. A.M.A., Mo. Med. Assn., Internat. Soc. Toxicology, Phi Beta Kappa, Sigma Xi. Publs. on descriptive epidemiology poisonous snakebites in U.S.; classification for severity snake venom poisoning; study risk factors asso. with coronary heart disease and strokes. Home: 1905 Dartmouth St., Columbia, Mo. 65201.*

PARROD, Jacques Auguste, French chemist; b. Paris, France, Oct. 23, 1901; s. François and Jeanne Parrod; Dr. es sc., Paris Sch. Physics and Indsl. Chemistry. Head organic synthesis services Physico-Chem. Biology Inst., Paris; chief chem. lab., med. clinic Saint-Antoine Hosp.; now head chem. service Center Macromolecular Research, also prof. Faculty of Scis., U. Strasbourg (France). Mem. French Chem. Soc., Animal Protection Soc., French Anti-Vivisection League. Publns. on organic chemistry allied with biology and macromolecular compositions. Address: 6, rue Boussingault, Strasbourg (Bas-Rhin), France.

PARROT, Jean Julien Louis, French physiologist; b. Paris, France, Apr. 7, 1908; s. Léon and Eugénie (Manson) P.; M.D., U. Paris; m. Marie-Rose Panisset, July 7, 1933; children—Marie-Louise, Geneviève, Annie. Intern Paris Hosps., 1933; agrégé in physiology U. Paris, 1946, prof. without chair Faculty of Medicine, 1955, prof. pathol. physiology Faculty of Medicine, 1960, hosp. biologist, service chief, 1963. Sec. Jour. Physiology, also Histamine Club; called to 1st Internat. Congress Angeiology, 1953. Decorated Legion of Honor, Palmes Acad.; laureate the Inst., Acad. Medicine, Faculty of Medicine. Author: Le mécanisme d'économie de la vitamine C, 1952; also thesis on manifestation of anaphylaxie and histamine, 1938. Research on capillary resistance, pharmacology, pathol. physiology, physiology; record of 1st synthetic antihistamine. Home: 27, bd. Pereire. Office: 45, rue des Saints-Pères, Paris, France.*

PARROT, Joseph, French physician; b. Exideuil, France, 1829; became prof. history medicine Paris Faculty Medicine, 1876, head physician Enfants-Assistés Hosp., 1877, prof. infantil clin. medicine, 1879. Mem. French Acad. Medicine. Author: la Syphilis héréditaire, 1886; l'Athrepsie, 1877. Described syphilitic pseudoparalysis in infants caused by epiphyseal separation (Parrot's disease), 1871, primary infantile atrophy or marasmus, 1877, nodes on outer table of frontal and parietal bones in syphilitic children (Parrot's nodes), 1879; 1st description of primary lesion of pulmonary Tb. in children (later described by Ghon, and called Ghon's primary focus), 1876; probably earliest description of pneumococcus, 1881. Died 1883.

PARRY, Caleb Hilliard, English physician; b. Cirencester, Eng., Oct. 21, 1775; M.D., Edinburgh, 1778; m. Miss Ridby, 1778; 4 sons, including Charles Henry, William Edward. Practiced medicine, Bath, Eng.; physician Bath Gen. Hosp.; fellow Royal Coll. Physicians. Author: Cases of Tetanus and Rabic Contagiosa or Canine Hydrophobia, 1814; Elements of Pathology, 1815; The Natural Cause and Varieties of the Arterial Pulse (established facts of pulse), 1816; Elements of Pathology and Therapeutics (unfinisted), 1925; Collected Works, 1825; also articles. Credited with observing exophthalmic goiter, 1786 (Parry's disease), 1786; described histaminic cephalalgia, 1792; established causes of angina pectoris, 1799; credited with being 1st to record case of congenital idiopathic dilatation of colon, also described perhaps 1st recorded cases of facial hemiatrophy, pub. 1825; wrote on rhubarb cultivation; tried to improve breeds of sheep for better wool. Died Mar. 9, 1822.

PARRY, Charles Christopher, botanist; b. Admington, Eng., Aug. 28, 1823; s. Joseph and Eliza (Elliott) P.; came to U.S., 1832; A.B., Union Coll., 1842; M.D., Columbia Coll., 1846; m. Sarah M. Dalzell, 1853; m. 2d, Emily R. Preston, 1859. Botanist under David Dale Owen in geol. survey of Wis., Ia. and Minn., 1848; apptd. botanist to U.S. and Mexican boundary survey, 1849; bot. explorer of western states and territories, 1850-79; 1st botanist in U.S. Dept. Agr.; organized plant collections brought back by govt. scientists and surveying expdns. at Smithsonian Instn., 1869-71. Author intro. Botany of the Boundary, for Report on U.S. and Mexican Boundary Survey, 1857-59. Discovered Picea Engeimannii spruce, new species Cal. manzanitas, lilium Parryi of southern Cal. mountains, lote bush of Colorado Desert, Ensenada Buckeye, many others; 1st investigator of these groups to study living plants in the field in connection with specimens in herbarium. Died Davenport, Ia , Feb. 20, 1890.

PARRY, Robert Walter, Am. chemist; b. Ogden, Utah, Oct. 1, 1917; s. Walter and Jeanette (Petterson) P.; B.S., Utah State Agr. Coll., 1940; M.S., Cornell U., 1942; Ph.D., U. Ill., 1946; m. Marjorie J. Nelson, July 6, 1945; children—Robert Bryce, Mark Nelson. Research asst. NDRC Munitions Devel. Lab., U. Ill., Urbana, 1943-45, teaching fellow U. Ill., 1946, asst. prof., 1949-54; faculty U. Mich., Ann Arbor, 1946-49, 54—, prof. chemistry., 1958—. Indsl. cons., 1952—. Mem. Am. Chem. Soc. (past chmn. inorganic div., award for distinguished service to inorganic chemistry 1965), A.A.A.S., Sigma Xi. Founding editor Inorganic Chemistry, 1961-63. Research, numerous publs. on some structural problems of inorganic chemistry, and incorporation results into theoretical models chemistry, synthesis new inorganic molecules. Home: 2588 Hawthorn St., Ann Arbor, Mich. 48104.*

PARRY, Sir William Edward, English Polar explorer; b. Bath, Eng., Dec. 19, 1790; s. Caleb Hillier Parry; D.C.L., Oxford, 1829; m. Isabella Louisa Parry, Oct. 1826; 2 daus., 2 sons; m. 2d, Catherine Edwards Hankinson, 1841; 2 daus. Entered navy, 1803, served in North Sea, Baltic, promoted to lt., 1810; protected whaling fisheries in Arctic regions, 1810-13; accompanied Capt. Ross in exploring voyage, 1818; obtained command of expdn. for discovery of N.W. Passage, 1819; made subsequent attempts to find N.W. Passage; comptroller Dept. of Navy, 1837, supt. of Haslar, 1846; made rear-adm., 1852. Author: Journal of a Voyage for the Discovery of a North-West Passage from the Atlantic to the Pacific, performed in the Years 1819-20 in H.M. Ships Hecla and Griper, 1921; Journal of a second Voyage for the Discovery of a North-West Passage . . . performed in the Years, 1821-3, 1824; Journal of a Third Voyage for the Discovery of a North-West Passage . . . performed in the Years 1824-5 in H.M. Ships Fury and Hecla, 1826; Narrative of an Attempt to reach the North Pole in Boats fitted for that purpose and attached to H.M. Ship Hecla, in the Year, 1827, 1828. Passed through Lancaster Sound, explored contiguous strait (he named it Barrow's Strait), penetrated beyond meridian of 110° W.; discoverer of Lancaster strait and Melville Island, 1819; made subsequent attempts to find N.W. Passage, 1821, 24 (without success), attempted to reach pole by boats and sledges from Spitzbergen, attained high latitude of 82° 45′ N., 1827. Died Ems, Hesse-Nassau (where he had been taken for med. treatment), July 8, 1855.

PARSEGIAN, Vozcan Lawrence, physicist; b. Van, Armenia, May 13, 1908; s. Sahag Hovsepian and Lillian (Krikorian) P.; came to U.S., 1916, naturalized, 1931; B.S. in Physics, Mass. Inst. Tech., 1933; postgrad. Washington U., St. Louis; Ph.D. in Nuclear Physics, N.Y. U., 1948; m. Varsenig Boyajian, Apr. 17, 1938; children—Vosken Adrian, Elsa Varsenig. Dir. research C.J. Tagliabue Mfg. Co., 1936-47; dir. instrument dept., dir. physics dept. Kellex Corp., N.Y.C., 1947-50; dir. research N.Y. Operations office AEC, 1950-54; dean Sch. Engring., Rensselaer Poly. Inst., Troy, N.Y., 1954-61, distinguished chair Rensselaer prof., 1961—; director of Cast Technology, Incorporated, also Radiation Applications, Inc. Mem. Gov. N.Y. Com. on Atomic Energy, 1958-59, N.Y. State Adv. Com. on Radiation Utilization, 1959—, U.S. C. of C. Com. on Comml. Utilization Atomic Energy, 1958—; cons. various industries N.Y. State, also U.S. Joint Com. on Atomic Energy; exchange scientist U.S. Nat. Acad. Scis. to Soviet Acad. Scis., 1962. Mem. Am. Nuclear Soc. (dir., chmn. Northeastern N.Y. sect. 1958-59),

A.A.A.S., Am. Assn. U. Profs., Am. Phys. Soc., Radiation Research Soc., Am. Soc. for Engring. Edn., Am. Soc. M.E., Sigma Xi, Sigma Pi Sigma, Tau Beta Pi. Author: Industrial Management in the Atomic Age, 1965; Introduction to Natural Science; also articles on instrumentation, edn., policy on atomic energy devel. and secrecy. Inventor nuclear, radiation, indsl. instruments. Home: Route 57, Brunswick Mills, Troy, N.Y. 12180.*

PARSHAD, Ram, Indian biologist, zoologist; b. Nurmahal, India, Mar. 31, 1933; s. Mela and Swarn (Kaur) P.; student D.A.V. Coll., Jullundur, 1950-58; B.Sc., M.Sc., Ph.D., Punjab U.; m. Nirmal, Dec. 11, 1958; children—Rakesh, Ranjeev. U. Grants Commn. research scholar, 1954-58; lectr. zoology Panjab U., Chandigarh, India, 1958-—. Recipient Roll of Honour, Punjab U. Coll., Hoshiarpur, India, 1958; USPHS, NIH Internat. Postdoctoral research fellow NIH, Bethesda, Md., 1965-67. Research and publs. on cytogenetics of animals, especially classification and evolution of karyotypes; cytogenetical and histochem. changes in mouse embryo cells during spontaneous neoplastic transformations in vitro. Home: EI/51, Sector 14, Chandigarh, Punjab, India.*

PARSHLEY, Howard Madison, Am. zoologist; b. Hallowell, Me., Aug. 7, 1884; s. John Howard and Julia Maria (Tuck) P.; student N.E. Conservatory Music, 1906-09; A.B., Harvard, 1909, A.M., 1910, Sc.D., 1917; m. Nancy Fredricson, June 28, 1910; children—Thomas Fredricson, Elsa Madison. Instr. zoology U. Me., 1911-14; research in zoology Bussey Inst. (Harvard), 1914-17; asst. prof. zoology Smith Coll., Northampton, Mass., 1917-19, asso. prof., 1919-25, prof., 1925-52, emeritus since 1952; tchr., summers at Biol. Lab., Cold Spring Harbor, N.Y., also U. Chgo. Mem. A.A.A.S., Am. Soc. Zoologists, Entomol. Soc. Am., Genetics Soc. Am., Am. Soc. Naturalists, Sigma Xi. Author: Bibliography of North America Hemitera Heteroptera, 1925; Science and Good Behavior, 1928; Science of Human Reproduction, 1933; Survey of Biology, 1940. Translator; The Second Sex (Simone de Beauvior); Life and Habits of the Mammals (F. Bourliere), 1953. Contbr. to books and jours. Died May 19, 1953.

PARSONS, Sir Charles Algernon, inventor; b. London, June 13, 1854; s. William Parsons (Lord Rosse); ed. Dublin, Cambridge. With Armstrong, Elswick, Newcastle, until 1884; founder C.A. Parsons Heaton works, 1889; pres. Parsons Marine Steam Turbine. Fellow Royal Soc., 1898. Developed steam turbine, 1st built as turbogenerator, 1884; inventor steam engine with planet slide valve for centrifugal pumps, also multi-celled reaction steam turbine, 1884; produced gearing with Parsons creep cutting (spiral cut); proved that erosion is a result of cavitation; built parabolic and elliptical-parabolic spotlights for navy, 1889; made test flights with a steam helicopter, 1893; patentee for application of steam turbine to ship propulsion, 1894 (1st used in ship Turbinia 1897); tested a steam monoplane; inventor valve amplifier which worked on a compressed airtube (before elec. amplifier of wireless), 1906; produced a geared turbine, 1910; inventor anti-skid motor chains; produced optical glasses after World War I. Died Kingston, Jamaica, Feb. 11, 1931.

PARSONS, Charles Lathrop, Am. chemist; b. New Marlboro, Mass., Mar. 23, 1867; s. Benjamin Franklin and Leonora (Bartlett) P., B.S., Cornell U., 1888; D.Sc., U. Me., 1911; D. Chem., U. Pitts., 1914; D.Sc., U.N.H., 1944; m. Alice Douglas Robertson, Dec. 29, 1887; children—Leonora Elizabeth, Charles Lathrop, Anna Guerard, Enith Alice, Priscilla Bartlett. Prif. chemistry N.H. Coll., 1890-1911; named chief chemist Bur. Mines, Washington, 1911; cons. practice since 1919. Mem Nitrate Commn.; mem. Adv. Bd. on Gas Warfare. v.p. for Am., Internat. Union Pure and Applied Chemistry, 1919-22. Recipient Nichol's medal for research on atomic weight of beryllium, 1904; Priestly medal for distinguished service, 1932, A.C.S. Spl. Gold medal of honor, 1946. Fellow A.A.A.S.;mem. Am. Chem. Soc. (sec. 1907, sec and bus. mgr. 1930), Am. Inst. Chemists (life), Chem. Soc. of Rumania, (hon.), Soc. of Chem. Industry (hon.) Chimique de France, (life), Sigma Xi. Author: (with A.J. Moses) Mineralogy, Chrystallography and Blow-pipe Analysis, 1895, 1901, 04, 09, 11, 16; Beryllium, Its Chemistry and Literaure, 1908; also many sci. papers in chem. jours. and govt. bulletins. Died Pocassett, Mass., Feb. 14, 1954.

PARSONS, Dennis Shirley, English physiologist; b. Aylesbury, Bucks, Eng., June 29, 1917; s. James Rainbow and Hester Cordelia (Fitz-Henry) P.; B.A. with 1st class honors, Oxford (Eng.) U., 1939, B.M., B.Ch., 1942, M.A., 1950, D.M., 1950; m. Margaret Elizabeth Floor, Oct. 28, 1942; children—Hannah, Sophie, Terence James. Med. Research Council scholar Oxford U., 1946-48; lectr., demonstrator biochemistry dept., 1948-50; lectr., fellow, tutor Merton Coll., Oxford, 1950-—, subwarden, 1955-57. Examiner for B. and D.'s degrees in physiol. scis. in various univs. in Gt. Britain. Mem. Biochem. Soc., Physiol. Soc. Author: (with others) Biochemistry in Relation to Medicine, 1959; also numerous publs. Editorial bd. Jour. physiology, 1966-—. Devel. and use of methods for investigation of absorptive functions of intestinal tract; research on mechanisms of processes underlying movement of substances into and out of cells of animal tissues. Home: 20 Holywell St., Oxford, Eng.*

PARSONS, Donald Frederick, biophysicist; b. Shoreham-by-sea, Sussex, Eng., Nov. 28, 1928; s. Reginald Alexander and Gladys (West) P.; B.Sc. in Chemistry, U. London, 1950, Ph.D. and D.I.C. in Phys. Chemistry (Ministry of Supply Sci. scholar), 1953, M.B., B.S. (St. Bartholomew's Hosp. scholar), 1957; m. Julieta Perez-Medrano, Aug. 1, 1958; children—Clare Olivia, Raymond Anthony, Francis Robin, Catherine Leslie. Research asso. dept. exptl. surgery Duke, 1957-59; biophysicist biology div. Oak Ridge Nat. Lab., 1959-61; asst. prof. med. biophysics U. Toronto (Ont., Can.), 1961-66; prin. cancer research scientist Roswell Park Meml. Inst., Buffalo, also research prof. biophysics State U. N.Y., Buffalo, 1966-—. Fellow Royal Micros. Soc.; mem. Biophys. Soc., Am. Crystallographic Assn., Am. Assn. Cancer Research, Electron Microscope Soc. Am., Internat. Soc. Cell Biologists. Editor Biochimica Biophysica Acta, 1966-—. Devel. of methods for analysis of structure of cells and viruses; research in electron microscopy, x-ray diffraction, electron diffraction. Home: 43 Ashland Av., Buffalo 14222. Office: 666 Elm St., Buffalo 14203.*

PARSONS, Elsie Clews, Am. anthropologist; b. 1875; d. Henry and Lucy Madison (Worthington) Clews; A.B., Barnard Coll. (Columbia), 1896, A.M., 1897, Ph.D., 1899; m. Herbert Parsons, Sept. 1, 1900. Pres. Am. Anthropol. Assn., 1940-41. Author: Educational Legislation and Administration of the Colonial Government (U.S.A.), 1899; The Family, 1906; The Old-fashioned Woman, 1913; Fear and Conventionality, 1914; Social Freedom, 1915; Social Rule, 1916; Notes on Zuni, 1917; Folk Tales of Andros Island, Bahamas, 1918; Winter and Summer Dance Series in Zuni, 1922; Laguna Genealogies, 1923; Folk Lore of the Sea Islands, S.C., 1923; Folk Lore from the Cape Verde Islands, 1924; Scalp Ceremonial of Zuni, 1924; Pueblo of Jemez, 1925; Pueblo Indian Journal, 1925; Tewa Tales, 1926; Kiowa Tales, 1929; Social Organization of the Tewa of New Mexico, 1929; Hopi and Zuni Ceremonialism, 1933; Folk Lore of the Antilles, French and English, 1933, 1936; Taos Pueblo, 1936; Mitla (Oaxaca, Mexico), 1936; Pueblo Indian Religion, 1939; Taos Tales, 1940; Notes on the Caddo, 1941. Editor: Notes on Cochiti (Dumarest), 1920; American Indian Life, 1922; Stephen's Hopi Journal, 1936. Died Dec. 19, 1941.

PARSONS, Sir John Herbert, English surgeon; b. 1868; M.B., B.S., D.Sc., London; D.Sc. (hon.), Bristol, Eng.; LL.D., Edinburgh, 1927; cons. surgeon Royal London Ophthalmic Hosp.; cons. ophthalmic surgeon Univ. Coll. Hosp.; fellow Univ. Coll., London; mem. departmental com. on sight tests Bd. Trade, 1910, on factory lighting Home Office, 1913, on causes and prevention of blindness Ministry of Health, 1920; mem. adv. med. council Air Ministry, 1919, Admiralty, 1922; mem. Med. Research Council, 1928-32. Fellow Royal Soc., 1921 (glass-workers cataract com. 1906), Royal Coll. Surgeons; pres. Illuminating Engring. Soc., 1921. Author: The Pathology of the Eye, 4 vols.; Diseases of the Eye, 10th edit., 1942; Elementary Ophthalmic Optics, 1901; Introduction to the Study of Colour Vision, 2d edit., 1925; Mind and the Nation, 1918; Introduction to the Theory of Perception, 1927. Died Oct. 7, 1957.

PARSONS, Laurence, Irish astrophysicist; b. Parsonstown, Ireland, Nov. 17, 1840; s. William Parsons; grad. Trinity Coll., Dublin, Ireland, 1864, LL.D., 1879; D.C.L., Oxford (Eng.) U., 1870; LL.D., Cambridge (Eng.) U., 1900. Asst. father's obs., Birr, Ireland; became sheriff, King's County, Ireland, 1867; named rep. peer of Ireland, 1868; lord lt. King's County, 1892-1908; chancellor Dublin U., 1885-1908. Fellow Royal Soc., 1867. Author: An Account of Observations of the Great Nebula in Orion, 1867. Magnetic observations, Valencia Obs., Ireland; research on heat radiation from moon. Died Birr, Aug. 30, 1908.

PARSONS, Talcott, Am. sociologist; b. Colorado Springs, Colo., Dec. 13, 1902; s. Edward Smith and Mary Augusta (Ingersoll) P.; A.B., Amherst, 1927, L.H.D., 1949; Dr.Phil., U. Heidelberg (Germany), 1927; Dr. Rer. pol., U. Cologne (Germany), 1963; LL.D., U. Chgo., 1967; m. Helen Bancroft Walker, Apr. 30, 1927; children—Anne (dec.), Charles Dacre, Susan Pendrell (Mrs. Alfred A. Cramer). Instr. econs. Amherst, 1926-27; faculty Harvard, 1927-—, prof. sociology, 1944-—; vis. prof. social theory U. Cambridge, Eng., 1953-54; fellow Center for Advanced Study, Stanford, Cal., 1957-58. Fellow Am. Acad. Arts and Scis. (pres. 1967-—); mem. Eastern Social. Soc. (pres. 1942), Am. Sociol. Assn. (pres. 1949, sec. 1961-65), Am. Philos. Soc. Author: Structure of Social Action, 1937; Essays in Sociol. Theory 1949; The Social System, 1951; Structure and Process in Modern Societies, 1960; Social Structure and Personality, 1964; Societies, 1966; Sociological Theory and Modern Sqciety, 1967; (with others) Toward a General Theory of Action, 1951; (with Smelsed) Economy and Society, 1955; (with Bales) Family Socialization and Interaction Process, 1955; (with Bales, Shils) Working Papers in Theory of Action, 1953. Editor: Marx Weber, Theory of Social and Economic Organization, 1949; (with others) Theories of Society, 1961. Elucidation and devel. gen. theory in sociology and related fields starting with critical comparative analysis of such central figures as Pareto, Weber, Freud, others; devel. and application of theoretical analysis in number of empirical fields such as professions and occupations, family and kinship, sociology of religion, higher edn., social evolution. Home: 62 Fairmont St., Belmont, Mass. 02178. Office: William James Hall, Harvard U., Cambridge, Mass. 02138.*

PARSONS, Usher, Am. surgeon; b. Alfred, Me., Aug. 18, 1788; s. William and Abigail (Blunt) P.; studied medicine under various physicians; M.D., Harvard Med. Coll., 1818; m. Mary Holmes, Sept. 23, 1822; 1 child. Licensed to practice medicine by Mass. Med. Soc., 1812; commd. surgeon's mate, 1812; in charge of sick and wounded at Black Rock, 1812-13; served with Perry at Battle of Lake Erie, 1813, distinguished at battle with operations on and treatment of wounded; promoted surgeon, 1814; became prof. anatomy and surgery Dartmouth, 1820; appt. prof. anatomy and surgery Brown U., 1822; apptd. prof. obstetrics Jefferson Med. Coll., Phila., 1831; pres. R.I. Med. Soc., several times; an organizer A.M.A., v.p., 1853; a founder R.I. Hosp.; recipient Boylston prize 4 times; papers collected and published in Boylston Prize Dissertations, 1839; recipient Fiske Fund prize, 1842; summary of larger surg. operations pub. in Am. Jour. of Med. Scis., 1848. Author: The Sailor's Physician (med. guide for use on mcht. vessels), 1820. Died Providence, R.I., Dec. 19, 1868.

PARSONS, Willard H., Am. surgeon; b. Brookhaven, Miss., May 3, 1898; s. William F. and Ophelia (Herring) P.; student Tulane U., 1914-17; M.D., Jefferson Med., Philadelphia, 1920; m. Edna Earl Sparks, Oct. 23, 1922; children—Edna Earl (Mrs. H. Thurston Whitaker), Ruth Lee (Mrs. Emmett C. Neil). Practice of medicine, Vicksburg, Miss., 1922-—; chief staff and dir. surgery Vicksburg Clinic, 1929-62; dir. surgery Vicksburg Hosp., Inc., 1929-62; civilian consultant surgery to Surgeon-General Army U. S. Far Eastern Theatre, 1962-63, to Surgeon Gen. of U.S.A., Dept. of Defense, 1964-—; former cons. thoracic surgery U. S. Vets. Hosp., Jackson; clin. asso. prof. surg. sch. med. U. Miss.; past dir. grad. tng. in surgery Vicksburg Hosp. and Vicksburg Clinic. Diplomate Am. Bd. Surgery Founders Group. Fellow A.C.S. (chmn. bd. govs. 1953-56, regent 1956-—, vice chmn. bd. regents 1963-65, 1st v.p. 1965-66), International Cardiovascular Soc.; mem. Soc. Surgey Alimentary Tact, Am. Cancer Soc. (Miss. Cancer pres. 1955-57, nat. director), Southeastern Surg. Congress (pres. 1960-61), Pan-Pacific Surgical Soc., Royal Soc. Medicine (affiliate), Southern Soc. Clin. Surgeons (pres. 1950-51), So. (v.p. 1950-51), Western, Am. surg. assns., New Orleans Surg. Soc., Societe Internationale de Chirurgie, Issaquena-Sharkey-Warren Counties Med. Soc. (past pres.), Miss. State Hosp. Assn. (past pres.), Miss. State Med. Assn. (past chmn. surgery sect.). Author: Cancer of Breast, 1960; also surg. publs. Adv. editorial bd. Cancer. Contbr. books and numerous clin. papers. Home: 2711 Confederate Av. Office: 1401 Adams St., Vicksburg, Miss. 39180.*

PARSONS, William (Earl of Rosse), Brit. astronomer; b. York, Eng., June 17, 1800; s. Laurence Parsons; ed. Trinity Coll., Dublin, Magdalen Coll., Oxford; children include Laurence, Charles Algeron. Mem. Parliament, 1821-34; elected peer, 1841; chancellor U. Dublin, 1862-67; built obs., Birr Castle, Parsonstown, Ireland, 1826. Fellow Royal Soc., 1831 (pres. 1849-54, Royal medal 1851). Author: Observations of Nebulae and Clusters of Stars made with the 6 foot and 3 foot Reflector at Birr Castle from 1848 to 1878; also articles. Discovered similarity of annular and planetary nebulae, also spiral configuration of nebulae; studied nebula of Orion; resolved some spiral nebulae into star groups; discovered binary and triple stars; research on methods of improving constrn. of spectrum of reflecting telescope; built 36 inch reflecting telescope, 1839; mounted speculum 54 feet in focal length, 6 feet in diameter, Birr Castle, 1845; discovered method of preventing cracking and warping of surface while cooling. Died Monkstown, Ireland, Oct. 31, 1867.

PARTRIDGE, Lloyd Donald, Am. physiologist; b. Cortland, N.Y., Dec. 18, 1922; s. Bert J. and Marian (Rice) P.; B.S., U. Mich., 1948, M.S., 1949, Ph.D., 1953; m. Jean M. Rutledge, Aug. 6, 1944; children—L(loyd) Donald, David L., Gayle Ann. Instr., U. Mich., 1953-56; research asso. Yale Med. Sch., New Haven, 1956-57, asst. prof. physiology, 1957-62; asso. prof. physiology U. Tenn., Memphis, 1962-—, dep. chmn. physiology and biophysics, 1965-—; vis. prof. U. Vt., 1965, 66. Cons. in neurophysiology Conn. State Hosp., 1956-62; lectr. Conn. Postgrad. Med. Edn. Program, 1958-62. Mem. I.E.E.E., Am. Physiology Assn., Biophysics Soc., Am. Acad Neurology, Assn. Am. Med. Colls., Sigma Xi. Research, publs. on communication and control accomplished by nervous system, including reflex, cerebellum, muscle and receptor studies, devel. necessary equipment and analytical procedures. Home: 3061 Dumbarton Rd., Memphis 38128.*

PARTRIDGE, William Maurice, English pharmacologist; b. Lincoln, Eng., June 5, 1913; s. Harold and Mary Ann Agnes (Stokes) P.; B.Sc., B.Pharmacy,

Ph.D., Nottingham (Eng.) U. Coll.; m. Monica McMain, June 30, 1940. Asst. in research Brit. Pharmacopoeia Commn.; now prof. pharm. chemistry U. Nottingham. Mem. Pharm. Soc., Chem. Soc., Am. Chem. Soc. Contbr. profl. jours. Research on structure, activity and mechanism of action of polyazapolycyclic compounds; synthesis of chemotherapeutic agts.; chemistry carcinogenesis cancer chemotherapy. Home: 3 Cambridge Rd., Wollaton Park. Office: Univ. Nottingham, Nottingham, Eng.

PARTSCH, Joseph, German geographer, glaciologist; b. Schreiberhau, Germany, July 4, 1851; prof., Breslau (now Wroclaw, Poland), also Leipzig, Germany. Author: Schlesien. Eine Landeskunde für das deutsche Volk, 2 vols., 1896-1911; Mitteleuropa, 1904. Died Bad Brambach, June 22, 1925.

PARTSCH, Karl Johann, German surgeon; b. Schreiberhau, Germany, Jan. 1, 1855; prof., Breslau, Germany (now Wroclaw, Poland); founder modern surgery of teeth and jaw. Died Breslau, Sept. 6, 1932.

PARZEN, Emanuel, Am. statistician, mathematician; b. N.Y.C., Apr. 21, 1929; s. Samuel and Sarah (Getzel) P.; A.B. in Math., Harvard, 1949; M.A., U. Cal. at Berkeley, 1951, Ph.D., 1953; m. Carol Tenowitz, July 12, 1959; children—Sara Leah, Michael Isaac. Research scientist Columbia, 1953-56, asst. prof. math. statistics, 1955-56; faculty Stanford, 1956-64, asso. prof. statistics, 1959-64, prof., 1964—; guest prof. Imperial Coll., London, Eng., 1961-62; vis. prof. Mass. Inst. Tech., 1964-65. Fellow Am. Statis. Assn., A.A.A.S., Royal Statis. Soc., Inst. Math. Statistics; mem. Am. Math. Soc., Soc. Indsl. and Applied Math., Math. Assn. Am., Profl. Group in Information Theory, Biometric Soc., Econometric Soc., N.Y. Acad. Scis., Phi Beta Kappa, Sigma Xi. Author: Stochastic Processes, 1962; Modern Probability Theory and its Applications, 1960; Empirical Time Series Analysis, 1967; Foundations of Time Series Analysis and Statistical Communications and Control Theory, 1967; also articles. Extended limit theorems probability theory to uniform convergence in a parameter, introduced reproducing kernel Hilbert space formulation founds. time series analysis; introduced Parzen window for statis. spectral analysis; developed approach to empirical time series analysis. Home: 820 Lathrop Dr., Stanford, Cal. 94305.*

PASAMANICK, Benjamin, Am. psychiatrist; b. N.Y.C., Oct. 14, 1914; s. Alex and Elizabeth (Moskalik) P.; A.B., Cornell, 1936; M.D., U. Md., 1941; m. Hilda Knobloch, May 1, 1942. Faculty, U. Mich. Med. Sch., 1946-47, L.I. Coll. Medicine, 1948-49, State U. N.Y., 1949-50, Johns Hopkins, 1950-55, Ohio State U., 1955-65; clin. prof. psychiatry U. Ill. Coll. Medicine, Chgo., asso. dir. research Ill. State Dept. Mental Health, 1965-67; staff Neuropsychiat. Inst., Ann Arbor,, Mich., 1946-47, Kings County Hosp., Bklyn., 1947-50; psychiatrist Phipps clinic Johns Hopkins Hosp., 1952-55, Harriet Lane Home, 1951-55; asso. N.Y. State Dept. Mental Hygiene Com. for Research, pres. and dean N.Y. Sch. Psychiatry, adj. prof. epidemiology Columbia Sch. Pub. Health and Administrv. Medicine, 1967—. Cons. coms. WHO, U.S. Office Econ. Opportunity, U.S. Nat. Com. Vital and Health Statistics, Council of State Govts. Mem. Am. Psychiat. Assn. (First Hofheimer Research prizes 1949, 67), Am. Psychopath. Assn. (Stratton award 1961, pres. 1967), Am. Orthopsychiat. Assn., Am. Acad. Child Psychiatry, Group Advancement Psychiatry, Soc. Biol. Psychiatry, Assn. for Research in Nervous and Mental Disease, Am. Am. Pub. Health Assn., Am. Assn. Mental Deficiency, Am. Coll. Psychiatrists, Am. Psychol. Assn., Am. Anthrop. Assn., Am. Social. Assn., Am. Coll. Neuropsychopharm., Soc. Research Child Devel., Soc. for Applied Anthropology, Soc. Psychol. Study Social Issues, N.Y. Acad. Sci., Am. Acad. Mental Retardation, Sigma Xi., others. Cons. editor Merrill-Palmer Quar., Community Mental Health Jour.; editorial staff Jour. Nervous and Mental Diseases; editorial bd. Am. Jour. Orthopsychiatry. Research, publs. on factors affecting devel. of intelligence and moral values; epidemiology, diagnosis, community treatment of mental disorders. Home: 140 Riverside Dr., N.Y.C. 10024. Office: N.Y. State Dept. Mental Hygiene, 15 Park Row, N.Y.C. 10038.*

PASCAL, Blaise, French mathematician, physicist and natural philosopher; b. Clermont (now Clermont-Ferrand), Auvergne, France, June 19, 1623; s. Étienne and Antoinette Bégon P.; ed. at home; lived in Paris, where he frequented scientific circles, from 1631; lived in Rouen, 1639-47; returned to Paris, 1647; experienced religious conversion, night of Nov. 23, 1654; withdrew to Convent of Port-Royal, 1654. Author: Essai sur les coniques, 1640; Experiences nouvelles touchant le vide, 1647; Récit de la grande experience sur l'equilibre des liqueurs, 1648; Traité du vide, 1651; Lettres provinciales, 1656-57; Lettre circulaire relative a la cycloïde, 1658; Traité de la pesanteur de la masse de l'air, 1663; Traité du triangle arithmetique, 1665; Pensées (notes for a projected Apologie de la religion chrétienne; pub. posthumously). Showed great precocity in mathematics and science; his sister claimed that when he was 12 years old, he discovered for himself the theorems of geometry up to the 32nd in Euclid's first book; made important contributions to conic sections, 1640; invented a calculating machine, 1642-44; with Fermat credited with founding modern theory of probability and combinatorial analysis; invented arithmetic triangle (Pascal's triangle); solved problem of quadrature of cycloid; in barometric experiments at the Puy-de-Dôme mountain in Auvergne, 1648 (conducted for Pascal by Florin Périer) and in tower of Saint-Jacques-de-la-Boucherie, Paris, increased knowledge of atmospheric pressure and showed phenomena attributed to horror vacui were caused by weight of air; peformed experiments in hydrostatics; considered founder of hydrodynamics; formulated law which now bears his name: pressure applied to a confined fluid at any point is transmitted undiminished through the fluid in all directions and acts upon every part of the containing vessel at right angles to its interior surfaces and equally upon equal areas. Died, Paris, France, Aug. 19, 1662.

PASCAL, Étienne, French mathematician; b. 1588; father of Blaise Pascal; worked with Mersenne. Conchoid of a circle is sometimes called Pascal's limacon; supported Fermat against Descartes. Died 1651.

PASCAL, Gerald Ross, Am. psychologist; b. Raritan, N.J., Aug. 3, 1907; s. Anthony and Mary (Ross) P.; B.A.., U. Cal. at Berkeley, 1940; M.A., Harvard, 1942; Ph.D., Brown U., 1948; m. Lalla Vincent Sullivan, Sept. 14, 1964; children—Walther Gerald, Lawrence Hiby, Christopher Biram, Roy Darby. Chief psychologist Butler Hosp., Providence, 1946-49; lectr. Brown U., 1948-49; asso. prof. dept. psychology, research psychologist Western Psychiat. Inst. and Clinic U. Pitts., 1949-51; prof. psychology, dir. tng. program in clin. psychology, dir. Psychol. Service and Research Center, U. Tenn., 1951-64; research prof. psychiatry, psychology, coordinator research U. Miss., 1964-66, cons. Social Sci. Research Center, 1964-66; prof. mental health Miss. State Jr. C. of C., 1964-66; cons. VA, 1951—. Mem. Tenn. Alcoholism Commn., 1957-59; chmn. VA Psychology Adv. Council, S.E. Area, 1963-64; cons. Oak Ridge Nat. Lab., 1951-54, Knox County Cerebral Palsy Center, 1951-56, Knox County Juvenile Ct., 1951-56, VA Hosp., Dept. Surgery, Atlanta, 1956—, Miss. State Vocational Rehab., 1964-66, Draper Correctional Inst., Elmore, Ala., 1964-66. Diplomate Am. Bd. Examiners in Psychol. Hypnosis. Fellow Am. Psychol. Assn.; mem. Southeastern, Miss. psychol. assns., Sigma Xi. Author (with B. J. Suttell) The Bender-Gestalt Test, 1951; Behavioral Changes in the Clinic, 1959; (with W. O. Jenkins) Systematic Observations of Gross Human Behavior, 1961; also numerous articles. Pioneer in quantifiable, systematic approach to psychotherapy which forms sci. basis to objective psychotherapy; devel. and use of systematic interview, demonstrated its use in research. Home: 3718 Kings Hwy. Office: 440 E. Woodrow Wilson St., Jackson, Miss. 39216.*

PASCAL, Paul Victor Henri, French chemist; b. Saint-Pol, France, July 4, 1880; s. Alfred and Marie (de Tape) P.; D.Sc., École Normale Supérieure; m. L. Couvieur, July 29, 1907. Prof. chemistry Lille (France) U., 1908-28, U. Paris, 1928-50, École Normale Supérieure, 1928-62, also École Centrale des Arts et Manufactures. Corr., French Acad. Scis., 1927, mem., 1945. Author: Poudres, explosifs et gaz de combat; Syntheses et catalyses industrielles; Metallurgie; Traité de chimie minérale, 12 vols., 2d edit., 30 vols.; Liaisons interatomiques, interioniques et intermoleculaires; Chimie générale, 1 6 vols.; Études magnétochimiques et chimie générale. Research on complex minerals, metaphosphates, synthetic prodn. of nitric acid and its derivatives. Home: Château du Mesnil-Soleil, Damblainville, Morteaux, France. Office: 6 place du Pantheon, Paris 5, France.

PASCH, Moritz, German mathematician; b. Breslau, Germany (now Wroclaw, Poland), Nov. 8, 1843; ed. Breslau, Berlin, Giessen (all Germany). Prof., Giessen, 1873-1911. Author: Vorlesungen über Neuere Geometrie (contained system of axioms for descriptive geometry), 1882. Research, publs. on variables, functions, differential and integral calculus, analysis; Pasch's axiom named after him; 1st after Euclid to present elements of geometry as abstract postulates of their relations; helped lay foundations of modern geometry. Died Hamburg, Sept. 20, 1930.

PASCHALSKI, Jerzy, Polish limnologist; b. Radom, Poland, Jan. 6, 1912; s. Eustachy and Romualda (Florianska) P.; M.Sc., Yagiellonian U., Cracow, Poland, 1937; Ph.D., Nicolaus Copernicus U., Torun, Poland, 1962. Faculty, dept. limnology Higher Sch. Agr., Olsrtyn, Poland, 1951-54, Lab. Exptl. Hydrobiology, Neneki Inst. Exptl. Biology, Polish Acad. Scis., Warsaw, 1954-66, dept. hydrobiology Inst. Ecology, Polish Acad. Scis., Warsaw, 1966—. With Inst. Inland Fisheries, Olsrtyn, 1952, State Agrl. Inst., Warsaw, 1953-54, Union of State Fish Farms, 1962-63, others. Mem. Polish Hydrobiol. Soc. Research, publs. on abiotic environment of different types of bodies of water. Home: 15/17 Deotymy, Warsaw. Office: 72 Nowy Swiat, Warsaw, Poland.*

PASCHEN, Friedrich, German physicist; b. Schwerin, Germany, 1865; prof. univs. Tübingen, Bonn, Berlin (all Germany). Research on fine structure of spectral lines; verified experimentally Sommerfield's relativistic theory of atoms by means of studies on structure of fine X-rays. Died Potsdam, Germany, 1940.

PASKIN, Arthur, Am. physicist; b. Bklyn., Feb. 15, 1924; s. Max and Pauline (Jacobs) P.; B.S., S.D. Sch. Mines and Tech., 1948; Ph.D., Ia. State U., 1953; m. Charlotte R. Lipson, Nov. 1, 1953; children—Judith Ann, Carol Joan, Amy Ruth. Physicist, Sylvania Elec. Products, Mineola, N.Y., 1953-55; solid state physicist U.S. Army Materials Research Agy., Watertown, Mass., 1955-63; asst. head metallurgy, materials sci. div. Brookhaven Nat. Lab., Upton, N.Y., 1963—. Mem. Am. Phys. Soc., Am. Soc. Metals, Research Soc. Am., Sigma Xi. Author: (with E. Hartnett) Atomics for Everyone, 1951. Co-editor jour. Physics and Chemistry of Liquids, 1966. Contbns., publs. to understanding of a wide range of phenomena; effects of vibrating atoms on diffraction, long-range atomic interactions and their effects on alloying, order-disorder and liquid state of metals, importance of surface effects in superconductivity, use of computer techniques to study liquid state and phenomena of under cooling and crystallization. Home: High Path, Belle Terre, N.Y. 11777. Office: Brookhaven Nat. Lab., Upton, N.Y. 11973.*

PASSARGE, Siegfried, German geographer; b. Königsberg, East Prussia, Feb. 26, 1866; studied medicine, later became interested in geomorphology. Prof. in Breslau and Hamburg, 1908-35; made several voyages to Cameroons, 1893-94, Kalahari, 1896-99, Venezuela, 1901-02. Author: Die Kalahari, 1904; Südafrika, 1908; Physiologische Morphologie, 1912; Die Grundlagen der Landschaftskunde, 3 vols., 1919-20; Landschaftskunde, 4 vols., 1921-30; Landschaft und Kulturenwicklung in unseren Klimatbreiten, 1922; Erdkundliches Wandenbuch, 1922; Die Erde und ihr Wirtschaftsleben, 1927; Geographische Völkerkunde, 1933; Einführung in die Landschaftskunde, 1933; Die deutsch Landschaft, 1936. Subdivided continent of Africa on basis of phys. and climatic factors. Died Bremen, Germany, June 26, 1958.

PASSERINI, Giovanni, Italian botanist; b. Picuc di Guastalla, Italy, June 16, 1816; s. Gaetano and Barbara (Allegretti) P.; diploma in philosophy and medicine, Parma, Italy. Asst. to curator Civic Mus. Natural History, Milan, Italy, 1843-44; head Bot. Gardens, Milan, 1843-93; rector sch. pharmacy and vet. medicine U. Parma, 1879-85. Mem. Royal Acad. Lincei, Superior Council Instrn. Author: Flora Italiae, 1844; Flora dei coutorni di Parma, 1852; also articles. Research in mycology, plant morphology and biology. Died Apr. 17, 1893.

PASSOW, Adolf, German otologist; b. 1859; dir. clinic and polyclinic for otology, Berlin (Germany) U.; made 1st attempt to form opening into labyrinth of ear to improve hearing in otosclerosis, circa 1897. Died 1926.

PASSY, Antoine-François, French botanist; b. Paris, Apr. 23, 1792; mem. French Acad. Scis., Soc. Agr. Author: Carte geologique 1857. Studied geol. structure of Eure region. Died Gisors, France, Oct. 8, 1873.

PASTEUR, Louis, French chemist, microbiologist; b. Dôle, Jura, France, Dec. 27, 1822; s. Jean-Joseph P.; ed. Royal College at Besançon, bachelor's degree, 1840; Ecole Normale Supérieure, from 1843, doctorate in chemistry, 1847; hon. degree U. Bonn (returned); m. Marie Laurent, 1849; prof. physics in Dijon, 1848-49; prof. chemistry, U. Strausbourg, 1849-54; prof. chemistry and dean of faculty of sciences (which he organized), U. Lille, 1854-57; dir. sci. studies, Ecole Normale Supérieure, Paris, 1857-67; also prof. geology, physics and chemistry, Ecole des Beaux Arts, 1863-68; prof. chem., Sorbonne, 1867-89; first dir., Pasteur Institute, Paris, 1889-95; mem. French Acad. Scis., 1862 (perp. sec., from 1887); elected to Académie de Médecine, 1873; Académie Française, 1888; Fellow, Roy. Soc., 1869 (Rumford Medal, 1856; Copley Medal, 1874); author of numerous sci. articles. Founder of microbiological sciences, germ theory of disease, and science of immunity; laid foundations of stereochemistry when he discovered isomerism by separating two optically different forms of tartatic acid; showed isomerism occurs in many organic substances; demonstrated lactic, alcoholic, and other fermentations caused by specific microorganisms (germs) acting on a specific type of substance to produce another substance (germ theory of fermentation); investigated diseases of wine and beer; demonstrated some microorganisms derive energy from reactions which do not make use of oxygen (i.e. discovery of anaerobiosis); investigated origin of microorganisms; showed by elegant and dramatic experiments that spontaneous generation does not occur, that each microbe derived from a pre-existing microbe; introduced heat treatment to destroy microbes in perishable products; discovered and eliminated diseases of silkworms (saved silk industry in France); demonstrated anthrax caused by bacillus which survives in form of resistant spores in bodies of dead animals or in soil; his work enabled Joseph Lister to develop antiseptic surgery; found innoculation with attenuated froms of microbes could be used for immunization against highly virulent cultures; demonstrated rabies caused by microscopically unobservable agent, thus revealed world of viruses; found technique to produce attenuated forms; developed effective treatment for hydrophobia; used it successfully on humans. Died, Villeneuve l'Etang, near Paris, G France, Sept. 28, 1895.

PASTINSZKY, Istvan, Hungarian physician, dermato-venereologist, allergist; b. Budapest, Hungary, June 24, 1910; s. Joseph and Elisabeth (Feyertag) P.; student U. Vienna (Austria), 1927, U. Geneva (Switzerland), 1929; student Faculty Medicine, U. Budapest, 1927-33, dermato-venereologist diploma, 1937; m. Elisabeth Bajnóczy, Oct. 17, 1942; 1 son, Istvan. Asst., U. Budapest, 1934-42; chief sects. dermato-venereology at various hosps. in Hungary, 1942-45; lectr. U. Debrecen, Budapest, 1946—, asso. prof. dept. dermato-venereology, 1964—. Recipient numerous decorations. Mem. Hungarian Dermatol. Soc. (asso. pres. 1951—), San. Sci. Council (vice chmn. sect. dermatology 1952—). Author: (with I. Rácz) Börapolás, kozmetika, börbetegségek Medicina, 1958; Belbetegségek börtünetei i Medicina, 1959; Hautveränderungen bei inneren Krankheiten. Verlag Volk und Gesundheit, 1965; also numerous articles. Research on assns. of skin manifestations and internal disorders; allergic diseases of genito-urinary tract. Address: Budapest II. Széher ut 78, Hungary.*

PATAI, Raphael, anthropologist; b. Budapest, Hungary, Nov. 22, 1910; s. Joseph and Edith (Ehrenfeld) P.; student U. Breslau, Germany (now Wroclaw, Poland) 1930-31; Ph.D., U. Budapest, 1933, Hebrew U. Jerusalem, 1936; children—Ofra-Jennifer, Daphne. Came to U.S., 1947, naturalized, 1952. Mem. faculty Hebrew U., Jerusalem, 1942-47; founder, dir. Palestine Inst. Folklore and Ethnology, 1944-48; prof. anthropology Dropsie Coll., Phila., 1948-57; dir. research Theodor Herzl Inst., N.Y.C., also editol Herzl Press, 1956—. Vis. prof. anthropology. Columbia, U. Pa., Princeton, New Sch. Social Research, N.Y. U., Ohio State U.; prof. anthropology Fairleigh Dickinson U., 1968—. Preparer report on social conditions Middle East, UN, 1952; dir. Syria-Lebanon-Jordan research project Human Relations Area Files, Inc., New Haven, 1955-56. Viking Fund fellow, 1947. Fellow Am. Anthrop. Assn., Am. Folklore Soc. Author: numerous books, including Israel Between East and West, 1953; The Kingdom of Jordan, 1958; Sex and Family in the Bible and the Middle East, 1959; Golden River to Golden Road: Society, Culture and Change in the Middle East, 1962, 2d edit., 1967; (with Robert Graves) Hebrew Myths, 1964; The Hebrew Goddess, 1967. Editor numerous books. including Jordan, 1957; Herzl Year Book, 6 vols., 1958-66; Women in the Modern World, 1967. Adv. editor Judaism, Ency. Americana, 1959—. Home: 39 Bow St., Forest Hills, N.Y. 11375. Office: 515 Park Av., N.Y.C. 10022.*

PATAI, Saul, educator; b. Budapest, Hungary, Aug. 2, 1918; s. Joseph and Edith (Ehrenfeld) P.; M.Sc., Hebrew U., Jerusalem, 191, Ph.D., 1947; m. Elise H. Schaner, May 23, 1944; children—Irith Biklah, Michael. Indsl. chemist Gordon & Co., Tel-Aviv, Israel, 1941-42; chemist Israel Def. Forces, 1947-48; with Hebrew U., Jerusalem, Israel, 1944—, lectr., 1953-57, asso. prof., 1957—; hon. research asso. U. Coll. London, 1954-56, 60. Mem. Chem. Soc. London, Israel Chem. Soc., N.Y. Acad. Sci. Author: Glossary of Organic Chemistry, 1962. Editor: The Alkenes, 1964; The Carbonyl Group, 1966; The Ether Linkage, 1966; The Amino Group, 1968; gen. editor series of advanced treatises Chemistry of the Functional Groups, Interscience, N.Y., London, 1964—; cons. editor Chem. Rubber Co. Research, numerous publs. on kinetics and mechanism of organic reactions, novel methods of organic reactions in solid and molten state, oxidative and nonoxidative decomposition of mono and polysaccharides. Home: 28 Radak St., Jerusalem, Israel.*

PATARO, Vincent Florence, Argentinian surgeon; b. Buenos Aires, Argentina, Mar. 2, 1907; s. Francis and Mary Rose (Lauria) P.; M.D., U. Buenos Aires, 1934; m. Susan Ocampo, Aug. 18, 1935; children—Edward, Mary Susan. Chief surgery dept. Avellaneda Hosp., Buenos Aires, 1954—; cons. surgeon Navy Hosp., Buenos Aires, 1950—; dir. Finochietto Clinic, 1961—. Fellow Internat. Coll. Surgeons; mem. A.C.S., Société Internationale de Chirurgie, Internat. Cardiovascular Surgery. Contbg. author: Surgical Technique, 1948; also numerous articles. Research on diaphragmatic anatomy, arterial hypertension, bowel exptl. anastomosis. Home: 4230 Navarro St. Office: 2678 Cordoba, Buenos Aires, Argentina.*

PATEK, Paul R., Am. anatomist; b. Cleve., May 1, 1908; s. Walter L. and Ernestine R. Patek; A.B., U. So.Cal., 1931, Ph.D., 1938; m. Esther B. Sturgeon, June 6, 1938; children—Patricia, Pamela, Paul. Asst. anatomy U. So. Cal., 1931-38, instr., 1938-39, asst. prof., 1939-44, asso. prof., acting head dept., 1945-46, prof., head dept. anatomy, 1946—; vis. asso. prof. Washington U., St. Louis, 1944-45; cons. for attending staff Los Angeles Country Hosp., 1946—; curator unclaimed dead So. Cal. 1942—. Research investigator OSRD, World War II. Commonwealth Fund fgn. fellow, 1964-65. Fellow Internat. Coll. Angiology; mem. Am. Assn. Anatomists, A.A.A.S., Med. Research Assn. Cal., Assn. Am. Med. Colls., N.Y. Acad. Scis., Reticuloendochelial Soc. Am., Sigma Xi. Home: 4123 Mantova Dr., Los Angeles 90008.*

PATEL, Chandra Kumar Naranbhai, physicist; b. Baramati, India, July 2, 1938; s. Naranbhai and Maniben (Patel) P.; B.E., Coll. Engring., Poona U., India, 1958; M.S., Stanford, 1959, Ph.D., 1961; m. Shela Rani Dixit, Aug. 20, 1961; 1 dau., Neela. Mem. tech. staff Bell Telephone Labs., Murray Hill, N.J., 1961—, head infrared physics and electronics dept. Recipient Adolph Lomb medal Optical Soc. Am., 1966. Fellow Am. Phys. Soc.; mem. I.E.E.E., Sigma Xi. Research, publs. in investigation of Laser action in gaseous discharges containing atomic and molecular gases; interaction of laser radiation with matter in regard to nonlinear properties of materials. Home: 575 Main St., Chatham, N.J. 07928. Office: Bell Telephone Labs., Murray Hill, N.J. 07971.*

PATEL, Jean, French surgeon; b. Nantes, France, June 21, 1900; s. Alphonse and Margaret (Renouard) P.; m. Nicole De Massary, June 24, 1930; children—Jean-Claude, Alain, Martine; M.D., U. Paris (France), 1931. Faculty, U. Paris, 1939—, prof. surg. technic, 1955—; prof. surg. clinic, chief surgeon Hotel-Dieu, Paris, 1959—. Decorated Officer de la Légion d'Honneur, War Cross. Mem. French Surg. Acad. Surgery (past pres.), Nat. Acad. Medicine, Surg. Acad. Author: Surgery of the Spleen, 1955; also numerous articles. Research on digestive, vascular and spleen surgery. Address: 6 Av. de Messine, Paris 8°, France.*

PATEL, Vithalbhai Lallubhai, physicist; b. Samathiara Guj State, India, Mar. 31, 1935; s. Lallubhai R. and Amrutben (Patel) P.; B.Sc. with distinction, U. Baroda (India), 1956; M.S., U. Md., 1960; Ph.D. in Physics, U. N.H., 1964; postgrad. Internat. Summer Sch., U. Grenoble (France); m. Pushpa N. Desai, May 19, 1952; children—Jyot, Vickas. Research fellow Phys. Research Lab., Ahmedabad, India, 1956-58; instr., research asso. U. N.H., Durham, 1964-65; sr. research asso. space sci. dept. Rice U., Houston, 1965-66; asst. prof. physics U. Denver, 1966—. Govt. scholar, Bombay State, India, 1952-56; recipient Grad. Sci. Soc. award U. N.H., 1964. Mem. Am. Phys. Soc., Am. Geophys. Union. Research, publs. on cosmic radiation and their variations with time, plasma physics properties of tenuous ionized gas in magnetosphere, solar wind; discovered hydromagnetic waves in space using earth satellites Explorer 12 and 14. Home: 1613 S. Ivy Way, Denver 80222.*

PATERNO, Emanuele (Marquis de Sessa), Italian chemist; b. Palermo, Italy, Dec. 12, 1847; named prof. gen. chemistry U. Turin (Italy), then U. Palermo, 1872; apptd. prof. analytical chemistry U. Rome, 1892; dir. Inst. Chemistry, Rome., from 1910. Mem. French Acad. Scis., 1919. Author works on analytical, phys. indsl. chemistry, photochemistry, "Cryométrie." Died Palermo, Jan. 18, 1935.

PATERSON, Hugh Edward, zoologist; b. Pietersburg, Transvaal, Dec. 26, 1926; s. Walter Norman H. and Vida (Lloyd) P.; B.Sc., U. Witwatersrand, Johannesburg, South Africa, 1949, B.Sc. with honors, 1951; m. Joan Shirley Martin, Feb. 21, 1953; children—Margaret Ann, Michael Norman, Entomolologist, S. African Inst. Med. Research, Johannesburg, 1952-55, Arbovirus Research unit Polio Research Found., Johannesburg, 1955-63; lectr. dept. biol. sci. U. Coll. of Rhodesia, Salisbury, 1963-66; sr. lectr. U. Western Australia Nedlands, 1967—. Cons. to WHO, Tanzania, 1961, Mauritius, 1962. Brit. Council scholar Oxford U., 1958-59; WHO fellow, Italy, 1964. Mem. Entomol. Soc. S. Africa, Royal Entomol. Soc., Soc. for Study of Evolution, A.A.A.S. Research, publs. in fields of speciation, medically important insects. Home: 32 Myers St., Nedlands, Western Australia 6009.*

PATERSON, Robert, Scottish physician; b. Scotland, 1814; M.D., Edinburg, 1836; practiced medicine, Edinburgh; became prof. clin. surgery U. Edinburgh, 1874. Author: Memorials of the Life of James Sepne. Described (independently of William Henderson) inclusion bodies in molluscum contagiosum (Paterson's, or Henderson-Paterson bodies), 1841. Died 1889.

PATETTA-QUEIROLO, Miguel Angel, physiologist, biophysicist; b. Montevideo, Uruguay, Aug. 2, 1921; s. Carlos Alberto and Maria (Queirolo) Patetta; M.D., Faculty Medicine, Montevideo, 1948; m. W. Carmen Montenegro, Jan. 10, 1949; children—Carmen, Maria, Juan, Ana, Miguel. Faculty, Faculty Medicine, Montevideo, 1942-59, prof. physiciis, 1949-59; prof. biophysics Faculty Medicine, U. Carabobo, Valencia, Venezuela, 1959-65, prof. physiology and physics, 1965—, dir. Sch. Medicine, 1961-64; dir. dept. radioisotopes Inst. Oncology, 1963—. Mem. A.A.A.S., Radiation Research Soc., Am. Nuclear Soc., Assn. Venezuelan Advance Ciencia. Author: Lectures in Radiobiology, 1955; also articles. Research on biol. actions of ultrasonics waves on elec. brain activity and poliomielitic contractures, action of X-rays and fast neutrons on small intestine, elec. activity of brain and retina in chick embryo. Home: Avda 99 No 156-14, Valencia, Venezuela.*

PATEY, David Howard, English physician; b. Monmouth, Eng., Oct. 25, 1899; s. Frank Walter and Anne (Davies) P.; M.B., U. London, 1923, M.S., 1927; m. Gladys Joyce Summers, Feb. 5, 1927; children—David Geoffrey Hamilton, Mark, Margaret Joyce (Mrs. McCullough). Cons. surgeon Middlesex Hosp., London, 1930-64, dir. surg. studies, 1952-64, emeritus cons., 1964—. Mem. sci. adv. com. Brit. Empire Cancer Campaign, 1946—; Hunterian prof. Royal Coll. Surgeons Eng., 1964—. Fellow Royal Coll. Surgeons Eng., Am. Surg. Assn. (hon.); mem. Royal Soc. Medicine (past pres. surg. sect.), Surg. Research Soc. (past pres.). Editor: An Introduction to Surgery, 1954. Research, numerous publs. on surgery of breast and salivary glands. Address; 65 Montagu Mansions, London W.1., Eng.*

PATHAK, A. N., soil scientist; b. Varanasi, India, June 20, 1927; s. Markandey P.; B.Sc. in Agr., Govt. Agrl. Coll., Agra U., Kanpur, India, 1946, M.Sc., 1948, Ph.D., 1952; m. Kusumlata P. Feb. 1942; children—Manjula, Ashutosh. Faculty, Govt. Agrl. Coll., 1950-52, prof. agrl. chemistry, 1959—; soil microbiologist Bihar, Sabour, India, 1952-53; asst. agrl. chemist Govt. of U.P., Kanpur, 1953-58. Mem. Indian Soc. Soil Sci. Research and publs. on soils, soil microbiology and soil fertility, especially major and minor plant nutrients, nutritional requirements of crop and deficiency symptoms caused by them, soil bacteria and other micro-flora related to soil fertility. Home: Vill.& P.O., Tanda, Varanasi, U.P., India. Office: Govt. Agrl. Coll., Kanpur, U.P., India.*

PATIALA, Jorma, Finnish physician; b. Luopioinen, Finland, Sept. 17, 1912; s. Kalle Kustaa and Maria (Isolukkari) P.; M.D., U. Helsinki, 1944; m. Aino Tellervo Koskinen, June 3, 1939; children—Erkki, Anneli. Prof. Tb and pulmonary diseases U. Helsinki (Finland), 1955—; chief physician dept. pulmonary diseases U. Central Hosp., Helsinki, 1955—. Cons. physician Mass X-ray Surveys Center; cons. mem. for med. damages Med. Bd. Finland. Fellow Am. Coll. Chest Physicians; mem. Am. Thoracic Soc. (corr.), Royal Swedish Med. Soc. (fgn.), Soc. Lung Physicians in Finland (chmn.), Finnish Anti-Tb Assn. (pres.). Research, numerous publs. on pulmonary Tb and carcinoma of lung, pleurisy, occupational diseases of lungs, pneumomediastinography, rehab. and social medicine. Home: 17 A 14 Temppelikatu, Helsinki 10. Office: Meilahti Hospital, Helsinki, Finland.*

PATIN, Guy, French physician; b. Hodenc-en-Bray (Oise), France, 1601; ed. Collège de France, M.D., 1624; prof. surgery faculty medicine, Collège de France, 1632; prof., Collège Royal, 1654. Author: Lettres; Traité de la conservation de la santé, 1632; Notes sur le livre de Galien; De la saignée; Editor: Apologie de Galien, by Gaspard Hoffmann, 1668. Enemy of novelties in medicine; opponent of chem. medicine and theories of Harvey; partisan of bloodletting, which he used widely; famous for his letters to physicians, which are full of details on social customs of France at time of Fronde. Died Paris, 1672.

PATON, Boris Evgeneévich, Russian metallurgist, technologist; b. Nov. 27, 1918; grad. Kiev Polytech. Inst., 1941. Engr., Krasnoe Sormovo Plant, 1941-42; with Inst. Electro-Welding, Ukrainian SSR Acad. Scis., 1942-53, dir. Paton Inst. Electro-Welding of inst., 1953—. Mem. Communist Party, 1952—, Central Com., 1961; dep. to 5th session Ukrainian Supreme Soviet; dep. to USSR Supreme Soviet, 1962. Recipient Stalin prize, 1950, Lenin prize, 1957. Corr. mem. Ukrainian SSR Acad. Scis., 1952, academician, 1958, pres., 1962; academician USSR Acad. Scis., 1962. Author: Selecting the Production System for Large-Gauge Straight-Seamed Welded Tubes; Large-Gauge Welded Tube Production in the German Federal Republic and France; Investigation Conditions and Control of Stable Combustion of Welding Arc, 1951; Electro-slag Rewelding of Metals, 1961; Production of Two-Layer Sheets for Electro-slag Welding, 1962; coauthor: Experimental Research in Automatic Welding Under a Flux Layer, 1944; Estimating the Chain and Apparatus of an Alternate Current for Arc Welding, 1953; Programmatic and Cybernetic Control of Welding Processes, 1960. Specialist in electrotech. metal welding; established basis process of closed arc welding, 1942-45; devised principles of automatic electric arc welding control and pipe welding methods, 1945-51; co-developer electro-slag welding of thick pieces of metal, 1952-57. Adress: ul. Gorkogo 69 Paton Inst. Electro-welding, Ukrainian SSR, Kiev, Ukrainian SSR.

PATON, Bruce Calder, surgeon; b. Coonoor, India, Aug. 28, 1925; s. William Calder and Marian (Williamson) P.; M.B., Ch.B., U. Edinburgh, Scotland, 1951; m. Patricia Allen Ryan, July 9, 1955; children—Peter W.C., Ian B., Allen Malcolm. Came to U.S., 1958, naturalized, 1962. Registrar cardiology Royal Infirmary, Edinburgh, 1955-56, sr. surg. house officer, 1957-58; faculty U. Colo. Med. Center, Denver, 1960—, asso. prof. surgery, 1966—, dir. Halsted Lab. for Exptl. Surgery, 1960—. Cons. VA Hosp., Denver, 1965—. Recipient Gold medal in surgery U. Edinburgh, 1951, William Leslie prize in medicine, 1956. Fulbright scholar, 1953-55. Diplomate Am. Bd. Surgery. Fellow Royal Coll. Surgeons; mem. Royal Coll. Physicians, Brit. Med. Assn., Royal Med. Soc., Internat. Cardiovascular Soc., Soc. U. Surgeons Am. Assn. U. Profs., Am., Colo. heart assns., Soc. for Vascular Surgery. Research, publs. on function American Association for Thoracic Surgery, Society for Vascular Surgery. Research, publs. on function and uses heart-lung machines; total body cooling for heart surgery; artificial heart valves; surg. treatment congenital and acquired heart disease; snake bite. Home: 260 Newport St., Denver 80220.*

PATON, Diarmid Noel, Scottish physiologist; b. Edinburgh, Mar. 19, 1850; s. Joseph Noel and Margaret (Ferrier) P.; B.Sc., Edinburgh U., 1881, M.B., C.M., 1882, LL.D., 1919; postgrad., Vienna, Paris, 1882-84; fellow under William Rutherford, Edinburgh, 1884; m. Agatha Balfour, 1898; 1 son, 1 dau. Became lectr., surgeon's hall, sch. medicine Royal Colls., Edinburgh, 1886; named supt. research lab. Royal Coll. Physicians, Edinburgh, 1889, also fellow; Regius prof. physiology Glasgow U., 1906-28. Mem. Royal Commn. on Salmon Fisheries, 1900; mem. Med. Research Council, 1918-23. Fellow Royal Soc., 1914, Royal Soc. Edinburgh. Author: (with Leonard Findlay) Poverty, Nutrition and Growth, 1926; Nervous and Chemical Regulators of Metabolism, 1913; The Physiology of the Continuity of Life, 1926. Pioneer in study of metabolism and nutrition in Gt. Britain; research on endocrine glands. Died Stobo, Scotland, Sept. 30, 1928.

PATON, Robert Frederick, Am. physicist; b. Calumet, Mich., Dec. 9, 1892; s. Andrew and Mary (Nowlin) P.; A.B., U. Mich., 1915, M.S., 1916, Ph.D., 1922; m. Agnes Spence, Sept. 9, 1919; children—Mary Ann (Mrs. John Vogt), Janet Spence (Mrs. Lawrence Corray), Martha Jean (Mrs. Allyn May). Instr. physics Western Res. U., 1918-19, U. Mich., 1920-22; asso. in physics U. Ill., Urbana, 1922-1956, Associate Professor 1933-1956, asso. prof. emeritus, since 1956—; sr. lectr. physics Fla. State U., Tallahassee, 1960—. Served with Signal Corps, U. S. Army, 1918-19. Fellow Am. Phys. Soc.; mem. Ill. Acad. Sci. (pres. 1948-49), Am. Assn. Physics Tchrs. (sec. 1950-56), Am. Inst. Physics (dir. 1956-59); sec. Fla., Am. Soc. Physics Tchrs. Author: (with Sheldon, Miller, and Kent) Physics for Colleges, 1924. Research on temperature effects on infrared absorption, temperatures in internal combustion engines, alpha ray disintegration phosphorus. Home: 2123 Jennette St., Tallahassee 32303.*

PATON, Stewart, Am. physician; b. N.Y.C., 1865; s. William and Anne Stavely (Agnew) P.; A.B., Princeton, 1886, A.M., 1889; M.D., Coll. Phys. and Surg. (Columbia), 1889; postgrad., Germany and Italy; m. F. Margaret Halsey, 1892; children—F. Evelyn, William, R. Townley. Asso. in psychiatry Johns Hopkins; dir. lab., Sheppard and Enoch Pratt Hosp., Balt., lectr. Princeton, Columbia; cons. mental hygiene, lectr. psychiatry Yale, 1926-28. Trustee Carnegie Instn., Washington. Fellow A.A.A.S.; mem. Am. Philos. Soc., Am. Neurol. Assn., N.Y. Acad. Medicine, Eugenics Research Assn. (pres. 1919). Author: Text-Book of Psychiatry for Use of Students and Practitioners of Medicine; Education in Peace and War, 1919; Human Behavior, 1921; Signs of Sanity and the Principles of Mental Hygiene, 1922; Prohibiting Minds, 1932. Died Jan. 7, 1942.

PATON, William Drummond Macdonald, English pharmacologist; b. Hendon, Eng., May 5, 1917; s. William and Grace (MacDonald) P.; B.A., Oxford (Eng.) U., 1938; B.M., B.Ch., U. Coll. Hosp. London (Eng.), 1942, M.A., 1948, D.M., 1953; m. Phoebe Margaret Rooke, Aug. 22, 1942. House physician U. Coll. Hosp., 1942-43; reader in applied pharmacology, 1952-54; pathologist Midhurst Hosp., 1943-44; sci. staff Nat. Inst. for Med. Research, 1944-52; prof. pharmacology Royal Coll. Surgeons, London, Eng., 1954-59; prof. pharmacology U. Oxford, 1959—. Justice of Peace, 1956—; mem. Med. Research Council, 1963-67. Fellow Royal Soc., 1956; mem. Physiol. Soc. Gt. Britain (past sec.), Brit. Pharmacological Soc., Med. Research Soc., Biometric Soc., Brit. Computer Soc. Recipient (with E.J. Zaimis) Cameron prize, 1956, Gairdner award, 1959; Campbell Oration, 1957. Research, numerous publs. on mechanisms of underwater breathing, decompression sickness, allergy, histamine release, introduced decamethonium and hexamethonium; developed kinetic theory drug action. Home: 13 Staverton Rd., Oxford, Eng.*

PATT, Harvey Milton, Am. physiologist; b. Chgo., Aug. 2, 1918; s. Jack and Minnie (Rosenthal) P.; B.S., U. Chgo., 1939, Ph.D., 1942; m. Marcia June Goldstock, June 4, 1965; children—Bruce E., Mark S., Emily C. Asso. physiologist Argonne Nat. Lab., 1946-52, sr. physiologist, 1952-64; prof. dir. lab. radiobiology U. Cal., San Francisco, 1964—. Mem. coms. NRC, USPHS, AEC; cons. govt. agys.; lectr. various univs. Recipient Ernest Orlando Lawrence Meml. award for radiobiology U.S. AEC, 1964. Fellow A.A.A.S.; mem. Am. Physiol. Soc., Soc. Exptl. Biology and Medicine, Am. Assn. Cancer Research, Radiation Research Soc. (past pres.), Sigma Xi. Research, numerous publs. on parathyroid control of blood calcium, chem. radiation protection, white blood calcium, chem. radiation protection, leukocyte balance, kinetics. Office: Lab. Radiobiology, U. Cal., San Francisco 94122.*

PATTEN, Bradley Merrill, Am. embryologist, educator; b. Milw., June 14, 1889; s. William and Mary Elizabeth (Merrill) P.; A.B. summa cum laude (Chamberlain fellow), Dartmouth, 1911; M.A., Harvard, 1912, Ph.D., 1914; m. Barbara Standish, June 13, 1914; 1 dau., Elizabeth (Mrs. Walter E. Garrey). Asst. in zoology Harvard, 1912-14; hydrographer Internat. Ice Patrol, 1914; faculty histology and embryology Western Res. U., 1914-34, asso. prof., 1921-34; asst. dir. med. scis. Rockefeller Found., 1934-36; prof., chmn. dept. anatomy U. Mich. Med. Sch., Ann Arbor, 1936-58, prof. emeritus, 1958—. Vis investigator Carnegie Embryological Inst., Balt., 1925, Pathologisches Institut, Vienna, Austria, 1927; vis. prof. U. P.R., 1952, U. Otago Med. Sch., Dunedin, New Zealand, 1954, U. Buenos Aires (Argentina), 1958, U. Miami, 1959, U. Adelaide (Australia), 1961; cons. embryology Stedman's Med. Dictionary, 1958—. Mem. Am. Soc. Naturalists, Am. Soc. Zoologists, Am. Assn. Anatomists (2d v.p. 1934-36), Sociedad de Biologa de Montivideo (hon. life), Mich. Med. Soc. (hon. life), Phi Beta Kappa, Sigma Xi, Alpha Omega Alpha (hon.). Author: Early Embryology of the Chick, 1920; Embryology of the Pig, 1927; Cardiovascular System in Morris' Anatomy, 1942; Human Embryology, 1946; Cardiac Development and Structure in Gould's Pathology of the Heart, 1953; Foundations of Embryology, 1958; also numerous articles. Asso. editor Am. Jour. Anatomy, 1941-54. Research on micro-moving picture methods of recording activities living embryos. Home: 2126 Highland Rd., Ann Arbor, Mich. 48104.*

PATTEN, William, Am. zoologist; b. Watertown, Mass., Mar. 15, 1861; s. Thomas and Mary Low (Bradley) P.; B.S., Lawrence Sci. Sch. (Harvard), 1883; A.M., Ph.D., Leipzig, Germany, 1884; m. Mary Elizabeth Merrill, June 28, 1883; 1 son, Bradley Merrill. Asst. in Lake Lab., Milw., 1886-89; prof. biology U. N.D., 1889-93; prof. zoology Dartmouth, 1893-1931. Trustee Marine Biol. Lab., Woods Hole. Mem. NRC for Biology and Agr. Pres. Sect. F (zoology) A.A.A.S., 1918. Author: The Evolution of the Vertebrates and Their Kin, 1912; The Grand Strategy of Evolution: The Social Philosophy of a Biologist, 1920. Known for research on evolution of vertebrates, fossil fishes, anatomy and embryology of invertebrates, especially mollusks and arthropods, for which he developed theory of color vision. Died Hanover, N.H., Oct. 27, 1932.

PATTERSON, Andrew, Jr., Am. chemist; b. Texarkana, Tex., July 23, 1916; s. Andrew and Nell (Young) P.; B.A., U. Tex., 1937, M.A., 1938, Ph.D., 1942; m. Elizabeth Chambers, June 6, 1940; children—Ellen Clifton, Andrew Muir, Elizabeth Lucinda, Katharine Chambers. Asst. prof. N. Tex. Agrl. Coll., 1941-42; acting asso. prof. U. N.C., 1942-43; research asso. Harvard Underwater Sound Lab., 1943-45; physicist USN Underwater Sound Lab., 1945-46; faculty Yale, New Haven, 1946—, asso. prof. chemistry, 1951—, research asso., Edwards St. Lab., 1951-53, dir., 1953-55. Mem. mine adv. com. Nat. Acad. Sci.-NRC, 1955-57, chmn., 1957-61. Mem. Am. Chem. Soc., Sigma Xi, Phi Lambda Upsilon, Alpha Chi Sigma. Author: (with H.C. Thomas) Quantitative Analysis, 1952; also numerous articles. Research on exptl. measurement of conductance of electrolytes at high elec. fields; phase separation phenomenon in alkali metal-liquid ammonia solutions. Home: 175 E. Rock Rd., New Haven 06511.*

PATTERSON, A(rthur) L(indo), physicist, x-ray crystallographer; b. Nelson, N.Z., July 23, 1902; s. Arthur Henry and Nellie Tweeddale (Slack) P.; B.Sc. (hons.), McGill U., 1923, M.Sc., 1924, Ph.D., 1928; m. Elizabeth Lincoln Knight, Sept. 14, 1935. Came to U.S., 1929, naturalized, 1945. Research worker Royal Instn., London, Eng., 1924-26, Kaiser Wilhelm Institut für Faserstoffchemie, Berlin-Dahlem, Germany, 1926-27; lecturer physics McGill U., 1928-29; asso. Rockefeller Inst. for Med. Research, 1929-31; lecturer Johnson Foundn. for Med. Physics, Phila., Pa., 1931-33; research worker Mass. Inst. Tech., 1933-36; asst. prof. physics Bryn Mawr Coll., 1936-40, asso. prof., 1940-49; research physicist Naval Ordnance Lab., Washington, 1944-45; sr. mem. and head dept. molecular structure Inst. for Cancer Research, Phila. Mem. of the U.S.A. National Committee on Crystallography 1948-55, 57—, chmn. com., 1948-50; member exec. com. Internat. Union Crystallography, 1948-54; mem. div. phys. scis. NRC, 1957-62. Fellow of the Am. Physical Soc., Physical Soc. London, Mineral. Soc. Am., N.Y. Acad. Sci.; mem. Am. Soc. for X-ray and Electron Diffraction (v.p. 1948, pres. 1949), Am. Crystallographic Assn. Author: (with W. C. Michels) Elements of Modern Physics; also papers on x-ray analysis of crystal structures; structure compounds biol. interest, theoretical papers on particle size determination, homometric structures. Devised Patterson synthesis (now basic method for applying x-ray diffraction to analysis of crystal structures), 1934. Died Nov. 6, 1966.

PATTERSON, Bryan, paleontologist; b. London, Eng., Mar. 10, 1909; s. Lt. Col. John Henry and Frances Helena (Gray) P.; student Malvern Coll., Eng., 1923-26, U. Chgo., 1927-33; M.A. (hon.), Harvard, 1955; m. Bernice Maurine Caine, Apr. 14, 1934; 1 son, Alan. Came to U. S., 1926, naturalized, 1938. Assistant paleontology to curator fossil mammals Chicago Natural History Museum, 1926-55; Alexander Agassiz prof. vertebrate paleontology Harvard, 1955—. Mem. adv. panel earth scis. Nat. Sci. Found., 1956-59. Served with 16th Regt., 1st Inf. Div., AUS, 1944-45. Carnegie Corp. grant-in-aid for travel, 1938; John Simon Guggenheim Found. fellow, 1951-53, 54-55. Mem. Am. Acad. Arts and Scis., Soc. Vertebrate Paleontology (sec.-treas. 1946-48, pres. 1948-49), Geol. Soc. Am., Soc. for Study Evolution, Soc. Systematic Zoology, Am. Soc. Mammalogists, Nat. Acad. Sci., Sigma Xi. Contbr. tech. papers sci. jours. Research on North and South America fossil mammals and birds; Tertiary crocodiles. Home: 234 Brattle, Cambridge 38, Mass.

PATTERSON, Edward Mervyn, Brit. chemist; b. Bangor, County Down, No. Ireland, Mar. 14, 1920; s. John Wilson and Dorothy Mary (Ekin) P.; B.Sc., Queen's U., Belfast, 1941, M.Sc., 1944, D.Sc., 1956; m. Violet Adams, Mar. 1953; 1 dau., Anna Dorothea. Tech. officer ICI Ltd., Nobel div., Stevenston, Ayrshire, Scotland, 1941-47, 1954—; lectr. geology and geochemistry U. St. Andrews, Scotland, 1947-54, warden Hamilton Hall, 1951-53. Fellow Royal Soc. Edinburgh, Geol. Soc. London; mem. Royal Irish Acad., Mineral. Soc. Gt. Britain, Geol. Soc. Glasgow (council 1942-45, 59-62, v.p. 1962-64, pres. 1964-67). County Donegal Hist. Soc., Stephenson Locomotive Soc., Irish Ry. Record Soc. Author books on Irish ry. history. Research, publs. on Tertiary igneous rocks of Gt. Britain; work on explosives, including 1st underground high-speed photography of blasting. Home: 25 Caldwell Rd., West Kilbride, Ayrshire. Office: ICI Ltd., Nobel Div., Stevenston, Ayrshire, Scotland.*

PATTERSON, Howard Alexander, Am. surgeon; b. Salem, N.C., Aug. 31, 1902; s. Andrew Henry and Eleanor Spurrier (Alexander) P.; A.B., U. N.C., 1921, postgrad., 1921-23; M.D., Harvard, 1925; m. Sarah Elizabeth Robertson, Nov. 22, 1930; children—Drew, Howard Alexander, Sarah Elizabeth. Asst. surgeon, chief surg. clinic Roosevelt Hosp., 1929-37, asso. surgeon, 1937-47, attending surgeon, 1947-52, chief surg. service, 1952—; cons. surgeon Southampton Hospital, United Hospital, Portchester, Paterson (N.J.) Gen. Hosp., Lawrence Hosp.; clin. prof. surgery Columbia. Diplomate Am. Bd. Surgery. Fellow A.C.S. (pres. 1965-66), Am., So. surg. assns.; Gastroent. Soc. (pres. 1952), mem. N.Y. Acad Medicine, Phi Beta Kappa, Sigma Alpha Epsilon, Phi Chi. Contbr. surg. jours. Home: 25 Sycamore St., Bronxville, N.Y. 10708.*

PATTERSON, John Miles, Am. chemist; b. Vineland, N.J., Nov. 5, 1926; s. James and Gladys (Barraclough) P.; B.S., Wheaton Coll., 1949; Ph.D., Northwestern U., 1953; m. E. Genevieve Christensen, Aug. 20, 1949; 1 son, John Miles. Teaching asst. Northwestern U., 1949-50, Comml. Solvents fellow, 1950-51, U.S. Rubber fellow, 1951-52, postdoctoral fellow, 1953; faculty U. Ky., 1953—, prof., 1967—. Mem. Am. Chem. Soc., A.A.A.S., Sigma Xi, Phi Lambda Upsilon. Contbr. articles in field to sci. jours. Research, publs. on heterocyclic compounds, study of high temperature reactions of organic compounds, photochem. reactions in solution, synthesis of organic compounds. Home: 726 Della Dr., Lexington, Ky. 40504.*

PATTERSON, John Ward, Am. physician; b. Baldwin, Kan., Dec. 6, 1916; s. John and Eleanor (Ward) P.; A.B., Ohio Wesleyan U., 1939, D.Sc. (hon.), 1965; M.Sc. in Organic Chemistry, Ohio State U., 1941, Ph.D., 1942; M.D., Western Res. U., 1949; m. Margaretta Elizabeth Crawford, Aug. 31, 1940; children—Eleanor Louise, Virginia Elizabeth, John Franklin, Margaret Ann. Instr. chemistry U. Vt., 1942-43; instr. anatomy Western Res. U., 1947-49, asst. prof., 1949-52, asso. prof., 1952-56, asso. dean med. edn., 1953-56; dean medicine, prof. anatomy U. B.C., 1956-58; dir. med. affairs, dean medicine Vanderbilt U., 1958-62, vice chancellor med. affairs, 1959-62, prof. physiology, 1962-63; prof. physiology, planning asso. new med. sch. U. Conn., 1963-65, dean Sch. Medicine, 1965—. Mem. Am. Chem. Soc., Assn. Research Ophthalmology, A.A.A.S., Am. Assn. Anatomists, Soc. Exptl. Biology and Medicine, Am. Diabetes Assn., Sigma Xi, Phi Beta Kappa, Alpha Omega Alpha, Phi Lambda Upsilon, Sigma Pi Sigma, Phi Mu Epsilon. Contbr. sci. articles on spectroscopy, exptl. diabetes, cataract formation, med. edn. Home: Lynwood Rd., Storrs, Conn. 06268.*

PATTERSON, Robert, mathematician; b. Hillsborough, County Down, Ireland, May 30, 1743; s. Robert Patterson; LL.D., U. Pa., 1819; m. Amy Ewing, May 9, 1774; 8 children, including Robert M. Came to U.S., 1768. Prin. of acad., Wilmington, Del., 1772-75; served in Revolutionary War; prof. math. U. Pa., 1779-1814, vice provost, 1810-13; mem. Select Council of Phila., pres., 1799; dir. Phila. mint (apptd. by Pres. Jefferson), 1805; elected mem. Am. Philos. Soc., 1783, pres., 1819-24. Author: Lectures on Select Subjects in Mechanics, 2 vols., 1806; Astronomy Explained upon Sir Isaac Newton's Principals, 1806-09; Newtonian System of Philosophy, 1808; A Treatise of Practical Arithmetic, 1818; also papers. Died Phila., July 22, 1824.

PATTERSON, Roy, Am. physician; b. Ironwood, Mich., Apr. 26, 1926; s. Donald Irving and Helmi (Lantta) P.; B.S., U. Mich., 1949, M.D. cum laude, 1953; m. Elaine Anita Gustafson, Aug. 28, 1949; children—Kathleen, David Roy, Thomas Donald. Instr. medicine U. Mich. Med. Sch., Ann Arbor, 1957-59, asst. prof., 1959, instr. U. Hosp., 1957-58, cons. U. Health Service, 1958-59; cons. VA Hosp., Ann Arbor, 1958-59; attending physician VA Research Hosp., Chgo., 1959-60; prof. medicine, chief allergy-immunology sect. Northwestern U. Med. Sch., Chgo., 1964—; attending physician Cook County Hosp., Chgo., 1963, Chgo. Wesley Meml. Hosp., 1964; civilian cons. Wilford Hall Hosp., Lackland AFB,

Tex. Recipient Research Career Devel. award USPHS, 1963. Fellow A.C.P., Am. Acad. Allergy; mem. Am. Assn. Immunologists, A.A.A.S., Am., Central socs. clin. investigation, Soc. Exptl. Biology and Medicine, A.M.A., Mich., Chgo. med. socs., Mich Allergy Soc., Alpha Omega Alpha, Phi Kappa Phi. Phi Eta Sigma. Research, publs. on characterization of antigens and antbodies and their interactions in animals. Home: 230 Latrobe St., Northfield, Ill. 60093. Office: 303 E. Chicago Av., Chgo. 60611.*

PATTIE, Frank A(cklen), Am. psychologist; b. Winchester, Tenn., Jan. 13, 1901; s. Frank Acklen and Lou (Gregory) P.; B.A., Vanderbilt U., 1922; A.M., Harvard, 1925; Ph.D., Princeton, 1925; m. Billie Whatley, Aug. 24, 1939; children—Robert Gregory, Frances (Mrs. Thomas C. Tanner), Margaret. Prof. dept. psychology U. Ky., Lexington, 1947—, head dept., 1947-49. Cons. Lexington VA Hosp., 1957—; mem. Ky. Bd. Examiners of Psychologists, 1948-56. Diplomate in clin. psychology Am. Bd. Examiners Profl. Psychology. Mem. Am. Psychol. Assn., Am. Soc. Clin. Hypnosis (pres. 1959-61, award for sci. contbns. 1965), Argentine Soc. Hypnotherapy (hon.), Phi Beta Kappa, Sigma Xi. Asso. editor Am. Jour. Clin. Hypnosis, 1958—; internat. editor Revista Latino-Americana de Hipnosis Clinica, 1960—. Research, publs. on hypnotic unilateral anesthesia of vision, hearing, touch; hypnotherapy of psychogenic deafness in children, history of hypnotism, audition, stuttering, animal learning. Home: 224 Shady Lane, Lexington, Ky. 40503.*

PATTINSON, Hugh Lee, English metall. chemist; b. Alston, Eng., Dec. 25, 1796; s. Thomas and Margaret (Lee) P.; ed. pvt. schs.; m. Phoebe Walton, Dec. 25, 1815. Became clk., asst. to soap-boiler, Newcastle, Eng., circa 1821; became assay master to lords of manor, Alston, 1825; became partner (with John Lee, George Burnett), mgr. chem. works, Felling, Eng., later Washington, Eng. Fellow Royal Soc., 1852, Geol. Soc., Royal Astron. Soc.; mem. Lit. and Philos. Soc. Newcastle. Discovered method of separating silver from lead ore, 1829, patented, 1833; discovered process for obtaining white lead which formed new compound oxychloride of lead, 1841, process for manufacturing magnesia alba, 1841. Died Nov. 11, 1858.

PATTISON, Granville Sharp, anatomist; b. nr. Glasgow, Scotland, 1791; s. John Pattison; attended U. Glasgow. Prof. anatomy, physiology, surgery in Andersonian Instn. (U. Glasgow); came to U.S., 1819; gave a series pvt. lessons in anatomy in Phila., 1819-20; prof. anatomy, physiology, surgery U. Md., Balt., 1820-26; prof. anatomy U. London, 1828-31; prof. anatomy Jefferson Med. Coll., Phila., 1832-41; prof. anatomy U. City of N.Y., 1841-51. Fellow Royal Coll. Surgeons; mem. Medico-Chirug. Soc., London. Author: Experimental Observations on the Operation of Lithotomy, 1820; "A Refutation of Certain Calumnies" in pamphlet Correspondence Between Mr. Granville Sharp Pattison and Dr. Nathaniel Chapman, 1821. Editor: Observations on the Surgical Anatomy of the Head and Neck (Allen Brun), 1924; Register and Library of Medicine and Surg. Sci., Washington, 1933-36; (with others) Am. Med. Library and Intelligence, Phila., 1936. Died N.Y.C., Nov. 12, 1851.

PATTON, Harry Dickson, Am. physiologist; b. Bentonville, Ark., Mar. 10, 1918; s. William Ernest and Nell (Dickson) P.; B.A., U. Ark., 1939; Ph.D., Yale, 1943, M.D., 1946; m. Barbara Martin Whitney, Aug. 14, 1943; children—Ann, Elizabeth. Faculty, U. Wash., Seattle, 1947—, prof. dept. physiology, biophysics, 1956—, chmn. dept., 1966—. Cons. U. Wash., VA hosps.; chmn. cultural exchange visit in neurol. sci. to USSR, 1962; chmn. physiology study sect. USPHS, 1965-66, mem. Nat. Adv. Neurol. Diseases and Blindness Council, 1966—. Mem. Nat. Bd. Med. Examiners, Am. Physiology Soc., A.A.A.S., Internat. Brain Research Orgn., Phi Beta Kappa, Sigma Xi, Alpha Omega Alpha. Author: Physiology and Biophysics, 1961; also numerous articles. Editorial bd. Jour. Neurophysiology, 1961-64. Research in role of brain in sensory and motor control. Home: 1717 Evergreen Pl., Seattle 98122.*

PATTON, John Barratt, Am. geologist; b. Marion, Ind., July 1, 1915; s. Baratt Marsh and Mary Frances (Kuntz) P.; A.B. in Chemistry, Ind. U., 1938, A.M. in Geology, 1940, Ph.D. in Geology, 1954 M. Jean Glenn, 1941; children—Baratt Marsh II, Roger Craig, Frank Jamison Campbell, Ian Alastair. Geologist, Magnolia Petroleum Co., 1940-47; head indsl. minerals sect. Ind. Geol. Survey, 1947-53, prin. geologist, 1951-59, state geologist, 1959—; asst. prof. geology Ind. U., 1948-52, asso. prof. econ. geology, 1952-55, prof. economic geology, 1955—, chmn. dept. geology, 1959—. Commr. Am. Commn. Stratigraphic Nomenclature, 1960-63, 64—, vice chmn., sec., 1962-63. Fellow Geol. Soc. Am., Ind. Acad. Sci.; mem. Am. Assn. Petroleum Geologists, Soc. Econ. Geologists, Soc. Econ. Paleontologists and Mineralogists, Am. Inst. Mining, Metall. and Petroleum Engrs., Ind.-Ky. Geol. Soc., Am. Assn. U. Profs., Assn. Am. State Geologists (pres. 1966-67), A.A.A.S., Interstate Cil Compact Commn., Phi Beta Kappa, Sigma Xi. Contributor of articles to profl. jours. Research on Midwestern Silurian and Devonian stratigraphy; sedimentary petrology of Paleozoic limestones; studies of basement complex; geology of dimension limestones. Home: 809 Sheridan Rd., Bloomington, Ind. 47401.*

PATTON, Stuart, Am. food biochemist; b. Ebenezer, N.Y., Nov. 2, 1920; s. George and Ina (Neher) P.; B.S., Pa. State U., 1943; M.S., Ohio State U., 1947, Ph.D., 1948; m. Colleen Cecelia Lavelle, May 17, 1945; children—John, Richard, Gail, Thomas, Mary, Patricia, Joseph. Chemist, Borden Co., 1943-44; research fellow Ohio State U., 1946-48; faculty Pa. State U., University Park, 1949—, prof. dept. dairy sci., 1959-66, Evan Pugh Research prof. agr., 1966—. Cons., U.S. Dept. Agr., 1958-61. Mem. Am. Chem. Soc. (Borden award chemistry of milk 1957), Am. Dairy Sci. Assn. (past dir.), Inst. Food Technologists, Am. Oil Chemists Soc., Sigma Xi. Author: (with Robert Jenness) Principles of Dairy Chemistry, 1959; also numerous articles. Research on chemistry of lipids and flavors of foods, chemistry and biosynthesis of milk. Home: 209 Circle Dr., State College, Pa. 16801. Office: 105 Borland Lab., University Park, Pa. 16802.*

PATTULLO, June Grace, Am. oceanographer; b. Newark, June 30, 1921; d. David Harry and Loraine Louise (Taylor) Pattullo; B.S., U. Chgo., 1948; M.S., U. Cal. at Los Angeles, 1950, Ph.D., 1957. Staff, Scripps Inst. Oceanography, La Jolla, Cal. 1948-60; faculty Ore. State U., Corvallis, 1960—, prof. oceanography, 1963—. Mem. Internat. Assn. Phys. Oceanography (mean sea level com.), Am. Soc. Limnology and Oceanography (sec.-treas. Pacific div. 1964-67), Am. Geophys. Union, A.A.A.S., Sigma Xi, Phi Beta Kapa. Contbns. on heat and water budgets of oceans; described temperature, salinity, and water mass variations; investigated current shear across oceanic fronts; explained large-scale seasonal variations in sea level. Home: 345 N. 23d St., Corvallis, Ore. 97330.*

PATZ, Arnall, Am. physician; b. Elberton, Ga., June 14, 1920; s. Samuel and Sara Berman; student U. Ga., 1940-41; B.S., Emory U., 1942, M.D., 1945; m. Ellen B. Levy, Mar. 12, 1950; children—William Michael, Susan Claire, David Samuel, Jonathan Alan. Research fellow in ophthalmology Armed Forces Inst. Pathology, Washington, 1950-51; practice medicine specializing in ophthalmology Balt., 1951—; opthalmologic surgeon Balt. Eye and Ear Hosp., 1955—, Sinai Hosp. Balt., 1953—; asst. prof. ophthalmology Johns Hopkins, 1961—. Adv. bd. Md. Med. Eye Bank, Balt., 1963—; dir. Md. Soc. for the Prevention of Blindness, Balt. Recipient Edward L. Holmes Research award in ophthalmology, 1954; Mead-Johnson award Am. Acad. Pediatrics, 1956; Albert Lasker award Am. Pub. Health Assn., 1956. Mem. Med. Research Club Balt., Phi Delta Epsilon. First controlled study proving role of oxygen in retrolental fibroplasia, exptl. prodn. of retrolental fibroplasia in animals; first demonstration of diabetic retinopathy in exptl. animals. Home: 2A Slade Av., Balt. 21208.*

PAUCTON, Alexis-Jean-Pierre, French physicist, mathematician; b. Lea Baroch-Gondouin, France, Feb. 10, 1732; prof. U. Strasbourg, France, later U. Döle, France; mem. French Acad. Scis. Author: Théorie de la vis d'Archimède, 1768; Traité des mesures poids et monnaies des anciens peuples, 1780. Publs. on helicopter with double propellors, weight and measurement systems of ancient civilizations; proposed use of Archimedian screw (propellor) to propel ships, 1768. Died Paris, June 15, 1798.

PAUL, Benjamin David, Am. anthropologist; b. N.Y.C., Jan. 25, 1911; s. Phillip and Esther (Kranz) P.; student U. Wis., 1928-29; A.B., U. Chgo., 1938, Ph.D., 1942; m. Lois Fleishman, Jan. 4, 1936; children—Robert Allen, Janice Carol. Lectr. to asso. prof. anthropology Harvard, 1946-62; prof. anthropology Stanford, 1963—, chmn. dept. anthropology, 1967—. Cons. NIH, 1957—. Social Sci. Research Council fellow, 1940-41; Center for Advanced Study Behavioral Scis. fellow, 1962-63. Mem. Am. Anthrop. Assn., A.A.A.S., Am. Assn. U. Profs., Phi Beta Kappa, Sigma Xi. Editor: Health, Culture and Community: Case Studies of Pub. Reactions to Health Programs, 1955; (with W.C. Gamson and Stephen Kegeles) Trigger for Community Conflict: The Case of Fluoridation, 1961. Research in significance of cultural differences in determining success or failure of health programs, sci. understanding of processes and factors involved in producing social and cultural change at local level. Home: 622 Salvatierra St., Stanford, Cal. 94305.*

PAUL, John Rodman, Am. physician; b. Phila., Apr. 18, 1893; s. Henry Neill and Margaret (Butler) P.; grad. St. George's Sch., 1911; A.B., Princeton, 1915; M.D., Johns Hopkins, 1919; A.M. (hon.), Yale, 1940; D.Sc. (hon.), U. Chgo., 1956; m. Mary Leita Harlan, Sept. 30, 1922. Faculty, Yale Med. Sch., New Haven, 1928—, prof. preventive medicine, 1940-61, prof. emeritus, 1961—; Dir. WHO Reference Serum Bank, 1961-66. Recipient Phillips award A.C.P., 1942; Ricketts medal U. Chgo., 1952; Kober medal Assn. Am. Physicians, 1963. Mem. Nat. Acad. Scis. Author: Epidemiology of Rheumatic Fever, 1930, Clinical Epidemiology, 1958. Researcher in infectious diseases. Home: Old Quarry Rd., Guilford, Conn. 06437. Office: 333 Cedar St., New Haven 06510.*

PAUL, Martin Ambrose, Am. phys. chemist; b. N.Y.C., June 29, 1910; s. Martin and Rosena (Sing) P.; B.A., Coll. City N.Y., 1930; M.A., Columbia, 1931, Ph.D., 1936; m. Genevieve Wells, June 28, 1935; children—Harriet (Mrs. J. Henry Jonquière), Dorothy (Mrs. Duvall A. Jones). Instr., Coll. City N.Y., 1930-42; research supr. Explosivies Research Lab., NDRC, Bruceton, Pa., 1942-45; asst. prof. chemistry Triple Cities Coll., Syracuse U., 1946-50; prof. chemistry Harpur Coll., State U., N.Y., Binghamton, N.Y., 1950-65, chmn. dept., 1950-60; exec. sec. div. chemistry and chem. tech. Nat. Acad. Scis.-NRC, Washington, 1965—; vis. prof. U. Cal. at Los Angeles, 1957-58, Columbia, 1960-61. Recipient Naval Ordnance Devel. award, 1946. John Simon Guggenheim fellow, 1957. Fellow A.A.A.S.; mem. Am. Chem. Soc., Am. Assn. U. Profs., Phi Beta Kappa, Sigma Xi, Phi Lambda Upsilon. Author: Principles of Chemical Thermodynamics, 1951; Physical Chemistry, 1962; co-author General Chemistry, 1962; also articles. Research on measurement of acidity in highly acid media, nature of indicators, catalysis by acids, thermodynamic properties of solutions. Home: 2030 F St. N.W., Washington 20006. Office: 2101 Constitution Av., Washington 20418.*

PAUL, Oglesby, Am. physician; b. Villanova, Pa., May 3, 1916; s. Oglesby and Laura Little (Wilson) P.; A.B. cum laude, Harvard, 1938, M.D. cum laude, 1942; m. Marguerite Black, May 29, 1943; children—Marguerite, Rodman Oglesby. Practiced medicine specializing in cardiology, Chgo., 1948—; clin. asso. prof. medicine U. Ill. Coll. Medicine, Chgo., 1952-54, clin prof., 1954-62; prof. medicine Northwestern U. Med. Sch., Chgo., 1963—; chief div. medicine Passavant Meml. Hosp., Chgo., 1963—; asso. attending physician Presbyn., St. Luke's Hosp., Chgo., 1954-59, attending physician, 1959-62. Cons. U.S. Naval Hosp., Gt. Lakes, Ill.; mem. Joint U.S.-U.K. Com. for Study Coronary and Pulmonary Diseases, 1959—; chmn. med. adv. com. Chgo. Bd. Health, 1960-64; v.p., sec. Mayor's adv. com. regional interests in President's program concerning heart, cancer, stroke, 1966—. Bd. dirs. Internat. Soc. Cardiology Found. Diplomate Am. Bd. Internal Medicine. Fellow A.C.P., Am. Coll. Chest Physicians; mem. Am. (Gold Heart award 1964, pres. 1960-61), Chgo. (pres. 1965-67) heart assns. Research on epidemiology of coronary heart disease. Home: 1012 Westmoor Rd., Winnetka, Ill. 60093. Office: 707 N. Fairbanks Ct., Chgo. 60611.

PAUL, William, physicist; b. Deskford, Scotland, Mar. 31, 1926; s. William and Jean (Watson) P.; M.A., Aberdeen U., 1946, Ph.D., 1951; A.M., Harvard, 1960; m. Barbara Anderson Forbes, Mar. 28, 1952; children—David, Fiona. Came to U.S 1952. Faculty, Harvard, Cambridge, Mass., 1952—, Gordon McKay prof. applied physics, 1963—; Guggenheim fellow, 1959-60; prof. Associe a l'Universite de Paris, 1966-67. Cons. indsl. firms. Fellow Am. Phys. Soc.; mem. A.A.A.S., Sigma Xi. Author: (with D.M. Warschauer) Solids Under Pressure, 1963; also numerous articles. Studies on properties of semiconductors, especially under conditions of high pressure. Home: 2 Eustis St., Lexington, Mass. 02173. Office: 229 Pierce Hall, Harvard, Cambridge, Mass. 02138.*

PAUL OF AEGINA (Paulos Aegineta), physician, surgeon; b. Aegina, Saronic Gulf; flourished Alexandria, circa 640. Author: De re medica (med. ency. in 7 parts, based on Galen and Oribasios), pub. by Günther of Andernach, Paris, 1532; also books on gynecology and toxicology ascribed to him by Muslim tradition, now lost. Last rep. of Greek medicine before domination of Muslims, whose medicine he influenced; attempted to summarize med. knowledge of his time and introduced many new procedures; described techniques of lithotomy, breast amputation, herniotomy, tonsillectomy, tracheotomy, ophthalmic surgery; treatment of fractures, aneurysms, ganglion; use of ligatures; gave 1st known description of lead poisoning.

PAUL OF ALEXANDRIA, Greek astrologer; author astrological intro. with oldest astrological chorography extant, circa 378.

PAULEY, Jean-Jacques, French botanist, physician; b. Anduze (Gard), France, Apr. 27, 1740; practice medicine; mem. French Acad. Scis. Author: Tabula plantarum fungosarum, 1791; Traité des champignons, 1793; Histoire de la petite vérole. Pioneered research in smallpox, epizoic diseases; studied mushrooms. Died Fontainebleau, France, Aug. 4, 1826.

PAULEY, Scott Samuel, Am. forest geneticist; b. Sault Ste. Marie, Mich., Dec. 21, 1910; s. John Livingood and Flossa Viola (Scott) P.; B.S., U. Minn., 1939, M.S., Mich. State Coll., 1942; Ph.D., Harvard, 1947; m. Fritzi Klawans, Dec. 28, 1937; 1 dau. Nan (Mrs. Roger P. Johnston). Forest ranger Wis. Conservation Dept., 1939-40; instr. forestry Mich. State Coll., East Lansing, 1942-43; asst. prof. Harvard, Cambridge, Mass., 1947-52, lectr., forest geneticist Maria Moore Cabot Found. Bot. Research, 1952-55; asso. prof. U. Minn., St. Paul, 1955-57, prof., 1957—. Vice pres., dir. Forest Genetics Research Found., San Francisco, 1953—; mem. Lake States Forest-tree Improvement Com., St. Paul, 1955—, tech. com. Quetice Superior Wilderness Research Center, Ely, Minn., 1956—; del. U.S.

Forestry Delegation to USSR, 1959. Mem. Soc. Am. Foresters, A.A.A.S. Am. Inst. Biol. Sci., Genetics Soc., Am. Soc. Study Evolution, Sigma Xi, Alpha Zeta, Gamma Sigma Delta, Xi Sigma Pi. Contbg. author: The Physiology of Forest Trees, 1958; Balsam Fir: A Monographic Rev. (E.V. Bakuzis, H.L. Hansen), 1965. Asso. Editor: Jour. Forestry, 1960-64. Home: 136 Iris St., Mathomedi, Minn. 55115.*

PAULI, Wolfgang, physicist, b. Vienna, Austria, Apr. 25, 1900; s. Wolfgang Joseph and Bertha (Schutz) P.; Ph.D., U. Munich, 1921; m. Franciska Bertram, Apr. 4, 1934. Asst. U. Göttingen (Germany), 1921-22, U. of Copenhagen (Denmark), 1922-23; docent U. Hamburg, 1923-28; prof. theoretical physics Eidgenössiche Technische Hochschule, Zürich, Switzerland, 1928——; vis. prof. theoretical physics Inst. for Advanced Study, Princeton, N.J., 1935-36, 40-45, 49-50, 54; lectr. U. Mich., summers 1931, 41, Purdue U., May-June, 1942. Recipient Lorentz medal, 1930; Nobel prize in physics, 1945; Franklin Medal, 1952; Max Planck medal, 1958. Fellow Royal Soc.; mem. Swiss Physics Soc., Am. Phys. Soc. A.A.A.S. Author: Theory of Relativity, 1920; Quantum Theory, 1926; Principles of Wave Mechanics, 1933. Contbr. tech. encys. and other reference works in many countries. Announced Pauli exclusion principle (1925), which states that only 2 electrons can occupy the same energy level (have same quantum numbers) at the same time in an atom; attempted to unite wave mechanics with theory of relativity while also considering magnetic moment of electron, 1927; 1st to apply matrices to spin of electron in hydrogen atom; showed relationship of quantum spin and distribution statistics of elementary particles; to explain energy anomalies in emission of beta particles from atoms, postulated existence of new subatomic particle, 1931 (it was named neutrino by Fermi in 1932 and detected in 1956). Died Dec. 15, 1958.

PAULING, Linus Carl, Am. chemist, physicist; b. Portland, Ore., Feb. 28, 1901; s. Herman William and Lucy Isabelle (Darling) P.; B.S. in Chem. Engring., Ore. State Agrl. Coll., 1922; Ph.D., Cal. Inst. Tech., 1925; 27 hon. degrees; m. Ava Helen Miller, June 17, 1923; children—Linus Carl, Peter Jeffress, Linda Helen (Mrs. Barclay Kamb), Edward Crellin. Tchg. fellow Cal. Inst. Tech., 1922-25, research fellow, 1925-27, asst. prof., 1927-29, asso. prof., 1929-31, prof. chem., 1931-64, chmn. div. chem. and chemical engring. and dir. of Gates and Crellin Labs. of Chemistry, 1936-58; mem. executive com., bd. trustees, 1945-48; research prof. Center for Study Dem. Instns., 1963-67; prof. chemistry U. Cal., San Diego, 1967——; George Eastman prof., Oxford Univ., 1948. Fellow Balliol College, 1948; lectr. chemistry several univs. Frequent cons. govtl. agencies and research groups. Fellow Nat. Research Council, 1925-26, John S. Guggenheim Meml. Found., 1926-27; mem. Nat. Acad. Scis., hon., corr., fgn. mem. numerous assns. and orgns. Numerous awards in field of chemistry, including Presdl. medal of merit, 1948, Nobel Prize in Chemistry, 1954, Nobel Peace Prize, 1962, Fermat medal, Paul Sabatier medal, Pasteur medal, medal with laurel wreath of Internat. Grotius Found., 1957. Author: The Nature of the Chemical Bond, and the Structure of Molecules and Crystals, 1939; General Chemistry, 1947; No More War!, 1958. Determined crystal structure of molecules by means of x-ray diffraction analysis; made important discoveries on nature of chem. bond; discovered atomic structure of many proteins, including hemoglobin; developed concept of molecular basis of disease. Home: Salmon Creek, Big Sur, Cal. 93920. Office: U. Cal. San Diego, La Jolla, Cal. 92038.

PAULINI, Ernest, chem. engr.; b. Szöllösardo, Hungary; Jan. 18, 1920; s. Armin and Helen (Korbai) P.; Chem. Engr., Royal U. Tech. Scis., Budapest, Hungary, 1942; San. Engr., Engring. Sch. Minas Geraes (Brazil), 1955, D.Sc., 1958; m. Livia Balog, Dec. 25, 1942; children—Erno Ivan, Helene Maria. Asst. prof. Royal U., 1942-45; chief chemist USAF Gen. Purpose Lab., Germany, 1946-47, Alianca Comercial de Anilinas, Rio de Janeiro, Brazil, 1948-51; head chem. lab. Instituto Nacional de Endemias Rurais, Belo Horizontee, Brazil, 1951——; prof. indsl. chemistry Sch. Engring., Belo Horizonte, 1964——. Mem. panel experts on insecticides WHO, 1953——. Mem. Academia de Ciencas de Minas Gerais, Soc. Brasil. Higiene, Soc. Brasil. Biologia, Soc. Brasil. Quimica, Soc. Brasil Progresso da Ciences. Author: Chemistry and Biological Activity of Organic Insecticides, 1958; Contribution to Fabrication of BHC, 1965; also numerous articles. Devel. new techniques for prevention of parasitic diseases including medicated salt against malaria, physico-chem. factors affecting pesticides; screening and field testing insecticides and molluscides. Home: C.P. 1166 Belo Horizonte M.G., Brazil.*

PAULUS, Max Pierre, French physicist; b. Drancy, France, Nov. 27, 1925; s. Eudore Jules and Emilienne (Prevot) P.; Engr. in Metallurgy, Conservatoire Nat. des Arts et Metiers, 1955, Dr. Engr., 1962, Maitre de Recherche, 1963; m. Alice Marie-Thérèse Wagner, August 9, 1947. Head dept. study and synthesis of microstructures Laboratoire de Magnetisme et de Physique du Solide Centre National de la Recherche Scientifique, Hauts de Seine, France, 1953——, mem. Commn. for Powder Metallurgy, 1964——. Mem. Société Francaise de Physique, Société Francaise de Mineralogie et de Cristallographie, Société Francaise de Metallurgie. Research, numerous publs. on connections between phys. properties magnetic and electric and microstructures, transformation of microstructures, reactive and simple sintering of spinal ferrites, thermodynamic equilibrium in solids; devel. techniques for study of microstructures, ceramic single crystal growth in the solid state, preparation of ferrite thin films by reactive sputtering, mechanism of formation of spinal ferrites. Home: 3 Residence des Quinconces - GIF-sur-Yvette, Essonne 91, France. Office: 1, Place Aristide Briand-Bellevue, Hauts de Seine 92, France.*

PAULY, John Edward, Am. anatomist; b. Elgin, Ill., Sept. 17, 1927; s. Edward John and Gladys (Myhre) P.; B.C., Northwestern U., 1950; M.S., Stritch Sch. Medicine, Loyola U., 1952, Ph.D., 1955; m. Margaret Mary Oberle, Sept. 3, 1949; children—Stephen John, Susan Elizabeth, Kathleen Ann, Mark Edward. Faculty, Chgo. Med. Sch., 1952-63, asst. prof. anatomy, 1959-63, asst. to pres. 1960-62; asso. prof. anatomy Tulane U. Sch. Medicine, New Orleans, 1963-67; prof., chmn. dept. anatomy U. Ark. Sch. Medicine, Med. Center, Little Rock, 1967——. Tech. adviser Ency. Brit. Films, Inc., 1956. Recipient certificates of merit A.M.A., 1953, 59; Bronze award Ill. State Med.. Soc., 1959; Lederle Med. Faculty award, 1966. Mem. Am. Assn. Anatomists, So. Soc. Anatomists, Am. Soc. Zoologists, Sigma Xi. Author: (with Hans Elias) Human Microanatomy, 1960, 62, 66; contbr. to A Handbook of Environmental Biology, 1966; also articles, sci. films. Research in electromyography of human musculature during routine movements and exercise, histoarchitecture of human adrenal cortex in health and disease, comparative histology of adrenal cortex, growth of muscle, temporal variations in susceptibility to drugs, daily rhythms in mammalian physiol. variables under normal and exptl. conditions. Home: 2800 Hudson Pl., New Orleans 70114.*

PAUNESCU, Eugeniu, Rumanian biochemist; b. Bucharest, Rumania, Dec. 24, 1925; s. Alexandru and Ortansa-Elena (Stefanescu) P.; M.D., U. Bucharest, 1952, prin. in microbiology, 1959. Chief lab. San. and Anti-epidemic Center, Bucharest, 1952-53; gen. san. insp. Ministry Food Industry, 1953-56; chief lab. Faculty Medicine, Dept. Tb, Bucharest, 1954-57; chief researcher Inst. Physiology, Bucharest, 1956-63, chief dept. biochemistry and bacteriology, 1963——. Tech. cons. pub. house Rumanian Acad., 1957——. Recipient medal for distinction in san. labor, 1956; Dr. V. Babes prize Rumanian Acad., 1960. Mem. Union Med. Socs. Rumania (sec. 1959-64), Biochem. Soc. (sec.). Author: (with Lydia Mesrobeanu) Bacterial Physiology, 1960; (with P. Georgescu) Biochemical Methods for Diagnosis and Research; also numerous articles. Research on enzymatic modifications and chem. compositions of Staphylococcus under action of penicilline, Mycobacterium under action of tuberculostatics, reduction of virulence under some chemotherapics actions, metabolic pathways of INH, PAS, ethionamide, role of microphages and histoid macrophages in Tb, presence of intracellular antibodies. Home: 42, Bd. Marasesti, Bucharest 53. Office: Inst. Tb Research, 90 Sos. Viilor, Bucharest 28, Rumania.*

PAUSANIAS, Greek geographer; b. Lydia; flourished 117-180, under Hadrian, Antonius Pius, Marcus Aurelius; traveled in Antioch, Joppa, Jerusalem, Greece, Egypt, Macedonia. Author: Description of Greece (on Greek art, folklore, superstitions), 10 vols., circa 160-74. Authority on Greek topography, legends, monuments.

PAUSESCU, Exacustodian, Rumanian physician; b. Buzau, Rumania, Nov. 20, 1928; s. Gheorghe and Elena (Negrila) P.; Physician with exceptional Diploma, Inst. Medicine and Pharmacy, Bucharest, Rumania, 1954, D.Med. Scis., 1966; m. Mioara Botez, Oct. 5, 1957. Asst. fellow, asst. chair of physiopathology Inst. Medicine and Pharmacy, 1955-60; head dept. exptl. pathology and artificial organs Clin. Hosp. Fundeni, Bucharest, 1960——. Mem. Union Socs. Med. Scis. (Bucharest), German Assn. Clin. Medicine, European Soc. Exptl. Surgery. Author: (with V. Marinescu, M. Ionescu) Circulatia extracorporeala si hipotermia profunda, 1962, Catecolaminele, 1965; also numerous articles. Research on method of differential hypothermia for open heart surgery, mechanism of cardiac insufficiency in arteriovenous fistula, original method for differential diagnosis in hypertensive syndrome, method for heparin neutralization, basic studies on mechanisms of heparin activity, biochem. aspects of homograft rejection. Home: 34A Bul. Magheru. Office: Clinical Hospital Fundeni, 37 Soseaua Fundeni, Bucharest, Rumania.*

PAUTHENET, René, French physicist; b. Saint-Aubin Du-Jura, France, Aug. 22, 1925; s. Georges and Marthe (Prost) P.; Elec. Engr., Inst. Poly., Grenoble, France, 1948; Ingénieur-Dr., Faculté des Sciences, Grenoble, 1951, Dr. ès Sciences, 1957; m. Claudine Moesch, July 25, 1957; children—Pierre, Yves, Jean. Research scholar electrostatics and physics of metals lab. Centre National de la Recherche Scientifique, 1948-51, engr., 1951-56; prof. Faculté des Sciences, U. Grenoble, 1957——. Sci. adviser Centre d'Études Nucléaires de Grenoble, 1960——. Recipient Médaille Blondel, 1961; silver medal Centre national de la recherche scientifique, 1961. Mem. Internat. Union Pure and Applied Physics (sci. adviser dept. research and investigation 1963——, sec. 1966——). Research and publs. (following Néel's theory of ferrimagnetism) on magnetic properties of spinulose ferrites; discovered magnetic properties of garnet ferrites; magnetostatic study of antiferrimagnetic or ferrimagnetic materials, alloying of rare earths with transition elements; prodn. intense magnetic fields. Home: Seyssins, 38, France. Office: Chemin des Martys-Grenoble, 38, France.*

PAUTHENIER, Marcel Frédéric, French physicist; b. Messigny, France, Mar. 14, 1887; s. Auguste and Marie (Grapin) P.; Agrégé des Sciences Physiques, Ecole Normale Supérieure, Paris, 1913; Doctorate ès-Sciences, Sorbonne, 1920; m. Germaine Bezard, Apr. 4, 1920; children—Geneviève (Mrs. Sorre), Anne-Marie (Mrs. Mariée), Michel. Lectr., U. Lille (France), 1921; dir. Physics Inst., Faculty Lille, 1927-30; faculty Sorbonne, 1930-58, titular prof. gen. electrochemistry, 1942-58; titular prof. l'Ecole Supérieure d'Electricité, 1942-58. Mem. French Soc. Physics (pres. 1953), French Soc. Electricians (pres. 1944), French Soc. Meteorology (pres. 1956). Numerous publs. on proof of theory of electrostriction; research in high tension generators, elec. study of fog, contra-emission of electrofilters; work in control of carbon fumes in France by laws and control of charge of spherical particles in ionized elec. fields. Home: 6, Place du Pantheon, Paris, France. Office: 44, Chemin de Ronde, Le Vesinet, France.*

PAV, Jaroslav, Czechoslovakian physician; b. Manetin, Czechoslovakia, Jan. 26, 1923; s. Jaroslav and Ludmila (Prochazkov) P.; m. Eva Spurna, July 2, 1949; children—Thomas, Henriette. Staff internist Hosp. Ceska Lipa,, Bohemia, 1949-53; research fellow Inst. Human Nutrition, Prague, Czechoslovakia, 1953-59; asst. Charles U., Faculty of Medicine, III Med. Clinic, Prague, 1960——. Recipient award for study insulin antibodies Czechoslovakian Med. Assn., 1963. Author: Diabetics Guide, 1959; also articles, chpt. in book. Editor: (with J. Wenkeova) Metabolism of Fat and Lipoprotein Lipase, 1962. Research on pathophysiol. aspects of diabetes mellitus, interrelationship of carbohydrate and lipid metabolism, action of lipoprotein lipase. Home: 5 Srobarova, Prague 3. Office: 1 Unemocnice, Prague 2, Czechoslovakia.*

PAVAN, Crodowaldo, Brazilian geneticist; b. Sao Paulo, Brazil, Dec. 1, 1919; s. Henrique and Rozaria (Fiorentino) P.; Ph.D., U. Sao Paulo, 1944, Livre Docente, 1951; m. Maria de Lourdes Vaz de Oliveira, Aug. 10, 1946; children—Octavio Henrique, Ricardo, Luciano. Faculty, U. Sao Paulo, 1942——, prof., head dept. gen. biology, 1952——; exchange visitor biology div. Oak Ridge Nat. Lab., 1965-66, cons., 1966——. Rockefeller Columbia U., U. Tex., 1945-46. Recipient 1st Brazilian award on genetics. Mem. Brazilian Acad. Sci. Author: (with A.B. da Cunha), Genetica, 1964; also articles. Research on genetic structure of natural populations of tropical Drosophila, morphology and physiology of polytene chromosomes of Drosophila and Sciaridae. Home: 320 Alvares Florence, Sao Paulo, Brazil.*

PAVEL, I. S., physician; b. Bucharest, Rumania, 1897; s. Sava and Elena (Pandele) P.; ed. Bucharest Faculty Medicine; m. Sanda Kiriacesco, July 2, 1933. Asst., Bucharest U. Med. Clinic; lectr., pof. Metabolic Diseases and Dietetics Clinic, Faculty Advanced Med. Study; dir. Bucharest Anti-Diabetic Center. Decorated Ordinul Muncii. Author: Les Ictères, 1948; Vésicule biliaire, 1958; Dietetica fiziopatologica et clinica, 1963; Diabetul, 1965; also numerous articles. Described functional trouble in bile ducts, 1924, jaundice caused by spasm of sphincter of Oddi, 1930, jaundice caused by bile pigment, 1956, relation between Disse spaces and bile channels; research in diabetic etiology and hereditary transmission of diabetes; codified diabetes prevention. Home: 6 Interaea Caragiale, Bucharest R.1 M. Office: 5 Ion Movila, Bucharest, Rumania.*

PAVENSTEDT, Eleanor, Am. psychiatrist; b. N.Y.C., Mar. 16, 1903; d. Edmund William and Caroline (Probst) P.; student U. Zurich, U. Geneva, 1921-28; M.D., U. Geneva, 1929. Psychoanalyst, Boston Psychoanalytic Soc. and Inst., 1940; J.J. Putnam Childrens Center, Boston, 1941-54; faculty Boston U. Sch. Medicine, 1949-65, dir. child psychiatry, 1949-65, clin. prof. psychiatry, 1965——; lectr. Tufts U. Med. Sch., 1965——; staff child psychiatrist Tufts-Columbia Point Health Center, 1966——. Mem. A.M.A., Am. Psychiat. Assn., Am. Psychoanalytic Assn., Am. Acad. Child Psychiatry. Editor: (with Lucie Jessner) Psychopathology of Childhood; editor, co-author The Drifters, 1967. Research, publs. on childhood psychosis, longitudinal study in child devel., children of seriously deprived and disorganized families, observations of 13 to 16 months old Japanese children. Home: 18 Hemlock Rd., Cambridge, Mass. 03138. Office: 300 Mt. Vernon St., Dorchester, Mass. 02150.

PAVLENKO, Stefan Makarovich, Russian pathophysiologist; b. 1900; grad. Med. Faculty, 1st Moscow U., 1923; D.Med. Sci., 1936. Asst., lectr. dept. path. physiology 1st Moscow U., 1923-33; head pathophysiology dept. Inst. Exptl. Endocrinology, 1925-34; prof., 1934——; head chair path. physiology Higher Med. Ednl. Establishment, Moscow Oblast Clin.

Inst., 1933-41, Soviet Army Mil. Vet. Acad., 1941-47; head chair path. physiology 1st Moscow Med. Inst., 1947—, dep. dir. for sci. work, 1956-59. Collegium mem., head main med. tng. establishments bd. USSR Ministry Health, 1947-50. Mem. All-Union (co-founder, bd. mem.), Moscow (chmn.) socs. pathophysiologists. Author numerous works including: Physiological Principles of Biotechnics, 1940; Natural Scientific Principles of Medical Genetics, 1964. Editor: Handbook of Pathological Physiology, 1961; mem. editorial bd. Path. Physiology and Exptl. Therapy. Address: 1st Moscow Med. Inst., B. Pirogovskaya ulitsa 2-6, Moscow, USSR.

PAVLOV, Igor Mikhailovich, Russian metallurgist; b. June 23, 1900; s. M.A. Pavlov; grad. Petrograd Poly. Inst., 1923. Worker in metall. plant, 1923-28; lectr., 1928-34; prof. Leningrad. Poly. Inst., 1934-42, Moscow Inst. Steel, 1943—; asso. Inst. Metallurgy USSR, 1953. Decorated Order of Lenin, Order Red Banner of Labor. Corr. mem. Moscow Acad. Scis., 1946. Author: Metall-rolling Theory, 1932; Charge Composition in Non-Ferrous Founding, 1932; The Rolling of Non-Ferrous Metals and Alloys, 1932; (with Y.A. Gallai) Forward Flow During Rolling, 1936; Grundlagen der Metall-Formung durch Druck, 1954; co-author: Pressure Treatment of Metals, 1955; The Relating of the Cold-Rolling Friction Factor to the Roughness of the Rollers, 1960; A Study of the Full Pressure Exerted on the Rollers During Cold Pipe Rolling, 1960; Axial Stresses During Cold Pipe Rolling, 1960; The Friction Factor in Cold Rolling and its Dependence on Shrinkage, 1962; Study of Certain Conditions of Hot Rolling of Molybdenum in a Vacuum in Argon Atmosphere and in Air, 1963. Contbn. to theory of metal processing, particularly metal rolling; effects of pressure and temperature. Home: ul. Gor'Kogo 8, Moscow. Office: A.A. Baykov Inst. Metallurgy, USSR Acad. Scis., Leninskii Prospekt 49, Moscow, USSR.

PAVLOV, Ivan Petrovitch, Russian physiologist; b. Ryazan, Russia, Sept. 14, 1849; student U. St. Petersburg; Ph.D., St. Petersburg Med. Acad., 1893; student in Germany, 1884-86. Conducted research in Breslau and Leipzig; returned to St. Petersburg Mil. Med. Acad.; in recognition of his research on nervous mechanisms controlling the secretion of digestive glands he was awarded professorial position at St. Petersburg Mil. Med. Acad., 1890; remained in Russia after revolution though anti-communist; Soviet govt. built him a lab., 1935. Recipient Nobel prize physiology and medicine, 1904; Copley medal Royal Soc., 1915. Author: Conditioned Reflexes, 1926. Major contbns. include discovery of the nerve fibres affecting action of the heart and the secretory nerves of the pancreas; work on physiology of circulation and digestion, especially the secretion of digestive glands; using live dogs made an exptl. study of conditioned or acquired reflexes associated with definite areas of brain cortex; established importance of autonomic nervous system; was a complete mechanist believing that all action depends on conditioned reflexes. Died Leningrad, Feb. 27, 1936.

PAVLOV, Mikhail Aleksandrovich, Russian metallurgist; b. Lenkoraw, Azerbaijan, 1860; prof. metallurgy; mem. Cem. Scis., Tech. Scis. brs. Acad. Scis. U.S.S.R.; recipient Stalin prize (2), 1943, 46; Order of Lenin, 1943; Order of Red Banner, 1945. Dean of Russian metallurgists; work significant in devel. of iron and steel industries; research on theory of blast-furnace process. Address: Acad. Scis. U.S.S.R., Moscow, U.S.S.R.

PAVLOV, Nikolay Nikiforovich, Russian astronomer; b. 1902. Asso., Main Astron. Obs., USSR Acad. Scis., Pulkovo, 1929-36, head dept. time service, 1936—; prof. Leningrad U., 1944—. Recipient Stalin prize, 1947. Research and publs. on photoelec. method to record stellar passages on transit instrument (excludes errors introduced by observer); designer new models of transit instruments. Address: Leningrad University, Universitetskaya n. 7-9, Leningrad, USSR.

PAVLOV, Todor Dimitrov, Bulgarian philosopher; b. Stip, Bulgaria, Feb. 14, 1890; s. Dimiter and Ekaterina (Nasteva) P.; grad. U. Sofia (Bulgaria), 1914; Ph.D., Communist Acad., Moscow, USSR, 1936; Dr.h.c., Sofia U., 1944, Karlovy U., Prague, Czechoslovakia, 1948, Humboldt U., Berlin, Germany, 1960; children—Vera, Assen. Tchr., 1910-22; editor, journalist, 1922-32, 36-40; prof. dialetic materialism Red U., Moscow, 1932-36; dean Faculty Philosophy, Inst. Philosophy, History and Lit., Moscow; cons. prof. Inst. Philosophy, Communist Acad., Moscow, 1932-36; regent of Bulgaria, 1944-46; hon. prof. Faculty History and Philosophy, Sofia U., 1945-48, prof., 1948—; mem. Presidum Nat. Assembly, 1947, 50, 62—; dir. Inst. Philosophy, Bulgarian Acad. Sci., 1949-52, 60—. Recipient Dimitrov prize laureate, 1950; Hero of Socialist Labour, 1960, 9th of Sept., 1946; Nat. Liberty, 1947; Golden medal Sci. and Arts, 1950; G. Dimitrov, 1956, 59; Kiril and Metodi, 1957; People's Sci. and Cultural Worker, 1949; People's medal of Labor—Golden, 1949; White Lion (Czechoslovakia), 1947; State Flag of Korean People's Democratic Republic, 1958; V.I. Lenin (USSR), 1964. Mem. Polish, Czechoslovakian, Serbian, Rumanian (hon.), acad. scis., USSR Acad. Arts, Acad. Sci. People's Republic Korea. Author: Selected Works, vols. I-IX, 1957-66; Theory of Reflection, 1936; Essential Problems of Aesthetics, 1949; Information Reflection, Creative Work, 1965; also numerous articles. Editor, in chief Filosofska Misal. Studied Lenin's theory of reflection as gen. property of matter which appears in forms of original quality due to its different stages of devel., interrelation between philosophy and other scis., reorgn. of history, archaeology, biology and physics on basis of dialectical materialism. Home: 2 Dobrudza. Office: 1 7 Noemvri, Sofia, Bulgaria.*

PAVLOVSKII, Evgenii, Nikanorovich, Russian parasitologist, zoologist; b. Voronezh Oblast, Mar. 5, 1884; grad. St. Petersburg Mil. Med. Acad., 1909; M.D., 1913; D.Zoology, 1917. Mus. curator Brain Inst., 1918-30; prof. zoology Stebutov Agrl. Inst., 1920-22; prof., 1921—; sr. zoologist, dep. head Zool. Inst. and Mus., USSR Acad. Scis., 1930-42, dir. dept. parasitology Tadzhikistan br., 1937-51, dir. Inst. Zoology, 1942-62, also chmn. Malaria Commn. of Zool. Inst. and Chmn. Ichthyological Commn.; head chair biology and parasitology Mil. Med. Acad., 1921-56; dir. dept. parasitology All-Union Inst. Exptl. Medicine, Leningrad, 1933-49; dir. dept. parasitology and med. zoology Inst. Epidemiology and Microbiology, USSR Acad. Med. Scis., 1946—. Mem. learned council USSR Ministry Health. Decorated Order of Lenin (5); recipient Stalin prize, 1941, 50. Mem. USSR (Mechnikov Gold medal 1949), Tadzhikistan (hon.), Polish acads. scis., USSR Acad. Med. Scis., All-Union (founder com. on med. geography, Presidium 1955), USSR (pres. 1952—, Grand Gold medal 1954) geog. socs., USSR Parasitological and Entomol. Soc. (chmn. 1952—), also other fgn. insts. and socs. Author: Poisonous Animals and Their Importance for Man, 1923; Poisonous Animals of the USSR, 1931; A Course in Human Parasitology, 2d edit., 1934; A Short Textbook on the Biology of Human Parasites, 1941; Pappataci Fever and Its Vectors, 1947; A Manual on Human Parasitology and Theory on the Vectors of Transmissible Diseases, 6th edit., 1951; Bloodsucking Diptera, Their Significance and Measures for Combatting Them, 1951; General Problems of Parasitology and Zoology, 1961; co-author: Human Diseases with Natural Foci of Infection, 1960. Co-editor: Large Medical Ency., 2d edit. Research and publs. on poisonous animals, cycles of parasite development, fauna of bloodsucking insects, methods of combating them, enteroprotozoan and worm invasions, parasite cenosis, regional and landscape parasitology. Died May 27, 1965.

PAVLVUCHENKO, Mikhail Mikhaylovich, Russian chemist; b. Glukhovichi, Russia, Mar. 17, 1909; s. Mikhail I. and Ulyana (Ivananko) P.; sci. grad. Moscow State U., 1933; m. Klavdia Vasilyevna Petrova, Nov. 2, 1938; children—Galya, Nina. Postgrad. research Acad. Scis. of B.S.S.R., 1934-37; prof. chemistry Byelorussian U., 1938-58; dir. Inst. Gen. and Inorganic Chemistry, Acad. Scis., Minsk, 1959—. Decorated Badge of Honor (2); hon. diploma Supreme Soviet of B.S.S.R. (3). Mem. Mendeleev All-Union Chem. Soc. Author: Heterogeneous Chemical Reactions, 1961; Potassium Salt in B.S.S.R., 1966; also numerous articles. Research on theory of heterogeneous chem. reactions with participation of a solid body, new methods of spectral determination of microelements in powdery mixtures. Home: 79 Lenin St., Minsk, B.S.S.R., USSR.

PAVONI, Nazario, Swiss geologist; b. Zurich, Switzerland, July 18, 1929; s. Franz and Elsa (Lezzi) P.; diploma in geology E.T.H. Zurich, 1952, Doctors degree in geol., 1956; m. Ruth Hörler, May 14, 1957; children—Leonhard, Anna, Christine, Verena. Tchr. grammar sch., Zürich, 1952-56; petroleum geologist, Turkey, 1956-58; sci. asso. Inst. Geophysics, Eidgenössische Technische Hochschule, 1959—, lectr. applied geophysics, 1964—; collaborator Seismol. Survey Switzerland, 1959—, Swiss Geol. Commn., 1958—. Mem. Swiss Assn. Petroleum Geologists and Engrs. (editor Bull. 1961-66), Swiss Geol. Society, Geol. Soc. Zurich (mem. directory 1966—, Geolog. Vereinigung, Am. Geophys. Union, A.A.A.S., Schweizerische Naturforsch. Gesellschaft. Research, publs. on geology upper fresh water molasse and geomorphology Swiss Plateau, molasse sedimentation, roll marks in flysch sediments, tectonics Jura Mountains, folding and wrench-fault tectonics, interpretation Cenozoic tectonics in Turkey and Nr. E., recent movements earth's crust, stress state earth's crust, geotectonic hypothesis, seismicity Switzerland, rock magnetism. Home: 15 Hofernweg, 8134 Adliswil ZH, Switzerland. Office: Inst. Geophysics, Eidgenössische Technische Hochschule, Hönggerberg, 8049 Zurich, Switzerland.*

PAVONI, Pietro, Italian physician; b. Rome, Italy, Apr. 29, 1928; s. Adalberto and Elena (Rinaldini) P.; M.D., U. Rome, 1954; m. Teresa Stallone, June 21, 1958; children—Alberto, Fabrizio. Vol. asst. in Inst. Med. Pathology, U. Rome, 1957-62, extraordinary asst., 1962—, tchr. nuclear medicine, 1958-63, tchr. applications of radioisotopes C.N.E.N., 1957-62, Sch. Nuclear Medicine, 1963, prof. nuclear medicine, 1964—. Mem. Italian Soc. Biology and Nuclear Medicine, Am. Nuclear Soc., A.A.A.S., Italian Soc. Endocrinology. Publs. on scanning techniques, analytical system for biol. studies with tracers, med. applications of radioisotopes. Home: 30 Via Cardinal Giovanni Bessarione, Rome, Italy.*

PAVY, Frederick William, English physician; b. Wiltshire, Eng., May 29, 1829; ed. Merchant Taylor's Sch.; Gulstonian lectr., 1862-63; Croonian lectr., 1878, 94; Harveian orator, 1886; staff Royal Coll. Physicians, London. Fellow Royal Soc., 1863; mem. Royal Med. and Chirurg. Soc. (pres.), Path. Soc. Author: On the Nature and Treatment of Diabetes, 1869; A Treatise on the Function of Digestion, its Disorders and their Treatment, 1869; A Treatise on Food and Dietetics, 1875; Physiology of the Carbohydrates, 1894; Carbohydrate Metabolism and Diabetes, 1906. Pioneered modern chem. pathology; developed Pavy's test for sugar, modern theory of diabetes; used sugar tests and albumin test in solid form; studied carbohydrate metabolism. Died Sept. 19, 1911.

PAVY, Octave, Am. physician, explorer; b. New Orleans, June 22, 1844; attended U. Paris, 1866; m. Lilla Stone, 1878. Organized (with Lt. Beauregard), equipped and financed ind. Zouave corps; surgeon and naturalist on H.W. Hawgate's expdn. to Greenland, 1800-81, sailed in ship Gulnare, explored country, collected specimens for Smithsonian Instn.; commd. surgeon of Lady Franklin Bay expdn. under Lt. A.W. Greely, 1881, naturalist and surgeon in Arctic region, 1881-84; discovered Pavy Valley and Pavy River between Cape Baird and Cape Ritter Bay. Died just before rescue ship arrived, Cape Sabine, Arctic, June 6, 1884.

PAWAN, Joseph Lennox, West Indian physician, bacteriologist; b. Trinidad, Brit. W.I., Sept. 6, 1887; ed. St. Mary's Coll., Trinidad; M.B., Ch.B., Edinburgh (Scotland) U.; D.P.H., Univ. Coll., London; postgrad., Germany; govt. bacteriologist, Trinidad; cons. on rabies WHO. Research and publs. on differentiation of B. coli in water analysis, pulmonary b, colon group of organisms, acute rabic-myelitis in Trinidad, transmission of paralytic rabies by vampire bat, other aspects of pub. health in Trinidad. Died Nov. 3, 1957.

PAWLIKOWSKI, Stefan, Polish chemist; b. Sanok, Poland, Dec. 9, 1899; s. Ferdynand and Stefania (Chitry) P.; Ed.E., Polytechnic Inst., Lwów, 1925, D.Sc., 1928; Docent, Poly. Inst., Gliwice, Poland, 1948; m. Anna Sheybal, June 6, 1925 (d. 1963); 1 dau., Janina (Mrs. Antoni Czubak); m. 2d, Danuta Pachulicz, Feb. 24, 1968. Asst., Poly. Inst., Lwów, 1925-27; engr., prodn. supr. Moscice State Factory Nitrogen Compounds, 1928-39; chemist pvt. co. for mfr. adhesives Ramnoza, Kraków, Poland, 1940-44; chief mgr. Oswiecim Chem. Works, 1945-47, head study and design office, 1947-49; faculty Silesian Poly. Inst., Gliwice, Poland, 1945—, prof., 1962—, chmn. chair tech. heavy inorganic industry, 1949—, head chemistry dept., 1957-58. Cons., Gliwice Design Office for Indsl. Constrn., 1950-61; mem. sci. bd. Central Lab. Gas Industry, 1965—; chmn. group chem. raw materials Com. for Sci. and Engring., 1965—. Recipient Golden Cross of Merit, 1931, 56; Gold medal for 15th Anniversary Silesian Poly. Inst., 1960; award 2d grade Minister Higher Edn., 1963. Mem. Trade Union Polish Tchrs., Polish Chem. Soc. Author: Air and Water Govern Chemistry, 1957; also numerous articles, chpts. in books. Research on nitrogen fertilizers and protection of materials for construction of chem. equipment from corrosion; developed method of producing neutral ammonium carbonate, method of using calcium cyanamide as corrosion inhibitor in asphalt masses for insulation of underground pipe-lines. Home: 11 ul. Szujskiego, Kraków, Poland. Office: 19 M. Strzody, Gliwice, Poland.*

PAXTON, Harold William, metall. engr.; b. Yorkshire, Eng., Feb. 6, 1927; s. John Wilfrid and Hilda Annie (Vasey) P.; B.Sc. with 1st class honours, U. Man., 1947, M.Sc., 1948; Ph.D., U. Birmingham (Eng.), 1952; m. Ann Dorothy Davies, May 13, 1953; children—Jane Elizabeth, Sally Patricia, Anthony Charles, Nigel John. Came to U.S., 1953, naturalized, 1961. Univ. fellow U. Birmingham, 1950-53; mem. faculty Carnegie Inst. Tech., 1953—, prof. metall. engring., 1962-65, Firth Sterling prof. metall. research, 1958-65, now prof., head dept. metallurgy and materials sci.; cons. to industry, 1953—; adj. sr. fellow Mellon Inst., Pitts., 1965-68. Chmn. Internat. Conf. High Velocity Deformation, 1960, Internat. Conf. Fracture, 1962. Mem. adv. council U.S. Army Materials Research Agy.; mem. adv. com. Pa. Tech. Assistance program. NSF sr. postdoctoral fellow Imperial Coll., London, Eng., 1962-63. Mem. A.A.A.S., Am. Inst. Mining, Metall. and Petroleum Engrs., Am. Soc. Metals (Bradley Stoughton Young

Tchrs. award 1960), Inst. Metals, Iron and Steel Inst. Author: Alloying Elements in Steel (with E. C. Bain), 3d edit., 1966; also numerous articles. Home: 115 Eton Dr., Pitts. 15215.*

PAYEN, Anselme, French chemist; b. Paris, Jan. 6, 1795; s. Jean Baptiste Pierre and Marie François (Jeanson) de C.; student École polytechnique; studied under Vanquelin, Fourcroy, Tremery; m. Zelie Charlotte Mélanie Thomas, 1821; 5 children. Managed several chem. factories founded by father; prof. indsl. and agrl. chemistry École centrale des arts et manufactures, 1835-71; prof. applied chemistry Conservatoire des arts et métiers, 1839-71. Mem. various govtl. commns. Recipient Gold medal Agrl. Soc. France, 1823. Mem. Central Agrl. Soc. France (permanent sec. 26 years), French Acad. Scis., French Acad. Medicine, Société d'Encouragement pour l'Industrie Nationale; Société d'Horticulture de la Seine, Council Hygiene and Pub. Health. Author: Traité de Chimie Industrielle, 1849; Traité de la distillation, 1858. Prepared borax from soda and boric acid; 1st to attempt to separate wood into its components; discovered and analyzed cellulose; discovered (with Persoz) enzyme, diatases, 1833-34; studied potato, starch, lignin, sugar, bitumen, decolorizing properties of animal charcoal; improved sugar refining; discovered dextrin, 1836, pektin, 1824. Died Paris, May 12, 1871.

PAYER, Jean-Baptiste, French botanist; b. Asfeld, France, Feb. 3, 1818; ed. U. Strasbourg (France); prof. Faculty Rennes (France), École normale, Paris. Mem. French Acad. Scis. Author: Organogénie de la fleur, 1854-59; Traité élémentaire de botanique. Died Paris, Sept. 5, 1860.

PAYER, Julius von, see von Payer.

PAYNE, Anthony Monck-Mason, epidemiologist, educator; b. London, Eng., Aug. 10, 1911; s. John Ernest and Sylvia May (Moore) P.; B.A. with honours, Trinity Coll., Cambridge U., 1933; M.B., B.Ch., Cambridge U. and St. Bartholomews Hosp., London, 1936; M.D., Cambridge U., 1946; M.A., Yale, 1960; m. Margaret Catherine Smith, Aug. 15, 1941. Came to U.S., 1960. Sr. epidemiologist Central Virus Reference Lab. and Oxford Regional Lab., Pub. Health Lab. Service, Med. Research Council, 1947-51; med. officer WHO, Geneva, Switzerland, 1952-55, chief med. officer charge endemo-epidemic diseases and virus diseases, 1955-60; asst. dir. gen., 1966—— (on leave from Yale); Anna M.R. Lauder prof. epidemiology and pub. health, chmn. dept. Yale Sch. Medicine, 1960——. Sec. expert adv. panel virus diseases, expert coms. hepatitis, influenza, poliomyelitis and respiratory virus diseases WHO, 1952-60, mem. expert adv. panel virus diseases, 1960——; sci. group virus research, 1960——; adv. com. med. research Pan-Am. Health Orgn., 1962——, vice chmn., 1964-66; asso. mem. commn. immunization. Armed Forces Epidemiological Bd., 1961-65; sub-com. geog. pathology Nat. Acad. Scis., 1962-65. Bd. dirs. Conn. Health League. Fellow Royal College of Physicians, Am. Pub. Health Assn.; mem. Assn. Tchrs. Preventive Medicine, N.Y. Acad. Scis., Am., Internat. epidemiological socs., Sigma Xi, Delta Omega (hon.). Contbr. numerous articles in field. Home: Maple Vale Dr., Woodbridge 15, Conn. Office: Yale Sch. Medicine, 333 Cedar St., New Haven*

PAYNE, Dwight Arthur, Am. chemist; b. Waterloo, Ia., Sept. 24, 1926; s. Dwight A. and Nina Nell (Mabie) P.; B.A., Ia. State Tchrs. Coll., 1950; Ph.D., State U. Ia., 1959; m. Ellen Colleen Goarcke, Oct. 25, 1946; children—Katherine Grace, Patricia Ellen, Julia Jeanne. Faculty, Tulane U., New Orleans, 1958——, adminstrv. officer dept. chemistry, 1964——. Mem. Am. Chem. Soc., A.A.A.S., Sigma Xi, Alpha Chi Sigma, Phi Lambda Upsilon. Contbr. articles to tech. jours. Pioneered isolation dimethyl compounds calcium, strontium, barium, syntheses tetrahedral silicon, germanium, tin, lead; research on optical activity on above series of compounds, flame analysis halogen, sulfur, nitrogen and phosphorus containing organic. Home: 838 Lowerline St., New Orleans 70118. Office: Dept. Chemistry, Tulane U., New Orleans 70118.*

PAYNE, James Patrick, Brit. anesthesiologist; b. Dalkeith, Scotland, Jan. 19, 1922; s. James and Mary (Cairney) P.; M.B., U. Edinburgh (Scotland), 1946, Ch.B., 1946, D.A., 1951; F.F.A.R.C.S., 1954; m. Alice Josephine McCorry, July 16, 1951; children—James Charles, Josephine Frances, Monica Louise, Jacqueline Mary. Registrar anesthetics Royal Infirmary Edinburgh, 1949-52; research asst. U. Manchester (Eng.), 1952-54; lectr. anesthetics Postgrad. Med. Sch. London, 1954-63; Brit. Oxygen Research prof. anesthetics Royal Coll. Surgeons Eng., London, 1963——. Cons. in anesthetics to hosps., RAF; Hunterian prof. Royal Coll. Surgeons Eng., 1963. Fellow Royal Soc. Medicine; mem. Assn. Anaesthetists Gt. Britain and Ireland, Scandinavian, Brazilian socs. anesthesiologists. Author: (with D.W. Hill) Oxygen Measurements in Blood and Tissues, 1966; also numerous articles. Research on factors modifying action of neuromuscular blocking agts., acid, base, and oxygen disturbances in blood during anesthesia, circulatory effects anesthetic drugs. Home: 36 Raymond Rd., Wimbledon, London S.W. 19, Eng.*

PAYNE, Joseph Frank, English physician; b. Camberwell, Eng., Jan. 10, 1840; s. Joseph and Eliza (Dyer) P.; student U. Coll., London; B.A., Magdalan Coll., Oxford (Eng.) U., 1862, M.B., 1867; B.Sc., London, 1865; M.D., 1880; m. Helen MacPherson, Sept. 1, 1882; 1 son, 3 daus. Became asst. physician St. Thomas Hosp., 1871, physician, 1887-1900, cons. physician, 1900-1910; mem. staff Hosp. for Skin Diseases at Blackfriars, over 30 years; Harveian librarian Coll. Physicians, from 1899; became chief med. witness Staunton Poisoning Case, 1877. Author: Life of Linacre, 1881; Manual of General Pathology, 1888; Observations on Rare Diseases of the Skin, 1889; Nomenclature of Diseases, 1896; Life of Thomas Sydenham, 1900; English Medicine in Anglo-Saxon Times, 1904. Editor works of Joseph Payne, 1880-92. Research on epidemiology, dermatology, pathology; reported on plague, Vetlanka, Russia, 1879. Died New Barnet, Eng., Nov. 16, 1910.

PAYNE, Robert Walter, psychologist; b. Calgary, Alta., Can., Nov. 5, 1925; s. Reginald William and Nora (Cowdery) P.; B.A., U. Alta., 1949; Ph.D., U. London (Eng.), 1954; m. Helen June Mayer, Dec. 1948; children—Raymond William, Barbara Joan, Margaret June. Lectr. psychology U. London, Inst. Psychiatry, 1952-59; prof. psychology Queens U., Kingston, Ont., 1959-65; prof. psychology, chpt. dept. behavioral sci. Temple U. Med. Sch., Phila., 1965——. Recipient Stratton Research award, 1964. Fellow Brit. Psychol. Soc.; mem. Am., Canadian psychol. assns. Research, numerous publs. devel. of objective techniques for assessing different kinds of thought disorder from which psychotic mental patients suffer. Home: 8011 Stenton Av., Phila. 19118. Office: Eastern Pa. Psychiat. Inst., Henry Av. and Abbottsford Rd., Phila. 19129.*

PAYNE-GAPOSCHKIN, Cecilia Helena, astronomer; b. Wendover, Eng., May 10, 1900; d. Edward John and Emma Leonora Helena (Pertz) Payne; A.B., Newnham Coll., Cambridge, Eng., 1923; Ph.D., Radcliffe College, 1925; D.Sc., Wilson College, 1942, Cambridge (England) University, 1950, Smith Coll., 1951, Western College for Women 1953; married m. Sergei Illarionowitsch Gaposchkin, March 6, 1934; children—Edward Michael, Katherine Leonora, Peter John Arthur. Came to U. S., 1923, naturalized, 1931. With Harvard Observatory since 1923, Phillips astronomer Harvard University, 1938——, professor of astronomy, 1956——. Member American Astronomical Soc., Royal Astron. Soc., Am. Philos. Soc., Am. Acad. Arts and Sciences, Phi Beta Kappa, Sigma Xi (pres. Radcliffe chapter, 1945-47). Author: Variable Stars, 1938; Stars in the Making, 1952; Variable Stars and Galactic Structure, 1954; Galactic Novae, 1957; numerous technical papers. Devised new techniques for determining stellar magnitudes from photog. plates; concentrated on study of variable stars, fom 1934.

PAYNTER, Raymond Andrew, Jr., Am. zoologist; b. N.Y.C., Nov. 29, 1925; s. Raymond Andrew and Hannah (Minch) P.; B.S., Bowdoin Coll., 1946; M.S., Yale, 1940, Ph.D., 1954; m. Elizabeth Storer, Mar. 12, 1960; children—Dorothy Storer. Field dir. Bowdoin Sci. Sta., 1946-48; leader Yale expdns., Mexico, 1948-49, Mexico, Brit. Honduras, 1950-51, Mexico, 1952; asst. curator birds Mus. Comparative Zoology, Harvard, 1953-56, asso. curator birds, 1956-60, curator, 1961——, mem. faculty arts and sci. Harvard, 1963——. leader Harvard expdn. Mexico, 1954, Ecuador, 1965; leader Yale-Harvard expdn. Nepal, Pakistan, India, 1957-59. Fellow Am. Ornithologists' Union, A.A.A.S. mem. Ecol. Soc. Am., Soc. Systematic Zoologists, Brit., Royal Australasian ornithol. unions, Wilson, Cooper ornithol. socs., Bombay Natural History Soc., Deutsche Ornithologen-Gesellschaft, Sigma Xi. Author: Ornithogeography of the Yucatan Peninsula, 1955; also numerous articles. Editor: (with E. Mayr) Check-list of Birds of the World, 1961. Asso. editor Am. Midland Naturalist, 1963—; editor Nuttall Ornithol. Club Pubs., 1966——. Research on systematics neotropical and Asiatic birds, population dynamics colonial birds. Home: 74 Sunset Rd., Weston, Mass. 02193. Office: Mus. Comparative Zoology, Harvard, Cambridge, Mass. 02138.*

PAYR, Erwin, surgeon; b. Innsbruck, Austria, Feb. 17, 1871; prof., Griefswald, Germany, Königsburg (now Kaliningrad, USSR), Leipzig, Germany. Author: Lehrbuch der speziellen Chirurgie, 2 vols., 1918-27; Die Klinik der Bösartigen Geschwülste, 3 vols., 1924-27. Authority on surgery of joints; described colonic stasis caused by kinking of colon at splenic or hepatic flexure, or by kinking of adhesion between transverse and descending parts of colon (Payr's disease), 1910. Died Leipzig, Apr. 6, 1946.

PAZ, Pedro, mathematician; accountant to Chpt. Mexico Cathedrals. Author: Arte para aprender todo el menor del arithmetica sin maestro (1st gen. arithmetic pub. in New World, based on Spanish texts), 1623.

PAZUR, John Howard, biochemist; b. Zubne, Czechoslovakia, Jan. 17, 1922; s. John and Mary (Bonko) P.; B.S., U. Guelph, 1944; M.S., McGill U., 1946; Ph.D., Ia. State U., 1950; m. Jean Josephine Glabais, Nov. 22, 1950; children—Robert Leslie, Barbara Jean, Beverly Ann, Carolyn Jo. Came to U.S., 1946, naturalized, 1961. Instr. chemistry Ia. State U., 1950-51; asst. prof. biol. chemistry U. Ill., 1951-52; faculty U. Neb., 1952-66, prof. biochemistry, 1959-66, chmn. dept., 1960-66; prof., head biochemistry Pa. State U., University Park, 1966——. Mem. Am. Chem. Soc., Am. Soc. Biol. Chemists. Research, publs. on chemistry of carbohydrates, structure of enzymes, mechanism of enzyme action, carbohydrate biosynthesis and degradation. Home: 542 W. Ridge Av., State College, Pa. 16801. Office: Biochemistry Dept., Pa. State U., University Park, Pa. 16802.*

PEACHEY, John Edwin, Brit. biologist; b. Nottingham, U.K., Apr. 8, 1935; s. John Bertram and Maud (Mallett) P.; B.Sc. with 1st class honors in Zoology, U. Durham (Eng.), 1955, Ph.D., 1958; m. Cherry Frances Gooding, May 29, 1965; 1 son, Stephen John. Officer in charge chem. control nematology dept. Rothamsted Exptl. Sta., Harpenden, Eng., 1958-64; officer in charge fisheries helminthology unit of the Natural Environment Research Council St. Albans, U.K., 1964——; asst. dir. Commonwealth Bur. Helminthology, St. Albans, 1965——. Cons. Interafrican Phytosan. Commn., Orgn. for African Unity, 1966——. Mem. Inst. Biology. Author: (with M.R. Chapman) Chemical Control of Plant Nematodes, 1965; also articles. Research on ecology of soil animals and on control of plant parasitic nematodes in soil, on parasites of economically important marine fish, information services in parasitology. Home: 142c St. Albans Rd., Sandridge, Herts. Office: 103 St. Peter St., St. Albans, Herts., U.K.*

PEACOCK, Erle Ewart, Jr., Am. physician; b. Durham, N.C., Sept. 10, 1926; s. Erle E. and Vera (Ward) P.; student U. N.C., 1943-45, U. N.C. Med. Sch., 1945-47; M.D., Harvard, 1949; m. Mary Lowrey, Apr. 17, 1954; children—James L., Susan L., Virginia Gayle. Faculty, U. N.C., Chapel Hill, 1956——, prof. surgery, 1965——. Cons. VA Hosp., Fayetteville, N.C., Womack Army Hosp., Ft. Bragg, N.C.; Mem. surgery study sect. NIH. Mem. Am., So. surg. assns., Am. Soc. Plastic and Reconstructive Surgery, A.C.S. Publs. on basic research in wound healing, tissue transplantation; clin. studies in gen. plastic surgery, reconstructive surgery of hand. Home: 322 W. University Dr., Chapel Hill, N.C. 27514.*

PEACOCK, George, English mathematician; b. Denton, Eng., Apr. 9, 1791; s. Thomas Peacock; B.A., Trinity Coll., Cambridge, 1812, M.A., 1816; m. Frances Elizabeth Selwyn, 1847. Apptd. lectr. math. Trinity Coll., Cambridge, 1815, tutor, 1823, sole tutor, 1835-39; Lowndean prof. astronomy, 1836-58; dean of Ely, 1839-59. Mem. Commn. on Weights and Measures. Fellow Royal Soc., 1818; founder (with Babbage and John Herschel) Analytical Soc. (devoted to intro. of Leibniz's notation in calculus, gen. revival of math. analysis in Eng.), 1812. Author: Collections of Examples of the Applications of the Differential and Integral Calculus, 1820; Arithmetic—Encyclopaedia Metropolitana, 1825-26; A Treatise on Algebra, 1830, 42; On the Recent Progress of Certain Branches of Analysis, 1834. Advocate of decimal coinage. Died Ely, Nov. 8, 1858.

PEACOCK, Lelon James, Am. psychologist; b. Brevard, N.C., May 25, 1928; s. L. J. and Dorothy (Barrett) P.; student Emory U., 1945-47; A.B., Berea Coll., 1950; M.S., U. Ky., 1952, Ph.D., 1956; m. Marian Davis, June 20, 1945; children—Lynn Barrett, Janice Davis, Timothy Lee. Psychophysiologist U.S. Army Med. Research Lab., Ft. Knox, Ky., 1954-56; research asso., acting dir. Yerkes Labs. of Primate Biology, Inc., Orange Park, Fla., 1956-59;; asso. prof. psychology U. Ga., 1959-65, prof., 1966——. Recipient U. Ga. M.G. Michael Research award. Mem. Am. Psychol. Assn., A.A.A.S., So. Soc. for Philosophy and Psychology, Animal Behavior Soc., Internat. Neuropsychology Soc., Sigma Xi, Phi Kappa Phi. Author (with R. V. Heckel): Textbook of General Psychology, 1966. Research, publs. on general activity of organisms, including circadian rhythms; effects of ionizing radiation on behavior; inventor of ultrasonic device for measuring animal motion. Home: 145 Woodland Way, Athens, Ga. 30601.*

PEACOCK, M(artin) A(lfred), mineralogist; b. Edinburgh, Scotland, Jan. 15, 1898; s. Alfred Norman and Antonie Ida (Fuller) P.; B.Sc., Glasgow U., 1922, Ph.D., 1925, D.Sc., 1932; A.M., Harvard, 1927; m. Katharine Louisa West, Apr. 2, 1937; children—Barbara Clendon, Nancy Bligh. Research asso. crystallography U. Heidelberg, 1929, 33; lectr.

geology and geography U. B.C. (Can.), 1929-30, asst. prof., 1930-31; research asso. mineralogy Harvard, 1932-37; asso. prof. mineralogy and petrography U. Toronto (Ont., Can.), 1937-42, prof. mineralogy, 1942-46, prof. crystallography and mineralogy since 1946. Mem. Canadian Nat. Com. on Crystallography. Mem. Royal Soc. Can., Walker Mineral. Club, Geol. Soc. Am. (v.p. 1948), Mineral. Soc. Am. (pres. 1948), Mineral. Soc. Gt. Britain (abstractor since 1941), Crystallographic Soc. of Am. Author and editor sci. papers on crystallography, mineralogy. Research on recognition of minerals by X-ray diffraction. Died Oct. 30, 1950.

PEACOCK, Thomas Bevill, English physician; b. York, Eng., Dec. 21, 1812; s. Thomas and Sarah (Bevill) P.; M.D., U. Edinburgh (Scotland), 1842; m. Cornelia Walduck, 1850. Became asst. physician St. Thomas's Hosp., 1849, lectr. materia medica, then on medicine; Croonian lectr., 1865; founder dispensary (became Victoria Park Hosp.), London. Licentiate, fellow Royal Coll. Physicians; mem. Path. Soc. (a founder 1846, sec. 1850, v.p. 1852-56, pres. 1865-66). Author: On the Influenza or Epidemic Catarrh of 1847-48, 1848; On Malformations of the Human Heart, 1866; On the Prognosis in Cases of Valvular Disease of the Heart, 1877; also papers. Described congenital anomaly of heart characterized by pulmonic stenosis, dextropostion of aorta, hypertrophy of right ventricle, interventricular septal defects, 1858. Died May 31, 1882.

PEALE, Albert Charles, Am. geologist, paleobotanist; b. Heckscherville, Pa., Apr. 1, 1849; s. Charles W. and Harriet (Friel) P.; A.B., Central High Sch., Phila., 1868, A.M., 1873; M.D., U. Pa., 1871; m. Emily W. Wiswell, Dec. 23, 1875. Mineralogist and geologist U.S. Geol. and Geog. Survey of Territories, 1871-79; geologist U.S. Geol. Survey, 1883-98; aid, sect. paleobotany U.S. Nat. Mus., 1898—. Author: Yellowstone National Park and Thermal Springs, 1882; The Classification of American Mineral Waters, 1887; Mineral Springs of the United States, 1886; The Natural Mineral Waters of the United States, 1895; Classification of Mineral Waters, 1902; Biographical Sketches of Charles Willson Peale and of Titian R. Peale, 1905; The Stratigraphic Position and Age of the Judith River Formation. Died 1913.

PEALE, Charles Willson, Am. naturalist; b. Queen Anne County, Md., Apr. 15, 1741; s. Charles and Margaret (Triggs Matthews) P.; m. Rachel Bruner, Jan. 12, 1762; m. 2d, Elizabeth De Peyster, 1791; m. 3d, Hannah Moore, 1805; 12 children, including Raphael, Rembrandt, Titian, Rubens, Franklin, Titian Ramsay. Apprenticed to saddler, Nathan Walters, 1754-61; began painting portraits, 1762; gained recognition which prompted several men to pay for his visit and study in Eng., 1766; studied under Benjamin West; returned to Annapolis, Md., 1769, moved to Phila., 1776; served in Revolutionary War; mem. Pa. Gen. Assembly, 1779-80; painted miniature portraits, 1780-90; established Phila. Mus.; founder Peales Mus. Natural History and Portraits, 1802; largely responsible for establishment Pa. Acad. Fine Arts, 1805. Author: An Essay on Building Wooden Bridges, 1797; Introduction to a Course of Lectures on Natural History, 1800; An Epistle to a Friend on the Mean of Preserving Health, 1803. Died Phila., Feb. 22, 1827.

PEALE, Titian Ramsay, Am. naturalist, artist; b. Phila., Nov. 17, 1799; s. Charles Willson and Elizabeth (De Peyster) P.; studied anatomy U. Pa.; m. Eliza Cecilia La Forgue, 1822; m. 2d, Lucy Mullen; 6 children. Apprenticed to spinning machine mfr., 1814; worked with brother Rubens (curator of mus., Phila.), 1816-18; joined William MacLure and Thomas Say in expdn. to coasts of Ga. and Fla. to study and collect fauna specimens, 1818-19; asst. naturalist under Stephen H. Long on U.S. Army expdn. to Upper Missouri, 1819-20; asst. mgr. Phila. Mus., 1821-24; exhibited 4 water color paintings Pa. Acad. Fine Arts, 1822; sent to Fla. by Charles Lucien Bonaparte to collect specimens and make drawings for book America Ornithology, 1824; mem. civil staff U.S. exploring expdn. to S. Seas under Charles Wilkes, 1838-42, made drawings which appeared in published accounts of expdn.; examiner U.S. Patent Office, Washington. Author: Lepidoptera Americana, 1833; Mammalia and Ornithology, vol. 8 of Reports of U.S. Exploring Expdn., 1838-42. Died Mar. 13, 1885.

PÉAN, Jules Émile, French surgeon; b. Chateaudun, France, Nov. 29, 1830; ed. Coll. of Chartres; began med. studies, Paris, 1849; became surgeon of hosps., Paris, 1868; apptd. chief of services St.-Louis Hosp., 1874; founder Internat. Hosp., 1892; mem. Acad. Medicine. Author: Splénotomie, 1860; Eléments de pathologie chirurgicale, 1875; De la forcipressure, 1875; Lecons de clinique chirurgicale, 1876-90; Du pincement des vaisseaux comme moyen d'hémostase, 1877. A founder of modern gynecol. surgery; devised techniques of vaginal hysterectomy, ovariotomy, gasterectomy; inventor hemostatic forceps, 1868; performed vaginal hysterectomy for carcinoma, 1890; credited with 1st operation for diverticula of bladder, circa 1895; practiced aseptic surgery during 1850's (before importance of sterilization was fully understood). Died 1898.

PEANO, Giuseppe, Italian mathematician, logician; b. Cuneo, Piedmont, Aug. 27, 1858. Prof. of infinitesimal calculus, U. Turin, 1890. Author: Calculo differenziale e principie di calcali integrale, 1884; Calcalo geometrico, 1888; I Principii de geometria logicamente esposti, 1889; Lezioni di anal. infinitesmale, 1893; others. Founder, Rivista de mathematica. A founder of symbolic logic; contributed to non-Euclidean geometry; devised Interlingua as an internat. auxiliary language (vocabulary consists of words common to Latin, French, German and English). Died Turin, Italy, Apr. 20, 1932.

PEARCE, George Whitenack, Am. chemist; b. State College, Pa., Mar. 21, 1907; s. Diemer Thomas and Cora (McMahon) P.; B.S., Pa. State U., 1929, M.S., 1930, Ph.D., 1946; m. Sara Minerva Mallory, June 20, 1931; children—Mallory Pearce, Peter Jon, Penelope Annette. With N.Y. State Agr. Expt. Sta., Cornell U., Geneva, N.Y., 1930-51, asso. prof., 1945-51; chief chem. sect. Tech. Devel. Labs. USPHS, Savannah, Ga., 1951-63, chief, 1963—, Cons. WHO, 1957—, mem. adv. panel on pesticides, 1957—. Recipient Distinguished Service award U.S. Dept. Health, Edn. and Welfare, 1955; Rockefeller Pub. Service award Princeton, 1956. Mem. Entomol. Soc. Am. (Gold medal 1940, 42, 47), Am. Chem. Soc., A.A.A.S., Research Soc. Am., Sigma Xi, Phi Lambda Upsilon. Research, numerous publs. on chemistry, analysis, formulations, specification and evaluation of insecticidal compounds; synthesis and toxicological studies of radioactive insecticides, chemistry fluorescein derivatives; co-discoverer dicholorvos. Home: 8 5th St., Savannah Beach, Ga. 31328. Office: P.O. Box 769, Savannah, Ga. 31402.*

PEARCE, Joseph Algernon, Canadian astrophysicist; b. Brantford, Ont., Can., Feb. 7, 1893; s. Joseph William and Clarissa Augusta (Rounds) P.; B.A., U. Toronto, 1920, M.A., 1922; Ph.D., U. Cal. at Berkeley, 1930; D.Sc., U. B.C., 1955; m. Esther Edith Mott, Dec. 8, 1917; children—Josephine Anna, Richard Mott; m. 2d, Elizabeth Mill Allan, Apr. 26, 1948. With Dominion Astrophys. Obs., Victoria, B.C., 1924—, dir., dominion astrophysicist, 1940-51, dir. emeritus, 1952—. Fellow Royal Soc. Can. (past pres.); mem. Royal Astron. Soc. Can. (Distinguished Service award 1962, pres.), Am. Astron. Soc. (life), A.A.A.S., Astron. Soc. Pacific (life). Research, numerous publs. on phys. characteristics of stars, orbital elements of binary stars and eclipsing variables, radial velocities of stars, rotation of the galaxy, distbn. of interstellar matter. Home: 1233 Union Rd., Victoria, B.C., Can.*

PEARCE, Louise, Am. physician; b. Winchester, Mass., Mar. 5, 1885; d. Charles Ellis and Susan Elizabeth (Hoyt) Pearce; A.B., Stanford, 1907; M.D., Johns Hopkins, 1912; Sc.D. (hon.), Wilson Coll., 1947. Bucknell U., 1950; LL.D., Skidmore Coll., 1950; Litt.D., Beaver Coll., 1948; D.M.S., Women's Med. Coll., 1952. Med. house officer Johns Hopkins, 1912-13; with Rockefeller Inst. Med. Research, from 1913, successively as fellow, asst., asso., asso. mem., 1923-51. Conducted African Sleeping Sickness Mission, Belgian Congo, 1920-21; vis. prof. medicine Peiping Union Med. Coll., China, 1931-32; mem. bd. corporators Women's Med. Coll., Phila., from 1941, pres., 1946-51; gen. adv. council Am. Social Hygiene Assn., 1924-44; NRC, 1931-33. Recipient Elizabeth Blackwell Citation, 1951, Woman's Med. Coll. Citation, 1952, Leopold II award, 1953. Mem. A.A.A.S., N.Y. Acad. Medicine, Harvey Soc., Am. Soc. Exptl. Pathology, Am. Assn. U. Women (dir. 1945-51), Am. Soc. Pharmacol. and Exptl. Therapeutics, Am. Assn. Pathologists and Bacteriologists, Soc. for Exptl. Biology and Medicine, Am. Assn. for Cancer Research, N.Y. Soc. Tropical Medicine, Johns Hopkins Surg. and Med. Assn., Am. Soc. Tropical Medicine, Coll. Physicians Phila., Am. Med. Women's Assn., Pathol. Soc. Gt. Britain and Ireland, Royal Soc. Tropical Medicine and Hygiene, Brit. Soc. for Study of Veneral Disease (hon.), Société belge de Médicine tropicale, Peiping Soc. Natural History, Phi Beta Kappa, Sigma Xi. Author: Treatment of Human Trypanosamiasis with Tryparsamide (monograph Rockefeller Inst.), 1930; also papers. Introduced use of tryparsamide in treatment of trypanosomiasis, circa 1921. Died Aug. 10, 1959.

PEARCE, Richard Mills, Jr., pathologist; b. Montreal, Que., Can., Mar. 3, 1874; s. Richard Mills and Sarah (Smith) P.; M.D., Harvard, 1897; postgrad. U. Leipzig (Germany), 1902, D.Sc., Lafayette Coll., 1915; m. May Harper Musser, Nov. 6, 1902; children—Agnes M., John M. Resident pathologist Boston City Hosp., 1896-99; instr. pathology Harvard, 1899-1900; demonstrator, later asst. prof. pathology U. Pa., 1900-03; prof. pathology, 1910-11, prof. research medicine, 1910-20; dir. Bender Hygienic Lab., Albany, N.Y., 1903-08; prof. pathology and bacteriology Albany Med. Sch., 1903-08; dir. bur. pathology and bacteriology N.Y. State Dept. Health, 1903-08; prof. pathology, Univ. and Bellevue Hosp. Med. Coll. (N.Y. U.), 1908-10; gen. dir. div. med. edn., Rockefeller Found., 1920—. Chmn. med. div. NRC, 1918. Author: Medical Research and Education, 1913; The Spleen and Anemia, 1917. Died Feb. 16, 1930.

PEARL, Raymond, Am. biologist; b. Farmington, N.H., June 3, 1879; s. Frank and Ida May (McDuffee) P.; A.B., Dartmouth, 1899, Sc.D., 1919; Ph.D., U. of Mich., 1902; U. of Leipzig, 1905, Univ. Coll., London, 1905-06, Carnegie Instn. Table, Naples Zoöl. Station, 1906; LL.D., University of Maine, 1919; Litt. D., from St. John's College, 1935; m. Maud M. DeWitt, June 29, 1903; children—Ruth DeWitt, Penelope Mackey. Asst. in zoölogy, 1899-1902, instr., 1902-06, U. of Mich.; instr. U. of Pa., 1906-07; biologist and head of dept. biology, Maine Agrl. Expt. Sta., 1907-18; prof. biometry and vital statistics, Sch. of Hygiene and Pub. Health, Johns Hopkins, 1918-25, research professor, 1925-30, prof. biology, Medical School, Johns Hopkins, 1923—; statistician, Johns Hopkins Hospital, 1919-35; director Inst. for Biol. Research, Johns Hopkins, 1925-30, professor of biology, School of Hygiene, 1930—. Engaged in biol. researches on variation in fishes, with Biol. Survey Great Lakes (U.S. Fish Commn.), 1901-02; awarded grants for research on variation in organisms from Carnegie Inst., 1904, 05, 06; expert, poultry breeding, U.S. Dept. Agr.; non-resident lecturer Grad. Sch. of Agr., Ames, Iowa, 1910, Lansing, Mich., 1912, Washington, 1939; Lowell lecturer, Boston, Mass., 1920; special lecturer, U. of London, 1927; Harrington lecturer U. of Buffalo, 1928; Health Clark lecturer U. of London, 1937; Patten Foundation lecturer, Indiana University, 1938. Member exec. com. and chmn. agrl. com. Nat. Research Council, 1916-18, and mem. exec. board, 1919-35; chief of statis. division U.S. Food Administration, 1917-19; pres. Internat. Union for Scientific Investigation of Population Problems, 1928-30. Mem. bd. visitors and govs. St. John's Coll., 1928-34; trustee Science Service, 1929-35. Editor Quarterly Review of Biology, Human Biology; asso. editor Biometrika, 1906-10, Journal Agrl. Research, 1914-18, Genetics, 1915—, Journal Exptl. Zoölogy, 1915—, Metron, 1920—, Biologia Generalis, 1923-27, Acta Biotheoretica, 1937—. Decorated Knight of the Crown of Italy, 1920, Officer, 1929. Author: Variation and Differentiation in Ceratophyllum, 1907; Variation and Correlation in the Crayfish (with A. B. Clawson), 1907; Poultry Diseases and Their Treatment (with F. M. Surface and M. R. Curtis), 1911; Modes of Research in Genetics, 1915; Diseases of Poultry (with F. M. Surface and M. R. Curtis), 1915; The Nation's Food, 1919; The Biology of Death, 1922; Introduction to Medical Biometry and Statistics, 1923; Studies in Human Biology, 1924; The Biology of Population Growth, 1925; Alcohol and Longevity, 1926; To Begin With, 1927; The Rate of Living, 1928; Constitution and Health, 1933; The Ancestry of the Long-lived (with Ruth D. Pearl), 1934; The Natural History of Population, 1939. Known for statis. studies on population growth, mortality, birth rates, human fertility; devel. theory that in absence of strong distbg. forces populations tend to increase in accordance with a logarithmic curve. Died Hershey, Pa., Nov. 17, 1940.

PEARLSTEIN, Edgar A(aron), Am. physicist; b. Pitts., Mar. 19, 1927; s. Charles and Sophia (Siegel) P.; B.S., Carnegie Inst. Tech., 1947, D.Sc., 1950; m. Laura Hyland (div.); children—Victoria, Leonard Eugene. Research asso. U. Ill., 1950-52; research physicist Carnegie Inst. Tech., 1952-56; physicist Westinghouse Electric Co., 1956; faculty U. Neb., Lincoln, 1956—, prof., 1963—. Fellow Am. Phys. Soc.; mem. A.A.A.S., Am. Assn. Physics Tchrs., Am. Assn. U. Profs. Research, publs. on elec. and optical properties insulating crystals, radiation effects in solids. Home: 632 Eastborough Lane, Lincoln, Neb. 68505.*

PEARSE, Anthony Guy Everson, English histochemist; b. Birchington, Eng., Aug. 9, 1916; s. Reginald Guy and Constance (Steels) P.; B.A. (Kitchener scholar) Trinity Coll., Cambridge, 1937, M.B., B.Chirurg., 1941, M.A., 1942, M.D., 1950; M.D. (hon.), U. Basel (Switzerland), 1960; m. Elizabeth Himmelhoch, Nov. 8, 1947; children—Mark Guy, Patricia, Elizabeth-Anne, Katharine. House physician St. Barts Hosp., 1940-41; asst. lectr. pathology Postgrad. Med. Sch., London, 1947-51; faculty U. London, 1951—, prof. histochemistry, 1965—; vis. prof. U. Ala., 1953-54. Guest instr. U. Kan., 1959-60, Vanderbilt U., 1967. Recipient Horton Smith prize, Cambridge U., 1950. Fulbright fellow, 1953. Fellow Royal Coll. Physicians, Coll. Pathologists, Royal Micros. Soc. (hon.), Royal Hort. Soc.; mem. Histochem. Soc., Path. Soc., Biochem. Soc., Arbg. Histochem. (Germany). Author: Histochemistry Theoretical and Applied, 1953; also numerous articles. Devel. histochem. and cytochem. techniques and application to physiology and pathology, especially endocrine (APUD cell series), muscular and neurol. fields; research on functional cytology of adenohypophysis, thyroid gland, ultimobranchial bodies; invented remote control cryostat, 1954, thermoelectric tissue freeze-dryer, 1963. Home: The Fortress, Letchmore Health, Hertfordshire, Eng. Office: Royal Postgrad. Med. Sch., London W. 12, Eng.*

PEARSE, Arthur Sperry, Am. zoologist; b. Crete, Neb., Mar. 15, 1877; s. Sherman L. and Sarah Louise (Gardner) P.; B.S., U. Neb., 1900, A.M., 1904, LL.D., 1941; Ph.D., Harvard, 1908; m. Mary Oliver Lehmer, Dec. 22, 1902; children—Frederick Deweese, Richard Lehmer, Frank, Elizabeth (Mrs. William Henry Caufman). Tchr. Omaha High Sch., 1900-04; asst. in zoology Harvard, 1904-07, teaching fellow in zoology, 1908; tchr. Lake High Sch., Chgo., 1907; instr. in zoology U. Mich., 1909-10, asst. prof., 1911; asst. prof. zoology U. Philippines, 3 mos., 1911; asso. prof. zoölogy St. Louis U. Sch. Medicine, 1911; asso. prof. zoology U. Wis., 1912; prof., 1919-26; mem. faculty grad. dept. Duke U., beginning 1927, dir. Marine Lab., 1938-45. Vis. prof. Keio U., Tokyo, Japan, 1929-30. Spl. investigator U.S. Bur. of Fisheries, 1913-25, 35-36, Internat. Health Bd., 1925-26, Carnegie Inst., 1928, 31, 32, 36. Fellow A.A.A.S. (v.p. sect. F 1933); mem. Am. Soc. Zoologists (pres. 1945), Am. Soc. Parasitologists, Am. Soc. Naturalists, Ecol. Soc. Am. (pres. 1925), S.E. Biol. Assn. (pres. 1942), N.C. Acad. Sci., Arts and Letters, Sigma Xi, Phi Beta Kappa. Author: General Zoology; Animal Ecology; Homoiothermism; Environment and Life; Migration of Animals from Sea to Land; Cenotes of Yucatan; Fauna of Caves of Yucatan; Hell's Bells; Introduction to Parasitology; Fauna in Encyclopedia Yucateca; also articles. Editor Ecology Monographs, 1931-51. Research on parasites, animal behavior, ecology, fisheries, crustacea, also animals of U.S., P.I., Africa, Japan, Yucatan, S.A. Died Dec. 11, 1956.

PEARSON, Anthony Augustus Jr., Am. anatomist; b. Greenville, S.C., Oct. 9, 1906; s. Anthony A. and Grace (Ellis) P.; B.S. Furman U., 1928; M.A., U. Mich., 1930, Ph.D., 1933. Instr., U. Chgo., 1933-37; asso. Loyola U. Med. Sch., 1937-43; asso. prof. Coll. Medicine, Baylor U., 1943-46; prof. med. sch. U. Ore., Portland, 1946—, chmn. dept. anatomy, 1953—. Mem. Am. Assn. Anatomists, Sigma Xi, Phi Beta Pi. Programmed textbook on The Development of the Eye, The Development of the Heart and Aortic Arches. Research on the comparative anatomy and devel. of the nervous system and the organs of spl. sense. Home: 1225 S.W. 6th St., Portland, Ore.*

PEARSON, Bjarne, Am. pathologist; b. Oslo, Norway (parents Am. citizens); s. Per and Katherine Pearson; B.S., U. Minn., 1927, M.B., 1929, M.D., 1930, M.S., 1932; m. Anne Pearson, 1928; children—Iday Kathryn (Mrs. Thomas Gillam), AnnaLee (Mrs. Douglas Ross), Bjarna Karen (Mrs. Alfred Prudhomme). Fellow, U. Minn., 1930-33; asst. prof. La. State U., 1935-42; asso. prof. Tulane U., 1942-45; prof. pathology, chmn. dept. U. Vt., Burlington, 1945-55; prof. pathology Wayne State U. Coll. Medicine, Detroit, 1954-64; chief pathologist U.S. Army Biol. Lab., Ft. Detrick, Frederick, Md., 1964—. Diplomate Am. Bd. Pathology. Mem. Am. Assn. Pathologists and Bacteriologists, Am. Assn. Cancer Research, Am. Soc. Exptl. Pathology, Internat. Acad. Pathology, Histochem. Soc., Soc. Exptl. Biology and Medicine, Sigma Xi, Alpha Omega Alpha, others. Research, numerous publs. on cancer of uterine cervix, lung, exptl. prodn. of breast cancer, syphilis of intestine in newborn, histochem. detection of enzymes in relation to pathology and cancer, devel. of series of enzyme tests in disease. Home: 200 Church St. Office: U.S. Army Biol. Lab., Ft. Detrick, Frederick, Md.*

PEARSON, Carl M., Am. physician; b. Seattle, Nov. 19, 1919; s. Maxwell Carl and Sophie (Gray) P.; A.B., U. Cal. at Los Angeles, 1943; M.D., Boston U., 1946; m. Gloria E. Mannix, July 23, 1949; children—Karen, Eric, Kyle. Faculty, U. Cal. at Los Angeles Sch. Medicine, 1956—, prof. internal medicine, 1965—, dir. div. rheumatology, dir. tng. program in arthritis, rheumatic diseases and disorders of muscular and neuromuscular system, 1958—. Cons. physician Wadsworth VA Hosp., Los Angeles, 1956—; mem. of coms. NIH, Arthritis Found.; med. adv. bd. Muscular Dystrophy Assn., 1958—. Fellow A.C.P., others; mem. Am. Soc. Clin. Investigation. Author: Diseases of Muscle: A Study in Pathology, 1953, 62; Corticosteroids in the Practice of Medicine, 1961; also numerous articles. Asso. editor Am. Jour. Medicine, 1962—, Rheumatism Revs., Annals Internal Medicine, 1963—, Arthritis and Rheumatism, 1967. Discovered exptl. model for prodn. of arthritis in animals called adjuvant arthritis; discovered new disease of muscle secondary to a deficiency of an enzyme called phosphorylase in muscle, subsequently called phosphorylase deficiency myopathy. Home: 1660 Roscomare Rd., Los Angeles 90024.*

PEARSON, Donald Emanual, Am. chemist; b. Madison, Wis., June 21, 1914; s. Gustav E. and Clara (Bjelde) P.; grad. U. Wis., 1936, U. Ill., 1940; m. Gwen Smiseth, June 5, 1950; children—Donald T., Jeanah C., Sam S. Chemist, Pitts. Plate Glass Co., Milw., 1940-42; tech. aide OSRD, 1942-45; chemist Mass. Inst. Tech., 1945-46; faculty Vanderbilt U., Nashville, 1946—, prof. chemistry. Mem. Am. Chem. Soc. Author: (with R.C. Elderfield) Phenazines; also numerous articles. Discovered new method of substitution in aromatic compounds, new rearrangement hydrazone; research on mechanisms of Beckmann rearrangement, reactions in polyphosphoric acid; structure of diuretic mercuhydrin; synthesis of barbiturates and anti-malarials. Home: 2111 Golf Club Lane, Nashville 37212.*

PEARSON, Egon Sharpe, English statistician; b. London, Eng., Aug. 11, 1895; s. Karl and Maria (Sharpe) P.; B.A., Trinity Coll., Cambridge, Eng., 1921, M.A., 1923; D.Sc., U. London, 1926; m. Dorothy Eileen Jolly, Aug. 31, 1934; children—Judith Reneé (Mrs. Ian Walker), Sarah Penelope; m. second, Margaret Scott Turner, January 1967. Faculty dept. statistics U. Coll., London, 1921—, prof., 1935-60, prof. emeritus, 1960—; staff Ordnance Bd., Ministry of Supply, 1939-46. Mem. statis. adv. com. Brit. Standard Instn., 1950—; mem. sci. adv. council War Office, 1960-63; mem. Nat. Council for Quality and Reliability, 1961-63. Decorated Comdr. Order Brit. Empire. Fellow Royal Soc., Royal Statis. Soc.; mem. Inst. Math. Statistics, Internat. Statis. Inst., Am. Statis. Assn., Biometric Soc., Inst. Actuaries (hon.). Author: The Application of Statistical Method to Industrial Standardization and Quality Control, 1935; Karl Pearson: An Appreciation of Some Aspects of His Life and Work, 1938; (with H.O. Hartley) Biometrika Tables for Statisticians, vol. 1, 1954; also numerous articles. Research on theory and application of statis. method, application statis. method to quality control, weapon devel. Home: 8 Cambanks, Union Lane, Cambridge, Eng. Office: Math. Dept., Univ. Coll., Gower St., London W.C.1, Eng.*

PEARSON, George, Brit. physician, chemist; b. Rotherdam, England, 1751; s. John Pearson; student medicine, Edinburgh, Scotland; studied under Joseph Black, chemist; M.D., 1733; student St. Thomas's Hosp., London; 2 daus. Travelled through France, Germany and Holland, 1775-77; settled in Doncaster, Eng., 1777; became physician St. George's Hosp., 1787, also lectr. chemistry, materia medica, practice physic. Fellow Royal Soc., 1791. Author: Observations and Experiments . . . on the Springs of Buxton, 1784; On the Nature of Gas produced by passing an Electric Discharge through Water, 1797. passing an Electric Discharge through Water, 1797. Translator: Nomenclature chimique, 1794. One of 1st to recognize value of vaccination; discovered calcium phosphide; introduced solution of sodium arsenate 1/10 the strength of usual Solution of Sodium Arsenate (Pearson's solution of sodium arsenate), 1774; discovered calcium phosphide; originated term, uri acid. Died Nov. 9, 1828.

PEARSON, Gerald Hamilton Jeffery, Am. psychiatrist; b. Key West, Fla., Sept. 21, 1893; s. George Lloyd and Frances (Baxter) P.; B.A., U. Western Ont., M.D., 1915; D.Sc. in Medicine (Commonwealth fellow), U. Pa., 1930; m. Mary Agnes Mackenzie, July 3, 1916; children—Frances Marion Hamilton (Mrs. Samuel E. Bucher), Lesley Agnes Joan (Mrs. Louis J. Fridenberg), George Robert Mackenzie. Faculty, Temple U., Phila., 1937-49; dean Inst. Phila. Assn. for Psychoanalysis, 1950-58; prof. emeritus child psychiatry Hahnemann Med. Sch. and Hosp., Phila., 1961—. Diplomate Am. Bd. Neurology and Psychiatry, Am. Bd. Child Psychiatry. Mem. Phila. Assn. Psychoanalysis, Am. Psychoanalytic Assn., Am. Psychiat. Assn., Am. Orthopsychiat. Assn., Am. Acad. Child Psychiatry, A.M.A., Am. Assn. Group Psychotherapy, Phila. Psychiat. Soc. (pres. 1948-49), Phila. Neurol. Soc., Phila. Pediatric Soc., Phila. County Med. Soc., Phila. Psychoanalytic Soc. (pres. 1943-46). Author: Emotional Disorders of Children, 1949; Psychoanalysis and the Education of the Child, 1954; Adolescence and the Conflict of Generations, 1958; (with Dr. O.S. English) Common Neuroses of Children and Adults, 1937; Emotional Problems of Living, 1945. Research, publs. in child devel., children's inability to learn acad. subjects, neuroses, psychoses and delinquency and behavior disorders in children. Address: Presdl. Apts., Phila. 19131.*

PEARSON, Gerald Leondus, Am. physicist; b. Salem, Ore., Mar. 31, 1905; A.B., Williamette U., 1926, Sc.D., 1956; M.A., Stanford, 1929; m. Mildred O. Cannoy, June 30, 1929; children—Ray L., Carl Ann (Mrs. Charles A. Parlette). Research physicist Bell Telephone Labs., Murray Hill, N.J., 1929-60; prof. elec. engring. Stanford, 1960—; vis. prof. electronics engring. U. Tokyo, 1966. Cons. Varian Assos., Palto Alto, Cal., 1963—, Douglas Aircraft Co., Santa Monica, Cal., 1963—, Thompson Ramo Wooldridge Space Tech. Labs., Redondo Beach, Cal., 1962—, Aerojet Gen. Nucleonics Corp., San Ramon, Cal., 1963—. Recipient John Scott award City of Phila., 1956; John Price Wetherill medal Franklin Inst., 1963; Golden Plate award Am. Acad. Achievement, 1964. Fellow Am. Phys. Soc., I.E.E.E. Contbr. articles to tech. jours. Study of semiconductor physics; photocells; transistors; diodes; contact erosion; noise in vacuum tubes and resistors; patentee solid state electronic devices. Home: 185 Golden Oak Dr., Menlo Park, Cal.*

PEARSON, Gustaf Adolph, Am. silvicultrist; b. Holdredge, Neb., Nov. 14, 1880; s. Anders Peter and Anna Christina (Arvidson) P.; A.B., U. Neb., 1906, B.S., 1906, A.M., 1907; m. May Perkins, June 14, 1910; children—Arthur Adolph, Margaret Angeline. In charge Ft. Valley Forest Expt. Sta., U.S. Forest Service, Flagstaff, 1909-29; dir. Southwestern Forest and Range Expt. Sta., 1930-35, sr. silviculturist in charge Ft. Valley Exptl. Forest, 1935-44; collaborator U.S. Forest Service, 1945. Trustee No. Ariz. Soc. Sci. and Art. Fellow A.A.A.S., Soc. Am. Foresters (award for best article on forestry 1944); mem. Ecol. Soc. Am., Sigma Xi. Author: Natural Reproduction of Western Yellow Pine, 1923; Forest Types in the Southwest as Determined by Climate and Soil, 1931; Timber Growing and Logging Practice (with R. E. Marsh), 1935; Forest Land Use, 1940; also articles, bulls. Pioneer in research on reforestation, forest meteorology and ecology, silviculture. Died Jan. 31, 1949.

PEARSON, Karl, English statistician; b. London, Mar. 27, 1857; s. William and Fanny (Smith) P.; B.A., King's Coll., Cambridge, 1879, also Math. Tripos, 3d Wrangler; LL.B., 1881, M.A., 1882, fellow, 1880-86; student Heidelberg, Germany, Berlin, Germany; LL.D., St. Andrew's, Scotland; D.Sc., London; m. Maria Sharpe, 1890; 1 son, 2 daus.; m. 2d, Margaret Child, 1929. Goldsmid prof. applied math. U. Coll., London, 1884-1911, Gresham prof. geometry, 1891-1911, Galtan prof. eugenics, dir. Galton Lab. Eugenics, 1911-33, emeritus prof., 1933-36. Named hon. fellow King's Coll., Cambridge, 1903, U. Coll. London; recipient medal Royal Anthropol. Inst., 1903, Rudolf Virchow medal Anthropol. Soc. Berlin, 1932. Fellow Royal Soc., 1896 (Darwin medal 1898), Royal Soc. Edinburgh (hon.), anthropol. socs. of Paris and USSR (hon.). Author: The Ethic of Free Thought, 1887; The Grammar of Science, 1892, 1900, 37; The Changes of Death and Other Studies in Evolution, 1897; National Life from the Standpoint of Science, 1901; The Life, Letters, and Labours of Francis Galton, 1915; (with Geoffrey M. Morant). The Portraiture of Oliver Cromwell, 1935. Editor: Tables of the Incomplete Gamma Function, 1923. Co-founder Biometrika, 1900, joint editor, 1900-06, mng. editor, 1906-36. Leader in application of statistics, especially to biology; introduced chi-square test, 1900. Died Coldharbour, Eng., Apr. 27, 1936.

PEARSON, Manuel Malcolm, Am. physician; b. Derby, Conn., Feb. 5, 1911; s. Samuel Sam and Lena (Lazar) P.; B.A., Brown U., 1932; M.D., Jefferson Med. Coll., 1936; m. Roslyn Lande, Oct. 1, 1937; children—Laurence Henry, David Michael. Rockefeller Found. fellow in psychiatry Inst. Pa. Hosp., 1940-41; practice psychiatry, Phila., 1941—; asst. prof. psychiatry U. Pa. Med. Sch., Phila., 1951-54, asso. prof. psychiatry Med. Sch. and Grad. Sch. Medicine, 1954-67, professor of clinical psychiatry, since 1967—; vis. chief U. Pa. sect. Phila. Gen. Hosp., 1960—; attending psychiatrist VA Hosp., Coatesville, Pa., 1946-55; sr. psychiatrist Inst. Pa. Hosp., 1964—. Cons. Valley Forge Army Hosp., 1955—. Mem. Phila. Psychiat. Soc. (pres. 1955), A.M.A., Am. Psychiat. Assn., Assn. Research and Nervous and Mental Diseases, Am. Psychosomatic Soc., Am. Group Psychotherapy Assn. Author: Strecker's Fundamentals of Psychiatry, 1963; also articles. Research on group psychotherapy, therapy drug addiction, mil. psychiatry. Home: 267 Meeting House Lane, Merion Station, Pa. 19066. Office: 11 N. 49th St., Phila. 19139.*

PEARSON, Olof Hjalmer, Am. physician; b. Boston, Feb. 7, 1913; s. Nils August and Esther (Peterson) P.; A.B. magna cum laude, Harvard, 1934, M.D. cum laude, 1939; m. Barbara Farr, Dec. 30, 1942; children—Jane, Alan, John, Thomas. Asst. physiology Harvard, 1935-37, instr., 1945-46; faculty Cornell U. Med. Coll., 1949-60, asso. prof. medicine, 1952-60; asso. mem. Sloan Kettering Inst., N.Y.C., 1948-60; professor medicine Western Res. U. Sch. Medicine, Cleve., 1968—, head sect. endocrinology dept. medicine, 1963—; dir. clin. labs. U. Hosps. Cleve., 1960-63, asso. attending physician, 1960—. Recipient Sloan award for cancer research Sloan Kettering Inst., 1955. Fellow A.C.P.; mem. Assn. Am. Physicians, Am. Assn. for Cancer Research, Am. Assn. for Clin. Research, Am. Soc. for Clin. Investigation, Harvey Soc., Central Soc. for Clin. Research. Author: (with C.C. Thomas), Hypophysectomy, 1957; also numerous articles. Editor: Dynamic Studies of Metabolic Bone Disease, 1964. Research in endocrinology and metabolism. Home: 1411 Larchmere Blvd., Shaker Heights, O. 44106. Office: 2065 Adelbert Rd., Cleve. 44106.*

PEARSON, Paul Brown, Am. biochemist, sci. adminstr.; b. Oakley, Utah, Nov. 28, 1905; s. Levi and Ada (Brown) P.; B.S., Brigham Young U., 1928; M.S., Mont. State U., 1930; Ph.D., U. Wis., 1937; m. Emma Snow, June 21, 1933; children—Paula (Mrs. Raymond Soller), Marilyn. Faculty, Mont. State U., 1929-32, U. Cal., Berkeley, 1933-35; faculty Tex. A. and M. Coll., 1937-49, dean. Grad. Sch., chmn. dept. biochemistry and nutrition, 1947-49; head biol. scis. AEC, 1949-58; prof. Johns Hopkins, Balt., 1952-58; scientist Ford Found., N.Y.C., 1958-63; pres. Nutrition Found., N.Y.C., 1963—. Cons. govt. agys., 1958—; U.S. del. to various internat. sci. congresses, 1952, 54, 55, 60. Contbr. articles to sci. lit., NRC bulls. Study of micronutrients and mineral metabolism; utilization of proteins and amino acids; metabolism of sulphur and enzymes; B vitamins. Home: 15 Mark Dr., Port Chester, N.Y. 10573. Office: 99 Park Av., N.Y.C. 10016.*

PEARSON, Ralph Gottfrid, Am. chemist; b. Chgo., Jan. 12, 1919; s. Gottfrid and Kerstin (Larson) P.; B.S., Lewis Inst., 1940; Ph.D., Northwestern U., 1943; m. Lenore Johnson, June 15, 1941; children—John Ralph, Barry Lee, Christie Ann. Faculty, Northwestern U., Evanston, Ill., 1946—, prof. chemistry, 1957—. Cons. Dow Chem. Co., Midland, Mich., 1957—. Mem. Am. Chem. Soc., Phi Beta Kappa, Sigma Xi. Author: (with A.A. Frost) Kinetics and Mechanism, 1961; (with F. Basolo) Mechanisms of Inorganic Reactions, 1958; also numerous articles. U.S.A. editor Theretica Chimica Acta, 1964—. Research on mechanisms of chem. reactions; discovered principle soft and hard acids and bases. Home: 2700 Bryant St., Evanston, Ill. 60201.*

PEARSON, William, English astronomer; b. Whitbeck, Eng., Apr. 23, 1767; s. William and Hanna (Ponsonby) P.; ed. grammar sch., Hawkshead, Eng.; LL.D., 1819; m. twice; 1 dau. by 1st marriage. Took orders and went to Lincoln, Eng.; became owner pvt. sch., Temple Grove, East Sheen, 1811, built obs. there. Fellow Royal Soc., 1819; mem. Royal Instn. (charter propr.), Astron. Soc. London (a founder). Author: Introduction to Practical Astronomy, 1824. Built astron clock and orrey, Lincoln, apparatus for demonstrating Jupiter's satellites, 1798; observed Halley's comet, 1835, and calculated value for obliquity of ecliptic, 1839; measured diameters of sun and moon during partial solar eclipse of 1820. Died Sept. 6, 1847.

PEART, William Stanley, English physician; b. South Shields, Eng., Mar. 31, 1922; s. John George and Margaret (Fraser) P.; M.D., St. Mary's Hosp. Med. Sch., London, 1945; m. Peggy Parkes, Mar. 10, 1947; children—Celia, Robert. Research fellow pharmacology dept. Edinburgh (Scotland) U., 1947-49, Nat. Inst. for Med. Research, London, 1952-54; faculty U. London St. Mary's Hosp., 1954—, prof. medicine, dir. med. unit, 1956—. Fellow Royal Coll. Physicians; mem. Med. Research Soc., Renal Assn. Contbg. author: Renal Disease, 1962; Biochemical Disorders in Human Disease, 1964; Cecil and Loeb Medicine, 1966. Research, publs. on 1st demonstration of release of noradrenaline into blood on sympathetic nerve stimulation; isolated angiotensin and showed its amino acid sequence; purified renin; showed release new humoral substances in certain cancers; carried out successful cadaveric renal transplantations. Home: 5 Fordington Rd., London N-6, Eng.*

PEARY, Robert Edwin, Am. arctic explorer; b. Cresson, Pa., May 6, 1856; s. Charles N. and Mary (Wiley) P.; C.E., Bowdoin Coll., 1877 (Sc.D., 1894, LL.D.; LL.D., Edinburgh and Tufts); m. Josephine Diebitsch, 1888. Entered U.S. Navy as civil engr., Oct. 26, 1881; asst. engr. Nicaragua Ship Canal under Govt. orders, 1884-85; engr. in charge of Nicaragua Canal Surveys, 1887-88; reconnaissance, 1886, of the Greenland inland ice-cap; chief of Arctic expdn. of Acad. National Sciences of Phila., June 1891-Sept. 1892, to N.E. angle of Greenland; another arctic voyage, 1893-95; made summer voyages, 1896-1897, bringing the Cape York meteorites to U.S.; comdr. Arctic expdn. under auspices of Peary Arctic Club of New York, 1898-1902; sailed north again, July 1905, in S.S. Roosevelt, specially built by Peary Arctic Club; returned Oct. 1906, having reached "highest north"; started on 8th Arctic expdn., July 1908; retired U.S.N., 1913. Promoted to rank of rear admiral, and given thanks of Congress by special act of Congress, Mar. 3, 1911. Spl. gold medals of Nat. Geog. Soc. (Washington); Royal Geog. Soc. (London); Phila. Geog. Soc., Peary Arctic Club and Explorers Club; awarded the Hubbard gold medal by the Nat. Geog. Society, Culver gold medal, Chicago Geog. Soc.; Kane gold medal, Phila. Geog. Soc.; Daly and Cullom gold medals, Am. Geog. Soc.; gold medal of Imperial German, Austrian, and Hungarian socs.; Royal, Royal Scottish, Italian, and Belgian socs.; Swiss, Paris, Marseilles, Normandy, and City of Paris. Pres. Am. Geog. Soc., 1903; pres. 8th Internat. Geog. Congress, Washington, 1904; hon. v.p. 9th Internat. Geog. Congress, Geneva, 1908, and 10th, at Rome, 1913; pres. Explorers Club, and Aerial League America. U.S. Govt. del. Internat. Polar Commn., Rome, 1913; sec. Internat. Polar Commn.; chmn. Nat. Aerial Coast Patrol Commn. Made Grand Officier d'Honneur, France, 1913; hon. mem. Philadelphia Geog. Soc., Am. Alpine Club, Nat. Geog. Soc., Am. Mus. Natural Hist., N.Y. Chamber of Commerce, and all prin. home and foreign geog. socs. Author: Northward Over the Great Ice, 1898; Nearest the Pole, 1907; The North Pole, 1910; Secrets of Polar Travel, 1917. Leader, organizer 1st expdn. to reach North Pole, Apr. 6, 1909; previously discovered and named Melville Land and Heilprind Land; determined insularity of Greenland, discovered Iron Mountain of meteorites, 1894; named Cape Morris F. Jesup, northernmost in the world; rounded Greenland archipelago; designed rolling-rock canal gates; adapted small amount to European equipment and technology to Eskimo equipment and customs in devel. successful methods of arctic exploration. Died Washington, D.C., Feb. 20, 1920.

PEASE, Francis Gladhelm, Am. astronomer; b. Cambridge, Mass., Jan. 14, 1881; s. Daniel and Katharine Bangs (James) P.; B.S., Armour Inst. Tech., 1901, M.S., 1924, D.Sc., 1927; D.Sc., Oglethorpe U., 1934; m. Caroline T. Furness, Apr. 20, 1905. Optician and observer Yerkes Obs., Williams Bay, Wis., 1901-04; instrument designer Mt. Wilson Obs., Pasadena, Cal., 1904-07, 1908-13, astronomer, 1911—; supt. Sci. Shop Works, Evanston, Ill., 1907-08. Chief draftsman NRC, Washington, World War, 1918. Made direct photographs and spectograms of nebulae and star clusters, of moon and planets; also interferometer measures of star diameters, measurement of velocity of light, ether drift; in charge of design of 100-inch telescope, 50-foot interferometer telescope; asso. in optics and instrument design, 200-inch reflector Cal. Inst. Tech. Died Feb. 7, 1938.

PEASE, Michel Stewart, English geneticist; b. Aberfeldy, Eng., Oct. 2, 1890; s. Edward and Mary Gammell (Davidson) P.; M.A., Trinity Coll., Cambridge (Eng.) U.; m. Helen Bowen Wedgwood, Feb. 24, 1920; children—Joanna, Sebastian, Richenda, Dora, Fabian. Research asst. Small Animal Breeding Inst., Cambridge, 1920-49; adminstr. Poultry Genetics Research Inst., Cambridge, 1950-57. Mem. Genetical Soc. Author articles in Jour. Genetics, 1922-50. Died July 27, 1966.

PEASE, Rendel Sebastian, English physicist; b. Cambridge, Eng., Nov. 2, 1922; s. Michael Stewart and Helen (Wedgwood) P.; B.A., Cambridge U., 1943, M.A., 1950, Sc.D., 1964; m. Susan Spickernell, Aug. 9, 1952; children—Rosamund, Sarah, Christopher, Roland, Rowan. Operational research R.A.F., 1942-46; sci. officer U.K. Atomic Energy Authority, Atomic Energy Research Establishment, Harwell, 1947-61; div. head Culham Lab., Abingdon, Berks., Eng., 1961-66, head lab., asst. dir. research group, 1967—. Mem. physics com. Sci. Research Council. Recipient Kelvin Premium award Inst. Elec. Engrs., 1959. Mem. Am. Phys. Soc., Inst. Physics, Phys. Soc. Editor: (with A.C. Kolb) Controlled Thermonuclear Reactions, 1964. Research, publs. on irradiation damage in solids, atomic displacement and disorder, neutron diffraction, controlled thermonuclear reactions. Home: The Poplars, W. Ilsley, Newbury, Berks. Office: Culham Lab., Abingdon, Berks, Eng.*

PEASLEE, David Chase, physicist; b. White Plains, N.Y., July 23, 1922; s. Arthur F. and Anita Q. (Clark) P.; A.B., Princeton, 1943; Ph.D., Mass. Inst. Tech., 1948; m. Virginia P. Close, Oct. 25, 1947; children—Anne C., Frank D., Graham F. Asst. prof. physics Washington U., St. Louis, 1950-51; research asso. Columbia, 1951-54; asso. prof. Purdue U., 1954-59; prof. Research Sch. Phys. Sci., Australian Nat. U., Canberra, 1959—. AEC postdoctoral fellow Eidgenössische Technische Hochschule, Zurich, Switzerland, 1949-50. Research on structure and interactions of elementary and nuclei particles. Home: 14 Hamelin Crescent Narrabundah, A.C.T., Australia. Office: Australian Nat. U., Canberra, Australia.

PEASLEE, Edmund Randolph, Am. physician; b. Newton, N.H., Jan. 22, 1814; s. James and Abigail (Chase) P.; grad. Dartmouth, 1836; postgrad. Dartmouth Med. Sch., 1837-39; M.D., Yale, 1840; m. Martha Kendrick, 1841, 2 children. Prof. anatomy and physiology Dartmouth, 1842-69, lectr. on diseases of women, 1868-70, prof. obstetrics and diseases of women, 1870-73, prof. gynecology, 1873-78, trustee, 1860-78; prof. surgery and anatomy Med. Sch. of Me., 1843-1860; prof. pathology and physiology N.Y. Med. Coll., 1852-56; prof. obstetrics, lectr. obstetrics Albany Med. Coll., 1872-74, prof. gynecology, 1874-78; prof. gynecology, 1856-60; Bellevue Hosp. Med. Coll., 1874-78; attending physician Demilt Dispensary, N.Y.C., 1958-65; private practice medicine, N.Y.C., 1858-78. Author: Necroscopic Tables for Postmortem Examinations, 1851; Human Histology in Its Relations to Descriptive Anatomy, Physiology, and Pathology, 1857; Ovarian Tumors; Their Pathology, Diagnosis, and Treatment, Especially by Ovariotomy, 1872. Authority on microscopy; pioneer in use of peritoneal lavage for prophylaxis against infection. Died N.Y.C., Jan. 21, 1878.

PEATMAN, John Gray, Am. psychologist, statistician; b. Centerville, Ia., Mar. 16, 1904; s. Clarence Albert and Binney (Gray) P.; student U. Colo., 1922-25; A.B., Columbia, 1927, M.A., 1928, Ph.D., 1931; m. Lillie Burling; children—Alice (Mrs. W.L. Dettmers, Jr.), John, William; m. 2d, Madeline Martin; 1 dau., Mary. Asst. psychology Columbia, 1927-28; instr. psychology St. Mark's Hosp., N.Y.C., 1928-29; faculty Coll. City N.Y., 1929—, prof., 1948—, sec. faculty liberal arts and sci., 1940-65, asso. dean, 1944-53, chmn. dept. psychology, 1952-63; grad. faculty, 1962—; vis. prof. Columbia, 1948. Pres., Office of Research Inc., N.Y.C., Norwalk, Conn., 1945-58, Research Consultant, Inc., Norwalk, 1950—. Diplomate Am. Bd. Examiners in Profl. Psychology. Fellow Am., Conn., N.Y. State (pres. 1946-47) psychol. assns.; mem. Met. N.Y., (dir. 1942-45), Internat. assns. applied psychology, Am. Statis. Assn., Psychonomic Soc., Am. Assn. U. Profs., Phi Beta Kappa, Sigma Xi (pres. City Coll. club 1962-63). Author: Descriptive and Sampling Statistics, 1947; (with Tore Hallonquist) Geographical Sampling in Testing Appeals of Radio Broadcasts, 1950; Introduction to Applied Statistics, 1963. Co-editor Festschrift for Gardner Murphy, 1960. Devel., application of sci. methods in area of statis. techniques, including sampling to research problems in psychology and related fields, especially market research. Home: Comstock Hill, R.D. 2, Silvermine, Norwalk 06850. Office: 83 East Av., Norwalk, Conn. 06851; also City U. N.Y., N.Y.C. 10031.*

PEAUCELLIER, A., engr.; flourished 19th century; capt. French army; engr., Nice, France; recipient Priz Montyou, Inst. France. Credited with giving 1st exact straight-line linkage, 1873, made diagram demonstrating solution (Peaucellier cell), which is important to inverse geometry and considered an improvement over Watt's parallelogram.

PEBAY-PEYROULA, Jean-Claude, French physicist; b. Chartres, France, Sept. 6, 1930; s. Jean-Pierre and Simone (Maréchaux) P.-P.; License es Sci., Ecole Normale Superieure Paris (France), 1949, Agregation de Scis. Physiques, 1953, Doctorat d'Etat de Scis. Physiques, 1958; m. Colette Bonnavaud, Feb. 21, 1955; children—Francois, Anne, Veronique. Staff, Laboratoire de Physique, Ecole Normale Superieure, Paris, 1955-60; faculty Faculte des Scis., U. Grenoble (France), 1960—, prof. physics, 1965—. Cons. Laboratoire Central de Télécommunications, Paris, 1960—. Mem. Société Francaise de Physique, Am. Phys. Soc. Research, publs. on atomic excited levels by resonance techniques using an excitation by slow electrons; determination of lifetimes, Zeeman structure and cross sect. of excitation; alloys and magnetic compounds using Mössbauer technique. Home: 9, Place St-Bruno, 38-Grenoble, France.*

PECHLIN, John Nicholas, anatomist; b. Leyden, Netherlands, 1646; M.D., Leyden, 1667; prof. physic, Kiel, Germany; chief physician to Duke of Holstein. Mem. l'Académie des curieux de la nature, Royal Soc., 1688. Author: De apoplexia, 1667; Exercitatio nova de purgantium, medicamentorum facultatibus, 1672; De epilepsia et remedius contra illam, 1676; Purgant med. facultatibus; Dissertations de fabrica et usu cordis; Habitu et colore aethiopum; also articles. Described villous coat of intestines, intestinal glands; discussed animals and men living without food or air for long periods. Died 1706.

PECHMANN, see von Pechmann.

PECK, Edson Ruther, Am. physicist; b. Evanston, Ill., Oct. 29, 1915; s. David Billings and Janet (Cameron) P.; B.A., Northwestern U., 1936, M.S., 1937; Ph.D., U. Chgo., 1945; m. Mary Catherine MacLean, Dec. 21, 1946; children—David Fraser, John Edson, Janet Willene, Keith MacLean, Kathryn Ann. Faculty, Northwestern U., 1942-62; prof. physics U. Ida., Moscow, 1962—. Cons. Argonne (Ill.) Nat. Lab., 1950-56, Bell & Howell Corp., 1955-57, Tech. Research Group, 1959; AAPT-AIP regional counselor in physics for Ida., 1963-66. Fellow A.A.A.S., Optical Soc. Am.; mem. Am. Phys. Soc., Am. Math. Soc., N.Y. Acad. Scis., Phi Beta Kappa, Sigma Xi. Author: Electricity and Magnetism, 1953. Research in modified Michelson interferants using corner reflectors, ind. of reflection angle, along with electronic methods of fringe counting and interpolation, applications to wave length comparisons, gaseous refractive index and dispersion, dilatometers, angle measurements, magnetostriction, microcreep, theory of corner-cube cavities. Home: 536 Ridge Rd., Moscow, Ida. 83843.*

PECK, Franklin Bruce, Sr., Am. physician; b. Remington, Ind., Sept. 28, 1898; s. Frank L. and May (Tedford) P.; A.B. in Physiology, Ind. U., 1920; M.D., Jefferson Med. Coll., Phila., 1923; m. Lisbeth C. Roeder, Sept. 18, 1924; children—Franklin Bruce, Elizabeth L. (Mrs. F. W. Schoeneman, Jr.). Asst. physiology, Ind. U., 1919-20; staff Lilly Research Labs., Indpls., 1936-40, asso. dir. med. div., 1940-46, dir., 1946-52, dir. med. research coop., 1953-58, dir. clin. research internat., 1958-61, spl. adviser diabetes dept. clin. research div., 1961—; asso. medicine Ind. U. Sch. Medicine, 1943-49, asst. prof., 1949-51, asso. prof., 1951—; physician Indpls. Gen. Hosp., 1942-50, cons. medicine, 1950—, pres. staff, 1951; cons. White Co. Meml. Hosp., 1961—. Organizing mem. Internat. Diabetes Fedn., Brussels, 1949, Am. Diabetes Assn. del., Dusseldorf, 1958. Recipient Davis prize Jefferson Med. Coll., 1923; named academician of honor Acad. Medicine and Surgery, Panama, 1957; Banting medal, 1961. Fellow A.C.P.; mem. Am. Diabetes Assn. (mng. editor jour. 1942-52, cons. com. on function and structure, pres. 1960-61), A.M.A.; hon. mem. numerous fgn. med. socs. and diabetes assns. Contbr. articles relating to diabetes and new insulin modifications. Address: 5738 Brockton Dr., Indpls. 46220.*

PECK, Merton Joseph, Am. economist; b. Cleve., Dec.17, 1925; s. Kenneth Richard and Charlotte (Harn) P.; A.B., Oberlin Coll., 1949; A.M., Harvard, 1951, Ph.D., 1954; A.M. (hon.) Yale, 1963; m. Mary McClure Bosworth, June 13, 1949; children—Richard, Katherine, Sarah, David. Instr. econs. Harvard, 1954-55, faculty, 1956-61, asst. and asso. prof. bus. adminstrn., 1956-61; dir. systems analysis Office of Asst. Sec. of Def., Washington, 1961-62; prof. econs. Yale, 1963—. Ford Faculty fellow, 1965-66. Mem. Am. Econ. Assn., Econometrics Soc. Author: Competition in the Aluminum Industry, 1961; (with Richard Nelson, Edward Kalachek) Technology, Economic Growth, and Public Policy, 1967. Analyzed (with others) role of science and tech. in econ. growth; statis. studies on r.r. coasts, econ. relationships between defense contracts and Dept. Def. Home: 415 Humphrey St., New Haven 06511.*

PECK, Raymond Elliott, Am. geologist; b. Hamilton, Mo., May 3, 1904; s. Frank Sherrard and Olive (Newton) P.; A.B., Park Coll., 1926; M.A., U. Mo., 1928, Ph.D., 1932; m. Vaona Olive Hedrick, Oct. 13, 1929. Asst. geologist Mo. Hwy. Commn., 1928-30; faculty geology U. Mo., Columbia, 1930—, prof., 1947—, chmn. dept. geology, 1950-59, acting dean Grad. Sch., 1959-60, asso. dean, 1961-63, dean research adminstrn., 1963—. Geologist, U.S. Geol.

Survey, intermittently 1950-56, 58——; contract geologist Shell Oil Co., summers 1956-58. Fulbright research fellow U. Paris (France), 1950-51. Recipient Distinguished Faculty award U. Mo. Alumni Assn., 1960. Fellow Geol. Soc. Am. (councilor 1963——), Paleontol. Soc. Am.; mem. Assn. Petroleum Geologists, Soc. Econ. Paleontologists and Mineralogists (sec.-treas. 1958-59), Mo. Geol. Assn. (pres. 1959). Research, publs. on evolution and systematics of Charophyta, Paleozoic and Mesozoic crinoids. Home: 703 W. Rollins Rd., Columbia, Mo. 65201.*

PECK, William Dandridge, Am. naturalist; b. Boston, May 8, 1763; s. John and Hannah (Jackson) P.; B.A., Harvard, 1782. Lived on farm, Kittery, Me., 1785-1805, engaged in bot. and zool. studies; 1st prof. natural history Harvard, 1805-22; established Botanic Garden, Cambridge, Mass.; 1st tchr. of entymology in U.S.; a founder Am. Antiquarian Soc., 1812. Author: The Description and History of the Canker-Worm (article), 1795, Natural History of the Slug Worm, 1799 (Gold medals for both from Mass. Agrl. Soc.). First to discover and describe an egg parasite in U.S. Died Oct. 3, 1822.

PECK, William Guy, Am. mathematician; b. Litchfield, Conn., Oct. 16, 1820; grad. U.S. Mil. Acad., 1844; A.M., Trinity Coll., 1853, LL.D., 1863; Ph.D., Columbia Coll., 1877; m. Miss Davis. Mem. survey of mil. fortifications of Portsmouth, N.H.; brevetted 2d lt. Topog. Engrs., U.S. Army; mem. John C. Fremont's 3d expdn. to Rocky Mountains; promoted lt. Topog. Engrs.; asst. prof. philosophy and math. U.S. Mil. Acad., 1846-55; prof. physics and civil engring. U. Mich., 1855-57; prof. math. Columbia, 1857-61; mem. bd. visitors U. S. Mil. Acad., 1868. Author: Elementary Mechanics, 1859; Manual of Arithmetic, 1874; Popular Astronomy, 1883. Died Greenwich, Conn., Feb. 7, 1892.

PECKER, Jean-Claude, French astrophysicist; b. Reims, Marne, France, May 10, 1923; s. Victor Noel and Nelly (Herrmann) P.; Agrégé des Sciences Physiques, Ecole Normale Supérieure, 1946, Docteur ès Sciences Physiques, 1950; div.; children—Martine, Daniel, Laure. With CNRS, 1945-52; faculty Faculté des Sciences, Clermont-Ferrand, France, 1952-55; staff Observatoire de Paris, 1955-65; dir. Observatoire de Nice (France), 1962——; prof. Coll. de France, Paris, 1962——. Decorated chevalier de la Légion d'Honneur; Officier des Palmes Académiques-Lauréat de l'Académie des Sciences de Paris et de L'Ac. Royale des Sciences d'Uccle (Belgium); Médaille d'argent du CNRS. Author: (with E. Schatzman) Astrophysique Générale, 1959; Le Ciel, 1962; also numerous articles. Research on stellar atmospheres including theory of curve of growth, introduction of saturation function, empirical determination of departures from local thermodynamical equilibrium, and cosmical abundances. Home: 2 r. Pierre Sémard, 92 Bagneux, France. Office: Observatoire de Nice le Mont-Gros, 06, Nice, France.*

PECKHAM, Ben Miller, Am. physician; b. Milw., Mar. 8, 1916; s. John William Gross and Mary Peckham; B.S., U. Wis., 1934-39, M.D., Northwestern U., 1942, M.S., 1947, Ph.D., 1949; m. Ann Faherty, Aug. 16, 1943; children—Mary, Roger, Gardner. Research asst. Northwestern U. Med. Sch., Chgo., 1946-47, instr., 1947-48, 51-53; clin. asst., 1950-51, asso., 1953-54, asst. prof., 1954-56; attending staff Chgo. Wesley Meml. Hosp., 1950-56; attending staff Chgo. Maternity Center, 1950-56; prof., chmn. dept. gynecology and obstetrics U. Wis. Med. Sch., Madison, 1956——, asso. dean for clinical affairs, since 1967——. Decorated Purple Heart, Bronze Star medal; recipient Centennial Merit award Northwestern U., 1959; Distinguished Teaching award U. Wis., 1963. Mem. Assn. Prof. Gynecology and Obstetrics (pres. 1964), Soc. for Gynecol. Investigation (pres. 1965), Am. Assn. Obstetrics-Gynecology, Am. Gynecol. Soc., Am. Gynecol. Club, Am. Coll. Obstetrics-Gynecology, Wis. Soc. Obstetrics-Gynecology, A.M.A., Wis., Madison med. socs. Research on pathology and physiology of female genital tract; cell population kinetics of placenta and vagina. Home: 3433 Crestwood Dr., Madison, Wis. 53705.

PECKHAM, John, English mathematician, physicist; b. Sussex, Eng., circa 1220; student, Oxford, Eng.; degree (under St. Bonaventure), U. Paris; became mem. Franciscan order, circa 1250; lived in Paris, circa 1250-circa 1270, also taught there; returned to teach in Oxford, circa 1270; lectr. in theology, Rome, sch. of papal palace; archbishop of Canterbury, 1279-92. Author: Perspectiva communis (on optics or perspective, reflection and refraction, includes description of eye and probably 1st diagram in print); Theorica planetarium; Tractatus sphaerae; De numeris; also religious works. Held astron. views similar to Roger Bacon's; opposed St. Thomas Aquinas; interested in meteorology and optics. Died Mortlake, Eng., Dec. 8, 1292.

PECKHAM, Robert Hamilton, Am. biophysicist; b. Youngstown, O., Mar. 15, 1908; s. Ray Clifford and Edith (Roake) Morse-Peckham; B.A., U. Rochester, 1930; Ph.D., Johns Hopkins, 1933; m. Neva Kempton Brown, Mar. 26, 1932; children—Henry Morse, John Kempton, Robert Smith. Faculty, Temple U. Sch. Medicine, Phila., 1934-52, asso. prof. research ophthalmology, 1946-52; dir. research Houze Glass Co., Point Marion, Pa., 1952-54, Pa. Optical Co., Reading, Pa., 1954-57; v.p. Penberthy Electromelt Co., Seattle, 1957; v.p. Eye Research Found., Bethesda, Md., 1957-66; research prof. physiol. optics U. Md., 1962-66, prof. physiol. optics So. Coll. Optometry, Memphis, 1966——. Cons. pvt. cos. NRC fellow in psychology U. Pa., 1933-34. Mem. Optical Soc. Am., Am. Psychol. Assn., Aero-Space Medicine Assn., Am. Ceramic Soc., Internat. Soc. Clin. Electroretinography, Sigma Xi, Phi Beta Pi, Epsilon Psi Epsilon. Research, numerous publs. on etiology squint, binocular vision, color vision, effect sunlight on retinal sensitivity, usefulness sunglasses, med. statistics, electroretinography frog eye, electroretinography, and visually evoked cortical responses in partially blind human eye, devel. galactose cataract in albino rat, effect retinal sensitivity in hwy. vehicular driving. Office: 1246 Union Av., Memphis 38104.*

PECLET, Jean-Claude-Eugène, French physicist; b. Besancon, France, 1793; ed. École Normale; prof., Marseille, France; founder, prof. École Centrale; head École Normale. Author: Cours de chimie, 1823-26; Cours de physique, 1823-26; Traité de l'éclairage, 1827; Traité de la chaleur et de ses applications aux arts et aux manufactures, 1829, English version with atlas, 1843. Determined (with Fourier) conductivity of various substances to 100 degrees Celsius, 1841; constructed a condensor and a galvanometer, 1838-41. Died 1857.

PECORA, William Thomas, Am. geologist; b. Belleville, N.J., Feb. 1, 1913; s. Cono and Ann (Amabil) P.; B.S., Princeton, 1933; Ph.D., Harvard, 1940; m. Ethelwyn Elizabeth Carter, Apr. 7, 1947; children—William Carter, Am. Stewart. Employed with the United States Geological Survey, Washington, 1939——, chief, geochemistry and petrology br., 1957-64, chief geologist, 1964-65, dir., 1965——. Chmn., Fed. Council Interagy. Com. Solid Earth Scis. Mem. adv. coms. various univs., NSF, NRC, numerous others. Fellow Am. Acad. Arts and Scis., Geol. Soc. Am., Mineral. Soc. Am.; mem. Nat., Washington, Brazilian (fgn.), acads. scis., Geol. Soc. Washington (pres. 1964), Am. Assn. Petroleom Geologists, Am. Inst. Profl. Geologists, others. Field and lab. investigations in western hemisphere including nickel deposits in U.S., Alaska, Brazil, Venezuela; mica and related pegmatite mineral deposits of Brazil and Colombia; rare mineral deposits in alkalic igneous rocks and carbonatite complexes; phosphate mineralogy; geologic field mapping and mineral evaluations. Home: 4572 Indian Rock Terrace N.W., Washington 20007. Office: U.S. Geol. Survey, Washington 20242.*

PECQUET, Jean, French physician; b. Dieppe, France, May 9, 1622; student Dieppe, with Jesuits, Rouen, France, Coll. Clermont, Paris; M.A., Acad. Paris; doctor's degree, Montpellier, France, 1652. Practice medicine, Montpellier; researcher, Paris. Mem. Royal Acad. Scis., French Acad. Scis. Author: Experimenta nova anatomica, 1651; Dissertation sur la circulation du sang et le mouvement du chyle; Traitez de l'équilibre des liquides; Nouvelle dissection sur les vaisseaux lactés, 1654. Discovered thoracic duct in dog (Pecquet's duct), 1651, cistern chyli (reservoir of Pecquet), course of lacteal vessels, termination of thoracic duct at opening into left subclavian vein. Died Paris, Feb. 1674.

PECSOK, Robert Louis, Am. chemist; b. Cleve., Dec. 18, 1918; s. Michael C. and Katherine (Richter) P.; S.B. summa cum laude, Harvard, 1940, Ph.D., 1948; m. Mary Bodell, Oct. 12, 1940; children—Helen (Mrs. Bent Hansen), Katherine (Mrs. Garrett Capune), Jean, Michael, Ruth, Alice, Sara. Prodn. foreman Procter & Gamble, Balt., 1940-43; prof. chemistry U. Cal., Los Angeles, 1948——. Panelist in chemistry NSF, 1961-63; sci. adviser U.S. FDA, Los Angeles Lab., 1966——. Guggenheim fellow, 1956-57; Am. Chem. Soc. Petroleum Research Fund Internat. fellow, 1963-64. Mem. Am. Chem. Soc. (chmn. So. Cal. sect. 1961), Am. Inst. Chemists, Phi Beta Kappa, Sigma Xi, Alpha Chi Sigma, Phi Lambda Upsilon. Author: Principles and Practice of Gas Chromatography, 1959; also articles. Research in electrochem. studies metal complexions, instrumental methods of chem. analysis. Home: 15321 DePauw St., Pacific Palisades, Cal. 90272. Office: Chemistry Dept., U. Cal., Los Angeles 90024.*

PEDERSEN, Asger, Danish cardiologist; b. Copenhagen, Denmark, Sept. 2, 1919; s. Gerhard P. and Margrethe (Jensen) P.; M.D., U. Copenhagen, 1957; m. Gertrud Galster, June 29, 1946; children— Torsten, Laust, Frederik. Staff, U. Hosp., Copenhagen, 1958-59, Bispebjerg Hosp., clin. fys. lab., Copenhagen, 1963-64; staff Copenhagen County Hosp. med. dept. C, 1959-63, 64——. Mem. various bds. for clin. pharmacology Danish Health Service; mem. bd. for med. instrumentation Copenhagen County Hosp., 1965——, mem. bd. for EDB, 1965——, chief physician, 1964——. Mem. Am. Coll. Chest Physicians. Author: Venous Pressure in Pulmonary Circulation, 1956; Drug Names and Formulas, 1965; Pharmacology for Nurses, 1966; also numerous articles. Research on pulmonary hemodynamics in acquired and congential heart disease, cardiac metabolism in cardiac arrest, defibrillation, renal hypertension, method for cardiac output determination; establishment and experiences for coronary care unit. Home: 82 Sandkrogen Farùm, Denmark. Office: Med. dept. C, Copenhagen County Hosp., Glostrup, Denmark.*

PEDERSEN-BJERGAARD, Kaj, Danish pharmacist; b. Maribo, Mar. 25, 1904; s. Lars and Sigrid (Bjergaard) Pedersen-B.; ed. Royal Danish Sch. Pharmacology, U. Copenhagen, Ph.D.; m. Grete Pedersen-Bjergaard, Apr. 21, 1931; children—Jens, Niels, Ole, Lars. Asst. pharmacology dept. Copenhagen Municipal Hosp.; asst. Univ. Inst. Pathology of Copenhagen; dir. dept. biology Lövens kemishe fabrik; head pharmacy dept. Bispebjerg Hosp.; pharmacist-in-chief Valby apotek of Copenhagen. Mem. Danish Pharmacopoeia Soc., Danish Endocrinology Soc. (treas.). Author: Comparative Studies Concerning the Strengths of Oestrogenic Substances, 1939; The Preparation of Solutions Isoosmotic with Blood, Tears and Tissue, 1947; A Graphic Method for use in the Adjustment of Isotonic Solutions, 1958. Editor, dir. Acta endocrinologica. Home: 114 B, Havdrupvej. Office: Valby apotek 49, Valby Langgade, Copenhagen, Denmark.

PEDOE, Daniel, mathematician; b. London, Eng., Oct. 29, 1910; B.A., Magdalene Coll., Cambridge, Eng., 1933, Ph.D., 1937; B.Sc., U. London, 1930; children—Dan S. Tunstall, Hugh D. Tunstall; m. Marilyn Joel Sjoberg, Aug. 29, 1966. Faculty, U. London, 1947-52; prof. U. Khartoum, Sudan, 1952-59, U. Singapore, 1959-62; prof. Purdue U., 1962-64; prof. geometry U. Minn., Mpls., 1964——. Mem. Am. Math. Soc., Am. Math. Assn. Author: (with Sir William Hodge) Methods of Algebraic Geometry, 3 vols. 1947-53; The Gentle Art of Mathematics, 1957; Circles, 1957; A Geometric Introduction to Linear Algebra, 1963; Introduction to Projective Geometry, 1963; also articles. Study of algebraic geometry; mathematics education. Office: Inst. Tech., U. Minn., Mpls. 55455.*

PEELE, Talmage Lee, Am. anatomist, physician; b. Goldsboro, N.C., Aug. 16, 1908; s. Ichabod and Nora (Spence) P.; A.B., Duke, 1929, M.D., 1934; Research fellow Johns Hopkins, 1937-38; instr. U. Rochester Sch. Medicine, 1938-39; faculty Duke, Durham, N.C., 1939——, asst. prof. pediatrics, 1958——, asso. prof. neurology, prof. anatomy, 1963——; temporary asst. prof. neurology Northwestern U., 1945; vis. lectr. in neurology U. N.C., 1947-50, vis. asst. prof. anatomy, 1955. Mem. N.C. Neurospsychiat. Soc., Am. Assn. Anatomists, Med. Soc. State N.C., A.M.A., Am. Acad. Neurology. Author: The Neuroanatomic Basis for Clinical Neurology, 1954, 2d edit., 1961; also articles. Research on descending cortical projections from parietal lobe monkey, limbic system anatomy and function. Home: University Apts., Durham, N.C. 27701.*

PEERY, Thomas Martin, Am. pathologist; b. Lynchburg, Va., Aug. 24, 1909; s. John Carnahan and Alean (Martin) P.; B.A., Newberry Coll., 1928; M.D., Med. Coll. S.C., 1932; m. Eleanor Bishop, Oct. 9, 1936; children—Eleanor Brooke (Mrs. Frank Joseph Russell), Sue Bishop (Mrs. William Wilson Moore), Linda Littleton. Instr. pathology Med. Coll., S.C., Charleston, 1934-38; asst. pathologist Roper Hosp., Charleston, S.C., 1934-38; faculty pathology George Washington U., Washington, 1938——, prof., 1950——, chmn. dept. pathology, 1954——, dir. postgrad. instrn., 1945-56, chmn. curriculum com., 1946-58, chmn. com. ednl. policy, 1958——, dir. ednl. facilities study, 1962-63; chief dept. pathology George Washington U. Hosp., 1947——. Dir. Cooke Engring. Co., Alexandria, Va., Technicon Instruments Corp., Chauncey, N.Y. Cons. in pathology, VA, 1950——, clin. center NIH, 1960——, Armed Forces Inst. Pathology, 1963——. Diplomate Nat. Bd. Med. Examiners, Am. Bd. Pathology. Mem. Am. Soc. Clin. Pathologists (exec. com. 1955-58, bd. schs. med. tech. 1958-61, v.p. 1961, gold award for original investigation 1958, spl. citation 1964), Coll. Am. Pathologists, A.M.A. (chmn. sect. pathology, physiology 1959, chmn. exhibit lab. Clin. Pathology Tests in Health Evaluation 1961-64), Am. Assn. Pathologists and Bacteriologists, Sigma Xi, Alpha Omega Alpha. Author: (with Frank N. Miller) Pathology, 1961; pathology cons. Am. Coll. Dictionary, 1964——. Bd. editors Med. Annals of D.C., 1956——. Study of heart disease and brucellosis; clinical pathology tests in health evaluation. Home: 2115 Belle Haven Rd., Alexandria, Va. 22307. Office: 901 23d St. N.W., Washington 20037.*

PEET, Max Minor, Am. surgeon; b. Iosco, Mich., Oct. 20, 1885; s. LaFayette and Eunice Ann (Minor) P.; A.B., U. Mich., 1908, A.M., 1910, M.D., 1910; M. Ed. (hon.), Mich. State Normal Coll., Ypsilanti, 1934; m. Grace Stewart Tait, Oct. 5, 1915; children—Max Minor, Stewart Tait, Martha Eunice Ann. Intern R.I. Hosp., Providence, 1910-12; Robert Robinson Porter fellow in research medicine U. Pa., 1912-13, asst. instr. surgery, 1913-15; asst. chief Surgeon Phila. Gen. Hosp., 1914-16; instr. in surgery U. Mich., 1916-17, asst. prof., 1918-27, asso. prof. of neuro-surgery, 1927-30, prof. surgery since 1930, chief, neurosurg. div. Univ. Hosp., since 1918. Ornithologist U. Mich. Museums expdns., 1904, 05, 32; mem. gen. adv. com., virus research, med. publs., chmn. ed. comm. Nat. Found. for Infantile Paralysis mem. Internat. Neurol. Congress, Berne, 1931, London, 1935; Internat. Surg. Congress, Brussels,

1938. Fellow A.M.A., A.C.S.; mem. Soc. Neurol. Surgeons, Am. Surg. Assn., Am. Bd. Surgery, Am. Neurol. Assn., Bd. Neurol. Internat. Soc. of Surgery, Harvey Cushing Soc., Sociedad Argentina de Cirujanos (hon.), other med. socs., Sigma Xi. Contbr. numerous articles on neurosurg.. problems to med. jours., chpts. on neuro-surgery in various surg. works. Used exercise of greater and lesser thoracic ganglia for relief of hypertension, from circa 1935. Died Ann Arbor, Mich., Mar. 25, 1949.

PEETERS, Georges Joseph, Belgian physiologist; b. Overwinden, Belgium, Nov. 25, 1919; s. Hubert and Seraphine (Boffin) P.; student Cath. U., Louvain, Belgium, 1937-39; D.Vet. Sci., Vet. Coll., U. Ghent (Belgium), 1943; m. Roza Vander Waeren, Oct. 2, 1946; children—Wim, Anne Marie, Magda, Joris. With Vet. Coll., U. Ghent, 1943—, prof. vet. physiology and pharmacology, 1946—. Recipient sci. awards, Belgium, 1947, 54, 61, Germany, 1961. Mem. Royal Flemisch Acad. Medicine Belgium, numerous other sci. socs. in Belgium, Eng., Germany, France. Research, publs. on lactation in ruminants, including mechanism of milk ejection, physiology of nerves of udder, synthesis of milk constituents, intermediate metabolism of mammary gland, release of oxytocin in male and female. Home: 71 Brusselse Steenweg, Melle, East Flanders. Office: 21 Casinoplein, Ghent, East Flanders, Belgium.*

PEGOLOTTI, Francesco Balducci, Italian geographer; b. Florence, Italy, flourished circa 1340; s. Balduccio Pegolotti; represented Florence for negotiation and ratification of treaty of 1311, Siena, Italy; with Bardi Banking House, 1310-46, conducted negotiations, Antwerp, Belgium, 1315, became dir., 1318; elected to city adminstrn., Florence, 1331. Author: Libro di divisamenti di paesi e di misure di mercatantie (earliest book on comml. practice; described trade routes, markets, merchandise, moneys, weights and measures, bus. methods; includes tables of calendar, alloys, interest), circa 1340.

PEGRAM, George Braxton, Am. physicist; b. Trinity, N.C., Oct. 24, 1876; s. William Howell and Emma Lenore (Craven) P.; A.B., Trinity Coll., Durham, N.C., 1895, Ph.D., Columbia, 1903; postgrad. U. Berlin, 1907, Cambridge U., 1908; D.Sc., Trinity (Duke), 1918, Columbia, 1929, George Washington, 1937, U. N.C., 1946, Northwestern U., 1946, Case Inst. Tech., 1951; LL.D., U. Denver, 1950; m. Florence Bement, June 3, 1909; children—William Braxton, John Bement. Asst. in physics Columbia U., 1901-03, tutor, 1903-05, instr., 1905-07, Tyndall fellow, 1907, asst. prof. physics, 1909-12, asso. prof., 1912-18, prof. since 1918, acting dean Sch. Mines, Engring. and Chemistry, 1917-18, dean, 1918-30, dean, Grad. Faculties, 1937-49, v.p., 1949-50, spl. adviser to pres. since 1950; chmn. Columbia Com. on War Research, 1941-45; sci., also ednl. cons. Oak Ridge Institute Nuclear Studies. Fellow A.A.A.S., Am. Phys. Soc. (treas. 1917-57), Am. Soc. M.E., Inst. Aero. Scis.; mem. Am. Inst. Physics (treas., 1939-55), Nat. Acad. Scis., Am. Philos. Soc., Inst. Radio Engrs., Sigma Xi (pres. 1949-51). Demonstrated transmutation of elements (sodium to magnesium) by splitting of atom with slow neutrons. Died Aug. 12, 1958.

PEGRUM, Dudley Frank, economist; b. Romford, Eng., May 28, 1898; s. Frank and Elizabeth (Keen) P.; B.A., U. Alta., 1922, M.A., 1924; Ph.D., U. Cal. at Berkeley, 1927; m. Marion Pheasey, July 29, 1924 (dec. Jan. 1966); 1 son, Douglas Frank. Came to U.S., 1925, naturalized, 1936. Faculty, U. Cal. at Los Angeles, 1927-65, prof. econs., 1943-65, prof. emeritus, 1965—. Cons. economist govt. agys., state agys., orgns., indsl. firms, pub. utilities. Mem. Am. (Distinguished mem. award transp. and pub. utilities group 1966), Western econ. assns., Artus, Transp. Research Forum, Am. Soc. Traffic and Transp., Alpha Kappa Psi, Pi Gamma Mu, Delta Mu Alpha. Author: Regulation of Industry, 1949; Public Regulation of Business, 1959, rev., 1965; Transportation: Economics and Public Policy, 1963; Rate Theories and the California Railroad Commission, 1932; also numerous articles. Formulation of theory of rate making; research on pub. policy in regulation of industry, application antitrust laws in U.S., formulation of publ. policy for transp. Home: 414 Denslow Av., Los Angeles 90049.*

PEGUY, Charles Pierre, French geographer; b. Bourg-la-Reine, Feb. 4, 1915; s. Charles and Charlotte (Baudouin) P.; ed. univs. Paris and Grenoble; Ph.D.; m. Suzanne Puget, Oct. 19, 1939; children—Marie-Claudine, Jean Franççois, Odile, Dominique, Jacques, Elizabeth, Vincent. Prof. agrégé at Lyceé de Gap and Prytanée; prof. phys. geography Rennes Faculty of Letters; dir. Nat. Center Sci. Research. Mem. French Nat. Geography Com., French Nat. Geodesy and Geophysics Soc. Author: Précis de Climatologie. Home: 27, rue de Fougères. Office: Faculté des lettres, pl. Hoche, Rennes (1 & V), France.

PEHL, Richard Henry, Am. physicist; b. Raymond, Wash., Nov. 27, 1936; s. Henry L. and Annabelle (Moyer) P.; B.S. in Chem. Engring., Wash. State U., 1958, M.S. in Nuclear Engring., 1959; Ph.D. in Nuclear Chemistry, U. Cal. at Berkeley, 1964. With Lawrence Radiation Lab., Berkeley, Cal., 1960—, mem. electronics engring. staff, 1965—. Mem. Sigma Xi. Research, publs. on multi-nucleon transfer nuclear reactions using semiconductor detectors, expts. determining masses of He8, C9, O13, Na20, Mg 21, characteristics of semiconductor detectors (silicon and germanium) for purpose of improving their usefulness in detecting nuclear radiation. Home: 2119D Russell St., Berkeley 94705. Office: Lawrence Radiation Lab., Berkeley, Cal. 94720.*

PEIERLS, Ronald Frank, physicist; b. Manchester, Eng., Sept. 8, 1935; s. Rudolf Ernst and Eugenia (Kannegiesser) P.; B.A., Gonville and Caius Coll., Cambridge (Eng.) U., 1956; Ph.D., Cornell U., 1959; m. Julie Mainwaring Wilson, Sept. 5, 1959; children—Timothy, Benjamin. Mem. Inst. for Advanced Study, Princeton, N.J., 1961-62; asst. prof. Cornell U., 1962-63; asso. physicist Brookhaven Nat. Lab., Upton, N.Y., 1963-66, physicist, 1966—. NATO fellow, 1959-61. Fellow Am. Phys. Soc. Theoretical investigations into nature and properties of interactions between elementary particles, especially involving strongly unstable particles. Home: R.F.D. 1, Box 92, Brookhaven, N.Y. 11719. Office: Physics Dept., Brookhaven Nat. Lab., Upton, N.Y. 11973.*

PEIRCE, Benjamin, Am. mathematician, astronomer; b. Salem, Mass., Apr. 4, 1809; s. Benjamin and Lydia (Nichols) P.; grad. Harvard, 1829; m. Sarah Hunt Mills, July 23, 1833, 5 children, including Charles S., James Mills, Herbert Henry Davis. Instr. Round Hill Sch., Northampton, Mass., 1829-31; tutor math. Harvard, 1831-33, prof. astronomy and math., 1833-42, Perkins prof. math. and astronomy, 1842-80, largely responsible for establishment of Harvard Obs., 1843; cons. astronomer for Am. Nautical Almanac (renamed Astron. Almanac for Use of Navigators 1860), 1849-67; dir. longitude determinations U.S. Coast Survey, 1852-67, supt., 1867-74, cons. geometer, 1874-80, dir. Am. expdn. to Sicily to observe eclipse of sun, 1870. An organizer Smithsonian Instn., 1847. Fellow Royal Soc., 1852, Am. Acad. Arts and Scis., U. St. Vladimir, Kiev, Russia (hon.), Royal Soc. Edinburgh (hon.); mem. Am. Philos. Soc., Royal Astron. Soc. (asso.), Brit. Assn. for Advancement Science (corr.), Royal Soc. Sciences, Göttingen, Germany (corr.); pres. A.A.A.S., 1853; a founder Nat. Acad. Scis., 1863, chmn. math. and physics class. Author: An Elementary Treatise on Sound, 1836; An Elementary Treatise on Algebra, 1837; An Elementary Treatise on Plane and Solid Geometry, 1837; An Elementary Treatise on Plane and Spherical Trigonometry, 1840; An Elementary Treatise on Curves, Functions, and Forces, vol. I, 1841, vol. II, 1846; Tables of the Moon, 1853; A System on Analytic Mechanics, 1855; Tables of the Moon's Parallax, 1856; Linear Associative Algebra (outstanding work), 1870; Ideality in the Physical Sciences, 1881. Asso. editor Am. Jour. Math., 1878; founder, editor Cambridge Miscellany of Math., Physics, and Astronomy, 1842-43. Examined basic concepts of algebra; contbd. to theory of matrices; proved there is no perfect number with fewer than 4 primes, 1832; formulated Peirce's criterion to solve practically important problem of probabilities in connection with series of observations; computed gen. perturbations of Uranus and Neptune; research on rings of Saturn. Died Cambridge, Mass., Oct. 6, 1880.

PEIRCE, Carleton Barnhart, physician, radiologist; b. Elgin, Ill., Apr. 4, 1899; s. Eugene Colfax and Amanda (Barnhart) P.; A.B., U. Mich., 1920, M.D., 1924, M.Sc., 1927; m. Esther Damaris Wood, Apr. 14, 1928; children—Patricia Jane (Mrs. W. Douglas Harper), Esther Katharine (Mrs. Charles Robert Scriver), Carleton Wood, Eugene Charles. Instr., U. Mich., 1925-27; asso. prof. U. Neb., 1927-30; faculty U. Mich., 1930-38, asso. prof., 1933-38; faculty McGill U., Montreal, Que., Can., 1938—, prof. radiology, 1947-64, emeritus prof., 1964—; radiologist-in-chief Royal Victoria Hosp., Montreal, 1938-64, hon. cons., 1964—. Cons. Royal Canadian Navy, 1941—, Med. and Life Scis., Canadian Corp. for 1967 World Exhbns., 1964-67, Canadian Armed Forces, 1954—, Montreal Neurol. Inst., Montreal Children's Hosp., U. B.C. Med. Faculty. Fellow A.C.P., Royal Coll. Physicians Can., Am. Coll. Radiology; mem. Am. Roentgen Ray Soc. (past mem. exec. council, chmn. 1937-38, first vice president 1964, 65), Canadian Cancer Society (board of dirs. 1945—), Internat. Congress Radiology (sec. gen. 1959-65). Research, numerous publs. on Hodgkin's disease, anatomy and diseases lung, nuclear medicine, radiation therapy in control nervous system tumors. Home: 472 Strathcona Av., Westmount, Montreal 6, Que., Can.*

PEIRCE, Charles Santiago Sanders, Am. logician, psychologist; b. Cambridge, Mass., Sept. 10, 1839; s. Benjamin Peirce; mother was d. U.S. Senator Mills, of Mass.; A.B., Harvard, 1859, Sc.B. in Chemistry summa cum laude, 1863; m. (Harriett) Melusina Fay, 1862; m. 2d, Juliette Froissy. Mem. staff U.S. Coast Survey, 1861-91; lectr. Johns Hopkins, 1897-84, Harvard, 1903, Lowell Inst., 1903-04; pvt. research in logic, Pike County, Pa., from 1887. Fellow Am. Acad. Arts and Scis. Author: Photometric Researches, 1878; Collected papers, 8 vols., 1931-58. Editor (with extensive additions): Studies in Logic (mems. of Johns Hopkins U.), 1883; Linear Associative Algebra (Benjamin Peirce), 1882. Founder pragmatism, 1878 (later developed by William James), also of pragmaticism; laid founds. of logic of relations, instrument for logical analysis of math., 1867-85; contbd. to theory of probability and to logic of sci. methodology; modified Booleian algebra to accomodate De Morgan's logic; distinguished 3-fold div. of predicates; elaborated triadic theory of meaning; demonstrated his father's algebra to be operational and matricular; represented Brassmann's system in logical notation; proved that bodies in 4-fold space must rotate about 2 axes at once or lose a dimension; made studies of pendulum; research in meteorology, psychophyscs, philology. Died Milford, Pa., Apr. 14, 1914.

PEIRCE, George James, botanist; b. Manila, P.I., Mar. 13, 1868; s. George Henry and Lydia Ellen (Eaton) P.; S.B., Lawrence Sci. Sch. (Harvard), 1890, fellow, 1892-94; postgrad. univs. Bonn, Leipzig, Munich, 1892-94; A.M., Ph.D., Leipzig, 1894; m. Anna Hobart, June 14, 1897; children—Elizabeth, Carolyn, Rosamond Hobart. Asst. prof. botany Ind. U., 1895-97; asst. prof. botany Stanford, 1897-1900, ass. prof., 1900-10, prof., 1910-33, emeritus. Collaborator U.S. Forest Service, 1909-10; spl. agt. Dept. of Justice, studying effects of smoke on vegetation, 1910-11; studied effect of cement dust on vegetation, 1911. Fellow Am. Acad. Arts and Scis., A.A.A.S., Bot. Soc. Am. (pres. 1932); mem. Brit. Assn. for Advancement Sci., Deutsche Botanische Gesellschaft. Author: Textbook of Plant Physiology, 1903; The Physiology of Plants—The Principles of Food-Production, 1925; Experimental Plant Physiology, 1931. Joint Author: General Biology. Contbr. articles to Dictionary Am. Biography, Sci. Monthly, Bot. Gazette. Died Oct. 15, 1954.

PEIRCE, James Mills, Am. mathematician; b. Cambridge, Mass., May 1, 1834; grad. Harvard 1853, A.M. Tutor, Harvard, 1854-58, 60-61, asst. prof. math., 1861-69, univ. prof., 1869-85, Perkins prof. math., 1885—, sec. academic council, 1872-90, dean of grad. sch., 1890-95, dean faculty arts and scis., 1895-98. Author: Text-Book of Analytic Geometry, 1857; Three and Four-Place Tables of Logarithmic and Trigonometric Functions, 1871; Elements of Logarithms, 1874; Mathematical Tables, Chiefly to Four Figures, 1879. Died 1906.

PEIRESC, Nicolas Claude Fabri de, see de Peiresc.

PEISINO, Giovanni, Italian astronomer; b. Somano, Apr. 16, 1890; s. Stefano and Christina (Durante) P.; ed. Faculty of Sci., U. Turin (Italy); m. Angiola Maria Griva, June 9, 1920. Named asst., 1925, then astronomer, 1929, then prof. U. Trieste (Italy), 1931-41; dir. Collurania Astron. Obs., 1941-59; retired after 40 years service. Decorated Order Merit Italian Republic. Mem. Italian Astronomy Soc. Address: Vicoforte (Cuneo) Italy.

PEISS, Clarence Norman, Am. physiologist; b. Ansonia, Conn., Jan. 3, 1932; s. Alexander and Rose (Pasternack) P.; A.B., Stanford, 1946, A.M., 1947, Ph.D., 1949; m. Evelyn Schwartz, July 17, 1949; children—Kathy Lee, Robert Laurence. Faculty, St. Louis U. Sch. Medicine, 1950-54, asst. prof., 1952-54; faculty Stritch Sch. Medicine, Loyola U., Chgo., 1954—, prof., physiology, 1958—. Liaison scientist Office Naval Research, London, Eng., 1965-66. Postdoctoral fellow Johns Hopkins, 1949-50; Markle scholar in med. sci., 1953-58; recipient Lederle Med. Faculty award, 1958-61. Mem. A.A.A.S., Chgo. Heat Assn., Am. Heart Assn., Am. Physiol. Soc. Research, numerous publs. on physiol. responses to heat stress in man, mechanisms by which brain and nervous system control heart. Home: 806 Leamington Av., Wilmette, Ill. 60091. Office: Loyola U. Med. Center, Hines, Ill.*

PEKAR, Solomon Isaklvich, Russian physicist; b. Kiev, 1917; grad. Kiev U., 1938; D.Physico-Math. Sci. Asso., Physics Inst., Ukrainian Acad. Scis., head dept. theoretical physics, 1941-46, 49-62, dir. inst., 1946-49, head dept. theoretical physics Inst. Semiconductors, 1962—; prof. Kiev U., 1944—. Mem. Ukrainian Acad. Scis. Formulated theories of semicondr. rectifiers, polarons, light absorption in crystals, additional electromagnetic waves in crystals, ultra sound amplification in semiconductors. Address: Ukrainian Acad. Scis., Vladimirskaya 54, Kiev, Ukrainian SSR, USSR.*

PEKAREK, Luděk, Czechoslovakian physicist; b. Prague, Czechoslovakia, Oct. 31, 1924; s. Ludvík and Bozéna (Schléeová) P.; student Charles U. Faculty Natural Scis., Prague, 1945-49; Ph.D., Moscow State U., 1953; m. Kveta Suchomelová, Dec. 10, 1954; 1 dau., Iva. Sci. worker Inst. Physics, Czechoslovakian Acad. Scis., Prague, 1954—, dir., 1955—; lectr. dept. physics Charles U., 1957—. Contbg. author: Little Modern Ency., 1959. Research, publs. on ionization waves in discharge plasmas, their artificial excitation by pulse disturbance, and explanation of their phys. nature. Home: 8 V luhu, Praha. Office: 1 Lumumbova, Prague, Czechoslovakia.*

PEKERIS, Chaim Leib, mathematician; b. Alytus, Lithuania, June 15, 1908; s. Samuel and Chaya (Rivel) P.; B.Sc., Mass. Inst. Tech., 1929, D.Sc., 1934; student Oslo U. Obs., 1932-33; m. Leah Kaplan, Jan. 9, 1933. Fellow Gen. Edn. Bd. Mass. Inst. Tech., 1934-35, Cambridge U., Eng., 1935-36; research asso. Mass. Inst. Tech., 1936-40, div. war research Columbia, 1941-45, dir. math.-physics group, 1945-47; mem. Inst. Advanced Study, Princeton, 1946-48; head dept. applied mathematics Weizmann Inst. Sci.,

Rehovot, Israel, 1949—. Mem. spl. com. on upper atmosphere NACA, 1948-53. Mem. Nat. Acad. Scis. (Israel), Am. Phys. Soc., Am. Geophys. Union, Am. Seismol. Soc. Contbr. profl. jours. Research on hydrodynamics; underwater sound; seismology; atmospheric waves; stellar hydrodynamics; electromagnetic wave propagation; applied mathematics. Home: American Com. for Weizmann Inst., 515 Park Av., N.Y.C. 22. Office: Weizmann Inst., Rehovot, Israel.

PEKKARINEN, Aimo Ilmari, Finnish physician, pharmacologist; b. Rääkkylä, Finland, Oct. 2, 1921; s. Pekka Vilho and Elli (Hirvonen) P.; candidate medicine U. Helsinki, 1945, licenciate medicine, 1950, docent, 1951, M.D., 1950; m. Anna Liisa Talvikki Kariki, June 2, 1948; children—Marja Elisa, Eeva Appeli, Juss Olavi. With U. Helsinki (Finland), 1945-52, lectr. med. chemistry, 1951-52; prof. pharmacology, chmn. dept. pharmacology Turku (Finland) U., 1952—, dir. endocrinology lab., 1954—. Med. adv. council med. mfg. co., Leiras, 1952—; mem. biol. standardization com. No. Pharmacopoea. Decorated Comdr. Order Lion Finland, 1962; Rockefeller fellow U. Pa., Phila., 1952-53; ASLA fellow, U. Pa., also U. Vt., Burlington, 1960-61. Mem. Soc. Clin. Chemistry and Physiology (past chmn.), Pharmacology Soc. (mem. council 1957—), Scandinavian Soc. Clin. Chemistry and Clin. Physiology (past mem. council), Internat. Study Group for Steroid Hormones, Internat. Com. of Standardization in Human Biology (mem. bur. 1959—), Am. Chem. Soc., Royal Soc. Medicine (affiliate) Editorial com. Acta Pharmacologica et Toxicologica, 1953—, European Jour. Steroids, 1966, Scandinavian Jour. Clin. and Lab. Investigation. Research and numerous publs. in adrenocortical and sympathetic activation in stress and neurogenic condition and effect of drugs; bioassay of corticotrophin in guinea pig; determination of adrenaline, noradrenaline, steroid hormones and their excretion in clin. condition; regulation (especially inhibition of adrenomedullary secretion) by drugs, by determination of adrenaline in urine and adrenal vein. Home: 6 A 12 Porthaninkatu, Turku, Finland. Office: Kiinamyllynkatu 10, Turku, Finland.*

PELACANI, Biagio, Italian physicist; b. Parma, Italy; doctorate U. Pavia, 1374; at least 1 son, Francesco. Author: Tractatus de ponderibus source of medieval mechanics for Da Vinci and Cardano); also commentaries on Aristotle's Organon, De anima, De coelo et mundo, De generatione et Corruptione, Meteora.

PELAGONIUS, Roman veterinarian; flourished 4th century. Author: Ars veterinaria (mainly prescriptions in vet. medicine).

PELCZAR, Michael Joseph, Am. microbiologist; b. Balt., Jan. 28, 1916; s. Michael and Josephine (Polek) P.; B.S., U. Md., 1936, M.S., 1938; Ph.D., U. Ia., 1941; m. Merna M. Foss, Aug. 28, 1941; children—Ann Foss, Patricia Mary, Michael Rafferty, Rita Margaret, Josephine Merna, Julia Foss, Instr., U. Ia., 1940-41, U. Md., 1941-42; bacteriologist, chief lab. service sta. Gen. Hosp., 1946-48; faculty U. Md., College Park, 1958—, prof. microbiology, 1950—, v.p. for grad. studies and research, 1966—. Councilor, Oak Ridge Affiliated U.; spl. cons. Am. Coll. Dictionary, Random House, N.Y.C., 1964—. Diplomate Am. Bd. Microbiology. Mem. Internat. Assn. Microbiol. Socs., Am. Soc. Microbiologists (council policy com. 1966 —, past pres., Washington br.), Am. Assn. U. Profs. (past pres. Md. chpt.), Brit. Soc. Gen. Microbiologists, A.A.A.S., Washington Acad. Sci., Am. Acad. Microbiology (bd. govs. 1966), Am. Inst. Biol. Sci. (vis. lectr. 1956—), Sigma Xi (past pres. Md. chpt.), Phi Kappa Phi, Phi Beta Kappa. Author: (with R.D. Reid) Microbiology, 1965. Lab Exercises Microbiology, 1958, 2d edit., 1965; (with P.A. Hansen, W.A. Konetzka) Quantitative Bacteriology Physiology Laboratory Experiments, 1955; also numerous articles. Research on nutritional requirements bacteria, media for growth of bacteria, bacteriological method for assay vitamins, classification bacteria related to meningococcus, and gonococcus. Home: 4318 Clagett Pineway, University Park, Md. 20782. Office: Office of Vice pres. for Grad. Studies and Research, U. Md., College Park, Md. 20742.*

PELESKA, Bohumil, physician; b. Zár, South Bohemia, Apr. 3, 1921; s. Jan. and Berta (Cejková) P.; M.D., Charles U., Prague, Czechoslovakia, 1951, Candidate Med. Scis., 1960, D.Med. Scis., 1963; m. Hana Kosárová, Dec. 29, 1956; children—Hana, Petr, Pavel. Staff, Faculty Surgery, Plzen, Czechoslovakia, 1951-53; chief exptl. dept. Inst. for Clin. and Exptl. Surgery, Prague, 1953-63; dir. Inst. for Med. Electronics and Modelling, Prague, 1963—; asst. prof. Med. Faculty, U. Brno, Czechoslovakia, 1965—. Recipient Grand Prix, World exhbn., Brussels, Belgium, 1958, Czechoslovak State Prize, 1965. Mem. Internat. Fedn. for Med. and Biol. Engring., Société francaise dánesthesie et de réanimation. Author: Present Trends of Medical Electronics in Surgery, 1964; also numerous articles. Research on reanimation, reanimation electronics, cardiac stimulation, cardiac electro-impulse therapy, artificial heart problems; evaluated optimal parameters of elec. impulses for defibrillation by condenser discharges; designed (with others) cardiac implantable pacemakers; introduced monitors and computing machines to medicine. Home: 9 U Keramické Skoly, Prague, 3. Office: 800 Budéjovická ulice, Prague 4, Czechoslovakia.*

PELETIER, Jacques (Peletarius), French philosopher, mathematician, poet; b. Mans, France, 1517; student law Coll. Navarra, Paris; later student medicine, Paris; head colls. of Bayeux and of Mans; bishop of Mans; practiced medicine in various French cities. Author: De constitutione horoscopi; De pest compendium (attributed plague to conjunction of Saturn and Jupiter in it), 1563; Arithmetic, 1549; Algebra, 1554; De l'usage de la géométrie, 1573. Editor: Elements (with commentary) (Euclid), 1st 6 books, 1559. Observed the root of an equation is divisor of last term; wrote equations with all terms on one side and equated to zero; criticized Cardan's ideas on angle of contingence. Died Paris, 1582.

PELIGOT, Eugene Melchior, French chemist; b. Paris, Feb. 24, 1811; ed. École Polytechnique, Paris; pupil of Dumas, whom he succeeded as prof. Conservatoire des Arts et Métiers; named dir. French Mint, 1846; mem. French Acad. Scis. Author: Traité élémentaire de manipulations chimiques, 1836; Recherches sur la nature et les propriétés chimiques des sucres, 1838. Research on cinnamic acid, cinnamyl radical, 1834; helped establish composition of alcohols; described methyl flouride; used phosphorus pentachloride for chlorination, 1836; isolated metallic uranium, 1841; showed that Klaproth's uranium was uranous oxide; found atomic weights of uranium and chronium; contbd. to tech. chemistry of beet sugar and sulfuric acid. Died Paris, Apr. 15, 1890.

PELIZAEUS, Friedrich, neurologist; b. Germany, 1850. Author: (article) Ueber eine eigenthümliche Formspastischer Lähmung mit Cerebraler-Scheinungen auf hereditärer Grundlage (Multiple Sklerose), 1885. Described a congenital, familial disease caused by atrophy of white matter of brain, marked by speech disturbances, poor coordination, mental retardation (Merzbacher-Pelizaeus disease, or aplasia axialis extracorticalis congenita), 1885. Died 1917.

PELL, John, English mathematician; b. Southwick, Sussex, Eng., Mar. 1, 1611; M.A., Cambridge U., 1630; tchr. in Sussex, Eng.; prof. math., Amsterdam, Netherlands, 1643-46; apptd. prof. philosophy and math., Breda, 1646; returned to Eng., 1652; diplomat for Cromwell in Switzerland, 1654-58; rector of Fobbing, 1661-81; vicar of Laindon, 1663-85. Fellow Royal Soc., 1663. Author: Description and Use of the Quadrant, 1628; An Idea of Mathematics, written to S.H. (Samuel Harliby), Philos. Collection, 1681-82; other writings on algebra, geometry, astronomy; revised and added to Rhonius algebra (translated by Branker); demonstrated 2d and 10th books of Euclid, most of Diophantus's 6 books on arithmetic, part of Archimedes. Worked on problems of nav. and applied math.; introduced division sign into Eng.; Pellian equation takes its name from solutions of an indeterminate equation which he presented. Died London, Eng., Dec. 12, 1685.

PELL, William Hicks, Am. mathematician; b. Lewisport, Ky., Oct. 15, 1914; s. William Clay and Beulah (Hicks) P.; B.S., U. Ky., 1936, M.S., 1938; Ph.D., U. Wis., 1943; m. Dorothy Small, Aug. 23, 1939. Instr., Brown U., Providence, 1943, faculty 1947-56, asso. prof., 1951-56; research aerodynamicist Bell Aircraft, Buffalo, 1943-47; chmn. dept. math. U. Ky., Lexington, 1952-53, prof. applied mechanics, 1959-60; aero. research engr. Nat. Bur. Standards, Washington, 1956-59, chief math. physics sect., 1960-65; program dir. applied math., statistics NSF, Washington, 1965-67, head mathematical sciences section, 1965—. Rockefeller Post-doctoral fellow applied math. Brown U., 1943; Inst. Fluid Dynamics and Applied Maths. Postdoctoral fellow U. Md., 1955-56. Fellow Washington Acad. Sci.; mem. Am. Math. Soc., Am. Soc. M.E., Math. Assn. Am., Soc. Indsl. and Applied Math. (asso. editor Jour 1961—), Soc. Natural Philosophy, Phi Beta Kappa, Sigma Xi. Research, publs. on behavior elastic and plastic plates and materials, dynamics fluid motion, aero-dynamic design. Home: 605 Muriel St., Rockville, Md. 20852. Office: 1800 G St. N.W., Washington 20550.*

PELLAT, Joseph-Solange-Henri, French physicist; b. Grenoble, France, 1850; became asst. prof. Obs., 1874; named prof. Rollin Coll., 1876, Louis-le-Grand Lycée, 1880; apptd. asst. prof. Sorbonne, 1899. Author: Cours de physique, 188-87; Lecons sur l'électricité, 1889; Électricité atmosphérique, 9 1890; Cours de physique générale, 1896-97; Thermodynamique, 1897; Cours d'électricité, 1901-07. Inventor electrodynameter; demonstrated that energy of 1 calorie transformed into elec. current in a cell can produce audible tones over telephone for thousands of years. Died Paris, 1909.

PELLEGRINO, Edmund Daniel, Am. physician; b. Newark, June 22, 1920; s. Michael J. and Marie (Catone) P.; B.S., St. John's U., 1941; M.D., N.Y. U., 1944; m. Clementine Coakley, Nov. 17, 1944; children—Thomas, Stephen, Virginia, Michael, Andrea, Alice, Leah. Fellow medicine N.Y. U., 1948, 49-50; supervising Tb physician Homer Folks Tb Hosp., Oneonta, N.Y., 1951-53; dir. internal medicine Hunterdon Med. Center, Flemington, N.J., 1953-59, med. dir., 1955-59; asst. prof. clin. medicine N.Y.U., 1953-59; asso. vis. physician Bellevue Hosp., N.Y.C., 1958-59; prof., chmn. dept. medicine U. Ky. Coll. Medicine, Lexington, 1959-66; dir. Health Scis. Center, prof., chmn. dept. medicine State U. N.Y., Stony Brook, 1966—; sr. vis. scientist Brookhaven Nat. Lab., 1966—; cons. Meadowbrook Hosp., East Meadow, N.Y., 1968—; Mem. sci. rev. com. for health research facilities NIH, 1964—. Diplomate Am. Bd. Internal Medicine. Fellow N.Y. Acad. Medicine, Am. Coll. Chest Physicians, Am. Coll. Cardiology, A.C.P.; mem. Am. Clin. and Climatol. Assn., A.A.A.S., N.Y. Acad. Scis., Am. Heart Assn. (council on circulation, council for high blood pressure research), Am. Assn. U. Profs., Am. Hosp. Assn. (com. on patterns hosp. services 1964-65), Metaphys. Soc. Am., Medieval Acad. Am., Am. Cath. Philos. Assn., Sigma Xi. Contbr. articles on cardiovascular renal disease, electrolyte metabolism, bone chemistry, med. edn. to profl. jours. Home: 8 Robinswood St., Setauket, N.Y. 11785. Office: State U. N.Y. at Stony Brook, Stony Brook, N.Y. 11790.*

PELLEGRINO, José, Brazilian parasitologist; b. Passagem, M. Gerais, Brazil, Nov. 11, 1922; s. Braz and Assunta (Magaldi) P.; M.D., Faculty of Medicine, Belo Horizonte, Brazil, 1946; m. Maria do Carmo Lins, Apr. 21, 1952; children—Marcus Lins, Sergio Lins, Bertha Lins. Research fellow Oswaldo Cruz Inst., Rio de Janerio, Brazil, 1947-51; prof. med. parasitology Cath. U., Belo Horizonte, 1952-55, head lab. schistosomiasis Nat. Inst. Endemic Diseases, 1955—; head div. immunology, biology of parasites Inst. Biology, U. M. Gerais, Belo Horizonte, 1952—; cons. F. Hoffmann - La Roche & Co., Basle, Switzerland. Cons. Pfizer Group, Sandwich, U.K., WHO, Geneva, Switzerland, 1964—, P.R. Nuclear Center, San Juan, 1966—. Recipient Pirajá da Silva award, 1958; Gaspar Vianna award, 1961; decorated Oswaldo Cruz medal, 1946. Mem. Brazilian, Royal socs. tropical medicine, Am. Soc. Parasitologists, Latin Am. Fedn. Parasitologists. Research, numerous publs. on Chagas' Disease, immunodiagnosis, prophylaxis, transmission through blood transfusion leshmaniasis, immunodiagnostic methods, Schistosomiasis, skin test, complement fixation, exptl. chemotherapy, devel. oogram method for drug screening. Home: 925 Washington St. Office: Instituto de Biologia, 288 Carangola St., C. Postal 253, Belo Horizonte, M. Gerais, Brazil.*

PELLETIER, Bernard Roderick, Canadian geologist; b. Toronto, Ont., Can., June 25, 1923; s. William and Mildred Pelletier; B.S., McGill U., 1950; M.S., McMaster U., Hamilton, Ont., 1953; Ph.D., Johns Hopkins, 1957; m. Judy Grace Sabine Lamb, Oct. 7, 1950; children—Carol, Margaret, David, Frederick, Catherine, Marianne. Party chief Peel Plateau Exploration Co., Yukon, 1953-54, 56, Geol. Survey Can., N.W.T., 1957, N.E. B.C., 1958-62; chief Marine Geology div. Canadian Dept. Energy Mines and Resources, Arctic Ocean and Archipelago, Hudson Bay and Approaches, Atlantic Ocean and Gulf St. Lawrence, 1960—; faculty U. Ottawa, 1958-61; panel mem., chmn. geology coms. Atlantic Provinces Interuniv. Com. Scis., Canadian Com. Oceanography, Internat. Quaternary Geology, 1963—. Mem. Canadian Inst. Mining and Metallurgy, Geol. Assn. Can., N.Y. Acad. Sci., Sigma Xi. Editor: Maritime Sediments, 1965—. Research in statis., graphical and pictorial models to demonstrate ancient routes of sedimentary transport from eroded area to site of deposition, organized permanent marine geology program for Canadian govt., outlined prospective oil and gas fields in So. Ont., Western Can. Home: 9 Dunolly St. Office: Bedford Inst. Oceanography, Box 1006, Dartmouth, N. S., Can.*

PELLETIER, Bertrand, French chemist; b. Bayonne, France, July 31, 1761; dir. pharmacy, Rouelle; became prof. Polytechnic Sch., Paris, 1795; mem. French Acad. Scis. Author: Memoirs and Observations (collected works), 1798. Studied formation of oxidized muriatic acid and metallic phosphorus. Died Paris, July 21, 1797.

PELLETIER, Pierre Joseph, French chemist; b. Paris, Mar. 22, 1788; s. Bertrand Pelletier; ed. by father; studied medicine; became prof. École de Pharmacie, dir., 1832; alkaloid pelletierine named by Tanret in his honor; mem. French Acad. Scis. (Montyon prize 1827). Author: Memoir on Quinine, 1821; other works and articles. Follower of Lavoisier; discovered strychnine, 1818, brucine, 1819, cinchonine, 1821, caffeine, 1821; discovered (with Joseph Caventou) quinine in cinchona bark, 1820, thus making possible its preparation in known strength for med. uses; obtained (with Walter) toluene by distilling pipe resin, 1836; named chlorophyll. Died Paris, July 19, 1842.

PELLETIER, S. William, Am. organic chemist; b. Kankakee, Ill., July 3, 1924; s. Antoine A. and Estella E. (Hayes) P.; B.S., U. Ill., 1947; Ph.D., Cornell U., 1950; m. Leona Jane Bledsoe, June 18, 1949; children—William Timothy, Jonathan Daniel, Rebecca Jane, Lucy Ruth, David Mark, Sarah Lynn. Instr., U. Ill., Urbana, 1950-51; faculty Rockefeller Inst., N.Y.C., 1951-62, asso. prof. chemistry, 1961-62; prof., head dept. chemistry U. Ga., Athens, 1962—. Gordon lectr., 1955, 59; Victor Coulter lectr. U. Miss., 1965; Am. Swiss Found. for Sci. Exchange lectr., 1960. Fellow Chem. Soc. London; mem. Am. Chem. Soc. (nat. symposium lectr. 1961, 65), A.A.A.S., Sigma Xi, Tau Beta Pi, Sigma Tau. Bd. editors Jour. Organic chemistry, 1966—. Research, numerous publs. in elucidation of chem. structures of complex products of plant origin such as alkaloids and terpenes; devel. of methods of synthesis of complex organic compounds. Home: 555 Glenwood Dr., Athens, Ga. 30601.*

PELOUZE, Théophile Jules, French chemist; b. Valognes, France, Feb. 26, 1807; studied under Gay-Lussac and Lasaigne, Paris. Became prof., Lille, France, 1830; apptd. assayer of mint, 1833; faculty L'École Polytechnique, Paris, also Collège de France, 1831-51; asso. with Salpetrèe, Paris; worked on chem. studies with Liebig, Germany, 1836. Became pres. mint commn., 1848. Became mem. French Acad. Scis., 1837. Author: (with Fremy) Traité de chimie générale, 7 vols., 1846-65; Traité de chimie analytique, 1847-50; Abrégé de Chimie, 1848; Notions générales de chimie, 1853; Discovered borneol, 1841, nitro-sulfates, cyanide acid of iron, that beet root and cane sugar are the same; developed process of producing tannin; improved plate glass manufacture, 1856; determined atomic weight of arsenic, phosphorus, nitrogen, silicon; prepared propionitrile, 1834; discovered composition of glycerine, 1836; studies in pyrogallic acid, terpene series and inorganic esters of alcohol, enamel, effect of sunlight on glass. Died Paris, May 31, 1867.

PELSENEER, Paul, Belgian biologist; b. Brussels, Belgium, June 26, 1863; prof., Ghent, Belgium; premanent sec. Belgian Royal Acad.; mem. French Acad. Scis. Set up a theory of devel. of molluscs. Died Brussels, May 5, 1945.

PELTIER, Jean Charles Athanase, French physicist; b. Ham, France, Feb. 22, 1785; watchmaker, Paris, until 1815; then engaged in sci. research. Author: Observations sur les multiplicateurs et sur les piles thermo-électriques, 1836. Discovered a thermo-electric reduction of temperature (Peltier effect); observed electrification of flowing steam jet from a boiler, 1840; measured temperature of water with heating, 1841. Died Paris, Oct. 27, 1845.

PELTONEN, Tuomas Eetvartti, Finnish physician; b. Virrat, Finland, May 3, 1924; s. Kalervo Ilmari and Tyyne (Kangaspeska) P.; Cand. medicine U. Helsinki (Finland), 1947; Med. Lic., U. Turku (Finland), 1951, Med. Sci., Dr., 1956; m. Leena Laesvuori, Aug. 20, 1949; children—Leena Kristiina, Juha Tuomas Kalervo. Staff, U. Turku, 1953—, prof. pediatrics, 1966—; staff Mil. Hosp. 2, Turku, 1957—, chief hosp., 1965-67; vis. asso. prof. U. Cal. at Los Angeles, 1959. Mem. Deutsche Geselschaft für Kinderheit, European, Internat. socs. hematology, Finnish Pediatric Assn., European Club Pediatric Cardiology, Scandinavian Paediatric Soc. (pres. 1967—). Author: Über die sog. funktionellen Störungen in Schulalter, 1956; (with Hirvonen) Experimental Studies in Foetal and Neonatal Circulation, 1965; also numerous articles. Research on vegetative dystonicity in childhood, neonatal physiology. Address: Yliopiston Lastenklinikka, Turku 3, Finland.*

PELZER, Karl Josef, geographer; b. Oberpleis, Germany, Feb. 23, 1909; s. Friederich and Christine (Broel) P.; Ph.D., U. Bonn (Germany), 1935; M.A. (hon.) Yale, 1952; m. Elizabeth A. Clark, Mar. 21, 1936; children—Christine (Mrs. D. Gordon White), Ingrid. Came to U.S., 1935, naturalized, 1940. Research asst. Council Fgn. Relations, 1936-37, Inst. Pacific Relations, 1938-39, 40-41; research asso. Johns Hopkins, 1939-41, asso. geography, 1942-44; sr. regional specialist OWI, 1944-45; agrl. economist Office Fgn. Agrl. Relations, U.S. Dept. Agr., 1945-47; faculty Yale, 1947—, prof. geography, 1952—, dir. Southeast Asia studies, 1958—; Fulbright research prof. Philippines, 1950. Ford fellow, Indonesia, 1955-56. Mem. Assn. for Asian Studies (pres. 1966-67), Asia Soc. (chmn. Indonesia council 1966—), African Studies Assn., Assn. Am. Geographers, Council for Fgn. Relations, Am. Geog. Soc. Author: Arbeiterwanderungen in Südostasien, 1935; Population and Land Utilization, 1941; Pioneer Settlement in the Asiatic Tropics, 1945; (with others) Indonesia, 1963; also articles, numerous revs. Pioneered tropical land use studies especially shifting cultivation. Home: 29 Cooper Rd., North Haven, Conn. 06473. Office: Yale, New Haven 06520.*

PEMBERTON, Henry, English physician; b. London, Eng., 1694; studied medicine with Boerhaave, Leyden, 1714; studied anatomy in Paris; clinical study, St. Thomas Hosp., London, M.D., Leyden, 1719. Pursued mathematical and scientific studies rather than practicing medicine; Gresham prof. of physics, 1728; lectr. on chemistry. Fellow Royal Soc., 1720. Author: A View of Sir I. Newton's Philosophy, 1728; Scheme for a course of Chemistry to be performed at Gresham College, 1731; Dissertatio Physico-Medicianalis Inauguratio de Facultate Oculi ad disversas Rerum Computatarum Distantias se accommodante, 1718; Epistulae ad Amicum de Rogeri Cotesii Inventis, 1722. Edited 3rd ed. of Newton's Principia, 1726; with Mead, edited Cowper's Myotomia Reformata, 1724; prepared 5th London Pharmacopoeia for Royal College of Physicians. Died Mar. 9, 1771.

PEMBERTON, John deJarnette, Am. physician; b. Wadesboro, N.C., May 3, 1887; s. John deJarnette and Emma (Lilly) P.; B.A., U. N.C., 1907, LL.D., 1932; M.D., U. Pa., 1911; M.S. in Surgery, U. Minn., 1918; m. Anna T. Hogeland, June 4, 1918; children—John deJarnette, Albert H., Henry W., Robert G., Elizabeth Anne (Mrs. Edgar Lovelace Lassetter). Practice surgery; head sect. surgery Mayo Clinic, Rochester, Minn., 1918-52; prof. surgery Mayo Found. Grad. Sch., 1936-52. Recipient Distinguished Alumni award U. Pa. Sch. Medicine, 1962. Diplomate Am. Bd. Surgery. Mem. Am. Assn. for Study Goiter (past pres.), A.C.S., A.M.A., Am., So. surg. assns., Soc. Med. Assn. Société Internationale de Chirugie, Soc. Clin. Surgery, Alumni Assn. Mayo Found., Sigma Xi, Alpha Tau Omega, Nu Sigma Nu, Alpha Omega Alpha; hon. mem. Chgo., St. Paul, Spokane, Central, Mpls. surg. socs. Research, publs. on gen. surgery, surgery thyroid gland. Home: 615 11th St. S.W. Office: 20 1st St. S.W., Rochester, Minn. 55901.*

PEMBERTON, Ralph, Am. physician; b. Phila., Sept. 14, 1877; s. Henry and Agnes (Williams) P.; B.S., U. Pa., 1898, M.S., 1899, M.D., 1903, Woodward fellow in physiol. chemistry, Pepper Lab., 1908-10; postgrad., Berlin, 1911, U. Strasbourg (now in France), 1912; m. Virginia Breckenridge Miller, May 23, 1911. Began practice, Phila., 1905; instr. medicine U. Pa., 1907-10, asso. prof. medicine Grad. Sch., 1928-31, prof. since 1931; asst. vis. physician U. of Pa. Hosp., 1908-10; asst. vis. neurologist Phila. Gen. Hosp., 1905-08; vis. physician, dir. dept. clin. chemistry Presbyn. Hosp., 1913-33; vis. physician Abington Meml. Hosp.; cons. physician Chester County Hosp. Nat. consultant in rheumatism and arthritis under Program of War-Time Grad. Med. Meetings since 1943; mem., officer various med. bds. and bodies, especially for control of rheumatism. Recipient meritorius service medal Commonwealth of Pa., 1939; Gold Key, Am. Congress Phys. Medicine, 1946. Fellow Coll. Physicians Phila.; mem. A.M.A., Acad. Phys. Medicine, Am. Soc. for Clin. Investigation, Am. Inst. Nutrition, Phila. County Med. Soc., Acad. Natural Scis., Franklin Inst., Internat. Soc. Med. Hydrology, Am. Rheumatism Assn. (pres. 1938-39), Sigma Xi. Author: Arthritis and Rheumatoid Conditions, 1929, translated into French, 1933, 2d edit., 1935; (with R. B. Osgood) Medical and Orthopedic Management of Chronic Arthritis, 1934. Contbr. to Nelson Loose Leaf System of Medicine, 1922, Bedside Diagnosis (by Am. authors), 1927, Textbook of Medicine (by same), 1928, Internat. Ency. of Medicine since 1931, Tice System of Medicine since 1934; also articles. Editor vol. on Medicine in Principles and Practice of Physical Therapy, 1932. Died June 17, 1949.

PÉNA, Jean, astronomer; pupil of Charles de l'Ecluse (Clusius); royal mathematician, Paris. Author: The Spherics of Theodosios (in Greek), 1558. Editor: Harmonicum introduction (Celeoneides, in Greek and Latin), 1557; Optica, 1557, Rudimenta (both Euclid, in Greek). Held that Ptolemy's astron. ideas were but math. abstractions; admired Copernicus but without formally adhering to his system; maintained on basis of optical reasoning that some comets exist beyond the moon, 1557.

PENAUD, Alphonse, French inventor; b. Paris, 1850. Staff, Ecole Navale. Theoretical research in air travel; patentee (with Paul Gauchot) amphibian single decker; proposed a helicopter design, 1870; built successful model airplane, 1871; proposed and built full-scale airplane design, 1873; developed tandem horizontal surfaces with longitudinal dihedral for airplanes, fixed vertical tail surface, set mainplanes at dihedral angle for lateral stability, 1871; developed kite-balloon with tail-fin, 1874. Committed suicide, Oct., 1880.

PENCHEV (Pentcheff), Nicola Petrov, Bulgarian chemist; b. Sofia, Bulgaria, July 31, 1901; s. Peter Benev and Rada (Petkova) P.; student U. Sofia, 1920-24, Dr.hab., 1958; student Collège de France, Paris, 1930-31, Institut d'Optique, Paris, 1931; m. Zdravka Eremieva Popova, Dec. 30, 1928; children—Elka Nicolova (Mrs. Dimiter Lubenov Yordanov), Rada Nicolova. Staff, U. Sofia, 1924—, prof. dept. analytical chemistry, 1947—, head dept., 1943—. Recipient award for civic service, 1939; Cyril and Methodi 2d degree, 1957, 1st degree, 1960; Dimitrov State prize for sci., 1959. Mem. Bulgarian Acad. Scis. (corr., mem. bur. and sci. council dept. chemistry), Union Sci. Workers Bulgaria, Société de Chimie-Physique (Paris). Author: (with B. Zagorchev) textbooks on analytical chemistry; also articles. Research on analysis, isotopic composition and geochemistry of noble gases in natural objects, geochronology, helium-bearing minerals at high pressures, trace elements in mineral waters, analytical principles and methods based on gas-producing processes in vacuum; 1st quantitative methods and geochem. data for neon; developed new macro- and micromethods for separation and determination of aluminum, chromium, and strontium; studies on aluminum hydroxides. Home: 35, General Parensov St., Sofia, Bulgaria.*

PENCK, Albrecht, German geographer; b. Leipzig, Germany, Sept. 25, 1858; prof., Vienna, Austria, also Berlin, Germany. Author: Die Vergletscherung der deutschen Alpen, 1882; Das Deutsche Reich, 1887; Die Morphologie der Erdoberfläche 2 vols., 1894; (with E. Brückner) Die Alpen in Eiszeitalter, 1901-09. Began thorough study of Alpine glaciation, 1882; reaffirmed Guettard's idea that erosion unopposed by upheavals leads to complete leveling of earth's surface, 1889. One of the founders of terrestrial morphology. Died Prague, Czechoslovakia, Mar. 7, 1945.

PENDE, Nicola, Italian physician; b. Noicattaro, Italy, Apr. 21, 1880; s. Angelo and Anna (Crapuzzi) P.; M.D.; hon. doctorate univs.Bordeaux, Aix-en-Provence, Marseilles (all France); m. Anna Bolacco, 1921. Prof. spl. med clinics and pathology Rome U., 1922—; organizer Bari U., 1924, served as rector; founder Biotypology and Orthogenesis insts., Genoa, Rome; organizer, dir. Instit. di Scienza della Costituzione, Rome. Mem. Italian Soc. Endocrinology (pres.), Internat. Bur. Differential Anthropology (pres.), various Italian and fgn. med. socs. Author: Patologia del simpatico (editor Pietro Castellino), 1915; Endocrinologica, Patologia e clinica degli organi a secrezione interna, 1916; Constitutional Inadequacies, an Introduction to the Study of Abnormal Constitutions, 1928; Trattato di biotipologia umana, 1939; La scienza della persona umana, 2d edit., 1940; Trattato di endocrinologia, 5th edit., 1949; Trattato di scienza dell, ortogenesi. Introduced term endocrinology, 1909; développed treatment for arterial hypertension, 1927. Address: 326 via Salaria, Rome, Italy.

PENDER, Harold, Am. elec. engr.; b. Tarboro, N.C., Jan. 13, 1879; s. Robert H. and Martha Wallace (Hanks) P.; A.B., Johns Hopkins, 1898, Ph.D., 1901; Sc.D., U. Pa., 1923; m. Alice Matthews, June 28, 1905; m 2d, Ailsa Craig MacColl, Dec. 22, 1934; 1 son, Peter Alexander. Tchr. McDonogh Sch., Md., 1901-02; instr. Syracuse U., 1902-03; elec. engr. Westinghouse Electric & Mfg. Co., 1903-04, N.Y.C. R.R., 1904-05; with Cary T. Hutchinson, elec. engr., N.Y., 1905-09; sec.-treas. McCall Ferry Power Co., 1905-09; prof. elec. engring. Mass. Inst. Tech., 1909-13, dir. research div., dept. of elec. engring., 1913-14; dir. dept. elec. engring. U. Pa., 1914-23, dean Moore Sch. Elec. Engring., 1923-49; cons. 1949—. Mem. Internat. Electrotech. Com. Fellow Am. Acad. Arts and Scis., Am. Inst. E.E.; mem. Am. Philos. Soc., Franklin Inst. Author: Principles of Electrical Engineering, 1911; Electricity and Magnetism for Engineers, 1918; Direct-current Machinery, 1921; Electric Circuits and Fields (with S. R. Warren, Jr.), 1943; also papers. Editor-in-chief Electrical Engineering Handbook (new edit.). Established beyond question existence of magnetic field around moving electrically charged body, at Sorbonne, Paris, 1903. Died Sept. 5, 1959.

PENDERGRASS, Eugene Percival, Am. physician; b. Florence, S.C., Oct. 6, 1895; s. Edward J. and E. Ethel (Smith) P.; student Wofford Coll., 1912-14, D.Sc., 1959; student U. N.C., 1914-16; M.D., U. Pa., 1918; D.Sc., Hahnemann Med. Coll., 1964; m. Rebecca Barker, Sept. 9, 1922; children—Henry P.; Jane B. (Mrs. William J.H. Hough, Jr.), Margaret B. (Mrs. William P. Huston, II). Faculty, U. Pa., Phila., 1919—, prof. radiology, 1936-61, chmn. dept., 1939-61, prof. emeritus, 1961-64, dir. bicentennial, 1962-65; dir. dept. radiology Jeanes Hosp., Phila., 1948-61, Dr. Matthew J. Wilson prof. research radiology, 1964—. Recipient Distinguished Service award U. N.C., 1958. Mem. A.M.A. (Gold medal 1929, del. sect. on radiology 1952-65, del. Nat. Health Council 1963—), Am. Cancer Soc., Phila. Tb Assn., Pa. Med. Com. for Better Govt., Phila. Art Alliance, Phila. Zool. Soc., Sigma Xi (pres. U. Pa. 1950-51), Alpha Omega Alpha, Alpha Tau Omega, Phi Chi. Author: (with H.K. Pancoast, J.P. Schaeffer) The Head and Neck in Roentgen Diagnosis, 1940, 56; The Pneumoconiosis Problem, 1958; also numerous articles. Research on pneumoconiosis; cancer; radiology, especially of head and neck. Home: 428 Owen Rd., Wynnewood, Pa. 19106. Office: 3400 Spruce St., Phila. 19104.

PENDLETON, Robert L(arimore), Am. soil scientist; b. Mpls., June 25, 1890; s. John Louis and Jessie (Larimore) P.; B.S., U. Cal., 1914, Ph.D., 1917; student N. India Hindustani Lang. Sch., 1918, N. China Lang. Sch., 1931-33; m. Anne Laurel Miltimore, June 10, 1917. Asst. in soil survey Cal. Soil Survey, 1914-15; asst. dir. Dept. Agrl., Gwalior State, India, 1918-20, dir., 1920-23; prof. soil technology Coll. Agrl. U. Philippines, Los Banos, Laguna, 1923-35, head, dept. soils, 1930-35; chief soil technologists Nat. Geol. Survey of China, 1931-33; soil scientist and agriculturist, dept. agr. Siamese Govt., Bangkok, 1935-42; soil scientist, office fgn. agrl. relations U.S. Dept. Agr., 1942-52; soil scientist S.T.E.M. to Siam, Mut. Security Agy., 1952-53; prof. topical soils and agrl. Isaiah Bowman Sch. Geography, Johns Hopkins University, 1946-55, emeritus, 1955; soil technologist to Ministry Agr., Thailand Govt., 1956—; field work in soils and land use in Central and S. Am., Philippines, Siam, China, India, Central and Brit. W. Africa. Mem. Mindanao Exploration Commn., Philippines, 1939; adviser FAO Mission to Siam, 1948; fellow Belgian Am. Ednl. Found., Belgian Congo, 1948-49; cons. War Dept., 1942-44, E.C.A., 1949-50. Recipient David Livingstone Centenary medal Am. Geog. Soc., 1950. Mem. Am. Geog. Soc. (hon.), Assn. Am. Geographers, Am. Chem. Soc., Am. Soc. Agronomy,

Soil Sci. Soc. Am., Am. Geophys. Union, A.A.A.S., Siam Soc., Bangkok, Sigma Xi. Author: Lateritte and Lateritic Soils (with J. A. Prescott), 1952. Translator (form Dutch); Soils of Equatorial Regions, 1944. Editor of Natural History Bull. of the Siam Soc., Bangkok. Died June 23, 1957.

PENFIELD, Wilder Graves, surgeon; b. Spokane, Wash., Jan. 26, 1891; s. Charles Samuel and Jean (Jefferson) P.; B.Litt., Princeton, 1913, D.Sc., 1939; Rhodes scholar, Oxford U., Eng., 1914-16, 19-20, B.A., 1916, M.A. and B.Sc., 1920, D.Sc., Oxon, 1935; M.D., Johns Hopkins, 1918; D.Sc. (hon.), Princeton, 1939, B.C., 1946, Toronto, 1953, Yale, Leeds, 1954, N.Y., Manitoba, 1955, Delhi, 1957, Wis., 1958, Dartmouth, 1960, McGill, 1960, McMaster, 1962, N.B., 1963; D.C.L., Bishops, 1944, Oxford, 1953; Docteur en Medicine, Montreal, 1945, Laval, 1951, Ottawa, 1966; LL.D., Wales, Dalhousie, 1953, St. Lawrence 1956, Queen's, 1957, Edinburgh, Saskatchewan, 1959, Vt., 1962, Melbourne, 1962; M.D. (hon.), Poland, 1960; D. de l'U. de Paris, 1954; D.U.C., Calgary, 1967; m. Helen Katherine Kermott, June 6, 1917; children—Wilder Graves, Ruth Mary, Priscilla, Amos Jefferson. Naturalized Canadian. Beit memorial research fellow, London, 1920-21; research and brain surgery Presbyterian Hospital, N.Y. C., 1921-28; surgeon Royal Victoria Hosp., Montreal Gen. Hosp.; dir. Montreal Neurol. Inst. until 1960, hon. cons.; apptd. Guggenheim fellow, 1961. Companion Order St. Michael and St. George, Order Merit (British); Medal of Freedom (U.S.); Chevalier Legion of Honor (France); gold cross Legion of George I (Greece); companion Order of Can. Recipient Lister Medal. Royal Coll. Surgeons, Eng., 1961. Fellow Royal Coll. Surgeons, Royal Soc. Can., Royal Soc., 1943, Royal Coll. Phys. and Surg. Can. (pres. 1940-44); mem. Am. Philos. Soc., Acad. Sci., U.S.S.R., Neurol. Society of Warsaw, Am. Neurol. Assn. (past pres.), Neurol. Soc. Estonia, Membre correspondent Societe de Neurologie de Paris, Academie des Sciences et des Lettres of Poland; fgn. asso. Nat. Acad. Scis., Am. Acad. Arts and Scis. Established Laboratory of Neurocytology, at Presbyterian Hospital, 1925; specializes in neurol. surgery. Author: Epilepsy and Cerebral Localization, 1941; Canadian Army Manual of Neurosurgery, 1941; Cerebral Cortex of Man, 1950; No Other Gods, 1954; Epilepsy and Functional Anatomy of Human Brain, 1954; Exaitable Cortex in Conscious Man, 1958; Speech and Brain-Mechanisms, 1959; The Torch, 1960; The Second Career, 1963; The Difficult Art of Giving, 1967; Man and His Family, 1967. Editor: The Cytology and Cellular Pathology of the Nervous System, 3 vols., 1932. Located several functional areas of human cerebral cortex in devel. of surg. treatment of epilepsy (motor, speech, vocalization control, gustatory, gastro-intestinal); discovered that stimulation of interpretive cortex activated neuronal record of past experience; developed centrocephalic hypothesis of memory control. Home: 4302 Montrose Av., Montreal 6. Office: Montreal Neurol. Inst. 3801 University St., Montreal 2, Que., Can.*

PENGELLY, William, English geologist; b. East Looe, Eng., Jan. 12, 1812; s. Richard and Sarah (Prout) P.; m. Mary Ann Mudge, 1837; 3 children; m. 2d, Lydia Spriggs, 1853; 2 daus.; opened sch., Torquay, circa 1836; pvt. tutor in math. and geology; lectr. math. and geology in various sects. of Gt. Britain; fellow Geol. Soc. (Lyell medal 1886), Royal Soc., 1863 Contbr. papers to jours. Studied geology of Devonshire, Eng.; collected fossils (given to mus. U. Oxford); examined plant-bearing deposits, Bovey Tracey, Brixham Cave, Kent's Hole at Torquay; his exploration of Brixham Cave proved man contemporary with several large extinct animals, 1858. Died Mar. 16, 1894.

PENHALLOW, David Pearce, botanist; b. Kittery Point, Me., May 25, 1854; s. Andrew Jackson and Ann Josepha (Pickering) P.; B.S., Mass. State Coll., 1873, Boston U., 1888; M.S., McGill U., 1896, D.Sc., 1904; m. Sarah A. Dunlap, May 4, 1876. Prof. botany and chemistry Imperial Coll. Agr., Sapporo, Japan, 1876-80, acting pres., 1879-80; botanist Houghton Farm Expt. State 1882-83; prof. botany, McGill U., 1883-—. Mem. Brit. Assn. com. on Canadian ethnology, 1897-1904 (chmn., 1902-04); chmn. Royal Soc. of Can. com. on ethnology, 1902-04; spl. commr. World's Indsl. and Cotton Centennial Expn., 1884. Trustee Marine Biol. Lab., Woods Hole, Mass.; dir. and sec. Biol. Stas. of Can.; dir. Atlantic Coast Biol. Sta., St. Andrews, N.B., 1907-—; chmn. Assn. Am. Biol. Research Stas., 1908-09; Brit. Assn. com. on pleistocene fauna and flora of Can., 1897-1901. Pres. Montreal Hort. Soc., 1888-92, Dominion Pomol. Soc., 1890, Soc. Plant Morphology and Physiology, 1899, Am. Soc. Naturalists, 1908-09, Natural History Soc. Montreal, 1904-—; fellow Royal Soc. Can. (pres. Sect. IV 1896-97), A.A.A.S. (v.p. Sect. G. 1908-09), Bot. Soc. Am., Geol. Soc. Am. Publs. on bot. subjects. Editor Canadian Record of Sci., 1888-90; asso. editor Am. Naturalist, 1897-1907; editor for paleobotany Botanisches Centralblatt, 1902-07. Died 1910.

PENICK, George Dial, Am. pathologist; b. Columbia, S.C., Sept. 4, 1922; s. Edwin A. and Caroline I. (Dial) P.; B.S., U. N.C. Sch. Medicine, 1944; M.D., Harvard, 1946; m. Marguerite M. Worth, Feb. 8, 1947; children—George Dial, H. Worth, David W., Anderson H., Marguerite W. Faculty, U. N.C. Sch. Medicine, Chapel Hill, 1949-—, prof. pathology, 1963-—, dir. Nat. Heart Inst. program-project on thrombosis and hemorrhage, 1961-—; staff N.C. Meml. Hosp., Chapel Hill, 1952-—, attending pathologist, 1963-—. Markle scholar in med. sci., 1953-58. Mem. A.M.A., Am. Assn. Pathologists and Bacteriologists, Am. Soc. for Exptl. Pathology, Soc. for Exptl. Biology and Medicine, Sigma Xi. Research, numerous publs. on properties plasma factor missing in blood of hemophiliacs, basis retarded coagulation in blood of hibernating rodents, changes in blood plasma asso. with thrombotic diseases. Home: Coker Dr., Chapel Hill, N.C. 27514.*

PENNAK, Robert William, Am. biologist; b. Milw., June 13, 1912; s. William Henry and Ella (Clemeson) P.; B.S., U. Wis., 1934, M.S., 1935, Ph.D, 1938; m. Alberta Vivian Pope, Sept. 7, 1935; children—Richard Dean, Cathy Ann. Faculty, U. Colo., Boulder, 1938-—, prof. biology, 1950-—, chmn. dept., 1963-66. Mem. Am. Micros. Soc. (pres. 1956), Soc. Systematic Zoology (pres. 1964), Am. Soc. Limnology and Oceanography (pres. 1962), Invertebrate Sect. Am. Soc. Zoologists (pres. 1963), Ecol. Soc. Am., Internat. Soc. Theoretical and Applied Limnology, Marine Biol. Assn. U.K. Author: Freshwater Invertebrates of the U.S., 1953; Collegiate Dictionary of Zoology, 1964; also numerous articles. Research on biology of mountain lakes and streams; biology of aquatic invertebrate animals. Home: Wild Horse Circle, Pine Brook Hills, Boulder, Colo. 80302.*

PENNANT, Thomas, Brit. naturalist; b. Downing, Flintshire, Eng., June 14, 1726; s. David and Araella Mytton) P.; student Queen's Coll., Oxford; m. Elizabeth Falconer, 1759; a dau., Arabella (Mrs. Edward Hanmer); m. 2d, Anne Mostyn, 1777; children—Sarah, Thomas. Traveled in Ireland, Scotland, and the continent. Became Fellow Royal Soc., 1767. Author: Tour in Scotland, 1771; Synopsis of Quadruped (later expanded to History of Quadrupeds), 1771; Second Tour of Scotland and a Voyage to the Hebrides, 2 vols., 1774-76; Arctic Zoology, 2 vols., 1784-85, Supplement, 1787; British Zoology, 4 vols., 1776-77; Genera of Birds, 1781; Indexes of the Orthologies of the Comte de Buffon and the Planches enluminees, 1786; Indian Zoology, 2d edit., 1790. Described classification of Crustacea, Arthropods, especially those of med. importance. Died Downing, Dec. 16, 1798.

PENNELL, Robert B., Am. biochemist; b. Berrien Springs, Mich., Mar. 27, 1909; s. Oscar Jude and Elethea (Brown) P.; B.S., Mich. State Coll., 1930, M.S., 1932, Ph.D., 1937; m. Naomi Jeanette Van Loo, July 31, 1937; 1 son, Robert Van Loo. Research asst. Mich. Agrl. Expt. Sta., U.S. Dept. Agr., 1934-39; research asso. Sharp & Dohme, Inc., 1939-50, asst. dir. immunochem. research, 1950-53; spl. asso. phys. chemistry Harvard Med. Sch., 1953-54; sr. investigator Protein Found., Jamaica Plain, Mass., 1953-—, dir. labs., 1963-—. Lectr. immunology Harvard Sch. Pub. Health, 1956-—; cons. div. biologic standards NIH, 1956-59; chmn. com. plasma NRC, 1960-—. Mem. Am. Chem. Soc., N.Y. Acad. Scis., A.A.A.S., Sigma Xi. Research, numerous publs. on chemistry of human blood components; on chemistry of brucella microorganisms; purification of toxoids and antitoxins; viruses and other antigens. Home: 35 Calvin Rd., Newtonville, Mass. 02160. Office: Blood Research Inst., Inc., 281 South St., Jamaica Plain, Mass. 02130.

PENNER, Samuel, Am. physicist; b. Buffalo, Oct. 3, 1930; s. Max and Rebecca (Cohen) P.; B.A., U. Buffalo, 1952; M.S., U. Ill., 1954, Ph.D., 1956; m. Beverly B. Shupe, May 25, 1952; children—Marcia Jo, Ellen Naomi. NSF fellow U. Ill., 1956-57; physicist Nat. Bur. Standards, Washington, 1957-—, chief accelerator physics sect., 1965-—; vis. physicist High Energy Physics Lab., Stanford, 1960; vis. physicist, Guggenheim Found. fellow Frascati Nat. Lab., Frascati, Rome, Italy, 1963-64; lectr. U. Md., 1959-61; Mem. Am. Phys. Soc., Phi Beta Kappa, Sigma Xi. Research, publs. on pi meson physics, photonuclear physics, ion optics, high energy electron scattering. Home: 11802 Milbern Dr., Potomac, Md. 20854. Office: Nat. Bur. Standards, Washington 20234.*

PENNER, Stanford Solomon, aero. engr.; b. Unna, Germany, July 5, 1921; s. Henry and Regina (Saal) P.; came to U.S., 1935, naturalized, 1942; B.S., Union Coll., 1942; M.S., U. Wis., 1943, Ph.D., 1946; m. Beverly Preston, Dec. 28, 1942; children—Merilynn Jean, Robert Clark. Research asso. Allegany Ballistics Lab., Cumberland, Md., 1944-45; research asso. Standard Oil Devel. Co., Esso Research Lab., Elizabeth, N.J., 1946; sr. research engr. jet propulsion lab. Cal. Inst. Tech., 1947-50, mem. faculty Guggenheim Jet Propulsion Center, 1950-64; prof., chmn. dept. aerospace and mech. engring. scis. U. Cal., San Diego, La Jolla, 1964-—. Cons. various mil. and govt. agys., 1954-—, aircraft, missile and chem. cos., 1952-—; dir. research and engring. support div. Inst. Def. Analyses, Washington (on leave from Cal. Inst. Tech.), 1962-64; vis. lectr., prof. various univs. Recipient OSRD award, 1945, certificate of Merit People-to-People Program, 1957. Fellow Am. Phys. Soc., Optical Soc. Am., Am. Inst. Aeros and Astronautics (mem. bd. 1964-66), A.A.A.S., N.Y. Acad. Scis.; mem. Am. Chem. Soc., Internat. Combustion Inst., Internat. Acad. Astronautics (engring. scis. sect.), Sigma Xi, Phi Lambda Upsilon. Author: Chemistry Problems in Jet Propulsion, 1957; Quantitative Molecular Spectroscopy and Gas Emissivities, 1959; (with D. B. Olfe) Radiation and Reentry, 1968; Thermodynamics, 1968; also monographs. Editor: Advanced Propulsion Techniques, 1961; Fundamental Problems in Rocket Propulsion, 1967; Advances in Tactical Rocket Propulsion, 1968. Mem. various editorial bds.; editor Internat. Jour. Quantitative Spectroscopy and Radiative Transfer, 1961-—, Jour. Missile Def. Research, 1963-67. Home: 6611 Avenida de las Pescas, La Jolla, Cal. 92037.*

PENNEY, William George, English physicist; b. Sheerness, Eng., June 24, 1909; s. William Alfred and Blanche Evelyn (Johnson) P.; B.S., Imperial Coll. London U., 1929, Ph.D., 1931, D.Sc., 1936; A.M. (Commonwealth Fund fellow), U. Wis., 1933; Ph.D., Cambridge U., 1936; LL.D., Melbourne U., D.Sc. (hon.), Durham U., 1957, Oxford U., 1959; m. Adele Minnie Elms, July 27, 1935 (dec.); children—Martin Charles, Christopher Charles; m. 2d, Eleanor Joan Quennell, Oct. 27, 1945. Asst. prof. math. Imperial Coll., 1936-45; atomic scientist Los Alamos, 1944-45; coordinator blast measurements Operation Crossroads, 1946; apptd. chief supt. armaments research Ministry of Supply, 1946, dir. Atomic Weapons Research Establishment 1953-59, dep. chmn. U.K. Atomic Energy Authority, 1961-64, chmn., 1964-—, mem. for weapons 1954, mem. for research, 1959. Seconded to Ministry Home Security, 1939-44, Dept. Sci. and Indsl. Research, 1944-45. Decorated Medal of Freedom with Silver Palm, 1946, Knight Comdr. Order Brit. Empire. Fellow Royal Soc., 1946; fgn. asso. Nat. Acad. Scis. Author: The Quantum Theory of Valency 1935; also papers. Research on nuclear weapons; greatly reduced size of Brit. atomic bomb by decreasing critical mass of uranium needed for explosion; work on solid state physics, molecular structure and hydrodynamics. Home: 1 Orchard House, East Hendred, Near Wantage, Berkshire, Eng. Office: 11 Charles II St., London, S.W.1, Eng.

PENNINGTON, William Alvin, Am. chemist, metallurgist; b. Halls, Tenn., Dec. 9, 1904; s. Marvin and Minnie (Abernathy) P.; B.S., Union U., 1925; D.Sc., 1962; Ph.D., Ia. State U., 1933; m. Jane Elizabeth Myer, May 25, 1954; 1 dau., Carolyn (Mrs. Christopher Fuller). Head dept. math. Union U., Jackson, Tenn., 1934-35; research engr. Armco Steel Corp., 1935-40; indsl. fellow Mellon Inst., Pitts., 1940-44; chief chemist and metallurgist Carrier Corp., Syracuse, N.Y., 1944-54; prof. chem. engring., metallurgy U. Md., College Park, 1954-63; phys. sci. Bur. Reclamation, Denver, 1963-—. Bd. govs. Acta Metallurgica, 1961-64. Fellow A.A.A.S., Am. Inst. Chemists, Am. Soc. Metals (pres. 1961, mem. coms., Howe medal 1947, Sauveur award 1962), Am. Chem. Soc., Nat. Assn. Chem. Engrs., Wash., N.Y. acads. sci., Sigma Xi, Alpha Chi Sigma, Alpha Sigma Mu (nat. pres. 1961), Phi Lambda Upsilon. Inventor 1st comml. azeotropic refrigerant; devel. of thermodynamics in decarbonization of steel; derivation of atmospheric corrosion equation; also publs. Home: 1705 Glen Moor Dr., Lakewood, Colo. 80215. Office: Bur. Reclamation, Denver 80225.*

PENNOCK, J(ames) Roland, Am. polit. scientist; b. Chatman, Pa., Feb. 4, 1906; s. James L. and Alice R. (Carter) P.; student London Sch. Econs., 1925-26; A.B., Swarthmore Coll., 1927; M.A., Harvard, 1928, Ph.D., 1932; m. Helen Sharpless, Jan. 24, 1931; children—Joan (Mrs. V. John Barnard), Judith C. (Mrs. Albert F. Lilley). Faculty, Swarthmore (Pa.) Coll., 1929-—, prof. polit. sci., 1946-—, dept. chmn., 1941-—; vis. prof. Columbia, 1950, Harvard, 1953. Guggenheim fellow, 1954-55. Mem. Social Sci. Research Council (chmn. com. on fellowship, polit. theory and legal philosophy 1954-64, dir. 1960-66), Am. Polit. Sci. Assn. (past mem. council, past v.p.), Am. Soc. Pub. Administrn., Internat. Polit. Sci. Assn., Am. Soc. Polit. and Legal Philosophy (editor Nomos, 1965-—), Pa. Assn. Polit. Sci., and Pub. Adminstrn., Phi Beta Kappa. Author: Administration and the Rule of Law, 1941; Liberal Democracy: its Merits and Prospects, 1950; (with David G. Smith) Political Science: an Introduction, 1964; contbr. to Democracy in the Mid-Twentieth Century: Problems and Prospects, 1960; also articles. Editor: Self Government in Modernizing Nations, 1964. Work in polit. theory especially theory democracy, definition and elucidation fundamental terms. Home: 3 Whittier Pl., Swarthmore, Pa. 19081.*

PENNY, Thomas, English botanist; ed. Cambridge; became clergyman; prebendary St. Paul's; asst. to Gesner; executed drawings used in Moffet's Insectorum Theatrum, 1734, thus may be considered a founder of entomology; discovered Hypericum balearicum (originally named Mytoeistus Pennaei by Clusius). Died 1589.

PENOTUS, George Bernard, chemist, alchemist, physician; b. Port-Sainte-Marie, Guienne, France, circa 1520; ed. Basel, Switzerland; follower of Paracelsus and supporter of Libavius; practiced alchemy and later attacked subject. Author of many pharmaceutical and chemical tracts including Theophrastisch Vade Mecum, 1596, and Tractatus Varii, de vera Praeparatione et Usu Medicamentorum Chemicorum, 1659. Strongly defended Paracelsus while attacking Galenists in his preface to Hundred and fourteene Experiments and Cures of the famous Physitian . . . Paracelsus, 1596 (Latin, 1582). Died in extreme poverty, Yverdun, Switzerland, circa 1620.

PENROD, Kenneth Earl, Am. physiologist; b. Blanchester, O., Mar. 30, 1916; s. William F. and Josie (Carman) P.; B.S., Miami U. (O.), 1938; Ph.D., Ia. State U., 1942; m. Virginia B. Hogue, June 29, 1942; children—Caroline L., Bruce H. Aviation physiologist USAAF, 1942-46; asst. prof. physiology Boston U. Sch. Medicine, 1946-50; asso. prof. physiology Duke Sch. Medicine, Durham, N.C., 1950-57, prof., 1957-59, asst. dean, 1952-59; prof. physiology W. Va. U. Sch. Medicine, Morgantown, 1959-65, v.p. W. Va. Med. Center, 1959-65; prof. physiology, provost Ind. U. Med. Center Indpls., 1965—. Spl. cons. on med. edn. in developing countries AID, 1963-65. Mem. Am. Physiol. Soc., A.A.A.S., Phi Beta Kappa, Sigma Xi, Phi Kappa Phi, Alpha Omega Alpha. Research, publs. on response of man to environmental stress, devel. of spl. protective equipment including that for outer space, diving medicine and effects of high pressures of gases on man. Home: 7645 Washington Blvd., Indpls. 46240.*

PENROSE, Edith Tilton, economist; b. Los Angeles, Nov. 29, 1914; d. George Albert and Hazel (Sparling) Penrose; B.A., U. Cal. at Berkeley, 1936; M.A., Johns Hopkins, 1950, Ph.D., 1951; m. David B. Denhardt, Aug. 25, 1934 (dec.); 1 son, David Tilton Denhardt; m. 2d, E.F. Penrose, Oct. 20, 1944; children—Perran, Trevear Tilton. Economist, social worker Cal. Relief Adminstrn., 1936-39; research asso. ILO, Geneva, Switzerland, 1939-41; spl. asst. to U.S. ambassador, London, Eng., 1941-46; asst. to adviser on econ. problems, U.S. del. to UN Econ. and Social Council, N.Y.C., 1946-47; lectr., research asso. Johns Hopkins, 1950-60; reader econs. U. London, 1960-64, prof., 1964—, head dept. econs. and politics Sch. Oriental and African Studies, 1962—; vis. fellow Australian Nat. U., 1955-56; asso. prof. U. Baghdad, Iraq., 1957-59. Mem. Brit. Com. to Investigate Pharm. Industry, 1964—. Mem. Am. Econ. Assn., Royal Econ. Soc., Assn. Brit. Orientalists. Author: Food Control in Great Britain, 1942; Economics of the International Patent System, 1951; Theory of the Growth of the Firm, 1959; also articles. First comprehensive analysis of theory of growth of bus. firm. significance of firm in internat. petroleum industry; studies on internat. firm, econs. of internat. patent industry. Home: 15 Chaldon Way, Coolsdon, Surrey, Eng. Office: Sch. Oriental and African Studies, U. London, Malet St., London WCI, Eng.*

PENROSE, Oliver, English theoretical physicist; b. London, Eng., June 6, 1929; s. Lionel Sharples and Margaret (Leathes) P.; B.Sc., U. Coll., London, 1949; Ph.D., Cambridge (Eng.) U., 1952; m. Joan Lomas Dilley, Oct. 17, 1953; children—Lucy Rebecca, Mathew David, Martin Sebastian. Math. physicist English Electric Co., Luton, Eng., 1952-55; lectr. asst. Yale, 1955-56; lectr. math. Imperial Coll., London, 1956-62, reader in applied math., 1963—; vis. prof. Belfer Grad. Sch. Sci., Yeshiva U., N.Y.C., 1962-63. Research, publs. on math. generalization of concept of Einstein condensation, criterion for stability of uniform plasma to electrostatic oscillations; analysis direction of time; proofs of convergence of Mayer's cluster expansion and related series; rigorous treatment of van der Waals-Maxwell theory of liquid vapor transition. Home: 42 Lebanon Park, Twickenham, Middlesex, Eng. Office: Imperial Coll., London S.W.7, Eng.*

PENSA, Antonio, Italian physician; b. Milan, Italy, Sept. 15, 1874; s. Michele and Giuseppina (Calzini) P.; M.D.; m. Maria Pergami, 1907; 1 dau., Adele. Prof. anatomy and histology; dir. Center for Study Nervous System; dir. History Museum, U. Pavia (Italy). Decorated Order of Merit for culture and teaching. Pontifical Academician. Mem. dei Lincei Acad., Lombard Inst. Sci. and Letters, Venetian Inst., Bologna Acad. Sci. Publs. on biology, morphology, comparative embryology, histology and cytology. Research in gen. histology, embryology, normal human anatomy, cytology, gen. biology. Address: viale 9 febbraio 22, Pavia, Italy.

PENSAK, Louis, Am. physicist; b. N.Y.C., June 16, 1911; s. Samuel and Ida (Feld) P.; B.S., L.I. U., 1932; postgrad. U. Pitts.; M.S., N.Y. U., 1936; m. Charlotte Markowitz, Mar. 7, 1943; children—Stephen, David, Martin. Devel. engr. tube div. RCA, Harrison, N.J., 1937-40, physicist RCA Research Labs., Princeton, N.J., 1941—; vis. asso. prof. Duke, 1962-63. Mem. I.E.E.E., Am. Phys. Soc. Designed oscilloscopes and kinescopes; patented aluminized kinescope, graphechon and metrechon storage tubes; contbd. to improvements in manufacture of transistors; discovered electron bombardment induced conductivity in thin insulating films; discovered high voltage photovoltaic effect in thin films of semiconductors. Home: 119 Random Rd. Office: RCA Labs., Princeton, N.J. 08540.*

PENTIMALLI, Luciano, Italian chemist; b. Messina, Italy, Feb. 28, 1923; s. Giuseppe and Teresa(de Stefano) P.; student U. Bologna (Italy), 1941-43, Laurea in Indsl. Chemistry, 1948; Laurea in Chemistry, U. Messina (Italy), 1946; m. Mariapia Peca, Sept. 16, 1964; 1 son, Carlo. Faculty, Facoltà di Chimica Industriale, U. Bologna (Italy), 1949-54, 56-64, prof. Facolta di Ingegneria, 1964—; asso. prof. U. Notre Dame (Ind.), 1954-55. Mem. Società Chimica Italiana, Am. Chem. Soc., Accademia Teatina della Scienze (hon.). Research, publs. in organic chemistry especially nitrogen and sulphur containing heterocyclics, organic dyestuffs. Home: via S. Stefano 75, 40125 Bologna. Office: Istituto Chimico Ingegneria, viale Risorgimento 2, 40136 Bologna, Italy..*

PENTSHEV, Iwan Georgiev, Bulgarian endocrinologist; b. Vratza, Bulgaria, No. 22, 1904; s. Georgi Pentshev and Iota (Atanasová) P.; grad. U. Sofia (Bulgaria) Med. Faculty, 1931; m. Savka Dimitrova, Nov. 19, 1944; children—Ivan, Julia. Practiced gen. medicine, 1931-33; staff therapeutic clinic Med. High Sch., Sofia, 1931-50, asst. prof., 1947-50; prof. Postgrad. Med. High Sch., I.S.U.L., 1950—. Chmn. med. sci. council Ministry Pub. Health; head Sci. Research Center Diabetes Mellitus, head research group endocrinology; mem. State Council Sci. and Tech. Progress. Decorated Red Flag of Labour, People's Republic of Bulgaria; Banting-Best decoration. Fellow Bulgarian Acad. Sci. (sec. med. dept.), Bulgarian Endocrine Soc. (chmn.), Czechoslovakian Med. Soc. Purkinje (hon. mem.). Author: Early Diagnosis of Internal Disease, 1954; Hormones and Hormonal Therapy, 1957; Endocrine and Metabolic Diagnosis, 1959; Endemic Goiter in Bulgaria, 1960; Current Progress in Endocrinology, 1963; Diabetes Mellitus, 1965; also numerous articles. Research in devel. of endocrinology in Bulgaria; initiated endemic goiter study and prophylaxis; pioneered study prediabetes; discovered insulin-precipitating antibodies. Home: 74, Patr. Ephtimii-str., Sofia, Bulgaria.*

PEPITONE, Albert, Am. psychologist; b. Bklyn., Jan. 22, 1923; s. Albert and Mabel (Davison) P.; A.B., N.Y. U., 1942; M.A., Yale, 1943; Ph.D., U. Mich., 1949; m. Emmy Angelica Berger, Oct. 3, 1949; children—Leslie Angelica, Jessica Lenore, Andrea Louise, Victor Albert. Teaching asst. dept. econs., social sci. Mass. Inst. Tech., 1946-48; teaching fellow dept. psychology U. Mich., 1948-49, project dir. Inst. for Social Research, 1949-51; faculty dept. psychology U. Pa., Phila., 1951—, prof., 1964—; Fulbright research prof. U. Groningen, Netherlands, 1960-61, Nat. Inst. Psychology, Rome, Italy, 1967-68. Mem. Am. Psychol. Assn., Soc. Exptl. Psychology. Author: Attraction and Hostility, 1964; also articles. Research in several areas of exptl. social psychology, determinants of distortions in social perception, role of cohesiveness in expression tion, threat and group cohesiveness, cognitive bases of aggression, conflict resolution in impression formation of interpersonal attraction and hostility, social determinants of decision-making. Office: Dept. Psychology, U. Pa., Phila. 19104.*

PEPPARD, Donald Francis, Am. chemist; b. Flanagan, Ill., Oct. 8, 1912; s. Charles Edward and Catherine (Schultheis) P.; B.S., U. Ill., 1937, Ph.D., 1939; m. Marvila Ann Aepinus, Mar. 27, 1946; children—Sean Bruce, Kevin Donal, Rory Jon, Brian Philip. Instr., Ill. Inst. Tech., 1939-41; research fellow organic chemistry U. Chgo., 1939-41, instr., 1941-43; research chemist thorium and rare earths Lindsay Chem. Co., West Chicago, Ill., 1945-46; group leader inorganic research and devel. Argonne (Ill.) Nat. Lab., 1946—. Recipient Scholarship medal U. Ill., 1937. Mem. Am. Chem. Soc. Contbg. author: Analytical Chemistry, 1961; Progress in the Science and Technology of the Rare Earths, 1964; also numerous articles. Patentee methods separating metals by liquid-liquid extraction. Home: 715 Thomas St., Oak Park, Ill. 60302. Office: 9700 S. Cass Av., Argonne, Ill. 60439.*

PEPPER, Bailey B(reazeale), Am. entomologist; b. Easley, S.C., Mar. 20, 1906; s. Bailey B. and Eugenia (Sheriff) O.; B.S., Clemson Coll., 1929; M.S., Ohio State U., 1931; Ph.D., Rutgers, 1934; m. Margaret M. Forgham, Oct. 9, 1937; children—James Bailey, Carl Forgham. Mem. faculty Rutgers U. since 1935, prof. entomology since 1945; staff N.J. Agrl. Expt. Sta. since 1935, research specialist, chmn. since 1945. Mem. Middlesex Co. Mosquito Extermination Commn., 1946—; sec. State Mosquito Control Commn. Mgr. Marlboro State Hosp. Mem. Am. Assn. Econ. Entomoloists, Entomological Society of Am. (chmn. Eastern branch 1963-64), A.A.A.S., New Jersey State Hort. Soc., N.J. Health and San. Assn., Am. Mosquito Control Assn., N.J. Mosquito Extermination Assn. (sec.), Sigma Xi. Contbr. articles sci. jours. Research on treefruit insects, 1931-34; vegetable insects with emphasis on processing props, 1935-50; pesticide residues and relationship between pesticide use and crop quality; insect ecology and insect control and its relationship to wild life. Home: 4 Lenox St., Edison, N.J. Office: Dept. Entomology, Rutgers U., New Brunswick, N.J.*

PEPPER, David Charles, Brit. chemist; b. Shipley, Yorkshire, Eng., Jan. 1, 1917; s. William Owen and Blanche (Tindall) P.; B.Sc. London, Tech. Coll. Bradford, London, Eng., 1937; Ph.D., Jesus Coll., Cambridge (Eng.) U., 1941, postgrad.; m. Deirdre Joan Donovan, Sept. 3, 1960; 1 son, William Eoin David. Extramural research scientist dept. colloid sci. Ministry Supply, Cambridge, 1940-45; faculty U. Dublin (Ireland), 1945—, prof. phys. chemistry, 1961—, fellow Trinity Coll., 1951—. Fellow Chem. Soc., Inst. Chemistry Ireland (past mem. council); mem. Royal Irish Acad. (mem. council 1965—), Faraday Soc. Contbg. author: Polymerization in Friedel-Crafts and Related Reactions, 4 vols., 1963-65. Research, numerous publs. on phys. chemistry macromolecules in solution and mechanisms their formation, polymer formation by cationic mechanisms. Home: 4 Brighton Vale, Monkstown, Dublin, Ireland.*

PEPPER, Paul Milton, Am. mathematician; b. Kendallville, Ind., May 16, 1909; s. Byron Jacob and Louisa (Dirrim) P.; A.B. with distinction, Ind. U., 1931, A.M., 1932; Ph.D., U. Cin., 1937; m. Nellie Margaret Jacqmain, June 9, 1934; 1 son, Ronald Paul. Instr., U. Cin., 1937-38; faculty U. Notre Dame, Ind., 1938-49; asso. prof. indsl. engring. Ohio State U., Columbus, 1949-51, 56-60, prof., 1960—, asso. prof. math., 1953-59, asst. to dir. U. Research Found., 1951-53, 57—, dir. Mapping and Charting Research Lab., 1953-57. Cons. USAF, 1958-59, Autometric Corp., 1957-63, Tech. Inc., 1964—, Chem. Abstracts Service, 1964—. Fellow A.A.A.S.; mem. Math. Assn. Am. (sec. Ind. sect. 1945-49), Am. Math. Soc., Soc. for Indsl. and Applied Math., Am. Astronautical Soc. (sr. mem.), Am. Soc. Tool and Mfg. Engrs., Assn. for Computing Machinery, Am. Geophys. Union, Am. Soc. Photogrammetry, Assn. for Machine Transplantation and Computational Linguistics, Ohio Acad. Sci., Am. Soc. for Engring. Edn., Phi Beta Kappa, Sigma Xi, Alpha Pi Mu. Author: Balanced Scheduling of Production Facilities, 1955; also articles. Patentee 5 significant figure slide rule with cooperating trigonometric functions including verniers for nonlinear scales; originator balanced cycle scheduling; research in distance geometry, photo grammetric math. Home: 517 E. Schreyer Pl., Columbus, O. 43214.*

PEPPER, William, Am. physician; b. Phila., Aug. 21, 1843; s. William and Sarah (Platt) P.; grad. U. Pa., 1862, Med. Dept., 1864; m. Frances Sargeant Perry, June 25, 1873, 4 children. Resident physician Pa. Hosp., Phila., 1864-65; pathologist, vis. physician Pa. Hosp., Phila. Hosp., 1865-68; lectr. morbid anatomy U. Pa., 1868-70, prof. clin. medicine, 1876-84, provost, 1880-94, prof. theory and practice of medicine, 1884-98, founder Univ. Hosp. (1st hosp. in U.S. directly associated with and staffed by faculty of univ. med. sch.), 1874, founded nurses' tng. sch., 1887, as provost founded Wharton Sch. Finance, vet., hygiene, biol. and archtl. schs., univ. extension lectures, Bennett Sch. (for grad. instrn. of women); med. dir. Phila. Centennial Exhbn., 1875-76; founder Am. Climatol. Soc., 1884, pres., 1886; pres Am. Clin. Assn., 1886, Assn. Am. Physicians, 1891; founded Coll. Assn. of Pa., 1886; mem. exec. com. A.M.A.; pres. 1st Pan-Am. Med. Congress, 1893; founded, endowed William Pepper Lab. of Clin. Medicine, U. Pa. (1st lab. in U.S. devoted to advanced clin. studies of causes of disease), 1894; directed establishment of 1st Phila. Free Library (endowed by his uncle George S. Pepper), 1890; founded Archaeol. Assn. of U. Pa., 1892, founded Univ. Museum. Author: The Morphological Changes of the Blood in Malarial Fever, 1867; Descriptive Catalogue of the Pathological Museum of the Pennsylvania Hospital, 1869; A Practical Treatise on the Diseases of Children,1870; Clinical Lecture on the Case of Progressive Muscular Sclerosis, 1871; Cases of Abdominal Tumor Attended with Profuse Sweating . . ., 1871; The Sanitary Relations of Hospitals, 1874; Higher Medical Education, the True Interest of the Public and the Profession, 1877, 94; A Contribution to the Clinical Study of Typhitis and Inflamation of the Bile-ducts, 1878; A Contribution to the Clinical Study of Exophthalmic Goitre, 1879; A Contribution to the Clinical Study of Typhities and Perityphitis, 1883; A Contribution to the Climatological Study of Phthisis in Pennsylvania; A System of Practical Medicine, 5 vols., 1885-86; Textbook of the Theory and Practice of Medicine, 2 vols., 1893-94; Two Addresses, 1894. Described bone marrow changes in prog. pernicious anemia, 1875. Died Pleasanton, Cal., July 28, 1898.

PEPPER, William, Am. physician; b. Phila., May 14, 1874; s. William and Frances Sergeant (Perry) P.; A.B., U. Pa., 1894, M.D., 1897, Sc.D., 1932; LL.D., Temple U., 1942; m. Mary Godfrey, Dec. 31, 1904 (dec. Oct. 1918); m. 2d, Phoebe S. (Voorhees) Drayton, Apr. 3, 1922. With med. dept. U. Pa., 1899; dean Sch. Medicine, 1912-45. Fellow Coll. Physicians of Phila., Assn. Am. Med. Colls. (pres. 1920-21); mem. A.M.A., Am. Philos. Soc. Described congenital sarcoma of adrenal and liver in infants (Pepper's type). Died Dec. 3, 1947.

PEPYS, William Hasledine, English inventor; b. London, Mar. 23, 1775; s. W.H. Pepys; philos. instrument maker. Fellow Royal Soc., 1808; a founder Akesian Soc. (led to Brit. Mineral. and Geol. socs., London Instn.), 1796, an original mgr., also hon. sec., 1821-24; pres. Royal Instn., 1816; treas., v.p. Geol. Soc.; early mem. Mineral. Soc. Contbr. papers to jours. Inventor modern mercury gasometer and water gasholder, other chem. apparatus. Died Aug. 17, 1856.

PEQUEGNAT, Willis E(ugene), Am. biol. oceanographer; b. Riverside, Cal., Sept. 18, 1914; s. Frank A. and Mary G. (Witherspoon) P.; A.B., U. Cal., 1936, A.M., 1938, Ph.D., 1942; m. Elizabeth Younghusband, 1937 (div. 1957); children—Willo Ann, John E.; m. 2d, Linda Lee Haithcock, 1957; children—Marina Lynn, William Gordon. Faculty mem. Pomona Coll., 1940-60, chmn. dept. zoology, 1944-60, prof. zoology, 1951-60; asso. program dir. NSF, 1960-61, program dir., 1962-63; prof., chmn. dept. biol. scis. State U. at Oyster Bay, 1961-63; prof. biol. oceanography Tex. A. and M. U., College Station, 1963—, acting head dept. oceanography and meteorology, 1964—; vis. prof. biology U. Chgo., 1947-48, U. Cal., summer, 1951; dir. Carnegie Conf. Progress and Horizons of Sci., 1953. Dir. summer session Marine Lab., Corona Del Mar, 1948-60; Ford fellow, research asso. Scripps Instn. of Oceanography, 1954-55; vis. prof. Cal. Inst. Tech., 1957, 58; Dir. Inst. Biology for NSF. Recipient research award in oceanography Office Naval Research, 1956, 64. NSF fellow, 1958-59. Fellow A.A.A.S.; mem. Am. Soc. Limnology and Oceanography, Western Soc. Naturalists, Sigma Xi. Contbr. articles sci. jours. Home: 2508 Willow Bend Dr., Bryan, Tex. 77801. Office: Dept. Oceanography, Tex. A. and M.U., College Station, Tex.*

PERCILE, Ninni Alessandro, Italian zoologist; b. Venice, Apr. 4, 1837; s. Christo and Maria (Polo) Ninni; student, Venice; made study trips to mus's., Switzerland, Austria, Germany, Greece; degree in natural sci. U. Modena (Italy), 1867; m. Emma Gasperini, 1863; children include Emilio. Research and publs. on vertebrates of Venice and Adriatic Sea, mollusks, insects, arachnids. Died 1892.

PERCIVAL, James Gates, Am. geologist; b. Kensington, Conn., Sept. 15, 1795; s. James and Elizabeth (Hart) P.; grad. Yale, 1815, Med. Instn., 1820; student U. Pa. Med. Sch., 1818-19. Apptd. asst. surgeon U.S. Army and prof. chemistry U.S. Mil. Acad., 1824; surgeon recruiting offices, Boston, 1824-25; editor Am. Atheneum, N.Y.C., 1825; Phi Beta Kappa poet Harvard, 1824; Phi Beta Kappa orator Yale, 1825; assisted Noah Webster in editing An American Dictionary of the English Language, 1827-28; state geologist of Conn., 1835-42; surveyor Am. Mining Co., Ill. and Wis., 1853; state geologist of Wis., 1854-56. Author: Zamor, 1815, Prometheus, 1821 (both single poems); Poems, 1821; Clio I and II, 1822; Prometheus Part II with Other Poems, 1822; Poems, 1823; Clio Number III, 1827; The Dream of a Day, and Other Poems, 1843; Report on the Geology of the State of Connecticut, 1842. Editor: Elegant Extracts (Vicesimus Knox), 6 vols., 1825; System of Universal Geography (Malte-brun), 3 vols., 1827-34. Distinguished crystalline rocks; determined crescent shape of trap dykes, also exemplified laws governing subterranean forces which form mountains, 1842. Died Hazel Green, Wis., May 2, 1856.

PERCIVAL, Thomas, English physician; b. Warrington, Eng., Sept. 29, 1740; s. Joseph and Margaret (Orred) P.; ed. Warrington Acad., Edinburgh U.; completed med. studies, Leiden, Netherlands, 1765; m. Elizabeth Basnett; 3 sons, including Percival. Moved to Manchester, Eng., 1767; helped form com. to enforce sanitation in Manchester; suggested city establish pub. baths; removed Warrington Acad. to Manchester, 1785. Fellow Royal Soc., 1765; a founder Manchester Literary and Philos. Soc., 1781. Author: De frigore, 1765; Essays on Medical Topics, 1767, 78, 79; Experiments and Observations on Water, particularly on the Hard Pump Water of Manchester, 1769. On the Disadvantages which attend the Inoculation of Children in Early Infancy, 1768; A Father's Instructions, 3 parts; Proposal to Establish More Accurate and Comprehensive Bills of Mortality in Manchester; Observations and Experiments on the Poison of Lead, 1774; Tables showing the Number of Deaths occasioned by the Small-pox, 1775; A Physical Inquiry into the Power and Operation of Medicine, 1786; Medical Ethics, 1803; also articles. Obtained census of Manchester, 1778; pioneer in epidemiol epidemiology; credited with introducing cod liver oil into therapeutics in Eng., 1782; early advocate of factory legislation. Died Manchester, Aug. 30, 1804.

PERCY, James Fulton, Am. surgeon; b. Bloomfield, N.J., Mar. 26, 1864; s. James and Sarah Ann (Fulton) P.; M.D., Bellevue Hosp. Med. Coll. (now Med. Dept. New York U.), 1886; postgrad. Chgo. Vet. Coll., 1895, in Germany, Switzerland, Belgium, 1897-98; visited clinics in Eng., France, Germany, Austria, 1914; A.M. (hon.), Knox Coll., 1914; m. Edna B. Post, 1925. Med. practice, Mazeppa, Minn., 1886-88, Galesburg, Ill., 1888-1917; practice limited treatment of cancer since 1917; surg. practice, San Diego, Cal., 1920-22, Los Angeles, since 1922; attending sr. surgeon, cancer service, and founder mem. Malignancy Bd. and Tumor Clinic, Los Angeles County Hosp.; clin. prof. surgery (neoplasms) Coll. Med. Evangelists Med. Sch.; attending surgeon French Hosp., 12 years; cons. surgeon Orthopedic Hosp. and Sch. for Crippled Children (all in Los Angeles). Fellow A.C.S. (founder mem.); mem. A.M.A., So. Cal. (hon.), Cal., Los Angeles (v.p. 1931) med. assns. Los Angeles Surg. Soc. (pres. 1929), Los Angeles Cancer Soc. (pres. 1939), Am. Bd. Surgery (founder mem.), Western Surg. Assn. (pres. 1918), Ill. Med. Soc. (pres. 1907; sec. Jud. Am. Assn. Obstet., Gynecol. and Abdominal Surgeons, Hollywood Acad. Medicine (hon. 1926), U.S. Mil. Surgeons Assn. Contbr. articles to med. jours., principally on actual cautery in treatment of cancer. Developed Percy actual cauteries for treatment of accessible cancer. Died Apr. 26, 1946.

PERCY, John, English metallurgist; b. Nottingham, Eng., Mar. 23, 1817; s. Henry Percy; studied chemistry, Nottingham; attended lectures by Gay-Lussac, Thenard, A. de Jussieu, in Paris; M.D., Edinburgh, 1838; m. Grace Piercy, 1839. Apptd. physician Queen's Hosp., Birmingham, Eng., 1839; worked (with David Forbes and William Hallowes Miller) on crystallized slags, 1846; named lectr. metallurgy Met. Sch. Sci. (now Royal Coll. Sci.), 1851, later prof.; lectr. metallurgy to arty. officers, Woolwich, 1864-89. Recipient Millar prize Inst. Civil Engrs., 1887, Albert medal Soc. Arts, 1889. Fellow Royal Soc., 1847, Geol. Soc.; mem. Iron and Steel Inst. (Bessemer medal 1876, pres. 1885, 86). Author: Experiments (on) Presence of Alcohol in the Ventricles of the Brain after Poisoning by that Liquid, 1839; On the Metallurgical Treatment and Assaying of Gold Ores, 1852; Iron and Steel, 1864; A Treatise on Metallurgy, 1864. Invented method for extracting silver from ore which led to other important metall. processes. Died London, June 19, 1889.

PERCY, Baron Pierre François, French surgeon; b. Montagney, France, Oct. 28, 1754; surgeon with Army of Napoleon; gen. insp. Army Health Services; prof. Paris Faculty Medicine. Mem. French Acad. Scis., Acad. Medicine, Acad. Surgery, Soc. Agr. Author: Manuel du chirurgien d'armée. Leading military surgeon of his time. Died Lagny, France, Feb. 18, 1825.

PERDRIEL, Georges, French ophthalmologist; b. Pontivy, France, Apr. 4, 1922; s. Francois and Marguerite (Cheny) P.; Docteur en médecine Hpitaux de Paris, 1946; Professor agrégé diploma Health Service, Dept. Army, 1960; m. Suzanne Monteil, July 27, 1944; children—Francoise, Evelyne. Lt. col. medic l'Armée de l'Air, 1964—; asst. Hosp. Val de Grace; adviser Centres du Personnel Navigant de l'Aeronautique de Paris (France), 1956-66; head ophthalmology dept., prof. l'Ecole d'Application du Service de l'Air, 1960—. Decorated medaille de'Aeronautique, medaille d'Honneur du Service de Sante des Armées, chevalier de la Legion de l'Honneu, chevalier de l'Instruction Publique. Mem. Académie Internautique de Med. Militaire, Soc. Internationale d'ERG. nationale de Med. Aéronautique, French Soc. Ophthalmology, Soc. Ophthalmology Paris, Soc. Med. Aéronautique de Med. Miltaire, Soc. Internationale d'ERG. Research, numerous publs. on ocular physio-pathology of pilot electroretinographic research, color perception, med. surg. ocular techniques, tng. apparatus for nocturnal vision. Home: 67 Boulevard Lefebvre, Paris 15°. Office: C.P.E.M.P.N. 5 bis, Av. de la Porte de Sèvres, Paris, 15, France.*

PEREGRINUS, Petrus (or Peter the Pilgrim, Peter de Maricourt), French engr., inventor; native of Picardy; flourished 2d half 13th century; engr., army of Louis IX; probably participated in Crusades; engaged in exptl. research; 1st to speculate magnetism might be converted to kinetic energy; 1st to determine poles of magnet, also stated laws of attraction between poles; inventor 1st practical compass, consisting of magnetic needle on pivot; introduced graduated scale for compass.

PEREIRA, Father, mathematician, astronomer; b. 18th century; mem. Soc. Jesus. Author: (with Father Koegler) Chinese treatise on astronomy based on Ptolemaic principles; contbg. author on Chinese and Western mus. instruments to Yü-Ting Li Hsiang Kao Chheng.

PERERA, George A(lfred), Am. physician, educator; b. N.Y.C., Dec. 29, 1911; s. Lionello and Carolyn (Allen) P.; A.B., Princeton, 1933; M.D., Columbia, 1937; Sc.D. in Medicine, 1942; m. Anna Paxson Rhoads, Dec. 22, 1934; children—Marcia (Mrs. Nicholas B. Van Dyck), David Rhoads. Asst. physiology N.Y.U., 1940-41; mem. staff Presbyn. Hosp. and Vanderbilt Clinic 1943—, attending physician, 1960—; dir. student health service Columbia Coll. Phys. and Surg., 1943-46; mem. faculty Columbia, 1946—, prof. medicine, 1958—, asso. dean, sec. faculty medicine, 1962—; mem. staff Columbia div. Goldwater Meml. Hosp., N.Y.C., 1947-48, cons., 1951-62. Mem. sect. endocrinology com. growth NRC, 1949-53; Lewis A. Conner meml. lectr. Am. Heart Assn., New Orleans, 1955; Pfizer lectr. Australian Postgrad. Fedn. Medicine, 1961; Lambie-Dew orator Sydney U. (Australia) Med. Soc., 1961. Trustee Riverdale Country Sch., N.Y.C. 1949—, Mary Imogene Bassett Hosp., Cooperstown, N.Y., 1961—, Columbia Univ. Press, 1961-64, 66—; bd. dirs. Intercollegiate br. N.Y.C. YMCA, 1956-65, Friends Med. Found., 1962—. Walter Belknap James research fellow, 1946-47. Diplomate Am. Bd. Internal Medicine (bd. dirs. 1957-64, vice chmn. 1963-64). Fellow A.C.P., A.A.A.S.; mem. Am., N.Y. (adv. council research 1951-58) heart assns., N.Y. Acad. Medicine (sec., then chmn. sect. medicine 1952-54), Friends Med. Soc. (adv. bd. 1953-55, chmn. 1955-59), A.M.A., Harvey Soc., Am. Soc. Clin. Investigation, Soc. Exptl. Biology and Medicine, Interurban Clin. Club, Assn. Am. Physicians, Practitioners Soc. (pres. 1961-63), Am. Clin. and Climatol. Assn., Alpha Omega Alpha. Contbr. profl. jours., chpts. textbooks. Mem. editorial bd. Circulation, 1953-57, Circulation Research, 1958-60, Annals Internal Medicine, 1961-65, Jour. Med. Edn., 1962—. Research on relationship adrenal hormones to high blood pressure, natural history and epidemiology of hypertension. Home: 5209 Sycamore Av., N.Y.C. 10471. Office: 630 W. 168th St., N.Y.C. 10032.*

PÉRÈS, Jean-Marie, French biologist, oceanographer; b. Montpellier, France, Oct. 8, 1915; s. Joseph and Robin P.; Doctorat és Sciences, U. Paris, 1943; m. Germaine Sidet, Sept. 17, 1937; children—Michéle (Mrs. Dominique Rain), Bernard. Asst., Faculty of Scis., Marseilles, France, 1940-43, lectr., 1947-51, prof., dir. Marine Sta., 1951—, also vice dean faculty; asst. dir. Musée Océanopraghique Monaco, 1943-44; asst. dir. malacological and marine labs. Nat. Mus. Natural History, Paris, 1944-47. Mem. all French coms. for oceanography and marine research; pres. directory com. French bathyscaphe; participant numerous oceanographic cruises. Author: Oceanographie Bilogique Biologi et Marine, 1951; (with L. Deveze) La Vie dans l'Océan, 1966; also articles Biol. observations with different deep-sea vehicles, including bathyscaphe, diving saucer; description of communities in various areas (temperate and tropical seals), especially flora and fauna of bottom. Address: Station Marine d'Endoume, Rue Batterie des Lions, Marseilles 7, France.

PÉRÈS, Joseph Jean Camille, French mathematician; b. Clermont-Ferrand, France, Oct. 31, 1890; s. Jean Peres; ed. École Normale Supérieure; agrégé de math., Docteur ès Scis.; became prof., Lycée, Montpellier, France, 1914, U. Strasbourg (France), 1919; named prof. fluid mechanics Faculty Scis. Paris, 1932; co-dir. Nat. Center for Sci. Research, from 1946. Mem. French Acad. Scis. Author: Les sciences exactes, 1930; Cours de mécaniques des fluides, 1936; (with V. Voiterra) Théorie générale des fonctionnelles, 1936; (with Comffignal and Brillouin) Les grandes machines mathématiques, 1950; Méchanique générale, 1953. Died Neuilly-sur-Seine, France, 1962.

PERETTI, Ettore Alex, Am. metall. engr.; b. Butte, Mont., Apr. 5, 1913; s. Charles L. and Mary (Crotto) P.; B.S., Mont. Coll. Mineral Sci. and Tech., 1934, M.S., 1935; Sc.D., Wuerttemberg (Germany) Inst. Tech., 1936; Met.E. (hon.), U. Mont., 1963; m. Pierina H. Giacoletto, June 6, 1937; children—Charlene Pier, D'Arcy Skinner, E.A. James, Martha Jayne Seats. Faculty, Mont. Coll. Mineral Sci. and Tech., 1936-40, asst. prof., 1939-40; asst. prof. Columbia, 1940-46; faculty U. Notre Dame (Ind.), 1946—, prof. metall. engring., 1949—, head dept., 1951—. Mem. Am. Inst. Metall. Engrs., Am. Soc. for Metals, Am. Soc. for Testing and Materials, Am. Soc. Engring. Edn., Am. Assn. U. Profs., Brit. Inst. Metals, Sigma Xi, Tau Beta Pi, Theta Tau, Alpha Sigma Mu (nat. trustee 1964—). Research, numerous publs. on nature chem. reactions in smelting operations and alloying behavior of metals and compounds. Home: 18295 Brightlingsea, South Bend, Ind. 466037. Office: Box E, Notre Dame, Ind. 46556.*

PEREVOZCHIKOV, Dmitri Matveevich, Russian mathematician, astronomer; b. Apr. 28, 1788; prof., Moscow; founder Moscow obs. Author: Teorija planet, 5 vols., 1863-68; Rucnaja matematiceskaja enciklopedija, 13 vols., 1826-27. Died St. Petersburg, Sept. 15, 1880.

PÉREZ, Charles, French zoologist; b. Bordeaux, France, May 19, 1873; s. Jean Perez; ed. École normal supérieure, Paris; docteur ès sciences, 1902; prof. zoology Bordeaux Faculty Scis.; dir. Roscoff Zool. Lab.; became prof. zoology Paris Faculty Scis., 1921. Mem. French Acad. Scis. Described numerous species marine fauna; research on parasitism, protozoa. Died Paris, Sept. 22, 1952.

PEREZ MASIA, Andrés, Spanish chemist; b. Requena, Valencia, Spain, Nov. 24, 1924; s. Andrés Pérez Laguna and Enriqueta Masiá López; Licentiate in Physico-chem. scis., Faculty Scis., U. Valencia, 1947, Licentiate in Chemistry, 1948; Dr. in Chem. Scis., Faculty Scis., U. Madrid (Spain), 1951; m. María Antonia Boada de la Cueva, Nov. 9, 1951; children—Andrés, Susana, Alejandro, Enrique, Margarita María. Staff, Spanish Superior Council for Sci. Research (Inst. Phys. Chemistry Rocasolano), Madrid, 1951—, chief dept. thermodynamics, 1954—, sci. research worker, 1956—, sec., 1957, adjoint sec. gen., 1962; prof. in charge chem. thermodynamics Faculty Scis., U. Madrid, 1960—. Recipient Torres Quevedo, Superior Council for Sci. Research, 1951, Order Mil. Merit with White Distinctive, 1945, Commendary with medal Order Alfonso el Sabio, 1964. Mem. Royal Spanish Soc. Physics and Chemistry (medal 1965). Research and publs. on physico chem. investigations of gaseous state using thermal conductivity, 2d virial coefficient and viscosity; dielec. behavior of solutions; interfacial phenomena in aqueous solutions of capillary inactive electrolytes; theoretical calculations of molecular structures. Home: 49, Avenida de América, Madrid-2. Office: 117, Serrano, Madrid-6, Spain.*

PEREZ-MENDEZ, Victor, physicist; b. Guatemala, Guatemala, Aug. 8, 1923; s. Moses and Rebecca (Gagin) P.; M.Sc., Hebrew U., Jerusalem, 1946; Ph.D., Columbia, 1951; m. Gladys E. Cobert, Aug. 5, 1949; children—David, Judith. Came to U.S., 1947, naturalized, 1951. Research asso. Columbia, 1951-53; staff physicist Lawrence Radiation Lab., Berkeley, Cal., 1953-59, sr. scientist, 1959—. Lectr. extension div. U. Cal., Berkeley, 1958—. Mem. Am. Phys. Soc., Sigma Xi. Research, numerous publs. on elementary particles to study meson-nuclear forces; developed electronic and nuclear instrumentation, automatic analysis nuclear data by computers. Home: 870 Santa Barbara Rd., Berkeley 94707. Office: Lawrence Radiation Lab., Hearst Av., Berkeley, Cal. 94720.*

PÉRIER, Auguste-Charles, French engr.; founder (with son Jacques-Constantin) workshop for manufacture of centrifugal pumps, weavers' looms, spinning apparatus, steam engines (1st large scale indsl. enterprise in France), Chaillot, circa 1770; built (with son) steam ship and tested it on Seine, 1775, also built 2 fire engines of Watt type (functioned 1781), 1st rotating machines (grinders driven by 2 steam engines), Fulton's submarine, then the engine for his steam boat, 1803.

PÉRIER, Floriu, physicist; b. 1605; m. Gilberte Pascal (sister of Blaise), 1641; children—Etienne, Jacqueline, Marquerite, Louis, Blaise. Carried out Blaise Pascal's expt. of comparing barometric pressure in Clermont, France, with that at top of mountains; instrumental in publ. of Pascal's Provincials and various sci. works. Died 1702.

PÉRIER, Jacques-Constantin, French engr.; b. Paris, Nov. 2, 1742; founder (with father Auguste) workshop for manufacture of centrifugal pumps, weavers' looms, spinning apparatus, steam engines (1st large scale indsl. enterprise in France), Chaillot, circa 1770; mem. French Acad. Scis., 1783. Built (with father) steam ship and tested it on Seine, 1775, also built 2 fire engines of Watt type (functioned 1781), 1st rotating machines (grinders driven by 2 steam engines), Fulton's submarine, then the engine for his steam boat, 1803. Died Aug. 16, 1818.

PÉRIER DE LAHITOLLE, Jules, French engr.; b. Eure, France, 1832; participant Mexican campaign, wounded 1870; dir. iron works, Bourges, France; developed 1st French field steel cannon (tested 1873), also 9.5 centimeter cannon. Died 1879.

PERIN, Charles Theodore, Am. psychologist; b. Hillsboro, O., Oct. 16, 1916; s. Charles Theodore and Faye (Dailey) P.; A.B., Miami U., Oxford, O., 1938; M.A., Yale, 1940, Ph.D., 1942; m. Jean Letcher Spargur, Dec. 27, 1941; children—Charles Theodore 3d, Nancy Lyne. With NACA, Cleve., 1945-46, USAF, Washington, 1945-46; asst. prof. to prof. psychology Miami U., 1947—. Cons. USAF, 1950-66, Office Sec. Def., 1950. Mem. Am., Midwestern psychol. assns., Psychonomic Soc., Am. Soc. Clin. Hypnosis, A.A.A.S. Research, publs. on delay-of-reinforcement gradient, multi-dimensional approach to habit strength measurement, hypnotherapy in phobic reactions. Home: 138 Hilltop Rd., Oxford, O. 45056.*

PERINAUD, Henri, ophthalmologist; b. France, 1844; described an infectious conjunctivitis believed to be caused by a leptothrix and passed from animals to man (Perinaud's conjunctivitis), 1889, (with Xavier Galezowski) form of Tb conjunctivitis occurring in both man and animals (Perinaud's oculo-glandular syndrome), 1889. Died 1905.

PERKIN, Sir William Henry, English chemist; b. Shadwell, London, Eng., Mar. 12, 1838; s. George Fowler and Sarah (Cuthbert) P.; student Royal Coll. Chemistry, South Kensington, Eng., 1853-57; hon. degrees from Royal College of Chemistry, Univs. of Würzburg, St. Andrews, Manchester, Oxford, Leeds, Heidelberg, Columbia, Johns Hopkins, others; m. Jemina Harriet Lissett, Sept. 13, 1859; 2 sons; m. 2d, Alexandrine Caroline Mollwo, Feb. 8, 1866; 1 son, 4 daus. Asst. to Hoffman, Royal Coll. Chemistry, 1855-57; indsl. chemist 1857-74; pvt. chem. research, from 1774. Recipient 1st Perkin medal Soc. Chem. Industry, N.Y.C., 1906. Fellow Royal Soc., 1866 (Royal medal, 1879, Davy medal 1889, v.p. 1883); mem. Chem. Soc. (pres. 1883), knighted, 1906. Author: A Course of Practical Chemistry, 1890; Organic Chemistry, 1894-95; Inorganic Chemistry, 1909-11. Developed process for producing artificial alizarin; produced, manufactured mauve, 1st synthetic dye (thus founding aniline dye industry); 1st to synthesize coumarin (beginning of synthetic perfume industry); originated Perkin reaction for making aromatic unsaturated acids such as cinnamic acid. Died Sudbury, Eng., July 14, 1907.

PERKIN, William Henry, Jr., Brit. organic chemist; b. Sudbury, Eng., June 17, 1860; s. Sir. William Henry and Jemina Lissett P.; ed. Royal Coll. Chemistry, South Kensington, Eng., 1877-80; student U. Würzburg (Germany), 1880-82, U. Munich, 1882; hon. degree, Cambridge, 1910, Edinburgh, 1910; St. Andrews, 1911; m. Mina Holland, 1887. Became prof. chemistry Heriot-Watt Coll., Edinburgh, 1887, prof. organic chemistry Owens Coll., Manchester, Eng., 1892; named Waynflete prof. chemistry Oxford U., 1912. Became fellow Royal Soc., 1890, recipient Davy medal 1904, Royal medal 1925; mem. French Acad. Scis., 1920, Chem. Soc. (Longstaff medal 1900, pres. 1913-15). Research on molecular structure of berberine, chemistry of brazilin and haemotoxylin, (with Baeyer) polyacetylenes; synthesized various organic compounds and plant products, including camphor and derivatives, alkaloids, terpines. Died Oxford, Eng., Sept. 17, 1929.

PERKINS, David Dexter, Am. biologist; b. Watertown, N.Y., May 2, 1919; s. Dexter M. and Loretta (Gardiner) P.; A.B., U. Rochester, 1941; Ph.D. in Zoology, Columbia, 1949; m. Dorothy Newmeyer, Aug. 1, 1952; 1 dau., Susan J. Faculty, Stanford, 1948—, prof., biology, 1963—. Research fellow U. Glasgow (Scotland), 1954-55, Columbia, 1951-52. Mem. genetics tng. com. NIH, USPHS, 1961-65. Mem. Genetics Soc. Am., Am. Soc. Naturalists. Editor: Genetics, 1963-67. Research, publs. on mechanism and chromosomal basis genetic recombination in Neurospora. Home: 345 Vine St., Menlo Park, Cal. 94026. Office: Dept. Biol. Scis., Stanford U., Stanford, Cal. 94305.*

PERKINS, Fred William, Am. mathematician; b. Malden, Mass., May 14, 1898; s. Fred William and Frances (Langille) P.; A.B., Harvard, 1921, A.M., 1922, Ph.D., 1928; Sheldon Travelling fellow U. Poitiers, U. Paris, U. Strasbourg; A.M. (hon.), Dartmouth, 1940. Instr. math. Harvard, 1923-26; faculty math. Dartmouth, Hanover, N.H., 1927—, prof., 1940-64, chmn. dept. math. and astronomy, 1939-43, 51-55, prof. emeritus, 1964—. Del., Internat. Congress Mathematicians, Cambridge, Mass., 1950, participant, Oslo, Norway, 1936, Amsterdam, Holland, 1954; Mem. Am. Math. Soc., Math. Assn. Am., Societe Mathematique de France, No. New Eng. Acad. Sci. (pres. 1962-63), Gamma Alpha. Co-author: An Introduction to Linear Analysis, 1966. Reviewer Mat.h. Revs., 1940-59. Research, publs. on theory of harmonic functions, domains with multiple boundary points. Home: 8 Prospect St., Hanover, N.H. 03755.*

PERKINS, Henry Augustus, Am. physicist; b. Hartford, Conn., Nov. 14, 1873; s. Edward H. and Mary E. (Dwight) P.; A.B., Yale, 1896; M.A., E.E., Columbia, 1899; postgrad. Yale, 1900-02, U. Paris, 1908-09, Coll. de France, 1921-22; Sc.D., Trinity, 1920; L.H.D., Gallaudet, 1944; m. Olga Flinch, April 8, 1903; children—Henry Augustus, Evelyn Ingeborg (Mrs. Amyas Ames). Prof. physics Trinity Coll., Hartford, 1902-42, ret. 1942, recalled 1943, ret. 1946; acting pres. 1915-16, 19-20. Mem. Am. Inst. E.E., Am. Phys. Soc., French Soc. Physics, Am. Alpine Club, Explorer's Club, phi Beta Kappa, Sigma Xi. Author: Thermodynamics, 1912; College Physics, 1938; College Physics, Abridged, 1941; Basic College Physics, 1949; also articles. Research in velocity of magnetism, discontinuous discharges, metallic conductivity, residual magnetism. Died July, 1959.

PERKINS, R(ichard) Marlin, Am. herpetologist, mammalogist; b. Carthage, Mo., Mar. 28, 1905; s. Joseph Dudley and Mynta Mae (Miller) P.; student U. Mo., 1924-26; m. Elise More, September 12, 1933 (div. Oct. 1953); 1 dau., Suzanne; m. 2d, Carol M. Cotsworth, August 13, 1960. Curator of reptiles St. Louis Zoo, 1926-38; curator Buffalo Zoo, 1938-44; dir. Lincoln Park Zoo, Chgo., 1944-62, St. Louis Zoo, 1962—. Sr. fellow Center for Biology of Natural Systems. Mem. Am. Soc. Ichthyologists and Herpetologists, Am. Soc. Mammalogists, Internat. Union Zool. Gardens Dirs. Author: Animal Faces, 1944; Marlin Perkins' Zooparade, 1954; (with Carol Perkins) I Saw You from Afar (The Bushmen of the Kalahari Desert), 1964; several sci. papers on herpetology. Originator of Zoo Parade, television program, 1949-57, winner of numerous awards including, Peabody, Look, and Sylvania; series Wild Kingdom. Office: St. Louis Zoo, St. Louis 10.*

PERKINS, Richard Wendell, chemist; b. Smithfield, Utah, Oct. 5, 1926; s. William Wendell and Jessie (Pierce) P.; B.S. in Chemistry, Utah State U., 1950, M.S., 1953; m. Billie Faye Shaw, Aug. 31, 1949; children—Darwin Richard, Kathy Lynn, Sharon, Melinda Jean, Jeanette Faye, Rebecca Ellen, Ginger, De Anna. Tchr. sci. Preston (Ida.) High Sch., 1948-49; chemist Cache Valley Dairy Assn., Smithfield, 1950-52; sr. research scientist radiol. chemistry research group Gen. Electric Co., Richland, Wash., 1952-65; sr. research scientist Battelle N.W. Lab., Richland, 1965-66, mgr. radiol. chemistry research group, 1966—. Mem. Am. Chem. Soc. Research, publs. on measurement, distbn., phys., chem., biol., geophys. behavior of radionuclides released to or present in environment; pioneer in devel. high sensitivity gamma-ray spectrometry; application of this technique to environmental tracing of numerous fallout and cosmic ray produced radionuclides. Home: 1413 Sunset St. Office: P.O. Box 999, Richland, Wash. 99352.*

PERL, Edward Roy, Am. physiologist; b. Chgo., Oct. 6, 1926; s. John I. and Blanche (Braun) P.; student U. Chgo., 1943-44; B.S., U. Ill., 1947, M.D., 1949, M.S., 1951; m. Marjorie Patricia Herdt, Dec. 23, 1953; children—Patricia Marie, Anne Elizabeth, John II. Asst. in physiology U. Ill., Chgo., 1948-49; fellow in physiology Johns Hopkins Sch. Medicine, 1950-52; asst. prof. physiology U. N.Y., 1954-57, asso. prof., 1957; asso. prof. physiology U. Utah Sch. Medicine, Salt Lake City, 1957-65, prof., 1965—, acting chmn. dept. physiology, 1964; vis. prof. Faculté des Sciences, Paris, 1965. Cons. USPHS, 1966—. USPHS fellow, 1951-52; NSF Sr. Postdoctoral fellow, Toulouse, France, 1962-63. Mem. Am. Physiol. Soc., Internat. Brain Research Orgn., A.A.A.S., Sigma Xi. Contbr. numerous articles to sci. jours. Develop. several devices for analyses of physiol. parameters; systematic analyses of physiol. mechanisms related to sensation from body, particularly those concerned with visceral sensibility and pain; demonstration and analyses of simplest reflex mechanisms for visceral control in mammals. Home: 4200 Mathews Way, Salt Lake City 84117.*

PERLIN, Irwin Earl, mathematician; b. Gomel, Russia, Dec. 1, 1911; s. Jacob and Lena Perlin; came to U.S., 1915, naturalized, 1923; B.S., Northwestern U., 1932, M.S., 1933; Ph.D., U. Chgo., 1935; m. Lilyan White, Dec. 5, 1939; 1 son, Martin Asst. instr. Northwestern U., 1932-33; asst. prof. Ill. Inst. Tech., 1937-42; faculty Ga. Inst. Tech., Atlanta, 1945—, research prof. math., 1955—, head math. analysis br. Rich Electronic Computer Center, 1963-67, chief, 1967—. Cons. Marshall Space Flight Center, NASA, 1962—. Mem. Am. Math. Soc., Math. Assn. Am., Soc. Indsl. and Applied Math., Assn. Computing Machinery, Phi Beta Kappa, Sigma Xi. Research, publs. on radio-frequency interference, random-walk problems, scheduling problems, numerical analysis, electronic computers, calculus of variations. Home: 5012 Carol Lane N.W., Atlanta 30327.*

PERLIS, Alan J., Am. computer scientist; b. Pitts., Apr. 1, 1922; s. Louis Philiip and Zelda Anne (Gilfond) P.; B.S. in Chemistry, Carnegie Inst. Tech., 1943; student Cal. Inst. Tech., 1946-47; M.S., Mass. Inst. Tech., 1950, Ph.D. in Math., 1950; m. Sydelle Gordon, Oct. 28, 1951; children—Mark Lawrence, Robert Gordon, Andrea Lynn. Asst. prof. math., dir. computer center Purdue U., 1952-56; mem. faculty Carnegie Inst. Tech., 1956—, prof. math., dir. computer center, 1960—, head dept. math., 1961-64, head dept. computer sci., 1965—. Mem. NSF computer nom., Nat. Joint Computer Com., 1954-56, Gov. Pa. Council Sci. and Tech., 1963—; com. computers research NIH, 1963—. Mem. Assn. Computing Machinery (pres. 1962-64, editor-in-chief jour. Communications 1958-62), Soc. Indsl. and Applied Math., Am. Math. Soc., Math. Assn. Am. Designed and supervised the development of the automatic programming language "IT", 1956; one of the inventors of Algol, 1958, 61, 62; designed the language formula Algol for programming symbolic mathematical manipulations on computers. Home: 5231 Forbes Av., Pitts. 17. Office: Carnegie Inst. Tech., Pitts. 15213.*

PERLIS, Sam, Am. mathematician; b. Maywood, Ill., Apr. 18, 1913; s. Adolph and Ida (Ziskind) P.; B.A., U. Chgo., 1934, M.S., 1936, Ph.D., 1938; m. Esther Rockoff, June 24, 1939; children—Donald, Robert. Instr., Ill. Inst. Tech., Chgo., 1938-42; engring. researcher Lockheed Aircraft Corp., Burbank, Cal., 1942-46; faculty Purdue U., Lafayette, Ind., 1946—, asso. prof., 1951—. Mem. Am. Math. Soc., Maths. Assn. Am. Author: Theory of Matrices, 1952; Introduction to Algebra, 1966; also articles. Described radical of an algebra in new terms; determined when two abelian groups have same group algebra; discovered various properties of fields and algebras. Home: 704 Sugar Hill Dr., West Lafayette, Ind. 47906. Office: Dept. Math., Purdue U., Lafayette, Ind. 47907. Office: Dept. math., Purdue U., Lafayette, Ind. 47907.*

PERLMAN, D(avid), Am. biochemist; b. Madison, Wis., Feb. 6, 1920; s. Selig and Eva (Shaber) P.; B.A., U. Wis., 1941, M.S., 1943, Ph.D., 1945. Biochemist, Hoffmann-LaRoche, Nutley, N.J., 1945; microbiologist Merck & Co., Inc., Rahway, N.J., 1945-47, Squibb Inst. for Med. Research, New Brunswick, N.J., 1947-67; Knapp Vis. prof. U. Wis. Sch. Pharmacy, Madison, 1958, prof. pharm. biochemistry, 1967—, dean Sch. Pharmacy, 1968—. Chmn., Intersci. Conf. on Antimicrobial Agts. and Chemotherapy, 1966, 67. Fellow N.Y. Acad. Scis.; mem. Am. Soc. for Microbiology, Tissue Culture Assn., Am. Chem. Soc. (chmn. div. microbial chemistry and tech. 1966), Am. Soc. Biol. Chemists, Biochem. Soc., Soc. for Gen. Microbiology, Am. Acad. Microbiology. Editor: Advances in Applied Microbiology, 1968—; editorial bd. Antimicrobial Agts. and Chemotherapy, 1961—. Research, numerous publs. on microbial transformation of steroids; devel. comml. processes for microbial prodn. penicillins, streptomycins, neomycins, tetracyclines, Vitamin B12; growth in vitro of mammalian tissues and cells. Home: P.O. Box 5072, Madison, Wis. 53705.*

PERLMAN, Henry Bernard, otolaryngologist; b. Russia, July 15, 1901; s. Hessel and Sarah (Dolgin) P.; B.A., U. Wis., 1923; M.D., U. Chgo., 1926; m. Dorothy Zimmerman, Aug. 9, 1931; children—Daniel, Robert. Practice medicine, specializing in otolaryngology, Chgo., 1931—; faculty U. Chgo., 1934—, prof., 1952—. Diplomate Am. Bd. Otolaryngology.

Mem. A.M.A., Am. Otol. Soc., Am. Laryngol., Rhinol., Otol. Soc., Soc. U. Otolaryngologists, Sigma Xi. Contbr. chpt. to Progress in Neurology and Psychiatry, 1947. Contbr. numerous articles to profl. jours. Research on eustachian tube, circulation of ear, acoustic trauma, vestibular system. Home: 5448 East View Park, Chgo. 60637. Office: 950 E. 59th St., Chgo. 60637.*

PERLMAN, Isadore, Am. chemist, educator; b. Milw., Apr. 12, 1915; s. Harry and Bella (Karpman) P.; student U. Cal. at Los Angeles, 1932-34, B.S., U. Cal. at Berkeley, 1936, Ph.D., 1940; m. Labelle Grinblat, July 26, 1937; children—Judith Ann (Mrs. Sanchez), Alice Lenore, Paula Jean. Research fellow U. Cal. at Berkeley, 1940-41, asso. prof. chemistry, 1945-49, prof., 1949—, chmn. dept. chemistry, 1957-58, asso. dir., head nuclear chemistry div. Lawrence Radiation Lab., 1958—; sr. chemist Manhattan Project, 1942-45. Recipient E.O. Lawrence Meml. award AEC, 1960. Guggenheim fellow 1955-56, 1963. Mem. Nat. Acad. Scis., Am. Chem. Soc. (Cal. sect. award 1953, award for nuclear applications in chemistry 1964), Am. Phys. Soc., A.A.A.S., Am. Assn. U. Profs., Phi Beta Kappa, Sigma Xi. Home: 1158 Glen Rd., Lafayette, Cal. Office: Lawrence Radiation Lab., U. Cal., Berkeley, Cal.*

PERLOFF, William Harry, Am. endocrinologist, physician; b. Phila., May 14, 1911; s. Harry B. and Esther (Brenner) P.; B.A., Swarthmore Coll., 1932; M.D., U. Pa., 1936; m. Dorothy Averett, Dec. 22, 1933; children—William Harry, Carol (Mrs. Robert Capper). Endocrine fellow Columbia Coll. Physicians and Surgeons, 1938-39; Henrietta Hecksher fellow in medicine U. Pa., 1939-42, George Keim fellow in medicine, 1945-47, instr. medicine, 1939-50; practice medicine specializing in endocrinology, Phila., 1939—; chief endocrine dept. Phila. Gen. Hosp., 1947-58; asso. prof. medicine, dir. dept. endocrinology Temple U. Sch. Medicine and Med. Center, 1950-59; chief div. endocrinology and reprodn. Research Labs., Albert Einstein Med. Center, Phila., 1961—. Fellow Acad. Psychosomatic Medicine, Sigma Xi; mem. A.A.A.S., Phila. (past pres.), Argentina, Mexican endocrine socs., Phila. County Med. Soc., N.Y. Acad. Sci., Phila. Psychoanalytic Inst. (affiliate), Am. Psychosomatic Soc., Phila. Coll. Physicians, Am. Soc. for Study Sterility, Internat. Fertility Assn., Pan Am. Med. Assn. (pres. endocrine sect. 1952—). Research, numerous publs. on endocrinology and human reprodn. Home: 1805 Laguna St., Santa Barbara 93105. Office: 320 W. Junipero St., Santa Barbara, Cal. 93101.*

PERLOFF, Robert, Am. psychologist; b. Phila., Feb. 3, 1921; s. Myer and Elizabeth (Sherman) P.; A.B., Temple U., 1949; M.A., Ohio State U., 1949, Ph.D., 1951; m. Evelyn Potechin, Sept. 22, 1946; children—Richard Mark, Linda Sue, Judith Kay. Instr. edn. Antioch Coll., 1950-51; with personnel research br. Dept. Army, 1951-55, chief statis. research and cons. unit., 1953-55; dir. research and devel. Sci. Research Assos., Inc., Chgo., 1955-59; vis. lectr. Chgo. Tchrs. Coll., 1955-56; mem. faculty Purdue U., 1959—, prof. psychology, 1964—, field assessment officer univ. Peace Corps Chile III project, 1962; cons. in field, 1959—. Diplomate Am. Bd. Examiners Profl. Psychology. Fellow A.A.A.S., Am. Psychol. Assn. (mem at-large exec. com., div. consumer psychology 1964—); mem. Am. Assn. U. Profs., Am. Marketing Assn., Emerson Soc., Interam. Soc. Psychology, Psychometric Soc., Soc. Psychol. Study Social Issue, Thoreau Soc., Sigma Xi. Contbr. profl. jours. Editor Indsl. Psychologist, 1963—, programs in rev. in Personnel Psychology, 1965—; book rev. editor Personnel Psychology, 1952-55; editorial bd. Inst. Personality and Ability Testing, 1962—. Pioneered in establishing consumer psychology as branch of academic psych., including research and teaching of survey research principles and procedures; questionnaire construction, analysis, interpretation; studies on parameters of taste and offensiveness in advertising. Home: 104 Wheeler Lane, West Lafayette, Ind. 47906. Office: Dept. Psychology, Purdue Univ., Lafayette, Ind. 47907.*

PERNOD, Jacques, physician; b. Düren, Germany, Oct. 31, 1923; s. Julien and Marie (Thiriet) P.; Externe, U. Lyon (France), 1946, Interne Suppl., 1948, M.D., 1948; m. Renée Marie Fischer, Feb. 26, 1949; children—Anne Christine, Philippe Jacques. Med. officer, Tunisia, 1949-51; charge gen. medicine br. hosps. of Tlemcen and Algiers, 1956-59; in charge thoracic med. br. Hopital Militaire d'Instruction Percy, Paris, 1959—; prof. medicine Val-de-Grâce, Paris, 1958—. Mem. Société Médicale des Hopitaux de Paris, Sociétés Française de Cardiologie, Pneumologie et Thérapeutique. Numerous publs. on epidemiologic studies in Algeria, including viral hepatitis, Reiter's syndrome, hepatic amoebiasis, Kala-Azar; research on sarcoidosis, hydatidosis, agammaglobulinemia, pulmonary fibrosis, ethionamid therapy for Tb, Tb meningitis, electrocardiographyof dyskaliemic disease, criteria of normal carotidogram. Home: 101 Av. Henri Barbusse, Clamart 92, France. Office: Val-de-Grâce 277 bis rue St-Jacques, Paris. V°, France.

PERNTER, Josef Maria, geophysicist; b. Neumarkt, Tyrol, Mar. 15, 1848; ed. Innsbruck, Austria, doctorate in physics, Vienna; mem. Soc. Jesus until 1877; became mem. staff Central Meteorol. Inst., 1878, later dir.; named prof., Innsbruck, 1890, Vienna, 1897. Author: Meteorologische Optik, 1902. Research on earth magnetism, higher layers of atmosphere, meteorol. optics. Died Arco, Italy, Dec. 20, 1908.

PERNY, Guy Charles-Marie, French physician; b. Sarreguemines, France, July 6, 1923; s. Charles-Marie and Florentine (Sandrino) P.; Licence-ès-Scis., Faculté des Scis. de Strasbourg (France), 1954, Doctorat d'Etat-es-Scis. physiques, 1957; m. Marylène Paulus, Aug. 2, 1949; children—Anne-Chantal, Genevieve-Noelle, Michel-Guy. Prof. Enseignement Technique, 1947-52; research staff Centre Nat. de la Recherche Scientifique, 1952-57; prof. Ecole Supérieure de Chimie de Mulhouse, 1957; master conf. Faculty Scis. de Strasbourg, 1959—, prof., 1964—; prof., dir. Centre Associé du Conservatoire Nat. des Arts et Metiers de Paris, 1959-60; dir. Laboratoire de Physico-Chimie des Couches Minces à Mulhouse, 1957—. Decorated Croix du Combattant, Palmes Academiques. Mem. Société Francaise de Physique, Société Francaise de Chimie-Physique, Am. Vacuum Soc. Author: Research on the Spectroscopy of Silver and Lead Iodide in the Crystalline State, 1960; also numerous articles. Research on absorption, luminescence and exciton spectra of thin films, synthesis of thin films by simultaneous projection of components, preparation of polymorphic varieties of thin films by vacuum heat treatment, condensation of cold plasmas in spite of thin films phys. chemistry, theoretical and exptl. investigations on reactive cathodic sputtering. Home: 10 Porte du Miroir. Office: 3 rue Werner, Mulhouse, France-68.*

PERON, François, French biologist; b. Cérilly, France, Aug. 22, 1775; participated in voyages of Naturalist and Geographer, 1800-03; mem. French Acad. Scis. Collected more than 2500 new species of animals on his voyages; studied ocean phosphorescent animals, jellyfish; enthnographic studies of Antilles Island, Australia, Tasmania. Died Cérilly, Dec. 14, 1810.

PEROT, Alfred, French physicist; b. Metz, France, 1863; ed. École Polytechnique; apptd. dir. confs. Sci. Faculty, Marseille, France, 1888; prof. indsl. physics, 1894; became dir. trial lab. Conservatory of Arts and Crafts, Paris, 1901; astron. physicist, Meudon, France, 1908; named prof. physics École Polytechnique, 1909. Studied latent heat of vaporization of water, 1887; measured (with Ch. Fabry) light wave length, experimented in star spectroscopy, developed (in 1896) Perot-Fabry inteferometer. Died Paris, 1925.

PEROV, Sergei Stepanovich, Russian biochemist; b. Aug. 22, 1889; grad. St. Petersburg U., 1913; prof. Vologda Dairy Inst., 1921-30, also from 1956; prof. Moscow Zoo-Veterinarian Inst., 1932-41, also Moscow Fur and Pelt Inst., 1930-34, 49-55; chief Protein Lab., All-Union (Lenin) Acad. Agrl. Scis., 1930-35; USSR Acad. Scis., 1935-42. Recipient Stalin prize, 1949. Author: Casein Proteic Protoacid. Methods of Production and Physiochemical Characteristics, 1947; The Proteic Protoacids of a Number of Seeds of Herbaceous, Fruticose and Ligneous Plants; The Collodial Characteristics of Super-Pure Casein Protoacid in an Activated Medium; The Physiochemical Characteristics of a Number of Protoacids in Relation to Concentration and Temperature, 1951. Basic research in field of biochemistry, including colloidal chemistry of protein and milk. Address: Vsesoyuznaya Selskokhozyaystvennaya Akademiya, B. Karitonevsky 21, Moskva, USSR.

PERRAULT, Claude, French architect, anatomist; b. Paris, Sept. 25, 1613; studied philosophy, natural history, mathematics, astronomy; M.D., U. Paris; one of 1st members of French Acad. Scis., 1666. Translated Vitruvius into French (Dix livres d'architecture de Vitruve, 1673). Author: Ordonnance des cinq espèces de colonne selon la méthode des anciens, 1683. Designed Colonnade (east facade of Louvre), 1667-70; also designed portions of south facade of Louvre; designed Paris Observatory, 1667-72, and decorations for the palace at Versailles. Proved existence of two types of sap, ascending and descending; dissected animals to study anatomy; wrote on lion, hedgehog, eagle, guinea fowl, ostrich, tortoise; discovered existence of spiral valves in intestine of shark; studied springs and evaporation of water. Died, Paris, France, Oct. 9, 1688.

PERRET, Frank Alvora, Am. volcanologist; b. Hartford, Conn., Aug. 2, 1862; ed. Bklyn. Poly. Inst. Worked under Edison; took up study volcanic eruptions, 1904; dir. Hawaiian expdn. Mass. Inst. Tech., 1911. Lived on Martinique Island to observe Mt. Pelée, 1929-39; founded Vocanological Mus. at St. Pierre, Martinique, 1933. Invented Perret electric motor; research in volcanology. Died Jan. 12, 1943.

PERRET, William Riker, Am. geophysicist; b. Newark, N.J., Feb. 29, 1908; s. Charles Edward and Katherine (Crane) P.; S.B., Mass. Inst. Tech., 1930, S.M., 1956; postgrad. U. at Berlin and Munich (Germany), 1930-31; m. Elizabeth DeGraffenreid, May 1938; children—William Edward, Robert Ferrand DeGraffenreid. Physicist, Waterways Exptl. Sta., Vicksburg, Miss., 1935-36, 38-43, 46-51, Schlumberger Well Survey Corp., Houston, 1936-37, Tenn.-Eastman Corp., Oak Ridge, 1943-46; physicist Sandia Corp., Albuquerque, 1951—. Cons. on ground motion Boeing Co., Seattle, 1963—; Edn. Council Mass. Inst. Tech., 1954—. Redfield Proctor Travelling fellow in physics, 1930-31, Charles A. Coffin fellow, 1932-33. Fellow Am. Phys. Soc.; mem. Am. Geophys. Union, Soc. Exploration Geophysicists, Seismological Soc. Am. Research on reaction soils and rocks in immediate vicinity very large explosions, detonated underground or in air; helped develop systems for observation of stress and motion. Home: 6116 Natalie N.E. St., Albuquerque, New Mexico 87110. Office: Div. 5232, Sandia Corp., Albuquerque, New Mexico 87115.*

PERRI, Guilio Cesare, Italian biochemist; b. Pavia, Italy, Apr. 12, 1916; s. Francesco A. and Olocco (Francesca) P.; student Ghislieri Coll., 1935-41; M.D., U. Pavia, 1941, Ph.D., 1950; m. Aurelio Imelda, Oct. 27, 1944; children—Rossella, Valeria, Giulia, Alessandro. Asst. in biochemistry U. Pavia Med. Sch., 1945-50, asso. prof., 1951-53; fellow Inst. Internat. Edn., Sloan Kettering Inst., N.Y.C., 1951, Sloan Found. fellow, 1952-53; asso. prof. U. Pavia, 1954-55; asst. div. chemotherapy Sloan Kettering Inst., 1955-57, asso., 1957-64; dir. research Richarson Merrel, Inc. Overseas Lab., Naples, 1964—. Author: History of Medicine, 1951; also numerous articles. With others isolated human myoglobin in crystalline form; determination of role of myoglobin in crush syndrome renal block, isolation and cystallisation of kidney lyosozime from rats and its role in tumor growth, significance of kidney lysozyme accumulation. Home: 2 Traversa Manzoni, Naples 80123. Office 111 P. Castellino, Naples, Italy 80100.*

PERRIA, Luigi, Italian surgeon; b. Cagliari, Apr. 7, 1909; s. Carolo and Maria (Murru) P.; M.D.; m. Valeria Mazzuchi, Sept. 18, 1947; 1 dau., Barbara. Dir. Neurosurg. Clinic, U. Genoa (Italy), also prof. Mem. Herbert Olivercrona Soc., Italian Neurosurgery Soc.; hon. mem. French Lang. Soc. Neurosurgery. Author numerous sci. papers. Research on surgery for epilepsy; endocranial tumors; endocranial aneurysms. Home: via Don G. Minzoni 10/1. Office: Clinica Neurochirurgica dell'Univ. Genova, Genoa, Italy.

PERRIER, (Jean Octave) Edmond, French zoologist; b. Tulle, France, May 9, 1844; prof. Agen Lyceum; became prof. Mus. Natural History, Paris, 1876, dir., from 1900; mem. French Acad. Scis., 1892, French Acad. Medicine. Author: Les colonies animales et la formation des organismes, 1881; Philosophie zoologique avant Darwin, 1884; Les explorations L'intelligence des animaux, 1887; Le transformisme, 1888; La vie des animaux, 4 vols., 1903-06; La terre avant l'historie, 1913; La vie en action, 1921. Supported theory of inheritability of acquired characteristics (Neo-Lamarckism); founder 2 marine biol. stas. in France, 1881. Died Paris, July 31, 1921.

PERRIER, François, French geodesist; b. Valleraugue, France, Apr. 18, 1833; brig. gen.; dir. army geog. services; mem. French Acad. Scis., Bur. Longitudes. Carried out geodetic work in Algeria and Corsica; worked (with Ibánez) to complete geodetic junction of Spain and Algeria, 1878; began new triangulation of France with instruments and methods which he contbd.; caused 1st repeating azimuth circle to be constructed. Died Feb. 20, 1888.

PERRIER, Georges-Antoine-François-Jacques-Justin, French geodesist; b. Montpellier, France, Oct. 28, 1872; s. François Perrier; ed. Ecole Polytechnique; prof. geodesy and astronomy Ecole polytechnique, 1929-42; mem. French Acad. Scis. (pres. 1940), Bur. Longitudes, Internat. Geodetic Assn. (sec.). Contbd. to modernization of French geodetic service; continued triangulation of France begun by father, also worked on measurement of meridian equatorial arc in Equador begun in 18th century, 1901-06. Died Paris, Feb. 16, 1946.

PERRIN, Francis Henri Jean Siegfried, French physicist; b. Paris, France, Aug. 17, 1901; s. Jean B. and Henriette (Duportal) P.; student Ecole normale Supérieure, U. Paris, 1918-22, Sc.D., 1928; m. Colette Auger, Mar. 27, 1926; children—Nils, David, Francoise (Mrs. Yves Chapuis). Asst. prof. U. Paris, 1933-46; vis. prof. Columbia, N.Y.C., 1941-43; prof. Coll. de France, Paris, 1946—; high commr. French Atomic Energy, 1951—. Decorated grand officer Legion d'Honneur, grand croix Ordre National du Mérite. Mem. French Acad. Scis., 1953, Société Francaise de Physique, Am. Phys. Soc., N.Y. Acad. Scis. (hon.). Author: Mecanique Statistique Quantique, 1939. Research on rotational Brownian motion of spheres, fluorescence and mean life of excited molecules in solution, 1929, thresholds for electron pair prodn. by interaction of photon or electron with electron, zero mass of neutrinos, 1933, rotational Brownian motion of ellipsoids, 1934, diffusion of ellipsoidal molecules in solutin, dielectric dispersion for such molecules, 1934, critical dimension for nuclear chain reactions, 1939. Home: 4 Rue Froidevaux, Paris, 14, France.*

PERRIN, Jean Baptiste, French physicist; b.Lille, France, Sept. 30, 1870; attended Ecole Normale Supérieure, Dr. ès Scis., 1897; hon. degrees Brussels, Liège, Ghent, Calcutta, Princeton, Manchester, Oxford, others; asst. in physics, Ecole Normale Supérieure, 1894-97; U. Paris, readership in Physical chemistry, 1897, prof. physical chemistry, 1910-40; left France for U. S. in 1938; mem. French Acad. Sciences, 1923 (La Caze Prize, 1914; pres. 1938); Fellow, Royal Soc. 1918 (Joule Prize, 1896); mem. many other sci. socs.; Nobel Prize for Physics, 1926; Commdr., Legion of Honor; Order of Leopold (Belgium). Author: Les atomes, 1913; Les éléments de la physique, 1930. Demonstrated cathode rays are negatively charged particles (not waves); worked on x-rays and conductivity of gases, fluorescence, radium decay, and sound; through research on colloids and Brownian motion of particles, demonstrated discontinuous structure of matter and discovered equilibrium of sedimentation, which permitted accurate calculation of size of atoms. Died, New York City, Apr. 17, 1942.

PERRIN, Maurice, French surgeon; b. Vézelise, France, Apr. 13, 1826; became doctor, Paris, 1851; entered army; mem. Acad. Medicine. Author: Traité pratique d'ophthalmoscopie et d'optométrie. Described snapping hip caused by movement of band of fascia lata over greater trochanter of femur, circa 1866. Died Aug. 31, 1889.

PERRIN, Nils Noël, French physicist; b. Paris, Mar. 20, 1927; s. Francis Henri and Colette (Auger) P.; grad. École de Physique et Chimie de Paris, 1949; Licencié, Sorbonne, Paris, 1953, D.Sc., 1958; m. Gricouroff, June 10, 1954; children—Antoine, Florence, Arnaud. Engr., Ecole de Physique et Chimie de Paris, 1949; staff Centre National de la Recherche Scientifique, Orsay, France, 1950—, maitre de recherches, 1961—. Mem. Société Francaise de Physique. Editor in chief Industries Atomiques. Research, publs. on effect of nuclear size on ratio of electrons to gamma in nuclear transition; measurement of short life time in isomeric transitions of recoil atoms following alpha emissions. Home: 2 Emile Faguet, Paris, France. Office: 15 Georsges-Clemenceau, Orsay 91, France.*

PERRIN, Paul Georges, French pathologist; b. Malo-les-Bains, Feb. 19, 1895; s. Georges and Hélène (Carton) P.; ed. Nantes (France) Acad. Medicine; m. Madeleine Barreau, Nov. 24, 1925; children—Jacqueline, Geneviève, Michel, Renëe,, Jean-Paul, Daniel,, Louis, Noël. Intern Nantes Hosps., 1920, mem. staff, 1933—; head med. clinic, also head pathol. anatomy works Nantes Sch. Medicine, 1925-38, prof. anatomy and histology, 1938—; prof. social medicine and labor medicine Faculty of Nantes. Founding dir. Revue de l'alcoolisme. Decorated Palms of Acad., Order Pub. Health; Bréant prize Acad. Sci. Mem. Med. Group Study Alcoholism (sec.). Author: Une croisade médicale contre l'alcoolisme, 1945; L'alcoolisme-Problèmes médico-sociaux, 1950; Alcoolisme, criminalité et délinquance, 1962. Research forensic medicine, alcoholism, sociology. Home: 31 rue Marzelle-de-Grillaud. Office: 14 bis, rue d'Alger, Nantes, France.

PERRIN-PAYOLLE, Max, French physician; b. Lyon, France, July 17, 1930; s. Louis and Gabrielle (Fayolle) Perrin; ed. U. Lyon Faculté de Médecine; Docteur en Médecine, 1960; m. Jaqueline Josserand, Feb. 2, 1957; children—Éric, Laurence, Olivier. Successively asst. head clinic, 1963, prof. agrégé in pneumophtisiology, physician hosps., 1965; asst. chief service Pneumophtisiology Clinic, U. Lyon. Author: (with J. Brun, F. Magnin, J. Gardère) Les Urgences cardio-respiratoires, 1966; also numerous articles. Research in pulmonary Tb., lung cancer, sarcoidosis, respiratory difficulties. Home: 21 cours Fr. Roosevelt, Lyon 6-69, France. Office: Hôpital Ste. Eugénie, Lyon-St. Genis-Laval 69, France.*

PERRINE, Charles Dillon, Am. astronomer; b. Steubenville, O., 1867; astronomer Lick Obs., 1895-1909; dir. Argentine Nat. Obs., S.Am., 1909-36. Discovered motion in nebulosity about new star in Persueus, 1901, also 6th and 7th satellites of Jupiter, 13 comets; studied solar eclipses. Died 1951.

PERRINE, Henry, Am. botanist; b. Cranburg, N.J., Apr. 5, 1797; s. Peter and Sarah (Rozengrant) P.; m. Ann Fuller Townsend, Jan. 8, 1822, 3 children. Practiced medicine, Ripley, Ill., 1819-24, Natchez, Miss., 1824-27; U.S. consul at Campeche, Mexico, 1827-37; made bot. collections in Mexico; introduced useful tropical plants to Fla., including henequen and sisal; built nursery, Indian Key, Fla., 1833; received grant of land in Fla. from Congress, 1838. Killed by Seminole Indians, Aug. 7, 1840.

PERRON, Oskar, German geometer; b. Frankenthal, Germany, May 7, 1880; Ph.D., Dr. ès sci. Prof. agrégé U. Munich; asso. prof. U. Tübingen; prof. U. Heidelberg, prof. emeritus, 1951—. Mem. various acads. scis. Author: Die Lehre von den Kettenbüchen; Irrationalzahlen; Algebra; Nichteuklidische Elementargeometrie; also numerous articles. Address: Friedrich-Hershcelstrasse 11, Munich 27, West Germany.

PERRONCITO, Edoardo, Italian parasitologist; b. Viale-d'Asti, Italy, Mar. 1, 1847; Agrégé in Parasitology, U. Paris, 1878; named prof. Faculty Medicine, Turin, Italy, 1888; prof. Turin Vet. Sch. Mem. French Acad. Scis., 1911. Proved simple hygienic measures could prevent disease; studied life of bees, prodn. of silk, especially treatment of silk worm diseases; parasitology of anemia in coal miners. Died Pavia, Italy, Nov. 4, 1936.

PERROS, Theodore Peter, Am. chemist; b. Cumberland, Md., Aug. 16, 1921; s. Peter G. and Christina (Sioris) P.; B.S., George Washington U., 1946, M.S., 1947, Ph.D., 1952; postgrad. Technische Hochschule, Munich, Germany. Analyst research div. U.S. Naval Ordnance Lab., 1943-46; faculty George Washington U., Washington, 1946—, prof. chemistry, 1960—; research chemist Bur. Ordnance, 1949; research dir. Air Force Office Sci. Research and Devel., 1958-59. Cons. U.S. Naval Ordnance Lab., 1953-56. NSF fellow, 1959; AEC grantee, 1951-53; Research Corp. grantee, 1953-54. Mem. Am. Chem. Soc., Chem. Soc. London, Gesellschaft Deutscher Chemiker, Washington Acad. Scis., Philos. Soc. Washington, Sigma Xi, Omicron Delta Kappa, Alpha Chi Sigma. Author: (with William F. Sager) Chemical Principles, 1961; (with C.R. Naeser, W. Harkness) Experiments in General Chemistry, 1961; College Chemistry, 1966; also articles. Research on stabilities of inorganic coordination polymers, preparation and characterization fluorine containing compounds transition metals. Home: 5825 3d Pl., N.W., Washington 20011.*

PERROT, Georges, French archaeologist; b. Villeneuve-Saint-Georges, Nov. 12, 1832; ed. l'Ecole normale supérieure, 1852; D.Litt., 1867. Taught at several lycées, then at lycée Louis-le-Grand, Paris, 1863; head sci. mission in Asia Minor, 1861; lectr. Ecole normale, 1872; prof. archaeology Faculté des lettres de Paris, 1875; dir. Ecole normale supérieure, 1888-1902. Mem. Académie des inscriptions (perpetual sec. 1904). Author: Exploration archéologique de la Galatie et de la Bithynie, 1862; Mémoire sur l'ile de Thasos, 1864; De l'état actuel des études homériques, 1864; L'Ile de Crète, 1866; Essai sur le droit public et privé de la Republic athénienne, 1867; L'Eloquence politique et judiciaire à Athènes; Les Précurseurs de Démosthène, 1873; Mémoires d'archéplogie d'épigraphie et d'histoire, 1875; Praxitèle, 1904; Histoire de l'art dans l'antiquité, 10 vols., 1882-1914. Translator: Nouvelles leçons sur la science du langage (Max Muller). Editor: Revue archéologique. Did archaeol. research in Asia Minor; wrote exact critical text of Testament of Augustus (Monument d'Ancyre). Died Paris, July 1, 1914.

PERROT, Marcel, French physicist; b. Paris, Aug. 9, 1908; s. Emile and Antoinette (Van der HaSselt) P.; Ingénieur, Ecole Nationale des Arts et Metiers, 1930; Docteur es Sciences, U. D'Aix-Marseille, 1944; m. Josette Valdy, 1937; children—Marie-Francoise (Mrs. Cosset), Anne-Marie (Mrs. Suaudeau), Elisabeth (Mrs. Merigoux), Marie-Pascale, Therese. Prof. physics and electrotechnics Ecole Nationale d'Arts et Métiers, Aix-en-Provence, France, 1937-46; research supr. Centre National de a Recherche Scientifique, 1946-48; prof. Faculté des Sciences, Algiers, Algeria, 1948-59; dir. Institut d'Energie Solaire, U. Algiers, 1959-62; now dir. Laboratories d'Electricité et d'Heliotechnique, Faculté des Sciences, Marseille, France. Internat. sec. gen. Cooperation Mediterraneenne pour l'Energie Solaire. Decorated laureat French Acad. Scis., 2 times; Officer Palmes Acad. Mem. French Assn. for Study and Devel. Solar Energy (v.p.), Solar Energy Soc. Author: Memorial des Sciences Physiques—Gauthier Villars, 1954; La Houille d'or, 1962; also numerous articles. Research on optical properties of thin laminae, including their elec. properties and evolutions; improved (with Blanc Lapierre) separating capacity of optical instruments by use of linear filters and theory of information; thermoelectric properties of various pulverizable metallic oxides (including discovery of evidence for particular thermoelectric effects); developed thermionic, thermoelectric and thermodynamic generators. Home: 42 Avenue Francoise Duparc, Marseille (13), France.*

PERROTIN, Henri Joseph Anastase, French astronomer; b. Saint-Loup, France, Dec. 19, 1845; dir. Nice (France) Obs.; mem. French Acad. Scis. Died Nice, Feb. 29, 1904.

PERRY, Clay Lamont, Am. applied mathematician; b. San Francisco, Feb. 26, 1920; s. Clay Lamont and Matie (Bishofberger) P.; A.B., U. Cal. at Los Angeles, 1942; M.A., U. So. Cal., 1946; Ph.D., U. Mich., 1949; m. Kathleen M. Kelly, Sept. 6, 1946; children—Virginia Gail, Carol Ann. Sr. mathematician Oak Ridge Nat. Lab., 1950-53; dir. computer center, prof. math. U.S. Naval Postgrad. Sch., Monterey, Cal., 1953-55; mgr. math. scis. dept. Stanford (Cal.) Research Inst., 1955-60; dir. Computer Center, prof. math. U. Cal. at La Jolla, 1960—. Mem. Co-op Computer Users Group (pres. 1963), Am. Math. Soc., Soc. for Indsl. and Applied Maths., Assn. for Computing Machinery, Math. Assn. Am. Author: (with G.E. Evans) Programming and Coding for Automatic Digital Computers, 1960; also articles; contbg. author: Handbook of Systems Engring., 1965. Developed new math. computing methods; designed math. method used in developing magnetic ink character recognition system, devel. system for automatic processing library serials records. Home: 8911 Nottingham St., La Jolla, Cal. 92037.*

PERRY, John Francis, Am., physician; b. Lubbock, Tex., Aug. 6, 1923; s. John F. and Avis (Frasier) P.; B.A., U. Tex., 1944, M.D., 1947; Ph.D., U. Minn., 1958; m. Genevieve Quist, July 12, 1958; 1 son, John Francis 3d. Faculty dept. surgery U. Minn. Med. Sch., Mpls., 1957—, prof., 1965—; chief of surgery St. Paul Ramsey Hosp., 1962—. Diplomate Am. Bd. Surgery, Bd. Thoracic Surgery. Fellow A.C.S.; mem. Soc. U. Surgeons, Central Surg. Assn., Am. Soc. for Thoracic Surgery. Research, publs. on intestinal obstruction, gastric physiology, radiation treatment of tumors, trauma. Home: 2 Red Fox Rd., St. Paul 55110. Office: 640 Jackson St., St. Paul 55101.*

PERRY, Joseph Earl, Jr., Am. physicist; b. Belmont, Mass., Oct. 29, 1917; s. Joseph Earl and Bessie (Stanford) P.; B.S. in Applied Physics, Mass. Inst. Tech., 1939; Ph.D. in Physics, U. Rochester, 1948; m. Marion Annette Brown, July 29, 1944; children—Joseph Earl III, Roger Jon. Physicist U. Cal., 1942-45, staff mem. Los Alamos Sci. Lab., 1950-58, group leader, 1958—; post-doctorate fellow Cal. Inst. Tech., 1947-50. Fellow Am. Phys. Soc. Research, publs. on nuclear physics and reactor controls; study of spectroscopy of light nuclei; monoenergetic neutron flux measurement; photoelectricity of semiconductors. Home: 126 Andanada St. Office: P.O. Box 1663, Los Alamos, 87544.

PERRY, Stephen Joseph, astronomer; b. London, Aug. 26, 1833; ed. Douai, Belgium, St. Acheul, Stonyhurst, London, Paris; became mem. Soc. Jesus, 1853; became head obs., tchr. physics and math., Stonyhurst, 1860; went to St. Beuno's, Wales, 1863; ordained, 1866; returned to Stonyhurst; made magnetic surveys of France and Belgium, 1868-71; observed transists of Venus, Kerguelen in Indian Ocean, 1874, Madagascar, 1882. Fellow Royal Soc., 1874. Died on expdn. to study solar eclipse, Salut Islands, off French Guiana, Dec. 27, 1889.

PERRY, Stuart, Am. inventor; b. Newport, N.Y., Nov. 2, 1814; grad. Union Coll., 1837; m. Amy Jane Carter, 1837; m. 2d, Jane W. Maxson, 1873; 1 child. In partnership with brother in wholesale dairy products business, Newport, 1840-60; invented internal combustion gas engine, patented, 1844; patented lock, key, and safe bolt, 1857, combination lock, 1858; also invented milk cooler, stereopticon, velocipede, hay tender and improved sawmill machinery. Died Feb. 9, 1890.

PERRY, Thomas Gregory, paleontologist; b. Toronto, Ont., Can., Nov. 5, 1919; s. Clarence G. and Minnie (Bennett) P.; B.A., U. Toronto, 1947, M.A., 1948, Ph.D., 1951; m. Lillian Elsie Hopkins, June 9, 1945; children—John, Michael, Peter, Paula. Faculty, Ind. U., Bloomington, 1950—. Mem. Geol. Soc. Am., Am. Assn. Petroleum Geologists, Paleontol. Soc., Soc. Econ. Paleontologists and Mineralogists (editor Jour. Paleontology 1964—), others. Research on internal features of fossilized invertebrate animals (Bryozoa) living in Paleozoic era, delineation of outcrop belts of late Mississippian rocks in S.Central Ind. Home: 4020 Stoutes Creek Rd., Bloomington, Ind. 47401.*

PERRYMAN, Charles Richard, Am. physician; b. Elliot, Ia., Sept. 21, 1916; s. Raymond H. and Mary Gertrude (Weir) P.; B.A., Dartmouth, 1938; M.D., Cornell U., 1942; m. Charlene E. Omen, Aug. 23, 1940; children—Charles A., Richard W. Practice medicine, specializing in radiology, Pitts., 1946—; dir. dept. radiology Mercy Hosp., 1951-65; faculty U. Pitts., 1951—, asso. prof. radiology, 1956—; dir. Mercy Sch. X-ray Tech., 1952-65; dir. Philip Murray Radiation Therapy Center, 1954-65; dir. dept. radiology Divine Providence Hosp., 1955—. Recipient sci. exhibit awards Pa. Radiol. Soc., 1953, 54, 56, Charles Luther Hinkel award, 1958. Diplomate Pan Am. Med. Assn., Nat. Bd. Med. Examiners. Mem. A.M.A., Am. Radium Soc., Am. Roentgen Ray Soc. (Gold medal award sci. exhibit 1947), Am. Coll. Radiology, Radiol. Soc. N.Am., Soc. Nuclear Medicine, Rocky Mountain, Inter Am. radiol. socs., Sigma Xi. Research, publs. on diagnostic cerebral angiography; reactions to radiation and their prevention; use of cobalt 60 teletherapy for cancer. Home: 640 Osage Rd., Pitts. 15216. Office: 1501 Locust St., Pitts. 15219.*

PERRYMAN, James Harvey, Am. physiologist; b. Kansas City, Mo., Aug. 18, 1918; s. Alfred Fey and Christine (Moore) P.; A.B., Stanford, 1941; M.A. in Anatomy, U. Cal. at Berkeley, 1943, Ph.D. in Physiology, 1955; m. Lucile Cooney, July 1, 1950. Research staff U. Cal. at San Francisco, 1949-56; asst. prof. U. San Francisco, 1954-56; faculty N.Y. U., 1956—, asso. prof. dept. physiology and pharmacology Dental Coll., 1960—; lectr. Bankstreet Coll., N.Y.C., 1961-64, Queens Coll., 1966—. USPHS grantee, 1956—. Mem. Am. Physiol. Soc., Am. Psychol. Assn., Psychonomic Soc., N.Y. Acad. Sci., Am. Assn. U. Profs., I.R.E., Harvey Soc., A.A.A.S., Sigma Xi. Research, publs. on nerve elec. activity of peripheral regeneration, reflexes and behavior brain function, central coordination, somatic and autonomic effect of pain and stress. Home: 330 3d Av., N.Y.C. 10010.*

PERSEUS, mathematician; flourished 2d century B.C.; lived before Heron and Geminos; credited with discovering spiric curves (plane sects. of surfaces generated by revolution of circle around axis in its plane not passing through center).

PERSHIN, Grigoriy Nikolaevich, Russian pharmacologist; b. 1908; grad. Med. Faculty, 2d Moscow Med. Inst., 1929; D.Med. Sci. Head dept. chemotherapy All-Union Chem. and Pharm. Research Inst., Moscow, 1932-—, dir. for sci. work, 1951-—. Recipient State prize. Mem. USSR Acad. Med. Scis. (corr.). Author over 270 works including: The Effect of Bactericidal and Chemotherapeutical Substances on Bacterial Enzymes, 1952; co-author, editor: Thiosemicarbazones, 1954; Phthivazide, 1954; Chemotherapy, 1956; Bactericides and Fungicides, 1959; Methods of Experimental Chemotherapy, 1959; co-author: Manual of Pharmacology, 1961. Chief editor: Pharmacology and Toxicology. Address: All-Union Chem. and Pharm. Research Inst., Zubovskaya ulitsa 7, Moscow, USSR.

PERSIDSKY, Konstantin Petrovich, Russian mathematican; b. Syzran, Russia, Oct. 23, 1903; grad. Kazan U., 1927; D.Physico-Math. Sci., 1947. Instr., Kazan U., 1927-34, prof., 1934-—, head chair math. analysis, 1940-46; head chair differential equations Kazakhstan U., 1946-—. Mem. Kazakhstan Acad. Scis. Author: One of Lyapunov's Theorems, 1937; The Theory of the Stability of Integrals in a Differential Equation System, 1938; A Certain Equation Calculation System with Partial Derivatives, 1950; The Spectrum of Characteristic Numbers, 1950; Some Theorems from Lyapunov's Second Method, 1960. Research and publs. on motion stability theory, infinite systems of differential equations, theory of probability, approximation analysis. Address: Kazakhstan University, ulitsa Kirova 136, Alma-Ata, Kazakhstan SSR, USSR.

PERSKY, Lester, Am. physician; b. Cleve., Mar. 4, 1919; s. Abraham E. and Sylvia (Meisel) P.; B.S., U. Mich., 1941; M.D., Johns Hopkins Sch. of Med., 1944; m. Mary E. Connor, June 25, 1949; children—David W., James M., Katherine S. Faculty, Western Res. U., Cleve., 1953-—, prof. urology Sch. Medicine, 1963-—. Mem. A.M.A., Am. Urol. Assn. Research, publs. on mechanisms of urinary obstruction, corrective techniques for urol., congenital disorders. Home: 14326 S. Park Blvd., Shaker Heights, O. 44120. Office: 2065 Adelbert Rd., Cleve. 44106.*

PERSON, Philip, Am. biochemist; b. N.Y.C., Aug. 6, 1919; s. Barney and Lena (Spindel) P.; B.S., Coll. City N.Y., 1940; D.D.S., N.Y. U., 1946; M.S., Rutgers U., 1951, Ph.D., 1952; m. Bertha Paula Kaufman, Mar. 14, 1953; children—Sarah, Naomi, Matthew. USPHS Postdoctoral fellow Rutgers U., 1949-52; chief dental research lab. VA Hosp., Bklyn., 1954-—; spl. research fellow USPHS Inst. Muscle Research, Woods Hole, Mass., 1961-62; vis. lectr. U. Pa. Grad. Sch. Medicine, 1960-—. Mem. corp. Marine Biol. Lab., Woods Hole, 1960-—. Fellow N.Y. Acad. Scis.; mem. Am. Soc. Biol. Chemists, Internat. Assn. for Dental Research. Cons. editor: The Metabolism of Oral Tissues, 1961; also articles. Research on intermediary metabolism of oral tissues, evolution of skeletal and oral tissues, chemistry of oxidative enzymes, origins of cartilage tissue in invertebrate animals, relationship between symptoms and metabolism of oral tissues; discovered new methods for extracting respiratory enzymes from heart muscle Home: 144-54 69th Av., Flushing, N.Y. 11367. Office: VA Hosp., Bklyn. 11209.*

PERSOON, Christian Hendrik, botanist; b. Cape of Good Hope, S. Africa, 1755; lived most of his life in Paris. Author: Observations mycologicae, 1795; Icones pictae, 1803-06; Species plantarum, 5 vols., 1817-27; Traité des Champignons comestibles, 1819; Mycologia Europae, 3 vols., 1822-23; Synopsis methodics fungorum, 2 vols., 1801. Linnaeus of mycology; classified fungi and introduced 71 genera; studied edible fungi. Died Paris, Feb. 17, 1837.

PERSOZ, Jean-Francois, chemist; b. Gex, Switzerland, 1805; became prof. chemistry Strasbourg (France) Faculty Scis., 1833; dir. Strasbourg Sch. Pharmacy; named prof. Sorbonne, 1850. Author: Introduction à l'étude de la chimie moléculaire, 1839; Traité pratique, 1846. Discovered (with Payen) enzyme diatase, 1834; responsible for classic preparation of methane from acetic acid; 1st to signal case of poisoning by phosphorus. Died Paris, 1868.

PERTHES, Georg Clemens, German surgeon; b. Sept. 17, 1869; prof., Leipzig, Tübingen, Germany. First to use deep X-ray therapy, and reported inhibitory effect of X-rays on neoplasms, especially carcinoma, 1903; described osteochondritis deformans juvenilis (Legg-Calvé-Perthes disease), 1910. Died 1927.

PERUTZ, Max Ferdinand, molecular biologist; b. Vienna, Austria, May 19, 1914; s. Hugo and Dely (Goldschmidt) P.; student U. Vienna, 1932-36; Ph.D., U. Cambridge (Eng.), 1940; Dr.Phil., U. Vienna, 1965; D.Sc., U. Edinburgh (Scotland), 1965; m. Gisela Peiser, Mar. 28, 1942; children—Vivien, Robin. Dir. unit for molecular biology Med. Research Council, Cambridge, 1947-62, chmn. Lab. Molecular Biology, 1962-—. Recipient (with J.C. Kendrew) Nobel prize in chemistry, 1962. Fellow Royal Soc., 1954. Author: Proteins and Nucleic Acids, 1962; also articles. Research on structure of crystalline proteins especially hemoglobin; (with J.C. Kendrew) discovered molecular structures of hemoglobin and myoglobin. Home: 42 Sedley Taylor Rd., Cambridge. Office: Lab. Molecular Biology, Hills Rd., Cambridge, Eng.*

PETAVEL, Sir Joseph Ernest, English physicist, engr.; b. London, Aug. 14, 1873; s. Emmanuel and Susanna (Ollift) P.; ed. Lausanne, Switzerland, also Univ. Coll., London. Research fellow Manchester U., 1901-03, prof. engring., dir. Whitworth Labs., 1908-19; dir. Nat. Phys. Lab., 1919-36; with Davy-Faraday Lab. Fellow Royal Soc., 1907. Author (with Sir J. H. Fleming) 1st paper on alternating current, 1896. Studied effect of high and low temperatures, pressure on gases and materials; designed Petavel gauge for measuring pressures set up in exploring gaseous mixtures; built and demonstrated reprodn. of plant for liquification of gases, St. Louis Internat. Expn.. Died Mar. 31, 1936.

PETCH, Howard Earle, Canadian physicist; b. Agincourt, Ont., Can., May 12, 1925; s. Thomas Earle and Edith (Painter) P.; B.Sc., McMaster U., 1949, M.Sc., 1950; Ph.D., U. B.C., 1952; m. Rosalind June Hulet, Aug. 13, 1949; children—Stephen, Patricia. Postdoctoral fellow McMaster U., Ont., 1952-53, faculty, 1954-67, prof. metallurgy and metall. engring., 1960-67, dir. research, 1961-67, prin. Hamilton Coll., 1963-67, chmn. interdisciplinary materials research unit, 1964-67; prof. dept. physics, v.p. acad. U. Waterloo, Kitchener, Ont., 1967-—. Mem. Sci. Council Can. Rutherford Meml. fellow Cambridge U., 1953-54. Research Council Ont. and NRC scholars; B.C. Acad. Scis. grantee; NRC fellow. Mem. Am. Crystallographic Assn., Am. Phys. Soc., Canadian Assn. Physicists, Canadian Research Mgmt. Assn., Internat. Union Crystallography, Royal Soc. Can. A developer fast coincidence circuit to measure lifetimes of excited nuclear states and other nuclear properties; pioneer in exptl. studies of the nuclear elec. quadrupole interaction in single crystals and applications of method to elucidating phys. properties of crystals; contbns. to understanding crystal chemistry of hydrated borates, atomic mechanism of order-disorder type of hydrogen-bonded ferroelec. Home: 11 Maple Hill Dr., Kitchener, Ont. Office: U. Waterloo, Waterloo, Ont., Can.*

PETER, Luther Crouse, Am. ophthalmologist; b. St. Clairsville, Pa., Feb. 14, 1869; s. Jacob and Harriet Jane (Crouse) P.; student Susquehanna U., 1887-88, Sc.D., LL.D.; A.B., Gettysburg Coll., 1891, A.M., 1894, Sc.D.; M.D., U. Pa., 1894; m. Carrie Chryistine Moser, June 20, 1916. Practiced medicine, Phila., from 1894; prof. diseases of eye Temple U., 1917-30, also grad. med. sch. U. Pa.; cons. ophthalmolodist Grad. Hosp., Rush Hosp. for Consumption and Allied Diseases, Roxborough Meml. Hosp. Fellow A.C.S., Am. Opthal. Soc., Coll. Physicians Phila., Oxford Congress Ophthalmology, Eng.; mem. Internat. Congress Ophthalmology (sec., treas. 1922), Nat. Bd. Ophthalmology (pres. 1929-37), A.M.A., Am. Acad. Ophthalmology and Oto-Laryngology (pres.), Phi Beta Kappa, Phi Gamma Delta. Author: The Principles and Practice of Perimetry, 4th edit., 1938; The Extra-Ocular Muscles, 3d edit., 1941. Died 1942.

PETER, Martin, Swiss physicist; b. Basel, Switzerland, July 12, 1928; s. Albin and Anna (Schmid) P.; diplôma with honor, Swiss Fed. Politechnic Sch., Zürich, 1952; Ph.D., Mass. Inst. Tech., 1955; m. Claudine Doll, June 14, 1954; children—Robin, Henry, Clarence. Staff, Mass. Inst. Tech., 1952-57, research asso., 1955-57; mem. staff, supr. Bell Telephone Labs., 1957-62; prof. physics, dir. Inst. for Exptl. Physics, U. Geneva, 1962-—, vice rector, 1966-—. Recipient Kern medal Swiss Fed. Politechnic Sch., 1952. Fellow Am. Phys. Soc.; mem. Société Hélvétique, Société Suisse de Physique Soc. Académique de Genève, Sigma Xi. Research, publs. on solid state physics, especially physics of metals, atomic clocks, microwaves. Home: 10 chemin Clos Alpestre, 1222 Vésenaz, Switzerland. Office: Sch. Physics, U. Geneva, Geneva, Switzerland.*

PETER OF SAINT OMER (or Petrus de Sancto Audemaro), French chemist; flourished 2d half 13th century; canon, St.-Omer, France. Author: Liber de coloribus faciendis (mentions oil painting, lists substances for painting, recipes for preparing pigments).

PETER OF SPAIN (Petrus Hispanus), philosopher, physician, psychologist; b. Lisbon, Portugal, 1210; s. Julian; student Compostela, Paris; Siena, Italy, 1246-50. Prof. Siena; became cardinal, 1273; elected pope under name of John XXI, 1276. Works ascribed to him include: Summa de conservanda sanitate; Regimen sanitatus; Regimen salutis per omnes menses; Liber de morte et vita et de causis longitudinis ac brevitatis vite; Diete super cyrurgian; Thesaurus pauperum; Liber de morbis oculorum (treatise on eye disease); also commentaries on works of Hippocrates, Galen, Theophilos, Hunain ibn Ishāq, Ishāq al-Isrā'ilī, Ibn al-Jazzār. Died 1277.

PETERING, Harold George, Am. biochemist; b. LaPorte, Ind., Oct. 8, 1910; s. George H. and Magdalena (Droege) P.; B.S., U. Chgo., 1935; Ph.D., U. Wis., 1938; m. Eva C. Petersen, Mar. 4, 1939; children—David H(arold), Marion Louise. Asst. prof. Mich. State U., East Lansing, 1938-41; research chemist E.I. DuPont de Nemours & Co., New Brunswick, N.J., 1941-45; research asso. U.S. Dept. Agr., Peoria, Ill., 1942-43; research asso., group leader Upjohn Co., Kalamazoo, 1945-66; asso. prof. environmental health and biochemistry U. Cin., 1966-—; adj. prof. biochemistry Mich. State U., East Lansing, 1965-66. Mem. A.A.A.S., Am. Soc. Biol. Chemists, Am. Chem. Soc., Am. Assn. for Cancer Research, Transplantation Soc. Research, publs. on devel. new drugs and methods for cancer, effects cancer drugs on immunity, biochemistry of trace metals, clin. implications of trace metal deficiencies; isolated and characterized new vitamins D; developed methods of analysis of chemistry of plant lipids and carotenoids, methods for studying respiration of biol. tissue, kinetics of photosynthesis and biochemistry of antimetabolites. Home: 1046 Clifton Hills Av., Cin. 45220.*

PETERLIN, Anton, physicist; b. Ljubljana, Yugoslavia, Sept. 25, 1908; s. Anton and Zofija (Pucnik) P.; Mr.Sc., U. Ljubljana, 1930; Ph.D., Humbold U., Berlin, 1938; m. Leopoldina Leskovic, Feb. 22, 1941; children—Tanja, Matija. Faculty, U. Ljubljana, 1939-60; head, J. Stefan Inst., Ljubljana, 1949-59; prof. physics, head Inst. Physics, Tech. U., Munich, W.Germany, 1960-61; dir. Camille Dreyfus Lab., Research Triangle Inst., Research Triangle Park, N.C., 1961-—. Fellow Am. Phys. Soc., Slovenian Acad. Scis.; mem. Am. Chem. Soc., German Phys. Soc., Kolloid Ges. (W.Germany), Sigma Xi. Author: (with H.A. Stuart) Accidental Birefringence, 1943. Research, publs. on solution properties of macromolecules, semicrystalline polymer solids. Home: 1212 Hill St., Durham, N.C. 27707. Office: P.O. Box 12194, Research Triangle Park, N.C. 27709.*

PETERMAN, Mynie Gustav, Am. physician; b. Merril, Wis., Mar. 5, 1896; s. Albert Frederick and Ida (Bratz) P.; B.S., U. Wis., 1918; A.M., Washington U., 1920, M.D., 1921; M.D. (hon) U. Madrid, 1960; m. Mildred Mackenzie, Sept. 29, 1924; children—Albert F., Mary Jean (Mrs. F.J. Harris). Fellow in surgery Washington U. Med. Sch., 1921; chief resident Ancker Hosp., 1922; fellow in pediatrics Mayo Clinic, 1922-23, cons. 1923-25; dir. labs. and research, Milw. Children's Hosp., 1925-33; prof., dir. dept. pediatrics, Marquette Med. Sch., 1928-38; practice medicine specializing in pediatrics, cons., Hosp., Balt.; cons. pediatrics USPHS, 1954-64; sr. cons. Bur. Medicine, Dept. Health, Edn. and Welfare, 1964-67. Mem. Sigma Xi, Phi Sigma. Introduced ketogenic diet, classified convulsions in childhood. Milw., Arlington, Va.; mem. staff Johns Hopkins Address: P.O. Box 360, Leesburg, Va. 22075.*

PETERMANN, Abdreas, Swiss physicist; b. Lausanne, Switzerland, Sept. 27, 1922; s. Felix and B. (Divorne) P.; Dr. es sci., Faculty of Sci. of Lausanne; m. E. Chochard, Nov. 19, 1952; children—Christian, Frank. Asst. to Prof. Perrier at Lausanne; asst. to prof. E. Stuekelberg at Geneva Atomic Energy Commn.; asst. to Prof. L. Rosenfeld, U. Manchester (Eng.); vis. scientist Copenhagen (Denmark) Inst. Theoretical Physics; work with Prof. N. Bohr and C. Moller for two years. Mem. French Alpine Club. Author: Découverte du groupe de renormalisation en théorie quantique, 1951; Découverte de la raison du désaccord entre l'expérience et la théorie du Lamb-Shift (dedicated to Prof. W. Heitler on his 60th birthday), 1963; Calcul des moments magnétiques anormaux de l'électron et du Muon, 1957. Home: 48, rue de Lausanne, Versoix, Geneva. Office: Cern, Geneva 22, Switzerland.

PETERMANN, August Heinrich, German geographer; b. Bleicherode, Germany, Apr. 18, 1822; founder a cartographic inst., London; later dir. Justus Perthes' Geog. Inst., Gotha. Author: An Atlas of Physical geography and Maps of Inner Africa and Transval. Founder jour. Petermanns Mitteilungen. Used new materials to improve maps; organized expds. to Africa and polar regions; drew up population maps of Brit. Isles which introduced use of density shading for distbn. by area and graduated dots for towns and cities; authority on geography of Africa and the Arctic. Died Gotha, Sept. 25, 1878.

PETERMANN, Mary Locke, Am. biochemist; b. Laurium, Mich., Feb. 26, 1908; d. Albert Edward and Anna (Grierson) P.; A.B., Smith Coll., 1929, D.Sc. (hon.), 1966; Ph.D., U. Wis., 1939. Postdoctoral fellow U. Wis., 1939-42, 44-45; profl. asst. com. on med. research NDRC, 1942-44; research chemist Meml. Hosp., 1945-46; Finney-Howell Found. fellow Sloan-Kettering Inst., N.Y.C., 1946-48, asso., 1948-60, asso. mem., 1960-63, mem., 1963-—; asso. prof. biochemistry Sloan-Kettering div. Cornell U. Med. Coll., 1951-66, prof. biochemistry, 1966-—. Recipient Sloan award in cancer research Alfred P. Sloan Found., 1963. Mem. Am. Soc. Biol. Chemists, Am. Assn. for Cancer Research, Biophys. Soc., Am. Chem. Soc. (Garvan medal 1966). Author: The Physical and Chemical Properties of Ribosomes, 1964; also numerous articles. Research on splitting of antibodies by enzymes, isolation animal cell nuclei, isolation and characterization animal ribosomes, cell structures where proteins are synthesized. Home: 315 E. 56th St., N.Y.C. 10022. Office: 425 E. 68th St., N.Y.C. 10021.*

PETERS, Christian August Friedrich, German astronomer; b. Hamburg, Germany, Sept. 7, 1806;

prof. Königsberg, Germany (now Kaliningrad, USSR); dir. Obs., Altona, Holstein; named prof. astronomy, Kiel, Germany, 1874. Mem. French Acad. Scis., 1857. Author: Numerus constans nutationis, 1842. Publs. on nutation, proper motion of Sirius; contributed to determination of stellar distances by parallax measurements at Pulkovo, Prussia, 1845; calculated orbit of Sirius's companion star (later discovered by Clerk, 1862), 1851. Died Kiel, May 8, 1880.

PETERS, Christian Henry Frederick, astronomer; b. Coldenbüttel, Germany, Sept. 19, 1813; Ph.D., U. Berlin (Germany), 1836. With expdn. surveying Mt. Etna, 1838-43; dir. trigonometrical survey in Sicily, 1843-48; came to U.S., 1854; with U.S. Coast Survey, 1854-58; dir. obs. Hamilton Coll., 1858-67, Litchfield prof. astronomy, dir. Litchfield Obs., 1867-90; led expdn. to observe solar eclipse, Des Moines, Ia., 1869; led U.S. expdn. to New Zealand to observe transit of Venus, 1874. Mem. Nat. Acad. Scis.; fgn. asso. Royal Astron. Soc. Author: Celestial Charts, 1882; Heliographic Positions of Sun Spots, Observed at Hamilton College from 1860 to 1870 (edited by E. B. Frost), pub. 1907; articles Contributions to the Atmospherology of the Sun, 1856, Corrigenda in Various Star Catalogues, 1886. Revised Ptolemy's Almagest, catalog stars' positions. Described apparent division of sun spots by bridges of luminous gas; discovered 48 new asteroids, 2 comets, 1846, 57; helped prepare charts locating all stars in zodiac. Died July 19, 1890.

PETERS, Dietrich, German virologist; b. Fulda, Germany, Nov. 5, 1913; s. Waldemar and Wilhelmine (Linnemann) P.; student U. Munich (Germany), 1932-35; Dipl.Chem., U. Leipzig (Germany), 1937; Dr.rer. nat., Kaiser Wilhelm-Inst. for Biochemistry, Berlin, Germany, 1939; m. Margrethe Backer Welhaven, Feb. 5, 1941; children—Jan., Henning, Sven. Asst., Tropical Inst. Hamburg (Germany), 1946-54, chief dept. for virus research, 1954—; privat dozent virology and molecular biology U. Hamburg, 1964—. Mem. German Soc. for Electron Microscopy, Soc. for Biol. Chemistry German Soc. for Hygiene and Microbiology. Editor: (with W. Bargmann, C. Wolpers) IV Internat. Congress for Electron Microscopy, vol. 2, 1960. Research, numerous publs. on protein and surface chemistry, ultrastructure of bacteria, rickettsiae, bartonellaceae and viruses especially pox viruses, ultracytochem. methods in electron microscopy including use of enzymes and staining reagts. substructures of genetical material of viruses.*

PETERS, Georges, pharmacologist; b. Mannheim, Germany, Jan. 25, 1920; s. Wilhelm and Anna (Siedentopf) P.; M.D., U. Istanbul, 1943; m. Editha E. Hofmann, Aug. 25, 1948; children—Claudia, Sylvia, Oliver. Clin. asst., chief lab. U. Istanbul, 1943-47; clin. asst. Municipal Hosp., Lubeck, Germany, 1947-49; med. cons., pharmacologist C.H. Boehringer Sohn, Ingelheim, Germany, 1949-53; research asst. in pharmacology U. Mainz, Germany, 1953-58, privat-docent, 1958-61; vis. scientist Lab. Kidney and Electrolyte Metabolism, Nat. Heart Inst., NIH, Bethesda, Md., 1959-60; pharmacologist Ciba Ltd., Basle, Switzerland, 1961-63; prof., chmn. dept. pharmacology U. Lausanne, Switzerland, 1963—. Mem. Swiss Soc. Physiology, Physiol. Chemistry and Pharmacology, Association des physiologistes de langue francaise, German Pharmacol. Soc., Swiss Assn. Pharmacologists, French Soc. Nephrology, Swiss Fedn. Physicians, others. Author: (with W. Graubner) Lobelin und Lobliaalkaloide, 1955; Nebennierenrinden-Inkretion und Wasser-Elektrolythaushalt, Befunde und Deutungen, 1960; also articles. Research on renal functions and water electrolyte exchanges as influenced by homeostatic mechanism, hormones and drugs; drugs interfering with carbohydrate metabolism, psychopharmacology. Home: 25, ch. de la Fontanettaz, CH-1012 Pully-La Rosiaz, Switzerland.*

PETERS, Gerd, German neuropathologist; b. Bonn, Germany, May 8, 1906; s. Josef and Margarete (Bachem) P.; M.D.; m. Dolfi Duensing, 1937; children—Helge, Karin. Asst., Kaiser-Wilhelm Inst.; named to chair neuropathology U. Bonn, 1952; dir. German Inst. Psychiat. Research, Munich, Germany, 1959—; now also prof. psychiatry and neurology U. Munich. Mem. numerous profl. societies. Author numerous sci. work, collaborator various books and manuals. Research on neuropathology, neurology, psychiatry. Home: Langenbachstrasse 13. Office: Kraepelinstrasse 2, Munich 23, West Germany.

PETERS, James L(ee), Am. ornithologist; b. Boston, Aug. 13, 1889; s. Austin and Frances Howie (Lee) P.; A.B., Harvard, 1912; m. Eleanor K. Sweet, May 28, 1932. Expdns. to Mexico, W.I., Central and S. Am., 1911-30; asst. curator birds Mus. Comparative Zoology, Harvard Coll., 1927-32, curator since 1932. Fellow Am. Ornithologists Union, Am. Acad. Arts and Scis.; mem. Washington Acad. Scis., Am. Soc. Mammalogists, Internat. Commn. Zool. Nomenclature, Deutsche Ornithologische Gesellschaft (corr.), Ornithologische Gesellschaft Bayern (corr.), Sociedad Ornithologica del Plata (corr.), Nutall Ornithol., Cooper Ornithol. socs., Biol. Soc. Wash. Author: Check-List of Birds of the World, vols. I-VII, 1931-48; also numerous articles. Died Apr. 19, 1952.

PETERS, John Emmett, Am. psychiatrist; b. Houston, Aug. 30, 1917; s. John Emmett and Mary (Joseph) P.; student Rice U., 1935-38; B.A., U. Cal. at Los Angeles, 1940; M.D., Johns Hopkins, 1950; m. Virginia Lee Otterback, Sept. 23, 1950; children—Mary Eguenia Lynn, Philip Hamblin. Faculty, U. Ark. Med. Center, Little Rock, 1954-62, prof. psychiatry, 1964—, dir. Psychiat. Clinic for Children, 1955-62, dir. child psychiatry, 1964—; asso. prof., asso. dir. Pavlovian Lab., Johns Hopkins Med. Sch., Balt., 1962-64; cons. neuropsychiat. research North Little Rock VA Hosp. Mem. Am. Bd. Psychiatry. Mem. A.A.A.S., Am. Phychiat. Assn., Am. Psychosomatic Soc., Pavlovian Soc., Sigma Xi. Research, publs. on age as factor in response to emotional trauma, behavior and learning problems in children, importance of inheritance in determination temperament in dogs, interpretation of Pavlov's canine typology. Home: 501 N. Elm St., Little Rock 72205. Office: 4301 W. Markham St., Little Rock 72205.*

PETERS, Karl Friedrich Wilhelm, German astronomer; b. Pulkowa, Prussia, Apr. 16, 1844; s. Christian August Friedrich Peters, prof., Kiel, Germany, also Königsberg (now Kaliningrad, USSR). Author: Astronomische Tafeln und Formein, 1871; Die Fixsterne, 1883. Died Königsberg, Dec. 2, 1894.

PETERS, Richard Morse, Am. physician; b. New Haven, Feb. 21, 1922; s. John Punnett and Charlotte (Hodge) P.; B.S., Yale, 1943, M.D. cum laude, 1945; m. Ann DeHuff, Oct. 12, 1946; children—Joan Forman, Deborah, Barbara Ann, Richard Morse. Faculty, U. N.C., Chapel Hill, 1952—, prof. surgery in charge thoracic and cardiovascular surgery, 1962—. Cons. thoracic surgery VA, 1960—. Fellow A.C.S.; mem. A.A.A.S., Am. Assn. Thoracic Surgery, Am. Assn. U. Profs., Am. Fedn. for Clin. Research, A.M.A., Am., So. surg. assns., Am., N.C. thoracic socs., N.C., Durham-Orange County med. socs., Elisha Mitchell Soc., N.Y. Acad. Scis., A.C.S., N.C. Rehad. Assn., Soc. Exptl. Biology and Medicine, Soc. U. Surgeons, So. Thoracic Surg. Assn., Soc. Thoracic Surgeons, Sigma Xi. Author: Pulmonary Physiology for Surgeons, 1968; also articles. Research in circulation of lung and its coordination with exchange of gas or ventilation, relationship between mech. properties of lung and alterations in blood gases, circulation of abdominal viscera in cirrhosis. Home: 421 Brookside Dr., Chapel Hill, N.C. 27514.*

PETERS, Robert Louis, Am. math. physicist; b. Boston, Dec. 7, 1925; s. George A. and Mary E. (Edwards) P.; B.A. in Math., U. Mass., 1950; postgrad. Boston U., Mass. Inst. Tech., Etudes U. Internat., Zurich, Switzerland; m. Elisabeth Jane Wheeler, June 20, 1959. Engr., Redwood Television, Glenwood, Cal., 1950-51; jr. engr. Stone & Webster Corp., Boston, 1951-52; design engr. Hamilton Standard div. United Aircraft Corp., Windsor Locks, Conn., 1952-53; chief engr., pres. Neo Co., N.Y.C., 1953—; devel. engr. Aerojet Gen. Corp. div. Gen. Tire Corp., Sacramento, 1959-60; research scientist NASA, Cleve., 1961-62; pres. Space Frontiers, Inc., Boston, 1962—. Prof. physics Etudes U. Internat., Switzerland, 1963—; cons. to industry. Mem. N.Y. Acad. Scis., Combustion Inst., Aerospace Elec. Soc., Ordnance Soc. Author: Materials Data Nomographs, 1965; Design of Liquid, Solid and Hybrid Rockets, 1965; Electronics Nomographs, 1966; Engineering with Plastics, 1966; also others. Inventor autophageous rocket, computer, aerospace devices. Research in automation, electronics, rocketry, space tech. and related fields. Home: 1112 Park Av., N.Y.C. 10025. Office: 520 Fifth Av., N.Y.C. 10036.*

PETERS, Robert William, Am. speech scientist; b. Boyden, Ia., Sept. 3, 1921; s. William Joseph and Janet Drucilla (Morris) P.; B.A., U. Minn., 1948; M.A., Ohio State U., 1950, Ph.D., 1953; m. Helen Abramson, Sept. 3, 1949; 1 son, Colin James. Research asso. Ohio Research Found., Pensacola, Fla., 1953-55; faculty U. So. Miss., Hattiesburg, 1955—, prof., chmn. dept. speech and hearing scis., 1964—, dir. office research and projects, 1965—. Fellow Am. Speech and Hearing Assn.; mem. Am. Psychol. Assn., Acoustical Soc. Am., N.Y. Acad. Scis. Research, publs. on factors contbg. to speech intelligibility, effects of auditory feedback on speech behavior, human auditory processing, studies relative to perceived temporal order of input signals. Home: 421 N. 37th Av., Hattiesburg, Miss. 39401.*

PETERS, Rudolph Albert, English biochemist; b. Kensington, London, Apr. 13, 1889; s. Albert Edward and Agnes Malvina (Watts) P.; student King's Coll., London; M.A., Gonville and Caius Coll., Cambridge, Eng., 1914; M.D., St. Bartholomew's Hosp., 1919; M.A., Oxon (by decree), 1923; M.D., Liege U., 1950; Doctor, U. Paris, 1952; D.Sc., U. Cin., 1953; M.D., Amsterdam University, 1953; D.Sc., London University, 1954, Leeds University, 1959, Australian Nat. U., 1961; LL.D., Glasgow U., 1963; m. Francis W. Vérel, 1917; children—Rudolph Vérel, Francis Raymond. Dunn lecturer, senior demonstrator biochemistry, Cambridge; Whitley prof. biochemistry Oxford U., 1923-54; past head biochemistry department agrl. research council, Institute Animal Physiology, Babraham, Cambridge; president Conseil Internat. des Unions Scientifiques, 1958-61; Dunham lecturer at Harvard, 1946-47; Herman Leo Loeb lecturer St. Louis U., 1947; Christian Herter lectr. N.Y.U., 1947; Croonian lectr. Royal Soc., 1952; Dohme lectr. Johns Hopkins Sch. Medicine, 1954; vis prof. Dalhousie University, Halifax, N.S., 1963. Mem. Med. Research Council, 1946-50; sci. adv. council Mil. Coll. of Sci., 1947-50, Ministry of Supply, 1950-53. Fellow, Trinity Coll., Oxford, 1925-54; hon. fellow Gonville and Caius Coll., Cambridge; recipient Thruston medal, 1918, Royal medal, Royal Soc., 1949; Cameron prize, Edinburgh, 1949; Medal of Freedom with silver palm (U.S.), 1947. Hon. mem. Biochem. and Physiol. Socs. Société Philomathique of Paris, Royal Netherlands Acad. Sci. and Letters, Belgian Royal Acad., Accademia Nazionale del Lincei of Rome, Am. Acad. Arts and Scis., Soc. Biological Chemists. Author: Biochemical Lesions and Lethal Synthesis, 1963. Discovered important aspects of tissue behavior which have increased understanding of mechanism of cancerous tissues. Home: 3 Newnham Walk, Cambridge. Office: Dept. of Biochemistry, Cambridge Univ., Cambridge, Eng.

PETERS, Theodore, Am. biochemist; b. Chambersburg, Pa., May 12, 1922; s. Theodore and Miriam (Lenhardt) P.; B.S., Lehigh U., 1943; Ph.D., Harvard, 1950; m. Margaret Campbell, June 9, 1945; children—Theodore D., James C., Melissa J., William L. Instr. physiol. chemistry U. Pa., 1950-51; biochemist VA Hosp., Boston, asso. in biochemistry Harvard Med. Sch., 1953-55; research biochemist M.I. Bassett Hosp., Cooperstown, N.Y., asso. in biochemistry Columbia Coll. Physicians and Surgeons, 1955—. Vis. sci. Carlsberg Lab., Copenhagen, 1958-59; cons. N.Y. State Dept. Health, 1965, 67, Albany Med. Coll., 1966—. Recipient Gold medal Electron Microscope Soc. Am., 1966. Fellow Am. Assn. Clin. Chemists; mem. Am. Soc. Biol. Chemists, Am. Soc. Cell Biology, Am. Chem. Soc., Sigma Xi. Research, publs. demonstrating that isolated slices of tissue can produce specific proteins; studies of mechanism of secretion of plasma proteins by liver cell, chem. structure of serum albumin; improved methods for clin. determination of iron and iron-binding capacity of blood serum and of free amino acids of serum. Home: 30 River St. Office: Atwell Rd., Cooperstown, N.Y. 13326.*

PETERSDORF, Robert George, physician; b. Berlin, Germany, Feb. 14, 1926; s. Hans Herbert and Sophie Petersdorf; B.A., Brown U., 1948; M.D., Yale, 1952; m. Patricia Horton Qua, 1951; children—Stephen Hans, John Eric. Practice medicine specializing in internal medicine, Seattle, 1960—; research fellow in medicine, Nat. Found. for Infantile Paralysis fellow Johns Hopkins Hosp., 1955-57, physician, 1958-59; chief resident in medicine U. Service Grace-New Haven Community Hosp., 1957-58; instr. medicine Yale, 1957-58; asst. prof. medicine Johns Hopkins, 1958-59, with U. Wash. Sch. Medicine, 1960—, prof. medicine, 1962—, chmn. dept., 1964—; physician-in-chief King County Hosp., 1960-64; attending physician Seattle VA Hosp., 1960—; with U. Wash. Hosp., 1960—, physician-in-chief, 1964—; cons. Surgeon Gen. U.S. Army, 1960—, U.S. Pub. Health Hosp., 1962—. Recipient Perkins prize, 1950; Campbell and Parker prize, 1952; Lederle Med. Faculty award, 1959. Fellow A.C.P.; mem. Am. Soc. Clin. Investigation, Am. Fedn. Clin. Research, Am. Assn. Immunologists, Soc. Exptl. Biology and Medicine, So. Soc. Clin. Investigation, Western Soc. Clin. Research (sec.-treas.), King County Med. Soc., Seattle Acad. Medicine, Wash., Am. med. assns., Am. Soc. Microbiology, Western Assn. Physicians, Assn. Am. Physicians, Am. Soc. Nephrology, Infectious Disease Soc. Am., Assn. Professors Medicine, Sigma Xi, Alpha Omega Alpha. Editor: The Prophylaxis of Infection, 1964; editorial bds. Jour. Chronic Diseases; Antimicrobial Agents and Chemotherapy; Nephron. Research, numerous publs. on bacterial infections including meningitis, urinary tract infections, pathogenesis of fever, antimicrobial therapy. Home: 3333 43d St. N.E., Seattle 98105.*

PETERSEN, Donald F., Am. pharmacologist; b. Brookings, S.D., May 20, 1926; s. Hans Valdemar and Sarah (Lione) P.; A.B., DePauw U., 1947; M.S., S.D. State U., 1950; Ph.D., U. Chgo., 1954; m. Lois Faith Ullman, Dec. 28, 1949; children—Hans, Christine, Sarah, Carl. Instr. pharmacology U. Chgo., 1954-56; mem. staff Los Alamos Sci. Lab., 1956—, leader cell biology sect. biomed. research group, 1963—; asso. prof. radiology Colo. State U., 1965—. Mem. Am. Soc. Pharmacology and Exptl. Therapeutics. Research, numerous publs. on sequence of biochem. events comprising life cycle of mammalian cells in tissue culture, effects of ionizing radiations, toxic materials on sequence of reactions regulating cell growth. Home: 2472 45th St. Office: P.O. Box 1663, Los Alamos 87544.*

PETERSEN, Gisli Fridrik, Icelandic physician; b. Reykjavik, Iceland, Feb. 21, 1906; s. Aage Lauritz and Gudbjorg (Gisladóttir) P.; M.D., U. Iceland, 1930, Dr.med., 1942; postgrad. internal medicine and radiology, Sweden, Denmark; m. Sigridur G. Brynjólfsdóttir, Nov. 3, 1934; children—Thóir, Aki. Physician radiology dept. U. Hosp., Iceland, 1934-48, chief physician, asso. prof. radiology, 1949—. Chmn. radiation Protection Com. Iceland, 1964—; cons. radiology Vifilsstadir State Sanatorium, 1962—. Mem. Cancer Soc. Reykjavik (mem. bd. 1949—), Scandinavian Radiology Soc. (mem. bd. 1949—), Icelandic Radiol. Soc. (mem. bd. 1957—), Nordic Soc. for Radiation Protection (mem. bd. 1964—).

Author: Röntgenologische Untersuchungen über Arteriosklerose, 1941; also articles. Radiol. research on arteriosclerosis, lung cancer in Iceland; orgn. radiol. protection in Iceland. Home: 16 Oddagata. Office: Radiology Dept., U. Hosp., Reykjavik, Iceland.*

PETERSEN, Magnus Christian, physician; b. Frederikshavn, Denmark, Feb. 6, 1893; s. Martin and Inger Marie (Mortensen) P.; student Dana Coll., 1915-17; B.S., U. Neb., 1924, M.D., 1925; m. Rosa C. Jensen, Feb. 26, 1926; 1 son, Allan Keith. Instr. psychiatry U. Colo., 1926-27; asst. supt. St. Peter (Minn.) State Hosp., 1928-35; supt. Willmar (Minn.) State Hosp., 1935-42, Rochester (Minn.) State Hosp., 1942-60; practice medicine specializing in psychiatry, Rochester, 1960-62, 65—; dir. tng. and research Mental Health Inst., Cherokee, Ia., 1962-65; asso. prof. psychiatry Mayo Found. for Med. Edn. and Research, U. Minn. Grad. Sch., Rochester, 1942-60, asso. prof. emeritus, 1960—. Fellow Central Neuropsychiat. Assn.; mem. A.M.A., Am. Psychiat. Assn., Soc. for Biol. Psychiatry, Minn. Soc. Neurol. Scis., Alpha Omega Alpha, Sigma Xi. Research, publs. on prefrontal lobotomy, especially in chronic schizophrenic patients, intracerebral electrography, tranquilizing drugs and psychic stimulating corrosion inhibition studies. Home: 11 Birch Hill Rd., Newtonville, Mass. 02160. Office: Dept. Chemistry, Simmons Coll., Boston 02115.*

PETERSEN, Quentin Richard, Am. chemist; b. Bridgeport, Conn., Mar. 10, 1924; s. Carl M. and Stella (Scharff) P.; B.S. in Chemistry, Antioch Coll., 1948; Ph.D., Northwestern U., 1952; m. Peggy Pottenger, Aug. 3, 1946; children—Jan Walker, Bardo Scharff. Faculty, Wesleyan U., Middletown, Conn., 1952-57, asst. prof., 1953-57; faculty Wabash Coll., Crawfordsville, Ind., 1957-66, prof., 1962-66; prof., chmn. chemistry dept. Simmons Coll., Boston, 1966—; vis. asst. prof. Trinity Coll., Hartford, Conn., 1953-54; Fulbright lectr. U. Barcelona, U. Valencia, Spain, 1961-62; vis. prof. Wesleyan U., 1964, U. Wis., 1966. Cons. dir. research and devel. Kelite Products, Inc., Chgo., 1950-56. Corr. Research fellow Inst. Chemistry, Academia Sinica, Taiwan, 1963—; Nat. Cancer Inst. fellow, 1951-52. Mem. Am. Chem. Soc., Chem. Soc. (London), A.A.A.S., Ind. Acad. Sci. (chmn. chemistry sect. 1964), Am. Assn. U. Profs., Sigma Xi. Designed Cenco-Petersen molecular model; research, publs. on alicylic compounds and steroids, corrosion inhibition studies. Office: Dept. Chemistry, Simmons Coll., Boston.*

PETERSEN, William, Am. sociologist; b. Jersey City, N.J., Aug. 3, 1912; s. Henry and Katherine (Gehrhardt) P.; A.B., Columbia, 1934, Ph.D., 1954; m. Renee Peller, Jan. 16, 1950; Vis. lectr. Smith Coll., Northhampton, Mass., 1953; faculty U. Cal. at Berkeley, 1953-56, 59-66, prof. sociology, 1963-66; faculty U. Colo., Boulder, 1956-59, asso. prof., 1957-59; research prof. Inst. Human Scis., Boston Coll., Chestnut Hill, Mass., 1966-67; Lazarus prof. social demography Ohio State U., 1967—. Spl. cons. state's population Gov. Colo., 1958-59. Hendrik Willem van Loon fellow, 1951-52; NSF Sr. fellow, 1958-59. Mem. Population Assn. Am. (Past dir.), Am., Netherlands sociol. assns., Am. Statis. Assn., Am. Eugenics Soc. Author: Planned Migration: The Social Determinants of Dutch-Canadian Movement, 1955; (with Renee Petersen) University Adult Education, 1960; Population, 1961, 2d edit., 1968; The Politics of Population, 1964; also articles. Editor, contbg. author: The Realities of World Communism, 1963; (with David Matza) Social Controversy, 1963; editor: American Social Patterns, 1956; asso. editor Am. Sociol. Rev., 1960-62. Research on population especially social policies governing migration and fertility. Home: 3501 Avigan Pl., Columbus, O. 43221.*

PETERSEN, William Ferdinand, Am. physician; b. Chgo., Mar. 25, 1887; s. Eduard and Wilhelmina Joanna (Klockziem) P.; student Armour Inst., 1904-06; B.S., U. Chgo., 1910; M.D., Rush Med. Coll. (U. Chgo.), 1912; m. Alma Catherine Schmidt, Sept. 16, 1919; children—Eduard Schmidt, Conrad William, William Otto. Instr. pathology Vanderbilt U., 1913, asst. prof. exptl. medicine and pathology, 1914-17; asso. in pathology and bacteriology U. of Ill. Coll. Medicine, 1919-24, prof., 1924-42; pres. Petersen Oven Co. Mem. A.M.A., Inst. Medicine, Soc. Exptl. Pathology, Am. Soc. Pathology and Bacteriology, Chgo. Pathol. Soc., Chgo. Soc. Internal Medicine, Am. Assn. Phys. Anthropology. Author: Protein Therapy and Non-Specific Reactions, 1922; Skin Reactions, Blood Chemistry and Physical Status of Normal Men and Clinical Patients (with S. A. Levinson), 1930; The Patient and the Weather (monographs with Margaret E. Milliken), 1934-38; Destiny—Lincoln-Douglas, 1943; Hippocratic Wisdom, 1945; Man-Weather-Sun, 1947. Introduced terms sympathetic status and parasympathetic status, circa 1923. Died Aug. 20, 1950.

PETERSON, Franklin Paul, Am. mathematician; b. Aurora, Ill., Aug. 27, 1930; s. Paul F. and Mildred (Umbriet) P.; B.S., Northwestern U., 1952; Ph.D., Princeton, 1955; m. Marilyn Lee Rutz, Aug. 8, 1959. Fellow, U. Chgo., 1955-56; Higgins lectr. Princeton, 1956-58; faculty Mass. Inst. Tech., Cambridge, 1958—, prof. math., 1965—. Alfred P. Sloan fellow Oxford (Eng.) U., 1960-61. Mem. Am. (editor Transactions 1966—), Mexican math. socs. Research, publs. on hotopy theory and cohomology operations, applications differential topology especially cobordism theory. Home: 14 Hammond St., Chestnut Hill, Mass. 02167. Office: Mass. Inst. Tech., Cambridge, Mass. 02139.*

PETERSON, George Earl, Am. physicist; b. Pitts., June 7, 1934; s. Harry and Lillian (Grabe) P.; B.S., U. Pitts., 1956, Ph.D., 1961. With tech. staff Bell Telephone Labs., Murray Hill, N.J., 1961—. Cons. univs., indsl. orgns. Mem. Am. Crystallographic Assn., Optical Soc. Am., Am. Phys. Soc., A.A.A.S., N.Y. Acad. Scis., Sigma Pi Sigma. Author: (with others) Transition Metal Chemistry, vol. 3, 1966; also articles. Research in nuclear quadrupole resonance spectroscopy, ferroelectric and other dielectric materials. Home: 806 Morris Turnpike, Short Hills, N.J. 07078. Office: Bell Telephone Labs., Murray Hill, N.J. 07971.*

PETERSON, Glen Ervin, Am. microbiologist; b. Jackson, Minn., July 12, 1926; s. Walter V. and Mabel (Petersen) P.; B.A., Luther Coll., Decorah, Ia., 1949; M.S., U. Minn., 1952, Ph.D., 1954; certificate in pharmacology Rutgers U., 1956; m. Joyce Tomasek, Nov. 24, 1949; children—Glenna, Brenda, Eric, Mark, Matthew. Faculty, U. Houston, 1953-55, 58-65, prof. biology, 1963-65; research asso. Squibb Inst. For Med. Research, 1955-58; cons., tchr. Biol. Scis. Curriculum Study U. Colo., Boulder, 1965—. NSF Summer Insts. adv. panel, 1964; cons. Lamar State Coll. Biology Inst., 1964; NSF cons. on Central Am., 1965—. Mem. Am. Soc. Microbiology, Am. Acad. Microbiology, Soc. Indsl. Microbiology, Inst. Food Tech., Nat. Assn. Biology Tchrs., Am. Assn. U. Profs., N.Y. Acad. Scis., Am. Inst. Biol. Scis., Nat. Sci. Tchrs. Assn., A.A.A.S., Sigma Xi. Research, publs. on antibiotic action, microbial action and biosynthesis; water purification in space capsules. Home: 12570 W. 31st Av., Denver 80215. Office: P.O. Box 930, Boulder, Colo. 80301.*

PETERSON, Jack Milton, Am. physicist; b. Portland, Ore., Apr. 25, 1920; s. Adolph Julius and Anna (Pearson) P.; B.S., Harvard, 1942; Ph.D., U. Cal. at Berkeley, 1950; m. Beverly Lael Begole, Aug. 31, 1946; children—Laelanne, Sharon, Diane. Staff mem. Mass. Inst. Tech. Radiation Lab., Cambridge, 1942-44; asst. dir. vacuum tube devel. group Columbia, N.Y.C., 1943-46; with U. Cal. Lawrence Radiation Lab., 1946—, div. leader, 1953-61, group leader, 1964—. Fulbright fellow Bohr Inst., Copenhagen, Denmark, 1960-61. Mem. Am. Phys. Soc. Author: Microwave Techniques, 1943; also numerous articles. Research field nuclear physics, accelerators; co-discoverer photo-mesons; theory of nuclear Ramsauer effect. Home: 350 Cordell Dr., Danville, Cal. 94526. Office: U. Cal. Lawrence Radiation Lab., Berkeley, Cal. 94720.*

PETERSON, John Booth, Am. soil chemist; b. Salem, Ore., July 18, 1905; s. Victor Allen and Bertha (Booth) P.; B.S., Ore. State Coll., 1928; M.S., Ia. State Coll., 1929, Ph.D. in Soil Chemistry, 1936; postgrad. phys. chemistry (NRC fellow) U. Cal. at Davis; m. Elizabeth L. Fogel, June 7, 1930; children—John Robert, Mary Jean (Mrs. Robert J. Brown). Faculty, Ia. State U., Ames, 1928-48, prof. agronomy, 1933-48; head dept. agronomy Purdue U., Lafayette, Ind., 1948—. Cons. Rockefeller Found., Latin Am., 1961-62, Greek Govt., 1963, Ford Found., Argentina, 1964—. Fellow Am. Soc. Agronomy (Stevenson award for research in soil sci. 1948, pres. 1958-59); mem. Sigma Xi. Research on soil fertility; soil microscopy; conservation; morphology and chemistry; genesis. Home: 201 Forest Hill Dr., West Lafayette, Ind. Office: Lilly Hall, Purdue U., Lafayette, Ind.

PETERSON, John Cyril, Am. physician; b. Lehi, Utah, Nov. 11, 1904; s. John Reynold and Johanna (Bauerle) P.; A.B., U. Utah, 1928; M.D., Vanderbilt U., 1931; m. Ruby May Clayton, May 9, 1933; children—John Reynold, Brent Marshall, Neill Clayton. Asst. in bacteriology and immunology Cornell Med. Coll., N.Y.C., 1935-38; faculty Vanderbilt U., Nashville, 1938-53, asso. prof. pediatrics, 1948-53; prof., chmn. dept. pediatrics Marquette U., Milw., 1953—; pediatrics-in-chief Milw. Children's Hosp., 1953—; dir. pediatrics Milw. City Gen. Hosp., 1963—; practice medicine specializing in pediatrics, Nashville, 1938-53, Milw., 1953—. Mem. Acad. Pediatrics, A.M.A., Am. Pediatric Soc., Soc. for Pediatric Research, Midwest Soc. Pediatric Research. Research, publs. on immunology, disease description and therapy. Home: 4068 N. Lake Dr., Milw. 53211. Office: 1700 W. Wisconsin St., Milw. 53233.*

PETERSON, Laurence Elmer, Am. physicist; b. Grantsburg, Wis., July 26, 1931; s. Elmer T. and Mary Estelle (Grant) P.; B.S. in Elec. Engring., U. Minn., 1954, Ph.D. in Physics, 1960; m. Mary Katherine LaBrie, Sept. 1, 1955; children—Mark Leo, Daniel Francis, Lynn Marie, Julianne. Research asso. U. Minn., Mpls., 1960-62; research physicist U. Cal. at San Diego, La Jolla, 1962-63, asst. prof. physics, 1963-67, associate professor of physics, since 1967—. Mem. solar physics subcom. space sci. steering com. NASA, 1964—. Mem. Am. Geophys. Union, Am. Phys. Soc., A.A.A.S., Am. Astron. Soc. Research, publs. on galactic cosmic rays, solar cosmic rays, modulation in interplanetary medium, auroral X-rays and magnetospheric effects, X-ray and gamma-ray astronomy and high energy astrophysics, balloon and satellite instrumentation. Home: Route 1, Box 52, Del Mar, Cal. 92014. Home: 345 Westbourne St. Office: Physics Dept., U. Cal. at San Diego, La Jolla, Cal. 92037.*

PETERSON, Lowell Niels, Am. periodontist; b. Stockton, Cal., Nov. 3, 1904; s. Niels C. and Lillie (Sperry) P.; D.D.S., U. Cal. Coll. Dentistry, 1928; m. Mildred Hutzen Neil, May 29, 1941; 1 son, Neil Bruce. Gen. practice dentistry, San Francisco, 1928-45, specializing in periodontics, 1946—; faculty U. Pacific Coll. Physicians and Surgeons Dental Sch., 1936—, clin. prof. periodontics, lectr. diet, nutrition, 1947—. Diplomate Am. Bd. Periodontology. Fellow Internat., Am. colls. dentists; mem. Am. Dental Assn., Am., Cal. acads. periodontology, Am. Soc. Periodontists, Western Soc. Periodontology, A.A.A.S., Am. Inst. Oral Biology, Am. Acad. Applied Nutrition, Omicron Kappa Upsilon. Research, publs. on dental occlusion, nutritional and dietary influences on periodontal health, oral hygiene and dental history. Home: 615 Harvard Rd., San Mateo, Cal. 94402. Office: 450 Sutter St., San Francisco 94108.*

PETERSON, Melvin Norman Adolph, Am. oceanographer; b. Evanston, Ill., May 27, 1929; s. Fred Gothard Walter and Norma (Johnson) P.; B.S., Northwestern U., 1951, M.S., 1956; Ph.D., Harvard U., 1960; m. Margaret Stewart Forbes, June 14, 1958; children—Katrina Elizabeth, John Frederick Forbes, Bruce Norman Stewart. Prof. oceanography Scripps Inst. Oceanography, La Jolla Cal., 1960—; chief scientist Joint Oceanographic Instns. Deep Earth Sampling-Deep Sea Drilling Project. Fellow Geol. Soc. Am. ; mem. A.A.A.S. Research on carbonate rocks, volcanic material in Pacific ocean; discoveries relating to dolomite and chert; invention of ultrasonic sieving. Home: 1221 Umatilla Rd., Del Mar, Cal. 92014. Office: Scripps Instn. of Oceanography, La Jolla, Cal. 92038.*

PETERSON, Merlin Dewey, Am. chemist; b. Provo, Utah, 1910; s. Joseph and Rhoda (Robinson) P.; A.B., Vanderbilt U., 1931; Ph.D., U. Cal. at Berkeley, 1936; m. Marian Sarquis, 1933; children—Carolyn (Mrs. Harry G. Carlson), Janice, Joseph. Instr. chemistry U. Cal. at Berkeley, 1936-37; chemist, ammonia dept. E.I. DuPont de Nemours & Co., Wilmington, Del., 1937-43; chemist, group leader metall. lab. U. Chgo., 1943, Clinton Labs., Oak Ridge, 1943-44; chief chem. process devel. Clinton Labs., 1944-48; dir. tech. div. Oak Ridge Nat. Lab., 1948-49; prof., head dept. chemistry Vanderbilt U., Nashville, 1949-55, prof. chemistry, 1955-57; dir. Indsl. Reactor Labs., Plainsboro, N.J., 1957-59; adj. prof. chem. engring. Columbia, N.Y.C., 1958-59; dep. asso. lab. dir. for edn. Argonne (Ill.) Nat. Lab., 1960—. Mem. Am. Chem. Soc., Am. Inst. Chem. Engrs., Am. Nuclear Soc., Am. Soc. Engring. Edn., A.A.A.S., Phi Beta Kappa, Sigma Xi. Research on high-pressure organic syntheses, ethylene polymerization, uranium reactor devel., separation and purification of inorganic materials; devel. separation and purification processes. Home: 45 Waverly Av., Clarendon Hills, Ill. 60514. Office: Argonne Nat. Lab., Argonne, Ill. 60439.*

PETERSON, Osler Luther, Am. physician; b. Cokato, Minn., May 28, 1912; s. Olaus Louis and Mathilda (Johnson) P.; A.B., Gustavas Adolphus Coll., 1934; M.B., U. Minn., 1938, M.D., 1939; M.P.H., Johns Hopkins, 1947; m. Delores M. Kealy, Sept. 28, 1939; children—Thomas Brooks, Osler Leopold. Asst. dir. med. edn. and pub. health Rockefeller Found., 1956—; vis. lectr. preventive medicine Harvard Med. Sch., 1959-62, vis. prof., 1962-66, prof., 1968—, faculty pub. adminstrn., 1961—, clin. asso. medicine Boston City Hosp., 1962-66, cons. physician 2d and 4th services, 1966—. Fellow American College of Physicians; member Am. Pub. Health Assn., Assn. Am. Med. Colls., Assn. Tchrs. Preventive Medicine, Mass. Med. Soc., Soc. for Exptl. Biology and Medicine. Author books, numerous articles on med. care research. Home: 1 Yale St., Winchester, Mass. 01890. Office: 25 Shattuck St., Boston 02115.*

PETERSON, Raymond D(ale) A(ugus), Am. physician; b. Mpls., Oct. 5, 1930; s. Elmer Ralph Theodore and Minnie (Groechel) P.; B.A. cum laude, U. Minn., 1952, B.S., 1953, M.D., 1955; m. Carol Ellen Johnson, Sept. 20, 1953; children—Mark Charles Edward, Craig Raymond Dale, Rose Paul Martin. Dermatologist, Tripler Army Hosp., Honolulu, 1956-58; research fellow Minn. Hosp., Mpls., 1960-63, established investigator Am. Heart Assn., 1963-65; established investigator Am. Heart Assn., U. Uppsala (Sweden), 1965-66; asst. prof. dept. pediatrics U. Minn. Hosps., Mpls., 1963-66; asso. prof. pediatrics La Rabida Inst., dir. research La Rabida Sanitarian, U. Chgo., 1966—; Guggenheim fellow, 1965-66. Mem. Am. Assn. Immunologists, Am. Soc. for Exptl. Pathology, Soc. for Pediatric Research, A.A.A.S. Research, numerous publs. on devel. lymphoid tissue of body, malignancy devel. in lymphoid tissue; discovered common leukemia of chickens can be prevented by surgically or chemically removing a small lymphoid organ in gut wall. Home: 823 Hutchison Rd., Flossmoor, Ill.*

PETERSON, Roger Tory, Am. ornithologist; b. Jamestown, N.Y., Aug. 28, 1908; s. Charles Gustav and Henrietta (Bader) P.; student Art Students League, 1927-28, N.A.D., 1929-31; D.Sc., Franklin and Marshall Coll., 1952, Ohio State U., 1962, Fairfield U., 1967; Allegheny Coll., 1967; m. Barbara Coulter, July 29, 1943; children—Tory Coulter, Lee Allen. Instr. sci. and art Rivers Sch., Brookline, Mass., 1931-34; adminstrv. staff Nat. Audubon Soc., N.Y.C., 1934-43; editor Field Guide Series, Houghton Mifflin Co., Boston, 1946—, Am. Naturalist Series, 1965— art dir. Nat. Wildlife Fedn., Washington, 1951—; scholar-in-residence Fallingwater, Western Pa. Conservancy, 1968. Mem. council Lab. Ornithology, Cornell U., Ithaca, N.Y., 1961—, Am. sect. Internat. Ornithol. Congress, 1954—; chmn. Am. sect. Internat. Bird Protection Com., 1962. Recipient John Burroughs medal for nature writing John Burroughs Assn., 1950; Geoffrey St. Hilaire gold medal Société National de'Acclimatation de France, 1958; Allen medal Lab. Ornithology, Cornell U., 1967. Fellow N.Y. Zool. Soc. (gold medal 1961), Linnaean Soc., Brit. Orithol. Soc. (corr.); mem. Nat. (sec. 1960-63, dir.), Mass. (hon. v.p.), N.J. (hon. v.p.), R.I. (hon. v.p.) Audubon socs., Am. Ornithologists Union (1st v.p. 1962, Brewster medal 1944), Atlantic Naturalist Soc. D.C. (hon. v.p.), Wilson Ornithol. Club (pres. 1964), Spanish Ornithol. Soc. (hon.). Author: A Field Guide to the Birds, 1934, 39, 42; The Junior Book of Birds, 1938; Field Guide to Western Birds, 1941; (with John Baker) The Audubon Guide to Attracting Birds, 1941; Birds Over America, 1948; How to Know the Birds, 1949; Wildlife in Color, 1951; (with Guy Mountfort, P.A.D. Hollom) A Field Guide to the Birds of Britain and Europe, 1952; Wild America, 1955; A Bird Watcher's Anthology, 1957; A Field Guide to the Birds of Texas, 1960; The Birds, 1963; (with James Fisher) The World of Birds, 1964; Wildflowers of Northeastern U.S., 1968. Contbr. articles to natural history and popular publs.; devel. sci. approach to field recognition. Address: Route 4, Box 131, Neck Rd., Old Lyme, Conn. 06371.*

PETERSON, Thurman Stewart, Am. mathematician; b. Mercur, Utah, May 5, 1905; s. Oscar Olaf and Dora (Stewart) P.; B.S., Cal. Inst. Tech., 1927; M.S., Ohio State U., 1928, Ph.D., 1930; m. Helen Mary Skowerski, June 16, 1932; children—Kay, Alan, William, Linda, Julie, Pamela. Instr., U. Mich., 1930-32; research fellow Inst. for Advanced Study, 1932-34; head sci. and math. Shipley Sch., 1934-38; with U. Ore., 1938-53, asso. prof., 1948-53; faculty Portland (Ore.) State Coll., 1953—, prof. math., 1955—. Mem. Am. Math. Soc., Math. Assn. Am., Am. Assn. U. Profs., Sigma Xi. Author: Intermediate Algebra, 3d edit., 1967; College Algebra, 1956; Calculus, 1958; Calculus with Analytical Geometry, 1958. Contbr. articles in field to sci. jours. Study of functional analysis; integral equations. Home: 1527 Lake Front Rd., Lake Oswego, Ore. 97034. Office: Div. Sci., Portland State Coll., Portland, Ore.*

PETISCUS, Bartholomaüs, see Pitiscus, Bartholomaüs.

PETIT, Alexis Thérèse, French physicist; b. Vesoul, France, Oct. 2, 1791; prof. physics Lycée Bonaparte, also L'École polytechnique, Paris. Developed (with P.L. Dulong) methods for determining thermal expansion and specific heat of solid bodies, also discovered that elements in solid state have nearly same atomic heat (law of Dulong and Petit), 1819. Died Paris, June 21, 1820.

PETIT, François Pourfour de, see de Petit.

PETIT, Georges, French chemist; b. Vireux-Wallerand, France, Sept. 15, 1909; s. Jean-Louis and Alice (Lemaire) P.; student Faculte de Pharmacie de Paris; D.Sc., Faculte des Sciences, Paris; m. Marié, June 16, 1946. Asst., Sorbonne, Paris; instr. l'Ecole de Medecinè de Reims, France; new head research Centre National de la Recherche Scientifique. Decorated Chevalier des palmes academiques; Laureat French Acad. Scis., Chem. Soc. Mem. Chem. Soc. France, Phys. Soc. France. Author: La cryometrie à haute température et ses applications, 1966; also articles. Research on osmotic effects especially high temperature cryometry, giving structure of substances dissolved in fusible salts at raised temperatures; originated theory of fusion in statis. thermodynamics. Home: 53, avenue de Neuilly, Neuilly/Sein (Seine), France. Office: Laboratoire de Physique générale, Faculté des Sciences, Paris, France.*

PETIT, Jean Louis, French surgeon; b. Paris, Mar. 13, 1674; became master surgeon, 1700. Fellow Royal Soc., 1729; mem. French Acad. Scis., 1716; a founder Acad. Surgery. Author: L'art de guérir les maladies des os, 2 vols., 1732; Dissertation sur les maladies des fosses nasales et leur sinus (1st authoritative work on nose diseases), 1804; Traité de Chirurgie (unfinished). Inventor screw-type tourniquet, circa 1705; credited with giving 1st description of osteomalacia, circa 1705; credited with observing anat. triangle in lumbar region, bounded by iliac crest, latissimus dorsi muscle, external oblique muscle, 1705, pub., 1774, also described lumbar hernia (Petit's hernia) within the triangle, 1774; believed to have performed 1st successful operation for mastoiditis, 1736; performed successful ganglionectomy for breast cancer. Died Paris, Apr. 20, 1750.

PETIT, Marc-Antoine, French surgeon; b. Lyons, France, Nov. 3, 1766; master surgeon Lyons Hôtel-Dieu; recipient Gold medal Paris Hosps., 1785; mem. French Acad. Scis. Author: Dissertation sur le phthisi laryngea, 1790; Essai sur la médecine du coeur, 1799. Pioneer in study of Tb of larynx; research on heart diseases. Died Villeurbanne, France, July 7, 1811.

PETIT, Pierre, French physicist, mathematician; b. Montlucon, France, Dec. 31, 1598; m. 1650; moved to Paris, 1633; pupil, friend, defender of Descartes; corresponded with Blaise and Etienne Pascal. Fellow Royal Soc., 1667. Author: Sur la nature des comètes, 1665; Sur la nature du chaud et du froid, 1671. Inventor machine for measuring stellar distances, also cylindrical adding machine; repeated Torricelli's barometric expts. before Pascal, in Rouen, France, 1646, also pub. detailed account of them. Died Paris, Aug. 20, 1677.

PETIVER, James, English botanist; b.Hillmorton, Eng., 1663; s. James and Mary Petiver; ed. Rugby free sch.; unmarried. Apprenticed to apothecary to St. Bartholomew's Hosp., London, 1683; practicing apothecary, 1692; demonstrator of plants Soc. Apothecaries, 1709. Fellow Royal Soc., 1695; mem. French Acad. Scis., 1699. Author: Museum Petiverianum, 8 vols., 1695-1703; Gazophylacium Naturae et Artis, 1702-09; The Monthly Miscellany, or Memoirs for the Curious, 3 vols., 1707-09; Plantarum Genevae Catalogus, 1709; Pterigraphia Americana, 1712; Aquat. Animalium Amboinae Catalogus, 1713; Herbarii Britannici clariss. D. Raii Catalogus cum Iconibus ad vivum delineatis, Vol. 1, 1713, Vol. II, 1715; Plantarum Etruriae rariorum Catalogus, 1715; Plantarum Italiae marinarum et Graminum Icones, 1715; Hortus Peruvianus medicinalis, 1715; Monspelii desideratarum Plantarum Catalogus, 1716; Proposals for the Continuation of an Iconical Supplement to Mr. John Ray his Universal History of Plants, 1716; Graminum, Muscorum, Fungorum . . . Condordia, 1716; Petiveriana sive Collectanea Naturae, iii, 1716-17; Plantae Silesiacae rariores, 1717; Plantarum Aegyptiacarum rariorum Icones, 1717; English Butterflies, 1717; Botanicum Anglicum; Hortus siccus Pharmaceuticus; Rudiments of English Botany; James Petiver his Book, being Directions for gathering Plants; Brief Directions for the easie making and preserving Collections; Plants engraved for Ray's English Herball, also articles in Philos. Trans., 1697-1717. Corresponded with naturalists in all parts of world forming large miscellaneous museum; his book and papers now in Brit. Mus.; furnished materials for Ray's History of Plants. Died London, Apr. 2, 1718.

PETRARCA, Francesco, Italian philosopher, humanist; b. Arezzo, Italy, July 20, 1304; student, sch. Carpentras, 1315-18; studied law, Montpellier, France, 4 years; student U. Bologna, 3 years; children—Giovanni, Francesca. Recipient laurel crown of poetry, Avignon, 1341. Influenced sci. thought through his anti-mediaeval, anti-Artistotelian, anti-Averroistic opinions; helped dethrone scholastic philosophy; criticized astrology to some extent. Died Arqua, nr. Padua, Italy, July 18, 1374.

PETREA, Ion, Rumanian physician; b. Teisani, Rumania, June 27, 1924; s. Ion and Aretina (Voicescu) P.; student Faculty Human Medicine, U. Bucharest (Rumania), 1943-49, endocrinologist primary doctor, 1962, doctor ex Med. Scis., 1964; m. Dina Gatulescu-Petrea, Dec. 24, 1954; children—Ioana, Dan. Preparator, Embryology and Anatomy Inst., Med. Faculty, Bucharest, 1945-47; external med. student clinics Health Ministry, Bucharest, 1947-48; staff Inst. Endocrinology, Rumanian Acad., Bucharest, 1948-51, main research worker, 1951-59, chief electron microscopic dept., 1959—; staff electron microscopic lab. Inst. Biology and Medicine, German Sci. Acad., Berlin, 1959. Eleanor Roosevelt Internat. Cancer fellow UICC, Radium Inst. (Fondation Curie), U. Paris, 1965-66. Mem. Rumanian Oncological Soc. (founder), Rumanian Electron Microscopy Soc. (founding, v.p.), Endocrinological Soc., Bucharest, Endocrinological Soc. Paris, Assn. Franc. de la Lutte contre le Cancer (corr.), French Electron Microscopy Soc. (corr.). Author: Experimental Cancerogenesis of the Thyroid Gland, 1964; (with C.I. Parhon, I. Potop, Babes) The Role of Thymus Gland in Experimental Cancer, 1955; Endocrinology of Salivary Glands, 1956; Elements of Electron Microscopy, 1962; also numerous articles. Sec. editorial staff Studii si Cercetárai de Endocrinologie, 1953-66, Revue Roumaine d'Endocrinologie, 1964-66; editorial staff Neoplasma Internat. Cancer Rev., 1961—. Developed exptl. research in endocrine cancerology in Rumania; studies of cytogenesis of hypophyseal and thyroid cancer, internal secretion of salivary glands, digestive tract; isolated metastatic transplantable sarcoma; developed techniques in exptl. endocrine surgery, heterotransplantation of human cancer, cytology. Home: 11, Bd. N. Balcescu. Office: 34, Bd. Aviatorilor, Bucharest, Rumania.*

PETRESCU, Arcadiu Tiberiu, Rumanian physician, neurologist; b. Calarasi, Rumania, Jan. 18, 1925; s. Valerian and Maria-Florica (Galiceanu) P.; physician Faculty Gen. Medicine, Bucharest, Rumania, 1949; m. Xenia Marinescu, Aug. 5, 1961. Lab. secondary physician Central Lab. Hygiene Bucharest, 1949-51; staff Inst. Neurology, Acad. Socialist Republic Rumania, Bucharest, 1951—, prin. researcher, 1958—, head neuropath. and histochem. lab., 1964—; secondary physician Clinic Neurology, Prof. Gh. Marinescu State Hosp., 1954-57, prin. neurologist in Policlin. Service, 1959—. Mem. Neurol. Soc., Histo-and Cytochem. Soc., USSM-Bucharest. Author: (with St. Draganescu, N. Draganescu) Human Viral Encephalitis, 1962; also numerous articles. Demonstrated histochemically 2 kinds of macrophages producing esterified cholesterol in demyelinating diseases; research on neuro-histochemistry, encephalitis, aphasia, cerebrovascular disease. Office: Inst. Neurology, 42 Str. Povernei, Bucharest, Rumania.

PETRI, Gabor, Hungarian surgeon; b. Budapest, Feb. 6, 1914; s. Endre and Flora (Pollak) P.; M.D., U. Pecs, 1937; surgeon U. Budapest, 1942, C.Sc., 1952; m. Margaret Pal, Sept. 1, 1948; children—Klara, Andras. Fellow in physiology U. Pecs, 1933-37; house officer St. Stieve Hosp., Budapest, 1937-39; registrar Postgrad. Med. Sch., Budapest, 1939-45; faculty U. Med. Sch., Szeged, 1945—, prof. surgery, 1958—, chief Inst. Exptl. Surgery, head 1st dept. surgery, prin., 1958-62. Mem. Internat. Soc. Surgery, Hungarian Soc. Surgeons, Hungarian Physiol. Soc., Austrian Soc. Surgery, others. Author: (with G. Kovacs) Metabolic Aspecs of Pre- and Postoperative Treatment, 1964; also numerous articles. Research on gastro-duodenal ulcer, hemorrhagic shock, operative trauma, hemodynamics in postoperative state. Home: 5 Maros. Office: 4 Pecsi, Szeged, Hungary.*

PETRI, Julius Richard, German bacteriologist; b. Barmen, Germany, 1852; asst. to Koch; became curator Hygiene Mus., Berlin, 1886; mem. Reichgesundheitsamt. Inventor Petri dish for cultivation of bacteria (possibly modification of vessel used by Koch), circa 1887; devised sand filter to separate bacteria from air, circa 1888, also test for kairin in urine and test for proteins. Died 1921.

PETRI, Nicolas, Dutch mathematician; b. Deventer, Netherlands; flourished circa 1567. Author: Arithmetica. Practicque omne cortelycken te lere chijphere . . . Door my Nicolaum petri F. Dauentriensem (work on solutions of spl. cases of quadratics), 1567.

PETRIDES, George Athan, Am. ecologist; b. N.Y.C., Aug. 1, 1916; s. George Athan and Grace Emmeline (Ladd) P.; B.S., George Washington U., 1938; M.S., Cornell U., 1940; Ph.D., Ohio State U., 1948; m. Miriam Clarissa Pasma, Nov. 30, 1940; children—George Henry, Olivia Ladd, Lisa Bonfoey. Naturalist, Nat. Park Service, Washington and Yosemite, Cal., 1938-43, Glacier Nat. Park, Mont., 1947, Mt. McKinley Nat. Park, Alaska, 1959; game technician W. Va. Conservation Commn., Charleston, 1941; instr. wildlife mgmt. Am. U., Washington, 1942-43; asst. leader Ohio Wildlife Research unit, instr. zoology Ohio State U., 1946-48; leader Tex. Coop. Wildlife unit, asso. prof. wildlife mgmt. Tex. A. and M. U., 1948-50; asso. prof. wildlife mgmt. and zoology Mich. State U., East Lansing, 1950-58, prof., 1958—, grantee, Nigeria, 1962. Research prof. U. Pretoria, Republic South Africa, 1965; participant, del. internat. confs. Fulbright research awardee East Africa, Royal Nat. Parks, Kenya, 1953-54, Uganda, 1956-57; N.Y. Zool. Soc. grantee, Ethiopia, Sudan, 1957; Inst. for Radiation Ecology postdoctoral research fellow U. Ga., 1964. Fellow World Acad. Art and Sci.; mem. Am. Ornithologists Union, Am. Soc. Mammalogists, Wildlife Soc. (exec. sec. 1953), Wilderness Soc., Am. Com. Internat. Wildlife Protection, Ecol. Soc., Fauna Preservation Soc., Kenya Wildlife Soc., Sigma Xi. Author: Field Guide to Trees and Shrubs, 1958. Editor wildlife mgmt. terrestrial sect. Biol. Abstracts, 1947—. Research, publs. on grassland ecology particularly African big game, ecol. problems in nat. park and land use planning, age determination in birds and mammals, resource ecology. Home: 4895 Barton Rd., Williamston, Mich. Office: Dept. Fisheries and Wildlife, Mich. State U., East Lansing, Mich.*

PETRIE, Sir (William Matthew) Flinders, English archeologist; b. Charlton, Kent, Eng., June 3, 1853; s. William and Anne (Flinders) P.; ed. by parents; hon. degrees from univs. Oxford (Eng.), 1892, Edinburgh (Scotland), 1896, Strasbourg (France), 1897, Cambridge (Eng.), 1900; m. Hilda Urlin, 1897; 1 son, 1 dau. Excavated ancient sites in Britain, 1875-1880, in Egypt, 1880-24, in Palestine, 1926-38; founder Egyptian Research Account (later British Sch. Archeology in Egypt), 1894; prof. Egyptology, Univ. Coll., London, Eng., 1892-1933, emeritus, 1933—. Fellow Royal Soc., 1902, Brit. Acad., 1904. Author: Inductive Metrology, 1875; Stonehenge, 1880; Pyramids and Temples of Gizeh, 1883; Naukratis I, 1886; Racial Portraits, 1888; Historical Scarabs, 1889; Ten Years' Digging, 1893; History of Egypt, 4 vols., 1894-1927; Decorative Art, 1895; Six Temples at Thebes, 1897; Religion and Conscience in Ancient Egypt, 1898; Royal Tombs of the First Dynasty, 1901; Methods and Aims in Archeology, 1904; Hyksos and Israelite Cities, 1906; Arts and Crafts of Egypt, 1909; Revolutions in Civilization, 1911; Corpus of Prehistoric Pottery, 1918; Prehistoric Egypt, 1920; Status of Jews in Egypt, 1922; Buttons and Design Scarabs, 1925; Descriptive Sociology of Ancient Egypt, 1926; Seventy Years in Archaeology, 1931; Ancient Ghaza, 5 vols., 1931-38; Egyptian Architecture, 1938; Egyptian Science, 1939; Wisdom of the Egyptians, 1940; also numerous reports of individual excavations. A major founder of Egyptology; expert on artifacts and 1st to stress importance of chronology of Egyptian miniature sculptures; most important discoveries were at Memphis, Naucratis, Daphnae, Abydos and Thebes; established enviable precedents in accurate measurement and immediate reporting of field work. Died Jerusalem, Palestine, July 28, 1942.

PETRIE, Robert Methven, astrophysicist; b. St. Andrews, Scotland, May 15, 1906; B.A., U. B.C., Can., 1928; A.M., U. Mich., 1929, Ph.D., 1932; m. Carlotta Joan Hallett (dec. 1959); children—Patricia Joyce (Mrs. Melvin Calkin), Robert Hallett; m. 2d, Jean Knox McDonald, May 9, 1960. Instr., U. Mich., 1930-35; staff Dominion Astrophys. Obs., Victoria, B.C., 1935-66, dominion astrophysicist, 1951-65, dominion astronomer, 1965-66, dir., 1951-66. Mem. senate, bd. govs. U. Victoria, 1963-66. Fellow Royal Soc. Can., Royal Astron. Soc.; mem. Am. Astron. Soc. (past v.p.), Royal Soc. Can. (past sect. pres., Tory medal 1961), Internat. Astron. Union (past v.p.), Am. Soc. Physicists (past pres.), A.A.A.S. (past sect. v.p.). Research, publs. on measurement line-of-sight stellar motions, structure galaxy, calculation orbits, masses, and dimensions double stars from spectroscopic observations; devel. spectroscopic methods estimating distances high-temperature stars. Home: Obs. Hill, Rural Route 7. Office: Dominion Astrophys. Obs., Rural Route 7, Victoria, B.C., Can.*

PETRLE, Miroslav, Czechoslovakian physician; b. Dolni Roven, Czechoslovakia, Jan. 27, 1923; s. Frantisek and Katerina (Schejbalova) P.; Dr., Charles U., Prague, 1950, C.Sc., 1960; m. Drahomira Kudrnkova, Aug. 1, 1950; children—Dagmar, Jana. Lectr. medicine Charles U., 1955; head cardiol. research group Research Inst. Exptl. Therapy, Prague, 1963—; lectr. Postgrad. Med. Sch., Prague. Mem. Med. Soc. J.Ev. Purkyne, Cardiol. Soc. Czechoslovakia. Author: (with J. Rosa) Phoncardiography; (with I. Kosmak, J. Endrys) Dye Dilution Curves; also numerous articles. Research on correlations of hemodynamics and phonocardiography, new methods for measuring end-diastolic vol. of ventricle, regurgitant vol. of ventricle. Home: 306 Unorova ul., Pardubice, Czechoslovakia. Office: 800 Budejovicka, VUET, Prague, Czechoslovakia.*

PETROV, Aleksandr Dmitrievich, Russian organic chemist; b. Aug. 28, 1895; grad. Petrograd U., 1922. Prof. Moscow Chemico-Tech. Inst., 1943—. Recipient Stalin prize, 1947; decorated Order Lenin, Order Red Banner of Labor. Corr. mem. USSR Acads. Scis. Author: The Developmental Course of Organic Synthesis, 1943; the Industry of Organic Aliphatic Compound Synthesis, 1943; Motor Fuel Chemistry, 1953; co-author: The Synthesis of Aromatic Ether Compounds and their Dehydrogenation, 1960. Research in organic synthesis of pure hydrocarbons in motor fuel; established relationship between composition and structure of hydrocarbons and their properties; investigated properties of silicon hydrocarbons. Address: Chemistry Dept., Moscow Chemico-Tech. Inst., Moscow, USSR.

PETROV, Boris Aleksandrovich, Russian surgeon; b. 1898; grad. Med. Faculty, Moscow U., 1922; D.Med. Sci., 1943. Asst., sr. asso. Moscow First Aid Research Inst., 1927-49, dir. for sci. work, chief surgeon, 1949—; prof., 1944—; head chair hosp. surgery 1st Moscow Med. Inst., 1944—. Recipient State prize, 1952. Mem. USSR Acad. Med. Scis., Internat. Soc. Surgeons, Moscow Surg. Soc. (hon.). Author numerous works including: Closed Plaster Cast, 1943; Treatment of Gunshot Wounds of the Knee Joint, 1945; Spinal Anesthesia, 1948; Liberal Skin Grafts in Major Defects, 1950; Cancer of the Large Intestine, 1960. Dep. editor Surgery jour., Surgery sect. Large Med. Ency., 2d edit. Address: Moscow First Aid Research Inst., B. Kolkhoznaya pl. 3, Moscow, USSR.*

PETROV, Boris Nikolaevich, Russian automation specialist; m. Mar. 11, 1913; grad. Moscow Inst. Energetics, 1939. Asso., Inst. Automation and Remote Control, USSR Acad. Scis., 1939-60; tchr. Moscow Aviation Inst., 1944-48; prof., head dept. Inst. Automation, USSR Acad. Scis., 1960—, sec. dept. mechanics and control processes, 1963—. Mem. Commn. for Lenin Prizes, 1964—. Decorated Order Red Banner of Labor, 1963. Corr. mem. USSR Acad. Scis., 1953, academician, 1960. Author: The Construction and Conversion of Structural Systems, 1945; Principles of Invariance and Conditions for its Application in Calculating Linear and Non-Linear Systems, 1960; co-author: Automatic Control of Linear Dimensions of Objects, 1947. Research in approximate integration of differential equations; theory and devel. of automatic control. Address: Moscow Aviation Inst., Moscow, USSR.

PETROV, Georgii Ivanovich, Russian engr.; b. May 31, 1912; grad. Moscow U., 1935. Staff various sci. research insts.; became prof. Moscow U., 1953; mem. bur. dept. tech. sci. USSR Acad. Scis., 1960— Recipient Stalin Prize, 1949. Mem. Acad. Scis. Author: On the Propagation of Oscillations in a Viscous Liquid and the Appearance of Turbulence, 1938. Research and publs. in aeromechanics and gas dynamics, stability of eddy layers, breakdown of laminar flow; proved covergence of Galerkin's method for characteristic value in equations, including non-conservative systems (especially equations of oscillation in a viscous liquid). Home: Prospekt Mira, 73, Moscow, USSR.

PETROV, Ioakim Romanovich, Russian pathophysiologist; b. 1893; grad. mil. med. orderlies' sch., 1912, Petrograd Mil. Med. Acad., 1922; D.Med. Sci. Asst. instr., instr. chair path. physiology Leningrad Mil. Med. Acad., 1922-38, head chair path. physiology, 1939-63, cons. prof., learned council, 1963—; asso. path. lab. 2d Leningrad Med. Inst., 1922-28; head lab. exptl. biology Inst. Hygiene and Occupational Diseases, 1926-36; head chair path. physiology 1st Leningrad Med. Inst., 1938; founder pathophysiology lab. Leningrad Inst. Osteotuberculosis, 1946; founder, head dept. exptl. pathology Leningrad Inst. Blood Transfusion, RSFSR Ministry Health. Decorated Order of Lenin. Mem. USSR Acad. Med. Scis., All-Union Soc. Pathophysiologists (chmn. 1960—), Internat. Blood Transfusion Soc., Leningrad Soc. Pathologists (chmn., bd. mem.). Author: Severe Loss of Blood and Its Treatment with Blood Substitutes, 1945; Shock and Collapse, 1947; Oxygen Starvation of the Brain, 1949; The Role of the Nervous System in Oxygen Starvation, 1952; co-author: Traumatic Shock, Its Pathogenesis, Prevention and Treatment, 1953; Pathological Physiology, 1957; Plasma Substitute Solutions, 1958. Developer blood substitutes, also antishock fluid; evolved method of complex therapy for shock; 1st to demonstrate that inner vascular membranes receive nutrition through walls of vessels. Address: Leningrad Mil. Med. Acad., ulitsa Lebedeva 6, Leningrad 9, USSR.

PETROV, Mikhaíl Platónovich, Russian botanist, geographer; b. Oct. 9, 1906; grad. Leningrad (USSR) U., 1930. Staff, All-Union Plant Inst., 1928-41; dir. Repetekskaia Sand Sta., Kara-Kumy, USSR, 1930-34, Turkmen Exptl. Sta., Kopet-Dag, USSR, 1937-41; dir. Biology Inst. Turkmen br. USSR Acad. Scis., 1941-44, dep. chmn., 1944-46; v.p., chmn. Biol., Agrl. Scis. sect., 1951-57; prof. Leningrad Pedagogical Inst., 1947-51, Turkmen U., Ashkhabad, USSR, 1956—. Author: Root Systems of Plants in the Kara-Kum Sand Desert, their Distribution and Interrelation in Connection with the Ecological Conditions, 1933; Shifting Desert Sands of the USSR and the Campaign Against Them, 1950; Agricultural and Forestry Sand Reclamation in the Deserts and Semi-deserts of the USSR, 1952; Iran, 1955. Specialist in plant life and geol. problems of arid and semi-arid regions. Office: Turkmensky gosudarstvenny universitet, 133 Pervomayskaya, Turkmen Soviet Socialist Republic, Ashkhabad, USSR.

PETROV, Nikolai Nikolaevich, Russian surgeon-oncologist; b. St. Petersburg, Russia, Dec. 14, 1876; grad. St. Petersburg Mil. Med. Acad., 1899, Dr.Med. Sci., 1902. Lectr. surgery Propedeutics Clinic, St. Petersburg Mil. Med. Acad., 1905-08, sr. asst., 1908-12, also lectr. gen. surgery Women's Med. Inst.; prof. Hosp. Surgery Clinic, Warsaw U., 1912-13; prof. surg. clinic St. Petersburg Postgrad. Med. Inst., 1913-14; cons. on Western Front, Red Cross, 1914-17; surgeon Armavir-Taupse R.R. Hosp., 1918-20; cons. surgeon 9th Kuban Red Army, 1920-21; prof. surgery Kuban U., Krasnodar; head surgery dept. Central Red Army Hosp.; prof. Hosp. Surgery Clinic, 1st Leningrad Med. Inst., 1921-26; dir. Leningrad Oncological Inst., 1926-42; cons. Leningrad mil. hosps., 1939-40, hosps. of Leningrad, Front-line Evacuation Center, 1941-42; head chair surgery Leningrad Postgrad. Med. Inst., 1921—; sci. dir. Leningrad Oncological Inst., 1942—; mem. sci. mission Pasteur Inst., Paris, also surg. clinics in Austria, Germany, France, Switzerland, 1903-05; an organizer Kuban U., Krasnodar, 1920-21; founder 1st Soviet Oncological Inst., Leningrad, 1926; founder Lab. Exptl. Cancer, Sukhumi Br. All-Union Inst. Exptl. Medicine (now Med. Biol. Sta., USSR Acad. Med. Sci.), Sukhumi. Named Hon. Sci. Worker of RSFSR, 1935; recipient Stalin Prize, 1942, Mechnikov prize, 1953, Order of Lenin, 3 times, Order of Red Banner of Labor, 2 times. Mem. USSR Acad. Med. Scis., USSR Acad. Scis. (corr.), Pirogov Surg. Soc. (chmn. 1924—), All-Union Soc. Surgeons (mem. bur.), Internat. Surg. Soc. (hon.). Author: General Tumor Theory, 1910; Free Osteoplasty, 1913; The Danger of Cancer and the Campaign Against It, 1930; Malignant Tumors, 2 vols., 1932-34, 3 vols., 1947-52; The Treatment of War Wounds, 1939; A Brief Outline of the Comparative Pathology of Tumors in Animals and Man, 1941. Editor, co-author: Gastric and Duodenal Ulcers and their Surgical Treatment, 1941. Research and publs. on origin, prophylaxis and treatment of malignant tumors and stomach and duodenal ulcers. Home: 5-Aya Cheremushkinskaya 3, Moscow. Office: Acad. Med. Scis., Solyanka 14, Moscow, USSR.

PETROVA, Maria Konstaninovna, Russian physiologist; b. 1874; Dr. Med. Sci., St. Petersburg Med. Inst. for Women; student and colleague of Ivan Pavlov; dir. lab. Pavlov Physiology Inst., Leningrad; recipient Order of Red Banner of Labor for service in field of physiology, Order of Lenin, 1945, Stalin prize first class, 1949. Research on conditioned reflex in animals, physiology of war neuroses and effects of air raids on nervous system; established bromides as curative for various neuroses; current research on nervous system disorders in relation to cancer, also nervous system in relation to old age. Address: Ivan Pavlov St., Leningrad, USSR.

PETROVIC, Alexandre Gabriel, physician, biologist; b. Belgrade, Yugoslavia, July 10, 1925; s. Gabriel and Maria (Miskovic) P.; M.D., Strasbourg (France) U., 1954; D.Sc., Faculté des Sciences de Strasbourg, 1961; certificate hematology Nat. Bd. France, 1957; m. Suzanne Durry, Feb. 25, 1956. Monitor histology Med. Sch., Strasbourg, 1950-54; research fellow Inst. Nat. d'Hygiène, Paris, France, 1953-61, Med. Research Postdoctoral fellow McGill U., Montreal, Que., Can., 1961-62; head research Nat.Inst. Health and Med. Research, Paris 1963—; chargé de cours Faculté des Scis., Strasbourg, 1966—; dir. tissue culture lab., asso. staff physician Chgo. Wesley Meml. Hosp., Chgo., 1964—; asst. prof. otolaryngology, lectr. anatomy Northwestern U. Med. Sch., Chgo., 1965—. Recipient Prize Vlès, Strasbourg U. Med. Sch., 1954; Prize Laborde, Biol. Soc., Paris, 1960. Mem. Soc. for Cryobiology (charter), Internat., Am. socs. for cell biology, European Tissue Culture Club, Tissue Culture Assn. Contbg. author: Physiologie, 1968; also numerous articles. Developed new method of organ culture in liquid medium; electron microscopic studies on cultures and homografts of endocrine glands, hypothalamus of guinea pig, effect of sodium fluoride on normal and otosclerotic bone, lack of rat mandibular condylar cartilage independent growth potential, hibernation. Home: 42, Boulevard d'Anvers, 67 Strasbourg, France. Office: 251 E. Chgo. Av., Chgo. 60611.*

PETROVITCH, Michel, Yugoslavian mathematician; b. Belgrade, May 7, 1868; s. Nikodic and Militza Petrovitch; Ph.D. in Math., U. Paris, 1894; became prof. math. U. Belgrade, 1895; editorial writer Politika (daily newspaper), Belgrade. Author: La mécanique des phénomènes, 1906; Éléments de phénoménologie mathématique; Mécanismes communs aux phénomènes disparates, 1921; Théorie des spectres mathématiques, 1928; also many articles on math. analysis, mechanics, geometry, math. analysis.

PETROVSKY, Boris Vasilevich, Russian surgeon; b. Yessentuki (now Stavropol Kray), 1908; grad. Med. Faculty, 1st Moscow U., 1930; D.Med. Sci., 1947. Asso., Moscow Oncological Inst., 1932-45; dep. dir. Inst. Exptl. and Clin. Surgery, USSR Acad. Med. Scis., 1945-47; lectr. dept. gen. surgery 1st Moscow Med. Inst., 1945-47; prof. chair hosp. surgery Budapest U., 1949-51; prof., 1949—; head chair faculty surgery Pediatric Faculty, 2d Moscow Med. Inst., 1951-56; head Martynov chair hosp. surgery Sechenov 1st Moscow Med. Inst., 1956—; chief surgeon 4th main bd USSR Ministry Health, until 1965, USSR minister health, 1965—. Del., 16th Internat. Congress Surgeons, Copenhagen, 1955, 18th, Munich, 1959. Decorated Order of Lenin; recipient Burdenko prize, 1950, Lenin prize, 1960. Mem. USSR Acad. Med. Scis., All-Russian (chmn. 1964—), All-Union (dep. chmn.), Polish (hon.) socs. surgeons, Hungarian Surg. Soc. (hon. chmn. 1950—), European Soc. Cardiovascular Surgery (v.p. 1959—). Author: Surgical Treatment of Vascular Lesions, 1949: Surgical Treatment of Cancer of the Esophagus and Cardia, 1950; Blood Transfusion in Surgery, 1954; Surgery of Mediastinum, 1960; co-author: Clinical Aspects and Surgical Treatment of Thyrotoxic Goiter, 1961. Editor: Surgery; mem. editorial bd. New Surg. Archives; co-editor Surgery sect. Large Med. Ency., 2d edit. Research and numerous publs. on oncology, blood transfusion, vascular surgery, surgery of thoracic organs, surg. treatment of congenital and acquired heart defects; 1st in USSR to operate on cancer of esophagus; developer new plastic operations including replacement of affected esophagus section in cases of cardiospasm by diaphragmal pedicle graft; developer new method of blood tranfusion into aorta for severe shock cases; devised modifications of operations on heart and major vessels including replacement of injured heart wall by diaphragmal graft in aneurysm of heart; one of 1st Russian surgeons to replace major arteries with plastic tubes. Address: USSR Ministry of Health, Rakhmanovsky p. 3, Moscow, USSR.*

PETROVSKY, Ivan Georgievich, Russian mathematician; b. Sevsk (now Bryansk Oblast), 1901; grad. Moscow U., 1927, postgrad., 1927-30; D.Physico-Math. Sci., 1935. Prof., Moscow U., dean mech. and math. faculty, 1940-44, rector, 1951—; prof., chair higher math. Moscow Machine-Bldg. Evening Inst., Peoples Commissariat of Heavy Machine-Bldg., 1930-41; sr. asso. USSR Acad. Scis., 1943-47, dep. dir. Math. Inst., 1947-49, acad. sec. dept. physico-math. sci., 1949-51, mem. editorial and publs. council, 1955—. Decorated Order of Lenin, 1961; recipient State prize, 1946. Mem. USSR Acad. Scis. (Presidium mem. 1953—). Author: Lectures on the Theory of Ordinary Differential Equations, 1939; Lectures on the Theory of Integral Equations, 1948; Lectures on Partial Differential Equations, 1950. Chief editor Math. Symposium, 1950—. Research and publs. on theory of partial differential equations, probability theory, qualitative theory of ordinary differential equations. Address: Moscow University, Leninskie gory, Moscow, USSR.

PETRUNKEVITCH, Alexander, zoologist; b. Pliski, Ukraine, Dec. 22, 1875; student U. Moscow, 1894-98; Ph.D., U. Freiburg, 1900; D.Sc., (hon.), U.

P.R., 1926, Ind. U., 1951; m.; 2 children. Pvt. docent U. Freiburg, 1901-03; lectr. Harvard, 1903-04; acting prof. zoology Ind. U., 1906; mem. expdn. to southern Mexico Am. Mus. Natural History, 1907, hon, curator Arachnida, 1909-11; prof. zoology Yale, 1917-44, emeritus, from 1944. Vis. prof. U. P.R., 1925-26. Mem. Nat. Acad. Scis., A.A.A.S., Entomol. Soc., Soc. Systematic Zoology, Soc. Naturalists, Soc. Zoology, Entomol. Soc. Belgium (hon.). Author: Free Will, 1905; Morphology of Invertebrate Types, 1916; An Inquiry Into the Natural Classification of Spiders, 1933; A Study of Paleozoic Arachnida, 1949. Research on amber spiders, anatomy of spiders, fossil Arachnids; compiled synonymic catalog of Am. spiders; conducted studies in cytology, physiology, exptl. zoology, paleontology, microphotography, morphology of invertebrates, heredity, psychology. Died New Haven, Conn., Mar. 9, 1964.

PETRUS, Alphonsus (Moses Sefordi), physician; b. 1062; physician to Alphonso VI, Spain; Author: Dialogi cum judaeo; Disciplina clericalis. Noted for treatises on astronomy, geography and medicine. Died 1110.

PETRUS APONENSIS, see D'Abano, Pietro.

PETRY, Gerhard, German surgeon; b. Ludwigshafen/Rhine, Germany, July 31, 1913; s. Ernst and Augusta (Neumayer) P.; M.D., U. Fribourg/Brisgau; m. Angelica May Cellista. Agrégé in anatomy, 1944; med. asst. intern and surgery div. Heidelberg Gen. Hosp.; instr. anatomy U. Fribourg; asso. prof. anatomy U. Marbourg (Germany), 1955-62, prof., 1962—; dir. Anatomy Inst. Marbourg/Lahn, 1963—. Publns. related to structure of ovaries, fibers, embryonic cells, alloplastic transplants of vessels, electronic microscopy. Research in epithelial domains. Home: Rentjof 30. Office: Robert Kochstrasse 6, Marburg/Lahn, West Germany.

PETRYANOV-SOKOLOV, Igór Vasilèvich, Russian phys. chemist; b. June 18, 1907; grad. Moscow U., 1930. Mem. staff Karpov Physico-Chem. Inst. USSR Acad. Scis., 1930—; prof. Moscow Chemico-Tech. Inst., 1947—. Recipient Stalin prize, 1941; decorated Order of Lenin. Corr. mem. USSR Acad. Scis., 1953. Co-author: Determining the Size and Charge of Particles in Fog; Formation of Aerosols During Condensation of Supersaturated Vapors; Edge Angles of Small Drops. Devel. method of studying aerosols with liquid dispersion phase (i.e. fog); studied occurrence of charges in fogs and effects of these charges on their stability. Address: Karpov Physicochem. Inst., Leninsky pr. 31, Moscow, USSR.

PETRYSHYN, Walter Volodymyr, mathematician; b. Lviv, Ukraine, Jan. 22, 1929; s. Vasyl and Maria (Pasirska) P.; B.A., Columbia, 1953, M.S., 1954, Ph.D., 1961; m. Arcadia Olenska, Sept. 2, 1956. Came to U.S., 1950, naturalized, 1956. Instr. math. Notre Dame Coll. S.I., N.Y., 1954-56; lectr. math. Columbia, 1956-59; lectr. math. Coll. City N.Y., 1959-61; mem. Courant Inst. N.Y. U., 1961-64; faculty U. Chgo., 1964—, asso. prof. math., 1966—; vis. asst. prof. Cal. Inst. Tech., 1965. Contbr. articles to sci. jours. Developed an abstract solvability theory and approximation theory of iterative and projectional methods for solution of K-p.d. and Non-K-p.d. linear and nonlinear operator equations in Hilbert spaces; proved a fixed point theorem for a gen. new class of nonlinear P-compact operators in Banach spaces which at same time includes number of important theorems for completely continuous, quasicompact and monotone operators; studied iterative methods for constrn. of fixed points of nonlinear mapping of contractive type; developed theory of projectionally solvable and approximation-solvable nonlinear operators. Home: 1401 E. 55th St., Chgo. 60637.*

PETSCHEK, Harry Ewald, physicist; b. Prague, Czechoslovakia, Sept. 12, 1930; s. Hans and Eva (Epler) P.; came to U.S., 1938, naturalized, 1945; B.Engring. Physics, Cornell U., 1952, Ph.D., 1955; m. Barbara Scaffidi, Nov. 7, 1959; children—Dell Irene, Bruce Irving, Kim Diane, Philip Edward. Faculty, Princeton, 1955-56; prin. research scientist Avco Everett Research Lab., Everett, Mass., 1956—. Mem. subcom. on fluid mechanics NASA, 1958-65. Mem. Am. Phys. Soc. (exec. com. plasma physics div. 1966), Am. Geophys. Union. Asso. editor Physics of Fluids, 1965-67. Ionization and radiation studies in high temperature shock tubes; magneto-hydro-dynamic flow calculations; dissipation by turbulence in collision-free plasmas; shock wave structure in plasmas; plasma propulsion; interaction of solar wind with magnetic field of earth and magneto-sphere tail; study of loss of trapped particles from earth's magnetic field. Home: 17 Preston Rd., Lexington, Mass. 02173. Office: 2385 Revere Beach Pkwy., Everett, Mass. 02149.*

PETTENKOFER, Max Joseph, see von Pettenkofer, Max Joseph.

PETTERSSEN, Sverre, meteorologist; b. Hadsel, Norway, Feb. 19, 1898; s. Edward Hildor and Petronella (Petersen) P.; B.S., Oslo, U., 1924, M.S., 1926, Ph.D., 1933; m. Daisy Bonner, Aug. 16, 1925; 1 dau., Eileen (Mrs. Anton Oerbeck); m. 2d, Lilian Bye, Sept. 25, 1941; 1 son, Bernt; m. 3d, Grace Beverly, Mar. 29, 1946. Came to U.S., 1948. Meteorologist Norwegian Weather Service, 1924-31, regional dir., 1931-39; vis. prof. Cal. Inst. Tech., 1935; prof., chmn. meteorol. dept. Mass. Inst. Tech., 1939-42; adviser Meteorol. Office Brit. Air Ministry, 1942-45; chief Norwegian Forecasting Service, 1945-48; adviser Dir. Gen. Observations, India, 1948; dir. sci. services U.S.A.F. Weather Service, Washington, 1948-52; prof. meteorology U. Chgo., 1952-63, chmn. dept., 1959-61, chmn. dept. geophys. scis., 1961-63. Decorated Comdr. Order Brit. Empire; Comdr. Order St. Olaf, Liberation Cross (Norway); recipient The Buys Ballot's Gold medal (Netherland Acad. Sci.), 1948; Letter commendation Gen. Dwight Eisenhower, 1944; Distinguished Service award US AF, 1953; N.Y. Bd. Trade Gold award, 1958; U. Helsinki Medal, 1958; Charles Franklin Brooks award, 1962; 10th ann. prize award, Internat. Meteorol. Orgn., 1965. Pres. Internat. Commn. Maritime Meteorology, 1939-46, Internat. Commn. Aerology, 1946-51; dir. Nat. Acad. Sci. Task Force for 10 yr. plans atmospheric scis., 1961—, chmn. com. internat. programs atmospheric scis. and hydrology, 1962-63; mem. panel of atmospheric scis., Pres.'s Sci. Adv. Com., 1962—. Mem. Norwegian, Finland (fgn.) acads. sci., Acad. Sci., Ill. Acad. Sci. and Letters, Peruvian, Royal (hon.), Am. (pres. 1958-60) meteorol. socs., Am. Geophys. Union, Acad. Polit. Scis., A.A.A.S., Nat. Geog. Soc. Author: Weather Analysis and Forecasting, 1940, 56; Introduction to Meteorology, 1941, 58. Co-author: The Meteorology of the Arctic, 1956. Home: 5700 Blackstone Av., Chgo. 60637.*

PETTERSSON, Sven Otto, Swedish chemist; b. 1848; prof. chemistry U. Stockholm (Sweden), 1881-1908; founder Internat. Council for Sea Investigations. One of 1st chemists to support Arrhenius' views of electrolytic dissociation; established (with Nilson) correct atomic weight of Beryllium, also studied titanium, germanium; gave exptl. proof of Rasult's law of depression of freezing point of solutions by finding value for ethylene bibromide had been correctly predicted; developed Pettersson-Nansen bottle for water sampling, also method for determining dissolved gases in sea water, various devices for ocean exploration. Died Göteborg, Sweden, Jan. 6, 1941.

PETTIGREW, James Bell, Scottish physician; b. Roxhill, Scotland, May 26, 1834; s. Robert and Mary (Bell) P.; ed. Glasgow U., 1850-55, LL.D., 1883; M.D., Edinburgh U., 1861; m. Elsie Gray, 1890. Croonian lectr. Royal Soc.; asst. curator Hunterian Mus., London, 1862-68; curator Mus. Royal Coll. Surgeons, Edinburgh, 1869, lectr. physiology, 1873; examiner in physiology Royal Coll. Physicians and Surgeons Edinburgh; Chandos prof. medicine and anatomy, 1875; examiner in anatomy U. Glasgow, 1883-87. Recipient Godard prize French Acad. Scis., 1874; named laureate Inst. France. Fellow Royal Soc., 1869, Royal Coll. Physicians; mem. Royal Med. Soc. (pres. 1860). Author: Arrangement of Muscular Fibres in Heart, and Bladder, 1864, 66; Structure and Function of Valves of Vascular System, 1864; Presumption of Survivorship, 1865; Mechanism of Flight, 1867; Physiology of Wings, 1870; Plants, Animals and Inorganic Matter, 1873; Animal Locomotion or Walking, Swimming, and Flying, with a Dissertation on Aeronautics, 1873; The Physiology of Circulation in Plants, in the Lower Animals and in Man, 1874; Flight: Natural and Artificial, 1879; Man's Place in Nature, 1882; The Phonograph or Speech Recorder, 1882; Crystals, Dendrites, and Spirals, in relation to growth and movements, especially rhythmic movements, 1901; Anatomical Preparation making at Edinburgh University and Royal College of Surgeons of England, 1901; Spiral Formations in Relation to Walking, Swimming and Flying, 1904; Design in Nature, 3 vols., 1908. Died St. Andrews, Jan. 30, 1908.

PETTIGREW, Thomas Fraser, Am. social psychologist; b. Richmond, Va., Mar. 14, 1931; s. Joseph Crane and Janet (Gibbs) P.; A.B. in Psychology, U. Va., 1952; M.A. in Social Psychology, Harvard, 1955, Ph.D., 1956; m. Ann Hallman, Feb. 25, 1956; 1 son, Mark Fraser. Research asso. Inst. Social Research, U. Natal, Republic of S. Africa, 1956; asst. prof. psychology U. N.C., 1956-57; asst. prof. social psychology Harvard, 1957-62, lectr., 1962-64, asso. prof., 1964-68, prof., 1968—. Chmn. Episcopal presiding Bishop's Adv. Com. Race Relations, 1961-63; v.p. Episcopal Soc. Cultural and Racial Unity, 1962-63; mem. Mass. Gov.'s Adv. Com. Civil Rights, 1962-64; cons. U. S. Commn. Civil Rights, 1966—. Fellow Am. Psychol. Assn., Am. Sociol. Assn.; mem. Soc. Psychol. Study Social Issues (council 1962-66, pres. 1967-68). Author: (with E. Q. Campbell) Christians in Racial Crisis: A Study of the Little Rock Ministry, 1959; A Profile of the Negro American, 1964. Mem. editorial bd. Jour. Social Issues, 1959-64; asso. editor Am. Sociol. Rev., 1963-65. Contbr. articles profl. publs. Applied social psychological principles to understanding of race relations in the U.S. and the Republic of South Africa with particular emphasis upon the processes of conformity and social comparison. Home: 5 Follen St., Cambridge, Mass. 02138.*

PETTIJOHN, Francis John, Am. geologist; b. Waterford, Wis., June 20, 1904; s. John J. and Elizabeth (Schenkenberg) P.; B.A., U. Minn., 1924, M.A., 1925, Ph.D., 1930; m. Dorothy M. Bracken, Aug. 9, 1930; children—Norma (Mrs. Alan Friedemann), Clare (Mrs. J. Glazer), Loren E. Instr., Oberlin Coll., 1925-29; faculty U. Chgo., 1929-52, prof., 1945-52; prof. geology Johns Hopkins, Balt., 1952—, chmn. dept. geology, 1963—; geologist U.S. Geol. Survey, Washington, 1943-53. Cons. Shell Devel. Co., 1954-64. Fellow Nat. Acad. Sci., Am. Acad. Arts and Sci.; mem. Soc. Econ. Paleontology and Mineralogy (past pres., hon. mem.), Geol. Soc. Finland (corr.), Geol. Soc. Am., Am. Assn. Petroleum Geologists, A.A.A.S., Geologische Verein., Sigma Xi. Author: (with W.C. Krumbein) Manual Sedimentary Petrography, 1938; Sedimentary Rocks, 1957; (with P.E. Potter) Paleocurrents and Basin Analysis, 1963, Atlas and Glossary of Primary Sedimentary Structures, 1964; also articles. Editor: Jour. Geology, 1947-52. Research on earliest Precambrian sedimentary rocks with spl. reference to conditions on earth in earliest times; discovered Precambrian glacial beds in Mich. Home: 512 Woodbine Av., Towson, Md. 21204. Office: Johns Hopkins, Balt. 21218.*

PETTINGILL, Olin Sewall, Jr., Am. zoologist; b. Belgrade, Me., Oct. 30, 1907; s. Olin Sewall and Marion Bradbury (Groves) P.; A.B., Bowdoin Coll., 1930, Doctor of Science (honorary), 1956; Ph.D., Cornell University, 1933; student University of Michigan Biological Sta., Summer, 1928; m. Eleanor Rice, Dec. 31, 1932; children—Polly-Ann, Mary-Ann. Instructor ornithology, Ithaca, N.Y. Night Sch., 1933; teaching fellow in biology, Bowdoin Coll., 1933-34; instr. ornithol., N.H. Nature Camp at Lost River, summers, 1934, 1936; instr. biology, Westbrook Jr. Coll., Portland, Me., 1935-36; mem. faculty, U. of Mich. Biological Station, summers 1938-45, 47-57, 59—; director Lab. of Ornithology, Cornell U., 1960—; faculty Carleton Coll., 1936-54, associate prof. zoology 1946-54; resigned 1954; leader expdn. Falkland Islands, Disney Prodns., 1953-54; lectr. Audubon Screen Tours, 1943—. Research Asso. Cranbook Inst. Sci., 1940-45; mem., dir. or leader ornithol. expdns. U. S., Can., Iceland and Mexico for various groups, 1929—; del. 12th Internat. Ornithol. Congress, Helsinki, Finland, 1958, 14th, Oxford, Eng., 1966. Member Wilson Ornithol. Society (past pres. and sec.), Nat. (director 1955-66, secretary 1957-59, 63-66), Maine (president 1959-60), Audubon societies, American Ornithologists Union (sec. 1946-51); member numerous profl. assns. and orgns. in field, Sigma Xi. Author ornithol. works including: The Am. Woodcock, 1936; Bird Life of the Grand Manan Archipelago, 1939; A Laboratory and Field Manual of Ornithology, 1946 (rev. 1956); A Guide to Bird Finding East of the Mississippi, 1951; A Guide to Bird Finding West of the Mississippi, 1953; author-photog., 9 motion pictures on birds (for Coronet Instructional Films), 1941, 3 for Walt Disney Prodns., 1951. Co-author: Birds of the Black Hills, 1965. Editor sect. on Aves on Biol. Abstracts, 1942-53, The Bird Watcher's America, 1965; rev. editor Wilson Bull., 1959—; columnist, contbg. editor Audubon Mag., 1957—; adv. editor Nature Books, McGraw-Hill Book Co., 1963—. Research on the distribution, life histories, and ecology of birds. Office: Lab. Ornithology, Cornell U., Ithaca, N.Y. 14850.

PETTIT, Edison, Am. astronomer; b. Peru, Neb., Sept. 22, 1890; B.Ed., Neb. Normal Coll., Peru, 1911; Ph.D., U. Chgo., 1920; LL.D., Carthage Coll., 1935; m., 1918; 2 children. High sch. tchr., Neb., 1911-14; tchr. astronomy and physics Washburn Coll., 1914-15; astronomy, 1915-18; asst. Yerkes Obs., Chgo., 1918-20; astronomer Mt. Wilson Obs., from 1920. Mem. eclipse expdns., Matheson, Colo., 1918, Point Loma, 1923, Middletown, 1925, Honey Lake, 1930, Lancaster, N.H., 1932; civilian AEC; with OSRD, 1944. Mem. Astron. Soc., Optical Soc., Internat. Astron. Union. Author: Forms and Motions of the Solar Prominences. Discovered laws governing movement of solar prominences; devised interference polarizing monochrometer, also a thermocouple for registering very small temperatures; other research on solar stars. Address: Mt. Wilson Obs., Pasadena 4, Cal.

PETTIT, George Robert, Am. chemist; b. Long Branch, N.J., June 8, 1929; s. George Robert and Florence (Seymour) P.; B.S., Wash. State U., 1952; M.S., Wayne State U., 1954, Ph.D., 1956; m. M. Jean Benger, June 20, 1953; children—William Edward, Margaret Sharon, Robin Kathleen, Lynn Benger, George Robert. Research fellow Wayne State U., 1954-56; sr. research chemist Norwich Pharmacal Co., 1956-57; faculty U. Me., 1957-65, prof. chemistry, 1965; vis. prof. chemistry Stanford, 1965; prof. chemistry Ariz. State U., Tempe, 1965—, chmn. organic div., 1966—. Cons. Nat. Cancer Inst., NIH. Mem. Am. Chem. Soc., Chem. Soc. (London), Sigma Xi, Phi Lambda Upsilon. Contbr. chpt. (with E. Van Tamelen) to Organic Reactions, 1962. Research, numerous publs. on design and synthesis of organic compounds for possible use in treatment of cancer, chemistry of natural products, in gen., organic chemistry. Office: Dept. Chemistry, Ariz. State U., Tempe, Ariz. 85281.*

PETTIT, Harvey Pierson, Am. mathematician; b. Chadron, Neb., Apr. 25, 1893; s. Lewie Durmont and Olive (Martin) P.; B.A., Kalamazoo Coll., 1914; M.A., U. Ky., 1919; Ph.D., U. Ill., 1922; m. Marcia Foster, Oct. 10, 1914; children—Marshall, Marcia Doris. High sch. math. tchr., Holland, Mich.,

1914-28; faculty U. Ill., Urbana, 1922-23; prof. math., head dept. Ill. Wesleyan U., Bloomington, 1923-26; prof. math. Marquette U., Milw., 1926—, chmn. dept. math., 1928-58. Fellow A.A.A.S.; mem. Am. Math. Soc., Math. Assn. Am., Phi Beta Kappa, Sigma Xi. Author: (with Luteyn) College Algebra, 1932, Analytic Geometry, 1942. Research on projective and algebraic geometry, gen. cylide, particularly quintic cylide, projective description some higher plane curves. Died 1966.

PETTY, Sir William, English political economist, statistician, physician; b. Romsey, Hampshire, Eng., May 26, 1623; went to sea as boy; abandoned on French coast, age 15; studied at Jesuit college, Caen, France; returned to Eng., entered Royal Navy; returned to continent, 1643; studied at Utrecht and Amsterdam, medicine at Leyden, 1644; traveled to Paris; returned Eng., 1646; became clothier at Romsey; Dr. Physics, U. Oxford, 1649 (Fellow of Brasenose Coll.). Prof. anatomy, Oxford, 1651; also organized informal sci. club there with John Wilkins and others; physician general to English army in Ireland, 1652 (reorganized medical services); commissioned to survey and map whole of Ireland (map completed 1673). Original mem. Royal Soc. Knighted, 1662. Author: History of the Down Survey (history of the survey of Ireland), 1851; A Treatise of Taxes and Contributions, 1662; Natural and Political Observations . . . made upon the Bills of Mortality of the City of London, 1662; The Political Anatomy of Ireland, 1672; Political Arithmetic concerning the growth of the city of London, 1683-89; Observations upon the Cities of London and Rome, 1687. Devised new kind of land carriage, a double bottomed ship, other practical inventions; held land and labor (not precious metals) constitute true wealth of nations; studied vital statistics of population of London (1st such work ever undertaken); studied population, agriculture, industry and commerce of Eng. and Ireland. Died London, Eng., Dec. 16, 1687.

PETTYJOHN, Elmore S(haw), Am. chem. engr.; b. Alma, Mich., Apr. 12, 1897; s. Elmore Sloan and Ada Ernst (Lozier) P.; A.B., U. Mich., 1918, B.S., M.S. in Chem. Engring. 1922, Chem. E., 1930; m. Isabel Bender Nairn, June 25, 1922; 1 dau., Nairn. Research engr. U. Mich., 1927-33, asst. prof. gas engring., 1927-30, asso. prof., 1930-33; sales and devel. engr. Leader Industry, Inc., Decatur, Ill., 1933-36; chief engr. Mervin Bldg. Corp., St. Charles, Ill., 1936-37; asst. prof. chem. engring. U. Mich., 1937-38, asso. prof., 1938-45 (on leave, 1941-45); dir. Inst. Gas Tech., 1945-55, v.p., 1951-55; pres. Gas Assos., 1956-57; research prof. chem. engring. Ill. Inst. Tech., 1946-51; prof. chem. engring. Fenn Coll., 1958-65, Cleve. State U., 1965—. Commd. ensign, USNRF, 1918, advanced through grades to capt., 1944; on active duty, Naval Overseas Transport Serv., 1918; Navy Liaison Selective Service, Mich., 1940-41; gunnery officer, U.S.S. Am. Legion, 1941-42; officer in charge amphor boat pool, Tulagi (Solomons and Noumea (New Caledonia), 1943, also spl. fuels sect., Research and Standards Br., Bur. of Ships, 1944-45; also oil sect., U.S. Naval Tech. Mission, Europe, 1945; ret. 1947. Awarded Navy and Marine Corps Medal, Victory medals (both World Wars), Am. Def. with star, Navy Reserve medal with 2 stars, and others. Registered profl. engr., Mich. Fellow A.A.A.S.; member Am. Assn. Cost Engrs. (dir. N.E. Ohio sect. 1964-65), Am. Gas Assn., Am. Inst. Chem. Engrs., Am. Soc. Engring. Edn., Am. Chem. Soc., Sigma Xi, Tau Beta Pi, Phi Lambda Upsilon, Iota Alpha, Phi Delta Theta. Research on distillation, evaporation, sedimentation, crystallization, manufactured and natural gas prodn., transmission, distbn. and utilization. Home: 3440 Avalon Rd., Shaker Heights 44120.*

PETUKHOV, Valentín Afanásievich, Russian physicist; b. 1907; grad. Leningrad Poly., 1934. With Kharkov Physico-Tech. Inst., 1934-54; dep. dir. Lab. High Energy, United Inst. Nuclear Research, 1954—; dr. physico-math. sci., also prof. Recipient Lenin prize for participation in creation of synchrocyclotron of 10 billion electronic volts. Author: A 680-Mev. Synchrotron, 1960. Research on nuclear physics, accelerations, also cosmogony. Address: Obedinenny Inst. yadernykh issledovaniy, Moskovskaya oblast, Dubna, A USSR.

PEUS, Fritz Ferdinand Christian, German zoologist; b. Siegen/Westphalia, Apr. 22, 1904; s. Hugo and Maria (Schlun) P.; Ph.D., univs. Münster and Rostock; m. Monika Jekelous, Mar. 14, 1942; children—Katharina, Stephen, Hans Michael. Collaborator, mem., sect. chief, later prof. Nat. Inst. Prussia, 1927-45, then prof. State Inst. of Wasser-Bodenund-Lufthgiene, Berlin-Dahlem; curator, head dept. zoology Museum of U. Humboldt, Berlin, 1947-61, dir., prof., 1958; prof., dir. Zoology Inst. Free U., Berlin, 1962—. Mem. German Acad. Sci., Leopoldina. Publns. on entomology, ecology, ornithology. Home: Am. Fischtal 2, Berlin 27. Office: Königin-Luise Strasse 1-3, Berlin 33, West Germany.

PEVSNER, Aihud, physicist; b. Haifa, Israel, Dec. 18, 1925; s. Yoshua and Esther (Benyeshaia) P.; came to U.S., 1928, naturalized, 1934; B.A., Columbia, 1947, A.M. in Math., 1958, Ph.D., in Physics, 1954; m. Lucille Wolf, June 19, 1949; children—Mark D., Laura R., Jonathan A. Instr., Mass. Inst. Tech., 1953-56; faculty Johns Hopkins, Balt., 1956—, prof. physics, 1963—. Guggenheim fellow, Sr. Fulbright fellow, 1963-64. Mem. Am. Phys. Soc. Contbr. articles to tech. jours. Research, publs. on properties of fundamental particles, using counter, nuclear emulsion, cloud chamber and bubble chamber techniques. Home: 5806 Stuart Av., Balt. 21215.*

PÉWÉ, Troy Lewis, Am. geologist; b. Rock Island, Ill., June 28, 1918; s. Richard and Olga (Pomrank) P.; A.B., Augustana Coll., 1940; M.S., U. Ia., 1942; Ph.D., Stanford, 1952; m. Mary Jean Péwé, Dec. 21, 1944; children—David Lee, Richard Hill, Elizabeth A. Instr., head dept. geology Augustana Coll., Rock Island, 1942-46; instr. geomorphology Stanford, 1946; geologist Alaskan Geol. br. U.S. Geol. Survey, 1946-58, staff geologist engring. geology, 1953-60, part time, 1958—; asso. prof. geology U. Alaska, 1953-58, head dept. geology, 1958-65; chmn. dept. geology, Ariz. State U., Tempe, 1965—. Chief glacial geologist USMC IGY, Antarctica, 1957-58; vis. lectr. Recipient Antarctic medal IGY. Fellow A.A.A.S. (exec. sec. 1954, pres. 1956 Alaska div.), Geol. Soc. Am. Arctic Acad. Sci.; mem. Assn. Geology Tchrs., Glaciological Soc., New Zealand Antarctic Soc., Am. Soc. Engring. Geologist, Internat. Geog. Union. Research, publs. on permafrost and ice wedges in polar areas, establishment of basic glacial chronology Antarctic. Office: Dept. Geology, Ariz. State U., Tempe.*

PEYER, Jean Conrad, Swiss anatomist; b. Schaffhausen, Switzerland, Dec. 26, 1653; ed. Paris, Basel; prof., Schaffhausen. Author: Exercitatio anatomico-medica de glandulis intestinorum, 1677; Methodus historiarum anatomico-medicarum, 1678. Studied history of illnesses; prepared way for studies in path. anatomy; noted raised areas of lymph nodules in mucous membrane of small intestine (aggregate follicles, agminated nodules, Peyer's patches or nodules), 1673. Died Schaffhausen, Feb. 29, 1712.

PEYNAUD, Emile, French biochemist; b. Bordeaux, France, June 29, 1912; s. Hermann and Alixine (Ferrier) P.; Diplôme d'études supérieures des sciences physique; Ingénieur; Docteur, Faculté des Sciences, Bordeaux; m. Yvonne Jameau, Sept. 3, 1938; children—Jean-Pierre, Daniele. Work in industry, first in collaboration with Dr. Jean Ribéreau, later as tech. dir., 1928-49; research chief Talence Agronomic and Cenological Sta., 1949—; under dir. Talence Inst. Oenology. Cons. engr. in winemaking industry. Decorated Order Agr. Merit, Palms Acad. Author: Analyses et contrôle des vins, 1947; Traité d'oenology, 1960, 61; also numerous articles. Research on biochemistry of fruit maturation and alcohol fermentation; quantitative analysis of plant constituents. Home: 26 ave. de Lattre de Tassigny. Office: 351 course de la libération, Talence (Gironde), France.*

PEYRILHE, Bernard, French physician; b. Perpignan, France, 1735; M.D., Acad. Surgery, Paris, 1769; became prof. materia medica l'École de Médecine, Paris, 1794. Author: (with Dujardin) Histoire de la chirurgie, 2 vols., 1760-74; Tableau d'historie naturelle des medicaments, 1800. Injected material from human breast cancer into dog in attempt to transmit cancer, circa 1774. Died 1804.

PEYSSONEL, Jean André, zoologist, botanist; b. Marseilles, France, June 19, 1694; practiced medicine, Marseilles; royal med. botanist, Guadeloupe. Mem. French Acad. Scis., 1723. One of 1st to recognize corals are animals rather than plants. Died Guadeloupe, Dec. 23, 1759.

PEYTON, Floyd Avery, Am. chemist; b. Charlestown, Ind., Feb. 2, 1905; s. James Avery and Birde (Gray) P.; A.B., Ind. U., 1928; M.S., Mich. Coll. Mining and Tech., 1929; D.Sc., U. Mich., 1933; m. Beatrice Stodden, Jan. 1, 1934; children—Floyd Terry, Keith Stodden. Research chemist Ames Dental Mfg. Co., Fremont, O., 1933-35; faculty U. Mich., 1935-45, asst. prof., 1941-45; prof. U. Tex. Dental Sch., 1945-48; prof. dentistry, head materials dept. U. Mich. Sch. Dentistry, Ann Arbor, 1948—. Cons. to orgns., govt. agys.; Fulbright lectr. in dentistry UAR, 1961, New Zealand, 1966. Recipient Distinguished Faculty Achievement award U. Mich., 1962; Acad. Plastics award in dentistry, 1959. Hon. mem. Am., Mich. State dental assns., Acad. Crown and Bridge Prosthesis; mem. Am. Assn. U. Profs., A.A.A.S., Am. Chem. Soc., Internat. Assn. Dental Research (Wilmer Souder award in dental research 1961), Am. Soc. Metals, Psi Omega, Phi Kappa Phi. Author: (with others) Restorative Dental Materials, 2d edit., 1964; also numerous articles. Research on chem., phys. and bioengring. characteristics restorative materials in dentistry; contbr. to devel. high speed rotary handpiece used in dental practice. Home: 10 Ridgeway, Ann Arbor, Mich. 48104.*

PEYVE, Aleksandr Vol'demarovich, Russian geologist; b. Feb. 9, 1909; grad. Moscow Geol. Survey Inst., 1930. Staff Sci. Inst. on Fertilizers; mem. Tadzhik-Pamir Expdn., USSR Acad. Scis., 1934-35, joined Geol. Inst. USSR Acad. Scis., 1952, became dept. dir., 1952, dir., 1961. Recipient Stalin prize, 1946. Corr. mem. USSR Acad. Scis. Author: Tectonics of the North Urals Bauxite Belt, 1947. Research and publs. in theoretical and regional tectonics, stratigraphy of magnas, metalogenesis; discovered deposits of bauxite, phosphorite, potassium salts; contributed to tectonic map of USSR, 1956. Home: 1-Aya Cheremushkinskaya 3, Moscow, USSR. Office: Inst. Geology, USSR Acad. Scis., Pyzherskii Pereulok, 7, Moscow, USSR.

PÉZÉNAS, Esprit, French mathematician, astronomer; b. Avignon, France, Nov. 28, 1692; mem. Soc. Jesus; became prof. hydrography, Marseille, France, 1728; placed in charge Marseille obs., 1749; mem. French Acad. Scis., Marine Acad. Author: Astronomie des marins, 1766; Histoire critique de la découverte des longitudes, 1775. Research on calculation of longitude; helped popularize science in 17th century France by his clearly and simply written treatises on astronomy and chemistry. Died Avignon, Feb. 4, 1776.

PFAFF, Christian Heinrich, German physician, chemist; b. Stuttgart, Germany, Mar. 2, 1773; prof., Kiel, Germany. Author: System der Materia medica nach chemischer Prinzipien, 7 vols., 1808-24; Handbuch der analytischen Chemie, 2 vols., 1821-22. Died Kiel, Apr. 24, 1852.

PFAFF, Franz, physician; b. Grafenort, Silesia, Mar. 16, 1860; studied natural sci., univs. Leipzig, Munich (both Germany); Ph.D., U. Zurich (Switzerland); postgrad. Würzburg, Germany, Basel, Switzerland; M.D., U. Strasbourg (now in France); postgrad. London hosps.; 1st asst. in chem. lab., pvt. docent U. Geneva; dir. chem. lab. Province of Amazonas, Brazil; studied med. plants in Brazil, 3 yrs.; physiology Harvard Med. Sch., 1894-95, instr. pharmacology, 1895-1900, instr. physiol. chemistry, 1898-1900, asst. prof. pharmacology and therapeutics, 1900-05, prof., from 1905. Licentiate Soc. Apothecaries, London. Died 1926.

PFAFF, Johann Friedrich, German mathematician; b. Stuttgart, Germany, Dec. 22, 1765; ed. univs. Göttingen, Berlin (both Germany); prof. math., Helstedt, 1788-1810, Halle, Germany, 1810-25; mem. French Acad. Scis., 1821. Author: Disquisitiones analyticae, 1797; Methodus generalis, aequationes differentiarum partialium . . . Abhandlungen Berlin Akademie, 1814-15; Observationes ad Euleri institutiones calculi integralis. Known for work on calculus, theory of series, solution of differential equations; forerunner of German sch. of Gauss (main determinant of course of 19th century math.); 1st to propose gen. method of integrating partial differential equations of 1st order, 1814-15; term Pfaffian problem introduced by Jacobi. Died Halle, Apr. 21, 1825.

PFAFF, Philippe, German dentist; b. 1715; dentist to Frederick the Great. Author book on dentistry, 1756. First to recognize need for curetting and drying dental cavity before filling; credited with 1st description of casting models for false teeth, 1756. Died 1767.

PFAFFMANN, Carl, Am. psychologist; b. Bklyn., N.Y., May 27, 1913; s. Charles and Anna (Haaker) P.; Ph.B., Brown U., 1933, M.A., 1935, B.A., Oxford U., 1937; Ph.D. (Rhodes scholar 1935-38), Cambridge U., 1939; D.Sc., Brown U., 1965, Bucknell U., 1966; married Hortense Louise Brooks, Dec. 26, 1939; children—Ellen Anne, Charles Broooks, William Sage. Research asso. Johnson Found., U. Pa., 1939-40; instr. psychology Brown U., 1940-42, asst. prof., 1945-48, asso. prof., 1949-51, prof., 1951-65, Florence Pirce Grant U. prof., 1960; vis. prof. Harvard, 1962-63, Nat. Sigma Xi lectr., 1963; v.p., professor Rockefeller University, 1965—, chmn. div. behavioral scis. NRC, 1962-64. Chmn. of 17th Internat. Congress Psychology, 1963. Trustee Butler Hospital, 1962-65. Recipient Howard Crosby Warren medal for research in psychology; Soc. for Exptl. Psychology, 1960; Distinguished Service Contbn. award Am. Psychol. Assn., 1963; Guggenheim fellow, 1960-61. Fellow Am. Psychol. Assn. (pres. div. exptl. psychology, 1956-57), A.A.A.S., Am. Acad. Arts and Scis.; mem. Am. Philos. Soc., Nat. Acad. Sci., Soc. Exptl. Psychologists, Am. Physiol. Soc., Eastern Psychol. Assn. (dir. 1953-57, pres. 1958-59). One of the 1st to record nerve impulses in single fibers of taste in mammals, 1939; helped provide sensory code for the different qualities of taste sensation; has studied species' difference in taste, central nervous system pathways, sensitivity in man, and specific taste preferences and cravings as well as neural mechanisms of olafaction. Home: 1161 York Av., N.Y.C. 10021.*

PFAHLER, George Edward, Am. radiologist; b. Numidia, Pa., Jan. 29, 1874; s. William H. and Sarah A. (Stine) P.; B.E., Bloomsburg State Normal Sch. (now Tchrs. Coll.), 1894; M.D., Medico Chirurg. Coll., Phila., 1898; Sc.D., Ursinus Coll., also LL.D., 1942; D.M. R.E., Cambridge U., Eng., 1926; m. Frances Simpson, Nov. 8, 1908 (dec. Mar. 1910); m. 2d, Muriel Bennett July 10, 1918. Intern Phila. Gen. Hosp., 1898-99, asst. chief resident physician, 1899-1902; clin. prof. roentgenology Medico-Chirurg. Coll., 1909-12, prof., 1912-16; prof. radiology U. Pa., 1916—; dir. radiol. dept. Misericordia Hosp., Phila. Hon. fellowship Faculty Radiologists, London, 1950. Mem. Am. Roentgen Ray Soc. (pres. 1910), Am. Electrotherapeutic Assn. (pres.

1912), Am. Radium Soc. (pres. 1922), Am. Coll. Radiology (pres. 1923), A.M.A., Pa. Med. Soc., Phila. Dermathol. Soc. (pres. 1956-57), Phila. County Med. Soc. (pres. aid assn. 1953—); hon. mem. Brit., French, German, Austrian, Scandanavian and Russian, Panama, Cuba, Peru radiol. socs., Radiol. sect. Royal Soc. Medicine London, England, and Mexico. Contbr. to med. jours. Pioneered in radium and X-ray treatment of cancer. Died Jan. 29, 1957.

PFAHLER, Gerhard, German psychologist; b. Freudenstadt, Germany, Aug. 12, 1897; prof., Giessen, Göttingen, Tübingen (all Germany). Author: System der Typenlehren, 1929; Vererbung als Schicksal, 1932; Warum Erziehung trotz Vererbung?, 1935; Der Mensch und sein Lebenswerkzeug, 1954. Tried to find relation between theories which considered character to be inherited and those which considered it an acquired trait. Deceased.

PFALZNER, Paul Michael, Austrian physicist; b. Vienna, Aug. 18, 1923; s. Ernest Arthur and Katherine (Hofman) P.; B.A. with honors in Math. and Physics, U. Toronto (Ont., Can.), 1946; M.Sc. in Physics, McGill U., Montreal, Que., Can., 1951; m. Anna Hauben, Oct. 7, 1950; 1 dau., Jenny Viviann. Research officer NRC Can., Ottawa, Ont., 1946-50; lectr. depts. physics and therapeutic radiology U. Western Ont., London, 1951-66; 1st officer dosimetry sect. div. life scis. dept. research and isotopes IAEA, Vienna, Austria, 1966—, expert adviser to Govt. Thailand, Bangkok, 1964-65; physicist Ont. Cancer Treatment and Research Found., London, 1952—. Mem. Canadian Assn. Physicists (past sec. med. physics div.), Hosp. Physicists' Assn. (Eng.), Am. Assn. Physics Tchrs., Brit. Inst. Radiology. Author: (with I.H. Smith, S.J. Lott) Cobalt-60 Teletherapy, 1964; also articles. Research on med. radiation dosimetry phys. aspects med. and biol. applications of radioisotopes, use computers in life scis. Home: Peter Jordan-Str. 127, A-1180, Vienna. Office: Internat. Atomic Energy Agy., Kaertnerring 11, A-1010 Vienna, Austria.*

PFANDER, William Harvey, Am. nutritionist; b. Lamar, Mo., Aug. 9, 1923; s. Edwin E. and Anna (Neely) P.; B.S., U. Mo., 1948; M.S., U. Ill., 1949, Ph.D., 1951; m. Margaret Alley, June 20, 1953; children—Michael Alan, James Eric, Geoffrey Ian. Fellow, U. Ill., Urbana, 1948-51; research asso. in biochemistry U. Wis., Madison, 1951; Fulbright fellow U. Aberdeen and Rowett Research Inst., Scotland, 1951-52; asso. prof. nutrition sect., animal husbandry dept. U. Mo., Columbia, 1952-54, prof., 1954—. NSF fellow U. Sydney (Australia), 1958-59; exec. sec. U. Mo. Nutrition Area Program, 1964-66; Fulbright fellow U. Alexandria (Egypt), 1966-67. Recipient Jr. Sci. award Gamma Sigma Delta, 1957. Fellow A.A.A.S., Am. Chem. Soc.; mem. Am. Inst. Nutrition (com. on exptl. animal nutrition 1962—), Am. Soc. Animal Sci., Am. Dairy Sci. Assn., Biochemistry Soc., Soc. Exptl. Biology and Medicine, Sigma Xi, Gamma Sigma Delta. Author: (with R. Seiden) Handbook of Feedstuffs, 1957; also numerous articles. Research in qualitative and quantitative requirements of nutrients including molybdenum and potassium for ruminants, vitamins and essential amino acids, determination of combination of minerals needed to insure proper utilization of high energy rations containing NPN, dietary factors related to metabolic disorders. Home: R.F.D. 2, Columbia, Mo. 65201.*

PFANNENSTIEL, Max Joseph, geologist; b. La Wanzenau, France, July 25, 1902; s. Hermann and Maria (Reinach) P.; Dr ès sci., U. Mainz; dr. honoris causa, U. Besançon; m. Christine Hermuth, Apr. 8, 1933; children—Marianne, Gerda. Librarian, U. Fribourg/Brisgau, also Ankara, Turkey; dir. library Ankara Coll. Agr., 1938-41; prof. geology U. Fribourg/Brisgau, 1947—. Decorated Order Palms of Acad. France. Mem. acads. Mainz, Heidelberg, Leopoldina, Halle, Italian Inst. Human Paleontology; hon. mem. Soc. Paleontology of Indies. Author books on The Mediterranean, Black Forest, paleontology, prehistory, history natural sci. Editor: Berichtel of Fribourg Natural Sci. Soc. Research on prehistory; history geol. sci.; geology; submarine geology. Home: Günterstalstrasse 32. Office: Hebelstrasse 40, Fribourg/Brisgau, West Germany.

PFANSTIEHL, Carl, Am. metallurgist, inventor; b. Columbia, Mo., Sept. 17, 1887; s. Albertus A. (Rev.) and Julia (Barnes) P.; spl. work Armour Inst. Tech.; m. Caryl Cody, June 24, 1915; children—Cody, Alfred, Rose-Caryl, Grace. Organized Pfantiehl Elec. Lab. (now Fansteel Metall. Corp.), 1907, pres. and dir. research until 1919; pres. and dir. research Pfanstichl Radio Co., 1922-28; v.p. and dir. research, chem. and metal divs., Pfanstiehl Chem. Co., since 1918; spl. research for War and Med. depts. Govt., World War I; cons. in the Government during World War; applied physics, metallurgy, radio. Recipient Modern Pioneer award, 1910. Mem. Am. Chem. Soc., Electrochem. Soc., A.A.A.S., N.Y. Acad. Sci., Am. Phys. Soc., Optical Soc. Am., Am. Inst. Mining and Metall. Engrs., Am. Soc. for Metals. Author: Ignition, 1912; also articles on radio theories and biochemical subjects. Patentee in elec., chem., metall. fields; research on prodn. metallic tungsten, molydenum, tantalum rhenium, osmium, ruthenium and their alloys, spl. anti-friction and hard tipping alloys, rare biol. chemicals, cold lighting, flourescent powders. Died Mar. 1, 1942.

PFANZAGL, Johann, mathematician; b. Vienna, Austria, July 2, 1928; s. Johann and Maria (Hoffmann); Ph.D., U. Vienna; m. Elvine Schlecht, Dec. 31, 1959; 1 dau., Beatrix. Statistician, Austrian C. of C.; asso. prof. U. Vienna; now prof., dir. Inst. Econ. and Social Statistics, U. Cologne, also dir. Inst. Math., 1964—. Mem. Internat. Inst. Statistics. Author: Die axiomatischen Grundlagen einer allgemeinen Theorie des Messens, 1959; Allgemeine Methodenlehre des Statistik, 2d edit., 1962. Research on foundational aspects of measurement theory, math. statistics, mainly abstract problems connected with hypothesis testing and estimation. Address: Albertus-Magnus-Platz 1, Cologne, West Germany.

PFAUNDER, Leopold von, see von Pfaunder.

PFEFFER, Richard Lawrence, Am. atmospheric scientist; b. Bklyn., Nov. 26, 1930; s. Lester Robert and Anna (Newman) P.; B.S. cum laude, Coll. City N.Y., 1952; S.M., Mass. Inst. Tech., 1954, Ph.D., 1957; m. Roslyn Ziegler, Aug. 30, 1953; children—Bruce, Lloyd, Scott, Glen. Meteorologist, atmospheric physicist Air Force Cambridge Research Center, Boston, 1955-59; sr. research scientist Columbia, N.Y.C., 1959-60, faculty, 1961-64, lectr. geophysics, 1961-62, asst. prof. geophysics, 1962-64; asso. prof. meteorology Fla. State U., Tallahassee, 1964-67, prof., dir. Geophys. Fluid Dynamics Inst., 1967—; vis. sr. research scientist NASA Inst. for Space Studies, N.Y.C., 1961-64. Broadcaster on tsunamis Voice of Am., 1961; cons. to pvt. cos. Mem. N.Y. Acad. Scis. (past chmn. planetary scis. sect.), Am. Meteorol. Soc. (past chmn. program com.), Am. Geophys. Union, Sigma Xi. Editor: Dynamics of Climate, 1960. Research, publs. on dynamics large-scale atmospheric processes, application concept available potential energy to rotating fluid systems, theory acoustic-gravity wave propagation in earth's atmosphere and atmospheres other planets. Home: 926 Waverly Rd., Tallahassee 32306.*

PFEFFER, Wilhelm, German botanist; b. Grebenstein, Germany, Mar. 9, 1845; Ph.D., U. Göttingen, 1865; prof. botany, Bonn, Germany, also Basel, Switzerland, Tübingen Germany; prof., dir. bot. inst. Leipzig (Germany) U. Fellow Royal Soc., 1897; mem. French Acad. Scis. Author: Handbuch der Pflanzenphysiologie, 2 vols., 1881; Studien zur Energetik der Pflanze, 1892. Research on osmosis; one of 1st to make semi-permeable membrane, 1877, used it to measure osmotic pressure of solutions; laid found. of modern theory of solutions; discovered that osmotic pressure for dilute sugar solutions is proportional to concentration; studied (with Julius Sachs) basic questions of plant physiology; experimented with artificial plant nutrients, 1900; found techniques of his time too limited to pursue studies of protein metabolism of plants; investigated plant growth and movements, 1904. Died Kassel, Germany, Jan. 31, 1920.

PFEIFER, Harry Erich, German physicist; b. Penig, Germany, Feb. 25, 1929; s. Erich and Johanne (Müller) P.; Diplom, U. Leipzig, 1952, Dr.rer.nat., 1954, Dr.rer.nat.habil., 1956; m. Christine Pfeifer, July 25, 1957; children—Sabine, Andreas. Asst., Phys. Inst., U. Leipzig, 1952-57, docent, 1958-60, prof., 1960—, dir. electronics dept., 1963—. Named Deserving Technologist of the People, 1965. Mem. E. German phys., chem. socs., Groupement Ampere. Author: Elektronisches Rauschen, part I, 1959, part II, 1967; Elektronik für Physiker I-VI, 1966-67; also numerous articles. Research on generation of harmonic oscillations by negative resistances; interaction between spin-systems and spin-detectors; nuclear magnetic resonance and relaxation in solutions of paramagnetic ions. Address: 163 Lenin-Strasse, 7027 Leipzig, East Germany.*

PFEIFER, Siegfried Artur, German pharm. chemist; b. Meerane, Germany, June 3, 1926; s. Arthur Albert and Else Martha (Gessner) P.; student Inst. Pharmacy, Jena (Germany), 1948-51; Dr.rer.nat., Inst. Pharmacy, Humboldt U., Berlin, East Germany, 1955, Dr. rer. nat. habil., 1961; m. Annelore Else Maria Gehrels, Jan. 24, 1958. Faculty, Humboldt U. Inst. Pharmacy, 1952—, prof., 1962—, dir., 1963—. Editor, Pharmazie, 1961—. Research, numerous publs. on analytical studies of drugs, analytical and physiol. studies of poppy alkaloids; isolation and structure elucidation of new alkaloids of genus papaver. Home: 82 Volkradstr., Berlin 1136 DDR.*

PFEIFFER, August, German bacteriologist; b. Germany, 1848; pupil of Koch and Flügge; med. officer health, Wiesbaden, Germany; pub. bacteriological observations on tubercle bacilli in lupus, 1883, typhoid bacilli in stools, 1885; discovered B. pseudotuberculosis rodentium, 1889. Died 1919.

PFEIFFER, Carl Curt, Am. physician, pharmacologist; b. Peoria, Ill., May 19, 1908; s. Curt Richard and Minnie D. (Meiers) P.; A.B., U. Wis., 1931, A.M., 1933, Ph.D., 1935; M.D., U. Chgo., 1937; m. Lillian H. Twenhofel, June 13, 1930; children—Helen Nancy (Mrs. Carl Nugent), Edward Carl. Instr. pharmacology U. Chgo., 1938-40; asso. prof. Wayne U. Coll. Medicine, Detroit, 1940-41; chief pharmacology Parke Davis & Co., Detroit, 1941-43; prof., head dept. pharmacology U. Ill. Coll. Medicine, Chgo., 1945-54; prof. chemistry Emory U., Atlanta, 1954-57, dir. div. basic health scis., 1956-60; head sect. pharmacology Bur. Research, N.J. Neuropsychiat. Inst., Princeton, N.J., 1960—. Sec.-treas. N.J. Mental Health Research and Devel. Fund, 1964—. Mem. Am. Schizophrenic Assn. (mem. sci. adv. bd. 1964—), A.M.A., Am. Soc. for Pharmacology and Exptl. Therapeutics (past pres.), Am. Soc. for Exptl. Biology and Medicine, A.A.A.S., Am. Chem. Soc., Am. Acad. Neurology, Soc. for Biol. Psychiatry, N.Y. Acad. Scis., Assn. for Research in Nervous and Mental Diseases, Sigma Xi. Co-editor Internat. Rev. Neurobiology, 1956—. Research, numerous publs. on physiology of pain, pharmacology of analgesic and antiepileptic drugs, chem. structure-activity relationships, drugs lowering normal body temperature, chemotherapy of schizophrenia, hydrazides as metabolic convulsants, pyridoxine as antidote to hydrazides, etiology of schizophrenia and chem. isomerism and drug action. Home: R.D. 4, 169 Cherry Hill Rd. Office: P.O. Box 1000, Princeton, N.J. 08540.*

PFEIFFER, Emil, physician; b. Germany, 1846; described infectious mononucleosis (Drüsenfieber), 1889. Died 1921.

PFEIFFER, Mildred C. J., Am. physician; b. Phila., Aug. 16, 1910; d. Emil and Natalie Marie (Doehler) Pfeiffer; B.S., U. Pa., 1933, M.D., 1936, M.P.H., 1945. Practiced medicine specializing in internal medicine, gastroenterology and proctology, Phila., 1938-52; faculty Woman's Med. Coll., Phila., 1939-53, dir. dept. oncology, 1948-52, vis. prof. pub. health, 1952-56; with Pa. Dept. Health, 1952—, dir. div. planning, 1957—. Mem. bd. Arthritis Found., numerous state vol. agys. Recipient citations and awards from numerous groups including Drexel Inst. Tech., Pa. Heart Assn., Central Pa. Arthritis Found., Woman's Hosp., others. Fellow A.C.P., Am. Coll. Chest Physicians; mem. Internat. Soc. Internal Medicine, Phila. Acad. Medicine, Am. Pub. Health Assn., A.A.A.S., Pub. Health Mgmt. Assn. Am., Geiatrics Soc., Proctol. Soc., A.M.A., Am. Soc. Planning Ofcls., Nat. Planning Assn., Am. Soc. Pub. Administrn., Am. Acad. Polit. and Social Sci., Gerontologic Soc., Nat. Rehab. Assn., Pa. Acad. Sci., Physics Club., World Med. Assn. (asso. mem.), Am. Soc. Internists, Am. Med. Womens Assn., Sigma Delta Epsilon, Alpha Omega Alpha, Delta Phi Alpha, others. Research, publs. on orgn. of health programs in adult heart disease, aging, arthritis, diabetes, expanded narcotic and alcoholism control, behavioral problems and science, chronic disease, home safety. Former editor Jour. Am. Med. Women's Assn., Biosciences Abstracts. Home: 358 Valley Rd., Merion Sta., Pa. 19066. Office: P.O. Box 90, Pa. Dept. Health, Harrisburg, Pa. 17120.*

PFEIFFER, Raymond Louis, Am. ophthalmic surgeon, roentgenologist; b. Lancaster, O., May 16, 1901; s. William F. and Louis K. (Metzger) P.; A.B., Wittenberg U., 1924, D.Sc., 1956; M.D., Ohio State U., 1928; M.Sc., Columbia, 1934; Sc.D. in Medicine, 1935; m. Gertrude B. Smith, Mar. 16, 1935; children—Nancy L. (Mrs. George D. Sauter), Robert W., Jean Bard, Raymond S. Staff, Inst. Ophthalmology, Presbyn. Hosp., N.Y.C., 1930—, fellow, 1932-33, roentgenologist, 1934—, attending ophthalmologist, 1945—; attending ophthalmologist Vanderbilt Clinic, N.Y.C., 1945-63; instr. Columbia Coll. Physicians and Surgeons, N.Y.C., 1932-36, asso., 1936-48, asst. prof. 1940-45, asso. prof., 1945—; practice medicine specializing in ophthalmology, N.Y.C. Cons. ophthalmologist, commn. for blind N.Y. State Dept. Social Welfare, 1936—, mem. med. adv. com., 1936—. Recipient Alumni Achievement award Coll. Medicine Ohio State U., 1948; award of merit Am. Acad. Ophthalmology and Otolaryngology, 1955, 65. Diplomate Am. Bd. Ophthalmology. Fellow A.C.S.; mem. N.Y. Acad. Medicine, N.Y., Am. ophthal. socs., Am., N.Y. State, County med. socs., Phi Kappa Psi, Phi Chi, Alpha Omega Alpha. Asso. editor N.Y. State Jour. Medicine, 1946—. Research, publs. on diagnostic techniques, procedures, new methods in roentgenology of eye and related structures. Home: 2 Rivercrest Rd., Riverdale, N.Y. 10471. Office: Inst. Ophthalmology, 635 W. 165th St., N.Y.C. 10032.*

PFEIFFER, Richard Friedrich Johann, German physician, bacteriologist; b. Zduny, Germany, Mar. 29, 1858; M.D., Berlin, 1892. Became asst. to R. Koch, Koch's Inst., 1888; named dir. Inst. Contagious Diseases, Berlin, 1891; named prof. hygiene, dir. Inst. Hygiene, Königsberg (now Kaliningrad, USSR), 1899; prof. U. Breslau (now Wroclaw, Poland), 1909-26. Discovered Pfeiffer's bacillus in cases of influenza, micrococcus catarrhalis; described reaction used to diagnose cholera (Pfeiffer's reaction or phenomenon); studies on serum against influenza. Died 1945.

PFLAUM, Ronald Trenda, Am. chemist; b. Webster, Minn., June 21, 1922; s. Clarence Carl and Josephine (Trenda); B.A., St. Olaf Coll., 1948; Ph.D., Purdue U., 1953, M.S., 1951; m. Avis Hanson, June 19, 1948; children—Ronald, Cynthia, Suzanne. Faculty, U. Ia., Iowa City, 1953—, prof., 1963—. Mem. Am. Chem. Soc., Sigma Xi, Alpha Chi Sigma. Research, publs. on analytical methods of analysis, particularly spectrophotometric methods of analysis of metal ions. Home: 706 Clark St., Iowa City 52240.*

PFLÜGER, Eduard Friedrich Wilhelm, German physiologist; b. Hanau, Germany, June 7, 1829; pupil of Ludwig, Johannes Müller; became prof., Bonn, 1859. Fellow Royal Soc., 1888. Author: Untersuchungen über der Physiologie des Electrotonus, 1859; (paper) Über die Diffusion d. Sauerstoffs, den Ort und die Gesetze der Oxydationsprozesse im thierischen Organismus, 1872; Die teleologische m Mechanik der lebendigin Natur, 1877; Wesen und Aufgabe der Physiologie, 1878; Die A allgemeine Lebenserscheinungen, 1889. Founder Archiv für die gesamte Physiologie des Menschen und der Tiere, 1869. Research on digestive and metabolic processes, intestinal nerves, sensory function of spinal cord, elec. stimulation of motor nerves; formulated law stating that reflexes spread toward head more quickly than in other directions (unchallenged until Sherrington), 1853; 1st to formulate laws of behavior of nerves in response to galvanic current, pub. 1859; helped build mercurial blood pump; conducted expts. on physiology of breathing; demonstrated that oxidation occurs in tissues rather than in blood and depends on requirements of organism; pointed out that some basic decomposition asso. with animal respiration occurs in absence of oxygen, 1875; applied physico-chem. techniques to study of respiration and nutrition; introduced respiratory quotient, 1877. Died Bonn, Mar. 16, 1910.

PFLUGFELDER, Otto, German physiologist; b. Rappoltshofen, Germany, Mar. 15, 1904; s. Christian and Magdalena (Weidner) P.; Dr. ès science naturelle, U. Tübingen (Germany); m. Luise Schneider, Mar. 19, 1938; children—Hartmut, Heidrun. Asst. U. Tübingen; prof. at large U. Jena (Germany); prof. Stuttgart (Germany) Agrl. Coll.; hon. prof. Stuttgart Technische Hochschule. Mem. German Zoology Assn., Deutsche Gesellschaft für Parasitenkunde, Gesellschaft deutscher Naturforscher und Arzte, others. Author: Vergleichende Anatomie und experimentell Untersuchungen über das Nerven-system der Hemipteren; Zooparasiten und die Abwehrreaktionen ihrer Wirte; Entwicklungsphysiologie der Insekten; others. Research on endocrinology, parasitology. Address: Egilofstrasse 35, Stuttgart, W. Germany.

PFOLSPEUNDT, Heinrich von, see von Pfolspeundt.

PFUETZE, Karl Hamilton, Am. physician; b. Manhattan, Kan., Sept. 14, 1908; s. Emil Carl and Rogene (Scott) P.; B.S., Kan. State U., 1930; M.D., U. Kan., 1934; m. Dorothy Constance Anderson, June 7, 1941; children—Douglas Phillip, Scott Hunter, Mary Vivian. Staff, Nopeming Sanitarium, Nopeming, Minn., 1938-40; med. dir., supt. Mineral Springs Sanitarium, Cannon Falls, Minn., 1941-51; med. dir., supt. Chgo. State Tb Sanitarium, 1951-62; gen. adminstr. Suburban Cook County Tb Sanitarium Dist., Hinsdale, Ill., 1962—; clin. prof. medicine U. Ill., 1951—; cons. pulmonary diseases VA Hosp., Hines, Ill., 1951—; instr. medicine Mayo Found., Rochester, Minn., 1947-51. Recipient Tb Inst. Chgo. and Cook County Tb Inst. medal, 1962. Mem. Am. Coll. Chest Physicians, Am. Thoracic Soc., Am. Pub. Health Assn., Inst. Medicine Chgo., A.M.A. Co-editor: (with David B. Radner) Clinical Tuberculosis: Diagnosis and Treatment, 1966. Research, numerous publs. on clin. investigation of various chemotherapy agts. in treatment of Tb; pioneer in adminstrn. of streptomycin; lectr. on control and treatment of Tb. Home: 5515 County Line Rd. Office: 55th and County Line Rd., Hinsdale, Ill. 60521.*

PHAEDIAS, astronomer; flourished 280 B.C.; 1 son, Archimedes; wrote treatise on diameters of sun and moon; believed diameter of sun was 12 times that of moon.

PHAER, Thomas (or Faier, Phayre), English pediatrician; b. Norwich, Eng., 1510; possibly s. Thomas Phaer; student Lincoln's Inn; M.D., Oxford U., 1559; soliciter, ct. of Welsh marches. Author: Natura Brivium, 1535, Newe Boke of Presidentes, 1543 (both law books); The Boke of Children (1st English work devoted entirely to pediatrics, possibly a transl. from French), 1545; The Regiment of Life (a version of Sanitatis Salerni), 1554; also an early English transl. of Aeneid. Died 1560.

PHAIR, John Joseph, Am. physician; b. Cin., Sept. 10, 1904; s. Robert James and Barbara (Arns) P.; B.S., U. Cin., 1926, M.B., 1928, M.S., 1931; M.P.H. (Rockefeller Found. fellow), Johns Hopkins, 1933, Dr. P.H., 1938; m. Phyllis Elizabeth Wolfe, Mar. 20, 1930; 1 son, John Phillip. Faculty, U. Cin., 1929-32, prof. preventive medicine dept. preventive medicine and indsl. health, 1949—; mem. field staff, internat. health div. Rockefeller Found., 1933-36; faculty Johns Hopkins, 1936-46; faculty Sch. Medicine, U. Louisville, 1946-49, lectr. community health, 1961-63; chmn. med. adv. com. to health Commr. City of Cin., 1961-67, commr., 1967—. Cons. to sec. war, 1941-46, USPHS, 1946—; also various brs. govt. WHO traveling fellow, 1951; Rockefeller Found. traveling fellow, 1958. Diplomate Am. Bd. Preventive Medicine. Fellow Am. Pub. Health Assn., A.C.P., A.A.A.S., Ohio Acad. Sci., Indsl. Med. Assn.; Am. Coll. Preventive Medicine (pres. 1962-63); mem. Am. Epidemiol. Soc., Am. Soc. Tropical Medicine and Hygiene, Soc. Exptl. Biology and Medicine, Assn. Tchrs. Preventive Medicine (pres. 1956-57), A.M.A., Am. Indsl. Hygiene Assn., Internat., Am. epidemiol. socs., Phi Chi, Delta Omega, Sigma Xi, others. Research and publs. in diphtheria and immunization; epidemiology of meningococcal, upper respiratory infections; air pollution and its prevention; and pulmonary dynamics. Home: 102 E. Shields St., Cin. 45220. Office: Cin. Health Dept., 2517 Burnet Av., Cin. 45219.*

PHALEN, Harold Romaine, Am. mathematician; b. Acton, Mass., Apr. 21, 1889; s. Edwin Anthony and Harriet Davis (Reed) P.; B.S., Tufts, 1912; M.S., U. Chgo., 1923, Ph.D., 1926; m. Lucie Hortense Snyder, Dec. 20, 1914 (dec. Aug. 1933); children—Carolyn Annette, Edward Snyder, m. 2d, Elizabeth Nagle Kinder, July 2, 1938. Draftsman Improved Paper Machine Co., Nashua, N.H., 1912, Am. Locomotive Co., Providence, 1913; instr. math. James Millikin U., 1913-15, Berea Coll., 1915-18, Armour Inst. Tech., 1918-26; prof. math. St. Stephen's Coll. (now Bard Coll.), Columbia, 1926—, provost, 1929-33; vis. prof. Brown U., 1939-40; asso. prof. math. Coll. William and Mary, 1941-46, prof. 1946-54, head dept., 1954—. Mem. Am. Math. Soc., Math. Assn. Am., A.A.A.S., Sigma Xi. Author: History of Action, 1954. Translator: Lezioni de Geometria Provittiva (Enriques). Died May 30, 1955.

PHAN, The Tran, Vietnamese biologist; b. Saigon, Oct. 30, 1933; s. Kien Khuong Phan and Thi Vang Le; B.S., Toulouse U., 1957, M.S., 1958, Ph.D., 1962; m. Ngoc Phuong Lan Nguyen, Apr. 24, 1960; children—The Trung, The Trinh. Research asso. Inst. Nuclear Research Dalat, Vietnam, 1958-64, on leave to Oak Ridge Nat. Lab., 1959-60; sr. research asso. Pasteur Inst., Saigon, 1964—; asst. prof. pharmacy Saigon U., 1964-66, asso. prof., 1966—. Laureate, U. Toulouse. Mem. Chem. Soc. Vietnam, Biol. Soc. Vietnam. Research, publs. on protection of bone marrow during freezing, bactericidal action of smoke and fumes, blood biochemistry of Vietnamese. Home: 290/10 Cong Ly, Saigon, Vietnam.*

PHARISEAU, Pierre, Belgian theoretical physicist; b. Aalst, Belgium, Nov. 5, 1931; s. Victor and Maria (Galle) P.; Dr.Sc., State U., Ghent (Belgium), 1959; m. Jeannine Van Vaerenbergh, Oct. 12, 1959; children—Marleen, Elsie, Ann. Sci. collaborator Inst. de Recherce Scientifique dans l'Industrie et l'Agriculture, comite de l'etat solide, 1958-64; aggregee Faculty Sci., State U. Ghent, 1964—; group leader theoretical group Lab. for Crystallography and Solid State Physics, 1964—. Research, publs. on diffraction of light by ultrasonic waves, theory of surface states in solids, quantum theory of disordered structures, electron states, conductivity theory, impurities in solids. Office: 105 Krijgslaan, Ghent, Belgium.*

PHEMISTER, James, Scottish geologist; b. Glasgow, Scotland, Apr. 3, 1893; s. John and Elizabeth G. (Crawford) P.; M.A., D.Sc., U. Glasgow; m. Margaret Clark, July 11, 1921; children—John Clark, Margaret R., Thomas G. Geologist, from 1921; petrographer, from 1935; curator Geology Mus., 1945-53, ret., 1953. Recipient Neill medal and prize Royal Soc. Edinburgh; mem. Mineral. Soc. Gt. Britain, Geol. Soc. London (Murchison medal), Am., Brit. (pres. 1952-56) mineral. socs., Mineral. Assn. Can., Geol. Soc. Glasgow. Author: Petrology of the Loch Ailsh Intrusion; Limestones of Scotland; Zoning in Plagioclase. Editor: Geol. Mineral. Abstracts, 1959—. Publs. on mineralogy, geophysics. Address: 39 A Fountainhall Rd., Edinburgh, Scotland.

PHEREKYDES OF SYROS, Greek natural scientist; flourished 6th century B.C.; considered 1st Greek prose writer; author Pentemychos (depicts cosmogeny cloaked in allegory and mythology according to 5 original elements, ether, earth, fire, water, air).

PHILBERT, Georges Marie Victor, French physician; b. Paris, Jan. 5, 1922; s. Victor Marie and Georgette (Dufaux) P.; Licencié es Sciences, Sorbonne, 1942, Diplome d'Etudes Supérieures de Physique, 1944; D. Science, 1953; 1 dau., Dominique-Chantal. Research physician Kaiser Wilhelm Inst. Physics, Herhinger (Germany), 1945-51, nuclear physics lab., College de France, Paris, 1951-58; asso. prof., dept. nuclear physics, U. Lyon, 1958-63, prof., 1963—. Recipient Ordre des palmes académiques. Mem. French Soc. Physics. Research, numerous publs. on electrodynamics of supraconductions, mass spectrometry, high energy exptl. and theoretical research. Address: 72 rue d'Auteuil, Paris, France.*

PHILINUS OF COS, Greek physician; b. flourished circa 250 B.C.; pupil of Herophilos; founder (with Serapion of Alexandria) of Empirical Sch. of Medicine; wrote on difficult words in Hippocratic works; rejected diagnosis based on pulse; adapted sceptic philosophy to med. thought; advocated use of clin. observations and tried remedies rather than med. hypotheses.

PHILIP, André, French economist; b. Pont St. Esprit, France, June 28, 1902; s. Louis and Gilberte (Vincent) P.; licence en philosophie, Sorbonne, agrege des sciences economique; m. Mirielle Conresson; children—Olivier, Jean, Christiane (Mrs. Geanson), Nicole (Mrs. Disdarine), Loic. Prof. econs., Lyon, France, 1926-36, Larubrück, 1950-55, U. Paris, 1955—; mem. Parliament, 1936-55; minister underground Free French Govt., London, 1942-45; minister economy and finance, 1945-47. Chmn., French delegation to GATT, 1950—. Mem. French Fedn. Youth and Culture Houses (chmn.). Author: Labor Problems in U.S.A., 1926; Modern Trends, 1928; Industrial Democracy, 1952; Europe and International Trade, 1953; History of Economics, 1963; also numerous articles. Research on labor problems, internat. trade, especially in developing countries. Home: 32 rue du Creivone S'Cloud (Set Oise), France. Office: Faculte des Sciences Economomiques, Paris, France.*

PHILIP, Cornelius Becker, Am. med. entomologist; b. Ft. Lupton, Colo., June 12, 1900; s. Smith Drury and Mattie (Shoemaker) P.; B. Sc., U. Neb., 1923, Sc.D., 1952; M.S., U. Minn., 1925, Ph.D., 1930; postgrad. Johns Hopkins, Harvard; m. Gladys Helen Hill, Nov. 11, 1922; children—Robert, Bonnie Dee (Mrs. Harold L. Holt), Jo Joyce (Mrs. William D. Dratz), Gordon. Asst. entomologist U. Neb., 1922-23, Expt. Sta., Mont. State Coll., 1926-27; staff Mont. Bd. Entomology, 1928; entomologist Rockefeller W. Africa Yellow Fever Comm., 1928-29; staff Rocky Mountain Lab., USPHS, Hamilton, Mont., 1930—, prin. med. entomologist, 1948—, dir., 1962-64. Mem. Internat. Commn. Bacterial Nomenclature 1958—, mem. jud. com., 1963—; expert panel tick-borne disease FAO/WHO, 1959—, chmn., 1966. Guggenheim fellow 1941-42. Recipient U.S. Am. Typhus Commn. medal, 1943; Outstanding Achievement award U. Minn., 1960. Mem. A.A.A.S., Am. Soc. Parasitology (past pres.), Entomol. Soc. Am. (bd. govs. 1959-62), Am. Soc. Tropical Medicine and Hygiene, Soc. Am. Microbiologists, Sigma Xi, Phi Beta Kappa. Research, numerous publs. including 1st proof of transmission of yellow fever by 2 natural vectors in Africa, tularemia in natural fauna in Alaska, murine typhus in Jamaica and P.I., tick typhus in Kasmir, Ethiopia, canine rickettsiosis in U.S., St. Louis encephalitis in horses, efficacy of protective and therapeutic value of chlormycetin in scrub Typhus, test for plague bacillus in dead rodents. Home: 908 S. 4th St., Office: Rocky Mountain Lab., Hamilton, Mont. 59840.*

PHILIP, Graeme Maxwell, Australian geologist; b. Ballarat, Victoria, Australia, Nov. 17, 1933; s. Percival Norman and Ruth (Osborne) P.; B.Sc., U. Melbourne (Australia), 1956, M.Sc., 1959; Ph.D., U. Cambridge, 1962; m. Judith Ann Sullivan, July 1, 1957; children—Michael Ross, Susan Vanessa, Timothy Alvin. Demonstrator geology U. Melbourne, 1955-59; faculty geology U. New England, Armidale, Australia, 1962—, prof. geology, 1966—, head dept. Mem. Geol. Soc. Australia, Geol. Soc. London, Paleontol. Soc., Paleontol. Assn., Australian Assn. for Advancement Sci. Research, publs. on relationship and evolution of echinoderms, Australian stratigraphical paleontology. Home: 14 Reginald Av., Armidale, N.S.W., Australia.*

PHILIP, John Robert, Australian physicist, mathematician; b. Ballarat, Australia, Jan. 18, 1927; s. Percival Norman and Ruth (Osborne) P.; B.C.E., Melbourne (Australia) U., 1946, D.Sc., 1960; m. Frances Julia Long, Apr. 30, 1949; children—Peregrine Paul, Julian Mark, Candida Jane. Research asst. Melbourne U., 1947; engr. Queensland Irrigation Commn., 1948-51; staff Commonwealth Sci. and Indsl. Research Orgn., Canberra, Australia, 1951—, sr. prin. research scientist, 1961-63, chief research scientist, asst. chief div. plant industry, 1963—, first dir. F.C. Pye Field Environment Lab., 1964—; research staff Cambridge (Eng.) U., 1954-55; research fellow Cal. Inst. Tech., 1957-58, Harvard, 1966-67; vis. prof. U. Ill., 1958, 61. Mem. Australian Nat. com. for I.H.D., 1965—. Nuffield Found. fellow Cambridge U., 1961-62; 1st recipient David Rivett medal for phys. scis., 1966. Fellow Royal Meteorol. Soc., Australian Acad. Sci., Australian Inst. Physics; mem. Am. Geophys. Union (Robert E. Horton award 1957). Research, publs. on math. physics of water movement in unsaturated porous media and soils, theory of local advection, theory of heat and mass transfer within vegetation canopies, theory of point quadrats in ecology and geometrical probability, theories of dynamics of osmotic cell and diffusion of tissue turgor in physiology; devel. theory of infiltration; numerical and analytical solution of non-linear diffusion equations; co-author Philip-deVries theory of thermally induced moisture movement in porous media; pioneered concept of soil, plant, atmosphere as thermodynamic continuum for water transfer. Home: 42 Vasey Crescent, Campbell, Canberra City. Office: Commonwealth Sci. and Indsl. Research Orgn., Div. of Plant Industry, P.O. Box 109, Canberra City, A.C.T. 2601, Australia.*

PHILIP, Sir Robert William, Scottish physician; b. Scotland, 1857; s. George Philip; ed. U. Edinburgh, also continental univs.; m. Elizabeth Motherwell, 1888; m. 2d, Edith Josephine McGaw, 1938. Prof. Tb, examiner in medicine U. Edinburgh; founder 1st Tb dispensary in Edinburgh, 1887; chief founder Royal Victoria Hosp. for Consumption, also farm colony, Edinburgh; extra physician to king in Scotland; cons. physician to King Edward VII. Recipient Trudeau medal Nat. Tb Assn., U.S., 1928. Mem. Assn. Physicians of Gt. Britain and Ireland; pres. Brit. Med. Assn., Royal Coll. Physicians, Edinburgh. Contbr. to med. publs. Research on prevention and cure of Tb; described enlarged supraclavicular glands in tuberculous children (Philip's glands), 1912. Died 1939.

PHILIPPI, Rudolf Amandus, botanist; b. Charlottenburg (Berlin), Germany, Sept. 14, 1808; prof., dir. Natural History Mus., founder Bot. Garden, Santiago, Chile. Described numerous new types of Chilean flora. Died Santiago, July 23, 1904.

PHILIPPOFF, Wladimir, rheologist; b. Peterhof, Russia, Aug. 26, 1907; s. George and Mary (Wonlarlarsky) P.; M.S., Tech. U., Berlin, Germany, 1932, Ph.D., 1934; m. Xenia Quade, May 3, 1939; children—Michael, Peter, George. Came to U.S., 1948, naturalized, 1953. With Franklin Inst. Research and Devel., Phila., 1951-59; sr. research asso. Esso Research & Engring. Co., Linden, N.J., 1959—; adj. prof. chem. engring. Bklyn. Poly. Inst., 1968—. Mem. Soc. Rheology (Bingham medal 1963), Am. Chem. Soc., Am. Phys. Soc. Author: Viskostaet der Kolloide, 1942. Research, numerous publs. in rheology, mainly on polymer solutions, vibrational testing, normal stresses, x-rays in soap solutions and flotation, polymers, asphalts. Home: 51 Nomahegan Ct., Cranford, N.J. 07016. Office: P.O. Box 51, Linden, N.J. 07036.*

PHILIPPOS OF MENDE, see Philippos of Opos.

PHILIPPOS OF OPOS (probably identical with Philippos of Mende), mathematician, astronomer; flourished circa 360 B.C.; pupil of Plato, at whose suggestion he carried out research; wrote on polygonal numbers, many astron. topics (no works extant); credited with editing Plato's Laws, also with writing Epinomis (may have been written by Plato); believed to have explained rainbow as phenomenon of refraction.

PHILIPPSON, Alfred, German geographer; b. Bonn, Germany, Jan. 1, 1864; prof., Bern, Switzerland, Halle, Bonn (both Germany). Author: Der Peloponnes, 1892; Das Mittelmeergebiet, 1904; Europa, 1905; Reisen und Forschungen im westlichen Kleinasien, 5 vols., 1910-15; Grundzüge der allegemeinen Geographie, 2 vols., 1921-24; Das byzantinische Land als geographische Erscheinung, 1939; (with E. Kirsten) Die griechischen Landschaften, 4 vols., from 1951. Studied Mediterranean area. Died Mar. 28, 1953.

PHILIPS, Frederick Stanley, Am. pharmacologist; b. Mt. Vernon, N.Y., Sept. 25, 1916; s. Alfred I. and Rose Ruth (Rehberger) P.; B.A., Columbia, 1936; Ph.D., U. Rochester, 1940; m. Clarinda May Burr, Oct. 26, 1940; children—Sally Burr, Susan Jane, John Frederick. Lab. asst. Seth Low Jr. Coll., Columbia, 1936; grad. asst. biology U. Rochester, 1936-40; Theresa Sessell fellow Osborn Zool. Lab., Yale, 1940-41, NRC fellow natural scis. Med. Sch., 1941-42, research asso. pharmacology, 1942-43; head department pharmacology, div. exptl. chemotherapy Sloan-Kettering Institute, 1946—, associate inst., 1946-56, mem. inst., 1956—, chief div. pharmacology, 1967—; research fellow biology Mass. Inst. Tech., 1946-47; instr. pharmacology Cornell U. Med. Coll., 1947, asst. prof., 1948-51; asso. prof. pharmacology Sloan-Kettering Grad. Div. Cornell U. Med. Coll., 1951-57, prof., 1957—. Mem. panel bioassay, com. growth NRC, 1947, panel chemotherapy, 1951-53; cons. USPHS, 1956-60, 62—, mem. pharmacolog. study sect. NIH, 1954-56; cons. USPHS, mem. panel pharmacology, cancer chemotherapy, Nat. Service Center, NIH, 1955-57; cons. USPHS, mem. cancer chemotherapy study sect. NIH, 1956-60; spl. cons. USPHS, mem. research career award com. Inst. Gen. Med. Sci., NIH, 1962—; research cons. research dept. Am. Cancer Soc., 1955-56. Received Alfred P. Sloan award cancer research, 1964. Fellow A.A.A.S.; mem. Soc. Exptl. Biol. and Med. (chmn. N.Y. chpt. 1962—), N.Y. State Soc. Med. Research (pres. 1951-52), Am. Soc. Pharmacology and Exptl. Therapeutics, Am. Assn. Cancer Research. Editorial bd. Pharmacol. Revs., 1957-63; editorial adv. bd. Cancer Research, 1958-60, asso. editor, 1963—. Studies of drugs used in cancer treatment with respect to their effects on normal cells and tissues, their distbn. among and metabolism by different tissues in intact animals, and to dangers likely to be encountered when given to human beings; (with A. Gilman) introduced nitrogen mustards in treatment of Hodgkin's disease, 1946. Home: 24 Mersereau Av., Mount Vernon, N.Y. Office: 410 E. 68th St., N.Y.C. 10021.

PHILIPS, Joseph-Pierre, see Durand de Gros, Joseph Pierre.

PHILIPSBORN, see von Philipsborn.

PHILIPSON, Lennart Carl, Swedish virologist, biochemist; b. Stockholm, Sweden, July 16, 1929; s. Carl Otto and Greta (Svanstrom) P.; B.Medicine, Uppsala U., 1952, M.D., 1957, D.M.Sc., 1958; m. Malin Kristina Jondal, June 21, 1954; children—Niklas, Andreas, Tomas. Research staff Inst. Bacteriology, Uppsala U., 1953-57, Inst. Biochemistry, also Inst. Virology, 1957-59; Sophie Fricke fellow Rockefeller Inst., N.Y.C., 1959-61; asso. prof. virology Swedish Med. Research Council, 1961—, also co-dir. cell biology unit. Fellow N.Y. Acad. Sci.; mem. Harvey Soc., Soc. for Exptl. Biology and Medicine, Sigma Xi. Research and numerous publs. in virology and biochemistry; developed methods for particle separation and purification; studies in early interaction between animal viruses and cells to elucidate mechanism by which a virus informs cell to shut off normal cellular activities and produce virus particles. Home: 15 Banergahan, Uppsala, Sweden.*

PHILISTON OF LOCRI, Greek physician; contemporary of Plato (whom he influenced); disciple of Empedocles; practiced medicine, Athens; said to have taught Eudoxus; main rep. of Sicilian Sch. of medicine; dogmatic med. writer; made accurate observations of living heart (probably based on expts.); practiced a dielectical therapeutics; equated 4 elements (fire, air, water, earth) with qualities hot, cold, moist, dry, held them responsible for processes of body; attributed disease to excess or lack of 4 elements, also to external causes, changes in constitution of body, especially breathing; accepted Plato's physiology of breathing; regarded respiration as cooling of innate heat.

PHILLIPS, David Chilton, English biophysicist; b. Ellesmere, Eng., Mar. 7, 1924; s. Charles Henry and Edith (Finney) P.; B.Sc., U. Wales, 1947; Ph.D., U. Coll., Cardiff, 1951; m. Diana Kathleen Hutchinson, Nov. 18, 1960; 1 dau., Sarah Anne. Postdoctoral fellow NRC Can., Ottawa, 1951-53; research officer Nat. Research Labs., Ottawa, 1953-55; research officer Davy-Faraday Lab., Royal Instn., London, 1956-66; prof. molecular biophysics Oxford U., 1966—. Mem. Med. Research Council, U.K. External Staff, 1960-66. Fellow Inst. Physics U.K., Royal Soc.; mem. Brit. Biophys. Soc., Biochem. Soc., Royal Instn. Contbr. articles to profl. jours. Devel. techniques for studying fine structure of biol. materials especially X-ray diffraction methods; research on protein structure in atomic detail of myoglobin, lysozyme enzyme structure. Home: 3 Fairlawn Rd., Upper Wolvercote, Oxford. Office: Molecular Biophysics Lab., Zoology Dept., Oxford, Eng.*

PHILLIPS, Edouard, French physicist, mathematician; b. Paris, May 21, 1821; became material adminstr. Cie. des Chemins de fer de l'Est, 1849; named prof. mechanics École Centrale, 1864, later at École Polytechnique; mem. French Acad. Scis. Author: Manual pratique sur le spiral réglant des chronometres et des montres. Determined by approximation the flexibility coefficient of metal bridge from sag when loaded with conveyence; solved problem of leaf springs on railroad cars, 1850; studied (with Jacob) chronometry, after 1858; laid math. and theoretical found. for constrn. of chronometer spring; pub. his improvement of terminal coile of spiral springs, 1861; studied watch spring, 1865. Died Chateau de Narmont, France, Dec. 14, 1889.

PHILLIPS, Edwin Allen, Am. botanist; b. Lowell, Fla., Mar. 18, 1915; s. William Henry and Jane (Goodman) P.; A.B., Colgate U., 1937; M.A., U. Mich., 1940, Ph.D., 1948; m. Margaret Ellen Knight, Jan. 16, 1942; children—Ellen Knight, Nancy Jane. Instr. botany Colgate U., 1946-48; faculty Pomona Coll., Claremont, Cal., 1948—, prof. botany, 1955—; vis. prof. plant ecology U. Mich. Biol. Sta., summers 1955-56, 58. Sci. cons. AID, India, summers 1964-65; adv. com. San Dimas Exptl. Forest, 1957—; supr. blue lab. exercises Biol. Sci. Curriculum Study Lab. Innovation Com., 1961—, steering com., 1965—, film com., 1964—. Mem. Bot. Soc. Am., Am., Brit. bryological socs., Ecol. Soc. Am., A.A.A.S., Phi Beta Kappa, Sigma Xi. Author: Methods of Vegetation Study, 1959; Ecology of Land Plants and Animals, 1961; Field Ecology, 1964; Introductory Biology, 1968; also articles. Editorial bd. Vegetatio, Internat. Jour. Plant Sociology, 1955—. Research in ecology, ecology of vegetation. Home: 1201 N. College Av., Claremont, Cal. 91711.*

PHILLIPS, George Elwood, Am. biochemist; b. Decatur, Ill., Feb. 28, 1913; s. Phillip D. and Maude E. (Wilkinson) P.; B.A., James Millikin Coll., 1937; Ph.D., U. Ill., 1944; m. Joy B. Burcham, May 30, 1939. Chemist, Decatur Sanitation Dist., Ill., 1937-40; biochemist Maltine Co. div. Chilcott Labs., Warner-Chilcott Labs., 1944-47; dir. biochemistry dept. Warner-Lambert Research Inst., Morris Plains, N.J., 1947—. Mem. cons. panel Sigma Xi, 1966—. Fellow Am. Inst. Chemists; mem. N.Y. Acad. Scis., A.A.A.S., Am. Chem. Soc., Am. Heart Assn., Soc. for Study Blood, Reticuloendothelial Soc. Research, numerous publs. on devel. new drugs especially for prevention and treatment of coronary thrombosis and hemorrhages, fate of drugs in body. Home: 51 MacKenzie Rd., Convent Station, N.J. 07961. Office: 170 Tabor Rd., Morris Plains, N.J. 07950.*

PHILLIPS, Gerald Cleveland, Am. physicist; b. Plainview, Tex., Feb. 27, 1922; s. Grover Cleveland and Evelyn (Claitor) P.; B.A. with honors in Physics, Rice Inst., 1944, M.A., 1947, Ph.D., 1949; m. Anne Reynolds, Dec. 24, 1946; 1 son, G(erald) Cleveland R. Faculty, Rice Inst., Houston, 1949—, prof. physics, 1958—, chmn. dept., 1961—. Cons. Ampex, Inc. 1952—; mem. adv. panel on physics NSF, 1962—; mem. adv. group nuclear cross sect. AEC, 1961—. Carnegie Inst. fellow, 1950-51, Guggenheim fellow, 1957-58; NSF Sr. Postdoctoral fellow, 1966-67. Fellow Am. Phys. Soc.; mem. Soc. Exploration Geophycists (del. to NRC), Am. Assn. Tchrs. Physics, A.A.A.S., Am. Geophys. Union, Council Oak Ridge Inst. Nuclear Studies. Author: Progress in Fast Neutron Physics, 1963. Contbr. articles to profl. lit. Research in nuclear physics and geophysics. Home: 5636 Longmont St., Houston 77021.*

PHILLIPS, Horatio F., English inventor; b. London, 1845; inventor Phillips entry wing constrn.; patentee series of curved shapes, prototypes of modern aerofoil, 1884, proved their superiority to flat shapes by using them in machine that flew 40 miles an hour, 1893. Died 1912.

PHILLIPS, John, English geologist; b. Marden, Eng., Dec. 25, 1880; brought up by uncle William Smith, accompanied him on geol. expdns.; keeper Yorkshire Muse., 1825-40; became prof. geology King's Coll., London, 1834, also in Dublin; named reader in geology Oxford U., 1853; 1st keeper New Science Mus., Oxford; with Geol. Survey; mem. commn. to inquire into coal-mine ventilation. Fellow Royal Soc., 1834, Geol. Soc. (Wollaston medal 1928); a founder Brit. Assn. for Advancement of Science. Author: Illustrations of the Geology of Yorkshire; Rivers, Mountains and Sea-coasts of Yorkshire; Geology of the Thames Valley; Vesuvius; Phlegraean Fields; also articles. Named the Cainozoic era; introduced term Mesozoic; studied contour of mountains of moon; investigated slaty cleavage. Died Oxford, Apr. 24, 1874.

PHILLIPS, John Charles, Am. naturalist; b. Boston, Nov. 5, 1876; s. John Charles and Anna (Tucker) P.; S.B., Lawrence Sci. Sch. (Harvard), 1899; M.D., Harvard, 1904; grad. Boston City Hosp., 1906; m. Eleanor Hyde, Jan. 11, 1908; children—John C., Jr., Madelyn, Eleanor, Arthur. Asso. curator of birds Mus. Comparative Zoology, Harvard. Trustee Boston Soc. Natural History; chmn. Mass. Conservation Council. Author of papers on birds, genetics, exptl. animal breeding, conservation. Died Nov. 14, 1938.

PHILLIPS, John Gardner, Am. astrophysicist; b. New Haven, Jan. 9, 1917; s. Ray E(dmund) and Dora (Larson) P.; B.A., Carleton Coll., Northfield, Minn., 1939; M.A., U. Ariz., 1942; Ph.D., U. Chgo., 1948; m. Margaret A. Butler, June 11, 1944; children—Mary Jane, Cynthia Ann, Gail Elizabeth. Instr., Yerkes Obs., U. Chgo., Williams Bay, Wis., 1948-50; faculty U. Cal., Berkeley, 1950—, prof. astronomy, 1961—; asst. dean Coll. Letters and Sci., 1960-64, chmn. dept. astronomy, 1964-67. Mem. Am. Astron. Soc., A.A.A.S., Astron. Soc. Pacific. Author: (with D. Alter, C. Cleminshaw) Pictorial Astronomy, 1963; (with S. Davis) The Red System of the CN Molencule, 1963. Lab. studies of molecules of interest to astrophysics, application of this information to study of spectra of cooler stars. Home: 1955: 1234 Lawrence St., El Cerrito, Cal. 94530. Office: Campbell Hall, U. Cal., Berkeley, Cal. 94720.*

PHILLIPS, John Hunter, Jr., Am. physician; b. Houston, Nov. 2, 1930; s. John Hunter and Ida Minta (Sholars) P.; B.S., Tulane U., 1952, M.D., 1955; m. Mary Jo Holland, Dec. 18, 1954; children—Cynthia Kay, John Hunter, Virginia Minta. Intern, Charity Hosp., New Orleans, 1955-56, mem. staff, 1956—; practice medicine, specializing in internal medicine and cardiology, New Orleans, 1959—, faculty Sch. Medicine Tulane U., 1956—, asso. prof. medicine, 1966—; research fellow USPHS, Nat. Heart Inst., 1957-60; med. cons. Lallie Kemp Charity Hosp., Independence, La., 1956-63, Huey P. Long Charity Hosp., Pineville, La., 1959-63; clin. asst. VA Hosp., New Orleans, 1958-61, staff physician, chief cardiology sect., 1961-63; attending physician VA Hosp., Alexandria, La., 1960-62; chief cardiovascular labs. U.S. Naval Hosp., Portsmouth, Va., 1963-65; med. cons. USPHS Hosp., New Orleans, 1967—. Recipient Querens-Rives-Shore award in cardiology Tulane U., 1955, Isador Dyer Coll. Chest Physicians; mem. A.A.A.S., Am. Heart Assn., Phi Beta Kappa, Sigma Xi. Asst. editor: Am. Heart Jour., 1959—; editorial bd. Am. Jour. Med. Scis., 1962—. Numerous publs. on studies of factors influencing flow of lymph, microscopy of human mast cells, racial differences in cardiovascular disease, relationship between palm prints and congenital heart disease, clin. cardiology and peripheral vascular diseases. Home: 3905 Courtland Dr., Metairie, La. 70002. Office: 1430 Tulane Av., New Orleans 70112.*

PHILLIPS, Owen Martin, geophysicist; b. Parramatta, New South Wales, Australia, Dec. 30, 1930; s. Richard Keith and Madeline (Lofts) P.; B.Sc., U. Sydney (Australia), 1951; Ph.D., U. Cambridge (Eng.), 1955; m. Merle Winifred Simons, Aug. 7, 1953; children—Lynette M., Christopher I., Bronwyn Ann, Michael S. Imperial chem. Industries fellow U. Cambridge, 1955-57, asst. dir. research dept. applied maths. and theoretical physics, 1962-64; faculty Johns Hopkins, Balt., 1957-63, prof. geophys. mechanics, 1963—; sr. scientist Hydronautics, Inc., Laurel, Md., 1960—. Cons. to pvt. cos. Recipient Adams prize U. Cambridge, 1965. Fellow, St. John's Coll., Cambridge, 1956-59. Fellow Cambridge Philos. Soc.; mem. Am. Geophys. Union, Sigma Xi. Author: The Dynamics of the Upper Ocean, 1966; also articles. Asso. editor Jour. Fluid Mechanics, 1964—; tech. reviewer Math. Revs., 1960—, Zentrallblatt fur Mathematik, 1962. Research on physics surface waves on ocean and air-sea interaction, theory turbulent motion, generation sound by turbulence, theory circulation sugar in plants. Home: 23 Merrymount Rd., Balt. 21210.*

PHILLIPS, Philip, Am. anthropologist; b. Buffalo, Aug. 11, 1900; s. Bradley H. and Ruth (Harnden) P.; A.B., Williams Coll., 1922; M.Archeology, Harvard, 1927, Ph.D., 1940; m. Ruth Schoellkopf; m. 2d, Wilhelmina Schoellkopf; children—Patricia (Mrs. John Ogier), Sayre (Mrs. Stanley Sheldon), Bradley. Curator, Peabody Mus., Harvard, 1941——; lectr. anthropology Harvard, 1947——; dir. archeol. field projects 1940-41, 47, 49-51, 54-55. Mem. Kappa Alpha. Author (with others) Archaeological Survey in the Miss. Alluvial Valley, 1951; Method and Theory in American Archaeology, 1958. Home: 22 Berkeley St., Cambridge, Mass. 02138.*

PHILLIPS, Ralph Wesley, Am. physiologist; b. Parsons, W. Va., Feb. 7, 1909; s. Elijah N. and Margaret (Auvil) P.; B.S., Berea Coll., 1930; M.A., U. Mo., 1931, Ph.D., 1934. Asst. animal husbandry U. Mo., 1930-33; instr. U. Mass., 1933-36; with U.S. Dept. Agr., 1936-39, 41-46, dir. internat. orgns. staff Office Asst. Sec. Internat. Affairs, Washington, 1957——; prof. animal husbandry, head dept. Utah State U., 1939-41; dep. dir. agr. div. FAO, UN, 1946-57. Sci. sec. for agr. UN Conf. on Sci. and Tech., Geneva, Switzerland, 1963. Recipient U.S. Dept. Agr. Superior Service award. Fellow Am. Soc. Animal Sci., A.A.A.S. (mem. council), Am. Genetic Assn. (pres., mem. council), Author: (with R. Johnson, R. Moyer) The Livestock of China, 1945; Breeding Livestock Adapted to Unfavorable Environments, 1948; (with I. Moskovits, F. Liniger) The Organization of Agricultural Research in Europe, 1953; (with N. Joshi) Zebu Cattle of India and Pakistan, 1953; (with N. Joshi, E. McLaughlin) Types and Breeds of African Cattle, 1957; also sci. papers, tech. bulls., chpts. in sci. books. Research in physiology of reprodn. in farm animals including estrous cycles in sheep and goats, semen physiology and factors affecting fertility; environmental physiology and breeding of animals adapted to unfavorable environments; inheritance of economically important traits in livestock; compiling of information on methods of prodn. and animal genetic resources available in various parts of world; orgn. agrl. sci. work; internat. agr. and inter-relationships between human population and food supplies. Home: 2401 S. Lynn St., Arlington, Va. 22202. Office: Office Asst. Sec. for Internat. Affairs, U.S. Dept. Agr., Washington 20250.*

PHILLIPS, Ralph Wilbur, Am. chemist; b. Farmland, Ind., Jan. 12, 1918; s. J. Stanley and Effie (Berger) P.; B.S., Ind. U., 1940, M.S., 1955; D.Sc., U. Ala., 1963; m. Dorothy McLeaster, Aug. 21, 1943; 1 dau., Cheryl D. Faculty, Ind. U. Sch. Dentistry, Indpls., research prof. dental materials, 1962——. Hon. v.p. Internat. Dental Congress, Rome, 1957; chmn. dental study sect., chmn. bio-materials research adv. com. Nat. Inst. Dental Research; mem. com. sci. manpower USPHS; cons. to govt. agys. Recipient Gold Medal Research award Alumni Assn. Columbia Dental Sch., 1957; Research awards Chgo. Dental Soc., 1948, 56; Distinguished citation Washington U., St. Louis, 1965; Ann. Recognition award Ind. U. Student Union Bd., 1950. Fellow Internat., Am. colls. dentists, A.A.A.S.; mem. Am. Chem. Soc., Am. Assn. Dental Schs., Am. Dental Assn., Am. Assn. Dental Editors, Internat. Assn. Dental Research (Wilbur Souder award 1959, president), Sigma Xi, many others. Author: (with Skinner) The Science of Dental Materials, 6th edit., 1967; Elements of Dental Materials for Dental Hygienists and Assistants, 1965; (with others) Modern Day Practice in Crown and Bridge, 1966; also numerous articles. Contbr. sects. to The Dentist and His Assistant, 1961; Review of Dentistry, 1961; Clinical Dental Hygiene, 1964. Editor: Dental Clinics of North America, 1958, 64; Proceedings from a Workshop on Adhesive Restorative Dental Materials, 1962, 66. Research on properties and behavior of materials used in dentistry, nature of tooth structure and effects of fluorides upon that structure. Home: 735 E. 70th Pl., Indpls. Office: 1121 W. Michigan St., Indpls. 46202.*

PHILLIPS, Richard, English chemist; b. London, 1778; s. James Phillips; studied chemistry and pharmacy under William Allen; became lectr. chemistry London Hosp., 1817; lectr. London Instn.; prof. chemistry Royal Mil. Coll., Sandhurst; lectr. chemistry Grainger's Sch. Medicine, Southwark; chemist, curator Mus. Practical Geology, 1839-51. Fellow Royal Soc., 1822; mem. Chem. Soc. (pres. 1849-50). Author: An Analysis of the Bath Water, 1806; An Experimental Examination of the latest edition of the Pharmacopoeia Londinensis, 1811; Remarks on the editio altera of the Pharmacopoeia Londinensis, 1816; transl. (with notes) of Pharmacopoeia, 1824; also papers. Discovered true character of uranite, 1823. Died May 11, 1851.

PHILLIPS, Robert Allan, physician, med. researcher; b. Clear Lake, Ia., July 16, 1906; s. Allen Blackmore and Agnes (Allan) P.; B.S., Ia. State U., 1927; M.D., Washington U., St. Louis, 1929; m. Hope N. Fuess, Jan. 11, 1946; children—Judy, Robin, Wilkie, Hope, Honor, Innes, Shan. Asst. prof. physiology Cornell U. Med. Coll., N.Y.C., 1936-40; commd. lt. M.C., USNR, 1940, advanced through grades to capt. USN, 1955; stationed at Rockefeller Inst. Med. Research Hosp., N.Y.C., 1942-46; comdg. officer med. research unit, Cairo, Egypt, 1947-49, Taipei, Taiwan, 1955-65; head medicine and surgery br. Research div. Bur. Medicine and Surgery, Navy Dept., Washington, 1952-55; dir. Pakistan-SEATO Cholera Research Lab., Dacca, E. Pakistan, 1965——. Bd. dirs. U.S. Ednl. Found. in Egypt, 1949-52, U.S. Ednl. Found. in Republic China, 1957-65, chmn. 1960-64; vis. lectr. Recipient Farouk Cholera medal Egypt, 1947, Stitt award for outstanding achievement in med. research, 1962, James D. Bruce award Am. Coll. Physicians, 1966. Fellow Rockefeller Inst. Med. Research, N.Y.C., 1941. Mem. Am. Physiol. Soc., Soc. Exptl. Biology and Medicine, Harvey Soc., N.Y. Acad. Scis., Sigma Xi, Alpha Omega Alpha. Research, publs. on diabetes insipidus, treatment, pathogenesis of cholera, also circulation, metabolism. Home: Brightlands, Rd. 20, Dhanmondi, Dacca, E. Pakistan. Office: Pakistan SEATO Cholera Research Lab., Mohakhali, Dacca 5, E. Pakistan.*

PHILLIPS, Ronald Edward, Am. soil physicist; b. Williamstown, Ky., Nov. 30, 1929; s. Stanley Young and Alberta (Hasank) P.; B.S., U. Ky., 1954, M.S., 1955; Ph.D., Ia. State U., 1959; m. Susan Conrad, Nov. 4, 1950; children—Gregory Conrad, Rhonda Ann, Barbara Sue. Research asso. Ia. State U., Ames, 1948-59; asst. prof. agronomy U. Ark., Fayetteville, 1959-64, asso. prof., 1964——. Mem. Am. Soc. Agronomy, Soil Sci. Soc. Am., Brit. Soil Sci. Soc., Soil Sci. Soc. Can., Sigma Xi, Gamma Sigma Delta. Research, publs. on relationships between soil compaction and plant growth, tillage of soil and plant growth, soil phys. properties and diffusion of cations and anions in soil. Home: 2200 Ora Dr., Fayetteville, Ark. 72701.*

PHILLIPS, Theodore Evelyn Reece, English astronomer; b. Eng., Mar. 28, 1868; s. Abel Phillips; M.A., St. Edmund Hall, Oxford; m. M.H. Kynaston, 1906; 1 son. Ordained, 1891; held various curacies; rector of Headley, Epsom, 1916-41; univ. extension lectr., 1903-22; lectr. under Gilchrist Ednl. Trust, 1935——; mem. League Nations Com. on Calendar Reform. Fellow Royal Astron. Soc. (Jackson-Gwilt medal and gift 1918, Walter Goodacre medal and gift 1930, sec. 1919-26, pres. 1927-28), Royal Meteorol Soc.; dir. Jupitor sect. Brit. Astron. Assn., 1901-33, Saturn sect., 1935-40, assn. pres., 1914-16; pres. commn. 16 Internat. Astron. Union, 1925-35; pres. Newcastle-upon-Tyne Astron. Soc.; Editor: Popular Guide to the Heavens (R. Ball), 4th edit; co-editor Splendour of the Heavens. Contbr. articles on planets to Ency. Brit., also papers to astron. jours. Made observations in pvt. obs. on planets, double stars, variable stars; noted for observations of Jupiter. Died May 13, 1942.

PHILLIPS, Wendell, Am. archeologist; b. Oakland, Cal., Sept. 25, 1921; s. Merley H. and Sunshine (Chrisman) P.; A.B. with honors, U. Cal. at Berkeley, 1943; Sc.D., Marietta Coll., 1952; Litt.D., U. Redlands, 1954; J.D., Calvin Coolidge Coll., 1954; LL.D., Coll. Emporia, 1961; L.H.D., Sterling Coll., 1961; D.C.S., Colo. State U., 1962; H.H.D., Trinity U., 1962; Dr. Comml. Law, Pacific U., 1963; Ph.D., Whitworth Coll., 1964. Mem. U. Cal. Mus. Paleontology Expdn., No. Ariz., 1940, Monument Valley, So. Utah, Grand Canyon, 1942; mem. Carnegie Inst. Paleobot. Expdn., Ore., 1941; leader U. Cal. African Expdn., 1947-49, Mt. Sinai Expdn., Library Congress and Alexandria (Egypt) U., 1950, Carnegie Mus. Arabian Expdn. I-II, 1950-51, Yemen Expdn., 1951-52, Oman Expdn. I-II, 1952-53, Oman-Sahar Expdn., 1958, Oman Geog. Expdn., 1960; pres., dir. Philpryor Corp., 1951-58; chmn. bd. P.T.P. Corp., Reno, 1962——, Phillips Pacific, Sacramento, 1960——; pres. Am. Found. for Study Man, Washington, 1949——. Dir. gen. antiquities Sultanate Oman, 1953, econ. adviser Sultan, 1956; lectr. N.Y. U., 1958. Fellow Egyptian, Royal geog. socs. London, Royal Anthrop. Inst. Gt. Britain and Ireland, Am. Geog. Soc., Royal Central Asian Soc., Royal Asiatic Soc.; mem. Am. Oriental Soc., A.A.A.S., N.Y. Acad. Sci., Royal Archeol. Inst. Gt. Britain and Ireland, Middle East, French insts., Soc. Vertebrate Paleontology, Soc. Study Evolution, Am. Anthrop. Assn., Soc. Am. Archeology, Inst. Internat. Chateaux Historiques Switzerland, Alpha Gamma Nu (hon.). Author: Qataban and Sheba, 1955; Unknown Oman, 1966; Oman, A History, 1967; also articles. Research on South Arabian archeology, Oman history; microfilmed ancient library Monastery St. Catherines, Mt. Sinai, for Library Congress, 1950; led 1st party Westerners to reach and excavate Queen Sheba's capitol city, Yemen, 1951-52. Home: Diamond Head Apts., 2969 Kalakaua Av., Honolulu 96815. Office: 2400 Kalakaua Av., Honolulu 96815.*

PHILLIPS, William Dale, Am. phys. chemist; b. Kansas City, Mo., Oct. 10, 1925; s. Elmer Ernest and Mabyl (Craft) P.; student U. Tex., 1943-45; B.A. in Chemistry, U. Kan., 1948; Ph.D. in Phys. Chemistry, Mass. Inst. Tech., 1951; m. Esther J. Parker, Jun. 20, 1950; children—Edward D., Katherine Ann. Chemist, central research dept. Expl. Sta., E. I. du Pont de Nemours & Co., Wilmington, Del., 1951-58, research supr., 1959-64, asso. dir. basic scis., 1965——. Mem. Am. Chem. Soc., Am. Phys. Soc. Biophys. Soc., N.Y. Acad. Sci., A.A.A.S. Editor: (with F.C. Nachod) Determination of Structures of Organic Molecules, 1962. Research, publs. on internal rotation in molecules, charge-transfer interactions in solution, electronic structures of organic solids, distbns. of unpaired electrons in paramagnetic species, structures and interactions of proteins and nucleic acids using techniques of nuclear magnetic resonance, electron spin resonance, Mössbauer and optical spectroscopies. Home: 50 Shellburne Dr. Office: Central Research Dept., Exptl. Sta., E.I. du Pont de Nemours & Co., Inc., Wilmington, Del. 19803.*

PHILOLAOS OF CROTON, Greek philosopher, mathematician, astronomer; b. Croton or Tarentum, Southern Italy; a contemporary of Socrates (according to Plato); lived in Thebes in last decades of 5th century B.C.; student of Archytas (disciple of Pythagoras); left Metapontum because of political disturbances, entered Lucania, later visited Thebes. Committed teachings of Pythagoras to writing for first time; distinguished sensory, animal, and vegetative functions localized in brain, heart, and navel respectively; held earth not center of universe, but that it, counter-earth, fixed stars, sun, moon, and 5 planets circle about central fire, which is hearth of the world.

PHILON OF BYZANTIUM, Greek mechanician, natural philospher; flourished early 2nd century B.C. Followed Ctesibius and preceeded Heron. Wrote on mechanics; engines of war, pneumatic machines; credited with solution of problem of finding two mean proportionals.

PHILPOT, Van Buren, Jr., Am. pathologist; b. Houston, Miss., Mar. 3, 1923; s. Van Buren and Lois (Atkinson) P.; student Miss. Coll., 1940-42, Southwestern Coll. at Memphis, 1942-43; M.D., Tulane U., 1950; m. Rachel Gene Eaves, Aug. 1, 1946; children——Marjorie Gene, Eloise, James George, Van Buren III, Rachel Eaves. Asst. prof. pathology U. Tenn., Memphis, 1957-58; pathologist, dir. lab. Cary Meml. Hosp., Caribou, Me., 1958-61, Mt. Auburn Hosp., Cambridge, Mass., 1961-62; pathologist Framingham (Mass.) Union Hosp., 1962-63, Carney Hosp., Boston, 1963-64, Houston (Miss.) Hosp., 1964-65; pathologist, dir. labs. N. Miss. Research Found., Houston, 1965——. Mem. Soc. Exptl. Pathology, A.M.A., N.Y. Acad. Sci., Internat. Acad. Pathology, Soc. Clin. Pathologists, A.A.A.S. Author: Battalion Medics, 1955; also articles. Research on activation and inhibition of snake venom hemolysis by mammalian sera, importance of bacterial counts on urine samples for diagnosis of urinary tract infections; discovered anti-proteases in serum of blood of snakes to neutralize snake venom proteases and fibrinolysin. Home: 480 Castle St. Office: 104 Huddleston St., Houston, Miss. 38851.*

PHILPOTT, Delbert Eugene, Am. molecular biologist; b. Loyal, Wis., Sept. 24, 1923; s. Lacey D. and Nettie (Goehring) P.; B.S., Ind. U., 1947, M.S., 1948; Ph.D., Boston U., 1963. Research asso. U. Ill. Med. Sch., 1949-51; head electron microscopy Inst. for Muscle Research and Marine Biol. Lab., Woods Hole, Mass., 1951-63; asst. prof. biochemistry U. Colo. Med. Sch., Denver, 1963-65; head dept. electron microscopy Inst. for Biomed. Research, Mercy Hosp., Denver, 1965-66; research scientist, head electron microscopy NASA, Ames Research Center, Moffett Field, Cal., 1966——. Mem. N.Y. Acad. Scis., Electron Microscope Soc. Am., A.A.A.S., Am. Soc. Cell Biology, Sigma Xi. Contbr. photographs to textbooks, covers of sci. jours. Research, numerous publs. on muscle structure and function, div. cells., ultrastructure and function in relation to space biology problems; photographed first protein crystal with electron microscope. Home: 1602 Kamsack Dr., Sunnyvale, Cal. 94087. Office: NASA, Ames Research Center, Moffett Field, Cal. 94035.*

PHILUMENOS OF ALEXANDRIA, physician; flourished circa, 180; Greek physician of eclectic sch.; probably younger contemporary of Galen; compiled works used extensively by Oribasios in 4th century and Aëtios of Amida in 6th century; author book on diseases of bowels (extant in part), another on gynecology (lost), also De Venenatis animalibus (on animal poisons basis of Aelian's 13th book).

PHINNEY, Bernard Orrin, Am. botanist; b. Superior, Wis., July 29, 1917; s. Bernard Orrin and Franc (Lawrence) P.; B.A., cum laude, U. Minn., 1940, Ph.D., 1946; m. Feb. 2, 1951 (div. 1962); children—William Scott, Kathlene Ann; m. 2d, Jean Swift, Dec. 11, 1965. Fellow, Cal. Inst. Tech., Pasadena, Cal., 1946-48; faculty U. Cal. at Los Angeles, 1947——, prof., botany, 1961——. NSF Sr. fellow Genetics Inst., U. Denmark, Copenhagen, 1959-60; vis. scientist NSF, U.S.-Japan Coop. Sci. Program, 1966-67. Mem. Am. Inst. Biol. Scis., A.A.A.S., Genetics Soc., Scandinavian, Am. socs. plant physiologists, Bot. Soc. Am., Am. Soc. Naturalists, Sigma Xi. Research, publs. on genetics and physiology of gibberellins in flowering plants and fungi; discovered gibberellins in flowering plants. Home: 671 Landfair St., Los Angeles 90024.*

PHLEGER, Fred B., Am. oceanographer; b. Kansas City, Kan., July 31, 1909; s. Fred B. and Norabelle (Elliott) P.; A.B., U. So. Cal., 1931; M.S., Cal. Inst. Tech., 1932; Ph.D., Harvard, 1936; m. Marjorie Temple, Oct. 20, 1933; children—Charles F. and Audrey P. (Mrs. Scott McElmury). Asst. paleontologist Harvard, 1934-36; Sheldon Traveling fellow, 1936-37; faculty Amherst Coll., 1937-49, asso. prof., 1944-49; vis. asso. prof. U. Cal. at La Jolla, 1949-51, asso. prof., 1951-57, prof. oceanography, 1957—; dir. Cushman Found. for Foraminiferal Research, 1952—; dir. Marine Foraminifera Lab., La Jolla, 1949—. Named Investigador Scientifico Honorario y Distinguido, Universidad Nacional Autonoma de Mexico, 1959. Mem. Geol. Soc. Am., Am. Assn. Petroleum Geologists, Paleontol. Soc., Soc. Econ. Palentologists and Mineralogists. Author: The Whales Go By, 1959; Ann Can Fly, 1959; Ecology and Distribution of Recent Foraminifera, 1960; Red Tag Comes Back, 1961; also articles. Research on ecology Foraminifera and their applications to problems oceanography and geology, history and processes devel. coastal lagoons, history oceans during Pleistocene ice ages. Home: 8593 La Jolla Dr. Office: Scripps Instn. Oceanography, La Jolla, Cal. 92037.*

PHOUPHAS, Chryssanthos, Greek botanist; b. Piraeus, Greece, Mar. 25, 1909; s. Alexios and Helene (Politou) P.; ed. univs. Athens and Paris; dr es sci. Successively asst., research chief, instr., prof. botany U. Athens. Participant 2d Internat. Congress Biochemistry, Paris, 1952. Decorated Order George I. Mem. Hydrobiology Council Greece, French Bot. Soc., French Biol. Chemistry Soc. Author: Dévelopment de l'embryon chez le géranium Robertianum; Action de l'hydrazide maléique sur la teneur en substances glucidiques des tissue de topinambours cultivs in vitro; Sur la présense de saponaroside chez cultivés in vitro; Sur la présense de saponaroside chez certaines aracées; Sur la distributions des tannoides dans les plants; Recheches sur certaines espèces du genre hibiscus. Home: Odos Pipinou 51 A. Office: Odos Solonos 104, Athens, Greece.

PHRIES (Phrisius, Phryes), Lorenz, see Fries, Lorenz.

PHYSICK, Philip Syng, Am. surgeon; b. Phila., July 7, 1768; s. Edmund and Abigail (Syng) P.; grad. U. Pa., 1785; studied medicine Windmill St. Sch., London, Eng., 1788. M.D., U. Edinburgh (Scotland), 1792; m. Elizabeth Emlen, Sept. 18, 1800, 7 children. House surgeon St. George's Hosp., London, 1790; physician during yellow fever epidemics, Phila., 1793, 98; clin. tchr. on staff Pa. Hosp., 1794-1816; surgeon Almshouse Infirmary, 1800; became lectr. U. Pa., circa 1800, prof. surgery, 1805-19; mem. English, French med. socs., Am. Philos. Soc.; 1st pres. Acad. Medicine. Called father of Am. surgery; responsible for many surg. advances, including manipulation instead of mech. traction in reduction of dislocations, new methods in treatment of hip-joint disease by immobilization, notable work in urinary tract and bladder-stone operations; pioneer in use of stomach tube; performer successful operation on arteriovenous aneurysm, 1804; introduced absorbable buckskin and kid ligatures, circa 1816, ceton in treatment of ununited fractures, circa 1822; inventor guillotine tonsillotome, needle forceps, snare for use in tonsillectomy, new type of catheters. Died Phila., Dec. 15, 1837.

PIA, Hans Werner, German neurosurgeon; b. Bochum, Germany, Jan. 26, 1921; s. Walther and Elisabeth (Wilhelmi) P.; ed. in medicine, U. Marbourg (Germany); m. Lore Koch, Sept. 26, 1943; children—Klaus, Bärbel. Asst., Univ. Clinic Cologne, Germany, 1946-53; head neurosurgery sect. U. Giessen (Germany), 1953, agrégé prof. neurosurgery, 1956, asso. prof., dir. Surgery Clinic, 1962, now prof. Recipient Langenbeck prize German Surgery Soc., 1958. Corr. mem. numerous societies. Publns. on neurosurgery, cerebro-spinal diseases, epilepsy, hemispherectomy. Home: Giessen Oberhof 6. Office: Klinikstrasse 37, Giessen, West Germany.

PIAGET, Jean, Swiss psychologist; b. Neuchatel, Switzerland, Aug. 9, 1896; studied biology, U. Neuchatel, B.A. 1915, Dr. ès Sci., 1918; also studied at U. Zürich and U. Paris; hon. degrees Harvard, 1936, Sorbonne, 1946, Brussels, 1949, Rio de Janeiro, 1949, Chicago, 1953; dir. studies, Institut J. J. Rousseau, Geneva, Switzerland, from 1921; prof. philosophy, U. Neuchatel, 1926-29; prof. child psychology and history of scientific thought, U. Geneva, since 1929, dir., Psychology Laboratory, since 1940; prof. gen. psychology, U. Lausanne, 1937-54; founded International Center for Genetic Epistemology, 1955; also associated with Bureau International d'Education (dir.); mem. exec. bd., UNESCO and Institute for Educational Sciences, Geneva; mem. Nat. Acad. Scis. and many other sci. socs. Author: Le language et la pensée chez l'enfant; Le jugement et le raisonnement chez l'enfant; La représentation du monde chez l'enfant; La causalité physique chez l'enfant; Le jugement moral chez l'enfant; Le devéloppement de la notion de temps chez l'enfant, 1946; Les notions de mouvement et de vitesse chez l'enfant; 1946; Traité de logique, 1949; Introduction à l'épistémologie génétique, 3 vols., 1950; Essai sur les transformations des operations logiques, 1952; Les mécanismes perceotifs: modèles probabilistes, analyse génétique, relations avec l'intelligence, 1961; many others and more than 200 articles. Using phenomenological approach, investigated all aspects of intelligence of children: their implicit philosophies, infant's construction of reality, the stages (related to age) of mental development; explored children's spontaneous ideas about physical world and their own mental processes; studied origins of children's mental growth in behavior of infants; conducted systematic theoretical and experimental work on perception and on problems of genetic epistemology; studied development of logical concepts; derived schema of developmental stages; also studied molluscs.

PIANA, Guiseppe, Italian nutritionist; b. Milan, Italy, Apr. 29, 1914; s. Giovanni and Maria (Cantaluppi) P.; ed. in agrl. sci. and vet. medicine; m. Isolde Lehmann, Oct. 15, 1941; children—Nigi, Gionni, Dani, Betti, Chicca. Prof., dean faculty, dir. Zootechnic Inst. of Nutrition Scis. of Agr., Faculty Catholic U. Milan. Decorated medal of merit in sci. Mem. Nat. Agr. Acad. Publns. on biochemistry, food, applied endocrinology. Address: via San Ambrogio 29, Milan, Italy.

PIANCIANI, Giambattista, Italian physicist; b. Spoleto, Italy, Oct. 27, 1784; became mem. Soc. Jesus, 1805; tchr. physics, various schs.; prof. Roman Coll., 25 years; tchr., Georgetown, 1849-50; returned to Rome, 1851; tchr. Roman Coll.; tchr. Collegio Filosofico, apptd. pres., 1860. Publs. on electricity and magnetism. Died Mar. 23, 1862.

PIAZZA, Giovanni, Italian physician; b. Vescovato-Cremona, Italy, Jan. 6, 1921; s. Charles and Catherine (Lupi) P.; student U. Bologna (Italy), 1940-44; D.Medecin, U. Parma (Italy), 1947; m. Sinelli Mina, Nov. 17, 1952; children—Charles, Daniel. Staff, Civil Hosp. Cremona, 1947-49; staff Inst. Radiology, U. Padua (Italy), 1950—, extraordinary asst., 1955-63, ordinary asst., 1963—, tchr. Sch. radiol. techniques Sch. Specialization Radiology, Specialization Respiratory Apparatus, 1955—, tchr. 1962—. Mem. Italian Soc. Radiology, Internat. Soc. for Study Bronchus. Author: Curved Tomography, 1960; Image Intensification in Radiology, 1966; also numerous articles. Radiol. studies of digestive tract, lead poisoning, respiratory system, morphological aspect of colon little transport, curved tomography on plane film, radiate tomography, mouldable filters in radiodiagnostics. Home: Via Vittorio Emanuele, Villafranca, Padova, Italy.*

PIAZZA, Marcello, Italian physician; b. Naples, Italy, Sept. 11, 1935; s. Raffaele and Carla (Kulhankova) P.; grad. with 1st class honors, U. Naples, 1959; m. Antonietta Caccuri, Apr. 30, 1962; children—Raffaele, Zaira. Staff, U. Naples, 1960—, dept. head infectious diseases clinic, 1965—, libero docente infectious diseases, 1966—, libero docente biochemistry, 1964—. Recipient award U. Naples, 1962, Carlo Erba Ministero pubblica istruzione, 1963, Dante De Blasi, 1964. Mem. Società italiana di biologia sperimentale. Author: Epatiti Sperimentali da Virus MHV-1 e MHV-3, 1965; Experimental Viral Hepatitis, 1968; also articles. Discovered biochem. changes during course of viral hepatitis, substance inactivating murine hepatitis virus in intestine of normal mice. Home: 63 Parco Comola, Napoli, Italia.*

PIAZZI, Giuseppe, Italian astronomer; b. Ponte, Valtellina, Italy, July 16, 1746; ed. Milan, Turin, Rome, Genoa, entered Theatine order, Milan, 1764; tchr., Genoa, Malta; became prof. theology, Rome, 1779; named prof. higher math. Palermo Acad., 1780; founder, dir. obs., Palermo, began observations, 1792; commd. to reform Sicily's weights and measures, 1812; in charge govt. obs., Naples, from 1817. Fellow Royal Soc., 1804. Author: Lezioni elementari di Astronomia, 1817. Piazzia (1000th planetoid discovered) named in his honor. Published catalogue of fixed stars, 1803, 2d catalogue listing 7646 stars, 1814; wrote on astronomy, length of tropical year, light aberration. Discovered 1st known planetoid (Ceres), 1801; demonstrated that proper motions detected by Halley are usual motions among stars; discovered star 61 Cygni. Died Naples, July 22, 1826.

PICARD, Charles Emile, French mathematician; b. Paris, France, July 24, 1856; ed. Ecole Normale; m. Mlle. Hermite. Master confs. U. Paris, 1878-79, tchr. mechanics Sorbonne, 1881-85, prof., 1886-1941; prof. U. Toulouse (France), 1879-81, Central Sch. Arts and Manufactures, Paris, 1893-1941. Vis. lectr. Clark U., Worcester, Mass., 1899. Mem. Royal Soc., 1909, Royal Soc. Edinburgh, French Acad. Scis. (v.p. 1908, pres. 1910, perpetual sec. math. scis. 1917), 1889 acads. Berlin (Germany), Leningrad (USSR), others. Author: Traité d'analyse, 3 vols., 1891-95; (with G. Simart) Théorie des fonctions algébriques de deux variables indépendantes, 2 vols., 1897-1906; Equations fonctionelles, 1929; Quelques applications analytiques de la théorie des courbes et des surfaces algébriques, 1931. Worked on theory of Fuchsian, hyperfuchsian functions, allied infinite discontinuous linear groups; proved Picard's theorem involving number of solutions of gen. equation $f(z)=A$, 1879; (in theory differential equations) contbd. analogue for linear differential equations of Galois theory of algebraic equations: helped develop Cauchy-Lipschitz method solving differential equations; made analytical study algebraic functions and their integrals (making possible fruitful cooperation between analytic and geometric methods), 1885. Died Paris, Dec. 11, 1941.

PICARD, Jean, French astronomer; b. La Flèche, France, July 21, 1620; became Roman Catholic priest; prior of Rille, Anjou; studied astronomy under Gassendi, succeeded him as prof. astronomy Coll. de France, 1655. Mem. French Acad. Scis., 1666. Author: La mesure de la terre, 1671. Traité du nivellement, 1684; Historie céleste, ou receuil de observations astronomiques faites par l'orore du roi, pub. 1741. Established Connaisance des temps, 1679, pub. 1st 5 ann. vols. Father of modern astronomy in France; observed (with Gassendi) solar eclipse of Aug. 25, 1645; determined longitude and latitude of Tycho Brahe's obs., Uraniborg; made 1st accurate measurement of degree of meridian, from Mahoisine to Sourdon, France (results used by Newton in work on gravitation), 1669-70; 1st to use telescope to measure angles; inventor transit instrument, micrometer, pendulum clock. Died Paris, Oct. 12, 1682.

PICARD, M. Dane, Am. geologist; b. Washburn, Mo., Aug. 7, 1927; s. Vincent Haynes and Velma (Stubblefield) P.; student Swarthmore Coll., 1945-46; B.S., U. Wyo., 1950; A.M., Princeton, 1962, Ph.D. (NSF fellow), 1963; m. Virginia Yvonne Reitz, July 5, 1958; children—Marion Velma, Jacqueline Ann, Dane Vincent, Bennet Verl. With Texaco, Inc., 1950, Shell Oil Co., 1950-56, St. Helens Petroleum Corp., 1956-57, Am. Stratigraphic Co., 1957-60, Pan Am. Petroleum Corp., 1964; faculty Ft. Lewis A. and M. Coll., 1957-59, Princeton, 1960-61; prof. geology U. Neb., Lincoln, 1963—. Fellow Geol. Soc. Am., A.A.A.S.; mem. Am. Assn. Petroleum Geologists, Soc. Econ. Paleontologists and Mineralogists, Internat. Assn. Sedimentology, Nat. Assn. Geology Tchrs., Sigma Xi. Research, numerous publs. on stratigraphy, sedimentary petrology, petroleum geology. Office: Dept. Geology, U. Neb., Lincoln, Neb. 68508.*

PICART, Théophile-Luc, French astronomer; b. La Hardoye, France, July 4, 1867; prof. Lille (France) Faculty Sci., also Bordeaux (France) Faculty Sci.; became dir. Bordeaux obs., 1906; mem. French Acad. Scis. Author: Astronomie générale, 1942; Sur le mouvement d'un corps de figure variable, 1895. Studied movement of comets; research in math. analysis. Died Floirac, France, Mar. 16, 1956.

PICCARD, Auguste, Swiss physicist; b. Basel, Switzerland, Jan. 28, 1884; D.Sc., Fed. Coll. Tech., Zurich, Switzerland. Became physics asst., later prof. physics Fed. Coll. Tech.; prof. Polytech. Inst., Brussels, Belgium, 1922-39, 46-47. Best known for his pioneer exploration of stratosphere; constructed balloon with air tight gondola attached; made balloon ascent (with Kipfer) of 51,793 feet, Ausburg, Germany, 1931, then (with Max Cosyns) ascent of 55,577 feet, Zurich, 1932; constructed Trieste bathysphere used for deep sea exploration and tested it off the Azores, 1948; probed ocean depths around Guam and Marianas, 1960; studied cosmic rays and other sci. phenomena. Died Lausanne, Mar. 25, 1962.

PICCARD, Jacques, oceanographic engr.; b. Brussels, Belgium, July 28, 1922; s. Auguste and Marianne (Denis) P.; Licence U. Geneva, 1946; Sc.D., hon., U. Mass., 1962; diploma Inst. Internat. Studies, Geneva, 1967; m. Marie-Claude Maillard, Nov. 14, 1953; children—Bertrand, Marie-Laure, Thierry. Asst. prof. U. Geneva, 1946-48; pvt. tchr., 1948-51; cons. underwater research high pressure studies, 1948—. Decorated Croix de Guerre, France, 1945; Distinguished Pub. Service award given by President Eisenhower, 1960; Theodore Roosevelt Distinguished Service award, 1960; Richard Hopper Day Meml. award Acad. Natural Scis. of Phila., 1960; Day award Drexel Inst. Engrs., 1960. Hon. mem. Swiss Inst. Naval Architects, Swiss Inst. Natural Scis.; life mem. Nat. Geog. Soc. Author: (with Robert S. Dietz) Seven Miles Down, 1961; also articles. With Auguste Piccard designed and operated 1st deep diving vessel, bathyscaph; designed, numerous dives in Trieste, mesoscaph Auguste Piccard; over 1100 dives; with Don Walsh made record descent to 35,800 feet In Marianas Trench, 1960. Home: 9, Chemin de Fontanettaz. Office: 15, Chemin de Fontanettaz, 1012 Lausanne, Switzerland.*

PICCARD, Sophie, mathematician; b. St. Petersbourg, Russia, Sept. 27, 1904; d. Eugene-Ferdinand and Eulalie (Güée) P.; Diploma in Math. and Phys. Scis., U. Smolensk, 1925; Lic., U. Lausanne (Switzerland), 1927, Dr., 1929. Faculty, U. Neuchatel (Switzerland), 1936—, prof. geometry and statistics, prof. math., dir. Institut Geometry. Lectr., Faculty Scis., Paris, 1948, Palais de la Découverte, Paris, 1956, Freie Universität Berlin, Technische U. Berlin, 1965. Internat. Fedn. U. Women fellow, 1941. Mem. math. socs. U.S., Austria, France, Poland and Switzerland, Assn. Française pour l'Avancement des Sciences, Société Helvétique des Sciences Naturelles. Author: Sur les ensembles de distances des ensembles de points d'un espace euclidien, 1939; Sur des ensembles parfaits, 1942; Sur les bases du groupe symétrique I et II, 1946, 48; Sur les bases des groupes d'ordre fini, 1958; Sur les groupes libres et quasi libres mod n., 1966; also numerous papers. Research on set theory, theory of groups of finite order, theorems about generators and relations of abstract groups; introduced notion of fundamental group and P-product of groups; discovered wide and various classes of P-groups; studies in linear algebra, theory of relations and history of math. Address: Poste restante Gare, 2002 Neuchätel, Switzerland.*

PICCIONI, Oreste, physicist; b. Siena, Italy, Oct. 24, 1915; s. Ubaldo and Calliope (Burali) P.; student

U. Pisa (Italy), 1934-36; Dr.Physics, U. Rome (Italy), 1938; m. Marina Aardema, Oct. 26, 1963; children—Robert, Richard, Gabriella. Came to U.S., 1946, naturalized, 1956. Asst. prof. U. Rome, 1938-45, prof. incaricato, 1945; vis. research scientist Mass. Inst. Tech., 1946-48; scientist Brookhaven Nat. Lab., 1948-60; vis. physicist Lawrence Radiation Lab., Berkeley, Cal., 1955-56, 58-59; prof. physics U. Cal. at San Diego, 1960—. Fellow Am. Phys. Soc.; Contbns. to study of elementary particles. Home: P.O. Box 1323, Rancho Santa Fe, Cal. 92067.*

PICCOLOMINI, Alessandro, Italian physicist; b. Siena, Italy, 1508; became prof. philosophy, Padua, 1540; named archbishop of Patras, 1574; coadjutor of Siena. Author: De la sfera del mondo (then the best Italian textbook of astronomy), 1540; De le stelle fisse; In mechanicas questiones Aristotelis paraphrasis paulo quidem plenior . . . (based on theory of impetus), 1542; other commentaries on Aristotle, 1562-75. Died 1578.

PICCONE, Antonio, Italian botanist; b. Albissola, Italy, Sept. 11, 1844; s. Francesco and Maria (Aprile) P.; grad. Atheneum, Genoa, Italy, 1864; became tchr. Genoa tech. schs., 1868; named tchr. Sch. for Men of Genoa, 1869; apptd. instr. natural history Christopher Columbus Licco, Genoa, 1874, prof., 1885-1901. Research and publs. on algae; mapped incidence of algae types in Italian sea waters and Red Sea. Died May 21, 1901.

PICENA, Juan Pedro, Argentinian pathologist; b. Rosario, Argentina, Jan. 1, 1906; s. Juan Luis and Rosa (Iula) P.; B.A., 1923; D.Medicine, U. Litoral, 1932; m. María Esther Diez Rodríguez, Oct. 15, 1934; 1 son, Juan Carlos. With Med. Sch., U. Ritaral, Rosario Argentina, 1924—, prof., 1955—, dir. dept. pathology, 1956—; dean Med. Sch., 1966—. Recipient Gold medal Med. Circle, Rosario Fellowship, 1935; Nat. Acad. Medicine Buenos Aires, Germany, 1934; Guggenheim fellow, Harvard, 1939. Mem. Med. Assn. Rosario, Argentine Med. Assn., Soc. Normal and Path. Anatomy Buenos Aires, N.Y. Acad. Scis., Internat. Soc. Hematology, Internat. Acad. Pathology. Author: La Biopsia de la Médula ósea, 1935; Las Linfomatosis, 1942; also numerous articles. Research on hematopoiesis, pathology of pancreas, lymph nodes. Home: 2190 San Lorenzo, Bosario-S.Fe, Argentina.*

PICHI-SERMOLLI, Rodolfo Emilio Giuseppe, Italian botanist; b. Florence, Italy, Feb. 24, 1912; s. Giuseppe and Maria (Del Rosso) P.-S.; Degree in Natural Sci., U. Florence, 1935; m. Carla Bernardini, Apr. 9, 1942; children—Christina, Niccolo (dec.). Asst., Bot. Inst., Florence U., 1935-58; prof. botany U. Sassari, (Sardina, Italy), 1958-59; prof., dir. bot. inst. and garden U. Genoa (Italy), 1959—. Mem. adv. com. arid zone UNESCO, 1954-56. Mem. Linnean Soc. London (fgn.). Research and numerous publs. on plant ecology, geography, flora and vegetation of tropical Africa, toxonomic botany. Home: Corso Dogali 1/c, Genoa, Italy.*

PICK, Ernest Peter, pharmacologist; b. Jaromer, Bohemia, May 18, 1872; s. David and Eleanor (Schick) P.; M.D., U. Prague, 1896; student U. Strasbourg, 1897-99; m. Margaret Rikea Janssen, Aug. 1, 1927. Research fellow Biochem. Inst. of Prof. Fr. Hofmeister, U. Strasbourg, 1897-99; asst., head biochem. dept. Austrian State Serum Inst., Vienna, 1899-1911; asst. prof. biochemistry, 1904-11; asst., asso. prof. exptl. pharmacology and toxicology U. Vienna, also head Inst. for Drug Investigation, Austrian Health Dept., 1924-38; clin. prof. pharmacology Columbia, N.Y.C., 1939-60; asso. pharmacologist Mt. Sinai Hosp., N.Y.C 1939-60. Author: Untersuchungen über Protein-Stoffe und peptische Spaltungsprodukte des Fibrins, 1897-1902; Zur Kenntis der Immunkörper, 1902; (with Fr. Obermayer) Über die Chemischen Grundlagen der Arteigenschaften der Eiweisskörper, 1906; Darstellung der Antigene mit chemischen und physikalischen Methoden, 1907; Biochemie der Antigene, 1912; Hypnotica, 1926; (with H. H. Meyer) Die Experimentelle Pharmakologie als Grundlage der Arzneibehandlung, 1933, 36; (with Fr. Silberstein) Biochemie der Antigene und Antikörper, 1938. Research on biochemistry, exptl. pathology and pharmacology, autonomous nervous system, diuresis, mechanism of bleeding in liver. Died N.Y.C., Jan. 15, 1960.

PICK, Ruth Holub, physician; b. Carlsbad, Czechoslovakia, Nov. 13, 1913; d. Arthur and Paula (Lenk) Holub; M.D., German U. Prague (Czechoslovakia), 1938; m. Alfred Pick, May 28, 1938. Came to U.S., 1949, naturalized, 1954. Research fellow cardiovascular dept. Michael Reese Hosp., Chgo., 1949-50, research asso., 1951-58, asst. dir. cardiovascular dept., 1958—; established investigator Am. Heart Assn., 1958-64. Mem. Am. Heart Assn. (mem. council arteriosclerosis), Am. Fedn. Clin. Research, Am. Physiol. Soc., A.A.A.S., Central Soc. Clin. Research, Chgo. Heart Assn. Contbr. articles in pub. sci. to profl. jours. Prodn., treatment exptl. hardening of arteries; research differences in this process in various vascular beds, also effect female sex hormones on coronary atherosclerosis. Home: 601 E. 32d St., Chgo. 60616. Office: 2900 S. Ellis St., Chgo. 60616.*

PICKARD, George Lawson, oceanographer; b. Cardiff, Wales, July 5, 1913; s. Harry Lawson and Phoebe (Crosier) P.; B.A., Oxford (Eng.) U., 1935; D.Phil., 1937, M.A., 1947; m. Lilian May Perry, Apr. 26, 1938; children—Rosemary Ann (Mrs. Roger McAfee), Andrew Lawson. Sci. officer Royal Aircraft Establishment., Farnborough, Eng., 1937-42; prin. ei. officer operational research sect. coastal command RAF, Eng., 1942-47; faculty U. B.C., Vancouver, Can., 1947—, prof. physics, 1954—, dir. Inst. Oceanography, 1958—. Mem. Fisheries Research Bd. Can., 1963—. Bd. mgmt. B.C. Research Council. Fellow Royal Soc. Can., A.A.A.S.; mem. B.C. Acad. Sci. (past pres.), Am. Soc. Limnology and Oceanography (past v.p.), Am. Geophys. Union. Author: Descriptive Physical Oceanography, 1964; also articles. Research on circulation in estuaries and coastal seas relating to anti-pollution studies. Home: 4546 W. 5th Av., Vancouver, 8, B.C., Can.*

PICKARD, Greenleaf Whittier, Am. elec. engr., inventor; b. Portland, Me., Feb. 14, 1877; s. Samuel Thomas and Elizabeth Hussey (Whittier) P.; ed. Westbrook Sem., Lawrence Sci. Sch., Harvard, Mass. Inst. Tech.; m. Miriam Watson Oliver, Apr. 5, 1902 (dec. Dec. 1912); m. 2d, Helen Liston, Apr. 27, 1914; children—Helen Liston, Elizabeth Whittier, Geraldine, Greenleaf Whittier, Mary Katherine, John. Engr. Am. Tel. & Tel. Co., 1902-06; cons. engr. Wireless Specialty Apparatus Co., 1907-30, RCA Victor Co. of Mass., 1930-31; cons., 1932-42; dir. research Am. Jewels Corp., Attleboro, Mass., 1942-45; pres. Pickard & Burns, Inc., 1945-52, chmn. bd., from 1952. Fellow Instn. Radio Engrs. (medal of honor 1926, pres.), Radio Club Am. (Armstrong medal 1941), Am. Inst. E.E., A.A.A.S., Am. Acad. Arts and Scis.; mem. Am. Meteorol. Soc. Patentee radio inventions; one of 1st to obtain successful transmission of speech by elec. waves; inventor crystal detector, the radio compass, the static eliminator. Died Jan. 8, 1956.

PICKART, Stanley Joseph, Am. physicist; b. Norway, Ia., May 12, 1926; s. Clemens J. and Edna (Storey) P.; m. Louise Salvio, Aug. 2, 1952; children—Cecile, Maureen, Andrea, Robert, David, George, Christopher. Research physicist U.S. Naval Ordnance Lab., Silver Springs, Md., 1951-54, 56—; guest physicist Brookhaven Nat. Lab., Upton, N.Y., 1955—. Asst. prof. physics State U. N.Y., Stony Brook. Recipient Navy Meritorious Civilian Service award, 1959. Fellow Am. Phys. Soc.; mem. A.A.A.S., Sigma Pi Sigma. Research, publs. on metals and alloys determining spin arrangements, spin density distbns. and spin wave excitation spectra, neutron diffraction research on magnetic oxides. Home: R.F.D. Number 1, Box 1-H, Brookeville, Md. 20729. Office: U.S. Naval Ordnance Lab., Silver Springs, Md.*

PICKAVANCE, Thomas Gerald, English physicist; b. St. Helens, Eng., Oct. 19, 1915; s. William and Ethel (Leyland) P.; B.Sc., U. Liverpool (Eng.), Ph.D. in Physics; m. Alice Isobel Boulton, Apr. 14, 1943; children—Margaret, John David, Robert Gerald. Lectr. physics, research staff nuclear energy project Directorate Tube Alloys, U. Liverpool, 1940-46; head cyclotron group Atomic Energy Research Establishment, Atomic Energy Authority, from 1946, dep. head gen. physics div., until 57; dir. Rutherford High Energy Labs., Chilton, Berkshire, Eng., 1957—. Mem. Am. Phys. Soc., Inst. Physics, Phys. Soc. London. Contbr. articles to tech. jours., chpts. in books. Research on nuclear physics and high energy physics; research and devel. on particle accelerators, detectors, instrumentation. Home: Craigellachie, Hinksey Hilltop, Oxford, Eng. Office: Rutherford Lab., Chilton, Berkshire, Eng.*

PICKERING, Charles, Am. physician, naturalist; b. Susquehanna County, Pa., Nov. 10, 1805; s. Timothy Pickering, Jr. and Lurena (Cole) P.; M.D., Harvard, 1826, A.B. (as of 1823), 1849; m. Sarah Hammond, 1851. Physician; mem. Acad. Natural Scis., 1827, librarian, 1828-33, curator, 1833-37; chief zoologist U.S. Exploring Expdn. S. Seas, 1838-42; traveled through Egypt, Arabia, East Africa, India, 1843-45. Mem. Am. Philos. Soc., Phila. Author: Races of Men and Their Geographical Distribution, 1848; The Chronological History of Plants: Man's Record of His Own Existence Illustrated Through Their Names, Uses and Companionship, 1879; Essay on the Invention of the Art of Writing; Notes on the Stinging Power of the Physalia; Geographical Distribution of Plants. Died Boston, Mar. 17, 1878.

PICKERING, Edward Charles, Am. astronomer; b. Boston, July 19, 1846; s. Edward and Charlotte (Hammond) P.; ed. Boston Latin Sch.; S.B., Lawrence Scientific Sch. (Harvard), 1865, A.M., (hon.), 1880; LL.D., univs. of Cal., 1886, Mich., 1887, Chgo., 1901, Harvard, 1903, Pa., 1906; Ph.D., Heidelberg U., 1903; D.Sc., Victoria U., Eng., 1900; L.H.D., Allegheny Coll., 1912; m. Lizzie Wadsworth Sparks, Mar. 9, 1874. Instr. math. Lawrence Sci. Sch., 1865-67; Thayer prof. physics Mass. Inst. Tech., 1867-76; prof. astronomy and dir. Harvard Coll. Obs., 1876—. Recipient Henry Draper medal for work on astron. physics, Bruce Gold medal, 1908, Gold medals Royal Astron. Soc., 1886, 1901. Fellow Royal Soc., 1907 (Rumford medal 1891), Am. Acad. Arts and Scis.; mem. Nat. Acad. Scis.; hon. mem. or mem. many Am. and fgn. sci. socs.; founder, 1st pres. Appalachian Mountain Club. Author: Elements of Physical Manipulation, 2 vols., 1874-76; also prepared work for Revised Harvard Photometry, 1908. Editor 70 vols. of annals, other publs. of Harvard Coll. Obs. Established 1st phys. lab. in U.S.; under his direction, invested capital and income of the obs. increased fourfold; study of light and spectra of stars spl. features of his work; devised meridian photometer and made 1,400,000 measurements of light of stars with it; by establishing an auxiliary sta. in Arequipa, Peru, Southern stars also observed, extending the work from pole to pole, in which 240,000 photographs included; mem. Nautical Almanac expdn. to observe total eclipse of sun, Aug. 7, 1869; mem. U.S. Coast Survey expdn. to Xeres, Spain, Dec. 22, 1870; clarified explanation of star Algol's periodicity as due to dark companion (confirmed by Vogel 1889), 1880; showed (with Vogel) that stars of type Beta Lyrae are eclipsing stars; used new spectroscopic techniques to obtain spectra of stars of entire No. hemisphere, 1886-89; devised method of establishing stellar magnitudes (basis of modern stellar photometry); discovered existence of spectroscopic binary systems, 1890. Died Cambridge, Mass., Feb. 8, 1919.

PICKERING, William Hayward, physicist; b. Wellington, N.Z., Dec. 24, 1910; s. Albert William and Elizabeth (Hayward) P.; B.S., Cal. Inst. Tech., 1932, M.S., 1933, Ph.D. in Physics, 1936; m. Muriel Bowler, Dec. 30, 1932; children—William B., Anne E. Mem. Cosmic Ray Expdn. to India, 1939; Mexico, 1941; faculty Cal. Inst. Tech., 1940—, prof. elec. engring., 1946—, dir. jet propulsion lab., 1954—. Mem. sci. adv. bd. USAF, 1945-48; chmn. panel on test range instrumentation Research and Development Bd., 1948-49; mem. U.S. nat. com. tech. Panel Earth Satellite Program, 1955-60; mem. Army Scientific Adv. Panel, 1960-64. Recipient James Wyld Meml. award Am. Rocket Soc., 1957; Columbus medal Genoa, 1964, Prix Galabert for Astronautics, Goddard award, Nat. Space Club, 1965. Fellow Am. Inst. Aeros. and Astronautics (pres. 1963); mem. Nat. Acad. Scis., I.E.E.E., Am. Assn. U. Profs. Head project Cal. Inst. Tech. which developed 1st U.S. artificial satellite (Explorer I), Ranger, 1964, Mariner, 1962, 64-65; developed microlock radio communications system; modified army Jupiter-C rocket to carry satellites into space. Home: 2514 N. Highland Av., Altadena, Cal. Office: 4800 Oak Grove Dr., Pasadena, Cal. 91103.*

PICKERING, William Henry, Am. astronomer; b. Boston, Feb. 15, 1858; s. Edward and Charles (Hammond P.; grad. Mass. Inst. Tech., 1879; m. Anne Atwood Butts, June 11, 1884; children— William T., Esther. Instr. Mass. Inst. Tech.; asst. prof. Harvard Obs.; head sxpdn. to observe total solar eclipses in Colo., 1878, Grenada W.I., 1886, Cal., 1889, Chile, 1893, Ga., 1900, N.H., 1932; expdn. to So. Cal. to make observations of moon, 1904; discovered Phoebe (9th satellite of Saturn), 1899, showed later why it revolved in a direction opposite to others, discovered 10th satellite (unconfirmed); predicted existence and gave location of 9th planet, Pluto, 1919; visited Hawaii, 1905, the Azores in 1907, in order to compare their crater formations with those in the moon; established temporary obs. in So. Cal., 1889, Arequipa Sta. of Harvard Obs., 1891; erected obs. and telescope for Dr. Lowell at Flagstaff, Ariz., 1894; established an astron. station for Harvard Obs., in Mandeville, Jamaica, W.I., 1900, reestablished it in 1911 and converted it into pvt. obs., 1925, with a new telescope; ascended Half Dome in Yosemite Valley, 1878, and El Misti in Peru (altitude 19,400 feet), besides over 100 other peaks. Fellow Am. Acad. Arts and Scis.; mem. Nat. Acad. Scis., Internat. Astron. Union and other astron. socs. Author: Walking Guide to Mt. Washington Range, 1882; Investigations in Astronomical Photography, 1895; Visual Observations of the Moon and Planets, 1900; An Atlas of the Moon, 1903; The Moon, 1903; Miscellaneous Papers, 1905; Lunar and Hawaiian Physical Features Compared, 1906; Researches of the Boyden Department, 1908; A Search for a Planet Beyond Neptune, 1909; A Statistical Investigation of Cometary Orbits, 1911; Mars, 1921. Contributor to Annals of the Harvard College Observatory. Began a series of Reports on Mars, in Popular Astronomy, 1914 (44th issued 1930). Died Jan. 17, 1938.

PICKETT, Morris John, Am. microbiologist; b. Beloit, Kan., Dec. 3, 1915; s. John A. and Mabel A. (Hall) P.; A.A., U. Chgo., 1936; B.A., Stanford, 1938, Ph.D., 1942; m. Marion Maurice Newcombe, Mar. 21, 1946; children—Linda N., David H., Charles H., Roger E. Faculty, U. Cal., Los Angeles, 1947—, prof., 1957—, acting chmn. bacteriology, 1957-59; cons. Olive View Hosp., 1963—, Space Gen. Corp., 1963-65, Consol. Electrodynamics Corp., 1966—, Hyland Labs., 1967—. Mem. Am. Soc. Microbiology, Am. Pub. Health Assn., N.Y. Acad. Sci., Soc. Gen. Microbiology (Eng.), Am. Acad. Microbiology, Sigma Xi. Author: (with C.R. Manclark) Laboratory Manual for Medical Bacteriology, 1966. Adv. editor: Lab. World, 1961—; Jour. Infectious Diseases, 1964—. Contbr. numerous articles to profl. jours. Developed, improved procedures for med. bacteriology; studies on nature of immunity in undulant fever and other chronic diseases. Home: 12213 Lawler St., Los Angeles 90066.*

PICKRELL, Kenneth LeRoy, Am. physician; b. Old Forge, Pa., June 6, 1910; s. Thomas and Anna May (Williams) P.; B.S., Franklin and Marshall Coll., 1931; M.D., Johns Hopkins, 1935; m. Katharine Council, June 11, 1935; children—Judith, Katharine Lee, Anna May, Elizabeth. Resident in surgery Johns Hopkins Hosp., Balt., 1935-44; chief div. plastic and reconstructive surgery Duke Med. Center, Durham, N.C., 1944-—; prof., chmn. div. plastic surgery Duke, 1944-—. Named N.C. Physician of Year, 1961. Diplomate Am. Bd. Plastic Surgery (chmn. 1960-61). Mem. A.C.S., Am. Soc. Plastic and Reconstructive Surgery (pres. 1959-60), Am., So. surg. assns., Soc. U. Surgeons, Internat. Soc., Soc. Head and Neck Surgeons, Soc. for Surgery of Hand. Contbr. over 100 articles, chpts. to sci. jours. Research on abnormalities in infancy and childhood, burns, cancer of head and neck. Home: 3 Sylvan Rd., Durham 27701. Office: Duke Medical Center, Durham, N.C. 27706.*

PICONE, Mauro, Italian mathematician; b. Palermo, Italy, May 2, 1885; s. Alfonso and Anna (Bongiovanni) P.; Ph.D., Superior Normal Sch., Pisa; hon. doctorate U. Sao Paulo (Brazil); m. Maria Agonigi, 1913. Asst., chair calculus Pisa U., 1908-03; became prof. infinitesimal analysis U. Cagliara, 1920; apptd. prof., Catane, 1921, Pisa, 1924, Naples, 1925; tchr. higher analysis U. Rome, 1932-41; founder Nat. Inst. for Application of Calculus, Naples, 1929, dir., from 1932. Mem. Lincei Acad., acads. scis. Poland, Peru, Madrid. Author: Corso di calcolo delle variazioni, 1923; Lezioni di analisi infinitesimale, 1923; Appunti di analisi superiore, 1940; (with T. Viola) Moderna teoria dell'integrazione delle funzioni, 1951; (with G. Fichera) Trattato di analisi matematica, 2 vols., 1953-56. Research in differential geometry, ordinary and partial differential equations, math. theory of elasticity; also functional analysis and calculus of variations, series expansions and approximations of functions. Address: 18 via delle Tre Madonne, Rome, Italy.

PICTET, Raoul Pierre, physicist; b. Geneva, Switzerland, Apr. 4, 1846; prof. physics, Geneva; went to Berlin, 1886, later to Paris. Author: Mechanische Theorie der Explosivstoffe, 1902; Die Entwicklung der Theorien und der Verfahrensweisen bei den Herstellund der flüssigen Luft, 1907. First to liquefy oxygen, 1877, also liquefied nitrogen, hydrogen, carbon dioxide; studied phenomena of low temperatures. Died Paris, July 27, 1929.

PICTET DE LA RIVE, François Jules, Swiss zoologist, paleontologist; b. Geneva, Sept. 27, 1809; prof. zoology, Geneva; mem. French Acad. Scis. Author: Traité de paleontologie, 4 vols., 1845-46; Matriaux pour paléontologie suisse, 6 vols., 1854-71; also (with Campiche) book on cretacious fossils, 5 vols., 1858-72. Died Geneva, May 15, 1872.

PICTET-TURRETIN, Marc-Auguste, Swiss physicist; b. Geneva, Switzerland, July 23, 1752; s. Charles Pictet de Rochemont; student of Saussare; became prof. philosophy Acad. Geneva, 1786; dedicated to sci. study, after 1814. Fellow Royal Soc., 1791; mem. French Acad. Scis., Author: Essai sur le feu (1st accurate research on vapor flow), 1791; Essais de physique, 1791; Voyage de trois mois en Angleterre, 1803. Founder (with father and F.G. Maurice) Bibliothèque britannique (became Bibliothèque universelle 1816). Died Geneva, Apr. 18, 1825.

PIDDUCK, Frederick Bernard, English mathematician; b. July 17, 1885; ed. Exeter Coll., Oxford, Tech. Hochsch., Charlottenburg, Germany; D.Sc., 1923; became fellow Queen's Coll., Oxford, 1907; tutor Corpus Christi Coll., 1920-50; univ. lectr., 1920; univ. reader in applied math., 1927-34. Author: A Treatise on Electricity, 1916; (paper) The Reform of Mathematics, 1937; Mathematical Theory of Electricity, 1937; Currents in Aerials and High-frequency Networks, 1946; (paper) The Semantic Discipline, 1947; (with R.K. Sas) The Metre-Kilogram-Second System of Electrical Units, 1947; also papers on physics and math. Died June 1952.

PIDINGTON, John Hobart, Australian physicist; b. Wagga, Australia, Nov. 6, 1910; s. Clarence and Constance (Millenet) P.; B.E., Sydney (Australia) U., 1929, M.Sc., 1933; PH.D., Cambridge U., 1938; m. Patricia Olive, Sept. 30, 1965. Sydney U. fellow, 1934-39; staff Commonwealth Sci. and Indsl. Research Orgn., Chippendale, Australia, 1939-—; vis. prof. Md. U., 1960. Australian rep. IGY, 1956. Recipient Sidey medal Royal Soc. New Zealand, 1959; Syme medal U. Melbourne (Australia), 1958. Fellow Australian Acad. Sci., Royal Astron. Soc. Author: Radio Astronomy, 1961; also articles. Designed radar set used in S. Pacific, World War II; designed and patented air radio navigation system; research on radio astronomy and astrophysics, theory of Earth magnetic storms, auroras, ionosphere; 1st to postulate magnetic tail of Earth, also magnetic streamers from sun. Home: 3 Churchill Crescent, Cammeray, N.S.W. Office: Radiophysics Lab., University Grounds, Chippendale, N.S.W., Australia.*

PIDOPLICHKO, Ivan Grigorevich, Russian zoologist, paleontologist; b. Kozatskoe (now Cherkassy Oblast), 1905; grad. higher pedagogical courses Korsun, 1924, Inst. Applied Zoology and Phytopathology, Leningrad, 1927; postgrad. Geology Research Inst., Ukrainian Acad. Scis., until 1933, Inst. Zoobiology, until 1935; D.Biol. Sci. Sr. asso. Inst. Archeology, Ukrainian Acad. Scis., 1934-41, 45-47, head dept. paleozoology Inst. Zoology, 1935-—, also chmn. Nature Conservation Commn; lectr. Kiev U., 1939-52, prof., 1952-59; prof., 1952-—. Mem. Ukrainian Acad. Scis. (corr.). Author over 250 publs. including: Material on the Study of Extinct Fauna of the Ukrainian SSR, 2 vols., 1938-56; The Ice Age, 4 vols., 1946-56; Short History of the World, 1958; Nature Conservation in the Ukraine, 1958. Dep. chief editor Ukrainian Soviet Ency. Address: Inst. Zoology, Ukrainian Acad. Scis., Vladimirskaya ulitsa 55, Kiev, Ukraine, USSR.

PIEKARA, Arkadiusz Henryk, Polish physicist; b. Warsaw, Poland, Jan. 12, 1904; s. Wincenty Jakob and Maria (Biala) P.; B.Sc., Warsaw U., 1927, Ph.D., 1930; m. Krystyna Chodzicka, Dec. 2, 1950; children—Andrzej Przemslaw, Marek Arkadiusz, Lidia Maria. Physics tchr. Exptl. Secondary Sch., Rydzyna, Poland, 1928-39; asst. prof. Jagiellonian U., Cracow, Poland, 1937-39; asst. prof. Underground U., Cracow, 1942-45; prof. physics, chmn. dept. Tech. U., Gansk, Poland, 1946-52, U. Poznan (Poland), 1952-65, U. Warsaw, 1965-—; head Dielectrics Lab., Inst. Physics, 1953-—; dir. Inst. Physics, Polish Acad. Scis., 1966-—. Recipient State award for sci. devel., 1955; Ministry award for sci. devel. and outstanding u. textbooks, 1963; medal U. Liege (Belgium), 1961. Mem. Polish Acad. Scis. (corr.), Polish Phys. Soc., Phys. Soc. London, Société Française de Physique. Author: Microwave Spectroscopy, 1953; Electricity and Structure, 1955; General Mechanics, 1961; also numerous papers. Editorial bd. Kolloid-Zeitschrift, 1930-39; Acta Physica Polonica, 1953-—, Physica Status Solidi, 1962-—. Discoverer or co-discoverer dielectric and magnetooptical anomalies due to moleculer fluctuations in vicinity of critical point, inverse saturation effect in dipolar fluids and its theory, effect of intense magnetic field, effect of laser beam on electric permittivity of dielectric liquids, and its theory, theory of dielectric polarisation and saturation of hydrogen bond systems. Home: 11 Ul. Opocznska, Warsaw 12. Office: 101 Al. Zwirki 1 Wigury, Warsaw 22, Poland.*

PIEPER, George Francis, Jr., Am. physicist; b. Boston, Jan. 1, 1926; s. George Francis and Katherine (Cross) P.; B.A., Williams Coll., 1946; M.S. in Engring., Cornell U., 1949; Ph.D., Yale, 1952; m. Barbara Ferguson, Dec. 27, 1950; children—Pamela, Lynell. Staff mem. Radiation Lab., Mass. Inst. Tech., Cambridge, 1944-45; instr. physics Williams Coll., 1946-47; asso. physicist Tracerlab, Inc., Boston, 1949-50; instr., asst. prof. physics Yale, New Haven, 1952-60; mem. prin. staff, project supr. Applied Physics Lab., Johns Hopkins, Silver Spring, Md., 1960-64; dep. asst. dir. for advanced research NASA Goddard Space Flight Center, Greenbelt, Md., 1964-65, asst. dir. for space scis., 1965-—. Carnegie Instn. fellow dept. terrestrial magnetism Carnegie Instn. of Washington, 1956-57; cons. to dir. geophysics and astronomy programs NASA Hdqrs., 1963-64. Mem. Am. Phys. Soc., Am. Geophys. Union, Phi Beta Kappa, Sigma Xi, Phi Sigma Kappa. Exptl. research in low energy nuclear physics, especially nuclear reaction mechanisms and instrumentation techniques, also exptl. research in space scis., especially measurement solar cosmic rays and Van Allen radiation belt particles. Home: 1109 Kathryn Rd., Silver Spring, Md. 20904. Office: Code 600, Goddard Space Flight Center, Greenbelt, Md.*

PIEPER, Heinz Paul, physiologist; b. Wuppertal-Barmen, Germany, Mar. 24, 1920; s. Heinrich Ludwig and Agnes (Koehler) P.; M.D., U. Munich, 1948; m. Rose Irmgard Hackl, Apr. 23, 1945. Came to U. S., 1957, naturalized, 1963. Med. service German Air Force, 1938-45; asst. prof. physiology U. Munich, 1950-57; asst. prof. Ohio State U., Columbus, 1957-60, asso. prof., 1960-—. Recipient Established Investigatorship award Am. Heart Assn., 1962. Mem. Am German physiol. socs., Am., Central Ohio heart assns., Ohio Acad. Scis., Biophysics Soc., Sigma Xi. Contbr. articles to profl. jours. Designed, used miniature instruments for exptl. research on heart function; used to study mechanisms of heart filling, heart ejection, and heart control. Home: 1563 Fishinger Rd., Columbus, O. 43221.*

PIERCE, Allan, Am. physicist; b. Clarinda, Ia., Dec. 18, 1936; s. Franklin D. and Ruth (Wright) P.; B.S., N.M. State Coll., 1957; Ph.D., Mass. Inst. Tech., 1962; m. Penelope Claffey, Oct. 28, 1961. Staff, RAND Corp., Santa Monica, Cal., 1961-63; sr. staff scientist AVCO Corp., Wilmington, Mass., 1963-66; lectr. Northeastern U., Boston, 1964-66; asst. prof. mech. engring. Mass. Inst. Tech., 1966-—. NSF fellow, 1957-58, 58-59, 59-60; Shell Oil fellow, 1960-61. Mem. I.E.E.E., Am. Inst. Aeros. and Astronautics, Am. Phys. Soc., Acoustical Soc. Am., Am. Geophys. Union, Am. Meteorol. Soc., Soc. Terrestrial Magnetism and Electricity Japan. Research and publs. in electron-vibration interaction in molecules and solids, acoustic waves in atmosphere, underwater acoustics, thunderstorm effects, upper atmosphere physics. Home: Autumn Lane, Carlisle, Mass. 01741.*

PIERCE, Benjamin Osgood, Am. physicist; b. Beverly, Mass., Feb. 11, 1854; s. Benjamin Osgood and M. (Seccomb) P.; A.B., Harvard, 1876; Ph.D., U. Leipzig, 1879; postgrad., Berlin, 1879-80; m. Isabella Turnbull Landreth, July 27, 1882. Tchr., Boston Latin Sch., 1880-81; instr. math. Harvard, 1881-84, asst. prof. math. and physics, 1884-88, Hollis prof. math. and natural philosophy, 1888-—. Fellow Am. Acad. Arts and Scis., Am. Philos. Soc., A.A.A.S. Author: Theory of the Newtonian Potential Function; Table of Integrals, Boston, 1899; Experiments in Magnetism. Research on thermal conductivity of stone; pioneer in sci. edn. in U.S. Died Jan. 14, 1914.

PIERCE, George Washington, Am. elec. engr.; b. Webberville, Tex., Jan. 11, 1872; s. George W. and Mary Elizabeth (Gill) P.; B.Sc., U. Tex., 1893, M.A., 1894; Harvard, 1898-1900, A.M., 1899, Ph.D., 1900; U. Leipzig, 1900-01; m. Florence H. Goodwin, Aug. 12, 1904 (dec. 1945); m. 2d, Helen Russell, Nov. 2, 1946. Asst. prof. physics Harvard, 1907-17, prof., 1917-21, Rumford prof., 1921-40, emeritus from 1940, Gordon McKay prof. communication engring., 1935-40, chmn. div. phys. scis., 1927-40, emeritus, from 1940; dir. Cruft High Tension Elec. Lab., 1914-40. Recipient medal Inst. Radio Engrs., 1928, Franklin medal Franklin Inst. Fellow Am. Acad. Arts and Sciences; mem. Nat. Acad. Scis., Am. Phys. Soc., Am. Inst. E.E., Inventors Guild, Philos. Soc. Tex., Radio Inst. (pres. 1918). Author: The Principles of Wireless Telegraphy, 1910; Electric Oscillations and Electric Waves, 1920; The Songs of Insects, 1948. Made inventions in elec. communications. Died Aug. 25, 1956.

PIERCE, G(ordon) Barry, pathologist; b. Westlock, Alta., Can., July 21, 1925; s. Gordon B. and Helen M. (Jones) P.; M.Sc. in Anatomy, U. Alta., 1950, M.D., 1952; m. Donna Jean Turner, Oct. 4, 1952; children—Donald B., C. Helen, Gordon B., Patricia L., Thomas A. Came to U.S., 1955. Sarah Mellow Scaife fellow, U. Pitts., 1955-59, asst. prof., 1958-61; faculty U. Mich., Ann Arbor, 1961-68, Am. Cancer Soc. prof. pathology, 1964-68; prof., chmn. dept. pathology, U. Colo. Med. Center, 1968-—. Markle scholar med. sci. Guiteras lectr. Am. Urol. Soc., 1962. Mem. Sco. Exptl. Pathology, Am. Soc. Cell Biology, Am. Assn. Cancer Research, Electron Microscope Soc. Am., A.A.A.S., Am. Assn. Pathologists and Bacteriologists, Soc. Exptl. Biology and Medicine, Internat. Acad. Pathology. Research, publs. on histogenesis, pathogenesis of testicular tumors, demonstrating that cancer cells differentiate and lose malignancy; epithial origins of basement membrane, chem. composition and role in disease. Home: 21 S. Kearney St., Denver 80222.*

PIERCE, John Robinson, Am. elec. engr.; b. Des Moines, Mar. 27, 1910; s. John Starr and Harriet Anne (Robinson) P.; B.S., Cal. Inst. Tech., 1933, M.S., 1934, Ph.D., 1936; D.Eng., Newark Coll. Engring., 1961; D.Sc., Northwestern U., 1961, Yale University, Polytech. Institute Brooklyn, 1963; D.Sc., Columbia U., 1965; E.D., Carnegie Institute Technology, 1964; married to Martha Peacock, Nov. 5, 1938 (div. Mar. 1964); children—John Jeremy, Elizabeth Anne; m. Ellen R. McKown, Apr. 1, 1964. With Bell Telephone Labs., Inc., 1936-—, mem. tech. staff, director electronics research, 1952-55, director research, communication principles, 1958-62, executive dir. research, communications principles and systems div., 1963-65, exec. dir. research, communications scis. div., 1965-—. Mem. Pres.'s Sci. Adv. Com., 1963-66, Pres.'s Com. of Nat. Medal of Sci., 1968-70. Trustee Battelle Meml. Inst. Recipient Morris Liebmann meml. prize, I.R.E., 1947; Stuart Ballantine medal Franklin Inst., 1960; H. H. Arnold trophy as aerospace man of yr. Air Force Assn., 1962; Golden Plate award, Acad. of Achievement, 1962; Arnold Air Society General Hoyt S. Vanenberg Trophy, 1963; Nat. Medal of Sci., 1963; Edison Medal Inst. of Elec. Engrs., 1963; Valdemar Poulsen Gold medal, 1963, H. T. Cedergren medal, 1964. Fellow Acoustical Soc. Am., Am. Phys. Soc., Inst. Elec. and Electronics Engineers; member Nat. Acad. Engring. (vice chmn. com. pub. engring. policy 1967-—), Am. Academy Arts and Sciences, Air Froce Assn., Nat. Acad. Scis. Author: Theory and Design of Electron Beams, rev. edit. 1954; Traveling Wave Tubes, 1950; Electrons, Waves and Messages, 1956; Man's World of Sound, published 1958; Symbols, Signals and Noise, 1961; (with A. G. Tressler) The Research State: A History of Science in New Jersey, 1964. Electrons and Waves, 1964; Quantum Electronics, 1966; Waves and Messages, 1967; articles on popular sci., short stories. Worked on important communications projects, on vacuum tubes, microwave oscillators and amplifiers, low voltage microwave reflex oscillator, high electron guns and travelling-wave amplifiers; work led to Echo I communication satellite program. Home: 16 Roberts Rd., Warren, N.J. 07060. Office: Bell Telephone Labs., Inc., Murray Hill, N.J. 07974.*

PIERCE, Madelene Evans, Am. biologist; b. Boston, Nov. 7, 1904; d. Frank Wendell and Bertha Evelyn (Bryant) Pierce; A.B., Radcliffe Coll., 1926, M.A., 1927, Ph.D., 1933. Instr., Smith Coll., 1927-29; mem. faculty Vassar Coll., 1931-—, prof. zoology, 1950-—, chmn. dept., 1962-—; instr. Marine Biol. Lab., Woods Hole, Mass., 1943-53; spl. research control aquatic weeds. Mem. corp. Marine Biol. Lab. Mem. Am. Soc. Zoologists, Am. Soc. Limnology and

Oceanography; Am. Inst. Biol. Scis., Am. Assn. U. Profs., Phi Beta Kappa. Contbr. articles in field. Research on effect of modern weedicides on exptl. areas of small ponds. Address: Vassar Coll., Poughkeepsie, N.Y. 12601.*

PIERCE, Richard Scott, Am. mathematician; b. Long Beach, Cal., Feb. 26, 1927; s. Robert Scott and Dorothea Stinson (Bloomfield) P.; B.S., Cal. Inst. Tech., 1950, Ph.D., 1952; m. Mary Ray, June 12, 1953. Mem. faculty U. Wash., 1955—, prof. math., 1960—, chmn. dept., 1962—. Office Naval Research fellow, Yale, 1952-53; Jewett fellow Harvard, 1953-55; NSF sr. postdoctoral fellow U. Cal. at Berkeley, 1961-62. Mem. Am. Math. Soc. (asso. sec. 1959—), Math. Assn. Am., Assn. Symbolic Logic, A.A.A.S., Sigma Xi. Home: 8714 26th St. N.E., Seattle 98115.*

PIERCE, Robert, English physician; b. Somerset, Eng., 1622; B.A., Lincoln Coll., Oxford (Eng.) U., 1642; M.A., M.B., 1650; M.D., 1661; m. dau. of David Pryme; 1 dau. Practiced medicine, Bath, Eng. Fellow Coll. Physicians. Author: Bath Memoirs, 1697; The History and Memoirs of the Bath, 1713. First to suggest condition now called trade palsies; 1st English writer to describe lympho-sarcoma of pericardium, also to note acute rheumatism following scarlet fever. Died June 1710.

PIERER, Johann Friedrich, German physician; b. Altenburg, Jan. 22, 1767; practiced medicine, Altenburg; later became book dealer. Author: Anatomisch-physiologisches Realwörterbuch, 9 vols., from 1816; Universal-Wörterbuch oder vollständiges enzyklopädisches Wörterbuch, 26 vols., from 1824. Died Dec. 22, 1832.

PIERI, Jean, French physician; b. Marseilles, France, Jan. 30, 1896; s. Marius and Clara (Guérin) P.; ed. faculties medicine Paris, Marseilles univs.; M.D.; m. Christiane Mounier, Oct. 25, 1926; children—Claude, Francoise, Charles, Chantal, Mireille, Christian, Jean-Philippe, Bertrand. Dir. clinic; hosp. physician; agrégé prof. therapeutics Faculty Medicine, U. Marseille. Recipient 7 prizes Acad. Medicine. Mem. Thermal Provence-Mediterranean-Corsica Fedn. ,pres.). Author: Les maladies méditerranéenes; La fievre exanthématique méditerranéene; Maunel d'hydroclimatologie clinique. Research, publs. on hydroclimatology, seawater therapy, clin. treatment, therapeutic medicine. Address: 67, Blvd. Longchamp, Marseilles, France.

PIERLOOT, Roland Alphonse, Belgian physician; b. St. Andries, Belgium, Feb. 23, 1924; s. Alois and Alphonsine (Braet) P.; M.D., U. Louvain, 1949; postgrad. U. Louvain, U. Amsterdam; m. Van Roey Raphaelle, Sept. 4, 1951; children—Myriam, Rita, Kristin, Kathleen. Cons. for psychosomatics U. Clinic, St. Raphael, Louvain, 1954—; med. dir. U. Clinic for Psychiatry, Kortenberg, 1961—; prof. med. psychology and psychotherapeutical methods U. Louvain, 1962—; head dept. psychopathology U. Clinic St. Raphael, 1964—. Mem. Internat. Psychoanalytical Assn., Royal Soc. Medicine, Belgian Royal Soc. Mental Medicine, Belgium Soc. Psychology. Author: General Problems of Clinical Psychosomatics, 1956. Research, publs. on outlook of psychosomatic, doctor-patient relationship. Home: 100 Leuvense baan. Office: 68 Leuvense baan, Kortenberg, Belgium.*

PIERONI, Antonio, Italian pharmacist, chemist; b. Viadana, Mantua, Italy, Sept. 20, 1885; s. Appolo and Emilia (Pasini) P.; Ph.D. in Chemistry and Pharmacy; m. Angela Pederelli, 1920. Prof. emeritus U. Parma (Italy); past sci. dir., chemotherapy dept. Milan (Italy) Inst. Serotherapy, various other pharm. insts. Mem. Nat. Research Council. Research, publs. on organic chemistry, on pyrrol and melanin, aribazossicarbonamides, syntheses of arsenobenzols. Home: via Renato Jucini 14, Milan, Italy. Office: via M. D'Aregeio 85, Parma, Italy.

PIERQUIN, B., French physician; b. Paris, Oct. 21, 1920; s. J. and F. (Vassal) P.; student Faulty Medicine, Paris, 1938-53; m. J. Touraine, Mar. 31, 1948; children—Odele, Claire, Anne, Beatrice. Asst., later chief service Institut Gustave Roussy, 1952—; radiologist Paris hosps., 1956; named head service de Curietherapie, 1961; became prof. agrégé Paris Faculty Medicine, 1965. Mem. French Radiology Soc., Assn. for Study Cancer, Internat. Radio Therapists' Club. Author: Précis de Curiethérapie, 1964; also numerous articles. Research on utilization of tomography in radiotherapy; techniques of Curietherapy and replacement of radium with artificial radio-elements, protection from its hazards. Home: 211 rue de l'université, Paris 7, France. Office: 16 bis avenue Voillon, Paris, France.*

PIERRE, William Henry, Am. soil scientist; b. Brussels, Wis., Aug. 2, 1898; s. Joseph and Mary (Leroy) P.; B.S., U. Wis., 1921, M.S., 1923, Ph.D., 1925; m. Alice Mary Oerkwitz, Oct. 16, 1928; children—Mary Frances (Mrs. Lloyd H. Morrisett), Louise Joanne (Mrs. Richard C. Connell), Nancy Jean (Mrs. Richard McKinney). Asst. soil scientist S.D. State U., Brookings, 1921-22; asst. soil chemist Auburn (Ala.) U., 1925-29, asso. prof., 1929-36; prof., head dept. agronomy and genetics W.Va. U., Morgantown, 1936-38; prof., Ia. State U., Ames, 1938—, head dept. agronomy, 1938-64. Mem. agrl. mission FAO, Uruguay, 1950; past mem. agr. bd. NRC; cons. U. Republic, Uruguay, 1964-65. Mem. Am. Soc. Agronomy (Nitrogen Research award 1931, past pres.), Soil Sci. Soc. Am. (past pres.), Internat. Soc. Soil Sci., Am. Assn. U. Profs. Co-editor books on soil sci. Research and publs. on effects of soil acidity, soluble aluminum and manganese, and percentage base saturation of soils on plant growth; effect of nitrogen and phosphorus fertilizers and of acid-base balance of plants on soil acidity; developed method for estimating physiol. acidity and basicity of fertilizers; determined minimum concentrations of soluble phosphorus and potassium necessary for optimum plant growth. Home: 3403 Oakland, Ames, Ia. 50010.*

PIERSON, Willard James, Jr., Am. oceanographer; b. N.Y.C., July 7, 1922; s. Willard James and Mary (Hand) P.; B.S., U. Chgo., 1944; Ph.D., N.Y. U., 1949; m. Joy Mary Kell, July 3, 1954; children—Mary Jean, Arthur Willard, Mark Lester. Faculty, N.Y. U., N.Y.C., 1949—, prof. oceanography, 1961—. Mem. Soc. Naval Architects and Marine Engrs. (mem. panel H7 1955—), Am. Geophys. Union (v.p. sect. on oceanography 1965-67), Am. Meteorol. Soc., Am. Astronautical Soc., Marine Tech. Soc., Sigma Xi. Author: (with G. Neuman) Principles of Physical Oceanography, 1966; also numerous articles. Research on motion of ships in waves, wave analysis by time series, and forecasting by computer techniques, use of radar on spacecraft to study oceans, turbulence, ocean currents. Home: 1641 Rosalind Av., Elmont, N.Y. 11003. Office: Dept. Meteorology and Oceanography, Sch. Engring. and Sci., N.Y. U., N.Y.C., 10053.*

PIETRO D'ABANO, see d'Abano, Pietro.

PIETRO DA TOSSIGNANO, see Da Tossignano, Pietro.

PIETSCH, Erich, German chemist; b. Berlin, Germany, May 6, 1902; s. Hermann and Auguste (Schultz) P.; ed. Friedrich Wilhelm U., Berlin; m. Gisela Hildebrandt, 1945. Sci. research worker, 1925; became acting head Gmelin Inst., 1927, dir., 1936; named hon. prof. Mining Coll. Clausthal, head inst. for gen. studies, 1947; apptd. hon. prof. Johann Wolfgang Goethe U., Frankfort/Main, Germany, 1957. Editor-in-chief: Gmelin Handbook, 8th edit. Research on surface chemistry and corrosion; contbr. publs. on history of chemistry, documentation, terminology. Office: 40-42 Varrentrappstrasse, Gmelin Inst., Frankfurt/Main, Germany.

PIETSCH, Paul Andrew, Am. molecular biologist; b. N.Y.C., Aug. 8, 1929; s. Elwood Paul and Bridget (McDonnell) P.; student Champlain Coll., 1949-50, Coll. City N.Y., 1951-52, U. Nevada, 1952; A.B., Syracuse U., 1954; postgrad. Ohio State U., 1954-56; Ph.D., U. Pa., 1960; m. Myrtle Evelyn Miller, Dec. 8, 1950; children—Samuel, Benjamin, Mary, Abigail. Asst. in anatomy Ohio State U., 1954-56; asst. instr. U. Pa., 1956-60; instr. Sch. Nursing, 1958; instr. Bowman Gray Sch. Medicine, Wake Forest Coll., 1960-61; asst. prof. anatomy U. Buffalo Sch. Medicine, 1961-63; research physiologist Biochem. Research Lab., Dow Chem. Co., Midland, Mich., 1963-65, sr. research molecular biologist, 1965—. Mem. Biophys. Soc., Soc. for Study Growth and Devel., Am. Soc. Cell Biology, Am. Assn. Anatomists, A.A.A.S. Contbr. articles to tech. jours. Investigated fate blastema in limb regeneration; studied relationship between growth in muscle cells and initiation their devel. as such; established obligatory relationship between onset of muscle differentiation and DNA synthesis; detected differences in DNA synthesis among different tissues; research on mechanism action antibiotic phleomycin, theoretical biology of replication and differentiation. Home: 1505 Airfield Lane, Midland, 48642. Office: Dow Chem. Co. Bldg. 1701, Midland, Mich. 48640.*

PIEZ, Karl Anton, Am. biochemist; b. Newton, Mass., Aug. 30, 1924; s. Karl Anton and Margaret (Snider) P.; B.S., Yale, 1947; Ph.D., Northwestern U., 1952; m. Glades Zambianchi, Apr. 18, 1948; children—Janet, Karl William, Barbara. Biochemist, Nat. Inst. Dental Research, NIH, Bethesda, Md., 1952—, chief protein chemistry sect., 1961—, chief Lab. Biochemistry, 1966—. Mem. Internat. Assn. for Dental Research, Am. Soc. for Biol. Chemists, A.A.A.S., Am. Chem. Soc., Sigma Xi, Phi Lambda Upsilon. Contbr. numerous articles to tech. jours. Research on structure proteins, relationship between structure and function connective tissue proteins. Home: 8313 Fenway Rd., Bethesda 20034. Office: Nat. Inst. Health, Bethesda, Md. 20014.*

PIGFORD, Robert Lamar, Am. chem. engr.; b. Meridian, Miss., Apr. 16, 1917; s. Lamar and Zula (Harrington) P.; B.S., Miss. State U., 1938; M.S., U. Ill., 1941, Ph.D., 1942; m. Marian Gray Pinkston, Aug. 30, 1939; children—Nancy Marie, Robert Harrington. Research engr. E.I. du Pont de Nemours & Co., Wilmington, Del., 1941-47; prof. chem. engring. U. Del., Newark, 1947-66; prof. chem. engring. U. Cal. at Berkeley, 1966—. Vis. prof., U. Cal., Berkeley, 1954, U. Wis., Madison, 1957, Cambridge (Eng.) U., 1959; chem. engring. cons. M. W. Kellogg Co., E. I. du Pont & Co. Mem. Am. Inst. Chem. Engrs. (Profl. Progress award, 1955, W. H. Walker award, 1958, mem. council 1965-68). Author: (with W. R. Marshall, Jr.) Application of Differential Equations to Chemical Engineering Problems, 1946, (with T. K. Sherwood) Absorption and Extraction, 1951. Editor: Fundamentals Quarterly, Indsl. and Engring. Chemistry. Work in gen. fields applied math., diffusion, solution thermodynamics. Home: 1510 Madera Ct., El Cerrito, Cal. 94530. Office: U. Cal., Berkeley, Cal. 94720.*

PIGFORD, Thomas Harrington, Am. nuclear engr.; b. Meridian, Miss., Apr. 21, 1922; s. Lamar and Zula (Harrington) P.; B.S., Ga. Inst. Tech., 1943; M.S., Mass. Inst. Tech., 1948, Sc.D., 1952; m. Catherine Kennedy Cathey, Dec. 31, 1948; children—Cynthia Thomas, Julie Catherine. Faculty, Mass. Inst. Tech., 1947-57; dir. engring., asst. lab. dir. Gen. Atomic, La Jolla, Cal., 1957-59; prof. nuclear engring. U. Cal. at Berkeley, 1959—, dept. chmn., 1959-64; cons. indsl. firms; cons. editor McGraw Hill Nuclear Engring. Series, 1962—. Mem. Am. Chem. Soc., Am. Nuclear Soc. Author: (with Manson Benedict) Nuclear Chemical Engineering, 1958. Research in fuel-cycle theory for nuclear power reactors, release of fission products from nuclear fuels, safety analysis of nuclear power reactors, direct energy conversion by thermionic emission, nuclear fuel elements for thermoelectric energy conversion. Home: 1 Garden Dr., Kensington, Cal. 94708. Office: U. Cal., Berkeley, Cal. 94720.*

PIGGOT, Charles Snowden, Am. chemist, geophysicist; b. Sewanee, Tenn., June 5, 1892; s. Cameron and Anne (Cockey) P.; B.A., U. of South, 1914, B.S., 1914, D.Sc. (hon.), 1926; grad. U. Pa., 1916, U. London, 1922; Ph.D., Johns Hopkins, 1920; m. Ruth Blaine, Aug. 10, 1927; children—Deboorne, Anne Marguerite (Mrs. Robert Woolfold Black). Research chemist U.S. Indsl. Chem. Co., Balt., 1920--24; with geophys. lab. Carnegie Instn. Washington, 1924-48; with NDRC, Washington, 1947-49; chief, sci. mission Am. embassy, London, 1950-52; insp. Sci. Research Instns. of India for NRC, 1954-56. Mem. Am. Chem. Soc., Phys. Soc., Am. Geog. Soc., Geophys. Union, Nat., Washington acads. sci., Royal Instn. Gt. Britain, Phi Beta Kappa, Sigma Xi. Research on radioactivity of rocks and determination of age of rocks; study of geology and radioactivity of ocean bottom using cores; determination of isotopes of lead (with Aston). Home: Route 2, Box 68, Lexington Park, Md. 20653.*

PIGGOTT, Stuart, Brit. archaeologist; b. May 28, 1910; s. G. H. O. Piggott; ed. St. John's College, Oxford; hon. D.Litt., U. Columbia, 1954. Staff mem., Royal Commission on Ancient Monuments, Wales, 1929-34; asst. dir., Avebury excavations, 1934-38; Abercromby prof. prehistoric archaeology, U. Edinburgh, 1946—; commissioner, Royal Commissions on Ancient Monuments, Eng. and Scotland. Fellow, British Acad., 1953; Royal Soc., Edinburgh; Soc. of Antiquaries; mem., German Arch. Inst., Am. Acad. Arts and Scis. Assoc. ed., Proceedings of Prehistoric Soc. Author: Some Ancient Cities of India, 1946; Fire Among the Ruins, 1948; British Prehistory, 1949; William Stukeley: an XVIII century Antiquary, 1950; Prehistoric India, 1950; (with G. E. Daniel) A Picture Book of Ancient British Art, 1951; Neolithic Cultures of British Isles, 1954; Scotland before History, 1958; Approach to Archaeology, 1959; The West Kennet Long Barrow, 1962; Ancient Europe, 1965; (with J. G. D. Clark) Prehistoric Societies, 1965; numerous papers. Conducted excavations and studied prehistory of southern England, Europe, India, and the Orient. Office: Edinburgh University, Edinburgh, Scotland.

PIGHINI, Giacomo, Italian psychiatrist; b. Parma, Italy, Dec. 18, 1876; s. Giuseppe and Enrichetta (Marenzoni) P.; ed. in medicine, psychiatry; m. Giuseppina Varoli Piazza, Feb. 1908; children—Raffaello, Giuseppe, Cesare. Prof. psychiatry U. Parma. Mem. Medico Psychol. Assn. London (corr.), Soc. Psychiatry and Endocrinology Bucharest (hon.). Author: Lo spirito che vince, 1950; Dove andiamo?, 1952; L'uomo d'oggi, 1955; Il piacere di confessarsi, 1963. Address: via Mazzini 52, Parma, Italy.

PIGMAN, W(illiam) Ward, Am. biochemist; b. Chgo., Mar. 5, 1910; s. James Ward and Olga (Chapel) P.; B.S., George Washington U., 1933, M.S., 1934; Ph.D., U. Md., 1936; m. Alice Wolfe, June 3, 1934; children—James Ward, Jean Louise (Mrs. Guy Lytle), John Charles. Research asst. Nat. Bur. Standards, 1930-36, research chemist, 1936-38, 40-43; group leader, dir. Berwyn Br. Lab., Corn Products Refining Co., Argo, Ill., 1943-46; group leader in organic chemistry Inst. Paper Chemistry, Appleton, Wis., 1946-49; asso. prof. biochemistry Med. Coll. Ala., 1949-60; asso. prof. U. Ala. Sch. Dentistry, Birmingham, 1949-60, Co.-dir. Arthritis and Rheumatism Research Lab., 1955-60; prof., chmn. biochemistry dept. N.Y. Med. Coll., N.Y.C., 1960—, asso. dir. Grad. Sch., 1962—. Recipient medal French Biochemistry Soc., 1963; medal U. Milan (Italy), 1965. Fellow A.A.A.S.; mem. Am. (C.S. Hudson award 1959), Swiss chem. socs., Ala., N.Y., Washington acads. sci., Am. Soc. Biochemists, Soc. for Exptl. Biology and Medicine, Internat. Assn. for Dental Research, Am. Rheumatism Assn., Biochem. Soc. (London, Eng.), Am. Fedn. for Clin. Research, Am. Soc. for Clin. Investigation, Harvey Soc., N.Y.

Acad. Scis. Editor: Advances in Carbohydrate Chemistry, vols. 1-4, 1954-49, editorial bd., 1949—. Research, numerous publs. on carbohydrate chemistry and biochemistry, enzyme action, composition of connective tissue in relation to arthritic diseases, chemistry of glycoproteins, autoxidations, nature and care of dental caries, devel. of artificial mouth and methods of rehardening softened tooth enamel; patentee, Niagara Starch. Home: 216 Malaga Av., Birmingham, Ala. 35209. Office: N.Y. Med. Coll., Fifth Av. at 106th St., N.Y.C. 10029.

PIGOTT, Edward, English astronomer; b. Middlesex, Eng.; s. Nathaniel Pigott; asst. in father's geodetic operations, Flanders, 1772; discovered nebula in Coma Berenices, 1779, comet named after him, at York, 1783; observed transit of Mercury, from Louvain, 1786; pub. catalog of 50 variable or suspected stars, 1786; early observer of great comet of 1807; attempted to establish artificial system of photometry; introduced John Goodricke to astronomy.

PIGOTT, Nathaniel, English astronomer; b. Whitton, Eng.; s. Ralph and Alethea (Fairfax) P.; m. Anna Mathurina de Beriol; kept metrorol. record, Caen, France, 1765-69; used Dolland's achromatic telescope to observe partial solar eclipse, from Caen, 1765; began project to determine geog. positions of prin. towns in Low Countries, 1772; became astronomer, obs. nr. London, 1777. Fellow Royal Soc., 1772; mem. French Acad. Scis., 1776, Acad. Scis. Caen (fgn.), Brussels Acad. Observed transit of Mercury, 1786, also eclipses of Jupiter's satellites; discovered some double stars. Died 1804.

PIHA, Runo Sakari, Finnish biochemist; b. Punkalaidun, Finland, Dec. 8, 1919; s. Anton Waldemar and Helma (Hellman) P.; B.Pharm., U. Helsinki (Finland), 1945, M.Sc. in Chemistry, 1950, Ph.D. in Biochemistry, 1957; m. Aini Elina Siintola, Aug. 5, 1950; children—Anna Helena, Riitta Inkeri. Faculty, U. Helsinki, 1947—, asst. prof. biochemistry, 1959—, reader physiology, 1964—; research asso. Nat. Inst. for Med. Research, London, Eng., 1953-54; research fellow Columbia, N.Y.C., 1962-64, N.Y. State Psychiat. Inst., 1965; prof. biochemistry U. Turku (Finland), 1964-66, U. Oulu, 1966—. Decorated Order Cross of Liberty (Finland). Mem. Finnish Biochem. and Biophys. Soc., Scandinavian Physiol. Soc., Biochem. Soc. (London), Finnish Physiol. Soc. (sec. 1961—), Internat. Brain Research Orgn. (mem. panel neurochemistry 1964—). Author: Studies on Blood Regeneration during Progressive Reticulocytosis due to Repeated Bleedings 1956; also articles. Research on blood regeneration, hemopoiesis and erythropoietic factor, role of kidney in erythropoiesis, amino acid and protein metabolism of central nervous system; characterization of basic proteins; comparison of turnover of deoxyribonucleic acid with respective histones, showing that histones are replaced concurrently with DNA. Home: Pihlajatie 52 B 27, Helsinki, 27. Office: Dept. Biochemistry, U. Oulu, Oulu, Finland.*

PIHL, Alexander, Norwegian biochemist, radiation biologist; b. Oslo, Norway, June 1, 1920; s. Oluf and Oddrun (Enge)) P.; M.D., U. Oslo, 1947, Dr. med. degree, 1954; m. Tove Agnethe Mohr, June 13, 1946; children—Joronn, Merete. Research fellow U. Rochester (N.Y.), 1947-48, U. Chgo., 1948-49; research asst. U. Oslo, 1949-52, prof. cancer biochemistry, 1963—; sr. biochemist Norwegian Radium Hosp., Oslo, 1956—, head clin. lab. and radioisotope lab., 1959—; sr. biochemist Norsk Hydros Inst. Cancer Research, 1956—, head dept. biochemistry, 1963—; vis. prof. Mass. Inst. Tech., 1966. Recipient Cancer Research award (with L. Eldjarn), Nordic Cancer Union, 1963. Mem. Fedn. European Biochem. Socs. (chmn. 1966-67), Norwegian Soc. Biochemistry and Physiology (chmn. 1964-67). Research, publs. on biochem. studies of cholesterol, uracil and cysteamine metabolism and of effects of adrenal steroids on mammalian cells in tissue culture, radiation inactivation of enzymes and electron spin resonance studies of radiation damage to amino acids, proteins and DNA constituents; discovered (in 1955, wtih L. Eldjarn) that sulfur containing radioprotective agts. become fixed as mixed disulfides in tissues and proposed on this basis a theory for their mechanism of action. Home: 19 Valleg. Office: Radiumhospitalet, Oslo, Norway.*

PIIPER, Johannes, physiologist; b. Tartu, Estonia, Nov. 11, 1924; s. Johannes and Elwine (Ounapuu) P.; M.D., U. Göttingen (Germany), 1954; m. Ilse Pfundt, Dec. 27, 1957; children—Hilja, Albrecht, Johanna-Elisabeth. Faculty, U. Göttingen, 1960—, prof. physiology, 1966—; staff Max Planck Inst. for Exptl. Medicine, 1953—, sci. mem. 1964—. Research, numerous publs. on physiology of respiration and circulation. Address: Hermann Rein Str. 3, Göttingen, West Germany.*

PIKE, Richard Bevis, marine biologist; b. Kobe, Japan, Oct. 2, 1911; s. Harold Bevis and Elisabeth Sidney (Parkin) P.; B.Sc., U. Reading (Eng.), 1939, Ph.D., 1945; m. Margaret Annette Hawkins, Apr. 30, 1946; children—Janice Elisabeth, Martin Brian, Hilary Anne. Asst. naturalist Scottish Marine Biologica Assn., 1945-48, sr. sci. officer, 1948-51, prin. sci. officer, 1951-61; prin. sci. officer to Marine Dept., Wellington, New Zealand, 1961-64; sr. lectr. invertebrate zoology Victoria U., Wellington, 1964—, biologist in charge Marine Sta., Island Bay, Wellington, 1966—. Author: Galathea, 1947; also articles. Biol. studies on prawns, spiny lobsters and crabs, hermit crabs, larvae of prawns and hermit crabs, parasites (Bopyridae) of crustacea. Home: 23 David Crescent, Karori, Wellington, New Zealand.*

PIKE, Robert Merrett, Am. microbiologist; b. Hiram, Me., Apr. 5, 1906; s. John Bennett and Cora (Hubbard) P.; A.B., Brown U., 1928, M.A., 1930, Ph.D., 1932; m. Mary Mather Brownell, June 17, 1932; children—Elizabeth (Mrs. Joe L. Dunlap), Mary Lusanna (Mrs. Ellery W. Sinclair), Robert Brownell. Bacteriologist, asst. dir. Otsego County and Bassett Hosp. Labs., Cooperstown, N.Y., 1932-43; faculty Southwestern Med. Sch., U. Tex., Dallas, 1943—, prof. microbiology 1951—. Diplomate Am. Bd. Microbiology. Charter fellow Am. Acad. Microbiology; mem. Am. Soc. Microbiology, Am. Assn. Immunologists. Research and publs. on properties of disease producing bacteria especially streptococci, mechanism and application of immunol. reactions in rheumatoid arthritis, nature of antibodies produced against microorganisms. Home: 5815 Elderwod Dr., Dallas 75230.*

PILAT, Albert, Czechoslovakian mycologist; b. Prague, Czechoslovakia, Nov. 2, 1903; s. Charles and Marie (Tauchen) P.; Rerum Naturalium D., 1926, D.Sc., 1967; m. Anny Pilat, Jan. 17, 1929; children—Stella (Mrs. M. Mares), Albert. With Nat. Mus., Prague, 1930—, head mycol. dept., 1965—. Mem. Czechoslovakian Acad. Scis., Czechoslovakian Mycol. Soc. (chmn.), Société Mycologique de France (hon.), Brit. Mycol. Soc. (hon.). Author: Atlas des Champignons d Europe: Polyporaceae 1936, Pleurotus, 1935, Lentinus, 1946, Crepidotus, 1948; also numerous articles. Research on wood-destroying fungi of Eurasia. Home: 12, Divcich hradu, Prague 5. Office: Nat. Mus., Václavske n. Prague 1, Czechoslovakia.*

PILATRE DE ROZIER, Jean-François, French aeronaut, chemist; b. Metz, France, Mar. 30, 1756. Prof. chemistry Athénée; Royal; supt. Natural History Collections under Louis XVI. Pub. memoirs in Journal de physique. Invented 1st gas mask; made 1st free balloon flight using balloon inflated with warm air, 1783; later made flight lasting 25 minutes, reached height of 1000 meters; crashed on flight across English Channel, 1785; founder Athénée Royal, 1781. Died nr. Boulogne-sur-Mer, June 15, 1785.

PILET, Charles, French microbiologist, immunologist; b. Coex, Vendée, France, Feb. 22, 1931; s. Charles and Bachel (Delcroix) P.; Baccalaureate, 1949; Doctorate in Vet. medicine, Vet. Sch. Lyon (France), 1955; Diploma Institut Pasteur in Bacteriology and Immunology, 1955; m. Christiane Fichet. Faculty Alfort (France) Vet. Sch., 1955—, dir. Lab. of Brucellosis Research, 1960—, prof. microbiology, immunology, exptl. medicine, 1964—. Expert for toxicology and pharmacology Ministry of Pub. Health, 1963—. Recipient Victor Lyon prize, 1966. Mem. Assn. for Vet. Microbiologists (gen. sec. 1966—), World Assn. Vet. Microbiologists (gen. sec. 1967—). Research and numerous publs. on immunology of Brucella, non agglutinogenic vaccine against Brucellosis, non specific resistance of body, antibacterial and antiviral drugs. Office: 7 av. Gen. de Gaulle Maisons Alfort, 94, France.*

PILET, Paul-Emile William, Swiss biologist; b. Lausanne, Switzerland, July 26, 1927; s. William and Berthe (Lemat) P.; Bachelor, Master, Ph.D., U. Lausanne; Docteur honoris causa U. Toulouse (France), 1966; m. Suzanne Gervaix, Aug. 15, 1962. Staff, Cal. Inst. Tech., Pasadena, London U., Sorbonne, Paris; faculty U. Lausanne, prof. plant biology and physiology, 1964—, dir. Inst. Plant Biology, 1964—. Mem. Swiss Soc. Plant Physiology (founding pres.). Author: les Phytohormones de croissance, 1961; Le cellule, 1963; also numerous articles. Analysis of plant cell growth; research on biochem. processes related to cell elongation and division, hormone metabolism. Home: 13 Av. Reymondiu, Pully, Switzerland. Office: Pala's de Rumine, Lausanne, Switzerland.*

PILGERAM, Laurence Oscar, Am. biochemist, physiologist; b. Great Falls, Mont., June 23, 1924; s. John Rudolph and Bertha (Phillips) P.; A.A., U. Cal. at Berkeley, 1948, B.A., 1949, Ph.D., 1953; m. Marilyn J. Heinrich, Feb. 11, 1951; children—Karl Erich, Kurt John. Postdoctoral fellow Life Ins. Med. Research Fund, U. Cal. at Berkeley, 1953-54; instr. physiology U. Ill. Coll. Medicine, Chgo., 1954-55; asst. prof. biochemistry Stanford 1955-57; dir. Arteriosclerosis Research Lab., U. Minn. Sch. Medicine, Mpls., 1957-65; dir. Center for Study of Aging, Sansum Clinic Research Found., Santa Barbara, Cal., 1965—. Recipient Ciba Found. award for studies on aging, 1958. Fellow Gerontol. Soc.; mem. Am. Soc. Biol. Chemists, Am. Chem. Soc., Council on Arteriosclerosis, Am. Heart Assn., Soc. Exptl. Biology and Medicine. Contbr. numerous articles to profl. jours. Discovered that biochem. process of aging causes arteriosclerosis, that biosynthesis of blood clot forming protein fibrinogen increases with aging and heart disease, and that synthesis of the protein fibrinogen is controlled by free fatty acids; discovered changes occurring in blood clot forming enzymes as a function of heart disease and aging. Home: 217 Morada Lane, Santa Barbara 93102. Office: 317 W. Pueblo St., Santa Barbara, Cal. 93102.*

PILGRAM, Kurt Hans, chemist; b. Hohenferehesar, Germany, July 6, 1927; s. Kurt Julius and Luise (Margenberg) P.; student Sch. Engring., Hamburg, Germany, 1951-54; Ph.D. in Chemistry, U. Würzburg (Germany), 1959; m. Elizabeth Maria Bäuml, June 28, 1958. Chem. technician Deutsche Erdöl A.G., Hamburg, 1951-54; research chemist Shell Grundlagen Forschung Gesellschaft m. b.H., Birlinghoven, Germany, 1960-65, research chemist, group leader Shell Devel. Co., Modesto, Cal., 1965—. Mem. Gesellschaft Deutscher Chemiker, Am. Chem. Soc. Patentee in field. Research, publs. on reactions and mechanisms of organophosphorus compounds, polychloro compounds, heterocyclic boron, nitrogen, sulfur containing compounds, oil additives, agrl. chems. Home: 2607 Warwick Lane. Office: P.O. Box 3011, Modesto, Cal. 95350.*

PILIERO, Sam Joseph, Am. pharmacologist; b. N.Y.C., Apr. 24, 1924; s. Joseph and Nancy (Rotondo) P.; A.B., N.Y. U., 1947, M.S., 1949, Ph.D., 1952; m. Catherine Aloise, July 7, 1956; children—Joseph, Anthony, Peter, Paul. Instr., St. Francis Coll., N.Y.C., 1948-50; faculty St. Johns U., N.Y.C., 1950-53, asst. prof., 1953; faculty N.Y. Med. Coll., N.Y.C., 1953-65, prof. anatomy, 1964-65, acting chmn. dept. anatomy, 1962-64; head endocrinology and hematology sects dept. pharmacology Geigy Research Labs., Ardsley, N.Y., 1965—. Mem. N.Y. State Bd. Examiners, N.Y. State Edn. Dept., 1963—. N.Y. State War Service scholar, 1949-52; NIH Research grantee, 1953-65. Fellow N.Y. Acad. Sci., A.A.A.S.; mem. Assn. Am. Anatomists, Am. Soc. Hematology, Endocrine Soc., Am. Physiol. Soc., Soc. for Exptl. Biology and Medicine, Soc. Zoologists, Am. Assn. U. Profs, Assn. Am. Med. Colls., N.Y. Zool. Soc., Sigma Xi. Author: (with M. Jacobs, S. Wishnitzer) Atlas of Histology, 1965; also numerous articles. Research on endocrine interrelations, hormones and hematopoieses, red blood cell formation, erythropoietic stimulating factor, drugs and inflammation; codeveloper microicterometer. Home: 548 Fowler Av., Pelham Manor, N.Y. 10803. Office: Ardsley, N.Y. 10502.*

PILLAI, Kunjan Sadasivan, Indian chemist; b. Kerala, India, June 11, 1922; s. Raman Kunjan and Devaki (Amma) P.; B.Sc., Travancore U., 1945, M.Sc., 1951; Ph.D., Kerala U., 1963; m. Bhargavi Ponnamma, Mar. 28, 1947; children—Girijadevi, Indrapal, Sobhanadevi, Rudrapal, Renukadevi. Sci. master Kottapuram High Sch., Quilon, Kerala, 1945-46; chem. asst. Govt. of Travancore State, 1946-52; sr. research scholar Govt. of India, Trivandrum, 1952-55; sci. asst. Norweigian Found., Cochin Sta., 1955-56; sr. chemist fisheries tech. sect. U. Kerala, Trivandrum, 1956—, part time tchr. marine chemistry, 1956—. Mem. Kerala U. Tchrs. Assn. (past sec.). Research, publs. on Indian sea weeds; developed method for extracting high polymer alginic acid from sargassum while recovering mannitol; discovered a diketo unsaturated fatty acid from balistes fish and determined its structure. Home: TC22/93, Chirakkulam Rd., Trivandrum, Kerala, India.*

PILLERI, Georg, neuroanatomist; b. Trieste, Italy, June 16, 1925; s. Mario and Ines (Zadnik) P.; student U. Padua (Italy), 1950-52, U. Vienna (Austria), 1952-55; doctor's diplome, U. Bern (Switzerland), 1958; m. Rosa Brand, Mar. 21, 1959; children—Olaf, Boris, Ralph. Staff, Neurol. Inst., H. Obersteiner, U. Vienna, 1952-55; research asst. Brain AnatomyInst., Psychiat. Clinic, U. Bern., 1955-62, dir., 1965—, cons. pathologist ophthalmic clinic, 1964—. Mem. Marine Biol. Assn. U.K. (life). Author: Beitrage sur vergleichenden Morphologie des Nagetiergehirnes, 1955; Beitrage zur Morphologie der Cetacea, 1962; also numerous articles. Research on instinctive behavioral patterns in psycho-organic deterioration, nervous systems of animals. Home: 1 Chrottesgassli, Bolligen/BE, Switzerland. Office: 3072 Wab au/BE Hirnanatomisches Inst., Switzerland.*

PILLSBURY, Walter Bowers, Am. psychologist; b. Burlington, Ia., July 21, 1872; s. William Henry Harrison and Eliza Crabtree (Bowers) P.; student Penn. Coll., 1888-90; A.B., U. Neb., 1892, Ph.D., 1896; postgrad. Cornell; m. Margaret M. Milbank, June 16, 1905; children—Margaret Elizabeth, Walter. Tchr., Grand Island (Neb.) Coll., 1892; mem. faculty U. Mich., from 1896, became dir. psychol. lab., 1901, prof., 1910-42, chmn. phychol. dept., 1929-42, prof. emeritus, from 1942. Non-resident lectr. psychology Columbia, 1909; vis. lectr. Sorbonne, 1922-23. Mem. Am. Physiol. Soc., Am. Psychology Soc., Mich. Sci. Club. Author: Attention, 1908; Psychology of Reasoning, 1910; Essentials of Psychology, 1911, rev., 1920, 28; Fundamentals of Psychology, 1916, 22, 34; Psychology of Nationality and Internationalism, 1919; Education as the Psychologist Sees It, 1925; (with C. L. Meader) Psychology of Language, 1928; History of Psychology, 1929; Elementary Psychology of the Abnormal, 1932; Psychology of Memory, 1938; (with L. A. Pennington) Handbook of General Psychology, 1942; Psychology as Science of Human Behavior (relates psychology to physiology). Mem. Cooperating bd. editors Am. Jour.

Psychology. Research on analysis, definition, identification of conditions resulting in psychol. phenomena of attention; opposed Hume's theory of causality. Died Ann Arbor, Mich., June 3, 1960.

PILSBRY, Henry Augustus, Am. conchologist, zoologist; b. Iowa City, Ia., Dec. 7, 1862; s. Dexter Robert and Elizabeth (Anderson) P.; student State U. Ia., Sc.D. (hon.), 1899; Sc.D., U. Pa. 1940, Temple U., 1941; m. Adeline Bullock Avery, 1890; children—Elizabeth, Grace P. Barcroft. Curator mollusks and other invertebrates Acad. Natural Scis., Phila., 1888-1957. Recipient Leidy medal, 1928. Fellow Am. Acad. Arts and Scis., Boston; mem. Malacological Soc. London, Am. Soc. Naturalists, A.A.A.S., Am. Conchological Soc. (1st pres. 1907), Am. Malacological Union (1st pres. 1931), Phila. Shell Club (hon. life pres.), Sigma Xi, various fgn. sci. socs. Author: The Manual of Conchology, 31 vols., 1888-1931; Marine Mollusks of Japan, 1895; Guide to the Study of Helices, 1907; Barnacles of the United States, 1916; Mollusks of the Belgian Congo, 1927; Land Mollusca of North America, 2 vols., 1939-47; other books and numerous articles on conchology, paleontology, zoology. Authority on land shells and other mollusks. Died Fla., Oct. 26, 1957.

PIMENTEL, David, Am. entomologist, ecologist; b. Fresno, Cal., May 24, 1925; s. Frank and Marion (Sylva) P.; B.S., U. Mass., 1948; Ph.D., Cornell U., 1951; Orgn. European Econ. Cooperation research fellow, Oxford U., Eng., 1961; m. Marcia Ruth Hutchins, July 16, 1949; children—Christina, Susan, Mark. Project leader Tech. Devel. Lab. USPHS, Savannah, Ga., 1954-55; faculty Cornell U., Ithaca, N.Y., 1955—, prof. ecology, head dept. entomology, limnology, 1963—; cons. pollution panel com. mem. Office Sci. and Tech., Exec. Office President. NSF Computer scholar Mass. Inst. Tech., 1966. Mem. Ecol. Soc. Am., Soc. For Study Evolution, Entomol. Soc. Am., Am. Soc. Zoologists, Am. Soc. Naturalists, A.A.A.S., Am. Assn. U. Profs. Contbr. numerous articles to profl. jours. Research, publs. on ecology and genetics of DDT-resistance in the housefly, mongoose control, ecology of snail of schistosomiasis, ecology and genetics of natural population regulation and competition, parasite-host relationships. Home: 147 N. Sunset Dr., Ithaca, N.Y. 14850.*

PIMENTEL, George Claude, Am. chemist; b. Fresno, Cal., May 2, 1922; s. Emile J. and Lorraine (Laval) P.; A.B., U. Cal. at Los Angeles, 1943; Ph.D., U. Cal. at Berkeley, 1949; m. Betty Anne Jeffrey, Oct. 4, 1942; children—A. Christine, Janice A., Tess L. Staff, Manhattan Project, 1943-44, Office Naval Research, 1944-46; faculty chemistry dept. U. Cal. at Berkeley, 1949—. Mem. Nat. Acad. Scis., Am. Chem. Soc. (Cal. sect. award, award in petroleum chemistry), Am. Phys. Soc. Author: (with A. L. McClelland) The Hydrogen Bond, 1960; also numerous articles. Editor: Chemistry, An Experimental Science, 1963; asso. editor Jour. Chem. Physics, Spectrochimica Acta. Research on molecular structure through infrared spectroscopy, hydrogen bonding, matrix isolation technique for free radical study, flash photolysis using infrared methods, chem. lasers. Office: Chemistry Dept., U. Cal. at Berkeley, Cal.*

PIMENTEL GOMES, Frederico, Brazilian mathematician; b. Piracicaba, Sao Paulo, Brazil, Dec. 19, 1921; s. Raymundo and Silvia (de Souza) P. G.; grad. in agronomic engring. Coll. Agr., U. Sao Paulo, 1943, Livre-Docent of Math., 1948; m. Mary Lee Fonseca de Bem; children—Marli, Valquiria, Vangri. Asst. prof. math. Coll. Agr., U. Sao Paulo, 1944-52, 53-56, 56-58, prof. physics and meteorology, 1956, prof. math., 1958-61, 61—, sec. substitute of grad. courses, chief dept. physics and math., 1965—. Vis. scholar U. N.C., 1952-53; vis. prof. statistics Instituto Nacional de Tecnologia Agropecuária, Argentina, 1961; pres. Rural U. Brazil, State of Rio de Janeiro, 1964; gen. dir. Dept. Agrl. Promotion, Rio de Janeiro, 1965. Recipient Gen. Rondon medal, 1965. Mem. Brazilian region Biometric Soc. (pres. 1958-59, 59-60, 66-67), Brazilian Soc. for Progress of Sci., Math. Soc. Sao Paulo, Paulist Agronomy Soc. Author: A Course on Experimental Statistics, 1st edit., 1960, 2d edit., 1963, 3d edit., 1966; (with E. M. Cardoso) The Fertilization of Sugar Cane, 1958; (with A. T. Vianna, M. Santiago) Breeding the Canchim cattle by Crossing Charolais and Brahma Animals, 1960; (with others) Cultivation and Fertilization of Sugar Cane 1964. Research and publs. on fractional differentation and integration, Mitscherlich's law (proposing new methods of analyzing response of crops to fertilizers), combined analysis of expts. with some common treatments (permitting better use of exptl. data), methods of statis. analysis of coffee-tasting expts. and comml. classification of coffee. Home: 575 Av. Carlos Botelho, Piracicaba, Sao Paulo. Office: Escola Sup. Agricultura Luiz de Queiroz, Piracicaba, Sao Paulo, Brazil.*

PINCHERLE, Salvatore, Italian mathematician; b. Trieste, 1853; ed. U. Pisa; (Italy); tchr., lycée Pavia, Italy, 1875-80; prof. algebra and analytical geometry U. Bologna (Italy), until 1912, prof. infinitesimal calculus, until 1928; organized, presided over Internat. Congress Mathematicians, Bologna, 1928. Author: Le operazioni distributive et le loro applicazioni all'analisi, 1901; Gli elementi della theoria delle funzioni analitiche, 1922; Lezioni di calcolo infinitesismale, 1926-27. A founder of modern functional calculus; elaborated theory of functional distributive operations (linear functions); developed concept of derived functional which led to devel. in parallel series to that of Taylor (systematic study of analytic functions considered as space functional). Died Bologna, 1936.

PINCHOT, Gifford, Am. conservationist, forester; b. Simsbury, Conn., Aug. 11, 1865; s. James W. and Mary (Eno) P.; A.B., Yale, 1889; studied forestry France, Germany, Switzerland, Austria; A.M. (hon.) Yale, 1901, Princeton, 1904; ScD., Mich. Agrl. Coll., 1907; LL.D., McGill, 1909, Pa. Mil. Coll., 1923, Yale, 1925 Temple, 1931; m. Cornelia Elizabeth Bryce, 1914; 1 son, Gifford Bryce. Began 1st systematic forest work in U.S. at Biltmore, N.C., 1892; mem. Nat. Forest Commn., 1896 forester, chief div. Bur. Forestry, also Forest Service, U.S. Dept. Agr., 1898-1910; prof. forestry, Yale, 1903-36; commr. forestry of Pa., 1920-22; gov. of Pa., 1923-27, 31-35; inspected forests of P.I., 1902; apptd. mem. Inland Waterways Commn., 1907; apptd. chmn. Nat. Conservation Commn., 1908; chmn. Joint Com. on Conservation, apptd. by the conf. of govs. and nat. orgns. at Washington, 1908; pres. Nat. Conservation Assn., 1910-25. Mem. Soc. Am. Foresters, Royal English Arboricultural Soc., Am. Mus. Natural History, Washington, Pa. acads. scis., Am. Acad. Polit. and Social Sci. Author: Biltmore Forest, 1893; The White Pine (with H. S. Graves), 1896; Timber Trees and Forests of North Carolina (with W. W. Ashe), 1897; The Adirondack Spruce, 1898; Report to the Secretary of the Interior on Examination of the Forest Reserves, 1898; A Study of Forest Fires and Wood Production in Southern New Jersey, 1899; A Primer of Forestry, Part I, Bull. 24, Div. of Forestry, 1899, Part 2, 1905; Recommendations on Policy, Organization and Procedure for the Bureau of Forestry of the Philippine Islands, 1903; The Fight for Conservation, 1909; The Country Church (with C. O. Gill), 1913; The Training of a Forester, 1914, 4th edit. (rewritten), 1937; Six Thousand Country Churches (with C. O. Gill), 1919; To the South Seas, 1930; Just Fishing Talk, 1936; Breaking New Ground, 1946. First Am. profl. forester; discovered fish juice for castaways at sea, 1942; secured placement of fishing tackle on all Am. lifeboats, 1943. Died Oct. 4, 1946.

PINCHOT, Gifford Bryce, Am. biochemist; b. N.Y.C., Dec. 22, 1915; s. Gifford and Cornelia Elizabeth (Bryce) P.; B.A., Yale, 1938; M.D., Columbia U., 1942; m. Sarah Huntington Richards, June 13, 1936; children—Marianna, Gifford III, Peter Cooper. Postdoctoral fellow dept. physiol. chemistry Yale, 1946-48, instr. dept. microbiology Yale, 1948-49, asst. prof., 1951-55, research asso., 1955-58; asso. prof. biology Johns Hopkins U., Balt., 1958-65, prof., 1965—; career investigator USPHS, Balt., 1964—. Trustee Conservation Found., Washington; bd. govs. Pinchot Inst. for Conservation Studies, Milford, Pa. Mem. Am. Soc. Biol. Chemists. Author: Giff and Stiff in the South Seas, 1931. Pioneered fractionation enzyme system catalyzing formation adenosinetriphosphate linked to oxidation diphosphopyridine nucleotide, isolation and characterization intermediate of reaction. Home: Mt. Zion Rd., Upperco, Md. 21155. Office: McCollum-Prat Inst., Johns Hopkins, Balt. 21218.*

PINCUS, Gregory, Am. biologist; b. Woodbine, N.J., Apr. 9, 1903; s. Joseph William and Elizabeth (Lipman) P.; B.S., Cornell U., 1924; M.S., Harvard, 1927, Sc.D., 1927; m. Elizabeth Notkin, Dec. 2, 1924; children—Alexis John, Laura Jane (Mrs. Michael Bernard). Faculty, Harvard, 1930-38, Clark U., 1938-45, Tufts Med. Sch., 1946-50; vis. investigator Cambridge U., 1937-38; dir. labs. Worcester Found. for Exptl. Biology, Shrewsbury, Mass., 1944-56, research dir., 1956—; research prof. biology Boston U. Grad. Sch., 1950—. Mem. numerous coms. USPHS, NIH, NRC; chmn. subcom. on program 1st Internat. Congress Endocrinology, 1960; mem. med. council Planned Parenthood Fedn. Am., 1961—; chmn. Laurentian Harmone Conf., Internat. Congress on Hormonal Steroids. Recipient Oliver Bird prize Soc. for Study Fertility (Eng.), 1957; Albert D. Lasker award in Planned Parenthood, 1960; Modern Medicine award for Distinguished Achievement, 1964; Cameron prize in practical therapeutics U. Edinburgh, 1966; Barren Found. medal, 1966. NRC fellow, 1928-30; Guggenheim fellow, 1939-41. Fellow Am. Acad. Arts and Scis., A.A. A.S., Nat Acad. Scis., Portuguese Endocrine Soc. (hon.); mem. Internat. Soc. for Research in Reproductive Biology (chmn.), Am. Assn. Anatomists, Am. Assn. for Cancer Research, Am. Genetics Soc., Am. Soc. Naturalists, Am. Physiol. Soc., Am. Soc. Zoologists, Société de Biologie France, Société d'Endocrinologie France, Soc. for Endocrinology Gt. Britain, Soc. for Study Growth and Devel., Royal Soc. Medicine Gt. Britain (affiliate); hon. mem. Sociedad Médica de Santiago, Chile, Société d'Endocrinologie d'Haiti, Sociedad Mexican de Nutricíon y Endocrinología. Author: The Eggs of Mammals, 1936; The Control of Fertility, 1965; also articles. Editor: Recent Progress in Hormone Research, The Hormones; cons. editor Life Sciences. Research, publs. in reproductive physiology, biochemistry of hormones and nervous responses to stress. Died Boston, Aug. 22, 1967.

PINCUS, Howard Jonah, Am. geologist, geophysicist; b. N.Y.C., June 24, 1922; s. Max O. and Gertrude (Janowsky) P.; B.S., City Coll. N.Y., 1942; A.M., Columbia, 1948, Ph.D., 1949; m. Maud Lydia Roback, Sept. 6, 1953; children—Glenn David, Philip Ethan. Mem. faculty Ohio State U., Columbus, 1949-67, prof. geology, 1959-67, chmn. dept., 1960-65; prof. geology U. Wis.-Milw., 1968—. Research asso. Columbia U., 1949-52; geologist Ohio Dept. Natural Resources, summers 1951-60; NSF sr. postdoctoral fellow Applied Physics Lab., U. S. Bur. Mines, 1962, research geologist, summers, 1963-67, research mgr., 1967-68; cons. engring. geology and geophysics, 1954-66. Fellow Geol. Soc. Am., Ohio Acad. Sci., Royal Astron. Soc. A.A.A.S.; mem. Am. Geophys. Union, Soc. for Exptl. Stress Analysis, Internat. Soc. Rock Mechanics, Assn. Engring. Geologists, Am. Inst. Profl. Geologists (pres. Ohio sect. 1965-66), Am. Assn. U. Profs. (pres. Ohio State U. chpt. 1955-56, council 1965-67), Soc. Mining Engrs., Phi Beta Kappa (pres. Ohio State U. Chpt. 1956-57), Sigma Xi. Research and publs. on statis. analysis applied to geol. problems, optical data processing, engring. geology, tectonics and rock mechanics. Address: Dept. Geology, Sabin Hall, U. Wis.-Milw., Milw. 53201.*

PINDELL, Merle Herbert, Am. pharmacologist; b. Bloomington, Ill., Jan. 22, 1926; s. Ira F. and Bertha E. (Shutt) P.; A.B., U. Ill., 1948, M.S., 1953; Ph.D., Med. Coll. Va., 1955; m. F. Arlene Kimler, Aug. 5, 1945; children—Terry Lee, Pamela Jo, James Lawrence. Pharmacologist, Mile Labs., Elkhart, Ind., 1949-53; instr. U. Colo. Med. Sch., Denver, 1955-56; with Bristol Labs., Syracuse, N.Y., 1956—, dir. pharm. research, 1959-65, dir. biol. research, 1965—. Mem. A.A.A.S., Am. Soc. Pharmacology and Exptl. Therapeutics, Am. Soc. Toxicology, European Toxicology Soc., Am. Chem. Soc. (medicinal div.), N.Y. Acad. Scis. Contbr. articles to sci. publs. Research on autonomic, central nervous system and cardiovascular pharmacology, antibiotics and cancer. Home: 26 Old Farm Rd., Fayetteville, N.Y. 13045. Office: Bristol Labs., Thompson Rd., Syracuse, N.Y. 13201.*

PINEAU, Severin, French surgeon; b. Chartres, France, circa 1550; M.D., Paris. Author: Opusculum tractans analytice primo notas integritis et corruptionis virginum, 1598. Noted as a lithotomist, also for his anat. description of virginity. Died Paris, 1619.

PINEL, Philippe, French physician; b. Saint-André, Tarn, France, Apr. 20, 1745; studied medicine at U. Toulouse, M.D. 1773; also studied at Montpellier and Paris; taught philosophy and mathematics, Paris; directing physician, Bicetre, 1793 and at Salpetriere, 1795; prof. of pathology, École de Médécine, Ca. 1803; member Institut de France, 1803. Author: Nosographie philosophique, 1788; Traité medico-philosophique sur l'aliénation mentale ou la manie, 1801. Advocated more humane treatment for mentally ill; also urged experimental study of mental disease; stressed role of passions in mental illness; established custom of maintaining accurate psychiatric case histories for research. Died, Paris, France, Oct. 26, 1826.

PINELLI, Paolo, Italian physician; b. Mantova, Italy, Dec. 16, 1921; s. Pietro and Beatrice (Biasi) P.; student U. Pavia (Italy) Faculty Medicine, 1939, M.D., 1954; postgrad. Institut Comparative Anatomy and Physiology, 1940; m. Lanzoni Maria Lavisa, Apr. 18, 1949; With U. Pavia, 1945-48, 50-54, prof. psychiatry, 1961-66; prof. U. Rome, 1966—; vis. scientist Institut Neurophysiology, U. Copenhagen (Denmark), 1948-50. Mem. Pavia Com. for EMG, 1961; mem. Italian Com. for EEG, 1959—. Author books, numerous articles. Electrophysiol. research in neuromuscular diseases; neurophysiol. processes elicited by E.S. therapy. Home: 67 Vittoro Montiglis, Rome, Italy.*

PINES, Herman, chemist; b. Lodz, Poland, Jan. 17, 1902; s. Isaac and Eugenie (Greenfield) P.; Ch. Eng., Institut de Chimie, Lyon, France, 1927; Ph.D., U. Chgo., 1935; m. Dorothy Mlotek, Aug. 13, 1927; 1 dau., Judith. Came to U.S., 1928, naturalized, 1932. Research chemist, coordinator exploratory research Universal Oil Products Co., Des Plaines, Ill., 1930-52; Vladimir Ipatieff prof. chemistry, dir. Ipatieff High Pressure and Catalytic Lab., Northwestern U., Evanston, Ill., 1941—.Mem., Am. Assn. U. Profs., Am. Chem. Soc. (Fritzsche award, Midwest award 1963), Sigma Xi. Contbr. numerous papers to tech. jours. Patentee approximately 140 petrochems. Co-discoverer with Ipatieff of alkylation and isomerization of alkanes used in W.W. II in aviation gasoline. Home: 827 Monticello Pl., Evanston, Ill. 60201.*

PINGRÉ, Alexandre Guy, French astronomer; b. Paris, Sept. 11, 1711; entered Augustinian order, circa 1727; ed. Augustinian coll., Senlis, tchr. theology, 1735-37; removed from position because of Jansenistic tendencies and objections to papal bull Unigenitus which he later abandoned; became prof. astronomy Rouen Acad., 1749; librarian Ste. Geneviève, also chancellor univ., built obs., Ste. Geneviève. Mem. French Acad. Scis. (corr. sec.), 1753. Author: Cométographie (calculations of 32 comet paths and solar eclipses 1000 B.C. to 1000 A.D.), 2 vols., 1783-84; Histoire de l'astronomie du XVIIe siècle, 1790; also compiled 1st nautical almanac for 1754. Observed transits of Venus from the Pacific, 1760, Haiti, 1769. Died Paris, May 1, 1796.

PINGS, Cornelius John, Am. chem. engr.; b. Conrad, Mont., Mar. 15, 1929; s. Cornelius John and Marjorie (O'Loughlin) P.; B.S., Cal. Inst. Tech., 1951, M.S., 1952, Ph.D., 1955; m. Marjorie Anna Cheney, June 25, 1960; children—John Cornelius, Ann Elizabeth, Mary Cathleen. Faculty, Stanford, 1955-59, asst. prof. chem. engring., 1956-59; faculty Cal. Inst. Tech., Pasadena, 1959—, prof., 1964—. Cons., Electro-optical Systems, Inc., 1966—. Mem. Am. Chem. Soc., Am. Phys. Soc., Am. Inst. Chem. Engrs. (Presentation award 1963, 65). Author: (with J.B. Opfell, B.H. Sage) Equations of State for Hydrocarbons, 1959; also articles. Research in applied chem. thermodynamics and liquid state physics, thermodynamic criteria for optimum design and operation practical chem. processes, liquid state physics especially testing and improvement theories of liquid state. Home: 393 S. Sierra Bonita Av., Pasadena 91106, Cal.*

PINHEY, Elliot Charles Gordon, entomologist; b. Knocke, Belgium, July 18, 1910; s. Alexander Fleetwood and Violet Beatrice (Gordon) P.; B.Sc., Imperial Coll. Sci., London U., D.Sc.; m. Nancy Tindal MacKenzie MacRae, Sept. 3, 1946; 1 dau., Rosalind Gordon. Sci. tchr., Sussex, Eng., Berlin, Germany, 1935-39, Rhodesia, 1939-42; econ. entomologist Rhodesia Dept. Agr., 1942-48; systematic entomologist Transvaal Mus., Pretoria, S. Africa, 1948-49, Coryndon Mus., Nairobi, Kenya, 1949-55, Nat. Mus., Bulawayo, S. Rhodesia, 1955—. Mem. Royal Entomol. Soc. London, Zool. Soc. London, Entomol. Soc. So. Africa. Author: Butterflies of Rhodesia, 1949; Dragonflies of Southern Africa, 1951; Survey of the Dragonflies of Eastern Africa, 1961; Hawk Moths (Sphinx Moths) of Central and Southern Africa, 1962; Butterflies of Southern Africa, 1965. Research, publs. on African dragonflies and butterflies including descriptions of new species and genera, ecology of African grasshoppers, sunnhemp beetles, olive bugs. Home: 10 Atterburg Rd., Kumalo, Bulawayo. Office: Nat. Mus., P.O. Box 240, Bulawayo, Southern Rhodesia.

PINKAU, Klaus, German physicist; b. Leipzig, Germany, Apr. 3, 1931; s. Werner and Anny (Hentschel) P.; student U. Tübingen, 1951-53; dipl. U. Hamburg Germany, 1956; Ph.D., U. Bristol, 1958; m. Ursula Frochtenicht, Dec. 30, 1958; children—Stephen, Christopher, Tobias. Research asst. U. Bristol, Eng., 1958-60; research asst. U. Kiel, 1960-64, lectr., 1963—; vis. prof. La. State U., 1964-65 cons., 1965—; sci. mem. Max-Planck Inst. fur Extraterrestrische Physik, Munich, Germany, 1966—; lectr. Tech. U. Munich, 1966—. Mem. Deutsche Physikalische Gesellschaft. Research, publs. on energy measurement on high energy cascades and nuclear interactions, understanding of these interactions and effects of cosmic rays in atmosphere, gamma ray astronomy, cosmic ray neutrons, use of high energy cosmic rays for study of high energy interactions. Home: 14 Rheinlandstrasse, Munchen 23 Germany 8000. Office: Max-Planck Inst., Garching, Germany 8046.*

PINKERTON, John Henry McKnight, Irish physician; b. Belfast, North Ireland, June 5, 1920; s. William Ross and Eva (Odgers) P.; M.B., B.Ch. B.A.O. with honors (Magrath scholar) Queen's U. Belfast, 1943, M.D., 1948; m. Florence McKinstry, Sept. 30, 1947; children—Mark Odgers, Charles Ross, John Robert, Paul William. Sr. lectr., cons. gynecology U. Coll. W.I., Jamaica, 1953-59, Rockefeller research fellow Harvard, 1956-57; prof. gynecology U. London, also cons. Queen Charlotte's Hosp. and Chelsea Hosp. for Women, 1959-63; mem. com. mgmt. Inst. Obstetrics and Gynecology, 1959-63; bd. govs. Queen Charlotte's Hosp., 1961-63; mem. governing body Brit. Postgrad. Med. Fedn., 1961-63; prof. gynecology U. Belfast 1963—; gynecologist Royal Maternity, Royal Victoria, Jubilee hosps., Belfast, 1963—. Fellow Royal Coll. Obstetrics and Gynecology (mem. council 1966—); mem. Royal Soc. Medicine (mem. council 1958-61, 66—), Soc. for Promotion Med. Edn. Overseas (found. mem), Blair Bell Research Soc. London (found. mem.), Family Planning Assn. (adv. council 1966—), Ulster Obstetrics and Gynecology Soc., Brit. Med. Assn. Author: Advances in Oxytocin Research, 1965—. Publs. on contbns. to knowledge of devel. of human ovary; clin. significance of bacteriuria; use of oxytocics. Home: Woodbank, Whiteabbey, County Antrim, No. Ireland.*

PINKUS, Hermann Karl Benno, physician; b. Berlin, Germany, Nov. 18, 1905; s. Felix and Elise (Etzdorf) P.; M.D., Friedrich Wilhelms Universität, Berlin, 1930; M.S. in Surgery, U. Mich., 1935; m. Hilda Marie Elizabeth Hensel, July 5, 1935; 1 son, Walter Hensel. Came to U.S., 1934, naturalized, 1939. Asst. U. Breslau (Germany), 1930-33; research fellow U. Mich., 1934-36, Wayne County Gen. Hosp., 1936-38; practice medicine specializing in dermatology, Detroit, 1938-39, Monroe, Mich., 1940-51, 53-56; research asso. Detroit Inst. for Cancer Research, 1951-53; chmn. dept., prof. dermatology Wayne State U. Detroit, 1957—. Hon. mem. Pacific, German (Karl Herxheimer medal 1965); Polish, Israeli, New Zealand dermatol. assns.; mem. Mich. Med. Soc., A.M.A., Am. Dermatol. Assn., Am. Assn. for Cancer Research, Soc. for Investigative Dermatology, Am. Soc. Dermatopathology, Detroit Dermatol. Soc., Am. Soc. Cell Biology. Contbg. author several books on dermatology including Handbuch der Haut- und Geschlechtskrankheiten, Ergänzungwerk vols. 1, 2, 1964. Research, numerous publs. on anatomy and biology skin by means tissue culture, microscopic histology, electron microscopy; interpretation path. processes in skin. Home: 12 E. 4th St., Monroe, Mich. 48161. Office: 1400 Chrysler Expressway, Detroit 48207.*

PINTNER, Rudolf, psychologist; b. Lytham, Eng., Nov. 16, 1884; s. William and Irma P.; M.A., Edinburgh U., 1906; Ph.D., U. Leipzig, 1913; L.H.D., Gallaudet Coll., 1931; m. Margaret M. Anderson, Aug. 15 1916; children—Irma Jane, Walter McKenzie. Came to U.S., 1912. Prof. psychology Toledo U., 1912-13; instr. psychology Ohio State U., 1913-14, asst. prof., 1914-17, prof., 1917-21; prof. edn. Tchrs. Coll., Columbia, since 1921. Fellow A.A.A.S.; mem. Am. Psychol. Assn., Am. Assn. for Applied Psychology. Author: A Scale of Performance Tests, 1917; The Picture Completion Test, 1917; The Mental Survey, 1918; Intelligence Testing, 1923, 2d edit., 1931; Educational Psychology, 1929; The Psychology of the Physically Handicapped (with Eisenson and Stanton), 1941. Translator: An Introduction to Psychology (Wundt), 1912; Experimental Psychology and Pedagogy (Schulze), 1912; The Idea of the Industrial School (Kerschensteiner), 1913. Mem. editorial bd. Psychol. Bull. Died Nov. 7, 1942.

PINTO, Fernao Mendes (or Francisco Mendez), Portugese explorer; b. Montemor-o-Velho, Portugal, circa 1509; 3 children; departed for E. Indies, 1537; became novice Soc. Jesus, Goa, India, 1554. Author: Peregrinacem, 1614. Introduced ancient Chinese technique of acupuncture into Europe; made geographical and ethnographical observations on his journey which have been confirmed in modern times. Died July 8, 1583.

PINTO-COELHO, Aristides, Brazilian chemist; b. Ponte Nova, Brazil, July 21, 1930; s. Antonio and Elvira (Pinto) P.-C.; Bachelor in Chemistry, U. Estado Guanabara, 1953, Licenciate in Chemistry, 1954; Ph.D. in Biochemistry U. Brazil, 1961, Radiochemist, 1957; children—Thales, Cristina. Tchr. math. SENAC, Rio de Janeiro, 1951-56; tchr. chemistry YMCA Coll., Rio de Janeiro, 1953-56; with Instituto La-Fayette, Rio de Janeiro, 1953-56; staff Instituto Biof-isica, U. Brazil, 1957-59, researcher, 1959; head radioisotopes lab. Instituto Nacional Cancer, Ministry Health, Rio de Janeiro, 1960—; prof. Brazilian Nuclear Energy Commn., 1957-63; asst. prof. biochemistry U. Estado Guanabara, Rio de Janeiro, 1957-63, prof., 1964—; researcher U. Cal., 1962-63. Fellow Brazilian Nuclear Energy Commn.; IAEA fellow. Recipient Prize Prof. Amadeu Fialho in cancerology, 1964. Mem. Brazilian Soc. Devel. Sci., Brazilian Soc. Phys. Medicine (sci. adviser), N.Y. Acad. Scis., A.A.A.S. Author: Introduction to Radiobiochemistry and Techniques of Radiobiochemistry, 1960; Introduction to Nuclear Physics, 1966; also articles. 1st S. Am. studies on artificial strontium 90, research on contamination on biol. materials and natural contamination on Brazilian biol. materials from high zones of radioactivity, scintillation scanning of organs, thyroid hormones and thyroid cancer. Home: 430/1503, Larranjeiras, Rio de Janeiro GB. Office: 23 (Inst. Nac. Cancer) Pa Cruz Vermelha, Rio-GB-Brazil.*

PIOBERT, Guillaume, French physicist; b. Lyons, France, Nov. 29, 1793; ed. École polytechnique; worker Lyons; arty. officer; became prof. artillery, Metz, France, 1831; mem. French Acad. Scis. Designed mountain gun, 1821; discovered method of breach shooting later used in siege of Constantinople; his research in inner ballistics showed necessity of decreasing charge density according to barrel length; studied resistance of soft and hard metals to projectiles, 1836, influence of rotation of movable bodies on their motion in resisting media, 1837. Died Beaujeu, France, June 9, 1871.

PIONTELLI, Roberto, Italian chemist; b. Lodi, Italy, May 11, 1909; s. Alfredo and Clotilde (Perego) P.; Dr.Engring., Poly. of Milan, 1931; Dr. Physics, U. Pavia, 1933, Dr.Chemistry, 1935; m. Edvige Tonolli, June 24, 1939; children—Maria Clotilde (Mrs. Giorgio Moro Visconti), Orestina (Mrs. Luigi Dacco), Alessandra (Mrs. Mauro Mancia), Vittoria (Mrs. Fabrizio Agustoni), Margherita. Faculty, U. Milan (Italy), 1937-48, prof. electrochemistry and chem. physics U. Milan, 1942-48; prof. chem. physics Polytechnic of Milan, 1948—, dir. dept. chem.-physics, electrochemistry and metallurgy. Recipient Gold medal Ministry Pub. Edn. Mem. M. Accademia Nazionale Lincei, Instituto Lombardo Scienze e Lettere, Accademia Scienze Turin. Author: Teoria della Corrosione dei Materiali Metallici, 1961; Lectures on Thermodynamics Electrochemistry; also numerous articles. Research on electrochemistry, including new methods of measurement, theory of electrochem. behavior of metals-anion influence-electrochemistry of metallic single crystals; discovery of electrometall. applications of sulfamic baths, overvoltage phenomena in fused salts, theory of aluminum electrolysis, chem. thermodynamics, new method for finishing of titanium. Home: 26, Viale Bianca Maria, Milan, Italy.*

PIORE, Emanuel Ruben physicist; b. Wilno, Russia, July 19, 1908; s. Ruben and Olga (Gegusin) P.; came to U.S., 1917, naturalized, 1924; A.B., U. Wis., 1930, Ph.D. 1935; m. E. Nora Kahn, Aug. 26, 1931; children—Michael Joseph, Margot Deborah, Jane Ann. Asst. instr. U. Wis., 1930-35; research physicist RCA, 1935-38; engr. in charge television lab. CBS, 1938-42; head spl. weapons group bur. ships U.S. Navy, 1942-44; head electronics br. Office Naval Research, 1946-47; dir. phys. sci., 1947-48, dep. for natural sci., 1949-51, chief scientist, 1951-55, v.p., dir. Avco Mfg. Corp., 1955-56; dir. research IBM, Armonk, N.Y., 1956-61, v.p. research and engring., 1961-63, v.p., group exec., 1963-65, v.p. chief scientist, 1965—, also dir.; physicist research lab. electronics Mass. Inst. Tech. 1948-49; dir. Sci. Research Assos., Inc. Mem. Nat. Sci. Bd., 1961—. Trustee, mem. corp. Woods Hole Oceanographic Instn.; mem. vis. com. to elec. engring. dept. Mass. Inst. Tech., 1956-57; mem. corp. Polytech. Inst. Bklyn.; trustee Sloan-Kettering Inst. Cancer Research. Mem. vis. com. Harvard Coll., 1958-59. Fellow Am. Phys. Soc., I.E.E.E., Am. Acad. Arts Scis.; mem. Sci. Research Soc. Am., Nat. Acad. Scis., Washington Acad. Sci., Am. Philos. Soc., Nat. Acad. Engring. Home: 115 Central Park W., N.Y.C. 10023. Office: IBM, Armonk, N.Y. 10504.*

PIORRY, Pierre Adolphe, French physicist; b. Poitiers, France, 1794; doctor's degree, Paris, 1816. Joined Spanish army as surgeon, 1812; mem. Med. Acad., 1823, faculty, 1826, doctor of central office, 1827, holder chair internal pathology, 1840; physician Charity Hosp., from 1836. Author: Sur le danger de la lecture des livres de médicine pour les gens du monde, 1816; De la percussion mediate, 1828; De l'irritation encéphalique des enfants, 1823; Du procédé opératoire à suivre dans l'exploration des organes pour la percussion médiate, 1831; Clinique médicale de la Pitié et de la Salpetrière, 1833; Traité de médicine pratique déduit des faits recueillis dans les hôpitaux, 1835-36; Des habitations et de l'influence de leur dispositions sur l'homme, 1838; De l'hérédité dans les maladies, 1840 (translated into German); Traité des alterations du sang, 1840; Traité de médicine pratique et de pathologie iatrique, 1841, also articles. Introduced term pyemia, 1828, septicemia, 1837; pioneer in mediate percussion; invented percussor and pleximeter, 1826; opposed to defining most illnesses by single word composed of Greek roots. Dec. 31, 1794.

PIPKIN, Allen Compere, Am. mathematician; b. Mena, Ark., May 21, 1931; s. Allen Compere and Leyland (Chambers) P.; student U. Ark., 1948-49; B.S., Mass. Inst. Tech., 1952; Ph.D., Brown U., 1959; m. Ann Brittain, May 26, 1956; children—Janet Louise, Lee Ann. Research asst. Brown U., 1955-58, faculty 1960—, prof., 1966—; research asso. U. Md. Inst. for Fluid Dynamics and Applied Math., 1958-60. Mem. Soc. Rheology, Soc. Natural Philosophy, Soc. Indsl. and Applied Math., Am. Math. Soc. Sigma Xi. Studies, publs. on symmetry principles in the math. description of the behavior of materials, fluid and solid mechanics, electromechanics, kinetics of macromolecules. Home: 87 Greenwood Av., Rumford, R.I. 02916.*

PIPKIN, Francis Marion, Am. physicist; b. Marianna, Ark., Nov 27, 1925; s. Larry Stewart and Augusta Pearl (Hill) P.; student U. Kan., 1943-44, Morningside Coll., Sioux City, Ia., 1946-47; B.A., State U. Ia., 1950; M.A., Princeton, 1952, Ph.D., 1954; m. Phyllis Burr, June 14, 1958; children—Jane, Augusta. Jr. fellow Soc. Fellows, Harvard, 1954-57, mem. faculty, 1957—, prof. physics, 1964—; spl. research atomic physics, nuclear orientation, high energy nuclear physics. Mem. Am. Phys. Soc., Am. Acad. Arts and Scis., Sigma Xi. Contbr. research papers. Home: 10 Kilburn Rd., Belmont, Mass. 02178. Office: Dept. Physics, Harvard Univ., Cambridge, Mass. 02138.*

PIPPARD, Alfred John Sutton, Brit. civil engr.; b. Yeovil, Somerset, Eng., Apr. 6, 1891; s. Alfred William and Mary Alice (Sutton) P.; M.B.E. LL.D., F.R.S., D.Sc., U. Bristol (Eng.), 1919; LL.D., 1966, D.Sc., U. Birmingham, 1966; m. Olive Tucker, Dec. 26, 1917 (dec. 1964); children—John Sutton, Alfred Brian. Articled asst. Cotterell & Carr, Bristol, 1911-13; engring. asst. Pontypridd & Rhondda Joint Water Bd., 1913-15; with Air Dept. and Air Ministry, 1915-19; partner Ogilvie & Partners, cons. aero. engrs., 1919-22; prof. engring. Cardiff Coll., U. Wales, 1922-28; prof. civil engring. U. Bristol, 1928-33, Imperial Coll., U. London, 1933-56; vis. prof. Northwestern U. Evanston, Ill., 1956-57; Fellow Imperial Coll., 1958. Chmn. Com. on Pollution of Thames, 1951-61; mem. various govt. coms. Fellow Royal Soc., 1954, Am. Soc. C.E., City and Guilds London Inst. (hon.); mem. Inst. Civil Engrs. (Ewing Gold medal 1963, past pres.). Author: (with J. L. Pritchard) Aeroplane Structures, 1919; Strain Energy Methods of Stress Analysis, 1928; Analysis of Engineering Structures, 1936 (with J. F. Baker); also numerous articles. Devel. theory of structure for use in airplane design, application of Castigliano's work to gen. structural problems, stress distbn. in arch Dams. Home: 1 Dorset House, St. John's Av., London, S.W. 15, Eng.*

PIPPARD, (Alfred) Brian, English physicist; b. London, Eng., Sept. 7, 1920; s. A. J. S. and Olive (Tucker) P.; ed. Cambridge (Eng.) U., M.A., 1945, Ph.D., 1949; m. Charlotte Dyer, July 2, 1955; children—Corinna, Deborah, Eleanor. With Ministry Provisions, 1941-45; research, from 1945; from instr. to prof. physics Cambridge U., 1949—; reader in physics, 1959-60; Plummer prof. physics, 1960—;

vis. prof. Inst. for Study of Metals, U. Chgo., 1955-56. Fellow Royal Soc., 1956 (Hughes medal 1959). Author: Elements of Classical Thermodynamics, 1957 Dynamics of Conduction Electrons, 1962. Contbr. articles to profl. publs. Home: 30 Porson Rd., Cambridge. Office: Cavendish Lab. Cambridge, Eng.

PIRAGINO, Guido, physicist; b. Moscow, USSR, Sept. 3, 1933; s. Renato and Alessandra (Gherasimenko) P.; grad. U. Turin, 1957; m. Mariarosa Canepa, Apr. 10, 1961; 1 son, Renato. Research, Nat. Inst. Nuclear Physics, Turin. Italy, 1957-58, Inst. Physics of Genoa, 1958-59, Nat. Labs. of Frascati, 1959-63; asso. prof. Inst. Physics, U. Turin, 1961——. Mem. Italian Soc. Physics. Research, publs. on hard component of cosmic radiation, photoprodn. of pimesons on light nuclei, influence of nuclear shell structure on photodisintegration of different nuclei, energy spectrum and polarization of photoneutron from different nuclei and structures in giant resonance of spherical nuclei. Home: 11 G. Matteotti, Turin, Italy.*

PIRANI, Conrad Levi, physician; b. Pisa, Italy, July 29, 1914; s. Mario Giacomo and Adriana (Coen) P.; M.D., U. Milan (Italy), 1938; m. Luciana Nahmias, Mar. 12 1955; children—Barbara Ann, Sylvia Joy, Robert John. Came to U.S., 1939, naturalized, 1945. Faculty, U. Ill. Coll. Medicine, Chgo., 1945——, prof. pathology, 1955——; chmn. dept. pathology Michael Reese Hosp. and Med. Center, Chgo. 1965——. Cons. to hosps., 1965——. Fellow Coll. Am. Pathology, Am. Soc. Clin. Pathologists; mem. A.M.A., A.A.A.S., Am. Assn. Pathologists and Bacteriologists, Am. Heart Assn., Am. Rheumatism Soc., Internat. Acad. Pathology (council), Chgo. Path. Soc. (pres. 1965-66), others. Research, numerous publs. on renal, connective tissue, cardiovasular diseases. Home: 2103 Orrington St., Evanston, Ill. 60201. Office: Michael Reese Hosp. and Med. Center, 29th and Ellis Av., Chgo. 60616.*

PIRANIAN, George, mathematician; b. Thalwil, Switzerland, May 2, 1914; s. Badwagan and Bertha (Walser) P.; came to U. S., 1929, naturalized, 1936; B.S., Utah State U., 1936, M.S., 1938; post-grad. Oxford (Eng.) U., M.A., Rice U., 1941, Ph.D., 1943; m. Joe Louise Mills, Dec. 22, 1941; children—Elizabeth, Margaret, Inga, Barbara, Deborah. Mem. applied math. group Columbia, N.Y.C., 1943-44, Northwestern U. Evanston, Ill., 1944-45; faculty U. Mich., Ann Arbor, 1945——, prof. dept. math., 1958——; mng. editor Mich. Math. Jour., 1954——. Mem. Am., Swiss math. socs., Math. Assn. Am. Research, publs. on summability theory, sets of convergence of power series, boundary behavior of analytic functions; solved Carathéodory's problem on topological distbn. of four kinds of prime ends of a simply connected domain. Home: 2105 Devonshire Rd., Ann Arbor, Mich. 48104.*

PIRET, Edgar Lambert, chem. engr.; b. Winnipeg, Manitoba, Can., July 1, 1910; s. Hubert and Maria Celine (Dutilleux) P.; came to U. S., 1922, naturalized, 1927; B. Ch.E., U. Minn., 1932, Ph.D., 1937; France-Am. fellow, 1935-36; Docteur de l'Université Lyon, France, in Biochemistry, 1936; m. Alice Moeglein, Sept. 4, 1945; children—Mary, Marguerite, Jacqueline, John, Robert, James. Instr. chem. engring. U. Minn., 1937-41, asst. prof., 1941-43, prof., 1945-65, dir. chemical products from peat project, 1954-59; chief chem. engr. Minn. Mining & Mfg. Co., 1943-45, cons., 1937-65; sci. attaché Dept. State, U. S. Embassy, Paris, 1959——; guest of inst., vis. prof. Mass. Inst. Tech., 1965-66; consultant for the U. S. Naval Research Lab., 1951-55; Fulbright research prof. univs. Nancy and Paris, 1950-51. Recipient Friedel medal U. Paris, 1951; medallist U. Liege, Belgium, 1951; Palms Acadèmiques, Officer d'Acadèmie, France, 1951; Bronze medal Swedish Assn. Engrs., 1954, lectr. Gothenburg, 1954, Royal Inst. Engrs., Holland, 1954; Chevalier Legion of Honor (France), 1957. Fellow A.A.A.S., N.Y. Acad. Scis.; mem. Chemists Club, Am. Inst. Chem. Engrs. (nat. program com. 1948-49-50; Walker award 1955), Am. Inst. Chem. Engineers (awards com. 1958-63), Am. Soc. Engring. Edn., Am. Assn. U. Profs., Sigma Xi. Author of numerous papers on continuous reactor theory and design, theory of crushing, leaching, heat and mass transfer. Editor: Chemical Engineering around the World, 1957. Mem. editorial bd. Jour. Am. Inst. Chem. Engrs., 1957——, Chem. Engring. Progress, 1957——, Jour. Internat. Chem. Engring., 1961——. Patents on dielectric drying; high vacuum transfer operations; study of the theory of crushing. Office: Am. Embassy, Paris, France.

PIRILÄ, Veikko Paavo, Finnish physician; b. Helsinki, Finland, May 25, 1915; s. Paavo Werner and Anna (Ikonen) P.; qualified physician U. Helsinki, 1943, M.D., 1947, specialized in dermatology, 1947; m. Louna Puranen, May 28, 1944; children—Pekka Väinämö, Hannu Antero, Laila Marjatta, Leena Tuulikki. Faculty, U. Helsinki, 1949—— prof. dermatology, 1958——; chief dermatologist Inst. Occupational Health, Helsinki, 1950——. Mem. sci. council State Med. Dept. Finland, 1959——; cons. dermatology State Accident Inst. Office, Helsinki, 1954——. Mem. Scandinavian Com. for Standardization of Patch Testing, 1962——. Mem. nat. and internat. socs. dermatology, occupational medicine, allergy. Author: On Occupational Diseases of the Skin Among Paint Factory Workers, Painters, 1947; also numerous articles. Research on specificity of contact allergy to metals, trace elements and oil of turpentine, cross-sensitivity between antibiotics of neomycin group, occupational dermatology and dermat. allergy. Home: 1 B, Pyhän Laurin tie, Helsinki 34, Finland.*

PIRLOT, Paul, zoologist; b. Mettet, Belgium, Mar. 17, 1920; s. Léon and Frumence (Quinet) P.; Licence Phil. Let Jury Central, Brussels, Belgium, 1942; Licence Zool., U. Louvain (Belgium), 1946, Candid. Geol., 1944, Agregation, 1949; Ph.D. Zool., U. London (Eng.), 1949; D.Sc., in Zoology, 1960; m. Renée Tarlost, Dec. 15, 1947; children— Marie-Antoinette Jean-Paul, Brigitte. Research asso. Inst. for Sci. Research, Central Africa, 1949-57; spl. advanced fellow Belgian-Am. Ednl. Found., N.Y.C., vis. prof. Carroll. Coll., Helena, Mont., 1955; postdoctoral fellow NRC Can., 1958; asso. prof. U. Montreal (Que., Can.), 1958-63, prof. zoology, 1963——; research asso. U. Zulia (Venezuela), 1964-66; vis. prof. U. Oriente (Venezuela), 1964. Mem. Canadian Soc. Zoologists (pres. 1964-65), Soc. for Study Evolution, Am. Soc. Mammalogists, Montreal Soc. Biology (v.p. 1965-66). Research, numerous publs. on comparative anatomy and ecology vertebrates, distbn. and biology African and S.Am. mammals. Home: 66 Merton Rd., Hampstead, Que. Office: Dept. Biology, U. Montreal, Montreal.

PIROFSKY, Bernard, Am. physician; b. N.Y.C., Mar. 27, 1926; s. Hyman and Yetta (Herman) P.; A.B., N.Y. U., 1946, M.D., 1950; m. Elaine Friedwald, June 19, 1953; children—Daniel Niles, Tandy Ellen. Am. Cancer Soc. research fellow, 1955-56; practice medicine, specializing in immunohematology, Portland, Ore., 1956——; dir. Pacific N.W. Blood Center, 1956-58; faculty Med. Sch. U. Ore., 1956——, prof. medicine, 1966——; cons. hematologist Portland VA Hosp., 1958—— St. Vincents Hosp., 1960——; cons. immunohematologist Pacific Regional Red Cross Center, 1958——, Barnes VA Hosp., 1963——. Mem. nat. med. adv. bd. Leukemia Soc., vis. prof. Nat. Inst. Nutrition, Mexico, 1966-67. Mem. A.C.P., Internat., Am. socs. hematology, Internat. Soc. Blood Transfusion, Western Soc. Clin. Research. Research, numerous publs. on devel. of new proteolytic enzyme techniques to demonstrate antibodies, new concepts of antibody structure and function; application of autoimmunization to human disease; exptl. studies of immune tolerance to human disease. Home: 10370 S.W. Ridgeview Lane, Portland 97219. Office: 3181 S.W. Sam Jackson Park Rd., Portland, Ore. 97201.*

PIROGOFF, Nicolai Ivanovich, Russian surgeon; b. Moscow, Russia, Nov. 15, 1810; grad. Moscow U., 1827; M.D., 1832; student, Dorpat, Estonia, Berlin. Prof., Dorpat, also St. Petersburg, Russia; went to Berlin, 1835; became prof. surgery, Dorpat, 1840; prof. surgery Petersburg Acad.; dismissed, 1867. Mem. med. coms. Ministry Edn. Author: Recherches pratiques et physiologiques sur l'etherisation, 1847; Anatomia topographica, sectionibus per corpus humanum congelatum, 4 vols., 1851-54; Principles of War Surgery, 1864. First in Russia to use ether by rectum as an anesthetic, 1847; introduced osteoplastic operation in which part of calcaneus is retained in amputation of foot (Pirogoff's amputation), 1854; introduced frozen sect. technique to Russia, circa 1852; angle formed by junction of internal jugular and subclavian veins named after him (Pirogoff's angle); developed Pirogoff's operation for hernia; classification of war injuries. Died Gut Visnja, Dec. 5, 1881.

PIRSSON, Louis Valentine, Am. geologist; b. N.Y.C., Nov. 3, 1860; s. Francis M. and Louise (Butt) P.; Ph.B., Sheffield Sci. Sch. (Yale) 1882, A.M., 1902; also studied at Heidelberg and Paris; m. Eliza Trumbull Brush, May 17, 1902. Asst. in analyt. chemistry Sheffield Sci. Sch., 1882-83, 1884-88, instr. geology and lithology, 1892-94, asst. prof. inorganic geology, 1894-97, prof. phys. geology, from 1897. Asst. and spl. expert U.S. Geol. Survey, 1893-1904, geologist, from 1904. Fellow Am. Acad. Arts and Scis. Author: (with Cross, Iddings, Washington) Quantitative Classification of Igneous Rocks, 1903; Rocks and Rock Minerals, 1908. Asso. editor Am. Jour. Science, from 1897. Developed new classification system and terminology for igneous rocks. Died Dec. 8, 1919.

PISA, Zbynek, physician; b. Pilsen, Czechoslovakia, Feb. 3, 1924; s. Joseph and Marie (Tagl) P.; grad. Faculty Medecine, Charles U. 1950, Candidate Scis., 1956; m. Milena Zeman, Aug. 8, 1953; children—Pavel, Irene. House physician dept. medicine Faculty Medicine, Pilsen, 1950-51; research fellow Inst. for Cardiovascular Research Prague, 1951-54, research worker, 1954-61, dep. dir., 1962-64; research fellow pneumoconiosis research unit Llandough, Penarth, South Wales, U.K., 1961-62; regional officer for chronic diseases European Office, WHO, Copenhagen, Denmark, 1964——; asso. prof. medicine Faculty Medicine, Prague, 1965——. Mem. Czechoslovakian Cardiological Soc. Research, publs. on relation between elec. and mech. events in heart muscle, counteraction of calcium on the depressive effect of pentobarbital on contractility of heart muscle, pathogenesis of cardiogenic shock, early diagnosis of coronary heart disease, heart efficiency; developed high-frequency myocardiography, methods of permanent or temporary coronary artery occlusion in closed-chest dogs.*

PISANO, Michael Anthony, Am. microbiologist; b. Bklyn., Mar. 31, 1923; s. Michael and Mary (Esposito) P.; B.A., Bklyn. Coll., 1946; M.S., St. John's U., Jamaica, N.Y., 1949; Ph.D., Mich. State U., 1953; m. Anne M. Salanitro, Nov. 6, 1948; children—Michael, Geralyn, Andrea, Lorraine. Bacteriologist, Mich. Dept. Health, Lansing, 1951-53; research microbiologist S.B. Penick & Co., Jersey City, 1953-55; faculty St. John's U., Jamaica, N.Y., 1955——, prof. biology, 1960——. Research asso. Mt. Sinai Hosp., N.Y.C., 1956——. Recipient Research Achievement award St. John's U., 1966; Spl. NIH Research fellow, Netherlands, 1968-69. Fellow Am. Acad. Microbiology; mem. Am. Soc. Microbiology, (pres. N.Y.C. br., 1966——), N.Y. Acad. Scis., Soc. Indsl. Microbiology (editorial bd. 1963-67), Mycol. Soc. Am., Soc. Gen. Microbiology, Sigma Xi. Research, publs. in microbial fermentations, antibiotics, chem. activities of fungi, microbial nutrition and cell morphology; isolation and purification of fibrinolytic enzymes; use of airborne sound for sterilization; chemiluminescence to detect airborne microorganisms. Home: 11 Lawrence Rd., Hempstead, N.Y. 11550.*

PISANTY, Jose O., physiologist; b. Sofia, Bulgaria, Sept. 20, 1924; s. Samuel D. and Buca (Ovadia) P.; M.D., Nat. Med. Sch. Mexico, 1947; m. Irma Marin, Mar. 21, 1949; children—Julieta, Ileana, Eugenia, Guillermo. Research asst. Mexican Inst. Cardiology, 1945-50; prof. physiology Mexican Nat. U., 1947-50; chmn. physiol. research Syntex Corp., 1950-51; head dept. physiology and pharmacology U. Guadalajara, 1952-57; head dept. physiology U. Nuevo Leon, Monterrey, Mexico, 1957——. Mem. N.Y. Acad. Sci., Am. Physiol. Soc. Author: Nociones de Fisiologia, 1958. Research, publs. on action of synthetic and natural digitalis, mechanism and treatment of shock. Home: FCO. FEDZ. Trevino 485, Monterrey, Nuevo Leon, Mexico.

PISCIOTTA, Anthony Vito, Am. physician; b. N.Y. C., Mar. 3, 1921; s. Andrew and Mary (Zinnanti) P.; B.S., Fordham U., 1941; M.D., Marquette U., 1944, M.S., Sch. Medicine, 1952; m. Lorraine Gault, June 15, 1951; children—Robert Andrew, Nancy Marie, Anthony Vito. Instr., Tufts Coll. Med. Sch., 1951-52; faculty Marquette U. Sch. Medicine, Milw., 1952——, professor medicine, 1966——; dir. bood research lab. Milw. County Gen. Hosp., 1952; lectr. hematology U.S. Naval Hosp., Great Lakes, Ill., 1960——. Recipient Encaenia award Fordham U., 1956; Phi Chi Teaching award Marquette U., 1959. Fellow A.C.P.; mem. Am. Fedn. for Clin. Research, A.A.A.S., Am. Assn. Immunologists, Am. Assn. Hematology, Am. Soc. Clin. Nutrition, Central Soc. Clin. Research, Soc. for Exptl. Biology and Medicine. Contbr. numerous articles to tech. jours. Demonstrated agglutinogens in normoblasts; described 2 hereditary methemoglobins; studied mechanism of chlorpromazine-induced bone marrow damage; showed that patients recovered from CPZ-induced marrow damage have imited DNA synthesis in marrow cells. Home: 12550 W. Grove Terrace, Elm Grove, Wis. 53122. Office: 8700 W. Wisconsin Av., Milw. 53226.*

PISHKIN, Vladimir, Am. psychologist; b. Belgrade, Yugoslavia, Mar. 12, 1931; s. Vasilije and Olga (Bartosn) P.; came to U. S., 1946, naturalized, 1951; B.A., Mont. State U., 1951, M.A., 1955; Ph.D., U. Utah, 1959; m. Dorothy L. Martin, Sept. 12, 1953; children—Gayle Ann, Mark Vladimir. Faculty, Mont. State U., 1954-55; clin. psychologist Mont. State Hosp., Warm Springs, 1955-56; dir. research labs. VA Hosp., Tomah, Wis., 1959-62; chief research psychologist, asso. dir. behavioral sci. labs. VA Hosp., Oklahoma City, 1962——; faculty U. Okla., 1963——, asso. prof. biol. psychology Sch. Medicine, 1966——; prof. psychology Oklahoma City U., 1963——. Fellow A.A.A.S., Am. Psychol. Assn.; mem. Psychonomic Soc., N.Y. Acad. Scis., Sigma Xi. Author: (with J. L. Mathis, C. M. Pierce) A Psychiatry Study Aid: A Three Part Program, 1967. Mem. editorial bd. Jour. Clin. Psychology, 1963——. Research in cognitive functioning and psychopathology as related to psychotropic drug effects in human decision making; animal research in sensory deprivation with operant conditioning; design math. model for prediction of information processing in schizophrenia. Home: 415 N.E. 15th St., Oklahoma City 73104. Office: VA Hosp., 921 N.E. 13th St., Oklahoma City 73104.*

PISO, Willem (or Le Pois), Dutch physician, botanist; b. Leyden, Netherlands, 1611; ed. Us. Leyden, Caen, M.D., 1630. Accompanied Maurice of Nassau to Brazil, 1637; began medical practice, Amsterdam, 1648. Author: Historia naturalis Brasiliae, 1638; De India utriusque re naturali et medica, 1648; De lue Indica, 1648. Co-introducer of Brazilian ipecacuanha into Europe, 1648; studied medicinal plants and snake venoms of Brazil; among 1st to become acquainted with tropical diseases; distinguished between yaws and syphilis, 1648. Died 1678.

PISTOLKORS, Aleksandr Aleksandrovich, Russian radio engr.; b. Oct. 10, 1896; student Officers Electrochem. Sch., World War I; grad. Moscow Tech. Coll., 1927. Worker at radio sta. on Caucasian front, World War I, in Nizhnii-Novgorod Radio Lab., 1926-28, at Central Radio Lab., Leningrad, 1931-45; tchr. Leningrad Electro Tech. Inst., then Leningrad Inst. Engrs. of Communication, 1931-45; prof. Moscow Inst. Communication Engrs., 1945-50. Recipient A.S.

Popov Gold medal, 1956. Corr. mem. USSR Acad. Scis., 1946. Author: Calculation of Radiation Resistance for Short-Wave Directional Antennas, 1928; The Theory of Asymmetrical Two-Wire Lines, 1937; The Problem of Contactless Electric Traction, 1938; The General Theory of Diffraction Atennas, 1944; Antennas, 1947; The Use of Mathieu Functions in Calculating the Distribution of Antenna Directivity Pattern, 1953; Oscillation of the Small Hydrotropshere in the Field of a Plane Wave, 1960; On the Line to Mars, 1963. Work on antennae and feeder Lines; two wire non symmetrical lines; slot antennae; developed TV and several other types of antennae. Home: ul. Gor'Kogo 43, Moscow. Office: USSR Acad. Sciences, Leninskii Prospeky 14, Moscow, USSR.

PITARD, Jean, French surgeon; b. Carentan, France, 1230-36; physician, courts of Philip the Fair, Philip V, Charles, Count of Artois; head, examining com. for practice of surgery in Paris (established 1311); improved conditions of surgery in Paris. Died after 1328.

PITCAIRN, David, Brit. physician; b. Fifeshire, Scotland, May 1, 1749; s. John Pitcairn; student U. Glasgow (Scotland), Edinburgh (Scotland) U.; M.B., Corpus Christi Coll., Cambridge (Eng.) U., 1779, M.D., 1784; m. Elizabeth Almack. Began practice medicine, London, 1779; physician St. Bartholomew's Hosp., 1780-93. Fellow Royal Soc., 1782, Coll. Physicians (Gulstonian lectr., Harveian orator 1780, censor 5 times). First to discover valvular disease of heart frequently results from rheumatic fever. Died Apr. 17, 1809.

PITCAIRNE, Archibald, Scottish biologist; b. Edinburgh, Scotland, Dec. 25, 1652; s. Alexander Pitcairne; M.A., U. Edinburgh, 1671; postgrad., Edinburgh, Paris; M.D., Rheims, France, 1680, U. Aberdeen, 1701; m. Margaret Hay; 1 son, 1 dau.; m. 2d, Elizabeth Stevenson; 1 son, 4 daus. Practiced medicine, Edinburgh, 1680-92, from 1693; prof. U. Leiden (Netherlands), 1692. Mem. Royal Coll. Physicians Edinburgh, French Acad. Scis. Author: Oratio qua ostenditur medicinam ab omni philosophandi secta essa liberam, 1692; De sanguinis circulatione . . ., 1693. Applied math. to med. theories and research; influenced by mech. theories in medicine; vindicated Harvey's claim of discovering circulation of blood. Died Oct. 20, 1713.

PITCHER, (Arthur) Everett, Am. mathematician, educator; b. Hanover, N.H., July 18, 1912; s. Arthur Dunn and Wilimina (Everett) P.; A.B., Western Res. U., 1932, D.Sc., 1957; M.A., Harvard, 1933, Ph.D., 1935; m. Sarah Hindman, July 2, 1936; children—Joan (Mrs. Joan P.McLean), Susan Sarah (Mrs. John L. Cooper). Asst., Inst. Advanced Study, Princeton, N.J., 1935-36, mem. 1945-46, 47-50; Benjamin Peirce instr. Harvard, 1936-38; faculty Lehigh U., Bethlehem, Pa., 1938—, prof., 1948-63, chmn. dept. math. and astronomy, 1960—, distinguished prof., 1963—. Guggenheim fellow, 1952-53. Fellow A.A.A.S. (council); mem. Am. Math. Soc. (sec. 1967—), Soc. Indsl. and Applied Math. (trustee 1960-63), Math. Assn. Am., Phi Beta Kappa, Sigma Xi, Pi Kappa Alpha. Study of calculus of variations; terminal and exterior ballistics; critical point theory. Home: 422 W. Broad St., Bethlehem, Pa. 18018.

PITELKA, Dorothy Riggs (Mrs. Frank Alois Pitelka), Am. zoologist; b. Marsovan, Turkey, Sept. 13, 1920 (parents Am. citizens); d. Theodore Dalzel and Winifred (Clark) Riggs; B.A. cum laude, U. Colo., 1941; postgrad. U. Cal., Berkeley; Ph.D., 1948; m. Frank Alois Pitelka, Feb. 5, 1943; children—Louis Frank, Wenzel Karl, Kazi Helen. Author, instr. corr. course gen. biology U. Cal. Extension, Berkeley, 1950-57; USPHS post-doctoral fellow Nat. Cancer Inst., Paris, 1957-58; asst. research zoologist U. Cal., Berkeley, 1952-60, asso. research zoologist, 1960-66, research zoologist, 1966—. Mem. Soc. Protozoologists (exec. com. 1964-67, pres. 1967-68), Electron Microscopy Soc. Am. Am. Micros. Soc., Am. Inst. Biol. Scis., Groupement des Protistologues de Langue Française (hon.), Sigma Xi, Phi Beta Kappa. Author: Electron-Microscopic Structure of Protozoa, 1963. Bd. editors Jour. Protozoology, 1960—, Jour. Morphology, 1961-64, Transactions of Am. Micros. Soc., 1966—. Electron-microscope studies of ultrastructure and devel. of cilia and flagella and asso. fibrillar systems in protozoa; studies of structure, distbn. and life cycle of mouse mammary tumor viruses; ultrastructure and devel. mammary gland and its adipose tissue stroma. Home: P.O. Box 9278, Berkeley, Cal. 94719.*

PITELKA, Frank Alois, Am. biologist; b. Chgo., Mar. 27, 1916; s. Frank Joseph and Frances (Laga) P.; student U. Mich., summer 1938; B.S., U. Ill., 1939; postgrad. U. Wash.; Ph.D., U. Cal., Berkeley, 1946; m. Dorothy Getchell Riggs, Feb. 5, 1943; children—Louis Frank, Wenzel Karl, Kazi Vlasta Helen. Faculty, U. Cal., Berkeley, 1946—, chmn. dept. zoology, 1963-66; Miller research prof., 1965—; asst. curator birds Mus. Vertebrate Zoology, 1945-49, curator, 1949-63, exec. chmn. Miller Inst. for Basic Scis., 1967—. Mem. NSF panel on environmental biology, 1959-62, panel biol. and med. sci. NAS Com. on Polar Research, 1960-65. Guggenheim fellow, 1949-50; NSF fellow Oxford U., 1957-58. Fellow A.A.A.S., Arctic Inst. N. Am., Am. Ornithologists Union, Cal. Acad. Scis.; mem. Cooper Ornithol. Soc. (hon.), Brit., Am. (Mercer award 1953) ecol. socs., Am. Soc. Mammalogists, Am. Inst. Biol. Sci., Phi Beta Kappa, Sigma Xi. Research, publs. in animal ecology, social systems, breeding and molt cycles populations various rodents and birds. Home P.O. Box 9278, Berkeley, Cal. 94719.*

PITISCUS (or PETISCUS), Bartholomaüs, German mathematician; b. Schlaume, nr. Grunberg, Silesia, Aug. 24, 1561; preceptor, later chaplain to Frederick IV. Author: Theasurus mathematicus, 1593. Improved Rheticus's tables of sines; noted for works on trigonometry. Died Heidelberg, July 3, 1613.

PITKIN, Walter Boughton, Am. psychologist; b. Ypsilanti, Mich., Feb. 6, 1878; s. Caleb S. and Lucy T. (Boughton) P.; A.B., U. Mich., 1900; postgrad. Sorbonne, Paris, U. Berlin, U. Munich, Hartford (Conn.) Theol. Sem., 1900-05; m. Mary B. Gray, 1903; children—Richard Gray, John Gray, David Bartholomew, Robert Bolter, Walter Boughton; m. 2d, Katharine B. Johnson. Lectr. in psychology Columbia, 1905-09, prof. journalism from 1912. Mem. editorial staff N.Y. Tribune, 1907-08, Evening Post, 1909-10, Parents' Mag., 1927-30; Am. mng. editor Ency. Brit., 1927-28; story supr. Universal Pictures Co., 1929; editorial dir. Farm Jour., 1935-38; cons. psychologist, adviser on teaching methods. Author: The Art and Business of the Short Story, 1913; Must We Fight Japan?, 1920; How to Write Stories, 1922; Seeing America—Farm and Field (with Harold Hughes), 1924; Seeing America—Mill and Factory (with Harold Hughes), 1926; The Twilight of the American Mind, 1928; The Art of Rapid Reading, 1929; The Psychology of Happiness, 1929; The Young Citizen, 1929; The Art of Sound Pictures (with William M. Marston), 1930; The Psychology of Achievement, 1930; Vocational Studies in Journalism, 1931; The Art of Learning, 1931; How We Learn, 1931; Short Introduction to History of Human Stupidity, 1932; Life Begins At Forty, 1932; The Consumer—His Nature and His Changing Habits, 1932; More Power To You, 1933; The Chance of a Lifetime, 1934; Take It Easy, 1935; Let's Get What We Want, 1935; Capitalism Carries On, 1935; Careers After Forty, 1937; Making Good Before Forty, 1939; Seeing Our Country (with Harold F. Hughes), 1939; The Art of Useful Writing, 1940; Escape from Fear, 1940; On My Own, 1944; The Best Years, 1946; Road to a Richer Life, 1949. Editor and contbr.: The New Realism, 1913; As We Are, 1923. Research into devel. of country-city population research, eugenics, large; scale farming, central mgmt. in agr. Died Jan. 25, 1953.

PITMAN, Sir Isaac, English inventor; b. Trowbridge, Eng., Jan. 4, 1813; s. Samuel Pitman; ed. privately; m. 1835; m. 2d, Isabelle Masters, Apr. 21, 1861; children—Alfred, Ernest. Founded instn. for shorthand, Bath, Eng., 1837. Knighted, 1894. Author: Stenographic Sound Hand, 1837; Phonography, 1840. Invented original system of shorthand based on phonetics rather than orthographic principles, 1837; proposed phonetic printing alphabet using new letters; advocated spelling reform. Died Bath, Jan. 22, 1897.

PITOT, Henri, French physicist; b. Aramon, France, May 29, 1695; dir. Languedoc Canal; mem. French Acad. Scis. Author: Théorie de la manoeuvre des vaisseaux, 1731. Inventor pressure jet, also tube named after him for determination of current speed and wakes of ships; studied machines driven by water falls, wheels of mills on rivers, theory of pumps, impulse inclined to surface area of striking fluid jets. Died Aramon, Dec. 17, 1771.

PITRE, Davide, Italian chemist; b. Como, Italy, Mar. 4, 1923; s. Pierro and Piera Vita (Durini) P.; Dott. in Chemistry, U. Pavia (Italy), 1947; m. Casali Simona, July 23, 1953. Fellow Inst. Chem. Pharms., U. Pavia, 1947-51, asst., 1951-52, docent, 1960—; faculty dept. chemistry Cilag Italiana, 1952-54; dir. dept. organic chemistry Bracco Industria Chimica, 1954-60. Mem. Societa Italiana di Chimica, Societa Italian Scienze Farmaceutische. Publs. bd. Il Farmaco, 1951—. Research, publs. on S-acyl-pantethine and homo-pantethine, new iodinated radiologic diagnostic agts, partial catalytic reduction of aromatic dinitro compounds by hydrazine, pyrazine; invented (with E. Felder) iodamide. Home: 24 Via Brocchi. Office: 50 Via E. Folli, Milan, Italy.*

PITT, Harry Raymond, Brit. mathematician; b. West Bromwich, Eng., June 3, 1914; s. Henry and Florence Pitt; B.A., Ph.D., Cambridge (Eng.) U.; m. Catherine C. Jacoby, Apr. 5, 1940; children—Matthew, John, Francis, David. Asst. U. Aberdeen (Scotland); prof. math. U. Belfast (Ireland), 1945-51, U. Nottingham (Eng.), 1951-64; rector U. Reading (Eng.), 1964—. Fellow Royal Soc., 1957. Author: Tauberian theorems; Integration; Measure and Probability. Research, publs. on Fourier analysis, probability. Office: U. Reading (Eng.).

PITTENDRIGH, Colin Stephenson, biologist; b. Whitley Bay, Eng., Oct. 13, 1918; s. Alexander and Florence Hemy (Stephenson) P.; B.Sc. with 1st class honors (Lord Kitchener nat. meml. scholar), U. Durham, Eng., 1940; A.I.C.T.A., Imperial Coll., Trinidad, B.W.I., 1942; Ph.D. (Univ. fellow), Columbia, 1947; m. Margaret Dorothy Eitelbach, May 1, 1943; children—Robin Ann, Colin Stephenson. Came to U.S., 1945, naturalized, 1950. Biologist internat. health div. Rockefeller Found., 1942-45; adviser Bromeliad-malaria, Brazilian govt., 1945; asst. prof. biology Princeton, 1947-50, asso. prof., 1950-57, prof., 1957—, Class of 1877 prof. zoology, 1963—, dean Grad Sch., 1965—;vis. prof. Rockefeller Inst., 1962; Phillips lectr. Haverford Coll., 1956; Timothy Hopkins lectr. Stanford, 1957; adviser Brit. colonial office on Bromeliad-malaria, 1958. Mem. Nat. Acad-NRC com. oceanography. Past v.p., trustee Rocky Mountain Biol. Lab. Guggenheim fellow, 1959. Fellow Am. Acad. Arts and Scis.; mem. Am. Inst. Biol. Scis., Am. Soc. Zoologists, Am. Soc. Naturalists (v.p. 1963), Nat. Acad. Scis., A.A.A.S. Soc. Study Evolution, Sigma Xi. Author: (with George Gaylord Simpson, L. H. Tiffany) Life, 1957. Contbr. books, profl. jours. Home: Wyman House Princeton, N.J.*

PITTINGER, Charles Bernard, Am. physician; b. Akron, O., Apr. 27, 1913; s. Charles B. and Mary (Leininger) P.; B.S. in Chemistry, U. Akron, 1935; M.D., U. Cin., 1949; M.S. in Anesthesiology, State U. Ia., 1952; m. Gertrude Cramer, June 20, 1949; children—Suzanne, Charles, Catherine. Faculty, State U. Ia., 1952-62, prof., acting chmn. div. anesthesiology, 1961-62; prof. anesthesiology, chmn. dept. Sch. Medicine, Vanderbilt U., Nashville, anesthesiologist-in-chief Vanderbilt U. Hosp., 1962—, asso. prof. pharmacology, 1963—; cons. VA Adminstrn. Hosp., Nashville, 1962—; staff Nashville Gen. Hosp., 1963—. Diplomate Am. Bd. Anesthesiology. Fellow Am. Coll. Anesthesiologists; mem. Am. Chem. Soc., A.A.A.S., A.M.A., Am. Soc. Pharmacology and Exptl. Therapeutics, Am. Soc. Anesthesiologists, Assn. U. Anesthetists, Central Soc. Clin. Research, Central Surg. Assn., Internat. Anesthesia Research Soc., N.Y. Acad. Scis., Soc. Exptl. Biology and Medicine, So. Soc. Anesthesiologists, Sigma Xi. Research, publs. on pharmacological and clin. effects of anesthetic drugs; xenon and theories of narcosis, adverse drug reactions during anesthesia, neuromuscular blocking agts., hyperbaric anesthesia and oxygenation. Home: 201 Vaughn's Gap Rd., Nashville 37205.*

PITTMAN, David Joshua, Am. sociologist; b. Rocky Mount, N.C., Sept. 18, 1927; s. Jay Washington and Laura (Edwards) P.; B.A., U. N.C., 1949, M.A., 1950; Ph.D., U. Chgo., 1956. Instr. sociology U. Rochester, 1950; Louis Asher fellow U. Chgo., 1953-55; asst. prof. sociology U. Rochester, 1955-58; faculty sociology Wash. U., 1958—, now prof., dir. Social Sci. Inst., 1963—. Chmn. 28th Internat. Congress on Alcohol and Alcoholism, 1968; cons. St. Louis Bd. Police Commnrs., 1959—, Pres.'s Commn. on Law Enforcement and Adminstrn. Justice. Nat. Inst. Mental Health fellow. Fellow Am. Sociol. Assn.; mem. Soc. for Study Social Problems, Population Assn. Am., Internat. Soc. Criminology, Internat. Council on Alcohol and Alcoholism (mem. exec. com.), Phi Beta Kappa. Author: (with C. Wayne Gordon) Revolving Door: A Study of the Chronic Police Case Inebriate, 1958; (with M. Glatt, D. Hills, D. Gillespie) Drug Scene in Great Britain: Journey Into Loneliness, 1967—; also numerous articles. Editor: Alcoholism: An Interdisciplinary Approach, 1959; (with Charles R. Snyder) Society, Culture, and Drinking Patterns, 1962; Alcoholism, 1967. Co-organizer of First Detoxification Center for Alcoholism; devel. mental health tng. procedures for law enforcement; sociol. principle in organizing treatment facilities for alcoholics and drug addicts. Home: 7417 Oxford Dr., Clayton, Mo. 63105. Office: Social Sci. Inst., Washington U., St. Louis 63130.*

PITTMAN, Frank King, Am. nuclear engr.; b. Sacramento, Sept. 30, 1914; s. Frank King and Julia (Coleman) P.; B.Engring., Vanderbilt U., 1936, M.S., 1937; Ph.D., Mass. Inst. Tech., 1941. Instr. chemistry Mass. Inst. Tech., 1941-43; chemist Corhart Refractories Co., Louisville, 1943-44; mgr. plutonium recovery Plutonium Prodn. Plant, Los Alamos Sci. Lab., 1944-48; with U.S. AEC, Washington, 1948-64, dept. dir. civilian application div., 1956-57, dir. indsl. devel. div., 1957-58, dir. reactor devel. div., 1958-64; spl. asst. to v.p. marketing N.Am. Aviation, El Segundo, Cal., 1964-67; mgr. nuclear marketing Kerr McGee Corp., Oklahoma City, 1967—. Recipient Fed. Career Service award Nat. Civil Service League, 1960; AEC Distinguished Service award, 1964. Fellow Am. Nuclear Soc. (chmn. pub. information com. 1966—); mem. N.Y. Acad. Scis., Atomic Indsl. Forum (asso.); Sigma Xi. Author: (with Zinn, Hogerton) Nuclear Power, 1964; also articles. Research on chemistry, metallurgy and prodn. techniques for plutonium, nuclear central sta. power, nuclear space power. Home: 333 N.W. 5th St., Oklahoma City 73102. Office: Kerr McGee Bldg., Oklahoma City 73102.*

PITTMAN, Margaret, Am. bacteriologist; b. Prairie Grove, Ark., Jan. 20, 1901; d. James (M.D.) and Virginia (McCormick) Pittman; A.B., Hendrix Coll., 1923, LL.D., 1954; M.S., U. Chgo., 1926, Ph.D., 1929. Acad. instr., prin. Galloway Coll., Searcy, Ark., 1923-25; asst. sci. Rockefeller Inst. Med. Research, N.Y.C., 1928-34; asst. bacteriologist N.Y. State Dept. Health, Albany, 1934-36; staff Nat.

Insts. Health, Bethesda, Md., 1936—, chief sect. haemophilus studies, 1954-60, chief lab. bacterial products, division of biologics standards, 1958—, chief sect. bacterial vaccines, 1960-65; study conjunctivitis Rio Grande Valley, 1949; del. 5th Internat. Microbiologist Congress, Rio de Janeiro, 1950; participant conf. on whooping cough Internat. Children's Center, Paris, 1957, Prague 1962; mem. biol. standardization groups WHO, Geneva, 1958, 59, 62, cons., 1962; mem. cholera adv. com. NIH, 1962—, organized pertussis vaccine symposium, 1963; mem. panel expert consultants to tech. com. for Pakistan, SEATO Cholera Research Lab., 1962. Recipient Superior Service award Dept. Health, Edn. and Welfare, 1963, Distinguished Service award, 1967. Diplomate Am. Bd. Microbiology. Mem. Washington Acad. Scis. (pres. 1955), Am. Soc. Microbiologists (pres. Washington br. 1949-50; councilor-at-large 1958-59), Academy Microbiology (member board of governors 1962—), Am. Association Immunologists, Soc. Gen. Microbiol., Society Exptl. Biology and Medicine, A.A.A.S., Harvey Soc., N.Y. Acad. Sci., Am. Assn. U. Women, Sigma Xi, Mu Sigma Chi. Author numerous papers med. subjects. Research on agt., serology, treatment of meningitis; standardization of pertussis, cholera and typhoid vaccines; prevention of neonatal tetanus; criteria for sterility testing. Home: 3133 Connecticut Av. N.W., Washington 20008. Office: National Institutes of Health, Bethesda, Md. 20014.*

PITTMAN, Melvin Amos, Am. physicist; b. Ft. Lawn, S.C., Aug. 4, 1905; s. William Amos and Nancy (Nunnery) P.; B.S., The Citadel, 1925; M.S., U. S.C., 1929; Ph.D., Johns Hopkins, 1936; m. Lorraine O'Hara, Oct. 2, 1943; children—Melvin Philip, Joyce Lorraine, William Pfund. Faculty, U. Md., 1929-37; head dept. physics Madison Coll., 1937-42, 46-55; head dept. physics Coll. William and Mary Williamsburg, Va., 1955-66, dir. sci. insts., 1966-67; dean Sch. Scis., Old Dominion Coll., Norfolk, 1967—. Mem. Grad. Studies Bd., Va. Asso. Research Center, 1964-66. Mem. Optical Soc. Am.,, Am. Phys. Soc., Va. Acad. Sci., Sigma Vi. Research in infrared dispersion, absorption and polarization, total internal reflection carbon dioxide lasers in connection with scattering, radiation effects on optical devices. Home: 6158 Powhatan Av., Norfolk, Va. 23508.*

PITTS, Robert Franklin, Am. physiologist; b. Indpls., Oct. 24, 1908; s. John Franklin and Estella (Coffin) P.; B.S., Butler U., 1929; Ph.D., Johns Hopkins, 1932; M.D., N.Y.U., 1938; Rockefeller Found. fellow, Neurol. Inst. Northwestern U., 1938-39, Johnson Found. Med. Physics, U. Pa., 1939-40; m. Marjorie Anna Wallace, Dec. 25, 1936; children—Robert Wallace, Marjorieann. Asst. physiology N.Y. U., 1932-33, instr. physiology, 1933-38, asst. prof. physiology, 1940-42; asst. prof. physiology Cornell U., N.Y.C., 1942-44, asso. prof. physiology, 1944-46, prof., chmn. dept. since 1950; prof., chmn. dept. physiology Syracuse U., 1946-50; mem. physiology study sect. NIH; mem. NRC adv. com. on grad. sch. Army Med. Services; chmn. conf. renal function Josiah Macy Found.; mem. adv. bd. Life Ins. Med. Research Fund, Lederle Med. Faculty Awards, NRC Med. Fellowships. Recipient Gail Borden Award, Assn. Am. Med. Colls., 1960, Med. Alumni Research award N.Y. U., 1962, Homer W. Smith award in renal physiology N.Y. Heart Assn., 1963. Fellow A.C.P., Am. Acad. Arts and Sciences; mem. Nat. Acad. Scis., Am. Physiol. Soc. (bd. publ. trustees, mem. council; pres. 1959-60), Soc. Exptl. Biology and Medicine (chmn. N.Y. sect), Harvey Soc. (pres. 1960—), Soc. Clin. Investigation, Phi Beta Kappa, Sigma Xi, Phi Kappa Phi, Phi Delta Theta, Alpha Omega Alpha. Author: The Physiological Basis of Diuretic Therapy, 1959, Physiology of the Kidney and Body Fluids, 1963. Editor Am. Lectures in Physiology since 1946. Home: 535 Stellar Av., Pelham Manor 65, N.Y. Office: 1300 York Av., N.Y.C. 10021.*

PITZER, Kenneth Sanborn, Am. chemist; b. Pomona, Cal., Jan. 6, 1914; s. Russell K. and Flora (Sanborn) P.; B.S., Cal. Inst. Tech., 1935; Ph.D., U. Cal., 1937; D.Sc., Wesleyan University, 1962; LL.D., University of Cal. at Berkeley, 1963; m. Jean Mosher, July 1935; children—Ann, Russell, John. Instr. chemistry U. Cal., 1937-39, asst. prof., 1939-42, asso. prof., 1942-45, prof., 1945-61, asst. dean letters and sci., 1947-48; pres., prof. chemistry Rice U., Houston, 1961—; tech. dir Md. Research Lab. for OSRD, 1943-44; dir. research U.S. AEC, 1949-51, mem. gen. adv. com., 1958—, chmn., 1960-62; dean coll. chemistry U. Cal., 1951-60. Mem. adv. bd. U.S. Naval Ordnance Test Sta., 1956-59, chmn., 1958-59. Mem. commn. chem. thermodynamics, Internat. Union Pure and Applied Chemistry, 1953-61. Member of board of trustees Rand Corp. Mem. Pres.'s Science Advisory Com., 1965—. Awarded Guggenheim Fellowship, 1950, Precision Sci. Co. award in petroleum chemistry, 1950; One of the 10 Outstanding Young Men, U.S. Jr. C. of C., 1950; Clayton Prize, Instn. Mech. Engrs., London, 1958; Priestley Memorial award Dickinson Coll., 1963. Mem. program com. for phys. scis. Sloan Found., 1955-60. Fellow Am. Nuclear Soc., Am. Inst. Chemists, Am. Acad. Arts and Scis., Am. Phys. Soc.; mem. Am. Chem. Soc. (award pure chemistry, 1943 Gilbert Newton Lewis medal, 1965), Faraday Soc., A.A.A.S., Nat. Acad. Sci., Am. Philos. Soc. Author: Selected Values of Properties of Hydrocarbons (with others), 1947; Quantum Chemistry, 1953; (with L. Brewer) Thermodynamics, rev. 1961. Editor: Prentice-Hall Chemistry series, 1955-61. Contbr. articles to profl. jours. Research in chem. thermodynamics, quantum theory and statis. mechanics applied to chemistry, and molecular spectroscopy, results made possible wide-ranging predictions of thermodynamic properties of various substances. Home: President's House, Rice U., Houston 70001.*

PIWNICA, Armand Hermann, French surgeon; b. Paris, July 16, 1927; s. Charles and Henrietta Piwnica; m. Huguette Lachter, Dec. 16, 1948; children—Dominique, Emmanuel. Surgeon, Hopitaux de Paris, 1965—; prof. agrégé thoracic and cardiovascular surgery Faculté de Médecine, Paris, 1965—. Fellow A.C.S.; mem. Soc. Thoracic Surgeons, Association Fre. de Chirurgie. Research, numerous publs. on open heart surgery, cardiac prosthesis, artifical pacemakers. Address: 47 rue de Courcelles, Paris, France.*

PIYP, Boris Ivanovich, Russian geologist, volcanologist; b. Nov. 6, 1906; grad. Leningrad. Mining Inst., 1931. Worker in Lab. Volcanology, USSR Acad. Scis., 1940-61; chief Kamchatka Volcanological Sta., 1940-46, 50-54; dir. Kamchatka Geol. and Geophys. Obs., USSR Acad. Scis., 1961-62; dir. Inst. Volcanology, Siberian dept. USSR Acad. Scis., 1962—. Mem. presidium of Siberian br. USSR Acad. Scis., also dir. Kamchatka Joint Expdn., 1961. Mem. Communist Party, 1945—. Recipient prize of presidium of USSR Acad. Scis., 1956; decorated Order Red Star, other medals. Corr. mem. USSR Acad. Scis. Author: The Thermal Springs of Kamchatka, 1937; Geological and Petrological Data on The Avachi, Rassoshina, Gavanka and Nalacheva River Areas in Kamchatka, 1941; The Klyuchevskaya Volcano and its Eruptions in 1944-45, and in the Past, 1956; The USSR Academy of Sciences Expedition to the Kurile Islands and Kamchatka, 1958. Authority on volcanos, hot springs and geology of Kamchatka; detailed study of eruptions of various types of volcanos; dir. field research in Urals and Kamchatka Penninsula. Home: Leninskii Prospekt 25, Moscow. Office: Lab. of Volcanology, Staromonetnyg Pereulok 35, Moscow, USSR.

PLAKSIN, Igor Nikolaevitch, Russian hydrometallurgist; b. Oufa (the Ural), Oct. 8, 1900; grad. Far Eastern U., Vladivostok, 1926; D.Tech.Sci., Moscow, 1937. Prof. extractive metallurgy and treatment complex ores, head chair metallurgy of precious metals Moscow Inst. Steel and Alloys (formerly Moscow Inst. Non-Ferrous Metals and Gold), 1930—; head sect. mineral processing and coal preparation Scotchinsky Inst. Mining (Ministry Coal Industry - USSR Acad. Scis.), Moscow. Participant internat. congresses mineral processing (mem. sci. com. 1960, 63, 64), surface activity, coal preparation, mining and metallurgy, others. Decorated Order of Lenin; recipient Stalin prize, 1951, 52, 2 prizes USSR Acad. Scis. 2 diplomas and medals Indsl. Exhbn., Moscow. Mem. USSR Acad. Scis. (corr., pres. sci. council, phys. and chem. principles of mineral processing 1963). Author: The Interaction of Alloys and Virgin Gold with Mercury and Cyanogen Solutions, 1937; The Metallurgy of Precious Metals, 1943; Assaying and Assay Analysis, 1947; (co-author) Hydrometallurgy, 1949; The Technological Equipment of Ore Enriching Plants, 1955; The Metallurgy of Precious Metals, 1958. Research, patents and many publs. on theory and tech. of hydrometallurgy, ore enrichment, applied chemistry, tech. physics. Address: Scotchinsky Institute of Mining, Luberzy 4, Moscow, USSR.*

PLANA, Baron Giovanni Antonio Amedeo, Italian astronomer, geometer; b. Voghera, Lombardy, Nov. 13, 1781; prof. astronomy U. Voghera, also dir. obs.; prof. analysis U. Turin (Italy), also dir. obs., Turin; mem. French Acad. Scis., 1826. Research on vibrations of elastic sheets, 1815, motion of pendulum in resisting medium, 1835, also specific heat of gases, distbn. of electricity of metal sphere, rotating magnetisms. Died Turin, Jan. 20, 1864.

PLANCHEREL, Michel, Swiss mathematician; b. Bussy, Switzerland, Jan. 16, 1885; s. Donat and Justine (Pamblanc) P.; ed. St. Michael Coll., Fribourg, Germany, U. Fribourg, U. Göttingen (Germany), U. Paris (France); dr. es sci. math Agrégé U. Geneva (Switzerland), 1910-11; asso. prof. U. Fribourg, 1911-13, prof., 1913-20; prof. Fribourg Ecole Polytechnique, rector, 1931-35; prof. Zurich Ecole Polytechnique, 1920-55. Mem. Geneva Soc. Phys. and Natural Scis. (hon.), Fribourg Soc. Natural Sci. (hon.), Swiss Math. Soc., Coimbre Inst. (corr.), Turin Acad. Sci. Publs. on orthogonal functions, Fourier transformations, asymptotic formulas. Address: Kraebuehlstrasse 64, 8044 Zurich, Switzerland.

PLANCHON, Jules-Émile, French pharmacist, botanist; b. Ganges, France, Mar. 21, 1823; Docteur ès scis., 1844; M.D., 1851; became asso. with Grand Bot. Inst., Belgium, 1850; apptd. prof. botany Montepellier (France) Faculty Scis., 1854; named dir. Montepellier Bot. Gardens, 1860; mem. French Acad. Scis. Author: Flore des serres et jardins de l'Europe. Introduced (with Riley) Australian plant louse which eliminated a plant lice plague attack on orange groves in Cal., 1886; organogenic research on plant ovule; identified phylloxera. Died Montpellier, Apr. 1, 1888.

PLANCK, Max Karl Ernst Ludwig, German physicist; b. Kiel, Schleswig, Germany, Apr. 23, 1858; s. Johann Julius Wilhelm and Emma (Patzig) P.; student U. Munich, 1874-77; Ph.D., U. Berlin, 1879 (under Helmholtz, Kirchoff); m. 1887; 1 son, 2 daus.; m. 2d, 1911; 1 son, Erwin. Privat dozent U. Munich, 1880-85; prof. extraordinary theoretical physics U. Kiel, 1885-89; prof. extraordinary theoretical physics U. Berlin, 1889-92, ordinary prof., 1892-28; retired; spent final years in pvt. research on problems of philosophy and causality. Mem. Prussian Acad. Scis., 1896, permanent sec., 1912-43; pres. Kaiser Wilhelm Soc. (later named Max Planck Soc.), 1930-37; chancellor Order of Merit, 1930; Recipient Nobel prize in physics, 1918. Fellow Royal Soc., 1926 (Copley medal 1929). Author: Vorlesungen über Thermodynamik, 1897; Einführung in die theoretische Physik, 5 vols., 1916-30; Wege sur physicalischen Erkenntnis, 1933; Wissenschaftliche Selbstbiographie, 1948; Where is Science Going, pub. 1959; The Universe in the Light of Modern Physics, pub. 1959; The Philosophy of Physics, pub. 1959. Originated and helped devel. quantum theory from 1901; this theory evolved from his work on black-body radiation problems; Planck's constant, h, the quantum of action, named for him; research in thermodynamics, especially theory of entropy; investigated mechanics, optical and elec. problems associated with radiation of heat and with quantum theory; his quantum theory had such an impact that physics before his time is now called classical physics. Died Göttingen, Germany, Oct. 4, 1947.

PLANE, Robert Allen, Am. chemist; b. Evansville, Ind.; s. Allen George and Altha (Warren) P.; A.B., Evansville, Coll., 1948; S.M., U. Chgo., 1949, Ph.D., 1951; m. Mary Moore, July 2, 1963; children—David Allen, Martha Lu, Ann, Jennifer. With Oak Ridge Nat. Lab. 1951-52; faculty Cornell U., Ithaca, N.Y., 1952—, prof. chemistry, 1962—. NIH Spl. fellow Nobel Med. Inst., Stockholm, Sweden, 1960, Oxford (Eng.) U., 1961. Mem. Am. Chem. Soc., Am. Assn. U. Profs., Sigma Xi. Author: (with M.J. Sienko) Chemistry, 1957, 61, 66, Experimental Chemistry, 1958, 61, 66, Physical Inorganic Chemistry, 1963, Chemistry Principles and Properties, 1966; (with R.E. Hester), Elements of Inorganic Chemistry, 1965; also numerous articles. Established formulas, structure and method of bonding in metal ion complexes by spectroscopic techniques; research mechanisms complex ion reactions, photochemistry, metal ions in biol. systems. Home: 108 N. Sunset Dr., Ithaca, N.Y. 14850.*

PLANELES, Khuan Khuanovich, pharmacologist, microbiologist, virologist; b. Jerez, Spain, 1900; grad. Med. Faculty, Madrid U., 1921; D.Med. Sci., 1923. With sci. instns., Germany, 1921-24, Inst. Pharmacotherapy, Amsterdam, 1924-25; sci. dir. pharm. factory, Spain, 1926-36; dir. Clin. Research Inst., Madrid, 1930-36; sr. physician 5th Regt., later state sec. health Spanish Republican Govt., 1936-39; lectr. dept. pharmacology Saratov Med. Inst., 1939-41; head lab. for chemotherapeutic agts. Sechenov 1st Moscow Med. Inst., 1942-44; head lab. expt. chemotherapy Gamaleya Inst. Epidemiology and Microbiology, USSR Acad. Med. Scis., 1943-49, head dept. infectious pathology and exptl. therapy of infections, 1943—, head dept. exptl. chemotherapy Research Inst. Pharmacology and Exptl. Chemotherapy, 1949—, prof., 1944—, also mem. com. for antibiotics. Recipient Gamaleya prize, 1956, 61. Mem. USSR Acad. Med. Scis. (corr.), Spanish Med. Acad. (corr., Gold medal 1926). Author: The Pathogenesis of Infectious Diseases, 1956; The New Antibiotic Mycerin, 1959; co-author: Side Effects in the Antibiotic Therapy of Bacterial Infections, 1960. Mem. editorial council Antibiotics, Jour. Microbiology, Epidemiology and Immunobiology; dep. editor Microbiology sect. Large Med. Ency., 2d edit. Research and numerous publs. on infectious pathology, exptl. therapy, pharmacology, pharmacotherapy. Address: Gamaleya Inst. Epidemiology and Microbiology, Uspensky p. 12, Moscow, USSR.

PLANO, Richard James, Am. physicist; b. Merrill, Wis., Apr. 15, 1929; s. Victor James and Minnie (Hass) P.; A.B., U. Chgo., 1949, B.S., 1951, M.S., 1953, Ph.D., 1956; m. Louise Sylvia Grevillius, July 3, 1956; children—Linda Sylvia, Robert James. Faculty, Columbia, 1956-60; faculty Rutgers U., New Brunswick, N.J., 1960—, prof., 1962—. Mem. Am. Inst. Physics, Am. Assn. Physics Tchrs., Am. Assn. U. Profs. Studies on properties of elementary particles using the bubble chamber technique. Home: P.O. Box 306, Middlebush, N.J. 08873. Office: Physics Dept., Rutgers U., New Brunswick, N.J. 08903.*

PLANTA VON SILS, Martin, Swiss inventor; b. Euoz, Switzerland, Mar. 1727; ed. Zurich, Switzerland; prof., Erlangen, Germany, also Zurich; tutor to family of Baron Seckendorf. Inventor Wimshurst electric machine, 1755; submitted his ideas on steam power to Choiseul. Died Marschlins, Switzerland, 1772.

PLANTE, Gaston, French physicist; b. Orthez, France, Apr. 22, 1834; lecture asst. for physics

Conservatoire des Arts et Metiers; prof. physics Assn. Polytechnique; built (on basis of Ritter's research) 1st accumulator, of rolled lead sheets in dilute sulphuric acid, 1860. Died Paris, May 21, 1889.

PLANTIN, Christophe, French printer; b. Tours, France, 1514-20; considered 1st indsl. printer; encouraged Dodonaeus to publish a Flemish herbal, 1554, Lobelius to publish new version of his Plantarum seu stirpium historia, 1581; used A. Nicolai's plates to illustrate Garcia da Orta's Coloquias. Died Anvers, France, circa 1589.

PLANTIN, Lars Olof, Swedish chemist; b. Stockholm, Sweden, Jan. 16, 1924; s. Ake Oskar and Ester (Steen) P.; Diplomaas chem. engr., Royal Inst. Tech., Stockholm, 1948; m. Ingeborg R.L. Enebratt, May 8, 1954; 1 son, Lars Mikael. Research asst. King Gustaf V Research Inst., Stockholm, 1948-56, first research engr., 1956—. Mem. Swedish Chem. Soc. Research and publs. on methodological studies on the chem. determination of 17-ketosteroids and corticosteroids in human urine; metabolic studies of same steroids in health and disease especially in cancer diseases; metabolic and kinetic studies of plasma proteins in humans with the aid of radioiodinated proteins; pathophysiology in severe burns; radioactivation analysis of trace elements in blood. Home: 32 Rastavägen, Solna, Sweden. Office: King Gustaf V Research Inst, Fack, Stockholm 60, Sweden.*

PLANUDES, Maximos, translator, mathematician; b. Nicomedia, circa 1260; Byzantine monk; sent by Andronicos II Palaeologos as ambassador to Venice, 1296; translated works from Latin into Greek, thus helping to lay found. for Greek renaissance of western Europe. Author an arithmetic after Hindu method, circa 1300, a commentary on 1st 2 books of Diophantos, 2 grammatical treatises; compiler collections of hist. and geog. extracts, of proverbs, of epigrams. Died circa 1310.

PLAS, F., French physician; b. Limoges, France, Oct. 10, 1912; s. Émile and Louise (Theullier) P.; ed. Paris hosps., Paris Faculty Medicine; m. Jacqueline Pecaut, Dec. 27, 1939; children—Alain, Marie Christine (Mrs. Dugardin), Jean Noël, Claudine. Agrégé prof. exptl. pathology Paris Faculty Medicine, since 1958—; dir. Paris Inst. Phys. Edn., 1963—; practice medicine, specializing in cardiology, 1942—. Courtroom expert on cardiology. Recipient Médaille de l'Aéronautique; Chevalier de la Légion d'Honneur; Chevalier des Palmes académiques. Mem. Soc. de Cardiologie, Soc. de Thérapeutique, Soc. de Médecine Aéronautique, Soc. de Pathologie comparée, Groupement Latin de Médecine Sportive, Société Nationale de Médecine Sportive (pres.). Author: (with Chailley Ber) Physiologie des activités physiques; also numerous articles. Research on minor disorders of repolarization, electrocardiographic disorders of athletes, cardiac condition tests and tele-recording of electrocardiogram, suprarenal gland of athletes and protidic metabolism, cardiac adaptation to changes in altitude. Address: 18, rue de Grenelle, Paris VII, France.*

PLASKETT, Harry Hemley, astronomer; b. Toronto, Ont., Can., July 5, 1893; s. John Stanley and Rebecca (Hemley) P.; B.A., U. Toronto, 1916; M.A., Oxford (Eng.) U., 1932; LL.D., St. Andrews U., Scotland, 1961; m. Edith Alice Smith, Jan. 4, 1921; children—Barbara (Mrs. Arthur L. Pidgeon), John Stanley. Research astronomer Dominion Astrophys. Obs., Victoria, B.C., 1919-27; prof. Harvard, 1928-32; Savilian prof. astronomy Oxford U., 1932-60. Recipient Gold medal Royal Astron. Soc., 1963. Fellow Royal Soc., 1936. Research, publs. on solar and stellar spectra, especially O-type stars, shell star and nebulae, solar granulation; interpretation of equatorial acceleration in sun's rotation. Home: 48 Blenheim Dr., Oxford, Eng.*

PLASKETT, John Stanley, Canadian astronomer; b. nr. Woodstock, Ont., Can., Nov. 17, 1865; B.A., U. Toronto, 1899, D.Sc., 1923; entered service astron. br. Dept. Interior, Ottawa, 1903; became astronomer Dominion Obs., Ottawa, 1905; dir. Dominion Astrophys. Obs., Victoria, B.C., 1918-35; supervised grinding and polishing of 82-inch telescope mirror for McDonald Obs., U. Tex., after 1935. Fellow Royal Soc., 1923, Royal Soc. Can. (Flavelle medal 1910). Contbr. to sci. publs. Discovered that Plaskett twins (now named) are 2 stars rather than 1, 1922; studied motion of Milky Way and of its stars, also rotation of galaxy. Died Esquimault, B.C., Oct. 17, 1941.

PLAT, Sir Hugh, (or Platt) English inventor, agriculturist; b. Garlickhythe, Eng., May 3, 1552; s. Richard and Alice (Birtles) P.; B.A., St. John's College, Cambridge U., 1572; m. 1, Margaret Jounge, 1573; m. 2nd., Judith Albany; 3 sons 1st m.; 2 sons, 3 daus., 2nd m. Mem., Lincoln's inn; resided at Bethnal Green and other places in London area where he carried out agricultural and other experiments. Author: The Floures of Philosophie, 1572; The Jewell House of Art and Nature, 1594; Sundrie New and Artificial Remedies against Famine, 1594; Newfounde Art of Setting Corne, circa 1596; Delights for Ladies to adorne their Persons . . . , 1602; Of Coal-Balls for Fewell wherein Seacoal is, 1603; Floraes Paradise beautified and adorned with sundry sortes and delicate Fruits and Flowers, 1608; others. Listed new inventions relating to mechanics, chemistry and agriculture; described brewing of beer without hops, preservation of food in hot weather and at sea; invented cheap fuel by kneading coal and clay into balls; investigated the making of wine from English grapes; translated Bernard Palissy's Des sels diverses and part of his De la marn, which were included in Jewell House of Art and Nature; in these translations and his own commentary stated that salt of manure was active part which resulted in increased agricultural yields; 1st attempt by English author to understand reason for beneficial results and manuring and marling in agriculture. Plat's works were popular and often reprinted. Died 1608.

PLATEARIUS, Matthaeus, Salernitan physician; probably s. Joannes Platearius the Younger. Author: De simplici medicina (also called Circa instans; one of prototypes of Western pharmacopoeias); also earliest commentary on Nicholas' Antidotarium. Died 1161.

PLATEAU, Joseph Antoine Ferdinand, Belgian physicist; b. Brussels, Belgium, Oct. 14, 1801; student Brussels; doctorate phys. and math. sci., Liège, Belgium, 1829. Became prof. physics, Ghent, Belgium, 1835; gave up teaching after he became blind, 1843; continued research aided by his son Félix and son-in-law Van der Mensbruyghe. Author: Statique expérimentale et théorique des liquids soumis aux seules forces molécules, 2 vols., 1873. Fellow Royal Soc., 1870; mem. French Acad. Scis. Developed stroboscopic method for study of vibratory motion; invented phenakiscope (using principles of motion picture camera), 1829; discovered 2d drop which always follows main drop of liquid falling from surface (named after him); research in physiol. optics, molecular physics, surface tension, properties of liquids; studies in light impression appearing on observation of colored objects, illumination, 1834-35. Died Ghent, Sept. 15, 1883.

PLATNER, Johann Zacharias, German surgeon; b. Chemnitz, Saxony, Aug. 16, 1694; ed. Leipzig and Halle, M.D., 1716; studied anatomy and surgery in Paris; children include Ernst Zacharias. Traveled and visited universities in Germany, Lyons, Switzerland, Savoy, Paris; visited Boerhaave and Albinus in Leyden; returned to Chemnitz, 1719; extraordinary prof. anatomy and surgery, Leipzig, 1721; prof. physiology, 1724; ordinary prof. surgery and anatomy, 1734; prof. pathology, 1737; sr. prof. medicine, 1736; perpetual dean of Med. Faculty, 1736; prof. therapeutics, 1747; councillor to Court of Saxony. Author: Institutiones Chirurgiae rationales, 1745. Took special interest in surgery on the eye. Died Dec. 19, 1747.

PLATO, Greek philosopher; b. Athens, Greece, 428/7 B.C.; s. Ariston and Perictione; had political ambitions, but abandoned them when family friend and mentor, Socrates, was executed, 399 B.C.; left Athens; visited Megara, Syracuse, Tarentum; returned to Athens, where he founded school for research (The Academy, founded 387 B.C.: the first university; continued in operation until 529 A.D.). Author of many dialogues: Lysis; Charmides; Laches; Euthythro; Apology; Crito; Gorgias; Protagoras; Euthydemus; Meno; Ion; Cratylus; Menexenus; Phaedo; Symposium; Phaedrus; Republic; Timaeus; Critias; Parmenides Theaetetus; Sophists; Politicus; Philebus; Laws; Epinomis; possibly others whose authenticity is questioned or difficult to determine. His influence on Western thought (in metaphysics, epistomology, ethical theory, political thought and science) cannot be over estimated; held that our senses are not reliable; attempted to grasp by reason the eternal and unchanging intelligible ideas or forms (a reality having no location in space or time) behind the ephermeral shifting and impermanent sense appearances; in science (see his Timaeus), stressed importance of mathematics, holding nature pervaded by mathematical harmony and mathematical structure. Ever since, Platonism in science has generally meant emphasis on a priori abstract mathematical thinking. Died Athens 348/7 B.C.

PLATO OF TIVOLI (Plato Tiburtinies), mathematician, astronomer; flourished in Barcelona, circa 1134-35; translated works from Arabic and Hebrew, including: De electinibus horarum (al-Imrant), 1133-34; De nativitatibus (Abu 'Ali al-Khaiyat), 1136; De revolutionibus nativatum (Abu Bakr al-Hasan ibn al-Khasib); Liber Abulcasim de operibus astrolabiae (Ibn al-Latin; Spherics (Theodosios); Liber embardorum (Abraham bar Hiyya, landmark in history of math., earliest Latin writing giving complete solution of a quadratic equation); also a work of al-Battani on trigonometry.

PLATT, John Rader, Am. biophysicist, educator; b. Jacksonville, Fla., June 29, 1918; s. Louis Walter and Jennie (Sharp) P.; B.S., Northwestern U., 1936, M.S., 1937; Ph.D., U. Mich., 1941; m. Ann Isabel Tammela, June 23, 1941; children—Terry, Christopher. Inst. physics U. Toledo, 1941; research asso., instr. physics U. Minn., 1941-43; instr. physics, OSRD research asso. Northwestern U., 1943-45; faculty U. Chgo., 1945—, prof. physics and biophysics, 1957-65; prof. physics and asso. dir. Mental Health Research Inst., U. Mich., 1965—. Mem. study sect. on biophysics and biophys. chemistry NIH, 1959-63. Recipient Career award USPHS, 1964—, ann. book award Nat. Assn. Ind. Schs., 1962; John Simon Guggenheim fellow U. London (Eng.), 1952-53. Mem. Am. Phys. Soc. (sec. div. chem. physics 1951-52, 55-58, vice chmn. (1959-60, chmn. 1960-61), Phi Beta Kappa, Sigma Xi, Phi Kappa Phi. Author: The Excitement of Science, 1962; Free-Electron Theary of Conjugated Molecules - A Source Book, 1964; Systematics of Electronic Spectra of Conjugated Molecules - A Source Book, 1964; New Views of the Nature of Man, 1965; The Step to Man, 1966. Asso. editor Jour. Chem. Physics, 1951-53, 59-61. Research on distbn. and spectroscopy of electrons in large organic molecules. Office: Mental Health Research Inst., U. Mich., Ann Arbor, Mich. 48104.

PLATT, Joseph Beaven, Am. physicist, educator; b. Portland, Ore., Aug. 12, 1915; s. William Bradbury and Mary (Beaven) P.; B.A., U. Rochester, 1937; Ph.D., Cornell U., 1942; m. Jean Ferguson Rusk, Feb. 9, 1946; children—Ann Ferguson, Elizabeth Beaven. Instr. physics U. Rochester (N.Y.), 1941-43; sect. chief, radiation lab. Mass. Inst. Tech., Cambridge, 1943-45; faculty U. Rochester, 1946-56, prof., 1953-56; chief physics br., research div. AEC, Washington, D.C., 1949-51; president of Harvey Mudd Coll., Claremont, Cal., 1956—. Cons. NSF, Nat. Acad. Sci.-NRC; alternate del., sci. adviser Nat. Acad. Sci.-NRC; mem. panel on internat. sci. Pres.'s Sci. Adv. Com.; alternate del., sci. adviser UNESCO Gen. Conf., 1960, 62; sci. adviser Republic of China, 1964. Fellow Am. Phys. Soc.; mem. Am. Optical Soc. (asso.), Am. Nuclear Soc., Am. Assn. Physics Tchrs., Sigma Xi, Phi Beta Kappa, Phi Kappa Phi Synchroclotron design, 1946-48; detection and preliminary measurements of pi-mesonic atoms, 1952-55. Home: 495 E. 12th St., Claremont, Cal. 91711.*

PLATT, Robert Baxter, Am. biologist; b. Knoxville, Tenn., Jan. 19, 1913; s. Robert Baxter and Nette (Lawson) P.; A.B., Emory and Henry Coll., 1933; M.A., Peabody Coll., 1935; Ph.D., U. Pa., 1948; m. Wilma Dean Sherrod, Nov. 19, 1941; children—Carolee Jeannette, Rosalind Louise. High sch. biology tchr., Roanoke, Va., 1936-39; prof. biology Armstrong Jr. Coll., Savanna, Ga., 1941-42; prof. biology Radford (Va.) Coll., 1941-42; instr., U. Pa., 1945-48; faculty Emory U., Atlanta, 1948—, prof. biology, 1960—. Cons. pvt. cos., govt. agys.; mem. council Oak Ridge Inst. Nuclear Studies, 1963—. Trustee Highland Biol. Station. Guggenheim fellow, 1959. Fellow A.A.A.S., Ga. Acad. Sci. (past chmn.); mem. Bot. Soc. Am. (past chmn. S.E. sect.), So. Appalachian Bot. Club (past v.p.), Ecol. Soc. Am. (v.p. 1965—), Am. Inst. Biol. Scis. (vis. lectr. 1962—), Internat. Soc. Biometeorology, Sigma Xi (award 1951). Author: (with J.F. Griffiths) Environmental Measurement and Interpretation, 1964; also numerous articles. Editorial bd. Ecol. Monographs, 1958-61. Research on natural environment and its interrelationship with life, effects ionizing radiation on population and ecosystem, effects nuclear war on man's environment, problems pollution. Home: 1811 E. Clifton Rd., N.E., Atlanta 30307.*

PLATTER, Felix, Swiss anatomist; b. Basel, Switzerland, Oct. 28, 1536; s. Thomas Platter; prof. medicine, Basel; city physician, Basel; founder bot. garden; founder, tchr. anat. theater. Author: De corporis humani structura et usu, 1583; Praxis medica, 1602-08; Observationum in hominis affectibus libri tres, 1614. Forerunner of path. anatomy; made observations on lesions of cadavers; 1st to classify illness according to symtomatology; rejected notions that demons cause disease; credited with being 1st to record a death from hypertrophy of thymus (in an infant), 1614. Died Basel, July 28, 1614.

PLATTES, Gabriel, English chemist, agriculturist; flourished 1638. Author: A Treatise of Husbandry, 1638; Discovery of Infinite Treasure, 1639; A Discovery of Subterraneal Treasure, 1639; Observations and Improvements in Husbandry, with twenty Experiments, 1639; Recreatio Agriculturae, 1640. One of earliest advocates in Eng. of improved system of agriculture, husbandry, conducted several practical expts.; gave 1st description in English of parting of gold and silver with nitric acid; 1st useful English text on metallurgy; gave directions for analysis of mineral waters; attempted to explain formation of earth's features by chemical analogy. Died London.

PLATZMAN, George W(illiam), geophys. scientist; b. Chgo., Apr. 19, 1920; s. Alfred and Rose I. (Kaufman) P.; B.S., U. Chgo., 1940, Ph.D., 1948; M.S., U. Ariz., 1941; m. Harriet M. Herschberger, Feb. 19, 1945. Instr., U. Chgo., 1942-45, research asso., 1947-48; mem. faculty, 1949—, head phys. scis. in coll., 1959-60, prof. geophys. scis., 1960—; hydrologic engr. C.E., U. S. Army, 1945-46; cons. Inst. Advanced Study, Princeton, 1950-53. Fellow Am. Geophys. Union, Am. Meteorol. Soc. (editor jour. 1948-49, chmn. publs. com. 1966—; Meisinger award 1966). Research, publs. on circulation and wave theory of atmosphere and ocean; also on storm surges.

PLATZMAN, Robert LeRoy, Am. chemist, physicist; b. Mpls., Aug. 23, 1918; s. Alfred and Rose (Kaufman) P.; B.S., U. Chgo., 1937, M.S., in Physics, 1940, Ph.D. in Chemistry, 1942; m. Eva Maria Platzman, 1947; children—Loren Kerry, Elena Marie, Kenneth Rainer. Asso. prof. Physics Purdue U., 1949-58; Fulbright prof. physics and chemistry U. Paris (France), 1959-61; sr. physicist Argonne (Ill.) Nat. Lab., 1958-65; prof. physics and chemistry, master phys. scis. collegiate div., asso. dean coll. U. Chgo.,

1966—. Cons. to govt. agys., labs.; chmn. subcom. on effects ionizing radiations com. on nuclear sci. NRC, 1958—. Guggenheim fellow U. Copenhagen (Denmark), 1946-48, U. Rome (Italy), 1948. Fellow Am. Phys. Soc., Phys. Soc. (London, Eng.); mem. Radiation Research Soc. (past councillor). Research, numerous sci. publs. on mechanism of action of ionizing radiations; first exposition of role of subexcitation electrons and superexcited states of molecules; prediction of formation and properties of hydrated electron. Office: U. Chgo., Chgo. 60637.*

PLAUDE, Karl Karlovich, Russian thermal engr.; b. Mar. 26, 1897; grad. Leningrad Inst. Civil Engring., 1926. Worker at Gidrovlika plant, 1926-36; lectr. Leningrad Inst. Civil Engring., 1928-34, Leningrad Inst. Indsl. Constrn. Engrs., 1932-38; worker Lengosproekstroi, 1937-41; chief Constrn. Directorate, Moscow, 1942-43; lectr. U. Latvia, 1941-53; dir. Latvian SSR Acad. Scis. Inst. Energetics and Electrotechnics, 1950—. Mem. Communist Party, 1946—; dep. of Supreme USSR; mem. Council Nationalities, 1962. Named Honored Scientist of Latvian USSR, 1955. Academician Latvian SSR Acad. Scis. (v.p. 1958-60, pres. 1960); corr. mem. USSR Acad. Scis. Author: Scheme of a Step System of Distant Heat Supplying, 1950; A Natural Ventilation Installation, 1952; System of Heat Supply According to a Two-Step Scheme, 1953; Characteristics of the Heating of Radiators in Central Water Heating in Increased Temperatures of the Heat-Carrier, 1956; Automatic Thermo-Regulator for Radiators of a Central Water Heating System, 1956; Calculated Temperature of Water in Radiators of a Central Heating System, 1957; Automatic Regulation of a Central Water Heating System, 1960; Automation of Subscriber Centres in District Heated Buildings, 1960; Applications of Water at High Temperatures in Heating Systems, 1962; Latvijas enerhetikas attistiba PRSA energosistema, 1961. Research on heat supply; developed two-step system of heat supply, elec. systems and automatic thermo regulators, control of radiators locally; studies on heat exchange using high temperature carrier; principles of heating automation. Address: Latvian SSR Acad. Scis., ul. Turgeneva 9, Riga, Latvia USSR.

PLAUT, Gerhard Wolfgang Eugen, biochemist; b. Frankfort, Germany, Jan. 9, 1921; s. Max and Clara (Mayer) P.; came to U.S., 1937, naturalized, 1944; B.S., Ia. State U., 1943; M.S., U. Wis., 1949, Ph.D., 1951. Asst. prof. biochemistry U. Wis., 1951-54; asso. prof. biochemistry N.Y. U., N.Y.C., 1954-58; asso. prof. biochemistry, asso. research prof. medicine U. Utah, Salt Lake City, 1958-60, prof. biol. chemistry, research prof. medicine, 1960-66; prof. biochemistry, chmn. dept. Rutgers U. Med. Sch., New Brunswick, N.J., 1966—. Established investigator Am. Heart Assn., 1954-58. Research Devel. award USPHS, 1959-62, Research Career award, 1962-66. Mem. Am. Chem. Soc., Am. Soc. Biol. Chemists, Biochem. Soc. Research, numerous publs. in mode of action of vitamins, biosynthesis of pteridines and riboflavin, phosphorylation reactions, mechanism of enzyme action. Office: Dept. Biochemistry, Rutgers Med. Sch., New Brunswick, N.J. 08903.*

PLAYFAIR, John, Scottish mathematician, geologist; b. Benvie, near Dundee, Forfarshire, Scotland, Mar. 10, 1748; s. James and Margaret Young P.; graduated U. St. Andrews, 1765; studied theology at St. Mary's College, licensed by presbytery, entered ministry, 1770, held various livings; joint prof. mathematics, U. Edinburgh, 1785-1805; chair natural philosophy, U. Edinburgh, 1805; toured France, Italy, Switzerland, 1815-16 and engaged in geological and mineralogical researches; one of original members Royal Soc. Edinburgh (gen. Sec.); Fellow, Royal Soc., 1807. Author: Elements of Geometry, 1795; Illustrations of the Huttonian Theory of the Earth, 1802; Outlines of Natural Philosophy, 2 vols., 1812-16; many articles. Proponent of Huttonian theory of the earth; helped create modern sci. of geology. Died, Edinburgh, Scotland, Jul. 20, 1819.

PLAYFAIR, Lyon (1st Baron of), English chemist; b. Chuar, Bengal, India, May 21, 1818; s. George and Janet (Ross) P.; student St. Andrews, studied chemistry under Thomas Graham, Glasgow; pupil of Leibig; Ph.D., Giessen, Germany; m. Margaret Eliza Oakes, 1846; 1 son, George James; m. 2d, Jean Ann Millington, 1857; m. 3d, Edith Russell, 1878. Asst. to Graham, Univ. Coll., London; hon. prof. chemistry Royal Instn., Manchester, Eng., 1842-45; chemist Geol. Survey, also prof. new Sch. Mines, 1845; helped organize Crystal Palace expn., 1851; sci. sec. Dept. Science and Art, 1853, sec. for sci. and art, 1855-58; prof. chemistry, Edinburgh, 1858-69; Liberal mem. Parliament for univs. Edinburgh, St. Andrews, 1868-85, for South Leeds, 1885-92; postmaster-gen., 1873. Fellow Royal Soc., 1848; pres. Chem. Soc., 1857-59. Author: Science in its Relations to Labour, 1853. Discovered nitroprusside, 1849; studied (with Joule) atomic volume and specific gravity of hydrated salts; studied (with Bunsen) gases of blast furnace; helped (as mem. commn. on health of towns) lay founds. of modern sanitation; increased pub. awareness of importance of science. Died London, Eng., May 29, 1898.

PLAYFAIR, William Smoult, Brit. physician; b. 1836; s. George Playfair; ed. St. Andrews, Edinburgh; M.D., LL.D.; m. Emily Kitson. Became asst. surgeon Bengal Army, 1857; prof. surgery Med. Coll. Calcutta, 1859-60; then practiced medicine, London; cons. physician diseases of women and children Kings' Coll. Hosp.; physician-accoucheur to duchesses of Edinburgh and Connaught; prof. obstetric medicine King's Coll. Fellow Royal Coll. Physicians, Royal Coll. Surgeons Edinburgh. Author: A Treatise on the Science and Practice of Midwifery, 2 vols.; The Systematic Treatment of Nerve Prostration and Hysteria; Handbook of Obstetric Operations; also papers. Editor: (with Clifford Allbutt) A System of Gynaecology. Died Aug. 3, 1903.

PLEIN, Elmer Michael, Am. pharmacist; b. Dubuque, Ia., Nov. 21, 1906; s. Michael Christopher and Clara (Esch) P.; B.S. in Pharmacy, U. Colo., 1929, Ph.C., 1929, M.S., 1931, Ph.D., 1936; m. Ellen Joy Bickmore, Aug. 30, 1952. Faculty, U. Colo., 1929-38; faculty U. Wash., Seattle, 1938—, prof. pharmacy, 1951—, dir. drug service dept., 1950—, coordinator pharm. services, 1959. Mem. revision com. U.S. Pharacopiea, 1950—. Recipient Lehn and Fink medal U. Colo., 1929. Mem. Am. Pharm. Assn. (Man of Year merit award Puget Sound chpt. 1965), Am. Chem. Soc., Am. Coll Apothecaries, Am. Soc. Hosp. Pharmacists, Sigma Xi. Author: Laboratory Manual: Dispensing Pharmacy, 1943, rev., 1947, 50, 55; American Pharmacy, 1948; Pharmaceutical Compounding and Dispensing, 1949; Prescription Pharmacy, 1963; also numerous articles. Research on stability of drugs, effectiveness dosage forms, dermatologic preparations, crystallography. Home: 5122 N.E. 75th St., Seattle 98115.*

PLENCIZ, see von Plenciz.

PLENTL, Albert Adolphe, physician, chemist; b. Cairo, Egypt., Oct. 31, 1913; s. Wolfgang and Albina (Seeger) P.; came to U.S., 1933, naturalized, 1940; student Amherst Coll., 1933-34; M.A., Princeton, 1935; Ph.D., Columbia, 1940; M.D., Cornell U., 1948. With E.I. duPont, 1935-36, Am. Cyanamid, 1936-38, Columbia, 1940-41, Eli Lilly & Co., Indpls., 1941-44; asst. prof. obstetrics and gynecology Columbia, Coll. Phys. and Surg., N.Y.C., 1953-55, asso. prof., 1955—; practice medicine specializing in gynecology, N.Y.C., 1953—; staff Columbia Presbyn. Med. Center, 1953—. Recipient Borden award Cornell U. Med. Coll., 1948. Diplomate Am. Bd. Obstetrics and Gynecology. Mem. A.M.A., Am. Chem. Soc., Soc. for Gynecologic Investigation, A.C.S. Research, numerous publs. on organic synthetic chemistry, isolation physiologically active compounds, application tracer isotopes to study metabolism, nutrition of fetus in utero, kinetics transfer mechanisms, clin. investigations drugs used in obstetrics, exptl. surgery, physiologic and chem. investigation fetus in utero, gynecologic cancer surgery. Home: 225 Central Park W., N.Y.C. 10024. Office: 180 Ft. Washington Av., N.Y.C. 10032.*

PLETSCHER, Alfred, Swiss physician; b. Altstätten S.G., Switzerland, Mar. 5, 1917; s. Alfred and Hermine (Rauch) P.; student med. faculties U. Geneva (Switzerland), 1935-36, U. Rome (Italy), 1938; Fed. Diploma, U. Zurich (Switzerland), 1942, M.D., 1942, Ph.D. Chemistry, 1948; m. Liselotte Gericke, Apr. 9, 1949; children—Martin, Rosmarie, Marianne. Chief sci. labs. Med. U. Clinic, Basle, Switzerland, 1953-54, lectr., 1952-60; guest worker Nat. Heart Inst., NIH, Bethesda, Md., 1955; chief dept. for exptl. medicine F. Hoffmann-La Roche & Co., Ltd., Basle, 1955-58, chief med. research dept., 1958-67, dir. research dept., 1967—; faculty U. Basle, 1960—, prof. pathophysiology, 1965—. Fellow N.Y. Acad. Scis.; mem. Swiss Soc. for Physiology and Pharmacology, Swiss Chem. Soc., Swiss Soc. for Biochemistry, European Soc. for Biochem. Pharmacology, Collegium Internationale Neuro-Psychopharmacologicum. Research, numerous publs. on metabolism, transport and distbn. of aromatic monoamines in central system and their role in physiology of brain, intermediary metabolism of fructose and its effect on ethanol metabolism. Home: 11 Am Hang 4125 Riehen, Switzerland. Office: 124 Grenzacherstrasse, 4002. Basle, Switzerland.*

PLETTA, Dan Henry, Am. engr.; b. South Bend, Ind., Dec. 31, 1903; s. John and Anna (Deutcher) P.; B.S., U. Ill., 1927, C.E., 1938; M.S., University Wisconsin, 1931; married Alice May Austin on June 13, 1931; children—Alice A. (Mrs. Thomas Richard Dyckman), Nancy (Mrs. Richard Lee Brehm). Various engineering positions, 1920-30; instructor Univ. Wisconsin, 1927-30; asst. prof. civil engring. U. S.D., 1930-32; asst. prof. applied mechanics Va. Poly. Inst., 1932-39, asso. prof., 1939-41, prof., 1941-48, prof., head dept. applied mechanics, 1948—, member bd. of directors of the Research Found., 1948-53; cons. engr. 1945—, stress analyst. Served as maj. Ordnance Dept. AUS, charge Richmond regional office Phila. Ordnance Dist., 1942-44, asst. dir. mech. dept. U. S. Mil. Acad., 1944-46. Recipient Certificate of Outstanding Service, Virginia Society of Professional Engineers, 1955. Registered profl. engr., N.Y., Va. Mem. Am. Soc. C.E.. (chmn. com. on exptl. analysis and analogues 1951-54, exec. com. engring. mechanics division 1956-60, chmn. 1960, president of Virginia section 1966), American Soc. Engring. Edn. (sec. and chmn. grad. studies div. 1958-61, council rep. 1962; sec. council on gen. divs. 1962-64), Soc. Exptl. Stress Analysis, Nat. (nat. dir. from Va. 1957-65, pres. Educational Found. 1966-68), Va. (pres. in Va. (pres. 1953) socs. profl. engrs., Am. Concrete Inst. (chmn. com. on rigid frame bridges 1940-42, 47-57), Sigma Xi. Author: Engineering Statics and Dynamics, 1951; Engineering Mechanics, 1964; also articles in field. Home: Highland Av., Blacksburg, Va.

PLIENINGER, Hans, chemist; b. Zurich, Switzerland, Jan. 1, 1914; s. Reginald and Adele (Hauck) P.; Dipl.Ing., Techn. Hochschule, Munich, Germany, 1939, Dr.Ing., 1942; m. Herta Freudenberg, Jan. 24, 1942; children—Thomas, Matthias Peter. Research chemist I.G. Farbenindustrie, Ludwigshafen, Germany, 1942-45, Knoll A.G. Pharms., Ludwigshafen, 1945-53; privatdozent Heidelberg (Germany) Chemisches Inst. of Univ., 1953—, extraordinarius organic chemistry, 1963—. Mem. Gesellschaft Deutscher Chemiker, Am. Chem. Soc. Author: (with K. Freudenberg) Organische Chemie, 1957, 1960. Research, publs. on synthesis of bile pigments; also synthetic and biosynthetic studies on ergot-alkaloids, synthesis of cyclohexadienones and cyclohexadienols as intermediates to prephenic acid; studies in indole and pyrrol-field; reactions under high pressure. Home: 1 Gotheinstrasse, Heidelberg, Germany.*

PLIMPTON, Calvin Hastings, Am. physician, coll. pres.; b. Boston, Oct. 7, 1918; s. George Arthur and Fanny (Hastings) P.; B.A. cum laude, Amherst Coll., 1939; M.D. cum laude, Harvard, 1943, M.A., 1947; Sc.D. in Medicine, Columbia, 1951; LL.D., Williams, 1960, Wesleyan, 1961, Doshioha U., Kyoto, Japan, 1962, St. Lawrence U., 1963; L.H.D., U. Mass., 1962; D.Sc., Rockford Coll., 1962, St. Mary's, 1963; Litt.D., Am. Internat. Coll., 1965; m. Ruth Talbot, Sept. 6, 1941; children—David, Thomas, George (dec.), Anne, Edward. Asst. attending physician Columbia-Presbyn. Med. Center, 1950-60; asso. medicine Coll. Phys. and Surg., 1950-59, asst. prof. clin. medicine, 1959-60; prof. medicine, chmn. dept. Am. U. Beirut, Am. U. Hosp., Beirut, Lebanon, 1957-59; pres. Amherst (Mass.) Coll., 1960—. Trustee Am. U., Beirut, World Peace Found., U. Mass., Phillips Exeter Acad., Commonwealth Fund. Decorated comdr. Order of Cedars (Lebanon). Diplomate Nat. Bd. Med. Examiners, Am. Bd. Internal Medicine. Fellow A.C.P.; mem. Soc. Mayflower Descs., Harvey Soc., Sigma Xi, Alpha Omega Alpha. Research, publs. in endocrinology. Home: 175 S. Pleasant St., Amherst, Mass.*

PLINY THE ELDER (Gaius Plinius Secundus), Roman naturalist; b. Novum Comum (now Como), Italy, 23 A.D.; ed. Rome, Italy. Cavalry officer, Germany, 47-57; explored many regions of Europe; studied law in Novum Comun, practiced law, Rome; became procurator, Spain, Gaul, Africa, circa, 65-70; prefect Roman fleet based at Misenum (now Miseno), from circa 74. Author: De iaculatione equestri; De vita Pomponi Secundi; Bellorum Germaniae libri XX; Studosi; Dubius sermo; A fine Aufidi Bassi; Naturalis Historia. Best known for his Natural History, an encyclopedic survey which contains 37 books, and cites 473 authors in summarizing ancient knowledge of astronomy, geography, ethnology, anthropology, physiology, botany, pharmacology, mineralogy, metallurgy, especially zoology. His work is often superficial, undiscriminating, inaccurate, and overly credulous, but is valuable storehouse of scattered facts, which was very influential in medieval and early modern Europe. Died Stabiae (now Castellammare di Stabia), nr. Mt. Vesuvius, Aug. 24, 79.

PLLATNER, Karl Friedrich, German metallurgist, mineralogist; b. Kleinwaltersdorf, nr. Freiburg, Saxony, Jan. 2, 1800; ed. Freiborger Bergschule, Frieburg Bergacademie; studied under Heinrich Rofe in Berlin, 1838-89; prof., Freiberg. Author: Probirkunst mit dem löthrohr; Die metallurgischen Köstprozesse theoretisch betrachtet. Described chem. properties of silver, copper, zinc, tin, other metals. Died 1858.

PLOT, Robert, English natural historian, chemist; b. Sutton-Barne, Eng., Dec. 13, 1640; s. Robert and Rebecca (Patenden) P.; B.A., Magdalen Hall, Oxford, 1661, M.A., 1664; 2 law degrees, 1671; m. Rebecca Burnam, Aug. 21, 1690; children—Robert, Ralph Sherwood. Apptd. 1st keeper of Elias Ashmole's mus., 1683; 1st prof. chemistry Oxford U., 1683; became sec. to earl marshal, 1687; named historiographer to James II, 1688; Mowbray herald extraordinary, also registrar of ct. of honor, 1695. Fellow Royal Soc., 1677 (sec. 1682, pub. Philos. Trans. numbers 143-166). Author: The Natural History of Oxfordshire, 1677; De Origine Fontium tentamen philosophicum . . ., 8 vols., 1684; The Natural History of Staffordshire, 1686. Died Apr. 30, 1696.

PLOTINOS, philosopher; said to have been born at Lyso or Lycopolis, Egypt, circa 204; began study of philosophy at age 28, worked under Ammonius Saccas, Alexandria, next 11 years; with Gordian's expdn. against Persia, 242-43; tchr. philosophy, Rome, from circa 245; among his students and admirers were Porphyry and the emperor Gallienus, who hoped to follow his suggestion of founding a Platonic Commonwealth. Author: Enneads, a series of 54 treatises which constitute the most detailed exposition of Neoplatonism; works constitute the application of Platonic metaphysical principles to Neo-Pythagorian concepts and the Oriental doctrine of Emanation; Enneads cover physics, cosmology, psychology, metaphysics, logic, epistemology. Died Campania, 269-70.

PLOTZ, Harry, Am. physician, bacteriologist; b. Paterson, N.J., April 17, 1890; s. Joseph and Ida

(Oedelson) P.; M.D., Columbus, 1913; m. Ella Sachs. Pathologist Mt. Sinai Hosp., N.Y.C., 1913-14; mem. Typhus Fever Commn. to Serbia, 1915; staff surgeon gen. U.S. Army, 1917-19; dir. med. com. to Poland for Joint Distbn. Com., 1920-21; became chief lab. Inst. Pasteur, Paris, France, 1921-39; returned to U.S., 1940; served on Typhus Commn., U.S. Army, dir. Army Med. Sch. virus div., 1941-46. Decorated by Serbian and Belgian govts. Officer Legion of Honor, France. Discovered bacillus which he believed caused disease in typhus-like fevers (Plotz's bacillus), 1914; proved Brill's disease was the same as typhus, 1914; studied measles and a serum for its treatment, 1938; research on virus disease; developed protective vaccine against typhus. Died Washington, D.C., Jan. 6, 1947.

PLUCKER, Julius, German math. physicist; b. Elberfeld, Germany, July 16, 1801; ed. univs. of Bonn, Heidelberg, Berlin, Germany, Paris. Prof. math., Halle, Germany, 1834-36; named prof. math. U. Bonn, 1836; also named prof. physics U. Bonn, 1847. Mem. French Acad. Scis., 1867. Author: Analytisch-geometrische Entwicklungen, 1821; Systeme der analytischen Geometrie, 1835; Theorie der algebraischen Kurven, 1839; System der Geometrie des Raumes, 1846; Neue Geometrie des Raumes, 1868. Research in diamagnetism, effect of magnetic field on elec. discharges in rarefied gases, optics of crystals and spectra of gases; laid founds. of modern analytical geometry; discovered cathode rays; developed 6 equations (named after him) which connected numbers of singularities in algebraical curves. Died Bonn, May 22, 1868.

PLUKENET, Leonard, English botanist; b. Jan. 4, 1642; s. Robert and Elizabeth P.; M.D. from Europe; studied under Dr. Busby, Westminster Sch.; m. Letitia; 13 children. Keeper, Royal Gardens, Hampton Court, Eng.; apptd. Royal prof. botany by Queen Mary. Author: Phytographia (describes new and rare plant species), 1691-92; Almagestum Botanicum, etc. (lists 8000 plants), 8 vols., 1696; Amaltheum botanical; Amalgesti Botanici Mantissa, etc., 4 vols., 1700; Amaltheum botanicum, etc., 1705; assisted John Ray in arrangement of 2d vol. of Historia Plantarum. Died Westminster, Eng., July 6, 1706.

PLUM, Fred, Am. physician; b. Atlantic City, Jan. 10, 1924; s. Fred and Frances (Alexander) P.; A.B., Dartmouth, 1944; M.D., Cornell U., 1947; m. Jean M. Houston, July 15, 1950; children—Michael F., Christopher N., Carol H. Faculty, U. Wash., Seattle, 1953-63, prof. neurology, 1961-63; faculty Med. Coll., Cornell U., N.Y.C., 1950-53, 63—, Anne Parrish Titzell prof. neurology, 1963—; neurologist in chief N.Y. Hosp., N.Y.C., 1963—. Mem. various study sects. NIH. Mem. Am. Neurol. Assn. (exec. bd. 1961-63), Am. Soc. Clin. Investigation, Am., Western assns. physicians, Am. Acad. Neurology. Author: (with J.B. Posner) The Diagnosis of Stupor and Coma, 1966; also numerous articles. Editor neurol. sect. Cecil's Textbook of Medicine, 12th edit., 1967; editorial bd. Archives of Neurology, 1959—. Research in normal and diseases of cerebral metabolism, how brain regulates external respiration, clarification of fundamental neurol. mechanisms for several abnormal forms of human breathing. Home: 45 Iden Av., Pelham, N.Y. 10803.*

PLUMIER, Charles, French botanist; b. Marseilles, France, Apr. 20, 1646; student botany under Boccone Trinità dei Monti monastery, Rome, Italy; studied under Joseph de Tournefort, France. Became a Franciscan Minim, at 16 years of age; accompanied Surian on govt. expdn. French W. Indies, 1689; royal botanist; head expdns. to W.I., C.Am., 1693, 95. Genus, Plumeria, named in his honor by Linnaeus and Tournefort. Author numerous works including: Description des plants de l'Amérique, 1693; Nova plantarum americanarum genera (described 700 species of 100 genera), 1703; Filicetum americanum, 1703; Plantarum americanarum fasciculi X, 1755-60; also numerous drawings of plants, birds, and fish. Died Puerto de Santa Maria, Spain, Nov. 20, 1704.

PLUMMER, Albert J., Am. pharmacologist; b. Somerville, Mass., Apr. 16, 1908; s. William and Anna (Mooney) P.; B.A., Boston U., 1929, M.A., 1930, Ph.D., 1935, M.D., 1949; m. Mary Rose Haverty, July 7, 1940; children—Albert, Mary Rose (Mrs. Robert Russo). Teaching fellow Boston U. Sch. Medicine, 1932-35, instr. pharmacology, 1935-39, asst. prof. pharmacology, 1939-49, lectr. pharmacology, 1950—; sr. pharmacologist CIBA Pharm. Co., Summit, N.J., 1949-51, dir. pharmacol. research, 1951-57, dir. macrobiology, 1958-66, dir. biol. research, 1966—; spl. lectr. pharmacology Woman's Med. Coll. Pa., 1959. Mem. Am. Soc. Pharmacology and Exptl. Therapeutics, A.M.A., Am. Chem. Soc., A.A.A.S., Am. Coll. Neuropsychopharmacology, Am. Coll. Clin. Pharmacology and Chemotherapy, N.Y. Acad. Scis., Summit Med. Soc. Contbr. numerous articles in field to sci., med. jours. Research, numerous publs. on devel. of new drugs for treatment of human disease; most important work has dealt with agts. used in lowering high blood pressure, including Serpasil, Ismelin, Ecolid, Esidrix. Home: 13 Harding Terrace, Morristown, N.J. 07960. Office: CIBA Pharm. Co., Summit, N.J. 07901.*

PLUMMER, Gayther Lynn, Am. ecologist; b. Indpls., Jan. 27, 1925; s. Conley L. and Rowena (Huber) P.; B.S., Butler U., 1948; M.S., Kan. State U., 1950; Ph.D., Purdue U., 1954; m. H. Eileen Barr, June 3, 1950. Faculty, Antioch Coll., 1954-55; faculty U. Ga., Athens, 1955—, prof. botany, 1967—. Mem. Ecol. Soc. Am., A.A.A.S., Agronomy Soc. Am., Am. Inst. Biol. Sci., Assn. So. Biologists, others. Research, numerous publs. on radioactive fallout in populations, watershed hydrology and agrl. practices, applications of remote sensing to ecol. problems. Office: Dept. Botany, U. Ga., Athens, Ga. 30601.*

PLUMMER, Henry Stanley, Am. physician; b. 1875; staff Mayo Clinic. Described Plummer-Vinson syndrome of dysphagia, glossitis and anemia, frequently accompanied by atrophy of oral and pharyngeal tissue and sphlenomegaly, 1912; reintroduced iodine for preoperative medication, 1923; proposed (with W.M. Boothby) that in Graves disease thyroid secretes excess normal thyroxine in addition to abnormal thyroxine deficient in iodine, 1924. Died 1936.

PLYLER, Earle Keith, Am. physicist; b. Greenville, S.C., Apr. 26, 1897; s. John Robert and Mary (Earle) P.; B.A., Furman U., 1917, M.A., 1918; M.A., Johns Hopkins, 1923; Ph.D., Cornell U., 1924; m. Eleanor Mays, Sept. 3, 1925; children—Mary Eleanor (Mrs. Gus Hellwig), Ruth Annie (Mrs. Robert Viall). Faculty, Furman U., 1920-22, U. N.C., Chapel Hill, 1924-41; research physicist U. Mich., 1941-45; sr. physicist Nat. Bur. Standards, Washington, 1945-52, chief, radiometry sect., 1952-58, chief, infrared spectroscopy sect., 1958-62; prof., head, dept. physics Fla. State U., Tallahassee, 1962—. Vis. lectr. Inst. Optics, Madrid, Spain, 1949; cons. Oak Ridge, 1947—; mem. phys. scis. div. NRC, 1958-64. Recipient Gold medal for exceptional service U.S. Dept. Commerce, 1961. Fellow Am. Phys. Soc., Indian Acad. Scis. (hon.), A.A.A.S., Optical Soc. Am. (asso. editor 1961—, dir. 1958-60), Soc. Applied Spectroscopy (ann. award N.Y. sect. 1963); mem. Philos. Soc. Washington, Coblentz Soc., Am. Chem. Soc. (bd. editors Jour. Chem. Physiology 1955-58). Author: (with Drs. R.C. Lord, R.N. Jones): Tables of Wavenumbers for the Calibration of Infrared Spectrometer, 1961. Research on optical properties of materials in infrared, pioneer research on infrared flame spectra, molecular constants of many molecules from analysis of high resolution infrared spectra; velocity of light from precise wavelengths of infrared spectroscopy, constrn. and design of high resolution infrared spectrometers, infrared measurements of spectra of solutions, and high precision wavelength measurements. Home: 1114 Waverly Rd., Tallahassee 32303.*

PNIEWSKI, Jerzy Maria, Polish physicist; b. Plock, Poland, June 1, 1913; s. Henryk and Amelia (Babecka) P.; M.Math., U. Warsaw (Poland), 1936, M. Physics, 1938, D.Physics, 1951; postgrad. U. Liverpool (Eng.); m. Maria Chojnacka, Sept. 23, 1953. Research worker Inst. Exptl. Physics, U. Warsaw, 1935—, dir., 1953—, prof. elementary particle physics, 1954—, chmn. exptl. physics, 1953-61, chmn. elementary particle physics, 1961—; prof. Nuclear Research Inst., Warsaw, 1955—; leader High Energy Lab., 1958—. Recipient state decorations, 1952, 54, 64, state prizes in physics (with M. Danysz), 1955, 1st order, 1964. Mem. Polish Acad. Scis., Polish, Italian phys. socs. Research, publs. on Rayleigh scattering of light, beta-spectroscopy, emulsion technique, elementary particle and high energy nuclear physics; fragmentation of heavy nuclei. (with M. Danysz) observation and interpretation of first hypernucleus and isomerism of hypernuclei; (with others first double hypernucleus). Home: 75/33 Koszykowa. Office: 69 Hoza, Warsaw, Poland.*

POBEDINSKY, Mikhail Nikolaevich, Russian roentgenologist, radiologist; b. 1900; grad. Med. Faculty, Moscow U., 1925; D.Med. Sci., 1935. Intern obstet. and gynecol. clinic Moscow U., 1926-39; intern Inst. Roentgenology and Radiology, RSFSR Peoples Commissariat Health, Moscow, 1926-30, asso., 1930-31, head radiology dept., 1932-43, prof., 1938—; head roentgenology and radiology dept. Moscow Inst. Obstetrics and Gynecology, 1944-48; head chair roentgenology and radiology Leningrad Postgrad. Med. Inst., 1951-52, head chair med. radiology, 1952—; dir. Central Research Inst. Med. Radiology, USSR Ministry Health, Leningrad, 1949—, chief radiologist, 1959—; founder, dir. Leningrad Inter-Inst. Seminar on Radiobiology and Physics of Penetrating Radiation, 1954—. Permanent USSR rep. Internat. Commn. for Protection against Radiation, 1956—; head Soviet delegation 8th World Congress Radiologists, Mexico, 1959, 9th, Munich. Decorated Order of Lenin. Mem. All-Union (bd. mem., dep. chmn.), Leningrad (bd. mem., dep. chmn.) socs. roentgenologists and radiologists, All-Union (bd. mem.), Leningrad. (dep. chmn.) socs. oncologists. Dep. editor Med. Radiology, 1956—; co-editor Radiobiology sect. Large Med. Ency., 2d edit.; mem. editorial council Problems of Oncology, Herald Roentgenology and Radiology. Research and numerous works on roentgenology, radiology, therapy, diagnosis and prophylaxis of precancerous states, cancer of female genitals. Address: Leningrad Postgrad. Med. Inst., ulitsa Saltykova-Shchedrina 41, Leningrad, USSR.

POCHIN, Edward Eric, English physician; b. Sale, Eng., Sept. 22, 1909; s. Charles Davenport and Agnes (Collier) P.; M.A., Cambridge U., 1935, M.D., 1945; m. Constance Tilly, Jan. 6, 1940; children—Charles William Davenport, Sally Teresa. Dir. Med. Research Council dept. clin. research U. Coll. Hosp. Med. Sch., London, 1946—, chmn. internat. commn. on radiol. protection, 1962—. Mem. UN Sci. Commn. on Effects of Atomic Radiation. Fellow Royal Coll. Physicians (councillor 1965—); mem. Med. Research Soc., Physiol. Soc., Brit. Inst. Radiology, Internat., Brit. radiation protection assns., Assn. Physicians, Brit. Med. Assn., Royal Soc. Medicine, Internat. Soc. Internal Medicine, Thyroid Club London, Am. Thyroid Assn. Research, numerous publs. on diseases of thyroid gland, use of radioactive nuclides in their study and treatment, exam. of problems involved in protection against ionizing radiation in clin. occupational and population exposure. Home: High Cross, Aldenham, Watford, Herts., Eng. Office: U. Coll. Hosp. Med. Sch., University St., London W.C.1, Eng.*

POCHON, J., French biologist; b. Paris, Nov. 30, 1907; s. H. and G. (Bamberger) Pochon; Docteur, Faculté de Medecine, Paris, 1933, Faculté des Sciences, Paris, 1936; Dr. honoris causa Faculté Sci. agronomique, Gembloux, Belgium, 1963, Grand, Belgium, 1965; m. S. Liberge, Feb. 27, 1933; children—François, Dominique, Jean, Luc, Jacqueline, Annick. With Institut Pasteur, Paris, 1939—, head lab., 1941-54, head soil microbiology dept., 1954—; became lectr. Faculté des Sciences, Paris, 1963. Decorated Chavalier Legion of Honour. Mem. French Soc. Microbiology. Author books including: Traité de microbiologie du Sol, 1948; also numerous articles. Research on soil biology, including role of bacteria in fertility and conservation of soil, effect of soil microorganism on nutrition and physiology of plants; analysis of mechanism of synthesis of humus. Home: 11bis E. Deschance, Paris 7, Office: Institut Pasteur, Paris 15, France.*

POCKER, Yeshayau, chemist; b. Kishineff, Rumania, Oct. 10, 1928; s. Ben Zion and Esther (Sudit) P.; M.Sc., Hebrew U. Jerusalem, 1949; Ph.D., U. Coll., London, 1953; D.Sc., U. London, 1960; m. Anna Goldenberg, Aug. 8, 1950; children—Rona, Elon I. Research asso. Weizmann Inst. Sci., Israel, 1949-50; Humanitarian Trust fellow U. Coll., London, Eng., 1951-52, asst. lectr., 1952-54, lectr., 1954-61; recognized tchr. U. London, Senate House, 1959; examiner sci. Russian U. London, 1960; vis. asso. prof. Ind. U., 1960-61; prof. chemistry U. Wash., Seattle, 1961—. Mem. Am. Chem. Soc., Chem. Soc. (London), A.A.A.S., N.Y. Acad. Scis. Research, publs. on chem. dynamics with particular reference to organic reaction mechanisms; rearrangements, catalysis, hydration and hydrolysis; high pressure effects in chem. and biol. systems, deuterium isotope effects; kinetics of fast reactions, enzyme reactions. Home: 3515 N.E. 42d St., Seattle 98105.*

PODDAR, Sailendra Nath, Indian chemist; b. Calcutta, India, Mar. 15, 1929; s. Nabadwip P. and Jyoti (Saha) P.; B.Sc., Scottish Ch. Coll., Calcutta, 1948; M.Sc., Calcutta U., 1950, D.Phil., 1958; m. Renu Saha, Aug. 8, 1950; children—Santi, Sarmistha, Debjani, Aniruddha. Research worker Indian Assn. for Cultivation Sci., Jadavpur, Calcutta, 1951-57; chemist explosives dept. Govt. of India, 1957-61; research officer in organic chemistry dept. Indian Assn. for Cultivation Sci., Calcutta, 1962—. Fellow Indian Chem. Soc.; mem. Indian Sci. News Assn., Indian Sci. Congress Assn., Indian Assn. for Cultivation Sci. Research, publs. on coordination complexes of transitional metals, synthesis of coordination polymers, devel. of new methods of inorganic analyses especially those of ore, minerals and alloys. Home: 4/2 Amherst St., Office: Jadavpur, Calcutta, West Bengal, India.*

PODOLSKY, Boris, physicist; b. Taganrog, Russia, June 29, 1896; s. Yakov and Elizabeth (Parnoch) P.; came to U.S., 1913, naturalized 1918; B.S. in Elec. Engring., U. So. Cal., 1918, M.A. in Math., 1926; Ph.D. in Physics, Cal. Inst. Tech., 1928; m. Bertha Polly Edelman, Aug. 3, 1937; 1 son, Robert Earl. Nat. Research fellow U. Cal., Berkeley, 1928; Internat. Research fellow, Leipzig, Germany, 1929-30; dir. research in theoretical physics Ukrainian Physico-Tech. Inst., Kharkov, Russia; fellow Inst. Advanced Studies, Princeton, N.J., 1934; prof. math. physics, fellow Grad. Sch., U. Cin., 1935-61; prof. theoretical physics Xavier U., Cin., 1961—. Gen. Electric Corp., Avco Corp., Remington Rand, Cin. Milling Machine Co., Convair, 1956-63. Fellow Am. Phys. Soc., Ohio Acad. Scis.; mem. N.Y. Acad. Scis., Sigma Xi, Sigma Pi Sigma. Author: A Generalized Electrodynamics, 1942; (with K.S. Kunz) The Fundamentals of Classical Electrodynamics, 1965. Research, publs. on founds. of quantum mechanics and electrodynamics, especially Einstein-Podolsky-Rosen Paradox in quantum mechanics. Home: 875 Lafayette Av., Cin. 45220.*

PODVINEC, Srécko, Yugoslavian surgeon; b. Osijek, Yugoslavia, Nov. 23, 1899; s. Lavoslav and Ruza (Eibenschütz) P.; Medic., U. Zagreb (Yugoslavia), 1922; M.D., U. Vienna (Austria), 1925; m. Marija Simonovic July 20, 1929; m. 2d, Marija Sepe, May 12, 1935; children—Radoje, Mihael; m. 3d, Ana Rajic, July 16, 1966. Asst. clinic ORL, U. Zagreb, 1927-41, faculty, 1948—, asst. prof., 1949-54; titulary prof. ORL, Med. Faculty U. Belgrade, Yugoslavia, hon. prof. Faculty Stomatology, head chair otolaryngology Med. Faculty, dir. univ. ENT clinic, 1954—. Recipient seventh July, Exec. Com. of Serbia, 1964. Mem. Serbian Med. Soc. (pres. sect ORL), Collegium Oto-Rhino-Laryngologicum Amicitiae Sacrum, French Soc. Bronchoesophagology (past pres.). Author: Surgery of the Jaws, 1936; (with A. Sercer, I. Cupar) Education of the Deaf, 1962; Textbook of Oto-Rhino-Laryngology, 1965; also numerous articles. Introduced surg. procedures for resection of duplicatures in middle nasal meatus to prevent recurrence of nasal polypi, pharyngoplasty for relief of open rhinophonia, plastic operation of nasal alae to prevent inversion; discovered high incidence of platybasia in regions of endemic goiter, indicating retardation of growth of central nervous system; through homologous transplantation of skin in animals, demonstrated transfer of immunological information; research on immunological treatment against cancer, physiology and pathology of deglutition. Home: 8 Smetanina, Beograd, Yugoslavia.*

POE, Charles Franklin (Hall), Am. chemist; b. Golden, Colo., Feb. 15, 1888; s. Jefferson Q. and Christie (Hoagland) Hall; B.A. in Chemistry, U. Colo., 1911, M.A. in Bacteriology, 1911, Ph.C., B.S. in Pharmacy, 1914; Ph.D. in Sanitary and Biol. Chemistry, Cornell U., 1926; grad. Army Indsl. Coll., Washington, 1941; M.P.H., U. Minn., 1957; m. Frances Elizabeth Woland, June 27, 1913; children—Emily Eleanor (Mrs. Frederick H. Belcher), Alice Louise (Mrs. William Ligtenberg), Frances Elizabeth (Mrs. Charles William Latshaw). Faculty, U. Colo., 1911-56, emeritus prof. chemistry, emeritus dean Coll. Pharmacy, 1956—, prof. pub. health, 1956-58; prof. chemistry Chapman Coll., Orange, Cal., 1963—; prof. bacteriology, co-dir. summer insts. NSF, 1960-61; chemist Colo. Dept. Health, 1911-50. Speaker, promoter Fed. Food and Drug Act, 1937-38; past pres. Colo. Bd. Examiners in Basic Scis.; mem. citizen com. to evaluate adequacy of enforcement fed. laws Sec. Health, Edn. and Welfare, 1955. Inst. Pasteur (France) Research fellow, 1929-30, 51; NIH grantee, 1958-63. Recipient Distinguished Service award Sch. Pharmacy U. Colo., 1960; 50 Year award Colo. Bd. Pharmacy, 1964. Fellow A.A.A.S.; mem. Am. Pharm. Assn. (life), Colo. Pharmacal Assn. (life); emeritus mem. Am. Pub Health Assn., Soc. Am. Microbiologists; mem. Sigma Xi. Research, publs. on food and drug analysis; studies on toxicity of various substances; contamination of foodstuffs; relationship of vitamin deficiency to learning ability in white rats. Home: 657 Milford St., Orange Cal. 92667.*

POECK, K., German neurologist; b. Berlin, Germany, Jan. 3, 1926; s. Erich and Elisabeth (Erdmann) P.; M.D., U. Heidelberg, 1953; m. Margrit Tobing, Sept. 7, 1963; 1 son, Karsten Axel. Research fellow Physiol. Inst., U. Pisa, Italy, 1958; head dept. neurology Neuropsychiat. Clinic, U. Freiburg, 1960—, tchr., 1961—. Mem. German Soc. Neurology, German Soc. for Psychiatry and Nervous Disease, German Soc. Arts and Sci., Internat. League Against Epilepsy. Author: Einfuhrung in die klinische Neurologie, 1966; also articles. Research on influence of transmitting substances on reticular activating system of brain stem, appearance of instinctive movements, abnormal behavior as neurol. symptoms, neuropsychology. Home: 16 Sonnenbergstrasse, Freiburg, Germany.*

POENARU, Valentin Alexandru, mathematician; b. Bucharest, Rumania, Oct. 5, 1932; s. Dumitru V. and Elena (Lupescu) P.; Licentiat in matematica, U. Bucharest, 1955; Docteur ès scis. mathematiques, U. Paris, 1963; m. Rigmor Lilian Dreyer Jensen, Oct. 3, 1964. Instr., U. Bucharest, 1955-58; research fellow Inst. Math., Buchaest, 1960-62; research fellow Inst. des Hautes Etudes Scientifiques, Paris, U. Lausanne, 1962-63; research fellow Harvard, 1963-64, vis. lectr., 1965-66; research fellow Inst. Advanced Study, Princeton, N.J., 1964-65; prof. Northeastern U., Boston, 1966—. Mem. Am., Swiss math. socs. Research, publs. on topology of low dimensional manifolds in connection with their cartesian products with an interval or a disk; partial differential equation. Home: 72 Wendell St., Cambridge, Mass. Office: Dept. Math., Northeastern U., Boston.*

POEVERLEIN, Hermann, physicist; b. Ludwigshafen a. Rh., Germany, Oct. 18, 1911; s. Hermann and Elisabeth (Stepp) P.; Diploma Physicist, Tech. U. Munich (Germany), 1936, D. in Physics, 1942; m. Lilly Applemann, May 14, 1955. Asso., Tech. U. Munich, 1937-49, dozent, apl. prof., 1963; 1949-53, research physicist Air Force Cambridge Research Labs., Bedford, Mass., 1953—; vis. lectr. Harvard, 1962. Recipient Marcus O'Day Meml. award Air Force Cambridge Research Labs., 1962. Mem. Am. Phys. Soc., Am. Geophys. Union, Sci. Research Soc. Am., Union Radio Scientifique Internationale. Research, publs. on graphical constn. of ray paths of radio waves in ionosphere, wave theory electromagnetic and hydromagnetic waves in upper atmosphere, four-dimensional geometric optics. Home: 33 Elm St., Wakefield, Mass. 01880. Office: Air Force Cambridge Research Labs., L.G. Hanscom Field, Bedford, Mass. 01730.*

POEY, Felipe, Cuban zoologist, entomologist; b. Havana, Cuba, 1799; student, Paris, France; prof., dir. Zool. Mus., Havana. Mem. French Entomol. Soc. (a founder), Zool. Soc. London (corr.). Author: Centurie de Lépidoptéres de l'isle de Cuba; Memorias Sobre la historie natural de l'isle de Cuba, 2 vols., 1851, 1858-61; also articles on Lepidoptera of Cuba. Father of Cuban zoology; studied ichthyology. Died 1891.

POFFENBERGER, Albert T., Am. psychologist; b. Dauphin, Pa., Oct. 23, 1885; s. Albert T. and Lillie (Umberger) P.; A.B., Bucknell U., 1909; A.M., Columbia, 1910, Ph.D., 1912; m. Florence V. Kauffman, Aug. 27, 1913; children—Dora Helen (Mrs. George A. Wilkens), John Roberts. Faculty, Columbia, 1914—, prof., head dept. psychology, 1927-41, prof. emeritus, 1950—; chmn. div. anthropology and psychology NRC, 1932-33. Pres., Social Sci. Research Council, 1946-48. Mem. Am. Psychol. Assn. (pres. 1935). Author: Psychology in Advertising, 1932; Principles of Applied Psychology, 1942. Editor: James McK. Cattell: Man of Science, 1947. Home: Station Rd., Montrose, N.Y. 10548.*

POGGENDORFF, Johann Christian, German physicist, chemist; b. Hamburg, Germany, Dec. 29, 1796; studied physics and chemistry; apothecary; prof. Berlin, 1834-77. Founder, editor, annalen der Physik und Chemie. Author: Biographisch-literarisches Handwörterbuch zur Geschichte der exacten Wissenschaften, 2 vols., 1863 (supplements through 1931). Inventor galvanometer, 1821, suspended mirror which reflects light beams on to a scale to magnify small deflections, 1827, instrument for measuring elec. polarization, 1843; developed method for determining elec. energy of battery; research in magnetism; made studies in biography and history. Died Berlin, Jan. 24, 1877.

POGORELOV, Aleksei Vasilevich, Russian mathematician; b. Belgorod Oblast, 1919; grad. Zhukovskii Air Force Acad., 1945; Dr. Physico-Math. Sci., 1948. Chief geometry sect., chmn. dept. Kharkov State U., 1947-59; head geometry sect. Inst. Math., also Inst. Physico- Tech., Ukrainian SSR Acad. Scis., 1959—. Recipient Stalin prize, 1950. Mem. Ukrainian SSR Acad. Scis., 1961; corr. mem. USSR Acad. Scis., 1960. Author: The Single-Valued Definitions of Closed Tubes, 1948; The Single-Valued Definition of Convex Surfaces, 1950; The Regularity of Convex Surfaces, 1951; The Rigidity of Convex Polyhedrons, 1952; The Outer Curvature of Plane Surfaces, 1953; Lecturs on Differential Geometry, 1955; A Surface of Limited Outer Curvature, 1956; Gauss' Work on Plane Geometry, 1956; Lecturs on Analytic Geometry, 1957; Geometric Imbedding in the Large of a Two-Dimensional Riemannian Manifold into a Tri-Dimensional One, 1957; Some Questions in Geometry in the Large in a Riemannian Space, 1957; On a Transformation of Isometric Surfaces, 1958; The Rigidity of General Convex Surfaces, 1969; Monge and Ampere's Highly Elliptic Equations, 1960; Elastic Deformations of Convex Shells in the Transcritical Region, 1960; The Regularity of Convex Surface with Regular Lobachevsky Metrics, 1961; The Rigidity of Closed Surfaces Non-Homeomorphic in Riemannean Space, 1961; Transcritical Deformations of Cylindrical Shells Under External Pressure, 1961. Address: Ukrainian Physico-Tech. Inst., Yumovskii Lupik 2, Karkov, Ukrainian, USSR.

POGREBNYAK, Petr Stepanovich, Russian sylviculturist; b. 1900; grad. Novo-Aleksandriysk Inst. Agr. and Forestry, Kharkov, 1924; head chair gen. forestry and pedology Kiev Forestry Inst., 1933-49; dir. Inst. Forestry, Ukrainian Acad. Scis., 1945-56; prof., 1935—; prof. Kiev U., 1954-56, head chair phys. geography, 1954-57. Chmn., Council for Study Prodn. Resources of Ukrainian SSR, 1948-52. Mem. Ukrainian Acad. Scis. (v.p. 1948-52), Ukrainian Nature Preservation Soc. (chmn.). Author: Principles of Forest Typology, 1955; The Conservation and Rational Use of Natural Resources, 1960. Research and publs. on forest geobotanics, forest cultures; developer method of forest typology by composition and productivity based on evaluation of soil and climatic conditions, 1926-28. Ukrainian Acad. Scis., Vladimirskaya 54, Kiev, Ukrainian SSR, USSR.

POGSON, Norman Robert, English astronomer; b. Nottingham, Eng., Mar. 23, 1829; s. George Owen Pogson; m. Elizabeth Ambrose, 1849; many children; m. 2d, Edith Louisa Stopford Sibley, Oct. 25, 1883; 3 children. Asst., S. Villa Obs., London; became asst. Radcliffe Obs., Oxford, 1852; apptd. dir. Hartwell Obs., 1859; govt. astronomer, Madras, from 1860. Recipient LaLande medal French Acad. Fellow Royal Astron. Soc. Contbr. to sci. publs. Calculated orbits of 2 comets, 1847; discovered minor planets Amphitrite, 1854, Isis, 1856, Ariadne, 1856, Hestia, 1857, Asia, 1861, 4 others, 1861-68; asst.. in George Airy's expts. for determining mean density of earth, 1854. Proposed that difference of 5 magnitudes correspond to stellar light ratio 1:100, 1856; discovered 7 variable stars. Died June 1891.

POHLAND, Erich Paul, German chemist; b. Wuppertal-Elberfeld, June 7, 1898; s. Emil and Maria (Wagener) P.; Ph.D., U. Berlin, 1923; m. Erna Söhle, May 20, 1926; children—Renate (Mrs. E. Feldmann), Hermann, Dorothea (Mrs. H.O. Hoelper), Ingrid (Mrs. R. Stein). Docent, Karlsruhe Inst. Tech., 1931-39; govtl. dir. Fed. Econ. Ministry, Bonn, 1951-56; Fed. Minister for Sci. Research, Bonn, 1951-63; ministerial counselor, 1957-63; gen. dir. Eurochemic, Mol, Belgium, 1960-63, cons., 1964—. Mem. German Chem. Soc., German Atom Forum. Numerous publs. in field. Research on boron-hydrides and halogenides, crystal structure of solidified gases, reprocessing of irradiated nuclear fuel elements. Address: 52 Germanenstrasse, 532, Bad Godesberg, West Germany.*

POILICI, Ionel, Rumanian physician, neurologist; b. Rumania, May 5, 1923; s. Iancu and Adela (Iancu) P.; Ph.D., Medicine U., Bucharest, Rumania, 1950, C.Sc., 1956, M.D., 1965; m. Ghizi Aroneanu, Apr. 15, 1953. Resident doctor clin. sect. Neurol. Inst. Bucharest, 1955—; prin. researcher Neurol. Inst., Rumanian Acad. Bucharest, 1956—; cons. neurologist, 1959—. Mem. Union Med. Scis. Socs. Author: Vascular Reactivity in Diseases of the Central Nervous System, 1959; also numerous articles. Research on vegetable disturbances in cerebrovascular diseases, role of cerebral formations in origin of cerebrovascular diseases, role of cerebral reticular formation in origin of cerebrovascular lesions. Home: 20 A. 13 Decembrie St., Bucharest, Rumania.*

POINCARÉ, (Jules) Henri, French mathematician and physicist; b. Nancy, France, Apr. 29, 1854; ed. Ecole Polytechnique, 1873; Ecole Supérieur des Mines, Dr. Scis., 1879; instr. math. analysis, U. Caen, 1880; lecturer, Faculté des Sciences, U. Paris, from 1881, prof. physical mechanics, 1885, prof. math. physics and calculus of probabilities, 1886, prof. celestial mechanics, 1896; mem. French Acad. Scis., 1887, Académie Française, 1908; Fellow, Royal Soc., 1894 (Sylvester Medal, 1901); Prix Ponçelet; Prix Reynaud; Medal, Roy. Astron. Soc., 1900; gold medal, Lobachevsky Fund; Prix Bolyai, 1905; gold medal, Association Française pour l'Avancement des Sciences, 1909. Author: Leçons sur la théorie mathematique de la lumière, 2 vols, 1889-92; Eléctricité et optique, 2 vols., 1890-91; Thermodynamique, 1892; Leçons sur la théorie d'élasticité, 1892; Les méthodes nouvelles de la méchanique céleste, 3 vols., 1829-99; Theorie des tourbillons, 1893; Capillarite, 1895; Théorie analytique de la propagation de la chaleur, 1895; Calcul des probabilities, 1896; Cinématique et mécanismes potentiels et mécaniques des fluides, 1899; Théorie du potential newtonien, 1899; La science et l'hypothese, 1906; Science et méthode, 1908; La valeur de la science, 1913; Dernières pen sées, 1913. Contributed to theory of functions, esp. automorphic, Abelian functions; introduced Fuchsian functions into mathematics; also research in differential equations and theory of astronomical orbits, esp. 3-body problem; studies electromagnetic theory of light and relativity theory; exponent of conventionalism in philosophy of science. Died Paris, France, July 17, 1912.

POINSOT, Louis, French mathematician; b. Paris, Jan. 3, 1777; ed. École Polytechnique; engr. for bridges and roads; became prof. math. Lyceum Bonaparte, 1804; prof. math. École Polytechnique, from 1809, named examiner, 1816; mem. Superior Council Pub. Instrn. Fellow Royal Soc., 1858; mem. French Acad. Scis., French Bur. Longitudes. Author: Éléments de statique, 1803; Nouvelle théorie de la rotation des corps, 1832, 1834. Developed a theory of couples and its applications; returned to synthetic methods and clarified their formal presentation; solved problem of permanent rotation of a body about privileged axes, created idea of inertia of ellipsoids, 1834. Died Paris, Dec. 5, 1859.

POIRIER, Jacques Charles, chemist; b. Mehun-sur-Yevre, France, Jan. 3, 1927 (parents Am. citizens); s. Anthony Joseph and Renee (Potier) P.; Ph.B., U. Chgo., 1947, S.B., 1948, S.M., 1950, Ph.D., 1952; m. Marsha London, June 22, 1951; children—Marc Raymond, Charles Joseph, Julia Eve. NSF post-doctoral fellow Yale, 1952-53; Corning Glass Works Found. postdoctoral fellow U. Cal. at Berkeley, 1953-55; faculty Duke, Durham, N.C., 1955—, prof. chemistry, 1967—; vis. asso. prof. Ind. U., 1961. Alfred P. Sloan fellow, 1959-63. Mem. Am. Chem. Soc., Am. Phys. Soc., Phi Beta Kappa, Sigma Xi, Phi Lambda Upsilon. Research, publs. on quantitative relations between bulk properties of fluids, especially ionic solutions, disordered solids and molecular properties of component molecules. Home: 210 W. Lavender Av., Durham, N.C. 27704.*

POISEUILLE, Jean Leonard Marie, French physiologist; b. Paris, France, Apr. 22, 1799; M.D., 1828; practice medicine, Paris. Recipient Gold medal French Acad. Scis. Author: Sur la force du coeur aortique, 1828; Le Mouvement des liquides dans des tubes de petits diamètres, 1844. First to use mercury manometer for measurement blood pressure, 1828; discovered law on velocity of flow of a liquid through capillary tube, 1843; studied flow of viscous liquids, 1846; invented hemodynamometer for measuring blood pressure inside arteries, viscosimeter. Died Paris, Dec. 26, 1869.

POISSON, Siméon Denis, French mathematician; b. Pithiviers, Loiret, France, June 21, 1781; studied medicine, chemistry, and mathematics at Ecole Polytechnique, 1798; examiner, from 1802, and prof., from 1808, Ecole Polytechnique; 1st prof. of mechanics, Sorbonne, from 1809 (mem. council, 1820); mem. French Acad. Scis., 1812; mem. Bureau des Longitudes; peer in 1837; Author: Traité de mecanique, 1811; Mémoire sur la theorie des ondes, 1826; Théorie nouvelle de l'action capillaire, 1831; Théorie mathématique de la chaleur, 1835; Recherches sur la probabilité des jugements en matière criminelle et en matière civile, 1837; others. Research, pubs. on definite integrals, Fourier series, calculus of variations, probability, heat, acoustics, capillarity, elasticity of materials; ratio between lateral and longitudinal strain in a wire named after him; applied mathematics to physical problems; discovered well-known equation introducing electric potential; studied mechanical properties of luminiferous ether; deduced certain phenomena from wave theory of light, which convinced him it incorrect, but which when later observed confirmed that theory. Died Paris, France, Apr. 25, 1840.

POIVILLIERS, Georges-Jean, French topographer; b. Drache, France, May 15, 1892; s. Florimond and Celeste (Lasmezas) P.; m. Yvonne Meneboode, Aug. 3, 1915; children—Alain, Henri-Albert. Prof. photogrammetry French Nat. Conservatory Arts and Trades; dir. Central Sch. Arts and Trades, 1952-62, hon. dir., 1962—. Mem. French Acad. Scis., 1946. Inventor instruments for photogrammetric restorations. Address: 11, boulevard de Levallois, Neuilly-sur-Seine, France.

POKORNY, Alex Daniel, Am. psychiatrist; b. Taylor, Tex., Oct. 18, 1918; s. John Robert and Olga (Susen) P.; B.A., U. Tex., M.D., 1942; m. Jeanice Allen, Mar. 13, 1948; children—Martha, Ross, Ellen, Sally. Asst. chief psychiatry Houston VA Hosp., 1949-55, chief psychiatry and neurology, 1955—. Instr. to prof. psychiatry Baylor U. Coll. Medicine, Houston, 1949—. Fellow Am. Psychiat. Assn., A.A.A.S.; mem. A.M.A., Soc. Psychophysiol. Research, Phi Beta Kappa, Sigma Xi, Alpha Omega Alpha. Research, publs. on suicide in relation to mental illness, social and meteorol. factors; psychiat. epidemiology, psychopharmacology. Home: 813 Atwell St., Bellaire, Tex. 77401. Office: 2002 Holcombe St., Houston 77031.*

POLAK, Feliks, Polish chemist; b. Lwow, Poland, Nov. 8, 1901; s. Franciszek and Maria (Bogdan) P.; chem.eng., Tech. U. Lwów, 1923, d. tech. sci., 1926; m. Amalia Baczynska, Jan. 11, 1930; 1 son, Lucjan. Adj., Techn. U., Lwów, 1921-29; research fellow Sugar Research Inst. Warsaw (Poland), 1929-32, Mil. Inst. Warszawa (Poland), 1932-45; faculty Jagiellonian U., Kraków, Poland, 1945—, prof. chemistry, 1948—, dir. ch. chem. tech., 1951—. Cons. adviser Inst. Petroleum Tech., Kraków, 1960—. Recipient Order Polonia Restituta, 1954; award Ministry Higher Schs., 1965. Mem. Polish Chem. Soc. Research, numerous publs. on structure of starch, glycerol fermentation, carbonation process in sugar industry, adsorbents as carbon active silica gel, synthetic zeolites, ion-exchange resins, applications of synthetic zeolites in petroleum industry. Home: 12 Garbarska, Kraków, Poland.*

POLANSKY, Norman Alburt, Am. social psychologist; b. Carbondale, Pa., Oct. 22, 1918; s. Joseph Joel and Celia (Kaplan) P.; A.B., Harvard, 1940; postgrad. U. Ia.; M.S. Social Administrn., Western Res. U., 1943; Ph.D., U. Mich., 1948-51; m. Nancy Gale Finley, Feb 7, 1964; children—Rachael Vaughn, Jonathan. Faculty, Wayne State U., 1948-53, also Smith Coll. Sch. Social Work, 1950-62; psychologist Riggs Center, Stockbridge, Mass., 1953-55; faculty Western Res. U., 1955-60; dir. social service, psychologist Highland Hosp., Asheville, N.C., 1960-64, cons., 1966—; prof. social work and sociology U. Ga., Athens, 1964—. Cons. to govt. and pvt. instns. Fellow Am. Psychol. Assn., Am. Orthopsychiat. Assn. (dir.), Soc. for Psychol. Study Social Issues; mem. Nat. Assn. Social Workers (dir.), Am. Sociol. Assn. Editor: Social Work Research, 1960; (with Grace Ganter, Margaret Yeakel) Retrieval from Limbo, 1967. Research, publs. on contbns. to our knowledge of use of groups in treatment of emotional disorders through studies of contagion of behavior and power-structures in such groups; how relationships are formed on inital contact; role of direct and open communication by patient in words, facilitating his recovery, studies on verbal accessibility. Home: 311 White Pine Dr., Asheville, N.C. 28805. Office: Sch. Social Work, U. Ga., Athens, Ga. 30601.*

POLANYI, Michael, phys. chemist; b. Budapest, Hungary, Mar. 12, 1891; s. Michel and Cecilia (Wohl) P.; ed. Budapest, Karlsruhe; M.D., D.Sc.; m. Magda Kemeny, 1920; 2 sons. Mem. Kaiser Wilhelm Inst. Phys. Chemistry, 1922-33 (resigned); prof. phys. chemistry Manchester (Eng.) U., 1933-48, prof. social studies, 1948-58; sr. research fellow Merton Coll., Oxford, Eng., 1959-61; fellow Center Behavioral Studies, Palo Alto, Cal., 1962-63; distinguished prof. religion Duke, 1964; senior fellow Center for Advanced Studies, Wesleyan U., 1965-66; lectr., Cambridge, Eng., 1960, Calcutta, India, 1961, Yale, 1962, U. Chgo., 1967, 68, various others univs. Recipient Am. Le Comte du Nouy award, 1959. Fellow Royal Soc., 1944; mem. Am. Acad. Arts and Scis. (hon.). Author: Atomic Reactions, 1932; U.S.S.R. Economics, 1935; (film) Money and Unemployment, 1939; The Contempt of Freedom, 1941; Full Employment and Free Trade, 1945; Science, Faith and Society, 1946; logic of Liberty, 1951; Personal Knowledge, 1958; The Study of Man, 1959; Beyond Nihilism, 1960; The Tacit Dimension, 1966. Research on reaction kinetics and crystal structure. Home: 22 Upland Park Rd., Oxford, Eng.

POLATIN, Phillip, Am. psychiatrist; b. N.Y.C., Aug. 25, 1905; s. Ira and Anne (Ruskin) P.; grad. N.Y. Med. Coll., 1929; M.D. Columbia Coll. Phys. & Surg., 1934; m. Esther C. Cott, July 2, 1927; 1 son, Peter B. Psychiatrist Pilgrim State Hosp., Brentwood, N.Y., 1934-38; practice psychiatry, N.Y.C., 1938—; asst. clin. psychiatrist in charge female div. N.Y. Psychiat. Inst., N.Y.C., 1938-59, clin. dir., 1959—; asso. attending psychiatrist Presbyn. Hosp., N.Y.C., 1938-59, attending psychiatrist, 1959—; prof. clin. psychiatry Columbia Coll. Phys. & Surg., N.Y.C., 1960—. Cons. psychiatry Barnert Meml. Hosp., Patterson, N.J., Gracie Sq. Hosp., N.Y.C. Fellow A.C.P., Am. Psychiat. Assn., N.Y. Acad. Medicine; mem. A.M.A., Am. Psychopath. Assn., Am. Psychosomatic Assn., Am. Assn. Med. Colls., Am., Internat. psychoanalytic assns. Author: (with Ellen C. Philtine) How Psychiatry Helps, 1949. The Well-Adjusted Personality, 1953, Marriage in the Modern World, 1956; Guide to Treatment in Psychiatry, 1966. Discovered vertebral fractures in shock therapy; devised ambulatory insulin treatment schizophrenia; (with Paul Hoch) orginated concept pseudoneurotic schizophrenia. Home: 5281 Independence Av., Riverdale, N.Y. 10471. Office: 722 W. 168th St., N.Y.C. 10032.*

POLE, William, English civil engr.; b. Birmingham, Eng., Apr. 22, 1814; Mus.D., St. John's Coll., Oxford, 1867; m. Matilda Gauntlett, 1846. Prof. civil engring. Elphinstone Coll., Bombay, 1844-47, Univ. Coll., London, 1859-67; lectr. Royal Engr. Establishment, Chatham; sec. to various royal commns. on sci. subjects. Fellow Royal Soc., 1861; mem. Instn. Civil Engrs. (hon. sec.). Author: Treatise on the Corning Pumping Engine, 1844; On the High-Pressure Steam Engine, 1848; On the Use of Iron in Construction, 1872; Scientific Chapters in the Lives of Robert Stephenson and I. K. Brunel, 1877; Life of Sir William Siemens, 1888; Evolution of Whist, 1895. Served govt. on many important sci. matters, including iron, armor, heavy arty., breech-loading rifles, met. water, gas and sewerage arrangements; cons. engr. for Japanese rys. many years. Died Dec. 30, 1900.

POLENI, Marquis Giovanni, Italian astronomer; b. Venice, Italy, Aug. 23, 1683; supt. rivers and waters, Venice; surveyed St. Peters, Rome, Italy on request of Pope Benedict; prof. astronomy, math., philosophy, Padua, Italy. Fellow Royal Soc., 1710; mem. French Acad. Scis., Acad. Berlin, Insts. Padua and Bologna. Author: De physices in rebus mathematicus, 1716; De mortu aquae mixto, 1717; De castellis per quae derivantur fluviorum aquae, 1718; Exercitationes vitruvianae, 1739-41; also articles. Attempted to build a mech. multiplying machine, 1709. Died Padua, Nov. 14, 1761.

POLGLASE, William James, Canadian chemist; b. Vancouver, B.C., Can., May 31, 1917; s. Joseph and Elizabeth Ann (Curnow) P.; B.A., U. B.C., 1943, M.A., 1944; Ph.D., Ohio State U., 1948; m. Rosemary Elaine Hadfield, Aug. 1, 1960; 1 son, John Hadfield. Research instr. U. Utah, 1948-51; research chemist Rayonier, Inc., Shelton, Wash., 1951-52; faculty U. B.C., Vancouver, 1952—, prof. biochemistry, 1962—. Vis. scientist Nat. Inst. Med. Research, London, 1966-67. Mem. Am. Chem. Soc., Am. Soc. Biol. Chemists, Am. Soc. Microbiology. Research in carbohydrates, peptides and specificity of aminopeptidase structure of glycogen from normal human liver and muscle and from glycogen storage disease; chemistry of wood cellulose; antibiotics, biochem. regulation. Home: 3995 W. 39th Av., Vancouver 13, B.C., Can.*

POLHEM, Christopher, Swedish inventor; b. Visby, Gotland Island, Dec. 18, 1661; became mining engr., 1690; traveled on continent, 1694-96; dir. mines, Falun, Sweden; built water-powered factory for mfr. tools, 1700; founder factory for iron and other metal products, Stjärasund, Sweden, circa 1704; built minting machine for George I. of Eng. Mem. Acad. Scis. (became pres. 1744). Author: Cogitationes mathematicae, 1714; Political Testament, 1746; also articles. Used rolls in metal working; advocated use of water power instead of human muscle, div. of labor, thus paving the way for indsl. revolution. Died Tingstäde, Sweden, Aug. 30, 1751.

POLI, Martino, inventor, chemist; b. Palagnana, Italy, Jan. 21, 1662; apothecary, Rome; royal engr., Paris; mem. French Acad Scis. Author: Il trionfo degli acido, 1706. Refuted popular conception of acids as cause of disease, showed their therapeutic qualities; inventor artillery used by Louis XIV. Died Paris, July 30, 1714.

POLITZER, Adam, physician; b. Alberti, Hungary, Oct. 1, 1835; grad. Vienna U., 1859, docent, 1861-70, chair of otology, 1870-1907; mem. Austrian Otol. Soc. (founder 1902). Author: Lehrbuch der Ohrenheilkunde, 1878; Anatomie und Histologie des menschl Gehörogans, 1889; Geschichte der Ohrenheilkunde, 2 vols., 1907-13; also paper On a Peculiar Affection of the Labyrinthine Capsule as a Frequent Cause of Deafness (description of otosclerosis as med. entity), 1893. Tchr. of otologists who praticed throughout the world; pioneer of modern methods in diagnosis, therapy and surgery of ear; established otology as specialty; devised Politzer bag for inflating middle ear, circa 1863; 1st to get pictures of tympanic membrane by direct illumination, circa 1865; brought perforated mirror into gen. use; developed method for achieving permeability of eustachian tube. Died Vienna, Sept. 10, 1920.

POLK, Charles, physicist; b. Vienna, Austria, Jan. 15, 1920; s. Henry P. and Amalie (Canar) P.; came to U.S., 1940, naturalized, 1943; B.S. in Elec. Engring., Washington U., St. Louis, 1948; M.S. in Physics, U. Pa., 1953, Ph.D., 1956; m. Dorothy R. Lemp, Apr. 27, 1946; children—Dean F., Gerald W. With antenna advance devel. staff RCA Victor div., Camden, N.J., 1948-52; faculty U. Pa. Moore Sch. Elec. Engring., Phila., 1952-57; mem. tech. staff RCA Labs., Princeton, N.J., 1957-59; adj. prof. elec. engring. Drexel Inst. Tech., Phila., 1957-59; prof. elec. engring., chmn. dept. U. R.I., Kingston, 1959—. Mem. I.E.E.E. (chmn. Providence 1964-65), A.A.A.S., Am. Geophys. Union, Am. Soc. for Engring. Edn., Sigma Xi, Tau Beta Pi, Phi Kappa Phi. Patentee, publs. in field of antennas, radio propagation, geophysics. Home: 21 Springhill Rd., Kingston, R.I. 02881.*

POLKANOV, Aleksándr Aleksétevich, Russian geologist, petrographer; b. May 25, 1888; grad. St. Petersburg U., 1911. Prof., Leningrad U., 1921—; dir. Lab. for Precambrian Geology, USSR Acad. Scis., 1950—. Mem. USSR Acad. Scis. Author: Geological-Petrological Outline of the Northwestern part of the Kola Peninsula, Part I, 1935; A Geological Outline of the Kola Peninsula; The Petrology of Gremyakha-Vyrmes Plut; (with N. A. Yeliseev) Kola Peninsula, 1941; Genetic Systematization of the Intrusion of the Platform—Cratogen, 1946; Pluto of Gabbro-Labradorite in Volynia, Ukranian SSR, 1948. Research on structure and genetic systematization of intrusive bodies; Precambrian formation, including their mineral resources.

POLKINGHORNE, John Charlton, English physicist; b. Weston-Super-Marc, Eng., Oct. 16, 1930; c. George Baulkwill and Dorothy (Charlton) P.; B.A., Trinity Coll. U. Cambridge, 1952, Ph.D., 1955, M.A., 1956; m. Ruth Isobel Martin, Mar. 26, 1955; children—Peter John Martin, Isobel Jane, Michael James. Fellow Trinity Coll. Cambridge U., 1954; lectr. math. physics U. Edinburgh, 1956-58; lectr. theoretical physics U. Cambridge, 1958-65, reader theoretical physics, 1965—. Mem. Am. Phys. Soc. Author (with R.J. Eden, P.V. Landshoff, D.I. Olive) The Analytic S-Matrix, 1966; also numerous articles. Research, publs. field analytic side of elementary particle physics, including the analytic and high energy properties of Feynman integrals and the foundations of S-Matrix theory. Home: 22 Rutherford Rd., Cambridge, Eng.*

POLLACK, Irwin, Am. psychologist; b. Bridgeport, Conn., Apr. 10, 1925; s. Benjamin and Mary (Beimel) P.; B.S., U. Fla., 1945; M.S., Harvard, 1947, Ph.D., 1948; m. Marcille Kaufman, Apr. 30, 1949; children—Sharron, Phyllis, Stanley. Research psychologist USN Electronics Lab., San Diego, 1949, U.S. Air Force Operational Applications Lab., Washington, Bedford, Mass., 1949-63; prof. psychology dept. psychology U. Mich., Ann Arbor, 1963—, research psychologist Mental Health Research Inst., 1963—. Mem. com. on hearing and bioacoustics Nat. Acad. Sci.-NRC, 1953—; mem. adv. com. on psychobiology NSF, 1965—. Recipient Civil Service Meritorious Achievement award, 1956—. Mem. Am., Midwestern psychol. assns., Psychonomic Soc., Acoustical Soc. Am., Am. Standards Assn. Research, numerous publs. on elementary auditory displays, speech communication in noise, short-term memory span. Home: 4068 Thornoaks Dr., Ann Arbor, Mich. 48104.*

POLLACK, Max, psychologist; b. Chernowitz, Rumania, Jan. 21, 1922; s. Morris and Lisa (Spector) P.; came to U.S., 1935, naturalized, 1944; A.B., N.Y. U., 1947, Ph.D., 1955; m. Yvette N. Fischer, July 13, 1944; children—Deborah Toni, Judith Mildred. Research psychologist N.Y. U. Coll. Medicine Lab. Psychophysiology, 1948-50, dept. neurology Mt. Sinai Hosp., N.Y.C., 1950-55; research asso. Henry Ittelson Center for Child Devel., N.Y.C., 1954-55; cons. dept. mental hygiene State N.Y., 1957-63; cons. psychologist League Sch. for Seriously Disturbed Children, N.Y.C., 1964; sr. research asso. Hillside Hosp., Glen Oaks, N.Y., 1964—; clin. asso. prof. State U. N.Y. Downstate Med. Center; asso. prof. psychology Queen's Coll., City U., 1966—. Recipient USPHS Research Career Devel. award. Research, numerous publs. on relation of early minimal brain damage syndromes to childhood devel., adult psychopath. conditions. Home: 65 Allenwood Rd., Great Neck, L.I., N.Y. 11023. Office: 75-59 263d St., Glen Oaks, N.Y. 11004.*

POLLACZEK, Felix, mathematician; b. Vienna, Austria, Dec. 1, 1892; s. Alfred and Marie (Gompers) P.; student U. Vienna; Ph.D., U. Berlin, 1922; student tech. univs. Vienna, Brno, Czechoslovakia; Elec. Engr., Brno, 1920; m, Vera Jacobowitz, Oct. 6, 1934; 1 dau., Magda (Mrs. Robert Buka). Sci. collaborator Reichspostzentralamt, Berlin, 1923-33; mem. Centre National de la Recherche Scientifique, Paris, 1939-59; vis. prof. U. Md., 1963, U. Pa., 1967. Mem. Société

Mathématique de France, Am. Math. Soc., Assn. Française d'Informatique et de Recherche Opérationnelle. Contbg. author: Mémorial des Sciences Mathématiques, 1956, 57, 61. Research and publs. in number theory, math. physics, math. analysis, theory of probability with applications to problems of telephone traffic. Home: 54 Rue du Point du Jour, 92-Boulogne-Billancourt, France.*

POLLAK, Herman, Belgian physicist; b. Brussels, Belgium, Oct. 5, 1933; s. Jacques and Alda (van Perk) P.; M.A. in Physics, U. Libre de Bruxelles, 1956, Ph.D., 1960; m. Claudine Vaneyck, Aug. 10, 1957; children—Eric, Alain, Andre, Bernard, Guy. Staff, Centre D'Etudes de L'Energie Nucleaire, Mol, Belgium, 1956-62; prof. theoretical physics U. Lovanium, Kinshasha, Congo, 1962—; gen. mgr. Centre Nucléaire, Trico, Kinshasa, 1965—. Mem. Société belge de physique, Société Francaise de Physique, Am. Phys. Soc. Author: Mécanique quantique I and II, 1966; Calcul tensoriel, 1966; also articles. Measurements of cross sect., calculations of nuclear spectrum, solid state research including impurity in crystals, ionic state, lifetime of ionic state, theoretical calculations of antishieldingfactors. Home: BP 171 Kinshasa XI, Congo, also 40 Kruisberg, Mol, Belgium.*

POLLAK, Leo Wenzel, Czechoslovakian physicist; b. Prague, Czechoslovakia, Sept. 23, 1888; s. Simon and Dorothea (Gluck) P.; student German U., Prague, 1906-10; Ph.D., U. Prague, 1912; Sc.D. honoris causa, Dublin (Ireland) U., 1963; m. Johanna Dittrich, 1920 (dec. Dec. 1958); m. 2d, Nessa Falconer, Dec. 27, 1962. Prof. geophysics German U., Prague, 1929-39; sr. meteorol. officer Irish Meteorol. Service, 1939-47; sr. prof. meteorology and geophysics Dublin Inst. for Advanced Studies, 1947-63, ret., 1963. Organizer 1st Internat. Symposium on Condensation Nuclei, Dublin, 1955; chief investigator on contracts U.S. Army and USAF; cons. Gen. Electric, Schenectady, 1958-64; adviser, research prof. in atmospheric scis. Research Center, State U. N.Y., 1963-64. Mem. Royal Meteorol. Soc. London, Royal Irish Acad. Author: Rechentafeln zur harmonischen Analyse, 1926; Harmonic Analysis and Synthesis Schedules for Three to One Hundred Equidistant values of Empiric Functions, 1947; All Term Guide for Harmonic Analysis and Synthesis, 1949; (with V. Conrad) Methods in Climatology, 1950; On the Systematic Influence in Series of Annual Rainfall Totals, 1951; Eight-Place Supplement to Harmonic Analysis and Synthesis Schedules for Three to One Hundred Equidistant Values of Empiric Functions, 1954. Research, publs. on devel. of meteorol. instruments; introduced punch card method for weather data; developed photo-electric condensation nucleus counter for measurement of fog density, air pollution. Died Nov. 24, 1964.

POLLAK, Otto, sociologist; b. Vienna, Austria, Apr. 30, 1908; s. Jacob and Elsa (Kohnberger) P.; came to U.S., 1938, naturalized, 1944; J.D., U. Vienna, 1930; Ph.D., U. Pa., 1947; m. Gertrude Kary, Mar. 24, 1939. Practice law, Vienna, 1930-38; faculty dept. sociology U. Pa., Phila., 1945—, cons. Family Service Phila., 1957-61, dept. psychiatry and neurology Valley Forge Gen. Hosp., 1964—. Mem. Am. Sociol. Assn., Am. Orthopsychiat. Assn., Nat. Assn. Social Workers. Author: Criminality of Women, 1950; Social Adjustment, in Old Age, 1948; Social Science and Psychotherapy for Children, 1952; Positive Experiences in Retirement, 1957; Integrating Sociological and Psychoanalytic Concepts, 1956; also numerous articles. Home: 13 Aldwyn Lane, Villanova, Pa. Office: Dietrich Hall, U. Pa., Phila.*

POLLARD, Cash Blair, Am. chemist, toxicologist; b. Hannibal, Mo., Feb. 22, 1900; s. William Braxton and Nannie Elizabeth (Robinson) P.; A.B., William Jewell Coll., 1921; M.S., Purdue U., 1923, Ph.D., 1930, D.Sc., 1954; postgrad. U. Wis., 1924; m. Ailene Atherton; 1 son, Thomas David. Grad. asst. chemistry Purdue U., 1921-23, instr. chemistry, 1923-30; with Graver Corp., 1923; asst. prof. chemistry U. Fla., 1930-35, asso. prof., 1935-37, prof., 1937—, chmn. organic div.; cons. chemist, toxicologist, 1927—; cons. chemist Fla. states attys. sci. crime detection, 1930—; expert witness Fla. and Fed. cts., on toxicology, blood stains and powder marks; mem. faculty Alachua Gen. Hosp. Nurses Tng. Sch.; lectr. physiol. and pathol. chemistry, 1945—; cons. toxicologist Alachua Gen. Hosp., Morton Plant Hosp., Clearwater, Fla., Munroe, Meml. Hosp., Ocala, Fla. Recipient Fla. Acad. Scis. Achievement Award (with John H. Pomeroy), 1945, for study of sensitivity of aldehyde reagents; USPHS Research grant, 1948, 49, research grants, Navy Dept. Office Naval Research 1948, 49, 50, Parke, Davis and Co., 1950—; research grant Dow Chem. Co., 1956. Fellow A.A.A.S., Am. Inst. Chemists; mem. Am. Chem. Soc. (Fla. award 1954), Fla. Acad. Sci., Am. Assn. Clin. Chemists, Sigma Xi. Author: Laboratory Manual and Study Outline of General Chemistry (with L. A. Test), 1928, rev. 1937; Bibliography of Animal Venoms (with Ralph W. Harmon), 1947; Problems in Organic Chemistry (with E. G. Rietz), 1951. Asst. editor: Outline of Organic Chemistry, 1937. Asso editor: Quadri-Service Manual of Organic Chemistry, 1938. Collaborator on Fundamental Organic Chemistry, 1940; The Work Book of Fundamental Organic Chemistry, 1941. Contbr. research articles to sci. jours. Died May 31, 1959.

POLLARD, Ernest Charles, physicist; b. Chaotong Fu, China, Apr. 16, 1906; s. Samuel and Emma (Hainge) P.; B.A., Cambridge U., 1928, Ph.D., 1932, M.S., Yale, 1950; m. Elizabeth Watson, Aug. 12, 1933; children—Anne, Carol, Stephen. Physicist Yale, 1933-60, prof. biophysics, 1950-61, chairman dept., 1954-61; chmn. com. on biophysics Pa. State U., 1961—, prof. biophysics, 1961—, head of the department of physics, 1963—. Fellow Am. Phys. Soc., A.A.A.S.; mem. Am. Phytopathological Soc., Radiation Research Soc., Biophys. Soc., Sigma Xi. Author textbooks; (with J.M. Sturtevant) Microwaves and Radar Electronics, 1948; (with W.L. Davidson, Jr.) Applied Nuclear Physics; (with R.B. Setlow) Physics of Viruses, 1953, Molecular Biophysics, 1962. Contbr. articles profl. jours. Research nuclear physics on energy levels, nuclear radius, radiation effects, biophysics on radiation action on viruses, cells, also design microwave radar. Home: 444 W. Fairmount Av., State College, Pa. Office: Pa. State U., University Park, Pa.*

POLLARD, William Grosvenor, Am. physicist; b. Batavia, N.Y., Apr. 6, 1911; s. Arthur Lewis and Ethel (Hickox) P.; B.A., U. Tenn., 1932; M.A., Rice U., 1934, Ph.D., 1935; D.Sc., Ripon Coll., 1951, U. of South, 1952, Kalamazoo Coll., 1955; D.D., Hobart Coll., Grinell Coll., 1957; LL.D., U. Chattanouga, 1958, Kenyon Coll., 1964; L.H.D., Keuka Coll., 1962; m. Marcella Hamilton, Dec. 27, 1932; children—W. Grosvenor Pollard III, A. Lewis Pollard II, Frank H. Fellow in physics Rice U., 1932-35, asst. 1935-36; faculty physics U. Tenn., 1936-47, prof., 1943-47; research scientist Columbia, 1944-45; exec. dir. Oak Ridge Asso. Univs., 1947—. Ordained to ministry Episcopal Ch. as deacon, 1952, priest, 1954; priest asso. St. Stephen's Ch., Oak Ridge, 1954—, priest in charge St. Alban's Ch., Clinton, Tenn., 1959-65; faculty grad. sch. theology U. of South, Sewanee, Tenn., 1956, 60, 61. Recipient Distinguished Service award So. Assn. Sci. and Industry, 1950; Semicentennial medal of Honor, Rice U., 1962. Fellow Am. Phys. Soc. (chmn. Southeastern sect. 1951-52), A.A.A.S., Am. Nuclear Soc. (dir. 1955-56, mem. by-laws and rules com. 1963-68), mem. Phi Beta Kappa, Sigma Xi, Beta Gamma Sigma, Phi Kappa Phi, Sigma Pi Sigma. Author: The Hebrew Iliad, 1957; Chance and Providence, 1958; Physicist and Christian, 1961. Research, publs. on radiation, nuclear physics, gases. Home: 191 Outer Dr., Oak Ridge 37830. Office: P.O. Box 117, Oak Ridge 37830.*

POLLENDER, Franz Aloys Antoine, German bacteriologist; b. Barmen, Germany, 1800; M.D., U. Bonn (Germany), 1824. Practiced first, Lindlar, then Sanitätsrath, until 1870; ret. to Schaerbeck, Brussels, Belgium. First clear description of rods (Bacillus anthracis) in blood and organs of animals dead from splenic fever, 1855. Died 1879.

POLLEY, Howard Freeman, Am. physician; b. Columbus, O., Nov. 12, 1913; s. David William Latimer and Mary (Lakin) P.; B.A., Ohio Wesleyan U., 1934, D.Sc., 1965; M.D., Ohio State U., 1938; M.S. in medicine, U. Minn., 1945; m. Georgiana Redrup, June 5, 1938; children—Alice Lynne, Marry Ann, William Redrup. Fellow in medicine Mayo Clinic, Mayo Found., Rochester, Minn., 1940-43, cons. in medicine, clinic, 1943—, faculty Grad. Sch., 1946-54; prof. medicine, 1960—, head sect. rheumatic disease, 1962—; practice medicine specializing in internal medicine and rheumatology, Rochester, 1943—. Recipient Alumni Achievement award Ohio State U., 1958. Fellow A.C.P.; mem. Am. Med. Assn., Central Soc. for Clin. Research, Am. Acad. Phys. Medicine and Rehab., Am. Rheumatism Assn. (spl. citation 1961, pres. 1964—), Uruguayo de Reumatologia (hon.), Sigma Xi. Author: (with others) Physical Examination of the Joints, 1965. Clin. investigations in treatment, diagnosis, other aspects of rheumatic and related diseases; codesigner Polley-Bickel punch biopsy. Home: 1015 Plummer Circle, Rochester, Minn. 55901.*

POLLI, Silvio, Italian geophysicist; b. Trieste, Italy, Dec. 26, 1904; s. Giuseppe and Amelie (Turco) P.; ed. in math. and geophysics; Dr ès sc.; m. Elfrieda Milessich; children—Elio, Gea, Talassia. Dir. Thalassography Inst., Trieste, also prof., instr. geophysics U. Trieste. Mem. com. maritime sci. UNESCO. Mem. Acad. Econ. Studies Trieste, Am. Geophys. Union, Italian Geophysics Union, Soc. Adriatic Scis., Italian Soc. Geophysics and Meteorology, Center Research and Oceanographic Studies. Publns. on meteorology, oceanography, Geophysics. Home: via Filippo Zamboni 1, Trieste. Office: viale Romolo Gessi 2, Trieste, Italy.

POLLOCK, George Howard, Am. psychiatrist, psychoanalyst; b. Chgo., June 19, 1923; s. Harry J. and Belle (Lurie) P.; student Lawrence Coll., 1939-40, U. Chgo., 1940-42; M.D., U. Ill., 1945, M.S., 1948, Ph.D. in Physiology (Commonwealth fellow), 1951; m. Beverly Yufit, July 3, 1946; children—Beth, Raphael, Daniel, Benjamin, Naomi. Faculty, U. Ill., 1955—, clin. prof. psychiatry, 1964—; mem. staff, diplomate Chgo. Inst. Psychoanalysis, 1956—, asst. dean edn., 1960-67, dir. research, 1963—, tng. analyist, 1961—, supervising analyst, 1962—. Diplomate Am. Bd. Neurology and Psychiatry. Fellow Am. Orthopsychiat. Assn., Am. Psychiat. Assn.; mem. Am. Anthrop. Assn., A.A.A.S., Am. Assn. U. Profs., Am. Electroencephalographic Soc., Am. Fedn. Clin. Research, Am. Heart Assn., Am. Inst. Biol. Scis., A.M.A., Am. Psychoanalytic Assn., Am. Psychol. Assn., Am. Psychosomatic Soc., Am. Sociol. Assn., Assn. Am. Med. Colls., Central Assn. Electroencephalographers, N.Y. Acad. Scis., Soc. Biol. Psychiatry, Soc. Exptl. Biology and Medicine, Soc. Gen. Systems Research, Am. Acad. Polit. and Social Science, World Med. Assn., Sigma Xi, others. Research, publs. on chem. poisons of nervous system; epilepsy and convulsions; psychosomatic medicine. Home: 5759 S. Dorchester Av., Chgo. 60637. Office: U. Ill., 180 N. Michigan, Chgo. 60601.*

POLLOCK, H(arry) E(velyn) D(orr), Am. archeologist; b. Salt Lake City, June 24, 1900; s. James Albert and Evelyn Prince (Dorr) P.; A.B., Harvard, 1923, A.M., 1930, Ph.D., 1936; m. Katherine Winslow, Aug. 15, 1940; 1 son, Harry Winslow. Excavations, exploration in Guatemala, Carnegie Instn. of Washington, 1928, 37, Yucatan and Mexico, 1929-32, 35-37, 40, 51-55, archeologist, 1931-50, dir. dept. archeology 1950-58, research asso. 1958—; research fellow archeology Harvard, 1953—. curator of Maya archeology, Harvard, 1963—. Fellow Am. Anthrop. Assn.; mem. Soc. Am. Archaeology, A.A.A.S. Author: (with J. E. Thompson and J. Charlot) Preliminary Study of the Ruins of Coba, 1932; Round Structures of Aboriginal Middle America, 1936; also archeol. articles. Home: 11 Berkeley Pl., Cambridge 38, Mass.*

POLLOCK, Herbert Chermside, Am. physicist; b. Staunton, Va., June 13, 1913; s. James King and Mabel (Chermside) P.; B.A., U. Va., 1933; Ph.D., Oxford U., Eng., 1937; m. Virginia E. Jones, Nov. 28, 1942; children—Herbert James, Martha Avery, Richard Chermside, Robert Alexander. Research physicist Gen. Electric Co., 1937-43, 44—; with radio research lab. Harvard, Eglin Field, Fla., 1943-44. Mem. adv. bd. Nucleonic, Chemistry and Electronic Shares, Inc. Dir. Schenectady Mus. Mem. bd. visitors U. Va., 1955—. Fellow Am. Phys. Soc.; mem. English-Speaking Union. Research in electron and plasma physics leading to inventions in the electrical field relating to electron accelerators, x-ray tubes, pulsed neutron tubes, high temperature plasma devices, nuclear power, and naval equipment. Home: 2147 Union St., Schenectady 9. Office: Gen. Electric Co., Schenectady.*

POLLOCK, Martin Rivers, Brit. biologist; b. Liverpool, Eng., Dec. 10, 1914; s. Hamilton Rivers and Eveline (Bell) P.; B.A., Trinity Coll., Cambridge (Eng.) U., 1936, M.B., B.Ch., 1940, M.A., 1947; m. Jean Ilsley Paradise, Mar. 12, 1941; children—Jessamy, Julian Rivers, Lisa Jane, Jonathan Ilsley. Staff, Emergency Pub. Health Lab. Service, 1941-45; staff Med. Research Council, 1945-65, head div. bacterial physiology Nat. Inst. for Med. Research, 1949-65; prof. biology U. Edinburgh (Scotland), 1965—. Mem. chemotherapy com. Med. Research Council, 1958-65. Fellow Royal Soc., 1962; mem. Soc. for Gen. Microbiology (past mem. council), Biochem. Soc., Assn. Sci. Workers. Research, numerous publs. on adaptation of micro-organisms to their environment by means of control of formation of specific enzymes. Home: 30 Saxe-Coburg Pl. Office: Dept. Molecular Biology, King's Bldgs., Edinburgh, Scotland.*

POLO, Marco, Italian explorer, traveler; b. Venice, circa 1254; s. Nicolo P.; m. Donata; daus.; Fantina, Bellela, Moreta. His father and uncle, Maffeo, made trading expedition to Constantinople, 1253-60; war blocked their return; journeyed east-ward to reach Kublai Khan's eastern capitol near Kaifeng, China, 1266; returned to Venice, 1269; left again, with Marco, for Kublai Khan's court, 1271; reached Cambulve (Peiping), 1275; Marco Polo became favorite of Khan; ruled Yangchow, China, for 3 years; left for Venice, 1292; returned 1295; joined Venetian forces fighting Genoa; taken prisoner, 1296-98; in captivity, dictated account of his travels; told of customs of inhabitants, paper currency, asbestos, coal (then unknown in Europe); described oriental splendor and wealth of the Indies; only Western source of information on the East during Renaissance. Died Venice, Jan. 9, 1324.

POLONOVSKI, Michel, French physician; b. Mulhouse (now France), 1889; Licentiate Phys. Scis., 1909; M.D., 1914; Tchr.'s Certificate, 1920. Became prof. chem. biology U. Lille (France), 1923; named prof. U. Paris, 1937. Research in organic chemistry, biol. chemistry. Died nr. Douai (formerly Doucey, France), 1954.

POLSON, Alfred, S. African phys. chemist; b. Brandfort, O.F.S., South Africa, Mar. 11, 1912; s. Gustaf and Paulina (Gous) P.; B.Sc., Stellen bosch U., 1932, M.Sc., 1934, D.Sc., 1937; postgrad. Uppsala (Sweden) U.; D.Sc., Cape Town (S. Africa) U., 1958; m. Suzanne le Roux, Feb. 11, 1941; children—Gottfried, Birgitta. Research asst. Inst. Phys. Chemistry, Uppsala, Sweden, 1934-37; research officer Onderstepoort Vet. Inst., S. Africa, 1938-45; fellow NIH, Bethesda, Md., 1946-48; sr. research officer Onderstepoort Vet. Research Inst., South Africa, 1948-51; sr. research officer virus research unit Council for Sci. and Indsl. Research, Med. Sch., Cape Town, 1951-62, chief research scientist, 1962—. Chmn.

exptl. biology group Univs. Cape Town and Stellenbosch, 1966—. NIH grantee for research on African viruses, 1961—. Mem. Soc. for Exptl. Biology and Medicine. Research, numerous publs. on macromolecules and viruses using diffusion, ultracentrifugation, viscosity, electrophoresis, gel filtration in granulated agar gel and quantitative gel precipitin reaction, methods for purification of macromolecules such as multi-membrane electrodecantation, use of polyethylene glycol, use cellophane membranes of prodn. of potent toxin from anaerobes. Home: 20 Redlands Rd., Milnerton, C.P., Home: Council for Sci. and Indsl. Research, Virus Research Unit, Med. Sch., Cape Town, South Africa.*

POLUBARINOVA-KOCHINA, Pelageia Iakovlevna; see Kochina, Pelageia Iakovlevna.

POLUNIN, Nicholas, ecologist; b. Checkendon, Eng., June 26, 1909; s. Vladimir and Elizabeth Violet (Hart) P.; B.A. with 1st class honors, Oxford U., 1932, M.A., 1935, D.Phil., 1935, D.Sc., 1942; M.S., Yale, 1934; m. Helen Eugenie Campbell, Jan. 3, 1948; children—April Xenia, Nicholas Vladimir Campbell, Douglas Harold Hart. Research Scientist U.K. Dept. Sci. and Indsl. Research, 1935-39; Fielding curator, keeper herbaria, lectr., then sr. research fellow Oxford U., 1939-47; Macdonald prof. botany McGill U., Montreal, 1946-52; Guggenheim fellow, research fellow Harvard, 1950-53; dir. USAF research project, lectr. Yale, 1953-55; prof. plant ecology and taxonomy, chmn. dept. botany, dir. U. Herbarium and Bot. Garden, Baghdad, Iraq, 1956-58; guest prof. U. Geneva, 1959-61; prof., head dept. botany U. Ife, Ibadan, Nigeria, 1962-66; editor Plant Sci. Monographs, 1954—; editor Biol. Conservation, 1967—. Past cons. C.E., U.S. Army, USAF, pub. cos. Recipient numerous research and exploration grants. Felow Linnean Soc. London, Royal Geog. Soc., Royal Hort. Soc., A.A.A.S., Arctic Inst. N.Am.; mem. numerous profl., sci. socs. Author numerous books and articles, including Botany of the Canadian Eastern Arctic, 3 vols. 1940-48; Arctic Unfolding, 1949; Circum-polar Arctic Flora, 1959; Introduction to Plant Geography, 1960; Eléments de Géographie botanique, 1967. Research on plant life and ecology of arctic, subarctic and high-altitude regions; collections for nat., univ. museums; demonstrated persistence of microbial life over North Pole; discovered arctic islands; proposed principle of reaction similarity. Home: 1249 Avusy, Geneva, Switzerland.*

POLVANI, Carlo, Italian physician; b. Pisa, Italy, Nov. 16, 1924; s. Giovanni and Rosa Ida (Catagnola) P.; degree in medicine, U. Milan (Italy), 1950, specialization in radiology, 1952, Libera Docenza, 1964; m. Kathjoucha Colombi, Aug. 30, 1956; children—Giovanni, Marino. Radiologist, Inst. Radiology, Milan U., 1950-57, lectr. radiation protection Postgrad. Sch. Nuclear Medicine; chief med. and health service Ispra Nuclear Study Center, 1957-61; dir. biology and health protection div. Nat. Com. for Nuclear Energy, Rome, Italy, 1957-64, dir. radiation protection div., 1965—. Mem. com. IV, Internat. Commn. Radiol. Protection, 1963—; mem. tech. commn. for nuclear plants Nat. Com. for Nuclear Energy, 1965—. Mem. Italian Assn. for Health Physics and Radiation protection (exec. council 1958—), Internat. Radiation Protection Assn. (exec. council 1966—), Italian Radiol. Soc. (sec. nat. com. for radiol. protection 1954—). Author: (with F. Fossati, P. Gallone, L. Parmeggiani, M. Scolari), Recommendations on Radiation Protection, 1956; Health Physics Control of Radioactive Contaminations (with Malvicini), 1958; (with L. Oliva and others) Protection of Patients in Medical Radiology, 1965. Research on models simulating expts. with tracers, 1958; analysis of survival of patients radiologically treated for lympho- and reticulo-sarcomatas, 1958-60; studies on intake of iodine 131 by Italian children following nuclear test explosions, 1962, statis. evaluation criteria for individual hematological examinations, 1963, concentration factors in fresh water fishes for cesium 137 and strontium 90, 1960, 67. Home: 299 Nomentana, Rome. Office: 15 Belisario, Rome, Italy.*

POLYA, Eugene (Jeno), Hungarian surgeon; b. 1876; student, Budapest, Hungary; studied under Dollinger, later under Herczel; lectr. anatomy and surgery U. Budapest, 1908-11; surgeon-in-chief St. Stephan Hosp., Budapest, 1910-40. Modified Billroth II operation for sub-total gastrectomy; studied pancreatic necrosis, surgery of gallstones, lymphatics, facial plastic surgery. Died 1944.

POLYA, George, mathematician; b. Budapest, Hungary, Dec. 13, 1887; s. Jacob and Anne (Deutsch) P.; Ph.D., Budapest U., 1912; postgrad. U. Göttingen (Germany), U. Paris (France); D.Sc., Swiss Fed. Inst. Tech., 1947; LL.D., U. Alta., 1961; m. Stella Vera Weber, Aug. 1, 1918. Came to U.S., 1940, naturalized, 1947. Faculty, Swiss Fed. Inst. Tech., Zurich, 1914-40, prof., 1928-40; vis. prof. Brown U., 1940-42; prof. math. Stanford U. 1946-53, prof. emeritus, 1953—. Cons. Edn. Research Council Greater Cleve., 1963—. Mem. Am. Math. Soc., Math. Assn. Am. (award for distinguished service to math. 1963), Acad. Scis. (corr. mem. Paris); hon. mem. London, Swiss, French math. socs. Author: (with G. Szegö) Analysis, 1925; (with G.H. Hardy, J.E. Littlewood) Inequalities, 1934; (with G. Szego) Isoperimetric Inequalities 1951; How to Solve It, 1945; Mathematics and Plausible Reasoning, 1954; Patterns of Plausible Inference, 1954; Mathematical Discovery, 1962; also numerous articles. Research in probability, complex variables, math. physics, number theory; pioneered modern treatment heuristics. Home: 2260 Dartmouth St., Palo Alto, Cal. 94306. Office: Stanford U., Stanford, Cal. 94305.*

POLYAKOV, Ilya Mikhaylovich, Russian biologist; b. Kharkov, 1905; grad. Kharkov Inst. Pub. Edn., 1926; D.Biol. Sci. Instr., later prof. Artem Communist U., Kharkov; prof. Kharkov U., 1932-41, 1944-48, Tomsk U., 1941-44; sr. asso. Inst. Genetics and Selection, Ukrainian Acad. Scis., 1947-56; head Lab. Plant Devel. and Insemination, dep. dir. Ukrainian Research Inst. Plant Growing, Selection and Genetics, 1956—. Mem. Ukrainian Acad. Scis. (corr.). Author: Lamarck and the Theory of Evolution of the Organic World, 1926; Course in Darwinism, 1941. Address: Ukrainian Research Inst. Plant Growing, Selection and Genetics, Kharkov, Ukrainian SSR, USSR.

POLYCHRONAKOS, Dimitros, Greek ophthalmologist; b. Corinth, Oct. 21, 1919; s. Jean and Henriette (Theodossiadis) P.; ed. U. Athens, Bonn Exptl. Inst.; M.D. agrégé; m. Mina Kokorigos, Oct. 26, 1950; children—Jean, Alexander. Asst. ophthalmol. clinic Red Cross Hosp.; head Univ. Clinic Athens, 1949-50; dir. ophthalmol. clinic Salonika Municipal Hosp. Mem. Ophthalmol. Soc. No. Greece (founder), Hellenic Biochem. Soc., Greek Ophthalmol. Soc., ophthalmol. socs. France, Germany, Austria, Great Britain. Author: Opérations plastiques sur l'ectropion cicatriciel des paupières; L'ophthalmodynamométrie et l'indice artérioveineux dans les troubles de la tension artérielle; also numerous others works. Home: Odos Patriarchou Ioakim 2. Office: Odos Aristote 10, Thessaloniki, Greece.

POMALES-LEBRON, Americo, Puerto Rican microbiologist; b. Guayama, P.R., Jan. 1, 1904; s. Matias Pomalos and Lucia Lebron; Ph.D., U. Mich., 1939; m. Angelina Saldana, Jan. 24, 1937; children—Americo, Lucy (Mrs. William Santana), Ina. Staff, Presbyn. Hosp., San Juan, 1927-33; faculty U. P.R. Sch. Medicine, San Juan, 1933—, prof., head microbiol. dept., 1959—. Guggenheim fellow, 1954, 63. Mem. World Acad. Art and Sci., Soc. Exptl. Biology and Medicine, Am. Soc. Microbiology, Am. Tissue Culture Assn., A.A.A.S., others. Author: (with others) Tissue Culture in the Study of Infectious Disease, 1963; also articles. Research in pathogenesis of brucellus, streptococcal infections, host-parasite relationship at cellular level. Home: 783 Guatemala St., Las Americas, Caparra Heights, P.R.*

POMERANCHUK, Isaak Yakovlevich, Russian physicist; b. 1913; grad. Leningrad Poly. Inst., 1936; D.Physico-Math. Sci. With various instns. of USSR Acad. Scis., 1936—, sect. head Inst. Theoretical and Exptl. Physics, 1953—; prof. Moscow Physics Engring. Inst., 1946—. Recipient Stalin prize, 1950. Research and publs. on theoretical physics, radiation theory, nuclear physics, low temperature physics, theory of heat conductivity of insulators, theory of neutron scattering in crystals. Address: Moscow Physics Engring. Inst., ulitsa Kirova 21, Moscow, USSR.

POMERANTZ, Martin Arthur, Am. physicist; b. Bklyn., Dec. 17, 1916; s. Joseph and Henrietta (Moses) P.; A.B., Syracuse U., 1937; M.S., U. Pa., 1939; Ph.D., Temple U., 1951; m. Molly Bernstein, Aug. 10, 1941; children—Jane, Martin Arthur. With Bartol Research Found., Swarthmore, Pa., 1939—, physicist, 1943-59, dir., 1959—; mem. leader numerous sci. expdns.; vis. prof. astronomy Swarthmore Coll. Mem. com. polar research Geophys. Research Bd., Space Sci. Bd., chmn. U.S. Com. for Internat. Year of Quiet Sun (all Nat. Acad. Sci.); v.p. spl. com. Internat. Years of Quiet Sun, Comite Internat. de Geophysique (all Internat. Council Sci. Unions). Fellow Am. Phys. Soc., A.A.A.S., Am. Geophys. Union; mem. Sigma Xi, Sigma Pi Sigma (hon.), Pi Mu Epsilon. Editor: Jour. Franklin Inst., 1963—; editorial bd. Space Sci. Revs. Research on cosmic rays, especially mesons and nucleons, primary cosmic radiation, solar particles and modulation effects; research on phys. electronics, solid state physics. Inventor thoria cathode. Home: 1322 Knox Rd., Wynnewood, Pa. 19096. Office: Bartol Research Found., Swarthmore, Pa. 19081.*

POMERANZ, Yeshajahu, biochemist; b. Tlumacz, Poland, Nov. 28, 1922; s. David and Rachel (Bildner) P.; B.Sc., Israel Inst. Tech., 1944, Chem.E., 1945; Ph.D., Kan. State U. 1962; m. Ada Weissberg, Oct. 22, 1948; children—Sol, David. Came to U.S., 1959, naturalized, 1967. Head, Food Testing Lab., Govt. Israel, 1948-59; research asso. Brit. Milling Industry Research Assn., 1954; research chemist U.S. Dept. Agr., Manhattan, Kan., 1961-66; prof. biochemistry Kan. State U., Manhattan, 1965—. Mem. A.A.A.S., Inst. Food Technologists, Am. Assn. Cereal Chemists, Sigma Xi, Phi Kappa Phi. Research, numerous publs. on cereal chemistry, biochemistry and nutrition of cereals, relations between chem. composition and functional properties of cereals. Home: 1715 Laramie St., Manhattan, Kan. 66502.*

POMEY, Jean Baptiste, French engr.; b. Paris, France, 1861; dir. Ecole supérieure des télégraphes, 1923. Studied capacity of 2 wire circuit, oscillation of distbn. network, 1902; gave representation of Thévenin's theorem and solution of applicable equation with complex numbers, 1911. Died 1943.

POMMERENKE, Christian Max Willi, mathematician; b. Copenhagen, Denmark, Dec. 17, 1933; s. von Haven-Theilade and Anne Marie Pommerenke; Diplom-Mathematiker, U. Göttingen (Germany), 1957, Dr.rer.nat., 1959; m. Renate H.M. Lippold, Mar. 1, 1960; children—Martin, David, Philipp. Asst., Math. Inst., Göttingen, 1958-61; asst. prof. U. Mich., Ann Arbor, 1961-62; research fellow Harvard, 1962-63; faculty U. Göttingen, 1963-65, privatdozent, 1963-65; reader Imperial Coll., London, Eng., 1966—. Mem. London Math. Soc. Research, publs. on complex function theory especially theory of unequivalent functions. Home: 41 Burton Lodge, Portinscale Rd., London S.W.15, Eng.*

POMPEIANO, Ottavio, Italian physiologist; b. Faenza, Ravenna, Italy, Sept. 29, 1927; s. Antonio and Maria (Padula) P.; M.D., U. Bologna (Italy), 1950; m. Stefi Monica Möller, Dec. 13, 1940; children—Maria Patrizia, Maria Christina, Lucia. Asst. prof. physiology U. Bologna, 1953-58; faculty U. Pisa (Italy), 1959—, prof. physiology Med. Faculty, 1967—; acting prof. physiology Med. Faculty, Cath. U. Rome (Italy), 1962-63; fellow Rockefeller Found., Anat. Inst., U. Oslo (Norway), 1956-57; research staff Nobel Inst. Neurophysiology, Karolinska Inst., Stockholm, Sweden, 1958-59; vis. prof. dept. physiology U. Göteborg (Sweden), 1964. Contbg. author: The Vestibular Nuclei and their Connections, Anatomy and Functional Correlations, 1962. Research, numerous publs. on physiology of cerebellum and vestibular nuclei, sleep mechanisms. Address: 31 Via S. Zeno, Pisa, Italy.*

POMPEIU, Dimitrie, Rumanian mathematician; b. 1873; univ. prof.; mem. Acad. Rumanian People's Republic. Author: On the Continuity of Functions with One Variable Complex, 1905. Research in math. analysis, theory of functions of complex variable, rational mechanics; introduced math. concept of areolar derivate and of distance between 2 crowds; constructed real, non-constant functions (Pompeiu functions) with derivates annulled at any interval. Died 1954.

POMPONIUS, see Mela, Pomponius.

POMPOWSKI, Tadeusz, Polish chemist; b. Lwow, Poland, Aug. 29, 1910; s. Edmund and Franciszka (Rusin) P.; grad. Politechinika, Lwow, 1937; dr.ing. tech. scis., Politechnika, Gdansk, Poland, 1948, prof. extraordinary, 1962; m. Helena Nowicka, Feb. 12, 1952; 1 dau., Lofia. Asst., Politechnika, Gwow, 1932-41; faculty Politechnika, Gdansk, 1945—, now chief chair chem. inorganic tech. and tech. analysis. Recipient cross Polinia Restituta; Medal of 10 years PRL; Medal 1000 years Polish State; Gold distinction of Supreme Tech. Orgn. Mem. Polish Acad. Scis. (commn. analytic chemistry), Sci. and Tech. Commn. Gdansk, Polish Chem. Assn., Sci. Assn. Gdansk, Polish Assn. Engrs. and Technicians. Research, publs., manuals in sphere of analytic chemistry and inorganic technology. Home: 25, Mickiewicsa, Sopot, Poland. Office: Majakowskiego, Gdansk, Poland.*

POMRANING, Gerald Carlton, Am. physicist; b. Oshkosh, Wis., Feb. 23, 1936; s. Carlton Chester and Lorraine (Volkman) P.; B.S. in Chem. Engring., U. Wis., 1957; certificate in chem. engring. Technische Hogeschool, Delft, Holland, 1958; Ph.D. in Nuclear Engring., Mass. Inst., Tech., 1962; m. eGayl Ann Burkitt, May 27, 1961; 1 dau., Linda Marie. Mgr. theoretical physics group Vallecitos Atomic Lab., Gen. Electric Co., Pleasanton, Cal., 1962-64; staff mem. theoretical physics dept. John Jay Hopkins Lab. for Pure and Applied Sci., Gen. Atomic Co., San Diego, 1964—. Vis. scientist Brookhaven Nat. Lab., 1966. Mem. Am. Nuclear Soc. (Mark Mills award 1963), Am. Phys. Soc., Soc. for Indsl. and Applied Math., A.A.A.S., Am. Math. Assn. Co-editor: Reactor Physics in the Resonance and Thermal Regions, vol. I, II, 1966. Research, publs. on reactor physics, transport theory, applied math., methods of approximation, variational techniques, deviser new calculational schemes for phys. problems, neutral particle transport and devel. new math. methods for solving problems. Home: 5916 Erlanger St., San Diego 92122. Office: Gen. Atomic Co., P.O. Box 608, San Diego 92112.*

PONCELET, Jean Victor, French mathematician, b. Metz, France, July 1, 1788; ed. lyceum at Metz, Ecole Polytechnique (under Monge), 1808-10, Ecole de l'Application, Metz, 1810-12. Became army engineer, served as lt. of engineers in French army during Napoleonic wars, 1812; captured during retreat from Russia, spent 2 yrs. in prison, 1813-14; military engineer, Metz, 1815-25; prof. mechanics, Ecole de l'Application, Metz, 1825-35; organized 1st courses in applied mechanics, U. Paris, 1838-48; commandant, Ecole Polytechnique, 1848-50; Mem. French Acad. Scis., 1834. Author: Traité des propriétés projectives des figures,

1822 (first systematic treatment of projective geometry). Laid foundations for modern projective geometry; formulated principle of continuity; invented new type of undershot water wheel, 1827; pub. 1st practicable theory of turbines, 1838. Died, Paris, Dec. 23, 1867.

PONCIN, Henri, French physicist; b. Poitiers, France, Oct. 1, 1904; s. Jean and Marie (Girardin) P.; ed. Lycée de Poitiers; Dr ès sc; m. Jeanne Fady, Aug. 21, 1935; children—Jean, Jacques, Daniel, Odile. Agrégé at Poly., later école normale supérieure; scholar Thiers Found.; research head tech. aero. service dept. Air Ministry; prof. mechanics Faculty of Sci. Poitiers; prof. mechanics Faculty of Sci., U. Paris (France), 1948—; dir. Nat. Sch. Mechanics and Aero., 1946-62; dir. Inter-Univ. Center Aerodynamic and Thermic Studies Poitiers-Biard, 1962—; prof. Nat. Inst. Sci. and Nuclear Tech., Sacley. Recipient Aero. medal; decorated Legion of Honor. Mem. French Mechanics Soc., French Thermics Soc. Research, publns. on cavitations in hydrodynamics; transfer of mass and heat; forced convection; treatises on gen. and energetic mechanics. Home: 7, rue de Lorraine, Cachan (Seine). Office: 61, av. du President-Wilson, Cachan (Seine), France.

POND, John, English astronomer; b. London, 1767; ed. Trinity Coll., Cambridge; m.; settled in Westbury, Eng., 1798, built an altazimuth; apptd. astronomer royal, 1811. Recipient Lalande prize French Acad. Scis., 1817. Fellow Royal Soc., 1807 (Copley medal 1823); mem. Astron. Soc. Author: Greenwich Observations, also papers. Detected error in Greenwich observations, circa 1782; revealed that the mutual quadrant at Greenwich had suffered deformations through age, 1806; substituted mercury-horizon for plumb-line and spirit-level, 1821; introduced system of observing objects alternately by direct reflection vision, 1825; catalogued 1113 stars with great accuracy, 1833. Died Greenwich, Eng., Sept. 7, 1836.

PONFICK, Clemens Emil, German pathologist; b. Frankfort/Main, Germany, Nov. 3, 1844; prof., Rostock, Poland, Göttingen, Germany, Breslau (now Cracow, Poland); discovered actinomycosis bacillus. Died Breslau, Nov. 4, 1913.

PONI, Petru, Rumanian chemist, mineralogist; b. 1841; mem. Rumanian Acad. Author: Recherches sur la composition chimique des pétroles roumains, 1902; also 1st textbook of chemistry in Rumanian lang., 1st monograph on Rumanian ores. Studied minerals and ores in Rumanian Carpathians; discovered brostenita, badenita. Died 1925.

PONNAMPERUMA, Cyril, chemist; b. Galle, Ceylon, Oct. 16, 1923; s. Andrew and Grace (Siriwardne) P.; B.A. in Philosophy, U. Madras (India), 1948; B.Sc., in Chemistry, Birbeck Coll., U. London (Eng.), 1959; Ph.D., U. Cal. at Berkeley, 1962; m. Valli Pal, Mar. 19, 1955; 1 dau., Roshini Manel. Came to U.S., 1959, naturalized, 1966. Sci. tchr. St. Aloysius' Coll., Galle, 1948-51; research chemist B.X. Plastics Ltd., also Bunzy & Biach, London, 1952-54; radiochemist radioisotopes lab. London Hosp., 1955-59; research asso. Lawrence Radiation Lab., Berkeley, Cal., 1960-61; Nat. Acad. Scis.-NRC fellow NASA-Ames Research Center, Moffett Field, Cal., 1962-63, research scientist exobiology div., 1963-64, chief chem. volution br., 1964—. Recipient award Ravenscroft Exhbn., 1956; Salter's Inst. award in chemistry, 1957; award Coll. Exhbn., U. London, 1958; Sustained Superior Performance award NASA, 1964. Asso. fellow Chem. Soc. (London), Royal Inst. Chemistry (London), mem. Am. Chem. Soc., Radiation Research Soc., Geochem. Soc., A.A.A.S., Photobiology Group. Research, publs. on chem. origin life, synthesis components of nucleic acids, ATP. Home: 3410 Janice Way, Palo Alto, Cal. 94303.* Office; Exobiology Div., NASA-Ames Research Center, Moffett Field, Cal. 94035.*

PONS, Jean Louis, French astronomer; b. Peyres, France, Dec. 24, 1761; mem. staff Obs. at Marseilles, 1789-1819; asst. dir. 1813-19; became dir. Lucca (Italy) Obs., 1819, Florence (Italy) Obs., 1825. Discovered 37 comets (several named after him), 1801-27; discovered comet with shortest period known, 1818. Died Florence, Oct. 14, 1831.

PONSETI, Ignacio Vives, orthopaedic surgeon; b. Ciudadela, Balearic Islands, Spain, June 3, 1914; s. Miguel and Margarita (Vives) P.; B.S., U. Barcelona, 1930, M.D., 1936; m. Mary Bell Newman, Sept. 19, 1943; 1 son, William Edward; married 2d, Helen Percas, June 21, 1961. Became naturalized U. S. citizen, 1948. Instr. dept. orthopaedic surgery State U. Ia., 1944-57, prof., 1957—. Served as capt., M.C., Spanish Army, 1936-39. Recipient Kappa Delta award for orthopaedic research, 1955. Mem. Assn. Bone and Joint Surgeons, Am. Acad. Cerebral Palsy, Soc. Exptl. Biology and Medicine, Internat. Coll. Surgeons, N.Y Acad. Sci., A.M.A., Am. Acad. Orthopedic Surgeons, A.C.S., Am. Orthopedic Assn., Ia. Med. Soc., Orthopedic Research Soc., Chgo. Orthopedic Soc., Midwest Orthopedic Club. La Sociedad Mexicana de Ortopedia, Sigma Xi; hon. mem. Asociacion Argentina de Cirugia, Sociedad de Cirujanos de Chile. Author papers on congenital and developmental skeletal deformities; experimental production of aortic aneurysm, scoliosis, and slipped epiphysis. Home: 315 Ellis Av. Office: Children's Hospital, Iowa City, Ia.

PONSIOEN, Johannes Antonius, Dutch sociologist; b. Schoonhoven, Netherlands, June 23, 1911; s. Cornelis and Geerdina (Wempe) P.; candidate, Catholic Sch. Econs., 1948, pre-doctoral, 1950, doctor, 1952. Ordained priest Roman Catholic Ch., 1936; port chaplain, Rotterdam, 1937-40; prof. social ethics Sem. Liesbosch, Breda, Netherlands, 1941-58; prof. sociology Inst. Social Studies, The Hague, Netherlands, 1954—, dep. rector, 1966—. Decorated Officer Oranje Nassau. Mem. Netherlands Sociol. Assn. (pres. 1964—), Assn. Cath. Intellectuals (pres. social sci. div. 1963—), Soc. Internat. Devel., Demographic Assn. Author: Symbols in Social Life, 1952; Philosophy of Social Life, 1952; (with G. Veldkamp) Today's Social Problems; Social Welfare Policy, I, Contributions to Theory, 1962, II Contributions to Methodology, 1963; The Analysis of Social Change Reconsidered, 1962; also articles. Editor: (with G. Veldkamp, W.K.N. Schmelzer) Welfare, Well-being and Happiness, 1958-63, I. Sociological Analysis, II. Social Reconstruction, III. Economic Reconstruction, IV. Cultural Policies, V. International Relations. Studies in social structure and policies in changing socs. especially in developing countries. Home: Treublaan 2, the Hague, Netherlands.*

PONTECORVO, Bruno Maksimovich, physicist; b. Italy, Aug. 22, 1913; grad. Pisa U., 1930, Rome U., 1933. Instr., Rome U., 1933-36; with sci. instns. in France, 1936-40; expert radiographic prospecting of oil deposits in U.S., 1940-43; with E. Fermi's team, Chalk River, Ont., Can., 1943-48; asso. Harwell Lab., Eng., 1948-50; in charge team Joint Nuclear Research Inst. Decorated Order of Lenin; recipient Lenin prize, 1963. Mem. USSR Acad. Scis. Author: Artificial Radioactivity Produced by Neutron Bombardment, 1935; Isomérie nucléaire produite par les rayons x du spectre continu, 1939; The Birth Processes of Heavy Mesons and Particles, 1955; Weak Reactions of Elementary Particles and Neutrinos, 1963; co-author: Hydrogen-Scattering of Pi-Mesons, parts 1-2, 1956. Research on neutron physics, 1943-48, birth of pi^{0}-mesons from neutrons, after 1950, interaction of pi-mesons with nucleons. Address: USSR Acad. Scis., Leninsky prospect 14, Moscow, USSR.

PONTECORVO, Guido P. A., geneticist; b. Pisa, Italy, Nov. 29, 1907; s. Massimo and Maria (Maroni) P.; D.Agrl. Sci., U. Pisa, 1928; Ph.D., U. Edinburgh (Scotland), 1941; D.Sc., Leicester U., 1968; m. Leonore Freyenmuth, Sept. 5, 1939; 1 dau., Lisa S. Staff, Ispettorato, Agrario Compart., Florence, Italy, 1930-38; research scholar Inst. Animal Genetics, Edinburgh, 1938-40; research scholar dept. zoology U. Glasgow (Scotland), 1941-43, faculty, 1945-68, prof. genetics, 1955-68; sci. staff Imperial Cancer Research Fund, 1968—. Lectr., Inst. Animal Genetics, Edinburgh, 1943-45; Jessup lectr. Columbia, 1956; Messenger lectr. Cornell U., 1958; vis. prof. Albert Einstein Coll. Medicine, N.Y.C., 1964-65; vis. lectr. Wash. State U., 1967. Recipient Christian Hansen prize for microbiology,1961. Fellow Royal Soc. Edinburgh, Royal Soc., 1955; mem. Am. Acad. Arts and Sci. (fgn. hon.), Danish Royal Acad. Sci. and Lit. (fgn.), Genetical Soc. (past pres.). Author: Trends in Genetic Analysis, 1958; also numerous articles. Developed idea that a gene is made up of several mutational sites separable by recombination but unitary in action, 1952; discovered processes of gene recombination (parasexual) other than sexual reprodn., 1954. Home: Cranfield House, 97 Southampton Row, London W.C.1. Office: Imperial Cancer Research Fund, Lincoln's Inn Fields, London W.C.2, Eng.*

PONTEDERA, Giulio, Italian botanist, classicist; b. Vicenza, May 7, 1688; s. Antonio and Lucia (Zenonate) P.; student college in Vicenzo with P. Somachi; entered U. Padua (studied Medicine and Philosophy), degree 1715. Prof. botany at Padua, dir. bot. garden, 1719-57; scholar in Greek and Latin, wrote treatises on various classical authors. Author: Compendium Tabularum botanicarum, 1718; Anthologia, sive De Floris Natura, 1720. Identified over 300 previously unresearched species of plants of Cisalpine region, organized these and others into scheme of classification developed by Tournefort; added much knowledge on flower morphology, explaining uses of parts of flowers in its reproductive process. Died Lonigo, Italy, Sept. 3, 1757.

PONTIERI, Giuseppe Mario, Italian pathologist; b. Nocera Terinese, Italy, Sept. 3, 1927; s. Ernesto and Maria (Mercurio) P.; M.D., U. Naples (Italy), 1950; m. Franca Nobili, Feb. 29, 1960; children—Francesco Ernesto, Maria Vittoria. Asst., Inst. Microbiology, 1951; asst. Inst. Gen. Pathology, U. Naples (Italy), 1952-63, libero docente in gen. pathology, 1958-61, in microbiology, 1961-64; prof. gen. pathology U. Palermo (Italy), 1964—, dir. Inst. Gen. Pathology, 1964—. Research, numerous publs. on ultrastructure of bacterial cells, immunology especially functions of complement and properdin in several path. conditions. Home: 74/B Prinicpe di Paterno, Palermo, Sicily, Italy.*

PONTONNIER, G., French physician; b. Narbonne, France, Nov. 23, 1932; s. Andre and Juliette (Lac) P.; M.D., U. Toulouse, 1962; m. Monique Nicoleau, June 31, 1956; children—Sylvie, Florence. Chief clin. medicine U. Toulouse, 1962—, also tchr. obstetrics and gynecology, chief research lab. dept. obstetrics and gynecology. Recipient award French Acad. Medicine, 1963. Mem. French Soc. Obstetrics and Gynecology, French Surgery Soc., Internat. Nephrologics Soc., Acad. Med. Scis. Spain. Author: Les Syndromes Varculo renaux de la grossesse, 1962; Les lesions renales de la toxemie, 1965; also numerous articles. Described a renal lesion of toxemia of pregnancy, method for measure of uterine blood flow in pregnant women. Home: 1 Pivoines, Toulouse, France.*

PONTRYAGIN, Lev Semenovich, Russian mathematician; b. Moscow, USSR, Sept. 3, 1908; grad. Moscow U., 1929. Prof., Moscow U., 1935—. Decorated Order of Lenin; recipient Stalin prize, 1941, Lenin prize, 1962. Mem. USSR Acad. Scis. Author: The General Topological Theorem of Duality for Closed Sets, 1934; Continuous Groups, 1938, 54; Principles of Combinatorial Topology, 1947; Characteristic Cycles of Differentiated Manifolds, 1947; Manifold Matrix Fields, 1949. Research and publs. on topology, theory of continuous groups; discovered gen. law of duality and formulated gen. theory of characteristics of commutative groups, 1932. Home: B. Kaluzhskaya 13. Office: Moscow University, Leninskie gory, Moscow, USSR.

PONZ, Francisco, Spanish biologist; b. Huesca, Spain, Oct. 3, 1919; s. Mariano and Francisca (Piedrafita) P.; Lic.Sci., U. Madrid (Spain), 1941, D.Sc., 1942. Sci. collaborator Instituto Cajal, Madrid, 1941-44; scholar biochemistry, Zurich and Freiburg, Switzerland, 1942-43; prof. animal physiology U. Barcelona (Spain), 1944-66; prof. animal physiology U. Navarra (Spain), 1966—; head animal physiology sect. C.S.I.C., 1945-62, head dept. physiology and biochemistry, 1962—; rector U. Navarra, 1966—. Mem. Spanish Higher Council Sci. Research, Royal Acad. Scis. and Arts Barcelona, Spanish Soc. Natural History, Spanish Soc. Physiol. Scis., Spanish Soc. Biochemistry. Reviewer, Espanola de Fisiologia, 1945—. Research, publs. on intestinal absorption of sugars, active transport by membranes, metabolism of intestine, action of radiation on physiology and biochemistry of intestine, enzymes. Office: Faculty Scis., U. Navarra, Navarra, Spain.*

POOL, Eugene Hillhouse, Am. surgeon; b. N.Y.C., June 3, 1874; s. John Hillhouse and Sophia (Boggs) P.; A.B. Harvard, 1895; M.D., Coll. Phys. and Surg. (Columbia), 1899; m. Esther Phillips Hoppin, Apr. 29, 1904; children— James Lawrence, Beekman; m. 2d, Kitty Lanier Lawrance, June 10, 1932; m. 3d, Frances Saltonstall, Dec. 12, 1940. Asst. demonstrator anatomy Coll. Phys. and Surg., 1901-04, instr. surgery, 1904-12, asso., 1912-15, prof. clin. surgery, 1915-38; clin. prof. surgery Cornell U. Med. Coll.; sr. attending surgeon N.Y. Hosp., surgeon in chief Ruptured and Crippled Hosp.; cons. surgeon to Presbyn., French, Harlem and Woman's hosps., N.Y. Infirmary for Women & Children, N. Country Community Hosp., Glen Cove, Elizabeth A. Horton Meml. Hosp., Middletown, N.Y., Monmouth Meml. Hosp., Long Branch, N.J., United Hosp., Portchester, N.Y., N.Y. Eye & Ear Infirmary, Central Islip State Hosp., Berwind Free Outdoor Maternity Clinic. Fellow A.C.S.; mem. A.M.A., Med. Soc. State of N.Y., Am., Internat. surg. assns., New York Surg. Soc. Died Apr. 9, 1949.

POOL, Ithiel de Sola, Am. polit. scientist; b. N.Y.C., Oct. 26, 1917; s. David de Sola and Tamar (Hirshenson) P.; A.B., U. Chgo., 1938, A.M., 1939, Ph.D., 1952; m. Jean MacKenzie, Mar. 5, 1956; children—Jonathan, Jeremy, Adam. Faculty, Hobart Coll., Wm. Smith Coll., 1942-49; asst. dir. research Internat. Studies Program Hoover Inst., Stanford, 1949-50, acting dir., 1950-53; faculty Mass. Inst. Tech., Cambridge, 1953—, prof., 1959—, chmn. polit. sci. dept., 1966—. Cons. Rand Corp., 1951—, Office Dir. Def. Research and Engring., 1959, Office of Sci. and Tech., 1965—; mem. research bd. Simulmatics Corp. Fellow Am. Acad. Arts and Scis.; mem. Am. Polit. Sci. Assn. (Woodrow Wilson award 1963), A.A.A.S., Am. Social. Assn. Author: (with others) The People Look at Educational Television, 1963; American Business and Public Policy, 1964; Candidates, Issue and Strategies, 1964. Research on techniques to study projection of pub. opinion by attitude simulation on a computer. Home: 105 Irving St., Cambridge, Mass. 02138.*

POOL, James Lawrence, Am. neurosurgeon; b. N.Y.C., Aug. 23, 1906; s. Eugene H. and Esther (Hoppin) P.; A.B., Harvard, 1928; M.D., Columbia, 1932, D.Sc. in Medicine, 1941; m. Angeline James, June 14, 1940; children—J. Lawrence, Eugene H., Daniel S. Research, Boston City Hosp., 1933--34; house surgeon Presbyn. Hosp., N.Y.C., 1934-36, resident, 1936-40, research, 1940-42, dir. Service Neurol. Surgery, 1949—; practice medicine, specializing in neurosurgery, N.Y.C.; prof., chmn. dept. neurol. surgery Columbia Coll. Phys. and Surg. Diplomate Am. Bd. Neurol. Surgery. Fellow A.C.S., mem. Harvey Cushing Soc., Am. Acad. Neurol. Surgery, others. Author: Paraplegia, 1951; Acoustic Neurinomas, 1956; Aneurysms of the Brain, 1965; also articles. Research, publs. on chemotherapy for malignant brain tumors; lobotomy for mental illness; cerebrovascular disorders, especially intracranial aneurysms; dynamics of cerebral vasospasm; etiology and therapy of tremor diseases. Home: Closter Dock Rd., Alpine, N.J. Office: 710 W. 168th St., N.Y.C. 10032.

POOL, Judith Graham, Am. physiologist; b. N.Y.C., June 1, 1919; d. Leon M. and Nellie (Baron) Graham; B.S., U. Chgo., 1939, Ph.D., 1946; m. Ithiel de Sola Pool, Mar. 24, 1938 (div.); children—Jonathan, Jeremy, Lorna. Instr., Hobart Coll., Geneva, N.Y., 1943-45; physiologist Toxicity Lab., U. Chgo., 1946; research asso. Stanford Research Inst., Menlo Park, Cal., 1950-53; postdoctral research fellow Stanford Sch. Medicine, Palo Alto, Cal., 1953-56, research asso. dept. medicine, 1957-60, sr. research asso., 1960—. Fulbright research scholar, Norway, 1958-59. Mem. Am. Physiol. Soc., N.Y. Acad. Sci., A.A.A.S., Western Soc. Clin. Research, Phi Beta Kappa, Sigma Xi. Research, publs. on synthesis of coagulation factors by tissue slices; invented capillary microelectrode for measurement of membrane potentials in single cells; developed quantitative assay for plasma antihemophilic globulin, simple method for concentrating antihemophilic globulin in blood banks. Home: 996 Altschul Av., Menlo Park, Cal. 94025.*

POOL, Raymond John, Am. botanist; b. Wabash, Neb., Apr. 23, 1882; s. William H. and Mary L. (Burrows) P.; A.B., U. Neb., 1907, A.M., 1908, Ph.D., 1913; postgrad. U. Chgo., 1908; m. Martha M. Stangland, June 30, 1909. Instr. botany, 1907, adj. prof., 1908-10, asst. prof. and curator Univ. Herbarium, 1910-11, asso. prof. and curator, 1911, prof., 1914, head dept. bot., 1915-48, emeritus U. Neb. Asst. prof., U. Mich., summer, 1910; vis. prof. So. Ill. U., 1955-56; investigator relations of indsl. operations to vegetation since 1917. Fellow A.A.A.S. (v.p., chmn. Sect. G, 1939), Bot. Soc. Am. (pres. syst. sect. 1925); asso. Am. Mus. Natural History; mem. Torrey Bot. Club, Ecol. Soc. Am., Neb. Acad. Sci. (pres. 1916), Am. Forestry Assn., Am. Phytopath. Soc., Am. Micros. Soc. (v.p. 1932; pres. 1941), Soc. Am. Foresters, Soc. Am. Naturalists, Am. Soc. Plant Taxonomists, Am. Inter-profl. Inst. (pres. Lincoln Chpt. 1940), Phi Beta Kappa (past pres. Neb. chpt.), Sigma Xi (past pres. Neb. chpt), Chi Phi, Phi Sigma. Author: Vegetation of Nebraska Sand Dunes; Experiments in Plant Physiology; Handbook of Nebraska Trees; First Course in Botany (with Dr. A. T. Evans); Flowers and Flowering Plants; Basic Botany for Colleges; Marching With the Grasses; also numerous articles on various phases of botany in bot. and other jours. Died Feb. 1, 1967.

POOLE, Sidman Parmelee, Am. geographer; b. Syracuse, N.Y., Oct. 19, 1893; s. Theodore Lewis and Carrie (Law) P.; B.S., Syracuse Univ., 1921, M.S., 1925; Ph.D., U. Chgo., 1932; postgrad. Cambridge (Eng.) U., 1925; m. Rachel Sumner, Aug. 31, 1922. Instr. Syracuse U., 1921-25, asst. prof. geography, 1925-32, asso. prof., 1932-39, prof., 1939-40; summer lectr. Cornell, 1932; prof., chmn. dept. geography U. Va., since 1946; dir. Va. Geog. Inst., from 1947. Geographer to Syracuse Andean Expdn., 1930-31, Syracuse Gaspe Expdn., 1933, dir. and geographer to Syracuse Yucatan Expdn., 1937-38; detailed field work in N.Y. State, Vt. (with Vt. geol. survey., upper Great Lakes region, Chgo. area, Eng., Brittany, Venezuela, Gaspe and Yucatan. Geographic adviser Air Command and Staff Sch., Air Univ., Maxwell Field, Ala. since 1946; mem. U.S. Bd. on Geog. Names, 1943-46. Fellow Am., Royal geog. socs.; mem. Nat. Council Geography Tchrs., Assn. Am. Geographers, Am. Soc. for Profl. Geographers (v.p. 1947), Phi Beta Kappa, Sigma Xi. Author: Manual for College Geography, 1933; chapter on Geography of Central New York (An Inland Empire by W. Freeman Galpin), 1941; chapter on Geography in America's Life (Twentieth Century America), 1947-51; History of Virginia (junior author). Contbg. editor: Econ. Geography. Contbr. articles to geog. publs., mags. and newspapers. Consultant editor Bobbs-Merrill Co. series of geog. texts and readers 1945—. Died Oct. 28, 1955.

POOR, Charles Lane, Am. astronomer; b. Hackensack, N.J., Jan. 18, 1866; s. Edward Eri and Mary Wellington (Lane) P.; B.S., Coll. City N.Y., 1886, M.S., 1890; Ph.D., Johns Hopkins, 1892; m. Anna Louise Easton, Apr. 19, 1892; children—Charles Lane, Alfred Easton, Edmund Ward. Asso. in astronomy Johns Hopkins, 1892-96, asso. prof., 1895-99; prof. astronomy Columbia, 1903-10, prof. celestial mechanics, 1910-44, prof. emeritus, 1944—. Fellow Am. Acad. Arts and Scis. (asso.), Royal Astron. Soc. Author: The Solar System, 1908; Simplified Navigation, 1918; Gravitation versus Relativity, 1922; Relativity and the Motion of Mercury, 1925; The Relativity Deflection of Light, 1926; Rules and Regulations for the Construction of Racing Yachts, 1928; What Einstein Really Did, 1930; Men Against the Rule, 1937. Inventor various navigational devices. Died N.Y.C., Sept. 27, 1951.

POP, Emil, Rumanian botanist; b. 1897; prof. Cluj (Rumania) U.; mem. Acad. Socialist Republic Rumania. Author: Cytophysiological Researches on the Movements of Protoplasma and the Relation Between the State of Plasma and Plant Resistance to Frost; Contributions to the History of Rumanian Botany; Peat-Bogs in Rumania, 1960; Protoplasmatic Currents in Labiatae, 1960; (with N. Salageanu) Natural Reserves in Rumania, 1965. Research on vegetal physiology, cytophysiology, history of Rumanian botany.

POPE, Clifford Hillhouse, Am. herpetologist; b. Washington, Ga., Apr. 11, 1899; s. Mark Cooper and Harriet (Hull) P.; student U. Ga., 1916-18; B.S., U. Va., 1921; m. Sarah Haydock Davis, Sept. 8, 1928; children—Alexander Hillhouse, Hallowell, Whitney. Explorer in China, Am. Mus. Natural History, 1921-26, asst. curator herpetology, N.Y.C., 1928-34; curator amphibians and reptiles Chgo. Natural History Mus., 1941-53; free lance writer, 1953—. Fellow N.Y. Zool. Soc.; mem. Am. Soc. Ichthyologists and Herpetologists (pres. 1935-36), Chi Phi. Author: The Reptiles of China, 1935; Snakes Alive and How They Live, 1937; Turtles of the United States and Canada, 1939; China's Animal Frontier, 1940; Amphibians and Reptiles of the Chicago Area, 1944; The Reptile World, 1955; Reptiles Round the World, 1957; The Giant Snakes, 1961; also articles on reptiles and amphibians of China; ecology, distbn., evolution of salamanders of So. Appalachians. Address: Route 3, Box 807A, Escondido, Cal. 92025.*

POPE, Franklin Leonard, Am. electrician, inventor; b. Great Barrington, Mass., Dec. 2, 1840; s. Ebenezer and Electa (Wainwright) P.; ed. Amherst Acad.; m. Sarah Amelia Dickinson, Aug. 6, 1873; 5 children. Edited and pub. small newspaper, Great Barrington; operator Am. Telegraph Co., Great Barrington, 1857-59, asst. engr., serving as circuit mgr. Boston & Albany R.R. telegraph lines, Springfield, Mass., 1862; reestablished communication between N.Y.C. and Boston during draft riots, 1863; asst. to chief engr. Russo-Am. Telegraph Co., 1864-66, made preliminary exploration and survey of B.C. and Alaska; N.Y. editor The Telegrapher, 1867-68; made valuable improvements in stock ticker, 1869; partner firm Pope, Edison & Co.,, 1869-70; devised system which made practicable automatic electric block signel for rys. (invented by Thomas S. Hall); in charge all patent interests of Western Union Telegraph Co., 1875-81; editor Electrician and Elec. Engineer and Engring. Mag., 1884-93; cons. engr. Great Barrington Electric Light Co., converted plant from steam to water power, 1893-95; charter mem. Am. Inst. E.E., an original v.p., pres., 1886-87. Author: Modern Practice of the Electric Telegraph, 1869; The Telegraphic Instructor, 1871; Life and Work of Joseph Henry, 1879; Evolution of the Electric Incandescent Lamp, 1889, 94. Died Great Barrington, Oct. 13, 1895.

POPE, Sir William Jackson, English chemist; b. London, Mar. 31, 1870; s. William and Alice (Hall) P.; ed. Finsbury Tech. Coll., Central Instn. Asst. to H.E. Armstrong, also to F.S. Kipping, Goldsmith's Inst., New Cross, 1897-1901; prof. chemistry Municipal Sch., Faculty Tech., U. Manchester, 1901-08; prof. chemistry U. Cambridge, also professorial fellow Syndey Sussex Coll., 1908-39. Recipient Davy medal, 1914. Fellow Royal Soc., 1902; mem. Chem. Soc. (pres. 1917-19), Soc. Chem. Industry (pres. 1920-21), Union internationale de Chimie (pres. 1922-25), French Acad. Scis. Developed (with William Barlow) valency-volume theory, 1906; changed the notion of stereoisomerism from its original narrow application to a broad and general concept; developed methods for producing mustard gas in quantity. Died Cambridge, Oct. 17, 1939.

POPENOE, Paul, Am. social biologist; b. Topeka, Kan., Oct. 16, 1888; s. Fred Oliver and Marion Amanda (Bowman) P.; student Occidental Coll., Los Angeles, Calif., 1905-07, Stanford U., 1907-08; hon. D.Sc., Occidental, 1929; m. Betty Lee Stankowitch, Aug. 23, 1920; children—Paul, Oliver, John, David. Newspaper work, Pasadena and Los Angeles, 1908-11; agrl. explorer, 1911-13; editor Jour. of Heredity, 1913-17; exec. sec. Am. Social Hygiene Assn., 1919-20; date growing, 1912-35. Lecturer in biology, U. of Southern Calif., 1933-47. Asst. to chief of med. sect. Council Nat. Defense, 1917. Sec. The Human Betterment Foundation, 1929-37; dir. Am. Inst. Family Relations, 1930-60, pres., 1960-63, 64—, adminstr., 1963—, also founder. Mem. Am. Genetic Assn., Soc. for Study of Evolution, Soc. Mex. de Eugenesia (hon.), Soc. Argentina de Eugenesia (hon.), A.A.A.S., Population Assn. of America. Author: Date Growing in the New and Old Worlds, 1913; Applied Eugenics (with Roswell H. Johnson), 1918, revised edit. 1933; Modern Marriage, 1925, revised edit., 1940; The Conservation of the Family, 1926; Problems of Human Reproduction, 1926; The Child's Heredity, 1929; Sterilization for Human Betterment (with E. S. Gosney), 1929; Practical Applications of Heredity, 1930; Marriage, Before and After, 1943; Sex, Love, and Marriage, 1963; (with Evelyn Duvall and David Mace) The Church Looks at Family Life, 1964. As an agrl. explorer brought 16,000 date palms to U.S.; pioneer study of results of surg. sterilizations performed on 6,000 persons in Cal. state instns. for mental defect, mental disease; adapted techniques of biol., psychol., sociol. counseling to provide integrated, effective technique for marriage counseling. Home: 2503 N. Marengo Av., Altadena, Cal. Office: 5287 Sunset Blvd., Los Angeles 27.*

POPENOE, Willis Parkison, Am. paleontologist; b. Topeka, Sept. 5, 1897; s. Edwin Alonzo and Carrie (Holcomb) P.; B.S., George Washington U., 1930; M.S., Cal. Inst. Tech., 1933, Ph.D., 1936; m. Kathryn Malay, Dec. 22, 1941; 1 son, Willis Parkison, IV. Curator invertebrate paleontology, lectr. Cal. Inst. Tech., 1930-45; faculty U. Cal. at Los Angeles, 1945—, prof. geology, 1958-65, prof. emeritus, 1965—; also geologist for various industries and U.S. Geol. Survey. Fellow Paleontol. Soc. Am., Soc. Systematic Zoology; mem. Geol. Soc. Am., Sigma Xi. Research, publs. contbg. to increased detailed knowledge of types of fossil invertebrate animals found in Cal. in later Cretaceous rocks; devised inferred course of evolution of a number of fossil stocks in the Cretaceous. Home: 15154 Clark St., Van Nuys Cal. 91401. Office: Dept. Geology, U. Cal., Los Angeles, 90024.

POPENOE, (Frederick) Wilson, Am. horticulturist, explorer; b. Topeka, Kan., Mar. 9, 1892; s. Fred O. and Marion (Bowman) P.; ed. Pomona Coll.; Sc.D., U. of San Marcos, Lima, Peru, 1925, Pomona College, 1947, University of Florida, 1950; m. Dorothy Hughes, Nov. 17, 1923 (dec. Dec. 1932); m. 2d, Helen Barsaloux, Jan. 10, 1939 (dec. Mar. 1961). With U.S. Dept. of Agr., 1913-25; conducted explorations in Central & S.Am. to obtain useful plants worthy of introduction into U.S. cultivation. With United Fruit Co., investigating cultural problems of bananas & other tropical crops throughout the Caribbean region, 1925—, founded Escuela Agricola Panamericana Tegucigalpa, Honduras, 1942, dir. 1942-57, now dir. emeritus; hon. prof., Univ. de San Carlos de Guatemala. Decorated Orden al Merito, Chile, Ecuador; George Robert White Medal Mass. Hort. Soc., 1950, Wilder medal Am. Pomological Soc.; Medalla Agricola Interamericana; Orden del Vasco Nunez de Balboa, Panama; also the Orden al Merito Agricola e Industrial, Cuba; Comendador, Orden al Mérito Agricola, Ecuador; Orden de Ruben Dario, Nicaragua; Orden de Morazan, Honduras; Orden del Quetzal, Guatemala. Hon. mem. Cal. Avocado Soc., Am. Soc. Foresters, Sociedad Geográfica de Lima (Peru); corr. mem. Soc. Venezolana de Ciencias Naturales; mem. Washington Acad. Scis. Author: Manual of Tropical and Subtropical Fruits, 1920; also many papers on avocados and other tropical and subtropical fruits. Authority on tropical and subtropical horticulture and pomology. Home: Gainesville, Fla. Address: Antigua, Guatemala, C. A.*

POPESCU, Georgeta, Rumanian virologist; b. Bucharest, Rumania, Feb. 18, 1925; d. Mihail and Elena (Niculescu) Danescu; grad. faculty Medicine, Bucharest, 1951; m. Alexandru Popescu, Mar. 31, 1950; 1 dau., Sanda. Research worker Inst. inframicrobiology, Bucharest, 1951—. Mem. Union Med. Socs. (Rumania). Research, publs. on diagnosis of viral diseases, antiviral immunity, viral diagnosis in cardiovascular diseases; pararickettsia. Home: 3 str. Bocsa. Office: 285 sos. Mihai Bravu, Bucharest 29, Rumania.*

POPESCU, Ioan Iovitzu, Rumanian physicist; b. Burila-Mare, Rumania, Oct. 1, 1932; s. Dumitru and Elvira (Iovitzu) P.; B.S., Faculty Physics, Bucharest (Rumania) U., 1955, doctor degree in Physics, 1962; m. Denisa Georgeta Chiru, Aug. 10, 1963. Asst. for optics and gaseous electronics Faculty Physics, Bucharest U., 1955-60; researcher, chief lab. elementary processes in plasma Physics Inst., Rumanian Acad., Bucharest, 1960; Humboldt Dozentenstipendium, Kiel (Germany) U., 1967-68. Recipient 3d class Labour Order of S.R. of Rumania, 1964, Rumanian Acad. Prize for Physics, 1965. Mem. Nat. Council Sci. Research S.R. of Rumania (mem. 1st commn. math. and physics). Author: (with Eugen Badareu) Ionized Gases—Basic Processes, 1963; Ionized Gases-Electrical Discharges in Gases, 1965; also articles. Research on collision and transport processes in ionized gases, elec. discharges in gases. Home: 7 Daniceni, Bucharest. Office: 114 Calea Victoriei, Bucharest, Rumania.*

POPHAM, Richard Allen, Am. botanist; b. Charleston, Ill., Sept. 29, 1913; s. Frank and Charlotte (Kluge) P.; B.Ed., Eastern Ill. U., 1936; M.S., Ohio State U., 1937, Ph.D., 1940. Faculty, Ohio State U., Columbus, 1940—, asso. prof., 1950—. Cons. Battelle Meml. Inst., 1958—; research collaborator Brookhaven Nat. Lab., 1964—. Recipient Alfred J. Wright award, 1966. Mem. Bot. Soc. Am., Am. Assn. U. Profs., Am. Soc. Plant Physiologists, Internat. Soc. Plant Morphologists, A.A.A.S., Gamma Alpha (past nat. pres.). Author: Developmental Plant Anatomy, 1952; Laboratory Manual for Plant Anatomy, 1966. Research in shoot apex orgn., variability, cytogenesis, flower initiation and devel., levels of tissue differentiation in roots, cell differentiation, seedling anatomy, reprodn. in seed plants, plant microtechniques. Home: 1519 Neil Av., Columbus, O. 43201.*

POPJAK, George Joseph, biochemist; b. Kiskundorozsma, Hungary, May 5, 1914; s. George and Maria (Mayer) P.; M.D. Sub auspiciis Regis, Royal Francis Joseph U., Szeged, Hungary, 1939; D.Sc., U. London (Eng.), 1962; m. Hasel Marjorie Hammond, Apr. 9, 1941. Brit. Council scholar Postgrad. Med. Sch., U. London, 1939-41; demonstrator dept. pathology St. Thomas's Hosp. Med. Sch., London, 1941-47; Beit Meml. fellow for med. Research, 1943-47; mem. sci. staff Med. Research Council, Nat. Inst. for Med. Research, London, 1947-53, dir. Exptl. Radiopathology Research Unit, 1953-62; joint dir. chem. enzymology lab. Shell Research Ltd., Sittingbourne, Kent, Eng.,

1962—. Recipient CIBA medal Biochem. Soc., 1965; Stouffer prize, 1967. Fellow Royal Inst. Chemistry, Royal Soc., 1961, Royal Soc. Medicine; mem. Royal Flemish Acad. Belgium, Biochem. Soc., Faraday Soc., Path. Soc. Gt. Britain and Ireland, Brit. Med. Assn. Editor (with E. LeBreton) Biochemical Problemso f Lipids, 1956; Biosynthesis of Lipids, 1962. Research, numerous publs. on mechanisms of biosynthesis of fats and cholesterol and application of isotopic tracers. Home: 341 Queen's Rd., Maidstone, Kent. Office: Broad Oak Rd., Sittingbourne, Kent, Eng.*

POPKEN, Jan, Dutch mathematician; b. Smilde, Dec. 14, 1905; s. Jan and Jantje (Hofman) P.; Dr ès sc. univs. Groningen and Göttingen; m. C. C. J. Houwinkten Cate; children—Jeannette H. E., Peter J. O., Marion A. H. Instr., U. Groningen, 1945-47; prof. math. U. Utrecht, 1947-55, U. Amsterdam, 1955—; vis. prof. U. Minn., 1953-54, U. Cal. at Berkeley, 1962-63. Mem. Royal Acad. Sci. Amsterdam. Author: Ueber Arithmetische Eigenschaften analytischer Funktion, 1935; also numerous articles. Research in history human culture, theory of numbers, analysis, algebra. Home: Dr. Boslaan 2, Amstelveen. Office: Math. Inst., Nw. Achtergracht 121, Amsterdam, Netherlands.

POPKOV, Valeri Ivanovich, Russian elec. engr.; b. Feb. 3, 1908; grad. Moscow Inst. Energetics, 1930. Worker at All-Union Electro-Tech. Inst., 1932-36; with Inst. Energetics USSR Acad. Scis., 1943—. Dep. chmn. All Union Znanie Soc. Mem. Communist Party, 1951—. Decorated Order Red Banner of Labor. Corr. mem. USSR Acad. Scis., 1953. Author: Ion Recombination Factor under Corona Discharge in Air, 1948; The Theory of Direct Current Unipolar Corona, 1949; The Electrical Field with a Transient Unipolar Corona, 1954; co-author: Theory of a Corona Under Constant Voltage, 1950; Determining Parameters in the Scheme of Replacing the Corona Lines, 1951; Theory of a Corona Under Constant Voltage, 1951; Reactive Effects of a Corona of Alternating Current, 1956; Methods of Evaluating Yearly Losses of Energy on the Corona, 1957; An Experimental Study of Volumetric Charge Movement in the Field of an Alternating Current, 1957; Results of a Study of the Stability of the Transcaucasian Power System, 1960. Specializes in high-voltage technology; elec. discharge in gases at high voltage; phys. processes in electric filters. Home: Novopeschanaya 21, Moscow. Office: Inst. of Energetics, USSR Acad. Scis., Moscow, USSR.

POPOFF, Methodi, Bulgarian biologist; b. Schumen, Bulgaria, Apr. 29, 1881; s. Athanas and Anastasia (Sawoff) P.; ed. Natural Sci. U., Sofia, U. Munich, Bacteriological Inst. Robert Koch, Berlin, Inst. Pasteur, Paris; Dr.Phil., Natural Sci. U., Munich, 1906; m. Franciska Roegels, 1914. Asst. in zoology U. Munich, 1906-09; prof. biology and comparative anatomy U. Sofia, 1911-54, rector, 1920-21. Mem. Agrl. Acad. (Prague), Leopold Acad. (Halle). Author: Die Zellstimulation in Beziehung zur Landwirtschaft und Medizin; (with Gleisberg) Zellstimulationsforschungen. Research on cellular morphology and cellular physiology, "cell stimulation." Died 1954.

POPOV, Alexander Stepanovich, Russian physicist, engr.; b. Turynskie Rudinki, Russia, Mar. 4, 1859; grad. U. St. Petersburg (Russia), 1882, postgrad., 1882-83. Joined staff Naval Sch., Kronstadt, Russia, 1883; tchr. elec. engring. and physics Naval Engring. Coll., Kronstadt, 1890-1900. Recipient Gold medal with diploma Internat. Tech. Congress, Paris, 1900. Mem. Russian Physico-Chem. Soc. (became v.p. 1904), Russian Tech. Soc. Studied use of electromagnetic waves to receive signals; research on atomospheric discharges, 1894, X-rays; attempted to use electromagnetic waves to demonstrate formation of thunderstorms; invented antenna; built instrument to record atmospheric electricity, 1896. Died St. Petersburg, Dec. 31, 1905.

POPOV, Nikolai Fédorovich, Russian physiologist; b. Oct. 18, 1885; grad. Med. Faculty, Kharkov U., 1912. Staff All-Union Livestock Inst., also All-Union Inst. Exptl. Medicine; prof. Moscow Vet. Acad., 1938—; in charge lab. for physiology central nervous system Brain Inst., 1945-53; became instr. physiology animals Dom Inst. Agr. and Melioration, 1921. Recipient Order of Lenin, Order of Red Banner of Labor, Lenin Prize; named Honored Science Worker of RSFSR. Mem. All-Union Acad. Agrl. Scis. Author: Peripheral Nerve Formations and their Role in Regulating Tissue Processes in Organism. Co-author textbook on physiology of farm animals. Research and publs. on devel. of method for suspending influence of nerve center on automatic processes.

POPOV, Yevgenii Pavlovich, Russian automation specialist; b. 1914; grad. Bauman Moscow Advanced Tech. Sch., 1939; Dr.Tech. Scis. degree, 1947. With A.F. Mozhaiskii Air Force Engring. Acad., Leningrad, USSR, 1943—, now chmn. dept. automation and remote control; also sr. sci. worker USSR Acad. Scis. Inst. Electromechanics, prof., 1948—. Recipient Stalin prize, 1949. Corr. mem. USSR Acad. Scis. Served with Soviet Army. Author: On the Approximate Study of Self and Forced Oscillations on Nonlinear Systems, 1954; Approximate Calculation of Self-Excited and Forced Vibrations in Nonlinear Systems of Higher Order on the Basis of the Harmonic Linearization of Nonlinearity, 1954; Approximate Determination of Auto-Oscillations and Forced Oscillations in Systems of Automatic Control, 1955; A Generalization of the Asmyptotic Method of N. N. Bogouliuboff in the Theory of Nonlinear Oscillations, 1956; Use of Harmonic Linearization Method in Automatic Control Theory, 1956; Isolation of Regions of Stability of Nonlinear Automatic Systems Based on Harmonic Lineation, 1959; The Effect of Vibrational Interference on the Stability and Dynamic Quality of Nonlinear Automatic Systems, 1959; Approximate Methods of Study of Non-linear Automatic Systems, 1960; Automatic Regulation and Control, 1962; On the Study of Auto-Oscillation Systems with Logic Devices, 1962; On Non-linear Laws of Controls in Automatics, 1962. Work on automatic, control, particularly nonlinear systems. Home: Inst. Electromechanics, USSR Acad. Scis., Dvortsovaya Naberezhnaya 18, Leningrad, USSR.

POPOVIC, Vojin, physiologist; b. Belgrade, Yugoslavia, Sept. 18, 1922; s. Pavle and Bojana (Pavloric) P.; M.Sc., U. Belgrade, 1949, Ph.D., 1951; m. Pava Jovanovic, Oct. 2, 1949; 1 son, Ray. Faculty, U. Belgrade, 1949-57; research asso. Nat. Center for Sci. Research, Paris, France, 1957; research asso. in physiology U. Rochester (N.Y.), 1957-58; asso. prof. U. Houston, 1958-60; asso. prof. physiology Emory U. Med. Sch., Atlanta, 1961-66, prof., 1966—. Fellow College de France, Paris, 1954-55; fellow Oceanographic Mus., Monaco, 1956; postdoctorate fellow NRC, Can., 1958-60. Recipient Career Devel. award NIH, 1962. Mem. Serbian Biol. Soc., Société Biologie France, Am. Physiol. Soc., Soc. Exptl. Medicine and Biology, Soc. Biol. Rhythm, Internat. Soc. Biometeorology, Soc. Cryobiology, Aerospace Med. Assn., Soc. Psychophysiol. Research. Re search, numerous publs. in hypothermia, hibernation, temperature regulation, cancer. Home: 2342 Street de Ville N.E., Atlanta 30329.*

POPP, Raymond Arthur, Am. biologist; b. Northport, Mich., Nov. 23, 1930; s. Clarence A. and Loretta (Reicha) P.; B.A., U. Mich., 1952, M.S., 1954, Ph.D., 1957; m. Diana Esther Marriott, Sept. 4, 1954; children—Carolyn, Raymond, David, Stevan. Sr. biologist biology div. Oak Ridge Nat. Lab., 1957—. Mem. Genetics Soc. Am., A.A.A.S., Am. Soc. Zoologists, Am. Genetic Assn., Soc. for Study Growth and Devel., Sigma Xi. Research, publs. on genetics and biochemistry mouse hemoglobins and esterases, metabolism mouse embryonic tissues, effects radiation in mice, regulation growth tissue transplants in mice. Home: Pennell Lane Route 20, Knoxville, Tenn. 37921. Office: Biology Div., Oak Ridge Nat. Lab., Oak Ridge 37831.*

POPPE, Erik, Norwegian radiologist; b. Oslo, Norway, Apr. 3, 1905; s. Trygve and Justine (Wiborg) P.; M.D., U. Oslo; m. Anne Catherine Hoy, 1936; children—Hedvig, Kunt, Iver, Anne Sofie. Asst. various Norwegian hosps.; 1st asst. radiology dept. Bergen Municipal Hosp.; head radiology dept. Troms and Tromso Hosp.; instr. Oslo Univ. Hosp.; now dir. dept. Norwegian Radium Hosp.; prof. therapeutic radiology U. Oslo. Mem. Norwegian Med. Soc., Norwegian Med. Radiology Soc., Nordic Med. Radiol. Soc. Author: Experimental Investigations of the Effects of Roentgen Rays on the Eye, 1942. Co-editor works on radiotherapy. Research on tumors of bone, clin. radiobiology. Home: Elisenberg veien 16. Office: Norwegian Radium Hosp., Oslo, Norway.

POPPEL, Maxwell Herbert, Am. radiologist; b. N.Y.C., Oct. 19, 1903; s. Samuel and Rose (Stern) P.; student Fordham U., 1920-23; M.D., U. and Bellevue Hosp. Med. Coll., 1927; postgrad. Harvard, 1934; m. Celia Bercow, Sept. 11, 1928. With N.Y.U., 1936—, advancing through grades to prof., chmn. dept. radiology N.Y.U.-Bellevue Med. Center, 1952—; cons. radiology Bellevue and University hosps.; cons. radiologist Elizabeth Horton Meml. Hosp., Middletown, N.Y., Mt. Vernon (N.Y.) Hosp., Beth Israel Hosp., N.Y.C., Bergen Pines Co. Hosp., Paramus, N.J., cons, roentgenologist VA Hosp., Bronx, Naval Hosp., St. Albans, L.I., St. Francis Hosp., Poughkeepsie, N.Y., St. Vincent's Hosp., N.Y.C., Nat. Naval Med. Center, Bethesda, Md., VA Hosp., Manhattan, Montefiore Hosp., Bronx (N.Y.) St. Barnabas Hosp., Bronx, N.Y., others; vice chmn. Med. Bd. U. Hosp., N.Y.C. Recipient Meritorious Medallion, N.Y.U., 1962. Diplomate Am. Bd. Radiology. Fellow Am. Coll. Radiology; mem. A.M.A., Radiol. Soc. N.Am., N.Y. Roentgen Soc. (historian, past pres.), Sigma Xi, Alpha Omega Alpha. Author: The Roentgen Manifestations of Pancreatic Disease, 1951; The Roentgen Aspects of the Papilla and Ampulla of Vater, 1953; The Lower Esophageal Vestibular Complex, 1963; Cardiac Calcifications, 1964; also sci. articles. Editor textbook. Address: 180 East End Av., N.Y.C. 10028.*

POPPEN, James L., Am. neurosurgeon; b. Drenthe, Mich., Feb. 27, 1903; s. John and Ann (Flotman) P.; student Hope Prep. Sch., Holland, Mich., 1916-20; grad. Hope Coll., 1926; D.Sc., 1953; M.D., Rush Med. Coll., Chgo., 1930; m. Nancy High, Apr. 10, 1933; children—Elizabeth, John. Neurosurgeon Lahey Clinic Found., Boston, 1933—, chief of neurosurgery, 1957—, now bd. govs. clin. div. Lahey Clinic Found., also exec. com. bd. trustees; neurosurgeon N.E. Deaconess, N.E. Bapt. and Boston psychopathic hosps., 1933—; cons. Taunton (Mass.) State Hosp., Sturdy Meml. Hosp., Attleboro, Mass., St. Luke's Hosp., Middleboro, Mass. Recipient Cross King Alfonso X, Spain, 1947; Sword of San Martin, Argentina, 1949; Decoration of Merit, 1954. Fellow A.C.S.; mem. A.M.A., Am. Neurol. Soc., Soc. Neurol. Surgeons (pres. 1960), Argentina Sociedad Neurologia Psiquiatria y Neurocirugia Honorari, Sociedad de Neurologia Neurocirugia Honerario Repubia Oriental del Uurguay, Soc. Neurology, Rio de Janeiro, Ardem Nacional de Cruzeiro de Sul, Brazil, Am. Surg. Assn.; hon. mem. faculty medicine U. Madrid; corr. mem. La Real Societa Italiana di Neuro-Chirurgia, Milano, Italy, Neurosurg. Soc. Brazil, Neurosurg. Soc. Venezuela; hon. mem. Nacional de Medicina, Spain; hon. mem. Societa Italiana di Neurochirurgia, Milano, Italy; hon. mem. Societe de Neuro-Chirurgie de Langue Francaise; mem. Harvey Cushing Society (pres. 1959), Sigma Nu. Author: Atlas of Neurosurgery, also articles. Home: 20 Sears Rd., Brookline, Mass. Office: 605 Commonwealth Av., Boston.*

POPPENDIEK, Heinz Frank, Am. applied physicist; b. Altona, Germany, Nov. 8, 1919 (parents Am. citizens); s. Frank and Helen (Bunsen) P.; B.S., U. Cal. at Berkeley, 1942, M.S., 1943, Ph.D., 1949; m. Elizabeth Secrest, Aug. 15, 1943; children—Niel Eugene, Carolyn, Mark Gregory. Research engr., asst. prof. U. Cal., Berkeley and Los Angeles, 1942-50; sect. chief Oak Ridge Nat. Lab., 1950-56; staff scientist Gen. Dynamics, San Diego, 1956-60; dir. applied research Geophysics Corp. Am., San Diego, 1960-61; pres. Geosci., Ltd., Solano Beach, Cal., 1961—. Tchr., cons. U. Cal., 1958-65; cons. Inst. for Def. Analyses, 1963-65. Recipient Am. Geophys. award for Hydrology, 1951; honorarium for heat transfer symposium U. Mich., 1953. Mem. Am. Nuclear Soc., Sigma Xi, Tau Beta Pi. Patentee circulating fuel neutronic reactor, salt water conversion system, electromagnetic electrolyte pump, volumetric blood heater. Research contbns. in heat flow meter devel.; heat, mass and momentum transfer in the atmospheric boundary layer; nuclear reactor safeguard analysis; exptl. and analytical liquid metal boiling and condensing; freezing and thawing of tissue and organs. Home: 7834 Esterel Dr., La Jolla, Cal. 92037. Office: 410 S. Cedros Av., Solana Beach, Cal. 92075.*

POPPENSIEK, George Charles, Am. veterinarian; b. N.Y.C., June 18, 1918; s. George Frederick and Emily (Miller) P.; V.M.D., U. Pa., 1942; M.S., Cornell U., 1951; m. Edith Marion Wallace, July 3, 1943; children—Neil Allen, Leslie Marion. Faculty, U. Pa. Sch. Vet. Medicine, 1942-43, U. Md., 1943-44; dept. head Lederle Labs. div. Am. Cyanamid Co., Pearl River, N.Y., 1944-49; dir. diagnostic lab. N.Y. State Vet. Coll., Cornell U., Ithaca, N.Y., 1949-51, research asso. Vet. Virus Research Inst., 1951-55, acting asso. prof. bacteriology, 1953-54, dean, prof. microbiology, 1959—; veterinarian animal disease and parasite research div. Agrl. Research Service, U.S. Dept. Agr., Plum Island Disease Lab., Greenport, N.Y., 1955-56, acting-in-charge (supervisory veterinarian) diagnostic investigations, 1956-58, in charge immunological investigations, 1958-59. Recipient certificate of merit U.S. Dept. Agr., 1958. Charter fellow Am. Acad. Microbiology; mem. Am., So. Tier vet. med. assns., U.S. Livestock San. Assn., Conf. Research Workers in Animal Disease N.Am., N.Y. State Assn. Professions (charter), Assn. Am. Vet. Med. Colls., Nat. Assn. Standard Med. Vocabulary, Am. Vet. Med. Assn., Am. Pub. Health Assn., Sigma Xi, Phi Zeta, Alpha Psi (pres. U. Pa. 1941-42), Phi Kappa Phi, others. Co-patentee vaccines. Contbr. articles to vet. jours. on virus diseases of domestic animals. Home: 143 Pine Tree Rd., Ithaca, N.Y. 14850.

POPPER, Daniel Magnes, Am. astronomer; b. Oakland, Cal., Aug. 11, 1913; s. William and Tess (Magnes) P.; A.B., U. Cal., 1934, Ph.D., 1938; m. Catherine May Salo, Jan. 12, 1940; 1 son, Roger David. Lick Obs. fellow U. Cal., 1936-38, Martin Kellogg fellow, 1938-39, physicist radiation lab., 1943-45, asst. prof. U. Cal. Los Angeles, 1947-49, asso. prof., 1949-55, prof., 1955—, past chmn. dept. astronomy; research asso. McDonald obs. U. Tex. and Yerkes obs. U. Chgo., 1939-40, instr. 1940-43, asst. prof., 1945-47; physicist Oak Ridge, 1943-45; guest investigator Mt. Wilson and Palomar obs's since 1949. NSF fellow, 1964-65. Mem. Am. Astron. Soc. (rep. to U.S. Nat Com. Internat. Astron. Union, 1964—), Astron. Soc. Pacific, Am. Assn. U. Profs., Phi Beta Kappa, Sigma Xi. Home: 1010 El Medio Av., Palisades, Cal. 90272. Office: U. Cal. Dept. Astronomy, Los Angeles 90024.*

POPPER, Hans, pathologist; b. Vienna, Austria, Nov. 24, 1903; s. Carl and Emma (Gruenbaum) P.; M.D., U. Vienna, 1928; M.S., U. Ill., 1944, Ph.D., 1944; M.D., Cath. U. Louvain, U. Bologna (Italy), 1965; Ph.D., U. Vienna, 1965; m. Lina Bilig, June 4, 1942; children—Frank, Charles. Dir. dept. pathology Cook County Hosp., Chgo., 1944-57; sci. dir. Hektoen Inst. for Med. Research, Chgo., 1943-57; faculty Northwestern U. Med. Sch., Chgo., 1946-57, prof. pathology, 1956-57; pathologist-in-chief Mt. Sinai Hosp., N.Y.C., 1957—; prof. pathology Columbia, 1957—; dean for acad. affairs, prof. pathology, head dept. pathology Mt. Sinai Sch. Medicine, 1965—. Cons. to hosps.; chmn. Surg. Gen's Adv. Com. on Gen. Medicine, U.S. Army, 1964—; chmn.

com. to rev. tolerances for aldrin and dieldrin FDA, 1965—. Named Hon. Prof., U. Arequipa, Perus, 1965. Mem. Internat. Assn. for Study Liver (pres. 1964—), Am. Assn. for Study Liver Diseases (past pres.), Chgo., N.Y. (past pres.), path. socs. Author: (with Kushner) Clinical Pathological Conferences of Cook County Hospital, 1948; (with Schaffner) Liver: Structure and Function, 1957, Clinical Pathological Conferences of Mount Sinai Hospital, 1963; also numerous articles. Editor: (with Schaffner) Progress in Liver Diseases, vol. I, 1961, vol. II, 1965. Research on structure and function of liver, distbn. vitamin A in tissues, endogenous creatinine clearance. Home: 3135 Johnson Av., Riverdale, N.Y. 10063. Office: 100th St. and Fifth Av., N.Y.C. 10029.*

PÖPPIG, Eduard, German naturalist; b. Plauen, Germany, July 16, 1798; prof., Leipzig, Germany. Author: Reise in Chile, Peru und auf dem Amazonenstrome, 2 vols., 1835; Illustrierte Kultergeschichte des Tierreichs, 4 vols., 1851. Noted for descriptions of nature. Died Wahren/Leipzig, Sept. 4, 1868.

PORADOVSKY, Karol, Czechoslovakian physician; b. Cadca, Czechoslovakia, Feb. 26, 1916; s. Robert and Margita (Kompanek) P.; M.D., Komensky U., Bratislava, 1940; C.Sc./candidat scientiae U. Brno; docent gynaecologiae U. Kosice; m. Milina Hurtová, Mar. 30, 1948; children—Juraj, Peter, Fedor. Asst. at midwife sch., 1940-48; asst. dept. gynecology and obstetrics U. Kosice, 1948-52; chief dept. gynecology and obstetrics, Zilina, 1952-60; regional chief gynecology-obstetrics, Kosice, 1960-63; chief dept. gynecology and obstetrics U. Kosice, 1963—. Author: Zena a prechod, 1964; Tehotnost' a co dalej, 1966; also articles. Studies on participation of parametrium by dilatation of cervix during delivery, coagulation of neonatal blood and its fibrinolytic activity, vaginal colpohysterectomy. Home: 25 Sturova ul. Office: 11 Moyzesova ul., Kosice, Czechoslovakia.*

PORATZ, Herbert August, Am. chemist; b. Sumner, Ia., July 2, 1902; s. August and Roslyn (Shumacher) P.; B.SZ., U. Ill., 1928; M.S., U. Colo., 1931, Ph.D., 1935; m. Vivian Leone Miller, Sept. 1, 1942; children—George August, Antoinette (Mrs. David B. Mathis). Instr., Blackburn Coll., 1928-29; faculty U. Colo., 1929-42, asst. prof., 1938-42; research asso. U. Chgo. Metall. Lab., 1942-43, sect. chief, 1943-44; group leader plutonium analysis and research Los Alamos Lab., 1944-46; asso. prof. chemistry Washington U., St. Louis, 1946-57, prof., 1958—; civilian AEC, OSRD, 1944. Cons. AEC, 1953-65. Mem. Am. Chem. Soc., A.A.A.S., Am. Assn. U. Profs., Sigma Xi, Omicron Delta Kappa. Research, publs. on analytical and geochemistry of indium, cobalt, thorium, protactinium uranium and plutonium, cosmic abundances of elements, chem. procedures for determining geologic age. Home: 9 Dwyer Pl., St. Louis 63124.*

PORCHER, Francis Peyre, Am. physician, botanist; b. St. John, S.C., Dec. 14, 1825; s. Dr. William and Isabella (Peyre) P.; A.B., S.C. Coll., 1844; grad. Med. Coll. S.C., 1847; m. Virginia Leigh; m. 2d, Margaret Ward. Established Charleston (S.C.) Prep. Med. Sch.; mem. Charleston Bd. Health; surgeon, physician marine and city hosps.; prof. clin. medicine, materia medica, therapeutics Med. Coll. State S.C.; opened hosp. for Negroes, 1855; surgeon Holcombe Legion, also naval hosp., Norfolk, Va., S.C. Hosp., Petersburg, Va., during Civil War. Asso. fellow Coll. Physicians Phila.; mem. S.C. Med. Assn. (pres. 1872), Elliot Soc. Natural History, A.M.A. (v.p. 1879), 10th Internat. Med. Congress, Berlin, 1890, Pan. Am. Congress (pres. sect. on gen. medicine 1892). Author: A Medico-Botanical Catalogue of the Plants and Ferns of St. John's Berkely, S.C., 1847; A Sketch of the Medical Botany of South Carolina, 1849; The Medicinal, Poisonous and Dietetic Properties of the Cryptogamic Plants of the United States, 1854; Illustrations of Disease with the Microscope, and Clinical Investigations aided by the Microscope and by Chemical Reagents, 1861; The Resources of the Southern Fields and Forests, 1863. Editor Charleston Med. Jour. and Rev., 1853-58, 73-76. Authority on diseases of chest and heart. Died Nov. 19, 1895.

PORFIZIER, Vladimir Borisovich, Russian geologist; b. July 8, 1899; grad. Leningrad Mining Inst., 1926; D.Geol. Sci. With Leningrad Petroleum Research Inst., 1929-39; with Inst. Geol. Sci., Ukrainian Acad. Scis., 1939-41, 44-50, dir., 1964—, dir. Inst. Mineral Resources Geology, 1950-64; prof. Lvov U., Lvov Poly. Inst., from 1945. Decorated Order of Lenin. Mem. Ukrainian Acad. Scis. Author: The Metamorphism of Coal, 1948; The Geological and Geochemical Conditions for the Formation of Oil, 1949; co-author: Menilite Shales of the Carpathians, 1963. Research on geology of oil and gas deposits. Address: Inst. Geol. Sci., Ukranian Acad. Scis., Kiev, Ukrainian SSR, USSR.

PORGES, Otto, physician; b. Brandeis/Elbe, Austria, 1879; s. Sam and Emilie (Nossal) P.; ed. univs. Prague (Czechoslovakia), Strasbourg (France); M.D.; postgrad., Vienna, Austria, Berlin, Germany; m. Marie Low, 1917. Asst., First Med. Univ. Clinic, Vienna, 1908-29, head, 1929-33; became lectr. U. Vienna, 1910, prof. internal medicine, 1920; head 2d dept. internal medicine, dep. dir. S.C. Childs Hosp. and Research Inst., Vienna, 1935-38; became prof. Northwestern U. Med. Sch., 1942, now prof. emeritus; staff Columbus Hosp., Chgo. Mem. N.Y. Acad. Scis., Vienna Soc. Internal Medicine, various Am. med. socs. Author: Magenkrankheiten; Darmkrankheiten, 1939; also articles. Discovered gamma globulin, 1903; devised methods of gastric photography, 1929. Home: 3200 Sheridan Rd., Chgo. 60613.

PORRETT, Robert, English chemist; b. London, Sept. 22, 1783; clk. War Office, 1795-1850; recipient medal Soc. Arts. Fellow Soc. Antiquaries, Royal Soc., 1848, Chem. Soc. (charter), Astron. Soc. (charter). Contbr. numerous articles to tech. jours. Discovered sulphocyanic acid between 1808 and 1814, ferro-cyanic acid, 1814, (independently) electric endosmosis, 1816. Died Nov. 25, 1868.

PORRO, Ignatio, Italian physicist; b. Pignerol, Italy, 1795. Invented anallatic lens (allowed subtended angle to be read directly without allowing for focal length of object-glass), 1823, prismatic binocular, 1850; built reflector micrometer with visible threads, helioscope, speedometer for mine use; participated in measurement of meridian arc. Died 1875.

PORSTMANN, Werner, German physician; b. Geyersdorf/Erzgebirge, Germany, Feb. 22, 1921; s. Theodor and Alma (Köhler) P.; student medicine U. Leipzig (Germany), 1939, U. Marburg (Germany), 1942; state exam U. Greifswald (Germany), 1946, M.D., 1946; m. Annemarie Röhrer, June 11, 1945; 1 son, Tomas. Physician-specialist in internal medicine, 1953; specialist in radiology, 1956; faculty Humboldt U., Berlin, 1960—, prof. radiology, 1965—, dir. dept. cardiovascular diagnostics, 1964—. Recipient Rudolph-Virchow prize, 1960. Research, numerous publs. on X-radiol. diagnostics of heart and blood vessels, specific methods of diagnosis of left half of heart, methods of non-surg. treatment of ductus arteriosus (Botallo's duct); permanent closure of open ductus arteriosus by spl. catheter technique. Home: 13 Mollstrasse, 102 Berlin, East Germany.

PORT, Sergio Pereira da Silva, physicist; b. Niteroi-Rio de Janerio, Brazil, Jan. 19, 1926; s. Eginete P.S. and Grimalda (Valle) P.; B.S. in Chemistry, U. Brazil, Rio De Janerio, 1946, Licensiado Chem., 1947; Ph.D., in Physics, Johns Hopkins, 1954; m. Hilta W. Cantanhede, Sept. 13, 1950; children—Sergio, Marcia, Ivan, Paulo. Instr., U. Brazil, 1947-49; faculty Aero. Inst. Tech., Sao Paulo, Brazil, 1954-60, asso. prof., 1956-60; mem. tech. staff, research supr. quantum electronics Bell Telephone Labs., Murray Hill, N.J., 1960—. Mem. Am. Phys. Soc., Optical Soc. Am., Brazilian Acad. Scis. (asso.). Research, publs. on early solid state lasers, pioneer work on applications of lasers to physics and chemistry especially to Raman effect, structure of solids, vibrations in solids using lasers. Home: 232 Belmont Av., North Plainfield, N.J. 07060. Office: Bell Telephone Labs., Murray Hill, N.J. 07971.*

PORTA, Giambattista della, Italian natural philosopher; b. Naples, probably between Dec. 7, 1534 and July 6, 1535; traveled widely in France, Italy, Spain. Author: Perspectiva, 1558; Magiae naturalis libri XX, 1558; De humana physiognomia, 1583; Villae libri XII, 1592; De refractione opices parte, 1593; De aeris transmutationibus, 1609; Ars reminiscendi. Brought together the Otiosi (men of leisure) and founded Accademia secretorum naturae (in Naples) to promote study of science; helped establish Accademia dei Lincei (v.p.); gave clear description of camera obscura; discussed use of a combination of concave and convex lenses for viewing objects far off or near at hand; discussed magnetism, optics, distillation, cosmetics, perfume, metallurgy, crystalography, farming, fortifications, palmistry, pyrotechny, saltiness of sea, existence of poisonous exhalations different from air; believed in occult forces and held doctrine of sympathies; also reported useful recipes and uncritically accepted myths and legends; set out first ecological grouping of plants according to their geographical locale and distribution; wrote on astrology. Died, Naples, 1615.

PORTAL, Antoine, French physician; b. Caillac, France, Jan. 5, 1742; M.D., Montpellier, France, 1764; prof. anatomy Coll. de France; became prof. anatomy Royal Garden, 1777; 1st physician to Louis XVIII, then to Charles X; mem. Acad. Medicine, French Acad. Scis. Author: Histoire de l'anatomie et de la chirurgie, 1770-73; Cours d'anatomie, 5 vols., 1803; also treatises on epilepsy and apoplexy. Died Paris, July 23, 1832.

PORTEN, Laurence, orthopedist; b. Hausham, Bavaria, Germany, Mar. 22, 1894; s. Hans and Walburga (Haberlander) Portenkirchner; B.S., Trade Sch., Mainz, Germany, 1911; M.S., Trade and Bus. Sch., Konstantz, Germany, 1924; Master of Orthopedic Mechanics and Surg. Mechanics, 1924; m. Saskia Stephani Nordheim, Aug. 7, 1933; children—Edith (Mrs. Everest McDade), Hedi (Mrs. Raymond Marvin). Pres., Union Artificial Limb & Brace Co., Inc., Pitts., 1944—. Fellow Soc. Cal. Orthotist and Prosthetist Assn. Los Angeles; mem. Translation and Abstract Service Am. Orthotics and Prosthetics Assn., German Orthopedic Assn. (life). Research, numerous publs. on suction socket artificial legs; designer adjustable air stump sockets, pads and cuff which are intended to fill gap in limb fitting for shrunken stumps; developed methods for cosmetic duplication, process of covering and padding leg and back braces without sewing or stitching; originator suction socket limbs, Germany, 1932-37, U. S., 1948. Home: 5836 Clark Av. Extension, Bethel Park, Pa. 15102. Office: Century Bldg., 130-134 7th St., Pitts. 15222.*

PORTER, Charles Walter, Am. chemist; b. Morgan, Utah, May 16, 1880; s. Charles Graves and Betsy (White) P.; B.S., Utah State Coll., 1905; A.M., Harvard, 1909; Ph.D., U. Cal. at Berkeley, 1915; m. Alberta Smith, June 26, 1901; children—Bessie Alberta (Mrs. Maurice Glenwood Adams), Verna (Mrs. John Amos Worley). Prof. chemistry Utah State Coll., 1907-17; faculty U. Cal. at Berkeley, 1917-46, prof. chemistry, 1926-46, prof. emeritus, 1946—, asst. dean Coll. Chemistry, 1926-40. Fellow A.A.A.S.; mem. Sigma Xi, Phi Lambda, Upsilon, Alpha Chi Sigma. Author: The Carbon Compounds, 1924; (with T.D. Stewart, G.E.K. Branch) The Methods of Organic Chemistry, 1927; Molecular; Molecular Rearrangements, 1928; (with L.E. Young) Gen. Chemistry, 1940; (with T.D. Stewart) Organic Chemistry, 1943; also numerous articles. Research on molecular rearrangement reactions. Home: 735 Eureka St., Redlands, Cal. 92373.*

PORTER, George, English chemist; b. Stainforth, Yorkshire, Eng., Dec. 6, 1920; s. John Smith and Alice Ann (Roebuck) P.; ed. U. Leeds (Eng.), also Emmanuel Coll., Cambridge (Eng.) U.; B.Sc., M.A., Ph.D., Sc.D.; m. Stella Jean Brooke, 1949; children—John B., Andrew C. G. Asst. research dir. physicochemistry Cambridge U., 1952-54; asst. dir. Brit. Rayon Research, 1954-55; prof. physico-chemistry U. Sheffield (Eng.), 1955-63, prof., dir. dept. chemistry, 1963—; prof. chemistry Royal Inst. (dir. 1967). Fellow Royal Soc., 1960, Royal Inst. Chemistry. Recipient (with M. Eigen and R. G. W. Norrish) Nobel prize in chemistry, 1967. Author articles in publns. Royal Soc., Faraday Soc., others. Research on fast chem. reactions, photochemistry, photosynthesis. Home: 451 Abbey Lane, Sheffield 7. Office: Dept. Chemistry, Sheffield Univ., Sheffield, Eng.

PORTER, James Pertice, Am. psychologist; b. Hillsboro, Ind., Sept. 23, 1873; s. Alfred and Elizabeth (Marksbury) P.; student Normal Sch., Terre Haute, Ind., 1890-91, 1892-93; A.B., Ind. U., 1898, A.M., 1901; hon. fellow Clark U., 1903-07, Ph.D., 1906; Sc.D., Waynesburg Coll., 1917; m. Myrta Wayne Brown, Dec. 24, 1895; children—Ernest C., Helen, Marjorie. Instr. psychology Ind. U., 1900-03; asst. prof. psychology Clark Coll., 1907-12, prof. psychology, 1912-22, dean faculty, 1909-22; prof. psychology Ohio U., 1922-43, prof. emeritus, 1943; instr. psychol. U. Ill. Extension, Danville, Ill., 1948. Lectr. ednl. psychology Columbia, 1913-14; mem. Internat. Congress Psychology and Psycho-technique, Paris, 1927; with Adj. Gen.'s Office, N.Y. 1944. Mem. Am., Midwestern (pres. 1941-42) psychol. assns., N.E.A., Internat. Congress of Zoology, Am. Assn. Applied Psychologists, Phi Beta Kappa. Editor Jour. Applied Psychology, 1920-43. Research on English sparrow, spiders, intelligence and imitation in birds, human intelligence and personality. Died Sept. 1956.

PORTER, Jarmain Gildersleeve, Am. astronomer; b. Buffalo, Jan. 8, 1852; s. John Jermain and Mary (Hall) P.; A.B., Hamilton Coll., 1873, A.M., 1876, Ph.D., 1888; postgrad. U. Berlin, Royal Obs., 1873-74; Sc.D., U. Cin., 1930; m. Emily Snowden, July 3, 1879; children—John Jermain, Ruth May, Harold Mitchel. Asst. prof. astronomy Hamilton Coll., 1875-78; mem. U.S. Coast and Geod. Survey, 1878-84; dir. Cin. Obs., also prof. astronomy U. Cin., 1884-1931. Observer Internat. Latitude Service, 1899-1905. Recipient Astron. Jour. Comet prize, 1894. Author: The Stars in Song and Legend, 1901; Catalogue of 4,280 Stars, 1905; Variation of Latitude, 1908; Catalogue of Nebulae, 1910; Catalogue of 3,164 Proper Motion Stars, 1918; All-American Time, 1918; How to Find the Stars and Planets, 1920; Catalogue of 5,000 Stars, 1925; Catalogue of Proper Motion Stars, 1930. Research in astronomy. Died Apr. 14, 1933.

PORTER, John Addison, Am. chemist; b. Catskill, N.Y., Mar. 15, 1822; s. Addison and Ann (Hogeboom) P.; grad. Yale, 1842, M.D. (hon.), 1854; postgrad. in agrl. chemistry, Germany, 1847-50; m. Josephine Sheffield, July 16, 1855, 2 sons. Prof. rhetoric Delaware Coll., Newark, 1844-47; prof. chemistry applied to arts Brown U., 1850-52; prof. analytical, agrl. chemistry Yale (later Sheffield) Scientific Sch., 1852-56, prof. organic chemistry, 1856-64, 1st dean Sheffield Sci. Sch., Yale, consol. depts. of instrn., established and extended course of study which emphasized science, made available reliable, useful information about agr. and nutrition; a founder Scroll and Key of Yale, 1842. Author: Plan of an Agricultural School, 1856; Principles of Chemistry, 1856; First Book of Chemistry and Allied Sciences, 1857; Outlines of the First Course of Yale Agricultural Lectures, 1860. Died New Haven, Aug. 25, 1866.

PORTER, John Roger, Am. microbiologist, educator; b. Alma, Neb., Aug. 14, 1909; s. Robert William and Celia Edith (Mitchell) P.; student Westminster Coll., 1928-30; B.S., Ia. State Coll., 1933, M.S.,

1935; Ph.D., Yale, 1938; m. Marjorie Ann Perkins, Sept. 17, 1934; children—Roberta (Mrs. Walter W. Barbee), Carol, Katherine Ann, John Roger. Faculty microbiology State U. Ia. Coll. Medicine, Iowa City, 1938—, prof., 1947—, chmn. dept., 1949—. Mem. microbiology panel Office Naval Research, 1951-59, chmn., 1957-59, cons. biology div., 1962—; mem. sci. information council NSF, 1961-65; mem. coms. NIH, 1962—, Nat. Acad. Scis.-NRC, 1964—. Trustee Biol. Abstracts, 1960-65. Recipient Pasteur award Ill. Soc for Microbiology, 1961. Diplomate Am. Bd. Microbiology. Mem. Am. Acad. Microbiology (gov. 1961—, chmn. 1962-63), A.A.A.S., (publs. com. 1964—), Am. Assn. Immunologists, Am. Chem. Soc., Am. Documentation Inst., Am. Inst. Biol. Scis. (governing bd. 1964, pres. elect 1966—), Am. Soc. for Microbiology (council 1951-61, 63—, del. Internat. Congress for Microbiology, Stockholm 1958, pres. 1963-64), Conf. Biol. Editors (rep. internat. Fedn. for Documentation, Rio de Janeiro 1960), Ia. Acad. Sci., Internat. Union Microbiol. Socs. (sec. U.S. nat. com. 1965—) Johnson County Med. Soc., Soc. for Exptl. Biology and Medicine (pres. Ia. sect. 1954), Soc. for Gen. Microbiology (Eng.), Sigma Xi (pres. Ia. chpt. 1951-52), Alpha Omega Alpha (hon.). Author: Bacterial Chemistry and Physiology, 1946. Chmn. publs. com.: Style Manual for Biol. Jours., 1960, 64. Editor-in-chief Jour. Bacteriology, 1951-61, compiler, editor 50 vol. index, 1953, 62; adv. editor Microbiology, 1960-64. Discovered functions of vitamins in bacterial nutrition, physiology; role of sulfonamides in certain bacterial animal infections. Home: 215 Ferson Av., Iowa City 52241.*

PORTER, John Willard, Am. biochemist; b. Mukwonago, Wis., June 12, 1915; s. George Willard and Josephine (Gunderson) P.; B.S., U. Wis., 1938, Ph.D., 1942; m. Helen Reynolds, June 12, 1941; children—John Willard, Mary Grace, Susan Ann. Asst. chemist Purdue U., 1942-45, asst. prof., 1945-49; asst. chemist Armed Services Med. Nutrition Lab., Chgo., 1945; sr. scientist Gen. Electric Co., Hanford, Wash., 1949-54; Enzyme Inst. fellow U. Wis., Madison, 1954-56, asst. dir., prin. scientist, asso. prof. dept. physiol. chemistry, 1956-64, chief Lipid Metabolism Lab., prof., 1964—. Chmn. subcom. Nat. Acad. Sci., 1964—. Mem. Am. Soc. Biol. Chemists, Am. Chem. Soc., Am. Soc. Plant Physiologists, Radiation Research Soc., A.A.A.S., N.Y. Acad. Scis., Sigma Xi. Author: (with E.T. Mertz) Laboratory Experiments in Biochemistry, 1948, Plant and Animal Biochemistry, 1949. Research, numerous publs. primarily in field of elucidation of pathways of biosynthesis of carotenes, cholesterol and fatty acids. Home: 1710 Baker Av., Madison. Office: VA Hosp., 2500 Overlook Terrace, Madison, Wis. 53705.*

PORTER, Keith Roberts, biologist, cytologist; b. Yarmouth, N.S., June 11, 1912; s. Aaron C. and Josephine (Roberts) P.; B.S., Acadia U., 1934; A.M., Harvard, 1935, Ph.D., 1938; m. Elizabeth Lingley, June 16, 1938. NRC fellow Princeton, 1938-39; research asst. Rockefeller Inst., 1939-45, asso., 1945-50, asso. member, 1950-56, mem., 1956-61; professor of biology Harvard University, 1961—; cons. Armed Forces Inst. Pathology, morphology and genetics study sect. Nat. Insts. Health. Recipient Passano Found. award, 1964. Mem. A. Soc. Zoologists, Am. Assn. Anatomists, Harvey Soc., Electron Microscope Soc. Am., N.Y. Soc. Electron Microscopists, N.Y. Acad. Scis., Tissue Culture Assn. (1st pres. 1946), Am. Soc. Cell Biology (1st pres. 1961), Am. Acad. Arts and Scis. Author: (with M. Bonneville) An Introduction to the Fine Structure of Cells and Tissues, 1963. Editor-in-chief Jour. Biophys. and Biochem. Cytology, 1955—. Home: 11 Willard St. Office: care Biol. Labs., Divinity Av., Cambridge, Mass.

PORTER, Richard Francis, Am. chemist; b. Fargo, N.D., Feb. 8, 1928; s. Richard C. and Alice (Gillies) P.; student N.D. State U., 1945-46; B.S., Marquette U., 1951; Ph.D., U. Cal. at Berkeley, 1954; m. Dolores L. Bolger, Sept. 1, 1955; children—Patricia, Thomas. Research asso. dept. physics U. Chgo., 1954-55; faculty Cornell U., 1955—, prof. dept. chemistry, 1964—; vis. prof. U. Fla., 1964. Alfred P. Sloan fellow, 1960-64; John Simon Guggenheim fellow, 1964. Mem. Am. Chem. Soc., N.Y. Acad. Scis., Sigma Xi. Research, numerous publs. in high temperature inorganic chemistry, mass spectroscopy, boron chemistry. Home: 940 E. State St., Ithaca, N.Y. 14850.*

PORTER, Richard Janvier, Am. immunologist; b. Swampscott, Mass., Oct. 22, 1913; s. Charles Irving and Ethel (Janvier) P.; B.A., U. Va., 1935, M.A., 1936; Ph.D., U. Chgo., 1941; m. Rachel Davidson, Kelly, June 12, 1937; children—Carol D., Richard Janvier, Anne W., Susan W., James S., Robert H. Faculty, U. Mich. Sch. Pub. Health, Ann Arbor, 1941—, prof. protozoology, 1954—. Mem. Am. Assn. Immunologists, Sigma Xi, Delta Omega, Phi Sigma. Research, publs. on life cycle malarial parasites, effectiveness amodiaquin in treatment malaria, effects X-ray on immunologic memory, characteristics booster response in immunology, automatic diluting machine for immunology. Home: 3051 Geddes Av., Ann Arbor, Mich. 48104.*

PORTER, Rufus, Am. inventor; b. Boxford, Mass., May 1, 1792; s. Tyler and Abigal (Johnson) P.; at least 1 child. Served as pvt. Me. Militia, War of 1812; popularized camera obscura (produced portraits in 15 minutes), 1820, cord-making machine, 1825; founder, editor Am. Mechanic (1st sci. mag. of its kind), circa 1840; established publ. Sci. Am., 1845; published Aerial Navigation, 1849; other inventions include horse-drawn flat boat, clock carriage, washing machine, fire alarm, portable house, revolving rifle, rotary plow, reaction wind wheel. Author: Aerial Navigation, New York and California in Three Days, 1849. Died New Haven, Conn., Aug. 13, 1884.

PORTER, Thomas Conrad, Am. botanist; b. Alexandria, Pa., Jan. 22, 1822; grad. Lafayette Coll., 1840; D.D., Rutgers Coll.; LL.D., F. & M. Coll.; grad. Princeton Theol. Sem., 1843; m. Susan Kunkel, Dec. 24, 1850. Prof. natural scis. Marshall Coll., (Franklin & Marshall Coll.), 1848-66; prof. botany, zoölogy and gen. geology Lafayette Coll., 1866-97, then curator bot. collections, emeritus prof. dean Pardee Sci. Sch. Pioneer (with John M. Coulter and Joseph Leidy) in botany of Rocky Mountains; earliest Am. champion of Finnish lit. Author: Synopsis of the Flora of Colorado; Botany of Pennsylvania. Died 1901.

PORTERFIELD, Austin Larimore, Am. sociologist, educator; b. Salem, Ark., Oct. 16, 1896; s. John Thomas and Mary Emily (Rodman) P.; B.A., Oklahoma City U., 1923; M.A., Drake U., 1924; B.D., Phillips U., 1926; Ph.D., Duke, 1936; m. Rose Ella McCollum, Mar. 14, 1917; children—Frances (Mrs. Clayton B. Willis), Vernon E., Rosella (Mrs. W. J. Chastant). Pastor, 1st Christian Ch., Okmulgee, Okla., 1927-28; prof. sociology Southeastern State Coll., Durant, Okla., 1928-37; prof., chmn. dept. sociology and anthropology Tex. Christian U., Ft. Worth, 1937—. Exec. dir. Leo Potishman Found., 1946—; vis. prof. Duke, summers 1938-39, 42; So. Meth. U., 1961, N. Tex. State U., 1962; research cons. Ft. Worth Fed. Housing Authority, 1939—. Mem. Am. Sociol. Assn. (exec. com. 1946-47), Southwestern Sociol. Soc. (past pres.), Southwestern Social Sci. Assn. Author: Creative Factors in Scientific Research, 1941; Youth in Trouble, 1946; Crime, Suicide and Social Well-Being, 1948; Wait the Withering Rain?, 1953; Mid-Century Crime in Our Culture, 1954; Mirror, Mirror: On Seeing Yourself in Books, 1957; Marriage and Family Living, 1962; Cultures of Violence: The Tragic Man in Society, 1965. Established Jour. Health and Human Behavior, 1960. editor 1960-63. Wrote numerous articles on recognition of similarities in deviant behavior among coll. students and juvenile deliquents and factors producing dissimiliar outcomes of such behavior. Home: 2900 W. Lowden St., Ft. Worth 76109.*

PORTES, Louis, French physician; b. Paris, 1891; Licentiate Sci., 1912; M.D., 1922; Tchr.'s Certificate, 1929; became gynecologist Hosps. Paris, 1927; prof. clin. obstetrics Med. Faculty Paris, until 1942. Mem. Order Physicians (became pres. nat. council 1946). Introduced classic cesarean section in which uterus is moved from abdominal cavity, emptied and replaced (Portes' operation), 1924. Died Paris, 1950.

PORTEUS, Stanley David, psychologist; b. Box Hill, Victoria, Australia, Apr. 24, 1883; s. David and Katherine (Hebden) P.; ed. Melbourne Ednl. Inst., and U. Melbourne, 1910-16; hon. D.Sc., U. Hawaii, 1932; m. Frances Mainwaring Evans, July 13, 1909; children—David Hebden, John Ruxton. Came to U.S., 1919, naturalized, 1932. Sch. tchr., Victoria, Australia, 1900-12; supt. spl. schs., Melbourne, 1912-16; govt. research scholar U. Melbourne, 1916-17, lectr. exptl. edn., 1917-19; dir. research, Psychol. Lab. Tng. Sch., Vineland, N.J., 1919-25; prof. clin. psychology, dir. psychol. and psychopathic clinic U. Hawaii, 1922-48, emeritus, 1948—, cons. Psychol. Research Center, contract research, 1958—; leader expdn. to N.W. Australia, 1929, expdn. to Kalahari Desert, S. Africa, 1934. Recipient Distinguished Contributions award, American Psychological Association. Fellow Internat. Inst. Arts and Lit.; hon. mem. Phi Beta Kappa; mem. Am. Psychol. Assn., Am. Assn. Cons. Psychologists, Sigma Xi. Author: Porteus Maze Tests, 1915; Studies in Mental Deviations, 1922; Temperament and Race (with M.E. Babcock), 1925; The Matrix of the Mind (with F. W. Jones), 1929; Race and Social Differences in Performance Tests (with D. M. Dewey), 1930; Psychology of a Primitive People, 1931; Maze Tests and Mental Differences, 1933; Primitive Intelligence and Environment, 1937; The Practice of Clinical Psychology, 1941; Mental Changes After Bifrontal Labotomy (with R. D. Kepner), 1944; Calabashes and Kings; An Introduction to Hawaii, 1945; And Blow Not the Trumpet, 1947; The Maze Test and Psycho-Surgery (with H. Peters, 1948); The Restless Voyage, 1948; The Maze Test and Intelligence, 1950; Providence Ponds, 1951; The Maze Test and Clinical Psychology, 1959; A Century of Social Thinking in Hawaii, 1962; Streamlined Elementary Education, 1964; A Psychology of Sorts, 1968. Devised Porteus Maze Tests for mental ability measurement, 1914; co-inventor Porteus-Diamond Learning Machine. Home: 2620 Anuenue St., Honolulu 96822.*

PORTEVIN, Albert-Marcel-Germain-René, French chemist, metallurgist; b. Paris, Nov. 1, 1880; s. Paul Albert and Marie Felicie (Ollivier) P.; Ingenieur des Arts et Manufactures, École Centrale de Paris; hon. doctorates from univs. Brussels, Liège, Louvain Quebec, Zurich; m. Madeleine Castillon, Feb. 2, 1929; children—Philippe, Jean Paul. Became engr. Metall. Soc. Booneville, 1904; named prof. metallurgy and metallography Central Sch. Arts and Manufacture, 1912; apptd. prof. Ecole Supérieure de Fonderie, 1929, dir., 1937; became Ecole de Soudure autogène, 1930; prof., then pres. sci. research council Ecole Centrale. Fellow Royal Soc.; mem. French Acad. Scis., 1942; pres. Institut de Soudure. Author: Précis de metallographie; numerous articles. Research on thermal treatment of steel, blending of ternary alloys, oxidation and corrosion of steel; studied thermal treatment for perfecting artillery and aero. equipment, 1914-18. Died Abano Terme, Italy, 1962.

PORTIER, Paul, French biologist; b. Bar-sur-Seine, France, May 22, 1866; s. Ernest and Laure Moreau-Thiesset P.; ed. faculty scis., faculty medicine, Paris; M.D., Ph.D. in Scis.; m. Louise Moiret, 1912; children—Andrée, Janine, Paulette. Prof. physiology Oceanography Inst.; became prof. Paris Faculty Scis., chair of physiology founded for him, 1923; mem. expdns. of Albert I of Monaco. Mem. French Acad. Scis., 1936, also Acad. Medicine. Author: Les symbiotes, 1918; Physiologie des animaux marins, 1938; Biologie des Lépidoptères, 1949. Discovered (with Charles Richet) phenomenon of anaphylaxis. Died Bourg-la-Reine, France, 1962.

PORTIS, Alan Mark, Am. physicist; b. Chgo., July 17, 1926; s. Lyon and Ruth (Libman) P.; Ph.B., U. Chgo., 1948; A.B., U. Cal. at Berkeley, 1949, Ph.D., 1953; m. Beverly Aline Levin, Sept. 5, 1948; children—Jonathan Marc, Stephen Robert, Lori Ann, Frederick Sean. Faculty, U. Pitts., 1953-56; faculty U. Cal., Berkeley, 1956—, prof. physics, 1964—, asst. to chancellor for research, 1966-67, asso. dean grad. div., 1967-68. Fellow Am. Phys. Soc.; mem. Am. Assn. Physics Tchrs. (Robert Andrews Millikan award 1966), A.A.A.S., Sigma Xi. Contbg. author: Laboratory Physics, 1964, 65, 66. Studies in devel. and use of magnetic resonance techniques in study of electronic properties of ordered magnetic materials. Home: 2723 Marin Av., Berkeley, Cal. 94708.*

PORTMANN, Adolf, Swiss zoologist, evolutionist; b. Basel, Switzerland, May 27, 1897; s. Adolf and Elizabeth (Rohr) P.; ed. univs. Basel, Geneva, Munich, Paris, Berlin; Ph.D., 1921; m. Geneviève Devillers, 1931. tng. in marine labs., 1924-29; prof., dir. Zool. Inst., U. Basel, from 1931. Author: Biologische Fragmente zu einer Lehre vom Menschen, 1951; Das Tier als soziales Wesen, 1956; Biologie und Geist, 1956; Von Vögeln und Insekten, 1957; Die Tiergestalt, 1960; Neue Wege der Biologie, 1961; Einführung in die vergleichende Morphologie der Wirbeltiere, 1964. Contbr. to Traité de Zoologie. Research on morphology of vertebrates, birds, insects. Office: Rheinfelderstrasse 14, Basel, Switzerland.

PORTOCALA, Radu-Constantin, Rumanian physician; b. Braila, Rumania, Mar. 30, 1915; s. Radu and Margareta (Olanescu) P.; M.D., Faculty Medicine, Bucharest, Rumania, 1941, D.Med. Scis., 1963, D. Docent in Med. Scis., 1966; m. Esmeralda-André Papudof, Nov. 23, 1947; 1 son, Radu-Constantin. Staff, Faculty Medicine, Bucharest, 1943—, chair of inframicrobiology, 1947-52; head lab. Inst. Inframicrobiology, Acad. Scis., Bucharest, 1951-55, head dept., 1955—, head viral biochem. and biophys. dept., 1956—. Decorated Ordinul Muncii, 1962, Galasecu award U. Iassy (Rumania), 1941, V. Babes award Acad. Scis., 1957. Mem. Union Soc. Med. Sci. Bucharest, Soc. Comparative Pathology (v.p. Bucharest sect.). Author: Microscopia electronica in biologie si inframicrobiologie, 1962; Acizii ribonucleici celulari si virali, 1966; also numerous articles. Virological research in viral hepatitis, rabies, herpes, silkworm-jaundice, influenza virus, cytomegalic virus, Herpes-zoster, electron microscopy, electrophoresis, 1940-56; studies in viral infectious nucleic acids, physico-chem. properties of viral nucleic acids, 1956—; discovered infectivity of influenza virus-ribonucleic acid (RNA), of adenovirus type 3 deoxyribonucleic acid (DNA), Home 11 Bd.N. Balcescu, Bucharest, Rumania.*

POSADA, Adolfo Gonzalez, Spanish sociologist; b. Oviedo, Spain,1 860; prof. law and polit. scit. at various Spanish and S. Am. univs. Author: Sociologia general,1903; Principios de sociologia, 2 vols., 1908. Founder of Spanish sociology; emphasized analysis of social reality. Died 1944.

PÖSCHL, Klaus, German mathematician; b. Prague, Czechoslovakia, Apr. 22, 1924; s. Theodor and Marta (Mitzky) P.; ed. U. Göttingen (Germany), Karlsruhe (Germany) Tech. Coll.; Dr ès sc.; m. Lore Auer, Dec. 23, 1948; children—Thomas, Wolfgang, Gertrud. Scholar, Deutsche Forschungsgemeinschaft, 1950-51; collaborator Siemens & Halske, Munich, Germany, 1951—; agrégé in math. Munich Tech. Inst., 1958—. Mem. German Math. Assn., Nachrichtentechnische Gesellschaft of German Union Electrotech., Soc. Math. and Applied Mechanics. Author: Mathematische Methoden in der Hochfrequenztechnik, 1956; Lauffeldröhren, 1958; also numerous articles. Home: 8024 Oberhaching, Holzstrasse 30. Office: Martinstrasse 76, Munich 8, West Germany.

POSER, Charles Marcel, physician; b. Antwerp, Belgium, Dec. 30, 1923; s. Maurice and Sadye (Gleitsman) P.; B.S., Coll. City N.Y., 1947; M.D., Columbia, 1951; m. Joan Doris Crawford, Sept. 3, 1950; children—William John, Nicholas Charles. Faculty, U. Kan. Med. Center, 1955-64, asso. prof. neurology, 1962-64; prof., head div. neurology U. Mo. Sch. Medicine, Kansas City, 1964—. Cons., U.S. Army Surgeon Gen., 1965—. Diplomate Am. Bd. Psychiatry and Neurology. Fellow A.C.P., Am. Acad. Neurology (past treas.), Royal Soc. Medicine, Japaneses Soc. Neurology; mem. Am. Neurol. Assn., Assn. for Research in Nervous and Mental Diseases, Am. Assn. Neuropathologists, Pan-Am. Med. Assn., World Fedn. Neurology (med. exec. officer 1959-60). Author: Syringomyelia), 1956; Dictionary of Drugs used in Psychiatry and Neurology, 1962; also numerous articles. Editor-in-chief World Neurology, 1959-61. Research on classification of diseases of myelin sheath, radiol. investigation of cerebrovascular diseases, clin. neurology. Home: 616 W. Meyer Blvd., Kansas City, Mo. 64113. Office: Kansas City Gen. Hosp., Kansas City, Mo. 64108.*

POSEY, Chesley Johnston, Jr., Am. civil engr.; b. Mankato, Minn., June 12, 1906; s. Chesley Justin and Maude (Johnston) P.; B.S. in Civil Engring., U. Kan., 1926; M.S., U. Ill., 1927; m. Mildred Darlene Misbach, June 9, 1940; children—James Bennett, Edith Ann. Faculty, State U. Ia., Iowa City, 1929-62, prof. civil engring., 1946-62, head dept., 1949-62; founder, dir. Rocky Mountain Hydraulic Lab., Allenspark, Colo., summers 1946—; prof. civil engring. U. Conn., Storrs, 1962—. Cons. engr. on structural and erosion protection problems. Fellow Am. Soc. C.E., mem. Am. Concrete Inst., Am. Geophys. Union, Am. Assn. U. Profs., Permanent Assn. Nav. Congresses, Internat. Assn. for Hydraulic Research, Am. Soc Engring, Edn. Author: (with S.M. Woodward) Hydraulics of Steady Flow in Open Channels, 1941; (booklet) Water Surface Profiles, 1961; also articles. Research on hydraulics of open channel flow, especially friction losses; devel. gen. methods of analyzing fluctuating records such as daily temperature, turbulence, roughness profiles; expts. with reinforced concrete, corners in tension, hooks, anchorage and bond. Home: Rural Route 2, Box 293, Storrs, Conn. 16268.*

POSIDONIOS OF APAMEA, Greek Stoic philosopher; b. Apamea, Syria, circa 135 B.C.; studied under Panaetios of Rhodes at Athens; traveled extensively to Egypt, Spain, Gaul, England; opened school at Rhodes, 97 B.C., where he taught Stoic philosophy (to Cicero 78 B.C., and to Pompey); supported Stoicism with contemporary learning; a monist, held world is hierarchy of grades of being, from lowest inorganic thing to God with the whole bound together in a universal harmony; held vital force emanating from sun permeated world; maintained supralunar world (imperishable) sustains sublunar world (perishable) by forces, but that the two worlds are bound together in man; ascribed tides to combined action of sun and moon (this an example of sympathy that unites all parts of cosmic system); recognized connection between tides and phases of moon; called attention to spring and neap tides; reaffirmed Stoic doctrine of divination and helped popularize astronomy; wrote many works history, astronomy, and geography, now lost; propounded a theory of cultural development that held for a golden age of the wise in past; composed a commentary on Plato's Timaeus; ascribed origins of atomic theory to Phoenecian, Mochos of Sidon (Moschus the Phoenecian); in math., wrote on parallels, on distinction between theorems and problems, and on existence theorems; studied ethnography, stressing influence of climate and natural conditions on character and way of life; observed earthquakes, volcanoes; visited and described mines; tried to improve Eratusthenes' estimate of circumference of earth, lowering Eratusthenes' value and claiming a man sailing west from the Atlantic could reach India (encouraged Columbus). Died after 51 B.C.

POSIDONIOS THE PHYSICIAN; Greek physician; b. 2d century A.D.; bro of Philagrios. First to try to localize functions in brain; described phrenitis, lethargy, coma, catalepsy, giddiness, nightmare, melancholy, hydrophobia (based on his own observations).

POSKANZER, Arthur M.; Am. chemist; b. N.Y.C., June 28, 1931; s. Samuel I. and Adele (Kerman) P.; A.B., Harvard, 1953; M.A., Columbia, 1954; Ph.D., Mass. Inst. Tech., 1957; m. Lucille Block, June 12, 1954; children—Deborah Rae, Jeffrey Alan, Harold Mark. Research asso., asso. chemist, chemist Brookhaven Nat. Lab., Upton, N.Y., 1957-66; chemist Lawrence Radiation Lab., Berkeley, Cal., 1966—. Mem. Am. Phys. Soc., Am. Chem. Soc. Research, publs. on high energy nuclear reactions, nuclear spectroscopy; discoverer many isotopes of light elements. Office: Lawrence Radiation Lab., Berkeley, Cal. 94720.*

POSNETTE, Adrian, English virologist; b. Birmingham, Eng., Jan. 11, 1914; s. Frank William and Enid (Webber) P.; B.A., Cambridge U., 1936, Sc.D., 1958; postgrad. Imperial Coll. Tropical Agr., Trinidad, W.I., 1936-37; Ph.D., U. London, 1952; m. Isabelle La Roche, July 25, 1937; children—Jane (Mrs. James M. Benson, Suzanne, John. Econ. botanist Dept. Agr., Gold Coast, Africa, 1937-44; head botany and plant pathology W. African Cacoa Research Inst., Gold Coast, 1944-49; staff E. Malling Research St., Kent, Eng., 1949—, head plant pathology dept., 1957—. Mem. Assn. Applied Biologists, Soc. for Gen. Microbiology. Editor: Virus Diseases of Apples and Pears, 1963; Research, numerous publs. on virus diseases of stone and pome fruit trees in Eng., strawberry and Ribes crops, including vectors and thermotherapy; pioneered research on virus diseases of cacao in W. Africa, including vector transmission, wild host plants, virus strains, resistance and control. Home: Ashurst, Sutton-Valence, Kent. Office: E. Malling Research Sta., Maidstone, Kent, Eng.*

POSPISIL, Leopold Jaroslav, anthropologist; b. Olomouc, Czechoslovakia, Apr. 26, 1923; s. Leopold and Ludmila (Petrlak) P.; J.U.C., Charles U., Prague, Czechoslovakia, 1948; student Masaryk's U., Ludwigsburg, Germany, 1948-49; A.B., Willamette U., 1950; M.A. in Anthropology, U. Ore., 1952; Ph.D. (Ford Found. fellow) Yale, 1956; m. Zdenka Smydova, Jan. 31, 1945; children—Zdenka, Miraslava. With Lawyer's Office, Olomouc, 1947-48; staff Peabody Mus., Yale, 1953—, curator anthropology, dir. div. anthropology, 1965—, faculty, 1956—, prof. anthropology, 1965—; prof. anthropology So. Conn. State Coll., New Haven, 1963—. Sr. Sterling fellow, 1955-56; Am. Philos. Soc. fellow, 1959, 62; John Simon Guggenheim fellow, 1962; NSF Research grantee, 1962-68; Social Sci. Research Council fellow, 1959, 62, 66; Fulbright fellow, 1966. Author: Kapauku Papuans and Their Law, 1958; Kapauku Papuan Economy, 1963; The Kapauku Pauans of West New Guinea, 1963; also articles. Research on comparative theory of law, formal analysis of legal systems, quantitative analysis of primitive economy, ethnology of New Guinea, Eskimo, Tyrol. Home: 554 Orange St., New Haven 06520.*

POSSEL, René de, French mathematician; b. Marseilles, France, Feb. 7, 1905; s. Raoul and Marthe (Seignon) P.; ed. École normale supérieure, Paris, France; Ph.D. agrégé; m. Yvonne Liberati, 1938; children—Yann, Maya, Daphné. Instr., Faculty of Sci. Marseilles, Clermont-Ferrand and Besançon; prof. differential and integral calculus at Besançon, later Clermont; prof. rational mechanics and analysis in Algiers; prof. numeric analysis at Paris; lectr. Paris Poly.; dir. Blaise-Pascal Inst., also Programming Inst.; pres. orgn. com. Internat. Calculus Center, Rome, Italy. Decorated Legion of Honor; several times laureate of Inst.; recipient prize Internat. Union Railroads. Author: Surfaces de Riemann et polygone fondamental de Poincaré; Dérivation abstraite des fonctions d'ensemble; Principes mathématiques de la mécanique classique; Bien-ordonnance effective de l'ensemble des parties d'un ensemble bien ordonné; Principe de Huyghens pour une onde electromagnétique. Home: 55, av. du Panorama, Bourg-la-Reine (Seine). Office: 23, rue du Marco, Paris 19, France.

POSSONY, Stefan Thomas, social scientist; b. Vienna, Austria, Mar. 15, 1913; s. Ernst and Hermine (Siller) P.; Ph.D., U. Vienna, 1935; LL.D., Lincoln U., 1965; Phil. degree (hon.), Chunghua Acad., 1967; m. Regina Golbinder, Nov. 1961; 1 dau., Andrea. Came to U.S., 1940, naturalized, 1945. Spl. adviser USAF, Washington, 1946-61; prof. internat. politics Grad. Sch. Georgetown U., 1946-61; asso. Fgn. Policy Research Inst. U. Pa., 1955—; dir. internat. polit. studies program Hoover Instn. Stanford, 1961—; vis. prof. U. Cologne, 1962-64. Recipient Exceptional Civilian award USAF, 1959. Author: Tomorrow's War, 1938; Strategic Air Power, 1949; (with Robert Strausz-Hupé) International Relations, 1950, 54; Century of Conflict, 1953; (with Friedrich Gentz) Three Revolutions, 1959; (with others) A Forward Strategy for America, 1961; (with Nathaniel Weyl) The Geography of Intellect, 1963; Lenin, The Compulsive Revolutionary, 1964; Strategie des Friedens, Sicherheit und Fortschritt im Atomzeitalter, 1964; Zum Bewacltigung der Kriegschuldfrage, 1968. Editor: Lenin, A Reader, 1966. Research on polit. reforms, adjustment of strategic thinking to facts of modern life and tech., various myths which are bedeviling social scis. Home: 23262 Montclair Way, Los Altos, Cal. 94022. Office: Hoover Instn., Stanford U., Stanford, Cal. 94305.*

POST, Howard William, Am. chemist; b. Syracuse, N.Y., Sept. 18, 1896; s. William Alexander and Edith (Van Wagenen) P.; B.S. in Chemistry, Syracuse U., 1919, M.S., 1921; Ph.D., Johns Hopkins, 1927; m. Clara Eleanor Knappenberger, Dec. 20, 1928; 1 dau., Martha Ann. Instr., Syracuse U., 1920-22; faculty State U. N.Y., Buffalo, 1923—, prof. chemistry, 1952—. Am. Philos. Soc. grantee, 1963; grantee various govt. agys. Mem. Am. Chem. Soc. (past chmn. Western N.Y. sect.), Am. Inst. Chemiste (past chmn. Niagara chpt.), Sigma Xi, Alpha Chi Sigma, Gamma Alpha, Chi Beta Phi, Pi Mu Epsilon. Author: The Chemistry of Aliphatic Orthoesters, 1943; Silicones and Other Organic Silicon Compounds, 1949; also numerous articles. Research on structure, synthesis, properties organic compounds containing silicon, germanium, tin, titanium. Home: 94 N. Ellicott St., Williamsville, N.Y. 14221. Office: 3435 Main St., Buffalo 14214.*

POST, Wright, Am. surgeon; b. North Hempstead, L.I., N.Y., Feb. 19, 1766; s. Jotham and Winifred (Wright) P.; studied at London Hosp.; M.D. (hon.) regents U. State N.Y., 1814; m. Mary Magdalen Bailey, 1790. Prof surgery med. dept. Columbia, 1792-1813, created anat. mus.; prof. anatomy and physiology Coll. Physicians and Surgeons, 1813, pres., 1821-26; attending surgeon N.Y. Hosp., 1792-1821, cons. surgeon, 1821; ligated subclavian outside scaleni muscles, 1817; helped introduce Hunterian principles of surg. thought and procedure in Am.; 1st in U.S. to perform operation for case of false aneurism of femoral artery. Died Throgg's Neck, N.Y., June 14, 1828.

POSTERNAK, Jean Marc, Swiss physiologist; b. Geneva, Switzerland, Sept. 29, 1913; s. Swigel and Rose (Kleiner) P.; ed. Coll. and U. Geneva; M.D.; m. Yvonne Gallia, Apr. 20, 1943; children—Laurence, Nicole. Intern Geneva U. Med. Clinic; asst. Inst. Microbiology, U. Lausanne (Switzerland), 1939-40, asst., later chief works Inst. Physiology, 1940-41; scholar Johnson Found. Med. Physics, U. Pa., 1941-46, 48-50, prof. pharmacology at univ., 1946-48; pvt. tutor U. Lausanne, 1950-51; now prof. physiology, dir. Inst. Physiology, U. Geneva. Decorated laureate Faculty of Medicine, U. Geneva. Mem. Swiss Nat. Council Sci. Research, Swiss Soc. Physiology, Assn. Physiologists, A.A.A.S., Am. Physiol. Soc. Author: La cirrhose pigmentaire, 1942; Les mécanismes élémentaires de la transmission synoptique, 1950; also publns. on neurophysiology about action of anesthetics on nervous system. Research on mechanisms of action of anesthetics; neurophysiology. Home: 22, av. Krieg. Office: Inst. Physiologie, Ecole de Médecine, Geneva, Switzerland.

POSTERNAK, Theodore, chemist; b. Paris, France, Sept. 28, 1903; s. Swigel and Rose (Kleiner) P.; ed. univs. Geneva (Switzerland), Munich (Germany) and London (Eng.); Dr ès sc.; m. Denise Bron, Sept. 18, 1937; children—Rose Françoise, Michel-Alexandre. Instr., U. Lausanne (Switzerland); prof. pharm. chemistry U. Basle (Switzerland); now prof. biol. chemistry U. Geneva. Mem. Swiss (Werner medal), Am. chem. socs., French Biol. Chem. Soc. Publns. on organic and biol. chemistry in Helvetica Chimica Acta, Jour. Am. Chem. Soc., Jour. Biol. Chemistry. Research on biochemistry of carbohydrates, of fungal pigments, of vitamines and antivitamins of micro-organisms. Home: 25, rue de l'Athenée. Office: Ecole de chimie, 22, bd des Philosophes, Geneva, Switzerland.

POSTGATE, John Raymond, English microbiologist; b. London, June 24, 1922; s. Raymond William and Daisy (Lansbury) P.; B.A. with 1st honors (Williams Exhibitioner, War Meml. student), Balliol Coll., Oxford (Eng.) U., 1945, M.A., 1950, D.Phil., 1950, D.Sc., 1965; m. Muriel Mary Stewart, Oct. 20, 1948; children—Selina, Lucy, Joanna. Sr. sci. officer Chem. Research Lab., Teddington, Eng., 1948-52, prin. sci. officer, 1952-59; prin. sci. officer Microbiol. Research Establishment, Poston, Eng., 1959-61, sr. prin. sci. officer, 1961-63; asst. dir. ARC unit Nitrogen Fixation, U. Sussex (Eng.), 1963—, prof. microbiology, 1964—. Fellow Inst. Biology, 1965. Mem. Soc. for Gen. Microbiology, Soc. for Applied Microbiology. Research, publs. on microbiology and biochemistry of sulfate-reducing bacteria, death and freezing damage in bacteria, biol. nitrogen fixation. Office: Chemistry Lab., U. Sussex, Falmer, Brighton, Eng.*

POSTLETHWAIT, Raymond Woodrow, Am. physician; b. New Martinsville, W. Va., Oct. 9, 1913; s. Melvin and Mina Mae (Butler) P.; B.S. in Medicine, W. Va. U., 1935; M.D., Duke, 1937; m. Mary Elizabeth Corbett, May 21, 1937; children—Margaret (Mrs. Joseph John Kalinowski), Mina B., Raymond Woodrow II, William E. Surg. instr. Bowman Gray Sch. Medicine, 1948-49; asst. surgery prof. S.C. Med. Coll., 1949-51; pvt. practice medicine, Kinston, N.C., 1951-55; chief surg. service VA Hosp., Durham, N.C., 1955-65, chief of staff, 1965—; asso. prof. surgery Duke Med. Center, 1955-57, prof. surgery, 1957—. Mem. A.M.A., Am. Assn. U. Profs., A.A.A.C., Soc. Exptl. Biology and Medicine, Southeastern Surg. Congress, So., N.C. surg. assns., Soc. for Surgery Alimentary Tract, Soc. U. Surgeons, A.C.S., Am. Bd. Surgery, Sigma Xi. Author: (with W. C. Sealy) Surgery of the Esophagus, 1961; (with J. C. Thoroughman) Results of Surgery for Peptic Ulcer, 1963; also numerous articles on healing of wounds. Home: 1513 Pinecrest Rd. Office: VA Hosp., Fulton St. and Erwin Rd., Durham, N.C. 27705.*

POSTLETHWAIT, Samuel Noel, Am. botanist; b. Wileyville, W. Va., Apr. 16, 1918; s. Frank and Etta (Mason) P.; A.B., Fairmont State Coll., 1940; M.S., W. Va. U., 1947; Ph.D., State U. Ia., 1949; m. Sara Madeline Cover, Mar. 22, 1941; children—John Harvey, Robert Neil. Tchr. pub. schs. W. Va., 1940-41; instr. U. Ia., 1947-49; faculty Purdue U., Lafayette, Ind., 1949—, prof. botany, 1963—. NSF fellow, 1957-58. Fellow Ind. Acad. Sci.; mem. A.A.A.S. (mem. coop. com. on teaching math. and sci. 1960—), Bot. Soc. Am., Internat. Soc. Plant Morphology, Am. Soc. Cell Biology, Torrey Bot. Club, Internat. Soc. Stereology, N.Y. Acad. Scis., Nat. Assn. Biology Tchrs., Sigma Xi. Author: Textbook of Intermediate Plant Science, 1957; (with H.T. Murray, J.D. Novak) An Integrated Experience Approach to Learning—With Emphasis on Independent Study, 1964; Plant Science—A Workbook with Audio-Programmed Approach, 1966; also articles, chpt. in books. Research on growth and devel. of corn with spl. emphasis on genetic mutants. Home: 3180 Soldiers Home Rd., West Lafayette, Ind. 47906. Office: Dept. Biol. Scis., Life Sci. Bldg., Purdue U., West Lafayette, Ind. 47907.*

POSTNIKOV, Mikhaîl Mikhâilovich, Russian mathematician; b. Shatura, Moscow Oblast, USSR, 1927; grad. Moscow State U., 1944; staff Steklov Math. Inst., USSR Acad. Scis., 1949—, sr. asso., 1954—; prof. Moscow U., 1954—. dr.math. sci. Author: The Structure of Intersecting Rings of Three Dimensional Manifolds, 1948; Homological Invariants of Continuous Mappings, 1949; A Link between Homologies and Homotopies, 1950; Classification of Continuous Mappings, 1951; Homotopic Theory, 1955; The Theory of Homologies of Smooth Manifolds and its Generalization, 1956; The Main Concepts of Algebraic Topology, 1960. Research and publs. on continuous mapping, theory of homotopies and homologies, manifolds. Address: USSR, Moskva, V-312, l-y Akademichesky pr. 28, Matematichesky Institut AN SSR.

POTAIN, Pierre Carl Édouard, French physician; b. Paris, July 19, 1825; M.D., Faculty of Medicine, Paris, 1853; head Clinic, Bauillaud, France; named physician Hosp. Ménarges, 1860, Saint-Antoine, 1865, Necker, 1866; named prof. pathology Faculty of Paris, 1876. Mem. French Acad. Scis., 1893. Author: De Sauffles vasculaires qui seuvent les hémorrhagies, 1853; Des lesions des ganglions lymphatiques viscéraiux, 1859; also articles. Invented portable air sphygmomanometer, 1889. Died Paris, Jan. 5, 1901.

POTIER, Alfred, French physicist, geologist; b. Paris, France, May 11, 1840; chief engr. Mines; prof. physics Ecole polytechnique, also Ecole des mines; mem. French Acad. Scis., 1891. Interpreted theories of Frenel as ether vibrations; applied Maxwell theory to electromagnetic phenomena; studied refraction of polarized light, elliptical and metallic reflection, magnetic rotation of polarization plane. Died Paris, May 8, 1905.

POTT, Johann Heinrich, German chemist; b. Halberstadt, Germany, 1692; studied under Hoffmann and Stahl, Halle, Germany; M.D., Halle, 1716 prof. theoretical chemistry, Collegium Medico-Chirurgicum, Berlin, later of practical chemistry, 1737; dir. Royal Hofapothek. Author: Chymische untersuchungen welche fürnhemlich . . . , 1746; Exercitationes chymicae, 1738; Observationum et animadversionum chymicarum, vol. 1, 1739, vol. II, 1741. Discovered manganese; worked out 30,000 mineral and earth analyses; studied Chinese porcelain techniques, zinc; differentiated earth into limestone, gypsum, clay and silica; prepared tables for various reactions of different kinds of earth; research on effects of high temperatures on mineral substances. Died Berlin, Mar. 20, 1777.

POTT, Percival, English surgeon; b. London, Jan. 6, 1714; s. Percival Pott; apprentice to Edward Nourse, 1729; m. Sarah Cruttenden, 1746; 5 sons, including Joseph Holden, 4 daus. Asst. surgeon St. Bartholomew's Hosp., London, 1745-49, chief surgeon, 1749-87. Fellow Royal Soc., 1764, Royal Coll. Surgeons Edinburgh (hon.); hon. mem. Royal Coll. Surgeons Ireland. Author: A Treatise on Ruptures, 1756; An Account of a Particular Kind of Rupture frequently attendant upon newborn Children, 1757; Fistula Lacrymalis, 1758; Remarks on that kind of Palsy of the Lower Limbs which is frequently found to accompany a Curvature of the Spine; also collected writings and books, 1775. Tchr. of John Hunter. Described type of spinal curvature (Pott's disease) now known to be caused by Tb of spine, 1779, also fracture of lower end of fibula (Pott's fracture) sometimes asso. with injury to medial malleolus and lower tibial articulation, 1765, and cancer of scrotum occuring in chimney sweeps (1st record of cancer caused by indsl. or occupational injury), 1775. Died Dec. 22, 1788.

POTTASCH, Stuart Robert, astrophysicist; b. N.Y.C., Jan. 16, 1932; s. Max and Juliette (Pollak) P.; B.Engring., Cornell U., 1953; student U. Leiden, 1955; M.S., Harvard, 1957; Ph.D., U. Colo., 1958; m. Anne Marie de Groot, Aug. 8, 1956; children—Carol, Edward. Physicist, U.S. Nat. Bur. Standards, Boulder, Colo., 1957-58; astronomer Obs. of Paris, France, 1959; vis. lectr. Princeton, 1960; mem. Inst. for Advanced Study, Princeton, 1961; prof. astronomy Ind. U., Bloomington, 1962-63; prof. astrophysics U. Groningen, The Netherlands, 1963—. Mem. astronomy sect. Dutch Acad. Council, 1963—. Mem. Internat. Astron. Union, Am., Dutch astron. socs. Research, numerous publs. on motion and radiation from gas between stars and initial stages of star formation, outer atmosphere of sun especially determination of chem. composition of sun, outburst and chem. composition of novae. Home: Zuidlaarderweg 1, Glimmen, The Netherlands. Office: Kapteyn Astron. Lab., Groningen, The Netherlands.*

POTTER, Andrey Abraham, engr., educator; b. Vilno, Lithuania, Aug. 5, 1882; s. Gregor and Rivza (Pelonsky) P.; came to U.S., 1897, naturalized, 1906; S.B., Mass. Inst. Tech., 1903; D.Eng., Kans. State U., 1925, Rose Poly. Inst., 1947, S.D. Mines and Tech., 1949, Purdue U., 1955, Mich. State U., 1956, Ind. Inst. Tech., LL.D., Norwich U., 1944; postgrad., D.Sc., Northeastern U., 1936, Alfred U., 1947, Northwestern U., 1953; m. Eva Burtner, June 28, 1906; children—James G., Helen C. With Gen. Elec. Co., Schenectady, Lynn, Mass., 1903-05, 13; faculty Kans. State U., Manhattan, 1905-20; faculty Purdue U., Lafayette, Ind., 1920-53, dean emeritus engring., 1953—. Pres., Bituminous Coal Research, Pitts., 1950-60; cons. to industry, utilities, govt. agys., 1913—. Hon. mem. Am. Soc. M.E. (pres. 1932-33), Soc. for Promotion Engring. Edn. (pres. 1924-25); mem. Am. Engring. Council (pres. 1936-38). Research covering thermodynamics; fuel; power plant engineering. Home: 517 Russell St., West Lafayette, Ind. 47406.

POTTER, D(avid) S(amuel), Am. physicist; b. Seattle, Jan. 16, 1925; s. William and Henrietta (Guernsey) P.; B.S., Yale, 1945; Ph.D. (AEC fellow) U. Wash., 1951; m. Juanita Mae Beck, July 14, 1945; children—Diana, Janice, Thomas, William. With Applied Physics Lab., U. Wash., Seattle, 1946-60, asst. dir., 1955-60; head sea operations dept. Def. Research Labs., Gen. Motors Corp., Santa Barbara, Cal., 1960-66, dir. AC Electronics div., 1966—. Chmn., Gov's. Adv. Commn. on Ocean Resources, 1967—. Mem Acoustical Soc., Nat. Oceanography Assn. (v.p. 1966—), Am. Phys. Soc., Am. Soc. Naval Engrs., Marine Tech. Soc. (charter, dir. 1964—). Research, publs. on acoustics and cosmic radiation; patentee acoustic instrumentation. Home: 1144 Estrella Dr., Santa Barbara, Cal. 93105. Office: 6767 Hollister Av., Goleta, Cal. 93017.*

POTTER, Edith Louise, Am. pathologist; b. Clinton, Ia., Sept. 26, 1901; d. William Harvey and Edna Rugg (Holmes) Potter; B.S., M.D., U. Minn., 1926, M.S., 1932, Ph.D., 1934; Douter Honoris Causa, U. Brasil, 1953; D.M.S., U. Pa., m. Alvin Meyer, June 17, 1944; one stepdaughter, Louisa. Intern Mpls. General Hospital, 1925; engaged in pvt. practice, 1927-31; instr., asst. prof. U. Chgo., 1934-47, asso. prof. pathology, dept. obstetrics and gynecology, 1947-56, professor, 1956-67, professor emeritus, 1967—; pathologist Chicago Lying-in Hospital; guest lecturer University Brasil, 1949; cons. Armed Forces Inst. Path., 1950—; cons. Surgeon Gen. Army, Far East, 1954. Member USPHS delegation to Russia, 1960. Distinguished scholar Nat. Assn. Retarded Children, 1960—. Recipient Award of Achievement, U. Minn. and Blackwell Citation, N.Y. Infirmary; Adair award Am. Gynecological Soc., 1963. Diplomate Am. Bd. Pathology. Mem. Terotology Soc., American Association Pathologists and Bacteriologists, Society Pediat. Research, Am. Pediatric Society, Am. Soc. Human Genetics, Chicago Pathologic Society, Academy Pediatrics, Am. Coll. Obstetricians and Gynecol. Am., Chgo. gynecol. socs., Sigma Xi. Hon. mem. 38th Parallel, Southern Honsh med. societies, Tex., Brazilian, Uruguayan, Brit., N.M. pediatric societies, Wash., N.M., Uruguayan obstet. assns., British Pediatric Pathology Club, Brazilian Society of Obstetricians and Gynecologists. Author: Fetal and Neonatal Death (with F. L. Adair), 1940, 49; Fundamentals of Human Reproduction, 1948; Rh, its relation to transfusion reactions and hemolytic disease of the newborn, 1947; Pathology of the Fetus and the Newborn, 1952, 2d edit. pub. as Pathology of Fetus and Infant, 1961. Contbr. articles med. jours. Home: Route 3, Box 658, Ft. Myers, Fla. 33901.

POTTER, James Leroy, Am. elec. engr.; b. Carthage, Mo., Dec. 4, 1905; s. Frank and Amanda Elizabeth (Hoover) P.; B.S., Kan. State U., 1928, M.S., 1930, E.E., 1939; m. Helen Trembley, Aug. 7, 1930; children—Donald Joseph, Mary Ann (Mrs. Richard Clinton Isaacs). With Western Electric Co., Chgo., 1928-29; grad. asst. Kan. State U., Manhattan, 1929-30; instr. U. Ia., 1930-39; asst. prof. Rutgers U., New Brunswick, N.J., 1939-42, asso. prof., acting chmn. 1942-46, prof., chmn. elec. engring. dept., 1946—. Mem. Soc. Profl. Engrs., Am. Soc. Engring. Edn., I.E.E.E. (sec. treas. Princeton sect. 1954-55, chmn. 1956-57, 1st nat. prize Winter conv. 1954), Sigma Xi, Phi Kappa Phi, Sigma Tau, Tau Beta Pi, Eta Kappa Nu. Author: (with Sylvan Fich) Theory of A.C. Circuits, 1958, Theory of Networks and Lines, 1963; also articles. Patentee electrical devices. Home: 415 Lincoln Av., Highland Park, N.J. 08904. Office: Rutgers The State U., New Brunswick, N.J. 08903.*

POTTER, Loren David, Am. biologist; b. Fargo, N.D., June 23, 1918; s. John Henning and Jessie Alice (Payne) P.; B.S., N.D. State U., 1940; postgrad. Ia. State U., 1941; M.A., Oberlin Coll., 1946; Ph.D., U. Minn., 1948; m. Elvira V. Heuer, June 10, 1941; children—Claudia Ann, Nancy Ellen, Audrey Louise, Janet Lynn. With N. D. Game and Fish Dept., 1941; research div. Goodyear Tire & Rubber Co., 1943-45; from asst. prof. to prof. biology N.D. State U., 1948-58; prof. biology, chmn. dept. U. N.M., 1958—, also Radiation Biol. Insts. Recipient numerous research grants plant ecology. Fellow A.A.A.S.; mem. Ecol. Soc. Am. (asso. bot. editor), Am. Soc. Range Mgmt., Sigma Xi, Phi Kappa Phi. Home: 1612 Lafayette Dr. N.E., Albuquerque 87106.*

POTTER, Roy Frank, Am. physicist; b. Portland, Ore., Dec. 3, 1919; s. R. Roy and Rachael (Van Fossen) P.; B.S., U. Wash., 1947; M.S., U. Md., 1951; m. Helen I. Muir, Jan. 2, 1943; children—Matthew M., Andrew R. Physicist gas ion physics field Nat. Bur. Standards, Washington, 1948-53, project leader solid state physics, 1953-56; head Solid State Physics br. Infrared div. Naval Ordnance Lab., Corona, Cal., 1957-58, head Infrared div. research dept., 1958—. Chmn. ad hoc com. for organizing Far Infrared Physics Symposium, Riverside, Cal., 1964; mem. staff Advanced Study Inst. on Optical Properties of Solids U. Freiburg (Germany), 1966. Mem. Am. Phys. Soc., Optical Soc. Am., Sigma Pi Sigma. Research on low energy ion-molecule charge transfer, elastic and electromech. studies of indium antimonide, multi-source techniques for evaporating compound semiconductors, optical properties of semiconductors with spl. emphasis on lattice vibrations and interband electronic transitions. Home: 3940 Melody Lane, Riverside, Cal. 92504. Office: U.S. Naval Ordnance Lab., Corona, Cal. 91720.*

POTTER, Van Rensselaer, Am. biochemist; b. Day County, S.D., Aug. 27, 1911; s. Arthur Howard and Julia Eva (Herpel) P.; B.S. with high honors, S.D. State Coll., 1933, D.Sc., 1959; M.S., U. Wis., 1936, Ph.D., 1938; m. Vivian Christensen, Aug. 3, 1935; children—Karin, John, Carl. Nat. Research fellow, U. Stockholm, 1938-39, Rockefeller fellow U. of Sheffield and Chgo., 1939-40, Bowman fellow U. of Wis., 1940-42, asst. prof. oncology, U. Wis., 1942-45, asso. prof., 1945-47, prof., 1947—; vis. prof. Andean Biology U., Lima, Peru, 1952-53. Recipient Paul Lewis award enzyme chemistry, 1947, Bertner award cancer research, U. Tex., 1961. Mem. Am. Soc. Biol. Chemists, Am. Assn. for Cancer Research (Clowes medal, lectr. 1964), Am. Chem. Soc., Am. Soc. Cell Biology (pres. 1964-65), Japan Biochem. Soc. (hon.), Am. Acad. Arts and Scis., Sigma Xi, Gamma Alpha. Author: Nucleic Acid Outlines, Vol. 1, 1960. Contbr. article, Advances in Enzymology; editor, Methods in Medical Research, 1948. Home: 167 N. Prospect Av., Madison 5, Wis.*

POTTER, Wilbur Furse, Am. physiologist; b. Cherryvale, Kan., Oct. 11, 1893; s. William Asahel and Nannie (Furse) P.; A.B., U. Kan., 1925, A.M., 1927, Ph.D., 1930; M.D., U. Chgo., 1939; m. Grace Vanetta Barton, June 16, 1920; 1 son, Paul Herbert. Instr. physiology U. Kan., 1925-30; asst. prof. U. Ga., 1930-33, acting prof., 1932-33; asso. prof. Georgetown U., 1933; prof. physiology U. Miss., 1933-48; dean, prof. dept. physiology U. N.D., Grand Forks, 1948-53, prof., 1953-64. Mem. N.Y. Acad. Sci., A.A.A.S., Phi Beta Kappa, Sigma Xi, Alpha Omega Alpha. Author: (with others) Laboratory Manual of Medical Physiology, 1948, Laboratory Manual of Physiology for College Students, 1953. Developed electronic stimulator, micro vol. pressure recorder; modification of Gibbs artificial heart; research in neuromuscular physiology. Home: 41 Minnebago Way, Mason City, Ia. 50401.

POTTS, Albert Mintz, Am. physician; b. Balt., June 8, 1914; s. Isaac and Leah (Mintz) P.; A.B., Johns Hopkins, 1934; Ph.D., U. Chgo., 1938; M.D., Western Res. U., 1948; m. Esther Topkis, June 14, 1938; children—William Topkis, Leah Susan, Deborah. Research asst. U. Chgo., 1938-42, research asso., instr., 1944, biochemist Metall. Lab. Manhattan Project, 1944-45; sr. instr. biochemistry Western Res. U., 1951-54, asso. prof. ophthalmic research, 1954-59; asst. ophthalmologist U. Hosps. Cleve., 1951-59; prof. opthalmology, dir. research in opthalmology dept. surgery U. Chgo., 1959—. Mem. sensory diseases study sect. NIH, 1958-62, mem. vision research tng. com., 1962-66, cons. surgeon gen.'s adv. com. on smoking and health, 1963-64; mem. NRC Com. on Vision Nat. Acad. Scis., Armed Forces. Mem. Assn. for Research in Ophthalmology (Jonas S. Friedenwald Meml. award 1961), Am., Chgo. opthal. soc., Am. Acad. Opthalmology and Otolaryngology, Am. Chem. Soc., Am. Soc. Biol. Chemistry, Soc. for Exptl. Biology and Medicine, A.A.A.S., Am. Assn. U. Profs. Contbr. chpt. to Physiological Pharmacology, vol. II, 1965. Research, publs. on chemistry of hormones of posterior lobe of pituitary gland, action of toxic substances on lung, transparency of ocular structures, nature of elec. phenomena produced by visual system, action of toxic substances on eye.*

POTTS, Laurence Holker, English physician; b. London, Eng., Apr. 18, 1789; s. Cuthbert and Ethelinda Margaret (Thorpe) P.; M.D., U. Aberdeen (Scotland), 1825; m. Ann Wright, 1820; children—John Thorpe, Benjamin L.F., 4 daus. Practice medicine, Truro, Eng.; named supt. county lunatic asylum, Bodmin, Eng., 1828; founder instn. for treatment spinal diseases, Blackheath, Eng., 1837, left instn., circa 1843. Recipient Gold medal Soc. Arts, 1848. Invented and patented sinking founds. under water for creating partial vacuum 1843. (leading to system of sinking founds. using compressed air), Died Mar. 23, 1850.

POTTS, Willis John, Am. surgeon; b. Sheboygan, Wis., Mar. 22, 1895; s. Horace and Hannah (Boeyink) P.; A.B., Hope Coll., Holland, Mich., 1918; S.B., U. of Chicago, 1920; M.D., Rush Med. Coll., 1924; interne Presbyn. Hosp., Chicago; Logan fellowship in surg. Rush Med. Coll., 1925-26; post grad. work, Frankfort, Germany, 1930-31; m. Henrietta Neerken, July 7, 1922; children—Willis John, Edward Eugene, Judith Eleanor. Began gen. practice, Oak Park, Ill.,

1925; specialized in surgery, 1931-65, ret.; author syndicated newspaper column, 1965—; professor of emeritus surgery Northwestern U. Med. sch., 1960—; cons. surgery, Children's Memorial Hospital, Chicago. Sergt. Chem. Warfare Service, 1917-18; 1 year in U. S. and 1 year in France; lt. col. & colonel A.U.S., serving in Southwest Pacific with 25th Evacuation Hosp., 1942-45. Fellow Am. Coll. Surgeons; certified by Am. Bd. of Surgery; mem. Am. Med. Assn., Ill. and Chicago Med. Socs., Chicago Surg. Soc., Western Surg. Soc., Inst. Med. of Chicago, Am. Assn. Thoracic Surgery, Am. Surg. Assn., Central Surg. Assn. Am. Heart Assn. (pres. Chgo. 1960-61). Unitarian. Author: The Surgeon and The Child, 1959; Your Wonderful Baby, 1966. Contbr. to med. jours. Research on surgical treatment of congenital heart disease; instrumental in devel. "blue baby" operation. Home: 524 Yawl Lane, Sarasota, Fla.

POUCHET, Félix Archimède, French naturalist; b. Rouen, France, Aug. 26, 1800; s. Louis-Ezechiel P.; studied medicine at Rouen, then Paris, until 1827; a son, Georges. Dir. Rouen Mus. Natural History, 1828, and Rouen Jardin des plantes; became prof. Sch. Medicine, Rouen, 1838. Mem. French Acad. Scis., 1849. Histoire naturelle des solanées, 1829; Traité élémentaire de botanique appliquée, 1835; Recherches sur l'anatomie et la physiologie des mollusques, 1842; Hetérogénie ou Traité de la génération spontanée (advocated doctrine of spontaneous generation), 1859; l'Univers, les infiniment grands et les infiniment petits, 1865. Studied night shade, applied botany, also anatomy and physiology of mollusks. Died Rouen, Dec. 6, 1872.

POUILLET, Claude Servais Matthias, French physicist; b. Cusance, France, Feb. 16, 1790; prof. École polytechnique, also Faculté des Scis.; leader Conservatoire des Arts et Metiers; became prof. physics Sorbonne, Paris, 1838. Mem. French Acad. Scis., 1837. Author: Éléments de physique expérimentale et de météorologie, 2 vols., 1827. Measured solar constant; used gas thermometer with platinum bulb to measure high temperatures, 1836; studied compressibility of gases, inner and outer resistance of elec. circuits and battery strength, 1837; invented magnetic pyrometer for high temperatures, 1837, tangent, sine galvanometer, pyrheliometer, 1839, sine magnetic compass, 1844; prepared color scale for platinum as it is heated to incandescence. Died Paris, June 14, 1868.

POULIQUEN, Y., French ophthalmologist; b. Mortain, Feb. 17, 1931; s. Jean-François and Renée Blanche (Merio) P.; M.D., U. Paris, 1963; m. Jacqueline Louise Brevet, Nov. 11, 1952; 1 dau., Muriel. Head clinic Med. Faculty, U. Paris, prof. ophthalmology, 1965—; ophthalmologist Tenon Hosp., Paris, 1965—. Mem. French, Paris ophthal. socs. Editorial sec. Archives of Opthalmology. Research, publs. on ocular anatomo-pathology, fine structures of normal and path. eye, electron microscopy, biol. problems of cornea development, transplantation; devel. silico-dessication method (with P. Payrau). Address: 22 rue Henri-Heine, Paris, France.*

POULSEN, Christian Henrik Otto, Danish paleontologist, stratigrapher; b. Copenhagen, Denmark, Mar. 8, 1896; s. Poulsen and Mary Rasmussen P.; M.Sc., U. Copenhagen, 1924, Ph.D., 1927; m. Lucie Hélène Perret-Gentil, May 14, 1924; children—Valdemar. With U. Copenhagen, 1921—, prof. paleontology and stratigraphy, 1945-66, curator Mineral.-Geol. Mus., 1936-45. Recipient Gold medal U. Copenhagen, 1922. Mem. Danish Geol. Soc. (past pres.), Internat. Paleontol. Union (pres. 1961-65), Paleontol. Soc. Am., Geol. Soc. London, Royal Danish Acad. Sci., World Acad. Art and Sci. (charter). Research, publs. on paleontology and stratigraphy (lower Palaeozoic) of Denmark, Greenland, Argentina. Home: 69 Egebjergvej, Ballerup, Denmark, 2750. Office: 7 Oster Voldgrade, Copenhagen, Denmark.*

POULSEN, Valdemar, Danish engr.; b. Copenhagen, Denmark, Nov. 23, 1869; ed. sch., Christiansharn, Denmark; numerous hon. degrees; began in machine shops, circa 1890; later joined engring. dept. Copenhagen Telephone Co. Poulsen Gold medal from Danish Acad. Tech. Scis. established in his honor. Originated method of magnetic wire recordings (telegraphone), arc generator for high-frequency continuous oscillations (Poulsen arc); achieved unattenuated high frequency vibrations using toning arc in hydrogen with electromagnetic spark extinguishing; pioneered (with Petersen) sound film. Died 1942.

POULSON, Donald Frederick, Am. biologist; b. Idaho Falls, Ida., Oct. 5, 1910; s. Christian Frederick and Esther (Johnson) P.; B.S., Cal. Inst. Tech., 1933, Ph.D., 1936; M.A., Yale, 1955; m. Margaret Judd Boardman, June 18, 1934; children—Donald Boardman, Christian Frederick. Teaching fellow biology Cal. Inst. Tech., 1934-36; research asst. dept. embryology Carnegie Inst. Washington, 1936-37; instr. biology Yale, 1937-40, asst. prof., 1940-46, asso. prof., 1946-55, prof., 1955—, chmn. dept., 1962-65; Gosney research fellow biology Cal. Inst. Tech., 1949; research collaborator Brookhaven Nat. Lab., 1951-55. Fulbright com. NRC, 1958-60; vis. lectr. developmental biology U. Chgo., 1959; vis. lectr. genetics U. Cal., Davis, 1963. Fellow Calhoun Coll., Yale, 1954—; Fulbright sr. research scholar Commonwealth Sci. and Indsl. Research Orgn., Canberra, Australia, 1957-58, 66-67; Guggenheim fellow, 1957-58. Mem. central qualifications bd. NIH, 1961-63, Fellow A.A.A.S. (council 1960-62); Internat. Inst. Embryology, Am. Assn. U. Profs., Am Soc. Naturalists (treas. 1951-53), Am. Soc. Zoologists, Genetics Soc. Am., Soc. Study Growth and Devel., Conn. Acad. Arts and Scis., Sigma Xi. Author: The Embryonic Development of Drosophila Melanogaster, 1937; contbr. Biology of Drosophila, 1950. Mem. editorial bd. Jour. Morphology. Contbr. articles sci. jours. Home: 96 Green Hill Rd., Orange, Conn. 06477. Office: Kline Biology Tower, Yale U., New Haven 06520.*

POULTER, Thomas Charles, Am. explorer, dir. research; b. Salem, Ia., Mar. 3, 1897; s. Micajah Lewis and Alberta Viola (Pool) P.; grad. Ia. Wesleyan Acad., 1918; B.S., Ia. Wesleyan Coll., 1923; hon. D.Sc., 1935; Ph.D., U. of Chicago, 1933; m. Helen Elizabeth Gray, Feb. 18, 1935; children—Howard, Glenn, Louis, Thomas Charles. Instr. of physics, Ia. Wesleyan Acad., 1916; asst. in biology, chemistry and physics, Iowa Wesleyan Coll., 1919-23; asst. in chemistry, U. of Chicago, 1923-25; head dept. of chemistry, Iowa Wesleyan Coll., 1925-27, head dept. of physics, 1927-30, head div. mathematics, physical science and astronomy, 1930-33; research Univ. of Ia., 1931; mem. of Arizona Meteor Expedition, 1932; John Simon Guggenheim fellow, 1933-35; with Byrd's Second Antarctic Expdn. as sr. scientist in charge of sci. program and second in command, 1933-35; asso. dir., Armour Research Found., Chgo., 1936-48; prof. Stanford, asso. dir. Research Inst. 1948—, dir. Poulter labs., 1954-60, sci. dir., gen. manager phys. and life scis., 1960—; cons. U.S. Antarctic Service Office Scientific Research and Development, U.S. Navy and Columbia U. Nat. Defense Laboratory. Designed Antarctic snow cruiser and was with U.S. Antarctic Service Expedition, 1939-40. Inventor of Poulter Seismic Method of Geophysical Exploration. Recipient of Spl. Congressional Medal, 1940, 47, World War Service Medal, spl. gold medal by Nat. Geographic Soc., 1937, Alumni medal U. Chgo., 1963. Fellow Am. Inst. Physics, Am. Phys. Soc., Cal. Acad. Scis.; life mem. National Geographic Society; mem. Society Exploration Geophysicists, Geophysical Union, A.A.A.S., Seismol. Soc. America, American Geographic Society, Am. Polar Soc. (sr. v.p.), The Explorers Club, Sigma Xi, Lambda Chi Alpha, Sigma Pi Sigma. Research, numerous U.S. and fgn. patents, and many publs. in chemistry, physics, geophysics, extreme pressure effects, terrestrial magnetism, Antarctic meteorology and auroral phenomena, seals of Antarctic, high explosive and shock pulse phenomena, biol. sonar and diving mammals. Address: 13054 La Creste Dr., Los Altos Hills, Cal. Office: Stanford Research Inst., Menlo Park, Cal.*

POULTON, Charles Edgar, Am. ecologist; b. Oakley, Ida., Aug. 2, 1917; s. Richard and Narrie Jane (Queen) P.; B.S. in Forestry, U. Ida., 1939, M.S., 1948; postgrad. Mont. State Coll.; Ph.D. in Ecology, Wash State U., 1955; m. Marcile B. McCoy, Sept. 29, 1939; children—Richard Charles, Robert John, Mary Jane, Betty Jean. Forest adminstr., dist. forest ranger U.S. Forest Service, 1937-47; field asst. Intermountain Forest and Range Expt. Sta., 1939-40; asst. prof. range mgmt. Mont. State Coll., 1946-47; instr. U. Ida., 1947-49; asso. prof., organizer new range mgmt. program Ore. State U., 1949-59, prof. range ecology, head range mgmt. program univ., also Agr. Expt. Sta., 1959—. Cons. in range ecology and range mgmt., 1958—. Recipient Merit certificate Am. Grassland Council, 1963. Mem. Soc. Am. Foresters (chmn. range mgmt. sect. 1955), Am. Soc. Range Mgmt. (charter pres. Pacific N.W. sect. 1962, nat. dir. 1966—), Ecol. Soc. Am., Soil Sci. Soc. Am., Sigma Xi, Phi Sigma, Xi Sigma Pi, Phi Eta Sigma. Asso. editor: Ecology, 1965—, Jour. Forestry, 1956-61. Research, publs. on phytosociology, vegetation-soil relationships in steppe and forest vegetation, multispectral remote sensing imagery as applied to ecology and range resource inventory. Home: 902 N. 36th St., Corvallis, Ore. 97330.*

POULTON, Sir Edward Bagnall, English zoologist, entomologist; b. Reading, Eng., Jan. 27, 1856; s. William Ford and Georgina Selina (Bagnall) P.; B.A., Oxford (Eng.) U., 1876; D.Sc., 1900; LL.D., Princeton; D.Sc. (hon.), Durham, Eng., Dublin, Ireland, Reading, Eng., Sidney, Australia; m. Emily Palmer, 1881; 2 sons, 3 daus. Lectr. natural sci., tutor Keble Coll., Oxford U., 1880-89, Hope prof. zoology, 1893-1933, became fellow Jesus Coll., 1898. Knighted, 1935. Fellow Royal Soc., 1889 (v.p. 1909-10); mem. Royal Entom. Soc. London (pres. 1903-04, 25-26), Linnean Soc. London (pres. 1912-16), Brit. Assn. (became pres. 1937). Author: The Colors of Animals, 1890; Charles Darwin and the Theory of Natural Selection, 1896; Essays on Evolution, 1908; Charles Darwin and the Origin of Species, 1909; also numerous articles. Studied entomology, including colors, marking and protective attitudes of caterpillars; supported Darwinian theory of organic evolution by natural selection and opposed Mendelian mutation theory. Died Oxford, Nov. 20, 1943.

POULTON, Edward Palmer, English physiologist; b. Eng., 1883; s. Edward Bagnall and Emily (Palmer) P.; B.A., Oxford, Eng., 1905, M.A., M.D.; m. Elfrida MacLean; 3 sons, 2 daus. Became demonstrator physiology Guy's Hosp., 1912, med. registrar, 1913, asst. physician, 1914, physician, 1926; sr. physician, cons. physician Lewisham Hosp.; prof. Oxford, 1893-1933. Mem. Internat. Soc. Med. Hydrology (chmn.), Brit. Assn. (pres. physiol, sect. 1937). Author: (with Agyll Campbell) Oxygen and Carbon Dioxide Therapy; Diets and Recipes and Treatment of Diabetes and Obesity; also articles on diabetes, dyspnea, goiter. Research (with Jack Joffa) on partition or distbn. of carbon dioxide between plasma and corpuscles in reduced and oxygenated blood, 1920. Died 1939.

POULTON, Eustace Christopher, English psychologist; b. Eng., Aug. 23, 1918; s. Edward Palmer and Elfrida (MacLean) P.; Exhibitioner, Trinity Hall, Cambridge, 1937-39; M.A., Guys Hosp. Med. Sch., 1942; M.B., B.Chirurg., U. London, 19—; m. Sept. 26, 1949; children—Philippa, Christopher, Joanna, Alison. Mem. applied psychology research unit Med. Research Council, Cambridge, 1947—, asst. dir., 1958—; research fellow psycho-acoustic lab. Harvard, 1953-54. Mem. Brit. Psychol. Soc., Exptl. Psychology Soc. Research, publs. on tracking displays showing absolute motion easier to use than displays showing only relative motion, motor skills, doing 2 things simultaneously, rapid reading, evaluation of printed alphabets and typographical layouts, short-term memory, assessment of sensory magnitudes, exptl. method and design, environmental stresses, including heat, cold, hypoxia, compressed air. Home: 65 Grantchester St., Office: 15 Chaucer Rd., Cambridge, Eng.*

POUND, Glenn Simpson, Am. plant pathologist, univ. dean; b. Hector, Ark., Mar. 7, 1914; s. Leroy and Maude (Bullock) P.; B.A., U. Ark., 1940; Ph.D., U. Wis., 1943; m. Daisy Ferrel Cole, June 29, 1932; children—Robert Arthur, Elizabeth Jane. With U.S. Dept. Agr., 1943-46; faculty plant pathology U. Wis., Madison, 1946—, prof., 1953—, chmn. dept. plant pathology, 1954-64, dean Coll. Agr., 1964—. Cons. Rockefeller Found., 1960—. Mem. Am. Phytopath. Soc. (pres. 1959), Phi Beta Kappa. Research, publs. on effects of environmental factors in plant pathology; devel. disease resistant vegetable crops. Home: 313 Cheyenne Trail, Madison, Wis. 53705.*

POUND, Guy Marshall, Am. phys. metallurgist; b. Portland, Ore., Apr. 2, 1920; s. Guy Aubrey and Besse (Marsh) P.; B.A., Reed Coll., 1941; M.S., Mass. Inst. Tech., 1943; Ph.D., Columbia, 1949; m. Barbara Ann Corbly, June 21, 1944; children—Stephen M., James W. Alexandra W., Guy Nichols, Howard K. Faculty staff metals research lab. Carnegie Inst. Tech., Pitts., 1949-60, Alcoa prof. phys. metallurgy, dir. lab., 1960-66; prof. materials sci. Stanford, 1966—; guest prof. U. Sheffield, 1959-60, Tech. U. Berlin, 1964-65. Guggenheim Found. fellow, Sr. Fulbright Research fellow, 1959-60. Mem. Am. Chem. Soc., Am. Inst. Metall. Engrs., Am. Soc. Metals. Author: (with J.P. Hirth) Condensation and Evaporation, 1963; also articles. Research in phase transition kinetics. Home: 3003 Country Club Ct., Palo Alto, Cal.*

POUND, James, English astronomer; b. Wiltshire, Eng., 1669; ed. St. Mary Hall, Oxford, Eng.; B.A., Hart Hall, 1694; M.A., Gloucester Hall, 1694; M.B., 1697; m. Sarah Farmer, Feb. 14, 1710; a dau., Sarah; m. 2d, Elizabeth Wymondesold, Oct. 1722. Went to Madras, India, as chaplain to mchts. at Ft. St. George, 1699; went to settlement on the Cambodia; held ecclesiastical preferments, Eng. Fellow Royal Soc., 1699, admittance deferred until 1713. Observed satellites of Jupiter and Saturn; phase determination of total eclipse, 1715; observed occultation of star by Jupiter, 1715. Died Nov. 16, 1724.

POUND, Robert Vivian, physicist; b. Ridgeway, Ont., Can., May 16, 1919; s. Vivian Ellsworth and Gertrude C. (Prout) P.; came to U. S., 1923, naturalized, 1932; B.A., U. Buffalo, 1941; A.M. (hon.), Harvard, 1950; m. Betty Yde Andersen, June 20, 1941; 1 son, John Andrew. Research physicist Submarine Signal Co., 1941-42; staff mem. radiation lab. Mass. Inst. Tech., 1942-46; jr. fellow Soc. Fellows, Harvard, 1945-48, asst. prof. physics, 1948-50, asso. prof., 1950-56, prof., 1956—; Fulbright research scholar Oxford, 1951. Recipient B. J. Thompson meml. award Institute Radio Engineers, 1948; John Simon Guggenheim fellow, 1957-58; Fulbright Lecturer, Paris, 1958; Eddington medal Royal Astronomical Society, 1965. Fellow of the American Physical Soc., Am. Acad. Arts and Scis.; mem. Nat. Acad. Scis., Soc. Franc. De Physique (membre du conseil 1958-61), Phi Beta Kappa, Sigma Xi. Author, editor: Microwave Mixers, 1948. Contbr. profl. jours. Research in microwave technique, frequency control; participated in original discovery of nuclear magnetic resonance; originated study of nuclear electric quadrupole interactions by radiospectroscopy in solids; helped explain effect of external fields on directional correlations of nuclear radiation and extended Mössbauer's discovery of nuclear gamma-ray resonance to high resolution and applied it to confirm Einstein's predicted red shift. Home: 87 Pinehurst Rd., Belmont, Mass. 02178. Office: 336 Lyman Lab., Harvard U., Cambridge, Mass., 02138.*

POUNDER, Elton Roy, Canadian physicist; b. Montreal, Que., Can., Jan. 10, 1916; s. Roy M. and Norval (McLeese) P.; B.Sc., McGill U., Montreal,

1934, Ph.D., 1937; m. Marion Crane Wry, Feb. 15, 1941; children—David Crane, Norval Gillian. Field engr. Bell Telephone Co. Can., Ottawa, Ont., Can., Montreal, 1937-39; faculty McGill U., 1945-—, prof. physics, 1959-—. Mem. Canadian Assn. Physicists (past pres.), Am. Phys. Soc., Am. Geophys. Union. Author: (with J.S. Marshall) Physics, 1957; The Physics of Ice, 1965; also numerous articles. Research on constrn. cyclotron, neutron prodn., doppler radar used mid-Can. line, geophysics sea ice. Home: 21 Windsor Av., Westmount 6, Que. Office: Physics Dept., McGill U., Montreal, Que., Can.*

POUNDS, Norman John Greville, geographer; b. Bath, Eng., Feb. 23, 1912; s. John Greville and Camilla M.M. (Fisher) P.; B.A., Cambridge U., 1934, M.A., 1940; B.A., London U., 1942, Ph.D., 1944; m. Dorothy Josephine Mitchell, July 30, 1938. Lectr., coll. tutor Cambridge U., 1944-50; vis. prof. U. Wis., 1950; faculty Ind. U., Bloomington, 1950-—, U. prof. geography, 1959-—, chmn. dept. geography, 1962-65; Rose Morgan prof. U. Kan., 1958-59. Mem. Assn. Am. Geographers, Royal Geog. Soc. Author: The Ruhr, 1952; (with W.N. Parker) Coal and Steel in Western Europe, 1954; The Upper Silesian Industrial Region, 1958; The Geography of Iron and Steel, 1959; numerous other books, articles. Research on hist. and polit. geography. Home: 2203 Moore's Pike, Bloomington, Ind. 47401.*

POUPART, François, French surgeon, naturalist; b. Le Mans, France, 1661; student medicine Paris; med. degree, Reims, France. Mem. French Acad. Scis., 1699. Research and publs. on anatomy, cantharis, leeches, mussels, osteology; described relation of inguinal ligament (Poupart's ligament) to hernia, 1695. Died Reims, Oct. 31, 1708.

POURBAIX, Marcel, electrochem. engr.; b. Myshega, Russia, Sept. 16, 1904; s. Louis and Adèle (Evers) P.; grad. with grande distinction U. Libre de Bruxelles, 1927; Agrégé de l'Enseignement Supérieur, U. Brussels (Belgium), 1945; Doctor in de Technische Wetenschap, U. Delft (Netherlands), 1945; m. Marcelle Trojan, Mar. 31, 1934; children—Etienne, Philippe, Antoine. With U. Brussels, 1927-28, researcher, 1934-—; staff tech. sect. research dept. Union Chimique Belge, 1928-34; with Centre Belge d'Etude de la Corrosion CEBELCOR, Brussels, 1950-—, adminstr., dir. Recipient prix Société Royale Belge des Ingénieurs et Industriels, 1933; prix Gijsberti Hodenpijl, U. Delft, 1940; Croix Civique; officier l'Ordre de la Couronne.. Mem. Comité International de Thermodynamique et de Cinétique electrochemiques (founder, past pres.), Internat. Union Pure and Applied Chemistry (founder, past pres. commn. on electrochemistry, past sect. phys. chemistry), Electrochem. Soc., Faraday Soc. Author: Thermodynamics of Dilute Aqueous Solutions with Application to Electrochemistry and Corrosion, 1945; (with others) Atlas of Electrochemical Equilibria in Aqueous Solutions, 1963; also numerous articles. Home: 115 Rue Gabrielle, Brussels 18. Office: 50 Av. F.D. Roosevelt, Brussels 5, Belgium.*

POURTALES, Louis François de, see de Pourtales.

POWELL, Alan, aero. engr., physicist; b. Buxton, Eng., Feb. 17, 1928; s. Frank and Gwendolen (Walker) P.; D.L.C. Engring., Loughborough U., 1948, D.L.C. Engring. with honors, 1949, B.Sc. Engring. with honors, 1949; Ph.D., Southampton (Eng.) U., 1953; m. June Sinclair, Mar. 28, 1953. Came to U.S., 1956, naturalized, 1962. Tech. asst. Percival Aircraft, 1949-51; lectr. Southampton U., 1951-56; research fellow Cal. Inst. Tech., 1956-57; prof. engring., head aerosonics lab. U. Cal. at Los Angeles, 1957-65; asso. tech. dir. David Taylor Model Basin, USN, Washington, 1965-66, tech. dir., 1966-67, tech. dir. Naval Ships Research and Devel. Center, 1967-—. Cons. Douglas Aircraft Co., Inc., 1956-65; mem. com. on hearing, bioacoustics and biomechanics Nat. Acad. Scis.-NRC, 1961-67, exec. council, 1963-—, chmn., 1965-66, mem. panel on submarine noise measurement com. on underseas warfare, 1965-66; panel Def. Sci. Bd., 1967-68. Recipient Baden-Powell prize Royal Aero. Soc., 1948, Wilbur Wright prize, 1953. Fellow Royal Soc., 1953, Acoustical Soc. Am., Phys. Soc. (London), Am. Inst. Aeros. and Astronautics (asso.), Royal Aero. Soc.; mem. Acoustical Soc. Am. (Biennial award 1962, asso. editor Jour. 1962-67, exec. council 1967-—, chmn. edn. com. 1964-66), Inst. Mech. Engring. (asso.), Am. Soc. Engring. Edn., Soc. Naval Architects and Marine Engrs., Brit. Acoustical Soc. Research, numerous publs. on acoustics and vibration, especially flow noise, jet noise, fundamental work on structural vibration and fatigue failure due to noise and their engring. application in aerospace and underwater systems. Office: Naval Ships Research and Devel. Center, Washington 20007.*

POWELL, Cecil Frank, English physicist; b. Tonbridge, Kent, Eng., Dec. 5, 1903; s. Frank and Elizabeth Powell; student Sidney Sussex Coll., Cambridge 1922-25; Ph.D., Cambridge U., 1928; D. Sc. (hon.), Univs. Berlin, Bordeaux, Padua, Warsaw. Research asst. U. Bristol, Eng., 1928, Melville Wills prof. physics H.H. Wills Physics Lab., 1948-63, Henry Overton Wills prof. physics, 1964-—, pro-vice-chancellor, 1964-—. Chmn. sci., policy com. of European Centre for Nuclear Research, Geneva, Switzerland, 1961-63. Recipient Nobel Prize for Physics, 1950. Fellow Royal Soc., 1949 (Hughes medal 1949, Royal medal 1961); mem. Acad. Scis. USSR (fgn). Most important work includes mem. expdn. Royal Soc. and Colonial Office to investigate seismic and volcanic activity in Monteserrat, B.W.I., 1935; dir. European expdns. for making high altitude balloon flights in Italy, 1952-61. Author: (with G. P. S. Occhialini) Nuclear Physics in Photographs, 1947; (with P.H. Fowler, D.H. Perkins) The Study of Elementary Particles by the Photographic Method, 1959; also numerous papers. Co-discoverer and investigator production of pi-mesons (pions); developed photographic methods for study of nuclear processes. Home: 12 Goldney Av., Clifton, Bristol 8, Eng.*

POWELL, Herbert Marcus, English chemist, crystallographer; b. Eng., Aug. 7, 1906; s. William Herbert and Henrietta (Powell); B.A. (scholar) St. John's Coll., Oxford, Eng., 1927, honours 1st class in Chemistry, 1928, M.A., 1931. Faculty, Oxford U., 1928-63, prof. chem. crystallography, 1963, head lab., 1944-63; professiorial fellow Herford Coll., Oxford, 1963-—. Fellow Royal Soc., 1953 (mem. nat. crystallography com.); mem. Chem. Soc. (council), Mineral. Soc. (council). Contbg. author: Non-Stoichiometric Compounds, 1964. Research on crystal structure determination by X-ray diffraction, including stereochemistry of elements, metal-metal bonding, nature of intermolecular compounds, optical activity; originated new class of molecular compounds (clathrates) in which one atom or molecule is enclosed in a cage formed by others; prepared stable molecular compounds of inert gas elements and devised method of separating mirror image molecules by trapping them in left or right handed molecular cavities. Office: Hertford Coll., Oxford, Eng.*

POWELL, John Wesley, Am. naturalist; b. Mt. Morris, N.Y., Mar. 24, 1834; s. Joseph and Mary (Dean) P.; attended schools in Ohio, Wis., and Ill., 2 yrs. each at Oberlin and Wheaton (Ill.) colls.; grad. Ill. Wesleyan, A.M., Ph.D. (LL.D., Columbian, 1882, Harvard, 1886, Ill. Coll., 1889; Ph.D., Heidelberg, 1886); m. Emma Dean, 1861. Served through Civil war reaching rank of maj.; Prof., Ill. Wesleyan Coll., 1865-67; Prof., Illinois Normal Univ., 1867. Apptd. dir. U. S. Bureau of Ethnology, 1879, and of U. S. Geol. Survey, 1880; resigned latter, 1894, retaining former. Author: Explorations of the Colorado River, 1875; Report on Geology of the Uinta Mountains; Report on Arid Regions of United States, 1878; Introduction to the Study of Indian Languages; Truth and Error, 1898; Studies in Sociology; Cañons of the Colorado; etc. Pioneered exploration of Green and Colo. rivers, especially Grand Canyon of the Colo., 1869; studied geology, geography and ethonolgy of western U. S. Died Haven, Me., Sept. 23, 1902.

POWELL, Percival Herbert, Brit. elec. engr.; s. James and E.M. Powell; ed. Liverpool Coll., U. Liverpool, Victoria U.; M.Sc., M.Eng.; m. Una K. Mayne; 1 son. Fellow Victoria U.; with Siemens Bros. Works; lectr. elec. engring. Canterbury Coll.; prof. elec. engring. Canterbury Coll., Christch., New Zealand; ret., 1946. Fellow Royal Soc. Australia; mem. Instn. Elec. Engr., Australian Instn. Elec. Engr. Author: The Air-gap Correction Coefficient; Electric Power in New Zealand; Circuits of an Electrical Engineering Laboratory; Measurement of Power; Graphical Construction for Impedance of Parallel circuits; Resistance of Induction Moror. Co-author: Hydrodynamical and Electromagnetic Investigation of Distribution of Magnetic Flux in Toothed Core Armature. Died Dec. 18, 1958.

POWELL, Ralph Waterbury, Am. engr.; b. nr. Ionia, Mich., Oct. 4, 1889; s. Herbert Ernest and Alice May (Waterbury) P.; B.S., Mich. State U., 1911; C.E., Cornell U., 1914; Ph.B., Yale, 1916; m. Maude Esther Nason, Sept. 10, 1914. Instr., Mich. State U., 1911-12, Cornell U., 1912-14; asst. Yale, 1914-16; asso. prof. Coll. of Yale-in-China, 1916-27; faculty Ohio State U., Columbus, 1927-57, prof. engring. mechanics, 1945-57; vis. prof. U. Ia., 1955-56, U. Kan., 1957-58; pres. Rocky Mountain Hydraulic Lab., Allenspark, Colo., 1963-—, sec., 1946-63. Mem. Am. Soc. C.E. (task force on friction factors in open channels 1957-64), Am. Soc. for Engring. Edn., Am. Geophys. Union, Internat. Assn. for Hydraulic Research, Sigma Xi, Tau Beta Pi, Pi Mu Epsilon. Author: Mechanics of Liquids, 1940; An Elementary Text in Hydraulics and Fluid Mechanics, 1951; also numerous articles. Pioneer in use of Reynolds Number for flow in pipes, use of dimensional analysis in study of hydraulic models, in investigation of resistance to flow in channels of definite roughness; inventor plotting paper for flood frequency studies. Home: 2153 Vine St., Berkeley, Cal. 94709.*

POWELL, Reginald Walter, physicist; b. Erith, Kent, Eng., June 8, 1903; s. William Martin and Lucy (Maynard) P.; B.Sc. with 1st class honors in Physics, U. Coll., London, Eng., 1924; Ph.D., London U., 1938, D.Sc., 1944; m. Jennie Elliot Saunders, Apr. 30, 1930; children—John Martin, Graham Reginald, Jennifer Lucy (Mrs. David Pfluger). With Nat. Phys. Lab., Teddington, Middlesex, Eng., 1924-64, sr. prin. sci. officer, head thermal properties sect., 1953-54; sr. scientist Thermophys. Properties Research Center, Purdue U., West Lafayette, Ind., 1965-—. Mem. Brit. Nat. Com. for Physics, 1960-63. Recipient Moulton gold medal Instn. Chem. Engrs., 1935. Fellow Inst. Physics (past sec. low temperature group); mem. Brit. Assn. for Advancement Sci. (past sect. sec.-recorder), Internat. Inst. Refrigeration (mem. com. 2 1959-—), Sigma Xi. Research, numerous publs. on thermal properties of materials, evaporation, flax processing, first to report anisotropic properties of gallium, thermal conductivity measurement and data assessment. Patentee thermal comparator. Home: 1010-2 N. Salisbury St., West Lafayette, Ind. 47906.*

POWELL, Richard, English physician; b. Oxfordshire, Eng.; baptised May 11, 1767; s. Joseph Powell; ed. Pembroke and Merton Colls., Oxford (Eng.) U., M.A., 1791, M.D., 1795. Physician St. Bartholomew's Hosp., London, 1801-24; Lumleian lectr., 1811-22; became Harveian orator, 1808. Author: Heads of Lectures on Chemistry, 1796; also numerous articles. A revised Pharmacopoeia Londinensis, 1809. Mem. Royal Coll. Physicians (became censor 1798, 1807, 20, 24). First description of simple facial palsy, 1813; described hematoma of dura mater, meningitis after necrosis of walls of inner ear, new growth of pituitary gland, 1814. Died Aug. 18, 1834.

POWELL, Wilson Marcy, Am. physicist; b. Litchfield, Conn., July 18, 1903; s. Wilson M. and Elsie (Knapp) P.; A.B., Harvard, 1926, Ph.D., 1933; m. Dorothy Johnson Gardner, June 29, 1956; children—Wilson M., David R., Fredrika (Mrs. James P. Spillman), Claire. Asst. physics Harvard, 1929-30, 34-35; instr. Conn. Coll., 1935-37; asst. prof. Kenyon Coll., 1937-46, dir. cosmic ray expdn. 1940, 41; research asso. Radiation Lab., Berkeley, Cal., 1943-46; faculty U. Cal., Berkeley, 1946-—, prof., 1952-—. Mem. Swarthmore eclipse expdn., Mexico, 1923, Sumatra, 1925. Guggenheim fellow, 1941-42, fellow Oceanographic Inst., Woods Hole, Mass., 1937. Fellow Am. Phys. Soc., N.Y. Acad. Scis. Research on optics in Schumann region, cosmic rays, use of cloud chamber, magnets, bubble chambers in high energy nuclear physics. Home: 1098 Grizzly Peak Blvd., Berkeley, Cal. 94708.*

POWER, Sir D'Arcy, English physician, surgeon; b. London, (Eng.) Nov. 11, 1855; s. Henry and Anna (Simpson) P.; B.A., Oxford (Eng.) U., 1878, M.A., 1881, M.B., 1882; m. Eleanor Forbroke, 1883 (dec. 1923), 2 sons, 1 dau. Hunterian prof. surgery and pathology Royal Coll. Surgeons, 1896-97; became Bradshaw lectr., 1919, Vicary lectr., 1920, Hunterian orator, 1925; vis. lectr. Johns Hopkins, 1931-32. Fellow Royal Coll. Surgeons (Eng.), Royal Faculty Physicians and Surgeons Glasgow (hon.), Am. Surg. Assn. (hon.); mem. Brit. Med. Assn. (pres. sect. surgery), Internat. Soc. History Medicine, Société Internationale de Chirurgie, Author: Surgical Diseases of Children, 1895; Some Points in the Anatomy, Pathology and Surgery of Intussusception 1897; A System of Syphilis, 6 vols., 1908-10; The Practitioners Surgery, 1919; Selected Writings, 1931; A Mirror for Surgeons, 1940; also articles on war surgery, syphilis, cancer, intestinal obstruction, surg. diseases of children, history of medicine. Died Northwood, Eng., May 18, 1941.

POWER, Frederick Belding, Am. chemist; b. Hudson, N.Y., Mar. 4, 1853; s. Thomas and Caroline P. (Belding) P.; grad. Phila. Coll. Pharmacy, 1874; Ph.D., U. Strassburg, 1880; LL.D., U. Wis., 1908; m. Mary Van Loan Meigs, Dec. 27, 1883 (died 1894); children—Mrs. Annie Louise Heimké, Donald Meigs. Asst. to prof. materia medica U. Strassburg, 1879-80; prof. analytical chemistry Phila. Coll. Pharmacy, 1881-83; prof. pharmacy and materia medica, organizer, dean sch. pharmacy U. Wis., 1883-92; dir. labs. Fritsche Bros., 1892-96, Wellcome Chem. Research Labs., London, Eng., 1896-1914; in charge Phytochem. Lab., Bur. Chemistry, Washington, 1916-27. Mem. com. revision U.S. Pharmacopoeia, 1890. Recipient Ebert prize Am. Pharm. Assn., 1877, 1902, 06; Gold medal, St. Louis Expn., 1904; Silver medal, Liège, 1905; Gold medal and diploma of honor, Milan, 1906; Gold medal Franco-British Exhbn., 1908; grand prize, Brussels, 1910; Gold medal and diploma of honor Turin, 1911; Hanbury Gold medal, 1913; Gold medal presented by Henry S. Wellcome, London, 1914, Flueckiger Gold medal, 1922. Editor: (with Fred Hoffman) Manual of Chemical Analysis, 1883. Research on constituents of plant products; studied chaulmoogra seeds (source of oil for treatment of leprosy). Died Mar. 26, 1927.

POWER, Harry Harrison, Am. petroleum engr., educator; b. Chassell, Mich., July 7, 1892; s. Elmer L. and Ivy Belle (Spencer) P.; B.S. in Chem. Engring., Wash. State U., 1919, Chem.E., 1936; M.S. in Mining, U. Cal., 1928; Ph.D., U. Pitts., 1946; m. Gladys Drach, Oct. 1, 1928. Jr. and asst. engr. Cities Service Co., Denver, Bartlesville, Okla., 1919-23; valuation engr. U.S. Treasury Dept., Washington, 1923-26; prodn. engr., chief prodn. engr. Gulf Oil Corp., Tulsa, 1926-36; prof. petroleum engring., chmn. dept. U. Tex., Austin, 1936-62, prof. emeritus, 1962-—. Fulbright lectr. Delft (Holland) Tech. U., 1956; Fulbright lectr., cons. U. Cairo (Egypt), 1962-63; chmn. Tex. Conf. for Advancement Sci. and Math. Teaching, 1960. Mem. Am. Inst. Mining, Metall. and Petroleum Engrs. (past chmn. petroleum div., chmn. mineral industries edn. div. 1951-52, Mineral Industry Edn. award 1964), Am. Petroleum Inst. (past chmn. com. on prodn. tech.), Am. Soc. for Engring. Edn. (chmn. mineral edn. div. 1951-52),

Sigma Xi, Tau Beta Pi, Sigma Tau. Contbr. numerous articles on oil, gas devel., prodn. to profl. publs. Home: 2704 San Pedro St., Austin, Tex. 78705.*

POWER, Henry, English physician; b. 1623; B.A., Christ's Coll., Cambridge, Eng. 1644, M.A., 1648, M.D., 1655. Practiced medicine at Halifax. Author: Experimental Philosophy, in three Books: containing New Experiments, Microsopical, Mercurial, Magnetical. With Some Deductions, and Probable Hypotheses, raised from them, in Avouchment and Illustration of the now famous Atomical Hypothesis, 1664. Made 1st English micros. observations; made observations using telescope, lodestone, Toricellian tube. Died New Hall, Dec. 23, 1668.

POWERS, Edwin B(ooth), Am. zoologist; b. Ellis County, Tex., Aug. 6, 1880; s. William Wilson and Evaline Crocia (Woods) P.; A.B., Trinity U., Waxahachie, Tex., 1906; M.S., U. Chgo., 1913; Ph.D., U. Ill., 1918; postgrad., Cambridge U., Eng.; m. Pauline Watkins, June 9, 1918; children—Edwine Watkins, Wilson Watkins. Instr. and prof. in biology Trinity U., 1908-15; research asst. Puget Sound Biol. Sta., U. Wash., summer, 1914, asst. in zoology, summer, 1918, prof. zoology, summers, 1919, 21, 22, 24, 27; asst. prof. zoology Colo. Coll., 1918-19; traveled and studied Imperial Inst. (London). Danish Biol. Sta., Naples Biol. Sta., 1919-20; instr. zoology U. Neb., 1920-22; asso. prof. anatomy and embryology U. Tenn. Med. Coll., Memphis, 1922-23; prof. zoology and acting head dept. U. Tenn., Knoxville, 1923-24, head dept. zoology, 1924-41, head dept. zoology and entomology, from 1941; at Marine Biol. Lab., Woods Hole, Mass., summer 1920; prof. limnology, Mt. Lake Biol. Sta., U. of Va., summer 1934; prof. physiology Franz Theodore Stone Lab., Put-in-Bay, Ohio State U., summers 1935, 36; stream pollution adv. to N.C. Pulp Co., 1940-41; at Solomon Island, Md. State Lab., 1945-46. Mem. A.A.A.S., Zool. Soc. Am., Ecol. Soc. Am. (pres. 1933-34), Limnol. Soc. Am., Am. Fish Soc., Entomol. Soc. Am., Tenn. Acad. Science, Sigma Xi. Asso. editor Ecology, 1925-32, Ecol. Monographs, 1935-39. Contbr. sci. papers and monographs on salmon migration, toxicities, physiology of respiration of fishes. Died Aug. 25, 1949.

POWERS, Grover Francis, Am. pediatrician; b. Colfax, Ind., Aug. 12, 1887; s. Francis William and Elizabeth Catherine (Shobe) P.; B.S., Purdue U., 1908, Sc.D., 1935; M.D., Johns Hopkins U., 1913; M.A., Yale, 1927; Sc.D. (hon.), Ind. U., 1949; m. Beatrice Farnsworth, Aug. 21, 1916; 1 son, Ross Farnsworth. Instr. and asso. in clin. pediatrics, Johns Hopkins U., 1916-21; med. dir. Babies Milk Fund Assn., Balt., 1916-21; asst., later asso. prof. pediatrics, Yale, 1921-27, prof. pediatrics, 1927-52, prof. emeritus since 1952; cons. pediatrician Grace-New Haven Community Hosp.; pediatrician in chief New Haven Hosp. since 1927; former cons. Mental Hygiene div., USPHS. Trustee Southbury Tng. Sch. Recipient Borden Award, Am. Acad. Pediatrics, 1947, John Howland award Am. Pediatric Soc., 1953; Jos. P. Kennedy Jr. award; 2d International award. Diplomate Am. Bd. Pediatrics. Fellow A.M.A., Am. Acad. Pediatrics; mem. Am. Pediatric Soc. (past pres.), Soc. for Pediatric Research, Am. Soc. for Clin. Investigation, Interurban Clin. Club, Nat. Assn. Retarded Children (chmn. sci. research adv. bd.), Brit. Pediatric Soc. (corr.) Sigma Xi, Alpha Omega Alpha. Mem. editorial bd. Pediatrics. Died Apr. 18, 1968.

POWERS, John Howard, Am. surgeon; b. Gardiner, Me., May 31, 1898; s. William Lincoln and Marion (Turner) P.; A.B., Bates Coll., 1919; B.A. (Rhodes scholar), Oxford (Eng.), 1923; M.D., Harvard, 1925; m. Emilie Deen Baldwin Simmons, June 28, 1946; children—Laura Emilie (Mrs. Samuel Wilson Smith, Jr.), Susan Deen Simmons. Arthur Tracy Cabot fellow Harvard Med. Sch., 1927-28, instr. surgery, 1926-30; staff Mary Imogene Bassett Hosp., Cooperton, N.Y., 1931-63, surgeon-in-chief, 1956-63; faculty Columbia, 1942-63, clin. prof. surgery, 1956-63. Diplomate Am. Bd. Surgery. Mem. Phi Beta Kappa. Research, numerous publs. on devel. cardiac surgery, early post operative ambulations, post operative complications, problems relative to surgery in old age, accidents on the farm, accident in city hwys., in home. Home: Lake Rd. Office: Mary Imogene Bassett Hosp., Cooperstown, N.Y. 13326.*

POWERS, Justin Lawrence, Am. pharm. chemist; b. Tekonsha, Mich., Mar. 12, 1895; s. Norman L. and Eva (Mills) P.; B.S., U. Mich., 1924, M.S., 1927, Sc.D., 1960; Ph.D., U. Wis., 1935; m. Gladys L. Laufman, Dec. 25, 1917. Faculty, Wash. State Coll., 1919-23, Ore. State Coll., 1924-26, U. Mich., 1926-40; dir. sci. div. Am. Pharm. Assn., Washington, 1940-60, dir. Nat. Formulary Revision, 1940-60, dir. lab., 1940-45; dir. Food Chems. Codex, Nat. Acad. Scis.-NRC, Washington, 1961-66, spl. cons., 1966—. Mem. U.S. Pharmacopoeial Revision Com., 1940-60; mem. adv. panel Internat. Pharmacopoea, WHO, Geneva, Switzerland, 1954—. Recipient Remington medal, 1959; cited for distinguished service to pharm. profession U. Wis., 1958. Mem. Am. Pharm. Assn. (editor Jour. 1942-60, Am. Chem. Soc.). Sigma Xi, Delta Tau Delta, Phi Delta Chi, Phi Lambda Upsilon, Phi Sigma. Contbr. articles to tech jours. on food chemicals' standards. Home: 36 Surf Dr., St. Augustine, Fla. 32084.* Office: 2101 Constitution Av. N.W., Washington 20418.*

POWERS, Philip N(athan), Am. nuclear engr.; b. Terre Haute, Ind., Aug. 17, 1912; s. Samuel Ralph and Eda (Olds) P.; B.A., Columbia, 1932, Ph.D., 1940; Gen. Edn. Bd. fellow U. Chgo., 1940-41; m. Eleanor Ritchie, Apr. 25, 1936 (div.); 1 son, Philip; married second, Evelyne Schanze, May 22, 1959; stepchildren—Jacques le Sourd, Liliane le Sourd. Associate sci. New College Tchrs. Coll., Columbia, 1932-35; asso. New College Community, 1937-38; instr. natural scis. Stephens Coll., 1939-42; physicist Naval Ordnance Lab., 1942-44, tng. dir., 1944-46; chief sci. edn. br. Office Naval Research, 1946; advisor sci. personnel Pres.' Sci. Research Bd., 1947, AEC, 1947-50; manpower specialist, sec. sci. manpower adv. com. NSRB, 1950-51, cons., 1953; dir. atomic project Monsanto Chem. Co., 1951-55; chmn. bd. Internuclear Co., Clayton, Mo., 1955-59, pres., 1955-60; prof., head dept. nuclear engring. Purdue University, 1960—, asso. dir. engring. expt. station, 1960-61, dir. engring. expt. station, 1961—, also vice president Purdue Research Found., 1961-62. Chmn. of council Asso. Midwest Univs., 1964-65, chmn. nuclear engring. edn. com. 1964-65; trustee, exec. com. Argonne Univs. Assn., 1965—, pres., 1966—; cons. com. on specialized personnel ODM, 1951-63, mem. com. manpower resources 1953 cons. Nat. Sci. Found., 1952-67, AEC, 1964-67, AEC Fellowship Bd., 1964-66, Select Com. Govt. Research, Ho. of Reps., 1964-65, also Pres.'s Com. Scientists and Engrs., 1956-58, Dept. of State, 1957; cons. N.Y. Univ. planning internat. activities, 1958-59. Mem. of Commn. on Survey of Dentistry; chmn. of atomic energy panel of Engineers Joint Council, 1956-58. Mem. Am. Nuclear Society (director 1963-66, mem. exec. com. aerospace div. 1963-65), Am. Soc. for Engring. Edn. (chmn. atomic energy edn. com. 1953-57; chmn. com. on space engring. 1962-66, chmn. grad. studies div. 1965-66), U. S. C. of C. (nat. def. com. 1953-59, com. on commercial uses of atomic energy, 1959-62, com. on new frontiers of tech. 1960-63), Am. Nuclear Soc., Am. Society Engineering Society, A.A.A.S., Am. Assn. Physics Tchrs., Am. Phys. Soc., Assn. Research Sci. Teaching, N.Y. Acad. Sci., Am. Inst. Chem. Engrs., Sigma Xi. Research on magnetic scattering of neutrons; underwater ordnance; nuclear reactor engineering; feasibility and planning of nuclear power plants. Home: 502 Calvert Lane, LaFayette, Ind. 47905.*

POWERS, Sidney, Am. geologist; b. Troy, N.Y., Sept. 10, 1890; s. Albert W. and Tillie (Page) P.; A.B., Williams Coll., 1911; M.S., Mass. Inst. Tech., 1913; Sheldon traveling fellow to Hawaiian Island, Harvard, 1915, A.M., Ph.D., 1915, research fellow, 1915-16; m. Dorothy Edwards Powers, Sept. 8, 1917; children—Deborah, Eleanor. Div. geologist Tex. Co., 1916-17; asst. geologist U.S. Geol. Survey, 1917-18; geol. officer, A.E.F., U.S. Army, 1918-19; chief geologist Amerada Petroleum Corp., 1919-26, cons. geologist from 1926. Fellow Geol. Soc. Am.; mem. Am. Assn. Petroleum Geologists. 1931-33). Recognized importance of unconformities in petroleum geology; brought functional importance of buried hills as structure-bldg. agys. into proper sci. perspective; studied age of folding and inter-relationships of Ouachita, Arbuckle, and Wichita mountains of Okla., Llano-Burnet and Marathon uplifts of Tex.; recognized petroleum possibilities of Crinerville anticline, Carter County, Okla.; contbd. to improvement of reflection (seismic) geophys. method as prospecting tool. Home: Tulsa, Okla. Died 1932.

POWERS, Wendell Holmes, Am. chemist; b. Richford, Vt., Mar. 30, 1915; s. Harry Oakley and Golda (Cowan) P.; B.S., Middlebury Coll., 1937; M.S., U. N.H., 1939; Ph.D., Columbia, 1943; m. A. Irene Douglass, Nov. 28, 1942; children—William H., Nancy Ann. Instr., U. N.H., 1938-39; asst. Columbia, 1939-42; faculty Wayne State U., Detroit, 1942—, prof. chemistry, 1956—. Exec. sec. Kresge-Hooker Sci. Library Asso., Detroit, 1956—. Mem. Am. Chem. Soc., A.A.A.S., Am. Assn. U. Profs., Sigma Xi, Phi Lambda Upsilon, Phi Beta Kappa. Author: (with J.H. Secrist) General Chemistry, 1966; also articles. Editor: Chemotherapy, 1945; (with David Krogmann) Biochemical Dimensions of Photosynthesis, 1965; Rec. Chem. Progress, 1952—. Research on molecular rearrangements, preparation and study of enzymes. Home: 16556 Greenlawn St., Detroit 48221.*

POWERS, William Edwards, Am. geologist, geographer; b. Florence, Ala., Mar. 25, 1902; s. William Edwards and Annie (Smith) P.; B.S., Northwestern U., 1925; M.A., 1928; Ph.D., Harvard, 1931; m. Marian Louise Drisko, Sept. 5, 1932; 1 son, William Edwards. Faculty, Northwestern U., Evanston, Ill., 1926—, prof. geography, 1945—. Geologist, Ill. Geol. Survey, summers, 1927-31, 37, U.S. Geol. Survey, summer 1942, N.D. Geol. Survey, summer 1945. U.S. Library of Congress fellow in geology, 1940-41; Fulbright research scholar, New Zealand, 1961. Mem. Geol. Soc. Am., Assn. Am. Geographers, Phi Beta Kappa, Sigma Xi. Author: Physical Geography, 1955; also numerous articles. Research on effects on landscape of former glaciers of Great Ice Age in Middle West, New Eng., Rocky Mountains, Hawaii, New Zealand. Home: 2141 Ewing Av., Evanston, Ill. 60201.*

POWLES, Jack Gordon, English physicist; b. Stroud, Eng., June 22, 1924; s. Jack Meredith and Stephanie (Newman) P.; B.Sc., U. Manchester (Eng.), 1945, M.Sc., 1946, Ph.D., 1948; D. ès Sci., U. Paris (France), 1951; m. Jill Redston, Aug. 4, 1951; children—Tamsin Jane, Simon Jack. Faculty, U. London, Imperial Coll., 1948-49; exchange fellow U. Paris, 1948-50; faculty U. Liverpool (Eng.), 1950-51, Princeton, 1951-52, U. Ill., 1952-54; Imperial Chems. Industries fellow U. Newcastle (Eng.), 1954-55; reader U. London physics dept. Queen Mary Coll., 1955-64; prof. physics, dir. physics labs. physics dept. U. Kent Canterbury, Eng., 1964—. Fellow Inst. Physics. Author: Particles, 1968; also numerous articles. Research in atomic and molecular motion in solids, liquids and gases particularly reorientation processes, initially dielectric absorption, nuclear magnetic resonance methods, scattering of laser light. Home: Beverley Farm House, St. Stephen's Green, Canterbury, Eng.*

POYNTING, John Henry, English physicist; b. Monton, Eng., Sept. 9, 1852; s. T. Elford and Elizabeth (Long) P.; B.Sc., London U., 1872; student Trinity Coll., Cambridge, 1872-76, fellow, 1878, Sc.D., 1887; m. Maria Adney Cropper, 1880; 1 son, 2 daus. Became demonstrator phys. lab. Owens Coll., 1876; became prof. physics Mason Coll., Birmingham, Eng., 1880, dean Sci. Faculty, U. Birmingham, 1902. Recipient Adams prize, 1893, Hopkins prize, 1903. Fellow Royal Soc., 1888 (v.p. 1910-11; Royal medal); mem. Brit. Assn. Author: On the Mean Density of the Earth, 1893; (with J.J. Thomson) Text-Book of Physics, 1899-1914; The Pressure of Light, 1910; The Earth: Its Shape, Size, Weight and Spin, 1913; Collected Scientific Papers, 1920; also articles on elec. phenomena, radiation, light pressure. Measured density of earth, 1891, constant of gravitation, 1893; studied motion of energy in electric field, including gen. law for transfer of energy (this theory has been employed in radio in calculating the power radiated from an antenna); demonstrated that for finely divided matter light-pressure could be greater than gravity (explained comet's tail pointing away from sun); studied (with W. H. Barlow) force of light, demonstrating that they produced tangential forces with light on absorptive surface and torque with light through a prism. Died Birmingham, Eng., Mar. 30, 1914.

POYNTON, Frederick John, English physician; b. 1869; s. F. J. Poynton; ed. Marlborough Coll., Eng., U. Coll., Bristol, Eng., St. Mary's Coll. Hosp., London; m. Alice Constance Campbell-Orde, 1904; 1 son, 1 dau. Cons. physician Heart Hosp., West Wickham, Heart Hosp., Lancing, Winford Orthopedic Hosp., U. Coll. Hosp. Recipient Dawson Williams prize for advance of study children's diseases, 1930. Hon. fellow Royal Inst. Pub. Health and Hygiene; mem. Nat. Children Adoption Assn., Cardiol. Soc. Gt. Britain and Ireland (hon.), Brit. Paediatric Assn. (became pres. 1931), Royal Coll. Physicians (became sr. censor 1930). Author: Researches on Rheumatism, Heart Disease in Childhood; Social Factors in Causation of Rheumatism, 1938. Concluded (with A. Paine) a streptococcus caused acute rheumatism, 1913. Died 1943.

POZZI, Samuel Jean, French surgeon; b. Bergerac, France, 1846; M.D., U. Paris, 1873. Joined faculty U. Paris, 1875; became surgeon Paris Hosp., 1877, Hosp. Broca, 1883. Mem. Congress Surgeons (founder), Acad. Medicine, Broca Soc. (became pres. 1888). Research in gynecology, connection of anatomy and anthropology; instituted ideas on hosp. constrn. and cleanliness. Died 1918.

PRAETORIUS, Jean, inventor, astronomer, mathematician; b. Joachimsthal, Bohemia, 1537. Tchr., Emperor Maximilian II; became prof. math., Wittenberg, Germany, 1571, Altof, 1576. Prepared astron. calendar; invented Praetorius' plane table, 1611, hydraulic balance. Died 1616.

PRAGER, William, civil engr., educator; b. Karlsruhe, Germany, May 23, 1903; s. Willy and Helen (Kimmel) P.; state certificate as civil engr., 1925; Dr. Engring. Scis., Inst. of Tech., Darmstadt, 1926; hon. degrees U. Liège (Belgium), 1962, U. Poitiers (France), 1962, Case Inst. Tech., 1963; Polytech. Inst. Milan, 1964; m. Gertrude A. Heyer, Sept. 16, 1925; 1 son, Stephen. Came to U.S., 1941. Instr. mechanics, Inst. Tech., Darmstadt, 1927-29; acting dir. Inst. Applied Mechanics, U. Göttingen, 1929-32; prof. mechanics, Tech. Inst., Karlsruhe, 1933; structural insp. German Airsport League, Berlin, 1929-33; sci. adviser Fieseler Aircraft Co., Kassel, Germany, 1933; prof. mechanics, U. Istanbul, 1934-41; mng. editor Revue de la Faculté des Sciences de l'Université d'Istanbul, 1935-41; prof. applied mechanics, Brown U., 1941-65, chmn. grad. div. of applied math., 1946-52, chmn. phys. scis. council, 1953-59; L. Herbert Ballou univ. prof. 1959-65; cons. IBM Research Lab., Zurich, Switzerland, 1963-65; now prof. applied mechanics U. Cal. San Diego, La Jolla. Recipient Worcester Reed Warner medal Am. Soc. M.E., 1957, Theodore Von Karman award Am. Soc. C.E., 1960. Fellow Am. Acad. Arts and Scis., Am. Soc. M.E. (Timoshenko medal 1966), Am. Soc. C.E.; mem. Inst. Mgmt. Scis., Am. Math. Soc., Nat. Acad. Engring., Nat. Acad. Scis. Mng. editor, Quar. Applied

Math., 1943-65. Author: Dynamik der Stabwerke (with K. Hohenemser), 1933 Mécanique des solides isotopes, 1937; Tersimi Hendese, 1937; Mihanik (with F. Gürsan), 1941; Theory of Perfectly Plastic Solids (with P. G. Hodge, Jr.), 1951; Probleme der Plastizitaetstheorie, 1955; An Introduction to Plasticity, 1959; Introduction to Mechanics of Continua, 1961; Introduction To Basic Fortran Programming and Numerical Methods, 1965. Office: Univ. of California, San Diego, La Jolla, Cal. 92037.*

PRAIN, Sir David, Brit. botanist; b. Fettercairn, Scotland, July 11, 1857; s. David and Mary (Thomson) P.; M.A., U. Aberdeen, 1878, M.B., C.M., 1833; ed. U. Edinburgh; m. Margaret Thomson, 1887; 1 son. Demonstrator anatomy Coll. Surgeons, Edinburgh, 1882-83, U. Aberdeen, 1883-84; entered Indian Med. Service, 1884; curator Calcutta Herbarium, 1887-98; prof. botany Med. Coll., Calcutta, 1895-1905; dir. Bot. Survey of India, supt. Royal Botanic Garden, Calcutta, 1898-1905; dir. Royal Botanic Gardens, Kew, 1905-22; dir. forest products research, 1922-25. Chmn. adv. council plant and animal products Imperial Inst., 1926-35. Recipient Barclay medal Asiatic Soc. Bengal, 1909, 42, Bruhl medal, 1938; Victoria medal Royal Hort. Soc., 1912, Veitch medal, 1932; Albert medal Royal Soc. Arts, 1925. Fellow Royal Soc., 1905, Linnean Soc. (Linnean medal 1935), Zool. Soc.; mem., hon. mem. numerous socs. Editor: Bot. Mag. 1907-20, also numerous monographs and articles on botany. Died Whyteleafe, Surrey, Eng., Mar. 16, 1944.

PRAKASH, Brahm, Indian metallurgist; b. Lahore, India, Aug. 21, 1912; s. Jodh and Khem (Kaur) Ram; B.Sc. with honors, Punjab U., 1933, M.Sc., 1934, Ph.D., 1942; Sc.D., Mass. Inst. Tech., 1948; m. Rajeshwari Prakash, Feb. 17, 1945; children—Suman, Arun, Maneesha. Asst. metallurgist Metall. Lab., Indian Rys., Ajmer, 1940-45; metallurgist Indian Atomic Energy Commn., 1949-50; prof., head dept. metallurgy Indian Inst. Sci., Bangalore, 1951-57; staff Bhabha Atomic Research Centre (formerly Atomic Energy Establishment Trombay), Bombay, India, 1957—, dir. metallurgy group, 1962—. Sci. sec. 1st Conf. on Peaceful Uses of Atomic Energy, UN, Geneva, Switzerland, 1955. Recipient Indian Nat. award Padma Shri, 1961; Shanti Swarup Bhatnagar Meml. award for engring. scis., 1963. Mem. Am. Inst. Mining and Metall. Engrs., Indian Inst. Metals. Designed and built fuel fabrication plant at Bhabha Atomic Research Centre; research, publs. on phys. chemistry, metallurgy especially extraction of zirconium. Home: 5C Anand Bhawan, Bhulabhai Desai Rd., Bombay 26. Office: Apllo Pier Rd., Bombay 1, India.*

PRAMER, David, Am. microbiologist; b. Mt. Vernon, N.Y., Mar. 25, 1923; s. Coleman and Ethel (Toback) P.; student St. Johns U., 1940-41, Tex. A. and M. Coll., 1941-42; B.Sc, Rutgers U., 1948, Ph.D., 1952; m. Rhoda Lifschutz, Sept. 6, 1950; children—Andrew, Stacey. Vis. investigator Imperial Chem. Industries, Ltd., Welwyn, Herts, Eng., 1952-54; faculty Rutgers U., New Brunswick, N.J., 1954—, prof. microbiology, 1960—, chmn. dept. biochemistry and microbiology, 1965—. Dir. New Brunswick Sci. Co., Inc. Cons., Am. Inst. Biol. Scis., 1961—. Mem. Am. Acad. Microbiology, Am. Soc. for Microbiology, Soc. for Gen. Microbiology, Soc. for Indsl. Microbiology, Am. Assn. U. Prfs., Phi Beta Kappa, Sigma Xi, Alpha Zeta. Author: (with E.L. Schmidt), Experimental Soil Microbiology, 1964; Life in the Soil, 1965; also numerous articles. Research on application antibiotics in agr., chem. transformation resulting from microbial activity, biochem. bases microbial interrelationships and morphogenesis, biol. control agrl. pests, nematode-trapping fungi. Home: 37 Grant Av., Highland Park, N.J. 08904. Office: Dept. Biochemistry and Microbiology, Rutgers —The State U., New Brunswick, N.J. 08901.*

PRANDTL, Ludwig, German physicist; b. Freising, Germany, Feb. 4, 1875; degree U. Munich, 1900; prof. applied mechanics, Göttingen, Germany, 1904-53; named dir. Kaiser Wilhelm Inst. for Fluid Mechanics, 1925; adviser to Ministry of Air Force, 1942-45; Founder, aerodynamics exptl. sta., Göttingen. Author: Vier Abhandlungen zur Hydrodynamik und Aerodynamik, 1927. A founder modern hydrodynamics and aerodynamics; proved percussion wave produced at sound barrier related to moving body and determines sudden change in aerodynamic power; discovered boundary layer on surface of body moving in water or air (led to understanding of friction drag and streamlining); research on supersonic flow, turbulence; developed wing theory explaining air flow over airplane wings; helped in devel. of wind tunnel; developed soap film analogy for torsion of noncircular sects. Died Göttingen, Aug. 15, 1953.

PRANKERD, Thomas Arthur John, English physician; b. Isle of Wight, Eng., Sept. 11, 1924; s. Prankerd Horace and Julia (Shorthose) P.; student St. Bartholomews Hosp. Med. Sch., 1942-47; M.B., B.S., U. London, 1947, M.D., 1949; m. Margaret Harrison-Cripps, Feb. 18, 1950; children—Richard, Henry, Nicola, Stephen. House physician, St. Bartholomews Hosp., 1947-49; Travelling fellow U. London, 1953-54; staff U. Coll. Hosp., London, 1957—, cons. physician, 1961—, prof. clin. hematology, 1964—, hon. cons. physician, 1964—. Dir. group on hemolytic anemias Med. Research Council, 1964—. Fellow Royal Coll. Physicians; mem. Physiol. Soc., Assn. Physicians. Author: The Red Cell, 1961; also articles. Research on metabolism of red cell and its application to clin. problems of hemolysis, dynamics of splenic circulation and its relationship to hemolysis. Home: Reeves Lodge, High Rd., Chipstead, Surrey, Eng. Office: U. Coll. Hosp., London WC1, Eng.*

PRASAD, Ananda Shiva, physician; b. Buxar, Bihar, India, Jan. 1, 1928; s. Radha Krishna Lall and Mahesha Kaur; B.Sc. (with honors in math.), Patna U., India, 1946, M.B., B.S., 1951; Ph.D., U. Minn., 1957; m. Aryabala Ray, Jan. 9, 1952; children—Rita, Sheila, Ashok, Nivedita. Instr., U. Minn., 1957-58; vis. faculty U. Shiraz (Iran), 1958-60, Vanderbilt U., 1961-63; asso. prof. medicine, chief hematology, dept. medicine Wayne State U., Detroit, 1963—; staff VA Hosp., Dearborn Mich., also Detroit Gen. Hosp. Mem. subcom. trace elements Nat. Acad. Scis.-NRC, Food and Nutrition Bd. Recipient Caldwell Meml. prize Patna Sci. Coll., 1944; Pfizer Resident scholar U. Hosp., Mpls., 1955-56, Research Recognition award Wayne State U., 1964; named hon. prof. medicine, U. Shiraz, 1960. Fellow A.C.P., Internat. Soc. Hematology; mem. Am. Fedn. Clin. Research, Central Soc. Clin. Research, Internat. Soc. Internal Medicine, Am. Inst. Nutrition, Am. Soc. Clin. Nutrition, Soc. Exptl. Biology and Medicine, A.M.A., Sigma Xi, others. Editor: Zinc Metabolism, 1966. Research, publs. on syndrome of hypogammaglobulinemia, splenomegaly and hypersplenism; zinc deficiency in man; internal medicine and metabolism. Home: 2212 W. Long Lake Rd., Orchard Lake, Mich. 48033. Office: Wayne State U. Sch. Medicine, 1400 Chrysler Expressway, Detroit 48207.*

PRASAD, Braham Govind, Indian physician; b. Orai, Nov. 14, 1915; s. Ram Govind and Ram (Kali) P.; M.B.B.S., Lucknow U., 1937, M.D., 1948; D.P.H., All India Inst. Hygiene and Pub. Health, Calcutta, 1939; D.T.M., Sch. Tropical Medicine, Calcutta, 1940; m. Shyama Jwala, Feb. 7, 1939; children—Vijay Kumar, Rachna (Mrs. Krishn Kumar Srivastava). Jr., sr. house physician King George's Med. Coll. Hosp., Lucknow U., 1937-38, prof., head dept. social and preventive medicine, King George's Med. Coll., 1958—; with Provincial Hygiene Inst., Lucknow, asst. dir. med. and health services, 1940-43, 46-58. Rockefeller Found. fellow Edinburgh U. Mem. expert adv. panel profl. and tech. edn. of med. and aux. personel WHO; mem. adv. com. on nat. health problems Indian Council Med. Research. Mem. Internat. Epidemiological Assn. (council). Research and numerous publs. on nutrition of sch. children, epidemiology of cholera, plague, kalazar, yaws, diphtheria, trachoma, hookworm, lathyrism, leprosy, diabetes and blindness in India, morbidity studies in children and mental health, family planning practices. Home: 8 Ram Mohan Raimars. Office: King George's Medical College, Lucknow, Uttar Pradesh, India.*

PRASAD, Jagdish, Indian physicist; b. Muzaffarnagar, India, July 25, 1933; s. Sumer and Yashodha (Devi) P.; M.Sc., U. Allahabad, 1953; D.Phil. in Sci., U. Calcutta (India), 1961; Diplome Optique Instrumentale et Technologie, ASTEF, Paris, 1963; m. Permila Rani Gupta, June 18, 1959; children—Sandeep, Sangeeta, Sumata. Lectr. physics Agra (India) U., 1953-54; staff Central Glass and Ceramic Research Inst., Calcutta, 1954-56, 57-65, sr. sci. officer, 1964-65; profl. asst. India Meteorol. Dept., Delhi, 1956-57; sr. sci. officer Central Sci. Instruments Orgn., Chandigarh, India, 1965—, scientist-in-charge optics div., since 1965—. Fellow Indian Phys. Soc. (prin. mem. optical and math. instruments sect. com.). Research and publs. on solid state physics, especially clay mineral problems, thermal conductivity of ceramic materials and applied optics; patentee improvements in electric incandescent filament lamps, focusing devices for projection lamps in paraboloidal or like shaped reflectors. Home: 3063 Sector 20-D. Office: Central Sci. Instruments Orgn., Sector 30, Chandigarh, Punjab, India.*

PRASAD, M.R.N., Indian endocrinologist; b. Bangalore, India, May 1, 1923; s. Iyengar and Yadugiriamma; M.Sc., U. Mysore, India, Ph.D.; Ph.D., U. Wis.; m. Nagarathna, June 25, 1944; children—Prema, Prakash, Raju. Lectr. zoology U. Mysore, 1945-49; faculty U. Delhi (India), 1959—, dir. tng. program physiology reprodn., 1963—, prof. zoology, 1964—, investigator in charge Ford Found. project. Recipient Gold medal U. Mysore, 1944. Mem. Brit. Soc. for Study Fertility, Indian Soc. for Study Reprodn., Internat. Planned Parenthood Fedn. (research com.), V Internat. Symposium on Comparative Endocrinology (chmn. internat. organizing com.). Research, publs. on comparative morphology and biochemistry of acecssory glands of reprodn. in mammals. Address; U., Delhi 7, India.*

PRASAD, Sarju, Indian chemist; b. Bahraich, India, July 1, 1907; s. Hanuman and Leclawati (Devi) P.; B.Sc., Banaras Hindu U., 1928, M.Sc., 1930, M.A., 1934, D.Sc., 1942; m. W. Rameshwari Devi, May 9, 1929; children—Kashi Prasad, Srivastara, Madhuri Lata, Srivastava, Savitri, Srivastara, Vijai Lakshm, Srivstavia, Krishna Srivastara, Hari Prasad Srivastra, Jagdish Prasad Srivastava. Teaching staff chemistry dept. Banaras Hindu U., 1932—, reader chemistry, 1951—. Recipient Cooper Meml. medal Instn. Chemists India, 1958. Fellow Indian Chem. Soc., Instn. Chemists India (mem. council), Nat. Acad. Scis. India. Research and numerous publs. on coordination compounds in non-aqueous media, micro-electroanalytical techniques, high temperature reactions, catalysis. Died, July 26, 1967.

PRASADA, R(aghubir), Indian botanist; b. Chandausi, India, Dec. 11, 1907; s. Jwala and Prem (Pyarie) P.; M.Sc., Agra (India) Coll., 1930; D.Sc., U. Agra., 1943; m. Chandra Gupta. Apr. 23, 1935; children—Vi-jay, Lata (Mrs. Surendra Kumar), Pushpa (Mrs. Virendra Mohan), Vikram. Research scientist Cereal Rust Research Lab., Simla, India, 1930-46; mycologist Indian Agrl. Research Inst., New Delhi, India, 1946-52. Mycologist, Wheat Improvement Program, 1952-67. Prof. plant pathology Udaipur U., Jobner, India, 1967—. Mem. Indian Phytopath. Soc. (past pres., editor-in-chief Indian Phytopathology 1962—), Indian Sci. Congress (past pres. agrl. scis. sect.). Research and publs. on epidemiology and physiologic specialization of rust diseases of cereals, oil-seeds and fibre crops, disease forecasting and their control; devel. disease resistance of varieties of plants; aerobiol. studies in relation to wind currents; discovered Alternaria blight of wheat, also new host of barberry rust. Home: Qr20 Agr. Coll., Jobner (Jaipur), India.*

PRAT, Henri Charles, French biologist; b. St. Germain-en-Laye, France, Aug. 20, 1902; s. Maurice and Marthe (Himmelspach) P.; Licencie ès Scis. Physiques, U. Paris, 1922, Lic. ès Sc. Naturelles, 1926, Agrégé Sc. Nat., 1926, Docteur és Sc., 1931; m. Germaine Bertrand, Sept. 30, 1930; children—Annie (Mrs. Janglius, Alain, Olivier. Prof. biology U. Montreal (Que., Can.), 1931-35, 45-60, chmn. dept. biology, 1931-35, 49-55; prof. botany Faculty Scis., Marseilles, France, 1936-46, 60—, chmn. dept. botany, 1961—; dir. Bot. Garden Marseilles, 1961—. Decorated Commandeur des Palmes académiques, Chevalier de la Légion d'Honneur. Mem. Société Botanique France, Société de Biologie, Société Teilhard de Chardin. Author: L'homme et le Sol, 1949; Microcalorimétrie, 1956; Métamorphose explosive de l'humanité, 1961; (with Calvert) Recent Progress in Micro-Calorimetry, 1963; Le champ unitaire en biologie, 1966; also numerous articles. Research on histology, organogenesis, anatomy of plants, plant and animal studies in thermogenesis and influence of phys. and chem. factors, ethnology, philosophy of sci. Home: 8 Cote d'Azur Av., Marseilles 8. Office: Faculté des Sciences, Place Victor Hugo, Marseilles 3, France.*

PRATA, Aluizio, Brazilian physician; b. Uberaba, Brazil, June 1, 1920; s. Joao and Delia Rosa Prata; M.D., U. Brasil, Rio de Janeiro, 1945; m. Martha Taubes Alonso, Apr. 2, 1938; children—Aluizio, Alvaro, Martha Maria. With Brazilian Navy Health Corps, 1946-58; faculty U. Bahia (Brazil), Med. Coll., 1957—, prof., chmn. dept. tropical and infectious diseases, 1958—; dir. fundaçao Gonççalo Moniz, 1961—; supr. research progrfam Ministry Health on Endemic Diseases, Bahia, 1965—; vis. prof. Cornell Med. Coll., 1965-66. Adviser on cardiomyopathies WHO, 1964, 66. Recipient Carlos Chagas, 1959; Gaspar Viana, 1962; Ordem de Mérito Naval, 1965. Mem. Associacao Brasileira de Medicina, Am., Royal socs. tropical medicine and hygiene. Author: (monograph) Biopsia Retal na esquistossomose, 1957; also articles. Research on chronology of maturation of Schistosoma mansoni eggs in vitro, evolution, prognosis and clin. forms of Brazilian endemic diseases, especially schistosomiasis and Chagas disease. Home: 2, Comendador Horacio Urpia-Salvador, Bahia, Brazil. Office: Cl. Dopenas Trop. Infec., Hosp. das Clinicas, Bahia, Brazil.*

PRATT, George K., Am. psychiatrist; b. Detroit, Dec. 17, 1891; s. George Oscar and Alice Elizabeth (Beedzler) P.; M.D., Detroit Coll. Medicine and Surgery, 1915; postgrad. State Psychopathic Hosp., U. Mich., also in Europe; m. Neva Emma MacArthur, Dec. 30, 1916; children—Shirley Jane (Mrs. Carleton W. Clark), Rodney George, Douglas MacArthur. Asst. physician Oak Grove Hosp., Flint, Mich., 1915-20; in pvt. practice, also asst. health officer, Flint, 1920-21; med. dir. Mass. Soc. for Mental Hygiene, 1921-25; also in out-patient dept. Boston Psychopathic Hosp., 1921-25; lectr. mental hygiene Smith Coll., 1923-25; asst. med. dir. Nat. Com. for Mental Hygiene, N.Y., 1925-33; med. dir. Mental Hygiene Com. N.Y., State Charities Aid Assn., 1930-35, Conn. Soc. for Mental Hygiene, 1936-42; cons. mental hygiene, U. of Vt., 1925-29; faculty New Sch. for Social Research, N.Y.C., 1930-33, also Bklyn. Inst. Arts and Scis.; cons. in psychiatry St. Christopher's Sch., Dobbs Ferry, N.Y., 1932-36, from 1942; asst. clin prof. psychiatry and mental hygiene Sch. of Medicine, Yale, 1936-43; psychiat. dir. Stamford Child Guidance Service, also Bridgeport Mental Hygiene Clinic, 1936-1947; instr. New Haven State Tchrs. Coll., 1939-42; asso. neuropsychiatrist Bridgeport Hosp.; med. dir. Hall-Brooke Sanitarium, 1948-54. Diplomate Am. Bd. Psychiatry and Neurology. Fellow Am. Psychiat. Assn., Royal Medico-Psychol. Assn. Gt. Britain, Conn. State, Fairfield County med. socs. Author: Your Mind and You, 1924. Why Men Fail (with others), 1928; Our Neurotic Age (with others), 1932; Morale; the Mental Hygiene of Unemployment, 1933; Three Family Narratives, 1935; Soldier to Civilian, 1944. Contbr. tech. articles. Died Dec. 11, 1957.

PRATT, George Woodman, Jr., Am. physicist; b. Boston, Aug. 13, 1927; s. George W. and Helen (Horton) P.; B.S., Mass. Inst. Tech., 1949, Ph.D., 1952; m. Ann Frances Polonsky, June 15, 1948; children—George Woodman III, Sandra Ann. Staff mem. Lincoln Lab., Mass. Inst. Tech., 1952-60, faculty, 1960-—, prof. elec. engring., 1965-—, dir. materials theory group, 1961-—; vis. prof. Brandeis U., 1957. Cons. labs., indsl. cos. Mem. Am. Phys. Soc., Phys. Soc. Japan. Research, publs. on semiconductor lasers, including pressure tuning and frequency modulation, relativistic effects in band structure of solids, theory of effective mass, g-factors and deformation potentials of semi-conductors, theory of ferromagnetism and anti-ferro magnetism, optical properties of antiferromagnetic salts; many body theory of dielectric function. Home: 67 Glezen Lane, Wayland, Mass. 01778. Office: Dept. Elec. Engring., Mass. Inst. Tech., Cambridge, Mass. 02139.*

PRATT, Gerald Hillary, Am. surgeon; b. Montello, Wis., Dec. 15, 1906; s. Martin Henry and Margaret Ann (Farr) P.; B.S., U. Minn., 1924; M.D., U. Ia., 1928; m. Mae D. Gargin, Mar. 19, 1935 (dec. 1954); children—Margaret Anne (Mrs. Thomas Taylor), Judith Mae, Gerald Hillary; m. 2d, Nancy H. Brown, Mar. 22, 1956; children—Dennis Gerald, James Martin. Practice gen. and cardiovascular surgery; staff St. Vincent's Hosp. and Med. Center, St. Clare's Hosp., N.Y. Poly Clinic, Doctor's Hosp. (all N.Y.C.); asso. clin. prof. surgery N.Y. U. Sch. Medicine, N.Y.C., 1948-—; instr. surgery Temple U., Phila., 1931-35; asst. prof. surgery Columbia, 1935-48. Decorated Order Don X Govt. of Spain, Sir Don Alphonso X. Diplomate Am. Bd. Surgeons. Fellow A.C.S.; mem. Surg. Soc. Cuba (hon.), Surg. Soc. (France), Surg. Soc. (U. Madrid, Spain), Surg. Soc. (U. Barcelona, Spain), Hon. Soc. Venezuela, N.Y. State, N.Y. County med. socs., A.M.A., Am. Therapeutic Soc. (past pres.), N.Y. Heart Assn. (dir. 1941-—), N.Y. Diabetes Soc. (dir. 1941-—), Am. Cardiovascular Soc. (founding). Author: Surgical Management of Vascular Diseases, 1948; Cardiovascular Surgery, 1956; also numerous articles, chpts. in textbooks. Research on surg. treatment of diseases of arteries, veins and lymphatics; devel. new techniques in replacement and repair of disease; open heart surgery in animals. Address: 58 E. 66th St., N.Y.C. 10021.*

PRATT, Harlan K(elley), Am. plant physiologist; b. Berkeley, Cal., Mar. 18, 1914; s. Burr B. and Louise (Kelley) P.; B.S. in Agr., U. Cal. at Los Angeles, 1939, Ph.D. in Plant Sci., 1944; postgrad. Cal. Inst. Tech., 1940-41; m. Anna Marie Martin, June 21, 1939; children—Robert M., Elizabeth Ann. Research fellow biology Cal. Inst. Tech., 1944-46; faculty dept. vegetable crops U. Cal. at Davis, 1946-—, prof., plant physiologist, 1963-—. Fulbright Research scholar plant physiology unit div. food preservation Commonwealth Sci. and Indsl. Research Orgn., Sydney, Australia, 1956; USPHS Spl. fellow Ditton Lab., Agr. Research Council, U.K., 1963-64. Mem. Am. Inst. Biol. Scis., Am. Soc. Plant Physiologists, Am. Soc. Hort. Sci., Bot. Soc. Am., Sigma Xi, Phi Beta Kappa, Alpha Zeta. Research, publs. on postharvest physiology of fruits and vegetables, especially ripening of fruits and role of ethylene as fruit ripening and growth hormone. Home: 753 Oeste Dr., Davis, Cal. 95616.*

PRATT, Henry Sherring, Am. zoologist; b. Toledo, Aug. 18, 1859; s. Charles and Catherine (Sherring) P.; A.B., U. Mich., 1882; Ph.D., Leipzig, 1892; postgrad. univs. Freiburg, Geneva (Switzerland), Harvard, Innsbruck, Graz (both Austria); m. Agnes Woodbury Gray, Sept. 1, 1894; 1 dau., Anna. Instr. biology Haverford Coll., 1893-98, asso. prof. 1898-1901, prof., 1901-29 (emeritus). Instr. comparative anatomy Cold Spring Harbor Biol. Lab., 1896-1926. Fellow A.A.A.S.; pres. Cambridge Entomol. Club, 1896; mem. Am. Soc. Naturalists, Am. Soc. Zoologists. Author: Invertebrate Zoology, 1902; Vertebrate Zoology, 1906, 3d edit., 1937; Manual of Common Invertebrates, 1916, rev. 1935; Manual of Vertebrates of the U.S., 1923, 2d edit., 1935; A Course in General Zoology, 1927; A Course in General Biology, 1927; General Biology—an Introductory Study, 1931; also zool. papers. Died Oct. 6, 1946.

PRATT, Ivan, Am. parasitologist; b. Navarre, Kan., Sept. 18, 1908; s. Walter T. and Lillie (Shank) P.; B.A., Coll. Emporia (Kan.), 1932; M.S., Kan. State U., 1935; Ph.D., U. Wis., 1938; m. Elizabeth L. Hartberg, Aug. 22, 1938; children—Charles Walter, Richard Edward, Katherine Elizabeth. Faculty, U. Ida., 1938-42, U. Kan. City Sch. Dentistry, 1942-44; asst. sanitarian USPHS, NIH, Bethesda, Md., 1944-45; faculty zoology Ore. State U., Corvallis, 1946-—. prof., 1952-—. NIH scholar U. Neuchatel, Switzerland, 1961-62. Mem. A.A.A.S., Western Soc. Naturalists, Sigma Xi, Phi Kappa Phi. Author: (with J.E. McCauley) Trematodes of the Pacific Northwest, 1961. Research on effects of parasite upon host and life cycles and ecol. factors of helminths, especially trematodes. Home: 3208 Johnson St., Corvallis, Ore. 97330.*

PRATT, Joseph Hyde, Jr., Am. surgeon; b. Chapel Hill, N.C., Mar. 9, 1911; s. Joseph Hyde and Mary (Bayley) P.; A.B., U. N.C., 1933; M.D., Harvard, 1937; M.S. in Surgery, U. Minn., 1947; m. Hazel Housman, Dec. 11, 1943; children—Judith Housman, Lisa Mary, Joseph Hyde III. Staff, Mayo Clinic, Rochester, Minn., 1943-—, head surg. sect., 1945-—; faculty Mayo Grad. Sch. Medicine, U. Minn., Rochester, 1947-—, prof. clin. surgery, 1963-—. Diplomate Am. Bd. Surgery, Am. Bd. Obstetrics and Gynecology. Mem. A.C.S. (gov. 1966-—), A.M.A., Am. Coll. Obstetricians and Gynecologists, Am. Assn. Obstetrics and Gynecology, Central Assn. Obstetricians and Gynecologists, Western Surg. Assn., Minn. Obstet. and Gynecol. Soc., Minn. Surg. Soc., Soc. Pelvic Surgeons, So. Surg. Assn., Sigma Xi. Research, numerous publs. on lesions and surg. problems of abdomen and female pelvis. Home: 721 12th Av., S.W., Rochester, Minn. 55901.*

PRATT, Parker Frost, Am. soil chemist; b. Verdin, N.M., Nov. 21, 1918; s. Ira W. and Mary (Merrill) P.; B.S., Utah State U., 1947, M.S., 1948; Ph.D., Ia. State U., 1950; m. Mary J. Skoro, July 19, 1945; children—Von, Craig, Koleen, Kathleen. Asst. prof. agronomy Ohio State U., 1951-55; faculty U. Cal., Riverside, 1955-—, prof. soil sci., chmn. dept. soils and plant nutrition, 1965-—. Cons. Boyle Engring., San Diego, 1963-64, IRI Research Inst., Inc., N.Y.C., Sao Paulo, 1964-65. Mem. Am. Soc. Agronomy, Soil Sci. Soc. Am., Internat., Western soil sci. socs., Sigma Xi. Author: (with H.D. Chapman) Methods of Analysis of Soils, Plants, and Waters, 1961; also articles. Contbns. to knowledge of cation-exchange processes in soils; availability of essential mineral elements to plants; long-term changes in irrigated soils. Home: 4198 Swain Ct., Riverside, Cal. 92507.*

PRATT, Robertson, Am. pharmacognosist; b. Bklyn., Oct. 28, 1909; s. Abram Johnston and Janet (Robertson) P.; B.A., Columbia, 1931, Ph.D., 1936; m. Jean Noack, Sept. 1, 1948; children—Robin Jean, David Walter. Faculty, U. Cal. Med. Center at San Francisco, 1938-—, prof. pharmacognosy and antibiotics, 1954-—; researcher Cutter Labs., 1943-45. Cons. on microbiologic and antibiotic problems, 1945-—. Fellow A.A.A.S.; mem. N.Y., Cal. acads. sci., Soc. Am. Bacteriologists, Am. Pharm. Assn. (Ebert prize for research on penicillin 1945), Am. Soc. Pharmacognosy, Sigma Xi, Rho Chi. Author: (with Jean Dufrenoy) Antibiotics, 1949; pharmacognosy (with H.W. Younken, Jr.), 1951; also numerous articles. Co-editor-in-chief U.S. Dispensatory, 1960-—. Research on devel. penicillin and understanding penicillin action, effect nutrition on growth Chlorella vulgaris and its vitamin content. Office: U. Cal. Med. Center, San Francisco, Cal. 94122.*

PRAVAZ, Charles Gabriel, French physician; b. Isere, France, Mar. 24, 1791; ed. Polytechnique Inst.; became physician, 1824; practiced in Paris, then, Lyons, France; founder Orthopedic Inst., Lyons. Invented 1st practical metal syringe with hollow needle which he used to inject aneurysms, 1853; invented procedures of modern galvanocautery, 1853. Died Lyons, June 24, 1853.

PRAXAGORAS OF COS, surgeon; b. Cos; flourished 2d half of 4th century B.C.; s. Nicharchos; mem. family of Asclepiades; tchr. of Herophilos. Defended humoral pathology; discovered arterial pulse and distinguished between veins and arteries, but considered arteries to be air-channels which led from heart and tapered into nerves; recognized connection of brain and spinal cord.

PRAY, Lloyd Charles, Am. geologist; b. Chgo., June 25, 1919; s. Allan Theron and Helen (Palmer) P.; B.A., Carleton Coll., 1941; M.Sc., Cal. Inst. Tech., 1943, Ph.D. (NRC fellow), 1952; m. Carrel Myers, Sept. 14, 1946; children—Lawrence Myers, John Allan, Kenneth Palmer, Douglas Carrel. Geologist, Magnolia Petroleum Co., summer 1942, U.S. Geol. Survey, 1943-44; from instr. to asso. prof. geology Cal. Inst. Tech., Pasadena, 1949-56; sr. research geologist Denver Research Center, Marathon Oil Co., Littleton, Colo., 1956-62, research asso., 1962-68; prof. geology U. Wis., Madison, 1968-—. Vis. prof. geology U. Tex., 1964, Colo., 1967. Fellow Geol. Soc. Am., A.A.A.S.; mem. Am. Petroleum Inst. (com. on carbonate research), Am. Assn. Petroleum Geologists (research com. 1958-61), Am. Inst. Mining, Metall. and Petroleum Engrs., Soc. Econ. Paleontologists and Mineralogists (sec.-treas. 1961-63, v.p. 1966-67), Colo. Sci. Soc., Rocky Mountain Assn. Geologists, Internat. Assn. Sedimentologists, Am. Geol. Inst. (council on geol. edn. 1964-66, com. on edn. 1966-—), Phi Beta Kappa, Sigma Xi. Contbr. articles to profl. jours. on stratigraphy and structural geology of N.M., Western Tex., carbonate facies, geology of sedimentary carbonates, rare earth mineral deposits. Home: 6777 Southridge Lane. Office: Marathon Oil Co., Littleton, Colo. 80121.*

PREBLE, Edward A., Am. naturalist; b. Somerville, Mass., June 11, 1871; s. Edward Perkins and Marcia (Alexander) P.; ed. high sch., Woburn, Mass., 1886-89; m. Eva A. Lynham, Dec. 29, 1896; children—Dorothy Marcia, Marjorie Elizabeth, Evelyn Morgan. With biol. survey U.S. Dept. Agr., 1892-1935. Established wild life sanctuary and library, Ossipee, N.H. Member Am. Ornithologists Union, Biol. Soc. Washington, Am. Soc. Mammalogist, Am. Soc. Ichthyologists and Herpetologists. Author: A Biological Investigation of the Hudson Bay Region (U. S. Govt. pub.), 1902, A Biological Investigation of the Athabaska-Mackenzie Region, 1908; The Fur Seals and other life of the Pribilof Islands, Alaska, in 1914 (in collaboration), 1915. Birds and Mammals of the Pribilof Islands, Alaska (in collaboration), 1923. Contbr. many sci. and popular articles. Asso. editor Nature Mag., Washington. Died Oct. 4, 1957.

PREDVODITELEV, Aleksandr Savvich, Russian physicist; b. Sept. 11, 1891; grad. Moscow U., 1915. Prof. Moscow U., 1930; lab. chief Inst. Energetics, USSR Acad. Scis., 1938-—. Recipient State prize, 1950; decorated Order Lenin. Corr. mem. USSR Acad. Scis., 1939. Author: The Relations Between the Thermal Conductivity, Thermal Capacity and Viscosity of Liquids, 1943; The Molecular-Kinetic Basis of Hydrodynamics Equations, 1948; Flucuations in Statistical Systems, 1948; Physical Gas Dynamics, 1961; Physics of Heat Exchange and Gas Dynamics, 1962; co-author: Thermodynamics Air Function Tables for Temperatures 12,000-20,000° and Pressures of 0.001 and 1.00 Atm. Worked Out Theory of heterogenous combustions relating chem. and phys. processed of carbon combustion; research in hydrodynamics, thermophysics, molecular physics. Address: Leninskiye Gory, Sekt. K, Moscow, USSR.

PREECE, Sir William Henry, Welsh elec. engr.; b. Bryn Helen, Wales, Feb. 15, 1834; s. Richard Mathias and Jane (Hughes) P.; ed. Kings Coll.; studied under Faraday at Royal Inst.; LL.D.; m. Agnes Pocock, 1864; 4 sons, 3 daus. Civil engr. for Edwin Clark, 1852; apptd. to Electric & Internat. Telegraph Co., 1853, supt. so. dist., 1856; engr. Channel Islands Telegraph Co., 1856-62; div. engr. Post Office, 1870, electrician, 1877, engr.-in-chief, electrician, 1892-99; cons. engr. to colonies, 1899-1904. Fellow Royal Soc., 1881; mem. Instn. Civil Engrs. (pres. 1898-99). Author: (with Sir James Sivewright) Textbook of Telegraphy, 1876, 15th edit., 1899, new edit., 1905; (with Julius Maier) The Telephone, 1889; (with Arthur J. Stubbs) A Manual of Telephony, 1893. Contbd. improvements and inventions in field of telegraphic communications, ry. signaling techniques, wireless telegraph; invented duplex telegraph, 1881, block system for r.r., 1873; discovered possibility of receiving in telephone through induction telegraphic signals which were transmitted through telegraph line running in certain interval, 1885; took out patent for reproducing by miniature signals in signal book the positions of actual signals and system of locking signals; introduced in Britain 1st telephone receivers as patented by Bell. Died Penrhos, Carnarvon, Wales, Nov. 6, 1913.

PREER, John Randolph, Jr., Am. biologist; b. Ocala, Fla., Apr. 4, 1918; s. John Randolph and Ruth (Williams) P.; B.S., U. Fla., Gainesville, 1939; Ph.D., Ind. U., 1947; m. Louise Bertha Brandau, Nov. 17, 1941; children—James Randolph, Robert William. Staff zoology U. Pa., 1947-—, prof. zoology, 1957-—, chmn. grad. group zoology, 1958-62. Mem Genetics Soc. Am., Soc. Zoologists, Am. Soc. Naturalists, Phi Beta Kappa, Sigma Xi (pres. local chpt. 1959). Contbr. articles profl. jours. Home: Box 132, R.D. 1, Glen Mills, Pa. Office: University of Pennsylvania, Phila. 4.*

PREFONTAINE, Georges, Canadian bacteriologist; b. Isle-Verte, Que., Can., May 26, 1897; s. Alexander and Louisa (Lindsay) P.; B.A., Joliette Coll., 1918; M.D., Montreal U., 1923, Licencie es Sciences, 1933; Doctor honoris causa U. Alger, 1944, U. Ottawa, 1960; m. Marie-Anna Dubreiul, Apr. 16, 1924; children—Francois Jean, Jacques, Claude, Yves. Dir., Inst. Biology, U. Montreal (Que., Can.), 1935-48; dir. dept. labs. St. Joseph Hosp., Rosemont, Montreal, 1948-—, head microbiology lab., 1950-67, mem. adminstrv. council, 1964. Mem Fisheries Research Bd. Can., 1931-52. Mem. Assn. Can. francaise pour Advancement de Sci. (past pres.), Canadian Soc. Natural History (past pres. Montreal), A.A.A.S., Am. Thoracic Soc., Arctic Inst., Soc. médicale de Montréal, Sigma Xi. Founder mem. Revue Canadienne de Biologie, 1941-—, mem. exec. Bur., 1941-51. Research, publs. on marine fauna of St. Lawrence Estuary, salmon of Gulf of St. Lawrence, freshwater biology of certain areas of St. Lawrence Basin, properties of certain new antibiotics. Home: 630 Davaar St., Montreal 8. Office: 5689 Rosemont St., Montreal 36, Que., Can.*

PREGL, Fritz, Austrian chemist; b. Laibach, Austria (now Ljubljana, Yugoslavia), Sept. 3, 1869; student Tübingen, Leipzig, Berlin (all Germany); Dr. med., U. Graz (Austria), 1893. Asst. physiology and histology, Graz, 1893-1904; traveled in Germany, 1905; staff, Graz, 1905-10, 13-30; prof., Innsbruck, Austria, 1910-13. Recipient Nobel prize in chemistry, 1923. Author: Die quantitative organischen Mikroanalyse, 1917. Founder microchem. analysis; developed micromethods for determining hydrogen, carbon, nitrogen and organic groups, 1911-18; studied biliary acids. Died Graz, Dec. 13, 1930.

PREISENDORFER, Rudolph William, Am. math. physicist; b. Bklyn., Dec. 30, 1927; s. William Joseph and Minna (Steube) P.; B.S., Mass. Inst. Tech., 1952; Ph.D., U. Cal., Los Angles, 1956; m. Eleanor

May Reynolds, Aug. 16, 1952; children—Russell, Lynn. Asst. research mathematician U. Cal., San Diego, 1956-60; research mathematician John Jay Hopkins Lab., San Diego, 1960-61; asso. research mathematician Visibility Lab. Scripps Inst. Oceanography, San Diego, 1961-65, research mathematician, 1965—. Fellow Optical Soc. Am.; mem. Math. Assn. Am., Am. Math. Soc. Author: Radiative Transfer on Discrete Spaces, 1965; also numerous articles. Research in physics and geometry of scattering and absorption of light in atmosphere and seas, lakes and other natural waters of earth; developed theoretical conections between observed light field in these media and their scattering and absorption parameters; devised purely math. found. of scattering theory similar to way Euclid developed geometry from axioms of geometry. Home: 8303 Prestwick Dr., La Jolla, Cal. 92037. Office: Sverdrup Hall, Scripps Inst. Oceanography, San Diego 92037.*

PREISS, Ekkehard, German mineralogist; b. Breslau, Oct. 10, 1908; s. Georg and Grete (Hasenjaeger) P.; Ph.D., State Sch. Pforta, 1935; m. Gudrun Herzog, May 7, 1940; four children. Prof., 1950—. Publns. on meteorites, mineralogy, spectral analysis. Home: Saint-Privatstrasse 11, Munich. Office: Kumpfmühlerstrasse 2, Regensburg, West Germany.

PREISWERK, Peter, Swiss physicist; b. Basel, Jan. 16, 1907; s. Wilhelm and Julia (Imhoff) P.; student univs. Basel, Berlin; Dr.phil.; m. Anna Gertrud Vischer, Sept. 6, 1937. Sci. asso. Radium Inst., Paris; sci. asso., inst. physics, Fed. Inst. Tech., Zurich, 1936-54, titular prof.; dept. dir., lab. group European Center Nuclear Research (CERN), Geneva, 1952-54, dir. site ad bldgs. div., 1954-61, dir. nuclear physics div., 1961—. Mem. Swiss, Am. phys. socs. Research, numerous publs. on artificial radioactivity, slow neutrons, nuclear spectroscopy. Home: 1242 Satigny, Switzerland. Office: CERN, Geneva 23, Switzerland.*

PRELOG, Vladimir, chemist; b. Sarajevo, Yugoslavia, July 23, 1906; s. Milan and Mara (Cettolo) P.; Ing.chem., Inst. Tech. Sch. Chemistry, Prague, Czechoslovakia, 1928, Dr., 1929; Dr.h.c., U. Zagreb (Yugoslavia), 1954, U. Liverpool (Eng.), U. Paris (France), 1963; m. Kamila Vitek, Oct. 31, 1933; 1 son, Jan. Chemist, Lab. G.J. Driza, Prague, 1929-35; docent U. Zagreb, 1935-40, asso. prof., 1940-41; faculty Swiss Fed. Inst. Tech., Zürich, 1942—, prof. chemistry, 1950—, head Lab. Organic Chemistry, 1957-65. Mem. bd. CIBA Ltd., Basel, Switzerland, 1963—. Recipient Werner medal, 1945; Stas medal, 1962; medal of honour Rice U., 1962; Marcel Benoist award, 1965. Fellow Royal Soc., 1962; mem. Am. Acad. Arts and Scis. (hon.), Nat. Acad. Scis. (fgn. asso.), Acad. dei Lincei (Rome, Italy, fgn.), Leopoldina, Halle/Saale, Acad. Scis. USSR (fgn.). Research, numerous publs. on constn. and stereochemistry alkaloids, antibiotics, enzymes, other natural compounds. Home: 41 Bellariastr. 8038, Zurich, Switzerland. Office: Universitätsstr. 8006, Zurich, Switzerland.

PREMONT, Michel, French surgeon; b. Paris, France, Apr. 11, 1927; s. Fernand and Jeanne Michelle (Avril) P.; M.D., Faculty Medicine, Paris, 1957; m. Jannine Courtecuisse, Apr. 8, 1953; children—Nicolas, François, Jean-Baptiste, Denis. With Faculte de Medecine de Paris, 1955—, prof. agrege, 1963—; with Hôpitaux de Paris, 1961—, surgeon, 1962—. Mem. anatomy Soc. Paris, French Soc. Orthopedics (asso.), French Assn. Surgery. Author: (with Lucien Leger) Semeiologie chirurgicale, 1964; also articles. Research in clin. and exptl. surgery, especially gastro-enterol. disease. Home: 20 Longchamp, Paris 16, France. Office: 30 Kilford, Courbevoie (92), France.*

PRENANT, Auguste, French anatomist; b. Lyons, France, 1861; prof., Nancy, France, also Paris. Author: Traité d'embryologie, 1896; Traité de cytologie et d'histologie, 2 vols., 1906-11. Pioneer in cytology; contbd. to improvement of methods of cytology. Died Paris, 1927.

PRENDERGAST, Joseph Benoit, Canadian geophysicist; b. Toronto, Ont., Can., Oct. 1, 1927; s. William Killoran and Nora Irene (Benoit) P.; B.A. in Physics and Geology, U. Toronto, 1950, M.A. in Geophysics, 1951; m. Shirley Joyce Stilwell; children—Katherine, Michael, Laura, Patrick, Matthew, Guillermo. Party chief seismologist Geophys. Service, Inc., 1951-54; sr. geophysicist Internat. Nickel Co., 1954-56; chief geophysicist Sulmac Exploration Services, 1956-59; cons. geophysicist mining areas Can., Ireland, Mexico, 1959-62; project mgr. mineral projects for UN, Mexico, 1962-65; cons. geophysicist-geologist, Mexico, Can.; pres. Dominion Geophysics, Ltd., Toronto, 1962—, United New Fortune Mines, Ltd., Toronto, 1965—, Tolteca Mines Ltd., Toronto, 1964—. Mem. Soc. Exploration Geophysicists, European Assn. Exploration Geophysicists, Am. Inst. Mining and Metall. Engrs. Research on gravitational study deep seated structural features in S. Ont., geophys.-geol. study Sudbury basin; devel. gravitation methods for iron ore exploration and devel.; discovered ore bodies in Can. and Ireland. Address: Monte Irazu 140, Mexico 10, D.F., Mexico; also office: Lancaster Bldg., Calgary, Alta., Can.*

PRENDERGAST, Kevin Henry, Am. astrophysicist; b. Bklyn., July 9, 1929; s. George A. and Nora (Sullivan) P.; A.B., Columbia, 1950, Ph.D., 1954; m. Jane Keston, Sept. 4, 1965. Faculty, U. Chgo, Yerkes Obs., 1954-61, asst. prof., 1956-61; mem. Inst. for Advanced Study, Princeton, N.J., 1961-62; Staff Inst. for Space Studies, NASA, N.Y.C., 1962-63; asso. prof. dept. astronomy Columbia, 1963-66, prof., 1966—. Mem. Am., Royal astron. socs. Research, publs. on photometry, rotation galaxies, stellar dynamics. Home: 750 Kappock St., Bronx, N.Y. 10463. Office: 1418 Pupin, Columbia, N.Y.C. 10027.*

PREOBRAZHENSKY, Boris Sergeevich, Russian otorhinolaryngologist; b. Moscow, 1892; grad. Med. Faculty, Moscow U., 1914; D.Med. Sci., 1926. With otorhinolaryn. clinic 2d Moscow U., 1921-35; dir. Semashko Clinic Ear, Nose and Throat Diseases, 1935-41; prof., head chair otorhinolaryngology Pirogov 3d Moscow Med. Inst., 1936-41; prof., 1936—; head chair otorhinolaryngology, dir. clinic ear, nose and throat diseases Pirogov 2d Moscow Med. Inst., 1941—. Chmn., Otorhinolaryn. Instrument Commn., Tech. Council, USSR Ministry Health; chmn. Otorhinolaryngology Problems Commn., Presidium, USSR Acad. Med. Scis., 1956—; mem. Internat. Com. of Otorhinolaryngologists, 1961—. Decorated Order of Lenin (5). Mem. USSR Acad. Med. Scis., All-Union (dep. chmn.), Moscow (chmn. 1943—) socs. otorhinolaryngologists. Author: Deaf-Mutism, 1933; War Injuries of the Ear, Nose and Throat, 1944; co-author: Ear, Nose and Throat Diseases, 5th edit., 1955; Angina and Its Prophylaxis. Editor: Herald of Otorhinolaryngology. Otorhinolaryngology sect. Large Med. Ency., 2d edit.; mem. editorial council Ear, Nose and Throat Diseases, Excerpta Medica (Holland). Research and publs. on theoretical and clin. otorhinolaryngology, pathogenesis and clin. aspects of deafness, deaf-mutism and baryecoia, clin. aspects, surg. treatment of chronic tonsillitis and its connection with other organic diseases, phlegmonous angina; one of 1st Soviet surgeons to adopt full tonsillectomy in chronic tonsillitis, improved operating techniques and anesthesia. Address: Pirogov 2d Moscow Med. Inst., Malaya Pirogovskaya 1, Moscow, USSR.

PRESCOTT, Benjamin, Am. biochemist; b. Fall River, Mass., Feb. 7, 1907; s. Isaac and Sarah (Abramski) P.; B.S., U. Chgo., 1930; postgrad. Columbia, Harvard; Ph.D., Georgetown U., 1941; m. Lillian Stein, June 22, 1932; children—Lawrence Malcolm, Elliot Jordan. Biochemist, Harvard Med. Sch., 1933-35, organic chemist, 1934-35; gen. inorganic chemist Franklin Union, Boston, 1934-35; immunochemist Johns Hopkins Med. Sch., 1935-38; biochemist NIH, Bethesda, Md., 1938—; instr. biochemistry U.S. Dept. Agr. Grad. Sch., 1954-57. Mem. Nat. capital area br. Animal Care Panel, 1961—. Mem. Am. Chem. Soc., Soc. for Exptl. Biology and Medicine, N.Y. Acad. Sci., Soc. for Indsl. Microbiology. Research, publs. on nitrogen and sulfur metabolism in Bright's disease, immunochemistry of pneumococcus, chemotherapy of bacterial, viral, fungal and parasitic diseases, detoxification mechanisms in these diseases, isolation of antimicrobials and antitumor substances from molluscs, isolation of protective antigen from mycoplasma pneumoniae. Home: 5215 Chevy Chase Pkwy. N.W., Washington 20015. Office: 9000 Rockville Pike, Bethesda, Md. 20014.*

PRESCOTT, James Arthur, agrl. chemist; b. Bolton, Lancashire, Eng., Oct. 7, 1890; s. Joseph Arthur and Mary Alice (Garsden) P.; B.Sc., U. Manchester (Eng.), 1911, M.Sc., 1919; D.Sc., U. Adelaide (Australia), 1932; D.Ag. Sci., U. Melbourne (Australia), 1956; m. Elsie Mason, Oct. 12, 1915; 1 son, John R. Research scholar Rothamsted Exptl. Sta., 1912-16; chief chemist Sultanic Agrl. Soc., Cairo, Egypt, 1916-24; prof. agrl. chemistry U. Adelaide, 1924-55, dir. Waite Agrl. Research Inst., 1938-55; chief div. soils Council for Sci. and Indsl. Research, 1929-48, chmn. oenological research com., 1938-55; mem. council Australian Wine Research Inst., 1955—. Named Comdr. Order Brit. Empire, 1947. Fellow Royal Soc., 1951, Australian Acad. Sci., Chem. Soc. London, Royal Australian Chem. Inst. (past pres.), Australian Inst. Agrl. Sci. (past pres.), Royal Soc. S. Australia (past pres., past editor); hon. mem. Internat. Soc. Soil Sci. Research, numerous publs. on soil factors, geography, classification of soils, climate and its influence. Home: 82 Cross Rd., Myrtle Bank, South Australia 5064.*

PRESCOTT, John Mack, Am. biochemist; b. San Marcos, Tex., Jan. 22, 1921; s. John Mack and Maude (Raborn) P.; B.S. in Chemistry, S.W. Tex. State Coll., 1941; M.S. in Biochemistry and nutrition, Tex. A. and M. U., 1949; Ph.D., U. Wis., 1952; m. Kathryn Ann Kelly, June 8, 1946; children—Stephen Michael Prescott, Donald Wyatt. Lab. asst. Dow Chem. Co., Freeport, Tex., 1942-43; staff Tex. A. and M. U., College Station, 1946-49, 52—, prof. biochemistry, 1959—; research asst. U. Wis., Madison, 1949-51, U. Tex., Austin, 1951-52. Cons. to indsl. orgns., 1956—. Mem. Am. Chem. Soc., Am. Soc. Biol. Chemists, Am. Soc. for Microbiology, Am. Inst. Nutrition, Soc. for Exptl. Biology and Medicine, Soc. for Gen. Microbiology (Eng.). Research, publs. on identification, purification and characterization of several proteolytic enzymes from microorganisms and snake venoms, bacterial nutrition and metabolism. Home: 1103 Walton Dr., College Station, Tex. 77840.*

PRESENT, Richard David, Am. physicist; b. N.Y.C., Feb. 5, 1913; s. David and Blanche (Wertheimer) P.; B.S., Coll. City N.Y., 1931; M.A., Harvard, 1932, Ph.D., 1935; m. Thelma Cohen, July 31, 1943; children—Irene Naomi, Constance Sarah. Instr. physics Purdue U., Lafayette, Ind. 1935-40; research asso. Harvard, Cambridge, Mass., 1940-41; asst. prof. physics N.Y. U., N.Y.C., 1941-43; mem. div. war research Manhattan Project, Columbia, N.Y.C., 1943-46; asso. prof. physics U. Tenn., Knoxville, 1946-48, prof., 1948—. Cons. Clinton Nat. Labs., 1946-48. Fellow Inst. Internat. Edn., Paris, 1937-38. Fellow Am. Phys. Soc. (chmn. S.E. sect. 1964—); mem. Am. Assn. U. Profs. (pres. chpt. 1957-58), Sigma Xi. Author: Kinetic Theory of Gases, 1958. Research in nuclear theory, molecular structure and kinetic theory. Home: 21 Circle Hill Dr., Knoxville, Tenn.*

PRESS, Frank, Am. geophysicist; b. Bklyn., Dec. 4, 1924; s. Solomon and Dora (Steinholz) P.; B.S., Coll. City N.Y., 1944; M.A., Columbia, 1946, Ph.D., 1949; m. Billie Kallick, June 9, 1946; children—William Henry, Paula Evelyn. Research staff Columbia, N.Y.C., 1946-49, asst. prof. geophysics, 1951-52, asso. prof., 1952-55; prof. geophysics Cal. Inst. Tech., Pasadena, 1955-65, dir. Seismol. Lab., 1957-65; head dept. geology and geophysics Mass. Inst. Tech., Cambridge, 1965—. Mem. seismology, polar research, interdisciplinary panels IGY, 1957-59; mem. Pres.' Sci. Adv. Com., 1961-64; cons. Dept. Def., 1958-61, AID, 1962-63, Arms Control and Disarmament Agy., 1962-63, Office Sci. and Tech., 1964—. Dir. United Electro Dynamics Corp. Recipient medal for excellence Columbia, 1959; Townsend medal Coll. City N.Y., 1962; named Cal. Scientist of Year, Cal. Mus. Sci. and Industry, 1960. Fellow Geol. Soc. Am. (councillor); mem. Internat. Union Geodesy and Geophysics (internat. geophysics com. 1959—), Acoustical Soc. Am., Am. Geophys. Union, Am. Phys. Soc., Seismol. Soc. Am. (pres. 1962), Nat. Acad. Scis. Author: (with Ewing, Jardetzky) Elastic Waves in Layered Media, 1957. Research on submarine geology and geophysics, crystal and mantle structure, ultrasonic model seismology, devel. long period seismography, excitation, dispersion and attenuation of long surface waves, geophys. studies in Gt. Basin, free oscillations of earth, lunar seismology, pulse propagation in atmosphere. Home: 26 Spring Valley Rd., Belmont, Mass.*

PRESS, Leon, Argentinian physican, nutritionist; b. Buenos Aires, Argentina, Dec. 3, 1917; s. Adolfo and Rebeca (Pobireski) P.; M.D., Nat. U. Buenos Aires, 1943, Physician Dietitian, 1949; m. Maria Barón, Aug. 24, 1944; children—Raul Edgardo, Eduardo Cesar. Physician, Rawson Hosp., Buenos Aires, 1943-45; 2d lt., physician Argentine Army, 1945-47; physician Model Inst. Nat. U., Buenos Aires, 1947-52, Nat. Nutrition Inst. Buenos Aires, 1952—; prof. U. Museo Social Argentino, 1952—. Mem. Argentine Assn. Nutrition and Diet, Argentine Med. Assn., Latin Am. Nutrition Soc. Research, publs. on problems of human nutrition. Office: 1074 Azcuénaga, Buenos Aires, Argentina.*

PRESSAT, Roland François Marcel, French demographer; b. Paris, France, June 28, 1923; s. François Adrien and Célénie (Mellet) P.; Licence de Mathematiques, Sorbonne, Paris, 1948; postgrad. Institut de Statistique, U. Paris; m. Cécile Bonino, July 15, 1949; children—François, Laurent. Chief dept. Institut Nat. d'Études Démographiques, Paris, 1953—; prof. Institut de Demographie de l'Université de Paris, 1958—. Mem. directory com. Centre de Sociologie et de Demographie medicale, Paris, 1961—. Mem. Union Internationale pour l'Etude scientifique de la Population, Société de Statistique Paris, Institut international de Statistique. Author: L'Analyse Démographique, 1961; Principes d'Analyse, 1966; Pratique de la Démographie, 1967; also articles. Research on population analysis. Home: 23 Av. de Bréteville, 92 Neuilly, France. Office: INED, 23 Av., Franklin Roosevelt, Paris (8°), France.*

PRESSMAN, David, Am. chemist; b. Detroit, Nov. 24, 1916; s. Jacob and Lottie (Frankfort) P.; B.S. in Chemistry, Cal. Inst. Tech., 1937, Ph.D., 1940; M.A., U. Cal. at Los Angeles, 1938; m. Reinie M. Epstein, July 11, 1940; children—Jeffrey L., Adele R. Sr. fellow research Cal. Inst. Tech., Pasadena, Cal., 1940-47; head sect. immunochemistry Sloan-Kettering Inst., N.Y.C., 1947-54; asso. prof. biochemistry Sloan-Kettering div. Cornell U. Med. Sch., N.Y.C., 1952-54; dir. cancer research biochemistry Roswell Park Meml. Inst., Buffalo, 1954—, asso. dir. sci. affairs, 1967—; research prof. chemistry Roswell Park div. State U. N.Y., Buffalo, 1955—. Mem. allergy and immunology study sect. USPHS, 1955-58, bd. sci. div. biologic standards NIH, 1967—; mem. com. on etiology cancer Am. Cancer Soc., 1958-61. Recipient Bertha Teplitz award Ann Langer Cancer Research Found., 1958. Fellow N.Y. Acad. Scis. A.A.A.S., mem. Am. Chem. Soc. (Schoellkopf medal Western N.Y. sect. 1965, Morley medal Cleve. sect. 1967), Am. Soc. Biol. Chemists, Am. Assn. for Cancer Research, Soc. for Exptl. Biology and Medicine (past pres. Western N.Y. sect.), Am. Assn. Immunologists. Author: (with I. Lucas, J. Howard) Principles and Practice in Organic Chemistry, 1949. Editorial adv. bd. Cancer Research, 1959-64, asso. editor, 1965—; asso. editor Jour. Immunology, 1959-63, 65—; editorial bd. Transplantation Bull., 1954-56; Jour. medicinal and Pharm. Chemistry, 1962-65; Immunochemistry, 1964—. Research, numerous publs. on elucidation of structure of antibody molecules; determined chem. nature of specific binding sites of antibody and enzymes, factors important for specific combination of antibodies with antigens; studied specific fixation of anti-tissue and anti-tumor antibodies in organs and tumors. Home: 76 Edge Park, Buffalo 14216. Office: 666 Elm St., Buffalo 14203.*

PRESTON, Frederick Willard, Am. physician; b. Chgo., June 27, 1912; s. Frederick Augustus and Margaret (Atwater) P.; B.A., Yale, 1935; M.D., Northwestern U., 1940, M.S. in Physiology, 1942; M.S. in Surgery, U. Minn., 1947; m. Gertrude Bradford, June 23, 1942 (div. Apr. 1960); children—Frederick Williard, David Eldred, William Blackmore; m. 2d., Barbara Gay Hess, July 30, 1961. Clin. asst. in surgery U. Ill., 1948-49; faculty Northwestern U. Med. Sch., Chgo., 1950—, prof. surgery, 1960—; staff Chgo. Wesley Meml. Hosp., VA Research Hosp., Chgo. Diplomate Am. Bd. Surgery. Mem. A.M.A., A.C.S. (pres. Chgo. met. chpt. 1965), Mayo Clinic Alumni Assn., Am. Fedn. for Clin. Research, Reticuloendothelial Soc., Am. Assn. for Cancer Research (past pres. Chgo. sect.), Chgo. Acad. Scis. (sec. 1961—), Chgo. Surg. Soc. (past sec.), Am., Western, Central surg. assns., Société Internationale de Chirurgie, Soc. for Surgery Alimentary Tract. Research, numerous publs. on surg. techniques, oncology, anti-cancer drugs, anticoagulants, portal hypertension, abdominal gastrointestinal physiology. Home: 3 Kent Rd., Winnetka, Ill. 60093. Office: 333 E. Huron St., Chgo. 60611.

PRESTON, Melvin Alexander, Canadian theoretical physicist; b. Toronto, Ont., Can., May 28, 1921; s. Gardener Alexander and Hazel (Melvin) P.; B.A., U. Toronto, 1942, M.A., 1946; Ph.D., U. Birmingham, 1949; m. Mary Whittaker, Aug. 16, 1947; children—Jonathan, Richard; m. 2d, Eugene Shearer, June 25, 1966. Asst. lectr. math. physics U. Birmingham, 1947-49, Nuffield fellow, 1957; faculty U. Toronto, 1949-53; faculty McMaster U., Hamilton, Ont., Can., 1953—, prof. physics, 1959—, prof. applied math., 1962—, dean grad. studies, 1965—. NRC fellow U. Institut for Teoretisk Fysik, Copenhagen, 1963-64. Mem. Am. Phys. Soc., A.A.A.S., Phys. Soc. London, Computer Soc. Can., Canadian Assn. Physicists, Royal Soc. Can. Author: Physics of the Nucleus, 1962; also articles. Studies in theoretical nuclear physics, particularly radioactive alpha-decay of non-spherical nuclei, weak interaction in nuclear beta-decay of nuclear structure, properties of nuclear matter, and force between nucleons. Home: 4058 Lakeshore Rd., Burlington, Ont. Office: McMaster U., Hamilton, Ont., Can.*

PRESTON, Reginald Dawson, English biophysicist; b. Leeds, Yorkshire, Eng., July 21, 1908; s. Walter Cluderay and Eliza (Dawson) P.; B.Sc., with 1st Class honors in Physics, Leeds U., 1929, Ph.D., in Botany, 1931, D.Sc., 1943; postgrad. (Rockfeller Found fellow) Cornell U.; m. Sarah Jane Pollard, Apr. 13, 1935 (dec.); children—David Roger (dec.), Maureen Ann (Mrs. Keith W. Roberts), Judith Margaret; m. 2d, Eva Frei, Oct. 19, 1962. Faculty, U. Leeds (Eng.), 1936—, prof. plant biophysics, 1953—, head Astbury dept. biophysics, 1962—.Mem. council John Innes Inst., 1961—. Fellow Inst. Physics, Linnean Soc., Inst. Wood Sci., Royal Soc., 1954; mem. Leeds Philos. and Lit. Soc. (editor 1950—), Brit. Biophys. Soc. (founder). Author: The Molecular Architecture Plant Cell Walls, 1952; also numerous articles, chpts. in books. Editor, Advances in Bot. Research, 1960—; co-editor Jour. Exptl. Botany, 1964—; editorial adv. bd. Biorheology, 1962—. Research on polysaccharide structure; discovered relation between crystallite orientation in cell walls and cell dimensions, microfibrillar structure plant cell walls; used structure adult cell walls to define nature biosynthetic mechanism of structural polysaccharides; demonstrated existence of plants in which structural polysaccharide is not cellulose. Home: 117 St. Annes' Rd., Leeds 6, Yorkshire, Eng.*

PRESTON, Richard Joseph, Jr., Am. forester; b. Rockford, Ill., Sept. 28, 1905; s. Richard Joseph and Margaret Amelia (Whitehall) P.; A.B. U. Mich., 1927, M.S. in Forestry, 1928, Ph.D., 1941; student Stanford 1930, 32, U. Chgo., 1932, 33; m. Bernice Boynton, June 1, 1937. With U. S. Forest Service, Laramie, Wyo., 1929; extension ranger Fla. Forest Service, Tallahassee, 1930-32; asst. prof. Principia Coll., Elsah, Ill., 1932-36; ranger naturalist U. S. Nat. Park Service, Yellowstone, Wyo., summers 1935-36; mem. faculty Colo. A. and M. Coll., 1936-42, 45-48, prof. forest mgmt. and utilization, head dept., 1945-48; technologist U. S. Forest Products Lab., Madison, Wis., 1942-45; dir. div. forestry N.C. State Coll., Raleigh, 1948-50, dean Sch. Forestry, 1950—. Mem. Gov. N.C. Adv. Com. Forestry, 1961—; governing bd. Agrl. Research Inst., NRC, 1961—; conf. bd. Asso. Research Council, 1960.—. Fellow of the Soc. Am. Foresters; member Am. Forestry Assn., N.C. Forestry Council, Sigma Xi. Author: Rocky Mountain Trees, 3d edit., 1966; North American Trees, 3d edit., 1967; also numerous articles. Research on wood preservation; gluing of wood; influence of shelterbelts; shear testing techniques; North American trees. Home: 3201 Churchill Rd., Raleigh, N.C.

PRESTON, Richard Swain, Am. physicist; b. Natick, Mass., Feb. 4, 1925; s. Arthur Charles and Esther (Swain) P.; B.A., Wesleyan U., 1949, M.A., 1950; M.S., Yale, 1952, Ph.D., 1954; m. Angela Louise Camurati, July 25, 1954; children—Claire, Mark. Asso. dir. Yale Goechronometric Lab., New Haven, 1954-55; asst., asso. physicist Argonne (Ill.) Nat. Lab., 1955—, Vis. physicist U.K. Atomic Energy Research Establishment, Harwell, Eng., 1965-66. Mem. Am. Phys. Soc. Editorial bd. Bull. Atomic Scienticts, 1965—. Precision measurements of masses of certain nuclei; radiocarbon dating of objects of archaeol. and biol. interest; search for new "subnuclear" particles; observation of annihilation of positrons in ferromagnets to demonstrate parity nonconservation in certain nuclear decays; use of Mossbauer effect to study nuclear properties of several isotopes, and to investigate properties of solids. Home: 725 Willow Rd., Naperville, Ill. 60540. Office: Physics Div., Argonne Nat. Lab., Argonne, Ill. 60440.*

PRESTON, Thomas Reginald, animal nutritionist; b. Windermere, Eng., Dec. 16, 1928; s. Thomas Nuttal and Beatrice (Graham) P.; B.Sc. with honors, U. Durham, 1952, Ph.D., 1955; m. Mary Braithwihte, 1954; m. Margaret May Duncan, 1961; children—Jane, Tanya. Head sheep sect. Rowett Research Inst., Aberdeen, Scotland, 1955-60, head cattle sect., 1960-65; dir. Inst. Animal Sci., Havana, Cuba, 1965—. Mem. Brit., Latin Am. socs. animal prodn., Brit. Nutrition Soc., Am. Soc. Animal Sci. Research, numerous publs. on influence of synthetic sex hormones on body composition of fattening sheep and cattle, devel. econ. calf rearing methods, system of intensive beef prodn. based on feeding all-concentrate diets (barley-beef). Home: Calle Norte 62, Esq. 35, Nuevo Vedado, Havana. Office: Institute of Animal Science, Km. 48 Carretera Central, Catalina de Guines, Havana, Cuba.*

PRESTON-THOMAS, Hugh, physicist; b. Havant, Eng., Dec. 26, 1923; s. Quintin Bernard and Marjorie (Rashleigh) P.-T.; B.Sc., Bristol (Eng.) U., 1945, Ph.D., 1951; m. Mary Elizabeth Bowler, Aug. 22, 1955; children—Timothy, Johnathan, Caroline, Peter. Sci. officer atomic energy project Brit. Ministry Supply, Birmingham, Eng., Montreal, Que., and Chalk River, Ont., Can., 1945-47; postdoctoral research Bristol U., 1950-51; postdoctoral fellow div. pure physics NRC Can., Ottawa, Ont., 1951-52, staff, 1952—, head heat and solid state physics lab. applied physics div., 1954—. Fellow Canadian Assn. Physicists. Research, publs. on superconductors, ion rockets and space ship orbits, heat transfer problems, specific heat, thermal conductivity, temperature metrology. Home: 1109 Blasdell Av., Ottawa 7. Office: NRC, Ottawa, Ont., Can.*

PRESTWICH, Sir Joseph, English geologist; b. Clapham, Surrey, Eng., Mar. 12, 1812; s. Joseph and Catherine (Blakeway) P.; student sci. and chemistry U. Coll., London; M.A., Christ Church, Oxford, 1874; D.C.L., 1888; m. Grace Anne M'Call Milne, 1870. Joined father's bus. as wine mcht., London; prof. geology, Oxford, 1874-88, Became mem. Water Commn., 1862. Knighted, 1896. Recipient Wollaston medal, 1849. Fellow Royal Soc., 1853 (Royal medal 1865), Geol. Soc.; mem. French Acad. Scis. Author: Geology, Chemical, Physical and Stratigraphical, 2 vols., 1886-88; The Tradition of the Flood, 1895; also numerous pamphlets, articles. Studied water bearing strata around London, 1851, Thames basin; classified English Tertiary deposits; theorized that man was co-existent with other Pleistocene mammals. Died Shoreham, Kent, Eng., June 23, 1896.

PRESTWOOD, Rene Jesse, Am. chemist; b. San Rafael, Cal., Oct. 18, 1920; s. Rodney W. and Louise (Cailleaud) P.; B.S., U. Cal. at Berkeley, 1942; Ph.D., Washington U., St. Louis, 1948; m. Sara Dowson, Feb. 5, 1944; children—Jennifer Louise, Linda Jane, Patti Anne. Staff, U. Cal. at Berkeley, 1942-43; staff Los Alamos Sci. Lab., 1943-46, 48—. Contbg. author: Radioactivity Applied to Chemistry, 1949; also articles. Research on absolute beta counting, excitation functions, radiochem. separation of elements from fission products, symmetry of fission in resonance region. Home: 683 47th St. Office: P.O. Box 1663, Los Alamos, 87544.*

PRETORIUS, Petrus Jacobus, South African pediatrician; b. Ugie, South Africa, Dec. 4, 1923; s. Jacobus Albertus and Hester (van der Walt) P.; M.B., Ch.B., U. Pretoria, 1947, M.D., 1957; m. Johanna Celestina de Klerk, July 2, 1949; children—Jacobus, Philippus. Clin. asst. dept. pediatrics Pretoria Hosp., 1952-55, chief pediatrician, 1965—; head Nutrition Clinic for Children, Nat. Nutrition Research Inst., Pretoria, 1956-64; prof. pediatrics U. Pretoria, 1965—. Mem. South African Pediatric Assn. (exec. council). Author: Infant Feeding and Nutrition, 1967; also numerous articles. Research on protein malnutrition in African infants, established that fish flour can replace milk in prevention of protein malnutrition, protein supplements to milk of no advantage in treating acute cases. Home: 92 Argyle. Office: P.O. Box 667, Pretoria, Transvaal, South Africa.*

PREUSS, Heinzwerner, German quantum chemist, physicist; b. Liegnitz, Germany, Sept. 12, 1925; s. Otto and Else (Pohl) P.; student U. Halle; Dr.rer.nat., U. Hamburg, 1954; m. Hannelore Röseler, May 19, 1956; 1 son, Glennfried. Asst., Max Planck Inst. Physics and Astrophysics, 1952-66, quantum chemistry group leader, 1966—; docent theoretical phys. chemistry U. Frankfort/Main, Germany, 1961—. Author: Integraltafeln zur Quantenchemie, 4 vols., 1956-60; Die Methoden der Molekülphysik und ihre Anwendungsbereiche, 1959; Grundriss der Quantenchemie, vol. 10 a,b,c, 1962; Quantentheoretische Chemie I, vol. 43, 1963, II, vol. 44 a,b, 1965; Quantenchemie für Chemiker 1966; also numerous articles. Research on theory of chem. bond, quantum chemistry, theoretical chemistry. Home: 31 Germaniastrasse. Office: Max Planck Inst. for Physics and Astrophysics, Föhringer Ring 6, Munich 23, West Germany.*

PREUSSMANN, Rudolf, German chemist; b. Stein/Nuremberg, Germany, Aug. 25, 1928; s. Werner and Claudia (Wolfarth) P.; Diplom-Chemiker, U. Munich (Germany), 1954; Doctor-Promotion, U. Freiburg (Germany), 1955; m. Erika Raabe, Dec. 1, 1955; children—Susanne, Roland. Head chem. dept. Forschergruppe Präventivmedicine Freiburg, 1955—. Research, publs. on carcinogenic action chem. substances, synthesis, chem. reactivity, analytical chemistry of carcinogenic organic N-Nitroso-compounds, hydrazines, azo-and azoxy-compounds; structure-activity relationship, biol. testing for carcinogenic activity in team research; synthesis and biol. evaluation of new cancer-chemotherapeutic agts. Home: 7, Rosastr. Office: 8, Stefan Meier-STr. 78, Freiburg/B, Germany.*

PREVOST, Charles Paul, French chemist; b. Champlitte, France, Mar. 26, 1899; s. Georges and Marie (Zimmermann) P.; Licence, Agrigation de Sciences Physiques, Ecole Normal Superieure, 1923; Pharmacien, Faculty de Pharmacie, Nancy, France, 1933; m. Eleonore Fumée, Oct. 6, 1923; children—Georges, Noemi (Mrs. Guern). Faculty, Ecole Normale Superieure, 1923-29; prof. Faculté de Pharmacie, Nancy, 1929-35, Faculté de Pharmacie, Lille, France, 1936-37; prof. Faculté de Sciences, Paris, 1937—, Ecole Centrale de Arts et Manufactures, 1955—. Decorated Officer Legion of Honour, Comdr. Palmes Academiques. Mem. Chem. Soc. France, Soc. Phys. Chemistry, Am. Chem. Soc. Author: Lecons de chimie organiques, 1951; Prêcis de chimie organique générale, 1960; Traité de chimique organique générale, 1967; also numerous articles. Research in organic chemistry, especially ethylene-glycol, kinetics of numerous heterolytic reactions. Home: 1, rue Clovis, Paris, France.*

PRÉVOST, Constant, French geologist; b. Paris, June 4, 1787; prof. geology Paris Faculty Scis.; mem. French Acad. Scis. Author: Historie des terrains tertiaires; Traité de géographie physique. Explained mountain formation as process of gradual retraction of earth's crust; opposed theory of successive acts of creation as explanation of prehistoric fossils; held that all life could be traced to one original act of creation; opposed opinion of his time by asserting that geol. causes of past operate in the present. Died Paris, Aug. 6, 1856.

PRÉVOST, Jean Louis, Swiss physician; b. Geneva, Switzerland, Sept. 1, 1790; student medicine, Paris; M.D., U. Edinburgh (Scotland), 1818; began practice medicine, Geneva, 1820. Research on physiol. and anat. relations between man and animals; used (with J.B.A. Dumas) defibrinated blood in animal transfusions (1st successful attempt to prevent coagulation in transfusion), 1821; discovered (with Dumas) Graafian follicles in dogs and rabbits, 1824; discovered method of fertilizing frog's eggs, also that filtration sterilizes frog semen; 1st description (with Dumas) of segmentation in frog's egg, 1827. Died Geneva, Mar. 14, 1850.

PRÉVOST, Pierre, Swiss physician, physicist, philosopher; b. Geneva, Switzerland, Mar. 3, 1751; student theology, law, medicine; prof. philosophy, Geneva, 1784—, physics, 1810—. Fellow Royal Soc., 1806. Author: Recherches physico-mécaniques sur la chaleur, 1792; Essai sur le calorique rayonnant, 1809. Translator works on physics and philosophy. Research and publs. on heat and magnetism; developed theory of exchanges of radiation from one body to another, 1792. Died Geneva, Apr. 18, 1839.

PRÉVOT, André Romain, French microbiologist; b. Douai, Nord, France, July 22, 1894; s. Romain and Alix P.; Docteur en Médecine, Faculty Paris, 1924; Docteur ès Sciences naturelles, U. Paris, 1933; m. Anna Soerson, Sept. 27, 1919; children—Johanne, (Mrs. Garein), Francois, Michel, Romain. Staff, Pasteur Inst., Paris, 1922—, head lab., 1929-39, departmental head, 1939-65, emeritus dept. head, 1965—; exchange prof. in numerous countries. Named officer Legion of Honor, 1952, Order Pub. Health, 1956. Mem. French Soc. Hematology (past pres.), French Soc. Microbiology (past pres.), Société géologique du Nord, Soc. Biology, Internat. Soc. Soil, Brussels Royal Acad. Medicine (emeritus), Acad. Scis.,

Nat. Acad. Medicine. Author: (with Lea, Fiebiger) Manual of Classification of Anaerobes, 13 vols., 1965; also numerous articles. Research on anaerobic bacteria; discovered gangrenous anatoxins, anaerobic corynebacterioses and their treatment especially ecology of Whipple's disease. Home: 6 Rue Gathelot Clamart 92, France. Office: 28 Rue du Dr Roux, Paris XV, France.*

PREVOT, François Gabriel, French physicist; b. Paris, July 7, 1923; s. Andre Romain and Anna (Sorensen) P.; Phys. Engr., Paris Inst. Physics and Chemistry, 1947; m. Micheline Hervouet, Sept. 24, 1947; children—Claude, Jean-Jacques, Alain, Eric. With Nat. Office. Aero. Research, 1947-49, AEC, Saclay and Fontenay aux Roses (France), 1949—. Mem. French, Am. phys. socs., French Soc. Engrs. and Technicians. Contbr. numerous articles to publs. Design and constrn. of particles accelerators; high vacuum and ultra-high vacuum devel.; ion sources and ion beam development; research on plasma physics for controlled fusion. Home: 27 Carnot, Antony 92. Office: BP6, Fontenay aux Roses 92, France.*

PREYER, Wilhelm Thierry, physiologist, psychologist; b. Moss Side/Manchester, Eng., July 4, 1841; prof., Jena, Germany; pvt. tchr., Wiesbaden, Germany. Author: Die Blausäure, 2 vols., 1868-70; Die Blutkrystalle, 1871; Die Entdeckung des Hypnotismus, 1881; Die Seele des Kindes, 1882; Das genetische System der chemischen Elemente, 1893; Darwin, sein Leben und Wirken, 1896. Research on physiology, blood, breathing, muscles, hypnotism. Supported Darwinism. Died Wiesbaden, July 15, 1897.

PRIANISHNIKOV, Dimitry, Russian agronomist; b. Kiakhta, USSR, Nov. 6, 1865; s. Nicolay and Alexandra (Lebedeva) P.; ed. Moscow U., Petrovskaia Acad. Agr.; Ph.D., 1899; m. Maria Terentieva, 1898; 3 children. Became prof. plant industry Petrovskaia Acad. Agr., 1895, later prof. agrl. chemistry; organizer Golitzyn's Coll. for women, 1908, dir. until merger with Petrovskaia Acad., 1917; organizer, mgr. agrochem. sect. State Research Inst. Fertilizers, 1919; mem. State Com. Planning of Nat. Economy (Gosplan), 1920-25; co-founder Com. of Chemization of Nat. Economy, 1927. Recipient Gold medal Com. Agrl. Exhbn. in USSR, 1940; Stalin prize, 1940; named Hero of Labor, 1945. Fellow Lenin's Acad. Agr. Scis.; mem. USSR, French acads. scis., Internat. Soc Soil Sci. (pres. soil fertility sect. 1930), Soc. Plant Physiology, U.S., Bot. Soc. Low Countries, Sci. Council Internat. Inst. Agronomy, Rome, Italy; hon. mem. Ceska Academia Zemedelska, Prague, Kgl. Landbruks Akademie, Stockholm, Naturforscher Akedemie zu Halle (Leopoldina), Deutsche Botanische Gesellschaft, Vereinigung für Angewandte Botanik, Physikalische Gesellschaft (Koenigsberg). Author: Course of Fertilizers; Agrochemistry; Special Plant Industry; Plant Chemistry; Die Einheitlichkeit der Prinzipen in Stickstoffwechsel bei Pflanzen und Tieren, 1928; The Result of Vegetation Experiments and Laboratory work, 17 vols., 1898-40; (monograph) Nitrogen in the Life of Plants. Address: Ivanovskaia 23, Moscow City 8, USSR.

PRIBRAM, Ernest August, physician; b. Prague, Bohemia, Feb. 2, 1879; s. Otto and Leonora (Popper) P.; grad. U. Prague (Czechoslovakia), 1903; postgrad. U. Strasbourg (France); m. Mrs. Maria H. Salsonson. Mar. 21, 1918; 1 son, Karl Harry. Came to U.S., 1925. With U. Vienna in various capacities and as asso. prof. pathology until 1925, also asst. and dir. State Serum Inst., 1907-25; mem. faculty Rush Med. Coll., Chgo., 1926-28; prof. bacteriology and preventive medicine Loyola U., Chgo., from 1928, dir. lab., from 1938; pathologist St. Elizabeth's Hosp., Chgo., St. Therese's Hosp., Waukegan, Ill. Owner and dir. microbiol. collection in Vienna, from 1911. Mem. numerous profl. socs. Author: Culture Media for Bacteria and for Fungi, 1925; Classification of Bacteria, 1933. Died Sept. 14, 1940.

PRIBRAM, Karl Harry, physician, psychologist; b. Vienna, Austria, Feb. 25, 1919; s. Ernest and Maria (Salmonson) P.; came to U.S., 1927, naturalized, 1933; S.B., U. Chgo., 1939, M.D., 1941; m. Amy Isle, June 25, 1960; children—John, Joan, Bruce (by previous marriage), Cynthia Ann, Karl Seward. Neurosurgeon, neurophysiologist Yerkes Labs. Primate Biology, Orange Park, Fla., 1946-48; individual practice neurology and neurol. surgery, Jacksonville, Fla., 1946-48; faculty Yale, 1948-58; chmn. neurophysiology Inst. of Living, Hartford, Conn., 1951-55, dir. research and labs., 1956-59, cons., 1959-60; faculty Stanford, 1958—, Nat. Inst. Mental Health, research prof. psychiatry and psychology, 1962—. Vis. lectr. univs., 1954, 56, 57, 61, 67, 68; mem. sci. adv. com. Mental Research Inst., Palo Alto, Cal., 1959—; mem. various coms. NRC, NIH. Diplomate Am. Bd. Neurol. Surgery. Mem. Acad. Arts and Scis., A.A.A.S., Am. Assn. Anatomists, Am. Electroencephalographic Soc., A.M.A., Am. Neurol. Assn., Am. Physiol. Assn. (chmn. com. on precautions and standards in animal experimentation), Am. Psychol. Assn., Harvey Cushing Soc., Internat. Brain Research Orgn., Pavlovian Soc. Am., Psychonomic Soc., Sigma Xi. Author: (with G. A. Miller, E. H. Galanter) Plans and the Structure of Behavior 1960. Research, publs. on analysis of brain-behavior relationships and specific areas of brain; theoretical systematizations of neuropsychological data pertaining to thinking, planning, emotion, sensory processing, memory. Home: 12499 Costello Ct. W., Los Altos, Cal. 94304. Office: Dept. Psychiatry, Stanford U. Sch. Medicine, Palo Alto, Cal. 94304.

PRICE, Albert Thomas, English math. physicist; b. Nantwich, Eng., Jan. 30, 1903; s. Albert T. and Marie L. (Light) P.; B.Sc., Manchester (Eng.) U., 1924, M.Sc., 1927, D.Sc., 1951; m. Ann Waterman, Mar. 29, 1947. Asst. math. Queen's U., Belfast, Ireland, 1925-26; faculty Imperial Coll., London, Eng., 1926-51, asst. prof., 1946-51; prof. math. Royal Tech. Coll., Glasgow, Scotland, 1951-52; vis. investigator dept. terrestrial magnetism Carnegie Inst., Washington, 1952; prof. applied math. Exeter (Eng.) U., 1952—, dean faculty sci., 1954-58. Mem. Nat. Adv. Council Edn. for Industry, 1957—; cons. RAND Corp., 1963. IGY Research fellow Coast and Geodetic Survey, Washington, 1961-62. Fellow Royal Astron. Soc. (past mem. council), Inst. Math. Applied; mem. London Math. Soc., Am. Geophys. Union, Soc. Exploration Geophysics, Soc. Terrestrial Magnetism and Electricity (Japan). Contbg. author: Physics of Geomagnetic Phenomena, 1966; The Earth's Mantle, 1966. Research, publs. on geomagnetism especially analysis and interpretation of magnetic variations, theory aeolotropic elasticity of wood, theory standing waterwaves of finite height; developed methods for finding elec. conductivity at various depths within earth. Home: 20 W. Garth Rd., Exeter, Devon, Eng.*

PRICE, Bartholomew, English mathematician; b. Coln St. Dennis, Gloucestershire, Eng., May 14, 1818; s. William Price; B.A. with 1st class in math., Pembroke Coll., Oxford, 1840, M.A., 1843, D.D., 1892; m. Amy Eliza Cole, Aug. 20, 1857; several sons and daus. Fellow, Pembroke Coll., 1844, tutor, 1846-57, master, 1892; visitor Royal Obs., Greenwich; canon of Gloucester; Sedleian prof. natural philosophy Oxford, 1853; pub. examiner, 1847-48, 53-55. Fellow Royal Soc., 1852, Royal Astron. Soc. Author: A treatise on the Differential Calculus, 1848; A Treatise on Infinitesimal Calculus, Vol. 1, Differential Calculus, Vol. II, 1852; Integral Calculus and Calculus of Variations, 1854; Statics and Dynamics of a Particle, Vol. III, 1856; Dynamics of Material Systems, Vol. IV, 1862, 2d edit., 1889. Died Dec. 29, 1898.

PRICE, Charles Coale, Am. chemist; b. Passaic, N.J., July 13, 1913; s. Thornton Walton and Helen Marot (Farley) P.; A.B., Swarthmore Coll., 1934, ScD. (hon.), 1950; M.A., Harvard, 1935, Ph.D., 1936; m. Mary Elma White, June 29, 1936; children—Patricia, Susanne, Sarah S., Judith S., Charles C. IV. Research asst. U. Ill., 1936-37, instr., 1937-39, asso., 1939-41, asst. prof., 1941-42, asso. prof., 1942-46; prof. chemistry U. Notre Dame, 1946-54, head dept., 1946-52; Blanchard prof. of chemistry, U. of Pa., dir. dept., 1954-65, Benjamin Franklin prof. chemistry, 1965—. Chmn. adv. council on coll. chemistry NSF, 1962-66. Pres. of United World Federalists, Inc., 1959-61. Recipient Am. Chem. Soc. award Pure Chemistry, 1946. Mem. Am. Chem. Soc. (pres.), A.A.A.S., Phi Beta Kappa, Sigma Xi, Phi Lambda Upsilon, Phi Sigma Kappa, Alpha Xi Sigma. Author: Brief Course in Organic Chemistry, 1940; Mechanism of Reactions at the Carbon-Carbon Double Bond, 1946; Sulfur- Bonding, 1962. Editor: Jour. Polymer Sci., Organic Synthesis; asso. editor, Chem. Revs. Contbr. numerous tech. papers to Jour. Am. Chem. Soc., Jour. Organic Chemistry, Jour. Polymer Sci. Home: 118 Hilldale Rd., Lansdowne, Pa. 19050. Office: Dept. Chemistry, U. Pa., Phila. 19104.*

PRICE, David Edgar, Am. pub. health physician; b. San Diego, July 5, 1914; s. Charles David and Pearl Mae (McCune) P.; A.B., U. Cal. at Berkeley, 1936, M.A., 1937, M.D., 1940; M.P.H., Johns Hopkins, 1945, Dr.P. H., 1946; m. Jean Shearer, June 5, 1936; children—William David, Janet Ruth. Commd. officer USPHS, 1941, epidemiologist San Diego Health Dept., 1941-42, staff physician Venereal Disease Center, Hot Springs, Ark., 1942-44, asst. to chief of research grants div. NIH, 1946-47, chief of cancer research grants, 1947-48, chief div. research grants, 1948-50, asso. dir. NIH, 1950-52, asst. surgeon gen. USPHS, Washington, 1952-57, dep. chief Bur. Med. Services, 1957-58, chief Bur. State Services, 1958-60, dep. dir. Inst., 1960-62, dep. surgeon gen. USPHS, 1962—. Mem. Phi Beta Kappa, Alpha Omega Alpha, Phi Chi. Research in exptl. and pituitary endocrinology; family planning. Home: 5215 Elsmere Av., Bethesda, Md. 20014. Office: Dept. Health, Edn. and Welfare, Washington 20201.*

PRICE, Don K., Jr., Am. sci. administrator; b. Middlesboro, Ky., Jan. 23, 1910; s. Don K. and Nell (Rhorer) P.; A.B., Vanderbilt U., 1931; B.A., Oxford U. (Rhodes scholar 1932), 1934, B.Litt., 1935; LL.D., Centre Coll. Ky., 1961, Syracuse U., 1962; m. Margaret Helen Gailbreath, Mar. 3, 1936; children—Don C., Linda G. Reporter and state editor Nashville Eve. Tennessean, 1930-32; research asst. H.O.L.C., and asst. to chmn. Central Housing Com., 1935-37; staff mem., com. on pub. adminstrn. Social Sci. Research Council, 1937-39; editorial asso. Pub. Adminstrn. Clearing House, 1939-41, asst. dir., 1941-43, asso. dir., 1946-53; staff mem. U. S. Bur. Budget, 1945-46; dep. chmn. Research and Development Bd., U. S. Dept. Def., 1952-53; lecturer political science University of Chicago 1946-53; associate director The Ford Foundation 1953-54, vice pres. 1954-59; professor of government, dean Graduate School Public Administration, Harvard, 1958—; member board of trustees Twentieth Century Fund; bd. trust Vanderbilt U., 1964—. Trustee Rand Corp., 1961—. Asst. to Herbert Hoover on study of U. S. Presidency under auspices Commn. on Orgn. Exec. Br. Govt., 1947-48; cons. Exec. Office of Pres., 1961—; dir. Social Sci. Research Counc., 1949-52, 64—; staff dir. Committee on Dept. Def. Orgn., 1953; mem. Pres.'s adv. com. on government orgn., 1959-61; adviser to the King of Nepal, Kathmandu, Nepal, 1960. Fellow American Academy Arts and Sciences member American Philos. Soc., A.A.A.S. (mem. bd. dirs. 1959-64, 66, pres. 1967), Phi Beta Kappa. Author: City Manager Government in the United States (with Harold and Kathryn Stone), 1940; U. S. Foreign Policy, Its Organization and Control (with W. Y. Elliott and others), 1952, also The Political Economy of American Foreign Policy, 1955; Government and Science, 1954; The Scientific Estate, 1965. Editor, co-author: The Secretary of State, 1960. Research, publs. on relations between science and government. Home: 114 Irving St., Cambridge 38. Office: Littauer Center, Cambridge, Mass.

PRICE, Glenn Albert, Am. physicist; b. Mpls., Feb. 9, 1923; s. Hugh Bruce and Jennie (Swab) P.; B.S., U. Ky., 1946; M.S., U. Ill., 1948, Ph.D., 1952; m. Charlotte Louise Jones, June 6, 1950; children—Beverly Jane, Daniel Jonathan, David Bruce. Sci. technician Los Alamos Sci. Lab., 1944-46; research asst. U. Ill., 1946-50, AEC Pre-doctoral fellow, 1950-52; physicist Brookhaven Nat. Lab., Upton, L.I., N.Y., 1952—. Mem. Am. Phys. Soc., Am. Nuclear Soc., Sigma Xi, Sigma Pi Sigma. Research, publs. on nuclear structure, nuclear reactors. Home: 256 Bayport Av., Bayport, N.Y. 11705. Office: Brookhaven Nat. Lab., Upton, N.Y. 11973.*

PRICE, Griffith Baley, Am. mathematician; b. Brookhaven, Miss., Mar. 14, 1905; s. Walter Edwin and Lucy (Baley) P.; B.A., Miss. Coll., 1925, LL.D., 1962; M.A., Harvard, 1928, Ph.D., 1932; m. Cora Lee Beers, June 18, 1940; children—Cora Lee, Griffith Baley, Lucy Jean, Edwina Clare, Sallie Diane and Doris Joanne (twins). Instr., Miss. Coll., 1925-26, 29-30, Union Coll., Schenectady, 1932-33, U. Rochester, 1933-36, Brown, 1936-37; faculty U. Kan., Lawrence, 1937—, prof. math., 1943—, chmn. dept. 1951—. Chmn. com. on regional devel. math. NRC, 1952-54; chmn. Conf. Bd. Math. Scis., 1959-60, exec. sec., 1960-62; mem. U.S. Nat. Commn. for UNESCO, 1961—. Guggenheim fellow, 1946-47. Mem. A.A.A.S., Am. Math. Soc. (editor Bull. 1950-57), Am. Statis. Assn., Inst. Math. Statistics, Math. Assn. Am. (gov. 1952—, pres. 1957-58), Nat. Council Tchrs. Math., Operations Research Soc. Am., Societe Mathematique de France, Kan. Acad. Sci., Sigma Xi. Author with others: Universal Mathematics, 1954, 58; Elementary Mathematics of Sets, with Applications, 1958; An Introduction to Mathematics, 1963; Linear Topological Spaces, 1963: Linear Equations and Matrices, 1966; also articles. Study of abstract spaces; functions of several real variables; integration; theoretical dynamics. Home: 1520 Barker Av., Lawrence, Kan. 66044.

PRICE, Henry Locher, Am. physician; b. Phila., Oct. 21, 1922; s. Henry Locher and Sara (Anderson) P.; A.B., Swarthmore Coll., 1945; M.D., U. Pa., 1946; m. Mary Lowe, Dec. 13, 1953; children—Susan Garrison, Kathryn Locher Anderson. NRC fellow in physiology Harvard Med. Sch., Boston, 1952-53; faculty U. Pa., Phila., 1953—, Wellcome prof. anesthesiology, 1960—. Cons. in anesthesiology U.S. Naval Hosp., Phila., 1960—; mem. com. on anesthesia NRC, 1961—, clin. research tng. com. Nat. Inst. Gen. Med. Scis., 1963-65, anesthesia tng. com., 1965-67. NIH fellow physiology U. Cal. Cardiovascular Research Inst., San Francisco, 1960-61. Mem. Am. Physiol. Soc., Am. Soc. Pharmacology and Exptl. Therapeutics, Am. Soc. Clin. Investigation, Am. Soc. Anesthesiology, Assn. U. Anesthetists, Sigma Xi, Nu Sigma Nu. Editor: Effects of Anesthetics on the Circulation, 1964. Author: Circulation during Anesthesia and Operation, 1967. Research, numerous publs. on effects of anesthetic agts. on cardiovascular system in man. Home: 510 Lynmere Rd., Bryn Mawr, Pa. 19010. Office: 3400 Spruce St., Phila. 19104.*

PRICE, Peter Jack, physicist; b. London, Eng., July 29, 1924; s. Mark Philip and Elizabeth (Cohen) P.; B.A., Oxford U., 1948; Ph.D., Cambridge U., 1951; m. Charlotte A. Alber, July 14, 1956. Came to U.S., 1951. With Royal Naval Sci. Service, Eng., 1944-46; Office Naval Research research asso. Duke, 1951-52, Inst. for Advanced Study, Princeton, N.J., 1952-53; physicist IBM Watson Lab., N.Y.C., 1953—. Fellow Am. Phys. Soc. Asso. editor Phys. Rev., 1964-67. Research, publs. on quantum statist. mechanics and theory of superfluid helium, physics of electrons in solids, especially theory of transport and related phenomena. Office: IBM Watson Lab., 612 W. 115th St., N.Y.C. 10025.*

PRICE, Philip Barbour, Am. surgeon; b. Sinchang, China, Mar. 7, 1897 (parents Am. citizens); s. Philip Frank and Esther (Wilson) P.; A.B., Davidson Coll., 1917, D.Sc., 1964; postgrad. U. Va., U. Nanking; M.D., Johns Hopkins, 1921; m. Octavia Duvall Howard, Nov. 3, 1925; 1 dau., Mary Greenwood (Mrs. Sidney M.B. Coulling). Faculty, Cheeloo U., Tsinan,

China, 1927-38, Johns Hopkins, 1938-43; prof. surgery U. Utah Coll. Medicine, Salt Lake City, 1943-62, head surgery, 1943-56, dean, 1955-62, prof., dean emeritus, 1962—, surgeon in chief Salt Lake Gen. Hosp., 1943-56; sr. surg. cons. Salt Lake VA Hosp., 1946-56, VA Hosp., Grand Junction, Colo., 1949-55. Recipient Distinguished Service award Utah Med. Assn., 1962. Mem. A.M.A., A.C.S., Southwestern Surg. Congress, Am., Western surg. assns., Internat. Surg. Soc., Soc. U. Surgeons. Research, numerous publs. on traumatic shock, burn wounds, wound healing, surg. techniques, radiation injury, antiseptics, measurement of physician performance, student selection. Address: 901 Bowyer Lane, Lexington, Va. 24450.*

PRICE, Vincent Edward, Am. physician; b. Battle Creek, Mich., July 11, 1920; s. George Gustav and Nora (Murley) P.; A.B., Oberlin Coll., 1942; M.D., U. Mich., 1945; m. Florence Rosalind Viancour, Dec. 25, 1943; children—Patricia Ellen, Robert Harold, James Vincent, Daniel Edward. Biochemist, head, enzymes and metabolism sect. Nat. Cancer Inst., Bethesda, Md., 1946-61; program adminstr. med. scientist tng. program, asso. chief sci. program, research tng. br., spl. asst. to dir. Nat. Inst. Gen. Med. Scis., NIH, Bethesda. Med. dir. USPHS. Mem. Am. Soc. Biol. Chemists, A.A.A.S. Research, publs. in enzymatic resolution of amino-acids; studies on anemia in cancer, tumor-host relationships, iron metabolism in tumor bearing animals, rates of synthesis and destruction of enzymes in vivo. Home: 4615 Edgefield Rd. Office: NIH, Bethesda, Md. 20014.*

PRICE, Willard, naturalist, explorer; b. Peterborough, Ont., July 28, 1887; s. Albert and Estella (Martin) P.; B.A., Western Res. U., 1909; M.A., Columbia, 1914, Litt. D., 1930; m. Jean Reeve, Aug. 4, 1914 (dec. Apr. 1929); 1 son, Robert; m. 2d, Mary Virginia Selden, May 28, 1932. Editorial staff Survey, 1912-13; editor World Outlook, other publs. on fgn. affairs, 1915-32; fgn. corr. Am., Brit. papers, 1933-46; ethnographic expdns. for Am. Mus. Natural History, Nat. Geog. Soc., 1915—. Author: Ancient Peoples at New Tasks, 1918; Pacific Adventure, 1936; Children of the Rising Sun, 1938; Barbarian, 1941; Japan Rides the Tiger, 1942; Japan's Islands of Mystery, 1944; Japan and the Son of Heaven, 1945; Key to Japan, 1946; Roving South—Rio Grande to Patagonia, 1948; I Cannot Rest from Travel, 1951; The Amazing Amazon, 1952; Journey by Junk, 1953; Adventures in Paradise—Tahiti and Beyond, 1955; Roaming Britain, 1958; Amazon Adventure, 1949; South Sea Adventure, 1952; Underwater Adventure, 1954; Volcano Adventure, 1956; Whale Adventure, 1960; Incredible Africa, 1962; The Amazing Mississippi, 1963; African Adventure, 1963; Rivers I Have Known, 1965; America's Paradise Lost, 1965; Elephant Adventure, 1964; Safari Adventure, 1966; also articles to nat. mags., Ency. Brit. Home: Verbena Lane, Cathedral City, Cal. 92234.*

PRICE, W(illiam) Armstrong, Jr., Am. geologist; b. Richmond, Va., Mar. 17, 1889; s. William Armstrong and Bessie (Lancaster) P.; A.B., Davidson Coll., 1909; Ph.D., Johns Hopkins, 1913; m. Evelyn Tyson Williams, Dec. 15, 1921; children—William Armstrong III, John Wilson; m. 2d, Lucie Harris Locke, Feb. 27, 1964. Paleontologist, W.Va. Geol. Survey, 1913-19; faculty, W.Va. U., 1913-15, asst. prof., 1915-18; with various oil cos., 1919-24; cons. geologist, Houston, 1925-30, Corpus Christi, 1930-51; prof. geol. oceanography and Pleistocene geology Tex. A. and M. Coll., 1950-54, research asso. dept. oceanography, 1955; cons. geologist, oceanographer, Corpus Christi, 1955—; oceanographer Corpus Christi Lab. S.W. Research Inst., 1967—. Chmn. submerged lands com. State Land Bd., Tex., 1963—; mem. internat. com. on Pleistocene shorelines Internat. Assn. Quaternary Research, 1957-61, sr. del. U.S. State Dept. to 4th Congress. Fellow Tex. Acad. Sci. (past prs.); mem. Am. Assn. Petroleum Geologists (past chmn. post-Cretaceous correlation com.), A.A.A.S., Geol. Soc. Am., Am. Geog. Soc., Am. Assn. Petroleum Geologists, Soc. Econ. Paleontologists and Minerologists, Corpus Christi Geol. Soc. (hon. life, past pres.). Research, numerous publs. on geomorphology and genesis coastal surface features S. Tex., mapping continental shelf, Gulf of Mexico, energy classification of shorelines and coasts, origin of barrier island, classification and genesis of oriented lakes of world including Carolina Bays, hydrodynamics and geomorphology of tidal inlet and delta, chem. basis of clay dune formation; discovered oil and gas fields Tex.; described Pa. invertebrate fauna. Home: 401 Southern St. Office: 1213 Ocean Drove, Corpus Christi, Tex. 78404.*

PRICE, Winston Harvey, Am. med. scientist; b. N.Y.C., Jan. 3, 1923; s. Julius Joseph and Florence (Cooper) P.; B.A., U. Pa., 1944; M.S., Ph.D., Princeton, 1949; m. Grace Gertrude Hartigan, Dec. 24, 1960. Spl. investigator Rockefeller Inst. for Med. Research, 1946-51; faculty Johns Hopkins Sch. Hygiene and Pub. Health, Balt., 1956—, prof. epidemiology, asso. prof. biochemistry, 1964—. Recipient Theobald Smith award in med. sci. A.A.A.S., 1954. Research, numerous publs. on epidemiology Rocky Mountain spotted fever, epidemic typhus; pioneered isolation common cold viruses; research epidemiology common cold, immunological relationships between arboviruses, vaccines against certain arboviruses. Home: 209 Edgevale Rd., Balt. 21210. Office: 615 N. Wolfe St., Balt. 21205.*

PRICHARD, Hesketh, Brit. naturalist; b. Nov. 1876; s. Hesketh and Kate Prichard; m. Elizabeth Grimston, 1908; 2 sons, 1 dau. Leader, Patagonian Expdn., 1900-01; became a.d.c. to Lord Lt. of Ireland, 1907; served in European War, 1914-18. Author: Hunting Camps in Wood and Wilderness, 1910. Studied natural history; traveled in Patagonia, Labrador, Can., Sardinia, Spain, Mexico. Died 1922.

PRICHARD, James Cowles, English physician, ethnologist; b. Ross, Herefordshire, Eng., Feg. 11, 1786; M.D., U. Edinburgh (Scotland), 1808; became physician St. Peter's Hosp., 1811, Bristol (Eng.) Infirmary, 1814; became commr. in lunacy, 1845. Fellow Royal Soc., 1827; mem. Ethnol. Soc. (became pres. 1848). Author: Researches into the Physical History of Man, 2 vols., 1813; Treatise on Diseases of the Nervous System, 1822; Eastern Origin of the Celtic Nations, 1831; Insanity and Other Disorders Affecting the Mind, 1835; Different Forms of Insanity in Relation to Jurisprudence, 1842; Natural History of Man, 1843; The Relation of Ethnology to Other Branches of Knowledge, 1847. Introduced concept of moral insanity (psychopathic personality) as a disease, 1835; suggested that civilization produced white varieties of man; assigned all mankind to single species; established Celtic lang. as Indo-European lang. Died London, Dec. 22, 1848.

PRICHARD, Marjorie Mabel Lucy, English physiologist; b. Oxford, Eng., Feb. 11, 1906; d. Harold Arthur and Mabel Henrietta (Ross) P.; B.A., St. Anne's Coll., Oxford (Eng.) U., 1928, M.A., 1938, D.Phil., 1950; D.Sc., 1968. Work in field archaeology and edn., 1928-36; personal asst. to hon. radiologist Nuffield Inst. for Med. Research, Oxford U., 1937-41, grad. asst., 1941-55, sr. research officer, 1955—, research fellow St. Anne's Coll., 1956—. Mem. Physiol. Soc., Anat. Soc. Gt. Britian and Ireland. Author: (with A.E. Barclay, K.J. Franklin) The Fetal Circulation and Cardiovascular System and the Changes that they Undergo at Birth, 1944; (with others) Studies of the Renal Circulation, 1947; also numerous articles. Research on fetal circulation and changes occurring at birth, blood vessels of and circulation through kidney, liver and pituitary gland, influence of pituitary gland on growth, on other endocrine glands, on pregnancy and lactation, on cancer of breast, arteriovenous anastomoses, exptl. studies on mammary cancer.*

PRIEN, Charles Henry, Am. chem. engr.; b. Lafayette, Ind., Oct. 9, 1916; s. Henry Carl and Alma (Wegner) P.; B.S. in Chem. Engring., Purdue U., 1938, Ph.D., 1948; m. Isabell L. Linn, July 3, 1954; children—(by previous marriage) Carol (Mrs. Ibrahim Ansari), David Manter. Faculty, U. Colo., Denver, 1941-48, asst. prof. chem. engring., asst. dir. chem. Engring. Practice Sch., 1944-48; faculty U. Denver, 1948—, prof., 1952—, research engr. Denver Research Inst., 1948-52, head chem. div., 1952—. Cons. UN Tech. Assistance Administrn., 1951; mem. Colo. Engring. Council, 1950-53. Mem. Am. Inst. Chem. Engrs. (past pres. Rocky Mountain sect.), Am. Chem. Soc., A.A.A.S., Am. Soc. for Engring. Edn., N.Y. Acad. Scis., Sigma Xi, Sigma Tau, Phi Lambda Upsilon. Author: (with J.E. Stepanek) Role of Rural Industries, 1953; also articles, chpts. in books. Research on chemistry of organic matter in oil shale and shale oil, chem. reaction oil shale organic matter and coal under hydrogenation, extraction, oxidation, relationship between fuel volatility and its distbn. in chem. engring. unit processes; devel. indsl. processes to produce shale oil. Home: 2266 Crabtree Dr., Littleton, Colo. 80210. Office: Denver Research Inst., U. Denver, Denver 80120.*

PRIESSNITZ, Vincenz, German agriculturist; b. Graefenberg, Germany, Oct. 5, 1799. Began as farmer; founder hydrotherapy establishment, Graefenberg. Discovered cold-water compress (Priessnitz's bandage); founder hydrotherapy; developed system for treatment of ailments with cold water. Died 1851.

PRIEST, Walter Scott, Am. cardiologist, physician; b. Denver, June 25, 1896; s. Walter Scott and Anna Ellen (Schaeffer) P.; A.B., U. of aKn., 1917; M.D., Washington U., 1920; M.S., Northwestern U., 1946; m. Ruth M. Biederman, June 12, 1923; 1 son, Walter Scott; m. 2d, Clara E. Rourke, June 7, 1939; children—David Hartzell and William Curtis (twins), Kenneth Lee. Asso. prof. medicine Northwestern U., 1949-62, prof., 1962-64, emeritus prof., 1964—; sr. attending physician Wesley Meml. Hosp., Chgo., 1940—; dir. Inst. Medicine, Chgo., 1966—. vis. physician Passavant Meml., Chgo. and Evanston (Ill.) Hosps. since 1940; cons. internal medicine, St. Francis Hosp., Evanston, since 1946; cons. cardiologist United Air Lines since 1930. Pioneer research work on pencillin and strepomycin in the treatment of subacute bacterial endocarditis; recipient gold medal for scientific exhibit of this work, Ill. State Med. Soc., Chicago, 1946; certificate of merit, A.M.A., San Francisco, 1946; Silver medal Miss. Valley Med. Soc., Sept. 1946. Diplomate Am. Bd. Internal Medicine, cardiovascular diseases. Fellow A.C.P., Am. Coll. Cardiology (trustee, pres. 1956); mem. A.M.A., Ill. State, Chgo. med. socs., Am. (gov.), Chgo. (v.p., bd. govs.) heart assns., Chgo. Soc. Internal Med., Sigma Xi, Sigma Nu, Nu Sigma Nu, Phi Mu Alpha, Alpha Omega Alpha. Club: Sigma Nu Alumni (past pres.). Editorial cons. Am. jour. Cardiology. Contbr. to Geriatric Medicine (edited by E. J. Stieglitz), 1943; numerous articles on internal medicine especially cardiovascular disease in med. jours. Home: 260 E. Chestnut St., Chgo. 60611. Office: 251 E. Chicago Av., Chgo. 60611.*

PRIESTER, Wolfgang, German astrophysicist; b. Detmold, Germany, Apr. 22, 1924; s. Wilhelm and Gertrud (Knauff) P.; Dr.Rer.Nat., U. Goettingen (Germany), 1953; Dr.Rer.Nat. Habil, U. Bonn (Germany), 1958; m. Gisela Preuss, Dec. 23, 1950; 1 son, Achim. Faculty, Inst. for Theoretical Physics, U. Kiel (Germany), 1953-55; with U. Bonn. Obs., 1955—, prof. astronomy, 1962—, dir. Inst. for Astrophysics and Space Research, co-dir. Astron. Insts., full prof., 1964—; sr. research asso. NASA Inst. for Space Studies, N.Y.C., 1961, 63-64, 65-66. Mem. Internat. Astron. Union, Internat. Radio Sci. Union, Com. on Space Research, Am. Astron. Soc., Am. Geophys. Union, Astron. Gesellschaft, A.A.A.S. Research, publs. on upper atmosphere physics, radioastronomy, astrophysics. Home: Luffrid Str. 2, Bonn 5300, Germany.*

PRIESTLEY, Joseph, English chemist; b. Fieldhead Leeds, Yorkshire, Eng., Mar. 13, 1733; s. Jonas and Mary (Swift) P.; attended Daventry Acad., Eng., 1751-54; LL.D., U. Edinburgh (Scotland), 1765; m. Mary Wilkinson, June 23, 1762; children—Joseph, Sarah, William, Henry. Pastor, Needham Market, Surrey, Eng., 1755; ordained to ministry Congl. Ch., 1762; tutor belles-lettres Warrington Acad., until 1767; preached, wrote, taught history, anatomy, botany, astronomy; pastor Mill Hill Ch., Leeds, Eng., 1767-72; librarian to Lord Shelburne, 1772-80; minister New Meeting, Birmingham, Eng., 1780-91; made citizen of France for Revolutionary sympathies, 1792; pastor, Hackney, Eng., 1792-94; came to U.S., 1794; settled in Northumberland, Pa.; chief early proponent of Unitarianism in U. S.; Fellow Royal Soc., 1766. Author: History and Present State of Electricity, 1767; Essay on the First Principles of Government, 1768; The History and Present State of Discoveries Relating to Vision, Light and Colours, 1772; Experiments and Observations on Different Kinds of Air, 3 vols., 1774, 75, 77; An History of the Corruptions of Christianity, 1782; A General History of the Christian Church, 4 vols., 1790-1802; Unitarianism Explained and Defended, 1796; The Doctrine of Phlogiston Established, 1803. Explained Priestly rings formed by elect. discharge on metallic surface, also proposed explanation of oscillatory nature of discharge from Leyden jar, 1767; began study of airs, 1767; announced discovery of dephlogisticated air (oxygen), 1774; still believed in theory of phlogiston; observed prodn. of fixed air (carbon dioxide) in fermentation; invented method of making soda water, 1772; isolated and described nitrous oxide, nitric oxide, nitrogen peroxide, ammonia, silicon flouride, sulpher dioxide, hydrogen sulphide, carbon monoxide; discovered decomposition of ammonia by electricity, 1781; discovered hydrochloric acid gas; recognized that oxygen is produced by green plants in sunlight, also that water forms during reduction of oxides by hydrogen. Died Northumberland, Feb. 6, 1804.

PRIESTLEY, Joseph Hubert, English botanist; b. Tewkesbury, Eng., Oct. 5, 1883; s. J.E. Priestley; ed. U. Coll., Bristol, Eng.; B.Sc., London; m. Marion E. Young, 1911. Head bot. dept., lectr. botany U. Coll., Bristol, 1905-11; prof. botany U. Leeds (Eng.), 1911—, vice chancellor, 1935-39. Cons. botanist to Bath, W. and So. Counties Soc., 1910-11. Recipient Distinguished Service Order, 1917. Fellow Linean Soc., Brit. Assn. for Advancement Sci. Author: (with Lorna I. Scott) An Introduction to Botany, 1938. Research on use of electric currents to promote plant growth, plant physiology and developmental anatomy. Died Oct. 31, 1944.

PRIESTLY, William Overend, English, gynecologist; b. Leeds, Eng., June 24, 1829; s. Joseph and Mary (Verend) P.; ed. Leeds, Eng., Kings Coll., London, Paris; M.D., U. Edinburgh (Scotland), 1853, also LL.D.; m. Eliza Chambers, Apr. 17, 1856; 2 sons, 2 daus. Pvt. asst. to James Young Simpson; came to London, 1856; became lectr. midwifery Middlesex Hosp., 1858; named prof. obstetric medicine Kings Coll., London, 1862; became cons. Kings Coll. Hosp., 1872; elected parliamentary rep. U. Edinburgh, 1896; obstetrician to H.R.H. Princess Louise of Hess. Recipient Senate medal for excellence in original work U. Edinburgh, 1893. Mem. Royal Coll. Surgeons (examiner), Obstet. Soc. London (pres.), Med. Soc. Paris (v.p.), Royal Coll. Physicians Edinburgh, Royal Coll. Physicians London. Author: Lecture on the Development of the Gravid Uterus; The Pathology of Intra-uterine Death. Editor: (with H.R. Storer) The Obstetric Writings and Contributions of Sir James Y. Simpson. First to apply modern sci. methods to midwifery; described intermenstrual pain (mittleschmerz), 1872. Died Apr. 11, 1900.

PRILLIEUX, Edouard-Ernest, French botanist, agronomist; b. Paris, Jan. 11, 1829; pupil of Duchartre; named prof. Agronomic Inst., Paris, 1883; founder 1st lab. of plant pathology in Europe, Paris, 1887; became nat. agronomic insp., 1907; mem. French Acad. Scis., Soc. Agr. Author: Traité des maladies des plantes agricoles, 1897. Introduced teaching of plant pathology into France; specialist in diseases of fruit and forest trees. Died Mondoubleau, France, Oct. 7, 1915.

PRIMAK, William Leo, Am. physicist; b. N.Y.C., June 4, 1917; s. Nathan and Elizabeth Primak; B.S., Coll. City N.Y., 1937; M.S., Bklyn. Poly. Inst., 1943, Ph.D., 1946; m. Dorothy Newfang, Oct., 1953; children—John Jefferson, Margaret Kay, Robert Carl. Jr. physicist Queens Coll., 1938-42; research asso. Bklyn. Poly. Inst., 1943-46; with Argonne (Ill.) Nat. Lab., 1946—, now sr. chemist. Mem. Am. Chem. Soc., Am. Phys. Soc., Sigma Xi, Phi Lambda Upsilon. Research, publs. on changes in phys. properties of solids such as graphite, diamond, silicon carbide, quartz, lithium fluoride, magnesium oxide, silicon, germanium, copper and nickel caused by exposure to energetic radiation, as x-rays, gamma rays, electrons, neutrons, accelerated ions; elucidation of laws governing rates at which irradiated solids alter on heating. Office: Argonne Nat. Lab., Argonne, Ill. 60439.*

PRIMAN, Jacob, anatomist, anthropologist; b. Riga, Latvia, Mar. 12, 1892; s. Jacob and Eva (Bernsons) P.; M.D., Mil. Med. Acad., Russia, 1918; M.D.Sc., U. Latvia, 1926; m. Margareta Raiska, Dec. 18, 1926; children—Ieva (Mrs. Joseph Cucinelli), Maija (Mrs. Philip Gresh). Came to U.S., 1948, naturalized, 1954. Faculty, U. Latvia, 1920-44, prof. anatomy, 1932-44; research worker U. Jena, Germany, 1945; prof. anatomy Baltic U., Hamburg, Germany, 1946-48; faculty U. Pitts., 1948—, prof. anatomy, 1958-62, prof. emeritus, 1962—. Sr. med. dr. for displaced persons, Germany, 1945-48. Mem. Biol. Soc. U. Pitts., Assn. Am. Anatomists, Assn. Latvian Physicians, Sigma Xi, Phi Beta Pi. Author: Nomina Anatomica, 1931, 44; Introduction in Anatomy (in Latvian), 1925. Research, publs. on anthropology of Latvians; devel. urinary organs, variations of blood vessels and skull, anat. terminology. Home: 1444 N. Euclid Av., Pitts. 15206.*

PRINCE, Albert Thomas, Brit. mathematician; b. Nantwich, Eng., Jan. 30, 1903; s. Albert and Marie Lavinia (Light) P.; student Monmouth (Eng.) Sch.; D.Sc., U. Manchester (Eng.); m. Rose Ann Waterman, Mar. 29, 1947. Asst., Queen's U., Belfast, Ireland, 1925-26; lectr. Imperial Coll., London, Eng., 1926-46; prof. Royal Tech. Coll., Glasgow, Scotland, 1946-51; prof. U. Exeter (Eng.), 1952—, dean Faculty of Sci., 1954-58; cons. RAND Corp. Mem. Nat. Adv. Com. Further Edn. Mem. Royal Astron. Soc., London Math. Soc., Am. Geophys. Union, Carnegie Instn., Nat. Acad. Scis. Publns. on applied math, geomagnetism; articles for Royal Soc. London, Jour. Geophys. Research. Research on devel. math. and computing techniques for dealing with theoretical problems and analysis of data; theoretical studies in geomagnetism and aeronomy. Home: Vermont, W. Garth Rd. Office: Univ. Exeter, Exeter, Eng.

PRINCE, David Chandler, Am. elec., mech. engr.; b. Springfield, Ill., Feb. 5, 1891; s. Arthur Edward and Charlotte (Hitchcock) P.; student U. Mich., 1908-10; B.S. in Elec. Engring, U. Ill., 1912, M.S. in Elec. Engring., 1913; postgrad. Lincoln Coll. Law; Sc.D., Union Coll., Schenectady, 1943; m. Winifred Notman, May 14, 1918; children—David C., George N., Winifred (Mrs. Charles D. Hyson), Edward. Test engr. Gen. Electric Co., Schenectady, 1913-14, mem. radio dept., 1919-23, staff research lab., 1923-31; chief engr. switch bd. dept., 1931-40, mgr. comml. engring., 1940-41, vice chmn. field devel. div. com. for econ. devel., 1941-43, v.p. in charge gen. engring. and cons. lab., 1945-51, v.p., mem. pres. staff, 1951; pvt. cons. practice Washington, 1951—. Recipient citation for especially meritorius service in the solution of engring. problems relating to aircraft U.S. Army; Modern Pioneer award Phila. Sesquicentennial, 1940. Fellow Am. Inst. Elec. Engring (pres. 1941-42, Lamme medal 1946); mem. Inst. Aero. Scis., Am. Soc. M.E., Am. Inst. Cons. Engrs., Soc. Automotive Engrs., I.R.E., Am. Rocket Soc., Am. Heliocopter Soc., A.A.A.S., Inst. Elec. Engrs. (Brit.), N.Y. State Soc. Profl. Engrs., Sigma Xi, Chi Psi, Eta Kappa Nu, Tau Beta Pi. Author: Vacuum Tubes as Oscillation Generators, 1920; (with others) Mercury Are Rectifiers and Their circuits; also numerous articles. Inventor of 3-cycle circuit breakers as part of first super, high-voltage, long-distance transmission lines, in Los Angeles, Calif., patentee in field. Home: 24 Hibiscus Way, Ocean Ridge, Fla.; 3009 P. St. N.W., Washington. Office: 3009 P. St. N.W., Washington.*

PRINCE, Helen Dodson, Am. solar astronomer; b. Balt., Dec. 31, 1905; d. Henry Clay and Helen (Walter) Dodson; A.B., Goucher Coll., 1927, Sc.D., 1952; M.A., U. Mich., 1932, Ph.D., 1933; m. Edmond Lafayette Prince, Oct. 24, 1956. Asst. statistician Md. Dept. Edn., 1927-31; faculty Wellesley (Mass.) Coll., 1933-45; asso. prof. math. and astronomy Goucher Coll., 1945-50; faculty U. Mich., Ann Arbor, 1947—, prof. astronomy, 1957—, asso. dir. McMath-Hulbert Obs., 1962—. Mem. Am. Astron. Soc. (Annie Jump Cannon prize 1954); Am. Geophys. Union, Internat. Astron. Union. Astron. Union (mem. com. 10 1946—), A.A.A.S., Phi Beta Kappa. Research, numerous publs. on observation and analysis solar activity as aspect solar physics and as possible cause certain geophys. phenomena. Home: 650 Lake Angelus Shores. Office: McMath-Hulbert Obs., 895 Lake Angelus Rd., Pontiac, Mich. 48055.*

PRINCE, Morton, Am. neuropsychiatrist; b. Boston, Mass., Dec. 21, 1854; s. Frederick Octavus and Helen Susan (Henry) P.; A.B., Harvard, 1875, M.D., 1879; LL.D., Tufts Coll., Mass., 1910; m. Fanny Lithgow Payson, Feb. 14, 1885; children—Mrs. Clara Morton Walcott, Morton Peabody. Practiced in Boston, 1880—; phys. for nervous diseases, Boston City Hosp., 1885-1913, consulting phys., 1914—; prof. nervous diseases, 1902-12, prof. emeritus, 1912, Tufts Coll. Med. Sch. Lecturer abnormal psychology, University of Calif., 1910; asso. prof. abnormal and dynamic psychology, Harvard University, 1926—. Editor Jour. Abnormal and Social Psychology, 1906—. Chmn. Boston Charter Assn. Commd. by the gov. as mgr. of Mass. Soldiers and Sailors Information Bur. in Paris, representing State of Mass. in France, 1918-19; instigated and organized the "Address (of the 500 Americans) to the Peoples of the Allied Nations," 1916; chmn. Serbian Distress Fund, 1915—; chmn. reception com. of Boston for Japanese Mission, Dec. 1917, and Serbian Mission, Jan. 1918; chmn. State and Boston reception committees for Marshal Foch, 1921, and State Com. for General Diaz, 1921. Decorated Order Chevalier of St. Sava (Serbia), 1916; Order Rising Sun (Japan), 1918; Cross Legion of Honor (France), 1919; Royal Order of Red Cross and Order of the White Eagle (Serbia), 1920. Author: Nature of Mind and Human Automatism, 1885; Dissociation of a Personality, 1906; The Unconscious, 1913; The Psychology of the Kaiser, 1915; The Creed of Deutschtum, 1918. Author of classic case study in abnormal psychology; integrated neurology and psychology in theory of abnormal psychology and mental dissociation. Died Brookline, Mass., Aug. 31, 1929.

PRINCE, William, Am. horticulturist; b. Flushing, L.I., N.Y., Nov. 10, 1766; s. William and Ann (Thorne) P.; m. Mary Stratton; 4 children, including William Robert. Founder Linnaean Botanic Garden and Nurseries, 1793. Mem. N.Y. Hort. Soc., Mass. Hort. Soc., Linnaean Soc. of Paris, Hort. Soc. of London and Paris, Imperial Soc. of the Georgofili, Florence, Italy. Author: A Short Treatise on Horticulture, 1828; (with son William Robert) A Treatise on the Vine, 1830, The Pomological Manual, 1831. Imported and introduced many varieties of fruits and plants, also exported Am. plants and trees to Europe; introduced Isabella grape (which he renamed), 1816; standardized name of Bartlett pear, others. Died Flushing, Apr. 9, 1842.

PRINCE, William Robert, Am. horticulturist; b. Flushing, L.I., N.Y., Nov. 6, 1795; s. William and Mary (Stratton) P.; m. Charlotte Goodwin Collins, Oct. 2, 1826; 4 children, including LeBaron Bradford. Botanized entire range of Atlantic states; importer 1st merino sheep to U.S., 1816; became mgr. (with brother) Linnaean Botanic Garden and Nurseries, circa 1835; pioneer (with father) in introducing silk culture, 1837; importer mulberry Morus multicaulis (became very popular.), from Tarascon, France; introduced culture of osiers and sorghum, 1854-55; importer Chinese yam, 1954; corr. mem. Mass. Hort. Soc., 1829; mem. Am. Pomological Soc.; greatest contbn. to Am. horticulture was advancement of grape-growing. Author: Prince's Manual of Roses, 1846; (with father) A Treatise on the Vine, 1830; The Pomological Manual, 1831. Contbr. to Gardners' Monthly and Rural New Yorker. Died Flushing, Mar. 28, 1869.

PRINGLE, Cyrus Guernsey, Am. botanist; b. Charlotte, Vt., May 6, 1838; s. George and Louisa (Harris) P.; classical edn. in Vt. and Can.; A.M., (hon.), Middlebury Coll., 1876; Sc.D. (hon.), U. Vt., 1906. Collected extensively in forestry and gen. botany in Ariz., Sonora, Cal., Ore., Wash., as collector for Am. Mus. Natural History, N.Y.C., 1881-84; bot. collector Harvard U., from 1885; keeper herbarium U. Vt.; engaged in thorough exploration of flora of Mexico, placing large collections (including many new plant species) in 50 or more of the most important herbaria of world. Asso. fellow Am. Acad. Arts and Scis. Contbr. to bot. publs. Died 1911.

PRINGLE, Sir John, Brit. physician; b. Roxburghshire, Apr. 10, 1707; s. John and Magdalen (Elliott) P.; ed. U. St. Andrews, U. Edinburgh, Leiden, Netherlands; M.D., 1730; practiced medicine, Edinburgh; joint prof. pneumatics (metaphysics) and moral philosophy Edinburgh U., 1734-44; apptd. physician-gen. to forces in Flander, 1744; settled in London, 1748; named physician to George III, 1774. Fellow Royal Soc., 1745 (pres. 1772, Copley medallist), Royal Coll. Physicians; one of 8 fgn. mems. French Acad. Scis. (successor to Linnaeus), 1778. Author: De Marcore Senili, 1730; Observations on the Nature and Cure of Hospital and Jayl Fevers, 1750; Observations on the Diseases of the Army (classic work), 1752; Six Discourses delivered at the Royal Society on occasion of the Annual Assignment of the Copley Medal, with Life of the Author by Andrew Kippis, D.D., 1783. Reformed mil. medicine and sanitation; considered founder of modern mil. medicine; credited with originating ideas of Red Cross, also with popularizing term influenza in English. Died Jan. 18, 1782.

PRINGLE, John James, Brit. dermatologist; b. Kirkudbrightshire, Scotland, 1855; s. Andrew Pringle; M.B.C.M., U. Edinburgh (Scotland), 1876; student dermatology Paris, Vienna, Berlin. Joined staff Middlesex Hosp., London, 1882, became asst. dermatologist, then head dept.; ret. at age of 65. Mem. Brit. Med. Assn. (pres. dermatol. sect.), Royal Soc. Medicine. A founder Brit. Jour. Dermatology. Contbr. articles to tech. jours. Described sebaceous adenoma (Pringle's disease or adenoma), 1890, tubercular lymphangitis, 1895, actinomycosis, 1895, lupus erythematosus, 1900, pemphigus vegetans, 1915; 1st description of granulosis rubra nasi, 1894. Died on voyage to New Zeland, Dec. 18, 1922.

PRINGLE, John William Sutton, English zoologist; b. Manchester, Eng., July 22, 1912; s. John and Dorothy (Beney) P.; M.A., King's Coll., Cambridge, Eng., 1938, Sc.D., 1955; m. Beatrice Gilbert-Carter, Aug. 17, 1946; children—John, Sally, Bridget. Demonstrator, Cambridge U., 1937-39, lectr., 1945-59, fellow King's Coll., 1938-45; fellow Peterhouse Coll., 1945-61, emeritus fellow, 1961—; staff Telecommunication Research Establishment, 1939-44, Ministry War Transport, 1944-45; reader in exptl. cytology, Oxford (Eng.) U., 1959-61, Linacre prof. zoology, fellow Merton Coll., 1961—. Mem. Order Brit. Empire; recipient Am. medal of Freedom with bronze palm, 1945. Fellow Royal Soc., 1954. Author: Insect Flight, 1957; also articles. Research on insect sensory physiology, dynamics, biophysics of insect flight muscle. Home: 437 Banbury Rd. Office: Dept. Zoology, Oxford U., Parks Rd., Oxford, Eng.*

PRINGSHEIM, Alfred, German mathematician; b. Ohlau, Silesia, Sept. 2, 1850; prof., Munich. Author: Vorlesungen über Zahlen- und Funktionenlehre, 2 vols., 1916-32. Studied theory of functions. Died Zurich, Switzerland, June 25, 1941.

PRINGSHEIM, Ernst, German physicist; b. Breslau, Germany (now Wroclaw, Poland), July 11, 1859; prof., Breslau. Author: Vorlesungen über die Physik der Sonne, 1910. Research on emission spectra of compounds in which he concluded line spectra and possibly fluted spectra are produced only when chem. changes are occurring in radiating substances; research (with Lummer) on black-body radiation (influenced Planck), 1899. Died Breslau, June 28, 1917.

PRINGSHEIM, Ernst Georg, German microbiologist, exptl. phycologist; b. Breslau, Germany, Oct. 26, 1881; s. Hugo and Hedwig (Heymann) P.; student U. Munich (Germany), 1901-02; Dr.phil. s.c.l., U. Leipsig (Germany), 1906; Dr.rer.nat. h.c., U. Goettingen, 1961; m. Olga Zimmermann, July 16, 1929; children—Ludwig, Wolfgang Leopold. Asst. prof. U. Halle, 1909-20; prof. Berlin U., 1920-23; prof. plant physiology Deutsche U., Prague, Czechoslovakia, 1923-39; curator cult collection algae and protozoa U. Cambridge (Eng.), 1942-51; guest prof. Ind. U., U. Cal. at Berkeley, Yale, Harvard; head div. algal. research U. Goettingen, 1953—. Mem. Academia Carolo-Leopoldina, Goettingen Acad. Sci. and Learning. Mem. Bose Research Inst. Calcutta, Bot. Soc. Am., Am. Soc. Protozoologists, Dutch Bot. Soc., Brit. Physological Soc. Author: Reizbewegungen der Pflanzen, 1912; Planzenphysiologische Uebungen, 1931; Julius Sachs, 1932; Pure Cultures of Algae, 1964; Farblose Algen, 1963; also numerous articles. Research on irritability of plants, cultivation and physiology of algae, flagellates and bacteria, exptl. elimination of chromatophores in Euglena gracilis, taxonomy and physiology of Beggiasoa; discovered acetate as specific nutritive substance, photo-assimilation of acetate. Home: 44 Pfingstanger, Goettingen, B.R.D., Germany.*

PRINGSHEIM, Nathanael, German biologist; b. Wziesko, Silesia, Nov. 30, 1823; ed. Breslau Germany (now Wroclaw, Poland), Leipzig, Berlin (both Germany); prof. botany U. Jena (Germany), then U. Berlin; later pvt. tchr. Mem. French Acad. Scis. Founded German Bot. Soc., also Jahrbücher für Wissenschaftlicher Botanik. Author: Geschichtliche Abhandlungen, 4 vols., 1895-96. Discovered (with Thunet) plant fertilization, 1853-55; 1st to observe spermatozoids enter plant female cells (used alga Vaucheria); studied algoid fungi, alternation of generations in mosses and the thallophytes. Died Berlin, Oct. 6, 1894.

PRINZLER, Helmut, German physicist; b. Magdeburg, Germany, Sept. 6, 1928; s. Otto and Hedwig (Bohme) P.; Diplomphysiker, Technische Universitat Dresden, 1960, Doctor, 1965; m. Margarete Bader, May 23, 1953; children—Klaus-Thomas, Hans-Martin. Lab. engr. Heinrich Hertz Inst., 1951—, group leader microwave plasma physics team, 1960—. Research, publs. on standard calibration of radio-astron. equipments, standard noise generators for u.h.f., plasma diagnostics with microwaves especially use of microwave radiation from plasmas for diagnostic, examination of emitted radiation from glow- and arc-discharges and its relation with the elementary processes in positive column and electrodezones. Home: 115 Volkswohl-Strasse, 1199 Berlin. Office: Heinrich Hertz Inst. der Deutschen Akademie der Wissenschaften, 1199 Berlin, DDR/Germany.*

PRIOR, John Alan, Am. physician, educator; b. Columbus, O., Apr. 17, 1913; s. John Clinton and Edna (Cones) P.; B.A., Ohio State U., 1935, M.D., 1938; m. Helen Eileen Zurmehly, Mar. 24, 1936; children—John Alan, Robert Lee, Richard Ray (dec.).

Instr. U. Cin., 1942-43; asst. prof. medicine Ohio State U., 1944-51, prof. medicine, 1951—, asst. dean Coll. Medicine, 1961-63, asso. dean, sec. faculty, 1963—; cons. internal medicine USAF, VA, USPHS. Recipient Distinguished Service Alumni award Ohio State U. Coll. Medicine, 1963; Alumni award Distinguished Teaching Ohio State U., 1965. Diplomate Am. Bd. Internal Medicine. Mem. A.C.P., Am., Ohio (pres. 1956-57) thoracic socs., Columbus Acad. Medicine, Am. Fedn. Clin. Research, Alpha Omega Alpha (hon.), Phi Delta Theta. Author: Physical Diagnosis, 3d edit., 1968; articles on pulmonary and fungus diseases. Home: 2650 Donna Dr., Columbus, O. 43221.*

PRIOR, John Thompson, Am. physician; b. St. Albans, Vt., Oct. 8, 1917; s. Thomas William and Pauline (Thompson) P.; B.S., U. Vt., 1939, M.D., 1943; m. Elizabeth Titus, July 24, 1948; children—Anne, Polly, John Thompson, Thomas, Jeffrey, Timothy. Faculty, Upstate Med. Center, Syracuse, N.Y., 1948—, prof. pathology, 1963—; lab. dir. Crouse Irving Hosp., Syracuse, 1950—. Mem. A.M.A., Am. Assn. Pathologists and Bacteriologists, Am. Soc. Clin. Pathology. Publs. on research on exptl. ateriosclerosis, pulmonary tumors, cardiac pathology. Home: 100 Lansdowne Rd., Dewitt, N.Y. 13214. Office: 766 Irving Av., Syracuse, N.Y. 13210.*

PRITCHARD, Charles, English astronomer; b. Alberbury, Eng., Feb. 29, 1808; s. William Pritchard; ed. Mchts. Taylors' Sch., Christ's Hosp., London; M.A., St. John's Coll., Cambridge, 1833; M.A. by decree, New Coll., Oxford, 1870, D.D., 1880; m. Emily Newton, Dec. 18, 1834; m. 2d, Rosalind Campbell, Aug. 10, 1858; children by both marriages. Headmaster Clapham grammar sch., 1834-62; Hulsean lectr. Cambridge, 1867; named Savilian prof. astronomy Oxford U., 1870; a founder new obs. in Parks; elected fellow New Coll., Oxford, 1883, hon. fellow St. John's Coll., Cambridge, 1886. Fellow Royal Soc., 1840. Author: Astronomical Observations made at the University Observatory, Oxford, 4 numbers, 1878-92; Occasional Thoughts of an Astronomer on Nature and Revelation, 1889; also papers. Pioneer in photog. measurements of stellar parallax; inventor wedge photometer, used it to measure stellar magnitudes of nearly 3000 stars (with inaccurate results), pub. 1885; determined libration of moon. Died May 28, 1893.

PRITCHARD, George Eric Campbell, Brit. physician; b. Isle of Wight, Sept. 1864; s. Charles Pritchard; B.A., Oxford, 1887, also M.A.; M.D., St. Mary's Hosp., London (Oxford U.) 1802; m. Marian Elizabeth Dawson, 1894; 1 son; m. 2d, Ellen Beatrice Hove. Med. insp. schs. for London City Council, 3 years; cons. physician Home for Blind Babies, Queen Charlotte's Maternity Hosp.; med. dir. Infants' Hosp., London, 1922-36; tchr. Queens Hosp.; postgrad. lectr. asso. with Fellowship of Medicine, London Pub. Med. Service. Fellow Royal Coll. Physicians; sec. Nat. Asn. for Prevention of Infant Mortality; mem. Brit. Med. Assn., Royal Soc. Medicine. Author: Physiological Feeding of Infants and Children; The Infant: a Handbook of Modern Treatment; also articles. Advocated preventive program in child welfare services and strict med. care of sick children, thus contbg. to decline in infant mortality. Died Oct. 20, 1943.

PRITCHARD, Huw Owen, chemist; b. Bangor, Wales, July 23, 1928; s. Owen and Venetia (McMurray) P.; B.Sc., Manchester U., 1948, M.Sc., 1949, Ph.D., 1951, D.Sc., 1964; m. Margaret B. Ramsden, Nov. 3, 1956; children—Karen Louise, David Huw. Asst. lectr., lectr. chemistry Manchester U., 1951-65; research fellow Cal. Inst. Tech., Pasadena, 1957-58; prof. chemistry York U., Toronto, Ont., Can., 1965—. Mem. Chem. Soc., Faraday Soc. Research, numerous publs. on structure and stability of simple chem. compounds, energy-transfer in thermal and photochem. reactions, application of electronic digital computers for problems in theoretical chemistry.*

PRITCHARD, Jack Arthur, Am. physician; b. Painesville, O., July 25, 1921; s. George Frederick and Marguerite (McKee) P.; B.S., Ohio No. U., 1942; M.D., Western Res. U., 1946; m. Signe M. Allen, Mar. 10, 1945; children—Jack Allen, David George, Allen Jeffrey. Fellow pharmacology Western Res. U., 1947-48, Oglebay fellow, 1952-54, asst. prof. dept. obstetrics and gynecology, 1954-55; prof., chmn. dept. obstetrics and gynecology U. Tex. Southwestern Med. Sch., Dallas, 1955—; chief obstetrics and gynecology Parkland Meml. Hosp., Dallas, 1955—. Mem. Am. Central assns. obstetricians and gynecologists, Am. Coll. Obstetricians and Gynecologists, Am. Gynecol. Soc., Soc. for Gynecol. Investigation, Central, So. socs. for clin. research. Research, numerous publs. on physiol. adaptation of mother to pregnancy. Home: 4336 Livingston St., Dallas 75216.*

PRITCHETT, Henry Smith, Am. astronomer; b. Fayette, Mo., Apr. 16, 1857; s. Carr Waller and Betty Susan (Smith) P.; A.B., Pritchett Coll., 1875; Ph.D., Munich, 1894; LL.D., Harvard, Yale, 1901, Johns Hopkins, 1902, U. Toronto 1906, McGill U., 1917; other hon. degrees; m. Eva McAllister, June 1900. Student with Asaph Hall in U.S. Naval Obs., 1876, asst. astronomer, 1878; astronomer Morrison Obs., Glasgow, Mo., 1880; asst. astronomer Transit of Venus expdn. to New Zealand, 1882; prof. astronomy and dir. obs. Washington U., St. Louis, 1883-97; supt. U.S. Coast and Geod. Survey, 1897-1900; pres. Mass. Inst. Tech., 1900-06; pres. Carnegie Found. for Advancement of Teaching, 1906-30 (emeritus). Trustee Carnegie Instn. of Washington. Died Santa Barbara, Cal., Aug. 28, 1939.

PRITIKIN, Roland I., Am. ophthalmologist; b. Chgo., Jan. 9, 1906; s. Edward and Bluma (Saval) P.; B.S., Loyola U., Chgo., 1928, M.D., 1930; Ophthalmology diploma U. Ill., 1938; m. Jeanne DuPre Moore, May 25, 1940; children—Gloria Anne, Karen. Faculty, Stritch Sch. Medicine Loyola U., 1930-32; practice medicine specializing in ophthalmology, Rockford, Ill., 1946—; faculty Rockford Meml. Hosp. Sch. Nursing, 1946—; vis. eye surgeon Henry Holland Hosps., India and Pakistan, 1939, 57, 60, 63; chief eye service Stark Gen. Hosp., Charleston, S.C., 1941—, mem. research and devel. com., 1949—. Fellow A.C.S., Internat. Coll. Surgeons, A.A.A.S., Soc. Acad. Achievement, Am. Geriatrics Soc.; mem. A.M.A., Assn. Mil. Surgeons U.S., Am. Acad. Ophthalmology and Otolaryngology, Assn. Am. Physicians and Surgeons, Pan-Am. Med. Assn., Internat. Assn. for Prevention Blindness, World Med. Assn., Ophthal. Found., Assn. Research Ophthalmology, Am. Assn. History Medicine, Internat. Orgn. Against Trachoma, N.Y. Acad. Scis., Soc. Nuclear Medicine, Ophthal. socs. U.K., Pakistan, Am. Nuclear Soc., Royal Soc. Health, Contact Lens Assn. Ophthalmologists, Sigma Xi, others. Author: Essentials of Ophthalmology, 1950; also contbr. sect. ophthalmology Ency. Americana supplements. Research, publs. on eye surgery of mass casualties; use of atomic and ultrasonic equipment, invention of diagnostic and surg. instruments for rapid eye surgery. Home: 3505 Highcrest Rd. Office: Talcott Bldg., Rockford, Ill. 61101.*

PRIVETT, Orville Samuel, biochemist; b. London, Ont., Can., June 6, 1919; s. Harry and Ruby (Essery) P.; B.S.A., U. Toronto, 1942; M.S., McGill U., 1944, Ph.D, 1947; m. Arlene Marion Larson, July 2, 1949. Came to U.S., 1949, naturalized, 1960. Postdoctoral fellow Purdue U., 1946-48; research fellow U. Minn., Hormel Inst., Austin 1944-52, mem. faculty 1952—, prof., biochemistry, 1960—. Recipient 1st award in basic research Glycerol Producers 1961. Mem. Am. Chem. Soc., Am. Oil Chemists Soc., N.Y. Acad. Sci., Am. Soc. Biol. Chemists. Research, publs. on fractionation and analysis of fatty acids and related lipids. Home: 1300 2d St. N.W., Austin, Minn. 55912.*

PROCHASKA, Georg, Czechoslovakian physiologist; b. 1749; prof. anatomy and ophthalmology Prague, Czechoslovakia, Vienna, Austria. Introduced concept of sensorium commune (section of cortex coordinating all impressions passing to individual nerve centers); believed animal movements could be produced without cerebral control and discarded mechanistic explanations. Author: De functionibus systematis nervosi commentatio, 1784. Died 1820.

PROCLOS DIADOCHOS (the Successor), Greek Neo-Platonic philosopher, mathematician, astronomer; b. Byzantium, 410; brought up at Xanthos, Lydia; studied in Alexandria with Olympiodoros the Elder and in Athens with Syrianos, whom he succeeded as head of the Academy, a position which he held until 485. Author: Commentaries on Plato's Alcibiades I, Timaeos, Parmenides, and Republic; on Ptolemy's Quadripartitum; on book 1 of Euclid's Elements; on Hesiod and on Aristotle's theory of motion (Istitutio physica sive de motu); Hypotyposis (introduction to astronomy of Hipparchos and Ptolemy); Platonic Theology; On Providence and Fate; On Ten Doubts About Providence, and on the Nature of Evil. Chief representative of late Neo-platonists, emphasized mystical and religious aspect of Plato; believed reality descends hierarchically from and imitates a single highest principle, the One; taught that the One is beyond grasp of human reason and can only be encountered in mystical experience; studied ancient Greek geometry; discussed Euclid's parallel-postulate and Ptolemy's attempt to prove it; studied higher curves, including the hippopede; made cinematical construction of curves; presents a proof of the geometrical equivalence of epicycles and eccentrics; mentioned an annular eclipse of the sun; denied the existence of the precession of equinoxes. Died Athens, Greece, April 17, 485.

PROCOPIU, Stefan, Rumanian physicist; b. 1890; prof. Iasi (Rumania) U.; mem. Acad. Socialist Republic Rumania. Author: Electricity and Magnetism, 2 vols., 1939; Thermodynamics, 1948. First to calculate magnetic moment of electron (Bohr magneton), 1913; studied Procopiu phenomenon of depolarization of ight in colloidal solutions and crystalline suspensions, 1921; discovered discontinuity of magnetization occurring at passing of alternative current through ferro-magnetic wire, 1929.

PROCTER, William, Am. zoologist; b. Cin., Sept. 8, 1872; s. Harley Thomas and Mary Elizabeth (Sanford) P.; grad. Phillips Exeter Acad., 1891; Ph.B., Yale, 1894; postgrad., Sorbonne (Paris), Columbia; D.Sc., U. Montreal, 1930; m. Emily Pearson Bodstein, Feb. 3, 1910. During early career specialized in railroad orgn. and securities; organized firm Procter & Borden, 1902, ret., 1929; established lab. on Mt. Desert Island, Me., 1921; established Biol. Survey of Mt. Desert Region, 1936; contbd. to curricula of univs. and state biol. depts. Trustee Am. Mus. Natural History, N.Y. C.; mem. bd. mgrs. Wistar Inst. of Anatomy and Biology (Phila.), 1928-40. Fellow Entomol Soc. Am., A.A.A.S.; mem. Acad. Natural Scis., Boston Soc. Natural History, Entomol. Soc. Am. (mem. edit. bd., 1940-47), Am. Micros. Soc., Acad. Natural Scis. of Phila., Ornithologists Union, Santa Barbara Natural History Soc., Genetic Soc., Ray Soc. London (Eng.), Plymouth Marine Assn. (Eng.), So. Cal. Acad. Scis., Entomol. Soc. Washington, Ecologists Union, Ecol. Soc. Am., Royal Canadian Inst., Am., Pacific Coast, Bklyn., N.Y., Cambridge entomol. socs., Sci. Research Soc. Am. (co-founder and gov.), Sigma Xi. Author of publs. on marine life and insect life of Mt. Desert Region. Died Apr. 19, 1951.

PROCTOR, Carlton Springer, Am. civil engr.; b. Washington, Sept. 18, 1893; s. Joseph Martin and Louise (Springer) P.; C.E., Princeton, 1915, Dr.Eng. Sci., 1947; Dr. Eng., Drexel Inst. Tech., 1952; C.E. (hon.), Renselaer Poly. Inst., 1952; Dr.Eng., Fenn Coll., 1955; m. Isabel Lucey, June 7, 1916; children—Richard Carlton, Robert Martin (dec.), Carol (Mrs. Wright) (dec.). With Moran, Proctor, Mueser & Rutledge (formerly Moran & Proctor),N.Y.C., 1919-64, sr. partner, 1964; dir. Am. & Fgn. Power Cos., Thiokol Chem. Co., Daystrom Inc., Phoenix Ins. Co. of N.Y., 1st Fed. Savs. & Loan Assn. Lectr. numerous univs., U.S., S.Am., U.S.S.R.; mem. U.S. com. to UNESCO, 1956-63, chmn. UNESCO 7th Internat. Conf. on Latin Am., 1958. Mem. Am. Soc. C.E. (pres. 1952), Am. Inst. Constrn. Engrs. (pres. 1955), Am. Soc. Moles (pres. 1949-50; Outstanding Engr. award 1963), Princeton Engrs. Assn. Contbr. numerous articles to tech. jours. Co-designer, co-inventor pneumatic false-bottom bridge caisson; designed 1st capping of deep limestone cavitations,controls for large landslides; cons. engr. dams, ports, harbors, tunnels, bridges, U.S., fgn. countries; cons. engr. Golden Gate Bridge, San Francisco Bay Bridge. Address: 9 Elm Rock Rd., Bronxville, N.Y. 10708.*

PROCTOR, Donald Frederick, Am. physician, physiologist; b. Red Bank, N.J., Apr. 19, 1913; s. Frederick R. and Gertrude (Chauncey) P.; A.B., Johns Hopkins, 1933, M.D., 1937; m. Janice Knighton Carson, June 10, 1937; children—Douglas Carson, Nan Knighton. Intern, Johns Hopkins, 1937-39; resident, Johns Hopkins, 1939-40; Baltimore City Hosps., 1940-41; asso. prof. laryngology and otology, Johns Hopkins, Balt., 1946—, prof. anesthesiology, 1951-55, asst. prof. physiology, 1955-64; asst. prof. environmental medicine, 1964-65; prof. 1965—. Mem., Sigma Xi. Author: Anesthesia and Otolaryngology, 1957; Tonsils and Adenoids in Childhood, 1960; Nose, Paranasal Sinuses and Ears, 1962. Contbr. articles on anesthesiology, otolaryngology, respiratory physiology to tech. jours. Research in conservation of hearing; respiratory air flow, breathing mechanics, physiology of singing; airborne disease including air pollution. Home: 1110 Hollins Lane, Balt. 21209. Office: Johns Hopkins School Hygiene, Balt. 21205.*

PROCTOR, Paul Dean, Am. geologist; b. Salt Lake City, Nov. 24, 1918; s. John and Emily Irene (Triptow) P.; B.A., U. Utah, 1942; M.A., Cornell U., 1943; Ph.D., Ind. U., 1949; m. Martha Facer, Mar. 5, 1945; children—Paul F., Lane F., Kirk F., Scot F. Staff, U.S. Geol. Survey, Western U.S., 1943-53, full geologist, 1952; faculty Brigham Young U., 1949-50, asso. prof., 1952-53; asso. prof. Ind. U., 1950-52; supr. exploration U.S. Steel Corp., Western states, 1953-57; prof., chmn. dept. geology Mo. Sch. Mines and Metallurgy, Rolla, 1957-64; dean, prof. Sch. Sci., U. Mo., Rolla., 1964—. Mem. earth sci. curricula com. State of Mo., 1965-66, UNESCO Spl. Geology, Middle East Tech. U., Ankara, Turkey, 1966-67. Fellow Geol. Soc. Am.; mem. Am. Inst. Mining, Metall. and Petroleum Engrs. (chmn. geol. engring. unit 1966—), Mo. Acad. Sci., Nat. Assn. Geol. Tchrs., Utah Geol. Soc., Sigma Xi, Phi Kappa Phi. Author: Laboratory Manual for Physical and Engineering Geology, 1953; Uranium—Where it is—How to Find It, 1954; also articles. Research on mineral deposits; co-designer field stereoscope. Home: P.O. Box 607, Rolla, Mo. 65401.*

PROCTOR, Richard Anthony, astronomer; b. Chelsea, Eng., Mar. 23, 1837; s. William Proctor; ed. King's Coll., London U., 1855, St. John's Coll., Cambridge; m. Mrs. Robert J. Crawley, 1881. Came to U.S., 1881; became tchr. Woolwich Mil. Sch., 1873; lived in Mo., 1881-87, Orange Lake, Fla., 1887-88. Founder popular sci. mag. Knowledge, 1881. Author: Saturn and His System, 1865; Handbook of the Stars, 1866; Half-hours with the Telescope, 1868; Other Worlds than Ours, 1870; Our Place Among Infinities, 1875; The Poetry of Astronomy, 1880; Mysteries of Time and Space, 1883; The Universe of Suns, 1884; Half Hours with the Stars, 1887; also numerous articles. Research on rotation of Mars; demonstrated resisting medium surrounding sun using its effect on trajectory of prominences. Died N.Y.C., Sept. 12, 1888.

PROGER, Samuel Herschel, physician; b. Jan. 21, 1906; s. Louis Proger; B.S., Emory U., 1925, M.D., 1928; D.Sc. (hon.), Tufts Coll., 1952; m. Evelyn Levinson, Sept. 8, 1929; children—Susan Jean, Nancy Lane. Mem. staff New Eng. Med. Center Boston, 1931—, physician-in-chief 1948—, chmn. adminstrv. bd. chief of staff Pratt Diagnostic Clinic since 1939; prof. medicine, Tufts Coll. Med. Sch. 1948—, chmn. dept. medicine. Pres. Bingham

Asos. Fund Med. cons. VA; pres. bd. trustees New Eng. Center Hosp.; trustee Wellesley Coll. Recipient Am. Design award, 1951. Mem. Am. Soc. for Clin. Investigation, Am. Acad. Arts and Scis., Am. Heart Assn., A.M.A., A.C.P., N.Y. Acad. of Medicine, New Eng. Cardiovascular Soc. (pres. 1958-59), Assn. Am. Physicians, Assn. Prof. Medicine, Phi Beta Kappa, Alpha Omega Alpha. Contbr. articles on cardio-vascular disease and med. edn. Home: 45 Willow Crescent, Brookline, Mass. 02146. Office: 171 Harrison Av., Boston 02111.*

PROHASKA, Fritz Georg, meteorologist; b. Vienna, Austria, May 16, 1914; s. Carl and Margarethe (Schmid) P.; Dr. phil., U. Vienna, 1937; m. Maria Amata Rodionoff, May 3, 1947; children—Juan-Miguel, Irene-Maria. Meteorologist, Zentralanstalt für Meteorologie, Vienna, 1937-39, Physikalisch-Meteorologisches Observatorium, Davos, Switzerland, 1939-42, 45-47, Servicio Meteorologic Nacional, Buenos Aires, Argentina, 1947-58; prof. U. Buenos Aires, 1957-58; prof. Jesuit U., Buenos Aires, 1959-60; meteorologist Instituto Nacional de Tecnologia Agropecuaria, Buenos Aires, 1959-64; tech. expert World Meteorol. Orgn., Guatemala, 1962-63; prof. meteorology U. Wis., Milw., 1964——. Mem. Am. Meteorol. Soc., Assn. Am. Geographers, Am. Geophys. Union, Oesterreichische Gesellschaft für Meteorologie, PanAm. Inst. Geography and History. Research, numerous publs. on sun and sky radiation in Alps, radiative qualities of snow, climate of Argentina, Paraguay and Uruguay with spl. emphasis on precipitation regime, semiarid and arid zones, problems of agrl. meteorology in semiarid regions, hydrometeorol. conditions in Guatemala.*

PROHASKA, John Van, surgeon; b. Prague, Bohemia, Sept. 6, 1904; s. Anthony and Teresa (Hampacher) P.; came to U.S., 1920, naturalized, 1925; B.S., U. Chgo., 1928, M.D., 1933; m. Astrid Paulson, Nov. 27, 1929; 1 son, Peter Van. Instr. surgery U. Chgo. Hosp., 1939; with Chgo. Meml. Hosp., 1940——, vice-chief of staff, trustee, 1951——; prof. surgery U. Chgo., 1954——. Mem. Soc. Surgery Alimentary Tract (sec.), Chgo. Surg. Soc. (pres., 1962), A.M.A., Am. Surg. Assn., A.C.S.; Soc. U. Surgeons, Internat. Soc. Surgeons, Chechoslovak Med. Soc., Sigma Xi. Author chpt. Adrenal Glands Lewis-Walters Practice of Surgery, vol. VII, 1965. Research, publs. on etiology, therapy of staphylococcal enteritis, surgery of ulcerative colitis, corticosteroid therapy, mammary carcinoma, suppression of gastric secretion. Home: 5830 Stony Island Av., Chgo. 60637. Office: 950 E. 59th St., Chgo. 60637.*

PROKHOROV, Aleksandr Mikhaylovich, Russian physicist; b. July 11, 1916; s. Mikhayl Ivanovich and Maria (Mikhaylova) P.; grad. Faculty Physics, Leningrad State U., 1939, Dr. Physics, Math., Sci., 1946; m. Shelepina Galina, Aug. 3, 1941; 1 son, Kirill. Jr. sci. worker, 1939-46, sr. sci. worker, 1946-54; head lab., FIAN, 1954——; prof. Moscow State U.; lab head Physics Inst., USSR Acad. Scis., 1954——. Chmn. council Nat. Com. U.R.S.I. Recipient Lenin prize, 1959, Nobel prize in physics (with N.G. Basov and C.H. Townes), 1964. Fellow acad. council FIAN; mem. USSR Acad. Scis. (corr.). Mem. editorial bd. Jour. Radioengring. and Electronics. Research on quantum radiophysics; molecular generators and amplifiers; with N. G. Basov, devised new method of amplifying electromagnetic radiation; patentee in field. Home: fl.85, N4 Gubkina, Moscow B333. Office: 53 Leninsky Prospect, Moscow, USSR.

PROKOP, Otto, physician; b. St. Pölten, Austria, Sept. 29, 1921; s. Ludwig and Elfriede (Worbs) P.; Dr.med., U. Bonn, 1948, Dozent, 1953; m. Wilhemine Cohnen, Apr. 1952; children—Uta, Eberhard. Asst., Inst. Forensic Medicine, Bonn, Germany, 1948-53, dozent, 1953-56; prof. forensic medicine Humboldt U., Berlin, Germany, 1957——, dir. Legal Medicine Inst. Mem. German Acad. Sci., Spanish (hon.), Yugoslav (hon.) socs. forensic medicine, Royal Soc. Medicine (London, affiliated); corr. mem. various socs. Author: Forensic Medicine, 1960; Medizinischer Okkultismus, 1962; (with Uhlenbruck) Lehrbuch der menschlichen Blutgruppen, 1963; (with Weimann) Picture Atlas of Forensic Medicine, 1965; others; also numerous articles. Research on methods for stain and trace investigation, death time determination, procedures for preparation of test sera, discovery of protectine (anti-Ahel, anti Bsal, anti-H per), different studies on heredity. Address: Hannoversche Strasse 6, Berlin 104, Germany.*

PROKOPCHUK, Andrey Yakovlevich, Russian dermatologist, venerologist; b. Chemery (now Grodno Oblast), 1896; grad. Med. Faculty, 2d Moscow U., 1924; D.Med. Sci., 1936. Head chair skin diseases Stalinabad Med. Inst., 1942-43. Prof., 1931——; dir. Belorussian State Research Inst. for Skin and Venereal Diseases, 1932——. Decorated Order of Lenin (2). Mem. Belorussian Acad. Scis. (mem. Presidium 1940-47). Mem. editorial council Herald Dermatology and Venerology; editor 7 vols. sci. works of Belorussian Inst. for Skin and Venereal Diseases, 6 symposiums of Practical Dermatology. Research and numerous publs. on treatment of skin diseases with radioactive phosphorus, use of radioactive isotopes in medicine; developer quinacrine (for treatment of Lupus vulgaris). Address: Belorussian State Research Inst. for Skin and Veneral Diseases, Miasnikowa 34/65, Minsk, Belorussia SSR, USSR.

PRONKO, Nicholas Henry, Am. psychologist; b. McKees Rocks, Pa., Feb. 28, 1908; s. Michael and Dora (Smarsh) P.; B.A. with distinction, George Washington U., 1941; M.A., Ind. U., 1941, Ph.D., 1944; m. Catherine Tippery, Mar. 28, 1930 (div. Aug. 1952); children—Michael John, Suzanne Kay; m. 2d, Geraldine Allbritten, Dec. 16, 1953. Instr., Ind. U., Bloomington, 1943-45; lectr. Shrivenham Am. U., Eng., 1945-46, Biarritz Am. U., France, 1946; asst. prof. psychology N.Y. City Coll., 1946-47; faculty U. Wichita (now Wichita State U.), Kan., 1947——, prof., head dept. psychology, Fulbright lectr. Istanbul U., 1952-53; vis. prof. Cracow U., 1959; vis. fellow, Rothschild fellow Inrael Inst. Tech., 1963. Fellow Conf. on Religion in Age of Sci., Star Island, N.H., 1957. Fellow Am., Kan., Midwestern psychol. assns., Am., Kan. socs. for clin. hypnosis; mem. A.A.A.S., Phi Beta Kappa, Sigma Xi. Author: (with J.W. Bowles, Jr.) Empirical Foundations of Psychology, 1951; (with F.W. Snyder) Vision with Spatial Inversion, 1952; Textbook of Abnormal Psychology, 1963; also numerous articles. Research on psychopathology perception and drug effects. Home: 1609 Fairmount St., Wichita, Kan. 67208.*

PRONY, Baron Gaspard-Clair-François-Marie Riche-de, see De Prony, Baron Gaspard-Clair-François-Marie Riche.

PROSKAUER, Johannes, botanist; b. Göttingen, Germany, Dec. 5, 1923; s. Walter P. and Margarete (Jacob) P.; B.Sc. (Gen.), U. London, 1943, B.Sc (Spl.) 1st class honors, 1944, Ph.D., 1947, D.Sc., 1964; m. Josephine Pia Schizzano, Jan. 25, 1951; children—Margaret, Bruno. Lectr., S.E. Essex Tech. Coll., Dagenham, Eng., 1945-48; faculty U. Cal. at Berkeley, 1948——, prof. botany, 1963——, prof. Miller Inst. for Research in Basic Sci., 1964-65. Guggenheim Found. fellow, 1954-55. Mem. Internat. Assn. Plant Morphologists, Internat. Assn. Plant Taxonomy, Am., Brit. bryological socs., Bot. Soc. Am. Research, publs. on gen. morphology, cytogenetics, biosystematics algae and bryophytes. Home: 132 Purdue Av., Kensington, Cal. 94708. Office: Dept. Botany, U. Cal., Berkeley, Cal. 94720.*

PROSSER, Charles Smith, Am. geologist; b. Columbus, N.Y., Mar. 24, 1860; s. Smith and Emeline A. (Tuttle) P.; B.S., Cornell, 1883, M.S., 1886; D.Sc., Union U., 1906; Ph.D., Cornell, 1907; m. Mary F. Wilson, Aug. 28, 1893. Instr. paleontology Cornell, 1885-88; asst. paleontologist U.S. Geol. Survey, 1888-92; prof. natural history Washburn Coll., Topeka, Kan., 1892-94; prof. geology Union Coll., 1894-99; asso. prof. hist. geology Ohio State U., 1899-1901, prof. geology, head dept., from 1901. Asst. geologist U.S. Geol. Survey, also state geol. surveys of Kan., N.Y., and Ohio; geologist Md. Geol. Survey. Fellow Geol. Soc. Am., A.A.A.S., Am. Paleontol. Soc. Author: The Devonian System of Eastern Pennsylvania and New York, 1895; The Classification of the Upper Palaeozoic Rocks of Central Kanas, 1895, 1902; The Upper Permian and Lower Cretaceous of Kansas, 1897; The Classification of the Hamilton and Chemung Series of New York, Part I, 1898, II, 1900; Cottonwood Falls (Kansas) Folio (with J. W. Beede), 1904; Revised Nomenclature of the Ohio Geological Formations, 1905; Anthracolithic or Upper Paleozoic Rocks of Kansas, 1910; Devonian and Mississippian Formations of Northeastern Ohio, 1912; Middle Devonian Deposits and Paleontology of Maryland (with Edward M. Kindle), 1913; Upper Devonian Deposits of Maryland (with Charles K. Swartz), 1913. Deceased.

PROSSER, Clifford Ladd, Am. physiologist, zoologist; b. Avon, N.Y., May 12, 1907; s. Clifford J. and Izora (Ladd) P.; A.B., U. Rochester, 1929; Ph.D., Johns Hopkins, 1932; m. Hazel Blanchard, Aug. 25, 1934; children—Jane Ellen (Mrs. John S. McReynolds), Nancy (Mrs. P.C. Ryan), Loring. Asst. prof. Clark U., 1934-39; faculty U. Ill., Urbana, 1939——, prof., 1949, head dept. physiology and biophysics, 1960——; asso. sect. chief Metall. Lab., U. Chgo., 1943-46; vis. lectr. U. Wash., Seattle, summer 1951, Stanford, 1954, U. Hawaii, Honolulu, 1957, U. Munich, Germany, 1963. Guggenheim fellow, 1963-64. Mem. Soc. Gen. Physiology (pres. 1958-59), Am. Soc. Zoologists (pres. 1961), A.A.A.S. (sect. chmn. 1965-66), Am. Physiol. Soc., Soc. Exptl. Biology. Author: Comparative Animal Physiology, 1950, 61. Research on smooth muscle and nerve function, animal evolution as related to molecular biology, adaptation of animals to cold. Home: 205 E. Michigan, Urbana, Ill. 61801.*

PROSSER, Reese Trego, Am. mathematician; b. Mpls., May 18, 1927; s. William Lloyd and Eleanor (Sewall) P.; A.B., Harvard, 1949; Ph.D., U. Cal. at Berkeley, 1955; m. Jeanne Elizabeth Melchior, June 30, 1962; children—Elizabeth Sewall, Charles Allen, James Montfort, David Lloyd. Numerical analyst Lawrence Radiation Lab., Berkeley, Cal., 1954-55; research instr. Duke, 1955-56; research instr. Mass. Inst. Tech., 1956-58, mathematician Lincoln. Lab., Lexington, Mass., 1958-62, group leader, 1962-65, project mgr. radar calibration spheres, 1965——. Mem. Am. Math. Soc., Am. Phys. Soc., I.E.E.E. (sr.), Assn. Computing Machinery, Soc. for Indsl. and Applied Math. Research, publs. on math structure underlying fundamental problems in phys. scis., especially classical and quantum mechanics particles and fields. Home: 64 Bloomfield St. Office: Lincoln Lab., Wood St., Lexington, Mass. 02173.*

PROTHRO, Edwin Terry, Am. psychologist; b. Robeline, La., Dec. 11, 1919; s. Edwin Thomas and Frances (Terry) P.; A.B., La. Coll., 1939; M.A., La. State U., 1940, Ph.D., 1942; m. Dorothy Kenworthy, Mar. 26, 1943; children—Martha Carol, Edwin Terry. Asst. prof. psychology La. State U., 1946-49; asso. prof. U. Tenn., 1949-51; faculty Am. U. Beirut (Lebanon), 1951——, prof. psychology, 1954——, dir. Center for Behavioral Research, 1963——, dean faculty arts and scis., 1965——. Mem. A.A.A.S., Am. Assn. U. Profs., Am. Sociol. Assn., Sigma Xi. Fellow Am. Psychol. Assn. Author: Child Rearing in the Lebanon, 1961; Psychology—A Biosocial Study of Behavior, 1950; also articles. Research on analysis of factors affecting racial attitudes, relation between child rearing patterns, family structure, and personality in Eastern Mediterranean cultures. Office: Am. U. Beirut, Beirut, Lebanon.*

PROTHRO, James Warren, Am. polit. scientist; b. Robeline, La., Apr. 15, 1922; s. Edwin Thomas and Frances (Terry) P.; B.A., N. Tex. State U., 1943; M.A., La. State U., 1948; M.A., Princeton, 1949, Ph.D., 1952; m. Mary Frances Harris, Oct. 17, 1943; children—Pamela, Barbara, Susan. Asst. prof. polit. sci. Fla. State U., 1950-53, asso. prof., 1953-57, prof., 1957-61; vis. prof. U. N.C., 1960-61, prof. polit. sci., research prof. Inst., for Research in Social Sci., 1961——, dir. Inst. for Research in Social Sci., 1967——. Vis. prof. Escuela Latinoamericana de Ciencia Politica y Adminstraction Publica, Flacso, Chile, 1966-67; book review editor Am. Polit. Sci. Rev., 1965——; mem. editorial bd. Pub. Opinion Quarterly, 1965——; mem. adv. bd. Journal of Politics, 1965——; cons. U.S. Commn. on Civil Rights, 1964-65; dir. Louis Harris Polit. Data Center, U. N.C., 1965——; editor Harbrace series in Political Behavior, 1966——. Rockefeller Found. grant, 1960-64. Mem. Am., So. polit. sci. assns., A.A.U.P. Author: Dollar Decade. Business Ideas in the 1920s, 1954; (with Marian D. Irish) The Politics of American Democracy, 1959, 2d edit., 1962, 3d edit., 1965, 4th edit., 1968; (with Donald R. Matthews) Negroes and the New Southern Politics, 1966. Contbr. articles in field to sci. jours. Research, publs. on Negro polit. participation in the South combining aggregate data, opinion survey data, depth community studies in a single explanatory model. Home: 306 Elliott Rd., Chapel Hill, N.C. 27514.*

PROTSENKO, Dmitriy Filippovich, Russian biologist; b. Maly Buzukov (now Kiev Oblast), 1899; grad. Kiev Inst. Pub. Edn., 1925; postgrad. Ukrainian Research Inst. Phytoculture, Kharkov, until 1931; D.Biol. Sci., 1940. Sr. instr. dept. botany Poltava Agrl. Inst., 1929-32, lectr., from 1932; head lab. plant physiology Ukrainian Horticulture Research Inst., 1933-36; acting head chair plant physiology and agrl. microbiology Azov and Black Sea Agrl. Inst., Novocherkassk, 1937-40; head chair physiology and agrl. microbiology Saratov Agrl. Inst., head chair plant physiology Saratov U., 1940-44; acting head dept. physiology and biochemistry Ukrainian Agrl. Research Inst., 1944-52; acting head chair microbiology and biochemistry Kiev Chem. Tech. Inst., 1944-55; prof., head chair physiology and biochemistry Kiev U., dir. Univ. Bot. Gardens, 1944——; acting head dept. plant resistance Ukrainian Research Inst. Plant Physiology, 1957——. Decorated Order of Lenin. Author: Practical Work on Plant Physiology and the Principles of Microbiology, 1951; The Anatomy and Morphology of Plants, 1953; Plant Physiology, 1958; co-author: Practical Work on Plant Anatomy, 1955. Address: Kiev University, Vladimirskaya ulitsa 64, Kiev, Ukrainian SSR, USSR.

PROTTER, Murray Harold, Am. mathematician; b. N.Y.C., Feb. 13, 1918; s. Aron and Bertha (Keller) P.; B.A., U. Mich., 1937, M.A., 1938; Ph.D., Brown U., 1946; m. Ruth Rotman, June 24, 1945; children—Barbara, Philip. Asst. prof. Syracuse U., 1947-51; mem. Inst. for Advanced Study, Princeton, N.J., 1951-53; faculty U. Cal. at Berkeley, 1953——, prof. math., 1958——, chmn. dept., 1962——. Mem. Am. Math. Soc., Math. Assn. Am., Soc. Indsl. and Applied Math. Author: (with C.B. Morrey) Calculus and Analytic Geometry, 1963, Modern Math Analysis, 1964; (with H. Weinberger) Maximum Principles, 1967; also articles. Research on partial differential equations and hydrodynamics. Home: 1016 Villa Nueva Dr., El Cerrito, Cal. 94530. Office: Math. Dept., U. Cal., Berkeley, Cal. 94720.*

PROUDFIT, William L., Am. physician; b. Connellsville, Pa., Feb. 16, 1914; s. John Lyle and Mary (Hanna) P.; B.S., Washington and Jefferson Coll., 1935; M.D., Harvard, 1939; m. Thelma Gladys Janaske, Aug. 31, 1940; children—John Paul, Ann Lyle, James Hanna. Practice medicine specializing in cardiology, Cleve., 1946——; staff Cleve. Clinic, 1946——, head clin. cardiology, 1965——. Fellow A.C.P.; mem. A.M.A., Am. Heart Assn. Research, publs. on clin. cardiology and electrocardiography. Home: 2607 N. Park Blvd., Cleve. 44106. Office: Cleve. Clinic, Cleve. 44106.*

PROUDHON, Pierre Joseph, French social, polit. theorist; b. Besançon, France, Jan. 15, 1809; ed. U. Bescançon; m. circa 1850. Working compositor, corrector for press; encountered socialistic ideas in Paris; owner small printing co., Besançon; mgr. comml. firm, Lyon, France; settled Paris, 1847 (gained nat. notoriety during revolution 1848); mem. for Seine dept. French Assembly; imprisoned for ideas in polit.

pamphlets, 1849-52; fled to Brussels, Belgium (to avoid prison for publ. of attack on ch., other instns.); returned to Paris, 1860-65. Author: Qu'est-ce que la Propriété?, 1840; Philosophie de la misère, 1846; Confessions d'un revolutionnaire, 1849; Le droit au travail et le droit de propriété, 1850; De la celebration du dimanche, 1851; Philosophie du progrès, 1853; De la justice dans la revolution et dans l'église, 3 vols., 1858; La guerreet la paix, 2 vols., 1860; others. Opponent pvt. property and the state; urged nat. bank for re-establishment of credit in interest of workers; wanted econ., not polit. innovation; believed ideal soc. would be union of order, anarchy; basic principle was that all are entitled to product of their work; developed idea of small, loosely federated groups with govt. limited by checks and balances as compromise between anarchism, state sovereignty. Died Passy, France, Jan. 16, 1865.

PROUDMAN, Joseph, English oceanographer; b. Lancashire, Eng., Dec. 30, 1888; s. John and Nancy (Blease) P.; B.Sc., U. Liverpool, 1910, M.Sc., 1915, D.Sc., 1916, LL.D., 1956; B.A., Trinity Coll., Cambridge, 1913, M.A., 1917; m. Rubina Ormrod, July 10, 1916; children—James, Nancy (Mrs. Kenneth G. Smith), Ian; m. 2d, Beryl G.W. Barker, May 29, 1961. Faculty, U. Liverpool, 1913-54, dir. Tidal Inst., 1919-46, prof. oceanography, pro-vice chancellor, 1940-46; fellow Trinity Coll., Cambridge, 1915-21. Recipient Smith prize, 1915; Adams prize Cambridge, 1923; Agassiz medal, 1946; Hughes medal Royal Soc., 1957; Comdr. Order Brit. Empire, 1952. Fellow Royal Soc., 1925; mem. Norske Videnskaps-Akademi i Oslo, Internat. Assn. Phys. Oceanography (sec. 1933-48, v.p. 1948-51, pres. 1951-54). Author: (with F.S. Carey) Elements of Mechanics, 1926; Dynamical Oceanography, 1953; also numerous articles. Research on dynamical theory of tides; temperature, salinity and turbulence of Irish Sea; dynamics of storm surges; Taylor-Proudman theorem of dynamics of rotating fluids. Home: Edgemoor, Verwood, Dorset, Eng.*

PROUST, Joseph Louis, French chemist; b. Angers, Maine-et-Loire, France, Sept. 26, 1754; trained as apothecary at his father's shop; then studied chemistry under Rouelle in Paris; settled in Spain; prof. chemistry at Real Seminario de Vergara; taught chemistry in Segovia; director, royal laboratory in Madrid, 1789-1808, when he returned to Paris; mem. French Acad. Scis., 1789. Established experimentally (in a famous controversy with Berthollet) Proust's law (law of definite proportions), viz. elements in any compound are present in a fixed proportion by weight, 1797; isolated grape sugar and discovered presence of sugar in some vegetables; mineral proustite named in his honor; recognized difference between oxides and hydroxides; developed use of hydrogen sulfide to precipitate heavy metals from solutions of their salts. Died, Angers, France, July 5, 1826.

PROUT, William, English chemist and physiologist; b. Horton, Gloucestershire, Eng., Jan. 15, 1785; M.D., Edinburgh, 1811; Licentiate of Royal College of Physicians, 1812; settled in London; delivered a course of lectures in chemistry, 1813. Fellow, Roval Soc., 1819; fellow, Royal Coll. Physicians, 1829. Author: Inquiry into the Nature and Treatment of Gravel, Calculus, and Other Diseases of the Urinary Organs, 1821; Chemistry, Meteorology, and the Function of Digestion, considered with reference to Natural Theology, 1834; many sci. papers. Pioneer in physiological chemistry; on Dec. 23, 1823, announced famous discovery of existence in gastric juice of stomach of pure hydrochloric acid; discovered alloxan, 1818; examined ink of cuttlefish; studied chemistry of blood and urine; discovered that excrement of boa constrictor contains 90 percent uric acid; prepared pure urea for first time, 1818; 1st to divide components of foodstuffs into carbohydrates, fats, and proteins, 1827; inventor of Prout's hypothesis, the view that the atomic weights of all the elements are multiples of the atomic weight of hydrogen and that elements are formed by a condensation or grouping of hydrogen atoms, 1815-16. Died, London, Eng. Apr. 9, 1850.

PROVANCHER, Léon Abel, Canadian naturalist; b. Bécancourt, Que., Can., Mar. 10, 1820; student, Nicolet, Que. Ordained, 1844; pastoral work for succeeding 25 years. Author: Flore du Canada, 1862; Faune entomologique du Canada, 3 vols., 1877-90. Founder, Naturaliste Canadien, 1868, editor, 1868-88. Research on Canadian flora and fauna. Died Cape Rouge, Que., Mar. 23, 1892.

PRUDDEN, T(heophil) Mitchell, Am. pathologist; b. Middlebury, Conn., July 7, 1849; s. George P. and Eliza A. Johnson (Prudden) P.; B.S., Sheffield Sci. Sch. (Yale), 1872, M.D., 1875 (LL.D., 1896); postgrad., Heidelberg, Vienna, Berlin. Instr. chemistry Yale, 1872-74; lectr. normal histology, 1880-86; asst. in pathology and histology Coll. Phys. and Surg. (Columbia), 1878-82, dir. pathol. and bacteriol. lab., 1882-91, prof. pathology, 1892-1909, emeritus prof., 1909. Dir. Rockefeller Inst. for Med. Research, from 1901. Fellow Am. Acad. Arts and Scis. Author: Manual of Normal Histology, 1881; Textbook of Pathology, 1885, 11th edit., 1919; Story of the Bacteria Dust and Its Dangers; Water and Ice Supplies; On the Great American Plateau, 1907. First to make diphtheria antitoxin in U.S.; studied remains of cliff dwellers in Am. Southwest. Died N.Y.C., Apr. 10, 1924.

PRUFER, Olaf Herbert, anthropologist; b. Berlin, Germany, Oct. 3, 1930; s. Curt Maximillian and Anneliese (Fehrmann) P.; came to U.S., 1954, naturalized, 1962; A.B., Harvard, 1956, M.A., 1958, Ph.D., 1961; m. Cynthia Munro, Sept. 20, 1958; children—Keith Malcolm, Diana Deirdre. Curator anthropology Cleve. Mus. Natural History, 1959-63; faculty Case Inst. Tech., Cleve., 1959-67, asso. prof. anthropology, 1963-67; prof. anthropology U. Mass., Amherst, 1967—. Fellow Am. Anthrop. Assn., Ohio Acad. Sci.; mem. Soc. for Am. Archaeology, N.Y. Acad. Scis., Phi Beta Kappa, Sigma Xi. Author: (with R. S. Baby) Palaeo Indians of Ohio, 1963; The McGraw Site, A Study in Hopewellian Dynamics, 1964; also numerous articles. Research on Old Stone Age India, Lebanon, Austria, Germany, Palaeo-Indian occupation Eastern U.S., formative Woodland cultures Ohio area. Home: 49 Kellogg Av., Amherst, Mass. 01002.*

PRUITT, Raymond Donald, Am. physician; b. Wheaton, Minn., Feb. 6, 1912; s. Lyman Burton and Ada (Brandes) P.; B.S., Baker U., 1933, D.Sc., 1956; B.A. (Rhodes scholar), Oxford (Eng.) U., 1936; M.D., Kan. U., 1939; M.S. in Medicine, U. Minn., 1945; m. Lillian Elaine Rasmussen, July 31, 1942; children—Virginia, Kristin, David, Charles. Fellow in medicine Mayo Found., 1940-43, faculty medicine U. Minn., Mpls., 1945-59, prof. medicine, chief medicine for med. edn. Grad. Sch., 1954-59, asso. dir. Mayo Found., 1954-57, cons. physician Mayo Clinic, 1943-59; prof., chmn. dept. medicine Baylor U. Coll. Medicine, Houston, 1959-68; v.p. med. affairs, chief exec. officer, 1966-68; dir. Mayo Grad. Sch. Edn., U. Minn., also dir. edn. Mayo Found., Rochester, Minn., 1968—. Mem. Nat. Adv. Heart Council, 1964-68, com. internships, residencies, and grad. med. edn. Assn. Am. Med. Colls., 1955-60; mem. tng. grants com. Nat. Heart Inst., 1961-64, adv. heart council, 1964—. Recipient Outstanding Achievement award U. Minn., 1964. Diplomate Am. Bd. Internal Medicine. Mem. Am. (research study com. Internal Medicine. Mem. Am. (research study com., council on clin. cardiology 1962-66), Tex. (dir. 1961-67), Houston (dir. 1959-68) heart assns., Am. Coll. Chest Physicians, A.C.P., Assn. Profs. Medicine, Assn. U. Cardiologists (council 1962-67, pres. 1967), Am. Coll. Cardiology, Am. Bd. for Cardiovascular Disease (sec. 1955-60), Central Soc. for Clin. Research, Sigma Xi, Alpha Omega Alpha. Editorial bd. Am. Heart Jour., 1962-68, Circulation, 1963—. Research, publs. on clin. and exptl. electrocardiography; med. edn. Address: Mayo Grad. Sch. Medicine, Rochester, Minn. 55901.*

PRUNIERAS, Michel Antoine, French cancer researcher; b. Boulogne-sur-Mer, France, Oct. 11, 1925; s. Aimé Stéphane and Marguerite (Mourraille) P.; M.D., Faculté Médecine, Lyons, France, 1949; m. Janine Goujon, Apr. 22, 1946; children—Catherine, Claude, Marie-Christine. Asst. dermatology, 1949-54; research fellow dermatology Mayo Clinic, Mayo Found., Rochester, Minn., 1954-55; asst. research tissue culture Lab. Hygiene, Lyons, 1955-58; asso. prof. dermatology U. Strasbourg (France), 1958-62; dir. of research, Unit for Research on Virus-Cancer Relationships, Lyons, 1963—. Mem. com. no. 3, French Nat. Inst. Health and Med. Research, mem. com. on cell cultures of permanent sect. on microbiol. standardization. Mem. Am. Dermatol. Assn. (corr.), French socs. dermatology, microbiology, electron microscopy. Author: La culture de l'épiderme, 1965; also articles. Research on long term cultivation of mammalian epidermal cells, chromosome studies on Shope's papilloma of rabbit skin cultured in vitro. Home: 109 Garibaldi, 69, Lyon 6. Office: Unité de recherches virus-cancer, Centre Léon Bérard 69, Lyons 8, France.*

PRUNTY, Francis Thomas Garnet, English chem. pathologist, physician; b. London, Jan. 5, 1910; s. Frank Hugh and Una (Marsden) P.; B.A. with 1st class honors in Natural Scis. Tripos, Trinity Coll., Cambridge U., 1932, M.A., 1936; M.B., B.Chir., Cambridge U., 1940, M.D., 1944; student St. Thomas's Hosp. Med. Sch., London, 1936-40; m. Rita Stobbs, Oct. 12, 1932; children—Una Kay (Mrs. Joachim Marroquin), John Garnet. Research asst. div. nutrition Lister Inst. Preventive Medicine, 1932-33; chemist Govt. Lab., London, 1933-36; house physician St. Thomas's Hosp., 1941; Rockefeller Travelling fellow, 1946-47; asst. medicine Peter Bent Brigham Hosp., Boston, 1946-47; research fellow medicine Harvard, 1946-47; reader chem. pathology U. London, 1947-54, prof. chem. pathology, 1954—; physician St. Thomas's Hosp., London, 1950—. Fellow Royal Coll. Physicians London, Royal Soc. Medicine (past pres. sect. endocrinology); mem. Assn. Physicians Gt. Britain and Ireland, Med. Research Soc., Soc. for Endocrinology (past hon. sec.), Biochem. Soc., Internat. Soc. Endocrinology (chmn.). Author: Chemistry and Treatment of Adrenocortical Diseases, 1964; (with R. McSwiney, J. Hawkins) Laboratory Manual of Chemical Pathology, 1959; also numerous articles. Research on chem. pathology and endocrinology especially of adrenal gland. Office: St. Thomas Hosp., London S.E.1., Eng.*

PRUSOFF, William Herman, Am. pharmacologist; b. N.Y.C., June 25, 1920; s. Samuel and Mary (Metrick) P.; B.S. cum laude, U. Miami, 1941; M.A., Columbia, 1947, Ph.D., 1949; m. Brigitte Auerbach, June 19, 1948; children—Alvin, Laura. Staff, Western Res. U., Sch. Medicine, Cleve., 1949-53; faculty Yale, 1953—, prof. pharmacology, 1966—; faculty Oxford (Eng.) U., 1959-60. Spl. Trainee award Nat. Inst. Neurol. Diseases and Blindness, 1959-60. Mem. Am. Assn. for Cancer Research, Am. Soc. Biol. Chemists, Am. Chem. Soc., Am. Soc. for Pharmacology and Exptl. Therapeutics. Research, numerous publs. on synthesis of anti-neoplastic and anti-viral compounds including 5-iodo-2-prime deoxyuridine. Home: De Forest Dr., North Branford, Conn. 06471. Office: 333 Cedar St., New Haven 06510.*

PRUXANSKY, Samuel, Am. orthodontist; b. N.Y.C., Sept. 10, 1920; s. Philip and Eva (Pasternak) P.; B.S., Coll. City N.Y., 1947; D.D.S., U. Md., 1945; M.S., Tufts U., 1949; postgrad. (USPHS fellow), U. Ill.; m. Donna Jean Schlarbaum, June 28, 1952; children—Elysa Louise, Marya Beth. Acting chief clin. investigation br. Nat. Inst. Dental Research, 1954-55, bd. sci. counselors, 1963—; faculty U. Ill., Chgo., 1956—, prof. dentistry, 1961—, asso. dir. Cleft Palate Center, 1956-67, dir. Center for Cranifacial Anomalies, 1968—. faculty Inst. Advanced Edn. Dental Research, 1963-65. Cons. Clin. Center, NIH, 1956—, mem. dental study sect., 1959-63, chmn. conf. maxillofacial prosthesis, 1965; project officer epidemiology and genetics study congenital malformations, Hebrew U., Israel, 1965. Diplomate Am. Bd. Orthodontists. Fellow A.A.A.S., Am. Coll. Dentists, Am. Pub. Health Assn.; mem. Am. Soc. Human Genetics, Am. Assn. Orthodontists (1st prize for research 1957), Teratology Soc., Am. Assn. Electromyography and Electrodiagnosis, Am. Cleft Palate Assn. (past pres.), Internat. Soc. Craniofacial Biology (sec.-treas., v.p. 1961-64, president 1967-68), Sigma Xi. on physiology of muscles of mastication, longitudinal growth studies of heads of children with congenital malformations and acquired deformities; epidemiologic and genetic studies of craniofacial malformations in man, exptl. pathology and surgery in animals relating to growth of face, orthodontic diagnosis and treatment, X-ray technics for studying growth of head in infants and children. Home: 425 Locust Rd., Wilmette, Ill. 60091. Office: 808 S. Wood St., Chgo. 60680.*

PRYLES, Charles Victor, Am. pediatrician; b. Atlanta, Oct. 4, 1920; s. Victor K. and Mary (Lucas) P.; A.B., Emory U., 1940; M.D. magna cum laude, U. Ga., 1947; m. Gloria Connell, Feb. 16, 1947; children—Victor, James, David. Research asso. Thorndike Meml. Lab., Harvard Med. Sch., 1955-59; faculty, Boston U. Sch. Medicine, 1954-65; prof. pediatrics N.Y. State Coll. Medicine, Downstate Med. Center, Bklyn., 1965—; asst. dir. pediatrics Boston City Hosp., 1960-65; dir. pediatrics Jewish Hosp. Bklyn., 1965—. Mem. Soc. Pediatric Research, Am. Acad. Pediatrics, Am. Bd. Pediatrics, N.Y. Acad. Sci., Infectious Disease Soc. Am., Sigma Xi, Alpha Omega Alpha. Research, numerous publs. in elucidation of diagnostic and therapeutic problems in pyelophites of infancy and childhood, infectious disease of infancy and childhood. Home: 5 Florence St., Rockland, Mass. Office: 555 Prospect Pl., Bklyn.*

PRYM, Friedrich, German mathematician; b. Düren/Rheinland, Sept. 28, 1841; prof., Würzburg, Germany; wrote on theory of functions. Died Bonn, Dec. 15, 1915.

PRZELECKA, Aleksandra, Polish cytologist, cytochemist; b. Lódz, Poland, Nov. 25, 1920; d. Aleksander Napiórkowski and Alicja (Wysznacka) P.; student Warsaw (Poland) Agrl. U., 1938-39; M.S. in Biology, U. Lódz, 1949, D.Sc. in Biology, 1961; postgrad. U. London (Eng.), Oxford (Eng.) U.; m. Marian, July 4, 1949. Asst. dept. botany U. Lódz, 1945-52; head lab. cytochemistry Nencki Inst.. Exptl. Biology, Warsaw, 1952—; lectr. U. Warsaw, 1961-62. Mem. Polish Histochem. and Cytochem. Soc., Polish Biochem. Soc. Research, publs. on lipid digestion in insects and vertebrates, correlation between increase in alkaline phosphatase activity and appearance of cellular phospholipids, cytochemistry and hormonal regulation of devel. insect ovarioles. Home: 16b Przasnyska. Office: 3 Pasteur str., Warsaw, Poland.*

PRZEWORSKA-ROLEWICZ, Danuta, Polish mathematician; b. Warsaw, Poland, May 25, 1931; d. Stefan and Janina (Rosen) Przeworski; M.A., U. Warsaw, 1956; Ph.D., Math. Inst., Polish Acad. Sci., 1958, docent, 1964; m. Stefan Rolewicz, Dec. 26, 1951; children—Jerzy, Jan. Asst., Politech. U. Warsaw, 1954-61; staff Math. Inst. Polish Acad. Scis., 1960—, docent, 1964—. Mem. Polish Math. Soc. Author: (with S. Rolewicz) Equations in linear Spaces, 1967; also articles. Research on approximation method for nonlinear singular integral equations, algebraic method of solving linear singular integral equations, linear operators. Home: 6/23 Sierpecka, Warsaw 32. Office: 8 Sniadeckich, Warsaw 10, Poland.*

PRZYBYLSKI, Antoni, astronomer; b. Rogozno, Poland, Dec. 29, 1913; s. Wladyslaw and Marie (Marciniak) P.; D.Tech. Scis., Fed. Inst. Tech., Zurich, Switzerland, 1950; Ph.D., Australian Nat. U., Canberra, 1954. Asst., Astron. Obs., Poznan, Poland, 1938-39; research scholar Mt. Stromlo Obs. (formerly Commonwealth Obs.), Australian Nat. U., 1950-55, sci. officer, 1955-57, fellow, 1957—. Research, publs. on computation of definitive orbits of comets, structure of stellar atmospheres, analysis of chem. composition of stars including 1st analysis of star outside Milky Way system; discovery and analysis of star HD 101065; determined radio velocities of stars; photometry of high velocity stars; search for subdwarf stars in so. hemisphere; discovered diurnal effect on structure of terrestrial atmosphere from acceleration of artificial satellites. Home: University House, Canberra, Australia.*

PSCHYREMBEL, Willibald Paul Bruno, German gynecologist; b. Jan. 1, 1901; s. Bruno and Claire Pschyrembel; M.D., Ph.D., U. Berlin (Germany); m. Ingrid Stiefel, May 16, 1960. Asst., chief doctor Municipal Clinic; later dir. clinic for women Berlin-Freidrichshain until 1961; prof. gynecology and obstetrics. Mem. German Soc. Gynecology. Author: Praktische Gynäkologie; Klinisches Wörterbuch. Address: Rüsternallee 45, 1 Berlin 19, West Germany.

PSELLUS, Michael Constantine (or Psellos), Byzantine natural philosopher; probably b. Nicomedia, 1018; ed. Athens; studied law, Constantinople; held judicial post, Philadelphia; became prof. philosophy Acad., Constantinople (under Constantine IX); reorganized faculty of philosophy, circa 1045; state sec., vestarch; various positions in govt. in Constantinople (except for 1054-55), until 1071; became prime minister under Michael VII (1071-78). Author: De omnifaria doctrina; De operatione daemonum; an account of topography of Athens; Categories; De quinque vocibus; commentary on Aristotle; treatise on Plato; Chronographia (history of 976-1077); On Agriculture; other sci. and philos. treatises on math., music, physics, astronomy, alchemy, medicine, other topics. Died 1078-79.

PTITSYN, Boris Vladimrovich, Russian inorganic chemist; b. 1903; grad. Leningrad State U., 1929, Dr. Chem. Scis., 1945. Successively asst., dozent, then chmn. dept. chemistry Naval Med. Acad., Leningrad, 1940-56; chmn. dept. gen. and analytical chemistry Leningrad Tech. Inst. of Food Industries, 1956-59; chmn. dept. complex compounds USSR Acad. Scis., Siberian br. Inst. of Inorganic Chemistry, also chmn. dept. gen. chemistry, Novosibirsk, 1959—. Corr. mem. USSR Acad. Scis., 1960. Coauthor: Oxidation Potential of Dichromate, 1956; Use of an Oxalate-Silver Electrode to Determine the Instability Constants of Complex Oxalates, 1957; Slightly Soluble Compounds of Quadrivalent Uranium Obtained with the Aid of Rongalite, 1957; Strontium Adsorption by Hydroxylapatite Crystals, 1958; A New Method of Obtaining Quadrivalent Uranium, 1958; The Zirconium Citrate Complex, 1959; Determination of the Instability Constants for the Uranyl Oxalate Complexes by the Equilibrium Displacement Method, 1959; Determination of Iodides in Presence of Bromides and Chlorides with the Aid of Radioactive Iodine, 1960; Determination of the Instability Constants of the Complex Oxalates of Magnesium and Uranyl with an Oxylate Silver Electrode, 1959. Contbn. in study of analytical chemistry. Address: Dept. Gen. Chemistry, Novosibirsk State U., Novosibirsk, Siberia.

PTOLEMY, Claudius, Alexandrian astronomer, mathematician, geographer; flourished 2nd century A.D. (probably b. Ptolemais, Egypt). Author: Almagest (Great Collection); Analemma; Planispherum; Optics; Geometry; Geography; others. Collated, expounded math. and astronomical systems developed by Greeks; relying on work of earlier astronomers, especially Hipparchus, and his own work, devised mathematical structures (epicycles and deferents) to predict future positions of planets; in Ptolemaic system, earth is at rest with sun, moon, planets and stars revolving about it; his Almagest claims only to enable prediction of future planetary positions, not necessarily to present a true picture of world; it was major work of math. astronomy until Copernicus; discovered irregularity in moon's motion (evection); his Geography founded on work of Marinus of Tyre; described earth; located places by latitude and longitude; also studied trigonometry, astrology, orthogonal and stereographic projection; attempted a theory of refraction; performed many optical experiments.

PTOLEMY I (or PTOLEMY SOTER), Macedonian noble and king of Egypt; b. Macedonia, 367; possibly s. Lagus; m. Ortacama; m. Eurydice, dau. of Antipater; 2 sons; at least 1 son (Ptolemy Philadelphus) by Berenice, dau. of Lagus; established Ptolemaic rule of nearly 300 years; patron of lit., science, art; founder of library and mus., Alexandria; gen. of Alexander and author narrative of his wars; one of 7 bodyguards of Philip. Died 283 B.C.

PUCHTLER, Holde, physician; b. Kleinlosnitz, Germany, Jan. 1, 1920; d. Gottfried and Gunda (Thoma) P.; student U. Wurzburg Sch. Medicine, 1942, 44, U. Halle Sch. Medicine, 1943; Dr. Med., U. Köln Sch. Medicine, 1951. Came to U.S., 1955, naturalized 1964. Research asso. dept. medicine U. Köln, Germany, 1949-51, resident dept. pathology, 1951-55; Damon Runyon research fellow McGill U., 1955-58, research fellow Nat. Cancer Inst. Can., 1958-59; with Ga. Med. Coll., Augusta, 1959—, asso. research prof. dept. pathology, 1962—, in charge histol., histochem. labs., 1959—. Mem. Biol. Stain Commn., Am. Histochem. Soc., Gesellschaft für Histochemie, Royal Micros. Soc. (London), Internat. Acad. Pathology, Am. Assn. Pathologists and Bacteriologists, Am. Soc. Exptl. Pathology, Am. Soc. Cell Biology, Am. Assn. Anatomists, So. Soc. Anatomists, A.A.A.S. Numerous publs., histochem. research on connective tissues, muscle, amyloid and arteriosclerosis; application of principles of textile and leather dyeing to histol. technique; devel. new staining methods for visible light, fluorescence and polarization microscopy, which are based on modern chem., electron microscopic and x-ray diffraction data but are inexpensive and technically convenient for gen. use in hosp. pathology. Address: Dept. Pathology, Med. Coll Ga., Augusta, Ga. 30902.*

PUCK, Theodore T(homas), Am. biophysicist; b. Chgo., Sept. 24, 1916; s. Joseph and Bessie (Shapiro) Puckowitz; B.S., U. Chgo., 1937, Ph.D., 1940; m. Mary Hill, Apr. 17, 1946; children—Stirling, Jennifer, Laurel. Mem. commn. airborne infections Army Epidemiological Bd., Office Surg. Gen., 1944-46; asst. prof. depts. medicine and biochemistry U. Chgo., 1945-47; sr. fellow Am. Cancer Soc., Cal. Inst. Tech., Pasadena, 1947-48; prof., chmn. dept. biophysics U. Colo. Med. Sch., 1948—; life-time research prof. American Cancer Soc. Recipient Albert Lasker award, 1958; Bordon award med. research, 1959. Fellow American Academy of Arts and Sciences; mem. Am. Chem. Soc., Soc. Exptl. Biology and Medicine, A.A.A.S., Am. Assn. Immunologists, Radiation Research Soc. Biophys. Soc., Genetics Soc. Am., Nat. Acad. Sci., Phi Beta Kappa, Sigma Xi. Research in physical chemistry; aerosol dynamics; bacterial and virus metabolism; human genetics; mammalian cell biochemistry; radiobiology; cancer. Office: 4200 E. Ninth Av., Denver, Colorado, 80220.*

PUCKETT, Allen Emerson, Am. aerospace engr.; b. Springfield, O., July 25, 1919; s. Roswell C. and Catherine C. (Morrill) P.; B.S., Harvard, 1939; M.S., Harvard, 1941; Ph.D., Cal. Inst. Tech., 1949; m. Marilyn Irene McFarland, May 17, 1963; children—Margaret Ann, (by previous marriage) Allen W., Nancy L., Susan E. Research asst. aerodynamics Cal. Inst. Tech., 1941-45, lectr., chief wind tunnel sect. Jet Propulsion Lab., 1945-49; staff Hughes Aircraft Co., Culver City, 1949—, v.p., dir. aerospace engring. div., 1959-61, v.p., group exec. aerospace group, 1962-65, exec. v.p., asst. gen. mgr., 1965—. Cons. govt. agys., pvt. cos.; mem. com. on aerodynamics NACA, 1953-55; mem. sci. adv. com. Army Ballistics Research Lab., 1958—; chmn. research adv. com. on control, guidance and nav., NASA, 1959-64, mem. research and tech. adv. council, 1967—; mem. exec. commn. and full bd. Def. Sci. Bd., 1962-66; mem. army sci. adv. panel Dept. Army, 1965—. Fellow Am. Inst. Aeros. and Astronautics, Inst. Aero. Scis. (Lawrence Sperry award 1949); mem. Nat. Acad. Engring., Am. Rocket Soc., Phi Beta Kappa, Sigma Xi. Author: (with Hans W. Liepman) Introduction to Aerodynamics of a Compressible Fluid, 1947; (with Simon Ramo) Guided Missile Engineering, 1959. Research, publs. on supersonic aerodynamic theory, especially wing theory; devel. techniques in supersonic wind tunnel design and operations; guided missile design and performance prediction. Home: 3572 Terrace View Dr., Encino, Cal. 91316. Office: Centinela and Teale Sts., Culver City, Cal. 90232.*

PUESTOW, Charles Bernard, Am. surgeon; b. Oshkosh, Wis., Jan. 10, 1902; s. Jacob and Adele (Fostre) P.; B.S., U. Wis., 1923; M.D., U. Pa., 1925; M.S., U. Minn., 1931, Ph.D., 1932; m. Lorraine Knowles, Dec. 17, 1949; children—Charles Bernard, Frances Fostre. Fellow surgery Mayo Found., 1928-31; practice medicine, specializing in surgery, Chgo., 1931—; chief surg. service VA Hosp., Hines, Ill., 1946—; sr. surgeon Henrotin Hosp. 1945—; attending surgeon Ill. Research and Ednl. Hosps., Chgo., 1931—; clin. prof. surgery U. Ill. Coll. Medicine, Chgo., 1945—. Author: (with Warren H. Cole) First Aid: Diagnosis and Management, 6th edit., 1965; Surgery of the Biliary Tract, Pancreas, and Spleen, 3d edit., 1964; also numerous articles. Research on physiology of biliary tract, intestinal motility, pancreatitis. Home: 999 Lake Shore Dr., Chgo. 60611. Office: 25 E. Washington St., Chgo. 60602.*

PUESTOW, Karver Louis, Am. physician; b. Oshkosh, Wis., Nov. 1, 1897; s. Jacob E. and Adele (Fosfre) P.; B.S., U. Wis., 1919; M.B., U. Minn., 1921, M.D., 1922; m. Mary Gertrude Sullivan, June 20, 1925; children—Sara A., Mary A., John C. Faculty, U. Wis., Madison, 1922—, prof. medicine, 1945—. Diplomate Am. Bd. Internal Medicine. Fellow A.C.P. (past gov., past regent, past v.p); mem. Am. Therapeutic Soc., Am. Soc. Gastro-intestinal Endoscopy. Contbr. articles to med. jours. Developed techniques for treatment of stenosing diseases of esophagus. Home: 2113 Adams St., Madison, Wis. 53711.*

PUFENDORF, Samuel, Baron von (pseud. Severinus de Monzambo), polit. theorist, jurist; b. Chemnitz, Saxony, Germany, Jan. 8, 1632; s. of Lutheran pastor; student univs. Leipzig and Jena, switching from theology to jurisprudence. Tutor in family of Swedish minister to Denmark, Copenhagen; imprisoned for eight months during hostilities between Sweden and Denmark; while a captive wrote a book dedicated to the elector of Palatinate; named to new professorship of natural and internat. law created by elector at U. Heidelberg (Germany) (1st such professorship in world), 1661-70; prof. U. Lund by invitation of king of Sweden, 1670-77; apptd. royal historgrapher, Stockholm, 1677; histographer to Frederick William elector of Brandenburg, 1686-94. Author: Elementa Jurisprudentiae Universalis, 1661; De Jure Naturaeer Gentium, 1672; De Officio Lominis et Civics, 1673; Einlitung zur Historie der Vornehmsten Reiche und Staaten, 1682; De Habita Religionis Christianae ed Vitam Civilem, 1687. Pioneer in theory of internat. jurisprudence; an influence on the Enlightenment; emphasized natural law as proper basis of all law; regarded internat. law as a form of natural law; argued that peace was more campatible with natural law than was war; conceived a theory of sovereignty; sought to define practical limits of polit. authority. Died Berlin, Germany, Oct. 26, 1694.

PUFF, Heinrich, German chemist; b. Mannheim, Germany, Nov. 1, 1921; s. Heinrich and Margarete (Beck) P.; Diplomchemiker, Heidelberg (Germany) U., 1952; Dr.rer.nat., Kiel (Germany) U., 1956; m. Sofie Reiss, Mar. 28, 1953; children—Heinrich Karl, Carola. Staff, Inst. fur Anorg. Chemie, Kiel, 1953-67, dir. Anorg.-Chem. Inst., Bonn, 1967—; prof. U. Kiel, 1965-67; prof. U. Bonn, 1967—. Mem. Gesellschaft Deutscher Chemiker, Bunsengesellschaft, Chem. Soc. (London). Research and publs. on manganese nitride, ternary compounds of mercury. Home: 7 Endenicher Allee. Office: 168 Meckenheimer Allee, Bonn, Germany.*

PUGH, Emerson Martindale, Am. physicist; b. Ogden, Utah, July 19, 1896; s. William and Hattie (Martindale) P.; B.S. in Elec. Engring., Carnegie Inst. Tech., 1918; M.S. in Physics, U. Pitts., 1927; Ph.D., Cal. Inst. Tech., 1929; m. Ruth Hazel Edgin, Sept. 18, 1920; children—George Edgin, Emerson William. Instr. physics Carnegie Inst. Tech., Pitts., 1920-27; Am. Petroleum Inst. fellow Cal. Inst. Tech., 1927-29, NRC fellow, 1929-30, faculty Carnegie Inst. Tech., 1930—, prof. physics, 1948-65, asso. head dept., 1961-65, chmn. com. on coordinating teaching in all depts., 1937-41, emeritus prof., 1965—. Dir. various research projects USNR, NDRC, OSRD, 1941-62; cons. industry and govt. agys. Mem. Sigma Xi, Tau Beta Pi, Pi Mu Epsilon, Phi Kappa Phi, Delta Tau Delta. Author: (with Emerson William Pugh) Principles of Electricity and Magnetism, 1960; (with George H. Winslow) Physical Measurements, 1965. Identified 2 Hall constants for ferromagnetic materials; produced band theory for transition alloys ordinary Hall constants; theoretical and exptl. explanation of lined cavity charges (shaped charges). Home: 1427 Walnut St., Pitts. 15218.*

PUGH, Hubert Lloyd David, Brit. metall. engr.; b. Port Talbot, Eng., May 11, 1914; s. Jenkin and Margaret (David) P.; B.Sc. with honors, U. Wales, 1935, diploma in edn., 1936; B.Sc. with honors U. London, 1947, D.Sc., 1966; m. Elsie May Durrans, Mar. 7, 1942; children—Carol Gwyneth, Gareth Alan David. With Rd. Research Lab., Dept. Sci. and Indsl. Research, Harmondsworth, Eng., 1938-48, prin. sci. officer, 1948; head plasticity div. Nat. Engring. Lab., Dept. Sci. and Indsl. Research, Ministry of Tech., 1948—. Republic Steel Distinguished vis. prof. Case Inst. Tech., Cleve., 1965-66; Bulleid Meml. Lectr. U. Nottingham, 1965; lectr., cons. Chmn. OECD Com., 1961—. Recipient Telford Premium, Inst. Civil Engrs.; W.H.A. Robertson medal Inst. Metals. Mem. Inst. Physics, Inst. Metallurgists, Inst. Mech. Engrs., Internat. Inst. Prodn. Engring., Sigma Xi. Research, publs. on application of hydrostatic pressure to metal working processes, pioneer work on hydrostatic extrusion and differential pressure extrusion, mech. properties of materials under hydrostatic pressure and at various rates of strain and temperatures, conventional metal working processes; vibrations, heat conduction, impact, wave propagation. Home: 3 Blacklands Rd., East Kilbride, Glasgow, Scotland. Office: Nat. Engring. Lab., East Kilbride, Scotland, U.K.*

PUGH, Lewis Griffith Cresswell Evans, English physiologist; b. Shrewsbury, Eng., Oct. 29, 1909; s. Lewis Evans and Adah E.S. (Chaplin) P.; M.A. in Law, Oxford (Eng.) U., 1931, B.Ch., 1939; m. Josephine Helen Cassel, Sept. 5, 1939; children—David, Simon, Harriet, Oliver. Med. officer Royal Army Med. Corps., 1939-45, physiologist Sch. Mountain Warfare, Lebanon, 1943-44; clin. researcher Dept. Medicine, Brit. Postgrad. Med. Sch., Hammersmith Hosp., London, Eng., 1945-50; mem. div. human physiology Med. Research Council, London, 1951—. Mem. Physiol. Soc., Med. Research Soc., Brit. Med. Assn. Research, publs. on physiol. effects of severe muscular exercise and altitude, thermal conductivity of human skin, muscle and fat; studies of channel swimmers and of fat and thin persons during immersion in cold water, physiol. problems of Himalayan mountaineering, cold stress and solar radiation, adaptability to high altitude life, causes of deaths from exposure among persons engaged in outdoor activities. Office: Med. Research Council Labs., Holly Hill, Hampstead, London, N.W.3, Eng.*

PUGH, Stanley Frederick, English metallurgist; b. Birmingham, Eng., July 7, 1922; s. Tom Stanley and Olive (Weston) P.; B.A., Christ's Coll., Cambridge (Eng.) U., 1943, M.A., 1948; m. Patricia Marie Wareham, July 12, 1952; children—Vyvyan Timothy, Jane Katherine, Owen Wareham. Tech. officer research dept. Imperial Metal Industries, Witton, Eng., 1942-51; staff Brit. Cast Iron Research Assn., Alvechurch, Eng., 1951; staff metallurgy div. Atomic Energy Research Establishment, Harwell, Eng., 1951—, head radiation br., 1960-66, head phys. metallurgy br., 1966—. Lectr. radiation damage materials sci. diploma course Oxford (Eng.) U., 1965—. Mem. Inst. Metals (chmn. Oxford local sect. 1965-67), Inst.. Physics (asso). Research and publs. on devel. powder metallurgy of copper and porous steel for bearings; rationalized annealing kinetics of lightly and heavily rolled copper and aluminum; studies in porosity of cast iron; dicoveries related to thermal cycling growth and irradiation growth and swelling of uranium metal, radiation damage in steele, choice of fuel element materials for power reactors, performance of ceramic nuclear fuels; studies in fracture, superconductivity and corrosion. Home: 89 Bath St., Abingdon, Berkshire, Eng. Office: Atomic Energy Research Establishment, Harwell, Berkshire, Eng.*

PUGLIESE CARRATELLI, Giovanni, Italian archeologist; b. Naples, Apr. 16, 1911. Prof. Greco-Roman history U. Florence; dir. Center for Studies, U. Naples. Mem. 'di Storia Patria' for Calabria and Lucania; corr. mem. German Inst. Archeology, Tosana Acad. Sci. and Letters, Inst. Etruscan Studies Acad. Archeology, Letters and Beaux-Arts Naples. Editor: La parola del passato. Address: via delle Farine 2, Florence, Italy.

PUGSLEY, Sir Alfred Grenvile, Eng. civil engr.; b. Wimbledon, Eng., May 13, 1903; s. Herbert William and Marian (Clifford) P.; B.Sc., London (Eng.) U., 1920, D.Sc., 1938; D.Sc. (hon.), U. Belfast (Ireland), 1965; m. Kathleen Mary Warner, May 24, 1928. Engring. apprentice Woolwich Arsenal, London, 1923-26; tech. officer Royal Airship Works Cardington, Bedfordshire, Eng., 1926-31; sci. staff Royal Aircraft Establishment Farnborough, Hants., 1931-45, head structural and mech. engring. dept., 1941-45; prof. civil engring. U. Bristol (Eng.), 1945-68. Chmn., Aero. Research Council, 1952-57; prof.-vice-chancellor U. Bristol, 1961-64. Named to Order Brit. Empire, 1944; created knight, 1956. Fellow Royal Aero. Soc. (hon.), Royal Soc., 1952; mem. Instn. Civil Engrs., Instn. Structural Engrs., Am. Inst. Civil Engrs. Author: Theory of Suspension Bridges, 1957; The Safety of Structures, 1966; also numerous articles. Research on theory structures especially airplane structures and suspension bridges, theory aeroelasticity, safety of structures. Home: 4 Harley Ct., Clifton, Bristol 8, Eng.*

PUGSLEY, Leonard Irving, Canadian chemist; b. Five Islands, N.S., Can., June 23, 1900; s. Irving Lorne and Elizabeth (Fulmer) P.; Acadia U., 1927; B.A., McGill U., Montreal, Que., Can., M.Sc., 1930, Ph.D., 1932; m. Elaine Bliss MacQuarrie, Oct. 17, 1937; children—William Irving, Janet Elaine. Biochemist, McGill U., 1932-36, Fisheries Research Bd., Prince Rupert, B.C., Can., 1936-39, Dept. Pensions and Nat. Health, Ottawa, Ont., 1939-45; pharmacologist food and drug directorate Dept. Nat. Health and Welfare, Ottawa, 1945-56, asso. dir., 1956-67, cons., since 1968—; professor of pharmacology Queen's University, since 1900—. Mem. Am. Chem. Soc., Am. Soc. Biol. Chemists, Endocrinologist, Chem. Inst. Can. Research, numerous publs. on calcium and phosphorus metabolism, vitamin A and D in fish liver oils, parathyroid hormone, biol. assay estrogens, androgens and insulin. Home: 11 Kippewa Dr., Ottawa. Office: Food and Drug Directorate, Dept. Nat. Health and Welfare, Ottawa, Ont., Can.*

PUISEUX, Pierre Henri, French astronomer; b. Paris, July 20, 1855; s. Victor Alexandre Puiseux; became astronomer Paris obs., 1893; named prof. Paris Faculty Scis., 1897; mem. French Acad. Scis. Research and publs. on kinematics; collaborated on map of moon; studied secular acceleration of moon's motion, also asteroids; determined constant of aberration. Died Fontenay, France, Sept. 28, 1928.

PUISEUX, Victor Alexandre, French astronomer; b. Argenteuil, France, Apr. 16, 1820; student L'École Normale; doctorate math., 1841. a son, Pierre Henri. Became asso. prof. math., 1840; tchr., Rennes, France; went to Besancon, France, 1845; rejoined L'École Normale, 1849; lectr. Sorbonne, Paris, Collège de France; asst. astronomer Paris Obs., 1855-59; became prof. astronomy Sorbonne, 1857. Mem. Bur. Longitudes, French Acad. Scis., 1871. Editor in chief new edit. Laplace's works. Editor, Connaissance des temps; became co-editor Annales scientifiques de l'école normale supérieure, 1864. Introduced new methods in algebraic functions; studied celestial mechanics. Died Frontenay, France, Sept. 9, 1883.

PULFRICH, Carl, German optician, physicist; b. Burscheid, Germany, Sept. 24, 1858. Invented stereo-comparer (named after him) which allows a stereoscopically photographed profile to be reproduced by dots; invented stereorangefinder with fixed mark, 1908; invented (with Zeiss) photometer. Died Timmendorf, Germany, Aug. 24, 1927.

PULLEN, Keats Abbott, Jr., Am. electronic engr.; b. Onawa, Ia., Nov. 12, 1916; s. Keats Abbott and Mabel (Faus) P.; student San Jose State Coll., 1934-36; B.S., Cal. Inst. Tech., 1939; Dr. Engring. Johns Hopkins, 1946; m. Phyllis Kouwenhoven, Jan. 6, 1945; children—Peter Kouwenhoven, Paul Vandercook, Keats Abbott III, Andrew, Victoria. With U.S. Army Ballistic Research Labs., Aberdeen Proving Ground, Md., 1946—, electronic research engr., 1956—; adj. prof. Drexel Inst. Tech. Fellow Am. Inst. E.E., I.E.E.E. (sect. chmn. Balt.); mem. Am. Ordnance Assn., Sigma Xi. Author: Conductance Design of Active Circuits, 1959; Handbook of Transistor Circuit Design, 1961; Theory and Application of Topological and Matrix Methods, 1962; Design of Communication Systems, 1963. Research, publs. on theory and application of active devices in networks, basic limitations to device behavior led to understanding of properties of electron tubes, bipolar transistors and field effect transistors; contbns. to study of nonlinear networks. Home: Route 1, Box 381, Kingsville, Md. 21087. Office: U.S. Army Ballistic Research Labs., Aberdeen Proving Ground, Md. 21005.*

PULLMAN, A. (Mrs. Bernard Pullman), French quantum chemist; b. Nantes, Aug. 26, 1920; D.Sc., U. Paris, 1946; m. Bernard Pullman, Mar. 3, 1945; children—Michel, Bertrand. With Nat. Center Sci. Research, Paris, 1945—, dir. research, 1962—; dir. research Inst. Phys. Chem. Biology, Paris. Author: (with B. Pullman) Les Theories Electroniques de la Chimie Organique, 1952, Cancerisation par les Substances Chimiques et Structure Moleculaire, 1956, Quantum Biochemistry, 1963; also numerous articles. Beginning research in quantum chemistry in France; elaboration of theory of carcinogenic activity of conjugated hydrocarbons; application of method of quantum chemistry to conjugated molecules; with B. Pullman started field of quantum biochemistry. Address: Inst. Biology, 13 Rue Pierre Curie, Paris, France.*

PULLMAN, Bernard, chemist; b. Wloclawek, Poland, Mar. 19, 1919; s. John and Helen (Wloclawska) P.; License-ès-scis., Sorbonne, Paris, France, 1946, Docteur-ès-scis., 1948; m. Alberte Bucher, Mar. 3, 1945; children—Michel, Bertrand. Staff, NRC France, 1946-54, maitre de recherches, 1953-54; faculty Sorbonne, 1954—, prof. quantum chemistry, 1958—, chief dept. quantum biochemistry, 1958—; dir. Institut de Biologie Physico-Chimique, Fondation Edmond de Rothschild, 1963—. Decorated Légion d'honneur, Palmes Académiques; recipient prize French Acad. Scis., 1958, 63; Prix Essec Union Against Cancer, 1956. Mem. Société Chimique de France, Société re Chimie Physique, Société de Chimie Biologique. Author: (with A. Pullman) Quantum Organic Chemistry, 1952, Chemical Carcinogenesis, 1955, Quantum Biochemistry, 1963; also numerous articles. Editor several books. Research on application of quantum-mech. theories to organic chemistry, biochemistry and biophysics; elucidation of electronic structure of fundamental biol. molecules and establishment of its significance for their properties; established relations between electronic structure and carcinogenic activity of molecules. Home: 6 rue Paul Appell, Paris 14e, France.*

PULLMAN, George Mortimer, Am. inventor; b. Brocton, Chautauqua County, N.Y., Mar. 3, 1831; s. James Lewis and Emily Caroline (Minton) P.; m. Harriet Sanger, June 13, 1867; 4 children. Contractor in Chgo., 1855-59, successful in raising level of some bldgs. and streets; store-keeper in mining town, Colo., 1859-63; contracted with Chgo. & Alton R.R. to remodel 2 day coaches into sleeping cars (incorporating his basic idea of upper berth hinged to side of car), 1858, constructed a 3d car, 1859; patentee (with Ben Field) folding berth, 1864, lower berth, 1865; completed 1st car The Pioneer, 1865, constructed other cars modeled after it; organized Pullman Palace Car Co., 1867 (became greatest car bldg. orgn. in world); established 1st mfg. plant, Palmyra, N.Y., then moved to Detroit; built Town of Pullman (Ill.) for accommodation of employees, completed 1881; responsible for combined sleeping and restaurant car, 1867, dining car, 1868, chair car, 1875, vestibule car, 1887; owner Eagleton Wire Works, N.Y.; pres. Met. Elevated R.R., N.Y.; contbd. bequest of $1,200,000 for establishment free manual tng. sch. at Pullman. Died Chgo., Oct. 19, 1897.

PULVARI, Charles Ferencz, elec. engr.; b. Karlsbad, Austria, July 19, 1907; s. Arpad M. and Anna Maria (Staniek) P.; M.S. in Elec. and Mech. Engring., U. Engring., Budapest, Hungary, 1929; m. Jozsa Maria Weber, Sept. 5, 1937; children—Ildiko Maria (Mrs. Frank McDonnell), Gunievere Gizella (Mrs. Joseph Beuchert), Csilla Margit. Came to U.S., 1949, naturalized, 1954. Research asso. Telephone Mfg. Co., Budapest, 1929-33; cons., adviser Hungarian Radio & Communication Co., Budapest, 1933-35; tech. mgr. labs. and factory Hungarian Film Co., 1935-45; founder Pulvari Elec-phys. Lab., 1943-49; cons. engr. Scophyny, Ltd., London, Eng., 1937; lectr. Tech. U., Budapest, 1943; faculty Cath. U., Washington, 1951—, prof. elec. engring., 1955—; head Solid State Electrocristal Corp., Inc., Washington, 1962—. Mem. I.E.E.E. (sr.), N.Y. Acad. Scis., Sigma Xi, Beta Pi. Patentee in field. Contbr. articles on solid state physics and elec. engring. to profl. jours. Discovered 2 crystal families ferrielectric materials and application to invention of type of asso. memory; discovered transpolarizer. Home: 2014 Taylor St. N.E., Washington 20018.*

PUMPHREY, Fred Homer, Am. elec. engr.; b. Dayton, O., July 31, 1898; s. Elgar Grant and Ella (Rhoades) P.; B.A., Ohio State U., 1920, B.E.E., 1921, E.E., 1926, D.Sc., 1962; m. Gladice Eno, June 20, 1922; children—Betty Ruth, Pattie Evelyn (dec.), James Eno, Margaret Elaine (Mrs. Joel N. Tobey). Instr., State U. Ia., 1927-28; prof., head dept. elec. engring. Rutgers U., 1928-45; engr. edn. service div. Gen. Electric Co., 1945-46; prof., head dept. elec. engring. U. Fla., Gainesville, 1946-58; Dean engring. Auburn (Ala.) U., 1958—. Head dept. elec. engring. Bihar (India) Inst. Tech., Tech. Cooperation Mission, 1955-56; cons. engring. summer sch. for So. India, P.S.G. Coll., Coimbatore, 1964. Fellow I.E.E.E.; mem. Nat., Ala. socs. profl. engrs., Am. Soc. for Engring. Edn. (project operating unit com. 1963—), Phi Beta Kappa, Sigma Xi (pres. Fla. chpt. 1953-54), Tau Beta Pi, Eta Kappa Nu, Phi Eta Sigma, Omicron Delta Kappa, Theta Xi. Author: (with others) Fundamentals of Radio, 1942, rev. edit., 1959; Electrical Engineering, Essential Theory and Typical Applications, 1946, 2d edit., 1953; Fundamentals of Electrical Engineering, 1951, 2d edit., 1959. Research on fundamentals of electronics and radio. Home: 706 Cary Dr., Auburn, Ala. 36830.

PUMPHREY, William Idwal, Brit. metallurgist; b. Flint, N. Wales, Feb. 10, 1922; s. Percy William and Mildred (Williams) P.; B.Sc., U. Birmingham (Eng.), 1942, M.Sc., 1944, Ph.D., 1948, D.Sc., 1956; m. Norah Marian Suckling, May 6, 1943; children—David Geraint, Martin Lloyd, Ann Jennifer. Jr. sci. officer Nat. Phys. Lab., Teddington, 1942-46; leader Aluminum Devel. Assn. Welding research team U. Birmingham, 1946-49; Harkness fellow Commonwealth Fund, U.S., 1949-51; research mgr. Murex Welding Processes Ltd., Waltham Cross, Eng., 1951—, dir., 1965—, mng. dir., 1967—; dir. Murex Positioning Equipment Ltd., Crawley, Eng. Mem. Inter-Services Metall. Research Council, 1964—. Liveryman, Worshipful Co. Blacksmiths, 1952, Freeman of City of London, 1952. Fellow Chem. Soc., Instn. Metallurgists (mem. council 1965—); mem. Brit. Welding Research Assn. (mem. research bd. 1961—, Iron and Steel Inst., Inst. Metals, Inst. Welding, Am. Welding Soc. Author: (with L. Aitchison) Engineering Steels, 1953; Researches into the Use of Aluminum and its Alloys, 1955; Using Steel Wisely, 1958; also articles. Research on structural changes occurring in steels during heating and cooling, basic metallurgy of welding, susceptibility of aluminum alloys to cracking during casting and welding; devised practical methods for prevention cracking during aluminum alloys. Home: The Croft, Low Hill Rd., Roydon, Harlow, Essex, Eng. Office: Murex Welding Processes Ltd., Waltham Cross, Herts., Eng.*

PUNGOR, Erno, Hungarian chemist; b. Vasszény, Hungary, Oct. 30, 1923; s. József and Franciska (Faller) P.; student U. Pázmany Péter, Budapest, 1943-48; Dr.rer.nat., U. Eötvös, Budapest, 1949, Cand.Sci., 1952, D.Sc., 1956; m. Erzsbet Lang, Oct. 27, 1950; children—Ernö, András, Katalin. With Inst. Inorganic and Analytical Chemistry, Eötvös U., Budapest, 1948-62, sr. lectr., 53-62; prof. analytical chemistry U. Veszprém (Hungary), 1962—, prorector, 1966—, dir. Inst. Analytical Chemistry, 1962—; faculty instrumental analysis U. Freiberg (Germany), 1965. Recipient Medal of Than, 1964, Medal of Hanus, 1966. Mem. Soc. Hungarian Chemists, Czechoslovak Chem. Soc. (hon.). Author: Theoretical Backgrounds of Flame Photometry, 1962; also numerous articles. Research in instrumental analysis, ion selective membrane electrodes, including devel. of high selective membrane electrodes; developed new methods in oscillometry, flame photometry, polarography, potentiometry. Home: 1 Bosnyák, Budapest, Hungary. Office: 8 Schönherz Veszprém, Hungary.

PUNNETT, Reginald Crundall, English geneticist; b. Tonbridge, Kent, Eng., June 20, 1875; s. George and Emily (Crundall) P.; B.A. with 1st class honors, Caius Coll., Cambridge, Eng., 1898, M.A., 1902. Faculty, St. Andrews U., 1899-1902; demonstrator zoology Cambridge U., 1902-05, Balfour student, 1905-09, supt. Mus. Zoology, 1909, prof. biology, 1910-12, prof. genetics, 1912-40, dir. Poultry Inst., 1921-40. Recipient Darwin medal, 1922. Fellow Royal Soc., 1912; hon. mem. Genetics Soc. Gt. Britain, Genetics Soc. Japan, Poultry Sci. Assn. Am. Author: Mendelism, 1905; Mimicry in Butterflies, 1915; Heredity in Poultry, 1923; also numerous articles. Co-discoverer linkage of characters in heredity of factorial interaction and phenomenon of sex-linkage; devised sex-linked method poultry breeding; created first auto-sexing breed; research on genetics rabbits and sweet peas. Office: Bilbrook Lodge, Nr. Minehcad, Somerset, Eng.*

PUPIN, Michael Idvorsky, physicist, inventor; b. Idvor, Banat, Yugoslavia, Oct. 4, 1858; s. Constantine and Olympiada P.; A.B., Columbia, 1883; Sc.D.,

1904; Ph.D., U. of Berlin, 1889; LL.D., Johns Hopkins U.; m. Sarah Katharine Jackson, 1888 (dec). Asst. teacher elec. engring., 1889-90, instr. math. physics, 1890-92, adj. prof. mechanics, 1892-1901, prof. electro-mechanics, 1901-31, Columbia, prof. emeritus. Mem. various societies. Awarded Washington medal (engring.), 1928; Pulitzer Prize, 1924. Author: Die Wirkung der Vakuumentladungen aufeinander, 1892; Elektrische Entladungen durch verdünnte Gase und koronaartige Entladungen, 1892; Über elektrische Oszillationen von geringer Frequenz und ihre Resonanz, 1893; Thermodynamics of Reversible Cycles in Gases and Saturated Vapors, 1894; Resonanz-Analyse automatische Quecksilberpumpe, 1895; Das Gesetz des elektromagnetischen Inductionsflusses, 1895; Versuche mit Kathodenstrahlen, 1896; From Immigrant to Inventor, 1924; The New Reformation, 1927; The Romance of the Machine, 1930. Applied automatic induction coils to decreasing capacity of telegraph for use in shallow depths; invented Pupin bobbins for use in telephone lines; studied electric phenomena in rarified gases and electric resonators; devised X-ray sensitive flourescent screen; improved radio transmitters; worked on gaseous discharge; devel. higher efficiency transmission lines by distbd. inductance. Died New York, N.Y., Mar. 12, 1935.

PUPPINI, Umberto, Italian hydraulic engr.; b. Bologna, Italy, 1884. Tchr. hydraulics U. Bologna, dir., then rector Engring. Faculty. Author: I fondamenti scientifici dell'idraulica, 1912; Idraulica, 1947. Research on currents of infiltration; discovered principle of reciprocity. Died Bologna, 1946.

PURBACH (Peuerbach, Peurbach), Georg von, Austrian mathematician, astronomer; b. Purbach, Upper Austria, May 30, 1423; studied under Nicholas of Cusa, gov. of Rome; learned Greek from Cardinal Bessarion so that he could read Ptolemy; lectr. math., Ferrara, Bologna, Padua (all Italy); prof. math. and astronomy, Vienna, Austria; taught Johann Müller (Regiomontanus). Studied mathematics and astronomy; began revision of Ptolemy's Almagest (completed by Regiomontanus); compiled table of sines; one of 1st Westerners to use sines in trigonometry. Died Vienna, Austria, Apr. 8, 1461.

PURCELL, Edward Mills, Am. physicist; b. Taylorville, Ill., Aug. 30, 1912; s. Edward A. and Mary Elizabeth (Mills) P.; B.S., Purdue University, 1933, D. Engring. (honorary), 1953; International Exchange student Technische Hochschule, Karlsruhe, 1933-34; A.M., Harvard, 1935, Ph.D., 1938; m. Beth C. Busser, Jan. 22, 1937; children—Dennis W., Frank B. Instr. in physics Harvard, 1938-40, asso. prof., 1946-49, prof. physics, 1949-58, Donner professor of science, 1958-60, Gerhard Gade University professor, 1960—; sr. fellow Soc. of Fellows 1949—; group leader Fundamental Developments Group, Radiation Lab., Mass. Inst. Tech., 1941-45. Member President's Science Advisory Com., 1957-60, 62—. Mem. sci. adv. bd. USAF, 1947, 48, 53-57. Co-winner (with Felix Bloch) Nobel Prize in Physics, 1952. Mem. American Philosophical Society, also National Academy of Sci. Mem. Am. Phys. Soc., Am. Acad. Arts and Scis., Sigma Xi. Contbr. author Radiation Lab. series, 1949, 50. Contbr. sci. papers on microwave phenomena, nuclear magnetism, radiofrequency spectroscopy. Developed method of nuclear resonance absorption for measurement of magnetic moments of nuclei and particles (method further proved valuable for investigation of molecular structures and chem. interactions of crystals and organic substances); research on relaxation phenomena, molecular structure, measurement of atomic constants, and low-temperature nuclear magnetic behavior. Home: 5 Wright St., Cambridge 38, Mass.

PURDIE, Thomas, Scottish chemist; b. Biggor, Scotland, Jan. 27, 1843; B.Sc., U. London, (Eng.), 1879; Dr.Phil., Wurzburg, Germany, 1881; LL.D., Aberdeen, Scotland, 1894. Became prof. chemistry St. Andrews, Scotland, 1903; asst. Sch. Mines, London, Eng. Author numerous works. Research on etherial and metallic salts and acids (optically active), alcohol. Died 1916.

PURDY, Laurence Henry, Jr., Am. plant pathologist; b. Miami, Ariz., Sept. 28, 1926; s. Laurence Henry and Winnie (Gibson) P.; student Mont. U., 1944-45, 46-47; B.S. in Chemistry, San Diego State Coll., 1949; Ph.D., U. Cal. at Davis, 1954; m. Barbara Ann Pershal, Feb. 14, 1948; children—Cynthia D., Laurence J., Timothy C., Paula E. Plant pathologist wheat investigations, cereal crops research br., crops research div. Agrl. Research Service, U.S. Dept. Agr., Pullman, Wash., 1953—. Recipient U.S. Dept. Agr. unit award for superior service, 1960. Mem. Am. Phytopath. Soc., Sigma Xi. Research, numerous publs. on hexachlorobenzene as seed-treatment fungicide for control wheat bunt; determined relationship between soil moisture and soil temperature in infection process of wheat bunt and flag smut; established presence of pathogenic races of stripe rust in natural population of fungus Puccinia striiformis Pacific N.W. Home: 709 State St. Office: Johnson Hall, Wash. State U., Pullman, Wash. 99163.*

PURDY, William Crossley, Am. chemist; b. Bklyn., Sept. 14, 1930; s. John Earl and Virginia (Clark) P.; B.A., Amherst Coll., 1951; Ph.D., Mass. Inst. Tech., 1955; m. Myrna Mae Moman, June 17, 1953; children—Robert Bruce (dec.), Richard Scott, Lisa Patrice, Diana Lori. Instr., U. Conn., Storrs, 1955-58; faculty U. Md., College Park, 1958—, prof. chemistry, 1964—; vis. prof. Institut für Ernährungswissenschaft, Justus Liebig-Universität, Giessen, Germany, 1965-66. Cons., Surg. Gen., U.S. Army, 1959—; sci. adviser Balt. dist. Food and Drug Adminstrn. Mem. Am. Chem. Soc., Soc. for Analytical Chemistry (London, Eng.), Sigma Xi. Author: Electroanalytical Methods in Biochemistry, 1965; also numerous articles. Research on application electroanalytical methods to biochem. systems, separations stereoisomers, nonaqueous titrations, complexes of transition metals. Home: 15211 Baughman Dr., Silver Spring, Md. 20906. Office: Dept. Chemistry, U. Md., College Park, Md. 20740.*

PURI, Balwant R., phys. chemist; b. Gujrat, West Pakistan, Feb. 6, 1909; s. Mathra Das and Ishwara (Sobti) P.; B.Sc., M.Sc., Forman Christian Coll., 1931; Ph.D., Punjab U., 1935; m. Lalita Puri, May 22, 1935; children—Sudarshan, Vijay. Asst. phys. chemist Punjab Irrigation Research, Lahore, 1935-42; lectr. phys. chemistry Punjab U., 1942-54, reader, 1954-62, prof., 1962—. Fellow Royal Inst. Chemistry (London), Chem. Soc. (London), Internat. Union Pure and Applied Chemistry, Internat. Soc. Soil Sci., Indian Chem. Soc., others. Author: (with L.R. Sherma) Principles of Physical Chemistry, 1963, Principles of Inorganic Chemistry, 1963; also over 200 articles. Editorial adv. bd. Internat. Jour. Carbons. Research on colloid and surface chemistry, absorption at solid/gas interface, Microcrystalline carbons and formation of carbon-oxygen, carbon-halogens and carbon-sulpher solid compounds. Home F/8, Sector 14, Chandigerh, Punjab, India.*

PURI, Harbans Singh, geologist; b. Kallar, Pakistan, Aug. 15, 1925; s. Hara Singh and Laj (Kochhar) P.; B.S., Lucknow U., 1945; Ph.D., La. State U., 1953; m. Martha Walker, Dec. 26, 1953; children—Edwin Everett, Nancy Gian, Jeannette Lucille. Came to U.S., 1949, naturalized, 1958. Geologist, Nat. Cement Mines and Industries, 1945; paleontologist Burmah Oil Co. Ltd., 1945-49; research cons. Fla. Geol. Survey, 1951-52, micropaleontologist, 1953-59, paleontologist in charge sect., Tallahassee, 1959—. Participating faculty in geology Fla. State U., 1956-65, investigator Oceanographic Inst., 1956—; cons. geologist. Fellow Geol. Soc. Am.; mem. Southeastern Geol. Soc. (sec.-treas. 1953-54, pres. 1959-60), Paleontol. Soc. India (fgn. sec. 1956—), Soc. Econ. Paleontologists and Mineralogists, Am. Assn. Petroleum Geologists, Am. Inst. Profl. Geologists, Sigma Xi. Author: Contributions to the Study of the Miocene of the Florida Panhandle, 1954; Stratigraphy and Zonation of the Ocala Group, 1957; (with R.O. Vernon) Summary of the Geology of Florida and a Guidebook to the Classic Exposures, 1964; also numerous articles. Editor: Ostracods as Ecological and Paleoecological Indicators, 1964. Research in structure, stratigraphy and econ. geology of Fla.; ecology, paleoecology and systematics of Ostracoda and Foraminifera; hydrography; sedimentation and environments of Gulf of Naples. Home: 305 Saratoga St., Tallahassee 32302. Office: 903 W. Tennessee St., Tallahassee, Fla. 32302.*

PURICA, Ionel Ion, Rumanian engr., physicist; b. Naipu, Rumania, Mar. 12, 1925; s. Ioan I. and Sofia (Necsulescu) P.; Ing.Dipl., Poly. Inst., Bucharest, Rumania, 1949; D.R.D. in Reactor Physics, U. Moscow, 1955, Dr. Engr., 1967; m. Elena Camarzan, 1950; 1 son, Ionut. With Ministry Power and Tech., 1949-55; chief reactor physics and tech. dept. Inst. for Atomic Physics, Acad. Socialist Republic of Rumania, Bucharest, 1955—. Mem. Commn. for Energy, Nat. Council Sci. Research Rumania, 1965—. Recipient Physics prize Acad. Socialist Republic Rumania, 1964. Mem. Brit. Nuclear Soc. Author: Neutron Gas Physics, 1967. Research, publs. on nuclear reactors, dynamic, theoretical and exptl. neutron thermalization by frequency characteristic method. Home: 4 Albotei. R. 30 Dec. Office: Inst. for Atomic Physics, P.O. Box 35, Bucharest, Rumania.*

PURKINJE, Jan Evangelista (or Jan Purkyne), Bohemian physiologist; b. Libochovice, Bohemia (now Czechoslovakia), Dec. 17, 1787; while preparing for priesthood, decided medicine was his true vocation; M.D., U. Prague, 1818. Prof. physiology, U. Breslau (now Wroclaw, Poland), 1823-50; prof. physiology, Charles U., Prague, 1850; representative in Parliament, 1861. Mem. French Acad. Sci., Order of Leopold, 1868. Author: Comentatio de examine physiologico organi visus et systematis cutanei, 1823; Beiträge zur Kenntnisse des Sehens, 1823; Symbolae ad ovi avium historium ante incubationem, 1830. A famous microscopist, discovered skin's sweat glands and excretory ducts; neurons in cortex of cerebellum (Purkinje cells); fiber network in cardiac muscle (Purkinje network, system, or tissue); introduced term protoplasm for formative material of young animal embryos, 1839; gave name "germinal vesicle" to egg nucleus, 1825; studied subjective visual figures and recurrent images (images on retina from shadows of blood vessels called Purkinje images); observed Purkinje phenomenon, that fields of different color and equal brightness become unequally bright when illumination is decreased in brightness; demonstrated importance of fingerprints and devised 1st classification, 1823; one of 1st to use microtome; translated poems of Goethe and Schiller into Czech; ardent propogandist for Czech nationalism. Died Prague, Bohemia, July 28, 1869.

PURMANN, Matthäus Gottfried, German surgeon; b. Lüben, Silesia, 1648; practice surgery Glogau (now Poland), Frankfort, Germany. Breslau (now Wroclaw, Poland). Studied exptl. surgery; performed 1st blood transfusion from animal to man, 1668. Died Breslau, 1711.

PUSHKOV, Nikolai Vasilievich, Russian geophysicist; b. 1903; student Moscow U., 1926-30, also Moscow Hydrometer Inst.; candidate's degree, physico-math. scis., 1934; lectr. Main Geo-Phys. Obs.; with magnetic obs., Pavlovsk, 1934—, dir., 1937—, continued as dir. Inst. Earth Magnetism (new name after reorgn.), Acad. Scis. USSR; head delegation to Internat. Geophysics Com., 1962. Recipient Lenin prize, 1960, for participation in discovery and study of magnetic pole of earth and moon.

PUSTOVALOV, Leonid Vasilevich, Russian petrographer; b. Aug. 8, 1902; grad. Moscow U., 1925. Prof., Moscow Petroleum Inst., 1934—; dir. Lab. Sedimentary Minerals; head dept. petrography of sedimentary Rocks Inst. Geol. Sci., USSR Acad. Scis., 1943-55, dep. chmn. Council for Study Prodn. Resources, 1953—. Recipient Stalin prize, 1941. Mem. USSR Acad. Scis. (corr.). Author: The Petrography of Sedimentary Rocks, parts 1-2, 1940; Basic Principles of Classifying Sedimentary Rocks, 1962. Research on geochem. facies, 1933, petrography, geochemistry of sedimentary rocks; founder hypothesis of differentiation of matter in zone of sedimentary accumulation, hypothesis of periodicity in formation of sedimentary rocks and minerals of sedimentary origin. Home: ulitsa Chaplygina la. Office: Moscow Petroleum Inst., Leninsky prospect 6, Moscow, USSR.

PUTMAN, John Laban, English physicist; b. Stockbridge, Eng., Jan. 15, 1919; s. John William and Margaret (Holmes) P.; B.A., Wadham Coll., Oxford U., 1940, M.A., 1944, D.Sc., 1958; m. Rhoda Pirie, Aug. 8, 1942; children—Lorna, Sylvia, Roderick. Staff, Telecommunications Research Establishment, Swanage, also Malvern, Eng., 1940-45, Atomic Energy/TRE, Malvern, 1946-47; with Atomic Energy Research Establishment, Harwell, 1947-60, staff Wantage (Eng.) Research Lab., 1960—, head isotope br. Fellow Inst. Physics. Author: (with W.J. Whitehouse) Radioactive Isotopes, 1953; Isotopes, 1960. Editor: Internat. Jour. Applied Radiation and Isotopes, 1956—. Research on properties and measurement of radioactive materials and radiations and their applications to industry and pub. works. Home: Tigh Na Gaoth, 3 Capel Close, Oxford, Eng. Office: Wantage Research Lab., Atomic Energy Research Establishment, Wantage, Berks., Eng.*

PUTNAM, Alfred Lunt, Am. mathematician; b. Dunkirk, N.Y., Mar. 10, 1916; B.S., Hamilton Coll., 1938; M.A., Harvard, 1939, Ph.D., 1942; m. Maryann Garbarino, April 29, 1966. Instructor in math. Yale, 1942-45; asst. prof. Univ. Chgo., 1945-50; assoc. prof., 1950-63; prof. 1963—. Mem. A.A.A.S., Am. Math. Soc., Math. Assn. Am., Phi Beta Kappa, Sigma Xi. Research in abstract algebra; history of math. Home: 5550 S. Dorchester Av., Chgo. 60637. Office: Dept. Mathematics, University of Chicago, Chicago, Ill. 60637.

PUTNAM, Calvin Richard, Am. mathematician; b. Balt., May 25, 1924; s. Calvin Sibley and Miriam May (Kolbe) P.; A.B., Johns Hopkins, 1944, M.A., 1946, Ph.D., 1948; m. Emogene Mae Ferrell, June 14, 1952; children—Bryan F., James E., Vicky L. Faculty, Johns Hopkins, 1948-50; mem. Inst. for Advanced Study, Princeton, N.J., 1950-51; faculty Purdue U., Lafayette, Ind., 1951—, prof. math., 1959—. Mem. Am. Math. Soc., Sigma Xi, Phi Beta Kappa. Research, publs. in singular boundary value problems, differential equations, Hilbert space. Home: 1715 Klondike Rd., West Lafayette, Ind. 47906.*

PUTNAM, Frederic Ward, Am. anthropologist, zoologist; b. Salem, Mass., Apr. 16, 1839; s. Ebenezer and Elizabeth (Appleton) P.; B.S., Harvard, 1862; (A.M., Williams, 1868; Sc.D., U. of Pa., 1894); m. Adelade Martha Edmands, 1864 (died 1879), three children; 2d, Esther Orne Clarke, 1882. Curator ornithology, 1856-64, curator vertebrata, 1864-66, supt. of mus., 1866-73, v.p., 1871-94; Essex Inst., and of East India Marine Soc., 1867-69, dir. of mus. Peabody Acad. Sciences, Salem, 1869-73; asst. in ichthyology, Mus. of Comparative Zoology, 1857-64 and 1876-78; asst. Geol. Survey of Ky., 1874, survey West of 100th meridian, U.S. engrs., 1876-79; prof. Am. archeology and ethnology, Harvard, 1886-1909, prof. emeritus, 1910; curator Peabody Museum of Harvard Univ., 1874-1909, honorary curator, 1909-13, hon. director in charge, 1913; prof. anthropology and dir. Anthrop. Mus. U. of Calif., 1903-09, prof. emeritus, 1909; state commr. inland fisheries, 1882-89; chief dept. ethnology, Chicago Expn., 1891-94; curator anthropology, Am. Mus. Natural History, 1894-1903. Decorated by French Govt. with Cross of Legion of Honor; awarded Drexel gold medal for archeol. research. Fellow Am. Acad. Arts and Sciences; mem. numerous Am. and foreign

Societies. Was originator and editor of Naturalists' Directory, 1865. One of founders of The American Naturalist, 1868. Has published over 400 papers on zoology and anthropology. From 1870 engaged in researches and explorations in Am. archeology; conducted archeol. field-work on Atlantic Coast, in Ohio Valley and Am. Southwest; responsible for acceptance and growth of archeology as sci. in U.S. Died Cambridge, Mass., Aug. 14, 1915.

PUTNAM, James Jackson, Am. neurologist; b. Boston, Mass., Oct. 3, 1846; s. Charles Gideon and Elizabeth Cabot (Jackson) P.; A.B., Harvard, 1866, M.D., 1870; postgrad., Leipzig, Vienna, London; m. Marian Cabot, Feb. 15, 1886. Lectr. Harvard, 1872-75; clin. instr. diseases nervous system, 1875-85, instr., 1885-93, prof., 1893-1912, prof. emeritus, 1912; neurologist Mass. Gen. Hosp., Boston, 1874-1907, founder one of 1st neurol. clinics in U. S., 1872, cons., from 1909. Author: Memoirs of Dr. James Jackson, His Father and His Brothers, 1905; Human Motives, 1915. Described dorsolateral sclerosis found in pernicious anemia and other conditions (Putnam-Dana syndrome), 1891. Died Nov. 4, 1918.

PUTNAM, Tracy Jackson, Am. neurologist; b. Boston, Mass., Apr. 14, 1894; A.B., Harvard, 1915, M.D., 1920; Instr. pathology Johns Hopkins Med. Sch., 1920-21; Moseley traveling fellow, Binnengasthuis, Amsterdam, 1925; Austing teaching fellow in surgery, Cabot fellow Harvard Med. Sch., 1926-27, asst., 1927-29, fellow neuropathology, 1929-31, faculty, 1931-39, prof. neurology, 1934-39, dir. service neurosurgery and neurology, neurol. inst., 1939-47; prof. neurology and neurosurgery Coll. Phys. and Surg., Columbia, 1939-47; dir. service neurosurgery Cedars of Lebanon Hosp., Los Angeles, 1947—. Pres., Fund for Research, Inc.; mem. com. psychiatry and neurology, com. neurosurgery NRC; dir. Health Ins. Plan, N.Y. Mem. A.A.A.S., Harvey Cushing Soc., Soc. Clin. Investigators, Soc. Neurol. Surgery, Epilepsy Soc., Soc. Exptl. Biology, Psychosomatic Soc. (past pres.), Soc. Neuropathology (past pres.), A.M.A., Neurol. Assn., Assn. Research Nervous and Mental Diseases, Nat. Epilepsy League, Acad. Cerebral Palsy, N.Y. Acad. Medicine. Research on anatomy of visual system, physiology of hypophysis, pathology and treatment of multiple sclerosis and of dyskisias, treatment of hydrocephalus; isolated (with H.H. Merritt) sodium dilantin from phenobarbitol for use as corrective for epilepsy, other nervous disorders. Address: 450 N. Bedford Dr., Beverly Hills, Cal. 90210.

PUTNINS, Pauls, meteorologist; b. Riga, Latvia, July 14, 1903; s. Heinrich P. and Ella (Faulbaums) P.; mag.math. U. Latvia, Riga, 1932, venia legendi, 1939; m. Aina M. Dzenis, Dec. 16, 1939; 1 son, Michael. Came to U.S., 1950, naturalized, 1956. With U. Latvia, Riga, 1926-44, asso. prof., chief meteorol. sect. Inst. Geophysics and Meteorology, chief Meteorol. Obs., 1940-44; research meteorologist Marineobservatorium Greifswald, Germany, 1944-45; asso. prof. Baltic U., Hamburg, Germany, 1946-50; meteorologist W.E. Howell Assos., Cambridge, Mass., 1951-54; research fellow Harvard, Blue Hill Meteorol. Obs., 1954-56; chief fgn. sect. Office Climatology, U.S. Weather Bur., 1956-61, chief investigation br., 1961-64, prin. project scientist synoptic climatology project Environmental Data Service, Environmental Sci. Services Adminstrn., Rockville, Md., 1964—. Fellow Washington Acad. Scis.; mem. Am. Meteorol. Soc., Am. Geophys. Union. Contbg. author: World Survey of Climatology, 1966-67. Research, publs. on halo appearances, some atmospheric circulation problems, climate of polar regions. Home: 10809 Georgia Av., Wheaton, Md. 20902. Office: Environmental Sci. Services Adminstrn., Environmental Data Service, Washington Sci. Center, Rockville, Md. 20852.*

PUVIS, Marc-Antoine, French agronomist; b. Cuiseaux, France, Oct. 26, 1776; ed. Ecole Polytechnique; mem. French Acad. Scis. Author: De l'emploi de la chaux en agriculture, 1878; also began a history of agrl. progress in 19th century France. Studied culture of silk worms; introduced use of lime to improve soils with slate content. Died July 30, 1851.

PYE, Sir David Randall, English engr.; b. Apr. 29, 1886; s. William A. Pye; ed. Trinity Coll., Cambridge; M.A., Sc.D.; m. Virginia Frances Kennedy, 1926; 2 sons, 1 dau. Engr. with Mather and Platt; lectr. engring. science Oxford U., 1909-14, also fellow New Coll.; fellow Trinity Coll., Cambridge, lectr. engring., 1919-25; dept. dir. sci. research Air Ministry, 1925-37, dir., 1937-43; mem. Aero. Research Council, 1943-46; provost Univ. Coll., London, 1943-51, also hon. fellow; Wilbur Wright lectr., 1936. Fellow Royal Soc., 1937, Royal Aero. Soc., Inst. Aero. Scis. (hon. U. S.); mem. Instn. Mech. Engrs. Author: The Internal Combustion Engine, 2 vols.; Heat and Energy; George Leigh Mallory (memoir); also articles. Editor: The Mummers Play (with memoir of R.J.E. Tiddy). Died Feb. 20, 1960.

PYE, Lucian Wilmot, polit. scientist; b. Fenchow, Shansi Province, China, Oct. 21, 1921; s. Watts Orson and Gertrude (Chaney) P.; B.A., Carleton Coll., 1943; M.A., Yale, 1948, Ph.D., 1951; m. Mary Toombs Waddill, Dec. 24, 1944; children—Evelyn, Christopher, Virginia. Faculty, Washington U., St. Louis, 1949-52, asst. prof., 1951-52; research asso. internat. relations, Yale, 1951-52; research asso. Center for Internat. Studies, Princeton, 1952-56; faculty Mass. Inst. Tech., Cambridge, 1956—, prof. polit. sci., 1960—. Mem. adv. com. to administr., AID, 1961—; mem. Ford Area Tng. Program, 1965—; cons. Dept. State, 1966—, RAND Corp., 1964—. Fellow Am. Acad. Arts and Scis.; mem. Social Sci. Research Council (chmn. com. on comparative politics, 1962—), Am. Polit. Sci. Assn. (mem. council 1963-65), Assn. Asian Studies (mem. S.E. Asia com.), Asia Soc. (bd. mem.), Council on Fgn. Relations (bd. mem. 1966—), Phi Beta Kappa. Center for Advanced studies in Behavioral Scis. fellow, 1963-64. Author: Guerrilla Communism in Malaya, 1956; Politics, Personality and Nation Building, 1962; Aspects of Political Development, 1966; Spirit of Chinese politics, 1968; also chpts. in books. Editor: Communications and Political Development, 1963; (with Sidney Verba) Political Culture and Political Development, 1965. Research in comparative polit. behavior and polit. psychology, especially in transitional societies of Asia. Home: 72 Fletcher Rd., Belmont, Mass. 02178. Office: Mass. Inst. Tech., Cambridge, Mass. 02139.*

PYE-SMITH, Philip Henry, English physician; s. Ebenezer Pye-Smith; ed. Univ. Coll., London, U. London, Guy's Hosp.; B.A., M.D.; M.D. (hon.), Dublin; m. Gertrude Foulger; 1 son. Cons. physician Guy's Hosp.; vice-chancellor U. London; joint-rep. Brit. govt. at Internat. Congress on Prevention Tb, Berlin, 1899; gov. Shrewsbury and St. Paul's schs. Fellow Royal Soc., 1886, Med. Soc. Phila. (hon.), sr. censor Royal Coll. Physicians. Died May 3, 1914.

PYLARINO, Giacomo (or James), Italian physician; b. 1659; inoculated 3 children with smallpox virus, Constantinople, 1701, thus considered by some the 1st immunologist. Died 1715.

PYM, Sir William, Brit. mil. surgeon; b. Edinburgh, Scotland, 1772; s. Joseph Pym; ed. U. Edinburgh. Med. officer, W.I., 1794-96; became insp. gen. army hosps., 1816, supt.-gen. quarantine, 1826; went to Gibraltar to control and supervise quarantine during outbreak of yellow fever, 1828. Named chmn. Central Bd. Health, Eng., 1832. Author: Observations upon Bulam Fever (contained 1st accurate description of yellow fever), 1815; Observations upon Bulam, Vomito-negro, or yellow fever, 1848. Died Mar. 18, 1861.

PYRRHON, Greek natural philosopher; b. Elis, Greece, circa 360 B.C.; s. Plistarchos; studied under Anaxarchos. Began as painter, later became philosopher; joined (with Anaxarchos) Alexander's expdn. to Persia, 334 B.C., and to India, 329-26 B.C.; founder Skeptical (or Pyrrhonian) Sch.; late in life taught philosophy of systematic skepticism in Elis. Believed happiness to be founded on suspension of judgement because certainty is unattainable; earliest Greek thinker influenced by India; supported atomic theory of Democritos and Anaxarchos. Died Elis, circa 270.

PYTHAGORAS OF SAMOS, Greek philosopher, mathematician; b. Samos, flourished 532 B.C.; s. Mnesarchos of Samos; became pupil of Pherekydes (soothsayer) on island of Syros; said to have been educated in Egypt and Babylonia; m. Theano of Croton; children—Arimnestus, Telauges, Damo. Left no writings, hard to determine what done by Pythagoras himself and what done by his followers; established secret brotherhood and sch. philosophy and religion at Croton (Greek colony in southern Italy), circa 531 B.C.; best known for theorem (Pythagorean theorem) on squares on sides of right-angled triangle; also said to have proved sum of 3 angles of triangle is equal to 2 right angles, and all angles inscribed in semi-circle are right angles; credited with discovery (probably by measuring lengths of string on a monochord) that chief musical intervals are expressible in simple numerical ratios between 1st 4 integers; held equation of things with numbers; attempted to interpret world through numbers; is said to have classified numbers into odd and even, to have defined perfect, defective, excessive and amicable numbers, to have defined arithmetic, geometric and harmonic proportions, introduced terms square and cube, shown incommensurability of diagonal and side of square (therby beginning theory of irrational quantities), constructed theory of 5 regular solids, and to have stressed deductive methods in geometry; invented (according to Aristotle) astron. system presupposing central fire around which celestial bodies circle including sun, earth and counter-earth; said to have invented harmony of spheres and to have recognized that morning star and evening star are in fact one and same star (Venus); held doctrine of metempsychosis and kinship of all living things; was vegetarian; also enjoined abstinence from beans; taught earthly life is only purification of soul.

PYTHEAS, Greek navigator, geographer; b. circa 310-306 B.C.; Supposedly sailed from Cadiz, past Cape Ortegal, Loire, N.W. France, Uxisame, to Cornwall, Eng. and tin depot, St. Michael's Mount; circumnavigated Britain; reported island, Thule, which is probably Norway or Iceland; sailed to the Vistula where he reported an estuary and an island containing much amber. Calculated latitude of Massilia; laid base for cartographic parallels through N. France and Britain; formulated theory of tides (including periodical fluctuation) and their relation to moon.

PYTTE, Agnar, Am. physicist; b. Kongsberg, Norway, Dec. 23, 1932; s. Ole and Edith (Christiansen) P.; came to U.S., 1949, naturalized, 1965; A.B., Princeton, 1953; A.M., Harvard, 1954, Ph.D., 1958; m. Anah Currie Loeb, Dec. 25, 1933; children—Anders, Anthony, Allyson. Faculty, Dartmouth, Hanover, N.H., 1958-59, 60—, professor physics, 1967—; research physicist theoretical div. Project Matterhorn, Princeton, 1959-60; NSF Faculty fellow U. Brussels, 1966-67. Member Am. Phys. Soc., Am. Assn. Physics Tchrs. Author: (with Robert W. Christy) The Structure of Matter, 1965; also articles. Research in theoretical nuclear physics and theoretical plasma physics. Home: Etna, N.H. 03750. Office: Wilder Lab., Hanover, N.H. 03755.*

Q

QUADT, Raymond Adolph, Am. metallurgist; b. Perth Amboy, N.J., Apr. 16, 1916; s. Adolph and Florence (McCracken) Q.; B.S., Rutgers U., 1939; M.A., Columbia, 1943; M.S. in Phys. Metallurgy, Stevens Inst. Tech., 1948; m. Helen Desmarais, Nov. 21, 1940; 1 son, Brian. Research metallurgist, mgr. aluminum dept. Am. Smelting & Refining Co., Barber, N.J., 1942-50; v.p. research Hunter Douglas Aluminum Corp., Riverside, Cal., 1950-56; dir. research, v.p. Bridgeport Brass Co. (Conn.), 1956-60; pres. Reactive Metals, Inc., Niles, O., 1960-62; v.p. def. devel. Bridgeport Brass Co. div. Nat. Distillers & Chem. Corp., 1962-63; v.p. aerospace and def. Nat. Distillers & Chem. Corp., 1963-65; v.p., gen. mgr. Pascoe Steel Corp., Pomona, Cal., since 1965—; dir. Wackenhut Corp. Recipient Meritorious Service citation USN, 1955. Mem. Phi Beta Kappa. Developed new comml. aluminum alloys, 1944-56; developed precision cold extrusion techniques for strong aluminum alloys and reactive metals, 1951-60. Home: 550 Bucknell Av., Claremont, Cal. 91711. Office: 1301 Lexington Av., Pomona, Cal.*

QUAGLIANO, James Vincent, Am. chemist; b. Bklyn., Nov. 9, 1915; s. John and Christina (Quagliano) Q.; B.S., Poly. Inst. Bklyn., 1938, M.S., 1940; Ph.D., U. Ill., 1946; m. Lidia Marisa Vallarino, July 3, 1961; children—John, Peter. Asst. prof. chemistry U. Md., 1946-48; asso. prof. chemistry U. Notre Dame, 1948-58; sci. adminstr. Office Naval Research, Chgo., 1954-55; prof. chemistry Fla. State U., Tallahasse, 1958—. Mem. Am. Chem. Soc. (chmn. St. Joseph Valley sect. 1953-54, councilor 1965-66), Am. Inst. Chemists, A.A.A.S., Sigma Xi, Phi Lambda Upsilon, N.Y. Acad. Scis. Author: Chemistry, 1958, 62; also numerous articles. Asso. editor Chem. Revs., 1965-67. Research on amino acids and related compounds with metals, fundamental work on basic problems of chemistry; determination of structure, properties and uses of newly synthesized complexes. Home: 1806 Westridge Dr., Tallahasse 32304.*

QUAIN, Richard, Brit. physician, surgeon; b. Fermoy, County Cork, Ireland, July 1800; s. Richard Quain; apprentice surgeon, Ireland; ed. Aldersgate Sch. Medicine, Univ. Coll., London; m. Ellen, Viscountess Midleton, 1859. Tutor under Richard Bennett, Paris, London; practiced medicine, London; prof. gen. antomy and physiology Royal Coll. Surgeons, 1832-50, pres., 1868, also fellow; spl. prof. clin. surgery Univ. Coll., 1848-66; surgeon extraordinary to Queen Victoria. Fellow Royal Soc., 1844. Author: Anatomy of the Arteries of the Human Body, 1844; The Diseases of the Rectum, 1854; Clinical Lectures, 1884. Editor: Elements of Anatomy (Jones Quain), 1848. Contbd. to improved standards of medicine and surgery; described fatty or fibrous degeneration of muscle of heart (Quain's fatty degeneration), 1849. Died Sept. 15, 1887.

QUAIN, Sir Richard, Brit. physician; b. Mallow, Ireland, Oct. 30, 1816; s. John Quain; M.B., Univ. Coll., London, 1840, M.D., 1842; became asst. physician Brompton Chest Hosp., 1848, physician, 1855-75, cons. physician, 1875; Harveian orator, 1885; apptd. physician extraordinary to Queen Victoria, 1890; pres. Gen. Med. Council, 1891-98. Fellow Royal Coll. Physicians (v.p. 1889), Royal Soc., 1871, Royal Coll. Physicians Ireland. Editor: Dictionary of Medicine (most successful med. publ. of its time), 1882. Contbr. articles to med. publs. Died Mar. 13, 1898.

QUANDT, Richard Emeric, economist; b. Budapest, Hungary, June 1, 1930; s. Richard F. and Elizabeth (Toth) Q.; came to U.S., 1949, naturalized, 1954; B.A., Princeton, 1952; M.A., Harvard, 1955, Ph.D., 1957; m. Jean H. Briggs, Aug. 6, 1955; 1 son, Stephen. Faculty, Princeton, 1956—, prof. econs., 1964—; cons. Alderson Assos., Phila., 1959-61; sr. cons. Mathematica, Inc., Princeton, N.J., 1961—. Guggenheim fellow, 1958-59; McCosh fellow, 1964; Mem. Am. Econ. Assn., Econometric Soc., Am. Statis. Assn., Regional Sci. Assn. Author: (with J.M. Henderson) Microeconomic Theory: a Mathematical Approach, 1958; (with W.L. Thorp) The New Inflation, 1959; also articles. Research on econometrics, math. econs. Office: Dept. Econs., Princeton U., Princeton, N.J. 08540.*

QUARE, Daniel, English clockmaker; b. Somerset, Eng., 1648; m. Mary Stevens, Apr. 18, 1676; children—Jeremiah, Anna, Sarah, Elizabeth. Admitted as brother of Clockmakers' Co., 1671, mem. court of assts., 1697, warden, 1705, 07, master, 1708. Credited with invented repeating watches, claimed that he adapted concentric minute hand; also received patent for portable barometer; said to have made watch for James II and William III; clock for William III went year without rewinding and did not strike (being made for bedroom), clock is at Hampton Court Palace, shows sundial time, latitude and longitude and course of sun. Died Croydon, Mar. 21, 1724.

QUARTON, Gardner Cowles, Am. psychiatrist; b. Des Moines, Feb. 6, 1918; s. Sumner and Bertha (Cowles) Q.; B.S., Harvard, 1940, M.D., 1944; grad. Boston Pschoanalytic Inst., 1960; fellow Center for Advanced Study-Behavioral Sci., Stanford, Cal., 1963-64; m. Frances Baxter, Oct. 9, 1942; children—Gardner Cowles, William, Thomas. Faculty, Harvard Med. Sch., Cambridge, Mass., 1948—, clin. asso. in psychiatry, 1966—; staff Mass. Gen. Hosp., Boston, 1951—, psychiatrist, 1960—; program dir. neurosci. research program Mass. Inst. Tech., Brookline, 1966—. Career investigator USPHS, 1959-64. Diplomate Nat. Bd. Med. Examiners, Am. Bd. Psychiatry and Neurology. Mem. A.M.A., Mass. Med. Soc., Am. Psychiat. Assn., Am. Psychosomatic Assn. Asso. editor Psychosomatic Medicine, 1957—, Daedelus, 1959—. Research, publs. primarily on hormones and drug effects on behavior, biol. mechanisms in cognition. Home: Weston Rd., Lincoln, Mass. 01773. Office: 280 Newton St., Brookline, Mass. 02146.*

QUASTEL, Judah Hirsch, biochemist; b. Sheffield, U.K., Oct. 2, 1899; s. Jonas and Flora (Itcovite) Q.; student Imperial Coll. Sci., London, 1919-21; Ph.D., Trinity Coll., Cambridge, Eng., 1924, D.Sc., 1926; m. Henrietta Jungman, Dec. 27, 1931; children—Michael, David, Barbara (Mrs. Norman Glick). Fellow Trinity Coll., 1924; lectr., demonstrator in biochemistry Cambridge U., 1924-29; dir. research Cardiff City Mental Hosp., Wales, 1929-41; dir. agrl. research council unit Soil Metabolism U.K., 1941-47; prof. biochemistry McGill U., 1947-66; prof. neurochemistry, hon. prof. biochemistry U. B.C., 1966—. Meldola Medallist, 1927; recipient Pasteur medal, 1955. Fellow Royal Soc., 1940, Royal Soc. Can., Royal Inst. Chemistry, Canadian Biochem. Soc. (past pres.), N.Y. Acad. Sci. (hon.); mem. Japanese Pharmacol. Soc., Can. Soc. Microbiologists, Biochem. Soc., Am. Soc. Biol. Chemistry, Canadian Biochem. Soc., others. Co-editor, co-author: Neurochemistry, 1955, 62; Metabolic Inhibitors, 1963; author: (with others) Chemistry of Brain Metabolism, 1961; also numerous articles. Discovered principle of competitive inhibition by subtrate analogues; co-discoverer of 2:4D; helped develop sci. of neurochemistry, biochem. pharmacology; introduced study of bacterial suspensions in biochemistry, new methods of soil biochem. studies; helped develop. synthetic soil conditioners. Home: 4585 Langara Av., Vancouver, 8. Office: Kinsmen Labs., U. B.C., Vancouver, B.C., Can.*

QUATREFAGES DE BREAU, Jean Louis Armand de, see de Quatrefages de Breau.

QUAY, Wilbur Brooks, Am. biologist; b. Cleve., Mar. 7, 1927; s. Paul Quilliam and Katharine (Brooks) Q.; A.B. magna cum laude, Harvard, 1950; M.S., U. Mich., 1952, Ph.D., 1952; m. Joan Hartt Franklin, Apr. 4, 1953; 1 dau., Karen Worthington. Instr. dept. anatomy U. Mich. Med. Sch., Ann Arbor, 1952-56; faculty dept. zoology U. Cal., Berkeley, 1956—, asso. prof. Miller Inst. for Basic Research in Sci., 1964-65. Editorial referee for sci. jours., 1962—. Fellow A.A.A.S., Gerontological Soc.; mem. Am. Assn. Anatomists, Arctic Inst. N. Am., Endocrine Soc., Histochem. Soc., Soc. for Exptl. Biology and Medicine. Chmn. editorial bd. U. Cal. Publs. in Zoology, 1959-61; asso. editor Jour. Mammalogy, 1962—. Research, numerous publs. on composition and function of mammalian tissues; demonstrated responses to light by mammalian pineal organ, also pineal involvement in homeostatis of brain composition, 24 hour rhythms in brain and pineal composition, differential methods for indolic compounds in tissues. Home: 624 Alvarado Rd., Berkeley, Cal. 94705.*

QUENEY, Paul, French meteorologist; b. Lyon, France, July 27, 1905; s. Auguste and Béatrix (Lacroix) Q.; Dr ès sc; m. Paule Thoumazou, Sept. 22, 1934; children—Pierre, Yvonne. Agrégé; asst. physicist in Algiers, Puy-de-Dôme Obs.; dir. Meteorol. Inst. Algeria; now prof. meteorol. Sorbonne, Paris, France. Decorated Palmes Acad., Order of Merit of Sahara, Legion of Honor. Author: Perturbations de relief; Circulation générale de l'atmopshère. Home: 1 rue Monticelli, Paris 14, France. Office: 1 Pl. Aristide-Briand, Bellevue (Seine & Oise), France.

QUENTIN, Karl-Ernst Hermann, German chemist; b. Treves, Nov. 17, 1918; s. Karl and Gertrud (Richter) Q.; Dr es sc, U. Munich (Germany), 1950; m. Friedl Mittelberger, Sept. 5, 1949; 1 dau., Angelika. Engaged as chemist, 1949; pres. chem. div. U. Munich Inst., 1952, mem. chem. council, 1959; instr. Munich Coll., 1959. Mem. European Orgn. for Coordination of Research on Fluorine and Prevention Dental Caries. Recipient ORCA-ROLEX prize, 1958. Mem. German Chemists Assn., Royal Soc. Medicine. Author: Die Heil und Mineralquellen Nordbayerns, 1963; also numerous articles. Home: Karl Theodorstrasse 104. Office: Leopoldstrasse 175, Munich, Bavaria, West Germany.

QUÉNU, Jean Augustin Edouard Eugène, French physician; b. Berck-sur-Mer, July 13, 1889; s. Edouard and Madeleine (Archambault) Q.; M.D., U. Paris; m. Madeleine Frémont, 1923; children—Louis, Jeanne Cécile, Jean-Baptiste, Catherine. Intern, 1911; prosector Paris Hosps., 1921, surgeon, 1926; agrégé Facilty Medicine Paris, 1926; surgeon l'Hôpital Notre-Dame du Bon Secours, 1925-42; prof. clin. surgery Cochin Hosp., 1942-60; mem. council U. Paris, 1956-60; hon. prof. Faculty of Medicine Paris; hon. chair Paris Hosps. Decorated War Cross World War II, Legion of Honor. Author: Chirurgie de l'abdomen, 1949; Traité de technique chirurgicale de l'abdomen, 1955-59; Nouvelle pratique chirurgicale illustrée; also articles. Address: 11, rue de l'Université, Paris 7, France.

QUER Y MARTINEZ, José, botanist; b. Perpignan, France, 1695. Traveled through Spain as surgeon for army regt., 1728; collected, observed plants; made study African plants, 1732; traveled to Italy with army, collected specimens which were lost at sea, 1733; went to Madrid, 1737; started bot. garden with aid of Duke of Atrisco, named cons. surgeon to army; returned to Italy, 1745; dir. El Botanico de Madrid, 1755-64; began series of sci. expdns., 1745-52; established new and larger bot. garden with assistance of Conde del Miranda, 1755. Author: Flora Española, o Historia de las Plantas que se crian en España, 6 vols., 1762-84. One of 1st Spaniards who pub. work on Spanish plants. Died 1764.

QUERCETANUS, Joseph, see Duchesne, Joseph.

QUERE, Maurice Alain, French physician; b. Brest, France, May 26, 1927; s. Rene and Marie (Salaun) Q.; m. Paule Plassart, July 1, 1952; children—Benedicte, Pascal, Marie-Pierre, Sibylle, Bertrand. Asst., Val de Grace, Paris, 1956-59; cons. ophthalmologist Alger, 1960-61; prof. Faculty Medicine, Dakar, 1961-66, Tours, 1967—; staff ophthalmologist Hosp. of Tours. Decorated chevalier Ordre National, Senegal; chevalier Palmes Academiques. Mem. Société française Ophtalmologie, Société Oto-Neuro-Ophtalmologie. Author: Cliniques Africaines, 1966; also articles. Research in neuro-ophthalmology, ocular histopathology, tropical ophthalmology, ocular parasitology, onchocerciasis, trachoma, glaucoma. Home: 248 Avenue de Grammont. Office: Hopital Bretonneau, Tours, France.*

QUERICIA, Italo Federico, Italian physicist; b. Rome, Italy, Jan. 25, 1921; s. Camillo and Ornella (Tanzarella) Q.; Laurea, U. Rome, 1943; m. Amessandra Perali, July 8, 1946; children—Piero, Francesca. Asst. prof. physics U. Rome, 1949-59; prof. physics. U. Catania (Italy), 1959—, dir. physics dept., 1965—; dir. Laboratori Nazionali di Frascati, CNEN, 1960-64, cons., 1964—. Vice pres. Comitato Regionale Ricerche Nucleari, 1967—. Bd. dirs. Am. phys. socs. Research and publs. in cosmic ray physics, 1943-53, constrn. of 1,2 GeV Frascati electronsyncrotron, 1953-59, solid and liquid state physics, also improvement of Frascati synchroton, and devel. 10: 12 MeV microtron. Home: 137 Viale Egeo, Rome. Office: Laboratori Nazionali di Frascati-Frascati, Rome, Italy; also Istituto di Fisica Università-Corso Italia 57 Catania, Italy.*

QUESNAY, François, French economist; b. Mérey, nr. Paris, France, June 4, 1694; apprenticed to surgeon, at age 16; studied medicine, Paris; qualified as master-surgeon; M.D., 1744; m. 1718; 1 son, 1 dau. Apptd. perpetual sec. Acad. Surgery, 1737; became surgeon in ordinary, physician in ordinary to king, 1744, later installed in palace, Versailles, as king's 1st cons. physician; devoted himself to study of econ. problems, after 1750. Author: Maximes générales de gouvernement économique d'un royaume agricoli, 1758; Tableau économique avec explication, ou extrait des economies royales de Sully, 1758; Dialogue sur le commerce et les travaux des artisans; also important articles, including, Fermiers, and Grains, in Encyclopédie. Founder of physiocratic sch. economics, first modern sch. of thinkers to call themselves economists and to regard their theory as objectively sci., also to develop a complete and self-contained view of economic order as a whole; attacked mercantilism and invented term and policy of laissez faire; stressed existence of a natural order, also that all wealth is derived from the land. Died Versailles, Dec. 16, 1774.

QUETELET, Lambert Adolphe Jacques, Belgian statistician, astronomer; b. Ghent, Belgium, Feb. 22, 1796; studied astronomy at Paris Observatory; studied under Laplace. Prof. math., Ghent, 1814; prof., math. Brussels Athenaeum and lectr., Museum of Sci. and Literature, 1828-34; head and founder of Royal Observatory, 1828-74; prof. astronomy, Military School, 1836. Helped organize 1st internat. statistics conference. Perpetual sec. of Belgian Royal Acad., 1834-74. Author: Astronomie élémentaire, 1826; Sur l'homme et le developpement de ses facultés, 1835; Du système scoial et des lois qui le regissent, 1848; L'Anthropométrie, 1871. Pioneer in statistics; helped develop uniformity and comparability in internat. statistics; applied math. methods of averages and probabilities to study of man; developed idea of "average man" as well as that of "vital statistics"; studied meteoric showers; developed methods of simultaneous observations of astronomical, meteorological, geodetic phenomena at various points through Europe. Died Brussels, Belgium, Feb. 17, 1874.

QUEVAUVILLER, André, French pharmacist; b. St. Aubin les Elbeuf, France, Sept. 20, 1910; s. Jules and Angélique (Lemercier) Q.; B.A., Faculté de Pharmacie, Paris, 1923, D. Pharmacy, 1938; D.Sc., Faculté des Sciences, Paris, 1943; m. Charlotte Paupert, June 15, 1935; children—Pierre,, Francois. With Faculté de Paris, 1941—, prof., 1957—; insp. gen. dangerous indsl. establishments Seine Dept.; prof. Facultés de Mauriance, Paris. Decorated chevalier Legion d'Honneur; officer l'Ordre des Palmes Academique. Mem. French Acad. Pharmacy, Vet. Acad., Dental Acad., Biology Soc. Author: (with Mercier et al) Les medicaments du systeme nerveux cerebrospinal; also numerous articles. Research on pharmacodynamics, local anesthesia, pathopharmacodynamics, environmental hygiene, pharmacology of alkaloids, civil protection. Home: 2, rue du Lt. Col. Deport, Paris 16, France.*

QUICK, Armand James, Am. hematologist; b. Theresa, Wis., July 18, 1894; s. Gabriel and Theresa (Meixensperger) Q.; B.S., U. of Wis., 1918, M.S., 1919; Ph.D., U. of Ill., 1922; student U. of Pa., 1924-26; M.D., Cornell U., 1928; m. Margaret Koll, July 2, 1937; children—Elizabeth (dec.), Edith Mary. Asst. in chem., U. of Wis., 1918-19; instr. chem., Vanderbilt U., 1919-20; asst. chemist, Philadelphia Gen. Hosp., 1921-22; instr. physiol. chemistry, Sch. of Medicine, U. of Pa., 1922-26; intern, Phila. Gen. Hosp., 1928-30; asso. in research surgery, Cornell U. Med. Coll.,, 1930-32; fellow, Fifth Av. Hosp., 1932-34; asso. prof. pharmacology, Sch. of Medicine, Marquette University, Milwaukee, Wisconsin, 1935-44, professor biochemistry, 1944-64, prof. emeritus, since 1964. Awarded gold medal for scientific exhibit, American Medical Association, 1944; Modern Medicine Award, 1953; Wisconsin State Medical Society Council Award, 1950; Beaumont lecturer, Wayne County Med. Soc., 1941; Lewis Linn McArthur lecture, Inst. Med., Chicago, 1946; John Phillips Meml. award A.C.P., 1961; Distinguished Service award Wis. Med. Soc., 1964. Mem. Am. Chem. Soc., A.M.A., A.A.A.S., Soc. Exptl. Biology and Med. Am. Soc. Biol. Chemists, Am. Society Pharmacology and Exptl. Therapeutics, Central Soc. Clin. Research, Milwaukee Acad. of Medicine, N. Mex. Clin. Soc. (hon.), Alpha Omega Alpha, Omicron Kappa Upsilon, Phi Beta Kappa, Sigma Xi, Phi Lambda Upsilon. Club: Tacoma Surgery. Author: The Hemorrhagic Diseases and the Physiology of Hemostasis, 1942; The Physiology and Pathology of Hemostasis, 1951; Hemorrhagic Diseases, 1957. Contbr. to Sahyun's Outline of the Amino Acids and Proteins, 1944; Tice's Practice of Medicine, 1947; Physiology and Pathology of Hemostasis, 1951. Publs. on devel. Quick's tests for blood coagulation and discovery of coagulation Factor V; work on detoxication mechanisms, liver function, hippuric acid test. Home: 3277 N. 46th St., Milw. 53216. Office: 561 N. 15th St., Milw. 53233.*

QUICK, Horace F., Am. biogeographer, population ecologist; b. Trenton, N.J., June 30, 1915; s. Horace Floyd and Hazel (Rickey) Q.; B.S., Pa. State U., 1937; M.S., U. Mich., 1940, Ph.D., 1956; m. Susanna McDermid, Dec. 21, 1961; children—Horace F. IV, Johanna, James. Staff, Pa. State Forestry Dept., U.S. Nat. Park Service, 1938; staff U.S. Fish and Wildlife Service, 1940-43, biologist, 1942-43, 45; asst. prof. forestry Colo. State U., Fort Collins, 1946-47, asso. prof. geography, 1965—; research asso. Office Naval Research and Arctic Inst. N.Am., 1947-49; faculty U. Me., 1950-62, asso. prof., 1956-62; research asso. Inst. Arctic and Alpine Research, Boulder, Colo., 1963-64, asst. dir. NSF program, 1966—; research supt. game div. State of Me., 1952-54. Recipient Wildlife Mgmt. Inst. award, 1950; Nat. Wildlife Fedn. award, 1950. Fulbright Sr. Research fellow, E. Africa, 1959-60. Mem. Assn. Am. Geographers, Am. Inst. Biol. Sci., Arctic Inst., Phi Sigma, Xi Sigma Phi. Author: Ecology of the African Elephant, 1965; also numerous articles. Research on application prins. demography to animal population control and mgmt., studies population-resource relationships in underdeveloped areas, natural resources, conservation. Home: 6893 Niwot Rd., Longmont, Colo. 80301. Office: Dept. Geography, U. Colo., Boulder, Colo. 80304.*

QUICKEL, Kenneth Elwood, Am. physician; b. Morgantown, Pa., July 4, 1909; s. Leroy and Eva Lena (Shertz) Q.; A.B., Johns Hopkins, 1930, M.D., 1934; m. Carolyn Reid Chick, June 25, 1909; children—Stephen Woodside, Kenneth Elwood, Carolyn Reid (Mrs. Dennis L. Gouse). Fellow in pathology Hartford Hosp., 1936; practice medicine specializing in internal medicine and cardiology, Harrisburg, Pa., 1936—; staff Harrisburg Hosp., 1936—, chief med. service, 1952—, cardiologist, 1954—, dir. med. edn.; clin. prof. medicine Hahnemann Med. Sch., Phila., 1959—. Recipient Seibert award Harrisburg Acad. Medicine, 1951; Dist. Service award Pa. Heart Assn., 1953; Dist. Achievement award, 19—. Fellow N.Y. Acad. Scis., A.C.P.; mem. Pa. (past pres.), Am. (past dir.), Tricounty (past pres.) heart assns., Am. Coll. Chest Physicians, Am. Coll. Cardiology, A.M.A.,

Contbr. articles on pneumonia, carotid sinus syndrome, treatment high blood pressure. Home: 400 Arlington Rd., Camphill, Pa. 17011. Office: 515 N. 2d St., Harrisburg, Pa. 17101.*

QUIGLEY, John Paul, Am. physiologist; b. Syracuse, N.Y., Apr. 26, 1896; s. John Ryan and Mary (Curtin) Q.; B.Sc., Syracuse U., 1918; M.Sc., U. Minn., 1921; Ph.D., U. Chgo., 1929. Teaching fellow in pharmacology U. Minn., 1919-21; instr. chemistry Ore. Agrl. Coll., 1922-24; instr. physiology and pharmacology U. Ga., 1924-25, U. Alta., 1925-28; instr. physiology U. Chgo., 1928-29; faculty Western Res. U., 1929-45, prof. gastro-intestinal physiology, 1943-45; prof. pharmacology, chief div. U. Tenn., Memphis, 1945-46, prof. physiology and pharmacology, chief div., 1946-47, prof. physiology, chief div., 1946—, prof. biophysics, chmn. dept., 1962—. Guest investigator physiology Northwestern U., 1932, Cambridge U., 1936; mem. com. on radiation studies USPHS, 1958—, physiology study sect., 1956—, gastro-intestinal cancer com., 1958—; cons. Oak Ridge Nat. Labs., John Gaston Hosp., Memphis. Mem. Am. Physiol. Soc., Assn. Am. Med. Colls., Soc. for Exptl. Biology and Medicine (editor physiology Proc.), Radiation Research Soc., Am. Gastroent. Assn., Am. Assn. U. Profs., Am. Soc. Pharmacology and Exptl. Therapeutics, A.M.A., Memphis and Shelby County Med. Soc., N.Y. Acad. Sci., Am. Assn. for Cancer Research, Am. Med. Writers Assn., Sigma Xi, Phi Lambda Upsilon, Alpha Chi Sigma, Gamma Alpha. Editor digestive system and digestive secretions sect. Biol. Abstracts; editorial council Am. Jour. Digestive Diseases, Gastroenterology; asso. editor Am. Jour. Physiology, Excerpta Medica, Ann. Revs. Physiology, Methods in Med. Research; Contbr. sects. to books. Research on gastrointestinal motility, hormonal regulation of pyloric sphincter activity, reflexes, food and relation to gastric evacuation; developed pyloric inductograph; first to study gut intralumen pressures. Home: 7454 McVay Rd., Germantown, Tenn. 38038. Office: 62 S. Dunlap St., Memphis 38088.*

QUIGLEY, Thomas Bartlett, Jr., Am. surgeon; b. North Platte, Neb., May 24, 1908; s. Daniel Thomas and Helen (Seyferth) Q.; A.B., Harvard, 1929, M.D., 1933; m. Ruth Elizabeth Pearson, Dec. 31, 1938; children—Jane Seyferth (Mrs. Robert Alexander), Thomas Bartlett, Pamela Pearson. Harvey Cushing fellow surgery Peter Bent Brigham Hosp., 1935-36, George Gorham Peter traveling fellow surgery, 1935-36, staff, 1938, surgeon, 1956—; Arthur Tracy Cabot fellow Harvard Med. Sch. 1936, faculty, 1938—, clin. prof. surgery, 1964—, surgeon univ. health services, 1945—; staff VA Hosp., West Roxbury, Mass., 1953—. Cons. to various hosps. Diplomate Am. Bd. Surgery. Mem. A.M.A. (chmn. com. on med. aspects sports), A.C.S. (mem. com. on trauma), New Eng., Eastern, Boston (v.p. 1964—) surg. socs., Mass. Med. Soc., Am. Assn. for Surgery Trauma, Boston Orthopaedic Club, Am. Coll. Health Assn., Am. Surg. Assn. Editorial bd. Am. Jour. Surgery. Research, publs. on surgery of knee and surgery; orgn. on nat. scale of med. care of sch. and coll. athletes. Home: 20 Hawthorn Rd., Brookline, Mass. 02146. Office: 319 Longwood Av., Boston 02115.*

QUIMBY, Edith Hinkley, Am. biophysicist; b. Rockford, Ill., July 10, 1891; d. Arthur Sealy and Harriet (Hinkley) Hinkley; B.S., Whitman Coll., 1912, Sc.D., 1940; M.A., U. Cal. at Berkeley, 1916; postgrad. Columbia; Sc.D., Rutgers U., 1956; m. Shirley Leon Quimby, June 9, 1915. Asst., asso. physicist Meml. Hosp., N.Y.C., 1919-42; asst. prof. radiology Cornell U. Med. Sch., N.Y.C., 1941-42; asso. prof. radiology Columbia Coll. Phys. and Surg., N.Y.C., 1943-54, prof., 1954-61, prof. emeritus, 1961—. Mem. Nat. Com. for Radiation Protection and Units, 1936—; mem. adv. com. med. uses of isotopes AEC, 1946-66. Recipient Gold medals Radiol. Soc. N. Am., 1941, Radiol. Soc. India, 1952, Ewing Soc. medal, 1953, Am. Cancer Soc. medal, 1957, Interam. Coll. Radiology, 1958, Am. Coll. Radiology, 1963. Mem. Am. Radium Soc. (pres. 1954, Janeway medal 1940). Author: (with others) Physical Foundations of Radiology, 1944, 3d edit., 1961; (with Fleitelberg, Silver) Radioactive Isotopes in Medicine and Biology, 1958, 2d edit., 1963; Safe Handling of Radioactive Isotopes in Medical Practice, 1960. Research, numerous publs. on radiation physics and radiobiology. Address: 630 W. 168th St., N.Y.C. 10032.*

QUIMBY, Shirley Leon, Am. physicist; b. San Francisco, Aug. 21, 1893; s. Leon Shirley and Mary E. (Carter) Q.; A.B., U. Cal. at Berkeley, 1915; Ph.D., Columbia, 1925; m. Edith Smaw Hinkley, June 9, 1915. Faculty physics Columbia, 1919—, prof., 1943-62, prof. emeritus, 1962—. Cons. operations evaluation group Office Chief Naval Operations, 1945—. Decorated Legion of Merit. Mem., Am. Phys. Soc. (treas. 1957—). Mason. Home: 302 W. 12th St., N.Y.C. 10014.*

QUIN, Louis DuBose, Am. chemist; b. Charleston, S.C., Mar. 5, 1928; s. Louis DuBose and Olga (Jatho) Q.; B.S., The Citale, 1947; M.A., U. N.C., 1949, Ph.D., 1952; m. Harriott Miller Johnson, Sept. 6, 1952; children—Gordon D., Howard R., Carol V. Research chemist Am. Cyanamid Co., Stamford, Conn., 1949-50; research project leader Westvaco Chem. div. Food Machinery & Chem. Corp., South Charleston, W.Va., 1952-54, 56; faculty Duke, Durham, N.C., 1956—, prof. chemistry, 1967—, dir. grad. studies in chemistry, 1965—. Mem. Am. Chem. Soc., N.Y. Acad. Scis., A.A.A.S., Sigma Xi. Research, publs. on synthetic methods organophosphorus chemistry, stereochemistry cyclic phosphorus compounds, occurrence compounds with carbonphosphorus bond in living systems, chemistry tobacco alkaloids and related heterocyclic compounds. Home: 2740 McDowell St., Durham, N.C. 27705.*

QUINCKE, Georg Hermann, German physicist; b. Frankfort/Oder, Germany, Nov. 19, 1834; brother of Heinrich Irenäus Quincke; prof., Berlin, also Würzburg, Heidelberg, Germany. Research on molecular forces of fluids, capillary phenomena, optical properties of metals, acoustics; built apparatus of measuring length of sound waves by interference. Died Heidelberg, Jan. 13, 1924.

QUINCKE, Heinrich Irenäus, German physician; b. Frankfort/Oder, Germany, Aug. 26, 1842; Author: Zur Physiologie der Cerebrospinalflussigkeit, 1872. Described alternate blanching and flushing of fingernails in aortic insufficiency (Quincke's sign, perceptible nail pulse (Quincke's pulse), 1868; observed aneurysm of hepatic artery, 1870; introduced (independently of Walter E. Wynter and others) lumbar puncture for diagnosis and treatment, 1891; differentiated (with Ernst Roos) between Entamoeba histolytica and Entamoeba coli, 1893; Quincke's disease (angioneurotic edema) named for him. Died Frankfort/Main, Germany, May 19, 1922.

QUINN, Alonzo Wallace, Am. geologist; b. Halltown, Mo., May 28, 1899; s. Llewellyn Wallace and Cora (Hubbard) Q.; B.S., Denison U., 1924, D.Sc., 1967; postgrad. Kan. U.; M.S., State U. Ia., 1926; Ph.D., Harvard, 1931; m. Alice Ripley, Sept. 18, 1929; children—Carolyn Elizabeth (Mrs. John Dennis Nixon), Judith (Mrs. Edwin A. Kartman). Instr., Williams Coll., 1926-28; faculty Brown U., Providence, 1929—, prof. geology, 1951—, chmn. dept., 1940-61. Recipient Erasmus Haworth Distinguished Alumni award Kan. U., 1956. Mem. Geol. Soc. Am., Mineral. Soc. Am., Am. Geophys. Union, A.A.A.S., Sigma Xi. Research on origins and relations igneous rock, devel. No. Appalachian Mountain System. Home: 21 Oriole Av., Providence 02906.*

QUINN, George Francis, Am. atomic engr.; b. Lawrence, Mass., July 29, 1920; s. Frank J. and Genevieve (McDermott) Q.; B.S., Mass. Inst. Tech., 1941, M.S., Columbia. m. Evelyn Mae Dow, Oct. 31, 1942; children—Kathleen M., David M., Jeffrey A., George T., Mark F., Elizabeth A., Mary J., Marcia J., Frances M., Michael J. Research asst. div. indsl. coop. Mass. Inst. Tech., Cambridge, 1941-42; engr. Manhattan Project contractors, N.Y., Chgo., Oak Ridge, 1942-46; instr. Columbia dept. chem. engring. Columbia, 1946-48; prodn. div. engr. AEC, 1948-53, asst. dir. operations, 1953-55, dep. dir., 1955-59, dir., 1959-61, asst. gen. mgr. plans and prodn., 1961—. Mem. Am. Inst. Chem. Engrs., Sigma Xi. Home: RFD 4, Mt. Airy, Md. 21771. Office: U.S. AEC, Washington.*

QUINN, Loyd Yost, Am. microbiologist; b. Cutler, Ind., June 16, 1917; s. Emory Loyd and Velda (Yost) Q.; B.S., Purdue U., 1941, M.S., 1947, Ph.D., 1950; m. Jeannette Cassell Bell, June 9, 1945; children—Annette, Terrence, Cynthia, Kevin. Faculty bacteriology dept. Ia. State U., Ames, 1949—, prof., 1963—. Mem. Am. Soc. for Microbiology, Am. Assn. U. Profs., Sigma Xi. Author: Modern Immunological Concepts, 1967; also articles. Research on immunology and ecology including detection of disease germs, methods of studying bacteria, tissue cultured lymphocytes (cells which produce antibodies against antigens). Home: 226 Stanton Av., Ames, Ia. 50010.*

QUINN, Robert William, physician, educator; b. Eureka, Cal., July 22, 1912; s. William James and Norma (McLean) Q.; student Stanford, 1930-33; M.D., C.M., McGill U., 1938; m. Julia Rebecca Province, Jan. 22, 1942; children—Robert Sean, Judith Dianne. Intern, Alameda County Hosp., Oakland, Cal., 1938-39; intern U. Cal. Hosp., San Francisco, 1939-40; research fellow in internal medicine, 1940-41; asst. resident in internal medicine Presbyn. Hosp., N.Y.C., 1941-42; research fellow preventive medicine Yale Sch. Medicine, New Haven, 1946-47, instr., 1947-49; asso. prof. preventive medicine U. Wis., Madison, 1949-52; prof., chmn. dept. preventive medicine Vanderbilt U. Sch. Medicine, Nashville, 1952—. Mem. Nashville Acad. Med., Tenn. Med. Assn., Am. Assn. Immunologists, Am. Soc. for Microbiology, Soc. Exptl. Biology and Med., Middle Tenn. Heart Assn. (pres. 1957). Research and numerous publs. on epidemiology, bacteriology, immunology, clin. aspects of hemolytic streptococcal infections, rheumatic fever, rheumatic heart disease, med. care. Home: 508 Park Center Dr., Nashville 37205. Office: Vanderbilt Hospital, Nashville 37203.*

QUINQUAUD, Charles Eugène, French physician; b. Lafat, France, 1843; ed. Paris; M.D., 1872; mem. Acad. Medicine. Author: Essai sur le puerpérisme infectieux chez la femme et chez le nouveau-né, 1875; Les affections du foie, 1879; Des métastases, 1880; Chimie pathologique, 1880; Traité technique de chimie biologique, 1882. Devised (with Nestor Gréhant) method for determining vol. of blood by means of carbon monoxide, 1882; described folliculitis decalvans (Quinquaud's disease), 1888. Died 1894.

QUINTUS, Roman anatomist, physician; flourished Rome, 117-38; pupil of Marinus; mem. eclectic sch.; founder med. sch. which included tchrs. of Galen, whom he influenced through his anat. teachings; banished from Rome; died in Pergamum.

QUISENBERRY, John Henry, Am. poultry scientist; b. Gainesville, Tex., June 25, 1907; s. Walter H. and Nancy (Lane) Q.; B.S., Tex. A. and M. U., 1931; M.S., U. Ill., 1933, Ph.D., 1937; m. Pearl Clara Buzy, June 6, 1936; children—Alex John, Judith Ann (Mrs. William L. Terry). Research asst. U. Ill., 1931-36; prof. genetics Tex. A. and M. U., College Station, 1936-45, prof., head poultry sci. dept., 1946—; head poultry dept. U. Hawaii, 1945-46. Fellow Nat. Poultry Sci. Assn., A.A.A.S., Tex. Acad. Sci., Am. Assn. U. Profs.; mem. Am. Genetics Assn., Soc. Exptl. Biology and Medicine, Poultry Sci. Assn., World Poultry Sci. Assn., Tex. Assn. Coll. Tchrs., Genetics Soc. Am., Am. Statis. Assn., Sigma Xi, Phi Kappa Phi. Research, numerous publs. on laying hen nutrition and mgmt., bird densities and environmental control housing, growing embryos in plastic shells, effects of irradiation upon reprodn., water metabolism. Home: 1006 Puryear Dr., College Station, Tex. 77840.*

QUSTA BEN LUGA AL-BA'LABAKKI, see al-Ba'labakki.

QUTB AL-DIN AL-SHIRAZI, see al-Shirazi.

R

RAAZ, Franz Friedrich, Austrian mineralogist; b. Neustadt, Oct. 28, 1894; s. Franz-Josef and Anna Amalia (Legler) R.; ed. univs. Vienna and Uppsala; Ph.D.; m. Valenta Stefanie Raaz, June 25, 1952. Asst., Mineral. Inst. Vienna, 1922; Rockefeller Found. scholar silicate research Kaiser Wilhelm Inst., Berlin/Dahlem, 1927-29; instr. U. Vienna, 1931, prof. at large, 1940, asso. prof., 1952, prof. mineralogy and crystallography, 1960—; prof., dir. Vienna Poly. Inst., 1957. Recipient Gold medal with crown and swords Austrian Order Merit. Mem. German, Austrian mineral. socs., Am. Geochem. Soc. Author: Sphär Trigonometrie für Naturwissenschaft und Technik, 1928; Mineralogie des Reichemberger Bezirkes, 1937; Polarisationsmikroskopie; Das Mikroskop und seine Nebensparate, 1950; Bau und Nildung der Kristalle, die Architektonik der stofflichen Welt, 1953; Geometrie und physikal. Kristallographie und deren Arbeitmethoden, 1939; also articles on crystallography, radiography, mineral. chemistry. Research on crystal growth and habitus; morphology of crystals; crystal optics; X-ray crystallography. Home: Hermanngasse 25, Vienna 7. Cffice: Inst. für Mineralogie, Kristallographie und angewandt Petrographie (TH), Getreidemarkt 9, Vienna 6, Austria.

RABANUS, Marus Magnentius (Hrabanus or Rhabanus), German theologian; b. Mainz, Germany, circa 776; ed. Fulda, Germany, Tours, France; studied under Alcuin. Became head Monastic Sch., Fulda, which became most famous in Germany under his leadership, 803, abbot, 822-42; ordained 814; archbishop of Mainz, 847-56. Author: De universo libri XXII, sive etymologiarum opus, circa 844; De institututione clericorum; also Latin-German glossary to Bible, commentaries on Scripture, treatise on edn. including 7 liberal arts. Important in Carolingian revival; opposed Gottschalk's theories of predestination. Died Winkel, Germany, Feb. 4, 856.

RABE, Eugene Karl, astronomer; b. Berlin, Germany, May 8, 1913; s. Hermann and Luise (Graetz) R.; Dr.phil., U. Berlin, 1937, Dr.phil.habil., 1940; m. Erika Marianne Schiller, Apr. 25, 1938; children—Ingrid Hannelore (Mrs. Anton Greiner), Karin Gudrun. Came to U. S., 1948, naturalized, 1955. Astronomer, Astronomisches Rechen-Institut, Berlin, Heidelberg, Germany, 1937-48; dozent astronomy U. Berlin, 1942-45; faculty U. Cin., 1948—, prof. astronomy, 1958—; visiting professor Yale University, 1967-68. Cons. Space Tech. Labs., 1958-59, Project Mercury, 1959-60, Aeronutronics, 1960-61. Fellow A.A.A.S.; mem. Am. Rocket Soc. (mem. flight mechanics com. 1959-60), Astronomische Gesellschaft, Internat. Astron. Union, Am. Assn. U. Profs. Contbg. author: The Solar System, vol. V; also numerous articles. Research on more accurate determination astron. unit and masses inner planets, dynamical implications Kuiper's theory origin solar system, celestial mechanics, theoretical and computational treatment orbital motions in solar system, periodic orbits asso. with Lagrange's equilateral libration centers and new results concerning stability librational motions. Home: 1016 Urbancrest Pl., Cin. 45226.*

RABELAIS, François, French physician; b. Chinon, France, circa 1494; s. Antoine R.; novice at Fontenay-le-Comte, 1520; took religious orders, 1523; studied at Maillezais Abbey, 1525; B.A., Montpellier, 1530; M.D., Montpellier, 1537; children—François, Junie, Théodule. Physician, Pont-du-Rhône Hosp.,

Lyons, 1532; private physician to Jean du Bellay, 1533; accompanied du Bellay to Rome, 1534, 47; private physician to Guillaume du Junie, Ghéodule. Physician, Pont-du-Rhône Hosp., 1546. Author: Pantagruel, 1532; Gargantua, 1534; published Aphorisms of Hippocrates, Ars Parva of Galen, Epitres de Giovanni Manardi, all 1531. Renaissance humanist considered in his time as philosopher, theologian, mathematician, physician, legal expert, musician, geometer and astronomer; today known for tales of Gargantua in which he satirizes scholasticism and superstition of Middle Ages and exalts Renaissance spirit of humanism and experimental science; invented the glottocomon, apparatus for setting fractures, 1537. Died Paris, France, Apr. 9, 1553.

RABI, Isidor Isaac, physicist; b. Rymanov, Austria, July 29, 1898; s. David Robert and Jennie (Teig) R.; B.Chem., Cornell U., 1919; Ph.D., Columbia, 1927; D.Sc., Princeton, 1947, Harvard, 1955, Williams Coll., 1958, U. Birmingham, 1960, Clark U., 1962, Adelphi Coll., 1962, Technion U., 1963, Franklin Marshall Coll., 1964, Brandeis U., 1965, U. Coimbra (Portugal), 1966; L.H.D., Hebrew Union Coll., 1958, Oklahoma City U., 1960; D.H.L., Yeshiva U., 1964; LL.D., Dropsie Coll., 1956; Litt.D., Jewish Theol. Sem. Am., 1966; m. Helen Newmark, Aug. 17, 1926; children—Nancy R. (Mrs. Immanuel Lichtenstein), Margaret R. (Mrs. Clement C. Beels). Tutor, Coll. City N.Y., 1924-27; faculty Columbia, N.Y.C., 1929—, prof., 1937—, exec. officer dept. physics, 1945-49, Univ. Prof., 1964—. Del., v.p. Internat. Conf. on Peaceful Uses of Atomic Energy, Geneva, 1955, 58, 64; Shreve fellow Princeton, 1961-62; Karl Taylor Compton lectr. Mass. Inst. Tech., 1962. Ernest Kempton Adams fellow, 1935; Sigma Xi Semicentennial prize, 1936; Elliot Cresson medal Franklin Inst., 1942; Nobel Prize in Physics, 1944; Barnard medal Nat. Acad. Scis., 1960. Fellow Am. Phys. Soc. (pres. 1950); mem. Council on Fgn. Relations, Am. Philos. Soc., Japanese Acad. (fgn.), Sigma Xi. Author: My Life and Times as a Physicist, 1960; also numerous articles. Asso. editor Phys. Rev., 1935-38, 41-44. Inventor molecular beam magnetic resonance method for study of atomic and molecular fine and hyperfine structure. Home: 450 Riverside Dr., N.Y.C. 10027.*

RABINOVICH, Isaak Moiseevich, Russian structural engr.; b. Jan. 23, 1886; grad. Moscow Tech. Coll., 1918. With Inst. Engring. Research, Sci. Tech. Com. of People's Commissariat in Rds. and Communication, 1918-32; also tchr. univs. and tech. colls., Moscow; became prof. Mil. Engring. Acad., 1932, Engr. Constrn. Inst., 1933. Named Honored Scientist of, R.S.F.-S.R. Corr. mem. USSR Acad. Scis., Acad. Constrn. and Architecture USSR. Author: Utilization of the Theory of Finite Differences in the Investigation of Continuous Beams, 1921; Investigation of Continuous Beams, 1921; Kinematic Method in Structural Mechanics in Connection with Graphic Kinematic and Static Plane Chains, 1928; On the Theory of Statically Indeterminate Trusses, 1933; Method of Calculating Frames, Part I-III, 1934-37; Course in Structural Mechanics of Rod Systems, Part I-II, 1938-40, 2d ed., 1950-54; Achievements of Structural Mechanics of Rod Systems in the USSR, 1949; The Bases of Dynamic Calculation of Structures on the Effects of Short-Term and Instantaneous Forces, Part I, 1952. Developed kinetic methods in structural mechanics, methods for calculation of complex; 1st USSR investigations of dynamic action on different loads on bridge spans, other engring. constructions; studies in statically indeterminate systems, theory of guy trusses. Home: Brusorskii, pr 7. Office: USSR Acad. Scis., Leninski Prospekt, 14, Moscow, USSR.

RABINOVITCH, B(enton) Seymour, chemist; b. Montreal, Can., Feb. 19, 1919; s. Samuel and Rachel (Schachter) R.; B.Sc., McGill U., 1939, Ph.D., 1942; m. Marilyn Werby, Sept. 18, 1949; children—Peter Samuel, Ruth Ann, Judith Nancy, Frank Benjamin. Came to U. S., 1946. Postdoctoral fellow Harvard, 1946-48; mem. faculty U. Wash., 1948—, prof. chemistry, 1957—. Cons. and/or mem. sci. adv. panels and coms. NSF, Nat. Acad. Scis.-NRC. Nat. Research Council Can. fellow, 1940-42; Royal Soc. Can. Research fellow, 1946-47; Milton Research fellow Harvard, 1948; Guggenheim fellow, 1961. Mem. Am. Chem. Soc. (past chmn. Puget Sound sect., now chmn. nat. exec. com. phys. div., also alternate councilor; mem. editorial bd. jour.), Am. Phys. Soc., Faraday Soc., Am. Assn. U. Profs., Sigma Xi. Editorial bd. Jour. Phys. Chemistry, Internat. Jour. Chem. Kinetics. Spl. research unimolecular gas phase reaction. Home: 4234 N.E. 125th St., Seattle 98125.*

RABINOWITCH, Eugene, biophysicist; b. St. Petersburg, Russia, Apr. 26, 1901; s. Isaak and Sinaida (Weinlud) R.; Ph.D., U. Berlin, Germany, 1926; D.H.L., Brandeis U., 1960; D.Sc., Dartmouth Coll., 1964; m. Anna Majersohn, Mar. 12, 1932; children—Alex and Victor (twins). Came to U. S., 1938, naturalized, 1943. Research asst. U. Göttingen, 1930-33, Rask Oested Found., Copenhagen, 1933; research asso. dept. chemistry U. Coll., London, 1934-37, Mass. Inst. Tech., Cambridge, 1937-45; sec. chief Manhattan dist. Metall. Lab., Chgo., 1944-46; prof. botany, biophysics U. Ill., Urbana, 1947—, mem. Center Advanced Studies, 1966—. Mem. internat. continuing com. Confs. on Sci. and World Affairs, 1957—. Recipient Immigrants League award, 1960; Calingar prize, 1966. Mem. Am. Biophys. Soc. (exec. com.). Author: Dawn of a New Age, 1963; many other sci. books, Photosynthesis and Related Processes, 3 vols., 1945-56. Editor, writer Bull. of Atomic Scientists, 1945—. Home: 1021 W. Church St., Champaign, Ill. 61820.*

RABINOWITSCH-KEMPNER, Lydia, bacteriologist; b. Kaunas, Lithuania, Aug. 22, 1871; colleague of Robert Koch; dir. bacteriol. dept. Berlin-Moabit Hosp.; 1st woman to receive Prussian professorial title; studied plague, human and animal Tb, sleeping sickness, other parasitic diseases. Died Berlin, Aug. 3, 1935.

RABINOWITZ, Joseph Loshak, biochemist; b. Odessa, Ukraine, Nov. 4, 1923 (parents Am. citizens); s. Laib and Rachel (Loshak) R.; B.S., Poly Inst. Mexico, 1940; B.S., Phila. Coll. Pharmacy and Sci., 1943; M.S., U. Pa., 1948, Ph.D., 1950; m. Josephine Feldmark, June 23, 1946; children—Malva E., Lois I., Martin D. Research asso. U. Pa., Phila., 1950-53; prin. scientist radioisotope service VA Hosp., Phila., 1953-54, acting chief, 1954-57, chief radioisotope research, 1958—; asso. dir. radioisotope service Meth. Hosp., Houston, 1957-58; asso. prof. Baylor U., Houston, 1957-58; asst. prof. U. Pa., 1958-65, asso. prof., 1965—. John L. Harrison fellow, 1949-50; Fulbright fellow, 1957-58. Mem. Am. Soc. Biochemistry, Soc. Exptl. Biology and Medicine, Soc. Nuclear Medicine, A.M.A., Soc. Nuclear Research, Radiation Research Soc., Am. Chem. Soc., Pa. Acad. Sci., N.Y. Acad. Sci., Sigma Xi. Author: (with Grafton Chase) Radioisotope Methodology, 4th edit., 1967. Editor: Topics in Medicinal Chemistry, 3 vols. Discovered several intermediaries in biosynthesis of cholesterol from acetate; intermediaries in biosynthesis from cholesterol to cholic acid; research in isotope effects, atherosclerosis, thyroid metabolism. Home: 127 Juniper Rd., Havertown, Pa. 19083. Office: VA Hosp., University and Woodland Avs., Phila. 19104.*

RABOTNOV, Yurii Nikolaevich, Russian physicist; b. Feb. 24, 1914; grad. Moscow U., 1935. Tchr., Moscow Inst. Energetics, 1935-46; named chief lab. strength of materials Inst. Mechanics, USSR Acad. Scis., 1946; became prof. Moscow U., 1947. Mem. USSR Acad. Scis. (chief editor tech. sci. sect. News). Author: Resistance of Materials, 1950; also articles. Research in theory of envelopes, theory of creep, theory of plasticity. Home: Lomonosovskii prospekt, 14, Moscow, USSR.

RABOTTI, Giancarlo Francesco, physician; b. Turin, Italy, Feb. 8, 1925; s. Carlo Alberto and Enrica (Maffioli) R.; Doctor in Medicine, Med. Sch., U. Parme (Italy), 1951. Asst. pathology Nat. Cancer Inst. Milan (Italy), 1951-56; asst. pathology Med. Sch., Milan, 1956-57; vis. scientist Nat. Cancer Inst., NIH, Bethesda, Md., 1958, sr. investigator Lab. Viral Oncology, 1963-66, mem. study sect. for contracts on cancer diagnosis NIH, 1963-66; instr. hematology Jefferson Med. Sch., Phila., 1959-61, asst. prof. pathology, 1962; staff Laboratoire de Médecine expérimentale, Collège de France, Paris, 1966—. Recipient Nat. award for cancer research Italian Council Research, 1957. Mem. A.A.A.S. Research and publs. on exptl. induction of cancer by viruses, especially avian sarcoma viruses in mammals and their immunology; induced cerebral tumors in rodents, tumors in dogs and cats; 1st demonstration of malignant transformation of tissues in monkeys. Home: 198 Rue de Rivoli, Paris, France.*

RACE, George Justice, Am. pathologist; b. Everman, Tex., Mar. 2, 1926; s. Claude Earnest and Lila (Bunch) R.; student Tex. Wesleyan Coll., 1942-43, Baylor U., 1943-44; M.D., U. Tex., 1947; M.S. in Parasitology, U. N.C., 1953; m. Anne Rinker, Dec. 21, 1946; children—George William Daryl, Jonathan Clark, Mark Christopher, Jennifer Anne, Elizabeth Margaret Rinker. Instr. pathology Duke, 1951-53, Harvard, 1953-54; staff pathologist Children's Med. Center, Dallas, 1955-59; faculty U. Tex. Southwestern Med. Sch., Dallas, 1955—, clin prof. pathology, 1964—; from asst. to asso. pathologist Parkland Meml. Hosp., Dallas, 1955-59; cons. pathologist VA Hosp., Dallas, 1955—; dir. labs. Baylor U. Med. Center, 1959—, pathologist-in-chief, 1959—, prof. pathology Coll. Dentistry, 1964—. Diplomate Am. Bd. Pathology. Fellow Coll. Am. Pathologists, Internat. Acad. Pathology, Am. Soc. Clin. Pathologists, A.A.A.S., N.Y. Acad. Scis.; mem. A.M.A., Am. Assn. Pathologists and Bacteriologists, Electron Microscope Soc. Am., Am. Soc. Cytology, Am. Soc. Tropical Medicine & Hygiene, Sigma Xi. Research, publs. on adrenal gland function in hypertension; immunological response of hosts to animal parasites and cancer nucleoproteins; effects of hydrogen peroxide systems on various disorders. Home: 3429 Beverly Dr., Dallas 75205. Office: 3500 Gaston Av., Dallas 75246.*

RACKER, Efraim, biochemist; b. Neu Sandez, Poland, June 28, 1913; s. Meier and Ella R.; M.D., U. Vienna, 1938; m. Franziska Weiss, Aug. 24, 1945; 1 dau., Ann. Naturalized Am. citizen, 1947. Joined Anderson Inst. for Biol. Research, U. Minn., 1941; asst. prof. microbiology N.Y. U. Sch. Medicine, 1944-52; asso. prof. biochemistry Yale, 1952-54; chief div. nutrition and physiology Pub. Health Research Inst. City N.Y., also adj. prof. microbiology N.Y. U.-Bellevue Med. Center, 1954-66; Albert Einstein prof., chmn. dept. biochemistry and molecular biology Cornell U., Ithaca, N.Y., 1966—. Chmn. biochemistry study sect. NIH, USPHS. Mem. Am. Soc. Biol. Chemists, Harvey Soc., Brit. Biochem. Soc., Nat. Acad. Scis., Am. Acad. Arts and Scis., A.A.A.S. Author: Mechanisms in Bioenergetics, 1965; also numerous articles. Research in multi-enzyme systems, function of regenerating systems of ATP, critical role of glyceraldehyde-3-phosphate dehydrogenase in energy metabolism; isolated enzyme which forms glyceraldehyde-3-phosphate from pentose phosphate and named it transketolase; studies on role of coupling factors in oxidative phosphorylation, including relationship between mitochondrial structure and function, metabolic control mechanisms. Home: 305 Brookfield Rd., Ithaca, N.Y. 14850.*

RACOVITA, Emil, biologist; b. 1868; naturalist Belgica Antarctic expdn.; prof. Cluj (Rumania) U., 1920-47; sub-mgr. marine zool. sta., Banyuls-sur-Mer, France, 1900-20, also at lab. comparative anatomy Sorbonne; Mem. Rumanian Acad. (chmn. 1927-29). Author: Essai sur les problèmes biospeologiques, 1907; Speology, 1927. Co-mgr. Archives de zoologie expérimentale et générale. Founder biospeology; founder 1st inst. of speology (at Cluj U.), 1920; made classic research studies of whales in Antarctic. Died 1947.

RACZYNSKI, Jan, Polish pediatrician; b. 1865; worked in Lvov, Poland, 1911-12. Author: Recherches experimentales sur la manque d'action du soleil comme cause rachitisme (paper presented at Congress of Pediatrics, Paris), 1912. Provided 1st exptl. proof of influence of sunlight on accumulation process of phosphorous and calcium in bones and its relationship to rickets. Died 1917.

RADAU, Jean-Charles-Rodolphe, astronomer, mathematician; b. Angerburg, Prussia, Jan. 22, 1835. Worked in acoustics; made geodetic studies of Ethiopia. Naturalized French citizen, 1874. Mem., French Acad. Scis., 1897; Bureau of Longitudes. Author: Le Spectre Solaire, 1863; L'Acoustique ou les Phénomènes du Son, 1867. Confirmed rotation of the sun on its axis is not the same when observed from different latitudes, 1864. Died Paris, France, Dec. 21, 1911.

RADCLIFFE-BROWN, Alfred Reginald, English social anthropologist; b. Birmingham, Eng., Jan. 17, 1881; fellow Trinity Coll., 1908-14; m. Wilifred Lyon (marriage dissolved 1933); 1 dau. Lectr. ethnology U. London, 1909-10; dir. edn. Kingdom of Tonga, 1916; prof. social anthropology U. Cape Town, 1921-25; prof. anthropology U. Sidney, 1925-31; prof. anthropology U. Chgo., 1931-37; fellow All Soul's Coll., prof. social anthropology Oxford U., 1937-46; prof. social science Farouk I U., Alexandria, Egypt, 1947-49. Mem. Royal Anthrop. Inst. (Rivers, Huxley medals, pres., 1939-40), Royal Acad. Scis. Amsterdam, (hon.) N.Y. Acad. Scis. Author: The Andamin Islanders, 1922; Social Organization of Australian Tribes, 1931; Structure and Function in Primitive Society, 1952; A Natural Science of Society, 1957; Method in Social Anthropology, 1958; also numerous papers. Did field work in Andaman Islands, 1906, western Australia, 1910; fostered devel. of social anthropology as science; contbd. to study of kinship, social, and family orgn.; developed systematic framework of concepts relating to social structures of simple socs.; thought soc. composed of interfunctioning and interdependent parts, also that usages and instns. in soc. function together to maintain whole. Died London, Oct. 24, 1955.

RADELEFF, Rudolph Durham, Am. toxicologist; b. Kerrville, Tex., Apr. 23, 1918; s. Fritz and Mary Mignonette (Durham) R.; A.A., Schreiner Inst., 1937; D.V.M., Tex. A. and M. U., 1941; m. Frances Lovic Bullard, Feb. 19, 1937; 1 dau., Sandra Sue. Research veterinarian in charge toxicological investigations Agrl. Research Sta., U. S. Dept. Agr., Kerrville, 1947-66, dir. Southwestern Vet. Toxicology and Livestock Insects Research Lab., College Station, Tex., 1966—. Recipient Distinguished Service citation Sec. of Agr., 1955, certificate of merit 1963. Mem. A.A.A.S., Am. Vet. Med. Assn., Am. Assn. Fed. Veterinarians, Sigma Xi. Author: Veterinary Toxicology, 1964. Research and numerous publs. on taenicides for ruminants, toxicology of pesticides in livestock, use of radioisotopes. Home: 115 Lee Av. S. Office: Drawer GE, College Station, Tex. 77840.*

RADEMACHER, Hans Adolph, mathematician; b. Hamburg, Germany, Apr. 3, 1892; s. Adolph Henry and Emma (Weinhover) R.; Dr. Phil., U. Gottingen, 1917; D.Sc., U. Pa., 1962; m. Irma Schoenberg, Sept. 10, 1949; children—Karin E. (Mrs. Ariel Loewy), Peter H. Came to U. S., 1934, naturalized, 1943. Prof., U. Hamburg, 1922-25, U. Breslau, 1925-34; faculty U. Pa., 1934-62, prof., 1939-62, prof. emeritus, 1962—; vis. prof. N.Y. U., 1962-64; vis. prof. Rockefeller Inst., N.Y.C., 1964-66, affiliate, 1966—; mem. Inst. Advanced Study Princeton, 1953, 60-61; vis. prof. Tata Inst. Fundamental Restarch, Bombay, India, 1954-55. Mem. Am., London math. socs., Math. Assn. Am., Soc. Indsl. and Applied Math., Société Mathematique de France. Author: (with O. Toeplitz) Von Zahlin und Figurea,

1930; Lectures on Elementary Number Theory, 1964; also numerous articles. Research primarily on theories of Lebesgue integral, modular functions, numbers, convex bodies. Home: 913 S. 48th St., Phila. 19143. Office: Rockefeller U., N.Y.C. 10021.*

RADEMAKER, John Adrian, Am. sociologist, anthropologist; b. Tacoma, Aug. 26, 1905; s. Arie Willem and Jannigje (de Jong) R.; A.B., Coll. Puget Sound, 1930; M.A., U. Wash., 1935, Ph.D., 1939; m. Elizabeth Dewey Spencer, Sept. 9, 1939; children—John Hendrick, Janice May. Asst. in sociology U. Wash., 1934-37; research asst. Wash. Emergency Relief. Adminstrn., 1933-34; asst. prof. sociology U. Ore., 1939; instr. Bates Coll., Lewistown, Me., 1939-43; community analyst U.S. War Relocation Authority, Amache Relocation Center, 1943-44; asst. prof. sociology and anthropology U. Hawaii, Honolulu, 1944-47; prof., chmn. dept. sociology and anthropology Williamette U., Salem, Ore., 1947—. Chmn. race relations com. Hawaii Assn. for Civic Unity, 1946-47; pres. Ore. Social Welfare Assn., 1963-65; mem. Ore. Joint Council for Social Welfare Legislation, 1949—, chmn., 1955-57; v.p. Ore. Mus. Anthropology and Natural History, 1965-67; sec. Ore. Natural History Soc. and Mus., 1967—. Recipient Freedom award Am. Vets. Com., Portland, 1958; award of merit Joint Council for Social Welfare Legislation, 1961. Mem. Am. Assn. U. Profs. (nat. bd. 1957-60, pres. Ore. 1954—, council 1954-68), Pacific Sociol. Soc. (sec. No. br. 1935-39, v.p. 1953-54), Am. Sociol. Assn., Soc. for Study Social Problems, Urban League, N.A.A.C.P., Fellowship of Reconciliation, Am. Civil Liberties Union, SANE. Author: (with Marion Hathway) Public Relief Administration in the State of Washington, 1934; (with James T. Lane) These Are Americans: The Japanese Americans in Hawaii in World War II, 1951; also articles. Home: 960 Shipping St. N.E., Salem, Ore. 97303.*

RADERMACHER, Jacques Corneille Matthieu, Dutch geographer, botanist; b. Netherlands, 1741; founder Soc. Scis. Batavia (Java), 1778, also library, mus., obs.; contbr. memoirs, papers, articles; studies on flora of Java. Died at sea, 1783.

RADEV, Tontscho Petrov, Bulgarian physiologist; b. Schumen, Bulgaria, Sept. 9, 1898; s. Peter Ivanov and Smaraida (Rascheva) R.; student chemistry U. Sofia (Bulgaria), 1917-20; Dr.vet. med. Higher Sch. Vet. Medicine, Berlin, Germany, 1924; m. Alice Aleksandrova, Aug. 1930. Staff, Higher Sch. Vet. Medicine, Sofia, 1931—, prof. physiology and biochemistry, 1951—; head dept. physiology Intitut Comparative Pathology, Bulgarian Acad. Sci., Sofia, dir. Central Sci. Research Lab., 1951—. Recipient several awards Govt. of Bulgaria. Mem. Soc. Bulgarian Physiologists and Biochemists, Bulgarian Acad. Sci. (sec. jour. Nature). Author: (with W. Roussev) Physiology of Domestic Animals, 1965; A Practicum of the Physiological, 1953; also numerous articles. Research on metabolism of calcium, phosphorus and iron in newborn and suckling animals, influence of rarified air on animals, movement of ruminant stomach, inheritance of hypocatalassemia in guinea pigs. Home: 37 Zar Ivan Assen, Sofia 4, Office: Bulgarian Acad. Scis., 1, 7 noemvri, Sofia, Bulgaria.*

RADFORTH, Norman William, biologist; b. Barrow-in-Furness, Lancastershire, Eng., Sept. 22, 1912; s. Walter Joseph and Kate Emma (Langley) R.; B.A., Riverdale Collegiate U. Toronto, 1936, M.A., 1937; Ph.D., Glasgow U., 1939; m. Isobel Limbert, June 30, 1939; children—John Robert, Janice Langley. Lectr. botany U. Toronto, 1939-46; prof., head botany dept. McMaster U., 1946-52; dir. Royal Bot. Gardens, Hamilton, Ont., Can., 1946-53; chmn. dept. biology McMaster U., 1960—, chmn. organic and asso. terrain research unit, 1961—, asso. mem. dept. geology, 1965—. Canadian nat. sec. Internat. Soc. for Terrain-Vehicle Systems, 1962—; chmn. subcom. NRC Can., 1948—; mem., asso. com. soil, snow mechanics; trustee, sci. adv. Ont. Waterfowl Research Found.; cons. U. S. Army C.E. Fellow Royal Soc. Can., Internat. Soc. Palaeobotanists (v.p.), Am. Inst. Biol. Scis., Am. Polar Soc., Arctic Inst. N.Am., Internat. Soc. Plant Morphologists, Internat. Soc. Plant Taxonomy, Palaeontol. Assn. Contbr. numerous articles to profl. jours. Home: Fell Croft, R.R. 1, Dundas, Ont. Office: McMaster U., Hamilton, Ont., Can.*

RADICATI DI BROZOLO, Luigi A., Italian physicist; b. Milan, Oct. 12, 1919; s. Giuseppe and Maria (Olliveri) R. di B.; Ph.D., U. Turin; m. Gianna Balbo di Vinadio, Feb. 8, 1947; children—Luca, Alessandro, Giuseppe, Paolo, Ottone. Research fellow U. Birmingham (Eng.); prof. physics U. Naples and Pisa; mem. Inst. Advanced Study, Princeton; now prof. physics Pisa Coll. Mem. Am. Phys. Soc., Italian Physics Soc., Lombard Inst. Sci. and Letters. Author articles in Phys. Rev., Nuovo Cimento, Annals of Physics, proc. Phys. Soc. Home: via del Capannone, Barboricina, Pisa. Office: Scoula Normale Superiore, Pisa, Italy.

RADNER, Roy, Am. economist, statistician; b. Chgo., June 29, 1927; s. Samuel and Ella (Kulansky) R.; Ph.B. with honors, U. Chgo., 1945, B.S., 1950, M.S., 1951, Ph.D. in Math. Statistics, 1956; m. Virginia Honoski, July 26, 1949; children—Hilary, Erica, Amy, Ephraim. Research asso. Cowles Commn. for Research in Econs., 1951-54; asst. prof. econs. Yale, also Cowles Found., 1955-57; faculty U. Cal. at Berkeley, 1957—, prof. econs. and statistics, 1961—. Cons. pvt. cos., govt. agys. Center for Advanced Study in Behavioral Sci. fellow, 1955-56; Guggenheim fellow, 1961-62, 65-66. Fellow Econometric Soc.; mem. Inst. Math. Statistics, Am. Statis. Assn., Inst. Mgmt. Scis., Am. Assn. U. Profs., Phi Beta Kappa. Author: Notes on the Theory of Economic Planning, 1963; (with D. W. Jorgenson, J. J. McCall) Optimal Replacement Policy, 1966; also articles. Research on theory of decision under uncertainty, theory of econ. growth and planning. Home: 2275 Eunice St., Berkeley, Cal. 94709.*

RADO, Sandor, psychiatrist; b. Hungary, Jan. 8, 1890; student U. Berlin, U. Bonn; Dr.Polit. Sci., U. Budapest (Hungary), 1911, M.D., 1915; Ednl. dir. N.Y. Psychoanalytic Inst., 1931-41; clin. prof. psychiatry, dir. psychoanalytic clinic for tng. and research Columbia U., N.Y.C., 1944-55; prof. psychiatry, dir. Sch. Grad. Psychiatry, State U. N.Y., Downstate Med. Center, 1956-58; pres., dean prof. psychiatry N.Y. Sch. Psychiatry, N.Y.C., 1958—. Mem. Am. Inst. Biol. Scis., Am. Psychiat. Assn., A.M.A., Am. Psychoanalytic Assn., N.Y. Acad. Scis., Am. Acad. Psychoanalysis. Fellow N.Y. Acad. Medicine. Author: Psychoanalysis of Behavior, Vol. I, 1922, Vol. II, 1956. Address: 25 E. 86th St., N.Y.C. 10028.*

RADOMSKI, Jack London, Am. pharmacologist; b. Milw., Dec. 10, 1920; s. Joseph Elwood and Evelyn (Hansen) R.; B.S., U. Wis., 1942; Ph.D., George Washington U., 1950; m. Margery Dodge, Feb. 1, 1947; children—Mark Stephen, Linda, Eric Parker, Janet. Chemist, Ozalid Products div. Gen. Aniline & Film Corp., Johnson City, N.Y., 1942-44; pharmacologist acute toxicity br. div. pharmacology FDA, Washington, 1944-52, acting chief, 1952-53; faculty U. Miami (Fla.) Sch. Medicine, 1953—, prof. pharmacology, 1959—. Mem. Am. Soc. Pharmacology and Exptl. Therapeutics, Am. Assn. Cancer Research, Soc. Toxicology. Discoverer that heptachor is converted to an epoxide in humans and animals, 1st demonstration of epoxidation in biol. systems; studies, numerous publs. on mechanism of toxic action of lithium chloride contbd. to understanding of carcinogenesis of aromatic amines and metabolism of azo-dyes. Home: 5350 S.W. 84th St., Miami, Fla. 33143.*

RADOSAVLJEVITCH, Paja, psychologist; b. Obrez, Srem, Yugoslavia, Jan. 11, 1879; ed. U. Zemun, U. Vienna, U. Jena; Ph.D., U. Zurich, Tchr. philosophy and pedagogical sci., Asonjam; prof. pedagogy and German, Yugoslavia; staff lab. N.Y. U.; faculty N.Y. U., now prof., tchr. exptl. psychology lab. Author: Experimentelle Grundlegenung der Gedächtnispsychologie, 1906; Das Behalten und Vergessen bei Kindern und Erwachsenen nach experimentellen Untersuchungen; Das Fortschreiten des Vergessens mit der Zeit, 1907; Entwürfe der Psychologiezweige der Wissenschaft; Der Entwurf der allgem. Psychologie für die Lehrer, 1908; Professor Boas' New Theory of the Form of the Head; a Critical Contribution to School Anthropology, 1911; New Moments in the Education, 1912; The Development of the Child Within the School Years, 1913; Who are the Slavs? a Contribution to Race Psychology, 1919; Modern Plans for the Instruction of Pupils, 1922; Positive Pedagogy of Ozist Kont, 1923; also numerous essays, articles. Research in psychology and sociology. Office: N.Y. U., N.Y.C.

RADWANSKI, Pierre Arthur, anthropologist; b. Latoszyn, Poland, Jan. 4, 1903; s. Henry and Marie (Kawecka) R.; Baccalaureat U. Krakow, Poland, 1922, Licentiate of Philosophy Jagellonian U., Krakow, 1926, D.Sc., 1931; m. Isabel Latoszynska, Aug. 2, 1937; 1 son, George. Asst. phys. anthropology Anthrop. Inst. Jagellonian U., 1922-31, prof. anthropology, ethnology Coll. Scis., 1931-39; research fellow anthrop. sect. Royal Inst. Natural Scis. Belgium, 1947-51; prof. phys., cultural anthropology U. Ottawa, Can., 1952-54; prof. anthropology, U. Montreal, Que., Can., 1952—, head Slavic Dept., 1962-68. Francqui Found. Belgium fellow, 1950; Queen Elisabeth of Belgium fellow, 1951; Dept. Resources and Devel. Can. fellow, 1953. Mem. Royal Anthrop. Soc. Belgium, N.Y. Acad. Scis., Polish Inst. Arts and Scis. (sec. gen. 1960-62), Polish Assn. Intellectuals, Veritas (pres. 1957—). Publs. on studies of synthesis of sci., sci. of man; anthrop. structure of various peoples; classification, systematics of culture. Home: 4855 Grosvenor Av., Montreal, Que., Can.*

RADWAY, Laurence Ingram, Am. polit. scientist; b. S.I., N.Y., Feb. 2, 1919; s. Frederick Stanley and Dorothy Radway; B.S., Harvard, 1940, M.A., 1946, Ph.D., 1950; M.P.A., (hon.), U. Minn., 1943; M.A. (hon.), Dartmouth, 1958; m. Patricia Ann Headland, Aug. 20, 1949; children—Robert Russell, Carol Sinclair, Michael Porter, Deborah Brooke. Tutor, teaching fellow Harvard, 1946-50; faculty Dartmouth, Hanover, N.J., 1950—, now prof. govt., chmn. dept., 1959-62, co-dir. Comparative Studies Center, 1963—; prof. fgn. affairs Nat. War Coll., 1962-63. Mem. bd. advisers Indsl. Coll. Armed Forces, 1959-62; civilian aide to sec. army, 1962—; cons. Office Sec. Def. Mem. Council on Fgn. Relations, Am. (nat. council 1965—), New Eng. (pres. 1964-65) polit. sci. assns., Am. Soc. Pub. Adminstrn. Author: (with John W. Masland) Soldiers and Scholars: Military Education and National Policy, 1957; also articles. Research in analysis and evaluation of service acads., war colls., analysis of mil. behavior in interservice and internat. instns., comparative and hist. analysis of variables that determine polit. power and policy orientations of mil. elites. Home: 22 Occom Ridge, Hanover, N.H.*

RAE, Bennet Birnie, Scottish biologist; b. Aberdeenshire, Scotland, Mar. 16, 1906; s. John and Christian (Lorimer) R.; M.A., Aberdeen (Scotland) U., 1926, B.Sc. with Honors in Zoology, 1929, Ph.D., 1938; m. Sophia Hutchison Kennedy, Apr. 15, 1936; children—Alison Fay, Richard Bennet, Colin Kennedy, Linden Christian. Kilgow scholar, Aberdeen U., also Carlsberg Lab., Copenhagen, Denmark, 1929-31; apptd. jr. naturalist Fishery Bd. for Scotland (now Scottish Home Dept.), 1931, prin. sci. officer, 1948, sr. prin. sci. officer, asst., dir., 1959. Fellow Royal Soc. Edinburgh. Author: The Lemon Sole, 1965; also numerous articles. Research on biology of lemon sole (Microstomus kitt), and other flatfishes including plaice, halibut, megnin and turbot; parasites of comml. fish, especially halibut, cod; predation of fish schools in Scottish waters, by common and grey seals food of seals and porpoises. Home: 31 Kingshill Av., Office: Marine Lab., Victoria Rd., Aberdeen, Scotland.*

RAE, John, physician, Arctic explorer; b. Stromness, Orkney Islands, Sept. 30, 1813; s. John Rae. Student, Edinburgh, Scotland, 1829-33; LL.D., U. Edinburgh; M.D., McGill Coll., Montreal, Que., Can.; m. Catharine Jane Alicia Thompson, 1860. Surgeon, Hudson Bay Co., 1833-45; mem. expdn. to Arctic, 1846; mem. Survey Shores of Committee Bay; mem. Richardson Expdn. in search of Franklin, 1848-49; became head similar expdn., 1850; made survey and mapped part of Wollaston Land, S. and E. coast of Victoria Land; joined expdn. to survey No. Coasts of Am., 1853; toured U. S., 1858; made survey for telegraph line to Am. from Eng. via Faroes, Iceland and Greenland, 1860; began telegraph survey from Winnipeg, Can. over Rockies to Pacific Coast, 1864; one of 1st dirs. Can. North-West Land Co.; dir. various bus. enterprises, Man., also B.C., Can. Recipient Founder's Gold medal Royal Geog. Soc., 1852. Fellow Royal Soc., 1880; mem. Royal Colonial Inst., Royal Geog. Soc., Imperial Inst. (gov.). Author: Narrative of an Expedition to the Shores of the Arctic Sea in 1846 and 1847, 1850; also articles. Proved King Williams' Land is an island; used solar heat to raise balloon. Died London, July 22, 1893.

RAE, Kenneth MacFarlane, oceanographer; b. Harwich, Eng., Jan. 15, 1913; s. John Macfarlane and Edith (Stephenson) R.; B.Sc., U. Coll., London, Eng., 1935; Ph.D., London, U., 1958; m. Ray W. Womer, Apr. 27, 1946; 1 son, Stephen B. Leverhulme fellow U. Coll., Hull, Eng., 1935-41, head dept. oceanography, 1946-50; dir. Scottish Marine Biol. Assn., Edinburgh, 1950-57; dir. marine labs. Tex. A. and M. Coll., 1957-61; dir. Inst. Marine Sci., U. Alaska, College, 1961-65, v.p. for research and advanced study, 1963—. Decorated officer Order of Brit. Empire. Fellow Royal Soc. Edinburgh; mem. Am. Inst. Biol. Scis., A.A.A.S., Am. Soc. Limnology and Oceanography, Challenger Soc., Sigma Xi. Editor, Bulls. Marine Ecology, 1948-57; Limnology and Oceanography, 1959-63. Research, publs. on planketon and its relation to fisheries, biology. Home: 520 Copper, Campus, U. Alaska, College, Alaska 99701.*

RAEKALLIO, Jyrki, Finnish forensic scientist, histochemist; b. Raahe, Finland, Mar. 27, 1929; s. Yrjö and Irja (Weiste) R.; M.D., U. Helsinki Sch. Medicine, 1954; postgrad. Medizinische Akademie Düsseldorft, Germany, 1958-59; m. Eeva Mustakallio, Dec. 12, 1953; children—Matti, Liisa, Pekka. With U. Helsinki (Finland) Sch. Medicine, 1955-63, sr. lectr. forensic medicine, 1963; Fulbright fellow Johns Hopkins, Balt., 1963; asso. prof. forensic medicine U. Turku (Finland), 1962, prof., chmn. forensic medicine, 1963—. Sci. councillor Finnish Central Med. Bd., 1962—. Fellow Royal Micros. Soc. (London); mem. Scandinavian Soc. for Forensic Medicine (past sec.), Finnish Medico-Legal Soc. (dir.), Deutsche Gesellschaft für gerichtliche Medizin, (corr.), Internat. Acad. Legal Medicine, Am. Soc. for Cell Biology, Histochem. Soc., Gesellschaft für Histochemie. Author: Histochemical Studies on Vital and Post-mortem Skin Wounds, 1961; Die Alterbestimmung mechanisch bedingter Hautwunden mit enzymhistochemischen Methoden, 1965; also articles. Discovered histochem. distinction between vital (inflicted during life) and post-mortem skin wounds; devised biol. time-table for dating of wounds based on enzyme reactions; discovered there is an immediate enzymatic response to injury; research on enzyme histochemistry of skin. Home: 13 Linnankatu, Turku, Finland.*

RAETHER, Heinz Arthur, German physicist; b. Nuremberg, Oct. 14, 1909; s. Arthur and Margarethe Raether; Ph.D., U. Munich; m. Hertha Daxenberger, 1937; children—Christoph, Joachim, Martin, Katharina, Stephen, Maria. Instr., U. Iena, 1937; prof. U. Hamburg, 1951—. Mem. Gottingen Akademie, Joachim Jungius-Gesellschaft Hamburg, French Soc. Electronic Microscopy (hon.). Author: Electron Avalanches and Breakdown in Gases, 1964; also articles on elec. discharge in gases. Research on interaction of electrons with solid bodies (diffraction of electrons, inelastic diffusion of electrons); gas discharges (elementary processes, electron avalanches, spark); interaction of electrons with crystals (elastic and inelastic scattering, plasma oscillation). Home: Siriusweg 1, Hamburg-Wellingbuttel. Office: Jungiusstrasse 11, Hamburg 36, W. Germany.

RAETTIG, Hansjürgen Bruno, German microbiologist; b. Stralsund, Oct. 12, 1911; s. Hans and Ilse (Schultze) R.; M.D., U. Greifswald; m. Ingrid Ripphoff; children—Johanna, Julia, Cornelia, Thomas, Christiane, Barbara. San. officer, 1939-44; asst. Greifswald Hygiene Inst., 1944-47, dir., 1946-48; sci. asst. Robert Koch Inst., Berlin, 1948—; sci. cons.; Bundesgesundheitsambt prof. Free U. Berlin. Mem. German Soc. Hygiene and Microbiology, Berlin Soc. Microbiology. Author: Gundlagen und Praxis der Seuchenstatistik; Typhusimmunität und Schutzimpfung; Bakteriophagie; Poliomyelitis Immunität; also articles on epidemiology, immunity and microbiology. Home: Senheimer Strasse 45a. Office: Nordufer 20, West Berlin, Germany.

RAFFEL, Sidney, Am. physician; b. Balt., Aug. 24, 1911; s. Saul and Leah (Katlin) R.; A.B., Johns Hopkins, 1930, Sc.D., 1933; M.D., Stanford, 1943; m. Yvonne Fay, Apr. 15, 1938; children—Linda, Gail (Mrs. Robert Drewes), Polly, Cynthia, Emily. Faculty, Stanford, Sch. Medicine, 1935—, prof. med. microbiology, 1948—, departmental exec., 1953—, acting dean, 1964-65. Cons., Va Hosp., Palo Alto, Cal., 1946—, NIH, 1953-63. Guggenheim fellow, 1950-51; Sr. fellow USPHS, 1961-62. Mem. Am. Assn. Immunologists, Am. Soc. Microbiologists, Soc. for Exptl. Biology and Medicine Am. Thoracic Soc., Alpha Omega Alpha. Author: Immunity, 1953; also numerous articles. Research on process of immunity in infections, in tissue grafts, constituents of body. Home: 770 Santa Ynez St., Stanford, Cal. 94305.*

RAFFLES, Sir Thomas Stamford, biologist; b. at sea off Port Morant, Jamaica, July 5, 1781; s. Benjamin R.; pvt. study langs. and natural scis.; LL.D.; m. Mrs. W. Fancourt, 1805; m. 2d, Sophia Hull, 1816; 5 children. Became asst. sec. E. India Co., Penang, 1805; named sec., 1807, also registrar recorder's ct.; named lt.-gov. of Java, 1811, of Bencoolen, Sumatra, 1818; lost notes for history of Sumatra, also large collection of plants and animals in ship fire on return to Eng., 1824. Founder, 1st pres. London Zoo. Fellow Royal Soc., 1817. Author: History of Java, 2 vols., 1817. Studies on products, history, and lang. of Java; Eastern sea regions; discovered flower, Rafflesia arnoldi (named after him), of a fungus parasite. Died July 5, 1826.

RAFINESQUE, Constantine Samuel, naturalist; b. Galata, Constantinople, Turkey, Oct. 22, 1783; s. G. F. Rafinesque; M.A., Transylvania U.; m. Josephine Vaccaro, 1809, 2 children, including Emily. Came to U. S., 1802; collected bot. specimens, So. N.J. and Dismal Swamp of Va., 1804; went to Palermo, Sicily, 1805; sec., chancellor to Am. consul, Palermo; exporter squills, medicinal plants, Palermo, 1808; returned to U. S., 1815; explored Hudson Valley, Lake George, L.I. and other regions, 1815-18; prof. botany, natural history, modern langs. Transylvania U., 1818-26; advocated Jussieu's method of classification. Wrote and published books and articles on botany, ichthyology, banking, econs. and other topics, including: Icthyologia Ohioensis, 1820; Medical Flora of the United States, 1828-30; (autobiography) A Life of Travels and Researches in North America and South Europe, 1836. Died Phila., Sept. 10, 1840.

RAFN (or RAVN), Carl Christian, Danish archeologist; b. Brahesborg/Fünen, Jan. 16, 1795; librarian, Copenhagen, Denmark. Author: Fornmanna Sögur, 12 vols., from 1825; Fornaldar-Sögur Nordrlanda, 3 vols., 1829-30; Antiquitates americanae, 1837. Proved that Vikings reached Am. under Leif Eriksson's leadership, circa 1000. Died Copenhagen, Oct. 20, 1864.

RAFTER, Thomas Athol, New Zealand chemist; b. Wellington, New Zealand, Mar. 5, 1913; s. Michael Edward and Grace (Clarkson) R.; B.Sc., Victoria U., Wellington, 1934, M.Sc., 1936; m. Ruby Valerie Organ, Dec. 2, 1939; children—Kenneth Reynold, Ian Lester, Janet Eleanor. Clerical officer New Zealand Govt., 1933-34; secondary sch. tchr. St. Patrick's Coll., Silverstream, New Zealand, 1934-35, Wellington, 1937-39; with Dept. Sci. and Indsl. Research, Wellingon, 1940—, dir. Inst. Nuclear Scis., 1959—. Fellow Royal Soc. New Zealand, New Zealand Inst. Chemistry (asso.), New Zealand Assn. Scientists. Research, publs. on use of carbon-14 and tritium in meteorology, hydrology, archaeology and geology; variations in sulphur isotopes in nature and their applications in volcanism ore genesis and biogeochemistry; oxygen isotope variations in sulphates and its application in isotope temperature studies. Home: 16 Simla Crescent, Wellington, New Zealand. Office: Inst. Nuclear Scis., Lower Hutt, New Zealand.*

RAFUSE, Robert Pendleton, Am. elec. engr.; b. Newton, Mass., Dec. 7, 1932; s. Albert E. and Faith E. (Pendleton) R.; B.S. summa cum laude in Elec. Engring., Tufts Coll., 1954; S.M., Mass. Inst. Tech., 1957, Sc.D., 1960; m. Mary Jane Lounsbury, June 13, 1959; 1 son, Robert Pendleton, II. Technician, jr. engr. Atomic Instruments Co., Cambridge, Mass., 1950-52, Elec. Engring. Research Lab., Tufts Coll., Medford, Mass., 1952-54; faculty Mass. Inst. Tech., 1957—, asso. prof. elec. engring, 1964—, staff Research Lab. Electronics, 1960—. Cons. to pvt. cos., govt. agys. Recipient Outstanding Paper award Internat. Solid State Circuits Conf., 1962, 66, Teaching award Mass. Inst. Tech., 1957. Mem. Am. Geophys. Union, Am. Ordnance Assn., Am. Inst. Aeros. and Astronautics, A.A.A.S., Sigma Xi, Tau Beta Pi, Sigma Pi Sigma. Author: (with P. Penfield, Jr.) Varactor Applications, 1962; also articles, chpt. in book. Developed unified theory for treating noise and energy conversion in varactor diode circuits; research on state-of-the art in solid-state power generation, amplification and conversion, microwave solid-state circuits, electromagnetic compatibility. Home: 77 Beaconsfield Rd., Brookline, Mass. 02146. Office: Mass. Inst. Tech., Cambridge, Mass. 02139.*

RAGATZ, Roland Andrew, Am. chem. engr.; b. Prairie du Sac, Wis., Dec. 1898; s. John Jacob and Anna (Tarnutzer) R.; B.S. in Chem. Engring., U. Wis., 1920, M.S., 1923, Ph.D., 1931; m. Nancy Gertrude Hansen, Mar. 15, 1930; children—Helen Karen, Andrew Roland. Faculty, U. Wis., 1920—, prof. chem. engring., 1942—, chmn. dept. 1941-46, 49-51, 55-64; research engr. A. O. Smith Corp., Milw., 1929-30. Recipient Benjamin Smith Reynolds Award, 1959. Mem. Am. Inst. Chem. Engrs., Am. Chem. Soc., Am. Soc. for Engring. Edn. (chmn. chem. engring. div. 1947-48, dir. 1948) Am. Soc. for Metals, Tau Beta Pi, Sigma Xi, Phi Lambda Upsilon, Alpha Chi Sigma. Collaborator: Applications of Chemical Engineering, 1940. Co-author: Chemical Process Principles, Vol. I, Material and Energy Balances, Vol. II, Thermodynamics; Chemical Process Principles Charts. Home: 4310 Cherokee Dr., Madison, Wis. 53711.*

RAGGATT, Harold George, Australian geologist; b. North Sydney, N.S.W., Australia, Jan. 25, 1900; s. Percy Claude and Annie (Barker) R.; B.Sc., Sydney (Australia) U., 1922, M.Sc., 1932, D.Sc., 1939; m. Edith Thora Raggatt, Jan. 22, 1927; 1 dau., Marcia Jean (Mrs. James Vane Lindesay). Staff, Geol. Survey New S. Wales, 1922-39; asst. Commonwealth geol. adviser, 1939; commonwealth geol. adviser, 1940-52; dir. Australian Bur. Mineral Resources, 1942-51; sec. Australian Dept. Nat. Devel., 1951-65; dep. chmn. Australian AEC, 1956-65, cons., 1965—; chmn. Snowy Mountains Council, 1956-65; geol. cons., Camberra, Australia, 1965—. Cons. Broken Hill Pty. Co. Ltd., Australian Mut. Provident Soc.; dir. Ampol. Expln. Ltd. Decorated comdr. Order Brit. Empire; created knight batchelor, 1963; recipient award of merit Profl. Officers Assn. Australia, 1965. Fellow Australian Acad. Sci.; mem. Australasian Inst. Mining and Metallurgy (life, medal 1964), Geol. Soc. Australia, Am. Assn. Petroleum Geologists, Soc. Econ. Geologists. Author: Mountains of Ore., A Survey of Australia's Mineral Resources, 1967; also numerous articles. Research on stratigraphy of Permian and Triassic rocks of Eastern and Western Australia especially resources of coal and petroleum; first to put geol. mapping and assessment of mineral resources in Australia on systematic basis. Home: 17 Glascow Pl., Hughes. Office: A.M.P. Bldg., Hobart Pl., Canberra, A.C.T., Australia.*

RAGHAVAN, Valayamghat, Indian plant biologist; b. Edavanakad, India, Mar. 19, 1931; s. P. Narayanan Kartha and Narayani Amma; B.Sc., U. Madras, 1950; M.Sc., Benares Hindu U., 1952; D.Phil., Gauhati U., 1957; Ph.D., Princeton, 1961; m. Lakshmi Menon, Dec. 21, 1962; 1 dau., Anita. Research fellow Gauhati U., Assam, India, 1952-57; editorial asst. Council Sci. and Indsl. Research, New Delhi, 1957-58; research asso. Harvard, Cambridge, Mass., 1961-63; lectr. botany U. Malaya, Kuala Lumpur, 1963—; guest investigator Rockefeller U., N.Y.C., 1966-67. Mem. Bot. Soc. Am., Soc. Developmental Biology, Am. Soc. Plant Physiologists, Japanese Soc. Plant Physiologists. Research, publs. on biology of arecanut palm; discovered balanced requirement of hormonal substances in regulating growth of very small plant embryos in culture; role of ribonucleic acid in differentiation of fern gametophytes. Home: 5 Lorong University, Petaling Jaya, Malaysia.*

RAGINSKY, Bernard Boris, physician; b. Ekatevindslav, Russia, Oct. 10, 1902; s. Abraham and Rena (Ratchevsky) R.; M.D., C.M., McGill U., 1927; m. Helen Theresa Steinkopf, May 31, 1938; 1 dau., Nina Barbara. Asso. physician Mayor of N.Y.C. Com. on Drug Addiction, 1929-30; demonstrator pharmacology McGill U., 1931-33; asso. physician Montreal (Can.) Jewish Gen. Hosp., 1934-62, hon. attending physician, 1962—; practice medicine, specializing in psychiatry and internal medicine, Montreal, 1930—. Dir., Inst. for Research in Hypnosis, N.Y.C., 1955—; sci. adviser Morton Prince Clinic Hypnotherapy, N.Y., 1965—; nat. sci. adv. council Inst. Comprehensive Medicine, 1966; also adv. editor various fgn. publs. Recipient Gold medal for Distinguished Profl. Contbns. to Sci. Hypnosis, Soc. Clin. and Exptl. Hypnosis, 1958; Diplomate Royal Coll. Physicians Can., Fellow Internat. Coll. Anesthesia, Am. Coll. Cardiology, A.A.A.S., Internat. Coll. Angiology, Acad. Psychosomatic Medicine (pres. 1957-58, Bronze award 1958), Soc. Clin. and Exptl. Hypnosis (pres. 1955-58; created Bernard B. Raginsky Annual award 1959), Internat. Soc. Clin. and Exptl. Hypnosis (Bronze plaque as founding pres. 1958), others. Author: Psychosomatic Medicine: Its History, Development, and Teaching, 1948; also articles. Research in devel. of tribromethynol and cyclopropane anesthetics; conditioned reflexes and devel. of neuroses in animals by this means; artificial heart expts.; temporary cardiac arrest through hypnosis; originator sensory hypnoplasty. Home: 2 Westmount Sq., Montreal 6. Office: 376 Redfern Av., Montreal 6, Que., Can.*

RAGOTZKIE, Robert Austin, Am. oceanographer; b. Albany, Sept. 13, 1924; s. Robert William and Edith (Van Wormer) R.; B.S., Rutgers U., 1948, M.S., 1950; Ph.D., U. Wis., 1953; m. Elizabeth M. Post, Aug. 25, 1949; children—Peter, Kim, Susan. Asst. prof., research coordinator Marine Biology Lab., U. Ga., Sapelo Island, 1954-57, asso. prof., dir. Marine Inst., 1957-59; faculty U. Wis., Madison, 1959—, prof., 1965—, chmn. dept. meteorology, 1964-67. Fellow A.A.A.S.; mem. Am. Meteorol. Soc., Am. Geophys. Union, Am. Soc. Limnology and Oceanography, Ecol. Soc. Am., Internat. Limnology Soc., Internat. Soc. Biometeorology. Research, publs. on biol. productivity and hydrography of estuaries, circulation and heat balance of lakes, infrared temperature measurement and circulation patterns of Great Lakes. Home: 2334 Tanager Trail, Madison, Wis. 53711.*

RAHN, Hermann, Am. physiologist; b. East Lansing, Mich., July 5, 1912; s. Otto and Bell S. (Farrand) R.; A.B., Cornell U., 1933; student U. Kiel, 1933-34; Ph.D., U. Rochester, 1938; hon. Docteur U. Paris, 1964; LL.D., Yonsci U., Korea, 1965; m. Katharine F. Wilson, Aug. 29, 1939; children—Robert F., Katharine B. NRC fellow Harvard, 1938-39; instr. physiology U. Wyo., 1939-41; asst. physiology Sch. Medicine U. Rochester, 1941-42, instr., 1942-46, asst. prof., 1946-50, asso. prof., vice chmn. dept., 1950-56; Lawrence D. Bell prof. physiology, chmn. dept. Sch. Medicine State U. N.Y., Buffalo, 1956—; vis. prof. Med. Faculty San Marcos U., Lima, Peru, 1955, Dartmouth Med. Sch., 1962. Mem. adv. com. biol. sci., Air Force Office Sci. Research and Devel., 1958-64; physiol. study sect. NIH, 1958-62, bd. sci. counsel Nat. Heart Inst., 1965—; working com. space sci. bd. Nat. Acad. Sci.-NRC, 1962-65. Fellow A.A.A.S., Am. Acad. Arts and Scis.; mem. Nat. Acad. Scis., Am. Physiol. Soc. (council mem. 1960-65, pres. 1963-64), Harvey Soc. (hon.), Internat. Union Physiol. Sci. (council 1965—), Soc. Exptl. Biology and Medicine, Am. Soc. Zoologists, Sigma Xi. Author (with W. O. Fenn) A Graphical Analysis of the Respiratory Gas Exchange, 1955, Handbook Physiology-Respiration, Vol. I, 1964, Vol. II, 1965. Editor: (with W. O. Fenn) Physiology of Breath-Hold Diving and the Ama of Japan, 1965. Mem. editorial bd. Am. Jour. Physiology, Jour. Applied Physiology, 1953—. Research on pulmonary physiology, especially ventilation-blood flow relationships in lung; effects of environmental stresses on man, including desert climates, high altitude, diving. Home: 75 Windsor Av., Buffalo 14209.*

RAHN, Johann Heinrich, Swiss mathematician, astronomer; b. Zurich, Switzerland, Mar. 10, 1622; s. Hans Heinrich Rahn; a son, Jean-Henri. Became town treas., Zurich, 1659. Author: Teutsche Algebra, 1659, translated into English by J. Pell, 1688; Philologischer Discurs über der Cometen Bedeutung, 1665; Algebra speciosa, 1667. First to use sign for division, 1659. Died 1676.

RAICHLE, Ludwig, German chem. engr.; b. Kirchheim, Teck, West Germany, Mar. 9, 1902; s. Hermann and Mathilde (Strobel) R.; M.Sc., Tech. U., Stuttgart, Germany, 1926, hon. Senator, 1964; m. Magda Waldmann, Feb. 14, 1931; children—Marianne (Mrs. Karl Kayser), Brigitte (Mrs. Heinz Schlage), Hermann, Irmgard. With planning and constrn. dept. Badische Anilin- & Soda-Fabrik AG, Ludwigshafen, Rhein, Germany, 1926—, chief engr., supr. all German hydrogenation plants, 1938-47, head central and prodn. plant workshops., 1947-58, head chem. engring. research dept., 1958-60, tech. dir., 1960—. Mem. Instn. German Engrs., S. German Plastic Centre. Research and publs. on chem. engring. and material problems of chem. prodn., devel. high pressure techniques. Patentee in field. Home: Weinbietstrasse 12, Limburgerhof, Pfalz, West Germany. Office: Badische Anilin- & Soda-Fabrik AG, Ludwigshafen, Rhein, West Germany.*

RAICU, Petre, Rumanian geneticist; b. Craiova, Rumania, Jan. 30, 1929; s. Constantin and Maria (Mita) R.; Dr.Genetics, U. Bucharest, 1956; m. Cristina Antoniu, Dec. 1, 1951. Asst., asst. chief Agronomical Inst. Bucharest (Rumania), 1951-57; prof. genetics U. Bucharest, 1957—; chief genetics lab. Biol. Inst., Rumanian Acad. Scis., 1958—. Mem. Nat. Council for Sci. Research, 1965—. Mem. Eucarpia. Author: New Methods in Genetics, 1962; Genetics in Actuality, 1966; Genetics, 1967; also articles. Research on cytogenetics of plants and animals, including polyploidy, evolution of karyotype, chem. mutagenesis. Home: 35, St.C. Sandu-Aldea. Office: 1-3, Aleea Portocalilor, Bucharest, Rumania.*

RAIMY, Victor Charles, Am. psychologist; b. Buffalo, Mar. 17, 1913; s. Christian and Alberta (Wiedman) R.; B.A., Antioch Coll., 1935; Ph.D., Ohio State U., 1943; m. Ruth Vendig, July 9, 1937; 1 son, Eric. Faculty, Ohio State U., 1941-43, 46-48; asst. prof. U. Pitts, 1946; prof. U. Colo., Boulder, 1948—, chmn. dept. psychology, 1954-62. Cons. govt. agys. Mem. Am. Psychol. Assn. (pres. div. clin. psychology 1962-63, dir. 1959-62, 64—). Contbr.

articles on research in psychotherapy, diagnosis, phrenophobia to tech. jours. Home: 1333 King Av., Boulder, Colo. 80302.*

RAINWATER, James, Am. physicist; b. Council, Ida., Dec. 9, 1917; s. Leo Jasper and Edna (Teague) R.; B.S. in Physics, Cal. Inst. Tech., 1939; M.A., Columbia U., 1941, Ph.D., 1946; m. Emma Louise Smith, Mar. 7, 1942; children—James Carlton, Robert Stephan, Elizabeth Ann (dec.) William George. Mem. faculty Columbia U., N.Y.C., 1939—, prof. physics, 1952—; dir. Nevis Cyclo Labs, 1951-53, 56-61; research sci. AEC, Office Naval Research projects, 1947—; adviser NSF, 1956-58, Oak Ridge Nat. Lab., 1962—. Recipient Ernest Orlando Lawrence award AEC, 1963. Fellow Am. Phys. Soc., A.A.A.S., N.Y. Acad. Sci., I.E.E.E.; mem. Optical Soc. Am., Am. Assn. Physics Tchrs. Research in neutron cross sections, 1942-49, 57—, neutron time of flight resonance spectroscopy, 1942—, angular distbn. of pion scattering by nuclei, 1954-62; 1st suggested spheroidal nuclear model, 1950; pioneered research on mu-mesonic X-rays, 1952—. Home: 342 Mount Hope Blvd., Hastings-on-Hudson, N.Y. 10706. Office: Dept. Physics, Columbia, N.Y.C. 10027.*

RAINWATER, Lee, Am. sociologist; b. Oxford, Miss., Jan. 7, 1928; s. Percy Lee and J. Tennis (McDowell) R.; student George Washington U., 1944-45, U. So. Cal., 1945-46; M.A., U. Chgo., 1950, Ph.D., 1954; m. Carol Lois Kampel, July 16, 1959; children—Jonathan Lee, Katherine Anne. Asso. dir. Social Research Inc., Chgo., 1950-63; asso. prof. sociology, anthropology Washington U., St. Louis, 1963-65, prof., 1965—, exec. dir. Pruitt-Igoe Research, Social Sci. Inst., 1963—; sr. editor Transaction Mag., 1963—. Mem. Soc. for Study of Social Problems (mem. exec. com. 1966-67), Am. Sociol. Assn. (chmn. family sect. 1967-68), Am. Psychol. Assn., Am. Anthropol. Assn. Author (with Richard Colemen, Gerald Handel) Workingman's Wife: Her Personality, World and Life Style, 1959; And the Poor Get Children: Sex, Contraception and Family Planning in the Working Class, 1961; Family Design: Marital Sexuality, Family Size and Contraception, 1965; (with William Yancey) The Moynihan Report and the Politics of Controversy, 1967; also articles. Investigated implications of research for Am. poverty for nat. anti-poverty-war policies and programs; analyzed dynamics of poverty and racial discrimination in Negro ghettoes; documented needs and problems of poor for family planning services. Home: 2 Ladue Hills, St. Louis 63132.*

RAINWATER, Leo James, Am. physicist; b. Council, Ida., Dec. 9, 1917; s. Leo Jasper and Edna (Teague) R.; B.S. in Physics, Cal. Inst. Tech., 1939; M.A., Columbia U., 1941, Ph.D., 1946; m. Emma Louise Smith, Mar. 7, 1942; children—James Carlton, Robert Stephen, Elizabeth Ann (dec. 1959), William George. Faculty, Columbia U., 1939—, prof. physics, 1952—, dir. Nevis Cyclo Labs. 1951-53, 56-61, research scientist AEC, Office Naval Research Projects, 1947—. Mem. adv. panel for phys. and elec. nuclear divs. Oak Ridge Nat. Lab., 1952-66, Argonne Nat. Lab., 1968—. Recipient Ernest Orlando Lawrence award in physics U. S. AEC, 1963. Fellow Am. Phys. Soc., A.A.A.S., N.Y. Acad. Scis., I.E.E.E.; mem. Optical Soc. Am., Am. Assn. Physics Tchrs. Contbr. author: Handder Physics, 1957; Annual Review Nuclear Science, 1957; also articles. Research on neutron time flight spectroscopy, spheroidal nuclear model, mu mesonic X-rays, theory multiple Coulomb scattering by extended nuclei, optical model for coherent scattering of pions by nuclei; pioneered establishment of nuclear charge radius. Home: 342 Mt. Hope Blvd., Hastings-on-Hudson, N.Y. 10706. Office: Physics Dept., Columbia U., N.Y.C. 10027.*

RAISZ, Erwin, geographer; b. Locse, Hungary, Mar. 1, 1893; s. Josephus and Rose (Balla) R.; diploma Royal Poly. Budapest, 1914; M.A., Columbia, 1924, Ph.D., 1929; m. Marika G. Patai, Dec. 24, 1924; 1 son, Lawrence Gideon. Came to U. S., 1923, naturalized, 1929. Faculty, Columbia, 1927-31, Harvard, 1931-50; vis. prof. Clark U., 1945-58; acting prof. U. Va., 1954-55; research prof. U. Fla., 1958-61; owner Cartographic Workshop, Cambridge, Mass., 1951—; staff mem. U. S. Nat. Atlas, 1962—. Recipient Gold Medal Soc. Geog. de Cuba, 1949; Meritorius Contbn. award Assn. Am. Geographers, 1955; Distinguished Service award Geog. Soc. Chgo., 1959. Mem. Assn. Am. Geographers, Royal Geog. Soc., Am. Congress Surveying and Mapping, Nat. Council Geog. Edn. (Goode prize 1947), many others. Author: General Cartography, 1938; Atlas of Global Geography, 1944; Mapping the World, 1956; Principles of Cartography, 1962; Atlas of Florida, 1964; also numerous articles. Research, publs. on methods of cartography; numerous maps of various kinds and locations. Address: 130 Charles St., Boston 02114.

RAJCHMAN, Jan Aleksander, elec. engr.; b. London, Eng., Aug. 10, 1911; s. Ludwik W. and Maria (Bojanczyk) R.; Diploma EE, Swiss Fed. Inst. Tech., Zurich, 1935, Dr.Sc., 1938; m. Ruth V. Teitrick, June 30, 1944; children—Alice R., John A. Came to U. S., 1935, naturalized, 1941. Staff, RCA Mfg. Co., 1935-42, RCA Labs., 1942-61, asso. dir. System Research Lab., Princeton, N.J., 1959-61; dir. Computer Research Lab., 1961-67, staff v.p. data processing research, 1967—. Recipient Levy medal Franklin Inst., 1947; RCA awards, 1948, 51, 54; Liebmann Meml. prize, 1960. Fellow I.E.E.E., Am. Phys. Soc., Nat. Acad. Engring., Assn. for Computing Machinery. Contbr. numerous articles to tech. jours. Pattentee in field. Developed electrostatically focused electron multiplier; invented and developed magnetic core memory; developed many digital computing circuits; invented transfluxor which started field of multiaperture magnetic elements. Home: 268 Edgerstoune Rd. Office: RCA Labs., Princeton, N.J. 08540.*

RAKE, Geoffrey William, physician, bacteriologist; b. Fordingbridge, Eng., Oct. 18, 1904; s. Herbert Vaughan and Rosemary (Satchell) R.; student Cliff House, Bournemouth, Eng., 1910-11, Christ Ch., Choir Sch., Oxford, Eng., 1911-19, King's Sch., Canterbury, Eng., 1919-21; Guy's Hosp. Medical School, London, Eng., 1922-28, M.B., B.S., 1928; came to U. S., 1928; became naturalized U. S. citizen, Dec. 14, 1942; m. Orpha May McNutt, July 1, 1932; 1 son, Adrian Vaughan; m. 2d, Helen Jones, March 23, 1946; children—Geoffrey, Juliet, James, Jane, Neave. House officer Guy's Hosp., London, 1926-28; asst. pathol., Johns Hopkins U., 1928-29, instr., 1929-30; asst. pathol., and bacteriol., Rockefeller Inst., N.Y., 1930-32, asso., 1932-36; research asso. Connaught Labs., U. Toronto, Canada, 1936-37; head div. microbiol., mem. Squibb Inst. Med. Research, New Brunswick, N.J., 1937-49; med. dir. E. R. Squibb & Sons, 1949-53; Consultant to pres., 1953-56; sci. dir. Internat. div. Olin Mathieson Chem. Corp., 1956—; director Squibb Inst. Med. Research, 1949-53; research prof. Sch. Medicine, U. Pa.; member Wistar Inst. Anatomy and Biology, Phila., 1953—. Fellow Royal Society Medicine (London), New York Academy of Medicine, American College of Physicians; licentiate Royal College Physicians (London); mem. Royal Coll. Surg. (Eng.), Harvey Soc., Am. Epidemiol. Soc. Awarded Hiltonprize in anatomy, 1923, Stokes prize in pathology, 1924, Beaney prize in pathology, 1927, gold medal in medicine, 1927—all Guy's Hosp.; Rettlinger prize, London, 1928, Stokes traveling scholarship in pathology, London, 1930. Research on nephritis, thyrotoxic heart, meningococcus, pneumococcus, measles and antibiotic substances. Died Apr. 20, 1958.

RAKESTRAW, Norris Watson, Am. chemist; b. Toledo, Jan. 16, 1895; s. Edwin and Clara (Norris) R.; A.B., Stanford, 1916, A.M., 1917, Ph.D. in Chemistry, 1921; m. Beatrice Brasefield, June 4, 1920; m. 2d, Hazel Stapp, Dec. 19, 1939; children—Douglas Cedric, Robert Norris. Instr. chemistry, Stanford, 1919-23, 24-25; asst. prof. Oberlin Coll., 1925-26, Brown U., 1926-46; prof. chemistry Scripps Instn. Oceanography, 1946—; dean grad. div. U. Cal., San Diego at La Jolla, 1961-62, acting dean, 1962—; research asso. Oceanographic Inst., Woods Hole, Mass., 1931-46; spl. research blood analysis, biochemistry, chem. oceanography. Mem. Nat. Adv. Com. Engring., Sci. and Mgmt. War Tng., 1943-46. Recipient James Flack Norris award N.E. sect. Am. Chem. Soc., 1956; award Sci. Apparatus Makers Assn., 1957. Mem. Am. Chem. Soc. (editor Jour. Chem. Edn., 1940-55), Am. Soc. Biol. Chemists, Geophys. Union, Phi Beta Kappa, Sigma Xi, Phi Lambda Upsilon. Home: 2611 Inyaha Lane, La Jolla, Cal.

RAKOFF, Abraham Edward, Am. physician; b. Phila., Feb. 2, 1913; s. Samuel and Esther (Pure) R.; A.B., U. Pa., 1933; M.D., Jefferson Med. Coll., 1937; m. Doris T. Michell, June 2, 1937; children—Jan D., Jed S., Todd D. Prof. obstetrics, gynecology Jefferson Med. Coll., Phila., 1938—, lectr. gynecic endocrinology Grad. Sch. Medicine, 1956—, prof. medicine, 1965—; practice medicine, specializing in endocrinology, Phila., 1939—; endocrinologist Jefferson Hosp. Labs., 1942—; cons. U. S. Naval Hosp., 1963—, Pa. Hosp., 1966, Meth. Hosp., 1967 (all Phila.). Mem. A.M.A., Soc. Exptl. Biology and Medicine, Endocrine Soc., Am., Internat. fertility socs., Am. Soc. Cytology (past pres.), Am. Cancer Soc. (dir. Phila. div.). Author: (with Bernstine) Vaginal Infections and Infestations, 1953; (with others) Clinical Endocrinology, 1967; also numerous articles. Research in metabolism of estrogens, biologic actions of estrogens and progestins, biologic characteristics of human vagina, effects of radiation on human ovary, influence of psychogenic factors on human reproduction, hormonal induction of ovulation. Home: 2028 Locust St., Phila. 19103. Office: 829 Spruce St., Phila. 19107.*

RAKOV, Aleksandr Ivanovich, Russian oncologist; b. Astrakhan, 1902; grad. Med. Faculty, Nizhniy Novgorod Med. Inst., 1926; postgrad. Leningrad Oncological Inst., 1932-34; D.Med. Sci., 1945. Head polyclinic, sr. asso. Leningrad Oncological Inst., 1934-50; asst., lectr. dept. oncology Leningrad Postgrad. Med. Inst., 1934-50, head chair oncology, 1953—, prof., 1949—; head surg. clinic Leningrad Oncological Research Inst., USSR Acad. Med. Scis., 1951—. Mem. Internat. Cancer League, also mem. Spl. Commn. on Tumor Classification; del. 6th, 7th, 8th Internat. Congresses Oncologists, Brazil, London, Moscow, 1962. Mem. USSR Acad. Med. Scis. (corr.), All-Union (bd. mem.), Leningrad (bd. mem.), Italian (corr.) socs. oncologists, Pirogov Surg. Soc. (bd. mem.). Mem. editorial bd. Problems of Oncology. Author numerous works including: Cancer of the lower Lip, 1952; The Principles of Anatomical Zones and Sheaths in Removal of Malignant Tumors, 1960; co-author: Thirty Years Experience of Treatment of Cancer of the Stomach, 1956. Address: Leningrad Postgrad. Med. Inst., ulitsa Saltykova-Shchedrina 41, Leningrad S-15, USSR.

RAKOVIC, Miloslav, Czechoslovakian chemist; b. Prague, Czechoslovakia, Sept. 28, 1935; s. Karel and Jana (Pumrova) R.; Ing., Inst. Chem. Tech., Prague, 1958, C.Sc., 1964; m. Eva Kovarova, Dec. 9, 1960. Chemist, dept. med. physics and nuclear medicine Charles U., Prague, 1958—, docent, 1967. Author: Activation Analysis; Nondestructive Methods of the Neutron activation Analysis of the Biological Materials; also articles. Research on use of nondestructive neutron activation analysis of biol. material without gamma spectrometer, selective activation analysis, separation procedures in activation analysis, analysis of decay curves for activation analysis; proposed new correction formula for activation analysis. Home: 548 U Kruharny Dobrichovice, Czechoslovakia. Office: 3 Salmovska, Prague, Czechoslovakia.*

RALL, David Platt, Am. pharmacologist; b. Aurora, Ill., Aug. 3, 1926; s. Edward Everett and Nell (Platt) R.; B.A., N. Central Coll., 1946; M.S. in Pharmacology (Baxter fellow), Northwestern U., 1948, Ph.D., 1951, M.D., 1951; m. Edith Levy, July 17, 1954; children—Jonathan David, Catharyn Elspeth. Research asso. pharmacology Northwestern U., 1950-51; with lab. chem. pharmacology Nat. Cancer Inst., 1953-55, with clin. pharm. and exptl. therapy service, 1955-63, head service, 1958-63, chief lab. chem. pharmacology, 1963—, asso. dir. for exptl. therapeutics, 1965—; lectr. physiology George Washington U. Sch. Medicine, 1958—; med. dir. USPHS, 1963—. Mem. chemotherapy panel Internat. Union Against Cancer, 1963—; chemotherapy study sect. NIH, 1964-65, chmn. pharmacology, toxicology, rev. com., 1965—. Trustee Mt. Desert Island Biol. Lab., Salsbury Cove, Me. Fellow A.A.A.S.; mem. Soc. Exptl. Biology and Medicine, Am. Soc. Pharmacology and Exptl. Therapeutics, Am. Assn. Cancer Research, Washington Acad. Scis. Editorial bd. proc. Soc. Exptl. Biology and Medicine, 1962-65. Research on drug distbn., especially in central nervous system, selected toxicity of anticancer agents. Home: 6601 Braeburn Pky., Bethesda, 20034. Office: Nat. Cancer Inst., Bethesda, Md. 20014.*

RALPH OF LAON, French mathematician; Lived at Laon (Aisne), early 12th Century. Succeeded brother, theologian Anselon (1030-1117), as head of cathedral school of Laon; sch. became famous under their direction. Wrote important work on the abacus containing hist. information on the subject; used Roman, also Hindu numerals; his work may be considered an abacus or an algorism. Died 1131.

RALSTON, Edgar Lee, Am. physician; b. Dubois, Pa., Dec. 27, 1911; s. Edgar Lee and Margaret J. (Camlin) R.; A.B., Muskingum Coll., 1933; M.D., U. Pa., 1937; m. Mary C. Black, Jan. 11, 1941. Mem. faculty U. Pa. Med. Sch., 1947—, prof. orthopaedic surgery, 1963—, also chmn. dept.; chief orthopaedic surgery service Hosp. U. Pa., 1963—. Mem. A.M.A., A.C.S., Am. Acad. Orthopaedic Surgeons, Am. Orthopaedic Assn., Internat. Soc. Orthopaedic Surgery and Traumatology, Am. Assn. Surgery Trauma, Alpha Omega Alpha. Home: 1105 Coventry Av., Cheltenham, Pa. 19012. Office: 3400 Spruce St., Phila.*

RAMACHANDRA, Rao B., Indian physicist; b. Visakhapatnam, A.P., India, Nov. 21, 1922; s. Barri Satyanaravana and Barri (Subadramma) R.; B.Sc. with 1st class honors, Andhra (India), 1944, M.Sc. with 1st class honors, 1945, D.Sc., 1949; m. Barri Susila, Apr. 10, 1951; 1 son, Barri Srikanth. Demonstrator, Andhra (India) U., 1945-48, faculty, 1955—, prof., 1958—, head dept. physics, 1961—; Commonwealth Council for Sci. and Indsl. Research Sr. Research fellow, 1948-50; Colombo Plan Sr. Research fellow Commonwealth Sci. and Indsl. Research Orgn. Labs., Canberra, Australia, 1951-52; vis. sr. research fellow Pa. State U., 1956. Recipient J. C. Bose Premium, 1958. Fellow Phys. Soc. U.K., I.E.E.E. (sr.), Brit. Inst. Elec. and Electronics Engrs., A.P. Acad. Scis. (founding). Research, numerous publs. on ultrasonics and iniosphere; discovered new phenomenon in diffraction of light and ionospheric drifts, absorption, and spread; developed new methods of measuring ultrasonic velocity and absorption in liquids and solids, new methods of measuring drifts. Home: 35/7 Siripuram Quarters, Visakhapatnam, Andhra, Pradesh, India.*

RAMACHANDRAN, Gopalasamudram Narayana Ayyar, Indian physicist, biophysicist; b. Ernakulam, Oct. 8, 1922; s. G. R. Narayana Ayyar and Lakshmi Ammal; I.Sc., Maharaja's Coll., Ernakulam, 1939; B.Sc. with honors U. Madras, M.A., 1942; Indian Inst. Sci., 1942-47, M.Sc., A.I.I.Ss., D.Sc.; U. Cambridge (Eng.), 1947-49, Ph.D.; m. Rajalakshmi Ramachandran, May 21, 1945; children—Ramesh Narayan, Vijaya Lakshmi, Hari Shankar. Lectr. in physics Indian Inst. Sci., Bangalore, 1946-47, asst. prof., 1949-52; exhibition scholar U. Cambridge (Eng.), 1947-49; prof., head dept. physics U. Madras, 1952—, dir. Centre of Advanced Study in Biophysics, 1964—; dean Faculty Sci., 1963—. Recipient Shanti Swarup Bhatnagar prize for phys. scis., 1961, Watumull Found. award for biophysics, 1964. Fellow Indian Acad. Scis. (mem. council, 1952, sec., 1957-58, v.p.,

1962-66), Nat. Inst. Scis. India; mem. Nat. Com. Biophysics, hon. Am. Soc. Biol. Chemists, chmn. Nat. Com. Crystallography. Editor, contbr. articles, chpts. to jours., books. Research on elucidation of molecular structure in form of triple helix, of protein collagen; developed methods for solution of conformation or molecular architecture of proteins and other biopolymers; contbd. theoretical and practical ideas for analysis of crystal structures by x-rays. Home: No. 3, Second Crescent Road, Gandhinagar, Madras-20. Office: Dept. physics, University Madras, A.C. College Buildings, Madras 25, India.*

RAMAGE, Colin Stokes, meteorologist; b. Napier, New Zealand, Mar. 3, 1921; s. James Gorman and Margaret (Stokes) R.; B.S., Victoria U., 1940; postgrad. U. Chgo.; Sc.D., U. New Zealand, 1961; m. Mary Louis Clements, Dec. 23, 1964; children—(by previous marriage) John, Kenneth. Came to U. S., 1956, naturalized, 1962. Meteorologist, New Zealand Meteorol. Service, 1940-41; meteorol. officer Royal New Zealand Air Force, 1942-46; with Royal Obs., Hong Kong, 1946-56, acting dir., 1955-56; faculty U. Hawaii, Honolulu, 1956—, prof. meteorology, 1958—, chmn. dept. geoscis., 1964—. Asso. dir. Hawaii Inst. Geophysics, 1964—; sci. dir. for meteorology, dir. U. S. Meteorology program Internat. Indian Ocean Expdn., 1961—. Fellow Commonwealth Fund of N.Y., 1953-54. Mem. Am. Meteorol. Soc., Am. Geophys. Union, Am. Assn. U. Profs. Research, publs. on various components of large-scale atmospheric circulation, devel. typhoons, steady state conditions in subtropical cyclones, origins and causes of monsoon rain systems. Home: 4959 Mana Pl., Honolulu 96816.*

RAMALINGASWAMI, V(ulimiri), Indian physician; b. India, Aug. 8, 1921; s. V. Gumpaswami and Sundaramma; M.B., B.S., Andhra U., India, 1944, M.D., 1946; D.Phil., Oxford U., 1951; m. Surya Prabha, June 13, 1947; children—Jagdish, Lakshmi. Sr. research officer, pathologist Nutrition Research Labs., Coonoor, S. India, 1951-54; dep. dir. Indian Council Med. Research, New Delhi, India, 1954-57; prof. pathology All India Inst. Med. Scis., New Delhi, 1957—. Mem. expert coms. in nutrition, anemia, cardiovascular research and med. edn. WHO. Recipient Watumull award for medicine, 1962. Mem. Path. Soc. Gt. Britain and Ireland, Am. Assn. Pathologists and Bacteriologists, Internat. Acad. Pathology. Contbg. author: Diseases of Children in Sub-tropics and Tropics, 1958; The Thyroid Gland, vol. I, 1964; Nutrition, A Comprehensive Treatise, vol. I, 1964. Research, publs. on human malnutrition, especially protein malnutrition (kwashiorkor), endemic goiter, cirrhosis of liver, nutritional anemias; experimentally produced protein malnutrition in primates; discovered cause of endemic goiter in Himalayas and contributed to its control. Home: CI-8, Med. Enclave, New Delhi-16, India.*

RAMAN, Chandrasekhara Vankata, Indian physicist; b. 1888; M.A., Ph.D., LL.D., D.Sc., Presidency Coll., Madras. Prof. physics Calcutta U., 1917-33; research asso. Cal. Inst. Tech., 1924; pres. Indian Sci. Congress, 1928; prof. Indian Inst. Sci., Bangalore, 1930; dir. Raman Research Inst. Hebbal. Recipient Mateucci medal, Rome, 1929, knighted, 1929; Hughes medal Royal Soc., 1930, Franklin medal, Philadelphia Institute, 1941; Nobel Prize in physics, 1930. Hon. fellow Royal Irish Acad., Hungarian Acad. Scis., Zurich Physics Soc., Indian Mathematical and Chem. Socs., Optical Soc. Am.; fgn. asso. French Acad. Scis., 1949; mem. Indian Acad. Scis. (pres. 1934—). Author: Theory of Bowed Strings and Violin Tone; Diffraction of X-rays; Theory of Musical Instruments; Molecular Diffraction of Light; Physics of Crystals. In important optics research, discovered that diffused light contained rays of other wave lengths (now called Raman effect), 1928; other research in light diffraction by sound waves, crystal dynamics, physiology of human vision and theory of musical instruments. Address: Raman Research Institute, Hebbal, Bangalore, India.

RAMANUJAN, Srinivasa, Indian mathematician; b. Erode, India, Dec. 22, 1887; ed. govt. coll. at Kumbakonam, India, Madras (India) U. Clk., Madras Port Trust, from 1909; research Trinity Coll., Cambridge (Eng.) U., 1914-19. Fellow Royal Soc., 1918. Author: Collected Papers, 1927, 2d edit., 1962. Original research on theory of numbers, theory of partitions, theory of continued fractions. Died Madras, Apr. 26, 1920.

RAMAZZINI, Bernardino, Italian physician; b. Carpi, Italy, Nov. 5, 1633; ed. Jesuits Sch., Carpi; M.D., Parma, Italy, 1659; student under Antonmariade Rossi, Rome, Italy. Practice medicine, Carpi, Modina, Italy; prof. medicine, Modena, 1682-1700, Padua, Italy, 1700-14. Author: De morbis artificium diatriba, 1700. Father of indsl. medicine; 1st systematic description of occupational diseases; described relationship between metals and symptoms of metal poisoning in metal workers, pneumoconiosis in miners, silicosis in stone masons, lead poisoning in painters and potters; believed cinchona bark (source of quinine) should be used only for treatment of malarial fevers and not for other types of fevers; described several plagues occuring in his area. Died Padua, Italy, Nov. 5, 1714.

RAMBERG, Hans, geologist; b. Trondheim, Norway, Mar. 15, 1917; s. Ragnvald and Olga (Petersen) R.; M.S. (Magister), Oslo U., 1943, Ph.D. (Dr. Phil.), Oslo U., 1946; m. Marie Louise Schistad, Apr. 25, 1942. Research asso. Sci. Council of Def., Oslo, Norway, 1946-47; expdn. leader, Greenland, 1946-51; prof. geology, U. Chgo., 1952-61; prof. mineralogy and petrology, U. Uppsala, Sweden, 1962—. Mem. Kungl. Vetenskap. Sweden, Kungl. Vetenskap. Soc. Uppsala, Kungl. Dansk Videnskap. Selskab, Köbenhavn Det Norske Vidensk. Akad. Oslo. Author: The Origin of Metamorphic and Metasomatic Rocks, 1952; Gravity, Deformation and the Earth's Crust, 1967; also articles. Editor: Bull. Geol. Inst., Uppsala, Tectomophysics, Elsevier, Holland. Research on origin of various rocks and ores; mineral chemistry, especially application of thermodynamics on mineral assemblages; geodynamics and structural geology. Home: 5 Murkleväg, 752-46 Uppsala, Sweden.*

RAMBERG, Walter, physicist; b. Florence, Italy, Feb. 16, 1904; s. Walter and Lucy (Dodd) R.; came to U. S., 1920; student Reed Coll., 1922-24; A.B., Cornell U., 1926; Dr.Tech.Sc., Technische Hochschule, Munich, Germany, 1930; m. Julia Elizabeth Lineberger, June 30, 1930; children—Walter Dodd, Julia Elizabeth (Mrs. W. V. Hall), Lucy Dodd. Grad. student engr. Westinghouse Elec. & Mfg. Co., Pitts., 1926-27; lab. asst. Westinghouse High Voltage Lab., Trafford, Pa., 1927-28; asst. physicist Nat. Bur. Standards, Washington, 1931-36, in charge aircraft structures group, 1937-46, chief engring. mechanics sect., 1946-47, chief mechanics div., 1947-59; sci. attaché Am. Embassy, Rome, Italy, 1959—. Mem. Washington Acad. Scis. (award 1942), Am. Soc. for Testing and Materials (Richard L. Templin award 1957), Soc. Exptl. Stress Analysts (past pres.), Philos. Soc. Washington (past pres.), Am. Soc. M.E. (past chmn. applied mechanics div.), Am. Phys. Soc., Am. Inst. Aeros. and Astronautics, Sigma Xi. Research, publs. and patents in strength of aircraft structures under static and dynamic loads, especially strength of plates, tubes and sheet stringer panels, vibration of propeller blades, response of aircraft structure to landing impact; studied technique of testing materials under static and dynamic loads, stress strain relations for structural materials. Home: Via delle Botteghe Oscure 32, Rome 00186. Office: Am. Embassy, Rome, Italy.*

RAMBO, Earle Kensington, Am. agrl. engr.; b. Ninety Six, S.C., Nov. 15, 1912; s. Benjamin T. and Lois Ann (Pratt) R.; B.S. in Agrl. Engring., Clemson Coll., 1936; M.S. in Agrl. Engring., Tex. A. and M. Coll., 1937; spl. study Purdue U., 1943, Stanford, 1955; m. Eberle Burge, Sept. 18, 1938; children—George E., Robert P., James E. Teaching fellow Tex. A. and M. Coll., 1936-37; successively instr., asst. prof., acting head agrl. engring. Dept. U. Tenn., 1937-41; asst. extension agrl. engr., S.C., summers 1936, 38, Tenn., summers 1939, 40, 41; extension agrl. engr. U. Ark., 1941-47, 48-50; vis. instr. Tex. A. and M. Coll., 1942; agrl. engring. adviser, acting head U. S. Agrl. Mission to Panama, 1947-48; agrl. engring. adviser to Govt. India, also engring. dir. agr. in India for Office Fgn. Agrl. Relations and TCA, 1950-52; engring. dir., agr. U. S. Operations Mission, Lebanon, 1952-57, Panama, 1957-58; food and agrl. officer with U. S. Operations Mission to Panama, 1960-63; water resources engr. U. S. AID, Laos, 1963—. Mem. Ark. Agrl. Machinery and Equipment Rationing Com., World War II. Decorated Order of Balboa, Comendador (Panama); Superior Service award, Silver Meritorious Service award (U. S. Govt.). Mem. Am. Soc. Agrl. Engrs., Indian Soc. Agrl. Engrs., Alpha Zeta. Home: Vientiane, Laos. Office: U. S. AID, APO 152, San Francisco.

RAMEAU, Jean-Philippe, French musician; b. Dijon, France, Oct. 23, 1683; ed. for career as magistrate; self taught in music; m. Marie-Louise Mangot, Feb. 1726. Conductor, southern France; organist Notre Dame d'Avignon, cathedral of Clermont-Ferrand; went to Paris, 1706, worked as organist; returned to Dijon, then to Lyons and Clermont-Ferrand; organist St.-Croix-de-la-Bretonnerie; returned to Paris, 1722; apptd. composer of king's chamber music. Author: Traité de l'harmonie reduite à ses principes naturels, 1722; Nouveau système de musique théorique, 1726; Plan abrégé d'une nouvelle méthode d'accompagnement; Génération harmonique, 1737; Demonstration du principe de l'harmonie, 1750; Code de musique practique, 1760; Origine des sciences, suivie d'une controverse sur le meme sujet, 1761; also operas, ballets, other music. Innovator in mus. theory and harmony; devised law of inversion in chords, system of chord bldg. by thirds upon basic common chord; conducted acoustical research. Died Paris, Sept. 12, 1764.

RAMEY, Estelle Rosemary (Mrs. James T. Ramey), Am. endocrinologist; b. Detroit, Aug. 23, 1917; d. Henry and Sarah White; M.A., Columbia, 1940; Ph.D. (Mergler scholar), U. Chgo., 1950; m. James T. Ramey, June 24, 1941; children—James North, Drucilla. Instr., N.Y. Queens Coll., 1938-41; professorial lecr. U. Tenn., 1942-47; USPHS fellow U. Chgo., 1950-51, asst. prof., 1952-56; prof. physiology Georgetown U. Med. Sch., Washington, 1956—. Mem. Am. Physiol. Soc., Am. Diabetes Assn., Endocrine Soc., A.A.A.S., Washington Heart Assn. (dir. 1964—). Author: (with D. O'Doherty) Electrical Studies on the Unanesthetized Brain, 1960; also articles. Research on relationship between adrenal cortical hormones and the nervous system in regulation blood vessel and metabolic responses to stresses. Home: 6817 Hillmead Rd., Bethesda, Md. 20034. Office: Dept. Physiology, Georgetown U. Med. Sch., Reservoir Rd., Washington 20007.*

RAMFJORD, Sigurd Peder, dentist; b. Kolvereid, Norway, June 6, 1911; s. Peter P. and Kristine (Landstrom) R.; L.D.S., Oslo U., 1934; M.S., U. Mich., 1948, Ph.D., 1951; m. Winifred Arnold, Dec. 22, 1956; 1 son, Per. Came to U. S., 1946, naturalized, 1956. Practice dentistry, Oslo, Norway, 1934-46; faculty U. Mich. Sch. Dentistry, Ann Arbor, 1950—, prof. oral pathology and periodontics, 1958—, chmn. periodontics, 1963—. Cons. VA Hosp., Ann Arbor, 1954—, Sch. Aviation Medicine dental div. Brooks AFB, 1958—. Fellow Am. Acad. Oral Pathology, A.A.A.S., Am. Coll. Dentists; mem. Am. Acad. Periodontology, Am. Dental Assn., Internat. Assn. for Dental Research, Mich. Path. Soc., Am. Acad. Periodontology (past pres.), Phi Kappa Phi, Omicron Kappa Upsilon. Author: (with M.M. Ash) Occlusion, 1966; also articles. Research in wound healing, epidermiology periodontal diseases, electromyography. Home: 393 Parklake St., Ann Arbor, Mich. 48103.*

RAMGREN, Olof Gustaf, Swedish physician; b. Stockholm, Sweden, Jan. 24, 1925; s. Erik Gustaf and Ingeborg (Hellberg) R.; M.B., Karolinska Inst. Med. Sch., 1951, M.D., 1962; m. Eva Maria Banck, Oct. 27, 1951; children—Katarina, Maria, Birgitta. Asst. chemistry dept. Karolinska Inst., 1945-53; physician City of Stockholm Blood Transfusion Centre, 1952-54, med. dir., 1955—; lectr. blood transfusion Karolinska Inst., Stockholm, 1962. Mem. Swedish Haemophilia Soc. (pres. 1964—). Author: A Clinical and Medicosocial Study of Haemophilia in Sweden, 1962; also numerous articles. Gen., statis. research in blood banking procedures; clin., medicosocial, genetical investigations in hereditary coagulation disorders; devel. of automatic lab. apparatures for cell washing, blood component separation and blood group serology. Home: 20 Leksandsvagen, Bromma, Sweden. Office: Sodersjukhuset Bloodcentre, Stockholm 38, Sweden.*

RAMIREZ, Jesús Emilio, Colombian geophysicist; b. Yolombó, Antioquia, Colombia, Apr. 25, 1904; ed. Colegio de San Ignacio, Medillín, Colombia; A.B., Colegio de San Bartolomé, 1927; M.A., Boston Coll., 1927; M.S., St. Louis U., 1931, Ph.D., 1939; D.D., Ignatius-kolleg, Valkenburg, Netherlands, 1935. Joined Soc. Jesus; prof. mineralogy, scis. Xavier Pontifical and Cath. U., Bogotá, Colombia, 1940-56, now rector; sub. dir. Geophys. Inst. Colombian Andes, 1941-43, now dir.; dir. Nat. Meteorol. Service, Colombia, 1949-50; prof. geophysics Nat. U. Colombia, Bogotá, 1951-60. Mem. permanent com. meteorology and hydrology, Bogotá. Mem. Colombian Acad. Phys. and Natural Exact Scis. (pres.), Inst. Nuclear Affairs Colombia (com.), Pan-Am. Inst. Geography and History (v.p. com. seismology), Colombian Physics Soc. (pres. 1957-59), Colombian Soc. Geology and Geophysics (1958-62), Am. Geophys. Union, A.A.A.S., Am. Meteorol. Soc., Am. Phys. Soc., Sigma Xi, others. Research, publs. on earthquakes, submarine volcanoes, aseismic housing design, radioactive materials in air. Office: Pontificia Universidad Javeriana, Carrera 7, 40-42, Bogota, Colombia.*

RAMMELKAMP, Charles Henry, Jr., Am. physician; b. Jacksonville, Ill., May 24, 1911; s. Charles Henry and Jeanette (Capps) R.; A.B., Ill. Coll., 1933, D.Sc. (hon.), 1958; M.D., U. Chgo., 1937; m. Helen Chisholm, Dec. 20, 1941; children—Charles Henry III, Colin Chisholm, Anne Capps. Johnson research fellow Boston U., 1941-43, instr. medicine, 1941-46; cons. sec. war, Washington, 1943-46; faculty medicine and preventive medicine Western Res. U., Cleve., 1946-48, prof., 1950—; dir. research labs. and medicine Cleve. Met. Gen. Hosp. Cons. Dept. Def., dir. strep. disease lab. Warren AFB, 1948-56, dir. commn. on strep. and staph. diseases Armed Forces Epidemiological Bd., 1954-57, 59—. Recipient Dist. Service award U. Chgo., 1953; Lasker award, 1954; James Bruce award 1963. Mem. Am. Assn. Immunologists, Am. Heart Assn. (Research Achievement award 1961, v.p. 1960-63), Am. Soc., Clin. Investigation (v.p. 1955), Assn. Am. Physicians, Central Soc. Clin. Research (pres. 1962), A.C.P. Important work includes treatment of infections, epidemiology of streptococcal and staphylococcal infections, prevention of rheumatic fever. Home: 3034 Berkshire Rd., Cleveland Heights, O. 44118. Office: 3395 Scranton Rd., Cleve. 44109.*

RAMMELSBERG, Karl Friedrich, German chemist, mineralogist; b. Berlin, Prussia, Apr. 1, 1813; B.A., medicine, U. Berlin, 1837. Prof., U. Berlin, 1841; prof. extraordinary of chemistry, 1845; prof. chemistry and mineralogy, Royal Industrial Institute, 1851; prof. inorganic chemistry and dir. of 2nd chem. lab., Berlin. Mem., Berlin Acad. Scis., 1874. Author: Handwörterbuch des chemischen Teils der Mineralogie, 1841; Handbuch der Mineralchemie, 1960; Grundriss der Chemie, 1881; Handbuch der Kristallographisch-physikalischen Chemie, 2 vol., 1881-82. Author of many handbooks; showed similarity of sulphur and selenium; his researches greatly improved inorganic and mineralogical chemistry. Died Grosslichterfelde, Germany, Dec. 29, 1899.

RAMO, Simon, Am. engring. exec.; b. Salt Lake City, May 7, 1913; s. Benjamin and Clara (Trestman) R.; B.S., U. Utah, 1933, D.Sc., 1961; Ph.D., Cal. Inst. Tech., 1936; D.Eng., Case Inst. Tech., 1960; D.Sc., Union Coll., 1963; m. Virginia May Smith, July 25, 1937; children—James Brian, Alan Martin. Various positions including sect. head gen. engring. lab., head, physics sect. electronics lab. Gen. Electric Co., Schenectady, 1936-46; with Hughes Aircraft Co., Culver City, Cal., 1946-53, dir. research electronics dept., dir. guided missile research and devel., v.p., dir. operations; exec. v.p., dir. Ramo-Wooldridge Corp., Los Angeles, 1953-58, pres. Space Tech. Labs., 1957-58; sci. dir. USAF ballistic missile program, Atlas, Titan, and Thor, 1954-58; exec. v.p., dir. Thompson Ramo Wooldridge, Inc., 1958-61, vice chmn. bd., 1961-64, dir., 1964—; pres., dir. Bunker-Ramo Corp., Canoga Park, Cal., 1964—. Research asso. elec. engring. Cal. Inst. Tech., 1946—; dir. Magna Corp., Union Bank; mem. various adv. coms. govt. Trustee Cal. Inst. Tech., Cal. State Colls., Am. Mus. Electricity, Case Inst. Tech., City of Hope, Aerospace Ednl. Found. Recipient numerous awards including citation Honor Air Force Assn., 1964. Fellow Am. Phys. Soc., Am. Acad. Arts and Scis., Am. Inst. Aeros. and Astronautics, I.E.-E.E.; mem. Nat. Acad. Engring. (founding mem.), Am. Rocket Soc. (dir. 1958-62), A.A.A.S., Sigma Xi, Tau Beta Pi, Phi Kappa Phi, Eta Kappa Nu, Theta Tau, Sigma Pi Sigma. Author: (with John R. Whinnery) Fields and Waves in Modern Radio, 1944; Introduction to Microwaves, 1945. Editorial bd. McGraw-Hill Electronic Sci. Series, 1966—. Editor and contbr. textbooks, articles, tech. publs. Patentee microwaves, electron optics, guided missiles, automatic controls. Home: 276 Tavistock Av., Los Angeles. Office: 8433 Fallbrook Av., Canoga Park, Cal.*

RAMON, Gaston, French bacteriologist; b. Bellechaume, France, Sept. 30, 1886; s. Leon and Clémence Ramon; ed. Vet. Sch., Alfort, France, also Pasteur Inst.; m. Martha Momont, 1917; 3 children. Became asst. Pasteur Inst., Paris, 1910, successively chief lab., chief service, sub-dir., dir. until 1940, became hon. dir., 1940. Mem. French Acad. Scis., French Acad. Medicine, Acad. Surgery. Research and numerous publs. on devel. antitoxins for vaccination against diphtheria, and tetanus; developed process which gave several immunities with single vaccination, 1923. Died 1963.

RAMON Y CAJAL, Santiago, Spanish histologist; b. Petilla de Avagon, Spain, May 1, 1852; licentiate in Medicine, Zaragoza, Spain, 1873; passed examination for doctorate in medicine, Madrid, Spain, 1877; m. 1879; 6 children. With Faculty Medicine, Zaragoza, 1877-84, mem., 1877-84; prof. anatomy, Valencia, Spain, 1884-87; prof. histology, Barcelona, Spain, 1887-92; prof. histology and pathologic anatomy, Madrid, 1892. Recipient (with C. Golgi) Nobel prize in medicine and physiology, 1906. Fellow Royal Soc., 1909; mem. French Acad. Scis., 1916. Author: Manual de anatomía pathológica general, 1890-92; Textura del sistema nervioso del hombre y de los vertebrados, 1899-1904, 3 vols.; Estudios sobre la degeneración y regeneración del sistema nervioso, 2 vols., 1913-14; Reglos y consejos sobre investagación científica, 1898. Research in structure and physiology of nerve cells especially their growth in relation to other nerve cells; developed neuron concept; investigated structure and connection of nerve cells in grey matter and spinal cord; described vertebrate retina. Died Madrid, Oct. 17, 1934.

RAMOND DE CARBONNIÈRES (le Baron Louis-François-Elisabeth), French geologist, politician; b. Strasbourg, Jan. 4, 1755; doctoral degrees in medicine and law. Counselor to Cardinal de Rohan; dep. to Legislative Assembly, Royalist Constl. Party; prof. natural history Ecole Centrale of Tarbes, 1796; dep. to Legislature, 1800-06; supported the transformation of the Consulate into the Empire; prefect of Puy-de-Dôme; master of petitions, 1818; state counsellor, Created baron, 1806; decorated comdr. Legion of Honor, 1804. Asso. French Acad. Sci., 1796, mem., 1802 (v.p. 1916, pres. 1817). Mem. Acad. Medicine. Author: Naturel et Legitime, 1804; Observations Faites Dans Les Pyrenes, 1789; Opinion sur les Lois Constitutionnelles, 1799; Voyage au Mont Perdu, 1801; Coup d'Oeil Generale et Comparative sur les Alpes et les Pyrenes, 1834. Sci. expdn. to Pyrenes to study their formation; promoted geol. study in France; did. hypsometric measurements, 1811, 13. Died Paris May 14, 1827.

RAMSAUER, Carl, German atomic physicist; b. Osternburg, Oldenburg, Germany, Feb. 6, 1879; ed. U. Munich, Tübingen, Berlin, Kiel, Ph.D., 1902; London, Breslau Us. Asst., Torpedo Laboratory, Kiel, 1902-06; asst., 1907-09, privatdocent, 1909-15, extraordinary prof., 1915-21, ordinary prof., 1921-28, U. Heidelberg; prof., Technische Hochschule, Danzig, 1928-45; dir., Allgemeine Elektrikalische Gesellschaft, Berlin, 1928-45; prof. U. Berlin, 1945—; rector, 1950-51. Studied slow electrons and protons of gas atoms; suspected wave nature of electrons in 1920. Died Berlin, Germany, Dec. 24, 1955.

RAMSAY, Sir Andre Crombie, Brit. geologist; b. Glasgow, Scotland, Jan. 31, 1814; s. William and Elizabeth (Crombie) R.; LL.D., Edinburgh, Scotland, 1866, Glascow, 1880; m. Louisa Williams, July 20, 1852; 4 daus., 1 son. Became mem. group conducting geol. survey of Gt. Britain, 1841, became a dir., 1845, sr. dir. for Eng. and Wales, 1862-72, dir. gen., 1872-81; prof. geology U. Coll., London, 1847-51; lectr. geology Govt. Sch. Mines, 1851-76. Recipient Cross of St. Maurice and St. Lazare, 1862, Neill Prize, 1866, Wollaston medal, 1871. Fellow Royal Soc., 1849, Royal medal, 1879; fellow Geol. Soc. (pres. 1862-64), Brit. Assn. (pres. geol. sect. 1856, 66, 81, assn. pres. 1880). Author: Physical Geology and Geography of Great Britain, 1863; Geology of North Wales, 1866; Geographical Map of British Isles, 1878; also articles. Research in dist. stratigraphy; drew up maps. Died Dec. 9, 1891.

RAMSAY, Sir William, British chemist; b. Glasgow, Scotland, Oct. 2, 1852; s. William and Catharine (Robertson) R.; ed. U. Glasgow, 1866-70; U. Heidelberg, 1871; U. Tübingen, Ph.D., 1872; numerous hon. degrees; m. Margaret Buchanan, 1881; 1 son, 1 dau. Asst. chemistry, Glasgow, 1872-80; prof. chem., U. College, Bristol, 1880, principal, 1881; held chair general chem., U. College, London, 1887-1913, emeritus from 1913. Knight Commander of the Bath, 1902; Nobel Prize for Chemistry, 1904; Prussian Order of Merit; Fellow Royal Soc., 1888 (Davy medal, 1895); mem. Brit. Assn. Advancement Sci. (pres. 1911); French Acad. Scis., 1895. Author: Elementary Systematic Chemistry, 1891; The Gases of the Atmosphere: the History of their Discovery, 1896; Modern Chemistry, 1900; Introduction to the Study of Physical Chemistry, 1904; Joseph Black, 1904; Elements and Electrons, 1913. Many contributions to organic, inorganic and physical chem.; discovered and investigated inert gases; discovered helium, (with Rayleigh) argon, (with Travers) krypton, neon, and xenon; studied molecular structure of pure liquids, radium emanation; determined density of gas radon. Died High Wycombe, Buckinghamshire, Eng., July 23, 1916.

RAMSAY, Sir William Mitchell, Brit. archeologist; b. Glasgow, Scotland, Mar. 15, 1851; s. Thomas and Jane (Mitchell) R.; ed. univs. of Aberdeen (Scotland), Oxford, Göttingen (Germany); D.C.L. (hon.), Oxford; LL.D., St. Andrews, Glasgow, Aberdeen (all Scotland); Litt.D., Cambridge; D.D., Edinburgh, N.Y.; Dr., U. Bordeaux (France); Dr.Theol. (hon.), Marburg, Germany, 1927; m. Agnes Dick Marshall, 1878 (d. 1927); 2 sons, 4 daus.; m. 2d, Phyllis Eileen Thorowgood, 1928. Made research trips to Asia Minor, 1880-1914, 20, 24-28; Lincoln and Merton prof. classical archeology and art, Oxford, 1885-86; prof. humanity Aberdeen U., 1886-1911; Levering lectr. Johns Hopkins, Balt., 1894; Rede lectr., Cambridge, 1906; Romanes lectr., Oxford, 1913. Recipient Gold medal Pope Leo XIII, 1893, Drexel Gold medal U. Pa for archaeol. research, 1905, Victoria Gold medal Royal Geog. Soc., 1906, medal Royal Scottish Geog. Soc., 1907. Mem. Inst. France (corr.), Imperial German Archeol. Inst., Archeol. Soc. Athens (hon.), Austrian Archeol. Inst. (hon.), Brit. Acad. (charter), Rumanian Acad. (hon.). Author: The Historical Geography of Asia Minor, 1890; The Church in the Roman Empire, 1893; The Cities and Bishoprics of Phrygia, vol. i, 1895, vol. ii, 1897; Impressions of Turkey, 1897; Was Christ Born at Bethlehem? 1898; Historical Commentary on Galatians, 1899; The Letters to the Seven Churches of Asia, 1905; Pauline and other Studies in Early Christian History, 1906; Studies in the History and Art of the Eastern Provinces of the Roman Empire, 1906; The Cities of St. Paul, 1907; The Revolution in Constantinople and Turkey, 1909; (with Gertrude Bell) The Thousand and One Churches, 1909; The Picture of the Apostolic Church, 1910; The First Christian Century, 1911; The Imperial Peace, an Ideal in European History, 1913; The Teaching of Paul in Terms of the Present Day, 1913; (with A. von Premerstein) Monumentum antiochenum, 1927; Asianic Elements in Greek Civilization, 1927; The Social Basis of the Permanence of the Roman Empire traced from the inscriptions of Sterrett and Other Travellers, in ann. parts, I, 1938; others; also articles. Made discoveries in geography and topography of Asia Minor and its ancient history; studies in Anatoliane. Died Bournemouth, Eng., Apr. 20, 1939.

RAMSAYER, Karl Heinrich, German geodesist; b. Schwäbisch Gmünd, Sept. 29, 1911; s. Karl and Wilhelmine (Seitzer) R.; ed. Stuttgart Tech. Coll.; m. Else Fischer, Dec. 14, 1938; 1 dau., Erika. Sci. collaborator German Research Establishment for Air Voyages, 1938-42, head dept. terrestrial navigation and astronomy, 1943-45; asst. Stuttgart Tech. Coll., 1946; prof. geodesy, dir. Geodesic Inst., Stuttgart Tech. Coll., 1949—, dir. Inst. Aerial Navigation, 1953—. Research, publs. on geodesy and aerial navigation. Home: Hasenbergstrasse 54. Office: Stuttgart Tech. Coll., Keplerstrasse 11, Stuttgart, W. Germany.

RAMSDEN, Jesse, English mechanic, optician; b. Salterhebble, Eng., Oct. 6, 1735; s. Thomas Ramsden; student math., from 1747; m. Sarah Dollond, 1775 or 76; 1 son, John. Apprentice to Burton, math. instrument maker, 1758-62; instrument engraver, from 1762; instrument maker, Haymarket, Eng., from 1775 or 76, later transferred to Picadilly. Fellow Royal Soc. (Copley medal 1795), 1786; mem. Imperial Acad. St. Petersburg. Author: New Universal Equatoreal, 1774; Description of an Engine for Dividing Mathematical Instruments, 1777. Invented dividing machine to grad. math. instruments; improved Hadley's quadrant or sextant, constn. theodite, barometer, refracting micrometer; celebrated for constn. five foot vertical circle for Palermo (Italy) Obs.; constructed electric machines with glass plates; investigated expansion of metals; designed optical equipment, screw-cutting lathe. Died Brighton, Eng., Nov. 5, 1800.

RAMSEY, Arthur Stanley, Brit. mathematician; b. Hackney, Eng., 1867; s. A. Averell Ramsey; ed. Magdalene Coll., Cambridge (Eng.) U.; m. Mary Agnes Wilson, 1902 (dec. 1927); 2 sons, 2 daus. Math. master Fettes Coll., Edinburgh, Scotland, 1890-97; fellow Magdalene Coll., Cambridge U., 1897, lectr., 1897-1934, tutor, 1912-27, bursar, 1904-13, pres., 1915-37, mem. univ. financial bd., 1912-34, univ. lectr. math., 1926-32; examiner U. London, 1915-18, 1932-34, Queen's U., Belfast, Ireland, 1942-44. Author: (with Richardson) Modern Plane Geometry; (with Besant) Hydromechanics; Hydrodynamics; Elementary Geometrical Optics; Dynamics; Statics; Hydrostatics; Electricity and Magnetism; Introduction to Newtonian Attractions. Died Dec. 31, 1954.

RAMSEY, Charles Eugene, Am. sociologist; b. Paragon, Ind., Apr. 24, 1923; s. Sarcefield Dodson and Stella (Goss) R.; B.S., Ind. State Tchrs. Coll., 1947; M.S., U. Wis., 1950, Ph.D., 1952; m. Alberta Mae Jordan, July 19, 1943; children—James D., Charles W., Jane E., Suzanne. Faculty, U. Wis., 1951-52, U. Minn., 1952-54, Cornell U., 1954-62, Colo. State U., 1962-65; prof. U. Minn., Mpls., 1965—. Vis. prof. Inter-Am. Instn. Agrl. Sci., Costa Rica, 1961, Exptl. Sta., U. P.R., 1961-62; research cons. to various univs., agys. Mem. Am. Sociol. Assn., Rural Sociol. Soc., Sigma Xi. Author: (with Lowry Nelson and Cooley Verner) Community Structure and Change, 1960; (with David Gottlieb) The American Adolescent, 1965, Understanding the Deprived Child, S.R.A., 1967; Problems of Youth, 1967; also articles. Developed and tested theory of variations in community power structure, types of sch. bds., and roles of sch. supt., developed method of comparative measurement of level of living for different countries. Home: 1162 Autumn St., St. Paul 55113.*

RAMSEY, Elizabeth Mapelsden (Mrs. H. A. Klagsbrunn), Am. physician, placentologist; b. New York, N.Y., Feb. 17, 1906; d. Charles Cyrus and Grace (Keys) R.; B.A., Mills College, 1928; M.D., Yale U., 1932; hon. D.Sc., Woman's College of Pa., 1965; m. Hans Alexander Klagsbrunn, Jan. 27, 1934. Asst. pathology, Yale U., 1933-34; guest, dept. embryology, Carnegie Inst., 1934-49; research assoc., 1949-51, staff mem., placentology, pathology, 1951—. Assoc., 1934-41; prof. lectr., George Washington, 1941-55; asst. information officer, Nat. Research Council, 1942-45. Mem., Assn. Anatomists. Research on placental circulation; myometrial activity in pregnancy; Kaposi's disease; vasa vasorum; vasculature of pregnant endometrium. Address: 3420 Quebec St. N.W., Washington, D.C. 20007.

RAMSEY, Glenn Virgil, Am. psychologist; b. Covington, Ind., Jan. 11, 1910; s. Arthur O. and Della (Dixon) R.; A.B., Ind. Central Coll., Indpls., 1932; M.A., U. Ill., 1937; D.Ed., Ind. U., 1941; m. Mary Gordon Steiner, July 5, 1947; children—Scott Dixon, Kim Erika. Faculty, Princeton, 1946-48; prof. psychology U. Tex., 1948-50; cons. psychologist, Austin, Tex., 1950—, cons. Tex. Commn. on Alcoholism, 1966—, chaplain's div. USAF, 1956—, Hogg Found. Mental Health, 1956—. Diplomate in Clin. Psychology, Am. Bd. Examiners in Profl. Psychology. Mem. Am., Southwestern psychol. assns., Am. Orthopsychiat. Assn., Soc. Sci. Study Sec. Author: The Sex Life of 289 Boys, 1941; (with R.R. Blake) Perception: An Approach to Personality. Research, publs. on sexual devel. and behaviour of boys; developed project for treatment of mentally ill. Address: 3301 Greenlee Dr., Austin, Tex. 78703.*

RAMSEY, Norman Foster, Am. physicist; b. Washington, Aug. 27, 1915; s. Norman F. and Minna (Bauer) R.; A.B., Columbia, 1937, Ph.D., 1939; B.A., Cambridge U. (Eng.), 1935, M.A., 1940, D.Sc., 1954; m. Elinor Jameson, June 3, 1940; children—Margaret (Mrs. Richard Kasschau), Patricia, Janet, Winifred. Asst. prof. physics Columbia, 1942-45, asso. prof., 1945-47; chmn. physics dept. Brookhaven Nat. Lab., 1946-47; asso. prof. Harvard, 1947-50, Higgins prof. physics, 1950—. Mem. gen. adv. com. AEC, Washington, 1961—, internat. sci. panel Pres.'s Sci. Adv. Com., Washington, 1960—. Recipient Lawrence award U. S. AEC, 1960; Guggenheim fellow Oxford U., 1954. Presdl. Certificate of Merit, 1947. Mem. Am. Phys. Soc., Nat. Acad. Sci., Phi Beta Kappa, Sigma Xi. Author: Nuclear Moments; Molecular Beams; Quick Calculus. Research on molecular beams, deuteron electric quadrupole moment, nuclear moment measurements, proton-nucleon and electron-nucleon scattering hydrogen maser research. Home: 55 Scott Rd., Belmont, Mass. 02178. Office: Lyman Lab., Harvard U., Cambridge, Mass. 02138.*

RAMSTAD, Egil, pharmacognosist; b. Namsos, Norway, July 4, 1911; s. Gustav Adolph and Petra (Devig) R.; B.S., U. Oslo, 1935; Ph.D., U. Liège, 1939; m. Petrine Rypdal, Aug. 22, 1936; children—Yngve, Tore, Liv May. Came to U. S., 1948, naturalized, 1954. With U. Oslo (Norway), 1936-48, asst. prof., 1939-40, examiner, tenant vacant chair pharmacognosy, 1940-48; vis. prof. Purdue U., Lafayette, Ind., 1948-49, prof. pharmacognosy, 1949-—. Mem. A.A.A.S., Am. Inst. Biol. Sci., Am. Soc. Pharmacognosy (pres. 1966-67), Internat. Pharm. Fedn., Sigma Xi, Rho Chi. Author: Drugs of Biological Origin, 1946; Modern Pharmacognosy, 1959; also numerous articles. Research on chemistry of plant drugs, biosynthesis of natural drug products especially ergot alkaloids. Home: 1712 Fernleaf Dr., West Lafayette, Ind. 47906. Office: Purdue U., Sch. Pharmacy, Lafayette, Ind. 47907.*

RAMUS, Petrus (De la Ramée, Pierre), French mathematician, philosopher; b. Cuts, France, 1515; ed. Collège de Navarre, Paris. Censored for his writings attacking Aristotle by Royal Council, 1544; became prin. Collège de Presles, 1546; became 1st prof. math. Collège de France, 1551; travelled to Germany, 1568; returned to Paris, 1570. Author: Aristotelicae animadversiones, 1543; Eloquentiae et philosophiae professoris regi arithmeticae libri tres, 1555; Les moeurs des anciens Gaulois, 1559; Scholarum mathematicarum libri XXXI, 1569; (with Schoner) Algebrae liber primus, 1586. Editor: Elements (Euclid). Sixteenth century reformer in scis. and humanities; studies in theoretical math., optics, geometry. Killed in St. Bartholomew's Day Massacre, Paris, Aug. 26, 1572.

RAMUSIO, Giambattista, Italian geographer, translator; b. Treviso, Italy, June 20, 1485; s. Paola and Tomyris (Macachio) R.; ed. Venice, Italy, Padua, Italy; m. Franceschina Navagero, 1524; 1 dau., Paola. Entered pub. service, 1505; became sec. of senate, 1515; named sec. Council of Ten; under Venetian State Govt. travelled to Rome, Switzerland, France. Author: Della navigationi e viaggi, 2 vols., 1554, 56. Geog. description with maps and illustrations of travels of various explorers, including Marco Polo and Magellan; publs. on astronomy, langs. Died 1557.

RANADIVE, Kamal Jayasing, Indian biologist; b. Poona, India, Nov. 8, 1917; d. Dinkar Dattatraya Samarth and Shantabai Dinkar Samarth; B.Sc., Fergusson Coll., Poona, 1938; M.Sc., Poona Coll. Agr., 1941; Ph.D., Indian Cancer Research Center, 1950; m. Jayasing Trimbak Ranadive, May 13, 1939; 1 son, Anil Jayasing. Chief, biol. labs. Indian Cancer Research Center, Bombay, 1952-62, chief biol. div., acting dir., 1962-—. Recipient Sr. Col. Basanti Devi Amirchand award, 1958; Silver Jubilee research award Med.Council India, 1964; Internat. Watumull Meml. award, 1964. Mem. Am., Japanese tissue culture assns., Internat, Cell Research Orgn. (UNESCO), Indian Council Med. Research (sci. adv. bd.), Council Sci. and Indsl. Research (biol. research com.), numerous nat., internat., univ. sci. assns. Editor: Proc. Cell Biology Meetings, Internat. Cell Research Orgn., 1965. Research, numerous publs. on breast cancer, etiology of cancer problems specific to India, mechanism of cell carcinogenesis — developed first human liposarcoma and myxofibrosarcoma cell strains, human leprosy — first successful isolation of acid-fast microorganism from lepromatous leprosy. Home: 14, Oval View, New Queen's Rd., Bombay 1. Office: Biology Div., Indian Cancer Research Centre, Bombay 12, India.*

RANCHIN, François, French physician; b. Montpellier, France, 1564; M.D., U. Montpellier, 1592; 1 son, 1 dau. Prof., U. Montpellier, 1605. Author: Questions Françaises sur la Chirurgie de Gui de Chauliac, 1904; Opuscula Medica Utili Jucundaque Rerum Varietate Referta, 1627. Account of Montpellier plaque which mentions the use of aromatic disinfectants, 1629. Died Montpellier 1641.

RAND, Austin Loomer, zoologist, mus. ofcl.; b. Kentville, N.S., Can., Dec. 16, 1905; s. Stanley Bayard and Carrie (Forsythe) R.; B.Sc., Acadia U., 1927, D.Sc., 1961; Ph.D., Cornell U., 1931; m. Rheua Medden, Aug. 15, 1931; children—Austin Stanley, William Medden. With Am. Mus. Natural History, N.Y.C., 1929-42, research asso., 1937-41; acting chief div. biology Nat. Mus. Can., 1942-47; curator birds Chgo. Natural History Mus., 1947-54, chief curator zoology, 1955-—. Mem. Am. (pres. 1962-64), Brit. ornithologists unions, Am. Soc. Mammalogy, Wilson, Cooper ornithol. socs., A.A.A.S., Author: Distribution and Habits of Madagascar Birds, 1936; (with R. Archbold) New Guinea Expedition, 1939; Mammals of Yukon, Canada, 1945; Mammals of Eastern Rockies and Western Plains of Canada; Development and Enemy Recognition of the Curve-billed Thrasher, 1941; Stray Feathers from a Bird Man's Desk, 1955; American Water and Game Birds, 1956; A Midwestern Almanac, 1961. Home: 403 W. Lincoln Av., Chesterton, Ind. Office: Chgo. Natural History Mus., Chgo., Ill. 60605.*

RAND, Gertrude, Am. psychologist; b. N.Y.C., Oct. 29, 1886; d. Lyman Fiske and Mary (Moench) Rand; A.B., Cornell U., 1908; M.A., Ph.D., Bryn Mawr Coll., 1911; D.Sc., Wilson Coll., 1943; m. Clarence E. Ferree, Sept. 28, 1918. (dec. 1942). Asso., Bryn Mawr (Pa.) Coll., 1913-27; research asst. com. on indsl. lighting NRC, 1925-27; asso. prof. physiol. optics John Hopkins Sch. Medicine, 1928-36; research asso. ophthalmology Columbia Coll. Phys. and Surg., 1943-57. Cons. Nat. Soc. Prevention Blindness, 1943-57. Sarah Berliner Research fellow, 1912-13. Recipient Edgar D. Tilyer medal Optical Soc. Am., 1959; Gold medal Illuminating Engring. Soc., 1963. Fellow Am. Acad. Ophthalmology and Otolaryngology (hon.). Patentee in field. Research, publs. in physiology vision; co-inventor Ferree-Rand perimeter, Hardy-Rand-Rittler plates for testing color vision. Home: 25 Erland Rd., Stony Brook, N.Y. 11790.*

RAND, Michael John, pharmacologist; b. Mildenhall, Eng., Aug. 19, 1927; s. Jackson Allan and Dora (White) R.; B.Sc., U. Melbourne, 1949, M.Sc., 1954; Ph.D., U. Sydney, 1957; m. Anne Stafford, Nov. 12, 1960; children—Madeleine, Joshua (by previous marriage). Demonstrator dept. pharmacology Oxford U., 1957-58; research fellow depts. pharmacology Australian and New Zealand Life Ins. Med. Research Fund, U. Oxford, U. Sydney, 1958-60; Wellcome research fellow dept. pharmacology U. London, 1960-61, lectr., 1961-65; prof. pharmacology U. Melbourne, Victoria, Australia, 1965-—. Mem. Brit., Australian physiol. socs., Brit. Pharm. Soc., Royal Soc. Medicine. Author: (with W. C. Bowman and G. B. West) Textbook of Pharmacology, 1967. Research and numerous publs. on source of serotonin from blood platelets, mode of action of cardiac glycosides, mode of action of sympathomimetic amines, role of acetylcholine in adrenergic mechanisms, mode of action of drugs affecting transmission in autonomic nervous system, pharmacology of tobacco smoke and nicotine. Home: 8.6/46, Lansell Rd., Toorak, Victoria. Office: Dept. of Pharmacology, U. Melbourne, Parkville, N.2, Victoria, Australia.*

RANDALL, Charles Chandler, Am. microbiologist, virologist; b. Cedar Rapids, Ia., Mar. 27, 1913; s. Frank Hall and Helen B. (Chandler) R.; B.S., U. Ky., 1936; M.D., Vanderbilt U., 1940; m. Virginia Gillette Smith, June 8, 1941; children—Gileitte Chandler, Stephen Hall. NIH, USPHS fellow Vanderbilt U. Sch. Medicine, Nashville, 1948-49, faculty, 1949-57, prof., acting head dept. microbiology, 1955-57; prof., chmn. dept. microbiology U. Miss. Sch. Medicine, Jackson, 1957-—. Diplomate Am. Bd. Pathology, Am. Bd. Microbiology. Mem. Am. Soc. Microbiology, Soc. for Exptl. Biology and Medicine, Tissue Culture Assn., Am. Assn. Pathologists and Bacteriologists, Am. Acad. Microbiology, Am. Assn. Immunology, Am. Soc. For Exptl. Pathology, N.Y. Acad. Scis., Electron Soc. Am. Research and numerous publs. on biochemistry of virus infection, nucleic acids of viruses, adaption of viruses and fungi to tissue culture. Home: 121 Ashcot Circle, Jackson, Miss. 39211.*

RANDALL, Clyde Lamb, Am. physician; b. Lyons, Kan., July 12, 1905; s. Harry Garfield and Ruth Ella (Lamb) R.; A.B., B.S., U. Kan., 1928, M.D., 1931; m. Vernie C. Theden, Aug. 17, 1929; children—Ruth Ann (Mrs. Rodney L. James), Harry Garfield II. Intern St. Margaret's Hosp., Kansas City, 1931-32; resident St. Luke's Hosp., Kansas City, then Buffalo Gen. Hosp., 1932-37; chief obstet. and gynecol. services Buffalo Childrens Hosp., 1961-—, Buffalo Gen. Hosp., 1942-—, E. J. Meyer Meml. Hosp., 1960-—; mem. faculty state U. N.Y., Buffalo, 1938-—, prof. obstetrics and gynecology, chmn. dept., 1942-—. Mem. Bd. Med. Examiners N.Y. State, 1948-57; mem. test com. obstetrics and gynecology Nat. Bd. Med. Examiners, 1957-59. Diplomate Am. Bd. Obstetrics and Gynecology (dir., 1961-—, sec.-tresas., 1964-—). Mem. Am. Assn. Obstetricians and Gynecologists (sec. 1960-62), A.M.A., Am. Coll. Obstetricians and Gynecologists (chmn. Dist. II, 1965-66), Dallas Southwestern clin. Soc., Am. Gynecol. Soc., Soc. Pelvic Surgeons, Beta Theta Pi, Nu Sigma Nu, Alpha Omega Alpha; hon. mem. Bklyn., Dallas, Kansas City, Pitts. obstet. and gynecol. socs., Kansas City Acad. Medicine, Central, S. Atlantic assns. obstetricians and gynecologists, Soc. Obstetricians and Gynecologists Can. Author articles med. jours. Mem. editorial bd. Obstetrics and Gynecology Jour., 1959-62. Home: 100 Meadow Rd., Buffalo 14216. Office: 100 High St., Buffalo 14203.*

RANDALL, Harrison McAllister, Am. physicist; b. Burr Oak, Mich., Dec. 17, 1870; s. Seth Cook and Ellen Louise (Plank) R.; Ph.B., U. Mich., 1893, Ph.M., 1894, Ph.D., 1902, LL.D., 1966; D.Sc., Ohio State U., 1956; postgrad. U. Tubingen (Germany); m. Ida May Muma, Aug. 24, 1898; children—John McAllister (dec.), Mary Foote (Mrs. S. H. Emerson), Esther McAllister (Mrs. David Miller), Robert DuBois, John Reed. Faculty dept. physics U. Mich., 1899-1940, prof., 1916-40, dir., 1918-40; ret., 1940. Conducted internat. symposiums in theoretical physics, between World Wars I and II. Recipient Ives medal, 1953, plaque Coblentz Soc., 1964. Russel lectr. U. Mich., 1940. Fellow A.A.A.S., Am. Phys. Soc. (past pres.), Optical Soc. (hon.), Phi Beta Kappa, Sigma Xi, Delta Upsilon. Author: (with others) College Physics, 1925, Infrared Determination of Organic Structure, 1948. Research on atomic and infrared spectra, invention of Randall prism grating spectrograph; studies combining column chromatography and infrared spectroscopy; discovery (with others) of mycosides. Home: 1208 Prospect St., Ann Arbor 48104. Office: Randall Lab., E. University St., Ann Arbor, Mich. 48104.*

RANDALL, Sir John (Turton), English biophysicist; b. Eng., Mar. 23, 1905; s. Sidney and Hannah (Cawley) R.; B.Sc., U. Manchester, 1925, M.Sc., 1926, D.Sc., 1938; m. Doris Duckworth, 1928; 1 son, Christopher John. Research physicist, research labs. GEC Ltd., 1926-37; Warren research fellow Royal Soc., U. Birmingham, 1937-43; lectr. Cavendish Lab., U.. Cambridge, 1943-44; prof. nat. philosophy St. Andrews U., Scotland, 1944-46; Wheatstone prof. physics King's Coll., U. London, 1946-61, prof. biophysics, 1961-—. Dir. biophysics research unit Med.. Research Council, 1947-—. Recipient John Scott award City of Phila., 1959; Duddell medal Phys. Soc. London, 1945. Fellow Royal Soc., 1946 (Hughes medal 1946), Inst. Physics. Author: Diffraction of X-rays by Amorph Solids, 1934. Editor: Nature and Structure of Collagen, 1953. Research, publs. on luminescence of solids, structure of collagen, biophysics of cilia; joint discoverer of cavity magnetron. Home: The Red Cottage, Highfield Lane, Liphook, Hants., Eng. Office: Dept. Biophysics, U. London, King's Coll., 26-29 Drury Lane, London W.C.2, Eng.*

RANDALL, John Ernest, Jr., Am. marine biologist; b. Los Angeles, May 22, 1924; s. John Ernest and Mildred (McKibben) R.; B.A., U. Cal. at Los Angeles, 1950; Ph.D., U. Hawaii, 1955; m. Helen L. S. Au, Nov. 9, 1951; children—Loreen Ann, Rodney Dean. Research fellow Office Naval Research, U. Hawaii, 1953-55; research fellow Yale, Bishop Mus., 1955-56; research asst. prof. Marine Lab., U. Miami (Fla.), 1957-61; prof. biology, dir. Inst. Marine Biology, U. P.R., Mayaguez, P.R., 1961-65; dir. Oceanic Inst., 1965-66; ichthyologist Bishop Mus., Honolulu, 1965-—; affiliate grad. faculty zoology dept. U. Hawaii, Honolulu, 1966-—, research prof. Hawaii Inst. Marine Biology, 1966-—. Mem. Assn. Island Marine Labs. Caribbean (past pres.), Am. Soc. Ichthyologists and Herpetologists, Am. Inst. Fishery Research Biologists, Hawaiian Acad. Scis., Phi Beta Kappa. Research, numerous publs. on classification and biology of tropical marine fishes, mimicry and protective resemblance in fishes, poisonous fishes, biology of queen conch, biology of echinoid Diadema, devel. artificial reefs. Home: 45-1033 Pahuwai Pl., Kaneohe, Hawaii 96744. Office: Bishop Mus., 1355 Kalihi St., Honolulu 96819.*

RANDALL, Lowell Orland, Am. pharmacologist; b. North Georgetown, O., Sept. 11, 1910; s. Cornelius Hall and Mary (Stoffer) R.; B.S., Mt. Union Coll., 1931, D.Sc., 1958; Ph.D., U. Rochester, 1935; m. Margaret V. Leary, Sept. 16, 1936; children—John C., Mary Margaret (Mrs. Robert J. Shackleton). With Mass. Civil Service Commn., 1935-39; pharmacologist Burroughs Wellcome Co., 1939-46; pharmacologist Hoffmann-LaRoche, Inc., Nutley, N.J., 1946-58, dir. pharmacology, 1958-—. Mem. Am. Soc. Pharmacology, Am. Soc. Biochemistry, Am. Chem. Soc., A.A.A.S., Am. Coll. Clin. Pharmacology. Author: Morphine and Allied Drugs, 1957; Physiological Pharmacology, 1963; Psychopharmacological Agents, 1964. Research and numerous publs. on pharmacology, toxicology and biochemistry of drugs in animals useful in cardiovascular, smooth muscle, skeletal muscle, central nervous system disorders, drugs for treatment infectious diseases, pharm. effects of central nervous system stimulants, depressants, hypnotics, sedatives, tranquilizers and antianxiety agts. Home: 161 Ridge Rr. Office: Hoffmann-LaRoche, Inc., Nutley, N.J. 07110.*

RANDALL, Walter Clark, Am. physiologist; b. Akeley, Pa., Dec. 12, 1916; s. Harry W. and Ruth N. (Wiggins) R.; A.B., Taylor U., 1938; M.S., Purdue U., 1940, Ph.D., 1942; m. Gwendolyn Ruth Niebel, Aug. 1, 1943; children—David, Marilyn, Douglas, Craig. Post-doctoral fellow physiology Western Res. U., 1942-43; faculty St. Louis U. Sch. Medicine, 1943-54; prof., chmn. dept. physiology Loyola U. Sch. Medicine, Chgo., 1954-—. Cons. heart program project com. NIH, 1963-67; mem. Nat. Bd. Med. Examiners. Fellow A.A.A.S.; mem. Am. Physiol. Soc. (chmn. circulation group and temperature regulation group), Am. Heart Assn., Soc. Exptl. Biology and Medicine, Sigma Xi. (pres. Loyola chpt.). Editor: Nervous Control of the Heart, 1965; bd. editors Circulation Research, Am. Jour. Physiology, Jour. Applied Physiology. Research, publs. on mechanisms and control of body temperature, nervous control of heart and blood vessels, regulation of sweating, autonomic nervous system outflows. Home: 624 N. Hamlin Av., Park Ridge, Ill. Office: 1400 S. 1st Av., Hines, Ill. 60141.*

RANDERS, Gunnar, Norwegian astrophysicist; b. Oslo, Norway, Apr. 28, 1914; s. Gunnar and Lubba (Bordtkorb) R.; Masters degree in Astrophysics, U. Oslo, 1937; m. Engelke R. Koren, Sept. 23, 1939; children—Jorgen, Karen, Jan G. Research asst. U. Oslo, 1937-39; research asso., Oslo U. fellow Mt. Wilson Obs., Pasadena, Cal., 1939-40; instr. astrophysics U. Chgo., 1940-42; dep. dir. Inst. Theoretical Astrophysics, U. Oslo, 1945-46, dir. research Def. Research Inst., 1946-48; dir. Joint Establishment for Nuclear Energy Research of Norway and Netherlands, 1951-59; managing director of Norwegian Inst. for Atomic Energy, 1959-—; chairman Norwegian Atomic Energy Council, 1954-—; spl. adviser atomic matters UN, 1954-56; adviser to dir. gen. International Atomic Energy Agency, Vienna, 1958-59; chmn. Norwegian Agy. for Internat. Devel., 1963-—. Decorated Comdr. Order St. Olave (Norway); Comdr. Order Orange-Nassau. Comdr. Order Ho. of Orange (Neth.); Comdr. Order of White Rose (Finland); comdr. Swedish Vasa

Order. Mem. Brit. Nuclear Energy Soc., Norwegian Acad. Scis., Royal Astron. Soc., Am. Nuclear Soc. (dir.), European Atomic Energy Soc. (v.p.). Author: Atom Energy, 1946; Atoms and Common Sense, 1950; also numerous sci. articles. Studies internal motion of stellar materials; cosmological research extra-galactic space; studies gen. relativity, atomic energy, heavy water reactors. Home: Trosterudstien 4. Slemdal, Oslo. Norway. Office: Institutt for Atomenergi, P.B. 40, Kjeller, Norway.*

RANDLE, P. J., English biochemist; b. Nuneaton, Eng., July 16, 1926; s. Alfred John and Nora (Smith) R.; M.A., U. Coll. Hosp., London, 1950, Ph.D., 1955; M.D., U. Cambridge, 1964; m. Elizabeth Ann Harrison, Sept. 27, 1952; children—Rosalind Jane, Peter John, Sally Elizabeth, Susan Penelope. House physician, house surgeon U. Coll. Hosp., 1951; research worker, lectr. biochemistry U. Cambridge, 1952-64, research fellow Sidney Sussex Coll. U. Cambridge, 1954-57, fellow, dir. med. studies Trinity Hall, 1957-64; prof., chmn. dept. biochemistry U. Bristol, 1964—. Research, numerous publs. on metabolism of carbohydrates and its control by hormones, causes of diabetes mellitus. Home: South Fall, Cadbury Camp Lane, Clapton in Eordano, Somerset, Eng.*

RANDOLPH, Jacob, Am. physician; b. Phila., Nov. 25, 1796; s. Edward Fitz and Anna Julianna (Steel) R.; M.D., U. Pa., 1817; m. Sarah Emlen Physick, 1822. Practiced as surgeon, circa 1820-35; asst. in Almshouse Infirmary, Phila.; surgeon Pa. Hosp., 1835-48; prof. clin. surgery U. Pa., 1847-48. Mem. Am. Philos. Soc., Phila. Coll. Physicians, Phila. Med. Soc. Author: A Memoir of the Life and Character of Philip Syng Physick, 1839; contbr. articles to Am. Jour. Med. Sci., N. Am. Med. and Surg. Jour., Med. Examiner. Introduced new method of removing stones from bladder (lithotripsy), 1831; performed amputation of lower jaw for osteosarcoma, ligation of external iliac aneurysm (both early radical operations). Died Phila., Feb. 29, 1848.

RANDOLPH, John (Adam) F(itz), Am. mathematician; b. Newton, Ia., May 2, 1904; s. Albert Clyde and Carrie (Engle) R.; B.S., W. Tex. Teachers, 1926; A.M., U. of Mich., 1929; Ph.D., Cornell U., 1934; m. Margaret Elizabeth Crowley, May 15, 1932; children—Peter Nelson Fitz, Rebecca Susan. Teacher mathematics and science, Canyon Tex. High Sch.; instr. math. and astronomy, Syracuse U., 1928-31; instr. Cornell U., 1931-36; asst. prof., 1938-44; mem. Inst. for Advanced Study, Princeton, N.J., 1936-38; prof. Oberlin Coll., 1944-48; prof. dept. math. U. Rochester, 1948—, chmn., 1948-59; vis. professor Am. U. of Beirut, 1955-56, 61-62. Mem. Am. Math. Soc., Math. Assn. of Am. Lloyd Green Allen Scholarship Society, also Sigma Xi, Phi Kappa Phi. Contributor to mathematical journals. Author: Analytic Geometry and Calculus (with Mark Kac), 1946; Primer of College Mathematics, 1950, Calculus, 1952, Trigonometry, 1952; Calculus and Analytic Geometry, 1961. Study of numerical analysis; metric properties of point sets; topology; measure theory; genetics. Home: 171 Highland Pkwy., Rochester, N.Y. 14620.

RANDOLPH, Malcolm Logan, Am. biophysicist; b. West Palm Beach, Fla., Oct. 11, 1920; s. Orrin and Elizabeth (Lockhart) R.; B.A., U. Va., 1943, M.S., 1946, Ph.D., 1947; m. Jeannette P. Caldwell, Dec. 18, 1949; children—Dorothy Elizabeth, Joseph Pendleton. With U. Va., 1942-47, NRC fellow, 1946-47; asst. prof., research asso. Tulane U., New Orleans, 1947-53; biophysicist biology div. Oak Ridge Nat. Lab., 1953—. Mem. Am. Phys. Soc., Radiation Research Soc., A.A.A.S., Biophys. Soc., Asso. editor Radiation Research, 1965—. Research, publs. on effects of inelastic neutron collisions in fast neutron dosimetry, measurement and analysis of RBE-LET relationships in radiobiology, kinetics of prodn. and decay of free radicals in crystallin amino acids by ionizing radiations. Home: 358 East Dr. Office: Biology Div., Oak Ridge Nat. Lab., Oak Ridge 37830.*

RANGE, Willard Edgar Allen, Am. polit. scientist; b. Pitts., June 24, 1910; s. Henry William and Caroline (Fritsch) R.; B.S., Harvard, 1933; M.A., U. Ga., 1939; Ph.D., U. N.C., 1958; m. Eula Pearl Ross, June 24, 1939; children—Peter, Franklin, Caroline. Faculty, Abraham Baldwin Agrl. Coll., Tifton, Ga., 1939-43; faculty U. Ga., Athens, 1946—, prof., 1963—. Fulbright lectr. Karnatak U., India, 1962-63. Mem. Am., So. polit. sci. assns., Internat. Studies Assn. Author: Rise and Progress of Georgia's Negro Colleges, 1865-1949, 1951; Century of Georgia Agriculture, 1850-1950, 1954; F.D. Roosevelt's World Order, 1959; J. Nehru's World View, 1961. Research in internat. relations, especially basic assumptions and attitudes of some of world's leading mgrs. of fgn. policies. Home: 276 Woodlawn Av., Athens, Ga. 30601.*

RANK, Otto, psychoanalyst; b. Vienna, Austria, Apr. 22, 1884; Ph.D., U. Vienna, 1912. Came to U. S., Apr. 22, 1884. Student and friend of Sigmund Freud; later turned away from Freudian theories; cofounder (with Hanns Sachs) psychoanalytic jour. Imago, 1912A Author: Der Künstler (Art and Artist), 1907; Der Mythus von der Geburt (Myth of the Birth of the Hero), 1909; Das Trauma de Geburt (The Trauma of Birth), 1924; Tecknik der Psychoanalyse (Will Therapy), 1926-31; Modern Education, 1932. Devel. theory of the birth trauma; new therapeutical method based on the will to achieve; books used in tng. psychotherapists and social workers. Died N.Y.C., Oct. 31, 1937.

RANKINE, Alexander Oliver, Brit. physicist; b. Guildford, Eng., 1881; s. John Rankine; B.Sc., Univ. Coll., London, D.Sc., 1910; m. Ruby Irene Short, 1907; 2 sons, 2 daus. Asst., dept. physics Univ. Coll., London, 1904-19, became fellow, 1912; chief research asst. Admiralty Exptl. Sta., Harwich, Eng. 1917-18; prof. physics Imperial Coll. Sci. and Tech., South Kensington, Eng., 1919-37, dir. dept. tech. optics, 1925-31; emeritus, 1937-56; chief physicist Anglo-Iranian Oil Co. Ltd., 1937-47. Fellow Royal Soc., 1934; fellow or mem. Brit. Assn. (pres. sect. A 1932), Phys. Soc. (pres. 1932-34), Inst. Physics (hon. sec. 1926-31), Optical Soc. (pres. 1931-32), Royal Instn. Gt. Britain (sec. 1945-52). Contbr. papers to sci. jours. Died Jan. 20, 1956.

RANKINE, William John Macquorn, Scottish engr., physicist; b. Edinburgh, Scotland, July 5, 1820; LL.D., U. Dublin, 1856; s. David and Barbara (Grahame) R.; ed. Superior Sch., Glasgow, Scotland; student U. Edinburgh, 1836-38; studied under Sir. John Benjamin MacNeill, surveyor of N. of Ireland; Worked on surveys for river improvements, water-works, also on Dublin and Drogheda Ry., 4 years; with various ry. projects Caledonian Ry. Co., intermittently, 1944-48; began research on molecular physics, 1948; became prof. civil engring. and mechanics U. Glasgow, 1855; named cons. engr. Highland and Agrl. Soc. Scotland, 1865. Recipient Keith medal, 1854. Fellow Royal Soc., 1853, Royal Soc. Edinburgh, 1849. Author: A Manual of Applied Mechanics, 1858; A Manual of the Steam Engine and Other Prime Movers, 1859; Manual of Civil Engineering, 1862; A Manual of Machinery and Millwork, 1869; A Mechanical Textbook, 1873; Miscellaneous Scientific Papers, pub. posthumously 1881; also numerous articles. Attempted to establish thermodynamics using principle of conservation of energy; research in thermodynamic theory of steam engines, hot air machine, equilibrium of elastic firm bodies, ship bldg., effect of friction on steam engine performance, air screw, water waves, including form, thermodynamic theory and effect on ship hull. Died Glasgow, Dec. 24, 1872.

RANSOM, Henry King, Am. surgeon; b. Jackson, Mich., Jan. 21, 1898; s. Fred C. and Gayle (King) R.; A.B., U. Mich., 1920, M.D., 1923, M.S., 1934. Faculty, U. Mich., Ann Arbor, 1929—, prof. surgery, 1950—, surgeon U. Hosp.; vis. surgeon St. Joseph Mercy Hosp.; cons. VA Hosp. Diplomate Am. Bd. Surgery. Mem. A.C.S. (life), A.M.A. (editorial bd. Archives of Surgery 1954-63), Am., Western, Central surg. assns., Am. Gastroent. Assn., Soc. Surgery Alimentary Tract, Internat., Frederick A. Coller surg. socs., Alpha Omega Alpha, others. Research, publs. on techniques, complications, observations of surgery on gastrointestinal tract, carcinoma of breast, rectum; vagotomy and gastric ulcer. Home: 721 S. Forest Av. Office: Dept. Surgery, U. Mich., Ann Arbor, Mich. 48104.*

RANSOM, John Paul, Am. microbiologist; b. Clarkston, Wash., Mar. 6, 1912; s. David H. and Ethel (MacVicar) R.; A.B., U. Cal., Berkeley, 1935, M.A., 1951, Ph.D., 1953; m. Madeline Smith, Oct. 19, 1940; children—John Paul, Frederick G. II. Commd. 2d lt. U. S. Army, 1935, advanced through grades to lt. col., 1954; asst. chief lab. service Ft. Lewis Sta. Hosp., Wash., 1940-45; bacteriologist, sanitation officer 251st Gen. Hosp., France, 1945-46; chief bacteriol. dept. 7th Army Med. Lab., Ft. McPherson, Ga., 1946-47; bacteriologist 406th Med. Gen. Lab., Tokyo, Japan, 1947-48; asst. chief lab. service 49th Gen. Hosp., 1948-49; chief bacteriol. dept. 6th Army Area Med. Lab., Ft. Baker, Cal., 1953-56; chief biol. and med. sci. br. Office of Chief Research and Devel., Washington, 1956-57; staff officer Army Res. Office, Arlington, Va., 1958-59; research bacteriologist Walter Reed Army Inst. Research, Washington, 1959-60; ret. 1960. Microbiologist New Eng. Inst. Med. Research, Ridgefield, Conn., 1960-65, 66—; project leader Bio-Sci., Labs., Los Angeles, 1965-66. Mem. A.A.A.S., Am. Soc. Microbiology, Am. Acad. Microbiology, Soc. Exptl. Biology and Medicine, N.Y. Acad. Scis., Sigma Xi. Research, publs. on pneumonic plague, immuno diffusion, host def. mechanisms, continously fed culture method. Home: 178 Danburg Rd. Office: New Eng. Inst. for Med. Research, Ridgefield, Conn. 06877.*

RANSOME, Frederick Leslie, geologist; b. Greenwich, Eng., Dec. 2, 1868; s. Ernest Leslie and Mary Jane (Dawson) R.; B.S. (teaching fellow), U. Cal., 1893, Ph.D., 1896; m. Amy Cordova Rock, May 25, 1899; children—Janet (Mrs. H. M. Baxter), Susan Clarkson (Mrs. E. D. Fry), Violet Jane (Mrs. H. Rodney Gale), Alfred Leslie. Asst. in mineralogy and petrography Harvard, 1896-97; asst. geologist U. S. Geol. Survey, 1897-1900, geologist, 1900-23; in charge; sects. of western areal geology, 1912-16, and of metalliferous deposits, 1912-23; prof. econ. geology U. Ariz., 1923-27, dean Grad. Coll., 1926-27; prof. econ. geology Cal. Inst. Tech., 1927—; cons. geologist U S. Bur. Reclamation and Met. Water Dist. of So. Cal., 1928—. Lectr. on ore deposits U. Chgo., 1907; Silliman lectr. Yale, 1913. Mem. Nat. Acad. Scis. (treas. 1919-24). NRC (treas. 1919-24), Soc. Econ. Geologists (pres. 1926-27), Washington Acad. Scis. (pres. 1918), Geol. Soc. Washington (pres. 1913). Author of numerous official monographs on geology of western mining dists. and papers in sci. jours. Asso. editor Econ. Geology. Specialist on econ. geology of copper, gold, other ore deposits, also geology applied to engring.; mapped geol. conditions, approved site for Boulder Dam, Madden Dam; investigated domestic resources in quicksilver during World War I. Died Oct. 6, 1935.

RANSON, S(tephen) Walter, Am. anatomist; b. Dodge Center, Minn., Aug. 28, 1880; s. Stephen William and Mary Elizabeth (Foster) R.; B.A., U. of Minn., 1902; M.S., U. of Chicago, 1903, Ph.D., 1905; M.D., Rush Med. Coll., 1907; m. Tessie Grier Rowland, Aug. 18, 1909; children—Stephen William, Margaret Jane, Mary Elizabeth. Fellow in neurology, U. of Chicago, 1904-06; intern Cook County Hosp., 1907-08; instr. anatomy, 1909-10, asst. prof., 1910-12, prof. and head of dept., 1912-24, Northwestern U. Med. Sch.; prof. neuroanatomy and head of Dept. of Neuroanatomy and Histology, Washington U. Med. Sch., 1924-27; prof. neurology and dir. Neurological Research Inst., Northwestern U. Med. Sch., since 1928. Fellow A.A.A.S.; mem. Am. Neurological Assn., Am. Assn. Anatomists (pres. 1938-40), Am. Physiol. Soc., Sigma Xi. Author: The Anatomy of the Nervous System, 6th edition, 1939. Contbr. results of investigations on structure of the peripheral nervous system of mammals, etc. Mem. editorial bd. Archives of Neurology and Psychiatry. With H. Kabat and H. W. Magoun, demonstrated presence of nerve pathways between the pituitary gland and the hypothalamus, 1935. Died Aug. 30, 1942.

RANVIER, Louis-Antoine, French histologist; b. Lyons, France, Oct. 2, 1835. Asst. dir. Histology Lab., Collége de France, Paris, 1872-75, dir., from 1875; prof. gen. anatomy, from 1886. Mem. Acad. Medicine, French Acad. Scis. Author: Traité Technique d'Histologie, 2 vols., 1875-77; Leçons sur l'Histologie du Systeme Nerveux, 1878; Leçons d'Anatomie Générale sur le Systeme Musculaire, 1880; Leçons d' Anatomie Général, 1880-81; (with Cornil) Manuel d'Histologie Pathologique, 1879-82. Founded exptl. histology; described structure of peripheral nerve, 1871. Died Thélys, France, Mar. 22, 1922.

RANYARD, Arthur Cowper, English astronomer; b. Swanscombe, Eng., 1845; s. Benjamin and Ellen Henrietta (White) R.; student Univ. Coll. London; M.A., Pembroke Coll., Cambridge, 1868. Barrister Lincoln's Inn, 1871; compiler systematic account of solar eclipses down to 1878 for Royal Astron. Soc., 1871-79, sec., 1874-80, also fellow, mem. council; asst. sec. expdn. for observing total solar eclipse, 1870. Editor Knowledge (in which he pub. his research on nebulae), 1888. Contbr. to Monthly Notices Royal Astron. Soc. Concluded that density of nebulae is low in comparison with earth's atmosphere, that star clusters show evidence of matter ejected from a center rather than of gradual condensation. Died Dec. 14, 1894.

RAO, Jivanna Thuljaram, Indian biologist; b. Coimbatore, India, May 5, 1919; s. Swaminatha Jivanna and Savithri Rao; m. Saroja Rao, Apr. 14, 1944; children—Uma, Sriram. With Sugarcane Breeding Inst., Coimbatore, India, 1943-57, botanist, 1959-61, dir., 1961—; agronomist Indian Inst. Sugarcane Research, Lucknow, India, 1957-59. Mem. Internat. Soc. Sugarcane Techologists, Indian Soc. Genetics and Plant Breeding. Author: (with N. L. Dutt) Coimbatore Canes in Cultivation, 1950, rev., 1956; (with U. Vijayalakshmi) Improved Canes in Cultivation, 1964, Indian Sugarcane Atlas, 1962; also articles. Devel. improved sugarcane varieties; morphological description of sugarcane varieties. Address: Sugarcane Breeding Inst., Coimbatore, Madras, India.*

RAO, M. Narayana, Indian nutritionist; b. Vizianagram, India, Feb. 27, 1929; s. M. C. and M. (Tayaramma) Satyanarayana; M.Sc., Andhra U. Coll., Waltair, India, 1949; Ph.D. in Biochemistry, Indian Inst. Sci., Bangalore, India, 1953; m. Suguna, Feb. 17, 1953; children—M. S. Prasad, M. V. Kamala. Scientist, Central Food Technol. Research Inst., Mysore, India, 1952—. NRC fellow Food and Drug Directorate, Canadian Dept. Nat. Health and Welfare, Ottawa, Ont., 1964-66. Mem. Assn. Food Technologists, Nutrition Soc. Contbg. author: World Reviews of Nutrition and Dietetics, vol. 7, 1966; Advances in Chemistry Series, number 57, 1966. Research, numerous publs. on protein-calorie relationship, amino acid availability, evaluation of processed protein foods in treatment and prevention of protein malnutrition. Home: 2947/1 IIIrd Main Rd., V.V. Mohalla, Mysore-2. Office: Central Food Technol. Research Inst., Mysore-2, India.*

RAO, Malempati Madhusudana, mathematician; b. Nimmagadda, Andhra, India, June 6, 1929; s. Sree Ramulu and Pitchamma (Korrapati) R.; B.A., Andhra U., 1949; M.A., Madras U., India, 1952, M.Sc., 1955; Ph.D., U. Minn., 1959; m. Durgamba Sept. 11, 1966. Came to U. S., 1959. Lectr. math. S.R.R. Coll., Vijayawada, India, 1952-53; research fellow U. Minn., 1958-59; with Carnegie Inst. Tech., 1959—, prof. math., 1966—. Mem. Am. Math, Soc., Inst. Math. Statistics. Global representation of continuous linear functionals on and smoothness properties of arbitrary orlicz spaces; interpolation theory of operators on such spaces; studies, publs. in function algebras; conditional expectations on function spaces and extension; inference, prediction and sampling of stochastic processes; theory of lower bounds in statist. estimation with convex loss functions. Home: Nimmagadda, Kistna Dist., Andhra, India. Office: Dept. Math., Carnegie Inst. Tech., Pitts. 15213.*

RAO, Nagaraja Adisheshappa, botanist; b. Channapatna, Mysore, India, Nov. 25, 1925; s. Nagappa Adisheshappa and Jayalakshamma Rao; B.Sc. with honors, Mysore U., 1949, M.Sc., 1952; Ph.D., State U. Ia., 1959; m. Shyamala Rao, Feb. 6, 1951; 2 sons. Lectr. botany U. Mysore, 1949-56, U. Malaya in Singapore, U. Singapore, 1960-64; research and teaching asst., research asso. dept. botany State U. Ia., Iowa City, 1956-59; sr. lectr. botany U. Singapore, 1964—, acting head dept. botany, 1965—, prof. botany, 1967—. Mem. various profl. and sci. assns. Research, publs. in plant embryology, anatomy and plant morphogenesis. Home: 84, Eng Neo Av., Singapore 11, Singapore.*

RAO, R.V.G., Indian chemist, physicist; b. Vizianagran, South India, Dec. 3, 1927; s. R. Joga and R. Manikyambha) R.; reader, M.Sc., Ph.D., Benares Hindu U., 1951; Ph.D., U. Cal. at Berkeley, 1957; postdoctoral fellow Pa. State U.; m. R. Jaya Lakshmi, Dec. 24, 1958; children—R. Sobha, R. Venkateswarulu. Lectr., M.R. Coll., Agra (India) U., 1951-53; James M. Goewry fellow, U. Cal. at Berkeley, 1953-56; postdoctoral research fellow Pa. State U., University Park, 1956-57; sci. officer Nat. Chem. Lab., India, 1958-64; reader, head chem. dept. S.V. U. Coll. Engring., Tirupati, India, 1964—; spl. research fellow Glasgow (Scotland) U., 1966-67. Fellow Indian Phys. Soc. Research, publs. on low temperature studies, adsorption of gases on solids, high temperature calorimetry, liquid state properties, viscoelastic properties of liquids; formulated several equations between properties of liquids. Home: 209/3RT Saidabad Colony, Hyderabad-36, India. Office: Dept. Chemistry, S.V.U. Coll. Engring., Tirupati, South India.*

RAOULT, François Marie, French phys. chemist; b. Fournes-en-Weppes, France, May 10, 1830; doctorate U. Paris, 1863. Became prof. chemistry Lycée, Sens, France, 1862; placed in charge chemistry classes, Grenoble, France, 1862; prof. chemistry, Grenoble, 1870-1901, became dean Faculty Scis., 1889. Recipient Davy medal, 1892. Corry. mem. French Acad. Sci., 1890. Developed Roult's Law (vapor pressure of solvent in solution is proportional to ratio of number of solvent molecules to solute molecules); demonstrated freezing points are depressed proportionally to concentration and molecular weight of dissolved substances; studies on heat of reaction, electromotive force of Galvanic cells; discovered law for reduction of compressibility of solutions with concentration. Died Grenoble, Apr. 1, 1901.

RAPANT, Vladislav, Czechoslovakian surgeon; b. Uh.Hradiste, Czechoslovakia, June 6, 1902; s. Jan and Marie (Zajicek) R.; M.D., U. Masaryk, Brno, Czechoslovakia, 1935, D.Sc., 1956; m. Jaroslava Podlipna, Mar. 21, 1912; 1 son, Vladislav. Staff, Surg U. Clinic, Brno, 1926-45; staff Surg. U. Clinic, Olomouc, Czechoslovakia, 1945—, dir., 1948—; prof. clin. surgery Palacky U., Olomouc, 1948—. Recipient Gold medal Palacky U., 1966, medal Vischnevski's Inst., Moskva, USSR, 1967. Fellow Internat. Coll. Surgeons; mem. Société Internat. de Chirurgie, Internat. Cardiovascular Soc., Jan. Evangelista Purkyne Soc. Author: Surgery of the Esophagus and Cardia, 1950; Cancer of the Stomach, 1956; Surgery of Massive Haemorrhage, 1959; also numerous articles. Research on surgery of benign diseases and cancer of esophagus, transthoracic esophagomyotomy in achalasia, retrovascular intracervical esophagogastroanastomosis, extramucous suture and resection of mucosa in treatment of bleeding esophageal varices, new technique of reconstrn. of esophagus using stomach, surgery for tumors of mediastinum and lungs, hiatal hernias. Home: 26 Polivkova Olomouc, Czechoslovakia.*

RAPAPORT, Felix Theodosius, surgeon; b. Munich, Germany, Sept. 27, 1929; s. Max W. and Adelaide (Rathaus) R.; came to U. S., 1945, naturalized, 1951; A.B. magna cum laude, N.Y. U., 1951, M.D., 1954. Intern, Mt. Sinai Hosp., N.Y.C., 1955-56; asst. resident surgeon, chief surg. resident Bellevue Hosp., N.Y.U. Hosp., 1958-62, asst. vis. surgeon, 1962-65, asso. vis. surgeon, 1965—; practice medicine, specializing in surgery, N.Y.C., 1962—; faculty N.Y. U., 1958—, asso. prof. surgery, 1965—, also dir. research Inst. Reconstructive Plastic Surgery, head transplantation, immunology unit dept. surgery. Recipient USPHS Research Career Devel. award Inst. Allergy and Infectious Diseases, 1961-62, Career Scientist award N.Y.C. Health Research Council, 1962—. Mem. A.M.A., A.A.A.S., N.Y. Acad. Sci., Transplantation Soc. (sec.), Infectious Disease Soc. Am., Am. Assn. Immunologists, Soc. Exptl. Biology and Medicine, Soc. U. Surgeons, Harvey Soc., Alpha Omega Alpha. Research, numerous publs. primarily in field of transplantation and microbiology. Home: 145 E. 15th St., N.Y.C. 10003. Office: 560 1st Av., N.Y.C. 10016.*

RAPAPORT, Howard G., Am. physician; b. N.Y.C., Oct. 25, 1908; s. Meyer and Fanny (Relkin) R.; B.S., N.Y. U., 1930; M.D., Bern (Switzerland) Med. Sch., 1934; m. Elizabeth Schnell, Sept. 24, 1939; children—Martin Seth, Lois Joyce, Richard Arthur. Practice medicine specializing in allergy, N.Y.C., 1938—; faculty Albert Einstein Coll. Medicine, Bronx, N.Y., 1955—, asso. clin. prof. pediatrics (allergy), 1955—; chief children's allergy, staff Abraham Jacobi Hosp., Bronx; med. dir. Convalescent Asthma Treatment Center, New Castle, N.Y., 1965—. Fellow Am. Coll. Allergists (pres. 1967-68), Am. Acad. Allergy; mem. Westchester Allergy Soc. (pres. 1965-67), Assn. Convalescent Homes and Hosps. for Asthmatic Children (past pres.), Nat. Allergy Seminars (exec. v.p.), A.M.A. (asst. sec. allergy sect.), N.Y. Allergy Soc. Contbg. author: Fundamental Aspects of Allergy, 1962; The Allergic Child, 1963; Somatic and Psychiatric Aspects of Childhood Allergies, 1963; also articles. Developed allergic index used to simplify recognition of allergic illness. Home: 14 Olmsted Rd., Scarsdale. Office: 16 E. 79th St., N.Y.C. 10021; also 14 Olmstead Rd., Scarsdale, N.Y. 10583.*

RAPAPORT, Samuel I., Am. physician; b. Los Angeles, Nov. 19, 1921; s. Hyman and Bertha (Krupnick) R.; student U. Cal. at Los Angeles, 1937-40; M.D., U. So. Cal., 1945; m. Joyce Mildred Cooperman, Oct. 3, 1951; children—Susan, Sally, Mark, Bruce. Chief hematology sect. VA Hosp., Long Beach, Cal., 1951-57; faculty U. Cal. at Los Angeles Sch. Medicine, 1957-58; faculty U. So. Cal. Sch. Medicine, 1958—, prof. medicine, 1964—. Cons. in medicine Long Beach Va. Hosps., 1962—; sr. attending physician, chief hematology sect. med. service Los Angeles County Hosp., 1958—. Fullbright Research scholar in hematology U. Oslo, Norway, 1953-54; Nat. Heart Inst. Spl. fellow, 1964-65. Fellow A.C.P.; mem. Am. Soc. Clin. Investigation, Western Assn. Physicians, Am. Soc. Hematology, Soc. for Exptl. Biology and Medicine, Western Soc. Clin. Research (councillor 1962-64), Am. Fedn. Clin. Research (past chmn. Western sect.). Research on blood coagulation and thrombosis. Home: 326 Glen Summer Rd., Pasadena, Cal. 91105. Office: U. So. Cal. Sch. Medicine, 2025 Zonal Av., Los Angeles 90033.*

RAPATZ, Gabriel Louis, Am. biologist; b. Browerville, Minn., May 11, 1925; s. Louis and Mary (Kmetz) R.; B.S., Coll. St. Thomas, St. Paul, 1949; M.S., St. Louis U., 1951, Ph.D. (grad. fellow), 1955; m. Eileen Lemm, July 22, 1948; children—Linda, Mary, Marcia, Jennifer, Michael, Gabrielle. Faculty, Webster Coll., Webster Grove, Mo., 1954-55, So. Ill. U., Carbondale, 1955-56; staff Am. Found. Biol. Research, Madison, Wis., 1956—, research prof., 1965—, asst. dir., 1960—. Mem. Am. Physiol. Soc., Biophys. Soc., Soc. Cryobiology, A.A.A.S., Sigma Xi. Research, chpts. in books, publs. on basic research in field of low temperature biology, with spl. emphasis effects of cooling rates and protective additives on preservation of living organisms; research is directed toward long term preservation of cells, tissues, organs and eventually entire organisms at cryogenic temperatures. Home: Route 1, Box 139, Madison, Wis. 53704.*

RAPER, John Robert, Am. biologist; b. Lexington, N.C., Oct. 3, 1911; s. William Franklin and Julia (Crouse) R.; A.B., U. N.C., 1933, M.A., 1936; M.A., Harvard, 1939, Ph.D., 1939; m. Ruth Scholz Dec. 20, 1936 (div. Oct. 1948); 1 son, William Thomas; m. 2d, Carlene Marie Allen, Aug. 9, 1949; children—Jonathan Arthur, Linda Carlene. NRC fellow Cal. Inst. Tech., 1939-41; instr. botany Ind. U., 1941-43; biologist Manhattan Dist., Chgo. also Oak Ridge, 1943-46; faculty botany U. Chgo., 1946-54, professor 1953-54; prof. botany Harvard, 1954—; Fulbright Research prof. U. Cologne, Germany, 1960-61. Guggenheim fellow, 1960-61. Mem. Nat. Acad. Scis., Am. Acad. Arts and Scis. (sec. 1962-64), Am. Soc. Naturalists, Bot. Soc. Am., Mycol. Soc. Am. (pres. 1957-58), Genetics Soc. Am. Incompatibility in Fungi, 1967; Genetics of Sexuality in Higher Fungi, 1967. Research in sexual hormones in aquatic fungi, biol. effects of external beta-irradiation, genetics of sexuality in higher fungi. Home: 2 Constitution Rd., Lexington, Mass. 02173. Office: Biol. Labs., Harvard, Cambridge, Mass. 02138.*

RAPER, Kenneth Bryan, Am. microbiologist, botanist; b. Welcome, N.C., July 11, 1908; s. William Franklin and Julia Selina (Crouse) R.; A.R., U. of N.C., 1929, D.Sc. (hon.), 1961; A.M., George Washington University, 1931; A.M., Harvard, 1935, Ph.D. (Austin Fellow 1935-36), 1936; m. Louise Montgomery Williams, July 18, 1936; 1 son, Charles Albert. With U. S. Dept. Agr. 1929-53, jr. mycologist Bur. Chemistry and Soils, 1929-36, asst. mycologist Bur. Plant Industry, Washington, 1936-40; microbiologist, sr. microbiologist and prin. microbiologist Northern Regional Research Lab. Peoria, Ill., 1941-53; vis. prof. botany, U. Illinois 1946-53, prof. bacteriology and botany U. Wis. since 1953. Chmn. Am. delegation to 13th gen. assembly, Internat. Union Biol. Scis., London, 1958, chmn. U.S. nat. com.; mem. exec. com. div. biology and agr. NRC, 1956-61. Fellow A.A.A.S.; mem. Nat. Acad. Sciences (mem. council 1961-64), Am. Acad. Microbiology, Am. Soc. Naturalists, Am. Philos. Soc., Soc. Gen. Microbiology, Am. Acad. Arts and Scis., Bot. Soc. Am. (microbiol. sect. chmn. 1949; award of merit 1961), Mycol. Soc. Am. (pres. 1951), Soc. Am. Bacteriologists (councilor 1954-58), Soc. Indsl. Microbiologists (past Pres.), Phi Beta Kappa, Sigma Xi. Author: A Manual of the Aspergilli (with Charles Thom), 1945; A Manual of the Penicillia (with Charles Thom), 1949; (with Dorothy Fennell) The Genus Aspergillus, 1965; also research papers. Trustee, Biol. Abstracts, 1963—. Research on isolation, selection superior penicillin producing molds, preparation definitive vols. on common molds of genera Aspergillus, Penicillium (important in industry, agr., medicine), isolation, study developmental processes in cellular slime molds (organisms nr. boundary between plants, animals). Home: 4110 Chippewa Dr., Madison, Wis. 53705.

RAPOPORT, Anatol, math. biologist; b. Lozovaya, Russia, May 22, 1911; s. Boris and Adel (Rapoport) R.; came to U. S., 1922, naturalized, 1928; S.B., U. Chgo., 1938, S.M., 1940, Ph.D., 1941; m. Gwen Goodrich, Jan. 29, 1949; children—Anya, Alexander, Charles Anthony. Instr. math. Ill. Inst. Tech., 1946-47; research asso. U. Chgo., 1947-49, asst. prof. math. biology, 1949-54; asso. prof. math. biology U. Mich., Ann Arbor, 1955-60, prof., 1960—, sr. research mathematician Mental Health Research Inst., 1960—. Fellow Center for Advanced Study in Behavioral Scis., 1954-55. Fellow A.A.A.S., Am. Acad. Arts and Scis.; mem. Am. Math. Soc., Math. Assn. Am., Biometric Soc. (charter), Internat. Soc. for Gen. Semantics (pres. 1953-55), Soc. for Gen. Systems Research (v.p. 1963—). Author: Science and the Goals of Man, 1950; Operational Philosophy, 1953; Fights, Games, and Debates, 1960; Strategy and Conscience, 1964. Editor: Gen. Systems, 1956—; asso. editor Review of Gen. Semantics, 1946, Bull. Math. Biophysics, 1949—; editorial bd. Behavioral Sci., Conflict Resolution, Information and Control. Home: 516 Oswego St., Ann Arbor, Mich.*

RAPPAPORT, Irving, Am. immunologist; b. N.Y.C., Sept. 4, 1923; s. Hyman and Sarah (Frock) R.; A.B., Cornell U., 1948; Ph.D., Cal. Inst. Tech., 1953; m. Helen Doris Magid, June 27, 1948; children—Glenn, Jeffrey, Paul. Research botanist U. Cal., Los Angeles, 1953-61; asst. prof. microbiology U. Chgo., 1961-64; prof. immunology N.Y. Med. Coll., N.Y.C., 1964-66, acting chmn. dept. microbiology, 1966—. Mem. Am. Assn. Immunologists, Genetics Soc. Am., Am. Assn. Microbiology, N.Y. Acad. Scis., N.Y. Acad. Medicine. Research, numerous publs. on reversible interaction of tobacco mosaic virus by rabbit antiserum, antigenic relationships between coat protein of wild type and defective mutants of TMV. Home: 200 Waverly Rd., Scarsdale, N.Y. 10583. Office: Fifth Av. at 106th St., N.Y.C. 10029.*

RAPPAPORT, Paul, Am. physicist; b. Phila., Mar. 25, 1922; s. Isadore and Ida (Braunstein); B.S., Carnegie Inst. Tech., 1948, M.S., 1947; m. Rose Goldblatt, Jan. 9, 1944; children—Edith Candace, Norma Wendy, Elycia Lois. Physicist, Naval Air Exptl. Sta., 1946-49; tech. staff RCA Labs., Princeton, N.J., 1949—, head energy conversion research, 1960-65, head device physics research, 1965-66, asso. lab. dir. applied materials research, 1967—. Mem. research adv. com. Elec. Energy systems NASA, 1959-61; mem. materials adv. bd. materials study for aux. power units Nat. Acad. Sci., 1960-62; cons. USAF, 1965—; indsl. adviser engring. dept. U. Pa., Phila., 1965—. Recipient Achievement award for outstanding research RCA Labs., 1953, 55, 66; Carnegie Int. Tech. Alumni award, 1955; IR100 award Indsl. Research, 1966. Fellow Am. Phys. Soc., Am. Inst. Aeros. and Astronautics (asso.); mem. I.E.E.E. (sr.), Pi Mu Epsilon, Sigma Xi. Research, publs. on discovery and devel. solar cell, radiation damage in semiconductors; first accurate measurement of damage thresholds in Germanium and Silicon. Home: 13 Broadripple Dr. Office: RCA Labs., Princeton, N.J. 08540.*

RAPPOPORT, Donald Aron, neurochemist; b. Kharkov, Russia, May 10, 1913; s. Aaron and Sonia (Sukotinsky) R.; brought to U. S., 1923, naturalized, 1936; B.S., U. Ill., 1935; M.A., U. Cal., Los Angeles, 1940, Ph.D., Berkeley, 1951; m. Voltairine Ehrlich, Dec. 9, 1964. Research biochemist Cancer Inst. Columbia, 1951-52; with Baylor U. Coll. Medicine, 1952-60, asst. prof. dept. biochemistry, 1953-60; asso. prof., dir. div. molecular biology dept. pediatrics U. Tex. Med. Br., Galveston, Tex., 1960—. Mem. A.A.A.S., Am. Chem. Soc., Am. Soc. Biol. Chemists, Biochem. Soc. (Eng.), Radiation Research Soc., Royal Soc. Tropical Medicine (London), Sigma Xi. Research, publs. on mechanism of information accrual in form of ribonucleic acid in brain of catfish following olfactory stimulation; elucidated detailed changes in biosynthesis of nuclear and cytoplasmic ribonucleic acid and proteins in growing rat brain. Home: 207 Barracuda St., Galveston, Tex. 77550.*

RASCH, Georg William, Danish statistician; b. Odense, Sept. 21, 1901; s. Vilhelm and Johanne (Duusgaard) R.; Ph.D., U. Copenhagen; m. Elna Nielsen, Feb. 8, 1928; children—Helga, Lotte. Successively instr., asst. prof., head statis. dept. Serology Inst. Denmark, 1923-56; prof. applied statistics and psychol. math., 1962—; cons. applied statistics and psychol. math. to Hygiene Inst., State Serology Inst., Danish Inst. Pedagogic Research, Research Group Mental Health. Mem. Internat. Inst. Statistics, Danish Math. Soc., Danish Soc. Actuaries, Internat. Biometry Soc. Author: Zur Theorie und Anwendung des Produktintegrals, 1934; Probablistic Models for Some Intelligence and Attainment Tests, 1960; On General Laws and the Meaning of Measurement in Psychology, 1961. Home: Skovmindej 14, Holte. Office: Statis. Inst., Sankt Peders Straede 19 1, Copenhagen K, Denmark.

RASCH, Philip John, Am. physiologist; b. Grand Rapids, Mich., Dec. 3, 1909; s. Walter J. and Elizabeth (Hilber) R.; B.A., U. So. Cal., 1947, M.A., 1951, M.Ed., 1954, Ph.D., 1956; m. Mary Kirk, Nov. 18, 1939; children—Philip E., Virginia M. Corrective therapist Brentwood Neuropsychiat. Hosp. VA, Los Angeles, 1947-54; dir. Biokinetics Research Lab., Cal. Coll. Medicine, Los Angeles, 1954-63; chief physiology div. Naval Med. Field Research Lab., Camp Lejeune, N.C., 1963—; head research and academic instrn. depts. USMC Phys. Fitness Acad., Quantico, Va., 1967—. Fellow Am. Coll. Sports Medicine (trustee), Assn. Phys. and Mental Rehab.; mem. A.A.H.P.E.R., Ergonomics Soc. Author: (with R. K. Burke) Kinesiology and Applied Anatomy, 3d edit., 1967; (with L. E. Morehouse) Sports Medicine For Trainers, 2d edit., 1963; also numerous articles. Research on phys. medicine and rehab., phys. fitness and heat acclimatization. Home: 405 Clyde Dr., Jacksonville, N.C. 28540. Office: Naval Med. Field Research Lab., Camp Lejeune, N.C. 28542.*

RASCHIG, Friedrich August, German chemist; b. Brandenburg, Germany, June 8, 1863; ed. U. Berlin, U. Heidelberg (both Germany); Asst. chem. lab. U. Berlin; chemist Badische Anilin und Sodafabrik, Ludwigshafen, Germany. Author: Schwefel- und Stick stoffstudien, P.E. 1924. Invented absorption device which is named after him; discovered nitroamide, chloramine; introduced new methods for producing hydroxylamine, hydrazine, phenol. Died Feb. 4, 1928, Duisburg, Germany.

RASETTI, Franco Dino, physicist; b. Castiglione del Lago, Italy, Aug. 10, 1901; s. Giovanni Emilio and Adele (Galeotti) R.; Ph.D. in Physics, U. Pisa (Italy), 1923; Ph.D. (hon.) Laval U., 1947; LL.D. (hon.), U. Glasgow (Scotland), 1957; m. Marie Madeleine Hennin, Sept. 9, 1949. Came to U.S., 1947, naturalized, 1952. Asst. prof. physics U. Florence (Italy), 1923-26; prof. physics U. Rome (Italy), 1927-30, Guggenheim fellow, 1959; prof. physics Laval U., Quebec City, Can., 1939-47; prof. physics Johns Hopkins, Balt., 1947—; vis. prof. Columbia, 1935, Cornell U., 1936, Washington U., 1946, U. Miami, 1958. Hon. research asso. paleobiology Smithsonian Instn., 1965—. Rockefeller Found. fellow Cal. Inst. Tech., 1928-29, Kaiser Wilhelm Inst., Berlin, 1931-32. Decorated Knight Order Ouissam Alaouite (Morocco); recipient Matteucci medal Italian Acad. Sci., 1931; Righi medal Acad. Sci. Bologna, 1932; Mussolini prize Italian Nat. Acad., 1938; Walcott medal Nat. Acad. Sci., 1952. Fellow Phys. Soc., Geol. Soc. Am., Paleontol. Soc.; mem. Accademia Nazionale dei Lincei (Rome), Pontifical Academi Sci., Phi Beta Kappa, Sigma Xi. Author: Elements of Nuclear Physics, 1936; Middle Cambrian Stratigraphy and Faunas of the Canadian Rocky Mountains, 1951; also articles on physics, geology, paleontology. Research in atomic and nuclear spectroscopy, nuclear physics, chiefly neutron induced reactions, cosmic ray; Collaborated with Enrico Fermi on early work on atomic energy, 1934-37. Home: 300 E. University Pkwy., Balt. 21218.*

RASHEVSKY, Nicolas, math. biologist; b. Chernigov, Russia, Sept. 20, 1899; s. Peter and Nadezhda (Konstantinovich) R.; grad. U. Kiev (Russia), 1919; m. Emily Zolotareva, Nov. 12, 1920; children—Emilie (Mrs. K. Strand), Nina (Mrs. C.O. Carlson), Nadezhda (Mrs. M.S. Pittman, Jr.). Came to U.S., 1924, naturalized, 1939. Instr. physics U. Kiev, 1919, Robert Coll., Constantinople, Turkey, 1920-21; prof. physics Russian U., Prague, Czechoslovakia, 1921-24; research physicist Westinghouse Research Labs., East Pittsburgh, 1924-34; Rockefeller fellow math. biophysics U. Chgo., 1934-35, faculty, 1935—, prof., 1946-47, prof. math. biology, chmn. com. on math. biology, 1947-65; prof. math. biology U. Mich. Mental Health Inst., Ann Arbor, 1965—. Lectr. mathbiology Am., fgn. univs.; cons. FDA, 1953-55; dir. internat. postdoctoral course physicomath. aspects biology, Varenna, Italy, 1960; research adviser U. Genoa (Italy), fall 1960; gen. chmn. 1st Internat. Symposium on Math. Theories Biol. Phenomena, N.Y.C., 1961. Fellow A.A.A.S.; mem. Internat. Biometric Soc. (charter), Biophys. Soc. Founder, editor Bull. Math. Biophysics, 1939. Author: Mathematical Biophysics, 1938, 3d edit., 1960; Advances and Applications of Mathematical Biology, 1940; Mathematical Theory of Human Relations, 1947; Mathematical Biology of Social Behavior, 1951, rev., 1960; Mathematical Principles in Biology and Their Application, 1960; Some Medical Aspects of Mathematical Biology; also numerous articles. Home: 1482 Bemidji Dr., Ann Arbor, Mich. 48103.*

RASHEVSKY, Petr Konstantinovich, Russian geometrician; b. Moscow, 1907; grad. Moscow U., 1928; D.Physico-Math. Sci., 1938. Instr., Moscow Power Engring. Inst., 1930-34; instr. Moscow Pedagogical Inst., 1931-34, prof., 1934-41; with Moscow Inst. R.R. Transport Engring., 1942-49; prof. Moscow U., 1938—. Author: Polymetric Geometry, 1941; A Course on Differential Geometry, 1956; Geometry and Its Axioms, 1960. Research and numerous works in Riemann geometry, geometry of affine connectivity, axioms of projective geometry, geometry of homogeneous spaces connected with Lie groups, developer polymetric geometry. Address: Moscow University, Leninskie gory, Moscow, USSR.

RASMUSSEN, John Oscar, Jr., Am. nuclear chemist; b. St. Petersburg, Fla., Aug. 8, 1926; s. John Oscar and Hazel (Ormsby) R.; B.S., Cal. Inst. Tech., 1948; Ph.D., U. Cal. at Berkeley, 1952; m. Louise Brooks, Aug. 27, 1950; children—Nancy L., Jane E., David B., Stephen J. Faculty, U. Cal. at Berkeley, 1952—, prof. chemistry, 1963—, research staff Lawrence Radiation Lab., 1952-66; vis. prof. Nobel Inst. Physics, Stockholm, Sweden, 1953. NSF Sr. fellow U. Inst. for Theoretical Physics, Copenhagen, Denmark, 1961-62. Recipient E. O. Lawrence award, 1967. Fellow Am. Phys. Soc.; mem. Am. Chem. Soc., Am. Assn. U. Profs., Fedn. Am. Scientists. Author: (with I. Perlman) Alpha Decay, 1958; also numerous articles. Expts. and calculations energy levels, transition rates heavy nuclei particularly on rate theory alpha emission. Home: 166 Ardmore Rd., Berkeley, Cal. 94707.*

RASMUSSEN, Knud Johan Victor, Danish explorer; b. Jakobshaven, Greenland, June 7, 1879; ed. U. Copenhagen (Denmark); m. Dagmar Andersen; 1 son, 2 daus. Dir. several expdns. to Greenland, beginning 1902; founder Thule Base, Cape York, Greenland, 1910; crossed from Greenland to Bering Strait, 1921-24. Recipient numerous medals from Scandinavian and fgn. geol. and anthropol. socs. Hon. mem. several geog. socs. Author: Greenland by the Polar Sea, 1921; Myths and Legends from Greenland, 1921-25; Across Arctic America, 1927. On his expdns. attempted to prove theory that Eskimos and N.Am. Indians were both descended from migratory tribes from Asia. Died Gentofte, Denmark, Dec. 21, 1933.

RASMUSSEN, Svend Erik, Danish chemist; b. Esbjerg, Denmark, Nov. 19, 1925; s. Svend Ove and Hilma (Hellsten) R.; Mag. scient., U. Copenhagen (Denmark), 1950, Dr. Phil., 1960; m. Clasy Dornt Holmlund, Dec. 9, 1950; children—Ulf Henrik Holmlund, Maja Holmlund. With dept. chemistry Royal Vet. and Agrl. Coll., 1950-51; asst., U. Copenhagen, 1951; research fellow Queen's Coll., Dundee, Scotland, 1951-52; asst. Tech. U. Denmark, 1952-59; prof. inorganic chemistry Aarhus (Denmark) U., 1959—. Mem. Lydsh Selskub for Fysikog Kemi, Chem. Soc., London. Research equilibria of complex ions in solution, X-ray and neutron crystallography, crystal chemistry. Home: 39 Hojkolvej, Aarhus C. Denmark.*

RASMUSSEN, Theodore Brown, physician; b. Provo, Utah, Apr. 28, 1910; s. Andrew Theodore and Gertrude (Brown) R.; B.S., U. Minn., 1934, M.B., 1934, M.D., 1935, M.S., 1939; m. Catherine Archibald, Dec. 28, 1947; children—Donald, Ruth, Mary, Linda. Neurol. fellow Mayo Clinic, 1936-39; neurosurg. fellow Montreal (Que., Can.) Neurol. Inst., 1939-42; lectr. McGill U., Montreal, prof. neurology and neurosurgery, 1954—; asso. neurosurgeon Montreal Neurol. Inst., 1946-47, dept. dir., 1954-60, dir., 1960—; prof. neurol. surgery U. Chgo., 1947-54. Recipient Outstanding Achievement award U. Minn., 1958, U. Chgo., 1963. Fellow Royal Coll. Physicians and Surgeons Can., N.Y. Acad. Scis., Am., Canadian med. assns.; mem. A.A.A.S., Sigma Xi, Alpha Omega Alpha. Author: (with Wilder Penfield) Cerebral Cortex of Man, 1950; also numerous articles. Research on localization of function in cortex of man, surg. treatment for focal epilepsy, cerebral circulation, radiation effects on nervous tissue. Home: 29 Surrey Dr., Montreal, 16, Que., Can.*

RASOOL, Ichtiaque, physicist; b. Lucknow, India, Jan. 16, 1933; s. Mairaj S. and Mushfiqua (Bano) R.; B.Sc., U. Luchnow (India), 1950; Ph.D. in Atmospheric Physics, U. Paris (France), 1956; m. Francoise Bouquet, Aug. 29, 1964; 1 dau., Elisa. Research asso. Centre National de la Recherche Scientifique, Paris, France, 1958-61; research physicist planetary scis. Goddard Inst. Space Studies, NASA, N.Y.C., 1961—. Vis. asst. prof. aeros. Polytech. Inst. Bklyn., 1963; adj. prof. meteorology N.Y. U., 1963—; asso. dir. summer faculty program space physics Columbia, 1966; vis. asso. prof. astro-geophysics U. Colo., 1966. Recipient Spl. Act award NASA, 1966. Mem. Am. Geophys. Union, Am., Royal (London) meteorol. socs. Constructed apparatus to measure ozone in upper atmosphere of earth, demonstrated use of meteorol. satellites in long range weather predictions; collaborated in theoretical calculations that early atmosphere of earth could have been composed of methane and ammonia. Home: 165 W. 66th St., N.Y.C. 10023. Office: 2880 Broadway, N.Y.C. 10025.*

RASORI, Giovanni, Italian physician; b. Parma, Italy, 1766; M.D., U. Parma, 1785; pupil of Girardi. Prof. U. Pavia (Italy), from 1796; went to Milan, 1800. Author several med. works. Devised new med. theory of counter-stimulus. Died 1837.

RASPAIL, François Vincent, French chemist; b. Carpentras, France, Jan. 29, 1794; s. Joseph and Marie (Laty) R.; student natural scis., Paris; children—Emile, Benjamin, Camille, Fought against Charles X, 1830; leader of Republicans; condemned to 6 years imprisonment for conspiracy against new regime, 1849. Author: Essai de chimie microscopique appliquée à la physiologie, 1831; Nouveau système de chimie organique, 1833; Nouveau système de physiologie végétale et de botanique, 19 vols., 1837; Histoire naturelle de la santé et de la maladie chez les végétaux et les animaux, 3 vols., 1839-43; le Médecin des familles, 1843; la Manuel de la santé, 1846. Used camphor as antiseptic, 1845. Died Paris, Jan. 8, 1878.

RASPE, Rudolph Erich, natural historian; b. Hanover, Germany, 1737; ed. univs. Göttingen, Leipzig (both Germany). Became clk. Univ. Library, Hanover, 1762; apptd. sec. Univ. Library, Göttingen, 1764; named prof. archeology Maurice Coll., Cassel, Germany, 1767, then curator antiquary mus.; went to Eng. to avoid arrest for jewel theft, 1775, became translator and publisher in London; assay master, storekeeper Dolcoath Mine, Cornwall, Eng., 1782-88; moved to Scotland, then Ireland. Hon. mem. Royal Soc. London, 1769, later dropped from membership. Author: Essay on the Origin of Oil-Painting, 1781; Baron Munchausen's Narrative of His Marvellous Travels and Campaigns in Russia (original version), 1785; Observations on the Work of M. Klotz, 1768; Documents on the Most Ancient History of Hesse-Cassel, circa 1774; An Account of Some German Volcanoes and Their Productions, 1776; A Descriptive Catalogue of a General Collection of Ancient and Modern Engraved Gems, 1791; also transls. of Leibnitz's philos. works. Died Muckross, Ireland, 1794.

RATCLIFF, Perry Albert, Am. periodontologist; b. Marion County, Ind., Aug. 25, 1915; s. Thomas Roscoe and Marie (Boardman) R.; D.D.S., U. Ind., 1939; m. Viola Ruth Hall, Aug. 20, 1939; children—Colleen Malcolm, James. Asso. prof. periodontology U. Soc. Cal., 1950-59; with U. Cal., San Francisco, 1959—, prof., 1964—, chmn. div. periodontology, 1960—; vis. asso. prof. Loma Linda U. 1960-62; vis. lectr. U. Neb., 1965. Cons. periodontics U.S. Army Letterman Gen. Hosp., 1962—; Cons. edn. in periodontics USPHS, 1963—. Mem. Am. Dental Assn., Am. Soc. Periodontists, Am. Acad. Periodontology, Western Soc. Periodontology, Assn. pour les Recherches sur les Periodontopathies, Internat. Assn. Dental Research, A.A.A.S. Co-author: Programmed Textbook on Gingivectomy-Gingivoplasty, 1966; Dimensions of Dental Hygiene, 1966. Editor: Periodontal Abstracts, 1956-62. Research and publs. on changes in periodontium and temporomandibular joint as affected by nutrition, endocrine imbalances. Home: 325 Buckingham Way, San Francisco 94132.*

RATCLIFFE, Gerald Alfred, Brit. chemist; b. London, Eng., Sept. 8, 1926; s. Alfred and Constance (Jones) R.; ed. Cambridge and Cornell univs.; M.A., Ph.D.; m. Kennen Kethley, Dec. 22, 1953; children—Heather, Kevin. Chemist-engr. Shell Oil Co., 1948-50, 54-56; instr. U. Cambridge, 1956-64; prof., dean faculty McGill U. Author articles on gas absorption, dispersion and solid-liquid separation pub. in Chem. Engring. Sci., Transactions Instn. Chem. Engrs., Analytic Chemistry. Address: McGill U., Montreal, Can.

RATEAU, Auguste, French engr.; b. Royan, France, Oct. 13, 1863; ed. Ecole Polytechnique. Engr., Corps Mines; prof. Ecole Nationale Saint-Etienne; prof., Paris. Mem. French Acad. Scis. Recipient Fourneyron and Poncelet prize French Acad Scis., 1911. Author: Considerations on Turbomachines and in particular on the Ventilators, 1892. Research on screw ventilators, compressors, pumps, turbines; invented gas turbine; improved diesel motor. Died Neuilly, France, Jan. 13, 1930.

RATHBUN, John Campbell, Canadian pediatrician; b. Toronto, Ont., Can., May 8, 1915; s. John Bell and Gladys (Jamieson) R.; M.D., U. Toronto, 1939; m. Catherine Coleman Moore, Jan. 10, 1942; children—Fredericka Adamson (Mrs. Inigo Adamson), Catherine (Mrs. Roy Brown), Flora, Elizabeth, Mary Kate, Amy. Practice medicine specializing in pediatrics, London, Ont., 1948—; faculty U. Western Ont., London, 1949—, prof., head dept. pediatrics, 1955—; staff War Meml. Children's Hosp., London. Mem. research sub-com. Canadian Dept. Nat. Health and Welfare, 1960; cons. Children's Psychiat. Research Inst., Byron, Ont., 1959—. Fellow Royal Coll. Physicians, Am. Acad. Pediatrics; mem. Canadian Paediatric Soc. (past pres.), A.A.A.S., Canadian Med. Assn., Am. Diabetic Assn., Canadian Soc. for Clin. Investigation. Contbg. author: Pediatrics, 1956; also articles. Research on hypophosphatasia, citrullinuria. Home: Rural Route 6, London, Ont., Can.*

RATHENAU, G. W., physicist; b. Berlin, Germany, June 25, 1911; s. Fritz M. and Sophie (Dannenbaum) R.; student U. Berlin; D.Physics, U. Göttingen (Germany), 1933; m. Johanna Huberta Van den Hoek, June 9, 1938; children—Jan Jacob, Frank Willem. Research asso. Physics Inst. U. Groningen, 1934-36, Lab. of Teyler's Found., Haarlem, 1936-38; with Philips Research Labs., Eindhoven, 1938-52, ending as chief physicist; prof. physics, dir. Natuurkundig Lab., U. Amsterdam, prof. extraordinary, 1963-66, spl. prof., 1967—; dir. research Philips Research Labs., 1967—. Research and publn. on molecular spectroscopy, spectroscopy of solids, metals, magnetism, also magnetic materials. Home: 7 Leeuwerikiaan. Office: Philips Research Lab., Eindhoven, The Netherlands.

RATHER, Lelland Joseph, Am. pathologist; b. College Station, Tex., Dec. 22, 1913; s. James Burness and Corinne (Carson) R.; A.B., Johns Hopkins, 1934, M.D., 1939; M.S., U. Chgo., 1936; m. Eleanor Edith Knight, June 20, 1940 (div. June 1958); children—Patricia, Leland, Noël; m. 2d, Ingeborg Gabrielle Arnold, June 19, 1959. Faculty dept. pathology Stanford Sch. Medicine, Palo Alto, Cal., 1946—, prof., 1957—. Cons. Surgeon Gen.'s Office, U.S. Dept. Agr., 1948-53; mem. study sect. NIH, 1961-65. Mem. Soc. for Exptl. Biology and Medicine, Am. Soc. Exptl. Pathology, Am. Assn. History Medicine, Internat. Soc. for History Medicine, Internat. Acad. for History Medicine, Hist. Sci. Soc. Author: Disease, Life and Man, 1958; Mind and Body in 18th Century Medicine, 1965; also numerous articles. Research on resorption protein by kidneys, experimentally induced hypertension, nuclear alterations in liver cells induced by various chems., exptl. nephritis, cardiac hypertrophy, liver cirrhosis and testicular atrophy, various aspects human pathology. Home: 1017 Bryant St., Palo Alto, Cal. 94301.*

RATHKE, Martin Heinrich, biologist; b. Danzig, Poland, Aug. 25, 1793; prof. Dorpat, Estonia, then Königsberg (now Kalingrad, USSR). Fellow Royal Soc., 1855; mem. French Acad. Scis. (corr.). Author: Abhandlungen zur Bildungs- und Entwicklungsgeschichte des Menschen und der Thiere, 2 vols., 1832-33. Discovered gill-slits and gill-arches in embryo birds and mammals (after Von Baer), 1829; Rathke's pocket (small pit on dorsal side of oral cavity which marks point of invagination of hypophysis in developing vertebrates) is named after him. Died Königsberg, Sept. 3, 1860.

RATLIFF, Floyd, Am. neurophysiologist; b. La Junta, Colorado, May 1, 1919; A.B., Colorado College, 1947; M.Sc., Brown U., 1949; Ph.D., 1950; m. 1942. Nat. Research Council fellow, Johns Hopkins, 1950-51; instr., Harvard U., 1951-52; asst. prof., Harvard, 1952-54; assoc. biophysics, Rockefeller U., 1954-58; assoc. prof., 1958-66; prof., Rockefeller, 1966—. Recipient Warren medal, Soc. Experimental Psychology, 1966. Mem., Nat. Acad. Scis., Optical Soc.; Psychology Assn. Research on retinal interaction, animal behavior; physiological nystagmus. Office: Rockefeller University, 66th St. and York Ave., New York, N.Y. 10021.

RATNER, Sarah, Am. biochemist; b. N.Y.C., June 9, 1903; s. Aaron and Hannah (Selzer) Ratner; B.A., Cornell U., 1924; M.A., Columbia, 1927, Ph.D., 1937. Geis fellow Columbia Coll. Phys. and Surg., 1934-36, faculty, 1937-46, asst. prof., 1945-46; faculty N.Y. U. Coll. Medicine, 1946-54, asso. prof., 1953-54, adj. asso. prof. biochemistry, 1954—; asso. mem. dept. biochemistry Pub. Health Research Inst. City N.Y., 1954-57, mem., 1957—. Schoenheimer lectr., 1956. Recipient Carl Neuberg medal Am. Soc. European Chemists, 1959. Fellow A.A.A.S., N.Y. Acad. Scis.; mem. Am. Soc. Biol. Chemists, Harvey Soc., Am. Chem. Soc. (Garvan medal 1961), Biochem. Soc. (Eng.), Sigma Xi. Editorial bd. Jour. Biol. Chemistry, 1958-63. Contbr. numerous articles to tech. jours. Pioneered research with nitrogen isotope leading to new concepts bodily protein and amino acid turn overs; established mechanisms for steps in intermediary nitrogen metabolism body concerned with biosynthesis of arginine and urea and worked out energy relationships; discovered argininosuccinic acid. Home 70 E. 10th St., N.Y.C. 10003. Office: Pub. Health Research Inst., 455 1st Av., N.Y.C. 10016.*

RATNER, Stanley, Am. psychologist; b. Pitts., Aug. 13, 1925; s. Harry and Lillian (Liberman) R.; student Stanford, 1943; B.S., U. Pitts., 1948; M.A., Kent State U., 1950; Ph.D., Ind. U., 1954; m. Anita Sikov, Sept. 5, 1950; children—Eric H., Keith A. Faculty, Ind. U., 1953-55; faculty Mich. State U., East Lansing, 1955-67, prof. psychology, 1963-67; prof., chmn. psychology Beloit (Wis.) Coll., 1967—. Vis. prof. U. Miami, 1964, U. Mont., 1966; cons. exptl. psychology VA, 1960—. NSF fellow U. Cambridge, 1959-60. Fellow Am. Psychol. Assn., Sigma Xi; mem. Animal Behavior Soc., Psychonomic Soc. Author: (with M.R. Denny) Comparative Psychology, 1964; (with W.C. Corning) Chemistry of Learning, 1967. Asso. editor Psychol. Record, 1959—. Research, publs. on animal learning, especially learning by worms. Home: 2207 Collingswood, Beloit, Wis. 53511.*

RATTE, see de Ratte, Estienne-Hyacinthe.

RATTI, Arduino, Italian physician; b. Milan, Italy, Jan. 7, 1901; s. Giovanni and Alice M. (Devizzi) R.; M.D., U. Pavia (Italy), 1924. Faculty, U. Milan, 1932-42, prof., Med. Sch., 1956—, dir. Postgrad. Sch. Radiology, 1959—; prof. radiology U. Pavia, 1942-56; chief radiol. service Istituto Nazionale Studio e Cura Tumori, Milano, 1956—. Sec. gen. XI Internat. Congress Radiology, 1965. Mem. Italian Soc. Radiology and Nuclear Medicine (editor La Radiologia Medica, 1960—, pres. 1958-60), Italian Soc. Radiology; hon. mem. Deutsche Röntgengesellschaft, Schweizerische Röntgengesellschaft, Tschekoslavakische Röntgengesellschaft. Author: Handbook of Radiology, 1965; also numerous articles. Research on roentgendiagnostics (lungs, arteriography, medullography), radium therapy of tumors. Home: 44/1 via Moscova, Milan, Italy.*

RATTRAY, Maurice, Jr., Am. hydrodynamicist, phys. oceanographer; b. Seattle, Sept. 16, 1922; s. Maurice and Margaret (Rae) R.; B.S., Cal. Inst. Tech., 1944, M.S., 1947, Ph.D., 1951; m. Mary Louise Wolsey, June 9, 1951; children—Julia Westwood, Maurice III, Gordon Winfield. Faculty, U. Wash., Seattle, 1950—, prof. oceanography, 1962—. Rossby fellow Woods Hole Oceanographic Instn., 1966-67. Mem. Am. Soc. Limnology and Oceanography (pres. 1965-66), Am. Geophys. Union, N.Y. Acad. Scis., Sigma Xi. Mem. editorial bd. Jour. Marine Research, 1966—. Developed theory of ocean internal tide generation by bathymetric coupling with surface tide at the continental shelf break; theoretical models and classification systems for estuary currents and salinity distbns.; unified theory for long period wave motions in oceans with applications of same. Home: 1315 Lexington Way E., Seattle 98102.*

RATTRAY, Sylvester, Scottish med. writer; b. Angus, Scotland; flourished 1650-66; s. Sylvester Rattray; credited with theol., med. degrees; m. Ingells, dau. King Gask, May 1652; 1 son, Sylvester. Author: Aditus Novus ad Occultas Sympathiae et Antipathiae Naturalis ex Fermentorum Artificiosa Anatomia Hausta Patefactus, 1658; Prognosis Medica ad Usum Praxeos Facili Methodo Digesta, 1666. Wrote treatises concerning sympathy and antipathy among animals, vegetables, minerals (definition of sympathy, antipathy was action at distance resulting from ferments).

RATZEL, Friedrich, German geographer, anthropologist; b. Karlsruhe, Germany, Aug. 30, 1844; prof., Munich, also Leipzig, Germany. Author: Vorgeschichte des europäischen Menschen, 1874; Die Vereinigten Staaten von Nordamerika, 1878-80; Anthropogeographie, 2 vols., 1882-91; Völkerkunde, 1885-88; Politische Geographie, 1891; Die geographische Verbreitung des Menschen, 1891; Deutschland, 1898; Die Erde und das Leben, 2 vols., 1901-02; Kleine Schriften, 2 vols., 1905-06. Editor Das Ausland. One of founders of anthropogeography; studied causes of human settlements and size and structure of such social groups; emphasized importance of phys. environment as a determining factor of human activity; opposed theory that primitive societies all underwent parallel developments. Died Ammerland/Starnberg Lake, Aug. 9, 1904.

RATZENHOFER, Gustav, Austrian sociologist; b. Vienna, Austria, July 4, 1842; ed. clock maker's apprentice; master's exam., 1859; self study philosophy and sociology. Entered mil. career, 1859; dir. archives, 1878; reached rank feldmar schalleutant, 1898; served as pres. superior mil. ct. Vienna, 1898-1901. Author: Wesen und Zweck der Politik, 3 vols., 1898; Die Soziologische Erkenntnis, 1898; Postive Ethik, 1901; Soziologie, 1908. A social-Darwinist, he conceptualized politics as conflict and adjustment of conscious interests of various groups in a state; devel. theory of accomodation that the individual adjusts to society by making changes in his behavior patterns; influenced Am. sociology. Died at sea returning from Am., Oct. 1904.

RAU, Charles, archeologist, mus. curator; b. Verviers, Belgium, 1826; attended U. Heidelberg (Germany); Ph.D. (hon.), U. Freiburg (Baden, Germany), 1882. Came to Am., 1848; resident collaborator in ethnology U. S. Nat. Mus., 1875, curator dept. archeology, 1881-87; assisted in preparation for Centennial Exposition of 1876; made contbns. to process of classification; 1st in Am. to recognize importance of study of aboriginal technology. Author: The Archeological Collection of the United States National Museum; Early Man in Europe, 1876; Prehistoric Fishing in Europe and North America, 1885. Translator: Account of the Aboriginal Inhabitants of the California Peninsula (Jacob Buegert), 1863. Died Phila., Jan. 25, 1887.

RAUBER, August Antinous, anatomist; b. Obermoschel, Germany, Mar. 9, 1841; prof. Leipzig, German, Dorpat (now Tartu, Estonia); described outermost layer of cells which form blastoderm of embryo (Rauber's layer), 1888. Died Feb. 6, 1917.

RAUCH, John Henry, Am. physician; b. Lebanon, Pa., Sept. 4, 1828; s. Bernard and Jane (Brown) R.; grad. Med. Sch., U. Pa., 1848. Established marine hosps., Galena, Ill., Burlingame, Ia.; aided in effecting passage of a bill providing for geol. survey of Ia., 1856; prof. materia medica Rush Med., Coll., Chgo., 1857-58; founder Chgo. Coll. Pharmacy, 1859, 1st prof. materia medica; served as surgeon during Civil War; helped reorganize Chgo. Bd. Health, 1867; 1st pres. Ill. Bd. Health, 1877-91, superintended adminstrn. of Med. Practice Act; assisted in establishing a quarantine sta. for cholera cases and suspects, 1892, Mem. Ia. Med. Soc. (1st del. to A.M.A. 1852), Ia. Hort. Soc., Ia. Hist. Soc., Ia. Geol. Soc., Am. Pub. Health Assn. (a founder, treas., pres.). Author: (article) Report on the Medical and Economical Botany of Iowa, 1851; The Smallpox Situation in the United States. Editor pub. health dept. Jour. of A.M.A., 1894. Made natural history collection from upper Mississippi and Missouri rivers, 1855-56; collected in Venezuela, 1870; made nation-wide study of prevalence and control of smallpox. Died Lebanon, Mar. 24, 1894.

RAUCH, Lawrence Lee, Am. aero. engr.; b. Los Angeles, May 1, 1919; s. James Lee and Mable (Thompson) R.; A.B., U. So. Cal., 1941; A.M., Princeton, 1948, Ph.D., 1949; m. Norma Ruth Cable, Dec. 15, 1961; 1 son, Lauren Lee. Radio broadcast engr. KMTR Radio Corp., Hollywood, Cal., 1936-39; research supr. NDRC telemetering project Princeton, 1943-46, instr., 1947-49; faculty U. Mich., Ann Arbor, 1950—, prof. aerospace engring. 1953—, chmn. instrumentation engring. program, 1952-63; supr. air blast telemetry Operation Crossroads; Bikini Atoll, 1946. Chmn. telemetering working group, panel on test range instrumentation, research and devel. bd. Dept. Def., 1952-53; exec. com. Nat. Telemetering Conf., 1960-63. Recipient Army and Navy award for contbn. to work of OSRD, 1947, Spl. I.R.E. award, 1957, Ann. award Nat. Telemetering Conf., 1960, Donald P. Eckman Edn. award Instrument Soc. Am., 1966. Fellow I.E.E.E., A.A.A.S.; mem. Am. Math. Soc., Soc. for Engring. Edn., Am. Inst. Aeros. and Astronautics, Soc. for Indsl. and Applied Math, Phi Beta Kappa, Phi Kappa Phi, Sigma Xi. Author: (with M.H. Nicholas) Radio Telemetry, 1956; also articles. Editor: (with others) Theory of Oscillations, 1949. Research on first high speed electronic time-div. multiplex telemetry system; research on aircraft, missile programs. Home: 11 Harvard Pl., Ann Arbor, Mich. 48104.*

RAUEN, Hermann Matthias Thomas, German biochemist; b. Offenbach am Main, Germany, Nov. 4, 1913; s. Peter Josef and Margarethe (Niemann) R.; Diplomchemiker, U. Frankfurt (Germany), 1939, Dr. rer.nat., 1941; m. Marianne Buchka, Feb. 18, 1957; 1 son, Florian Thomas. Faculty physiol. chemistry U. Frankfurt, 1950-55; faculty U. Münster (Germany), 1955—, prof., 1957—. Mem. N.Y. Acad. Scis., Gesellschaft für Biologische Chemie Deutschland, Internat. Institut der Görres-Gesellschaft. Author: (with W. Stamm) Gegenstromverteilung, 1953; Biochemisches Taschenbuch, 1956, 2d edit. 1964; (with K.G. Ober, J. Schoenmakers, J. Zander) Symposion über Krebsproblems, 1961; also numerous articles. Research on intermediate metabolism, cancer problems. Home: 32 von Bodelschwinghstrasse, Münster/Westfalen, Germany.*

RAUP, David Malcolm, Am. paleontologist; b. Boston, Apr. 24, 1933; s. Hugh Miller and Lucy (Gibson) R.; S.B., U. Chgo., 1953; M.A., Harvard, 1955, Ph.D., 1957; m. Susan Creer Shepard, Aug. 25, 1956; 1 son, Mitchell D. Instr., Cal. Inst. Tech., 1956-57; faculty Johns Hopkins, 1957-65, asso. prof., 1963-65; faculty U. Rochester (N.Y.), 1965—, prof. geology, 1966—; geologist U.S. Geol. Survey, parttime 1959—; vis. prof. U. Tübingen (Germany), 1965. Dir. In-Service Inst. in Geobiology, NSF, 1960-61. NSF grantee, 1960-66; Cal. Research Corp. grantee, 1955-56; Am. Assn. Petroleum Geologists grantee, 1957; also Am. Philos. Soc. grantee 1957; Am. Chem. Soc. grantee, 1965—. Fellow Geol. Soc. Am.; mem. Paleontol. Soc., Soc. Econ. Mineralogy and Paleontology, Soc. for Systematic Zoology, Soc. for Study Evolution, N.Y. Acad. Scis., A.A.A.S., Sigma Xi. Editor: (with B. Kummel) Handbook of Paleontological Techniques, 1965. Research, publs. on crystallography skeletons of living and fossil invertebrates, systematics and evolution of tertiary echinoderms, geometry biol. forms, applications digital and analog computer techniques to paleontol. problems, dynamics skeletal growth in plated echinoderms. Home: 97 Woodland Rd., Pittsford, N.Y. 14534. Office: Dept. Geology, U. Rochester, Rochester, N.Y. 14627.*

RAUP, Hugh Miller, Am. botanist; b. Springfield, O. Feb. 4, 1901; s. Gustavus Philip and Fannie (Mitchell) R.; A.B., Wittenberg Coll. 1923, D.Sc., 1968; A.M. U. Pitts., 1925, Ph.D., 1928; (hon.) A.M., Harvard, 1945; m. Lucy Catherine Gibson, June 20, 1925; children—Karl Alfred, David Malcolm, Instr. biology Wittenberg Coll., 1923-25, asst. prof. 1925-32; research asst. Arnold Arboretum, Harvard, 1932-34, research asso., 1934-38; asst. prof. plant ecology Harvard prof. forestry, 1960-67, emeritus, 1967—; dir. Harvard Forest, 1946—; vis. prof. plant geography Johns Hopkins, 1967—; field botanist Nat. Mus. Can., summers 1928-30, bot. survey expdns. in Mackenzie basin, summers 1926-30, 32, 35, 39, Alaska Hwy., 1943-44, bot. survey of Black Rock Forest, N.Y., 1936-37. Nat. Research fellow 1930-32. Mem. A.A.A.S., Am. Acad. Arts and Scis., Assn. Am. Geographers, Ecol. Soc. Am. Geol. Soc. Am., Arctic Inst. N. Am., New Eng. Bot. Club. Contbr. articles to jours. Address: Dept. Geography, Johns Hopkins U., Balt. 21218.*

RAURAMO, Lauri, Finnish gynecologist, obstetrician; b. Viipuri, Finland, June 14, 1917; s. Mauno and Margit (Becker) R.; med.lic., U. Helsinki (Finland), 1945, M.D., 1947; m. Salme Edit Anttilainen, Dec. 4, 1941; children—Mauno, Pekka Mauno, Eeva Marketta. Docent obstetrics and gynecology U. Turku (Finland), 1950, asso. dir. dept. obstetrics and gynecology, 1952-61, gynecol. radiologist, 1960-64, prof. obstetrics and gynecology, dir. dept. obstetrics and gynecology, 1964—; dir. Radiumhemmet, Turku, 1956—; dir. dept. obstetrics Turku City Hosp., 1963-64. Recipient Silver badge of merit for preventive work against cancer, 1961. Mem. Med. Assn.

Duodecim Turku (past pres.), Finnish Gynecol. Soc. (pres. 1967——). Author: (with Lea Harpa) Textbook of Obstetrics and Gynecology for Nurses, 1955; also numerous articles. Research on toxemia, improvement of foetal prognosis, gynecol. cancer, glucuronation. Address: Lääkäriasunnot, Keskussairaala, Turku 3, Finland.*

RAUSCH, Ludwig, German radiobiologist; b. Berlin, Feb. 11, 1922; s. Ernst and Charlotte (Steiff) R.; U. Berlin, U. Marburg; Dr.med., 1945; m. Annelise Meyer, Sept. 28, 1955; children—Thomas, Bettina, Peter. Head lab. dermatol. clinic U. Münster, 1946-51, head radiation dept., 1952-55; asst. dept. radiobiology and isotopes U. Marburg, 1956-62, docent, 1959-64; unscheduled prof., head dept. exptl. radiology and radiation protection Röntgen Clinic, U. Giessen (West Germany), 1965——. Mem. German X-ray Soc., German Soc. for Biophysics, German Soc. for Biol. Chemistry, Assn. German Radiation Protection Physicians, European Soc. for Radiation Biology. Research and numerous publs. on analytical methods in clin. chemistry, application to dermatol. problems; radiation protection and therapy. Home: 19 Im Loh, Wehrda (Kreis Marburg), W. Germany. Office: 25 Friedrichstrasse, Giessen, W. Germany.*

RAUTENSTRAUS, Roland Curt, Am. civil engr.; b. Gothenburg, Neb., Feb. 27, 1924; s. Christian and Emma (Steine) R.; student Colo. State Coll., 1942-43; B.S. in Civil Engring., U. Colo., 1947, M.S., 1950; m. Willie A. Atler, June 30, 1946; 1 son, Curt Dean. Faculty U. Colo., Boulder, 1946——, prof., chmn. dept. civil engring., 1958-64, asso. dean faculties, 1964——. Lectr. to local and state groups; chmn. local, state and nat. coms. Recipient Faculty Appreciation award for outstanding teaching U. Colo., 1964, Bronze medal award Lincoln Arc Welding Found.; Robert L. Stearns medal U. Colo. Registered profl. engr., Colo. Mem. Am. Soc. Photgrammetry, Am. Soc. for Engring. Edn., Am. Soc. C.E., Nat. Com. on Accreditation Colls. Engring., Chi Epsilon, Tau Beta Pi. Author: Plane Surveying, 1956; Surveying and Mapping, 1958; City Planning, 1962. Study of highway engring., photogrammetry, altimetry. Home: 3130 Jefferson St., Boulder, Colo. 80302.

RAVDIN, Isidor Schwaner, Am. surgeon; b. Evansville, Ind., Oct. 10, 1894; s. Marcus and Wilhelmina (Jacobson) R.; B.S., Ind. U., 1916, Sc.D., 1959; M.D., U. Pa., 1918, LL.D., 1953; L.H.D., N.Y. Med. Coll., 1951; LL.D., Temple U., 1956, Franklin and Marshall U., 1958, Hahnemann Med. Coll., 1961; Sc.D., Phila. Coll. Pharmacy and Sci., 1958, Northwestern U., 1959, Lehigh U., 1960, Bucknell U., 1960; M.D. (hon.), U. Oslo, 1961; m. Elizabeth Glenn, June 2, 1921; children—Robert, Elizabeth (Mrs. Donald Clayton Bergus), William Dickie. Comd. lt. col. U.S. Army, 1941, advanced through grades to maj. gen., 1954, ret., 1956; dir. Harrison dept. surg. research U. Pa., 1935-59, John Rhea Barton 59, prof surgery Sch. Medicine, U. Pa., Phila., 1959-65, professor emeritus of surgery, since 1965——, v.p. for med. affairs, 1959-65. Cons. in gen. surgery Monmouth Med. Center, Long Branch, N.J., 1953——, VA Hosp., Phila., 1953——, Doylestown (Pa.) Hosp., 1960——, Phila. Gen. Hosp., 1959——. Mem. Nat. Adv. Health Council; sr. civilian cons. Surgeon Gen.; v.p. Internat. Fedn. Surg. Colls. Decorated, Legion of Merit with one oak leaf cluster, Order of the Cloud and Banner (China), First Order (China), Order of Ranking (China). Mem. A.C.S. (bd. regents 1954-60, pres. 1960-61), Am. Acad. Arts and Scis., Am. Philos. Soc., Am. Surg. Soc. (pres. 1958-59), Am. Cancer Soc. (pres. 1962-63), Am. Physiol. Soc. Contbr. articles to tech. jours. Editor, Am. Surgery, 1951——; mem. editorial bds. numerous jours. Research on cancer chemotherapy; gall bladder function; liver and pulmonary physiology; metabolism during anesthesia; nutritional edema in surgical patients. Home: 2015 Delancey Pl., Phila. 19103.

RAVEN, Peter Hamilton, Am. botanist; b. Shanghai, China, June 13, 1936 (parents Am. citizens); s. Walter Francis and Isabelle (Breen) R.; student U. San Francisco, 1953-55; A.B. with highest honors U. Cal. at Berkeley, 1957; Ph.D., U. Cal. at Los Angeles, 1960; m. Sally Barrett, Aug. 16, 1958 (dec.); children—Alice Catherine, Elizabeth Marie. NSF postdoctoral fellow in botany Brit. Mus. Natural History, 1960-61; botanist Rancho Santa Ana Bot. Garden, Claremont, Cal., 1961-62; asst. prof. biol. sci. Stanford (Cal.) U., 1962-66, asso. prof., 1966——. Fellow Cal. Acad. Scis.; mem. Am. Soc. Naturalists, Bot. Soc. Am., Soc. for Study Evolution, Am. Soc. Plant Taxonomists. Editor, Brittonia, 1963-66. Research and numerous publs. in evolution of higher plants, especially those of western N.Am., eveningprimrose family; taxonomic theory. Home: 1100 Palo Alto Av., Palo Alto, Cal.*

RAVENEL, Henry William, Am. botanist, agrl. writer; b. Berkely, S.C., May 19, 1814; s. Henry and Catherine (Stevens) R.; grad. S.C. Coll., 1832; LL.D., U. N.C., 1886; m. Elizabeth Gaillard Snowden, 1835; m. 2d, Mary Huger Dawon, 1858; 10 children. Planter, circa 1832-65; sent by fed. govt. to investigate a cattle disease in Tex., 1869; corr. Acad. Natural Sci., Phila.; mem. Zoologische Botanische Gesellschaft, Vienna. Author: The Fungi Cardiniai Exsiccati, 5 vols., 1853-60 (1st published series of named specimens of Am. fungi) Published (with English botanist Prof. M. C. Cooke) 2d series Fungi Americani Exsiccati, 8 parts, 1878-82. Agrl. editor Weekly News and Courier, 1882-87. Collected and classified herbarium and fungi, mosses and lichens, later sold to Brit. Mus. and Converse Coll., Spartanburg, S.C. Died Aiken, S.C., July 17, 1887.

RAVENEL, St. Julien, Am. physician, agrl. chemist; b. Charleston, S.C., Dec. 19, 1819; s. John and Anna Elizabeth (Ford) R.; grad. Charleston Med. Coll., 1840; studied medicine in Phila. and Paris, France; m. Harriett Horry Rutledge, Mar. 20, 1851; 9 children. Established 1st stone lime works in S.C. at Stoney Landing on Cooper River, 1857; surgeon in charge Confederate hosp., Columbia, S.C., also in charge Confederate lab. (where much medicine used for Confederate Army was made) during Civil War: designer torpedo cigarboat Little David; originated a process which rendered phosphate rocks readily soluble, produced an ammoniated fertilizer, produced phosphate fertilizer without use of ammonia (acid fertilizer), developed process of adding marl to acid fertilizers to counteract free acid; chemist Charleston Agrl. Line Co.; discovered that planting and plowing of leguminous plants restored the worn out soil properties which made it produce large crops; proposed artesian well system for Charleston. Died Charleston, Mar. 17, 1882.

RAVENEL, Mazyck Porcher, Am. bacteriologist; born Pendleton, S.C.; s. Henry Edmund and Selina E. R.; grad. Univ. of the South; studied medicine, Med. Coll. State of S.C.; m. Jennie Carlile Boyd, Oct. 1898. Bacteriologist State Live Stock Sanitary Bd. of Pa., 1896-1904; asst. med. dir. Henry Phipps Inst. for Study, Treatment and Prevention of Tuberculosis; chief of laboratory, Henry Phipps Inst., 1904-07; prof. bacteriology, Univ. of Wis., since 1907. Mem. Nat. Assn. for Study and Prevention of Tuberculosis (1st v.p., 1907-08), Coll. Physicians, Phila., Am. Philos. Soc., Am. Pub. Health Assn., Am. Med. Assn., Am. Assn. Pathologists and Bacteriologists, Phila. Pathol. Soc., S. C. Huguenot Soc. Author numerous published papers. Research on med. and bacteriol. subjects, especially on tuberculosis and rabies. Died Jan. 14, 1946.

RAVENTOS, Antolin, Am. radiologist; b. Wilmette, Ill., June 3, 1925; s. Enrique Antolin and Juanita (Gillespie) R.; student Northwestern U., 1941-44; S.B., U. Chgo., 1945, M.D., 1947; M.Sc. in Medicine, U. Pa., 1955; m. Phyllis Joye Wilson, June 11, 1946. Faculty, Sch. Medicine, U. Pa., Phila., 1951-52, 54——, prof. radiology, 1963——. Fellow Am. Coll. Radiology, Coll. Physicians of Phila.; mem. A.M.A., Pa. State Med. Soc., Am. Roentgen Ray Soc., Am. Radium Soc., Radiol. Soc. N.Am., Radiation Research Soc., Am. Assn. Cancer Research, John Morgan Soc., Soc. Nuclear Medicine, others. Research, publs., in radiobiology, treatment cancer. Home: 2218 Rittenhouse Sq., Phila. 19103.*

RAVICH, Abraham, urologist; b. Biten, Poland, June 10, 1889; s. Isadore and Ida (Karelitz) R.; came to U.S., 1893, naturalized, 1898; A.B., Coll. City N.Y., 1908; A.M., Columbia U., 1912, M.D., 1912; m. Martha L. Zuckert, Sept. 3, 1918; 1 son, Robert. Asst. urologist Jewish Hosp., Bklyn., 1914-18, asso. urologist, 1918-25, attending chief urologist, 1925-34; asso. urologist Beth Moses Hosp., Bklyn., 1920-23; attending urologist Beth-El Hosp., Bklyn., 1920-24, chief urologist, 1936-39; chief urologist Bklyn. Home and Hosp. for Aged, 1925-28, dir. urology, 1938-49; cons. urologist Menorah Home for Aged, Bklyn., 1930-52; attending urologist Israel-Zion Hosp., Bklyn., 1922-34; attending urologist Adelphi Hosp., Bklyn., 1940-49, cons. urologist, 1950-60; dir. Ravich Urol. Inst., Bklyn., 1934-40; instr. postgrad. urology L.I. Coll. Medicine, 1930-34, asst. prof. clin. urology, 1934. Exec. dir. Cancer Research and Hosp. Found., 1947-58, sec.-treas., 1958-63; founder, pres., exec. dir. Inst. Applied Biology, 1947-57; founder Trafalgar Hosp., N.Y.C., 1956; bd. visitors Creedmoor State Hosp., 1946-57. Served as 1st lt. M.C., U.S. Army, 1918-19. Named Hon. Surgeon, N.Y.C. Police Dept.; recipient Townsend Harris medal for achievement Asso. Alumni of Coll. City N.Y., 1950. Fellow A.C.S.; mem. A.M.A., Am. Urol. Assn. (sr. mem.). Mason (32°, Shriner). Mem. editorial bd. Clin. Abstracts, 1934-37. Contbr. articles to med. jours. Pioneered research in renal hypertension and urinary stone formation, also urol. instrumentation; 1st to advocate early circumcision to prevent transmissible cancer factor of genitourinary tract; invented cystoscope, cystoscopic lithotrite. Home: 1135 103rd St., Miami Beach, Fla. 33154; summers 47 E. 88th St., N.Y.C. 10028.*

RAVIN, Abe, Am. physician; b. Denver, Sept. 9, 1908; s. Hyman and Lena (Rapkin) R.; B.A., U. Colo., 1929, M.D., 1932, M.A., 1938; m. Rose Steed, Dec. 27, 1937; children—Thomas Hyne, Lenore Rose. Instr. medicine U. Colo. Sch. Medicine, 1935-39; NRC fellow Harvard, 1939-41; with U. Colo., 1941-59, asso. clin. prof. medicine, 1950——; dir. Cardio Pulmonary Lab., Rose Meml. Hosp., Denver, 1959——; pvt. practice cardiovascular disease consultation, Denver, 1952——. Mem. Am. Coll. Cardiology (trustee, v.p.), Am. Heart Assn., Colo. Heart Assn. (pres.), A.C.P., Phi Beta Kappa, Sigma Xi, Alpha Omega Alpha. Author: Auscultation of the Heart, 1958, 66. Research and numerous publs. on coronary artery injection, cardiac arrhythmias, cardiac auscultation; inventor heart sound simulator. Address: 45 S. Dahlia St., Denver 80222.*

RAVIN, Arnold Warren, Am. geneticist; b. N.Y.C., Aug. 15, 1921; s. Max and Rae (Levinson) R.; B.S., Coll. City N.Y., 1942; M.A., Columbia U., 1948, Ph.D., 1951; m. Sophie Mary Brody, June 11, 1956; children—Sonia, Andrea. Faculty, U. Rochester (N.Y.), 1953-68, chmn. dept. biology, 1957-60, dean coll. arts and sci., 1961-63, prof. 1962-68; prof. dept. biology, master collegiate div. biology, asso. dean div. biol. sci., 1968——; visiting prof. genetics U. Cal., Bergeley, 1963. Served to 1st lt., USAAF, 1942-46. Pre-doctoral fellow, Nat. Cancer Inst., 1949-50, post-doctoral fellow, 1951-53; spl. research fellow, NIH, 1960-61. Fellow, A.A.A.S.; mem. Genetics Soc. Am., Am. Soc. for Microbiology, Soc. for Gen. Microbiology (Eng.), Am. Soc. for Cell Biology, Am. Soc. Naturalist, Sigma Xi. Author: Evolution of Genetics. Research, publs. on genetic basis of antibiotic bacterial resistance, transfer of bacterial DNA, effects of DNA mutation and evolution. Address: Dept. Biology, U. Chgo. 60637.*

RAVITCH, Mark Mitchell, Am. surgeon; b. N.Y.C., Sept. 12, 1910; s. Mitchell M. and Annette (Manevitch) R.; A.B., U. Okla., 1930; M.D., Johns Hopkins, 1934; m. Irene Ravitch, Feb. 27, 1932; children—Nancy (Mrs. Edwards Park Schwentker), Michael Mark, Mary Robin. Practice medicine, specializing in surgery, Balt., 1934-52, 56-66, N.Y.C., 1952-56, Chgo., 1966-; faculty Sch. Medicine Johns Hopkins, 1946-52, 56-66, Coll. Phys. and Surgs. Columbia, 1952-55; surgeon-in-chief Mt. Sinai Hosp., 1952-55, Balt. City Hosps., 1956-66; prof. surgery, prof. pediatric surgery, dir. div. pediatric surgery U. Chgo., 1966——; vis. prof. numerous univs.; 11th Annual Barney Brooks lectr. Vanderbilt U., 1964, 15th Annual Alfred A. Strauss lectr. U. Wash., 1964, Heineman Found. lectr. Mecklenburg Med. Soc., Charlotte, N.C., 1965, Gardner lectr. Duke Med. Center, 1965. Mem. Am., So. surg. assns., Am. Assn. Thoracic Surgery, Am. Acad. Pediatrics, Soc. Vascular Surgery, Soc. U. Surgeons, A.C.S. Author: Intussusception in Infants and Children, 1960; (with James Hitzrot) The Operations for Inguinal Hernia, 1960. Editor: (with others) Pediatric Surgery, 1962; Current Problems in Surgery, 1964; Pediatric Surgery Monographs, 1960——. Asso. editor: Surgery, 1956——. Mem. editorial bd. Rev. of Surgery, 1950——. Research, publs. on surgery of colon; pulmonary surgery; mech. suture methods; childhood surgery. Home: The Cloisters, Dorchester St., Chgo. 60637. Office: 950 E. 59th St., Chgo. 60637.*

RAVN, Carl Christian, see Rafn, Carl Christian.

RAWER, Karl Maria Aloys, German physicist; b. Neunkirchen, Saar, Germany, Apr. 19, 1913; s. Peter A. and Luise (Menzinger) R.; student U. Freiburg (Germany), 1931-32; Diplom-Physiker, Technische Hochschule München, Germany, 1937, Dr.rer.nat., 1939; m. Waltraut A.L. Hien, July 28, 1939; children—Thomas, Bernhard, Johanna (Mrs. Frieder Stoll), Adelheid, Martin, Hedwig, Katharina. Research worker Hochschule, München, 1938-39, Erprobung-Stelle Rechlin, 1939-43, Zentralstelle für Funkberatung Bad Vöslau, 1943-45; sci. dir. Service de Prévision Ionosphérique Militaire (France), Freiburg, 1946-56; dir. Ionosphären-Institut, Breisach, Germany, 1956——; faculty U. Freiburg, 1955——, asso. prof., 1961——; asso. prof. Sorbonne, Paris, 1958-60, exchange prof., 1961-64. Mem. Deutsche Geophysikalische Gesellschaft, Am. Geophy. Soc., Nachrichtentechnische Gesellschaft. Author: Die Ionosphäre, 1952; The Ionosphere, 1957; (with W.R. Piggott) Handbook of Ionogram Interpretation and Reduction, 1961; also numerous articles; contbr. to Ency. of Physics, 1967. Established prediction system for radio wave propagation via ionosphere; with internat. group established reduction system for ionospheric sounding records; research on radio wave propagation with satellites, ionosphere using satellite observations; different layers of ionosphere and exosphere; direct research on space with rocket expts., 1954——. Home: 43 Herrenstr., Hugstetten 7801, F.R. Germany. Office: Ionosphären-Institut, Breisach 7814, F.R. Germany.*

RAWLINS, William Arthur, Am. entomologist; b. Geneva, N.Y., Dec. 5, 1908; s. Thomas Henry and Elizabeth (Baxter) R.; B.S., Cornell U., 1930, Ph.D., 1936; m. Miriam Alma Dolan, Aug. 6, 1932; children—Elizabeth (Mrs. Allen Thomsen) (dec.), Phyllis L. (Mrs. Alson Sherman). Mem faculty Cornell U., Ithaca, 1936——, prof. entomology, 1949——. Fellow Indian Entomol. Soc., A.A.A.S.; mem. Entomol. Soc. Am., Am. Inst. Biol. Scis., Ecol. Soc. Am., N.Y. Acad. Scis., Sigma Xi, Alpha Zeta. Research and numerous publs. on applied entomology, insect control, insect vectors of plant viruses. Home: 115 James St., Ithaca, N.Y. 14850.*

RAWLINSON, Sir Henry Creswicke, English Assyriologist; b. Chadlington, Oxford, Eng., Apr. 11, 1810; s. Abram and Eliza (Creswicke) R.; LL.D., Oxford U., 1850, Cambridge (Eng.) U., 1862, Edinburgh (Scotland) U.; m. Louisa Seymour; 2 sons. Mil. cadet E. India Co., 1826, service in India, 1827-33, crown dir., from 1855; helped reorganize Persian army, 1833-39; polit. agt., Kandahar, Afghanistan, from

1840; consul gen., from 1851; mem. Ho. Commons, 1858, rep. Frome dist., 1865-68; mem. India Council, 1858-59, 68-95; minister plenipotentiary to Persia, 1859-60. Trustee Brit. Mus., 1876-95. Fellow Royal Soc., 1850; mem. Royal Asiatic Soc. (dir. from 1862, pres. 1878-81), Royal Geog. Soc. (pres. 1871-72, 74-75), London Oriental Congress (pres. 1874), French Academie des Inscriptions (corr., fgn.). Author: The Persian Cuneiform Inscription at Behistun, 1846-51; A Commentary on the Cuneiform Inscriptions of Babylon and Assyria, 1850; Outline of the History of Assyria, 1852; Notes on the Early History of Babylonia, 1854; (with others) Cuneiform Inscriptions of Western Asia, 6 vols., 1861-80. Copied, translated cuneiform inscription of Darius I (inscribed on practically inaccessible rock at Behistun); largely responsible for successful deciphering Class III Persian cuneiform. Died Mar. 5, 1895.

RAWSON, Rulon Wells, Am. physician; b. Idaho Falls, Ida., Sept. 28, 1908; s. Wilford Woodruff and Eugenia (Lefgren) R.; student Weber Coll., 1926-27, U. Utah, 1930-32; M.B., Northwestern U., 1937, M.D., 1938; m. Jane Young, Aug. 21, 1940; children—James R. Y., Elizabeth Jane, Daniel Y. Research fellow Harvard Med. Sch., 1938-42, instr. and Henry P. Walcott fellow clin. medicine, 1942-45, asso. medicine, 1945-47, asst. prof. medicine, 1947-48; asst. physician Mass. Gen. Hosp., 1946-47, asso. physician medicine, 1947-48; asso. prof. medicine Cornell U. Med. Coll., 1948-51, prof., 1955-67; v.p. N.J. Coll. Medicine and Dentistry, Jersey City, 1967—. Attending physician Meml. Hosp., N.Y.C., 1948-58, chief dept. medicine, Meml. Sloan-Kettering Inst., 1958—. Mem. adv. com. med. uses of isotopes AEC, 1959—; mem. com. on personnel for research American Cancer Soc., 1958-63, chmn. 1959-63; mem. adv. com. sr. clin. traineeship program Dept. Health, Edn. and Welfare, 1962—. Recipient award, Am. Goiter Assn., 1959. Centennial award Northwestern U., 1959. Fellow N.Y. Acad. Medicine; mem. A.C.P., A.M.A., Am. Physiol. Soc., A.A.A.S., Am. Goiter Assn. (council 1950-53, pres. 1955-56), Endocrine Soc. (v.p. 1957-58; chmn. publs. com. 1957-63), Am. Assn. Cancer Research, Am. Soc. Clin. Investigation, N.Y. State Med. Soc., Harvey Soc. (sec. 1955-58), Assn. Am. Physicians. Author sci. papers, chpts. in texts on physiology and diseases of the thyroid and endocrinology of cancer. Developed microhistometric method for assaying thyroid stimulating hormone; demonstrated that thyroid stimulating hormone in exerting its action is inactivated by thyroid thymic and lymphoid tissues; developed methods studying thyroid physiology with radioactive isotopes of iodine; demonstrated that cancer of thyroid can be forced to assume functions of normal thyroid by removing normal thyroid and administering thyroid stimulating hormone or goiterogenic agts. which augment action of thyroid stimulating hormone, also evidence that thyroid physiology with radioactive isotopes of iodine; immune mechanisms are optimum; showed cancerolytic effects of ACTH, adrenal steroids on certain lymphoid tumors, dissociation of thyroid hormonal effects by altering chem. structure thyroxine molecule. Home: 131 Corona Av., Pelham, N.Y. Office: 24 Baldwin Av., Jersey City.*

RAY, Ernest Clark, Am. physicist; b. St. Joseph, Mo., Feb. 23, 1930; s. Ernest L. and Helen (Tilbury) R.; B.A., State U. Ia., 1951, M.S., 1953, Ph.D., 1956; m. Mary Alice Simmering, Nov. 21, 1951; children—Brandon David, Nigel Kent. Asst. prof. physics State U. Ia., Iowa City, 1956-61; cons. Rand Corp., Santa Monica, Cal., 1961; Nat. Acad. Sci. fellow Goddard Space Flight Center, Greenbelt, Md., 1962-63; sr. research asso. Cornell U., Ithaca, N.Y., 1963-65; aerospace technologist Goddard Space Flight Center, 1965—. Mem. Am. Phys. Soc., Am. Geophys. Union. Research and publs. on theoretical studies of motion of cosmic rays and trapped radiation in earth's magnetic field; participated in discovery of Van Allen radiation belts. Home: 9505 2d Av., Silver Spring, Md. 20910. Office: Goddard Space Flight Center, Greenbelt, Md.*

RAY, John, English naturalist; b. Black Notley, nr. Braintree, Essex, Eng., Nov. 29, 1627; s. Roger Ray; ed. Braintree grammar sch.; entered Catharine Hall, Cambridge U., 1644; changed to Trinity Coll., Cambridge, 1646, B.A., 1647, M.A., 1661; m. Margaret Oakeley, 1673; 4 daus. Apptd. Greek lectr. Cambridge U., 1651, math. lectr., 1653, humanity reader, 1655, praelector, 1657, jr. dean, 1658, coll. steward, 1659-60; ordained deacon and priest, 1660; frequently toured Britain collecting plant and animal specimens, 1658-62; left coll. life, became tutor to sons of his former student Francis Willughby, 1662; toured Holland, Germany, Switzerland, Italy, Sicily and Malta with Willughby, collected plant and animal specimens, 1663-66; fellow Royal Soc., 1667; resided at Middleton Hall, Warwickshire, from 1666, Sutton Coldfield, 1676, Falkbourne Hall, nr. Witham, Essex, 1677, Dewlands, Black Notley, 1679-1705. Author: Catalogus plantarum circa Cantabrigiam nascentium, 1660; A Collection of English Proverbs, 1670; Catalogus plantarum Angliae, 1670; Observations made on a Journey through Part of the Low Countries, Germany, Italy and France, with a Catalogue of Plants not Native of England, 1673; A Collection of English Words not Generally Used . . . with Catalogues of English Birds and Fishes, with an Account of the Preparing and Refining of such Metals and Minerals as are gotten in England, 1674; Methodus plantarum nova, 1682; Historia plantarum, 1686-1704; Fasciculus stirpium Britannicarum, 1688; Synopsis methodica stirpium Britannicarum, 1690; The Wisdom of God Manifested in the Works of Creation, 1691; Synopsis methodica animalium quadrupedum et serpentini generus, 1693; De variis plantarum methodis dissertation, 1696; Methodus insectorum, 1705; Historia insectorum, 1710; Synopsis methodica avium et piscium, 1713, several others. Sometimes called founder of systematic biology and natural history in Britain; defined species by distinctive structural qualities thought to be transmitted from parent to offspring without change and unaffected by change of habitat; held to fixity of species; arranged plants according to their structural form, using characters taken from root, stem, leaves, flowers and fruit, this found. of natural system of classification; divided flowering plants into monocotyledons and dicotyledons, thus naming them; catalogued and described 18,600 different plant species; studied quadrupeds, reptiles and birds; tried to systematize animal kingdom; arranged animals by hoofs, toes and teeth; brought order to study of insects; performed expts. on movement of plants and on ascent of sap in trees; suggested erosion would eventually reduce land to sea level; thought fossils petrified remains of extinct creatures; Ray Soc. (founded 1844) named for him. Died Dewlands, Black Notley, Essex, Eng., Jan. 17, 1705.

RAY, Louis Lamy, Am. geologist; b. St. Louis, July 26, 1909; s. Louis L. and Roy (Kercheval) R.; A.B., Washington U., St. Louis, 1930, M.S., 1932; M.A., Harvard, 1937, Ph.D. (Austin Teaching fellow), 1938; m. Eleanor Bennett, Apr. 28, 1948; children—Deborah, Victoria. Teaching asst. Harvard, 1938-39; asst. prof. geology Mich. State U., 1939-42; geologist U.S. Geol. Survey, Washington, 1942—; vis. prof. geology Cornell U., Ithaca, N.Y., 1959-60; cons. United Nations, Argentina, 1967-68; member of geology Cornell U., Ithaca, N.Y., 1959-60; mem. Wissenschaflichen Mitarbeiter Archiv für Polarforschung, Kiel, Germany, 1951. Mem. Instituto Italiano di Paleontologica Umana (corr. mem. Rome, Italy), Geol. Soc. Am., Am. Assn. Geographers, Soc. Econ. Geologists, Am. Geog. Soc., Am. Geophys. Union, A.A.A.S., Sigma Xi. Author: Geomorphology and Quarternary Geology of the Owensboro Quadrangle, 1965; also numerous articles. Research on relating human cultural remains to glacial chronology western U.S., frozen ground conditions in Alaska, geomorphic history no. Ky., Alaskan glaciers, mil. geology. Home: 2230 California St. N.W., Washington 20008. Office: U.S. Geol. Survey, Washington 20242.*

RAY, Robert Durant, Am. orthopaedic surgeon; b. Cleve., Sept. 21, 1914; s. Clifford Arthur and Edna (Durant) R.; B.A. cum laude, U. Cal. at Berkeley, 1936, M.A., 1938, Ph.D., 1948; M.D., Harvard, 1943; m. Genevieve Triau, Dec. 19, 1953; children—Frances Carol, Robert Triau, Eston Bernard, Gisele Antoinette, Charles Alexander. Instr. anatomy U. Cal. at Berkeley, 1948-51; asst. prof. orthopaedic surgery U. Wash., 1948-51, asso. prof., head dept. othopaedic surgery, 1954-56; prof., head dept. orthopaedic surgery U. Ill. Coll. Medicine, Chgo., 1956—; chmn. dept. orthopaedic surgery Presbyn.-St. Luke's Hosp., Chgo., 1956—; attending orthopaedic surgeon research and ednl. hosps. Spl. cons., chmn. orthopaedic tng. com. USPHS, 1961-65, also mem. surg. study sect. NIH; mem. spl. med. adv. com., cons. VA, 1961—. Carnegie research fellow U. Cal. at San Francisco, 1938-40. Recipient nat. award for research in orthopaedic surgery Kappa Delta, 1954. Diplomate Am. Bd. Surgery. Mem. Am. Acad. Orthopaedic Surgeons, Am. Orthopaedic Assn., Orthopaedic Research Soc., A.C.S., A.M.A., Ill., Cook County med. socs., Soc. Nuclear Medicine, A.A.A.S., Sigma Xi, Phi Sigma. Research and publs. on isotope studies of bone, metabolic and bone circulation, exptl. studies of bone grafts and implants; use of decalcified bone in healing of bone defects. Home: 227 Dempster St., Evanston, Ill. 60201. Office: 1753 W. Congress Pkwy., Chgo. 60201.*

RAY, Verne Frederick, Am. anthropologist; b. Enfield, Ill., Mar. 13, 1905; s. Arden H. and Della (Duncan) R.; B.A., U. Wash., 1931, M.A., 1932; Ph.D., Yale U., 1939; m. Julia C. Wright, Jan. 1, 1927 (div. 1951); 1 dau., LaVerne Charlotte (Mrs. Gerald Fromberg); m. 2d Dorothy Jean Tostlebe, Feb. 2, 1955. With U. Wash., 1930-66, prof. anthropology, 1947-66, asso. dean Grad. Sch., 1947-51; vis. prof. Yale U., 1951-54; vis. prof. N.Y. U., 1939, U. Cal., Berkeley, 1946, U. Cal., Los Angeles, 1948, Whittier Coll., 1947; research dir. Human Relations Area Files, New Haven, 1951-54; cons. anthropology, Washington, 1966—; mem. editorial bds. Pacific Northwest Quar., Am. Anthropologist. Mem. Am. Ethnol. Soc. (pres. 1953), Seattle Anthropol. Soc. (pres. 1946). Author: The Sanpoil and Nespelem: Salishan Peoples of Northeastern Washington, 1932; Lower Chinook Ethnographic Notes, 1938; Cultural Relations in the Plateau of Northwestern America, 1939; Cultural Element Distributions: Plateau, 1942; Primitive Pragmatists: The Modoc Indians of Northern California, 1963. Ethnol., social anthropol. research, numerous publs. on the Indians of Western Am.; devel. of theories of primitive religion and polit. orgn.; devel. techniques for measurement of human color perfection. Home: 3001 Veazey Terrace N.W., Washington 20008. Office: 1908 Que St. N.W., Washington 20009.*

RAYCHAUDHURI, Amal Kumar, Indian physicist; b. Barisal, India, Sept. 14, 1923; s. Sures C. and Surbala (Roy) R.; B.Sc., Presidency Coll., 1942; M.Sc., U. Coll. Sci., Calcutta, India, 1944; D.Sc., Indian Assn. for Cultivation Sci., Calcutta; m. Nomita Sen, Jan. 30, 1958; children—Amlanava, Madhukshara, Asimava, Pupu. Research scholar Indian Assn. for Cultivation Sci., 1943-49; lectr. Asutosh Coll., Calcutta, 1949-52; research officer Indian Assn. for Cultivation Sci., 1952-61; prof. Presidency Coll., 1961—; vis. prof. U. Md., 1964-65. NSF Sr. Fgn. Scientist fellow, 1964-65. Mem. Indian Assn. for Cultivation Sci., Calcutta Math. Soc. Publs. on deduction of an equation in gen. relativistic dynamics of dust (known as Raychaudhuri equation) from which it follows that an expanding universe without rotation must necessarily originate from a collapsed state of infinite density in finite past. Home: 102/D Ballygunge Pl., Calcutta, India.*

RAYER, Pierre-François-Olive, French physician, pathologist; b. St. Sylvain, France, Mar. 7, 1793; M.D., Paris, 1818; physician Central Bur. Paris Hosp.; became physician St. Antoine Hosp., 1825; named chief physician Charity Hosp., 1832; apptd. physician to Napoleon III, 1852; prof. medicine Faculty of Paris, named dean faculty, 1882; founder Soc. Biology; 1st pres. Assn. Medicine of France; mem. Acad. Medicine, French Acad. Scis. (v.p. 1850, pres. 1851). Author: Traité théorique et pratique des maladies de la peau, 2 vols., 1826-27; Traité des maladies des reins, 3 vols., 1839-41. Described pituitary obesity, 1823; named and described xanthoma and xanthoma multiplex (Rayer's disease), 1826; 1st to distinguish acute from chronic eczema, 1826; gave 1st description of glanders in man and proved it is not a form of Tb, 1837; described and named condition hydronephrosis, 1839. Died Paris, Sept. 10, 1867.

RAYET, Georges-Antoine-Pons, French astronomer; b. Bordeaux, France, Dec. 12, 1839. Became tchr. physics Lycée d'Orleáns, France, 1862; named asst. astronomer Observatoire de Paris, 1863; prof. astronomy Faculty Scis., Marseilles, France later at Bordeaux, France; dir. Floirac (France) Obs., 1879-1906. Mem. French Acad. Scis., 1892. Author: L'Astronomie pratique et les observatiores en Europe et en Amérique depuis le milieu du XVIIe Siècle. Discovered (with G. T. E. Wolf) 3 stars in constellation Cygnus with brilliant rays indicating hydrogen, helium (Wolf-Rayet stars). Died Floirac, June 14, 1906.

RAYLEIGH, Lord, see Strutt, John William.

RAYMOND, Arthur E(mmons), Am. aeronautical engineer; b. Boston, Mass., Mar. 24, 1899; B.S., Harvard U., 1920; M.S., Mass. Inst. Technology, 1923; hon. D.Sc., Polytechnical Inst., Brooklyn, 1947; m. 1921. Engr., Douglas Aircraft Co., 1925-34; vice-pres. engrg., Douglas Aircraft, 1934-60; consultant, Rand Corp., 1960—; Mem., Nat. Advisory Committee Aeronautics, 1946-56; consultant, NASA, 1962—. Mem., Nat. Acad. Sic., Nat. Acad. Engrg.; Soc. Automotive Engrg.; hon. fellow Inst. Aeronautics and Astronautics. Work in development of aeronautics and astronautics. Office: Rand Corp., 73 Oakmont Dr., Los Angeles, Calif. 90049.

RAYMOND, Fulgence, French neurologist; b. St. Christophe, France, 1842; ed. Ecole d'Alfort. Became departmental mgr. anatomy and physiology Ecole d'Alfort, 1866, after med. studies, apptd. staff physician, 1878, prof. agrégé, 1880, then prof. nervous illnesses; mem. Acad. Medicine. Author: Etudes anatomiques, physiologiques et clinique de l'hémichorée, de l'hémemianesthésie, et des tremblements symptomatiques, 1874; Clinique des maladies du système nerveux, 1894-1903; Atrophies musculaire et maladies amyotrophiques, 1899. Most important work in nerve and mental pathology; studies on dorsalis, poliomyelitis, muscular atrophy, syphilis, neurasthenia. Died Vienne, France, 1910.

RAYMOND, Harry, Am. astronomer; b. La Grange, O., Feb. 17, 1876; A.B., Harvard, 1905; m. 1907; 2 children. Asst., Dudley Obs., Union, N.Y., 1905-07; staff dept. meridian astronomy Carnegie Inst., 1907-18, astronomer, 1918-37, research asso., 1938-40; ret.; asst. physicist Signal Lab., Camp Evans, Belmar, N.Y. Recipient Gold medal Royal Danish Acad. Sci., 1926. Mem. Am. Acad. Scis. Research in meridian astronomy, sidereal motions, radio direction finding, star positions. Home: 318 N. Scotland Av., Albany 8, N.Y.

RAYMONT, John Edwin George, English biol. oceanographer; b. Exeter, Gt. Britain, Apr. 6, 1915; s. Walter and Ellen (Isaac) R.; Ext. Degree in Zoology with 1st class honours, U. London, 1936; A.M. (Henry fellow) Harvard, 1938; D.Sc., U. Exeter, 1960; m. Joan Katharine Brigit Sloan, 1945; children—Carol, Michael, Catriona. Asst. lectr. zoology U. Coll., Exeter, 1938-39; lectr. zoology U. Edinburgh, 1939-46; prof. zoology U. Southampton (Eng.), 1946-64, prof. oceanography, 1964—. Cons. Atomic Energy Authority, 1959—. Fellow Indian Acad. Scis.; mem. Soc. Exptl. Biology, Challenger Soc. Brit. Ecol. Soc. Author: Plankton and Productivity in the Oceans, 1963; also articles. Research on feeding relations between marine phytoplankton and zooplankton, reprodn., metabolism and biochemistry of zooplankton, zooplankton cycles in the sea, bottom fauna, marine area in relation to fish farming and clam culture, mass cultures of phytoplankton; pollution studies. Home: 12 Russell Pl., Southampton, Hampshire, Eng.*

RAYNAUD, Maurice, French physician; b. Paris, France, Aug. 10, 1834. Prof. in Paris. Author: De l'asphysie locale et de la gangrène symétrique des extremitiés. Described pallor of extremities, especially toes and fingers, (called Raynaud's disease); experimented with transfusions of his own blood. Died June 29, 1881.

RAYNE, John Albert, physicist; b. Sydney, Australia, Mar. 22, 1927; s. Albert Wallace and Ethel (Inglis) R.; B.S., U. Sydney, 1948, B.E., 1950; Ph.D. (Fulbright scholar, Coffin fellow), U. Chgo., 1954; m. Ann Middleton, July 31, 1954; children—Susan, Julie, Brian. Came to U.S., 1956, naturalized, 1962. Research officer Commonwealth Sci. and Indsl. Research Orgn., Sydney, 1954-56; research physicist Westinghouse Electric Corp., Pitts., 1956-61, adv. physicist, 1961-64; asso. prof. physics Carnegie Inst. Tech., Pitts., 1964-65, prof. physics, 1965—. Fellow Am. Phys. Soc. Research, numerous articles on low temperature properties of metals and alloys with particular emphasis on their electronic structure. Home: 419 Burlington Rd., Pitts. 15221.*

RAYNOR, Geoffrey Vincent, English phys. metallurgist; b. Nottingham, Eng., Oct. 2, 1913; s. Alfred Earnest and Florence (Champion) R.; grad. with 1st class honors Keble Coll., Oxford, Eng., 1936, M.A., D.Phil., 1940, D.Sc., 1948; m. Emily Jean Brockless, July 29, 1943; children—John Campion, Peter Campbell, Duncan Hope. Imperial Chem. Industry Research fellow U. Birmingham (U.K.), 1945-47, faculty, 1947—, prof. metal physics, 1949-54, Feeney prof. phys. metallurgy, head dept., 1954—. Mem. adv. panel Brit. Council, 1965—. Recipient Beilby Meml. award, 1947, Rosenhain medal Inst. Metals, 1951, Heyn Meml. medal Deutsche Gesellschaft für Metallkunde, 1956. Fellow Royal Soc., 1959, Royal Inst. Chemistry, Inst. Physics, Instn. Metallurgists; mem. Assn. Commonwealth Univs. (mem. adv. panel 1965—), Inst. Metals, Iron and Steel Inst., Chem. Soc., Royal Soc. Arts, N.Y. Acad. Scis. Author: Instruction to Electron Theory of Metals, 1953; Structure of Metals and Alloys (with W. Hume-Rothery), 1962; Physical Metallurgy of Magnesium and its Alloys, 1959; also numerous articles. Research on formation of alloys, with spl. reference to solid solubilities, structures and compositions of intermediate phases, and phys. and mech. properties. Home: 94 Gillhurst Rd., Birmingham 17, Eng.*

RAYNOR, George Emil, Am. mathematician; b. San Francisco, Feb. 27, 1895; s. Charles Edmund and Louise (Huptcher) R.; B.S., U. Wash., 1918; A.M., Princeton, 1920, Ph.D., 1923; m. Amy Becraft, Aug. 24, 1920; 1 dau., Georgia Emily. Instr., Princeton, 1919-23; asst. prof. Wesleyan U., 1923-27; asso. prof. U. Okla., 1927-31; faculty LeHigh U., Bethlehem, Pa., 1931—, prof. math., 1946—. head dept. math., 1948-60. Mem. Math. Assn. Am., Am. Math. Soc., Phi Beta Kappa, Sigma Xi, Pi Mu Epsilon. Author: (with J.B. Reynolds) Elementary Mechanics, 1943; also articles. Research in potential theory, definite integrals, math. logic. Home: 349 8th Av., Bethlehem, Pa. 18018.*

RAYSKI, Jerzy, Polish physicist; b. Warsaw, Poland, Apr. 6, 1917; s. Stefan and Wiktoria (Janota) R.; M.S., Jagiellonian U., Krakow, 1946; Ph.D., U. Warsaw, 1947; m. Joanna Stec, Dec. 31, 1944; children—Jacek, Malgorzata. Asst., U. Warsaw, 1946-47; faculty U. Torun, 1947-57; prof. physics Jagiellonian U., Krakow, 1958—; vis. prof. U. Bern, 1962-63, Middle East Tech. U., Ankara, 1966-67. Mem. Polish Phys. Soc., Polish Math. Soc. Research, numerous publs. on regularization of quantum electrodynamics, quantization of non-local fields, unified field theory, gen. relativity theory, dynamical theory of quarks and hadrons. Home: 4 Pl. Wolnosci, Krakow, Poland.*

RAZAK, Charles Kenneth, Am. engr.; b. Collyer, Kan., Sept. 5, 1918; s. Charles and Dorothy (Brown) R.; B.S., U. Kan, 1939, M.S., 1942; m. Lillian Blanche McCall, July 4, 1940; children—Nancy Louise, Jeanne Marie. Successively instr., asst. prof. aero. engring., coordinator civilian pilot tng. program U. Kan., 1939-43; asso. prof. aero. engring. U. Wichita, 1943-45, prof., head dept. aero. engring, 1945-47, prof., dir. sch. engring., 1947-50, dean sch. engring., 1954-64, prof., 1950-66, research prof. in aero. engring., acting dean coll. bus. and industry 1950-54, responsible for orgn. and administrn. sch. engring., design, constrn. 7x10 wind tunnel, direction aero. research program; dir. Kan. Indsl. Extension Service, prof. engring. Kan. State U., Manhattan, 1966—; cons. local aircraft cos. Pres., Aerial Distributers, Inc., Mgrs., Inc., dir. Kan. Investment Co., KARD-TV (both Wichita). One of 3 dels. to work on problem of shortage of engring. and tech. manpower Engring. Manpower Commn. of Kan., 1951; adviser pest control div. U. S. Dept. Agr. on operation analysis, U. Cal. at Davis on agrl. aviation. Tech. dir. research contract held by U. Wichita under Office of Naval Research to investigate problem of low speed flight. Mem. subcom. low speed aerodynamics, mem. of com. on aircraft aerodynamics NASA. Fellow Inst. Aero. Scis. (asso.); mem. Am. Soc. Engring. Edn. (mem. gen. council 1961-62), Tau Beta Pi, Pi Tau Sigma, Sigma Gamma Tau. Author articles on agrl. aviation engring. report. Aerodynamic research, involving principles of boundary layer control to agrl. airplane. Home: 2415 Mesita Dr., Wichita, Kan. 67211. Office: Kan. Indsl. Extension Service, Kan. State U., Manhattan, Kan. 66502.*

RAZOUK, Rashad Ilias, Egyptian chemist; b. Ras-El-Bar, Damietta, Egypt, Aug. 22, 1911; s. Elias A. and Martha A. (Israfil) R.; B.Sc., Cairo (Egypt) U., 1933, M.Sc., 1936, Ph.D., 1939; postgrad. U. Exeter (Eng.), 1947-48, U. So. Cal., 1965; m. Emilie Selim Habib, Aug. 24, 1946; children—Reda, Rami. Faculty, Faculty Sci., Cairo, 1933-50, asso. prof., 1947-50; prof. phys. chemistry, chmn. chemistry dept. Faculty Sci., Ain Shams U., Cairo, 1950-66, vice dean, 1954-60; prof. phys. chemistry Am. U., Cairo, 1966—. Research dir. Nat. Research Centre, Dokki, Cairo, 1950—. Decorated Cross Saint Marc Apostle, Greek Orthodox Patriarch Christoforous II, 3d order, 1964, 2d order, 1966. Mem. Egyptian Acad. Scis., Inst. Egypt. Editorial bd. Jour. Chemistry U.A.R., 1958—. Research, publs. on heat of wetting and wetability of solids, adsorptive properties of carbon, oxides and mixed oxides, endothermic decomposition of solids, roasting of sulfides. Home: 11, Sharia Botros Ghali, Heliopolis, Egypt.*

RAZRAN, Gregory, psychologist; b. nr. Slutsk, Russia, June 4, 1901; s. Solomon and Rebecca (Ongeyber) R.; came to U. S., 1920, naturalized, 1927; B.S., Columbia, 1927, A.M., 1928, Ph.D., 1933; m. Elna Bernholz, Sept. 15, 1939; 1 dau., Lydia. Univ. scholar, Columbia, 1929-30, lectr. psychology, 1930-38, research asso., 1938-40; statistical cons. (expert), C.A.A., 1939-40; instr. to professor Queens Coll., N.Y.C., 1940—, chmn. dept. psychol. since 1945, Guggenheim fellow, 1948-49. Psychol. cons., O.S.S., 1941-44. Fellow (divs. exptl., physiol. and gen. psychology) Am. Physchol. Assn. (pres. div. gen. psychology), A.A.A.S.; member N.Y. Acad. Scis., (chmn. div. psychology), Eastern Psychol. Assn. Author monographs, contbr. articles on conditioning and learning theory. Home: 67-15 A 186 Lane, Fresh Meadows 65, N.Y.

RAZUVAEV, Grigory Alekseevich, Russian organic chemist; b. Moscow, USSR, Aug. 11, 1895; s. Alecsey and Ekaterina (Chernobrovkin) R.; grad. Leningrad U., 1925; m. Elena Kchodkova, Aug. 30, 1946; children—Olga, Alecsey, Vladimir. Staff, Research Inst. Acad. Scis., Leningrad, 1925-34; head chair Leningrad Inst. Chemistry and Tech., 1931-46; head organic chemistry chair Gorky (USSR) State U., 1946—, dir. Chemistry Research Inst., 1956-62; dir. Polymer Stablzn. Lab., Acad. Scis. USSR, 1963—. Recipient Lenin prize, 1958, Order of Lenin, 1961-, Zasluzenny Deyatel Nauki, 1966. Mem. Acad. Sci. of USSR, All-Union Mendeleev Chem. Soc. USSR. Author: Radicals in Organic Chemical Reactions, 1947; Free Radicals in Organic Reactions, 1949; The Reaction of Free Radicals in the Liquid Phase: The Reaction Capacity of Oxyacyl Radicals, 1955. Research and numerous publs. on free radical reactions in solutions on base thermo- and photo-reactions of organometallic compounds and peroxides. Address: Dept. Chemistry, Gagarina prospect 17, Gorky 22, USSR.*

RAZZAK, Muhammad Abdel, physician, b. Cairo, Egypt, Apr. 23, 1928; s. Abdel and Fathia Razzak; M.B., B.Ch., Cairo U., diploma medicine, Doctorate Medicine; m. Laila A. Hassaballa, Sept. 6, 1958; children—Omar M.A., Rawia M.A. Faculty, Cairo U., 1954—, asso. prof. internal medicine and nuclear medicine, 1966—; head med. research Atomic Energy Establishment Egypt, 1964-65; staff med. unit Regional Radioisotope Center for Arab, 1963-66. Mem. Egyptian Med. Assn., Soc. Nuclear Medicine. Research, publs. on use of radioisotopes in study hemodynamics and blood flow to different regions in normal states and in hepato-splenic bilharziasis; effect drugs on hemodynamics of liver and kidneys, photoscanning of liver in bilharziasis.*

RAZZELL, Wilfred Edwin, Canadian microbiologist; b. St. Boniface, Man., Can., May 10, 1931; s. Wilfred James and Vera (Razzell) Deeley; B.A. with honors U. B.C., 1952; Ph.D., U. Ill., 1957; m. Mary Catherine Slinn, Sept. 22, 1951; children—Daniel Edwin, Robin Catherine, James Wilfred. Research fellow B.C. Research Council with H. G. Khorana, U. B.C., 1956-59, research scientist applied biology, 1959-61; head enzymology sect. Syntex Inst. Molecular Biology, Palo Alto, Cal., 1961-64, asst. dir., 1962-64; asso. prof. microbiology U. B.C., 1964-66; prof., head microbiology dept. U. Alta., Edmonton, Can., 1966—. Mem. Am. Soc. Biol. Chemists, Am. Chem. Soc., Am. Soc. Microbiology, Canadian Soc. Microbiology, Biochem. Soc., Sigma Xi. Contbr. articles in field to sci. jours. Developed methods for enzymatic analysis of sequences in nucleic acids; established general occurrence in animal tissues of various enzymes which degrade nucleic acids and their absence from cultured cells; various methods and procedures for purifying enzymes. Home: 4203 123d St., Edmonton, Alta., Can.*

READ, Clark Phares, Am. biologist; b. Ft. Worth, Feb. 4, 1921; s. Clark P. and Helen (Chaudoin) R.; student Tulane U., 1943-45; B.A., Rice Inst., 1948, M.A., 1948, Ph.D., 1950; m. Leota A. Wolff, Oct. 24, 1944; children—Jo-Hanna G., Victoria H. Thomas Jefferson, Cathleen E. Faculty, U. Cal., Los Angeles, 1950-54; asso. prof. parasitology Johns Hopkins, 1954-59; prof. biology Rice U., Houston, 1959—, chmn. biology dept., 1964—; prof. parasitology Med. Sch. Baylor U., 1961—; chmn. study sect. tropical medicine USPHS, 1960-65; head zoology sect. Marine Biol. Lab., Woods Hole, Mass., 1960-63. Recipient Henry B. Ward medal, 1959; NRC-AEC fellow, 1949; Guggenheim fellow, 1960; Career prof. USPHS, 1963—. Mem. Soc. Gen. Physiologists, Am. Soc. Parasitologists. Author: (with Chandler) Introduction to Parasitology, 1960; Parasitism and Symbiology, 1967. Research and numerous publs. primarily in field of biochemistry, physiology of parasites of man and animals, diseases caused by parasites. Home: 1819 Dunstan Rd., Houston 77005. Office: 6100 Main St., Houston 77001.*

READ, Herbert Harold, English geologist; b. Whitstable, Kent, Eng., Dec. 17, 1889; s. Herbert and Caroline (Keam) R.; B.S., Imperial Coll., U. London (Eng.), 1912, D.Sc., 1924; D.Sc., Columbia U., 1954, U. Dublin (Ireland), 1956; m. Edith Browning, June 21, 1917; 1 dau., Marguerite (Mrs. Peter Alexander Colligan). With H.M. Geol. Survey, Scotland, 1914-31; George Heidman prof. geology U. Liverpool (Eng.), 1931-38; prof. Imperial Coll., U. London, 1939-55, prof. emeritus, 1955—, pro-rector, 1952-55, dean Royal Sch. Mines, 1943-45; Walker-Ames vis. prof. U. Wash., Seattle, 1949. Alexander du Toit Meml. lectr. Johannesburg, S. Africa, 1951. Fellow Royal Soc., 1939 (Royal medal 1963), Royal Soc. Edinburgh, Geol. Soc. London (Wollaston medal 1952, Bigsby medal 1955); fgn. mem. Geol. Soc. Portugal; corr. mem. Societe Geologique de Belgique (A. Dumont medal 1950), Geol. Soc. Am. (Penrose medal 1967), Societe geologique de France, French Acad. Scis., Associe de l'Academie Royale de Belgique (priz Fourmarier 1960); mem. Royal Coll. Sci. London (asso.) Norwegian Acad., Geol. Soc. Edinburgh (past pres.), Geol. Soc. Liverpool (past pres.), Geologists' Assn. (past pres.), Geol. Soc. London, Internat. Geol. Congress. Author: Geology, 1949; The Granite Controversy, 1954; (with J. Watson) Introduction to Geology, Vol. I., 1962, Beginning Geology, 1966; also numerous articles. Research on origin granite, natural history of metamorphic rocks, study migmatites formed by mixture of granite material with other rocks, timing in rock-formation and mountain-bldg., differentiation in basic magnas. Office: Geol. Dept., Imperial Coll., London, S.W. 7, Eng.*

READ, John, Brit. chemist; b. Maiden Newton, Eng., Feb. 17, 1884; s. John and Bessie (Gatcombe) R.; student Finsbury Tech. Coll., London, 1901-05; Ph.D., U. Zurich (Switzerland), 1907; postgrad. U. Manchester (Eng.), 1907-08; M.A., U. Cambridge, 1912, Sc.D., 1934. Became asst. chemist Thames Conservatory Lab., London, 1905; became asst. to prof. chemistry U. Cambridge, 1908; apptd. prof. organic chemistry pure and applied U. Sydney (Australia). 1916-23; prof. chemistry, dir. chemistry research labs. United Coll. St. Salvadore and St. Leonard, U. St. Andrews, Scotland, 1923—. Charter mem. Australian Nat. Research Council. Fellow Royal Soc., 1935, Australian Inst. Chemistry (charter); mem. Soc. Chem. Industry, Chem. Soc. Author: A Textbook of Organic Chemistry, 1926; An Introduction to Organic Chemistry, 1931; Prelude to Chemistry, 1936; Explosives, 1942; Humor and Humanism in Chemistry, 1947; The Alchemist in Life, Literature, and Art, 1947; also numerous articles. Research on high explosives, for Ministry Munitions, 1915. Deceased.

READ, John, radiation biologist, physicist; b. Hendon, Eng., Mar. 31, 1908; s. John William and Evelyn (Wilmot) R.; student Derby Tech. Coll., 1924-29; B.Sc., London (Eng.) U., 1929, B.Sc., Nottingham U. Coll., 1931; Ph.D., Cal. Inst. Tech., 1934; m. Henrietta Russell McCree, Mar. 4, 1939; children—Ian (dec. Jan. 1967), Anne (Mrs. William Alexander Grant), Barbara, Marion. Asst. physicist Radium Beam Therapy Research, London, 1934-35; research physicist Mt. Vernon Hosp., Northwood, Eng., 1936-43, chief physicist, 1946-50; chief physicist London Hosp., 1943-45; dir. radiation biology research Cancer Soc. New Zealand, Dunedin, 1950—. Created scheme for exchange of information between radiotherapy centres. Recipient Röntgen award Brit. Inst. Radiology, 1947-48, Anderson Berry prize and Gold medal Royal Soc. Edinburgh, 1953. Mem. Brit. Inst. Radiology, New Zealand Med. Physicists Assn. Author: Radiation Biology of Vicia faba, 1959; also numerous articles. Research on relative biol. efficiency of gamma rays, X-rays, fast neutron and alpha rays, effect of oxygen in potentiating radiation damage; demonstrated importance of chromosome damage by radiations. Home: 121 Oakwood Av. Office: Wakari Pub. Hosp., Dunedin, New Zealand.*

READ, John Robert, Australian physician; b. Sydney, Australia, Feb. 16, 1929; s. William Robert and Emily (Watt) R.; M.B., B.S., U. Sydney, 1952, M.D., 1959; m. Barbara Stewart Cope, Sept. 16, 1953; children—Lesley Ann, Robyn Elizabeth, Jennifer Rosalind. Research fellow U. Sydney, 1956-57, faculty, 1959—, prof., 1966—; research fellow Postgrad. Med. Sch., London, 1958; asst. physician Sydney Hosp., 1957-60, Royal Prince Alfred Hosp., 1961—. Fellow Royal Australasian Coll. Physicians (E.L. Susman prize 1962); mem. Physiol. Soc.

Australia Cardiac Soc. Australia and New Zealand, Thoracic Soc. Australia. Research and publs. on etiology of interstitial pulmonary fibrosis; physiol. studies of factors controlling distbn. of lung ventilation and blood flow, mech. function of lung in asthma. Home: 12 Rickard Rd., Strathfield, N.S.W., Australia. Office: Dept. Medicine, U. Sydney, Australia.*

READ, Nathan, Am. iron mfr., inventor; b. Warren, Mass., July 2, 1759; s. Rueben and Tamsin (Meacham) R.; grad. Harvard, 1781; m. Elizabeth Jeffrey, Oct. 20, 1790. Tchr., Beverly and Salem, Mass., 1781-83; tutor Harvard, 1783-87; devised double-acting steam engine, 1788; invented manually-operated paddle-wheel propelled boat, 1789; granted patents on portable multitubular boiler, improved double-acting steam engine, chain wheel method of propelling boats, 1791; organized Salem Iron Factory, 1796-1807; patented rail cutting and heading machine, 1798; mem. U. S. Ho. of Reps. (Federalist) from Mass., 6th-7th congresses, 1800-03; apptd. spl. justice Ct. of Common Pleas for Essex County (Mass.), 1803; chief justice Hancock County (Me.) Ct. of Common Pleas, 1807; mem. Am. Acad. Arts and Scis., 1791; hon. mem. Linnaean Soc. New Eng., 1815. Died Belfast, Me., Jan. 20, 1849.

READ, Thomas Albert, Am. physicist; b. Montclair, N.J., Oct. 21, 1913; s. Thomas Thornton and Mary (Peck) R.; A.B., Columbia, 1934, Ph.D., 1940; student Technische Hochschule, Munich, Germany, 1932-33; m. Doris Pascal, Nov. 4, 1935; 1 son, Thomas Thornton II. Research fellow Westinghouse Research Lab., East Pitts., 1939-41; principal physicist Franford Arsenal, Phila., 1941-47; Oak Ridge Nat. Lab., 1947-48; asso. prof. metallurgy Columbia, 1948-54; prof. metallurgy, head dept. mining, metallurgy and petroleum engring. U. Ill., Urbana, 1954——. Cons. for govt. agys., private cos. Fellow Am. Phys. Soc.; mem. Am. Inst. Mining and Metall. Engrs., Am. Soc. Metals, Inst. Metals (London), Phi Beta Kappa, Sigma Xi, Tau Beta Pi. Research on phase changes in crystals, internal friction of metals. Home: 507 E. Harding Dr., Urbana, Ill. 61801.*

READ, Willard Oliver, Am. physiologist; b. Clay Center, Kan., Nov. 1, 1916; s. Robert Randolph and Elizabeth (Wilson) R.; B.A., U. Mo., 1941, M.A., 1947, Ph.D., 1949; m. Virginia Ann Root, Aug. 18, 1947; children—John R., Catherine A., James M., Melinda C., Robert R., Sandra C., Jennifer. Instr., U. Mo., 1947-49; asso. in physiology U. S.D., Vermillion, 1949-53, faculty, 1953——, prof., chmn. dept., 1960——. Mem. Am. Physiol. Soc., Soc. Exptl. Biology and Medicine, S.D. Acad. Sci., Sigma Xi. Research, publs. on cause of cardiac arrhythmias, cause of muscular dystrophy. Home: 306 Lewis St., Vermillion, S.D. 57069.*

READE, Joseph Bancroft, English chemist, microscopist; b. Leeds, Eng., Apr. 5, 1801; s. Thomas Shaw Bancroft and Sarah (Paley) R.; M.A., Trinity and Caius colls., Cambridge (Eng.) U., 1828; m. Charlotte Dorothea Farish; 3 children. Rector of Stone, Eng., 1839-59, Ellesborough, Eng., 1859-63, Bishopsbourne, Eng., 1863-70. Fellow Royal Soc., 1838; mem. Micros. Soc. (charter), Royal Micros. Soc. (pres.). Contbr. articles to tech. jours. Discovered method of separating heat rays from light rays; improved photography; invented hemispherical condenser for microscope (Reade's kettledrum), 1861. Died Dec. 12, 1870.

REAM, Howard William, agronomist; b. Maywood, Ill., Dec. 20, 1908; s. Charles Lennox and Emma (Pogensee) R.; B.S., U. Wis., 1930, M.S., 1939; m. Eunice Ann Horn, Dec. 25, 1931; children—Lois Ann (Mrs. Harry O. Higgins), Robert Ray, David William. Agronomist, soil conservationist Soil Conservation Service, U.S. Dept. Agr., Fennimore, Milwk., Wis., 1935-48, Washington, 1948-51, research liason rep. Agrl. Research Service, also Soil Conservation Service, Washington, 1952-54; agronomy adviser, dep. dir. rural devel. div. AID, Thailand, 1951-52, P.I., 1954-56, Taiwan, Taipei, 1957-61, Brazil, 1962——; Cons., Land Mgmt., Office Chief Engrs., U.S. Army, Washington, 1947-48. Fellow A.A.A.S.; mem. Am. Soc. Agronomy, Soil Sci. Soc. Am., Am. Inst. Biol. Scis. Research in forage crops; selected disease resistant strain Pangola grass to introduce to Taiwan, U.S. and Brazil; introduced high yielding corn variety into Thailand. Home: US AID/ARD APO N.Y.C. 09676. Office: care of Am. Embassy, Rio de Janeiro, Brazil.*

RÉAUMUR, René Antoine Ferchault de, French naturalist, physiologist, physicist; b. La Rochelle, Charente-Maritime, France, Feb. 28, 1683; ed. Jesuit colls., Poitiers and Bourges, later Paris. In charge of ofcl. description of useful arts and manufactures of France 1710; mem. French Acad. Scis., 1708. Author: l'Art d'convertir le fer forge en acier, 1722; Mémoires pour servir a l'historie naturelle des insectes, 1734-42; Sur la digestion des oiseaux, 1762. Studied malleability of metals, tensile strength of cables, chem. differences between iron and steel; invented method of tinning iron; worked on prodn. of steel and improvements in manufacture of iron, 1722; invented Réaumur's thermometer with scale reading from 0 to 80 degrees to indicate freezing and boiling point, 1730; investigated chem. composition of Chinese porcelain; invented

Réaumur's procelain, 1740; best known for showing that gastric juices of birds have solvent effect on foods, thereby demonstrating that digestion is chem. and not mech. grinding process, 1752; studied regeneration of lost limbs in crustaceans, locomotion of starfish, action of electric ray, marine phosphorescence, artificial incubation of eggs, effect of heat on insect devel., auriferous rivers; demonstrated that zoophytes were animals, not plants, also that turquoise beds were fossil teeth of extinct animals; opposed theory of spontaneous generation. Died La Bermondière, Maine, France, Oct. 18, 1757.

REAVES, Gibson, Am. astronomer; b. Chgo., Dec. 26, 1923; s. Hart Walker and Helen (Gibson) R.; B.A., U. Cal. at Los Angeles, 1947; Ph.D., U. Cal., at Berkeley, 1952; m. Mary Craig Kerr, Apr. 2, 1955; 1 son, Benjamin Kerr. Lick Obs. fellow, 1950-52; faculty U. So. Cal., Los Angeles, 1952——, prof. astronomy, 1965——. Mem. Am. Astron. Soc., Astron. Soc. Pacific, Royal Astron. Soc., Sigma Xi. Study dwarf galaxies in clusters of galaxies, supernovae. Home: 3830 S. Sycamore Av., Los Angeles 90008.*

REBER, Elwood Frank, Am. biochemist, nutritionist; b. Reading, Pa., June 24, 1919; s. Aquilla Peter and Irene (Ringler) R.; A.B., Berea Coll., 1944; M.Sc. in Nutrition, Cornell U., 1948; M.S. in Chemistry, Okla. State U., 1950, Ph.D., 1951; m. Alta Mae Davis, Dec. 18, 1942; children—Ruth Ann, Margaret Beth, Rebecca, Irene. Research Chemist Swift & Co., 1950-51; faculty Ill. Agr. Expt. Sta. and Coll. Vet. Medicine, U. Ill., Urbana, 1952-64, prof. vet. research, vet physiology and pharmacology, 1952-64; prof., chmn. food and nutrition Sch. Home Econs., U. Mass., Amherst, 1964-68; prof. head dept. foods and nutrition Purdue U., 1968——. Mem. Am. Inst. Nutrition, Am. Chem. Soc., Inst. Food Technologists, Am. Soc. Animal Sci., Ill. Acad. Sci. Research, numerous publs. on interrelationship nutrition and disease; Vitamin K, protein and amino acid metabolism; wholesomeness of irradiation preserved food. Address: Dept. Foods and Nutrition, Purdue U., Lafayette, Ind. 47907.

REBIKOFF, Dimitri, oceanographic engr.; b. Paris, France, May 6, 1921; s. Dimitri I. and Eugenie (Savici) R.; student Sorbonne, Paris, 1940-42; m. Ada Niggeler, Apr. 23, 1952. Came to U.S., 1961, naturalized, 1966. Developer, Pegasus Photogrammetry System for Undersea Mapping for USN, Ft. Lauderdale, Fla., 1964——. Recipient Niepce medal Societe de Photographie. Mem. Am. Soc. Photogrammetry, Soc. Motion Picture and Television Engrs., Soc. Photographic Scis. and Engrs., Optical Soc. Am., Inst. Navigation, Marine Tech. Soc. Author: Underwater Photography Amphoto, 1955; Free Diving, 1956, Aviation Sous Marine, 1961. Research in underwater exploration and photomapping of large areas by integrated camera and vehicle system; sci. applications of electronics to photography; developed first underseas television system, first portable electronic flash. Home: 2426 Sea Island Dr. Office: 245 S.W. 32d St., Ft. Lauderdale, Fla. 33315.*

REBINDER, Petr Aleksandrovich, Russian phys. chemist; b. St. Petersburg, Oct 2, 1898; grad. Moscow U., 1924. Asso., Moscow Inst. Physics and Biophysics, 1923——; prof. chair physics Moscow Pedagogical Inst., 1929——; head dept. disperse systems Inst. Phys. Chemistry, USSR Acad. Scis., 1934——; prof. Moscow U., 1942——. Cons. various research insts.; del. 2d Internat. Congress on Surface Activity, London, 1955. Decorated Order of Lenin; recipient Stalin prize, 1942. Mem. USSR Acad. Scis. Coauthor: The Physical Chemistry of Flotation Processes, 1933; The Physical Chemistry of Detergent Action, 1935; Hardness-Reducing Agents in Drilling, 1944; author: Research on the Applied Physical Chemistry of Surface Phenomena, 1936; The Physical Chemistry of Flotation Processes, 1937; Physicochemical Studies of Deformation Processes in Solids, 1947; Some Results of the Development of Colloidal Chemistry, 1959. A founder colloidal chemistry in USSR; developer theory of flotation ore concentration and theory of thixotropy; discovered relaxation of deformation and drop in mech. strength of solids due to action of adsorbents (Rebinder effect); founder disciples of physiochem. mechanics. Address: Inst. Phys. Chemistry, USSR Acad. Scis., Leninsky prospect 31, Moscow, USSR.

REBOUL, Georges Scipion Antoine, French physicist; b. Calvisson, France, June 1, 1879; D.Sc., U. Paris (France), 1908. Lectr. physics Faculty Sci. U. Nancy (France), 1910-14; prof. physics U. Montpellier (France), from 1927. Lectr. meteorology Nat. Agronomical Inst., Paris. Corr. mem. French Acad. Sci. Research, publs. on radiation, meteorology.

RÉCAMIER, Joseph Claude Anthelme, French physician, surgeon; b. Rochefort, France, 1774; became physician to Hôtel Dieu, Paris, 1801, also lectr., path. demonstrator; prof. medicine Coll. de France (despite preference of French Acad. Scis. for Magendie), dismissed for refusing to pledge allegiance to Louis Philippe, 1831. Author: Recherches sur le traitement du cancer par la compression méthodique, 2 vols., 1829. Introduced term metastasis to describe spread of cancer, 1829; performed 1st colpo-hysterectomy, 1829. Died Paris, 1856.

RECHCIGL, Miloslav, Jr., biochemist; b. Mlada Boleslav, Czechoslovakia, July 30, 1930; s. Miloslav and Marie (Rajtr) R.; came to U.S., 1950, naturalized, 1955; B.S., Cornell U., 1954, M. Nutritional Sci., 1955, Ph.D., 1958; m. Eva Edwards, Aug. 23, 1953; children—John Edward, Karen Marie. Research asso. dept. biochemisty Cornell U., Ithaca, N.Y., 1958; USPHS research fellow Nat. Cancer Inst., NIH, Bethesda, Md., 1958-60, research biochemist, 1960-64, sr. investigator Lab. Biochemistry, 1964——, mem. council Assembly Scientists, 1963-65.USPHS Research fellow, 1958-59, 59-60. Fellow A.A.A.S.; mem. Czechoslovak Soc. Arts and Scis. in Am. (dir.-at large 1962——, dir. publs. 1962——, mem. presidium 1964——), Am. Chem. Soc., Am. Assn. for Cancer Research, Soc. for Exptl. Biology and Medicine, N.Y. Acad. Scis., Am. Inst. Biol. Scis., Am. Inst. Nutrition, Am. Soc. Animal Sci., Am. Soc. for Cell Biology, Fedn. Am. Socs. for Exptl. Biology, Soc. for Biol. Rhythm, History of Sci. Soc., Am. Assn. for Advancement Slavic Studies, Sigma Xi, Phi Kappa Phi. Author: Czechoslovakia and its Arts and Sciences: a Selective Bibliography in the Western European Languages, 1964; Ten Years of the Czechoslovak Society of Arts and Sciences in America, 1966; also articles, chpts. in books. Editor: The Czechoslovak Contribution to World Culture, 1964; Czechoslovakia Past and Present, 2 vols., 1967. Abstractor, Chem. Abstracts. Research and publs. on mechanisms and rates of action, genetic regulation of enzymes; amino acid metabolism and nutrition; interaction of normal and neoplastic cells; history, historiography, and bibliography of science. Home: 1703 Mark Lane, Rockville, Md. 20852. Office: Lab. Biochemistry, Nat. Cancer Inst., Bethesda, Md. 20014.*

RECHNITZER, Andreas Buchwald, Am. oceanographer; b. Escondido, Cal., Nov. 30, 1924; s. Ferdinand M. and Dagmar C. (Buchwald) R.; B.S., Mich. State U., 1947; MA., U. Cal. at Los Angeles, 1951, Ph.D., 1956; m. Martha Jean Mitchell, Aug. 18, 1946; children—David F., Andrea J., Martin A., Michael J. Cons. oceanographer, La Jolla, Cal., 1950-56; research staff U. Cal., Scripps Inst. Oceanography, La Jolla, 1952-56; phys. oceanographer U. S. Naval Electronics Lab, San Diego, 1956-61; program devel. mgr. autonetics div. N.Am. Aviation, Anaheim, Cal., 1961-65, chief scientist deep submergence systems, 1965-67, dir. ocean scis., ocean systems operations N.Am. Rockwell, Long Beach, Cal., 1967——. Sci. adviser Cal. State Coll., Fullerton, 1963——. Recipient Navy Dept. Distinguished Civilian Service award, 1960; Richard Hopper Day Meml. award, Phila., 1960; Internat. Underwater Film Festival award, 1961; Geog. Soc. Chgo. Gold medal, 1961. Mem. Am. Soc. Naval Engrs., Marine Tech. Soc., Underwater Photog. Soc., Am. Soc. Ichthyologists and Herpetologists, Am. Soc. Oceanography, CEDAM Internat., World Fedn. Underwater Activities (co-pres. 1961——). Contbr. articles to tech. jours. pioneered devel. techniques for manned exploration of sea; scientist in charge of expdn. leading to deepest manned dive with bathyscaphe, 1960; established world record dive, 1959; immunological investigation of taxonomic relationships of viviparous sea perches. Home: 532 Miguel Pl., Fullerton, Cal. 92632. Office: 350 S. Magnolia Av., Long Beach, Cal. 90802.*

RECHT, Pierre, Belgian physician; b. Sept. 21, 1915; M.D., spl. degree hygiene U. Brussels; m., 1929; 2 children. Dir., Ministry of Pub. Health, Belgium; now gen. dir. health and safety br. EURATOM, Brussels; prof. U. Brussels. Mem. Conseil Supérieur Hygiene de Belgique. Mem. various sci. socs. Contbr. articles profl. jours. Research in radiobiology and radioprotection; active in establishment radiation standards and norms in radioprotection field. Home: Boulevard General Jacques 30. Office: 51 Rue Belliard, Brussels, Belgium.*

RECKLINGHAUSEN, Friedrich Daniel von, see von Recklinghausen, Friedrich Daniel.

RECLUS, Jean Jacques Elisee, French geographer; b. Sainte-Foy la Grande, Gironde, Mar. 15, 1830; s. of Protestant pastor; ed. Protestant coll. of Montauban; degree from U. Berlin. Left France because of polit. views, 1851; travelled to British Isles, U. S., C.A. and Colombia doing geog. work, 1952-57; returned to France; sentenced to perpetual banishment from France, 1872; settled in Switzerland; prof. geography U. Brussels (Belgium), 1892-1905. Author: Voyage à la Sierra Neveda de Ste Marthe, 1861; Histoire d'un Ruisseau, 1869; La Terre: Description des phénomènes de la vie du globe, 2 vols., 1867-68; Les Phénomènes terrestres: Les Mers et les Météores, 1873; La Nouvelle Géographic Universelle, 19 vols., 1875-94; Histoire d'un Montagne, 1880; L'Homme et la Terre, 6 vols., 1905-08. Known especially for his La Nouvelle Géographie, which won Gold medal Paris Geog. Soc., 1892; one of outstanding geographers of 19th century; his works on Western hemisphere popularized its geography in Europe. Died Thourout, nr. Bruges, Belgium, July 4, 1905.

RECORDE, Robert, English mathematician, b. Tenby, Pembroke, circa 1510; studied Oxford, elected fellow of All Souls Coll., Oxford, 1531; studied math., medicine, Cambridge U. Probably taught math. and medicine, Cambridge U., later instr. in many subjects, Oxford U.; physician to Edward VI and Queen Mary; general surveyor of mines and money, 1551. Author:

The grounde of Artes, 1540-1699; The Urinal of Physick, 1547; The Pathway to Knowledge, or first principles of geometry, 1551-1602; The Whetstone of Witte, or the second part of arithmeticke, 1557; The Castle of Knowledge, a treatise on astronomy and the sphere, 1551. His math. works used as textbooks in English schools for more than 100 years; 1st to use sign for equality and to extract root of multinomial algebraic expression; considered skillful physician and accomplished philologist. Died in debtor's prison, London, Eng., 1558.

REDDY, Duvur Jaganatha, Indian pathologist, med. adminstr.; b. Veeranathur, Madras, India, Aug. 5, 1915; s. Duvur Venkata Subba and Munimma Reddy; student Presidency Coll., Madras, 1933-34; M.B., B.S., Madras Med. Coll., 1940, M.D., 1959; m. Lakshmi Reddy, Dec. 6, 1941; 1 son, Ramachandra. Prof. forensic medicine Andhra Med. Coll., Visakhapatnam, India, 1949-54; prof. pathology, prin. Gunter (India) Med. Coll., 1952-62; prof., dir. upgraded dept. pathology Andhra Med. Coll., 1962-64; prof. pathology, prin. Jawaharlal Inst. Postgrad. Med. Edn. and Research, Pondicherry, India, 1964——. Graded pathologist Med. Corps, Indian Army, 1941-47. Fellow Acad. Med. Scis., Coll. Am. Pathologists; hon. mem. Internat. Assn. Hydatidi Logia, S.Am.; mem. London (Eng.) Coll. Pathologists, Assn. Pathologists India, Indian Med. Assn., Internat. Acad. Pathology, Indian Assn. Advancement Med. Edn. Author: Clinical Laboratory Guide, 1960; also numerous papers. First to describe occurrence of sarcoidosis in India; research in etiology and pathology of still births and neonatal deaths in S. India, of carcinoma of liver in India; etiology of carcinoma of cervix in S. India, of carcinoma of oropharynx in S. India; induction of carcinoma of cervix in mice by human smegma, of atherosclerosis by cooking oils and egg yolk in rabbits by variation of serum cholesterol levels with exercises and stress in rabbits; sickle-cell anemia in Andhra Pradesh.; spontaneous lesions in cardio-vascular system with special reference to atherosclerosis in fowls; contbr. to geographic pathology in India. Home: Principal's Bungalow, Dhanovantari Nagar. Office: Jawaharlal Inst. Postgrad. Med. Edn. and Research, Pondicherry 6, India.*

REDDY, William John, Am. biochemist; b. Boston, Aug. 10, 1926; s. Neil and Margaret (Doherty) R.; A.B., Harvard, 1949, M.S., 1957, Sc.D., 1960; m. Elizabeth Ann Enaire, Jan. 1, 1955; children—Margaret, Elizabeth, Ann, William. Chemist endocrine lab. Peter Bent Brigham Hosp., 1949-55, dir. endocrine lab., 1955-64, cons. Psychiat. Research Group, 1960——, cons. Research Center, 1961-64; research asso. medicine Med. Sch. Harvard, Boston, 1960-61, research asso. biochemistry, 1961-64, asso. in biochemistry, 1964——. Mem. Am. Soc. Biol. Chemists, Am. Physiol. Soc., Am. Chem. Soc., Endocrine Soc., N.Y. Acad. Scis. Numerous Publs. Developed first method for measurement of urinary corticosteroids in the diagnosis of adrenal dysfunction; physiol. hormonal effects and biochem. mechanisms. Home: 45 Kilsyth Rd., Brookline, Mass. 02146. Office: 185 Pilgrim Rd., Boston 02215.*

REDFIELD, Alfred Clarence, Am. oceanographer; b. Phila., Nov. 15, 1890; s. Robert S. and Mary (Guillou) R.; grad. Haverford Sch., 1909; student Haverford Coll., 1909-10; B.S., Harvard, 1914, Ph.D., 1917; Ph.D. (hon.), U. Oslo (Norway), 1956, Lehigh U., 1965, U. Newfoundland, 1967; m. Martha Putnam, June 8, 1914; children—Elizabeth (Mrs. Charles R. Marsh), Martha W. (Mrs. George Koch, Jr.), Alfred Guillou. Asst. prof. U. Toronto, 1919-20; mem. faculty Harvard, Cambridge, Mass., 1921——, prof. physiology, 1931-56, prof. emeritus, 1956——, chmn. dept. biology, 1935-38. Sr. biologist Woods Hole (Mass.) Oceanographic Inst., 1930-53, sr. oceanographer, 1953-56, emeritus, 1956——, trustee, 1936-64; vis. prof. Stanford, 1930, U. Wash., 1948; pres. Bermuda Biol. Sta., 1963-66. Trustee Marine Biol. Lab., 1930-53. Mem. Ecol. Soc. Am. (pres. 1946), Am. Soc. Limnology And Oceanography (pres. 1956), Boston Soc. Natural History (hon.), Nat. Acad. Scis. (Agassiz medal 1956), Am. Acad. Arts and Scis., Marine Biol. Assn. U.K. (hon.). Contbr. articles in biology, biol. chemistry, geology, oceanography to profl. publs. Home: Maury Lane, Woods Hole. Office: P.O. Box 106, Woods Hole, Mass.*

REDFIELD, Robert, Am. anthropologist; b. Chicago, Ill., Dec. 4, 1897; s. Robert and Bertha (Dreier) R.; Ph.B., U. of Chicago, 1920, J.D., 1921, Ph.D., 1928; L.H.D. (hon.), Fisk University, 1947; m. Margaret Park, June 17, 1920; children—Lisa Redfield Peattie, Robert (dec.), Joanna (Mrs. David Gutmann), James Michael. Instructor sociology, University of Colorado, 1925-26; fellow Social Science Research Council, 1926-27; instructor in anthropology. University of Chicago, 1927-28, asst. prof., 1928-30, asso. prof., 1930-34, prof. and dean of Div. of Social Sciences, 1934-46, chmn. dept. of anthropology, 1947-49, Robert Maynard Hutchins Distinguished Service prof. since 1953; Messenger lectr. Cornell University, 1952; Gottesman lecturer at Upsala University, 1954; research associate Carnegie Institution of Washington, in charge of ethnol. and sociol. fieldwork, 1930-46; research in Yucatan and Guatemala, 1930-48. Member Social Science Research Council, 1935-43; dir. Am. Council Race Relations, 1947-50; dir. Social Sci. Found. Recipient Viking medal, 1954; Huxley Meml. medal, 1955. Fellow A.A.A.S., Am. Philos. Society; member American Anthrop. Association (pres. 1944), Phi Beta Kappa, Sigma Xi. Author: Tepoztlan, A Mexican Village, 1930; (with Alfonso Villa) Chan Kom, A Maya Village, 1934; The Folk Culture of Yucatan, 1941; A Village That Chose Progress, 1950; The Primitive World and its Transformations, 1953; The Little Community, 1955; Peasant Society and Culture, 1956. Developer ideal type concepts folk culture and folk society; contrasted integration and solidarity folk culture to modern soc.; adapted sociol. concepts to anthropology; in studies attempted to bring together social scis. and humanities. Died 1958.

REDFIELD, William C., Am. meteorologist; b. Middletown, Conn., Mar. 26, 1789; s. Peleg and Elizabeth (Pratt) R.; m .Abigail Wilcox, Oct. 15, 1814; m. 2d, Lucy Wilcox, Nov. 23, 1820; m. 3d, Jane Wallace, Dec. 9, 1828; 4 children, including Charles Bailey. Apprenticed to saddle and harness maker, Middletown, 1804; a founder, 1st pres. A.A.A.S., 1848. Author: Observations on the Hurricanes and Storms of the West Indies and the Coast of the United States (meteorol. classic), 1833; Remarks on the Prevailing Storms of the Atlantic Coast of the North American States (paper which introduced correct and fundamental concept of rotary motion of hurricanes), 1831; other papers on meteorology, geography, geology. Demonstrated that direction of revolution in gales and hurricanes is uniform, also that velocity of rotation increases toward center; devised practical rules by which mariners might determine their position during hurricanes. Died N.Y.C., Feb. 12, 1857.

REDHEFFER, Raymond Moos, Am. mathematician; b. Chgo., Apr. 17, 1921; s. Raymond L. and Elizabeth (Moos) R.; S.B., Mass. Inst. Tech., 1943, S.M., 1946, Ph.D., 1948; m. Heddy Gross Stiefel, Aug. 25, 1951; 1 son, Peter Bernard. Research asso. Radiation Lab., Mass. Inst. Tech., 1942-45, Research Lab. of Electronics, 1946; instr. Harvard, Radcliffe, 1946-48; faculty U. Cal. at Los Angeles, 1950——, prof. math., 1960——; guest prof. Tech. U. Berlin, 1962, Inst. for Angewandte Math., Hamburg, 1966. Pierce fellow Harvard, 1948-50; sr. postdoctoral fellow NSF, Göttingen, Germany, 1956; Fulbright research scholar Vienna, 1957, Hamburg, 1961-62. Mem. Am. Math. Soc., Math. Assn. Am., I.E.E.E. (sr. mem.), Sigma Xi. Author: (with Ivan Sokolnikoff) Mathematics of Physics and Modern English, 1958; (with Charles Eames) Men of Modern Mathematics, 1966; also numerous articles. Research in algebraic, differential, integral inequalities; designed, constructed first free-space microwave interferometer. Home: 176 N. Kenter Av., Los Angeles 90049.*

REDI, Francesco, Italian physician and poet; b Arezzo, Tuscany, Italy, Feb. 19, 1626; ed. Florence and Pisa, Italy. Became physician to dukes of Tuscany; mem. Accademia del Cimento. Author: Osservazioni intorno alle vipere, 1664; Esperienze intorno alla generazione degli insetti, 1668; Osservazione intorno agli animali viventi, 1684. As poet best known for Bacco in Toscana, 1685; tested scientifically long-held theory of spontaneous generation; proved experimentally that maggots do not form in decaying meat exposed to air, as was widely held; showed instead that if meat is covered with gauze maggots are formed from eggs laid by flies and trapped on gauze; however he continued to believe gall insects were spontaneously generated. Died Pisa, Mar. 1, 1667.

REDI, Rodolfo, Italian surgeon; b. Florence, Italy, Mar. 16, 1896; s. Alberto and Pia Ferri; M.D.; m. Lemmi Blankite, Sept. 20, 1920; children—Francesco, Paolo. Prof. at Bari Surg. Clinic. Decorated Gold medal of merit in teaching. Mem. Internat. Surgery Soc. Author numerous articles. Home: Policlinico, Bari. Office: Chirurgica Università, Bari, Italy.

REDL, Fritz, child psychologist; b. Klaus, Austria, Sept. 9, 1902; s. Gustav and Rosa (Schwarz) R.; Ph.D., U. Vienna (Austria), 1925; postgrad. Wiener Psychoanalytisches Institut, 1925-36; m. Helen Burstein, Feb. 5, 1959. Came to U. S., 1936, naturalized, 1943. Tchr. pub. schs., clin. couselling Wiener Volkschochschulen, Vienna, 1925-36; research asso. Gen. Edn. bd. Rockefeller Found., Progressive Edn. Assn., N.Y.C., 1936-38; lectr. mental hygiene, U. Mich., 1938-41; cons. guidance dept. Cranbrook Sch. for Boys, Bloomfield, Mich., 1938-41; research asso. div. for child devel. and tchr. personnel, Am. Edn. Assn., U. Chgo., 1939-40; prof. social work Sch. Social Work, Wayne State U., Detroit, 1941-53, prof. behavioral scis., 1959——; chief Child Research br. NIH, Bethesda, Md., 1953-59. Founder, dir. Detroit Group Project Summer Camp, 1942——; dir. Pioneer House, 1946-47; cons. div. mental hygiene USPHS, 1946-49, prin. investigator research project on use group medium for clin. work with disturbed children, 1948-51. Ford Found. fellow Center for Advanced Study in Behavioral Scis., Stanford (Cal.), 1960-61. Mem. Am. Psychoanalytic Assn., Am. Orthopsychiat. Assn., Am. Psychol. Assn., Am. Assn. Social Workers, Am. Assn. Group Therapy, Am. Assn. Applied Anthropology, A.A.A.S., Nat. Probation Assn., Phi Delta Kappa. Author: (with William Wattenberg) Mental Hygiene in Teaching, 1951; (with David Wineman) Children Who Hate, 1951; (with David Wineman) Controls From Within, 1952. Contbr. articles to profl. jours., chpts. to books. Lect. edn., mental hygiene, group work, group therapy, other subjects, throughout U. S., 1936——. Home: 20001 Warrington Dr., Detroit 48221.*

REDMAN, John, Am. physician; b. Phila., Feb. 27, 1722; s. Joseph and Sarah Redman; M.D., U. Leyden, July 15, 1748; m. Mary Sobers; at least 3 children. Strenuously advocated saline purgatives rather than emetics and bleeding in yellow fever epidemics in Phila., 1762, 93; a founder Coll. of Physicians, Phila., 1786; educated young physicians, including John Morgan, Benjamin Rush, Caspar Wistar; cons. physician Pa. Hosp., 1751-80; 1st pres. Coll. of Physicians, Phila., 1786-1804; trustee Coll. of N.J. (now Princeton), Coll. of Phila.; mem. Phila. Common Council, 1751; mem. Am. Philos. Soc. Author: A Defense of Inoculation (advocating direct inoculation for smallpox), 1759. Died Phila., Mar. 19, 1808.

REDMAN, Lawrence V.; chemist; b. Oil Springs, Ont., Can., Sept. 1, 1880; s. Richard and Mary Jane (Monteith) R.; A.B., U. of Toronto, 1908; fellow 1908-10, D.Sc., 1931; LL.D., U. of West Ontario, 1930; m. Ellen Blossom Corey, Dec. 22, 1909 (dec.); children—Alice Blossom, Lawrence Truman. Came to U. S., 1910. Asso. prof. industrial chemistry, U. of Kan., 1910-1913; pres. Redmanol Chem. Products Co. since 1914; v.p. Bakelite Corp., 1922-40, also dir. research. Dir. Industrial Research Assn.; mem. Am. Inst. Chem. Engrs., Am. Inst. Chemists, A.A.A.S., Am. Chem. Soc. (ex-pres.; chmn. Chicago sect., 1918-19, N.J. sect., 1930-31), Soc. Chem. Industry (chmn. Am. and N.Y. sects., 1926-27), Farm Chemurgic Council (plastics com.), Sigma Xi. Awarded Grasselli medal, Soc. Chem. Industry, 1931. National (Can.). Author: (with A. V. H. Mory) The Romance of Research (Century of Progress Series). Contbr. papers on tech. subjects. Research on phenol condensation products; invented Redmanol. Died Nov. 25, 1946.

REDMAN, William Charles, Am. physicist; b. Washington, June 19, 1923; s. Lewis D. and Eva (Ridgell) R.; B.S., Georgetown U., 1943; M.S., Yale, 1947, Ph.D., 1949; m. Eileen Keenan, June 26, 1948; children—Timothy, Brian, Maribeth. Jr. physicist Carnegie Inst. Washington, 1943, Nat. Bur. Standards, 1943, Metall. Lab., U. Chgo., 1944-46; staff Argonne (Ill.) Nat. Lab., 1949-54, exptl. physics sect. head reactor engring. div., 1955-63, asso. dir. reactor physics div., 1963——; lectr. Fournier Inst. Tech., Lemont, Ill., 1952-54, pres., 1954-55. Recipient Naval Ordnance Devel. award, 1945. Fellow Am. Nuclear Soc. (vice chmn. physics div. 1967——), Am. Phys. Soc.; mem. Sci. Research Soc. Am. Patentee radiation detection and reactor control; research, publs. on nuclear physics, particularly charged particle reactions, radioactive decay and neutron interactions; devel. nuclear reactors for power, propulsion and research applications; instrumentation for detection of neutrons and radiation. Home: 608 S. Monroe St., Hinsdale, Ill. 60521. Office: 9700 S. Cass Av., Argonne, Ill. 60439.*

REE, Alexius Taikyue, phys. chemist; b. Yesan, Choong-Nam, Korea, Jan. 26, 1902; s. Yong Kyoon and Sakum (Park) R.; B.S., Kyoto Imperial U., 1927, Sc.D., 1931; Sc.D., Seoul Nat. U., 1962; m. Cathleen Pak, Aug. 16, 1932; children— Bernadette Juhye, Francis Heyin, Teresa Shinhye, Joan Junghye. Came to U.S., 1948. Instr., asst. prof., prof. Kyoto (Japan) Imperial U., 1931-45; dean, prof. Seoul (Korea) Nat. U., 1945-48; research prof. chemistry U. Utah, Salt Lake City, 1948-65, prof. chemistry, 1965——. Mem. Am., Korean chem. socs., Am. Phys. Soc., Nat. Acad. Scis. of Korea (award 1960). Research and numerous publs. on generalized equation of flow, theory of liquids to calculate thermodynamic properties and transport properties of liquids. Home: 228 Douglas St., Salt Lake City 84102.*

REEB, Georges Henri, French mathematician; b. Saverne, France, Nov. 12, 1920; s. Theobald and Caroline (Engel) R.; Docteur ès Sciences, Faculté des Sciences, U. Strasbourg (France), 1948; m. Gertrude Siefert, Jan. 12, 1945; children—Christiane, Elizabeth. Faculty, Faculté Science, U. Grenoble (France), 1951-63, prof., 1956-63; prof. U. Strasbourg, 1963——, dir. Institut Recherche Mathematique Avancèe. Mem. Societe Mathematique, Am. Math. Soc., Assn. Européenne des Enseignants. Author: (with Wu Wen Tsun) Propriétés Topologiques des varietes feuiletées et fibrees, 1952; also articles. Research on qualitative behavior of trajectories of a dynamical system, including ordinary differential equations, foliations, actions of groups. Home: 3 Bd Gambetta, Strasbourg, France 67, France.*

REECE, Ralph Parlette, Am. animal husbandry scientist; b. Bucyrus, O., Mar. 10, 1909; s. John Frank and Rose (Price) R.; B.S., Ohio State U., 1929; M.S., U. Conn., 1932; Ph.D., U. Mo., 1937; m. Janet Coerper Works, July 28, 1938; 1 son, Alan John. Asst. dairy husbandry N.J. Agrl. Expt. Sta., Rugers U., New Brunswick, N.J., 1937-38, faculty, 1938——, prof., research specialist, 1949——. Recipient medal N.Y. Farmers, 1957. Mem. A.A.A.S., Am. Anat. Assn., Am. Dairy Sci. Assn., Am. Soc. Animal Sci., Endocrine Soc., Soc. for Exptl. Biology and Medicine, N.Y. Acad. Scis., N.J. Acad. Scis., Sigma Xi. Contbg. author: The Artificial Insemination of

Farm Animals, 1960; The Endocrinology of Reproduction, 1958; Reproduction in Farm Animals, 1968; also numerous articles. Research on hormonal control mammary gland growth and function, effects growth substances on farm animals. Home: 43 S. Main St., Cranbury, N.J. 08512. Office: Dept. Animal Sci., Rutgers—The State U., New Brunswick, N.J. 08903.*

REED, Charles Allen, Am. anthropologist, zoologist; b. Portland, Ore., June 6, 1912; s. Charles Allen and Gladys (Donohoe) R.; B.S., U. Ore., 1937; Ph.D., U. Cal. at Berkeley, 1943; m. Lois Wells, Aug. 18, 1951; children—Charles Allen, III, Robert Morris, Brian Wells. Faculty U. Ore., 1936-37, Reed Coll., Portland, 1943-46, U. Ariz., 1946-49; faculty U. Ill. Profl. Colls., Chgo., 1949-54, 55-61, asso. prof. zoology, 1957-61; faculty U. Chgo., 1954-55; asso. prof. biology, curator mammalogy and herpetology Peabody Mus., Yale, New Haven, 1961-66; prof. anthropology and biology Chgo. Circle Campus, U. Ill., 1966—. Dir. prehistoric expdn. to Nubia, 1962-65; zoo-archeologist, Iraq-Jarmo project Oriental Inst., U. Chgo., 1954-55, Iranian Prehistoric project, 1960. Mem. A.A.A.S., Am. Anthrop. Assn., Am. Assn. Phys. Anthropology, Am. Assn. Anatomy, Assn. Current Anthropology, Am. Soc. Mammalogy, Soc. for Study Evolution, Sigma Xi. Research and numerous publs. on evolution burrowing adaptations in mammals with emphasis on fossil and living moles; clarification of origins of animal domestication in prehistoric Nr. E., relations of such domestication to successive human cultures; environmental changes, human cultures and animal life in Egyptian Nubia for past 60,000 years. Office: Dept. U. Ill. at Chgo. Circle, Box 4348, Chgo. 60680.*

REED, Eugene Clifton, Am. geologist; b. Holdrege, Neb., Dec. 12, 1901; s. Eugene W. and Della E. (High) R.; A.B., U. Neb., 1923, M.S., 1933; m. Kathryn M. Slaughter. Dec. 28, 1940. Field geologist Cia, de Petroleo El Aguila, So. Mexico, 1923-25; field geologist, chief exploration geologist Lago Petroleum Corp., Venezuela, 1925-31; mem. Neb. Geol. Survey, 1933—, asst. state geologist, 1938-44; asso. dir. conservation and survey div. U. Neb., 1940-54, dir., 1954-67, research geologist, 1967—, prof. conservation, 1940—, also mem. grad. faculty; asso. state geologist. State of Neb., 1944-54, state geologist, 1954-67. Mem. Neb. Soil and Water Conservation Commn., 1957-67; acting dir. Neb. Water Resources Research Inst., 1964—; mem. numerous coms. on devel. and conservation of Neb. natural resources. Mem. Am. Assn. State Geologists (hon.), Am. Assn. Petroleum Geologists, Geol. Soc. Am., Soc. Econ. Mineralogists and Paleontologists, Neb. Acad. Scis., Neb. Irrigation Assn. (dir.), Neb. Reclamation Assn. (dir.), Sigma Xi, Sigma Gamma Epsilon. Author: Neb. Geol. Survey Bulls. on geology, stratigraphy, and ground water of Neb. contbg. author: The Quaternary of the United States; Am. Inst. M.E. Petroleum Development and Technology for Neb., 1920—. Research on stratigraphy of Pennsylvanian deposits of Neb. and Eastern Wyo., oil and gas possibilities in Neb.; summary of ground water potentials in Neb.; evaluation of changes in ground water storage; stratigraphy of pleistocene deposits in Neb. Home: 2210 Smith St., Lincoln, Neb. 68502.*

REED, Hugh Daniel, Am. zoologist; b. Hartsville, N.Y., Mar. 4, 1875; s. Charles Hart and Sarah (Acker) R.; B.S., Cornell, 1899, Ph.D. (fellow), 1903; postgrad., Freiburg, 1909-10; m. Madeline Kingsley Church, Aug. 20, 1919; 1 dau., Sarah Acker. With Cornell 1900—, successively instr., asst. prof., prof., head dept. zoology, 1910-24. Research and publs. on sound transmitting organs in Amphibia, poison organs and skin of fishes, dermal rays of fishes, fauna of Cayuga Lake Basin, melanosis in fishes, biol. significance of the family. Died Aug. 23, 1937.

REED, J(ohn) Fielding, Am. soil scientist; b. Baton Rouge, Dec. 15, 1912; s. Albert Granberry and Margaret (McDearmon) R.; B.S. in Chem. Engring., La. State U., 1932, M.S. in Chemistry, 1934, Ph.D., 1937; student Washington U., St. Louis, 1929, Duke, 1930; Rockefeller postdoctorate fellow, Cornell U., 1939-40; m. Olive Mae Macdonald, May 13, 1933; children—Martha (Mrs. John C. Moore), Jeannie (Mrs. R. A. Summerlin). Instr., then asst. prof. soil chemistry La. State U., 1934-39; asso. prof., then prof. agronomy N.C. State Coll., 1942-49; dir. soil testing div. N.C. Dept. Agr., 1946-49; So. dir. Am. Pctash Inst., 1949-62, exec. v.p., 1963-64, pres., 1964—, also chmn. bd.; pres., chmn. bd. Found. Internat. Potash Research, 1964—; dir. Kali Kenkyu Kai (Japan), Inst. Brasileiro de Potassa (Brazil), Sadan Birbin Kali Yeun Koo Hwae (Korea), Taiwan Potash Research Found. Sec.-treas., Ga. Plant Food Ednl. Soc., 1952-62, hon. life mem., 1964. Recipient Nat. Peanut Council award, 1947. Fellow A.A.A.S. (v.p. 1956), Am. Soc. Agronomy (pres. So. sect. 1957; Ann. Achievement award 1962); mem. Soil Sci. Soc. Am. (pres. Ga. 1955), Am. Chem. Soc., Sigma Xi, Phi Kappa Phi, Pi Beta Tau, Sigma Chi, Alpha Chi Sigma. Author: Los Suelos de las Regiones Caneras, 1953; also articles. Developed methods for evaluating soil fertility needs by chem. analysis; proved essentiality of calcium in fruiting zone of peanut for pod filling; established worldwide programs to educate developing nations in use of fertilizers for food prodn. Home: 5613 McLean Dr., Bethesda, Md. 20034. Office: 1102 16th St. N.W., Washington 20036.*

REED, Lester James, Am. biochemist; b. New Orleans, Jan. 3, 1925; s. John T. and Sophie (Pastor) R.; B.S., Tulane U., 1943; Ph.D., U. Ill., 1946; m. Janet Louise Gruschow, Aug. 7, 1948; children—Pamela, Sharon, Richard, Robert. Research asst. NDRC, Urbana, Ill., 1944-46; research asso. biochemistry Cornell U. Med. Coll., 1946-48; faculty U. Tex., Austin, 1948—, prof. biochemistry, 1958—, research scientist Clayton Found. Biochem. Inst., 1949—, asso. dir., 1962-63, dir., 1963—. Mem. Am. Soc. Biol. Chemists, Am. Chem. Soc. (Eli Lilly & Co. award in biol. chemistry 1958), Phi Beta Kappa, Sigma Xi. Research and numerous publs. on isolation, chemistry and biol. function lipoic acid; pioneered research on organized enzyme systems; studied resolution, reconstn. and macromolecular orgn. of the multienzyme pyruvate and alpha-ketoglutarate dehydrogenase complexes. Home: 3502 Balcones Dr., Austin, Tex. 78731.*

REED, Richard John, Am. meteorologist; b. Braintree, Mass., June 18, 1922; s. William Amber and Gertrude (Volk) R.; student Boston Coll., 1941, Dartmouth, 1943-44; B.S., Cal. Inst. Tech., 1945; Sc.D., Mass. Inst. Tech., 1949; m. Joan Murray, June 10, 1950; children—Ralph Murray, Richard Cobden, Elisabeth Ann. Staff, Mass. Inst. Tech., 1949-54; faculty U. Wash., Seattle, 1954—, prof. meteorology, 1963—. Cons. to govt. agys. Recipient Meisinger award Am. Metorol. Soc., 1964. Mem. Am. Metrol. Soc., Am. Geophys. Union, A.A.A.S., Royal Metorol. Soc. Asso. editor, Jour. Meteorology, 1957-62, Jour. Atmospheric Scis., 1962-66; editor Jour. Applied Meteorology, 1966—. Contbr. articles to tech. jours. Co-discoverer 26-month oscillation winds in tropical stratosphere; research on ozone-weather relationships, large-scale tropospheric-stratospheric exchange processes, frontogensesis, polar meteorology, numerical weather prediction, stratospheric meteorology. Home: 7338 49th Av. N.E., Seattle 98115.*

REED, Stanley F(oster), Am. engr., b. Bogota, N.J., Sept. 28, 1917; s. Morton H. and Beryl (Turner) R.; m. Stella Swingle, Sept. 28, 1940; children—Nancie, Beryl Ann, Alexandra. With Bethlehem Steel Corp., Balt., 1940-41, cons., 1942-44; founder, pres. Reed Research, Inc., Washington, 1945-62; pres., founder Reed Research Inst. for Creative Studies, 1951—; founder, chmn. LogEtronics, Inc., Washington, 1955; founder, pres. Tech. Audit Corp., 1962; editor, pub. Mergers and Acquisitions, Jour. Corporate Venture, 1965—; dir. Books, U.S.A. Lectr., George Washington U., 1954, 60, U. Pa., 1962, Pa. State U., 1963, 64, Am. U., 1958, 62, U. Colo., 1963, 64, 65, 66, 67, Claremont Grad. Sch., 1964, Union Theol. Sem., 1966, Georgetown U., 1965, 66, 67. Wilton Park (Eng.) fellow, 1967. Mem. Soc. Naval Architects and Marine Engrs., Am. Phys. Soc., A.A.A.S., Soc. I.E.E.E., N.E. Coast Instn. Engrs. and Shipbuilders (Brit.), Am. Econ. Assn., Soc. for Internat. Devel. Contbr. author: Automation, Education, and Human Values, 1965. Research, publs. and numerous patents on cognitive devices, acoustic materials, cast-printing process, creative and learning process in sci. and edn., information systems, technol. balance sheet. Home: 1621 Brookside Rd., McLean, Va. Office: RCA Bldg., 1725 K St. N.W., Washington. 20006.*

REED, Thomas Edward, Jr., geneticist; b. Gadsden, Ala., Nov. 12, 1923; s. Thomas Edward and Katherine (Houlden) R.; student Cal. Inst. Tech., 1941-43, 46; A.B., U. Cal., Berkeley, 1948; Ph.D., U. London, 1952; m. Patricia Fay Doty, Sept. 10, 1949; children—Christopher, Thomas. Faculty, U. Mich., 1952-60; asso. prof. zoology, pediatrics U. Toronto, 1960—; vis. prof. Sch. Pub. Health U. Cal., 1965-66. Mem. Am. Soc. Human Genetics (past dir.), Genetics Soc. Am., A.A.A.S., Soc. Study Evolution, Sigma Xi. Research, publs. on distbn., mutation rates, and natural selection in rare hereditary diseases of man, natural selection and inheritance in human blood groups. Home: 21 Dale Av., Toronto 5. Office: Dept. Zoology, U. Toronto, Toronto 5, Ont., Can.*

REED, Walter, Am. physician; b. Belroi, Va., Sept. 13, 1851; s. Lemuel Sutton and Pharaba (White) R.; M.D., U. Va., 1869, Bellevue Hosp. Med. Coll., 1870; A.M. (hon.), Harvard, 1902; LL.D., U. Mich., 1902; m. Amelia Laurence, 1876; 2 children. Commd. asst. surgeon with rank of lt. M.C., U. S. Army, 1875; stationed at Ft. Lowell, Ariz., 1876-87; attending surgeon and examiner of recruits, Balt., 1890-93; promoted maj., 1893; curator Army Med. Mus., prof. bacteriology and clin. microscopy Army Med. Sch., Washington, 1893-1902; prof. pathology and bacteriology Columbian U., Washington, 1901-02. Elected to Hall of Fame for Gt. Americans, 1945. Author: The Contagiousness of Erysipelas, 1892. Made extensive studies of bacteriology of erysipelas and diphtheria; apptd. chmn. com. to study typhoid fever among U. S. soldiers in Cuba, 1898, proved that disease is transmitted by dust and flies; head commn. to study yellow fever, Cuba, 1900, proved (with James Carroll, Aristide Agramonte, Jesse W. Lazear) that the disease is caused by a virus and transmitted by mosquito Aedes aegypti. Died Washington, Nov. 23, 1902.

REEDER, John Raymond, Am. botanist; b. Grand Ledge, Mich., July 29, 1914; s. Ray and Hazel (Ingersoll) R.; B.Sc., Mich. State U., 1939; M.Sc., Northwestern U., 1940; M.A., Harvard, 1946, Ph.D., 1947; m. Charlotte Goodding, Aug. 15, 1941. Sheldon Travelling fellow Harvard, 1946-47; faculty Yale, 1947-68; prof. botany, curator Rocky Mountain Herbarium, acting head botany dept. U. Wyo., Laramie, 1968—; vis. prof. Central U. Venezuela, 1964, hon. prof., 1966. Mem. Bot. Soc. Am. (past chmn. systematics sect.), Internat. Assn. Plant Taxonomists, Evolution Soc., Am. Soc. Plant Taxonomists, Linnean Soc. London, Sigma Xi. Editor: Brittonia, 1967—. Contbr. articles to tech. jours. Discovered and named numerous new species plants, 3 new genera one tribe; research on grass embryo in terms of utility in systematics, systematics and evolution gramineae. Address: Dept. Botany, U. Wyo., Laramie, Wyo. 82070.*

REEDS, Chester Albert, Am. geologist, paleontologist; b. La Cygne, Kan., July 20, 1882; s. Armstead Madison and Elizabeth (Barrett) R.; B.S., U. Okla., 1905; M.S., Yale, 1907, Ph.D., 1910; m. Marie Louise Haase, Feb. 3, 1916; 1 son, Marcel Chester. Instr. mineralogy and petrology U. Okla., 1908; with Bryn Mawr Coll., 1908-12; asst. curator geology and invertebrate paleontology Am. Mus. Natural History, N.Y.C., 1912-17, asso. curator, 1917-27, curator, 1927-38, observer in charge seismograph, 1921-38, sec. sci. staff, 1913-24, editor div. I publs., 1925-35, ret., 1938; instr. geology Columbia U. Extension, 1914-17. Fellow Geol. Soc. Am., Paleontol. Soc. Am., Seismol. Soc. Am., Nat., Am. geog. socs., N.Y. Acad. Scis., N.Y. Zool. Soc.; mem. Sociedad de Geografico de Columbia (hon.), Am. Mus. Natural History (life). Author: The Earth, 1937 (placed in Time Capsule, N.Y. World Fair ground 1938). Editor: Natural History of Central Asia, 1925-35. Research on water resources of East St. Louis Dist., Hunton formation, Arbuckle Mountains, Okla., use of sulphur to mount rock and fossil specimens, geologic history of P.R., geology of N.Y.C. and vicinity, clays of Little Ferry, N.J., glaciers, deserts of Northwestern Nev., 1927. Address: Brook Cove Estate, P.O. Box 142, Ghent, N.Y. 12075.*

REEMTSMA, Keith, Am. physician; b. Madera, Cal., Dec. 5, 1925; B.S., Ida State Coll.; M.D., U. Pa., 1949; Med.Sci.D., Columbia, 1958; m. Ann Katharine Pierce; children—Lance Brewster, Dirk Van Horn. Asst. surgery Columbia Coll. Phys. and Surg., 1957; faculty Tulane U. Sch. Medicine, 1957-66, prof. surgery, 1966; prof., head dept. surgery U. Utah Coll. Medicine, Salt Lake City, 1966—. Mem. Soc. Clin. Surgery, Am. Surg. Assn., Soc. U. Surgeons, A.C.S., So. Soc. for Clin. Research, Am. Fedn. for Clin. Research, Am. Assn. for Thoracic Surgery, Soc. for Vascular Surgery, Internat. Cardiovascular Soc., So. Thoracic Surg. Assn., New Orleans Surg. Soc., Surg. Assn. La., A.M.A., La. State, Orleans Parish med. socs., Alpha Omega Alpha. Contbr. articles to profl. jours. Home: 2795 Comanche Dr., Salt Lake City.*

REES, Albert, Am. economist; b. N.Y.C., Aug. 21, 1921; s. Hugo R. and Rosalie (Landman) R.; B.A., Oberlin Coll., 1943; M.A., U. Chgo., 1947, Ph.D., 1950; m. Candida Kranold, July 15, 1945 (div. Feb. 1962); 1 son, David Kranold; m. 2d, Marianne Russ, June 22, 1963; children—Daniel Ira, Jonathan Hugo. Instr., Roosevelt Coll., 1947-48; faculty U. Chgo., 1948-66, prof. econs., 1961-66; prof., Princeton (N.J.) U., 1966—; research asso. Nat. Bur. Econ. Research, N.Y.C., 1953-54; staff mem. Council Econ. Advisers, Washington, 1954-55. Mem. Am. Econ. Assn., Indsl. Relations Research Assn. Author: Real Wages in Manufacturing, 1890-1914, 1961; The Economics of Trade Unions, 1962; also articles. Editor, Jour. Polit. Economy, 1954-59; asso. editor Internat. Ency. Social Scis., 1962—. Improvement of measurement of wage movements, wage differentials, consumer prices, and unemployment, improvement understanding of forces underlying wage determination. Home: 32 Turner Ct., Princeton, N.J. 08540.*

REES, Carl John, Am. mathematician; b. Millersville, Pa., Nov. 23, 1896; s. Charles John and Helena (Rapp) R.; A.B., Franklin and Marshall Coll., 1918, D.Sc., 1951; A.M., U. Chgo., 1925; Ph.D., U. Pa., 1940; m. Eleanor Adams Knight, June 21, 1930. Faculty U. Del., Newark, 1920—, prof. math., 1940—, chmn. dept. math., 1940-50, chmn. interdept. statistics, 1947-50, dir. div. grad. studies, 1945-50, dir. summer sch. and univ. extension, 1950-55, dean Sch. Grad. Studies, 1950-62, provost, 1955-62. Mem. Gov. Del.. Adv. Council on Natural Resources, 1954-58, Del. Commn. on Mental Health, 1955-59; mem. adv. com. Wilmington Trust Co., Newark, 1961—; cons. liberal arts Millersville State Coll., 1962-64. Mem. Am. Math. Soc. (chmn. com. on printing contracts 1947-56), Operations Research Soc. Am., Math. Assn. Am., N.Y. Acad. Scis., Inst. for Math. Statistics, Soc. for Engring. Edn., Phi Beta Kappa, Sigma Xi, Kappa Alpha Order, Phi Kappa Phi, Pi Mu Epsilon, Sigma Pi Sigma. Author: (with Walling and Hill) Nautical Mathematics and Marine Navigation, 1944; Graduate Education in Land-Grant Colleges and Universities, 1962. Research and publs. on statistics; developed basic theory of elliptic orthogonal polynomials. Home: 9 Georgian Circle, Newark, Del. 19711.*

REES, Don Merrill, Am. biologist; b. Wales, Utah, Sept. 9, 1901; s. Thomas Davis and Elizabeth (Rees) R.; B.S., U. Utah, 1926, M.S., 1929; Ph.D., Stanford, 1936; m. Norma Anderson, Apr. 30, 1924; children—Thomas Dee, Joseph Richard. Mem. faculty U. Utah, Salt Lake City, 1929—, prof. zoology and entomology, 1948—, head dept. zoology and entomology, 1954-63, chmn. div. biol. sci., 1951-65, organizer, dir. Inst. Environmental Biol. Research, 1960-67. Mem. staff Cal. Med. Sch., U. Indonesia, 1957-58; cons. U.S. Army, trustee Salt Lake City Mosquito Abatement Dist., 1938—, pres., 1938, 42, 46, 50, 54, 58, 62. Fellow A.A.A.S., mem. Entomol. Soc. Am., Am. Pub. Works Assn., Am. (dir. 1952-56, pres. 1952), Utah (organizer, pres. 1948) mosquito control assns., Am. Assn. U. Profs., Entomol. Soc. Washington, Utah Acad. Sci., Sigma Xi, Phi Kappa Phi, Phi Sigma. Research, numerous publs. on arthropods of medical importance and their control, particularly mosquitoes. Home: 717 11th Av., Salt Lake City 84103.*

REES, John Krom, Am. astronomer; b. N.Y., Oct. 27, 1851; grad. Columbia, 1872, Ph.D., 1895; E.M., Columbia Sch. Mines, 1875, A.M., 1875; m. Louise E. Sands, Sept. 7, 1876. Asst. in math. Sch. Mines, 1873-76; prof. math. and astronomy Washington U., St. Louis, 1876-81; dir. obs. Columbia, 1881—, instr. geodesy and practical astronomy, 1881-82; chmn. bd. editors, Sch. Mines Quar., 1883-90, prof. astronomy, 1892—. Vice pres. Am. Math. Soc., 1890-91; pres. N.Y. Acad. Scis., 1894-96; sec. Am. Metrol. Soc., 1882-96; v.p., 1896—; sec. Univ. Council of Columbia, 1892-98; Fellow Royal Astron. Soc., London; Legion of Honor, 1901. Research on variation of terrestrial latitudes and constant of aberration; contbd. to celestial photography. Died Mar. 9, 1907.

REES, Manfred Hugh, physicist; b. Nuremberg, Germany, June 29, 1926; s. Alfred and Magda (Glueck) R.; B.S. in Elec. Engring., W.Va. U., 1948; M.S. in Physics, U. Colo., 1956, Ph.D., 1958; m. Marjorie A. Grant, Aug. 28, 1949; children—Eric M., Peter M. Elec. engr. NACA, Langley Field, Va., 1948-49; faculty U. Alaska, College, 1958-60, 65-66, asso. prof., head physics dept., 1965-66; research physicist Lab. for Atmospheric and Space Physics, U. Colo., Boulder, 1966—. sr. research fellow Queen's U., Belfast, Ireland, 1960-61; research asso. U. Colo., 1961-65. Mem. Optical Soc. Am., A.A.A.S., Am. Geophys. Union. Research, publs. in phys. processes upper atmosphere of earth; airglow phenomena and aurora; atomic and molecular collision processes.

REES, Richard John, English pathologist; b. Wimbledon, Eng., Aug. 11, 1917; s. William and Gertrude (Smith) R.; B.Sc., U. London, 1939, L.R.C.P., M.R.C.S., 1941, M.B., B.S., 1942, F.C. Path., 1964; m. Kathleen Harris, Aug. 21, 1942; children—Lorna, Hazel, Diana. Asst. clin. pathologist Guy's Hosp., London, 1946-47, lectr. dept. exptl. pathology, 1947-49; sr. sci. staff Med. Research Council, Nat. Inst. for Med. Research, London, Eng., 1949—, sec. Leprosy com., 1961—. Chmn. med. com. Brit. Leprosy Relief Assn.; mem. expert adv. panel on leprosy WHO, 1964—; cons. leprosy panel U. S.-Japan Coop. Med. Sci. Program. Mem. Royal Soc. Medicine (hon. sec. 1965-67), Path. Soc. Gt. Britain, Brit. Soc. Immunology, Internat. Acad. Pathology, Internat. Leprosy Assn. Research, numerous publs. on leprosy, Tb, application of tissue culture for growing rat and human leprosy bacilli, recognition of dead leprosy bacilli on basis of morphology, discovery that exptl. human leprosy can be reproduced in mice by depressing their immunological response (thymectomy plus irradiation). Home: Highfield, Highwood Hill, Mill Hill. Office: Nat. Inst. for Med. Research, Mill Hill, London N.W.7, England.*

REES, William Linford Llewellyn, Brit. psychiatrist, educator; b. Burry Port, Wales, Oct. 24, 1914; s. Edward Parry and Mary (John) R.; B.Sc., Welsh Nat. Sch. Medicine, Maudsley Hosp. U. London (Eng), 1935, M.B., B.Ch., 1938, M.D., 1943, D.P.M., 1940, M.R.C.P., 1942, F.R.C.P., 1960; m. Catherine Magdalen Thomas, June 15, 1940; children—David, Angharad, Vaughan, Cathrin. Asst. med. officer Worcester City and County Mental Hosp., Powick and Claybury Hosp., London, Eng., 1938; psychiatrist Mill Hill War Emergency, Maudsley Hosp., 1940-42, specialist, 1942-44, dep. physician supt., 1944-45; specialist Prisoner of War Neurosis Unit, Dartford, 1945-47; dep. physician supt. Whitchurch Hosp., Cardiff, Wales, 1947; regional psychiatrist Wales and Monmouthshire, Cardiff, 1948-54; cons. physician Maudsley and Bethlehem Royal hosps., 1954-59; physician in charge dept. psychol. medicine St. Bartholomew's Hosp., London, 1959-66, prof. psychiatry U. London, 1966—. Recipient Alfred Sheen prize; David Hepburn medal in anatomy; Alfred Hughes medal; Howell Rees scholar; John MacLean prize and medal. Mem. World (treas. 1966—), Swedish psychiat. assns., Royal Medico-Psychol. Assn., Royal Coll. Physicians, Med. Research Council, Soc. for Psychosomatic Research (pres. 1955-56), Asthma Research Council, Royal Soc. Medicine (hon.). Author: A Short Textbook of Psychiatry, 1967; also numerous articles. Research into causes of asthma, vasomotor rhinitis, urticaria, psychiat. illness following childbirth, sci. evaluation of new methods of treatment in psychiatry, study of phys. constn. in relationship to disease and personality. Home: 34 Oakwood Av., Purley, Surrey, Eng. Office: St. Bartholomew's Hosp., London E.C. 1, Eng.*

REESE, Albert Moore, Am. zoologist; b. Lake Roland, Md., Apr. 1, 1872; s. Henry and Mary Anna (Miller) R.; A.B., Johns Hopkins, 1892, Ph.D. 1900; m. Nelle Summers, June 29, 1927; 1 son, Albert Moore, Lectr., Homeopathic Med. Coll., Balt., 1893-98, Pa. Coll., Gettysburg, 1897; prof. biology Allegheny Coll. Meadville, Pa., 1901-02; asso. prof. Syracuse U., 1903-07; prof., head dept. zoology W.Va. U., Morganstown, 1907-46, prof. emeritus, 1946; collaborator Smithsonian Instn., 1913, mem. expdn. to Orient, 1913; mem. expdn. to S.Am., Carnegie Instn., 1919. Recipient Order of Vandalia award for distinguished service to W.Va. U., 1964, Presdl. citation W.Va. Acad. Sci., 1965, Emma H. Clark award Crippled Children Soc., 1962. Fellow A.A.A.S.; mem. Am. Soc. Zoologists, Soc. Icthyologists and Herpetologists, W.Va. Acad. Sci. (pres. 1939-40, exec. bd. 1938-41, editor publ. 1947-61), Phi Beta Kappa, Beta Theta Pi. Author: Introduction to Vertebrate Embryology, 1904; The Alligator and its Allies, 1915; Outlines of Economic Zoology, 1919; Wanderings in the Orient, 1919; also numerous articles. Research on Am. alligator. Died Dec. 30, 1965.

REESE, Algernon Beverly, Am. ophthalmologist; b. Charlotte, N.C., July 28, 1896; s. Algernon Beverly and Mary Cannon (Wadsworth) R.; B.S., Davidson Coll., 1917, D.Sc. (hon.), 1946; M.D., Harvard, 1921; student U. Vienna, 1925-26; M.D. (hon.), U. Melbourne, 1952; LL.D., Duke, 1957; m. Joan Leeds, Sept. 26, 1942; children—Algernon Beverly, III, Rigdon Leeds, Jonathon Wadsworth. Cons. ophthalmologist, pathologist Eye Inst., N.Y.C.; cons. ophthalmologist to hosps. in N.Y. met. area; clin. prof. ophthalmology, emeritus Coll. Phys. and Surgs., Columbia; former mem. com. on Ophthalmology, Div. of Med. Sci., NRC.; DeSchweinitz lectr. 1946; Proctor lectr., 1949; Jackson lectr., 1954; Bedell lectr., 1955; Gifford lectr., 1955, Snell lectr., 1956, Schoenberg lectr., 1959, Montgomery (Royal Coll. Surgeons, Dublin) lectr., 1962. Cons. to Surgeon General, U. S. Army; mem. bd. sci. counselors Nat. Inst. Neurological Diseases and Blindness, Dept. Health, Edn., and Welfare. Bd. dirs. Nat. Soc. for Prevention Blindness. Received Hon. Key, Am. Acad. Ophthalmology, 1955; Citation of Merit, Davidson Coll., 1948; Howe Medal, Am. Ophthal. Soc., 1950; Howe Medal, U. Buffalo, 1956. Fellow A.C.S., A.M.A. (chmn. eye sect. 1956-57, Lucien Howe medal eye sect. 1961); mem. N.Y. Cancer Soc., Am. Ophthal. Soc. (pres. 1960), Am. Acad. Ophthalmology and Otolaryngology (pres. 1954-55), Phi Beta Kappa; hon. mem. Greek, Cuban, Mexican, Australian, Chilean, Panamanian, French, New Zealand ophthal. socs. Author: Tumors of the Eye, 1950, 62; Tumors of Eye and Adnexa (Fascicle 38, Armed Forces Inst. Pathology), 1956; also articles; contbg. author: The Eye and Its Diseases; The Treatment of Cancer and Allied Diseases; Systematic Ophthalmology. Mem. editorial staff and dir. Am. Jour. Ophthalmology; asso. editor Survey of Ophthalmology; editorial bd. Sight-Saving Rev. Home: 1088 Park Av., N.Y.C. 10028. Office: 73 E. 71st St., N.Y.C. 10021.*

REESE, Charles Lee, Am. chemist; b. Balt., Nov. 4, 1862; s. John S. and Arnoldina O. (Focke) R.; grad. U. Va., 1884; Ph.D., Heidelberg, 1886; Sc.D. (hon.), U. Pa., 1919, Colgate U., 1919, U. of Delaware, 1928, Wake Forest Coll., 1934, Heidelberg, 1936; m. Harriet S. Bent, April 10, 1901; children—Charles Lee, John Smith, David Meredith, Eben Bent, William Fessenden. Asst. in chemistry Johns Hopkins, 1886-88, instr., 1896-1900; prof. chemistry Wake Forest (N.C.) Coll., 1888, S.C. Mil. Acad., 1888-96; chief chemist N.J. Zinc Co., 1901-02, Eastern Dynamite Co., and dir. Eastern Lab., 1902-06; in charge chem. div. high explosive operating dept. E. I. du Pont de Nemours Powder Co., 1902-11, chem. dir., 1911-June 1, 1924, cons., until 1931. Asso. mem. Naval Cons. Bd., also chmn. Del. sect.; mem. Nat. Indsl. Conf. Bd.; mem. vis. com. Bur. Standards, 1930—; v.p. Internat. Union Pure and Applied Chemistry, 1929-34; founder, chmn. Bd. Indsl. Research. Mem. Am. Chem. Soc. (chmn. bd. 1930—, pres. 1934), Am. Inst. Chem. Engrs. (pres. 1923-25), Mfg. Chemists' Assn. (pres. 1923-25). Contbr. to chem. jours. Pioneer in devel. of modern blasting explosives; improved dye industry, also contact process for manufacture of sulfuric acid; developed successful theory to explain formation of Carolina phosphates. Died Apr. 12, 1940.

REESE, Hans H., physician; b. Bordesholm, Holstein, Germany, Sept. 17, 1891; s. H.F. and Dora (Kaak) R.; M.D., U. Kiel (Germany), 1917; M.D., U. Kyushu (Japan), 1965; married Theresa Schmidt children—Sibyl W. (Mrs. Wayne Millner), Ernst S., Alma P. (Mrs. Paul Gray). Came to U.S., 1924. Asst., U. Hamburg, 1918-23; asst. research neuropsychiatry U. Wis., Madison, 1924-25, faculty, 1925—, prof. neuropsychiatry, 1927-63, emeritus prof., 1963—, chmn. dept. neuropsychiatry, 1945-53. Mem. A.M.A., Am. Neurol. Assn., Central Neuropath. Soc., A.C.P.; hon.. mem. German, Japanese, Chgo., N.Y. neurol. socs. Author: Year Book of Neuropsychiatry, 1934-43; also numerous articles. Research on muscular and brain-spinal cord-nerve disorders; head injuries. Home: 3421 Circle Close, Madison, Wis. 53705.*

REEVE, Edward Wilkins, organic chemist; b. Westmont, N.J., July 31, 1913; s. Edward White and Roxana (Wilkins) R.; B.S. in Chem. Engring., Drexel Inst. Tech., 1936; Ph.D. in Chemistry, U. Wis., 1940; m. D. Lotus Marsh, Feb. 3, 1940; children—Roslyn Diane, Sheryl Anne, Donald Wilkins. Instr., U. Md., College Park, 1940-41, faculty, 1945—, prof. chemistry, 1951—; tech. aide OSRD, Washington, 1941-45, cons., 1949-52. Recipient Alumni citation Drexel Inst. Tech. for outstanding achievements, 1956. Mem. Am. Chem. Soc., A.A.A.S., Washington Acad. Scis., Sigma Xi, Alpha Chi Sigma, Phi Kappa Phi, Tau Beta Pi. Research, publs., patents on chemistry of aryltrichloromethylcarbinols, pinacol rearrangement, catalytic hydrogenation organic compounds. Home: 4708 Harvard Rd., College Park, Md. 20740.*

REEVE, Ernest Basil, physiologist; b. Liverpool, Eng., May 5, 1912; s. William Ernest and Irene (Gill) R.; B.A., Oxford U., 1935, B.M., B.Ch., 1938; M.R.C.P., London, 1940; m. Jeanne Knight Schenck, July 23, 1953; children—Jonathan, Sarah. Came to U.S., 1953. Mem. sci. staff Med. Research Council Longdon Clin. research unit Guy's Hosp., 1940-52; faculty U. Colo. Med. Sch., Denver, 1953—, prof. medicine, 1962—, head div. lab. medicine, 1961—; vis. prof. physiology Columbia U., 1953. Mem. com. on shock NRC, 1957—. Mem. Central Soc. for Clin. Research, Am., Brit. physiol. socs., Western Assn. Physicians, Med. Research Soc. (London). Author: (with R.T. Grant) Observations on the General Effects of Injury on Man with Special Reference to Wound Shock, 1951; also numerous articles. Research on wound shock, methods and significance of blood volume measurements, control levels plasma albumin and fibriniogen, systems analysis methods applied to physiol. function. Home: 420 Cherry St., Denver 80220.*

REEVES, Leonard Wallace, chemist; b. Bristol, U.K., Feb. 8, 1930; s. Wallace Edward and Violet Edith (Smith) R.; B.Sc., U. Bristol, 1951, Ph.D., 1954, D.Sc., 1965; m. Beryl Evelyn Usher, Nov. 9, 1957; 1 dau., Wendy Jane. Research asso. U. Cal., Berkeley, 1954-56; post-doctoral fellow NRC Can., 1956-57; jr. fellow Mellon Inst., Pitts., 1957-58; faculty, U. B.C., Vancouver, 1958—, prof., 1966—. Ford Found vis. prof. U. Sao Paulo, Brazil, 1967-68. Fellow Canadian Inst. Chemistry; mem. Am. Chem. Soc., Am. Phys. Soc., Faraday Soc., Chem. Soc. London, Sigma Xi. Research in application of nuclear magnetic resonance studies to problems in chemistry; hydrogen bonding, intermolecular forces, chem. exchange reactions and studies of NMR signals of heavier nuclei. Home: 3745 Puget Dr., Vancouver 8, B.C., Can.*

REEVES, Robert James, Am. physician; b. Matador, Tex., Dec. 28, 1898; s. Walter E. and Henrietta (Bryant) R.; A.B., Baylor U., 1920, M.D., 1924; m. Gipsie Proctor, Aug. 30, 1935; children—Elizabeth (Mrs. Roger Leverton), Judy Bryant. Instr. radiology and medicine Columbia Med. Center, N.Y.C., 1926-30; prof., chmn. dept. radiology Duke U. Med. Center, Durham, N.C., 1930—. Cons. Oak Ridge Inst. Nuclear Sci., Armed Forces Inst. Pathology, VA Hosps.; mem. teletherapy bd. Oak Ridge Inst. Nuclear Sci.; mem. administrv. com., coll. of radiology, Duke U. Bd. dirs. Am. Cancer Soc. Fellow Am. Coll. Radiology; mem. A.M.A., Am. Roentgen Ray Soc., Radiol. Soc., So. Med. Soc., Nuclear Medicine Soc., So. Radiol. Soc., Pan Pacific Surg. Soc., N.C. Med. and Radiol. Soc., Durham-Orange Med. Soc. (pres. 1964), Sigma Xi, Sigma Nu. Research, publs. on radium, Cobalt 60, Cesium 137. Home: 920 Anderson St., Durham, N.C.*

REEVES, Roy Franklin, Am. mathematician; b. Warrensburg, Mo., July 8, 1922; s. Archie Ray and Lucie (John) R.; B.S. in Elec. Engring., U. Colo., 1947; Ph.D., Ia. State U., 1951; m. Priscilla L. La Vanway, Mar. 17, 1951; children—Dennis, James, Terrence, Sandra, Timothy. Instr. elec. engring. Ia. State U., 1949-51; prof. math., dir. computer Ohio State U., Columbus, 1951—. Cons. to govt. and pvt. orgns. Mem. Am. Math. Soc., Math. Assn. Am., Assn. Computing Machinery. Contbr. articles to tech. jours. Design and devel. programming langs. and operating systems, numerical solution differential equations. Home: 1640 Sussex Ct., Columbus, O. 43221.*

REEVES, William Carlisle, Am. epidemiologist; b. Riverside, Cal., Dec. 2, 1916; s. William Claude and Alice (Brant) R.; B.S., U. Cal., at Berkeley, 1938, Ph.D., 1943, M.P.H., 1949; m. Mary Jane Moulton, July 6, 1940; children—William C., Robert F., Terrence M. Entomologist, research asst., research asso. Hooper Found., U. Cal. Med. Sch., San Francisco, 1941-49; faculty Sch. Pub. Health, U. Cal. at Berkeley, 1946—, prof. epidemiology, 1956—, dean Sch. Pub. Health, 1968—. Cons. to govt. agys.; chmn. arboviruses com. research reference reagts. br. NIH, USPHS, 1966—; mem. comm. on viral infections Armed Forces Epidemiological Bd., 1960—. Mem. A.A.A.S., Am. Pub. Health Assn., Am. Soc. Tropical Medicine and Hygiene, Am. Entomol. Soc., Am. Epidemiological Soc., Am. Mosquito Assn. Author: (with W. McD. Hammon) Epidemiology of Arthropod-Borne Viral Encephalitides in Kern County, California, 1943-52, 1962; also numerous articles. Research on biol. factors controlling spread viral disease by arthropod vectors. Home: 2846 Kinney Dr., Walnut Creek, Cal. 94598.*

REGAMEY, Robert-Henri, pathologist; b. Henniez, Sept. 30, 1907; s. Jules and Alice (Delafontaine) R.; ed. univs. Lausanne, Zurich, Berlin and Hamburg; M.D.; m. Elisabeth Diserens, July 16, 1938; children—Claude, Gilles, Fabienne. Asst. in pathol. anatomy, internal medicine, dermatology, bacteriology; service chief, later tech. dir. Serotherapeutic Inst., Berne, Switzerland; hon. prof. U. Berne; prof. U. Geneva (Switzerland), also dir. Med. Microbiology Inst. Publs. on immunity, tetanus, mil. medicine, allergy, standardization. Home: 15B ch. de Conches. Office: Inst. d'hygiène, Geneva, Switzerland.

REGAN, Francis, Am. mathematician; b. nr. Terre Haute, Ind., Jan. 10, 1903; s. Patrick M. and Ella (Curley) R.; A.B., Ind. State U., 1922; LL.B., La Salle Extension U., 1926; A.M., Ind. U., 1930; Ph.D., U. Mich., 1932. Tchr., Ind. High Schs., 1923-25; prof. commerce Columbus Coll., Sioux Falls, S.D., 1925-29; asst. prof. math. Colo. State U., 1929-30; faculty St. Louis U., 1932—, prof., math., 1945—, chmn. dept. maths., 1950—. Mem. Math. Assn. Am., Am. Math. Soc., Inst. Math. Statistics, Phi Beta Kappa, Sigma Xi, Pi Mu Epsilon (editor in chief Jour. 1957-62, mem. council 1962-66). Contbr. articles to tech. jours. Research in foundations of probability; testing assumptions made in time series; analysis of infinite series. Home: 378 N. Taylor St., St. Louis 63108.*

REGAN, Peter Francis, III, Am. psychiatrist educator; b. Bklyn., Nov. 11, 1924; s. Peter Francis and Veronica (Tierney) R., Jr.; M.D., Cornell U., 1949; m. Laurette O'Connor, June 18, 1949; children—Peter Francis, IV, Stephen, William, Elizabeth Susan, John, Carol. Asst. prof. psychiatry, Med. Coll. Cornell U., 1956-58; prof., head dept. psychiatry, Coll Medicine U. Fla., Gainesville, 1958-64; v.p. for health affairs State U. N.Y. at Buffalo, 1964-67, exec. v.p., 1967—, also prof. psychiatry, Sch. Medicine, U. psychiatrists Buffalo Gen. Hosp. Diplomate Am. Bd. Psychiatry and Neurology. Fellow Am. Psychiatric Assn.; mem. Am. Bd. Psychiatry and Neurology, A.M.A., Erie County Med. Soc., Alpha Omega Alpha. Author: (with Frederick Flach) Chemotherapy in Emotional Disorders, 1959; (with E. Pattishall) Behavioral Science Contributions to Psychiatry, 1965; also articles. Research on physiological correlates of emotions; social psychiatry. Home: 59 Chapin Pkwy., Buffalo, N.Y. 14209.

REGENER, Victor H., physicist; b. Berlin, Germany, Aug. 25, 1913; s. Erich and Victoria (Mincin) R.; Dr.-Ing. in Physics, Tech. Hochschule, Stuttgart, Germany, 1938; m. Birgit Hamilton, Aug. 23, 1941; children—Eric, Vivian. Came to U.S., 1940, naturalized, 1947. Research fellow U. Padova, Italy, 1938-40; research fellow U. Chgo., 1940-42, instr., 1942-46; asso. curator Mus. Sci. & Industry, Chgo., 1945-46; faculty U. N.M., Albuquerque, 1946—, research prof. physics, 1957—, chmn. dept., 1946-57, 1962—. Hon. prof. U. Mayor de San Andres, La Paz, Bolivia, 1958. Fellow Am. Phys. Soc., N.Y. Acad. Scis. Research in cosmic radiation, atmospheric ozone, zodiacal light, electronics. Office: U. N.M., Albuquerque 87106.*

REGGE, Tullio Eugenio, physicist; b. Turin, Italy, July 11, 1931; s. Michele and Lidia (Petrini) R.; Laurea in Fisica U. di Torino, 1948-52; Ph.D., Rochester U., 1956; m. Rosanna Cester, Feb. 27, 1957; children—Daniele, Marta, Anna-Marzia. Asst. prof. Turin U., 1957-62, prof. theory of relativity, 1962—; prof. Inst. for Advanced Study, Princeton, 1965—. Premio Della Societa Italiana Di Fisica, 1962; recipient Dannie Heineman award, 1946 Premio Francesco Somaini (cittá di Como), 1967. Author: (with D. E. Alfaro) Potential Scattering, 1965. Contbr. articles in field to sci. jours. Introduction into particle theory the idea of analytic continuation and of moving poles in the complex angular momentum; discovered new symmetries of Racah and Clebsch Gordan coefficients. Home: 45 Veblen Circle Princeton 08540. Office: Inst. Advanced Study, Olden Lane, Princeton, N.J. 08540.*

REGGIO, Niccolo da, see da Reggio, Niccolo.

REGIOMONTANUS (originally Johann Müller), German astronomer, mathematician; b. Königsberg, Franconia, Germany, June 6, 1436; ed. U. Leipzig (Germany), 1447-50; studied under Peurbach, Vienna, 1450. Accompanied his patron, Cardinal Bessarion to Italy, 1461; learned Greek, collected Greek manuscripts; computed eclipses and dates for Easter; apptd. keeper of King Matthaias Corvinus' library, 1468; went to Nürnberg, became tchr. at univ. where Bernhard Walther provided him with obs., tool shop and printing works. Author: De doctrina triangularum, 1463; De quadratura circuli, 1463; De triangulis omnimodis libri quinque, 1464; Tabella sinus, 1531; Sphaera mundi, 1531; also translator Conics of Apollonios, mech. works of Hero and works of Archimedes; pub. (with Walther) Peurbach's Theoriae planetarum novae and his own Calendarium novum, 1473, Ephemerides ab anno 1475-1506. Advanced study of algebra and trigonometry in Germany; among 1st to apply algebra to solution of geometric problems; made observations of comet of 1472 (later called Halley's comet); completed Peurbach's translation of Ptolemy's Almagest and wrote commentary on it which presented many later famous arguments against motion of earth; invented method of lunar distances for determination of longitude at sea (used by Columbus to reach Am.); invited by Pope Sixtus IV to help reform Julian Calendar, 1475; apptd. titular bishop of Ratisbon, 1475, but before occupying chair, died Rome, July 6, 1476.

RÉGIS, Pierre-Sylvain, French philosopher; b. Salvetat-de-Blanque Fort, France, 1632. Apptd. first asso. geometrician by Louis XIV, 1699. Author: Cours entier de philosophie ou système général selon les principes de Descartes, 1690. Popularized Cartesianism in lectures at Toulouse, France, 1665, Montpellier, France, 1671, also Paris. Died Paris, Jan. 11, 1707.

REGISTER, Ulma Doyle, Am. biochemist; b. West Monroe, La., Feb. 4, 1920; s. John W. and Lillian (Reagen) R.; B.S., Madison Coll., 1942; M.S., U. Vanderbilt, 1944; Ph.D., U. Wis., 1950; m. Helen L. Hite, June 18, 1942; children—Rebecca, Dorothy, Deborah. Postdoctoral fellow Tulane U. Sch. Medicine, 1950-51; mem. faculty Loma Linda (Cal.) U., 1951—, prof., chmn. dept. nutrition, since 1967—. Member American Public Health Association, Am. Chem. Soc., Am. Inst. Nutrition, Soc. Exptl. Biology and Medicine, Sigma Xi. Research, publs. on effect of diet factors on consumption of alcohol; biol. value of vegetable protein combinations in various dietaries; obesity; Vitamins B12, B6, Niacin. Home: 11452 Iris Av., Loma Linda, Cal. 92354.*

REGLER, Friedrich Maria, Austrian physicist; b. Vienna, Austria, Mar. 9, 1901; s. Karl and Emilie (Praschek) R.; Dr.phil., U. Vienna, 1924; m. Melanie Müller, July 12, 1926; children—Waltraud (Mrs. Wolfgang Pexa), Roderich, Dietburga (Mrs. Gerhard Huber), Meinhard, Gerlinde (Mrs. Helmut Bachmayer), Adelheid, Norbert. Owner, dir. state authorized Exptl. Inst. X-ray Materials Testing, Vienna, 1929-38; faculty Mining Acad., Freiberg/Saxony, Germany, 1942-47, prof., 1945-47, chancellor, 1945-46; prof. physics, chmn. inst. exptl. physics Vienna Inst. Tech., 1947—, dean faculty applied math. and physics, 1952-54, chancellor, 1958-59. Austrian dep. del. European Center Nuclear Research, Geneva, Switzerland, 1961-66; chmn. Atomic Insts. Austrian univs., 1961—. Recipient Silver Service medal Republic Austria; Wilhelm Exner medal. Mem. Austrian Phys. Soc. (pres. 1960), Chem. Phys. Soc. (pres. 1950), Austrian Acad. Scis. Author: Grundzüge der Röntgenphysik, 1937; Verformung und Ermüdung metallischer Werkstoffe im Röntgenbild, 1939; Röntgenfeinstruktur von Metallen, 1954; Atomphysik, 1956; Einführung in die Physik der Röntgen- und Gammastrahlen, 1967. Coeditor Acta Physica Austriaca. Research, publs. on physics of X-rays and gamma rays, X-ray fine structure investigation, X-ray spectroscopy. Address: Arsenal, Obj. 14, A-1030 Vienna, Austria.*

REGNAULT, Henri Victor, French chemist, physicist; b. Aix-la-Chapelle, France, July 21, 1810; student Ecole Polytechnique, Paris, France, 1829-32, Ecole des Mines, 1832; student under Justus von Liebig, Giessen, Germany; 1 son, Alexandre Georges Henri. Prof. chemistry U. Lyons (France), Ecole Polytechnique, from 1841; prof. physics College de France; dir. porcelain factory, Sèvres, France, from 1854; prof. Ecole Centrale des Arts et Manufactures; chief engr. mines, from 1847. Recipient Rumford medal, 1869, Copley medal Royal Soc. Mem. French Acad. Scis. (v.p. 1854, pres. 1855), 1840. Author: Relation des Expériences Entreprises, pour Déterminer les Lois et les Données Physiques Nécessaires au Calcule des Machines a Feu, 3 vols., 1847-70; Treatise on Chemistry, 4 vols., 1847. Measured co-efficients of expansion, specific heats, vapor pressures of mixtures; designed several pieces of measuring apparatus (including Regnault's hygrometer); discovered carbon tetrachloride, participated in adjustments of earlier phys. constants and laws, especially with regard to Boyle's law; important research on halogens, other derivatives unsaturated hydrocarbons; (with Jules Reiset) made 1st determination of respiratory quotient. Died Auteuil, France, Jan. 19, 1878.

REHAK, Svatopluk, Czechoslovakian physician; b. Cesky Brod, Czechoslovakia, Jan. 20, 1926; s. Svatopluk and Olga (Ossobova) R.; M.D., Charles U., Prague, 1951, D.Sc., 1964; m. Jirina Svobodova, Dec. 1, 1951; children—Svatopluk, Jiri. Staff, Charles U., Hradec Kralove, 1951—, prof. ophthalmology, 1966—, dean Med. Faculty, 1963—. Mem. Med. Commn., Ministry of Culture, 1963—; regional expert for ophthalmology, 1957-58. Recipient Deyl's prize for research in ophthalmology, 1961; Meml. medal Charles U., Prague, 1965; Hus's Meml. medal, 1965. Mem. Czechoslovakian Cuban (corr.) Ophthal. Socs. Research, numerous publs. on mechanisms of regulation of intra-ocular pressure, exchange of intra-ocular fluids in exptl. animals, regularity of reactive hypertension, discovered secondary reactive hypertension of eye. Home: 1057 Nalepkova, Hradec Kralove, Czechoslovakia.*

REHDER, Harald Alfred, Am. zoologist; b. Jamaica Plain, Boston, Mass., June 5, 1907; s. Alfred and Anneliese (Schrefeld) R.; A.B., Bowdoin Coll., 1929; M.S., Harvard, 1933; Ph.D., George Washington U., 1934; m. Lois Fleming Corea, Oct. 15, 1938; children—Anne Fleming, Alfred Luis. Sci. aid div. mollusks U. S. Nat. Museum, Smithsonian Instn., 1932-34, asst. curator, 1934-42, asso. curator, 1942-46, curator, 1946-65, sr. zoologist, 1965—; mem. Smithsonian-Bredin expdns. to French Polynesia, 1957, Yucatan, 1960; Pacific Sci. Bd. Expdn. to Jaluit, Marshall Islands, 1960; expdns. to S. Pacific, 1963, 64, 65. Del. 8th Am. Sci. Congress, 1940. Fellow A.A.A.S., Cal. Acad. Sci.; mem. Am. Malacological Union (pres. 1941), Washington Acad. Sci., Biol. Soc. Washington, Paleontological Soc., Soc. Systematic Zoology, Malacological Soc. London. Author: (with P. Bartsch) Marine Pelecypod Mollusks of Hawaiian Islands, 1938; also sci. papers. Editor, Jour. Washington Acad. Scis., 1944-46; co-editor Indo-Pacific Mollusca, 1959—. Research on taxonomy and geog. distbn. of mollusks, especially marine mollusks of tropical Indo-Pacific region; ecology of shore and reef mollusks of tropical Pacific. Home: 5620 Ogden Rd., Washington 20016. Office: U. S. Nat. Museum, Smithsonian Instn., Washington 20560.*

REHFELD, Carl Ernest, Am. vet. pathologist; b. Warner, S.D., July 1, 1918; s. John William and Ida F. (Angerhofer) R.; student Pasadena City Coll., 1938-40, Colo. State U., 1941-42; B.S., S.D. State U., 1944; D.V.M., Kan. State U., 1947; M.S., U. Minn., 1955; m. Sue Marie Theis, July 26, 1951; children—Jean Suzanne, Ann Kathryn, Patricia Ruth, Richard Carl. Asst. prof. vet. sci. S.D. State U., Brookings, 1947-48; asst. prof. vet. pathology Kan. State U., Manhattan, 1948-50; instr. U. Minn., St. Paul, 1950-55; head clinic sect. Radiobiology Lab., Coll. Medicine, U. Utah, Salt Lake City, 1955-61; asso. veterinarian div. biol. and med. research Argonne Nat. Lab. (Ill.), 1961—. Mem. A.A.A.S., Am. Vet. Med. Assn., Am. Coll. Lab. Animal Medicine, Am. Coll. Vet. Toxicologists, Am. Vet. Radiology Soc., Animal Care Panel, Radiation Research Soc., Conf. Research Workers in Animal Diseases. Research and publs. on clin. and pathol. effects of ionizing radiation in dogs and other mammals, electronic sizing and computer methods in blood cell population studies, dental radiology and periodontal disease; recessive anomalies in inbred dogs, dosimetry in dental radiology.*

REHN, Ludwig Mettler, German physician, surgeon; b. Allendorf, Germany, Apr. 13, 1849; prof. Frankfort, Germany. Performed 1st recorded operation for removal of thyroid in exophthalmic goiter (Basedow's disease), 1880; 1st operation for diseases of esophagus previously considered inoperable, 1884; discovered anilin cancer of bladder in anilin-dye worker, 1895; 1st successful heart operation (cardiac suture of stab wound), 1896. Died Frankfort/Main, May 29, 1930.

REICH, Edgar, mathematician; b. Vienna, Austria, June 7, 1927; s. Jonas and Luna (Lunenfeld) R.; B.E.E., Poly. Inst. Bklyn., 1947; M.S. Mass. Inst. Tech., 1949; Ph.D., U. Cal. at Los Angeles, 1954; m. Phyllis Masten, June 10, 1949; children—Eugene, Frances. Mathematician, Rand Corp., Santa Monica, Cal., 1949-54, 55-56; mem. Inst. for Advanced Study, Princeton, N.J., 1954-55; prof. math., U. Minn., Mpls., 1956—; vis. prof. Technion U., Haifa, Israel, 1965-66. Guggenheim fellow, 1960-61; Fulbright Research scholar Aarhus, Denmark, 1960-61. Mem. Am. Math. Soc., Eta Kappa Nu. Research, publs. on convergence approximate methods for solving problems by computing machines, investigation waiting time on waiting lines, theory functions complex variable. Home: 1961 East River Rd., Mpls. 55414.*

REICH, Ferdinand, German physicist, mineralogist; b. Bernburg, Germany, Feb. 19, 1799; ed. Leipzig, Freiberg, Göttingen (all Germany), Paris; studied under Strohmeyer, Göttingen. Went to Paris where he learned of metric system, 1823; prof physics Freiberg (Germany) Sch. Mines. Discovered (with Richter) element, indium, using spectroscopic analysis; Introduced metric system to Germany; studied temperature of rocks at different depths. Died Freiberg, Apr. 27, 1882.

REICH, Herbert Joseph, Am. elec. engr.; b. Staten Island, N.Y., Oct. 25, 1900; s. Jacques and Caroline (Bellinger) R.; M.E., Cornell U., 1924, Ph.D. in Physics, 1928; m. Anne Elizabeth Evans, Apr. 3, 1926; children—Robert J., Donald E. Instr. physics Cornell U., 1926-29; faculty U. Ill., 1929-46, prof., 1939-46; spl. research asso. Radio Research Lab., Harvard, 1944-46; prof. engring. and applied sci. Yale, New Haven, 1946-63, prof. engring. and applied sci., 1963—. Mem. adv. group on electron tubes Office Sec. Def., 1950-59; mem., tech. com. 39 Internat. Electrotech. Commn., 1960—, chmn. subcom. SC39A, 1965—. Fellow I.E.E.E., Am. Phys. Soc.; mem. Sigma Xi, Tau Beta Pi. Author: Theory and Applications of Electron Tubes, 1939; Principles of Electron Tubes, 1941; (with others) Ultrahigh-Frequency Techniques, 1942; Very High Frequency Techniques, 1947; (with others) Microwave Theory and Techniques, 1953, Microwave Principles, 1957; Functional Circuits and Oscillators, 1961; (with others) Theory and Applications of Active Devices, 1966; also numerous articles. Research on electron devices, electron-device circuits, microwaves, communications. Home: 1021 Benham St., Hamden, Conn. 06514. Office: Dunham Lab., Yale U., New Haven 06520.*

REICH, Nathaniel Edwin, Am. physician; b. N.Y.C., May 19, 1907; s. Alexander and Betty (Feigenbaum) R.; B.S., N.Y. U., 1927; postgrad. Marquette U., 1927-29; M.D., U. Chgo., 1931; m. Joan Finkel, May 23, 1943; children—Andrew, Mathew. Practice medicine specializing in cardiology, Bklyn., 1933—; attending cardiologist Jewish Chronic Disease Hosp., Bklyn., 1949—, Unity Hosp., Bklyn., 1934—; Cons. various hosps.; clin. asso. prof. medicine State U.

N.Y. Downstate Med. Center, 1963—; lectr. numerous univs. U.S., Europe, Asia, Latin Am. Diplomate Am. Bd. Internal Medicine, Am. Bd. Legal Medicine. Fellow A.C.P., Am. Coll. Chest Physicians (sec. cardiovascular rehab. and coronary disease coms. 1960—), Am. Coll. Cardiology, Am. Coll. Angiology (recipient award, 1956, 59); mem. Am. Geriatrics Soc., A.M.A., Kings County Med. Soc., N.Y. Heart Assn., Am. Heart Assn., N.Y. State Soc. Internal Medicine, N.Y. Cardiol. Soc. (pres. 1965—). Author: Diseases of the Aorta, 1949; The Uncommon Heart Diseases, 1954; (with R.E. Fremont) Chest Pain: Systematic Differentiation and Treatment, 1959; also articles. Introduced 2-ethyl hexanol for treatment of left heart failure. Home: 1620 Av. I, Bklyn. Office: 135 Eastern Pkwy., Bklyn.*

REICH, Robert Claude, French physicist, metallurgist; b. Paris, France, Nov. 2, 1929; s. Félix and Nelly (Bélestin) R.; Engr., Ecole Nationale Superieure de Chimie Paris, 1953; Licence ès Sciences Physiques, U. Paris, 1954, Docteur ès Sciences Physiques, 1965. Attaché de recherche Centre Nat. de La Recherche Scientifique, Centre d'Etude de Chimie Métallurgique, Vitry, France, 1953-54, 58-65; chargé de recherche Service de Physique des Solides, U. Orsay, Essonne, France, 1965—. Mem. Société Française de Physique, Société Française de Metallurgie. Research and publs. on purification of metals by zone-melting method, study of thermal conditions of this process, perfected stablzn. of zone-melting length during motion, study of size-effect at low temperature and residual resistivity of metals of various purities; ideal resistivity: temperature squared term in nonmagnetic metal; determination of characteristic Debye temperatures, supraconducting transition by resistivity measurement. Home: 25 St. Emile Zola, Villejuif, 94-Val de Marne, France. Office: Service de Physique des Solides, Bâtiment 210, Faculté des Sci., Orsay, 91-Essone, France.*

REICH, Wilhelm, psychologist, psychoanalyist; b. Austria, 1897; ed. Sch. Medicine, U. Vienna, Vienna Neuropsychiatric Inst. Became asso. Freud's Psychoanalytic Polyclinic, Vienna, worked in Berlin, then came to U. S., 1939; taught at New Sch. for Social Research, 1939-41; founder Orgone Inst., 1942. Author: Der Triebhafte Charakter, 1925; Der Funktion des Orgasmus, 1927; Uber Charakteranalyse, 1928; Character Formation and the Phobias of Childhood, 1930; Characteranalysis, 1933; The Mass Psychology of Fascism, 1933; The Discovery of the Orgone, 1942; Listen, Little Man!, 1948. Disagreed with Freud's conclusions that destructiveness is biol. instinct, and that sexuality drains the individual's energies for other pursuits; stressed methodical treatment of whole personality; made early study applying psychoanalytic approach to sociol. interpretation of national character and personality; developed concept of armoring; discredited through attempt to create an energy restoring device. Died Nov., 1957.

REICHARDT, Werner Ernst, German biologist; b. Berlin, Germany, Jan. 31, 1924; s. Wilhelm and Hedwig (Schütz) R.; dr. engring., Tech. U. Berlin; m. Barbara Lüdecke, May 17, 1958; 1 dau., Andrea. Asst., Fritz-Haber Inst., Berlin/Dahlem, 1952-54; postdoctoral fellow Cal. Inst. Tech., 1954-55; asst. Max Planck Inst. Phys. Chemistry, Göttingen, Germany, 1955-58, Max Planck Inst. Biology, Tübingen, Germany, 1958-60; sci. mem., div. chief, dir. Biol. Inst. Tübingen, 1960—. Mem. Deutsche physikalische Gesellschaft, Gesellschaft für physikalische Biologie, Internat. Brain Research Orgn. Author: Autocorrelation, a Principle for the Evaluation of Sensory Information by the Central Nervous System, 1961; Nervöse Verarbeitung optischer Nachrichten im Facettenauge, 1962. Home: Im Schönblick 42. Office: Spemanstrasse 34, Tübingen, W. Germany.

REICHELDERFER, Francis Wilton, Am. meteorologist; b. Harlan, Ind., Aug. 6, 1895; s. Francis Allen and Mae (Carrington) R.; A.B., Northwestern U., 1917, Sc.D. (hon.), 1939; postgrad. Norway Geophys. Inst., 1931; m. Beatrice Coralyn Hoyle, June 19, 1920; 1 son, Bruce Allen. Chemist, Calumet Co., Chgo., 1918-19; commd. ensign USN, 1918, advanced through grades to comdr., 1938, sci. assignments as meteorol. officer; ret., 1938; dir. U.S. Weather Bur., Washington, 1938-63, cons. internat. meteorology programs, 1963—. Vice chmn. NACA, 1940-58. Recipient Civilian Meritorious Service award USAF, 1947; Internat. Service awards Cuba, 1945, Japan, 1960. Fellow A.A.A.S., Am. Meteorol. Soc., Am. Inst. Aeros. and Astronautics; mem. World (pres. 1951-55, Internat. Meteorology award 1964), N.Am. (pres. 1940-51, 59-63) meteorol. orgns., Am. Meteorol. Soc. (pres. 1939-40, Abbe award 1964), Nat. Acad. Scis., Washington Philos. Soc., Inst. Aero. and Aerospace Scis. (chmn. Washington chpt.), Am. Geophys. Union, Internat. Union Geophysics. Pioneer in design and devel. methods and equipment for atmospheric soundings, especially for naval purposes and in research into air turbulence and winds aloft; responsible, with others, for advancement of space satellites for photographing global cloudiness and storms; weather prediction using modern electronic computers. Home: 3031 Sedgwick St. N.W., Washington 20008. Office: U.S. Weather Bur., Silver Spring, Md. 20910.

REICHENBACH, Georg von, see von Reichenbach, Georg.

REICHENBACH, Hans, mathematician, philosopher; b. Hamburg, Germany, Sept. 26, 1891; s. Bruno and Selma (Menzel) R.; ed. Coll. Engring., Stuttgart, Germany, univs.. Berlin, Munich, Göttingen, Erlangen (all Germany); Ph.D., 1915. Engr. in radio industry until 1920; instr., then prof. Coll. Engring., Stuttgart, 1920-26; prof. philosophy of physics U. Berlin, 1926-33; prof. philosophy U. Istanbul, 1933-38; U. Cal. at Los Angeles, 1938-53. Mem. Am. Philos. Soc., Am. Phys. Soc., Internat. Com. for Unity of Sci. Author: Relativitätstheorie und Erkenntnis a priori, 1920; Axiomatik der relativistischen Raum-Zeit-Lehre, 1924; Pilosophie der Raum-Zeit-Lehre, 1928; Atom and Cosmos, transl. from German, 1931; Zeile und Wege der heutigen Naturphilosophie, 1932; Wahrscheinlichkeitslehre, 1935; Experience and Prediction, 1938; From Copernicus to Einstein, transl. from German, 1942; philosophic Foundations of Quantum Mechanics, 1944; Elements of Symbolic Logic, 1947. Early member of Vienna circle of logical positivists. Died 1953.

REICHENBACH, Baron Karl von, see von Reichenbach, Baron Karl.

REICHENOW, Anton, German ornithologist; b. Charlottenburg (Berlin), Germany, Aug. 1, 1847; dir. zool. mus., Berlin; founder Rossitten bird sanctuary. Author: Vogelbilder aus fernen Zonen. Abbildungen und Beschreibungen der. Papageien, 1878-83; Die Vögel Afrikas, 3 vols., 1900-05; Die Vögel. Handbuch der systematischen Ornithologie, 2 vols., 1913-14. Contb. to ornithol. nomenclature. Died 1941.

REICHENOW, Johann Eduard, German zoologist; b. Berlin, Germany, July 7, 1883; s. Anton and Marie (Cabanis) R.; ed. U. Heidelberg, U. Berlin, U. Munich; Ph.D.; m. Lilly Mudrow, 1946. Became govt. zoologist, Cameroons, Africa, 1913; named dept. chief Tropical Inst, Hamburg, Germany, 1921; dir., Guatemala, Mexico, 1932-33, E. Africa, 1936-37. Mem. German Acad. Natural Scientists (Leopoldina), German Zool. Soc., German Soc. for Hygiene and Microbiology, Soc. Protozoologists (hon.). Author: Grundriss der Protozoologie, 3d edit., 1952; Lehrbuch der Protozoenkunde, 6th edit., 1953. Research in sleeping sickness and parasites of man and domestic animals of Africa. Office: Address: 10 Friedr.-Bayer-Str., Wuppertal-Vohwinkel.

REICHERT, Benno, German pharm. chemist; b. Rathenow, Germany, Apr. 3, 1906; s. Gustav and Martha (Werbelow) R.; Dr.Phil., U. Berlin (Germany), 1932, Dr.Phil.habil., 1935; m. Ursula Schwebs, Mar. 21, 1956; 1 dau., Ingrid Grude. Prof., U. Berlin, 1943-56; prof. dept. pharmacy and food chemistry U. Munich (Germany), 1956—. Recipient Silver medallion Hagen-Bucholz-Stiftung, 1928; prizes Meurer-Stiftung, 1926, 27. Mem. German Chem. Soc., German Pharm. Assn. Author: Harnanalytisches Praktikum, 1944; Hagers Handbuch d. Pharm. Praxis, 1944; Ergänzungsband I, 1944; Die Mannich-Reaktion, 1959; also numerous articles. Editor: Arzneimittelforschung, 1950—. Research in drugs, including mescaline, arbutine, derivatives of khellin, oxy-croton-lactom, oka-stilbene, acid tectronic oxidopyrolidine, nitrostilbene, catalytic reductive of aromatic nitro compounds; addition reactions with hexamethylenetetramine. Home: 15 Richard Strauss, Munich, Bayern, Germany.*

REICHERT, Philip, Am. physician; b. N.Y.C., Mar. 29, 1897; s. Nathaniel and Charlotte (Fischel) R.; A.B., Coll. City N.Y., 1918; M.D., Cornell U., 1923; m. Helen Faith Keane, Sept. 29, 1939; Staff Rockefeller Inst. Med. Research, 1923-26; dir. Witkin Found. Cardiological Research, 1938-40; v.p., dir. profl. div. Doherty, Clifford, Steers & Shenfield Corp., N.Y.C., 1947-58; nat. sec., trustee, gov. Am. Coll. Cardiology, 1949-59, exec. dir., historian, 1959—; dir. research Yorktown Products Corp., N.Y.C., 1956—; dir. course med. advt. and promotion N.Y. U., N.Y.C., 1956-61; practice medicine specializing in cardiology, N.Y.C., 1926—. Exhibit, History of Diagnostic Instrumentation in Cardiology, Smithsonian Instn., Washington, 1951-64. Named Distinguished fellow Am. Coll. cardiology, 1965. Diplomate Nat. Bd. Med. Examiners. Fellow A.C.P. (life), Am. Coll. Legal Medicine, Am. Acad. Compensation Medicine, Am. Coll. Clin. Pharmacology, N.Y. Acad. Medicine. Research and publs. on non-viability of viruses, 1926; lung and heart function test; collection of historic diagnostic med. instruments. Address: 480 Park Av., N.Y.C. 10022.*

REICHSTEIN, Tadeus, chemist; b. Wloclawek, Poland, July 20, 1897; s. Isidor and Gustava (Brokmann) R.; Dr.Ing.Chem., Fed. Poly. Sch. (ETH), Zürich, Switzerland, 1921, Dr.h.c., 1967; Dr.h.c., U. Genf, 1967, U. Abidjan, 1967; U. Paris; m. Louise Henriette Quarles v. Ufford, July 21, 1927; 1 dau., Ruth (Mrs. Straumann Bruno). Became asst. to Prof. H. Staudinger, 1921; began work in industry, 1925; joined faculty Fed. Poly. Inst., 1929, asso. prof., 1937-38; prof., head Pharm. Inst., U. Basle (Switzerland), 1938—, head Inst. Organic Chemistry, 1946-60. Recipient Nobel prize for physiology and medicine, 1950, Marcel-Benoist prize, 1947. Fellow Royal Soc., 1952; mem. Chem. Soc. London (hon.), Nat. Acad. Scis. (fgn. hon.), India Acad. Scis. (hon.), Museum d'Histoire Naturelle Paris (corr.), Royal Irish Acad. (hon.), Am. Acad. Arts and Scis. (fgn.), l'Acad. Royal de Médecin de Belgique (hon. fgn.), N.Y. Acad. Scis., Pharm. Soc. Japan (hon.), Am. Soc. Biol. Chemists, Deutsche Akad. d. Naturforscher Leopoldina. Research and numerous publs. on heterocyclic compounds, sugars, vitamin C, pantothenic acid, hormones of adrenal glands, steroids, especially cardiac glycosides and pregnane derivatives of plant origin. Home: 22 Weissensteinstrasse Ch-4000, Basel, Switzerland.*

REID, Allen Francis, Am. biophysicist; b. Deer River, Minn., July 31, 1917; s. Allen Roy and Rose (Seidel) R.; B. Chemistry, U. Minn., 1940; A.M., Columbia U., 1942, Ph.D., 1943; M.D., U. Tex., 1959; m. Dorothy Mary Cullen, May 31, 1943; children—Sally Anne, David Mark. Dir. radioactivity labs. Columbia U., 1942-46; staff Manhattan Project, 1947-50; prof., chmn. dept. biophysics and phys. chemistry Baylor U. Grad. Research Inst., Dallas, 1947-51; faculty Southwestern Med. Sch., U. Tex., Dallas, 1947-60, prof., 1951-60, chmn. dept. biophysics, 1947-60; prof., dept. biology U. Dallas, 1960—; con. biophysicist, 1947—. Mem. Am. Chem. Soc., Am. Phys. Soc., Am. Physiol. Soc., A.M.A., A.A.A.S., Am. Assn. for Cancer Research, Soc. for Exptl. Biology and Medicine, N.Y. Acad. Scis., Sigma Xi, Phi Lambda Upsilon. Contbr. numerous articles to tech. jours. Patentee in field. Codiscoverer I-125; pioneered demonstration of chem. synthesis by nuclear recoil; invented processes for fractionating proteins, isotopes, desalting water; research in physiol. mechanisms. Home: 3145 Spur Trail, Dallas 75234.*

REID, Charles Edward, Am. chemist; b. Amite, La., Nov. 17, 1917; s. Columbus and Hope (Bidez) R.; student So. La. Coll., 1934-35; B.S., Tulane U., 1939; M.S., La. State U., 1947, Ph.D., 1948; m. Barbara Louise Mylius, Jan. 24, 1948; children—Charles Edward, Jean Barbara, Helen Ann. Chemist, E.I. duPont de Nemours & Co., Baton Rouge, 1940-45; with dept. chemistry, U. Fla., Gainesville, 1948—, asso. prof., 1955—. Mem. Am. Chem. Soc., Sigma Xi, Phi Kappa Phi. Author: Principles of Chemical Thermodynamics, 1960. Publs. on devel. of reverse osmosis for desalination of water, employing cellulose acetate; calculation in quantum mechanics. Home: 11 S.W. 43d Terrace, Gainesville, Fla. 32601.*

REID, Emmet, Am. chemist; b. Fincastle, Va., June 27, 1872; s. Thomas Alfred and Virginia (Ammen) R.; M.A., U. Richmond, 1892, LL.D., 1917; Ph.D., Johns Hopkins, 1898; m. Margaret Kendall, Dec. 28, 1915; children—Emmet Kendall, Alfred Gray, Martha (Mrs. Charles C. Hudson). Faculty, Johns Hopkins, Balt., 1914—, prof. chemistry, 1917-37, prof. chemistry emeritus, 1937—; cons. chem. warfare Dupont Co., 1919-59, Hercules Powder Co., 1926-56, Thiokol Corp., 1936-62, Socony Vacuum Co., 1937-54. Recipient Herty medal, 1947. Mem. Am. Chem. Soc., Am. Inst. Chemists, Phi Beta Kappa. Author: Introduction Organic Research, 1924; College Organic Chemistry, 1929; Invitation to Chemical Research, 1961; Organic Chemistry of Bivalent Sulfur, 1958; (with P.H. Emmett) Catalysis Then and Now, 1965. Contbr. numerous articles to profl. jours. Research, publs. on organic sulfur compounds; identification of acids; phthallic esters as plasticizers; chloroacetophenone, chief constituent of tear gas. Home: 203 E. 33d St., Balt. 21218.*

REID, Evans Burton, chemist; b. Brock Twp., Ont., Can., Mar. 29, 1913; s. William Thomas and Ethel Elizabeth (Burton) R.; B.Sc. with first class honors, McGill U., 1937, Ph.D. cum laude, 1940; m. Isabel Sue Lewin, Apr. 2, 1942 (deceased November 3, 1962); 1 son, Nicholas Evans David; m. second, to Dorothy Pearson, on Aug. 12, 1963. Came to United States, 1941, naturalized, 1944. Grad. asst. chem. engring. McGill U., 1937-38, demonstrator organic chemistry, 1938-40; research chemist Dominion Tar & Chem. Co., Montreal, 1940-41; instr. chemistry Middlebury Coll., 1941-43, asst. prof., 1943-46; asst. prof. chemistry Johns Hopkins, 1946-54; cons. Tainton Products, Balt., 1951-54; Merrill prof. chemistry, chmn. dept. Colby Coll., since 1954—, acting dean of the faculty, since 1967—, director College National Science Foundation Summer Science Inst., 1959—; consultant to National Sci. Found., 1963—. Smith-Mundt vis. prof. chemistry U. Baghdad, 1960-61. Member of screening panel for National award in pure chemistry, 1953. Recipient J. Shelton Horsley award Va. Acad. Scis. (with Albert W. Lutz), 1955. Mem. Am. Chem. Soc. (exec. com. Md. sect. 1951-54, chmn. Me., 1956-57, 62). Chem. Soc. London, Sigma Xi. Contbr. profl. jours and encys. Research on plant growth hormones; chemistry of tetronic acids, neurospora, sesquiterpenes, cyclobutanediones and dimeric ketenes. Home: 11 Highland Av., Waterville, Me. 04901.

REID, George Colvin, physicist; b. Edinburgh, Scotland, Sept. 13, 1929; s. George Colvin and Elizabeth (Gibb) R.; B.Sc. with honors in Physics, Edinburgh U., 1950, Ph.D., 1954; m. Joan Alison Tingley, Sept. 1, 1956; children— Colin, Ingrid, Brian. Came to U.S., 1963. NRC fellow, Ottawa, Ont., Can., 1953-54; sci. officer Def. Research Telecommunications Establishment, Ottawa, 1954-58, cons. radio physics, 1960-62; asso. prof. geophysics Geophys. Inst., U. Alaska, 1958-60; with Nat. Bur. Standards (now

ESSA), Boulder, Colo., 1963—, asst. dir. research Space Disturbances Lab., 1966—. Mem. U.S. Commn. III, Internat. Sci. Radio Union, 1965—. Mem. Am. Geophys. Union, Canadian Assn. Physicists. Contbns. to interpretation of physics of the high-latitude ionosphere, especially concerning the phenomenon of polar-cap absorption; contbns. to theories of aurora of interplantary diffusion of solar protons, and electrodynamics of the ionosphere. Home: 971 Crescent Dr. Office: ESSA, Boulder, Colo. 80302.*

REID, George Kell, Am. biologist; b. Fitzgerald, Ga., Mar. 23, 1918; s. George K. and Pauline (Bowles) R.; B.S., Presbyn. Coll. S.C., 1940; M.S., U. Fla., 1949, Ph.D., 1952; m. Eugenie Chazal, July 23, 1949; children—Deborah Louise, George Philip. Mem. faculty Texas A. and M. U., 1953-56, Rutgers U., 1956-60; prof. biology Fla. Presbyn. Coll., St. Petersburg, 1960—; cons. to various state and pvt. orgns., 1952—. Fellow A.A.A.S.; mem. Fla. Acad. Scis. (pres. 1965), Ecol. Soc. Am., Am. Soc. Limnology and Oceanography, Internat. Soc. Theoretical and Applied Limnology, Sigma Xi, Phi Sigma (award 1952). Author: Ecology of Inland Waters and Estuaries, 1961; (with R.B. Platt) Bioscience, 1967; Pond Life, 1967. Research and publs. on the relationships of organisms and environment in Tex. bays and Fla. lakes; enumerated the fishes and outlined their ecology in shallow Gulf of Mexico at Cedar Key, Fla.; investigated and described how salinity changes in an estuary relate to certain fishes. Home: 2928 DeSoto Way S., St. Petersburg, Fla. 33712.*

REID, Harry Fielding, Am. geologist; b. Baltimore, Md., May 18, 1859; s. Andrew and Fanny Brooke (Gwathmey) R.; A.B., Johns Hopkins, 1880; K.E., Pa. Mil. Acad., 1876; Ph.D., Johns Hopkins, 1835; studied in Germany and England, 1884-86; m. Edith Gittings, Nov. 22, 1883; children—Francis Fielding, Doris Fielding. Prof. mathematics, 1886-89, physics, 1889-94, Case School of Applied Science, Cleveland; lecturer Johns Hopkins, 1894-96; asso. prof. physical geology, U. of Chicago, 1895-96; asso. prof., Johns Hopkins U., 1896-1901, prof. geol. physics, 1901-11, prof. dynamical geology and geography, 1911-30, prof. emeritus since 1930. Chief of highway div., Md. Geol. Survey, 1898-1905; spl. expert in charge of earthquake records, U. S. Geol. Survey, 1902-14. Mem. Commn. Internationale des Glaciers; rep. of U. S. in the Internat. Seismol. Assn. since 1906; hon. mem. Société Helvétique des Sciences Naturelles; corr. mem. Phila. Acad. of Natural Sciences; fellow Geol. Soc. America, Am. Phys. Soc., Washington Acad. Sciences; mem. Nat. Acad. Sciences. Am. Philos. Soc., Seismol. Soc. America (pres. 1913), Am. Geophys. Union (chmn. 1924-26). Author: Parts vi, vii, viii of Highways of Maryland, 1899. Joint author: (with A. N. Johnson) Second Report on the Highways of Maryland, 1902; Vol. II of Report of Calif. State Earthquake Investigation Commn., 1910; also several reports and articles on glaciers, earthquakes, etc. Mem. com. Nat. Acad. Sciences apptd. at request of the President to report on the possibility of controlling the Panama slides, 1915. Authority on dynamic geology, seismology, and glaciology. Died June 18, 1944.

REID, John Thomas, Am. nutritionist; b. Cumberland, Md., Mar. 14, 1919; s. Frank Ernest and Sarah (Tipton) R.; B.S., U. Md., 1941; M.S., Mich. State U., 1943, Ph.D., 1946; m. Alice Evelyn Smalley, June 8, 1945; children—J. Douglas, Sarah R.; Nancy S., Gwen L., Cindy A. Asst. prof. Mich. State U., 1944-45; asso. prof. Rutgers U., 1945-48; faculty Cornell U., Ithaca, N.Y., 1948—, prof. nutrition, 1951—, head div. nutrition and physiology, 1963—; vis. scientist Reading (Eng.) U. 1955-56. Cons., U.S. Dept. Agr., FAO, Nat. Acad. Sci. coms. Guggenheim fellow, 1955-56. Mem. Am. Dairy Sci. Assn. (nutrition award 1950, Borden award 1957; dir. 1960-63), Am. Soc. Animal Sci. (Morrison award 1967), Am. Inst. Nutrition, Brit. Soc. Nutrition. Contbr. numerous articles to profl. jours., chpts. in books. Discovered methods of measuring chem. composition, digestibility, amount of herbage ingested by grazing ruminants; quantified, systematized chem. composition of body of pigs, sheep, cattle, rats; demonstrated dietary characteristics affecting metabolism and energetic efficiency of animals, nutritional effects on reprodn. Home: 105 Sheldon Rd., Ithaca, N.Y. 14850.*

REID, Roger Delbert, Am. microbiologist; b. Lamont, Ia., July 31, 1905; s. George Sawyer and Florence (McCormack) R.; A.B., U. S.D., 1927; M.S., Pa. State U., 1931, Ph.D., 1935; m. Erma House, June 7, 1933; children—Donald House, Barry House. Prin., tchr. Gann Valley (S.D.) High Sch., 1927-29; instr. sci. U. S.D. High Sch., 1929-30; mem. faculty Pa. State Coll., 1931-36, U. Ida., 1936-37, Johns Hopkins, 1937-45; research biologist Hynson, Westcott and Dunning, Inc., Balt., 1946-48; head, microbiology br. Office Naval Research, Washington, 1948-56, dir. biol. scis. div., 1956-67; prof. biology U. W. Fla., Pensacola, 1967—. Mem. div. biology and agr. NRC, 1959—; biology cons. The Sci. Tchr.; mem. biology panel World Book Ency. Diplomate Am. Bd. Microbiology in Pub. Health and Med. Lab. Microbiology. Fellow A.A.A.S., Am. Acad. Microbiology (sec. bd. govs. 1958-60); mem. Am. Soc. for Microbiology (pres. Md. br. 1947), Soc. for Indsl. Microbiology, Am. Inst. Biol. Scis. (governing bd. 1958—), Soc. for Gen. Microbiology, Sigma Xi, Phi Sigma, Gamma Sigma Delta, Lambda Chi Alpha. Author: Laboratory Manual in Bacteriology, 1933; (with M.J. Pelczar, Jr.) Microbiology, 1958. Editor: (with M.G. Sevag) Resistance to Toxic Agents, 1955. Mem. editorial bd. Biosci., 1962-66. Pioneer in penicillin research in U.S., 1930—; subsequent work on chemotherapeutic drugs and antibiotics; patentee aluminum penicillin. Home: 6219 Vicksburg Dr., Oakfield Acres, Pensacola, Fla. 32503.*

REID, Rolland Ramsay, Am. geologist; b. Wilbur, Wash., Nov. 12, 1926; s. Max S. and Hazel (Covert) R.; B.S., U. Wash., 1951, M.S., 1953, Ph.D., 1959; m. Eileen Harris, Sept. 2, 1947; children—Robin Anne, Rolland Robert. Instr., Mont. Sch. Mines, 1953-55; mem. faculty U. Idaho, Moscow, 1955—, prof. geology, dean Coll. Mines, dir. Ida. Bur. Mines and Geology, 1965—. Fellow Geol. Soc. Am., Geol. Soc. London; mem. Soc. Econ. Geologists, Sigma Xi. Publs. on geologic mapping; research in igneous and metamorphic petrology and structural geology. Home: Route 1, Box 43, Moscow, Ida. 83843.*

REID, Walter Phillip, Am. physicist; b. Cleve., Aug. 8, 1916; s. Sidney Leonard and Edith Mary (Steele) R.; B.S., Carnegie Inst. Tech., 1938, M.S., 1939; Ph.D., U. Pitts., 1942; m. Elizabeth Randall MacKay, Nov. 25, 1948; children—Brian Keith, Russell MacKay, Bruce Harold, Harvey Douglas, Glenn Charles, Lucille Jean. With U.S. Rubber Co., Detroit, 1942-43; faculty N.Y. U., 1944-45, Purdue U., 1943-44, 45-49, USAF Inst. Tech., Wright Field, O., 1949-50, U.S. Naval Ordnance Test Sta., China Lake, Cal., 1950-55; staff theoretical div. Los Alamos Sci. Lab., 1955-56; asso. prof. math. Mich. State U., 1956-61; research mathematician U.S. Naval Ordnance Lab., Silver Spring, Md., 1961—; vis prof. math Am. U., Washington, 1966—; also at various univs. summers. Mem. Am. Assn. Physics Tchrs., Am. Math. Soc., Math. Assn. Am., Am. Soc. M.E., Soc. Indsl. and Applied Math., Sigma Xi, Phi Kappa Phi, Sigma Pi Sigma, Pi Mu Epsilon. Research and publs. on solutions to a number of problems in heat conduction: an analysis of stability of a towed object, the motion of a missile under water; test for validity of sedimentation results. Home: 9210 Willow Lane, Adelphi, Md. 20783. Office: Math. Dept., U.S. Naval Ordnance Lab., Silver Spring, Md. 20910.*

REID, Willard Malcolm, Am. parasitologist; b. Ft. Morgan, Colo., Oct. 9, 1910; s. Willard and Caroline (Riggs) R.; B.S., Monmouth Coll., 1932, D.Sc., 1959; postgrad. Heidelberg (Germany) U., 1933; M.S., Kan. State Coll., 1937, Ph.D., 1941; postgrad. Brown U., 1937-38, U. Mich., 1939; m. Janet Helen Sharp, Sept. 4, 1937; children—Caroline (Mrs. Ted R. Ridelhuber), Donald Malcolm, Willard Sharp, Nancy Jane, Judith Ann. Instr., Assiut (Egypt) Coll., 1933-35; prof. biology, dept. head Monmouth Coll., 1937-51; Fulbright Research prof. U. Cairo (Egypt), 1951-52; head poultry unit U.S. State Dept. AID, 1952-55; parasitologist poultry sci. dept. U. Ga., Athens, 1955—, Distinguished Alumni Found. prof., 1963—. Mem. Am. Soc. Parasitologists, Am. Soc. Zoologists, Poultry Sci. Assn., Am. Assn. Avian Pathologists, Am. Soc. Protozoologists, Contbr. numerous articles to tech. jours. Research on morphology and physiology of poultry cestodes, parasite control programs in U.S. and Egypt; developed immunity and poultry flock parasite control; growth and devel. of individual species of parasites under germfree conditions. Home: 240 Burnett St., Athens, Ga. 30601.*

REID, Sir William, Brit. mil. engr.; b. Kinglassie, Scotland, Apr. 25, 1791; s. James and Alexandrina (Fyers) R.; student Royal Mil. Acad., 1806-10; studied surveying under William Mudge; m. Sarah Bolland, Nov. 5, 1818; 5 daus., including Charlotte Cuyler (Lady Neville Chamberlain). Served under Wellington in Portugal, 1810-14; advanced from lt. to majgen., 1856; with Packenham expdn. against New Orleans, 1815, occupation of Paris, 1815; adj. Royal Sappers and Miners, Woolwich, Eng., 1816; with Exmouth expdn. against Algiers, 1816; with ordnance survey of Ireland, 1824; sent to W.I., 1831, helped rebuild govt. bldgs. destroyed during hurricane; fought in Spain, 1835-36; gov. Bermuda Islands, 1839-46; gov.-in-charge Windward W.I. Islands, 1846-48; commanding royal engr., Woolwich, 1849; gov, comdr.-in-chief, Malta, 1851. Chmn. exec. com. Gt. Exhbn. of 1851, London, 1850. Fellow Royal Soc., 1839, v.p., 1849; Inst. Civil Engrs. Author: (pamphlets) Defence of Fortresses, 1823, Defence of Towns and Villages, 1823; An Attempt to Develop the Law of Storms by Means of Facts, Arranged According to Place and Time, and Hence To Point out a Cause for the Variable Winds, 1838, 3d edit., 1850; the Progress of the Development of the Law of Storms and of the Variable Winds, with the Practical Application of the Subject to Navigation, 1849; also papers. Editor: Narrative, written by Sea-Commanders, illustrative of the Law of Storms and of its Practical Application to Navigation, 1851. Research on storms, gave gen. rules (law of storms) for seamen; as colonial gov. established schs., pub. libraries, introduced improved agrl. methods. Died London, Oct. 31, 1858.

REID, William Hill, Am. mathematician; b. Oakland, Cal., Sept. 10, 1926; s. William MacDonald and Edna (Hill) R.; B.S., U. Cal. at Berkeley, 1949, M.S., 1951; Ph.D., U. Cambridge (Eng.), 1955; A.M., Brown U., 1961; m. Elizabeth Mary Emily Kidner, May 26, 1962; 1 dau., Margaret Frances. Lectr., Johns Hopkins 1955-56; NSF fellow Yerkes Obs., Williams Bay, Wis., 1957-58; faculty Brown U., 1958-63, asso. prof. applied math., 1961-63; faculty U. Chgo., 1963—, prof. applied math., 1965—. Cons., Gen. Motors Research Labs., 1960—. Fellow Cambridge Philos. Soc.; mem. Am. Math. Soc., Am. Phys. Soc., Am. Astron. Soc., Sigma Xi. Contbr. articles to tech. jours. Research in applied maths., fluid mechanics, hydrodynamic stability and turbulence; applications to astrophysics and geophysics. Home: 5619 S. Dorchester Av., Chgo. 60637.*

REID, William Thomas, Am. mathematician; b. Grand Saline, Tex., Oct. 4, 1907; s. David Garfield and Laura (Burton) R.; B.A., Hardin-Simmons U., 1926, Sc.D., 1943; M.A., U. Tex., 1927, Ph.D., 1929; m. Frances Idalia Steere, July 8, 1929. Nat. Research fellow U. Chgo., 1929-31, faculty, 1931-44, asso. prof., 1942-44; prof. Northwestern U., 1944-59; prof., head dept. math. State U. Ia., Iowa City, 1959-64; Phillips prof. math. U. Okla., Norman, 1964—; vis. prof. U. P.R., 1946, U. Cal., Los Angeles, 1957-58; staff mem. Sandia Corp., Albuquerque, 1952. Cons., USAF, 1943-44. Fellow A.A.A.S.; mem. Am. Math. Soc. (asso. editor Trans. 1944-49, councilor 1950-53), Math. Assn. Am., Soc. Indsl. and Applied Math. (vis. lectr. 1964-66), Soc. Math. de France, Sigma Xi. Research and publs. on calculus of variations and differential equations, especially interrelations between variational principles and qualitative nature of solutions of differential equations. Home: 1213 Barkley Av., Norman, Okla. 73069.*

REIF, Frederick, physicist; b. Vienna, Austria, Apr. 24, 1927; s. Gerschon and Clara (Gottfried) R.; came to U.S., 1941, naturalized, 1946; B.A., Columbia, 1948; M.A., Harvard, 1949, Ph.D., 1953. Staff, Lincoln Lab., Mass. Inst. Tech., 1952-53; faculty U. Chgo., 1953-60, asst. prof., 1955-60; faculty U. Cal. at Berkeley, 1960—, prof. physics, 1964— Alfred P. Sloan Research fellow, 1955-59; Miller Research prof., 1964-65. Fellow Am. Phys. Soc.; mem. A.A.A.S. Author: Fundamentals of Statistical and Thermal Physics, 1965; also articles. Research in solid state and low temperature physics. Home: 60 Codornices Rd., Berkeley, Cal. 94708.*

REIFER, Ignacy, Polish biochemist; b. Chrzanow, Poland, June 30, 1909; s. Henryk and Matylda (Reifer) R.; Dipl. engr. chemistry, Politech. Coll., Brna, Czechoslovakia, 1931, D.Sc., 1932; m. Erna Ulreich, Aug. 13, 1937; children—Martin, Peter. Asst., Inst. Exptl. Pathology, Vienna, 1933-37; biochemist Dept. Sci. and Indsl. Research, Palmerston North, New Zealand, 1938-49; prof. plant biochemistry U. Agr., Warsaw, Poland, 1949-63; prof. plant biochemistry Inst. Biochemistry and Biophysics, Polish Acad. Sci., Warsaw, 1953—, dep. dir., 1963—. Decorated Officers Cross Polonia Restituta; recipient 3d State prize, 1953, 2d prize Atomic Energy Commn., 1963. Fellow New Zealand Inst. Chemistry; mem. Polish Acad. Sci. (corr.), Biochem. Soc. Poland. Research and numerous publs. on alkaloids of ryegrass and lupines, ornithine cycle in plant material, alternative path of pyrimidine nucleotides in plant material; discovered new alkaloid in ryegrass and lupines, 2 natural enzyme inhibitors. Home: ul. Wiejska 15/3, Warsaw. Office: ul. Rakowiecka 36, Warsaw, Poland.*

REIFF, Ferdinand, German chemist; b. Cologne, Germany, June 25, 1897; s. Franz and Agathe (Decker) R.; Ph.D, univs. Leipzig (Germany) and Bonn (Germany); m. Leni Erb, Sept. 26, 1936; children—Helmut, Ursel, Gisela, Günther, Klaus, Birgit, Ferdinand, Peter. Chemist, Goldschmidt Labs., Essen Germany, 1922-25, Staatshüttenlaboratorium, Hamburg, Germany, 1925-26; successively asst., instr, prof. univs. Konigsberg (Germany) and Marbourg (Germany), 1927-41; dir. exptl. labs. synthetic matter Zellstoffabrik Waldhor, Mannheim, Germany, 1941-60; dir. Central Benckiser Labs., Ludwigshafen, Germany, 1960—; asso. prof. U. Berlin (Germany), 1941-45, U. Heidelber (Germany), 1947—. Mem. German Chem. Soc., Soc. Naturalists and German Doctors, Soc. Physiol. Chemistry, Deutsches Atomforum, T.A.P.P.I. Author articles on inorganic chemistry, wood, cellulose, paper, decontamination. Editor: Die Hafen, 1960; coeditor: Chemie-Labor-Betrieb; collaborator Enzyklopädie der technischen Chemie, Handbuch der Lebensmittelchemie. Home: Leinpfad 30, Mannheim-Sandhofen, W. Germany. Office: Chem. Fabrik J. A. Benckiser, Ludwigshafen, W. Germany.

REIFFEL, Leonard, Am. physicist; b. Chgo., Sept. 30, 1927; s. Carl and Sophie (Miller) R.; B.S., Ill. Inst. Tech., 1947, M.S., 1948, Ph.D., 1953; m. Judith Eve Blumenthal, June 20, 1952 (div. Feb. 1962); children—Evan Carl, David Lee. Physicist, Perkin-Elmer Corp., 1948, U. Chgo., 1948-49; physicist IIT Research Inst. (formerly Armour Research Found.), 1949-51, supr. nuclear physics, 1951-56, dir. physics research, 1956-63, dir. Astro Scis. Center, v.p., 1963-65; cons. Apollo Program Office, NASA Hdqrs., 1965—; chmn. bd. Instrnl. Dynamics Inc.,

1964—; sci. editor WBBM radio sta., Chgo., 1962—; network sci. cons. CBS, 1967—. Cons., Parliamentarians' Confs., NATO, 1962-64; mem. adv. com. on isotope and radiation devel. AEC, 1959-66. Fellow Am. Phys. Soc.; mem. Nat. Acad. Sci. (mem. com. on research reactors 1960-62), Council for Study of Mankind (dir. 1962—), Am. Inst. Aeros. and Astronautics, Am. Nuclear Soc., Am. Geophys. Union, Am. Astron. Soc., Marine Tech. Soc. Research, patents and publs. in nuclear physics, solid state physics, high altitude weapons effects, space physics; invented isotopic low energy X-ray sources, imaging detectors, radiation gauging devices; responsible for devel. 1st indsl. nuclear research reactor. Home: 62 E. Division St., Chgo. 60610. Office 166 E. Superior St., Chgo. 60611.

REILLEY, Edward Michael, Am. physicist; b. Ellwood City, Pa., Apr. 15, 1919; s. Edward Michael and Jeanne Lena (Blandine) R.; student Geneva Coll., 1936-37; B.Sc. in Physics, Carnegie Inst. Tech., 1940; Ph.D., U. Pitts., 1951; m. Mary Elizabeth Davidson, Apr. 7, 1946; children—Edward Michael, IV, Kathleen Patricia, Linda Maureen, Laura Lee. Research asst. Mellon Inst., Pitts., 1940-42; staff U. Pitts., 1946-51, research asso., 1949-51; con. research and devel. bd. U.S. Dept. Def., 1949-51; with Signal Corps. Engring. Lab., Ft. Monmouth, N.J., 1951-58, chief math. and metrology br., 1954-55, asst. dir. for research, 1955-58; dir. Inst. for Exploratory Research, Ft. Monmouth, 1958-64; asst. dir. def. research and engring. Office of Sec. of Def., 1964-66; dir. research and devel. Post Office Dept., 1966—. Chmn., Army Sci. Adv. Group on Physics, 1958-64; mem. adv. bd. Sci. Information Exchange, 1964—. Recipient Army Meritorious Service award, 1964. Mem. Am. Phys. Soc., A.A.A.S., Am. Geophys. Union, Am. Geol. Inst., Sigma Xi, Sigma Pi Sigma, Alpha Tau Omega. Designed automatic radar controls, plate glass measuring instruments, cyclotron apparatus, nuclear resonance devices for measuring magnetic fields, aircraft flame detection system; research on nuclear scattering by light elements, nuclear detection, computers, quantum electronics. Home: 8420 Wendell Dr., Alexandria, Va. 22308. Office: Post Office Dept., Washington 20004.*

REIMANN, Hobart Ansteth, Am. physician; b. Buffalo, Oct. 31, 1897; s. George and Ottillia (Ansteth) R.; M.D., U. Buffalo, 1921; m. Cecilia DeMise, Sept. 5, 1950; children—George A., William P. Asst., Rockefeller Inst., 192-26; NRC fellow, Prague, Czechoslovakia, 1926-27; asso. prof. medicine Peking Union Med. Coll., 1927-30; prof. medicine U. Minn., 1930-36, Jefferson Med. Coll., 1936-51, U. Indonesia, 1956-57, Pahlevi Med. Sch., Shiraz, Iran, 1958-60; vis. prof. medicine Am. U., Beirut, 1952-54; prof. Hahnemann Med. Coll., Phila., 1960—. Area cons. in medicine VA, 1963-; field dir. A.M.A. Vietnam Med. Sch. Project, 1967. Decorated Order of Cedars (Lebanon); recipient Charles V. Chapin medal City of Providence, 1950; cited by U. Buffalo, 1950. Mem. Assn. Am. Physicists, Am. Soc. Clin. Investigation, Am. Soc. Tropical Medicine, Soc. Exptl. Pathlogy, A.M.A., A.C.P., Soc. Biol. Rhythm, Sigma Xi, Alpha Omega Alpha. Author: Treatment in General Medicine, 4 vols., 1936; Pneumonias, 1938; Pneumonias, 1954; Periodic Disease, 1963. Pioneered description of viral pneumonia, viral dysentery, periodic diseases, crystallosis; pioneered description of microbial genetics of Pneumococcus, Tetragenus. Home: 125 Old Gulph Rd., Wynnewood, Pa. 19096. Office: 235 N. 15th St., Phila. 19102.*

REINACH, Salomon, French archeologist; b. St. Germain-en-Laye, France, Aug. 29, 1858; s. Hermann Joseph Reinach; ed. Ecole Francaise D'Athènes. Sec., Archaeol. Commn. Tunis, Tunisia, 1882-85; joined Mus. Nat. Antiquities, St.-Germain-en-Laye, 1886; asst. prof. archeology École de Louvre, 1890-92; named asso. curator Nat. Museums, 1893, dir., 1902. Mem. Acad. des Inscriptions et Belles-Lettres (became pres. 1906), Brit. Acad. Author: Manuel de philólogie classique, 1880-84; Grammaire Latin; Exploration scientifique de la Tunisie, 3 vols., 1884-88; (with E. Pottier) La Necoropole de Myrina, 2 vols., 1886-88; Chronique d'Orient, 1891-96; Description raisonnée du musée de St.-Germain, 1894; Repertoire de la statuaire grecque et romaine, 3 vols., 1897-1904; Catalogue illustré du musée de St.-Germain, 1899; Répertoire des vases peints grecs et étrusques, 2 vols., 1899-1900; Apollo: Histoire générale des arts plastique, 1904; Cultes mythes et religious, 5 vols., 1905-23; Orpheus des relief grecs et romains, 3 vols., 1909-12; Chronologie de la guerre, 10 vols., 1915-19; A Short History of Christianity, 1922; Ephé merides de glozel, 2 vols., 1928-30; also numerous articles. Important archeol. discoveries include Myrina, nr. Smyrna, 1880-82, Cyme, 1881, Thasos, Imbros, Lesbos, 1882, Carthage, Meninx, 1883-84, Odessa, 1893. Died Paris, Nov. 4, 1932.

REINBERG, Alain Elie, French physiologist; b. Paris, Nov. 4, 1921; s. Arthur and Genia R.; Docteur ès Sciences, U. Paris, 1952, Docteur en Medecine, 1954; m. Marie-Anne Spanjaard, Sept. 12, 1945; children—Olivier, Agnes. Research head Centre National de la Recherche Scientifique, Paris, 1961—. Mem. French socs. for endocrinology (gen. sec. 1963—), allergy, physiology; N.Y. Acad. Scis. Author: (with J. Ghata) Biological Rhythms (English transl.), 1964; also articles. Study of several circadian (about 24 hours) rhythms in man during and after isolation (free-run expts.); extension to man of concept of hours of changing resistance or susceptibility (Franz Halberg); circadian changes in responsiveness to drugs or agts. in adult men and women. Home: 78 Avenue de Versailles (75) Paris 16. Office: Laboratoire de Physiologie, Fondation A. de Rothschild, 23 rue Manin (75), Paris, France.*

REINEKE, Ezra Paul, Am. physiologist; b. Waseca, Minn., Jan. 5, 1909; s. Henry William and Alvina (Fehmer) R.; B.S. with distinction, U. Minn., 1934; M.A., U. Mo., 1936, Ph.D., 1942; m. Lillian Elizabeth Vedness, Aug. 17, 1935; children—Barbara Joan (Mrs. Norman A. Spring), Richard Paul. Asst. instr. U. Mo., 1934-36, instr., 1938-42, asst. prof., 1942-45; asst. prof. Mich. State U., East Lansing, 1945-46, asso. prof., 1946-47, prof., 1948—. Recipient Borden award for research in dairy prodn., 1946, Sr. award for meritorious research Sigma Xi, 1959. Research in synthesis of thyroactive iodinated proteins, thyroid physiology, energy metabolism, body composition; designed methods for measuring thyroid secretion in living animals. Home: 680 Gunson St., East Lansing, Mich. 48823.*

REINER, Irving, Am. mathematician; b. Bklyn., Feb. 8, 1924; s. Max and Mollie (Bolotin) R.; B.A., Bklyn. Coll., 1944; M.A., Cornell U., 1945, Ph.D., 1947; m. Irma Ruth Moses, Aug. 22, 1948; children—David, Peter. Mem. Inst. for Advanced Study, Princeton, N.J., 1947-48, 54-56; faculty U. Ill., Urbana, 1948-54, 56—, prof. math., 1958—; with U. Paris (France), 1962-63. Guggenheim fellow, 1962-63; recipient Distinguished Alumnus award Bklyn. Coll., 1963. Mem. Am. Math. Soc. (editor procs. 1966—), Math. Assn. Am., Sigma Xi, Pi Mu Epsilon (councillor). Author: (with C.W. Curtis). Representation Theory of Finite Groups and Associative Algebras' 1962; also numerous articles. Research on integral group representations, auto-morphisms of matrix groups, matrix theory, homological algebra. Home: 101 E. Michigan Av., Urbana, Ill. 61801.*

REINER, Leopold, physician; b. Leipzig, Germany, Jan. 22, 1911; s. Joel and Golda (Michlewitsch) R.; student U. Freiburg (Germany), 1930, U. Leipzig, 1930-33; M.D., U. Vienna (Austria), 1936; m. Lillian Irene Myers, Sept. 22, 1946. Came to U.S., 1939, naturalized, 1945. Dir. Pathology Bronx Hosp., N.Y.C., 1956-65; dir., chmn. dept. pathology Bronx-Lebanon Hosp. Center, N.Y.C., 1956—. instr. pathology Harvard Med. Sch., Boston, 1950-53, clin. asso. pathology, 1953-56; vis. asso. prof pathology Albert Einstein Coll. Medicine, 1956—. Fellow N.Y. Acad. Medicine; mem. Am. Assn Pathologists and Bacteriologists, A.A.A.S., Internat. Acad. Pathology, Harvey Soc., Histochem. Soc., N.Y. Acad. Scis., A.M.A. Research and publs. on weight of human heart in normal and path. states, interarterial coronary anastomoses in neonatal hearts of man and animals, mesenteric arterial circulation, anatomy and pathology including technique of postmortem examination. Home: 277 Old Colony Rd., Hartsdale, N.Y. 10530. Office: 1276 Fulton Av., Bronx, N.Y. 10456.*

REINER, Markus, civil engr.; b. Czernowitz, Austria, Jan. 5, 1886; s. Ephraim and Rose (Altmann) R.; Dr. techn., U. Vienna, (Austria), 1914; m. Rebecca Schoenfeld, 1929 (dec. Nov. 1962); children—Ephraim, Hannah Reicher, Dorit Galil, Shlomit Ilan. With Austrian State Rys., Dist. Czernowitz, Austria, 1911-19, Roumanian State Rys., Dist. Cernauti, Roumania, 1919-22, pub. works dept. Govt. of Palestine, Jerusalem, 1922-47, Standard Instn., Tel Aviv, Palestine, 1947-49; prof. mechanics Israel Inst. Tech., Haifa, 1947—; research prof. Lafayette Coll., Easton, Pa., 1930-32; lectr. rheology Princeton, 1931-32, Hebrew U., Jerusalem, 1941, Bklyn. Inst. Tech., N.Y.C., 1963-64. Recipient Weizmann prize. Twp. of Tel Aviv, 1956, Israel prize, 1960, Rothschild-prize, 1962. Mem. Israel Acad. Scis. and Humanities, Israel Soc. Theoretical and Applied Mechanics. Author: Lectures on Theoretical Rheology, 1960; Deformation, Strain, and Flow, 1960; also numerous articles. Formulated equations for plastic flow in tubes and between rotating cylinders; assisted Bingham in establishing rheology as separate branch of physics; inaugurated considerations of second order effects in elasticity and fluid dynamics. Home: 21 Kirat Sefer St. Office: Technion City, Haifa, Israel.*

REINERT, Karl-Ernst Wilhelm, German physicist; b. Gräfenhain, Germany, Feb. 12, 1929; s. Alfred Martin and Emma (Schaller) R.; diplom U. Jena, 1955, Dr. rer. nat., 1961; m. Gertrud Wotzel, June 10, 1955; 1 son, Hans-Martin. Asst., Phys. Inst., U. Jena (Germany), 1955-61; with dept. phys. chemistry, Inst. Microbiology and Exptl. Therapeutics, German Acad. Scis., Jena, 1961—. Mem. Phys. Soc., Biophys. Soc., Soc. German Scientists and Physicians. Research, publs. on microwave spectroscopy of aromatics, reation kinetics and polarography, phys. chemistry of DNA, influence of polydispersity on mean values of phys. parameters of macromolecules. Home: 34 Leo-Sachse-Strasse. Office: 11 Beuthenbergstrasse, Jena, East Germany.*

REINES, Frederick, Am. physicist; b. Paterson, N. J., Mar. 16, 1918; s. Israel and Gussie (Cohen) R.; M.E., Stevens Inst. Tech., 1939, M.S., 1941; Ph.D., N.Y. U., 1944; Sc.D. (hon.), U. Witwatersrand (ohannesburg, S. Africa); m. Sylvia H. Samuels, Aug. 30, 1940; children—Robert G., Alisa K. Staff mem., group leader Los Alamos Sci. Lab., 1944-59, dir. operation greenhouse, 1951; prof., head physics dept. Case Inst. Tech., Cleve., 1959-66; prof. physics, dean phys. scis. U. Cal. at Irvine, 1966—. Mem. adv. com. on electrophysics NASA, 1962—; mem. Fulbright physics screening com., 1962—. Guggenheim fellow, 1958-59; Sloan fellow, 1959-63. Fellow Am. Phys. Soc., A.A.A.S. Contbr. articles to tech. jours. Research on effects and physics of nuclear weapons, of detections of free antineutrino interactions, scintillation detection techniques for elementary particles, cosmic rays, whole body counting; co-discoverer of neutrinos. Home: 2655 Basswood St., Newport Beach, Cal. 92660. Office: Physical Sciences Bldg., U. Cal., Irvine, Cal.

REINHARDT, William O(scar), Am. physician, anatomist; b. Colorado Springs, Colo., Apr. 18, 1912; s. Herman O. and Mildred (Davis) R.; A.B., U. Cal. at Berkeley, 1932, M.D., 1938; m. Elizabeth E. Parker, Mar. 16, 1940; children—William P., Susan Shelley (Mrs. Frank Frisch). Staff dept. anatomy U. Cal. Sch. Medicine, San Francisco, 1939—, prof. anatomy, 1952—, chmn. dept., 1956-63, dean Sch. Medicine, 1963-66; spl. research fellow Nat. Heart Inst., 1954-55; vis. prof. The University, Bristol, Eng., 1954-55, U. Indonesia, Djakarta, 1955-56, Kyoto (Japan) U., 1961, Nat. Def. Med. Center, Taipai, Taiwan, 1966-67. NIH Spl. fellow, 1954-55; Alan Gregg Traveling fellow in med. edn. China Med. Bd. N.Y., Inc., 1961. Assn. Anatomists, The Society for Experimental Biology and Medicine, A.A.A.S., N.Y. Acad. Sci., Am. Assn. U. Profs., Am. Physiol. Soc., Sigma Xi, Alpha Omega Alpha. Research on exptl. biology of endocrine-lymphatic system interrelationships, especially thymus, thoracic duct lymph and lymphocytes. Office: U. Cal. Sch. Medicine, San Francisco 94122.*

REINHART, Bruce Lloyd, Am. mathematician; b. Wernersville, Pa., Oct. 20, 1930; s. Russell Lloyd and Ruth (Snyder) R.; B.A., Lehigh U., 1952; M.A., Princeton, 1954, Ph.D., 1956; m. Virginia May Seems, Jan. 29, 1955; children—Gail, Davis, Hugh. Instr., Princeton, 1955-56, U. Chgo., 1956-58; research asso. U. Mich., Ann Arbor, 1958-59; faculty U. Md., College Park, Md., 1959—, prof. math., 1965—; research scientist Research Inst. for Advanced Studies, Balt., 1959-64. NATO fellow U. Strasbourg (France), 1961-62; Fulbright fellow U. Pisa (Italy), 1965-66. Mem. Am. Math. Soc., Math. Assn. Am., Société Mathématique de France, A.A.A.S. Publs. on geometric study of ordinary differential equations on a surface using asso. family of curves and points; gave necessary conditions for existence of closed curve, for curves on torus (no points). Home: 6604 44th Av., Hyattsville, Md. 20782. Office: Math. Dept., U. Md., College Park, Md. 20742.*

REINHOLD, Erasmus, German astronomer; b. Saalfeld, Germany, Oct. 22, 1511; prof. math. U. Wittenberg (Germany). Author: Tabulae prutenicae coelestium motuum, 1551; Primus liber tabularum directionum, 1554. One of 1st to adopt Copernican theory; calculated 1st astron. tables based on Copernican system; discovered dual nature of orbits of Mercury and the moon. Died Wittenberg, Feb. 19, 1553.

REINHOLD, John Gunther, Am. biochemist; b. Milw., Oct. 29, 1900; s. Rudolph and Louise (Peterson) R.; B.S., U. Wis., 1924; M.S., Yale, 1926; Ph.D., U. Pa., 1933; m. Sally Priscilla Edwards, Nov. 17, 1924; children—John Edwards, Peter Edwards, Gretel Edwards (Mrs. Richard Leed). Chief biochemist Phila. Gen. Hosp., 1927-48; faculty U. Pa., Phila., 1948-64, prof. clin. chemistry, 1955-64; vis. prof. biochemistry Pahlavi U., Shiraz, Iran, 1962-64; prof. biochemistry Am. U., Beirut, Lebanon, 1964—. Cons. Surgeon Gen., U. S. Army, 1952—, Clin. Center, NIH, 1952-56. Fellow A.A.A.S.; mem. Am. Assn. Clin. Chemists (past pres., Bischoff award, 1954, John Gunther Reinhold award 1964), Internat. Congress Clin. Chemistry (past sec.), Am. Chem. Soc. (past chmn. com. on clin. chemistry), Am. Soc. Biol. Chemists, Biochem. Soc. (London), Research, numerous publs. on chem. evaluation of liver function, especially as related to alterations in plasma proteins, zinc deficiency in humans, experimentally produced zinc deficiency in rats. Home: Gennaoui Apt., Maamari St., Beirut, Lebanon.*

REINKE, Johannes, German botanist, natural philosopher; b. Ziethen, Germany, Feb. 3, 1849; prof., Göttingen, Germany, Kiel, Germany. Author: Lehrbuch der allgemeine Botanik, 1880; Einleitung in der theoretischen Biologie, 1901; Kritik der Abstammungslehre, 1920; Naturwissenschaft, Weltanschauung, Religion, 1923. Research on sea algae, roots of phanerogams; opposed Darwin's and Haeckel's theories on origin of species; advocated vitalism. Died Preetz, Germany, Feb. 25, 1931.

REINOSO-SUAREZ, Fernando, Spanish neuroanatomist; b. Granada, Spain, Mar. 17, 1927; s. Francisco Reinoso-Lopez and Remedios Suarez-Cobo; M.D., U. Granada, 1950; Ph.D., U. Madrid, 1951; m. Maria Luisa Barbero-Morales, Sept. 29, 1955; children—Fernando, Maria Luisa, Maria Isabel, Maria Rosa, Francisco, Maria Lourdes. Faculty, U. Granada, 1950-57, prof., chmn. dept. anatomy, 1960-62; prof. chmn.

dept. U. Salamanca, 1957-60; prof., chmn. dept. U. Navarra, Pamplona, Spain, 1962—, dean faculty medicine, 1966—. Mem. I.B.R.O. Commn. Comparative Neuroanatomy. Recipient Cajal prize Spanish Research Council, 1956; Franco prize, 1958. Mem. World Fedn. Neurology, Spanish Soc. Anatomy, others. Author: El sistema reticular ascendente de activacion, 1960; Topographischer Hirnatlas der Katze, 1961; also articles. Research on devel. and connections of subthalamus, mesencephalic and pontine structures, brain centers of sleep. Home: 45 Bergamin, Pamplona Navarra, Spain.

REIS, Johann Philipp, German physicist; b. Gelnhausen, Jan. 7, 1834. Taught in Friedrichsdorf. Experimented with fricition electricity and galvanoplastics; in working on hearing aids, thought of transforming air vibrations into electrical impulses leading to construction of 1st rude telephone, 1861. Died Friedrichsdorf, Germany, Jan. 14, 1874.

REIS, Ralph A., Am. physician; b. Chgo., Dec. 18, 1895; s. Ignace J. and Nannie (Ashenheim) R.; B.S., M.D., Northwestern U., 1919; m. Rose Frances Kramer, Apr. 11, 1922; children—Jory Graham, Ruth (Mrs. Mark M. Pomaranc), Herbert K. Practice medicine, specializing in obstetrics and gynecology, Chgo., 1921—; mem. staff Passavant Meml. Hosp.; prof. emeritus obstetrics and gynecology Northwestern U. Med. Sch., Chgo., 1964—. Hon. fellow Obstet. and Gynecol. Soc. Chile, Obstet. and Gynecol. Soc. Uruguay, Peruvian Soc. Fertility, Am. Brit. Cowdray Med. Soc. (Mexico City), Soc. Obstetricians and Gynecologists Buenos Aires, New Eng., N.W. Pacific, Ind., S.W., Pitts. obstet. and gynecol. socs., S. Atlantic Assn. Obstetricians and Gynecologists, N.D., New Orleans, La. obstet. socs., Dallas Post-Grad. Soc., Wash. State Obstet. Assn.; fellow Am. Coll. Obstetricians and Gynecologists (past v.p.), Royal Coll. Obstetricians and Gynecologists, Am. Gynecol. Soc., Am., (past pres.), Central (past pres.) assns. obstetricians and gynecologists, A.C.S. (life), Chgo. Gynecol. Soc. (past pres.), Inst. Medicine Chgo.; mem. Pan Am. Cytology Soc. (co-chmn. for U.S.). Author: (monograph) Diabetes and Pregnancy; also numerous articles. Editor-in-chief: Obstetrics and Gynecology, 1952-67. Research on human reprodn., pregnancy, diabetes. Home: 215 E. Chicago Av., Chgo. 60611. Office: 707 N. Fairbanks Ct., Chgo. 60611.

REISER, Morton F., Am. psychiatrist; b. Cin., Aug. 22, 1919; s. Sigmund and Mary (Roth) R.; B.S., U. Cin., 1940, M.D., 1943; grad. N.Y. Psychoanalytic Inst., 1960; m. Jane Lomas, Mar. 18, 1945; children—David E., Barbara, Linda. Practice medicine specializing in psychiatry, Cin., 1947-52, Washington, 1954-55, N.Y.C., 1959—; faculty Cin. Gen. Hosp., also U. Cin. Coll. Medicine, 1949-52, Washington Sch. Psychiatry, 1953-55; faculty Albert Einstein Coll. Medicine, Yeshiva U., N.Y.C., 1955-58, prof. psychiatry, 1958—, dir. research dept. psychiatry, 1958-65; chief div. psychiatry Montefiore Hosp. and Med. Center, N.Y.C., 1965—. Cons., Walter Reed Army Inst. Research, 1957-58, High Point Hosp., Port Chester, N.Y., 1957—, com. WHO, 1963; also lectr. Recipient Stella Feis Hoffheimer Meml. prize U. Cin. Coll. Medicine, 1943. Diplomate Am. Bd. Psychiatry and Neurology. Fellow A.A.A.S., Am. Psychiat. Assn.; Mem. Am. Soc. Clin. Investigation, Am. Psychosomatic Soc. (pres. 1960-61), Am. Fedn. Clin. Research, A.M.A., Am. Psychoanalytic Assn., Internat. Psycho-Analytical Assn., Assn. for Psychophysiol. Study of Sleep, Psychiat. Soc. A. Graeme Mitchell Undergrad. Pediatric Soc., Phi Eta Sigma, Pi Kappa Epsilon, Alpha Omega Alpha, Sigma Xi, others. Editor-in-chief, Psychosomatic Medicine, 1962—; mem. editorial bd. A.M.A. Archiv. of Gen. Psychiatry, 1961—. Research and publs. on factors and progress of hypertension and psychol. implications; psychosomatic diseases; doctor-patient relationship. Home: 735 Kappock St., N.Y.C. 10463. Office: Montefiore Hosp. and Med. Center, 111 E. 210 St., N.Y.C. 10467.*

REISET, Jules, French chemist, agronomist; b. Bapaume, France, Oct. 6, 1818; corr. mem. French Acad. Scis., 1857, mem., 1884; mem. Agrl. Soc. Author: Recherches chimiques sur la respiration des animaux, 1849; research in agrl. chemistry; measured (with H. V. Regnault) chem. reactions asso. with respiratory process. Died Paris, Feb. 5, 1896.

REISNER, George Andrew, Am. Egyptologist; b. Indianapolis, Nov. 5, 1867; s. George Andrew and Mary Elizabeth (Mason) R.; A.B., Harvard Univ., 1889, A.M., 1891, Ph.D., 1893, hon. Litt.D, 1939, grad. courses in Semitic langs.; m. Mary Putnam Bronson, Nov. 22, 1892. Asst. Egyptian dept., Royal Museum, Berlin, 1895-96; instr. Harvard, 1896-97; mem. Internat. Commn. on Catalogue Khedivial Museum, Cairo, 1897-99; Hearst lecturer in Egyptology and dir. Hearst Egyptian Expdn. from U. of Calif., 1899-1905; asst. prof. Semitic archeology, 1905-10, asst. prof. Egyptology, 1910-14, prof. since 1914 Harvard U.; dir. Egyptian expdn. of Harvard and Boston Museum Fine Arts since 1905; curator Egyptian dept. Boston Museum Fine Arts since 1910. Delegate to Archeological Congress, Cairo, Egypt, 1909; fellow Am. Acad. Arts and Sciences; corr. mem. Sächsische Akademie der Wissenschaften (Phil. Hist. Klasse), 1929. Author: Sumerisch-Babylonische Hymnen, nach thontafeln Griechischer Zeit, 1896; Tempelurkunden aus Telloh, 1901; Hearst Medical Papyrus, 1905; The Early Dynastic Cemeteries of Naga-ed Der, Part 1, 1907; First Annual Report, Nubian Archeol. Survey, 1910; Models of Ships and Boats, 1913; Excavations at Kerma (2 vols.). 1923; Harvard Excavations at Samaria (2 vols.), 1924; Mycerinus, The Temples of the Third Pyramid at Giza, 1931; A Provincial Cemetery of the Pyramid Age, Naga-ed-Der III, 1932; The Development of the Egyptian Tomb, 1935. Archeologist in charge of excavations of Egyptian Govt. in Nubia in preparation to flooding lower Nubia by raising the Assuan dam, 1907-09; dir. Harvard Palestinian Expdn., conducting excavations at Samaria, 1907-10; has excavated at Bersheh, Girga, Giza Pyramids, Samaria, in Lower Nubia, in Halfa, Dongola and Berber Provinces in the Sudan and tomb of Hetep-heres I, mother of Cheops. Died June 6, 1942.

REISS, Albert John Jr., Am. sociologist; b. Cascade, Wis., Dec. 9, 1922; s. Albert J. and Erma (Schueler) R.; Ph.B., Marquette U., 1944; M.A., U. Chgo., 1948, Ph.D., 1949; m. Emma Lucille Hutto, June 11, 1953; children—Peter Clemens, Paul Wetherington, Amy Susan. Faculty, U. Chgo., 1947-52; asso. dir. Chgo. Community Inventory, 1947-52; faculty Vanderbilt U., Nashville, 1952-58, prof., 1954-58, chmn. dept. sociology, 1952-58; prof. State U. Ia., Iowa City, 1958-60, chmn., 1959-60; prof. sociology, dir. Survey Research Lab., U. Wis., Madison, 1960-61; prof. U. Mich., Ann Arbor, 1961—, chmn. 1964—, dir. Center for Research on Social Orgn., 1961—. Mem. behavioral scis. div. NRC, Nat. Acad. Sci., 1963-64; mem. adv. panel for sociology and social psychology NSF, 1963-65; mem. mental health small grants com. Nat. Inst. Mental Health, 1960-64. Author: Cities and Society, 1951; rev., 1957; (with O.D. Duncan) Social Characteristics of Urban and Rural Communities, 1956; Louis Wirth on Cities and Social Life, 1956, rev., 1964; Occupations and Social Status, 1961. Research and publs. on causes of crime and juvenile delinquency and variation in their prevalence and incidence; devel. of scale of prestige of occupations, a typology of cities, time budgets for rural-urban status groups. Home: 2900 Provincial Dr., Ann Arbor, Mich. 48104.*

REISS, Frederick, dermatologist; b. Edelsthal, Austria, Oct. 12, 1891; s. Sigismund and Helena (Knopfmacher) R.; B.A., Hungarian Royal Catholic Gymnasium, 1909; postgrad. Royal Hungarian U., U. Vienna; M.D., U. Budapest, 1914; postdoctoral research various skin clinics, 1918-23; student mycology several univs. Came to U.S., 1941, naturalized, 1944. Chief dermatology service Montefiore Hosp., N.Y.C., 1944-64, now emeritus, cons. dermatologist, dir. mycology lab., 1964—; asso. prof. clin. dermatology and syphilology N.Y. U., 1946-61; consulting dermatologist Manhattan Eye, Ear and Throat Hosp., N.Y.C., 1950—. Vis. lectr. N.Y. U. Dental Coll., 1962. Recipient George Washington award Am. Hungarian Studies Found., 1965; cited by Med. Faculty U. Vienna, 1965. Diplomate Am. Bd. Dermatology and Syphilology. Fellow N.Y. Acad. Scis., A.M.A., N.Y. Acad. Medicine, Am. Coll. Allergists, Royal Soc. Medicine (London), Am. Acad. Dermatology and Syphilology; mem. Chinese Med. Assn. (life), Soc. Investigative Dermatology, Am. Acad. Compensation Medicine, New Eng. Dermatol. Soc., Internat. Soc. Tropical Dermatology (sec. gen. 1960—), Sigma Xi; corr. or hon. mem. various fgn. dermatol. socs. Mem. editorial bd. Dermatologica, 1940—, Mycologia and Mycopathologica applicata, Mykosen, Excerpta Medicia; chmn. editorial bd. Dermatologia Internationalis. Contbr. chpts. to books. Research, publs. on leprosy, syphilis, psoriasis, various tropical skin diseases and treatment, especially mycoses. Home: 1225 Park Av., N.Y.C. 10028. Office: 19 E. 80th St., N.Y.C. 10021.*

REISSMAN, Leonard, Am. sociologist; b. Cleve., June 10, 1921; s. Jacob and Lillian (Star) R.; B.A., Wayne State U., 1942; M.A., U. Wis., 1948; student Princeton, 1948-49; Ph.D., Northwestern U., 1952; m. Ethel Banner, June 10, 1950; children—Alison Freya, Carla Jennifer. Faculty, Tulane U., 1951—, prof., 1957—; vis. mem. London Sch. Econs. 1961-62; sr. staff mem. The Brookings Instn., 1965-67; fellow Center for Advanced Study in Behavioral Scis., 1963-64. Mem. Am., So. social. assns., Soc. for Study Social Problems, Am. Assn. U. Prof. Author: Class in American Society, 1959; The Urban Process, 1964. Studies and publs. on social stratification in Am. society; process of urbanization and social change in U.S., Latin Am. countries; research on dynamics and character in urban environment in So. U.S. Home: 2600 Calhoun St., New Orleans 70118.*

REISSMANN, Kurt Rudolph, physician; b. Schoenebeck, Germany, Feb. 25, 1912; s. Franz and Mina (Ehlert) R.; student U. Halle, U. Innsbruck, 1930-35; M.D., Berlin U., 1936; m. Hildegard Weiss, Apr. 10, 1950; children—Ronald Andreas, Peter Thomas. Came to U.S., 1947, naturalized, 1953. Research fellow U. Berlin, 1942-45; head biology div. Helmholtz Inst., Nussdorf, Germany, 1946-47; research scientist USAF, 1947-51; faculty U. Kan. Med. Center, Kansas City, 1951—, prof. medicine, 1959—; vis. prof. physiology U. Philippines, 1958; cons. VA, 1953—. Fellow A.C.P.; mem. Am. Physiol. Soc., Internat. Soc. Hematology, Am. Soc. Hematology, Central Soc. Clin. Research, Am. Fedn. Clin. Research, Alpha Omega Alpha. Research and publs. on erythropoietin, iron metabolism. Home: 9629 Lee Blvd., Leawood, Kan. 66206. Office: 39th and Rainbow Blvd., Kansas City, Kan. 66103.*

REISSNER, Max Erich, applied mathematician; b. Aachen, Germany, Jan. 5, 1913; s. Hans and Josephine R.; Dipl. Ing., Technische Hochschule, Berlin, 1935, Dr. Ing., 1936; Ph.D., Mass. Inst. Tech., 1938; D. Eng. (hon.), U. Hanover (Germany), 1964; m. Johanna Siegel, Apr. 19, 1938; children—John E., Eva M. Came to U. S., 1936, naturalized, 1945. Mem. faculty Mass. Inst. Tech., 1936—, prof. math., 1949—; vis. prof. U. Mich., summer 1949, U. Cal. San Diego, 1967; aero. research scientist NACA, Langley Field, summers 1948, 1951; cons. to indsl. and govtl. engrng. and research orgn. Dir. Weston Land Co. Recipient Clemens Herschel award, Boston Soc. Civil Engrs., 1956; Theodore von Karman medal Am. Soc. C.E., 1964. Guggenheim fellow, 1962-63. Mem. Am. Soc. M.E. (exec. com. applied mechanics div. 1958-63), Am. Math. Soc., Am. Acad. Arts and Scis. (council 1957-61). Mng. editor Jour. Math. and Physics, 1945-67; asso. editor Quar. Applied Math., 1946—, Jour. Applied Physics, 1950-53, Jour. Applied Math. and Physics, 1965—; cons. editor Math. and Mechanics, Addison-Wesley Pub. Co., 1949-60. Contbr. to books and numerous articles in periodicals. Formulation of variational principles of theory of elasticity, linear and non-linear theories of elastic plates and shells and asymptotic expansions for solution of equations of these theories. Office: Mass. Inst. Tech, Cambridge 39, Mass.*

REITAN, Ralph Meldahl, Am. psychologist; b. Beresford, S.D., Aug. 29, 1922; s. John O. and Anna (Meldahl) R.; B.A., Central YMCA Coll., 1944; Ph.D., U. Chgo., 1950; m. Lucille Kirsch, Feb. 15, 1952; children—Ellen, Jon, Ann, Richard. Psychometrist, Armed Forces Induction Sta., 1944-45; psychologist dept. neurology Mayo Gen. Hosp., 1945-46; faculty U. Chgo., 1946-51; asst. prof. psychology Roosevelt U., 1951; faculty Ind. U. Med. Center, Indpls., 1951—, prof. psychology, dir. sect. neuropsychology, 1960—. Mem. space med. adv. group NASA, 1964—; mem. perinatal research com. Nat. Inst. Neurol. Disease and Blindness, 1965—. Recipient Gordon Barrows Meml. award, 1964. Mem. Am., Midwestern psychol. assns., Am. Acad. Neurology, Gerontol. Soc. Research and numerous publs. on brain behavior relationships in human beings, especially psychol. significance of damage to various areas of brain. Home: Rural Route 18, Box 376, Indpls. 46234. Office: Ind. U. Med. Center, 1100 W. Michigan St., Indpls. 46207.*

REITER, Elmar Rudolf, atmospheric physicist; b. Wels, Austria, Feb. 22, 1928; s. Rudolf and Maria (Huber) R.; student U. Innsbruck (Austria), 1947-52, Ph.D., 1953, docent, 1959; student U. Chgo., 1952-53; m. Gabriella J. Paar, Aug. 5, 1954; children—Bernadette G., Reihold R., Christa M. Came to U.S., 1961, naturalized, 1966. Instr., NATO Officers' Sch., Fuerstenfeldbruck, Germany, 1954; research asso., instr. U. Chgo., 1954-56; research asso., asst. prof. meteorology and geophysics U. Innsbruck, 1956-61; faculty Colo. State U., Ft. Collins, 1961—, prof. atmospheric sci., 1956—. Cons. govt. agys., aerospace cos.; dir. research project U.S. AEC, U.S. Weather Bur., Environmental Scis. Service Admnstrn., U.S. Navy, FAA, 1961—. Research fellow German Weather Service, 1959; recipient award for advancement sci. and cultural Fed. Govt. of Upper Austria, 1962. Mem. Am., Royal, meteorol. socs., Am. Geophys. Union, Nat. Geog. Soc., Meterol. Soc. Japan, Am. Inst. Aeros and Astronautics (panelist), Sigma Xi. Author: Meteorologie der Strahlstroeme, 1961; Jet Stream Meteorology, 1963; also numerous articles. Research on clear air turbulence measurement, jet streams in gen. circulation of atmosphere, transport of radioactive debris; formulated theory on generation of clear air turbulence. Home: 1513 Hillside Dr., Ft. Collins, Colo. 80521.*

REITER, Melchior, German pharmacologist; b. Berlin, Germany, Apr. 2, 1919; s. Casper and Anna (Hütt) R.; M.D., U. Berlin, 1944; m. Veronica Brantl, July 22, 1947; children—Christoph, Michael, Susanne. Asst. internal medicine and pharmacology U. Munich (Germany), 1945-48, med. faculty, 1955—, prof., 1961—, head div. dept pharmacology 1966—; asst. dept. pharmacology Harvard Med. Sch., Boston, also dept. medicine chem. div. U. Chgo., 1949-50. Mem. German Soc. Pharmacology. Research and numerous publs. on cellular pharmacology, physiology and pharmacology of cardiac function, substances enhancing contractility of heart muscle.*

REITER, Reinhold, German physicist; b. Munich, Nov. 17, 1920; s. Gustav and Maria (Irlbeck) R.; Dr.rer.nat., Technische Hochschule München; Promotion, U. Munich, 1953; m. Mirjam David, Nov. 25, 1954; 1 dau., Franziska. Dir., Physikalische-Bioklimatische Forschungsstelle, Fraunhofer-Gesellschaft zur Förderung der angewandten Forschung, Skistadion, West Germany, 1954—. Mem. Internat. Soc. Biometeorology (sec. working group on influences of electric fields), Am. Meteorol. Soc., Am. Geophys. Union. Author: Meteorobiologie und Elektrizität der Atmosphäre, 1960; Felder, Ströme und Aerosole in der unteren Troposphäre, 1964; also numerous articles. Research on statistics on effects of weather on human beings, influences of meteorol. parameters on behaviour of elements of atmospheric electricity, atmospheric fallout, vertical distbn. of natural and artificial radioactivity of atmosphere; devel. model of human respiratory tract as retention simulator for inhaled aerosol particles. Home: 38 a Partenkirchenerstrasse, 8105 Farchant, West Germany. Office: Skistadion, 81 Garmisch-Partenkirchen, West Germany.*

REIZENSTEIN, Peter Georg, Swedish physician; b. Feb. 25, 1928; s. Max L. and Berta R.; M.D., Karolinska Inst., 1954, D.Sc., 1959; m. Anna Lisa, 1952; children—Johan Anders, Elisabet, Margareta. Instr. dept. pathology Karolinska Inst., Stockholm, Sweden, 1953-56, attending physician dept. internal medicine, 1956-59, asst. head, 1961-66, sr. physician, 1967——, head dept. internal medicine Faculty Dentistry, 1963——; research asso. Brookhaven Nat. Lab. Med. Research Center, Upton, L.I., N.Y., 1959-61. Mem. Soc. for Exptl. Biology and Medicine, Internat. Soc. Hematology, Föreningen Skärgardsbaten (mem. bd.). Author: Vitamin B12 Metabolism, 1959; Clinical Nutrition, 1967; The Patient and the Organization of Medical Care, 1967; also articles. Studied enterohepatic circulation and human requirement of vitamin B12; diagnostic whole body radioactivity counting; folic acid metabolism and cellular growth rate in leukemia. Home: 9 Villavägen, Stocksund, Sweden. Office: Karolinska Hosp., Stockholm, 60, Sweden.*

REKVELD, Johan, Dutch physicist; b. Kampen, Netherlands, Nov. 2, 1902; s. Henri Jean and Anna (Botter) R.; Dr. ès sc., U. Utrecht (Netherlands); m. Charlotte Kallabis, Dec. 5, 1929; children—Anna V. C., Hedwig E. J. Johan C. Joined Holy Trinity Coll., 1929, U. Leyden (Netherlands), 1947; lectr. U. Tenn., 1956-57; with U. Delft (Netherlands), 1957-. Mem. Dutch Physics Assn., Sci. Masters Assn. Author textbooks, manuals. Research on Raman effect. Address: Prins-Hendriklaan 110 Overveen, Netherlands.

RELMAN, Arnold Seymour, Am. physician; b. N.Y.C., June 17, 1923; s. Simon and Rose (Malach) R.; A.B., Cornell U., 1943; M.D., Columbia U., 1946; m. Harriet Morse Vitkin, June 26, 1953; children—David Arnold, John Peter, Margaret Rose. House officer New Haven Hosp., also asst. in medicine Yale, 1946-49; NRC fellow Evans meml. Dept. Clin. Research, Boston, 1949-50; faculty Boston U. Sch. Medicine, 1950——, prof., 1961——; chief renal sect. Univ. Hosp., Boston; cons. in medicine Boston Vets. Hosp. Diplomate Am. Bd. Internal Medicine. Fellow Am. Acad. Arts and Scis., A.C.P.; mem. Am. Fedn. for Clin. Research (past pres.), Am. Soc. for Clin. Investigation, Assn. Am. Physicians, A.M.A., Physiol. Soc. Editor: (with F.J. Ingelfinger, M. Finland) Controversy in Internal Medicine, 1966. Editor-in-chief, Jour. Clin. Investigation, 1962——. Contbr. numerous research articles, revs. to tech. jours., chpts. to textbooks, monographs. Research on effects of kidney disease on physiol. processes in body, effect of certain biochem. disorders on kidney function, acid base physiology. Home: 4 Bennington Rd., Lexington, Mass 02173. Office: 750 Harrison Av., Boston 02118.*

REMAK, Robert, German physiologist, neurologist; b. Posen (now Poznan, Poland), July 26, 1815; ed. under Johannes Müller at U. Berlin; M.D., 1838; prof. U. Berlin. Author: Observationes anatomicae et microscopicae de systematis nervosi structura, 1838; Galvanotherapie der Nerven- und Muskelkrankheiten, 1858. A founder of electrotherapy; pioneer in use of electric currents to treat nerve disorders; discovered non-medullated nerve fibers (fibers of Remak), chiefly in sympathetic nervous system, also described axon (Remak's band), 1838; described groups of nerve cells nr. superior vena cava in heart (Remak's ganglion), 1844; conducted studies in embryology; reduced Baer's 4 germ layers to 3 primary layers and named them ectoderm, mesoderm, endoderm, 1845; 1st to state that growth of new tissue begins with div. of existing cells, 1852. Died Kissingen, Germany, Aug. 29, 1865.

REMANE, Adolf, German zoologist, oceanographer; b. Krotoschin, Aug. 10, 1898; s. Adolf and Martha (Heinze) R.; Dr.phil., U. Berlin, 1921; hon. dr.rer.nat., U. Hamburg 1963; m. Martha Borck, Feb. 21, 1923; children—Reinhard, Jürgen. Prof., U. Halle, 1934-36; faculty U. Kiel, 1925——, prof. zoology, oceanography, 1936——, dep. dir. inst. oceanography, 1937-44, dir. zool. inst. and mus., mus. ethnology, 1936-. Mem. Leopoldina, Mainz Acad. Scis. and Lit., Helsinki Soc. Scis., Royal Swedish Acad., Vienna Zool. and Bot. Soc. Author: Ökologie der Nord- und Ostsee, 1941; Die Grundlagen des Natürlichen Systems der Vergleichenden Anatomie und der Phylogenetik, 1952; (with C. Schlieper) Die Biologie des Brackwassers, 1958; Das Sozialleben der Tiere, 1960; also numerous articles. Research on problems of phylogeny. Home: 11 Prinzenstrasse, Plön. Office: Zool. Inst. and Mus. U. Kiel, 3 Hegewischstrasse, Kiel, West Germany.*

REMICK, Arthur Edward, Am. chemist; b. Chgo., June 5, 1899; s. Andrew Edward and Maude (Coe) R.; B. Chemistry, Cornell U., 1922; Ph.D., U. Chgo., 1928; m. Marie Louise Whiddit, Sept. 1, 1928; 1 dau., Sylvia Louise (Mrs. Harold N. Reid). Chemist, Kirkman & Sons, Bklyn., 1920-21, Nat. Aniline and Chem. Co., Buffalo, 1922-23; research chemist E.R. Squibb and Sons, Bklyn., 1923-24, The Todd Co.,

Rochester, N.Y., 1927-29; faculty Wayne State U., Detroit, 1929——, prof. chemistry, 1949-66, now emeritus professor, since 1966——. Mem. Am. Chem. Soc., Electrochem. Soc., Am. Assn. U. Profs., Sigma Xi. Author: Electronic Interpretations of Organic Chemistry, 1943, 49; also articles. Patentee safety paper for bank checks; research on electroorganic chemistry. Home: 14425 Archdale Rd., Detroit 48227.*

REMICK, Forrest Jerome, Am. mech. engr.; b. Lock Haven, Pa., Mar. 16, 1931; s. Forrest Jerome and Ruth (Saiers) R.; student Lycoming Coll., 1949-50; B.S., Pa. State U., 1955, M.S., 1958, Ph.D., 1963; diploma Oak Ridge Sch. Reactor Tech., 1956; m. Grace Louise Grove, June 7, 1953; 1 dau., Beth Ann Remick. Mechanical engineer with Bell Telephone Research Labs., Whippany, N.J., 1955-56; research asso. Nuclear Reactor Facility, Pa. State U., 1956-59, acting dir., 1959-63, dir., 1963-65; acting dir. Curtiss-Wright Nuclear Research Lab. of Pa. State U., Quehanna, Pa., 1960-63, dir., 1963-65, asso. prof. nuclear engring., 1965——, asst. to v.p. for research, dir. Inst of Sci. and Engring., 1967——; chief tng. sect. internat. AEC, Vienna, Austria, 1965-67. Indsl. cons. U.S. AEC, 1964——, NSF, 1964——; mem. reactor safeguards com. Saxton Nuclear Exptl. Corp., 1963-65; Mem. Am. Nuclear Soc. (exec. com. reactor operations div. 1963-66, vice chmn. 1965-66, editorial adv. bd. Nuclear Applications 1964——), Am. Soc. M.E., Am. Soc. for Engring. Edn., Phi Kappa Phi, Tau Beta Pi, Pi Tau Sigma. Author: (with Donald E. Kline, Alan M. Jacobs, D. Van Nostrand) Basic Principles of Nuclear Science and Reactors, 1960; also articles. Research on operation and utilization of research reactors. Home: 455 E. Foster Av., State College, Pa. 16801. Office: Old Main Bldg., Pa. State U., University Park, Pa. 16802.*

RE MINE, William Hervy, Am. surgeon; b. Richmond, Va., Oct. 11, 1918; s. William Hervey and Mable (Walthall) Re M.; B.S., U. Richmond, 1940, D.Sc., 1965; M.D., Med. Coll. Va., 1943; M.S. in Surgery, U. Minn., 1952; m. Doris Irene Grumbacher, June 9, 1943; children—William Hervey III, Steven Gordon, Walter James, Gary Craig. Mem. staff Mayo Clinic, Mayo Found., Rochester, Minn., 1947——, head sect. surgery, 1952——; faculty U. Minn., 1954——, asso. prof. surgery Mayo Grad. Sch., 1965——; spl. cons. to surgeon gen. U.S. Army, 1959——. Diplomate Am. Bd. Surgery. Fellow A.C.P.; mem. Minn., St. Paul, Venezuelan (hon.) surg. socs., Western, Central, So. surg. assns., Soc. Head and Neck Surgeons, Soc. Surgery Alimentary Tract, Priestley Soc. (sec.-treas. 1965——), Am. Med. Writers Assn. Soc. History Medicine, Assn. Mil. Surgeons. U.S., Soc. for Med. Consultants Armed Forces, Colombian Coll. Surgeons (hon.), Internat. Soc. Surgery, Sigma Xi, Omicron Delta Kappa, Alpha Omega Alpha. Author: (with J.T. Priestley, Joseph Berkson) Cancer of the Stomach, 1964. Research and publs. on surgery of the stomach, biliary tract and pancreas with spl. emphasis on cancer and gastrointestinal physiology which includes development, assessment and evaluation of surg. procedures for lesions both benign and malignant. Home: 800 12th Av. S.W., Rochester 55901. Office: 200 First St., Rochester, Minn. 55902.*

REMINGTON, Charles Lee, Am. evolutionary geneticist; b. Reedville, Va., Jan. 19, 1922; s. Pardon Sheldon and Maud (Skoglund) R.; B.S., Principia Coll, 1943; A.M., Harvard, 1947, Ph.D., 1948; m. Jeanne R. Ehrenborg, June 7, 1946; children—Eric Elliott, Janna Lee, Sheldon Tertius Carpenter. Faculty, Yale, New Haven, 1948——, asso. prof. biology, 1956——, asso. curator entomology Peabody Mus., 1953——, fellow Pierson Coll., 1950——. Guggenheim fellow Oxford (Eng.) U., 1958-59. Hon. fellow Societas Entomologica Fennica, Lepidopterists' Soc. (life); mem. Soc. for Study Evolution, Genetics Soc. Am., Royal Entomol. Soc. London, Entomol. Soc. Am., Am. Soc. Naturalists, Soc. for Systematic Zoology, Sigma Xi. Research and publs. on evolutionary processes in organisms especially evolutionary genetics, interspecific hybridization, mimicry, chromosomes, classification of Lepidoptera and primitive insects; formulated theory of suture zones of natural hybridization. Home: 170 Forest Hill Rd., Hamden, Conn. 06518.*

REMINGTON, Eliphalet, Am. inventor; b. Suffield, Conn., Oct. 27, 1793; s. Eliphalet and Elizabeth (Kilbourn) R.; m. Abigail Paddock, May 12, 1814; at least 5 children, including Philo, Samuel, Eliphalet. Made gun barrel out of scrap metals, subsequently received orders for gun barrels; was forging barrels and rifling, stocking, lock fitting for guns, by 1828; erected new gunshop, 1828; purchased entire gun-finishing machinery of Ames & Co., Springfield, Mass., 1845, assumed unfinished contract for several thousand carbines for U. S. Govt.; procured contract in own right for 5,000 Harper's Ferry rifles; marketed Remington pistol, 1847; began mfg. agrl. implements, beginning with cultivator tooth, later plows, mowing machines, wheeled rakes, horse hoes; received large contract from U. S. Govt. for rifles, carbines, pistols during Civil War; 1st pres., one of 1st dirs. Ilion Bank (N.Y.). Died Ilion, Aug. 12, 1861.

REMINGTON, John Wood, Am. physiologist; b. Fargo, N.D., Dec. 19, 1914; s. Roe E. and Jessie (Jepson) R.; B.S., Coll. Charleston, 1935; M.S., N.Y. U., 1937, Ph.D., 1939; m. Emily Henderson, Aug. 13, 1938; children—Olina Porter, Martha Lynn. Research asso. Princeton, 1939-43; faculty Med. Coll. Ga., Augusta, 1943——, prof. physiology, 1951——; Minn. Heart Assn. research fellow Mayo Found., 1955; Spl. fellow NIH, U. Munich, Germany, 1962-63. Mem. Am. Physiol. Soc., Am. Heart Assn., Sigma Xi. Author: Tissue Elasticity, 1957. Research and publs. on the hydraulics of aorta, cardiac output, blood vol. regulation, shock, hypertension, cardiodynamics, phys. characteristics of smooth and cardiac muscle, adrenal cortex, thyroid. Home: 1904 Flintwood Dr., Augusta, Ga. 30904.*

REMINGTON, Philo, Am. inventor, mfr.; b. Litchfield, N.Y., Oct. 21, 1816; s. Eliphalet and Abigail (Paddock) R.; ed. Cazanovia Sem.; m. Caroline Lathrop, Dec. 28, 1841; at least 2 children. Took charge of father's factory, 1861, reorganized firm, separating agrl. implements from armory; manufactured and aided in developing over 50 types of pistols; then manufactured Remington Breechloader rifle; organized armory as E. Remington & Sons, pres. until 1889; manufactured sewing machines (1st marketed), 1870; sole owner Sholes & Glidden typewriter after 1873; began manufacturing Remington typewriter, 1873, introduced to public at Centennial Exhbn., Phila., 1876. Died Silver Springs, Fla., Apr. 4, 1889.

REMLER, Johann Christian Wilhelm, German chemist; b. Oberbösa, 1759; pharmacist in Naumberg, Germany. Author: Neues chemisches Wörterbuch und Handlexikon, 1793; Tabelle über die chemische Nomenklature, 2 vols., 1793; Meusel, Gelehrtes Deutschland, 1798; also tables on water loss. Research on hormone content in plants; isolated various substances from aconite; extracted grape sugar from raisins. Died circa 1810.

REMLEY, Marlin Eugene, Am. physicist; b. Walcott, Ark., Apr. 25, 1921; s. Aubrey James and Kate (Clarida) R.; A.B., Southeast Mo. State Coll., 1941; postgrad. Ia. State Coll., 1941-42; M.S., U. Ill., 1948, Ph.D., 1952; m. Ruth Neoma Evens, Apr. 4, 1943; children—Carol Sue, Nancy Ann, Barbara Jean. Instr., Southeast Mo. State Coll., Cape Girardeau, 1946-47; research asst. U. Ill., Urbana, 1947-52; with N.Am. Aviation, 1952——, dir. tech. dept. Atomics Internat. Div., Canoga Park, Cal., 1959——. Mem. Am. Nuclear Soc. (standards com.), Atomic Indsl. Forum (reactor safety com.), Am. Phys. Soc., Nat. Mgmt. Assn., Sigma Xi, Phi Kappa Phi, Pi Mu Epsilon. Research and publs. in exptl. determination of response of scintillation crystals to nuclear particles, nuclear scattering, nuclear reactor kinetics and reactor safety, research and pulse reactors, sodium graphite and organic moderated reactors for central sta. power, compact reactor systems for space power plants. Home: 19112 Halsted St., Northridge, Cal. 91324. Office: 8900 DeSoto Av., Canoga Park, Cal. 91304.*

REMLINGER, Paul Ambroise, French bacteriologist; b. Bertrange, France, Dec. 29, 1871. Dir., Pasteur Inst., Tangiers. Mem. Fench Acad. Scis. (corr.), Académie de médicine. Remlinger's sign or difficulty in protuding tongue and presence of fine tremors in it when protuded, as seen in typhus, is named after him. Address: Pasteur Institute, Tangiers, Morocco.

REMMER, Herbert K. H., German pharmacologist; b. Berlin, Germany, Mar. 16, 1919; s. Henry W. and Margarete (Piech) R.; Dr.med., U. Berlin, 1945; m. Ingeborg Flemming, Mar. 6, 1950; children—Maria, Hildegard, Johannes. Docent, Pharmacological Inst. Free U. Berlin, 1950-58, unscheduled prof., 1958-64; prof., chmn. dept. toxicology, U. Tübingen, Germany, 1964——. Mem. German Soc. Pharmacology, German Soc. Biol. Chemistry, Royal Soc. Medicine. Research, publs. on standardization of hemoglobin determinations, oxidation of hemoglobin to methemoglobin, tolerance to drugs by increased breakdown, adaptation to drugs by increased synthesis of drug-metabolizing enzymes combined with structural changes in the liver cell. Address: 60 Hartmeyerstrasse, Tübingen, West Germany.

REMSEN, Ira, Am. chemist; b. N.Y.C., Feb. 10, 1846; A.B., Coll. City N.Y., 1865; M.D., Coll. Phys. and Surg. (Columbia), 1867; Ph.D., U. Göttingen, 1870; LL.D., Columbia, 1893, Princeton, 1896, Yale, 1901, Toronto, 1902, Harvard, 1909, Pa. Coll., 1910, U. Pitts., 1915; D.C.L., U. of South, 1907; m. Elisabeth H. Mallory, Apr. 5, 1875; children—Ira Mallory, Charles Mallory. Prof. chemistry Williams College, 1872-76; prof. chemistry Johns Hopkins, 1876-1913, dir. Chem. Lab., 1876-1908, sec. Academy Council, 1887-1901, pres., 1901-12, pres. and prof. emeritus, 1913. Recipient Priestly medal Am. Chem. Soc., 1923. Mem. Soc. Chem. Industry (medal 1904, pres. 1910-11), Nat. Acad. Scis. (pres., 1907-13). Author: The Principles of Theoretical Chemistry, 1876; An Introduction to the Study of the Compounds of Carbon, or Organic Chemistry, 1885; Introduction to the Study of Chemistry, 1887; The Elements of Chemistry, 1888; Inorganic Chemistry, 1889; A Laboratory Manual, 1889; Chemical Experiments, 1895; also many scientific articles and addresses. Founder, editor Am. Chem. Jour., 1879-27. Research in organic and inorganic chemistry; discovered saccharin, 1879; founder grad. research in chemistry in U. S. Died Carmel, Cal., Mar. 5, 1927.

REMY, David Carroll, Am. chemist; b. Waco, Tex., July 17, 1929; s. Theron P. and Ono (Carroll) R.; B.S., U. Cal. at Los Angeles, 1951, M.S., 1952; Ph.D., U. Wis., 1958; postgrad. U. Wis., 1960-62; m. Nancy Joan Wagner, Feb. 23, 1963. Chemist E.I. DuPont de Nemours & Co., Wilmington, Del., 1958-60; McArdle Meml. Labs., U. Wis. Med. Sch, Madison, 1960-62; chemist Merck, Sharp and Dohme, West Point, Pa., 1962—. Mem. Am. Chem. Soc., Phi Beta Kappa, Sigma Xi, Alpha Chi Sigma, Phi Lambda Upsilon. Research, publs. on fluointed momomers for synthesis thermally stable elastomers, medicinal chemistry. Home: Jenkins Lane, Mounted Route 1, North Wales, Pa., 19454. Office: Merck, Sharp and Dohme, Research Lab., West Point, Pa. 19486.*

RENALDINI, Carlo, Italian engr.; b. Ancona, Italy, 1615; engr. Papal army. Mem. Accademia del Cimento. Author: Philosophia Naturalis, 1694. Proposed that fixed points for thermometric scale be freezing and boiling points of water, 1693. Died 1698.

RENARD, Alphonse François, Belgian geologist, mineralogist; b.Renaix, Flanders, Belgium, Sept. 27, 1842; student Jesuit Tng. Coll., Eifel, Germany, from 1870. Supt. Collège del la Paix, Namur, Belgium, 1866-69; prof. chemistry, geology Louvain (Belgium) Jesuit Coll.; curator Royal Natural History Mus., Brussels, Belgium, 1877; ordained priest Roman Catholic Ch., 1877; prof. mineralogy, geology U. Ghent (Belgium), 1888-1903, founder mineralogy lab.; renounced clerical vows, 1901. Recipient Bigsby medal Geol. Soc. London, 1885. Author: (with Charles de la Vallee Poussin) Mémoire sur les Caractères Mineralogiques et Stratigraphiques des Roches Dites Plutoniennes de la Belgique et de l'Ardenne Française, 1876; Report on the Scientific Results of the Voyage of H.M.S. Challenger; (with John Murray) Deep Sea Deposits, 1891; Observations Géologiques sur les iles Volcaniques Explorées par l'Expedition du Beagle. Wrote on rock structure, mineral composition, metamorphism; research on cosmic dust, zeolitic crystals, ocean deposits. Died Brussels, July 9, 1903.

RENARD, Charles, French aviation pioneer; b. Damblain, France, 1847; ed. Ecole Polytechnique. Dir. air sta. workshops, Chalais-Mendon, France, 1877-1905; mem. commn. for commerce through air passage, 1874. (With Krebs and Paul Renard) built 1st dirigible, La France; worked on problems involving maneuvering, travel with free and captive balloons, better constn. materials for balloons, hydrogen; developed light motor for dirigible, also ventilation principle; predicted airplane travel. Died Paris, France, 1905.

RENARD, Fernand Michel, French physicist; b. Sardy, Nievre, France, Nov. 16, 1937; s. Fernand and Marie-Rose (Gauthier) R.; Licence-ès-Sciences Paris, 1958, Doctorat de Physique Theorique, 1961, Doctorat-ès-Sciences Physiques, 1965; m. Yvette Villachon, Oct. 16, 1965. Staff, Laboratoire de Physique Theorique et des Hautes Energies, Centre Nat. de la Recherche Scientifique, Orsay, France, 1960—, chargé de recherche, 1966—. Mem. Comite National de la Recherche Scientifique, 1963—. Research and publs. on high energy physics of elementary particles, relativistic description of nuclear structure of deuteron and determination of electromagnetic structure of nucleons by inelastic electron scattering. Home: 7 rue des Lauriers Roses. Office: Dept. de Physique Mathematique, Faculté des Sciences, 34 Montpellier, France.*

RENAULT, Louis, French engr.; b. Paris, France, Jan. 12, 1877; research for Delaunay-Belleville; (with bro.) developed largest automobile factory in France. Built auto with high gear ratio and drive shaft; invented guide frame for gearshift, 1899.

RENAUT, Joseph-Louis, French physician; b. La Haye-Descartes, France, Dec. 7, 1844; head med. clinic, Paris, France, circa 1869; lab. dir. charity clinics, from 1875; prof. gen. anatomy, histology Faculty Lyons (France), from 1877; hosp. physician, Lyons, from 1883. Mem. Acad. Medicine, French Acad. Scis. Author: Traité d'Histologie. Prin. work in normal and path. anatomy. Died Lyons, Dec. 26, 1917.

RENDEL, Jan, animal geneticist; b. Mjölby, Sweden, Apr. 29, 1927; s. Kurt E. and Hagar (Richardson) R.; B.S. in Agronomy, Royal Agr. Coll., Uppsala, Sweden, 1952, Agr.lic., 1957, Agr.D., 1958; M.S., U. Wis., 1953; m. Ingmarie Alund, Oct. 21, 1950; children—Johan, Henrik. Asst. Royal Agr. Coll., 1953-58, asst. prof. animal breeding, 1958-63, asso. prof., 1964—; animal breeding officer FAO, UN, Rome, Italy, 1968—; vis. serologist U. Cal. at Davis, 1963-64; vis. prof. U. Göttingen (Germany), 1966. Mem. Genetics Soc. Uppsala (chmn. 1962-66), European Soc. Animal Blood Group Research (mem. bd. 1962—). Author: (with Ivar Johansson) Genetics and Animal Breeding, 1963; also numerous articles. Editorial adviser Immunogenetics Letter, 1965—. Research on animal blood groups, chimerism in cattle twins, protein and enzyme polymorphism in farm animals, application blood groups to animal breeding, genetic influence prodn. traits in cattle based on monozygous and dizygous twins. Address: Animal Prodn. Br., FAO, V. della Terme di Caracalla, Rome, Italy.*

RENDLE, Alfred Barton, English botanist; b. Lewisham, Eng., Jan. 19, 1865; s. John Samuel and Jane Barton (Wilson) R.; student St. John's Coll., Cambridge (Eng.) U.; D.Sc., U. London (Eng.); m. Alice Maud Armstrong, 1892 (dec. 1896); 2 sons, 1 dau.; m. 2d, Florence Brown, 1898; 5 sons, 1 dau. Asst. botany Brit. Museum, 1888; head botany dept. Birkbeck Inst., 1894; keeper botany Brit. Mus., 1906-30; hon. prof. botany Royal Hort. Soc. Pres. confs. of delegates of Corr. Socs., 1936. Recipient Victoria medal of honour, 1917, Veitch Meml. medal. Fellow Royal Soc., 1909. Mem. Brit. Assn. (pres. bot. sect. 1916), Linnean Soc. (bot. sec. 1916-23, pres. 1923-27), Quekett Microscopial Soc. (pres. 1919-21), S. Eastern Union Sci. Socs. (pres. 1927), S. Western Union Naturalists (pres. 1931-32), S. London Bot. Inst. (pres. 1911). Author: Classification of Flowering Plants, 2 vols., 2d edit., 1930; (with W. M. Fawcett) Flora of Jamaica, 7 vols.; also sect. monocotyledons and symnosperms in Catalogue of the African Plants, collected by Dr. Friedrich Qelwitsch in 1853-61, 1899. Worked on descriptions of plants, plant nomenclature, classification and systematic botany. Died Leatherhead, Eng., Jan. 11, 1938.

RENDU, Henri-Jules-Louis, French physician; b. Paris, 1844; became intern, 1873, staff physician, Paris, 1877, agrégé, 1878; mem. Acad. Medicine; described form of hysterical tremor, known as Rendu's tremor, 1888, also Rendu-Osler-Weber disease, hereditary hemorrhagic telangiectasis. Died Paris, 1902.

RENGARTEN, Vladimir Pavlovich, Russian geologist; b. July 24, 1882; grad. Mining Inst. Petersburg, 1908. Staff, Geologic Com. (now All-Union Geology Inst.); staff USSR Acad. Scis., 1941—. Author: Geological Structure of the Murga-Istyk Region of Eastern Pamires. Geology and Paleontology of South Eastern Pamires, 1935. Editor: Geology of the USSR, 9th and 10th vols., 1941-47. Recipient Stalin Prize, 1948. Research and publs. on geology of Caucasus, Eastern slope of Urals, Pamir and Amur ter., tectonics of Caucasus, stratigraphy of chalk deposits, also paleontology. Office: USSR Acad. Scis. Leninskii Prospekt, 14, Moscow, USSR.

RENN, Charles Easterday, Am. biologist; b. Frederick County, Md., Mar. 27, 1905; s. Eli Charles and Adah (Easterday) R.; B.S., Columbia, 1928; M.A., N.Y. U., 1932; Ph.D., Rutgers, 1935; M.A. (hon.), Harvard, 1944; m. Elisabeth Sheffield, May 30, 1940; children—Eli C., Mary E. Marine bacteriologist Woods Hole Oceanographic Instn., 1932-38; biology Harvard, 1935, asso. prof. san. biology, 1942-44; biologist Mass. Dept. Health, 1944-46; asso. prof. san. engring. Johns Hopkins, 1952-54, prof., 1954—; pres. Charles E. Renn, Inc. Cons. chem. and oil producing industries in indsl. waste treatment, water supply problems. Research OSRD, NRC. Mem. Am. Pub. Health Assn., A.A.A.S., Marine Biol. Assn., Geophys. Union, Am. Water Works Assn. Studies of biodegradation of detergents and organic synthetics. Home: Route 2, Hampstead, Md. 21074. Office: Ames Hall, Johns Hopkins, Balt. 18.*

RENNELL, James, Brit. geographer; b. Chudleigh, Eng., Dec. 3, 1742; s. John and Anne (Clarke) R.; m. Jane Thackeray, 1772; children—Thomas, William Jane (Mrs. Rodd). Joined Navy, 1756; marine surveyor for E. India Co., later became surveyor-gen. Co.'s lands, Bengal, India, 1764, later geog. cons.; named maj. Bengal Engrs., 1776. Recipient Copley medal Royal Soc., 1791, Gold medal Royal Soc. Lit., 1825. Fellow Royal Soc., 1781; asso. mem. French Acad. Scis. Author: Bengal Atlas, 1779; Herodotus, 2 vols.; Observations on the Topography of the Plain of Troy, 1814; Illustrations of the Retreat of the Ten Thousand, 1816; A Treatise on the Comparative Geography of Western Asia, 1831. Made 1st survey of Bengal, 1st nearly correct map of India, current charts of Atlantic; drew up map of No. Africa, 1790; 1st to explain Rennell's Current (occasional northerly set South of Scilly Islands). Died London, Mar. 29, 1930.

RENNELS, Edward Gerald, Am. anatomist; b. Charleston, Ill., May 7, 1920; s. Ivory F. and Lucy (Gossett) R.; B.Ed., Eastern Ill. State Coll., 1947; M.A., Harvard, 1948, Ph.D. (AEC fellow in biology), 1950; Pd.D. (hon.), Eastern Ill. State Coll., 1956; m. Ruth Allison, Oct. 15, 1942; children—Virginia Beth, Douglas Edward. Faculty, Med. Br., U. Tex., San Antonio, 1950—, prof. anatomy, 1961-66, prof., chmn. dept. anatomy Med. Sch., 1966—. Mem. Am. Assn. Anatomists, Endocrine Soc., Am. Soc. Zoologists, N.Y. Acad. Scis., A.A.A.S., Soc. for Exptl. Biology and Medicine, Sigma Xi. Research and numerous publs. on devel. and histochemistry of ovarian interstitial tissue in immature rat; correlated neurosecretion and posterior lobe hormones in rat with respect to lactation; described FSH and LH gonadotrophs in rat pituitary; studies on function and cytology of pituitary glands of rat after transplantation; electron microscope studies of pituitary cells and luteal cells during altered states of secretory activity. Home: 104 Bent Oak St., San Antonio 78231.*

RENNIE, John, Scottish engr.; b. Phantassie, Scotland, June 7, 1761; s. James Rennie; student Edinburgh, Scotland, 1780-83; apprenticed to Andrew Meikle; m. Martha Mackintosh; children—George, John. Worked under Boulton and Watt; responsible for series of bridges, canals, land drainage works, docks, harbor works, including Southwark Bridge, Old Waterloo Bridge, Plymouth Breakwater, London, E. and W. India docks; designed (built by his son, John) London Bridge. Fellow Royal Soc., 1798. Improved flour-milling machinery. Died Oct. 4, 1821.

RENNINGER, Mauritius Karl, German physicist; b. Ravensburg, Germany, June 8, 1905; s. Albert and Alwine (Haubennestel) R.; Dipl.-Phys., Technische Hochschule, Munchen, Germany, 1928, Dr.rer. Techn., 1930; m. Friedl Brücklmeier, Sept. 19, 1932; children—Luitgard (Mrs. Dietrich), Wolfgang; Birgit (Mrs. Striepens), Michaela (Mrs. v. Greve-Dierfeld), Ulrich. Asst., Technische Hochschule, Stuttgart, Germany, 1930-37; physicist in industry, 1937-46; head dept. for interference optics kristallograph. Institut, U. Marburg (Germany), 1946—, prof., 1952—. Research, publs. on X-ray interference optics, Umweganregung (Renninger effect). Home: 9 Waldweg, Wehrda BRD, Hessen, Germany. Office: 18 Gutenbergstr., Marburg/Lahn, Germany.*

RENOLD, Albert Ernst, physician, biologist; b. Karlsruhe, Germany, July 10, 1923; s. Ernst J. and Agnes (Wunderlich) R.; M.D., U. Zurich (Switzerland), 1947; m. Jacqueline A. Reverdin, Aug. 4, 1951; children—Frederic Kersting, Marc-Andre Jean. Came to U.S., 1948, naturalized, 1958. Staff, Peter Bent Brigham Hosp., Boston, 1949-62, asso., 1953-58, sr. asso. in medicine, 1958-62; faculty Harvard Med. Sch., Boston, 1951-62, asst. prof. medicine, 1959-62, dir. Baker Clinic Research Lab., 1957-62; professeur ordinaire ad personam Faculty Medicine, U. Geneva (Switzerland), 1963—, dir. Institut de Biochimie Clinique, 1963—. Cons., USPHS, 1960-62. Fellow A.C.P. (hon.); mem. Swiss Acad. Med. Sci. (mem. senate 1964—), European Assn. for Study Diabetes (sec. 1965—), Am. Diabetic Assn. (Lilly award 1960), Am. Physiol. Soc., Am. Soc. for Clin. Investigation, Swiss Soc. Physiology, Physiol. Chemistry and Pharmacology. Editor, author: Handbook of Physiology, 1965. Research and numerous publs. on pathogenesis of human diabetes mellitus and spontaneous diabetes in animals, metabolism and obesity, immunological activities of insulin in several species, adrenocortical physiology and pathology. Home: 8 Cours des Bastions 1200 Geneva, Switzerland.*

RENOUX, Gerard Eugène, French microbiologist; b. Paris, Mar. 23, 1915; s. Joseph Jean and Andree (Bechofer) R.; Dr.Méd., U. Montpellier (France), 1939; Lic.Sciences, U. Marseille (France), 1939; Agrégation microbiologie, U. Paris (France), 1949; m. Micheline Dehais, Feb. 19, 1961; children—Jean, Catherine, Marianne, Frank. Staff, U. Montpellier, 1935-52, prof. microbiology, 1960-67; lab. chief Institut Pasteur, Tunis, Tunisia, 1952-55, dir., 1955-60; prof. immunology U. Tours (France), 1967—; chief Brucellosis Lab., Nat. Inst. Agronomic Researches, Nouzilly, France, 1967—. Expert Brucellosis; WHO/FAO, 1950; Brucellosis sr. cons. FAO, 1952; expert trachoma WHO, 1955; cons. Pub. Health Labs., 1964-66. Decorated Legion D'Honneur, Medaille des Epidémies, Palmes Academiques, Laureat Faculté de Medecine, Montpellier. Mem. Italian Acad. Medicine, Argentine, French Am. socs. microbiologists, N.Y. Acad. Scis., A.A.A.S., Assn. Francaise Vet. Microbiol., Soc. roy. belge Med. Tropicale, Research and numerous publs. on brucellosis including discovery of vaccine for livestock, new techniques of study, new species, new diagnostic methods; techniques and antigens discovery in immunology; studies on antibodies and antigens in human diseases, lab. procedures for diagnostics, trachoma, rickettsial diseases. Home: 250 Av. de Grammont, Tours 37, France. Office: Institut de Physiologie de la Reproduction, Nouzilly, Indreet Loire 37, France.*

RENSCH, Bernhard Carl Emmanuel, German zoologist; b. Thale, Germany, Jan. 21, 1900; s. Carl and Lisette (Siebenhuner) R.; Dr.phil., U. Halle, 1922; Dr.phil.h.c., Uppsala U., Sweden, 1957; m. Ilse Maier, May 20, 1926. Asst., Inst. for Plant Breeding, U. Halle, 1923-25; asst., leader dept. Zool. Mus., U. Berlin, 1925-37; dir. Mus. Natural Sci., Munster, 1937-54; reader in zoology U. Munster, 1938, prof., 1943—; dir., prof. zoology Zool. Inst. Munster, 1947-68. Recipient Leibniz medal Prussian Acad. Sci., 1938; Darwin-Wallace medal Linnean Soc. London, 1958; Darwin placque Acad. Scis. of Germany. Mem. Soc. Natural Sci. Madrid (hon.), Linnean Soc., Physiographical Soc. Lund, Am. Zoologists Soc., Am. Ornitohlogists Union, Arbeitsgemeinschaft fur Forschung. Author: Das Prinzip geographischer Rassen Kreise, 1929; Die Geschichte des Sundabogens, 1930; Evolution Above the Species Level, 1947, 60; Homo Sapiens, 1959; Bibliosophie, 1968. Publs. on studies of rules of climatic speciation in animals; fauna of Sunda Islands; juvenile and phylogenetic growth ratios in animal organs; animal psychology; natural philosophy. Home: 16 Moellmannsweg, 44 Munster. Office: 9 Bade-Strasse, 44 Munster, West Germany.*

RENSE, William Alphonsus, Am. physicist; b. Massillon, O., Mar. 11, 1914; s. Joseph and Rose (Luther) R.; B.S., Case Sch. Applied Sci., 1935; M.S., Ohio State U., 1937, Ph.D., 1939; m. Wanda Evelyn Childs, May 5, 1942; children—William C., John A.H., Charles E.C. Asst. prof. Rutgers U., 1941-42;

faculty La. State U., 1943-49, asso. prof., 1947-49; faculty U. Colo., Boulder, 1951—, prof., 1957—, co-dir. Lab. for Atmospheric and Space Physics, 1957—. Mem. Am. Phys. Soc., Am. Astron. Soc., Am. Geophys. Soc., Sigma Xi, Tau Beta Pi. Author: Physical Science, 1965; also articles. Discovered solar hydrogen Lyman alpha line, solar lines in far ultraviolet; devel. rocket and satellite optical instruments for solar ultraviolet measurements. Home: 601 10th St., Boulder, Colo. 80301.*

RENSON, Jean Félix, Belgian biochemist; b. Liége, Belgium, Nov. 11, 1930; s. Louis and Laurence (Crahay) R.; M.D., U. Liége, 1957; NIH fellow, 1960-63; Ph.D., George Washington U., 1966; m. Giselle Bouillenne, Sept. 9, 1956; children—Marc, Dominique, Jean-Luc. Asst. prof. physiopathology U. Liége, 1957-60; research asso. Nat. Heart Inst., 1963-66; dir. research FNRS, 1966—; with U. Liége, 1966—. Recipient Laureat du Concours, Universitaire de Belgique, 1959. Mem Belgian socs. physiology, pharmacology, biochemistry, European Soc. Radiobiology. Research, publs. on radiobiology of indolealkylamines; pharmacology of catecholamines, serotonin, mechanism of biol. hydroxylations. Home: 18 pl. Sainte Gertrude, Blegny, Belgium. Office: 153 Boulevard de la Constitution, Liége, Belgium.*

RENTSCHLER, Harvey Clayton, Am. physicist; b. Hamburg, Pa., Mar. 26, 1881; s. Joseph F. and Rebecca (Ritzman) R.; B.A., Princeton, 1903, M.A., 1904; Ph.D., Johns Hopkins, 1908, hon. D.Sc., Princeton University, 1941; LL.D., honorary, Ursinus College, 1942; m. Margaret Bender, 1904; 1 son, Lawrence Bender. Instr. physics, U. of Mo., 1908-10, asst. prof., 1910-13, asso. prof., 1913-17; dir. of research lamp div. Westinghouse Electric & Mfg. Co., 1917-47. Fellow A.A.A.S.; mem. Am. Optical Soc., Am. Physical Soc., Am. Inst. Elec. Engrs., New York Elec. Soc. (past pres.), Am. Inst. of Science (New York), Sigma Xi. Contbr. to tech. publs. Research on ultraviolet rays, preparation and properites of ductile thorium and uranium; developed ultraviolet lamp (with Robert F. James) to kill microorganisms. Died March 23, 1949.

RENTSCHLER, Walter, German physicist; b. Tübingen, Germany, Mar. 19, 1911; s. Christian and Hanne (Fritz) R.; Dr.ès.s., U. Tübingen; children—Hannelore, Gisela, Rolf, Werner. Asst, Stuttgart-Hohenheim, 1935-36; physicist, Robert Bosch Gesellschaft mit beschrankter Haftung, 1936-46; asst. prof. physics, 1946-54; joined faculty Agr. Inst., Stuttgart-Hohenheim, 1954, became asso. prof., 1957, prof., 1963. Mem. Physikalische Gesellschaft, Deutsche Gesellschaft für Elektronen-mikroscopie. Author: Aufbau der Materie, 1948; Physikalische Grundlagen der Naturwissenschaft und Technik, 1952. Research in aerosols, radioactivity, mass spectroscopy, radioactive isotopes as tracers, electron microscopy, micrometeorology. Home: Bitzerweg 4, Stuttgart, W. Germany. Office: Institut für Physik, Stuttgart-Hohenheim, W. Germany.

RENWICK, James, engr.; b. Liverpool, Eng., May 30, 1792; s. William and Jane (Jeffery) R.; grad. Columbia, 1807, A.M., 1810, LL.D., 1829; m. Margaret Anne Brevoort, 1816; children—Henry Brevoort, Edward Sabine, James, Laura K. Lectr. natural philosophy Columbia, 1812, trustee, 1817-20, prof. natural philosophy and exptl. chemistry, 1820-53, 1st emeritus prof., 1853-63; topog. engr. with rank of maj. U. S. Army, 1814; commd. col. of engrs. N.Y. Militia, 1817; authority in every branch of engring. of his day; his suggestions for uniting Hudson and Delaware rivers resulted in Morris Canal, a system of inclined planes or railways for transporting canal boat in cradle up or down the incline (awarded medal from Franklin Inst. for cradle innovation 1826); commd. to test usefulness of inventions to improve and render safe boilers of steam-engines against explosions, 1838; a commr. to survey Northeastern boundary of disputed territory between U. S. and New Brunswick, 1840; Author: Outlines of Natural Philosophy (1st extensive treatise on subject from Am. writer), 2 vols., 1822-23; Treatise on the Steam Engine, 1830; Applications of the Science of Mechanics to Practical Purposes, 1840. Editor: (Am. editions with notes) Rudiments of Chemistry (Parke), 1824, Chemical Philosophy (Daniell), 2 vols., 1840. Translator; (from French) Treatise on Artillery (Tallemand), 2 vols., 1820. Contbd. biographies including those of David Rittenhouse (1839), Robert Fulton (1845), Count Rumford (1848) to Sparke's Library of American Biography. Died N.Y.C., Jan. 12, 1863.

RENZEMA, Theodore Samuel, Am. physicist; b. Grand Rapids, Mich., July 12, 1912; s. Richard J. and Dora (Dreyer) R.; A.B., Hope Coll., 1934; M.S., Rutgers U., 1937; Ph.D., Purdue U., 1948; m. Tena H. Havinga, Aug. 22, 1939 (dec.); children—Marc William, Susan (dec.). Instr., Purdue U., 1943-48; faculty Clarkson Coll. Tech., Potsdam, N.Y., 1948-65, prof. physics, chmn. physics dept., 1950-65; research asso. Dudley Obs., Albany, N.Y., 1965-66; vis. prof. physics State U. N.Y. at Albany, 1965-66, prof., 1966—. Cited for contribution to crystal rectifier research Nat. Def. Research Council - Purdue, 1946. Mem. Am. Phys. Soc. (past chmn. N.Y. State sect.), Am. Crystallographic Assn., Am. Assn. Physics Tchrs., Electron Probe Analysis Soc. Am., Sigma Xi, Sigma Pi Sigma. Research on corrosion, micrometeorites. Home: 81 Sycamore St., Albany, N.Y. 12208.

RENZI, Alfred Arthur, Am. physiologist; b. Rochester, N.Y., July 20, 1925; s. Gennaro Hector and Linda (Insogna) R.; B.S., Fordham U., 1947; M.S., Syracuse U., 1949, Ph.D., 1952; m. Martha Louise Gilman, Sept. 4, 1954; children—Linda, David, Richard, Stephen, Maura Lee. With CIBA Pharm. Co., Summit, N.J., 1952-55, 58—, asso. dir. physiology, 1960-63, head endocrine-pharmacology sect., 1963-67; head dept. pharmacology Dow Human Health Research and Devel. Labs., 1967—. Mem. Am. Physiol. Soc., Endocrine Soc. Research, publs. on endocrine pharmacology drugs particularly action drugs and hormones on adrenal and renal physiology. Home: 5220 74th Pl., Indpls. 46250. Office: Box 10, Zionville, Ind. 46077.*

REOENER, Erich, German physicist, meteorologist; b. Bromberg, Germany, Nov. 12, 1881; s. Amandus and Anna (Urban) R.; ed. U. Berlin; m., 1906; 1 son, 1 dau.; m. 2d, Gertrud Heiter, 1949. Became pvt. docent U. Berlin, 1909, titular prof., 1912, prof. physics and meteorology, 1914; named prof. physics U. Stuttgart, 1920, dismissed for polit. views, 1937, reinstated, 1946-55; dir. Max Planck Inst. for Physics of Stratosphere. Fellow Am. Phys. Soc.; mem. German Acad. Sci., Kaiser Wilhelm Soc. for Advancement of Sci.; v.p. Max Planck Soc. for Advancement of Sci. Pioneer in cosmic radiation research; research and publs. on cosmic rays of great depths in water and in stratosphere; determined accurately ionization growth curve of cosmic radiation as function of height; found energy of cosmic rays equal to that which all stars (except the sun) emit; determined oxygen and ozone contents of stratosphere. Died Stuttgart, Feb. 27, 1955.

REPCIUC, Emil, Rumanian physician, biologist; b. Chernovitz, Rumania, Nov. 26, 1911; s. Isidor and Helen (Lewandowski) R.; Bachelor, Faculty Medicine, Jassy, Rumania, 1928, Doctor, 1936; m. Elena Atanasiud, Apr. 12, 1939; 1 dau., Rodica (Mrs. Stefan Puiu). Asst. human anatomy Faculty Medicine, Jassy, 1931-42; asst. U. Bucharest (Rumania), 1942-43, lectr., 1943—, prof. human anatomy, 1947—. Recipient Medalia Muncii, Rumania, 1957, Hufeland Medaille in Silver, Berlin-East, 1964. Mem. Societatea de Morfologie, European Soc. Pathology, N.Y. Acad. Scis., Soc. for Biol. Rhythm, Internat. Soc. for Stereology. Author: Treatise of Anatomy, 1951; also numerous articles. Research on spiral structure of minute chromosomes in mammals, musculo-elastic cushions in arteries of kidney, spleen and testis, an autonomous ganglion testiculare in human and vertebrates, eliciting and inhibiting factors of liver regeneration, arteries of pars distalis of hypophysis, proximal end of portal vessels of hypophysis, genetics of metabolic disorders. Home: 21 Comdor Eug.Botez, Bucharest, Rumania.*

REPKE, Kurt Robert Hermann, German biochemist, pharmacologist; b. Friesack, Germany, June 7, 1919; s. Karl Richard and Elisabeth (Fiebig) R.; student Med. Sch., U. Jena (Germany), 1940, U. Greifswald (Germany), 1941-44; M.D., U. Rostock (Germany), 1945; m. Käthe Berta Auguste Sasz, Jan. 8, 1951; children—Heinrich, Ingeborg, Wolfgang, Reinhardt. Asst., med. clinic U. Greifswald, 1946-50, head physician Pharmacologic Inst., 1950-55; research fellow Inst. Biochemistry German Acad. Scis., Berlin, 1955-58, dept. chief, 1959-64, dir., 1964—. Recipient Fraenkel award German Soc. for Circulation Research, 1961. Fellow Leopoldina Acad.; mem. Internat. Soc. Biochem. Pharmacology, Biochem. Soc. GDR, German Pharmacologic Soc. Research and numerous publs. on digitalis biochemistry; elucidation of metabolic transformations in animal body; digitalis receptor enzyme; designed pengitoxin as first therapeutically useful partial synthetic digitalis derivative. Home: 14 Roentgentaler Weg, Berlin-Buch 1115. Berlin. Office: 70 Lindenberger Weg, Berlin-Buch, 1115 GDR, Germany.*

REPLOH, Heinrich Maria, German bacteriologist; b. Bochum, Germany; s. Anton and Bernhardine (Stuckstette) R.; ed. univs. Fribourg, Munich, Vienna, Munster; M.D.; m. Ida-Elisabeth Lucas, Apr. 17, 1934; children—Hans-Dieter, Ursula, Karl, Georg, Heiner, Marigret. Asst., instr. hygiene and bacteriology; dir. Hygiene Inst, Acad. Medicine, Dusseldorf, Germany, 1940-42; mil. hygienist, 1943-45; dir. Bielefeld Bacteriological Hygiene Inst., 1946-55; prof. hygiene and bacteriology, dir. Hygiene Inst., U. Munster (Germany), 1955—. Author books on microbiol. medicine, hygiene. Research and numerous publs. on microbiology, hygiene. Home: Schlüterstrasse 8, Munster/Westphalia, W. Germany.

REPPE, Walter, German chemist; b. Göringen, Germany, July 29, 1892; prof., Mainz, also Darmstadt, Germany; research dir. firm Badische Anilin-und Soda-Fabrik (BASF). Author: Neue Entwicklungen auf den Gebiete der Chemie des Acetylens und Kohlenoxyds, 1949. Developed process for obtaining vinyl ethers from alcohol and acetylene using increased temperatures and pressure in presence of basic compounds.

REPSOLD, Johann George, German instrument maker; b. Wremen, Germany, Sept. 19, 1770; children—Georg, Adolf; engr., chief fire brigade, Hamburg, Germany. Used microscopes instead of verniers on median circles; furnished meridian circles to obs. in Hamburg, Königsberg (now Kaliningrad, USSR), Pulkova,; designed pendulum for determination of force of gravity (named after him); constructed transit circles, equatorial mountings. Died Hamburg, Jan. 14, 1830.

RESAL, Henri-Amé, French engr.; b. Plombières, France, Jan. 27, 1828; ed. l'Ecole polytechnique, Ecole des mines; Sc.D., 1854. Mining engr., Besancon, France; became prog. mechanics, Besancon, 1855; moved to Paris, 1870; supr. railroads; became prof. Ecole polytechnique, also Ecole des mines, 1872. Mem. French Acad. Scis., 1873. Author: Eléments de mécanique, 1851; Traité de cinématique pure, 1862; Traité élémentaire de mécanique céleste, 1865; Traite de mécanique générale, 1873-79. Research on hydraulics, thermodynamics, dynamics, application of thermodynamics to study of pressure in firearms; orginated theory of transmission of motion through cable, 1874; established interior ballistics. Died Annemasse, France, Aug. 22, 1896.

RESNICK, Oscar, Am. neuropsychopharmacologist; b. Bayonne, N.J., Apr. 27, 1924; s. Samuel and Rebecca (Rubinstein) R.; A.B., Clark U., 1944; A.M., Harvard, 1945; Ph.D., Boston U., 1955; m. Janice Zelda Ravitz, July 13, 1949; children—Sandra, Scott Alan. Research fellow U. Ia. Sch. Medicine, 1945-46, U. Kans. Sch. Medicine, 1947-49; faculty St. Petersburg (Fla.) Jr. Coll., 1946-47, U. Minn., 1949-50; editorial asst. Biol. Abstracts, U. Pa., 1950-51; scientist Nat. Drug Co., Phila., 1951-53, Worcester Found. for Exptl. Biology, Shrewsbury, Mass., 1953—, sr. scientist, 1963—. Lectr. hosps. and univs.; cons. staff Medfield State Hosp. (Mass.), 1958—, Norwich State Hosp (Conn.), 1964—. Mem. A.A. A.S., Am. Coll. Neuropsychopharmacology Mass. Soc. Research in Psychiatry, Inc., Psychodelic Research and Study Group, Sigma Xi, Phi Sigma. Contbr. articles to profl. jours., textbooks. Research in mechanisms of action of insulin on sugar uptake; metabolism of adrenaline in man; mechanisms of action of psychoactive drugs and LSD-25 in animals and man; psychoactive properties of anticonvulsants in man. Home: 5 Meadow Lane, Worcester, Mass. 01602. Office: 222 Maple Av., Shrewsbury, Mass. 01545.*

RESNICK, Robert, Am. physicist; b. Balt., Jan. 11, 1923; s. Abraham and Anna (Dubin) R.; A.B., Johns Hopkins U., 1943, Ph.D., 1949; m. Mildred Saltzman, Oct. 14, 1945; children—Trudy, Abby, Regina. Physicist Nat. Adv. Com. for Aero., U.S. Govt., Cleve., 1944-46; faculty U. Pitts., 1949-56; with Rensselaer Poly. Inst., 1956—, prof. physics, 1958—, mem. Commn. on Coll. Physics, 1960—; research fellow Harvard, 1964-65. Fellow Am. Phys. Soc.; mem. Am. Assn. Physics Tchrs. (Distinguished Service citation 1967), Am. Soc. Engring. Edn., Phi Beta Kappa, Sigma Xi. Author (with D. Halliday) Physics for Students for Science and Engineering, 1960, Physics, 2d edn., 1966; Introduction to Special Relativity, 1968. Contbr. articles in field to sci. jours. Research, publs. in nuclear physics, aerodynamics, and upper atmosphere physics; dir. course-content improvement programs leading to new lab. expts., lecture demonstration apparatus, films, texts; advisor and cons. on nat. programs in sci. edn. Home: 13 Oxford Rd., Troy, N.Y. 12180.*

RESPIGHI, Lorenzo, Italian astronomer; b. Cortemaggiore, Italy, Oct. 7, 1824; student, Parma, Bologna, Italy. Tchr., U. Bologna; apptd. prof. astronomy, 1851; named dir. Bologna Obs., 1855-65; became astronomer Vatican Obs., Rome, named dir., 1866; prof. physics, Sapienza; mem. Brit. expdn. to Indies to observe solar eclipse, 1871. Compiled catologue of declinations of 2534 boreal stars. Discovered 3 comets; 1st systematic observations of protuberances on sun; discovered new methods of measuring diameter of sun. Died Rome, Dec. 10, 1890.

RESSER, Charles Elmer, Am. geologist; b. East Berlin, Pa., Apr. 28, 1889; s. George M. and Sallie R.; grad. Blue Ridge Coll., 1908; B. Pedagogy, Pa. State Tchrs.' Coll., 1912; B.A., Franklin and Marshall Coll., 1913, D.Sc. (hon.), 1934; M.A., George Washington U., 1917; Ph.D., 1917. Fellow Geol. Soc. Am., Paleontol, Soc.; mem. Geol. Soc. Washington, Sigma Gamma Epsilon. Research and publs. on Cambrian paleontology and stratigraphy; photographed paleozoic fossils. Died Sept. 14, 1943.

RESTLE, Frank Joseph, Am. psychologist; b. Weehawken, N.J., Mar. 2, 1927; s. Frank Joseph and Isabel (Alexander) R.; A.B., Lafayette Coll., 1950; M.A., Stanford U., 1952, Ph.D., 1954; m. Barbara June Blackledge, Sept. 3, 1950; children—Kathleen, Phillip, Andrea. Research asso. Human Resources Research office, Washington, 1953-55; fellow Center for Advanced Study in Behavioral Scis., Stanford, Cal., 1955-56; faculty Mich. State U., 1956-61; vis. asso. prof. Ind. U., Bloomington, 1961-62, prof. psychology, 1962—. Fellow A.A.A.S.; mem. Am., Midwestern psychol. assns., Psychonomic Soc., Psychometric Soc. Author: Psychology of Judgment and Choice, 1961. Contbns. include math. theory of concept-formation, memory, and discrimination learning; analysis of group problem-solving, performance of arithmetic, and gen. theory of behavior; discovery that rate of learning depends on proportion of relevant cues, and formulation of theory of selection of strategies. Home: Rural Route 10, Bloomington, Ind. 47401.*

RETHMAN, Heinz, German dentist; b. Ibbenbüren, Germany, Apr. 25, 1913; s. August and Theresia (Tenambergen) R.; ed. U. Munster, U. Kiel; M.D. in Dental Medicine; m. Helia Donath, Mar. 18, 1954; children—Bodo, Thilo, Cornelia, Helia-Simone. Instr., dir. maxillary orthopedics dept. Dental Inst., U. Kiel. Mem. Fed. Instrs.' Assn. Author: manuals on maxillary orthopedics. Research and publs. on evolution. Home: 235 Neumunster, Altonaer Strasse 173. Office: Weimarer Strasse 8, Haus 3,230 Kiel (Kiel/Wik), Holstein, W. Germany.

RETTIG, Hans Moritz, German physician; b. Darmstadt, Germany, June 25, 1921; s. Hans Georg and Anna (Guntermann) R.; M.D., U. Erlangen (Germany), 1947; student U. Frankfurt/Main, 1940, U. Bonn, 1942-44; m. Maria Letzel, Feb. 6, 1954; children—Anna Maria, Thomas Nikolaus, Barbara Juliane. Asst. pub. welfare hosp. Bad Tölz, 1948-54; head physician univ. orthopedic clinic Oskar-Helen Heim, Berlin, 1954-60; venia legendi Free U. Berlin, 1957-61; dir. orthopedic clinic U. Giessen (Germany), 1961—, prof. orthopedics, 1961—, chair orthopedics, 1961—. Mem. Deutche orthopädische Gesellschaft, Deutsche Gesellschaft für Chirurgie, Österreichische Gesellschaft für Chirurgie und Traumatol. Author: Frakturen in Kindesalter, 1957; Patholphysiologie angeborener Fehlbildung Lendenwirbelsäule, 1959. Research and publs. on deformities, childhood injuries, damages and diseases of spinal column. Home: 14 Oberhof, 63 Giessen, Germany.*

RETZIUS, Anders Adolf, Swedish anatomist, anthropologist; b. Lund, Sweden, Oct. 13, 1796; s. Anders Jahan Retzius; M.D., U. Lund, 1819; postgrad. U. Copenhagen (Denmark), Carolinian Inst., Stockholm, Sweden; 1 son, Gustaf Magnus. Prof. anatomy Carolinian Inst., from 1824, also prof. physiology. Invented cranial or cephalic index (ratio of width to length of skull x 100), 1842 (concept used in various attempts at racial classification of humans); described ciliary ganglion, 1840, retroperitoneal or Retzius veins, preperitoneal space or space of Retzius or cave of Retzius, also ligaments of Retzius, gyri of Retzius. Died Stockholm, Apr. 18, 1860.

RETZIUS, Gustaf Magnus, Swedish anatomist, anthropologist; b. Stockholm, Sweden, Oct. 17, 1842; s. Anders Adolph Retzius. Prof. histology Sch. Medicine, Carolinian Inst., Stockholm, from 1877, prof. anatomy, 1889-91. Mem. French Acad. Scis. (corr.). Author: Studien zur Anatomische des Nerven Systems und Bindegewebes, 2 vols., 1875-76; Biologische untersuchungen, 21 vols., 1881-1921; Das Gehörorgan der Wirbeltiere, 2 vols., 1881-84; Menschenhirn, 2 vols., 1896; Das Affenhirn, 1906. Described microscopic lines in tooth enamel denoting layers enamel formation (lines of Retzius), 1873; anthrop. studies ancient Swedish, Finnish crania; studied brains of apes, also highly endowed people in order to localize special gifts in separate areas of brain; cerebral convolution named after him. Died Stockholm, July 21, 1919.

REUBI, François Charles, Swiss physician; b. Neuchatel, Switzerland, July 10, 1917; s. Charles Henri and Marie (Grisel) R.; student U. Neuchatel, 1935-36, U. Berne, 1937-41; M.D., U. Geneva, 1943; m. Claudine Petitpierre, Apr. 12, 1944; children—Jean Claude, Monique. Faculty, U. Berne, 1951—, prof. medicine, dir. outpatient dept., 1954—. Mem. Internat. Soc. Nephrology (v.p. 1963—), French Renal Assn. (pres. 1965—), German Renal Assn. (past pres.), Swiss Acad. Med. Sci., Soc. méd. Hôp. Paris (corr.). Author: Renal Diseases, 1960; Clearance Tests in Clinical Medicine, 1963; also numerous articles. Described vascular lesions in neurofibromatosis; measurement of renal blood flow in normal and diseased subjects; research on mechanism of renal glycosuria, high blood pressure, renal diseases. Home: 14 Roschistrasse, Berne, Switzerland.*

REULING, George, opthalmologist; b. Romrod, Germany, Nov. 11, 1839; s. Robert and Amalie (Vogler) R.; M.D., Giessen, Germany, 1865; studied opthalmology, Berlin, Vienna; postgrad., Paris, 1867-68; m. Elisa Külp, Sept. 21, 1871. Surgeon in Prussian army during war with Austria; asst. Eye Hosp., Wiesbaden, Germany, 1866-67; became chief physician Md. Eye and Ear Infirmary, 1869; named prof. eye and ear surgery Washington U., Balt., 1870; apptd. prof. ophthalmology U. Balt., 1881; prof. eye and ear diseases Balt. Med. Coll., 1886-1910, emeritus, 1910—. Mem. A.A A.S. Contbr. papers to med. publs. Inventor apparatus for eye and ear surgery. Died Nov. 25, 1915.

REUTER, Heinz, Austrian meteorologist; b. Vienna, Jan. 22, 1914; s. Fritz and Marianne (Nikola) R.; Ph.D., U. Graz (Austria); m. Paula Krech, July 29, 1943; children—Brigitte, Karin, Claudia. Became meteorologist Reichswetterdienst, 1938-45; observer Zengralanstalt für Meteorologie, Vienna, 1945-62; asso. prof. theoretical meteorology U. Vienna, 1962—. Mem. Austrian Physics Soc., Austrian Meteorol. Soc. Author: Theorie des Wärmehausalts einer Schneedecke, 1947; Methoden und Probleme der Wettervorhersage, 1954; Synoptische Aerologie, 1961. Home: Kupelwiesergasse 5/4, Vienna 8, Austria.

REUTHER, Walter, horticulturist; b. Kaitaia, New Zealand, Sept. 21, 1911; s. A.W.G. and Martha (Krüger) R.; came to U.S., 1919, naturalized, 1937; B.S., U. Fla., 1933; Ph.D., Cornell U., 1940; m. Flora Astbury Nelson, Aug. 10, 1935; children—David Walter, Charles Arthur. Asst. prof. pomology Cornell U., 1940; from asso. to prin. horticulturist U.S. Dept. Agr., Indio, Cal. and Orlando, Fla., 1941-55, mem. hort. crops research adv. com., 1963—; head dept. horticulture U. Fla., Gainesville, 1955-56; chmn. dept. hort. sci., prof. horticulture U. Cal., Riverside, 1956-66, prof. horticulture, 1966—. Mem. team adv. Greek Govt., 1963-64; cons. Rockefeller Found., 1966-67. Mem. Am. Soc. for Hort. Sci. (pres. 1962-63, chmn. bd. 1963-64), Internat. Com. on Plant Analysis and Fertilizer Problems (pres. 1959-62), Soil Sci. Soc. Fla. (v.p. 1954-55), Am. Inst. Biol. Sci., A.A.A.S., Am. Soc. Plant Physiology, Fla. State Hort. Soc., Rio Grande Valley Hort. Soc., Cal. Avacado Soc., Date Growers Inst., Sigma Xi, Gamma Sigma Epsilon. Editor: Plant Analysis and Fertilizer Problem, 1961; The Citrus Industry, volumes I and II, 1967. Contbg. author: Fruit Nutrition, 1954; others. Research, publs. on mineral nutrient, environmental effects on tree growth and fruiting. Home: 5002 Rockledge Dr., Riverside, Cal.*

REUTOV, Oleg Aleksandrovich, Russian organic chemist; b. Sept. 5, 1920; grad. Moscow U., 1941. Instr. Moscow U., 1945-54, prof., 1954—. Mem. USSR Acad. Scis. Author: Fundamentals of Theoretical Organic Chemistry, 1967; Rection Mechanisms of Organometallic Compounds, 1968. Research and publs. on mechanism of organometallic compound synthesis via diazo compounds, mechanism of homolytic and electrophilic substitution reactions in hydrocarbon atom, rearrangement of free alkyl radicals and carbonium ions in solution; developer methods to synthesize organometallic compounds. Address: Dept. Chemistry, Moscow University, Moscow B-234, USSR.*

REVELLE, Roger, Am. oceanographer; b. Seattle, Mar. 7, 1909; s. William Roger and Ella Robena (Dougan) R.; A.B. at Pomona College, 1929, D.Sc., 1957; student Claremont Coll., 1929-30; Ph.D., U. Cal., 1936; m. Ellen Virginia Clark, June 22, 1931; children—Anne (Mrs. Shumway), Mary (Mrs. Paci), Carolyn, William Roger. Teaching asst. Pomona Coll., 1929-30; teaching asst. U. Cal., 1930-31, research asst. Scripps Instn. Oceanography, 1931-36, instr., 1936-41, asst. prof., 1941-46, asso. dir., 1948-50, prof. oceanography, since 1948, acting dir., 1950-51, dir., 1951—, director LaJolla campus, 1958-64; head Center for Population Studies, Harvard U.; civilian cons. Office Naval Research, 1948; chmn. spl. commn. oceanic research Internat. Council Sci. Unions, 1957—; mem. for N.A. of UNESCO, internat. adv. com. on marine scis. Adv. bd. Mus. Arts Soc., La Jolla, 1947—; trustee La Jolla Town Council, 1949—. Dir. First Nat. Bank of San Diego, Scripps Clinic and Research Found.; dir. Theatre and Arts Found. San Diego County. Served with USN, 1941-47, naval officer Radio and Sound Lab., San Diego, Cal., 1941-42, with Bur. Ships, Navy Dept., 1942-46, staff Joint Task Force No. 1, 1946; Office Naval Research, 1946-47. Fellow San Diego Soc. Natural History, Geol. Soc. Am.; mem. Am. Assn. Petroleum Geologists, Am. Geophys. Union (v.p. oceanography sect. 1950-53, pres., 1956-59), Nat. Acad. Sci. (commn. oceanography, 1957—), Nat. Sci. Found. (commn. math., phys. and engring. scis.), American Academy of Arts and Sciences, A.A.A.S., American Philosophical Society, National Research Council, Am. Meteorol. Soc., Soc. Limnology and Oceanography, Geol. Soc. Washington, Western Soc. Naturalists. Contbr. articles profl. publs. Research in phys. oceanography, geology of sea floor; population studies and natural resource analysis. Home: 7348 Vista del Mar, La Jolla, Cal.

REVERCHON, Julien, naturalist, botanist; b. Diemoz, France, 1837; s. J. Maximilien and Florine (Pete); ed. privately; m. Marie Henry, 1858. Came to U. S., 1856; farmer nr. Dallas; prof. botany Baylor Coll. Medicine and Pharmacy, Dallas. Various bot. species named in his honor. Mem. Torrey Bot. Club. Contbr. numerous articles to bot. and hort. jours. Collected numerous species of Tex. flora, including numerous new species. Died 1905.

REVERDIN, Auguste, Swiss surgeon; b. Geneva, Switzerland, Dec. 2, 1848; M.D., Strasbourg, France, 1874. Surgeon, Hosp. Geneva (Switzerland); prof. Faculty Medicine, Geneva; prof. surgery Polyclinic, 1899. Recipient Laborie prize Acad. Medicine, 1895. Mem. Soc. Surgery Paris, Soc. Obstetrics Paris, French Acad. Medicine. Author: L'Antisepsie et l'asepsie chirurgicals; Souvenire gynécologues, 1906. Improved and invented various surg. instruments including modification of Jacques Reverdin's suturing needle. Died Geneva, June 18, 1908.

REVERDIN, Frédéric, Swiss chemist; b. Cologny, Switzerland, 1849; ed. Fed. Polytechnicum, Zurich, hon. doctorate, 1924; hon. doctorate U. Geneva, 1924; research chemist for dyestuff factory (later named Societé Chimique des Usines du Rhone), nr. Geneva, 1872-91; transferred to Meister Lucius Brüning, Höchst, 1891-1906; with l'École de Chimie, Geneva, 1907-30. Swiss del. 7th Consress Applied Chemistry, London, 1909, 8th congress, N.Y., 1912. Author: (with Noelting) Tables of Naphthalene Derivatives, 1880, rev., 1894, 1927; also articles. Translator: (into French with de la Harp) Textbook of Organic Chemistry (Fittig), 1878. Research on triphenglmethane colors and manufacture of alkyl chlorides from alcohols with hydrochloric acid under pressure; discovered migration of iodine during nitration of p-iodoanisole; made series of studies of nitration of substituted phenols and of behavior of nitramines with sulphuric acid. Died 1931.

REVERDIN, Jacques-Louis, surgeon; b. Geneva, Switzerland, Aug. 28, 1842; B.A. in Arts and Scis., Paris, 1862. Became intern, 1865; prof. external pathology and operative medicine Faculty Medicine, Geneva; also prof. surg. polyclinic, Geneva. Recipient Gold medallion, 1869; Prize of Civiale, 1870; Bronze medallion Faculty of Paris, 1870; Amusart prize Acad. Medicine, 1872. Mem. Anatomy Soc. (sec., v.p.), Surgery Soc. Paris. Author: Greffe épidermique experiénce faite dans le service de M. le Docteur Guyon, à l'hôpital Necker; Leçons de chirurgie de guerre. Performed epidermic graft named after him, 1869; described operative myxedema (disease caused by removal of thyroid called Reverdin's disease); invented suture needle. Died Jan. 9, 1929.

REVERS, Wilhelm Josef, German psychologist; b. Mulheim/Cologne, Germany, Aug. 18, 1918; s. Heinrich and Margarete (Lommersum) R.; Ph.D., U. Bonn (Germany); m. Erni Haupt, July 24, 1946; children—Sigrid, Rainer-Claus, Peter. With U. Bonn, 1945-46, became sci. collaborator Psychol. Inst., 1946; became sci. asst. Psychol. Inst., U. Würzburg (Germany), 1948, now prof. psychology. Mem. Wiener Arbeitskreis für Tiefenpsychologie, Profl. Union German Psychologists, Görresgesellschaft, Allgemein Gesellschaft für Philosophie in Deutschland, German Psychol. Soc. Author: Psychologie der Langeweile; Charakterprägung und Gewissensbildung; Det Thematische Apperzeptionstetst; Ideologische Horizonte der Psychologie; Editor: Jahrbuch für Psychologie, Psychotherapie und Medizin Anthropologie. Home: Schiessausstrasse 21, Würzburg, W. Germany.

REVESZ, Laszlo, physician, radiobiologist; b. Nagykaroly, Hungary, Mar. 31, 1926; s. Joseph and Lilly (Halasz) R.; student Kolozsvar (Hungary) Med. Sch., 1945-46, Med. Sch., Budapest, Hungary, 1947-48; Dr.med.univ., U. Innsbruck (Austria), 1950; Med.lic., U. Stockholm (Sweden), 1956, Med.dr., 1958. Asst. physician neurosurgery, Stockholm, 1951-52; research fellow Karolinska Institutet, Stockholm, 1952-58, asso. prof. tumor biology, 1958—; asst. physician radiotherapy, Radiumhemmet, Stockholm, 1959. Research and numerous articles in radiobiology and radiopathology of tumors; developed methods for quantitive estimation of tumor growth and cellular radiosensitivity, for immunizing with heavily irradiated cells; discovered growth stimulating effect of radiation killed cells; established role of naturally occurring sulphydryls in determining radiosensitivity and role of oxygen for cellular recovery. Home: 10 Olaus Magnusvag, Johanneshov-Stockholm. Office: Karolinska Institutet, Stockholm 60, Sweden.*

REVUZ, André, French mathematician; b. Paris, France, Mar. 15, 1914; s. Edouard and Suzanne (Sabouret) R.; ed. Ecole Normale Supérieure, Paris, 1934-37; m. Chazottes, Apr. 25, 1935; children; Daniel, Jean, Jacqueline Authier, François, Christine. Asst. prof., Technical U., Istanbul, Turkey, 1945-49; research assoc., Centre National de Recherche Scientifique, Paris, 1950-52; chef de travaux, Faculty of Sci., Paris, 1952-55; lectr., Faculty of Sci., Bordeaux, 1955-56; prof., Faculty of Sci., Poitiers, 1956-67; prof. Faculty of Sci., Paris, 1967—. Mem., French Math. Soc.; Assn. of Math. Profs., France. Author: Mathematique Moderne. Mathematique Vivante, 1963; numerous additional publications. Investigations in measure theory, general topology; study in mathematical education. Home: 16 rue de Rome 78 - Les Essarts le Roi, France. Office: 11 rue Pierre Curie, Parie 5e, France.*

REX, Daniel Ferrell, Am. meteorologist; b. Wichita, Kan., Dec. 4, 1916; s. Loren Edgar and Leda (Ferrell) R.; B.S., Mass.Inst. Tech., 1939; M.S., U. Tenn., 1941; Ph.D., U. Stockholm, Sweden, 1951; m. Mary Emory Hill, Oct. 13, 1942; children—Lloyd Ferrell, Aline Beverley, Anne Loren. Commd. ensign USN, 1941, advanced through grades to capt., 1959, officer in charge Navy Weather Research Facility, 1958-60, dir. meteorology Supreme Allied Command Atlantic, 1960-62; asso. dir. Nat. Center for Atmospheric Research, Boulder, Colo., 1963—. Asst. chief for tech. services U.S. Weather Bur., 1959-60. Mem. Am. Meteorol. Soc. (councilor 1952-54, 57-59), Am. Geophys. Union, Royal Meteorol. Soc., U.S. Naval Inst. Editor: The World Survey of Climatology, vol. 3, 1966; also articles in profl. jours. Identified and described a kind of harmonic instability affecting multi-phase power rectifiers; pioneer in use of radar as weather observing tool; described relationships between large-scale atmospheric circulation patterns and regional climate. Home: "Stonecroft", Lee Hill Rd., Boulder 80302. Office: Nat. Center for Atmospheric Research, Boulder, Colo. 80302.*

REX, Robert Walter, Am. geochemist; b. N.Y.C.; s. Frederick and Olga R.; A.B. cum laude Harvard, 1951; M.S., Stanford, 1953; Ph.D., U. Cal. Scripps Instn. Oceanography, 1958. Groundwater geologist U.S. Geol. Survey, Ida., 1951-52; oceanographer U.S. Navy Electronic Lab., San Diego, 1953; research geologist Scripps Instn. Oceanography, U. Cal., San Diego, 1954-57; sr. research asso. geochemistry Chevron Research Co., La Habra, Cal., 1958-67; prof. geology U. Cal. at Riverside, 1967—. Asso., Woods Hole (Mass.) Oceanographic Instn., 1962—; research asso. Hawaii Inst. Geophysics, 1967—. Fellow Geol. Soc. Am.; mem. Clay Minerals Soc. (councilor 1967—), Am. Geophys. Union, Arctic Inst., Mineral. Soc. Am. Contbg. author: The Sea, 1962; Papers in Marine Geology, 1964. Research and publs. on eolian origin of portion red clays of Pacific Ocean; discovered role of stable free radicals in formation of humic acids, peat, and lignin decomposition products; demonstrated low temperature phase relationships for portions of system kaolinite-K-mica-chlorite-montmorillonite-phillipsite-K-feldspar. Home: 2780 Casalero Dr., La Habra, Cal. 90631.*

REY, Jean, French physician, chemist; b. Bugue, France, circa 1582-83; M.A., Montaubon; M.B., U. Montpellier (France), 1607, M.D., 1609. Practice medicine, Perigord, France, also Bugue. Author: Essais sur la recherche de la cause pour laquelle l'étain et le plomb augmentent de poids quand on les calcine, 1630. Attributed increase of weight of calcinated tin to its absorption of air, 1630. Died 1645.

REY, Jean-Alexandre, engr.; b. Ouchy, Switzerland, Aug. 11, 1861; student Ecole des mines, 1883-86. Mem. staff geol. chart of France; became dir. plant specializing in constrn. lighthouses, 1915. Mem. French Acad. Scis., 1930, Marine Acad. Author: Note sur les feux-éclairs à l'hule et à l'électrcité, 1896; Sur l'incandescence par la vapeur su pétrole appliquéz à l'éclair age des phares, 1901; Notice sur un nouveau système de phares à réflecteurs metalliques, 1903; De la portée des projecteurs de lumière électrique, 1905. Developed (with Blondel) search light with parabolic sectors, alumnium bronze, reflector, projector with non-blinding light; perfected (with Rateau) turbine, 1893; invented 1st thermocompressor with liquid vaporization at normal pressure, 3-phase apparatus, 1902, submarine motor parts, 1904, 1st Rateau-type tubocompressor, 1908; studies on laying of underwater mines. Died Paris, Dec. 25, 1935.

REYERSON, Lloyd Hilton, Am. chemist; b. Dawson, Minn., May 1, 1893; s. John Emil and Lydia (Hilton) R.; A.B., Carleton Coll., 1915, D.Sc., 1956; A.M., U. Ill., 1917; Ph.D., Johns Hopkins, 1920; m. Nelle Nickell, Mar. 7, 1918; children—Jean Elizabeth (Mrs. Albert H. Moseman), James Hilton (dec.). Faculty, U. Minn., Mpls., 1919—, prof. phys. chem., 1930-61, prof. emeritus, 1961—, adminstrv. head Sch. Chemistry, 1937-54; staff scientist New Eng. Inst. for Med. Research, Ridgefield, Conn., 1962—. Chmn., Minn., Dakota Resources Devel. Commn., 1943-54. Recipient Knight's Cross 1st class, Royal Order St. Olaf (Norway), 1950, Distinguished Alumnus award, Carleton Coll., 1955; Guggenheim Found. fellow, Berlin, 1927-28, Copenhagen, 1958. Fellow A.A.A.S.; mem. Am. Chem. Soc. (mem. council policy com. 1952-54, 56-61), Am. Inst. Chemists (hon., councillor at large 1958-61, pres. 1965-66), Phi Beta Kappa, Sigma Xi (recipient Distinguished Service award Minn. chpt. 1962), Phi Lambda Upsilon, Alpha Chi Sigma. Patentee in field. Research in colloid and surface chemistry; discovered super magnetism in very finely divided para magnetic metals, deuterium exchange from adsorbed D20 on very dry proteins; discovered action of gaseos hydrochloric acid on dry nylon. Home: 5 Huckleberry Lane. Office: New Eng. Inst. for Med. Research, Grove St., Ridgefield, Conn. 06877.*

REYES, Teodoro Flores, Mexican geologist; b. Mexico City, 1873; s. José Conrado and Jerónima (de los Reyes Niño Ladrón) Labastida; ed. Nat. Engring. Sch., U. Mexico, graduated as assayer of metals, 1894, surveyor and hyrrographic engr., 1895, mining and metall. engr., 1900; worked in mining dists. of Pachuca, Real del Monte, Hidalgo, Guanajuato, Zacatecas; successively head exploration sect., chief research geologist, research geologist emeritus, dir. Geol. Inst., 1903-55 (except for 2 years as mining cons. in Cuba); faculty Sch. Engring., U. Mexico, 41 years; mem. Internat. Geologic Congress, Mexico, 1906, Toronto, 1913, London, 1948; mem. exec. organizing com. 20th session Internat. Geologic Congress, 1956. Fellow Geol. Soc. Am. mem. Am. Inst. Mining and Metall. Engrs., Soc. Engrs. and Architects of Mexico, Sociedad de Geografía y Estadística. Contbr. to geol. publs. Studied mineral deposits in Guanajuato, Oaxaca, Guerrero, Texcoco Lake, El Oro, Tlalpujahua, Hidalgo, Sonora; investigated Paricutin and other volcanoes. Died 1955.

REYHER, Samuel, German mathematician; b. Schlensingen, Germany, 1635; student, Leipzig, Germany, Leiden, Holland. Prof. math., Keil, Germany; preceptor to Prince of Gotha; counsellor to Duke of Saxe-Coburg. Author: Mathesis Mosaika Biblica, 1679. Translated to German with algebraic demonstrations, Elements by Euclid. Studies on human voice, including fact that each sound has basic note plus overtones, different vowels are asso. with different tones; developed math. explanations for passages of Pentateuch. Died Kiel, 1714.

REYNAUD, Emile, French inventor; b. Montreuil, 1844; studied movement; invented praxinoscope (1877), perfected it, then invented optical theatre, a device for projecting movements onto screen (1888), used in Musée Grévin (1892-1900). Died 1918.

REYNBERG, Samuil Aronovich, Russian roentgenologist; b. 1897; grad. 1st Leningrad Med. Inst., 1924; D.Med. Sci. Head chair child roentgenology Leningrad Pediatric Med. Inst., 1927-30; head chair roentgenology Leningrad Postgrad. Med. Inst., 1930-41; head 1st chair roentgenology and radiology Central Postgrad. Med. Inst., Moscow, 1943—. Sci. cons. 1st Moscow Clin. Hosp., RSFSR Ministry Health; mem. learned med. council USSR Ministry Health. Mem. All-Union (Presidium mem.), All-Russian (Presidium mem.) socs. roentgenologists and radiologists, also other socs. Author over 350 works including: X-Ray Diagnosis of Bone and Joint Diseases, 1929; Pulmonary Tuberculosis and Stoppages of the Bronchial Passages, 1937; How To Work on a Medical Thesis, 1940; Essays on Military Roentgenology, 1942; Detection of Pulmonary Tuberculosis with the Aid of Group X-Ray Examinations, 1942; X-Ray Diagnosis of Cancer of the Stomach, 1952. Editor Roentgenology sect. Large Med. Ency., 2d edit.; mem. editorial council Herald of Roentgenology and Radiology, Clin. Medicine. Address: Central Postgrad. Med. Inst., pl. Vosstaniya 1-2, Moscow, USSR.

REYNEAU, Charles-René, French mathematician; b. Brissac, France, 1656. Oratorian, tchr. philosophy Toulon, France, later Pezenas; tchr. math. Angers Coll., 1683-1705. Mem. French Acad. Scis., 1716. Author: Analyse démontrée, 2 vols., 1736; La Science du calcul, 2 vols., pub. 1739. Died Paris, Feb. 24, 1728.

REYNOLDS, Charles Albert, Am. chemist; b. Colorado Springs, Colo., Apr. 1, 1923; s. Charles Albert and Lulu (Highsmith) R.; A.B., Stanford, 1944, M.A., 1946, Ph.D., 1947; m. Priscilla Anne Lynch, Feb. 20, 1953; children—Marcia Noel, Victoria Anne, Thomas Allen, Amy Christine, John William, Joseph Patrick. Faculty, U. Kan., Lawrence, 1947—, prof. chemistry, 1959—, asso. chmn. dept. chemistry, 1962—; mem. operations research group Army Chem. Corps., 1951-53. Mem. Am. Chem. Soc., Phi Beta Kappa, Sigma Xi, Phi Lambda Upsilon, Alpha Chi Sigma. Author: Principles of Analytical Chemistry, 1966; also articles. Developed new analytical methods for determination weak acids, aldehydes, ketones, amines and amides; research on complexions in aqueous and non-aqueous solutions. Home: 2209 Hill Ct., Lawrence, Kan. 66044.*

REYNOLDS, David H., Am. physician; b. East Liverpool, O., Apr. 13, 1919; s. Oscar A. and Mina (Daniels) R.; student Ohio State U., 1937-47; M.D., Duke, 1951; m. Marjorie Flower, Aug. 26, 1944. Asst. resident neurol. surgery Duke Hosp., 1955-56, instr. neurosurgery, 1956-57; practice medicine, specializing in neurosurgery, Miami, Fla., 1957—; chief div. neurosurgery Jackson Meml. Hosp., 1957—, acting chmn. dept. surgery, 1961-63; cons. neurosurgery VA Hosp., Coral Gables, Fla., 1957-58, sr. cons., 1958—; asso. prof., chief div. neurosurgery U. Miami. Sch. Medicine, 1958-62, prof. neurosurgery, 1962—; hon. cons. surgeon to Ministry Health Govt. Bahamas. Diplomate Am. Bd. Neurol. Surgery. Fellow A.C.S.; mem. Congress Neurol. Surgeons, Harvey Cushing Soc., Am. Soc. Neurol. Surgeons, A.M.A., Am. Acad. Neurol. Surgery, Neurol. Soc. Am., Forum U. Neurosurgeons, Neurol. Soc. Am., Sigma Xi. Research, publs. on convulsive disorders in children; hypothermia, hypotension in neurosurgery, cranial fractures. Home: 1701 Espanola Dr. Office: 1700 N.W. 10th Av., Miami, Fla.*

REYNOLDS, Herbert Hal, Am. psychologist; b. Frankston, Tex., Mar. 20, 1930; s. Herbert Joseph and Avanell (Taylor) R.; B.Sc., Trinity U., Tex., 1952; M.Sc., Baylor U., 1958, Ph.D., 1961; m. Joy Myrla Copeland, June 17, 1950; children—Kevin Hal, Kent Andrew, Rhonda Sheryl. Mem. mil assistance adv. group to Japan, USAF, 1954-56; asst. prof. air sci., instr. psychology Baylor U., 1956-59, teaching and research fellow, 1959-61; chief research and tng. comparative psychology div. Aeromed. Research Lab., Holloman AFB, N.M., 1961-63, chief comparative psychology div., 1963-65, dir. research, dep. comdr., 1965—; commd. 2d lt. USAF, 1951, advanced through grades to lt. col., 1966; faculty U. N.M., 1962—; asso. prof. Baylor U., 1962—. Decorated Commendation medal for meritorious achievement, Commendation medal for meritorious service. Mem. Am., N.M. psychol. assns., A.A.A.S., Am. Assn. U. Profs., Sigma Xi. Research and publs. on space and space-related variables which are hazardous to man, early social-developmental influences on later response to stressful situations; formulation of comparative program in neurosciences. Home: 2845 Quay Loop. Office: Aeromed. Research Lab., Holloman AFB, N.M. 88330.*

REYNOLDS, James Emerson, Brit. chemist; b. Gooterstown County, Dublin, Ireland, Jan. 8, 1844; s. James Reynolds; M.D., Dublin U.; D.Sc. causa honoris; m. Janet Elizabeth Finlayson, 1875; 1 son, 1 dau. Practice medicine, few years; became keeper minerals Nat. Mus. Dublin, 1867; analyst Royal Dublin Soc., 1868-75; prof. chemistry Royal Coll. Surgeons, Dublin, 1870-75; prof. chemistry Trinity Coll., Dublin, 1875-1903; research staff Davy-Faraday Labs., London. Fellow Royal Soc., 1880; mem. Royal Coll. Physicians Dublin, Royal Coll. Physicians Edinburgh, Soc. Chem. Industry (became pres. 1891), Brit. Assn. (became pres. chem. sect. 1893), Chem. Soc. London (pres. 1902-08), Royal Soc. London (became v.p. 1902). Author: Experimental Chemistry for Junior Students, 4 vols., 1882; also articles. Discovered thiocarbamide or thiourea, 1868; 1st colloidal derivation of mercury, (basis of Reynold' acetone test), 1871; originated new teaching technique for exptl. chemistry; research on silicon compounds; discovered various groups of organic derivatives of silicon. Died Kensington, Eng., Feb. 18, 1920.

REYNOLDS, John Hamilton, Am. physicist; b. Cambridge, Mass., Apr. 3, 1923; s. Horace Mason and Catharine (Coffeen) R.; A.B., Harvard, 1943; M.S., U. Chgo., 1948, Ph.D., 1950; m. Helen Genevieve Marshall, Mar. 18, 1950; children—Amy, Horace Marshall, Brian Marshall. Research asst. Electroacoustic Lab., Harvard, 1941-43; asso. physicist Argonne Nat. Lab., Chgo., 1950; physicist U. Cal., Berkeley, 1950—; dir. Lidseen of N.C.; cons. Nuclide Corp. Guggenheim fellow, 1956-57; N.S.F. fellow, 1963-64; recipient Franklin Inst. Wetherill medal, 1965; J. Lawrence Smith medal Nat. Acad. Scis., 1967. Research, publs. on mass spectroscopy; isotope abundance anomalies in nature; meteorite studies; extinct radioactivity; geochronology. Office: Dept. Physics, U. Cal., Berkeley, Cal. 94720.*

REYNOLDS, Joseph Benson, Am. mech. engr.; b. New Castle, Pa., May 17, 1881; s. Peter Specard and Lydia (Kemp) R.; B.A., Lehigh U., 1907, M.A., 1910, Ph.D., 1919; m. Chloey Bessie Graham, July 2, 1908; children—Peter Graham, Jane Niblock (Mrs. W.A. Parsons), Joseph Benson. Mem. faculty Lehigh U., Bethlehem, Pa., 1907-48, prof. math. and theoretical mechanics, 1927-45, head dept. math. and astronomy, 1945-48; ret., 1948. Dir. spring research Am. Soc. M.E., 1934—. Mem. Am. Math. Assn., Pa. Acad. Sci., A.A.A.S., Phi Beta Kappa, Sigma Xi, Pi Mu Epsilon. Author: Analytic Mechanics, 1929; Elementary Mechanics, 1928, 34; also numerous articles. Research on applications of math. to engring.; authority on minor planet Lehigh. Home: 2075 Andover Lane, Erie, Pa. 16509.*

REYNOLDS, Joseph Melvin, Am. physicist; b. Woodlawn, Tenn., June 16, 1924; s. James J. and Frances (Shelby) R.; student David Lipscomb Coll., 1942-44; B.A., Vanderbilt U., 1946; M.S., Yale, 1947, Ph.D. (Sheffied-Loomis fellow), 1950; m. Ruth Anna Heise, Sept. 2, 1950; children—Molly Elizabeth, John Shelby, Wendy Lee. Asst. prof. physics La. State U., 1950-54, asso. prof., 1954-58, prof., 1958-62, head dept., 1962-65, Boyd prof. physics, 1962—, v.p. grad. studies and research devel., 1965—; vis. prof. (Guggenheim fellow), Kammerlingh-Onnes Lab., U. Leiden, The Netherlands, 1959. Mem. Nat. Sci. Bd. NSF; v.p. Gulf Univs. Research Corp. Fellow Am. Phys. Soc.; mem. A.A.A.S., Am. Ordnance Assn., Nat. Council Univ. Research Adminstrs., Sigma Xi. Contbr. articles in field to profl. jours. Research on low temperature physics; liquid helium; magnetic properties of metals; superconductivity in pure metals and alloys; transport effects and nuclear magnetic resonance in metals. Home: 998 W. Lakeview Dr., Baton Rouge.

REYNOLDS, Lloyd George, Am. economist; b. Wainwright, Alta., Can., Dec. 22, 1910; s. George Franklin and Dorothy (Carl) R.; B.A., U. Alta., 1931, LL.D., 1958; M.A., McGill U., 1933; Ph.D., Harvard, 1936; m. Mary C. Trackett, June 12, 1937; children—Anne (Mrs. James F. Skinner), Priscilla, (Mrs. Kermit Roosevelt, Jr.), Bruce. Came to U.S., 1934, naturalized, 1940. Mem. faculty Harvard, 1936-39, Johns Hopkins, 1939-45; chief economist War Manpower Commn., 1942-43; pub. mem. appeals com. Nat. War Labor Bd., Washington, 1943-45; faculty Yale, New Haven, 1945—, Sterling prof. econs., 1952—, dir. Econ. Growth Center, 1961—. Cons., mem. adv. bds. univs., govt. agys.; dir. program in econs. and bus. adminstrn. Ford Found., 1955-57. Bd. dirs. Nat. Bur. Econ. Research. Rockefeller fellow, 1951; Guggenheim fellow, 1955, 67; Ford fellow, 1959, Fellow Am. Acad. Arts and Scis.; mem. Am. Econ. Assn. (v.p. 1959), Indsl. Relations Research Assn. (pres. 1955), Royal Econ. Soc., Am. Statis. Assn., Econometric Soc. Author numerous books, including: Labor Economics and Labor Relations, 1949, The Structure of Labor Markets, 1951, Economics: A General Introduction, 1963, also articles. Analyzed operation of labor markets and the determination of relative wage rates for various occupations; analyzed wage and employment behavior, and optimal wage policy in early industrialization of underdeveloped countries. Home: 75 Old Hartford Turnpike, Hamden, Conn. 06517.*

REYNOLDS, Osborne, Brit. engr., physicist; b. Belfast, Ireland, Aug. 23, 1842; s. Osborne and Jane (Hickman) R.; ed. Queen's Coll., Cambridge (Eng.) U., fellow, 1867; m. Charlotte Chadwick, 1868; m. 2d, Annie Charlotte Wilkinson, 1881; 3 sons, 1 dau.

Worked in office C.E., 1867-68; prof. engring. Owens Coll., Manchester, Eng., 1868-1905, ret., 1905. Fellow Royal Soc. (Gold medal 1888), 1877. Author: Papers on Mechanical and Physical Subjects, 3 vols., 1900-03. Contbd. laws of flow of water in pipes, recognized critical velocity at which flow changes character (Reynold's number), ratio characterizing dynamic state of fluid; studied water flow in rivers; designed turbines, centrifugal pumps; determined mech. equivalent of heat; developed theory of lubrication; studied atmosphere refraction of sound and group-velocity of water waves; developed theory of radiometer. Died Somerset, Eng., Feb. 21, 1912.

REYNOLDS, S(amuel) A(llen), Am. chemist; b. Stithton, Ky., Dec. 5, 1921; s. M. Melvin and Evelyn (McFarland) R.; A.B., Emory U., 1942; M.S., U. Tenn., 1953; m. Mary Louise Brown, Oct. 5, 1942; children—Thomas M., David A. Explosives chemist Ala. Ordnance Works, Sylacauga, 1942-44; staff Oak Ridge (Tenn.) Nat. Lab., 1944—, group leader radiochemistry group, analytical chemistry div., 1946-61, radiochem. and isotope specialist, analytical and isotopes divs., 1961—; instr. Oak Ridge Sch. Reactor Tech., 1950-58. Cons., USPHS, 1960—; mem. radioactivity standards subcom. NRC, 1955—. Mem. Am. Pub. Health Assn. (subcom. on radiol. methods 1966—), Am. Chem. Soc., Am. Nuclear Soc., Am. Soc. for Testing and Materials, Research Soc. Am., A.A.A.S., Sigma Xi. Contbg. author: Progress in Nuclear Energy, Process Chemistry, Vol. 2, 1958; Handbook of Analytical Chemistry, 1963; Analysis of Essential Nuclear Reactor Materials, 1964. Research and publs. on analytical radiochemistry, activation analysis, radioisotope devel. and quality control, environmental radioanalysis. Home: 110 Kingfisher Lane, Oak Ridge 37830. Office: Oak Ridge Nat. Lab., Box X, Oak Ridge, Tenn. 37830.*

REYNOLDS, Samuel Godfrey, Am. inventor; b. Bristol, R.I., Mar. 9, 1801; s. Greenwood and Mary (Caldwell) R.; m. Elizabeth Anthony, 1823; m. 2d, Catherine Ann Hamlin, Nov. 18, 1845; 5 children. Invented machine for making wrought-iron nails and rivets, patented 1829, patented improvements to original machine, 1835; patented a spike-making machine, went directly to Eng., secured financial backing for manufacture of machines from Coates & Co., bankers, also obtained patents in Eng., Holland, Belgium, France; patented machinery for heading and pointing pins, 1845; returned to U. S. 1850; invented horse-nail machinery, patented 1852, patented improvements, 1866, 67; perfected steam plow, invented a rotary plow. Died Bristol, R.I., Mar. 1, 1881.

REYNOLDS, Samuel Robert Means, Am. physiologist; b. Swarthmore, Pa., Dec. 9, 1903; s. Walter Doty and Elizabeth Brown (Means) R.; A.B., Swarthmore Coll., 1927, M.A., 1928, D.Sc., 1950; Ph.D., U. Pa., 1931; Dr. hon. causa, Cath. U., Chile; prof. honoris causa Faculty Medicine, Montevideo, Uruguay; m. Mary Elizabeth Curtis, Aug. 18, 1931; children—Nancy Tupper, Harriet Jeffers (Mrs. Fred C. Clark). Mem. faculty Swarthmore Coll., 1927-30, U. Pa., 1929-30, Western Res. U., 1932-33, L.I. Coll., 1933-41, Carnegie Inst., Washington, 1941-56; prof., head dept. anatomy U. Ill., Chgo., 1956—; vis. prof. fgn. and U.S. univs.; cons. USPHS. NRC fellow, 1931-32; Guggenheim fellow, 1937-38, 50; Commonwealth Fund fellow, 1965. Fellow A.A.A.S., N.Y. Acad. Sci., Am. Gynecol. Soc. (hon.), Am. Coll. Obstetrics and gynecology (hon. aso.); mem. Am. Physiol. Soc., Am. Assn. Anatomists (2d v.p. 1962-64), Soc. Exptl. Biol. Medicine (council 1955-56), HarveySoc., Am. Soc. Naturalists, Centro Americana de Cardiologia (hon.), fgn. biol. and gynecol. socs., Phi Beta Kappa, Sigma Xi, others. Author: Physiology of the Uterus, 1939, 2d edit., 1949; Physiological Bases of Obstetrics and Gynecology, 1952; Clinical Measurements of Uterine Forces in Pregnancy and Labor, 1954. Research and publs. on the nature of the uterus; effects of hormones and uterine growth, contractions; mechanisms of pregnancy; placental functional anatomy; fetal circulation and effects of gravity. Home: 2052 Lincoln Park West, Chgo. 60614.*

REYNOLDS, William Norman, English physicist; b. London, Nov. 27, 1925; s. Henry and Mary (Almond) R.; student King's Coll., London, 1943-45, U. Coll., London, 1948-51; Ph.D., U. London, 1952; m. Hilda Mary Sullivan, Aug. 23, 1958; children—Gregory Hugh, Isabel Mary Clare. Research physicist U. Reading (Eng.), 1951-54; sr. sci. officer Royal Naval Sci. Service, 1954-58; prin. sci. officer Atomic Energy Research Establishment, Harwell, Eng., 1958-68. Research and publs. on semi-conducting materials, including germanium, silicon and indium phosphide; developed theory of radiation damage in graphite, especially for its behavior as moderator in nuclear power reactors. Home: Pentagon, East Hendred, Wantage, Eng. Office: Atomic Energy Research Establishment, Harwell, Berkshire, Eng.*

RHABDAS, Nicholaos Artabasdos (or Artavasdan), Byzantine mathematician; b. Smyrna; flourished 1341; children include Paul. Author of 2 letters on arithmetic; interest in numeration and computation; made use of new numerals, Greek letters and finger notation; edited Planudes' work on Hundu arithmetic.

RHAZES (or Razi, al-Razi, Abu-Bakr Muhammad ibn Zakariyya), Arbian physician; b. Ray, nr. Teheran, Iran, 850-60; pursued philosophy; turned to study of medicine (under Abu al-Hasan ben Sahl ben Rabbin) at age 30; became head, provincial hosp., Ray; named head of hosp. in Baghdad; later moved to other courts. Author: Liber continens (med. ency. which may have been compiled posthumously by students), also 100 to 200 other documents (mostly lost). Famed as a clinician and a writer; accepted Galenist view of medicine; classified all matter into animal, vegetable or mineral; 1st to report smallpox, circa 910, also measles; 1st to suggest blood as cause of infectious diseases; contbd. to gynecology, obstetrics, ophthalmic surgery; wrote on med. innovations, including use of animal gut for sutures, use of plaster of paris for casts, chem. remedies such as mercurial ointments and sulphuric acids; also wrote on children's diseases, alchemy. Died 923-32.

RHEITA, Anton Maria Schyrlaus von, see von Rheita, Anton Maria Schyrlaus.

RHETICUS (or RHAETICUS), see Joachim, Georges.

RHIJN, Pieter Johannes van, see van Rhijn, Pieter Johannes.

RHINE, Joseph Banks, Am. parapsychologist; b. Waterloo, Pa., Sept. 29, 1895; s. Samuel Ellis and Elizabeth (Vaughan) R.; B.S., U. Chgo., 1922, M.S., 1923, Ph.D., 1925; postgrad. Harvard, 1926-27, Duke, 1927-28; m. Louisa Ella Weckesser, Apr. 8, 1920; children—Robert Eldon, Sara Louise (Mrs. Ben W. Feather), Elizabeth Ellen, Rosemary (Mrs. Richard R. Trevarthen). Asst. plant physiologist Boyce Thompson Inst., Yonkers, N.Y., 1923-24; instr. botany W.Va. U., 1924-26; instr. psychology Duke, Durham, N.C., 1928-34, asst. prof., 1934-37, asso. prof., 1937-42, prof., 1942-50, dir. parapsychology lab., 1935—, exec. dir. Found. for Research on Nature Man, 1964—. Mem. A.A.A.S., Parapsychol. Assn., Phi Beta Kappa, Sigma Xi. Author: Extrasensory Perception, 1934; New Frontiers of the Mind, 1937; The Reach of the Mind, 1947; New World of the Mind, 1953; (with J.A. Greenwood, J.G. Pratt, B.M. Smith, C.E. Stuart) Extrasensory Perception After Sixty Years, 1940; (with J.G. Pratt) Parapsychology, Frontier Science of the Mind, 1957. Cons., Jour. Parapsychology, 1964—. Home: R.D. 3, Hillsboro, N.C. 27278. Office: Box 6847, College Sta., Durham, N.C. 27708.*

RHINE, Louisa Ella, Am. parapsychologist; b. Sanborn, N.Y., Nov. 9, 1891; B.S., U. Chgo., 1919, M.S., 1921, Ph.D., 1923; m. Joseph Banks Rhine; 3 daus., 1 son. Staff mem. Parapsychol. Lab., Duke, 1948-62; now research dir. Inst. for Parapsychology research br. Found. for Research into Found. in Nature of Man. Mem. Parapsychol. Assn. (charter). Author: Hidden Channels of the Mind, 1961; also articles. Co-editor, Jour. Parapsychology. Research in spontaneous cases of psi phenomenon, psychokinesis.

RHINELANDER, F(rederic) W(illiam), Am. physician; b. Middletown, Conn., May 25, 1906; s. Philip M. and Helen (Hamilton) R.; A.B., Harvard, 1928, M.D., 1934; B.A., Oxford (Eng.) U., 1931; m. Julie C. Hale, Sept. 12, 1947; children—Eric Hale, Clare Hamilton, Laura Pierson. Asst. orthopedic surgery Harvard Med. Sch., 1941-47; asst. clin. prof. U. Cal. Sch. Medicine, San Francisco, 1947-55; faculty Sch. Medicine, Western Res. U., Cleve., 1955—, prof., 1964—. Cons., U.S. Army, 1947-48, USAF, 1951-55. Diplomate Am. Bd. Orthopaedic Surgery. Mem. Orthopaedic Research Soc., A.M.A., Am. Rheumatism Assn., Am. Acad. Orthopaedic Surgeons, A.A.A.S. Research and publs. on healing of fractures and bone grafts. Office: Cleve. Met. Gen. Hosp., Cleve. 44109.*

RHINES, Frederick Nims, Am. metallurgist; b. Toledo, July 25, 1907; s. George Volney and Annie (Nims) R.; B.S.E. in Chem. Engring., U. Mich., 1929; Ph.D., Yale, 1933; m. Janet Adair Clark, July 5, 1941; children—Margaret Clark, Walden Clark. Research metallurgist Aluminum Co. Am., Cleve., 1929-31; contract metallurgist Naval Research Lab., Washington, summer 1931; instr. Yale U., 1933-34; mem. staff Metals Research Lab., Carnegie Inst. Tech., Pitts., 1934-59, Alcoa Prof. light metals, 1946-59; prof. metallurgy, head metallurgy lab., U. Fla., Gainesville, 1959—, chmn. dept. metallurgy, materials engring., 1963—. Cons. on non ferrous physical metallurgy, light metals, nuclear metallurgy, phase equilibria. Mem. Am. Inst. Mining and Metall. Engers. (chmn. Inst. Metall. div. 1949, Mathewson award 1932, 42), Am. Soc. Metals (Howe medal 1957, 60), Am. Soc. Testing Materials (chmn. subcom. 1943-65), (service award 1962), Inst. Metals (Gt. Britain), Iron and Steel Inst. (Gt. Britain), Internat. Soc. for Stereology, Am. Daffodil Soc., Sigma Xi, Gamma Alpha, Alpha Sigma Mu, Phi Eta Sigma, Scabbard and Blade. Author: Phase Diagrams in Metallurgy, 1956. Research and publs. on phase equilibria, oxidation of metals, diffusion in metals, sintering, fracture in metals, quantitative metallography, grain boundaries, freezing of alloys, gas-metal interaction. Home: 1540 NW 37th Terr., Gainesville, Fla. 32601.*

RHOADS, Cornelius Packard, Am. pathologist, physician; b. Springfield, Mass., June 20, 1898; s. George Holmes and Harriet (Barney) R.; A.B., Bowdoin Coll., 1920, D.Sc. (hon.), 1944; M.D. cum laude, Harvard, 1924; D.Sc. (honorary), Williams College, 1952; married Katherine S. Bolman, Sept. 9, 1936, Interne, dept. of surgery, Peter Bent Brigham Hosp., Boston, 1924-25, Trudeau fellow, Trudeau Sanatorium, N.Y.C., 1925-26; instr. in pathology Harvard Med. Sch., and asst. pathologist Boston City Hosp., 1926-28; asso. Rockefeller Inst. for Med. Research, 1928-33; asso. mem. in charge service for study hematologic disorders, 1933-39; pathologist Hosp. of Rockefeller Inst. for Med. Research, 1931-39, dir. Memorial Center for Cancer and Allied Diseases, N.Y.C., 1940-52, scientific dir., 1953—; dir. of The Sloan-Kettering Inst. for Cancer Research, 1945—, James Ewing Hosp., N.Y.C., 1950—; prof. pathology, dept. pathology, Cornell U. Med. Coll. 1940-52; prof. pathology, dept. biology and growth, Sloan Kettering Div., Cornell U. Med. Coll., 1952—. spl. cons. U. S. Public Health Service, Nat. Adv. Cancer Council, 1947—; cons., med. div., Chem. Corps. Army Chem. Center, Md., 1948-50. Col., M.C., A.U.S.; chief med. div., Chem. Warfare Service, 1943-45. Awarded Legion of Merit. Mem. com. to visit dept. of chemistry, Harvard. Trustee Kettering Foundation. Member National Research council (member sub. com. on blood substitutes 1940-42; chmn. blood procurement 1941-42; mem. com. for treatment of war gas casualties 1941-43; mem. com. on veterans med. problems 1945-47; chmn. com. on growth 1945-48; mem. com. on atomic bomb casualties, 1946; chmn. exec. com. of com. on growth, 1946-47; mem. adv. com. of chem.-biol. coordinator center 1946-47; mem. at large, div. of med. sciences 1946-49), Office of Sci. Research and Dev. (mem. com. on insect and rodent control 1945-46). Fellow A.C.P., N.Y. Acad. Medicine (mem. com. on public health relations 1942-43; v.p. 1943-45), A.A.A.S. (v.p. 1953, chmn. sect. med. sci. 1953), Am. Geriatrics Soc.; mem. Am. Pub. Health Assn., Am. Soc. Control Cancer, Am. Cancer Soc. (bd. dirs. 1941-46; exec. com. 1944-46; mem. N.Y.C. cancer com. 1943-51) Blood Transfusion Assn. (bd. 1940-51), Harvey Soc., Soc. Exptl. Biology and Medicine, Soc. of Med. Jurisprudence, Am. Assn. Pathologists and Bacteriologists, Am. Assn. for Cancer Research, Am. Indsl. Hygiene Assn., A.M.A., Am. Radium Soc., Am. Soc. for Clin. Investigation, Am. Soc. Exptl. Pathology, Am. Soc. Tropical Med., Armed Forces Chem. Assn., Assn. Am. Physicians, Med. Soc. of State of N.Y., Med. Soc. of County of N.Y. N.Y. Soc. Tropical Medicine, N.Y. Acad. Scis., N.Y. Zool. Soc. Contbr. med. articles in profl. jours. Died Aug. 13, 1959.

RHOADS, Paul Spotswood, Am. physician; b. Terre Haute, Ind., Mar. 12, 1898; s. Harry B. and Mary (Spotswood) R.; B.S., U. Chgo., 1922; M.D., Rush Med. Coll., 1924; m. Hester C. Chapin, Dec. 24, 1925; children—Goerge, Paula (Mrs. Robert Menary), Hester Mary (Mrs. Robert Bradbury), Emily (Mrs. Tom Johnson). Physician, Scarlet Fever Commn., 1926-35; faculty Rush Med. Coll., 1926-35; attending physician, head infectious disease service Evaston Hosp., 1933-41; attending physician Presbyn. Hosp., 1929-35, Cook County Contagious Hosp., 1938-48; attending physician, chief dept. medicine Chgo. Wesley Meml. Hosp., 1941—; faculty Northwestern U. Med. Sch., 1933—, prof. medicine, 1947—. Fellow A.M.A. (mem. com. on medicine and religion 1962—); mem. A.C.P., Chgo. Soc. Internal Medicine, Chgo. Med. Soc., Inst. Medicine Chgo., Soc. Exptl. Biology and Medicine, Am. Assn. Pathologists and Bacteriologist, Central Soc. for Clin. Research, Sigma Xi. Chief editor, Archives Internal Medicine, 1950 62. Contbr. articles to med. jours. on scarlet fever, diphtheria, focal infection pneumonia, rheumatic fever, antibiotic therapy, urinary tract infections. Home: 814 Roslyn Pl., Evanston, Ill. Office: 251 E. Chicago Av., Chgo. 60611.*

RHODES, Alan, biologist; b. Madras, India, July 19, 1915; s. Norman and Jessy (Horsfall) R.; B.Sc. with 1st class honors in botany, Leeds (Eng.) U., 1937, D.Sc., 1964; Ph.D., Cambridge (Eng.) U., 1946; m. Renée Taylor, July 2, 1945; children—Linda, Cherry, Jan-Marie. Dep.-head botany and plant physiology dept. Jealott's Hill Research Sta., Imperial Chem. Industries, Ltd., Bracknell, Berks., Eng., 1946-51; head biology dept. Glaxo Research Ltd., Sefton Park, Stoke Poges, Bucks., Eng., 1951—. Fellow Inst. Biology; mem. Soc. Chem. Industry, Soc. for Gen. Microbiology, Assn. Applied Biologists, Phytochem. Group, Soc. for Exptl. Biology, Royal Hort. Soc. Author: (with Derek L. Fletcher) Principles of Industrial Microbiology, 1966; also numerous articles. Research on plant growth regulators, gen. microbiology including antibiotics in plant protection, relationship between structure and biol. activity in griseofulvin; analogues biosynthesis and fermentation of griseofulvin; studies on submerged prodn. of griseofulvin and new antibiotic, venturicidin. Home: Linden Lea, Bagshot Rd., Bracknell, Berks. Office: Glaxo Research Ltd., Sefton Park, Stoke Poges, Bucks, Eng.*

RHODIN, Johannes Arne Gösta, electron microscopist; b. Lund, Sweden, Sept. 30, 1922; s. Johannes and Alma (Carlson) R.; M.D., Karolinska Inst., Stockholm, Sweden, 1945; Ph.D., U. Stockholm, 1954; m. Gunvor Thorstenson, Aug. 9, 1947; children—Anders, Erik. Came to U.S., 1958. Asst. prof. anatomy Karolinska Inst., 1954-57; faculty dept. anatomy N.Y. U. Sch. Medicine, N.Y.C., 1958-63; prof., chmn. dept. anatomy N.Y. Med. Coll., N.Y.C.,

1964—. Mem. Am. Assn. Anatomists, Am. Soc. Cell Biology, A.A.A.S., A.M.A., N.Y. Acad. Scis., N.Y. Soc. Electron Microscopists (past pres.), Reticuloendothelial Soc. Author: An Atlas of Ultrastructure, 1963; also articles. Research on ultrastructure of cells and tissues, ultrastructure of kidney, skin, microcirculation, devel. of tissues in human fetus by electron microscopy. Home: 50 Midland Av., Bronxville, N.Y. 10708.*

RHYNE, Willen Ten, Dutch physician, naturalist; b. DeVenter, circa 1640. Physician to Dutch settlement, Batavia, Java; assisted Van Rheede in writing Hortus Malabaricus. Author: Meditaticnes in Hippocrates Textum; also several dissertations, articles on E. Indian plants in Breyne's publn. on rare exotics. Collected information on Javanese plants; famous 17th Century naturalist.

RIBAUD, Gustave Marcel, French physicist; b. Conflans/Lanterne, France, Jan. 9, 1884; D. Natural Scis., Ecole Normale Supérieure de Paris (France). Prof. sci. U. Strasbourg (France), 1919-33, U. Paris, 1933-54, later emeritus. Mem. French Acad. Sci. Author: Four à Haute Fréquence; Les Hautes Températures; Etude des Flammes; Convection de la Chaleur. Developed high frequency induction furnace, 1919, patented similar furnace, 1926.

RIBBERT, Hugo, German antomist, pathologist; b. Hohenlimburg, Germany, Mar. 1, 1855; ed. Bonn, Germany, Berlin, Strassburg, France; prof. pathology, Zurich, Switzerland, Marburg, Germany, Göttingen, Germany, Bonn. Author new and improved edit. of Lehrbuches der Pathologie, 1901. Research and numerous publs. on bacteriology, cancer; discovered and introduced term chordoma for malignant tumor originating from embryonic remains of notochord, 1894. Died Bonn, Nov. 10, 1920.

RIBEREAU-GAYON, Jean Félix André, French agronomist; b. Bordeaux, France, Feb. 13, 1905; s. Daniel and Madeleine (Gayon) R.-G.; Dr.ès.sc., Faculty Sci., Bordeaux, France; dr.honoris causa, Tech. U. Lisbon (Portugal); m. Marie-Gabrielle Vallantin-Dulac, Oct. 17, 1927; children—Bernard, Pascal, Gilles, Marielle, Luc. Various positions in pvt. industry, 1927-48; prof. Faculty Sci., Bordeaux; dir Agronomic and Oenological Sta. Decorated Legion of Honor, Palms of Acad.; Order of Merit (Agr.). Research and numerous publs. (with E. Peynaud) in biochemistry and oenology. Home: 45, rue Théodore-Gardère, Bordeaux, France. Office: 341 avenue de la Libération, Talence (Gironde), France.

RIBNER, Herbert Spencer, physicist; b. Seattle, Apr. 9, 1913; s. Herman Joseph and Rose (Goldberg) R.; B.S., Cal. Inst. Tech., 1935; M.S., Washington U., St. Louis, 1937, Ph.D. in Physics, 1939; m. Lelia Carolyn Byrd, Oct. 29, 1949; children—Carol Anne, David Byrd. With Brown Geophys. Co., 1939-40; with NACA, Langley Lab., Va., 1940-49, sect. head, 1946-49, staff NACA Lewis Flight Propulsion Lab., Cleve., 1949-54; research asso. Inst. for Aerospace Studies, U. Toronto (Ont., Can.), 1955—, prof., since 1959—; vis. prof. U. Southampton (Eng.), 1960-61. Cons. to pvt. cos., govt. agys. Fellow Am. Inst. Aeros. and Astronautics (asso.), Acoustical Soc. Am., Canadian Aero. and Space Inst., also Am. Phys. Soc. Research, numerous publs. on X-rays, cosmic rays, propeller theory, wing theory, aerodynamics, turbulence, shock waves, jet noise, acoustics; devel. gravity meter; invented transonic propeller, differential flaps device, airfoil anemometer probe for turbulence. Home: 24 Hawksbury Dr., Willowdale, Ont., Can. Office: Inst. for Aerospace Studies, U. Toronto, Toronto, Ont., Can.*

RIBOT, Théodule Armand, French psychologist; b. Guingamp, France, 1839; Ph.D., l'Ecole normale; became prof. exptl. psychology Sorbonne, Paris, 1885, Collège de France, 1888; Mem. Académie des sciences morales et politiques. Author: Psychologie anglaise contemporaine, 1870; l'Hérédité, 1873; la Philosophie de Schopenhauer, 1874; la Psychologie allemande contemporaine, 1879; les Maladies de la mémoire, 1881; les Maladies de la volonté, 1883; les Maladies de la personnalité, 1885; Psychologie de l'attention, 1888; la Psychologie des sentiments, 1896; L'evolution des idées générales, 1897; Essai sur l'imagination gréatrice, 1900; Essai sur les passions, 1906; Problèms de psychologie affective, 1910; La vie inconsciente et les mouvements, 1913. Founder Revue philosophique, 1876. Pioneered in exptl. psychology. Died 1916.

RICAMO, Renato, Italian physicist; b. Trieste, Italy, Aug. 21, 1914; s. Vittorio and Olga (Giorovich) R.; Master degree in Physics, Bologna (Italy) U., 1936, Ph.D., 1949; m. Elena Bellingeri, Dec. 16, 1945. Research asst. physics dept. Zürich (Switzerland) Institut Tech., 1950-53; prof. physics Catania (Italy) U., 1954-64, founder, dir. Centro Sciiliano di Fisica Nucleare; prof. L'Aquila (Italy) U., 1965—, head physics dept. Mem. Am. Phys. Soc., Am. Assn. Physics Tchrs., Soc. Suisse de Physique, Societa' Italiana di Fisica. Author: Esperimentazioni di fisica, 1964; also articles. Research in nuclear physics, electronics and electrophysiology; measured total and neutron-proton reaction cross section for many light nuclei; one of earliest studies on polarization of d-d neutrons; studied nuclear photoeffect. Home: 1 Nesazio, Rome, Italy. Office: Istituto Fisico-Universitá, L'Aquila, Italy.*

RICARD, Jerome Sixtus, astronomer; b. Plaisians, Drome, France, Jan. 21, 1850; s. Leger and Marianne (Eyssartel) R.; ed. high sch., Turin, Italy; came to U. S., 1873; student Woodstock Coll., Md.; Ph.D., U. Santa Clara, 1887. Became mem. Soc. Jesus, 1872; observed and studied sunspots and faculae, 1900-23, discovered a method of using them in forecasting weather long in advance; pub. forecast for Santa Clara (Cal.) County, monthly long range weather forecast for U. S., The Sunspot, also occasional seismographic record. Contbr. to Popular Astronomy (mag.). Died Dec. 8, 1930.

RICARDO, David, English economist; b. London, Eng., Apr. 19, 1772; early edn. Eng., Holland; m. Priscilla Anne Wilkinson, Dec. 20, 1793. Employed in father's brokerage firm, 1786-97; independent broker, stock exchange, 1797-98, ret. 1798; mem. Parliament, 1819-23. Author: The Price of Bullion, A Proof of the Depreciation of Bank Notes, 1809; Proposals for an Economical and Secure Currency, 1816; Principles of Political Economy and Taxation, 1817; On Protection to Agriculture, 1822; Plan for the Establishment of a National Bank, 1824; Essay on the Infulence of a Low Price of Corn on the Profits of Stock, 1895. Student polit. econs. (after reading Adam Smith's Wealth of Nations); founder classical polit. economy; contbd. important econ. theories on rent, value, wages (points out that rent for highly productive land is unearned because it is based on ownership of valued and scarce natural resource; rent result, not cause of price); Iron Law of Wages asserts that because of demand and supply on labor market, wages can never be much more than level necessary to maintain minimal subsistence; asserted that on nat. level there must be balance between outgoing and ingoing payments; favored laissez faire govt., abolition of protective tariffs (theory influenced repeal of Corn Laws in Eng., change which helped make Eng. largely mfg. nation). Died Gloucestershire, Eng., Sept. 11, 1823.

RICARDO, Sir Harry Ralph, English mech. engr.; b. London, Jan. 26, 1885; s. Halsey Ralph and Catherine Jane (Rendel) R.; B.A., U. Cambridge, 1906; m. Beatrice Bertha Hale, 1912; children—Cicely Kate (Mrs. Bertram), Angela Edith (Mrs. Hughesdon), Camilla Bertha (Mrs. Bosanquet). Cons. engr. Rendel Palmer & Tritton, 1907-14; cons., adviser Brit. Govt. depts., 1914-17; founder, chmn. bd. Ricardo & Co., Cons. Engrs., 1918-64, pres., 1964—. Mem. Brit. War Cabinet Tech. Com. Knighted; recipient James Alfred Ewing medal for engring. research, 1940; James Watt medal, 1953; Horning medal, 1955; medal, Comdr. Order of Merit, Soc. for Advancement of Research and Invention, Paris, 1964. Fellow Royal Soc., 1929 (Rumford medal, 1944); mem. Instn. Mech. Engrs. (Clayton medal), Instn. Civil Engrs., Soc. Automotive Engrs. Author: The Internal Combustion, 1922; Engines of High Output, 1926; The High Speed Internal Combustion Engine, 1931; also articles. Research on motor fuels, gasoline engines, diesel engines; inventor of internal combustion devices. Home: Woodside, Graffham, Petworth, Sussex. Office: Bridge Works, Shoreham-by-Sea, Sussex, Eng.*

RICCATI, Francesco, Italian mathematician; b. Castelfranco, Italy, Nov. 28, 1718; s. Jacopo Francesco; brother of Vincenzo and Giordano Riccati. Author works on geometry applied to architecture. Died July 18, 1791.

RICCATI, Giordano, Italian mathematician, physicist; b. Castelfranco Treviso, Italy, Feb. 25, 1709; s. Jacopo Francesco Riccati; ed. Coll. of Jesuits, Bologna, Italy; collected and published (with bro. Vincenzo): Opere del Conte Jacopo Riccati, 4 vols., 1758; Delle corde ovvero delle fibre elastiche, 1767. Research and publs. on oscillation of elastic strings, Newtonian philosophy, cubic equations, geometry, phys. properties of drums. Died Treviso, Italy, July 20, 1790.

RICCATI, Count Jacopo Francesco, Italian mathematician; b. Venice, Italy, May 28, 1676; student, Padua, Italy; grad. 1696; children including Vincenzo, Giordano, Francesco. In latter part of life lived in Venice and Treviso, Italy. Author: Opere del Conte Jacopo Riccati (collected by sons after his death), 4 vols., 1758. Publs. on philosophy, differential equations, mensuration, physics; contributed to spread of Newton's theories; studied Riccati equation, 1724 and gave solutions for spl. cases. Died Treviso, Apr. 15, 1754.

RICCATI, Vincenzo, Italian mathematician; b. Castelfranco, Treviso, Italy, Jan. 11, 1707; s. Jacopo Francesco Riccati. Mem. Soc. Jesus; prof. math. Collegio di San Francesco Saverio, Bologna, Italy. Author: Delle Forze vive e dell' azione delle forze morte, 1749; Opuscula ad res physicas et mathematicas pertinentia, 2 vols., 1757-62. Also treatise on infinitesimal calculus. Collected and pub.: (with brother Giordano) Opere del Conte Jacopo Riccati, 4 vols., 1758. Introduced hyperbolic function to trigonometry, 1757; research in differential equations, quadratic problems, series. Died Jan. 17, 1775.

RICCI, Curbastro Gregorio, Italian mathematician; b. Lugo, Italy, Jan. 12, 1853; prof. math. physics U. Padua (Italy), 1880-1925. Author: Lezioni sulla Teoria delle Superficie, 1898; Lezioni di Algebra Complementare, 1900; Origine e Sviluppo dei Moderni Concetti Fondamentali sulla Geometria, 1902; Lezioni di Analisi Algebrica e Infinitesimale, 1918. Research in math. physics, differential calculus; combined work of Riemann, others to formulate theory of tensors (T. Levi-Civita, Ricci's pupil, later contbd. to theory); work provided math. found. for theory of relativity. Died Aug. 7, 1925.

RICCI, Enzo, radiochemist; b. Buenos Aires, Argentina, Nov. 8, 1925; s. Juan and Rosa (Bagnasco) R.; Licensee in chem. sci. U. Buenos Aires, 1952, Ph.D., 1954; m. Blanca Esther Iacobucci, Sept. 21, 1957; children—Steve Caesar, Carlos Albert. Came to U.S., 1962, naturalized, 1967. Mem. staff Hickethier & Bachmann Lab., Buenos Aires, 1949-51; mem. staff Buenos Aires Forestry Research Bd., 1951-53; with Argentinian AEC, 1953-62, head activation analysis and tracer applications group, 1961-62; mem. staff Oak Ridge (Tenn.) Nat. Lab., 1962—; Internat. Atomic Energy Agy. fellow Atomic Energy of Canada Ltd., Chalk River, 1959-61; faculty U. Buenos Aires, 1950-62, prof. applied radio-chemistry and radioisotopes to industry and engring., 1961-62. Mem. Am. Chem. Soc., Am. Nuclear Soc., Argentine Chem. Assn. Author: chpt. in Guide to Activation Analysis, 1964. Discovery, determination of nuclear properties of radioisotope Fe-61; investigation of errors in activation analysis; errors in measurement of fast neutron fluxes; interferences due to 2d order nuclear reactions; study of charged-particle activation analysis; math. formulation; sensitivity determinations for fifteen low-z elements; excitation functions for He-3 reactions. Home: 996 W. Outer Dr., Oak Ridge, Tenn. 37830. Office: Oak Ridge Nat. Lab., P. O. Box X, Oak Ridge, Tenn. 37831.*

RICCI, John Ettore, Am. chemist; b. N.Y.C., Jan. 1, 1907; s. Analito and Ida (Vitale) R.; B.S., N.Y. U., 1926, M.S., 1928, Ph.D., 1931; m. Ada Iuliani, Mar. 21, 1964. Faculty, dept. chemistry, N.Y. U., N.Y.C., 1931—, prof., 1946—. Cons., Oak Ridge Nat. Lab., 1953—. Recipient Gt. Tchrs. Award, N.Y. U., 1959. Mem. Am. Chem. Soc. Author: The Phase Rule and Heterogeneous Equilibrium, 1951; Hydrogen Ion Concentration, 1952; also numerous articles. Asso. editor Jour. Am. Chem. Soc., 1938-48; editorial bd. Jour. Phys. Chemistry, 1961-65. Research in phase equilibria in aqueous salt systems. Home: 76 Wakefield Av., Yonkers, N.Y. 10704. Office: N.Y. U., University Heights, N.Y.C. 10453.*

RICCI, Matteo, Italian mathematician, geographer; b. Macerata, Oct. 6, 1552; student law, Rome. Joined Soc. Jesus, 1571; missionary, India, 1577-83; opened China to Jesuit missionaries, 1583; worked in China, 1583-1600; entered Peking, China, 1600; became ct. mathematician, astronomer, Peking, 1601. Author: Tung-Wen-Suan-Ki; True Doctrine of God. Introduced Western math. and astronomy to China; made 1st modern maps of China available to Europeans. Translated first 6 books of Euclid's Elements into Chinese. Died Peking, May 11, 1610.

RICCO, Annibale, Italian astronomer; b. Modena, Italy, 1844; dir. Catania Obs., Mt. Etna. Editor, contbg. author: Memoires of the Spectroscopists. Founder (with Hale), Astrophys. Jour. Contributed to devel. astrophysics; research on sun spots. Died 1911.

RICE, Alexander Hamilton, Am. geographer, explorer; b. Boston, Aug. 29, 1875; s. John Hamilton and Cora Lee (Clark) R.; A.B., Harvard, 1898, M.D., 1904, A.M. (hon.), 1915; D.Sc. (hon.), Hamilton, 1951; surgical intern Mass. Gen. Hosp., 1903-05; certificate Royal Geog. Soc. Sch. of Geog. Surveying and Field Astronomy, 1908-10; m. Mrs. Eleanor Elkins Widener, Oct. 6, 1915 (died July, 1937); m. 2d, Mrs. Dorothy Farrington Upham, Nov. 5, 1949. Surgeon, dir. Hosp. No. 72, Société de Secours aux Blessés Militaires, Paris, France, 1914-15; surgeon Am. Ambulance, Neuilly, 1914-15; commd. lt. U. S. N.R., 1917; instr. navigation and nautical astronomy, 2d Naval Dist. Training Sch., Newport, R.I., 1917; dir. same sch., 1917-19; lectr. Lowell Inst., 1922; instrumental in establishing Indian Sch., hosp. and créche at Sao Gabriel, Brazil; established Inst. of Geog. Exploration, Harvard, dir., prof. same; hon. curator S.A. sect. Peabody Mus. Archaeology and Ethnology; lectr. diseases of tropical S. America, Dept. Tropical Medicine, Harvard Med. Sch. Fellow Royal Astron. Soc., mem. Royal Geog. Society, London, Am. Geog. Soc, Reale Società Geografica Italiana, Société de Geographie de Paris, Royal Institution of Great Britain, numerous other fgn. geog. and sci. socs. Awarded Patron's gold medal, Royal Geog. Soc., 1914; David Livingstone centenary medal, Am. Geog. Soc., 1920; Elisha Kent Kane medal, Phila. Geog. Soc., 1920; gold medal Société Royale de Geographie d'Anvers, 1931; Contbr. numerous articles to Geog. Jour. Quito to Iquitos via River Napo, Apr. 1903; The River Uaúpes, with Map, June 1910; Further Explorations in Northwestern Amazon Valley, Map, Aug. 1914; Notes on the Rio Negro (Amazonas), Map, Oct. 1912; The Rio Negro, the Casiquiare Canal and the Upper Orinoco (map), Nov. 1921; Rio Branco-Uraricuera-Parima (maps), 1928; El rio Negro (Amazonas) y sus grandes afluentes de la Guayana

Brasileña, 1934; Exploration en Guyane Brésilienne, Paris, 1937. Exploration, research in tropical S.Am.; organized and conducted 7 expdns. into Colombian Caqueta, Brazilian Amazonas and Venezuelan Guayana, exploring, surveying and mapping an area of 500,000 sq. miles and carrying out studies of geol., biol., anthropol., ethnol. character; expdn. 1924-25 to Rio Branco-Uraricuera-Parima, hydroplane successfully used in reconnaissance and air photography and showed efficiency of short wave in such explorations. Died July 23, 1956.

RICE, Francis Owen, chemist; b. Liverpool, Eng., May 20, 1890; s. James and Mary (Byrne) R.; D.Sc., Liverpool U., 1916; postgrad. Princeton, 1919; m. Katherine K. Kempner, May 23, 1930; children—Monica Ellen, Cecilia Joan. Came to U.S., 1919, naturalized, 1953. Instr., N.Y. U., 1919-20, asst. prof., 1920-24; asso. Johns Hopkins, 1924-26, asso. prof., 1926-38; prof., head dept., Cath. U., Washington, 1938-59; prof. chemistry dept. Georgetown U., Washington, 1959-62; prin. research scientist U. Notre Dame, South Bend, Ind., 1962—. Author: The Mechanism of Homogeneous Organic Reactions, 1928; the Aliphatic Free Radicals, 1935; (with E. Teller) The Structure of Matter, 1949. Research on free radicals. Home: 1704 Bader Av., South Bend, Ind.*

RICE, John Rischard, Am. mathematician; b. Tulsa, June 6, 1934; s. John C. Kirk and Margaret (Rischard) R.; B.S., Okla. State U., 1954, M.S., 1956; Ph.D., Cal. Inst. Tech., 1959; m. Nancy Bradfield, December 19, 1954; children—Amy Lynn, Jenna Margret. NRC-Nat. Bur. Standards fellow Nat. Bur. Standards, Washington, 1959-60; sr. research mathematician Gen. Motors Research Labs., Warren, Mich., 1960-64; prof. maths. and computer sci. Purdue U., West Lafayette, Ind., 1964—. Nat. lectr. in numerical maths. Assn. Computer Machinery, 1966. Mem. Assn. Computing Machinery, Am. Math. Soc., Soc. for Indsl. and Applied Maths. Author: The Approximation of Functions, Vol. 1, 1964, Vol. 2, 1968; also articles. Research on approximation of functions using non-linear methods, partial difference equations and error analysis. Home: 112 E. Navajo St., West Lafayette, Ind. 47906.*

RICE, Oscar K(nefler), Am. phys. chemist; b. Chgo., Feb. 12, 1903; s. Oscar Guido and Thekla (Knefler) R.; student San Diego Jr. Coll., 1920-22; B.S., U. Cal., 1924, Ph.D., 1926; NRC fellow, Cal. Inst. Tech., 1927-29, U. Leipzig, 1929-30; m. Hope Ernestyne Sherfy, Dec. 23, 1947; 2 daus., Margarita S., Pamela S. Asso. in chemistry U. Cal., 1926-27, research asso. chem., 1935-36; instr. chemistry Harvard, 1930-35; asso. prof. U. N.C., 1936-43, prof. 1943-59, Kenan prof. chemistry, 1959—; Stewart lectr. U. Mo., 1948; Reilly lectr. U. Notre Dame, 1957; Barton lectr. U. Okla., 1967; prin. chemist, Clinton Labs., Oak Ridge, 1946-47. Adv. panel chemistry NSF, 1958-61. Mem. Solvay Congress, Brussels, 1962. Recipient Am. Chem. Soc. award in pure chemistry, 1932, So. Chemist Award, 1961; N.C. award in sci., 1966. Mem. Réunion Internationale de Chimie Physique, Paris, 1928. Fellow Am. Phys. Soc., A.A.A.S.; mem. Faraday Soc., Am. Chem. Soc. (sec. 1942, chmn. 1944, div. physical and inorganic chem., chmn. N.C. Sect., 1946), Fedn. Am. Scientists (mem. adv. panel), Nat. Acad. Scis., N.C. Acad. Scis., Am. Assn. Univ. Profs., Am. Forestry Assn., Sigma Xi. Author: Electronic Structure and Chemical Binding, 1940; Statistical Mechanics, Thermodynamics, and Kinetics, 1967. Asso. editor: Jour. Chem. Physics, 1934-36, 1945-47. Mem. adv. bd. Philosóphy of Sci. Jour., 1934-54. Contbr. articles to sci. periodicals. Research in theoretical and exptl. chem. kinetics, photochemistry, quantum mechanics of chem. reactions and related phenomena, thermodynamics and statis. mechanics of liquids, solids and phase transitions; critical phenomena, especially in binary liquid systems, surface chemistry, irreversible thermodynamics, theory of liquid helium. Home: 311 Clayton Rd., Chapel Hill, N.C. 27514.*

RICE, Paul LaVerne, Am. entomologist; b. Bancroft, Neb., Dec. 28, 1906; s. Eugene Taylor and Fannie (Dalton) R.; student Coll. Ida., 1925-27; B.S., U. Ida., 1931, M.S., 1932; Ph.D., Ohio State U., 1937; m. Elfriede Neufeld, Feb. 4, 1939; children—Stephen Paul, Glen Eugene, Dennis Henry. Instr. entomology U. Ida., also asst. entomologist Ida. Agrl. Expt. Sta., 1931-33; asst. entomologist Del. Agrl. Expt. Sta. and Extension Div., 1936-37, asso. entomologist, acting head dept. entomology, 1942-45; prof. biology Alma (Mich.) Coll., 1937-42, prof. biology, dean faculty, 1945-50; prof. biology Whittier (Cal.) Coll., 1950-58; on leave as sr. entomologist, div. internat. health USPHS, Ethiopia, 1955-57; med. entomologist, asso. dir. Malaria Eradication Tng. Center, ICA, Jamaica, 1958-62; sci. adminstr. NIH, Bethesda, Md., 1962-64; asst. chief vector-borne disease sect., tng. br. Communicable Disease Center USPHS, Atlanta, 1964—. Commd. sr. scientist, comdr. USPHS Res., 1955, sci. dir., capt., 1960. Mem. Mich. Bd. Basic Sci. Examiners, 1940-42. Fellow A.A.A.S., Royal Soc. Tropical Medicine and Hygiene; mem. Am. Assn. Tropical Medicine and Hygiene, Am. Mosquito Control Assn., Commd. Officers Assn. USPHS, Entomol. Soc. Am., Sigma Xi, Phi Beta Kappa, Alpha Zeta, Tau Kappa Epsilon. Contbg. author: The Caribbean—Its Health Problems, 1965. Research and publs. on application of entomology to world-wide malaria eradication. Home: 2373 Burnt Creek Rd., Decatur, Ga. 30033. Office: Training Branch, Communicable Disease Center, 1600 Clifton Rd., Atlanta 30333.*

RICE, Robert Vernon, Am. biophysicist; b. Barre, Mass., Aug. 13, 1924; s. Laurence Vernon and Edith (Middlemiss) R.; B.S., Northeastern U., 1950; M.S., U. Wis., 1952, Ph.D., 1955; m. Betty Marts, July 13, 1945; children—Lee Robert, Paul Vernon. Asst., Godfrey L. Cabot, Inc., 1946-49; biochemist U. Wis., 1951-54; fellow chem. physics Mellon Inst., Pitts., 1954-57, sr. fellow, 1957—; prof. biochem istry Carnegie-Mellon U. Lectr., Grad, Sch. Pub. Health, U. Pitts., 1961—; vis. prof. U. Cal. at San Francisco, 1962-63. Mem. Am. Soc. Biol. Chemists, Biophys. Soc., A.A.A.S., Am. Chem. Soc., Electron Micros. Soc. Am. Research and publs. on position nucleic acid in tobacco mosaic virus; demonstrated return of normal conformation of collagen after its denaturation, earthworm cuticle collagen in polymerized form was the same as ordinary collagen; described shape muscle protein myosin; determined sense of helix of F-actin to be right-handed. Home: 1401 N. Negley Av., Pitts. 15206. Office: 4400 5th Av., Pitts. 15213.*

RICE, Stuart Alan, Am. phys. chemist; b. N.Y.C., Jan. 6, 1932; s. Laurence Harry and Helen (Rayfield) R.; B.S., Bklyn. Coll., 1952; A.M., Harvard, 1954, P.D., 1955; m. Marian Ruth Coppersmith, June 1, 1952; children—Barbara Ellen, Janet Ann. Jr. fellow Soc. Fellow, Harvard, 1955-57; faculty U. Chgo., 1957—, prof. phys. chemistry, 1960—, dir. Inst. for Study Metals, 1962—. Alfred P. Sloan fellow, 1958-62; Guggenheim fellow, 1960-61; recipient Alumni award Honor, Bklyn. Coll., 1961; NSF Sr. fellow, 1965-66. Mem. Nat. Acad. Scis., Am. Chem. Soc. (award in pure chemistry 1962), Am. Phys. Soc., Faraday Soc., N.Y. Acad. Scis. (A. Cressy Morrison prize in natural scis. 1955), A.A.A.S., Phi Beta Kappa, Sigma Xi. Author: (with Mitsuru Nagasawa) Polelectrolyte Solutions, 1961; (with Peter Gray) Statistical Mechanics of Simple Liquids, 1965; also numerous articles. Devel. theory of liquids; optical properties of crystals; mass transport in solids, biol. macromolecule. Home: 5421 Greenwood Av., Chgo. 60615.*

RICE, Stuart Arthur, Am. sociologist, statistician; b. Wadena, Minn., Nov. 21, 1889; s. Edward M. and Ida Emeline (Hicks) R.; A.B., U. Wash., 1912, A.M., 1915; Ph.D., Columbia, 1924; m. Sarah Alice Mayfield, May 29, 1934; 1 son, Stuart Arthur. Faculty, Dartmouth Coll., 1923-26, U. Pa., 1926-40 (on leave 1927-28, 32-40); asst. dir. Bur. Census, Washington, 1933-35; chmn. Central Statis. Bd., Washington, 1935-40; asst. dir. statis. standards Bur. Budget, Exec. Office President, Washington, 1940-54; pres. Surveys & Research Corp., Washington, 1955-65, cons., adviser, 1965—. Chmn., UN Statis. Commn., 1946, U.S. mem., 1947-55; 1st exec. v.p. Inter-Am. Statis. Inst.; pres. Internat. Statis. Inst., 1947-53, hon. pres.; U.S. rep. Inter-Am. Com. Improvement Nat. Statistics, 1950-55, also many others. Fellow A.A.A.S. (v.p. social scis. 1937, 57); mem. Am. Sociol. Assn., Population Assn. Am., Internat. Union for Sci. Study Population, Washington Acad. Sci., Am. Statis. Assn., Phi Beta Kappa. Author: Quantitative Methods in Politics, 1928. Editor: Methods in Social Science, 1931. Pioneer in applications of quantitative methods to social research and polit. sci.; created agys. and procedures for coordination of Fed. statistics and devel. internat. statistics. Home: 1870 Wyoming Av. N.W., Washington 20009. Office: 1030 15th St. N.W., Washington 20005.*

RICE-WRAY, Edris, physician; b. Newark, Jan. 21, 1904; d. Theron Canfield and Mabel (Simon) Rice-Wray; B.A., Vassar Coll., 1927; M.D., Northwestern U., 1932; M.P.H., U. Mich., 1950; m. Robert Carson, Mar. 30, 1929 (div. 1943); children—Lynn (Mrs. Gustavo Lopez), Barbara (Mrs. Alejandro Cervantes). Practice medicine, Chgo., 1935-47; dir. P.R. Health Dist., 1948-49; dir. Pub. Health Tng. Center, Rio Piedras, P.R., 1950-56; prof. preventive medicine U. P.R., 1950-56; med. dir. P.R. Planned Parenthood Assn., 1950; founder, dir. Asociación Pro-Salud Maternal, Mexico City, Mexico, 1959—. Diplomate Am. Bd. Preventive Medicine. Fellow Am. Coll. Preventive Medicine; mem. Asociación Médica de P.R., A.M.A., Am. Pub. Health Assn., Sociedad Mexicana de Nutrición y Endocrinología, International Fertility and Sterility Assn., Population Assn. Am., N.Y. Acad. Scis. Research and publs. on first field study in P.R. of oral contraceptives, comparisons, availability, accentability, hazards of oral contraceptives. Home: 59-103 Gral. Cano, Mexico City 18, D.F., Mexico. Office: Apartado Postal 7-1050, Mexico City 7, D.F., Mexico.*

RICH, Alexander, Am. molecular biologist; b. Hartford, Conn., Nov. 15, 1924; s. Max and Bella (Shub) R.; A.B., Harvard, 1947, M.D., 1949; m. Jane Erving King, July 5, 1952; children—Benjamin, Josiah, Rebecca, Jessica. Postdoctoral fellow Cal. Inst. Tech., Pasadena, 1949-54; chief sect. on phys. chemistry NIH, Bethesda, Md., 1954-58; vis. scientist Cavendish Lab., Cambridge, Eng., 1955; asso. prof. biology dept. Mass. Inst. Tech., Cambridge, 1958-61, prof. biophysics, 1961—. Guggenheim fellow, 1964. Fellow Am. Acad. Arts and Scis.; mem. Am. Chem. Soc., Am. Crystallographic Soc., A.A.A.S., Biophys. Soc. (mem. council). Editorial bd. Sci., 1963—; mem. editorial adv. bd. Jour. Molecular Biology, 1958—, Biopolymers, 1962—. Research and numerous publs. on molecular structure of nucleic acids and proteins, with X-ray diffraction techniques, study mechanism protein synthesis. Home: 12 Linnaean St., Cambridge, Mass. 02138.*

RICH, Arnold Rice, Am. pathologist; b. Birmingham, Ala., Mar. 28, 1893; s. Samuel and Hattie (Rice) R.; A.B., U. of Va., 1914, M.A., 1915; M.D., Johns Hopkins University, 1919; M.D. (honorary) University Zurich; married Helen Elizabeth Jones, June 3, 1925; children—Adrienne Cecile, Cynthia Marshall Asst. in pathology, Johns Hopkins U., 1919-20, instr., 1920-21, asso., 1921-23, asso. prof., 1923-44, prof., 1944-47, Baxley prof., dir. dept., 1947-58, Baxley professor emeritus, 1958—; resident pathologist, Johns Hopkins Hosp., 1920-26, asso. pathologist, 1929-44, pathol.-in-chief, 1947-58, now hon. cons.; expert cons. to the Surg. Gen., U. S. Army; mem. sci. adv. bd. Armed Forces Inst. of Pathology (chmn. 1951); special cons. USPHS; consultant in pathology Veterans Administration; consultant in med. research, Chem. Warfare Service, since 1943; adv. consultant Tuberculosis Control Div., U.S.P.H.S.; Nat. Research Council (com. on pathol., 1947-52). U. S. State Dept. del., 1st Internat. Allergy Congress, 1951; mem. Comité d'Honneur, 50th anniversary celebration of discovery of anaphylaxis, Paris, 1952. Decorated Chevalier Legion of Honor (France); awarded Charles Mickle hon. fellowship, U. Toronto Faculty of Medicine, 1956; Kober medal, Assn. of Am. Physicians, 1958; Gordon Wilson medal, Am. Clin. and Climatol. Assn., 1960; Trudeau medal, Nat. Tb Assn., 1960; Gairdner Found. (Can.) Internat. award, 1960; honorary Plaque, Japanese Soc. for Tb, 1960; medal A.C.P., 1963; Seaman award Assn. Mil. Surgeons U. S., 1963. Trustee Roland Park Co. Sch., 1944-51. Fellow A.A.A.S., Internat. Assn. Allergists, Royal Soc. Medicine London (hon.); honorary member Pathological Society of Great Britain and Ireland, Am. Clin. and Climatol. Assn., Harvey Soc., Soc. Française d'Allergie; fgn. corr. mem. Soc. Med. des Hôpitaux de Paris; fgn. mem. Soc. Aregentina de Anat. Norm. y Patol.; corr. mem. Soc. Brasileira de Tuberc., Tb Soc. of Scotland; asso. mem. Soc. Anat. de Paris. Dir. Md. Tb Assn., 1947-51. Mem. Nat. Acad. Scis., Assn. Am. Physicians, Soc. Exptl. Biology and Medicine (editorial bd. Proc. 1943-47), Soc. Exptl. Pathology, Am. Assn. Pathologists and Bacteriologists, Phi Beta Kappa, Sigma Xi. Author: The Pathogenesis of Tuberculosis 1944, rev. 1951, Spanish edit., 1946, Japanese edit., 1954. Mem. editorial bd. Bull. of the Johns Hopkins Hosp., 1925-63, Internat. Archives Allergy and Immunology, Internat. Review. Experimental Pathology, Contributor articles in field. Research on tuberculosis; hypersensitivity; diseases of liver; mechanisms of immunity. Home: 14 Edgevale Rd., Baltimore, Md. 21210. Office: Johns Hopkins Hosp., Balt.

RICH, Claudius James, archaeologist; b. Dijon, France, Mar. 28, 1787; raised, ed. Bristol, Eng.; m. Miss Mackintosh, Jan. 22, 1808. Entered E. India Co., 1803, counsel in Baghdad, Iraq, from 1808. Author: Memoir on the Ruins of Babylon, 1815; Narrative of a Residence in Koordistan and on the Site of Ancient Nineveh, 2 vols., 1836. Visited, described various sites ancient Mesopotamian cities, collected antiquities, manuscripts, provided 1st descriptions of some areas; considered forerunner field archaeology. Died Shiraz, Iran, Oct. 5, 1820.

RICHARD, Dean Boyd, Am. wood technologist; b. Baca County, Colo., Feb. 20, 1920; s. Dean Willard and Kate (Taylor) R.; student Park Coll., 1937-40; B.S., Colo. State U., 1944; M.S., Syracuse U., 1947, Ph.D., 1950; m. Alice Frances Skillman, July 16, 1944; children—Nicola D., Lindsay A., Jennifer H., Michael T., Karl E., Veronica J. Airport mgr. Otsego Aviation Service, Coopertown, N.Y., 1945-46; wood technologist, lacquer chemist Lilly Co., High Point, N.C., 1950-51; prof. forestry and wood tech. Auburn (Ala.) U., 1951-65; chmn. dept. forestry, prof. U. Ky., Lexington, 1965—. Cons. on end jointing and corner jointing wood, devel. So. pine plywood industry to various cos. in S.E. and Pacific N.W. Mem. Soc. Am. Foresters, Forest Products Research Soc. (past chmn. Fla.-Ga.-Ala.), Sigma Xi, Xi Sigma Pi, Gamma Sigma Delta. Research and publs. on devel. and design high strength end-joints for wood; invented high strength corner joints, new atlas system for aerial navigation; identification and study importance of stress concentration at tips of finger joints; determination of progressive phys. and chem. changes in wood decayed by various fungi. Home: 3257 Lansdowne Dr., Lexington, Ky. 40502.*

RICHARD, Louis-Claude-Marie, French botanist; b. Versailles, France, Sept. 19, 1754; prof. botany Ecole de Medecine, Paris; mem. French Acad. Scis., 1795; Author: Demonstrations de Botanique, 1808; made trip to Guiana and Antilles for bot. study, returned to France with herbarium of some 3,000 plants, 1781-89. Died Paris, June 6, 1821.

RICHARD OF SALERNO (or Nicholas II of Salerno), Salernitan physician; flourished 2nd half of 12th century. Author: Anatomia Ricardi (Salernitani). Wrote important early Latin treatise on anatomy (a similar treatise is entitled Anatomia Nicolai; both texts could have been taken from the lectures of A. Salernitan Master called Richard or Nicholas).

RICHARD OF WALLINGFORD, English mathematician; b. Wallingford, Eng., circa 1292; s. William and Isabella of Wallingford; B.D., Oxford (Eng.) U. Ordained priest Roman Catholic Ch., joined Soc. St. Benedict, abbot St. Albans, from 1327; lectr. liberal arts Oxford U. Author: Canones de Instrumento . . . Albion Dicto; Albion est Geometricum Instrumentum de Arte Componendi Rectangulum; Rectangulum in Remedium, 1326; Ars Operandi cum Rectangulo; Quadripartitum de Sinubus Demonstratis; De Sinubus et Arcubus in Circulo Inveniendis; Exafrenon Prognosticorum Temporis; De Opimetris; De Eclipsibus Solis et Lunae; Decretales et Constitutiones Capitulorum Provincialium et Concernentium; Super Prologum Regulaes Benedicti; Privilegia Monasterii sui; De Rebus Arithmeticis; De Computo. Invented clock which showed time, seasons, course of sun, moon and planets; skilled in liberal scis. and mech. arts; wrote on trigonometry, arithmetic. Died St. Albans, May 23, 1336.

RICHARD OF WENDOVER (Richard Anglicus), physician; b. Eng.; physician to Gregory IX, 1227-41; then moved to Paris; later canon St. Paul's, London. Author: Practica sive Medicamenta Ricardi; Micrologus Magistri Ricardi Anglici (med. ency. using Greek and Arabic knowledge which was available in Latin translations); Anatomia (based on Avicenna's Canon); De signis prognosticis; De modo conficiendi et medendi; De Phebotomia; De urinis; Correctorium alchymiae. Died London, 1252.

RICHARDS, Albert Glenn, Am. zoologist; b. Lake Forest, Ill., May 29, 1909; s. Albert Glenn and Grace (Nettleton) R.; A.B., U. Ga., 1929; Ph.D., Cornell U., 1932; m. Sara L. Mende; children—Jonathan Ian, Glenn Mende, Conrad Steven; m. 2d, Patricia Ann Wiegand, Aug. 10, 1966. Acting head dept. entomology Ward's Natural Sci. Establishment, Rochester, N.Y., 1932; asst. dept. zoology U. Rochester, 1933-36; research asst. Am. Museum Natural History, N.Y.C., 1936-37; instr. biology Coll. City N.Y., 1937-39; with U. Pa., 1939-45, asst. prof., 1944-45; asso. prof. dept. entomology and dept. zoology U. Minn., 1945-49, prof., 1949—. Guggenheim fellow, Fulbright Research scholar Max Planck Inst. for Biology, Tubingen, Germany, 1957-58; guest prof. U. Munich (Germany), guest investigator Max Planck Inst. for Behavioral Psychology, Munich, 1966-67. Mem. Am. Soc. Zoologists (program dir. 1963-65), Entomol. Soc. Am. (v.p. 1949), Am. Entomol. Soc. (sec. 1942-45), Histochem. Soc., Marine Biol. Lab., Soc. Exptl. Biology, Soc. Gen. Physiologists, A.A.A.S., Am. Inst. Biol. Sci., Am. Assn. U. Profs., Sigma Xi. Author: The Integument of Arthropods, 1951. Publs. on pioneer work on application of electron microscopy to biology; descriptive and exptl. insect morphology and physiology; studies on acclimatization and the physiol. bases of temperature threshholds. Home: 1415 N. Cleveland Av., St. Paul, Minn. 55108.*

RICHARDS, Albert Gustav, Am. dental researcher; b. Chgo., Jan. 7, 1917; s. John Young and Celia (Kuehl) R.; student Wright Jr. Coll., 1934-36, Northwestern U., 1936-37; B.S. in Chem. Engring., U. Mich., 1940, M.S. in Physics, 1943; m. Marian Ruth Kauffman, Feb. 13, 1942; children—Jean Diane, Kathleen Anne, Susan Patricia, Joanne Ruth, Nancy Louise. Faculty U. Mich. Sch. Dentistry, Ann Arbor, 1940—, prof. dentistry, 1959—. Cons., VA Hosp., Ann Arbor, 1953—, Dental health project U.S. Dept. Health, Edn., and Welfare, 1962; pvt. cos. Fellow A.A.A.S.; mem. Am. Dental Assn. (cons. council on dental research 1957—, editorial cons. 1953—), Am. Acad. Oral Roentgenology, Michigan State Dental Assn., Am. Standards Assn. (mem. Nat. Com. on Radiation Protection 1963—). Editorial cons. Jour. Oral Surgery, Oral Medicine, and Oral Pathology, 1962—, abstractor, 1962—; abstractor Excerpta Medica, 1953—. Patentee in field. Pioneered investigation human teeth with electron microscope; publs., research on prodn. erythema human skin with dental X-ray machine, scintillation spectroscopy, floral radiography; dental X-ray machine design, radiation hygiene. Home: 395 Rock Creek Dr., Ann Arbor, Mich. 48104.

RICHARDS, Alfred Newton, Am. pharmacologist; b. Stamford, N.Y., Mar. 22, 1876; s. Rev. Leonard E. and Mary E. (Burbank) R.; B.A., Yale, 1897, M.A., 1899; Ph.D., Columbia, 1901; hon. Sc.D., U. of Pa., 1925, Western Reserve, 1931, Yale, 1933. Harvard University, 1940, Columbia U., 1942. Williams College, 1943, Princeton University, 1940, New York University, 1955, Rockefeller Institute, 1960, Oxford University, 1960; honorary M.D., University of Pa., 1932, University of Louvain, 1949; LL.D., Univ. of Edinburgh (Scotland), 1935; Johns Hopkins University, 1949; married Lillian L. Woody, Dec. 26, 1908. Instr., physiol. chemistry Columbia, 1898-1904, pharmacology, 1904-08; prof. pharmacology, Northwestern U., 1908-10; prof. pharmacology U. Pa., 1910-46, emeritus prof., 1946—, vice president charge medical affairs, 1939-48; Herter lectr. New York University, Bellevue Hosp. Med. Coll., 1926; Beaumont lecturer, 1929; Croonian lecturer, Royal Soc. London, 1938. Chmn. Com. on Med. Research, Office Scientific Research and Development, U. S. Govt., 1941-46. Mem. Com. on Federal medical services of Hoover Commn. on Orgn. Exec. Branch Govt., May-Dec. 1948. Member Scientific staff, British Medical Research Committee, London, 1917-18; maj. Sanitary Corps, U. S. Army, attached to Chem. Warfare Service, Chaumont, France, July-Dec. 1918. Fellow Royal Soc., 1942; mem. Nat. Acad. of Sciences (pres. 1947-50), Am. Philos. Soc., (v.p. 1944-47), Am. Physiol. Soc., Harvey Soc., Sigma Xi, Phi Beta Kappa (hon.), corr. mem. Gesellsch. der Aertze in Wien; Royal Danish Acad. Science, Royal Soc. of Edinburgh; fellow Am. Acad. Arts and Sciences, fgn. corresponding member British Medical Assn. Awarded Gerhard medal, 1932; Kober medal, 1933; Keys medal, 1933; John Scott medal, 1934; Procter medal; Guggenheim Cup Award; Lasker Award, 1946; Kovalenko medal Nat. Acad. Sci., 1953; Abraham Flexner award Assn. Am. Med. Colls., 1959. Decorated Medal for Merit, 1946; Hon. Comdr., Order British Empire, 1948. Trustee Rockefeller Found., 1937-41. Successively asst., assoc. and mng. editor Jour. Biol. Chem., 1905-14. Research and numerous publs. on action of chloroform and histamine, kidney action, chemistry of connective tissues, adrenalin secretion, and salivary digestion. Died Mar. 24, 1966.

RICHARDS, Aute, Am. zoologist; b. Tekamah, Neb., Oct. 31, 1885; s. John Fletcher and Silvia (McNabb) R.; B.A., U. Kan., 1908; M.A., U. Wis., 1909; Ph.D., Princeton, 1911; D.Sc., Marietta Coll., 1939; m. Mildred Albro Hoge, Dec. 19, 1917; children—James Hoge, Ernest John. Tchr., Sumner County High Sch., Wellington, Kan., 1907-08; faculty U. Tex., 1911-16, Wabash Coll., Crawfordsville, Ind., 1916-20; faculty U. Okla., Norman, 1920—, prof. zoology, 1920-50, prof. emeritus, 1950—, dir. Sch. Applied Biology, 1924-42; dir. Okla. Biol. Survey, 1927-49; dir. Okla. Mus. Zoology, 1924-42. Mem. A.A.A.S., Am. Soc. Zoologists, Am. Micros. Soc., Am. Soc. Naturalists, Sigma Xi, Beta Beta Beta (past editor Bios.). Author: Laboratory Guide in General Zoology, 1925; Outline of Comparative Embryology, 1931; also numerous articles. Research on cellular biology, comparative embryology. Home: 2950 E. Mabel St., Tucson 85716.*

RICHARDS, Dickinson W., Am. physician; b. Orange, N.J., Oct. 30, 1895; s. Dickinson W. and Sally (Lambert) W.; A.B., Yale 1917, D.Sc., 1957; A.M., Columbia University, 1922, M.D., 1923, Doctor of Science (hon.), 1966; m. Constance B. Riley, Sept. 19, 1931; children—Ida E., Gertrude W., Ann H., Constance L. Research fellow Nat. Inst. for Med. Research, London, Eng., 1927-28; research on problems of pulmonary and cardiac physiology, Coll. Physicians and Surgeons, Columbia U., since 1928; Lambert prof. medicine, 1947-61, emeritus, 1961—; director first medical division Bellevue Hospital, N.Y.C., 1945-61. Decorated Chevalier Legion of Honor (France). Recipient Nobel Prize in medicine and physiology (with Courmand and Forssmann) 1956. Fellow American Academy of Arts and Sciences; mem. Assn. Am. Physicians (pres. 1962), Nat. Acad. Scis. Editor: (with A. P. Fishman) Circulation of the Blood: Men and Ideas, 1964. Research in cardiac and pulmonary physiology; developed, with Andre Cournand, methods of studying action of heart and lungs in health and disease, including cardiac catheterization; study of heart and lung function in traumatic shock. Home: Woodland Dr., Lakeville, Conn. Office: 620 W. 168th St., N.Y.C. 10032.*

RICHARDS, Ellen Henrietta, Am. san. chemist; b. Dunstable, Mass., Dec. 3, 1842; d. Peter and Fanny Gould (Taylor) Swallow; A.B., Vassar, 1870, A.M., 1873; S.B., Mass. Inst. Tech., 1873; m. Robert Hallowell Richards, June 6, 1875. Instr. in woman's lab. Mass. Inst. Tech., Lab., 1876-84, san. chemistry, 1884—; as chemist of Mfrs.' Mutual Fire Ins. Co. had much to do with oils, in reference to safety from spontaneous combustion, explosion; specialist in water analysis. An organizer, 1st pres. Am. Home Econs. Assn., 1908. Author: Air, Water and Food, 1900; First Lessons in Minerals; The Cost of Food, 1900; Cost of Shelter, 1905; First Lessons in Food and Diet; The Art of Right Living, 1905; Sanitation in Daily Life, 1907; Cost of Cleanness, 1908; Chemistry of Cooking and Cleaning, 1908; Industrial Water Analysis, 1909; Euthenics, 1910. Pioneer in application of chemistry to nutrition; leader in survey of Mass. waters. Died 1911.

RICHARDS, Francis Asbury, Am. oceanographer; b. Newton, Ill., May 26, 1917; s. Larkin A. and Eva May (Eck) R.; B.S., U. Ill., 1939; M.S., U. Nev., 1942; Ph.D., U. Wash., 1950; m. Mary Jane Brady, Feb. 12, 1944; children—Francis Asbury, Christian R., Peter L., John B. Chem oceanographer Woods Hole (Mass.) Oceanographic Inst., 1950-59; faculty U. Wash., Seattle, 1959—, prof., oceanography, 1964—, asst. chmn. dept., 1967—. Lectr., Mass. Inst. Tech., 1958. Mem. panel on ocean wide surveys sci. com. on oceanography Nat. Acad. Sci., 1966—. Guggenheim fellow U. Oslo (Norway), 1956-57. Mem. Am. Soc. Limnology and Oceanography (pres. Pacific sect. 1963-64, editor Limnology and Oceanography 1963-67), Am. Geophys. Union, Oceanographic Soc. Japan. Research, publs. on nutrient and gas cycles in sea, oceanic circulations of chem. constituents of sea water, oxygen deficient and sulfidebearing marine environments, developed spectrophotometric method of plankton pigment analysis. Home: 3914 48th Pl. N.E., Seattle 98105.*

RICHARDS, Hugh Taylor, Am. physicist; b. Baca County, Colo., Nov. 7, 1918; s. Dean Willard and Kate (Taylor) R.; B.A., Park Coll., 1939; M.A., Rice U., 1940, Ph.D., 1942; m. Mildred Paddock, Feb. 11, 1944; children—David Taylor, Thomas Martin, John Willard, Margaret Paddock, Elizabeth Nichols, Robert Dean. Fellow, Rice U., 1939-41, asst., 1941-42; scientist U. Minn., 1942-43, Los Alamos Sci. Lab., 1943-46; research asso. U. Wis., 1946-47, asst. prof., 1947-48, asso. prof., 1948-52, prof., 1952—, chmn. dept. physics, 1961-63, asso. dean Coll. Letters and Sci., 1963—. Research on neutron spectra, neutron measurements at first A-bomb test, nuclear reaction energy measurements, scattering and reaction. Home: 1902 Arlington Pl., Madison, Wis. 53705.*

RICHARDS, Lorenzo A(dolph), Am. physicist; b. Fielding, Utah, Apr. 24, 1904; s. Calvin Willard and Louisa (Madsen) R.; B.S., Utah State U., 1926, M.A., 1927; Ph.D., Cornell U., 1931; D.Tech. Sci. (hon.), Israel Inst. Tech., Haifa, 1952; m. Zilla Linford, Sept. 3, 1930; children—Lorenzo Willard, Paul Linford, Mary (Mrs. Robert L. Armstead). Instr. physics Cornell U., 1927-34; research physicist Battelle Meml. Inst., 1935; asst. prof., then asso. prof. physics Ia. State U., 1935-39; sr. physicist U. S. Salinity Lab., 1939-42; Nat. Def. Res. fellow, group supr. Cal. Inst. Tech. 1942-45; chief physicist U. S. Salinity Lab., Dept. Agr., Riverside, Cal., 1945-66; physics cons., 1966—. Tech. observer rocket ordnance, staff comdr. chief S.W. Pacific Area, 1944. Vice pres., dir. Moistomatic, Inc., 1954-62. Lectr., Internat. Symposium Desert Research, Jerusalem, 1952; cons. Ministry Agr., Egypt, 1952; cons. Ford Found., 1962, also lectr. U. Alexandria (Egypt), 1962. Recipient Ordnance Devel. award Navy Dept., 1945, Presdl. Certificate of Merit, 1948, Superior Service award Dept. Agr., 1959. Fellow A.A.A.S., Am. Soc. Agronomy (pres. 1965; Stevenson award 1949); mem. Am. Phys. Soc., Soil Sci. Soc. Am. (charter, chmn. div. 1, 1938, pres. 1952), Am. Geophys. Union (chmn. com. physics soil water 1947-50), Internat. Soc. Soil Sci. (del., gen. lectr. 4th Internat. Congress 1950, Western Soc. Soil Sci. (pres. 1950), Sigma Xi, Phi Kappa Phi. Author numerous research papers, sect. in books. Editor: Diagnosis and Improvement of Saline and Alkali Soils, 1954. Clarified relation of energy of soil water to growth and yield of agrl. crops; studied physics of soil water, including unsaturated flow theory; 1st methods and measurements of capillary conductivity; invented and patented membrane methods for determining soil water content in relation to matric water potential, thermocouple psychrometer for determining thermodynamic condition of water and soil, elec. conductivity methods for determining salinity and energy status of water in field soil, comml. systems for autonomic irrigation; patentee vacuum tube control circuits, barrage rocket launchers. Home: 4455 5th St., Riverside, Cal. 92501.*

RICHARDS, Oscar W(hite), Am. biologist; b. Butte, Mont., Jan. 4, 1902; s. Frank E. and Henrietta L. (White) R.; B.A., U. Ore., 1923, M.A., 1925; Ralph Sanger scholar, Harvard, 1826-27; Ph.D., Yale, 1931; m. Cecilia Rosser, Aug. 19, 1923; children—Adrian F., Richard E. Faculty, Yale, 1931-37; research biologist Spencer Lens Co., Buffalo, 1937-45; with Am. Optical Co., 1945—, chief biologist, Southbridge, Mass., 1953-67; lectr. Pacific U., Forest Grove, Ore., 1967—. Cons. Armed Forces Inst. Pathology, 1952—; also lectr. Fellow Biol. Photographers Assn. (pres. 1949-51, editor jour. 1951-52, Schmidt award 1953), A.A.A.S., Optical Soc. Am., Royal Micros. Soc. (hon.); mem. Am. Micros. Soc. (pres. 1953), Am. Assn. Anatomists, Am. Soc. Zoologists, Soc. Exptl. Biology and Medicine, Am. Physiol. Soc., Biometric Soc., Soc. Photographic Engr., Sigma Xi, others. Author: Phase Microscopy, 1951; also instn. manuals for microscopy. Research and publs. on analysis of growth of organisms; interference, fluorescence, and phase microscopy; night driving vision. Home: 2832 B St., Forest Grove, Ore. 97116.*

RICHARDS, Owain W., English zoologist; b. Croydon, Eng., Dec. 31, 1901; s. Harold Meredith and Mary C. (Todd) R.; B.A., Brasenose Coll., Oxford (Eng.) U., 1924, M.A., 1927, D.Sc., 1934; m. Maud Jessie Norris, June 27, 1931; children—Gillian Meredith, Jane Meredith (Mrs. R.W. Smithers). With Imperial Coll. Sci. and Tech., London, Eng., 1927—, prof. zoology and applied entomology, head dept., 1953—. Fellow Royal Soc., 1959; hon. fellow Royal Entomology Soc. London, Entomol. Soc. Egypt, Entomol. Soc. Netherlands. Author: (with G.C. Robson) Variations of Animals in Nature, 1936; Social Insects, 1954; also numerous articles. Research on taxonomy, biology, and ecology of insects especially hymenoptera and diptera. Home: 89 St. Stephens Rd., Ealing, London W. 13, Eng.*

RICHARDS, Paul Westmacott, Brit. botanist; b. Walton-on-the-Hill, Eng., Dec. 19, 1908; s. Harold Meredith and Mary Cecilia (Todd) R.; student U. Coll., U. London, 1925-27; B.A., Trinity Coll., Cambridge (Eng.) U., 1930, Ph.D., 1934, Sc.D., 1954; m. Sarah Anne Hotham, Dec. 22, 1935; children—Catherine Meredith (Mrs. Paul Whittle), Martin Paul Meredith, Mary Cecilia Meredith, Sarah Anne Meredith. Coutts Trotter student Trinity Coll., Cambridge, 1931-33, fellow, 1933-37, demonstrator, 1938-45, lectr. botany, 1945-49; prof., head dept. botany U. Coll. N.

Wales, Bangor, 1949—, vice prin., 1965-67. Mem. Nat. Parks Commn., 1955-59; chmn. com. for Wales, Nature Conservancy, 1954-67. Bullard fellow Harvard, 1964-65. Fellow Linnean Soc. London; mem. Brit. Ecol. Soc. (past pres.), Brit. (past pres.), Am. bryological socs., Assn. for Tropical Biology (mem. council). Author: The Tropical Rain Forest, 1952, rev., 1957; A Book of Mosses, 1950; also articles., Editor: Jour. Ecology, 1956-63. Research on autecology of Brit. plants (studies of ecology of individual species), bryology (taxonomy and ecology of mosses and liverworts), tropical forest ecology. Home: Talmaen, 30 Ffriddoedd Rd., Bangor, Caernarvonshire, U.K.*

RICHARDS, Rex Edward, English phys. chemist; b. Colyton, Devon, Eng., Oct. 28, 1922; s. Harold W. and Edith (Humphries) R.; B.A., Oxford (Eng.) U., 1945, M.A., Ph.D., 1948; m. Eva E. Vago, July 2, 1948; children—Jill Anne, Frances Janet. Sr. demy Magdalen Coll., Oxford, 1946; fellow Lincoln Coll., Oxford, 1947; lectr. Oxford U., 1948, reader in phys. chemistry, 1960-64, Dr. Lee's prof. chemistry, 1964—; research fellow Harvard, 1955. Fellow Royal Soc., 1959; mem. Chem. Soc. (past mem. council, Corday-Morgan medal 1954), Faraday Soc. (mem. council 1963—), Am. Phys. Soc. Author: (with E.E. Richards, J.H. Wolfenden) Advanced Problems in Physical Chemistry, 1964; also numerous articles. Research on chem. applications infra-red spectroscopy, phys. properties clathrate compounds, all aspects nuclear magnetic resonance and its chem. applications. Home: 8 Hawkswell Gardens, Summertown, Oxford, Eng.*

RICHARDS, Richard Davison, Am. physician; b. Grand Haven, Mich., Mar. 10, 1927; s. Marshall Foster and Alice (Davison) R.; A.B., U. Mich., 1948, M.D., 1951; M.Sc., State U. Ia., 1957; m. Alice Marion Durham, June 15, 1950; children—William O., Margaret Ann, Robert F. Practice medicine specializing in ophthalmology, Waterloo, Ia., 1957-58; asst. prof. ophthalmology State U. Ia., 1958-60; prof., head dept. ophthalmology U. Md. Med. Sch., Balt., 1960—; asso. dir. Eye Research Found., Bethesda, Md., 1964—. Chmn. med. bd. dirs. Med. Eye Bank Md. Mem. Md. Ophthal. Soc. (past pres.), Am. Acad. Ophthalmology and Otolaryngology, Assn. for Research in Ophthalmology, A.A.A.S., A.C.S., Am. Bd. Ophthalmology, A.M.A., N.Y., Md. acads. scis. Contbr. articles to tech. jours. Research on radiation effects on rabbit lenses. Home: 206 Skyline Dr., 1204.*

RICHARDS, Richard Kohn, physician, pharmacologist; b. Lodz, Poland, June 16, 1904; s. Stanley K. and Johann (Nachmann) R.; student U. Hamburg, 1923-25, M.D., 1931; student U. Berlin, 1925-30; m. Erika Nord, Apr. 14, 1946; 1 dau., Evelyn Jean. Came to U.S., 1935, naturalized, 1941. Asst. pharmacology U. Hamburg (Germany), 1930-33, asso. in medicine, 1934-35; research fellow U. Chgo., 1935-36; research pharmacologist Abbott Labs., Chgo., 1936-57, dir. exptl. therapy, 1957—, research adviser, 1963-66; med. faculty orthwestern U., Chgo., 1945—, prof. pharmacology, 1959—; prof. pharmacology emeritus U. Hamburg, 1964—; med. cons. St. Francis Hosp., Evanston, Ill., 1959—. Cons. to research and govt. instns. Recipient Research medal Internat. Coll. Anesthetists, 1938. Fellow A.C.P.; mem. Am. Pharmacol. Soc., Am. Physiol. Soc., A.M.A., Am. Soc. Anesthetists, French Therapeutic Soc. Contbg. author: Therapeutics in Internal Medicine, 1952; also numerous articles. Research in hypnotics, local anesthetics, exptl. and clin. toxicology and therapy, side effects of drugs, convulsive and anticonvulsive drug action; discovered novel drugs for treatment epilepsy. Home: 1580 Wakefield errace, Los Altos, Cal. 94022. Office: Dept. Pharmacology, Stanford U. Med. Sch., Palo Alto, Cal. 94305.*

RICHARDS, Theodore William, Am. chemist; b. Germantown, Pa., Jan. 31, 1868; s. William T. and Anna (Matlack) R.; S.B., Haverford Coll., 1885, LL.D., 1908; A.B., Harvard, 1886, A.M., Ph.D., 1888; postgrad. univs. Göttingen, Leipzig, Tech. Sch., Dresden; Sc.D., Yale, 1905; Ph.D., Royal Bohemian U., Prague, 1909; Sc.D., Harvard, 1910; M.D., Berlin U., 1910; D.Sc., Cambridge (Eng.), Oxford, 1911; other hon. degrees; m. Miriam Stuart Thayer, May 28, 1896; children—Grace Thayer (Mrs. James Bryant Conant), William Theodore, Greenough Thayer. Asst. prof. Harvard, 1894-1901, prof. chemistry, 1901—, chmn. chem. dept., 1903-11, dir. Gibbs Meml. Lab., 1912—. Exchange prof. from Harvard to Berlin U., 1907; Lowell lectr. 1908; mem. Internat. Com. on Elements and Atomic Weights; adviser Carnegie Instn., 1902, research asso., 1902—; mem. NRC. Recipient Davy medal Royal Soc., 1910; medal (also Faraday lectr.) Chem. Soc., 1911; Willard Gibbs medal Am. Chem. Soc., 1912; Nobel prize in chemistry, 1914 (awarded 1915); Franklin medal Franklin Inst., 1916; Lavoisier and Le Blanc medal (Paris), 1922. Fellow Am. Acad. Arts and Scis. (pres. 1919-21), A.A.A.S. (pres. 1917); mem. French Acad. Scis. (corr.), Contbr. numerous papers to sci. publs. Improved chem. determination of atomic weights; made (with students) very accurate revision of atomic weights of some 60 elements; reaffirmed existence of isotopes and led way to measurement of atomic weights by electromagnetic methods by discovery that atomic weight of lead from radioactive minerals is lower than that of ordinary lead, 1913; research on atomic volume, compressibility, molecular forces, thermochemistry, thermodynamics, electrochemistry; inventor calorimeter for precise measurement of heat quantities, instrument for measuring low concentrations of precipitates by light scattering. Died Cambridge, Mass., Apr. 2, 1928.

RICHARDS, Trevor Lloyd, Brit. physicist, metallurgist; b. Bangor, N. Wales, Dec. 20, 1910; s. Gwilym Arthur and Anne (Thomas) R.; B.Sc. with Honors in Physics, U. Coll. N. Wales, 1932, Ph.D., 1936; D.Sc., U. Wales, 1964; m. Katharine Mairgretta Jones, Jan. 4, 1940; children—Gwen, Mair. Temp. geophysicist Anglo-Iranian Oil Co., Lincoln, U.K., 1936; sect. head research dept. metals div. Imperial Chem. Industries Ltd., Witton, Birmingham, Eng., 1937-59; reader phys. metallurgy U. Aston, Birmingham, 1959—. Fellow Inst. Physics, Instn. Metallurgists; mem. Brit. Soc. Rheology, Brit. Non-Ferrous Research Assn., Inst. Physics. Research and publs. on application of X-ray diffraction and other methods to study of structure of metals after deformation and annealing, control of structure and properties of wrought products, particulary sheet; earthquakes. Home: 45 Maxstoke Rd., Sutton Coldfield, Warwickshire, U.K. Office: U. Aston, Birmingham 4, U.K.*

RICHARDS, Victor, Am. surgeon; b. Fort Worth, June 4, 1918; s. Jules Kelly and Minnie (Certain) R.; A.B., Stanford, 1935, M.D., 1939; m. Jennette O'Keefe, June 7, 1941; children—Lane Jennette, Victoria Michel, Victor Frederick, Peter Cromwell. Mem. faculty Sch. Medicine Stanford U., San Francisco, 1943—, prof. and chmn. dept. surgery, 1954-59, clin. prof. surgery, 1959—; clin. prof. surgery U. Cal. Sch. Medicine, 1964—; Commonwealth Research fellow in physiology Harvard, 1950-51; chief of surgery Presbyn. Med. Center, San Francisco, 1960—; chief of surgery Children's Hosp., San Francisco, 1960—; cons. USPHS. Hosp., Travis AFB, Hamilton AFB. Bd. dirs. San Francisco Heart Assn., Am. Cancer Soc. (San Francisco chpt.), Thomas Dooley Found. Mem. surg. study sect. B, USPHS; mem. bd. examiners Am. Bd. Surgery. Mem. Am. Thoracic Assn., A.C.S. (pres. No. Cal. chpt. 1952-63), Am. Surg. Assn., Soc. U. Surgeons, Pacific Coast Surg. Assn., Internat. Surg. Soc., San Francisco Surg. Soc., A.M.A., Cal. Med. Assn., Sigma Xi, Alpha Omega Alpha. Author: Surgery for General Practice, 1953; Abdominal Pain, 1955; also articles. Mem. editorial bd. Cal. Medicine, Am. Jour. Surgery, 1956—. Research in fields of cryobiology, tissue transplantation, cancer immunology, cardiovascular research, hyperbaric oxygenation. Home: 2714 Broadway. San Francisco 94115. Office: Children's Hosp., 460 Cherry St., San Francisco 94116.*

RICHARDSON, Archibald Read, mathematician; b. Eng., 1881; B.Sc., London; m. Margaret Harris, 1922. Lectr., U. London; asst. prof. math. Imperial Coll., 1912-14; prof. aero. sci. Cadet Coll., Cranwell, Eng., 1920; prof. math. U. Coll., Swansea, Wales, 1920-40; ret. to Capetown, S. Africa where he continued research. Fellow Royal Soc., 1946; asso. Royal Coll. Surgeons. Research and publs. in algebra, theory of invariants. Died 1954.

RICHARDSON, Sir Benjamin Ward, English physician; b. Somerby, Leicestershire, Eng., Oct. 31, 1828; s. Benjamin and Mary (Ward) R.; M.D., Glasgow, 1850; M.A., M.D., St. Andrews U., 1854; m. Mary J. Smith, Feb. 21, 1857; 2 sons, 1 dau. Became physician Royal Infirmary for Diseases of Chest, City Rd., 1856, London Temperance Hosp., 1892. Recipient Fothergillian Gold medal Med. Soc. London, 1856; Astley Cooper prize (for new concept that ammonia keeps blood fluid and escapes to permit coagulation), 1857. Fellow Royal Soc., 1867, Royal Coll. Surgeons, Soc. Antiquaries; pres. Med. Soc. London, 1868; hon. mem. Philos. Soc. Am. Contbr. to Med. Times and Gazette. Author: Diseases of Modern Life, 1876; National Health, 1890. Introduced methylene bichloride and 13 other anesthetics; used ether spray as local anesthetic; described amyl nitrite, 1863; introduced bromides of quinine, iron and strychnia, ozonized ether, styptic and iodized collloid, hydrogen peroxide, soda ethylate. Died Nov. 21, 1896.

RICHARDSON, Charles Clifton, Am. biochemist; b. Wilson, N.C., May 7, 1935; s. Barney Clifton and Elizabeth (Barefoot) R.; B.S.M., Duke, 1959, M.D., 1960; A.M. (hon.), Harvard, 1967; m. Ute Ingrid Hanssum, July 29, 1961; 1 son, Thomas Clifton. Research fellow Stanford Med. Sch., Palo Alto, Cal., 1961-63; asst. prof. biol. chemistry Harvard Med. Sch., Boston, 1964-67, asso. prof., 1967—. Recipient Am. Chem. Soc. award biol. chemistry, 1968. Mem. Am. Soc. Biol Chemists, Phi Beta Kappa, Pi Mu Epsilon, Alpha Omega Alpha. Research and publs. on identification and characterization of enzymes involved in deoxyribonucleic acid metabolism, structure of viral DNA's. Home: 78 Chestnut Hill Rd., Newton, Mass. 02167. Office: 25 Shattuck St., Boston 02115.*

RICHARDSON, Charles Henry, Am. geologist; b. Topsham, Vt., Sept. 26, 1862; s. Robert Fletcher and Rosetta (Dexter) R.; student Bates Coll., 1887-89; A.B., Dartmouth, 1892, A.M., 1895, Ph.D., 1898; postgrad. U. Chgo., 1902, Johns Hopkins, 1906; m. Katharine May Davis, June 16, 1892. Prin. schs. in Vt. to 1895; fellow and instr. chemistry Dartmouth, 1895-1905; in life ins. business, Manchester, N.H., 1905-06; asst. prof. geology and mineralogy Syracuse (N.Y.) U., 1906, asso. prof., 1907-09, prof. mineralogy, 1909—. Mem. Vt. Geol. Survey, summers, 1895—. Fellow Geol. Soc. Am. A.A.A.S.; mem. advisary council Simplified Spelling Bd.; mem. council Nat. Resources Assn.; mem. 8th and del. 12th Internat. Geologic Congress; mem. gen. com. of science, arts and edn. Paris Expn., 1901. Author works on econ. geology, 1911, on bldg. stones and clay, 1918. Research on econ. geology of Ky.; studied terrains of metamorphosed rocks of Green Mountains in Vt., substantiated theory that the belt is of Paleozoic age. Died Sept. 19, 1935.

RICHARDSON, Edward Glick, English physicist; b. Watford, Eng., 1896; s. Edward and Ada Elizabeth Richardson; ed. Cooper's Co.'s Sch., Univ. Coll., London; B.A., Ph.D., D.Sc. With RAF, 1918-19; asst. master Kilburn Grammar Sch., 1919-22; lectr. Univ. Coll., London, 1923-31, King's Coll., Newcastle, Eng., 1931-40; sci. adviser, mine design dept. Admiralty, 1940-42, Royal and Marine Aircraft Establishments, 1943-45; prof. acoustics, reader in physics U. Durham (Eng.); became Leverhulme research fellow, 1956. Mem. Phys. Soc., Acoustical Soc. Author: Sound, a Physical Textbook, 1927; Acoustics of Orchestral Instruments and of the Organ, 1929; Introduction to Acoustics of Buildings, 1933; Physical Science in Modern Life, 1938; Physical Science in Art and Industry, 1940; Dynamics of Real Fluids, 1949; Relaxation Spectrometry, 1957; Ultrasonic Physics, 1959. Editor Acustica; Technical Aspects of Sound, 1953, 56. Contbr. papers on sound and fluid dynamics to jours. Died Mar. 31, 1960.

RICHARDSON, Frank Howard, Am. physician; b. Bklyn., July 1, 1882; s. William James and Mary Carrington (Raymond) R.; A.B., Cornell U., 1904, M.D., 1906; certificate U. Vienna, 1909; m. Clara Louise Dixon, Sept. 8, 1915; children—Mary Faison (Mrs. Victor A. Gauthier, Jr.), Howard, Raymond Moseley, Dixon, Ruth Cadbury (Mrs. Tom C. Innes). Practice medicine, specializing in pediatrics, Black Mountain, N.C., Asheville, N.C., 1918-62; co-founder, vice dean So. Pediatric Seminar, Saluda, N.C.; founder So. Parenthood Inst., Black Mountain, 1930. Diplomate Am. Bd. Pediatrics. Fellow A.C.P., Am. Acad. Pediatrics, A.M.A.; mem. Nat. Writers Club. Author: Simplifying Motherhood, 1925; Parenthood and the Newer Psychology, 1926; Rebuilding the Child, 1927; The Nervous Child and His Parents, 1928; A Doctor's Letters to Expectant Parents, 1929; The Preschool Child and His Posture, 1930; Feeding the Child, 1931; For Boys Only, 1952; For Girls Only, 1953; The Nursing Mother, 1953; How to Get Along With Children, 1954; For Teenagers Only, 1957; For Young Adults Only, 1961; For Parents Only, 1962; A Christian Doctor Talks With Young Parents, 1963; Grandparents and Their Problems: A Guide to Three Generations, 1964; also numerous articles. Research on protection of children physically and emotionally; promotion of breast feeding. Home: "Hilltop", P.O. Box 965, Black Mountain, N.C. 28711.*

RICHARDSON, George Burr, American geologist; b. N.Y.C., 1872; s. George Wentworth and Emma (Breck) R.; student Coll. City N.Y.; B.S., Harvard, 1895, M.S., 1897; Ph.D., Johns Hopkins, 1901; m. Irene Dashiell, June 23, 1904; 1 dau., Mrs. Edward Russell True, Jr. with Geol. Survey, 1900-42, specializing in coal, oil and gas; geologist charge petroleum resources World War II Mem. A.A.A.S., Am. Assn. Petroleum Geologists, Geol. Soc. Am., Phi Beta Kappa. Author several books, numerous articles on mineral resources. Extensive research in geol. surveying of U. S., including descriptions of stratigraphy of S.W. which formed a basis for all subsequent work in N. and W. Tex., also Southeastern N.M. Died 1949.

RICHARDSON, Sir John, Scottish explorer, naturalist; b. Ninth Place, Scotland, Nov. 5, 1787; s. Gabriel and Anne (Mundell) R.; ed. U. Edinburgh; LL.D., Dublin, 1857; m. Mary Stiven, June 1, 1818 (dec. 1831); m. 2d, Mary Booth, Jan. 1833 (dec. 1845); 4 sons, 2 daus.; m. 3d, Mary Fletcher, Aug. 4, 1847. Became house-surgeon Dumfries (Scotland) Infirmary, 1804; named asst. surgeon Royal Navy, 1807; apptd. surgeon, naturalist Franklin's polar expdn., 1819, passed winter in Sask., reached Ft. Providence, 1821; mem. Franklin's 2d expdn. to mouth of Mackenzie, 1825, continued alone to explore coast to Coppermine River and Gt. Slave Lake, 1826; chief med. officer Melville Hosp., Chatham, Eng., 1828-38; apptd. physician Royal Hosp., Haslar, 1838; became insp. hosps., 1840; conducted search expdn. for Franklin, 1847-49; ret., 1849. Fellow Royal Soc., 1825 (Royal medal 1856); mem. Royal Coll. Surgeons. Author: Fauna boreali-Americana, 4 vols., 1829-37; Arctic Searching Expedition, 2 vols., 1851. Journal, 1851; also works on ichthyology, fossil mammals. Died Lancrigg, Scotland, June 5, 1865.

RICHARDSON, John Marshall, Am. physicist; b. Rock Island, Ill., Sept. 5, 1921; s. George Warren and Eloise (Hanna) R.; B.A., U. Colo., 1942; M.A., Harvard, 1947, Ph.D., 1951; m. Barbara Bennett, July 22, 1944; children—J. Jeffrey, Gregory B., Wendy L., Nancy B. Asst. research physicist Denver Research Inst., 1950-52; physicist Nat. Bur. Standards, Boulder, Colo., 1952-60, chief radio standards lab., 1960-67; dir. Office of Standards and Rev., U. S. Dept. Commerce, 1967-68; exec. sec. com. on telecommunications Nat. Acad. Engring., 1968——. Mem. com. for definition of second Internat. Com. Weights and Measures, 1961-67. Recipient gold medal Dept. Commerce, 1964. Fellow Am. Phys. Soc., A.A.A.S.; I.E.-E.E.; mem. Internat. Sci. Radio Union (chmn U. S. Commn. 1, 1964), Sci. Research Soc. Am., Phi Beta Kappa, Sigma Xi. Research, publs. on gaseous electronics, microwave physics, frequency and time standards, radio measurements and standards, sci. adminstrn. Home: 4616 Harrison St., Chevy Chase, Md. 20015. Office: 2201 Constitution Av. N.W., Washington 20418.*

RICHARDSON, John Reginald, physicist; b. Edmonton, Alta., Can., Oct. 31, 1912; s. John Thomas and Lilly (Reay) R.; came to U. S., 1923, naturalized, 1934; A.B., U. Cal. at Los Angeles, 1933; Ph.D., U. Cal. at Berkeley, 1937; m. Louise Edna Frisbie, July 31, 1938; children—Pamela, Barbara. Nat. Research fellow U. Mich., Ann Arbor, 1937-38; asst. prof. physics U. Ill., Urbana, 1938-42; chief physicist U. Cal. Radiation Lab., Berkeley, 1942-46; faculty U. Cal. at Los Angeles, 1946——, prof. physics, 1952——. Cons. Lawrence Radiation Lab., 1948——; mem. tech. adv. com. Oak Ridge Nat. Lab., 1958-62; mem. com. sr. reviewers U. S. AEC, 1952——. Fellow Am. Phys. Soc.; mem. Phi Beta Kappa, Sigma Xi. Research, publs. on nucleon-nucleon and nuclear interaction, beta and gamma radioactivity; discovered 12 new radioactive isotopes; devel. of frequency modulated cyclotron, sector-focusing cyclotron. Home: 20505 Big Rock Dr., Malibu, Cal. 90265. Office: Physics Dept., U. Cal., Los Angeles, 90024.*

RICHARDSON, Lewis Fry, English physicist, psychologist; b. 1881; s. David and Catherine Richardson; ed. Durham Coll. Sci., Newcastle, Eng., King's Coll., Cambridge; D.Sc. in Physics, 1926; B.Sc. in Psychology, 1929; m. Dorothy Garnett, 1909; 2 sons, 1 dau. With Nat. Phys. Lab., 3 years; in charge chem. and phys. lab. in tungsten lamp factory, 3 years; with geophys. obs., including period as supt. Eskdalemuir, more than 4 years; in charge physics dept. Westminster Tng. Coll., London, 1920-29; prin. Paisley Tech. Coll., 1929-40. Fellow Royal Soc., 1926, Brit. Psychol. Soc. Author: Weather Prediction by Numerical Process, 1922; Generalized Foreign Politics, 1939; Arms and Insecurity, 1949; (microfilm) Statistics of Deadly Quarrels, 1950. Research in physics, math, meteorology, psychology.

RICHARDSON, Sir Owen Williams, English physicist; b. Dewsbury, Eng., Apr. 26, 1879; s. Joshua and Charlotte Maria (Willans) R.; B.A., U. of Cambridge, Eng., 1900, M.A., 1904; B.Sc., U. of London, 1900, D.Sc., 1903. Fellow of Trinity Coll., Cambridge, 1902-8; m. Lilian Maud Wilson, June 12, 1906. Prof. physics, Princeton, 1906-13; dir. research physics Kings Coll., 1924-44. Recipient Nobel Prize in physics, 1928. Fellow Royal Soc., 1913, Cambridge Philos. Soc., Phys. Soc. London; mem. Am. Philos. Soc., Am. Phys. Soc., Soc. Française de Physique (council). Author: The Ionisation Produced by Hot Platinum in Different Gases, 1906; The Electron Theory of Matter, 1914; The Emission of Electricity from Hot Bodies, 1916; Molecular Hydrogen and its Spectrum, 1934. Originated study of thermionics; developed Richardson equation governing thermionic emission; other work on photoelectric action, gyromagnetic effect, electronic emission by chem. action, electron and quantum theory, molecular hydrogen spectra, also fine structure of alpha particles. Died Feb. 15, 1959.

RICHARDSON, Thomas, English chemist; b. Newcastle-on-Tyne, Eng., Oct. 8, 1816; ed. Glasgow; studied chemistry under Thomas Thomson; Ph.D., Giessen, Germany; studied under J. Pelouze, Paris; M.A. (hon.) U. Durham (Eng.), 1856; engaged in chem. manufacture, Newcastle; began to remove impurities from hard lead to convert it to soft lead, Blaydon, 1840, to manufacture superphosphates (upon Liebig's suggestion), 1844; became lectr. chemistry U. Durham (Eng.), 1856. Fellow Royal Soc., 1866; mem. Royal Irish Acad. Author: (with Armstrong and James Longridge) Steam Coals of Hertley District of Northumberland in Steam Boilers, 1857-58; also papers. Patentee various processes. Died July 10, 1867.

RICHARDSON, William Springer, Am. oceanographer; b. Providence, Oct. 27, 1923; s. Oscar S. and Lillian (Demerest) R.; Sc.B., Brown U., 1947; Ph.D., Harvard, 1950; m. Ruth Pangburan, Nov. 26, 1947; children—James Steven, David William, Janet, Kathy. Jr. fellow Mellon Inst., 1950-51, sr. fellow, 1951-52; staff Woods Hole (Mass.) Ocenaographic Instn., 1952-53; asso. prof. oceanography Inst. Marine Sci., U. Miami (Fla.), 1963-66; prof. Nova U., Ft. Lauderdale, Fla., 1966——. Mem. Am. Geophys. Union, Marine Technol. Soc. Research and publs. on phys. oceanography, thermal structure, current in seas, instrumentation for oceanographic research. Home: 1015 Cordova Rd., Ft. Lauderdale, Fla. 33316.*

RICHER, Jean, French astronomer; b. France, circa 1630; elected astronomer French Acad. Scis., 1666; astronomer Obs. Paris; sent to Cayenne to determine parallax of Mars at opposition French Acad. Scis., 1672-73. Author: Observations astronomiques et physiques faites en l'isle de Cayenne, 1679. Discovered beat of a pendulum is slower in Cayenne that at Paris which is in different latitude and concluded Cayenne was further from center of earth than Paris (Newton and Huygen used this to prove the earth is oblate spheroid such as would be required by theory of gravitation). Died Paris, 1696.

RICHERT, Dan Arnold, Am. biochemist; b. Goessel, Kan., Nov. 6, 1915; s. Henry and Mary (Richert) R.; A.B., Bethel Coll., 1939; B.S., Kan. State U., 1938, M.S., 1939; Ph.D., St. Louis U., 1944; m. Esther Irene Beamer, May 1, 1943; children—John, Eric, Evan, Mary. Research asso. Harvard Med. Sch., 1944-45; faculty Syracuse U. Med. Sch., 1946-50, asst. prof., 1948-50; asso. prof. dept. biochemistry State U. N.Y. Upstate Med. Coll., Syracuse, N.Y., 1951-61, prof., 1962——. Cons., Hosp. Clin. Chemistry labs. Mem. Am. Soc. Biol. Chemists, Am. Inst. Nutrition, Soc. for Exptl. Biology and Medicine, Sigma Xi, Phi Lambda Upsilon. Contbr. numerous articles to tech. jours. Co-discoverer relationship dietary molybdenum and iron to enzyme xanthine oxidase, role vitamin B6 cofactor in hemoglobin synthesis; research on biol. assay procedure for thyroid hormones based on enzyme analysis. Home: 116 Westminster Av., Syracuse, N.Y. 13210.*

RICHERT, Hans-Egon, German mathematician; b. Hamburg, Germany, June 2, 1924; s. Johann and Agnes (Hinsch) R.; m. Irmgard Schaefer, Mar. 12, 1951; children—Manfred, Ranko. Successively asst. agrégé, 1st asst., prof. at large U. Göttingen; prof. U. Syracuse, U. Marburg (Germany). Mem. Union German Mathematicians, Am. Math. Soc. Author: Verschärfung der Abschätzung beim Dirichletschen Teilerproblem; Über quadratfreie Zahlen mit genauen Primfaktoren in einer arithmetischen Progression; Über Dirichletreihen mit Funktionalgleichung; Einführung in die Theorie der starken Rieszchen Summierbarkeit von Dirichletreihen. Research on theory of numbers. Home: Wehrda Kr. Marburg Im Loh 14, West Germany.

RICHET, Charles Robert, French physiologist; b. Paris, France, Aug. 26, 1850; s. Alfred Richet; M.D., U. Paris, 1877. Faculty, Collège de France, until 1887; prof. U. Paris, 1887-1927. Founder Annals des Sciences Physiques, 1891, editor until 1929. Recipient Nobel prize in physics and medicine, 1913. Mem. French Acad. Scis. (v.p. 1932, pres. 1933), 1914, Acad. Medicine, Institut Metaphysique International (founder, 1st pres.). Author: Les Poisons de L'Intelligence, 1877; Recherches Ekpérimentales et Cliniques sur la Sensibilité, 1877; Structure des Circonvolutions Cérébrales, 1878; Du Suc Gastrique Che L'Homme et Les Animaux, 1878; Physiologie des Muscles et des Nerfs, 1882; L'Homme et L'Intelligence, 1884; Essai de Psychologie Générale, 1887; La Physiologie et la Médecine, 1888; La Chaleur Animale, 1889; Programme, 1890; Physiologie, Travaux du Laboratoire, 3 vols., 1892-95; Dictionnaire de Physiologie, 1895; Les Guerres et la Paix, 1899; L'Anaphylaxie, 1911; Abrege D'Histoire Generale, 1919; L'Homme Stupide, 1919; Traite de Metapsychique, 1923; Le Savant, 1923; L'Homme Impuissant, 1927; Apologie de la Biologie, 1929. Research on serums and antigens producing immunity; discovered and described phenomenon of anaphylaxis; studied digestion, discovered hydrochloric acid base of gastric juice; studied nervous system, respiration, epilepsy, Tb; investigations in psychology, parapsychology. Died Paris, Dec. 4, 1935.

RICHEY, Herman Glenn, Am. psychologist, educator; b. Cory, Ind., Nov. 4, 1897; s. George Willard and Susan May (Bennett) R.; A.B., Ind. State Tchrs Coll., 1920; Ph.M., U. Wis., 1924; A.M., U. Chgo., 1927, Ph.D., 1930; m. Marie Magdalena Rousseau, Aug. 9, 1923; 1 son, Herman Glenn. High sch. prin., North Terre Haute, Ind., 1920-26; research sec. Lab. Schs., U. Chgo., 1928-33; instr. edn. U. Chgo., 1930-35, asst. prof. from 1935, asso. prof. until 1948, prof., 1948——, sec. dept., 1942——, dean students Grad. Sch. Edn., 1959——. Departmental editor Ency. Brit., 1951——; cons. President's Adv. Com. on Edn., 1937; mem. Commn. to Survey Higher Edn. Facilities in Ill., 1944; mem. Pa. Com. on Post-High Sch. Edn., 1948-49. Trustee, Morgan Park Acad., 1955-61. Mem. Nat. Soc. Study Edn. (bd. dirs. 1959——, sec.-treas., 1959——, editor yearbooks, 1960——), Am. Ednl. Research Assn., N.E.A., Am. Assn. Coll. Tchrs. Edn., Comparative Edn. Soc., Phi Delta Kappa. Author: (with Newton Edwards) The School in the Social Order, 1947; Editor: Child Psychology, 1963; Improvement and Impact of School Testing Programs, 1963; Theories of Learning and Instruction, 1964; Behavioral Science and Educational Administration, 1964; Art Education, 1965; Vocational Education, 1965. Home: 5801 S. Dorchester Av., Chgo. 60637.*

RICHEY, James Ernest, geologist; b. Cookstown, N. Ireland, Apr. 24, 1886; s. John and Susanna (Best) R.; B.A., Trinity Coll., U. Dublin (Ireland), 1908, B.A. in Engring., 1909, Sc.D., 1932; m. Henrietta Lily McNally, June 23, 1924; children—Henrietta Mary Joyce, Angela Margaret Grace (Mrs. John Sinfield), Jennifer Alison Suzanne (Mrs. Robert Evans). Demonstrator, Oxford U. (Eng.), 1910-11; with Geol. Survey, Scottish office, 1911-46, dist. geologist, 1926-46; lectr. geology Queen's Coll., U. St. Andrews (Scotland), 1946——; cons. in engring. geology, Edinburgh, 1946——. Recipient Lyell medal Geol. Soc. London, 1933; Baker medal Soc. Engrs., 1956; Clough medal Geol. Soc. Edinburgh, 1964; Neill prize Royal Soc. Edinburgh, 1965. Fellow Royal Soc., 1938, Royal Soc. Edinburgh (past gen. sec.), Geol. Soc. Am. (hon.), Soc. Engrs.; mem. Internat. Assn. Volcanology (past v.p.), Inst. Civil Engrs. (asso.). Author: Elements of Engineering Geology, 1964; also articles. Research on tertiary volcanic dists. of Western Scotland and N.E. Ireland.*

RICHMAN, Chaim, physicist; b. Poland, Jan. 18, 1918; s. Jacob and Frieda (Goren) R.; B.A., So. Meth. U., 1938; M.A., U. Okla., 1939; Ph.D., U. Cal. at Berkeley, 1953; m. Betty Stekoll, July 3, 1942; children—Steve, Cathey, Ann, Jeffrey. Came to U.S., 1930, naturalized, 1955. With Los Alamos (N.M.) Sci. Lab., 1943-46; lectr., asst. prof., asso. prof., staff mem. Lawrence Radiation Lab., U. Cal. at Berkeley, 1946-56; dir. research Stekoll Petroleum Corp., Dallas, 1956-60; cons. physicist Microwave Power Lab., Varo Inc., Dallas, 1960-61; prof. physics So. Meth. U., Dallas, 1960-62; prof. Southwest Center for Advanced Study, Dallas, 1961——. Mem. Phi Beta Kappa, Sigma Xi. Research in production and application of pions to radiotherapy of cancer. Home: 6749 Prestonshire Lane, Dallas. Office: 1500 Dallas N. Pkwy., Plano, Tex.*

RICHMANN, Georg Wilhelm (Rickman), physicist; b. Livonia, 1711; prof. natural history, St. Petersburg, Russia, 1745. Studies on law of cooling (refrigeration); built instrument for measurement of water evaporation. Killed by lightning while trying to duplicate Franklin's expt., 1753.

RICHMOND, Donald Everett, Am. mathematician; b. Meriden, Conn., May 23, 1898; s. Elmer Ellsworth and Elizabeth (Kinne) R.; A.B., Cornell U., 1920, M.S., 1922, Ph.D., 1926. Engr., Am. Tel. & Tel., 1922-23; Nat. Research fellow, Harvard, 1926-27; faculty Williams Coll., Williamstown, Mass., 1927——, prof. math., 1940——, chmn. dept., 1941-63, acad. co-ord. V-5 program, 1942-43; mem. Inst. for Advanced Study, Princeton, N.J., 1944-45; operations analyst, USAF, 1945; lectr., Mass. Inst. Tech., 1951-52; vis. prof. Dartmouth, 1957-58. Dir., NSF Inst., 1956, 1959-60. Mem. Math. Assn. Am. (bd. govs. 1962-64), Sch. Math. Study Group, Sigma Xi, Kappa Phi, Gamma Alpha. uthor: Fundamentals of alculus, 1950, Calculus with Analytic Geometry, 1959. Research, publs. on Laplace transforms, complex numbers, vectors; developed treatment of areas and volumes without using limits. Home: Sweet Brook Rd., Williamstown, Mass.*

RICHMOND, Jonas Edward, Am. biochemist, biophysicist; b. Prentiss, Miss., July 17, 1929; s. Benjamin F. and Bessie (Collins) R.; M.S., U. Rochester, 1950, Ph.D., 1953; postgrad. (Commonwealth fellow) Oxford (Eng.) U., 1955-56; m. Mattie Lee Humes, Aug. 2, 1957; 1 son, Perry Keith. Instr., U. Rochester (N.Y.), 1956-57; established investigator Harvard Med. Sch., Boston, 1957-63; research biochemist U. Cal. at Berkeley, 1963——. Cons., NIH, 1965——, AEC, 1960——, Ross Labs., 1963——. Mem. Am. Soc. Biol. Chemists, Biophysicist Soc., Biochem. Soc., Am. Chem. Soc., A.A.A.S., Radiation Research Soc. Research and numerous articles on characterization of reactions involved in biosynthesis of hemin and hemoglobin, quantitation of rate of uptake and utilization amino acids for protein synthesis and for catabolic reactions, characterization, structure, biosynthesis, metabolism, and function of glycoproteins, and glycolipids, chemistry of morphogenesis and aging. Home: 6870 Charing Cross Rd., Berkeley, Cal. 94705.*

RICHOU, Rémy Jean Henry, French immunologist; b. La Pommeraye, Jan. 4, 1905; s. Rémy and Désirée (Coutant) R.; M.D. in Vet. Medicine, Alfort Vet. Sch.; m. Henriette Thaubin, Aug. 4, 1930. Lab. chief Pasteur Inst.; dir. research Nat. Hygiene Inst. Decorated Legion of Honor; Order of Pub. Health; Crder of Merit (Agr.); laureate Acad. Sci., also Acad. Medicine. Contbr. numerous articles on immunology to tech. jours Home: 36 allée des Haras, Vaucresson (Seine and Oise), France. Office: 22 rue Pierre-Curie, Alfort, (Seine) France.

RICHTER, August Gottlob, German surgeon; b. Zörbig, Apr. 13, 1742; prof. Göttingen, Germany. Author: Chirurgische Bibliothek, 15 vols., 1771-97; Abhandlung von der Brüchen, 1777; Anfangsgründe der Wundarzneikunst, 7 vols., 1782-1804, Attempted to unify surgery and medicine; studies in intestinal ruptures (hernias), including description of hernia involving only part of lumen (Richter's hernia). Died Göttingen, July 23, 1812.

RICHTER, Charles Francis, Am. seismologist; b. nr. Hamilton, O., Apr. 26, 1900; student U. So. Cal., 1916-17; A.B., Stanford, 1920; Ph.D., Cal. Inst. Tech., 1928; m. Lillian Brand, July 18, 1928. Asst., Seismol. Lab., Carnegie Instn. Washington, Pasadena, Cal., 1927-36; asst. prof. seismology Cal. Inst. Tech., 1937-47, asso. prof., 1947-52, prof., 1952——. Fulbright research scholar Tokyo U., Japan, 1959-60. Fellow Geol. Soc. Am., Royal Astron. Soc.; mem. Am. Geophys. Union, Seismol. Soc. Am., Sigma Xi. Author: (with B. Gutenberg) Seismicity of the Earth, rev. edit., 1954; Elementary Seismology, 1958. Contbr. to Internal Constitution of the Earth, rev. edit., 1951. Research on earthquake magnitude scale, geography and statistics. Home: 1820 Kenneth Way, Pasadena 91103.*

RICHTER, Curt Paul, Am. psychologist; b. Denver, Feb. 20, 1894; s. Paul Ernest and Martha (Dressler) R.; B.S., Harvard, 1917; Ph.D., Johns Hopkins U., 1921; m. Leslie Prince Bidwell, Apr. 11, 1937; children—Ann (Mrs. William A. Roy), Peter, Martha (Mrs. Donald Bailey). Dir. psychobiol. lab. Phipps Psychiat. Clinic, Johns Hopkins Hosp., Balt., 1921—, prof. emeritus of psychobiology Johns Hopkins U. Med. Sch., 1960—. Mem. coms. on neurology, food habits, rodenticides OSRD, Washington, 1942-47; lectr. Harvey Soc., 1943; Thomas Salmon lectr. 1959; mem. Inst. for Advanced Study, Princeton, N.J., 1957-58. Named Distinguished Citizen, Denver, 1958; Warren medal Soc. Exptl. Psychology, 1950. Mem. Nat. Acad. Scis., Am. Philos. Soc., Am. Acad. Arts and Scis., Halsted Soc., entury Assn., Phi Beta Kappa. Author: Biological Clocks in Medicine and Psychiatry, 1965. Research on behavior in fields of neurology, endocrinology, physiology, nutrition. Home: 14 Meadow Rd., Balt. 21212. Office: John Hopkins Hosp., Balt. 21205.*

RICHTER, Derek, Brit. neurochemist; b. Bath, Eng., Jan. 14, 1907; s. Charles Augustus and Frances (Mann) R.; M.A., B.Sc., Magdalen Coll., Oxford U., 1929; D.phil., U. Munich (Germany), 1932; M.R.C.S., L.R.C.P., London, 1945; m. Beryl Ailsa Griffiths, Dec. 27, 1937, (div. Sept., 1953); children—Sally (Mrs. Michael Festing), John, Polly; m. 2d, Winifred Molly Bullock, Dec. 18,1953. Asst. demonstrator Biochem. Lab., Cambridge, Eng., 1933-37; biochemist Maudsley Hosp., London, 1938-39; biochemist Mill Hill Emergency Hosp., London, 1940-45; house physician Addenbrookes Hosp., Cambridge, 1945-46; dir. research Whitchurch Hosp., Cardiff, Wales, 1946-49; med. cons., dir. Neuropsychiat. Research Centre, Cardiff, 1950-58; dir. neuropsychiat. research unit Med. Research Council, Carshalton, also Epsom, Eng., 1959—. Hon. cons. W. Park, Queen Mary's hosps., Carshalton, 1960—; hon. lectr. Inst. Neurology, U. London, 1941—; hon. sec. Mental Health Research Fund, 1949—. Recipient Chapman Research prize, Oxford U., 1933. M.R.C.P. Fellow Inst. Biology; mem. Internat. Brian Research Orgn. (mem. central council since 1966), Internat. Soc. for Neurochemistry (treas. since 1966—). Royal Medico-Psychol. Assn., Biochem. Soc., Physiol. Soc., Sci. Publs. Council, Collegium Internat. Neuro-Psychopharm., Royal Soc. Medicine. Author: Perspectives in Neuropsychiatry, 1950; Schizophrenia: Somatic Aspects, 1957; Metabolism of the Nervous System, 1957; Aspects of Psychiatric Research, 1962; Comparative Neurochemistry, 1964; Aspects of Learning and Memory, 1965; also numerous articles. Determined chem. constitution of several plant pigments including rose madder glycoside, 1933-35; research on enzymes inactivating adrenaline, identified monamine oxidase, isolated and discovered chem. constitution of adrenochrome, showed esterification of adrenaline in body, 1936-37; showed occurrence of chem. changes in brain under physiol. conditions (with R. M. C. Dawson), 1946-49; measured turnover of proteins in vivo in brain (with M. K. Gaitonde), 1951-56; (with others) studied turnover of amino acids in brain and showed changes with age in brain metabolism; studied chem. pathology of different forms of mental illness, 1957—. Home: Deans Cottage, Walton-on-the Hill, Surrey. Office: Med. Research Council, Neuropsychiat. Research Unit, Carshalton, Surrey, Eng.*

RICHTER, Edward, Austrian glaciologist; b. Mannersdorf, nr. Vienna, Oct. 3, 1874; prof., Graz, Austria. Author: Das Gletscherphänomen, 1877; Die Gletscher der Ostalpen, 1888; Die Erschliessung der Ostalpen, 3 vols., 1892 94. Died Graz, Feb. 6, 1905.

RICHTER, Goetz Wilfried, pathologist; b. Berlin, Germany, Dec. 19, 1922; s. Werner and Ursula (Meyer) R.; came to U.S., 1939, naturalized, 1944; B.A., Williams Coll., 1943; M.D., Johns Hopkins U., 1948; postgrad. Cornell U., 1948-51; m. Lucretia Henry, Dec. 28, 1946; children—James Wilfrid, Elizabeth Margot, Marianne. Mem. faculty Med. Coll., Cornell U., 1948-53; chief of labs. U.S. Naval Hosp., Corona, Cal., 1953-55; asso. attending pathologist N.Y. Hosp., 1958—; faculty Cornell U., 1955-58, asso. prof. pathology, 1958—; vis. investigator Rockefeller Inst., 1956-57; career sci. Health Research Council, N.Y.C., 1961-67; prof. pathology U. Rochester Med. Sch., 1967—. Cons. USPHS, U.S. Army. Fellow A.A.A.S.; mem. N.Y. Soc. Electron Microscopists (pres. 1965-66), Am. Assn. Pathologists and Bacteriologists, Am. Soc. Exptl. Pathology, Am. Soc. Clin. Pathology, Internat. Acad. Pathology, Am. Soc. Cell Biology, Electron Microscopy Soc. Am., Harvey Soc., Soc. Exptl. Biology, Phi Beta Kappa. Founder, editor Internat. Rev. Exptl. Pathology, 1961—. Research and publs. in field of cellular iron metabolism; protein synthesis; cell fine structure in disease; discovered isoferritins in cancer cells; demonstrated characteristic, ultrastructural changes in hypertrophied heart muscle; traced various compounds of iron in cells at molecular level by electron microscopy. Home: 155 Avon Rd., Rochester, N.Y. 14625.*

RICHTER, Henry John, chemist; b. Germany, May 29, 1910; B.S., Hamline U., 1933; M.S., U. Minn., Ph.D., 1937; m. Dorothy M. Schroeder, Aug. 20, 1938; children—Judith Anne, Henry John, Jay. Research chemist DuPont Chem. Dept., Wilmington, 1937-44; reserved chemist A.B. Dick Co., Chgo., 1945-49; asso. prof. Hamline U. 1948-53; prof. chemistry U. Colo., Boulder, 1953—. Mem. Am. Chem. Soc. (chmn. Colo. sect. 1960), Sigma Xi, Phi Lambda Upsilon. Research, publs. on preparation, modification, use of vinyl chloride polymers, emulsions and plasticizers, acenaphthene chemistry, heterocyclics. Home: 844 Iris Av., Boulder, Colo. 80301.*

RICHTER, Henry Leopold, Jr., Am. space scientist; b. Long Beach, Cal., June 14, 1927; s. Henry L. and Catherine (Moore) R.; B.S., Cal. Inst. Tech., Ph.D., 1956; m. Marilyn Wood, July 19, 1946; children—Gail, Norman, David, Steven, Jan. With Jet Propulsion Lab., Pasadena, Cal., 1955-60, chief space instruments sect., 1959-60; v.p. Electro-Optical Systems, Inc., Pasadena, Cal., 1960—. Mem. Am. Chem. Soc., I.E.E.E., Am. Inst. Aeros. and Astronautics, N.Y. Acad. Scis., A.A.A.S. Contbr. articles to tech. jours. Designed, developed 1st Am. earth satellite; developed space probe instruments. Home: 401 Patrician Way, Pasadena 91105. Office: 300 N. Halstead St., Pasadena, Cal. 91107.*

RICHTER, Hieronymus Theodor, German chemist, mineralogist; b. Dresden, Germany, Nov. 21, 1824; asst. to Reich, Freiberg (Germany) Sch. Mines, became dir., 1875. Discovered (with Reich) indium in zinc-blende using spectroscopic analysis, 1863. Died Freiberg, Sept. 25, 1898.

RICHTER, Jeremias Benjamin, German chemist; b. Hirschberg, Silesia, Mar. 10, 1762; studied under Kant at Königsberg, grad. 1789. Assayer in Breslau, 1794; 2nd chemist in porcelain factory in Berlin, from 1798; assessor to dept. of mines and chemist to Royal Porcelain Factory, Berlin, 1800. Author: Anfangsgründen der Stöchiometrie oder Messkunst chemischer Elemente, 3 vols., 1792-4; Über die neueren Gegenstände in der Chemie, 1792-1802 Made early determinations of quantities by weight in which acids saturate bases and bases which can saturate the same quantity of a certain acid are equivalent to each other; held that chemistry is branch of applied math.; attempted to find regularities among combining proportions; generalized that "elements must have among themselves a certain fixed proportion of mass," determined several combining proportions and decomposition ratios. Died Berlin, Prussia, Apr. 4, 1807.

RICHTER, Maurice Nathaniel, Am. physician; b. Chgo., Aug. 26, 1897; s. Emanuel and Cecelia (Meyer) R.; student U. Chgo., 1914-15, Coll. City N.Y., 1916-17; B.S., Columbia, 1919, M.D., 1921; m. Brina H. Kessel, June 30, 1928; children—Maurice Nathaniel, Marcel K., Wayne H. Faculty Columbia, N.Y.C., 1924-47, N.Y. U., N.Y.C., 1947-62, prof. pathology, 1948-62; mem. staff U. Hosp. N.Y. U. Med. Center, 1948-62. Cons. pathologist Good Samaritan Hosp., Phoenix, Ariz., 1962—, Armed Forces Inst. Pathology 1951—. Diplomate Am. Bd. Pathology. Fellow A.M.A., A.A.A.S., Biol. Photog. Assn., N.Y. Acad. Scis., Internat. Soc. Hematology, Coll. Am. Pathologists (gov. 1960-63); mem. Am. Assn. Pathologists and Bacteriologists, Am. Soc. for Clin. Pathology, Am. Soc. Exptl. Pathology, Am. Soc. Hematology, Harvey Soc., Internat. Acad. Pathology, N.Y. Acad. Medicine, Med. Soc. County N.Y., N.Y. Path. Soc. (pres. 1940), N.Y. State Soc. Pathologists, (v.p. 1961-62), Soc. Exptl. Biology and Medicine, Soc. Study Blood. Contbr. articles to med. jours. Pioneered successful transmission leukemia in mice, 1928; research exptl., path. studies blood and blood forming organs. Home: 6837 N. 2d St., Phoenix 85012. Office: 1033 E. McDowell Rd., Phoenix, Ariz. 85012.*

RICHTER, Richard Biddle, Am. physician; b. La Porte, Ind., May 9, 1901; s. Harry Walter and Elizabeth (Biddle) R.; S.B., U. Chgo., 1922; M.D., Rush Med. Coll., 1925; m. Gudrun Anderson, Jan. 27, 1926; children—Tor, Anders. Grad. work in neurology, Rush Med. Coll., 1925-30, asst. clin. prof. neurology, 1932-39; asst. prof. medicine, U. Chgo., 1936-45, asso. prof. medicine, 1945-46, prof. neurology, 1946—; attending neurologist, Albert Merritt Billings Hosp., 1936—. Mem. Am. Neurol. Assn. (pres. 1963), Am. Assn. Neuropathologists (pres. 1960-61), Assn. for Research Nervous and Mental Diseases, Am. Acad. Neurology, A.A.A.S., A.M.A., Chgo. Neurol. Soc. (past pres.), Chgo. Med. Soc., Phi Beta Kappa, Alpha Omega Alpha, Sigma Xi. Mem. editorial adv. bd. Jour. Neuropathology and Exptl. Neurology, 1950—, Neurology, 1950—. Home: 5550 Dorchester Av., Chgo. 37*

RICHTERICH, Roland, Swiss clinical chemist; b. Basel, Switzerland, Feb. 27, 1927; s. Werner and Mary (Bürgi) R.; ed. Medical Faculty, U. Basel, 1949-55; m. Britta van Baerle, Aug. 12, 1953. Intern, Mass. Memorial Hosp., Boston, 1955-56; instr. pharmacology, Boston U., 1956-58; oberassistent biochemistry, U. Berne, 1959-65; head, Central Lab., Inselspital Bern, 1965—; prof. clinical chemistry, U. Berne, 1966—. Author: Enzymopathologie, Enzyme in Klinik und Forschung, 1958; Klinische Chemie, Theorie und Praxis, 1965; over 100 published articles. Research on enzymes in muscular dystrophy, particularly as applied to heterozygote detection; diagnostic use of serum enzymes; work on inborn errors of metabolism (Refsum's disease, Lysine intolerance). Home: 6 Arvenweg, Spiegel/Bern, Switzerland. Office: 28 Bühlstrasse, Bern, Switzerland.*

RICHTHOFEN, Ferdinand Paul Wilhelm (Baron) von, see von Richthofen, Ferdinand Paul Wilhelm (Baron).

RICK, Charles Madeira, Jr., Am. geneticist; b. Reading, Pa., Apr. 30, 1915; s. Charles M. and Miriam C. (Yeager) R.; B.S., Pa. State U., 1937; A.M., Harvard, 1938, Ph.D., 1940; m. Martha Elizabeth Overholts, Sept. 23, 1938; children—Susan Charlotte (Mrs. Eugene Baldi), John Winfield. Asst. plant breeder W. Atlee Burpee Co., Lompoc, Cal., 1937-40; faculty U. Cal. at Davis, 1940—, prof., geneticist, 1955—. Carnegie vis. prof. U. Hawaii, 1963; vis. prof. U. Sao Paulo, Piracicaba, Brazil, 1965; Gen. Edn. Bd. vis. lectr. N.C. State U., 1956; vis. scientist Agr. Expt. Sta., U. P.R., 1968. Chmn. coordinating com. Tomato Genetics Coop., 1950—. Guggenheim fellow, 1948, 50; NSF grantee, 1953-68; Rockefeller Found. grantee, 1956-57; NIH grantee, 1959—. Fellow A.A.-A.S. (Campbell award 1959), Cal. Acad. Sci.; mem. Am. Soc. Hort. Sci. (Vaughan Research award 1946), Bot. Soc. Am., Am. Soc. Naturalists, Am. Genetics Assn., Genetics Soc. Am., Am. Inst. Biol. Scis., Soc. for Study Evolution, Nat. Acad. Scis. Research, publs. on sex determination and polyembryony in asparagus, X-ray induced deletions in chromosomes, hybridization between chicory and endive and its consequences, tomatoes, prodn. of complete set of primary trisomics, compatability and other relations between species, mechanisms of transfer of germ plasm from wild to cultivated species; established all linkage groups and relations with chromosomes. Home: 8 Parkside Dr., Davis, Cal. 95616.*

RICKART, Charles Earl, Am. mathematician; b. Osage City, Kan., June 28, 1913; s. Charles Day and Ola (Brewer) R.; student Emporia State Tchrs. Coll., 1932-34; B.A., U. Kan., 1937, M.A., 1938; Ph.D., U. Mich., 1941; m. Annabel E. Erickson, Mar. 31, 1942; children—Mark Charles, Eric Alan, Thomas Melvin. Teaching fellow U. Mich., 1940-41; Benjamin Pierce instr. Harvard, 1941-43; mem. faculty Yale, New Haven, 1943—, prof., chmn. dept. math., 1959-65. Author: General Theory of Banach Algebras, 1960. Research on algebra, theory of Banach algebras, theory of integration; groups of linear transformations. Home: 131 Whitney Ridge Terrace, North Haven, Conn. 06518.

RICKELS, Karl, psychiatrist; b. Wilhelmshaven, Germany, Aug. 17, 1924; s. Karl E. and Stephanie (Roehrhoff) R.; M.D., U. Muenster, 1951; m. Rosalind Wilson, June 27, 1964; 1 son, Laurence. Practice medicine, specializing in psychiatry, Phila., 1957—; fellow psychiatry Hosp. U. Pa., 1955-57, faculty, 1957—, asso. prof., 1964—; dir. psychopharmacology Phila. Gen. Hosp. Fellow Am. Coll. Clin. Pharmacology and Chemotherapy, Am. Coll. Neuropsychopharmacology, Acad. Psychosomatic Medicine; mem. Am. Psychiatric Association, Collegium Internat. Neuro-Psychopharmacologicum, A.M.A., Am. Soc. Pharmacology and Exptl. Therapeutics, American Psychosomatic Society, Am. Therapeutic Society. Research, numerous publs. on non-drug factors influencing psychiatric drug therapy; devel. of clin. methodology for psychiatric drug evaluation. Home: 1518 Sweet Briar Rd., Gladwyne, Pa. 19035. Office: Piersol Bldg., U. Hosp., 3400 Spruce St., Phila. 19104.*

RICKER, Norman Hurd, Am. physicist; b. Galveston, Tex., Oct. 11, 1896; s. John Romaine and Julia Hurd (Shaw) R.; B.A. with honors in physics, Rice U., 1916, M.A., 1917, Ph.D., 1920; m. Sallie Lee St. Louis, June 18, 1923; children—Florence Harris (Mrs. William Sandford Pottinger), Norman Hurd, Sallie Lee (Mrs. James Byron Snow, Jr.). Instr., Rice U., 1919-21; lab. engr. Western Electric Co. Research Lab. (now Bell Telephone Lab.), 1921-23; head dept. geophysics Humble Oil & Refining Co., Houston, 1923-25; cons. physicist, Houston, 1925-38; sr. research physicist Carter Oil Co., also Jersey Prodn. Research Corp., Tulsa subsidiaries Standard Oil Co. N.J., 1938-58; prof. physics U. Okla., Norman, 1959-65, prof. emeritus, 1965—. Cons., Nat. Acad. Scis.-NRC, 1955—. Recipient Naval Ordnance Devel. award, 1945. Mem. Am. Phys. Soc., Am. Math. Soc., Acoustical Soc. Am., Am. Inst. Physics, A.A.A.S., Soc. Exploration Geophysicists (editor 1955-57), Am. Geophys. Union, Nat. Geog. Soc., Geophys. Soc. Tulsa (hon.), Sigma Xi. Patentee in field. Research in basic laws and prins. regulating the universe; invented paper cone loud speaker; developed wavelet theory seismogram structure. Home and office: 501 Terrace Pl., Norman, Okla. 73069.*

RICKER, William Edwin, Canadian biologist; b. Waterdown, Ont., Can., Aug. 11, 1908; s. Harry Edwin and Rebecca (Rouse) R.; B.A., U. Toronto, 1930, M.A., 1931, Ph.D., 1936; m. Marion Torrance Cardwell, Mar. 30, 1935; children—Karl Edwin, John Fraser, Eric William, Angus Clemens. Sci. asst. Biol. Bd. Can., 1931-37; jr. scientist Internat. Pacific Salmon Fisheries Commn., 1938; with Ind. U., 1939-50, prof. zoology, 1945-50; with Fisheries Research Bd. Can., Nanaimo, B.C., Can., 1950—, chief scientist, 1966—. Recipient Profl. Inst. Pub. Service Can. Gold medal, 1966. Fellow Royal Soc. Can., numerous others. Author: Handbook of Computations for Biological Statistics of Fish Populations, 1958; also numerous articles. Contbr. procedures for studying dynamics, ecology of prodn. sockeye salmon, mgmt. small warm-water lakes, systematics and distbn. Plecoptera. Home: Rural Route No. 1, Wellington, B.C., Can. Office: Fisheries Research Bd. Can., Nanaimo, B.C., Can.*

RICKETTS, Howard Taylor, Am. pathologist; b. Findlay, O., Feb. 9, 1871; s. Andrew Duncan and Nancy Jane (Taylor) R.; B.A., U. Neb., 1894; M.D., Northwestern U., 1897; m. Myra Tubbs, Apr. 18, 1900; children—Henry, Elizabeth. Fellow cutaneous pathology Rush Med. Coll., 2 years; joined dept. pathology U. Chgo., 1902; accepted position as prof. pathology U. Pa., 1910. Author: Infection, Immunity, and Serum Therapy, 1906. Discovered tick infected by microorganisms (later called Rickettsia) is carrier of Rocky Mountain Spotted Fever (important for control of typhus epidemics which are caused by similar microorganisms). Died Mexico City, Mexico, May 3, 1910.

RICKHAM, Peter Paul, pediatric surgeon; b. Wannsee, Germany, June 21, 1917; s. Otto Louis and Susan (Huldschinsky) R.; student St. Bartholomew's Hosp. Med. Sch.; M.B., B.S., U. London, 1948, M.S., 1949, D.C.H., 1950; m. Elizabeth Marie Hartley, May 10, 1938; children—David Nicholas, Susan, Diana, Mary Anne. First house surgeon, Chester, Eng., 1943; surg. 1st asst. St. George's Hosp., London, 1947; surg. registrar Hosp. for Sick Children, London, 1948; sr. surg. registrar Alver Hey Children's Hosp., Liverpool, Eng., 1949-51; sr. cons. pediatric surgeon Royal Liverpool Children's Hosp., Birkenhead Children's Hosp., 1952—; lectr. pediatrics surgery U. Liverpool. Examiner for diploma child health U. London; external examiner pediatric surgery U. Glasgow (Scotland); examiner pediatric surgery U. Liverpool. L.R.C.P. Fellow Royal Coll. Surgeons (Hunterian prof. 1963, 67), Brit. (pres.), French (hon.), German assns. pediatic surgeons, Brit. Pediatric Assn. Author: The Metabolic Response to Neonatal Surgery; also articles, chpts. in textbooks. Devel. and orgn. of surgery of new born on regional basis; established first neonatal surg. unit; research on numerous surg. conditions of childhood, especially treatment of congenital malfcrmations of newborn, metabolic response to surgery in newborn, treatment of hydrocephalus and spina bifida, urinary incontinence. Home: The Bield, Mt. Rd., Upton Mirral, Cheshire, Eng. Office: Alder Hey Children's Hosp., Liverpool, 12, Eng.*

RICKOVER, Hyman George, engr.; b. Jan. 27, 1900; s. Abraham and Rose Rickover; grad. U. S. Naval Acad., 1922; m. Ruth D. Masters; 1 son, Robert. Commd. ensign, U.S.N., 1922, advanced through grades to vice admiral; qualified submariner, 1930; assigned to atomic submarine project with A.E.C., Oak Ridge, Tenn., 1946-47, continues in development atomic submarine Bur. of Ships since 1947, now in charge nuclear propulsion div.; chief naval reactors br. AEC. Recipient Egleston Medal, Columbia Engring. Alumni Assn., 1955, Pupin med-1958; Enrico Fermi award for contbn. to atomic sci., 1965. Author: Education and Freedom, 1958; Swiss Schools and Ours, 1961; American Education—A National Failure. Leader in devel. nuclear propulsion systems for submarines and other naval ships; responsible for launching of Nautilus (1st atomic submarine), 1952. Address: care Bur. Ships, Dept. Navy, Washington.*

RICORD, Philippe, physician; b. Balt., Dec. 10, 1800; M.D., med. sch. Paris, France, 1826. Became surgeon to Central Bur., Paris, 1828; surgeon-in-chief for syphilis Hopital du Midi, 1831-61; became ofcl. surgeon to Prince Napoleon, 1852, later to Emperor Napoleon III; of cons. surgeon to imperial troops; dir. Lazaretto (instn. for care of needy and ailing poor) during Siege of Paris; founder sch. at Hosp. Venereal Diseases, Paris. Mem. Acad. Medicine (pres. 1868). Author: De la Blennorrhagie de la Femme, 1834; Monographic du Chancre, 1837; Traite pratique des maladies veneriennes (treatise on venereal disease) 1838; Lettres sur la Syphilis, 1851; Leçons sur le Chancre, 1857. Established rational theraphy of syphilis; gave laws of transmission of syphilis in precise terms, 1834, divided it into primary, secondary and tertiary stages, 1838; differentiated between hard and soft chancres, 1838; devised new method of curing varicocele and a spl. technique in urethroplasty (Monthyon prize). Died Paris, Oct. 22, 1889.

RIDDELL, John Leonard, Am. physician; b. Leyden, Mass., Feb. 20, 1807; s. John and Lephe (Gates) R.; grad. Rensselaer Sch., Troy, N.Y., 1829; M.D., Cin. Med. Coll., 1836; Adj. prof. chemistry and botany Cin. Med. Coll., 1835-36; published Synopsis of the Flora of the Western States, 1835; prof. chemistry Med. Coll. of La., 1836-65; published Catalogus Florae Ludovicianae (catalogue of La. plants), 1852; melter and refiner in branch of U. S. Mint, 1838/39-49; mem. commn. to devise means of protection for New Orleans against Mississippi River, 1844; later discovered microscopical characteristics of blood and black vomit in yellow fever; devised binocular microscope, 1851, instrument finished and sent to him, 1854; mem. 1st La. State Med. Soc., New Orleans Physico-Medico Soc. Died New Orleans, Oct. 7, 1865.

RIDDELL, Robert James, Jr., Am. physicist; b. Peoria, Ill., June 25, 1923; s. Robert James and Hazel (Gwathmey) R.; B.S., Carnegie Inst. Tech., 1944; M.S., U. Mich., 1947, Ph.D., 1951; m. Kathryn Jane Gamble, Aug. 12, 1950; children—Cynthia, Stephen Louis, James Duncan. Faculty U. Cal., Berkeley, 1951-55; physicist Lawrence Radiation Lab., Berkeley, 1955-58, sr. scientist, 1960—; with dept. research AEC, Washington, 1958-60. Mem. Am. Phys. Soc., Research, publs. on theory of condensation, plasmas, scattering, mu-mesonic molecules. Home: 98 Acacia Av., Berkeley 94708. Office: Lawrence Radiation Lab., Berkeley, Cal. 94720.*

RIDDLE, Oscar, Am. biologist; b. Cincinnati, Ind., Sept. 27, 1877; s. Jonathan and Amanda Emiline (Carmichael) R.; A.B., Ind. U., 1902, LL.D., 1933; Ph.D., U. Chgo., 1907; D.H.C., Catholic U. Chile, 1945; m. Leona Lewis, June 3, 1937. Mem. natural history expdn. to Orinoco River, 1901, Western Cuba, 1902; asst. in zoology U. Chgo., 1904-07, instr., 1908-11; research asso. Carnegie Instn. Washington, Chgo., 1912-14, staff mem. dept. genetics, Cold Spring Harbor, N.Y., 1914-45; lectr. biology, U.S. Dept. State, Latin Am., 1946-47; guest lectr., India, 1947. Mem. Am. Inst. City N.Y. (trustee 1923-44, v.p. 1942, gold medal 1933), A.A.A.S. (v.p. 1935), Endocrine Soc. (pres. 1929), Am. Euthanasia Soc. (dir. 1935-45), Am. Humanist Assn. (humanist of year 1958), Nat. Assn. Biology Tchrs. (founder, Distinguished Service award 1958, hon. mem.); Sigma Xi, Gamma Alpha, Pi Gamma Mu; hon. fellow or mem. 10 fgn. acads. Author: Endocrines and Constitution in Doves and Pigeons, 1947; The Unleashing of Evolutionary Thought, 1954. Correlated metabolic, developmental sexual processes with genetics; directed isolation, identification of chief functions of prolactin (pituitary hormone) in vertebrates. Address: Rural Route 4, Box 576, Plant City, Fla. 33566.*

RIDEAL, Sir Eric Keightley, English phys. chemist; b. London, Eng., Apr. 11, 1890; s. Samuel and Elizabeth (Keightley) R.; M.A., Cambridge U., 1911; Ph.D., Bonn (Germany) U., 1912; D.Sc., U. London, 1918; Hon. Dr., U. Dublin (Ireland), 1960, Queen's U. (Belfast, Ireland), U. Birmingham (Eng.), U. Turin (Italy), 1962, U. Bonn., 1961, U. Brunel (Eng.), 1967; m. Margaret Jackson, June 21, 1924; 1 dau., Mary (Mrs. Peter Oliver Chichester). Lectr., U. Coll., U. London, 1920, prof. King's Coll., 1950-55; sr. research fellow Imperial Coll., 1955—; prof. U. Ill., 1921; faculty U. Cambridge, 1922-45, prof., 1930-45; dir. Royal Instn., London, 1945-50. Decorated Mem. Brit. Empire; created knight, 1951; recipient Gold medal Soc. Chem. Engrs., 1913, Gold medal oc. Chem. Industry, 1923. Fellow Royal Soc., 1930 (Davy Medal, 1951); mem. Chem. Soc. (past pres.), Soc. Chem. Industry, Faraday Soc. Author: (with H.S. Taylor) Catalysis in Theory and Practice, 1918; Ozone, 1920; Electrometallurgy, 1918; (with S. Rideal) Water Supplies, 1914; Surface Chemistry, 1927; (with J. T. Davies) Interfacial Phenomena, 1961; Concepts in Catalysis, 1968; also numerous articles. Research on colloids and surface chemistry especially liquid surfaces, adsorption on solid surfaces, catalytic actions at solid surfaces. Home: 22 Westbourne Park Rd, London W. 2, Eng.*

RIDER, Paul Reece, Am. mathematician, statistician; b. Independence, Mo., Oct. 14, 1888; s. Walter and Alwilda (Reece) R.; A.B., William Jewell Coll., 1909, A.M., 1910; A.M., Yale, 1914, Ph.D., 1915, postgrad. (fellow), 1928-29; postgrad. (fellow) U. London (Eng.), 1935-36; m. Madeline Elizabeth Bostian, Oct. 5, 1918; children—Paul Reece, William Bostian, Margaret Louise (Mrs. David Graham Moore). Instr. math. Yale, 1915-16; instr. to prof. math. Washington U., St. Louis, 1916-54, prof. emeritus, 1954—. Exchange prof. Nat. U. Mexico, 1942-43; prof. U.S. Army U., Shrivenham, Eng. and Germany, 1945-46; chief statistician Aerospace Research Labs., Wright-Patterson AFB, O., 1951-64; vis. prof. U. P.R., Rio Piedras, 1965; cons. Southwestern Bell Telephone Co. Fellow Inst. Math. Statistics (pres. 1939), Am. Statis. Assn (past mem. council); mem. Am. Math. Soc., Math. Assn. Am., Phi Beta Kappa, Sigma Xi. Author: (with Alfred Davis) Plane Trigonometry, 1923; Essentials of the Mathematics of Investment, 1938; An Introduction to Modern Statistical Methods, 1939; College Algebra, 1940, also fgn. edits.; Plane and Spherical Trigonometry, 1942; (with Charles Hutchinson) Navigational Trigonometry, 1943; Analytic Geometry, 1947; Intermediate Algebra for College, 1949; First-Year Mathematics for Colleges, 1949, 2d edit., 1962; (with Carl H. Fischer) Mathematics of Investment, 1951; Plane Trigonometry, 1953. Research papers in calculus of variations, mathematical statistics, applications of mathematical statistics. Address: 422 Harmon Blvd., Dayton, O. 45419.*

RIDGEWAY, Sir William, Brit. archeologist, anthropologist; b. King's County, Ireland, Aug. 6, 1853; s. John Henry and Marianna R.; M.A., Sc.D., Cambridge, Eng.; hon. D.Litt., Dublin, Ireland, 1902, Manchester, Eng., 1906; LL.D., Aberdeen, Scotland, 1908, Edinburgh, Scotland, 1921; m. Lucy Samuels, 1880; 1 dau. Became fellow Caius Coll., Cambridge, 1880; prof. Greek, Queen's Coll., Cork, Ireland, 1883-94; prof. archaeology, Cambridge, 1892-1926, Brereton reader in classics, 1907-26, a founder depts. anthropology and architecture; Gifford lectr. natural religion U. Aberdeen (Scotland), 1909-11; became Stokes lectr. Irish archaeology, Dublin, 1909, Hermione lectr. art, 1911. Fellow Brit. Assn. (pres. anthropol. sect. 1908-10), Zool. Soc.; corr. mem. Archaeol. Soc. Athens, French Assn. pour les sciences anthropologiques; fgn. mem. Sociétés d'anthropologie Paris and Brussels, Deutsche Gesellschaft für Anthropologie; hon. mem. Bithar and Orissa Research Soc.; mem. Royal Anthrop. Inst. (pres. 1908-10), Classical Assn. Eng. and Wales (became pres. 1914), Cambridge Philological Soc. (pres.), Cambridge Antiquarian Soc. (pres.), Cambridge Classical Soc. (pres.), Cambridge Anthrop. Soc. Author: Origin of Metallic Currency and Weight Standards, 1892; The Origin and Influence of the Thoroughbred Horse, 1905; Homeric Land System; Who were the Romans?, 1907; The Oldest Irish Epic, 1907; The Origin of Tragedy, 1910; Minos the Destroyer, 1910; The Differentiation of the Chief Species of Zebras, 1909; The Dramas and Dramatic Dances of Non-European Races in Special Reference to the Origin of Tragedy, 1915; also articles. Studies and new interpretations in Greek history and anthropology. Died Fen Ditton, Aug. 12, 1926.

RIDGWAY, Robert, Am. ornithologist; b. Mt. Carmel, Ill., July 2, 1850; s. David and Henrietta James (Reed) R.; M.S. (hon.), Ind. U., 1884; m. Julia E. Perkins, Oct. 12, 1875; 1 son, Audubon Wheelock. Zoologist, U. S. Geol. Exploration of 40th Parallel (under Clarence King), in Cal., Nev., southern Idaho, Utah, 1867-69; in charge bird collections Smithsonian Instn., 1869-80; curator, div. of birds U. S. Nat. Mus., 1880—. Mem. permanent ornithol. com. First Internat. Congress, Vienna, 1885; hon. mem. 2d Congress Ornithologique Internat., Budapest, 1891; mem. com. of patronage Internat. Congress of Zoology, London, 1897. Mem. various societies. Author: A History of North American Birds (with Prof. Spencer F. Baird and Dr. Thomas M. Brewer) 5 vols., A Manual of North American Birds; A Nomenclature of Colors for Naturalists and Compendium of Useful Information for Ornithologists; Color Standards and Color Nomenclature; The Ornithology of Illinois, 2 vols.; The Birds of North and Middle America, 16 vols. Devised Ridgway color system for describing bird coloration. Died Mar. 25, 1929.

RIDLEY, Henry Nicholas, Brit. botanist, geologist; b. Dec. 10, 1855; s. Oliver Matthew and Louisa (Pole) R.; m. Lilly Eliza Doran. Joined botany dept. Brit. Mus., 1880; explored Island Fernando de Noronha, Brazil, 1887; dir. gardens Straits Settlements, 1888-1911; began research on cultivation para rubber, 1889; founder rubber industry, 1895; made several expdns. to parts of Malaysia, Indonesia. Recipient Gold medal Rubber Planters Assn., 1914; Frank Meyer medal Fgn. Plant Introduction, 1928; Linnean medal, 1950. Fellow Royal Soc., 1907, Linnean Soc., Royal Hort. Soc., Instn. Rubber Industry, (hon.); corr. mem. Pharm. Soc., Zool. Soc., Moscow Anthrop. and Ethnol. Soc., Mass. Hort. Soc., Royal Asiatic Soc. (sec., editor Jour. Straits br. 1889-1911). Editor, Agrl. Bull. Straits Settlements and Federated Malay States. Research and numerous publs. on botany, geology, zoology, tropical agr., especially of Malay Peninsula. Died Oct. 24, 1956.

RIDLEY, Humphrey, Brit. physician; anatomist; b. 1653; s. Thomas Ridley; grad. Merton Coll., Oxford (Eng.) U., 1671; M.D., U. Leiden (Netherlands), 1679; M.D., Cambridge (Eng.) U., 1688. Gulstonian lectr., 1694. Mem. Coll. Physicians. Author: The Anatomy of the Brain, 1695; Observationes Quaedam Medico-Practicae et Physiologicae, 1703. Gave 1st description of circular sinus, 1695, 1st English description of a sarcoma; dissected venous supply of corpora striata; observed lymph vessels in brains of men who had been hanged. Died Apr. 1708.

RIEBSOMER, Jesse Leroy, Am. chemist; b. Connersville, Ind., June 6, 1905; s. Anthony Lewis and Alice (Chenoweth) R.; B.A., DePauw U., 1928; student Cornell U.; m. Mildred M. Tate, July 12, 1928; 1 son, James Lewis. Mem. faculty DePauw U., Greencastle, Ind., 1932-45; prof. chemistry U. N.M., Albuquerque, 1945—, chmn. dept., 1948-63. Recipient award for outstanding teaching Mfg. Chemists Assn., 1958. Mem. Am. Chem. Soc., N.M. Acad. Sci., Phi Beta Kappa, Sigma Xi. Asso. editor Jour. Hetero-Cyclic Chemistry. Research and many publs. in field of natural products and synthesis of hetero-organic compounds as potential pharm. interest. Home: 2160 Don Andres Rd. S.W., Albuquerque 87105.*

RIECHENBACH, Heinrich Gottlieb Ludwig, German biologist; b. Leipzig, Germany, Jan. 8, 1793; a son, Heinrich Gustav. Became prof. natural history, Dresden, Germany, 1820. Author: Iconographia botanica seu plantae criticae, 1823-32; Flora exotica, 5 vols., 1834-36; Icones florae germanicae et helveticae, 25 vols., 1834-1911; Handbuch des Naturlichen Pflanzensystems, 1837; Deutschlands Fauna, 2 vols., 1842; Vollständigste Naturgeschichte des In- und Auslands, 9 vols., 1845-54. Developed system of plant classifications based on relationships between plants. Died Dresden, Mar. 17, 1879.

RIECHERT, Traugott Karl, German neurosurgeon; b. Lyck, Germany, Oct. 29, 1905; s. August and Mathilde (Pichler) R.; ed. univs. Heidelberg, Fribourg/Brisgau, Konigsberg, Vienna, Rostock, Breslau; M.D.; agrégé, 1940; m. Mar. 9, 1937; children—Bernd-Traugott, Christa-Maja. Asst. ophthalmol. clinic U. Königsberg (now Kaliningrad, USSR); staff neurology clinic U. Frankfort; with neurosurgery clinic (Luitpoldkrankenhaus), Würzburg; staff surgery div. Magdebourg Municipal Hosp.; head mil. hosp. Eastern Front, World War

II; dir. U. Clinic Neurosurgery, Fribourg/Brisgau, West Germany, 1946——. Mem. German-Portuguese (hon.), Italian (hon.) socs. neurosurgery, Leopoldina Acad. Sci., Surg. Acad. Peru., Neuro-psiquiatria y Medicina Legal de Lima, Soc. Neurosurgery French Lang. Paris. Author: Die Arteriographie der Hirngefässe; Kopfverletzungen; Beitrag zum Taschenbuch der praktischen Medizin. Research in irradiation; cerebral and spinal medulla surgery; inducoagulation; stereotactic operations. Home: Sonnhalde Strasse 10. Office: Hugstetter Strasse 55, Freiburg/Brisgau, West Germany.

RIECKE, Carl Victor Eduard, German physicist; b. Stuttgart, Germany, Dec. 1, 1845; prof., Göttingen, Germany. Author: Lehrbuch der Experimental Physik, 2 vols., 1896. Research in crystallography; application electron theory to conductivity in metals, electric heat, hydrodynamics. Died Göttingen, June 11, 1915.

RIED, Walter, German chemist; b. Frankfort/Main, Germany, May 5, 1920; s. Karl and Amalie (Leger) R.; Dipl.chem., U. Frankfort, 1942, Dr.phil.nat., 1942; m. Hildegard Moos, Mar. 28, 1953; children—Matthias, Sibylle, Walter-Antonius. Asst., Inst. Inorganic and Organic Chemistry, 1941-46, Paul Ehrlich Inst., Frankfort, 1947-49; staff Inst. Organic Chemistry, 1946-55, docent, 1952-55; prof. organic chemistry U. Frankfort 1955——. Wissenschaftlicher Rat, 1958——. Mem. Gesellschaft Deutscher Chemiker, Gesellschaft Deutscher Ärzte und Naturforscher. Author: (with H. Mengler Zur präpaparativen Chemie der Diazocarbonylverbindungen Fortschritte Chem., 1965; also numerous articles. Research on preparation organic chems., chemistry of heterocycline, peptide chemistry. Home: 27 Arndtstrasse. Office: Robert Mayerstrasse 7-9, 6 Frankfort/Main, Germany.*

RIED, William Wharry, Am. surgeon; b. Argyle, N.Y., 1799; A.B., Union Coll., Schenectady, 1825; studied medicine under Dr. A. G. Smith, Rochester, N.Y., 1826-28; m. Elizabeth Manson, Oct. 4, 1830. Practiced medicine, Rochester, 1828-64, N.Y.C., 1864-66; pres. Monroe County (N.Y.) Med. Soc., 1836, 49; 1st to perfect and publish flexion method of reducing hip dislocations in article Dislocation of the Femur on the Dorsum Ilis Reducible without Pulleys or any other Mechanical Means, Buffalo Med. Jour., 1851. Died N.Y.C., Dec. 9, 1866.

RIEDEL, Bernhard Mortiz Carl Ludwig, German surgeon; b. Laage, Germany, Sept. 18, 1846; prof. Jena, Germany. First surg. reposition of hip luxation; described tongue-shaped part of liver attached to right lobe (Riedel's lobe), 1888, chronic infection of thyroid (Riedel's struma), 1896. Died Jena, Sept. 12, 1916.

RIEDEL, Oswald Paul, German physicist; b. Ratibor, Germany, Aug. 24, 1921; s. Josef and Margarete (Stanjek) R.; student Technische Hochschule Hannover (Germany), 1940, 41-43; Diplom-Ingenieur, Technische Hochschule Breslau (Germany), 1944; D. rerum naturalium, U. Freiburg (Germany), 1949; m. Elena Bothe, July 23, 1949; children—Agnes, Reinhold. Sci. asst. Technische Hochschule Breslau, 1944-45, Kaiser-Wilhelm-Institut für Physik, Hechingen, Germany, 1945-49, Kaiser-Wilhelm-Institut für Chemie (Otto-Hahn-Institut), 1949-55; phys. analyst Badische anilin- & Soda-Fabrik AG., Ludwigshafen am Rhein, West Germany, 1955——; vis. N.C. State Coll., Raleigh, 1956; vis. Argonne Nat. Lab., Lemont, Ill., 1957. Author: (with F. Bopp) Die physikalische Entwicklung der Quantentheorie, 1950; (with others) Physikalisches Wörterbuch, 1952; also articles. Research on radiometric analyses, gas-chromatographic assay of deuterated compounds, tracers in process tech. and reaction mechanisms; theory of nuclear converter; theoretical studies on relativistic Doppler shifts; reaction calorimetry. Office: Badische Anilin- & Soda-Fabrik AG., Bismarckstrasse 83, 67 Ludwigshafen am Rhein, West Germany.*

RIEDER, Hermann, German internist, radiologist; b. Rosenheim, Germany, Dec. 3, 1858; prof., Munich, Germany. Studies on use of X-ray examination in lung diseases; became founder radiol. gastro-intestinal diagnostics by introducing use of bismuth subnitrate in contrast meal (Rieder's meal) for gastro-intestinal examinations. Died 1938.

RIEDESEL, Marvin LeRoy, Am. physiologist; b. Iowa City, Ia., Nov. 8, 1925; s. Elmer V. and Augusta (Stankee) R.; B.A., Cornell Coll., Mount Vernon, Ia., 1949; M.S., State U. Ia., 1953, Ph.D., 1955; m. Rosemary P. Dorothy, Aug. 28, 1949; 1 son, David Lee. Research asso. U. Pitts., 1955-59; asst. prof. biology U. N.M., Albuquerque, 1959-65, asso. prof., 1965——. Cons. NASA. Mem. Am. Assn. U. Profs., N.Y. Acad. Sci., Internat. Soc. Biometerology, N.M. Soc. for Exptl. Biology and Medicine, Am. Soc. Zoology, Am. Soc. Mammalogy, Am. Physiol. Soc., Sigma Xi, Phi Sigma. Research and publs. on effect of temperature on biol. distbrn. of radioisotopes and tolerance of mammalian hibernators to environmental stress; described increased serum magnesium concentration during hibernation of many mammals, work-rest cycles of men working in hot environments, sweating rates of men in hot water, tolerance of mammalian hibernators to dehydration. Home: 1826 Arizona St. N.E., Albuquerque 87110.*

RIEGEL, Byron, Am. chemist; b. Palmyra, Mo., June 17, 1906; s. William Henry and Elena (Beagle) R.; A.B., Central Meth. Coll., Fayette, Mo., 1928, D.Sc., 1963; postgrad. Princeton, 1929-30; A.M., U. Ill., 1931, Ph.D., 1934; m. Belle M. Huot, Aug. 25, 1934; 1 son, Byron William. NRC fellow Tech. Inst., Danzig, 1935-36, Harvard, 1936-37; research asso. George Washington U., 1934-35; faculty Northwestern U., 1937-59, prof., 1948-51, lectr., 1951-59; dir. chem. research and devel. Searle & Co., Chgo., 1951——. Mem. A.A.A.S., Am. Chem. Soc., Am. Soc. Biochemists, Am. Assn. Cancer Research, Sigma Xi, Phi Lambda Upsilon, Alpha Chi Sigma. Directed chem. research on cancer, steroids and drugs; pioneered first oral contraceptive. Home: 2500 Sheridan Rd., Evanston, Ill. 60201. Office: P.O. Box 5110, Chgo. 60680.*

RIEGER, Herwigh, Austrian ophthalmologist; b. Mödling, Vienna, Austria, May 2, 1898; s. Ludwig and Anna Elisabeth (Flatz) R.; M.D., Faculty Medicine, U. Vienna; m. Marianne Kerschbaum, Jan. 5, 1929; children—Gerhild, Ludwig, Margund, Waltraut, Hiltrud, Gebhard. Asst., Ophthalmol. Clinic, Vienna, 1923-26; staff Hygiene Inst., U. Vienna, 1927-28; staff Ophthal. Clinic of Prof. Linder, Vienna, 1929-39; dir. Ophthal. Clinic, German U., Prague, Czechoslovakia, 1940-45; dir. Municipal Ophthal. Clinic, Linz/Danube, Austria, 1950——. Mem. German, Viennese ophthal. socs. Author: Der Augenarzt; also articles. Home: Museumstrasse 15, Linz/Donau, Austria.

RIEHL, Herbert, meteorologist; b. Munich, Germany, Mar. 30, 1915; s. Maximilian and Olga (Bach) R.; came to U. S., 1933, naturalized, 1939; M.S., N.Y. U., 1942; Ph.D., U. Chgo., 1947; m. Jamia Cone, Mar. 27, 1952; children—Natalie Ann, Herbert Ernst. Instr., U. Wash., 1941-42; faculty U. Chgo., 1942-60; prof., head dept. meteorology Colo. State U., Ft. Collins, 1960——; dir. Inst. Tropical Meteorology, U. P.R., 1945-46. Cons. USN, U. S. Weather Bur. Mem. Am. Meteorol. Soc. (Messinger award 1947), Am. Inst. Aeros. and Astronautics (Losey award 1962), Am. Geophys. Union, Royal Meteorol. Soc., Sigma Xi. Author: Tropical Meteorology, 1954; Introduction to the Atmosphere, 1965; also numerous articles. Description and analysis dynamics and energetics weather disturbances in tropics including hurricanes, upper airjet streams, gen. circulation tropics, interaction between low and high latitudes. Home: 1601 Sheeley Dr., Ft. Collins, Colo.*

RIEHL, Nikolaus, physicist; b. St. Petersburg, Russia, May 24, 1901; s. Wilhelm and Helen (Kahan) R.; Dr. Phil., U. Berlin (Germany), 1927; Dr. phil. habil., Technische Hochschule Berlin-Charlottenburg, 1938; m. Ilse Przybyla, June 9, 1933; children—Ingeborg (Mrs. Erich Hahne), Irene. Head radiol. dept., head light tech. dept. Auer-Co., Berlin, 1927-39, dir. sci. dept., 1939-45; head German sci. group in USSR, 1945-55; prof. Technische Hochschule, Munich, Germany, 1957——, dir. physics dept., 1966——, dean faculty for gen. scis., 1965-66; head Nuclear Reactor Sta., Garching, Germany, 1963——; vis. prof. N.Y. U., 1965; Recipient Stalin prize, 1949, Heroes of Socialist Work, 1949, Lenin Decoration, 1949. Mem. Bavarian Phys. Soc. (mem. bd. 1960——). Author: Physik und technische Anwendungen der Lumineszenz, 1941; Energy Migration in Non-living and Living Systems, 1950; also numerous articles. Developed first luminescent lamp for illumination, 1st application of gamma rays for material testing; discovered radiationless energy migration in luminescent crystals; research on luminescence in crystals, protonic semiconductors. Home: 20 Lochhamer Str., Lochham nr. Munich. Office: 21 Arcisstr., Munich 8, West Germany.*

RIEKE, William Oliver, Am. anatomist; b. Odessa, Wash., Apr. 26, 1931; s. Henry William and Hutoka S. (Smith) R.; B.A. summa cum laude, Pacific Lutheran U., 1953; M.D. with honors U. Wash., 1958; m. Joanne Elynor Schief, Aug. 22, 1954; children—Susan Ruth, Stephen Harold, Marcus Henry. Instr., U. Wash., Seattle, 1958-61, asst. prof., 1961-64, asso. prof., 1964-66, adminstrv. officer, 1963-66; prof., chmn. dept. anatomy U. Ia., Iowa City, 1966——. Named one of Outstanding Young Men in Nation, U.S. Jr. C. of C., 1964. Mem. Am. Assn. Anatomists, Am. Soc. for Cell Biology, A.A.A.S., Sigma Xi. Contbg. author: Functional Neuroanatomy, 1965; also articles. Research on mechanisms lymphocyte activity, life span, growth, interactions with fgn. lymphoid cells in vitro. Home: Rural Route 1, Iowa City, Ia. 52240.*

RIEMANN, Georg Friedrich Bernhard, German mathematician; b. Breselenz, nr. Dannenberg, Hanover, Germany, Sept. 17, 1826; s. Friedrich Bernhard and Charlotte (Ebell) R.; student math. U. Göttingen, 1846, U. Berlin, 1947; Ph.D. (approved by Gauss), U. Göttingen, 1854. Served with Prussian King Frederick William IV against revolutionaries, 1848. Prof. math. U. Göttingen. Fellow Royal Soc., 1866; mem. French Acad. Scis., 1866. Deepened ideas of Bolyai and Lobachevsky and devel. new non-Euclidean system of geometry (Riemannian geometry) and a theory of space which provided a geometric found. for modern physical theory (for example, Einstein's work); introduced idea of finite but unbounded space; bldg. on Cauchy, devel. theory of analytic functions of a complex variable; devised method of representing such functions on coincident planes or sheets (Riemann surfaces); studied potential theory; devel. definition of definite integral. Died of Tb, Lake Maggiore, Italy, July 20, 1866.

RIEMSCHNEIDER, Randolph, German chemist; b. Hamburg, Germany, Nov. 17, 1920; chem. diploma U. Gottingen, 1941, Dr.rer.nat., 1943. With Free U. Berlin (Germany), 1947——, prof. biochemistry, 1958——, dir. Biochem. Inst., 1962——; dir. Chem. Inst., U. Santa Maria (Brazil), 1964——. Author several books; also numerous articles. Research on insecticide of diene group, hexachoro-cyclopentadiene as diene components, configuration and effect. Home: Postfach 136, 1 Berlin 19, Germany.*

RIES, Elias Elkan, inventor; b. Baden, Germany, Jan. 16, 1862; s. Elkan Elias and Bertha (Weil) R.; brought to U. S. at age of 3; ed. Md. Inst., Johns Hopkins; m. Helen Hirshberg, Apr. 21, 1895. Elec., mech. and tech. engr.; pres. Am. Audioscope Co.; holder over 250 patents. Fellow A.A.A.S. Principal pioneer inventions: The underground electric ry. conduit or sub-trolley system; the modern urban and long-distance alternating-current system of generation, transmission and conversion of electricity for operating electric railways, by which earlier restrictions of 500-volt trolley systems were successfully overcome, and which has made possible operation of rapid-transit elevated, subway and tunnel systems in N.Y.C. and elsewhere, and electrification of suburban and long distance steam rys.; the original automatic electric motor-starters; the 1st practical device for turning down light of incandescent lamps without wasteful resistance; the controller system used on electric elevators; original methods and apparatus for electric welding, riveting, soldering, metal working, methods and appliances for electric heating and cooking; telephone, phonograph and tele-phonograph systems; original processes and apparatus for mfg. iron and steel tubes from hot billets in one continuous operation; original methods and apparatus for producing talking motion pictures directly from film; new methods and appliances for detecting presence of unseen vessels, icebergs, in fog, as well as locating and following position of hostile aircraft and submarines and also for precise location and salvage of sunken vessels directly from the surface, by means of a novel electro-acoustic range-finder, indicating and signalling apparatus, which effectively accomplished for the ears, in the domain of sound, what the binocular telescope has done for the eyes; submarine detector and other inventions offered to U. S. Govt., 1917. Died Apr. 20, 1928.

RIES, Heinrich, Am. geologist; b. Bklyn., Apr. 30, 1871; s. Heinrich and Caroline Bowman (Atkins) R.; Ph.B., Columbia, 1892, A.M., 1894, Ph.D., 1896; Sc.D. (hon.) Alfred U., 1945; m. Millie Timmerman, July 1, 1893; (dec.); children—Victor H., Donald T.; m. 2d, Adelyn Halsey Gregg, June 7, 1948. On N.Y. geol. survey, 1891-92; asst. geologist summer of 1895; lectr. pub. schs. N.Y., 1895-97; asst. in mineralogy, Columbia, 1896-97; instr. econ. geology, Cornell U., 1898-1902, asst. prof., 1902-05, prof., 1905-39, head dept., 1914-37, emeritus prof. 1939——. Asst. dir. N.Y. sci. exhibits. Chgo. Expn., 1893; mem. jury of awards, Cotton States and Internat. Expn., 1896, Buffalo Expn., 1901, St Louis Expn. 1904. Del. Geol. Congress, St. Petersburg, 1897, Internat. Geol. Congress Paris, 1900, Mexico, 1906, Toronto, 1913, Washington, 1933; Fellow. Am. Geog. Soc., Am. Mineral. Soc., A.A.A.S., Seismol. Soc., British Ceramic Soc., Soc. Econ. Geologists; mem. Am. Ceramic Soc. (hon., pres. 1910-11); Am. Foundrymens Assn., (tech. dir., hon.), Rochester Acad. Sci.; Ky. Acad. Sci.; Geol. Soc. Am. (1st v.p. 1925-26, pres. 1928-29); Am. Inst. Mining Engrs. (life mem.); Can. Mining Inst. (life mem.). Author: Economic Geology; Clay, Occurrence, Properties and Uses; History of Clay Working Industry of United States (with H. Leighton); Building Stones and Clay Products; Engineering Geology (with T. L. Watson); Elements of Engineering Geology (with same); Elementary Economic Geology; Conservation in United States (with others). Has published reports on clays of N.Y., Ala., La., Fla., N.C., Mich., Md., Colo., N.J., Va., Tex., Wis., Can., in state geologists' and govt. reports. Contbr. articles on geology and mining, Internat. Year-book, 1898-1901 articles on geology, New International Encyclopaedia. Specialist on occurrence, uses, properties and geol. distbn. of clay. Died Apr. 11, 1951.

RIES, Herman Elkan, Jr., Am. phys. chemist; b. Scranton, Pa., May 6, 1911; s. Herman Elkan and Henrietta (Brenner) R.; B.S., U. Chgo., 1933, Ph.D., 1936; m. Elizabeth Hamburger, Aug. 17, 1940; children—Walter Elkan, Richard Alan. Lab. instr. U. Chgo., 1934-36; asst. dir. catalysis div., research labs. Sinclair Refining Co., Harvey, Ill., 1936-51; sr. research asso., research labs. Standard Oil Co. of Ind., Whiting, 1951——. Lectr. Institut Francais du Petrol, Paris, 1951, 57; vis. scientist Cavendish labs. U. Cambridge (Eng.), 1964; chmn. A.A.A.S. Gordon Research Conf. on Interfaces, 1961; chmn. Nat. Bur. Standards Com. on Monolayer Critical Tables, 1964——; mem. exec. com., colloid div. Am. Chem. Soc., 1963——. Bd. dirs. South Park Improvement Assn., Chgo., 1960——. Recipient Ipatieff prize in Catalysis, Am. Chem. Soc., 1950. Mem. Am. Inst. Chemists, Faraday Soc. (London, Eng.), Am. Soc. Lubrication Engrs., Phi Beta Kappa, Sigma Xi. Contbr. chpts. on structure and sintering of catalysts, 1950, physical adsorption, 1952, monolayers, 1965. Research in electron microscope studies of monolayers, 1955—, radiotracer adsorption, 1957—. Home: 5721 Blackstone Av., Chgo. 60637. Office: Box 431, Whiting, Ind.*

RIESE, Adam (or Ries), German mathematician; b. nr. Bamberg, Germany, circa 1489; tchr. arithmetic, Erfurt, Germany, also Annaberg, 1525. Author: Rechnung auff der Linihen..., 1518; Rechnung auff der Linihen und Federn... (gave rise to phrase nach Adam Riese to indicate correct calculation), 1522; Ein Gerechent Büchlein, 1536; Rechenung nach der lenge..., 1550. Credited with originating radical sign still in common use. Died Mar. 30, 1559.

RIESE, Walther, physician; b. Berlin, Germany, June 30, 1890; s. Emil and Anna (Rosenthal) R.; student U. Berlin, 1910-14; M.D., U. Koenigsberg, Germany, 1914; Ph.B., U. Lyon, France, 1937; m. Hertha R.I. Pataky, Aug. 7, 1915; children—Renee (Mrs. Judd Hubert), Beatrice (Mrs. Willie Riese). Practice medicine, specializing in psychiatry, Richmond, Va., 1941—; faculty Med. Coll. Va., 1941—, emeritus asso. prof. neurology, psychiatry, history of medicine, 1960—; cons. neuropathologist Dept. Mental Hygiene and Hosps. Commonwealth Va., 1942-60; cons. Dept. Health, Edn. and Welfare Nat. Inst. Mental Health, Bethesda, Md., 1960-62; ofcl. debater First Internat. Congress Neuropathology, Rome, 1952; lectr. various hosps., schs, instns. Mem. A.M.A., Am. Assn. Neuropathologists, Am. Assn. History of Medicine, Acad. Psychoanalysis, Am. Soc. Mammalogists, Am. Geriatrics Soc. Author: Vincent van Gogh in der Krankheit, 1926; Das Sinnesleben eines Dichters, 1928; (with others) Die Unfallneurose als Problem der Gengenwartsmedizin, 1929; (with O. Rothbarth) Die Unfallneurose und das Reichsgericht, 1930; Das Triebverbrechen, 1933; (with A. Réquet) L'idée de l'homme dans la neurologie contemporaine, 1938; La pensée causale en médecine, 1950; Principle of Neurology in the Light of History and Their Present Use, 1950; The Conception of Disease, Its History, Its Versions and Its Nature, 1953; La pensée morale en médecine, Premiers Principes d'une Ethique Medicale, 1954; History of Neurology, 1959; also numerous articles. Research on inter-relation of mind and brain using anatom., physiol., clin., pathol., psychol. and hist. observations; analysis, clarification, unification of med. concepts in hist.-philos. framework. Home: Box 397, Route 2, Francis Rd., Glen Allen, Va. 23060.*

RIESEN, Austin Herbert, Am. psychologist; b. Newton, Kan., July 1, 1913; s. Emil Richert and Rachel (Penner) R.; B.A., U. Ariz., 1935; Ph.D., Yale, 1939; m. Helen Charlotte Haglin, July 29, 1939; children—Carol Aneda, Kent Murdoch. Staff, Yale, 1935-49, asst. prof., 1941-49; cis research prof. U. Rochester, 1951-53; faculty U. Chgo., 1949-62, prof., 1956-62; prof. psychology, U. Cal. at Riverside, 1961—, chmn. dept. psychology, 1963—; research staff Yerkes Labs. Primate Biology, Orange Park, Fla., 1940-55. Cons. Nat. Inst. Mental Health, 1963—. Mem. Am. Psychol. Assn., A.A.A.S., Soc. for Research in Ophthalmology, Phi Beta Kappa, Sigma Xi. Author (with E.F. Kinder) Postural Development of Infant Chimpanzees, 1952; also articles. Research on relation between stimulation and growth nervous system, devel. behavior in chimpanzees, effects visual deprivation. Home: 5159 Brockton Av., Riverside, Cal. 92506.*

RIESMAN, David, physician; b. in Saxe-Weimar, Germany, Mar. 25, 1867; s. Nathan and Sophie (Eismann) R.; M.D., U. Pa., 1892, Sc.D.; LL.D; m. Eleanor L. Fleisher, Jan. 20, 1908; children—David, John Penrose, Mary. Practiced at Phila., 1893—; prof. clinic medicine, Phila. Polyclinic, 1900-18; asst. prof. medicine U. Pa., 1908-12, prof. clin. medicine, 1912-33 (emeritus), prof. history of medicine, 1933—, prof. clin. medicine Grad. Sch., 1933—; physician Phila. Gen. and Univ. hosps.; cons. phys. to Women's, Chestnut Hill, Jewish and Kensington hosps.; pres. med. bd. Phila. Gen. Hospital. Fellow Coll. Physicians Phila., A.A.A.S.; mem. numerous socs. Editor: (with Dr. Ludwig Hektoen) American Textbook of Pathology, 1901; Life of Thomas Sydenham; Medicine in the Middle Ages; Medicine in Modern Society. Described a degenerative fibrosis of myocardium (Riesman's myocardosis), a form of chronic bronchopneumonia (Riesman's pneumonia), Riesman's sign for diagnosis of exophthalmic goiter, diabetic coma, gall bladder disease. Died June 3, 1940.

RIESMAN, David, Am. sociologist; b. Phila., Sept. 22, 1909; s. David and Eleanor L. (Fleisher) R.; A.B., Harvard, 1931, LL.B., 1934; LL.D., Marlboro Coll., 1954, Grinnell Coll., 1957; Litt. D., Wesleyan U., 1960, Temple U., 1962; Ed.D., R.I. Coll., 1960; D.C.L., Lincoln U., 1962; m. Evelyn Hastings Thompson, July 15, 1936; children—Paul Hastings, Jennie Burdsal, Lucy Leathers, Michael deKay. Practice law, Mass., 1935-37; faculty law U. Buffalo, 1937-41; vis. research fellow Columbia Law Sch., N.Y.C., 1941-42; dep. asst. dist. atty. N.Y. County, 1942-43; contract termination dir., asst. to treas. Sperry Gyroscope Co., N.Y.C., 1943-46; vis. prof. Yale, New Haven, Conn., social scis. staff U. Chgo., 1946-58; Henry Ford II prof. social scis. Harvard, Cambridge, Mass., 1958—. Author: The Lonely Crowd, 1950; Faces in the Crowd, 1952; Thorstein Veblen: A Critical Interpretation, 1953; Individualism Reconsidered, 1954; Constraint and Variety in American Education, 1956; Abundance for What: and Other Essays, 1964. Analysis of contemporary Am. mass soc.; developed concepts of inner-directed and other-directed personality; studied sociol. basis of individual isolation. Home: 49 Linnaean St., Cambridge, Mass. 02138.

RIESS, Peter Theophil, German physicist; b. Berlin, June 27, 1804; doctorate, U. Berlin, 1831; hon. doctorate U. Bavia, 1872. Mem. Acad. Petersburg, Acad. Göttingen, Acad. Munich. Author: Lehre von der Reibungselektrizität, 1853; Abhandlunge, 1867-78. Research in electricity, especially static electricity; determined electricity quantity in Leyden jar is proportional to surface area, 1859. Died Berlin, 1883.

RIGAS, Demetrios Anton, biochemist; b. Andros, Greece, Feb. 2, 1921; s. Anton D. and Vasiliki (Kritikou) R.; Ph.D., Nat. U. Engring. Sci., Athens, Greece, 1941; m. Chrysiis K. Eginitou, Apr. 28, 1955; children—Anna Maria, Vasiliki Diane. Came to U.S., 1946, naturalized, 1946. Faculty U. Ore. Med. Sch., Portland, 1947—, prof. biochemistry and exptl. medicine, 1961—. Fellow Am. Inst. Chemists, Internat. Soc. Hematology; mem. Am. Chem. Soc., A.A.A.S., Am. Nuclear Soc., N.Y. Acad. Scis., Soc. Nuclear Medicine, Am. Soc. Hematology, Am. Fedn. for Clin. Research, Western, N.W. socs. for clin. research, Pacific Slope Biochem Soc., Portland Acad. Medicine, Biophys. Soc., Sigma Xi. Contbr. articles to tech. jours. Research kinetics of cell proliferation and differentiation, abnormal human hemoglobins, lymphocyte growth factors, macromolecular aging. Home: 6420 S.W. 90th St., Portland, Ore. 97223.*

RIGAULT, Germano, Italian chemist; b. Biella, Italy, June 6, 1930; s. Raoul and Aldo (Viretti) R.; Laureato in chimica, 1953. With U. Turin (Italy), 1953—, prof. Sch. Mineralogy, 1962—, dir. Istituto di Mineralogia, Petrografia i Geschimica. Author: Elementi di ottica cristallografica; Introduzione alla Cristallografia; also articles. Research in mineralogy, crystallography, geochemistry. Home: 49 via Magenta, Turin, Italy.*

RIGBY, Malcolm, Am. meteorologist; b. Hartford, Wash., Oct. 26, 1909; s. William Edward and Laura (McFadden) R.; student Wash. State U., 1927, U. Wash. 1928-41, Alaska U., 1930-31, Great Falls Coll., 1941-42, U.S. Dept. Agr. Grad. Sch., 1942-46; m. Marian Kingdon Smith, June 17, 1933; children—David John, Carolyn Louise. Cooperative weather observer, Everett, Wash., 1927-28; observer, climatologist, meteorologist U.S. Weather Bur., Seattle, Fairbanks, Alaska, 1930-31, Spokane, Wash., 1932-38, North Head, Wash., 1938-40, Great Falls, Mont., 1940-42, Chgo., 1940, Washington, 1942—, sci. information specialist U.S. Weather Bur., 1964—; analyst USAF, 1942-45. Chmn. working group on hydrological classification World Meteorol. Orgn., 1965—. Mem. Am. Documentation Inst. (chmn. Potomac Valley chpt. 1962-63, 65—), Internat. Fedn. for Documentation (chmn. U.S. nat. com. 1961-64) exec. bd. 1965—), Am. Standards Assn. (com. on classifications 1963—). Editor, Meteorol. and Geoastrophys. Abstracts, 1949—; founder, editor Meteorol. Abstracts, 1949—. Research and publs. on applications of computers to problems of information retrieval, indexing of sci. lit. and internat. classification schedules, Home: 5816 22d St. N., Arlington, Va. 22205. Office: U.S. Environmental Sci. Services Administrn., Rockville, Md. 20852.*

RIGDON, Raymond Harrison, Am. pathologist; b. Musella, Ga., July 30, 1905; s. William C. and Mary E. (Harrison) R.; B.S., Emory U., 1929, M.D., 1931; m. Margaret Britt, June 17, 1944; children—Mary Virginia, Anita Louise (Mrs. E.J. Clarke, Jr.). Faculty, Vanderbilt U. Sch. Medicine, Nashville, 1935-38; asso. prof. pathology Med. Sch., U. Tenn., Memphis, 1939-44; prof., chmn. dept. pathology Med. Sch., U. Ark., Little Rock, 1944-47; prof. pathology U. Tex. Med. Br., Galveston, 1947—. Sr. cons. M.D. Anderson Hosp., Houston, 1957—; cons. USPHS Hosp., Galveston, 1957-65 Diplomate Am. Bd. Pathology. Fellow Am. Coll. Legal Medicine; mem. A.M.A., Am. Assn. Pathologists and Bacteriologists, Am. Soc. Exptl. Pathologists, Tex. Soc. Pathologists, Soc. for Exptl. Biology and Medicine. Research and numerous publs. on exptl. study effects staphylococcus toxin in man and animal, inflammation, malarial infection in White Pekin duck, muscular dystrophy in duck, chicken, and turkey, carcinogensis in duck. Home: 1328 Ball St., Galveston, Tex. 77551.*

RIGGS, Lorrin Andrews, Am. psychologist; b. Harput, Turkey, June 11, 1912 (parents Am. citizens); s. Ernest Wilson and Alice (Shepard) R.; A.B., Dartmouth, 1933; M.A., Clark U., 1934, Ph.D., 1936; m. Doris Robinson, Aug. 28, 1937; children—Douglas Rikert, Dwight Alan. NRC fellow in biol. scis. U. Pa., 1936-37; instr. U. Vt., 1937-38, 39-41; faculty Brown U., Providence, R.I., 1938-39, 41—, prof. psychology, 1951—, L. Herbert Ballou Found. U. Prof., 1960—. Mem. exec. bd. Armed Forces-NRC Vision Com., 1946-64; mem. sensory diseases study sect. USPHS, 1962-63. Mem. Soc. Exptl. Psychologists (Howard Crosby Warren medal 1957), Am. (council reps. 1953—, pres. div. exptl. psychology 1962-63), Eastern (nat. com. to combat blindness) psychol. assns., Am. Acad. Arts and Scis., A.A.A.S. (v.p., chmn. sect. I 1964), Psychonomic Soc., Optical Soc. Am., Internat. Soc. for Clin. Electroretinography, Am. Physiol. Soc., Nat. Acad. Scis., Sigma Xi. Contbr. chpts. to Vision and Visual Perception edited by C.H. Graham, 1965; also numerous articles. Research in human and animal vision; developed method of studying human electroretinogram by means of electrodes imbedded in contact lens, methods of studying involuntary motions of eye showing a series of rhythmic tremors which play a significant role in vision, method of stabilizing retinal image. Home: 9 Diman Pl., Providence, R.I. 02906.*

RIGHI, Augusto, Italian physicist; b. Bologna, Italy, Aug. 27, 1850; engr.; prof. univs. Palermo, Bologna. Fellow Royal Soc., 1907; mem. French Acad. Scis., 1913. Author: L'ottica delle oscillazioni electtriche, 1897; La moderna teoria dei fenomeni fisici, 1904. La materia radiante e i raggi magnetici, 1910; Da rotazioni ionomagnetiche, 1915. Research on hysteresis of dielectrics; built microphone with conducting power; built spherical oscillator for centimetric waves, 1892; investigated reflection of electric waves by mental mirrors, 1893, propagation of electricity in gases through which X-rays fall, 1896; showed that radio waves differ from light waves in wave length, but not in their nature. Died Bologna, June 8, 1921.

RIGLER, Leo George, Am. radiologist; b. Mpls., Oct. 16, 1896; s. Harris and Rose (Rabinowitz) R.; B.S., U. Minn., 1917, B.M., 1919, M.D., 1920; m. Matyl Sprung, Sept. 8, 1920; children—Stanley Paul, Nancy Judith (Mrs. Eugene Saxon), Ruth Margaret (Mrs. J. Danby Olincy). Gen. practice medicine, New Eng., N.D., 1921-22; faculty U. Minn., 1924-57; practice medicine, specializing in radiology, Mpls., 1925-26; chief, dept. radiology Mpls. Gen. Hosp., 1927-57; exec. dir. Cedars of Lebanon-Mt. Sinai Hosps., Los Angeles, 1957-63; prof. radiology Center for Health Scis., U. Cal., Los Angeles, 1963—. Cons. radiology Cedars of Lebanon, San Diego Naval, Wadsworth VA, Long Beach VA hosps.; cons., mem. numerous govt. coms. Recipient numerous awards including Gold medal Am. Coll. Radiology, 1960, Medal of Distinction, Internat. Coll. Surgeons, 1963, Grubbe Medal, Chgo. Med. Soc., 1964. Fellow Am. Coll. Radiology (former chancellor), Am. Coll. Chest Physicians; mem. Am. Roentgen Ray Soc., A.M.A. (past chmn. sect. radiology), Radiol. Soc. N.Am. (pres. 1958), Am. Assn. Thoracic Surgery, A.A.A.S., Brit. Inst. Radiology, Sigma Xi, Alpha Omega Alpha (past pres. Minn. chpt.); hon. mem. No. Assn. Med. Radiology, Swedish Soc. Radiologists, Indian, Colombian, Mexican radiol. socs., Indian Soc. Chest Physicians. Author: Outline of Roentgen Diagnosis, 1938, The Chest, 1946; (with C.M. Nice, Jr. and A.R. Margulis) Roentgen Diagnosis of Abdominal Tumors in Childhood, 1957; also articles. Asso. editor Radiology; asst. editor Diseases of Chest; editorial bd. New Physician, Med. Tribune. Studies in roentgen diagnosis, including determinations of mode of origin, devel. and duration cancer of lung, physiology of pulmonary disease, signs of acute abdominal conditions. Home: 10633 Kinnard Av., Los Angeles 90024.*

RIJLANT, Pierre-Bernard, Belgian med. engr.; b. Antwerp, Belgium, Apr. 28, 1902; s. Michel and Louise-Adelaide (Velle) R.; M.D., Brussels U., 1927, 30; m. Odette Jeanne Hoquet, Aug. 1, 1927. Advanced fellow C.R.B. Ednl. Found., 1927-31; prof. human physiology U. Brussels Sch. Pedagogy and Sch. Commerce, 1929—; prof. gen. physiology Faculty Medicine, also Faculty Sci., 1934-65; prof. spl. physiology Faculty Medicine, Faculty Sci., dir. physiol. lab., 1937—; dir. SOLVAY Inst., 1941—; prof. Psychophysiol. Faculty Philosophy, of Letter, of Law, of Psychology, Brussels, Belgium, 1945—. Recipient prix Internat. Cardiologie, San Remo, 1961. Mem. Royal Belgian Acad. Medicine, European (hon. pres.), Internat. (hon. councillor), Belgian (hon. pres.) socs. cardiology, Belgian, French German socs. physiology; hon. mem. Italian, French, German, Spanish, Brit., East German, Polish socs. cardiology. Author, Cours de physiologie experimentale générale, 1941; Cours de physiologie expérimentale spéciale, 1941; Elements de Psychophysiologie, 1945; also numerous articles. Research on automatism and conduction in hearts, comparative physiology of heart, heart of limulus polyphemus; electrocardiography, vectorcardiography, analog computation and simulation in cardiology, respiratory centers, extensive three dimensional resistive network, on line analog computation of multidipolar equivalents, solution of black box problems. Home: 27 Avenue de la Floride, Bruxelles 18. Office: 115 Blvd. de Waterloo, Bruxelles 1, Belgium.*

RIKER, Albert Joyce, Am. plant pathologist; b. Wheeling, W. Va., Apr. 3, 1894; s. Albert Birdsall and Mary Edith (Davis) R.; A.B., Oberlin Coll., 1917; A.M., U. Cin., 1920; Ph.D. (fellow in plant pathology), U. Wis., 1922; m. Regina Emma Stockhausen, Dec. 26, 1922 (dec.); m. 2d, Helen Burgoyne, Apr. 18, 1953 (dec.). Asst., instr. botany U. Cin., 1917-20; instr. U. Wis., Madison, 1922-25, asst. prof., 1925-29, asso. prof., 1929-31, prof. plant pathology, 1931-64, prof. emeritus, 1964—. Spl. investigator Imperial Coll. Sci. and Tech., London,

Eng., 1926, Pasteur Inst., Paris, France, 1927; v.p. Forest Genetics Research Found., 1952-53; U.S. del. Atoms for Peace Conf., Geneva, Switzerland, 1955; mem. Latin Am. Sci. Bd., 1963——. Recipient medal 8th Internat. Bot. Congress, Paris, 1954. Fellow Crop Protection Inst., 1923-24, NRC, 1924-25, Internat. Edn. Bd., 1926-27; Haight Travelling fellow, 1959-60, 62-63. Mem. A.A.A.S., Am. Acad. Microbiology, Am. Assn. Cancer Research, Am. Assn. U. Profs., Am. Phytopath. Soc. (pres. 1947), Am. Soc. Naturalists, Am. Soc. Plant Physiologists, Bot. Soc. Am., Soc. Am. Microbiologists, Soc. Am. Foresters, Soc. Devel. and Growth, Wis. Edn. Assn., Am. Documentation Inst., Nat. Acad Scis. (chmn. sect. botany 1959-61), Conf. Biol. Editors, Sigma Xi, Gamma Alpha, Phi Sigma. Author: (with Regina S. Riker) Introduction to Research on Plant Diseases, 1936. Editor, Phytopathology, 1929-32, 43-47. Research, publs. on causes of plant disease, breeding of resistant trees, studies of plant culture and growth. Home: 211 N. Spooner St., Madison, Wis. 53705.*

RIKER, Walter Franklin, Am. pharmacologist; b. Bronx, N.Y., Mar. 8, 1916; s. Walter Franklyn and Eleanore Louise (Scafard) R.; B.S., Columbia, 1939; M.D., Cornell U., 1943; m. Virginia Helene Jaeger, Nov. 28, 1941; children—Donald Kay, Walter Franklyn III, Wayne Scaford. With Cornell U. Med. Coll., 1941——, beginning as research fellow Pharmacology, successively instr. pharmacology, instr. medicine, asst. prof. pharmacology, asso. prof., 1941-56, prof. pharmacology, 1956——, also chmn. dept. Research adv. bd. United Cerebral Palsy, 1954-58, USPHS, 1956——. Recipient John J. Abel prize Am. Pharmacologic Soc., 1951. Mem. N.Y. State Soc. Medical Research (past president), American Society Pharmacology and Experimental Therapeutics (councillor), A.A.A.S., Harvey Society, Society Experimental, Biology and Medicine, Sigma Xi. Associate editor Biol. Abstracts, 1948——, Jour. Pharmacology and Exptl. Therapeutics, 1950-58. Author sci. papers. Analyzed mechanisms operating in transmission of nerve impulses to muscle, also analysis, devel. drugs enhancing and reducing this function. Home: 3 Horizon Rd., Ft. Lee, N.J. Office: 1300 York Av., N.Y.C. 21.*

RIKITAKE, Tsuneji, Japanese geophysicist; b. Tokyo, Japan, Mar. 30, 1921; s. Tokuji and Toku (Takayama) R.; M.S., Tokyo Imperial U., 1942; m. Kiyoko Nagata, Apr. 24, 1947; 1 son, Kenji. Faculty, Earthquake Research Inst., U. Tokyo, 1942——, prof. geophysics, 1962——. Mem. Am. Geophys. Union, Soc. Terrestrial Magnetism and Electricity (Tanakadate prize 1949). Author: Electromagnetism and the Earth's Interior, 1966; also numerous articles. Research on distbn. elec. conductivity in earth by analysis geomagnetic variation; discovered a geomagnetic variation anomaly in Japan; pioneered geothermal work in Japan. Home: 10-302, I 4-chome, Sakuratosui, Setagaya-ku, Tokyo, Japan.*

RILEY, Gordon Arthur, oceanographer; b. Webb City, Mo., June 1, 1911; s. James Arthur and Georgia (Radcliff) R.; B.S., Drury Coll., 1933; M.S., Washington U., 1934; Ph.D., Yale, 1937; m. Lucy Madana Fuller, Sept. 14, 1940; children—Louise Georgia, Grace Fuller, Mildred Kathryn. Sterling fellow Yale, New Haven, 1937-38, instr. Bingham Oceanographic Lab., 1938-42, asso. prof., 1948-52, asso. dir., 1952-65; marine physiologist Woods Hole Oceanographic Instn., 1942-48; dir. Inst. Oceanography and research prof. marine biology Dalhousie U., Halifax, N.S., Can., 1965——. Trustee Bermuda Biol. Sta., 1946——, pres., 1954-61. Mem. Nat. Acad. Scis. Com. on Oceanography, 1957-62. Mem. Am. Soc. Limnology and Oceanography (pres. 1961-62), Am. Geophys. Union, Conn. Acad. Arts and Scis. Publs., descriptive and theoretical studies of marine plankton and oceanic productivity; effects of phys. oceanographic processes and chem. environment on plankton productivity; relationship of non-living particulate and dissolved organic matter and living populations in sea. Home: 1658 Vernon St., Halifax, N.S., Can.*

RILEY, Harris DeWitt, Am. physician; b. Clarksdale, Miss., Nov. 12, 1925; s. Harris DeWitt and Louise (Allen) R.; B.A., Vanderbilt U., 1945, M.D., 1948; student Emory U., 1943-44; m. Margaret Estelle Barry, Sept. 16, 1950; children—Steven, Mark, Margaret. Chief pediatric service, chief infectious disease service U.S. Air Force, 1951-53; with Vanderbilt U. Sch. Medicine, 1953-57 research fellowship, 1953-54, instr. pediatrics, vis. pediatrician, 1953-57; pediatrician-in-chief, chief infectious disease div. Children's Meml. Hosp., U. Okla. Med. Center, Oklahoma City, 1958——; prof., chmn. dept. pediatrics U. Okla. Sch. Medicine, 1958——. Mem. Soc. Pediatric Research, Am. Pediatric Soc., Infectious Disease Soc., numerous others. Contbr. articles to profl. jours. Work in infectious diseases, immunology in infants and children. Home: 606 N.E. 17th St., Oklahoma City, Okla. 73104. Office: Children's Memorial Hosp., U. Okla. Med. Center, 800 N.E. 13th St., Oklahoma City, Okla. 73104.*

RILEY, Herbert Parkes, Am. botanist; b. Brooklyn, N.Y., June 28, 1904; s. William Parkes and Amelia C. (Goergens) R.; A.B., Princeton, 1925, A.M., 1929, Ph.D., 1931, Procter fellow, 1931-32; student Cornell U., summers 1930, 31, 36; Mass. Inst. Tech., summer 1935; Nat. Research Council fellow Bussey Instn. of Harvard U., 1932-34; m. Agnes Graham Sanders, Aug. 21, 1935; 1 son, William Parkes, 2d. Asst. prof. biology Tulane U., 1934-38; asst. prof. botany U. of Wash., 1938-42, asso. prof., 1942; prof. of botany and head dept. of botany U. of Ky., 1942-56, distinguished prof., 1956——, head dept., 1956-65; (on leave) vis. investigator, Oak Ridge Nat. Lab., 1949, summers, 1950-51; cons., 1949-55; Fulbright professor, dept. genetics U. Pretoria, Union S. Africa, 1955-56, dept. botany, U. Cape Town, 1956; external examiner University Calcutta; vis. professor U. Cal. at Irvine, 1967-68. Fellow A.A.A.S.; mem. Genetics Society Canada, Biological Stain Commn., American Society Naturalists, Am. Genetic Assn., Genetics Soc. Am., Bot. Soc. America, Soc. for Study of Evolution, Torrey Bot. Club, Assn. S.E. Biologists, Ky. Acad. Sciences, American Institute Biological Scis., Bot. Soc. South Africa, Sigma Xi, Phi Beta Kappa, Author: An Introduction to Genetics and Cytogenetics, 1948; The Families of Flowering Plants of Southern Africa, 1963. Contributor of numerous articles in genetics and cytogenetics to scientific journals. Editor of the plant genetics sect., Biological Abstracts, editorial bd., Jour. Heredity. Fgn. adv. bd.: The Nucleus. Demonstrated chromosomal factors affecting speciation in Haworthia and Gasteria; described plant families in Southern Africa; demonstrated the mechanism of self-incompatability in Capsella; showed that natural hybridization occurs frequently in Iris and Tradescantia in Louisiana; showed (with N. H. Giles) that radiation damages chromosomes less in absence of oxygen. Home: 1023 E. Copper Dr., Lexington, Ky.*

RILEY, John Winchell Jr., Am. sociologist; b. Brunswick, Me., June 10, 1908; s. John Winchell and Marjorie Webster (Prince) R.; A.B., Bowdoin Coll., 1930; M.A., Harvard, 1933, Ph.D., 1936; m. Matilda White, June 19, 1931; children—John Winchell III, Lucy Ellen. Faculty, Marietta Coll., 1933-35, Wellesley Coll., 1935-37, Douglass Coll., 1937-42; faculty Rutgers U., 1945-60, prof. sociology, chmn. dept., 1945-60; v.p., dir. social research The Equitable Life Assurance Soc. of U. S., N.Y.C., 1960——; faculty Harvard, summer 1955; cons. Market Research Co. Am., also CBS; editorial adviser Henry Holt & Co. Vice chmn. adv. panel spl. operations Research and Development, Dept. Def. Mem. Sociol. Research Assn. (pres. 1964-65), Am. (sec. 1949-54), Eastern sociol. socs., Nat. Acad. Scis. (commn. human resources), Am. Assn. Pub. Opinion Research (pres. 1961-62), Market Research Council. Author: (with Bryce Ryan, Marcia Lifshitz) The Student Looks at His Teacher, 1950; (with Wilbur Schramm) The Reds Take a City, 1951; (with Matilda W. Riley, Jackson Toby; Sociological Studies in Scale Analysis, 1954. Editor: The Corporation and Its Publics, 1963. Pioneered in social research relevant to instn. of life insurance; studied man's relationship to time and death, sociology of communication. Home: 270 Riverside Dr., N.Y.C. 10025. Office: 1285 Av. of the Americas, N.Y.C. 10019.*

RILEY, Ralph, English cytogeneticist; b. Scarborough, Eng., Oct. 23, 1924; s. Ralph and Clara (Urmson) R.; B.Sc., U. Sheffield (Eng.), 1950, Ph.D., 1955, D.Sc., 1964; M.A., U. Cambridge, 1966; m. Joan E. Norrington, July 21, 1949; children—Susan Elizabeth, Jennifer Ann. Head cytogenetics dept. Plant Breeding Inst., Cambridge, Eng., 1952——. Nuffield Found./NRC Can. lectr., 1966. Inst. Biology; mem. Genetical Soc. Britain (sec. 1960-66). Editor: (with K. R. Lewis) Chromosome Manipulations and Plant Genetics, 1966; also articles. Research on crop plant evolution, genetic and agronomic effects of transfers of single chromosomes from related species to wheat, regulation of meiatic chromosome pairing with genetic variants; discovered genetic regulation of meiotic chromosome pairing in wheat, and its use in wheat breeding. Home: 19 Priam's Way, Stapleford, Cambridge. Office: Plant Breeding Inst., Trumpington, Cambridge, Eng.*

RILEY, Vernon Todd, Am. biologist; b. Idaho Falls, Ida., Sept. 2, 1914; s. Francis Vernon and Nellie Mabel (Todd) R.; student Spokane Coll., 1933-35, U. Wash., 1936-40, George Washington U., 1943-50; D.Sc., Sorbonne, Paris, France, 1966; m. Jeannine Marcel Roy, June 24, 1961; children—Elizabeth Anne, Dennis Vernon, Michael Francis, Christine. Asst. bacteriologist Hollister Steir Labs., Spokane, Wash., 1935-36; biologist Nat. Cancer Inst., Bethesda, Md., 1943-53; biologist Sloan-Kettering Inst., Rye, N.Y., 1954-67; chmn., div. microbiology Pacific Northwest Research Found., Seattle, 1967——. Fellow N.Y. Acad. Scis.; mem. Am. Assn. for Cancer Research, A.A.A.S., Soc. for Exptl. Biology and Medicine, Soc. Gen. Physiologists, Harvey Soc., Am. Soc. Cell Biologists, Am. Soc. Microbiology. Author: The Pigment Cell, 1963; the Enzyme-Elevating Viruses, 1966, also numerous articles. Research on nature cell and life processes in normal and malignant states; discovered new virus causing no known disease but widespread in tumor-bearing mice. Office: Pacific Northwest Research Found., 1102 Columbia St., Seattle 98104.*

RIMINGTON, Claude, biochemist; b. London, Eng., Nov. 17, 1902; s. George Garthwaite and Hilda (Klyne) R.; B.A., U. Cambridge (Eng.), 1924, M.A., 1928, Ph.D., 1928; B.Sc., U. London, 1924, D.Sc., 1947; m. Soffi Anderson, Feb. 23, 1929; 1 dau., Greta. Biochemist, Wool Industries Research Assn., Leeds, Eng., 1928-30; research officer div. vet. services Govt. of Union of S. Africa, Onderstepoort Vet. Research Lab, Pretoria, 1931-37; biochemist Nat. Inst. Med. Research, Med. Research Council, London, 1937-45; prof. chem. pathology U. London at U. Coll. Hosp. Med. Sch., 1945-67, emeritus, 1967——, dir. Graham Labs., 1964-67, chmn. bd. mgmt. clin. sci., 1965-68. Fellow Royal Soc., 1954; mem. Royal Soc. Medicine, Biochem. Soc., Endocrinol. Soc., Zool. Soc., Assn. Clin. Pathologists. Author: (with A. Goldberg) Diseases of Porphyrin Metabolism, 1962; also numerous articles. Research on biochem. formation of hemoglobin and related pigments; elucidation of chem. abnormalities in porphyrias, chemistry of poisonous plants; discovered drug for treatment disease of thyroid gland.*

RINDGE, Frederick Hastings, Am. entomologist; b. Los Angeles, Jan. 1, 1921; s. Samuel K. and Agnes (Hole) R.; B.S., U. Cal. at Berkeley, 1942, Ph.D., 1949; m. Phyllis Jane Denton, Apr. 9, 1943; children—Janet, Barbara, Marguerite. Staff, Am. Mus. Natural History, N.Y.C., 1949——, curator dept. entomology, 1963——. Mem. The Lepidopterists' Soc. (past pres.), A.A.A.S., N.Y., Pacific Coast entomol. socs., So. Cal. Acad. Scis., Soc. Evolution. Research and publs. on New World moths family Geometridae, and other groups Lepidoptera. Home: Buckingham Dr., Alpine, N.J. 07620. Office: Dept. Entomology, Am. Mus. Natural History, Central Park W. at 79th St., N.Y.C. 10024.*

RINEHART, John Sargent, Am. physicist; b. Kirksville, Mo., Feb. 8, 1915; s. Rupert Lloyd and Gertrude (Upright) R.; B.S., N.E. Mo. State Tchrs. Coll., 1934, A.B., 1935; M.S., Cal. Inst. Tech., 1937; Ph.D., State U. Ia., 1940; m. Marion Sladky, Aug. 10, 1940; children—Margot, Eric. Instr. physics Wayne U., Detroit, 1941-42; exec. sec. div. 4, NDRC, 1942-45; supr. N.M. Exptl. Range, N.M. Sch. Mines, Albuquerque, 1945-49; research sci. U.S. Naval Ordnance Test Sta., China Lake, Cal., 1949-55; asst. dir. Smithsonian Astro-physical Obs., Cambridge, Mass., 1955-58; research asso., Harvard, Cambridge, 1955-58; dir. mining research lab., prof. mining engring. Colo. Sch. Mines, Golden, 1958-64; dir. research and devel. U.S. Coast and Geodetic Survey, Washington, 1964-65; dir. sci. and engring. Environmental Sci. Services Adminstrn., Washington, 1965-66, sr. research fellow, dir. univ. relations Insts. for Environmental Research, Boulder, Colo., 1966——. Cons. to industries; founder Rinehart and Assos.; founder, pres. Hyperdynamics, 1963——. Recipient Presidential Certificate of Merit, 1948; Exceptional Service awards U.S. Navy, U.S. Army, 1947; Fulbright lectr. Egypt, 1962-63. Mem. Am. Phys. Soc., Explorers Club, Meteoretical Soc., Am. Astron. Soc., Am. Geophys. Union, Am. Soc. for Metals, Sigma Xi. Author: (with J. Pearson) Behavior of Metals under Impulsive Loads, 1954, Explosive Working of Metals, 1963; On Fractures Caused by Explosions and Impacts, 1960; also numerous articles. Research on transients, stress waves, shock waves, breakage, impacts, explosions, ballistics, hypervelocity flight, meteorities, metal crystals, seismology, geysers, applications of explosives. Mome: 756 6th St. Office: Insts. for Environmental Research, Boulder, Colo. 80302.*

RINEHART, Robert Fross, Am. mathematician; b. Springfield, O., May 31, 1907; s. Ezra C. and Jessie I. (Fross) R.; B.A., Wittenberg Coll., 1930; M.A., Ohio State U., 1932, Ph.D., 1934; Sc.D., Wittenberg U., 1960; m. Lillian Mae Grob, Dec. 25, 1942; children—Robert E., Carla Jeanne, Ronald W., Kim B. Prof. math. Ashland (O.) Coll., 1934-37; mem. faculty Case Inst. Tech., Cleve., 1937-65, prof. math., 1947-65; academic dean Naval Postgrad. Sch., Monterey, Cal., 1965——. Exec. sec. research and devel. bd. Dept. Def., 1948-50; dir. spl. research and operations research Duke U., 1958-60; dir. Weapons Systems Evaluation Div., Inst. for Def. Analyses, 1962-64; mem. Sci. Adv. Bd., Nat. Security Agy., 1960-65; mem. Adv. Bd. Air Tng. Comd., USAF, 1961-65, 67——. Recipient Ordnance Devel. award USN, 1945, Medal for Merit, 1946, Medal of Freedom, 1946. Mem. Operations Research Soc. Am. (pres. 1953), Soc. for Indsl. and Applied Math. (pres. 1962), Am. Math. Soc., Math. Assn. Math, A.A.A.S., Am. Assn. U. Profs., Sigma Xi, Alpha Tau Omega. Research and publs. in mathematics, complex variables and intrinsic functions; operations research. Home: Naval Postgrad. Sch., Monterey, Cal. 93940.*

RINGBOM, Anders Johan, Finnish chemist; b. Abo, Finland, July 21, 1903; s. Lars and Elin (Granit) R.; Diplomingenieur, Abo Akademi (Swedish U. in Finland), 1925, Dr.Techn., 1934; postgrad. U. Berlin (Germany), 1928; m. Astrid Signild Sonck, Aug. 3, 1930; children—Gunnel Elisabeth (Mrs. Pavel Hjalmar Carpelan), Hakan Erik. Faculty, Abo Akademi, 1928——, prof. chemistry, 1952——, v.p., 1954-57; research fellow U. Minn., 1950-51. Mem. various Finnish socs. and acads., N.Y. Acad. Scis., K. Vetenskapssocieteten i Uppsala. Author: Complexation in Analytical chemistry, 1963; also numerous articles. Research on phys.-chem. and complexometric methods of analysis applying theoretical principles to analytical problems. Home: 8 C Vardbergsgatan, Abo, Finland.*

RINGER, Robert Kosel, Am. physiologist; b. Ringoes, N.J., Feb. 21, 1929; s. Louis and Louise (Kosel) R.; B.S., Rutgers U., 1950, M.S., 1952, Ph.D., 1955; m. Joan L. Harwick, Aug. 18, 1951; 1 son, Kevin J. Asst. prof. Rutgers U., New Brunswick, N.J., 1955-57; mem. faculty Mich. State U., East Lansing, 1957-—, prof. physiology, 1964-—. Mem. Am. Physiol. Soc., Am. Assn. Avian Pathologists, Poultry Sci. Assn., Sigma Xi, Alpha Zeta. Contbr. chpt. to Avian Physiology by P.D. Sturkie, 1965; also numerous articles. Research in field of cardiovascular avian physiology. Home:2336 Rockwood Dr., East Lansing, Mich. 48823.*

RINGER, Signey, English physician, physiologist; b. Norwich, Eng., 1835; s. John M. and Harriet R.; M.B., Univ. Coll., London, 1860, M.D., 1863; m. Ann Darley; 2 daus. Became asst. physician U. Coll. Hosp., 1863, physician, 1865-1900, cons. physicians, 1900; prof. materia medica, pharmacology and therapeutics U. Coll., London, 1862-78; asst. physician Hosp. for Sick Children, 1864-69; prof. prins. and practice medicine, U. Coll., London, 1878-87, Holme prof. clin. medicine, 1887-1900. Fellow Royal Soc., 1885, Royal Coll. Physicians; mem. Acad. Medicine Paris (corr.), N.Y. Med. Soc. (hon.). Author: The Temperature of the Body as a Means of Diagnosis of Phythisis, Measles and Tuberculosis, 1865; A Handbook of Therapeutics, 1869; also numerous articles. Research on effect of organic salts on circulatory system, clin. medicine. Died Lastingham, Eng., Oct. 14, 1910.

RINGOLD, Howard Joseph, Am. chemist; b. Seattle, Sept. 7, 1923; s. Morris and Rose (Band) R.; B.Sc., U. Wash., 1946, Ph.D., 1935; m. Vivian Davidson, Sept. 8, 1946; children—Gordon Mark, Leslie Jean, Jeffrey David. With Syntex, S.A., Mexico, 1951-60, dir. research, 1957-60; sr. scientist Worcester Found. for Exptl. Biology, Shrewsbury, Mass., 1960-—. Cons. Syntex Corp., 1960-—. Fellow N.Y. Acad. Scis.; mem. Endocrine Soc., Am. Soc. for Biol. Chemists, Am. Chem. Soc., London Chem. Soc., Sigma Xi, Phi Lambda Upsilon. Research, patents, numerous publs. on synthesis numerous useful steroid drugs, oral contraceptives, topical and systemic anti-inflammatory agts., anti-tumor agt., protein bldg. agt., mechanism hormone action, mechanism enzyme action, mechanism organic chem. reactions. Home: 4 Beaver Dr. Office: 222 Maple Av., Shrewsbury, Mass. 01545.*

RINGS, Roy Wilson, Am. entomologist; b. Columbus. O., Aug. 15, 1916; s. John J. and Blanche (Tipton) R.; B.Sc., Ohio State U., 1938, M.Sc., 1940, Ph.D., 1948; m. Virginia M. Thalgott, Nov. 26, 1942; children—Gary Wayne, Steven Edward, Cynthia Marie. Faculty, Ohio Agr. Expt. Sta. (now Ohio Agrl. Research and Devel. Center), 1947-—, prof., asso. chmn. dept. zoology and entomology, Wooster, 1961-—; professor Academic Faculty of Entomology Ohio State U., Columbus, 1961-—. Fellow Ohio Acad. Sci., A.A.A.S.; mem. Entomol. Soc. Am., Am. Soc. Zoology, Am. Mosquito Control Assn., Entomol. Soc. Can. Research, numerous publs. on biology and control fruit insects; pioneered research with radioisotopes to tag insects for behavior studies. Home: 2438 Christmas Run, Wooster 44691. Office: Agrl. Research and Devel. Center, Madison Av., Wooster, O. 44691.*

RINGWOOD, Alfred Edward, Australian geochemist; b. Melbourne, Australia, Apr. 19, 1930; s. Alfred E. and Wilhelmena (Robertson) R.; B.Sc., Melbourne U., 1950; M.Sc., 1953, Ph.D., 1956; m. Gur Ivor Carlson, Aug. 29, 1960; children—Kristina, Peter. Research officer, CSIRO, 1956; research fellow geochemistry Harvard, 1957-58; sr. research fellow Australian Nat. U., Canberra, 1959-60, sr. fellow, 1960-63, personal prof. geochemistry, 1963-—. Mem. Bur. Internat. Upper Mantle Com., 1963-—. Fellow Australian Acad. Sci.; mem. Geol. Soc. Australia, Am. Geophys. Union, Mineral. Soc. Gt. Britain, Geochem. Soc. Research and numerous articles on phase transitions under very high pressures and their bearing on internal constitution of earth, particularly constn. of earth's mantle., geochem. studies on origin of earth, and terrestrial plants; originator of single stage theory of earth and inner planets. Home: 37 Gawler Crescent, Deakin, Canberra, A.C.T., Australia.*

RINKENBACH, William Henry, Am. chemist; b. Carbon County, Pa., Mar. 17, 1894; s. Leopold and Ellamanda (Oplinger) R.; grad. Perkiomen Sch., Pennsburg, Pa., 1911; A.B., Cornell, 1915; M.S., U. Pitts., 1922; m. Ruth M. Allender, Feb. 22, 1933. Chemist, E.I. du pont de Nemours & Co., 1915-18; asst. explosives chemist, U.S. Bur. Mines, 1919-27; asst. chief chemist Picatinny Arsenal, U.S. Dept. Army, Dover, N.J., 1927-29, chief, head chemist, 1929-48, asst. chief tech. div., 1948-50. Lectr. industrial chemistry U. Pitts., 1921-23; instr. George Sch. Social Scis., 1940-48; cons. chemist explosives to industry. Recipient, Exceptional Civilian Service award, U.S. Dept. Army, 1944. Fellow A.A.A.S., Am. Inst. Chemists; mem. Am. Chem. Soc., Am. Ordnance Assn., Sigma Xi, Phi Lambda Upsilon. Research, patents and publs. on comml. and mil. explosives and related process, ichthyology, polit. economy. Home: 2010 Cypress Av., R.D. 60, Allentown, Pa. 18103.*

RINNE, Friedrich Wilhelm Berthold, German mineralogist, crystallographer; b. Osterode, Hanover, Mar. 16, 1863; Ph.D., U. Göttingen, 1883; hon. dr. engrng., Hanover, 1926. Asst., 1883-85, lectr. 1885-86, mineralogical-petrographic institute, U. Göttingen; asst. and lectr., U. Berlin, 1887-94; prof., Technical U., Hanover, 1894-1904; prof., U. Giessen, 1904; prof. U. Königsberg, 1908; prof., U. Kiel, 1908-28; prof. U. Leipzig, 1929; emeritus, 1928; hon. prof. U. Freiburg, 1928. Mem., Saxon Acad. Scis., Leipzig; Göttingen Soc. of Sci.; Heidelberg Acad. Scis. Author: Praktische Gesteinskunde, 1901; Kristallographische Formenlehre, 1919; Das feinbaulische Wesen der Materie nach dem Vorbilde der Kristalle, 1920. Research in practical and theoretical mineralogy of potassium salts; crystallography; petrology. Died Freiburg im Breisgau, Baden, Germany, Mar. 12, 1933.

RIOLAN, Jean, French physician; b. Paris, 1577; s. Jean Riolan; qualified to practice medicine, 1604; apptd. royal prof. anatomy and botany, 1613; chief physician to Marie de Medici. Author: Schola anatomica, 1607; Anatome corporis humani, 1610; Osteologia ex veterum et recentiorum praeceptis descripta, 1614; l'Anatomica seu anthropographia, 1926. Well-known anatomist and adversary of Harvey's theory on circulation of blood. Died Paris, Feb. 19, 1615.

RIOPELLE, Arthur Jean, Am. psychologist; b. Thorp, Wis., Apr. 20, 1920; s. Wilfred G. and Ann Marie (Schroeder) R.; B.S., U. Wis., 1941, M.S., 1948, Ph.D., 1950; m. Mary Jane Astell, May 2, 1942; children—Mary Ann, James, Jean. Faculty, Emory U., 1950-57, asso. prof., 1954-57; dir. psychology div. U.C. Army Med. Research Lab. Ft. Knox, Ky., 1957-59; dir. Yerkes Labs. Primate Biology, Orange Park, Fla., 1959-62; dir. Delta Regional Primate Research Center, Tulane U., Covington, La., 1962-—, prof. psychology, 1963-—. Mem. Internat. Primatological Soc., Am., La. psychol. assns. Research and numerous publs. on learning, brain functioning, primate behavior. Home: Riverside Dr., Covington, La. 70433.*

RIOS GARCIA, Sixto, Spanish mathematician; b. Pelaustan, Spain, Jan. 4, 1910; s. José Maria and Maria Cristina (García) Rios; Dr.ès.sc., U. Madrid (Spain); m. Maria Jesus, July 26, 1950; children—Maria Jesus, Maria Cristina, Sixto, Maria Valeria, David. Prof., U. Madrid; dir. Sch. Statistics and Statis. Research, Consejo Superior de Investigaciones Cientificas. Fellow Inst. Math. Statistics; mem. Royal Acad. Sci. Spain, Statis. Inst. Hague, Soc. Econometry and Operations Research (asso.). Author: Teoría de la integral; Sucesiones de funciones analiticas; Representacion analitica de functiones; Series ortogonales; Analisis funcional; Fundamentos del calculo de probabilidades. Home: calle Ministro Ibanez Martin 5, Madrid. Office: calle Serrano 123, Madrid, Spain.

RIOUX, Marcel, Canadian anthropologist; b. Amqui, Que., Can., Jan. 19, 1919; s. Louis Philippe and Febronie (Roy) R.; B.A., U. Laval, 1939; M.A., U. Montreal, 1942; Ph.D., U. Paris, 1952; m. Helene Barkeau, Jan. 5, 1942; children—Aude, Morency, Solange, Francois. Research anthropologist Nat. Mus. Can., 1947-59; asso. prof. sociology Carleton U., 1959-61; prof. sociology U. Montreal, 1961-—. Recipient Parizeau medal, 1956. Mem. l'Association des Sociologues du Can. Français (past pres.), l'Association Internationale des Sociologues de Lanque Française (gov.). Author: Description de la Culture de l'Isle Verte, 1954; Belle-Ause, 1956; French-Canadian Society, 1962; also numerous articles. Research on culture French Can. Home: 5077 Victoria Av., Montreal, Que., Can.*

RIPAN, Raluca, Rumanian chemist; b. 1894; Doctor Honoris Causa, Torum Nicolaus Copernicus; prof. Cluj (Rumania) U.; mem. Acad. Socialist Republic Rumania. Author (with others) textbook of analytical chemistry. Discovered new classes of complex combinations (metallic selenocyanates, tellurates); drew up macro-and micro-chem. methods for titration of cations and anions; research on iso- and heteropolycombinations, chemistry of compounds of sellenium and tellurite.

RIPLEY, George, English alchemist; b. Ripley, Yorkshire, Eng., circa 1415; traveled and studied in France, Germany, Italy (especially Rome). Augustinian; canon of Bridlington; made chamberlain by Innocent VIII while living in Rome, 1477; returned to England, 1478, supposedly with secret of transmutation; continued alchemical studies. Author: The Compound of Alchymy (one of most popular treatises on the subject), 1471; Medulla Alchimiae, 1476; Concordantiae Guidonis et Raimundi (Lullii), after 1471. Probably 1st to popularize works attributed to Raymond Lully; greatly influenced English alchemical revival. Died 1490.

RIPLEY, Herbert Spencer, Jr., Am. psychiatrist; b. Galveston, Tex., June 29, 1907; s. Herbert Spencer and Caroline (Kenison) R.; A.B., U. Mich., 1929; M.D., Harvard, 1933; certificate tng. in psychoanalytic medicine Columbia, 1949; m. Elizabeth Stuart Smith, July 20, 1940; children—Caroline Bird, Nancy Stuart. Instr. psychiatry Cornell U. Med. Coll., 1937-46, asst. prof., 1946-49; prof., chmn. dept. psychiatry U. Washington Sch. Medicine, Seattle, 1949-—. Mem. Am. Psychiatric Assn., Am. Psychoanalytic Assn., Am. Psychosomatic Soc., Alpha Omega Alpha, Sigma Xi. Research, publs. on psychopathology of schizophrenia, relationship between personality and visceral function; use of barbiturates, hypnosis in psychotherapy. Author numerous articles and chpts. in books on psychiatry. Home: 3707 47th Pl., N.E., Seattle 98105.*

RIPLEY, Sidney Dillon II, Am. zoologist, mus. dir.; b. N.Y.C., Sept. 20, 1913; s. Louis Arthur and Constance Baillie (Rose) R.; grad. St. Paul's Sch., 1932; B.A., Yale, 1936, M.A. (hon.), 1961; Ph.D., Harvard, 1943; m. Mary Moncrieffe Livingston, Aug. 18, 1949; children—Julie Dillon, Rosemary Livingston, Sylvia MacNeill. Staff, Acad. Natural Sci., Phila., 1936-39; vol. asst. Am. Mus. Natural History of N.Y., 1939-40; teaching asst. Harvard, 1941-42; asst. curator birds Smithsonian Inst., 1942, sec., 1964-—; lectr., asso. curator Yale, 1946-52, curator, 1952-64; asst. prof., 1949-55, asso. prof. zoology, 1955-64, prof. biology, 1961; dir. Peabody Mus. Natural History, Yale, New Haven, 1959-64. Pres., Internat. Council Bird Preservation, London, Eng. 1958-—; mem. U.S. Nat. Commn. UNESCO; dir. Research Corp., 1965-—; exec. bd. Internat. Union Conservation Nature, Morges, Switzerland, 1964-—. Dir. World Wildlife Fund: trustee White Meml. Found., Henry Francis Du Pont Winterthur Mus., 1965-—. Decorated Freedom medal (Thailand); recipient Order White Elephant, (Thailand), 1947; Fulbright fellow, 1950, Guggenheim fellow, 1954, NSF fellow, 1954. Fellow A.A.A.S., Soc. Ornithological de la Plata (hon.); mem. Nat. Acad. Scis., French Inst., Am. Ornithologists Union, Zool. Soc. India, Brit. Ornithol. Union, Soc. Systematic Zoology, Bombay Natural History Soc., Soc. Study Evolution, Wilson Soc., Council Fgn. Relations; corr. mem. South African, New Zealand, Cooper ornithol. socs., Sigma Xi. Author: Trail of the Money Bird, 1942; Search for the Spiny Babbler, 1952; A Paddling of Ducks, 1957; A Synopsis of the Birds of India and Pakistan, 1961; (with L. Schribner) Land and Wildlife Southeast Asia, 1964. Expdns. to Pacific, S.E. Asia, India, Nepal. Research on evolution and speciation in vertebrate zoology, particularly ornithology. Home: 2324 Massachusetts Av., N.W., Washington 20008. Office: Smithsonian Inst., Washington 20060.*

RIPPEL, August (Rippel-Baldes), German microbiologist; b. Birkenfeld/Nahe, Germany, Nov. 1, 1888; s. Ludwig and Berta (Baldes) R.; Ph.D.; Dr. agron. honoris causa; m. Anna Rickes, Mar. 21, 1921; children—Johann Karl, Georg Wilhelm. Became asst., 1913; agrégé Breslau (now Wroclaw, Poland); prof., dir. Microbiol. Inst., U. Göttingen (Germany), 1923. Mem. Gottingen Acad. Sci., Assn. for Applied Botany, German Soc. Agrl. Sci., German Bot. Soc. Author: Wachtsumsgesetze bei höheren und niehderen Pflanzen, 1925; Vorlesungen über theoretische Mikrobiologie, 1927; Vorlesungen über Boden-Mikrobiologie, 1933; Grundriss der Mikrobiologie, 1957, became editor, founder Archiv für Mikrobiologie, 1930. Address: Albrechtstrasse 6, 34 Gottingen, W. Germany.

RIS, Hans, cell biologist; b. Bern, Switzerland, June 15, 1914; s. August and Martha (Egger) R.; came to U.S., 1938, naturalized, 1945; student U. Bern, 1933-38; Ph.D., Columbia U., 1942; m. Hania Wislicka, Dec. 26, 1947; children—Christopher Robert, Annette Margot. Lectr. zoology Columbia U., 1941-42; Seessel fellow Yale, 1942; instr. biology Johns Hopkins, 1942-44; staff Rockefeller Inst., 1944-49, asso. physiology, 1946-49; faculty U. Wis., Madison, 1949-—, prof. zoology, 1954-—. Mem. Am., Internat. socs. cell biologists, Genetics Soc. Am., Am. Soc. Naturalists, Am. Soc. Zoologists, A.A.A.S., Sigma Xi. Research and publs. on chromosome movement in cell div., coiled structure of chromosomes, chem. analysis of isolated nuclei and chromosomes; determination of DNA on single nuclei using cytophotometry; molecular organization of chromosomes with electron microscope; presence DNA in chloroplasts and mitochondria. Home: 1102 Dartmouth Rd., Madison, Wis. 53705.*

RISING, Jesse David, Am. physician; b. Kansas City, Mo., Oct. 20, 1914; s. Dean Stelle and Zula Z. (Reynolds) R.; A.B., U. Kan., 1935, M.D., 1938; m. Myra Elizabeth Wildish, Nov. 25, 1937; children—Dean Robert, James Davis, John Edward. Practice medicine, 1939-53; mem. faculty U. Kan. Sch. Medicine, 1940-—, prof. medicine, chmn. dept. postgrad. med. edn., 1960-—, prof. pharmacology, 1966-—. Mem. U. S. Pharmacopeial Conv., 1950, 60. Mem. A.M.A. (internship rev. com. 1960-—, chmn. commn. med. practice 1964-65), Am. Therapeutic Soc. (com. med. edn. and research 1962-—), Am. Acad. Gen. Practice, Assn. Am. Med. Colls., Am. Therapeutic Soc., Am. Coll. Clin. Pharmacology and Chemotherapy, Phi Beta Kappa, Sigma Xi, Alpha Omega Alpha. Demonstrated possibility of electroshock modification in exptl. animals for application in human electrotherapy. Home: 4406 Sunrise Dr., Kansas City, Mo. 64123.*

RISING, Louis Wait, Am. chemist, pharmacist; b. Rising, Ill., May 11, 1902; s. Louis John and Olive (Wait) R.; B.S., Ph.G., Ore. State U., 1924; M.S., U. Wash., 1926, Ph.C., 1926, Ph.D., 1929; m. Lois Edwina Fendal, Apr. 28, 1929. Faculty U. S.C., 1929-30, Rutgers U., 1930-34; faculty U. Wash., Seattle, 1934-—, prof., chmn. dept. pharmacy

1940—; tech. dir. Western Pacific Chem. and Dermetics, Inc., Seattle, 1945-52. Pres. R and S Chem. Co., Portland, Ore., 1956-60. Mem. Am. Pharm. Assn., Am. Coll. Apothecaries, Am. Inst. Chmists. Author: (with Burlage, Burt, and Lee) Fundamental Principles and Processes of Pharmacy, 1949; Introduction to Pharmacy, 1954; (with Burlage and Lee) Orientation to Pharmacy, 1959. Editor Squibb Review, 1962—. Research, publs. on dermatol. preparations, toxicology, rheology, emulsification. Home: 6835 18th Av. N.E., Seattle 98115.*

RISING, Willard Bradley, Am. chemist; b. Mechlenburg, N.Y., Sept. 26, 1839; grad. Hamilton Coll., N.Y., 1864; M.E., U. Mich., 1867; Ph.D., Heidelberg, Germany, 1871. Instr. chemistry Coll. of Cal., 1866-67, prof. natural sciences, 1867-69; prof. chemistry U. Cal., from 1872. State analyst of Cal., 1885—, also adviser and chemist bds. of viticulture and of health. Specialist in thermal chemistry; contbd. to chemistry of explosives. Died 1910.

RISLEY, Sir Herbert Hope, Brit. anthropologist; b. Akeley, Eng., Jan. 4, 1851; s. John and Frances (Hope) R.; B.A., New Coll., Oxford, 1872; m. Elsie Julie Oppermann, June 17, 1879; children—Crescent Gebhard, Sylvia. Became asst. collector Indian Civil Service, Midnapur, 1873, asst. to William Wilson Hunter (compiler of Gazetteer of Bengal), 1875, under-sec., Bengal, 1878, under-sec., imperial secretariat, home dept. Govt. of India, 1879, supt. Ghatwali Survey, 1880, mem., sec. police commn., 1890, acting financial sec. Govt. of India, 1898, census commr., 1899, dir. ethnography for India, 1901, sec., home dept. Govt. of India, 1902, sec. pub. and judicial dept. India office, London, 1910. Pres. Royal Anthrop. Inst., 1910; officer French Acad. Scis., 1891. Author: Tribes and Castes of Bengal, 1891-92; Anthropometric Data, 2 vols., 1891; Ethnographical Glossary, 2 vols., 1892; Gazeteer of Sikhim: Introductory Chapter, 1892; The People of India, 1908; also chpt. on Chota Nagpur in Gazetteer of Bengal; chpt. on ethnology and caste in Imperial Gazetteer of India, Vol. I, 1907. Expert on processes by which non-Aryan tribes are admitted into Hinduism; proved that Kolarians south of Bengal cannot be distinguished from neighboring Dravidians; compiled statistics on castes and occupations of residents of Bengal, 1885; applied Popinard's system of anthrop. research; advocated inclusion of ethnology in tng. of civil servants for India. Died Wimbledon, Eng., Sept. 30, 1911.

RISSER, Jacob Rutt, Am. physicist; b. Lancaster, Pa., Aug. 3, 1910; s. Jacob Stauffer and Anna (Rutt) R.; A.B., Franklin and Marshall Coll., 1931; M.A., Princeton, 1935, Ph.D., 1938; m. Marguerite Bard, June 29, 1937; children—Thomas Bard, William Leigh. Instr. physics Purdue U., Lafayette, Ind., 1937-41, asst. prof., 1946; staff Radiation Lab., Mass. Inst. Tech., Cambridge, 1941-46; asst. prof. physics Rice U., Houston, 1946-48, asso. prof., 1948-58, prof. physics, 1958—. Cons. Radioisotopes Service, Houston VA Hosp. Fellow Am. Phys. Soc.; mem. A.A.A.S., Am. Assn. U. Profs., Phi Beta Kappa, Sigma Xi. Research, publs. on nuclear physics, nuclear structure from neutron and charged particle scattering and reactions. Home: 5106 Pocahontas St., Bellaire, Tex. 77401. Office: Box 1892, Rice U., Houston 77001.*

RISSO, Giovanni Antonio, naturalist; b. Nice, France, 1777; prof. Lycée, also École de médecine, Nice. Author: The Ichthyology of Nice, 1810; Natural History of the Principal Productions of Southern Europe, 5 vols., 1826; Histoire naturelle des productions de l'Europe méridionale; Histoire naturelle des Orangers, 1818-22. Various discoveries in zoology of Mediterranean. Died 1845.

RISTIC, Miodrag, microbiologist; b. Zagubica, Serbia, Yugoslavia; s. Miodrag S. and Vera (Dordevic) R.; D.V.M., Coll. Vet. Medicine, Hanover, Germany, 1950; M.S., U. Wis., 1953; Ph.D., U. Ill., 1959; m. Ingrid Otte, May 12, 1951; 1 dau., Patricia. Asst. prof. Inst. for Microbiology, Hanover, Germany, 1950-51; research asso. dept. vet. sci. U. Wis., Madison, 1951-53; prof. vet. medicine U. Fla., Gainesville, 1953-59; prof. vet. research and vet. pathology, sr. staff mem. Center for Research, Coll. Vet. Medicine, U. Ill., Urbana, 1959—. Am. Vet. Med. Assn. Postdoctoral fellow. Mem. Am., Ill. vet. med. assns., Am. Soc. Protozoologists, Am. Soc. for Tropical Medicine and Hygiene, Research workers in Animal Disease, Am. Soc. Electron Microscopy, U. S. Livestock San. Assn. Contbg. author: Advances in Veterinary Science, 1960; Immunity to Parasitic Animals, 1966; Biology of Parasites, 1966; Infectious Blood Diseases of Man and Animals, 1967. Research, numerous publs. on microbiology, immunology, and electron microscopy; developed capillary tube agglutination test for diagnosing anaplasmosis; co-developer vaccine for transmissible gastroenteritis of swine and attenuated vaccine for anaplasmosis. Home: 410 Buekwood Ct. W., Urbana, Ill.*

RITCHEY, George Willis, Am. astronomer; b. Tuppers Plains, O., Dec. 31, 1864; student U. Cin., 1883-84, 86-87. Staff, Yerkes Obs., 1896-1904; supt. instrument constrn., 1899-1904; tchr. astronomy U. Chgo., 1901-05; staff solar obs. Carnegie Instn., 1905-09; dir. astrophotog. lab. Observatoire de Paris, 1924-30. Designed and built reflecting telescopes, including 60 inch and 100 inch telescopes, Mt. Wilson Obs., 40 inch U. S. Naval Obs., Washington, 1931; invented (with Crétun) aplanatic reflecting telescope; invented fixed vertical universal type of reflecting telescope, cellular type optical mirror; discovered nova in galaxy NGC 6946, 2 novae in Andomeda Galaxy m31, 1917. Died 1945.

RITCHEY, Harold W., Am. chem. engr.; b. Kokomo, Ind., Oct. 5, 1912; s. Glen Robert and Mabel Ann (Wilson) R.; B.S., Purdue U., 1934, M.S. in Phys. Chemistry, 1936, Ph.D., 1938, D. Eng., 1960; M.S. in Chem. Engring., Cornell U., 1945; D.Sc., Utah State U., 1961; m. Helen Louise Hively, Aug. 29, 1941; children—Stephen Robert, David Allen. Tech. dir., asst. mgr. Redstone div. Thiokol Chem. Corp., Huntsville, Ala., 1949-58, v.p., tech. dir., 1958-59, v.p. rocket operations, Ogden, Utah, 1959-63, exec. v.p., 1963-64, pres., Ogden, Bristol, Pa., 1964—; cons. space tech. and propulsion panels USAF Sci. Adv. Bd., 1958-62; mem. propulsion com. NACA, 1959-60; defender review panel inceptor design, tech. Inst. Def. Analyses, 1962—. Mem. Am. Ordnance Assn., Aerospace Industries Assn. Am., Am. Inst. Aeros. and Astronautics (past com. chmn.), Am. Rocket Soc. (C.N. Hickman award 1954, pres. 1961), Sigma Xi, Phi Lambda Upsilon. Contbr. chpts to McGraw-Hill Ency. Sci. and Tech., 1960, Advances in Space Sci., 1963. Research on lubricating oil detergents, oil field prodn., nuclear reactor tech., solid propellant rockets, solid propellants. Home: 911 Queen's Dr., Yardley, Pa. 19067. Office: Thiokol Chem. Corp., P.O. Box 27, Bristol, Pa. 19007.*

RITCHIE, Alexander Charles, pathologist; b. Auckland, New Zealand, Apr. 2, 1921; s. Percy C. and Olive M. (Hodge) R.; M.B., Ch.B., U. Otago, New Zealand, 1944; D.Phil., U. Oxford, 1950; m. Susan Liszauer, Nov. 10, 1956. Vis. fellow div. oncology Chgo. Med. Sch., 1951-52; asst. resident in pathology Mass. Gen. Hosp., 1952-54; faculty McGill U., 1955-61; prof., head dept. pathology U. Toronto, head, div. pathology Toronto Gen. Hosp., 1962—. Cons. Canadian Tumour Registry, Wellesley Hosp., Hosp. for Sick Children, Sunnybrook Hosp., Ont. Dept. Health, Women's Coll. Hosp., Ont. Cancer Inst. Fellow Coll. Am. Pathologists, Royal Coll. Physicians Can.; mem. Coll. Pathologists, Brit., Ont. med. assns., Am. Assn. for Cancer Research, Am. Assn. Pathologists and Bacteriologists, Internat. Acad. Pathology, Path. Soc. Gt. Britain and Ireland, Canadian Assn. Pathologists. (pres.), Nat. Cancer Inst. Can. (clin. adv. com.). Studies, publs. on nature of epidermal carcinogenesis, reaction of carcinogens on skin, pathology. Home: 500 Avenue Rd., Toronto 7. Office: 100 College St., Toronto 2, Ont., Can.*

RITCHIE, Donald Dirk, Am. botanist; b. Atlanta, Mar. 28, 1914; s. William Fair and Jennie (Heiserman) R.; B.A. summa cum laude, Furman U., 1934; postgrad. U. Va., summers 1936-40; M.A., U. N.C., 1937, Ph.D., 1947; m. Nancy Rose Vawter, Apr. 27, 1940; children—Nancy Rose, Douglas Dirk, Alice Vawter. Instr., Furman U., 1934; asst. U. N.C., 1935-38; agt. U.S. Dept. Agr., 1938; instr. W. Va. U., 1938-42, asst. prof., 1946-48; faculty Barnard Coll., Columbia, N.Y.C., 1948—, prof., 1962—, head botany dept., 1951—. Scientist-in-charge Naval Research lab. Tropical Exposure Sta., Panama, 1953-54; microbiologist submarine habitability program U.S. Navy, 1957; microbiologist Pershing Missile Tropical Test, U. S. Army, 1963; Fulbright lectr. in marine biology Univ. Coll., Galway, Ireland, 1967-68. Mem. A.A.A.S., Bot. Soc. Am., Am. Soc. for Microbiology, Soc. for Indsl. Microbiology, Mycol. Soc. Am., Torrey Bot. Club., Sigma Xi, Pi Kappa Phi. Research on cellular structure fungi, cultivation algae, distbn., growth and evolution of marine fungi, osmophilism in microorganisms. Home: 191 Brookside Av., Cresskill, N.J. 07626.*

RITCHIE, Lawrence S., Am. med. parasitologist; b. Glenvil, Neb., Dec. 9, 1906; s. Garfield R. and Bertha (Starr) R.; B.A., Grand Island Coll., 1928; M.A., Northwestern U., 1930, Ph.D., 1936; m. Ethel Allen, Jan. 2, 1932; children—Glennys (Mrs. Joseph E. Weiss), Carolyn (Mrs. Philip F. Spelt), Ann. Instr., Grand Island Coll., 1930-31; part-time instr. Northwestern U., 1932-36; asst. prof. Women's Coll., U. N.C., 1936-46; parasitologist U. S. Army 406th Med. Gen. Lab., Tokyo, Japan, 1947-56, Army Tropical Research Med. Lab., San Juan, P.R., 1956-66, mem. overseas field team Walter Reed Army Inst. Research in Nigeria, 1967—; lectr. U. P.R. Sch. Pub. Health, 1956-66. Cons. WHO, Geneva, 1962—. Recipient Isaac Gonzalez Martinez award P.R. Com. for Bilharzia Control. Mem. A.A.A.S., Am. Soc. Parasitologists, Am., Royal socs. tropical medicine and hygiene. Contbr. numerous articles to profl. jours. Developed formalin-ether concentration technique for diagnosing parasitic diseases; research on human parasites in Japan, chems. used against snails transmitting bilharziasis, biology of snails, biology of bilharziasis in different hosts. Office: U. Ibadan, Nigeria.*

RITCHIE, Rufus Haynes, Am. physicist; b. Blue Diamond, Ky., Sept. 24, 1924; s. Rufus and Eula (Haynes) R.; B.S., U. Ky., 1947, M.S., 1949; Ph.D., U. Tenn., 1959; m. Dorothy Estes, Dec. 2, 1944; children—Susan Alice (Mrs. Paul Witkowski), David. Instr., U. Ky., 1948-49; physicist Oak Ridge Nat. Lab., 1949-58, sr. physicist, 1959—; faculty U. Tenn., Knoxville, 1965—, asso. prof., 1966—. Ford Found. prof., 1966—. Mem. Am. Phys. Soc., Radiation Research Soc., Health Physics Soc., Sigma Xi. Contbg. author: Nuclear Instruments and their Uses, 1962. Research and publs. on devel. modern methods of measuring fast neutron energy deposition in matter, concepts in quantum plasma physics, interaction of charged particles and photons with quantum plasmas; postulated existence of surface plasmon in metals. Home: 109 Euclid Pl., Oak Ridge 37830. Office: P.O. Box X, Oak Ridge Nat. Lab., Oak Ridge 37830.*

RITLAND, Richard Martin, Am. paleontologist; b. Grants Pass, Ore., July 3, 1925; s. Martin C. and Mae E. (Puckett) R.; B.A., Walla Walla Coll., 1946; M.S., Ore. State U., 1949; Ph.D., Harvard, 1954; m. Juanita A. Hansen, Sept. 29, 1946; children—Lynda Beth, Stephen L., Stanley M., John M., Forrest R. Mem. faculty Atlantic Union Co., South Lancaster, Mass., 1947-52, Loma Linda U., 1954-60; faculty Andrews U., Berrien Springs, Mich., 1960—, prof. paleontology, 1964—, also faculty Geo-sci. Research Inst., 1960—, dir., 1967—. Fellow Geol. Soc. London; mem. Paleontol. Soc., Soc. Vertebrate Paleontology, Soc. for Study of Evolution, Am. Geophys. Union, Sigma Xi. Author: Meaning in Nature, 1966. Research and publs. on comparative morphology of amphibia—osteology and myology; vertebrate ecology, paleoecology, sci. and religion. Home: Box 162, Coll. Sta., Berrien Springs, Mich. 49104.*

RITT, Joseph Fels, Am. mathematician; b. New York, N.Y., Aug. 23, 1893; son of Morris and Eva (Steinberg) R.; student Coll. of City of N.Y., 1908-10; A.B., George Washington U., 1913. D.Sc., 1932; Ph.D., Columbia, 1917; m. Estelle Fine, June 29, 1928. Instr. in mathematics, Columbia, 1910-21, asst. prof., 1921-27, asso, prof., 1927-31, prof. since 1931, executive officer, department of mathematics, 1942-45, Davies Prof. Mathematics since July 1, 1945. Master computer ordnance dept., 1918-19. Mem. Nat. Acad. Sciences, Nat. Research Council, 1938-41. Am. Math. Soc. (v.p. 1938-40). Author: Differential Equations from the Algebraic Standpoint, 1932; Theory of Functions, 1947; Integration in Finite Terms, 1948; Differential Algebra, 1950. Studies of algebra and analysis, theory of functions, and functional equations; developed theory of differential algebraic equations. Died Jan. 5, 1951.

RITTENBERG, David, Am. biochemist; b. N.Y.C., Nov. 11, 1906; s. Joseph and Sadie (Bloch) R.; B.A., Coll. City N.Y., 1929; Ph.D., Columbia, 1934; m. Sara Merson, June 30, 1930; 1 son, Stephen. Prof. biochemistry Coll. Physicians and Surgeons. Columbia, 1934-56, exec. officer dept. biochemistry, 1956-64, chmn. dept. biochemistry, 1964—. Bd. govs. Weizmann Inst. Sci., Rehovoth, Israel. hon. fellow. Member Nat. Acad. Scis., Am. Acad. Arts and Scis. Application of isotope techniques to the study of intermediary metabolism; the study of the biogenesis of cell constituents from small molecules. Home: 192 N. Woodland St., Englewood, N.J. 07631.*

RITTENBERG, Sydney Charles, Am. bacteriologist; b. Chgo., Dec. 19, 1914; s. William and Lena (Goldman) R.; A.B. in Chemistry, U. Cal. at Los Angeles, 1935, M.A., 1936; Ph.D. in Microbiology, U. Cal. at Berkeley, 1941; m. Beatrice Prushan, Sept. 18, 1941; children—William Lloyd, Stephen Alan. With research div. Technicolor Motion Picture Corp., 1941-42; with contaminated wounds project Tulane U. Sch. Medicine, 1942-43; research div. S. B. Penick & Co., 1943-47; mem. faculty dept. bacteriology U. So. Cal., 1947-62; prof. bacteriology, chmn. dept. U. Cal. at Los Angeles, 1962—; cons. NSF; sci. adv. bd. St. Jude Meml. Hosp., Memphis; spl. research bacterial physiology and metabolism, autotrophic bacteria, biogeochemistry. Mem. Am. Soc. Microbiology (councilor at large), Am. Chem. Soc., Soc. Gen. Microbiology (Great Britain), Phi Beta Kappa, Sigma Xi. Home: 2502 St. George St., Los Angeles 90027.*

RITTENHOUSE, David, Am. inventor, astronomer, mathematician; b. Paper Mill Run, nr. Germantown, Pa., Apr. 8, 1732; s. Matthias and Elizabeth (Williams) R.; m. Eleanor Colston, Feb. 20, 1766; m. 2d, Hannah Jacobs, 1772; 3 children. Conducted boundary survey for William Penn, 1763-64; engaged in boundary surveys and commns. involving Pa., Del., Md., Va., N.Y., N.J., Mass.; conducted canal and river surveys; served on coms. to test specimens of flint glass, to inspect the 1st steam-engine in U. S.; supervised casting of cannon, manufacture of saltpeter; engr. Council of Safety, 1775, v.p., 1776, pres., 1777; mem. Pa. Gen. Assembly, Pa. Constl. Conv., 1776; trustee loan fund (Pa. loan to Continental Congress); mem. Bd. of War created by Continental Congress; prof. astronomy, trustee U. Pa.; mem. commn. to organize U. S. Bank; 1st dir. U. S. Mint (apptd. by George Washington), 1792-95; curator, librarian, sec., v.p., pres. (1791-96) Am. Philos. Soc.; Fellow Royal Soc., 1795. Author: Easy Method of Deducing the True Time of the Sun's Passing the Meridian, 1770; Method of Raising the Common Logarithm of any Number, 1795; To Determine the True Place of a Planet, in an Elliptical Orbit, 1796. Designed his orrery (represents motions of bodies of the solar system and illustrates solar and lunar eclipses and other phenomena for a period of 5000 years either forward or back), 1767; experimented on compressibility of water; invented metallic thermometer; made calculations on transit of Venus that was to occur in 1796; said to have made 1st telescope in Am.; invented collimating telescope (introduced spider threads in eyepiece), 1785; measured grating intervals, deviations of several orders of spectra; experimented on magnetism and electricity; measured barometric effect on a pendulum clock rate and expansion of wood by heat; constructed compensating pendulum, wooden hygrometer; solved math. problem of finding sum of several powers of sines, 1792. Died Phila., June 26, 1796.

RITTER, Eugene Kerfoot, Am. mathematician, research adminstr.; b. Blackstone, Va., Mar. 28, 1909; s. Lacy Milton and Georgia Kerfoot (Laws) R.; B.A. (Lawrence fellow), U. Richmond, 1930; M.A. U. Va., 1934; Ph.D. (duPont fellow), 1949; postgrad. U. Pa., 1941-42; m. Lucille New, Dec. 29, 1936; children—Eugene Kerfoot, Martha Lucille. Dial switchman Chesapeake & Potomac Telephone Co., 1930-32; instr. U. Va., 1932-35; asst. prof. math. The Citadel, 1935-37, U. Richmond, 1937-41; instr. U. Pa., 1941-42; asst. prof. math. and mechanics U. S. Naval Postgrad. Sch., 1946-47, asso. prof., 1947-49; head simulation and computation dept., lectr. applied math. U. Mich. Engring. Research Inst., Willow Run Research Center, 1949-52; dir. computation and ballistics dept. U. S. Naval Proving Ground, 1952-55; dir. Rich Electronic Computer Center, Engring. Expt. Sta., Ga. Inst. Tech., 1955-57; mgr. math. analysis dept. Lockheed Aircraft Corp. (Ga. div., 1957-63; cons. scientist, acting mgr. Computation Center, Lockheed Missiles & Space Co., Sunnyvale, Cal., 1963-65, mgr. Programming and Analysis div., 1965-67, chief math. dept. Naval Ordnance Lab., Silver Spring, Md., 1967—. Mem. Am. Math. Soc., Math. Assn. Am., Am. Ordnance Assn., Soc. for Indsl. and Applied Mathematics, Phi Beta Kappa, Sigma Xi, Omicron Delta Kappa, Pi Mu Epsilon, Sigma Phi Epsilon, Pi Delta Epsilon. Research and publs. in exterior ballistics, trajectories of missiles, theory of ordinary differential equations. Home: 15413 Carrolton Rd., Rockville, Md. 20853. Office: Naval Ordnance Lab., Silver Spring, Md.*

RITTER, Howard L(ester), Am. chemist; b. Washington, Feb. 12, 1916; s. George Howard and Ruby (MacFarlane) R.; B.S., U. S. Naval Acad., 1938; Ph.D., Pa. State U., 1943; m. Marie Louise Kruecke, July 12, 1941; children—Howard Lester, Carl Peter, Mary Ann, Margaret Jo, Mark MacFarlane. Research chemist Socony-Vacuum Oil Co., 1943-46; instructor, asst. prof. chemistry U. Wis., 1946-52; professor, head dept. chemistry Miami U., 1952-62, research professor, 1962—. Member Am. Chem. Soc., Am. Phys. Soc., Pi Kappa Phi, Pi Mu Epsilon, Phi Lambda Upsilon, Sigma Pi Sigma, Sigma Xi. Author: An Introduction to Chemistry, 1955; also articles chem. jours. Investigations concerning the spatial arrangement of atoms in crystals and especially in liquids. Home: 1025 Cedar Dr., Oxford, Ohio. 45056.*

RITTER, Johann Wilhelm, German physicist; b. Samitz, Silesia, Dec. 16, 1776. Apothecary, Liegnitz, Jena, Munich (all Germany); research expts. in electricity. Author: Beweis, Dass Ein Beständiger Galvanismus den Lebensprozess im Tierreich Begleitet, 1798; Beiträge zur Näheren Kenntnis des Galvanismus, 1800-05; Physikalisch-Chemische . . . , 3 vols., 1806. Das Electrische System des Körpers, 1805. Discovered ultraviolet rays, 1802 (also discovered by Wollaston); invented dry pile; 1st to propose chem. theory of electricity; discovered principle of accumulator, 1803; originated 1st electro-chem. series; collected oxygen, hydrogen separately by electrolysis, 1800; repeated isolation of potassium; demonstrated attraction and repulsion electric charges. Died Munich, Jan. 23, 1810.

RITTER, Karl, German geographer; b. Quedlinburg, Prussia, Aug. 7, 1779; ed. Schnepfenthal Inst., U. Halle, 1796-98, also U. Göttingen. Tutored children of banker, Bethmann Hollweg, while student at U. Göttingen, 1798-1815; travelled through Europe; prof. history Frankfort (Germany) Gymnasium, 1819; prof. extraordinarius history U. Berlin (Germany), 1920-59; lectr. mil. coll. in Berlin. Mem Berlin Acad. Scis., Royal Soc., 1848. Author: Europa, ein geographisch-historisch-statistiches Gemälde, 1807; Die Erdkunde im Verhältnis zur Natur and zur Geschichte des Menschen, 2 vols., 1817-1818 (expanded to 19 vols. 1822-59); Die Stupas, oder die architektonischen Denkmäler an der indobaktrischen Königstrasse und die Kolosse von Bamyan, 1838. Laid foundations for modern study of geography and a founder of human geography; brought new conception into field, stressed its importance in other disciplines; believed geography could be considered the physiology and comparative anatomy of earth; helped define scope of geography, emphasised influence of natural environment on man; known as great sci. compiler. Died Berlin, Sept. 29, 1859.

RITTER, Kurt, German chemist; b. Salzhof-Spandau, Germany, June 10, 1891; s. Heinrich and Gertrud (Wersche) R.; student U. Königsberg, Berlin, also Berlin Tech. Coll.; Ph.D.; m. Margarete Streitz, May 20, 1917; children—Barbara, Ursula. Asst., Inorganic Chemistry Inst., Berlin-Charlottenburg, Germany, 1912-14, 18-23; chemist-in-chief in industry, 1925-42; dir. dept. German Inst., 1942-45; cons. chemist, 1945—. Recipient M. P. Neumann medal, 1954; Cross of Merit, German Republic, 1961. Mem. Tech. Sci. Commn. German Community of Milling Trade, Soybean Assn. (mem. sci. commn.), Internat. Cereal Chemistry Soc., Soc. German Chemists, German Com. Cereal Chemists (pres.). Author: Zur Methodik der Bestimmung der diastatischen Kraft, 1928; Enzyme und Konditionierung, 1938; Der Aufbau der Eiweisskorper und die Eiweisstoffe des Weizenmehl, 1939; Über ein einfaches Verfahren zur Feststellung des Auswuchsgrades, 1942; Zur Methodik der Teigmikroskopie, 1950. Research on microscopy; physiology of nutrition; documentation. Address: Eitorferstrasse 1, Köln-Deutz, W. Germany.

RITTNER, Edmund Sidney, Am. physicist; b. Boston, Mass., May 29, 1919; s. Philip and August (Beich) R.; B.S., Mass. Inst. Tech., 1939, Ph.D., 1941; m. Marcella Weiner, Oct. 6, 1942; 1 dau., Leona. Project leader, Laboratory Insulation Research, Mass. Inst. Tech., 1942-45; sect. chief Philips Labs., Briarcliff Manor, N.Y., 1946-62, dir. dept. physics, 1962—. A.D. Little Post-docorate fellowship, Mass. Inst. Tech., 1941-42. Fellow Am. Phys. Soc.; mem. Sigma Xi. Patentee, cathodes, electron bombardment, induced conductivity, infrared photoconductors, apparatus for measuring relative humidity. Contbr. approximately 40 articles to tech. jours. Research on infrared, photoconductivity, cathodes, transistors, solar batteries, thermoelectric heat pumps and generators, thermionic converters, theory of solids. Home: 23 Winslow Rd., White Plains, N.Y. 10606. Office: Philips Labs., 345 Scarborough Rd., Briarcliff Manor, N.Y. 10510.*

RITTS, Roy Ellot Jr., Am. physician, microbiologist; b. St. Petersburg, Fla., Jan. 16, 1929; s. Roy Ellot and Dorothy (Bliss) R.; A.B. (U. scholar), 1948, M.D., 1951; m. Hilda Joan Stump, June 19, 1953; children—Leslie Sue, Graham Douglas, Ian Christopher. Intern, D.C. Gen. Hosp., 1951-52; fellow in medicine George Washington U. Washington, 1952-53, resident in medicine U. Hosp., 1953-54; research fellow in medicine Harvard Med. Sch.; asst. in medicine Peter Bent Brigham Hosp., Boston, 1954-55; vis. investigator, research asso. Rockefeller Inst., N.Y.C., 1955-58; prof. microbiology, chmn. microbiology, professorial lectr. medicine Georgetown. U. Sch. Medicine, Washington, 1958-64; professorial lectr. microbiology U. Chgo. Sch. Medicine, 1964—; dir., mem. A.M.A. Inst. Biomed. Research, Chgo., 1964—. Cons. in clin. pathology NIH, 1959-64; cons. immunology USN Hosp., Bethesda, 1963-64; spl. lectr. USN Dental Sch. Bethesda, 1963-64, Walter Reed Army Med. Research Inst., Washington, 1963-64. Life Ins. Med. Research Fund fellow, 1954-56. Diplomate Am. Bd. Microbiology, A.M.A. Fellow Am. Acad. Microbiology; mem. Am. Assn. Pathology and Bacteriology, Am. Soc. Tropical Medicine and Hygiene, Am. Fedn. Clin. Research, Am. Rheumatism Assn., Soc. Exptl. Biology and Medicine, Soc. Gen. Microbiology (Eng.), Sigma Xi. Research in cellular mechanisms in delayed-type hypersensitivity, immunologic tolerance. Home: 630 N. Aberdeen Rd., Inverness, P.O. Palatine, Ill. 60067. Office: 535 N. Dearborn St., Chgo. 60610.*

RITZ, Walter, Swiss physicist; b. Sion, Switzerland, Feb. 22, 1878; docent Zurich (Switzerland), Göttingen (Germany) univs. Author: Gesammalte Werke, 1911. Research in spectral physics; devised theory called Ritz's combination principle (stated there is simple math. scheme for distbn. spectral lines of atoms), confirmed by later research. Died Göttingen, July 7, 1909.

RITZEL, Guenther Nicolaus, physician; b. Frankfort/Main, Germany, Oct. 11, 1924; s. Heinrich Georg and Elisabeth (Lack) R.; M.D., U. Basel (Switzerland), 1950, Specialist Internal Medicine, 1956; m. Doris Sylvia Friedrich, Oct. 7, 1956; children—Rainer, Lukas. Asst. pharmacological inst. U. Basel, 1950-53, asst. biochem. inst., also Swiss Inst. on Vitaminology, 1951-53, chief officer Sch. med. services, 1957—; asst. Med. U. Clinic Basel, 1953-57; lectr. social and preventive medicine Tchr.'s Tng. Coll. Basel, 1959—; lectr. biochemistry and nutrition research Med. Faculty, U. Basel, 1963—. Recipient 1st award Swiss Research Assn. for Nutrition, 1966. Mem. Nutrition Research Assn. (chmn. 1960—), Biochem. Assn., Physiol. and Pharmacological Assn., Assn. for Preventive Medicine, Assn. Sociology. Research and publs. on metabolic changes especially in nutritive aspects, vaccinations, dental caries in sch. children before and after use of fluorides; neuroimmunological studies on multiple sclerosis. Home: 7, Buttertalstrasse, 4106 Therwil, Switzerland. Office: 19, St. Albanvorstadt, 4052 Basel, Switzerland.*

RIUST, Paul, French anthropologist; b. Wasigny, France, May 7, 1876; s. Gustav and Marie (Lajoux) R.; ed. Sch. Mil. Health Service, Lyons, France; M.D.; m. Mercédès Andrade, Apr. 1, 1922. Became mil. physician, 1898; physician Geodetic Mission to Equator, 1901-06; sub-dir. Anthrop. Lab., Mus. Natural History, 1906-28, prof., 1928—; sec. gen. Ethnol. Inst., U. Paris, 1926—; mem. Municipal Council, Paris, 1935—; mem. 1st and 2d Constituent Assembly, 1945—, Nat. Assembly, 1946—. Research and numerous publs. on linguistics, ethnology and ethnography, anthropology. Office: Musée de l'homme, Palais de Chaillot, Paris XVI, France.

RIVA SANSEVERINO, Eugenio, Italian physiologist; b. Palermo, Italy, Dec. 23, 1934; s. Ferdinando and Maria Felice (Boscogrande) R.S.; M.D., U. Palermo, 1959; m. Corrada Fanasi, Jan. 30, 1961; children—Ferdinando, Luisa. Asst. prof. U. Catania, 1959-63; asst. prof. U. Bologna (Italy), 1963—, libera docenza physiology, 1965—. NATO-Nat. Council Italy fellow Lab. Physiology, Sch. Medicine, Johns Hopkins, Balt., 1966-67. Mem. Soc. ital. Fisiologia, Soc. ital. EEG and Neurofisiologia, Boll. Soc. ital. Biol. sper. Research and publs. on cerebellum, including patterns of EEG convulsive activity caused by sensorial peripheral stimulation following local strychninization of various cerebellar areas, functional projections of cerebellar paraflocullus on cerebral cortex, histological and electrophysiol. observations on chronically isolated cerebellar cortex. Office: Istituto Fisiologia-Piazza Porta S. Donato, 2-Bologna, Italy.*

RIVER, Louis Philip, Am. surgeon; b. San Francisco, Mar. 12, 1901; s. Louis Philip and Amy (Hopkins) R.; B.S., U. Chgo., 1922, M.D., 1925; m. Elizabeth Lambert, Sept. 18, 1928; children—Louis, George, Amy (Mrs. Teodolo Valenzuela), Valerie (Mrs. Aubrey W. Vaughan). Practice medicine, specializing in surgery, Oak Park, Ill., 1925—; faculty Stritch Med. Sch. Loyola U., Chgo., 1925—, clin. prof. surgery, 1952—; prof. surgery Cook County Post-Grad. Med. Sch., Chgo., 1937; chief, breast tumor service Cook County Hosp., Chgo., 1947—. Diplomate Am. Bd. Surgery. Fellow A.C.S.; mem. A.M.A., Am. Med. Writers Assn., Ill., Chgo. surg. socs., Internat. Coll. Surgeons (gov. U.S. sect.). Research, publs. on surgery of tumors of breast, intestine. Home: 165 N. Kenilworth Av., Oak Park 60301 Office: 715 Lake St., Oak Park, Ill. 60301.*

RIVERO, Juan Arturo, Puerto Rican biologist; b. Santurce, P.R., Mar. 5, 1923; s. Agustin and María Leticia (Quintero) R.; B.S.A., U. P.R., Mayaguez, 1945; M.A., Harvard, 1951, Ph.D., 1952; m. Eneida J. Bordallo, Apr. 27, 1945; children—Juan Augustín, Hedrick J. Faculty, U. P.R., Mayaguez, 1947—, prof., 1958—, dir. Inst. Marine Biol. and Zool. Garden, 1954-63, dir. biology dept., 1959-60, dean arts and scis., 1962-66. Mem. Acad. Arts and Scis. P.R.; corr. mem. LaSalle Soc. Scis. Natural, also Sociedad Venzolana de Ciencias Naturales (both Venezuela). Research and publs. on taxonomy, ecology, geography and behavior of amphibia. Home: 345 Harvard St. Office: Museum of Comparative Zoology, Harvard U., Cambridge, Mass. 02138.*

RIVERS, William Halse, Brit. anthropologist; b. Kent, Eng., Mar. 12, 1864; s. H. F. Rivers; M.A., M.D., LL.D., St. Andrews, Scotland; D.Sc., Manchester, Eng. Became lectr., Cambridge, Eng., 1887; made expdns. to study preliterate peoples, Torres Straits, 1898, Toda tribes in S. India and Melanesia, 1908; fellow, praelector in natural scis. St. John's Coll., Cambridge. Fellow Royal Soc., 1908, Royal Coll. Physicians; mem. Royal Anthrop. Inst. (pres.), Folklore Soc. (pres.), Brit. Assn. (pres. sect. for anthropology 1911). Recipient Royal medal Royal Soc., 1915. Author: The Todas, 1906; The History of the Melanesian Society, 2 vols., 1914; Instinct and the Unconscious, 1920; Psychology and Politics, 1923; Conflict and Dream, 1923; Medicine, Magic and Religion, 1924; Psychology and Ethnology, 1926; also articles. Founder, Cambridge Sch. Exptl. Psychology. Pioneered use of geneal. method in sociol. investigation; research on influence of drugs and alcohol on mental fatigue. Died June 4, 1922.

RIVIER, Dominique-Casimir, Swiss physicist; b. Jouxtens-près-Lausanne, Switzerland, Nov. 12, 1918; s. Louis and Julie (De Rham) R.; student U. Lausanne, U. Geneva, Princeton; Dr.ès.sci.; m. Florence Matthey, July 11, 1918; children—Francine, Léonore, Augustin, Constance, Louis. Asst., U. Lausanne, 1940-44, U. Geneva, 1944-49; vis. fellow Princeton, 1950-52; research fellow Nat. Canadian Research Council, Ottawa, Ont., 1952-53; became asso. prof. U. Lausanne, 1953, prof., 1957. Mem. Swiss Physics Soc. (pres. 1961-63), Am. Phys. Soc., Sigma Xi. Research and publs. on theory of fields, ferromagnetic transport, statis. mechanics. Home: L'Oche, Jouxtens (VD), Switzerland. Office: 3 place du Chateau, Lausanne, Switzerland.

RIVIÈRE, André, French sedimentologist; b. Piriac, France, Oct. 2, 1904; s. Eugéne and Cecile (Boucher) R.; Ed. Ecole Normale Superieure, also Faculté de Science, U. Paris; Agrége des Sciences Naturelle, 1929; m. Suzanne Meneux, July 2, 1936; children—Marc, Claude, Colette. Prof., Faculté des Sciences, U. Téheran (Iran), 1929-35; in charge research Centre National de la Recherche Scientifique, 1935-38; lectr. Sorbonne, Paris, 1938-45; faculty U. Paris, 1945—, became prof., 1952; later prof. Faculty Sci., Orsay, France. Mem. French Geol. Soc., Nat. Sci. Soc. Lausanne (hon.), Internat. Assn. Sedimentology. Research and publs. on gen. sedimentological geology; tectonic and stratigraphic research on Central Elburz; sedimentologic research, including screen analysis chemistry of sediments, mechanics of sedimentation; use of radioactive substances in littoral sedimentology; sedimentology of lagoons; works on maritime defense. Home: 28 Avenue du Chateau, Bourg-la-Reine, 92, France. Office: Labor Sédimentologie, Faculté des Sciences, Orsay, 91, France.*

RIVLIN, Ronald Samuel, applied mathematician; b. London, Eng., May 6, 1915; s. Raoul and Bertha (Aronsohn) R.; B.A., Cambridge (Eng.) U., 1937, M.A., 1939, Sc.D., 1952; m. Violet LaRusso, June 16, 1948; 1 son, John Michael. Came to U.S., 1952, naturalized, 1955. Research physicist Gen. Electric Co., Eng., 1937-42; sci. officer Telecommunications Research Establishment, Ministry of Aircraft Prodn., Eng., 1942-44; staff Brit. Rubber Producers Research Assn., 1944-52, supt. research, 1950-52; head research group Davy Faraday Lab., Royal Instn., London, Eng., 1948-52; cons. Naval Research Lab., Washington, 1952-53; prof. Brown U., Providence, 1953—, chmn. div. applied mathematics, 1958-63, prof. applied math. and engring. sci., L. Herbert Ballou U. prof., 1963—. Cons. to indsl. corps.; co-chmn. 4th Internat. Congress on Rheology, 1963. Guggenheim fellow, 1961-62. Mem. Soc. Rheology (exec. com. 1957-59, recipient Bingham medal, 1958), Soc. for Natural Philosophy (chmn. 1963-64), Am. Acad. Arts and Scis. Editorial com., Jour. Rational Mechanics and Analysis, 1952-57, Archive for Rational Mechanics and Analysis, 1957—, Jour. Math. Physics, 1960, Jour. Applied Physics, 1960-63, Acta Rheologica, 1963—. Contbr. articles to tech. jours. Formulation of non-linear theories in mechanics and physics. Home: 1 Drowne Parkway, East Providence, R.I. 02916.*

RIZNICHENKO, Yuriy Vladimirovich, Russian geophysicist; b. Sept. 28, 1911; grad. Kiev Mining and Geol. Inst., 1935. Asso., Geophys. Inst., USSR Acad. Scis., 1938-47, prof., 1947-56, asso. Inst. Geophysics, 1956—, bur. mem. dept. physico-math. sci., 1960—. Del., 13th gen. assembly Internat. Union Geodesy and Geophysics, U.S., 1963. Mem. USSR Acad. Scis. (corr.). Author: Seismic Hodograph Theory, 1939; The Seismic Properties of Parmafrost Strata, 1942; Seismological Speeds in Stratified Media, 1947; The Simulation of Seismic Waves, 1951; A Seismic Impulse Methods of Studying Rock Pressure, 1955; Mass Methods of Determining the Coordinates of the Foci of Near Earthquakes and the Speed of Seismic Waves in the Area of the Location of the Foci, 1958; Development of the Principles of a Quantitative Method of Seismic Zoning, 1960; co-author: A Correlation Method of Refracting Waves, 1952. Chief editor News of the USSR Acad. Scis.: Geophys. Series. Research and publs. on propagation of seismic waves, developer seismic prospecting methods. Address: Inst. Geophysics, USSR Acad. Scis., B. Gruzinskaya 10, Moscow, USSR.

ROACH, Franklin Evans, Am. physicist; b. Jamestown, Mich., Sept. 23, 1905; s. Richard F. and Belle (Torgersen) R.; student Wheaton (Ill.) Coll., 1923-26; B.S., U. Mich., 1927; M.S., U. Chgo., 1930, Ph.D., 1934; m. Eloise Blakslee, Jan. 25, 1930; children—John R., Gerard, Janet, Charlotte (Mrs. Georg Vedeler). Physicist, Naval Ordnance Test Sta., China Lake, Cal., 1945-54; physicist Nat. Bur. Standards/Inst. Telecommunications Sci. and Aeronomy, Boulder, Colo., 1954—; Fulbright research scholar Inst. d'Astrophysique, Paris, France, 1951-52; specialist E-W Center Hawaii, 1963-64. Recipient Gold Medal award Dept. Commerce, 1961. Mem. Am. Astron. Soc., Am. Geophys. Union, Internat. Assn. Geomagnetism and Aeronomy, Internat. Astron. Union, Research Soc. Am., Sigma Xi. Contbr. numerous articles to profl. jours. Research in stellar spectroscopy, photoelectric photometry of eclipsing binaries, propellant and explosive engring., photometry of the light of night sky. Home: 2132 Bluebell Av. Office: NBS/ESSA, Boulder, Colo. 80302.*

ROAF, Herbert Eldon, physiologist; b. Toronto, Ont., Can., 1881; s. James R. Roaf; ed. U. Toronto; D.Sc., Liverpool, Eng. Asst. lectr. physiology U. Liverpool (Eng.), 1906-11, lectr. chem. physiology, 1909-11, George Holt prof. physiology, 1932-44, then prof. emeritus; lectr. St. Mary's Hosp. Med. Sch., 1911-20; prof. London Hosp. Med. Coll., U. London, 1920-32. Johnston Colonial fellow U. Liverpool, 1902-05; Brit. Med. Assn. Research scholar, 1905-06; licentiate Royal Coll. Physicians London. Mem. Royal Coll. Surgeons Eng. Author: Biological Chemistry, 1921; Text Book of Physiology, 2d edit., 1936; also numerous articles. Studied color vision. Died Sept. 21, 1952.

ROAF, Robert, Brit. physician; b. London, Eng., Apr. 25, 1913; s. Herbert Eldon and Beatrice (Herdman) R.; ed. Oxford (Eng.) U., 1931-34, Liverpool (Eng.) U., 1934-37; m. Ceinwen Roberts, Oct. 25, 1939; children—David, Sarah, Margery, John. Dir. clin. research Orthopaedic Hosp., Oswestry, Eng., 1955-65, cons., 1947—; prof. orthopaedic surgery U. Liverpool, 1963—. Cons., Royal Liverpool United Hosp., 1946—, Nat. Inst. for Blind, 1946—; vis. cons. Irwin Hosp., New Delhi, India, 1952-54. Fellow Royal Soc. Medicine, Assn. Surgeons E. Africa (hon.), Brit. Orthopaedic Assn., Internat. Soc. Orthopedics. Author: Potts Paraplegia, 1956; Textbook of Orthopaedic Nursing, 1963; Bone and Joint Tuberculosis, 1960; Scoliosis, 1967; also numerous articles. Research on diseases and deformities of spine; injuries to peripheral nerves; Tb. of bones and joint; spinal injuries. Home: 215 Speke, Liverpool, U.K.*

ROBBINS, Frederick C(hapman), Am. microbiologist, physician; b. Auburn, Ala., Aug. 25, 1916; s. William J. and Christine (Chapman) R.; A.B., U. Mo., 1936, B.S., 1938; M.D., Harvard, 1940; D.Sc. (hon.), John Carroll U., 1955, U. Mo., 1958; m. Alice Havemeyer Northrop, June 19, 1948; children—Alice, Louise. Sr. fellow virus diseases Nat. Research Council, 1948-50; staff research div. infectious diseases Children's Hosp., Boston, 1948-50, asso. physician, asso. dir. isolation service, asso. research div. infectious diseases, 1950-52; instr., asso. in pediatrics Harvard Med. Sch., 1950-52; dir. dept. pediatrics and contagious diseases Cleve. Met. Gen. Hosp., 1952—; associate pediatrician U. Hospitals, Cleve., 1952—; professor pediatrics Western Reserve U., 1952—; vis. scientist Donner Lab., U. Cal., 1963-64. Served as maj. AUS, 1942-46; chief virus and rickettsial disease sect. 15th Med. Gen. Lab. Decorated Bronze Star, 1945; received 1st Mead Johnson prize application tissue culture methods to study of viral infections, 1953; co-recip. (with Enders and Weller) Nobel prize in physiology and medicine, 1954. Diplomate Am. Bd. Pediatrics. Mem. American Epidemiol. Soc., Am. Academy Arts and Scis., Central Soc. Clinical Research (emeritus member), American Academy Pediatrics, Soc. Pediatric Research (president 1961-62, emeritus mem.), Am. Assn. Immunologists, Am. Soc. Clin. Investigation, Soc. Exptl. Biol. and Medicine, Am. Pediatric Soc. Research on infectious hepatitis, typhus and Q fever, and immunology of mumps; discovered that poliomyelitis virus could be grown in test-tube cultures of various human tissues. Home: 2469 Wellington Rd., Cleveland Heights 18, O. Office: 3395 Scranton Rd., Cleve., Ohio 44109.*

ROBBINS, Laurence Lamson, Am. physician, radiologist; b. Burlington, Vt., Mar. 26, 1911; s. George E. and Grace (Lamson) R.; B.S., U. Vt., 1935, M.D., 1937; m. Ruth Dick, Feb. 18, 1932; children—Dick, Carol, Janice. Intern, Mary Fletcher Hosp., 1937-38; instr. pathology, U. Vt., 1938-39; resident in radiology Mass. Gen. Hosp., Boston, 1939-41, radiologist, 1941-46, radiologist-in-chief, 1946—; asst. radiology Harvard Med. Sch., 1941, clin. prof., 1960—. Dir., James Picker Found., 1962—, pres., 1964—. Mem. Am. Bd. Radiology (trustee 1954—, pres. 1960-63), Am. Roentgen Ray Soc., Radiological Soc. N. Am. (pres. 1959), Am. Coll. Radiology, New Eng. Roentgen Ray Soc. (pres. 1953-54), New Eng. Cancer Soc., Am. Med. Soc. Editor: Golden's Diagnostic Roentgenology, 1959; Roentgen Interpretation, 1955. Research on clinical radiology; reducing radiation hazards to patients; radiologic problems of the lung. Home: 98 Cambridge St., Winchester, Mass. 01890. Office: Mass. Gen. Hosp., Boston. 02114.

ROBBINS, Mary Louise, Am. microbiologist; b. St. Paul, Sept. 26, 1912; d. Orison Benjamin and Mary (Fiske) Robbins; B.A., Am. U., 1934; M.A., George Washington U., 1940, Ph.D., 1944. Faculty, George Washington U. Sch. Medicine, Washington, 1944—, prof. microbiology, 1958—, acting chmn. dept. microbiology, 1961-62; vis. scientist U.S. Naval Med. Research Unit Number 3, Cairo, Egypt, 1955; vis. prof. Baghdad (Iraq) U., 1963; USPHS spl. research fellow, Japan, 1968-69. Diplomate Am. Bd. Microbiology. Fellow Am. Acad. Microbiology, A.A.-A.S., Washington Acad. Scis.; mem. Am. Assn. Immunologists, Am. Soc. for Microbiology, N.Y. Acad. Scis., Sigma Xi, Sigma Delta Epsilon. Contbr. articles to tech. jours. Research on mycobacteriophages, interactions between microorganisms, also between microorganisms and viruses, antigenic analysis bacteria, inhibition of multiplication viruses by chems., serological methods in virology. Home: 825 New Hampshire Av. N.W., Washington 20037. Office: 1339 H St., N.W., Washington 20005.*

ROBBINS, Stanley Leonard, Am. pathologist; b. Portland, Me., Feb. 27, 1915; B.S., Mass. Inst. Tech., 1936; M.D., Tufts U., 1940; m. Eleanor L. Peskin, June 20, 1940; children—Janet (Mrs. Stephen Schachter), Jonathan H., Jeffrey M. Staff, Mallory Inst. Pathology, Boston City Hosp., 1940—, asso. pathology, asso. dir., 1953-66, dir., 1966—; prof. pathology Boston U. Sch. Medicine, 1957—, chmn. dept. pathology, 1964—. Mem. Mass., N.E. socs. pathologists, Am. Heart Assn. Author: Textbook of Pathology, 3d edit. Research, publs. on gross and finer circulation of heart in patients with coronary artery disease. Home: 175 Brookline St., Chestnut Hill, Mass. Office: Mallory Inst., Boston City Hosp., Boston.*

ROBBINS, William Jacob, Am. botanist; b. North Platte, Neb., Feb. 22, 1890; s. Frederick Woods and Clara (Federhoof) R.; A.B., Lehigh U., 1910, D.Sc., 1937; Ph.D., Cornell, 1915; D.Sc., Fordham U., 1945; m. Christine F. Chapman, July 15, 1915; children—Frederick Chapman, William Clinton, Daniel Harvey. Instr. biology, Lehigh U., 1910-11; instr. plant physiology, Cornell U., 1912-16; asst. in plant physiology, Marine Biol. Lab., Woods Hole, Mass., summers 1912, 13; prof. botany, Ala. Poly. Inst., also plant physiologist Agrl. Expt. Sta., 1916-17; soil biochemist, Bur. Plant Industry, U. S. Dept. Agr., 1919; prof. botany, U. of Mo., 1919-37, dean Grad. Sch., U. of Mo., 1930-37, actg. pres., U. of Mo., 1933-34; prof. Columbia, 1937-58, prof. emeritus, 1958—; dir. N.Y. Bot. Garden 1937-58; pres. Fairchild Tropical Garden, 1963—; asso. dir. Internat. Sci. Activities Nat. Sci. Found., 1962-63. Trustee Rockefeller Inst., 1956—. Spl. investigator Citrus Expt. Sta., University of California, summer 1924. Connected with European Office, Rockefeller Foundation, 1928-30. Chmn. Nat. Research Fellowship Bd. in Biol. Sciences, 1931-37. Fellow Am. Acad. Arts and Scis., New York Academy of Sciences, A.A.-A.S.; member Nat. Acad. Scis. (treas.), Am. Philos. Soc. (pres. 1956-59, exec. officer 1959-60), Bot. Soc. Am. Author: (with H. W. Rickett) Botany, 1929. Work with tissue culture of higher plants, antibiotics, plant tumors and other abnormal growth; discovered a specific soil bacteria capable of destroying certain organic compounds toxic to higher plants. Home: 301 E. 66th St. Office: Rockefeller Inst., 66th and York Av., N.Y.C.

ROBEL, Robert Joseph, Am. animal ecologist; b. Lansing, Mich., May 21, 1933; s. Joseph J. and Loretta (Pung) R.; B.S., Mich. State U., 1956; M.S., U. Ida., 1958; Ph.D., Utah State U., 1961; m. Anice Marie Blanc, Aug. 27, 1960. Research fellow U. Ida., 1956-57; biologist Ida. Dept. Fish and Game, 1956-57; research asst. Utah State U., 1958-61; faculty Kan. State U., Manhattan, 1961—, asso. prof. zoology, 1966—. Council mem. Acad. Conf., 1964—. Fellow A.A.A.S.; mem. Wildlife Soc., Ecol. Soc. Am., Brit. Ecol. Soc., Animal Behavior Soc., Am. Inst. Biol. Scis., Am. Soc. Mammalogists, Kan. Acad. Sci. (sec. 1964—). Research and publs. in animal behavior, population dynamics, census techniques, and avian dispersal patterns, bobwhite quail, prairie chickens, elk. Home: 1822 Fairchild St., Manhattan, Kan. 66502.*

ROBERT GROSSETESTE, see Grosseteste, Robert.

ROBERT OF CHESTER (or Robertus Cataneus or Robertus Retinensis), English translator; flourished circa 1143; student astrology, Spain, probably studied with Plato of Tivoli, Barcelona, 1136. Lived in Spain, circa 1141-47; archdeacon of Pampeluna; hired by Abbot of Cluni to translate Arabic works into Latin. First translator of Koran, circa 1143; translated Al-Khowarizmi's Algebra into Latin, circa 1145; prepared many astron. tables; translated 1st alchem. work into Latin, 1144.

ROBERT OF YORK (or Robertus Perscrutator or Robertus Eboracensis), English meteorologist; flourished 1313-48; D. Divinity degree. Dominican monk, studied medicine, astrology and alchemy. Author: De Impressionibus Aeris, 1325; Corruptorium Alchimiae; De Elementorum Mixtione, Musica; De Moralibus Caeremoniali; De Mysteriis Secretorum. Observations on natural phenomena; used the theory of elements and the influences of the seven planets to explain variations in the weather.

ROBERTS, Arthur, Am. physicist; b. N.Y.C., July 6, 1912; s. Abraham and Rose (Bloch) R.; B.S., Coll. City N.Y., 1931; diploma in piano Manhattan Sch. Music, 1933; M.A., Columbia U., 1933; Ph.D., N.Y. U., 1936; m. Janice Banner, May 21, 1935; children—Judith Anne (Mrs. John Henry Neale), Richard Martin. Faculty, Mass. Inst. Tech., 1937-46, State U. Ia., 1946-50, U. Rochester, 1950-60; sr. physicist Argonne (Ill.) Nat. Lab., 1960-67, Nat. Accelerator Lab., Oak Brook, Ill., 1967—. Cons. to labs. and U.S. Govt. agys.; liaison officer U.S. Office Naval Research, London, 1954-55. Recipient Presdl. Certificate of Merit, 1948. Fellow Am. Phys. Soc.; mem. I.E.E.E. Editor: Radar Beacons, Vol. 3, McGraw Hill Radar Lab. series, 1947. Research in radioactive tracers in biology and medicine, first use of radioiodine; studies in beta and gamma spectroscopy; microwave spectroscopy; high-energy particle physics; contbns. to instrumentation in nuclear and high-energy physics. Home: 5512 S. Everett Av., Chgo. 60637. Office: Nat. Accelerator Lab., 1301 W. 22d St., Oak Brook, Ill. 60521.*

ROBERTS, Daniel Altman, Am. plant pathologist; b. Micanopy, Fla., Jan. 8, 1922; s. Simon and Pearle (Altman) R.; B.S., U. Fla., 1943, M.S., 1948; Ph.D., Cornell U., 1951; m. Ruth Remsen, Feb. 1, 1944; children—Peter R., Kathleen E., Stephen B. Asst. prof. plant pathology Cornell U., 1951-55, asso. prof., 1955-59; with U. Fla., 1959—, prof., 1964—. Guggenheim fellow, 1958. Mem. Am. Phytopath. Soc., Sigma Xi, Phi Kappa Phi, Phi Sigma, Gamma Sigma Delta, Alpha Zeta. Research, publs. on plant virology; effects of viruses on reduced photosynthesis; translocation of viruses. Home: 3662 N.W. 7th Av., Gainesville, Fla. 32601.*

ROBERTS, Eliot Collins, Am. agriculturist; b. Camden, N.J., Apr. 25, 1927; s. Benjamin Jenkins and Ruth (Collins) R.; B.S., U. R.I., 1950; M.S., Rutgers U., 1952; Ph.D., 1955; m. Beverly Mae Cruickshank, June 30, 1951; children—Eliot Collins, Mary Alice, William Cruickshank. Asst. prof. agronomy U. Mass., 1954-57; faculty Ia. State U., Ames, 1959-67, prof. horticulture and agronomy, 1964-67; prof., chmn. ornamental horticulture U. Fla., Gainesville, 1967—. Mem. Weed Soc. Am. (past chmn. div. 7), Am. Soc. Agronomy (past chmn. div. C-5), Am. Soc. for Hort. Sci. Asso. editor Agronomy Jour., 1966—. Research and numerous publs. on physiology of closely clipped grasses, effects of microclimate and plant nutrition on disease, incidence in grasses; developed improved cultural practices for maintenance of sports turf and lawns. Home: 3535 N.W. 14th Av., Gainesville, Fla. 32601.*

ROBERTS, Harold Selig, economist; b. Zdunska-Wola, Poland, Mar. 14, 1911; s. Selig Wolf and Adela (Erlich) R.; came to U.S., 1921, naturalized, 1926; B.S. in Social Sci. cum laude, Coll. City N.Y., 1934; M.A., Columbia, 1938, Ph.D., 1944; m. Margery Prescott, Apr. 4, 1939; children—Joyce Donald Kennedy), Pamela Jean. Analyst, U.S. Senate com. on interstate commerce, subcom. on r.r. finance, 1936-38; economist div. econ. research NLRB, 1938-41; co-editor Am. Labor Year Book, Bur. Labor Statistics, Washington, 1941-42; adminstrv. officer, cons. Nat. War Labor Bd., 1942-45; chief collective bargaining div. Bur. Labor Statistics, U.S. Dept. Labor, Washington, 1945-47; faculty U. Hawaii, Honolulu, 1947—, dean Coll. Bus. Adminstrn., 1950-59, sr. prof. bus. econs. and indsl. relations, dir. Indsl. Relations Center, 1948—. Tech. adviser U.S. Govt. dels. at confs., Mexico City, Seattle, 1946; adminstrv. asst. Gov. of Hawaii, 1952-53; mem. Hawaii Jud. Council, 1961—. Recipient Wills Meml. award in indsl. relations Wills Meml. Fund, U. Hawaii, 1959; named Outstanding Citizen of Year, Hawaiian Govt. Employees Assn., 1966-67. Mem. Am. Econ. Assn., Acad. Polit. and Social Sci., Indsl. Relations Research Assn., Am. Arbitration Assn. (nat. panel arbitrators, cons. editor Jour.), Fed. Mediation & Conciliation Service Arbitration Panel, Phi Kappa Phi. Author: The Rubber Workers, 1944, Roberts' Dictionary of Industrial Relations, 1966; (with Paul F. Brissenden) Challenge of Industrial Relations in Pacific-Asian Countries, 1965; also articles. Research in methods of labor-mgmt. dispute settlement; intensive study of govt. health and welfare services. Home: 2646 Oahu Av., Honolulu 96822.*

ROBERTS, Harry Vivian, Am. statistician; b. Peoria, Ill., May 1, 1923; s. Harry V. and Mary D. (Pickels) R.; B.A., U. Chgo., 1943, M.B.A., 1947, Ph.D., 1955; m. June H. Hoover, Nov. 19, 1943; children—Andrew H., Mary D. Market research analyst McCann-Erickson, Inc., 1946-49; faculty U. Chgo., Grad. Sch. Bus., 1949—, prof. statistics, 1959—. Cons. on bus. and mil. applications statistics, 1949—. Fellow Am. Statis. Assn.; mem. Am. Econ. Assn., Royal Statis. Soc., Econometric Soc., A.A.A.S., Am. Marketing Assn. Author: (with J. H. Lorie) Basic Methods of Marketing Research, 1951; (with W. Allen Wallis) Statistics: A New Approach, 1956; also articles. Research on Bayesian decision theory and its application to problems of bus., econs., and sci. research. Home: 1353 Burr Oak Rd., Homewood, Ill. 60430. Office: University Chgo. Grad. Sch. Bus., Chgo. 60637.*

ROBERTS, Henry Stoutte, Am. cytologist; b. Macon, Ga., Nov. 14, 1913; s. Henry Stoutte and Annie (Baldwin) R.; A.B., Mercer U., 1934; Ph.D. in Zoology, Duke, 1948; m. Clotilda Anne Houle, Nov. 8, 1943; children—Henry S., Roberts, III, Frederick Eugene. High sch. biology tchr. Ga. Pub. Schs., 1934-37; biology tchr. Gordon Mil. Coll., 1937-41; asso. prof. zoology Duke, 1948-64; prof., head dept. biology Washington and Lee U., Lexington, Va., 1964—. Mem. Am. Soc. Cell Biology, Am. Soc. Zoologists, Va. Acad. Sci., Assn. Southeastern Biologists, Sigma Xi. Contbr. articles to tech. jours. Research on mitochondria, mechanisms of cell division. Home: 913 Shenandoah, Lexington, Va. 24450.*

ROBERTS, Hyman Jacob, Am. physician; b. Boston, May 29, 1924; s. Benjamin and Eva (Sherman) R.; M.D. cum laude, Tufts U., 1947; m. Carol Antonia Klein, Aug. 9, 1953; children—David Barry, Jonathan Stuart, Mark Elliott, Stephen Bennett, Scott Fleming, Pamela Beth. Practice medicine specializing in internal medicine, cons. in diagnosis and metabolic diseases, West Palm Beach, Fla., 1955—; staff Good Samaritan Hosp., St. Mary's Hosp., West Palm Beach, 1955—; dir. Palm Inst. for Med. Research, West Palm Beach, 1964—; instr., research fellow Tufts U. Med. Sch., 1948-49, Georgetown Med. Sch., 1949-50. Named one of Fla.'s 5 Outstanding Young Men Fla. Jr. C. of C., 1958. Mem. Am. Fedn. for Clin. Research, A.M.A., Fla., So. med. assns., A.C.P., Am. Coll. Chest Physicians, Am. Coll. Angiology, Am. Heart Assn., Am. Diabetes Assn., Am. Geriatrics Soc., Alpha Omega Alpha. Author: Difficult Diagnosis: A Guide to the Interpretation of Obscure Illness, 1958; Mimics in Medicine, 1967; also chpts. in texts, numerous articles. Research on diabetes, hyperinsulinism, coronary disease, narcolepsy, obesity, myasthenia gravis, porphyria, reading disability in children, prostatic tumors, gout, multiple sclerosis, gout, parkinsonism; originated afternoon glucose tolerance testing, intravenous betamethasone-glucose tolerance testing, postglucose electroencephalography, basic clin.-pharmacologic researches on synthetic oxytocin, betamethasone (systemic and local), estrogens (oral and intravenous), nandrolone, phenformin and formula diets. Home: 6708 Pamela Lane, West Palm Beach 33401. Office: 300 27th St., West Palm Beach, Fla. 33407.*

ROBERTS, James Edwin, biophysicist; b. Glendale, Cal., Aug. 22, 1922; s. Merle Ellison and Dorothy (Briggs) R.; student U. Colo., 1941-45, M.D., 1950; m. Erna May Sargent, Aug. 13, 1952; children—Kirk Foster, Cynthia Lee, Jay, Jeffrey. Head anatomy dept. Lovelace Found., Albuquerque, 1952-53; research asso. Radiation Lab., U. Cal. at Berkeley, 1953-55; research asso. U. Colo. Med. Sch., Denver, 1958-59; head kinetics Sch. Aerospace Medicine, Brooks AFB, 1960-61; dep. chief plans and analysis SAM, Brooks AFB, 1962-63; dir. bioastronautics Aerospace Corp., El Segundo, Cal., 1963-66; mem. sr. staff adv. Systems Lab., TRW Systems, Redondo Beach, Cal., 1966—. Mem. Am. Physiol. Soc. Research, publs. on nuclear weapons effects, effects high energy radiation on pituitary and central nervous system, math. models biol. system, bioengring. Manned space flight. Home: 624 Paseo Lunado, Palos Verdes Estates, Cal. 90275. Office: 1 Space Park, Redondo Beach, Cal. 90278.*

ROBERTS, James H., Am. physicist; b. Tucson, Oct. 27, 1915; s. James and Annie (Armitage) R.; B.S., U. Ariz., M.S., 1938; Ph.D., U. Chgo., 1946; postgrad. Dallas Theol. Sem., 1940-41; m. Mary Elizabeth Kinne, June 6, 1942; children—James Norman, Jean Elizabeth, Judith Ann, John Philip. Lectr.-demonstrator Mus. Sci. and Industry, Chgo., 1939-40, 41; research physicist Metall. Lab., U. Chgo., 1942-44; research physicist Los Alamos Sci. Lab., 1944-46, asso. group leader, 1946-48; faculty Northwestern U., Evanston, 1948—, prof. physics, 1958—; vis. prof. physics Macalester Coll., St. Paul, 1963-65. Fellow Am. Phys. Soc.; mem. Am. Assn. Physics Tchrs., Phi Beta Kappa, Sigma Xi, Phi Kappa Phi. Research and publs. on measurements in phys. quantities dealing with nuclear reactions and decay; contributed to devel. of use of nuclear emulsions and other solid particles tract detectors to study neutron speeds, energy with which strange particles are bound to atomic nuclei and characteristics of nuclear fission. Home: 601 Appletree Lane, Deerfield, Ill. 60015. Office: Physics Dept., Northwestern U., Evanston, Ill. 60201.*

ROBERTS, Job, Am. agriculturist; b. Whitpain, Pa., Mar. 23, 1756; s. John and Jane (Hunk) R.; m. Mary Naylor, May 22, 1781; m. 2d, Sarah (Williams) Thomas, Oct. 12, 1820; at least 2 children. Apptd. justice of peace, 1791. Began experimenting in better farming methods, 1785; published results of expts. as The Pennsylvania Farmer, 1804; experimented with fertilizers, use of lime, plaster, various barnyard manures, deep ploughing of land; built improved harrow, devised new roller, 1792; attached water wheel to dairy churn making it possible to churn 150 pounds of a butter a week, 1797; invented machine for planting corn, 1815; advanced growing season of corn by soaking before planting; introduced Merino sheep into Pa.; interested in cultivation of mulberry for silk culture; substituted green fodder for his cattle for grazing. Died Whitpain, Aug. 20, 1851.

ROBERTS, John Alexander Fraser, Brit. geneticist; b. Denbighshire, Wales, Sept. 8, 1899; s. Robert Henry and Elizabeth (Fraser) R.; B.Sc., U. Coll. N. Wales, 1920; M.A., Gonville and Caius Coll., Cambridge, Eng., 1925; D.Sc., U. Edinburgh (Scotland), 1933, M.D., 1943; m. Doris Breamer Hare, Mar. 15, 1941; children—Susan Jane (Mrs. Frederick John Griffith), Catherine Ann Fraser. Research asst. Inst. Animal Genetics, U. Edinburgh, 1922-28; biologist Wood Industries Research Assn., Leeds, Eng., 1928-31; Macaulay fellow U. Edinburgh, 1931-33; dir. Burden Mental Research Dept., Bristol, Eng., 1933-57; lectr. med. genetics London Sch. Hygiene and Tropical Medicine, 1946-57; dir. clin. genetics research unit Med. Research Council, 1957-64; geneticist pediatric research unit Guy's Hosp., London, 1964—. Fellow Royal Soc., 1963, mem. Royal Anthropol. Inst. Great Britain and Ireland; mem. Royal Soc. Medicine, Royal Statis. Soc. Author: An Introduction to Medical Genetics, 1940, 14th edit., 1967; also numerous articles. Research on animal and human genetics, psychometrics and mental deficiency, inherited diseases. Home: 13 Ruvigny Mansions, Embankment, London S.W. 15, Eng. Office: Pediatric Research Unit, Guy's Hosp. Med. Sch., London S.E.1, Eng.*

ROBERTS, John D., Am. chemist; b. Los Angeles, June 8, 1918; s. Allen Andrew and Flora (Dombrowski) R.; B.A., U. Cal. at Los Angeles, 1941, Ph.D. in Organic Chemistry, 1944; Dr. rer. nat. h. c., U. Munich (Germany), 1962; D.Sci., Temple U., 1964; m. Edith M. Johnson, July 11, 1942; children—Anne Christine, Donald William, John Paul, Allen Walter. Instr. U. Cal. at Los Angeles, 1945; Nat. Research fellow in chemistry Harvard, 1945-46, instr., 1946; instr. Mass. Inst. Tech., Cambridge, 1946-47, asst. prof., 1947-50, asso. prof., 1950-53; prof. organic chemistry Cal. Inst. Tech., Pasadena, 1953—, chmn. div. chemistry and chem. engring., 1963—. Cons. E.I. du Pont de Nemours Co., 1949—, Union Carbide and Carbon Corp., Oak Ridge Nat. Lab., 1949-62; mem. adv. panel for chemistry NSF, 1957-60; chmn., 1959-60, adv. com. for math., phys. and engring. scis., 1961—, chmn., 1962—, adv. com. for math. and phys. scis., 1964—, chmn., 1964—; vis. prof. Harvard, 1959-60; Am. Swiss Found. for Sci. Exchange lectr., vis. prof. U. Munich, 1962; dir., editorial cons. W.A. Benjamin, Inc., 1960—. Mem. Am. Chem. Soc. (exec. com. organic div. 1953-57, chmn. organic chemistry div. 1956-57, award in pure chemistry 1954, Harrison Howe award, 1957), Nat. Acad. Scis., Am. Acad. Arts and Scis., Sigma Xi, Alpha Chi Sigma, Phi Lambda Upsilon. Author: Nuclear Magnetic Resonance, 1959, Russian edit., 1961, Japanese edit., 1962; Spin-Spin Splitting, 1961, Russian edit., 1963, Polish edit., 1964; Molecular Orbital Calculations, Russian edit., 1963; (with Marjorie C. Caserio) Basic Principles of Organic Chemistry, 1963. Editor; Organic Syntheses, Vol. 41, 1961. Editorial bd. Organic Syntheses, 1956-63. Extensive research, publs. in structure and uses of the Grignard reagent; original discoveries and syntheses of certain cyclic compounds; studies and application of nuclear magnetic resonance (NMR) spectroscopy. Ofifce: Gates and Crellin Labs., Cal. Inst. Tech., Pasadena, Cal. 91109.*

ROBERTS, John Edwin, Am. chemist; b. Laconia, N.H., Mar. 6, 1920; s. Edwin Jay and Grace (Moore) R.; B.S., U. N.H., 1942, M.S., 1944; Ph.D., Cornell U., 1947; m. Martha Holt, Apr. 29, 1944; children—Christine Lee, Jay Timothy. Faculty, U. Mass., Amherst, 1946—, prof. chemistry, 1963—; Postdoctorate fellow U. Wash., Seattle, 1958-59; vis. prof. chemistry Cairo (Egypt) U., 1961-62. Mem. Am. Chem. Soc., New Eng. Assn. Chemistry Tchrs., Sigma Xi. Contbr. articles to tech. jours. Research on chemistry of rare earth metals and flourine. Home: 18 Red Gate Lane, Amherst, Mass. 01002.*

ROBERTS, John Eric, English physicist; b. Leeds, Eng., Sept. 5, 1907; s. James John and Minnie (Land) R.; B.Sc., U. Leeds, 1928, Ph.D., 1930, D.Sc., 1944; m. Sarah Raybould, Nov. 12, 1932; children—Gwendoline Nora (Mrs. Richard L. Moore), Susan Margaret (Mrs. Peter R. Woodthorpe). Research asst. U. Leeds, 1930-32; asst. physicist Royal Cancer Hosp., London, Eng., 1932-36; faculty Middlesex Hosp. Med. Sch., U. London, 1937—, Joel prof. physics applied to medicine, 1946—. Cons. adviser physics Ministry Health, 1960—. Mem. Hosp. Physicists' Assn. (pres. 1950-51), Inst. Physics and Phys. Soc., Brit. Inst. Radiology (pres. 1951-52). Author: (with Bishop of Chichester) Nuclear War and Peace, 1950; also articles. Editor, Brit. Jour. Radiology, 1964-67, Physics in Medicine and Biology, 1956-61. Research on radiation dosimetry and other aspects of radiol. physics. Home: 72 Lampton Rd., Hounslow, Middlesex, Eng. Office: Middlesex Hosp., Med. Sch., London, W.1., Eng.*

ROBERTS, John Henderson, Am. mathematician; b. Raywood, Tex., Sept. 2, 1906; s. John and Alice (Percival) R.; A.B., U. Tex., 1927, Ph.D., 1929; m. Doretta von Boeckmann, Aug. 29, 1928; 1 son, John Edward. NRC fellow U. Pa., 1929-30; adj. prof. U. Tex., 1930-31; faculty Duke, Durham, N.C., 1931—, prof. math., 1947—; mng. editor Duke Math. Jour., 1950-61; vis. lectr. Princeton, 1937-38. Mem. Am. Math. Soc. (past asso. sec.), Math. Assn. Am. (vis. lectr. 1964—), A.A.A.S. Contbr. articles to tech. jours. Research in fields dimension theory, continuous transformations, integral equations. Home: 2813 Legion Av., Durham, N.C. 27709.*

ROBERTS, John Milton, Am. anthropologist; b. Omaha, Dec. 8, 1916; s. John Milton and Ruth (Kohler) R.; A.B., U. Neb., 1937; postgrad. U. Chgo., 1937-39; postgrad. Yale, 1939-41, Ph.D., 1947; m. Marie Louise Kotouc, May 22, 1941 (dec.); children—Tania Marie Roberts, Andrea Louise, James Barton, John Milton; m. 2d, Joan Marilyn Skutt, Oct. 22, 1961. Asst. prof. U. Minn., 1947-48, Harvard, 1948-53; faculty U. Neb., 1953-58, prof., 1955-58; prof. dept. anthropology Cornell U., Ithaca, N.Y., 1958—. Fellow Center for Advanced Study in Behavioral Scis., Palo Alto, Cal., 1956-57. Mem. Am. Ethnol. Soc. (past pres.), Am. Anthropol. Assn. Contbr. articles to tech. jours., monographs. Research on nature small group cultures, study values, ethnography Zuni and Navaho, understanding expressive models in culture. Home: 205 Ridgedale Rd., Ithaca, N.Y. 14850.*

ROBERTS, Louis Douglas, Am. physicist, govt. ofcl.; b. Charleston, S.C., Jan. 27, 1918; s. Louis Wigfal and Evelyn (Douglas) R.; A.B. with honors, Howard Coll., 1938; postgrad. Johns Hopkins, 1938-39; Ph.D. in Phys. Chemistry, Columbia, 1941; postgrad. (NRC fellow), Cornell U., 1941-42; m. Marjorie Lawson, Aug. 31, 1942; 1 dau., Joyce Carol. Physicist research lab. Gen. Electric Co., Schenectady, 1942-46; with Oak Ridge Nat. Lab., 1946-68, sr. physicist; prof. physics U. Tenn., 1963-68; prof. physics U. N.C., Chapel Hill, 1968—. Fulbright and Guggenheim fellow Oxford (Eng.) U., 1958-59. Fellow Am. Phys. Soc.; mem. Am. Phys. Soc. (chmn. Southeastern sect. 1963-64). Research and publs. on phys. chemistry, ultrasonics, electron optics, neutron physics, nuclear fission and low energy nuclear physics, magnetron devel., gas discharge studies, semiconductors, cryogenics, hyperfine structure coupling, elec. resistance of metals and alloys. Home: Towne House Apts., Chapel Hill, N.C. 27514.*

ROBERTS, Paul Harry, Brit. physicist; b. Aberystwyth, Wales, Sept. 13, 1929; s. Percy Harry and Ethel (Mann) R.; student U. Coll. Wales, 1946-48; B.A., Gonville and Caius Coll., Cambridge, Eng., 1951, M.A., Ph.D., 1954, Sc.D., 1966; m. Joyce Atkinson, Aug. 24, 1959. Research asso. astronomy dept. U. Chgo., 1954-55, asso. prof. astronomy, 1961-63; sci. officer Atomic Weapons Research Establishment, Aldermaston, Eng., 1955-56; Imperial Chem. Industries fellow physics dept. U. Newcastle-Upon-Tyne, 1956-59, lectr. physics, 1960-61, prof. math., 1963—; research asso. Inst. Math. Sci. N.Y. U., N.Y.C., 1960; Fellow Royal Astron. Soc., Inst. Math. and Its Applications (pres. N.E. br. 1965). Author: An Introduction to Magnetohydrodynamics, 1967. Research, numerous publs. on hydrodynamics, particularly problems involving stability, non linear effects. Home: 12A Woodlands Pk., Newcastle-Upon-Tyne 3, U.K.*

ROBERTS, Richard Brooke, Am. biophysicist; b. Titusville, Pa., Dec. 7, 1910; s. Erastus Titus and Helen (Chambers) R.; A.B., Princeton, 1932, A.M., 1933, Ph.D., 1937; m. Josephine Taggart, Jan. 7, 1967; children (from previous marriage)—Richard Furness, Helen Juliette, Edward Thomas. Nuclear physicist Carnegie Instn. Washington, 1937-41, biophysicist, 1947—; weapon devel. applied physics lab. Johns Hopkins, 1942-46. Recipient Medal of Merit, Pres. Truman, 1947. Mem. Biophys. Soc. (past pres.), Nat. Acad. Sci., A.A.A.S. (chmn. zoology sect., v.p.). Research, publs. on reactions of light elements, scattering, uranium fission; first observations of radioactivity of Be7; emission of neutrons from uranium fission; proximity fuse, fire control, guided missiles; biosynthesis of small molecules and ribosomes. Home: 4430 Linnean Av., Washington 20008. Office: 5241 Broad Branch Rd., Washington 20015.*

ROBERTS, Richard William, Am. chemist; b. Buffalo, Jan. 12, 1935; s. William B. and Eugenia (Pratt) R.; B.S., U. Rochester, 1956; Ph.D., Brown U., 1959; m. Carol Jean Elmer, Aug. 18, 1956; children—Beth Carol, William Charles. Nat. Acad. Scis. Postdoctoral fellow Nat. Bur. Standards, Washington, 1959-60; chemist Gen. Electric Research Lab., Schenectady, 1960-65, mgr. structures and reactions br., 1965-67. mgr. phys. chemistry lab., 1968—. Mem. Am. Chem. Soc., Am. Phys. Soc., Am. Vacuum Soc., N.Y. Acad. Scis., Phi Beta Kappa. Author: Ultrahigh Vacuum and its Application, 1963; also articles. Research on phys. and chem. properties of atomically clean metal and semiconductor surfaces, ultrahigh vacuum tech., boundary lubrication of space age metals; inventor iodine lubricants for space age metals. Home: 1917 Townsend Rd., Schenectady 12309. Office: P.O. Box 8, Schenectady 12301.*

ROBERTS, Royston Murphy, Am. chemist; b. Sherman, Tex., June 11, 1918; s. Charles Stanly and Leska (Murphy) R.; B.A., Austin Coll., 1940, D.Sc., 1965; M.A., U. Ill., 1941, Ph.D., 1944; postgrad. U. Cal. at Los Angeles, 1946-47; m. Phyllis Arlene Benson, Sept. 18, 1943; children—Richard Royston, David Merrill, Charles Stanly, III, Jean Ellen. Fellow, U. Ill., 1944-45; research chemist Merck & Co., Inc., 1945-46; fellow U. Cal. at Los Angeles, 1946-47, faculty, 1947-61, asso. prof., 1951-61; prof. chemistry U. Tex., Austin, 1961—. Cons., Continental Oil Co., Ponca City, Okla., 1963—, Tracor, Inc., Austin, Tex., 1957—. Am. Chem. Soc.-Petroleum Research Fund Internat. fellow U. Zurich (Switzerland), 1959-60. Mem. Am. Chem. Soc., Chem. Soc. (London, Eng.), Sigma Xi, Alpha Chi Sigma, Phi Lambda Upsilon. Contbr. numerous articles to tech. jours. Research in organic chemistry, synthesis of antimalarial drugs, synthesis of aromatic hydrocarbons, mechanisms of rearrangements of complex organic molecules induced thermally or by acidic catalysts. Home: 841 E. 38th St., Austin, Tex. 78705.*

ROBERTS, Shepard, Am. elec. engr.; b. New Rochelle, N.Y., Apr. 12, 1915; s. John Willard and Edith (Shepard) R.; student Antioch Coll., 1932-35; Sc.B., Mass. Inst. Tech., 1938, Sc.M., 1939, Sc.D., 1946; m. Muriel Beatty, June 28, 1941; children—Stephen, James. Research asso. Radiation Lab., Mass. Inst. Tech., 1940-45, Lab. for Insulation Research, 1945-46; physicist Research Lab. and Research and Devel. Center, Gen. Electric Co., Schenectady, 1946—. Fellow Am. Phys. Soc.; mem. I.E.E.E., A.A.A.S. Co-editor Digest Lit. on Dielectrics, 1951, editor, 1952. Research, publs. on dielectrics, electroluminescence, optical properties of metals, far infrared spectroscopy; discovered piezoelectric effect in ceramic barium titanate. Home: 821 Sanders Av., Scotia, N.Y. 12302. Office: Gen. Electric Research and Devel. Center, Box 8, Schenectady 12301.*

ROBERTS, Walter Orr, Am. solar astronomer, geophysicist; b. West Bridgewater, Mass., Aug. 20, 1915; s. Ernest Marion and Alice Elliott (Orr) R.; A.B., Amherst Coll., 1938, D.Sc., 1959; M.A., Harvard, 1940, Ph.D., 1943; D.Sc., Ripon Coll., 1958, Colo. Coll., 1962, C.W. Post Coll., L.I. U., 1964, Carleton Coll., 1966, Southwestern at Memphis, 1968, U. Colo., 1968; m. Janet Naomi Smock, June 8, 1940; children—David Stuart, Alan Arthur, Jennifer, Jonathan, Orr. Observer-in-charge Fremont Pass Sta., Harvard Coll. Obs., 1940-46; dir. High Altitude Obs., U. Colo., Boulder, 1946-60, prof. astro-geophysics, 1957—; dir. Nat. Center for Atmospheric Research, Boulder, 1960-68; pres. Univ. Corp. for Atmospheric Research, Boulder, 1967—; Inter-Union Commn. on Solar-Terrestrial Relationships, 1964-66, U. S.-Japan Com. on Sci. Coop., 1968—. Fellow A.A.A.S. (dir., v.p., chmn. sect. on astronomy 1964-68, pres. 1968), Am. Acad. Arts and Scis., Royal Astron. Soc., Am. Geophys. Union; mem. Nat. Acad. Scis. (geophysics research bd. 1960—, Pacific Sci. bd. 1964—), Aspen Soc. Fellows, Internat. Acad. Astronautics, Am. Inst. Aeros. and Astronautics, Am. Meteorol. Soc. (council 1960-63), Internat. Astron. Union (del. congresses Zurich 1948, Dublin 1955, Moscow 1958, Berkeley 1961), Am. Astron. Soc. (council 1960-63), Century Assn., Phi Beta Kappa (nom. com. 1961—, vis. scholar 1963-64), Sigma Xi (nat. lectr. 1951-53). Contbr. articles to profl. jours., symposia, texts. Editorial adv. bd. Jour. Planetary and Space Sci., 1958—; asso. editor Jour. Geophys. Research, 1960-64. Research on solar corona, solar prominences, origin geomagnetic disturbances, influences of variable solar activity on terrestrial atmosphere; assembled and operated 1st solar coronagraph in Western hemisphere; 1st detected and measured solar spicules. Home: 1829 Bluebell Av., Boulder 80302. Office: Nat. Center for Atmospheric Research, Boulder, Colo. 80302.*

ROBERTS, William, English physician; b. Bodedern, Eng., Mar. 18, 1830; s. David and Sarah (Foulkes) R.; B.A., U. Coll., London, 1851, M.B., 1853, M.D., 1854; student Paris, Berlin; m. Elizabeth Johnson, 1869; 1 son, 1 dau. Physician, Manchester Royal Infirmary, 1855-83; lectr. anatomy and physiology Royal Sch. Medicine, Manchester, Eng.; became lectr. on pathology Owens Coll., 1859, lectr. principles and practice medicine, 1863; 1st joint prof. medicine Victoria U., 1873-76. Represented London U. on Gen. Med. Council, 1896-99. Fellow Royal Soc., 1877, Royal Coll. Physicians; mem. Royal Coll. Surgeons. Author: An Essay on Wasting Palsy, 1858; On Peculiar Appearances exhibited by Blood-corpuscles under the Influence of Solutions of Magenta and Tannin, 1863; A Practical Treatise on Urinary and Renal Diseases, 1865; On the Digestive Ferments and the Preparation and Use of Artificially Digested Food, 1880; On the Chemistry and Therapeutics of Gout and Uric-Acid Gravel; Collected Contributions on Digestion and Dietetics. Research on renal diseases, biogenesis, sterilization by heat; 1st to feed invalids with predigested food; supported germ theory of disease. Died Apr. 16, 1899.

ROBERTS-AUSTIN, Sir William Chandler, English metallurgist; b. London, Mar. 3, 1843; s. George and Maria Louisa (Chandler) Roberts; ed. Royal Sch. Mines; chemist, assayer Royal Mint, from 1870; prof. metallurgy Royal Sch. Mines, from 1880; Fellow Royal Soc., 1875; pres. Iron and Steel Inst.; hon. gen. sec. Brit. Assn. for Advancement of Sci.; mem. Chem. Soc., Phys. Soc., Soc. Arts. Author: An Introduction to the Study of Metallurgy. Designed automatic recording pyrometer with thermocouple for high temperature work; demonstrated that diffusion can occur between sheet of gold and block of lead. Died London, Nov. 22, 1902.

ROBERTSON, Douglas Moray Cooper Lamb Argyll, Scottish ophthalmic surgeon; b. Edinburgh, Scotland, 1837; s. John Argyll Robertson; ed. Edinburgh Instn., Neuwied, Germany, U. Edinburgh; M.D., St. Andrews (Scotland) U., 1857; LL.D., U. Edinburgh, 1896; m. Carey Fraser, 1882. House surgeon Royal Infirmary, Edinburgh, 1857, ophthalmic surgeon, 1867-97, later cons.; asst. to prof. John Hughes Bennett, U. Edinburgh, conducted 1st course practical physiology. Pres. Internat. Ophthal. Congress, Edinburgh, 1894; surgeon oculist in Scotland to Queen Victoria, King Edward VII. Fellow Royal Coll. Surgeons Edinburgh (pres. 1886-87); mem. Ophthal. Soc. Gt. Britain (pres. 1893-95), Edinburgh Medico-Chirurg. Soc. (pres. 1896). Proved that physostigmine (eserin) led to constriction of pupil of eye (therefore it acted as a miotic); showed disease of spinal cord is sometimes connected with loss of light reflex of pupil (called Argyll Robertson pupil); introduced several new procedural methods such as trephining sclerotic for relief of glaucoma. Died Gondal, India, Jan. 3, 1909.

ROBERTSON, Etienne-Gaspard (Robert), physicist; b. Liège, Belgium, 1763; prof. physics, Liège; later moved to Paris. Author: Mémoires récréatifs et anecdotiques, 1830-34. Made numerous balloon ascents including one in Hamburg, Germany, 1803 which was 1st ascension to nearly 8,000 n meters; invented phantascipe (projection XXXX lantern); sometimes credited with invention of parachute (rather than Garnerin). Died Paris, 1837.

ROBERTSON, Eugene Corley, Am. geophysicist; b. Tucumcari, N.M., Apr. 9, 1915; s. Ernest Neilson and Maple (Neafus) R.; B.S., U. Ill., 1936; M.A., Harvard, 1948, Ph.D., 1952; m. Olivia Cooney, May 22, 1950; children—Andrew Whitmore, James Gregory. Mining geologist Anaconda Copper Co., Butte, Mont., 1936-42; research fellow Harvard, Cambridge, Mass., 1951-55; geophysicist U.S. Geol. Survey, Silver Spring, Md., 1955—. Mem. Geol. Soc. Am., Am. Geophys. Union, Am. Inst. Mining Engrs., Am. Mineral. Soc., Royal Astron. Soc. Contbr. articles to tech. jours. Research on mech. fracture and plastic flow rocks in lab., mineral composition of interior earth; measurement of phys. properties of minerals and rocks, amount of heat flowing from interior of earth. Home: 1905 Hanover St. Office: 8001 Newell St., Silver Spring, Md. 20910.*

ROBERTSON, James David, Am. physician; b. Tuscaloosa, Ala., Oct. 13, 1922; s. Floyd Earl and Gladys (Williams) R.; B.S., U. Ala., 1942; M.D., Harvard, 1945; Ph.D. (Am. Cancer Soc. fellow), Mass. Inst. Tech., 1952; m. Doris Elizabeth Kohler, Oct. 21, 1946; children—Karen Lee, Elizabeth Ann, James David. Practice internal medicine, Lowndes County, Ala., 1948; faculty U. Kan. Med. Sch., Kansas City, 1952-55; hon. research asso. dept. anatomy U. Coll. London (Eng.), 1955-60; asso. biophysicist Research Lab., McLean Hosp., Belmont, Mass., 1960-63, biophysicist, 1964—; faculty Harvard Med. Sch., 1960-64, asso. prof. neuropathology dept. neurology and psychiatry, 1964-66; prof., chmn. dept. anatomy Sch. Medicine, Duke, Durham, N.C., 1966—. Sci. cons. NSF, 1963—. Recipient W.B. Saunder's award in pathology and bacteriology U. Ala. Sch. Medicine, 1943. Mem. Phi Beta Kappa, Sigma Xi. Editorial bd. Jour. Cell Biology, 1963—, Jour. Lipid Research, 1963-65. Contbr. numerous articles to tech. jours. Research on cell membranes, synaptic junction of nerve tissue and neuro-muscular junctions with electron microscope, nerve myelin and retinal rod receptors by electron microscopy and X-ray diffraction; originated unit membrane concept of molecular orgn. of cell membranes. Home: 32 Oak Dr., Durham, N.C.*

ROBERTSON, James Ian Summers, Brit. physician; b. Welbeck, Eng., Mar. 5, 1928; s. James Charles and Sybil (Summers) R.; B.Sc. with 1st Class Honors, St. Mary's Hosp. Med. Sch., London, 1949, M.B., B.S. with Honors, Distinction in Pathology and Medicine, 1952; m. Maureen Patricia Doherty, Sept. 10, 1955; children—Fiona Elizabeth, James Andrew. House physician St. Mary's Hosp., London, 1952-53, lectr. medicine med. unit, 1956-63, sr. lectr. therapeutics, 1963-66, hon. cons., 1963-67; house physician Brompton Hosp., London, 1953-54; research fellow Leverhulme Nephrological unit St. Mary's Hosp., 1966-67; sci. staff blood pressure research unit Med. Research Council; cons. physician Western Infirmary, Glasgow (Scotland), 1967—. Fellow Zool. Soc. London; mem. Royal Coll. Physicians, Med. Research Soc., Physiol. Soc., Internat. Nephrological Soc., European Soc. for Clin. Investigation Contbg. author: Recent Advances in Medicine, 1967; Recent Advances in Endocrinology, 1967; Research and publs. in pathogenesis and treatment of carcinoid disease, renin-angiotensin-aldosterone system in regulation of sodium balance in normal subject and various diseases, pathogenesis and treatment of high blood pressure; devel. techniques for measurement of hormones serotonin and renin in blood. Home: Elmbank, Manse Rd., Bowling, Dunbartonshire. Office: Med. Research Council Blood Pressure Research Unit, Western Infirmary, Glasgow W.1, Scotland.*

ROBERTSON, James Sydnor, Am. med. physicist; b. Richmond, Va., Nov. 27, 1920; s. Paul Augustus and Beth (Whitacre) R.; B.Sc., U. Minn., 1943, M.B., 1944, M.D., 1945; Ph.D. in Physiology, U. Cal. at Berkeley, 1949, postgrad., 1949-50; m. Ruth Elizabeth Henrici, Jan. 15, 1944; children—Kathleen Mary, John Paul, Marion Adelle. Practice gen. medicine, Eureka, Mont., 1946; staff med. dept. Brookhaven Nat. Lab., Upton, N.Y., 1950—, sr. scientist, sr. physician, 1955—. Cons. in nuclear medicine, M.D. Anderson Hosp., Houston, 1960—; lectr. radiobiology U.S. Naval Sch. Submarine Medicine, Groton, Conn., 1958—. Mem. A.A.A.S., Am. Physiol. Soc., Radiation Research Soc., Soc. for Nuclear Medicine, Am. Nuclear Soc., N.Y. Scis., Sigma Xi. Contbr. numerous articles. Research in math. basis for analysis tracer kinetics, in membrane transport, exptl. therapy with neutrons in treatment of brain tumor. Home: 28 Gnarled Hollow Rd., East Setauket, N.Y. 11733. Office: Med. Physics Div., Med. Dept., Brookhaven Nat. Lab., Upton, N.Y. 11973.*

ROBERTSON, John Gow, physician; b. Calcutta, India, Oct. 19, 1927; s. John and Winifred (Gow) R.; M.B., Ch.B., U. Edinburgh (Scotland), 1950, M.D., 1963; m. Winifred Barlow, Jan. 2, 1961; children—Iain, Neil, Gillian Elizabeth. Sr. registrar in obstetrics and gynecology Royal Infirmary, Edinburgh, 1960-66, cons. obstetrician and obstetrician, 1966—; lectr., clin. tutor obstetrics and gynecology U. Edinburgh, 1958—. Lectr., examiner Central Midwives Bd., Scotland, 1961—. Wellcome Trust Research fellow Columbia-Presbyn. Med. Center, N.Y.C., 1964-65. Fellow Royal Coll. Surgeons Edinburgh, Edinburgh Obstet. Soc.; mem. Royal Coll. Obstetricians and Gynaecologists, Simpson Obstet. Club (gen. sec. 1966—), Blair Bell Research Soc., Brit. Med. Assn. Research and publs. on prevention, diagnosis and assessment of severity of Rh isoimmunization and hemolytic disease of newborn, problems of twin pregnancy, perinatal mortality, effect of significant asymptomatic bacteriuria in pregnant patient. Home: 15 Ravelston House Rd., Office: 39 Chalmers St., Edinburgh, Midlothian, Scotland.*

ROBERTSON, John Monteath, Scottish chemist; b. Perthshire, Scotland, July 24, 1900; s. William and Jane (Monteath) R.; B.Sc., Glasgow U., 1923, M.A., 1925, Ph.D., 1926, D.Sc., 1933; LL.D., Aberdeen (Scotland) U., 1963; m. Stella Kennard Nairn, June 2, 1930; children—Kennard, Keith, Patricia. Harkness fellow, 1928-30; staff Davy Faraday Lab., Royal Instn., London, Eng., 1930-39; sr. lectr. phys. chemistry U. Sheffield (Eng.), 1939-41; sci. advisers Hdqrs. Bomber Command, RAF, 1941-42; Gardiner prof. chemistry, dir. chem. labs. U. Glasgow, 1942—. Named Comdr. Brit. Empire; recipient Longstaff medal, 1966. Fellow Royal Soc., 1945 (Davy medal 1960), Royal Soc. Edinburgh, Royal Inst. Chemistry, Inst. Physics; mem. Chem. Soc. (past pres.). Author: Organic Crystals and Molecules, 1953; also numerous articles. Research on structure determinations of organic molecules by x-ray crystallography; discovery, devel. phase determining methods using heavy atoms, isomorphous replacement. Home: 42 Bailie Dr., Bearsden, Glasgow, Scotland.*

ROBERTSON, Malcolm Slingsby, mathematician; b. Brantford, Ont., Can., July 18, 1906; s. Malcolm and Bertha (Slingsby) R.; B.A., U. Toronto (Ont., Can.), 1929, M.A., 1930; Ph.D., Princeton, 1934; m. Margaret Grant, Sept. 1, 1934; children—John Malcolm, Alan Grant. Came to U.S., 1931, naturalized, 1941. NRC fellow U. Chgo., 1934-35; instr. Yale, 1935-37; faculty Rutgers U., New Brunswick, 1937-66, prof. math., 1950-66, dir. grad. studies in math., 1959-66; Unidel prof. math. U. Del., Newark, 1966-——. Recipient Outstanding Research award adv. bd. for research and grad. edn. Rutgers U., 1964. Mem. Am. Math. Soc., Am. Math. Assn., Sigma Xi, Phi Beta Kappa. Contbr. articles, numerous revs. to tech. jours. Research on complex analyses; theory of functions of complex variable; univalent functions; multivalent and typically real functions; conformal mapping. Home: 104 Hullihen Ct., Newark, Del. 19711.

ROBERTSON, Nat Clifton, Am. phys. chemist; b. Atlanta, July 23, 1919; s. Henry Booker and Eura (Williams) R.; A.B., Emory U., 1939; Ph.D., Princeton, 1942; m. Elizabeth Bates Peek, Nov. 29, 1946; children—Henry Bartlett, Mary Amanda, Paul Edward. Staff, OSRD, Princeton, 1942-43; staff Standard Oil Devel. Co., 1943-47; group leader Celanese Corp., 1947-51; dir. petro. chems. dept. Nat. Research Corp., 1951-55; dir. research, v.p. Escambia Chemi. Corp., Cambridge, Mass., 1955-58; v.p. research and devel. Spencer Chem. div. Gulf Oil Corp. (formerly Spencer Chem. Co.), Kansas City, Mo., 1958-——. Pres. Research Found. Kan., 1963-64. mem. Gov's Sci. Adv. Com., Mo., 1961-63. Trustee Midwest Research Inst., Kansas City; bd. dirs. Marion Labs., Kansas City. Mem. Phi Beta Kappa, Sigma Xi. Patentee, catalysis, hydrocarbon oxidation and phys. methods separation. Home: 6821 Tomahawk Rd., Mission Hills, Kan. 66208. Office: Dwight Bldg., Kansas City, Mo. 64105.*

ROBERTSON, Oswald Hope, physician; b. Woolwich, Eng., June 2, 1886; s. Theodore and Kathleen (Conlan) R.; came to U.S., 1888, naturalized, 1920; B.S., U. Cal., 1910, M.S., 1911; M.D., Harvard Med. Sch., 1913; m. Ruth Allen, Nov. 30, 1916; children—Alan Morley, Donald Irwin, Robert Conlan. Fellow in pathology Mass. Gen. Hosp., Boston, 1914-15; asst. bacteriologist, pathologist Rockfeller Inst., N.Y.C., 1915-17; asso. prof. medicine Peking Union Med. Coll., 1919-23, prof., head dept., 1923-27; prof. U. Chgo., 1927-51, prof. emeritus, 1951-——; lectr. biology Stanford, 1950-——. Mem. U. S. NRC, 1937-40; dir. commn. on air-borne infections U. S. Army, 1941-45. Mem. Nat. Acad. Scis., Assn. Am. Physicians (pres., 1952), Am. Soc. Clin. Investigation. Publs. on establishment of 1st blood bank in World War I, studies on exptl. pneumonia in dog, disinfection of air, similarity of a condition in salmon to Cushing's disease. Home: 9150 Los Gatos Hwy., Santa Cruz, Cal. 95062.*

ROBERTSON, Rutherford Ness, Australian plant physiologist, biochemist; b. Melbourne, Australia, Sept. 29, 1913; s. Joshua and Josephine (Hogan) R.; B.Sc., U. Sydney, 1934, D.Sc., 1961; Ph.D., St. Johns Coll., Cambridge (Eng.) U., 1939; D.Sc., U. Tasmania, 1965; m. Mary Helen Bruce Rogerson, Sept. 9, 1937; 1 son, Robert James. Linnean Macleay fellow Sydney U., 1934-36, faculty, 1939-46, lectr., 1942-46; scholar Exhbn. 1851, Cambridge U., 1936-39; staff div. food preservation Commonwealth Sci. and Indsl. Research Orgn., 1946-59, chief research officer, 1957-59, mem. exec., 1959-62; prof. botany U. Adelaide (Australia), 1962-——; vis. prof. U. Cal., 1958-59. Chmn., Australian Research Grants Com., 1965-——. Recipient Clarke Meml. medal Royal Soc. New South Wales, 1954. Fellow Australian Acad. Sci., Royal Soc., 1961; mem. Nat. Acad. Scis. (fgn. asso.), Australian and New Zealand Assn. for Advancement Sci. (past pres., Farrer Meml. medal 1963, medal 1968), Australian Soc. Plant Physiologists, Am. Hort. Soc. Author: (with G. E. Briggs, A. B. Hope) Electrolytes and Plant Cells, 1961; also articles. Research on plant ion absorption, plant respiration and behaviour of mitochondria; physiology and biochemistry of fruit devel. and ripening. Home: 2 Martin Av., Prospect, South Aus. Australia. Office: Dept. Botany, U. Adelaide, Australia.*

ROBERTSON, William Donald, metallurgist; b. Montreal, Que., Can., Dec. 23, 1913; s. James McCallum and Grace (Brecken) R.; B.Sc., Mass. Inst. Tech., 1942, D.Sc., 1948; m. Liza Tencer, Apr. 18, 1938; children—James Duncan, Eric Clifton. Came to U.S., 1948, naturalized, 1954. Research metallurgist Aluminum Labs. Ltd., Kingston, Can., 1942-45; research asso. Inst. for Study Metals, U. Chgo., 1948-50; asst. prof. Yale, 1950-53, asso. prof., 1953-57, prof. 1957-——, chmn. dept. metallurgy, 1958-63, prof. applied sci. 1963-——. Chmn., Gordon Research Conf. on Corrosion, 1957, on Metallurgy, 1962; cons. Brookhaven Nat. Lab., Upton, L.I., 1962-——, Chase Brass and Copper Co., Waterbury, Conn., 1953. Recipient Willis Rodney Whitney award for contributions to sci. of corrosion Nat. Assn. Corrosion Engr., 1964, 65; Overseas fellow Churchill Coll., Cambridge, Eng. 1964-65; Fulbright Sr. Research scholar U. Cambridge, 1964-65. Research and publs. on electrical properties of metals and metallic compounds, corrosion of metals and alloys, plasticity and phase transformations of metal crystals; surface physics, analysis of surface structure by low energy electron diffraction; identified 1st intermetallic-compound semi-conductor. Home: 164 Linden St., New Haven 06511.*

ROBERTSON, William van Bogaert, Am. biochemist; b. N.Y.C., Sept. 15, 1914; s. William Merry and Wilhelmina (VanBogaert) R.; M.E., Stevens Inst. Tech., 1934; Ph.D., Albert Ludwig U., Freiburg, Germany, 1937; m. Alice Freeman Slemmons, May 21, 1941; children—William Van Bogaert, Wilhelmina Louise. Research chemist Mass. Gen. Hosp., Boston, 1938-41; research fellow Nat. Cancer Inst., Bethesda, Md., 1941-44; research asso. U. Chgo., 1944-45; faculty U. Vt., Burlington, 1945-60, prof. biochemistry, 1952-60; faculty Stanford, 1961-——, prof. biochemistry-pediatrics, 1963-——; dir. research and edn. Stanford Childrens Convalescent Hosp., Palo Alto, Cal., 1962-——. Mem. Am. Soc. Biol. Chemistry, A.A.A.S., Am. Chem. Soc., Soc. Gen. Physiologist, Sigma Xi. Contbr. numerous articles to tech. jours. Research on metabolism of connective tissue, intracellular cardiac electrolyte concentrations.*

ROBERTUS CATANEUS, see Robert of Chester.

ROBERTUS EBORACENSIS, see Robert of York.

ROBERTUS PERSCRUTATOR, see Robert of York.

ROBERTUS RETINENSIS, see Robert of Chester.

ROBERVAL, Gilles Personne (Personier) de; French mathematician; b. Roberval, nr. Beauvais, France, Aug. 8, 1602; Prof. philosophy Collège de Maistre Gervais, Paris, France, 1631; prof. math. Collège de France, from 1632. Original mem. French Acad. Scis., 1666. Author: Traité des indivisibles, 1634; Aristarchi Samii de mundi systemate, 1644. Developed method of indivisibles and applied it to method of drawing tangents and for determining area of cycloid and volume generated by it; invented Roberval's balance; proposed universal attraction between smallest particles of matter. Died Paris, Oct. 27, 1675.

ROBEV, Stefan Kirov, Bulgarian chemist; b. Sofia, Bulgaria, Jan. 9, 1931; s. Kiro Vassilev and Stefana (Vodeva) R.; Specialist degree of Organic Chemistry, Sofia U., 1955; m. Tatjana Sumerska, Sept. 6, 1956; children—Anna, Raina. Chief central plant's lab I. Lilov, Sliven, Bulgarian; research asst. Inst. Radiology, Sofia, 1956-59, sr. scientist, chief dept. radiation bio-chemistry, 1961-——; sr. scientist, chief chemistry dept. Bulgarian Acad. Scis., 1959-61. Mem. Sci. Union Bulgaria. Research, publs. on new applications of sodium amide in organic synthesis; discovered transformation of arylhydrazones into amidines, radioprotective properties of amidine compounds; described radiosensitivity of messenger RNA-synthesis; introduced mRNA/DNA hybridization in radiobiology. Home: 43-W kompl. Lenin. Office: Inst. Radiology, P.O. Box 673, Sofia, Bulgaria.*

ROBIN, Albert-Charles-Edmon, French physician; b. Dijon, France, 1847; asst. chemist to Thénard; Became intern Hosps. of Paris, became physician, 1881, prof. medicine, 1883, prof. clin. therapeutics, 1906. Mem. French Acad. Medicine. Author: Traité des maladies de l'estomac; Thérapeutique usuelle des practiciens, 3 vols. Popularized concept of clinics and therapeutics; research in biochemistry, physiology, pathology, hot mineral springs, gastric pathology; introduced to medicine glycerinephosphoric compounds, (with Bardet) colloidal metals. Died Dijon, France, 1928.

ROBIN, Charles, French anatomist; b. Jasseron, France, June 4, 1821; agrégé, 1847; tchr. anatomy, also organized lab. of comparative anatomy, after graduation; became prof. histology Faculty Medicine, Paris, 1862; elected senator from Ain, 1876. Mem. French Acad. Scis., Acad. Medicine. Author: Anatomie pathologique des cataractes en général, 1866 1856; Les theories des mouvements du coeur, 1864; (with Littré) Dictionnaire de médecine et de chirurgie; Leçons sur les humeurs normales et morbides du corps de l'homme, 1867; Leçons sur les vaisseaux capillaires et l'inflammation, 1867; Anatomie microscopique, 1868; Des tissus et des secretions, 1869; Anatomie et physiologie cellulaires ou des cellules animales et végétales, 1873; numerous other works. Described osteoclasts in regard to bone formation (Robin's myeloplaxes), 1851. Died Jasseron, Oct. 6, 1885.

ROBIN, Eugene Debs, Am. physician, biologist; b. Detroit, Aug. 23, 1919; s. Benedict and Anna (Cooper) R.; A.A., U. Chgo., 1939; B.S., George Washington U., 1946, M.S., 1947, M.D., 1951; m. Evelyn Beatrice Cowen, Aug. 23, 1942; children—Anna, Donald. Mem. faculty Harvard Med. Sch., 1955-59; faculty U. Pitts. 1959-——, prof. medicine, 1963-——. Trustee Mt. Desert Island Biol. Lab., Salisbury, Mem. Am. Fedn. Clin. Research, Am. Thoracic Soc., A.C.P., Am. Physiol. Soc., Am. Soc. Clin. Investigation, Archives Environmental Health (basic sci. editor), Am. Assn. Physicians, Alpha Omega Alpha, Sigma Xi. Research and publs. on treatment of syphilis; pulmonary embolism, fibrosis, ventilation; myocardiopathies; acid-base buffers and ammonia; electrolyte and fluid metabolism. Home: 5411 Northumberland St., Pitts. 15217.*

ROBIN, Gordon de Quetteville, geophysicist; b. Melbourne, Australia, Jan. 17, 1921; s. Reginald James and E. Mabel (Berryman) R.; B.Sc., U. Melbourne, 1940, M.Sc., 1942; Ph.D., U. Birmingham (U.K.), 1956; m. Jean Margaret Fortt, June 10, 1953; children—Caroline Margaret, Elizabeth Clare. Lectr., research fellow U. Birmingham, 1948-56; meteorologist Brit. Sta. Signy Island, South Orneys, Antarctic, 1947-48; physicist, sr. Brit. mem. Norwegian-Swedish Antarctic Expdn., 1949-52; sr. fellow dept. geophysics Australian Nat. U., Canberra, 1957-58; dir. Scott Polar Research Inst., U. Cambridge (Eng.), 1958-——. Back grantee Royal Geog. Soc., London, 1953; recipient Bruce medal Royal Soc. Edinburgh, 1953. Fellow Inst. Physics. Author: Annals of the IGY-Glaciology, 1967; also articles. Co-editor: Antarctic Research, 1964. Research on ice sheets especially measurement of thickness, temperature distbn. in ice sheets, seismic waves in ice, application of radio echo methods to study ice sheets. Home: Lordship House, Swaffham Bulbeck, Cambridge, Eng.*

ROBINEAU-DESVOIDY, Jean-Baptiste, French physician, naturalist; b. St. Sauveur, France, 1799; collected geol. and entomol. specimens which contbd. to progress in those scis. Author: Les crustacés fossiles trouvés dans un terrain, néocomien de Saint-Sauveur; Essai sur la tribu des culicides; L'histoire naturelle des diptères des environs de Paris, 1863. Died 1862.

ROBINET, Jean Baptiste René, French philosopher, grammarian; b. Rennes, France, 1735; sec. to minister Amelot; royal censor, 1780. Author: De la Nature, 1761; Grammaire Anglaise, 1764; Grammaire Francaise, 1768; Recueil Philosophique, 1769; Des Lettres sur les Debats de l'Assemblee Nationale Relatifs a la Constitution en Vers, 1814; (with others) Histoire Universelle, 46 vols. Developed materialistic theories (he disavowed later); described principles of continuity as being new manner of contemplating nature; speculative transformist. Died Paris, France, 1820.

ROBINS, Benjamin, English mathematician, mil. engr.; b. Bath, Eng., 1707; s. John Robins; self-educated. Taught math. and phys. sci.; became engr. in mil. projects; apptd. sec. com. to examine H. Walpole; invited to improve fortifications, by Prince of Orange; repaired and designed forts of East India Co., also designed Fort St. David (both India). Fellow Royal Soc., 1927 (Copley medal 1747). Author: New Principles of Artillery, 1742; Mathematical Track, 1761; also wrote polit. phamplets, papers on rockets and flight, gunnery. Invented ballistic pendulum, 1742; gave logical presentation of fluxions, rejecting all infinitely small quantities, 1735; investigated velocity of projectiles and resistance of air. Died Madras, 1751.

ROBINS, Charles Richard, Am. ichthyologist, biol. oceanographer; b. Harrisburg, Pa., Nov. 25, 1928; s. Claude R. and Helen (Ayres) R.; A.B., Cornell U., 1950, Ph.D., 1954; m. Catherine Hale, Sept. 3, 1965. Faculty, U. Miami (Fla.), 1956-——, prof., 1964-——, curator fishes Marine Lab., 1956-——, chmn. dept. marine sci., 1961-63. Mem. systematic biology panel NSF, 1966-——. Jessup scholar Acad. Nat. Sci., Phila., 1960. Mem. A.A.A.S., Am. Inst. Biol. Scis., Am. Fisheries Soc., Am. Inst. Fishery Research, Am. Ornithologists Union, Am. Soc. Ichthyologists and Herpetologists, Am. Soc. Zoologists, Biol. Soc. Washington, Fla. Acad. Sci., Herpetologists League, Marine Biol. Assn. India, Soc. for Study Evolution, Soc. for Systematic Biology, Wilson Ornithol. Soc., Phi Kappa Phi, Sigma Xi. Editor: Bull. Marine Sci. Gulf and Caribbean, 1961-62; editorial bd. Copeia, 1963-64, Bull. Marine Sci., 1963-——, Tulane Studies in Zoology, 1959-——, Studies in Tropical Oceanography, 1965-——. Research, publs. on marine and freshwater fishes, especially N. of N.Am. and tropical Atlantic Ocean from Africa to C.Am.; systematics, ecology, morphology and behavior; mem. oceanic expdns. to Gulf of Guinea, 1964, 65, numerous expdns. to tropical western Atlantic. Home: 6840 S.W. 76th Terrace, South Miami, Fla. 33143. Office: 10 Rickenbacker Causeway, Miami, Fla. 33149.*

ROBINS, Eli, Am. psychiatrist, neurochemist; b. Houston, Feb. 22, 1921; B.A., Rice U., 1940; M.D., Harvard, 1943; m. Lee Nelken, Feb. 22, 1946; children—Paul, James, Thomas, Nicholas. Asst., Harvard Med. Sch., 1944-45; faculty Washington U. Sch. Medicine, St. Louis, 1951-——, prof. psychiatry, 1958-66, head dept. psychiatry, 1963, Wallace Renard prof. psychiatry, 1966-——; psychiatrist-in-chief Barnes and Renard Hosps., St. Louis, 1963-——. Editor: (with Korey, Pope) Ultrastructure and Metabolism of the Nervous System, 1962; also numerous articles. Research on neuroses, suicide, chemistry of brain; discovered screening test for newborns to detect presence or absence of preventable type of mental deficiency. Home: 1 Forest Ridge, Clayton, Mo. 63105. Office: 4940 Audubon Av., St. Louis 63110.*

ROBINS, Roland Kenith, Am. chemist; b. Scipio, Utah, Dec. 13, 1926; s. Kenith R. and Florence (Cropper) R.; B.A., Brigham Young U., 1948, M.A., 1949; Ph.D., Ore. State U., 1952; m. Lessa Rasmussen, June 1, 1948; children—Renee, Leon, Rhonda, Corinne, Rochelle, Roy Lynn. Research asso. Wellcome Research Labs., N.Y.C., 1952-53; faculty N.M. Highlands U., 1953-57, asso. prof., 1957-60; prof. Ariz.

State U., 1957-64; prof. chemistry U. Utah, Salt Lake City, 1965—. Mem. Am. Chem. Soc. Cons., Nat. Cancer Inst., 1958—, Merck, Sharpe and Dohme, 1957—. Contbr. numerous articles to tech. jours. Research on synthesis nitrogen heterocycles, structural elucidation and synthesis of heterocyclic nucleoside antibiotics, aromaticity in nitrogen heterocyclic systems, synthesis of drugs as antiviral and antitumor agts.; synthesis of organic compounds of biochem. interest. Home: 1141 Alton Way, Salt Lake City, 84108.*

ROBINSON, Abraham, mathematician; b. Waldenburg, Germany, Oct. 6, 1918; s. Abraham and Hedwig (Bahr) R.; student Hebrew U., Jerusalem, Israel, 1936-40, M.Sc., 1946; Ph.D., London (Eng.) U., 1949, D.Sc., 1957; m. Renée Kopel, Jan. 30, 1944. Asst., sci. officer Royal Aircraft Establishment, Farnborough, Eng., 1942-46; sr. lectr. math., dep. head dept. aerodynamic Coll. Aero., Cranfield, Eng., 1946-51; faculty U. Toronto (Ont., Can.), 1951-57, asso. prof., 1951-56, prof. math., 1956-57; prof. Hebrew U., Jerusalem, 1957-62; prof. math. and philosophy U. Cal. at Los Angeles, 1962-67; prof. math. Yale, 1967—; vis. prof. at univs. of Princeton, Paris (France), Rome (Italy), Tübingen (Germany); vis. fellow St. Catherine's Coll., Oxford (Eng.) U., 1965. Mem. Am. Math. Soc., London Math. Soc., Assn. for Symbolic Logic. Author: On the Metamathematics of Algebra, 1951; Theorie métamathematique desidéaux, 1955; (with J.A. Laurmann), Wing Theory, 1956; Complete Theories, 1956; Introduction to Model Theory, 1963; Numbers and Ideals, 1965; Non-Standard Analysis, 1966; also articles. Research on math. logic especially model theory, applications of logic to algebra and analysis, functional analysis, partial differential equations, also aerodynamics, supersonic wing theory. Office: Math. Dept., Yale U., New Haven 06520.*

ROBINSON, Brian John, Australian physicist; b. Melbourne, Australia, Nov. 4, 1930; s. Raymond John and Ellen (Guernin) R.; B.Sc., U. Sydney, 1952, M.Sc., 1953; Ph.D., Cambridge U., 1958; m. Judith Ogilvie White, June 28, 1956. Research officer Commonwealth Sci. and Indsl. Research Orgn., Radiophysics Lab., Sydney, Australia, 1953-54, prin. research scientist, 1962—; research scientist Netherlands Found. for Radio Astronomy, Leiden, 1958-61. Mem. Australian Nat. Com. for Radio Sci., 1966—. Recipient award Instn. Elec. Engrs., Engrs., London, 1962. Fellow Royal Astron. Soc. (London), Australian Inst. Physics; mem. I.E.E.E. (sr.), Astron. Soc. Australia, Internat. Astron. Union. Contbr. articles to profl. jours. Research on radio astronomy, radiation from hydrogen atoms in external galaxies, radiation from molecules in interstellar space, devel. of semiconductor parametric amplifiers for high-sensitivity radio astronomy receivers. Home: 44 Willowie Rd., Castle Cove, New South Wales, Australia. Office: Radiophysics Lab., P.O. Box 76, Epping 2121, Australia.*

ROBINSON, Charles Franklin, Am. physicist; b. Vernon, Tex., June 21, 1915; s. Charles B. and Bervia (McElreath) R.; B.A., S.W. Mo. State Coll., 1935; B.S., Drury Coll., 1936; M.S., Cal. Inst. Tech., 1938, Ph.D. (NRC predoctoral fellow 1945-47), 1949; m. Virginia Coke, Sept. 11, 1943; children—Karen Robinson, Palmer, Control operator Springfield Broadcasting Corp. (Mo.), 1936-37; staff physicist Cedars of Lebanon Hosp., also Clyde Emery Med. Group, 1939-41; tech. staff OSRD Project, Cal. Inst. Tech., 1941-45, resident dir. Goldstone Lake Rocket Weapon Test Center, 1942-44; cons. Naval Ordnance Test Sta., Inyokern, Cal., 1941-45; with Consol. Electrodynamics Corp., 1947-51, successively staff physicist, sr. physicist, chief research physicist, asso. dir. research, 1947-60, v.p. research, 1960-61; with Bell & Howell Co., 1960—, v.p., 1961—. Mem. governing bd. Cal. Inst. Tech. YMCA, 1955-58. Recipient Naval Ordnance Devel. citation, 1945. Mem. Am. Phys. Soc., So. Cal. Optical Soc., A.A.A.S., N.Y. Acad. Sci., Am. Soc. Testing Materials, Am. Ordnance Assn. Research in mass spectroscopy and fine structure; ultra-high frequency gas discharges; holder numerous patents. Home: 3340 Calvert Rd. Office: 360 Sierra Madre Villa, Pasadena, Cal.

ROBINSON, Edwin James, Jr., Am. biologist; b. Wilkes-Barre, Pa., Feb. 7, 1916; s. Edwin James and Edna (Lewis) R.; A.B., Dartmouth, 1939; M.S., N.Y. U., 1941, Ph.D., 1949; m. Marie Josette Bentivoglio, Apr. 11, 1948; children—Nancy Ellen, James Lewis, Loretta Jean, Margaret Ann, Catherine Marie, Edwin Anthony. Instr., Cornell U. Med. Sch., 1948-51; sr. asst. scientist USPHS, 1951-54; faculty Kenyon Coll., Gambier, O., 1954-63, prof., 1961-63; chmn. dept., 1962-63; prof., chmn. biology dept. Macalester Coll., St. Paul, 1963—. Fellow A.A.A.S.; mem. Am. Soc. Parasitologists, Am. Soc. Zoologists, Am. Micros. Soc., Soc. Protozoologists, Am. Inst. Biol. Scis., Sigma Xi. Research, publs. on taxonomy, life histories and ecology of parasites, both digenetic flukes and filarial worms. Home: 5122 Garfield Av. S., Mpls. 55419. Office: Grand and Snelling Avs., St. Paul 55105.*

ROBINSON, Francis Pleasant, Am. psychologist; b. Danville, Ind., Dec. 21, 1906; s. Plezsant S. and Grace (Huron) R.; B.A., U. Ore., 1929; M.A., State U. Ia., 1930, Ph.D., 1932; m. Carolyn G. Bostwick, Aug. 15, 1931; children—Mary Frances (Mrs. Kemp Edward Dietiker), Carol Lynette (Mrs. Raymond F. Hopkins). Mem. faculty State U. Ia., 1932-33, Stout Inst., Menomonie, Wis., 1933-37; mem. faculty Ohio State U., Columbus, 1937—, prof. psychology, 1945—. Mem. Am. Psychol. Assn. (pres. div. counseling psychology 1954-55), Am. Personnel & Guidance Assn., Sigma Xi, Phi Beta Kappa. Author: Effective Study, 1941; Principles & Procedures in Student Counseling, 1950; Effective Reading, 1962; (with S.L. Pressey and J.E. Horrocks) Psychology in Education, 1959. Editor, Jour. Counseling Psychology, 1964—. Designed SQ3R study method (a higher-level study skill); emphasized communications analysis of counseling interactions. Home: 167 Rustic Pl., Columbus, O. 43214.*

ROBINSON, Frederick Byron, Am. surgeon, anatomist; b. Hollandale, Wis., Apr. 11, 1857; s. William and Mary Robinson; B.S., U. Wis., 1878; M.D., Rush Med. Sch., 1882; postgrad., Berlin, Germany, Vienna, Austria, Birmingham, Eng.; m. Lucy Waite, 1894. Schooltchr.; practiced medicine, Grand Rapids, Wis., 1882-89; named prof. anatomy and clin. surgery Toledo Med. Coll., 1889; became prof. gynecology Chgo. Post-grad. Med. Sch., 1891; prof. gynecology and abdominal surgery Ill. Med. Sch., Chgo. Coll. Medicine and Surgery; mem. staff Woman's Hosp., Mary Thompson Hosp. Author: Experimental Intestinal Surgery, 1889; Practical Intestinal Surgery, 1891; Landmarks in Gynecology, 1894; The Peritoneum—Histology and Physiology, 1897; Abdominal and Pelvic Brain, 1906; also numerous articles. Studied abdominal and pelvic anatomy, especially that of peritoneum, seminal vesicles, biliary and pancreatic ducts, abdominal sympathetic nervous system; investigated relation of psoas muscle to appendicitis; described celiac ganglion or ganglia of abdomen (abdominal brain), 1899. Died Mar. 23, 1910.

ROBINSON, George Wilse, Am. phys. chemist; b. Kansas City, Mo., July 27, 1924; s. George Wilse and Elizabeth (Millett) R.; B.S., Ga. Inst. Tech., 1947, M.S., 1949; Ph.D., U. Ia., 1952; m. Ellen Elizabeth Johnson, June 5, 1950. Research asst. phys. chemistry U. Ia., Iowa City, 1950-52; research fellow U. Rochester, 1952-54; asst. prof. chemistry Johns Hopkins, 1954-59; faculty Cal. Inst. Tech., 1959—, prof. phys. chemistry, 1961—. Mem. Am. Chem. Soc., Am. Phys. Soc. Editorial bd. Jour. Molecular Spectroscopy, 1966—, Chem. Physics Letters, 1966—, Ann. Revs. Phys. Chemistry, 1966—, Jour. Chem. Physics, 1963-65. Research and publs. on microwave, vibrational and electronic spectroscopy, low temperature chemistry, trapped-free-radical spectroscopy, energy transfer phenomena and photosynthesis. Home: 8000 N. Grand St., Glendora, Cal. 91740. Office: 1201 E. California Blvd., Pasadena, Cal. 91109.*

ROBINSON, Gilbert de Beauregard, Canadian mathematician; b. Toronto, Ont., Can., June 3, 1906; s. Percy J. and Esther (de Beauregard) R.; B.A., U. Toronto, 1927; Ph.D., U. Cambridge (Eng.), 1931; m. Joan Howard, Sept. 1, 1936; children—John, Nancy. Faculty, U. Toronto, 1931—, now prof. math.; with NRC Can., 1941-45; vis. prof. Mich. State U., 1953, U. B.C., 1963; v.p. research U. Toronto, 1965—. Decorated Mem. Brit. Empire. Fellow Royal Soc. Can.; mem Canadian Math. Congress (past pres.), Am. Math. Soc. (mem. council), London Math. Soc., Australian Math. Soc. Author: Foundations of Geometry, 1940; Representation Theory of the Symmetric Group, 1961; Vector Geometry, 1962; also articles. Mng. editor Canadian Jour. Maths., 1949—. Research in theory of group representations, especially symmetric group. Home: 1 Neville Park Blvd., Toronto 13, Ont., Can.*

ROBINSON, Hamilton B(urrows) G(reaves), Am. oral pathologist; b. Phila., Feb. 16, 1910; s. William J. and Kathryn (Burrows) R.; D.D.S., U. Pa., 1934; Rockefeller fellow U. Rochester, 1934-37, M.S., 1936; m. Katherine Long, Oct. 1, 1929; children—William Edward, Marian Kay (Mrs. P. W. Burnside), Peter Jay Robinson. Assistant professor of pathology Washington University, St. Louis, 1937-41, asso. prof., 1941-44; prof. dentistry Ohio State U., 1944-58, dir. postgrad. dental div. 1947-52, dir. clin. teaching, 1950-52, asso. dean, 1952-58, chief dental staff Univ. Hosp., 1952-58; dean, prof. oral pathology and diagnosis Sch. Dentistry, U. of Mo. at Kansas City, 1958-67, acting chancellor, 1967-68, dean, 1968—. Webb Johnson lectr. Royal Coll. Surgeons (London), 1961; cons. Surgeon-Gen. U. S. Army, Surgeon Gen. USN, Kansas City VA Hosp., Wadsworth (Kan.) VA Hosp.; staff Kansas City Gen. Hosp., Mercy Hosp.; editor Jour. Dental Research, 1935-58, Jour. Ohio State Dental Soc., 1951-58; asso. editor Oral Surg., Oral Medicine, Oral Pathology, 1948-56; former cons. Surgeon Gen. USAF, U. S. Naval Med. Center, Bethesda, Md.; cons. Surgeon Gen. U. S. Army and Navy; nat. adv. com. community health USPHS, 1963-66, nat. adv. council health services, 1966-67; bd. sci. councillors, Nat. Inst. Dental Research, 1963-67. Pres. Columbus Bd. Health, 1946-54. Recipient Tufts University award for leadership oral pathology, 1959; Callahan medal, 1964; Jarvis-Burkhart medal, 1967. Hon. fellow Internat. Coll. Dentists; fellow A.A.A.S., Royal Soc. Medicine (Eng.), Am. Soc. Exptl. Pathology; mem. Am. Cancer Soc., Pierre Fauchard Acad. (past pres.; Fauchard medal, 1950), Am. Acad. Dental Medicine (hon.), Fedn. Dentaire Nationale Francais (hon.), Am., Mo. dental socs., Internat. Assn. Dental Research (p.p.), Am. Soc. Oral Surgeons (hon.), Am. Bd. Oral Pathology (past pres.), American Academy Oral Pathology (past pres.), Am. Assn. Dental Schs. (past pres.), Am. Acad. Dental Sci. (hon.), Am. Acad. Gen. Dentistry (hon.), Sigma Xi, Omicron Kappa Upsilon, Omicron Delta Kappa. Author: Oral Diagnosis and Treatment Planning (with K. H. Thoma), 1960; Tumors of the Oral Regions, 1958; Color Atlas of Oral Pathology (with R. A. Colby and D. A. Kerr), 1961; Yearbooks of Dentistry, 1949—. Research and numerous publs. on periodontal diseases, oral lesions, dental edn. Home: 4311 Rockhill Rd., Kansas City 64110. Office: 1108 E. 10th St., Kansas City, Mo. 64106.*

ROBINSON, Harold Roper, English physicist; b. Nov. 27, 1889; s. James Robinson; Ph.D., Cambridge U.; D.Sc., Manchester, Eng.; m. Marjorie Eve Powell, 1920 (dec. 1939); 1 son, 1 dau.; m. 2d, Madeleine Jane Symons, 1940. Lectr., asst. dir. phys. labs. U. Manchester; reader, Carnegie teaching fellow U. Edinburgh; prof. physics Univ. Coll., Cardiff; prof. physics Queen Mary Coll., U. London, 1930-53, Coll. vice-princ. 1946-53, univ. vice-chancellor, 1954-55, fellow coll., 1955. Fellow Royal Soc., 1929, Indian Assn. for Cultivation of Scis. (hon.); mem. Vic-Wells Assn. (v.p., hon. life mem.). Contbr. papers on radioactivity, Röntgen rays, atomic structure to sci. publs. Died Nov. 28, 1955.

ROBINSON, Harry Maximilian, Jr., Am. physician; b. nr. Balt., Dec. 15, 1909; s. Harry M. and Verna (Wilson) R.; B.S., U. Med., 1931, M.D., 1935; postgrad. Johns Hopkins, 1937-40; m. Maurice Hardin, July 3, 1937 (dec. Dec. 1963); children—Emily Ann, Harry Maximilian; m. 2d, Elizabeth S. Rehm, June 6, 1965. Faculty, U. Md., Balt., 1938-42, 45—, prof. dermatology, head dept., 1953—; Mem. A.M.A. (award 1953, 62), Am. Acad. Dermatology (award 1951, 62, past dir.), So. Med. Assn. (award 1959, 61), Balt. Dermatologic Soc. (past pres.), Balt. City Med. Soc. (past pres., past dir.), Am. Dermatol. Assn., Assn. Mil. Dermatologists. Author: (with Williams and Wilkins) Clinical Dermatology, 1961; also numerous articles. Research on fungus infections, antibiotic therapy, tropical diseases, syphilis, fluorescence fungi. Inventor biopsy tool. Home: 216 Northway St., Balt. 21218.*

ROBINSON, Howard Addison, Am. physicist; b. Rotterdam, N.Y., July 30, 1909; s. Ralph Chandler and Eva (Mitchell) R.; S.B., Mass. Inst. Tech., 1930, Ph.D., 1935; exchange student U. Munich (Germany), 1930-31; Am.-Scandinavian fellow, U. Uppsala (Sweden), 1935-36; m. Kira Volkoff, June 4, 1935; children—Marina Chandler, Peter Holbrook, Holbrook Chandler. Instr. physics Ohio State U., 1936-37; chief physicist, mgr. phys. research Armstrong Cork Co., Lancaster, Pa., 1937-52; attache Am. embassy, Stockholm, Sweden, 1948-49; 1st sec. embassy, spl. asst. to U. S. ambassador, Paris, France, 1952-56; prof. physics, chmn. dept. Adelphi U., 1958—, dir. Inst. Sci. and Math., 1960-62. Adviser, State Dept. at Geneva Atomic Energy Conf., 1955; cons. to sec. state, 1948-52; U. S. dep. del. Orgn. European Econ. Conf., 1956. Fellow Am. Phys. Soc. (chmn. high polymer div. 1945); mem. Optical Soc. Am., Am. Assn. Physics Tchrs. Editor: High Polymer Physics, 1945. Editor, translator: (with Mrs. Robinson) Lecture Demonstrations in Physics as Given at the University of Moscow, 1964; sci. translation editor Soviet Jour. Optical Tech. Research on phys. properties and chem. structure of silicate glasses. Home: 73 Roxbury Rd., Garden City, L.I., N.Y. 11530.*

ROBINSON, James, Brit. physicist; b. Seghill, Eng., Sept. 9, 1884; s. Robert Robinson; ed. Armstrong Coll., Newcastle on Tyne, U. Göttingen, (Germany), U. Durham (Eng.); D.Sc., Ph.D.; m. Beatrice W. Buckley, 1914; 2 daus. Lectr. physics Armstrong Coll., Sheffield U., London U., 1909-14; in charge research and devel. in wireless and photography RAF, 1918-25; tech. adviser Brit. Radiostat Corp. Mem. Instn. Elec. Engrs., Brit. Inst. Radio Engrs. (v.p.). Contbr. papers on photoelectricity, acoustics, and their applications to radio to jours. Pioneer in devel. wireless compass for airplanes; introduced stenode system, also quartz crystal gate, into radio. Died Oct. 21, 1956.

ROBINSON, John Thomas Romney, Irish astronomer and math. physicist; b. Dublin, Ireland, Apr. 23, 1792; s. Thomas and Ruth (Buck) R.; B.A., Trinity Coll., Dublin, 1810; D.D., LL.D. (hon.), Dublin and Cambridge (Eng.) univs.; D.C.L. (hon.), Oxford (Eng.) U.; m. Eliza Isabelle Rambaut, 1821 (dec. 1839); 3 children; m. 2d, Lucy Jane Edgeworth, 1843. Dep. prof. natural philosophy Trinity Coll.; became astronomer Armagh Obs., 1823. Fellow Royal Acad. Sci., 1830 (pres. 1851-56), Royal Soc., 1856 (Royal medal 1862); mem. Brit. Assn. at Birmingham (pres. 1849); hon., corr. mem. fgn. socs. Author: Systems of Mechanics, 1820; Places of 5,345 Star Observed at Armagh from 1828-54, 1859; also several papers on astronomy and physics. Inventor cup anemometer, 1846; observed and recorded many stars not previously observed; conducted various experiments in physics. Died at Armagh, Feb. 28, 1882.

ROBINSON, Laurence Charles, Australian physicist; b. Chiltern, Victoria, Australia, June 10, 1926; s. James and Rita (O'Brien) R.; B.Sc. with honors, Sydney (Australia) U., 1951; Ph.D., 1964; M.Sc., Adelaide (Australia) U., 1958; m. Kathleen Connolly, Nov. 29, 1952; children—Marcus, Pia, Lisa, Ema. Higher research officer Weapons Research Establishment, So. Australia, 1952-53, 56-58; vis. scientist Services Electronic Research Lab., Harlow, Eng., 1954-55; fellow Cambridge Electron Accelerator Harvard, 1959-61; sr. lectr. physics Sydney U., 1962—. Research and publs. on plasma physics through innovations in microwave interferometry, magneto-microwave interactions at a plasma surface, physics of microwave generations and particle accelerators. Home: 75 Shadforth Mosman, N.S.W., Australia. Office: Sydney U., Sydney, N.S.W., Australia.*

ROBINSON, Lawrence Baylor, Am. physicist; b. Tappahannock, Va., Sept. 14, 1919; s. William Harvey and Fannie (Pollard) R.; B.S. summa cum laude, Va. Union U., Richmond, 1939; M.A., Harvard, 1941, Ph.D., 1946; m. Laura Gwendolyn Carter, June 15, 1956; children—Lyn Adrian, Gwendolyn Harvey, Lawrence Baylor. Faculty, Howard U., 1946-47, 48-51; faculty U. Chgo., 1947-48; research engr. N.Am. Aviation, Inc., Downey, Cal., 1951-53; physicist Naval Research Lab., Washington, 1953-54; faculty Bklyn. Coll., 1954-55; mem. tech. staff Ramo-Wooldridge Corp., Space Tech. Labs., Los Angeles, 1956-60; asso. prof., prof. engring. U. Cal., Los Angeles, 1960—; vis. prof. Technische Hochschule Aachen (Germany), 1966-67. Mem. Am. Phys. Soc., Soc. Engring. Sci., Sigma Xi, Tau Beta Pi, Sigma Pi Sigma. Research on correct interpretation of surface tension phenomena, basic information regarding effect of inter-atomic spacing in solids on Curie temperature and Neel temperature. Home: 1120 Roberto Lane, Los Angeles 90024.*

ROBINSON, Margaret King, Am. oceanographer; b. Provo, Utah, Feb. 23, 1906; d. Samuel Andrew and Maynetta (Bagley) King; B.A. with honors, U. Utah, 1928; student U. Cal. at Berkeley, 1928-30; M.A., Scripps Instn. Oceanography, 1951; m. Arthur G. Robinson, May 24, 1937; children—Renan (Mrs. Ell E. Sercarz), Creighton Homer. Tool designer Consol.-Vultee Aircraft Co., San Diego, 1943-45; tchr. Pacific Beach Jr. High Sch., San Diego, 1945-46; mem. staff Scripps Instn. Oceanography, 1946—, head bathythermograph data processing and analysis sect., div. oceanography, 1957—. Cons. to industry, 1952—; UNESCO oceanography expert to Bangkok, Thailand, 1962-63; participant U. S.-Japan Coop. Sci. Program Phys. Oceanography, 1964. Grantee, NSF, 1960-61, 61-63, 63-64. Mem. Arctic Inst. Am., A.A.A.S., Am. Geophys. Union, Am. Soc. Limnology and Oceanographys, Phi Kappa Phi. Research and numerous publs. on phys. oceanography, temperature distbn., statis. analysis variability temperature in oceans based on bathythermograph and reversing thermometer temperature data. Home: 7721 Hillside Dr., La Jolla, Cal. 92037.*

ROBINSON, Marsh Edward, Am. surgeon; b. Sioux City, Ia., Oct. 30, 1916; s. James Edward and Edna (Jepson) R.; A.B., U. Cal., at Los Angeles, 1938; D.D.S., U. So. Cal., 1942, M.D., 1946. Practice oral, maxillofacial and plastic surgery, Los Angeles, 1949-54; prof., head dept. oral surgery U. So. Cal., Sch. Dentistry, Los Angeles, 1954—; chief oral surgery Los Angeles County Gen. Hosp., Los Angeles, 1954—. Cons. to hosps. Fellow Am. Coll. Dentists, A.C.S.; mem. A.M.A., Am. Dental Assn., Am. Soc. Maxillofacial Surgeons, Am. Soc. Oral Surgeons, So. Cal. Soc. Oral Surgeons, Acad. Oral Pathology. Contbr. articles to tech. jours. Proposed theory of how jaw works; developed new ways to correct surgically congenital and acquired deformities of jaws. Home: 3003 Santa Monica Blvd., Santa Monica, Cal. 90404. Office: 925 W. 34th St., Los Angeles 90404.*

ROBINSON, Raphael Mitchel, Am. mathematician; b. National City, Cal., Nov. 2, 1911; s. Bertram H. and Bessie (Stevenson) R.; A.B., U. Cal. at Berkeley, 1932, Ph.D., 1935; m. Julia Bowman, Dec. 22, 1941. Instr., Brown U., 1935-37; faculty U. Cal. at Berkeley, 1937—, prof. math., 1949—. Mem. Am. Math. Soc., Math. Assn. Am. Author: (with A. Tarski, A. Mostowski) Undecidable Theories, 1953. Publs. on functions of complex variable, theory of numbers, founds. of math. and geometry. Home: 243 Lake Dr., Berkeley, Cal. 94708.*

ROBINSON, Rex Julian, Am. chemist; b. Maxwell, Ind., Nov. 15, 1904; s. John Porter and Ethel (Webb) R.; B.A., DePauw U., 1925; M.A., U. Wis., 1927, Ph.D., 1929; m. Ruth E. (Clark), Aug. 18, 1932; children—Richard C., Neal C., R. Clark. Asst. dept. chemistry U. Wis., 1925-29; faculty U. Wash., Seattle, 1929—, prof. chemistry, 1945—. Mem. Am. Chem. Soc., Am. Soc. Liminology and Oceanography. Contbr. numerous articles to tech. jours. Research on analysis of fresh and ocean waters, methods for analysis of small quantities. Home: 7324 16th Av. N.E., Seattle 98115.*

ROBINSON, Sir Robert, English chemist; b. Chesterfield, Derbyshire, Eng., Sept. 13, 1886; s. W. B. Robinson; ed. U. of Manchester; D.Sc., Univs. of Liverpool, London, Wales, Sheffield, Belfast, Delhi, Cambridge, Nottingham, Bristol, Oxford, Sydney; Doctor of Laws (honorary), Universities of Birmingham, Edinburgh, St. Andrews, Glasgow, Liverpool, Manchester; hon. Dr., Univs. Paris, Madrid, Zagreb; m. Gertrude Walsh, 1912 (dec. 1954), 1 son and 1 daughter; m. 2d, Stearn S. Hillstrom, 1957. Prof. organic chemistry, U. of Sydney, N. S.W., 1912-15, U. of Liverpool, 1915; dir. research, Brit. Dyestuffs Corp., 1920; prof. chemistry, St. Andrews U., 1921; prof. organic chemistry, Manchester U., 1922-28, Univ. Coll., London, 1928-30; Bakerian lectr., 1929; fellow Magdalen Coll. and Waynflete prof. of chemistry, Oxford U., 1930-55; hon. fellow Magdalen; prof. emeritus Oxford U.; dir. and consultant Shell Chemical Co. Ltd. Pres. Chem. Soc., 1939-41; pres. Royal Soc., 1945-50. Longstaff medalist, Chem. Soc.; Davy medalist, Royal Soc., 1930; Royal medalist, 1932; Paracelsus medalist, Swiss Chem. Soc., 1939; Copley medalist, 1942; Albert gold medal, Royal Soc. Arts, 1947; Franklin medal, Franklin Inst., Phila., 1947; Nobel prize for chemistry, 1947; Priestley medal, American Chem. Soc., 1952, 53; Hofmann Meml. medal, German Chem. Soc.; 2d Order Rising Sun with Broad Rays (Japan), 1964. Decorated Order of Merit, 1948. Fellow Royal Soc., 1920; comdr. Legion of Honor; Fellow Royal Inst. Chemistry. Corr. mem. French Acad. Scis., 1947. Foreign mem. Nat. Acad. Sciences (Washington), Am. Acad. Arts and Sciences (Boston), Am. Philos. Soc. (Phila.), Soc. Chem. Industry (pres. 1958-59), British Assn. Advancement Sci. (president 1955); hon. mem. Chemists Club of N.Y., N.Y. Acad. of Sciences. Awarded Medal of Freedom (U. S.). Author: Structural Relations of Natural Products, 1955. Early research on plant products having biol. importance, especially alkaloids; synthesized brazilin and haematoxylin substances, 1906; developed theory of alkaloid biogenesis, 1917; synthesized steroid hormones (with J. W. Cornforth), 1933-35; suggested an important qualitative electronic theory in organic chemistry, and a dual biological-abiological theory of origin of petroleum; elucidated penicillin structure. Home: Grimm's Hill Lodge, Gt. Missenden, Bucks, Eng. Office: Shell Centre, Downstream Bldg., London, S.E.1, Eng.

ROBINSON, Robert Alexander, Am. surgeon, microanatomist; b. Rochester, N.Y., Jan. 9, 1914; s. Robert Clarence and Anna (Bill) R.; M.D., Columbia, 1939; m. Beatrice Robinson, June 21, 1941; children—Anne Clark (Mrs. Ross Brian Hooker), Barbara Bill, Robert Alexander, Elizabeth Wells. Faculty, U. Rochester, 1949-53; prof. orthopaedic surgery Johns Hopkins Med. Sch., Balt., 1953—; surgeon-in-charge Johns Hopkins Hosp., 1953—. Cons. to hosps.; mem. coms. Nat. Acad. Sci.-NRC, 1963—; mem. tng. grant study sect. in orthopaedic surgery NIH, 1963—. Recipient Kappa Delta award for research in orthopaedic surgery. Mem. Orthopaedic Research Soc. (past pres.), Soc. U. Surgeons, Am. Acad. Orthopaedic Surgery (rep. to NRC 1965—), Am. Orthopaedic Assn., Soc. Internat. Chirurgy, Orthopaedics, Traumatology, A.M.A., A.C.S., Research on ultra-structure of bone and its relation to bone analysis, physiology; also bio-phys. properties and mechanism of bone matrix mineralization. Home: 314 Broxton Rd., Balt. 21212. Office: 601 N. Broadway, Balt. 21205.*

ROBINSON, Robin, Am. mathematician; b. Clinton, Conn., May 11, 1903; s. Charles Frederick and Florence (Pringle) R.; A.B., Dartmouth, 1924; A.M., Harvard, 1925, Ph.D., 1929; m. Ellen Martha Newsome, July 13, 1927; children—Peter, Julia Ann (Mrs. Robert A. Walkling). Instr. math. Harvard, 1926-27, Sheldon fellow, 1927-28; faculty Dartmouth, Hanover, N.H., 1928—, prof. math., 1942—, chmn. dept. math., 1943-46, chmn. div. scis., 1951-55, registrar, 1958—. Mem. Am. Math. Soc., Math. Assn. Am., Am. Assn. U. Profs., Am. Assn. Collegiate Registrars and Admissions Officers. Author: Analytic Geometry, 1949; also articles. Research in non-Euclidean differential geometry. Home: 16 Allen St., Hanover, N.H. 03755.*

ROBINSON, Russell Lee, Am. physicist; b. Louisville, July 30, 1931; s. Russell Bates and Margaret (Fulton) R.; B.A., U. Louisville, 1953; M.S., Ind. U., 1955, Ph.D., 1958; m. Velda June Flener, Aug. 1, 1953; 1 dau., Marjorie Jannelle. Physicist, Oak Ridge Nat. Lab., 1958—. Mem. Research Soc. Am., Sigma Xi. Determination of properties of low-lying nuclear states and investigation of nuclear structure, including such techniques as beta and gamma-ray spectroscopy, Coulomb excitation, inelastic neutron and proton scattering and charged particle reactions. Home: 104 E. Morningside Dr. Office: Oak Ridge Nat. Lab., P.O. Box X, Oak Ridge 37830.*

ROBINSON, Thane Sparks, Am. zoologist; b. Kansas City, Kan., Apr. 8, 1928; s. Harry E. and Ethyl (Cox) R.; diploma Kansas City Jr. Coll., 1949; B.A., U. Kan., 1951; Ph.D., U. Kan., 1956; m. Eunice Elizabeth Davis, Feb. 5, 1954; children—Amy Beth, Martha Suzanne. Instr. zoology U. Kan., 1956-57; faculty Western Mich. U., 1957-63, asso. prof. biology, 1962-63; asso. prof. biology U. Louisville, 1963—, prof., 1967, head dept., 1966—; dir Adams Center for Ecol. Studies, Kalamazoo, 1959-63, editor publs. series, 1960-63. Mem. A.A.A.S., Am. Assn. U. Profs., Am. Soc. Zoologists, Am. Ornithologists Union, Wilson Soc., Kan., Mich. acads. scis., Ky. Soc. Natural History (pres. 1964—), Sigma Xi, Beta Beta Beta. Contbr. articles to tech. jours. Research on environmental requirement of bobwhite quail, micrometerology of forest stands, geog. distbn. of small mammals, physiology and ecol. stress in natural population of vertebrate animals, breeding biology in birds and amphibians. Home: 202 Nob Hill Lane, Louisville 40206.*

ROBINSON, Thomas Romney, Irish astronomer, math. physicist; b. Dublin, Ireland, Apr. 23, 1792; s. Thomas and Ruth (Buck) T.; B.A., Trinity Coll., Dublin, 1810; D.D., LL.D., Dublin, Cambridge; D.C.L., Oxford; m. Eliza Isabelle Rambaut, 1821; 3 children, including Mary Susanna Stokes; m. 2d, Lucy Jane Edgeworth. Became fellow Trinity Coll., 1814, also dep.-prof. natural philosophy; placed in charge Armagh Obs., 1823; rector Garrickmacross, 1824-82; prebendary St. Patrick's, Dublin, 1872; mem. nautical almanac com., 1830. Fellow Royal Astron. Soc., Royal Soc., 1856 (Royal medal 1862); pres. Royal Irish Acad., 1851-56. Author: Places of 5,345 Stars observed at Armagh from 1828 to 1854, 1859; also numerous papers. Inventor cup-anemometer, completed 1846. Died Feb. 28, 1882.

ROBINSON, William Dodd, Am. physician; b. Hoosac, N.Y., Aug. 9, 1911; s. William Goodwin and Catherine (Dodd) R.; A.B., Albion Coll., 1931; M.D., U. Mich., 1934; m. Anna Louise Ebert, Sept. 8, 1932; children—William Harvey, David Wells, Peter Dodd. Fellow, mem. internat. health div. Rockefeller Found., 1940-43; instr. Vanderbilt U. Med. Sch., 1943-44; faculty U. Mich. Med. Sch., Ann Arbor, 1944—, prof. internal medicine, 1952—, chmn. dept. internal medicine, 1958—, head Rackham Arthritis Research unit, 1944-53. Mem. Nat. Adv. Arthritis Metabolic Diseases Council, 1957-61, 63-67. Diplomate Am. Bd. Internal Medicine. Mem. Am. Rheumatism Assn. (past pres.), A.M.A. (mem. council food and nutrition 1964-67), Phi Beta Kapa, Sigma Xi, Sigma Chi, Phi Rho Sigma, Phi Kappa Phi, Alpha Omega Alpha. Editor, Jour. Lab. and Clin. Medicine, 1955-60. Research and publs. on vitamin and protein nutrition, metabolic diseases, biochemistry connective tissue diseases. Home: 1616 E. Stadium Blvd., Ann Arbor, Mich. 48104.*

ROBIQUET, Pierre-Jean, French chemist; b. Rennes, France, Jan. 13, 1780; student of Vauquelin; pharmacist in French army, then at Vauquelin's pvt. lab.; founder chem. factory; named prof. chemistry Sch. of Pharmacy, 1812, became treas. Mem. French Acad. Scis., 1833, also Acad. Medicine. Author: Nouvelles expériences sur la les amandes amères et sur l'huile volatile qu'elles fournissent, 1830; Nouvelles expériences sur la semence de moutarde, 1831. Conducted expts. on chem. contents of plants; discovered codeine in opium, 1833; studied meconic acid, uric acid from leaf-eating insects, bitter almonds and their volatile oils; discovered (with Vauquelin) asparagine. Died Paris, Apr. 29, 1840.

ROBISON, John, Scottish natural philosopher; b. Boghall, nr. Glasgow, Scotland, 1739; s. John Robison; M.A., Glasgow, 1756, LL.D.; m. Rachel Wright, 1777; children—John, Euphemia, Hugh, Charles. In charge of John Harrison's chromometer on trial voyage to Jamaica for bd. longitude, 1762; became lectr. chemistry, Glasgow, 1766; apptd. to math. chair (with rank of col.), sea cadet corps, St. Petersburg, 1772; prof. natural philosophy, Edinburgh, from 1783. Mem. Royal Soc. Edinburgh (gen. sec. 1783). Author: A System of Mechanical Philosophy, 1882; Outlines of a Course of Lectures on Mechanical Philosophy, 1797; Elements of Mechanical Philosophy, 1804; also articles in Ency. Brit. Anticipated Mayer's elec. discovery that law of force is approximately an inverse square, 1769. Died Edinburgh, Jan. 30, 1805.

ROBISON, Robert, Brit. biochemist; b. Newark-on-Trent, Eng., 1883; s. Robert and Jessie (Thomson) R.; ed. U. Coll., Nottingham, Eng., U. Leipzig, Germany; D.Sc., London, Eng.; Ph.D., Leipzig; m. Ethel Ray Walker, 1910; 1 dau. Became Exhbn. Research scholar, 1851; named lectr., demonstrator chemistry U. Coll., Galway, Ireland, 1909, U. Coll., Nottingham, Eng., 1910; became asst. biochemist Lister Inst., 1913; head dept. biochemistry Lister Inst. Preventive Medicine, 1931-41. Herter lectr. N.Y. U., 1931. Recipient Baly medal Royal Coll. Physicians, 1933. Became fellow Royal Soc., 1930. Author: The Significance of Phosphoric Esters in Metabolism, 1932; also numerous articles. Research in glucopyranose-6-monphosphate which is involved in chem. process of muscle contraction (Robison ester). Died June 18, 1941.

ROBITZEK, Edward Heinrich, Am. physician; b. N.Y.C., Dec. 12, 1912; s. Arthur Harrison and Kate (Heinrich) R.; A.B., Colgate U., 1934; M.D., Columbia, 1938; m. Katherine Robertson, Nov. 9, 1940 (dec. Feb. 1945); children—Arthur Scott, John Edward; m. 2d, Christine Baldwin, June 4, 1952. Med. resident Sea View Hosp., S.I., 1941-42, x-ray resident, 1942-43, chief med. resident, 1943-44, pathologist, 1944-46; dir. med. services Sea View Hosp. & Home, 1961—; attending in medicine Sailors Snug Harbor; pathologist Richmond Meml. Hosp., 1944-46, now cons. staff; pvt. practice as lung specialist, S.I., 1946—; attending staff S.I. Hosp.; cons.

staff St. Vincent Hosp., Richmond Meml. Hosp, USPHS Hosp., Clifton, S.I. Recipient Albert and Mary Lasker award, American Public Health Association, 1955. Diplomate Am. Bd. of Internal Medicine. Fellow A.C.P., Am. Coll. Chest Physicians; mem. Am. Soc. Internal Medicine (pres. S.I. chpt. 1962-65), N.Y. Tb and Health Assn. (dir. 1950-60), Am. Thoracic Soc., A.M.A., N.Y. State Med. Soc., Lambda Chi Alpha. Contbr. articles med. jours. With assos. conducted pioneer trials of various derivatives of the hydrazides of isonicotinic acid in treatment of human tuberculosis, resulting in determination of effectiveness, toxicity and dosage of isoniazid (basic drug responsible for world-wide decline in significance of Tb), 1951-52. Home: 37 Windsor Rd., Staten Island 10314. Office: 100 Central Av., Staten Island, N.Y. 10301.*

ROBLIN, Richard Owen, Am. chemist; b. Rochester, N.Y., Dec. 11, 1907; s. Richard Owen and Minnie (Jeffs) R.; B.A., U. Rochester, 1930; M.A., Columbia U., 1931, Ph.D. (Pfieiffer fellow, U. fellow), 1934; m. Jane Milham Andrews, Oct. 10, 1936; children—Richard Owen, Linda R. (Mrs. Spencer A. Darby), William Murray. Research chemist Am. Cyanamid Co., Linden, N.J., 1934-36, group leader Stamford, Conn., 1936-42, dir. chemotherapy div., 1942-54, asst. gen. mgr. research div., N.Y.C., 1954-56, asst. gen. mgr. pigments div., 1956-57, gen. mgr. comml. devel. div., 1957-59, v.p., 1959—; pres. Cyanamid European Research Inst., Geneva, Switzerland, 1960—. Bd. dirs. Jefferson Chem. Co., Houston. Mem. Mfg. Chem. Assn. (mem. research adv. com. 1959—), A.A.A.S., Am. Chem. Soc. (past chmn. Western Conn. sec., medicinal chem. div.), Am. Soc. Biol. Chemists, Chem. Soc. (London), N.Y. Acad. Sci., Nat. Assn. Mfrs. (mem. research com. 1963), Soc. Chem. Industry (hon. treas. Am. sect. 1964—), Dirs. Indsl. Research, Sigma Xi, Phi Lambda Upsilon, Alpha Delta Phi. Patentee in field. Contbr. numerous articles and revs. to tech. jours. Discoverer sulfadiazine, diamox and other therapeutic agents. Home: 821 Scioto Dr., Franklin Lakes, N.J. 07417. Office: American Cyanamid Co., Wayne, N.J. 07470.*

ROBSON, Douglas S(herman), Am. biometrician; b. St. John, N.D., July 30, 1925; s. Charles Gordon and Myra (Ewart) R.; student U. N.D., 1943, U. Mont., 1945-47; B.S., Ia. State Coll., 1949; M.S., Cornell U., 1952, Ph.D., 1955; m. Anne Thoen Marie Gronna, Apr. 6, 1949; children—Parry Arthur, Suzanne Lorraine, Ricky Charles. Biometrician, Cornell U., Ithaca, 1949-54, Livingston-Farrand Sr. fellow, 1954-55, faculty, 1955—, prof. biol. statistics, 1963—; research asso. U. Wash., Seattle, 1960; asst. dir. statis. techniques research group Princeton, 1961-62; research asso. Colo. State U., 1962. Recipient NIH Career Devel. award, 1963. Mem. Am. Statis. Assn., Biometrics Soc., Am. Fisheries Soc., A.A.A.S. Contbr. articles to tech. jours. Research on statis. techniques applicable to biol. research particularly in vet. virology, fishery biology, statis. genetics, quantitative ecology. Home: R.D. 2, Freeville, N.Y. 13068. Office: Warren Hall, Cornell U., Ithaca, N.Y. 14850.*

ROBSON, Geoffrey Robert, geologist; b. Eng., Jan. 24, 1929; s. Robert and Mary (Darbyshire) R.; B.Sc., U. Durham (Eng.), 1949, Ph.D., 1956; m. Dorothy E. S. Jewitt, Aug. 26, 1953; children—James A. J., Robert C. Officer in charge seismol. rec. stas. Colonial Research Service, Windward and Leeward Islands, W.I., 1952-60; head seismic research U. W.I., 1960-68, reader, 1967-68; econ. affairs officer UN Secretariat, N.Y.C., 1968—. Mem. UNESCO volcanological Mission to Mexico and C.Am., 1966, UNESCO earthquake reconnaisance mission to Venezuela, 1967. Fellow Geol. Soc. London; mem. Internat. Assn. Volcanology (working group on volcano physics), Am. Geophys. Union, Mineral. Soc., Geochem. Soc., Volcanological Soc. Japan. Contbg. author: Catalogue of the Active Volcanoes of the World. Research, publs. on description of seismicity and volcanism of W.I.; orgn. eruption warning system in W.I., recognition of extension failure as earthquake mechanism in volcanic areas. Office: UN, N.Y.C. 10017.*

ROBSON, George Bernard, Am. physician; b. Campbell, Cal., June 16, 1909; s. George S. and Louise (Blenkinsop) R.; A.B., Stanford U., 1930, M.D., 1934; m. Elizabeth McCoy, Sept. 18, 1939; step-children—Betty Walters, L.F. Fenster. Mem. faculty Stanford (Cal.) U., 1936—, clin. prof. medicine, 1954—; clin. prof. medicine U. Cal. at San Francisco, 1961—; asso. dean Stanford U., 1954-57. Mem. A.C.P., Cal. Acad. Medicine. Contbr. articles to profl. jours. Home: 65 5th Av., San Francisco 94118. Office: 2410 Clay St., San Francisco 94115.*

ROCAMORA, Enrique Balcellis, Spanish zoologist; b. Barcelona, Spain, Mar. 31, 1922; s. Francisco Balcellis and Asunción (Rocamora) Vallbona; Licenciado en Cienclas Naturales, U. Barcelona, 1943; D.Naturals Scis., U. Madrid (Spain), 1950. Scientist, Consejo Superior de Investigaciones Científicas, Barcelona, also prof. Barcelona U., 1944-64; dir. Centro Pirenaico de Biología Experimental, Jaca, Spain, 1963—. Research, publs. on Insects and vertebrates. Home: 87 Vía Layetana, Barcelona, Spain. Office: 64 Apartado, Jaca, Spain.*

ROCHE, Marcel, Venezuelan physician; b. Caracas, Venezuela, Aug. 15, 1920; s. Luis and Beatriz (Dugand) R.; B.S., St. Joseph's Coll., 1942; M.D., Johns Hopkins, 1946; Hon. D.Sc., Case Inst., 1963; m. Maria Teresa Rolando, May 20, 1947; children—Antoinette, Noelle, Christian, Diana. Research fellow Sch. Medicine, Harvard, 1948-50; voluntary investigator N.Y. Pub. Health Dept., 1950-51; investigator, dir. Instituto Investigaciones Médicas, Caracas, 1952-58; investigator, dir. Instituto Venezolano de Investigaciones Cientificas, Caracas, 1958—; chief medicine I, Hosp. Vargas, Caracas, 1955-59. Recipient José Maria Vargas prize for sci. research, 1954, Orden Andrés Bello, Venezuela, 1961, Odre de la Couronne de Belgique, 1959. Mem. Council Higher Edn. in Am. Republics (co-pres. 1965—). Author: Bitacora, 1963; Ediciones IVIC, 1963; also numerous articles. Application of techniques of endocrinology and metabolism to environmental medicine especially endemic goiter and anemia asso. with hookworm; research on iodine metabolism in affected regions; measured factors influencing synthesis and fate of erythocytes in subjects with iron deficiency anemia in tropics. Home: 41 Avenida Avila, Caracas. Office: Apartado 1827, Caracas, Venezuela.*

ROCHEFORT, Michel, French geographer; b. Joux-La-Ville, France, Jan. 1, 1927; s. Henri and Agnes (Lecourt) R.; Agrégé de l'Université; Docteur es lettres; m. Miss Espindola, Apr. 22, 1952. Asst., Faculté des Lettres, Strasbourg, France, 1952-57; research attaché Centre National de la Recherche Scientifique, 1957-58; lectr. Faculté des Lettres, Strasbourg, 1958-64; prof., Sorbonne, Paris, 1964—. Mem. Orgn. for Cultural and Tech. Aid to Brazil, 1960-61, 66; cons. doing research on French city structure Minister of Material. Mem. Union for Sci. Study Population. Author: L'organisation urbaine de l'Alsace, 1958; Géographie de l'Amérique du Sud, 1966; also co-author books; articles. Definitions of various conceptions of urban transp. networks and of urban structuring of a colony; importance of regional role of a city and importance of power of polarization of activities of tertiary econ. sector. Home: 32 Croulebarbe, Paris XIII. Office: 191 Saint-Jacques, Paris V, France.*

ROCHESTER, George Dixon, English physicist; b. Wallsend, Eng., Feb. 4, 1908; s. Thomas and Ellen (Dixon) R.; B.Sc., U. Durham (Eng.), 1930, M.Sc., 1932, Ph.D., 1937; m. Idaline Bayliffe, Apr. 18, 1938; children—Dorothy Margaret, Anthony John. Earl Grey Meml. scholar Armstrong Coll., Durham U., 1926-29; Earl Grey fellow, Stockholm U., 1934-35; Commonwealth Fund fellow U. Cal. at Berkeley, 1935-37; faculty U. Manchester, 1937-55, reader, 1953-55; head dept., prof. physics Durham U., 1955—. Sci. adviser in civil def. N.W. and N.E. Regions, 1952-55; founder mem. Council for Nat. Acad. Awards, 1964. Recipient C.V. Boys prize Phys. Soc., 1956; Symons lectr., 1962. Fellow Royal Soc., 1958, Inst. Physics. Author: (with J.G. Wilson) Cloud Chamber Photographs of the Cosmic Radiation, 1952; also numerous articles. Research on cosmic rays; discovered (with C.C. Butler) new elementary particles. Home: 18 Dryburn Rd., Durham, Eng.*

ROCHON, Alexis Marie de, see de Rochon, Alexis Marie.

ROCHOW, Eugene George, Am. chemist; b. Newark, Oct. 4, 1909; s. Theodore Charles and Mary (Lieberman) R.; B.Chem., Cornell U., 1931, Ph.D., 1935; M.A. (hon.), Harvard, 1948; Dr. rer.nat.hic., U. Braunschweig (Germany), 1966; m. Priscilla Ferguson, Sept. 28, 1935 (dec. July 1950); children—Stephen, Jennifer; m. 2d, Helen Louise Smith, Sept. 8, 1951; 1 son, Eugene George. Research asso. Gen. Electric Research Lab., Schenectady, N.Y., 1935-48; asso. prof. Harvard, Cambridge, Mass., 1948-52, prof., 1952—. Recipient Baekeland medal, 1949 Kipping award in silicon chemistry, 1965. Am. Chem. Soc.; Myer award in ceramic chemistry Am. Ceramic Soc., 1951; Perkin medal Soc. Chem. Industry, 1962. Mem. Am. Acad. Arts and Sci. (council 1956-58), Am. Chem. Soc., Am. Inst. Chemists, Acad. Law and Sci., Chem. Soc. of London, Societe de Chemie Industrielle. Editor, Inorganic Syntheses, 1960; regional editor Inorganic and Nuclear Chemistry, 1965—. Author: Chemistry of the Silicones, 1946, 51; (with D.T. Hurd and R.N. Lewis) Organometallic Chemistry, 1952; (with M.K. Wilson) General Chemistry, 1954; Organometallics, 1963; Metalloids, 1965; also many articles. Research in high-temperature insulation for elec. equipment, silicone Holymers; developed method for making organic derivatives of metals, new compounds of element silicon. Home: 37 Squire Rd., Winchester, Mass. 01890.*

ROCK, John, Am. physician; b. Marlborough, Mass., Mar. 24, 1890; s. Frank Sylvester and Ann Jane (Murphy) R.; S.B., Harvard, 1915, M.D., 1918, LL.D., 1966; D.honoris causa, U. Nacional Mayor de San Marcos de Lima, 1962; Sc.D., Amherst Coll., 1965; m. Anna Thorndike, Jan. 3, 1925; children—Rachel Sherman (Mrs. Hart Achenbach), John (dec.), Ann Jane (Mrs. Henry Levinson), Martha (Mrs. Martha LeFevre), Ellen (Mrs. J. Gerald Phillips). Practice medicine specializing in gynecology, Boston, 1921—; clin. prof. gynecology Harvard Med. Sch., 1947-56, emeritus, 1956—; dir. fertility and endocrine clinic Free Hosp. for Women, Brookline, Mass., 1926-56; dir. Rock Reproductive Clinic, Inc., Brookline, Mass., 1956—. Recipient Lasker award Planned Parenthood Fedn. Am., 1948, Ortho awards Am. Gynecol. Soc., 1949, Am. Soc. for Study Sterility, 1963; Oliver Bird Trust (Eng.) medal, 1963, Modern Medicine Publs. award for distinguished achievement, 1966. Author: (with David Loth) Voluntary Parenthood, 1949; The Time Has Come—A Catholic Doctor's Proposals to End the Battle of Birth Control, 1963; also numerous articles. Documented (with Marshall K. Bartlett) cyclic endometrial changes, thus providing test of preceding ovulation in infertile women; demonstrated (with Miriam F. Menkin) human ovum could be fertilized in vitro; developed (with William J. Mulligan) technique for use of plastic material in repair of occluded oviducts; collected (with Arthur T. Hertig) series of dated embryos during 1st 2 weeks of human devel.; proved (with Celso-Ramón García) contraceptive efficacy of progestational steroids and developed concept that these agts. can serve as natural method of fertility control; utilized hyperthermia of testes to induce subfecund oligospermia. Home: 35 Allerton St., Brookline 02146. Office: Rock Reproductive Clinic, Box 623, Brookline, Mass. 02147.*

ROCKERT, Hans Otto Ernst, Swedish histologist; b. Sundsvall., Sweden, Jan. 1, 1932; s. Ernst O.W. and Henny (Pettersson) R.; M.D., U. Gothenburg (Sweden), 1958; m. Lise Lotte Odelram, Oct. 31, 1958; children—Fredrika, Anders, Bengt. Mem. faculty U. Gothenburg, 1956—, asso. prof. histology, 1966—; naval doctor, 1962—. Research and numerous publs. on X-ray microscopic analyses of mineralised tissues as teeth and bone, elementary analyses of single cells, also single muscle cells and extracellular fluids of importance for shock- and postoperative treatments. Home: 17 Bronsaaldervagen, Hovas, Sweden.*

ROCKMORE, Ronald Marshall, Am. physicist; b. N.Y.C., Aug. 10, 1930; s. Nat and Bessie (Pavenick) R.; B.S., Bklyn. Coll., 1951; Ph.D., Columbia, 1957; m. Miriam Miller, Mar. 13, 1960; children—Daniel Nahum, Adam Jeffrey. Mem. Inst. Advanced Study, Princeton, 1957-58; research asso. Brookhaven Nat. Lab., 1958-60; asst. prof. Brandeis U., 1960-63; asso. prof. Rutgers U., New Brunswick, N.J., 1963—; cons. Rand Corp., Santa Monica, Cal., 1961—, Inst. Def. Analyses, Washington, 1962-64. Fellow Am. Phys. Soc.; mem. Phi Beta Kappa, Sigma Xi. Research, publs. on pion-deuteron scattering cross sections in terms of impulse approximation, properties of normal many fermion sysetms at low temperature using field theoretic techniques. Home: 60 Rose St., Metuchen, N.J. 08840. Office: Rutgers U., New Brunswick, N.J. 08903.*

ROCKSTEIN, Morris, physiologist; b. Toronto, Ont., Can., Jan. 8, 1916; s. David and Mina (Segal) R.; came to U. S., 1923, naturalized, 1934; A.B. magna cum laude, Bklyn, Coll., 1938; M.A., Columbia, 1941; Ph.D. (NRC postdoctoral fellow 1946-48), U. Minn., 1948; student Oak Ridge Inst. Nuclear Studies, 1950; widower; children—Susan M., Madelaine Jo. Research asst. U. Minn., 1941-42; asst. prof., then asso. prof. zoophysiology Washington State U., 1948-53; asso. prof. physiology N.Y. U. Sch. Medicine, 1953-61; instr. Marine Biol. Lab., Woods Hole, Mass., summers 1954-59, trustee, 1962-65; cons. Am. Pub. Health Assn., 1961—; prof. physiology U. Miami (Fla.) Sch. Medicine, 1961—, acting chmn. dept. physiology Sch. Medicine, 1967—. Cons., study sect. tropical medicine NIH, 1962-66; adviser Nat. Inst. Child Health and Human Devel., 1964—; mem. research adv. council Bklyn. Hebrew Home Aged, 1958-61; del. President's White House Conf. Aging, 1961. Recipient Distinguished Alumnus Achievement award Bklyn. Coll., 1959. Fellow A.A.A.S., Gerontological Soc. (v.p. biol. scis. 1962-63, pres. 1965-66), Entomol. Soc. Am. (bd. govs. 1958-61); mem. Am. Physiol. Soc., Harvey Soc., Soc. Gen. Physiologists, Soc. Exptl. Biology and Medicine, Am. Soc. Zoologists, Internat. Assn. Gerontology (council), Phi Beta Kappa, Sigma Xi. Author articles in field. Editor: Physiology of Insecta, 3 vols., 1964; Bannerstone Lecture Series, 1964—; editorial bd. Thomas Say Found., 1962-66; abstractor Excerpta Medica, 1960—. Investigation of the underlying mechanisms of the aging process. Home: 335 Fluvia Av., Coral Gables, Fla. 33134. Office: P.O. Box 875. Biscayne Annex, Miami, Fla. 33152.*

ROCKWELL, Alphonso David, Am. physician; b. New Canaan, Conn., May 18, 1840; s. David S. and Betty R.; A.M., Kenyon Coll., 1868; M.D., Bellevue Hosp. Med. Coll. (New York U.), 1864; m. Susie Landon, Oct. 7, 1868. Prof. electro-therapeutics, N.Y. Post-Grad. Sch. Medicine, 1888-92; neurologist and electro-therapeutist, Flushing Hosp., 1904-12. Member commn. to aid in establishing method of executions by electricity. Author: Relation of Electricity to Medicine and Surgery; Treatise on the Medical and Surgical Uses of Electricity (with G. M. Beard), 1871, 9 edits.; Nervous Exhaustion, 1901; Rambling Recollections (autobiography), 1920. Pioneer (with George M. Beard) in electro-therapeutics. Died Apr. 12, 1933.

ROCKWELL, Theodore, III, Am. engr.; b. Chgo., June 26, 1922; s. Theodore and Paisley (Shane) R.; B.S., Princeton, 1943, Chem. Engr., 1944; Sc.D. (hon.) Tri-State Coll., Angola, Ind., 1960; m. Mary Compton, Jan. 25, 1947; children—Robert C., William T., Lawrence E., Juanita C. Research fellow OSRD, Princeton, 1944; process improvement engr.

Clinton Engr. Works, Oak Ridge, Tenn., 1944-45; nuclear engr. Oak Ridge Nat. Lab., 1945-49; nuclear engr., naval reactors program AEC-USN, 1949-55; tech. dir. naval reactors program, 1955-64; dir., prin. officer MPR Assos., Inc., Washington, 1964—. Research asso. Johns Hopkins Sch. Advanced Internat. Studies, 1965—; mem. reactor safety task force U.S. Atomic Indsl. Forum, 1965—; mem. ad hoc panel on artificial heart program NIH, 1966—. Registered profl. engr., D.C. Recipient Distinguished Service medal AEC, 1960; Distinguished Civilian Service medal USN, 1960. Editor: Reactor Shield Design Manual, 1956. Author: (with H.G. Rickover and others) Some Problems in Application of Nuclear Propulsion to Naval Vessels, 1957; also articles. Editor: The Shippingport Pressurized Water Reactor, 1958. Patentee neutron-absorbing material in metal matrix and process for making; helped develop criteria, procedures and facilities for safe operation of nuclear powerplants on land, at sea, in ports; helped systematize analysis of radiation and radiol. problems. Home: 4317 Chestnut St., Bethesda, Md. 20014. Office: 815 Connecticut Av. N.W., Washington 20006.*

RODBERTUS, Johann Karl, German polit. economist; b. Griefswald, Germany, Aug. 12, 1805. With Prussian Civil Service, 1829-32; became mem. Prussian Nat. Assembly, 1848; minister edn., 2 weeks, 1848; carried Frankfurt Constn., 1849. Author: Overproduction and Crises, 1850-51. Founder sci. socialism, however believed in change through legal methods as compared to Marx's revolutionary beliefs; developed theory of falling wage quota in an increasing nat. income. Died Jagetzow, Dec. 6, 1875.

RODECK, Heinrich F. J., German pediatrician; b. Gladbeck, Nov. 1, 1920; s. Franz and Ottilie (Hahne) R.; student univs. Halle, Münster, Würzburg; state exam., U. Münster, 1946; M.D., U. Munich, 1946; m. Käthe Dreses, Apr. 25, 1952; children—Wolfgang, Burkhard, Ulrich, Ortwin, Egbert, Ute. Sci. asst. Children's Clinic, U. Düsseldorf (Germany), 1950-60, faculty, 1956—, asst. prof. pediatrics; vis. asst. U. Kiel (Germany), 1955-59; vis. lectr. U. Montreal (Can.), Harvard, Albert Einstein Coll. Medicine, N.Y.C., NIH, Bethesda, Md., 1959; dir. Children's Hosp., Datteln (Germany), 1960—. Recipient Mem. German Soc. Pediatrics (Moro prize 1959), Rhenish-Westphalian Assn. Pediatrics, Med. Soc. Düsseldorf. Author: Neurosekretion und Wasserhaushalt bei Neugeborenen und Säuglingen, 1958; Untersuchungen über den Einfluss der Dehydration auf die postnatale Entwicklung der Regulationszentren des Wasserhaushaltes, 1962; also numerous articles. Research on water metabolism and neurosecretion in fetus and newborn, diabetes insipidus, new antibiotics in pediatrics, various fields of metabolism in childhood, stress and adaptation in neurosecretion. Home: 17 am Rosengarten, Recklinghausen. Office: 5 Lloydstrasse, Datteln, West Germany.*

RODEWALD, Martin, German meteorologist; b. Hänigsen, Germany, Jan. 12, 1904; s. Ferdinand and Hermine (Voigts) R.; student U. Göttingen 1922-25; Dr.rer.nat. in Geophysics, U. Hamburg, 1939; m. Anni Vollmer, Sept. 3, 1929; children—Wulf Wittekind, Guda. Meteorologist, Deutsche Seewarte, Hamburg, 1925-47, regierungsrat, 1940, regierungsdirektor, 1966, head synoptic dept., 1945-47; chief br. for applied maritime meteorology and overseas meteorology Met. Office for N.W. Germany (now Seewetteramt), Hamburg, 1948-58, head dept. for maritime meteorology, 1958—. Mem. commn. for maritime meteorology WMO, 1955—. Mem. Deutsche Meteorol. Gesellschaft, Deutsche Geophysikalische Gesellschaft. Author: Klima und Wetter der Fischereigebiete Bäreninsel, 1949; Island, 1951; West-u. Südgrönland, 1955; also numerous articles. Founder, Der Wetterlotse, 1949; co-editor Der Seewart, 1953—. Discovered Dreimasseneck (triple point between three different air masses) and its cyclogenetic effect; developed forecasting rules from upper air charts. Home: 78 Rantzaustrasse, Hamburg 70. Office: Seewetteramt, 76 Bernhard Nocht-Strasse, Hamburg 4, Germany.*

RODGERS, Eric, Am. physicist; b. Goshen, Ala., Oct. 17, 1904; s. Ernest and Eva (Carter) R.; B.S., U. Ala., 1931, M.A., 1932; Ph.D., U. Chgo., 1937; m. Sarah Ella Haughton, May 23, 1933; children—Linda, Anne, Joseph Eric, Sarah. With U. Ala., 1932—, prof. physics, 1944—, chmn. dept., 1947-58, dean grad. sch., 1958—; asso. editor Am. Jour. Physics, 1949-52. Mem. Am. Phys. Soc., Am. Assn. Physics Tchrs., A.A.A.S., Phi Beta Kappa, Sigma Xi. Author: (with F. H. Mitchell) Experimental Physics, 1948. Research on X-rays, statistics, theoretical mechanics. Home: 2903 16th Av., Northport, Ala. 35486.*

RODGERS, John, Am. geologist; b. Albany, N.Y., July 11, 1914; s. Henry Darling and Louise Woodward (Allen) R.; B.A., Cornell U., 1936, M.S., 1937; Ph.D., Yale, 1944. With, Cornell U., 1935-37; field geologist U. S. Geol. Survey, 1939-46, part-time geologist, 1946—; faculty Yale, New Haven, 1946—, Silliman prof. geology, 1962—, sr. fellow NSF, 1959-60; vis. lectr. Coll. de France, Paris, 1960. Sec. gen. Commn. on Stratigraphy, Internat. Geol. Congress, 1948-56; commr. Conn. Geol. and Natural History Survey, 1960—; sci. cons. U. S. Army Engrs., 1944-46. Mem. Geol. Soc. Am. (councillor 1962-65), A.A.A.S., Am. Geophys. Union, Am. Assn. Petroleum Geologists, Am. Acad. Arts and Scis., Société géologique de France (v.p. 1960), others. Author: (with C. O. Dunbar) Principles of Stratigraphy, 1957; also articles. Editor: El sistema cambrico, su paleogeografia y el problema de su base, 3 vols., 1956-63. Asst. editor Am. Jour. Sci., 1948-54, editor, 1954—. Research on stratigraphy and structural geology of sedimentary rocks; Appalachian Mountains; regional geology. Office: Dept. Geology, Yale U., New Haven, Conn. 06520.*

RODHE, Wilhelm Carl Olof, Swedish limnologist; b. Stockholm, Oct. 26, 1914; s. Olof and Willy (Herdin) R.; fil.mag. U. Uppsala, 1940, fil.lic., 1942, fil.dr., 1948; m. Kerstin Collin, Aug. 31, 1943; children—Andreas, Caroline, Barbro. Faculty, U. Uppsala, 1949—, prof. limnology, 1959—, dir. Inst. Limnology, 1949—. Convenor freshwater sect Internat. Biol. Programme, 1962-64. Mem. Internat. Assn. Limnology (past gen. sec.). Research and publs. on conditions for growth of plankton algae in lakes. Home: 6 Bergagatan, Uppsala, Sweden.*

RODIN, Alvin Eli, physician; b. Winnipeg, Can., Mar. 25, 1926; s. Paul and Bessie (Oretsky) R.; M.D., U. Man., 1950, M.Sc., 1960; F.R.C.P., Royal Coll. Physicians and Surgeons of Can., 1960; m. Bernice Eta Block, Jan. 14, 1951; children—Beverly Ann, Paula Jacqueline, Mindy Claire, Lisa Phyllis. Came to U.S., 1963. Practice medicine, Minitonas, Man., 1951-54; asso. pathologist Royal Alexandra Hosp., Edmonton, Alta., 1959-60; dir. labs. Misericordia Hosp., Edmonton, 1960-63; asst. prof. pathology U. Tex. Med. Br., Galveston, 1963-65, asso. prof., 1965—, dir. undergrad. teaching of pathology, 1964—. Mem. History of Medicine Soc. (pres. chpt. 1966—), A.M.A., A.A.A.S., N.Y. Acad. Sci., Am. Soc. Exptl. Pathology, Can. Assn. Pathology, Tex. Med. Assn., Assn. Am. Med. Colls., Am. Assn. Pathologists and Bacteriologists, Am. Soc. Clin. Pathologists, Internat. Acad. Pathology, Coll. Am. Pathologists. Contbr. articles to profl. jours. Demonstrated that mercuric chloride is not toxic to mitochondria of kidney tubule cells but to other components, that cancer grows more quickly, spreads more and kills more rapidly in absence of pineal gland, that pineals in humans are larger in cancer cases than non-cancer cases; clarification of electron microscopic appearance of pineal gland. Home: 2908 Beluche St., Galveston, Tex. 77550.*

RODIONOV, Vladimir Mikhailovich, Russian chemist; b. 1878. Prof., Mendeleyev Inst. of Industrial Chem.; prof. organic chem., Moscow Medical Inst. Awarded Stalin Prize. Mem., U.S.S.R. Acad. Scis., chmn., Moscow branch, All-Union Mendeleyev Chem. Soc.; mem., Am. Chem. Soc.; Société Chimique de France. Synthesized alkaloids, expecially light-proof dyes; developed original synthesis technique; isolated and studied ipianic acid; research on tautomerism of aldehyde acids and ethers; obtained beta-amino acids by condensation of aldehydes of malonic acid; founder and developer of Russian dye industry. Address: Mendeleyev Inst. of Industrial Chemistry, Moscow, U.S.S.R.

RODRIGUES DE CASTRO, Estevan (Stephanus Rodericus Castrensis), physician; b. Lisbon, Portugal, circa 1550; prof. medicine U. Pisa (Italy). Author: De meteoris microcosmi, 1621; De complexu morborum tractatus, 1624; Medicae consultationes, 1644; others. Attacked 4 elements of Aristotle and argued for atoms. Based medical views on macrocosm-microcosm analogy and discussed formation of meteors in body. Died Pisa, 1627.

RODRIGUEZ-MOLINA, Rafael, Puerto Rican physician; b. San Juan, P.R., Sept. 23, 1901; s. Manuel and Encarnacion (Molina-Cifredo) Rodriguez Serra; B.S., U. P.R., 1923; student St. Johns Coll., 1918-19; M.D., Med. Coll., Va., 1926; D.M.S., Columbia U., 1935; m. Mirian Alberta Mehrhof, Aug. 25, 1937; children—Mirian (Mrs. John P. Lew, Jr.), Rafael Alberto, Manuel Eugenio, Ana Margarita. Practiced medicine specializing in tropical medicine, San Juan, 1927-30; faculty Sch. Tropical Medicine, U. PR., 1930-42, asst. prof., 1935-42; faculty Columbia U., 1930-42, asst. prof., 1935-42; ward physician to chief Med. Service, Rodriguez U. S. Army Hosp., San Juan, 1942-47; chief Med. Service, VA Hosp., San Juan, 1947-55, asso. chief staff-adminstr. gen. med. research program, 1955-65; clin. prof. medicine, lectr. clin. parasitology U. P.R. Sch. Medicine, San Juan, 1950—. Recipient Bronze plaque for distinguished service as trustee Pan Am. Med. Assn., 1960, Bronze plaque for meritorious service in combating Schistosomiasis, P.R. Bilharzia Com., 1963, Fellow A.C.P. (past gov. P.R.), Royal Soc. Tropical Medicine and Hygiene; mem. P.R. Med. Assn., A.M.A., Am. Soc. Tropical Medicine, A.A.A.S. Author: Americanization of Manuel de Rosas, 1967. Research and numerous publs. on anemia and blood pressure, hypersensitiveness to hookworm proteins, hematology and treatment of hookworm disease and of schistosomiasis mansoni, immunology of trichiniasis, ascaris infestation, hematology, clin. aspects and treatment of sprue, treatment of schistosomiasis mansoni, malabsorption and effect folic acid in tropical sprue, evaluation of circumoval precipitin test as diagnostic tool in schistosomiasis mansoni. Home: 12 M Rodriguez-Serra, Santurce, P.R.*

ROE, Anne (Mrs. George Gaylord Simpson), Am. psychologist; b. Denver, Aug. 20, 1904; d. Charles Edwin and Edna (Blake) Roe; B.A., U. Denver, 1923, M.A., 1925; Ph.D., Columbia, 1933; L.H.D., Lesley Coll.; M.A. (hon.), Harvard, 1963; m. George Gaylord Simpson, May 27, 1938. Grad. asst. U. Denver, 1923-25, Columbia, Tchrs. Coll. 1925-27; research psychologist Commonwealth Fund grant to T. H. Weisenberg for research on aphasia, 1931-33; asst. psychologist Worcester State Hosp., 1933-34; research psychologist WPA Project, N.Y. Infirmary for Women and Children, 1935-36; pvt. cons. psychologist, 1936-38; asst. editor Research Council Problems of Alcohol, 1940-41; cons. Foster Child Study, Carnegie Grant under Social Sci. Research Council, 1941-42; dir. study alcohol edn. in U. S., Yale Sch. Alcohol Studies and Research Council Problems of Alcohol, 1941-42; research asst., asst. prof. lab. applied physiology Yale, 1943-46; chief clin. psychologist charge research VA, br. 2, 1946-47, chief psychology tng. unit Franklin Delano Roosevelt VA Hosp., Montrose, N.Y., 1955-57; dir. Study of Scientists, research grant USPHS, 1947-51; Guggenheim fellow, 1952-53; adj. prof. psychology N.Y. U., 1957-59; lectr. edn. and research asso. edn. Harvard Grad. Sch. Edn., 1959—, prof. edn., 1963-67, emerita, 1967—, dir. Center Research in Careers, 1963-66, emerita, 1966—; lectr. psychology U. Ariz., 1967—. Recipient Lifetime Career award Nat. Vocational Guidance Assn., 1967. Diplomate Am. Bd. Examiners in Profl. Psychology. Fellow Am. Psychol. Assn. (pres. div. clin. psychology 1957-58), Am. Acad. Arts and Scis.; mem. Eastern, Mass., New Eng. (pres. 1965-66), psychol. assns. Contbr. to sci. books. Research and publs. on intellectual functions in normal, aphasic, and mentally disordered adults, adjustment of foster children from different backgrounds, alcohol edn., personalities of scientists, and artists; psychology of creativity, psychology of occupations, origin of interests, behavior and evolution. Home: 5151 E. Holmes St., Tucson 85711.*

ROE, Benson Bertheau, Am. surgeon; b. Los Angeles, July 7, 1918; s. Hall and Helene (Bertheau) R.; A.B., U. Cal. at Berkeley, 1939; M.D. cum laude, Harvard, 1943; m. Jane F. St. John, Jan. 20, 1945; children—David Benson and Virginia St. John. NRC fellow Harvard Med. Sch., 1948-49; Moseley Traveling fellow U. Edinburgh (Scotland), 1950-51; faculty U. Cal. at San Francisco, 1951—, prof. surgery, 1967—, chief cardiac surgery, 1958—, chief cardiotheracic surgery, 1966—. Cons., U.S. VA Hosp., San Francisco, 1958—, San Francisco Gen. Hosp., 1951—. Mem. Am., Pacific Coast surg. assns., Am. Assn. for Thoracic Surgery, Soc. U. Surgeons, Soc. Thoracic Surgery, Soc. Vascular Surgery (v.p.), Western Soc. Clin. Research, A.C.S., Am. Coll. Cardiology, Internat. Soc. Cardiovascular Surgery, Alpha Omega Alpha. Contbr. numerous articles to tech. jours. Research on cardiac function in relation to stress, cold, interruption blood supply, ventricular fibrillation, coronary artery flow patterns, hemodilation in extra corporal circulation. Home: 3647 Washington St., San Francisco 94118.*

ROE, Francis John Caldwell, English pathologist; b. London, Eng., Aug. 16, 1924; s. Stanley Bernard and Ivy Olive May (Caldwell) R.; B.A., U. Oxford, 1943, B.M., B.Ch., 1948, M.A., 1950, D.M., 1957; D.Sc., U. London, 1965; m. Brenda Joan Beckett, Aug. 28, 1948; children—Sarah Jane, Jonathan Caldwell, Catherine Elizabeth. Pathologist, cancer research dept. London Hosp. Med. Coll., 1951-61; with McArdle Meml. Labs., U. Wis., Madison, 1956-57; U. London reader exptl. pathology Chester Beatty Research Inst., Inst. Cancer Research, London, 1961—; asso. pathologist Royal Marsden Hosp. Mem. pharmacology subcom. Com. Med. Aspects Food Policy, Ministry Health. Mem. Coll. Pathologists, Brit. Med. Assn., Path. Soc. Gt. Britain, Med. Assn. Prevention War, European Soc. Study Drug Toxicity, Med. Research Club (London), Brit. Postgrad. Med. Fedn. (mem. central acad. council), Brit. Assn. Cancer Research, Lab. Animal Sci. Assn., Marie Curie Found. (mem. sci. com.), Internat. Com. Lab. Animals, numerous others. Author: (with E. J. Ambrose) The Biology of Cancer, 1966; also numerous articles. Editorial bd. Brit. Jour. Cancer, Internat. Jour. Cancer, Food and Cosmetics Toxicology, Excerpta Medica (cancer sect.). Research on carcinogenesis, chronic toxicity and carcinogenicity of food, drugs, air pollutants, tobacco smoke, metals, other minerals, pathology of lab. animals, animal husbandry. Home: 19 Marryat Rd., Wimbledon, London. Office: 237 Fulham Rd., London, Eng.*

ROE, Joseph Hyram, Am. biochemist; b. Winchester, Va., Dec. 27, 1892; s. Joseph Ashby and Julia (Winkfield) R.; A.B., Roanoke Coll., 1916; AM., Princeton, 1917; Ph.D., George Washington U., 1923, Yale, 1934; m. Clara Grace Lauck, Aug. 19, 1922; 1 son, Joseph Hyram. Prof. biochemistry George Washington U., Washington, 1922-59, chmn. biochemistry dept., 1922-32, 38-59, asst. dean, 1929-31, prof. emeritus biochemistry, 1959, sr. scientist research staff Med. Sch., 1963—. Recipient Alumni Achievement award George Washington U., 1955, Ernst Bischiff award Am. Assn. Clin. Chemists, 1956, Eloy Alfaro award, 1961. Mem. Alpha Chi Sigma, Alpha Omega Alpha. Author: Principals of Chemistry, 9th edit., 1963; A Laboratory Guide in Chemistry, 4th edit., 1963. Contbr. research articles to tech. jours. Work in biochem. analytical procedures for ascorbic acid, fructose, glucose, pentoses, inulin, glycogen, dextran, amylase, lipase, fructose phosphate esters. Home: 1352 Jefferson St. N.W., Washington 20011.*

ROEBLING, John Augustus, civil engr.; b. Mühlhausen, Thuringia, Germany, June 12, 1806; s. Christoph Polycarpus and Friederike (Mueller) R.; studied architecture, engring, bridge constrn., hydraulics, langs., also philosophy (under Hegel) Royal Poly. Inst., Berlin, Germany, civil engr. degree, 1826; m. Johanna Herting, May 1836; m. 2d, Lucia Cooper; 9 children, including Washington Augustus, Ferdinand W., Josephine. Roadbuilder for Prussian govt. in Westphalia, 1826-29; made spl. study of chain suspension bridge, Bamberg, Bavaria; came to U. S., 1831, naturalized, 1837; engr. working on constrn. dams and locks on Beaver River, Pa., 1837; conceived idea of twisted wire rope (to replace hempen cables), devised equipment to manufacture wire rope, produced 1st wire rope made in U. S. in his factory, Saxonburg, Pa., 1841; built wooden aqueduct for Pa. Canal, 1844-45; completed hwy. bridge over Monongahela River at Pitts. (his 1st suspension bridge), 1846; constructed 4 suspension aqueducts for Del. and Hudson Canal; built pioneer railroad suspension bridge at Niagara Falls, 1851-55; built suspension bridge over Over River between Cincinnati and Covington, Ky. (completed 1867), bridge over Allegheny River at Pitts., 1858-60; apptd. chief engr. for bridge over East River between Lower Manhattan and Bklyn., drew up plans (approved 1869), died before constrn. started on Bklyn. Bridge (completed by his son Washington Augustus, 1883); an early advocate of railroad transp., trans-Atlantic telegraph. Author: Diary of My Journey from Mühlhausen in Thuringia via Bremen to the United States of North America in the Year 1831, Written for My Friends, published 1931; Long and Short Span Railway Bridges, 1869. Died Bklyn., July 22, 1869.

ROEBUCK, John, English physician, inventor; b. Sheffield, Eng., 1718; s. John Roebuck; student, Edinburgh; M.D., Leyden, Netherlands, 1741; m. Ann Ward, circa 1746; at least 3 sons, including Ebenezer. Founder chem. lab., Birmingham; began manufacture of sulphuric acid, Pretonpans, 1749; founder co. to manufacture iron, on river Carron, Sterlingshire, circa 1760, also produced carronade; patron of James Watt. Fellow Royal Soc., 1764, Royal Soc. Edinburgh. Contbr. papers to jours. Improved methods of refining precious metals and of producing sulphuric acid and other chemicals; patentee iron mfg. process using pit coal, 1762. Died July 17, 1794.

ROEDER, Kenneth David, zoologist; b. Richmond, Eng., Mar. 9, 1908; s. Carl David and Grace (Phillips) R.; B.A. with honors, Cambridge U., 1929, M.A., 1933; D.Sc. (hon.), Tufts U., 1951; m. Sonja von Cancrin, July 9, 1931; children—Peter Ludwig, Stephanie Eve. Research asst. Toronto (Can.) U., 1930-31; faculty Tufts U., Medford, Mass., 1931—, prof. physiology, 1948—, chmn. dept. biology, 1959-64. Mem. study sects. NIH, codification com. NRC, 1950-53. Fellow Am. Acad. Arts and Scis., Royal Entomol. Soc. (hon.); mem. Nat. Acad. Scis., A.A.A.S., Am. Soc. Zoologists, Am. Physiol. Soc., Entomol. Soc. Am., Sigma Xi. Mem. editorial bd. Jour. Insect Physiology, 1959—, Ann. Rev. Entomology, 1964—. Author: Insect Physiology, 1953; Nerve Cells and Insect Behavior, 1963; also many articles. Research on insect nerve physiology and behavior, insect flight mechanisms, insect neuropharmacology, ultrasonic detection in moths, ultrasonic interaction of moths and bats, evasive behavior mechanisms. Home: 84 Monument St., Concord, Mass. Office: Dept. Biology, Tufts U., Medford, Mass. 02155.*

ROEDERER, Johann Georg, German obstetrician; b. Strasbourg (now in France), May 15, 1726; prof., Göttingen, Germany. Author: De morbo mucoso, 1762. Described typhus abdominalis, 1862; founder 1st German obstet. clinic, Göttingen; research in astronomy. Died Apr. 4, 1763.

ROEDERER, Juan Gualterio, physicist; b. Trieste, Italy, Sept. 2, 1929; s. Ludwig Alexander and Ana (Lohr) R.; Ph.D., U. Buenos Aires, 1952; m. Beatriz Susana Cougnet, Dec. 20, 1952; children—Ernesto, Irene, Silvia, Mario. Group leader Argentine AEC, 1955-64; prof. physics Technol. Sch. Argentine Army, 1956-60; prof. physics U. Buenos Aires, 1959-66; dir. Argentine Nat. Cosmic Ray Center, 1964-66; sr. research asso. U. S. Nat. Acad. Scis. NASA Goddard Space Flight Center, 1964-66; prof. physics U. Denver, 1967—, dir. Center for Study of Planetary Radiation Environment, 1968—. Mem. American Geophysical Union, Argentine Association Geophysicists, Author: Mecanica Elemental, 1962; also numerous articles. Research in modulation of galactic cosmic rays by interplanetary magnetic fields, propagation of energetic particles from the sun to the earth, precipitation of Van Allen electrons into the atmosphere, motion of trapped particles in the atmosphere, motion of trapped particles in the outer magnetosphere, diffusion of Van Allen particles. Home: 2437 S. Fillmore St., Denver 80210.*

ROEDIG, Alfred Heinrich, German chemist; b. Düsseldorf, Germany, Oct. 2, 1910; s. Christian and Johanna (Albers) R.; student U. Bonn, Freiburg, Würzburg; Dr. ès. sc. chem.; Dr. ès. sc. nat. Became asst. U. Würzburg, 1939; sci. collaborator, dept. head Tech. and Chem. Inst. Berlin, 1939; became prof., also prof. at large U. Würzburg, 1950, asso. prof., 1960—. Author: Herstellung von Brom-und-Jodverbindungen in Houben-Weyl Methoden der organischen Chemie, 1960; also articles on organic combinations of halogen elements. Home: Winterleitenweg 65 a, Würzburg, W. Germany.

ROELSEN, Einar, Danish physician; b. Fredericia, Denmark, Mar. 23, 1904; s. Carl Vilhelm and Ellen Elisabeth Roelsen; grad. U. Copenhagen, 1929, M.D., 1937; m. Karen Malling, Oct. 15, 1938; children—Niels Christian, Karen Elisabeth. Sr. registrar, several depts. internal medicine, Copenhagen, 1937-45; practice medicine, specializing in internal medicine, 1939—; docent U. Copenhagen, 1942-45; chief physician, dept. internal medicine Central County Hosp., Silkeborg, Denmark, 1945—. Cons. Hammel County Hosp., 1945—. Mem. Orgn. Danish Internists (chmn. 1950-58), Med. Soc. Jutland (pres. 1954-60), Danish Med. Soc. (pres. 1956-59), Norwegian Soc. Internal Medicine (corr.). Author: Fractional Analysis of Alveolar Air, 1937; also numerous articles. Research on heart, lung and metabolic diseases. Home: Skovhuset, pr. Kjellerup, Denmark. Office: Central County Hosp., Silkeborg 1600, Denmark.*

ROEMER, Elizabeth, Am. astronomer; b. Oakland, Cal., Sept. 4, 1929; d. Richard Quirin and Elsie (Barlow) Roemer; B.A. with honors, U. Cal. at Berkeley, 1950, Ph.D., 1955. Tchr. adult class Oakland Pub. Schs., 1950-52; lab. technician U. Cal. at Mt. Hamilton, 1954-55; grad. research astronomer U. Cal. at Berkeley, 1955-56; research asso. Yerkes Obs., U. Chgo., 1956; astronomer U. S. Naval Obs., Flagstaff, Ariz., 1957-66; asso. prof. lunar and planetary lab., dept. astronomy U. Ariz., Tucson, 1966—. Chmn. working group on orbits of periodic comets, commn. 20, Internat. Astron. Union. Bertha Dolbeer scholar U. Cal., 1948-50; recipient Dorthea Klumpke Roberts prize U. Cal., 1950; Mademoiselle merit award, 1959. Fellow A.A.A.S., Royal Astron. Soc., Am. Geophys. Union; mem. Am. Astron. Soc. (program vis. profs. astronomy 1960—, council), Astron. Soc. Pacific (publs. com.), Internat. Astron. Union, Brit. Astron. Assn., Sigma Xi. Research and numerous publs. on astrometry and astrophysics of comets including recovery of 37 returning periodic comets; visual and spectroscopic binary stars, computation of orbits of comets and minor planets, photographic astronometry.

ROEMER, (Karl) Ferdinand, German geologist; b. Hildesheim Hanover, Germany, Jan. 5, 1818; s. Friedrich and Charlotte (Lüntzel) R.; student law at U. Göttingen, Germany, 1836-39; doctorate in science, U. Berlin (Germany), May 10, 1842; m. Katharina Schäfer, spring 1869. Contbr. to Neues Jahrbuch für Mineralogie Geologie und Palaeontologie, until circa 1887; sailed for Am. on funds provided by Soc. for Protection of German Emigrants in Tex. and Berlin Acad. Science, 1845, undertook mission to study conditions of colonists in Tex. and to report on natural resources of country; became pvt. docent in mineralogy and paleontology at U. Bonn (Germany), 1848; became dir. mineral cabinet at U. Breslau (now Wroclaw, Poland), 1855, also prof. geology and paleontology; recipient Murchison medal Geol. Soc. London, 1855. Author: Texas—Mit besonderes Rücksicht auf deutsche Auswandesung und die physischen Verhältnisse des Lands nach eigener Beobachtung geschildert, 1849, Die Kreidbildungen von Texas and ihre organischen Einschlüsse, 1852, Die Silurische fauna des Westlichen Tennessee, 1860; Geologie von Oberschlesien, 3 vols., with maps and plates, 1870. Died Breslau, Dec. 14, 1891.

ROEMER, Gerd-Benno, German bacteriologist; b. Düsseldorf, Germany, Mar. 24, 1909; s. Josef and Gerda (Giesen) R.; student univs. Cologne, Marburg Düsseldorf; M.D.; LL.D.; agrégé, 1948. Prof. agrégé hygiene and microbiology, Düsseldorf; became physician in chief Hygiene Inst., Acad. Medicine, Düsseldorf, 1950; prof. at large, 1955; became asso. prof., dir. Inst. Clin. Bacteriology and Serology, U. Hamburg (Germany), 1955; prof., 1963—. Research and publs. on serodiagnosis and immunology; antibacterial chemotherapy; clin. bacteriology. Home: Innocentiastrasse 7, Hamburg 13, W. Germany.

ROENTGEN, see Röntgen.

ROESS, Dieter Adolf Friedrich, German physicist; b. Wurzburg, Apr. 6, 1932; s. Eduard F. and Berta (Ebert) R.; Dipl.-Phys., U. Wurzburg, 1956, Dr.rer.-nat., 1958; m. Doris H. E. Hoffmann, Nov. 23, 1958; children—Joachim, Monika. Asst. U. Wurzburg, 1957-60; research physicist Central Lab., Siemens AG, Munich, 1960-62, head lab. phys. research, 1962—. Mem. Soc. Communications Tech. (award 1965). Author: Laser-Lichtverstärker und-Oszillatoren, 1966; also numerous articles. Research on x-rays, masers, lasers: new pump systems for solid state lasers, continuous ruby lasers, laser resonator and transient phenomena, lasers as logical elements. Home: 2 Zeismeringerstrasse. Office: Hofmannstrasse, Munich, West Germany.*

ROESSLER, Edward Biffer, Am. mathematician; b. Redlands, Cal., Aug. 28, 1902; s. Edward E. and Annie (Biffer) R.; A.B., U. Cal. at Berkeley, 1924, M.A., 1925, Ph.D., 1929; m. Elizabeth Stevenson, Aug. 1, 1931 (dec.); children—Bruce, Jean (Mrs. Robert Long); m. 2d, Joan Bugbee, June 21, 1960. Faculty, U. Cal. at Berkeley, 1929-33; faculty U. Cal. at Davis, 1933—, prof., statistician, 1949—, chmn. dept. math., 1953-58, 63—, dir. U. Extension No. area, 1961-63, asso. dean acad. affairs, 1963-67, acting dean, 1967—. Mem. bd. govs. Pacific Jour. Math. Mem. A.A.A.S., Math. Assn. Am., Am. Math. Soc., Am. Statis. Assn., Am. Biometric Soc., Am. Soc. for Hort. Sci., Am. Inst. Food Technologists, Am. Soc. Enology, Am. Extension Assn. Author: (with Henry L. Alder) Introduction to Probability and Statistics, 1960; (with Maynard A. Amerine, Rose Marie Pangborn) Principles of Sensory Evaluation of Food, 1965; also articles. Design and anlysis of experiments especially in food tech. Home: 68 College Park, David, Cal. 95616.*

ROESSLER, Robert Louis, Am. physician; b. Neillsville, Wis., Sept. 2, 1921; s. Julius Otto and Anna (Perrwitz) R.; B.Ph., U. Wis., 1942; M.D., Columbia U., 1945; m. Eleanor Ramsdell, Nov. 6, 1946; children—Ronald, Christina, Erich. Instr. psychiatry, Strong Meml. Hosp., U. Rochester, (N.Y.), 1949-50; faculty U. Wis. Med. Sch. 1950-63, prof., 1960-63, chmn. dept. psychiatry, 1956-61; prof psychiatry Baylor U. Med. Sch., Houston, 1964—; practice medicine specializing in psychiatry, Rochester, 1949-50, Madison, 1950-63, Houston, 1963—. Dir., Wis. Psychiatric Inst., Madison, 1960-61. Served to capt., M.C., AUS, 1946-48. Fellow Am. Psychiatric Assn.; mem. A.A.A.S., Am. Psychosomatic Soc. Group for Advancement Psychiatry, N.Y. Acad. Scis., Soc. for Psychophysiol. Research, Wis. Psychiat. Assn. (pres. 1957-58), Sigma Xi. Editor: (with N.S. Greenfield) Physiological Correlates of Psychological Disorder, 1962. Research in area of stress tolerance, relating physiolog. response to ego strengths and weaknesses and concomitant susceptibility to disease. Home: 2615 Pemberton Dr., Houston 77005. Office: 1300 Moursund Av., Houston 77025.*

ROESSLIN, Eucharius, German physician; b. circa 1490; student, Frankfort, Germany; municipal physician, Frankfort. Author: Der swangeren Fraven und Hebammen Rosegarten, 1513; De partu hominis, 1532. First to deal with midwifery independently of surgery. Died Frankfort/Main, Germany, 1526.

ROFFENI, Giovanni Antonio, Italian astronomer; b. 16th century. Lived in Bologna, Italy; corresponded with Kepler, Galileo. Author: Epistola Apologetica contra Coecam Cuiusdam Martini Horchii Peregrinationem circa Sydereum Nuntium excellentissimi Galilae (defending Galileo), 1611; De Laudibus Verae Astrologiae et Adversus Eiusdem Calumniatores, 1614; Discorso Astrologico delle Mutationi De' Tempi e D'Altri Accidenti dell'Anno 1642. Made astrological predictions, meteorol. explanations. Died 1643.

ROGER, Frédéric, French mathematician; b. Abbeville, Somme, France, Sept. 1, 1910; s. Michel and Madeleine (Neullies) R.; licence agrégé, Superior Normal Sch., Paris, France, 1933, D.Sc., 1938; m. Hélène Chevet, July 21, 1934; 1 dau., Michèle (Mrs. Gerard Lubet-Moncla). Scholarship, Nat. Center Sci. Research, 1936-38, attaché, 1938-40; lectr. math. faculty of scis. U. Aix-Marseille (France), 1942-44; titular prof. chair rational mechanics faculty scis. U. Bordeaux (France), 1944—, dir. Center of Third Cycle of Corpuscular and Theoretical Physics, 1956-62; prof. physco-physics, 1966—. Laureat, French Acad. Sci., 1938, 42; recipient Peccot prize Coll. of France, 1940. Mem. Math. Soc. France, Am. Math. Soc. Studies of tangential properties of Euclidean ensembles of points; matrix method of coupling constituents in mollecular spectroscopy; alpha and beta system of heavy radioelements; methods of factorial analysis in psychology; teaching psychophysics. Home: 51 rue du Parc, 33 Bordeaux Cauderan, France. Office: Faculté des Sciences, 351 cours de la Libération 33 Talence, France.*

ROGER, Georges-Eugène-Henri, French bacteriolgist, physiologist; b. Paris, 1860; intern, 1883, doctor, 1887; became doctor of hosps., prof. Sch. Medicine, Paris, 1892, dir. lab. gen. pathology, 1895, prof. exptl. pathology, 1905, prof. physiology, 1925, dean., 1917-30. Author: Introduction a l'étude de la médecine, 1898; also articles. Research on bacteriology, self-intoxication (poisoning through improper elimination of body wastes), reaction of liver against poisons, destruction of greasy substances by lungs. Died St. Leu-la-Foret, France, 1946.

ROGER, Henri-Louis, French physician; b. Paris, France, 1809. Physician, Sainte-Eugenie Hosp. Mem. Acad. Medicine, 1862. Pub. description of abnormal congenital opening between ventricles of the heart, 1897; prin. work in pediatrics; 1st to give systematic clin. instrn. in this field; research on myelitis. Died Paris, 1891.

ROGER OF HEREFORD, mathematician, astronomer; b. Herefordshire, Eng.; flourished 12th century. ed. probably at Cambridge; reputed author tracts: Theorica Planetarum Rogeri Herefordensis; Introductorium in artem judiciariam astorum; Liber de quatuor partibus astronomiae judiciorum editus a magistro Rogero de Herefordia; De ortu et occasu signorum; Collectaneum annorum omnium planetarum; De rebus metallicis; Compotus (1176). Studied math., astrology, natural philosophy; authority on mines and metals; made astron. table, 1178.

ROGER OF SALERNO (Roger Salenitanus), Italian physician; flourished 12th century; s. Frugardi; student, Salerno, Italy; probably prof., Salerno; chancellor U. at Montpellier, France. Editor: Practica chirurgiae (earliest Western surg. treatise; used as surg. text at Salerno).

ROGERS, Arthur William, English geologist; b. nr. Taunton, Eng., 1872; s. George and Emma (Mills) R.; ed. Clifton Coll., Bristol; student Christ's Coll., Cambridge, 1891-95; D.Sc. (hon.), univs. Cape Town, Witwatersrand; m. Hester J. van der Riet, 1902. Asst., Geol. Commn., Cape Town, S. Africa, 1896-1916, dir., Pretoria, 1916-32; mem. Vernay-Lang expdn. to Kalahari, 1930; ret., 1932. Fellow or mem. Geol. Soc. London (Bigsby medal 1907, Wollaston medal 1931), Royal Soc., 1918, Geol. Soc. S. Africa (Draper Meml. medal 1936, pres. 1915), S. African Assn. (S. African medal 1913, pres. 1933), Internat. Geol. Congress (pres. 1929), Royal Soc. S. Africa (pres. 1934-35). Author: Introduction to the Geology of the Cape Colony, 1905; Handbuch der regionalen Geologie, 1929; The Pioneers in South African Geology and Their Work, 1937; also papers, addresses, reports. Explored and mapped little known geology of S. Africa; specialized in geology of Heidelberg area and its gold fields. Died 1946.

ROGERS, Carl R(ansom), Am. psychologist; b. Oak Park, Ill., Jan. 8, 1902; s. Walter A. and Julia (Cushing) R.; B.A., U. Wis., 1924; M.A., Columbia U., 1928; Ph.D., 1931; L.H.D., Lawrence Coll., 1956; m. Helen Elliott, Aug. 28, 1924; children—David, Natalie (Mrs. L. H. Fuchs). Psychologist in child study dept., Soc. for Prevention of Cruelty to Children, Rochester, N.Y. 1928-30, dir., 1930-38; dir., Rochester (N.Y.) Guidance Center, 1939; prof. clin. psychology, Ohio State U., 1940-45; dir. counseling services USO, 1944-45; prof., psychology, exec. sec. Counseling Center, U. Chicago, 1945-57; Knapp prof. U. Wis., 1957, prof. psychology and psychiatry, 1957-63; resident fellow Western Behavioral Scis. Inst., La Jolla, Cal. 1964—. Recipient Distinguished Sci. Contbn. award Am. Psychol. Assn., 1956; fellow Center for Advanced Study in Behavioral Scis., 1962-63; named Humanist of Year, Am. Humanist Assn., 1964. Fellow Am. Acad. Arts and Scis.; mem. Am. Acad. Psychotherapists (pres. 1956-57), Am. Orthopsychiat. Assn. (v.p. 1944-45), Am. Assn. for Applied Psychology (pres. 1944-45), Am. Psychol. Assn. (pres. 1946-47), Phi Beta Kappa. Author: Measuring Personality Adjustment in Children, 1931; Clinical Treatment of the Problem Child, 1939; Counseling and Psychotherapy, 1942; Counseling with Returned Servicemen (with J. Wallen), 1946; Client-Centered Therapy, 1951; Psychotherapy and Personality Change (with others), 1954; On Becoming a Person, 1961; also articles. Devel. of practice and theory of client-centered therapy; research in process and outcomes of psychotherapy; devel. of theory of personality; application of therapeutic principles to edn. Home: 2311 Via Siena, La Jolla, Cal. 92037.*

ROGERS, David Elliott, Am. physician; B. N.Y.C., Mar. 17, 1926; S. Carl R. and Helen Martha (Elliott) R.; student Ohio State U., 1944; M.D., Cornell U., 1948; m. Cora Jane Baxter, Aug. 13, 1946; children Anne Baxter, Gregory Baxter, Julia Cushing. Research fellow medicine Cornell U. Med. Coll., 1950-51; vis. investigator Rockefeller Inst., 1954-56; asst. prof. medicine, then asso. prof. N.Y. Hosp.-Cornell U. Med. Center, 1954-59; prof. medicine, chmn. dept. Vanderbilt U. Sch. Medicine, 1959—; chief staff Vanderbilt U. Hosp. Chmn. staphylccoccal sect. Armed Forces Epidemology Bd., 1958—; adv. bd. Nat. Bd. Med. Examiners, 1960-65; cons. surgeon gen. USPHS; cons. Nashville VA Hosp., Ft. Campbell Army Hosp.; mem. A-1 study sect. NIH, 1962-65. Named one of ten outstanding young men U. S., Jr. C. of C., 1961. Fellow N.Y. Acad. Scis., A.C.P.; mem. Am. Assn. Physicians, Soc. Clin. Investigation, Am. Clin. and Climatol. Assn., Harvey Soc., So. Soc. Clin. Research, Am. Fedn. Clin. Research. Research and publs. on dynamics of staphylococcal infection, influenza, botulism. Home: 1163 Gateway Lane, Nashville 4.*

ROGERS, Donald Phillip, Am. botanist; b. Toledo, Feb. 5, 1908; s. Philip John and Ella (Johnston) R.; student Toledo U., 1925-26; B.A., Oberlin Coll., 1929; postgrad. U. Neb., 1929-30; Ph.D., U. Ia., 1935; m. Alpha Mae Looney, Dec. 25, 1934; 1 dau., Helen Patricia. NRC fellow Harvard, 1935-36, research, 1940-41; instr. Ore. State Coll., 1936-40; instr. Brown U., 1941-42; asso. prof. biology Am. Internat. Coll., 1942-45; asst. prof. botany U. Hawaii, 1945-47; curator N.Y. Bot. Garden, 1947-57; prof. botany, curator mycol. collections. U. Ill., Urbana, 1957—. Mem. Mycol. Soc. Am. (past pres.), Internat. Assn. for Plant Taxonomy (past sec. spl. com. for fungi and lichens), Brit. Mycol. Soc., Torrey Bot. Club, Bot. Soc. Am., A.A.A.S. Mng. editor Mycologia, 1948-57, editor-in-chief, 1957-60. Research and publs. on taxonomy, comparative morphology, cytology of Basidiomycetes; the basidium; philosophy of taxonomy; history of mycology. Home: 407 W. Delaware Av., Urbana, Ill. 61801.*

ROGERS, Edward Saunders, Am. epidemiologist, human ecologist; b. Fall River, Mass., June 18, 1905; s. Alfred Paul and Georgina (Crosby) R.; A.B., Colgate U., 1927; M.D., Harvard, 1930; M.P.H., Johns Hopkins, 1939; m. Lucinda A. Sutton, June 14, 1932; children—Richard Crosby, Jane Sutton (Mrs. Jack Stergeon), Elizabeth Page. Practice medicine specializing in internal medicine, Boston, 1934-35; dir. bur. pneumonia control, New York State Dept. Health, Albany, 1935-41; asst. prof. medicine Albany Med. Coll., Union U., N.Y., 1937-46; asst. commr. for med. administrn. New York State Dept. Health, 1941-46; prof. pub. health and med. health and med. adminstrn., U. Cal., Berkeley, 1946—, dean Sch. Pub. Health, 1946-51; sr. scholar Inst. Advanced Projects, East-West Center, Honolulu, 1963-64. Chmn. human ecology study sect., NIH, 1960-62; mem. expert adv. panel on health statistics WHO, 1956—. Mem. Am. Pub. Health Assn., Am. Epidemiological Soc. Author: Human Ecology and Health: Introduction for Administrators, 1960. Home: 6 Oak Dr., Orinda Cal.*

ROGERS, George Ernest, Australian biochemist; b. Melbourne, Australia, Oct. 27, 1927; s. Percy and Bertha (Baxter) R.; B.Sc., U. Melbourne, 1948, M.Sc., 1951; Ph.D., U. Cambridge (Eng.), 1957; m. Alison Phoebe Wright, June 8, 1951; children—Jonathan Caroline Mary, Sarah Elizabeth. Sr. research scientist Commonwealth Sci. and Indsl. Research Orgn., 1951-62; reader dept. biochemistry U. Adelaide (Australia), 1963—. Mem. nat. com. for electron microscopy Australian Acad. Sci., 1965—. Mem. Biochem. Soc. (Eng.), Australian Biochem. Soc., Med. Scis. Club S. Australia. Research, publs. on fibrous proteins, especially keratin, using electron microscopy and allied techniques; biochemistry of skin and hair growth; discovered occurrence of citrulline (amino acid) in certain proteins. Home: 212 Cross Rd., Adelaide, S. Australia.*

ROGERS, Hartley, Jr., Am. mathematician; b. Buffalo, July 6, 1926; s. Hartley and Margaret (Kinsey) R.; B.A., Yale, 1946, M.S., 1950; M.A., Princeton, 1951, Ph.D., 1952; M.A., Cambridge (Eng.) U., 1968; postgrad. (Henry fellow) Trinity Coll., Cambridge (Eng.) U., 1946-47; m. Adrianne Thorine Ellefson, Aug. 6, 1954; children—Hartley Raymond, Campbell David Kinsey, Caroline Rebecca. Faculty, Harvard, 1952-55, Benjamin Peirce instr., 1953-55, Eliot House tutor, 1952-55; faculty Mass. Inst. Tech., Cambridge, 1955—, prof. math., 1964—. Guggenheim fellow, 1960-61; NSF sr. postdoctoral fellow, 1967-68; vis. fellow Clare Hall, Cambridge, 1967-68. Mem. Assn. for Symbolic Logic (v.p. 1965—), Am. Math. Soc., Math Assn. Am., Phi Beta Kappa, Sigma Xi. Author: Theory of Recursive Functions and Effective Computability, 1967; also articles. Editor, Jour. Symbolic Logic, 1963—. Research in math. logic and probability especially recursive function theory; work on teaching maths. Home: 19 Lakeview Rd., Winchester, Mass. 01890.*

ROGERS, Henry Darwin, Am. geologist, educator; b. Phila., Aug. 1, 1808; s. Patrick Kerr and Hannah (Blythe) R.; LL.D., U. Dublin, 1857; m. Elza Lincoln, Mar. 1854; 1 child. Lectr. chemistry Md. Inst., Balt., 1828; prof. chemistry and natural philosophy Dickinson Coll., Pa., 1830-31; accompanied socialist Robert Dale Owen to London, Eng., 1832; lectr. geology Franklin Inst., Phila., circa 1833-35; prof. geology and mineralogy U. Pa., 1835-46; dir. N.J. Geol. Survey, 1835-38, Pa. Geol. Survey, 1836-42; Regius prof. natural history U. Glasgow (Scotland), circa 1855-66; published Geology of Pennsylvania (a report on the Pa. Geol. Survey which was one of the most important geol. documents of its time in Am.) 2 vols., 1858. Fellow Royal Soc., 1858; asso. Geol. Soc. London. Author: Description of the Geology of New Jersey, 1840. Advanced noteworthy ideas (with brother William) regarding structure of Appalachian Mountains. Died Glasgow, Scotland, May 29, 1866.

ROGERS, James Blythe, Am. chemist, educator; b. Phila., Feb. 11, 1802; s. Patrick Kerr and Hannah (Blythe) R.; attended Coll. William and Mary, 1820-21; M.D., U. Md., 1822; m. Rachel Smith, 1830, 3 children. Supt. chem. works of Tyson and Ellicott, 1827; named prof. chemistry Washington Med. Coll., Balt., circa 1827; lectured at Md. Inst.; prof. chemistry med. dept. Cincinnati Coll., 1835-39; worked on Va. survey with his brother William, 1837; helped brother Henry, who was conducting Pa. Geol. Survey; prof. chemistry Med. Inst. Phila., 1841, Franklin Inst., Phila., 1844; became prof. chemistry U. Pa., 1847; mem. Am. Philos. Soc. Author: (with broter Robert) A Text Book on Chemistry, 1846, (article) On the Alleged Insolubility of Copper in Hydrochloric Acid . . . , 1848; (article with George W. Andrews and William R. Fischer) Minutes of an Analysis of Soup Containing Arsenic, 1834; (article with James Green) Experiments with the Elementary Voltaic Battery, 1835. Died Phila., June 15, 1852.

ROGERS, J(ames) Harris, Am. inventor; b. Franklyn, Tenn., July 13, 1850; s. James Webb and Cornelia (Harris) R.; ed. St. Charles Coll., London, Eng.; D.Sc., Georgetown U., 1919, U. of Md., 1919. Settled at Hyattsville, Md., 1895; devoted life to scientific work, especially elec. research; awarded many patents relating to multiplex and rapid printing telegraphy, electric lights, the telephone and radio telegraphy; discoverer visual synchronism; the secret telephone and underground and underwater radio communication (enabling U. S. Govt. to carry on uninterruptedly during World War I, communication with allied govts., also with submarines when submerged, and with battleships and airplanes). Hon. fellow Med. Acad. Scis., 1919, also Inventor's Medal; extended thanks by Md. Legislature, 1919, for distinguished contribution to science, also by Md. Legion of Honor. Died Dec. 12, 1929.

ROGERS, Leonard James, English mathematician; b. Oxford, Eng., Mar. 30, 1862; s. James Edwin Thorold and Ann S. C. (Reynolds) R.; B.Mus., Balliol Coll., Oxford, 1884. Became prof. math. Yorkshire Coll., 1888; prof. math. Leeds (Eng.) U., 1888-1919. Fellow Royal Soc., 1924; mem. London Math. Soc. (mem. council). Research and publs. on theory of reciprocals, elliptic functions, theta function, hypergeometric series, G-series. Died Sept. 12, 1933.

ROGERS, Leonard, Brit. physician; b. Plymouth, Eng., Jan. 1868; s. Henry and Jane R.; student Plymouth Coll., St. Mary's Hosp., London; M.B., B.S., London U., 1892, M.D., 1897; LL.D., Glasgow U., 1936, St. Andrews U., 1939; m. Una Elsie North, Sept. 19, 1914; 3 children. Joined Indian Med. Service, 1893; dir. Vet. Bacteriological Lab., Mukhtesar, India; prof. pathology Calcutta (India) Med. Coll., 1900-20; mem. med. bd. India Office, London, 1922-27; pres. Med. Bd., med. adviser to sec. state for India, 1927-33; physician London Hosp. for Tropical Medicine; Croonian lectr. Royal Coll. Physicians, 1924; founder Calcutta Sch. Tropical Medicine. Recipient Moxon Gold medal Royal Coll. Physicians, 1924, Fothergillian Gold medal Med. Soc. London, 1920; Patrick Manson medal Royal Soc. Tropical Medicine and Hygiene, 1938, Cameron prize in therapeutics Edinburgh U., 1929, Laveran Gold medal, Paris, 1956. Fellow Royal Soc., 1916, Royal Soc. Medicine (hon.); mem. Am. Soc. Tropical Medicine (hon.), Am. Climatol. Assn. (corr.), Cambridge Philos. Soc. (hon.), Asiatic Soc., Bengal India Sci. Congress, Brit. Empire Leprosy Relief Assn. Author: Fevers in the Tropics, 1907, 3d edit., 1919; Cholera and its Treatment, 1911; Dysenteries, 1913; Recent Advances in Tropical Medicine, 1929, 2d edit., 1930; (with E. Muir) Leprosy, 1925; (with Sir Joh Mewgaw) Tropical Diseases; The Truth about Vivisection; Happy Toil, Fifty-Five Years of Tropical Medicine. Used anesthesia to distinguished leprosy from similar diseases; demonstrated use of emetine in treatment of amebiasis, 1912; a discoverer of pellagra-preventive factor. Died Sept. 16, 1962.

ROGERS, Lloyd Sloan, Am. physician; b. Waukegan, Ill., Apr. 23, 1914; s. Irvin Lloyd and Maude (Sloan) R.; B.S., Trinity Coll., Hartford, Conn., 1936; M.D., U. Rochester, 1941, Asst. chief surg. service VA Hosp., Cleve., 1951-52; chief surg. service VA Hosp., Syracuse N.Y., 1953—; faculty State U. N.Y. Upstate Med. Center, Syracuse, 1955—; coordinator Upstate Med Center. Surg. cons. VA, NIH; surg. cons.Community, Crouse-Irving, St. Joseph hosps. Diplomate Am. Bd. Surgery. Fellow A.C.S., A.M.A.; mem. Assn. VA Surgeons (pres. 1968—), Soc. Surgery Alimentary Tract, Central N.Y. Surg. Soc., N.Y. Acad. Sci., N.Y. State, Onondaga County med. socs. Research and publs. in fields of gastro-intestinal, cardiac surgery and physiology; cancer chemotherapy; physiology in area of tissue oxygenation including hyperbaric medicine. Home: Staff Apts., VA Hosp. Office: VA Hosp., Syracuse, N.Y. 13210.*

ROGERS, Martha Elizabeth, Am. nursing educator; b. Dallas, May 12, 1914; d. Bruce Taylor and Lucy Mulholland (Keener) Rogers; student U. Tenn., 1931-33; R.N., Knoxville Gen. Hosp. Sch. Nursing, 1936; B.S., George Peabody Coll., 1937; M.A., Columbia, 1945; M.P.H., Johns Hopkins, 1952, Sc.D., 1954. Rural pub. health nurse Children's Found. Mich., 1937-39; successively staff, asst. supr., acting edn. dir. Hartford Vis. Nurse Assn., 1940-45; exec. dir. Vis. Nurse Service of Phoenix, 1945-51; vis. lectr. Cath. U. Am., 1951-52; research fellow Johns Hopkins, 1953-54; prof., chmn. nurse edn. N.Y.U., 1954—. Mem. Ariz. Bd. Nurse Examiners, 1947-51; adviser in nursing edn. Meml. Center for Cancer and Allied Diseases, mem. research com. Montefiore Hosp.; adv. com. nursing program Bklyn. Coll. Recipient award for outstanding contbn. to nursing N.Y. U. Sch. Edn. Nurse Alum. Mem. N.Y. State, Am. (com. on functions, standards and qualifications for pub. health nursing) nurses assns., N.Y. State League for Nursing (pres.), Vis. Nurse Service N.Y., Lower East Side Neighborhood Assn. of N.Y.C. (cons health com.), Nat. League for Nursing, Am. Pub. Health Assn. Author: Educational Revolution in Nursing, 1961; Reville in Nursing, 1964; also articles. Chmn. editorial bd. League Lines; chmn. editorial bd., contbr. Primed for Progress; editor Nursing Sci. Theoretical research in nursing. Home: 50 E. 8th St., N.Y.C. 10003.*

ROGERS, Robert Emple, Am. chemist; b. Balt., Mar. 29, 1813; s. Patrick Kerr and Hannah (Blythe) R.; M.D., U. Pa., 1836; m. Fanny Montgomery, Mar. 13, 1843; m. 2d, Delia Saunders, Apr. 30, 1866.

Connected with ry. surveying parties in New Eng., 1831-32; chemist 1st Pa. Geol. Survey; made independent analysis of limestones (with Martin H. Boye); prof. gen. and applied chemistry U. Va., 1842; prof. chemistry Med. Sch., U. Pa., 1852, dean, 1856; asst. surgeon West Phila. Mil. Hosp.; made study of petroleum, circa 1864; investigated waste silver in Phila. mint, made suggestions about refining, 1872; prepared plans for equipment of refinery of mint, San Francisco, 1875; prof. med. chemistry and toxicology Jefferson Med. Coll., Phila., 1877- circa 1884; an original mem. Nat. Acad. Scis.; an organizer Assn. Am. Geologists and Naturalists (now A.A.A.S.). Author: Textbook on Chemistry, 1846. Devised (with brother William) new process for preparing chlorine; improved processes for making formic acid, and aldehyde, perfected method of determining carbon in graphite, studied volatility of potassium and sodium carbonates, decomposition of rocks by meteoric water, absorption of carbon dioxide by liquids; studied (with brother James) alleged insolubility of copper in hydrochloric acid. Died Phila., Sept. 6, 1884.

ROGERS, William Augustus, Am. mathematician, astronomer, physicist; b. Waterford, Conn., Nov. 13, 1832; s. Daivid Potter and Mary (Rogers) R.; M.A., Brown U., 1857; m. Rebecca Titsworth, 1857; 3 children. Instr. and tutor math. Alfred Acad., 1857; prof. math. Alfred U., 1859, prof. indsl. mechanics, 1860-circa 1870, built and equipped astron. obs., 1865; pursued advanced mechanics Yale; studied practical astronomy (under Prof. Bond) at Harvard Obs., became asst., 1870, asst. prof. astronomy, 1877-86; prof. physics and astronomy Colby. U., 1886-98; sent to Europe by Am. Acad. Arts and Scis. to obtain copies of imperial yard and French meter; organized phys. lab. with model for accurate measurements Colby U. Fellow Royal Soc. London (hon.), Royal Micros. Soc., A.A.A.S.; mem. Am. Micros. Soc. (pres. 1887), Am. Acad. Arts and Scis., Nat. Acad. Scis. Contbr. papers to Harvard Annals. Made important changes in value of yard and meter; made observations of exact positions of catalogued stars between 50° and 55° north declination. Died Waterville, Me., Mar 1, 1898.

ROGERS, William Barton, Am. geologist; b. Phila., Dec. 7, 1804; s. Patrick Kerr and Hannah (Blythe) R.; grad. Coll. William and Mary, 1822; LL.D., Harvard, 1866; m. Emma Savage, June 20, 1849. Conducted sch. (with brother Henry), Windsor, Md.; lectr. Md. Inst., 1827; prof. natural philosphy and chemistry Coll. William and Mary, 1828-35; prof. natural philosophy U. Va., 1835; state geologist Va., 1835-48; state insp. gas meters Mass., 1861; responsible for act incorporating Mass. Inst. Tech., 1861, 1st pres., 1862-70, pres., 1878-81; prof. emeritus of geology and physics. 1878-82; Chmn. Assn. Am. Geologists and Naturalists, 1845, 47; corr. sec. Am. Acad. Arts and Scis., 1863-69; an original mem. Nat. Acad. Scis., pres., 1878-82. Author: Strength of Materials, 1838; Elements of Mechanical Philosphy, 1852; A Reprint of Annual Reports and Other Papers on the Geology of the Virginias. Noted for work (with brother Henry) on structure of Appalachian Mountain chain; devised (with brother Robert) new process for preparing chlorine. Died Boston, May 30, 1882.

ROGERS, William Percy, Australian biologist; b. Katanning, West Australia, Nov. 23, 1914; s. Percy Nunn and Agnes (Bishop) R.; B.Sc., U. W. Australia, 1936, M.Sc., 1938; Ph.D., U. London (Eng.) Sch. Hygiene and Tropical Medicine, 1940, D.Sc., 1959; m. Lillian Readhead Taylor, Sept. 10, 1940. Research worker London Sch. Hygiene and Tropical Medicine, 1940-42, Molteno Inst., U. Cambridge (Eng.), 1942-46, Commonwealth Sci. and Indsl. Research Orgn., 1947-52; prof. zoology U. Adelaide (S. Australia), 1952-64, prof. parasitology, 1964——. Fellow Australian Acad Sci.; mem. Physiol. Soc., Biochem. Soc. Author: The Nature of Parasitism, 1963; also numerous articles. Research on biochemistry of parasitism especially biochemistry of interaction of parasite and host during process of infection. Home: Lirra Lirra, Oakbank, S. Australia. Office: Waite Agrl. Research Inst., Glen Osmond, S. Australia.*

ROGET, Peter Mark, English physician; b. London, Jan. 18, 1779; s. John and Catherine (Romilly) R.; M.D., Edinburgh, 1798; also studied medicine, London; m. dau. of Jonahthan Hobson, 1824; 2 children, including John Lewis; became physician to infirmary, Manchester, Eng., 1805, No. Dispensary, 1810, Spanish embassy, 1820, Milbank penitentiary, 1823; commd. by govt. to study water supply of metropolis, 1827-28; Gulstonian lectr. Royal Coll. Physicians, 1831, censor, 1834, 35, also fellow, licentiate; 1st Fullerian prof. physiology Royal Instn., 1833-36. Fellow Royal Soc., 1815, sec., editor Proc., 1827-49. Author: Animal and Vegetable Physiology considered with reference to Natural Theology, 1834; Physiology and Phrenology, 1838; Theasaurus of English Words and Phrases (1st pub. collection of synonyms and antonyms), 1852; also papers. Inventor new sliding rule, 1815; a founder U. London, 1837. Died West Malvern, Worcestershire, Sept. 12, 1869.

ROGINSKII, Simon Zalmanovich, Russian phys. chemist; b. Mar. 12, 1900; grad. Dnepropetrovsk U., 1922. Instr., Dnepropetrovsk Mining Inst., 1923-28; asso. Ukraine Inst. Phys. Chemistry, 1925-28; with Inst. Chem. Physics, USSR Acad. Sci., 1928-41; later at Leningrad Poly. Inst.; asso. Inst. Chem. Physics, USSR Acad. Scis., 1941——. Recipient Stalin Prize, 1941. Corr. mem. USSR Acad. Scis. Author: The Kinetics of Chemical Type Reactions, 1938; Theoretical Principles of Catalyst Production, 1944; Adsorption of Heterogeneous Surfaces, 1948; The Principles of Catalyst Theory, 1949; Theoretical Principles of Heterogeneous Catalysis, 1950; Theoretical Principles of the Use of Isotope Methods for the Study of Chemical Reactions, 1956; Semiconductor Catalysis, 1957; Electron Microscope Studies of Catalysis, 1958; The Presorption Effect and Some Other Anamalies of Catalytic Oxidation in Oxide Semiconductors, 1960; An Express Chromatographic Method of Measuring the Adsorption Isotherms of Gases and Vapors, 1960. Research on kinetics of heterogeneous reactions and isotopes, catalysis; formulated theory of adsorption and catalysis for heterogeous surfaces, microchem. theory for active surfaces; 1st in USSR to write on isotope exchange and use of artificial radioactive isotopes in study of chem. reactions. Home: Leninskii Prospekt, 13. Office: Inst. Phys. Chemistry, USSR Acad. Scis., Leninskii, Prospekt 31, Moscow, USSR.

ROGLER, George Albert, Am. agronomist; b. Matfield, Green, Kan., Apr. 5, 1913; s. Henry William and Maud (Sauble) R.; B.S., Kan. State U., 1935; M.S., U. Minn., 1942; m. Vera Thompson, Mar. 14, 1936; children—Joan (Mrs. James Schieffer), Susan (Mrs. Ronald Troop). Agronomist, Soil Conservation Service, Manhattan, Kan., Mandan, N.D., 1935-37; research agronomist Agrl. Research Service, No. Gt. Plains Research Center, Mandan, 1937——. Forage research cons. in Peru for N.C. State U., 1958. Recipient Superior Service award U.S. Dept. Agr., 1955, Merit award Am. Grassland Council, 1964. Fellow Am. Soc. Agronomy; mem. Am. Soc. Range Mgmt., Am. Grassland and Forage Council, N.D. Acad. Sci. Research and numerous publs. on devel. grasses; pioneered concept of use of seeded grasses in combination with native range and fertilization of No. Plains rangelands. Home: 1701 Monte Dr. Office: P.O. Box 459, Mandan, N.D. 58554.*

ROGOFF, Julius M., physiologist; b. Riga, Latvia, Nov. 17, 1884; Ph.G., Ohio No. U., 1900, Sc.D., 1933; M.D., Western Res. U., 1908; L.H.D., Yeshiva U., 1952. Mem. faculty Vanderbilt U., 1911-15, Western Res. U., 1915-34, U. Chgo., 1934-39; prof. endocrinology, dir. lab. expt. endocrinology U. Pitts. Sch. Medicine, 1939-50; vis. prof., hon. gov. Hebrew U.; vis. prof., fellow Brandeis U.; also lectr. to med., sci. and lay audiences. Trustee Ohio No. U.; founder, dir. G.N. Stewart Meml. Lab. and Library, Rogoff Found.; hon. pres. Am. Physicians Fellowship for Israel Med. Assn.; bd. dirs. Am. Friends of Hebrew U. Recipient Brandeis U. medal, 1965, Bronze medal Weizmann Inst. Sci., Israel, 1965, Earl of Balfour certificate. Fellow A.A.A.S., N.Y. Acad. Sci.; mem. Am. Physiol. Soc., Am. Pharm. Soc., Am. Diabetes Assn., Am. Endocrine Soc., Am. Soc. Exptl. Biology and Medicine, Sigma Xi, Kappa Nu, Phi Lambda Kappa. Author: Comprehensive Plan for Medical and Health Education, A New School of Medicine. Pioneer in med. research of the adrenal gland and discoverer of the indispensable life-sustaining hormone of the adrenal cortex, interrenalin; originator of successful treatment of Addison's disease and adrenal cortical insufficiency; authority on diagnosis of Addison's disease. Died Norwalk, Conn., June 22, 1966.

ROGOFF, Stanley Myron, Am. physician; b. Auburn, N.Y., Jan. 30, 1922; s. Charles and Anna (Glasser) R.; B.A. summa cum laude, Coll. of City of N.Y., 1943; M.D., U. Rochester, 1946; m. June B. Goldman, June 17, 1945; children—Harry Edward, Kenneth Saul, Richard David. Mem. faculty U. Rochester (N.Y.) Med. Center, 1951——, prof. diagnostic radiology, 1961——, chief div. diagnostic radiology, 1960——. Recipient Honors Achievement award Angiology Research Found., 1964-65. Mem. A.M.A., Assn. U. Radiologists, Am. Coll. Radiology, Radiol. Soc. N. Am., Phi Beta Kappa, Alpha Omega Alpha. Research and publs. on X-ray studies of peripheral arteries and veins, particularly arteriosclerosis and venous thrombosis; contrast liquids injected into arteries and veins. Address: U. Rochester Med. Center, Rochester, N.Y. 14620.*

ROGOFF, William Milton, Am. entomologist; b. N.Y.C., Mar. 15, 1916; s. Michael and Etta (Rogoff) R.; B.S. with distinction in Zoology, U. Conn., 1937; postgrad. Yale, 1937-40; Ph.D., Cornell U., 1943; m. Esther Johana Petersen, Oct. 14, 1947; children—Barbara Lynn, James David. Asso., Citrus Expt. Sta., U. Cal. at Riverside, 1946-47; faculty S.D. State U., Brookings, 1947——, prof. entomology, 1953-62; research entomologist entomology research div. U.S. Dept. Agr., Corvallis, Ore., 1962-68, investigations leader, Fresno, Cal., 1968——. Fulbright Sr. Research scholar U.S. Ednl. Found., div. entomology Commonwealth Sci. and Research Orgn., Canberra, Australia, 1955-56. Fellow A.A.A.S.; mem. Entomol. Soc. Am. (past chmn. sect. D), Am. Inst. Biol. Scis., Am. Soc. Zoologists, Am. Mosquito Control Assn., Am. Micros. Soc., Sigma Xi. Author: (with others) Fundamentals of Applied Entomology, 1962; also numerous articles. Research on metamorphosis nerve tracts within mosquito central nervous system, insecticides involving efficacy against specific pests, repellency and resistance, systemic insecticides, insect mating behavior, oviposition behavior, sex attractants; developed insecticide treatments cattle for control horn flies and cattle grubs. Home: 1027 E. Alamos Av., Fresno 93704. Office: Entomol. Research Div., U.S. Dept. Agr., 5544 Air Terminal Dr., Fresno, Cal. 93727.*

ROHAULT, Jacques, French physicist; b. Amiens, France, 1620; m. Mlle. Clerselier; author: Traité de physique, 1671; Entretiens sur la philosophie, 1671; devoted to Cartesianism; worked to correct and complete Descartes' proofs; wrote one of most important textbooks of Cartesian physics; conducted studies on surface tension and cosmogony based on Descartes' three elements. Died Paris, 1675.

ROHEIM, Geza, psychoanalyst, anthropologist; b. Budapest, Hungary, Sept. 12, 1891; student Berlin (Germany) U., 1909-10, U. Leipzig (Germany), 1911; Ph.D., U. Budapest, 1914; m. 1919. Came to U. S., naturalized, 1944. Mem. folklore orgn. dept. Hungarian Nat. Mus., 1917-19; field work central Australia, Normanby Island, Somali, Yuma, 1929-32; tchr. psychoanalysis, anthropology Budapest Psychoanalytical Inst. 1932-38; analyst Worcester (Mass.) State Hosp., 1938-39; practicing psychoanalyst, 1940——. Lectr. N.Y. Psychoanalytical Inst.; field work with Navaho, 1947. Author: The Origin and Function of Culture, 1943. Studied mythology, magic in cultural anthropology; strongly Freudian, stressed sexuality in culture and psychoanalysis. Died 1953.

ROHRBACH, Hans Joachim, German mathematician; b. Berlin, Feb. 27, 1903; s. Paul and Clara (Muller) R.; Ph.D., U. Berlin; m. Rose Gadebusch, July 12, 1932; children—Gerhild, Waltraut, Gisela, Dagmar, Helmut. Asst., U. Berlin, 1929-35; became asst. U. Gottingen, Germany, 1936, prof., 1937; named asso. prof. German U. Prague, Czechoslovakia, 1941, prof., 1942; became vis. prof. U. Mainz (Germany), 1946, prof., 1951, named dir. Inst. Math., 1958; vis. prof. U. N.C., Chapel Hill, 1957-58. Mem. German, Am. math. socs., Dutch Christian Assn. Physicians and Doctors. Author: Einführung in die höhere Mathematik, 1953. Research and publs. on number theory, algebra, natural sci., cryptography, relation between sci. and religion. Home: Pfeiffer-Weg 7, Mainz, W. Germany.

ROHRER, John Harrison, Am. psychologist; b. Bourbon Mo., Mar. 26, 1914; s. John and Etta (Harrison) R.; A.B., Westminster Coll., 1937; A.M., U. Denver, 1940; Ph.D., Ia. State U., 1942; m. Anne Conway, May 5, 1935 (div. 1960); children—Agatha, Lucinda. Dir. personnel research G. L. Martin Co., Omaha, 1942-43; personnel investigator Western Electric Co., Chgo., 1943-44; asso. prof. psychology U. Okla., 1945-48, prof., dir. social research, 1948-49, prof., asst. dean Coll. Arts and Scis., 1949-50; prof. psychology, dir. Urban Life Research Inst., Tulane U., 1950-57; prof. psychology Georgetown U. Med. Sch., 1957——. Cons., Surg. Gen., U. S. Navy, 1958——, Research and Development Bd., U. S. Army, 1958-59, Nat. Inst. Mental Health, 1958-59; mem. Operation Deep Freeze to Antarctica, 1957, 58. Social Sci. Research Council fellow anthropology, Inst. Human Relations, Yale, 1948-49. Fellow Am. Psychol. Assn., A.A.A.S.; mem. Acad. Psychoanalysis (sci. asso.) Soc. Applied Anthropology, Am. Statis. Assn., Sigma Xi, Kappa Alpha (nat. scholarship officer). Co-editor: Social Psychology at the Crossroads, 1951; Change and Dilemma in the Nursing Profession, 1957; The Eighth Generation, 1959. Research and publs. on learning process, personality development, and adjustment to polar isolation. Deceased.

ROHRLICH, Fritz, physicist; b. Vienna, Austria, May 12, 1921; s. Egon and Illy (Schwarz) R.; Chem.E., Israel Inst. Tech., 1943; A.M., Harvard, 1947, Ph.D., 1948; m. Beulah Friedman, June 24, 1951; children—Emily H., Paul E. Came to U.S., 1946, naturalized, 1956. Mem. Inst. for Advanced Study, Princeton, 1948-49; faculty Cornell U., Ithaca, N.Y., 1949-51, Princeton, 1951-53, State U. Ia., Iowa City, 1953-63; prof. physics Syracuse (N.Y.) U., 1963——. Vis. prof. Johns Hopkins, 1958-59. Fellow Am. Phys. Soc.; mem. A.A.A.S., Sigma Xi. Author: (with J.M. Jauch) Theory of Photons and Electrons, 1955; Classical Charged Particles, 1965. Research and publs. in theoretical physics; electomagnetic interactions, quantum theory of fields, relativity, theory of atomic spectra. Home: 226 Ridgecrest Rd., Dewitt, N.Y. 13214. Office: Physics Dept., Syracuse U., Syracuse, N.Y. 13210.*

ROHRMAN, Frederick Alvin, Am. chem. engr., govt. ofcl.; b. Pendleton, Ore., Jan. 10, 1904; s. Charles Albert and Ella (Kloepzig) R.; B.Sc., Ore. State U., 1926; M.Sc., U. Minn., 1928; Ph.D., Columbia, 1931; m. Velma Elizabeth Birdwell, June 8, 1938; 1 son, Douglass Frederick. Asst. prof., asso. prof. Mich. Technol. U., 1931-41; prof., head dept. chem. engring. Kan. State U., 1946-47; exec. dir., chmn. research U. Colo., Boulder, 1947-55; superintending scientist U.S. Navy Mine Def. Lab., Panama City, Fla., 1955-60; prin. scientist Oak Ridge Inst. Nuclear Studies, 1960-64; staff adviser div. air pollution Lab. Engring. and Phys. Scis., USPHS, Cin., 1964——. Registered profl. engr., Colo. Fellow A.A.A.S.; mem. Am. Chem. Soc., Am. Inst. Chem. Engrs., Am. Soc. Mech. Engrs., Am. Soc. Engring. Edn., Sigma Xi, Phi Lambda Upsilon, Sigma Tau, Sigma Phi Epsilon. Author: Materials of Construction, 1939; Introduction to Nuclear Engineering, 1939. Contbr. numerous articles on engring. and sci. problems to profl. publs.

Patentee in fields of corrosion, constn. materials, underwater ordnance, atomic energy. Home: 1813 Mears Av., Cin. 45230. Office: 4676 Columbia Pkwy., Cin. 45200.*

ROHRMANN, Charles Albert, Am. chem. engr.; b. Pendleton, Ore., Dec. 17, 1911; s. Charles and Ella (Kloepzig) R.; B.S., Ore. State U., 1934; postgrad. Mich. Technol. U., 1934-36; Ph.D., Ohio State U., 1939; m. Elva Ann Chamblin, July 21, 1939; children—Charles A., George F., Virginia L., Joanna E. With E.I. duPont de Nemours & Co., Cleve., Phila., 1939-48; chem. engr., sect. chief supr., mgr. tech. specifications Hanford Atomic Products Operation Gen. Electric Co., Richland, Wash., 1948-65; sr. research asso. Battelle-N.W., Richland, 1965—. Mem. Am. Chem. Soc., Am. Nuclear Soc., Am. Inst. Chem. Engrs. Research on recovery of chlorine from gas mixtures, purification of sulfamates, radioistopes useful as heat and radiation sources, by product values in spent fuels from nuclear power reactors; described processes for recovery of plutonium and uranium from spent fuels; mineral industries investigations. Home: 4707 W. 7th Av., Kennewick, Wash. 99336. Office: Battelle-N.W., Richland, Wash. 99352.*

ROHSENOW, Warren Max, Am. mech. engr.; b. Chgo., Feb. 12, 1921; s. Fred A. and Selma (Gorss) R.; B.S. in Mech. Engring., Northwestern U., 1941, M.Engring., Yale, 1943, D.Eng., 1944; m. K. Towneley Smith, Sept. 20, 1946; children—John, Brian, Damaris, Sandra, Anne. Instr. mech. engring. Yale, New Haven, 1941-44; asst. prof. mech. engring. Mass. Inst. Tech., Cambridge, 1946-52, asso. prof., 1952-56, prof., 1956—, dir. Heat Transfer Lab., 1954—. Cons. in heat transfer, fluid mechanics to numerous indsl. orgns., 1946—. Chmn. bd. dirs. Dynatech Corp., Cambridge, 1957—. Recipient award for advancement basic and applied sci. Yale Engring. Assn., 1952. Registered profl. engr., Mass. Fellow Am. Acad. Arts and Scis.; mem. Am. Soc. M.E. (Jr. award 1951, Pi Tau Sigma Gold medal 1952), Sigma Xi, Tau Beta Pi, Pi Tau Sigma. Author: (with H.Y. Choi) Heat, Mass. and Momentum Transfer, 1961. Editor: Developments in Heat Transfer, 1964. Research on thermodynamics, heat transfer, gas turbines. Home: 47 Windsor Rd., Waban, Mass. Office: Mass. Institute of Technology, 77 Massachusetts Av., Cambridge, Mass. 02139.

ROKITANSKY, Baron Karl von, see von Rokitansky, Baron Karl.

ROKKAN, G(eorg) Stein, Norwegian sociologist; b. Vagan, Norway, July 4, 1921; s. Georg Emil and Charlotte (Arntzen) R.; Mag.Art., U. Oslo (Norway), 1948, Cand.Philol., 1948; postgrad. Am. univs., London Sch. Econs.; m. Elizabeth Clough Harris, Oct. 14, 1950; children—Siriol, Bendik. Research asso. UNESCO, Paris, 1948-49; lectr. U. Oslo, 1951-56; dir. research Oslo Inst. Social Research, 1951-60; dir. research Michelsen Inst., Bergen, Norway, 1958-68; prof. U. Bergen, 1966—. Cons. UNESCO, 1956, 60, 65. Center for Advanced Study Behavioral Scis. fellow, Stanford, 1959-60, 67. Mem. Internat. Sociol. Assn. (past sec., v.p. 1966—), Internat. Soc. Sci. (exec. council 1964—). Author: Approaches to the Study of Political Participation, 1962; (with J. Meyriat) International Guide to Electoral Statistics, 1968; also numerous articles. Editor: (with R. McKeon) Democracy in a World of Tensions, 1950; (with R. L. Merritt) Comparing Nations, 1966; Data Archives for the Social Sciences, 1966; (with S. M. Lipset) Party Systems and Voter Alignments, 1967; Comparative Research across Cultures and Nations, 1968. Orgn. large-scale projects on elections, comparative polit. sociology; studies in comparative theory of elections and party systems. Home: 34 Sudmansvei, Bergen, Norway.*

ROLAND OF PARMA (or Rolando Capelluti, or Rolandus Parmensis), Italian physician; flourished 2d half of 13th Century; pupil of Roger of Salerno. Prof., U. Bologna; brought Salernitan surgery to Lombardy. Author: Chirurgia Rolandina, which was Roger of Salerno's Surgery, or Practica re-edited with additions (which showed Arabic influences). Roger of Salerno and Roland were subjects of a commentary (probably French), Glossulae Quater Magistrorum Super Chirurgiam Rogerrii et Roland, which was a very successful mediaeval textbook on surgery.

ROLANDO, Luigi, Italian biologist, anatomist; b. Turin, Italy, June 20, 1773; apptd. prof. medicine, Sassari, 1804, then anatomy U. Turin, 1814; author: Saggio Supra la Vera Struttura del Cervello, 1809; Humani Corpis Fabricae, 1817. Research on structure and function of nervous system; described club-shaped portion of posterior part of lateral funiculus of medulla oblongata, Rolando's gelatinous substance, 1809, and rolandic fibers, area, fissure. Died Apr. 20, 1831.

ROLDAN ROMAN, Enrique, Mexican neurophysiologist; b. Fed. Dist., Mexico, Dec. 27, 1933; s. Enrique Roldán Santillán and Maria Román; Ph.D. in Psychology, 1957; m. Martha Roldán, Feb. 7, 1958; children—Hugo, Arturo, Gabriel, Sara. Psychologist, Fed. Dist. Pentitentiary, Mexico, 1953-54, Gen. Bd. Rehab., 1953-56, Guidance Clinic, 1955-56; prof. psychology UNAM, 1959—, prof. neurophysiology, 1961—, prof. neuroanatomy, 1963—, part time asst. researcher, 1963-65. Vis. researcher Inst. Physiology, Czechoslovakia, 1962; sci. research Cerebral Investigation Unit, 1964-66. Mem. Am. Psychol. Soc., Mexican Soc. Physiol. Scis. Research, publs. on facilitating and inhibiting factors of convulsion, visual parareceptors, physiol. characteristics and origin of dream cycles. Home: 68 Tebas, México, D.F. 16, México. Office: 3877 Insurgentes Sur, México, D.F., México.*

ROLFINCK, Guerner, German physician, anatomist, chemist; b. Hamburg, Nov. 15, 1599; studied medicine at Wittenberg under Sennertus, 1616; at Leyden, 1618; at Oxford, 1621; at Paris and Padua, M.D. 1625. Taught anatomy in Venice; prof. of anatomy, surgery, botany, U. Jena, 1629; prof. practical medicine and chemistry, 1641; founded lab. and botanic garden at Jena; dir. botanical garden, 1630-38; physician to William IV, Duke of Saxe-Weimar. Author: Chimia in artis Formam Redacta, Sex Libris comprehensa, 1661. 1st prof. of chemistry in Germany; refuted Thurneysser's supposed transmutation of the nail. Died Jena, Saxe-Weimar, May 6, 1673.

ROLLE, Michel, French mathematician; b. Auvergne, France, Apr. 21, 1652; moved to Paris, circa 1675, began work as copy-boy, studied math.; elected mem. French Acad. Scis., 1685, 2d geometrical-pensionary, 1699-1719; recipient govt. pension for answering an arithmetic problem posed by Ozanam. Author: Traité d'algebre, 1692; Methode pour la résolution des problèmes indéterminés, 1699; Sur la question inverse des tangentes, 1704; other works. Formulated theorem of Rolle (important in calculus); known for disputes with de Gua, Varignon, Saurin. Died Paris, Nov. 8, 1719.

ROLLER, Duane Emerson, Am. physicist; b. Minerva, O.; Oct. 9, 1894; s. Urban Duane and Etta (Eyster) R.; student U. Wis., 1914-15, Ohio State U., 1915-17; A.B., U. Okla., 1923, M.S., 1925; Ph.D., Cal. Inst. Tech., 1928; D.Sc., Hamline U., 1952; m. Doris Della DuBose, June 15, 1919; children—Duane Henry DuBose, Richard Eyster (dec.). Faculty U. Okla., 1924-37, prof., 1934-37; faculty Hunter Coll., 1937-44, prof., 1943-44; prof. Wabash Coll., 1944-52; asst. dir. Hughes Aircraft Research and Devel. Labs., 1952-53; editor Science, also Sci. Monthly, 1953-55; sr. staff Ramo-Woolridge Corp., Los Angeles, 1955-57; prof. physics Harvey Mudd Coll., Claremont, Cal., 1957-64, prof. emeritus, 1964—; vis. lectr. Harvard, 1948-49. Tech. staff NDRC, 1941-44. Ford Found. fellow, 1951-52. Fellow Am. Phys. Soc., A.A.A.S.; mem. Am. Assn. Physics Tchrs. (pres. 1951 Oersted medal 1947), Sigma Xi, Phi Beta Kappa. Author: Physical Terminology, 1930; (with R.A. Millikan and E.C. Watson) Mechanics, Molecular Physics, Heat and Sound, 1937; Early Development of the Concepts of Temperature and Heat, 1950; (with D. H. D. Roller) Early History of the Concept of Electric Charge, 1954. Editor: Am. Jour. Physics, 1933-49. Contbr. numerous articles on exptl. physics and history, language and teaching of physics. Office: Harvey Mudd Coll., Claremont, Cal. 91716.*

ROLLESTON, George, English physician; b. Maltby, Eng., July 30, 1829; s. George Rolleston; B.A., Pembroke Coll., Oxford, 1850; M.A., 1853, M.B., 1854; M.D., St. Bartholomew's Hosp., London, 1857; m. Grace Davy, Sept. 21, 1861; 7 children. Physician, Brit. Civil Hosp., Smyrna, 1855-57; asst. physician Hosp. for Sick Children, London; became physician Radcliffe Infirmary, 1857; also Lee's reader anatomy; Linacre prof. anatomy and physiology, 1860-81. Became fellow Merton Coll., 1872, Fellow Royal Soc., 1862, Royal Coll. Physicians. Author: Forms of Animal Life (first zool. text to organize material into series of types system), 1876; also articles. Research on brain classification; urged improvement of sanitary conditions, especially in Oxford. Died Oxford, June 16, 1881.

ROLLESTON, Sir Humphrey Davy, English physician; b. Oxford, Eng., June 21, 1862; s. G. and Grace (Davy) Rolleston; ed. St. John's Coll., Cambridge; student medicine St. Bartholomew's Hosp.; M.D., Cambridge, 1891; D.Sc. (hon.), Oxford U., U. Pa.; D.C.L., Durham, Eng.; LL.D., Glasgow, Scotland, Edinburgh, Scotland, Bristol, Eng., Birmingham, Eng., Nat. U. Ireland, Jefferson; Doctor honoris causa, Padua, Italy, Dublin, Ireland, Bordeaux, France, Paris; m. Lisette Eila Ogilvy; 2 sons (both dec.). Demonstrator anatomy St. Bartholomew's Hosp.; became asst. physician St. George's Hosp., 1893, Met. Hosp., Victoria Hosp. for Children; named cons. physician Imperial Yeomanry Hosp., Pretoria S. Africa, 1901, Royal Navy, 1914-18; physician in ordinary to King George V, 1923-32, physician extraordinary, 1932-36; Regius prof. physics Cambridge U., 1925-32; mem. Royal commns. on lunacy and mental disorder, also on nat. health ins.; mem. med. consultative bd. Navy, Med. Adminstrv. Com., RAF, Colonial Office Com. Med. Services, chmn. coms. on vaccination and med. records Ministry Health, com. on workmen's compensation Home Office. Fellow Royal Coll. Physicians Ireland, Royal Faculty Physicians and Surgeons, Royal Coll. Surgeons Eng., Faculty Radiologists (hon.), N.Y. Acad. Medicine (hon.), mem. Brit. Med. Assn. (v.p.), Royal Coll. Physicians (pres. 1922-26), Royal Soc. Medicine (pres. 1918-20), Assn. Physicians Gt. Britain (became pres. 1925, 29), Brit. Inst. Radiology (pres.), Acad. de Médecine de Paris (corr. fgn.), Assn. Am. Physicians (hon.), Roentgen Soc. (hon., past pres.), other fgn. acads. Author: Some Medical Aspects of Old Age, 1922; (with J. W. McNee) Diseases of the Liver, 3d edit. 1929; Memoir of Sir Clifford Allbutt, 1929; The Cambridge Medical School, 1932; The Endocrine Organs, 1936. Editor: British Ency. of Medical Practice; (with Sir Clifford Allbutt) A System of Medicine, 11 vols., 2d edit. Research and publs. in clin. medicine, pathology, internal medicine, med. biography. Died Haslemere, Eng., Sept. 23, 1944.

ROLLET, (Martin-Pierre) Joseph, French physician; b. Lagnieu, France, Nov. 12, 1824; surgeon-maj., Lyon; prof. hygiene Faculty Medicine, Lyon; mem. French Acad. Scis., 1893; Acad. Medicine. Author: Traité des Maladies Vénériennes, 1865. Conceived idea of mech. glass blowing; proved existence of Rollets's disease, mixed lesions resulting from syphilis and other infection. Died Lyon, Aug. 2, 1894.

ROLLET, Pierre Antoine, French chemist; b. Belfort, France, May 18, 1902; s. Prosper and Jeanne (Ringenbach) R.; Ingénieur chimiste, Faculté des Sciences, Strasbourg, France, 1922, Licencié ès Sciences, 1926, Doctorate ès Sciences, 1929; m. Marguerite Feltz, Sept. 15, 1932; children—Henry, Marie-Claire, Jacques, Anne-Lise (Mrs. Samson). Asst., Faculté des Sciences, Strasbourg, 1930-34; instr. Faculté des Sciences, Besançon, France, 1934-36; titular prof. Faculté des Sciences, Alger, Algeria, 1937-57; titular prof. mineral chemistry Faculté des Sciences, Paris, 1958—, Decorated Legion of Honor, Officer Pub. Instrn., Laureat French Acad. Scis. Contbg. author: Chimie minérale. Research and publs. on inter-phase equilibrium (solid, liquid, gas); devel. techniques of study under pressure, including solubility measurements, direct of differential thermal analysis; dehydration under pressure, identification of inferior hydrates; monovalent positive ions of borates (lithium, potassium) by establishment of binary and teritiary systems; oxides, hydroxides and alkaline carbonates. Home: 127 avenue Philippe Auguste, Paris 11, France.*

ROMAGNOLI, Aldo, Italian vet. physician; b. Guardistallo, Mar. 22, 1924; s. Emidio and Anna Maria (Bastelli) R.; ed. Pisa and Cornell univs.; M.D. in Vet. Medicine, 1946; M.S., 1952; m. Rosarita Volpi, Jan. 3, 1957; children—Stefano, Anna. Prof. spl. pathology and clin. vet. medicine U. Messina, 1956—, also dir. Vet. Med. Clinic. Mem. Soc. Italian Scienze Veterinarie, Soc. Italian di Biologia Sperimentale, Acad. Peloritana di Sci. Mat. Fis. e Fis. e Nat. Author numerous work on diseases of domestic animals. Research on small and large animals; vet. clin. pathology and medicine. Home: via Vincenzo D'Amore 162. Office: via S. Cecilia 30, Messina, Italy.

ROMAGNOSI, Giovanni Domenico, Italian physicist, philosopher; b. Salsomaggiore, Italy, Dec. 11, 1761; student Jesuit Sch., Borgo di Donnino, 1772-75, Alberoni Coll., Piacenza, Italy, 1775-81. Became advocate, Trent, Italy, 1791; named prof. law, Parma, Italy, 1802, Milan, Italy, 1806; tried and acquitted of revolutionary activities, 1821; gave pvt. lessons until his death. Author: La genesi del diritto penale, 1791; Introduzione allo studio del diritto pubblico universale, 2 vols., 1805. Assisted in writing criminal code, Milan; performed expts. with electricity resulting in deflection of magnetic needle, 1802. Died Milan, June 8, 1935.

ROMAN, Nancy G(race), Am. astronomer; b. Nashville, May 16, 1925; d. Irwin and Georgia Frances (Smith) Roman; B.A. (Joshua Lippincott Meml. fellow), Swarthmore Coll., 1946; Ph.D., U. Chgo., 1949. Asst., Sproul Obs., Swarthmore Coll., 1943-46; asst. Yerkes Obs., U. Chgo., at Williams Bay, Wis., 1946-48, research asso., 1949-52, instr. stellar astronomy, 1952-55, asst. prof., 1955; research asso. Warner and Swasey Obs., Case Inst. Tech., Cleve., summer 1949; physicist radio astronomy by. U. S. Naval Research Lab., Washington, 1955-56, astronomer, head microwave spectroscopy sect., 1956-58, astronomer cons., 1958-59; head observational astronomy program Office Space Flight Devel., NASA, Washington, 1959-60, chief astronomy and solar physics, geophysics and astronomy programs, 1960-63, chief astronomy programs, physics and astronomy programs, 1963—. Recipient Fed. Woman's award, 1962. Mem. Am., Royal astron. socs., Internat. Astron. Union (editor symposia 1956, 58), Astron. Soc. Pacific, Internat. Sci. Radio Union, Am. Assn. U. Women. Contbr. articles sci. periodicals. Research on stellar clusters, including Ursa Major cluster, high velocity stars, radio astronomy, F-type subdwarfs, spectra of components of eclipsing binaries, bright high velocity stars; use of open clusters to calibrate spectroscopic parallaxes; statis. study of apparently bright stars. Home: 722 Hunting Pl., Balt. 21229. Office: NASA Hdqrs., 400 Maryland Av. S.W., Washington 20546.*

ROMANES, George John, English biologist, physiologist; b. Kingston, Eng., May 20, 1848; s. George and Isabella Gair (Smith) R.; grad. Gonville and Caius Coll., Cambridge, 1870; M.A., Oxford; postgrad. Univ. Coll., London, 1874-76; LL.D., Aberdeen, 1882; me. m. Ethel Duncan, Feb. 11, 1879; 5 sons, 1 dau. Prof., Edinburgh, 1886-90; Fullerian prof. physiology Royal Instn., 1888-91; founder Romanes lecture Oxford, 1891; became hon. fellow, Cambridge, 1892. Fellow Royal Soc., 1879; mem. Linnean Soc. (zool. sec.). Author: Candid Examination of Theism, 1878; Animal Intelligence, 1881; Scientific Evidences of Organic Evolution, 1882; Mental Evolution in Animals, 1883; Jelly-fish, Starfish, and Sea Urchins, 1885; Mental Evolution in Man, 1888; Darwin and after Darwin, 1892; Examination of Weismannism, 1892; Mind and Motion and Monism, 1895. Demonstrated parallel devel. of mental faculties of man and animals, 1881; applied Darwin's theory of evolution to devel. of mind, 1888; maintained that acquired characteristics can be inherited, 1892; set forth theory of physiol. selection pertaining to influence of environment on evolution of distinct species from isolated groups of original species, 1892. Died May 23, 1894.

ROMANO, Antonio Harold, Am. microbiologist; b. Penns Grove, N.J., Mar. 6, 1929; s. Antonio and Maria (Del Borello) R.; B.Sc., Rugers U., 1949, Ph.D., 1952; m. Marjorie Jean Backus, Aug. 22, 1953; children—Stephen Anthony, James William, Charles Paul. Research microbiologist Ortho Research Found., Raritan, N.J., 1952-54; instr. Rutgers U., New Brunswick, N.J., 1954-56; biochemist USPHS, Cin., 1956-59; faculty U. Cin., 1959-—, prof. bacteriology, 1964-—. NSF Sr. Posdoctoral research fellow, 1967-68. Mem. Am. Acad. Microbiology, Am. Soc. Microbiology, A.A.A.S., Sigma Xi. Research and publs. on biochemistry and physiology of microorganisms. Home: 2586 Bonnie Dr., Cin. 45230.*

ROMANO, John, Am. psychiatrist; b. Milw., Nov. 20, 1908; s. Nicholas Vincent and Frances Louise (Notari) R.; B.S., Marquette U., 1932; M.D., 1934; m. Miriam Modisett, May 13, 1933; 1 son, David Gilman. Commonwealth Fund fellow psychiatry U. Colo. Sch. Medicine, 1935-38; Rockefeller fellow neurology Harvard Med. Sch., 1938-39, asst. in medicine, 1939-40, instr., 1940-42; fellow neurology Boston City Hosp., 1938-39; Sigmund Freud fellow psychoanalysis Boston Psychoanalytic Soc., 1939-42; prof., chmn. dept. psychiatry U. Cin. Coll. Medicine, 1942-46; dir. dept. psychiatry Cin. Gen. Hosp., 1942-46; prof., chmn. dept. psychiatry U. Rochester (N.Y.) Sch. Medicine and Dentistry, 1946-—; psychiatrist-in-chief Strong Meml. Hosp., 1946-—; cons. surgeon gen. U.S. Army, Europe, 1955; cons. various hosps. Mem. Nat. Adv. Mental Health Council, 1946-49, chmn. research study sect., 1946-48, chmn, tng. com. on psychiatry, 1949-52; chmn. adv. com. on human growth and emotional devel. Social Sci. Research Found.-Ford Found., 1953; mem. adv. com. Behavioral Scis. div. Ford Found., 1954-58; chmn. mental health career investigator selection com. USPHS, 1956-61; sci. adv. council Nat. Found., 1959-60; mem. bd. psychiat. examiners N.Y. State Dept. Mental Hygiene, 1963-—; vis. prof., lectr. various univs. in U.S. and fgn. countries. Commonwealth Fund Advanced fellow for European study, 1959-60. Diplomate Am. Bd. Psychiatry and Neurology. Fellow Am. Psychiat. Assn., Am. Acad. Arts and Scis.; mem. Internat. League Against Epilepsy, Am. Soc. for Psychosomatic Medicine, Am. Neurol. Assn., A.A.A.S., Am. Soc. for Clin. Investigation, Group for Advancement Psychiatry, Assn. for Research in Nervous and Mental Diseases, Rochester Acad. Medicine, Monroe County Med. Soc., Sigma Xi, Alpha Omega Alpha. Editor: Adaptation, 1949. Co-editor sect. psychiatry Interne Mag., 1945-47; editorial bd. Jour. Psychiat. Research, 1961-—. Research, publs. on physiology, psychology of delirium; teaching of psychiatry; decompression sickness; syncope; epidemiology of mental illness. Home: 240 Chelmsford Rd., Rochester 14618. Office: Strong Meml. Hosp., 260 Crittendent Blvd., Rochester, N.Y. 14620.*

ROMANOFF, Alexis Lawrence, embryologist; b. St. Petersburg, Russia, May 17, 1892; s. Lawrence Mercury and Daria (Kondratieff) R.; came to U.S., 1921, naturalized, 1927; Diploma, St. Petersburg Tchrs. Coll., 1910-15; student Acad. Fine Arts, St. Petersburg, 1912-14; student Tomsk (Russia) U., 1918-19, Vladivostok Poly. Inst., 1919-20; B.S., Cornell U., 1925, M.S., 1926, Ph.D., 1928; m. Anastasia J. Sayenko, Sept. 1, 1928. Research asst. Cornell U., Ithaca, N.Y., 1924, 26-28, research instr., 1928-32, research asst. prof., 1932-43, asso. prof., 1943-48, prof. chem. embryology, 1948-60, prof. emeritus, 1960-—; research fellow biology Harvard, 1939-40; research fellow Yale, 1940-41; research asso. physics U. Fla., 1942. Lectr., cons. in U.S. and fgn. countries, 1932-—. Fellow N.Y. Acad. Scis., A.A.A.S., Poultry Sci. Assn. (Borden award 1950); mem. Am. Physiol. Soc., Soc. Am. Zoologists, Am. Chem. Soc., Sigma Xi. Author: The Avian Egg, 1949; The Avian Embryo, 1960; Contbr. numerous articles to sci. jours. on physiology of reprodn.; structure and chemistry of eggs; chem. embryology; embryo growth and metabolism; influence of environment on devel.; embryonic mortality. Patentee sci. equipment. Home: Belleayre Apts., 700 Stewart Av., Ithaca, N.Y. 14850.

ROMANS, Bernard, civil engr., naturalist; b. Netherlands, 1720; m. Elizabeth Whiting, Jan. 28, 1779; at least 1 child, Hubertus. Sent to N. Am. by Brit. Govt. for engring. work, 1757; apptd. dep. surveyor Ga., 1766; went to East Fla. to survey Lord Edgmont's estates on Amelia Island and St. John's River; apptd. prin. dep. surveyor Fordham dist.; explored Fla. and Bahama banks, and West Coast as far as Pensacola, Fla.; assisted in survey of West Fla., also in map preparation, circa 1771; made bot. discoveries, became king's botanist in Fla.; mem. Conn. com. to take possession of Ticonderoga and its outposts, 1775; with N.Y. Com. of Safety to construct fortifications on the Hudson nr. West Point, N.Y., 1775; gave allegiance to N.Y. Provincial Congress, 1775; served in war. Mem. N.Y.C. Marine Soc., Am. Philos. Soc. Author: A Concise Natural History of East and West Florida, Vol. I, 1775; The New American Military Pocket Atlas, 1776; Annals of the Troubles in the Netherlands, Vol. I, 1778, Vol. 2, 1782; also printed maps, including plans of Pensacola Harbor, Mobile Bay, Tampa Bay; an article on indigo, 1774. Died at sea, 1784.

ROMAS, Jacques de, see de Romas, Jacques.

ROMBERG, Moritz Heinrich, German pathologist, neurologist; b. Meiningen, Germany, Nov. 11, 1795; M.D., U. Berlin, 1817; practiced medicine, Berlin, 1817-73; named prof. U. Berlin, 1838. Author: Lehrbuch der Nervenkrankheiten (classic work), 1840. Founder of sci. neurotherapy in Germany; described symptoms of tabes dorsalis (Romberg's syndrome), 1840, progressive facial hemiatrophy (Romberg's disease), 1846. Died Berlin, June 16, 1873.

ROMER, Alfred Sherwood, Am. zoologist, paleontologist; b. White Plains, N.Y., Dec. 28, 1894; s. Harry Houston and Evalyn (Sherwood) R.; A.B., Amherst Coll., 1917, D.Sc., 1952; Ph.D., Columbia, 1921; D.Sc., Harvard, 1949, Dartmouth, 1959, U. Buffalo, 1960, Lehigh U., 1966; m. Ruth Hibbard, Sept. 12, 1924; children—Sally Hibbard (Mrs. Paul R. Evans), Robert Horton, James Henry. Instr. anatomy Bellevue Med. Coll., N.Y. U., 1921-23; asso. prof. vertebrate paleontology U. Chgo., 1923-31, prof., 1931-34; prof. zoology, curator vertebrate paleontology Harvard, Cambridge, Mass., 1934-65, Alexander Agassiz prof. zoology, 1947-65, dir. biol. labs., 1944-46, dir. Mus. Comparative Zoology, 1946-61. Pres. XVI Internat. Zool. Congress, Washington, 1963. Mem. Nat. Acad. Sci. (Thompson medal 1956, Elliot medal 1960), Am. Assn. Anatomists, Paleontol. Soc. (v.p. 1939), Soc. Vertebrate Paleontol. (pres. 1940), Am. Soc. Zoologists (pres. 1950), A.A.A.S. (v.p. 1948, dir. 1958-64, pres. 1966), Royal Soc. Edinburg (fgn. mem.), Soc. Animal Morphologists and Physiologists (India, hon.), Soc. Systematic Zoology (pres. 1952), Soc. Study Evolution (pres. 1953), Am. Philos. Soc., Geol. Soc. London, Soc. Naturalists, Geol. Soc. Am. (Penrose medal 1962), Am. Acad. Arts and Scis., Am. Soc. Mammalogists, Am. Soc. Ichthyologists and Herpetologists, Linnean Soc. (London), Zool. Soc. London, Acad. Nac. Cien. Córdoba (fgn.), Bayer Akad. Wiss. (corr.), Phila. Acad. Nat. Sci. (Hayden medal 1962), Sigma Xi. Author: Vertebrate Paleontology, 1933; Man and the Vertebrates, 1933; The Vertebrate Body, 1949; Osteology of the Reptiles, 1956; The Vertebrate Story, 1959. Contbr. articles to sci. jours. on vertebrate paleontology and anatomy, especially amphibians and reptiles. Home: 38 Avon St., Cambridge, Mass. 02138.

RöMER, Olaus (or Ole), Danish astronomer; b. Aarhus, Jutland, Sept. 25, 1644; studied under Erasmus Bartholinus at U. Copenhagen. Assisted J. Picard to determine geographical position of Tycho Brahe's observatory (Uraniborg), 1671; assistant to J. Picard at royal observatory, Paris, 1672-79; summoned to Denmark by King Christian V, 1681; royal mathematician and prof. of astronomy, U. Copenhagen, 1681-1710; dir. Copenhagen Observatory, 1681-1710; mayor of Copenhagen, 1705; chief of police; privy councillor; built "Tusculan" observatory, nr. Copenhagen, 1704. All of Römer's observations destroyed in conflagration of Oct. 21, 1728. While at Paris discovered velocity of light, which he determined by observing eclipses of satellites of Jupiter, 1672-79; constructed 1st good transit instrument, 1689; and meridian circle, 1690. Died Copenhagen, Denmark, Sept. 19, 1710.

ROMETSCH, Rudolf, phys. chemist; b. Basel, Switzerland, Nov. 9, 1917; s. Emanuel and Elisabeth (Ofteringer) R.; Doktorate in phys. chemistry U. Basel, 1943; m. Adelheide Galli, 1946; children—Eva, Michael, Sibylla. Asst., U. Basel, 1943-45; staff pharm. research dept. CIBA, 1945-59, head div. phys. chemistry, 1955-59; lectr. Fed. Inst. Tech., Zurich, Switzerland, 1956-59; dir. research dept. Eurochemic Co., Mol Donk, Belgium, 1959-64, mng. dir., 1964-—. Mem. Sté Suisse des Spécialistes du Génie Nucléaire, Schweizerische Physikalische Gesellschaft, Schweizerische Chemische Gesellschaft, Schweizerischer Chemiker Verband. Research and publs. on determination of absolute configuration of assymetric diphenyl compounds; application of tracer technic in pharm. research, fuel cycle studies; fabrication of uranium-IV-nitrate and application as plutonium reductant; radioactive waste solidification. Home: Kastelsebaan Mol, Belgium. Office: Eurochemic, 200 Boeretang, Mol Donk, Belgium.

ROMINGER, Carl Ludwig, geologist; b. Schnaitheim, Würtemberg, Dec. 31, 1820; s. Ludwig and Johanna Dorothea (Hoecklin) P.; ed. in Latin sch., then became apprentice in drug store and prepared for U. of Tübingen, where studied 1839-44, and was grad. M.D.; asst. in chem. laboratory of univ., 1844-47, and received from State annual gift of money to prosecute geol. studies, traveling over large portions of Germany, France, Austria, Switzerland, etc.; m. Frederica Mayer, Nov. 30, 1854. Came to U. S., 1845, to continue geol. studies; practiced medicine over 25 yrs., but afterward devoted attention exclusively to geology and palaeontology; State geologist of Mich., 1870-84. Received medal of merit from Royal Acad., Munich, 1892. Extensive contbr. to geol. reports (wrote entire 3d and 4th vols. Mich. Reports and parts of others), 1869-83. Authority on geology of Mich.; important writings on fossil corals. Died 1907.

ROMINGER, Celestin Jules, Am. physician; b. Phila., May 10, 1925; s. Jules Alfred and Marcella (Culin) R.; grad. LaSalle Coll., Phila., 1944; M.D., Jefferson Med. Coll., Phila., 1948; m. Martina Henry, June 14, 1948; children—John, Marcella, Robert, Martina, David. Practice medicine, specializing in radiology, Phila., 1949-—; asso. radiologist Misericordia Hosp., 1954-59, chmn. isotope com., 1955-—, dir. dept. radiology, 1959-—, chmn. research com., 1964-—; asso. therapeutic radiologist Am. Oncologic Hosp., 1955-—, chmn. isotope com., 1957-—, chmn. conf. com., 1960-—; dir. Mercy Radiation Therapy Center, 1965-—. Diplomate Am. Bd. Radiology, Am. Bd. Therapeutic Radiology. Mem. Am. Coll. Radiology, A.M.A., Inter-Am. Coll. Radiology, Soc. Nuclear Medicine, Am. Radium Soc., Radiol. Soc. N.Am., Alpha Omega Alpha. Contbr. numerous articles to profl. jours. Research in field of cancer, including application of radiation alone and with surgery and chems. in treatment of cancer, application of x-rays and radioactive isotopes in diagnosis of disease states, especially cancer. Home: 320 Strathmore Dr., Rosemont, Pa. 19010. Office: 5301 Cedar Av., Phila. 19143.*

ROMME, (Nicholas-) Charles, French mathematician; b. Riom, France, Dec. 8, 1745. Prof. navigation Ecole de la marine, Rochefort, France. Mem. French Acad. Scis., 1801. Author navigational books including: Nouvelle méthode pour déterminer les longitudes en mer, 1777; L'Art de la marine, 1787; Dictionnaire de la marine anglaise, 1804; Tableau des vents des marées et des courant sur toutes les mers du globe, 1806. Died Rochefort, Mar. 14, 1805.

ROMPP, Hermann, German chemist; b. Weiden-bei-Horb, Feb. 18, 1901; s. Christian and Dorothea (Reich) R.; Dr. ès sc., U. Tübingen. Author: Chemie-Lexikon, vol. 1-3, 1962; Chemie des Alltags; Rezeptbuch des Alltags; Chemische Zaubertränke; Chemie der Metalle; Anorganische Chemie; Wunder-welt der Atome; Atom-Lexikon; Chemie der Zukunft; Spurenelemente; Wuchsstoffe; Under täglich Brot; Isotope; also 7 other books translated into 17 langs., and numerous articles; collaborator on Enzyklopädie der medezinischen Fachausdrucke. Address: 7241 Weiden-über-Horb, Baden-Württemberg, W. Germany.

RONA, George, physician; b. Budapest, Hungary, Mar. 8, 1924; s. Adolph Arona and Ilona (Steiner) R.; M.D., Med. U. Budapest, 1949; Ph.D., Hungarian Sci. Acad., 1954; m. Agnes Szilas, Jan. 13, 1950; children—Zoltan, Gabriel. Faculty, dep. dir. 1st Inst. Pathology and Exptl. Cancer Research, Med. U. Budapest, 1951-56; asso. pathologist St. Mary's Hosp., Montreal, Que., Can., 1963-64; sr. and cons. pathologist Ayerst Research Labs., Montreal, 1957-—; dir. labs., pathologist Lakeshore Gen. Hosp., Pointe Claire, Que., 1965-—; lectr. pathology McGill U., 1962-64, asst. prof., 1964-—. Fellow Coll. Am. Pathologists, Royal Coll. Physicians (Can.); mem. Canadian Fedn. Biol. Socs., Canadian Assn. Pathologiss, Am. Assn. Pathologists and Bacteriologists, Coll. Pathology (Eng.), Fedn. Biol. Socs. (Can.), Montreal Physiol. Soc. Author: (with H. Jellinek) Autopsy Technic, 1955. Contbr. over 100 articles to med. jours. Described pathology of antibiotic treated Tb meningitis, vascular changes in human and exptl. hypertension, discoverer isoproterenol myocardial necrosis; research on pathology and methods of prevention isoproterenol cardiac necrosis, human coronary heart disease. Home: 80 Henley Ave., Town of Mt. Royal, Que. 16. Office: 160 Stillview Rd., Pointe Claire, Que., Can.*

RONALDS, Sir Francis, English meterologist; b. London, Feb. 21, 1788; s. Francis and Jane (Field) R.; studied practical electricity under Jean André de Luc; hon. dir., supt. Meteorol. Obs., Kew, Eng., 1843-52; lived on continent. Fellow Royal Soc. 1844. Author: Descriptions of an Electric Telegraph and some other Electrical Apparatus, 1823; Mechanical Perspective, 1828; also articles on his inventions. Conducted expts. in 1816 which led to invention of telegraphic instrument based on principle of synchronously revolving discs (described in article, 1823); patentee perspective tracing instrument, 1825; developed photog. system of automatic registration of meteorol. instruments, 1844-45. Died Aug. 8, 1873.

RONDELET, Guillaume, French naturalist, physician; b. Montpellier, France, Sept. 27, 1507; prof. medicine Montpellier; friend co-worker of Rabelais. Author: Methodus de materia medicinali et compositione medicamentorum; Libri de piscibus marinis (includes descriptions of about 250 fish species), 2 vols., 1554-55; Universe aquatilium historiae pars altera, 1555; Liber de ponderibus. Laid found. for modern ichthyology. Died Réalmont, France, July 30, 1566.

RONNE, Finn, Antarctic explorer, and geographer; b. Horten, Norway, Dec. 20, 1899; s. Martin Richard and Maren (Gulliksen) R.; M.E. and naval architect,

Horten Tech. Coll., 1923; m. Edith X. Maslin, March 18, 1944; one daughter, Karen. Came to the United States, 1923, naturalized, 1929. Mem. Adm. Byrd's 2d Antarctic Expdn., 1933-35; 2d in Command, East Base, U. S. Antarctic Service, 1939-41; apptd. 1st Am. postmaster in the Antarctic, 1946; leader of Ronne Antarctic Research Expedition 1946-48, claimed 250,000 sq. miles of new land, including Edith Ronne Land for the U. S. Wintered four times on the Antartic Continent; expdn. to Antarctica with Argentine Navy, 1958-59. Capt. USNR, now consultant U. S. Dept. of Defense; comdg. officer, Weddell Sea Sta., Antarctica, sci. director IGY, 1956-58; mil. comdr., scientific leader Ellsworth Sta., Edith Ronne Land, Antarctica. Decorated knight 1st class Royal Order St. Olav (Norway). Recipient Elisha Kent Kane gold medal Phila. Geog. Soc., 1966; gold medal Explorers Club, 1968. Mem. Am. Geophysical Union (honorary), American Polar Society (honorary). Author: Antarctic Conquest; Hellhole of Antarctic; Antarctic Command; also articles on Polar regions. Decorated Congl. Silver Medal, 1935, Gold Medal, 1943, 48, 61; Argentine Govt. Decoration, 1959; Legion of Merit, 1964. Lectr. on polar subjects. Early exploration and mapping of Wendell Sea Area (Antarctica) and coastline; discovered glacier and valley on Alexander Island; ascertained that Antarctica was definitely one continent. Address: 6323 Wiscasset Rd., Washington 20216.

RONSSEUS, Balduinus, physician; b. Ghent, Belgium, 1525; pupil of Dryvère, Louvain, Belgium, 1541, then studied in Paris; described use of orange and lemon juices by Dutch sailors for prevention and treatment of scurvy, 1564. Died 1597.

RÖNTGEN, Wilhelm Konrad, German physicist; b. Lennep, Rhenish Prussia, Mar. 27, 1845; student mech. engring., Netherlands, also physics, Zurich, Switzerland; doctorate, 1869. Privatdozent, Strasbourg, 1874-75; extraordinary prof. math. and physics, 1876-79; prof. Agrl. Acad., Hohenheim, 1875-76; prof., dir. Phys. Inst., Giessen, 1879-85; prof., Würzburg, 1885-1900; prof. Ludwig-Maximilian U., Munich, 1900-23. Recipient Rumford medal Royal Soc., 1896, first Nobel prize for physics, 1901. Best known for his discovery of X-rays (frequently called Röntgen rays), 1895; performed many expts. to study their fundamental properties; also investigated specific heat of gases, 1870-73; absorption of infrared rays by gases, 1884, influence of pressure on viscosity in fluids, 1884, compressibility of solutions, 1886, capillarity, elasticity, and influence of pressure on refractive index of various fluids; conducted research on pyroelectricity, piezoelectricity, optical and elect. properties of quartz, and magnetic rotation of plane of polarization on light. Died Munich, Feb. 10, 1923.

ROOD, Ogden Nicholas, Am. physicist; b. Danbury, Conn., Feb. 3, 1831; s. Anson and Aleida Gouverneur (Ogden) R.; grad. Princeton, 1852; postgrad. univs. Munich and Berlin, 1854-58; m. Matilde Prunner, 1858. Prof. chemistry Troy U., 1858-63; prof. physics Columbia, 1863——. Mem. Nat. Acad. Sci. Author: Modern Chromatics, 1879; also papers. First to apply stereoscopic photography to microscope, to make quantitative expts. on color-contrast, to measure duration of flashes of lightning, to make a photometer ind. of color; improved Sprengel air pump to produce higher vacuums; inventor process for measuring great elec. resistances; research in acoustics. Died 1902.

ROOKE, Lawrence, English astronomer; b. Deptford, Eng., Mar. 13, 1622; s. George and Mary (Burrell) R.; M.A., King's Coll., Cambridge, 1647; m. Barbara Heyman; 4 daus., 5 sons, including Heyman, James. Became fellow commoner Wadham Coll., Oxford, 1650; prof. astronomy Gresham Coll., London, 1652-57, prof. geometry, 1657-62; asst. to Robert Boyle. Author: Observationes in cometam qui mense Decembri anno 1652 apparuit, 1653; On the Effect of Radiant Heat on the Height of Oil in a Long Tube; other papers. Died June 27, 1662.

ROOMEN, Adriaen van, see van Roomen, Adriaen.

ROONWAL, Mithan Lal, Indian zoologist; b. Jodhpur, India, Sept. 18, 1908; s. Mool Chand and Jamnabai Roonwal; B.Sc., Lucknow U., 1930, M.Sc., 1931; Ph.D., Cambridge U., 1935, Sc.D., 1962; m. Mrs. Premvati; children—Mrs. Shanti Negi, Dalpat S.; Mrs. Vimla Jwala, Mrs. Sarla Jhingran. Asst. prof., head zoology dept. Govt. Coll., Ajmer, India, 1935; asst. locust entomologist Indian Council Agrl. Research, Pasni, Baluchistan, India, 1935-39; in charge bird and mammal sect. Zool. Survey of India, Calcutta, 1939-49; with Forest Research Inst., Dehra Dun, 1949-56, Zool. Survey of India, Calcutta, 1956-65; with Jodhpur U., 1965——. Fellow Zool. Soc. India (pres., Tata medal), Nat. Inst. Sci. India; mem. Human Tropics Research Soc. (internat. adv. com. 1956-65). Author 2 books (with P. K. Sen-Sarma), 1960, (with O. B. Chhotani), 1962. Contbr. numerous articles to sci. jours. Several discoveries on insect embryology including new theory of multiphased gastrulation; ecology and morphology of locusts, especially population dynamics; ecology and taxonomy of mammals of Assam; new evolutionary phenomenon of increased variation with minimum populations in locusts, systematics and ecology of oriental termites. Home: House 182, Polo No. 2, Jodhpur, Rajasthan, India.*

ROOSEVELT, Theodore, Am. explorer, conservationist, 26th U. S. pres.; b. New York, Oct. 27, 1858; s. Theodore (1831-78) and Martha (Bulloch) R.; A.B., Harvard, 1880; LL.D., Columbia, 1899, Hope Coll., 1901, Yale, 1901, Harvard, 1902, Northwestern, 1893, U. of Chicago, 1903, U. of Calif., 1903, U. of Pa., 1905, Clark U., 1905, George Washington U., 1909, Cambridge U., 1910; D.C.L., Oxford U., 1910; Ph.D., U. of Berlin, 1910; m. Alice Hathaway Lee, Oct. 27, 1880 (died 1884); m. 2d, Edith Kermit Carow, Dec. 2, 1886. Mem. N.Y. Legislature, 1882-84; del. Rep. Nat. Conv., 1884; resided on ranch in N.D., 1884-86; candidate for mayor of New York, 1886; U. S. civil service commr., 1889-95; pres. New York Police Bd., 1895-97; asst. sec. of the navy, 1897-98; resigned to organize, with Surgeon (later Maj.-Gen.) Leonard Wood, 1st U. S. Cav.; was lt. col. of regt. in Cuba; promoted col.; mustered out Sept. 1898. Gov. N.Y., Jan. 1, 1899, to Dec. 31, 1900; elected Vice-President of the United States, Nov. 4, 1900, for term, 1901-05; succeeded to the presidency on death of William McKinley, Sept. 14, 1901; elected President of the U. S., Nov. 8, 1904, for term 1905-09, by largest popular majority ever accorded a candidate; Progressive Party candidate for president of U. S., 1912. Awarded the Nobel Peace Prize ($40,000), 1906. Spl. ambassador of U. S. at funeral of King Edward VII, 1910. Long a contbr. to leading mags. and reviews; known for yrs. as advocate of civil service and other reforms, nat. and municipal; contbg. editor The Outlook, 1909-14. Hon. fellow Am. Mus. Natural. History, 1917. Author: Winning of the West, 1889-96; History of the Naval War of 1812, 1882; Hunting Trips of a Ranchman, 1885; Life of Thomas Hart Benton, 1886; Life of Gouverneur Morris, 1887; Ranch Life and Hunting Trail, 1888; History of New York, 1890; The Wilderness Hunter, 1893; American Ideals and Other Essays, 1897; The Rough Riders, 1899; Life of Oliver Cromwell, 1900; The Strenuous Life, 1900; Works (8 vols.), 1902; The Deer Family, 1902; Outdoor Pastimes of an American Hunter, 1906; American Ideals and Other Essays; Good Hunting, 1907; True Americanism; African and European Addresses, 1910; African Game Trails, 1910; The New Nationalism, 1910; Realizable Ideals (the Earl lectures), 1912; Conservation of Womanhood and Childhood, 1912; History as Literature, and Other Essays, 1913; Theodore Roosevelt, an Autobiography, 1913; Life Histories of African Game Animals (2 vols.), 1914; Through the Brazilian Wilderness, 1914; America and the World War, 1915; A Booklover's Holidays in the Open, 1916; Fear God, and Take Your Own Part, 1916; Foes of Our Own Household, 1917; National Strength and International Duty (Stafford Little lectures, Princeton Univ.), 1917. Influenced interest in conservation of natural resources in U. S.; brought many bird and mammal specimens to U. S. from Africa, 1909-10, and S.Am.; headed expdn. that explored 600 miles of tributary (Reo Teodoro) of Madeira River, Brazil, 1914. Died Jan 6, 1919.

ROOZEBOOM, Hendrik Willem Bakhuis, Dutch phys. chemist; b. Alkmaar, Netherlands, Oct. 24, 1854; began career as indsl. chemist; asst. to Bemmelen, Leiden, Netherlands; named prof. chemistry, Amsterdam, Netherlands, 1896. Developed applications of phase rule (deduced by Gibbs), 1904, thus contbg. to modern chemistry of alloys; studied triple points; classified chem. equilibria and solid solutions. Died Amsterdam, Feb. 8, 1907.

ROQUETAILLADE, Jean de, see Rupescissa, Johannes de.

RORSCHACH, Hermann, Swiss psychiatrist; b. Zurich, Switzerland, Nov. 8, 1884. Author: Psychodiagnostik, 1921. Originator of psychiatric test based on interpretation of free forms in ink blots (Rorschach test); developed systematic analysis of patient's attention to ink blots as wholes or details, color, shading, and other factors. Died Herisau, Switzerland, Apr. 2, 1922.

ROSA, Edward Bennett, Am. physicist; b. Rogersville, N.Y., Oct. 4, 1861; s. Edward David and Sarah G. (Rowland) R.; B.S., Wesleyan U., 1886, Sc.D., 1906; Ph.D., Johns Hopkins, 1891; m. Mary Evans, Mar. 22, 1894. Prof. physics Wesleyan U., 1891-1902; physicist Nat. Bur. Standards, 1901-10, chief physicist, 1910——, in charge elec. div. including investigations on standards of service and safety for public utilities; sec. Internat. Com. on Elec. Units and Standards. Mem. Nat. Acad. Scis. Developed (with W. O. Atwater) Atwater-Rosa respiration calorimeter for nutrition study; inventor Rosa curve tracer for use in alternating electric currents; pioneer in determining absolute value of ampere; contbd. to establishment of units and standard nomenclature in photometry; studied flame standards; established laws of electrolytic corrosion; developed war instruments, including sound ranging device for locating big guns, geophone for detecting mining operations, aircraft radio apparatus, radio direction finders for locating craft; conceived of Nat. Electric Safety Code. Died May 17, 1921.

ROSA, Umberto, Italian chemist; b. Turin, Italy, Aug. 13, 1933; s. Alfonso and Laura Bassani (Romano) R.; D. in Chemistry (magna cum laude), Chem. Inst., U. Turin, 1957; m. Lucia Pavesio, Jan. 3, 1960; 1 son, Carlo. With high activity lab. Commissariat Energie Atomique, Saclay, France, 1959-60; head radioisotope research group Nuclear Research Center, Societa Richerche Impianti Nucleari, Saluggia, Vercelli, Italy, 1964——, also radioisotope prodn. rep. to tech. com. of Societa Richerche Impianti Nucleari, Commissariat Energie Atomique, Centre d'Etude de l'Energie Nucleaire Belgium Assn., 1962——. Asst. prof. radiochemistry Sch. Medicine U. Turin, 1965——. Mem. Italian Soc. for Nuclear Biology and Medicine. Mem. editorial bd. Jour. of Labeled Compounds, Jour. Biology and Medicine. Research and publs. in prodn. radioisotopes, developer of analytical procedure in radiochemistry and original labelling procedure for proteins. Home: 162 Cso. Galileo Ferraris, Turin, Italy. Office: Nuclear Research Center, Saluggia, Vercelli, Italy.*

ROSAHN, Paul Dolin, Am. pathologist; b. N.Y.C., Dec. 21, 1903; s. Kalman and Anna (Dolin) R.; A.B., Columbia U., 1924; M.D., N.Y. U., 1928; D.Sc. (hon.), U. Med. Scis., Bangkok, Thailand, 1960; m. Jean Maclean Wight, Aug. 8, 1959. Asst., Rockefeller Inst., 1930-35; asst. dir. research Pub. Health Inst., Chgo., 1935-36; Josiah Macy Jr. Found. fellow Yale Med. Sch., 1936-37; pathologist-in-chief New Britain Gen. Hosp., (Conn.), 1937——; pathologist New Britain Meml. Hosp., 1940——; dir. tumor clinic New Britain Gen. Hosp., 1939; asso. clin. prof. pathology Yale, 1947——; asso. prof. pathology Washington U. Sch. Medicine, in Thailand, 1952-53; vis. prof. pathology U. Med. Scis., Bangkok, Thailand, 1959-60. Recipient 1st prize essay contest Am. Dermatol. Assn., 1952. Fellow A.A.A.S., Coll. Am. Pathology; mem. New Britain Med. Soc. (pres. 1966——, certificate merit 1961), Harvey Soc., Am. Assn. Cancer Research, Am. Assn. Pathologists and Bacteriologists, Am. Soc. Exptl. Pathology, Internat. Acad. Pathology, Am. Soc. Clin. Pathologists. Contbr. numerous articles to tech. jours. Research on biology syphilis, constl. factors in disease, hematology, biology cancer, biometrics, Letterer-Siwe's Disease, hyaline membrane disease, weight normal heart. Home: 206 Ellwood Rd., Kensington, Conn. 06037. Office: 100 Grand St., New Britain, Conn. 06050.*

RÖSCH, Jean, French astronomer; b. Sidi-Bel-Abbès, Jan. 5, 1915; s. Gabriel and Lucile (Forgues) R.; Dr. ès sc., Ecole normale superieure; m. R. Postel Sept. 23, 1937. Astronomer, Bordeaux Obs., 1940-47; dir. Pic-du-Midi Obs., 1947——; prof. astronomy U. Paris, 1963——. Mem. Ramond Soc. Publs. on stereoscopic vision, astron. instruments, solar physics. Address: Obs. du Pic-du-Midi, Bagnères-de-Bigorre; also Inst. d'astrophysique, 98 bis, blvd. Arago, Paris 14, France.

ROSCOE, Henry Enfield, English chemist; b. London, Jan. 7, 1833; s. Henry and Marie (Fletcher) R.; B.A., U. Coll., London, 1852; Ph.D., Heidelberg, Germany, 1854; studied under R. W. von Bunsen; m. Lucy P., 1863; 1 son, 2 daus. Prof. chemistry Owens Coll., Manchester, Eng., 1857-85; mem. Parliament, 1885-95; vice chancellor U. London, 1896-1902; became privy councillor, 1909. Mem. Royal Commn. on Tech. Edn., 1882-84; gov. Lister Inst.; a founder Victoria U.; became exec. Carnegie Trust, 1901. Mem. Commn. Reporting on Pasteur's Treatment of Hydrophobia. Fellow Royal Soc., 1863, recipient Royal medal, 1873; mem. French Acad. Sci., 1893, Soc. Chem. Industry (a founder 1880), Soc. Chem. Industry (1st pres.), Chem. Soc. (pres. 1881-83), Brit. Assn. (became pres. 1887). Author: Lessons in Elementary Chemistry, 1866; Spectrum Analysis, 1868; (with Carl Schorlemmer) Treatise on Chemistry, 6 vols., 1877-89, The Hydrocarbons and Their Derivatives, 1881-92; A New View of the Origin of Dalton's Atomic Theory, 1896; Life and Experiences, 1906; Investigations on the Chemical Action of Light. Research and publs. on sewage filtration, niobium, tungsten, uranium, perchloric acid, (with Bunsen) photochemistry; originated reciprocity law; co-inventor of actinometer; 1st to prepare pure vanadium and show it belongs to phosphorus-arsenic family. Died Leatherhead, Eng., Dec. 18, 1915.

ROSE, Albert, Am. physicist; b. N.Y.C., Mar. 30, 1910; s. Simon and Sarah (Cohen) R.; A.B., Cornell U., 1931, Ph.D., 1935; m. Lillian Loebel, Aug. 25, 1940; children— Mark L., Jane R. With R.C.A., 1935——, dir. research, Zurich, Switzerland, 1955-58, fellow tech. staff, Princeton, N.J., 1958——; vis. prof. Princeton, 1961-62, Cornell U., 1967. Recipient Morris Liebman award Inst. Radio Engrs., 1946, Sarnoff award Soc. Motion Picture and TV, 1958, also awards from TV Broadcasters Assn., 1946, Soc. Motion Picture and TV, 1947. Fellow Am. Phys. Soc., I.E.E.E.; mem. Societe Suisse de Physique. Author: Concepts in Photoconductivity, 1963. Inventor the Orthicon and Image Orthicon, also dir. program leading to Vidicon (TV camera tubes now in gen. use); research and publs. on theories of human vision as ltd. by quantum flucuations, photoconductivity, solid state triodes and various transport phenomena in solids. Home: 292 Stockton Rd. Office: RCA Labs., Princeton, N.J. 08540.*

ROSE, Arnold Marshall, Am. sociologist; b. Chgo., July 2, 1918; s. Frank A. and Ruth (Wilansky) R.; B.A., U. Chgo., 1938, M.A., 1940, Ph.D., 1946; m. Caroline Baer, Dec. 24, 1942; children—Richard, Ruth (Mrs. Michael Parker), Dorothy. Research asso.

Carnegie Corp., N.Y., 1940, 1941-43; statistician U.S. War Dept., Washington, 1943; lectr. Bennington, Coll., Vt., 1946-47; asso. prof. Washington U., St. Louis, 1947-49; prof. sociology, U. Minn., Mpls., 1949—. Cons. HHFA, 1961—, Nat. Inst. Child Health and Human Devel., Washington, 1964—; chmn. Minn. planning com. White House Conf. on Aging, 1960-61. Pres, Marshall High Sch. P.T.A., 1960-66; nat. co-chmn. Centennial Celebration, Wayne State U., Detroit, 1963. Mem. Minn. Ho. of Reps., 1943-45. Recipient 1st award in sociol theory, A.A.A.S., 1953; named Fulbright research scholar, France, 1951-52, Fulbright prof., Italy, 1956-57. Mem., Soc. for Study Social Problems (pres. 1955-56), Midwest Sociol. Soc. (pres. 1961-62), Internat. (com. chmn. 1956—), Am. (pres. 1968—) sociol. assns., Am. Assn. U. Profs., Citizens League Minn., Com. for Cultural Freedom. Author: (with Gunnar Myrdal and R. Sterner) An American Dilemma, 1944; (with C.B. Rose) America Divided, 1948; The Negro in America, 1948; Theory and Method in the Social Sciences, 1954; Sociology: The Study of Human Relations, 1956, 65. Editor: Race Prejudice and Discrimination, 1951; Mental Health and Mental Disorder, 1955; Institutions of Advanced Societies, 1958; Human Behavior and Social Processes, 1962; Aging in Minnesota, 1963; Assuring Freedom to the Free, 1964; Minority Problems, 1965; The Power Structure, 1967; Libel and Academic Freedom, 1968. Studies, publs. on analysis and prediction of development of race relations; theories, methods of social science; contbr. concepts of group identification, covert and world culture, pseudo-mores, the aged. Home: Malcolm Av., S.E., Mpls. 55414.*

ROSE, Bruce, Canadian geologist; b. Mountain, Ont., Can., Jan. 5, 1885; s. Charles and Eliza Jane (Keys) R.; student U. Toronto (Ont.), 1905-06; B.Sc., Queen's U., Kingston, Ont., 1909; Ph.D., Yale, 1913; m. Minnie Olive, Jan. 12, 1921; children—Kenneth Campbell, Margaret Ailsa. Asst. prof. geology Queen's U., 1909-10, prof., 1922-51; conducted Geol. Survey Can., 1913-20; staff Whitehall Petroleum Corp.; cons. geologist, 1920—. Mem. Royal Soc. Can. (past pres. sect. IV). Research and publs. on structural and stratigraphic geology; proved existence of additional coal regions around Crownest pass in Alta, B.C., Can. Died 1956.

ROSE, David John, physicist; b. Victoria, Can., May 8, 1922; s. David A. and Nora (Birkett) R.; B. Applied Sci., U. B.C. (Can.), 1947; Ph.D., Mass. Inst. Tech., 1950; m. Constance Vivienne Fox, Feb. 6, 1948; children—Elizabeth, Victoria, Hugh, Andrew. Came to U.S., 1951, naturalized, 1958. Staff, B.C. Research Council, Vancouver, 1950-51; mem. tech. staff, supr. gas tube devel. Bell Telephone Labs., Murray Hill, N.J., 1951-58, cons., 1959-60; asso. prof. Mass. Inst. Tech., Cambridge, 1948-60, prof., 1960—. Bd. dirs. Magnion Corp., Burlington, Mass. Cons., Oak Ridge Nat. Labs., 1952—, Cornell Aero. Labs, Buffalo 1959-60, Gen. Motors Research Labs., Warren, Mich., 1963—; mem. numerous U.S. Govt. coms. on plasma physics and related topics. Fellow Am. Phys. Soc. (chairman div. plasma physics 1965), Am. Acad. Arts and Scis.; mem. Sigma Xi. Author: (with M. Clark Jr.) Plasmas and Controlled Fusion, 1961. Contbr. articles to tech. jours. Research on plasma physics, gaseous electronics, electron physics. Home: 45 Ridge Rd. Waban, Mass. 02168.*

ROSE, Francis Leslie, English chemist; b. Lincoln, Eng., June 27, 1909; s. Frederick William and Elizabeth (Watts) R.; B.Sc. with 1st honors in Chemistry, Nottingham (Eng.) U., 1930, Ph.D., 1934, D.Sc., 1950; m. Ailsa Buckley, June 15, 1935; 1 son, Peter Francis. With Imperial Chem. Industries, 1932—, sect. head dysstuffs div., 1942-54, research mgr. pharms. div., 1954—. Hon. reader in organic chemistry Faculty Tech., U. Manchester (Eng.), 1959—, mem. ct. govs., 1960—; mem. ct. govs. Manchester Coll. Sci. and Tech. Decorated Order Brit. Empire. Fellow Royal Soc., 1957, Royal Inst. Chemistry; mem. Biochem. Soc., Pharmacological Soc., Chem. Soc. London, Soc. Chem. Industry. Research and numerous publs. in medicinal chemistry; discovery drugs for treatment bacterial infections, malaria, cancer. Home: 26 Queensway Heald Green, Cheshire, Eng. Office: Imperial Chem. Industries Ltd., Pharmaceuticals Div., Alderley Park, Macclesfield, Eng.*

ROSE, George Gibson, Am. tissue culture scientist; b. Liberty, Ind., Aug. 7, 1922; s. Joseph Sims and Dorothy (Gray) R.; B.A., U. Tex., 1947, M.D., 1951; m. Jeanne W. Roco, Dec. 24, 1945; children—Mary Dianne, George Glenn. Research asso. M.D. Anderson Hosp. and Tumor Inst., U. Tex., Houston, 1955-60, asso. biologist, 1966—, asso. prof. medicine Dental Br., 1960—. Cons., Pasadena (Cal.) Found. for Med. Research, 1961-65. Fellow Royal Micros. Soc. (London); mem. Am. Assn. for Cancer Research, A.A.A.S., Am. Soc. Cell Biology, Tissue Culture Assn., Internat. Assn. Dental Research. Author: Cimemicrography in Cell Biology, 1963; also articles. Developed chamber for cultivation living cells which was incorporated into a multichamber miniature heart-lung circulating system; research by time-lapse movie techniques. Home: 5306 Rotherglenn St., Houston 77035.*

ROSE, Gustav, German mineralogist; b. Berlin, Germany, Mar. 28, 1798; studied under Berzelius, Stockholm, Sweden. Given charge mineral. collection U. Berlin, 1822, became asso. prof. mineralogy, 1826, prof., 1839; travelled around No. Asia, 1829. Fellow Royal Soc., 1866; mem. French Acad. Scis., 1832. Author: Elemente der Krystallographie, 1833; Reise nach dem Ural dem Altai und dem Kapische Merre, 1837-42; Ueber das Krystallisationsystem des Quarzes, 1846; Das Krystalochemische Mineralsystem, 1852; Beschreibung und Einteilung der Meteoriten, 1864; also essays. One of 1st to describe and classify rocks; developed system of crystallography. Died Berlin, July 15, 1873.

ROSE, Heinrich, German chemist; b. Berlin, Germany, Aug. 6, 1795; s. Valentin Rose the Younger; student pharmacy, Berlin; D.Sc., U. Kiel, 1820; studied under Berzelius, Stockholm, Sweden. Became extraordinary prof. chemistry, Berlin, 1823, prof., 1835. Fellow Royal Soc., 1842; mem. French Acad. Scis., 1843. Author: Handbuch der analytischen Chemie, 2 vols., 1829; Ausführliches Handbuch der analytischen Chemie, 2 vols., 1851; also articles. Analyses in inorganic chemistry; pioneered research on mass action in chem. transformation; proved niobium (columbium) and tantalum were different metals, 1844; discovered antimony pentachloride; developed new lab. methods. Died Berlin, Jan. 27, 1864.

ROSE, Jerzy Edwin, neurophysiologist; b. Buczacz, Poland, Mar. 5, 1909; s. Henryk and Regina (Deiches) R.; M.D., Jagiellon U., 1937; m. Annelies Argelander, Mar. 6, 1939. Came to U.S., 1939, naturalized, 1941. Research brain anatomy and brain physiology; faculty Johns Hopkins, 1940-60; prof. neurophysiology U. Wis., 1960—. Mem. Soc. Exptl. Biology, Physiol. Soc., Anat. Assn., Neurol. Assn. Mem. editorial bd. Jour. Neurophysiology, 1964—, Jour. Comparative Neurology, 1966—. Contbr. numerous articles to profl. jours. Publs. on the anatomy and neurophysiology of the mammalian central nervous system. Home: 814 Oneida Pl., Madison, Wis. 53711.*

ROSE, Morris Erich, Am. physicist; b. Bklyn., May 9, 1911; s. Samuel and Dora (Lipset) R.; A.B., Wayne U., 1931; M.A., U. Mich., 1933, Ph.D., 1935; m. Alice Moskowitz, Dec. 25, 1934, (dec. Feb., 1965); children—Barbara Susan (Mrs. Irwin Beitch), Joel Samuel; m. 2d Olive Plunkett, Oct. 11, 1965. Research fellow Inst. for Advanced Study, Princeton, N.J., 1935-36, Cornell U., 1936-39; Sterling fellow Yale U., 1939-40; with Bartol Research Found., Swarthmore, Pa., 1940-43, physicist, 1940-43; asst. prof. Princeton U., 1943-45; asso. prof. Ill. Inst. Tech., 1945-46; chief physicist Oak Ridge (Tenn.) Nat. Lab., 1945-61; with U. Va., 1961—, Taylor prof. to present; chmn. scientific advisory bd. Va. Asso. Research Center. Fellow Am. Phys. Soc., mem. Phi Beta Kappa, Sigma Xi. Author: Multipole Fields, 1955, Elementary Theory of Angular Momentum, 1957, Internal Conversion Coefficients, 1958, Relativistic Electron Theory, 1961. Theoretical research, publs. in fields of nuclear physics and quantum theory with emphasis on electromagnetic interactions and processes involved in nuclear spectroscopy. Home: Midstream, R.R. 3, Charlottesville, Va. 22901.*

ROSE, Sylvan Meryl, Am. biologist; b. Camden, N.J., Sept. 29, 1912; s. Horace Lewis and Gertrude (English) R.; A.B., Amherst Coll., 1933, M.A., 1933; Ph.D., Columbia, 1940; m. Florence Estelle Cracauer, Dec. 16, 1933; children—Marguerite (Mrs. Marc Shulman), Charles Leland, John Kenneth, Carol Anne (dec.). Faculty, U. Ill., 1949-60; instr.-in-charge embryology Marine Biol. Lab., Woods Hole, Mass., 1950-55, trustee, 1954-61; 63—; Frank B. Weeks vis. prof. biology Wesleyan U., Middletown, Conn., 1960-61; prof. exptl. embryology Med. Sch. Tulane U., New Orleans, 1961—. Recipient Career Devel. award NIH, 1961—; Guggenheim fellow, 1955-56. Mem. A.A.A.S., Am. Soc. Zoology, Assn. Anatomists, Am. Soc. Naturalists, Soc. Devel. Biology, Soc. Exptl. Biology and Medicine, Sigma Xi. Contbr. chpt. to Physiology of the Amphibian, 1964. Asso. editor: Growth. Research, publs. on induction of limb regeneration in naturally non-regenerating frogs; demonstrated that tumor agts. and tissue differentiation agts. are related and transmutable, that differentiation is controlled by specific inhibitors from already differentiating regions, that inhibitors are positively charged and move in polarized bioelectric fields. Home: 2118 Diana St., Gretna, La. 70053; also 34 High St., Woods Hole, Mass. 02543. Office: Lab. Developmental Biology, Tulane U., Riverside Research Labs., F. Edward Herbert Center, Belle Chasse, La. 70037.*

ROSE, Valentine the Younger, German chemist; b. 1762; s. Valentin Rose the Elder; ed. by Martin Heinrich Klaproth; children—Heinrich, Gustav; worked with Klaproth in researches, verified his analyses before publ.; discovered bicarbonate of soda, insulin; demonstrated presence of chronium in a kind of serpentine. Died 1807.

ROSE, Walter Deane, Am. chem. engr.; b. Liberty, Ind., Jan. 10, 1920; s. Joseph Sims and Dorothy (Gray) R.; B.S., U. Chgo., 1944; m. Edith Schroder, Apr. 1, 1959; children—Dixie (Mrs. Paul Rexroat), Walter Deane, John, Bonnie, Deane Michael, James, Elisabeth. Grad. faculty U. Colo., 1954-56; prof. petroleum engring. U. Ill., Urbana, 1956—; research affiliate Ill. Geol. Survey, 1956-58; sci. counselor French Petroeum Inst., 1962-63. Mem. Am. Geophys. Union. Author: Reflexions sur la description et l'explitation des reservoirs souterrains, 1965. Contbr. numerous articles to profl. jours. Research, publs. on surface chemistry, applied geophysics, study of oil recovery processes, transport processes in porous solid systems, inventions of lab. instrumentation and oil field equipment and processes. Home: 409 E. Springfield Av., Champaign, Ill. 61820. Office: M & M Bldg., U. Ill., Urbana, Ill. 61803.*

ROSE, William Cumming, Am. biochemist; b. Greenville, S.C., Apr. 4, 1887; s. John McAden and Mary Evans (Santos) R.; B.S., Davidson (N.C.) Coll., 1907, Sc.D., 1947; Ph.D., Yale, 1911, Sc.D., 1947; postgrad. U. Freiburg (Ger.), 1913; Sc.D., U. Chgo., 1956; Doctor of Science, University of Illinois, 1962; m. Zula Franklin Hedrick, Sept. 3, 1913. Asst. physiol. chemistry Yale, 1908-11; instructor in physiol. chemistry, Univ. of Pa., 1911-13; asso. prof. biol. chemistry, Coll. of Medicine, U. of Tex., 1913-14, prof. and head of dept., 1914-22; prof. physiol. chemistry, U. of Ill., 1922-36, prof. of biochemistry, 1936-53, research professor biochemisstry, 1953-55, emeritus, 1955—, acting head chemistry dept., 1942-45; vis. lecturer, U. of Mich., summer 1938; Fourteenth Hektoen lecturer Inst. Medicine of Chicago, 1938. Sci. award Grocery Mfrs. Am., 1947; Osborne-Mendel award American Institute of Nutrition, 1949; Willard Gibbs Medal of Am. Chem. Soc., 1952; Charles F. Spencer award Am. Chem. Soc., 1957. Associate fellow A.M.A. (member Council on Pharmacy and Chemistry, 1936-43); mem. adv. bd. Wistar Inst., 1936-40; mem. food and nutrition bd., chmn. com. Protein Foods, Nat. Research Council, 1940-47, Nat. Adv. Health Council, 1944-49; mem. adv. com., Nutrition Found., 1943-57. Fellow A.A.A.S.; mem. Biochemistry Society of Great Britain, Am. Soc. of Biological Chemists (councilor 1931-34, 1936-37; v.p. 1937-39; pres. 1939-41); Am. Chem. Soc., Soc. Exptl. Biology and Med., Am. Inst. Nutrition (pres. 1945-46, hon. fellow), Ill. Acad. Sci., Harvey Soc. (hon.), Nat. Acad. of Sci, Phi Beta Kappa, Sigma Xi. Mem. editorial bd. Jour. of Nutrition (1935-39), Jour. of Biol. Chemistry, 1936-49. Contbr. papers on nutrition and intermediary metabolism and biochem. of the amino acids, to Jour. Biol. Chemistry, Jour. of Nutrition, Jour. Pharmacology and Exptl. Therapeutics, Am. Jour. Physiology and Physiol. Reviews. Discovered and identified the amino acid threonine; research on pepsin, creatine and creatinine, origin of uric acid, mucic acid metabolism, nephrotoxic action of dicarboxylic acid, and replacement of amino acids by synthetic products. Home: 710 W. Florida Av., Urbana, Ill.

ROSEAU, Maurice Edmond Adolphe, French mathematician; b. Asnières, France, Nov. 3, 1925; s. Marcel Simon and Cecile (Debuc) R.; Agrégé, Ecole Normale Supérieure, Paris, 1948, Doctorate es sciences mathématiques, 1951; m. Marie Francois Louët, Sept. 1, 1960; children—Christine, Michel. Chargé de recherches Centre National de la Recherche Scientifique, 1949-52; maitre de conférences Facultè des Sciences de Caen, Poitiers, France, 1952-57; temp. mem. Courant Inst., N.Y. U., 1957-58; prof. Faculté des Sciences de Lille, France, 1958-62; prof. Faculté des Sciences, Paris, 1962—. Lauréat Acad. des Sciences, 1951-66. Mem. Société Mathématique de France. Author: Vibrations non linéaires et théorie de la Stabilité 1966; also articles. Research in water waves, diffusion theory, diffraction theory, non-linear problems in theory of oscillation and stability. Home: 144bis Avenue de Général Leclerc, Sceaux 92, France. Office: 11 Pierre Curie, Paris, France.

ROSEBURY, Theodor, bacteriologist; b. London, England, Aug. 10, 1904; s. Aaron R. and Emily (Dimesets) R.; brought to U.S., 1910, naturalized, 1916; D.D.S., U. Pa., 1928; m. Amy Pearl Loeb, Nov. 21, 1949; children—Joan R. (Mrs. Myron Lubow), Celia R. (Mrs. Stephen Lighthill). Asst. histopathologist U. Pa., 1927-28; Wm. J. Gies fellow biol. chemistry Columbia, 1928-30, instr. dept. bacteriology Coll. Physicians and Surgeons, 1930-35, asst. prof., 1935-44, asso. prof., 1944-51; prof. bacteriology Wash. U., 1951—, emeritus, 1967—; bacteriologist; chief airborne infection project Camp Detrick, Md., 1943-45; cons. U.N.A.E.C., UN Secreteriat, 1950; mem. Nat. Bd. Dental Examiners, 1954-58. Fellow A.A.A.S.; mem. Internat. Assn. Dental Research (chmn. N.Y. Sect. 1937-39), Am. Soc. Microbiology, Am. Soc. Immunology, Soc. Exptl. Biology and Medicine, Harvey Soc., Am. Pub. Health Assn. Author: Experimental Air Borne Infection, 1947; Peace or Pestilence, 1949; Microorganisms Indigenous to Man, 1962. Research on identification of protein in tooth enamel; tooth decay, role of bacteria, impacted starch particles, of Vitamin D, fatty oils; role of infection in periodontal disease, man's normal microbiota, nature, concentration, ecology, distbn., significance in health and disease; cultivation, classification of indigenous anaerobic bacteria, especially actinomycetes, spirochetes; significance of airborne infection for biol. warfare; biol. warfare in relation to disarmament and peace; literary and sociol. implications of man's indigenous microbiota. Home and Office: 6837 S. Bennett Av., Chgo. 60649.*

RÖSEL VON ROSENHOF, August Johann, German naturalist; b. nr. Arnstadt, 1705; ed. as engraver and artist; Author: History of Fresh Water Polyps; periodical on insects, Insecten-Belustigen (contained illustrations considered best of period), 4 vols., 1746-61. Made and collected drawings of various insects; engraved over 300 plates. Died 1759.

ROSEMAN, Ephraim, Am. neurologist; b. Balt., Jan. 1, 1913; s. Adolph Joseph and Sarah (Levitzsky) R.; A.B., Johns Hopkins, 1933; M.D., U. Md., 1937; m. Wilma Beryl Shaw, Sept. 23, 1960. Teaching fellow neurology McGGill U., Montreal, Can., 1938-39, Harvard, 1940-41; teaching fellow U. Cin., 1939-40, instr. 1942-46; mem. faculty, Rockefeller fellow U. Ill., 1941-42, faculty U. Cal. at San Francisco, 1946-47; faculty U. Louisville Med. Sch., 1947—. prof. neurology, 1949—; dir. neurologic clinic also lab. electroencephalography Louisville Gen. Hosp., 1947—; cons. also dir. lab. electroencephalography Nichols Gen. Hosp., Louisville, 1947—; mem. staff Children''s Free Hosp., Norton Meml. Infirmary, Jewish Hosp., Kosair Crippled Children's Hosp. (all Louisville); cons. to surgeon gen., others. Diplomate Am. Bd. Psychiatry and Neurology, Am. Electroencephalography Soc. Mem. Am. Neurol. Assn., Am. Psychiat. Assn., Am. Soc. Electroencephalography, Assn. Research in Nervous and Mental Disease, A.M.A., Acad. Neurology, Internat. League Against Epilepsy, Assn. Mil. Surgeons U.S., Nat. Multiple Scelerosis Assn., Nat. Rehab. Assn., Alpha Omega Alpha, others. Research and publs. on encephalopathies following drug therapy; intracranial circulation in pathologic states; use of the electroencephalogram; epilepsy. Home: 6302 Transylvania Beach, Prospect, Ky. 40059. Office: 323 E. Chestnut St., Louisville 40202.*

ROSEMAN, Saul, Am. biochemist; b. Bklyn., Mar. 9, 1921; s. Emil and Rose (Markowitz) R.; B.S., Coll. City N.Y., 1941; M.S., U. Wis., 1942, Ph.D., 1948; m. Martha Ozrowitz, Sept. 9, 1941; children—Mark Alan, Dorinda Ann, Cynthia Bernice. Instr., asst. prof., research asso. U. Chgo., 1948-53; asst. prof., then prof. bio-chemistry, U. Mich., chemist Rackham Arthritis Research Unit, 1953-65; prof. dept. biology, also McCollum-Pratt Research Inst., Johns Hopkins, Balt., 1965—. Mem. study sects. NIH, 1958—; mem. research coms. for various founds. Mem. Am. Soc. Biol. Chemists, Am. Chem. Soc., Am. Soc. for Microbiology. Editorial bd. Jour. Biol. Chemistry, 1962—. Research and publs. on intermediary metabolism, especially biosynthesis, of simple and complex carbohydrates, including glycoproteins and glycolipids. Home: 8206 Cranwood Ct., Balt. 21208.*

ROSEN, J(udah) Ben, Am. computer scientist; b. Phila., May 5, 1922; s. Ben and Susan (Hurwitch) R.; B.S., Johns Hopkins, 1943; Ph.D. in Applied Math. (NRC fellow), Columbia, 1952; m. Harriet Lee Sussman, June 25, 1943; children—Susan Beth, Lynn Ruth. Jr. engr. Gen. Electric Co., 1943-44; devel. engr. Manhattan Project, 1944-47; with Brookhaven Nat. Lab., 1947-48; research asso. Princeton, 1952-54; head applied math. dept. Shell Devel. Co. subsidiary Shell Oil Co., 1955-62; vis. prof. computer sci. Stanford, 1962-64; prof. computer sci. dept. Math. Research Center, U. Wis., Madison, 1964—, chmn. computer sci. dept., 1965—. Cons. IBM, 1962-64, Shell Devel. Co., 1962—. NASA Research grantee, 1963-66; NSF grantee, 1966—. Mem. Am. Math. Soc., Assn. for Computing Machinery, Soc. Indsl. and Applied Math. (asso. editor Jour. on Control 1964—). Asso. editor Jour. Computer and System Sci., 1966—. Research, publs. on nonlinear optimization (math. programming) methods, new digital computer techniques for optimal control, game theory, ordinary and partial differential equations; developed theory and computational methods for certain chem. reactor and combustion problems. Office: 1210 W. Dayton St., Madison, Wis. 53706.*

ROSEN, Louis, Am. physicist; b. N.Y.C., June 10, 1918; s. Jacob and Rose (Lipionski) R.; B.A., U. Ala., 1939, M.S., 1941; Ph.D., Pa. State U., 1944; m. Lillah Terry, Sept. 5, 1941; 1 son, Terry Leon. Instr. physics U. Ala., 1940-41, Pa. State U., 1943-44; staff Los Alamos Sci. Lab., 1944—, group leader, 1947—, alternate physics div. leader, 1963—, dir. Los Alamos Meson Physics Facility, 1964—, head medium energy physics div., 1966—. Mem. adv. com. nuclear cross sect. AEC, 1962—. Charter mem. Los Alamos County Planning Commn., 1964. Recipient Ernest Orlando Lawrence Meml. award, 1963, Nat. Acad. Achievement award, 1964; Guggenheim fellow, 1959 Fellow Am. Phys. Soc. (exec. com.); mem. A.A.A.S., Sigma Pi Sigma. Editorial bd. Nuclear Data. Research on neutron interactions, polarization of protons elastically scattered by nuclei, nuclear structure. Home: 1170 41st St. Office: P.O. Box 1663, Los Alamos.*

ROSEN, Milton William, Am. engr., physicist; b. Phila., July 25, 1915; s. Abraham and Regina (Weiss) R.; B.S. in Elec. Engring., U. Pa., 1937; student U. Pitts., 1937-38, Cal. Inst. Tech., 1946-47; m. Josephine H. Haar, Feb. 28, 1948; children—Nancy Elizabeth, Deborah Anne, Janet Suzanne. Engr., Westinghouse Electric & Mfg. Co., 1937-38; engr., physicist Naval Research Lab., Washington, 1940-58, sci. officer Viking rocket, 1947-55, head rocket devel. br., 1953-55, tech. dir. Project Vanguard (earth satellite), 1955-58; engr. NASA, Washington, 1958—, chief rocket vehicle devel. programs, 1958-59, asst. dir. for vehicles, 1959-60, dir. launch vehicle and propulsion, 1961-63, sr. scientist office def. affairs, 1963—. Recipient James H. Wyld meml. award for application of rocket power Am. Rocket Soc., 1954. Fellow Am. Rocket Soc. (dir.). Author: The Viking Rocket Story, 1955. Research and patents on radar and radio control systems for guided missiles and ceramic liners for rocket motors; designed and directed devel. of Viking upper-atmosphere research rocket, Delta launch vehicle for NASA; designed Vanguard satellite launch vehicle. Home: 5610 Alta Vista Rd., Bethesda, Md. 20034. Office: 400 Marland Av., Washington 20025.*

ROSEN, Philip, Am. physicist; b. N.Y.C., Oct. 31, 1922; s. Louis and Gussie (Brazaza) R.; B. Chem. Engring., City Coll. N.Y., 1944; M.S., Yale, 1946, Ph.D., 1949; m. Rose Taubenblatt, Aug. 21, 1966; children—Gary M., Robert M., Ellen S. Asst. prof. Renssalaer Poly. Inst., 1948-51; sr. physicist Johns Hopkins Applied Physics Lab., 1951-54; faculty U. Conn., 1954-57; prof. physics U. Mass., Amherst, 1957—; vis. prof. U. Wis., summer 1965; vis. fellow Yale, 1965-66; also cons. bus. orgns. Fellow Am. Phys. Soc.; mem. Biophys. Soc., Am. Astro. Soc., Sigma Xi. Research and publs. on propagation of electromagnetic waves in ionized gas; discovered that chem. potential energy is not additive; used variational principles in irreversible phenomena of thermodynamics; calculated entropy flux of electromagnetic radiation and probability of ionization for neutral colliding atoms; explained twin paradox of relativity. Home: 87 Harlow Dr., Amherst, Mass. 01002.*

ROSEN, Sam, Am. microbiologist; b. N.Y.C., Apr. 14, 1923; s. Jacob and Minnie (Safran) R.; B.A., Bklyn. Coll., 1948; M.S., U. Ill., 1949; Ph.D., Mich. State U., 1953; m. Mindla Weinrich, Sept. 26, 1954; children—Paul, Grant. With Pyridium Corps., Yonkers, N.Y., 1949-50; faculty Mich. State U., 1952-61; faculty Ohio State U., Columbus, 1961—, asso. prof. dentistry, 1964—. Mem. Am. Soc. Microbiology, Internat. Assn. Dental Research, A.A.A.S., Ohio Acad. Sci. Research and publs. on inherited resistance to dental caries in rats, discovered that salivary and extra salivary factors contribute to resistance and that Lactobacillus casei can cause dental caries in gnotobiotic rats. Home: 3625 Ridgewood Dr., Columbus, O. 43221.*

ROSENBACH, Anton Julius Friedrich, German bacteriologist, surgeon; b. Grohnde/Weser, Germany, 1842; ed. Heidelberg, Göttingen, Vienna, Paris; prof. surgery, Göttingen, Germany. Noted as early and careful worker in surg. bacteriology; studied microorganisms of tetanus and suppuration; credited with isolating Streptococcus pyogenes, circa 1881; divided Staphylococcus pyogenes into aureus and albus, circa 1884; credited with 1st use of term erysipeloid to describe dermatitis caused by Erysipelothrix rhusiopathiae, 1887. Died 1923.

ROSENBAUM, Ottomar, physician; b. Krappitz, Germany, Jan. 4, 1851; grad. U. Breslau, Germany (now Wroclaw, Poland), 1873; med. dir. Allerheilgen Hosp., 1887-93; prof. U. Breslau, 1888-96. Author 14 books, numerous monographs and articles on physiology, pathology, hygiene, diagnosis, clin. medicine. Pioneer in devel. of concepts of functional disease and diagnosis; studied cardiac neuroses, psychotherapy, recurrent laryngeal nerves; developed Rosenbach's bile test, 1876, Rosenbach's digestive reflex, 1879, Rosenbach's law that in lesions of nerve centers paralysis appears in extensor muscles before in flexor muscles, 1880, Rosenbach's sign for hemiplegia, 1880; described Rosenbach's disease, characterized by nodes on ends of finger phalanges, 1890. Died 1907.

ROSENBAUM, Robert Abraham, Am. mathematician; b. New Haven, Nov. 14, 1915; s. Joseph and Goldey (Rostow) R.; A.B., Yale, 1936, Ph.D., 1947; m. Louise Johnson, Aug. 1, 1942; children—Robert J., Joseph G., David W. Faculty, Reed Coll., Portland, Ore., 1939-42, 46-50, 51-53, prof., 1949-53; vis. prof. Swarthmore (Pa.) Coll., 1950-51; prof., 1953—, dean scis., 1963-65, provost Wesleyan U., Middleton, Conn., 1965—67, academic vice president, sin e 1967—. Mem. Math. Assn. of America (bd. govs. 1965-67, sect. chmn. 1954-55, editor 1966—), Am. Math. Soc., Nat. Council Tchrs. Math., Am. Assn. U. Profs. Author: Introduction to Projective Geometry and Modern Algebra, 1963; also articles. Research in field of sub-additive functions in n-dimensions; math. edn. Home: 29 Long Lane, Middletown, Conn. 06457.*

ROSENBERG, Jerome Laib, Am. chemist; b. Harrisburg, Pa., June 20, 1921; s. Robert and Mary (Katzman) R.; A.B., Dickinson Coll., 1941; M.A., Columbia U., 1944, Ph.D., 1948; m. Shoshana Gabriel, Sept. 15, 1946; children—Jonathan, Judith. Research asso., asst. prof. Inst. Radiobiology and Biophysics, U. Chgo., 1950-53; faculty dept. chemistry U. Pitts., 1953—, prof., 1961—; research chemist S.A.M. Labs., N.Y.C., 1944-46. NSF Sr. fellow Technion, Israel Inst. Tech., 1962-63; AEC fellow phys. sci. U. Chgo., 1948-50. Author: Photosynthesis, 1965; also articles. Editor, reviser: Outline Theory and Problems of College Chemistry (Schaum), 1949, 58, 66. Research on biosynthetic pathways for carbon compounds and interrelationships between photosynthesis and respiration, photosynthetic energy transfer pathways, nature of two photochem. reactions of photosynthesis, nature of photosensitization using dyes, delocalized electronic structure crystalline iodine. Home: 1029 S. Negley Av., Pitts. 15217.*

ROSENBERG, Mihai Serban, Rumanian physicist; b. Bucharest, Rumania, Apr. 21, 1927; s. Iohan S. and Cecilia (Hirsch) R.; M.D. in Physics, Bucharest U., 1950; Ph.D. in Physics, Moscow (USSR) U., 1955; m. Louisette Grünwald, Feb. 12, 1949; 1 dau., Ileana. Faculty, Faculty Physics, U. Bucharest, 1951—, asso. prof., 1955—; leader magnetic group Inst. Physics, Acad. Socialist Republic Rumania, 1960—. Mem. Rumanian Soc. Physics and Chemistry. Author: The Magnetic Properties of Solid State, 1963; also articles. Discovered new antiferromagnetic type of ordering in manganites; research on pecularities of magnetic domain structure in ferrimagnetics with a high anisotropy field and in single crystals with preferred plane of magnetization. Home: 6 Gheorghiu-Dej, Bucharest, Rumania.*

ROSENBERG, Milton J., Am. psychologist; b. N.Y.C., Apr. 15, 1925; s. Jacob and Rae (Dumbrowitz) R.; B.A., Bklyn., 1946; M.A., U. Wis., 1948; Ph.D., U. Mich., 1953; m. Marjorie Anne King, Sept. 5, 1954; 1 son, Matthew. Instr., U. Wis., Milw., 1948-49; instr., U. Mich., 1952-54; asst. prof. Yale, 1954-61; asso. prof. Ohio State U., 1961-63; prof. Dartmouth, 1963-65; prof. psychology U. Chgo., 1965—. Mem. Am. Psychol. Assn., Soc. for Psychol. Study Social Issues. Author: (with others) Attitude Organization and Change, 1960; also articles. Research on cognitive process and dynamics attitude, pub. opinion in relation to policy, devel. of consistency theory attitude. Home: 5421 S. Cornell Av., Chgo. 60615.*

ROSENBERG, Paul, Am. physicist; b. N.Y. City, Mar. 31, 1910; s. Samuel and Evelyn (Abbey) R.; A.B., Columbia, 1930, M.A., 1933, grad. study, 1933-40; m. Marjorie S. Hillson, June 12, 1943; 1 dau., Gale B. E, Chemist, Hawthorne Paint & Varnish Corp., N.J., 1930-33; grad. asst. physics Columbia, 1934-39, lectr., 1939-41; instr. Hunter Coll., N.Y. City, 1939-41; research asso. elec. engring. Mass. Inst. Tech., Cambridge, 1941; staff mem. Radiation Lab., Nat. Def. Research Com., 1941-45; pres. Paul Rosenberg Assos., cons. physicist, Pelham, N.Y., 1945—; pres. Inst. Nav., 1950-51. Mem. war com. radio Am. Standards Assn., 1942-44; gen. chmn. joint meeting Radio Tech. Commn. for Aeronautics, Radio Tech. Commn. for Marie Services, and Inst. of Nav. 1950; co-chmn. Nat. Tech. Development Com. for upper atmosphere and interplanetary nav., 1947-50. Mem. maritime research adv. com. National Academy of Sciences, Nat. Research Council, 1959-60. Talbert Abrams award, Am. Soc. Photogrammetry, 1955. Fellow A.A.A.S. (council 1961—, vice president 1966—), I.E.E.E. ,American Institute Aeronautics and Astronaugics (asso.); mem. Am. Phys. Soc., Am. Chem. Soc., Inst. Aeronautical Scis., Acoustical Soc. Am., Armed Forces Communication Assn., Polar Soc., Sci. Research Soc. Am., New York Academy of Sciences, Am. Soc. Photogrammetry, Am. Assn. Physics Teachers, Sigma Xi. Author sci. and tech. articles in prof. jours. Editorial Com. Journal of Aero-Space Scis., 1952-60. Holder patents. Research in electronics; physics; aeronautics; radar; photogrammetry; ultrasonics; medicine; molecular beams; automation; and engineering. Home: 53 Fernwood Rd., Larchmont, N.Y. Office: 330 Fifth Av., Pelham, N.Y. 10803.*

ROSENBLATT, Alfred, Polish mathematician; b. Cracow, Poland, 1880; prof. U. Cracow, 1910-36, U. Lima (Poland), from 1936; studied fundamentals of calculus, analytic geometry, hydrodynamics of viscous fluids; attempted to unite means of restoring ordinary differential equations and partial differential equations. Died 1947.

ROSENBLATT, Murray, Am. mathematician; b. N.Y.C., Sept. 7, 1926; s. Hayman and Esther (Goldberg) R.; B.S., Coll. City N.Y., 1946; M.S., Cornell U., 1947, Ph.D., 1949; m. Adylin Isabelle Lipson, July 17, 1949; children—Karin, Daniel. Research asso. Cornell U., 1949-50; with U. Chgo., 1950-55, asst. prof., 1951-55; asso. prof. Ind. U., 1956-59; prof. Brown U., 1959-64; prof. U. Cal., San Diego, 1964—. Recipient Office Naval Research Research grant, 1949-50; Guggenheim fellow, 1965-66. Fellow Inst. Math. Statistics; mem. Am. Math. Soc., Soc. for Indsl. and Applied Math. Author: (with U. Grenander) Statistical Analysis of Stationary Time Series, 1956; Random Processes, 1962. Research, publs. in probability theory, math. statistics, time series analysis. Home: 2717 Hidden Valley Rd., La Jolla, Cal. 92037.*

ROSENBLITH, Walter Aalter, biophysicist; b. Vienna, Austria, Sept. 21, 1913; s. David A. and Gabriele (Roth) R.; Ingenieur Radiotelegraphiste, U. Bordeaux, 1936; Ingenieur Radioelectricien, École Supericeure d'Electricite, Paris, 1937; m. Judy Olcott Francis, Sept. 27, 1941; children—Sandra Yvonne, Ronald Francis. Came to U.S., 1939, naturalized, 1946. Lowry scholar dept. physics U. Cal. at Los Angeles, 1940-43; faculty S.D. Sch. Mines and Tech., 1943-47; research fellow Psycho-Acoustic Lab.. Harvard, 1947-51, research asso. otology Med. Sch., 1957—; faculty Mass. Inst. Tech., Cambridge, 1951—, prof., 1957—, staff mem. research lab.

electronics, 1951—, Center For Communications Scis., 1958—; chmn. various coms. Nat. Acad. Sci.-NRC; mem. exec. com. Internat. Brain Research Orgn., UNESCO, 1960—, hon. treas., 1962—; pres. Commn. on Biophysics Communication and Control Processes, Internat. Orgn. For Pure and Applied Biophysics, 1961—, also mem. council; Weizmann lectr. Tata Inst. Inaugural lectr., 1962; vis. prof. Tech. U., Berlin, 1965, 66. Fellow Am., World acads. arts and scis., Acoustical Soc. Am., N.Y. Acad. Scis., A.A.A.S., mem. I.E.E.E., Biophs. Soc., Soc. Exptl. Psychologists, Psychonomic Soc., Am. Inst. Biol. Scis., Sigma Xi. Author: (with K.N. Stevens) Noise and Man, 1953. Editor: Processing Neuroelectric Data, 1959, 62; Sensory Communication, 1961. Research, numerous publs. on quantification of electric activity of brain; sensory communication; role of engring. in study of living systems. Home: 164 Mason Terrace, Brookline, Mass. 02146. Office: Mass. Inst. Tech., Cambridge, Mass. 02139.*

ROSENBLUETH, Arturo Stearns, Mexican phyisiologist; b. Chihuahua, Mexico, Oct. 2, 1900; s. Julio G. and María (Stearns) R.; Student Nat. Sch. Medicine, Mexico City, Mexico, 1918-21, Sch. Medicin, Berlin, Germany, 1923; M.D., Ecole de Médecine, Paris, France, 1924-27; m. Virginia Thompson, Sept. 5, 1931. Faculty, U. Mexico, 1928-30, prof., 1929-30; J.S. Guggenheim Found. fellow Harvard, 1930-32, fellow in physiology, 1932-33, asst. prof., 1934-44; head dept. physiology Nat. Inst. Cardiology, Mexico City, 1944-60; head dept. physiology, dir. Center for Research and Advanced Studies, Nat. Poly. Inst., Mexico City, 1960—. Mem. El Colegio Nacional, Instituto Nacional de La Investigación Científica. Author: (with W. B. Cannon) Autonomic Neuroeffector Systems, 1937, The Supersensitivity of Denervated Structures. A Law of Denervation, 1949; Transmission of Nerve Impulses at Autonomic Neuro-Junctions and Peripheral Synapses, 1950; also numerous articles. Research on transmission nervous impulses in autonomic and synaptic junctions, physiology central nervous system and of heart. Home: 514-201 Schiller St., Mexico City, Office: P.O. Box 14-740, Centro de Investigación del I P N, Mexico City 14, Mexico.*

ROSENBLUTH, Marshall Nicholas, Am. physicist; b. Albany, N.Y., Feb. 5, 1927; s. Robert and Margaret (Sondheim) R.; B.S., Harvard, 1945; M.S., U. Chgo., 1947, Ph.D., 1949; m. Arianna Wright, Jan. 26, 1951; children—Alan, Robin, Jean, Mary. Instr., Stanford, 1949-50; staff mem. Los Alamos Sci. Lab., 1950-56; sr. research adviser Gen. Atomic Co., La Jolla, Cal., 1956-67; prof. U. Cal., San Diego, 1960-67; with Inst. Advanced Study Princeton, N.J., 1967—. Mem. adv. com. on fluid mechanics NASA, 1961—; dir. Internat. Plasma Summer Schs., Riso, Denmark, 1960, Varenna, Italy, 1962, Trieste, 1964. Served with USNR, 1944-46. Recipient E.O. Lawrence award AEC, 1964. Research on electron nucleon scattering, plasma stability theory. Home: 284 Mercer St. Office: Inst. for Advanced Study, Princeton, N.J. 08540.*

ROSENBUSCH, Karl Harry Ferdinand, German geologist, mineralogist; b. Einbeck/Hannover, Germany, June 24, 1836; Ph.D., U. Freiburg; named prof. petrography U. Strasbourg (now in France), 1873; apptd. mem. commn. for geol. survey of Alsace-Lorraine, 1873; prof. mineralogy and geology U. Heidelberg (Germany), from 1878. Corr. mem. French Acad. Scis. Author: Mikroskopische Physiographie, 2 vols., 1873-77; Elemente der Gesteinslehre, 1898. Advanced (with others) the then new sci. of mineralogy, from 1880's; emphasized field studies; observed rock masses, classified igneous rocks as plutonic rocks, dike rocks, eruptive lavas. Died Heidelberg, Jan. 20, 1914.

ROSENFELD, Arthur Hinton, Am. physicist; b. Birmingham, Ala., June 22, 1926; s. Arthur H. and Pauline (Lewy) R.; B.S., Va. Poly. Inst., 1944; Ph.D., U. Chgo., 1954; m. Roselyn Bernheim, Oct. 2, 1955; children—Margaret, Anne. Research asso. U. Chgo., 1954-55; research asso. Lawrence Radiation Lab., U. Cal. at Berkeley, 1955-57, asst. prof, 1957-60, asso. prof., 1960-63, prof. physics, 1963—. Mem. panel on univ. computing facilities NSF, 1963-66, chmn, 1965; mem. com. on uses computer Nat. Acad. Scis., 1962-65. Fellow Am. Phys. Soc.; mem. Fedn. Am. Scientists (exec. com. 1964—). Author: (with Fermi, Orear, Schluter) Nuclear Physics, 1949; also numerous articles. Asso. editor Jour. Computational Physics, 1964—. Research in high energy physics using hydrogen bubble chambers and digital computer; discoverer or co-discoverer several elementary particles. Home: 160 Southhampton St., Berkeley, Cal. 94707.*

ROSENFELD, Leon, physicist; b. Charleroi, Belgium, Aug. 14, 1904; s. Leon and Jeanne (Pierre) R.; D.Sc. in Physics and Math., U. Liège (Belgium), 1926; Dr.honoris causa, U. Brussels (Belgium), 1965, U. Copenhagen (Denmark), 1965; m. Yvonne Cambresier, July 8, 1933; children—Andrée, Jean. Lectr., prof. U. Liège, 1931-40; prof. U. Utrecht (Netherlands), 1940-47, U. Manchester (Eng.), 1947-58; prof. Nordic Inst. for Theoretical Atomic Physics, Copenhagen, Denmark, 1958—. Recipient Prix Francqui, Fondation Francqui, Brussels, 1949. Mem. Royal Danish Acad. Scis. and Letters, Royal Belgian Acad. Scis., Letters, and Fine Arts, Internat. Acad. History Sci. Author: Nuclear Forces, 1948; Theory of Electrons, 1950; other books, numerous articles. Chief editor Nuclear Physics, 1955—. Research on quantum electrodynamics, epistemology of quantum mechanics, nuclear physics, especially nuclear reactions, statis. mechanics, history of sci. Home: 9II Carl Plougsvej, 1913 Copenhagen V. Office: Blegdamsvej 17, 2100 Copenhagen O, Denmark.*

ROSENFELD, Sheldon, Am. physiologist; b. N.Y.C., Dec. 28, 1921; s. Abram and Mary (Simowitz) R.; D.V.M., Middlesex Sch. Vet. Medicine, 1945; B.S., Bklyn. Coll., 1948; M.S., U. Cal. Sch. Vet. Medicine, Davis, 1955; Ph.D., U. So. Cal., 1964; m. Rosalie Weisberger, July 10, 1949; children—Diane, Barbara, Mitchel, Gary. Asst. prof. U. So. Cal., Los Angeles, 1950, instr., research asso., 1950-53; research asso. Cedars of Lebanon Hosp., Los Angeles, 1953-62; sr. research asso. Cedars-Sinai Med. Center, Los Angeles, 1962—. Mem. Am. Physiol. Soc., Am. Heart Assn., Western Soc. for Clin. Research, Soc. for Exptl. Biology and Medicine, Nat. Kidney Disease Found., Med. Research Assn. Cal., So. Cal. Vet. Med. Assn., Sigma Xi. Contbr. articles to tech. jours. Pioneered development of method to keep kidney alive and functioning while isolated from body and perfused with artificial heart-lung system; developed method to produce exptl. hypertension in rabbit by occluding blood supply to brain and correcting by denervating kidneys. Home: 2819 Hutton Dr., Beverly Hills, Cal. 90210. Office: 4751 Fountain Av., Los Angeles 90029.*

ROSENHEAD, Louis, English applied mathematician; b. Leeds, Eng., Jan. 1, 1906; s. Abraham and Ellen (Nelson) R.; B.Sc., U. Leeds, 1926, Ph.D., 1928, D.Sc., 1935; Ph.D., Cambridge (Eng.) U., 1930; m. Esther Brostoff, July 13, 1932; children—Martin David, Jonathan Vivian. Asst. lectr. applied math. U. Coll. Wales, Swansea, 1931-33; prof. applied math. U. Liverpool, 1933-40, 46—, dean, Faculty Sci., 1945-47, council, 1956-65, pro-vice chancellor, 1961-65; sci. war service Ministry of Supply, U.K., 1940-46. Adviser to govt. coms. on various sci. and ednl. projects, 1946—; sci. visitor Brit. Hydromechanics Research Assn., 1949-54; vis. prof. U. Toronto (Ont., Can.), 1950, Israel Inst. Tech., Haifa, 1956-60. Fellow Royal Soc., 1946 (council 1956-58), London Math. Soc., Cambridge Philos. Soc., Inst. Math. and Its Applications. Author: (with A. Fletcher, J. C. P. Miller) Index of Mathematical Tables, 1st edit., 1946; (with A. Fletcher, J. C. P. Miller, L. J. Comrie) Index of Mathematical Tables, 2 vols., 1946; (with others) A Selection of Tables for Use in Calculations of Compressible Airflow, 1952; others; also several translations. Mem. editorial bd. Quarterly Jour. Mechanics and Applied Math., 1947—. Research on flow of fluids, motion of surface of earth; math. tabulations. Home: 30 Wheatcroft Rd., Liverpool 18, Eng.*

ROSENHEIM, Sigmund Otto, chemist; b. Würzburg, Germany, 1871; Ph.D., U. Würzburg; postgrad. U. Bonn, also with Graebe in Geneva, then at U. Manchester (Eng.), 1894-95; m. Mary Christine Tebb, July 1910. Analytical and cons. chemist, Chancery Lane, London, 1896-1901; research student of pharmacological chemistry King's Coll., 1901-04, named lectr. chem. physiology, 1904, then reader in biochemistry, 1915-20, researcher in dept. physiology, from 1920, also asst. prof. physiology. Mem. accessory food factors com. Med. Research Council. Fellow Royal Soc., 1927, Linnean Soc. Contbr. articles to jours. Research in chemistry of brain; isolated spermine from pancreas and other tissues, discovered new base, spermidine, determined their constitutions and synthesized both; discovered that more vitamin A occurs in fish other than cod; studied sterols; discovered (with H. King) structure of cholestorol and ring-system of bile acids. Died 1955.

ROSENKRANTZ, Harris, Am. biochemist; b. Bklyn., Mar. 23, 1922; s. Abraham and Mary (Heller) R.; A.B., Bklyn. Coll., 1943; M.S., N.Y. U., 1946, Cornell U. Med. Sch., 1948; Ph.D., Tufts U. Med. Sch., 1951; m. Natalee F. Scheiner, May 19, 1951; children—Elliot Dale, Mark Steven. Technician dept. metabolism N.Y. (N.Y.C.) Hosp., 1943-46, research asso., 1948-51; research fellow dept. physiology Cornell U. Med. Sch., 1946-48; research fellow Worcester Found. for Exptl. Biology, Shrewsbury, Mass., 1951-52, staff scientist, 1952-59; prof. biochemistry Clark U., Worcester, Mass., 1959—; dir. biochemistry Mason Research Inst., Worcester, 1959—. Recipient Adm. Ralph Earle award Worcester Engring. Soc., 1956. Mem. Am. Soc. Biol. Chemists, Soc., A.A.A.S., Endocrine Soc., N.Y. Acad. Scis., Coblentz Soc., Sigma Xi. Research and numerous publs. on biochemistry of vitamin E. and muscular dystrophy, infrared spectroscopic analysis biol. substances, metabolism of steroid hormones, biochemistry prostatic tissue and fluid, cancer. Home: 136 S. Flagg St., Worcester 01602. Office: 21 Harvard St., Worcester, Mass. 01608.*

ROSENLICHT, Maxwell, Am. mathematician; b. Bklyn., Apr. 15, 1924; s. Martin and Julia (Dalinski) R.; A.B., Columbia U., 1947; Ph.D., Harvard, 1950; m. Carla Zingarelli, 1953; children—Nicholas, Alan, Joanna. NRC scholar, 1950-52; asst. prof. Northwestern U., Evanston, Ill., 1952-55, asso. prof., 1955-59; prof. U. Cal. at Berkeley, 1959—; vis. prof. U. Mexico, summers 1952, 65; mem. Institut des Hautes Etudes Scientifiques, France, 1962-63. Organizing com. Woods Hole Summer Inst., 1964. Fulbright scholar U. Rome, 1953-54; Guggenheim fellow, Holland, 1957-58; Cole prize in algebra Am. Math. Soc., 1960. Mem. Am. Math. Soc., Sigma Xi. Chmn. gov. bd. Pacific Jour. Math., 1964—. Research and publs. in algebra and algebraic geometry, especially theory of algebraic groups. Home: 263 Forest Lane, Berkeley, Cal. 94708.*

ROSENMÜLLER, Johann Christian, German anatomist; b. Hessberg, Germany, May 25, 1771; became prosector, Leipzig, Germany, 1794, prof. anatomy and surgery, 1800. Author: Quaedam de ovariis embryonum et foetuum humanorum, 1802; Handbuch der Anatomie, 1808. Described palpebral part of lacrimal gland (Rosenmüllers gland), 1797, epoophoron or parovarium, 1802, tensor tarsi muscle, 1805, lateral pharyngeal recess (fossa of Rosenmüller), 1808. Died Leipzig, Feb. 28, 1820.

ROSENOW, Edward Carl, Am. bacteriologist; b. Alma, Wis., 1875; M.D., U. Chgo., 1902; LL.D., Park Coll., 1925; m. 1906; 2 children. Fellow pathology Rush Med. Coll., Chgo., 1902-04, fellow medicine, 1904-09, instr., 1907-15; mem. McCormick Inst., 1904-15; prof. exptl. biology Mayo Found., Minn., 1915-44, prof. emeritus, 1944—; mem. staff Cal. Inst. Tech., 1944-45, Longview Hosp., 1945—. Recipient Stacey award Coll. Medicine, Cin., 1937. Mem. Soc. Exptl. Biology, Soc. Immunology, Soc. for Exptl. Pathology, A.M.A. (Gold Medal 1912), Internat. Assn. Allergologists. Research in transmutation of pneumococci and streptococci, localization of bacteria, lobar pneumonia, poliomyelitis, influenza, focal infection, epilepsy, schizophrenia; in vitro prodn. of antibodies to streptococci; 1st to point out focal infection of teeth, 1921. Office: Longview Hosp., Paddock Rd., Cin. 16.

ROSENQVIST, Ivan Thoralf Koss, mineralogist, petrologist; b. Vienna, Austria, May 17, 1916; s. Einar and Maria (Koss) R.; student Vienna U.; M.S., Oslo U., 1940, Dr.Philos., 1945; m. Anna Magdalene Sommerfeldt, Mar. 22, 1941; children—Einar, Christine. Mineralogist, Norwegian Hwy. Research Lab., 1941-46; head materials sect. Norwegian Def. Research Inst., 1946-49; sr. lectr. Bergen (Norway) U., Inst. Geology, 1950-52; staff Norwegian Geotech. Inst., 1952-61; faculty Oslo U., 1956—, prof., 1961—, chmn. Inst. Geology. Fellow Norwegian Acad. Sci., Norwegian Acad. Tech. Scis.; mem. Assn. Internationale pour l'Etude des Argiles, Norsk Kjemisk Selskap, Norsk Geoteknisk Forening affiliate Internat. Soc. Mechanics and Found. Engring., Norsk Geologisk Forening, Geochem. Soc. Am., Clay Minerals Soc. Author: Subsoil Corrosion of Steel, 1961; also numerous articles. Research on border field between phys. chemistry and geology, especially clays and clay minerals; formulated theories for high sensitivity clays (quick clays), 1946. Home: 19 Anton Schjodts gate, Oslo, Norway.*

ROSENSTEIN, Nils Rosen, Swedish physician; b. Gothland, Sweden, 1706; ed. Acad. Lund, also with physician at Uppsala, Sweden; at least 1 son, Nils; physician to king; assessor Coll. Medicine. Author: Compendium anatomicum, 1771; Pharmacie domestique et de voyage; Traité sur les maladies des enfants, 1774. A founder of modern pediatrics. Died 1773.

ROSENSTEIN, Solomon Nathan, Am. dentist; b. N.Y.C., Sept. 1, 1906; s. Israel S. and Sara (Tat) R.; B.S., Coll. City N.Y., 1929; D.D.S., Columbia U., 1930; postgrad., 1951; postgrad. Cornell U., 1948; m. Beverly B. Gutterman, Dec. 19, 1943; children—Roger G., Dwight J., Frederick L.,, Elliott M. Practice dentistry, N.Y.C.; faculty Sch. Dental and Oral Surgery Columbia, 1930—, prof. dentistry, 1953—, dir. div. pedodontics, 1952—, dir. Cerebral Palsy Dental Program, 1952—, dir. Tng. Program Dentistry For Handicapped Children, 1962—, sec. faculty, 1963—; attending dental surgeon Presbyn. Hosp.; cons. Guggenheim Dental Clinic; chmn. dental com. United Cerebral Palsy N.Y. State; mem. exec. com., dental adv. com. Am. Assn. Help Retarded Children. Mem. Am. Dental Assn., Am. Pub. Health Assn., Internat. Assn. Dental Research, Am. Acad. Pedodontics, Am. Acad. Cerebral Palsy, Am. Soc. Dentistry for Children, Am. Assn. Dental Editors, Am. Coll. Dentists, Sigma Xi, Alpha Omega Alpha, Omicron Kappa Upsilon. Contbr. chpt. to Pediatrics, 1967; Dental Problems in Cerebral Palsy, 1967. Editor, Bull. Dentistry Guidance Council for Cerebral Palsy, 1961—. Research in conservation of children's teeth, preventive dentistry, dentistry for handicapped, devel. of new tng. programs on grad. level in dentistry for handicapped. Home: 32 Saddlewood Dr., Hillsdale, N.J. 07642. Office: 630 W. 168th St., N.Y.C. 10032.*

ROSENSTOCK, Henry Meyer, phys. chemist; b. Mannheim, Germany, Feb. 15, 1928; s. Isidore and Rose (Rennert) R.; came to U.S., 1936, naturalized, 1944; B.S. magna cum laude, Coll. City N.Y., 1949; Ph.D., U. Utah, 1952; m. Shirley Dolores Jonap, July 14, 1952; children—Stephen, Paul, Peter. Chemist, Oak Ridge Nat. Lab., 1952-56; physicist Operations Research, Inc., Silver Spring, Md., 1956-58; sr. scientist William H. Johnston Labs., Lafayette, Ind., 1958-60; chemist Nat. Bur. Standards, Washington, 1960-62, chief mass spectrometry sect.,

1962—. Recipient Silver medal U.S. Dept. Commerce, 1965. Fellow Washington Acad. Sci.; mem. Am. Soc. Testing and Materials (chmn. com. mass spectrometry), A.A.A.S. Author: (with Norman Desrosier) Radiation Technology, 1960; also numerous articles. Originated quasiequilibrium theory of mass spectra of polyatomic molecules; devel. of coincidence mass spectrometry, theoretical aspects of formation and behavior of polyatomic gaseous ions; devel. operations research techniques for r.r. systems, air def. Office: Nat. Bur. Standards, Washington 20234.*

ROSENTHAL, Friedrich Christian, German anatomist; b. Greifswald, Germany, 1779; M.D., Jena, Germany, 1802; pupil of Pierre Frank, Vienna; prof. anatomy and physiolcgy, dir. zootomic mus. U. Greifswald. Author: De organo olfactus quorumdam animalium, 1902; Desquisitio anatomica de organo olfactus quorumdam, animalium, 1807; Epistola de Baloenopteris quibus dam ventre sulcata distinictis, 1824. Described vena basalis (Rosenthal's vein). Died 1829.

ROSENTHAL, Ira Maurice, Am. physician; b. N.Y.C., June 11, 1920; s. Abraham Leon and Jean (Kalotkin) R.; student Coll. City N.Y., 1936-38; A.B. with honors, Ind. U., 1940, M.D., 1943; m. Ethel June Ginsburg, Oct. 17, 1943; children—Anne Margaret, Judith Lucille. Faculty, U. Ill. Coll. Medicine, Chgo., 1953—, prof. prediatrics, 1963—; dir. pediatrics Cook County Hosp., 1967—. Member of the American Pediatric Society, Midwest Soc. for Pediatric Research (secretary-treasurer 1962-65), Soc. for Pediatric Research, Endocrine Soc., Acad., Pediatrics. Author: (with Beulah Bosselman, Marvin Schwarz) Introduction to Developmental Psychiatry, 1965; also numerous articles. Research in endocrine, metabolic and genetic disorders of childhood, gout in children, fructose intolerance, leucine sensitive hyoglycemia, intersexuality; co-discoverer new form glucose-6-phosphate dehydrogenase deficiency and hemoglobin M. disease. Home: 5535 S. Harper Av., Chgo. 60637. Office: 700 S. Wood St., Chgo. 60612.*

ROSENTHAL, Isidore, German physiologist; b. Labischin, Posen, July 16, 1836; student med. and natural scis., Berlin; apptd. prof. physiology, Berlin, 1867, of physiology and hygiene, Erlangen, 1872. Author: (in German) Respiratory Movements and their Relation to the Vagus Nerve, 1852; Lessons on Medical Electricity, 1862; The Regulation of Heat in Warm-Blooded Animals, 1872; General Physiology of Muscles and Nerves, 1877; Clinical Treatise on Diseases of the Nervous System, 1878. Described spiral canal around modiolus of cochlea (Rosenthal's canal), 1890; research on nervous system. Died Erlangen, 1915.

ROSENTHAL, Robert, Am. psychologist; b. Giessen, Germany, Mar. 2, 1933; s. Julius and Hermine (Kahn) R.; came to U.S., 1940, naturalized, 1946; A.B., U. Cal., Los Angeles, 1953, Ph.D., 1956; m. Mary Lu Clayton, Apr. 20, 1951; children—Roberta, David, Virginia. Clin. psychology trainee VA, Los Angeles, 1954-57; mem. faculty U. So. Cal., 1956-57, U. Cal., Los Angeles, 1957, U. N.D., 1957-62; lectr. clin. psychology Harvard, Cambridge, Mass., 1962-67, professor of social psychology, since 1967—. Vic. asso. prof. Ohio State U., 1960-61; cons. U.S. Dept. Def., Mass Dept. Pub. Health. Diplomate Am. Bd. Examiners in Profl. Psychology. Mem. Am., Eastern Midwest, Mass. psychol. assns., A.A.A.S. (recipient Socio-psychol. prize 1960), Am. Assn. U. Profs., Soc. Projective Techniques (past treas.), Sigma Xi, Phi Beta Kappa. Author: Experimenter Effects in Behavioral Research, 1966; (with L. Jacobson) Pygmalion in the Classroom, 1968; also articles. Research in methods of behavioral research. Home: 12 Phinney Rd., Lexington, Mass. 02173. Office: William James Hall, Harvard, Cambridge, Mass. 02138.*

ROSENTHAL, Sol Roy, physician; b. Tiktin, Russia, Sept. 6, 1903; s. Harry and Sarah (Kahn) R.; came to U. S., 1908, naturalized, 1917; B.S., U. Ill., 1925, M.D., 1927, M.S., 1930, Ph.D., 1934; postgrad. U. Freiburg (Germany), Pasteur Inst. (France); m. Dorothy Bobinsky, May 26, 1950; children—Anthony J., Wendy Elizabeth. Practice medicine specializing in exptl. pathology and immunology, Chgo., 1934—; dir. Tice Lab. Chgo. Municipal Tb. Sanitarium, 1934-48; dir. Inst. Tb. Research, 1948—, med. dir. Research Found., 1948—; faculty U. Ill., 1929—, prof. preventive medicine, 1965—; cons. USPHS, 1947-53. Diplomate Am. Bd. Pathology and Clin. Pathology. Mem. Am. Soc. Pathology and Bacteriology, Soc. Exptl. Biology and Medicine, Am. Physiol. Soc., Am. Thoracic Soc., Am. Soc. Exptl. Pathology, Sigma Xi, Alpha Omega Alpha. Author: The General Tissue and Humoral Response to an Avirulent Tubercle Bacillus, 1938; (with others) BCG Vaccination Against Tuberculosis, 1957. Contbr. numerous articles to profl. jours. Publs. on origination of theory of histamine as chem. medicator for cutaneous pain; prodn. of antitoxins in vitro to "Burn Toxin" in burn patients; anesthetic ability of antihistamines; introduction of tuberculin tine test; first demonstration of animal to animal transmission of coccidioidomycosis. Home: 230 E. Delaware Pl., Chgo. 60611. Office: 1853 Polk St., Chgo. 60612.*

ROSENZWEIG, Mark Richard, Am. psychologist; b. Rochester, N.Y., Sept. 12, 1922; s. Jacob Z. and Pearl (Grossman) R.; B.A., U. Rochester, 1943. M.A., 1944; Ph.D., Harvard, 1949; m. Janine S.A. Chappat, Aug. 1, 1947; children—Anne J., Suzanne J., Philip M. Faculty, U. Cal. at Berkeley, 1950—, prof. psychology, 1960—. Research prof. Miller Found. for Basic Research in Sci., U. Cal. at Berkeley, 1958-59, 65-66. Fulbright fellow, 1955-56; Social Sci. Research fellow, 1955-56. Fellow Am. Psychol. Assn.; mem. Am. Physiol. Soc., Internat. Brain Research Orgn., Soc. Exptl. Psychologists. Research, publs. on discovery that brains of animals can be modified anatomically and chemically by tng. and experience. Home: 470 Michigan Av., Berkeley, Cal. 94707.*

ROSENZWEIG, Saul, Am. psychologist; b. Boston, Feb. 7, 1907; s. David and Etta (Tuttle) R.; A.B. summa cum laude Harvard, 1929, M.A., 1930, Ph.D., 1932; m. Louise Ritterskamp, Mar. 21, 1941; children—Julie Maya, Ann Gradiva. Research asso. Harvard Phychol. Clinic, 1929-34, Worcester State Hosp., 1934-43; affiliate prof. Clark U., 1938-43; chief psychologist Western State Psychiat. Inst. and Clinic, Pitts., 1943-48; lectr. psychology U. Pitts., 1943-48; chief psychologist Community Child Guidance Clinic, Washington U., St. Louis, 1949-59, faculty, 1949—, prof. psychology, 1951—. Mem. study sect. on history life scis. NIH, 1964-68. Fellow Am. Psychol. Assn. (mem. council reps. 1964—); mem. Am. Psychopath. Assn., Sigma Xi, Phi Beta Kappa. Author: (with Kate L. Kogan) Psychodiagnosis, 1949; also numerous articles. Adv. editor Jour. Cons. Psychology, 1959-64, Jour. Abnormal Psychology, 1965-67; asso. editor Zeitung für Psychologie and Persönlichkeitsforschung, 1953-58, Diagnostica, 1959—. Research on reactions to frustration, effects of sibling deaths in dynamics schizopherenia, theory of projective techniques, idiodynamics; developed psychol. test, the Rosenzweigh Picture-Frustration Study. Home: 8029 Washington St., St. Louis 63114.*

ROSETT, Joshua, neurologist; b. Ekaterinburg, Russia, Jan. 22, 1875; s. Marcus and Sara Lea (Mackin) R.; came to U. S., 1891, naturalized, 1896; M.D., U. Md., 1903; m. Louise Carey, 1914; Blanche (Brownell) Grant, 1937. Mem. Md. Tb. Commn., 1902-05; officer Balt. Health Dept., 1903-07; practice gen. medicine, 1903-14, then specialized in neurology; faculty neurology Columbia U. Coll. Phys. and Surg., 1919—. Mem. Commn. Pub. Information. Mem. A.M.A. N.Y. Neurol. Soc., Assn. for Research Nervous and Mental Diseases, N.Y. Assn. Clin. Psychiatry, N.Y. Acad. Medicine, Am. Neurol. Soc. Author: The Intercortical Systems of the Human Cerebrum, 1934; also articles. Research in anatomy of human brain, neurology, epilepsy; used technique of exploding brain to study long cerebral tracts. Died Mexico, Apr. 1, 1940.

ROSKIN, Grigoriy Iosifovich, Russian histologist; b. 1893; grad. natural sci. dept. Moscow U.; D.Biol. Sci. Researcher, asst., lectr., prof. chair histology Moscow U., 1915—, now head chair cytology and histology Biopedological Faculty, head lab. cytology and cytochemistry of cancer. Author over 180 works including Data on the Protofauna of the USSR, 1930; The Therapeutic Effect of Protozoan Endotoxins on Cancer, 1938; Cytodiagnosis of Malignant Cells, 1938; History of Histology in Moscow University, 1940; Biotherapy of Malignant Tumors, 1946; Microscope Techniques: A Manual, 1946; Plastic and Regeneration Processes, 1959; Key Problems of Cytology, 1959. Co-author: Skeletal and Retractile Apparatus of Protozoa, 1930; Microscope Techniques, 1957; Antibiotics for Use against Cancer, 1957. Address: Moscow University, Leninskie gory, Moscow, USSR.

ROSOFF, Betty, Am. physiologist; b. N.Y.C., May 28, 1920; d.. Saul and Aranka (Blum) Greenebaum; B.A., Hunter Coll., 1942, M.A., 1960; Ph.D., City U. of N.Y., 1966; m. Human Rosoff, Oct. 30, 1942; 1 son, David. Research chemist Nat. Aniline Co., Buffalo, 1943-48; research asso. Montefiore Hosp., Bronx, N.Y., 1952-62; lectr. Hunter Coll., 1961-64, 65-67; instr. Bronx Community Coll., 1964-65; asso. prof. Paterson State Coll., 1967—. Mem. Sigma Xi, Phi Sigma. Author (with others) Chelation Therapy, 1964. Contbr. articles in field to sci. jours. Studies of the metabolism of zinc and of rare earth isotopes in man and animals; developed methods of removing rare earth and other isotopes after accidental contamination; effects of various hormones on the zinc content and growth of the prostate gland; developed (with others) the method of injecting Yttrium as a treatment for pleural effusions resulting from cancer. Home: 3472 Knox Pl., Bronx, N.Y. 10467. Office: Paterson State Coll., Wayne, N.J.*

ROSS, Douglas Allen, med physicist; b. Westmount, Que., Can., Jan. 5, 1907; s. Robert Baldwin and Kate E. (Paterson) R.; B.Sc., McGill U., Montreal, Que., Can., 1929, M.Sc., 1931, Ph.D., 1934; postgrad. U. Chgo.; M.D., Harvard, 1938; children—John Robert R., Duncan R. Came to U.S., 1943, naturalized, 1956. Demonstrator, lectr. McGill U., 1930-34, 38-43; spl. research asso. Psycho-Acoustic Lab., Harvard, 1943-45; with Sanborn Co., 1945-53; faculty U. Tenn. Med. Sch., 1953-56; chief med. physics Med. Div., Oak Ridge Inst. Nuclear Studies, 1956-63; research staff mem. Oak Ridge Nat. Lab., 1963—. Cons. on thyroid uptake measurements IAEA, 1960. Mem. A.A.A.S., Sci. Research Soc. Am., Soc. Nuclear Medicine (past trustee, council mem. S.E. sect.). Author: (with others) Principles of Nuclear Medicine. Contbr. articles to profl. jours. Research on hearing and balancing organs, muscle spindles, speech transmission, acoustics of auditory canal, design problems in electrocardiographs, metabolism testers, ballistocardiographs, rec. of biol. pressures, med. gamma-ray spectrometry, thyroid uptake, instrumental problems in scanners, rate recorders, whole-body counters. Home: 214 N. Purdue Av., Oak Ridge 37830. Office: Oak Ridge Nat. Lab., Bldg. 9201-2, Box Y, Oak Ridge 37830.*

ROSS, Edward Alsworth, Am. sociologist; b. Virden, Ill., Dec. 12, 1866; s. William Carpenter and Rachel (Alsworth) R.; A.B., Coe Coll., Ia., 1886; U. of Berlin, 1888-89; Ph.D., Johns Hopkins, 1891; LL.D., Coe, 1911; m. Rosamond C. Simons, June 16, 1892; children—Frank Alsworth, Gilbert, Lester Ward; m. 2d, Helen Forbes, Sept. 29, 1940. Prof. economics, Ind. U., 1891-92; asso. prof. polit. economy and finance, Cornell, 1892-93; prof. sociology, Stanford, 1893-1900, U. of Neb. 1901-06. U. of Wis., 1906-37. Lecturer on sociology Harvard, 1902, U. of Chicago, 1896, 1905; dir. of edn., "The Floating University," 1928-29, Northwestern U., 1930. Pres. Am. Sociol. Soc., 1914 and 1915; sec. Am. Econ. Assn., 1892-93; advisory editor Am. Journal of Sociology, since 1895; mem. Institut. International de Sociologie, Phi Beta Kappa. Author: Honest Dollars, 1896; Social Control. 1901; The Foundations of Sociology, 1905; Sin and Society, 1907; Social Psychology, 1908; Latter Day Sinners and Saints, 1910; The Changing Chinese, 1911; Changing America, 1912; The Old World in the New, 1914; South of Panama, 1915; Russia in Upheaval, 1918; What Is America?, 1919; The Principles of Sociology, 1920, 30, 38; The Russian Bolshevik Revolution, 1921; The Social Trend, 1922; The Social Revolution in Mexico, 1928; The Outlines of Sociology, 1923; The Russian Soviet Republic, 1923; Roads to Social Peace, 1924, Reports on the Employment of Native Labor in Portuguese Africa, 1925; Civic Sociology, 1925, 33. Standing Room Only?, 1927; World Drift, 1928; Seventy Years of It, 1936; New Age Sociology, 1940. Part author: Changes in the Size of American Families in One Generation, 1924; Readings in Civic Sociology, 1926. Contbr. of numerous articles to econ. and sociol. jours. and lit. periodicals. Early contbr. to popularizing and systematizing of sociology in U. S.; studies on social order, social control, collective behavior; believed sociology should investigate products of social processes and relations; took early sociol. interest in 1917 Russian Revolution. Died Madison, Wis., July 22, 1951.

ROSS, Frank Elmore, Am. astronomer; b. San Francisco, Calif., Apr. 2, 1874; s. Daniel Walter and Katherine (Harris) R.; B.S., U. of Calif., 1896, fellow, 1897-98; at Lick Obs., 1898-99; Ph.D., U. of Calif., 1901; m. Margaret J. Benton, May 5, 1904; 1 son, Robert D.; m. 2d, Elizabeth Bischoff, June 10, 1913; children—Alan K., Barbara H.; m. 3d, Anna Lee, Aug. 21, 1939. Teacher mathematics and physics, Mt. Tamalpais Mil. Acad., Calif., 1896-97; asst. prof. mathematics, U. of Nev., 1900; asst. Nautical Almanac Office, 1902-03; research asst. Carnegie Inst., 1903-05; dir. Internat. Latitude Obs., 1905-15; physicist Eastman Kodak Co., 1915-24; prof. astronomy, Yerkes Observatory, 1924-39, prof. emeritus, 1939—. Mem. National Academy, American Astronomical Society. Has specialized in math. astronomy, variation of latitude, physics of the photographic plate, planetary and stellar photography, development of wide-angle high-speed lenses, corrector lenses for mirrors. Died Sept. 21, 1966.

ROSS, Herbert Holdsworth, Brit. entomologist; b. Leeds, Eng., Mar. 3, 1908; s. Jonathan and Jessie (Holdworth) R.; B.S.A., U. B.C. (Vancouver, Can.), 1927; M.S., U. Ill., 1929, Ph.D., 1933; m. Jean Alexander, Feb. 3, 1932; 1 son Charles A. Asst. entomologist Canadian Dept. Agr. 1926-27; with Ill. Natural History Survey, 1927—, head sect. faunistic surveys and insect identification, 1955—, prin. scientist, 1956—, acting chief, 1962-63, asst. chief 1963—; prof. entomology U. Ill., Urbana, 1947—. Panelist, NSF, 1958-61. Guggenheim fellow, 1951-52. Mem. Entomol. Soc. Am. (past sec.-treas., past pres.), Soc. for Study Evolution (past sec., pres. 1966—), A.A.A.S. Author: A Textbook of Entomology, 1948; Evolution and Classification of Mountain Caddisflies, 1956; A Synthesis of Evolutionary Theory, 1962; Understanding Evolution, 1966; also articles. Editorial bd. Ann. Rev. Entomology, 1956-61. Research on classification of insects, inter-relations of animals and plants in natural communities. Home: 208 W. Iowa St., Urbana 61801. Office: Ill. Natural History Survey, Natural Resources Bldg., Urbana, Ill. 61801.*

ROSS, James Alexander, Scottish surgeon; b. Edinburgh, Scotland, June 25, 1911; s. James and Bessie (Flint) R.; M.B., Ch. B., Edinburgh U., 1934, M.D., 1947; m. Catherine Elizabeth Curtis, Sept. 27, 1940; children—Elizabeth (Mrs. William Yuill), David, Mary, Jane. House surgeon, physician Edinburgh Royal Infirmary, 1934-35; asst. surgeon Leith (Scotland) Hosp., 1946-50, surgeon, 1950-61; asst. surgeon Edinburgh Royal Infirmary, 1947-61; surgeon in administrv. charge Eastern Gen. Hosp., Edinburgh, 1961—; sr. lectr. surgery Edinburgh U., part time

1961—. External examiner surgery St. Andrews (Scotland) U., 1962-64, Liverpool (Eng.) U., 1966-67. Decorated Order Brit. Empire. Fellow Royal Coll. Surgeons Edinburgh (hon. sec. 1960—), Royal Coll. Surgeons Glascow, Assn. Surgeons Gt. Britain and Ireland, Brit. Assn. Urol. Surgeons; mem. Internat. Soc. Urologists, Urologists Corr. Club U.S.A. Author: Memoirs of an Army Surgeon, 1948; Manual of Surgical Anatomy (with J. Bruce, R. Walmsley), 1964; also numerous articles. Research on abdominal surgery and surg. anatomy, arterial patterns of intestine, renal arteries, arteries of liver, pancreas, surgery and physiology of genitourinary system especially activity of ureter in health and disease and the effects of drugs on it. Home: 5 Newbattle Terrace, Edinburgh. Office: 5 Newbattle Terrace, Edinburgh, Scotland.*

ROSS, Sir James Clark, English polar explorer; b. London, Apr. 15, 1800; s. George Ross; D.C.L., Oxford, 1844; m. Anne Coulman, 1843. Joined Navy, 1812; joined expdn. with uncle Sir John Ross to find N.W. Passage, 1818; mem. Sir William Edward Parry's 4 Arctic expdns. 1819-27; commd. lt., Navy, 1822, comdr., 1827; mem. Felix Booth Expdn., 1829-33; became comdr. expdn. to Baffin's Bay, 1836; completed magnetic survey of Gt. Britain, 1838; named comdr. Antarctic Expdn., 1839; last expdn., 1848-49. Recipient Gold medal London and Paris geog. socs., 1842. Fellow Royal Soc., 1828; mem. French Acad. Scis., 1852. Author: A Voyage of Discovery in the Southern and Antarctic Seas, 2 vols., 1847. Discovered magnetic pole, 1831, Victoria Land, 1841, Ross Sea, 1843; authority on Arctic navigation. Died Aylesbury, Eng., Apr. 3, 1862.

ROSS, Sir John, Brit. Arctic explorer; b. Inch, Scotland, June 24, 1777; s. Andrew and Elizabeth (Corsane) R.; m. twice; 1 son. Joined E. India Co., 1794; commd. midshipman Brit. Navy, 1799, advanced to lt., 1805, comdr., 1812; in Baltic and N. Sea regions, 1812-17; apptd. comdr. expdn. looking for N.W. passage, 1818; commd. post rank, 1818; comdr. expdn. surveying Boothia peninsula, King William Land, and Gulf of Boothia, 1829-33; consul of Stockholm, Sweden, 1839-46; made expdn. in search of Franklin, 1850. Recipient gold medal geog. socs. of London and Paris, 1834. Author: A Voyage of Discovery Made Under the Orders of the Admiralty in His Majesty's Ships Isabell and Alexander for the Purpose of Exploring Baffin's Bay, and Inquiring into the Probability of a North-West Passage, 1819; A Treatise on Navigation by Steam, 1828; A Narrative of a Second Voyage in Search of a North West Passage and of a Residence in the Arctic Regions during the Years 1829-1833, with Appendix, 2 vols., 1835; Memoirs and Correspondence of Admiral Lord de Saumarez, 2 vols., 1838; On Steam Communication to India, 1838; Observations on a Work entitled Voyages of Discovery by Sir John Barrow, 1846; A Short Treatise on the Deviation of the Mariner's Compass, 1849; On Intemperance in the Royal Navy, 1852; Rear-Admiral Sir John Franklin: A Narrative of the Circumstances and Causes which led to the Failure of the Searching Expeditions sent by Government and others for the Rescue of Sir John Franklin, 1855. Discovered Boothia Peninsula, King William Island; explored Smith, Jones, Lancaster sounds. Died London, Aug. 30, 1856.

ROSS, John, chemist; b. Vienna, Austria, Oct. 2, 1926; s. Mark and Anna (Krecmar) R.; B.S., Queens Coll., 1948; Ph.D., Mass. Inst. Tech., 1951; m. Virginia Franklin, Aug. 26, 1950; children—Elizabeth, Robert. Came to U.S., 1940, naturalized, 1945. Research Asso. Mass. Inst. Tech., Cambridge, 1950-52, prof., chmn. dept. chemistry, 1966—; research fellow Yale, 1952-53; faculty Brown U., 1953-66, prof., 1963-66. Cons., AEC, 1961—. NSF fellow, 1952-53; Guggenheim fellow, 1959-60; Sloan fellow, 1960-64; vis van der Waals prof. U. Amsterdam, 1966. Mem. Am. Chem. Soc., Am. Phys. Soc., A.A.A.S., Am. Acad. Arts and Scis. Author: Molecular Beams, 1966. Contbr. articles to profl. jours. Research on molecular interactions and chem. reactions by molecular beam techniques, irreversible processes in reacting and nonreacting systems.*

ROSS, Joseph Foster, Am. physician b. Azusa, Cal., Oct. 11, 1910; s. Verne Ralph and Isabel Mills (Bumgarner) R.; A.B., Leland Stanford Jr. U., 1933; M.D., Harvard, 1936; m. Eileen Sullivan, Dec. 19, 1942; children—Louisa, Elisabeth, Joseph, Jeanne, Marianne. Asst. topographical anatomy Harvard, 1934-37, research fellow biochemistry, 1943-46; asst. pathology U. Rochester Sch. Medicine, 1939-40; physician, dir. hematology and radioisotope divs. Mass. Meml. Hosp., 1940-54; instr., asst. prof., asso. prof. Medicine Boston U.; 1940-54; dir. radioisotope unit Cushing VA, Boston VA hosps., 1948-54; prof. medicine U. Cal. at Los Angeles, 1954—, prof. radiobiology, 1954-59, asso. dean, 1954-58, chmn. dept. nuclear med. and radiation biology, 1958-60, dir., Lab. of Nuclear Med. and Radiation Biology, 1958-65; prof., chmn. dept. biophysics and nuclear medicine, 1960-65; chief staff U. Cal. Hosp., Los Angeles, 1954-58, attending physician, 1954—. U. S. del. Internat. Conf. Peaceful Uses Atomic Energy, Geneva 1955; mem. U. S. Atoms for Peace mission to Latin Am., 1956; mem. U. S. AEC Life Scis. Mission to Greece and Turkey, 1961; mem. Cal. Adv. Council on Cancer, 1959-68, chmn., 1963-66; mem. CENTO Sci. Mission Iran, Turkey, Pakistan, 1963. Research preservation of whole blood OSRD, World War II, Mem. nat. adv. cancer council Nat. Cancer Inst., 1956-60. Recipient certificate of merit Pres. of U. S., 1948; Gordon Wilson lectr. and medal Am. Clin. and Climatol. Assn., 1964. Diplomate Am. Bd. Internal Medicine. Mem. Am. Soc. Exptl. Pathology, Am. Soc. Clin. Investigation, Assn. Am. Physicians, Radiation Research Soc. (council 1964-65), A.M.A. Internat. Soc. Hematology, Am. Soc. Hematology (pres. 1961-62), Western Assn. Physicians (pres. 1962-63), Soc. Nuclear Medicine (trustee 1962-68, pres. Soc. Cal. chpt. 1964-65), Phi Beta Kappa, Sigma Xi. Editorial bd. Blood, Jour. Hematology, 1946—, Annals Internal Medicine, 1960—, med. book div. Little, Brown & Co., 1958—, med. book series U. Cal. Press, 1961—. Contbr. articles profl. jours. Pioneered in application of radioisotope tracers to study of metabolic and physiologic processes in health and disease, especially iron metabolism, blood formation, and destruction in neoplastic diseases; developed methods of preservation of whole blood for transfusion; studies of environmental radiation and human radioecology. Home: 11246 Cashmere St., Los Angeles 90049. Office: U. of Cal. Sch. Medicine, Los Angeles 90024.*

ROSS, Marion Amelia Spence, Scottish physicist; b. Edinburgh, Scotland, Sept. 4, 1903; d. William Baird and Marion (Thomson) Ross.; M.A., Edinburgh U., 1925, Ph.D., 1943; Faculty, Edinburgh U., 1928-41, 46—, reader, 1957—; with Mine Design Dept., Admiralty, 1942-46. Cons. archtl. acoustics. Fellow Royal Soc. Edinburgh; mem. Phys. Soc. London, Brit. Acoustical Soc. Research, publs. on X-rays especially yields from L-shells of heavy atoms, nuclear physics, underwater acoustics, low-speed wind tunnel study of boundary layers. Home: 24 Belford Gardens, Edinburgh 4, Scotland.*

ROSS, Robert Alexander, Am. physician; b. Morganton, N.C., July 18, 1899; s. Charles Ellis and Kate (Chambers) R.; B.S., U. N.C., 1920; M.D., U. Pa., 1922; m. Rosalie Walter, Oct. 3, 1933; children—Robert Alexander, Charles Allen, Rosalie W. Practice medicine specializing in obstetrics and gynecology, Durham, N.C., 1926-30; prof. obstetrics and gynecology, Duke Sch. Medicine, 1930-52; prof., chmn. dept. obstetrics and gynecology U. N.C., Chapel Hill, 1952-66. Cons., Watts Hosp., Durham, 1930—. Diplomate Am. Bd. Obstetrics and Gynecology (examiners 1952—). Mem. A.C.S., Am. Coll. Obstetricians and Gynecologists, N.Y. Acad. Scis., Am. Gynecol. Soc., Am. Assn. Gynecologists, Sigma Xi, Alpha Omega Alpha. Author: (with Gladys Groves) The Married Woman, 1934; also numerous articles. Research on nutrition and pregnancy, maternal and perinatal mortality, malignancy in the female, blood clotting mechanism, female endocrinology.*

ROSS, Sir Ronald, physician; b. Almora, India, May 13, 1857; s. Campbell Claye Grant and Matilda Charlotte (Elderton) R.; M.D., St. Bartholomew's Hosp. Med. Sch., 1879; D.P.H., LL.D., D.Sc.; m. Rosa Bessie Bloxam, 1889; 2 sons, 2 daus. With Indian Med. Service, 1881-99; leader expdn. which found malaria bearing mosquitoes, W. Africa, 1899; became lectr. tropical medicine, Liverpool (Eng.) Sch. Tropical Medicine, 1899, prof., 1902-12; prof. tropical sanitation, Liverpool, 1912-17; became physician for tropical diseases King's Coll. Hosp., 1913; named dir. in chief Ross Inst. and Hosp. for Tropical Diseases, 1926. Became cons. in malaria War Office, 1917, Ministry Pensions, 1925; mem. adv. bd. Indian Research Fund. Recipient Parke Gold medal 1895; Nobel prize for physiology and medicine, 1902; Royal medal Royal Soc., 1909. Fellow Royal Soc., 1901, Royal Coll. Surgeons, Royal Soc. Edinburgh, Royal Soc. Sci. Uppsala; mem. Acad. Royal de Belgique (corr.), acads. medicine Paris and Turin (asso.), A.C.P. (asso.), Soc. Apothecaries London (hon. freeman). Author: The Prevention of Malaria, 1910; Memoirs, 1923; Studies on Malaria, 1928; also books on math., novels. Began research on malaria, 1892; discovered malaria is transmitted by bite of female Anopheline mosquito, 1897-99, life history of malaria parasites in mosquitoes. Died London, Sept. 16, 1932.

ROSS, Sherman, Am. psychologist; b. N.Y.C., Jan. 1, 1919; s. Max and Rachel (Khoutman) R.; B.S., Coll. City N.Y., 1939; A.M., Columbia, 1941, Ph.D., 1943; m. Jean Goodwin, Aug. 18, 1945; children—Norman Kimball, Claudia Lisbeth, Michael Lachlan. Faculty, Bucknell U., 1946-50, asso. prof., 1946-50; research fellow N.Y. Zool. Soc., 1948; guest investigator, sci. asso. R.B. Jackson Lab., Washington, 1947—; faculty U. Md., 1950-60, prof. psychology, 1956-60. Spl. cons. psychopharmacology service center Nat. Inst. Mental Health, 1956-64; cons. to govt. agys., pvt. cos.; mem. Md. Bd. Examiners Psychology, 1957-58. Mem. Am. Psychol. Assn. (exec. sec., edn. and tng. bd. 1960—), Washington Acad. Scis., Aerospace Med. Assn., A.A.A.S., Am. Soc. Zoologists, Ecol. Soc., Ergonomics Research Soc., Sigma Xi (past pres. U. Md. chpt.), Phi Kappa Pi, Psi Chi (nat. pres. 1964—). Author: Laboratory Manual for Experimental Comparative Psychology (with C. J. Warden, G. S. Klein), 1942; Graduate Education in Psychology, 1959; Principles of Comparative Psychology, 1960; also numerous articles, chpts in books. Research in psychopharmacology, genetics, instinctual and social basis of behavior in animals, studies endocrine factors in behavior, placebo effects in human behavior. Home: 24 Wessex Rd., Silver Spring, Md. 20910; also 23 Glen Mary Rd., Bar Harbor, Me. 04609. Office: 1200 17th St. N.W., Washington 20036.*

ROSS, Sidney David, Am. chemist; b. Lynn, Mass., Jan. 31, 1918; s. Samuel and Dora (Pross) R.; B.A., Harvard, 1939; M.A., Boston U., 1940; Ph.D., Harvard, 1944; m. Rhoda R. Revman, Nov. 15, 1942. Spl. research asso. Harvard-OSRD, 1940-45, Pitts. Plate Glass fellow, 1945-46; dir. organic lab., research asso. Sprague Electric Co., North Adams, Mass., 1946—; vis. prof. U. N.H., 1966. Mem. Am. Chem. Soc., Electrochem. Soc., Chem. Soc. (London, Eng.), Sigma Xi. Patentee in field. Research and numerous publs. on mechanisms of organic reactions including charge transfer complexes, displacement reactions and anodic oxidations. Home: Colonial Av., Williamstown, Mass. 01267. Office: Sprague Electric Co., Marshall St., North Adams, Mass. 01248.*

ROSS, Walter Charles Joseph, English chemist; b. London, Eng., Feb. 15, 1918; s. Walter C. and Edith (Pepper) R.; student Chelsea Poly. London, 1934-39; B.Sc., Imperial Coll. London, 1940, Ph.D., 1943, D.Sc., 1951. Research chemist Organon Labs., London, 1943-46; research chemist Inst. Cancer Research, London, 1946—, reader in chemistry, 1953-66, prof. organic chemisty, 1966—. Sir Halley Stewart Research fellow, 1946; Brit. Empire Cancer Campaign Research fellow, 1950. Fellow Royal Inst. Chemistry, Chem. Soc. (London); mem. Brit. Assn. for Cancer Research. Author: (with Sir John Simonsen) The Terpenes, vols. IV and V, 1957; Biological Alkylating Agents, 1962; also numerous articles. Research on preparation, properties of potential chemotherapeutic agts. for treatment of cancer; discoverer Chlorambucil Leukeran, drug widely used for treatment of certain forms of cancer. Home: 31, Cumbernauld Gardens, Sunbury-on-Thames, Middlesex, Eng. Office: Inst. Cancer Research, Fulham Rd., London S.W. 3, England.*

ROSSBY, Carl-Gustaf Arvid, meteorologist; b. Stockholm, Sweden, Dec. 28, 1898; s. Arvid and Alma Charlotta (Marelius) R.; studied U. of Stockholm, 1917-18, 1922-25; student Geophysical Inst., Bergen, Norway, 1918-19. U. of Leipzig, 1920; hon. D.Sc., Kenyon Coll., 1939; m. Harriet Marshall Alexander, Sept. 2, 1929; children—Stig Arvid, Hans Thomas, Carin. Came to U. S., as fellow Swedish-Am. Found., 1926. Research asso. in meteorology Daniel Guggenheim Fund for Promotion of Aeronautics, and chmn. com. on aero meteorology, 1927-28; prof. meterology Mass. Inst. Tech., 1931-39; research asso. Woods Hole Oceanographic Inst. since 1931; asst. chief for research and edn. U. S. Weather Bureau, 1931-41; distinguished service prof. meteorology U. of Chicago since 1943, on leave to serve as prof. meteorology U. of Stockholm, 1947-48. Expert cons. to Office of Sec. of War, and cons. on weather problems to comdg. gen. of A.A.F., during World War II. Hon. fellow Royal Meteorol. Soc., London; mem. Inst. Aero. Sciences, Am. Meteorol. Soc. (pres. 1944-45), Am. Philos. Soc., Nat Acad. Sciences, Det Noske Videnokaps-Akademi (Oslo). Received (with H. C. Willett) the Sylvanus Albert Reed award from Inst. Aero. Scis., 1934; received the Robert Losey award, 1947. Organizer model aero-weather service for Guggenheim Fund. 1928. Contbr. articles to sci. jours. in U. S. and abroad. Studies of application of fluid mechanics to meteorology, general circulation of atmosphere, and atmospheric wave motions; developed important theories of large-scale air movements and techniques for effective long-range weather prediction. Died Stockholm, Sweden, Aug. 19, 1957.

ROSSE, Earl of, see Parsons, William.

ROSSEL, Elizabeth-Paul-Edouard, Chevalier de, French naval officer, hydrographer; b. Sens, France, Sept. 11, 1765. Joined Marine Guard, at age of 15; left for Indies with D'Entrecasteaux, 1785; searched for La Pérouse, 1791; named head of expdn., 1794; captured by English, 1795; returned to Paris after peace of Amiens, 1802; joined Marine Depot; became asst. dir.-gen., 1814; named dir.-gen. Naval Charts and Plans Depository, 1826; became rear-adm., 1828. Mem. French Acad. Scis., 1812, Bur. Longitudes. Author: La voyage de D'Entrecasteaux envoyé à la recherche de la Pérouse, 2 vols., 1808; also publs. on naut. astronomy. Died Paris, Nov. 20, 1829.

ROSSELAND, Svein, Norwegian astrophysicist; b. Kvam, Norway, Mar. 31, 1894; s. Isak and Ragna Rosseland; Dr.Philos., Oslo (Norway) U., 1927; hon. degree Copenhagen (Denmark) U., 1946, Stockholm (Sweden) U., 1961; m. Ragna Michelsen, Aug. 26, 1924; 1 son, Hallvard. Rockefeller Found. fellow Mt. Wilson Obs., Pasadena, Cal., 1924-26; prof. astronomy Oslo U., 1928-65; prof. astrophysics Princeton, 1941-46; sci. war service Brit. Admiralty, Am. OSRD, 1943-45; dir. Inst. Theoretical Astrophysics, Olso U., 1934-65, dir. Oslo Solar Obs., 1954-65, dean Sci. Faculty, 1950-53. Norwegian Govt. del. NATO Sci. Com., 1958-65. Decorated Royal Norwegian Order St. Olaf, comdr. with star, 1957. Mem. Royal Norwegian Acad. Sci. and Letters Oslo (past pres.), Royal acads. at Oslo, Copenhagen, Stockholm, Uppsala, Leige, Belgium, Royal Instn. Author: Astrophysik auf atomtheoretischer Grundlage, 1931; Theoretical Astrophysics, 1936; Pulsation Theory of Variable Stars, 1949;

also articles. Research on collision of 2d kind, internal constn. of stars, stellar constns. such as pulsation problem of variable stars. Home: 18C Ovenbakken, Oslo 7. Office: P.O. Box 1029, Blindern, Oslo 7, Norway.*

ROSSER, J(ohn) Barkley, Am. mathematician; b. Jacksonville, Fla., Dec. 6, 1907; s. Harwood and Ethel (Merryday) R.; B.S., U. Fla., 1929, M.S., 1931; Ph.D., Princeton, 1934; m. Annetta Louise Hamilton, Sept. 7, 1935; children—Edwenna Merryday, John Barkley. Faculty math. dept. Cornell U., 1936-63, chmn. dept., 1961-62; dir. Math. Research Center, U. S. Army, prof. math., computer scis. U. Wis., Madison, 1963-—; dir. research Inst. Numerical Analysis, Nat. Bur. Standards, 1949-50; mem. Stewart Com. monitoring U. S. space satellite, 1955-58; dir. Focus Project Inst. Def. Analyses, 1959-61; chmn. Math. div. NRC, 1960-62; chmn. Conf. Bd. Math. Scis., 1963-65. Recipient Presidential certificate of merit for rocket work, 1948, certificate of commedation Sec. Navy for work on Polaris missile, 1960. Guggenheim and Fulbright research fellow U. Paris, 1953-54. Mem. Am. Math. Soc., Math. Soc. Am., Assn. Computing Machinery, Soc. Indsl. and Applied Math. (pres. 1964-66), Assn. Symbolic Logic (past pres.), Sigma Xi, Phi Kappa Phi. Author: (with R.R. Newton and G.L. Gross) Mathematical Theory of Rocket Flight, 1947; Theory and Application of Various Integrals, 1948; (with A.R. Turquette) Many-Valued Logics, 1952; Logic for Mathematicians, 1953; Deux Esquisses de Logique, 1955. Contbr. numerous articles to profl. jours. Research on devel. in logic where there are other possibilities besides true and false, math theory of rockets, devel. of Polaris missile and the current space activity; pioneer devel. of computing machines. Home: 4209 Manitou Way, Madison, Wis. 53711.*

ROSSI, Bruno Benedetto, physicist; b. Venice, Italy, Apr. 13, 1905; s. Rino and Lina (Minerbi) R.; student Univ. of Padua, 1923-25, University of Bologna, 1925-27; married Nora Lombroso, April 10, 1938; children—Florence S., Frank R., Linda L. Asst. physics dept., Univ. of Florence, 1928-32; prof. of physics, Univ. of Padua, 1932-38; research asso., Univ. of Manchester, Eng., 1939; research asso. in cosmic rays, U. of Chicago, 1939-40; asso. prof. of physics, Cornell U. 1940-45; prof. of physics, Mass. Inst. Tech. since 1945. Mem. Am. Acad., Nat. Acad. Sci., Am. Phys. Soc., Am. Assn. Physics Tchrs., Accademia dei Lincei, Bolivian Academy of Sciences (corr. member), A.A.-A.S., American Philosophical Society, Italian Physical Society, American Geophysical Union. Author: Rayons Cosmiques, 1935; Lezioni di Fisica Sperimentale Elettrologia, 1936; Lezioni di Fisica Sperimentale Ottica, 1937; Ionization Chambers and Counters, 1949; High Energy Particles, 1952; Optics, 1957; Cosmic Rays, 1964. Showed capacity of cosmic rays to traverse great thicknesses of matter and demonstrated that collisions of individual cosmic rays with atoms may generate large numbers of secondary particles (now called showers), 1931; proved that primary cosmic waves are positively charged; demonstrated that some particles are capable of producing nuclear reactions and some are not; experiments on meson decay; first to measure absorption curves. Address: 7 Scott St., Cambridge, Mass.

ROSSI, Corrado, Italian chemist; b. Florence, Italy, Apr. 4, 1905; s. Giuseppe and Egle (Saccenti) R.; Ph.D. Prof. phys. chemistry U. Milan (Italy); prof. indsl. chem., also instr. phys. chemistry U. Genoa (Italy); cons.-chemist Italian Electrochemistry Soc.; collaborator various indsl. projects, especially electricity. Corr. mem. Ligure Acad. Sci. and Letters. Author textbooks, publs. on colloids, viscosity of gases and liquids, kinetic chem. reactions. Research on physicochem., structural and tech. applications of radiation chemistry to processes involving macromolecules, polymers and polymerization. Address: via Moretto 42, Milan, Italy.

ROSSI, Giovanni, Italian physician; b. Savona, Italy, July 12, 1925; s. Ferdinando and Teresa (Chiappella) R.; M.D. U. Pisa (Italy), 1950; postgrad. E.N.T., U. Turin (Italy), 1958; m. Giovanna Astuti, Oct. 26, 1959; children—Ferdinando, Elena. Faculty ear, nose throat dept. U. Turin (Italy), 1953-—, asst. prof., 1958-—, research asso. in neuroanatomy sensory organs, 1961-—. Recipient Citelli award, 1961, 64, Philips Internat. award French Soc. Otorhino-laryngology, 1965. Author books, numerous articles. Research on ototoxicity of streptomycin, di-hydrostreptomycin and neomycin, synthesis of new salts, efferent innervation of inner ear; discovered interposed vestibular nucleus and two bundles of efferent vestibular fibers. Home: Corso Massimo d'Azeglio 49, Turin, Italy.*

ROSSI, Girolamo (or Rebeus, or De Rubeis), Italian historian, physician; b. Ravenna, 1539; M.D., Ph.D., U. Padua, 1561. Physician to Pope Clement VII, 1604. Author: Hieronymi Rubei Ravenna, 1585; De Distillatione, 1582 (description of Persian method of preparing distilled oil of roses); De Melonibus, 1607. Died Apr. 22, 1607.

ROSSI, Harald Herman, physicist; b. Vienna, Austria, Sept. 3, 1917; s. Oswald J. and Hedwig (Braun) R.; student U. Vienna, 1935-39, U. Bristol (Eng.), 1939; Ph.D., Johns Hopkins, 1942; m. Ruth M. Gregg, June 22, 1946; children—Gerald C., Gwendolyn C., Harriet May. Faculty, Johns Hopkins, Balt., 1940-42; research scientist AEC contract radiol. research lab. Coll. Phys. and Surg., Columbia, 1946-60, faculty, 1949-—, prof. radiology, 1960-—, dir. AEC contract, 1960-—; physicist, radiation protection officer Presbyn. Hosp., N.Y.C., 1954-60. Chmn. radiation study sect. NIH 1965-67; mem. main commn. Internat. Commn. on Radiol. Units and Measurements, 1959-—; mem. adv. com. on med. use isotopes U. S. AEC, 1964-—. Mem. Radiation Research Soc. (mem. council 1964-—), Radiol. Soc. N.Am., Sigma Xi. Research and publs. on radiation dosimetry, radiobiology, radiation protection. Home: Undercliff Dr., Upper Nyack, N.Y. 10960. Office: 630 W. 168th St., N.Y.C. 10032.*

ROSSI, Lino, Italian pathologist; b. Milan, Italy, Dec. 31, 1923; s. Emilio and Anna Visco (Gilardi) R.; Dr. in Medicine, U. Milan (Italy), 1947; specialist in blood disease U. Pavia (Italy), 1951; specialist in cardiology U. Turin (Italy), 1959, in endocrinology, 1961, Lib.Doc.Pathol., 1956; m. Graziella Belluschi, June 25, 1951; Research staff dept. pharmacology U. Bristol (Eng.), 1951, dept. anatomy U. Hamburg (Germany), 1953; head lab. and pathology Civic Hosp., Gallarate, Italy, 1964-—; faculty dept. pathology U. Milan, 1956-—. Recipient medal for distinguished service Fatebenefratelli Hosp., Milan, 1963. Mem. Italian Soc. Pathology and Cardiology, N.Y. Acad. Scis., Ist. Di Studi Romani, Soc. for Roman Studies (London). Author: Sistema di Conduzione e Nervi nel Cuore Dell'Uomo, 1954; (with C. Cavallero) Iperparatiroidismo Renale, 1950; (with G. Tusini) Neoformazioni Epit. Linf. Tiroidee, 1956; also numerous articles. Research and publs. on impulse conducting system and nerves of heart, exptl. pathology of arrhythmias, pathology of endocrine glands, mil. history of Rome. Home: 23 Via Annunciata, Milan, Italy.*

ROSSIER, Paul Louis, Swiss astronomer; b. Crans, Apr. 24, 1895; s. Emile and Aline (Michaud) R.; Dr. ès sc., Zurich Fed. Poly.; m. Jeanne Delarue, Mar. 22, 1920; children—Claude, Georges, Yvette. Asst. at Geneva; astronomer Geneva Obs.; instr., later prof. Faculty of Sci., Geneva. Mem. Geneva Soc. Physics and Natural History, Helvetian Nat. Sci. Soc., Swiss Math. Soc. Author: Géométrie synthétique moderne; Perspective; Géographie mathématique; Surfaces usuelles; Mathématiques générales. Research on history and philosophy of the exact scis.; synthetic geometry; astrophysics. Address: 8 blvd. de la Tour, Geneva, Switzerland.

ROSSINI, Frederick Dominic, Am. chemist; b. Monongahela, Pa., July 18, 1899; s. Martino and Costanza (Carrara) R.; B.S., Carnegie Inst. Tech., 1925, M.S., 1926, D.Sc. (hon.), 1948; Ph.D., U. Cal., 1928; D.Eng. Sci., Duquesne Univ., 1955; D.Sc., U. Notre Dame, 1959, Loyola U., Chicago, 1960, U. Portland, 1965; Litt.D., St. Francis Coll., 1962; m. Ann Kathryn Landgraff, June 29, 1932; 1 son, Frederick Anthony. Phys. chemist Nat. Bur. Standards, Washington, 1928-36, chief sect. thermochemistry and hydrocarbons, 1936-50; Silliman prof., head dept. chemistry and dir. chem. and petroleum research lab. Carnegie Inst. Tech., Pitts., 1950-60; dean coll. sci., asso. dean Graduate sch. U. Notre Dame, 1960-67, vice president for research, 1967-—. Chairman commission chemical thermo-dynamics Internat. Union Pure and Applied Chemistry, 1953-61; member of committee on physical chemistry National Research Council, 1948-62, chmn. div. chemistry and chem. tech., 1955-58, exec. com. office critical tables, 1957-—, chmn., 1963-—. Cons., cooperating expert for tech. adv. com. Petroleum Industry War Council, Office Rubber Res., Office Sci. Research and Development, Atomic Energy program during World War II; cons. National Science Found., 1952-62; dir. Am. Petroleum Inst. research projects 1935-60, Manufacturing Chemists, Assn. Research Projects 1955-60; mem. policy adv. bd. Argonne Nat. Lab., 1958-66; pres. permanent council World Petroleum Congresses, 1967-—. Awarded Hillebrand prize award Chem. Soc. Washington, 1934; Gold medal Exceptional Service award by Department of Commerce, 1950; Pittsburgh Jr. Chamber of Commerce award in chemistry for 1957; Pitts. award American Chem. Soc., 1959; Laetare medal U. Notre Dame, 1965; John Price Wetherill medal Franklin Inst., 1965; William H. Nichols medal N.Y. sect. Am. Chem. Soc., 1966. Fellow Am. Inst. Chemists, A.A.A.S. (past counsilor), Am. Phys. Soc.; mem. Am. Chemical Society (past councilor), Chemical Society Washington (president 1950), American Institute Chemical Engineers, Am. Petroleum Inst., Am. Soc. Testing Materials (Marburg lecturer, 1953), Faraday Soc., Washington Acad. Scis. (pres. 1948), Philos Soc. Washington, Cath. Commn. Intellectual and Cultural Affairs (sec, 1946-47, chmn., 1958-59), American Society of Engineering Edn., Am. Acad. Arts and Scis., Albertus Magnus Guild (past pres.), Nat. Acad. Scis., Sigma Xi (exec. com. 1953-58, pres. 1963-64). Author: (with F. R. Bichowsky) Thermochemistry of the Chemical Substances, 1936; (with others) Selected Values of Properties of Hydrocarbons, 1947; Chemical Thermodynamics, 1950; Chemical Thermodynamics, Fractioning Processes, Hydrocarbons from Petroleum, 1950; (with others) Selected Value of Chemical Thermodynamic properties, 1952; (with others) Selected Values of Physical and Thermodynamic Properties of Hydrocarbons and Related Compounds, 1953; (with others) Hydrocarbons from Petroleum, 1953; (with others) Properties of Titanium Compounds and Related Substances, 1957; numerous papers in field. Research in chemical thermodynamics; thermochemistry; composition of petroleum; properties of hydrocarbons; critically selected numerical data for sci. and technology. Home: 411 N. Ironwood Dr., South Bend, Ind. 46615. Office: U. Notre Dame, Notre Dame, Ind. 46556.*

ROSSITER, Richard Alfred, Am. astronomer; b. Oswego, N.Y., Dec. 19, 1886; A.B., Wesleyan U., 1914; A.M., U. Mich., 1920, Ph.D., 1923; m. 1915; 2 children. Tchr. math. Genesee (N.Y.) Wesleyan Sem., 1914-19; telescope asst., astronomy lab. U. Mich., 1919-20, instr., 1920-23, asst. prof., 1923-26, asso. prof., astronomer in charge Lamont-Hussey Obs., 1925-52, emeritus asso. prof., 1953-—. Fellow Royal Astron. Soc.; mem. Internat. Astron. Union (commn. on double stars), Astron. Soc., S. African Astron. Soc. (pres. 1940). Sought and measured new double stars in So. hemisphere; studied bota lyrae rotation effect, orbital motion.

ROSSITER, Roger James, biochemist; b. Adelaide, S. Australia, July 24, 1913; s. James L. and M.S.K. (Jacobs) R.; B.Sc., U. Western Australia, 1934; B.A., U. Oxford (Eng.), 1938, D.Phil., 1940, B.M., 1941, B.Ch., 1941, M.A., 1942, D.M., 1946; m. Helen M. Randell, Mar. 16, 1940; children—James R., Margaret R., George M. Demonstrator biochemistry, tutor U. Oxford, 1938-46; prof. biochemistry U. Western Ont., London, Can., 1947-—, dean faculty grad. studies, 1965-—. Cons. Westminster Hosp., London, Victoria Hosp., London, St. Joseph's Hosp. Mem. Royal Soc. Can. Contbr. numerous articles to tech. jours., chpts. to numerous books, monographs. Research on chemistry brain; chemistry, biosynthesis, metabolism, and function lipids. Home: 504 Colborne St., London, Ont., Can.*

RöSSLE, Robert, German pathologist; b. Augsburg, Germany, Aug. 19, 1876; prof., Jena, Germany, Basel, Switzerland; Berlin; studied diseases of liver; contbd. to knowledge of constitution pathology; formulated own theory of infection. Died Berlin, Nov. 21, 1956.

ROST, Rudolf, Czechoslovakian mineralogist; b. Libusín, Czechoslovakia, July 7, 1912; s. Rudolf and Maria (Holub) R.; RNDr., Charles U., Prague, Czechoslovakia, 1936; m. Jirina Kukelka, Sept. 28, 1944; children—Jiri, Adéla. Tchr. high sch. gymnasium, 1938; scientist dept. mineralogy Nat. Mus., Prague, 1939-46; faculty Mining U., Ostrava, Czechoslovakia, 1946-51, prof. mineralogy, 1948-51; prof. mineralogy and geochemistry Charles U., Prague, 1951-—, dean Faculty Scis., 1952-57. Mem. Mineral. Soc. Am. Czechoslovakian Soc. for Mineralogy and Geology (treas. 1952-—), Soc. Nat. Mus. (vice chmn. 1967-—;. Author: Heavy Minerals, 1956; Microchemical Determination of Minerals, 1961; (with M. Kocar), Atlas of Minerals, 1964; also articles. Discovered minerals kladnoite and kratochvilite; research on topographical mineralogy of Czechoslovakia, meteorites, tektites, heavy minerals; discovered test for diamond. Home: 29, Jecná, Prague, 2, Czechoslovakia.*

ROSTAN, (Louis) Léon, French physician; b. St.-Maximin, France, 1790; student, Marseille, France; intern, Paris, 1809; pupil of Lallement and Pinel; completed med. studies, 1812; became prof. medicine, Paris, 1833; mem. Acad. Medicine (pres. 1854). Author: Ramollissement cérébral, 1819; Traité élémentaire de diagnostic, 1826; L'organicisme, 1826. Precursor of doctrine of organicism later formulated by Bérard; described cardiac asthma (Rostan's asthma), 1817. Died Paris, 1866.

ROSTAND, Jean, French biologist; b. Paris, Oct. 30, 1894; s. Edmond and Rosemonde (Gerard) R.; ed. Paris U.; m. Andrée Mante, Apr. 20, 1920; 1 son, François. Mem. superior council R.T.F. Recipient literary prize City of Paris, 1951; prize Foundation Singer-Polignac, 1955; prize Acad. Scis.; prize Palais de la Decouverte; Kalinga prize, 1959. Mem. French Acad. Scis., 1959, Soc. Biology, Internat. Acad. History of Sci. Author: La loi des riches, 1919; Ignace ou l'ecrivain, 1923; Deux angoisses: la mort, l'amour, 1924; De l'amour des idées, le mariage, valère ou l'exaspère, 1927; La formation de l'etre de la mouche à l'homme, 1930; L'état présent du transformisme, journal d'un caractère, 1931; L'évolution des espèces, historie des idées transformistes, 1932; Les problèmes de l'heredité et du sexe, la vie des crapauds, 1933; L'aventure humaine, 1933-35; La vie des libellules, 1935; (with Cuenot) Introduction à la genetique, 1936; La nouvelle biologie, 1937; La parthenogénèse des vertebres, 1938; Biologie et médecine, Hérédité et racisme, moeurs nuptiales des betes; Pensées d'un biologiste, La vie et ses problemes, 1939; La génèse de la vie, 1943; La vie des vers à soie, 1944; Esquisse d'une histoire de la biologie, 1945; Hommes de verité, 1942-48; Ce que je coris, 1953; Les chromosomes; Les idées nouvelles de la genetique; Science et génération; La biologie et l'avenir humain; Science fausse et fausses sciences; Bestiaire d'amour; Cahier de notes; Peut-on modifier l'homme?; Carnet d'un biologiste; also articles. Research on parthenogenesis, artificial ovulation, effect of cold on reproductive system, anomalies of batrachians.

ROSZKOWSKI, Ireneusz, Polish physician; b. Lapy, Poland, Mar. 24, 1909; s. Franciszek and Natalia (Wnorowska) R.; grad. Warsaw U., 1935, M.D., 1945; m. Halina Obrodzka, Dec. 27, 1949; children—Elisabeth, Peter, Barbara, Katherine. Obstetrician, Lying-In Hosp., Warsaw, 1936-42, prof., dir., 1955—; surgeon Town Hosp., Warsaw, 1944; asst. prof. Lying-In Clinic, Gdansk, 1944-51, prof., dir., 1951-55; dean med. dept. Warsaw Med. Sch., 1958-62; state specialist of obstetrics, 1951-60. Chief editor Polish Gynecology, 1963; chmn. fetal malformations com. Polish Acad. Sci., 1960-65. Recipient Golden Cross of Merit, 1951; Distinguished Health Service Worker, 1952; cavalier Cross of Polonia Restituta, 1954, officer, 1959. Mem. Polish Soc. Obstetrics and Gynecology, Polish Soc. Endocrinology, Polish Soc. Friends of Sci. Author: Obstetrics and Gynecology for Students, 1948; Obstetrics and Gynecology for Midwives, 1966; also numerous articles. Research on operative techniques in cases of genital cancer, physiopathology of fetus, congenital malformations, ovarian function, obstetric procedures. Home: Klonowa 12, Warsaw, Poland.*

ROTBLAT, Joseph, physicist; b. Warsaw, Poland, Nov. 4, 1908; s. Zygmunt and Sonia Rotblat; M.A., U. Warsaw, 1932, D.Physics, 1936; Ph.D., U. Liverpool (Eng.), 1950; D.Sc., U. London (Eng.), 1953. Asst. dir. Atomic Energy Inst., Warsaw, 1937-39; faculty U. Liverpool, 1940-49, sr. lectr., 1945-49; dir. research in nuclear physics, 1945-49; atomic bomb research Los Alamos Lab., 1943-45; prof. physics U. London, St. Bartholomews Hosp. Med. Coll., 1950—, physicist, 1950—. Sec.-gen. Pugwash Confs. on Sci. and World Affairs, 1959—. Fellow Inst. Physics; mem. Brit. Inst Radiology, Polish Acad. Scis., Am. Inst. Physics, Radiation Research Soc., Hosp. Physicists. Assn. Author: (with Chadwick) Radioactivity and Radioactive Substances, 1953; Atomic Eenergy, A Survey, 1954; Atoms and the Universe (with Jones, Withrow), 1956; Science and World Affairs, 1962; also numerous articles. Research in nuclear physics especially discovery of inelastic scattering of neutrons, spontaneous fission, and radioactive isotopes; research on applications of nuclear physics to medicine and radiobiology. Home: 8 Asmara Rd., London N.W. 2, Eng.*

ROTCH, A(bbott) Lawrence, Am. meteorologist; b. Boston, Mass., Jan. 6, 1861; s. Benjamin Smith and Annie Bigelow (Lawrence) R.; pvt. schs. and tutor, Paris, Florence, Berlin, Boston, 1875-80; S.B., Mass. Inst. Tech., 1884; (hon. A.M., Harvard, 1891); m. Margaret Randolph Anderson, Nov. 22, 1893. In 1885 established and has since maintained the Blue Hill Meteorological Obs., nr. Boston, famous for its investigations of clouds, and for 1st use of kites to record meteorol. data; prof. meteorology, Harvard, 1906—. Mem. Internat. Jury Awards, Paris Expn., 1889, and then made Chevalier Legion of Honor; received Prussian Orders of the Crown, 1902, and Red Eagle, 1905, in recognition of efforts to advance knowledge of atmosphere; mem. various Am. and foreign scientific socs. and coms. Asso. editor of Am. Meteorological Journal, 1886-96; lectured before Lowell Inst. of Boston, 1891, 98; librarian Am. Acad. Arts and Sciences; trustee several ednl. instns., Boston. Editor: Observations and Investigations at Blue Hill, pub. in Annals Harvard Coll. Observatory. Author: Sounding the Ocean of Air, 1900; The Conquest of the Air, 1909; Charts of the Atmosphere for Aeronauts and Aviators (with A. H. Palmer), 1911. Obtained 1st observations high above Atlantic Ocean with kites, 1901, 1st observations 5 to 10 miles above Am. Continent with registration balloons, 1904; 1st trigonometrical measurements of pilot balloons in U. S., 1909; collaborated with Teisserenc de Bort in sending a steam yacht to explore the tropical atmosphere, 1905-06; has taken part in scientific expdns. in U. S., S. America, Europe and Africa. Died Apr. 7, 1912.

ROTCH, Thomas Morgan, Am. physician; B. Philadelphia, Pa., Dec. 9, 1849; s. Rodman and Helen (Morgan) R.; A.B., Harvard, 1870; M.D., 1874; m. Helen d. William J. Rotch, of New Bedford, June 4, 1874. Med. house officer, Mass. Gen. Hosp., 1873-74; studied in European hosps., 1875-76; lecturer, 1878-88, asst. prof. diseases of children, 1888-93, prof., 1893-1903, prof. pediatrics, 1903—, Harvard; visiting phys. Children's Hosp. and Thomas Morgan Rotch, Jr., Memorial Hosp. for Infants, Boston; apptd. consulting phys., St. Francis' Hosp. for Infants, London, Eng., 1903. Author: Pediatrics (textbook); The Roentgen Ray in Pediatrics (text-book). Pioneer in field of pediatrics; founded 1st infants' hospital in Am.; introduced use of scientifically determined diets for infants. Died Mar. 9, 1914.

ROTH, Etienne Georges Alfred, French nuclear chemist; b. Strasbourg, France, June 5, 1922; s. Georges Jules and Marguerite (Neymarck) R.; Sc.D., Ecole Polytechnique, Paris, 1942; m. Francoise Hirsch, June 3, 1949; children—Catherine, Elisabeth, Marianne, Brigitte. With French AEC, 1946—, head stable isotope service, Saclay, 1959—; vis. chemist Brookhaven (N.Y.) Nat. Lab., 1957-58; lectr. in nuclear chemistry Conservatoire National des Arts et Metiers, Paris, 1961-63, prof., 1963—. Mem. Conseil National de la Recherche Scientifique, 1967—. Mem. Société de Chimie Physique (council), Société de Chimie, Am. Nuclear Soc., Faraday Soc. Research publs. on isotope separation and analysis, phys. chemistry of isotopes. Home: 103 rue Brancas, Sevres, France. Office: Cen-Saclay, B.P. n. 2, Gif sur Yvette, France.*

ROTH, George Byron, Am. pharmacologist; b. Mt. Eaton, Ohio, May 22, 1879; s. Charles Conrad and Magdelene (Miller) R.; student Western Res. U., 1902-04; A.B., U. Mich., 1906, M.D., 1909; m. Dorothea Payne, Sept. 14, 1912; 1 dau., Dorothy (Mrs. Wilson). With U. Mich., 1907-12, instr. pharmacology, 1909-12; pharmacologist USPHS, 1913-21; asst. prof. pharmacology Western Reserve U., 1921-24; with George Washington U., 1924—, prof. emeritus pharmacology, 1945—. Fellow Internat. Coll. Anesthetists; mem. A.M.A., D.C. Med. Soc., Am. Soc. Exptl. Pharmacology and Exptl. Therapeutics, Am. Physiol. Soc., Soc. Exptl. Biology and Medicine, Sigma Xi, others. Author: Practical Pharmacology. Contbr. numerous articles in field to profl. jours. Died May 23, 1967.

ROTH, Grace Marguerite, Am. clin. investigator; b. Rochester, Minn., June 8, 1894; d. Edmond C. and Nellie (Conway) Roth; B.S., U. Minn. Med. Sch., 1931, M.S., Ph.D. (Mayo Found. fellow), 1936. Chief vascular lab., dept. clin. physiology Mayo Clinic, 1937-40; faculty Mayo Found., U. Minn., 1940-59, prof. clin. physiology, chief vascular lab., 1953-59; chief vascular lab. Lovelace Clinic, Albuquerque, 1959—. Mem. Central Soc. Clin. Research, Am. Heart Assn. Silver medal 1963), Am. Congress Phys. Medicine, Am. Coll. Chest Physicians, World Med. Assn., Alumni Assn. Mayo Found., La Agrupacion Medica Femenina de Chile (hon.), Sigma Xi. Research and publs. on cardiovascular physiology of blood and circulation. Home: Apt. A, 1017 Palomas S.E. Office: 1017 Palomas Dr. S., Albuquerque 87108.*

ROTH, Jay Sanford, Am. biochemist; b. N.Y.C., June 10, 1919; s. Harold C. and Beatrice (Brown) R.; B.S., Coll. City N.Y., 1940; M.S., Cornell U., 1941; Ph.D., Purdue U., 1944; m. Lorraine Anne Blake, May 13, 1951; children—Diana, Robbie, Katherine, Christopher. Asst. prof. U. Ida., Moscow, 1944-47; asst. prof. Rutgers U., New Brunswick, N.J., 1947-50; asso. prof. Hahnemann Med. Coll., Phila., 1950-60; prof. U. Conn., Storrs, 1960—. Cons. Hartford Hosp. (Conn.), 1965—. Recipient Research Career award Nat. Cancer Inst., NIH, 1962—. Fellow N.Y. Acad. Scis., A.A.A.S.; mem. Am. Soc. Biol. Chemists, Cell. Biology Soc., Biochem. Soc. London, Am. Assn. Cancer Research, Radiation Research Soc., Soc. for Exptl. Biology and Medicine, Soc. Gen. Physiologists, Sigma Xi. Research, numerous publs. on properties, distbn. and function of enzymes concerned with breaking down of nucleic acids in normal and tumor tissues, mechanism of control of gene action in higher animals in normal and tumor tissues. Home: 5 Meadowood Rd., Storrs, Conn. 06268.*

ROTH, John Paul, Am. mathematician; b. Detroit, Dec. 16, 1922; s. Peter G. and Mary L. (Regan) R.; B.Mech. Engring., U. Detroit, 1946; M.S. in Math., U. Mich., 1948, Ph.D. in Math., 1953; m. Dorothy Dolentz, Jan. 31, 1948; children—Erich Paul, Dana Jonathan. Research engr. Continental Motors, Detroit, 1946-47; adjunct prof. math. Wayne State U., 1946-47; research asso. U. Mich., 1948-53; Pierce instr. math. U. Cal. at Berkeley, 1953-55; mem. electronic computer project Inst. for Advanced Study, Princeton, 1955-56; mathematician IBM Research, Yorktown, Heights, N.Y., 1956—. Cons., Shell Devel. Co., Cal., 1951 55. OOR fellow, 1953; vis. asst. prof. math. Princeton, 1957-58. Mem. Am. Math. Soc., A.A.A.S., I.E.E.E., Am. Automatic Control Council (chmn. automation and edn. com. 1962-64), Sigma Xi. Contbr. articles to tech. jours. Research on methods in math. analysis elec. and phys. networks, topology manifolds, combinatorial theory algorithms, automation design computers. Home: 26 Underhill Rd., Ossining, N.Y. 10562. Office: IBM Research Center, Yorktown Heights, N.Y. 10598.

ROTH, Robert Steele, Am. applied mathematician; b. Phila., July 3, 1930; s. Clyde Christian and Gertrude (Steele) R.; A.B., Kenyon Coll., 1953; M.S., Carnegie Inst. Tech., 1954; Ph.D. in Applied Maths., Harvard, 1962; m. Micheline M. Mathews, May 13, 1966. Physicist, Aberdeen Proving Grounds (Md.), 1954-56; head structures research group AVCO Corp., 1961—. Mem. Am. Inst. Aeros. and Astronautics, Sigma Xi. Research and publs. on theory of thin shell structures including static plastic buckling of shallow spherical shells, criterion of buckling of thin shell structures, effect imperfections on dynamic instability of shells, theory of data unscrambling, analysis of complex biol. phenomena. Office: 201 Lowell St., Wilmington, Mass. 01887.*

ROTH, Walter Lester, Am. chemist; b. Salt Lake City, Jan. 18, 1917; s. George and Mary (Klein) R.; B.A., U. Utah, 1936; Ph.D., U. Cal. at Berkeley, 1941; m. Julia Goldman, Sept. 16, 1949; children—Robert Irwin, Irene Roberta. Lab. asst. Shell Devel. Co., Emeryville, Cal., 1937-38; asso. chemist TVA, Wilson Dam, Ala., 1941-43; research scientist, sect. head Manhattan Project, Columbia, 1943-45; phys. chemist Gen. Electric Co. Research and Devel. Center, Schenectady, 1945—, instr. molecular and atomic structure Advanced Tng. Program, 1957-50; guest scientist Brookhaven Nat. Lab., Upton, N.Y., 1953. Mem. Am. Crystallographic Assn., Am. Chem. Soc., Sigma Xi. Research and publs. on utilization of nuclear fission; determination of atomic arrangement in solids by X-ray diffraction; magnetic structures of solids by neutron diffraction, magnetism, photochemistry; reactions in solids, relation of elec. magnetic properties to structure. Home: 1552 Baker Av., Schenectady 12309. Office: Gen. Electric Research Lab., Schenectady 12301.*

ROTHBALLER, Alan Burns, Am. physician; b. N.Y.C., May 15, 1926; s. Albert Floyd and Helen (Burns) R.; student Princeton, 1943-44; M.D., U. Pa., 1948; M.Sc., McGill U., 1955; postgrad. Montreal (Que., Can.) Neurol. Inst., 1951-56; faculty, Albert Einstein Coll. Medicine, 1956-65, asso. prof. anatomy and neurol. surgery, 1960-65; prof., chmn. dept. neurosurgery, research prof. physiology N.Y. Med. Coll., N.Y.C., 1965—. Fellow A.C.S.; mem. Am. Physiol. Soc., Harvey Cushing Soc. Translator: (with J. Olszewski) Brain Tumors (K. Zülch), 1965. Research and publs. on hypothalamus and pituitary inter-relations, especially brain control of neurohypophysis and vasopressin. Home: 411 E. 53d St., N.Y.C. 10022. Office: 1249 Fifth Av, N.Y.C. 10029.*

ROTHBART, Harold A., Am. mech. engr., coll. dean; b. Newark, Dec. 17, 1917; s. Edward and Jeanette (Colt) R.; B.S., Newark Coll. Engring., 1939; M.S., U. Pa., 1942; D.Eng., Technische Hochschule, Munich, Germany, 1959; m. Florence Hollander, Dec. 14, 1943; children—Daniel, Ellen, Jane. Marine engr. Phila. Navy Yard, 1939-43; chief engr. Trought Assos., Newark, 1943-46; prof. mech. engring. Coll. City N.Y., 1946-61; dean Coll. Sci. and Engring., Fairleigh Dickinson U., 1961—; cons. high-speed machinery; lecture tours Scotland, Eng., Germany, France, Switzerland, Italy, Netherlands, 1958, 59, 64; vis. prof., lectr. Japanese univs. and indsl. firms, 1966. Mem. Am. Soc. for Engring. Edn., Sigma Xi, Tau Beta Pi, Pi Tau Sigma. Author: Cams—Design, Dynamics and Accuracy, 1956. Editor-in-chief: Mechanical Design and Systems Handbook, 1964. Developed math. concept in field of complex machine elements, high speed machinery. Home: 315 Sherman Av., Teaneck, N.J. 07666.*

ROTHCHILD, Irving, Am. physician; b. N.Y.C., Dec. 2, 1913; s. Phillip and Yetta (Fisch) R.; B.A., U. Wis., 1935, M.A., 1936, Ph.D., 1939; M.D., Ohio State U., 1954; m. Rose Kanet, Sept. 27, 1935 (div. 1956); 1 dau., Susan Elisabeth; m. 2d, Ellen Newman, Sept. 6, 1958. Biologist, Hosp. Liquids, Inc., Chgo., 1940-41; chemist Michael Reese Hosp., Chgo., 1941-43; asst. physiologist U. S. Dept. Agr., Beltsville, Md., 1943-48; asst. prof. physiology U. Md., Balt., 1948-49; asst. prof. obstetrics and gynecology Ohio State U., Columbus, 1949-53; asso. prof. obstetrics and gynecology Western Res. U., Cleve., 1955-66, prof. reproductive physiology, 1966—. asso. obstetrician, gynecologist U. Hosps., Cleve., 1955—; vis. asst. Cleve. Meth. Gen. Hosp., 1962—; cons. St. Ann's Hosp. Mem. Am. Soc. Zoologists, Endocrine Soc., Am. Physiol. Soc., Soc. Exptl. Biology and Medicine, Soc. Gynecologic Investigation, N.Y. Acad. Scis., Royal Soc. Medicine. Research on physiology of reprodn. Home: 2441 Kenilworth Rd., Cleveland Heights, O. 44106. Office: 2065 Adelbert Rd., Cleveland. 44106.*

ROTHÉ, Edmond Ernest Antoine, French physicist; b. Paris, France, Oct. 13, 1873; D.Sc., U. Paris. Prof. faculties sci. Paris, Grenoble, Nancy (all France) univs.; hon. dean Faculty Sci., U. Strasbourg (France), until 1939. Prof., dir. Institut de Physique du Globe; pres. French and internat. seismic offices. Mem. French Acad. Scis. (corr.). Author: La Polarisation des Electrodes; Cours de Physique, 1914-28; Les Applications de la T.S.F., 1921; Le Tremblement de Tierre, 1925; Les Methodes de Prospection du Sous-Sol, 1930. Died Lezoux, France, 1942.

ROTHE, Erich Hans, mathematician; b. Berlin, Germany, July 21, 1895; s. Wilhelm and Else (Horwitz) R.; Ph.D., U. Berlin, 1926; m. Hildegard Ille, Mar. 20, 1928 (dec. Oct. 1942); 1 son, Erhard William. Came to U. S., 1937, naturalized, 1944. Faculty, Engring. Coll., Breslau, Germany (now Wroclaw, Poland), 1927-35, U. Breslau, 1931-35; prof. math. William Penn Coll., Oskaloosa, Ia.., 1937-43; prof. U. Mich., Ann Arbor, 1943-65, prof. emeritus, 1965; vis. prof. Western Mich. U, 1965-66. Mem. Am. Math. Soc., Math. Assn. Am., Ia. Acad. Sci, A.A.A.S., Am. Assn. U. Profs., Sigma Xi. Author: (With others) Differential und Integralgleichungen, 1925; (with others) Handbook for Automation, Computation and Control, 1958; also articles. Research in fields of boundary value problems, functional analysis, topology in function spaces. Home: 413 S. Forest Av., Ann Arbor, Mich. 48104.*

ROTHÉ, Jean-Pierre Edmond, French seismologist; b. Nancy, France, Nov. 16, 1906; s. Edmond and Marguerite (Tilly) R.; Dr. ès sc., univs. Strasbourg (France) and Paris (France); m. Marguerite Méjan, Dec. 8, 1942; children—Christine, Olivier, Jean-Louis, Lucile. Lectr., U. Strasbourg, 1937-45, titular prof., 1945—; seismotectonic work, France and N. Africa. Sec.-gen. Internat. Assn. Seismology and Inner Earth Physics. Decorated Palms of Acad., Mil. Vol. Service cross; hon. pres. Alpine Ski Club Lower Rhine. Author: Séismes et volcans; Prospection géophysique; also a study on the anomalies of magnetic earth fields. Research on protection against earthquakes; surveying methods; gen. geophysics; seismicity and seismotectonics. Home: 77, rue du Gén. Conrad. Office: 38, blvd. d'Anvers, Strasbourg, France.

ROTHE, Rudolf, German mathematician; b. Berlin, Germany, Oct. 15, 1873; ed. U. Berlin; m. Johanna Binner, 1899. Asst., Physikalische Technische Reichanstalt, 1897-1905; asst., then privatdocent, Mathematische-Technische Hochschule, Charlottenburg, 1908-13; prof., Mechanische Bergakademie, Calusthal, from 1913. Author: Darstellende Geometrie des Geländes, 1914; Aufgaben der Technischen Hochschulen auf dem Gebiete der Geisteskultur, 1921; Höhere Mathematik, 1924. Published and edited works of Karl Weierstrass. Died 1942.

ROTHMAN, Milton A., Am. physicist; b. Phila., Nov. 30, 1919; s. Isadore and Goldie (Glazer) R.; B.S., Ore. State Coll., 1944; M.S., U. Pa., 1948, Ph.D., 1952; m. Doris Weiss, Apr. 22, 1950; children—Anthony, Lynne. Research physicist Bartol Research Found., Swarthmore, Pa., 1952-58; research staff Plasma Physics Lab., Princeton, 1958-68; prof. physics Trenton State Coll., 1968—. Mem. Am. Phys. Soc., Sigma Xi. Author: The Laws of Physics, 1963; Men and Discovery, 1964; also articles. Research in nuclear physics, measurement of neutron scattering cross-sections, plasma physics, heating of plasma by absorption of ion cyclotron waves using model C stellarator to confine plasma. Home: 1687 Lawrenceville Rd., Trenton, N.J. 08638.*

ROTHMANN, Christopher, German astronomer; b 16th century. Astronomer to Landgrave William IV of Hesse-Kassel. (With Landgrave William IV and Jost Bürgi) made observations to correct positions of fixed stars and motion of planets, 1564; disciple of Copernicus but disagreed with Copernican theory of earth's motion.

ROTHNER, Jacoby Theodore, Am. dentist; b. Bklyn., Aug. 30, 1902; s. Joseph H. and Fanny (Karfiol) R.; D.D.S., Temple U., 1925; m. Leah Margolis, Dec. 27, 1926; children—Phyllis Leah (Mrs. Robert L. Kirson), Doris Jean (Mrs. Robert L. Horwitz). Practice dentistry specializing in periodontics, Phila., 1931—; prof. periodontics Temple U. Dental Sch., 1946—. Cons. U. S. Army, VA. Fellow Internat., Am. colls. dentists, A.A.A.S.; mem. Eastern Dental Soc. (past pres.), Phila. Soc. Periodontology (past pres.), Acad. Dental Medicine (past pres. Phila.), Am. Dental Assn., Am. Acad. Periodontology, Am. Soc. Periodontists, Pan Am. Odontological Soc. Author: (with others) Periodontics, 1958; also articles. Developed set of periodontal instruments. Home: 714 N. Park Town Pl., Phila. 19130. Office: Med. Arts Bldg., Phila. 19102.*

ROTHSTEIN, Aser, biophysicist; b. Vancouver, B.C., Can., Apr. 29, 1918; s. Samuel and Etta (Wiseman) R.; B.A., U. B.C., 1938; student U. Cal., Berkeley, 1938-40; Ph.D., U. Rochester, 1942; m. Evelyn Paperny, 1940; children—Sharon, David, Steven. Came to U. S., 1938. With U. Rochester Med. Sch., 1946—, prof. dept. radiation biology, 1959—, vice-chmn. dept. radiation biology, 1961-65, co-chmn. dept. radiation biology and biophysics, 1965—; with AEC Project, 1943—, asso. dir. atomic energy project 1961—, co-dir., 1965—. Mem. A.A.-A.S., Am. Physiol. Soc., Soc. Pharmacology and Exptl. Therapeutics, Soc. Cell Biology, Soc. Gen. Physiology, Biophys. Soc. Author: Enzymology of the Cell Surface Protoplasmatologia Band II, E.4, 1954. Contbr. numerous articles in field to sci. jours. Research, publs. in areas of membrane transport; heavy metal toxicology; biophysics; electrolyte metabolism; permeability. Home: 1415 Clover St., Rochester N.Y. 14610. Office: Crittenden Blvd., Rochester, N.Y. 14620.*

ROTINI, Orfeo Turno, Italian chemist; b. Fauglia, Pisa, Feb. 15, 1903; s. Giuseppe and Isola (Panizzi) R.; Dr. agrl. chemistry; m. Carmen Formaggia, Nov. 18, 1937; children—Paola, Andrea, Sandro. Dir. agrarian chemistry U. Pisa. Mem. Internat. Soil Sci. Soc., Accademia dei Georgofili, Bologna Nat. Acad. Agr., Académie de la vigne et du vin, Am., French chem. socs., Deutsche Chemische Gesellschaft. Author numerous articles in field. Dir. Agrochimica mag. Research on enzymology, plant physiology, soil sci., biochemistry. Home: via Crispi 68. Office: via S. Michele degli Scalzi 2, Pisa, Italy.

ROTTER, Hans, Austrian dermatologist; b. Vienna, Austria, Jan. 21, 1910; s. Ludwig and Helene (Haas) R.; Dr. med., U. Vienna, 1933; m. Herta Kornherr, Nov. 25, 1939; children—Manfred, Elisabeth, Gertraud. With Prof. Priesel at Inst. for Pathol. Anatomy, Histology and Seriology, 1934; asst. physician clinic cutaneous and venereal diseases U. Vienna, 1934-38; asst. physician U. Pharmacological Inst., 1935-38; exptl. sci. studies with Prof. E. P. Pick, 1938; specialist cutaneous and venereal diseases, Vienna, 1938; war service as med. specialist, France, Russia, Finland, Norway, 1939-45; med. supt. cutaneous sect. Landeskrankenanstalten, Salzburg, Austria, 1945—; mgr. prodn. artificial eyes of plastic, devel. complete procedure treatment and prophylaxis of hypostatic venous diseases of legs; engaged in regulation of sci. information all over world; devel. tech. aids, especially reading and reproduction equipment for registration of documents, equipment for shaping of writing desk and hand library of scientist, furniture, card indexing and filing aids. Decorated Silver Cross of Merit (Republic Austria); recipient silver badge for achievements Austrian Union Inventors, 1954. Author essays. Patentee in field. Home: 6 Ernest Thunstrasse. Office: Landeskrankenanstalten, Salzburg, Austria.*

ROTZSCH, Wolfgang, German physiol. chemist; b. Meissen, Mar. 2, 1930; s. Erhard and Margarete (Kuhnert) R.; Dr.med., U. Leipzig, 1954; m. Rosemarie Rotzsch, Aug. 25, 1956; 1 son, Cornelius. Faculty, Inst. Physiol. Chemistry, U. Leipzig, 1961—, prof., 1966—. Mem. Soc. Biochemistry. Author: (with V. Bayerl, M. Quarg) Taschenbuch des Verfahrens-chemikers, 1965; (with H. Aurich) Physiologisch-chemisches Institut in Leipzig, Geschichte und Forschungsergebnisse, 1962; also numerous articles. Research on biochemistry of thyroid hormones, anuran metamorphosis, cellular and energy metabolism, metabolism and functions of carnitine, aminotransferases, application of isotopics in biochemistry. Home: 4 Ernst-Thälmann-Strasse, 7113 Markleeberg. Office: 16 Liebigstrasse, Leipzig 701, East Germany.*

ROUARD, Pierre, French physicist; b. Marseilles, France, Oct. 25, 1908; s. Adolphe and Rose (Petit) R.; Agrégé des sciences physiques, Faculty Scis., Marseilles, 1930, Docteur es-sciences, 1936; m. Lucrèce Kohler, Apr. 12, 1946. With Faculty Scis., Marseilles, 1935-42, head works, 1939-42, prof., 1944—, dean, 1958—; lectr. Faculty Scis., Clermont-Ferrand, France, 1942-44. Mem. sci. council Commissariat à l'Energie Atomique; dir. Centre d'étude des couches minces. Lauréat, French Acad. Scis.; Parville prize, 1951. Mem. French Soc. Physics. Author: Propriâtés optiques des lames minces solides, 1952; Applications optiques des lames minces solides, 1952; Electroacoustique, 1960; also articles. Editorial bd. Thin Solid Films. Research on antireflecting layers, optical properties of thin solid films; determination of optical constants of thin metallic films; variation with wave-lengths of optical constants of thin metallic films. Home: 58 Tellène, 13-Marseilles, France.*

ROUBAUD, Emile (-Charles-Camille), French biologist; b. Paris, Mar. 2, 1882; s. Charles Albert and Catherine (Posler) R.; Lic.Sc., Sorbonne, Paris, 1901, Sc.D., 1909; m. Suzanne Veillon, 1916; children —Louise (Mrs. P. de Sablet), Gabrielle (Mrs. A. Baron), Geneviève (Mrs. Word). Became head mission French Equatorial Africa, 1906; sci. missions to colonial Africa, Pasteur Inst., 1909-12, dir. lab., Paris, 1912-20, prof., 1920—. Decorated Commandeur de la Légion d'Honneur. Mem. Acad. Colonial Scis. (became pres. 1943), Soc. Entomology, Soc. Tropical Pathology, Acad. Agr., Geog. Soc., Biol. Soc., Biog. Soc., Rumanian Acad. Scis., Belgian Soc. Tropical Medicine (corr.), Acad. Colonial Sci. (founding), French Acad. Scis., 1938. Research on disease carrying insects.

ROUCH, Jules (Alfred-Pierre), French geographer, oceanographer, explorer; b. Marseille, France, May 24, 1884; student Ecole Navale; children—Genevieve, Jean. Capt., French Navy; mem. Charcot's Antarctic Expdn., 1908-10; chief Meteorol. Service, French Armies, World War I; prof. l'Ecole navale, 1919-23; naval attaché, 1933-37; prof. Oceanographic Inst., Paris, 1937-45; dir. Oceanographic Mus., Monaco, 1945-57. Mem. French Acad. Scis., 1946, Académie de Marine Paris (past pres.), Overseas Acad. Sci. Author narratives on his voyages, also numerous articles. Research in phys. geography, oceanography, meteorology, geography.

ROUCHÉ, Eugène, French mathematician; b. Sommières, France, Aug. 18, 1832; docteur ès scis. École polytechnique; prof. math. Lycée Charlegmagne, École centrale; admissions examiner École polytechnique. Mem. French Acad. Scis., 1896. Author: (with Comberousse) Traité de géométrie; also works on descriptive geometry, algebra, math. analysis. Studied functions in series, theory of equal roots, decomposition of rational numbers, definite integrals, probability theory. Died Lunel, France, Aug. 19, 1910.

ROUGNON, Nicolas François, French physician; b. Morteau, Franche-Comté, Apr. 19, 1727; M.D., Besancon, France; physician Hôtel de Dieu, Paris, then in Noyon; prof. medicine and botany U. Besançon, from 1759; chief physician Besançon hosp.; mem. Acad. Besançon. Author: Codex physiologicus, 1776; Considerationes pathologico-semeioticae de omnibus corporis functionibus, 1786-88; Médecine et curative générale et particulière, 1799. Gave 1st description of angina pectoris. Died Aug. 5, 1799.

ROULE, Louis, French zoologist; b. Marseille, France, 1861; lectr., then prof. faculty scis., Toulouse, France; named prof. ichthyology and herpetology Mus. Natural History, Paris, 1910. Recipient 2 Grand prizes in phys. scis. French Acad. Scis. Author: Études ichtyologiques. Research in embryology and comparative anatomy, later on migratory and deepwater fish; expounded main problems of modern biology. Died 1942.

ROULSTON, Kenneth Irwin, physicist; b. Armagh, No. Ireland, Aug. 4, 1916; s. Thomas and Evelyn (Irwin) R.; B.A. with honors, Dublin (Ireland) U., 1938, M.Sc., 1940; Ph.D., U. Man., Winnipeg, Can., 1952; m. Doris M. Felgate, Aug. 17, 1944; children—Enid Maureen, Clifford K. Asst. lectr. Trinity Coll., Dublin, 1938-39; physicist Standard Telephones & Cables, Ltd., 1940-42; electronic engr. Elliott Bros., Ltd. (London) 1947-48; faculty U. Man., 1948—, prof. physics, 1959—; dir., v.p. Nuclear Enterprises, Ltd., Winnipeg, 1951—; dir. Nuclear Enterprises (GB) Ltd., Edinburgh, Scotland. Fellow I.E.E.E., Am. Phys. Soc., Inst. Physics; mem. N.Y. Acad. Scis., Canadian Assn. Physicists. Research, publs. on scintillation counter and gamma ray spectroscopy, applications to geophys. problems, including surveying for radioactive minerals and petroleum deposits; co-inventor geophys. surveying instrument. Home: 314 Park Blvd., Winnipeg 29, Man., Can.*

ROUNSEFELL, George Armytage, Am. marine biologist; b. Ketchikan, Alaska, Feb. 11, 1905; s. George Forsyth and Lillian (Thompson) R.; A.B., Stanford, 1926, Ph.D., 1931; M.S., U. Wash., 1929; m. Velma Kathleen Lawler, Mar. 7, 1942; children—Eric, Thane, Jennifer, Karen. Mem. staffs various govtl. fishery agys., 1924-50; sci. editor Fish and Wildlife Service, 1951-52, 54-56; leader fishery mission FAO, Turkey, 1953-54; dir. Fisheries lab U. S. Bur. Fisheries Galveston, Tex., 1957-63; dir. Ala. Marine Resources Lab., 1963-64; faculty U. Ala., 1963—, prof. marine biology, 1964-65, dir. Marine Sci. Inst., Bayou La Batre, 1966—. Recipient Distinguished Service award U. S. Dept. Interior, 1964. Mem. Am. Soc. Limnology and Oceanography, Am. Fisheries Soc., Am. Inst. Fishery Research Biologists, Wildlife Soc., Gulf and Carribbean Fisheries Inst., Ala., Miss. acads. sci., Sigma Xi, Phi Sigma. Author: (with W. H. Everhart, John Wiley) Fishery Science, 1953. Research and publs. on various fish, reservoirs, fishery edn. and techniques, estuarine research. Address: P.O. Box 667, Marine Sciences Inst., Bayou La Batre, Ala. 36509.*

ROUS, (Francis) Peyton, Am. physician, virologist; b. Baltimore, Md., Oct. 5, 1879; s. Charles and Frances Anderson (Wood) R.; B.A., Johns Hopkins, 1900, M.D., 1905; (hon.) Sc.D., Cambridge Univ., U. of Mich., 1938. Yale, U. of Birmingham, McGill U., 1949, U. Chgo., 1954, Rockefeller Inst., 1959; M.D. (hon.), Univ. of Zurich, 1946; LL.D. (honorary), St. Lawrence University, 1963; m. Marion Eckford de Kay, June 15, 1915; children—Marion (Mrs. Alan Hodgkin), Ellen DeKay, Phoebe (Mrs. Thomas J. Wilson). Resident house officer, Johns Hopkins Hosp., 1905-06; instr. in pathology, U. of Mich., 1906-08; asst., asso., asso. mem., Rockefeller Inst. for Med. Research, 1909-20, mem. in pathology and bacteriology, 1920-45, mem. emeritus 1945; mem. bd. scientific consultants Sloan-Kettering Inst. Cancer Research, New York City, 1957—; Linacre Lecturer Cambridge U., 1929, honorary fellow Trinity Hall, same. Fellow A.A.A.S.; mem. Nat. Acad. Sciences, Am. Philos. Soc., Assn. Am. Physicians, Am. Assn. Pathologists and Bacteriologists, Am. Soc. for Exptl. Pathology, Soc. for Experimental Biology and Medicine, Harvey Soc., N.Y. Pathological Soc., N.Y. Acad. Medicine, Am. Assn. for Cancer Research, Phi Beta Kappa, Sigma Xi. Foreign mem. Royal Society; hon. fellow Royal Soc. Medicine and Coll. Pathologists, London; Weizmann Inst. Science, Israel; hon. mem. British Physiol. Soc., Pathol. Soc. of Great Britain and Ireland, American Society for Microbiology, New York Pathological Society, foreign correspondent mem. British Med. Association, Académie de Médecine, Paris; mem. Royal Acad. Sciences of Denmark, Norwegian Academy of Science and Letters. Awarded John Scott medal and award, 1927; Walker prize, Royal Coll. Surgeons, England, 1941; Anna Fuller Award, 1952; Kober Medal, Assn. Am. Physicians, 1953; Bertner medal and award, U. Tex., 1954; Kovalenko Medal, Nat. Acad. Scis., 1956; distinguished service award Am. Cancer Soc., 1957; Lasker award Am. Pub. Health Assn., 1958; Landsteiner award, Am. Assn. Blood Banks, 1958; N.Y. Acad. Medicine medal, 1959; Judd award, Memorial Center for Cancer, 1959; Gold Medal, Royal Soc. of Medicine (London), 1962, United Nations Prize for cancer research, 1962; Gold-headed Cane award American Assn. of Pathologists and Bacteriologists, 1964. Co-editor the Journal of Experimental Medicine. Research on cancer, viruses, liver pathology and physiology, blood and blood vessels; important studies with chickens of relationship of virus to cells containing it and organism supporting it; showed that influences changing normal cells into tumor cells differ from those encouraging growth; built first blood banks during World War I. Home: 122 E. 82d St. Address: Rockefeller Inst. Med. Research, 66th St. and York Av., N.Y.C. 21.

ROUSE, George E., Am. geochemist; b. Chugwater, Wyoming, May 4, 1934; s. Elverton Frank and Lucy (Wall) R.; geo. engr., Colorado School of Mines, 1961, D.Sc. 1968; m. Sue Ellen Werber, June 4, 1961; children—Dorthea Lyn, Valerie Sue. Mining geologist, Anglo American Corp. of South Africa, Ltd., 1961-64; geological engr., project mgr., Earth Sciences, Inc., Golden, Colorado, 1968. Mem., Am. Inst. Mining Engrs., Sigma Xi. With R. E. Bisque, developed theory that stresses at interface of earth's interior (globally) control distribution of surface features, such as mountains, deep oceanic trenches, mineral belts, island arcs, etc.; theory asserts that quakes, volcanoes, rifts fall into ordered pattern and are not chance happenings; provides possibility for prediction of earthquakes; asserted that intergalactic and interplanetary magnetic fields may react sufficiently with earth's magnetic field to cause core to rotate a little differently than earth does on regular turning around axis. Office: Earth Sciences, Inc., 1101 Washington Ave., Golden, Colorado 80401.

ROUSE, Hunter, Am. hydraulician, univ. dean; b. Toledo, Mar. 29, 1906; s. Henry Esmond and Jessie (Hunter) R.; student Toledo U., 1924-25; S.B., Mass. Inst. Tech., 1929, S.M., 1932; Dr.-Ing., Technische Hochschule, Karlsruhe, 1932; Doctorat ès Sciences Physiques, U. Paris, 1959; m. Dorothee Hüsmert, July 7, 1932; children—Richard Hunter, Allan Hüsmert, Patricia Mary. Began as instrumentman Lucas County (O.) Surveyors, 1925; Mass. Inst. Tech. traveling fellow in hydraulics, 1929-31; asst. in hydraulics, Mass. Inst. Tech., 1931-33; instr. civil engring. Columbia, 1933-35; asst. prof. fluid mechanics Cal. Inst. Tech., also asso. hydraulic engr. Soil Conservation Service, Pasadena, Cal., 1936-39; prof. fluid mechanics U. Ia., 1939-—, dean Coll. Engring., 1966-—; cons. engr. Ia. Inst. Hydraulic Research, 1939-42, asso. dir. in charge, 1942-44, dir., 1944-65; Fulbright Research scholar, France, 1952-53; Professeur d'Èchange, Faculté des Sciences, U. Grenoble, 1952-53. Tech. rep. OSRD, 1943-45; cons. Office Naval Research, 1948-—, Waterways Expt. Sta., 1949-—; Smith-Mundt cons. Egyptian univs., 1960; organizer Soviet-Am. exchange hydraulic lab. dirs., 1961-62, Japanese-Am. exchange, 1965; adv. com. ESSA, 1966-—. Sr. post-doctoral fellow NSF at univs. Göttingen, Rome, Cambridge, Paris, 1958-59. Recipient Norman medal Am. Soc. C.E., 1938, shared Karl Emil Hilgard prize, 1951, 61, Theodore von Karman award, 1963; George Westinghouse award Am. Soc. Engring. Edn., 1948, Vincent Bendix award, 1958. Fellow Am. Acad. Arts and Scis., Am. Soc. C.E.; mem. Nat. Acad. Engring., A.A.A.S., Internat Assn. for Hydraulic Research, Gesellschaft für angewandte Mathematik und Mechanik, Am. Soc. for Engring. Edn., Sigma Xi, Sigma Chi, Tau Beta Pi, Chi Epsilon. Author: Fluid Mechanics for Hydraulic Engineers, 1938; Elementary Mechanics of Fluids, 1946; Basic Mechanics of Fluids (with J. W. Howe) 1953; History of Hydraulics (with S. Ince), 1957; also numerous articles. Editor, co-author: Engineering Hydraulics, 1950; Advanced Mechanics of Fluids, 1959. Research on fluid mechanics including similitude, efflux and overflow, cavitation, jet diffusion, free convection, turbulence, boundary roughness and sediment suspension. Home: 701 Templin Rd., Iowa City, 52240.*

ROUSE, Irving, Am. anthropologist; b. Rochester, N.Y., Aug. 29, 1913; s. Benjamin Irving and Louise (Bohachek) R.; B.S., Yale, 1934, Ph.D., 1938; m. Mary Uta Mikami, June 24, 1939; children—Peter Mikami, David Christopher. Asst. curator anthropology Peabody Mus. Natural History, Yale, New Haven, 1938-47, asso. curator, 1947-54, research asso. in anthropology, 1954-62; instr. anthropology Yale, New Haven, 1939-43, asst. prof., 1943-48, asso. prof., 1948-54, prof., 1954-—, dir. grad. studies in anthropology, 1953-57, chmn. dept., 1957-63. Recipient A. Cressy Morrison prize in natural sci. N.Y. Acad. Scis., 1948, medal in anthropology Viking Fund, 1960. Mem. Nat. Acad. Scis. Archaeol. fieldwork in Haiti, 1935, P.R., 1936-38, 62, Cuba, 1941, Fla., 1944, 49, Trinidad, 1946, 53, Venezuela, 1946, 50, 55-57, 61-62, Mass. 1952, N.Y., 1953, Conn.; studied Caribbean prehistory, also archeol. method, theory. Home: 12 Ridgewood Terrace, North Haven, Conn. 06473. Office: Box 2114, Yale Station, New Haven 06520.*

ROUSI, Arne Henrik, Finnish botanist; b. Jamsa, Finland, Sept. 1, 1931; s. Arvo Sulo and Karin (Sahlberg) R.; Ph.D., U. Helsinki, 1956; m. Liisa Maria Sipila, Apr. 28, 1956; children—Maarit Kirsti Kaarina, Laura Maria, Martti Arvo Henrik. Asst., U. Turku, Finland, 1957-60, docent, 1962-66, asso. prof. botany, 1966-—; research asso. U. Cal. at Berkeley, 1959-60; prin. researcher dept. horticulture Agrl. Research Center, Piikkio, 1960-66; docent U. Helsinki, 1963-—. Publs. on research on chromosomes and evolution of higher plants; race formation of plant species with a wide distbn. Home: Kupittank 9C, Turku. Office: U. Turku, Finland.*

ROUSSEAU, Jean Jacques, social philosopher, writer; b. Geneva, Switzerland, June 28, 1712; s. Isaac and Suzanne (Bernard) R.; mother died in childbirth, spent rough childhood; m. Thérèse la Vasseur. Footman to a Madame de Vercellis, Turin; worked for Madame de Warens at Annecy, 1729-31; studied classics under seminarists of St. Lazare; also studied under a music master; "amant en titre" to Madame de Warens, 1732-38; second to Vintzenreid as "amant en titre", 1738-40; tutor to M. de Mably's children, Lyons, 1740; traveled to Paris, 1741; sec. of M. de Montaigu, ambassador to Venice; sec. to Madame Dupin and her son-in-law, M. de Fracueil, Paris, 1745; copied music for money, Paris; associated with literary group headed by Diderot; contributed to Encyclopédie; brought out an operetta "Devin du village" at Fountainbleau, 1752; quarreled with Diderot; returned to cottage nr. Montmorency, 1756; established himself at Montlouis, winter 1757-58; Duke and Duchess of Luxembourg his patrons, 1758-62; fled to Motiers in Neuchatel after Paris parlement condemned Émile; expelled from Bern, 1765; accepted David Hume's offer to live at his house in Wotton, Derbyshire; spent time in Eng., 1766-67; returned to France, 1767; spent time with Mirabeau (father of the great Mirabeau), then Prince de Conti at Tyre; copied music at Paris, 1770-78; accused of being (nearly) insane in his later years; shortly before his death moved to cottage at Ermonville. Author: La Nouvelle Héloïse, 1760; Contract social, 1762; Émile (traces principles of modern education), 1762; Lettres de la montagne, 1763; Confessions; Dialogues; Reveries du promeneur solitaire. Supported concept of social contract among men for attaining civil liberty; stressed idea of general will as being important in decision making; asserted society corrupted man's native goodness; called for return to "natural man"; attacked unnatural and social inequalities; stressed life sciences in his work; believed in expression rather than repression in education; mentioned importance of fresh air and hygiene in child's upbringing; considered father of romanticism in literature. Died Ermonville, nr. Paris, France, July 2, 1778.

ROUSSELOT, Jean Pierre, French linguist; b. St. Cloud, France, 1846; sem. student, Angoulême, France, 1865-70; degree in arts, 1878, doctorate, 1892; ordained priest; began course in exptl. phonetics Catholic Inst., 1889; founder Soc. Langs. of France, 1893; founder phonetics lab. Coll. de France, 1897, became lab. dir., later prof. exptl. phonetics. Author: Principes de phonétique expérimentale, 1897-1909; (with F. Laclotte) Précis de prononciation français, 1902. Studied Spanish, English, German, French provincial dialects; called father of instrumental phonetics and geog. linguistics; inventor recording apparatus based on research of physiologist Marey. Died Paris, 1924.

ROUSSIN, Baron Albin-Reine, French naval officer, hydrographer; b. Dijon, France, Apr. 21, 1781; a son, Albert-Edmond-Louis. Mem. D'Irlande expdn. under Hoche; successively became lt.-comdr., 1807, comdr., 1810, capt., 1814, rear adm., 1822, adm., 1840; captured by English, 1808; later returned to France, named capt. Antilles Squadron, 1821; ambassador to Constantinople, 1832-34; named minister of Navy, 1840, 43. Mem. French Acad. Scis., 1830, Bur. Longitudes. Made hydrographic explorations of African and Brazilian coasts, 1816-20. Died Paris, Feb. 21, 1854.

ROUSSIN, François-Zacharie, French chemist; b. Vieux-vy, France, 1817; named prof., Val-de-Grâce, France, 1858. Author: (with Dr. Tardieu) Étude médico-légale et clinique sur l'empoisonnement, 1867. Developed dyeing processes and materials used in industry. Died Paris, 1894.

ROUSSY, Gustave, neurologist, pathologist; b. Vevey, Switzerland, Nov. 24, 1874; became French citizen, 1906; named rector U. Paris, 1937; organized sci. and lit. lectures Sorbonne, during World War II; dismissed by Vichy govt. for protesting closing of French univs., reinstated, 1944; hon. prof. path. anatomy, hon. dean Faculty Medicine, Paris; founder, dir. Cancer Inst. Mem., French Acad. Scis., Acad. Medicine (gen. sec.). Research on prevention and cure of cancer, also on structure of eye, cortex, medulla, endocrine glands. Died Paris, Sept. 30, 1948.

ROUTH, Edward John, mathematician; b. Que., Can., Jan. 20, 1831; s. Sir Randolph Isham and Marie Louise (Taschereau) R.; B.A., London, 1849; M.A., 1853; student Peterhouse, Cambridge, grad. sr. wrangler, 1854, Sc.D., Cambridge, 1883; LL.D., Glasgow, 1878; Sc.D., Dublin, 1892; m. Hilda Airy, Aug. 31, 1864; 5 sons, 1 dau. Joined Peterhouse, Cambridge, 1850, fellow, 1855, hon. fellow, 1883, lectr. math., 1855-1904, had unprecedented success as pvt. coach, 500 out of 700 pupils became wranglers, 27 sr. wranglers, 43 Smith's prizemen; examiner Univ. Coll., London, Cambridge. Fellow Royal Soc., 1872, Geol. Soc., Royal Acad. Sci. Author: An Analytical View of Newton's Principia, 1855; Rigid Dynamics, 1860; Treatise on the Stability of a Given State of Motion (recipient Adams prize, treatise greatly advanced knowledge of dynamics), 1877; Statics, 1891; Dynamics of a Particle, 1898, also numerous articles. Greatest contbn. was as math. tutor. Died Cambridge, Eng., June 7, 1907.

ROUTH, Joseph Isaac, Am. biochemist; b. Logansport, Ind., May 8, 1910; s. William Arthur and Ethel (Etnire) R.; B.S., Chem.E., Purdue, 1933, M.S., 1934; Ph.D., U. Mich., 1937; m. Dorothy Frances Hayes, Sept. 4, 1937; children—Joseph Hayes, John Michael. Faculty State U. Ia., Iowa City, 1937-—, prof. biochemistry, 1951-—, dir. clin. biochemistry lab., 1952-64. Cons. clin. chemistry VA Hosp., Iowa City, 1952-—. Diplomate Am. Bd. Clin. Chemistry (pres. 1959-—). Fellow Am. Inst. Chemists, Am. Assn. Clin. Chemists (past mem. exec. com., past pres.), Am. Chem. Soc. (past pres. Ia. sect.), Am. Soc. Biol. Chemistry, Soc. Exptl. Biology and Medicine, Ia. Acad. Sci., Sigma Xi, Phi Lambda Upsilon, Alpha Chi Sigma Gamma Alpha. Author: Twentieth Century Chemistry, (1st edit.) 1953, 3d edit., 1963; Fundamentals of Inorganic, Organic and Biological Chemistry, 1945, 5th edit., 1965; Laboratory Manual of Chemistry, 1st edit., 1945, 5th edit, 1965; also numerous articles. Editorial bd. Clin. Chemistry, 1954-60. Research on phys. chemistry, chem. composition and nutritional properties finely powdered keratins, nutritional requirement hamster, electrophoretic analysis plasma, serum and other body fluids in normal individuals and patients with various diseases, chemistry, metabolism and therapeutic effect analgesic drugs. Home: 2 Oak Park Ct., Iowa City 52240.*

ROUTIEN, John Broderick, Am. microbiologist; b. Mt. Vernon, Ind., Jan. 23, 1913; s. William Evert and Frances Lolita (Broderick) R.; B.A., DePauw U., 1934; M.A., Northwestern U., 1936; Ph.D., Mich. State Coll., 1939; m. Helen Harrison Boyd, Mar. 11, 1944 (dec. Sept. 1965); m. Constance C. Connolly, Feb. 22, 1967. Instr. botany U. Mo., 1939-42; mycologist Charles Pfizer & Co., Inc., Bklyn., 1946-—. Mem. adv. com. on fungi Am. Type Culture Collection. Recipient Comml. Solvents award in antibiotics, 1950. Mem. Mycol. Soc. Am., Bot. Soc. Am., Soc. Am. Bacteriologists, N.Y Acad. Scis. (chmn. div. mycology 1955-57), Soc. Indsl. Microbiology. Editorial bd. Applied Microbiology, 1964-—. Research, publs. and patents on identification, distbn. and genetics of fungi; co-inventor several antibiotics. Home: Grassy Hill Rd., Lyme, Conn. 06371.* Office: Groton, Conn. 06340.*

ROUTLY, Paul McRae, Am. astrophysicist, assn. exec.; b. Chester, Pa., Jan. 4, 1926; s. James Lawrence and Adelaide (Russell) R.; B.Sc., McGill U., Montreal, Que., Can., 1947, M.Sc., 1948; A.M., Princeton, 1950, Ph.D., 1951; m. Angelina M. Catanese, Aug. 5, 1951; children—Pamela Lynn, Paula Elizabeth. Fellow, NRC Can., 1951-53; research fellow Cal. Inst. Tech., 1953-54; faculty Pomona Coll., Claremont, Cal., 1954-63, prof., 1962-63, chmn. dept. astronomy, dir. Obs., 1954-63; exec. officer Am. Astron. Soc., Princeton, N.J., 1962-—; vis. prof. Rutgers U., 1967. Mem. Am. Astron. Soc., Am. Mus. Natural History, Nat. Geog. Soc., Internat. Astron. Union. Contbg. author: Galactic Astronomy, 1968. Research, publs. on astrophysics, molecular spectroscopy, absolute and relative atomic transition. Home: 184 Shady Brook Lane. Office: 211 FitzRandolph Rd., Princeton, N.J. 08540.*

ROUX, Edward Rudolph, S. African plant physiologist; b. Pietersburg, Transvaal, Apr. 1903; s. Philip Rudolph and Edith May (Wilson) R.; B.Sc. with honors U. Witwatersrand, 1926, M.Sc., Ph.D., Cambridge U., 1929; m. Winifred Mary Lunt, Nov. 1933; 1 dau., Alison (Mrs. Luigi Pineschi). Research pathologist U. Capetown, 1937-39; research with Dr. C. J. Mattero, 1940-45; sr. lectr. plant physiology U. Witwatersrand, 1946-62, prof. botany, 1963-64. Mem. S. African Assn. Advancement Sci., Bot. Soc. S. Africa, Tree Soc., others. Author: Harvest and Health in Africa, 1942; Cattle of Kiamalo, 1943; The Veld and the Future, 1946; Botany for Medical Students, 1951, others. Research on conservation of veld grasses and soil. Died Johannesburg, S. Africa, Mar. 1, 1966.

ROUX, Philibert Joseph, French surgeon; b. Auxerre, France, Apr. 26, 1780; studied medicine, Paris; pupil of Bichat; m. Miss Boyer, circa 1810; became asst. surgeon Hopital de la Charité, Faculty Medicine, Paris, 1810, prof. clin. surgery, 1820; succeeded Dupuytren at Hôtel Dieu, 1835. Mem. French Acad. Scis., Acad. Medicine (pres. 1828). Author: Treatise on Resection, 1812; other surg. treatises. Credited with 1st staphylorrhaphy in France, 1819; sutured ruptured female perineum, 1832; suggested ether per rectum for anesthesia, 1847. Died Paris, Mar. 23, 1854.

ROUX, Pierre Paul Emile, French bacteriologist, physician; b. Confolens, France, Dec. 17, 1853; studied medicine at Clermont-Ferrand and U. Paris. M.D. 1881. Clinical aid, Faculty Medicine, U. Paris, 1874-78; asst., Pasteur Institute, from 1888, dir. 1904-18; pres., Higher Council of Public Hygiene, Paris. Mem. Fr. Acad. Scis., 1899. Recipient Prix Osiris, Acad. Scis., 1903. Research on tetanus, anthrax, hydrophobia; discovered pneumonia microbe (with Edmond Nocard), 1880; his research on the diphtheria baccillus (and those of Emil von Behring) resulted in the discovery of a diphtheria vaccine, 1894; worked on syphilis (with Élie Metchnikoff). Died Paris, France, Nov. 3, 1933.

ROUX, Wilhelm, German zoologist; b. Jena, Germany, June 9, 1850; studied under Haeckel; student sci., Jena, Berlin, Strasbourg, France. Become asst. Inst. Hygiene, Leipzig, Germany, 1879; named prof. Inst. Embryology and Mechanics Devel., U. Breslau (now Wroclaw, Poland), 1886, dir., 1888; became prof. anatomy Innsbruck, Austria, 1889, Halle, 1895-1921. Author: Geschichtliche Abhandlung über Entwicklungsmechanik, 2 vols., 1895; Der Kampf der Teile im Organismus, 1881; Über die Entwicklungsmechanik der Organismen, 1890; Die Entwicklungsmechanik, 1905; Terminologie der Entwicklungsmechanik, 1912. Founder, Archiv für Entwicklungsmechanik, 1894. Founder exptl. embryology; research on early devel. of fertilized egg; developed technique for development mechanics; studied functional differentiation (organ and tissue changes caused by use or disuse). Died Halle, Sept. 15, 1924.

ROVELSTAD, Gordon Henry, Am. dentist; b. Elgin, Ill., May 19, 1921; s. Henry Randolph and Margot (Greenhill) R.; student St. Olaf Coll., 1939-41; D.D.S., Northwestern U., 1948, Ph.D., in Pathology, 1960; m. Barbara Jean Johnson, Apr. 8, 1945; children—Craig, Martha, Andrew. Faculty, Northwestern U., 1943-53, asst. prof. pedodontics, 1948-53, also cons. cleft palate inst., 1949-53; commd. ensign, 1942, USN, advanced through grades to capt., 1957-—, head research and scis. div., also research coordinator dental sch., Nat. Naval Med. Center, Great Lakes, Ill., 1960-65, dir. dental research facility, dental dept., adminstrv. command U. S. Naval Tng. Center, Great Lakes, 1965-66, officer in charge Naval Dental Research Inst., 1966-—. Lectr. Georgetown U., 1961-65; cons. grants div. NIH, 1960-64; cons. presdl. planning com. Project Headstart, 1965. Diplomate Am. Bd. Pedodontics (chmn. 1964). Fellow Am. Coll. Dentists, A.A.A.S.; mem. Am. Dental Assn. (sec.-treas.), Internat. Assn. Dental Research, Am. Soc. Dentistry Children, Am. Acad. Pedontics, N.Y. Acad. Sci., Am. Assn. Mil. Surgeons, Omicron Kappa Upsilon, Sigma Xi. Research and publs. on dental caries and the effect of fluorides in prevention; electrophoresis of saliva; dental surveys of naval personnel. Home: 1897 Enterprise St. Office: Naval Dental Research Inst., U. S. Naval Tng. Center, Great Lakes, Ill. 50088.*

ROVENSTINE, E(mery) A(ndrew), Am. physician; b. Atwood, Ind., July 20, 1895; s. Cassius Andrew and Lulu (Massena) R.; A.B., Wabash Coll., Crawfordsville, Ind., 1917, D.Sc., 1948; M.D., Ind. U., 1928; student Grad. Sch., U. of Wis., 1930-34; m. Jewel Sonya Gould, 1939. Asst. prof. of anesthesia, U. of Wis., 1934; asst. prof. surgery, Coll. of Medicine, New York U., 1935-36, prof. of anesthesia since 1937, prof. of anesthesia Coll. of Dentistry since 1938; dir. of div. of anesthesia, Bellevue Hosp. since 1935; guest dir. anesthesia, Oxford (Eng.) U., 1938; guest prof. anesthesia, U. Rosario (Argentina), 1939; mem. med. teaching mission to Czechoslovakia, 1946. Cons. anesthetist—Beth Israel Hosp., Hosp. for Spl. Surgery, Goldwater Meml. Hosp., Knickerbocker Hosp., Horace Harding Hosp., Gouverneur Hospital (N.Y.C.). Sr. cons. in anesthesia Veterans Hosp. (Bronx, N.Y.); director of anesthesia, University Hospital, 1949—. Served as 1st lt. Engrs., A.U.S., 1917-19. Fel. A.M.A., N.Y. Acad. Med., N.Y. Acad. Science; mem. Am. Bd. Anesthesiology (pres. 1948), Am. Soc. Anesthetists (past pres.), Am. Soc. Regional Anesthesia (past pres.), Nat. Research Council (med. sci. div., 1941-46), Soc. Exptl. Biology and Medicine, Soc. Pharmaceutical and Experimental Therapy, International Anesthesia Research Society, Sigma Xi; honorary member French and Mexican socs. of anesthesia, honorary member South African Society Anesthesia. Recipient Internat. Anesthesia Research Award, 1938. Decorated Order of the White Lion (Czechoslovakia). Asso. editor: Anesthesiology, Geriatrics. Contbr. to numerous med. and dental publs. Conducted clinical research in cyclopropane anesthesia and spinal anesthesia, devising endotracheal airway and laryngoscope; investigation of therapeutic and diagnostic nerve blocking; prepared system for collecting statis. data of anesthesia and surgery. Home: 320 E. 57th St. Office: 550 First Av., New York 16.

ROVSING, Niels Thorkild, Danish surgeon; b. Flensborg, 1862; s. Vinhandler Hans Georg and Villumine Caroline Anoreia (Nielsen) R.; M.D., Copenhagen; m. Marie Emilie Raaschou, Apr. 30, 1890. Pres. Danish Surg. Soc. Author: Inflamation of Urinary Bladder, 1889; Surgical Dissection of Urinary Organs, 1897; Abdominal Surgery, 1910. Research on nitrate of silver, antiseptics, cystitis, carbonic treatment of tubercle, treatment of arthritis by vaseline injections, methods of gastropexis and hepopexis. Died 1927.

ROWE, Chandler William, Am. anthropologist, univ. pres.; b. Torrington, Conn., Oct. 8, 1917; s. Clarence and Bessie Margaret (Hotchkiss) R.; B.A., Beloit Coll., 1939; M.A., U. Chgo., 1947, Ph.D., 1951; m. Margaret Shirley Grubbs, May 31, 1941; children—Cynthia Margaret, Chandler William, Sara Lea. Staff archaeologist U. Tenn., 1941-42; tng. supr. Hercules Powder Co., 1942-43; prof. anthropology, chmn. dept. Lawrence U. (formerly Lawrence Coll.), 1946-61, dean coll., 1961-64, dean acad. affairs, 1964-65; pres. Hawaii Loa Coll., Oahu, 1965—. Dir. restoration Aztalan State Park, Wis., prehistoric Indian village, 1949-64. Mem. planning com. Inst. Indian Life and Community Resources for Indians Wis., 1962—. Fellow Am. Anthrop. Assn.; mem. Soc. Am. Archaeology, Am. Assn. Phys. Anthropologists, Central States Anthrop. Society (past pres.), Wis. Archaeol. Survey (dir., past pres.; Lapham Research medal 1959). Research and publs. in archeology of N.Am. Indian and Hawaiian prehistory, effigy mound culture of Wis., cultural change in Japan. Home: 1974 Halekoa Dr., Honolulu 96821.*

ROWE, George Giles, internist; b. Vulcan, Alta., Can., May 17, 1921; s. James Giles and Cora (Blotz) R.; came to U. S., 1923, naturalized, 1929; B.A., U. Wis., 1943, M.D., 1945; m. Patricia Barnett, Sept. 10, 1947; children—George, James, Jane. Instr. Washington U., St. Louis, 1948-50; mem. research staff U. Wis., Madison, 1952-57, faculty, 1957—, USPHS, sr. research fellow, 1961—, prof. medicine, 1964—. Mem. sect. NIH, since 1965—. Recipient Markle scholar U. Wis., 1955-60, Distinguished teaching award Wis. Med. Alumni Assn., 1963-64, Bardeen award Phi Kappa Phi, 1943. Research fellow Am. Heart Assn., 1952-54. Diplomate Am. Board Internal Medicine. Fellow A.C.P.; mem. Am. Physiol. Soc., Am. Soc. Pharm. and Exptl. Therapy, Am. Fedn. Clin. Research, Am. Soc. Clin. Investigation, Sigma Xi, Phi Beta Kappa, Alpha Omega Alpha, A.M.A., Am. Heart Assn., others. Research and publs. on the diagnosis and treatment of congenital and acquired heart disease, and drug treatment of heart and circulation; primary interest in coronary circulation and drugs which affect it. Home: 5 Walworth Ct., Madison, Wis. 53705.*

ROWE, John Howland, Am. anthropologist; b. Sorrento, Me., June 10, 1918; s. Louis Earle and Margaret Talbot (Jackson) R.; A.B., Brown U., 1939; M.A. in Anthropology, Harvard, 1941, Ph.D. in Latin Am. History and Anthropology, 1947; Litt.D. (hon.), U. Nacional del Cuzco (Peru), 1954; student U. Paris (France), 1945-46; m. Barbara Bent Burnett, June 6, 1942; children—Ann Pollard, Lucy Burnett. Field supr. So. Peru, Inst. Andean Research, 1941-42; prof. de arqueologia, dir. Seccion Arqueologica, U. Nacional del Cuzco, 1942-43; rep. in Colombia, Inst. Social Anthropology, Smithsonian Instn., tchr. U. del Cauca, Popayan, 1946-48; mem. faculty U. Cal. at Berkeley, 1948—, prof. anthropology, 1956—, chmn. dept., 1963—, curator S.Am. archaeology Mus. Anthropology, 1949—; field research in Me., 1938, 40, Mass., 1939-41, Fla., 1940, Guambia and Popayan, Colombia, 1946-48, Peru, 1939, 41-43, 46, 54, 58, 59, 61-65. Hon. prof. U. del Cauca 1947; research prof. Mil. Inst. Basic Research Sci., U. Cal. at Berkeley, 1964-65. Recipient Diploma de Honor, Soc. Cientifica del Cuzco, 1954, Prémio de Honor, Concejo Provincial de Inca, Peru, 1958; Guggenheim fellow, 1958. Mem. Archaeol. Inst. Am., Soc. Am. Archaeology, Soc. des Americanistes de Paris, Am. Anthrop. Assn., Soc. Peruana de Historia, Soc. History Tech., Inst. Andean Studies (pres. 1960—). Author: An Introduction to the Archaeology of Cuzco, 1944; Max Uhle, 1954; Chavin Art, 1962; (with Menzel and Dawson) The Paracas Pottery of Inca, 1964. Editor Nawpa Pacha, 1963—. Research on dating and cultural interpretation in Peruvian archeology, Inca history and instns., archeol. theory, early history of anthropology; developed new methods of qualitative stylistic analysis for ancient pottery; contributed to theory of use of archeol. assns. Home: 2137 Rose St., Berkeley, Cal. 94709.*

ROWE, Joseph Everett, Am. elec. engr.; b. Highland Park, Mich., June 4, 1927; s. Joseph and Lillian May (Osbourne) R.; B.S. Engring., 1951, B.S. Engring. in Math., 1951, M.S. Engring., 1952, Ph.D., 1955; m. Margaret Anne Prine, Sept. 1, 1950; children—Jonathan Dale Rowe, Carol Kay. Faculty, U. Mich., Ann Arbor, 1953—, prof. elec. engring., 1960—, dir. electron physics lab, 1958-68, chmn. dept. elec. engring., 1968—. Mem. adv. group on electron devices U. S. Dept. Def., 1966—. Cons. to pvt. cos. Fellow I.E.E.E. (chmn. adminstrv. com. group on electron devices 1968—; mem. Am. Phys. Soc., Am. Math. Soc., Am. Soc. Engring. Edn. (Curtis McGraw Research award), Sigma Xi, Phi Kappa Phi, Tau Beta Pi, Eta Kappa Nu. Author: Nonlinear Electron-Wave Interaction Phenomena, 1965; also articles. Research on nonlinear theory of microwave electron devices and plasma systems, solid-state masers, noise in microwave and quantum systems, solid state materials and devices. Home: 2745 Bedford Rd., Ann Arbor, Mich. 48104.*

ROWE, Robert Seaman, Am. civil engr., univ. dean; b. Wilmington, Del., Jan. 31, 1920; s. Everett Whittemore and Frances (Seaman) R.; B.S., U. Del., 1942; M.S., Columbia, 1949; M.Eng., Yale, 1950, D.Eng., 1951; m. Emma Jean Mote, Apr. 4, 1942; children—Robert Seaman, William E. Designing engr. Triumph Explosives, Inc., Elkton, Md., 1937-42; from instr. to asso. prof. civil engring. Princeton, 1946-55, research asso. rivers and harbors sect., 1952, dir., 1953-56; vis. prof. civil engring. U. Del., 1949; asst. in instrn. Yale, 1950; prof. civil engring., chmn. dept. Duke, 1956-60, J. A. Jones prof. engring., 1957-60; dean Sch. Engring., Vanderbilt U., 1960—. U. S. del. Internat. Navigation Congress, Rome, Italy, 1953; asst. engring. and metall. div. Office: Ordnance Research, 1957; chief investigator research project and design and analysis structural elements U. S. Army, 1958. Recipient Morris K. Blumberg Meml. prize U. Del., 1942. Mem. Am. Soc. C.E., Am. Concrete Inst., Am. Soc. Engring. Edn., Soc. Am. Mil. Engrs., Nat. Soc. Profl. Engrs., Permanent Internat. Assn. Navigation Congresses, Yale Engring. Assn., Tau Beta Pi, Phi Kappa Phi, Sigma Phi Epsilon. Author: Bibliography of Rivers and Harbors and Related Fields in Hydraulic Engineering, 1953; French-English Translation Guide for River and Harbor Engineering, 1954. Contbr. articles to profl. jours. Invented hyperbolic paraboloidal mortar base plate, tangent measuring device cables, guywires. Home: 4914 Roselawn Circle, Nashville 37215.*

ROWE, Thomas Dudley, Am. pharmacist; b. Missoula, Mont., June 25, 1910; s. Jesse P. and Anna (Richards) R.; student U. Mich., 1927-29; B.S., U. Mont., 1932, M.S., 1933; Ph.D., U. Wis., 1941; m. Georgia Stripp, Aug. 11, 1934; 1 son, Thomas Dudley. Instr., U. Neb., 1934-35; faculty Sch. Pharmacy, Med. Coll. Va., 1935-45, asso. prof., asst. dean, 1941-45; prof. pharmacy, dean Rutgers U., 1946-51; prof., dean Coll. Pharmacy, U. Mich., Ann Arbor, 1951—. Cons. to surgeon gen. U. S. Army., 1959—. Mem. Am. Assn. Colls. Pharmacy (past pres.), Am. (past 1st v.p.), Mich. State (Distinguished Service award 1956, exec. com. 1967—) pharm. assns., Am. Council on Pharm. Edn., Phi Beta Kappa (hon.), Sigma Xi, Phi Kappa Phi, Rho Chi, Kappa Psi, Sigma Chi. Author: (with R. A. Deno, D. C. Brodie) The Profession of Pharmacy, 1966; also articles. chpt. in book. Phytochem. and pharmacological study fresh Aloe vera and its effect on radiation burns, antiseptic value of iodine in decoloraized iodine tincture. Home: 1336 Glendaloch Circle, Ann Arbor, Mich.*

ROWE, Verald K., Am. toxicologist; b. Warren, Ill., Oct. 5, 1914; s. Earl James and Edna (Leverington) R.; A.B., Cornell Coll., Mt. Vernon, Ia., 1936; M.S., State U. Ia., 1937; m. Mary Gardner, Sept. 6, 1937; children—James K., Karen K. Staff biochem. research lab. Dow Chem. Co., Midland, Mich., 1937—, dir. toxicological research, 1954-64, asst. dir., 1964—. Mem. com. on toxicology NRC-Nat. Acad. Sci., 1964—. Diplomate Am. Acad. Indsl. Hygiene. Mem. Soc. Toxicology (pres. 1966-67), Am. Indsl. Hygiene Assn. (dir. 1956-57), Am. Soc. Pharmacology and Exptl. Therapy, U. S. Permanent Commn. Internat. Assn. Occupational Health. Research, chpts. in books, publs. on physiologic properties of chems. for purpose of evaluating their safety, determining the consequences of excessive exposure, prescribing proper uses and handling procedures. Home: 709 Columbia Rd. Office: Biochem Research Lab., Dow Chem. Co., Midland, Mich. 48640.*

ROWLAND, Frank Sherwood, Am. chemist; b. Delaware, O., June 28, 1927; s. Sidney A. and Margaret L. (Drake) R.; A.B., Ohio Wesleyan U., 1948; M.S., U. Chgo., 1951, Ph.D., 1952; m. Joan E. Lundberg, June 7, 1952; children—Ingrid, Jeffrey. Instr., Princeton, 1952-56; faculty U. Kan., Lawrence, 1956-64; prof., chmn. dept. chemistry U. Cal. at Irvine, 1964—. Mem. Am. Chem. Soc., Am. Phys. Soc. Contbr. numerous articles to sci. jours. Research on chem. kinetics of energetic atomic reactions, radiochemistry. Home: 4807 Dorchester Rd., Corona del Mar, Cal. 92625. Office: Dept. of Chemistry, University of Cal., Irvine, Cal. 92664.*

ROWLAND, Henry Augustus, Am. physicist; b. Honesdale, Pa., Nov. 27, 1848; C.E., Rensselaer Poly. Inst., 1870 (Ph.D., Johns Hopkins, 1880); LL.D., Yale, 1895; LL.D., Princeton, 1896; m. Henrietta Harrison, June 4, 1890. Engaged in railroad surveys, 1871; taught in Wooster Univ., 1871-72; instr., 1872-74, and asst. prof. physics, 1874-75, Rensselaer Poly. Inst.; studied abroad a year; prof. physics, Johns Hopkins, from 1876. Was member Electrical Congress at Paris, 1881; served on jury of electrical exhibition there, 1881; made chevalier, 1881, and in 1896 Officer Legion of Honor; recipient Rumford, Draper, and Mattenci medals for his discoveries. Mem. French Acad. Scis., 1893; Fellow Royal Soc., 1889. Author: On Magnetic Permeability, 1873; Research on the Absolute Unit of Electrical Resistance, 1878; On the Mechanical Equivalent of Heat, 1880; Elements of Physics, 1900. Principal scientific work was discovery of magnetic action due to electric convection; the exact determination of the mechanical equivalent of heat and of the ohm; the discovery of concave grating and the machine for ruling gratings by which spectrum analysis has been revolutionized. Died Balt., Apr. 16, 1901.

ROWLAND, Lewis Phillip, Am. neurologist; b. N.Y.C., Aug. 3, 1925; s. Henry A. and Cecile (Coles) R.; B.S., Yale, 1945, M.D., 1948; m. Esther Edelman, Aug. 31, 1952; children—Andrew Simon, Steven Samuel, Judith Mora. Clin. asso. NIH, 1953-54; asst. neurologist Montefiore, N.Y.C., 1954-57; faculty Columbia, 1957-67, co-dir. neurol. research center Columbia Presbyn. Med. Center, 1961-67; prof., chmn. dept. neurology U. Pa., Phila., 1967—. Sec. med. adv. bd. Myastenia Gravis Found., 1965—. Diplomate Nat. Bd. Med. Examiners, Am. Bd. Psychiatry and Neurology. Mem. Am. Neurol. Assn., Am. Acad. Neurology, Assn. Research Nervous Mental Disease, A.M.A., Alpha Omega Alpha, others. Mem. editorial bd. Archives of Neurology, 1968—. Research and publs. on biochem. and clin. studies of muscle disease and inherited diseases affecting the nervous system. Home: 5125 Woodbine Av., Phil. 19131.*

ROWLAND, Theodore Justin, Am. physicist; b. Cleve., May 15, 1927; s. Thurston Justin and Lillian (Nesser) R.; B.S., Western Reserve U., 1948; M.A., Harvard U., 1949, Ph.D., 1954; m. Janet Claire Millar, June 28, 1952; children—Theodore Justin, Dawson Ann, Claire Millar. Research physicist Union Carbide Metals Co., Niagara Falls, N.Y., 1954-61; prof. phys. metallurgy U. Ill., 1961—; cons. physicist, 1961—. Mem. Am. Phys. Soc., Am. Inst. Mining, Metall. and Petroleum Engrs., A.A.U.P., A.A.A.S., Sigma Xi. Contbr. articles in field to sci. jours. Applied nuclear magnetic resonance techniques to the study of metals and alloys. First demonstrated effects of defects in metals; demonstrated experimentally the nature of the electron charge distbn. around impurity atoms in solid solutions, indirect exchange, anisotropy of Knight shift. Office: Metallurgy Dept. U. Ill., Urbana, Ill. 61801.*

ROWLEY, Howard Holmes, Am. chemist; b. Medina, N.Y., Nov. 27, 1905; s. Harry Eglin and Kathrine (Taylor) R.; B.S. with highest honors, Northwestern U., 1927, M.S., 1928; Dr. phil. nat., Goethe U., Frankfurt, Germany, 1932; m. Elizabeth Marie Hayes, Aug. 23, 1939. Organic research Dow Chem. Co., 1928-30; instr. Northwestern U., 1933-37; asso. in chemistry U. Ia., 1937-42; asso. prof. Lawrence Coll., 1942-46; prof. chemistry U. Okla., Norman, 1946—, cons. U. Research Inst. Research participant Oak Ridge Nat. Lab., 1952, Ozark-Mahoning Co., 1954. Fellow Okla. Acad. Sci.; mem. Am. Chem. Soc., Phi Beta Kappa, Sigma Xi, Phi Lambda Upsilon, Alpha Chi Sigma. Author: (with James E. Belcher and James C. Colbert) Properties and Numerical Relationships of Common Elements and Compounds, 1961; Lecture Outlines, 1963; also articles. Research on preparation of inorganic compounds, phase studies of aqueous and non-aqueous systems. Home: 611 Morningside Dr., Norman, Okla. 73069.*

ROWLINSON, John Shipley, English chemist; b. Handforth, Cheshire, Eng., May 12, 1926; s. Frank and Winifred (Jones) R.; B.A., Trinity Coll., Oxford (Eng.) U., 1947, B.Sc., 1948, M.A., D.Phil., 1950; m. Nancy Gaskell, Aug. 3, 1952; children—Paul John, Stella Margaret. Research asso. Naval Research Lab., U. Wis., 1950-51; Imperial Chems. Industry Research fellow U. Manchester (Eng.), 1951-54, lectr., 1954-57, sr. lectr., 1957-60; prof. chem. tech. Imperial Coll., U. London (Eng.), 1961—; vis. prof. U. Wis., 1963. Cons., U. Cal. at Los Angeles, 1965. Recipient Meldola medal Royal Inst. Chemistry, 1954, Marlow medal Faraday Soc., 1957; named hon. asso. City and Guilds London Inst., 1966. Fellow Royal Inst. Chemistry; mem. Faraday Soc. (past mem. council), Chem. Soc. London (mem. council 1964—). Author: Liquids and Liquid Mixtures, 1959; The Perfect Gas, 1963; also numerous articles. Research on phys. properties of gases and liquids, interpretation of those properties in terms of properties of constituent molecules by methods of statis. thermodynamics. Home: 19 Burdett Av., London S.W.20., Eng.*

ROWNTREE, Leonard George, physician; b. London, Ont., Can., Apr. 10, 1883; s. George and Phoebe (Martindale) R.; M.D., Western Med. Coll., Can., 1905; D.Sc. (hon.), Western Ont., 1916; m. Katherine Campbell, July 9, 1914; 2 children. Asst. medicine Johns Hopkins, 1907-08, instr. pharmacology, 1908-10, asso. exptl. therapeutics, 1910-13, asso. prof., 1913-14, prof. medicine, 1914-16; prof. medicine, chief dept. Med. Sch., U. Minn., 1916-20; prof. medicine, sr. med. cons. Mayo Found., 1920-32; dir. Phila. Inst. Med. Research, 1932-45; chief med. adviser Am. Legion, 1944—; chief med. div. Nat. Hdqrs., SSS, 1940-45. Recipient Stritmatter award. Mem. A.M.A., A.C.P., Am. Physicians, Soc. Pharmacology, Soc. for Exptl. Pathology, Soc. Biol. Chemistry, Soc. for Clin. Investigation, Endocrine Soc. Research in clin. medicine, including water intoxication, renal and hepatic functional tests, Addison's disease, thymus and pineal glands, exptl. tumors, water balance, water intoxication; absorption from pleural and peritoneal cavities, vividiffusion, plasmaphaeresis, radium, antimonials in trypanosomiasis, subcutaneous purgatives, Addison's disease, exptl. tumors; developed (with others) method for determination volume of blood and plasma. Home: 1342 du Pont Bldg., Miami, Fla.

ROWSON, Lionel Edward Aston, English vet. surgeon; b. Stafford, Eng., May 28, 1914; s. Lionel Frederick and Maude (Aston) R.; M.R.C.U.S., Royal Vet. Coll. London (Eng.), 1938; m. Audrey Kathleen Foster, Jan. 8, 1942; children—Peter Aston, John Aston, Kathleen Ann, Mary Audrey. Practiced vet. medicine, 1939-41; sterility investigation officer, Cambridge, Eng., 1942; dir. Cambs and Dist. Cattle Breeders Ltd., 1942—; dep. dir. A.R.C. unit reproductive physiology and biochemistry, Cambridge, Eng., 1956—. Dir. several comml. cos. Recipient Thomas Barton award, 1956. Fellow Zool. Soc.; mem. Royal Coll. Vet. Surgeons, Soc. for Study Animal Breeding (past pres.). Co-author: The Semen of Animals and Artificial Insemination, 1962; Reproduction of Domestic Animals, 1959; (with J. Hammond, J. Edwards, A. Walton) Artificial Insemination of Cattle, 1947; also articles. Research on artificial insemination, physiology of repdn., relationships between function and anat. parts in animal reprdn. Home: Water Lane, Histon. Office: 307 Huntingdon Rd., Cambridge, Eng.*

ROXBY, Percy Maude, English geographer; b. Nov. 21, 1880; s. Rev. H. M. Roxby; B.A., Christ Church, Oxford (Eng.) U., 1903; m. Marjorie Peers Howden, 1941. Lectr. geography U. Liverpool (Eng.), 1904-12, prof., 1917-44; vis. prof. Egyptian U., Cairo, 1930. Mem. China Edn. Commn., 1921-22; chief rep. in China of Brit. Council, 1945-47. Albert Kahn travelling fellow, 1912-13. Mem. Brit. Assn. (pres. sect. E, 1930). Author: Henry Gratton Gladstone; The Far Eastern Question in its Geographical Setting, 1920; Report to Albert Kahn Trustees; China (Oxford Pamphlets on World Affairs), 1942. Contbr. Ency. Brittanica, various geog. jours. revs. Expert on China; stressed importance of studying geography of region as a framework for understanding its characteristics. Died Feb. 17, 1947.

ROY, Louis-Maurice, French physicist; b. Troyes, France, June 21, 1882; ed. Lille, Paris (both France) faculties scis.; M.Sc., Sc.D. Became asst., then prin. l'Ecole supérieure d'electricité, 1904, lectr., 1914; prof. rational and applied mechanics, faculty scis. U. Toulouse (France), 1914-52, emeritus, 1952—. Mem. French Acad. Scis., 1927, also French Phys. Soc. French Astron. Soc., Acad. Sci., Inscriptions and Letters (past pres. Toulouse). Author: Cours de mécanique rationelle, 4 vols., 1946; l'Electrodynamique des milieux isotropes en repos d'après Helmholtz et Duhem; Cours de statique, graphique et de résistance des matériaux; Cours d'électricité générale; also papers. Home: 9, rue Leo-Lagrange, Toulouse (Haute-Garonne), France.

ROY, Maurice, French mech. engr.; b. Bourges, France, Nov. 1, 1899; ed. Ecole Polytech., Paris, 1917-19, Ecole Nle Sre des Mines, Paris, 1920-22; Dr. Es-Sciences, U. Strasbourg, 1923; m. Nebout-Maritchu, Oct. 4, 1932. Chief engr. Tech. Control French Rys., 1922-35; dir. mech. industry France, 1935-46; sci. cons. to SNECMA and ONERA, Paris, 1946-49; gen. dir. ONERA, Paris, 1946-62; pres. Com. on Space Research, Paris, 1962—; prof. Ecole Poly., Paris, 1947—; gen. sec. Internat. Union Theoretical and Applied Mechanics, 1956—; pres. Comite de Coordination de la Recherche Industrielle, 1959—. Hon. fellow Inst. Aerospace Scis., fellow Royal Aero. Scis., French Acad. Scis., 1949; mem. Nat. Acad. Scis., Societé Française des Mecaniciens (chmn.), Assn. Tech. Maritime et Aeronautique (vice chmn.). Author: Mécanique des milieux contraires et déformables, 1950. Pioneer work theory and devel. jet propulsion; research work thermodynamics, mechanics, fluid dynamics. Home— 86 Av. Niel, Paris 17. Office: 55 Blvd. Malesherbes, Paris 8, France.

ROY, Rustum, chemist; b. Ranchi, India, July 3, 1924; s. Narendra Kumar and Rajkumari (Mukherjee) R.; B.Sc. with honors, Patna U., 1942, M.Sc., 1944; Ph.D., Pa. State U., 1948; m. Della Martin, June 8, 1948; children—Neill Rathan, Ronnen Andrew, Jeremy Rustum. Research asso. Pa. State U., 1948-49; sr. sci. officer Central Glass and Ceramic Research Inst., Calcutta, India, 1950; faculty Pa. State U., 1950—, prof. chemistry, 1957—, dir. material research lab., 1962—. Chmn. materials adv. panel Commonwealth of Pa., 1965—; nat. adv. bd. Coms. on Characterization, Ceramic Processing, 1964—; mem. Gov.'s Sci. Adv. Com., 1965—; mem. NRC; mem. com. mineral sci. and tech. Nat. Acad. Scis.-NRC, 1966—. Founder, dir. Tem-Pres Research, Inc., State College, Pa., 1957—. Mem. Am. Chem. Soc., Phys. Soc., Sigma Xi. Fellow Ceramic Soc., Mineral. Soc. Am. (award 1957). Editor, Materials Research Bull., 1966—. Research and numerous publs. synthesis and preparation of new inorganic materials, very high pressures; characterization of crystalline and non-crystalline solids. Home: 528 S. Pugh St., State College, Pa. 16801. Office: 1-112 Research Bldg., University Park, Pa. 16802.*

ROY, William, Brit. mil. engr.; b. Lanarkshire, Eng., May 4, 1726; s. John Roy. Assisted Watson prepare Duke of Cumberland's Map in survey of Scotland, 1747; made practioner-engr., Brit. army, 1755, sub-engr., lt., 1759, engr., capt., 1759, dept. quartermaster gen., Germany, 1760-61, later in south Britain, returned to Germany, 1762, promoted to lt.-col., surveyor gen. for mil. surveys, 1765, suggested improvements for defense of Gibraltar, 1768, served to maj. gen., from 1781; dir. Royal Engrs., from 1783. Fellow Royal Soc., 1767, Soc. Antiquaries. Author: Military Antiquities of the Romans in Britain, 1793; also papers: Experiments and Observations made in Britain in order to Obtain a Rule For Measuring Heights with the Barometer, 1778; An Account of the Mode Professed to be Followed in Determining the Relative Situations of the Royal Observations of Greenwich and Paris, 1787. Made sci. and archeol. studies; accurately measured base line at Hounslow Heath for triangulation to determine relative positions of observatories at Paris and Greenwich; drew several maps, 1752-66. Died London, 1790.

ROYCE, Sir Frederic Henry, English engr.; b. Petersborough, Eng., 1863; s. James Royce. Engring. apprentice Gt. No. Ry.; worked at Gun Machinery, Leeds, Eng.; chief elec. engr. in charge lighting Liverpool (Eng.) stas., 1882-83; founder Mech. & Elec. Engring. Co., Manchester, Eng., 1884; dir., cons. engr. Royce, Ltd.; co-founder (with C. S. Rolls) Rolls-Royce, Ltd., 1907, also dir., chief engr. Mem. Instn. Mech. Engrs., Instn. Elec. Engrs., Designer, builder automobile and airplane engines. Died Apr. 22, 1933.

ROYCE, James E(mmet), Am. psychologist; b. Spokane, Wash., Oct. 20, 1914; s. James Emmet and Lucie F. (Reilly) R.; B.A., Gonzaga U., 1939, M.A., 1940; Ph.D., Loyola U., Chgo., 1945, S.T.L., 1948. Instr., Gonzaga U., 1940-41; faculty Seattle U., 1948—, prof., head dept. psychology, 1960—; pres. Notre Dame Coll., Nelson, B.C., Can. 1954-55, vis. prof. various colls., summers. Chmn. Wash. State Examining Bd. in Psychology, 1964-66; mem. Gov.'s Adv. Bd. on Alcoholism, 1958—. Recipient Gov.'s Spl. award for contbrn. to alcoholism edn., 1965; Catholic Author's award N.W. Cath. Library Assn., 1962. Fellow Am. Psychol. Assn.; mem. Wash., Cath. psychol. assns., Acad. Religion and Mental Health. Author: Man and His Nature, 1961; Personality and Mental Health, 1964; Man and Meaning, 1968; also articles, revs. Address: Seattle U., Seattle 98122.*

ROYCE, Joseph R(ussell), psychologist; b. New York, N.Y., Aug. 19, 1921; A.B., Denison, 1941; Ph.D., U. Chicago, 1951; studied Ohio State; m. Apr. 3, 1944; children—Christopher J., Janet L. Asst. prof., Drake U., 1949-51; asst. then assoc. prof., Redlands, 1951-60; prof. and head dept. psych., U. Alberta, Canada, 1960—; dir., Center for Advanced Study in Theoretical Psychology, 1965—. Mem., Canada Nat. Research Council, 1962-65; Canada Council Sr. Research Fellow; Distinguished Visiting Lectrs. Program; Educational Testing Service, Princeton, N.J.; consultant, U. S. Air Force, 1951-55. Mem. Am. Psychology Assn., Soc. Multivariate Experimental Psychology, Canadian Psychology Assn. Author: The Encapsulated Man, Van Nostrand, 1964; (ed.) Psychology and the Symbol, 1965; Toward the Unification of Psychology, 1966. Work in theoretical unification of psychology; philosophical and methodological issues in psychology; general experimental psych.; application of multiple factor analysis to problems in comparative physiological psychology; research in behavior genetics; autokinetic phenomenon; development of psychological tests. Address: Dept. of Psychology, University of Alberta, Edmonton, Alberta, Canada.*

ROYCE, Josiah, Am. philosopher, psychologist; b. Grass Valley, Calif., Nov. 20, 1855; A.B., U. of Calif., 1875; Leipzig and Göttingen, 1876; Ph.D., Johns Hopkins, 1878; (LL.D., U. of Aberdeen, 1900, Johns Hopkins, 1902, Yale University, 1911, St. Andrew's Univ., 1911; Litt.D., Harvard, 1911; D.Sc., Oxford, 1913); m. Katharine Head, Oct. 2, 1880—3 sons. Instr. English literature and logic, U. of Calif., 1878-82; instr. philosophy, 1882-85, asst. prof., 1885-92, prof. history of philosophy, 1892-1904, Alford prof. natural religion, moral philosophy, and civil polity, Mar. 1914—, Harvard University. Fellow Am. Acad. Arts and Sciences. Author: A Primer of Logical Analysis, 1881; Religious Aspect of Philosophy, 1885; History of California (in Am. Commonwealth series), 1886; The Spirit of Modern Philosophy, 1892; The Conception of God (joint author), 1897; Studies of Good and Evil, 1898; The World and the Individual (2 vols.), 1900, 01; The Conception of Immortality, 1900; Outline of Psychology, 1903; Herbert Spencer, 1904; The Relation of the Principles of Logic to the Foundations of Geometry (Trans. Am. Math. Soc., July 1905); The Philosophy of Loyalty, 1908; Race Questions, Provincialism and Other American Problems, 1908; William James and Other Essays on the Philosophy of Life, 1911; Bross Lectures on The Sources of Religious Insight, 1912; The Problem of Christianity (2 vols.), 1913; War and Insurance, 1914; The Hope of the Great Community, 1917; Fugitive Essays, 1920. Leading Am. exponent of philosophical idealism, held monistic view of consciousness as a universal principle; believed that truth could be proven, that an absolute mind exists and that man as part of this mind can perceive truth; stressed importance of community in individual fulfillment. Died Cambridge, Mass., Sept. 14, 1916.

ROYDEN, Halsey Lawrence, Am. mathematician; Phoenix, Sept. 26, 1928; s. Halsey Lawrence and June (Slavens) R.; B.S., Stanford U., 1948, M.S., 1949; Ph.D., Harvard U., 1951; m. Virginia Voegeli, June 15, 1948, children—Leigh Handy, Halsey, Lawrence III, Constance Slavens. Asst. prof. Stanford U., 1951-53, asso. prof., 1953-58, prof., 1958—, asso. dean Sch. Humanities and Scis., 1962-65; NSF fellow Swiss Federal Inst. Tech., Zurich, 1958-59; editor Pacific Jour. Math. 1955-58. Mem. Am. Math. Soc., Math. Assn. Am., A.A.A.S., Finnish Math. Soc., Am. Inst. Archeology, Phi Beta Kappa, Sigma Xi. Author: Real Analysis, 1963. Contbns. to the theory of functions of a complex variable, Rieman surfaces, conformal mapping, foundations of geometry. Home: 13466 Three Forks Lane, Los Altos Hills, Cal. 94022. Office: Mathematics Dept., Stanford Univ., Stanford, Cal.*

ROYER, Clemence Augustine, French naturalist; b. Nantes, France, 1830; recipient (with Proudhon) prize for essay on polit. economy. Translator: Origines de l'homme et des sociétés (1st French transl. of Darwin's Origin of the Species), 1869. Author: Deux hypothèses sur l'hérédité, 1877; Nouveaux principes de philosophie naturelle, 1900; also articles on anthropology and archaeology. Died Neuilly-sur-Seine, France, 1902.

ROYER, Louis Michael, French mineralogist; b. Metz, France, Apr. 6, 1895; s. Eugene and Josephine (Petit) R.; D.Sc., U. Strasbourg (France); m. Mlle. Offenstein, Aug. 27, 1928. Lectr. mineralogy Faculty Scis., Montpellier, 1928-30; prof. mineralogy and crystallography, Faculty Scis.; Algiers, 1930-61, dean, 1939-61; apptd. full prof. U. Aix-Marseille, 1961; corr. mem. French Acad. Scis. Research on regular joining of different types of crystals.

ROYLE, John Forbes, Brit. naturalist; b. circa 1800. Physician with med. service of India Co.; holder chair med. material Royal Coll. London (Eng.). Fellow Royal Soc., 1837. Author: Manuel de Matière Médicale. Publs. on flora of India. Died Acton, Eng., 1864.

ROYSTER, Wimberly Calvin, Am. mathematician; b. Robards, Ky., Jan. 12, 1925; s. Fred and Ruth (Denton) R.; B.A., Murray State Coll., 1946; M.A., U. Ky., 1948, Ph.D., 1952; m. Betty Jo Barnett, July 1, 1950; children—David Calvin, Paul Barnett. Asst. prof. math. Auburn U., 1952-56; mem. faculty U. Ky., 1956—, prof. math., 1961—, chmn. dept. math., dir. Sch. Math. Scis., 1963—; mem. Inst. Advanced Study, Princeton, 1962. Mem. Am. Math. Soc., Math. Assn. Am., Sigma Xi. Contbr. articles profl. jours. Contbn. in field of complex analysis, mainly study of univalent (1 to 1) conformal mappings of canonical domains such as circle; ellipse, half-plane, also theory of series, polynomials. Home: 133 Vanderbilt Dr., Lexington, Ky. 40503. Office: U. Ky., Lexington, Ky. 40506.*

ROYTER, Vladimir Andreevich, Russian phys. chemist; b. Nizhnedneprovsk (now Dnepropetrovsk Oblast), 1903; grad. Dnepropetrovsk Oblast, 1926. Sr. asso. Inst. Phys. Chemistry, Ukrainian Acad. Scis., 1929—, prof., 1934—. Mem. Ukrainian Acad. Scis. Author numerous works including: Development of the Theory of Heterogeneous Catalysis, 1950; Introduction to the Theory of Kinetics and Catalysis, 1962. Address: Inst. Phys. Chemistry, Vladimirskaya ulitsa 55, Kiev, Ukrainian SSR, USSR.

ROZE, Janis Arnold, biologist; b. Tukums, Latvia, Oct. 31, 1926; s. Bernhard and Konstance Roze; student Albert Ludwigs U., Freiburg, Germany, 1946-48; B.S., Central U. Venezuela, 1950, Licenciado, 1954, Ph.D., 1958; m. Binka Yanakieva, Apr. 9, 1950; children—Ingrid, Bernhard, Peter. Faculty, Central U. Venezuela, 1951-66, asso. prof., 1962-66, head dept. zoology, 1958-63, chmn. ecology and zoology invertebrates, 1953—; research asso. Am. Mus. Natural History, N.Y.C., 1962—; vis. prof. biology and philosophy sci. Manhattan Coll., also Coll. Mt. St. Vincent, N.Y.C., 1966—. Curator zoology Mus. Biology, Central U. Venezuela, 1954—, curator herpetology, 1956—; prof. zoology Pedagogical Inst.., Venezuela, 1960-62; prin. investigator pub. health service Dept. Health, Edn., and Welfare, 1963-66; dir. research Internat. Center for Integrative Studies, N.Y.C., 1965—. Cons. NIH, 1966—. Recipient Nat. Sci. Research award Venezuela, 1961; Gold medal merit Sci. Faculty, Central U. Venezuela, 1962. Mem. Am. Inst. Biol. Scis., Ecol. Soc. Am., Am. Soc. Ichthyologists and Herpetologists, Herpetologists League, Am. Soc. Zoologists, Herpetologists Soc. Phila. (hon.), N.Y. Acad. Scis., Venezuelan Soc. Natural Sci. Author: (with A. Gomero) Textbook of General Zoology, 1958; Taxonomy and Zoogeography of the Snakes of Venezuela, 1966; New World Poisonous Coral Snakes, 1966; also articles. Research in new World herpetology, especially snakes; revisions of new world poisonous coral snakes; biology, poisons, pathology, ecol. studies of neotropical fauna; mimicry; philosophy of science, especially evolution. Home: 4 Washington Sq. Village, N.Y.C. 10012. Office: Am. Mus. Natural History, N.Y.C. 10024.*

ROZEBOOM, Lloyd Eugene, Am. med. entomologist; b. Orange City, Ia., Oct. 17, 1908; s. William and Antoinette (Vandergyp) R.; student Morningside Coll., 1927-29; B.S., Ia. State Coll., 1931; M.D., Johns Hopkins, 1926, Sc.D., 1934; m. Mae Thompson, Aug. 5, 1939; children—Kenneth, Carol A. (Mrs. David A. Wike). Med. entomologist Gorgas Meml. Lab., Panama, 1934-37; asso. prof. entomology Okla. A. and M. Coll., Stillwater, 1937-39; faculty Sch. Hygiene and Pub. Health, Johns Hopkins, Balt., 1939—, prof. med. entomology, 1958—; exchange prof. parasitology U. Philippines, 1954-55. Mem. parasitology and malaria commns. Armed Forces Epidemiological Bd., 1953—; cons. div. research grants USPHS, 1948-53, mem. microbiology rev. panel, 1960-62. Mem. A.A.A.S., Am. Soc. Parasitologist, Am. Soc. Tropical Medicine (Bailey K. Ashford award 1941), and Hygiene, Entomol. Soc. Am. Research and numerous publs. on epidemiology of malaria in Panama and Trinidad, epidemiology of filariasis in Republic of Philippines and India, taxonomy, biology and genetics of mosquitoes, mechanisms of disease transmission by mosquitoes. Home: 3635 Campfield Rd., Balt. 21207.*

ROZHKOV, Ivan Sergeevich, Russian geologist; b. 1908; grad. Leningrad Mining Inst., 1933; Dr. Geolog.-Mineral Scis., 1952. Worked in gold-platinum industry, 1933-57; became chmn. Yakutsk br. Siberian br., USSR Acad. Scis., 1957, dir. Geology Inst., 1958. Recipient Stalin Prize, 1950, 51. Corr. mem. USSR Acad. Scis. (mem. presidium Siberian br. 1961). Coauthor: Geological Cross-Section of the Urals from Zlatovita to Chelyabinsk, 1957. Research in metallurgy of gold and platinum, methods for survey and prospecting of ore deposits, geology and geomorphology of ore deposits. Office: Inst. Geology, Siberian Br., USSR Acad. Scis., Yakutsk, Siberia.

RUARK, Arthur Edward, Am. physicist govt. ofcl.; b. Washington, Nov. 9, 1899; s. Oliver Miles and Margaret Gordon (Smith) R.; B.A., Johns Hopkins, 1921, M.A., 1923, Ph.D., 1924; m. Sarah Grace Hazen, Mar. 19, 1927; children—Margaret Ann, Helen (Mrs. Nicholas Van Laer), Mary Lee (Mrs. John Fennel), Patricia (Mrs. Neal Obert). Mem. atomic structure sect. Nat. Bur. Standards, 1922-26; asst. prof. Yale, 1926-27; physicist Gulf Oil Co., Mellon Inst., 1927-29, chief physics div. Gulf Research Lab., 1930; prof. physics U. Pitts., 1930-34; Kenen prof., head dept. physics U. N.C., 1934-44; cons. Naval Research Lab., 1944-46; head research lab. div. Applied Physics Lab., Johns Hopkins, 1946-47, asst. dir. Inst. for Coop. Research, 1947-49, research contract dir., 1949-52; Temerson prof. physics U. Ala., 1953-56; chief controlled thermonuclear br. research div. AEC, 1957-61, asst. dir. research controlled thermonuclear research program, Washington, 1961-66, sr. asso. dir., 1966—. Mem. Am. Phys. Soc. (council 1944-45, chmn. Southeastern sect. 1956), Philos. Soc. Washington, Phi Beta Kappa, Sigma Xi. Author: Multiple Electron Transitions and Primed Spectral Terms, 1925; (with Harold C. Urey) Atoms, Molecules and Quanta, 1930, 63; (with others) Atomic Physics, 1933, 55; also articles. Research on atomic physics, physics fundamental particles, wave mechanics, interaction electrons with matter, engring. properties metals; discovered double electron jumps in atoms, selection rule governing total angular momentum, limits of accuracy in measurement of a single dynamical variable. Home: 4101 Byeforde Ct., Kensington, Md. 20795. Office: Atomic Energy Commn., Washington 20545.*

RUBEL, Lee Albert, Am. mathematician; b. N.Y.C., Dec. 1, 1928; s. Arthur John and Marcella (Crohn) R.; B.S. cum laude, Coll. City N.Y., 1950; M.S., U. Wis., 1951, Ph.D., 1954; m. Nina Marcus, Sept. 1, 1954; children—Mark Burrill, Natasha Audrey. Scientist, U. S. Naval Proving Ground, Dahlgren, Va., 1954; instr. Cornell U., 1954-56; mem. Inst. for Advanced Study, Princeton, N.J., 1956-58, 67-68; faculty U. Ill., Urbana, 1958—; vis. asso. prof. Columbia, 1960-62; mem. Center for Advanced Study, U. Ill., 1964-65; vis. prof. Princeton, 1967-68; Inst. Math. Scis., Madras, India, 1967-68. Prin. investigator Air Force grant, 1963—. NSF Postdoctoral fellow, U. Paris, 1965-66. Mem. Am. Math. Soc., Société Mathematique de France, Österreichische Mathematische Gesellschaft. Research, publs. on theory of analytic functions, theory of functions of real variable, harmonic analysis, and theory of numbers. Home: 807 W. Daniel St., Champaign, Ill. 61822. Office: Math. Dept., U. Ill., Urbana, Ill. 61801.*

RUBEN, Laurens Norman, Am. biologist; b. N.Y.C., May 14, 1927; s. Samuel and Rena (Koch) R.; A.B., U. Mich., 1949, M.S., 1950; Ph.D., Columbia, 1954; m. Judith Marion Starr, Aug. 29, 1950; children—Bruce Leonard, Ellen Patti, Barbara Jo. Faculty, Reed Coll., Portland, Ore., 1955-62, 63—, prof. biology, 1967—, chmn. biology dept., 1959-60, 67-68; vis. researcher Sta. de Zoologie Experimentale, Geneva, Switzerland, 1962-63. Mem. Soc. Developmental Biology, Am. Soc. Zoologist, Am. Soc. Anatomists, N.Y. Acad. Scis., A.A.A.S., Am. Insts. Biol. Sci., Sigma Xi, Phi Kappa Phi. Research, publs. on two kinds of developmental systems in adult organisms: regeneration and cancer; some approaches have dealt with their interrelationships, others have been solely within confines of one or other system. Home: 3108 S.E. Crystal Springs Blvd., Portland, Ore. 97202.*

RUBENCHIK, Lev Iosifovich, Ukrainian microbiologist; b. Odessa, 1896; grad Odessa Inst. Pub. Edn., 1922. Prof. microbiology Grain Flour Tech. Inst., 1927-30, Odessa U., 1932-41, Kiev U., 1946-50; head dept. soil and gen. microbiology Inst. Microbiology, Ukrainian Acad. Scis., 1944—. Mem. Ukrainian Acad. Scis. (corr.). Research and publs. on sewage purification, gen., aquatic, soil and indsl. microbiology, microbiology of saline lakes and therapeutic muds, biol. corrosion of bldg. materials, vital activity of microorganisms. Address: Inst. Microbiology, Ukrainian Acad. Scis., Vladimirskaya ulitsa 55, Kiev, Ukrainian SSR, USSR.

RUBENS, Heinrich, German physicist; b. Wiesbaden, Germany, Mar. 30, 1865; student Frankfort (Germany), Strasbourg (France) univs.; Ph.D., U. Berlin (Germany), 1889; D.Sc. (hon.), Leeds, Cambridge (both Eng.) univs. Prof. Technische Hochschule, Charlottenburg, Germany, from 1895, U. Berlin; dir. Physics Inst., U. Berlin. Recipient Rumford medal Royal Soc. Fellow Berlin Acad. Scis., Gottingen Acad.; hon. mem. Royal Inst. (London). Built galvonometer, dynamo bolometer; studied stationary waves in conductors, long heat waves, infrared spectrum; measurements of black body radiation; confirmed quantum theory. Died Berlin, Feb. 17, 1922.

RUBESKA, Ivan, Czechoslovakian chemist; b. Prague, June 16, 1931; s. Milos and Vera (Krizenecka) R.; student Charles U., Prague, 1952-54; candidate sci. U. Chem. Tech., 1962; m. Dana Exlerova, Apr. 12, 1956; 1 dau., Monica. Research chemist Central Agrl. Control and Examination Inst., Prague, 1954-56; with Central Geol. Survey, Prague, 1956—, chief flame photometry atomic absorption group. Lectr., U. Prague, 1966. Mem. Soc. for Spectroanalytical Research (sec. atomic spectroscopy group 1958-62). Author: Laboratory Praxis of Flame Photometry, 1966; Atomic Absorption Spectrophotometry, 1967; also articles. Research in emission spectrography, flame photometry, atomic absorption spectrometry. Home: 64 Vacluske N., Prague 1. Office: 26 Kostelni, Prague 7, Czechoslovakia.*

RUBEUS, see Rossi, Girolamo.

RUBEY, William Walden, Am. geologist; b. Moberly, Missouri, Dec. 19, 1898; s. Ambrose Burnside and Alva Beatrice (Walden) R.; A.B., U. of Missouri, 1920, D.Sc., 1953; student Johns Hopkins, 1922, Yale, 1922-24, D.Sc., 1960; D.Sc., Villanova U. 1959; m. Susan Elsie Manovill, November 27, 1919; children—Susie Lee Putnam (Mrs. Edwin Randolph Middleton, dec.), Jean Manovill (Mrs. Francis Joseph Eisenman, Jr.), Elizabeth Walden (Mrs. Thomas A. Dean). Asst. valuation engr., Johnson Huntley, Pittsburgh, Pa., 1920; geologic aid, United States Geol. Survey, 1920-21; asst. geologist, 1921-22; instr. in geology, Yale, 1922-24; with U. S. Geol. Survey, 1924-60, successively asso. geologist, geologist, sr. geologist, prin. geologist, geologist in charge div. area geology and basic scis., staff geologist, research geologist; guest scientist Inst. Geophysics U. Cal. Los Angeles, 1954, professor geology and geophysics, 1960—; vis. prof. geology Cal. Tech., 1955, Johns Hopkins, 1956; Silliman lectr. Yale, 1960. Mem. com. on geophysics and geog. Research and Devel. Bd., 1947-50; com. math., physics, engring. sci. Nat. Sci. Found., 1951-55; dir., v.p. Am. Geol. Inst., 1950-51, 58-59; member of the National Science Board, 1960—. Recipient Award of Excellence, U. S. Department of Interior, 1943, Distinguished Service award, 1958. Mem. Nat. Acad. Scis., Am. Philos. Soc. (council 1956-59), Am. Acad. Arts and Scis., Acad. Med. Wash.; A.A.A.S. (dir.), Am. Assn. Petroleum Geologists, Am. Geophys. Union, Geol. Soc. Washington (pres. 1948), Washington Acad. Sci. (pres. 1957), Geol. Soc. Am. (councilor 1941-43, v.p., 1948-49, pres., 1949-50), Seismol. Soc. Am., Geochem. Soc. (dir. 1955-57), Am. Ornithologists Union. Author govt. reports and tech. articles in sci. jours. Investigations of origin of sedimentary rocks and tectonic history of folded mountain ranges; authority on structural geology, geomorphology, sedimentation, and geochemistry. Home: 1678 San Onofre Dr., Pacific Palisades, Cal. Office: care Dept. Geology, U. Cal., Los Angeles 24.

RUBIN, Alan, Am. physician; b. Phila., Nov. 10, 1923; s. Hyman and Miriam (Magil) R.; student U. Pa., 1941-43, M.D., 1947; m. Helen Metz, May 1, 1947; children—Alan Magil, Stephen Curtis, Blake Douglas. Nat. Cancer Inst. trainee, instr. obstetrics and gynecology U. Pa. Med. Sch., Phila., 1951-52, research asso. 1952—; practice medicine specializing in obstetrics and gynecology, Phila., 1952—; asso. U. Pa. Hosp., 1952—. Mem. med. adv. com. Planned Parenthood Assn., Phila., 1953—, chmn., 1966—. Diplomate Am. Bd. Obstetrics and Gynecology. Fellow A.C.S., Am. Coll. Obstetricians and Gynecologists, Royal Soc. Medicine (Eng.), Coll. Physicians Phila.; mem. Pa. (mem. commn. on maternal and Child health 1962-64), Phila. County (past br. dir.) med. socs., Am. Fertility Soc., Am. Geriatric Soc., N.Y. Acad. Scis., Am. Soc. Human Genetics, Obstet. Soc. Phila., A.M.A., U. Pa. Med. Alumni Soc., W. Area Health and Welfare Council (exec. com. 1960—), Southeastern Pa. Food and Nutrition Council, Sigma Xi, Alpha Omega Alpha. Editor: Handbook of Congenital Malformations, 1967. Research, numerous publs. on cancer in women, treatment of infertility, pharmacology of drugs used for relief of birth injuries, treatment of common diseases of women. Home: 5900 Woodbine Av., Phila. 19131. Office: 255 S. 17th St., Phila. 19103.*

RUBIN, Eli Hyman, physician; b. Russia, Apr. 20, 1899; s. Michael and Lena (Kantrowitz) R.; Ph.B., Yale, 1921, M.D., 1925; m. Adena Lipschitz, Aug. 9, 1942; children—Leonard, Susan (Mrs. Bertram M. Gesner), Isaac Michael. Practice medicine, specializing in chest diseases, N.Y.C., 1926—; attending physician pulmonary div. Montefiore Hosp., 1942—, pres. sr. med. bd., 1955-57; 1st prof. medicine Albert Einstein Coll. Medicine, Yeshiva U., 1954—, prof. clin. medicine, 1955—; vis. physician Chest Service Bronx Municipal Hosp. Center, 1955—; cons. physician pulmonary disease Morrisania City, Lebanon, Union hosps. Gibson Student in Tb,Montefiore Hosp., 1928-29, Goldschmidt fellow, 1929-30, Bd. Trustees fellow, 1930-32. Fellow A.C.P., Am. Coll. Chest Physians (pres. N.Y. State chpt. 1967); mem. A.M.A., Am. Thoracic Soc., N.Y. Acad. Medicine. Author: (with Dr. Morris Rugin) Diseases of the Chest Emphasizing X-ray Diagnosis, 1947; The Lung as a Miror of Systemic Disease, 1956; (with Relationships, 1961; The Lungs in Systemic Diseases, 1968; also numerous articles. Research on pulmonary diseases including Tb, cancer and systemic diseases. Home: 1749 Grand Concourse, Bronx 10453. Office: 2021 Grand Concourse, Bronx, N.Y. 10453.*

RUBIN, Leonard Sidney, Am. psychologist; b. N.Y.C., Aug. 27, 1922; s. Hyman H. and Tillie (Schneider) R.; B.S., Coll. City of N.Y., 1943; Ph.D., N.Y. U., 1951; m. Blanche Bailer, Mar. 30, 1950; children—Joshua Tarbut, Matthew Maccabee, Beth Sabra. Asst. instr. N.Y. Med. Coll., 1944-45; instr. N.Y. U., N.Y.C., 1947-50, research asso., 1950-53; chief psychology br. Chem. Corps Med. Labs., 1953-57; head psychobiology Eastern Pa. Psychiat. Inst., Phila., 1957—; asso. prof. U. Pa., Phila., 1965—. Mem. A.A.A.S., N.Y. Acad. Sci., Am. Psychol. Assn., Sigma Xi. Research, publs. in elucidation of role of autonomic nervous system in neurotic disorders; formulated approach to pharmacotherapy in psychotic behavior employing pupillographic measurements as index of autonomic neurohumoral dysfunction. Home: 706 Powder Mill Lane, Phila. 19151.*

RUBIN, Meyer, Am. geologist; b. Chgo., Feb. 17, 1924; s. Abe and Esther (Fleischer) R.; B.S., U. Chgo., 1947, M.S., 1949, Ph.D., 1956; m. Mary Louise Tucker, June 13, 1944; children—Ronald, Robert, Mark. Asst., U. Chgo., 1948-50; with mil. geology br. U.S. Geol. Survey, Washington, 1950-53, in charge radiocarbon dating lab. geochemistry and geology br. (now isotope geology br.), 1953—. Vis. geophysicist Am. Geol. Inst., 1958-61, Am. Geophys. Union, 1961-66. Mem. A.A.A.S., Geol. Soc. Am., Geol. Soc. Washington, Washington Acad. Sci. (Outstanding Young Phys. Scientist 1958), Am. Geophys. Union. Research on geol. and archeol. dating, especially last stage of continental glaciation of U.S., and rise of sea level for past 10,000 years. Home: 8215 Cottage St., Vienna, Va. 22180. Office: 18th and F Sts. N.W., Washington 20242.*

RUBIN, Mitchell Irving, Am. physician; b. Charleston, S.C., Apr. 1, 1902; s. Abraham and Eva (Feintuch) R.; student Coll. Charleston, 1919-21; M.D., Med. Coll. State of S.C., 1925; m. Maizie-Louise Cohen, Sept. 1, 1934; children—Henry Park, Eve. Instr. pediatrics Johns Hopkins Hosp., U. Sch. Medicine, 1930-31; asso. prof. pediatrics U. Pa. Sch. Medicine, 1931-45; prof., chmn. dept. pediatrics State U. N.Y. at Buffalo, 1945—; pediatrician-in-chief Children's Hosp., Buffalo, 1945—. Mem. N.Y. State Bur. Handicapped Children, Handicapped Children's Council; adv. council N.Y. State Assn. Crippled Children, Nat. Kidney Found.; cons. pediatrics Erie County Dept. Health, Buffalo. Mem. Am. Pediatric Soc., Soc. Pediatric Research (past sec., pres.), Soc. Exptl. Biology and Medicine, Am. Acad. Pediatrics, A.M.A., Sigma Xi, Alpha Omega Alpha. Contbg. author: Nelson's Textbook of Pediatrics; Handbook of Biological Data; Practice of Pediatrics; The Child in Health and Disease. Editorial staff Biol. Revs., Quar. Rev. Pediatrics; editorial bd. Pediatrics; 1950-56. Research and publs. on kidney function in childhood devel., especially in premature infant and newborn baby, nature of kidney disease in children and functional alteration produced by them. Home: 703 W. Ferry St. Office: Children's Hosp., 219 Bryant St., Buffalo 14222.*

RUBIN, Saul Howard, Am. chemist; b. N.Y.C., Oct. 15, 1912; s. Maxwell L. and Evelyn (Sussman) R.; B.S., Coll. City N.Y., 1931; M.S., N.Y. U., 1932; Ph.D., in Biochemistry, N.Y. U. Coll. Medicine, 1939; m. Anne Bissom, Aug. 2, 1932; children—Barbara, Richard. Staff, N.Y. U. Med. Coll., 1932-41; dir. nutrition Labs., Hoffmann-La Roche, Inc., Nutley, N.J., 1941-49, dir. product devel, 1949—. Mem. sci. adv. com. Nat. Vitamin Found., 1958—. Mem. Assn. Research Dirs. N.Y.C. (past pres.), Am. Soc. Biol. Chemists, Am. Inst. Nutrition, Soc. for Exptl. Biology and Medicine, Am. Chem. Soc., N.Y. Acad. Scis., Am. Pharm. Assn. Contbg. author: Vitamin Methods, 1955; Physiological Chemistry, 1965; also numerous articles. Research on metabolism and nutrition vitamins, enrichment foods, drug formulations; patentee drug forms and stablzn.; inventor process for enrichment corn grits and rice. Home: 62 Beech St. Office: Hoffmann-La Roche, Inc., Nutley, N.J. 07110.*

RUBIN DE LA BORBOLLA, Daniel Fernando, Mexican anthropologist, museum dir.; b. Puebla, Puebla, Mexico, May 20, 1907; s. Juan Rubin de la Borbolla and Trinidad Cedillo; student State Coll., Puebla, Mexico; grad. U. Mexico, 1930; Ph.D., Northwestern U.; m. Sol Arguedas, Oct. 1944; children—Daniel, Sol, Maria de la Paz. Chief anthropologist Nat. Mus. Anthropology, 1931-36; prof. human biology Nat. Prep. Sch., U. Mexico, 1931-36, prof. faculty philosophy and letters at univ., 1931-37; explored Monte Alban, Oaxaca, other anthrop. research, Tzintzuntzan, Michoacan, Teotihuacan, Chupicuaro, Guanajuato, also Tlatilco, Mexico, 1933-53; founder, dean Nat. Sch. Anthropology and History, Mexico City, 1940-45; dir.-gen. Nat. Mus. Anthropology, 1947-53; chmn. U. Mus., Nat. Autonomous U. of Mexico, counselor for rector, dir. gen. of univ., 1951—; dir. U. Mus. Sci. and Art, Nat. U.; dir. Nat. Mus. Popular Arts and Crafts, 1949—; exec. sec. Nat. Council Arts and Crafts, 1951—; tech. adviser cultural matters Ministry Fgn. Relations, 1950—. Mem. bd. trustees arts, crafts Nat. Coop. Bank Mex. Decorated comendador de la órden del Sol (Peru); Cabellero de la 'Legión de Honor (France); Comendador de la Orden de León (Finland). Mem. Mexican Royal, Am. anthrop. socs., Mexican Folklore Soc., Am. Acad. Natural Scis., Am. Assn. Phys. Anthropologists, Am. Mus. Assn. Contbr. articles anthrop. publs. Home: Galeana 115, San Angel Inn., Villa Obregon 20. Office: Nat. Museum of Popular Arts and Crafts, Avenida Juarez 44, Mexico City, Mexico.*

RUBINOWICZ, Adalbert Wojciech, Polish physicist; b. Sadágora, Bukowina, Feb. 22, 1889; s. Damian and Malgorzata (Brodowska) R.; Ph.D., Czernowitz U., 1914, venia legendi, Czernowitz U., 1918; studied Munich, Copenhagen; hon. dr. Humboldt U., Berlin, and Jagellonian U., Cracow; m. Elzbieta Norst, July 21, 1921; son, Jan. Asst., Czernowitz U., 1912; asst. to Sommerfled, Munich U., 1917; docent, Czernowitz U., 1918; prof. Ljubljana U., 1920; prof. Lwów Polytechnic U., 1922; prof. Lwów U., 1937; prof. Warsaw U., 1946-60. Chmn. Scientific Advisory Board, Physical Inst., Polish Acad. Scis., 1960—. Recipient State Scientific Prize, 1st class, 1951. Mem., Polish Physical Soc. (pres. 1949-51, 1961); IUTAM; Polish Acad. Scis.; Polish Math. Soc. Author: Ursprung und Entwicklung der älteren Quantentheorie, 1933; Vectors and Tensors, 1950; (with W. Królikowski) Theoretical Mechanics, 1955; Quantum Theory of Atom, 1954; numerous published articles. Discovered selection and polarization rules for azimuthal and magnetic quantum numbers for electric dipole radiation, 1918; theory of electric quadrupole radiation on basis of which equivalence of Young's and Fresnel's views in Kirchhoff diffraction theory was discovered. Home: 74 Hoza, Warsaw, Poland. Office: Physics Dept., Warsaw University, 69 Hoza, Warsaw, Poland.*

RUBINSTEIN, Hyman Solomon, psychoanalyst, neuro-psychiatrist; b. Leeds, Eng., Mar. 16, 1904; s. Myer David and Rose (Schneederman) R.; came to U.S., 1904, naturalized, 1909; Ph.G., U. Md., 1924, M.D., 1928, B.S., 1932, Ph.D., 1934, postgrad. (Weaver fellow); postgrad. Washington-Balt. Psychoanalytic Inst.; diploma in psychiatry Washington Sch. Psychiatry, 1952; certificate in psychoanalysis William Alanson White Inst., 1958; m. Ellen Steinhorn, July 21, 1929; children—Madelyn Hope (Mrs. Herbert Shapiro), Roberta Faith (Mrs. John Herts). Dir., Lab. for Neuroendocrine Research, Sinai Hospital, Balt., 1935-45, dir. Alfred Ullman Lab. for Neuropsychiat. Research, 1945-51, head neuro-psychiat. div., 1947-49, asso. phys. in neurology and psychiatry, 1951-64; practice psychiatry and psychoanalysis, Balt., 1936—; attending psychiatrist Seton Psychiat. Inst., Balt., 1955—. Faculty, Washington Sch. Psychiatry, 1952-56; cons. Jewish Big Bro. League, 1955—. Named Big Bro. of Year for Md., Big Bro. League, 1963. Diplomate Am. Bd. Psychiatry and Neurology. Fellow A.M.A., Am. Psychiat. Assn., A.A.A.S., Am. Acad. Psychoanalysis; mem. Balt. Med. Soc., William Alenson White Psychoanalytic Soc., Soc. Med Psychoanalysts, Am. Physicians Fellowship of Israel Med. Assn., Sigma Xi, Phi Lambda Kappa, others. Author: (with C.L. Davis) Stereoscopic Atlas of Neuroanatomy, 1947; The Study of the Brain, 1953; also numerous articles. Research on growth hormone, its effects on brain-body relations; sex hormones, their effect on genital devel.; gen. body growth, spermatogenesis, ovulation, sexual urges, structure and function brain and spinal cord, body build and behavior, dural reflexes, cerebellar agenesis, treatment behavior-problem children, hypnosis as diagnostic aid, interpersonal conflict, dynamics of behavior. Address: 3900 N. Charles St., Balt. 21218.*

RUBNER, Max, German physiologist, hygienist; b. Munich, Germany, June 2, 1854; studied under Voit; prof., Marburg, Germany, Berlin. Author: Gesetze des Energieverbrauchs, 1902; Kraft und Stoff im Haushalt der Natur, 1909. Research on effects of climate on man, caloric requirements of infants; proved law of conservation of energy could be applied to animal metabolism and nutrition, 1894; showed metabolism of body is proportional to surface area; discovered law of constant growth quotient (energy for growth is constant fraction of organism's total energy). Died Berlin, Apr. 27, 1932.

RUBY, Lawrence, Am. physicist; b. Detroit, July 25, 1925; s. Irving and Rose (Markin) R.; B.A., U. Cal.. at Los Angeles, 1945, M.A., 1947, Ph.D., 1951; m. Judith Friedberg, Apr. 8, 1951; children—Jill Alice, Peter Albert, Frederick Carl. Staff physicist Lawrence Radiation Lab., U. Cal. at Berkeley, 1950—, asso. prof. nuclear engring., 1962-66, prof., 1966—. Mem. Am. Phys. Soc., Am. Assn. Physics Tchrs., Am. Nuclear Soc., Sigma Xi. Research, publs. in accelerator design, nuclear spectroscopy, neutron physics, reactor dynamics. Home: 54 Cowper Av., Berkeley, Cal. 94707.*

RUBY, Stanley Lawrence, Am. physicist; b. N.Y.C., July 19, 1924; s. Walter and Selma (Ratner) R.; B.A., Columbia, 1947; m. Helga Fanny Ringel, June 7, 1947; children—Walter, Daniel, Joanne. Research asst. Brookhaven Nat. Lab., Upton, N.Y., 1951-52; physicist IBM, N.Y., 1953-55, Westinghouse Electric Corp., 1955-63; physicist Argonne (Ill.) Nat. Lab., 1964—, vis. scientist Israel AEC, Rehovoth, 1961-62. Vis. prof. Technion U., 1962, Tsing Hua U., Hsincha, Taiwan, 1966. Mem. Am. Phys. Soc., Sigma Xi. Research, publs on neutrinos, cybernetics and electrostatic copying, Mössbauer effect to illuminate chemistry, solid-state and nuclear physics; inventor measurement technique (including signal averaging) utilizing equipment previously specialized for nucleonics. Home: 649 Hill Av., Glen Ellyn, Ill. 60137. Office: Argonne Nat. Lab., Argonne, Ill. 60440.*

RUCH, Theodore Cedric, Am. physiologist; b. Guthrie Center, Ia., May 22, 1906; s. William Wallace and Libby (Young) R.; B.A., U. Ore., 1927; M.A., Stanford, 1928; B.A., Oxford U., 1930, B.Sc., 1932; Ph.D., Yale, 1933; m. Helen B. Hembroff, Sept. 5, 1940; 1 dau., Libby Olive. Prof., physiology and biophysics U. Wash. Sch. Medicine, Seattle, 1946—, chmn. 1946-64, dir. Regional Primate Research Center, 1961—; mem. sci. adv. com. Nat. Council to Combat Blindness, 1956—, Fels Research Inst. For Study Human Devel., 1965—; mem. med. student research tng. grant com. NIH, 1960-64; mem. nat adv. council, 1963-64, mem. med. com. Nat. Inst. Gen. Med Scis. NIH, 1964-66; bd. advisers Nat. Inst. Mental Health. Mem. Am. Physiol. Soc., Internat. Brain Research Soc., Internat. Primatological Soc., Phi Beta Kappa, Sigma Xi, Alpha Omega Alpha. Author: Bibliographia Primatologica, 1941; Bibliography of Primate Malaria, 1944; Diseases of Laboratory Primates, 1959; also numerous articles. Editor: (with J.F. Fulton) Medical Physiology and Biophysics, 1960; (with H.D. Patton) Physiology and Biophysics, 1965. Research on apparatus for inclined plane discrimination, location of descending spinal pathways inhibiting flexion reflex, nonsegmental origin of and reciprocal innervation in Schiff-Sherrington phenomenon, retinal basis of elec. method of rec. eye movements, taste localization in thalmus and cerebral cortex, muscle representation in motor cortex of monkey brain, neural control of bladder, disease, care and housing of primates. Home: 1014 34th Av. E., Seattle 98102. Office: Cancer-Primate Bldg., Regional Primate Research Center, U. Wash., Seattle 98105.*

RUCHKOVSKY, Sergey Nikiforovich, Ukrainian microbiologist, epidemiologist; b. Tarashcha (now Kiev Oblast), 1888; grad. Med. Faculty, Kiev U., 1914; D.Med. Sci. Epidemiologist, infectionist charge bacteriological labs., 1914-25; head epidemiology dept. Kharkov Bacteriological Inst., 1926-29, Kiev Health Bacteriology Inst., 1929-39; lectr. Kiev Postgrad. Med. Inst., 1929-38, head chair epidemiology, 1939-58; head chair epidemiology Kiev Med. Inst., 1939-48; head virus dept. Ukrainian Inst. Epidemiology and Microbiology, 1958—. Mem. USSR Acad. Med. Scis. (corr.). Author numerous works including: The Mechanism of Botulin-Type Poisoning Caused by Fish, 1928; Typhoid, Paratyphoid and Dysentary, 1936. Research on epidemiology of Volhynia fever, use of opthalmic test in rabbits to diagnose cancer in man, immunological diagnosis of infective hepatitis. Address: Ukrainian Inst. Epidemiology and Microbiology, ulitsa S. Razina 4, Kiev, Ukrainian SSR, USSR.

RUCKENSTEIN, Eli, Rumanian chem. engr., chemist; b. Botosani, Rumania, Aug. 13, 1925; s. Beno and Dina (Steinberg) R.; Chem.E., Poly. Inst., Bucharest, Rumania, 1949, Ph.D., 1966; m. Velina Rotstein, May 15, 1948; children—Andrei, Lelia. Faculty chem. engring. Poly. Inst., Bucharest, 1949—, asso. prof. chem. engring., 1960—; sr. research worker Inst. Physics, Rumanian Acad., 1950-57, sr. research worker colloid dept. Research Centre, 1964—. Recipient Ministry Edn. award for physics, 1958, Gheorgne Space award for chemistry Rumanian Acad., 1963, Ministry Edn. award for chemistry, 1964. Research and numerous contbns. on turbulent heat or mass transfer, boiling, condensation, heat transfer in two phase systems, hydrodynamics of two phase systems, elementary phenomena of separation processes, effect surface phenomena on mass transfer, regulation and synthesis in living cells. Home: Drumul Taberei 60, Raion16 Februarie, Bloc F2, sc C et 1, Ap. 48, Bucharest, Rumania.*

RÜCKER, Sir Arthur William, English physicist; b. Clapham, London, Oct. 23, 1848; ed. Brasenose Coll., Oxford, fellow, 1871-76. Became prof. physics and math., Leeds, Eng., 1874; prof. physics Royal Coll. Sci., London, 1886-1901; prin. U. London, 1901-08. Fellow Royal Soc., 1884, sec., 1896-1901, recipient Royal medal, 1902. Author: On the Expansion of Sea Water by Heat, 1876; Properties of Liquid Films, 1880-92; Magnetic Surveys of the British Isles for for the Epochs, 1886, 1891, 2 vols., 1890-96; also articles. Research on properties of liquid films; made magnetic survey of Gt. Britain; measured thickness of soap film using interferometer. Died Newbury, Eng., Nov. 1, 1915.

RUD, Finn Torgeir, Norwegian psychiatrist; b. Fredrikstad, June 14, 1907; s. Karl and Josefine (Rade) R.; M.D., U. Oslo; m. Sigrid Olsen, Oct. 30, 1935; children—Hilde Elizabeth, Finn Torgeir. Prof. psychiatry U. Bergen; dir. Neevengarden Sykehus, Bergen. Mem. Assn. Mental Health (sec.), Assn. Pub. Health (sec.), Norwegian Psychiatry Assn., Norwegian Social Work Assn., Norwegian Gerontology Assn., Soc. Promotion Sci., Medico-Psychol. Soc. Neuilly-sur-Marne, Bergen Med. Soc. (v.p.). Author: The Eosinophil Count in Health and Mental Disease, 1948; The Social Psychopathology of the Schizophrenic State, 1951; Psychiatric Activities and Instruction at a General Hospital, 1953; Euthanasia, 1953; also report on psychiatry service for Bergen and Sogn og Fjordane regions. Editor Folkehelsen mag. Home: Helleveien, Bergen. Office: Neevengarden Sykehus, Bergen, Norway.

RUDBECK, Olof, Swedish naturalist; b. Westeras, Sweden, Dec. 12, 1630; student medicine U. Uppsala (Sweden). Prof. botany, anatomy U. Uppsala, prof. anatomy, botany, chemistry, math. Med. Sch., founder Poly. Inst., later one of 3 curators. Author: Nova Exercitatio Anatomica, Exhibens Ductus Hepaticos Aquosis, et Vasa Glandularum Serosa Västeras, 1653; 3 vols, 1675-98; Campus Elysii, 2 vols., 1701-02. Tchr. of Linnaeus; credited with discovering lymphatic system, especially of intestine and its connection with thoracic duct, 1651; believed and tried to prove that Sweden was site of Plato's Atlantis and cradle of culture. Died Uppsala, Sept. 17, 1702.

RUDBERG, Erik Gustaf, Swedish physicist; b. Stockholm, Sweden, Nov. 17, 1902; s. Ture A. and Emma (Munck) R.; B.Sc., Stockholm U., 1923, Ph.D., 1930; student U. Göttingen, 1924, Kings Coll., London U., 1925-27; m. Birgit Holmgren, Sept. 7, 1930. Asst. chemistry U. Stockholm, 1922-24; asst. phys. chemistry Nobel Inst., Stockholm, 1924-31; lectr. physics U. Stockholm, 1929; instr. physics Mass. Inst. Tech., Cambridge, 1931-32, asst. prof., 1932-36; resident physicist Swedish Gen. Electric Co., 1936-40; prof., head dept. physics (Chalmers fund) Inst. Tech., Gothenburg, Sweden, 1940-45; dir. Swedish Inst. Metals Research, Stockholm, 1945-59; permanent sec. Royal Acad. Sci., Stockholm, 1959—. Fellow Am. Phys. Soc., Swedish Acad. Engring. Sci.; mem. Royal Acad. Sci., Société Francaise de Metallurgie (hon.), others. Research, publs. on electron emission, electron theory of metals, electron collisions, vapour pressure, metallography, reaction velocities, solid state physics. Home: Royal Acad. Sci., Stockholm, Sweden.*

RUDBERG, Fredrik, Swedish natural philosopher; b. Norrkjöping, Sweden, 1800; ed. U. Uppsala (Sweden), 1815; student exptl. physics, Paris, 1828. Became prof. physics, Uppsala, 1828. Research and publs. in physics and philosophy. Inventer goniometer for measurement angles, 1826; determined expansion coefficient of air (rate of expansion of air with heat); 1st measurement of double refraction in uni- and biaxial crystals, 1828-29. Died 1839.

RUDD, Jacob Louis, Am. physician; b. Boston, May 5, 1902; s. Simon and Fannie (Baron) R.; B.A., Harvard, 1924, M.D., 1928; m. Dorothy Wiseman, Sept. 4, 1936; children—Edward J., Elaine J. Gen. practice medicine, Cambridge, Mass., Boston, 1930-60; chief phys. medicine and rehab. Boston City, West Roxbury VA, Brockton VA hosps., 1946-60; chief phys. medicine and rehab. service VA Outpatient Clinic, Boston, 1960—, chmn. med. research com., 1964—. Cons. dept. rehab. and spl. edn. Northeastern U., 1966. Diplomate Am. Bd. Phys. Medicine and Rehab. Mem. Am. Acad. Phys. Med. and Rehab., Am. Assn. for Rehab. Therapy (chmn. registry), New

Eng. Soc. Phys. Med. and Rehab. (past pres.), N.Am. Acad. Manipulation Medicine (founder, 1st pres.), Nat. Geriatrics Soc. (hqn. life), A.M.A., Mass. Med. Soc. Author: (with Reuben J. Margolin) Maintenance Therapy for Geriatric Patients, 1967. Contbr. numerous articles to profl. jours. Mem. editorial bd. Am. Archives of Rehab. Therapy, 1950—; asso. editorial staff Rehab. Rev., 1954—; abstractor Exerpta Medica, 1964—. Research on methods of cervical traction, maintenance therapy, phys. fitness for hypertensives and cardiacs, passive stretching or manual mobilization for back and neck conditions. Home: 50 Massapoag Av., Sharon, Mass. 02067. Office: 17 Court St., Boston 02108.*

RUDENBERG, Reinhold, elec. engr.; b. Hanover, Germany, Feb. 4, 1883; s. Georg and Elsbeth (Herzfeld) R.; diploma Inst. Tech., Hanover, 1906, Dr. Engring., 1906; m. Lily Minkowski, Sept. 7, 1919; children—H. Gunther, Angelika (Mrs. Robert T. Howard), F. Hermann. Came to U.S., 1938, naturalized, 1944. Instr., U. Göttingen, 1906-08; with Siemens Schuckertwerke, 1908-36, chief elec. engr., 1923-36; cons. engr. Gen. Electric Co. Ltd., London, Eng., 1936-38; Gordon McKay prof. elec. engring. Harvard, Cambridge, Mass., 1939-52, emeritus, 1952-61. Vis. prof. Brazil, Uuruguay. Recipient 1st Montefiore prize, Belgium; Stevens medal Stevens Inst. Tech., 1946; Gold Cedergreen plaquette, Sweden, 1949; Grand Cross of Order of Merit, Germany, 1957; Elliott Cresson medal Franklin Inst., 1961. Fellow Am. Acad. Arts and Scis., I.E.E.E., A.A.A.S.; mem. Am. Inst. Physics, Electron Microscope Soc., Sigma Pi. Author: Transient Performance of Electric Power Systems; Shock-wave Phenomena in Electric Power Systems; also numerous articles. Research on theory of electric machinery, high voltage transmission, electric and mech. transient phenomena; patentee electron microscope, electrostatic lenses, radar scanning devices, superspeed cathode ray tubes, others. Died Boston, Dec. 15, 1961.

RUDERMAN, I(rving) Warren, Am. physicist; b. N.Y.C., Jan. 7, 1920; s. Jack and Mollie (Ettin) R.; Ph.D. (teaching fellow), Columbia, 1949; m. Carol Carver Schmied, June 16, 1945; children—Barbara Lee, Clifford Eric, William Brandon, Genevieve Kathryn. Lectr. chemistry Columbia, 1947-48, research scientist, 1949-51, cons., 1953-57; guest physicist Brookhaven Nat. Lab., 1952-53; founder, pres. Isomet Corp., Palisades Park, N.J., 1956—, also dir. Fellow N.Y. Acad. Scis.; mem. Am. Chem. Soc., Am. Phys. Soc., Electrochem. Soc., Am. Nuclear Soc., Am. Optical Soc., Aerospace Med. Assn., Sigma Xi, Tau Beta Pi, Phi Lambda Upsilon. Research, publs. on exptl. verifications of scattering of neutrons by paramagnetic ions, devel. of new techniques for measuring radioactivity, devel. new crystals for diffraction of soft x-rays, devel. methods for large scale growth of single crystals. Home: 45 Duane Lane, Demarest, N.J. 07627. Office: 433 Commercial Av., Palisades Park, N.J. 07650.*

RUDERMAN, Malvin Avram, Am. physicist; b. N.Y.C., Mar. 25, 1927; s. Aaron and Hanna (Cohen) R.; A.B. Columbia, 1945; Ph.D. Cal Tech., 1951; m. Paula Rosenberg, Aug. 30, 1953; children—Peter, Robert. Physicist Radiation Lab., Berkeley, 1951-52; NSF fellow Columbia, 1952-53; faculty dept. physics U. Cal., Berkeley, 1953-59, prof., 1959; prof. N.Y. U., 1964—. Mem. Janson div. Inst. for Def. Analyses, Washington, 1960—. Guggenheim fellow, 1957-58. Home: 29 Washington Sq., W., N.Y.C.*

RUDNEV, Georgiy Pavlovich, Russian infectionist; b. 1899; grad. Med. Faculty, Rostov-on-Don U., 1923; D.Med. Sci., 1936. Lectr. infectious diseases clinic Rostov U., 1932-34; founder, head chairs propedeutics of internal diseases Faculty Therapy and Infectious Diseases, Daghestan Med. Inst., Makhachkala, 1934-37, prof., 1937—; head chair infectious diseases Rostov Med. Inst., 1937-41; prof. Central Postgrad. Med. Inst., Moscow, 1944—. Acad. sec., Presidium mem. dept. clin. medicine, USSR Acad. Med. Scis., 1953-57. Recipient Stalin prize, 1940. Mem. USSR Acad. Med. Scis., All-Union (bd. mem., dep. chmn.), Moscow (dep. chmn.) socs. microbiologists, epidemiologists and infectionists. Author: Clinical Aspects of Plague, 2d edit., 1940; Clinical Aspects of Brucellosis, 1949; Zoonoses, 1950; Brucellosis: Clinical Aspects, Diagnosis and Treatment, 1955; Clinical Aspects, Diagnosis and Treatment of Tularemia, 1960; Treatment of Infectious Diseases, 1960; Hormone Therapy in Infectious Diseases; co-author, editor: Treatment of Infectious Patients, 1950-60. Dep. editor Epidemiology and Infectious Diseases sects. Large Med. Ency., 2d edit.; mem. editorial bd. Jour. Microbiology, Epidemiology and Immunobiology. Research and numerous publs. on plague, brucellosis, tularemia, hematology, malaria, intestinal infections, antibiotic therapy. Address: Central Postgrad. Med. Inst., pl. Vosstaniya 1-2, Moscow, USSR.

RUDNEY, Harry, biochemist; b. Toronto, Ont., Can., Apr. 14, 1918; s. Joshua and Dina (Gorback) R.; B.A., U. Toronto, 1947; M.A., 1948; Ph.D., Western Res. U., 1952; m. Bernice Dina Snider, June 25, 1946; children—Joel David, Paul Robert, Came to U. S., 1948, naturalized, 1956. Faculty, Western Res. U., Cleve., 1952-67, prof. biochemistry, 1965-67; prof., dir. dept. biol. chemistry U. Cin. Coll. Medicine, 1967—; vis. prof. Case Inst. Tech., Cleve., 1965-66. Mem. research adv. panel N.E. Ohio Heart Assn., 1966—. Recipient USPHS Research Career award, 1963-67; Am. Cancer Soc. scholar, 1954-56; NSF Sr. Research fellow U. Amsterdam (Netherlands), 1957. Mem. Am. Soc. Biol. Chemists, Am. Chem. Soc., Am. Soc. Microbiology, Biochem. Soc. (Eng.), Sigma Xi. Editorial bd. Archives of Biochemistry and Biophysics, 1965—. Research, publs. on biosynthesis of cholesterol, biosynthesis and function of naturally occurring quinones. Office: Dept. Biol. Chemistry, U. Cin. Coll. Medicine, Cin. 45219.*

RUDOLF, Christoff, mathematician; b. Javer, circa 1500; studied under Grammateus; lived in Vienna, Austria. Author: Behend und Hübsch Rechnung Durch die Künstreichen Regeln Algebras Gemeinicklich die Coss Genennt Werden, 1525. Invented radical sign for square root; pub. book in colloquial German on whole numbers and fractions (1st of its kind in Germany), also 1st German algebra textbook. Died 1545.

RUDOLPH, Abraham Morris, physician; b. Johannesburg, S. Africa, Feb. 3, 1924; s. Chone and Sarah (Feinstein) R.; M.B.B.Ch. summa cum laude, U. Witwatersrand, Johannesburg, 1946, M.D., 1951; M.R.C.P., Royal Coll. Physicians London, Royal Coll. Physicians Edinburgh (Scotland), 1949; m. Rhona Sax, Nov. 2, 1949; children—Linda, Colin, Jeffrey. Instr., Harvard Med. Sch., 1955-60, asso. pediatrics, 1957-60; asso. cardiologist in charge cardiopulmonary lab. Children's Hosp., Boston, 1957-60; prof. pediatrics, asso. prof. physiology Albert Einstein Coll. Medicine, N.Y.C., 1960-66; dir. pediatric cardiology, vis. pediatricain Bronx Municipal Hosp. Center, N.Y.C., 1960-66; prof. pediatrics U. Cal. San Francisco Med. Center, 1966—; practice medicine specializing in pediatric cardiology, San Francisco. Mem. cardiovascular study sect. NIH, 1961-65; established investigator Am. Heart Assn., 1958-62; career scientist Health Research Council, City N.Y., 1962-66. Mem. Am. Acad. Pediatrics (E. Mead Johnson award for research in pediatrics 1964, past chmn. sect. on cardiology), Am. Phys. Soc., Soc. for Clin. Investigation, Am. Coll. Chest Physians, Soc. for Pediatric Research (mem. council 1961-64), N.Y. (dir. 1965—), Am. heart assns. Editorial bd. Pediatrics, 1964—; asso. editor Circulation Research, 1966—, Circulation, 1966—. Research, numerous publs. on cardiopulmonary physiology. Home: 758 Contra Costa Av., Berkeley, Cal. 94707. Office: Cardiovascular Research Inst., U. Cal. Med. Center, San Francisco 94122.*

RUDOLPH, Philip S., Am. chemist; b. Syracuse, N.Y., May 10, 1912; s. Simon and Flora (Kliman) R.; B.A., Syracuse U., 1933; Ph.D. (Dupont fellow in ohemistry), 1951; m. Mary Kemzura, May 7, 1942; children—Simon Alan, Florence Ann. Social worker Dept. Pub. Welfare, Syracuse, N.Y., 1934-37; faculty Syracuse U., 1944-46; chemist Oak Ridge Nat. Lab., 1951—. Mem. Am. Chem. Soc., A.A.A.S., N.Y., Tenn. acads. sci., Radiation Research Soc., Sigma Xi, Phi Lambda Upsilon, Sigma Pi Sigma. Research and publs. on radiation chemistry, chem. kinetics, mass spectrometry; co-inventor of alpha-particle ionization source and electric field radiolysis source for mass spectrometers. Home: 106 E. Damascus Rd., Oak Ridge 37830. Office: P.O. Box X, Oak Ridge 37830.*

RUDOLPHI, Karl Asmund, Swedish naturalist; b. Stockholm, Sweden, June 14, 1771. Physician, prof. vet. medicine U. Greifswald (Germany); prof. anatomy, physiology U. Berlin (Germany), from 1810; co-founder Mus. Anatomy and Zoology, Berlin. Mem. Berlin, French (corr.) acads. scis. Author: Anatomy of Plants, 1807; Entozoorum siehe Vermium Intestinalium Historia Naturalis, 3 vols., 1808-10; Documents for Anthropology and for Universal Natural History, 1812; Principles of Physiology, 3 vols., 1821-28. Said to have introduced term echinococcus, 1808; research, publs. on intestines, dentition, anatomy of plants, animals. Died Berlin, Nov. 29, 1832.

RÜDORFF, Walter, German chemist; b. Berlin, Germany, Oct. 3, 1909; s. Fritz and Gertrud (Flügeel) R.; dr. engring., agrégé, Berlin Tech. Coll.; m. Gerda Fischer, Sept. 22, 1939. Agrégé, 1942; instr., 1942; prof. at large Berlin and Vienna (Austria) tech. colls., 1945; asso. prof. U. Tübingen (Germany), 1949-52, prof., 1952—. Mem. Assn. German Chemists. Author: Chemie des Graphits und Graphitverbindungen; Struktur, Magnetismus und Leitfähigkeit von Doppeloxiden und sulfiden; Thio und Selenoverbindungen der Uebergangsmetalle; Flouride; also articles. Collaborator (with K. Hofmann) Anorganische Chemie. Research on X-ray structure determination; magnetism; graphite intercalation compounds; ternare oxides, flourides and oxyflourides of transition metals. Home: Correnstrasse 7. Office: Wilhelmstrasse 31-33, Tubingen, W. Germany.

RUEDEMANN, Rudolf, geologist, paleontologist; b. Georgenthal, Germany, Oct. 16, 1864; Ph.D., U. Jena, 1887, Strassburg U., 1889. Asst. geology Strassburg U., 1887-92; tchr. high schs., N.Y., 1892-99; asst. state paleontologist, N.Y., 1899-1926, state paleontologist, 1926-27; curator paleontology N.Y. State Mus., 1913-37. Recipient Walker prize Boston Soc. Natural History, 1901. Mem. Nat. Acad. Sci., Am. Geol. Soc. (v.p. 1916), Paleontological Soc. (pres. 1916). Author: Graptolites of New York, 1904-08; Cephalopodo of Champlain Basin, 1906; Euryterida of New York, 1912; Existence and Configuration of Precambrian Continents, 1922; Paleozoic Plankton of North America, 1934, others. Authority on graptolites; other studies of cephalopods, eurypterids, radiolarians, inliers, origin of chert, Precambrian continents, and of fossils and geology of the Ordovician and Silurian of N.Y. Died July 19, 1956.

RUEDI, Luzius, Swiss physician; b. Thusis, Switzerland, Dec. 2, 1900; s. Thomas and Anna (Siegrist) R.; student med. sch. U. Geneva (Switzerland), U. Zurich (Switzerland); m. Lotti E. Rudolph, Dec. 12, 1931; children—Thomas, Monica (Mrs. Trümpler). Staff otorhinolaryngology Zurich U. Hosp., 1937-41; prof. otorhinolaryngology U. Bern (Switzerland), 1941-48; prof. U. Zurich, 1948—, chief ENT dept. Recipient Guyot prize Groningen, Germany; Shambaugh prize in otology, Chgo. Hon. mem. Am. Otol. Soc., Triological Soc. Research, publs. on diseases of ear. Home: 50 Kurhausotr., Zurich, Switzerland.*

RUEGAMER, William Raymond, Am. biochemist; b. Huntingron, Ind., Dec. 15, 1922; s. C. Raymond and Edna (Roberts) R.; B.S., Ind. U., 1943; M.S., U. Wis., 1944, Ph.D., 1947; postgrad. U. Colo.; m. Arlene Frankenberg, Apr. 5, 1946. Biochemist, Swift & Co., 1947-48; instr. biophysics U. Colo., 1950-51; asst. dir. Radiosotope Labs., VA Hosp., Syracuse, N.Y., 1952-60; asso. prof. biochemistry State U. N.Y., Syracuse, 1960-67; prof., acting chmn. biochemistry dept., 1967—. Cons. Warner Lambert Pharm. Co., Morris Plains, N.J., 1960—. Mem. Am. Soc. Biol. Chemists, Am. Chem. Soc., Am. Soc. Exptl. Biology and Medicine, A.A.A.S., Sigma Xi, Phi Sigma, Alpha Chi Sigma. Research, numerous articles in thyroid metabolism, cholesterol and ground substance metabolism in relation to atherosclerosis. Home: 105 Brockton Lane, De Witt, N.Y. 13214. Office: Biochemistry Dept., State U. N.Y., Upstate Med. Center, Syracuse, N.Y. 13210.*

RUEGER, Lauren John, Am. physicist; b. Archbold, O., Dec. 30, 1921; s. Edwin and Hazel (Fisher) R.; B.S., Ohio State U., 1943, M.S., 1947; m. Florence Marian Scott, July 30, 1944; children—Carol Ann, Beth Marlene, Lauren Allen, Mary Lynn. Mem. staff Radiation Lab., Mass. Inst. Tech., Cambridge, 1943-45; instr. Ohio State U., 1946-47; research asso., electronics dept. Battelle Meml. Inst., 1947-49; project leader, microwave frequency standards Nat. Bur. Standards, Washington, 1949-53; group surp. Applied Physics Lab., Johns Hopkins, Balt., 1953—; project scientist Navy Satellite Nav. System. Mem. I.E.E.E. Am. Phys. Soc., Washington Philos. Soc., Sigma Xi. Developed microwave ferrites, varactor frequency multiplication, techniques for precision satellite time and frequency measurements; extended Nat. Bur. Standards calibration service to 70,000 Mc.; discovered microwave spectrum of deuterated ammonia; patentee printed microwave circuits, ionospheric refraction correction instrumentation. Home: 1415 Glenallan Av., Silver Spring 20902. Office: 8621 Georgia Av., Silver Spring, Md. 20910.*

RUEGG, Werner, geologist; b. Zürich, Switzerland, Apr. 3, 1902; s. Werner and Emma (Gyseler) R.; student U. Heidelberg (Germany), 1922-23, U. Berne (Switzerland), 1924-26; Dr.rer.nat. magna cum laude, U. Fribourg (Switzerland), 1928; m. Epifanía Santos Vasquez, June 26, 1937; 1 dau., Erica Carmen. Geologist, party and expdn. chief geol. investigations Royal Dutch Shell Oil Co., Dutch East Indies, Egypt, Central and S. Am., 1928-40; chief expdn. and exploration various pvt. cos. in Upper Amazon basin, No. and So. Peru, No. Chile, 1941-46; geologist, head non-metallics Geol. Inst. Peru, 1947-54; prof. geology San Marcos U., also Universidad Nacional de Ingenería, Lima, Peru, 1948-63; dir. geology dept., investigation of splty. U. Sci. and Tech., Lima, 1965—; head geol. services Nat. Corp. Fertilizers, Lima, 1964—. Fellow Geol. Soc. Am., A.A.A.S.; mem. Geochem. Soc., N.Y. Acad. Scis. Research, publs. on mapping of Peruvian coast, mapping and exploitation of mineral deposits, hist. geology and comparative tectonics; introduced 1st revised vocabulary of Spanish tech. terms used in geology and side branches. Home: Los Manzanos 230, Lima-Orrantia, Peru, S. Am.*

RUEHLMANN, Klaus, German chemist; b. Neumark, Germany, Nov. 8, 1929; s. Alfred and Johanna (Hebecker) R.; Dr., Martin Luther U., Halle, Germany, 1956, Dr.habil., 1961; m. Brigitte Schwinger, Aug. 13, 1960; children—Susanne, Christine. Prof., Hochschule für Verkehrswesen, Dresden, Germany, 1961; prof. chemistry Humboldt U., Berlin, Germany, 1962—. Mem. Chemische Gesellschaft in der DDR. Research, publs. on chemistry of organosilicon compounds. Home: 1 Alexanderstrasse, Berlin 2, DDR, 102. Office: Hessischestrasse 1-2, Berlin 4, DDR 104.*

RUEL, (Ruellius) Jean, French physician, botanist; b. Soissons, France, 1479; widower; canon of Notre-Dame; physician to Francis I. Author: De natura stirpium, 1536. Compiled Greek and Latin works on botany. Died Paris, Sept. 24, 1537.

RUESCH, Jurgen, psychiatrist; b. Naples, Italy, Nov. 9, 1909; s. Oscar and Vera (Meissner) R.; M.D.,

U. Zurich (Switzerland), 1935; m. Annemarie Jacobson, Dec. 11, 1937; 1 son, Jeffrey. Came to U. S., 1939, naturalized, 1945. Fellow Rockefeller Found., asst. psychiatry Mass. Gen. Hosp., Boston, 1939-41; research fellow neuropathology Harvard Med. Sch., 1941-43; faculty U. Cal. Sch. Medicine, San Francisco, 1943—, prof. psychiatry, 1956—; research psychiatrist, Langley Porter Neuropsychiat. Inst., San. Francisco, 1943-58, dir. treatment research center, 1958-64, dir. sect. social psychiatry, 1965—. Cons. to various fed., state, community, pvt. agys. Diplomate Am. Bd. Psychiatry and Neurology. Mem. Internat. Fedn. for Med. Psychotherapy (dir. 1964—), A.M.A., San Francisco County Med. Soc., No. Cal. Psychiat. Soc., Am. Psychiat. Assn., Am. Coll. Psychiatrists, Assn. for Research in Nervous and Mental Disease (past v.p.), A.A.A.S. Author: (with F.L. Wells) Mental Examiners Handbook, 1942; (with G. Bateson) Communication, 1951-68; (with Kees) Noverbal Communication, 1956; Disturbed Communication, 1957; Therapeutic Communication, 1961; (with Brodsky, Fischer) Psychiatric Care, 1964; also numerous articles. Editor: Archives Gen. Psychiatry, 1961—, Psychotherapy and Psychosomatics, 1964—, Jour. Nervous and Mental Disease, 1966—; adv. bd. Internat. Jour. Social Psychiatry (London), Social Psychiatry (Heidelberg). Research on social class, culture change, disturbances of communication, social factors in mental disease. Address: Langley Porter Neuropsychiat. Inst., U. Cal Med. Center, San Francisco 94122.*

RUFFIN, Julian Meade, Am. physician; b. Norfolk, Va., Aug. 26, 1900; s. Edmund Sumter and Cordelia (Byrd) R.; A.B., U. Va., 1921, A.M., 1922, M.D., 1926; m. Lucy Landon Noland, June 22, 1929; children—Lucy Landon (Mrs. Henry H. Sprague), Jane Byrd (Mrs. Robert I. Ayerst), Judith Meade (Mrs. David G. Simpson). Mem. faculty Duke Sch. Medicine, Durham, 1930—, prof. medicine, 1949—, dir. med. clinic, 1930—. Area cons. internal medicine and gastroenterology VA, 1946—; mem. gen. medicine study sect. NIH, 1957-62. Diplomate Nat. Bd., Am. Bd. Internat Medicine, Am. Bd. Gastroenterology. Master A.C.P.; mem. So. Med. Assn., A.M.A., Am. Gastroenterol. Assn. (pres. 1953-54), Am. Clin. and Climatol. Assn., Am. Soc. Clin. Investigation, Assn. Am. Physicians, Phi Beta Kappa, Alpha Omega Alpha, Sigma Xi, others. Mem. editorial bd. Gastroenterology, 1955-60. Research and publs. on pellagra, other vitamin deficiencies; diagnosis, mgmt. of peptic ulcer and ulcerative colitis; malabsorption syndrome. Home: 816 Anderston St., Durham N.C. 27706.*

RUFFINI, Paolo, Italian mathematician; b. Valentano, Italy, Sept. 22, 1765; diploma D. in medicine and surgery U. Modena (Italy), 1788. Prof. math., medicine U. Modena, 1787-91, rector, prof., from 1814; mem. Milan (Italy) legislature, 1796 (lost position for refusal to take republican oath imposed by invading France), reinstated, 1799; prof. math. mil. sch., 1806 (when Modena lost rank as univ.). Recipient Gold medal for improvements in solution of numerical equations Italian Sci. Soc., 1804, Gold medal for work on immaterial nature of souls Pope Pius, 1806. Mem. Italian Soc. (pres.), Nat. Napoleonic Inst. Author: Teoria Generale dell' Equazioni, 1799; Reflessioni Intorno alla Soluzione dell'Equazioni Algebriache, supra la Determinazione delle Radici nell'Equazioni numbriche di Qualunque Grado, 1804; Dell' Immaterialita Dell'Anima, 1806; Memoria sul Tifo Contagioso; Opere Matematiche, 1915-54 (3 vols. pub. posthumously). Made 1st noteworthy attempt to prove no algebraic solution of quintic equation, 1803, 05; developed theory of transforming one equation into another whose roots are diminished by certain constant; research on equations anticipated algebraic theory groups. Died Modena, May 9, 1822.

RUFFO, Jordan (Giordano Ruffo, Jordanus Ruffus, Jourdain Rufi), veterinarian; flourished at ct. Frederick II Hohenstaufen; author treatise on veterinary art, especially on med. care horses (1st western treatise of its kind), basis of medieval vet. medicine, circa 1252. Died after 1250.

RUFUS OF EPHESUS, Greek biologist, physician, anatomist; flourished early 2d century; probably lived during reign of Trajan and studied in Alexandria; knew Egypt well, visited Caria and Cos; practiced at med. center, Ephesus. Author: De Appellationibus Partium Corporis Human; De Renom et Vesicae Morbis. Treatise on anat. nomenclature contained good description of eye, heart and nerves, the last divided into 2 classes corr. to sensory reception and movement (earliest attempt to base pathology on anatomy); studied single diseases; made reference to bubonic plague. Work had greater influence in Orient.

RUGH, Roberts, Am. radiobiologist; b. Springfield, O., Apr. 16, 1903; s. Arthur and Gertrude (Roberts) R.; B.A., Oberlin Coll., 1926, M.A., 1927; Ph.D., Columbia, 1935; m. Harriette C. Sheldon, July 24, 1926; children—Mary Elizabeth (Mrs. A.J. Downs), William Arthur. Instr. biology Lawrence U., 1927-28, Hunter Coll., 1928-38; asst. prof. biology N.Y. U., 1938-48; asso. prof. radiology Columbia Coll. Phys. & Surg., 1948—, dir. radiobiol. research, 1948—; cons. Armed Forces Radiobiol. Research Inst., 1966—. Mem. Radiol. Soc. N. Am., Roentgen Ray Soc., Brit. Inst. Radiology, Harvey Soc., Soc. Exptl. Biology and Medicine, Am. Soc. Neuro-pathologists, Tissue Culture Soc., Soc. Cell Biology, Radiation Research Soc., Am. Cancer Soc., A.A.A.S., Sigma Xi. Author: Manual of Vertebrate Embryology, 1962; The Frog, Its Reproduction and Development, 1949; Experimental Embryology, 1962; Vertebrate Embryology: The Dynamics of Development, 1964. Editorial bd. Biol. Bull. Atompraxis (Germany). Research, numerous publs. in biology and radiation biology, especially in normal devel. but recently in effects of ionizing radiations on biol. systems, principally fetus. Home: 110 Morningside Dr., N.Y.C. 10027.*

RUHENSTROTH-BAUER, Gerhard, German biochemist; b. Troppau, June 2, 1913; s. Rudolf and Margarete R-B.; M.D., Dr. ès sc.; m. Renate von Hase, Apr. 18, 1943; 1 son, Eberhard. Qualified in exptl. medicine U. Tübingen, 1951; mem. Max Planck Inst. Biochemistry, 1962—. Publs. on exptl. medicine. Home: Spitzelbergerstrasse 11, Gräefling. Office: Goethestrasse 31, Munich 15, W. Germany.

RUHMKORFF, Henirich Daniel, inventor, mechanic, electrician; b. Hanover, Germany, Jan. 15, 1803. Precision instrument maker, Paris, France, from 1840; opened workshop (firm later named Carpentier). Inventor Ruhmkorff induction coil which produced high tension current in thin secondary armature winding through sudden interruption of direct current in primary armature winding (displayed at Paris Exhbn. 1855), 1851, used in radio, prodn. x-rays, other areas. Died Paris, Dec. 20, 1877.

RULAND, Martin (The Elder), German physician, philologist; b. Freisingen, Germany, 1532; 1 son, Martin. Physician to Emperor Rudolph II, also Count Palatine Philip Ludwig; tchr. medicine Acad. Lavingen (Germany). Author: De Lingua Graeca Ejusqie Dialectis Omnibus, 1556; Clavis Scripturae, 1564; Medicina Practica Nova, 4th edit., 1564; Hydriatice, 1568; The Saurus Rulandinum, circa 1679. Valued health giving properties of spas and mineral springs. Died Lavingen, Feb. 2, 1602.

RULAND, Martin (The Younger), German physician; b. Lavingen, Germany, Nov. 11, 1569; s. Martin Ruland; doctor's degree, U. Basle. A follower of Paracelsus; physician in Ratisbonne; physician to Emperor Rudolph II, Prague, Czechoslovakia, 1607; physician to Count Palatine Philip Ludwig. Author: De Perniciosa Luis Hungaricae Curatione, 1600; Progymnasmata Alchemiae, 1607; Problematica Medica Physica, 1608; Propugnaculum Chyiatriae, 1608; Alexicaeau Chymiatricus, 1611; Lexicon Alchemoiae, 1607. Used secret remedies, especially antimony compounds. Died Prague, Apr. 23, 1611.

RULAND, Wilhelm Otto, physicochemist, crystallographer; b. Stolberg, Germany, Oct. 25, 1925; s. Josef and Elisabeth (Fell) R.; Diplom chemiker, Technische Hochschule, Aachen, Germany, 1954, Dr.rer. nat., 1957; m. Dieta Susanne Hoehne, May 2, 1959; children—Guido Alexander, Alice Verena. Staff, Union Carbide European Research Assos., Brussels, Belgium, 1957—, project leader, 1962-64, team leader, 1964-67, gen. mgr., 1967—. Research, publs. on molecular structure and state of order in coal, crystallinity of polymers, structure on non-graphitic carbons, lattice imperfections in graphitic carbons, X-ray small-angle scattering, Compton scattering, scattering theory of two dimensional lattices. Home: 61, Av. Kersbeek, Brussels, 19. Office: 95, Rue Gatti de Gamond, Brussels 18, Belgium.*

RUMFORD, Count, see Thompson, Benjamin.

RÜMKER, Karl Ludwig Christian, astronomer; b. Neubrandenburg, Germany, May 28, 1788. Worked for East Indies Co.; later prof. nav. for Brit. Navy; dir. Sch. Nav. Hamburg, Germany, 1817-21; went to Australia with Sir Thomas Brisbane, 1822; dir. Obs., Paramatta, Australia; became dir. Obs., Hamburg, 1831. Fellow Royal Soc., 1855. Author: Mittlere Örter von 12,000 Fixsternen, 1843-59; Handbuch der Schiffahrtskunde, 1818; Längenbestimmungen durch den Mond, 1849. Compiled several stellar catalogues. Made observations at Paramatta, 1822-31. Died Lisbon, Portugal, Dec. 21, 1862.

RUMPF (or Rumph), George Eberhard, naturalist; b. Hanau, Hesse-Nassau, 1627. Consul, Dutch East India Co. Author: Thesaurus imaginum piscium, testaceorum ut et cochlearum, quibus accedunt conchylia, 1711; Herbarium Amboinense, 6 vols., 1741-55. Described tropical plants. Died Amboina, Dutch East Indies, June 13, 1701.

RUMSEY, James, Am. inventor; b. Bohemia Manor, Cecil County, Md., Mar. 1743; s. Edward and Anna (Cowman) R.; 1st wife unknown; m. 2d, Mary Morrow; 3 children. Began operating grist mill, Sleepy Creek, Md., 1782; opened (with a friend) gen. store, engaged in bldg. trade, Bath (now Berkley Springs), W.VA., 1783-84; accepted position as supt. constrn. of canals Potomac Navigation Co., 1785; began to experiment with steam engine, 1785, experimented in bldg. steamboat, 1783; exhibited boat propelled by streams of water forced out through stern, steam engine being employed to operate the force pump on Potomac River, nr. Sherpherdstown, W.Va., 1787; Rumseian Soc. formed to promote Rumsey's projects which included improved saw mill, improved grist mill, improved steam boiler, 1781; sent by Rumseian Soc. to Eng. to patent his improvements and to interest English capital; secured English patents on boiler and steamboat, 1788, U. S. patents 1791. Author: A Plan wherein the Power of Steam is fully Shewn, 1788; A Short Treatise on the Application of Steam, 1788. Died (shortly before his 2d steamboat was completed) London, Eng., Dec. 20, 1792.

RUNCORN, Stanley Keith, English geophysicist; b. Southport, Eng., Nov. 19, 1922; s. William Henry and Lily Idena (Roberts) R.; student Gonville and Caius Coll., Cambridge, Eng., 1941-43; M.A., Cambridge U., 1948, Sc.D., 1963; Ph.D., Manchester (Eng.) U., 1949. Radar Research Devel. Establishment, Ministry of Supply, 1943-46; faculty Manchester U., 1946-49; fellow Gonville and Caius Coll., 1948-55; asst. dir. research in geophysics Cambridge U., 1950-55; dir. Sch. Physics, U. Newcastle upon Tyne (Eng.), 1956—. Mem. Natural Environment Research Council, 1965—. Recipient Napier Shaw prize Royal Meteorol. Soc., 1959. Fellow Royal Soc., 1965, Inst. Physics, Royal Astron. Soc. (v.p. 1967—); mem. Am. Geophys. Union, Am. Phys. Soc. Editor: (with Press, Urey, Ahrens) Physics and Chemistry of Earth, vols. 1-6, 1956—. Research, publs. on ancient geomagnetic field by paleomagnetism, continental drift, convection currents in earth's mantle, paleowind directions, earth's rotation, theory of geomagnetic field, planetary interiors. Home: 16 Moorside Ct., Newcastle upon Tyne, Eng.*

RUNGE, Carl David Tolmé, German mathematician, physicist; b. Bremen, Germany, Aug. 30, 1856; ed. Munich, Berlin; prof. math., Hanover, later Göttingen, Germany. Author: Praxis der Gleichungen, 1900; Theorie und Praxis der Reihen, 1904; Graphische Methoden, 1915; Vorlesungen über numerisches Rechnen, 1924. Research on magnetic resolution in spectroscopy; studied theory of functions, differential equations. Died Göttingen, Jan 3, 1927.

RUNGE, Friedlieb Ferdinand, German chemist; b. Hamburg, Feb. 8, 1795. Prof. chemistry at Breslau; lectr. in Berlin; later dir. chem. factory in Oranienburg. Author: Farbenchemie, 3 vols., 1834-50; Chemischtechnische Monographie des Krapps, 1835; Der deutsche Guano in Oranienburg, 1838; Der Bildungstrieb der Stoffe, 1855; Das Gift der deutschen Sprache, ausgetrieben durch Runge, 1857. Discovered carbolic acid, pyrrole, rosolic acid and aniline in coal tar, 1834; investigated dry distillation and composition of matter; discovered mydriatic effect of belladonna, 1816; considered father of paper chromotography; discovered process of obtaining sugar from beet juice; 1st to prepare phenol by destructive distillation of coal, 1834. Died Oranienburg, Prussia, Mar. 25, 1867.

RUNKLE, John Daniel, Am. mathematician, astronomer; b. Root, N.Y., Oct. 11, 1822; s. Daniel and Sarah (Gordon) R.; ed. public schools, acads. at Canajoharie, Ames and Cortland, N.Y.; grad. Lawrence Scientific School, Harvard, 1851 (A.M., Harvard; Ph.D., Hamilton; LL.D., Wesleyan); m. Catharine Robbins Bird, 1862. Apptd. asst. on Am. Ephemeris and Nautical Almanac, 1849; resigned, 1884; originated, 1858, Mathematical Monthly, and edited it until outbreak of Civil war; prof. of mathematics, Mass. Inst. Tech., 1865-68 and 1880-1902; acting pres. Mass. Inst. Technology, 1868; pres. same, 1870-78; introduced manual training into U. S. from Russia, 1876. Author: Analytic Geometry. Wrote: New Tables for Determining the Values of the Coefficients in the Perturbative Function of Planetary Motion, Smithsonian Contributions, 1856; The Manual Element in Education; Report on Industrial Education; The Elements of Plane Analytical Geometry, 1888. Introduced physical laboratory work for studies in mining and metallurgy. Died Southwest Harbor, Maine, July 8, 1902.

RUNNER, Meredith Noftzger, Am. biologist; b. Schenectady, Jan. 7, 1914; s. Claude C. and Gladys (Noftzger) R.; student Union Coll., Schenectady, 1932-34; A.B., Ind. U., 1937, Ph.D., 1942; m. Helen Falacy, Jan. 31, 1941; children—Carol, Charles, Alan, Marilyn, Peter, Suzanne. Eigenmann Research fellow, 1941-42; faculty U. Conn., Storrs, 1942-46, asst. prof., 1946-48; research asso. Roscoe B. Jackson Meml. Lab., 1948-57, staff scientist, 1957-62; research biologist Roswell Park Meml. Inst., 1955-56; program dir. developmental biology NSF, 1959-62; prof. U. Colo., Boulder, 1962—, chmn. dept. biology, 1962-63, dir. Inst. for Developmental Biology, 1966—. Cons. cell biology study sect. NIH, 1963-68, sci. facilities NSF, 1966—. Mem. A.A.A.S., Am. Assn. Anatomists, Am. Inst. Biol. Sci., Am. Soc. Cell Biology, Genetics Soc. Am., Soc. for Developmental Biology, Teratology Soc. (pres. 1966-67), Internat. Inst. Embryology, Internat. Soc. Cell. Biologists, Congenital Anomalies Research Assn. (Japan). Research, numerous publs. on transplantation and explantation of embryos, genetic and environmental factors causing abnormal devel., interaction of agts. causing abnormal devel., mechanisms by which environmental agts. produce congenital deformity. Home: 715 Willowbrook St., Boulder, Colo. 80302. Office: Inst Developmental Biol., Univ. Colo., Boulder, Colo.*

RUPE, Hans, Swiss chemist; b. Basel, Switzerland, Oct. 9, 1866; s. Johannes and Mathilde (Fischer) R.; ed. U. Basel, 1885-87, U. Strasbourg, 1888, U. Munich, 1888-95; m. Marguerite Hagenbach; m. 2d, Marg. Lutz, 1939. Prof., Sch. Chemistry, Mulhouse,

Alsace; privat docent U. Basel, 1899, prof. chemistry, from 1903. Author: Chemie der natürlichen Farbstoffe, 2 vols., 1900-09; Anleitung zum Experimentieren in der Vorlesung über Organische Chemie, 2d edit., 1930. Research on stereochemistry; a hydrogenation catalyzer bears his name. Died 1951.

RUPEL, I(saac) Walker, Am. dairy scientist; b. Walkerton, Ind., May 3, 1900; s. David Edmund and Daisie Flora (Snethen) R.; B.S., U. Ill., 1923; M.S., U. Wis., 1924, Ph.D., 1932; m. Ruth Mabel Peterson, Sept. 6, 1924, (dec. 1956); children—John W., Joan (Mrs. Don P. Hegi); m. 2d, Cora Bradley Davies, Aug. 17, 1957. Tchr. rural sch., Wyatt, Ind., 1917-18; dairy herdsman, also test supr. in Ill., 1918-19; instr. dairy husbandry U. Wis., 1924-28, asst., then asso. prof., 1929-45; exchange prof. agr. U. Hawaii, 1930; prof. dairy sci., head dept. A. and M. Coll. Tex., College Station, 1945-65, now emeritus, chief party tech. advisers under AID contract to E. Pakistan Agrl. U., 1965-67. Ofcl. type classification judge Am. Jersey Cattle Club, 1945—; ofcl. judge livestock Nat. Fair Guatemala, 1941, Brown Swiss Cattle XIII Expn., Girardot, Colombia, 1961; mem. U. S. delegation XVth World's Dairy Congress, London, 1959, XIVth Congress, Copenhagen, 1962; del. World Conf. on Animal Prodn. 1963. Dir. Lilly Ice Cream Co., Bryan, Tex. Fellow A.A.A.S., Tex. Acad. Sci.; mem. Am. Dairy Sci. Assn. (exec. bd. 1955-58, pres. 1962), Am. Dairy Assn. Tex. (dir. 1945-60), Tex. Jersey Cattle Club (dir. 1950-53), Am. Soc. Animal Sci., Am. Soc. Exptl. Biology and Medicine, Am. Assn. U. Profs., Sigma Xi, Phi Kappa Phi, Phi Sigma, Alpha Zeta. Author, co-author research and extension bulls. Asso. editor Jour. Dairy Sci. 1944-51; editorial bd. Jour. Animal Sci., 1951-54. Demonstrated susceptibility of calves to rickets and protective and curative action of vitamin D; confirmed role of cobalt in anemia and anorexia in cattle; introduced reconstituted dry buttermilk as semen extender for artificial insemination; contributed to crossbreeding work with Zebu and European stocks of cattle. Home: 305 College View St., Bryan, Tex. Office: Tex. A. and M. U., College Station, Tex.*

RUPESCISSA, Joannes de (Jean de Roquetaillade), French alchemist; flourished 14th century; studied philosophy at Toulouse. Franciscan monk; lived at Aurillac in Aquitaine; imprisoned by Innocent VI, circa 1356, for criticizing clergy and pope and for making prophecies about kings and states; may have been freed by Urban VI in 1378, or burned 1362, or perhaps died in prison. Author: Coelum philosophorum, 1548; De consideratione Quintae essentiae rerum omnium opus sane egregium . . . nunc primum in lucem data . . . ; Epistola, 1661; Liber de confectione veri Lapidis philosophorum; Liber lucis (modifed version of preceeding), 1579. Wrote on quintessence of which 1 part could transmute 100 parts mercury into gold; applied alchemical and astrological knowledge to solution of medical problems; described 7 stages leading to complete transmutation; described construction of alchemical furnace.

RUPP, Charles, Am. physician; b. Phila., Oct. 12, 1908; s. Charles and Helena Christina (Bauer) R.; A.B., Harvard, 1929, M.D., 1933; m. Jane M. Dubbs, Jan. 3, 1954. Practice medicine specializing in neurology, Phila., 1939—; faculty U. Pa. 1936—, asso. prof. neurology, 1960—, chmn. dept. neurology, 1960—; chief neurology Phila. Gen. Hosp., 1944—; chief Neurology and Psychiatry Michael Lambrian Hosp., Phila., 1954—. Diplomate Am. Bd. Psychiatry and Neurology (dir. 1961—). Fellow Am. Acad. Neurology, Am. Psychiat. Assn.; mem. Am. Neurol. Assn. (past 1st v.p. and sec.), A.A.A.S., A.M.A. Asso. editor Progress in Neurology and Psychiatry, 1946—. Contbr. numerous articles to tech. jours. Research in neurology and psychiatry; tumors; degenerations of the central nervous system. Home: 237 S. 18th St., Phila. 19103. Office: 133 S. 36th St., Phila. 19104.*

RUSAKOV, Mikhail Petrovich, Russian geologist; b. Nov. 20, 1892; grad. Petrograd Mining Inst., 1921; employed in Kazakhstan, 1920—; dir. prospecting, copper-porphyry ore deposits, 1925-30. Recipient Stalin prize, Lenin prize; named hon. sci. worker of Kazakh SSR, 1945. Mem. Kazakh Acad. Scis., Inst. Geog. Sci. (asso.). Research and publs. on geology of Kazakhstan; discovered ore deposits at Kounrad, Karganda Oblast.

RUSBY, Henry Hurd, M.D., Am. botanist; b. Franklin, N.J., Apr. 26, 1855; s. John and Abigail Holmes) R.; Mass. State Normal Sch., 1872-74; M.D, Univ. Med. Col. (New York U.), 1884; hon. Pharm.M., Phila. Coll. Pharmacy and Science, 1923; Sc.D., Columbia, 1930; m. Margaretta Saunier Hanna, 1887. Awarded medal at Centennial Exhibition, 1876, for herbarium of plants of Essex Co., N.J.; made bot. explorations, N.M. and Ariz., 1880-81, and 1883, as agt. Smithsonian Instn., and S. America, 1885-87, interest of med. botany, crossing the continent; also on lower Orinoco River, 1896, and the Republic of Colombia, 1917; Bolivia and Brazil, 1921-22. Prof. of botany, physiology and materia medica, Dept. Pharmacy, Columbia, 1888-1930, and dean of faculty; prof. materia medica, U. and Bellevue Hosp. Med. Coll., 1897-1902. Hon. curator Economic Mus., New York Bot. Garden (chmn. bd. scientific directors, 1908-17, and mem. board of managers); Revision Com. 7th, 8th and 9th revisions, U. S. Pharma-

copoeia; mem. Revision Com. of Nat. Formulary; chmn. Commn. Pan-Am. Med. Congress for study of Am. medicinal flora; hon. mem. Pharm. Soc. of Great Britain; hon. mem. Instituto Medico Nacional of Mexico; pres. Torrey Bot. Club, 1905-12; pres. Am. Pharm. Assn., 1909-10. Expert in drug products in Bur. Chemistry, U. S. Dept. Agr., 1907-09; then pharmacognosist in same bur., 1912-17. Secured the vindication of Dr. Wiley and associates from charges, 1911. Fought successfully to stop the common use of decomposed ergot in those medicinal preparations for use in childbirth, from 1929 to 1935. Awarded Hanbury medal, Brit. Pharm. Assn., 1929; Flueckiger medal, German Apothecaries Assn.; hon. mem. Brit. Pharm. Assn., 1930. Author: Essentials of Pharmacognosy, 1895; Morphology and Histology of Plants, 1899; Materia Medica of Buck's Reference Handbook of the Medical Sciences (8 volumes), 1899; National Standard Dispensatory, 1905; Wild Vegetable Foods of United States, 1906; Fifty Years of Materia Medica, 1907; Manual of Botany, 1911; Three Hundred New Species of South American Plants; A Guide to the Economic Collections of the New York Botanical Garden; Properties and Uses of Drugs, 1930; Jungle Memories, 1933. Introduced important drugs to Am. Materia Medica, among them pichi, cocillana, miré and caäpi. Died Nov. 18, 1940.

RUSCH, Harold Paul, Am. oncologist; b. Merrill, Wis., July 15, 1908; s. Henry A. and Olga (Brandenburg) R.; B.A., U. Wis., 1931, M.D., 1933; m. Clara Lenore Robinson, Aug. 6, 1940; children—Carolyn E. (Mrs. George R. Schlotthauer), Judith A. Faculty, U. Wis., Madison, 1935—, prof. oncology, 1945—, dir. McArdle Lab. for Cancer Research, 1946—. Mem. Nat. Adv. Cancer Council, 1954-58, Commn. on Cancer Research of Internat. Union Against Cancer, 1958-66, Pres.'s Com. on Heart Disease and Cancer, 1961. Fellow Am. Acad. Arts and Scis.; mem. Am. Cancer Soc. (research adv. council 1962-65, dir. 1965—), Am. Assn. Cancer Research (pres. 1954), A.A.A.S., Am. Soc. Exptl. Pathology, Alpha Omega Alpha, Sigma Xi. Editor in chief Cancer Research, 1950-65; editorial bd. Perspectives in Biology and Medicine, 1959—; editorial com. Acta Unio International Contra Cancrum, 1953-65. Research, numerous publs. in carcinogenic action of ultraviolet irradiation, influence of diet on devel. of hepatic cancer, effect of caloric restriction on tumor formation, stages in tumor formation, biochemistry of growth and differentiation in Physarum polycephalum. Home: 3511 Sunset Dr., Madison, Wis. 53705.*

RUSE, Harold Stanley, Brit. mathematician; b. St. Leonards on Sea, Eng., Feb. 12, 1905; s. Frederick and Lydia (Backshall) R.; M.A., D.Sc., Oxford (Eng.) U. Scholar, U. Edinburgh (Scotland), 1928-36; Rockefeller fellow Princeton, 1932-33; prof. math. Univ. Coll. Southhampton (Eng.), 1937-46; prof. pure math., head dept. math. U. Leeds (Eng.), 1946—. Mem. Royal Soc. Edinburgh, London; Edinburgh, Am. math. socs., Assn. Math., Tensor Soc. Japan. Author: (with A. Walker and T. Willmore) Harmonic Spaces, 1962; also articles on differential geometry and relativity. Research on algebra, relativity, differential geometry. Home: Oak Bank, Shaw Lane, Leeds 6. Office: Univ. Leeds, Leeds 2, Eng.

RUSH, Benjamin, Am. physician; b. Phila., Jan. 4, 1746; s. John Harvey and Susanna (Hall) R.; A.B., Coll. of N.J. (now Princeton), 1760; studied medicine under Dr. John Redman, 1761-66; attended 1st lectures of Dr. William Shippen and Dr. John Morgan in Coll. of Phila.; M.D, U. Edinburgh (Scotland), 1768; m. Julia Stockton, Jan. 11, 1776, 13 children including James, Richard. Returned to Phila., 1769, began practice of medicine; prof. chemistry Coll. of Phila., 1769-91, also prof. theory and practice, 1789; published A Syllabus of A Course of Lectures on Chemistry (1st Am. text on chemistry), 1770, reissued 1773; published anonymously Sermons to Gentlemen upon Temperance and Exercise (one of 1st Am. works on personal hygiene) 1772; mem. Am. Philos. Soc.; published An Address to the Inhabitants of the British Settlements in America, upon Slave-Keeping, 1773; an organizer Pa. Soc. for Promoting the Abolition of Slavery, 1774, pres. 1803; elected to Pa. Provincial Conv., 1776; mem. Continental Congress, 1776-77, signer Declaration of Independence; apptd. surgeon gen. Armies of the Middle Dept. Continental Army, 1777; became lectr. U. State of Pa., 1780; mem. staff Pa. Hosp., 1783-1813; established 1st free dispensary in Am., 1786; recognized as the "instaurator" of the Am. temperance movement; persuaded the Presbyns. to found Dickinson Coll., 1783, served as trustee; mem. Pa. Conv. which ratified U. S. Constn. 1787, with James Wilson led successful fight for adoption; with James Wilson inaugurated a campaign which secured a more liberal and effective constn. for Pa., 1789; apptd. treas. U. S. Mint by Pres. John Adams, 1797-1813; became prof. the Institutes Medicine and Clin. Practice, U. Pa., 1792, prof. theory and practice, 1796; a founder Phila. Coll. Physicians, 1787. Author: Medical Inquiries and Observations, initial vol., 1789; An Account of the Bilious Remitting Yellow Fever, As It Appeared in the Essays, Literary, Moral and Philosophical, 1798; Medical Inquiries and Observations upon the Diseases of the Mind, 1812. A pioneer worker in exptl. physiology in U. S.: 1st Am. to write on cholera infantum, 1st to recognize focal infection of teeth; regarded as father of Am. psychiatry; chief interest lay in corre-

lating phys. and mental aspects in medicine; attempted to prove insanity was arterial disease; his views on the causes of yellow fever epidemic of 1793 caused controversy. Died Phila., Apr. 19, 1813.

RUSH, Benjamin Franklin, Jr., Am. physician; b. Honolulu, Jan. 14, 1924; s. Benjamin Franklin and Vera (Marston) R.; B.A., U. Cal. at Berkeley, 1944; M.D., Yale, 1948; m. Norah Grant, June 5, 1948; children—Stephen Hughes, Christopher Marston. Instr. surgery Johns Hopkins Med. Sch., 1957-59, asst. prof., 1959-62; asso. prof. surgery U. Ky. Coll. Medicine, Lexington, 1962-66, prof., 1966—. Mem. Soc. U. Surgeons, A.C.S., Central, Ky. surg. assns. Lab., clin. research, numerous publs. on shock and treatment of cancer. Home: 1862 Parkers Mill Rd., Lexington, Ky. 40504.*

RUSH, Joseph Harold, Am. physicist; b. Mt. Calm, Tex., Apr. 11, 1911; B.A., Texas, 1940, M.A., 1941; Ph.D., Duke, 1950; m. Juanita Erickson 1936; 1 dau., 2 sons. Instr. physics Tex. Tech. Coll., 1941-42, prof. physics, 1954-56; asst. prof. physics and astronomy Denison U., Granville, O., 1942-44; asso. physicist Clinton Lab., Oak Ridge, Tenn., 1944-46; sec.-treas. Fedn. Am. Scientists, Washington, 1946-47; research physicist High Altitude Obs., Boulder, Colo., 1949-54; cons., 1956-62; research physicist Nat. Center Atmospheric Research, 1962—. Lectr. U. Colo., 1950-51; Mem. A.A.A.S., Astron. Soc., Optical Soc., Fedn. Am. Scientists, Parapsychology Assn., Phi Beta Kappa, Sigma Xi.

RUSHTON, John Henry, Am. chem. engr.; b. New London, Pa., Nov. 25, 1905; s. Edward Wester and Daisy Rich (Garber) R.; B.S. in Chem. Engring., U. Pa., 1926, Ph.D., 1933; m. Ellott May McLellon, Dec. 9, 1933; 1 son, Edward Wester. Chem. engr. Royal Electrotype Co., Phila., 1926-28; Kieckheffer Container Co., Delaire, N.J., 1928-29; asst. prof. Drexel Inst. Tech., Phila., 1929-36; asst. prof. U. Mich., 1936-37; prof. chmn. sch. of chem. engring. U. Va., 1937-46; prof., dir. dept. chem. engring. Ill. Inst. Tech., Chgo., 1946-55; prof. chem. engring. Purdue U., 1955—; cons. in chem. engring. to equipment mfrs. and process industries, 1933—; tech. aide and sect. chief OSRD, NDRC, 1942-46; dir. thermodynamics research lab. U. Pa., 1945-46. Expert cons. Research and Development Bd., Dept. Def. Mem. Am. Chem. Soc. (chmn. div. indsl. and engring. chemistry 1952-53, mem. council), Engrs. Council Profl. Devel. (exec. com.), Am. Inst. Chem. Engrs. (pres. 1957, treas. 1958-62, recipient William H. Walker award 1952; recipient Founders award 1962; dir.), Am. Soc. M.E., Am. Soc. Engring. Edn. (dir., chmn.), Soc. Chem. Ind., Sigma Xi, Alpha Chi Sigma, Sigma Tau, Theta Xi. Author: (with H. McCormack) Applications of Chemical Engineering, 1938; (with H. C. Hesse) Process Equipment Design, 1946. Contbr. articles on subjects including mixing, extraction, fluid flow, low-temperature processes, oxygen, design, and teaching in chem. and tech. jours. Home: 5 Hitching Post Rd., West Lafayette, Ind. 47906.*

RUSHTON, Martin Amsler, English physician, dentist; b. London, Mar. 29, 1903; s. William and Alice (Amsler) R.; M.A., Cambridge U., 1928, M.B., 1932, M.D., 1946; student Guy's Hosp., 1924-31; LL.D., U. Toronto (Ont., Can.), U. Belfast (Ireland), 1965; D.Odont., U. Stockholm (Sweden), U. Copenhagen (Denmark); m. Dorothy Whiteside, Apr. 27, 1949; 1 dau., Elisabeth. Practice dentistry, London, 1932-39; dental surgeon Guy's, St. Thomas's hosps.; oral surgeon Emergency Med. Service, 1939-46; prof. dental medicine London U., 1946-66. Chmn. dental com. Med. Research Council; dean Faculty Dental Surgery, Royal Coll. Surgeons. Decorated comdr. Brit. Empire. Recipient John Tomes prize Royal Coll. Surgery, Elmer Best award P. Fauchard Acad., 1967. Fellow Dental Soc., Royal Coll. Surgeons, Royal Soc. Medicine (hon.); mem. Internat. Assn. for Dental Research (pres.), Brit. Dental Assn. (pres.). Author: (with B. E. D. Cooke) Oral Histopathology, 1958; also numerous articles. Research on minute structure of dental and oral tissues including effects of heredity, age and disease. Home: 4 Kippington Rd., Sevenoaks, Kent, Eng.*

RUSHWORTH, John, English surgeon; b. Northamptonshire, 1669; s. Thomas Rushworth; m. Jane Danvers; 10 children; practiced surgery, Northampton; suggested (with Sir Samuel Garth) found. of local infirmaries and dispensaries. Author: A Proposal for the Improvement of Surgery, 1732; Two letters showing the great advantage of the Bank in Mortifications, 1732. Discovered usefulness of cinchona bark in treatment of gangrene, 1721. Died Dec. 6, 1736.

RUSINOV, Mikhail Mikhaylovich, Russian optician; b. 1909; grad. Leningrad Technicum Precision Mechanics and Optics, 1927, Leningrad Inst. Precision Mechanics and Optics, 1931. Designer, All-Union Assn. Optical Mechanics Industry, 1929-32; with Leningrad br. Central Research Inst. Geodesy, Aerial Survey and Cartography, 1932-42; instr. Moscow High Tech. Sch., 1943-44, prof., 1944—; prof. Leningrad Inst. Precision Mechanics and Optics, 1946—. Decorated Order of Lenin; recipient Stalin prize, 1941, 49, 50. Author: Optics of Aerial Survey Instruments, 1936; Optics of Instruments for Recording Vibrations, 1939; Wide-angle Lenses, 1949; Distortion in Dual Anastigmats, 1949; Technical Optics, 1957. Research and publs. on theory for designing optical lenses, developed Russar wide-angle and extra wide-angle lenses for aerial survey; designer lenses and condensers for extra wide-angle multiplexes (basic instrument for processing aerial survey photographs), and reprodn. apparatus. Address: Leningrad Inst. Precision Mechanics and Optics, Grivtsova p. 14, Leningrad, USSR.

RUSINOV, Vladimir Sergeevich, Russian physiologist; b. 1903; grad. Biology Faculty, Leningrad U., 1926; D.Biol. Sci. Asso. physiol. lab. USSR Acad. Scis., Leningrad, from 1926, later sr. asso. electrophysiol. labs. Inst. Higher Nervous Activity, dir., 1960-—, head lab. gen. physiology; former sr. asso. electro-physiol. labs. Inst. Neurosurgery, USSR Acad. Med. Scis., now head electro-physiol. lab. Burdenko Inst. Neurosurgery. Co-author: Theory and Practice of Electroencephalography in Focal Disorders of the Brain, 1954. Co-author, editor: Experimental Study of the Higher Nervous Activity of Man and Animals, 1960. Author: Experimental Study of the Functioning of the Cerebral Cortex, 1960. Address: Inst. Higher Nervous Activity, Pyatnitskaya 48, Moscow, USSR.*

RUSK, Howard A., Am. physician; b. Brookfield, Mo., Apr. 9, 1901; s. Michael Yost and Augusta Eastin (Shipp) R.; A.B., U. of Mo., 1923, LL.D., 1947; M.D., U. of Pa., 1925; D.Sc. (hon.), Boston University, 1949, Lehigh U., 1956, Middlebury Coll., 1957, Woman's Medical College, 1962; LL.D. (honorary), Westminster College, 1950, Hahnemann Med. Coll., 1952, Chungang U., Korea, 1956, L.I.U., 1957; D.Sc., Trinity Coll., 1961; Litt. D., Ithaca College, 1961; LL.D., Missouri Valley College, 1965; married Gladys Houx, October 20, 1926; children —Martha (Mrs. Preston Sutphen, Jr.), Howard A., John Michael. Engaged in practice of internal medicine, St.L., 1926-42; instr. med., Washington U., St.L., 1929-42; asso. chief of staff St. Luke's Hosp.; St.L. Col., M.C., chief of convalescent tng. div., Office of Air Surgeon, 1943-45. Asso. editor, N.Y. Times; prof. and chmn., dept. of rehabilitation and phys. med., N.Y. U. Coll. of Med., 1946-—. Consultant in rehabilitation, Baruch Com. on Physical Medicine Vets. Adminstrn.; also secreteriat United Nations; consultant in rehabilitation, N.Y. City Dept. Hosps.; chmn. health resources adv. com. O.D.M., 1950-57, chmn. nat. adv. com. S.S.S., 1950-57. Chmn. Am.-Korean Found.; pres. World Rehabilitation Fund, Internat. Soc. for Welfare of Cripples; director Chemical Fund, Inc. Trustee U. Pa. Decorated D.S.M.; Nat. Medal of Republic Korea; Order Jose Fernandez Madrid (Colombia); Chevalier, French Legion of Honor; recipient Dr. C. C. Criss award, 1952; Lasker award Am. Pub. Health Assn., 1952; Medal of Honor, Nat. Assn. Women Artists, 1953; Gold Medal, Inst. Soc. Sci., 1954, International Benjamin Franklin Soc., 1955, Albert Lasker award for services to physically disabled, 1957, 60, Gold Key award, American Congress Physical Medicine and Rehabilitation, 1958, others. Diplomate American Board Internal Medicine; American Bd. Phys. Medicine and Rehabilitation. Fellow American College of Physicians, Royal College of Physicians; mem. N.Y. Acad. Medicine, A.M.A., Internat. Soc. Welfare of Cripples (president, 1954), Phi Beta Kappa. Author: New Hope for the Handicapped (with Eugene J. Taylor), 1949; Living with a Disability, 1953; Rehabilitation Medicine, 1958; co-author Cardiovascular Rehabilitation, 1958; Rehabilitation of the Cardiovascular Patient, 1958; also contbr. med. articles to Ency. Britannica, sci. publs. and papers presented before med. assns. Home: 330 E. 33d St., N.Y.C. Office: Inst. Rehab. Medicine, 400 E. 34th St., N.Y.C. 10016.*

RUSK, Rogers D., Am. physicist; b. Washington, Nov. 17, 1892; s. James Madison and Annie (Rogers) R.; B.S., Ohio Wesleyan U., 1916; M.A., Ohio State U., 1917; Ph.D., U. Chgo., 1925; m. Sarah Florence Eldredge, Aug. 19, 1931; children—Susannah Margaret, James Rogers. Prof. physics, head dept. North Central Coll., Naperville, Ill., 1919-28; asso. prof. physics Mt. Holyoke Coll., 1928-40, prof., 1941-58, chmn. dept. 1940-46; head dept. physics Holyoke Jr. Coll., 1946-58, 1960-61; John Hay Whitney prof. Furman U., Greenville, S.C., 1948-59; vis. prof. Wesleyan U. Middletown, Conn., 1959-60; chmn. dept. physics State U. N.Y., Oswego, 1962-63. Cons. Induction Heating Corp., N.Y.C., 1942-46, Plastic Coating Corp., Holyoke, Mass., 1959-62. Author: Atoms, Men and Stars, 1937; Forward with Science, 1943; College Physics, rev., 1960; Atomic and Nuclear Physics, rev., 1964. Research in conduction in gases, cosmic rays. Home: 70 Woodbridge St., South Hadley, Mass.*

RUSKA, Ernst August Friedrich, German elec. engr.; b. Heidelberg, Germany, Dec. 25, 1906; s. Julius Ferdinand and Elisabeth (Merx) R.; student Tech. U. München, 1925-27; certified engr. Tech. U. Berlin, 1931, Dr.-Engr., 1934, Dr.-Engr. recognized as lectr., 1944; Dr. honoris causa in Medicine, Kiel U., 1958; Dr. honoris causa in Physics, Moderna U., 1963; m. Irmela Ruth Geigis, May 15, 1937; children—Ulrich, Irmtraud, Jürgen. Devel. engr. Television, Berlin, 1934-36; dept. head, head clk. Siemens & Halske AG, Berlin, 1937-55; prof. Free U. Berlin, 1949-—; asst. prof. Tech. U. Berlin, 1959-—; sci. mem. Fritz-Haber Inst. of Max Planck Soc. Promotion Scis., 1954-—, dir. Inst. Electron Microscopy, 1957-—; prin. research in invention, exptl. testing, tech. devel. electron microscope, 1931-—. Recipient Senckenberg prize U. Frankfurt a Main, 1939, Silver Leibniz medal Prussian Acad. Scis., 1941, Albert Lasker award med. research Am. Pub. Health Assn., 1960. Hon. fellow Royal Micros. Soc., socs. for electron microscopy of Japan, Germany, France. Research and numerous publs. on electron optics, electro-magnetic and permanent-magnetic electron lenses; designed 1st electron microscope; improved transmission and reflection electron microscope with electro-magnetic lenses; devel. of simplified electron microscopes with permanent-magnetic lenses. Home: 7 Falkenried. Office: 4-6 Faradayweg, Berlin-Dahlem, Germany.*

RUSKA, Helmut, German biophysicist; b. Heidelberg, Germany, June 7, 1908; s. Julius F. and E. (Merx) R.; M.D., U. Heidelberg; m. C. C. Menze, June 23, 1945; 1 son, Erdmann Amadeus. Asst., Berlin Charity Hosp.; dir. elec. microscopy labs. Siemans & Halske, Berlin, Germany, 1938-45; asst. insp. N.Y. State Dept. Health, 1952-58; now prof., dir. Inst. Biophys. and Electronic Microscopy, Acad. Medicine. Recipient Aronson prize. Mem. Dusseldorf Med. Soc., German Electron Microscopy Soc., Max Planck Soc. Devel. Sci. Author: Elektronenmikroskopie in der Virusforschung, 1950. Research on electron microscopy in biology and medicine. Home: Brehmstrasse 82. Office: Moorenstrasse 5, Dusseldorf, W. Germany.

RUSKIN, Richard Alan, Am. physician; b. New Rochelle, N.Y., Oct. 1, 1922; s. Randall and Dorothy (Prussak) R.; B.A., Duke, 1940, M.D., 1943; m. Claralee Johnson, Feb. 18, 1946; 1 son, Steven Johnson. Research asso. biochemistry Cornell U. Med. Sch., N.Y.C., 1952-54, asso. prof., obstetrics and gynecology Cornell U. Med. Sch., 1955-—; prof. obstetrics and gynecology N.Y. Polyclinic Med. Sch., N.Y.C., 1965-—; attending physician obstetrics and gynecology N.Y. Polyclinic Hosp.; asso. attending N.Y. Lying-In Hosp. N.Y. State Workmen's Compensation specialist obstetrics and gynecology. Diplomate Am. Bd. Obstetrics and Gynecology. Fellow A.C.S., Am. Coll. Obstetrics and Gynecology; mem. Am. Geriatric Soc. Research, publs. in use of oxytoxin and vasopressin in problems of labor and breast feeding, analysis of premature separation of placenta cases. Home: 415 E. 52d St., N.Y.C. 10022. Office: 737 Park Av., N.Y.C. 10021.*

RUSSEL, John Edward, Brit. chemist; b. Oct. 31, 1872; s. Edward T. and Clara Angel (Hallet) R.; D.Sc., univs. Aberystwyth and Manchester; D.Sc. (hon.), Reading U.; m. Elnor Oldham, Sept. 15, 1903; children—Edward W., John L., Maurice H., Elnor R., Clare. Asst. chem. research, 1896-97; demonstrator chemistry U. Manchester, 1897-1900; instr. chemistry Wye Agrl. Coll., 1901-07; chemist at Goldsmith, 1907-12; dir. Rothamsted Exptl. Sta., 1912-43; now retired. Created knight; decorated Crder British Empire; Order Crown Belgium; Order Merit in Agr. (France). Mem. Royal Soc. London (past v.p.), Acad. Sci. Paris; hon. mem. Royal Swedish Acad., agr. acads. Poland, Finland, London Agrl. Club. Author: Soil Conditions and Plant Growth; Report on Agriculture in India, 1936; World Population and Food Supplies; The Land Called Me. Address: Woodstock House, Woodstock, Oxfordshire, Eng.

RUSSEL, Richard Doncaster, Canadian geophysicist; b. Toronto, Ont., Can., Feb. 27, 1929; s. Richard Douglas and Gwen (Doncaster) R.; B.A. with honors in Physics and Chemistry, U. Toronto, 1951, M.A., 1952, Ph.D., 1954; m. Virginia Ann Clippingdale, Aug. 11, 1951; children—Linda Jean, Morna Ann, Mary Joyce. Faculty U. Toronto, 1954-58, prof. physics, 1962-63; faculty U. B.C., Vancouver, Can., 1958-62, prof. geophysics, 1963-—. Fellow Royal Soc. Can.; mem. Canadian Assn. Physicists, Am. Geophys. Union. Author: (with J.A. Jacobs, J.T. Wilson) Physics and Geology, 1959; (with R.M. Farquhar) Lead Isotopes in Geology, 1960; also numerous articles. Research on applications mass spectrometry to geology especially as concerned with lead. Home: 2061 Allison Rd., Vancouver 8, B.C., Can.*

RUSSELL, Bertrand Arthur William (3d Earl Russell), English mathematician, philosopher; b. Trelleck, Eng., May 18, 1872; s. John (Viscount Amberley) and Katherine (Stanley) R.; M.A., Trinity Coll., Cambridge University; m. Alys W. Smith, 1894 (div.); m. 2d, Dora Winifred Black, 1921 (div.); children—John Conrad, Katharine Jane; m. 3d, Patricia Helen Spence, 1936 (div.); 1 son, Conrad Sebastian Robert; m. 4th, Edith Finch, 1952. Began as fellow and lecturer Trinity College, Cambridge; temporary professor Harvard University, Lowell lecturer, 1914; professor philosophy, National Univ. of Peking, 1920-21; lecturer, U. of Chicago, 1938; professor philosophy, University of California at Los Angeles, 1939. Awarded Nicholas Murray Butler medal, 1915; Sylvester medal of Royal Society, England, 1934; received Order of Merit, 1949; recipient of the Nobel Prize for literature, 1950. Fellow Royal Society; fellow Trinity College. Author: Proposed Roads to Freedom, 1918; Introduction to Mathematical Philosophy, 1919; The Analysis of Mind, 1921; The ABC of Atoms, 1923; The ABC of Relativity, 1925; Philosophy, 1927; Sceptical Essays, 1928; Marriage and Morals, 1929; Mysticism and Logic, 1918; Our Knowledge of the External World as a Field for Scientific Method in Philosophy, 1915; Principia Mathematica (with Dr. A. N. Whitehead), 1910-13; The Conquest of Happiness, 1930; The Scientific Outlook, 1931; Education and the Good Life, 1926; Education and the Social Order, 1932; In Praise of Idleness, 1935; Which Way to Peace?, 1936; The Amberley Papers (with Patricia Russell), 1937; Power, 1938; An Inquiry Into Meaning and Truth, 1940; A History of Western Philosophy, 1946; Human Knowledge, its Scope and Limits, 1948; Authority and the Individual, 1949; Unpopular Essays, 1950; New Hopes for a Changing World, 1951; The Impact of Science on Society, 1952; Satan in the Suburbs, 1953; Nightmares on Eminent Persons, 1954; Human Society in Ethics and Politics, 1954; Portraits from Memory, 1956; Why I Am Not A Christian, 1957; Common Sense and Nuclear Warfare, 1958; Wisdom of the West, 1959; My Philosophical Development, 1959; Fact and Fiction, 1961; Has Man a Future?, 1961; Unarmed Victory, 1963; Autobiography, Vol. I, 1967, Vol. II, 1968; War Crimes in Vietnam, 1967. Pioneer in study of mathematical logic (with A. N. Whitehead); made profound studies of the logical foundations of mathematics; calls his own philosophy logical atomism; is generally considered one of the great philosophers of the 20th century. Home: Plans Penrhyn, Penrhyndeudraeth, Merioneth, Wales. Office: Bertrand Russell Peace Found., 3 and 4 Shavers Pl., Haymarket, London S.W.1, Eng.

RUSSELL, Findlay Ewing, Am. neurologist, toxicologist; b. San Francisco, Sept. 1, 1919; s. William and Mary Jane (Findlay) R.; B.A., Walla Walla Coll., 1941; M.D., Loma Linda U., 1950; m. Janet Louise Thiel, May 4, 1950; children—Christa, Sharon, Robin, Constance, Mark. Research fellow Cal. Inst. Tech., 1951-53; physiologist Huntington Inst. Med. Research, Pasadena, Cal., 1953-55; dir. lab. neurology research Los Angeles County Gen. Hosp., 1955-—; faculty Loma Linda U., 1955-66; prof. neurology (pharmacology) U. So. Cal. Sch. Medicine, 1966-—; cons. med. and biol. orgns., including WHO, NSF, Office Naval Research, Office Naval Intelligence, A.M.A. Recipient Walter Reed Soc. award, 1955; citation Los Angeles County Bd. Suprs., 1960. Mem. Internat. Soc. Toxinology (Francisco Redimedal 1967, pres. 1962-66), Am. Physiol. Soc., A.C.P., Am. Coll. Cardiology, Royal Soc. Tropical Medicine, Sigma Xi. Author: Bibliography of Snake Venoms and Venomous Snakes, 1964; Marine Toxins and Venomous and Poisonous Marine Animals, 1965: Animal Toxins, 1967. Research, publs. on chem. properties of venoms, how they produce changes in tissues and how their effects can be neutralized and treated. Home: 1086 Armada Dr., Pasadena, Cal. 91103. Office: 1200 N. State St., Los Angeles, Cal. 90033.*

RUSSELL, Frederick Fuller, Am. physician; b. Auburn, N.Y., Aug. 17, 1870; M.D., Columbia U., 1893; Sc.D., George Washington U., 1917; Sc.D. (hon.), Rochester. Commd. lt., USMC, 1898, capt., 1903, maj., 1909, col., 1917; ret., 1920; curator Army Med. Mus., 1908-20; instr. Army Med. Sch., 1908-20; prof. pathology and bacteriology George Washington U., 1909-20; dir. pub. health lab service internal health div. Rockefeller Found., 1920-23, gen. dir., 1923-27, dir., 1927-36; prof. preventive medicine and epidemiology Harvard Med. Sch., also Sch. Pub. Health, 1936-39; emeritus prof., 1939-—; dir. Lab., Bd. Health, Ancon, C.Z., 1915-17. Recipient Sedgwick Meml. medal Am. Pub. Health Assn., 1936; Hartley medal Nat. Acad. Sci., 1936; Buchanon medal Royal Soc. Arts. Mem. Am. Assn. Pathologists and Bacteriologists, Soc. for Exptl. Pathology, A.M.A., Soc. Tropical Medicine, Assn. Mil. Surgeons. Research on effect of anti-typhoid vaccination in U. S. Army, 1910; studies in ancylostomiasis, dysentery, yellow fever, malaria. Office: Harvard Med. Sch., Boston 15.

RUSSELL, Frederick Stratten, English zoologist; b. Bridport, Dorset, Eng., Nov. 3, 1897; s. William and Lucy (Newman) R.; B.A., Gonville and Caius Coll., Cambridge, Eng., 1921; LL.D., Glasgow (Scotland) U., 1957; D.Sc., Exeter (Eng.) U., 1960, Birmingham U., 1966; m. Gweneth Kate Moy Evans, Sept. 25, 1923; 1 son, William Moy Stratten. Asst. dir. fisheries research Govt. Egypt, 1922-23; sci. staff Plymouth (Eng.) Lab., Marine Biol. Assn., 1924-65, dir., 1945-65, sec. Marine Biol. Assn., 1945-65; mem. Gt. Barrier Reef Expdn., 1928-29. Mem. Nat. Oceanographic Council, 1950-65. Decorated Comdr. Order Brit. Empire; knighted, 1965; hon. fellow Caius Coll., 1965-—. Fellow Royal Soc., 1938, Inst. Biology; mem. Royal Danish Acad. (fgn.), Physiol. Soc. (hon.), Challenger Soc. (hon.). Author: The Medusae of the British Isles, 1953; (with C.M. Yonge) The Seas, 1928; also numerous articles. Research on plankton and their migration, systematic, biology, and description of new species of medusae. Home: Wardour, Derriford, Plymouth. Office: The Lab., Citadel Hill, Plymouth, Eng.*

RUSSELL, Glen Allan, Am. chemist; b. Jo ville, N.Y., Aug. 23, 1925; s. John Allen a ion (Cottrell) R.; B.S., Rensselaer Poly. In M.S., 1948; Ph.D., Purdue U., 1951; Ellen Havill, June 6, 1963; childre June Ellen. Research asso. Gen. F Lab., Schenectady, 1951-58; a State U., Ames, 1958-61, pro adv. panel Petroleum Research prof. U. Wurzburg (Germany) Notre Dame, 1966. Alfred 1959-63. Mem. Am. Che chemistry 1965), Am. aday Soc., Chem. S

1460

publs. on directive effects in aliphatic substitutions occurring via free radicals, photochlorination, oxidation, bromination, solvent effects, application of E.S.R. spectroscopy to problems of structure and conformation, mechanism of autoxidation of organic substances in basic solution. Home: 1014 Murray Dr., Ames, Ia. 50010.*

RUSSELL, Harry Luman, Am. bacteriologist; b. Poynette, Wis., Mar. 12, 1866; s. E. Fred (M.D.) and Lucinda Estella (Waldron) R.; B.S., U. of Wis., 1888, M.S., 1890; post-grad. studies U. of Berlin, Pasteur Inst., Paris, Zoöl. Sta., Naples; Ph.D., Johns Hopkins U., 1892; hon. Sc.D., U. of Wis., 1934; m. H. May Delany, Dec. 20, 1893 (dec.); children—Gertrude E. (dec.), Eldon Babcock; m. 2d, Susanna Cocroft Headington, July 27, 1932 (dec.). Assistant prof. bacteriology, 1893-97, prof., 1897-1907, dean College Agriculture, director Experimental Station, 1907-31, U. of Wisconsin; dir. Wis. Alumni Research Foundation since 1931. Dir. Wis. State Hygiene Lab., 1903-08; pres. advisory board Wis. Tuberculosis Sanatorium; mem. Wis. com. U. S. War Finance Corp.; apptd. mem. staff U. S. Food Adminstrn., Jan. 1918; mem. agrl. advisory com. Am. Bankers Assn., 1922-40; rep. Internat. Edn. Bd. in making survey of ednl. instns. of Far East, 1925-26. Author: Outlines of Dairy Bacteriology, 1894-1905; Agricultural Bacteriology, 1898; Public Water Supplies (with Prof. F. E. Turmeaure), 1939; Experimental Dairy Bacteriology (with Prof. E. G. Hastings), 1909; Agricultural Bacteriology (with same), 1921; Dairy Bacteriology (with same), 1919; also reports and bulls., Wis. Expt. Sta., 1893-1930. Contbn. to research in dairy bacteriology, pasteurization of milk and cheese curing, Tb in cattle; bacterial diseases of plants. Died 1954.

RUSSELL, Henry Norris, Am. astronomer; b. Oyster Bay, N.Y., Oct. 25, 1877; s. Alexander Gatherer and Eliza Hoxie (Norris) R.; A.B., Princeton, 1897, A.M., 1898, Ph.D., 1899; research student, King's Coll., Cambridge U., Eng., 1903-03; Docteur, Lauvain, 1927; D.Sc. (hon.), Dartmouth, 1923, Harvard, 1929, U. Chgo., 1941, Michoacau, Mexico, 1942, Yale, 1951, Princeton, 1954; married Lucy May Cole, Nov. 24, 1908; children—Lucy May, Elizabeth Hoxie, Henry Norris, Emma Margaret. Research asst. Carnegie Instn., Washington, stationed at Cambridge, Eng., 1903-05; instr. astronomy Princeton, 1905-08, asst. prof., 1908-11, prof. 1911-27, dir. of obs., 1912-47, research prof. 1927-47, emeritus 1947; research associate Harvard College Observatory, 1947-52, Mt. Wilson Observatory, 1922-42. Fellow A.A.-A.S. (pres. 1933); mem. Nat. Acad. Sciences, Am. Philos. Soc. (pres. 1931-32), Am. Acad. Arts and Sciences, Am. Astron. Soc. (pres. 1934-37), Am. Phys. Soc.; fgn. asso. Royal Astron. Soc., London (gold medalist 1921); fgn. mem. Royal Soc., 1937; hon. fellow Royal Soc. of Edinburgh; asso. Belgian Royal Acad.; corr. French Acad. of Sciences, 1939; fgn. asso. Academia dei Lincei, Rome, 1948. Awarded Henry Draper medal Nat. Acad. Sciences, 1922; Lalande medal, French Acad., 1922; Bruce medal Astron. Soc. Pacific, 1925; Rumford medal, Am. Acad. Arts and Sciences, 1925; Franklin medal, 1934; Janssen medal French Acad., 1936. Author: Determinations of Stellar Parallax, 1911; Probable Order of Stellar Evolution, 1914; Astronomy, 1926; Fate and Freedom, 1927; (with R. S. Dugan and J. Q. Stewart) vol. 2 of Astronomy: Analysis of Stellar Spectra, 1928; The Solar System and Its Origin, 1935; The Masses of the Stars, 1940. Contbr. on astron. topics to sci. jours. Authority on sidereal astronomy and astrophysics; developed theory of stellar evolution from giant red to dwarf stars; (with A. R. Hinks) developed procedure for photographic determination of parallaxes; determined orbits of double stars. Died Princeton, N.J., Feb. 18, 1957.

RUSSELL, Israel Cook, Am. geologist; b. Garrattsville, N.Y., Dec. 10, 1852; s. Barnabas and Louisa (Cook) R.; grad. New York U., 1872; post-grad. studies School of Mines, Columbia Coll.; M.S., C.E., LL.D., New York U.; m. J. Augusta Olmsted, Nov. 27, 1886. Mem. U. S. Transit of Venus expdn. to New Zealand, 1874-75; asst. prof. of geology, School of Mines, Columbia, 1875-77; asst. geologist U. S. Geog. and Geol. Survey West of 100th Meridian, 1878; traveled in Europe; geologist U. S. Geol. Survey, 1880—; prof. geology, U. Mich., from 1892. Author: Lake Lahontan, 1885; The Newark System, 1892; Lakes of North America, 1895; Glaciers of North America, 1897; Volcanoes of North America, 1897; Rivers of North America, 1898; A View of the World in 1900—North America. Authority on N.Am. geography, particularly of northwestern areas and Alaska; studies of dynamic geology, especially glaciology. Died Ann Arbor, Mich., May 1, 1906.

RUSSELL, James, Scottish surgeon; b. Edinburgh, 1754; s. James and Margaret (Balfour) R.; ed. Edinburgh; m. Eleanor Oliver, Sept. 21, 1798; 5 sons, including Francis, 4 daus. Surgeon, Royal Infirmary, Edinburgh; clin. lectr. in practical surgery, Edin-[illegible]h, 1786-1803; 1st prof. clin. surgery Edinburgh [illegible] 1803-34. Fellow Royal Coll. Surgeons (pres. 1796-[illegible]), Royal Soc. Edinburgh (original fellow, v.p.); mem. [illegible]os. Soc. Edinburgh. Author: Practical Essay on [illegible]rtain Disease of the Bones termed Necrosis, 1794; [illegible]he Morbid Affections of the Knee-joint, 1802; [illegible]tise on Scrofula, 1808; A System of Surgery. [illegible]ug. 14, 1836.

RUSSELL, James Samuel Risien, Brit. physician; b. 1863; M.D., (Gold medal) Edinburgh; student St. Thomas Hosp., also Berlin, Paris; m. Ada Clement Hartley, 1924. Cons. physician for diseases nervous system, bd. mgmt. Nat. Hosp., London; physician U. Coll. Hosp.; prof. clin. medicine and med. jurisprudence U. Coll. Research scholar Brit. Med. Assn. Fellow Royal Coll. Physicians London; mem. Nat. Soc. for Lunacy Law Reform (chmn.), Royal Soc. Medicine (vice pres., pres. neurol. sect.), Canadian Med. Assn. (hon.), N. Staffordshire Med. Soc. (hon.), Soc. de Neurologie de Paris (corr.). Research and publs. on diseases of nervous system; described (with F. E. Batten, James S. Collier) subacute combined degeneration of spinal cord and its relationship to pernicious anemia, 1900. Died Mar. 20, 1939.

RUSSELL, John Albert, Am. astronomer; b. Ludington, Mich., Mar. 23, 1913; s. Albert James and Orra (Woodruff) R.; A.B., U. Cal. at Los Angeles, 1935, M.A., at Berkeley, 1937, Ph.D., 1943; m. Phyllis Mae Rock, June 13, 1936; children—Carolyn Frances, Stanton James. Instr., Pasadena City Coll., 1939-41; fellow Lick Obs., 1941-42; faculty U. So. Cal., Los Angeles, 1946—, prof. astronomy, 1956—, chmn. dept., 1946—, asso. dean natural sci. and math., 1963—. Faculty Research lectr., 1957; moderator Elementary Astronomy (tv course), 1962—. Recipient First U. (So. Cal.) Assos. Teaching Excellence award, 1960. Fellow A.A.A.S.; mem. Meteoritical Soc. (past pres.), Inst. Nav. (past mem. exec. com.), Astron. Soc. Pacific (past dir.), Am. Astron. Soc. Research, publs. on spectra meteors and very bright meteors. Home: 5654 Coliseum St., Los Angeles 90016.*

RUSSELL, John Scott, Scottish engr.; b. Parkhead, Scotland, May 8, 1808; s. David Russell; student univs. of Edinburgh, St. Andrews, Glasgow, (all Scotland), also engring. workshops. Tchr. natural philosophy U. Edinburgh; mgr. shpbldg. works, Greenoch; moved to London, 1844; shipbuilder on Thames; built iron steamship Gt. Eastern, 1854-57; built Dome of Vienna Exhbn., 1873. Fellow Royal Soc., 1849, recipient Gold medal, 1837; mem. Inst. Civil Engrs. (v.p.), Soc. Arts (sec. 1845-50), Inst. Naval Architects (founding, v.p.). A founder Royal Sch. Naval Architecture. Author: The Fleet of the Future: Iron or Wood? Containing a Reply to Some Conclusions of General Sir H. Douglas in favour of Wooden Walls, 1861, (2d edit.) The Fleet of the Future in 1862, or England without a Fleet, 1862; Very Large Ships, their Advantages and Defects, 1863; The Modern System of Naval Architecture for Commerce and War, 3 vols., 1864-65; Systematic Technical Training for the English People, 1869; The Wave of Translation in the Ocean of Water, Air, and Ether, 1885; The Conical Forum as the Maximum in Resistance. Invented wave-line system of ship construn.; studied wave of translation and devel. law of resistance for ship's hull in water, 1835; designed boilers, ship engines, cellular double bottom for iron ships, bridge across Thames; built steam carriages running between Paisley and Glasgow. Died Ventnor, Isle of Wight, June 8, 1882.

RUSSELL, Loris Shano, palaeontologist; b. Bklyn., Apr. 21, 1904; s. Milan Winslow and Matilda (Shano) R.; B.Sc., U. Alta., Can., 1927, LL.D., 1958; M.A., Princeton, 1929, Ph.D., 1930; m. Grace Evelyn Le Feuvre, June 11, 1938. With Geol. Survey of Can., 1930-37; faculty U. Toronto, 1937-50, 63—, prof. geology, 1963—; asst. dir. Royal Ont. Mus. Palaeontology, Toronto, 1937-45, dir., 1945-50, chief biologist, 1963—; chief zoologist Nat. Mus. Can., Ottawa, Ont., 1950-56, dir., 1956-63. Fellow Royal Soc. Can. (Willett G. Miller medal 1959, pres. sect. IV 1958-59), Geol. Soc. Am., Paleontol. Soc.; mem. Soc. Vertebrate Palaeontology (pres. 1958-59), Canadian Museums Assn. (pres. 1961-63). Author: (with R.W. Landes) Geology of the Southern Alberta Plains, 1940; also numerous articles. Research in geol. history of Western Can. in Cretaceous and Tertiary times, description of new fossil mollusks and vertebrates, studies on biology and evolution of fossil reptiles and mammals. Home: 26 Carluke Crescent, Willowdale, Ont. Office: Royal Ont. Museum, 100 Queen's Park, Toronto 5, Ont., Can.*

RUSSELL, Paul Snowden, Am. surgeon; b. Chgo., Jan. 22, 1925; s. Paul Snowden and Carroll (Mason) R.; Ph.B., U. Chgo., 1944, B.S., 1945, M.D., 1947, M.A. (hon.), 1962; m. Allene Lummis, Sept. 25, 1952; children—Katherine Swift, Paul Snowden, Allene Gilbert, Laura Rice. USPHS Research fellow, 1954-55; asst. surgery Mass. Gen. Hosp., Boston, 1957-60, chief gen. surg services, 1962—; faculty Harvard Med. Sch., Boston, 1957-60, John Homans prof. surgery, 1962—; asso. prof. dept. surgery Columbia Coll. Physicians and Surgeons, 1960-62; staff Presbyn. Hosp., Francis Delafield Hosp., 1960-62. Mem. allergy immunology study sect. div. research grants NIH, 1963-65, chmn. allergy and immunology study sect. B, 1965; mem. coms. on trauma and tissue transplantation div. med. scis. Nat. Acad. Scis.-NRC, 1963—. Diplomate Am. Bd. Surgery, Bd. Thoracic Surgery. Fellow Royal Soc. Medicine London, A.C.S.; mem. N.Y. Acad. Scis., Mass. Med. Soc. Boston Surg. Soc., A.A.A.S., Soc. U. Surgeons, Soc. for Exptl. Biology and Medicine, A.M.A., Halsted Soc., Internat. Soc. Surgery, Am. Surg. Assn., Am. Acad. Arts and Scis., Am. Assn. Immunologists. Editorial bd. Surgery, 1963—, Archives of Surgery, 1963—, Monographs in the Surgical Scis., 1963—, Transplantation, 1965—, Annals of Surgery, 1966—. Research, publs. on immunology of tissue transplantation and on wound healing. Home: 201, Suffolk Rd., Chestnut Hill, Newton, Mass. 02167. Office: Mass. Gen. Hosp., Boston 02114.*

RUSSELL, Richard Joel, Am. geographer; b. Hayward, Cal., Nov. 16, 1895; s. Frederick James and Nellie (Brennan) R.; A.B., U. Cal. at Berkeley, 1920, Ph.D., 1926; m. Dorothy King, Jan. 1, 1924 (dec. 1936); 1 son, Benjamin James; m. 2d, Josephine Burke, Aug. 20, 1940; children—Robert Burke, Charles Douglas, John Walter, Thomas William. Asso. prof. geology Tex. Tech. Coll., 1926-27; faculty La. State U., Baton Rouge, 1928—, dean grad. sch., 1949-62, Boyd prof. geography, 1963—, dir. Coastal Studies Inst., 1954—. Chmn. earth scis. div. NRC, 1955-59; cons. Council on Research of Indonesia, 1959. Recipient Vega medal Royal Swedish Anthropol. and Geog. Soc., 1961; Cullum medal Am. Geog. Soc., 1962. Mem. Nat. Acad. Scis., Acad. Scis. of Göttingen, Assn. Am. Geographers (pres. 1948, outstanding achievement award 1960), Geol. Soc. Am. (pres. 1959), Royal Dutch Geog. Soc. (hon.), Sigma Xi (nat. lectr.), Theta Tau (nat. pres. 1928-32). Author: (with Fred B. Kniffen) Culture Worlds, 1950; also many articles. Research in climates of Cal., Tex., dry climates of U.S., morphology of deltas, flood plains and sea coasts. Home: 4575 Highland Rd., Baton Rouge 70808.*

RUSSELL, Roger Wolcott, Am. psychologist; b. Worcester, Mass., Aug. 30, 1914; s. Leonard W. and Sadie (Stanhope) R.; A.B., Clark U., 1935, M.A., 1936, grad. student, 1936-38; Ph.D., U. Va., 1939; D.Sc., U. London, 1954; m. Kathleen Sherman, Mar. 13, 1945; children—Leonard Walker II, Gillian S. Du Pont research fellow U. Va., 1938-39; instr. psychology U. Neb., 1939-41; asst. prof. psychology Mich. State Coll., 1941; asso. prof. psychology U. Pitts., 1946-49; Fulbright advanced research fellow Inst. Psychiatry U. London, 1949-50; prof., head dept. psychology U. Coll., London, 1950-56; exec. sec. Am. Psychol. Assn., Washington, 1956-59; prof., chmn. dept. psychology, Ind. U., 1959—. Mem. sci. grants com. Dept. Sci. and Indsl. Research, Britain, 1954-56; dir. med. research council Group for Research Indsl. Psychology, Britain, 1951-56, Group for Exptl. Investigation of Behavior, Britain, 1955-57; mem. adv. com. psychopharmacology, Nat. Inst. Mental Health, 1956-63; sec. gen. Internat. Union Sci. Psychology, 1960—; mem. U. S. Army sci. adv. panel, 1958—; Australian Am. Ednl. Found. vis. prof. U. Sydney, 1965-66; mem. profl. council Am. Child Guidance Found. Livermore scholar, 1934-35; Bruce R. Payne scholar, 1936-37; Jonas G. Clark fellow, 1937-38; Vis. Erskine fellow U. Canterbury (New Zealand), 1966. Fellow Am., Brit. (council 1950-55) psychol. assns.; mem. Am. Psychol. Found. (sec. 1956-59), NRC, Am. Coll. Neuropsychopharmacology (charter mem.), Psychonomics Soc., Midwestern, Eastern psychol. assns., Sigma Xi, Phi Delta Kappa, Psi Chi (nat. pres.). Editor: Library of Psychology, 1952-60; The American Psychologist, 1957-59. Research on interactions between biochem. events within body and behavior of living organism, processes by which information about environmental changes are encoded and transmitted to response mechanisms involved in organism's adjustment to its environment, modes of action of drugs influencing biochem. events and thus affecting behavior. Home: 2233 Moores Pike, Bloomington, Ind.*

RUSSELL, William Low, Am. geologist; b. New Haven, Dec. 20, 1897; s. Talcott Huntington and Geraldine (Low) R.; B.A., Yale, 1920, M.S., 1922, Ph.D., 1927; m. Leonore Kathryn Schuppert, Jan. 11, 1936; children—William Hart, Philip Low. Geologist, state surveys, oil firms, 1922-35, Tidewater Oil Co., 1935-38, Standard Vacuum Oil Co., 1939-40, Well Surveys, Inc., 1940-43; research Stanolind Oil & Gas Co.,, 1943-46; asso. prof. Tex. A. and M. U., College Station, 1946-51, prof., 1951-63, ret., 1963. Fellow Geol. Soc. Am.; mem. Soc. Econ. Geologists, Am. Assn. Petroleum Geologists, Phi Beta Kappa, Sigma Xi, Tau Beta Phi, Phi Kappa Phi. Author: Principals of Petroleum Geology, 1951, 60; Structural Geology for Petroleum Geologists, 1955; also many articles. Research in hydrodynamics, effects of regional alteration of sediments, structural problems. Home: 400 Brookside St., Bryan, Tex. 77801.*

RUSSWORM, see von Gleichen, Friedrich Wilhelm.

RUST, John Howard, Am. pharmacologist; b. Many, La., Sept. 29, 1909; s. Milbern James and Lucile (Osborn) R.; D.V.M., Kan. State U., 1932, D.Sc., 1963; postgrad. Duke; Ph.D., U. Chgo., 1956; m. Mary Jo Cortelyou, Sept. 29, 1932; children—Mary Van Zandt (Mrs. J.T. Townsend), Milbern James, John Henry, Joan Helen (Mrs. David W. Johnson). Pvt. practice vet. medicine, Concord, N.H., Wellesley, Mass., 1932-35; lectr., project dir. Mass. Inst. Tech., 1958-59; prof. pharmacology and radiology University of Chicago, since 1959—, Member of the agrl. research program AEC, Oak Ridge, 1950-54; cons. atomic energy U. Tenn., 1954-63; com. path. effects atomic radiation NRC, 1957—, com. effects on livestock, 1960—; chmn. life sci. exhibits U.S. del. Atoms for Peace Conf., Geneva, Switzerland, 1958; mem. com. radiation WHO, 1959—. Mem.

A.A.A.S., Soc. Exptl. Biology and Medicine, Am. Soc. Exptl. Pathology, Radiation Research Soc., Am. Vet. Medicine Assn., Am. Pub. Health Assn., Am. Coll. Vet. Pathologists, Internat. Acad. Pathology, Sigma Xi, Phi Kappa Phi, Sigma Phi Epsilon. Research on effects of exptl. variations of radiation on animals, radioisotopes, radiol. experimentation, radioisotope methodology. Home: 5715 S. Kenwood, Chgo. 60637.*

RUSTIGIAN, Robert, Am. microbiologist; b. Boston, July 26, 1915; s. Nicholas Karnig and Rose (Avadanian) R.; B.S., U. Mass., 1938; Ph.D., Brown U., 1943; m. Gloria L. Florentino, May 17, 1956; children—Robert, Gabrielle, Christina. NRS fellow Harvard Med. Sch., 1946-49, faculty, 1949; asst. prof. U. Chgo., 1949-55; faculty Tufts U. Sch. Medicine and Dental Medicine, Boston, 1955—, asso. prof. microbiology, 1961-67; chief microbiologist virology and tissue culture lab. VA Hosp., Brockton, Mass., 1967—. Fellow Am. Pub. Health Assn.; mem. Am. Assn. Immunologists, Soc. Am. Microbiologists, N.Y. Acad. Scis., Sigma Xi. Research, publs. on biochem. and serologic relationships enteric bacteria, rickettsial diseases, Q fever, pathology and serology animal viruses, persistent infection with animal viruses at cellular level. Home: 5 Lincolnshire Way, Winchester, Mass. 01890. Office: VA Hosp., Brockton, Mass. 02401.*

RUSTON, Walter Ralph, chem. engr.; b. Vienna, Austria, Jan. 25, 1916; s. Victor Charles and Elisabeth (von Maltzan) R.; Dipl. Ing. Chem., Poly. U. Vienna, 1943; m. Edda Adeline Keinert, Apr. 12, 1945; 1 dau., Béatrice. Research chemist Association pour l'Etude de la Paléontologie et de la Stratigraphie Houillère, Brussels, 1945-46, Association pour les Etudes Texturales, Brussels, 1946-55; mgr. Soc. Indsl. Study, Research, Application, Brussels, 1955-58, gen. mgr., 1958—, also bd. dirs.; bd. dirs. Profl. Group Nuclear Industry, Brussels, BELGOSPACE Assn. Mem. Indsl. Commn. Nuclear Energy Research Center, 1964—. Mem. A.A.A.S., Am. Chem. Soc.,Brit. Nuclear Energy Soc., French Astronautical Soc., DECHEMA, German Ceramic Soc., Iron and Steel Inst. Japan, Belgian Phys. Soc., Royal Belgian Soc. Engring. and Industry. Research, publs. on inorganic chemistry, X-ray crystallography, electron microscopy, heterogeneous catalysis, corrosion of steels in high temperature water and superheated system, gas-metal reactions at high temperatures, nuclear tech. and nuclear testing facilities, fuel cell devel. Home: 64 Avenue Emile Duray, Brussels 5. Office: 1091 Chaussée d'Alsemberg, Brussels 18, Belgium.*

RUSTOW, Dankwart Alexander, polit. scientist; b. Berlin, Germany, Dec. 21, 1924; s. Alexander and Anna (Bresser) R.; student U. Istanbul, 1943-45; B.A., Queen's Coll., 1947; M.A., Yale, 1949, Ph.D., 1951; m. Tamar Gottlieb Aiello, June 28, 1962; children—Stephen L., Janet S., Timothy D. Came to U.S., 1946, naturalized, 1951. Asso. prof. citizenship Oglethorpe U., 1950-52; faculty Princeton, 1952-59, asso. prof., 1957-59; faculty Columbia, 1959—, prof. internat. social forces, 1963—; vis. lectr. Hunter Coll., 1955-56, Yale, 1961, U. Heidelberg, Germany, 1961, Sch. Advanced Internat. Studies, Washington, 1963-64, London Sch. Econs., 1965, Mass. Inst. Tech., 1966-67, N.Y. U., 1967-68. Cons. U. S. Dept. State, Rand Corp.; mem. bd. govs. Inst. Current World Affairs, N.Y.C. Guggenheim fellow, 1965-66. Founding fellow Middle East Studies Assn.; mem. Am. Polit. Sci. Assn., Council Fgn. Relations. Author: The Politics of Compromise, 1955; Politics and Westernization in the Near East, 1956; (with G. Almond, others) Politics of the Developing Areas, 1960 (with R. E. Ward, others) Political Modernization in Japan and Turkey, 1964; A World of Nations, 1967. Editor: America's Role in World Affairs Series, 1967—; Philosophers and Kings: Studies in Leadership, 1968. Devel., publs. on empirical and dynamic theory of politics; contbd. to expansion of comparative politics to non-Western areas and theory of polit. modernization. Home: 239 Central Park West, N.Y.C. 10024.*

RUTHERFORD, Daniel, Scottish physicist, botanist; b. Edinburgh, Scotland, Nov. 3, 1749; s. John and Anne (Mackay) R.; M.A., U. Edinburgh, M.D., 1777; m. Harriet Mitchelson, Dec. 13, 1786. Practice medicine Edinburgh, from 1775; prof. botany, keeper Royal Bot. Garden, Edinburgh, from 1786; physician to Royal Infirmary, from 1791. Fellow Royal Coll. Physicians Edinburgh (pres. 1796-98), Philos. Soc. Edinburgh; mem. Linnean Soc. Author: De aere mephitico, 1772; Characteres Generum Plantarum, 1793. Contbr. papers to Trans., Royal Soc. Edinburgh. First to distinguish between carbon dioxide and nitrogen, 1772; inventor maximum and minimum thermometer, 1794. Died Edinburgh, Nov. 15, 1819.

RUTHERFORD, Sir Ernest, physicist; b. Spring Grove (later Brightwater), nr. Nelson, New Zealand, Aug. 30, 1871; s. James and Martha (Thompson) R.; ed. Nelson Coll., 1887-89. Canterbury Coll., Christchurch; B.A., 1892, M.A., 1893, B.Sc., 1894; worked under J. J. Thomson, Trinity Coll., Cambridge, 1895-98, Coutts Trotter Studentship Trinity Coll., 1897; numerous hon. degrees; m. Mary Georgina Newton, 1900; 1 dau. Macdonald research prof. physics McGill U., Montreal, Quebec, Canada, 1898-1907; Langworthy prof. physics, dir. phys. labs. Victoria U., Manchester, 1907-19; Cavendish prof. exptl. physics, dir. Cavendish lab. Cambridge U., 1919-37, fellow Trinity Coll., 1919-37, chmn. advisory council Dept. Sci. and Indsl. Research, 1930-37. Recipient Nobel prize for chemistry, 1908. Fellow Royal Soc., 1903 (Rumford medal, 1904, Copley medal, 1922, mem. council 1910, pres., 1925-30), (hon.) Royal Coll. Physicians; mem. French Acad. Scis., 1921, pres. Brit. Assn. for Advancement of Science, 1923; also hon. or asso. mem. numerous others. Author: Radioactivity, 1904; Radioactive Transformations, 1906; Radioactive Substances and their Radiations, 1912; (with J. Chadwick and C. Ellis) Radiations from Radioactive Substances, 1930; The Newer Alchemy, 1937. Built detector for electromagnetic waves (held for time world record for detection at greatest distance, 2 miles); studied conduction of electricity through gases, first by using X-rays, then by rays from radioactive substances; proved rays from radioactive substances different in nature from X-rays and other kinds of ionizing radiation (found them to consist of at least 2 distinct types of radiation: alpha and beta); discovered thorium emanation and its subsequent radioactive parts; showed (with Frederick Soddy) that radioactivity is phenomenon accompanying spontaneous disintegration of atoms of radioactive elements into intermediate elements; unravelled successive radioactive changes in radium, thorium, and actinium families; determined that alpha rays were in fact alpha particles and that they were minute particles of matter projected with great velocity; found alpha particles to be positively charged, then deduced that they must be positively charged atoms of helium (helium atoms with electrons removed, later known as helium nuclei), later experimentally showed (with Thomas Royds) this definitely the case; made calculations of earth's source of heat, also age of uranium-bearing minerals; devised (with Hans Geiger) method for counting alpha particles; accurately determined number of molecules in gram-molecule of any gas (Avogadro's number) and the elementary charge of electron; found value for e about 40% greater than old assumed value, but which agreed closely with value Max Planck deduced theoretically from his quantum ideas; from expts. on scattering of alpha particles by thin sheets of metal, showed that atom is mostly empty space; developed nuclear theory of atom, arguing that nearly all mass of an atom is concentrated in small positively charged central nucleus, and that positive charge is balanced by negative charges on electrons which revolve around nucleus in orbit's radius, which is very large compared to size of nucleus; deduced that positive charge on nucleus had numerical value approximating to half the atomic weight; made expts. bombarding nuclei of atoms of lighter gases with alpha particles; found nucleus of nitrogen atom disintegrated under intense forces developed in close collision with alpha particle, and that nucleus of hydrogen atom liberated, and thus found that nucleus of hydrogen atom (later called proton) must be constituent part of nitrogen nucleus (first observed case of deliberate artificial transmutation). Died Cambridge, Eng., Oct. 19, 1937.

RUTHERFORD, William, Scottish physiologist; b. Ancrum, Scotland, 1839; M.D. Prof. physiology Edinburgh (Scotland) U., from 1874. Fellow Royal Soc., 1876. Research, publs. on influence of vagus nerve on circulation, action of drugs on bile secretion, role of cochlea of ear in perception of pitch. Died Edinburgh, Feb. 21, 1899.

RUTHERFURD, Lewis Morris, Am. astrophysicist; b. Morrisania, N.Y. Nov. 25, 1816; s. Robert Walter and Sabina (Morris) R.; grad. Williams Coll., 1835; m. Margaret Stuyvesant Chanler, July 22, 1841; 7 children. Admitted to N.Y. bar, 1837; practiced law, N.Y.C., 1837-49; began work in astron. photography and spectroscopy, 1856; invented a photographic telescope, 1858; made his 1st attempt at classification of stellar spectra, 1862; made many photographs of sun and moon and various star fields; designed ruling engine, 1870, used it to construct interference gratings finer than any made earlier, for use in spectroscopy; responsible for establishing dept. of geodesy and astronomy Columbia, 1881; donated all his equipment to Columbia. Died Tranquility, N.J., May 30, 1892.

RÜTIMEYER, Ludwig, Swiss comparative anatomist; b. Biglen, Feb. 26, 1825. Prof. in Bern. Author: Über die Herkunft unserer Thierwelt, 1867; Über Thal-und Seebildung, 1869; Gesamte Kleine Schriften, 1898. Compared anat. structure of living animals, traced individual animal types back to fossil ancestors; recognized significance of teeth in classification of fossil mammals. Died Basel, Nov. 25, 1895.

RUTKOWSKI, Antoni, Polish food chemist; b. Poznan, Poland, Nov. 13, 1920; s. Witold and Wanda (Pokrzywnicka) R.; M.Sc., U. Poznan, 1948, D.Sc., 1951; m. Zofia Kazmierczak, Mar. 30, 1948; children—Halina, Aleksandra. Research asst. dept. agrl. tech. U. Poznan, 1946-54; dir. Inst. Fat Industry, Warsaw (Poland), 1954-59; dir. research Inst. Gen. Chemistry, Warsaw, 1960—; head dept. food tech. and storage Agrl. U. Olsztyn (Poland), 1960—. Decorated Officer Cross of Polonia Restituta, 1965. Mem. Polish Acad. Sci. (mem. presidium 1966—), Internat. Union Pure and Applied Chemistry, Am. Oil Chem. Soc., Internat. Soc. for Fat Research, Deutsche Chemische Gesellschaft, Polski Towarzystwo Chemiczne. Research, publs., patents on influence of tech. process on quality of foods, especially rapeseed oil and meal. Home: 2/18 m 25 Chodna, Warsaw, Poland. Office: Blok 31, Kortowo, Olsztyn, Poland.*

RUTLEDGE, L(ester) T., Am. physiologist; b. Big Sandy, Mont., June 12, 1924; s. Lester Thomas and Jennie (Weller) R.; B.A., with honors, Mont. State U., 1950; M.A., U. Utah, 1951, Ph.D., 1953; m. Erma Erwina Heydt, Sept. 4, 1949. Instr. physiology U. Utah, 1953-56; faculty U. Mich. Med. Sch., Ann Arbor, 1956—, professor physiology, 1965—, mem. Med. Sch. Animal Care Com., 1961—, mem. dept. physiology grad. com., 1961—; vis. scientist U. Wash. Med. Sch., 1958-59, Nobel Inst. for Neurophysiology, Stockholm, Sweden, 1959-60. Recipient Research Career Devel. award USPHS, 1958—. Mem. A.A.A.S., Am. Physiol. Soc. Research, publs. on higher brain functions, exptl. surg. methods to study epilepsy and prevent seizures, integrate and associative functions of cerebral cortex neurons using micromethods, regulative mechanisms Home: 11964 Greenway Circle, Route 2, South Lyon, Mich. 48178. Office: Dept. Physiology, U. Mich. Med. Sch., Ann Arbor, Mich. 48104.*

RUTSCH, Rolf Friedrich, Swiss geologist; b. Berne, Switzerland, Aug. 11, 1902; s. Friedrich and Nelly (Siegfried) R.; Ph.D., univ. Berne and Basel (Switzerland); m. Ida Guggisberg, Apr. 23, 1929; 1 son, Bernhard. Curator geology Basel Natural History Museum, 1927-37; prof. agrégé U. Basel, 1935-40; prof. agrégé, U. Berne, 1940-48, prof. geology and paleontology, 1948—; collaborator geol. map Switzerland; cons. geologist. Mem. Swiss Assn. Geologists and Oil Engrs. Publs., research Swiss geog. maps. Address: Melchenbühlweg 75, Berne, Switzerland.

RUTSTEIN, David Davis, Am. internist; b. Wilkes-Barre, Pa., Feb. 5, 1909; s. Harry and Nellie (Davis) R.; B.S. cum laude, Harvard, 1930, M.D. cum laude, 1934; m. Mazie E. Weissman, Feb. 27, 1935; children—Catherine Ann, David Davis; m. 2d. Ruth E. Rickel, Aug. 11, 1951. Research fellow, asst. Harvard Med. Sch., Boston, 1936-37, prof., head dept. preventive medicine, 1947—, Ridley Watts prof., 1965—; mem. staffs Mass. Gen., Peter Bent Brigham, Boston City, Boston Lying-in hosps., Mass. Eye and Ear Infirmary, Children's Hosp. Med Center, Beth Israel Hosp.; faculty Albany (N.Y.) Med. Coll., 1937-43, Columbia Coll. Phys. and Surg., 1943-47; staff N.Y. State Dept. Health, 1937-46; nat. dir. gas protection sect., med. div. USPHS, 1942-43, commd. med. dir. Res., 1951; med. dir. Am. Heart Assn., Am. Council on Rheumatic Fever, 1946-47. Am. chmn. U.K.-U.S. Coop. Rheumatic Fever Study, 1950-65; chmn. com. WHO, 1956, mem. coms., 1958, 64-65; mem. research adv. com. United Cerebral Palsy Research and Ednl. Found., 1956—, chmn., 1963-66; mem. com., med. dept. Mass. Inst. Tech., 1964—; cons. Office Surgeon Gen., USPHS, NIH, founds., various brs. govt. Recipient Benjamin Franklin Mag. award, 1956; chevalier Legion d'Honneur de France, 1959; Jubilee medal Swedish Med. Soc., 1966. Diplomate Nat. Bd. Med.Examiners, Am. Bd. Internal Medicine, Am. Bd. Preventive Medicine and Pub. Health. Fellow Am. Acad. Arts and Scis., A.A.A.S., N.Y. Acad. Scis.; mem. Am. Assn. U. Profs., Internat., Am. (pres. 1966-67) epidemiological socs., Am. Fedn. Clin. Research, Am. Genetic Assn., Am. Heart Assn. (Award of Merit 1956, Gold Heart award 1959), Am. Pub. Health Assn., Am. Rheumatism Assn., Am. Soc. Clin. Investigation, Assn. Tchrs. Preventive Medicine, Boylston Med. Soc. (pres. 1960-61), Harvey Soc., World Commn. Cerebral Palsy, Alpha Omega Alpha, Delta Omega, Sigma Xi, others. Author: Lifetime Health Record, 1958. Research, publs. on cardiovascular changes in pneumococcal pneumonia; pathogenesis of rheumatic fever and heart disease; atherosclerosis. Home: 98 Winthrop St., Cambridge, Mass. 02138. Office: Dept. Preventive Medicine, Harvard Med. Sch., 25 Shattuck St., Boston 02115.*

RÜTTIMANN, Alois Martin Franz Xaver, Swiss radiologist; b. Hitzkirch, Switzerland, Sept. 20, 1922; s. Xaver Martin and Josephine (Kottmann) R.; grad. Faculty Medicine, U. Zürich, 1949, M.D., 1951; m. Margarete Dedial, Oct. 17, 1955; 2 children. Docent med. radiology U. Zurich, 1959—; staff central X-ray dept. Kantonsspital Zurich. Organizer, 1st Internat. Symposium on Lymphology, Zurich, 1966. Mem. Internat. Soc. Lymphology (founder, 1st pres.). Author: (with H. R. Schinz, R. Glauner) Ergebnisse der medizinischen Strahlenforschung, 1964; (with P. Ricklin, M. S. Del Buono) Die Meniscuslaesion, 1964; also articles. Research on arthrography of knee joint, exams. of colon, selective angiography, lymphatic system, cholangiography. Home: 18, Hintere Rietstrasse, Unterengstringen ZH, Switzerland. Office: Kantonsspital, Rämistrasse 100, Zürich, Switzerland.*

RUZICKA, Jiri A., Czechoslovakian chemist; b. Kolin, Czechoslovakia, Apr. 15, 1918; s. Vaclav and Karolina (Kubikova) R.; Dipl.-Ing., Tech. U., Prague, Czechoslovakia, 1947, Ph.D., 1962; m. Olga Sixtova, Jan. 29, 1949; Chemist, Yeast Factory, Kolin, 1940-43, Lab. for Vitamin and Hormone Chemistry, Prague, 1943-45; research worker dept. phys. chemistry Tech. U., Prague, 1948-51; head lab. phys. chemistry Inst. Hematology, Prague, 1951-55; head chem. dept. Inst. Dental Research, Prague, 1955—. Mem. Czechoslovak Com. for Fluoride Dental Caries Prevention, 1959—. Mem. Czechoslovak, Am. chem. socs., Czechoslovak

Biochem. Soc., Czechoslovak Med. Soc. Author: Chemical Dictionary, 1953; (with C. Michalec, J. Korinek, J. Musil) Electrophoresis on paper and Other Carriers, 1959; also articles. Research on analytical chemistry of fluorine, metabolism of fluorides; adv. for fluoridation of drinking water. Home: 45 Severozapadni V., Prague 4. Office: 48 Vinohradska, Prague 2, Czechoslovakia.*

RUZICKA, Leopold, chemist; b. Vukovar, Croatia (now Yugoslavia), Sept. 13, 1887; ed. Tech. High Sch., Karlsruhe, Germany; M.D. honoris causa, U. Basel, Switzerland; D.Sc. (hon.), Harvard. Became prof. organic chemistry Zurich Fed. Inst. Tech., 1923; lectr. U. Utrecht (Netherlands), 1926-29; prof. U. Zurich, 1929—. Recipient Nobel prize in chemistry (with Adolph Butenandt), 1939. Fellow Royal Soc., 1942. Research in organic synthesis, especially polymethylenes and higher terpenes; 1st to synthesize musk; synthesized androsterone and testosterone (male sex hormones) from cholesterol. Home: 101 Freudenbergstrasse, Zurich 44, Switzerland.

RYABCHIKOV, Dmitri Ivanovich, Russian chemist; b. Shcholkovo, nr. Moscow, USSR, Nov. 6, 1904; s. Ivan Alexeevich and Kharitina (Ivanova) R.; grad. Plekhanov Inst. Pub. Industry, Moscow, 1930; candidate chemistry degree Acad. Scis. USSR, 1935; doctorate in sci., 1943; m. Valentina Grigorievna Silnichenko, July 26, 1936; children—Igor, Olga. Researcher, inst. gen. and inorganic chemistry Acad. Scis. USSR, 1935-41, dep. dir. Vernadski inst. geochemistry and analytical chemistry, 1944-65, also corr. mem. acad.; dep. dir. Inst. Chemistry Uzbek SSR, 1941-44; lectr., head chair Moscow Pedagogical Ednl. Inst., from 1935, prof., 1944-60. Recipient State prize; named hon. scientist RSFSR. Mem. research tech. council State Com. on Coordination of Research Work, also of State Com. on Research Equipment of USSR; mem. Superior Attestation Commn. USSR. Author: General Chemistry, 1949; The Analytical Chemistry of Thorium, 1963; also papers. Editorial bd. Analiticheskaya Khimia. Research on chemistry of noble metals and rare earths complex compounds, application of ion exchange and paper chromatography in analytical methods and industry; his work on rare earths became basis of methods of their separation and analysis. Died Sept. 18, 1965.*

RYAN, Harris Joseph, Am. electrical engr.; b. Powell's Valley, Pa., Jan. 8, 1866; s. Charles W. and Louisa (Collier) R.; Baltimore City Coll., 1880-81; Lebanon Valley Coll., Annville, Pa., 1881-83; M.E., Sibley Coll. (Cornell U.), 1887; m. Katharine E. Fortenbaugh, Sept. 12, 1888. Instr. physics, 1888-89, prof. elec. engring., 1889-1905, Cornell; prof. elec. engring., Leland Stanford Jr. U. (now Stanford U.), 1905-31, prof. emeritus, 1931—, also hon. dir. Harris J. Ryan High Voltage Lab., Stanford U. Cons. engr. Los Angeles Aqueduct Power Bur., 1909-24. Mem. Jury of Awards, dept. electricity, Chicago Expn., 1893; del. Internat. Elec. Congress, St. Louis Expn., 1904; mem. jury, Panama-P.I. Expn., 1915. Fellow Am. Inst. E.E. (v.p., 1896-98; hon. v.p. representing Inst. at P.-P.I. Expn., 1915; pres. 1923-24). Author: (with H. H. Norris and G. L. Hoxie) Text-Book of Electrical Machinery, Vol. I, 1903. Contbr. to Trans. Am. Inst. Elec. Engrs.; numerous monographs on elec. subjects. Dir. Anti-submarine Supersonics Lab., of Nat. Research Council, Pasadena, Calif., 1918-19. Studies on compensation winding and potential distbn. in an insulator chain. Died July 3, 1924.

RYAN, John Donald, Am. geologist; b. Norristown, Pa., July 9, 1921; s. Stephen Joseph and Grace Anna (Swenker) R.; B.A. in Geology, Lehigh U., 1943, M.S., 1948; Ph.D., Johns Hopkins, 1952; m. Marjorie Heebner Bean, Oct. 30, 1943; children—Elizabeth Ann, Barbara Kathleen, Judith Marjorie. Geologist, U. S. Geol. Survey, 1948-49; faculty Lehigh U., 1952—, prof. geology, chmn. dept., 1962—. Served to 1st lt., inf., AUS, 1943-46. Fellow Geol. Soc. Am.; mem. Geol. Soc. Finland, Soc. Econ. Mineralogists and Paleontologists, Nat. Assn. Geology Tchrs. (pres. Eastern sect. 1959-60), Pa. Acad. Sci., Sigma Xi. Investigation of Rocky Mountain stratigraphy; estuarine sedimentation; geology of the moon. Home: 305 Walnut St., Hellertown, Pa. 18055. Office: Lehigh U., Bethlehem, Pa. 18015.*

RYAN, Kenneth John, Am. steroid chemist, physician; b. N.Y.C., Aug. 26, 1926; student Northwestern U., 1946-48; M.D. magna cum laude, Harvard, 1952; m. Elizabeth Marion Kinney, June 7, 1948; children—Alison, Kenneth John, Christopher. Resident in medicine Mass. Gen. Hosp., 1953-54; research fellow, resident medicine Harvard, 1955-56, Columbia Presbyn. Hosp., 1956-57; resident obstetrics-gynecology Boston Lying-In Hosp., Free Hosp. for Women, 1957-60; instr. obstetrics-gynecology Harvard, 1960-61, dir. Fearing Research Lab., 1960-61; prof., chmn. obstetrics-gynecology dept. Western Res. U., Cleve., 1961—; dir. obstetrics-gynecology U. Hosp., Cleve., 1961—. Recipient Soma Weis award; Schering award, 1951; Borden award Harvard, 1962. Diplomate Am. Bd. Obstetrics and Gynecology. Mem. Am. Soc. Biol. Chemists, Am. Coll. Obstetrics and Gynecology, Endocrine Soc. (Ernst Openheimer award 1964). Contbr. articles to tech. jours. Demonstrated estradiol dehydrogenase in human placenta and other tissues, adrenal steroid 21-hydroxylase, carbon monoxide inhibited biol. oxidation in adrenal, biol. aromatization system in human placenta, pathway for estriol formation in pregnancy, elucidation steroid biosynthesis in human ovarian follicle. Home: 2368 Tudor Dr., Cleve. 44106. Office: 2065 Adelbert Rd., Cleve. 44106.

RYAN, Robert Frank, Am. plastic surgeon; b. Hoquiam, Wash., June 23, 1922; s. Andrew B. and Zylma (Upson) R.; A.B., Stanford, 1944, M.D., 1947; M.S. (Gen. Surgery fellow), Mayo Found., 1956; m. Yvonne Arnoult, Mar. 21, 1964. Staff, Tulane U., New Orleans, 1956-58, Sch. Medicine, 1958—, prof. surgery, 1967—; cons. VA, USAF. Diplomate Am. Bd. Gen. Surgery, Am. Bd. Plastic Surgery. Mem. A.M.A. (Hoektan gold medal 1959), A.C.S., Soc. U. Surgeons, Am. Soc. and Am. Assn. Plastic and Reconstructive Surgery. Research, publs. on contbns. to devel. tech. drug dosage in isolated perfusion treatment for malignancy. Home: 1524 7th St., New Orleans.*

RYAN, Thomas Arthur, Am. psychologist; b. Batavia, N.Y., Sept. 15, 1911; s. Thomas Francis and Ruby (Barber) R.; A.B., Cornell, 1933, Ph.D., 1937; m. Mary McElheny Shaw, Sept. 12, 1935; children—Thomas A., Adelaide. Mem. faculty, Cornell U., since 1937, prof. of psychology since 1949, chairman of department of psychology, 1953-61. Fellow Am. Psychol. Assn.; mem. Am. Statis. Assn., Psychometric Soc., Inst. of Math. Statistics, Phi Beta Kappa, Sigma Xi, Phi Kappa Phi. Author: Work and Effort: the Psychology of Production, 1947; Principles of Industrial Psychology (with P. C. Smith), 1954. Contbr. psychol. publs. Studies of measurement of fatigue and effort in indsl. work, also motivational factors in perception and learning. Home: 210 Mitchell St., Ithaca, N.Y.*

RYANS, David Garriott, Am. psychologist; b. Jeffersonville, Ind., July 29, 1909; s. John Bryon and Carrie Maude (Griffith) R.; B.A., DePauw U., 1932; M.A., U. Minn., 1933, Ph.D., 1937; m. Rosann Greco, Aug. 5, 1935; children—David Garriott, Richard Frank. Instr. psychology, dir. testing Eveleth (Minn.) Jr. Coll., 1934-37; instr. ednl. psychology U. Minn., summer 1937; dean faculty, prof. psychology William Woods Coll., 1937-39; exec. sec. coop. test service Am. Council Edn., 1939-42, exec. sec. nat. tchr. exams., 1939-46, asso. dir., then dir., 1946-48; prof. ednl. psychology and research U. Cal. at Los Angeles, 1948-58; prof. ednl. psychology, chmn. dept., grad. adviser ednl. psychology U. Tex., 1958-61; sr. research leader, human factors scientists sr. head center for research in systems devel., head edn. research and devel. System Devel. Corp., 1961-64; lectr. ednl. research U. Cal. at Los Angeles, 1961-64; dir. ednl. research and devel. center, dir. improvement of edn. in Hawaii project (Ford Found. grant), prof. ednl. psychology and research U. Hawaii, Honolulu, 1964—; vis. prof. ednl. psychology U. N.H., summer 1938, Mich. State Coll., summer 1942. Dir. tchr. characteristics research Grant Found. and Am. Council Edn., 1948-56; mem. Cal. Adv. Council Ednl. Research, 1949-58; cons. Mental Health in Tchr. Edn., Human Talent program U. Tex. Recipient travel grant Social Sci. Research Council, 1958, grant Hogg Fund., 1960. Diplomate Am. Bd. Examiners in Profl. Psychology. Fellow Am. Psychol. Assn. (council rep. 1954-57, pres. div. ednl. psychology 1954-57, sec.-treas. 1959-60), Am. Ednl. Research Assn. (v.p. 1960, pres. 1961-62, exec. com. 1960-64), Psychonomic Soc., Psychometric Soc., Internat. Assn. Applied Psychology, Nat. Soc. for Study of Measurement in Edn., Research Soc. Am., A.A.A.S., Phi Beta Kappa, Sigma Xi. Author: Characteristics of Teachers: A Research Study, 1960; also numerous research papers, monographs, chpts., revs. Editorial bd. Rev. Ednl. Research, 1958-62; research editor Jour. Tchr. Edn., 1959-65; ednl. psychology editor Chandler Pub. Co.; editorial cons. Contemporary Psychology, 1961—. Research and devel. in methodology and instrumentation for identifying basic classes (sets) of human behavior involved in occupational activities; devel. of applications of human information processing models and tech. to theory and study of learning and instruction; research on human behavior measurement, theory of research, research methodology applied to human behavior. Home: 281 Poipu Dr., Honolulu 96821; also 3726 Weslin Av., Sherman Oaks, Cal. Office: Edn. Research Center, U. Hawaii, 1776 University Av., Honolulu 96822.

RYCHLAK, Joseph Frank, Am. psychologist; b. Cudahy, Wis., Dec. 17, 1928; s. Joseph Walter and Helen (Bieniek) R.; B.S. in Psychology, U. Wis., 1953; M.A., Ohio State U., 1954, Ph.D., 1957; m. Lenora P. Smith, June 16, 1956; children—Ronald Joseph, Stephannie Dianne. Faculty, Fla. State U., 1957-58, Wash. State U., 1958-61; faculty St. Louis U., 1961—, prof. psychology, 1964—. Research cons. mgmt. progress study Am. Tel. & Tel., 1957—. Fellow Am. Psychol. Assn., Soc. Projective Techniques; mem. Sigma Xi. Author: A Philosophy of Science for Personality Theory, 1968. Asso. editor Psychotherapy, 1965—. Contbns. to theory of projective phantasy, including dreams, Rorschach contents, and free assns; devel. term "reinforcement value" to demonstarte how personally determined affective factors influence rate of learning in human subjects. Home: 3841 Flora Pl., St. Louis 63110.*

RYCKMAN, Raymond Edward, Am. entomologist, systematist; b. Shullsburg, Wis., June 19, 1917; s. Edward Thomas and Jennie E. (Wright) R.; B.S., U. Cal. at Berkeley, 1950, M.S., 1957, Ph.D., 1960; m. Evelyn Ethel Larson, July 29, 1943; children—Ruby K., Albert E., Joseph V. Chmn. dept. entomology Sch. Tropical and Preventive Medicine, Loma Linda, Cal., 1950-59; faculty Loma Linda U., 1959—, asso. prof. microbiology and biology, 1965—. research asso. Los Angeles County Mus., 1964—; project dir. grants-in-aid USPHS, Office of Surgeon Gen. U.S. Army, 1951—. Fellow A.A.A.S.; mem. Entomol. Soc. Am., Am. Soc. Mammalogists, Sigma Xi. Research, publs. on description of new species of bedbugs, kissing bugs, and mites; maj. research on determination of species of various bugs, which has led to a genetic evaluation of these insect populations, epizootiology of Chagas's disease. Office: Dept. Microbiology, Loma Linda U., Loma Linda, Cal. 92354.*

RYDBECK, Olof Erik Hans, Swedish physicist; b. Greifswald, Sweden, Feb. 12, 1911; s. Thure and Gertrud (Bendz) R.; M.E.E., M.Sc., Royal Inst. Tech., Stockholm, Sweden, 1937; Sc.D., Harvard; m. Kerstin Modin, Aug. 4, 1939; children—Hans, Gustaf, Erik, Anders, Bo. Became Gordon McKay fellow Harvard, 1939; asst. prof. Chalmers U. Tech., Gothenburg, Sweden, 1941-45, prof. radiophysics, 1945-48, prof. electron physics, also, dir., founder research lab. electronics, 1948—; univ. prof., dir., founder Onsala Space Research Obs., Gothenburg, 1963—. Recipient Polhem Gold medal Swedish Soc. Engrs.; named Comdr. Royal Order North Star, 1945. Mem. Royal Soc. Sci. and Letters (past pres.), Royal Acad. Engring. Sci., Royal Acad. Sci., Sigma Xi. Developed theories on radio wave propagation, microwave electronic devices, physics of ionosphere. Home: Fjaras, Sweden. Office: Research Lab. Electronics, Chalmers U., Gibraltargatan 5, G, Gothenburg S, Sweden.*

RYDBERG, Johannes Robert, Swedish physicist; b. Halmstad, Sweden, Nov. 8, 1854. Prof., Lund, Sweden, 1901-19. Fellow Royal Soc., 1919. Developed formula for spectral lines showing relation between light spectra and various elements, 1890; demonstrated existence of a constant (named after him). Died Lund, Dec. 28, 1919.

RYDBERG, Per Axel, botanist; b. Odh, Sweden, July 6, 1860; s. Adolf Fredrik and Elfrida (Otterstrom) R.; grad. Royal Gymnasium, Skara, Sweden, 1881; B.S., U. of Neb., 1891, M.A., 1895; Ph.D., Columbia U., 1898; m. Alfrida Amanda Rydberg, Nov. 11, 1903; children—Arthur Alfred, Elsa Margreta (dec.), Lilly Irene, Linnea Astrid. Prof. natural sciences and mathematics, Luther Acad., Wahoo, Neb., 1884-90, 1891-93; at Upsala Inst., Brooklyn, 1895-96 and 1897-98; Upsala Coll., New Orange, N.J., 1898-99; fellow in botany, Columbia, 1896-97; asst. Bot. Lab., U. of Neb., 1894-95; field agt. dept. agr., divs. of botany and agrostology, summers of 1891-96; curator New York Botanic Garden, Bronx Park, 1899—. Wrote monographs on Potentilleae, Physalis, Saxifragaceae, Rosaceae, Carduaceae and Fabaceae. Author: Flora of the Sand Hills of Nebraska, 1895; Catalogue of the Flora of Montana and the Yellowstone Park, 1900; Flora of Colorado, 1906; Flora of the Rocky Mountains and Adjacent Plains, 1917; Key to the Rocky Mountain Flora, 1919; Flora of the Prairies and Plains of North America, 1932. Died July 25, 1931.

RYGE, Gunnar, dentist; b. Copenhagen, Denmark, Dec. 15, 1916; s. Axel and Guntha W. (Clausen) R.; student U. Copenhagen, 1934-36; D.D.S., Royal Danish Dental Sch., 1939; M.S., Marquette U., 1957; m. Elin Hillersted Larsen, June 27, 1939; children—Peter, Svend A., Michael. D. Came to U. S., 1950, naturalized, 1957. Instr., Royal Danish Dental Sch., 1939-41, asst. prof., 1941-43, asso. prof., 1946-47, research asso., 1947-50; pvt. practice dentistry, Copenhagen, 1939-43, Stuer, Denmark, 1943-46; faculty Marquette U Sch. Dentistry, Milwk., 1950-64, prof., chmn. dept. dental materials, 1958-64; chief materials and tech. br., div. dental health USPHS, San Francisco, 1964—. Mem. Bio-Materials Research Adv. Com., 1961—; com. on teaching Am. Assn. Dental Schs., 1963-68. Fellow A.A.A.S., Am. Coll. Dentists; mem. Am., Danish dental assns., Federation Dentaire Internationale (sec. commn. on dental materials, instruments, equipment and therapeutics 1957—), Internat. Assn. for Dental Research (chmn. dental materials group 1958-59, member of specifications committee 1963-68, chairman 1967-68), Sigma Xi, Sigma Pi Sigma, Omicron Kappa Upsilon. Author: (with R.W. Phillips) Adhesive Restorative Dental Materials, 1961; Materials—A Programmed Review of Selected Topics (with W.J. O'Brien), 1965; also numerous articles. Research in physics, chemistry and metallurgy of dental materials, clin. research. Home: 49 Graceland Dr., San Rafael, Cal. 94901. Office: USPHS, 14th Av. and Lake St., San Francisco 94118.*

RYLANDER, Paul N., Am. chemist; b. N.Y.C., May 28, 1920; s. Paul Nels and Grace (Bull) R.; B.Ch.E., Johns Hopkins, 1942; M.A., Ph.D, Ind. U., 1948; m. Rosemary Reupert, Apr. 26, 1962. Control chemist Gen. Chem. Co., Pt. Pleasant, W.Va., 1942-43; research asst. U. Ill., 1943-45; research fellow U. Rochester, 1948-49, Harvard, 1949-51; group leader Standard Oil Co. of Ind., Whiting, Ind., 1951-56; sect. head Engelhard Inc., Newark, 1956—. Mem. Am. Chem. Soc., Chem. Soc. (London), N.Y. Catalysis Club, N.Y. Acad. Sci. Author: Catalytic Hydrogenation Over Platinum Metals, 1967. Research, publs. on applications of precious metals and other catalysts in field of organic chemistry, particularly in hydrogenation and oxidation reactions, reactions of molecules under electron impact and in organic reaction mechanisms. Home: 25 Clifton Av. Office: 497 Delaney St., Newark.*

RYNEARSON, Edward Harper, Am. physician; b. Pitts., Apr. 27, 1901; s. Edward and Rosetta Ann (Harper) R.; A.B., Ohio Wesleyan U., 1922; M.D., U. Pitts., 1926; M.S. in Medicine, U. Minn., 1931; Sc.D., U. Pitts., 1961, Sc.D., Ohio Wesleyan U., 1952; m. Lida Brickell Repp, June 26, 1928; children—Jean Harper (Mrs. Franklin Michaels), Robert Repp, Lida Ann (Mrs. Ben Cummins), Edward King. Fellow Mayo Clinic, Mayo Found., Mayo Sch. Medicine, U. Minn., Rochester, 1927-30, cons. div. medicine, 1947-55, chmn. sects. metabolic diseases, 1955-62, sr. cons. sects. metabolic diseases, 1962-66, asso. prof. Medicine, Mayo Found., 1944-50, prof., 1950-66, emeritus, 1966—. Diplomate Am. Bd. Internal Medicine. Mem. A.M.A., A.C.P., Am. Diabetes Assn., Endocrine Soc. (past pres.), Central Soc. for Clin. Research, Min. State, Zumbro Valley County med. assns. Author: (with C. F. Gastineau) Obesity, 1949; also numerous articles on endocrinology; metabolism; internal medicine.*

RYNN, Nathan, Am. physicist; b. N.Y.C., Dec. 2, 1923; s. Meyer and Rose (Wolkerwiczer) Rynkowsky; B.E.E., Coll. City N.Y., 1944; M.S., U. Ill., 1947; Ph.D., Stanford U., 1956; m. Ruth Anisfield, Sept. 3, 1954; children—Jonathan Michael, Margaret Ellen. Research engr. RCA Labs., Princeton, N.J., 1947-52; research asst. Stanford, 1952-56, research asso., 1958; mem. tech. staff Ramo Woolridge Corp., Los Angeles, 1957; research staff physicist Princeton Plasma Physics Lab., 1958-66; prof. elec. engring., physics U. Cal., Irvine, 1966—; vis. lectr. elec. engring. U. Cal., Berkeley, 1965-66. Cons. Curtiss-Wright Corp., 1964. Asso. fellow Am. Inst. Aeros. and Astronautics (com. on plasma dynamics 1965); mem. I.E.E.E. Contbr. articles to sci. jours. Inventor Q-Machine, a device for generating a manageable lab. plasma suitable for basic investigations in plasma physics; contbns. in fields of travelling wave tubes and picture tubes for color TV; patentee field color tv. Home: 1635 Eleanor Dr., Laguna Beach, Cal. 92651. Office: Sch. Engring., U. Cal., Irvine, Cal. 92644.*

RYSER, Herbert John, Am. mathematician; b. Milw., July 28, 1923; s. Fred G. and Edna (Huels) R.; B.A., U. Wis., 1945, M.A., 1947, Ph.D., 1948. Faculty, Ohio State U., 1949-62, prof., 1956-62; prof. Syracuse U., 1962-67; prof. Cal. Inst. Tech., Pasadena, 1967—; cons. Rand Corp. Mem. Math. Assn. Am., Am. Math. Soc. Research on modern algebra, matrix theory, finite geometries, combinatorial math. Home: 960 San Pasqual, Pasadena, Cal. 91106.*

RYZHKOV, Vitaliy Leonidovich, Russian microbiologist, virologist; b. June 30, 1896; postgrad. in microbiology; D.Biol. Scis. With Artem Communist U., Kharkov, 1922-30; instr. microbiology Crimean Pedagogical Inst., Simferopol, head lab. phytopathogenic viruses Ukrainian Inst. Plant Protection, 1930-33; prof., 1933—; prof., head chair gen. biology and genetics Kharkov U., 1933-37; sr. asso. Inst. Microbiology, USSR Acad. Scis., 1936—; head chair gen. biology med. inst. RSFSR Ministry Health, 1943-48. Recipient Stalin prize, 1946. Mem. USSR Acad. Scis. (corr.). Author over 200 works including Mutations and Diseases of the Chlorophyll Grains, 1933; Plant Virus Diseases, 1935; Genetics of Sex, 1936; The Principles of Plant Virus Disease Theory, 1944; Phytopathogenic Viruses, 1946; The Basic Concepts of Genetics, 1956. Co-editor Microbiology sect. Large Med. Ency., 2d edit.; mem. editorial bd. Problems of Virology. Developer (with Ye. P. Gromyko) new method to obtain purified tobacco mosaic virus, determined nature of changes in metabolism of higher plants caused by viruses, 1938. Address: Inst. Microbiology, USSR Acad. Scis., Leninsky prospect 33, Moscow, USSR.

RZESZOT, Tadeusz, Polish physicist; b. Warsaw, Poland, Nov. 28, 1927; s. Tadeusz and Stanislawa (Michalska) R.; M.Philosophy in Physics, Warsaw U., 1952; m. Barbara Ziebart, Mar. 26, 1952; 1 son, Peter. Asst., U. Warsaw, 1951-55; head lab. F.S.O. Factory, Warsaw, 1955-56, ZOPAN Factory, Warsaw, 1956-58; head lab. Inst. Nuclear Research, Swierk nr. Warsaw, 1958—. Recipient Sci. State prize of IIIth order, 1964. Research, publs. on elaboration of temperature measurement of thermal neutron flux using boron filters and measured distbn. neutron temperature across reactor core using this method; using fast chopper as pulsed neutron source of low energy neutron pulses neutron diffusion parameters in very small samples/large buckling of organic glass, epoxy resin graphite and beryllium were measured. Home: 1 A Polna St., Warsaw. Office: I.N.R. Dep. IX-A, Swierk nr. Warsaw, Poland.*

RZEWUSKI, Jan., Polish physicist; b. Lodz, Poland, Dec. 19, 1916; s. Feliks and Irma (Heinzel) R.; M.Sc., U. Warsaw (Poland), 1946, dr.sci., 1947; m. Alicja Wallner, Oct. 27, 1949; children—Marek, Maria. Asst. prof. U. Torun (Poland), 1948-52; prof. physics U. Wroclaw (Poland), 1952—, dir. Inst. Theoretical Physics, 1960—. Decorated Polonia restituta; recipient numerous awards for sci. research. Mem. Polish Acad. Scis., Sci. Soc. Torun, Sci. Soc. Wroclaw, Polish Phys. Soc., Polish Math. Soc. Author: Theoretical Mechanics, 1950; Field Theory, part I, 1958; 64, part II, 1967; also numerous articles. Research in theoretical physics, especially structure and scattering of elementary particles, math. methods especially reduction of integro-differential equations, functional methods in quantum field theory. Home: 34 Lewartowskiego, Wroclaw. Office: 36 Cybulskiego, Wroclaw, Poland.*

S

SAATY, Thomas Lorie, Am. mathematician; b. Mosul, Iraq, July 18, 1926 (parents Am. citizens); s. David M. and Dola (Hayali) S.; Ph.D., Yale, 1953; children—Linda, Michael, Emily, John. Staff, Melpar, Inc., 1953-54, Mass. Inst. Tech., 1954-57; mathematician Navy Dept., 1957-58; sci. liason officer Office Naval Research, U.S. Embassy, London, Eng., 1958-59; dir. advanced planning, Washington, 1959-61, head math. br., 1961-63; mathematician Arms Control and Disarmament Agy., Washington, 1963—; exec. dir. conf. bd. Math. Scis., 1965-67; prof., lectr. George Washington University, 1962—, Catholic U. Am., 1961—, Am. U., 1960, U. Cal., at Los Angeles, 1963-68. Ford Found. cons. Nat. Planning Council, Cairo, Egypt, 1959-64. Mem. Am. Math. Soc. Math. Assn. Am., Operations Research Soc., French Operations Research Soc. Author: Mathematical Methods of Operations Research, 1959; Elements of Queueing Theory, 1961; Nonlinear Mathematics (with J. Bram), 1964; (with R. Busacker) Finite Graphs and Networks, 1965; Modern Nonlinear Equations, 1967; also articles. Editor: Lectures on Modern Mathematics, 3 vols., 1964-65. Devel. linear programming, queueing theory, theory graphs, basic methods arms control and disarmament. Home: 7814 Marion Lane, Bethesda, Md. 20014. Office: Arms Control and Disarmament Agy., Washington 20451.*

SABATIER, Armand, French anatomist; b. Ganges, France, Jan. 14, 1834; intern Lyons Hosp., 1858; agrégé, prof. Acad. Medicine, Montpellier; prof. zoology and medicine U. Montpellier; founder, dir. zool. sta., Cette, France; corr. French Acad. Scis., 1895. Author: An Anatomical, Physiological and Clinical Study of Ausculation of the Lungs of Children, 1863; A Comparison of the Mid-Regions and the Anterior and Posterior Membranes in the Series of the Vertebrates, 1880; Collection of Writings on the Morphology of the Sexual Organs and on the Nature of Sensuality, 1887; Philosophy of Effort, 1904. Died Montpellier, Dec. 22, 1910.

SABATIER, Paul, French chemist; b. Carcassonne, France, Nov. 5, 1854; ed. École Normale; agrégé in physics, 1877; D.Sc., 1880; tchr., secondary sch., Nimes, France, 1877; became lab. asst. to Berthelot, Coll. de France, 1880; lectr., Bordeaux, France; became mem. faculty, Toulouse, France, 1882, titular prof. chemistry, 1884, dean, 1905; mem. Inst. Recipient (with Grignard) Nobel prize in chemistry, 1912. Fellow Royal Soc., 1918 (Davy medal 1915). Author: Thermal Research on Sulfur, 1880; Elementary Lessons in Agricultural Chemistry, 1889; La Catalyse en chimie organique, 1913. Research on catalytic behavior of oxides as oxidizing catalysts, also as dehydrating and dehydrogenating agts.; discovered (with Senderens) process for catalytic hydrogenation of oils to solid fats; developed theory of formation of a temporary unstable combination of catalysts in gen. Died Vichy, France, Aug. 16, 1941.

SABATIER, Raphael-Bienvenu, French surgeon; b. Paris, Oct. 11, 1732; prof., anat. demonstrator Coll. de France; tchr. of Dessault; pvt. surgeon to Napoleon; mem. French Acad. Scis., 1733, (pres. 1799; mem. Acad. Surgery, Acad. Marine. Author: Traité d'anatomie, 13 vols., 1764-81; De la médecine operatoire, 3 vols., 1796. Died Noisy-le-Roi, July 19, 1811.

SABIN, Albert B(ruce), microbiologist, physician; b. Bialystok, Russia, Aug. 26, 1906; s. Jacob and Tilly (Krugman) S.; came to U. S., 1921; naturalized, 1930; student Coll. of Dentistry, New York U., 1924-26; B.S., N.Y. U., 1928, M.D., 1931, D.Sc., 1959; L.H.D., Hebrew Union Coll. and College-Conservatory of Music, Cin., 1960; D.Sc., Temple U., Ohio State U., 1961; Bowling Green State University, 1961, Miami University, Oxford, Ohio, 1963; Dr. Hon. Causa, U. Brazil, 1961; Ph.D. honoris causa, Hebrew U., Jerusalem, 1964; D.Sc., University of Newcastle (England), 1965, Albert Einstein College of Medicine, 1967; hon. degrees U. Gama Filho, Rio de Janeiro, U. Nat. de la Plata, Argentina, U. Sao Paulo, 1967; married Sylvia Tregillus, September 12, 1935; (deceased August 1966); children—Deborah, Amy. m. second, Jane Blach Warner, June 19, 1967. Research asso. bacteriology N.Y.U. Coll. Med., 1926-32; house physician Bellevue Hosp., N.Y.C., 1932-34; fellow in med., Nat. Research Council, Lister Inst., Eng., 1934; asst. Rockefeller Inst., N.Y.C., 1935-37, associate, 1937-39; asso. prof. of pediatrics, Univ. of Cincinnati Coll. of Medicine, 1939-46, professor of research pediatrics, 1946-60, Distinguished service professor since 1960; professor (honorary) University of Chile, U. Buenos Aires. Fellow Children's Hospital Research Foundation since 1939. Consultant to secretary of war on epidemic diseases, and mem. of U. S. Army Commn. for neurotropic Virus Diseases, 1941-61; member Armed Forces Epidemiological Bd., 1963—. Bd. govs. Hebrew U., Jerusalem, Weizmann Inst. Science, Israel; mem. bd. of trustees N.Y. U. Commd. maj., Medical Corps, Army of United States, to serve with bd. for Investigation and Control of Epidemic Diseases in the Army, Preventive Medicine Division, Office of Surgeon General, 1943; promoted to lieutenant colonel, 1944. Decorated Legion of Merit; recipient of the Theobald Smith award in medical science ($1,000 and medal) from A.A.A.S. chiefly for research in pneumonia and infantile paralysis, 1939; E. Mead Johnson award for research on virus diseases of nervous system, Am. Acad. of Pediatrics, 1941; T. Ricketts award, U. of Chgo., 1959; Sachs award, Cin. Inst. Fine Arts, 1960; James D. Bruce Meml. award A.C.P., 1961; Robert Koch medal, West Germany, 1963; Oscar B. Hunter award Am. Therapeutic Soc., 1963; Order of Merit, Bavaria, 1964; Liceago gold medal, Mexico, 1964; Antonio Feltrinelli prize ($40,000), Accademia dei Lincei, Rome, 1964; Gold Medal for merit in pub. health, Italy, 1965; Lasker award, 1965; Dag Hammarskjold award Internat. Assn. Diplomatic Corps, 1965; Scopus award American Friends Hebrew U., 1965; Gran Ofcl. Orden. Hipdito Unanue, Peru, 1966; Order Yugoslav Flag with ribbon, 1966; Matricola d'Onore, University of Pavia, 1967; also Murray-Green award from the AFL-CIO, 1967, Fellow Am. Arts and Sci., Royal Soc. Health Eng.; member National Acad. of Sciences, American Society Tropical Medicine and Hygiene, Society Pediatric Research, Soc. Am. Bacteriologists, Soc. Exptl. Biology and Medicine, Soc. Clin. Investigation, Am. Pediatric Soc., Am. Epidemiol. Soc., Am. Assn. Immunologists, Harvey Soc., Alpha Omega Alpha, Sigma Xi; corr. mem. Brit. Paediatric Assn., Royal Acad. Medicine (Belgium); hon. mem. Soc. Microbiologists, Epidemiologists and Infectionists (USSR), Brazilian Pediatric Soc., Med. Soc. Santiago, Academia Nacional de Medicina (Brazil). Contbr. numerous papers to jours. on pneumococcus infection, infantile paralysis, virus diseases of nervous system, and other topics relating to various infectious diseases. Discoverer B virus, reoviruses and many enteroviruses and their role in human diseases; isolation and cultivation of sandfly fever and dengue viruses and preparation of vaccines against them; studies hemagglutination by neurotropic viruses; elucidation of natural history of encephalitis and poliomyelitis; devel. oral polio vaccine; dye test for toxoplasmosis; new phenomena in cancers experimentally produced by DNA viruses. Home: 7420 E. Aracoma Dr., Clin. 45237. Office: The Children's Hospital Research Foundation, Elland and Bethesda Av., Cin. 45229.*

SABIN, Florence Rena, Am. anatomist; b. Central City, Colo., Nov. 9, 1871; d. George Kimball and Rena (Miner) Sabin; B.S., Smith Coll., 1893, Sc.D., 1910; M.D., Johns Hopkins, 1900; Sc.D., U. Mich., 1926, Mt. Holyoke, 1929, N.Y. U., 1933, Syracuse U., 1934; other hon. degrees; Tchr. math. Wolfe Hall, Denver, 1893-94; asst. in zoology, Smith Coll., 1895; intern Johns Hopkins, 1900-01; fellow Balt. Assn. for Advancement of Univ. Edn. of Women, 1901; asst. in anatomy Johns Hopkins, 1902, asso., 1903-05, asso. prof., 1905-17, prof. histology, 1917-25; mem. Rockefeller Inst. for Med. Research, 1925-38; mem. emeritus since 1938. Recipient Nat. Achievement award, 1932; M. Carey Thomas prize, 1935; Lasker award, 1951; Bronze Statue placed in her honor in Statuary Hall, Washington. Mem. Am. Assn. Anatomist (v.p. 1908-09; president 1924-26), Am. Assn. Physiologists, Soc. Exptl. Biology and Medicine (councilor 1932), Nat. Tb Assn., Nat. Acad. Scis.; hon. life member N.Y. Acad. Scis., Harvey Soc., A.A.A.S. Author: An Atlas of The Medulla and Mid-Brain, 1901; Biography of Franklin Paine Mall, 1934; also articles on lymphatic system, on blood vessels and origin of blood cells, on blood and bone marrow, on Tb. Discoverer origin and process of lymphatic system; developed method of studying living cells. Died Oct. 3, 1953.

SABINE, Sir Edward, Brit. physicist, astronomer; b. Dublin, Ireland, Oct. 14, 1788; s. Joseph and Sarah (Hunt) S.; mil. edn.; D.C.L., Oxford, 1855; LL.D., Cambridge; m. Elizabeth Juliana Leeves, 1826; career as soldier; astronomer to N.W. passage expdns. of John Ross, 1818, William Edward Parry, 1818-20; Fellow Royal Soc., 1818 (pres. 1861-71), Linnean Soc., Royal Astron. Soc.; mem. French Acad. Scis., 1875. Author: Observations on Days of Unusual Magnetic Disturbance, 1843; On the Cosmical Features of Terrestrial Magnetism, 1862; also articles; editor various works. Carried out pendulum expts. to determine shape of earth, Spitsbergen, also tropical Africa; discovered interrelation between magnetic disturbances on earth and periodic variation on sunspots, 1852. Died Richmond, Eng., June 26, 1883.

SABINE, Joseph, Brit. naturalist; b. Tewein, Eng., 1770; s. Joseph and Sarah (Hunt) S.; ed. in law, practiced until 1808; insp. gen. of assessed taxes until 1835. Recipient Gold medal Hort. Soc. 1816; leguminous genus Sabinea named in his honor. Fellow Royal Soc., 1779, Linnean Soc. (original fellow); hon. sec. Hort. Soc., 1810-30. Contbr. appendix to Narrative (Sir John Franklin), 1823, list of plants to History of Hertfordshire (Clutterbuck), 1815, also papers to trans. of socs. Contbd. to establishment of gardens of Hort. Soc., helped found various local socs. in connection with the soc.; sent David Douglas, others as collectors; authority on Brit. birds, their habits, migration, moulting. Died Jan. 24, 1837.

SABINE, Wallace Clement, Am. physicist; b. Richwood, O., June 13, 1868; s. Hylas and Anna (Ware) S.; A.B., Ohio State U., 1886; A.M., Harvard, 1888, D.S., 1914; D.Sc., Brown U., 1907; m. Jane Downs Kelly, Aug. 22, 1900; 2 daus. Asst. physics, Harvard, 1889-90, instr., 1890-95, asst. prof., 1895-1905, prof., 1905-—, became dean Lawrence Sci. Sch., 1906, dean Grad. Sch. Applied Sci., 1908; prof. agrégé U. Paris, 1916-17. Fellow Am. Acad. Arts and Scis., A.A.A.S.; mem. Nat. Acad. Scis., A.I.A. (corr.). Author: Architectural Acoustics; Collected Papers on Acoustics, 1922. Discovered Sabine's law of acoustics (basis of archtl. design of acoustically useful rooms). Died Jan. 10, 1919.

SABISTON, David Coston, Jr., Am. surgeon; b. nr. Jacksonville, N.C., Oct. 4, 1924; s. David Coston and Marie (Jackson) S.; B.S., U. N.C., 1943; M.D., Johns Hopkins, 1947; m. Agnes Barden, Sept. 24, 1955; children—Anne Barden, Agnes Foy, Sarah Coston. Practice medicine, specializing in surgery, Balt., 1955-64, Durham, N.C., 1964-—; faculty Johns Hopkins, 1955-64, Howard Hughes investigator, 1955-60; Fulbright Research scholar U. Oxford (Eng.), 1960; research asso. Hosp. For Sick Children U. London, 1961; Career Research award NIH, 1962-64; prof., chmn. dept. surgery Duke Med. Center, 1964-—. Mem. Soc. U. Surgeons, Am., So. surg. assns., A.C.S., Am. Assn. Thoracic Surgery, Soc. Clin. Surgery, Internat. Soc. Cardiovascular Surgery, Soc. Vascular Surgery, Halsted Soc., Surg. Biology Club II, Soc. Thoracic Surgery, Soc. for Surgery Alimentary Tract, Soc. Thoracic Surgeons Gt. Britain and Ireland. Editor: Monographs in Surg. Scis., 1964. Editorial bd. Lewis-Walters Practice of Surgery, 1963-—; Jour. Cardiovascular and Thoracic Surgery, 1964-—; Annals of Surgery, 1965-—; Circulation, 1966-—. Research, publs. on cardiovascular physiology, cardiac surgery and pulmonary circulation. Home: 1528 Pinecrest Rd. Office: Duke Med. Center, Durham, N.C. 27706.*

SABOURAUD, Raymond Jacques-Adrien, French physician, dermatologist; b. Nantes, France, 1864; intern, later dir. lab. St.-Louis Hosp., Paris; used X-rays to determine origin and treatment of several scalp diseases; suggested x-ray treatment of ringworm, 1904; studied seborhea, baldness, eczema. Died Paris, 1938.

SABRA, Fuad Amin, Lebanese physician; b. Beirut, Lebanon, Feb. 13, 1919; s. Amin Elias and Asma (Richa) S.; B.S., Am. U. Beirut, 1938, M.D.; m. Ellen Habib Badr, June 23, 1949; children—Amin, George, Ramzi. Faculty, Am. U. Beirut, 1945-46, 1948-—, asso. prof. medicine, 1952-62, acting chmn. dept. internal medicine, 1963-—, head div. neurology and E.E.G., 1949-—; Rockefeller fellow med. research Neurol. Inst., Columbia U., N.Y.C., 1947-48, 44-46, Neurol. Inst., McGill U., Montreal, Que., Can., also Mass. Gen. Hosp., Boston, 1955-56. Named Knight Order Cedar. Fellow A.C.P.; mem. Lebanese Soc. Neurology and Psychiatry (pres. 1962-—), World Fedn. Neurology (permanent del. 1962-—), Am. EEG Soc. (hon.), Assn. Research in Nervous and Mental Disease, Internat. Coll. Surgeons, Fedn. Neuro. Rehab. (founding), N.Y. Acad. Scis., Brit. Med. Assn., Am. Neurol. Assn., Alpha Omega. Contbg. author: Text Book of Medicine, 1966. Research and publs. on symptomatology of cerebrovascular malformations, clin. diagnosis of inclusion body encephalitis, heredofamilial cerebellar disease simulating clin. findings of multiple sclerosis. Home: Makdisi, Beirut, Lebanon.*

SABROSKY, Curtis Williams, Am. entomologist; b. Sturgis, Mich., Apr. 3, 1910; s. Herman August and Jennie (Curtis) S.; A.B., Kalamazoo Coll., 1931, Sc.D. (hon.), 1966; postgrad. U. Mich.; M.S., Kan. State U., 1933. Faculty, Mich. State U., 1936-44, asst. prof. entomology, 1944; research entomologist U. S. Dept. Agr., Washington, 1946-—. Mem. Internat. Commn. on Zool. Nomenclature, 1963-—, Permanent Com. for Internat. Congresses Entomology, 1960-—. Recipient Superior Service award U.S. Dept. Agr., 1962; Distinguished Service award Kan. State U., 1965. Fellow A.A.A.S.; mem. Am. Inst. Biol. Scis., Entomol. Soc. Am. (life), Am., Kan. entomol. socs., Entomol. Soc. Can., Soc. Systematic Zoology (past pres.), Entomol. Soc. Washington, Sociedade Brasileira de Entomologia. Research, numerous publs. on classification and recognition of flies of certain families, zool. nomenclature. Home: 5426 Lincoln St., Bethesda, Md. 20034. Office: Systematic Entomology, U.S. Dept. Agr., U. S. Nat. Mus., Washington 20560.*

SACCARDO, Francesco, Italian botanist; b. Trevist, Italy, 1845; titular prof., Padua, Italy. Author: Sylloge fungorum, 22 vols., to 1913; also iconography of 1500 varieties of microscopic mushrooms, 1877-78. Died 1920.

SACCHERI, Girolamo, Italian mathematician; b. San Remo, Italy, Sept. 5, 1667; mem. Soc. Jesus; tchr. theology, logic, metaphysics, math. in Jesuit schs. at Turin, Pavia, Milan, Italy. Author: Quaesita geometrica, 1693; Logica demonstrativa, 1701; Euclides ab omni naevo vindicatus, 1733. First to discuss consequences of denying parallel axiom (Euclid's 5th postulate), and to suggest constrn. of non-Euclidean geometry independent of it. Died Milan, Oct. 25, 1733.

SACHER, Joseph Albert, Am. plant physiologist; b. Irvington, N.J., Dec. 27, 1918; s. George Albun and Margaret (Reilly) S.; B.S., N.Y. State Coll. Forestry, Syracuse U., 1941; Ph.D., U. Cal. at Berkeley, 1953; m. Mary Virginia Gayman, Mar. 7, 1946; 1 son, David Alan. Instr., U. Ill., Urbana, 1953-55; faculty Cal. State Coll., Los Angeles, 1955-—, prof. botany, 1962-—, chmn. dept., 1964-—; research fellow div. biology Cal. Inst. Tech., Pasadena, 1956-59. Mem. Am. Bot. Soc., Am. Soc. Plant Physiologists, A.A.A.S., Sigma Xi. Research, publs. on histology of shoot and leaf devel., mechanism of enzyme action, role extracellular enzymes, regulation of sugar uptake and accumulation, physiology of tissue senescence, hormonal control of nucleic acid and protein synthesis. Home: 4155 Aralia Rd., Altadena, Cal. 91001. Office: 5151 State College Dr., Los Angeles 90032.*

SACHS, Allen Maxwell, Am. physicist, educator; b. N.Y.C., July 13, 1921; s. George M. and Elsa (Shenfield) S.; B.A., Harvard, 1942, M.A., 1947, Ph.D., 1949; m. Judith Morrison, Dec. 19, 1949; children—Carolyn, George, Marjorie. Faculty physics dept. Columbia, N.Y.C., 1949-—, prof., 1960-—, chmn. dept., 1968-—. Decorated Bronze Star. Fellow Am. Phys. Soc.; mem. Am. Assn. Physics Tchrs., Phi Beta Kappa. Research in high energy particle physics. Home: 51 Belden Av., Dobbs Ferry, N.Y. Office: Columbia, N.Y.C. 10027.*

SACHS, Joseph, Am. engr., inventor; b. N.Y.C., Aug. 17, 1870; s. Louis Von and Bertha (Sanger) S.; student Coll. City of N.Y.; m. Caroline Norman, June 5, 1895; children—Kelvin N., Margaret N. (Mrs. J. J. Bissell). Began with Sprague Elec. Motor Co. prior to 1890, later with Edison Machine Works (now Gen. Electric Co.), Schenectady; developed trolleyless elec. ry., nonarching enclosed fuse, elec. ry. appliances, 1892-94; cons. patent expert, also engaged in constrn. and operation engring., work and pioneer magnetic ore separating plant, 1892-98; developed electric fire alarm signalling system, elec. fire engine system, canal boat haulage, cable and elec. ry. signals, elec. type-setting machinery, electric metal heating and melting, 1893-98, elec. drive and control for motor vehicles, electrically controlled carburetor, elec. primer and other automobile accessories, 1900-15; cons. and chief engr. Johns-Pratt Co. (later elec. div. Colts Patent Fire Arms Mfg. Co.), mfrs. elec. devices, Hartford, Conn.; pres. and mgr. The Sachs Co., and v.p. and mgr. Sachs Labs., Inc.; identified with devel., mfr. and sale of Sachs inventions, including elec. protective devices, enclosed fuses, cutouts, switches, sockets, circuit breakers, enclosed safety switches, meter service switches, meter testing devices, standardized service installations, 1898-1937; research and invention, also engring. and devel. cons. for several mfg. cos. since 1939; awarded over 250 U. S. patents. Recipient John Scott Legacy medal Franklin Institute for pioneer invention electric fuse protective devices, 1903. Fellow Am. Inst. E.E.; mem. Am. Soc. M.E., Nat. Elec. Mfrs. Assn. Author: (with T. C. Martin) Electrical Boats and Navigation, 1894. Contbr. articles to tech. publs. Died Nov. 15, 1946.

SACHS, Julius von, see von Sachs, Julius.

SACHS, Leo, biologist; b. Leipzig, Germany, Oct. 14, 1924; s. Elijah and Louise (Lichtblau) S.; Ph.D., Cambridge U., 1951; m. Rachel Eliash, Dec. 12, 1950; children—Shlomit, Judith, Mordechai, Naomi. Research scientist John Innes Instn., 1951-52; mem. sci. staff Weizmann Inst. Sci., Rehovoth, Israel, 1952-—, prof., head sect. genetics, 1962-—. Research, numerous publs. on mechanism of carcinogenesis and of normal cell differentiation, cytogenetics. Address: Weizmann Inst. Sci., Rehovoth, Israel.*

SACHS, Robert Green, Am. physicist; b. Hagerstown, Md., May 4, 1916; s. Harry Maurice and Anna (Green) S.; Ph.D., Johns Hopkins, 1939; m. Jean K. Woolf, Dec. 17, 1950 (dec.); children—Judith, Joel, Rebecca, Jennifer, Jeffrey. Fellow, Washington U., 1939-41; instr. physics Purdue U., 1941-43; research fellow U. Cal. at Berkeley, 1941, cons. radiation lab., 1955-59; sect. chief Ballistic Research Lab., Aberdeen (Md.) Proving Ground, 1943-45; dir. theoretical physics div. Argonne (Ill.) Nat. Lab., 1945-47, cons., 1947-50, asso. dir. lab., 1964-—; asso. prof. physics U. Wis., 1947-48, prof., 1948-64; prof. U. Chgo., 1964-—; vis. prof. Princeton, 1955-56, U. Paris 1959-60. Mem. adv. panel physics NSF; high energy physics adv. panel AEC. Guggenheim fellow, 1959-60. Fellow Am. Acad. Arts and Scis.; mem. A.A.A.S., Am. Phys. Soc., Phi Beta Kappa. Author: Nuclear Theory, 1953. Research, publs. in theoretical high energy physics, field theory, theoretical nuclear physics. Home: 5490 S. Shore Dr., Chicago 60615. Office: Enrico Fermi Inst. Nuclear Studies, U. Chgo., Chgo. 60637.

SACHS, Volkmar Robert, German physician; b. Küstrin, Germany, Dec. 7, 1922; s. Ernst Emil and Gertrude (Siegert) S.; student medicine U. Berlin, 1941, U. Hamburg, U. Kiel, 1943-44; student medicine and philosophy U. Heidelberg, 1945-48; approbation in medicine, grad. in medicine U. Kiel, 1949; m. Ursula Keara Luise Straube, Oct. 16, 1948; children—Heinke Hildburg-Elfriede, Tilmann Ernst-Georg. Asst. dept. medicine U. Kiel, 1949-52, asst. Inst. Forensic Medicine, 1952-59, research staff Inst. Hygiene, 1961-—, dir. dept. blood transfusion, 1961-—, lectr. on blood group serology and blood transfusion, 1963-—, privat-docent, 1967-—; asst. Central Inst. for Blood Transfusion Hamburg, 1959-60. Mem. Internat., German socs. blood transfusion, German Soc. Scientists and Physicians, Inst. Nuclear Energy (mem. sci. com. nuclear hematology 1964-—). Contbg. author: Nuclear Hematology, 1965. Research, publs. on hematology, alcohol metabolism, chromometric determination of carbon monoxide in blood, determination of tissue alchol, genetics of blood and serum groups, blood transfusion and gen. hemotherapy, math. statistics and immunohematology. Home: 73a Forstweg. Office: Brunswiker Strasse 2-6, 23 Kiel, Germany.*

SACKLER, Arthur Mitchell, Am. physician; b. Bklyn., Aug. 22, 1913; s. Isaac and Sophie Sackler; B.S., N.Y. U., 1933, M.D., 1937; m. Else Jorgensen, 1935; children—Carol Ingrid, Elizabeth Ann; m. 2d, Marietta Lutze, 1949; children—Arthur Felix, Denise Marica. House physician and pediatrician Lincoln Hosp., N.Y.C., 1937-39; resident psychiatry Creedmoor State Hosp., Queens Village, L.I., 1944-46; dir. research Creedmoor Inst. Psychobiol. Studies, 1949-54; with William Douglas McAdams, Inc., pharm. advt., N.Y.C., 1941-—, pres., 1942-57, chmn. bd., 1957-—; dir. Labs. for Therapeutic Research, Bklyn. Coll. Pharmacy, L.I. U., 1957-—; chmn. bd. Med. Press, Inc., 1954-—; pres. Physicians News Service, Inc., 1955-—, Med. Radio and Television Inst., Inc., 1955-—, Medimetric Inst., Inc., 1955-—; Asso. chmn. internat. com. research First Internat. Congress Psychiatry, 1950. Recipient awards for sci. research exhibits Med. Soc. State N.Y., 1950, 52; named officer Carlos J. Finlay Nat. Order Merit (Cuba), 1958. Fellow Am. Psychiat. Assn., Am. Geriatrics Soc.; mem. Gerontological Soc., A.A.A.S., N.Y. Acad. Scis., World Med. Assn. (U. S. com.), Assn. Med. Advt. Agencies (pres. 1956), Chinese Art Soc. (bd. govs.). Co-author: Great Physiodynamic Therapies, 1956. Editor in chief Jour. Clin. and Exptl. Psychopathology, 1950-62. Pioneer in development of histamine, sex steroids, other biochemotherapies; predicted psychosis resulting from use of cortisone and ACTH. Home: Searingtown Rd., R.F.D. Roslyn, L.I., N.Y. Office: 130 E. 59th St., N.Y.C. 10022.*

SACKS, Jacob, Am. biochemist; b. Chgo., Sept. 27, 1901; s. Mendel and Libby (Braude) S.; S.B., U. Chgo., 1922, S.M., 1924; Ph.D., U. Ill., 1926; M.D., Northwestern U., 1932; m. Wilma R. Cohn, Sept. 5, 1926. Instr., U. Tenn., 1922-24; research instr. Northwestern U., 1928-31; instr. pharmacology U. Mich., 1932-37, asst. prof., 1937-44; head pharmacology lab. Endo Products, Inc., 1944-47; physiologist Brookhaven Nat. Lab., Upton, N.Y., 1947-52; asso. prof. U. Ark., Fayetteville, 1952-55, prof. chemistry, 1955-—; vis. prof. U. El Salvador, 1960-61. Recipient Distinguished Research award U. Ark., 1965. Mem. Am. Chem. Soc. (S.W. regional award 1963), A.A.A.S., Am. Soc. Physiology, Am. Soc. Pharmacology and Exptl. Therapeutics, Soc. Exptl. Biology and Medicine. Author: The Atom at Work, 1951; Isotopic Tracers in Biochemistry and Physiology, 1953. Research in field of histamine and gastric secretory hormone, mechanism of insulin action, chemistry of muscular contraction. Home: 873 California Blvd., Fayetteville, Ark. 72701.*

SACKTOR, Bertram, Am. biochemist; b. N.Y.C., May 11, 1922; s. Elias and Mollie (Miller) S.; B.S., Cornell U., 1943; M.S., Rutgers U., 1947, Ph.D., 1949; m. June Dale Charlton, Dec. 27, 1942; children—Todd C., Ned C. Research fellow Rutgers U., 1946-49; chief biochemist, dir. med. research Edgewood Arsenal, Md., 1949-67; chief sect. on intermediary metabolism NICHD, Nat. Insts. Health, Balt. City Hosps., 1967-—. Research asso. Johns Hopkins, 1964-—; prof. biochemistry U. Md., University College 1963-65. Recipient U. S. Army Superior Performance award, 1959; Travel award Internat Congress Biochemistry, Vienna, Austria, 1958, Moscow, Russia, 1961, Tokyo, 1967. Fellow A.A.A.S.; mem. Am. Soc. Biochemistry, Am. Chem. Soc., Soc. for Cell Biology, Soc. Gen. Physiologists, N.Y. Acad. Scis. Research, numerous publs. on metabolic pathways and control factors in tissues in transition from resting to physiologically active state. Home: 6502 Old Pimlico Rd., Balt. 21209. Office: Gerontology Research Center, NICHD, Nat. Insts. Health, Balt. City Hosps., Balt. 21224.

SACREZ, Robert, French physician; b. Fumay, Ardennes, France, Oct. 20, 1907; s. Louis and Marguerite (Renard) S.; Med. doctorate Strasbourg (France) U., 1937, Licence ès Sciences, 1937; Prof. pediatrics, dir. Children's Clinic, physician, dir. pediatric studies Strasbourg Hosp., 1947-—. Named Officer d'Academie, 1949. Mem. French socs. biology, biol. chemistry and pediatrics, Royal Soc. Medicine, Internat. Soc. Internal Medicine, Swiss Soc. Pediatrics (corr.). Author: (with others) Traite de Pathologie Infantile, 1945; also numerous articles. Research on vitamin c, digestive disorders, enzymic illness, jaundice; anat. and clin. studies of agenesia of intrahepatic bile ducts, reticulo-endotheliosis of newborn infant, fibro-elastosis of endocardium; research in dietetics. Home: 2 Quai Fustel de Coulanges, Strasbourg. Office: Clinque de Pediatrie et Puériculture, Hospices Civils de Strasbourg 1, Place de l'Hôpital, France.*

SACROBOSCO, Johannes de, see John of Hollywood.

SADE, Albert J. V., French algebraist; b. Pertuis, France, May 3, 1898; s. Victor and Claire (Guerin) S.; hon. prof.; m. Marie-Louise David, July 1927. University Prof. Recipient, Order of Public Instruction. Mem., Am. Math. Soc.; Italian Math. Union; Brussels Sci. Soc.; Palermo Math. Club; Austrian Math. Soc. Author numerous published articles and papers. Work in study of general theory of quasi-groups, morphisms, autopies, demosian systems, parastrophics and half-symmetry. Home: 364 Cours de la République, Pertuis 84, France.*

SADEK, Hussein, Egyptian chemist; b. Cairo, Egypt, Jan. 15, 1917; s. Mohamed and Fatma Sadek; B.Sc. with honours, Cairo U., 1938, M.Sc., 1941; Ph.D., Princeton, 1948; m. Soad Khaled, Aug. 14, 1941; children—Ahmed Rafig, Mohamed Hani, Hafez El-hamy, Amira. Demonstrator chemistry Cairo U., 1938-43; faculty Alexandria (Egypt) U., 1943—, prof. phys. chemistry, 1956—. Cons. Ministry Sci. Research, 1964; mem. Com. of Nat. Prizes in Chemistry, 1965. Decorated Order of Merit in Scis. and Arts, 1962. Mem. Faraday Soc., Am. Chem. Soc. Author: Freshman's Notes in Physical Chemistry, 1956, rev. (with Th.F. Tadros), 1966; also numerous articles. Research on ionic investigations of weak acid solutions, stability of oxides and hydrolysis of acetates in acids. Home: 62 Moustafa Kamel, Alexandria, Egypt.*

SADHU, D(ulal) P(ada), Indian physiologist; b. Khulna, India, May 9, 1916; s. Abinash Chandra and Induprabha S.; student N.R.S. Med. Coll., Calcutta; Licentiate, Med. Faculty, Bengal, India, 1937; B.Sc. with Honors, Pres. Coll., Calcutta, 1942, M.Sc. in Physiology, 1944; Ph.D., Mo. U., 1947; m. Lekha Saha, Feb. 24, 1949; children—Bani, Nachiketa, Anirundha. House physician, clin. asst. N.R.S. Med. Coll., 1937-38, 39-40; practice gen. medicine, 1938-39; lectr. physiology Pres. Coll., 1944-45, asst. prof., 1948-49, 49-53; asst. research officer Indian Council Med. Research Scheme, 1948, sr. scientist Office Def. Sci., Kanpur, India, 1948; prof. physiology and nutrition Bengal Vet. Coll., Calcutta, 1953-61; prin. P.G. Coll. Animal Sci., Izatnagar, India, 1961—. Recipient Silver medals for proficiency in medicine N.R.S. Med. Coll., 1937, Gold medal U. Calcutta, 1942. Fellow Nat. Acad. Scis. India, Nat. Inst. Sci. India; mem. Societe de Chimie biologique, Physiol. Soc. India, Zool. Soc., Soc. Biol. Chemists India, Indian Soc. for Study Reprodn., Nutrition Soc. India. Research and numerous publs. on rumen and reproductive physiology, especially effect of thermal stress on semen qualities; discovered role of deamination and transamination in mechanism of protein action in rat and ruminants, stenothermal stage in evolution of thermoregulation in animals; studied thyroid and vitamin A physiology, cholesterol metabolism, milk sugar in nutrition. Home: Dak Bungalow Rd., Katwa, Bengal, India. Office: P.G. Coll. Animal Scis., I.V.R.I., Izatnagar, U.P., India.*

SADOVE, Max Samuel, Am. physician; b. Balt., Mar. 8, 1914; s. Harry and Rebecca (Must) S.; B.S., Balt. City Coll., 1932; Pharm.B., U. Md., 1937, M.D., 1939; m. Ethel Segall, Apr. 6, 1941; children—Ellen Rose, Alan Michael, Richard Craig. Intern St. Agnes Hosp., Balt., 1939-40; resident Hines (Ill.) VA Hosp., 1946-47; mem. faculty U. Ill. Coll. Medicine, 1947—, prof. anesthesiology, 1951—, chmn. dept., 1949—; head dept. anesthesiology U. Ill. Research and Ednl. Hosps., 1950—; dir. anesthesiology dept. W. Side VA Hosp., Chgo., 1954—, Hines VA Hosp., 1954—; cons. numerous hosps. Chgo. area VA; hon. chmn. dept. anesthesia U. Philippines Coll. Medicine, 1957—; member adv. bd. Chgo. Narcotic Rehab. and Research Center; adviser Med. Council Civil Def., Chgo. Bd. Health; med. lectr. Great Lakes Naval Hosp. Diplomate Am. Bd. Anesthesiologists. Fellow A.A.A.S., Hines Surg. Assn., Soc. U. Profs., Am. Soc Anesthesiologists (chmn. sub. com. motion pictures 1958—), Am. Coll. Clinical Pharmacology and Chemotherapy; member Society of Med. Consultants to Armed Forces, Internat. Coll. Surgeons (vice regent), Chgo. Soc. Anesthesiologists (pres. 1950-51), Walter Reed Soc. (pres. 1954), Am. Soc. Assn. Inhalation Therapists (bd. govs., adv. bd. 1956-57), Am. Coll. Chest Physicians (chmn. sect. resuscitation 1957—), Am. Fedn. Clin. Research, A.M.A., Ill. Med. Soc., Chgo. Med. Soc., Ill. Soc. Anesthesiologists, Soc. Med. History Chgo., Ill. Soc. Med. Research, Inst. Medicine Chgo., Am. Med. Authors; hon. mem. American Dental Society of Anesthesiologists, also Flint, Hollywood acads. medicine; corr. fellow Soc. Mexicana de Anesthesiology, Cuban Soc. Anesthesiologists. Co-author: Recovery Room, 1956; Cardiovascular Collapse in the Operating Room, 1958; Halothane, 1962; also numerous articles. Contbr. to books, cons. editor med. jours. Basic research analgesia (pain relief), anesthesia, shock, resuscitation, post operative techniques. Home: 1021 Lathrop St., River Forest, Ill. 60305. Office: 840 S. Wood St., Chgo. 60612.*

SADOVSKII, Mikhail Aleksandrovich, Russian physicist; b. Nov. 6, 1904; grad. Leningrad Poly. Inst., 1928. Staff, Seismol. Inst., USSR Acad. Scis., 1930-41, staff presidium, 1941-46, joined staff Inst. Chem. Physics, 1946, apptd. dir. Shmidt Inst. Terrestrial Physics, 1961. Decorated Hero of Socialist Labor, 1949; Recipient Stalin Prize, 1948. Corr. mem. USSR Acad. Scis. Research and publs. on mechanics of explosions; developed devices for studying blast effects; confirmed law of similarity during explosion. Office: Inst. Chem. Physics, USSR Acad. Scis., Vorob'evskoye, Shosse 2, Moscow, USSR.

SADRON, Charles Louis, French biophysicist; b. Cluis, France, May 12, 1902; s. Jacques and Louise Marie (Bougault) S.; ed. U. Poitiers (France), U. Strasbourg (France); Licencié-ès-Sciences, Agrégé Université, 1926; Docteur-ès-Sciences, 1934; m. Marie-Louis Eck, Dec. 28, 1931 (dec.); 1 dau. (dec.); m. 2d, Geneviève Aubel, Mar. 3, 1959. Prof., Lycée Troyes, 1927-38; prof. Lycée Kléber Strasbourg, France, 1928-31; with Coll. Services, Rech. Aéronautique, 1931-33; boursier Rockefeller Inst. Tech., Pasadena, 1933-34; staff Caisee Nat. Recherche Scientifique, 1934-43; prof. U. Strasbourg, 1943-60; prof. Muséum Nat. d'Histoire Naturelle, Paris, 1961—; dir. Centre de Recherches sur les Macromolécules, Strasburg, 1947-67, Centre de Biophysique Moléculaire, Orléans, France, 1967—. Decorated Officer Ordre Nat. Legion Honor, Comdt. Order Palmes Acad., Lauréat Inst. (Fond. Hirn); recipient Prix Holweck, Phys. Soc. London. Fellow N.Y. Acad. Sci.; mem. Société de Physique, Société de Chimie Physique, Société Chimie Biologique (v.p.), Sté Fr. Microscopie Electronique. Research and numerous publs. on magnetism, phys. chemistry of macromolecules, biol. macromolecules, biophysics. Address: Centre de Biophysique Moleculaire, La Source (45) Orleans, France.*

SADYKOV, Abid Sadykovich, Soviet organic chemist; b. Sept. 30, 1913; grad. Central Asian U., 1937. Instr., Tashkent Textile Inst., 1937-39, Central Asian U., 1939-47; head lab. cotton chemistry Usbek Acad. Scis., 1947—, dir. Inst. Chemistry, 1946-50. Mem. Uzbek Acad. Scis. Author: Chemistry of Anabasis aphylla Alkaloids, 1956; co-author: Results of Chemical Studies of Anabasis aphylla Growing in the Turkmen SSR; The Structure of Aphyllidine, 1955; The Structure of Oxiaphillidine, 1956. Research in chemistry and content of alkaloids, carbohydrates in wild and cultivated plants, organic acids. Office: A. USSR, Tashkent, Uzbek, SSR. Marksa 32, Sredneaziatsky gosudarstvenny universitet.

SAELEN, Odd Henrik, Norwegian oceanographer; b. Bergen, Norway, June 8, 1920; s. Odd and Hansia (Carstens) S.; cand. real. U. Oslo (Norway), 1945; Dr.philos., U. Bergen, 1963; m. Mary Jane Dahm, Nov. 12, 1948; children—Astri, Marianne, Kari, Odd. Amanuensis, U. Bergen, 1947-60, 1st amanuensis, 1960-65; prof. oceanography U. Oslo, 1965—. Research, publs. on hydrography of Norwegian Fjords, currents and circulation in Norwegian Sea. Home: Otto Ruges vei 84, Oslo 7, Norway.*

SAENGER, Eugen, Rocket engr.; b. Bréznice, Bohemia, Sept. 22, 1905; ed. U. Graz (Austria); grad. in aero. sci. U. Vienna, 1929; m. Ireen Bredt, 1946. Asst. to Prof. Ludwik, Vienna Tech. High Sch.; worked in Germany; later adviser to French Ministry Armaments, 1946; became head Inst. Jet Propulsion Physics, Stuttgart (Germany) Tech. High Sch., 1954. Author: The Technique of Rocket Flight, 1933; Rocket Propulsion of Long-Range Bombers, 1944. Founder (with wife) study high-altitude aerodynamics; pioneered in liquid-propellent rockets, rocket combustion chamber design; studies in turbulence-injection nozzles, thermo-nuclear reaction; originated idea of long-range rocket bomber, photon.

SAENGER, Eugene Lange, Am. radiologist; b. Cin., Mar. 5, 1917; s. Eugene and Therese (Lange) S.; A.B. cum laude, Harvard, 1938; M.D., U. Cin., 1942; m. Sue Reis, Jan. 18, 1941; children—Katherine C., Eugene Lange. Faculty, U. Cin., 1949—, prof. radiology, 1962—, dir. radioisotope lab. Gen. Hosp., 1950—; radiotherapist Children's Hosp., Cin.; cons. Brooke Army Hosp., Ft. Sam Houston, Tex., Lackland (Tex.) AFB., U. S. AEC, USPHS. Bd. dirs. Nat. Council Radiation Protection and Measurements. Diplomate Am. Bd. Radiology. Mem. A.M.A., Am. Roentgen Ray Soc., Radiol. Soc. N. Am., Am. Coll. Radiology, Health Physics Soc., Soc. Nuclear Medicine, Am. Radium Soc., Alpha Omega Alpha, Sigma Xi. Asso. editor Jour. Nuclear Medicine, 1960—. Research, publs. on effects of acute whole body radiation in humans, radiation epidemiology, computers in medicine. Home: 9160 Given Rd., Cin. 45243.*

SAEZ, Francisco Alberto, Uruguayan biologist; b. Montevideo, Uruguay, Mar. 10, 1898; s. Francisco and Luisa (Sanchez) S. D.Sc., U. La Plata (Argentina), 1927; m. Rosa Gershanik; children—Federico Carlos, Jorge Alberto, Graciela Luisa, Guillermo Francisco. Asst. zoology Museo La Plata, Argentina, 1924-33; asst. med. faculty U. Buenos Aires, Argentina, 1934-43; head cytology lab., agrl. faculty U. La Plata, 1938-47, prof. biol. sci., 1938-47, prof. zoology, 1941-47; prof., head dept. cytogenetics Inst. Biol. Research, Montevideo, 1950—. Recipient Nat. prize biol. scis., 1936; numerous fgn. research fellowships. Hon. mem. Linnean Soc. Montevideo, Chilean, Argentine biol. socs., Genetic Soc. Am., Histochem. Soc., Zoöl. Soc.; corr. mem. Brazilian Genetic Soc.; mem. Sigma Xi. Author: (with DeRobertis, Nowinski) Cell Biology. Research, publs. on general cytology, cytogenetics, cytochemistry, particularly karyotype alterations in insects. Home: 6031 Caramuru. Office: 3318 Avenida Italia, Montevideo, Uruguay.*

SAFAR, Peter, physician; b. Vienna, Austria, Apr. 12, 1924; s. Karl and Vinca (Landauer) S.; M.D., U. Vienna, 1948; m. Eva Kyzivat, July 6, 1950; children—Elizabeth, Philip. Came to U.S., 1949, naturalized, 1959. Chief anesthetist Balt. City Hosps., 1955-61; asst. prof. Johns Hopkins, clin. asso. prof. U. Md., 1954-61; prof., chmn. dept. anesthesiology U. Pitts., 1961—. Research contractor Surgeon Gen., U. S. Army, 1957—, USPHS, 1963-64; mem. or chmn. nat. and internat. resuscitation coms. and symposia, 1958—; mem. MEDICO surg. team to Indo China, 1959. Diplomate Am. Bd. Anesthesiology. Fellow Am. Coll. Anesthesiologists; mem. Am. Soc. Anesthesiologists, Am. Physiol. Soc., Assn. U. Anesthetists, Am. Coll. Chest Physicians, Am. Heart Assn., N.Y. Acad. Sci., Sigma Xi. Author: Resuscitation, Controversial Aspects, 1963; (with M. McMahon) Resuscitation of Unconscious Victim, 1959; also numerous articles. Editor: Respiratory Therapy, 1965. Research on airway obstruction, resuscitation in drowning, hemorrhage, and cardiac arrest, anesthetic drugs and techniques; developed and proved superiority mouth-to-mouth methods of resuscitation; developed modern methods mech., prolonged artificial ventilation and intensive care: Home: 243 Hoodridge Dr., Pitts. 15234. Office: Presbyn.-Univ. Hosp., Pitts. 15213.*

SAFFORD, Truman Henry, Am. astronomer; b. Royalton, Vt., Jan. 6, 1836; grad. Harvard, 1854; Ph.D., Williams, 1878; m. Elizabeth Marshall Bradbury, Mar. 8, 1860. Prof. astronomy U. Chgo., 1866-76; Field meml. prof. astronomy Williams Coll., 1876—. Fellow Am. Acad. Arts and Scis., A.A.A.S.; Royal Astron. Soc. of Eng. Author: Mathematical Teaching and Its Modern Methods; The Development of Astronomy in the United States, 1888. Compiled catalog of stars; discovered new nebulae; calculated orbits of planets, comets. Died June 13, 1901.

SAFFORD, William Edwin, Am. botanist; b. Chillicothe, O., Dec. 14, 1859; s. William Harrison and Pocahontas (Creel) S.; grad. U. S. Naval Acad., 1880; postgrad. Yale, 1883-85, Harvard, 1885; Ph.D., George Washington U., 1920; m. Clare Wade, Sept. 14, 1904; children—Decius Wade, Bernice Galpin. Served in USN, 1880-1902; served in Spanish-Am. War, collecting for U. S. Nat. Mus. in depts. of ethnology and ethnobotany; asst. botanist Dept. Agr., 1902-15, economic botanist, 1915—. Commr. for Chgo. Expn. to Peru and Bolivia, 1891-92, for Chicago Expn. Recipient medal Chgo. Expn., 1893. Author: A Year on the Island of Guam, 1902-04; Useful Plants of the Island of Guam, 1905; The Chamorro Language of the Island of Guam, 1905; Cactaceae of Northeastern and Central Mexico, 1909; the classification of the genus Annona with descriptions of new and imperfectly known species, 1913; An Aztec Narcotic, 1915; Lignum nephriticum, 1916; Narcotics and Stimulants of the Ancient Americans; Chenopodium nuttaliae, a food-plant of the Aztecs, 1918; Cosmos sulphureus, the Xochipalli, or Flower-paint of the Aztecs, 1918; Notes on the genus Dahlia, with descriptions of two new species from Guatemala, 1919, Natural History of Paradise Key and the nearby Everglades of Florida, 1919; Synopsis of the Genus Datura; Datura, an Inviting Genus for the Study of Heredity, 1921; Peyote, the narcotic Mescal Button of the Indians, 1921; Daturas of the Old World and New, 1922; Ant Acacias and Acacia Ants of Mexico and Central America, 1923; The Potato of Romance and of Reality, 1925. Died Jan. 10, 1926.

SAFFRAN, Murray, Canadian biochemist, educator; b. Montreal, Que., Can., Oct. 30, 1924; s. Isador and Reva (Elimelech) S.; B.S., McGill U., 1945, M.S., 1946, Ph.D., 1949; m. Judith Cohen, June 8, 1947; children—David, Wilma, Arthur, Richard. With McGill U., 1943—, demonstrator organic chemistry, 1943-46, demonstrator biochemistry, 1946-48, asso. prof., 1959-65, prof., 1965—, dir. McIntyre Med. Scis. Bldg., 1963-65. Cons. USPHS. Mem. Montreal Physiol. Soc. (sec. 1954-56), Canadian Soc. Biochemistry, Canadian Soc. Physiology, Am. Soc. Biol. Chemistry, Endocrine Soc., Internat. Brain Research Orgn., A.A.A.S., Sigma Xi. Editorial bd. Endocrinology. Research, numerous publs. on prodn. hormones by adrenal cortex tissue outside body, control of hormone prodn. by pituitary gland by a hormone produced in brain. Home: 5140 Mayfair St., Montreal 29, Que., Can.*

SAGAERT, Lucien Léonard Camille, French physicist; b. Asnières, May 28, 1923; s. Arthur and Zoé (Corneillie) S.; ed. Vauben Tech. Coll., Paul Langevin Lycée, Sch. Pub. Works, St. Andrew's U., U. London; Dr ès sc; Dr. Littl. and ès sc. hon. causa; m. Catherine Michelet, Apr. 13, 1952; children—Martine, Michel, Marc-Joel, Jacques, Alain. Engr. of pub. works; now physicist-engr. Nat. Research Center, Paris. Decorated Order of Merit, Order Merit for Research and Invention; Belgian Order Merit for Invention; Order Pub. Encouragement; Oder Philanthropic Merit. Mem. Phys. Chemistry Soc., Biol. Chemistry Soc., Nat. French Assn. Sci.

Writers, Imperial Philo-Byzantine U. Madrid (Spain), Acad. Human Sci. and Relations Mexico; corr. mem. Classical Acad. Arts and Letters Madrid. Author: L'ingénieur Voyer; Le Guide pratique de l'Inventeur; also numerous articles. Constrn. of various bridges and indsl. bldgs. Home: Villa Vent de Solitude, 101-101 bis rue des Bas, Asnières (Seine), France. Office: 13 rue Pierre-Curie, Paris, France.

SAGAN, Carl (Edward), Am. astronomer; b. N.Y.C., Nov. 9, 1934; s. Samuel and Rachel (Gruber) S.; A.B., U. Chgo., 1954, B.S., 1955, M.S., 1956, Ph.D., 1960. Research asst. genetics Ind. U., 1952, lectr. U. Chgo., 1957; asst. physicist Armour Research Found., Chgo., 1958-59; Miller Research fellow Inst. Basic Research U. Cal. at Berkeley, 1960-62; vis. asst. prof. genetics Med. Sch. Stanford, 1962-63; asst. prof. astronomy Harvard, Cambridge, Mass., 1962—; staff mem. Smithsonian Astrophys. Obs., 1962—; mem. numerous coms. Nat. Acad. Sci. and NASA; cons. govt. agys., univs. Recipient A. Calvert Smith prize Harvard, 1964. NSF fellow, 1955-60, Alfred P. Sloan Found. Research fellow Harvard, 1963-65. Fellow Brit. Interplanetary Soc., Am. Inst. Aeros. and Astronautics, A.A.A.S.; mem. Genetics Soc. Am., Soc. For Study Evolution, Internat. Astron. Union, Am. Astron. Soc., Am. Phys. Soc., Am. Geophys. Union, Astron. Soc. Pacific. Author: (with W.W. Kellogg) The Atmospheres of Mars and Venus, 1961; (with I.S. Shklovsky) Intelligent Life in the Universe, 1966. Asso. editor: Icarus, 1962—. Contbr. numerous articles to profl. jours. Research on planetary atmospheres and surfaces, especially Venus and Mars; studies of origin of life on earth; studies of exobiology. Office: 60 Garden St., Cambridge, Mass. 02138.*

SAGANE, Ryokichi, Japanese physicist; b. Nagasaki Prefecture, Japan, Nov. 27, 1905; s. Hantaro Nagaoka; grad. Tokyo U., 1929; studied under Prof. E. O. Lawrence, U. Cal. Prof., Tokyo U.; mng. dir. Nippon Atomic Generation of Electric Power Co. Author books including: Devices for Experiments in Nuclear Physics. Research in nuclear physics.

SAGE, Balthazar-Georges, French chemist, mineralogist; b. Paris, May 7, 1740; ed. by Abbé Nollet and Rouelle; aide-maj. Hôtel des Invalides; royal censor; became prof. exptl. mineraology, Mint, 1778; Prin. founder Ecole des Mines, Paris, 1783, and the Museum of Mineralogy; mem. French Acad. Scis. Author numerous treatises on chemistry, mineralogy, electricity including: Mémoires de Chimie, 1773; Exposé des effets de la Contagion nomenclative, et réfutation de Paradoxes qui dénaturent la Physique, 1810. Research emphasized assaying and analysis, galvanism and natural history; showed importance of chemistry for industry; improved methods of extracting metals from their ores; rejected the new chemistry of Lavoisier and published against it as late as 1818; known as "le Fossile" because of his conservative scientific views. Died Sept. 9, 1824.

SAGER, Günther, German oceanographer; b. Rostock, Germany, June 4, 1923; s. Karl and Elly (Wegener) S.; student Tech. U. Stuttgart (Germany), 1942-43, U. Rostock, 1946-50; D. rer.nat., Leipzig (Germany) U., 1961, D. (rer. nat.) habil, 1964; m. Brigitte Korsing, Mar. 10, 1954; children—Iduna, Guido. Staff, Inst. für Meereskunde Warnemünde, German Acad. Scis., Warnemünde, Germany, 1952—, dept. leader, 1954—, lectr. oceanography, 1965—. Mem. Nat. Geophys. Com., 1966—. Author: Gezeitenvoraussagen und Gezeitenrechenmaschinen, 1955; Gezeiten und Schiffahrt, 1959; Ebbe und Flut, 1959; Atlas der Elemente des Tidenhubs und der Gezeitenströme, 1963; Atlas der Tidewasserstände für die Nordsee, den Kanal und die Irische See (with Rudolf Sammler), 1964; also numerous articles. Improvement, devel. charts with elements of tidal range and tidal streams for N. Sea, English Channel and Irish Sea; research on theory of damping in tide gauges. Home: 9 Grüner Weg. Office: 15 Seestrasse, Warnemünde, Germany.*

SAGER, Ruth, Am. geneticist; b. Chgo., Feb. 7, 1918; d. Leon B. and Deborah (Borovik) Sager; B.S., U. Chgo., 1938; M.S., Rutgers U., 1944; Ph.D., Columbia, 1948. Merck fellow NRC, 1949-51; staff Rockefeller Inst. for Med. Research, 1951-53; sr. research scientist Columbia, 1955-65; prof. biology City U. N.Y. Hunter Coll., N.Y.C., 1966—. Non-resident fellow U. Edinburgh (Scotland) Inst. Animal Genetics, 1962—. Mem. Internat. Soc. for Cell Biology, Am. Soc. for Cell Biology, Genetics Soc. Am., Am. Soc. Naturalists, Phi Beta Kappa, Sigma Xi. Author: (with F. J. Ryan) Cell Heredity, 1961; also numerous articles. Research in molecular genetics, genetics systems not carried on chromosomes; pioneered devel. exptl. material for analysis non-chromosome heredity by combination genetic, biochem. and electron microscope techniques.*

SAH, Peter Pen Tieh, bio-organic chemist; b. Foochow, China, Feb. 3, 1900; s. Chung Lu and Haw Lin (Young) S.; student Tsing Hua Coll., China, 1912-20, Worcester Poly. Inst., 1920-22; B.S., U. Wis., 1924, M.S., 1924, Ph.D., 1926; postgrad. Yale, Halle U., Germany; m. Ethel Hsia, Feb. 2, 1932; children—Louise Amy, Peter Pen Tieh, Jenghis John, Joanne Dorothy, Hamilton James, Emil Daniel. Prof. bio-organic chemistry Nat. Tsing Hua U., China, 1929-37, Cath. U. Peking, China, 1935-45; vis. research prof. U. Göttingen (Germany), 1934, Oxford (Eng.) U., 1935; chmn. dept. chemistry Nat. Peking (China) U., 1945-46; faculty U. Cal. at San Francisco, 1946-53, at Davis, 1953—, prof. comparative pharmacology Sch. Vet. Medicine, 1959—, pharmacologist, 1961—. Dir. prodn. Fu Min Pharm. Labs., Peking, 1943-45; dir. research China Biol. and Chem. Labs., Peking and Shanghai, China, 1944-46. Recipient China Found. Research award, 1940, Ebert prize Am. Pharm. Assn., 1949; research fellow Yale, 1926-27, U. Halle (Germany), 1927-28. Mem. Chinese Chem. Soc. (founder), numerous internat. Socs., Kaiserlich Deutsche Akademie der Naturforscher, Sigma Xi, Phi Lambda Upsilon, Rho Chi. Research, publs. on micro reagents; synthesis vitamin C from carbohydrates; discoverer vitamin K, new drugs for lung diseases; discovery INHG for human Tb, vitamin C derivative of sulfone for leprosy, chemotherapy of psoriasis with vitaminized dye solutions, tetrahalogenated quinones for ameba and fungal infestions, new antihypertensives related to quanethidine, new antivirus agents related to thiosemicarbazones. Home: 1404 Colgate Dr., Davis, Cal. 95616.*

SAHA, Meghnad N., Indian astrophysicist; b. Dacca, Bengal, India, Oct. 6, 1893; s. Jagannath and Bhubaneswari (Devi) S.; ed. Dacca Coll., Calcutta Presidency Coll.; m. Radharani Roy, June 1918; 7children, including A.K. Saha. Prof. Allahabad U., 1923-38; prof., head dept. physics Calcutta U., 1938-56; founder, dir. Inst. Nuclear Physics, Calcutta; founder various Indian sci. orgns.; Fellow Royal Soc., 1927, Am., French astron. socs. (hon.). Author: (with B.N. Srivastava) Treatise on Heat, 4th edit., 1958; Treatise on Modern Physics; Theory of Thermal Tonization of Gases; (papers) On Measurement of the Pressure of Radiation, Physicial Theory of Stellar Spectra, Physical Theory of Solar Corona. Research on radiation and ionization; contbd. to theory of spectra; derived equation (named for him) connecting rates of ionization with partition function for ionization states, temperature and electron pressure. Died Feb. 16, 1956.

SAHADE, Jorge, Argentinian astronomer; b. Alta Gracia, Cordoba, Argentina, Feb. 23, 1915; s. Nallib Jorge and Maria (Kassab) S.; Surveyor, U. Cordoba, 1937; D.Astronomy, U. La Plata (Argentina), 1943; m. Miriam Stella Elkin Font, Dec. 29, 1948; children—Patricia Adriana, Carlos Alberto. Staff, Observatorio Astronomico, La Plata, 1941-46, head dept., prof. astrophysics, 1958—; staff Observatorio Astronomico, Cordoba, 1946-55, acting dir., 1953-55; prof. practical astronomy U. Cordoba, 1948-55; Vol. research asst. Yerkes Obs., U. Chgo., 1943-46; research fellow U. Cal. at Berkeley, 1955-57, staff, 1957-58, research astronomer, III, 1960; Am. Astron. Soc.-NSF Fgn. vis. prof. Ind. U., 1965; guest investigator Mount Wilson Obs., 1955-61, 63; mem. Argentine Commn. for Space Research, 1966—. Guggenheim fellow, 1955-57. Mem. Internat. Astron. Union (pres. commn. 29 1964—), Am., Argentine astron. socs. Contbg. author: Stars and Stellar Systems, vol. 6, 1960; also numerous articles. Editor: Information Bull. for So. Hemisphere, 1962—. Research on spectrographic investigation and interpretation close binary systems and Wolf-Rayet stars in framework stellar evolution. Discovered spectral peculiarities in several systems. Home; 47848, La Plata. Office: Observatorio Astronómico, La Plata, Argentina.*

SAHLI, Hermann, Swiss physician; b. Bern, Switzerland, May 23, 1856; grad., Bern; prof., internal medicine, Bern, 1888-1929. Author: Lehrbuch der klinishen Untersuchungs-methoden (most comprehensive work of its kind), 1894. Research on diseases of blood, digestion, lungs, also infectious diseases; introduced Shali's test of ascertaining functional activity of stomach, 1891; gave 1st description of acute rheumatism as attenuated form of generalized staphyloccic infection; devised Sahli hemoglobinometer, micrometer for estimation of blood sugar; inventor arteriometer for measuring changes in caliber of arteries, pulsometer for measuring rate of blood flow, blood pressure apparatus; built mercury manometer, 1904; introduced methyl violet method for estimating free hydrocholoric acid in stomach; described various anemias; introduced drugs salol and guaiacol into therapeutics; suggested use of drug pantopen by injection, 1909. Died Bern, Apr. 28, 1933.

SAHLINS, Marshall David, Am. anthropologist; b. Chgo., Dec. 27, 1930; s. Paul Aaron and Bertha (Skuo) S.; A.B., U. Mich., 1951, A.M., 1952; Ph.D., Columbia, 1954; m. Barbara Vollen, Feb. 12, 1951; children—Julia, Peter, Elaine. Lectr. anthropology Columbia, 1955-56; faculty U. Mich., Ann Arbor, 1956—, prof. anthropology, 1964—; field research, Fiji Islands, 1954-55, New Guinea, 1964. Fellow Center for Advanced Study in Behavioral Scis., 1963-64; Guggenheim fellow, 1967-68. Mem. Am. Anthrop. Assn., Am. Ethnol. Soc., Royal Anthrop. Inst., A.A.A.S. Author: Social Stratification in Polynesia, 1958; (with E. R. Service) Evolution and Culture, 1960; Moala, 1962; Tribesmen, 1967; also articles. Described native life in Moala, Fiji Islands; research on theory of cultural evolution, analysis of primitive economies, analysis of primitive social structures. Home: 2466 Newport Rd., Ann Arbor, Mich. 48103.*

SAHNI, Birbal, Indian botanist; b. Bhera, Punjab, India, Nov. 14, 1891; s. RuchiRam and Shrimati Ishwar Devi Anand; ed. Govt. Coll., Lahore, Ind.;M.A., Sc.D., Cambridge (Eng.) U.; D.Sc., U. London, (Eng.); postgrad. U. Munich (Germany); D.Sc. (hon.) Patna (India) U.; m. Savitri Suri, 1920. Prof. botany univs. Benares (India), 1919-20, (hon.) from 1936, Punjab, 1920-21, Lucknow (India), from 1921, dean sci. faculty Lucknow U. from 1933; spl. lectr. Lahore, 1932, extension lectr. Lahore, Rohtak, India, 1936; Sukhraj Rai reader natural sci. Panta U., 1937; Adharchandra lectr. natural sci. Calcutta (India) U., 1938; Gaekwad lectr. at Baroda, India, 1944-45; founder, hon. dir. Inst. Palaeoboany, Lucknow; v.p. Nat. Inst. Scis., India, 1935, also fgn. sec.; mem. Jubilee Session India Sci. Congress pres. session Madras, India, 1940; v.p. palaeobotanical Section, Internat. Bot. Congress, Cambridge, 1930, Amsterdam, Netherlands, 1935; del. govt. India 18th Internat. Geol. Congress, London, 1948; mem. Sci. Manpower Commn., Sci. Consultative Commn. Govt. India. Recipient Barclay medal Asiatic Soc. Bengal, 1936. Fellow Royal Soc., 1936 (del. ex-officio India sci. conf. London, 1946); mem. India Bot. Soc. (founder, pres. 1924), Nat. Acad. Sci. India (pres. 1937-38, 42-44), (fgn. corr.) Bot. Soc. Am., (fgn. hon.) Am. Acad. Arts and Scis., Lahore Philos. Soc. (pres. 1921). Author: (with Lowson) Text Book of Botany. Research on extinct plants, fossil floras and their geological bearings. Died Apr. 10, 1949.

SAHOVIC, Ksenofon, pathologist; b. Ohrid, 1898; s. Alexander and Natalija (Snegarovic) S.; ed. Coll. Bitolj, France, also Lyons, France; D.Sc., M.D.; m. Julijana Trickovic. Asst., Lyons, 1922; became lectr., Belgrade (Yugoslavia) Med. Faculty, 1923; prof. U. Belgrade, from 1926; chmn. path. inst., from 1932; mem. San. Council; sec. Soc. against Cancer, Yugoslavia. Editor book on cancer. Exptl. research and publs. on cancer, basal metabolism, termo-regulation, metabolism of carbon hydrates, shriveled goiter.

SAHRHAGE, Dietrich Wilhelm, German fishery biologist; b. Hamburg, Germany, Oct. 21, 1926; s. Heinrich and Elly (Rowoldt) S.; Doktor Rer. Nat., U. Hamburg, 1953; m. Eike Beuck, Aug. 21, 1954; children—Ute, Gerlind. Fishery biologist Bundesforschungsanstalt für Fischerei, Hamburg, 1953-65; fisheries officer FAO, Rome, Italy, 1965—, chief marine biology sect. dept. fisheries, 1966—. Supr., German Bilateral Fisheries-AID projects, Guinea, Togo, Libya, Africa, 1959-65; mem. German Sci. Commn. for Exploration of Sea, 1962—. Research, publs. on distbn. of fish species in N. Sea, herring, haddock and whiting stocks in N. Sea, population dynamics, fisheries devel. in W. Africa. Home: 12/u Via del Poggio Fiorito, Rome. Office: FAO, Viale delle Terme di Caracalla, Rome, Italy.*

SAHU, B. N., Indian agronomist; b. Kalantira, Aug. 1, 1910; s. Mani and Bajamoni (Devi) S.; student Ravenshaw Coll., 1928-30; B.Agr., Agrl. Coll., Nagpur, India, 1935; postgrad. Ont. (Can.) Agrl. Coll., Guelph; M.S. In Agr., U. Toronto (Ont.); Ph.D., Mich. State Coll., 1949; m. Kubuma Kumari Dei, Feb. 1945; children—Bibhuti Bhusan, BiChitranand, Bibekanand, Bipinbehari, Bimalkant, Bishnu Charan, Vijayalakshmi (Mrs. D. Lenka). With Dept. Agr., Bihar and Orissa, India, 1935-60, agronomist to Govt. Orissa, 1954-59, dep. dir. agr., 1959-60; faculty Utkal Kushi Mahavidyalaya, Bhulaneswar, India, 1954—, prof., head dept. agronomy, 1960—. Recipient 2d prize for sci. writings Orissa Sahitya Acad., 1962. Mem. Indian Sci. Congress (pres. agrl. scis. sect. 1966-67), Internat. Soil Sci. Soc., Indian Soc. Agronomy, Indian Soc. Soil Sci., Bigyan Prachar Samiti, Author: Land Utilization in Orissa, 1950; Vegetable Cultivation, 1953; Fruit Cultivation, 1954; Flower Garden, 1955; Go-Mangal O Gochikitsa, 1955; Our fish wealth, 1957; Forage Crops, 1959; Agricultural Sphorism, 1965; also articles. Research, publs. on crop prodn. and environment, manurial, cultural irrigation and weed control of rice; evolved a manurial schedule for different crops on basis of soil types of Orissa. Home: Village-Kalantina, Post-Somepur, Dust-Cuttock, Orissa, India.*

SAIBEL, Edward, Am. mathematician; b. Boston, Dec. 25, 1903; s. Robert and Martha (Ulin) S.; S.B., Mass. Inst. Tech., 1924, postgrad. Cornell U.; Ph.D., Mass. Inst. Tech., 1928; m. Lillian Howe, Aug. 8, 1930; children—Patricia (Mrs. Rolf Winter), Mahlon. Instr., U. Minn., 1927-30; faculty Carnegie Inst. Tech., 1930-57; prof., chmn. mechanics dept. Rensselaer Poly. Inst., Troy, N.Y., 1957-67; prof. applied mechanics Carnegie-Mellon U., 1967—. Fellow Am. Soc. M.E.; mem. Am. Math. Soc., Math. Assn. Am., Soc. for Natural Philosophy, Soc. Engring. Sci., Am. Soc. Lubrication Engrs., Sigma Xi, Phi Kappa Phi, Pi Tau Sigma. Author: (with W.W. Lai) Elements of Mechanics of Elastic Solids, 1965; also numerous articles. Research on vibration and stability, mech. properties materials, viscous fluids, numerical methods, stress analysis. Home: 5460 Northumberland St., Pitts. 15217.*

SAILA, Saul Bernhard, Am. oceanographer; b. Providence, May 23, 1924; s. John A. and Hannah (Koskinen) S.; student Purdue U., 1942-43, The Citadel, 1943; B.S., U. R.I., 1949; M.S., Cornell U., 1950, Ph.D., 1952; m. Mary Ann Hartikka, Jan. 29, 1949 (div. Dec. 1966); children—Sylvia, Eric, Anne Marie. Biologist, N.Y. State Conservation Dept., Ithaca, 1951-52; research asso. zoology dept. Ind.

U., Bloomington, Ind., 1952-54; fishery biologist R.I. Dept. Nat. Resources, Providence, 1954-56; faculty U. R.I. Grad. Sch. Oceanography, 1956—, prof., 1967—, dir. computer lab., 1959-61, dir. marine expt. sta., 1967—. Cons. Gen. Dynamics Corp., 1966—. Mem. Am. Soc. Limnology and Oceanography, Am. Fisheries Soc., Marine Tech. Soc., Nat. Shell Fisheries Assn. Research, publs. in population dynamics of marine organisms, methods of estimating vital statistics and computer simulation models of living systems. Home: R.F.D. Hope Valley, R.I. 02832. Office: Grad. Sch. Oceanography, U. R.I., Kingston, R.I. 02881.*

SAILER, Reece Ivan, Am. entomologist; b. Roseville, Ill., Nov. 8, 1915; s. Ethan Calvin and Millicent (Coghill) S.; student Western Ill. State Tchrs. Coll., 1934-36; A.B., Kan. U., 1938, Ph.D., 1942; m. Jessie Marie Bradbury, Sept. 7, 1939; children—Sigrid Eileen (Mrs. David J. Matson), Enid Louise (Mrs. Jay Lagree). Asst. state entomologist, Kan., 1942; asso. entomologist Bur. Entomology and Plant Quarantine USDA, Washington, 1942-48, entomologist, 1948-53; entomologist Entomol. Research div. Agrl. Research Service, Washington, 1953-57, asst. dir. Insect Identification and Parasite Intro. Research br. labs., 1957-60, research entomologist in-charge European Parasite Lab., Nanterre, France, 1960-66, asst. chief Insect Identification and Parasite Intro. Research br., 1966-67, acting chief of br., 1967—. Mem. Ecol. Soc. Am., Arctic Inst. North Am., Soc. Systematic Zoology, Washington Entomol. Soc. (pres. 1958), Entomol. Soc. Am. (chmn. Eastern br. 1959), Kan. Entomol. Soc., Entomol. Soc. Washington, Washington Acad. Scis., A.A.A.S., Sigma Xi. Contbr. numerous articles in field to sci. jours. Research, publs. on the classification of insects called true bugs, especially chinch bugs, stink bugs, bed bugs, plant bugs; research on cross breeding species of stink bugs demonstrated that natural species could arise from inter-specific hybridization; research on ecology of Alaskan mosquitoes made useful predictions of summer abundance possible; research which resulted in intro. of natural enemies to aid in control of pests such as alfalfa weevil, European elm bark beetle, European pine shoot moth, balsam woolly aphid; mgmt. of research programs in insect classification and biol. control of insect pests and weeds. Home: 10401 46th Av., Beltsville, Md. 20705. Office: Agrl. Research Service, Entomology Research div. USDA, Plant Industry Sta., Beltsville, Md. 20705.*

SAILOR, Vance Lewis, Am. physicist; b. Springfield, Mo., June 28, 1920; s. Vance Laird and Madge (Lewis) S.; A.B., DePauw U., 1943; M.S., Yale, 1947, Ph.D., 1949; m. Marguerite Erdman, Oct. 2, 1943; children—Richard, Nancy, John. Physicist, group leader Brookhaven Nat. Lab., Upton, N.Y., 1949—. Chmn. nuclear cross sects. adv. group AEC, 1961-63; adv. Turkish Nuclear Research and Tng. Center, 1962-65. Mem. Am. Phys. Soc., Am. Assn. Physics Tchrs., Phi Beta Kappa. Research, publs. in exptl. aneutron and low temperature physics. Home: 100 Durkee Lane, East Patchogue, N.Y. 11774. Office: Physics Dept., Brookhaven Nat. Lab., Upton, N.Y. 11973.*

SAINBEL, Charles Vial (or Saint Bel), vet. surgeon; b. Lyons, France, Jan. 28, 1753; student of M. Pean, vet. sch., also of Claude Bourgelat; asst. surgeon, pub. demonstrator, vet. coll., Lyons, 1773; noted for work during epizootic among horses in France, 1774; asst. prof. Royal Vet. Coll., Paris; vet. surgeon, physician, Lyons; equerry to Louis XVI; chief of manege, acad. of Lyons; went to Eng., 1788; founder Vet. Coll. London, became prof., 1791. Recipient prize Royal Soc. Medicine, 1772. Author: Essai sur les proportions géométrales de l'éclipse, 1791, 2d edit., 1795; Lectures on the Elements of Farriery, 1793; collection of 4 essays, pub. 1695. Founder of sci. vet. practice in Eng. Died Aug. 21, 1793.

ST. AMAND, Pierre, Am. geophysicist; b. Tacoma, Feb. 4, 1920; s. Cyrias Zephure and Mable (Berg) St. A.; B.S. in Elec. Engring. and Physics, U. Alaska, 1948; M.S. in Geophysics, Cal. Inst. Tech., 1951, Ph.D., 1953; m. Marie Magdalene Poss, Dec. 9, 1945; children—Gene Pierre, Barbara Michelle, Denali Marie, David Cyrias Humberto. Observer, Carnegie Inst. Washington, dept. terrestrial magnetism Coll. Obs., College, Alaska, 1940-41; observer, head ionospheric research Coll. Geophys. Obs. and Inst., Coll., 1945-49; research asso. div. geol. scis. Cal. Inst. Tech., Pasadena, 1949-53; physicist U. S. Naval Ordnance Test Sta., Pasadena, 1950-54, head optics br., China Lake, Cal., 1955-58, head div. earth and planetary scis., 1961—; Fulbright Research fellow, France, 1954-55; fgn. service res. officer U. S. State Dept., prof. earth scis. U. Chile, Santiago, 1958-61; prof. extension dept. U. Cal. at Los Angeles, 1955-57. Cons. to fgn. govts., state govs.; cons. engring. geology and seismology; mem. spl. panel on Latin Am., Pres.'s Sci. Adv. Com., 1961—. Fellow Geol. Soc. Am.; mem. Seismol. Soc. Am., Am. Geophys. Union, A.A.A.S., Earthquake Engring. Research Assn., Weather Research Control Assn., Soc. Geologica de Chile, (Diploma of Honor 1965), Soc. Phys. de Chile, Sigma Xi. Contbr. numerous articles to tech. jours., chpts. in books. Research on seismology and mountain bldgs., maj. fault systems, rotation Pacific basin, geologic maps, ionosphere, photometric instruments, pyrotechnics in cloud seeding. Home: 602 A Essex Circle, China Lake 93556. Office: Code 502, U. S. Naval Ordnance Test Sta., China Lake, Cal. 93557.*

SAINT-ANGE, Gaspard-Joseph Martin, French naturalist; b. Nice, France, 1803. Author: Circulation du sang chez l'homme et les animaux, 1830; Etude de l'appareil reproducteur dans 5 classes d'animaux vertebrates, 1954. Research on embryology and animal physiology. Died Paris, 1888.

SAINT BEL, Charles Vial, see Sainbel, Charles Vial.

SAINTE-CLAIRE DEVILLE, Charles, geologist; b. St. Thomas, W.I., Feb. 26, 1814; ed. L'Ecole des Mines, Paris; named insp. gen. meteorol. service, 1872; asst. to Elie de Beaumont, then succeeded him as prof. geology Coll. de France, 1875; founder dir. obs., Montsouris; founder other observatories in France and Algiers; mem. French Acad. Scis., 1857. Author: Geological Voyage to the Antilles and to the Islands of Teneriffe and Fogo, 1847; Actual Eruptions of the Volcano of Stromboli, 1858; Research on the Principal Meteorological and the Terrestrial Features of the Antilles, 1861; On the Periodical Variations in Temperature, 1866. Research and publs. on allo-tropic forms of sulfur and composition of volcanic gases; studied changes in molecular states; developed theory that volcanic disturbances are caused by passage of sea water through earth fissures onto hot rocks, on basis of studies in Antilles and at Vesuvius and Stromboli. Died Paris, Oct. 10, 1876.

SAINTE-CLAIRE DEVILLE, Henri Étienne, chemist; b. St. Thomas, B.W.I., Mar. 11, 1818; student Institution Sante Barbe; M.D., U. Paris, France, 1843, D.Sc. Prof. chemistry, dean, sci. faculty, Besançon, France, 1845-51; prof. École Normale, 1851-80; substitute lectr. to Dumas, U. Paris, 1853-66, named titular prof., 1866. Recipient Prix Jean Reynaud, French Acad. Mem. French Acad. Scis., 1857. Research on turpentine (discovered hydrobromide, terpineol), resins (discovered dinitrobenzene), creosote, toluene; isolated aluminum from alum (developed indsl. method for obtaining aluminum); devised very high temperature lab. furnaces; discovered nitrogen pentoxide, 1849; worked on preparation of sodium, silicon, boron (with Wöhler), silicides (with Caron); revived theory that a catalyst acts as porous body. Died Boulogne, France, July 1, 1881.

SAINT-FOND, see Faujas de Saint-Fond, Barthélemy.

ST. JOHN, Charles Edward, Am. solar physicist; b. Allen, Mich., Mar. 15, 1857; s. Hiram A. and Lois A. (Bacon) S.; grad. Mich. Normal Coll., 1876, Pd.M. (hon.), 1906; B.S., Mich. Agrl. Coll., 1887; postgrad. U. Mich., 1890-92; A.M., Harvard, 1893, Ph.D., 1896; postgrad. U. Berlin, 1894-95; Sc.D., Oberlin College, 1931. Tchr. physics Mich. Normal Coll., 1885-92; instr. physics U. Mich., 1896-97; asso. prof. physics and astronomy Oberlin Coll., 1897-99, prof., 1899-1908, dean Coll. of Arts and Scis., 1907-08; astronomer, Mt. Wilson Obs., Carnegie Instn., 1908—. Mem. internat. commns. on solar rotation and on standards of wavelength; pres. Internat. Commn. on Solar Physics. Author of Revision of Rowland's Preliminary Table of Solar Spectrum Wave-lengths. Research on motion and circulation in sun spots, gen. circulation and levels in solar atmosphere; made spectroscopic measurements of sun's rotation period; studied wave lengths and displacement of terrestrial and solar lines with reference to theory of generalized relativity; applied results of solar studies to planetary and stellar spectra. Died Apr. 26, 1935.

ST. JOHN, Harold, Am. botanist, explorer; b. Pittsburgh, Pa., July 25, 1892; s. Charles Elliott and Martha Elizabeth (Everett) St. J.; A.B., Harvard University, 1914, A.M., 1915; Ph.D., 1917; student Sorbonne, Paris, 1919; m. Elizabeth Chandler, June 24, 1922; children—Charles Elliott, Robert Pierce, Mary Merrill, Martha Everett. Asst. in botany, Harvard and Radcliffe, 1913-15; asst. botanist, Canadian Geol. Survey, 1915, 17; asst. in botany Univ. Extension, 1917; Sheldon traveling fellow. Harvard University, 1917; asst. at Gray Herbarium, 1913-17, 1919-20; asst. prof. of botany, State College of Wash., 1920-23, asso. prof. botany and curator of Herbarium, 1923-29; prof. of botany, U. Hawaii, 1929-50, sr. prof., 1950-57, Wilder professor, 1957-58, chairman of the department, 1929-40, 1943-54; asso. director Manoa Arboretum 1953-57, dir., 1957-58; J. H. Whitney visiting professor Chatham College, Pittsburgh, 1958-59; Smith-Mundt vis. prof. Université de Saigon, also Université de Hue Vietnam, 1959-61; Fulbright professor of botany Cairo U., 1963; botanist B. P. Bishop Mus., Honolulu, 1929—, curator Herbarium, 1965—; vis. prof. Yale, 1939-40. Botanist, exploring for quinine, Andes of Colombia, F.E.A., 1943-44. Served as pvt. to 2nd lt., Inf., U. S. Army, 1918-19; with A.E.F., 1 yr. Fellow A.A.A.S., Linnean Society of London; mem. Botanical Society of America, also New England Bot. Club, Northwest Scientific Soc., Torrey Bot. Club, Acad. Nat. Sciences of Phila., Hawaiian Bot. Soc. (pres. 1932-33), Hawaiian Acad. Science (pres. 1947-48), Am. Society Plant Taxonomists, Botanical Society of Japan (honorary member), Société Internationale de Phytosociologie et de Geobotanique, Am. Legion, Phi Beta Kappa (hon.), Phi Kappa Phi (pres. Univ. Hawaii chapt. 1945-46), Sigma Xi. Author: Weeds of the Pineapple Fields of the Hawaiian Islands (with E. Y. Hosaka), 1932; Flora of Southeastern Washington and of Adjacent Idaho, 1937, 3d edit., pub. 1963; Nomenclature of Plants, 1958. The Founder, first editor Research Studies, State Coll. of Wash., bd. editors Pacific Science, 1946-53, 55-58. Contbr. botanical publs. Made botanical explorations or expdns. to Newfoundland and maritime provinces of Can., Hawaiian Islands, Eastern Polynesia, Melanesia, Micronesia, S.E. Asia, Indian Ocean, Africa, Northeastern and Northwestern U. S., etc. Home: 2365 Hoomaha Way, Honolulu 96822. Office: Bishop Museum, Honolulu 96819.*

SAINT PIERRE, Jacques, Canadian math. statistician; b. Three-Rivers, Que., Can., Aug. 30, 1920; s. Oscar and Lucie (Landreville) St. P.; L.Sc., U. Montreal, 1948, M.Sc., 1951; Ph.D., U. N.C., 1954; m. Marguerite Lachaine, July 15, 1947; children—Marc, Guy, Andre, Louis, Francois, Mireille. Faculty, U. Montreal, Que., 1947—, prof., 1961—, vice dean faculty of pure scis., 1961-64, dir. Computing Centre, 1964—, dir. computer science department since 1966—. Mem. Inst. Math. Statistics, Biometric Soc., Assn. for Computing Machinery, Can. Operational Research Soc., Can. Math. Congress, Can. Assn. U. Tchrs. Research, publs. in statis. decision functions, applied math. statistics in biology. Home: 1795 Sauve Croissant, Duvernay, Que. Office: P.O. Box 6128, Montreal, Que., Can.*

SAINT-SIMON, Claude Henri de Rouvroy (comte de), French social philosopher; b. Paris, Oct. 17, 1760, of aristocratic family claiming descent from Charlemagne; ed. under direction of philosopher-mathematician Jean d'Alembert; briefly married. Entered mil. service at age 17; served under Comte de Grasse in Yorktown campaign of Am. Revolution, served afterwards in Mexico where he suggested constrn. of canal to unite Atlantic and Pacific oceans; returned to France with rank of col. but resigned; became a radical but took no active part in French Revolution; imprisoned in Luxembourg during Reign of Terror; speculated in confiscated ch. lands and amassed fortune; devoted himself to study, discussion groups, expts. and writing; reduced to poverty during his final years subsisting on laborer's job and generosity. Author: Letters of a Resident of Geneva, 1802; The Reorganization of European Society, 1814; The Industrial System, 1821; The New Christianity, 1825. Regarded as founder French socialism but this designation debatable; advocated orgn. of society upon sci. principles (physico-politics), rooted in study of history and natural scis.; believed leaders in this new sci. should control human affairs; believed that universal peace maintained by supra-nat. parliament needed for this sci. world; exercised influence on later thinkers including his associate Auguste Comte. Died in Paris, May 19, 1825.

SAINT-VINCENT, Grégoire de, see de Saint-Vincent, Grégoire.

SAISSY, Jean Antoine, French physician; b. nr. Grasse, France, 1756; ed. Paris; intern l'Hôtel-Dieu, Lyons, France; surgeon-maj. Royal Co., Africa; practiced medicine, specializing in ear diseases, Lyons; fellow Coll. Surgeons. Author: Recherches experimentales, anatomiques et chimiques sur la physiologie des animaux mammifères, hibernants . . . , 1808; Essais sur les maladies de l'oreille interne, 1827. Inventor, 1st to use bougie for Eustachian tube, 1829. Died Lyons, 1822.

SAITO, Kazuo, Japanese chemist; b. Tokyo, Japan, July 26, 1923; s. Yoshisuke and Mitsuko (Okumura) S.; B.Sc., Tokyo Imperial U., 1945; D.Sc., 1953; m. Michiko Araki, Oct. 25, 1952; children—Hiroyuki, Takako. Research asst. Tokyo Imperial U., 1946-53, U. Coll., London, Eng., 1953-55; asst. prof. Inst. for Nuclear Studies, Tokyo U., 1956-63; prof. coordination chemistry Tohoku U., Sendai, Japan, 1963—. Abstractor, Analytical Abstracts, London, 1954—. Fellow Chem. Soc. Japan, Chem. Soc. (London), Am. Chem. Soc. Asst. editor Bull. Chem. Soc. Japan, 1957-59. Contbr. articles to profl. jours. Research on complex formation of germanium in solution, isotopic exchange reaction kinetics with radioisotopes in solution (organic and aqueous), metal ions of typical elements. Home: 12-55, Kawauchi-Ohashidori, Sendai. Office: 75, Katahiracho, Sendai, Japan.*

SAITO, Kendo, Japanese bacteriologist; b. Ishikawa Prefecture, Japan, June 28, 1878; grad. botany dept. Tokyo U., 1900; studied in Germany; D.Sc.; childred—Hiroshi, Mrs. Tomonaga Nakagawa, Mrs. Mutsuo Hioki, Mrs. Toshio Shimizu; chief, central research inst. S. Manchuria Ry.; named prof. Osaka U. Tech., 1929, Osaka U., 1933; recipient prize (for research on prodn. of vitamin 2 complex) Imperial Invention Assn. Pres. Osaka Zymological Soc. Author: Introduction to Science of Ferment Bacillus; Science of Bacillus for Brewing of Refined Sake. Died Oct. 14, 1960.

SAITO, Nobufusa, Japanese chemist; b. Tokyo, Japan, Sept. 28, 1916; s. Tsunesaburo and Mitsu (Tominaga) S.; student U. Tokyo, 1936-40; D.Sc., U. Tokyo, 1949; m. Haruko Umeda, Sept. 6, 1944; children—Hiroko, Manabu, Mari. Asst. prof. Seoul

(Korea) Imperial U., 1941-46; asst. prof. Kyushu (Japan) U., 1946-49; faculty U. Tokyo, 1949—, prof. chemistry, 1956—. Dir. div. isotopes Internat. Atomic Energy Agy., Vienna, Austria, 1963-65; chief researcher Inst. for Phys. and Chem. Research, Tokyo, 1959—. Mem. Am. Chem. Soc., Chem. Soc. Japan (editor bull. 1968—), Atomic Energy Soc. Japan, Japan Soc. for Analytical Chemistry, Japan Radiation Research Soc. Author, editor: Radiochemistry, 1956; also numerous articles. Research on radiochem. separations, hotatom chemistry, radiation chemistry. Home: 5-12-9, Koshigoe, Kamakura, Japan. Office: Dept. Chemistry, Faculty of Sci., U. Tokyo, Bunkyo-ku, Tokyo, Japan.*

SAITO, Yukimasa, Japanese geophysicist; b. Masuda, Shimane, Japan, Dec. 4, 1916; s. Masayoshi and Ima S.; grad. Hokkaido (Japan) U., 1937, dept. physics Faculty Sci., Osaka (Japan) U., 1940; D.Sc., Tokyo (Japan) U., 1949; m. Chizuko, Nov. 26, 1941; 1 child, Masashi. With oceanographical sect. Kobe (Japan) Marine Obs., 1940, chief 1946-49; wartime research, 1943-45; asst. prof. physics Osaka City U., 1949—, prof. geophysics, 1954—. Mem. Deep-sea Research Com. Japan, 1956—; mem. Com. Pub. Injury, City of Osaka, 1963—. Fellow Royal Astron. Soc. (Eng.); mem. Phys. Soc. Japan, Oceanographical Soc. Japan (councilor 1956—), Am. Geophys. Union, Am. Seismol. Soc. Author: Sea water, Current and Tide, 1959; also articles. Research on movement of sea water, structure of oceanic crust. Home: 92-4 Umemoto-cho, Hyogoku, Kobe, Japan. Office: Osaka City U. 12, Sugimoto-cho, Sumiyoshi-ku, Japan.*

SAITOH, Noburu, Japanese chemist; b. Nagaoka, Japan, Feb. 1, 1916; s. O. and K. (Akira Saitoh) S.; Dr.Eng., Tokyo (Japan) Inst. Tech., 1953; m. Akira Saitoh, Dec. 3, 1964. Staff, Toshiba Elec. Co. Central Lab., 1928-66, sr. researcher, 1950-66, chief vacuum tube compression sect., 1960-65; prof. Faculty Engring., Niigata U., Nagaoka, 1966—. Mem. Japanese Vacuum Soc., Japanese Mass Spectroscopy, Japanese Inst. Metals. Author: (with others) Metals for Electron Tubes, 1950; (with S. Asao, K. Okamoto) Materials for Vacuum Technology, 1965; also articles. Research on evolution mechanism of gas from metals and other materials in vacuum tubes, mass spectroscope. Home: Gakocho 1 chome, Nagaoka City, Japan.*

SAJOUS, Charles Euchariste de' Médici, physician; b. at sea, nr. coast of France, Dec. 13, 1852; s. Count Charles E. and Marie Pierrette (Curt) de'Médici; steps. Charles Sajous; M.D., Jefferson Med. Coll., 1878; LL.D., St. Joseph Coll., 1909; Sc.D., Temple U., 1915; m. Emma Christine Bergner, Jan. 30, 1884; 1 son, Louis Theodore de'Médici. Clin. lectr. Jefferson Med. Coll., Phila., 1881-90; dean and prof. laryngology Medico Chirurg. Coll., 1897-98; prof. anatomy and physiology. Wagner Inst. Sci., 1880-82; prof. therapeutics Temple U., Phila., 1909-22; prof. endocrinology U. Pa., 1921—. Fellow Coll. Physicians, Phila., A.C.P.; mem. 22 sci. bodies; pres. Am. Assn. Study of Internal Secretions, 1917-18, Am. Therapeutic Soc., 1919-20. Author: Curative Treatment of Hay Fever, 1884; Diseases of the Nose and Throat, 1885, Editor: Sajous' Analytical Cyclopaedia of Practical Medicine 8 vols. (10 edits.), 1898-1925, Sajous' Annual of the Universal Medical Sciences (5 vols. annually), 1888-96 (45 vols); The Internal Secretions and the Principles of Medicine, 2 vols., 1903-22 (10 edits.); Strength of Religion as Shown by Science, 1926. Editor N.Y. Med. Jour., 1911-19. First to suggest defensive role of endocrine glands in diseases. Died Apr. 27, 1929.

SAKABE, Kohan (called Yuzaemon, originally named Toda), Japanese mathematician; b. 1759; Shogunate policeman then became tchr. math.; follower of Naonobu Ajima of the Seki sch. Author: Sampo Tenzan Shinan-roku 1810. Established series for determining value of pi divided by 4; discovered method of calculating circumference of oval. Died 1824.

SAKAI, Shoichiro, mathematician; b. Kanuma, Japan, Jan. 2, 1928; s. Hajime and Moto (Kojima) S.; B.S., Tohoku U., Sendai, Japan, 1953, Ph.D., 1961; m. Yoshiko Sasaki, July 10, 1958; children—Masato, Kiyoshi. Research asst. Tohoku U., 1953-60; asst. prof. Waseda U., Tokyo, Japan, 1960-64; vis. lectr. Yale, 1962-64; asso. prof. U. Pa., Phila., 1964—. Research, publs. on operator algebras; characterization of W-algebras, type I C*-algebras; derivations of W*-algebras. Home: 315 Gypsy Lane, Wynnewood, Pa. 19096. Office: U. Pa., Phila. 19104.*

SAKAI, Tetsuro, Japanese surgeon; b. Tokyo, Japan, Mar. 18, 1911; s. Zenzaburo and Kane (Hasegawa) S.; Igakushi, Niigata U. Sch. Medicine, 1934, Igakuhakase, 1939; m. Teiko Yahta, Oct. 19, 1934; children—Toshiharu, Noriko (Mrs. Ken Hashimoto), Akiko (Mrs. Daishiro Kasahara), Miyuki. Asst. prof. dept. surgery Niigata (Japan) U. Sch. Medicine 1944-53, dir. dept. surgery, prof., 1953—; surgeon in chief Niigata Cancer Hosp., 1962—. Author: Nutrition and Surgery, 1952; also numerous articles. Research on abdominal surgery, especially cancer of stomach, secretory function of stomach. Home: 1940-30 Kobari, Niigata, Japan.*

SAKAMI, Warwick, Am. biochemist, educator; b. Phila., Nov. 21, 1916; s. Isami and Toraye (Haraguchi) S.; B.A., Swarthmore Coll., 1938; postgrad. Yale; Ph.D., U. Pa., 1944; m. Adalyn Berardi, July 17, 1952. Faculty, U. Pa. Sch. Medicine, 1944-46; faculty dept. biochemistry Western Res. U. Sch. Medicine, Cleve., 1946—, prof., 1963—. Vis. scientist U. Paris, 1955-56; mem. biochemistry test sect. Nat. Bd. Med. Examiners, 1959-62. Mem. Am. Soc. Biol. Chemists, Am. Chem. Soc., A.A.A.S., Am. Assn. Med. Colls., N.Y. Acad. Scis., Sigma Xi. Research, numerous publs. on metabolism of amino acids and one-carbon compounds, metabolic functions of folic acids. Home: 2700 Edgehill Rd., Cleveland Heights, O. 44106. Office: 2109 Adelbert Rd., Cleve. 44106.*

SAKAMOTO, Yoshiyuki, Japanese physicist; b. Nara, Japan, June 6, 1932; s. Yoshitaka and Yoshino (Kita) S.; B.Sc., Kyoto U., 1956, M.Sc., 1958, D.Sc., 1961; m. Kimiko Adachi, Mar. 19, 1966; 1 son, Kazayuki. Research fellow Nordita Gustaf Werner Inst., U. Uppsala, Sweden, 1961-62; NRC postdoctorate fellow Ottawa, Ont., Can., 1962-64; lectr. nuclear physics dept. nuclear engring. Kyoto (Japan) U., 1964—; research fellow Inst. de Physique Nucleaire, Orsay, France, 1967—; lectr. physics Chubu Inst. Tech., Nagoya, Japan, 1966—. Mem. Phys. Soc. Japan, Atomic Energy Soc. Japan. Contbr. numerous articles to sci. jours. Research on theory of high energy reactions, elastic and inelastic scatterings of nucleons by nuclei, quasi-free scattering and interactions between elementary particles and nuclei, designed to give a unified description of nuclear structure. Home: 18 Minaguchicho, Shimogamo, Sakyo-ku, Kyoto, Japan.*

SAKATA, Shoichi, Japanese physicist; b. Tokyo, Jan. 14, 1911; s. Kanta Sakata; grad. physics Kyoto U., 1933; D.Sc.; m. Nobuko Kakiuchi. Prof., Nagoya U., 1942—. Recipient Japan Acad. prize, 1949. Author: Physics and Methods; Theory of Two Mesons; (with Yukawa) The Theory of the Atomic Nucleus and Cosmic Rays. Assisted (with Mitsuo Taketani) Hideki Yuakawa in devel. atomic nuclear theory; originated theory of 2 kinds of meson, 1942 and proved their existence using photog. plate registering cosmic rays, 1948.

SAKAZAKI, Riichi, Japanese bacteriologist; b. Mie, Japan, Aug. 21, 1920; s. Toshizo and Kinu (Shibata) S.; student Nippon Vet. Coll., 1938-40; Ph.D., Hokkaido (Japan) U., 1958; m. Shigeko Nishiyama, Mar. 1, 1926; children—Shinichiro, Akemi. Staff, Pub. Health Lab., Nakataki Pharm. Co., Tokyo, Japan, 1950-53, Nat. Inst. Animal Health, Tokyo, 1953-58; chief div. phage typing Nat. Inst. Health, Tokyo, 1958—; mem. Japan Monkey Center, Aichi, Japan, 1961—. Recipient Asahi prize for culture, Asaki News, 1965. Mem. Japaneses Assn. Bacteriologists. Author: Isolation and Identification of Enteric Bacteria, 1958; Bacterial Media, 1965; also articles. Clarified ecology of Salmonella and pathogenic coliform organisms in food poisoning; discovered several new pathogenic coliform bacteria; research on Vibrio parahaemolyticus (most important causative agt. of food poisoning in Japan). Home: 1561 Tsuda- machi, Kodaira-shi, Tokyo. Office: 2-chome, Kamiosaki, Shinaguwa-ku, Tokyo, Japan.*

SAKEL, Manfred Joshua, physician; b. Nadvorna, Austria, June 6, 1900; s. Mayer and Judith Golde (Friedman) S.; student 1st State College, Brno, Czechoslovakia, to 1920; med. degree U. Vienna, 1925; hon. D.Sc. Colgate U., 1938. Asso. physician Vienna (Austria) Hosp., 1925-27; research fellow Urban Hosp., Berlin, 1927; psychiatrist in chief Lichterfelde Hosp., Berlin, 1927-33; asso. in Neuro-psychiatric Univ. Clinic, Vienna, 1933-36; on the invitation of Commr. Frederick Parsons, of New York, taught method of treatment he had devised circa 1929 for schizophrenia and other nervous and mental diseases by shock with insulin and metrazol to various doctors of U. S., 1936; came to U. S., 1936, since practiced privately in N.Y.C.; med. dir. Manfred Sakel Research Found. Fellow Am. Psychiatric Assn.; life mem. N.Y. Acad. Scis. Author: Theory of Addiction; New Treatment of Addiction (Morphinism, Alkoholism); New Method of treating Nervous and Mental Ailments; The Pharmacological Shock Treatment of Mental Diseases; The Results of Shock Therapy; The History of the Origin of Shock Terapy; Medical Psychiatry and Psychological Medicine; An Approach to the Causative Treatment of Idiopathic Epilepsy; Schizophrenia; Epilepsy. Died Dec. 2, 1947.

SAKHAROV, Andrei Dmitrievich, Russian physicist; b. May 21, 1921; grad. Moscow U., 1942. Joined Inst. Physics USSR Acad. Scis., 1945. Decorated Order of Lenin; recipient Stalin prize. Author: The Generation of the Hard Component of Cosmic Rays, 1947; Electron-Positron Interaction during Pair Production, 1948; The Excitation Temperature in Gas Discharge, 1948. Research and publs. in theoretical physics; proposed (with I. E. Tamm) application of elec. discharge in plasma in magnetic field to obtain controlled thermonuclear reaction. Home: Luxhnikovskaya 1/7. Office: A. N. Lebedev Inst. Physics, USSR Acad. Scis., Levinskii Prospekt, 53, Moscow, USSR.

SAKIZ, Edvart, biologist; b. Istanbul, Turkey, Apr. 17, 1926; s. Migirdic and Adrine (Nahnikian) S.; M.D., Faculty Medicine Paris (France), 1953; Ph.D., Faculty Scis. Paris, 1964; m. Jacqueline Oxombze, July 30, 1964; 1 son, Alex. With French Nat. Center Sci. Research, 1954-61, charge, 1959-61; asso. dir. Ecole des Hautes Etudes, Sorbonne, Paris, 1961—; vis. asst. prof. Baylor U. Coll. Medicine, Houston, 1963-66; dir. dept. physiology and biochemistry Roussel-Uclaf, Paris, 1966—. Recipient Silver medal Faculty Medicine Paris, 1953, award Eugenie de Rosemont, U. Paris, 1956, award Godard French Biol. Soc., Mem. French Endocrine Soc., Research and publs. on adrenal-gonadal relationship, isolation and purification of hypothalamic hormones, application of math. analysis in exptl. biology by use of computers. Home: 26 Clos d'Orléans, Fontenay sous Bois 94, France. Office: 111 Route de Noisy, Romainville 93, France.*

SAKS, Vladimir Nikolaevich, Russian geologist; b. Apr. 22, 1911; grad. Leningrad Mining Inst., 1933. With All Union Arctic Inst., 1935-40, 44-48, mem. mining geol. dept. Main Directorate of No. Seaway, 1940-44; mem. staff Inst. Arctic Geology, 1948—. Decorated Order Banner of Labor. Corr. mem. Acad. Scis. USSR. Mem. Communist Party. Author: Conditions in the Formation of Bottomset Beds in Arctic Seas of the USSR, 1952; Quarternary Period of the Soviet Arctic, 2d edit., 1953; (with Z. Z. Ronkin) Jurassic and Cretaceous Deposits of the Ust-Enisei Depression, 1957; Problems in Arctic Geology, 1960. Research in quaternary geology, paleography, stratigraphy of Soviet Arctic and in geology of its formation; presented gen. scheme on separation of quaternary deposits of Arctic; worked out stratigraphy of Mesozoic deposits of oil bearing territories of No. Siberia; quarternary Arctic sea fauna and history of its formation. Address: Leninskii Prospekt 14, Acad. Scis. USSR, Moscow, USSR.

SAKURADA, Ichiro, Japanese chemist; b. Kyoto, Japan, Jan. 1, 1904; s. Bungo and Masa (Yamada) S.; grad. Faculty Engring., Kyoto Imperial U., 1925, Dr. Eng., 1931; m. Chiyoko Okumura, May 10, 1931; children—Yutaka, Mariko (Mrs. Ei-Ichiro Mori), Toshiko (Mrs. Koichi Takakura). Researcher, Inst. Phys. and Chem. Research, Tokyo, Japan, 1925-33; faculty Kyoto Imperial U., 1934—, prof. chemistry, 1935—, dean Faculty Engring., 1965—; dir. Osaka labs. Japan Radiation Research on Polymers, 1958—. Recipient Japan Acad. prize, 1955; Purple Ribbon prize Japanese Govt., 1956. Mem. Soc. High Polymers (pres. 1960—). Research, numerous publs. on invention of boiling water insoluble polyvinyl alcohol fiber vinylon, polymer and fiber chemistry. Home: 49 Higashi-senouchi, Kitashirakawa, Kyoto, Japan.*

SAKURAI, Joji, Japanese chemist; b. Kanazawa, Ishikawa Prefecture, Japan, 1858; s. Jintaro Sakurai; ed. Kaisei Sch., also London (Eng.) U.; became prof. Tokyo U., 1882; elected to House of Peers, 1920; mem. Imperial Acad. (pres. 1926); made study trips to Europe and Am.; improved Beckmann's methods of determining boiling points of various solutions; raised teaching of chemistry in Japan to level of Europe and Am. Died 1939.

SALA, Angelus, physician, chemist; b. Vicenza, Italy, probably 1576; probably studied chemistry, Venice, from circa 1593; at least 1 dau. (Mrs. Anthor Gunther Billich); lived at the Hague, 1613-17, at Hamburg, 1620-25; physician to Count of Oldenburg, 1617-30, to dukes of Mencklenburg-Güstrow, 1625-36; to Gustavus Adolphus, 1636-37. Author: Tractatus duo de variis tum Chymicorum, tum Galenistarum erroribus, in praeparationibus medicinali commissis, 1602; Emetologia du enarration . . . , 1613; Anatomia antimonii . . . , 1617; Aphorismorum chymiatricorum synopsis, 1620; Chrysologia, seu examen auri Chymicum, 1622; Essentiarum vegetabilium anatome . . . , 1630; De auro potabili . . . , 1631; Spagyrische schatzkammer, 1634; Saccharalogia, 1637; Septem planetarium terrestrium . . . , 1641; also treatises on study of tartaric acid, on oil of vitrol; his Opera medico-chymica . . . omnia went through several editions from 1647-1712. Studied chem. aspects of sugar, brandy and fermentation, also distillation of oils, antimony, some salts and acids; discovered process for refining sugar by means of heat and egg whites; 1st to mention of salts of sorrel in a chem. work; rejected universal medicine; described preparation of metallic antimony and antimonic acid, also of crystalline silver by dissolving silver in aqua fortis and evaporation; described preparation of sal ammoniac from spirit of salt and ammonia; declared extraction of mercury from minerals impossible; rejected the possibility of metallic transmutation; observed that metals have different affinities for acids; noted that sulphur takes something from air to burn; an advocate of iatrochemistry. Died Bützow/Mecklenburg, Germany, Oct. 2, 1637.

SALA, Oscar, Italian surgeon; b. Marostica, Italy, June 3, 1922; s. Ardeo and Emma (Santini) S.; Doctor's degree cum laude, U. Padua (Italy), 1945; Ear-Nose-Throat Specialist cum laude, U. Bologna (Italy), 1948; m. Maria Poli, Dec. 27, 1950; children—Tito, Annalisa. Staff, Ear-Nose-Throat Clinic, Padua U., 1942—, asso. prof. audiology, 1966—. Mem. Oesterreichische Otolaryngologische Gesellschaft (corr.).

Author: The Cortison and Cortison-life Drugs in Otorhinolaryngology Experimental and Clinical Researches, 1959; also numerous articles. Research in vestibular apparatus, cortisonic therapy in otorhinolaryngology, especially hypoglotic laryngitis of children, efferent vestibular system, immunoglobulins in ENT diseases. Home: 20 Negrelli. Office: U. Policlinic of Padua, Padua, Italy.*

SALADIN, Nicolas Joseph, French mathematician; b. La Bassée, France, 1743; gave up practice of medicine after many years to study math. at Acad. Lille (France); became prof. math. and physics, Douai, France, 1792, Strasbourg, France, 1803; mem. many sci. socs. Author: Traité d'algèbre, d'arithmétique et de géométrie practiques, 1775; Grammaire francoise, 1794; also works in manuscript. Collaborator Pharmacopée de Lille, 1772. Died 1829.

SALADINO DI ASCOLI (or da Ascoli, de Asculo, de Esculo), Italian physician; physician to Prince of Tarentum. Author: Compendium Aromatariorum (classic work intended to upgrade pharmacy, includes discussion of medicines used then), 1488.

SALAM, Abdus, physicist; b. Jhang, Pakistan, Jan. 29, 1926; M.A., U. Panjab, Pakistan, 1946; B.A., Cambridge (Eng.) U., 1948, Ph.D., 1952, D.Sc., 1957. Mem. Inst. for Advanced Study, Princeton, N.J., 1951; prof., head math. dept. U. Panjab, Lahore, 1951-54; lectr. Cambridge U., 1954-56, fellow St. John's Coll., 1951-56; prof. theoretical physics Imperial Coll., London, Eng., 1957—; dir. Internat. Centre Theoretical Physics, Trieste, Italy, 1964—. Mem. AEC, Pakistan, 1958—; chief sci. adviser Pres. Pakistan, 1961—; mem. Nat. Sci. Council, Pakistan, 1963—; mem. adv. com. on sci. and tech. UN, 1964—. Recipient Hopkins prize Cambridge U., 1957, Adams prize, 1958; Maxwell medal Phys. Soc. London, 1961; Pride of Performance award Pres. Pakistan, 1959. Fellow Royal Soc., 1959 (Hughes medal 1964); mem. Pakistan Assn. Advancement Sci. (pres. 1961). Research, numerous publs. on physics of elementary particles, especially two-component neutrino theory of parity violation in weak interactions, symmetry properties of elementary particles. Office: Imperial Coll., London, Eng.; also Internat. Centre for Theoretical Physics, Trieste, Italy.*

SALAMAN, Redcliffe Nathan, English pathologist; b. London, Eng., Sept. 12, 1874; s. Myer and Sarah (Solomon) S.; B.A., Trinity Hall, Cambridge (Eng.) U., 1896, B.Ch., 1900, M.B., 1901, M.A., M.D., 1904; postgrad. London Hosp. and Path. Labs., Urban Krankenhaus, Berlin, Germany; m. Nina Davis, 1901 (dec. 1925); m. 2d, Gertrude Lowy, 1926; children—Myer, Arthur, Edward, Raphael, Ruth, Esther. Dir. London Hosp. Path. Inst., 1901-04; hon. pathologist Zool. Soc. London, 1902-03; research on genetics of potato, other subjects, Barley, 1906-26; dir. Potato Virus Research Sta., Cambridge U. Gov. Hebrew U., Jerusalem. Fellow Royal Soc. (Snell medal), 1935, Linnean Soc., Nat. Inst. Agrl. Botany (hon., v.p.); licentiate Royal Coll. Physicians; m. Royal Coll. Surgeons, Genetical Soc. (v.p.). Author: Palestine Reclaimed, 1920; Potatoe Varieties, 1926; The History of the Potato and its Social Influence. Contbr. articles on virus genetics, diseases, problems related to potato. Initiated bldg. of virus free stocks of seed potatoes. Died Royston, Hertfordshire, Eng., June 12, 1955.

SALANAVE, Leon Edward, Am. geophysicist; b. San Francisco, Nov. 19, 1917; s. Edward Jean and Ethel (Allen) S.; A.B., U. Cal., Berkeley, 1940, M.A., 1948; m. Marilyn C. Sydenstricker, July 29, 1949 (div. June 1960); children—Camille Anna, Jonathan Lewis. Instr., Sacramento Coll., 1948-51; asso. astronomer Cal. Acad. Scis., San Francisco, 1952-56; research asso. Aura, Inc., Tucson, 1956-58, Inst. Atmospheric Physics, U. Ariz., Tucson, 1959—; design cons. Morrison Planetarium, San Francisco, 1950-53. Mem. Am. Astronom. Soc., Am. Geophys. Union, Internat. Union for Geodesy and Geophysics, Phi Beta Kappa. Research, publs. on application of slitless spectroscopy to observation of lightning, first determinations of temperature of lightning, ultraviolet and infrared spectra.*

SALANKI, János, Hungarian biologist; b. Debrecan, Hungary, May 11, 1929; s. János and Ilona (Sarkadi) S.; grad. Med. U. Debrecen, 1954; Ph.D., Moscow (USSR) State U., 1959; m. Katalin Rózsa, Dec. 29, 1960; children—Zsuzsanna, Katalin. Staff, Med. U. Debrecen, 1954—, lectr. physiology dept., 1959—; dir. Biol. Research Inst., Hungarian Acad. Scis., 1962—. Mem. Hungarian Physiol. Soc., Hungarian Biol. Soc., Soc. for Biol. Rhythm. Research and publs. on role of nerves and central nervous system in regulation of rhythmic and periodic activity of Pelecypoda; discovered periodic activity in Pelecypoda is not independent of outside factors, but a special regulation mechanism is suggested; described polymorphic morphological structure and some physiol. properties of abductor muscles in Pelecypoda. Home: 12. Fürdötelep, Office: Biol. Research Inst., Tihany, Hungary.*

SALDANHA, Pedro Henrique, Brazilian geneticist; b. Rio de Janeiro, Brazil, Sept. 1, 1929; s. Alcino Barbosa and Annita (Gonçalves) S.; B.S., U. Brazil, 1953; D.Sc., U. Sao Paulo, 1959; m. Sônia Guinsburg, Jan. 11, 1955; children: Frederico G., Henrique G., Suzana G. Tchr. Sao Paulo high schools, 1954-57; Rockefeller Found. fellowship, 1957-59, and 1960-61; lectr. genetics, Faculty of Medicine, U. Sao Paulo, 1959-65; founder and dir., Laboratory for Medical Genetics, U. Sao Paulo, 1959—; privat-docent, medical genetics, U. Sao Paulo, 1965—; asst. prof., physiological chemistry, U. Sao Paulo. Mem., Sociedade Brasileira de Genética; Sociedade Brasileira para o Progresso da Ciencia; Biometrical Society. Author: Hereditariedade Humana, 1964; ed., Aspectos Modernos da Genética Médica, 1966; fifty published articles. Research in medical genetics, including population biochemical and clinical; serological genetics; human cytogenetics; human biology and anthropological genetics. Home: 175, Rua Oquira, Sao Paulo, Brazil. Office: 455, Avenida Dr. Arnaldo, Sao Paulo, Brazil.*

SALEMINK, Cornelis August, Dutch chemist; b. Isgravenhage, Netherlands, Apr. 27, 1918; s. August Andreas and Eliza Salemink; Candidaats Examination, U. Utrecht (Netherlands), 1942, honors degree, 1946, doctoral exam. honors degree, Ph.D., 1949; Asst. organic chem. lab. U. Utrecht, 1946-58, sr. reader, 1958—. Recipient study award Shell Co., 1947-48. Mem. Provincial Utrecht Soc. Arts and Scis. Author: (with E. Veen) Introduction into Organic Chemistry, 1965; (with M. T. Bos) Chemistry and Safety, 1965; also articles. Research on purine analogues, quaternary nitrogen compounds related to muscarine (one of main alkaloids of fly agaric); plant toxins including elm disease toxin, raspery, disease toxin; entomol. research, including juvenile hormone. Home: 46, Jos. Hayducaan 46, Utrecht, Holland.*

SALFI, Mario, Italian zoologist; b. Cosenza, Italy, Nov. 11, 1900; s. Enrico and Giuseppina (De Marinis) S.; Dr.ès.sc.; m. Maria Borrea, Apr. 2, 1930; Asst., prof., Naples, 1922-36, prof. zoology and comparative anatomy, 1942—; instr. Cagliari, Italy, 1937-38; asso. prof., Genoa, Italy, 1939-41. Mem. Acad. Sci., Letters and Arts Genoa, Naples Acad. Sci., Nat. Entomologia Acad. Florence, Pontaniana Acad. Naples. Author: Elementi di morfologia dei vertebrati, 1950; Zoologia, 1957; Entomologia, 1960. Research and publs. on orthoptera. Home: corso Umberto 118, Naples, Italy.

SALIN, Jarl Paavo, Finnish engr.; b. Helsingfors, Oct. 5, 1898; s. Jeremias and Hulda Salin; civil engr., dr. tech., Helsingfors Tech. U.; m. Elsa Lindroos, Aug. 4, 1904. Engr., 1922-29; head mgmt. Finnish Hydraulic Force Services, 1929-32; mem. Technico-Chem. Faculty, Abo Acad., 1933—, prof., 1937—, dean, 1954-62. Decorated Order White Rose; recipient Gold decoration Swedish Assn. Tech. Sci. in Finland. Mem. Royal Swedish, Finnish acads. tech. sci. Author: Definitionen der diffusionskronstanten, 1935; Wahrscheinlichkeitsberechungen in der Kraft-und Warmetechnik, 1936; also chpt. in tech. manual, articles. Research in heat and mass transfer. Home: Slottsgaten 36 c. Office: Abo Acad., Abo, Finland.

SALING, Erich Zeno, physician; b. Stanislawow, Poland, July 21, 1925; s. Henrich and Emma (Hoffmann) S.; student U. Jena (Germany), 1946-49; doctor medicine Free U. Berlin (Germany), 1952; m. Hella Weymann, June 24, 1952; children—Peter, Michael. Clin. staff Women's Host., Berlin-Neukoelln, 1954—, chief research group for perinatal care, 1967—; faculty Free U. Berlin, 1963—, privat dozent, 1963—, lectr. obstetrics and gynecology, 1963—. Mem. German, Berlin socs. obstetrics and gynecology, German Soc. for Perinatal Medicine (chmn.). Author: Das Kind im Bereich der Geburtshilfe, 1966; also articles. Research on aortic catheterization of new born for blood exchange, fetal blood sampling, amniscopy, fetal and neonatal pathophysiology. Home: 233 Spandauer Damm, 1 Berlin 19. Office: 2838 Mariendorfer Weg 1 Berlin 44, Germany.*

SALIO OF PADUA, Italian translator; b. Italy; cannon, Padua, Italy; translator (from Arabic into Latin) Liber de nativitatibus (astrological treatise of Abu Bakr), 1218, 44, or 48, also De stellis fixis (attributed to Hermos Trismegistos).

SALISBURY, Edward James, English botanist; b. Harpenden, Eng., Apr. 16, 1886; s. James Wright and Elizabeth (Stimpson) S.; B.Sc., U. Coll. London, 1909, D.Sc., 1913, LL.D. (hon.), 1949; m. Mabel Elwin-Coles, 1918 (dec. 1956). Faculty, U. Coll. London, 1918-43, Quain prof., 1929-43; dir. Royal Bot. Gardens, Kew, 1943-56; prof. physiology Royal Instn., 1947-52. Vice chmn. Agrl. Improvement Council, 1944-56. Decorated comdr. Brit. Empire, 1939; knighted, 1946; recipient Gold medal Royal Soc., 1945, Veitchian Gold medal, 1936, Victoria medal of honor, 1953. Fellow Royal Soc., 1933 (biol. sec. 1945-55), Inst. Biology (hon.), Bot. Soc. Edinburgh (hon.); mem. Brit. Ecol. Soc. (hon.). Author: The East Anglian Flora, 1933; The Living Garden, 1935; Plant Form and Function, 1938, subsequent edits.; The Reproductive Capacity of Plants, 1942; Downs and Dunes, 1952; Weeds and Aliens, 1961; The Biology of Garden Weeds, 1962. Studies, publs. on soil influence on woodlands, sand dunes; plant germination and significance of seed output. Home: Croindene, Strandway Felpham, Sussex, Eng.*

SALISBURY, Frank Boyer, Am. plant physiologist; b. Provo, Utah, Aug. 3, 1926; s. Frank M. and Catherine (Boyer) S.; B.S., with high honors U. Utah, 1951, M.S., 1952; Ph.D. in Plant Physiology, Cal. Inst. Tech., 1955; m. Lois Marilyn Olson, Sept. 1, 1949; children—Frank Clark, Steven Scott, Michael James, Cynthia Kay, Phillip Boyer, Rebecca Lynn. Asst. prof. botany Pomona Coll., Claremont, Cal., 1954-55; faculty Colo. State U., Ft. Collins., 1955-66, prof. plant physiology, 1961-66, plant physiologist, Exptl. Sta., 1961-66; head plant sci. dept., prof. plant physiology Utah State U., Logan, 1966—. Trustee Colo. State U. Research Found. Fellow A.A.A.S.; mem. Am. Soc. Plant Physiologists, Ecol Soc. Am., A.A.A.S., Colo.-Wyo. Acad. Sci., Astron. Soc. Am., Sigma Xi, Phi Kappa Phi, Gamma Sigma Delta. Author: The Flowering Process, 1963; (with R.V. Parke) Vascular Plants, Form and Function, 1964; Truth by Reason and Revelation, 1965; also articles. Research on process plant flowering, physiol. ecology plants, speculations on life on Mars. Home: 2020 N. 13th East, North Logan, Utah 84321. Office: Utah State U., Logan, Utah.*

SALISBURY, John William, Am. space scientist; b. Palm Beach, Fla., Feb. 6, 1933; s. John William and Mary (Bates) S.; B.A., Amherst Coll., 1955; M.S., Yale, 1957, Ph.D., 1959; m. Lynne Trowbridge, June 7, 1957; children—John William, Matthew Trowbridge. Served as lt. USAF, 1959-62, in lunar and planetary research Air Force Cambridge Research Labs., Bedford, Mass., civilian employee, chief lunar planetary research br., 1962—. Vis. prof. Purdue U., 1962-64. Named Outstanding Mil. Scientist, 1962. Mem. A.A.A.S., Geol. Soc. Am., Am. Geophys. Union. Author: The Lunar Surface Layer, 1965; also articles. Research on nature and origin of Lunar and Martian surface features; (with G.R. Hunt) 1st detected compositional differences on lunar surface, using infrared spectroscopic techniques. Home: 12 Paul Revere Rd., West Action, Mass. 01720. Office: Air Force Cambridge Research Labs., Bedford, Mass. 01730.*

SALISBURY, Peter F., physiologist; b. Dresden, Germany, Apr. 12, 1913; s. Friedrich and Greta Ann (Gerstle) S.; B.A., M.A., U. Cambridge, Eng., 1938; M.D., U. Rome, Italy, 1937; Ph.D., U. Minn., 1941; June Downey, June 24, 1944; children—Ann Pope, Karin Eve. Came to U. S., 19—, naturalized, 1943. Practice medicine restricted to consultations and spl. procedures field cardiovascular, renal diseases, 1953-64; research asso. Inst. Med. Research, Cedars of Lebanon Hosp., 1948-58; mem. dept. med. research St. Joseph Hosp., Burbank, Cal., 1958-64; cons. several teaching hosps., space medicine Cal. Inst. Tech. Mem. Am. Soc. Artificial Internal Organs (founder, past pres.), Am. Physiol. Soc., Instrument Soc. Am., Found. Med. Tech., Am., Aerospace, Cal., Los Angeles County med. assns., Cal. Soc. Internal Medicine, Am. Soc. Internal Medicine, Western Soc. Clin. Research, Soc. Exptl. Biology and Medicine, Western Soc. Clin. Research, Soc. Exptl. Biology and Medicine, Biophys. Soc., Am. Heart Assn.; Fellow Am. Coll. Cardiology. Editor: (with W. E. Murray) Biomedical Sciences Instrumentation, 1944; Transactions of the American Society for Artificial Internal Organs, 1955-56. Research, publs. cardiac physiology; inventor cross transfusion machine, heart-lung machine, continuous indirect blood pressure recorder; 1st implantation artificial kidney into live animal; 1st prototypes for artificial hearts and kidneys. Died Nov. 5, 1964.

SALISBURY, Rollin D., Am. geologist, geographer; b. Sring Prairie, Wis., Aug. 17, 1858; s. Daniel and Lucinda (Bryant) S.; Ph.B., Beloit Coll., 1881, A.M., 1884 (LL.D., 1904); Instr. geology and biology, 1883-84, prof. geology, 1884-91, Beloit Coll.; student in Europe (chiefly at Heidelberg), 1887-88; prof. general and geographic geology, U. of Wis., 1891-92; prof. geographic geology, 1892—, dean of Univ. Colls., 1894-96, dean of Ogden (grad.) Sch. of Science, 1899—, head of dept. of geography, 1903-19, head of dept. of geology, 1919—, U. of Chicago. U. S. geologist, glacial div., 1882-94, geologist, 1894—, geologist in charge of Pleistocene geology of N.J., 1891-1915. Fellow Geol. Soc. America, Assn. Am. Geographers, A.A.A.S., Ill. Acad. Sciences. Author: Geology (3 vol., with Thomas C. Chamberlin), 1904-06. Organized Geography Dept., Univ. Chgo., drew distinction between phys. and human geography; expert on deposits of glacial and Pleistocene Ages. Died Aug. 15, 1922.

SALISBURY, Winfield Wyman, Am. physicist; b. Carthage, Ill., Dec. 27, 1903; s. Herbert Spencer and Leona (Scott) S.; student Tex. A. and M. Coll., 1919-20; A.B., U. Ia., 1926; postgrad. U. Cal., 1927-28, 36; D.Sc., Cornell Coll., 1950; m. Elma M. Stone, Dec. 22, 1928; children—Winfield Wyman, Sylvia Leona (Mrs. John W. Myers), Alan William, April Carina. Research asst. physics dept. U. Ia., 1926-27; teaching asst. physics U. Cal., 1927-28, ultrasonics, chem. dept., 1936-37, research asst. cyclotron development Radiation Lab., 1937-40, Mackay prof. elec. engring., 1951-52; cons. engr. sound movies and phonograph recording, 1928-30; chief recording engr. Hollywood Film Enterprises, 1930-31; cons. engr. radio, sound recording, 1931-36; group leader radio frequency group Mass. Inst. Radiation Lab., 1940-41; group co-ordinator, mem. gov. bd. high-powered radar countermeasures Harvard Radio Research Lab., 1941-45; dir. research Collins Radio Co., Cedar Rapids, Ia., 1945-51; chief physicist Cal. Research & Development Co., Livermore, Cal., 1952-54; mgr. Gray

Sci. div. Remler Co., 1954-55; dir. research Salisbury Lab., Zenith Radio Research Corp., 1956-58; technical director microwave power lab., chief scientist Varo, Inc., Garland, Tex., 1958-65; research asso. Harvard Coll. Obs., also sr. scientist Astrophys. Obs., Smithsonian Instn., 1966—. Recipient certificates of merit, Pres. Truman, 1948, Mass. Inst. Technology, and Harvard Fellow I.E.E.E., American Physical Society, A.A.-A.S.; mem. Am. Astronautical Soc., Am. Astron. Soc., Optical Soc. Am., N.Y. Acad. Scis., Sigma Xi. Contbns. include design and constrn. cyclotrons, research and design high power vacuum tubes, devel. radio telescope, investigation of lunar surface by microwaves (leading to proposal of dust on lunar surface), constrn. and operation large accelerators, generation of light from free electrons and effect of simulated lightning on meteoric dust. Home: 11 Glenn Rd., Belmont, Mass. 02178. Office: 60 Garden St., Cambridge, Mass. 02138.*

SALK, Jonas Edward, Am. physician, virologist, b. N.Y.C., Oct. 28, 1914; s. Daniel B. and Dora (Press) S.; B.S., Coll. City of N.Y., 1934; LL.D., 1955; fellow chemistry, N.Y. U., 1935-37, exptl. surgery, 1937-38, bacteriology, 1939-40, M.D., 1939, Sc.D., 1955; NRC fellow, sch. pub. health U. Mich., 1942-44, Sc.D., 1955; LL.D., U. of Pitts., 1955; Ph.D., Hebrew U., 1953; LL.D., Roosevelt U., 1955; Sc.D., Turin U., 1957; Sc.D., U. Leeds, 1959; Sc.D., Hahneman Med. Coll., 1959; D.H.L., Yeshiva U., 1959; Sc.D., Franklin and Marshall, 1960; m. Donna Lindsay, June 8, 1939; children—Peter Lindsay, Darrell John, Jonathan Daniel. Intern Mt. Sinai Hosp., N.Y.C., 1940-42; research asso. epidemiology, sch. pub. health U. Mich., 1944-46; asst. prof. epidemiology, 1946-47; asso. research prof. bacteriology, head virus research lab., sch. medicine U. Pitts., 1947-49, research prof. bacteriology, 1949-54, prof. preventive medicine, chmn. dept., 1954-56; Commonwealth prof. preventive medicine, 1956-57, Commonwealth prof. experimental medicine, 1957-63; fellow, dir. Salk Inst. for Biol. Studies, 1963—. Cons. epidemic diseases Sec. War, 1944-46, Sec. Army 1946-54; mem. commn. on influenza Army Epidemiol. Bd. since 1944, acting dir. commn. on influenza 1944; expert cons. virus diseases WHO. Recipient Criss award, 1955, Lasker award, 1956. Fellow A.A.A.S., Am. Pub. Health Assn.; asso. fellow Am. Acad. Pediatrics (hon.); mem. Soc. Am. Bacteriologists, Am. Coll. Preventive Med., Am. Acad. Neurology, Assn. Am Physicians, Soc. for Exptl. Biology and Medicine, Am. Soc. Clin. Investigation, Am. Assn. Immunologists, Am. Epidemiol. Soc. A.M.A., Am. Assn. Univ. Profs., Sigma Xi. Contbr. sci. articles profl. jours. Developed effective poliomyelitis vaccine, 1954; research on immunization and immunological properties of influenza virus, immunological problems of poliomyelitis, and mechanisms of delayed hypersensitivity. Home: 6397 LaJolla Scenic Dr., La Jolla, Cal. Office: Salk Institute for Biological Studies, P.O. Box 9499, San Diego. 92009.*

SALKOVITZ, Edward Isaac, Am. metal physicist; b. Braddock, Pa., Sept. 3, 1917; s. Sam and Pauline (Taub) S.; B.S., Carnegie Inst. Tech., 1939; M.S., Carnegie Inst. Tech., 1948, D.Sc. in Physics, 1950; m. Suzanne Feld, Nov. 6, 1946; children—Lisa, Judith, Daniel David. Tchr., Pitts. Bd. Edn., 1939; jr. physicist Bacharach Indsl. Inst., Pitts., 1939-40, Navy Dept., Bur. Ordnance Washington, 1940-41; asst. physicist Am. Soc. Heating and Ventilating Engrs., Pitts., 1941; physics instr. Carnegie Inst. Tech., 1941-42; asst. physicist U. S. Army Signal Corps, Balt., Washington, 1942; staff Naval Research Lab., Washington, 1942-60, head metal physics br., 1958-60, acting asso. supt. metallurgy div., 1958-60; head metallurgy br. Office Naval Research, Washington, 1960-65; dir. materials sci. office Advanced Research Project Agy., Washington, 1965; prof. metall. engring. and physics, U. Pitts., 1965—; instr. optics and phys. metallurgy U. Md., 1950-53. Head cons. ARPA, 1965—; lectr. physics Howard U., 1958-61. Recipient USN Meritorious Civil Service award, 1959, Civil Service performance awards, 1955, 57-60, 62. Buhl Found. scholar, 1935; Carnegie Honor scholar, 1935-39. Fellow Am. Phys. Soc., Washington Acad. Sci.; mem. Am. Inst. Mining Engrs., Am. Soc. Metallurgists, A.A.A.S., Philos. Soc. Washington, Sigma Xi. Research on single crystal plasticity; radiation effects in magnetic materials; x-ray studies of deformed metal and nonmetal crystals. Home: 5031 Bayard St., Pitts.

SALLER, K(arl) F(elix), German anthropologist; b. Kempton, Germany, Sept. 3, 1902; s. G. (Stieve) Saller; Ph.D., 1924; M.D., 1926; Ph.D. (hon.), U. Munich (Germany), 1962; m. Herta Gross, 1930; 3 daus., 1 son; m. 2d, Nina Elsner, 1948; 2 daus., 1 son. Became lectr. anthropology and anatomy Univs. of Kiel and Gottingen until 1935; practice medicine; sanatorium physician, Badenweiler, 1936-39; dir. Robert Bosch Hosp., Stuttgart, Germany, 1945-48; prof. anthropology and genetics, Munich, Germany 1949—. Senator, German European Acad. Author: Lehrbuch der Anthropologie, 1955; Zivilisationund Sexualität, 1956. Pub., Jour. Die Heilkunst. Research in parapsychology, especially related to age and racial differents, suggestion; growth of Nazi socialism in Germany, including effect of propaganda on devel. of Nazism.

SALLEY, John Jones, Am. dentist, univ. dean; b. Richmond, Va., Oct. 29, 1926; s. Thomas Raysor and Kathryn (Josey) S.; D.D.S., Med. Coll. Va., 1951; Ph.D., U. Rochester, 1954; m. Jean Gordon Cunningham, Dec. 21, 1950; children—Katharine Gordon, John Jones, Martha Cunningham. Research fellow U. Rochester, 1951-54; from instr. to prof., chmn. dept. oral pathology Med. Coll. Va., 1954-63; prof. pathology, dean Sch. Dentistry, U. Md., 1963—; cons. div. research grants NIH, 1962—, spl. cons. Nat. Inst. Dental Research, 1957-64; cons. USPHS Hosp., Balt., U. S. Naval Hosp., Portsmouth, Va., VA Hosp., Balt. Recipient Outstanding Civilian Service medal Dept. Army, 1961. Fellow A.A.A.S.; mem. Am. Dental Assn., Internat. Assn. Dental Research (Novice award 1953), Am. Acad. Oral Pathology, Sigma Xi, Sigma Zeta, Omicron Kappa Upsilon. Research and publs. on oral pathology, predisposing factors to devel. oral malignant disease; 1st to induce carcinoma in oral epithelium of exptl. animals. Home: 901 Greenleigh Rd., Balt. 21212.*

SALMASIUS, see Saumaise, Claude.

SALMI, Ernest William, Am. physicist; b. Detroit, Dec. 28, 1922; s. Toivo William and Elsie (Linn) S.; B.A., Wayne State U., 1946; M.S., U. Mich., 1947, Ph.D., 1950; m. Alice Nelly Hellsten, Aug. 20, 1949; children—Nadine Ann, Ernest William, Douglas R., Karen L. Staff mem. Los Alamos Sci. Lab., 1950-56, alternate group leader, 1956—. Cons. Euratom, Ispra, Italy, 1962-63. Mem. Am. Phys. Soc., Sigma Xi. Developed 1st digital code to calculate atomic bomb expansion; one of many inventors termionic diode converter. Home: 2920 Arizona St. Office: P. O. Box 1663, Los Alamos 87544.*

SALMOIRAGHI, Gian Carlo, neurophysiologist; b. Gorla Minore (Varese), Italy, Sept. 19, 1924; s. Giuseppi Carlo and Dina (Rinetti) S.; M.D., U. Rome, 1948; Ph.D., McGill U., 1959; m. Boyka S. Akraboff, Oct. 23, 1954. Came to U.S., 1952, naturalized, 1958. Med. officer-sr. med. officer Internat. Refugee Orgn., Italy, 1949-52; fellow, research asso. Cleve. Clin. Found., 1952-56; lectr. dept. physiology McGill U., 1956-58; neurophysiologist Nat. Inst. Mental Health, NIH, Bethesda, Md., 1959—, chief, sect. on neurophysiology Clin. Neuropharmacology Research Center, 1961-67, acting chief, 1965-66, chief center, 1966-67, chief lab. neuropharmacology div. spl. mental health research programs, dir. div., dir. research St. Elizabeths Hosp., Washington, 1965—. Mem. Am. Physiol. Soc., Am. Soc. for Pharmacology and Exptl. Therapeutics, Internat. Brain Research Orgn., A.A.A.S. Research, publs. on electrophysiology and neuropharmacology. Home: 8216 Hamilton Spring Ct., Bethesda, Md. 20034. Office: St. Elizabeths Hosp., Washington 20032.*

SALMON, Daniel Elmer, Am. veterinarian; b. Mt. Olive, N.J., July 23, 1850; s. Daniel Landon and Eleanor (Flock) S.; D.V.S., Cornell, 1872, D.V.M., 1876; m. Mary Thompson Corning, Oct. 17, 1872 (dec. 1902); m. 2d, Agnes Christina Dewhurst, Nov. 15, 1904. With U. S. Dept. Agr. as investigator, 1879-84, chief Bur. Animal Industry, 1884-1906; dir. Nat. Vet. Sch., Montevideo, Uruguay, 1906-12, organized dept. vet. medicine. Fellow A.A.A.S.; pres. U. S. Vet. Med. Assn., 1898; hon. asso. Royal Coll. Vet. Surgeons of Gt. Britain; hon. mem. Royal Agrl. Soc. Eng., Epidemiol. Soc. London; fgn. asso. Société Centrale de Médicine Vétérinaire of France. Demonstrated that dead virus or filtered products of microorganisms may be used to provide immunity, 1886; discovered (with Theobald Smith) causative organism of swine plague, Salmonella cholerae-suis, 1886; brought under control cattle diseases contagious pleuro-pneumonia and Tex. fever; studied cause and prevention of fowl cholera, also of nodular disease in sheep; responsible for system of meat inspection, quarantine system for imported livestock, inspection of exported cattle and ships transporting them. Died Butte, Mont., Aug. 30, 1914.

SALMON, George, Irish mathematician; b. Dublin, Ireland, Sept. 25, 1819; D.D., Trinity Coll., Dublin; D.C.L. (hon.), Oxford (Eng.) U.; LL.D., Cambridge (Eng.) U. Regius prof. divinity Trinity Coll., Dublin, 1866-88, provost, from 1888. Fellow Royal Soc. (Royal medal, Copley medal), Royal Soc. Edinburg, French Acad. Scis., 1884, Royal Irish; Brit. acad.; mem. Brit. Assn. (pres. math. and physics sect. 1878). Author: Conic Sections, 1847; Higher Plane Curves, 1852; Geometry of Three Dimensions, 1862; Modern Higher Algebra, 1863; The Reign of Law, 1873; Non-Miraculous Christianity, 1881; Historical Introduction to New Testament, 1885; Gnosticism and Agnosticism, 1887; Infallibility of the Church, 1888; Thoughts on Textual Criticism of New Testament, 1897; Cathedral and University Sermons, 1900. Studied properties of cubic surfaces and determined (with Cayley) their straight lines; defined Steiner's hexagrammum mysticum. Died Dublin, Jan. 22, 1904.

SALMON, Peter Alexander, Canadian surgeon; b. Victoria, B.C., Can., Aug. 5, 1929; s. Arthur L. and Ina (Gordon) S.; came to U.S., 1947, naturalized, 1960; B.S., U. Wash., 1951, M.D., 1955; M.S. in physiology, U. Minn., 1961, Ph.D. in Surgery, 1961; m. Janet Nancy Zeeb, Dec. 28, 1953; children—Steven Alexander, David James, Peter Gordon. Asst. prof. surgery U. Minn., 1962—; dir. surg. edn. and research Mt. Sinai Hosp., Mpls., 1963-66; asso. prof. surgery U. Alta. Hosps., Edmonton, Can., 1966—. Mem. Soc. Acad. Surgeons, A.M.A., Canadian Med. Assn., Soc. for Exptl. Biology and Medicine, Sigma Xi. Research, publs. on gastric hypothermia for hemorrhage, carcinoma of pancreas, isolated gastric perfusion, gastric homotransplantation, microsurgery; invented pneumatic needle holders. Home: 84 Quesnell Crescent. Office: Dept. Surgery, U. Hosps., Edmonton, Alta., Can.*

SALMON, Robert, Brit. inventor; s. William Salmon; ed. in law; clerk of works under Henry Holland, circa 1746-1906; worked on rebuilding of Carlton House; employed at Woburn Abbey, Bedfordshire, 1790; architect to Francis Russell, 5th duke of Bedford, 1794-1821. Recipient Silver medals Soc. Arts, for surg. instruments, a canal lock, a weighing machine, a humane man-trap, a system of earthwalls. Author: An Analysis of the General Construction of Trusses, 1807; also several papers. Devised and improved agrl. instruments; inventor 1st haymaking machine, 1814. Died Oct. 6, 1821.

SALMON, William Davis, Am. animal nutritionist; b. Cork, Ky., Feb. 12, 1895; s. Thaddeus Robert and Lucy (Demumbrun) S.; B.S., U. Ky., 1920, D.Sc., 1958; M.S., U. Mo., 1921; m. Helen Bowman, Aug. 7, 1924; children—William Davis, Joseph Thaddeus, Jane Helen (Mrs. Robert DeVan Jones), Robert Bruce, Charles Richard. Asst. animal husbandsman Clemson (Ala.) U., 1921-22; animal nutritionist Auburn (Ala.) U., 1922-50, prof., head dept. animal sci., 1950-57, prof., 1957-65, prof. emeritus, 1965-66. Bd. dirs. Auburn Research Found. Named Distinguished Alumni U. Ky., 1964. Fellow Am. Inst. Nutrition; mem. A.A.A.S., Am. Assn. Cancer Research, Am. Soc. Animal Sci., Am. Soc. Biol. Chemists, Soc. Exptl. Biology and Medicine, N.Y., Ala. acads. scis., Am. Chem. Soc., Sigma Xi, Gamma Sigma Delta, Alpha Zeta, Phi Kappa Phi. Asso. editor: Jour. Nutrition, 1954-58. Contbr. numerous articles to profl. jours. Research, publs. on mineral and vitamin nutrition, lipids, proteins, amino acids, nutrition to carcinogenesis; pproved that vitamin B was a group of vitamins, parakeratosis was a zinc deficiency, prolonged deficiency of choline can lead to cancer; research exptl. atherogenesis. Home: 455 Brookside Dr., Auburn, Ala. 36830. Died Feb. 5, 1966.

SALMON-LEGAGNEUR, François, French organic chemist; b. Paris, June 1, 1897; s. Raymond and Marguerite (Danloux du Mesnil) S.-L.; Ecole Superieure de Physique et Chimie, Paris, 1921; Docteur ès sciences, Ingenieur EPCI, Faculté des Sciences, Paris, 1927; m. Elisabeth Ribadeau-Dumas, Feb. 7, 1927; children—Maud, Emmanuel, Pascale (Mrs. Carné-Carnavalet), Francine (Mrs. Francois Bergot). Faculty, Faculté de Scis. Paris, 1923-33; faculty Faculté des Sciences, Rennes, France, 1933—, prof., 1937—, dir. Ecole Nationale Supérieure de Chimie, 1960-64, hon. dir., 1964—. Decorated chevalier Legion of Honor; officer Order Nat. Merit; comdr. Order Acad. Palms; officer Agrl. Merit; recipient medal Soc. Liberated France, 1949. Mem. Chem. Soc. France, Soc. Biol. Chemistry, Sci. Soc. Biology, French Chem. Soc. Contbg. author: Traité de Chimie organique, 1937. Research, numerous publs. on organic chemistry, including camphoric acids and diacids, acids homologous to camphoric acid and monocamphoric acid, mixed acids and diacides, acid-aldehydes alpha alpha di and monosubstitutions. Home: 32 Blvd. de Metz, Rennes 35, France.*

SALO, Torsti Päiviö, Am. biochemist; b. Virginia, Minn., Jan. 24, 1916; s. Victor E. and Anna (Kujala) S.; B.S. with distinction, U. Minn., 1938; Ph.D., U. Mich., 1943; m. Evelyn Whitney, Aug. 23, 1941; children—Lawr Victor, Ann Whitney, Margaret Claire. Research asso. dept. biology Mass. Inst. Tech., 1942-49; asso. prof. biochemistry and chemistry U. Tenn., Knoxville, 1949—. Mem. Sigma Xi, Phi Lambda Upsilon, Gamma Sigma Delta. Contbr. articles to tech. jours. Co-inventor process for purification collagen and its extrusion into continuous monofilament for suture manufacture; research on collagen structure, characterization components melanoma tissue, biosynthesis protein in melanoma tissue. Home: 1527 Ramsay St., Alcoa, Tenn. 37701. Office: Dept. Biochemistry, U. Tenn., Knoxville, Tenn. 37916.*

SALOMAA, Pentti, Finnish chemist; b. Tuusula, Finland, June 28, 1922; s. Jalmari Edward and Aili (Prins) S.; M.Sc., U. Turku (Finland), 1947, Ph.D., 1953; postgrad. U. Coll., London, Eng., 1955; m. Raili Vuorinen, Dec. 20, 1947; children—Martti Antero, Sinikka Anneli. Supr. analytic Lab., Outokumpu Oy, Finland, 1947-50; faculty U. Turku, 1950—, prof. chemistry, 1961—. Mem. bd. Nat. Research Council for Scis., 1966—. Recipient Komppa award Finnish Chem. Soc., 1954. Mem. Finnish Chem. Soc. (past mem. bd.), Am. Chem. Soc. Contbg. author: The Chemistry of the Carbonyl Group, 1966; Chemical Physics of Ionic Solutions, 1966; also numerous articles. Research on kinetics and mechanisms of various ethers, esters, 5-membered ring compounds, applications of deuterium solvent isotopes effects. Home: Koulukatu 9 A, Turku, Finland.*

SALPETER, Edwin Ernest, physicist; b. Vienna, Austria, Dec. 3, 1924; s. Jakob L. and Friedericke (Horn) S.; B.Sc., Sydney (Australia) U., 1941, M.Sc., 1945; Ph.D., Birmingham U., 1948; m. Miriam M. Mark, June 11, 1950; children—Judy

G., Shelley R. Dept. Sci. Indsl. Research research fellow Birmingham U., 1949; research asso. Cornell U., Ithaca, N.Y., 1949-53; prof., U., 1954—; vis. prof. Australian Nat. U., 1953-54. Editorial bds. Astrophys. Jour., Jour. Math. Physics. Fellow Am. Acad. Arts and Scis.; mem. Nat. Acad. Scis., Internat. Astron. Union, Am. Phys. Soc., Am. Astron. Soc., Union Radio Sci. Internat. Author: (with H. A. Bethe) Quantum Mechanics of Atoms, 1957. Research, numerous publs. on quantum field theory, nuclear astrophysics, stellar evolution, ionospheric physics atomic theory. Home: 116 Westbourne Lane, Ithaca, N.Y. 14850.*

SALT, George, English biologist; b. Loughborough, Eng., Dec. 12, 1903; s. Walter and Mary (Hulme) S.; B.Sc., U. Alta., Edmonton, Can., 1924; S.M. (Atkins scholar) Harvard, 1925, S.D. (Anna C. Ames scholar), 1927; Ph.D., Cambridge (Eng.) U., 1933, Sc.D., 1941; m. Joyce Laing, July 27, 1939; children—Michael, Peter. Entomologist, United Fruit Co., Republic of Colombia, 1927; NRC fellow, Harvard, 1928; entomologist Imperial Inst. Entomology, 1929-31; Fellow Royal Soc., 1956. Moseley Research student Cambridge U., 1932, lectr., 1937-65, reader, 1965—; fellow King's Coll., Cambridge, 1933—, dean, 1939-45, tutor for research students, 1945-51. Researcher, E. African Agr. Research Orgn., 1948-49, Pakistan Dept. Plant Protection, 1958-59; vis. prof. U. Cal. at Berkeley, 1966. Fellow Royal Soc. London; mem. Soc. for Exptl. Biology, Brit. Soc. Parasitologists, Brit. Ecol. Soc., Royal Entomol. Soc. Research, numerous publs. on behaviour and physiology parasitic insects as a basis for biol. control pests, reactions insect blood cells to parasites and other fgn. bodies, anthropod fauna soil, ecol. studies. Address: King's Coll., Cambridge, Eng.*

SALTER, William Thomas, Am. pharmacologist; b. Boston, Mass., Dec. 19, 1901; s. William Thomas Hall and Frances B. F. (Patten) S.; grad. Roxbury Latin Sch., 1918; A.B., Harvard, 1922, M.D., 1925; m. Eleanor Vallandigham, June 27, 1935; children—Frances V., Eleanor C., Katharine C. Intern Mass. Gen. Hosp., 1925-27, resident, 1927-28; Moseley traveling fellow Harvard, 1928-29, research fellow in medicine, 1929-32; instr. biochem. sciences, Med. Sch., 1928-39; asst. prof., 1934-41; asso. physician Huntington Meml. Hosp., 1929-39; jr. asso. in medicine Peter Bent Brigham Hosp., 1937-39; research fellow in biochemistry Harvard U. Cancer Commn., 1929-39; faculty instr. medicine Harvard, 1932-34; prof. pharmacology Yale U. Sch. Medicine, since 1941; asso. physician, Thorndike Meml. Lab., 1939-41. Recipient Iodine Ednl. Bur. Inc. award for research in pharm. chemistry of iodine. Mem. A.C.P., Am. Assn. for History of Medicine, Am. Assn. for Cancer Research, Am. Inst. of Nutrition, Am. Soc. Clin. Investigation, Assn. for Study of Internal Secretions, Biochem. Soc. (Eng.), Am. Chem. Soc., Sigma Xi, Phi Beta Kappa. Died July 30, 1952.

SALTZMAN, Herbert Aaron, Am. physician; b. Phila., Nov. 27, 1928; s. Maurice and Sophie (Aaron) S.; student Ursinus Coll., 1946-48; M.D., Jefferson Med. Coll., 1952; m. Charlotte Wolf, Aug. 15, 1954; children—Ann Louise, Patricia Gay, Stephen George. Cardiologist, asst. pulmonary physiology Sampson Air Force Hosp., 1955-57; cardiologist, internist Lackland Air Force Hosp., 1956-57; chief pulmonary diseases Durham (N.C.) VA Hosp., 1958-63; asso. prof. medicine, dir. hyperbaric unit Duke Med. Center, Durham, 1963—. Diplomate Nat. Bd. Med. Examiners. Mem. A.M.A., Am. Fedn. for Clin. Research, So. Soc. for Clin. Research, N.C. Thoracic Soc. (past pres.), Am. Physiol. Soc. Diplomate Am. Bd. Internal Medicine. Research, publs. on biol. response to changing environments and gas concentrations, response of animals, man and certain patients to very high pressures of oxygen. Home: 2728 McDowell St., Durham, N.C. 27705.*

SALUTSKY, Murrell L., Am. chemist; b. Goodman, Miss., July 16, 1923; s. Rubin and Fay (Goldberg) S.; B.S., U. Ky., 1944; postgrad. civil engring. Ohio State U.; Ph.D. with distinction Mich. State Coll., 1950; m. Mary Ellen Butler, Mar. 20, 1948 (dec. June 1959); 1 dau., Laura Lee; m. 2d, Phyllis (Perlman) Robin, June 26, 1966; stepchildren—Susan Robin, Ilene Robin. Staff, Mound Lab., Monsanto Chem. Co., Miamisburg, O., 1950-55, sr. research chemist Monsanto Chem. Co., Dayton, O., 1955-57; supr. chem. research div. W.R. Grace & Co., Clarksville, Md., 1957-65; dir. research Dearborn Chem. div., Lake Zurich, Ill., 1965—. Cons. AEC, Mound Lab., Miamisburg, O., 1955-57. Commd. Ky. col., 1964. Mem. Am. Chem. Soc. (certificate of merit div. water and waste chemistry 1963), Am. Inst. Chemists, Marine Tech. Soc., A.A.A.S. Author: (with H.W. Kirby) The Radiochemistry of Radium, 1965; also articles, chpts. in books, ency. Research on rare earths, properties of precipitates, radiochemistry, phosphorus and comounds of it, fertilizers, desalination and mineral recovery from water. Home: 442 Elm Pl., Highland Park, Ill. Office: 320 Genesee St., Lake Zurich, Ill. 60047.*

SALVADORI, Mario Giorgio, civil engr.; b. Rome, Italy, Mar. 19, 1907; s. Riccardo and Ermelinda (Alatri) S.; D.Civil Engring., U. Rome, 1930, D. Math, 1932; m. Giuseppina Tagliacozzo, June 30, 1935; 1 son, Vieri R. Asst. to libero docente theory of structures U. Rome, 1930-38; cons. engr. Istituto Naz di Calcolo, Rome, 1933-38; time and motion engr. The Lionel Corp., Irvington, N.J., 1939; v.p. Indsl. Products Trading Corp., N.Y.C., 1939-42; from lectr. to prof. civil engring. and architecture Columbia U., 1940—, chmn. div. architectural tech., 1965—, chmn. session on sci. and tech. U. Bicentennial, 1954; partner Paul Weidlinger, cons. engr., N.Y.C., 1955—; v.p. Advanced Computer Techniques Corp., N.Y.C. Mem. Assembly Internat. Union Theoretical and Applied Mechanics, Am. Soc. C.E. (com. on applied mechanics 1944-50, com. on thin shells, structure div. 1949-52; chmn. com. on math. methods engring. mechanics div. 1952-54), Am. Soc. M.E., Sigma Xi, Tau Beta Pi. Fellow N.Y. Acad. Scis. Author: (with K. S. Miller) The Mathematical Solution of Engineering Problems, 1948; (with M. L. Baron) Numerical Methods in Engineering, 1952; (with R. J. Schwartz) Differential Equations in Engineering Problems, 1954; (with R. A. Heller) Structure in Architecture, 1963; (with John N. McCormick) Numerical Methods in Fortran, 1964; (with M. Levy) Structural Design in Architecture, 1967; Mathematics in Architecture, 1968; also articles in Am., English, and Italian jours. Home: 2 Beekman Pl., N.Y.C. 10022.*

SALVEMINI, Giovanni, see Castillon, Jean.

SALVIN, Samuel Bernard, Am. immunologist; b. Boston, July 10, 1914; s. Benjamin and Lena (Smith) S.; B.A., Harvard, 1935, Ed.M., 1937, Ph.D., 1941; m. Helen Elizabeth Huleatt, Apr. 21, 1920; children—Helen Thalia, Carol Marjorie, Donald Samuel. Instr., Harvard, 1941-43; staff NIH, Bethesda, Md., 1946-50; head immunology and mycology Rocky Mountain Lab., Hamilton, Mont., 1950-63; vis. prof. depts. microbiology and pediatrics U. Minn., Mpls., 1963-64; head immunology research div. CIBA Pharm., Summit, N.J., 1964-67; prof. dept. microbiology U. Pitts. Sch. Medicine, 1967—. Mem. Am. Assn. Immunologists, Am. Soc. Microbiologists, Mycology Soc., Am. Acad. Microbiology, N.Y. Acad. Sci., Internat. Soc. Med. Mycology, Sigma Xi. Research, numerous publs. on immunology of mycotic infections, delayed hypersensitivity, immunologic tolerance, allergic thyroiditis, allergie encephalomyelitis. Office: Dept. Microbiology, U. Pitts. Sch. Medicine, Pitts. 15213.*

SALZANO, Francisco Mauro, Brazilian geneticist; b. Cach. doSul, Brazil, July 27, 1928; s. Francisco and Onelia (Pertille) S.; B.Sc., U. R.G. doSul, Porto Alegre, Brazil, 1950, Sc.Lic., 1952; Ph.D., U. Sao Paulo (Brazil), 1955; m. Thereza Praseires Torres, Mar. 20, 1952; children—Felipe, Renato, Faculty, U. R.G. doSul, 1952—, asso. prof. genetics, 1966—, head genetics sect. Instituto de Clencias Naturals, 1963, dep. dir., 1963—. Mem. N.Y. Acad. Scis., Am. Soc. Human Genetics, Am. Assn. Phys. Anthropologists, Soc. Bras. Genet., Soc. Bras. Prog. Ciencia. Research, numerous publs. on population genetics of fruit fly, demographic and genetic structure of primitive Brazilian Indian populations. Home: 1092 Venancio Aires, Caixa Postal 1953, P. Alegre, R.G. do Sul, Brazil.*

SALZBERG, Paul Lawrence, Am. chemist; b. Galesburg, Ill., Mar. 2, 1903; s. Karl and Clara (Seastedt) S.; B.S., Knox Coll., 1925; Ph.D. in Chemistry (Carr fellow), U. Ill., 1928; D.Sc., Knox Coll., 1958; m. Grace Ella Johnson, June 25, 1929; children—Robert Harris, John Paul, Claire Grace. With E. I. du Pont de Nemours & Co., 1928-67, successively chemist, research supr., lab. dir., asst. dir., dir. central research dept., 1953-67. Recipient Knox Coll. Alumni Achievement Award, 1949. Mem. Am. Chem. Soc., Indsl. Research Inst. (medal 1966), Del. Astron. Soc., A.A.A.S., Phi Beta Kappa, Sigma Xi, Alpha Chi Sigma, Phi Lambda Upsilon. Club: Wilmington (Del). Research and numerous patents on plastics, synthetic fibers, insecticides, pharmaceuticals, weed killers, lubricants, chem. intermediates. Home: 1525 Foulk Rd., Wilmington, Del. 19803.*

SALZMAN, Leon, Am. physician; b. N.Y.C., July 10, 1915; s. Morris and Sarah (Wolff) S.; B.S., Coll. City N.Y., 1935; M.D., Royal Coll. Phys. and Surg., Edinburgh, Scotland, 1940; Ann Bailin, Oct. 30, 1950; children—Carol, Sarah. Practice medicine specializing in psychoanalysis, Washington, 1945—; faculty Washington Sch. Psychiatry, 1948—; prof. clin. psychiatry Georgetown U. Med. Sch., 1963—; vis. lectr. Cath. U. Grad. Sch., 1950—. Cons. St. Elizabeth's Hosp., Washington, 1957—, Nat. Inst. Mental Health, 1960—. Diplomate Am. Bd. Neurology and Psychiatry. Mem. Acad. Psychoanalysis (past pres.), Am. Psychoanalytic Assn., Am. Psychiat. Assn. Author: (with Masserman) Modern Concepts of Psychoanalysis, 1962; Developments in Psychoanalysis, 1962; also articles. Research on theory and therapy of obsessional states, problems of sexuality especially homosexuality and female sexuality, relationship of religion and psychiatry. Home: 7402 Barra St., Bethesda, Md. 20014. Office: 1610 New Hampshire Av., Washington 20009.*

SAM, Joseph, Am. med. chemist; b. Gary Ind. Aug. 15, 1923; s. Andrew and Flora (Toma) S.; student Drake U., 1942-43; B.S., U. S.C., 1948; Ph.D., Kan. U., 1951; m. Frances Adickes, Sept. 11, 1945; children—Sherrie, Joseph A., Suzanne F. Sr. research chemist McNeil Labs., Phila., 1951-54; research group leader Bristol Labs., Syracuse, N.Y., 1955-57; sr. scientist E. I. du Pont de Nemours & Co., Camden, S.C., 1957-59; faculty U. Miss., 1959—, prof. pharm. chemistry, 1961—, chmn. dept., 1963—, asso. dir. Research Inst. Pharmaceut. Scis., 1966—. Vis. scientist Am. Assn. Colls. Pharmacy. Fulbright fellow Cairo U., 1965-66. Mem. Am. Chem. Soc., Am. Pharm. Assn., Sigma Xi, Rho Chi, Phi Lambda Upsilon, Phi Kappa Phi. Research, publs. in medicinal chemistry, organic syntheses, mechanism of drug action. Home: Box 351, University, Miss.*

SAMADDAR, Surendra Nath, elec. engr.; b. Kashipur, Bengal, India, Feb. 1, 1926; s. Bhava Ranjan and Umashashi (Banerjee) S.; B.Sc. with honours in math., U. Calcutta, 1948, M.Sc. in Applied Physics, 1950; Ph.D. in Elec. Engring., U. Mich., 1961. Came to U. S., 1957. With Calcutta Engring. Coll., 1951-52, Ericsson Telephone Co., Calcutta, India, 1953-57; sr. research fellow Microwave Research Inst., Poly. Inst. Bklyn., 1957-60; sr. scientist Raytheon Co., Sudbury, Mass., 1961—. Mem. I.E.E.E., Sigma Xi. Contbr. articles to profl. jours. Research in diffraction and scattering of electromagnetic waves in anisotropic media, interaction of electromagnetic waves with plasma, wave phenomena in plasma in presence of boundaries, propagation of electromagnetic waves through inhomogeneous media. Home: 58A Edgehill Rd., Waltham, Mass. 02154. Office: 528 Boston Post Rd., Sudbury, Mass. 01776.*

SAMARIN, Aleksandr Mikhailovich, Russian metallurgist; b. Aug. 14, 1902; grad. Moscow Inst. Steel, 1930. Became prof. Moscow Inst. Steel, 1938; named dep. dir. USSR Acad. Scis. Inst. Metallurgy, Moscow, 1955. Dep. chmn. State Com. for Coordination of Sci. Research, 1962. Recipient Order Red Banner of Labor, 1962. Corr. mem. USSR Acad. Scis.; hon. mem. Hungarian Acad. Scis. Author: Electrometallurgy. Steel Production, 1943; Structure and Properties of Liquid Methods, 1960. Research in electrometallurgy of steel, feroalloys; history sci. and tech. Home: Leningradskii prospekt, 13, Moscow, USSR. Office: A. A. Baykor Inst. Metallurgy, USSR Acad. Scis., Leninskii Prospekt, 49, Moscow, USSR.

SAMBUCUS, Johannes, physician; b. Tyrnau, 1531; prof., Bologna, Italy; physician, historian to Kaiser Maximilian II. Author: Icones veterum aliquot ac recentium medicorum philosophorumque elegiolis suis editae, 1574; commentary on Bonfini's Rerum ungaricarum decades quatuor. Collected manuscripts; pub. classics, also portrait work of physicians and philosophers. Died Vienna, 1584.

SÄMISCH, Edwin Theodor, German ophthalmologist; b. Luckau, Germany, Sept. 30, 1833; prof., Bonn, Germany. Editor: (with Graefe) Handbuch der gesamten Augenheilkunde, 7 vols., 1874-80. Gave 1st description of lucus serpens. Died Bonn, Nov. 29, 1909.

SAMITZ, Morris Harriss, Am. physician; b. Phila., Dec. 18, 1909; s. Philip H. and Rachel (Rabinowitz) S.; student Temple U., 1926-29, M.D., 1933; M.Sc. in Medicine, U. Pa., 1942; postgrad. N.Y. U.; m. Doris Rubin, Nov. 4, 1945; children—Phyllis, Joel. Faculty, U. Pa., Phila., 1940—, prof. Sch. Medicine, 1964-67, dir. dermatology Grad. Div. of Medicine, 1967—, chief. dept. dermatology Grad. Hosp., 1949-53; attending chief Skin and Cancer Hosp., Phila., 1949-53, med. dir., 1953-54; vis. prof. dermatology Pa. Coll. Podiatry, Phila., 1965—; vis. prof. medicine Hahnemann Med. Coll., 1967—, U. S. Naval Hosp., 1963—, (both Phila.). USPHS Research grantee, 1958—. Diplomate Am. Bd. Dermatology (preceptor grad. tng. in dermatology 1950-64). Mem. Am. Acad. Dermatology, Am. Coll. of Allergists, American College of Physicians. Soc. Investigative Dermatology, College of Physicians Phila., N.Y. Acad. Scis., A.A.A.S., Argentina (corr.), Uruguay (corr.) dermatol socs. Research, numerous publs. on metal sensitivity, dermatologic-gastrointestinal relationships. Home: 1715 Pine St., Phila. 19103. Office: Duhring Labs., U. Pa., Phila. 19104.*

SAMPEY, John Richard, Am. chemist; b. Louisville, Aug. 5, 1896; s. John Richard and Annie (Renfroe) S.; S.B., U. Chgo., 1920, S.M., 1921, Ph.D., 1923; LL.D., Furman U., 1965; m. Ida Jewell Cheatham, Sept. 4, 1925; children—John Richard III, Jane Renfroe. Postdoctoral research Johns Hopkins U., 1923-24, 30-31; asso. prof. Howard Coll., 1924-26, prof., 1926-34; prof. Furman U., Greenville, S.C., 1934-64, chmn. faculty, 1945-48. Inspector gen. U.S. Army, 1941-45. Recipient Alumni award Furman U., 1963; Tchr. award Mfg. Chemists Assn., 1961. Fellow A.A.A.S.; mem. Am. Chem. Soc. (Charles H. Herty medal 1954), Ala. (pres. 1927-28), S.C. (pres. 1940-41) acads. sci., So. Assn. Sci. and Industry (v.p., trustee), Sigma Xi, Chi Beta Phi, Kappa Phi Kappa. Research and many publs. on polarity of carbon-halogen bond, mechanisms of organic reactions, determination of organic sulfur compounds, chemotherapy of carcinogenic and anticancer agts., mgmt. of malignant blood diseases. Home: 4 Jones Av., Greenville, S.C. 29601.*

SAMPSON, Edward, geologist; b. Oxford, Eng., May 31, 1891 (parents Am. citizens); s. Alden and Mary Agnes (Yarnall) S.; C.E., Princeton, 1914, M.Sc. in Geology, 1915, D.Sc., 1920; m. Alfreda

Cope Lewis, June 5, 1917 (dec. 1932); children—Edward Sampson, Harold Y., Agnes (Mrs. Donald Lunn Appleby Sawyer); m. 2d, Katharine Westerlo van Rensselaer Arnold, Oct. 20, 1940 (dec. 1943); m. 3d, Eleanor Rodman Townsend, Jan. 15, 1955 (dec. 1965). Instr., Rutgers U., 1919-20; prin. geologist U. S. Geol. Survey, 1920-35, 43-46; faculty Princeton, 1925-—, prof. geology, 1934-59, chmn. dept., 1934-36, prof. emeritus, 1959-—. Recipient medal of Freedom, SCAP (Tokyo), 1946. Mem. geol. socs. Am., S. Africa, Philippines, Mineral. Soc. Am., Mineral. Soc. (London, Eng.), Soc. Econ. Geology, Am. Inst. Mining and Metallurgy, Geochem. Soc., Canadian Inst. Mining and Metallurgy. Research, publs. on mineral econs., origin chromite, use polarized light in study of ore sects.; co-inventor isodynamic separator; inventor polishing machine. Home: 102 Lafayette Rd., Princeton, N.J. 08540.*

SAMPSON, Ralph Allen, Brit. astronomer, mathematician; b. Skull, Ireland, June 25, 1866; s. James and Sarah Anne (Macdermott) S.; B.Sc., St. John's Coll., Cambridge, 1888; hon. degrees, Durham, Glasgow; m. Ida Binney, 1893; 1 son, 4 daus. Named fellow St. John's Coll., Cambridge. 1890; lectr. math., King's Coll., London; 1st Isaac Newton student in astronomy and phys. optics, 1891; became prof. math. Durham Coll., Sci., Newcastle-upon-Tyne, Eng., 1893; apptd. prof. math., 1896, also prof. astronomy, 1908; prof. astronomy Edinburgh U., from 1910; astronomer royal for Scotland; ret., 1937. Recipient 1st Smith's prize, 1890; Gold medal Royal Astron. Soc, 1928; Fellow Royal Soc., 1903. Author: Tables of the Four Great Satellites of Jupiter, 1910. Studied Stokes' current function and astron. spectroscopy. Died Bath, Eng., Nov. 7, 1939.

SAMSON, Frederick Eugene, Jr., Am. physiologist; b. Medford, Mass., Aug. 16, 1918; s. Frederick E. and Annie (Pratt) S.; Ph.D., U. Chgo., 1952; m. Camila Albert, Mar. 5, 1945. Faculty, U. Kan., Lawrence, 1952-—, prof., chmn. dept. comparative biochemistry and physiology, 1963-—. Resident scientist Mass. Inst. Tech. Neuroscis. Research Program, 1965-—. Mem. Am. Physiol. Soc., A.A.A.S., Am. Inst. Biol. Scis., N.Y. Acad. Scis., Biochem. Soc. (Gt. Britain), Am. Chem. Soc., Am. Cell Biology Soc., Assn. Neurol. and Mental Diseases. Research, publs. on energy metabolism brain. Home: 3401 Tam O Shanter St., Lawrence, Kan. 66044.*

SAMSON, James Alexander, physicist; b. Kilmarnock, Scotland, Sept. 9, 1928; s. James and Agnes (Ross) S.; came to U. S., 1953, naturalized, 1959; B.Sc. with honors, Glasgow (Scotland) U., 1952; M.S., U. So. Cal., 1955, Ph.D., 1958; m. Mary Simpson Richardson, Apr. 17, 1954; children—Ross, Scott. Research asso. U. So. Cal., 1958-60, Harvard, 1960-61; prin. scient'st GCA Corp., Bedford, Mass., 1961-—. Mem. Am. Phys. Soc., Optical Soc. Am. Author: Techniques of Vacuum Ultraviolet Spectroscopy, 1967; also articles. Research on mechanism by which extreme ultraviolet radiation is absorbed by atoms and molecules, prodn. of photelectrons by such absorption processes. Home: 210 Doty St., Waltham, Mass. 02154. Office: GCA Corp., Burlington Rd., Bedford, Mass. 01730.*

SAMSON, Paul Curkeet, Am. thoracic and cardiac surgeon; b. Emporia, Kan., June 12, 1905; s. Paul Bryant and Rena Mae (Curkeet) S.; student Municipal U., Toledo, 1921-23; B.S., U. Mich. 1928, M.D., 1928, M.S. in Surgery, 1935; m. Marion Doris Smith, July 13, 1940; foster children—Robert Bruce, Arthur Montell Atterbury. James Herrick fellow Rush Med. Coll., asst. dean, 1930-32; Mary Bissell fellow thoracic surgery U. Mich., 1934-36; instr. thoracic surgery U. Mich. Hosp., 1934-36; instr. to adj. clin. prof. surgery Stanford U. Sch. Medicine, 1937-—; practice thoracic and cardiac surgery, Oakland, Cal., 1936-—; chief thoracic and cardiac surgery service Highland-Alameda County Hosp., 1946-58, cons., 1958-—; chief thoracic and cardiac surgery service Children's Hosp. of East Bay, 1946-62; chief surg. service Samuel Merritt Hosp., 1950-55, pres. med. staff, 1960-61; area cons. thoracic and cardiac surgery VA; cons. thoracic and cardiovascular surgery Travis AFB Hosp., Letterman Army Hosp., San Joaquin Hosp., Stockton, Cal., Peralta Hosp., Oakland, Oak Knoll Naval Hosp.; Luis Guerera Meml. lectr. Santo Tomas U., Manila, P.I., 1960; Stuart Graves lectr. Beta chpt. Nu Sigma Nu, U. Ore., 1962; John Alexander Meml. lectr., Ann Arbor, Mich., 1964. Diplomate Bd. Thoracic Surgery (founders group, mem. bd. representing A.C.S. 1955-60). Fellow A.C.S. (case records prize 1936; chmn. bd. govs. 1960-63; mem. bd. regents 1963-—), Am. Coll. Chest Physicians (Cal. chpt. award for outstanding services 1960); mem. A.M.A., Soc. Thoracic Surgeons (founding, 1st pres. 1963-65), Am. Thoracic Soc. (past pres., mem. exec. com.), Cal. Tb and Health Assn. (past pres.), Am. Assn. Thoracic Surgery (past pres.), Am. Broncho-Esophagological Assn., Pacific Coast, Am. surg. assns., San Francisco, Excelsior (pres. 1962-63), Reno (hon.) surg. socs., John Alexander Soc., Nat. Tb Assn., Soc. Air Force Clin. Surgeons (hon.). Mem. adv. editorial bd. Surgery, Gynecology and Obstetrics, others. Research and publs. on endoscopy and surg. diseases of lungs, esophagus, diaphragm, intrathoracic gt. vessels and heart; contributed to clin. and bronchoscopic recognition of tuberculous tracheobronchitis, acute and chronic thoracic trauma; devel. techniques for pulmonary decortication. Home: 15 LaSalle Av., Piedmont, Cal. 94611. Office: 459 30th St., Oakland, Cal. 94609.*

SAMSON, Sten Otto, chemist; b. Stockholm, Sweden, Mar. 25, 1916; s. Lage and Maria (Lode) S.; Fil.kand. U. Stockholm, 1953 Fil.lic, 1957, Fil. Dr., 1968; m. Lalli Sandström, July 23, 1948; children—Karl Otto, Karin Helena. Research fellow chemistry Cal. Inst. Tech., Pasadena, 1953-56, 57-61, sr. research fellow, 1961-—. Cons. Nuclear Corp. Am., 1958-61. Mem. Am. Crystallographic Assn., Sigma Xi. Co-editor: Crystal Data Determinative Tables, 1964-—. Research, publs. on very complex crystal structures intermetallic compounds by X-ray diffraction techniques: developed methods for structure determination, instrumentation for X-ray diffraction. Home: 351 S. Parkwood Av., Pasadena, Cal. 91107.*

SAMSONOV, Pavel Fedorovich, Russian microbiologist; b. Margelan, 1892; grad. Med. Faculty, Moscow U., 1916; D.Med. Sci. Sr. asst. dept. infectious diseases, lectr. microbiology Med. Faculty, Turkestan U., 1921-32; head epidemiology and microbiology depts. Uzbek Inst. Epidemiology and Microbiology, 1932-35; head Uzbek Brucellosis Sta., Tashkent, 1935-39, sci. cons., 1939-—; head chair microbiology Tashkent Med. Inst., 1939-—. First dep. chmn. learned med. council Uzbek Ministry Health. Decorated Order of Lenin. Mem. All-Union Soc. Microbiologists, Epidemiologists and Infectionists (chmn. Uzbek br.). Mem. editorial council Jour. Microbiology, Epidemiology and Immunobiology. Research and numerous publs. on etiology, epidemiology and microbiology of infectious diseases in Uzbekistan, bacteriology of wounds taking long time to heal. Address: Tashkent Med. Inst., ulitsa Karla Marksa 88, Tashkent, Uzbekistan SSR, USSR.

SAMTER, Max, physician; b. Berlin, Germany, Mar. 3, 1908; s. Paul and Claire (Rawicz) S.; student U. Freiburg (Germany), 1926; U. Innsbruck (Austria), 1928; M.D., U. Berlin (Germany), 1933; m. Virginia Svarz Ackerman, Oct. 17, 1947; 1 dau., Virginia-Claire. Came to U. S., 1937, naturalized, 1943. Practice medicine, Berlin-Karow, 1933-37; faculty Johns Hopkins, 1937-38, U. Pa., 1938-43; faculty U. Ill., Chgo., 1946-—, prof. medicine, 1960-—. Area cons. U. S. VA, 1962-—. Diplomate Am. Bd. Internal Medicine (past chmn. subsplty. bd. allergy). Mem. A.M.A., A.C.P., Am. Acad. Allergy (past pres.), Internat. Assn. Allergology (pres. elect), Histamine Club, Sigma Xi, Alpha Omega Alpha. Author: (with Oren C. Durham) Regional Allergy, 1954; (with Harry L. Alexander) Immunological Diseases, 1965; also numerous articles. Research on function eosinophils, pathophysiology bronchial asthma, drug reactions particularly aspirin. Home: 645 Sheridan Rd., Evanston, Ill. 60202. Office: 840 S. Wood St., Chgo. 60612.*

SAMUEL, Arthur L., Am. engr.; b. Emporia, Kan., Dec. 5, 1901; s. John R. and Lee Ella (McHenry) S.; A.B., Coll. Emporia, 1923; S.B., S.M., Mass. Inst. Tech., 1926; postgrad. Columbia; m. Bernice Deborah Crawford, May 23, 1931; children—Margaret Lee (Mrs. Frank H. Finch), Donna Elizabeth (Mrs. Matsood Hussain). Instr. elec. engring. Mass. Inst. Tech., 1926-28, vis. prof., 1963-—; mem. tech. staff Bell Telephone Co., 1928-46; prof. elec. engring. U. Ill., 1946-49; research cons. I.B.M., Yorktown Heights, N.Y., 1949-66; with computer sci. dept. Stanford, 1966-—. Recipient Exceptionally Meritorious Civilian Service award U. S. Dept. Def. Fellow Am. Phys. Soc., I.E.-E.E. Patentee in field. Contbr. articles to tech. jours. Research on ion and electron dynamics; design of high vacuum and gas filled electron tubes; solid state physics, digital computers; microwave circuit theory. Home: Lake Dr. S., Candlewood Isle, Box 53, New Fairfield, Conn. Office: Polya Hall, Stanford U., Stanford, Cal. 94305.*

SAMUEL, David, Israeli phys. chemist; b. Jerusalem, Israel, July 8, 1922; s. Edwin and Haddassah (Goor) S.; M.A., (hons.), Balliol Coll., Oxford (Eng.) U., 1948; Ph.D., Hebrew U., Jerusalem, 1953; m. Rinna Grossman, Dec. 14, 1960; children—Judith, Naomi. Faculty isotope dept. Weizmann Inst. Sci., Rehovoth, Israel, 1949-—, Sherman prof. phys. chemistry, 1967-—; vis. fellow chemistry dept. Harvard, 1957-58; research fellow biodynamics lab. U. Cal. at Berkeley, Lawrence Radiation Lab., 1965-66; vis. prof. U. Warwick, 1967-—. Chmn., Israel Com. for Reform of High Sch. Chemistry Curriculum, 1964-—; cons. Yeda Research & Devel. Corp., 1960-—. Fellow Chem. Soc., Faraday Soc.; mem. Israel Chem. Assn. Author: (with F. Steckel) Bibliography of the Stable Isotopes, 1959; (with I. Dostrovsky) Syntheses of Inorganic Compounds with Oxygen-18, 1962; also numerous articles, chpt. in book. Research on reaction mechanisms (way chem. reactions occur) especially those containing phosphorus; use of oxygen isotopes in various fields; chem. basis of memory and learning. Home: Neve Wolfson, Rehovoth, Israel.*

SAMUEL, Eric, Brit. radiologist; b. Cwmllynfell, Wales, Aug. 26, 1914; s. William and Miriam (Jones) S.; B.Sc., U. Wales, 1933; M.B., B.S., U. London (Eng.), 1936, M.D., 1939; diploma in med. radiology and electrology Cambridge (Eng.), U., 1939; m. Vera Eileen Meredith, May 6, 1942; children—William Meredith, Lesley Mary. Hon. radiologist Middlesex Hosp., London, Central London E.N.T. Hosp., 1944-46; radiologist in charge Royal Infirmary, Edinburgh, Scotland, 1958-—; reader radiology U. Edinburgh, 1958-—. Adviser in radiology to Brit. Army, 1944-45, Scottish Home and Health Dept., 1961-—; Hunterian prof. Royal Coll. Surgeons, 1953. Decorated Bronze Star (U.S.); Order of White Lion; recipient Roentgen award Brit. Inst. Radiologists, 1945. Fellow Royal Soc. Medicine, Royal Coll. Physicians Edinburgh, Faculty Radiologists; mem. Internat. Soc. Radiology (sec.-gen. 1966-—.). Author: Clinical Radiology of Ear, Nose and Throat, 1952; (with Rodney Smith) Radiology of Acute Abdomen, 1946; also numerous articles, chpts. in books. Research on radiology of ear, nose, throat and acute abdomen. Address: 2 Clinton Rd., Edinburgh, Scotland.

SAMUEL, Iosif Alexandru, Rumanian physician; b. Bucharest, Rumania, Feb. 4, 1928; s. Alexandru I. and Victoria (Klapper) S.; grad. Medico-Pharm. Inst., Bucharest, 1933, D.Med. Scis., 1966; m. Elisabeta Nastac, Nov. 5, 1961; 1 dau. Anca Ruxandra. Staff, Inst. Inframicrobiology, R.S.R. Acad., Bucharest, 1953-—, research worker, 1954-62, head research worker, 1962-—. Recipient Evidentiat munca medicosanitara, 1963. Mem. Union Socs. Med. Scis. Contbg. author: Inframicrobiologie speciala, 1962. Research, numerous publs. on encephalitis, mumps, Coxsackie viruses, infectivity of nucleic acids from influenza virus, adenovirus 3, leukomogenic viruses, enzymatic biosynthesis in vitro of infective ribonucleic and deoxyribonucleic acids. Home: 78 Bd. Muncii, Bucharest. Office: 285 Bd.Mihai Bravu, Bucharest, 29, Rumania.*

SAMUEL, Paul, physician; b. Janoshaza, Hungary, Feb. 17, 1927; s. Adolf and Magda (Zollner) S.; M.D., très honnorable, U. Paris, France, 1953; m. Gabriella Relly Zeichner, Mar. 27, 1954; children—Robert Marc, Adrianne Jill. Came to U.S., 1954, naturalized, 1960. Research fellow Columbia research div. Coll. Phys. & Surg., Goldwater Meml. Hosp., N.Y.C., 1957-58; instr. medicine N.Y. U., 1958-60; dir. arteriosclerosis research div., mem. staff L.I. Jewish Hosp., New Hyde Park, N.Y., 1961-—, Queens Hosp., 1965-—. Fellow Am. Soc. Study Arteriosclerosis; mem. Harvey Soc., A.C.P. Research, publs. on cholesterol and bile acid metabolism; effect of different fats on serum cholesterol and exptl. arteriosclerosis, genetic aspects of hypercholesterolemia, discovered that neomycin and other antibacterial drugs reduce serum cholesterol levels in man; demonstrated that bile acids are absorbed from large bowel in man. Home: 25 Nassau Dr., Great Neck, N.Y. 11021. Office: L.I. Jewish Hosp., New Hyde Park, N.Y. 11043.*

SAMUEL, Pierre, French mathematician; b. Paris, France, Sept. 12, 1921; s. Raymond and Jacqueline (Dalmeyda) S.; Ph.D., Princeton, 1947; Doctorat ès Scis., U. Paris, 1949; m. Nicole Herrmann, July 23, 1948. Prof., U. Clermont-Ferrand, 1949-61; prof. math. U. Paris, 1961-—; vis. prof. Cornell U., 1952-53, Harvard, 1955, U. Mexico, 1955, 64, U. Ill., 1959-60, U. Cal. at Berkeley, 1960, U. Rio de Janeiro (Brazil), 1958, Tata Inst., Bombay, India, 1963, 66, Brandeis U., 1964. Mem. Societe Mathematique de France, Def. de la Jeunesse Scolaire, Am. Math. Soc. Author or co-author several books, numerous articles. Research on commutative algebra applied to geometry; discoveries concerning intersect. multiplicities, singular points, unique factorization, algebraic curves. Home: 3 Av. du Lycee Lakanal, Bourg-La-Reine 92, France. Office: ENS, 48 Bd, Jourdan, Paris (14), France.*

SAMUELS, George, Am. plant physiologist; b. Phila., July 7, 1922; s. Philip and Dora (Spitalnick) S.; B.S. with honors, U. Del., 1946; Ph.D., Rutgers U., 1949; m. Mollie Freed, May 27, 1944; children—Lynn, Sharon. Research asst. soils Rutgers U., 1946-49; plant physiologist U. P.R., 1949-52, agronomist Agr. Expt. Sta., 1952-—, lectr. Foliar diagnosis for sugarcane and soil fertility, 1955-—. Research cons. on tropical agr., 1960-—. Mem. Am. Soc. Agronomy, Soil Sci. Soc. Am., Am. Soc. Hort. Sci., Sigma Xi, Gamma Sigma Delta. Author: Foliar Diagnosis of Sugarcane, 1966; also numerous articles. Devel. foliar diagnosis of plant to determine its fertilizer needs especially tropical crops, mineral deficiency studies in tropical crops, evaluation of fertility requirements of pineapples, sugarcane, tobacco, other tropical crops. Home: 260 Himalaya St., Rio Piedras 00926. Office: Agr. Expt. Sta., U. P.R., Rio Piedras, P.R. 00928.*

SAMUELS, Leo Tolstoy, Am. endocrinologist; b. Indpsl., Nov. 15, 1899; s. John M. and Lillie Alice (Davisson) S.; B.A., Emanuel Missionary Coll., 1925; Ph.D., U. Chgo., 1930; m. Barbara Katherine Petri, Mar. 11, 1935. Instr., U. So. Cal., 1934-37; faculty U. Minn., 1937-44, asso. prof., 1943-44; prof. biochemistry U. Utah, Salt Lake City, 1944-—, head dept. biol. chemistry, 1944-64. Recipient Am. Acad. Achievement award, 1965. Mem. Soc. Endocrinology Gt. Britain, Biochem. Soc. Gt. Britain, Endocrine Soc. (Fred Konrad Koch award 1964), Am. Pharmacological Soc., N.Y. Acad. Scis., Soc. for Exptl. Biology and Medicine, Am. Soc. Biol. Chemists, Western Assn. Physicians, Western Soc. for Clin. Investigation, Sigma Xi. Research, numerous publs. on endocrine system and control physiology, intermediate metabolism, determination, metabolism hormones, and enzyme influence. Home: 850 18th Av., Salt Lake City 84103.*

SAMUELSON, Bengt Ingemar, Swedish biochemist; b. Halmstad, Sweden, May 21, 1934; s. Anders G. and Stina (Nilsson) S.; D.Med.Sc., Karolinska Institutet, Stockholm, Sweden, 1960, M.D., 1961; m. Karin Bergstein, Aug. 19, 1958; children—Bo, Elisabet, Astrid. Asst. prof. med. chemistry Karolinska Institutet, 1960-66; prof., head dept. med. chemistry Royal Vet. Coll., Stockholm, 1967—. Research, publs. on biochemistry of bile acids and prostaglandins. Home: 7 Vikinga Vägen, Danderyd, Sweden. Office: Dept. Med. Chemistry, Royal Vet. Coll., Stockholm, Sweden.*

SAMUELSON, Paul A., Am. economist; b. Gary, Ind., May 15, 1915; s. Frank and Ella (Lipton) S.; B.A., U. Chgo., 1935, LL.D., 1961; M.A. (Social Sci. Research Council fellow), Harvard, 1936, Ph.D. (David A. Wells prize), 1941; LL.D., Oberlin Coll., 1961, Boston Coll., 1964; D.Litt., Ripon Coll., 1962; m. Marion E. Crawford, July 2, 1938; children—Jane Kendall, Margaret Wray, William Frank, Robert James, John Crawford, Paul Reid. Prof. econs. Mass. Inst. Tech., Cambridge, 1940—. Cons. Nat. Resources Planning Bd., 1941-43, WPB, 1945, U.S. Treasury, 1945-52, Rand Corp., 1949—; mem. Nat. Task Force on Econ. Edn., 1960-61; Com. Econ. Devel. Research Adv. Bd., 1960—; adviser Pres.'s Nat. Goals Commn., 1960. Guggenheim fellow. 1948-49; Ford Faculty Research fellow, 1958-59. Fellow Brit. Acad.; mem. Internat. (pres.), Am. (John Bates Clark medal 1947, past pres.) econ. assns., Econometric Soc. (past pres.), Am. Acad. Arts and Scis., Am. Philos. Soc., Soc. Fellows Harvard. Author: Foundations of Economic Analysis, 1947; Economics, An Introductory Analysis, 1948; (with Robert Dorfman and Robert Solow) Linear Programming and Economic Analysis, 1958; Collected Scientific Papers,1 1966; also numerous articles. Editor: (with others) Readings in Economics, 1952. Home: 75 Clairemont Rd., Belmont, Mass. Office: Dept. Econs., Mass. Inst. Tech., Cambridge, Mass.*

SANARELLI, Giuseppe, Italian bacteriologist; b. San Savino, Italy, Sept. 24, 1864; prof. hygiene, Rome; mem. French Acad. Scis., 1934; isolated bacillus (Salmonella) icteroides, or Sanarelli's bacillus, 1894. Died Rome, Apr. 6, 1940.

SANBORN, Richard Carl, Am. biologist, physiologist; b. Cleve., Jan. 13, 1921; s. Jay Frederick and Lottie (Kreinbring) S.; A.B., Harvard, 1943, M.A., 1948, Ph.D., 1950; m. Katherine Schaff, July 31, 1943; 1 son, Keith Jay. Faculty, Harvard, 1950, U. Ill., 1950-52, Mass. Inst. Tech., 1952-57; faculty Purdue U., Lafayette, Ind., 1957—, asso. prof., 1963—; mem. corp. Marine Biol. Lab., Woods Hole, Mass., 1959-64. Mem. Am. Soc. Zoologists, Entomol. Soc. Am., Soc. Gen. Physiologists. Author: (with M.X. Zarrow, J.L. Yochim, J.L. McCarthy) Experimental Endocrinology: a Source Book of Basic Techniques, 1964; also numerous articles. Research on biochemistry insect devel., physiology and biochemistry regulation rate heart-beat in invertebrates. Home: 148 Seneca Lane, West Lafayette, Ind. 47906. Office: Dept. Biol. Scis., Purdue U., Lafayette, Ind. 47907.*

SANCHEZ, Francesco, physician; b. Braga, Portugal, 1552; student, Toulouse, France; M.D., Montpellier, France, 1574; became prof. medicine, Montpellier, 1576; then physician, royal prof. medicine, Toulouse. Author: De multum nobili et prima universali . . . , 1581. Opponent of scholastic method; precursor of Descartes and Francis Bacon. Died Toulouse, 1632.

SANCHEZ-CASCOS, Andres, Spanish physician; b. Trujillo, Spain, Feb. 27, 1935; s. Andrés Sanchez-Marcos and Rafaela Cascos; M.B., U. Madrid (Spain), 1959, M.D., 1960; m. Rosario Pernaute Montreal, May 6, 1963; children—Rosario, Andrés, Olga. Internal physician Fundacion Jiménez-Diaz, Madrid, 1959-61, asst. cardiac dept., chief dept. human genetics, 1962—; cardiac research fellow Guy's Hosp., London, Eng., 1961-62. Mem. Spanish Soc. Cardiology (mem. sci. com.). Author: Aetio-pathogenesis of Congenital Hearth Disease, 1967; also numerous articles. Description of forms of right atrial overloading, physiol. consequences of congenital heart diseases and genetic implications. Home: 13 Alc. Lopez Casero, Madrid. Office: 2 Av. Reyes Catolicos, Madrid, Spain.*

SANCHEZ DEL RIO, Carlos, Spanish physicist; b. Borja, Spain, Aug. 16, 1924; s. Carlos and Asuncion (Sierra) S. del R.; M.Physics, U. Madrid (Spain), 1946, D.Physics, 1948; m. Lilliana Skorel, Sept. 23, 1953; children—Andrés, Ana, Cristina. Prof. nuclear physics U. Madrid, 1953—; dir. physics dept. Spanish Atomic Energy Commn., 1958—; dir. div. reactors LAEA, 1961-62. Author: Introduccion a la Interferometria, 1949; Fundamentos Teoricos de la Fisica Atomica y Nuclear, 1960; also articles. Research on optics and nuclear physics especially neutron physics. Home: 38 Urquijo, Madrid 8, Spain.*

SANCIER, Kenneth M., Am. chemist; b. N.Y.C., June 21, 1920; s. Martin M. and Melitta (Straus) S.; B.S., Bklyn. Poly. Inst., 1942; M.A., Johns Hopkins, 1947, Ph.D., 1949. Chemist, Linde Air Products Co., Tonawanda, N.Y., 1942-46; instr. Johns Hopkins, 1948, U. Conn., 1948; research chemist Brookhaven Nat. Lab., Upton, N.Y., 1949-54, Stanford Research Inst., Menlo Park, Cal., 1954—; vis. scientist U.S.-Japan Coop. Sci. Program, U. Tokyo, Japan, 1966-67. Mem. Am. Chem. Soc., N.Y. Acad. Scis., Phi Beta Kappa, Sigma Xi, Phi Lambda Upsilon. Research, publs. on low temperature optical absorption spectroscopy of weak molecular complexes, quantum states of rare earth ions and photochemistry, heterogeneous catalysis on semiconductor catalysts by electron spin resonance, luminescence of solids by surface atom recombination, ion pairing by ESR, hydrogen bonding; photochemistry chlorophyll. Home: 65 Spanish Trail. Office: Stanford Research Inst., Menlo Park, Cal. 94025.*

SANCTORIUS, Sanctorius, see Santorio, Santorio.

SANDAGE, Allan Rex, Am. astronomer; b. Iowa City, June 18, 1926; s. Charles Harold and Dorothy (Briggs) S.; A.B., U. Ill., 1948; Ph.D. Cal. Inst. Tech,. 1953; D.Sc., Yale, 1966, Univ. of Chgo., 1967, Univ. of Ill., 1967; m. Mary Lois Connelley, June 8, 1959; children—David Allan, John Howard. Astronomer, Mt. Wilson and Palomar Obs., Pasadena, Cal., 1952—. Vis. lectr. Harvard, 1957, Haverford Coll., 1958; research asso. Australian Nat. U., 1968—. Recipient gold medal Royal Astron. Soc., 1967; Rittenhouse medal Rittenhouse Astron. Society, 1968. Mem. Am. Astron. Soc. (councilor Soc. (Eddington medal 1963), A.A.A.S., Nat. Acad. Scis., Am. Acad. Arts and Scis., Sigma Xi. Author: The Hubble Atlas of Galaxies, 1961; also numerous articles. Contbns. to theory stellar evolution especially regarding age of oldest stars; found main sequence of globular clusters establishing them as oldest bodies in the galaxy; participated in discovery of quasi-stellar radio sources; devised methods and gave 1st determination of rate of change of expansion rate of the universe. Home: 701 Santa Barbara St. Office: 813 Santa Barbara St., Pasadena, Cal. 91106.*

SANDER, Bruno, Austrian geologist; b. Innsbruck, Austria, Feb. 23, 1884; s. Max and N. (Rizzoli) S.; Ph.D., U. Innsbruck (Austria), 1907; Ph.D. (hon.) U. Gottingen (Germany), 1937; m. Elizabeth Holzknecht, 1920. Asst., Tech. U., Vienna, 1908-09; with U. Innsbruck, 1909-13, lectr. geology, 1912-14; with U. Vienna, 1914-22; geologist Geol. Inst. Vienna, 1913-22; prof. mineralogy and petrography U. Innsbruck, 1922-56, prof. emeritus, 1956—. Recipient Gustav Steinmann medal German Geol. Assn., 1950, Penrose medal Geol. Soc. Am., 1957. Mem. Nat. Acad. Scis., acads. sci. of Vienna, Bologna, Upsala, Berlin, Halle, Geol. Soc. Am. (corr.), Geol. Soc. London (fgn.). Author: Gefügekunde der Gesteine, 1930, Einführung in die Gefügekunde der Geologischen Korper, 2 vols., 1948-50. Developed descriptive terminology for studies in petrology, tectonics; studied both scaler and vectorial properties of fabric elements; used axes distbn. analysis in mapping and microscopic analysis. Home: Alte Universitat, Innsbruck, Austria.

SANDERMANN, Wilhelm, German chemist; b. Selsen/Lippe, Dec. 6, 1909; s. Hermann and Juliana (Reese) S.; ed. univs. Leipzig and Giessen; Ph.D.; m. Charlotte Streubel, Dec. 9, 1939; children—Heinrich, Margarete, Günther. Chemist in industry; researcher Univ. and Fed. Inst. Agr. and Forestry Re-research, 1947—. Mem. German Chem. Soc. Author: Holzchemie; Naturharze; Biochemie; Der Güttervogel ruft nicht mehr; also articles on organic and wood chemistry. Address: Kuckallee 34, Reinbek-Hamburg, West Germany.

SANDERS, Benjamin Elbert, Am. biochemist; b. Bowersville, Ga., Oct. 19, 1918; s. Clifton L. and Christine (Hogsed) S.; B.S., Wofford Coll., 1939; M.S. in Chemistry, U. Ga., 1942; Ph.D., Purdue U., 1949; m. Dorothy McIntyre, Sept. 7, 1946; children—Lamar, Marcia Jean. Am. Chem. Soc. fellow Purdue U., 1946-49; research biochemist Henry Ford Hosp., Detroit, 1949-51; research asso. immunochemistry dept. Sharp & Dohme, 1951-53, research asso. protein chemistry Research Labs., 1953-58; dir. protein chemistry dept. Merck Inst., West Point, N.Y., 1958-61; professor biochemistry State U. N.Y., Buffalo Sch. Medicine, 1967—; research collaborator med. dept. Brookhaven Nat. Labs., Upton, N.Y., 1958-64. USPHS grantee, 1962—; Diplomate Am. Bd. Clin. Chemists. Fellow A.A.A.S.; mem. Am. Soc. Biol. Chemists, Am. Chem. Soc., Sigma Xi. Research, publs. on purified plasma proteins especially on antibodies, coagulation, dissolution blood clots, biochemistry schizophrenia; patentee in field. Home: 52 Rosedale Blvd., Buffalo 14226.*

SANDERS, Clarence Elmer, Am. physician; b. Roodhouse, Ill., May 15, 1885; s. George Francis and Laura (Hughes) S.; M.D., U. Kan., 1907; m. Martha Beamer, June 30, 1909. Intern Research Hosp., Kansas City, Mo.; mem. staff St. Mary's Hosp., Kansas City, Mo., 1909-49; divisional surgeon M.-K.-T. R.R. Contbr. to med. jours. Specialist in diseases of heart and blood disorders due to circulatory diseases; devised method of pressure necrosis amputation by tightening silver wire attached to screw mechanism, mainly for amputation of gangrenous digits; inventor Sanders oscillating bed for treatment of circulatory diseases, 1932. Died Apr. 28, 1949.

SANDERS, James Glossberger, Am. entomologist; b. Louisville, O., July 17, 1880; Ph.B., Otterbein Coll., 1901; Sc.D. (hon.), 1948. Joined U. S. Bur. Entomology, 1905; became prof. entomology U. Wis., 1910; joined Wis. Dept. Agr., 1915; became state zoologist Pa. Dept. Agr., 1916; apptd. head Bur. Plant Industry, Pa., 1919; entomologist Comml. Solvents Corp., 1919-22; entomologist, mgr. spray oil dept. Sun Oil Co., 1923-38; tropical collections, 1939-41; chief Amazon party food div. Office of Coordinator Inter-Am. Affairs, 1942-43; gen. mgr. chinchina plantation Bd. Econ. Welfare, Peru, 1943-44; with food div. Fgn. Econ. Adminstrn., Persia, 1944-45; entomologist Comml. Solvents Corp., 1945—. Mem. Am. Assn. Econ. Entomologists (became pres. 1922), Entomol. Soc. D.C. (corr. sec. 1906-09). Research in Homoptera, Coccidae, miscible oils and oil emulsions, dustless treatments of coal and coke, benzene-hexachloride; pioneered devel. benzene hexachloride; originated 3-5-40 dust combination for cotton dusting; invented Kwell ointment. Office: Comml. Solvents Corp., 17 E. 42d St., N.Y.C. 32.

SANDERS, James Meadows, Am. zoologist; b. Knoxville, Tenn., Nov. 18, 1905; s. Frederick O. and Frances (Meadows) S.; A.B., William Jewell Coll., 1926; A.M., U. Ill., 1927, Ph.D., 1931; m. Mary Lou Pfister, Sept. 9, 1952; children—Mark, Peter. Faculty, U. N.H., 1929-30, Ark. A. and M. Coll., 1931-32, Cumberland U., 1934-37, Sioux Falls Coll., 1937-39; faculty Ill. Tchrs. Coll., Chgo. S., 1939-43, 46—, prof. zoology, 1959—; faculty U. Mo. Med. Sch., 1944-45. Mem. Am. Inst. Biol. Scis., A.A.A.S., Am. Assn. U. Profs., N.Y. Acad. Scis. Research and publs. on animal behavior, embryology, conservation, lab. techniques in zoology. Home: 10,000 W. 127th St., Palos Park, Ill. 60464. Office: 6800 Stewart Av., Chgo. 60621.*

SANDERSON, Donald Eugene, Am. mathematician; b. Oskaloosa, Ia., Feb. 4, 1926; s. Arlo Melvin and Laura (Goudy) S.; B.A., Cornell Coll., Mt. Vernon, Ia., 1949; M.S., Cal. Inst. Tech., 1951; Ph.D., U. Wis., 1953; m. Carol Mary Shaw, June 24, 1949; children—Robert Charles, Mark Alan, Karen Ann. Faculty, Ia. State U., Ames, 1953—, prof. math., 1964—; vis. prof. Mich. State U., 1962-63. Mem. Math. Assn. Am. (vice chmn. Ia. sect. 1964-65, chmn. 1965-66), Am. Math. Soc., Am. Assn. U. Profs., Phi Beta Kappa, Sigma Xi. Contbg. author: Topology of 3-manifolds, 1962. Research, publs. on theory isotopic deformations in 3-dimensional spaces, structure space homeomorphisms, theory classes function more gen. than continuous functions. Home: 1811 Coolidge Dr., Ames, Ia. 50010.*

SANDERSON, Sir John Scott Burdon, Brit. pathologist, physiologist; b. Newcastle-on-Tyne, Eng., 1828; grad. U. Edinburgh; studied in Paris; prof. physiology Univ. Coll., London, 1874-82; 1st Waynflete prof. physiology Oxford U., 1882-95, Regius prof. medicine, 1895-1903; med. officer of health for Paddington, research on infection and contagion for med. dept. Privy Council. Fellow Royal Soc., 1867. Author: Hand-Book of the Sphygmograph; also papers on physiology, cattle diseases, pub. health. Supported view that cause of infection was corpuscular rather than fluid at first, later accepted germ theory; made physiol. and path. studies on living animals. Died 1905.

SANDGROUND, Jack Henry, microbiologist; b. Johannesburg, South Africa, Feb. 2, 1899; s. Charles S. and Sarah (Freeder) S.; M.Sc., U. S. Africa, 1919; D.Sc., Johns Hopkins, 1925; m. Rose Plotler, Nov. 29, 1929; 1 son, Mark B. Came to U.S., 1922, naturalized, 1931. Asst. prof. Harvard Med. Sch., 1925-38, curator Harvard Mus. Zoology, 1925-38; chief parasitology, chemotherapy Lilly Research Labs., Indpls., 1939-47; lectr. preventive medicine N.Y. U. Med. Sch., 1948-53; prof. zoology Dartmouth, 1953-54; research asso. Haskins Lab., N.Y.C., 1954—. Guggenheim Found. fellow, 1938-39. Recipient S. African Biol. Soc. Scott medal, 1921. Contbr. numerous articles to sci. jours. Home: 390 1st Av., N.Y.C. 10010.*

SANDIFORT, Eduard, Dutch anatomist, physician; b. Dortrecht, Netherlands, 1742; doctorate, Leiden, Netherlands, 1763; prof. anatomy, Leiden, 1772-1814. Author: Observations anatomico-pathologicae (contains 1st known illustration of vegetative endocarditis in patient with pulmonary stenosis and interventricular septal defect), 1777-81. Research in path. anatomy; called father of path. iconography. Died 1814.

SANDLER, Stanley R., Am. chemist; b. N.Y.C., Jan. 24, 1935; s. Meyer and Bertha Sandler; B.S. cum laude, Bklyn. Coll., 1956; Ph.D., Pa. State U., 1960. Sr. research chemist Central Research Lab., Borden Chem. Co., Phila., 1959-64, group leader, 1964-66, head lab., 1966—; mem. evening staff chem. engring. tech. dept. Temple U., 1964—. Mem. Am. Chem. Soc., Phi Lambda Upsilon. Research, publs. on reactions of 1, 1-dihalocyclopropanes; synthesized fluorescent additives and polymers capable of detecting nuclear particles; synthesis and polymerization of epoxy, urethane and 2-oxazolidines; synthesis of new heterocyclic compounds; patentee in field.*

SANDLER, Yehuda Ludwig, phys. chemist; b. Kudowa, Germany, May 14, 1913; s. Aron and Adele (Straus) S.; B.S., Manchester (Eng.) U., 1936; Ph.D., Hebrew U., Jerusalem, 1940; m. Rivka Black, May

23, 1947; children—David, Judith Ada. Came to U.S., 1952, naturalized, 1959. Sr. lectr., acting chmn. dept. phys. chemistry Hebrew U., 1940-50; research fellow Nat. Research Labs., Ottawa, Ont., Can., 1950-52; sr. research asso. Poly. Inst. Bklyn., 1952-53; sr. staff mem. div. indsl. cooperation Mass. Inst. Tech., 1953-55; sr. research physicist, fellow research scientist Westinghouse Research Labs., Pitts., 1955—. Mem. Am. Phys. Soc., Am. Chem. Soc., Electrochem. Soc., Internat. Congress Catalysis, Sigma Xi. Author: (with M. Schiffer) Quantum Chemistry, 1945. Publs. on research on hydrogen molecule, its modifications (ortho, para, heavy); mechanisms of gas adsorption, catalysis, and electrode reactions; fuel cells. Home: 6368 Burchfield Av., Pitts. 15217. Office: Westinghouse Research Labs., Pitts. 15235.*

SANDON, Harold, Brit. zoologist; b. London, Dec. 15, 1894; s. Robert and Louisa Rudkins (Watts) S.; ed. Queen's Coll., Cambridge (Eng.) U., also Rutgers U.; M.A., Ph.D.; m. Suzanne Galinier, Sept. 3, 1928; 1 son, Oliver. Protozoologist, Rothamsted Agrl. Exptl. Sta., 1920-28; instr. Fouad U., Cairo, Egypt, 1928-34, U. Cape Town (S. Africa), 1934-46; prof., dir. Zool. Faculty, U. Khartoum (The Sudan), 1946-60, dean Faculty Sci., 1954-57; cuartor Museum Natural History The Sudan, 1957-60; now prof. zoology Bedford Coll., London. Mem. Zool. Soc. London, Inst. Biology, Soc. Protozoologists, Eugenics Soc. Author: The Protozoan Fauna of the Soil, 1927; The Food of Protozoa, 1932; Essays on Protozoology; also articles. Home: 53 Dartmouth Park Hill, London N.W. 5. Office: Zoology Dept., Bedford Coll., Regent's Park, London N.W. 1, Eng.

SANDORFY, Camille, chemist; b. Budapest, Hungary, Dec. 9, 1920; s. Kamill and Paula (Fenyes) S.; B.A., Cistercian Coll., Budapest, 1939; B.Sc., Ph.D., U. Szeged (Hungary), 1946; D.Sc., Sorbonne, Paris, France, 1949. Research officer Centre Nat. de la Recherche Scientifieque, Paris, 1946-51; postdoctoral fellow NRC Can., Ottawa, Ont., 1951-53; faculty U. Montreal (Que., Can.), 1954—, prof. chemistry, 1959—. Recipient 1st prize sci. Province of Que., 1964. Fellow Royal Soc. Can.; mem. Canadian Inst. Chemistry, Chem. Soc. France, Soc. Phys. Chemistry (France). Author: Les spectres électroniques en chimie théorique, 1959; Electronic Spectra and Quantum Chemistry, 1964; also articles, chpt. in book. Research in quantum chemistry, ultraviolet spectroscopy and infrared spectroscopy. Home: 4800 Cote-des Neiges, Montreal, Que., Can.*

SANDRITTER, Walter, German physician; b. Frankfort/Main, Germany, July 7, 1920; s. Karl and Paula (Zwirner) S.; student medicine, Erlangen, Kiel, Germany, M.D., U. Frankfort, 1947, Habilitation, 1952; m. Erika Gerber, Mar. 18, 1948; children—Gundula, Bettina. Faculty, U. Giessen (Germany), 1957—, prof., dir. dept. pathology Iustus Liebig, 1961—. Recipient Ludwig Schunk Prize, Med. Faculty Giessen. Mem. N.Y. Acad. Scis. Author: 100 Years of Histochemistry in Germany, 1963; (with J. Schorn) Histopathologie; also numerous articles. Demonstrated nucleic acid content of tumors is high in comparison to normal cells, carcinoma in situ has a DNA distbn. pattern like carcinomas; studies in stochiometry between nucleic acids and basic dye stuff; theoretical and practical background of cytophotometry; growth of single mammalina cells. Home: Am alten Friedhof 27, 63 Giessen, Germany.*

SANDSTRÖM, Ivar Victor, Swedish anatomist; b. Stockholm, Mar. 22, 1852; M.D., U. Uppsala (Sweden), 1887; tchr. histology U. Uppsala, 1881-86. Gave 1st full, systematic description of parathyroid glands, 1880. Died June 2, 1889.

SANDUS, Oscar, Am. chemist; b. N.Y.C., July 29, 1924; s. Samuel and Helen (Zwei) S.; B.S., U. Ky., 1949; M.S., U. Chgo., 1950; Ph.D., Ill. Inst. Tech., 1955; m. Esther Bernice Herman, Dec. 22, 1946; children—Deborah Sharon, Samuel Ira, Aaron Joel. Asst. chemist Argonne Nat. Lab., Lemont, Ill., 1955-58; asso. research phys. chemist U. Mich. Radiation Lab., Ann Arbor, 1958-61, asso. research chemist Infrared Physics Lab., 1964—; research phys. chemist Conductron Corp., Ann Arbor, 1962-63; sr. chemist Chemotronics, Inc., Ann Arbor, 1963-64. Mem. Am. Chem. Soc., A.A.A.S., Ill. Acad. Sci., Phi Beta Kappa, Sigma Xi, Phi Lambda Upsilon. Research, publs. in thermodynamics and properties of nonelectrolytic solutions, uranium fuel and feed materials process devel., electromagnetic materials research, electroplating process devel., plastic foam process devel., fundamental aspects of physics and chemistry relating to missile reentry. Home: 2540 Pamela Av., Ann Arbor, Mich. 48103.*

SANDWEISS, Jack, Am. physicist; b. Chgo., Aug. 19, 1930; s. Charles Ray and Florence (Hymovitzj) S.; B.Sc., U. Cal. at Berkeley, 1952, Ph.D., 1956; m. Letha Ann Boeck, Jan. 16, 1956; children—Daniel Howard, Anne Florence, Benjamin Lewis. Physicist, U. Cal. at Berkeley Radiation Lab., 1957; faculty Yale, 1958—, prof. physics, 1964—. Cons. Brookhaven Nat. Lab., Upton, L.I., N.Y., 1962—, Argonne (Ill.) Nat. Lab., 1966. Mem. Am. Phys. Soc. Contbg. author: Bubble Chamber and Spark Chamber Techniques. Research, publs. on devel. high energy separated particle beams, anti-proton-proton interaction, K-meson interactions and decays. Home: 122 Bedford Av., Hamden, Conn. 06517. Office: Sloane Lab., Yale, New Haven 06520.*

SANÉ, Jacques Noël, French naval engr.; b. Brest, France, Feb. 18, 1740; entered navy, Brest, circa 1755; became maritime dir., Brest, 1793; helped organize 1st fleet of Republic; mem. Marine Acad., French Acad. Scis., 1796. Contbd. to improvement of travel in sailing ships; called Vauban of the French navy. Died Paris, Aug. 22, 1831.

SANFORD, Arthur Hawley, Am. physician; b. New Albin, Ia., Jan. 12, 1882; s. Alcimcre Mead and Amanda Elizabeth (Gilbert) S.; A.B., Northwestern U., 1904, A.M., 1907, M.D., 1907; m. Margaret Loretta Seager, Aug. 23, 1906; children—Hawley Seager, Raymond Arthur, Gertrude Loretta Elizabeth. Asst. prof. physiology Med. Sch., Marquette U., 1907-08, asso. prof., 1908-09, prof. 1910-11; bacteriologist Mayo Clinic, 1911; head div. of clin. pathology, 1915; apptd. asso. prof. pathology, 1915, prof. pathology Mayo Found., U. Minn., 1921-50; dir. lab. Rochester State Hosp., 1950—. Spl. cons. USPHS, Pres. Am. Bd. Pathology. Fellow A.C.P.; mem. A.M.A., Am. Soc. Clin. Pathology (pres. 1927-28; Ward Burdick medal 1933), A.A.A.S., Am. Soc. Immunologists, Soc. Exptl. Biology and Medicine, Phi Beta Kappa, Sigma Xi. Author: Clinical Diagnosis by Laboratory Methods (with Dr. J. C. Todd), 11th edit., 1948. Developed (with Charles Sheard) Sheard-Sanford photelometer, (Photo-electric colorimeter for determination of hemoglobin, circa 1930. Died Apr. 28, 1959.

SANFORD, Fernando, Am. physicist; b. Taylor, Ill., Feb. 12, 1854; s. Faxton and Mariah (Bly) S.; B.S., Carthage (Ill.) Coll., 1879, M.S., 1882, Sc.D., 1920; studied with Helmholtz in Berlin, 1886-88; m. Alice E. Crawford, Aug. 12, 1880; children—Burnett, Alice. Prof. phys. sci. Mt. Morris Coll., 1879-82; county supt. schs., Ogle County, Ill., 1882-86; instr. in physics and chemistry Englewood High Sch., Chgo., 1888-90; prof. phys. sci. Lake Forest U., 1890-91; prof. physics Leland Stanford Jr. U., 1891-1919. Fellow A.A.A.S., Am. Phys. Soc. Author: Elements of Physics, 1902; A Physical Theory of Electrification Charges of Atoms and Ions, 1919; A Diurnal Variation in the Electrical Potential of the Earth, 1920; Terrestrial Electricity, 1931; also numerous monographs pertaining to original investigations in physics. Engaged in investigations in terrestrial electric obs. which he established at Palo Alto, Cal., 1920. Died May 21, 1948.

SANFORD, Jay Philip, Am. physician; b. Madison, Wis., May 27, 1928; s. Joseph A. and Arlyn (Carlson) S.; student U. Mich., 1945-48, M.D., 1952; m. Lorraine Burklund, Apr. 9, 1950; children—Jeb, Nancy, Sarah, Philip, Catherine. Research asst. bacteriology sect. dept. medicine U. Hosp., Ann Arbor, Mich., 1950-52; chief bacteriology sect. dept exptl. surgery Walter Reed Army Med. Center, Washington, 1954-56; faculty U. Tex. Southwestern Med. Sch., Dallas, 1957—, prof. internal medicine, 1965—. Cons. to hosps., div. health moblzn. USPHS; mem. bacteriology and mycology study sect. NIH, 1962-66; dist. dir. Disaster Med. Care, Tex, 1963—. Recipient Pfizer award merit U.S. Civil Def. Council, 1965. Mem. Am. Fedn. for Clin. Research, Am. Soc Microbiology, Central Soc. for Clin. Research, Am., So. socs. for clin. investigation, So. Soc. for Pediatric Research, A.M.A., Sigma Xi, Alpha Omega Alpha. Research, numerous publs. on infectious-communicable diseases especially immunity in infections and mechanisms of spread and control of infections in hospitals. Home: 3516 St. John's Dr., Dallas 75205.*

SANFORD, Katherine Koontz, Am. biologist; b. Chgo., July 19, 1915; s. William James and Alta (Koontz) S.; B.A., Wellesley Coll., 1937; M.A., Brown U., 1939, Ph.D., 1942. Instr., Western Coll., 1941-42, Allegheny Coll., 1942-43; asst. dir. Johns Hopkins Nursery Sch., 1943-47; biologist tissue culture sect. Lab. Biology, Nat. Cancer Inst., NIH, Bethesda, Md., 1947—. Recipient Ross Harrison award for work in tissue culture, 1954. Mem. Tissue Culture Assn., Am. Assn. for Cancer Research, Internat., Am. socs. for cell biology, Am. Soc. for Exptl. Pathology. Research, numerous publs. on neoplastic transformation of cells in tissue culture, nutrition of cells in tissue culture, chemically defined culture media for cells in vitro; pioneered single cell isolation and growth procedure. Home: 3831 Rodman St. N.W., Washington 20016. Office: Nat. Cancer Inst., Bethesda, Md. 20014.*

SANFORD, Paul Everett, Am. nutritionist; b. Milford, Kan., Jan. 14, 1917; s. Charles R. and Ina (Kneeland) S.; B.S., Kan. State U., 1941; M.S., Ia. State U., 1942, Ph.D., 1949; m. Helen Louise Crenshaw, Oct. 31, 1942; children—Paula Louise, Patricia Kathleen, Carolyn Ruth. Asso. prof. dept. poultry sci. Ia. State U., Manhattan, 1949—, prof., 1960—, nutritionist, 1949—. Fellow A.A.A.S.; mem. Poultry Sci. Assn., World's Poultry Sci., Broilers Soc. Japan, Am. Inst. Biol. Scis., Sigma Xi, Phi Kappa Phi, Gamma Sigma Delta, Alpha Zeta. Research and publs. on poultry nutrition and physiology, field hemmorrhagic syndrome, flavor and aroma of poultry meat, shell quality of eggs, ultization for sorghum grain protein, dehydrated alfalfa for poultry. Home: 343 N. 14th St., Mahattan, Kan. 66502.*

SANG, James H., Brit. geneticist; b. Aberdeen, Scotland, Nov. 4, 1912; s. James and Helen (Henderson) S.; B.Sc., Aberdeen U., 1935; Ph.D., Cambridge U., 1941; m. Pauline Caddy, Dec. 27, 1941; children—Robert Henderson, David Caddy, Helen Mary. Asst. dir. aircraft equipment Ministry of Aircraft Prodn., London, Eng., 1942-48; prin. sci. officer Animal Breeding and Genetics Research Orgn., Edinburgh, 1948-51; dep. dir. Poultry Research Centre, Edinburgh, 1951-65; prof. genetics U. Sussex, Eng., 1965—. Kilgour Research scholar Aberdeen U., 1934-38; Hutchison research scholar St. John's Coll. Cambridge U., 1938-39. Mem. Inst. Sci. Tech. (v.p. 1966—), Royal Soc. Edinburgh, Soc. Exptl. Biology, Genetical Soc. Contbr. numerous articles to sci. jours. Exptl. research on determinants of insect population growth; devel. germ-free culture techniques for rearing Drosophila on chemically defined diets; studies of gene action during devel., and of methods of assessing these; examination of genetic and physiol. basis of hybrid vigor in poultry and exptl. animals. Home: 60 Surrenden Crescent, Brighton, Eng. Office: U. Sussex, Falmer, Sussex, England.*

SANGER, Frederick, English biochemist; b. Rendcombe, Gloucestershire, Eng., Aug. 13, 1918; s. Frederick and Cicely (Crewdson) S.; B.A., St. John's Coll., Cambridge U., 1939, Ph.D., 1943; m. Margaret Joan Howe, Dec. 28, 1940; children—Robin, Peter, Sally. Research, 1940—, research metabolism of amino acid lysine, 1940-43, study chemistry of proteins, structure of insulin, Beit Meml. fellow for medical research, 1944-51, external staff Medical Research Council, 1951—; now head div. protein chemistry Medical Research Council, Laboratory Molecular Biology, Cambridge, England. Recipient Corday-Morgan medal and prize Chem. Soc., 1951; Nobel prize for chemistry, 1958; fellow King's Coll. Fellow Royal Soc., 1954; mem. Am. Soc. Biochem. Chemists (hon.), Nat. Acad. Scis., Acads. Sci. of Argentina and Brazil, Japanese Biochem. Soc. (hon.), World Acad. Art and Sci.; hon. fgn. mem. Am. Acad. Arts and Scis. Research on protein structure; determined amino acid sequence of the protein insulin. Office: Med. Research Council, Lab. Molecular Biology, Hills Rd., Cambridge, Eng.

SANGER, Ralph G(rafton), Am. mathematician; b. Cin., Apr. 10, 1905; s. Andrew Lewis and Louise (Pomeroy) S.; S.B., U. Chgo., 1925, S.M., 1926, Ph.D., 1931; student U. Wis., 1926-28. Instr. mathematics U. Wis., 1926-28; asst. U. Chgo., 1928-30, instr., 1930-40, asst. prof., 1940-46, asst. dean of students div. phys. scis., 1943-44, dean, 1944-46; prof. math. Kan. State Coll., 1946—, head dept., 1946-67, acting dean sch. arts and scis., summer 1955, acting associate dean sch. arts and sciences; 1955-56. Mem. Am. Math. Soc., Math. Assn. Am. (asso. editor Monthly 1931-40, gov. 1944-46, 49-51), Nat. Council Tchrs. Mathematics, Phi Beta Kappa, Sigma Xi. Author: Synthetic Projective Geometry, 1939. Study of calculus of variations; projective geometry. Died Mar. 13, 1968.

SANGER, Vance LaVerne, Am. veterinarian; b. Fayetteville, W.Va., Aug. 11, 1917; s. Frank P. and Matilda (Quelhorst) S.; A.B., Manchester Coll., 1943; D.V.M., Ohio State U., 1949, M.S., 1953; Ph.D., Mich. State U., 1963; m. Anna F. Clopper, Sept. 3, 1941 (dec. Nov. 1961). Practice vet. medicine, Kirkwood, Ill., 1949-51; asst. prof. Ohio State U., Columbus, 1952-55; prof. Ohio Agrl. Expt. Sta., Wooster, O., 1955-65; asso. prof. vet. medicine Mich. State U., East Lansing, 1965-67, professor since 1967—. Member of the American Veterinary Med. Assn., Am. Coll. Vet. Pathologists, Internat. Acad. Pathology, Conf. Research Workers in Animal Diseases, Sigma Xi. Contbg. author: Diseases of Wild Birds, 1968. Research, publs. on infectious and nutritional diseases of swine, sheep, cattle, poultry including avian leukosis. Home: 2937 Kenwick Circle, Lansing 48912. Office: Dept. Pathology, Mich. State U., East Lansing, Mich. 48823.*

SANI, Guelfo, Italian physician, obstetrician, gynecologist; b. Ferrara, Italy, Dec. 14, 1924; s. Gino and Renata (Zanardi) S.; Degree, Bologna (Italy) U., 1948; m. Maria Teresa Brusarosco, Oct. 24, 1959. Faculty, U. Ferrara, 1949-60, asst. prof. dept. obstetrics and gynecology, 1955-60; faculty Bologna U., 1960—, prof. obstetric and gynecol. pathology, 1964—. Recipient Premio Rachele Paolucci, 1948; Premio Merletti, 1955; Prix Internat., Soc. Royale Belge Ost. Gyn., 1958. Author: Fluorescence Microscopy in the Cytodiagnosis of Cancer, 1964; also numerous articles. Research on alkaline phosphatase and secretion of female genital tract, early diagnosis of tumors by fluorescent microscopy, hormonal influences of sexual devel., diagnosis and therapy of infertility. Home: 31 Strad Maggiore, Bologna, Italy.*

SANIELEVICI, Alexander, physicist; b. Bucharest, Rumania, May 20, 1899; s. Simon and Raluca (Crainic) S.; M.S., U. Jassy (Rumania), 1921; D.Sc. with honors, U. Paris (France), 1936; m. Elisabeth Fuhn, Sept. 11, 1945; 1 son, Sergiu. Research fellow French Nat. Council for Research, Radiuminstitue, Paris, 1931-39; tchr. physics secondary sch., Bucharest, 1940-45; faculty U. Bucharest, 1945—, prof., 1955—; head dept. nuclear spectroscopy Inst. Atomic Physics, Bucharest, 1950—; dep. dir. dept. research and labs. I.A.E.A., Vienna, Austria, 1958—. Decorated Order of Labour, First Class, 1956. Mem. Acad. Rumanian Socialist Republic, Rumanian Sci. Assn. Engrs. and Technicians. Author: Radioactivity, 1956; Introduction into Radioactivity, 1957; Nuclear Structure and Radioactive Transitions, 1958;

also articles. Research and devel. in application of calorimetry to radioactivity, and dosimetry of high energy ionizing radiations; devel. chem. dosimetry systems. Home: Vienna, 1190, Chimanistrasse 30/II-/8, Austria. Office: I.A.E.A., Vienna 1010, Kaerntnerring 11, Austria.*

SANKALÉ, Marc Henri Christain, physician; b. Sainto-Louis, Senegal, Feb. 7, 1921; s. Louis and Anne (Colas) S.; Docteur en Médecine, Faculty Medicine, Montpellier, France, 1944, aggregate prof., 1961; m. Yvette LePelletier, Nov. 24, 1944; children—Louis, Michèle, Joële, Hervé. Joined M.C., 1944; in Senegal, 1944-48, French Guiana, 1948-50, Indo-China, 1952-54, Mali, 1956-57; ret. as lt. col., 1962; asst. Med. Clinic, Dakar (Senegal) Faculty Medicine at Le Dantec Hosp., 1957-61; faculty U. Dakar, 1961—, prof. clin. medicine, 1966—; dir. Applied Tropical Medicine Inst. Dakar. Tech. adviser Ministry of Health and Social Affairs, Republic of Senegal, 1961—; mem. WHO, 1963—; specialist for testing new pharm. products, 1963—. Decorated chevalier Legion of Honor; commdr. Nat. Order Republic of Senegal; chevalier de l'Ordre des Palmes Académiques. Author: (with P. Pène) Médecine Sociale au Sénégale, 1960; (with M. Payet, P. Pène) Cliniques Africaines, 1966; also numerous articles. Research on tropical diseases especially intestinal diseases, cosmopolitan diseases of Africa (hepatology and hematology, splenomegaly, athersclerosis, diabetes); problems of pub. health and question of univs. in Black Africa. Home: 81, Boulevard de la République, Dakar, Senegal, also 82 Boulevard Michelet, Marseille, France. Office: Hôpital Le Dantec, Avenue Pasteur, Dakar, Senegal.*

SANKAR, Siva Devarakonda Venkata, biochemist; b. Vizianagram, India, Apr. 7, 1927; s. D.V. Jagannadham and Pullama (Neti) S.; M.Sc., U. Madras (India), 1949, Ph.D., 1951; postgrad. Mass. Inst. Tech., Johns Hopkins, Tex. A. and M. U.; m. Barbara Cossano, June 4, 1959; children—Priscilla, Jason. Asst. prof. biochemistry Adelphi U., Garden City, N.Y., 1956-58, adj. prof., 1961—; sr. research scientist Creedmoor State Hosp., N.Y. State Dept. Mental Hygiene, Jamaica, 1958-63, asso. research scientist, 1963—; adjunct professor Brooklyn Coll. City U. N.Y., 1960-62, M.J. Lewi Coll. Podiatry, 1965-66, State U. N.Y. Coll. Agr. and Tech., Farmingdale, 1965, L.I. U., 1964—, Hofstra U., 1965—. Perla scholar, 1940-41; State scholar, 1941-46; Lady Tata fellow, 1947-49; Fulbright fellow, 1953-54; Damon-Runyon fellow, 1954-55; NSF fellow, 1956—. Recipient Quinn Gold medal M.R. Coll., 1943; Das Gupta medal Indian Chem. Soc., 1953; Citys and Guilds Silver medal, London, 1948. Mem. Am. Chem. Soc., A.A.A.S., N.Y. Acad. Scis., N.Y. Acad. Medicine, Am. Assn. Clin. Chemists, Am. Soc. Microbiology, Soc. Biol. Psychiatry, Am. Fedn. for Clin. Research. Editor: Some Biological Aspects of Schizophrenic Behavior, 1962. Research, publs. on effects of biotin deficiency in rat, fungi and insects, biosynthesis of folic acid and derivatives in bacteria from purines, effects of psychoactive drugs especially lysergic acid diethylamide on animals and humans, growth in animals and children, schizophrenias; synthesized adenylobiotin. Home: 10 Oakdale Dr., Westbury, N.Y. 11590. Office: Creedmoor State Hosp., Jamica, N.Y. 11427.*

SANNERSTEDT, (Nils Axel) Rune, Swedish physician; b. Munkfors, Sweden, Oct. 21, 1931; s. Gunnar S. and Ellen (Skoglund) S.; Candidate Medicine, U. Göteborg (Sweden), 1952, M.D., 1957, Ph.D., 1966; m. E. Barbro E., Dec. 31, 1956; children—Gunilla, Torbjön, Lena, Jörgen. Instr. medicine, asst. physician dept. medicine I, Sahlgrenska Hosp., U. Göteborg, 1959—; physician Royal Swedish Air Force, 1962—; research asso. U. Mich., Ann Arbor, 1967-68. Author: (with E. Varnauskas, L. Werkö) Hjärtarytmier. Göteborgssymposiet, 1963; (with G. Cramér, O. Thulesius, L. Werkö) Cardiac arrhythmias and Quinidine, 1965; Hemodynamic Response to Exercise in Patients with Arterial Hypertension, 1966. Research on cardiovascular diseases, especially on hemodynamics in human arterial hypertension; clin. pharmacodynamic studies especially hypotensive agts. Home: Björkedal, Billdal, Sweden. Office: Dept. Medicine I, Sahlgrenska Hosp., Göteborg SV, Sweden.*

SANO, Keiji, Japanese neurosurgeon; b. Shizuoka Prefecture, Japan, June 30, 1920; s. Takeo and Haru (Sase) S.; M.D., Imperial U. Tokyo, 1945; Dr.Med. Sci., U. Tokyo, 1951; m. Sumako Nawata, June 10, 1957. Faculty, U. Tokyo, 1956—, prof. neurosurgery, dir. dept. neurosurgery, 1962—. Mem. Japan Neurosurg. Soc. (past pres.), Japanese Assn. for Research in Stereoencephalotomy (pres. 1966—), Harvey Cushing Soc. (corr.), Deutsche Gesellschaft für Neurchirurgie (corr.). Author numerous books, articles. Established and designated pulseless disease as clinicopath. entity, 1948; research on sedative neurosurgery, including stereotaxic surgery of hypothalamus (posteromed. hypothalamotomy), stereotaxic surgery of pain (thalamalaminotomy), tissue culture continuous infusion, radiosensitizer in brain tumors. Home: Den-enchofu 6-22, Oota-ku, Tokyo, Japan.*

SANO, Shizuo, Japanese physicist; b. Tokyo, 1872; grad. Tokyo U.; D.Sc.; became prof. Mil. Acad., 1906; named asst. prof. Tokyo U., later prof. Developed theories on relations between energy within substances and elec. and magnetic fields. Died 1925.

SANOTSKY, Vladimir Antonovich, Russian toxicologist; b. 1890; grad. Petrograd Mil. Med. Acad., 1914; D.Med. Sci. Head lab., dep. dir. for sci. work, dir. Inst. Pathology and Treatment of Intoxications, head course on toxicology of radioactive substances Central Postgrad. Med. Inst., Moscow, 1934-52; head toxicology lab. Inst. Biophysics, USSR Ministry Health. Chmn. bur., dept. med. and biol. sci. USSR Acad. Med. Scis. Decorated Order of Lenin. Mem. USSR Acad. Med. Scis., All-Union Soc. Physiologists (central council), Moscow Soc. Physiologists, Biochemists and Pharmacologists (chmn. pharmacology and toxicology sect. 1950—). Co-author: Pathology, Therapy and Prophylaxis of Poisoning by Animal Chemical Warfare Agents, 1940; author: Toxicology of Radioactive Substances, 1957; General Principles for Treating Injuries Caused by Radioactive Substances, 1957; Thiolic Compounds in the Treatment of Injuries Caused by Radioactive Substances, 1959. Mem. editorial bd. Pharmacology and Toxicology, Bull. Exptl. Biology and Medicine. Address: Inst. Biophysics, Profsoyuznaya ulitsa 7, Moscow, USSR.

SANSON, André, French veterinarian; b. Matha, 1826; mil. veterinarian; dir. chem. work Vet. Sch., Toulouse, France; later prof. zootech. Nat. Agr. Sch., Grignon, France, also Nat. Inst. Agronomy. Author: Treatise on Zootechnology, 1866-74. Introduced some empirically-based ideas into stock-farming which were later proved incorrect. Died Palais-sur-Mer, France, 1902.

SANSON, Louis Joseph, French surgeon, physician; b. Paris, Jan. 24, 1790; M.D., Paris, 1817; succeeded Dupuytren as prof. clin. surgery Hôtel Dieu, 1836. Author: New Elements of Medico-Surg. Pathology, 4 vols., 1825; Des Hemorrhagies traumatiques, 1836; also papers. Eminent med. practitioner; described 3 pairs of images of single object reflected from anterior and posterior surfaces of lens and from cornea, seen when looking into pupil (Purkinje-Sanson images), circa 1823. Died Nogent-sur-Seine, France, Aug. 2, 1841.

SANSONE, Giovanni, Italian mathematician; b. Port-Empedocle, Sicily, May 24, 1888; s. Giuseppe and Carmella (Li Fonti) S.; ed. U. Pisa; student Superior Normal Sch., Pisa, Italy, 1906-10; m. Emma Galli, Dec. 27, 1913. Asst. algebra and analytical geometry U. Pisa, 1910-13; math. tchr. Tech. Inst., Florence, Italy, 1913-26; prof. math. analysis, acting prof. higher analysis U. Florence, 1927—. Mem. Superior Council Pub. Instrn., 1945—; mem. Nat. Com. Math. and Physics, Nat. Research Council. Recipient Gold medal Societá Italiana dei XL. Mem. Nat. Acad. Lincei (corr.), Societá Italiana Mathesis (pres. 1937-41). Author: Lezioni di Analisi Matematics, vol. I, vol. II; Svliuppi in serie di Funzioni ortogonali; Equazioni differenziali nel campo reale, 2 vols.; Lezioni sulle funzioni di una variable complessa, 2 vols. Research on number theory, discontinuous groups, differential geometry, differential equations. Home: Via Francesco Crispi 6, Florence, Italy.

SANTE, LeRoy, Am. radiologist; b. St. Louis, Mar. 31, 1890; s. August Henry and Laura (Woodrow) S.; M.D., Washington U., St. Louis, 1913; grad. U. S. Army Sch. of Roentgenology, Ft. Oglethorpe, Ga., 1918; m. Jewel Hartt, July 2, 1914; children—Elsie Eleanor (Mrs. Carl Edward Weaver), Le Roy (dec.), Henry. Physician to St. Louis pub. schs., 1915-16; in gen. med. practice, Ellendale, N.D., 1916-18; chief radiologist St. Louis City Hosp., 1919-45, cons. radiologist, 1945-60; radiologist St. Louis U. hosp. group: St. Mary's Hosp., Firmin Desloge Hosp., Mount St. Rose Sanitorium, 1920—; cons. radiologist Koch Hosp. for Tb., St. Louis Tng. Sch., and Mo. State Sanitorium, 1919—; prof. radiology, dir. dept., St. Louis U. Sch. Medicine, 1919-60. Diplomate, charter mem. Am. Bd. Radiology. Fellow A.C.P., Am. (gold medal 1959), Inter Am. colls. radiology, Am. Trudeau Soc.; mem. Radiol. Soc. N. Am., Am. Radium Soc. (1st v.p.), A.M.A., Mo. State, St. Louis (pres.) med. socs., St. Louis Soc. Radiol. (pres.), Am. Roentgen Ray Soc. (1st v.p., rep. on Am. Bd. Radiol. 10 yrs.), Am. Registry X-ray Technicians (pres.), Denver, Panamanian (both hon.) radiol. socs., Brit. Inst. Radiology (hon.), Mexican Radiol. and Physiotherapy Soc. (hon.), Alpha Omega Alpha, Sigma Xi. Author: Lobar Pneumonia, a Roentgenological Study, 1928; The Chest Roentgenically Considered, 1929 (4 eds.); Manual of Roentgenological Technique, 1934 (20th rev. edit., 1962); Principles of Roentgenological Interpretation, 1937 (12th edit. 1961); Radiology For Nurses, 1936 (4th edit., 1945); Atlas of Miniature Roentgenograms, 1952. Contributor numerous articles on radiolog. subjects to profl. jours. Home: 308 Orchard Av., Webster Groves, Mo. 63119. Office: Mo. Theatre Bldg., St. Louis. Died Oct. 27, 1964.*

SANTINI, Giovanni Sante Gaspero, Italian astronomer; b. Caprese, Italy, Jan. 30, 1787; ed. Prato, Pisa, Milan; became asst. to dir. Padua obs., 1806, dir. obs., 1814; named prof. astronomy U. Padua, 1813, rector, 1924, 56, dir. math. studies, 1845-72. Research on comets; calculated orbital disturbances caused by comet of Biela, 1832-52; catalogued stars between declination plus 10° and minus 10°, 1847-57. Died Padua, June 26, 1877.

SANTIROCCO, Raymond Anthony, Am. physicist; b. Rochester, N.Y., July 19, 1930; s. Anthony and Eleanor (Rotonno) S.; B.S., U. Rochester, 1952, Ph.D., 1959; m. Bertha L. Meisel, Aug. 27, 1955; children—Laura, Paula, Carla. Sr. physicist, research dept., electronics div. Gen. Dynamics, Rochester, 1958-63, mgr. communications scis. lab., 1963-65, research specialist, 1965—; instr. astronomy U. Rochester, 1962—. Dir. Rochester Council Sci. Socs., 1962-64. Mem. Am. Phys. Soc., Am. Astron. Soc., Am. Geophys. Union. Research, publs. in fields of antisubmarine warfare, natural electromagnetic noise and wave propagation in very low frequency region of radio spectrum, geomagnetism and geomagnetic fluctuations. Home: 51 Kevin Dr., Rochester 14625. Office: 1400 N. Goodman St., Rochester, N.Y. 14609.*

SANTO DA BARLETTA, Marino, Italian surgeon; b. Barletta, Italy, 1490. Author: Compendium in chirurgia, 1514; Libellus aureus de lapide ex vesica per incisionem extrahendro, 1522. First to perform John de Romani's operation for removal of urinary calculi (forerunner of modern lateral lithotomy), 1531; eponym of metodo Mariano (Marion method of operation for lithotomy); recommended lateral incision which became popular with Colot and other French lithotomists. Died 1550.

SANTORIN(I), Giovanni Domenico, Italian physician, anatomist; b. Florence, Italy, 1681; ed. Pisa; practiced medicine, also prof. anatomy, Florence, Italy; prof. anatomy Physico-Med. Coll., Venice. Pioneer in obstetrics; described small nodule at tip of arytenoid cartilages (Santorini's cartilages, or corniculate cartilages), 1724, accessory pancreatic duct (Santorini's duct), 1724, also risorius of Santorini, plexus of Santorini in pubio-prostrate region. Died Venice, 1737.

SANTORIO, Santorio (or Sanctorius, Sanctorius), Italian physician; b. Capodistria, Italy, 1561; med. degree U. Padua, 1582; physician to Polish court, 1587-1611; became prof. theoretical medicine, Padua, 1611-29; began pvt. practice, Venice, 1629. Author: Methodus vitandorum errorum omnium, qui in arte medica, contingunt, 1602; Ars de statica medicine, 1614; De Remed. inventione, 1631; also commentaries on Hippocrates, Galen, Avicenna. Introduced balance thermoscope with scale, pulsimeter into med. practice; 1st to apply Galilean principles of measurement to biol. matters; laid founds. of modern study of metabolism; used scales to determine weight given off by body beyond excrement; inventor string hygrometer, pulsimeter, new trocar, instrument for extracting stones from bladder. Died Venice, Feb. 24, 1636.

SANTOS, Francisco O., Philippine chemist; b. Calumpit, Bulacan, Philippines, June 3, 1892; s. Miguel O. and Maria (Alvarez) S.; A.B., U. Philippines, 1914, M.S., 1919; Ph.D., (U. Philippines fellow), Yale, 1922; postgrad. U. Minn., Columbia, Cornell U.; m. Adela Paz de Guzman, July 29, 1923; children—Fernando, Celso, Orlando, Ruben, Adela (Mrs. Teodor M. Katigbak). Faculty, U. Philippines Coll. Agr., 1915-59, head dept. agrl. chemistry, 1929-55, dean, 1943-45, prof. emeritus, 1959—; pres. Los Banos Rural Bank, Inc., Laguna, Philippines, 1961—. Recipient Distinguished Service medal Republic Philippines, 1955; Col. Andres Soriano award in chemistry U.P. Chem. Soc., 1956; Gold medal of merit, Diploma of Honor as Distinguished Alumnus, U. Philippines, 1961; named Rural Banker of Year, Rural Bankers Assn. Philippines, 1961. Fellow A.A.A.S.; mem. Los Banos Biol. Club (past pres.), Soc. for Avancement Research (past pres.), Philippine Assn. Nutrition (past 1st pres.), Am. Inst. Nutrition, Inst. Nutrition Bd., Philippines Assn. Agriculturists, Philippine Acad. Scis. and Humanities, Am. Chem. Soc., N.Y. Acad. Scis., Sociedad Española de Bromatologia (hon.), Sigma Xi, Phi Kappa Phi, Phi Sigma Delta. Research, publs. on nutrition, defects of Filipino common diet, suggested methods of improving it. Address: Coll., Laguna, Philippines. 40219.*

SANTOS-DUMONT, Alberto, Brazilian inventor; b. Sao Paulo, Brazil, July 20, 1873; ed. France; began expts. with dirigible airships, Paris, 1898; built, flew cylindrical balloon with gasoline engine, 1898; built prize-winning airship which made 1st flight from St.-Cloud, around Eiffel Tower and back, 1901; built (with clockmaker Tatin) other airships, one of which made numerous flights over Paris; built 1st airship sta., Neuilly, 1903; flew 715 feet in box-kitelike airplane, 1906; built 260-pound monoplane, 1909. Author: My Airships: A Story of My Life, 1904. Died July 23, 1932; buried as nat. hero, Rio de Janerio, Brazil.

SANYAL, Ranajit Kumar, Indian pharmacologist; b. Ranchi, India, Aug. 2, 1922; s. Surendra Nath and Bimla (Bagchi) S.; M.B., B.S., Patna Med. Coll., 1947, M.D., 1951; Ph.D., London Sch. Pharmacy, 1958; m. Sujaya Lahiri, Nov. 25, 1951; children—Ajanta, Arun. Tutor in medicine Patna Med. Coll., 1950-51; demonstrator, lectr. physiology Darbhanga Med. Coll., 1951-54, lectr. pharmacology, 1954-60; prof. pharmacology Maulana Acad. Med. Coll., New Delhi, India, 1960—; pres. Indian Coll. Allergy and Applied Immunology, 1968—. Recipient Bombay Med. Union Gold medal. Mem. Internat. Coll. Allergy, European Acad. Allergy, Brit. Physiol. Soc., Royal Soc. Medicine, British Pharmacol. Soc. Contbr. numerous articles to sci. jours. Fundamental work on various aspects of exptl. allergy; demonstrated that species variation in allergy is due to different sensitivity to toxic substances produced during process; allergic manifestations have been shown to be mild if blood sugar level is high and vice versa; anticoagulants have been shown to possess antiallergic action. Home: D-1, 212 Satya Marg, New Delhi-11, India.*

SANZAEMON, see Aida, Yasuaki.

SAPIR, Edward, anthropologist, linguist; b. Lauenburg, Pomerania, Jan. 26, 1884; s. Jacob David and Eva (Sigel) S.; came with parents to U. S., 1889; A.B., Columbia, 1904, A.M., 1905, Ph.D., 1909, Sc.D., 1929; m. Florence Delson, 1911 (died 1924); children—Herbert Michael, Helen, Philip; m. 2d, Jean V. McClenaghan, 1926; children—Paul Edward, James David. Research assistant in anthropology, U. of California, 1907-08; instr. in anthropology, U. of Pa., 1908-10; chief of division of anthropology, Can. Nat. Museum, 1910-25, asso. prof. anthropology, U. of Chicago, 1925-27, prof. anthropology and gen. linguistics, 1927-31; Sterling prof. anthology and linguistics, Yale, 1931. Mem. Am. Acad. Arts and Sciences; mem. Am. Anthropol. Assn. (pres.). Author: Wishram Texts, 1909; Takelma Texts, 1909; Yana Texts, 1910; The Takelma Language of S.W. Oregon, 1912; Time Perspective in Aboriginal American Culture, 1916; Language, an Introduction to the Study of Speech, 1921; (with Marius Barbeau) Folk-Songs of French Canada, 1925; The Southern Paiute Language, 1931. Studied several groups of N.Am. Indians; also descriptive linguistics; stimulated co-ordination of anthropology and psychiatry in studies of personality within a cultural milieu; developed techniques of fieldwork in cultural anthropology; contbd. to theory of language as means of communication reflecting a culture. Died New Haven, Conn., Feb. 4, 1939.

SAPOZHNIKOV, Leonid Mikhailovich, Ukranian engr.; b. Apr. 29, 1906; grad. Dnepropetrovsk Coal and Chem. Inst., 1930. Joined staff Dnepropetrovsk Coal-Chem. Inst., 1930; joined faculty Dnepropetrovsk Chem.-Tech. Inst., 1930, prof., 1935-37; named lab. chief Inst. Fuel Minerals, USSR Acad. Scis., 1937. Mem. Commn. for Evaluation Fuel and Power Balance for Period, 1959-65. Corr. mem. USSR Acad. Scis. Author: A Study of the Coking Process, Coal Classification and Calculation of Charges by the Plastometric and Other Methods, 1955; Coals and Metallurgical Coke, 1941; The Baykal Meridian, 1961; co-author: An Examination of the Modern Principles of Coal Coking, 1953; A New Technique of Coking and Coal Enrichment, 1956. Research on coal coking process. Office: Dnepropetrovsk Chem.-Technol. Inst., Dnepropetrovsk, USSR.

SAPPENFIELD, Bert R(eese), Am. psychologist; b. Ind., Jan. 2, 1912; s. Alonzo Seth and Lena (Miller) S.; B.A., DePauw U., 1935; M.A., N.Y. U., 1938, Ph.D., 1941; m. Louise Pile, Oct. 2, 1937; children—Elizabeth Anne, Mary Jean. Grad. asst. N.Y. U., 1935-39, research asst., 1940-41; sec. com. on tests Life Office Mgmt. Assn., 1939-40; instr. psychology U. Mont., 1941-46, asst. prof., 1946-48, asso. prof., 1948-53, prof. psychology, 1953-—, chmn. dept., 1956-60; psychologist Utah State Hosp., 1954-55. Fellow Am. Psychol. Assn.; mem. Mont. (pres. 1957-59), Rocky Mountain (pres. 1961-62) psychol. assns., Phi Beta Kappa, Psi Chi (v.p. Rocky Mountain region 1959-—). Author: Personality Dynamics, 1954. Developed neo-Freudian theory of repression, anxiety, mechanisms of defense which views repression as conflict-solving process, anxiety as consequence of conflict solution; research on stereotypical perception of personality traits, especially perception of masculinity-feminity. Home: 3032 Queen St., Missoula, Mont. 59801.*

SAPPEY, Marie Philibert Constant, French anatomist; b. Cerdon, France, Aug. 10, 1810; doctorate, Paris, 1843; prof., Paris, 1868-86; mem. French Acad. Scis., 1886, Acad. Medicine. Author: Recherches sur l'appareil respiratoire des oiseaux, 1847; Traité d'anatomie descriptive, 1847-63; Anatomie, physiologie, pathologie des vaisseaux lymphatiques chez l'homme et les vertebrés, 1874; Atlas d'anatomie descriptive, 1879; Les éléments figurés du sang sans la série animale, 1881; Description et iconographie des vaisseaux lymphatiques, 1886. A leading 19th century anatomist, noted for work on lymphatic system; described accessory portal veins (Sappey's veins), 1859. Died Paris, Mar. 13, 1896.

SARADZHISHVILLI, Petr Mikhaylovich, Russian neuropathologist; b. 1894; grad. Med. Faculty, Odessa U., 1917; D.Med. Sci. Lectr. dept. nervous diseases Tbilisi Postgrad. Med. Inst., 1933-37, head chair nervous diseases, 1937-—; dir. Inst. Neurology, Georgian Acad. Scis., 1958-—. Mem. USSR Acad. Med. Scis. Author numerous works including: Tricresolphosphate Poisoning of the Nervous System, 1935; Injuries to the Peripheral Nerves of the Limbs, 1945; co-author: Clinical Aspects of the Functional Importance of the Reticular Formation of the Brain Stem, 1958; Clinical Syndromes of Injuries (Obstructions) of the Vessels of the Brain, 1960. Address: Tbilisi Postgrad. Med. Inst., Lunacharskogo 12, Tbilisi, Georgian SSR, USSR.

SARAGEA, Marcel Daniel, Rumanian physician; b. Ramnieu Sarat, Rumania, Apr. 21, 1914; s. Daniel and Maria (Zosmer) S.; M.D., Med. Inst. Bucharest, 1938; M.D., Leningrad Med. Inst., 1952; m. Alice Tauber, Dec. 25, 1940. Asst., lectr. Clinic Internal Medicine, Bucharest, Rumania, 1948-52; prof. physiopathology Med. Inst. Bucharest, 1952-—; dep. minister health for Rumania, 1953-57. Mem. Soc. Normal and Pathol. Physiology, Soc. Comparative Pathology. Author: Fundamental Principles of Nervous System Activity, 1953; Problems of Physiopathology, 1954; Normal and Pathological Physiology of Billiary Passages, 1954; also numerous articles. Research on physiopathology of biliary passages, exptl. allergic encephalitis, neuro-viscereous correlations. Home: 10 Galati St., Bucharest, Rumania.*

SARAJAS, H(eikki) S(eppo) Samuli, Finnish physiologist; b. Loppi, Finland, Apr. 11, 1927; s. Lauri Heikki and Sylvi (Wathén) S.; M.Sc., U. Helsinki (Finland), 1951, M.D., 1961, Ph.D., 1965; m. Kerstin Maria Soderlund, Jan. 7, 1956; children—Pia Maria, Sylvia Kristina, Henrik Hjalmar Samuli, Anna Maija. Research fellow Wenner-Gren Cardiovascular Research Lab., Stockholm, Sweden, 1953-56; asst., lectr. Inst. Physiology U. Helsinki (Finland), 1958-63, asso. prof. physiology, 1966-—; established investigator Finnish State Med. Research Council, 1962-63; acting prof. physiology Coll. Vet. Medicine, Helsinki, 1964-66. Mem. Hibernation Information Exchange, Finnish Physiol. Soc. Research and publs. on cardiovascular, endocrine and metabolic changes in induced hypothermia and natural hibernation, physiology of exercise, role of hemodynamics in altering circulating leucocyte and platelet pools, platelet adhesiveness; demonstrated that induced hypothermia in nonhibernating mammals seriously affects heart muscle and gives rise to different adrenal cortical reaction patterns depending on type and depth of precooling anesthesia. Home: 5 B 11, Niittykumpu, Finland. Office: Inst. Physiology, U. Helsinki, Helsinki, Finland.*

SARASIN, Paul, Swiss naturalist; b. Basel, Switzerland, Dec. 11, 1856; founder Swiss nature sanctuaries and nat. parks; promoted prehistoric collection, Basel. Died Apr. 7, 1929.

SARATEANU, Dan Emil, Rumanian virologist; b. Bucharest, Rumania, Mar. 11, 1924; s. Florian E. and Elena (Dumitrescul) S.; Grad., Medium Sch., Bucharest, 1942; M.D., U. Bucharest, 1948, D.Med. Scis., 1966; div.; 1 son, Serban. Researcher, Inst. Inframicrobiology, Rumanian Acad., Bucharest, 1950-—, head dept., 1957-—; asst. lectr. dept. virology Faculty Medicine, 1955-65. Recipient Babes prize for med. research Rumanian Acad. Mem. Internat. Soc. Infectious Pathology. Research, numerous publs. in viral respiratory diseases, ornithosis, viral immunology, poliomyelitis. Home: 2 str. Capt. Demetriade, Bucharest 3. Office: Inst. Inframicrobiology, 283 sos. Mihai Bravu, Bucharest 29, Rumania.*

SARATIKOV, Albert Samoilovich, Russian pharmacologist; b. Kharkov, Ukraine, USSR, Oct. 10, 1921; s. Samuil B. and Vera (Rivkina) S.; M.Sc. in Medicine, Tomsk State Med. Inst., 1946, D.Sc., 1953; m. Evgenia Dumenova, Nov. 16, 1943; 1 dau., Natalia. With pharmacology dept. Tomsk (Siberia) Med. Inst., 1946-—, prof. pharmacology, 1954-—, dep. dir., 1959-—, dep. dir. Tomsk Inst. Physiotherapy, 1952-59. Mem. Pharmacologists' Soc. USSR. Author: Bile Formation and Choleretics, 1962; Camphor, 1966; also numerous articles. Research on mechanism of camphor cardiotonic action, bioenergy of bile-formation and action of choleretics, pharmacology and biochemistry of central nervous system, vegetable stimulators and pharmacology of brain blood circulation. Home: 16 Savinikh, Tomsk, Siberia.*

SARAUX, Henry Camille Amédée, French physician; b. Niort, France, 1927; s. Jean Henri and Yvonne (Chaumier) H.; m. Geneviève Cluzeau, Feb. 16, 1953; children—Jean Luc, Bertrand, Nathalie. Agrégé ophthalmology, ophthalmologist Paris hosps., 1958-—. Cons. ophthalmologist Société nationale des chemins de fer francais. Author: Abrégé d'ophtalmologie, 1959; Anatomie de l'oeil, 1965; Exploration neuro-radiologique en ophtalmologie, 1966; Ophalmologia fondamental, 1966; also numerous articles. Research on pediatric ophthalmology, genetics, allergology. Home and office: 38 Niel, Paris 17, France.*

SARAZIN, Armand, French physicist; b. Rans (Jura), France, Aug. 8, 1924; s. Louis and Magdeleine (Paupard) S.; license in sci., engring. degree, Sorbonne, Paris, 1949, D.Sc., 1953; m. Monique Biolley, June 1955; children—Arielle, Francoise, Laurent. With Nat. Center Sci. Research, Paris, 1949-52; physicist European Orgn. Nuclear Research, Geneva, Switzerland, 1952-56; prof. nuclear physics Faculty Scis., U. Algiers (Algeria), also asst. dir. Inst. Nuclear Studies, 1956-60; prof. nuclear physics Faculty Scis., U. Lyons (France), also dir. Inst. Nuclear Physics, 1960-—. Sci. counselor to Commissariat Atomic Energy, Paris. Decorated chevalier Order Sci. and Tech. Merit; officer Nat. Order Merit; chevalier Palmes Acad. Mem. physics socs. France, Can., Am., Great Britain, Japan, I.E.E.E. Author: (with J. J. Samueli and J. Pigneret) L'Instrumentation Electronique en Physique Nucléaire, 1968; also articles. Research in particle acceleration, also in area of nuclear instrumentation and rapid electronics; studies of nuclear structure, life levels excited in the domaine of nanosecond. Home: 3 chemin de la Cigaline, Caluire (69), France. Office: 43 Blvd. du 11 novembre 1918, Villeurbanne (69), France.*

SARD, Arthur, Am. mathematician; b. N.Y.C., July 28, 1909; s. Frederick N. and Maria (Belloch) S.; B.S., Harvard, 1931, M.A., 1932, Ph.D., 1936; postgrad. U. Paris; m. Marguerite Cochran, Oct. 9, 1933. Instr. Harvard, 1933-36; faculty Queens Coll., City U. N.Y., Flushing, 1937-—, prof., 1958-—; research mathematician applied math. group div. war research Columbia, 1943-45, dir., 1945. Mem. Am. Math. Soc., Assn. for Symbolic Logic, Math. Assn. Am. Author: Linear Approximation, 1963; also articles. Research on measure of critical images, function spaces, remainders in approximation, efficient approximation, minimal response to error, spline formulas. Home: 146-19 Beach Av., Flushing, N.Y. 11355.*

SARGENT, Charles Sprague, Am. dentrologist; b. Boston, Apr. 24, 1841; s. Ignatius and Henrietta (Gray) S.; A.B., Harvard, 1862, LL.D., 1901; m. Mary Allen Robeson, Nov. 28, 1873. Prof. horticulture Harvard, 1872-73, dir. Botanic Garden, 1873-79, dir. Arnold Arboretum, 1872-—, prof. aboriculture, 1879-—. Planned Jesup collection of N. Am. woods for Am. Mus. Natural History, N.Y.; chmn. commn. for preservation of Adirondack forests, 1885; chmn. commn. apptd. by Nat. Acad. Scis. upon a forest policy for forestry lands of U. S., 1896-97. Fellow Am. Acad. Arts and Scis.; pres. Mass. Soc. Promotion Agr., 1890-—; trustee Mass. Hort. Soc. Author: Catalogue of the Forest Trees of North America; The Woods of the United States; The Forest Flora of Japan; Silva of North America; Report of the Forests of North America; Manual of the Trees of North America, 1905, 1922; Trees and Shrubs. Editor Garden and Forest, 1887-97. Died Mar. 22, 1927.

SARGENT, James, Am. inventor; b. Chester, Vt., Dec. 5, 1824; s. William and Hannah S.; ed. dist. schs.; m. Angeline M. Foster, Apr. 29, 1847. Worked on farm, later foreman in woolen mill; traveling daguerreotyper, 1848-52; manufactured and sold an automatic apple-parer, 1852-57; partner Yale & Greenleaf Lock Co., 1857-65; invented burglar-proof locks and established in business, Rochester, N.Y., 1865; invented Sargent timelocks, 1873; later added many styles of patented locks; pres. Sargent & Greenleaf Co., mfrs.; pres. Pfaudler Vacuum Fermentation Co., inventor their glass enameled steel tanks and vacuum pump; inventor, automatic electric semaphore ry. signals, pres. Gordon Railway Signal Co.; inventor automatic smoke consumer; pres. Waterloo Gold Mining Co., Cal. Died 1910.

SARIO, Leo Reino, mathematician; b. Lieksa, Finland, May 18, 1916; s. Walde and Katri (Gertzin) S.; Ph.D., Helsinki (Finland) U., 1948; postgrad. U. Zürich (Switzerland), Paris (France) U. Came to U.S., 1950, naturalized, 1955. Lectr., Helsinki U., 1945-48; research asso. Finland Acad., 1948-50; vis. mem. Inst. for Advanced Study, Princeton, 1950-52; research fellow Harvard, 1952-54; faculty U. Cal., Los Angeles, 1954-—, prof. math., 1958-—; prin. investigator project on theory Riemann surfaces U.S. Army Research Office, 1955-—. Decorated Cross of Comdr. Order of Finland's Knighthood. Mem. Am. Math. Soc., Société Mathématique de France. Author: (with L. Ahlfors) Riemann Surfaces, 1960; (with K. Noshireo) Value Distribution Theory, 1966; also numerous articles. Initiated and developed classification theory of Riemann surfaces; introduced capacity of boundaries and boundary components; introduced, developed theory of prin. functions on Riemann surfaces and in Riemannian spaces; developed value distbn. theory for analytic mapping. Home: 521 Georgina Av., Santa Monica, Cal. 90402. Office: 405 Hilgard Av., Los Angeles 90024.*

SARJEANT, William Antony Swithin, English geologist; b. Sheffield, Eng., July 15, 1935; s. Harold and Margaret (Cantrell) S.; B.Sc. with Honors in Geology, U. Sheffield (Eng.), 1956, Ph.D., 1959; m. Anne Margaret Crowe, Apr. 2, 1966. Research demonstrator U. Sheffield (Eng.), 1956-59; faculty U. Coll. N. Staffordshire, 1960-61; univ. research fellow U. Reading (Eng.), 1961-62; faculty U. Nottingham (Eng.), 1963-—, lectr. geology, 1964-—, adult edn. tutor geology, 1963-—; vis. prof. U. Okla., Norman, 1967-68. Fellow Geol. Soc. London, Mineral. Soc.; mem. Société Géologique de France, Palaeontol. Assn., Geologists' Assn., Marine Biol. Assn., Commn. Internationale sur la Microflore du Paléozoique, Peak Dist. Mines Hist. Assn. (founder, chmn. 1959). Author: (with C. Downie) Bibliography and Index of Fossil Dinoflagellates and Acritarchs, 1954; (with G. Norris) A Descriptive Index of Genera of Fossil Dinophycene and Acritarcha, 1965; (with R. J. Davey, C. Downie, G. L. Williams) Studies on Mesozoic and Cenozoic Dinoflagellate Cysts, 1966; also articles. Editor, Sorby Record, 1958-59, Mercian Geologist, 1964-67. Research on cysts of dinoflagellates, acritarchs; pioneered research on Jurassic, Cretaceous and Post-Pleistocene assemblages, described many new genera and species, contributed to devel. classifications schemes; mineralogy of Derbyshire, fibrous chlorites and gypsum, Triassic vertebrate footprints, history of geology in English Midlands; rediscovered (with T. D. Ford) lost stalactitic berytes. Home: 47 Milton Rd., Nether Edge, Sheffield S7IHP, Eng.*

SARKANY, Imrich, physician; b. Svaty Jur, Czechoslovakia, Jan. 7, 1923; s. Edmund and Maria (Pollitzer) S.; grad. St. Thomas' Hosp., London, Eng., 1952; M.R.C.S., L.R.C.P., London, 1952,

M.R.C.P., 1956; m. Helen Pomerance, Dec. 12, 1956; children—Elizabeth, Robert, Andrew. Cons. physician for diseases of skin Royal Free Hosp., 1960—, Hampstead Gen. Hosp., London, 1961—; research fellow dept. dermatology U. Miami (Fla.), 1960-61. Med. Research Council grantee, 1962—. Licentiate Royal Coll. Physicians. Fellow Royal Soc. Medicine; mem. Royal Coll. Surgeons, Royal Coll. Phyicians London, Brit. Assn. Dermatology. Author: (with G. Hildick-Smith, Harvey Blank) Fungus Diseases and their Treatment, 1964; also articles. Showed that erythrasma was bacterial in nature; clin. and exptl. research in fungus diseases; jointly first used griseofulvin in human fungus infections; organ culture of skin and tissue culture of lymphocytes; lymphocytes transformation used as test for drug reactions. Home: 34 Northwick Circle, Kenton, Middlesex, Eng. Office: Royal Free Hosp., Gray's Inn Rd., London, W.C.1., Eng.*

SARKER, Abdul Qayyum, Pakistani physicist; b. Rangpur, East Pakistan, Feb. 28, 1936; s. Maher Uddin and Kalsuma Khatun (Miah) S.; B.Sc. with honors in Physics, Rajshahi U., 1956; M.Sc., Dacca (E. Pakistan) U. 1958; Ph.D., Birmingham (Eng.) U., 1963; m. Niloufar Naseem, Aug. 11, 1963; 1 dau., Kashfi. Research asso. U. Cal. at San Diego, 1963-65, Rockefeller U., N.Y.C., 1965-66; reader in physics U. Islamabad, Rawalpindi, Pakistan, 1966—. Mem. Am. Phys. Soc. Research, publs. on use of dispersion technique to devel. theory of exchange currents in nuclei; predicted and related electromagnetic mass differences of higher resonances (decouplet) to stable baryons; explained certain electromagnetic decay modes of eta meson in relation to its other decay modes. Home: P.O. Tulshighat, Dist. Rangpur, East Pakistan. Office: U. Islamabad, Rawalpindi, Pakistan.*

SARKES, Louis Anthony, Am. physicist; b. New Bedford, Mass., June 15, 1925; s. Thomas and Josephine (Thomas) S.; B.S. in Physics, Boston Coll., 1951, M.S., 1953; m. Anne C. Thomas, June 17, 1955; children—Timothy, Louis, Jane. Radioisotope physicist Boston VA Hosp., 1952-55; research asst. Brown U., 1955-56; cons. radioisotopes Lemuel Shattuch Hosp., 1955-57; research and devel. chief applied physics lab. Metals & Controls, Inc., Attleboro, Mass., 1956-63, mgr. quality control-nuclear core tech., 1963-65; mgr. advanced devels. and basic research Am. Gas Assn., N.Y.C., 1965—. Mem. Am. Nuclear Soc., Soc. Nondestructive Testing, Am. Soc. for Testing Materials, Boston Med. Physics Soc., Am. Soc. Metallurgists. Research, publs. on low level alpha counting techniques for detection trace amounts uranium 235 on fuel elements, use iodine 131 for thyroid uptake measurements; developed variety nondestructive measuring devices to determine bond integrity uranium-235 homogeneity and surface condition in of fuel elements nuclear reactor cores, manage direct conversion research and basic research program for natural gas industry. Home: 36 Crane Av., White Plains, N.Y. 10603. Office: 605 3d Av., N.Y.C. 10016.*

SARKISOV, Semen Aleksandrovich, Russian neurologist, neuromorphologist; b. 1895; grad. Med. Faculty, Moscow U., 1923; D.Med. Sci. An organizer, sr. asso. Brain Inst., USSR Acad. Med. Scis., Moscow, 1927, dir., sci. dir., 1928—, acad. sec., 1948-51. Del., 20th Internat. Congress Physiologists, Brussels, 1956; head Soviet delegation 2d Internat. Congress Physicians, Cannen, France, 1957. Decorated Order of Lenin. Mem. USSR Acad. Med. Scis. (Presidium mem. 1957-59, 62—), Royal Soc. Medicine (London), Internat. Health Assn. for Study Living and Health Conditions, All-Union Soc. Neuropathologists and Psychiatrists (bd. mem.). Author numerous works including: Biological Phenomena of the Cortex and Problems of Localization, 1938; co-author, editor: Cytoarchitectonic Atlas of the Human Cortex. Co-editor Neurology sect. Large Med. Ency., 2d edit.; mem. editorial council Archives Anatomy, Histology and Embryology; mem. editorial bd. Korsakov Jour. Neuropathology and Psychiatry. Helped introduce encephalography into Soviet med. practice. Home: Novoslobodskaya 57-65. Office: USSR Acad. Med. Scis., Solyanka 14, Moscow, USSR.

SARLES, Henri Louis, French physician; b. Ermont, France, Feb. 6, 1922; s. Roger and Andrée (Wiart) S.; Docteur en Médecine, U. Marseille (France), 1949, Docteur es-Sciences, 1952; m. Janine Simon, Mar. 22, 1945; children—Alain, Agnes, Jacques, Marie, Isabelle. Research staff medicine and biochemistry U. Marseille, 1940-52; physician, Marseille, 1953; prof. agrége Faculte de Médecine, 1958—; dir. research unit digestive pathology Institut National de la Sante et dela Recherche Médicale, Marseille, 1965—. Author: les Pancreatites chroniques, 1962; Le Pancreatites aigues, 1964. Research in pancreatic pathology, etiology of gall stones, chem. composition of human bile in normal and diseases people, exptl. pathology of bile ducts and abdominal lymph ways. Home: 155 Av. du Prado 13, Marseille 8. Office: Hospital Sainte Marguerite, 13, Marseille 8b, France.*

SARNAT, Bernard George, Am. plastic surgeon; b. Chgo., Sept. 1, 1912; s. Isadore M. and Fanny (Sidran) S.; B.S., U. Chgo., 1933; M.D., 1936; M.S., U. Ill., 1940; m. Rhoda Elaine Gerard, Dec. 25, 1941; children—Gerard, Joan. Asst. to Dr. Marshall Davison, gen. surgery, Chgo., 1942-43; asst. to Drs. Vilray P. Blair and Louis T. Byars, plastic and reconstructive surgery, St. Louis, 1943-46; pvt. practice plastic surgery, Chgo., 1946-56, Beverly Hills, Cal., 1956—; asst. histology, U. Ill. Coll. Dentistry, 1937-40; asst. dept. surgery, Washington U. Sch. Medicine, St. Louis, 1944-46; prof., dir. dept. oral and plastic surgery St. Louis U. Coll. Dentistry, 1945-46; prof., head dept. oral and maxillofacial surgery Coll. Dentistry, U. Ill., 1946-56, clin. asst. prof. surgery (plastic surgery), Coll. Medicine, 1949-56; mem. staff Cedars-Sinai Med. Center, Los Angeles; cons. in gen. plastic and maxillofacial surgery VA Regional Office, Chgo., to 1956. Co-winner Joseph A. Capps prize for med. research, 1940; received Certificate of Merit and 2d prize, J. E. Kerbs award for research in plastic and reconstructive surgery, 1950, 1st prize, sr. award Found. Am. Soc. Plastic and Reconstructive Surgery, 1957; nat. achievement award Phi Epsilon Pi, 1964. Fellow A.C.S., A.A.A.S., Am. Bd. Plastic Surgery; mem. Am., Cal., Los Angeles med. socs., Am. Soc. Plastic and Reconstructive Surgery, Plastic Surgery Research Council, Cal. Soc. Plastic Surgeons, Am. Assn. Plastic Surgeons, Beverly Hills Acad. Medicine (pres. 1962-63), Sigma Xi. Sr. author: Oral and Facial Cancer (with Dr. Isaac Schour) 2d edit., 1957. Author: (with Dr. Daniel Laskin) Surgery of the Temporomandibular Joint, 1962; also articles, chpts. in textbooks. Editor, contbg. author: The Temporomandibular Joint, 1951. Research on normal and abnormal growth of bones; cartilage and cartilage implants. Home: 616 N. Maple Dr. Office: 435 N. Roxbury Dr., Beverly Hills, Cal. 90210.*

SARNGADHARA, Hindu physician; flourished 13th century or earlier. Author: Sarngadharasamhita (early Sanscrit work of its kind, treats calcination of mercury, other mercurial and metallic preparations and their therapeutic uses, refers to opium and bertram root as irritants).

SARNOFF, Irving, Am. psychologist; b. Bklyn., May 5, 1922; s. Nathan and Rose (Gelfand) S.; B.A., Bklyn. Coll., 1946; M.A., U. Mich., 1949, Ph.D., 1951; m. Suzanne Fischbach, Nov. 28, 1946; children—David, Sara. Research asso. Mich. Attitude Change Project U. Mich., 1951-54, lectr. Extension Service, sr. psychologist Student Health Service, 1952-54; Fulbright advanced research scholar U. Coll., London, 1954-55; faculty Yale, 1955-60; dir. research, prof. psychology, social work Western Res. U. Sch. Applied Social Scis., 1960-62; prof. psychology N.Y. U., N.Y.C., 1962—. Mem. Am. Psychol. Association. Author: Personality Dynamics and Development, 1962; Society With Tears, 1966. Research, publs. primarily in field of psychoanalytic hypotheses; developed social psychol. theory to account for maj. social pathologies of Am. and similar socs. Office: 4 Washington Pl., N.Y.C. 10003.*

SARRAU, (Jacques Rose Ferdinand) Émile, French engr., physicist; b. Perpignan, France, June 24, 1837; student, later prof. math. and mechanics École Polytechnique; became prof., dir. Sch. Explosives and Saltpetres, 1878; mem. French Acad. Scis., 1886. Author: Recherches théoriques sur les effets de la poudre et les substances explosives, 1874-75; Formules pratiques des vitesses et des pressions dans les armes, 1877-78; Theory of Explosives, 1895. Set forth, on basis of hypothesis for ether waves, law of polarization in 2-axis crystals (contractual with electromagnetic light theory); registered origination pressures in studies on inner ballistics; devised characteristic equation of Clausius of different gases; used simple assumptions about thermal conditions, relations to derive thermodynamic principles. Died Saint-Yrieix, France, May 10, 1904.

SARRAZIN, Michel, botanist, physician; b. Nuits, France, Sept. 5, 1659; went to Can., 1685; royal physician to Quebec; mem. French Acad. Scis., 1699; discoverer, eponym of a Canadian plant. Died Quebec, Sept. 8, 1734.

SARRE, Hans, German physician; b. Neubalbelsberg, Germany, Mar. 25, 1906; ed. U. Berlin, U. Heidelberg, U. Freiburg; M.D.; m. Irmgard Jacobi, 1953; 4 children. Became privat docent U. Frankfort/Maine, Germany, 1939; named prof., dir. clinic, U. Freiburg, 1948. Author: Akutes Nierenversagen, 1962; also numerous articles. Research in allergy, asthma, kidney diseases, hypertonicity.

SARRUS, Pierre Frédéric, French mathematician; b. St. Affrique, France, 1798; agrégé in scis., 1823; became prof. analysis, Strasbourg, 1823, later dean. Author: Memoire sur la determination des orbites des comètes, 1843; Recherches sur le calcul des variationes (prize French Acad. Scis. 1845); Methode pour trouver les conditions d'integralité d'une fonction differentielle, 1847. Solved problem of extending method of variations for determining maxima to multiple integrals, thus making perhaps most significant original contbn. to calculus of variations in period of 1800-60; inventor straight line motion (allowing surfaces to execute straight line motion). Died St. Affrique, 1861.

SARTAIN, Carl Clinton, Am. physicist; b. Coker, Ala., Jan. 16, 1913; s. Searcy Brown and Ceba (Rosser) S.; B.S. in Elec. Engring., U. Ala., 1935, M.A., 1937; Ph.D., U. Va., 1949; m. Mildred Hollis, Sept. 29, 1941; children—Cynthia Joanna, Robert Lawrence. Mem. faculty U. Ala., 1939-41, 46-54; sr. physicist Oak Ridge Nat. Lab., 1954-58; physicist U. Cal. at Livermore, 1958-62, Bourns, Inc., Riverside, Cal., 1962—. Cons., Oak Ridge, 1951-56. Mem. Am. Phys. Soc., Am. Assn. Physics Tchrs., Am. Ordnance Assn., Soc. Glass Technologists, Am. Vacuum Soc., So. Cal. Industry-Edn. Council, A.A.A.S., Sigma Xi (sec. U. Ala. chpt. 1950-51), Phi Beta Kappa, Tau Beta Pi, Sigma Pi Sigma, Pi Mu Epsilon. Research, publs. on solid state physics, nuclear properties, cryogenics, radiation effects, thermoelec. and thermionic properties of metals, nuclear weapons effects, nuclear explosive design; inventor variable resistor and element. Home: 3943 Chapman Pl., Riverside 92506. Office: 1200 Columbia Av., Riverside, Cal. 92507.*

SARTON, George (Alfred Leon), science historian; b. Ghent, Belgium, August 31, 1884; s. Alfred and Léonie (Van Halmé) S.; B.Sc., U. Ghent, 1906, Sc.D., 1911; hon. degrees Brown U., Harvard, Goethe U., Frankfurt-am-Main, Kenyon Coll., U. Chgo., Technion Hafia; m. Eleanor Mabel Elwes, June 22, 1911; 1 dau., Eleanor May. Came to U. S., 1915, naturalized, 1924. Lectr. history of sci. George Washington U., 1915-16, Harvard, 1916-18, 20—, Radcliffe, 1927—; prof. history sci. Harvard, 1940-51; Lowell lectr., Boston, 1916; asso. Carnegie Instn., Washington, 1918-48; Colver lecturer Brown U., 1930; in Near East and N. Africa for study of Arabic and Islam, 1931-32; Hitchcock prof. U. Cal., 1932; Elihu Root lectr., Washington, D.C., 1935; Fielding H. Garrison lectr., 1941; lectr. in London, Paris, Bruxelles, Liége, Geneva univs., 1948; Kaiser lectr. Library of Congress, 1950; pres. 3d Internat. Congress History of Science, Portugal, 1934. Recipient Prix Binoux, Acad. Scis., Paris, 1915, 35. Fellow A.A.A.S. (v.p.), Am. Acad. Arts and Sciences, Am. Philos. Soc., Am. Oriental Soc., Mediaeval Acad. (Haskins medal 1949), Am. Assn. History of Medicine (hon.), Royal Asiatic Soc. (London), Royal Soc. of Edinburgh (hon.), Swedish Soc. for History of Astronomy (hon.); mem. Am. Hist. Assn., Internat. Acad. History of Sci., Paris, Internat. Union History of Sci. (pres. 1950—). History of Science Soc. (hon. pres. 1938—), Swedish Soc. Hist. Astronomy, India and Pakistan Soc., Newcomen Soc., Deutsche Akademie der Naturforscher, Gesellschaft für Gesschichte d. Med., Soc. Statistique Paris, Soc. Histoire Médecine (Paris), Société asiatique (Paris), Academia de la historia (Madrid), Swedish Linnaean Soc., Gesellschaft für Geschichte der Pharmazie, Soc. italiana di storia d. scienze; Institut de philologie et d'histoire orientales (Bruxelles), Am. Philol. Assn., Belgian Com. History of Science (Bruxelles), S.E. Asia Inst. (N.Y.), Hakluyt Soc. (London), Turkish Oriental Soc. (Ankara); hon. mem. history of sci. socs. in Belgium, Eng., Germany, Holland, Italy. Author: Introduction to the History of Science (Vol. 1, From Homer to Omar Khayyam, 1927; Vol. 2, From Rabbi Ben Ezra to Roger Bacon, 1931; Science and Learning in the Fourteenth Century, Part I, 1947, Part II, 1948); The History of Science and the New Humanism, 1931; The Study of the History of Science, 1936; The Study of the History of Mathematics, 1936; The Life of Science, 1948; Science and Tradition (with bibliography of history of sci.), 1951; Horus. A Guide to the History of Science, 1952; Ancient Science through the Golden Age of Greece, 1952; Ancient Science and Modern Civilization, Galen of Pergamon, 1954; Appreciation of ancient and medieval sci. during Renaissance, 1955; Six wings—men of science in the Renaissance, 1957; A history of science Hellenistic science and culture in the last three centuries B.C., 1959; papers on history and philosophy of sci., New Humanism, Asiatic Art, Arabic culture. Founder Isis, 1912, also editor; founder Osiris, 1936, also editor. Died Mar. 22, 1956.

SARTOR, J(ames) Doyne, Am. meteorologist; b. McAlester, Okla., Jan. 31, 1919; s. Robert Lee and Mabel (Wood) S.; B.A., U. Cal., Los Angeles, 1946; M.S., Cal. Inst. Tech., 1952; S.M., Mass. Inst. Tech., 1953; m. Mary Elizabeth Carruth, Jan. 25, 1942; children—Leslie Ann, James Robert. Phys. scientist Rand Corp., Santa Monica, Cal., 1956-62; program scientist Nat. Center For Atmospheric Research, Boulder, Colo., 1962—. Mem. Am. Meteorol. Soc., Am. Geophys. Union, Sigma Xi. Research, numerous publs. on growth of precipitation through hydrodynamic and electrostatic interaction, importance of particle interaction in electrification of clouds and particles in them. Home: 727 10th St. Office: P.O. Box 1470, Boulder, Colo. 80302.*

SARWER-FONER, Gerald Jacob, physician; b. Volkovsk, Poland, Dec. 6, 1924; s. Morris Mischa and Ronia (Kaplan) Sarwer-F.; B.A., U. Montreal, 1945, M.D., 1950; D. Psychiat., McGill U., 1955; qualified psychoanalyst Canadian Inst. Psychoanalysis, 1962; m. Ethel Sheinfeld, May 28, 1950; children—Michael, Gladys, Janice, Henry, Brian. Cons. psychiatry, dir. psychiat. research Queen Mary Vets. Hosp., Montreal, Que., Can., 1955-60; asso. psychiatrist, dir. research, clin. investigation unit Jewish Gen. Hosp., Montreal, 1955-66, asso. psychiatrist, 1965—; cons. psychiatry Notre Dame Hosp., Montreal, 1964—; dir. dept. psychiatry Queen Elizabeth Hosp., 1966—; faculty psychiatry McGill U., 1955—, asso. prof., 1966—; vis. prof. psychiatry, psychotherapy Laval U., Quebec City, Que., 1964—. Mem. adv. panel psychiatry Def. Research Bd. Can., 1958-62. Fellow Am. Psychiat. Assn. (pres. Que. dist. br. 1960-61,

65-66), A.A.A.S., Am. Coll. Neuropsychopharmacology, Canadian Psychiat. Assn. (dir. 1958-62, chmn. sect. psychotherapy 1960-64), Canadian Psychoanalytic Soc.; mem. Internat. Psychoanalytic Assn., Canadian Med. Assn., Coll. Internat. Neuropsychopharmacoligicum, Royal Coll. Physicians and Surgeons Can. (mem. com. on psychiat. edn., 1958-64), Que. Psychiat. Assn. (treas. 1959-66, pres. 1966-68). Author: The Dynamics of Psychiatric Drug Therapy, 1962; The Depressive Group of Illnesses, 1966; also numerous articles. Research, publs. in psychoanalysis and psychotherapy particularly on depressive illnesses and treatment of psychosis as well as on marriage and marital discord; pioneered studies on psychodynamic and psychoanalytic aspects of psychotropic drugs, as well as many papers on social psychiatry and milieu therapy, including studies of adaptive difficulties of Hungarian immigrants to Can. Address: 613 Cote St. Antoine Rd., Montreal 6, Que., Can.*

SARYMSAKOV, Tashmukhamed Alievich, Russian mathematician; b. Sept. 4, 1915; grad. Central Asian U., Tashkent, 1936; Dr.physico-math. sci., 1942. Prof., Central Asian U., 1943-44, rector, 1952-59; pres. Uzbek Acad. Sci., 1946-52; chmn., State Com. for Higher and Secondary Spl. Edn. Recipient Order of Lenin, twice; Order Red Banner of Labor; Stalin Prize, 1947. Mem. Uzbek Acad. Scis. Author: Elementary Arithmetical Properties of the Chain Law, 1940; The Use of the Probability Methods to Determine Certain Function, 1941; Sequences of Stochastic Matrices, 1945; The Law of Multiple Logarithm as Applied to Markov's Project, 1945; Sequences of Polynomials with Regular Distribution of Zeros, 1946; The Roots of Hermite and Laguerre Polynomials, 1947; Statistical Methods and Problems in Geophysics, 1949; The Basic Theories of Markov's Processes, 1954; Contributions to Markov's Ergodic Theorem for Heterogeneous Chains, 1957. Research in theory of probability and theory of optoganal multiples; introduced new methods for compiling short- and long-term weather forecasts. Address: Soviet Ministry, ulitsa Tukaeva 3, Tashkent, Uzbekistan SSR, USSR.

SASAKI, Teiichi, Japanese biochemist; b. Otaru, Japan, Dec. 16, 1928; s. Ryosuke and Tsugeko (Kobayashi) S.; B.Sc., Hokkaido U., 1951, Dr. Med., 1957, Ph.D., 1962; m. Michiko Kitano, Mar. 18, 1955; children—Kiminori, Noriko. Research asso. Sapporo Med. Coll., 1951-58, asst. prof., 1958-63; research asso. dept. chemistry, chem. engring. U. Ill., 1958-60; asso. prof. Central Clin. Labs. Sapporo Med. Coll., Hosp., 1960-63, head biochem. sect., urinalysis sect., 1963—; lectr. Sch. Nurses Hokkaido. Mem. Assn. Japanese Bacteriologists, Japanese Biochem. Soc., Hokkaido Med. Soc., Japanese Soc. Clin. Pathology, Am. Soc. Microbiology. Abstractor, Chem. Abstracts. Contbr. numerous articles to sci. jours. Chem. analysis of diphtherial toxin and toxoid; chem. analysis of staphylococcus aureus; chem. analysis of O-antigenic polysaccharides of Salmonellae and Pasteurellae; studies on analytical procedures in clin. chemistry; enzyme activities in serum and its distbn. in tissues. Home: 203, Yamanoto 6-6, Kotoni-cho, Sapporo. Office: S. 1, W. 17, Sapporo, Hokkaido, Japan.*

SASLAW, Samuel, Am. physician; b. Cleve., Apr. 26, 1915; s. Louis and Fannie (Burstein) S.; A.B., Transylvania Coll., 1937, D.Sc., 1966; M.S., U. Ky., 1939; Ph.D., Ohio State U., 1942, M.D., 1946; m. Eloise White, July 26, 1947; children—Anne Louise, Karen Lee, Louis Newton. Grad. asst. bacteriology U. Ky., 1938-39; grad. asst. Ohio State U., Columbus, 1940-41, research fellow, 1941-42, Muellhaupt research scholar, 1942-43, asst. prof. medicine, bacteriology, 1950-52, asso. prof., 1952-57, prof., 1958—, dir. div. infectious diseases U. Hosp., 1952—; with Walter Reed Inst. Research, 1947-50; asst. prof. medicine Georgetown U., 1949-50; cons. Dayton VA Hosp., 1952—, Walter Reed Hosp., 1950-56, Wright Patterson AFB Hosp., 1960—. Project supr. Ohio State Research Found., 1955—; chmn. program project com. Nat. Inst. Health Allergy and Infectious Diseases, 1961-65. Recipient Borden Undergrad. Research award, 1946; Distinguished Teaching award Ohio State U. Alumni, 1963; Patriotic Civilian Service award Army, 1962; Coll. of Medicine Ohio State U. Alumni Achievement award, 1966. Diplomate Am. Bd. Internal Medicine. Fellow A.C.P.; mem. Am. Assn. Immunologists, Soc. Exptl. Biology and Medicine, Central Soc. Clin. Research, Infectious Disease Soc. Am., Am. Fedn. Clin. Research, A.M.A., N.Y. Acad. Scis., Am. Bd. Microbiology, Sigma Xi, Alpha Omega Alpha. Research, numerous publs. on problems related to infectious diseases, bacterial, viral, fungal, rickettsial; studies in immunology, antibiotic therapy, nutrition and diagnosis. Home: 425 Walhall Dr., Columbus 43202. Office: 410 W. 10th Av., Columbus, O. 43210.*

SASS, Ronald L., Am. phys. chemist; b. Davenport, Ia., May 26, 1932; s. Edwin L. and Flora (Puck) S.; A.B., Augustana Coll., 1954; Ph.D., U. So. Cal., 1957; m. Joyce Ruth Moorhead, Sept. 9, 1951; children—Denise Anne, Andria Lynn. Research fellow Brookhaven Nat. Lab., 1957-58; faculty Rice U., Houston, 1958—, prof. phys. chemistry, 1966—, master Hanszen Coll., 1966—. Guggenheim fellow, Cambridge, Eng., 1965. Recipient Rice U. Outstanding Tchr. award, 1964; Nicholas-Salgo Outstanding Tchr. award Salgo-Noren Found., 1966-67. Mem. Am. Chem. Soc., Am. Crystallographic Assn., Phi Beta Kappa, Sigma Xi. Research, publs. on crystal and molecular structure using crystal X-ray diffraction techniques. Home: Hanszen House, Rice U., Houston 77001.*

SASSER, Joseph Neal, Am. nematologist; b. Goldsboro, N.C., May 19, 1921; s. John A. and Minnie Gertrude (Neal) S.; B.S., N.C. State U., 1943, M.S., 1950; Ph.D., U. Md., 1953; m. Laura Elizabeth Long, Nov. 3, 1945; children—Anita Gail, Joseph Neal, Betty Louise, Laura Ann. Asst. nematologist U.S. Dept. Agr., Beltsville, Md., 1951-53; faculty N.C. State U., Raleigh, 1953—, prof. plant pathology, 1964—. Tech. cons. Rockefeller Found., Chile, 1963-64. Mem. Soc. Nematologists (past v.p., pres.), Am. Phytopath. Soc. (past editor), Helminthological Soc. Washington, Am. Inst. Biol. Scis., Sigma Xi, Kappa Phi Kappa. Editor: (with W.R. Jenkins) Nematology—Fundamentals and Recent Advances with Emphasis on Plant Parasitic and Soil Forms, 1959. Research, numerous publs. on etiology and control of plant-pathogenic nematodes, interrelationships between nematodes and other soil-inhabiting organisms in causing plant disease, chem. control. Home: 628 Grove Av., Raleigh, N.C. 27606.*

SASTRY, Bhamidipaty Venkata Rama, pharmacologist; b. Rayavaram, East Godavary Dist., Andhra, India, Oct. 21, 1927; s. B.V. and B.S. (Lakshmi) Chandrasekharam; B.Sc. with honors, Andhra U., 1949, M.Sc., 1950, D.Sc., 1955; M.S., Emory U., 1959; Ph.D., Vanderbilt U., 1962. Came to U.S., 1956, naturalized, 1965. Demonstrator dept. pharmacy Andhra U., 1951-52, lectr., 1952-57; research asst. Emory U., Atlanta, 1957-58, fellow, 1958-59; faculty Vanderbilt U., Nashville, 1959—, asso. prof. pharmacology, 1965—. Fellow Royal Inst. Chemistry (London, Eng.); mem. Am. Chem. Soc., A.A.A.S., N.Y. Acad. Scis., Am. Soc. for Pharmacological and Exptl. Therapeutics, Soc. for Exptl. Biology and Medicine, Soc. Toxicology, Am. Assn. U. Profs., Sigma Xi. Research, publs. on isolation, characterization and testing plant insecticides, physiology, pharmacology and toxicology radionuclides, synthesis and screening psychotherapeutic agts., pharmacology and structure-activity drugs active on autonomic nervous system. Office: Dept. Pharmacology, Vanderbilt U., Nashville 37203.*

SATCHELL, Derek Peter Norman, English chemist; b. Dulwich, London, Eng., Sept. 24, 1928; s. Frank Ernest and Ethel (Geiss) S.; B.Sc., U. London, 1952, Ph.D., 1955, D.Sc., 1963; m. Rosemary Sally Smith, Aug. 23, 1960. Faculty, U. London, King's Coll., 1955—, reader in phys.-organic chemistry, 1966—; research asso. Brookhaven Nat. Lab., L.I., N.Y., 1957-58. Fellow Chem. Soc. London. Research, numerous publs. on discoveries in fields of acidity, acylation, and hydrogen isotope reactions. Home: Spinney Cottage, The Glade, Kingswood, Surrey, Eng. Office: Chemistry Dept., King's Coll., Strand, London, Eng.*

SATCHLER, George Raymond, physicist; b. London, Eng., June 14, 1926; s. George Cecil and Georgina (Strange) S.; B.A., Oxford U., Eng., 1951, M.A., 1951, D.Phil., 1955; m. Margaret Patricia Enid Gibson, Mar. 26, 1948; children—Patricia Ann, Jacqueline Helen. Research fellow Oxford U., 1954-59; research asso. U. Mich., Ann Arbor, 1956-57; physicist Oak Ridge Nat. Lab., 1959—. Fellow Am. Phys. Soc. Author: Angular Momentum, 1962. Research on devel. of theory of direct nuclear reactions and its application to exptl. measurements for purpose of extracting information about shell and collective effects in nuclear structure. Home: 973 W. Outer Drive, Oak Ridge 37830. Office: Oak Ridge Nat. Lab., Oak Ridge 37830.*

SATHER, J(ohn) Henry, Am. biologist; b. Presho, S.D., July 12, 1921; s. Anton and Anna (Imster) S.; B.Sc., U. Neb., 1943, Ph.D., 1952; M.A., U. Mo., 1948; m. Shirley M. Johnson, Aug. 21, 1948; children—Kristi, Signe, Ingrid. Research biologist Neb. Game, Forestation and Parks Commn., Lincoln, 1948-55; prof. biol. scis. Western Ill. U., Macomb, 1955—, dean Grad. Sch., 1964—. Edward K. Love fellow, 1946-48. Mem. Wildlife Soc. (pres. Ill. chpt. 1965, v.p. N. Central sect. 1967), A.A.A.S., Ecol. Soc. Am., Soc. Mammalogists, Sigma Xi. Research, publs. on life history and ecology of Great Plains muskrat in Neb., life history and ecology studies of ring-necked pheasant in Neb., directed radio-telemetry studies of movements of white-tailed deer and muskrats, conducted survey of effects of water level fluctuations of Great Lakes upon wildlife populations. Home: Rural Route 1, Macomb, Ill. 61455.*

SATHYANESAN, A. G., Indian endocrinologist; b. Neyyoor, India, Oct. 3, 1926; s. T. J. and Agnes Sathyanesan; B.Sc., Annamalai U., 1948, B.Sc. with honors, 1950, M.A., 1952, M.Sc., 1955; Ph.D., Banaras Hind U., 1957; m. Oct. 17, 1957; children—Kumar, Arun Prakash. Lectr. zoology Andhra Christian Coll., Guntur, India, 1950-51; fellow Indian Council Agrl. Research, Annamalai U., 1951-54; research fellow Govt. of India Manpower Tng. Scheme, 1954-57; lectr. zoology Banaras Hindu U., Varanasi, India, 1957—. Postdoctoral research fellow Population Council, Rockefeller Inst., N.Y.C., 1962-64. Mem. Indian Sci. Congress Assn., Indian Zool. Soc., Indian Zoolomical Soc. Research, publs. on endocrinology of fish, neuroendocrine structure and regeneration of neuro secretory axons in fish, effects of irradiation on neuroendocrine system in fish. Home: G/12 Beyond Ladies Colony, Banaras, Hindu U., Varanasi-5, India.*

SATO, Daihachiro, mathematician; b. Fujinomiya, Shizuoka-ken, Japan, June 1, 1932; s. Risaburo and Teiko (Tsuyuki) S.; B.S. in Physics, Tokyo (Japan) U. Edn., 1955; A.A., Los Angeles City Coll., 1956; M.A., U. Cal. at Los Angeles, 1957, Ph.D. in Math., 1963; m. Yoriko Okamoto, Sept. 28, 1956; 1 dau., Mika Elaine. Asso., U. Cal. at Los Angeles, 1960; asst. prof. San Fernando Valley State Coll., 1961; instr. Tokai U., Tokyo, 1962; instr. U. Sask. (Can.), Regina, 1962, lectr., asst. prof., 1963-66, asso. prof., 1966—; vis. asst. prof. U. Cal. at Los Angeles, 1964; research fellow Summer Research Inst., Queen's U., Kingston, Ont., Can., 1965; U. Alta. (Can.), Edmonton, 1966. Mem. Math. Soc. Japan, Phys. Soc. Japan, Canadian Math. Congress, Soc. for Indsl. and Applied Math., Am. Math. Soc., Math. Assn. Am., Sigma Xi, Pi Mu Epsilon. Author: Integer Valued Entire Functions, 1961; also articles. Research on entire functions infinite order and generalized interpolation, integer valued entire functions, rate of growth of entire functions, generalized interpolation by analytic functions and prime-representing functions, utterly algebraic valued transcendental entire functions, rate of growth of Hurwitz functions of complex or p-adic variables. Home: 12 Falcon Bay, Regina, Sask., Can.*

SATO, Hidemi, biologist; b. Fukuoka, Japan, Sept. 17, 1926; s. Daiyu and Kozue (Sato) S.; D.Sc., Kyoto U., 1963; m. Yukiko Umeda, Apr. 29, 1952; children—Hideo, Masahiko, George Haruki. Came to U.S., 1959, naturalized, 1963. Instr. biology, Japan, 1953-59; faculty Dartmouth Med. Sch., Hanover, N.H., 1959-66, asst. prof. cytology, 1962-66; sr. staff asso. Marine Biol. Lab., Woods Hole, Mass., 1962-66; asso. prof. biology U. Pa., Phila., 1966—. Mem. Soc. Protozoologists, Am. Soc. Zoologists, Soc. Gen. Physiologists, Am. Microscopists Soc. Fellow A.A.A.S. Research, publs. in application of polarized light to living biol. system, cytology of amitosis, dynamic structure of mitotic spindle; research in devel. of high sensitivity, high resolution polarizing microscope. Home: 610 Penfield Av., Havertown, Pa. Office: Dept. Biology, U. Pa., Phila.

SATO, Yasuo, physicist; b. Tsingtan, China, Sept. 1, 1918; s. Tetsuichi and Chiyo (Kitamura) S.; grad. U. Tokyo (Japan), 1941, Ph.D., 1956; m. Miyoko Masuko, Sept. 24, 1945; 1 son, Jun. Faculty, U. Tokyo, 1941—, prof. geophysics, 1961—; vis. prof. Columbia, N.Y.C., 1960. Mem. Am. Geophys. Union, Seismol. socs. Am., Japan. Author: (in Japanese) For Fortran Users, 1964; also numerous articles. Theoritical studies of propagation of seismic waves especially study of seismic surface waves and free oscillation of earth. Home: 19-196 Goko-Mutsumi, Matsudo, Japan. Office: Earthquake Research Inst., U. Tokyo, Tokyo, Japan.*

SATOSKAR, Rajaninath Shantaram, Indian physician; b. Chiplun, Bombay, India, May 9, 1928; s. Shantaram Raghunath and Kamala (Weling) S.; M.B., B.S., Seth G.S. Med. Coll., 1951, B.Sc., 1954; Ph.D., U. Sheffield, 1957; m. Shashiprabha Gandbhir, Mar. 11, 1958; children—Rejeev, Raghunandan. Research asst. in pharmacology Seth G.S. Med. Coll., 1953-55, lectr., 1958, asst. prof., 1959-63, asso. prof., head radioisotope lab., 1963-65, prof. pharmacology, head radioisotope lab. Coll. and K.E.M. Hosp., Bombay, 1965—. Mem. Assn. Physiologists and Pharmacologists India. Asso. editor Jour. Postgrad. Medicine, 1960—. Research, publs. on serum proteins, cholesterol, vitamin B12, folic acid and blood vols. in Indians with particular reference to veganism and nutrition; nutritional status of newborn infants and Hypoproteinaemia in children, sodium metabolism in Indians from warm and humid climates, metabolism of sodium in bones, tetanus, thyroid disease. Home: 8 K.E.M. Hosp. Quarters, Dr.Rau Rd., Bombay 12, Maharashtra, India.*

SATPAEV, Kanysh Imantaevich, Russian geologist; b. Apr. 11, 1899; grad. Tomsk Technol. Inst., 1926; Dr.Geol. and Mineral. Scis. Tchr., 1917-21; dir. prospecting work, Dzhezkazgan, Uspensky mines Atbasar Metals Trust, Kazakhstan, 1926-29; dir. Geol. prospecting dept., chief geologist, 1929-41; chmn. presidium Kazakhstan br. USSR Acad. Scis., (now Kazakhstan Acad. Sci.), 1942-46, pres., 1946-52, 55—, chmn. presidium, 1957—; del. to Supreme Soviet from Kazakh SSR. Recipient Stalin Prize, 1942; Lenin Prize; Order of Lenin, 3 times. Mem. USSR Acad. Sci. (mem. presidium 1961), Tadzhik SSR Acad. Scis. (hon.). Author: Main Features of Geology and Metallogenesis of the Dzhezkazgansk Copper Ore Region, Greater Dzhezkazgansk, 1935; also articles. Research on geology of Central Kazakhstan; supr. expdn. which discovered copper deposits and other minerals in Ulutau-Dzhezkazgan Region, also compiled maps of area. Office: Pres., Acad. Scis., Kazakh SSR, Shevchenko Ulitsa, 28, Alm-Alta, Kazakh SSR.

SATTEN, Robert Arnold, Am. physicist; b. Chgo., Aug. 4, 1922; s. Frank and Mabel Satten; B.S., U. Chgo., 1944; M.A., U. Cal., Los Angeles, 1947, Ph.D., 1951; m. Erica Semone, Oct. 19, 1946; children—Corey, Glen. Faculty, U. Cal., Los Angeles, 1951—, prof., 1963—; asst. prof. Mass. Inst.

Tech., 1952-53. Cons. Argonne (Ill.) Nat. Lab., 1959-—, Hughes Research Labs., Malibu, Cal., 1959-—. Fulbright Research scholar, France, 1961-62. Fellow Am. Phys. Soc.; mem. Western Spectroscopy Assn. (chmn. 1962), Phi Beta Kappa. Interpretation of rare earth and actinide ion spectra in crystalline fields, vibronic spectra in crystals; optical double resonance method for measuring spin-lattice relaxation. Home: 1948 Parnell Av., Los Angeles 90025.*

SATTERFIELD, G(eorge) Howard, Am. nutritionist; b. Mt. Airy, N.C., Dec. 2, 1893; s. George Andrew and Mary (Callaway) S.; A.B., Duke, 1920, A.M., 1920; student Northwestern U., 1922-24; B.S., N.C. State U., 1934; m. Alleece Voss Sapp, Dec. 28, 1926; children— G(eorge) Howard II, Benton Sapp. Tchng. asst. Duke, 1919-20, indsl. fellow, 1920-21, instr., 1921-22; teaching fellow Northwestern U., 1922-23; research fellow Pub. Health Inst., 1922-24; with N.C. State U., 1924-—, prof., 1935-61, prof. emeritus, 1961-—; lectr. nutrition civic clubs, med. organizations, P.T.A.s, chem. socs. Fellow A.A.A.S.; mem. Am. Inst. Nutrition, Am. Soc. Biol. Chemistry, Soc. Exptl. Biology and Medicine, Am. Inst. Chemists, Inst. Food Technologists, A.A.U.P., Scientific Research Soc. Am., N.C. Acad. Sci., Sigma Xi, Phi Kappa Phi, Alpha Chi Sigma, Pi Gamma Mu, Sigma Chi. Co-editor (with W. J. Dann) and co-author Vol. XIII Biological Symposia, entitled Estimation of the Vitamins, 1947; author (with others) Gen. Chemistry Lab. Manual; Editions 3-14, 1931-63. Research, publs. on determination, source, content of Vitamin C and certain minerals in various substances. Home: 407 W. Park Dr., Raleigh, N.C. 27605.*

SATTERTHWAITE, Thomas Edward, Am. physician; b. N.Y., Mar. 26, 1843; s. Thomas Wilkinson and Ann Fisher (Sheafe) S.; A.B., Yale, 1864; postgrad. Harvard Med. Sch., 1864-65; M.D., Coll. Phys. and Surg. (Columbia), 1867; postgrad., Vienna, 1869-70; LL.D., U. Md., 1908; Sc.D., St. John's Coll., Md., 1912; m. Isabella Banks, 1884. Intern N.Y. Hosp., 1867-69; served in Franco-German War as asst. surgeon and surgeon Prussian Army microscopist, later pathologist St. Luke's Hosp., N.Y.C., 1872-82; pathologist Presbyn. Hosp., 1873-88; a founder Postgrad. Med. Sch., sec. 2 yrs., prof. pathol. anatomy, 1 yr., prof. gen. medicine, 7 yrs., v.p., 1890, prof. medicine, 1904-08; lectr. comparative pathology Columbia Vet. Coll., 1881-82; organized med. and surg. staff Chambers St. House of Relief (now Hudson St. Hosp.), 1875; a founder, pres., Babies' Hosp. 1894-95; cons. physician Post-Graduate, Orthopedic and Manhattan State hosps., N.Y.; ret., 1923. Pres. N.Y. Pathol. Soc., 1880-81; a founder pres., Am. Therapeutic Soc., pres., 1902-03; a founder A.C.P.; fellow N.Y. Acad. Medicine. Author: Manual of Histology, 1881; Practical Bacteriology, 1887; Diseases of the Heart and Aorta, 1905; Cardiovascular Diseases, 1912; Diseases of the Heart and Blood Vessels, 1918. Died Sept. 19, 1934.

SAUBERLICH, Howerde Edwin, Am. biochemist; b. nr. Appleton, Wis., Jan. 23, 1919; s. George H. and Ann Gertrude (Ray) S.; B.A. summa cum laude, Lawrence U., 1944; M.S., U. Wis., 1946, Ph.D., 1948; postgrad. U. Tenn.; m. Irene Kathrine Cartwright, Sept. 15, 1945; children—Melissa Kay, Howerde Edwin II. Prof. animal husbandry and nutrition Auburn (Ala.) U., 1948-58; prof. animal nutrition and microbiology U. Ky., U. Indonesia, 1957-59; asso. prof. animal husbandry and food tech. Ia. State U., Ames, 1959; chief chemistry div. U.S. Army Med. Research and Nutrition Lab., Denver, 1960-—. Cons. on biochemistry, 1944-—. Recipient Meade-Johnson award, 1952; McLester award, 1965; Meritorious Civilian Service award U.S. Army, 1964. Mem. Am. Soc. Biol. Chemists, Am. Inst. Nutrition, Am. Soc. Microbiology, Soc. Exptl. Biology and Medicine, Am. Chem. Soc., Am. Assn. Cancer Research, Am. Soc. Clin. Nutrition, Am. Soc. Animal Scis., N.Y. Acad. Scis., Phi Beta Kappa, Sigma Xi, Gamma Sigma Delta. Research, numerous publs. on protein and amino acid metabolism, imbalances, toxicities, methodology and value as related to animal, human and microorganisms, lipid and carbohydrate metabolism, vitamin B-complex, nutritional edema, cancer research, antibiotics and growth factors, radioisotopes and food irradiation preservation, microbial nutrition and metabolism, internat. nutrition, human requirements for vitamins and other nutrients. Home: 306 Moline St., Aurora, Colo. 80010. Office: U.S. Army Medical Research and Nutrition Lab., Fitzsimmons Gen. Hospital, Denver 80240.*

SAUCEROTTE, Nicolas, French surgeon; b. Lunéville, France, June 10, 1741; master surgeon, 1761; became mil. surgeon; insp. Army Med. Services; recipient prize from Convention for proposal on phys. edn. program for pub. schs.; mem. Acad. Surgery (1st prize 1769), French Acad. Scis. Author: Théorie des lésions de la tete mélanges de chirurgie, 2 vols., 1801. Noted for operations for gallstones; studied lesions of head. Died Luneville, Jan. 5, 1814.

SAUCIER, Walter Joseph, Am. meteorologist; b. Moncla, La., Oct. 5, 1921; s. Louis Edmond and Sidonie (Moncla) S.; B.S., U. Southwestern La., 1942; S.M., U. Chgo., 1947; Ph.D., 1951; m. Helen Nobles, May 8, 1943; children—Walter, Susanne, Diane, Janine, Gerard, Laurence, Loraine. Faculty, U. Chgo., 1946-52, Tex. A. and M. Coll., 1952-60; prof. meteorology, chmn. dept. U. Okla., Norman, 1960-—, dir. Atmospheric Research Lab., 1961-—. Cons. World Book Ency., USAF, U.S. Army, Am. Council on Edn. Mem. Am. Meteorol. Soc., Am. Geophys. Union, A.A.A.S., Am. Assn. U. Profs., Okla. Acad. Sci., Sigma Xi. Author: Principles of Meteorological Analysis, 1955. Research, contbns. in atmospheric circulations, convective to synoptic scale. Home: 848 Elm Av., Norman, Okla. 73069.*

SAUDINOS, Jean, French physicist; b. Montauban, France, Nov. 9, 1933; s. Emile and Lucienne (Theau) S.; Ingenieur, U. Toulouse (France), 1956; Docteur es Sciences, U. Paris, 1962; m. Dominique Schaad, July 24, 1965; 1 dau., Laure. Physicist, Service de Physique Nucleaire Moyenne Energie, Centre D'Etudes Nucleaires de Saclay, Giv/Yvette, France, 1957-—; staff Lawrence Radiation Lab., Berkeley, Cal. Research, publs. on nuclear spectroscopy by inelastic scattering of 44 Mev alpha particles, especially double nuclear excitation; spectroscopy of heavy nuclei by inelastic scattering of 24 Mev Protons. Home: 98 bis Rue de Lozere, Orsay-91-France. Office: S.R.N.M.E., C.E.N. Saclay, B.P. No2, Gif/Yvette-91-France.*

SAUER, John A(nthony), Am. physicist; b. Elizabeth, N.J., Oct. 26, 1912; s. Alfred Francis and Lucy M. (Blatz) S.; B.S., Rutgers U., 1934, M.S., 1936; student Harvard, 1939-40; Ph.D., U. Cambridge, Eng., 1942; Guggenheim fellow U. Oxford, 1959-60; m. Marian Short; children—Virginia, Pamela. Instr. math. engring. dept. Union Jr. Coll., 1934-38; instr. gen. engring. dept. Rutgers U., 1936-38; instr. mechanics dept. Carnegie Inst. Tech., 1940-41; lectr. physics, U. Pitts., also sr. fellow Mellon Inst. Indsl. Research, 1941-44; asst. to pres. Elastic Stop Nut Corp., Union, N.J., 1944, dir. research and engring., 1944-45; prof., head dept. engring. mechanics Pa. State Coll., 1946-52; vis. prof. Oxford U., Eng., 1952-53 (on leave); prof., head physics dept. Pa. State U., 1953-63; prof., chmn. dept. mechanics Rutgers-The State U., 1963-—. Guggenheim fellow Clarendon Lab., Oxford U., 1959-60. Fellow A.A.A.S., Am. Phys. Soc.; mem. N.Y. Acad. Sci., Am. Inst. of Physics, Am. Soc. Engring. Edn., Phi Kappa, Sigma Xi, Sigma Pi Sigma, Tau Beta Pi. Author: A Study of Magnetic Phenomena at Low Temperatures, 1942; co-author (with J. Marin) Strength of Materials, 1954. Research and publs. on solid state physics, mech. and magnetic properties of materials, polymer sci., relation of atomic and molecular structure to behavior of plants. Home: 31 Patton Dr., East Brunswick, N.J. Office: Rutgers-The State U., New Brunswick, N.J. 08903.*

SAUER, Robert, German mathematician; b. Pommersfelden, Sept. 16, 1898; s. Hans and Anna (Falch) S.; Dr ès sc, Univ. and Polytech., Munich; dr. engring. hon. causa; m. Hanni Winter, Oct. 6, 1930. Prof., Aix-la-Chapelle Poly.; prof. math. Munich Poly., 1932, 48-ular Formation of the Brain Stem, 1958; Clinical Syn-—, rector, 1954-56, 56-58, 61-62. Mem. Bavarian Senate, 1962-—. Decorated Order Bavarian Merit. Mem. Bavarian Acad. Sci., Cath. Acad. Bavaria, Bologna, Leopoldina acads. Publns. on math. Editor and/or co-editor math. jours. Research on differential geometry, theory of characteristics, applied math., gas dynamics, partial differential equations of hyperbolic type. Address: Tech. Hochschule, Arcisstrasse 21, Munich, West Germany.

SAUERBRUCH, Ernst Ferdinand, German surgeon; b. Barmen, Germany, July 3, 1875; prof., Marburg, Germany, also Zurich, Switzerland, Munich, Berlin. Author: Die Chirurgie der Brustorgane, 2 vols., 1920-25; Über die physiologischen und physikalischen Grundlagen bei intrathorakalanen Eingriffen (in reports German Surg. Soc.), 1904. Founder thorax surgery through intro. of pneumatic cabinet, 1904; introduced artificial paralization of diaphragm in treatment of Tb of lungs; introduced phrenicotomy in treatment of pulmonary Tb, circa 1913; developed (with Max Bernard Gerson and A. Herrmannsdorfer) salt restricted diet in treatment of Tb, circa 1926; advocated thymectomy for relief of myasthenia gravis; 1st to operate for heart aneurysms; devised artificial limbs which permit movement when attached to muscles of amputation stump. Died Berlin, July 2, 1951.

SAUGRAIN DE VIGNI, Antoine François, physician; b. Paris, France, Feb. 17, 1763; s. Antoine and Marie (Brunet) Saugrain; m. Genevieve Rosalie Michau, Mar. 20, 1793, 6 children. Went to Mexico to examine mines and mineral prodn., 1785-86; sailed for U. S., 1787; taken captive by Indians in attack, escaped, 1788; apptd. post surgeon by Spanish lt. gov. Delassus, St. Louis, 1800; army surgeon (apptd. by Jefferson), 1805-11; only practicing physician in St. Louis when Upper Louisiana was transferred to U. S.; made and sold ink, thermometers, phosphoric lights for hunters, barometers; conducted experiments in electricity and had electric battery; introduced 1st smallpox vaccine virus brought to St. Louis and publicly offered to vaccinate free of charge all indigent persons, paupers and Indians, 1809. Died circa May 19, 1820.

SAUKOV, Aleksandr Aleksandrovich, Russian geochemist; b. Aug. 15, 1902; grad. Leningrad Poly. Inst., 1929. Staff, USSR Acad. Scis., head dept. geochemistry Inst. Geol. Scis.; prof. Moscow U., 1952-—. Recipient Stalin Prize, 1947, 52. Corr. mem. USSR Acad. Scis. Author: Geochemistry of Mercury, 1946; Geochemistry, 2d edit., 1951; also articles. Research on geochemistry of rare elements; devel. method for determining small quantities of mercury; originated method of prospecting for mercury deposits. Home: B. Kommunisticheskaya, 24, Moscow, USSR.

SAUMAISE, Claude (or Salmasius), scholar; b. Semur, Apr. 15, 1588; studied philosophy, Paris; pupil of Casaubon; studied law under Denis Godefroy, Heidelberg; m., 1623. Author: De annis climactericis et antiqua antrologia diatribae, 1648; various other works. Attacked astrology, but maintained belief in marvelous secrets of chemistry and medicine.

SAUNDERS, Dorothy Chapman, Am. biologist; b. Hackensack, N.J., Apr. 2, 1912; d. Adelbert and Frances (Thompson) Chapman; A.B., cum laude, Syracuse U., 1934; M.S., U. Mich., 1935, Ph.D. with high honors, 1937; m. George Bradford Saunders, June 19, 1947. Prof. biology Cedar Crest Coll., Allentown, Pa., 1937-41; research anatomist Office of Fgn. Agrl. Relations, U. S. Dept. Agr., 1944-47; collaborator biology U. S. Fish and Wildlife Service, 1948-52, 1965-67; asst. prof. biology U. Fla., Gainesville, 1953-54; research asso. Mote Marine Lab., Sarasota, Fla., 1955-—. Emma J. Coles fellow U. Mich., 1936-37; Fulbright Postdoctoral Research scholar to Egypt, 1955-56; NSF grantee, 1963-64, 65-66. Mem. Am. Micros. Soc., Am. Soc. Ichthyologists and Herpetologists, Soc. Woman Geographers. Contbr. articles to profl. jours. Gen. surveys of incidence and identity of blood parasites in game birds and marine fishes of parts of Latin Am. Caribbean Sea, Gulf of Mexico and Red Sea; identified and described several new species; studied types of blood cells present in sharks and other marine fishes. Home: P.O. Box 295, Englewood, Fla. 33533. Office: 9501 Midnight Pass Rd., Sarasota, Fla. 33581.*

SAUNDERS, Francis Joseph, Am. endocrinologist; b. Dubuque, Ia., May 9, 1911; s. Charles J.W. and Loutilda (Traut) S.; B.S. in Animal Husbandry, U. Cal. at Davis, 1933, Ph.D. in Comparative Physiology, at Berkeley, 1937; m. Helen H. Embury, July 17, 1937; children—Francis Joseph, Donald Roy, Gertrude H. (Mrs. T.E. Scott), Philip A. Instr., U. Cal. at Davis, 1937-38; staff G.D. Searle & Co., Chgo., 1938-—, chief endocrinologist div. biol. research, 1947-60, head endocrine dept., 1960-—. Mem. Endocrine Soc., Am. Physiol. Soc., Soc. Exptl. Biology and Medicine, N.Y. Acad. Sci., Chgo. Chemists Club. Research, publs. on pituitary gland and relation of adrenal and sex hormones to reprodn., control fertility, effects synthetic steroids on reproductive function and glands of reproductive tract. Home: 8506 Kedvale, Skokie, Ill. 60076. Office: P.O. Box 5110 Chgo. 60680.*

SAUNDERS, John Bertrand de Cusance Morant anatomist; b. Grahamstown, South Africa, July 2, 1903; s. Frederick Anastasius and Lucy (Meiklejohn) S.; student St. Andrews Coll., Grahamstown, South Africa, 1909-17, Rhodes U. Coll., 1919-20; M.B., Ch.B., U. Edinburgh, Scotland, 1930; came to U. S., 1931; m. Alison Jean Maxwell Wood, July 2, 1930; children—Alison B. (Mrs. Alyn Duxbury), Margery H. (Mrs. Donald Hellmann). Demonstrator anatomy U. Edinburgh, 1925-30, in physiology, 1925; lectr. applied anatomy, 1928-30, anatomy Dunfermline Coll. Phys. Edn. and Hygiene, 1925-30; hon. surgeon. Settlers Hosp., Grahamstown, S. Africa, 1930; asst. prof. anatomy, U. Cal. Med. Sch., 1931-35, asso. prof., 1935-38, prof., 1938-—, dean Sch. Medicine, 1956-63, provost, 1958-64, chancellor, 1964-—, lectr. med. history and bibliography, 1933-—, cons. U. Cal. Hosp., 1931-—. Librarian, Med. Central Library, 1943-—. Medallist in Medicine, Royal Coll. of Physicians (Edinburgh); Comdr., Most Venerable Order Hosp. St. John Jerusalem. Fellow Royal Soc. Tropical Medicine and Hygiene, Am. Coll. Dentists, Royal Coll. Surgeons; hon. mem. Hollywood Acad. Medicine, Am. Orthopaedic Assn.; mem. A.M.A., N.Y. Acad. Sci., Am. Acad. History Dentistry (hon.), Reno Surg. Soc. (hon.), Anat. Soc. Gt. Britain, Ireland, Brit. Med. Assn., Edinburgh Med. and Chirurg. Soc., History of Sci. Soc., Am. Assn. Anatomists, Am. Assn. Phys. Anthropologists, A.A.A.S., Cal. Med. Soc., Soc. Exptl. Biology and Medicine, m. Assn. History Medicine, Western Orthopaedic Assn., Am. Acad. Orthopaedic Surgeons, Cal. Acad. Medicine, Sigma Xi, Sigma Kappa Theta (hon.), Alpha Omega Alpha. Asso. editor for Isis, quar. of History of Sci. Soc. Author: Andreas Vesalius, the Venesection Letter of 1539, 1947; Illustrations from the Works of Andreas Vesalius, 1950; Leonardo da Vinci on the Human Body, 1952; Ancient Egyptian and Cnidian Medicine, 1958; The Transitions from Ancient Egyptian to Greek Medicine, 1963. Contbr. sci. publs. Home: 1412 Willard St., San Francisco 94117.*

SAUNDERS, John Cunningham, Eng. surgeon; b. Eng., Oct. 10, 1773; s. John Cunningham and Jane Saunders; apprentice to John Hill, surgeon of Barnstaple, 1790-95; student St. Thomas, Guy's hosps.; m. Jane Louisa Colkett, Apr. 7, 1803. Became demonstrator anatomy St. Thomas Hosp., 1797; a founder Royal London Ophthalmic Hosp. Author: Anatomy of the Human Ear, with a Treatise of Deafness and their Treatment, 1806, 17, 29; A Treatise on some Practical Points relating to Diseases of the Eye, 1811. One of 1st surgeons to advocate incision of membrana tympani for relief of acute middle ear suppuration, 1806. Died Feb. 9, 1810.

SAUNDERS, Joseph Francis, Am. biochemist; b. Mt. Pleasant, Pa., Apr. 2, 1927; B.S., Duquesne U., 1950; M.S., Georgetown U., 1955, Ph.D., 1960; m. Pauline Claire Dugan, Nov. 23, 1950; children—Joseph Francis, William Paul. Asst. to head medicine and dentistry br. Office Naval Research, 1952-57, sci. project officer, 1957-59, asst. head, 1959-60, apptd. head medicine and dentistry br., 1960; now chief environmental biology biosci. programs office Space Sci. and Applications, NASA; organized research program on transplantation of tissues and cryobiology, program for research on indefinite preservation of whole blood at ultra-low temperatures. Instr. hematology and lab. techniques Bus. Tng. Coll. Pitts.; guest scientist Naval Med. Research Inst., 1958-60. Served with USNR, 1945-46, 51-52. Recipient Arthur S. Flemming award in sci., 1962. Fellow Am. Inst. Chemists; mem. Am. Chem. Soc., U. S. Naval Inst., A.A.A.S., Alpha Chi Sigma. Author: Forms of Water in Biologic System; numerous tech. reports. Patentee blood preservation process. First to show choline (a constituent of nerve) is taken up by nerve and turned into cholesterol; publs. on cryobiology, space biology. Home: 8131 Greeley Blvd., Springfield, Va. 22150. Office: 400 Maryland Av., Washington 20546.*

SAUNDERS, Sir Owen Alfred, English mech. engr.; b. London, Eng., Sept. 24, 1904; s. Alfred George and Margaret (Jones) S.; B.Sc., Birkbeck Coll., London U., 1923; M.A., Trinity Coll., Cambridge (Eng.) U., 1926; B.Sc. with honors in Math., U. London, 1928, M.Sc., 1930, D.Sc. in Engring., 1938; m. Marion McKechney, July 27, 1935; children—Jill Margaret (Mrs. Harry Diack Johnstone), Patricia Joan (Mrs. John Francis Lees), Owen Lyn David. Sci. officer fuel research Dept. Sci. and Indsl. Research, 1926-32; faculty Imperial Coll., U. London, 1932-42, prof., head dept. mech. engring., 1946—; dep. dir. turbine engine research Ministry of Supply, 1942-46. Dir. Internat. Research & Devel. Co., 1964—. Mem. Ind. Television Authority, 1964—. Created knight, 1965. Fellow Royal Soc., 1958, Phys. Soc., Royal Aero. Soc., Inst. Fuel; hon. mem. Instn. Mech. Engrs. (past pres.), Am. Soc. M.E., Japan Soc. Mech. Engrs. Author: (with M. Fishenden) The Calculation of Heat Transmission; An Introduction to Heat Transfer; also articles. Founds. of sci. of heat transfer; research in fluid mechanics and internal combustion turbine. Home: 11 Heath Dr., Sutton, Surrey, Eng. Office: Imperial Coll., London S.W. 7, Eng.*

SAUNDERS, Robert Mallough, elec. engr.; b. Winnipeg, Can., Sept. 12, 1915; s. Robert and Mabel Grace (Mallough) S.; B.Elec. Engring., U. Minn., 1938, M.S., 1942; m. Elizabeth Lenander, June 24, 1943. Design engr. Electric Machinery Co., Mpls., part-time, 1938-42; teaching asst. elec. engring. U. Minn., 1938-42, instr., 1942-44; faculty U. Cal. at Berkeley, 1946—, prof. elec. engring., 1957—, chmn. dept., 1959-63; asst. to chancellor for engring. U. Cal. at Irvine, 1964-65, dean School of Engineering, 1965—; visiting associate prof. Mass. Inst. Tech., 1954-55. Cons. Gen. Motors Def. Lab., Santa Barbara, Apollo Support Dept., Gen. Electric Co., Daytona Beach, Fla. Simon fellow elec. engring. Manchester (Eng.) U., 1960. Fellow I.E.E.E.; mem. Am. Soc. Engring. Edn. (sec. elec. div. 1963), Sigma Xi, Tau Beta Pi, Eta Kappa Nu. Co-author: Analysis of Feedback Control Systems, 1956. Research, publs. on elec. machinery theory, applications of digital computers to elec. machine design, theory electromech. energy converters, system simulation and optimization. Home: 2407 Bunya St., Newport Beach, Cal. 92660. Office: U. Cal., Irvine, Cal. 92650.*

SAUNDERS, William, Brit. physician; b. Banff, Scotland, 1743; s. James Saunders; M.D., U. Edinburgh, 1765; practiced medicine, London; physician Guy's Hosp., London, 1770-1802; named physician to prince regent George, 1807; Gulstonian lectr., 1792; Harveian orator, 1796. Licentiate Coll. Physicians. Fellow Royal Soc., 1793, Royal Coll. Physicians (censor 1791, 98, 1805, 13); 1st pres. Royal Med. and Chirurg. Soc., 1805. Author: Compendium medicinae practicum, 1767; A Treatise on the Structure, Economy, and Diseases of the Liver. First physician in Eng. to observe that liver becomes enlarged then contracted in some forms of cirrhosis (then called scirrhosity); wrote extensively on mercury, antimony, mephitic acid, red Peruvian bark, mineral waters. Died May 29, 1817.

SAUNDERS, William, horticulturist; b. St. Andrews, Scotland, Dec. 7, 1822; attended U. Edinburgh (Scotland); m. Martha Mildwaters, 1848. Came to U. S., 1848. In partnership with Thomas Meehan in landscape gardening, Phila., 1854-62; designed parks and cemeteries, including Rose Hill Cemetery, Chgo.; supt. exptl. gardens U. S. Dept. Agr., 1862-1900; designed grounds of Dept. Agr., Washington, nat. cemetery, Gettysburg, grounds for Lincoln monument in Springfield, Ill., and U. S. Dept. Agr. exhibits for Centennial Exhbn., Phila., 1876, New Orleans Expn., 1884, Paris Exhbn., 1889; established conservatories and greenhouses for study of economically important plants; mem. Parking Commn. of Washington, 1871; introduced Washington Navel orange from Brazil, 1871, several hundred varieties of Russian apples, 1870, eucalyptus globulus from Australia, 1866; co-founder Patrons of Husbandry, 1867, pres., 1867-73. Died Sept. 11, 1900.

SAUNDERS, William Hundley, Jr., Am. chemist; b. Pulaski, Va., Jan. 12, 1926; s. William Hundley and Vivian (Watts) S.; B.S., Coll. William and Mary, 1948; Ph.D., Northwestern U., 1952; m. Nina Velta Plesums, June 25, 1960; children—Anne Michele, Claude William. Research asso. Mass. Inst. Tech., 1951-53; faculty U. Rochester (N.Y.), 1953—, prof. chemistry, 1964—, chmn. dept., 1966—. Guggenheim fellow, 1960-61; Sloan Found. fellow, 1961-64. Mem. Am. Chem. Soc., Chem. Soc. (London, Eng.), Phi Beta Kappa, Sigma Xi, Phi Lambda Upsilon. Author: Ionic Aliphatic Reactions, 1965; also articles, chpts. in monographs, revs. Research on mechanisms organic reactions, isotope effects in organic reactions. Home: 15 Parkwood Av., Rochester, N.Y. 14620.*

SAUNDERSON, Nicholas, English mathematician; b. Thurlston, Eng., Jan. 1682; lost his sight due to smallpox at 1 year of age; M.A., Christ's Coll., Cambridge, 1711, D. Laws, 1728; became lectr. Newtonian philosophy Cambridge U., 1707, prof. math., 1711; also lectured on hydrostatics, mechanics, sounds, astronomy, tides, optics. Fellow Royal Soc., 1718. Author: Elements of Algebra, 2 vols., 1740-41; On Fluxions, 1751. Inventor computing board. Died Boxworth, Eng., Apr. 19, 1739.

SAUNOVICS, Jean, astronomer, linguist; b. 1733; of Hungarian origin; mem. Soc. Jesus; with obs., Bude; sent to observe movement of Venus, Holl, Norway. Author: Demonstratio idioma ungarorum et lapodum idemesse (established relation between langs. of Finland and Hungary), 1772. Died 1785.

SAUPE, Alfred Otto, German physicist; b. Badenweiler, Germany, Feb. 14, 1925; s. Alfred and Luise (Henssler) S.; M.Sc. (Dipl. Phys.), U. Freiburg (Germany), 1955, Dr.rer.nat., 1958; m. Brigitte J. Wiedemann, May 31, 1963; children—Anja Sabine, Welf Alfred. DFG Research fellow Phys. Inst. Freiburg, 1958-61; co-worker (with Prof. R. Mecke) Inst. for Phys. Chemistry, Freiburg, 1962; sci. asst. U. Freiburg, 1963—, dozent phys. chemistry, 1968—. Research, publs. on theoretical and exptl. work on liquid crystalline states and intermolecular forces in liquid crystals, high resolution nuclear magnetic resonance spectroscopy on oriented molecules in liquid crystalline solvents. Home: 2 Bismarckstr. Badenweiler/Baden 7847, Germany. Office: 31 Stefan-Meier-Str., Freiburg i Br. 78, Germany.*

SAURIN, Joseph, French geometer; b. Courtheton, France, Sept. 1, 1655; in politics and govt.; fled to Switzerland; returned to France, 1689; became mem. French Acad. Scis. Author memoirs on geometry; collaborator Jour. des Savants, 1702-08. First to solve problem of tangents at multiple points of algebraic curves, 1720. Died Paris, Dec. 29, 1737.

SAUSSURE, Nicholas Théodore de, Swiss botanist, chemist; b. Geneva, Switzerland, Oct. 14, 1767; s. Horace Benedict de Saussure; prof. mineralogy and geology Geneva Acad. Mem. French Acad. Scis., 1808. Fellow Royal Soc., 1820. Author: Recherches chimiques sur la végétation, 1804; Carbonic Acid Content of the Atmosphere and its Variation, 1816, 28, 30. Pioneered in exptl. physiology; research in fermentation, germination, structure of alcohol, transformation of sugar from starch, nutrition and respiration of plants; demonstrated that carbon dioxide is essential to green plants, that water is absorbed into plants and they are dependent on nitrogen in soil. Died Geneva, Apr. 18, 1845.

SAUTER, Johann Nepomuk, German physician; b. Reichenau Island, June 29, 1766; municipal physician, Constance, performed 1st successful complete extirpation of uterus through vagina, 1822. Died Thurgau, Switzerland, Nov. 30, 1840.

SAUTTER, Jay Howard, Am. vet. pathologist; b. Waynesburg, O., Nov. 11, 1912; s. John Frederick and Daisy (Shearer) S.; D.V.M., Ohio State U., 1944; Ph.D., U. Minn., 1948; m. Margaret Elizabeth Bentley, Mar. 7, 1937; children—Marjorie Ann, David Ashley, Thomas Howard. Research fellow U. Minn., 1944-47, faculty, 1947—, prof. vet. pathology, 1953—, head dept., 1953-65. Tech. adviser AID, Peru, 1965. Recipient award for contbn. to student edn. Minn. Student Assn., 1961. Diplomate Am. Coll. Vet. Pathologists. Mem. Am., Minn. vet. med. assns., Internat. Acad. Pathology, Twin City Clin. Pathologists Assn., A.A.A.S., Sigma Xi. Research, publs. on aplastic anemia of cattle, leukemia, cardiovascular diseases of animals. Home: 1550 Fulham St., St. Paul 55108.*

SAUVAGE, François Clement, French engr., geologist; b. Sedan, France, 1814; ed. Polytechnic Sch.; mining engr. on sci. voyages to Spain and Greece; later mil. engr.; active in Franco-Prussian War. Author: Cartes geologiques des Ardennes et de la Marne; Description geologique des Ardennes; Description geologique de la Grèce continentale et de l'Ile de Milo. Died Paris, 1872.

SAUVAGE, Frédéric, French inventor; b. Boulogne-sur-Mer, France, 1785; shipbuilder, Boulogne; patentee single stroke propellor, before Smith and Ericsson's patent of 1836; inventor mech. device (physionotype) for forming figures in relief, also apparatus for reducing size of statues. Died 1857.

SAUVAGEAU, Camille-François, French biologist; b. Angers, France, May 12, 1861; Docteur ès scis. naturelles; prof. botany Bordeaux (France) Faculty Scis.; mem. French Acad. Scis. Author: Sur les feuilles de quelques monocotyledones aquatiques, 1891; Utilisation des algues marines, 1920. Continued work of Bornet; studied plant anatomy; specialized in marine algae, especially brown algae, after 1892; also studied reproduction of kelp family. Died Vitrac, France, Aug. 5, 1936.

SAUVEUR, Albert, metallurgist; b. Louvain, Belgium, June 21, 1863; s. Lambert and Hortense (Franquin) Sauveur; ed. Athénée Royal, Brussels, Belgium, also Sch. Mines Liège, 1881-86; S.B., Mass. Inst. Tech., 1889; Sc.D., Case Sch. Applied Sci., 1921; Sc.D., U. Grenoble, (France), 1924, U. of San Marcos, Peru, 1925, Harvard, 1935; D.Eng., Leigh U., 1926; m. Mary Prince Jones, June 4, 1891; children—Hortense (Mrs. Romeyn Taylor), Mary Isabella (Mrs. George C. Eaton), Albert. Chemist and metallurgist various steel cos., 1889-97, instr. metallurgy, Harvard, 1889-1900, asst. prof. metallurgy and metallography, 1900-05, prof. metallurgy, 1905, Gordon McKay prof., 1924-35, emeritus, 1935—. Metall. expert French govt., also Am. Aviation Commn. in France, 1917-19. Recipient Elliott Cresson Gold medal, Franklin Inst., Phila., 1913; Bessemer medalist, Brit. Iron and Steel Inst., 1924; first recipient of the Albert Sauveur Achievement medal, Am. Soc. for Metals. Fellow Am. Acad. Arts and Scis., A.A.A.S.; hon. mem. Am. Inst. Mining and Metall. Engrs. (v.p. 1910-12), Am. Soc. for Metals, Soc. Engrs. Liège Sch. Mines, Soc. Ingénieurs Civils ce France, Soc. de l'Industrie Nationale; corr. mem. Soc. Encouragement Nat. Industry, France. Author: The Metallography of Iron and Steel; Metallurgical Dialogues. Editor Metallographist, 1898-1903, Iron and Steel Mag., 1903-06. Pioneer in technique of photo micrographs, also developed methods of improving steel through study of crystal structure of metals, before 1900; inventor pipeless process for casting ingots, magnetic deflector for rolling steels; advanced heat treatment of metal. Died Jan. 26, 1939.

SAUVEUR, Joseph, French mathematician, physicist; b. La Flèche, France, Mar. 24, 1653; named prof. math. Coll. de France, 1686; apptd. tutor to pages of dauphiness, 1680; mem. French Acad. Scis., 1696. Author: Traité de fortification; Géo métrie élémentaire; various writings on acoustics, 1710-13. Founder of musical acoustics; determined nodes, antinodes, frequency by beats; established higher harmonics in simple numerical ratio to basic tone; determined hearing limits of human ear; credited with 1st sci. explanation of human ear; credited with 1st sci. explanation of overtones, also 1st calculation of absolute vibration; assisted in Mariotte's expts., 1681; developed principles of fortification tested at siege of Mons, 1691. Died Paris, July 10, 1716.

SAUVY, Alfred-Bernard-Marie-Louis, French economist; b. Ville-neuve-de-la-Raho, France, 1898; ed. Ecole polytechnique; Dr. honoris causa, U. Brussels, U. Geneva. Dir., Institut de Conjoncture, 1937-45; became sec.-gen. for family and population, 1945; dir. Institut National d'Etudes Démographiques; 1945—; prof. econs. and opinion Inst. Polit. Studies, 1942-58; prof. demography College de France, 1959—. Became mem. statis. com. UN, 1946, pres. population com., 1950-53. Mem. Société Statistique de Paris (became pres. 1947), Nat. Social and Econ. Council, Inst. Statis. Inst., Internat. Population Union. Author: Essai sur la conjoncture et la prévision économiques, 1938; Richesse et population, 1943; Le pouvoir et l'opinion, 1949; Théorie générale de la population, 1952; L'Europe et sa population, 1953; L'Opinion publique, 1956; La Bureaucratie, 1956; La Nature social, 1957; De Malthus à Mao Tsé Toung, 1958; La Montée des Jeunes, 1959; Le plan Sauvy, 1960; Les Limites de la vie humaine, 1961; Fertility and Survival, 1961; Population Explosion, 1961; La Prévention des Naissances, 1962; Marx contre Malthus, 1963. Application of statis. methods to solution of econ. and population problems. Home: 76 rue Lepic, Paris 18, France.

SAVAGE, Albert Wilcox, Jr., Am. chemist; b. N.Y.C., May 24, 1923; s. Albert Wilcox and Faye (Immick) S.; grad. Hotchkiss Sch., 1941; A.B., Oberlin Coll., 1948; M.S., Yale, 1949, Ph.D., 1951; m. Joanna Griffies Corley, May 5, 1944; children—Antoinette Corley, Albert Wilcox III, Carroll Immick. Mem. staff Los Alamos Sci. Lab., 1951—. Mem. Am. Chem. Soc., Sigma Xi. Research, publs. on fundamental uranium chemistry, particularly with respect to uranium-flourine-chlorine compounds and water solutions of UF4. Home: Route 1, Box 346, Espanola, N.M. 87532. Office: Box 1663, Los Alamos 87544.*

SAVAGE, Hiram Newton, Am. civil engr.; b. Lancaster, N.H., Oct. 6, 1861; s. Hazen Nelson and Laura Ann (Newton) S.; B.S., N.H. Coll. Agr. and Mechanic Arts, 1887; C.E., Thayer Sch. Civil Engring. (Dartmouth), 1891; D.Sc., U. N.H., 1913; m. Linna Belle Clough, Dec. 1891 (dec. 1897); children—Lucy Eunice (Mrs. Robert L. Colthart), Laura Ada (Mrs. Laurence W. Hope); m. 3d, Eugenia Hurlock, 1927. Asst. engr. E. Tenn., Va. & Ga. R.R., Nashville & Tellico R.R. and Athens (Tenn.) Improvement Co., 1888; chief engr. Hydraulic Mining & Irrigation Co., N.M., 1889-90; engr. and supt. constrn. Billings Park and race track, White River Junction, Vt., also designed and located sewerage system for W. Randolph, Vt., 1890; chief engr. San Diego (Cal.) Land & Town Co., Sweetwater dam system, San Diego, 1891-1903; chief engr. Sweetwater Park and Race Track, 1893-94; cons. engr. San Diego & Cuyamaca R.R., San Diego & La Jolla R.R., Coronado Beach R.R., 1893-1903, Cuyamaca (Cal.) Water Co., 1896; chief engr. to contractor for U. S. Govt. San Diego Bay-Zuninga Shoals Jetty, 1894-98; cons. engr. So. Cal. Mountain Water Co., Lower Otay Dam, Upper Otay Dam, Barrett Dam, Morena Dam & Carrying System to San Diego, 1893-1903; cons. engr. U. S. Govt. Reclamation Service, 1903—; also supervising engr. Northern Div. in charge investigation, designs, construction, operation 12 irrigation projects and storage features, including Shoshone Dam (highest in world), Corbett Tunnel, 1904-15; cons. and constrn. engr. Sweetwater Water Co., enlargement extension Sweetwater Dam and Water Carrying System, 1916-17; hydraulic engr. City of San Diego Municipal Water System impounding and carrying features; designed and constructed with city forces, Lower Otay Dam, Barrett Dam, Morena Dam and Spillway enlargement, Rapid High Pressure Filtration Plant; directed investigation of all additional Water Supply resources in San Diego County and vicinity; field location and design all impounding and carrying features for fifty years future water requirements, 1917-23; toured world 3 times for engring. and archtl. research; reported on tech. adminstrv. matters to Pres. of U. S., 1925-26; cons. engr. Research Service (internat.), Washington, also Newell, Corse & McDaniel, Washington; recalled to San Diego, 1928, to take charge of Municipal Bur. of Water Devel. Deceased.

SAVAGE, Jay Mathers, Am. biologist; b. Santa Monica, Cal., Aug. 26, 1928; s. Jesse Mathers and Mary Louise (Bird) S.; A.B., Stanford, 1950, M.A., 1954, Ph.D., 1955; m. Ruth Louise Brynes, June 28, 1952; children—Nancy Diane, Charles Richard. Asst. prof. zoology Pomona Coll., Claremont, Cal., 1954-56; faculty U. So. Cal., Los Angeles, 1956—, prof. biology, 1964—, dir. biosystematics program Hancock Found., 1964—. Chmn. conf. tropical biology, 1962. NSF grantee, 1957—; USPHS grantee, 1960-64; John Simon Guggenheim fellow, 1963-64. Fellow So. Cal. Acad. Scis. (pres. 1966-68); mem. Am. Soc. Ichthyologists and Herpetologists (bd. govs., mem. editorial bd. 1959-60, treas. 1960-64), Brit. Herpetological Soc., Heretologists League, Soc. Study Evolution, Soc. Systematic Zoology, Sigma Xi, Orgn. for Tropical Studies (mem. bd.), others. Author: Evolution, 1963. Research, publs. on ecology, biogeography, evolution of lower vertebrates, osteology, tropical biology. Office: Dept. Biol. Scis., U. So. Cal., Los Angeles 90007.*

SAVAGE, Leonard Jimmie, Am. statistician; b. Detroit, Nov. 20, 1917; s. Louis and Mae (Rugawitz) S.; B.S., U. Mich., 1938, Ph.D. in Math., 1941; D.Sc. (hon.), U. Rochester; m. Jean Strickland, July 10, 1964; children by previous marriage—Sam Linton, Frank Albert. Rackham fellow Inst. Advanced Study, Princeton, 1941-42; instr. math. Cornell U., 1942-43; research mathematician Brown U., 1943; research asso. Columbia, 1944-45, N.Y.U., 1945-46; Rockefeller fellow Marine Biol. Lab., Woods Hole, Mass. and U. Chgo., 1946-47; research asso. U. Chgo., 1947-49, mem. faculty, 1949-50, prof., 1954-60, chmn. dept., 1956-59; prof. U. Mich., 1960-64; Eugene Higgins Institute Arts and Letters, Nat. Soc. Mural Painters, Painters' and Sculptors' Gallery Assn., Am. Inst. Architects. Study of mathematical and applied statistics; foundations of statistics; application of mathematics to biology. Home: Woodbury, Conn.

SAVARD, F. G. Kenneth, biochemist; b. Quebec City, Can., Feb. 26, 1918; s. Joseph D. and Jane (Davis) S.; B.S., Laval U., 1939, D.Sc., 1946; M.S., McGill U., 1943; m. Lorraine Kelly, 1943; children —Christopher, Julia. Research chemist Ayerst, McKenna and Harrison Ltd., Montreal, 1943-46; asst. staff Cleve. Clinic Found., 1948-51; sr. staff scientist Worcester Found. for Exptl. Biology, 1956-57; investigator Howard Hughes Med. Inst., 1957—; prof. biochemistry U. Miami, Fla., 1961—, prof. medicine, 1961—, chief endocrinology lab., 1957—. Former mem. reproductive biology study sect. NIH. Mem. A.A.A.S., Am. Chem. Soc., Canadian Physiol. Soc., Am. Soc. Biol. Chemists, Endocrine Soc., Sigma Xi. Research, numerous publs. on steroid hormone synthesis and metabolism, mechanism of control by pituitary gonadotropin, function of male and female gonads. Home: 3614 Bayview Rd., Coconut Grove, Fla. 33133. Office: P.O. Box 875, Miami, Fla. 33152.*

SAVART, Félix, French physicist; b. Mézières, France, June 30, 1791; began as mil. surgeon; moved to Metz, France; became physician, Strasbourg, France, 1816; curator physics collection then prof. physics College de France. Mem. French Acad. Scis., 1827. Fellow Royal Soc., 1839. Author: Mémoire sur les instruments à cordes, 1819. Studied hearing, including limits of audibility; invented acoustical instrument to determine number of vibrations per second in a tone (Savart's wheel), quartz plate for study of polarized light (Savart plate); studied resonance, especially in stringed instruments; 1st measurement (with Biot) of effect of force of elec. current on magnetic pole and originated law named after them; studied behavior of jets of liquid coming out of openings in thin walls or short nozzles, whirling motion of cyclones; built polariscope. Died Paris, Mar. 16, 1841.

SAVARY, Félix, French astronomer, physicist; b. Paris, Oct. 4, 1797; student Ecole polytechnique. Became mem. Central Astron. Office, 1830; prof. astronomy geodesy, founder studies on surveying and machines Ecole polytechnique. Mem. French Acad. Scis., 1832, French Bur. Longitudes. Author: Mémoire sur l'application du calcul aux phenomènes électro-dynamiques, 1823; Mémorie sur la détermination des orbites que décrivent autour de leur centre de gravité deux étoiles très rapprochées l'une de autre, 1827. Developed theorem on curvature of curve created by movement of point lying on a curve rolling on a fixed curve (named after him); applied law of gravity to reckoning of course of double stars; studied refraction of electro-dynamic phenomena, 1823, intensity of magnetization through elec. discharge, 1827. Died Estagel, July 15, 1841.

SAVIANO, Mario, Italian biochemist; b. Sarno, May 31, 1909; s. Leonardo and Giuseppina (Nunziante) S.; M.D.; m. Clara Modugno, 1944; children—Massimo, Leonardo. Asst., then prof. human physiology U. Naples; dir. Human Physiol. Inst., U. Modena; prof. alimentation sci. and biochemistry Faculty Pharmacology; dir. research group for physiopathology of work and sports Nat. Council Research, Inst. Human Physiology, U. Modena. Decorated Order Saint Agatha (Republic San Marino). Mem. Surg., Chem. and Human Physiol. Soc., Italian Biology Soc., Am. Chemistry Soc. Author: Ricerche sulla poliglobulia da cobalto; Ricerche istiofisiologiche sul pancreas di Ofidi e Cheloni; also articles on physiology and biochemistry. Home: via Selmi 33. Office: via S. Eufemia 19, Modena, Italy.

SAVIC, Pavle Petar, Yugoslavian phys. chemist; b. Solun, Jan. 10, 1909; s. Petar and Ana (Trpkovic) S.; Dipl. physicochemistry U. Beograd (Yugoslavia); Hon. doctorate chemistry, U. Beograd; m. Branka Bozinovic, Sept. 1, 1934; 1 dau., Ana. Asst. U. Beograd, 1934-39, prof. phys. chemistry, 1929-41; with Nat. Liberation Front, 1941-44; vis. research prof. Inst. Phys. Problems, Moscow Acad. Scis., 1944-46; coworker (with I. Joliot Curie) Inst. Radium, 1935-39; chmn. Inst. Boris Kidricin, Vinca, Sci. Council, 1948-60. Recipient award of Avnoj, 1966. Mem. Serbian Acad. Scis., Acad. Scis. USSR, Serbian Chem. Soc. Research and publs. on nuclear physics, low temperatures; discovered (with Joliot-Curie) 1st fission product; discovered method for obtaining low temperatures; elaborated original theory on origin of rotation of system of particles and behavior of materials under high pressures. Home: 22 Kicevska, Beograd, Yugoslavia.

SAVIGNY, (Marie-) Jules-César, French biologist; b. Provins, France, Apr. 5, 1777; studied botany under Lamarck; prof. École Centrale de Seine-Inférieure; mem. Napoleon's expdn. to Egypt; mem. French Acad. Scis., Egyptian Inst. Author: L'histoire naturelle et mythologique de l'Ibis; Histoire naturelle. Studied birds of Egypt and Syria, golden carp of China; his work led to more detailed studies of lower animals. Died Gally, France, Oct. 5, 1851.

SAVILE, D(ouglas) B(arton) O(sborne), botanist; b. Dublin, Ireland, July 19, 1909; s. Hugh Osborne and Kathleen (Barton) S.; B.S.A., McGill U., 1933, M.Sc., 1934; Ph.D., U. Mich., 1939; m. Constance Eleanor Cole, July 29, 1939; children—Harold Alan, Mary Elizabeth (Mrs. David F. Rhodes). Staff, Can. Dept. Agr., 1932-41, 43—, now prin. mycologist. Fellow Royal Soc. Can., A.A.A.S. (council 1962, 64-65), Arctic Inst. N.Am.; mem. Canadian Phytopath. Soc. (past mem. editorial bd.), Ottawa Field Naturalists Club, Canadian Bot. Assn., Mycol. Soc. Am., Am. Soc. Plant Taxonomists, Internat. Assn. Plant Taxonomists, Soc. for Study Evolution (past mem. editorial bd.), Am. Ornithologists Union. Author: Collection and Care of Botanical Specimens, 1962; also numerous articles. Arctic research on taxonomy, distbn., breeding behavior and gen. ecology parasitic fungi, ecology, evolution, bio-geography, glacial history arctic flowering plants, gen. arctic ecology including breeding bird density, relationships insects to plants, taxonomy and ecology arctic fungi. Home: 357 Hinton Av., Ottawa 3, Ont. Office: Plant Research Inst., Central Exptl. Farm, Ottawa, Ont., Can.*

SAVILE, Sir Henry, English mathematician; b. Bradley, nr. Halifax, England, Nov. 30, 1549; s. Henry and Elizabeth (Ramsden) S.; B.A., Merton Coll., Oxford, 1566, M.A., 1570; m. Margaret Dacres Gerrard, 1592; 1 dau., Elizabeth. Apptd. Greek and math. tutor to Queen Elizabeth, 1578; warden of Merton Coll., 1585-1621; named provost of Eton, 1596; founder Savilian professorships of geometry and astronomy at Oxford U. Published folio edit. of St. Chrysostom, 8 vols., 1610-13, an edit. of Cyropaedia, 1613. Author: Praelectiones tresdecem in principium elementorum Euclidis, 1621; translated 4 books of Tacitus' Historiae, 1591; apptd. to prepare authorized version of Bible, including parts of Gospels, Acts, Book of Revelation. Reinstituted study of geometry in Eng.; studied Ptolemy's Almagest. Died Eton, Eng., Feb. 19, 1622.

SAVILL, Thomas Dixon, English physician; b. Eng., Sept. 7, 1856; s. Thomas Choate and Eliza (Dixson) S.; student St. Thomas's Hosp., St. Mary's Hosp., La Salpêtrière, Paris, Hamburg, Vienna; M.B., U. London, 1881, M.D., 1882; m. Agnes Forbes Blackadder, 1901. Asst. physician, registrar, pathologist W. London Hosp.; med. supt. Paddington Infirmary, pres. Supts. Soc.; began med. teaching in poor law infirmaries, 1890; med. officer Royal Commn. on Vaccination, 1891-94; physician W. End Hosp. for Diseases of Nervous System, Hosp. for Diseases of Skin, Leicester Sq.; examiner in medicine U. Glasgow, Soc. Apothecaries, London; v.p., a founder Brit. Coll. Phys. Edn. Author: A New Form of Epidemic Skin Disease, 1893; Reports to the Royal Commission on Vaccination, 1893-94; Lectures on Neurasthenia, 1899, 4th edit., 1908; Lectures on Hysteria, 1909; A System of Clinical Medicine, 2d edit., 1909; also papers on diseases of nervous system, skin, arteries. Credited with 1st description of epidemic exfoliative dermatitis (Savill's disease), 1891. Died Jan. 10, 1910.

SAVILLE, Marshall Howard, Am. archeologist; b. Rockport, Mass., June 24, 1867; spl. student in anthropology, Harvard, 1889-94; Sc.D., U. of San Marcos, Lima, Peru; m. Annie W. Lyon, June 14, 1893. Did field work under direction of Prof. Putnam; explorations in Yucatan, 1890; Honduras, 1891-92, for Peabody Mus., Harvard, and for the Am. Museum Natural History, 1897-98; 4 expeditions to ruins of Mitla and Oaxaca, 1899, 1902, 04; numerous expdns. to Peru, Ecuador, Colombia, Honduras, Guatemala, Central America; hon. curator of Mexican and Central Am. archeology, Am. Museum of Natural History, New York, 1908-10; prof. Am. archeology, Columbia, 1903—; staff of museum of the Am. Indian, Heye Foundation, 1918-32, Am. Museum of Natural History, New York, 1933; hon. prof. Museo Nacional, Mexico, 1933. Had charge of Central Am. exhibit, Chicago Expn., 1893. Founder of the Cortes Soc.; Officier de l'Academie Française; corr. mem. Royal Acad. History of Spain; hon. mem. Soc. de Geog. y Estadistica, Mexico, Instituto Historico, Peru; mem. Am. Anthrop. Assn. (pres. 1926-28). Made important discoveries among the remains of mound-builders of So. Ohio, among Maya ruins of Copan and Palenque, among Aztec and Zapotec ruins in Mexico. Died N.Y.C., May 7, 1935.

SAVILLE, Thorndike, Am. engr.; b. Malden, Mass., Oct. 3, 1892; s. Caleb Mills and Elizabeth (Thorndike) S.; A.B., Harvard, 1914, M.S., 1917; B.S., Dartmouth, 1914, C.E., 1915; M.S., Mass. Inst. Tech., 1917; hon. degrees include E.D., Clarkson Coll., Syracuse U., D.Sc., N.Y. U.; m. Edith Stedman Wilson, Sept. 10, 1921; 1 son, Thorndike. Faculty, U. N.C., 1919-32, also chief engr. N.C. Dept. Conservation and Devel., 1920-32; prof. hydraulic and san. engring. N.Y. U., 1932-57, dean Coll. Engring., 1936-57, now emeritus; dir. sci. and engring. center study U. Fla., 1958-60, cons. to pres., 1960-62, cons. numerous govt. groups. Mem. beach erosion bd. Office Chief Engr. U.S. Army, 1930-63, costal engring. research bd., 1963—. Recipient Jubilee medal Am. Soc. M.E., 1955—; Distinguished Civilian Service decoration Dept. of Army, 1964. Fellow Am. Pub. Health Assn., A.A.A.S., Am. Soc. C.E. (hon.); mem. Am. Soc. Engring. Edn. (Lamme medal 1954), Water Pollution Control Fedn., Am. Water Works Assn., Am. Inst. Cons. Engrs., Am. Meterol. Soc., Am. Geophys. Union, Am. Acad. San. Engrs., Nat. Soc. Profl. Engrs., Internat. Assn. Hydraulic Research, Phi Beta Kappa, Tau Beta Pi, Sigma Xi, others. Research, publs. on water treatment, hydrology, municipal water supply, coastal erosion, water pollution; participated in govt. adminstrn. of water resources devel. and control. Home: 2225 N.W. 2d Av., Gainesville, Fla. 32601.*

SAVIN, Guriy Nikolaevich, Russian engineer; b. 1907; grad. Dnepropetrovsk U., 1932. Instr., Dnepropetrovsk Civil Engring. Inst., Dnepropetrovsk U., 1932, prof., from 1941; dir. Inst. Mining Mechanics, Ukrainian Acad. Scis., Lvov, 1945-48, dir. Inst. Mechanics, 1958-59, head dept. elesticity theory, 1959—; rector Lvov U., 1948-52; prof. Kiev U., 1951—. Recipient Stalin prize, 1952. Mem. Ukrainian Acad. Scis. (v.p. 1952-57). Co-author: A Course of Theoretical Mechanics, 1957. Research and publs. on elasticity theory, concentration of stresses near apertures in machine parts and structural elements, chamfering of aperture angles to take concentrated stresses. Address: Inst. Mechanics, USSR Acad. Scis., Vladimirskaya 54, Kiev, Ukrainian SSR, USSR.

SAVITSKY, Aleksandr Ivanovich, Russian surgeon, oncologist; b. 1887; grad. Med. Faculty, Moscow U., 1911; D.Med. Sci., 1936. Asst. faculty surgery clinic

1st Moscow U., 1921-31, sr. asst. propedeutic surgery clinic, 1931-39; dep.dir. Moscow Oncological Inst., 1931-39; prof., 1939—; head chair surgery Moscow Stomatological Inst., 1939-44; prof. chair hosp. surgery 1st Moscow Med. Inst., 1942-43; dir. Gertsen Oncological Inst., 1944-53, sci. dir., 1953—; head chair oncology Central Postgrad. Med. Inst., 1946—. Chief oncologist, dep. chmn. cancer com. USSR Ministry Health; mem. learned council dept. clin. medicine USSR Acad. Med. Scis.; mem. com. for cancer control Internat. Cancer League. Decorated Order of Lenin; recipient Spasokukotsky prize. Mem. USSR Acad. Med. Scis., All-Union (chmn.), Moscow (chmn.) socs. oncologists, All-Union (bd. mem.), Moscow (bd. mem.) socs. surgeons. Author numerous works including: Clinical Esophagoscopy, 1940; Lung Cancer, 1957; Limits of Operative Interference in Treatment of Cancer, 1960; co-author, editor: Short Manual on Early Diagnosis of Cancer, 1948. Mem. editorial bd. Problems of Oncology, Surgery; co-editor Surgery sect. Large Med. Ency., 2d edit. Address: Central Postgrad. Med. Inst., pl. Vosstaniya 1-2, Moscow, USSR.

SAVITSKY, J(erome) Philip, Am. physiologist, physician; b. N.Y.C., May 18, 1929; s. Nathan and Anna (Schiff) S.; A.B., Columbia, 1948; M.D., N.Y. U., 1952; m. Annette Coopersmith, June 17, 1951 (div. 1958); children—Daniel Elliott, Grace Ann. Research fellow May Inst., Cin., 1953-55; research asso. Montefiore Hosp., N.Y.C., 1957—; asso. in medicine Albert Einstein Coll. Medicine, Yeshiva U., N.Y.C., 1966—; established investigator Am. Heart Assn., 1960-65. Mem. Am. Physiol. Soc., Soc. for Exptl. Biology and Medicine, Am. Fedn. for Clin. Research. Contbr. articles to tech. jours. Research, publs. on human blood platelet physiology; discovered prolonged non-genetic effects of injected DNA on mammalian lipid metabolism. Home: 2630 Kingsbridge Terrace, N.Y.C. 10463. Office: Montefiore Hosp., N.Y.C. 10467.*

SAVORNIN, Justin Eugène Célestin, geologist; b. Pertuis, Dec. 4, 1876; Célestin and Eugénie (Cazal) S.; D.Sc., Faculty Scis., Marseille, France; m. Thérèse Jaumont; children—Mireille, Odelte, Jean, Violette; m. 2d, Gabrielle Canuët, 1919; 1 dau., Josette. Became asst. Faculty Algiers (Algeria), 1920, prof., 1920; named tchr. dir. Geol. Services Algeria and So. Tys., 1923; became sci. cons. Service Hydraulics and Rural Equipment, 1944. Corr. mem. French Acad. Scis., 1927—. Decorated Officer Palmes Acad., Officer Sahara merit. Author: Geologie du Hodna et du plateau Setifien, 1920; Geologie et Hydrologie des territories du Sud; Geologie de l'Algergie et de l'Afrique du Nord. Drew up (with others) detail geol. map of France, 1898, Algeria, 1900; studied geology of Algeria, Sahara, Morocco, Tunisia. Home: 14, rue d'Alembert, Algiers, Algeria.

SAVOT, Louis, French physician, chemist; b. Saulien, France, 1579; physician to Louis XIII; author treatises on medicine and numismatics; described use of nitric acid (aqua fortis) for separating gold from silver, also methods of refining gold by cementation and by antimony and silver by cupellation with lead. Died Paris, 1640.

SAVULESCU, Alice, Rumanian mycologist, plant pathologist; b. Oltenita, Rumania, Oct. 16, 1905; d. Jancu and Jeannette Savulescu; ed. U. Biology, Bucharest, Rumania, 1924-29; Ph.D., Columbia, 1931-34; m. Traian Savulescu. Asst. plant pathology Inst. Tobacco Cultivation and Fermentation, 1929-31; prin. research worker plant pathology sect. Inst. Agrl. Research, 1934-39, head lab., 1939-49, head plant pathology sect., 1949-57; asst. dir. Inst. Biology, Rumanian Acad. Scis., also head sect. microbiology and gen. phytopathology; postgrad. tchr. to Ph.D. students, 1950—. Recipient Meritul Stuntofic, Municci Clasa Ta, Steaua Republicii Clasa II a. Corr. mem. Rumanian Acad. Scis., 1952, mem., 1963. Mem. Soc. Natural Scis. and Geography Roumania, N.Y. Acad. Scis. Author: Plant Pathology, a Guide to Crop Zonation; Diseases of Corn; also numerous articles. Research on black spot of roses, rusts of cereals and their control; diseases of potatoes especially black wart disease; diseases of ornamental plants and fruit trees; problems concerning fungicides; host parasite relationships; virus diseases of plants. Home: 24 Sos Kiselef. Office: 296 Splainl Nidependentii, Bucharest, Rumania.*

SAVULESCU, Traian, Rumanian botanist; b. 1889; mem., chmn. Acad. Rumanian People's Republic. Author: Protection des plantes et organisation phytopathologique en Roumanie, 1935; Contribution à la connaissance des Ustilaginées en Roumanie, 1936; L'immunité aux maladies bactériennes des plantes, 1936; Uredinalia in Rumania, 1953; Ustilagineae in Rumania, 2 vols., 1957; Maize—monographic study, 1957. Chief editor Flora in the Socialist Republic of Romania, 8 vols., 1952-61. Founder phytopathology in Rumania; research in biology and systematics of microscopic fungi. Died 1963.

SAWADA, Ryukichi, Japanese physicist; b. Tokyo, Japan, Jan. 10, 1917; s. Katsutaro and Yori (Sakuragi) S.; student Higher Meteorol. Sch., Tokyo, 1937-40, 44-48; M.S., N.Y. U., 1952, Ph.D., 1954; Sc.D., Tokyo U., 1956; m. Aiko Takagri, Apr. 29, 1946; children—Mari, Kenji. Prof. atmospheric physics Kyushu U., Fukuoka, Japan, 1959—; vis. prof. Tech. U. Istanbul (Turkey), 1957-59, 60-61; vis. scientist Nat. Center for Atmospheric Research, Boulder, Colo., 1964-65. Recipient Japan Meteorol. Soc. prize for work on atmospheric tides, 1956. Research, publs. on theory of atmospheric oscillations due to effects of moon and sun. Home: 14, 1-chome Shimo-Nakahama-cho, Fukuoka, Japan.*

SAWAGUCHI, Kazuyuki, Japanese mathematician; flourished late 17th century. Author: Kokon Sampoki (included illustrations showing approximation to circle area measurement by crude integration of thin rectangles). Developed a rough calculus similar to Cavalieri's; recognized plurality of roots.

SAWAMOTO, Hachie, Japanese metallurgist; b. Osaka, Japan, Dec. 18, 1901; s. Taminojo and Ren (Okuda) S.; V.Engring., Kyoto (Japan) U., 1927, Dr.Engring., 1941; m. Sawako Sawamoto, Oct. 15, 1928; children—Yoshie, Yasuko (Mrs. Masami Tsutsui), Toru, Kazuko (Mrs. Takahiko Asayama), Mamoru. Engr., Nippon Seiko-Sho, Hiroshima, Japan, 1928-33; lectr. Kyoto U. Inst. Chemistry, 1933-38; prof. Niihama (Japan) Technol. Coll., 1938-42; prof. nonferrous metallurgy Nagoya (Japan) U., 1942-65, hon. prof., 1965—. Vice dir. Mining Pub. Peace Conf., Ministry of Trading and Industry, 1950—. Mem. Inst. Metals, Iron and Steel, Mining and Metallurgy (councilor 1920—). Author: Light Alloys, 1942; Light Alloys for Airplanes, 1943. Publs. on Investigation of new method for prodn. zinc, mechanisms of sinstering ores, electrolytic smelting of complexed metallic sulfide ores. Home: 1973 Nishijin-Cho, City Tsu, Prefecture Mie, Japan. Office: 1 Furo-Cho, Chikusa-Ku, City Nagoya, Japan.*

SAWICKI, Eugene, Am. chemist; b. New Bedford, Mass., July 11, 1916; s. Victor and Magdalene (Dubinska) S.; B.S., U. Cin., 1949, M.S., 1950; Ph.D., U. Fla., 1952; m. Dorothy Eleanor Hanly, Mar. 6, 1943; 1 son, Charles Anthony. With U. Fla., 1952-56, Sloan Kettering research fellow Cancer Research Lab., 1952-53, asst. prof., 1953-56; with Dept. Air Pollution, USPHS Taft San. Engring. Center, Cin., 1956—, chief airborne particulates research, 1960—. Mem. Am. Chem. Soc., Air Pollution Control Assn., Research Engring. Soc. Am., Phi Beta Kappa, Phi Lambda Upsilon. Co-editor: Symposium on Analysis of Carcinogenic Air Pollutants, Nat. Cancer Inst. monograph 9, 1962. Contbr. numerous articles to sci. jours. Devel. microanalytical techniques for determination of trace amounts of air pollutants, carcinogens and other cytotoxic agts. present in human environment; devel. colorimetric, room- and low-temperature fluorimetric and phosphorimetric methods for trace analysis of organic compounds; introduction of quenchofluorimetric and quenchophosphorimetric methods of functional group analysis. Home: 2315 Kenlee Dr., Cin. 45230. Office: 4676 Columbia Pkwy., Cin. 45226.*

SAWICKI, J(erzy) Jozef, physicist; b. Warszawa, Poland, July 14, 1931; s. Jerzy Kazimierz and Zofia (Kunicka) S.; M.S., U. Warsaw, 1954, Ph.D. in Theoretical Physics, 1957; m. Valena Settimi, Oct. 15, 1964. Faculty, U. Warsaw, 1953-57, Princeton, 1954-59, U. Cal. at Berkeley, 1959-61, univs. Bologna, Rome (Italy), 1961-63; asso. prof. U. Paris (France), 1963-66; physicist CERN, Geneva, Switzerland, 1966—; invited prof. Summer Sch., Conv. Intercantonale en Suisse Romande, 1966. Mem. Italian Phys. Soc. (prize 1962). Contbr. articles to tech. jours. Developed theories of nuclear reactions, especially photonuclear, optical model potential, structure of finite spherical and deformed nuclei, many body problem of mixed systems of bosons and fermions, nuclear absorption of kaons. Home: 56 Taro, Roma, Italy. Office: Internat. Center for Theoretical Physics, Trieste, Italy.*

SAWIN, Paul Baldwin, Am. geneticist, anatomist; b. Canton, O., Jan. 14, 1900; s. Farwell Lyman and Carrie (Baldwin) S.; B.S., Cornell U., 1924; M.S., Kansas State Coll., 1925; postgrad. Grinnell Coll. Inst. Zoology and Genetics; M.S., Sc.D., Harvard, 1931; m. Hallie Laughlin, June 8, 1926 (dec. 1946); children—Farwell Laughlin, Sylvia Primrose; m. Vera Mosley, Oct. 1, 1949. Faculty, Brown U., 1931-47, asso. prof., 1946-47; research asso. Jackson Lab., Bar Harbor, Me., 1947-57, staff scientist, 1957-65, emeritus, 1965—. Mem. A.A.A.S., Soc. Naturalists, Soc. Zoologists, Genetics Soc., Am. Genetics Assn., Assn. Phys. Anthropologists, Am. Assn. Anatomists, N.Y. Acad. Scis. Research, numerous publs. on genetics domestic rabbit, coat color, growth, morphology, maternal behavior. Home: R.D. 3, Box 19A, Ellsworth, Me. 04605. Office: Hamilton Sta., Jackson Lab., Box 257, Bar Harbor, Me. 04609.*

SAWYER, Charles Henry, Am. neuroendocrinologist; b. Ludlow, Vt., Jan. 24, 1915; s. John Guy and Edith (Morgan) S.; A.B., Middlebury Coll., 1937; postgrad. Cambridge (Eng.) U.; Ph.D., Yale, 1941; m. Ruth Schaeffer, Aug. 23, 1941; 1 dau., Joan Eleanor. Instr., Stanford, Palo Alto, Cal., 1941-44; faculty Duke, 1944-51, prof. anatomy, 1950-51; prof. U. Cal. at Los Angeles, 1951—, chmn. dept. anatomy, 1955-63; cons. VA Hosp., Long Beach, Cal., 1951—. Mem. neurology study sect. NIH, 1963-67, W. Coast regional dir. neuroanatomy vis. scientists program, 1964-68. Mem. Internat. Brain Research Orgn. (mem. council 1964-68), A.A.A.S., Am. Assn. Anatomists, Am. Soc. Zoologists, Am. Physiol. Soc., Soc. for Exptl. Biology and Medicine (editorial bd. Procs. 1959-62), Endocrine Soc., Histochem. Soc., Am. Zool. Soc., Phi Beta Kappa, Sigma Xi. Editorial bd. Endocrinology, 1955-59; co-editor Exptl. Brain Research, 1965—. Research, numerous publs. on disbn. and functions cholinesterases, effects of hormones on enzyme synthesis, nervous control pituitary gland and ovulation, reproductive function and control steroid synthesis by ovary, feedback action steroid hormones on central nervous system, gonadotropins and hypothalamic releasing factors, circadian rhythms and sex cycles, nervous control lactation and milk ejection. Home: 466 Tuallitan Rd., Los Angeles 90049.*

SAWYER, Constance Bragdon, Am. astronomer; b. Lewiston, Me., June 3, 1926; d. William Hayes and Beatrice (Burr) Sawyer; A.B., Smith Coll., 1947; A.M., Radcliffe Coll., Ph.D., 1952; m. James W. Warwick, Sept. 6, 1947; children—Sarah Haskell, David Irwin, Rachel Joan, Joel Howard. Asst., Sacramento Peak Obs., Sunspot, N.M., 1952-55; staff High Altitude Obs., Boulder, Colo., 1955-58; astronomer Space Disturbance Lab., Environmental Sci. Services Adminstrn., Boulder, 1958—; vis. lectr. U. Colo., 1963, 67. Mem. Am. Astron. Soc., Am. Geophys. Union, Internat. Geophys. Union, Internat. Sci. Radio Union. Publs. and research on solar-terrestrial relations. Home: 1900 Baseline Rd., Boulder 80302. Office: Environmental Sci. Services Adminstrn., Boulder, Colo. 80302.*

SAWYER, John Stanley, English meteorologist; b. Wembley, Middlesex, Eng., June 19, 1916; s. Arthur Stanley and Florence E. (Frost) S.; M.A., Cambridge (Eng.), 1937; m. Betty Tooke, Sept. 22, 1951; 1 dau., Jane Elizabeth. Weather forecaster Brit. Meteorol. Office, 1938-46, research staff, 1946—, dir. research, Bracknell, Eng., 1965—. Fellow Royal Soc., 1962, Royal Meteorol. Soc. (past pres.). Author: The Ways of the Weather, 1957; also numerous articles. Research on dynamical meteorology and introduction of numerical weather prediction in U.K., structure of atmospheric fronts. Home: 8 Sherring Close. Office: Meteorol. Office, London Rd., Bracknell, Berkshire, Eng.*

SAWYER, Philip N., Am. physician; b. Bangor, Me., Oct. 25, 1925; s. Frank Saad and Linda (Makanna) S.; student Harvard, 1942-45; M.D., U. Pa., 1949; m. Grace Makla, June 13, 1953; children—Margaret Ann, Elizabeth Lynn, Susan Jean, Philip Michael. Asst., Harrison dept. surg. research U. Pa., 1949, surg. research, surg. fellow, 1953-56; Nat. Cancer Inst. Research fellow, 1950; staff Naval Med. Research Inst., 1951-53; chief resident St. Luke's Hosp., N.Y.C., 1956-57; faculty surgery State U. N.Y. Downstate Med. Center, 1957—, Markle scholar, 1959-64, professor, 1966—, head vascular surg. services, biophys., electrochemistry labs. dept. surgery, 1964—; vis. surgeon Kings County Hosp., U. Hosp. State U. N.Y.; asso. attending physician St. John's Episcopal Hosp.; asso. attending surgeon Meth. Hosp., Bklyn. Mem. A.A.A.S., Am. Heart Assn., A.M.A., Am. Soc. Artificial Internal Organs, N.Y., Bklyn. surg. socs., Internat. Cardiovascular Soc., I.E.E.E., Soc. U. Surgeons, Soc. for Vascular Surgery, Am. Chem. Soc., Am., Phila. physiol. socs., N.Y. Soc. for Cardiovascular Surgery, Kings County Med. Soc. Editor: Biophysical Mechanisms in Vascular Homeostasis and Intravascular Thrombosis, 1965; editorial rev. bd. Med. Research Engring.; editorial bd. Jour. Biomed. Materials Research. Contbr. numerous articles to profl., med. jours. Home: 606 3d St., Bklyn. 11215. Office: 450 Clarkson Av., Bklyn. 11203.*

SAWYER, Ralph Alanson, Am. physicist; b. Atkinson, N.H., Jan. 5, 1895; s. George Alanson and Lillie Elvira (Noyes) S.; A.B., Dartmouth, 1915, Sc.D., 1947; Ph.D., U. Chgo., 1919; LL.D., Wayne State U., 1954; Sc.D., Mich. Coll. Mining and Tech., 1959; m. Martha Green, Apr. 22, 1919 (dec. 1957); children—George Alanson II, Rosalind (Mrs. George S. Springsteen); m. 2d, Frances Tracy Hay, 1964. Faculty U. Mich., Ann Arbor, 1919-65, prof. physics, 1930-65, v.p. for research, 1959-65, dean Horace Rackham Sch. Grad. Studies, 1946-65. Civilian tech. dir. Joint Task Force One atom bomb tests, Bikini Atoll, 1946; mem. sci. adv. bd. Naval Weapons Lab., 1952—; mem. U.S. Nat. Com. Internat. Commn. on Optics, 1954-56; cons. Goddard Space Flight Center, NASA, 1964—. Recipient award Spectroscopy Soc. Pitts., 1961. Guggenheim Meml. fellow, Berlin, Germany, 1926-27. Fellow Optical Soc. Am. (asso. editor 1941-57, pres. 1955-57, Frederic Ives medal 1963); mem. Am. Inst. Physics (chmn. governing bd. 1959—, acting dir. 1964-65), A.A.A.S. (v.p. sect. B 1964), Am. Phys. Soc., Am. Assn. Physics Tchrs., Phi Beta Kappa, Sigma Xi, Phi Kappa Phi, Gamma Alpha. Author: Experimental Spectroscopy, 1944, 3d edit., 1963. Contbr. articles to sci. jours. Work on optics, determination of atomic structure by classification of spectra, chem. analysis by spectra, ordnance engring. Home: 1208 Wells St., Ann Arbor, Mich. 48104.*

SAWYER, Robert Tom, Am. mech. engr.; b. Schenectady, June 20, 1901; s. Willits Herbert and Georgina (Case) S.; B.E.E., Ohio State U., 1923;

M.E., 1930; m. Ruth L. Ennis, Sept. 20, 1934; children—Holly Ruth (Mrs. K. Richard Augenbaugh), Joan Ennis (Mrs. Dick M. Miller). With Gen. Electric Co., 1923-30; with Am. Locomotive Co., 1930-56, mgr. research, 1941-56; editor Gas Turbine Internat., Ho-Ho-Kus, N.J., 1960—. Recipient numerous awards. Fellow Am. Soc. M.E. (award gas turbine div. 1947, 66), I.E.E.E.; mem. Am. Inst. Aeros. and Astronautics, Inst. Mech. Engring. London, Soc. Automotive Engrs., Sigma Xi. Author: The Modern Gas Turbine, 1945; Applied Atomic Power, 1946; Gas Turbine Construction, 1947. Research and numerous publs. on gas turbines; designer 1st gas turbine locomotive with mech. drive. Address: Box 188, Ho-Ho-Kus, N.J. 07423.*

SAWYER, Sylvanus, Am. inventor; b. Templeton, Mass., Apr. 15, 1822; s. John Sawyer. Sent to work in gunsmith shop, Augusta, Me., 1839; invented several things, including small railroad car operated by foot; employed in coppersmith's shop, Boston, 1844; employed by mfr. locks and house trimmings, 1845; patentee machinery for splitting and dressing rattan, 1849, for cutting rattan, 1851, additional rattan machinery, 1854, 55; supt. Am. Rattan Co., Fitchburg, Mass., circa 1852-55; patented improvements in rifled cannon and projectiles, 1855, dividers and calipers, 1867, steam generator, 1868, shoe-sole machine, 1876, centering lathe, 1882. Died Fitchburg, Oct. 13, 1895.

SAWYER, Wilbur Augustus, Am. hygienist; b. Appleton, Wis., Aug. 7, 1879; s. Wesley Caleb and Minnie Edmea (Birge) S.; student U. Cal., 1898-99, LL.D., 1945; A.B., Harvard, 1902, M.D., 1906; m. Margaret Henderson, Oct. 14, 1911; children—Margaret (Mrs. J. Wallace Carroll), Gertrude (Mrs. R. W. Howell), Ruth Henderson (Mrs. D. P. Yeuell, Jr.), Wilbur Henderson. Intern Mass. Gen. Hosp., 1906-08; med. examiner U. Cal., 1908-11, lectr., 1914-16, clin. prof., 1916-19; dir. hygienic lab. Cal. Bd. Health, 1910-15, sec. and exec. officer, 1915-18; apptd. state dir. Internat. Health Bd., N.Y., 1919; dir. Australian hookworm campaign, 1919-22; adviser in pub. health, Australian Ministry of Health, 1922-24; asst. regional dir. for the East, internat. bd. Health Rockefeller Found., 1923-24, dir. Public Health Lab. Service, 1924-27, also asso. dir. Internat. Health, 1927-35, dir., 1935-44; dir. health U.N.R.R.A., 1944-47; mem. W. African Yellow Fever Commn. Rockefeller Found., 1926-27, in charge Yellow Fever Lab., 1928-35; cons., adviser, mem. various govt. and pvt. instns. and bodies. Decorated Grand Ofcl. Order Carlos J. Finlay, Cuba; recipient Leon Bernard prize League of Nations, 1939; Richard P. Strong medal, 1949. Fellow Am. Pub. Health Assn., A.A.A.S., N.Y. Acad. Medicine, Washington Acad. Medicine, Royal Soc. Tropical Medicine and Hygiene (hon.); mem. Am. Found. Tropical Med., Assn. Am. Physicians, Am. Acad. Tropical Medicine (pres. 1936-37), Am. Soc. Tropical Medicine, (pres. 1943-44), Am. Epidemiol. Soc., Am. Soc. Exptl. Pathology, Sigma Xi; hon. or corr. mem. other profl. assns. Contbr. on yellow fever, internat. health. Died Nov. 12, 1951.

SAWYER, Wilbur Henderson, physiologist; b. Brisbane, Australia, Mar. 23, 1921; s. Wilbur Augustus and Margaret (Henderson) S.; A.B., Harvard, 1942, M.D., 1945, Ph.D., 1950; m. Marian Gholson Kittredge, Nov. 14, 1942; children—Wilbur Kittredge, Robert Kittredge, Thomas Kittredge, Richard Kittredge. Instr. biology Harvard, 1950-53; asst. prof. physiology N.Y. U. Coll. Medicine, N.Y.C., 1953-57; asso. prof. pharmacology Columbia, N.Y.C., 1957-64, prof., 1964—. Mem. Am. Soc. Zoologists, Am. Physiol. Soc., Soc. Gen. Physiologists, Soc. Exptl. Biology and Medicine, Endocrine Soc., Am. Soc. Pharmacology, Soc. for Endocrinology. Asso. editor Endocrinology, 1963-68. Research, numerous publs. on comparative renal physiology and neurohypopophysial endocrinology. Home: 12 Warnke Lane, Scarsdale, N.Y. 10583. Office: 630 W. 186th St., N.Y.C. 10032.*

SAX, Adolphe Antoine Joseph, inventor; b. Dinant, Belgium, 1814; became prof. music, conservatory, Paris, 1857; constructed (with son) prototypes of mus. instruments, Paris; inventor saxophone, 1840; also other brass instruments of mil. music. Died Feb. 9, 1894.

SAX, Karl, Am. biologist; b. Spokane, Wash., Nov. 2, 1892; s. William L. and Minnie (Morgan) S.; B.S. in Agronomy, Wash. State Coll., 1916; M.S., Harvard, 1917, Sc.D., 1922; Sc.D. (hon.), U. Mass., 1965; m. Hally Mary Delilia Jolivette, Sept. 22, 1915; children—Karl J., William P., Edward A. Instr. genetics U. Cal. at Berkeley, 1919; biologist Riverbank Lab., 1919-20, Me. Agrl. Expt. Sta., 1920-23; cytologist, prof. botany Harvard, 1928-59; vis. prof. U. Fla., Yale, N.C. State Coll., Oxford U., U. Cal., U. Tenn., U. Ga., Cornell U., Cranbrook Inst. Sci., 1959—. NIH fellow U. Ga., 1964-67; dir. Arnold Arboretum, 1947-54; nat. lectr. Sigma Xi, 1962. Recipient Colman award Am. Assn. Nurserymen, 1961; Distinguished Alumnus award Wash. State U., 1966. Mem. Genetics Soc. Am., Bot. Soc. Am. (certificate of merit 1956), Am. Naturalists, Nat. Acad. Scis. Author: Standing Room Only, 1955. Research, numerous publs. on effect of x-rays on chromosomes, methods of producing dwarf fruit trees, prodn. of hybrid ornamental cherry, apple, forsythia, magnolia plants, radio mimetic effects of beverages, drugs and food additives. Home: 95 Bishop Hollow Rd., Media, Pa. 19063.*

SAXEN, Lauri O., Finnish embryologist; b. Helsinki, Finland, July 27, 1927; s. Arno and Katri (Palmroth) S.; M.D., U. Helsinki, 1954, Ph.D., 1962; m. Leena Pelkonen, May 8, 1951; children—Harri, Meri, Anu, Minna. Research asso. State Research Council, Helsinki, 1954-56; research asso. in pathology U. Helsinki, 1956-61, asso. prof., 1965-66, prof. exptl. pathology, 1966—; sr. sci. State Med. Research Council, Helsinki, 1962-65. Mem. sci. adv. bd. Finnish Army, 1961-65; sr. Lalor fellow Marine Biol. Lab., Mass., 1965. Mem. Acad. of Finland (sec. gen. 1959—), Internat. Acad. Pathology, Am. Assn. Cancer Research, Internat. Inst. Embryology. Author: (with Sulo Toivonen) Primary Embryonic Induction, 1962. Numerous publs. on exptl. analysis of devel. of endocrine system in lower vertebrates; acquired viral resistance; teratogenic actions of drugs in tissue cultures; devel. of two-gradient theory in primary induction. Home: 28 Tiilimaki, Helsinki. Office: Haartmaninkatu 3, Helsinki, Finland.*

SAXENA, Devendra Bahadur, Indian zoologist; b. Fatehgarh, India, Mar. 20, 1928; s. Balbir and Shiam (Kumari) S.; B.Sc., D.A.V. Coll., Kanpur, India, 1947; M.Sc., Agra (India) U., 1951; Ph.D., Lucknow (India) U., 1959; m. Saroj Rani, Apr. 30, 1954; children—Anshu, Aruna. Lectr. zoology D.A.V. Coll., 1949-63; sr. research fellow J. and K. U., Srinagar, Kashmir, 1963-65; fisheries research officer dept. fisheries J. and K. Govt., Srinagar, 1965—. Fellow Acad. Zoology India, Internat. Soc. Ichthyology and Hydrobiology (gen. sec. 1960—), Zool. Soc. India (life); mem. Nat. Acad. Scis. India (Gold medal for best research publs. of year 1946, life), Japanese Soc. Ichthyology, Zool. Soc. Czechoslovakia, Indian Sci. Congress Assn., Indian Sci. News Assn., Zootomical Soc. India. Mng. editor Ichthyologica, 1962—. Research, publs. on circulatory modifications in fishes due to air breathing habit, functional anatomy of digestive organs, swim-bladder, cardiovascular and respiratory system of fishes, osteology of some freshwater fishes, fisheries mgmt. problems of Jammu and Kashmir State; estimated gill area of air-breathing fishes and influence of oxygen and carbon dioxide on them studied; proposed polyphyletic polyniche theory. Home: Wazir Bagh, P.O.B. 64. Office: Fisheries Research Lab., Harwan, Srinagar, Kashmir, India.*

SAXENA, Satish Chandra, physicist, educator; b. Lucknow, India, June 24, 1934; s. Raja Ram and Vidya (Saxena) Sinha; B.Sc., Lucknow U., 1951, M.Sc., 1953; D.Phil., Calcutta U., 1957; m. Asha Saxena, June 20, 1960; children—Alka, Alok, Anup. Research asso. Inst. for Molecular Physics, College Park, Md., Columbia, Yale, 1956-59; research officer Atomic Energy Establishment Trombay, Bombay, India, 1959-61; reader, head physics dept. Maharaja's Coll., Rajasthan U., Jaipur, India, 1961-66; asso. research prof. Thermophys. Properties Research Center, Purdue U., West Lafayette, Ind., 1966—68; prof. dept. energy engring. U. Ill. at Chgo. Circle, 1968—. Mem. Indian Sci Congress. Research, numerous publs. on molecular physics, solid state crystal physics covering various bulk macroscopic properties. Home: 216 Sheetz St., West Lafayette, Ind. 47906. Office: U. Ill. at Chgo. Circle, Box 4348, Chgo. 60680.*

SAXL, Erwin Joseph, physicist; b. Vienna, Austria, May 7, 1904; s. Richard and Charlotte Saxl; M.D., Ph.D., U. Vienna; postgrad. U. Berlin (Germany), U. Freiburg/Baden (Germany); m. Lucretia Lawrence Hildreth, Sept. 9, 1944; children—Richard, Ellen, Stanley, Mark. Came to U.S., 1929, naturalized, 1936. Staff, Carnegie Inst., 1929-30; dir. research, v.p. Waypoyset Mfg. Co., 1933-35; pres. Saxl Instrument Co., 1935-53; pres., gen. mgr. Tensitron, Inc., Harvard, Mass., 1953—. Cons. physicist, 1932—. Fellow A.A.A.S.; mem. Am. Wire Assn. (medal 1959), Soc. for Exptl. Stress Analysis, N.Y. Engrs. Club, Soc. Photog. Scientists and Engrs., U.S. Textile Inst. (chmn. com. fibers div. 1934—). Contbg. author: Rayon Handbook, 1935. Research, numerous publs. on tension engring., gravitation and electricity, ultraviolet microscopy; discovered micro-wave reception from interstellar space; developed first power tube for micro-wave transmissions and reception at centimeter wave lengths. Patentee in field, especially measurement, rec. and control of phys. tension. Home: Pin Hill. Office: P.O. Box 185, Harvard, Mass. 01451.*

SAXL, Paul, Austrian physician; b. Vienna, Nov. 2, 1880; 1st to treat typhus with milk injections, 1916; credited with introducing injections of mercurial diuretic (Novasurol) to treat cardiac failure, circa 1920; studied oligodynamic effect of metals. Died Rax Alps, Austria, Mar. 13, 1932.

SAXON, David S(tephen), Am. physicist; b. St. Paul, Feb. 8, 1920; s. Ivan and Rebecca (Moss) S.; B.S., Mass. Inst. Tech., 1941, Ph.D., 1944; m. Shirley Goodman, Jan. 6, 1940; children—Margaret E., Barbara S., Linda C., Catherine L., Victoria J., Charlotte M. Staff mem. Mass. Inst. Tech. Radiation Lab., 1943-46; research physicist Phillips Research Labs., 1946-47; prof. physics U. Cal., Los Angeles, 1947—, chmn. dept. physics, 1963-66, dean phys. scis., 1966—. Cons. Project Sorrento, 1959, Convair Astronautics, 1960-63. Guggenheim fellow, 1956-57, 61-62; Fulbright award, 1961-62. Mem. Am. Phys. Soc., Am. Assn. U. Profs., Am. Assn. Physics Tchrs., Am. Inst. Physics, Sigma X, Sigma. Research in quantum theory; nuclear physics; electromagnetic theory; contbd. to devel. of optical model of atomic nucleus, which accounts for scattering and polarization of neutron and proton beams when they impinge on the nucleus. Home: 1008 Hilts Av., Los Angeles 90024. Office: Dept. Physics, U. Cal., Los Angeles 90024.*

SAXOV, Svend Erik, Danish geophysicist; b. Aalborg, Denmark, Mar. 20, 1913; s. Peter Dusinus and Christine (Christenson) Jensen; M.Sc. in Math., Copenhagen (Denmark) U., 1937, M.Sc. in Phys. Geodesy, 1957; m. Ina Piculell, Oct. 30, 1937; children—Svend-Erik, Karsten Ulrik. Jr. geodesist Geodetic Inst. Denmark, 1933-56, sr. geodesist, 1956-64; prof. geology Aarhus (Denmark) U., 1964—; pres. NOFTIG, 1966—. Spl. lectr. Copenhagen U., 1951-64; geophys. cons., 1960—. Decorated Knight of Dannebrog. Mem. European Assn. Expln. Geophysicists (past pres.), Soc. Expln. Geophysicists, Danish, Norwegian, Swedish geol. socs., Danish, German geophys. socs., A.A.A.S., Danish Geol. Commn. Editor: Geoexploration, 1962—, Geoexploration Monograph, 1965—. Research, publs. on gravimetry especially gravimetric applicability to geol. problems, groundwater and mining geophys. prospecting, petrophys. properties of rocks. Home: 11 Elmevej, Skanderborg 8660, Denmark. Office: 28 Carl Blochsgade, Aarhus, 8000, Denmark.*

SAXTON, Joseph, Am. inventor; b. Huntington, Pa., Mar. 22, 1799; s. James and Hannah (Ashbaugh) S.; m. Mary Abercrombie, 1850, 1 child. Watchmaker, Phila., 1817-28; made clock for belfry of Independence Hall; in Eng., 1828-37; invented magneto-electric machine, 1833, also invented a fountain pen, locomotive differential pulley; constructor and curator of standard weighing apparatus U. S. Mint, Phila., 1837-43; designed standard balance used in govt. assay and coining offices; supt. weights and measures for U. S. Coast Survey, Washington, D.C., 1843-73; invented reflecting pyrometer, hydrometer, fusible metal seal, eversharp pencil; mem. Nat. Acad. Scis., Am. Philos. Soc. Died Oct. 26, 1873.

SAY, Thomas, Am. entomologist, conchologist; b. Phila., June 27, 1787; s. Benjamin and Ann (Bonsall) S.; m. Lucy Way Sistaire, Jan. 4, 1827. Apptd. zoologist to accompany expdn. to Rocky Mountains under Maj. Stephen H. Long, 1819; accompanied Long's 2d expdn. which explored sources of the Minnesota River, 1823; curator Am. Philos. Soc., 1821-27; prof. natural history U. Pa., 1822-28; went to New Harmony, Ind., 1825; fgn. mem. Linnaean Soc. of London; original mem., also bequeathed his library and collections to Phila. Acad. Natural Scis. after his death. Author: American Entomology; or Descriptions of the Insects of North America, 3 vols., 1824, 25, 28; American Conchology, 6 vols., 1830-34; prepared for publ. American Ornithology, or the Natural History of Birds Inhabiting the United States (Charles Bonaparte), 1825; works collected in The complete Writings of Thomas Say on the Conchology of the United States (W. G. Binney), 1858; The Complete Writings of Thomas Say on the Entomology of North America with a biographical memoir by George Ord (edited and pub. by J. L. LeConte), 2 vols., 1859. Called father of descriptive entomology in Am. Died New Harmony, Oct. 10, 1834.

SAYCE, Archibald Henry, English philologist, Orientalist; b. Shirehampton, Eng., Sept. 25, 1845; s. Henry Samuel and Mary Anne (Cartwright) S.; D.Litt., LL.D., D.D., Queen's Coll., Oxford (Eng.) U. Ordained, 1870; fellow Queen's Coll., 1869—, tutor, 1870-79; mem. O. T. Rivision Com., 1874-84; dep. prof. comparative philology Oxford U., 1876-90, prof. Assyriology, 1891-1919. Hon. fellow Brit. Acad.; corr. mem. Inst. de France, 1919. Author: Assyrian Grammar for Comparative Purposes, 1872; The Principles of Comparative Philology, 1874; Elementary Assyrian Grammar, 1874; Translations in Records of the Past, 1874-77; Lectures on the Assyrian Language and Syllabary, 1877; Babylonian Literature, 1877; Introduction to the Science of Language, 1879; The Monuments of the Hittites, 1881; The Inscriptions of Van Deciphered, 1882; Herodotos I-III, 1883; Fresh Light from the Monuments, 1883; The Ancient Empire of the East, 1884; The Inscriptions of Mal-Amir, 1885; Introduction to Ezra, Nehemiah and Esther, 1885; Assyria, 1885; Hibbert Lectures on Babylonian Religion, 1887; The Hittites, 1889; The Races of the Old Testament, 1891; The Higher Criticism and the Verdict of the Monuments, 1894; Patriarchal Palestine, 1895; The Egypt of the Hebrews and Herodotos, 1895; Early History of the Hebrews, 1897; Israel and the Surrounding Nations, 1898; Babylonians and Assyrians, 1900; Genesis in Temple Bible, 1901; Commentary on Tobit, etc., 1903; The Archaeology of Cuneiform Inscriptions, 1907; Reminiscences, 1923. Primarily known as an Assyriologist, contb. important studies of Hittites; deciphered the Cuneiform of Van; was among 1st to appreciate the importance of Schliemann's discoveries at Mycenae and Troy. Died Bath, Eng., Feb. 4, 1933.

SAYERS, James, Brit. physicist; b. Sept. 2, 1912; s. J. Sayers; ed. U. Belfast (Ireland); St. John's Coll.,

Cambridge; M.Sc., Ph.D.; m. Diana Alina Joan Montgomery, 1943; 2 sons, 1 dau. Research staff for Admiralty, U. Birmingham, 1939-43; mem. Brit. group atomic scientists on U. S. Manhattan Project, 1943-45. Brit. del. Internat. Sci. Radio Union, Zurich, Switzerland, 1950. Recipient award Royal Commn. on awards to Invents, 1949. Mem. Royal Soc., Physics Soc. Research in upper atmosphere physics, physics ionized gases; participated in devel. cavity magnetron valve, 1940, devel. of radar. Home: 25 Twatling Rd., Barnt Green, Worcestershire, Eng.

SAYLES, Robert Wilcox, Am. geologist; b. Pawtucket, R.I., Jan. 29, 1878; s. Frederick Clark and Deborah Cook (Wilcox) S.; A.B., Harvard, 1901; m. Adelaide K. Burton, June 1, 1904; children—Deborah W., Robert W. Began geol. work in Mont., 1899; curator Harvard Geol. Mus., 1907-28; research asso. Division of Geology, Harvard, since 1928. Pres. Baltic Mills Co., Baltic Ct. Fellow Geol. Soc. Am., A.A.A.S., Am. Acad. Arts and Scis., Am. Geog. Soc.; mem. Seismol. Soc. Am., Am. Meteorol. Soc., Boston Soc. Natural History, Geol. Soc. Boston (pres. 1928-29), Washington Acad. Scis., Conf. Geophys. Union. Author of monographs and papers on glacial geology and seismology. Research work on origin of Bermuda Islands, on seasonal banding in rocks and on glacial geology of So. Maine and Cape Cod; discovered 1st Permian tillite to be identified in U. S., in Mass., 1908, thus proving glaciers had existed in New Eng. in Permian period. Died Oct. 23, 1952.

SAYLOR, Charles (Hamilton) Proffer, Am. chemist; b. Camden, N.J., June 14, 1901; s. Harvey Reigner and Annabella (Young) S.; student Pratt Inst., 1918-20; B.Chemistry, Cornell U., 1923, Ph.D., 1928; m. Zella Annis Proffer, June 15, 1930; 1 son, Dwight P. With De Laval Separator Co., 1920-21, Pub. Service Commn. N.Y., 1923-24, Div. Trade Waste, Cleve., 1928-30; chemist Nat. Bur. Standards, Washington, 1931-50, chief pure substances sect., 1950-60, cons. analytical chemistry div., 1960—. Mem. commn. on physico-chem. data and standards Internat. Union Pure and Applied Chemistry, 1957—; mem. com. on physico-chem. standards NRC, 1959-63, President's Com. on Prevention Polution Seas by Oil, 1960-64. Recipient certificate of merit OSRD, 1946, Silver medal Dept. Commerce, 1961. Mem. Am. Chem. Soc., A.A.A.S., Optical Soc. Am., Am. Crystallographic Assn., Am. Soc. for Testing and Materials (com. on microscopy), Washington Acad. Sci., Philos. Soc. Washington, Sigma Xi. Research and publs. on crystal form as influenced by chem. environment during growth; co-discoverer of secondary elongation upon crystallization of oriented elastic materials, color phase-contrast microscopy, developer methods of accurate measurement with microscope, refractive index, crystal orientation, temperature, size, methods for determining purity of very pure substances, chem. micros. analysis; inventor freezing staircase method of ultrapurification, epicycloidal phonograph pick-up. Home: 10001 Riggs Rd., Adelphi, Md. 20783. Office: Nat. Bureau of Standards, Washington 20234.*

SAYLOR, John Henry, Am. chemist; b. Lamar, Mo., July 22, 1904; s. James Clyde and Mary (Shackelford) S.; student S.W. Mo. State Coll., 1921-23; A.B., So. Meth. U., 1925; A.M., Duke, 1928, Ph.D., 1930; m. Lettie McLane, Apr. 3, 1926; children—Letty Lois (Mrs. H. Julian Lewis), John Henry. Faculty, Duke, Durham, N.C., 1927—, prof. chemistry, 1946—, exec. officer dept., 1948-51, dir. undergrad. studies, 1946-54, chmn. chemistry dept., 1954—. Sci. adviser Office Ordnance Research, U.S. Army, 1951-61, Army Research Office, Durham, 1961. Mem. Am., Royal Dutch chem. socs., Am. Assn. U. Profs., A.A.A.S., Sigma Xi, Phi Lambda Upsilon. Author: (with Hill, Vosburg, Wilson) Elementary Chemistry, 1937; also articles. Research in solubility of non-electrolytes, polarography of organic compounds, spectrophotometry and fluorometry of chelates. Home: 2500 Perkins Rd., Durham, N.C. 27706.*

SAYRE, Lewis Albert, Am. physician; b. Battle Hill (now Madison), N.J., Feb. 29, 1820; grad. Transylvania U., 1839; Coll. Phys. and Surg., N.Y., 1842; m. Eliza Ann Hall, Jan. 25, 1849. Prosector in surgery Coll. Phys. and Surg., 1842-52; surgeon Bellevue Hosp., 1853-73, Charity Hosp., Blackwell's Island, 1859-73; cons. surgeon to both, 1873—; prof. orthopaedic surgery, later of clin. surgery Bellevue Hosp. Med. Coll.; emeritus prof. orthopaedic surgery Univ. and Bellevue Hosp. Med. Coll.; resident physician City of N.Y., 1860-66; lectured in hosps. in Gt. Britain, 1877. Author: Practical Manual of the Treatment of Club-Foot; Lectures on Orthopaedic Surgery; Spinal Curvature and Its Treatment. Leading Am. orthopedic surgeon; credited with introducing plaster of paris bandages, in jacket form, in treatment of Pott's disease, scoliosis, other spinal column diseases, circa 1877; 1st Am. surgeon to operate successfully for hip joint disease; introduced new methods of treatment in various diseases. Died 1900.

SAYRE, Mortimer Freeman, Am. mech. engr.; b. Newark, July 14, 1885; s. Joseph M. and Ella G. (Browne) S.; M.E., Columbia, 1907, A.M., 1911; m. Grace S. McKinney, Sept. 25, 1912; children—Harrison S., Penelope C. (Mrs. Edward S. Setchko). Engaged in engring. work in N.Y., Pa., Ariz., 1907-14; instr. Union Coll., 1914-18, asst. prof., 1918-25, asso. prof. 1925-37, prof. applied mechanics, 1937-52, dir. extension, 1945-49, head dept. mech. engring., 1952-55, now emeritus prof.; dir. Mech. Tech., Inc.; cons. engr. Alco Products Corp., 1940-63. Pres., Schenectady Bur. Municipal Research, 1931-35, 39-47. Fellow Am. Soc. M.E.; mem. Am. Inst. Mining Metall. and Petroleum Engrs., Am. Soc. Metals, Am. Soc. Engring. Edn., Am. Soc. Testing Materials, Am. Assn. U. Profs., Sigma Xi, Tau Beta Pi. Research and publs. on elastic and pseudo-elastic behavior in metals, theory of mech. springs, creep in aluminum and magnesium alloys, transient stresses in thermonuclear and other power plant equipment, low cycle fatigue failures; designed accurate spring scales; proved aluminum alloys differ in resistance to sea water. Home: 1169 Parkwood Blvd., Schenectady 12308.*

SCADDING, John Guyett, English physician; b. London, Eng., Aug. 30, 1907; s. John William and Jessima (Guyett) S.; M.B., B.S., U. London, 1930, M.D., 1932; m. Mabel Pennington, Aug. 30, 1940; children—Jessica Jane, John William, Sarah Elizabeth. Faculty, Postgrad. Med. Sch., London, 1935—, sr. lectr., 1946-63; physician Brompton Hosp., London, 1939—; prof. medicine Inst. Diseases of Chest, Brompton, London, 1963—. Mem. Assn. Physicians, Med. Research Soc., Brit. Tb Assn. (pres. 1959-60), Thoracic Soc. Author: Sarcoidosis, 1967; also articles. Research on pneumonias, including those asso. with viral infections, therapy of Tb, sarcoidosis, pulmonary fibrosis, med. semantics and logic. Home: 19 South Sq., London N.W. 11. Office: Inst. Diseases of Chest, Brompton, London S.W. 3, Eng.*

SCAFF, Alvin Hewitt, Am. sociologist; b. Dallas, Dec. 27, 1914; s. Alvin Gillam and Ina Vera (Phipps) S.; B.A. with highest honors, U. Tex., 1936; B.D., Chgo. Theol. Sem., 1940; M.A. (grad. fellow 1946), U. Tex., 1946, Ph.D., 1949; m. Marilee Kone, June 17, 1938; children—Lawrence Alvin, Charles Guy, Marilyn. Ordained to ministry Conglist. Ch., 1940; missionary Silliman U., Philippines, 1941-45; instr. sociology U. Tex., 1946; mem. faculty Pomona Coll., 1947-66, prof. sociology, chmn. dept. sociology and anthropology, 1956-66, Henry Snyder prof., 1960-66; asso. dean Grad. Coll., U. Iowa, Iowa City, 1966—; chief social research sect. UN Econ. Commn. Africa, 1960-61; lectr. Africa and Ethiopia, Peace Corps, Washington, 1962. Pres. Claremont Coordinating Council, 1955-56, Claremont Civic Housing Assn., 1950-52, Claremont Community Welfare Assn., 1952-53; co-chmn. Claremont Am. Field Service, 1962-63; head of United Nations team United Nations Technical Assistance Board, Kampala, Uganda, 1963-64; member Commn. Review Voting Procedures in Los Angeles County, 1959-60, Claremont Planning Commn., 1957-59. Fulbright research grantee, Philippines, 1953-54; Ford Found. pub. affairs research grantee, 1957, 58, 62. Mem. Am., Pacific (v.p.), Philippine sociol. assns., Soc. Study Social Problems, Phi Beta Kappa. Author: (with others) Our Needy Aged, 1954, Recommendations for Urban Development in Kampala and Mengo, 1964; The Philippine Answer to Communism, 1955. Editor Sociol. Inquiry, 1957-60. Research on identification of social factors of econ. devel., revolutionary movements. Home: 421 Lee St., Iowa City 52240.*

SCALIGER, Joseph Justus, scholar; b. Agen, France, Aug. 5, 1540; s. Julius Caesar Scaliger; ed. Paris; prof. Geneva Acad.; lived in France 20 years, then prof., Leiden, Netherlands. Laid found. for modern chronology through study of ancient chronology; devised 1st completely sci. time system, based on cycle of 7980 years (named Julian Period after his father). Died Leiden, Jan. 21, 1609.

SCALIGER, Julius Caesar, physician; b. Apr. 23, 1484; probably s. Benedetto Bordone; probably studied medicine, Padua, Italy; at least 3 sons, including Joseph Justus Scaliger; became French citizen, 1528; settled in Agen, France. Author: De subtilitate ad Hieronymum Cardanum, Exotericarum exercitationum libri XV (includes attack on Cardan's theory of motion), 1557; also works on Latin cases, on Theophrastus, Aristotle, Hippocrates, an attack on Erasmus; proposed classification of plants by unique characteristics. Died Oct. 21, 1558.

SCANLON, Jane Cronin, Am. mathematician, educator; b. N.Y.C., July 17, 1922; d. John Timothy and Janet (Murphy) Cronin; B.S., Wayne State U., 1943; M.A., U. Mich.; 1945, Ph.D., U. Mich., 1949; m. Joseph C. Scanlon, Mar. 5, 1953; children—Justin, Mary, Emer, Edmund. Mathematician, USAF Cambridge (Mass)Research Center, 1951-54; instr. math. Wheaton Coll., Norton, Mass., 1954-55; faculty math. Rutgers, the State U., New Brunswick, N.J., 1965—. Mem. Am. Math. Soc., Soc. Indsl. and Applied Math., Phi Beta Kappa. Author: Fixed Point and Topological Degree in Nonlinear Analysis, 1964; Advanced Calculus, 1967. Research, publs. in nonlinear differential equations, topology celestial mechanics. Home: 846 Monroe Av., Elizabeth, N.J. 07201. Office: Dept. Math., Rutgers, The State U., New Brunswick, N.J. 08903.*

SCARAMUZZI, Giovanni, Italian plant pathologist; b. Fidenza, Jan. 19, 1922; s. Donato and Alberta (Rovida); ed. engring. agronomy; m. Gianna Comelli, Sept. 3, 1953. Dir. Plant Pathology Inst., also Faculty Agronomy, U. Catana. Decorated Cross of Merit, World War II. Publns. on plant pathol. research. Address: via Valdisavoia 5, Catana, Italy.

SCARBOROUGH, James Blaine, Am. mathematician; b. Mt. Gilead, N.C., June 22, 1885; s. Isham Wilson and Jane (Haywood) S.; A.B., U. N.C., Chapel Hill, 1913, A.M., 1914; Ph.D., Johns Hopkins, 1923; m. Lessie Neville, June 30, 1915 (dec.), children—Lucile Elizabeth, James Blaine (dec.) Ernest Neville; m. 2d, Julia Kauffman, Aug. 18, 1930; 1 son, William Kauffman. Instr. math. N.C. State Coll., Raleigh, 1914-18; instr. math. U. S. Naval Acad., Annapolis, Md., 1918-21, asst. prof. math., 1921-28, asso. prof., 1928-37, prof. 1937-50, prof. emeritus, 1950—. Cons. in numerical analysis Naval Ordnance Lab.; sr. cons. Trident Engring. Assos., 1961—. Fellow A.A.A.S.; mem. Am. Math. Soc., Math. Assn. Am., Soc. for Indsl. and Applied Math., Phi Beta Kappa. Recipient Cain Math. medal U. N.C., 1912. Author: Numerical Mathematical Analysis, 1930, 50, 55, 58, 62; The Gyroscope: Theory and Applications, 1958; Differential Equations and Applications, 1965. Co-author: Fundamentals of Statistics, 1948. Research and publs. on numerical analysis, statistics, applied math., ballistics, gyrodynamics. Home: Ferry Farms, Annapolis, Md.*

SCARBURGH, Sir Charles, English physician, mathematician; b. London, 1616; s. Edmond Scarburgh; B.A., Caius Coll., Cambridge, 1637, M.A., 1640; M.D., Oxford, 1646; studied math under William Oughtred; at least 1 son, Charles. Elected anat. reader Barber-Surgeons' Coll., 1649; while gov. Christ's Hosp., founded (with Christopher Wren, Jonas Moor) math. sch., 1673; physician to Charles II after Restoration, also to James II, Queen Mary, Prince George of Denmark; mem. Parliament for Camelford, Cornwall, 1685-87. Fellow Coll. Physicians (censor 1655, 64, 65, lectr. 1656), Royal Soc., 1663 (original fellow). Author: Syllabus musculorum (guide to human dissection), also elegy on Cowley; left materials for English edit. of Euclid pub. by his son. Worked (with William Harvey) on generation of animals; furthered math. studies in Eng. Died Feb. 26, 1693/94.

SCARDIGLI, Gualfredo, Italian physician; b. Firenze, Italy, Nov. 24, 1922; s. Ermanno and Gisa (Calvani) S.; student U. Rome (Italy), 1941-47; m. Maria Teresa Ortalli, Apr. 3, 1952; 1 dau., Barbara. Faculty, U. Florence (Italy), 1948—, univ. tchr. climatology, 1955-57, clin. medicine, 1957-59, gerontology and geriatrics, 1959—, head dietetics, 1959—; dir. Instituto Geriatrico Comunale de Firenze, 1957-60. Mem. Societa Italiana di Gerontologia e Geriatria (gen. sec. 1951—), Internat. Assn. Gerontology (treas. 1957-60), Associazione Medici Geriatri Italiani (dir. 1965—), Soc. Italiana Medicina Interna, Soc. Italiana di Ematologia, Am. Gerontological Soc., Med. Idrology. Research, numerous publs. in hemocoagulation, dietetics, geriatrics, metabolism, thrombosis, anticoagulant therapy. Home: Via B., Varchi 38, Firenze, Italy.*

SCARFF, John Edwin, Am. neuro-surgeon; b. Bellefontaine, O., Mar. 19, 1898; s. Edwin Curl and Margaret Gorton (Riddle) S.; student Ohio-Wesleyan U., Delaware, O., 1916-18; B.S., Princeton U., 1920; M.D., Johns Hopkins, 1924; D.Sc., U. Brazil, 1953; m. Ellen Backland, Apr. 14, 1934; children—John Edwin, Timothy Backland. Staff, Neurol. Inst. N.Y., 1937—, attending neurosurgeon, 1947-63, cons., 1963—; faculty Coll. Phys. and Surg. Columbia U., N.Y.C., 1937—, prof. clin neurol. surgery, 1947-63, emeritus prof., 1963—; became sr. cons neurosurgery VA Hosp., Bronx, N.Y., 1946, also founder dir. residency tng. program in neurosurgery. Diplomate Am. Bd. Neurol. Surgery. Fellow A.C.S.; mem. Harvey Cushing Soc., Soc. Neurol. Surgeons, Soc. Brit. Neurol. Surgeons, Am. Electroencephalographic Soc., med. cons. World War II, several fgn. socs., A.M.A., N.Y. Acad. Medicine, N.Y. Neurol. Soc., N.Y. Acad. Scis., Am. Neurol. Assn., Assn. Research in Nervous and Mental Diseases, Am Neurol. Assn., Phi Delta Theta, Phi Beta Kappa, Alpha Omega Alpha, Key and Seal. Contbr. chapters to Pediatric Surgery by E. C. Brenner, 1938; chapters to Treatment of Cancer and Allied Diseases by G. T. Pack, 1939. Military Surgical Manual, 1942; Management of Battle Injuries of Brain, Spinal cord and Peripheral Nerves; (with L. Davis) The Causes and Treatment of High Altitude Frost Bite. Research on expl. prodn. of pulmonary abscess, primary cortical motor areas of human brain, treatment of cystic craniopharyngiomas by intra-ventricular drainage; developed technique for Aseptic end-to-end anastomosis of intestines; originated (with Rahm) 1st pre-set calibrated thyrotron for controlled cortical stimulation; pioneered electro-corticography in man; originated (with Stookey) puncture of lamina terminalis for treatment of obstructive hydrocephalus. Home: 761 W. 231 St., N.Y.C. 10463. Office: Neurol. Inst. N.Y., 710 W. 168th St., N.Y.C. 10032.*

SCARPA, Antonio, Italian anatomist, surgeon; b. Motta, Italy, June 13, 1747; student of Morgagni, Padua, Italy; became prof. anatomy, Modena, Italy, 1772; apptd. prof. anatomy and surgery, chmn. anatomy U. Pavia (Italy), 1783, lost post upon found. of Cisalpine Republic, reinstated by Napoleon, 1805; became dir. faculty medicine, Pavia, 1814, resigned when his reforms were not carried out. Mem. French Acad. Scis. Fellow Royal Soc., 1791. Author: Anatomicae disquisitiones de auditu et olfactu, 1779; Tabulae neurologicae, 1794; Saggio di osservazioni sulle principali malattie delgi occhi, 1801; Sull'erni, 1809; Opere, 2 vols., 1836-39. Credited with discovering membranous labyrinth of ear, circa 1772; described ganglion of vestibular nerve at junction with facial (Scarpa's ganglion), 1779; gave 1st good description of labyrinth of ear, 1789; described 2 foramina behind middle incisors, transmitting nasopalatine nerves (Scarpa's foramen of or foramina), 1799; credited with 1st full description of morbid anatomy of clubfoot, 1803; described femoral triangle formed by inguinal ligament, sartorius muscle, adductor longus muscle (Scrapa's triangle), 1809, part of deep layer of superficial fascia of anterior abdominal wall (Scarpa's fascia), 1809; carried out research on aneurism, sense of smell. Died Bonasco, nr. Pavia, Oct. 31, 1832.

SCATCHARD, George, Am. chemist; b. Oneonta, N.Y., Mar. 19, 1892; s. Elmer Ellsworth and Fanny (Harmer) S.; B.A., Amherst Coll., 1913, Sc.D., 1948; Ph.D. (Goldschmidt fellow), Columbia, 1917; m. Willian Watson Beaumont, July 28, 1928. Asso. prof. Amherst Coll., 1919-23; NRC fellow Mass. Inst. Chemistry, 1923-24; faculty Mass. Inst. Tech., Cambridge, 1924—, prof. phys. chemistry, 1937-57, emeritus prof., hon. lectr., 1957—. Cons. Oak Ridge Nat. Lab., 1950—, Brookhaven Nat. Lab., Upton, N.Y., 1954-55, Los Alamos Sci. Lab., 1955-58. Mem. U.S. Nat. Acad. Scis., Am. Acad. Arts and Scis., A.A.A.S., Am. Chem. Soc. (T.W. Richards medal 1954, Kendall Co. award in colloid chemistry 1962), Biophys. Soc., Am. Soc. U. Prof., Phi Beta Kappa, Sigma Xi, Phi Lambda Upsilon. Research, numerous publs. on phys. properties solutions, from simple hydrocarbons, metals or solutions alkali halides in water to ion-exchange resinds and proteins. Home: 984 Memorial Dr., Cambridge, Mass. 02138.*

SCELLINCK, Thomas, Flemish surgeon; flourished 1st half 14th century. Practiced medicine, circa 1317. Author: Boeck Van Surgien (important surg. treatise), 1343.

SCHAAF, S(amuel) A(lbert), Am. mathematician; b. Fort Wayne, Ind., Jan. 26, 1918; s. Albert H. and Bertha (Hart) S.; A.B., U. Cal. at Berkeley, 1939, Ph.D. in Math., 1944; m. Phyllis Burleson, Dec. 2, 1943. Lectr. math. N.Y. U., 1944-47; asst. prof., mech. engr. U. Cal., Berkeley, 1947-51, dir. Rarefied Gas Dynamics Research Lab., 1951—, prof. engring. sci., 1957—, chmn. aeronautics, 1957-60, chmn. mech. engring, 1960—. Cons. Rand Corp. Santa Monica, Cal., 1955—, NSF, 1960—, Office Naval Research, 1950—, Air Force Office Sci. Research, 1955—, various corps.; mem. sci. adv. bd. Vidya-Itek Corp., 1959-65; mem. research adv. com. on fluid mechanics NASA, 1957-62. Fellow Am. Phys. Soc. (chmn. fluid dynamics div. 1956); mem. Am. Math. Soc., Engring. Sci. Soc., Phi Beta Kappa, Tau Beta Pi, Lambda Chi Alpha. Author: Flow of Rarefied Gases, 1961. Bd. editors Physics of Fluids, 1957-65. Research, publs., on fluid mechanics, combustion, heat transfer and applied math. Home: 2 Greenwood Common, Berkeley, Cal. 94708.*

SCHAAFFAUSE, Hermann, German anthropologist; b. Coblenz, Germany, July 18, 1816; prof., Bonn; 1st to describe Neanderthal skull correctly as belonging to earlier human race (in opposition to Virchow), 1858. Died Bonn, Jan. 26, 1893.

SCHAAFFS, Werner Heinrich Albert, German physicist; b. Münster, Westfalen, Germany, Sept. 6, 1910; s. Albert Hermann and Olly (Schumacher) S.; student math., physics, phys. chemistry U. Tübingen (Germany), 1929, U. Munich (Germany), 1929-30, U. Halle (Germany), 1932; Ph.D., U. Göttingen (Germany), 1936; m. Edith Pflughaupt, July 17, 1937; children—Johannes, Christian Werner, Elisabeth (all dec.). Research staff firm Siemens, 1936-64; faculty Technische U. Berlin, 1950—, prof. exptl. physics, 1957-64, prof., sci. councillor, 1964—. Mem. Phyikalische Gesellschaft, Chemische Gesellschaft, Kolloid Gesellschaft, Gesellschaft Deutscher Naturforscher und Arzte. Author: Molekularakustik, 1963; Christus und die physikalische Forschung, 1966; also numerous articles. Research on relationships between acoustics and atomic physics, X-ray flashes of physics of highest pressures; generation and interpretation of rhythmic structures in colloid chemistry and biology. Home: Im Heidewinkel 3, Berlin 13, Germany.*

SCHAAP, Ward Beecher, Am. chemist; b. Holland, Mich., Sept. 15, 1923; s. John C. and Gertrude (Hemkis) S.; B.S., Wheaton Coll., 1944; M.S., U. Ill., 1948, Ph.D., 1950; m. Mary Womack, Oct. 28, 1944; children—Jeanne, Shelley, Julie. Chemist, Manhattan Project, U. Chgo., 1944, Tenn. Eastman Corp., 1944-45, Carbide and Carbon Chem. Corp., 1946-47; faculty dept. chemistry Ind. U., Bloomington, 1950—, prof., 1963—, asso. dean Coll. Arts and Scis., 1966—. Cons. Union Carbide Nuclear Corp., 1954-60; cons. explosives dept. E.I. du Pont de Nemours & Co., Inc., 1957-60. Recipient Ulysses S. Weatherly award for distinguished teaching Ind. U., 1964. Mem. Am. Chem. Soc., Sigma Xi, Phi Lambda Upsilon, Alpha Chi Sigma. Research, publs. on electrochemistry in nonaqueous solvents of low dielectric constant, thermodynamics and kinetics of reactions of coordination compounds, ligand field theory, applications of radioisotopes in plarography, improvements in polarographic instrumentation. Home: 819 S. Jordan St., Bloomington, Ind. 47401.*

SCHACHMAN, Howard Kapnek, Am. biochemist; b. Phila., Dec. 5, 1918; s. Morris H. and Rose (Kapnek) S.; B.S., Mass. Inst. Tech., 1939; Ph.D., Princeton, 1948; m. Ethel Lazarus, Oct. 20, 1945; children—Marc, David. Research asst. Rockefeller Inst., Princeton, N.J., 1947-48; faculty U. Cal. at Berkeley, 1948—, prof. biochemistry and molecular biology, 1959—; instr. physiology Marine Biol. Lab., Woods Hole, Mass., 1956-61. Recipient John Scott award City Phila., 1964; Warren Triennial prize Mass. Gen. Hosp., 1965; Guggenheim Found. fellow Washington U. Sch. Medicine, 1957-58. Mem. Nat. Acad. Scis., Am. Chem. Soc. (Cal. sect. award 1958, E. H. Sargent & Co. award for chem. instrumentation, 1962), Am. Soc. Biol. Chemists, A.A.A.S., Am. Assn. U. Profs., Am. Acad. Arts and Scis., Sigma Xi. Author: Ultracentrifugation in Biochemistry, 1959; also numerous articles. Research on structure, function, and interaction macromolecules biol. interest; invented cells, techniques and optical systems for ultracentrifuge and its application to study biol. problems. Home: 7705 Ricardo Ct., El Cerrito, Cal. 94530. Office: Molecular Biology-Virus Lab., U. Cal., Berkeley, Cal. 94720.*

SCHACHTEL, Ernest George, psychologist; b. Berlin, Germany, June 26, 1903; s. Franz Jacob and Flora (Isaacsohn) S.; J.D., U. Heidelberg, 1925; m. Zeborah Suesholtz, Jan. 11, 1952. Came to U.S., 1935, naturalized, 1943. Research asso. Inst. Social Research, Columbia, 1935-38; pvt. practice psychoanalysis and diagnostic testing, 1936—; lectr. New Sch. for Social Research, 1943-58; tng. and supervising analyst William Alanson White Inst. Psychiatry, Psychoanalysis and Psychology, N.Y.C., 1958—; adj. prof. psychology N.Y. U., 1961—. Recipient Helen Sargent award for outstanding contbn. to clin. psychology Menninger Found., 1967. Fellow Am. Psychol. Assn., Soc. for Projective Techniques and Personality Assessment, A.A.A.S.; mem. W.A. White Psychoanalytic Soc. Author: Metamorphosis: On the Development of Affect, Perception, Attention and Memory, 1959; Experiential Foundations of Rorschach's Test, 1966. Research, publs. on memory perception, attention, affect, repression, Rorschach's test. Address: 299 Riverside Dr., N.Y.C. 10025.*

SCHACHTER, Mendel, pedopsychiatrist; b. Zalutza, Rumania, July 13, 1903; s. Israel and Myriam (Liebster) S.; M.D., Med. Faculty of Nancy (France), 1930; m. Denise Nedler, Dec. 1932. Former asst. Faculty of Medicine Marseilles (France); now chief-physician Comité de l'Enfance Déficiente de Marseille. Decorated chevalier Ordre des Palmes Academiques. Mem. Soc. Médico-psychologique Paris, Internat. Soc. Rorschach, Soc. Psychologie de France, Soc. Psychiatry du Sud-Est, Royal Soc. Medicine (London, Eng. affiliate). Author: (with S. Cotte) L'Enfant Enurétique, 1941; El Mongolisma, 1943; (with A. Cremieux, S. Cotte) L'Enfant devenu Délinquant, 1945; (with L. Cornil, J. Vague) Les Maigreurs, 1945. Research, publs. on mongoloid idiocy, problems of infantile encephalopathias, enuresia in children, prematurity and mental devel., Rorschach study in pedopsychiatry. and in gen. adult psychiatry. Address: 40a Bd. Voltaire, Marseille, France.*

SCHADÉ, Johannes Petrus, Dutch neurobiologist, physician; b. Amsterdam, Netherlands, May 12, 1931; s. Johannes P. and Johanna M. (Bos) S.; B.Sc., U. Amsterdam, 1950, M.D., 1955, Ph.D., 1956; m. Johann Cornelia Zwaan, Aug. 17, 1957; children—Marcus Cornelis, Annemarie Cornelia, Marjolein Janine. Instr. neuroanatomy Central Inst. for Brain Research, Amsterdam, 1954-56, asso. neurobiology, 1956-57, head dept. neurophysiology, 1957, asso. dir., 1960—; vis. prof. comparative neurology U. Mich., 1957, Cal. Inst. Tech., 1957-59; vis. prof. neurphysiology U. Rome (Italy), U. Naples (Italy), 1964—; Ann. Cajal Meml. lectr., 1957. Mem. adv. panel radiation research Netherland Govt. Recipient A. E. Bennett award, U. S., 1959. Mem. Internat. Brain Research Orgn., Am. Soc. Biol. Psychiatry, Internat. Orgn. Neurobiologists, Netherlands Orgn. Anatomy, Physiology and Neurology. Author: Basic Neurology; The Peripheral Nervous System; Atlas of the Human Brain; also numerous articles. Analysis of developmental patterns in mammalian brain; devel. new methods for quantitative analysis of brain, teaching aids for neurol. scis. Home: Gooilustraat 37, Amsterdam-N. Office: Ijdijk 28, Havens-Oost, Amsterdam, Netherlands.*

SCHAEBERLE, John Martin, astronomer; b. Württemburg, Germany, 1853; s. Anton and C. Catherine (Voegele) S.; came to Ann Arbor, Mich., 1854; apprentice in Chgo. machine shop, 1868-71; studied at Ann Arbor High Sch.; C.E., U. Mich., 1876; LL.D., U. Cal., 1898. Pvt. asst. to Prof. Watson, 1876-78; asst. Ann Arbor Obs., instr. in astronomy, acting prof. of astronomy U. Mich., 1878-88; astronomer Lick Obs., Mt. Hamilton, Cal., 1888-97, acting dir., 1897-98; in charge of eclipse expdns. of Lick Obs., 1889, 93, Cayenne and Chile, and in 1896 to Japan. Contbr. extensive to astron. jours. Discovered 3 comets; constructed long-focus telescope camera. Died Sept. 17, 1924.

SCHAEFER, Hermann Joseph, biophysicist; b. Krefeld, West Germany, Aug. 2, 1905; s. Joseph H. and Maria (Camp) S.; Ph.D. summa cum laude, U. Frankfurt Main, Germany, 1929; m. Emily L. Schmitz, July 21, 1937. Came to U.S., 1948, naturalized, 1955. Asst. prof. U. Frankfurt Inst. Biophysics, 1924-30; head electromed. devel. lab. Koch & Sterzel, Inc., Dresden, Germany, 1930-32; first asst. Max Planck Inst. for Biophysics, Frankfurt, 1932-37, lectr., 1937-44, asst. prof., 1944-48; chief phys. scis. div. Naval Aerospace Med. Inst., Pensacola, Fla., 1948—; vis. prof. pathology Emory U., 1961—; lectr. pathology Georgetown U. Sch. Medicine, 1962—. Mem. Radiation Research Soc., Aerospace Med. Assn. (Arnold D. Tuttle award), Am. Inst. Aeros. and Astronautics Assn., Am. Geophys. Union, Health Physics, A.A.A.S., Internat. Acad. Astronautics. Contbr. numerous articles in field to sci. jours. Demonstration of inhaled hot particles in human lung tissue, research on biol. significance of background ionization and radiation in space, radiation measurements on manned space missions. Home: 525 S. 1st St., Warrington, Fla. 32507. Office: Naval Aerospace Med. Inst., Pensacola, Fla. 32512.*

SCHAEFER, Milner Baily, Am. oceanographer; b. Cheyenne, Wyo., Dec. 14, 1912; s. Heinrich Gottlieb and Kate Rosse (Baily) S.; B.S. magna cum laude, U. Wash., 1935; Ph.D., 1950; m. Isabella Long, May 25, 1949; children—Kate Baily, Kurt Milner, Patrick Joseph. Sci. asst. Internat. Fisheries Commn., 1934-35; asst. biologist, then biologist Wash. Dept. Fisheries, 1935-39; scientist Internat. Pacific Salmon Fisheries Commn., 1939-42; instr. Sch. Fisheries, U. Wash., 1946; with U. S. Fish and Wildlife Service, 1946-50, chief research and devel. Pacific Oceanic fishery investigation, Honolulu, 1948-50; dir. investigations Inter-Am. Tropical Tuna Commn., La Jolla, Cal., 1951-63, sci. cons., 1963—; mem. staff Scripps Instn. Oceanography, La Jolla, 1951—, prof. oceanography, dir. Inst. Marine Resources, 1962—; sci. adviser U. S. Dept. Interior, 1967—. Mem. com. effects atomic radiation oceanography and fisheries, Nat. Acad. Scis.-NRC, 1956-63, com. oceanography, 1957-68, chmn. 1964-67, mem. Latin Am. sci. bd., 1963—, sci. adv. com. marine protein resources devel., 1963—; expert fisheries, secretariat Internat. Conf. Law of Sea, Geneva, Switzerland, 1958; cons. spl. fund UN, 1960-65; chmn. standing com. marine sci. Pacific Sci. Assn., 1962-66; mem. expert panel tuna research FAO, 1964—, chmn. 1964-66, mem. IWP working group marine resources appraisal, 1966—; mem. Gov. Cal. Adv. Council Marine Resources, 1965—, chmn. 1965-66; adv. com. fisheries and oceanography State Dept., 1965—; adv. com. marine resources devel. Dept. Interior, 1967; cons. Nat. Council Marine Resources and Engring. Devel., 1967—. Served as officer USNR, 1942-46. Recipient Diploma de Reconocimiento (Costa Rica), 1967. Founding fellow Am. Inst. Fishery Research Biologists; fellow Cal. Acad. Scis.; mem. Am. Soc. Ichthyologists and Herpetologists, Nat. Oceanography Assn. (bd. dirs. 1966-67), Pacific Fishery Biologists (pres. 1939-40), Am. Fisheries Soc., Am. Geophys. Union, Am. Statis. Assn., Biometrics Soc., Am. Soc. Limnology and Oceanography (pres. Western div. 1956-57), Marine Tech. Soc., Phi Beta Kappa, Sigma Xi. Research in biological oceanography; population dynamics and marine ecology; dynamics of exploited animal populations. Home: 3337 Elliott St., San Diego 92106. Office: Office of Science Adviser, Dept. Interior, Washington 20240.*

SCHAEFER, Vincent Joseph, Am. chemist, meteorologist; b. Schenectady, N.Y., July 4, 1906; s. Peter A. and Rose A. (Holtslag) S.; student Union Coll.; grad. Davey Inst. of Tree Surgery, 1928; Sc.D. (honorary), University of Notre Dame, 1948; m. Lois K. Perret, July 26, 1935; children—Susan, Katherine Rose, James Michael. Grad. Gen. Elec. Apprentice System, 1926; research asst. to Dr. Irving Langmuir, Gen. Elec. Research Lab., 1933-38, research asso., 1938-54; dir. research Munitalp Found., Inc., 1954-59, cons. in N.Am., 1959-64; prof. State U. N.Y. at Albany, also dir. Atmospheric Scis. Research Center; dir. research Nat. Scis. Institute; research asso. Mus. No. Ariz.; cons. Gen. Electric Co., and other manufacturing companies; collaborator U. S. Forest Service. Fellow Woods Hole Oceanographic Institute, 1959-60; distinguished lecturer sci. State U. N.Y. Coll. Edn., Albany, 1959-60; Fulbright lecturer, Australia, 1960-61. Member mus. council State Univ. New York; mem. fire research council NRC. Trustee Schenectady Museum, Mohawk-Caughawaga Mus. Recipient Losey award, Institute Aero. Scis., 1953, award American Meteorol. Society, 1957. Fellow A.A.A.S., Rochester Museum Arts and Scis., Am. Meteorol. Soc., Am. Geophys. Union; mem. Ecol. Soc., Am. Chem. Soc., N.Y. Archeol. Assn., Schenectady Co. Hist. Soc., A.A.A.S., Am. Meteorol. Soc., Wilderness Soc., Friends of Forest Preserve, Royal Meteorol. Soc., Nat. Speleol. Soc., Soc. Am. Archeology, Sigma Xi. Discoverer of cloud seeding methods. Research in experimental meteorology, on precipitation, surface chemistry of monolayers, thin films and small crystals, atmospheric physics, and on time lapse photography, sci. edn. Contbr. sci. jours. Home: R.F.D. 3, Schermerhorn Rd., Schenectady, N.Y.*

SCHAEFFER, Bobb, Am. vertebrate paleontologist; b. New Haven, Sept. 27, 1913; s. Jacob Parsons and Mary Mabel (Bobb) S.; B.A., Cornell U., 1936; M.A. in Zoology, Columbia, 1937, Ph.D. in Zoology, 1941; m. Elizabeth Chapin White, Oct. 11, 1941; children

—Elizabeth Bobb, Richard White. Demonstrator histology and embryology Jefferson Med. Coll., 1941-42; asst. curator vertebrate paleontology Am. Mus. Natural History, N.Y.C., 1946-49, asso. curator, 1949-55, curator, 1955—, chmn. dept. vertebrate paleontology, 1967—; prof. zoology Columbia, 1959—, field work in Western U. S. Adv. panel NSF, 1961-64. Fellow Geol. Soc. Am., A.A.A.S.; mem. Am. Soc. Zoologists, Soc. Study Evolution, Soc. Vertebrate Paleontology (past pres.), Paleontol. Soc., Am. Soc. Ichthyologists and Herpetologists, Soc. Systemic Zoology. Research and publs. on morphology, functional anatomy and evolution of fishes, both extinct and living types, evolutionary problems, especially in relation of maj. organizational levels. Home: 174 Elm St., Tenafly, N.J. 07670. Office: Am. Mus. Natural History, Central Park West at 79th St., N.Y.C. 10024.*

SCHAEFFER, Clemens, German physicist; b. Remscheid, Germany, Mar. 24, 1878; ed. U. Bonn, U. Berlin. Became asst., Berlin Tech. High Sch., 1900; named lectr. physics Breslau (now Wroclaw, Poland), 1917, asst. prof., 1910, prof. natural philosophy, 1917; apptd. prof. exptl. physics Marburg (Germany) U., 1920; became prof., dir. Phys. Inst., Breslau U., 1926; prof. gen. physics Cologne (Germany) U., 1946—. Author: Principles of Dynamics; 1919; Letters between C. F. Gauss and C. L. Gerlin, 1927; Introduction to Maxwell's Theory of Electricity and Magnetism, 1929; Introduction to Theoretical Physics, 4 vols., 1929; The Physical Work of C. F. Gauss, 1929; (with Matossi) Infra-red Spectrum, 1930; (with Bermann, Kliefoth) Practical Physics, 1943; (with Bergmann) Text-Book of Experimental Physics, Vol. I, 1943, Vol. II, 1950, Vol. III, 1956. Research in infrared rays, structure of molecules and crystals. Office: Mehlemerstr. 21, Cologne, German Fed. Republic.

SCHAEFFER, Harold Franklin, Am. chemist, educator; b. Phila., Sept. 21, 1899; s. Franklin P. and Margaretta (Morgan) S.; B.Sc., Muhlenberg Coll., 1922; M.Sc., U. N.H., 1926; postgrad. U. Pitts., W.Va. U.; m. Katherine Arlene Fisher, Apr. 4, 1932. Faculty chemistry Waynesburg Coll., 1926-42, U. Mo., 1942-43, Valparaiso U., 1948-52, Grove City Coll., 1953-55; Coll. Emporia, 1955-59; faculty Westminster Coll., Fulton, Mo., 1959—, prof. chemistry, 1960—. Abstractor Chem. Abstracts Service, 1948—; participant Vis. Scientist Program Mo. Acad. Sci. Mem. Am. Chem. Soc., N.Y. Acad. Scis., Am. Microchem. Soc., A.A.A.S., Sigma Xi. Author: Microscopy for Chemists, 1953; Standard Methods of Chemical Analysis, 1966; also publs. on methods, uses of chem. microscopy. Contbg. editor Isotopics, 1954-55. Home: 15 E. Chestnut St., Fulton, Mo. 65251.*

SCHAEFFER, Jacob Christian, biologist; b. Querfurt, Germany, May 30, 1718; became prof., Ratisbonne, Bohemia, 1779; fellow Royal Soc.; mem. French Acad. Scis., 1762, Acad. Berlin. Author: Polypes d'eau douce, 1755; Fungorum qui in Bavaria nascuntur icones, 1765. Gave detailed and accurate descriptions of flora and fauna of Ratisbonne area; authority on fungi of Bavaria; built optical instruments. Died Regensburg, Bohemia, Jan. 5, 1790.

SCHAEFFER, Karl Ernst, physiologist; b. Bad-Hauheim, Germany, Apr. 19, 1912; s. Karl Ernst and Christine (Grunewald) S.; student U. Frankfort, Germany, 1930, U. Marburg, Germany, 1931, U. Innsbruck, Austria, 1932; M.D., U. Kiel, Germany, 1935; m. Ursula Kolbe, Nov. 26, 1938; children—Michael, Christopher, Barbara Maria (Mrs. Robert M. Quinn), Ernst John. Came to U. S., 1949, naturalized, 1957. Physiologist U. S. Naval Med. Research Lab., New London, Conn., 1949-51, head physiology br., 1951—; adj. prof. U. R.I., 1965—; mem. adv. com. NASA biotechnology, human research, 1962. Fellow Am. Coll. Cardiology, A.A.A.S.; mem. Aerospace Med. Assn., Am. Physiol. Soc., Am. Inst. Aerospace and Astronautics, Internat. Biometeorology, Internat. Soc. Biol. Rhythms, N.Y. Acad. Scis. Author: Man's Dependence on the Earthly Atmosphere, 1962; Environmental Effects on Consciousness, 1962; Bioastronautics, 1964; also numerous articles. Studies on CO2 toxicity; circadian cycles in confinement; diving physiology, submarine and space physiology, hyperbaric medicine. Home: Neck Rd., Old Lyme, Conn. 06371. Office: Physiology Br., Submarine Med. Research Lab., Naval Submarine Med. Center, Box 600, Code 413, Groton, Conn. 06340.*

SCHAEFFER, Oliver Adam, Am. geochemist; b. Fleetwood, Pa., Feb. 20, 1919; s. Charles B. and Mary (Heffner) S.; B.S., Pa. State U., 1941; M.S., U. Mich., 1942; Ph.D., Harvard, 1946; m. Viola A. Long, Oct. 21, 1944; children—Mary, Oliver G., Nancy, George, Clare, Alice. Asst. chemist TVA, Wilson Dam, Ala., 1942-44; research asso. Harvard, 1946-47; asso. chemist Brookhaven Nat. Lab., Upton, N.Y., 1947-51, chemist, 1951-63, sr. chemist, 1963-65; prof. geochemistry, chmn. dept. earth and space scis. State U. N.Y., Stony Brook, N.Y., 1965—, cons. in earth sci., 1962-65; vis. prof. Max-Planck Inst. für Kernphysik, Heidelberg, Germany, 1961-62. Recipient (with R. Davis) Boris Pregel prize N.Y. Acad. Scis., 1957. Mem. Am. Geophys. Union, Meteorol. Soc., Geochem. Soc., Am. Chem. Soc., N.Y. Acad. Scis. Editor: (with J. Zäringer) Potassium Argon Dating, 1966; asso. editor Jour. Geophys. Research, 1965-68. Research, publs. on motions molecules by electrons and radiation, cosmic rays, meteorites, cosmic dust in ocean, Helium-3 in solar flares; discovered Chlorine 36 in nature. Home: 16 Childs Lane, Setauket, N.Y. 11785. Office: State U. N.Y., Stony Brook, N.Y. 11790.*

SCHAERF, Carlo, Italian physicist; b. Rome, Italy, May 2, 1935; s. Samuele and Adriana (Modigliani) S.; Dr. in Physics, U. Rome, 1958; m. Mirella Casini, June 30, 1960; children—Marco, Andrea. Research asso. Nat. Inst. Nuclear Physics, Rome, 1958-60, High Energy Physics Lab., Stanford, 1960-63; research asso., asso. dir. Project Leale, Nat. Lab. of Frascati of Comitato Nazionale Energia Nucleare, Rome, 1963—; prof. exptl. techniques U. Rome Grad. Schs., 1964—, U. Turin (Italy), 1965—. Mem. Italian, Am. phys. socs. Research, publs. on pion photoprodn. and 2d pion-nucleon resonance, structure of deuteron and deuteron magnetic moment, electromagnetic interactions and tests of Born approximations, nuclear structure at high energy. Home: 172 Viale della Tecnica, 00144 Rome. Office: C.P. 70, 00044 Frascati, Rome, Italy.*

SCHÄFER, Klaus, German physicist; b. Cologne, Germany, Aug. 23, 1910; s. Otto and Adèle Schäfer; Ph.D., agrégé, U. Göttingen; m. Liselotte Thomas, Dec. 19, 1942. Asst., Gottingen Inst. Phys. Chemistry, 1936-39; instr. Gottingen, 1939-46; dir. Heidelberg (Germany) Inst. Phys. Chemistry, 1947—; rector U. Heidelberg, 1955-56. Mem. German Leopoldina Acad., Assn. German Chemists, Heidelberg Acad. Sci., Bunsen Assn., Real Sociedad Fisica y Animica de Madrid (hon.), Bunsen Assn. Home: Mozartstrasse 32. Office: Tiergartenstrasse 5, Heidelberg, West Germany.

SCHAFER, Richard Donald, Am. mathematician; b. Buffalo, Feb. 25, 1918; s. Edward J. and Ruth (Stone) S.; B.A., U. Buffalo, 1938, M.A., 1940; Ph.D., U. Chgo., 1942; m. Alice E. Turner, Sept. 8, 1942; children—John Dickerson, Richard Stone. Instr., U. Mich., 1945-46; mem. Inst. for Advanced Study, 1946-48; asst. prof. math. U. Pa., 1948-53; prof., head math. U. Conn., 1953-59; prof., dep. head dept. math. Mass. Inst. Tech., Cambridge, 1959—. Mem. Sci. Manpower Commn., 1959-63. NSF Sr. Postdoctoral fellow, 1958-59. Mem. Am. Math. Soc. (past asso. sec.), Math. Assn. Am. (past vis. lectr.), Sigma Xi, Phi Beta Kappa. Author: An Introduction to Nonassociative Algebra, 1966; also articles. Research in nonassociative algebras; proved several structure theorems for alternative algebras; discovered several relationships between Lie algebras and alternative and Jordan algebras. Home: 60 Spring Valley Rd., Belmont, Mass. 02178. Office: Mass. Inst. Tech., Cambridge, Mass. 02139.*

SCHÄFER, Werner, German virologist; b. Wanne-Eickel, Germany, Mar. 9, 1912; s. Wilhelm and Margarete (Buchlo) S.; student U. Giessen (Germany), 1932, 37, Dr.med.vet., 1938, Hon. Prof., 1964; Hon. Prof., U. Tubingen (Germany), 1965; m. Emmi Schafer, May 27, 1939; children—Irmgard (Mrs. Horst Weissenborn), Werner, Margret. Asst., Tierseuchen Inst., Giessen, 1937-44; collaborator Reichsforschungsanstalt, Insel Riems, 1944-45; chief dept. Kaiser Wilhelm-Inst. für Biochemie, Tübingen, 1948-54; staff Max-Planck-Institut für Virusforschung, Tübingen, 1954—, dir., 1956—; virology dept. U. Giessen, 1952—, U. Tübingen, 1966—. Recipient Carus Medaille, Deutsche Akademie d. Naturforscher Leopoldina, 1957, Emil v. Behring-Preis, 1962, Schunk-Preis, U. Giessen, 1965. Mem. Royal Soc. Medicine (affiliate, London, Eng.). Research and numerous publs. on fine structure of animal viruses, multiplication mechanisms of animal viruses action animal viruses on normal metabolism of host cells, immunological studies, devel. new prins. for prodn. vaccines against virus diseases. Home: 18, Wolfgang Stock-Str., Tübingen. Office: Max-Planck-Inst. für Virusforschung, 35/III Spemannstr., Tübingen, Germany.*

SCHAFFER, Norwood Korter, Am. chemist; b. Seattle, June 23, 1905; s. John W. and Martha (Korter) S.; B.S., U. Wash., 1926, 1927; postgrad. (research fellow), Cal. Inst. Tech., 1927, fellow) Johns Hopkins Med. Sch.; Ph.D., Harvard, 1936; M.D., Western Res. U., 1943; m. Martha Denny, Aug. 23, 1939; children—Robert Denny, Martha Storrow (Mrs. Frederick John Kluth), Elizabeth Denny (Mrs. William Earl Tamblyn, Jr.). Biochemist Meml. Found. for Neuro- Endocrine Research, Boston, 1930-33, Phila. Inst. for Med. Research, 1936-39; asso. in obstetrics Cornell U. Med. Coll., N.Y.C., 1941-43; fellow in pharmacology Yale Med. Sch., 1944-45; faculty Boston U. Med. Sch., 1945-49; biochemist, chief, asst. chief, biochemistry br. physiology dept. Research Labs. Edgewood Arsenal, Md., 1949—, physician dispensary, 1952, asst. U. Md. Extension Service, 1949-52. Recipient Sustained Superior Performance award Edgewood Arsenal, 1960. Mem. Am. Soc. Biol. Chemists, Am. Chem. Soc., Research Soc. Am., A.A.A.S., Am. Assn. U. Profs., N.Y. Acad. Scis., Sigma Xi, Phi Beta Kappa. Research, publs. on structure of active site of enzymes, action of nerve gases, anterior pituitary growth hormone, toxemia of pregnancy. Home: 8 Beechdale Rd., Balt. 21210. Office: Biochemistry Br. Research Labs., Edgewood Arsenal, Md. 21010.*

SCHÄFFLE, Albert Eberhard Friedrich, German sociologist, polit. economist; b. Nürtingen, Württemberg, Feb. 24, 1831; ed. U. Tübingen, 1848. Mem. editorial staff Schwabische Merkur, Stuttgart, 1850-60; prof. polit. economy U. Tübingen, 1860; prof. polit. sci. U. Vienna (Austria), 1868; editor Zeitschrift fur die gesamte Staatswissenschaft, 1892-1901. Mem. Württemberg Diet, 1862-64; received mandate to German Zollparlament, 1868; minister of commerce for Austria, 1871. Author: Das Gesellschaftliche System der menschlichen Wirthschaft, new edit., 1873; Die Nationalokonomische Theorie der ausschliessenden Absatzverhaltnisse, 1867; Bau und Leben des socialen Körpers, 4 vols., 1875-78; Gesammelie Aufsatze, 1885-87; Ein Votum gegen den neuestan Zolltarif, 1901; Dir agrarische Gefahr, 1902; Abriss der Soziologie, 1906. Drew elaborate parallels between biology and social instns.; compared social phenomena to biol. tissue; described group life as the unit of conflict, mut. aid and survival; important contbns. to theory of taxation; believed socialism would evolve out of capitalism; tried to unite all social and natural scis. into a system based on conception of society as an organism and his philos. idealism. Died Stuttgart, Germany, Dec. 25, 1903.

SCHAFFNER, Fenton, Am. physician; b. Chgo., Dec. 8, 1920; s. Samuel and Rose (Cannon) S.; B.S. in Anatomy, U. Chgo., 1941, M.D., 1943; M.S. in Pathology, Northwestern U., 1949; m. Lillian Levin, May 23, 1943; children—Roberta, John, Andrea, Marjorie. Asst. pathologist Northwestern U., 1948-53, instr. medicine, 1955-57; asso. in medicine, asst. prof. pathology Columbia Coll. Phys. and Surg., 1958-66; prof. pathology, asso. prof. medicine Mt. Sinai Sch. Medicine, N.Y.C., 1966—; asso. attending physician and pathologist Mt. Sinai Hosp., N.Y.C., 1966—. Diplomate Am. Bd. Internal Medicine. Mem. A.M.A., A.C.P., A.A.A.S., Am. Gastroent. Assn., Am. Assn. Study Liver Diseases (sec.), Am. Soc. Exptl. Pathology, Am. Assn. Pathologists and Bacteriologists. Author: (with H. Popper) Liver: Function and Structure, 1957, Progress in Liver Diseases, vol. 1, 1961, vol. 2, 1965; Clinical Pathological Conferences of the Mt. Sinai Hospital, 1966. Editorial bd. Gastroenterology, Med. Letter, Digestion (Basel). Research, numerous publs. on structure and ultra structure of liver in various diseases of that organ, especially jaundice, classification and mechanisms of liver abnormalities produced by adverse reactions to drugs. Home: 12 Wynmor Rd., Scarsdale, N.Y. 10583. Office: 1176 Fifth Av., N.Y.C. 10029.*

SCHAIRER, George Swift, Am. aeronautical engr., b. Pittsburgh, May 19, 1913; s. Otto Sorg and Elizabeth Blanche (Swift) Schairer; B.S. Swarthmore College, 1934, Doctor of Engineering (honorary), 1958; M.S., Mass. Inst. Tech., 1935; m. Mary Pauline Talbox, June 20, 1935; children—Mary Elizabeth, George Edward, Sally Helen, John Otto. With Bendix Aviation Corp., South Bend, Ind., 1935-37, Consol. Vultee Aircraft Corp., San Diego, Cal., 1937-39; joined Boeing Airplane Co., Seattle, 1939, successively chief aerodynamist, staff engr. aerodynamics and powerplant, 1948-51, chief tech. staff, 1951-56; assistant chief engr. Seattle div. Boeing Airplane Co., 1956-57, dir. research, 1957-59, vice president research and development, 1959—. Science adv. group USAAF, 1945-46; mem. com. on aerodynamics Nat. Adv. Com. Aeronautics; mem. Nat. Acad. Scis., Tech. adv. panel on aeros. Dept. Def. 1954-61; sci. adv. bd. USAF, 1955-60; cons. operations evaluation group, USN, 1961; panel sci. and tech. manpower Pres.'s Sci. Adv. Com., 1962—. Recipient Spirit of St. Louis medal Am. Soc. M.E., 1959. Fellow Am. Inst. Aeros. and Astronautics (Sylvanus Albert Reed award 1950, tech. dir.); mem. Sigma Xi, Sigma Tau. Author tech. papers sci. jour. Head of group which designed Boeing 707, 880, and DC-8 commercial jet transports which revolutionized passenger transportation, providing efficient load carrying over long distances. Home: 4242 Hunts Point Rd., Bellevue, Wash. Office: Boeing Co., Seattle 98124.

SCHAIRER, John Frank, Am. geologist, research scientist; b. Rochester, N.Y., Apr. 13, 1904; s. John George and Josephine Marie (Frank) S.; B.S. in Chemistry, Yale, 1925, M.S. in Geology, 1926, Ph.D. in Chemistry, 1928; m. Ruth Naylor, July 20, 1940; children—John Everett and Jeanne Evelyn (twins). Physical chemist, geophysical lab. Carnegie Instn. of Washington, 1927—; spl. asst. div. one Nat. Def. Research Com., 1942-45. Recipient Hillebrand award, Chem. Soc. Washington, 1942; President's Certificate of Merit, 1948; Medal of Honor (Eng.), 1948, Arthur L. Day medal, Geol. Soc. Am., 1953. Mem. Nat. Acad. Scis., Mineral. Soc. Am. (pres. 1943, recipient Roebling medal 1963), Geological Soc. Am. (v.p. 1944), Am. Chem. Society, Am. Geophys. Union, Nat. Capital Orchid Soc. (p.p.; editor bull. 1951—), Geochem. Soc. (p.p.), Internat. Assn. Volcanology (v.p. 1957-60), Sigma Xi. Research in physical chemistry; mineralogy; origin of igneous rocks; stability relations of minerals; phase equilibria in silicates at high temperatures. Home: 4617 Chevy Chase Blvd., Chevy Chase, Md. 20015. Office: Geophys. Lab., 2801 Upton St., Washington 20008.

SCHALLER, Karl Friedrich, dermatologist; b. Luederitzbucht, S.W. Africa, Feb. 4, 1914; s. Karl Justus and Auguste (Faber) S.; student Edinburgh U., 1938-39; M.D., U. Berlin, 1939; Diploma in Pub. Health, U. Hamburg, 1947; m. Helga de Vries, Aug. 7, 1942; children—Arend, Dieter, Jörg, Angela, Christian. With Ministry of Health, Addis Ababa,

Ethiopia, 1952-63, chief pub. health adv. group, 1960-63; lectr. tropical dermatology Hamburg (Germany) U., 1964—. Guest lectr. Tropical Inst., Hamburg, 1964—, Gondar Health Coll., 1958-63; cons. on tropical diseases, 1952—. Mem. Internat. Soc. Tropical Dermatology, Soc. Italiana di Medicina Sociale (hon.), Ethiopian Med. Assn., German Soc. for Dermatology, German Soc. for Tropical Medicine, German Soc. Hygiene. Research on leprosy, tropical dermatology, mycology, med. pathology of Ethiopia, epidemiological studies of tropical diseases. Home: 34 Parkstrasse, Hamburg, Germany.*

SCHALTENBRAND, Georges Otto, German physician; b. Oberhausen, Germany, Nov. 26, 1897; s. Eugen and Adele (Pastor) S.; student univs. Breslau, Göttingen, Munich; postgrad. U. Hamburg, Harvard; m. Luise Kleinwort, Aug. 9, 1928; children—Jan-Peter, Hans-Jürgen, Inge-Lu (Mrs. Mintzel), Else Li. Docent, U. Hamburg, 1928; asso. neurology Peking (China) Union Med. Coll., 1928-30; prof. U. Wurzburg (Germany), 1938—, dir. neurol. dept., 1950—. Recipient Martini award, 1928; Röntgen award, 1943; Erb medal, 1952; Nonne medal, 1965. Mem. Am. Neurol. Assn.; hon. mem. Am. Acad. Neurology, Brit., French neurol. socs., Chilean Soc. Neurology, Psychiatry and Neurosurgery; corr. mem. Am. Nat. Multiple Sclerosis Soc., Seville Acad. Medicine and Surgery, Swiss Neurol. Soc. Author: Die Multiple Sklerose, 1943; Lehrbuch der Nervenkrankheiten, 1951; Zeit in nervenärztlicher Sicht, 1963; also numerous contbns. to handbooks and profl. lit. Editor: (with P. Bailey) Atlas der stereotatkischen Operationen. Research on motor function, spinal fluid analysis, multiple sclerosis, stereotaxy. Address: 4 Lerchenweg, Wurzburg, West Germany.*

SCHAMBERG, Jay Frank, Am. dermatologist; b. Phila., Nov. 6, 1870; s. Gustav and Emma (Frank) S.; A.B., Central High Sch., Phila., 1888; M.D., U. Pa., 1892; postgrad., Vienna, Paris, Berlin and London; m. May Ida Bamberger, Oct. 11, 1905; children—Elizabeth, Ira Leo. Practiced medicine, Phila., 1892—; prof. dermatology and syphilology Grad. Sch. of Medicine, U. Pa., 1919—; dir. Dermatol. Research Inst., Phila. Apptd. mem. Pa. Vaccination Commn., 1912. Mem. Am. Dermatol. Assn. (pres. 1920-21). Author: Compendium of Diseases of Skin, 1898; Acute Contagious Diseases (with Dr. Welch), 1905; Skin Diseases and Eruptive Fevers, 1911. Described prog. pigmentary dermatosis (Schamberg's disease), 1901; asso. with Drs. George W. Raiziss and John A. Kolmer, in research which early in World War I led to 1st elaboration in U. S. of slavarsan, (previously made only in Germany), donated profits from sale as permanent endowment for Research Inst. of Cutaneous Medicine; eponym of Schamberg's ointment for grain itch. Died Mar. 30, 1934.

SCHAMP, Homer Ward, Jr., Am. physicist; b. St. Marys, O., June 23, 1923; s. Homer Ward and Estella (Long) S.; A.B., Miami U., Oxford, O., 1944; M.Sc., U. Mich., 1947, Ph.D., 1952; m. Juliana Reese, July 17, 1948; children—Joseph Brough, David Julian. Physicist, Monsanto Chem. Co., 1951-52; faculty U. Md., College Park, 1952-65, Balt., 1965—, prof. molecular physics, 1960-66, dir. Inst. for Molecular Physics, 1964-66, dean faculty, 1965—. Fellow Am. Phys. Soc.; mem. Phi Beta Kappa, Sigma Xi. Author: (with J.T. Vanderslice, E.A. Mason), Thermodynamics, 1966. Research, publs. on diffusion in solids and properties of gases at high pressures. Home: 521 Overdale Rd., Balt. 21229.*

SCHANKER, Lewis Stanley, Am. pharmacologist; b. Kansas City, Mo., Sept. 23, 1930; s. Herman H. and Florence (Fishman) S.; student Jr. Coll. Kansas City, 1947-48; B.S., U. Mo., 1951, M.S., 1953; Ph.D., U. Wis., 1955; m. Joyce Ann Minkin, May 27, 1953; children—Neil B., Lynn S. Asst. instr. pharmacology U. Mo., Kansas City, 1951-53; research fellow pharmacology U. Wis., Madison, 1953-55; staff Nat. Heart Inst., NIH, USPHS, Bethesda, Md., 1955-66, head sect. on cellular pharmacology, 1957-66, head sect. on biochemistry of drug action lab. chem. pharmacology, 1959-66; trustee prof. pharmacology, prof. dentistry, coordinator devel. environmental health research U. Mo., Kansas City, 1966—; Mem. pharmacological panel U. S. Bd. Civil Service Examiners, 1959-66. Recipient Lehn-Fink medal award U. Mo., Kansas City, 1951, Outstanding Alumni Achievement award, 1967. Fellow A.A.A.S.; mem. Am. Soc. Pharmacology and Exptl. Therapeutics (John J. Abel prize 1966), Am. Physiol. Soc., Soc. for Exptl. Biology and Medicine, (editorial bd. Proc. 1965-68), Sigma Xi. Editorial bd. Jour. Pharmacology and Exptl. Therapeutics, 1962-65, Internat. Jour. Clin. Pharmacology, Therapy and Toxicology, 1967—. Research, publs. on mechanisms of absorption of drugs from gastro-intestinal tract absorption of substances from lung, passage of drugs from blood into brain and cerebrospinal fluid, into red blood cells and body tissues, excretion of drugs in bile and formation of bile, uptake of drugs by heart, transport of drugs by choroid plexus, protein binding of drugs, cellular binding of drugs, physiology of membranes. Home: 8409 Juniper Lane, Prairie Village, Kan. 66207. Office: U. Mo., Kansas City, Mo. 64110.*

SCHARFF-GOLDHABER, Gertrude (Mrs. Maurice Goldhaber), physicist; b. Manneheim, Germany, July 14, 1911; d. Otto and Nelly (Steinharter) Scharff; Ph.D., U. Munich (Germany), 1935; m. Maurice Goldhaber, May 24, 1939; children—Alfred S., Michael H. Came to U.S., 1939, naturalized, 1944. Research asso. Imperial Coll., London, Eng., 1935-39; physicist U. Ill., Urbana, 1939-48, asst. prof., 1948-50; staff Brookhaven Nat. Lab., Upton, N.Y., 1950—, physicist, 1958-62, sr. physicist, 1962—, cons. Standard Reference Data System, 1965—. Cons. Argonne (Ill.) Nat. Lab., 1946-56, Los Alamos Sci. Lab., 1953—. Fellow Am. nuclear data project NRC, 1955—. Fellow Am. Phys. Soc.; mem. Sigma Xi. Research on identity beta-particles with electrons, research on ferromagnetism, neutron physics, photoneutrons, neutrons from spontaneous fission uranium, nuclear isomers, systematics nuclear levels. Home: 91 S. Gillette Av., Bayport, L.I., N.Y. 11705. Office: Physic Dept., Brookhaven Nat. Lab., Upton, L.I., N.Y. 11973.*

SCHARMANN, Arthur Otto, German physicist; b. Darmstadt, Germany, Jan. 26, 1928; s. Arthur Ludwig and Gustel (Darmstädter) S.; student U. Marburg; Dipl.-Phys., U. Giessem, 1951, Dr.rer.nat., 1955; m. Irmgard Hanle, Aug. 23, 1957; children—Albrecht, Alexander, Martin. Faculty, U. Giessen, 1959—, prof., sci. adviser, 1965—, head radiation physics dept., I. phys. inst., 1966—. Mem. German, Am. phys. socs. Research, numerous publs. on scintillation process, exoelectron emission, radiation damage in isolators and crystal phosphors, sputtering, reciprocal effect of radiation, dosimetry. Address: 17 Bergstrasse, 63 Giessen, West Germany.*

SCHARRER, Berta Vogel, biologist; b. Munich, Germany, Dec. 1, 1906; d. Karl Phillip and Johanna (Greis) Vogel; Ph.D., U. Munich, 1930; m. Ernst Albert Scharrer, Mar. 1, 1934 (dec. 1965). Faculty, U. Colo. Sch. Medicine, 1947-55; prof. anatomy Albert Einstein Coll. Medicine, N.Y.C., 1955—, prof. anatomy, acting chmn., 1965-66. Mem. Nat. Acad. Scis., Am. Soc. Zoologists, Am. Assn. Anatomists. Author: (with Ernst Scharrer) Neuroendocrinology, 1963. Editorial bd. Zeitschrift für Zellforschung, Jour. Gen. and Comparative Endocrinology. Jour. Neurochemistry. Research, numerous publs. on interactions of hormones with nervous system in various animals, elucidation of ultrastructure of endocrine glands and brains in animals. Home: 1240 Neill Av., Bronx, N.Y. 10461.*

SCHAUDINN, Fritz Richard, German zoologist; b. Röseningken, Germany, Sept. 19, 1871; expdn. to Arctic Ocean, 1898; became dir. dept. protozool. research Inst. Tropical Diseases, Hamburg, Germany, 1906. Author: Fauna arctica, 6 vols., 1900-33; Vorläufiger Bericht über das Vorkommen von Spriochaeten in syphilitischen Krankheitsprodukten, 1905; Geschichtliche Arbeiten, 1911. Discovered (with Hoffmann) cause of syphilis in Spirochaeta pallida, 1905; revealed amoebic nature of tropical dysentery; studied tryponosomes, other organisms and malaria; discovered alternation of generations in Coccidae and Foraminifera. Died Hamburg, June 22, 1906.

SCHAWLOW, Arthur Leonard, Am. physicist; b. Mt. Vernon, N.Y., May 5, 1921; s. Arthur and Helen (Mason) S.; B.A., U. Toronto (Ont., Can.), 1941, M.A., 1942, Ph.D., 1949; m. Aurelia Keith Townes, May 19, 1951; children—Arthur Keith, Helen Aurelia, Edith Ellen. Postdoctoral fellow, research asso. Columbia, N.Y.C., 1949-51, vis. asso. prof., 1960; research physicist Bell Telephone Labs., Murray Hill, N.J., 1951-61, cons., 1961-62; prof. physics Stanford (Cal.) U., 1961—. Cons. Varian Assos., Palo Alto, Cal., 1962—. Dir. Optics Tech., Inc., Palo Alto, 1963—. Recipient Stuart Ballantine medal Franklin Inst., 1962, Thomas Young medal and prize Phys. Soc. and Inst. Physics, 1963. Fellow Am. Phys. Soc., Optical Soc. Am., I.E.E.E. (Morris N. Liebmann Meml. prize 1964). Author: (with Charles H. Townes) Microwave Spectroscopy, 1955. Editorial bd. Phys. Rev., 1963-65, Jour. Applied Physics, 1962-64, Physics, 1964—, Laser Abstracts, 1963—. Research on radiofrequency, microwave and optical spectroscopy, quantum electronics, lasers, superconductivity. Home: 849 Esplanada Way, Stanford, Cal. 94305.*

SCHECHTER, Martin, Am. mathematician; b. Phila., Mar. 10, 1930; s. Joshua and Rose (Shames) S.; B.S., Coll. City N.Y., 1953; M.S., N.Y. U., 1955, Ph.D., 1957; m. Naomi Deborah Kirzner, Dec. 23, 1957; children—Sharon Libby, Arthur Irving, Isaac David. Faculty, N.Y. U., 1958-66, prof. math., 1965-66; prof., chmn. math. dept. Belfer Grad. Sch. Sci., Yeshiva U., N.Y.C., 1965—; mem. Inst. for advanced Study, Princeton, N.J., 1965-66. NSF Sr. fellow, 1965-66. Mem. Am. Math. Soc., Phi Beta Kappa, Pi Mu Epsilon. Author: (with L. Bers, F. John) Partial Differential Equations, 1964; also articles. Solutions of boundary value problems for certain types of partial differential equations and proofs of related estimates, theorems concerning certain classes of operators, function spaces, methods of interpolation between Banach spaces. Home: 1677 51 St., Bklyn. 11204.*

SCHEELE, Carl Wilhelm, Swedish chemist; b. Stralsund, Sweden (now Germany), Dec. 9, 1742; apprenticed to apothecary, Goteborg (Sweden), where he began to teach himself chemistry, 1757-65; m. 1786. Pharmacist in Malmo, 1765-67; pharmacist in Stockholm, 1768-69; in Uppsala, where he met Torbern O. Bergman, 1770-1774; in Koping (where he was proprietor of a pharmacy), 1775-86. Refused offers of Frederick the Great to chair of chemistry at Berlin and a similar offer in Eng. Mem. Stockholm Academy of Sci., 1775. Author: Air and Fire, 1777; Collected Papers of Carl Wilhelm Scheele (trans. by L. Dobbin), 1931. Discovered many new chemical substances, including oxygen, 1771, and chlorine, 1774; among organics, he isolated tartaric acid, 1770, gallic acid, oxalic acid, citric acid, and malic acid; among inorganics, isolated toxic gases; hydrogen fluoride, hydrogen cyanide, hydrogen sulfide, 1778; discovered copper arsenite (Scheele's green), 1776; 1st to demonstrate presence of calcium phosphate in bone; remained in favor of phlogiston theory; obtained molybdic acid from mineral molybdena nitens (molybdenite), which he was 1st to distinguish from ordinary molybdena (graphite), 1778; proved that acidity of sour milk due to what was afterwards called lactic acid, 1778; obtained tungstic acid from mineral now called scheelite (calcium tungstate), 1781; experimented on ether, 1782; investigated properties of glycerine, 1783; described properties, composition, and compounds of prussic acid. Died Koping, Sweden, May 21, 1786.

SCHEER, Bradley Titus, Am. biologist; b. Los Angeles, Dec. 17, 1914; s. William Bradley and Mabel (Titus) S.; B.S., Cal. Inst. Tech., 1936; Ph.D., U. Cal. at Berkeley, 1940; m. Marlin Ann Ray, Aug. 14, 1936. Research asst. Scripps Inst. Oceanography, 1936-37, 38-40, Columbia Coll. Phys. and Surg., 1942; faculty U. Cal. at Berkeley, 1937-38, W. Va. U., 1940-41, Cal. Inst. Tech., 1943-45, U. So. Cal., 1945-48, U. Hawaii, 1948-50; asso. prof. biology U. Ore., Eugene, 1950-54, prof., 1954—, head dept., 1956-57, 58-64. Mem. A.A.A.S., Am. Physiol. Soc., Soc. Gen. Physiologists, Am. Soc. Zoologists, Western Soc. Naturalists, Sigma Xi. Author: Comparative Physiology, 1948; General Physiology, 1953; Animal Physiology, 1963. Research, numerous publs. on fertilization of invertebrate eggs in relation to gen. theories of excitation, effects of fat level in diet on phys. condition of rats, hormonal control of carbohydrate metabolism in crustaceans, salt transport across biol. membranes and control by hormones. Home: 1825 Longview St., Eugene, Ore. 97403.*

SCHEFFER, Henric Theophil, Swedish chemist; b. Stockholm, 1710; grandson of Jean Scheffer; pupil of André Celsius at Uppsala, of Brandt at Stockholm; contbr. papers to Swedish Acad. Sci.; research on platinum, calcite, comml. potash, preparation of a gold-like zinc and copper alloy (pitch-beck); obtained chemically pure silver by precipitating solution of silver nitrate with salt and reducing silver chlorine by smelting with potassium carbonate; asserted that metals gain or lose weight as phlogiston is removed from, or added to them, 1757. Died Stockholm, 1759.

SCHEFFER, Victor Blanchard, Am. biologist; b. Manhattan, Kan., Nov. 27, 1906; s. Theophilus and Celia E. (Blanchard) S.; B.S., U. Wash., 1930, M.S., 1932, Ph.D., 1935; m. Mary Elizabeth MacInnes, Oct. 12, 1935; children—Brian M., Susan Elizabeth (Mrs. Robert D. Irvine), Ann Blanchard (Mrs. Roy W. Carlstrom). Biologist, U.S. Fish and Wildlife Service, Seattle, 1936—. NSF grantee, 1956-57; recipient Interior Dept. Distinguished Service medal, 1965. Mem. Am. Soc. Mammalogists Wilderness Soc., Am. Inst. Biol. Scis. Author: Seals, Sea Lions and Walruses, 1958; also numerous articles. Research on marine mammals especially fur seals. Home: 14806 S.E. 54th St., Bellevue, Wash. 98004. Office: Bldg. 192, Sand Point Naval Air Sta., Seattle 98115.*

SCHEIBE, Arnold, German agronomist, plant breeder; b. Greiz, Germany, Oct. 20, 1901; s. Johannes and Katharina (Schilbach) S.; student natural sci. and agr. U. Munich (Germany), 1923-24, U. Göttingen (Germany), 1924-26; LL.D., U. Cal. at Davis, 1965; m. Sophie-Charlotte Geest, May 15, 1931; children—Reinhild, Erdmuth, Hans-Christoph, Gabriele. Prof. Tech. U., Munich, 1941-48; prof. U. Giessen (Germany), 1951-55; dir. Max-Planck-Inst. Breeding Research, Gut Neuhof, nr. Giessen, 1948—; prof. plant cultivation and plant breeding Göttingen U., 1955—, dir. Institut für Pflanzenbau und Pflanzenzüchtung, 1955—. Mem. Deutsche Akademie der Naturforscher Leopoldina, German Soc. Agrl. Sci., German Bot. Soc., Soc. Applied Botany. Author: Einführung in die allgemeine Pflanzenzüchtung, 1951; Handbuch der Landwirtschaft, 5 vols., 2d edit., 1952-54; also articles. Research on physiology and biochemistry on germination problems in cereal seeds, problem of developmental physiology in cereals, phylogenetic problems on wild and cultivated plants, breeding problems on cultivated leguminous and oil plants, breeding fasciated new varieties of Pisum, content of growth substances in fasciated varieties of pisum. Home: 61 Merkelstrasse, Göttingen, Germany.*

SCHEIBEL, Arnold Bernard, Am. physician; b. N.Y.C., Jan. 18, 1923; s. William and Ethel (Greenberg) S.; B.A., Columbia, 1943, M.D., 1946; M.S., Ill., 1951; m. Madge E. Ragland, Mar. 3, 1950. USPHS fellow U. Ill. Med. Center, Chgo., 1950-51; faculty U. Tenn. Med. Center, Memphis, 1952-55, asso. prof., 1955; faculty U. Cal. at Los Angeles Med. Center, 1955—, professor of anatomy and psychiatry, 1967—; cons. Cal. State, Brentwood VA, Sepulveda VA hosps. Guggenheim fellow, 1953-54. Mem. A.A.A.S., Am. Psychiat. Assn., Am. Acad. Neurologists, Soc. Biol. Psychiatry, Psychiat. Research

Soc., Am. Assn. Anatomists. Research, publs. in reticular formation of brain, structural and physiol. substrates of Ego function, circuit orgn., models and individual neuron behavior in reticular system. Home: 16231 Morrison St., Encino, Cal. 91316.*

SCHEIBL, Adolf, German physicist; b. Zeulenroda, Germany, Mar. 9, 1895; sect. dir. Phys.-Tech. Inst., Brunswick, Germany; Author: Piezoelektrizität des Quarzes, 1938. Research on high frequency physics, piezo electricity, precision time measurement; developed the quartz clock. Died Brunswick, Apr. 20, 1958.

SCHEIBLER, Charles, German chemist; b. Gemehret, nr. Eupen, 1827; named asst. prof, Königsberg (now Kaliningrad, USSR), 1853; became chemist, Stettin, Poland, 1858. Founder lab. for sugar industry, Berlin, 1861. Author: Documents pour l'histoire de la fabrication du sucre en Allemagne, 1875. Research on beet sugar; discovered dextrose, arabinose, also method of extracting sugar from molasses; a founder indsl. chemistry in Germany. Died Berlin, 1899.

SCHEID, Francis, Am. mathematician; b. Plymouth, Mass., Sept. 24, 1920; s. John J. and Rose (Bergdoll) S.; B.S., Boston U., 1942, M.A., 1943; Ph.D., Mass. Inst. Tech., 1948; m. Barbara Paty, June 2, 1944; children—Betsy, Lisa, Sarah. Faculty, Boston U., 1948—, prof., 1958—, chmn. math. dept., 1957—. Fulbright lectr. Rangoon U., 1961-62; guest mathematician U.S. Nat. Bur. Standards, 1958, Swiss Fed. Inst. Tech., 1962; cons. Instrumentation Lab., Mass. Inst. Tech., 1954-61. Mem. Soc. for Indsl. and Applied Math., Math. Assn. Am., Nat. Council Tchrs. Math., Assn. Tchrs. Math. in New Eng. Author: Elements of Finite Mathematics, 1962; 1 plus 1 equals ?, 1966; Theory and Problems on Numerical Analysis, 1967. Research, publs. on concepts of modern math. Home: Elm St., Kingston, Mass. 02360. Office: 270 Bay State Rd., Boston 02215.*

SCHEIDEGGER, Adrian Eugen, geophysicist; b. Basel, Switzerland, May 2, 1905; s. Eugen M. and Mathilde (Inwiller) S.; diploma in Physics, Swiss Fed. Inst. Tech., Zurich, 1948; Ph.D., U. Toronto, 1950; m. Mary Ammann, Apr. 12, 1952; children—Katherine, Theodor, Paul. Lectr., Queens U., Kingston, Ont., Can., 1950-52; research engr. Imperial Oil Co., Calgary, Alta., Can., 1952-54, 57-61; seismologist Dominion Obs., Ottawa, Ont., 1954-57; asso. prof. U. Alta., 1961-62; prof. U. Ill., Urbana, 1962—; vis. prof. Cal. Inst. Tech., 1959; vis. sr. lectr. U. Sydney (Australia), 1962; vis. prof. Munster Westphalia; 1964, 65. Mem. Am. Geophys. Union, Am. Inst. Mining, Metall. Engrs., Seismol. Soc. Am., Geophys. Soc. Am., Canadian Inst. Mining and Metallurgy, Deutsche Geophys. Gesellschaft. Author: Physics of Flow Through Porous Media, 1960; Principles of Geodynamics, 1963; Theoretical Geomorphology, 1963; also numerous articles. profl. jours. Originator stochastic theory of dispersion in porous media; contbns. in mech. theory of earth's surface features. Office: Metall. and Mining Bldg., U. Ill., Urbana, Ill. 61803.*

SCHEIE, Harold Glendon, Am. physician; b. Brookings County, S.D., Mar. 24, 1909; s. Lars T. and Ella (Ware) S.; B.S., U. Minn., 1931, M.B., M.D., 1935; D.Sc., U. Pa., 1940; m. Mary Anne Tallman, 1951; children—Eric, Nancy. Faculty U. Pa. Med. Sch., 1946—, prof ophthalmology, 1956—, chmn. dept., 1960; pvt. practice, Phila., 1960—; chief ophthalmology service Phila. Gen., Childrens, VA hosps.; mem. courtesy staff Presbyn. Hosp. Lectr. U.S. Naval Hosp. Phila.; mem. nat. com. MEDICO; mem. nat. adv. com. Eye-Bank for Sight Restoration; cons. numerous hosps., govt. agys.; hon. lectr. Recipient Lindback Teaching award U. Pa., 1964; Howe award A.M.A., 1964; Gold medal U. Buffalo, 1964; certificate of achievement Assn. U.S. Army, 1964. Diplomate Am. Bd. Ophthalmology. Fellow A.C.S. (bd. regents); mem. Am. Acad. Ophthalmology and Otolaryngology (Honor award), A.A.A.S., Am. Med. Writers Assn., Am. Ophthalmology Soc, Nat. Med. Found. for Eye Care, Nat. Soc. For Prevention Blindness, Ophthal. Soc. U.K., Pan-Pacific Surg. Assn., Pan-Am. Ophthal. Soc., World Med. Assn, Pa. Acad. Ophthalmology and Otolaryngology (pres. 1968), Sigma Xi, Alpha Omega Alpha, others. Author: (with others) Surgery of the Eye, 1952; Glaucoma, 1955-56, 58; Pediatric Clinics of North America, 1957; What the General Practitioner Should Know About Ocular Surgery, 1957; Symposium on Glaucoma, 1959; Ophthalmology Lecture Notes, 1960; chpts. to Ophthalmic Plactic Surgery, 1961, Plastic and Reconstructive Surgery of the Eye and Adnexa, 1962; also numerous articles. Research on various ophthalmologic diseases especially glaucoma, heritable diseases of connective tissue. Ardmore, Pa. 19003.*

SCHEIER, Ivan Henry, Am. psychologist; b. Plattsburgh, N.Y., Jan. 7, 1926; s. Joel Henry and Melba (Gottlob) S.; A.B. in Philosophy, Union Coll., Schenectady, 1948; M.A., McGill U., Montreal, Que., Can., 1951, Ph.D., 1953. Research asso. Human Resources Research Office, Washington, 1953-55; research asso. Lab. Personality assessment and Group Behavior, U. Ill., Urbana, 1955-60; test editor, asso. dir. Inst. for Personality and Ability Testing, Champaign, Ill., 1960—; project dir. Boulder County Juvenile Delinquency Project, Office Juvenile Delinquency and Youth Devel., U.S. Dept. Health, Edn. and Welfare, 1965—. Cons. psychologist Boulder County (Colo.) Juvenile Ct., 1963—. Mem. Am. Psychol. Assn., N.Y. Acad. Scis., Soc. Multivariate Exptl. Psychologists, Phi Beta Kappa, Sigma Xi. Author: (with Raymond B. Cattell) The Meaning and Measurement of Neuroticism and Anxiety, 1961; Not Without Love, 1964; also articles, personality tests. Research on mil. tng. techniques, nature and structure of human personality especially anxiety and neurosis; action research on effective use of community vols. in local juvenile ct. programs; co-prin. investigator on research on effects of drugs on human personality; personality of med. doctors. Home: Salina Star Route, Boulder, 80304. Office: Hall of Justice, Juvenile Probation Dept., Boulder, Colo. 80301.

SCHEIN, Marcel, physicist; b. Trstena, Czechoslovakia, June 9, 1902; s. Henry and Hermina (Messinger) S.; student U. Wurzburg (Germany), 1921-23; Ph.D., U. Zurich (Switzerland), 1927; m. Hilde Schoenbeck, June 2, 1927; 1 son, Edgar H. Came to U. S., 1938, naturalized, 1943. Asst. physics dept. U. Zurich, 1926-29, docent, 1931-35; prof. exptl. physics U. Odessa, 1935-37; fellow Rockefeller Found. U. Chgo., 1929-30; research in cosmic rays, 1938-42, asst. prof., 1942-45, prof. physics 1946—; cons. Research Lab., Gen. Elec. Co., 1945-46, Fermi Inst. Nuclear Studies, 1946; vis. prof. Princeton U., spring 1946, Stanford, summer 1948; cons. Manhattan Dist., 1945-46; dir. Task Order 18, Office of Naval Research project, U. Chgo., 1947—; charge cosmic ray research dept. of physics; vis. prof. Brazilian Center Phys. Research, 1951. Numerous expdns. on cosmic rays to Colo. (Mt. Evans, Climax); research on Mt. McKinley, Alaska, 1947, in B-29 planes, Aircraft Carrier (balloon flights); balloon ascensions from Chgo. to study cosmic-rays close to top of our atmosphere, 1940—; expdn. to Guam, 1957; worked with Berkeley Bevatron on K-mesons and hyperons, 1956—. Recipient prize, Schnyder von Wartensee Stiftung, Zurich, 1928. Fellow Am. Phys. Soc., N.Y. Acad. Sci., A.A.A.S.; mem. Ill. Acad. Sci., Swiss Phys. Soc., Brazilian Acad. Sci. Author: Problems in Cosmic Ray Physics, 1946. Died Feb. 20, 1960.

SCHEIN, Martin Warren, Am. biologist; b. N.Y.C., Dec. 23, 1925; s. Jacob A. and Anna (Kawesch) S.; A.B., U. Ia., 1949; Sc.D., Johns Hopkins, 1954; m. Genevieve Madaleine Jauneau, Sept. 13, 1946 (div. June 1954); children—Lonnie G., Michael R.; m. 2d, Maxine M. Kennedy, Jan. 30, 1961; 1 son, Josh L. With USPHS, 1947-48; asso. animal climatologist La. Agrl. Expt. Sta., Jeanerette, La., 1952-55; faculty Pa. State U., 1955-68, prof. zoology, 1965-68; Centennial prof. biology W. Va. U., 1968—; vis. prof., dir. commn. on undergrad. edn. in biol. scis. George Washington U., Washington, 1965-68; vis. lectr. dept. zoology So. Ill. U., 1964. Mem. Animal Behavior Soc., Am. Soc. Zoologists, Ecol. Soc. Am., Am. Soc. Mammalogists, Am. Inst. Biol. Scis., A.A.A.S., Sigma Xi. Research, numerous publs. on behavior cattle, chickens, turkeys, behavioral ecology polar bears, edn. in biology. Office: Dept. Biology, W. Va. U., Morgantown, W. Va. 26505.*

SCHEINBERG, Israel Herbert, Am. physician; b. N.Y.C., Aug. 16, 1919; s. Joseph Reuben and Rachel (Rosenmond) S.; A.B., Harvard, 1940, M.D., 1943; m. Tess J. Levine, June 22, 1952 (dec. May 1953); 1 dau., Anne; m. 2d, Denise Mangravite, Jan. 26, 1957; children—David, Cynthia. Faculty, Harvard, 1950-53; prin. research scientist internal medicine N.Y. State Psychiat. Inst., N.Y.C., 1953-55; prof. medicine Albert Einstein Coll. Medicine, Yeshiva U., Bronx, N.Y., 1958—; vis. prof. U. Cal., La Jolla, 1963-64. Mem. med. adv. bd. Nat. Found. for Neuromuscular Diseases, 1962—, chmn. nat. genetic alert program, 1965—; cons. Pan Am. Health Orgn., WHO, 1964—. Recipient Research award Assn. for Research in Nervous and Mental Diseases, 1959; Ralph E. Miller lectr. Dartmouth Med. Sch., 1961. Commonwealth Fund fellow, 1963-64. Mem. Am. Soc. Clin. Investigation (v.p. 1964-65), Assn. Am. Physicians. Research, publs. in increased understanding of chem. nature of Wilson's disease, phys. chemistry of various proteins. Home: 5447 Palisade Av., Bronx, N.Y. 10471.*

SCHEINBERG, Labe Charles, Am. physician; b. Memphis, Dec. 11, 1925; s. Jacob and Ardie (Cohen) S.; student Southwestern U., Memphis, 1943-44; A.B., U. N.C., 1945; M.D., U. Tenn., 1948; m. Louise Goldman, Jan. 6, 1952; children—Susan, David, Ellen, Amy. Asst. in neurology Columbia, 1952-56; faculty Albert Einstein Coll. Medicine, N.Y.C., 1956—, prof., co-chmn. dept. neurology, 1964—. Guest investigator Rockefeller Inst., 1959—; career scientist Health Research Council, 1962—; scholar, Sister Elizabeth Kenny Found., 1959-61; dir. neurology tng. program Bronx Municipal Hosp. Center, Montefiore Hosp. Med. Center, 1960—. Fellow Am. Acad. Neurology; mem. N.Y. Neurol. Soc., Assn. for Research in Nervous and Mental Disease, N.Y.C. Med. Soc., Am. Soc. Exptl. Pathology, Am. Neurol. Assn., Med. Adv. Bd. Nat. Multiple Sclerosis Soc. Research, publs. on exptl. brain tumors involving immunological mechanisms and tumor growth studies, immune mechanisms of brain in response to tumor grafts and skin homografts. Home: 9 Oak Lane, Scarsdale, N.Y. 10583. Office: 1300 Morris Park Av., N.Y.C. 10461.*

SCHEINER, Christoph, astronomer; b. Wald, Swabia, July 25, 1575; became mem. Soc. Jesus, 1595; studied math., Ingolstadt; tchr., Dillingen; began astron. studies, Ingolstadt, 1610; went to Innsbruck, Austria, 1616; became rector Jesuit Coll., Neisse, Silesia, 1626; prof. math., Rome. Author: De maculis solaribus tres epistolae, 1613; Exegesis fundamentorum gnomonices, 1616; Oculus sive fundamentum opticum, 1619; Rosa ursina, 1626-30. Discovered (independently of Galileo) sunspots, 1611; devised Scheiner's test for demonstrating refraction by a card with a pinhole, 1619; credited with showing that curvature of crystalline lens changes during accomodation, 1619; inventor refracting telescope, parallactic machine, a pantograph, improved helioscope; supported theory of stable earth with moving sun; research on physiology of vision; connected the retina with vision. Died Neisse, July 18, 1650.

SCHEINER, Julius, German astronomer, astrophysicist; b. Cologne, Germany, Nov. 25, 1858; dir. obs., Potsdam, Germany. Author: Populäre Astrophysik, 1908; Spektralanalytische und photometrische Theorien, 1909. Conducted spectroscopic studies to determine motion of stars in line of sight; developed sensitometer to measure sensitivity of photog. plates; eopnym of Scheiner grade in light measurement; determined temperatures of stars. Died Potsdam, Dec. 20, 1913.

SCHELER, Werner Ehrenhold, German pharmacologist; b. Coburg, Germany, Sept. 12, 1923; s. Karl and Elise (Vogel) S.; M.D., U. Jena, 1951; m. Ingeborg Fischbach, Dec. 31, 1960; children—Karin, Ute, Heike. Chief asst. Inst. Medicine and Biology, German Acad. Scis., Berlin-Buch, 1954-59; faculty Humboldt U., Berlin, 1956-59, prof., 1959; prof. U. Greifswald, 1959—, prof. with chair, 1962—. Recipient E. German distinguished service medal, 1959, honor medal, U. Jena, 1964. Mem. European Soc. for Study Drug Toxicity, German socs. biochemistry, pharmacology. Editor Acta biologica et medica germanica, 1958—. Research and numerous publs. on molecular pharmacology and toxicology of blood and respiration poisons; pharm. development of new drugs. Address: 9a Apfelweg, Greifswald, East Germany.*

SCHELL, Irving Israel, meteorologist; b. Szczuczyn, Poland, Oct. 26, 1906; s. Harry and Freida (Rabinowicz) S.; student Mass. Inst. Tech., Harvard; m. May Magriel, Feb. 6, 1938; children—Louisa Faith, Lawrence Magriel. Came to U.S., 1923, naturalized, 1931. Research asst. Blue Hill Meteorol. Obs., Harvard, 1934-46; meteorologist Woods Hole (Mass.) Oceanographic Instn., 1948-54; research asso., prin. investigator dept. geology Tufts U., Medford, Mass., 1954-64; sr. scientist, dir. Ocean-Atmosphere Research Inst., Cambridge, Mass., 1965—. Mem. Am. Geophys. Union, Am. Meteorol. Soc., Arctic Inst. N. Am., Glaciological Soc. Research, publs. on prins. and concepts relating to largescale changes in circulation of atmosphere and ocean and their interaction. Home: 82 Pine Grove, Newton, Mass. 02162. Office: 1640 Massachusetts Av., Cambridge, Mass. 02138.*

SCHELLING, Friedrich Wilhelm Joseph von, German philosopher, epistemologist; b. Leonberg, Würtemberg, Jan. 27, 1775; s. Joseph Frederick Von Schelling; student theology, philosophy Tübingen, Leipzig (both Germany) univs.; m. Caroline Michaelis; m.2d, Pauline Gotter, 1810; several sons, daus. Prof. U. Jena (Germany), from 1798, U. Würzburg (Germany), 1803-08, U. Erlangen (Germany), 1820-26, U. Berlin (Germany), 1841-46; employed Munich, 1827-41. Mem. Royal Acad. Arts Munich (sec. 1808-20). Author: Ideen zu einer Philosophie der Natur, 1797; Von der Weltseele, eine Hypothese der Höhern Physik, 1798; Erste Entwurf Eines Systems der Naturphilosophie, 1799; System des Trancendentalen Idealismus, 1800; Bruno . . . , 1802; Philosophie und Religion, 1804; Über das Verhältniss Philosophischen Untersuchungen über das Wesen der Menschlichen Freiheit, 1809. Mem. naturphilosophie sch. (German transcendentalist reaction against 18th century French materialism); criticized post-Kantian idealism; shifted from interpretation of nature, history in rational dialectic to existential dialectic which sought to ground conditions of thought in datum of experience; taught that nature and spirit are linked by series of developed potencies or powers which in union form one gt. organism in which nature is dynamic (in medicine, this preceded theory of cells within tissues, theory of repetition of parts; prepared way for Heglian absolute idealism; studied epistomology; believed understanding of real world (ultimate truth) is derived by rational means; attempted to develop metaphys. empiricism; early in life supported idealistic pantheism, later turned to voluntarism. Died Ragaz, Switzerland, Aug. 20, 1854.

SCHELLING, Thomas C., Am. economist; b. Oakland, Cal., Apr. 14, 1921; s. John Martin and Zelda Maud (Ayres) S.; B.A., U. Cal. at Berkeley, 1943; Ph.D., Harvard, 1951; m. Corinne Tigay Saposs, Sept. 13, 1947; children—Andrew Crombie, Thomas Ayres, Daniel David, Robert John. With ECA, Denmark, 1948-49, Paris, 1949-50, Office Spl. Asst. to Pres. for Fgn. Affairs, Office Dir. for Mut. Security, Washington, 1951-53; faculty econs. Yale, 1953-58; prof. econs. Harvard, Cambridge, Mass., 1958—, faculty Center for Internat. Affairs, 1958—, on leave to RAND Corp., 1958-59, Inst. for Strategic Studies, London, 1965. Cons. Dept. State, Dept. Def., Arms Control and Disarmament Agy., RAND Corp., other nat. security orgns.; mem. sci. adv. bd. USAF, 1960-63; mem. research adv. bd. Com. for Econ. Devel., 1961-64; mem. Def. Sci. Bd., 1966—. Mem. Am. Econ. Assn., Council on Fgn. Relations, Inst. for

Strategic Studies of London. Author: National Income Behavior, 1951; International Economics, 1958; The Stategy of Conflict, 1960; (with Morton H. Halperin) Strategy and Arms Control 1961; Arms and Influence, 1966; also articles. Home: 20 Oakland St., Lexington, Mass. 02173.*

SCHENCK, Benjamin Robinson, Am. surgeon; b. Syracuse, N.Y., Aug. 19, 1872; s. Adrian A. and Harriet (Robinson) S.; A.B. Williams, 1894; M.D., Johns Hopkins, 1898; m. Jessie McCallum, Aug. 17, 1904. Resident gynecologist Johns Hopkins Hosp. 1898-1903, instr. in gynecology, 1901-03; moved to Detroit, 1903; gynecologist Harper Hosp.; cons. obstetrician Woman's Hosp., Detroit; asso. prof. gynecology Detroit Coll. Medicine. Contbr. numerous articles to med. lit., also to several systems of surgery. Described Schenck's disease, marked by ulcerating nodules in skin, lymph nodes, subcutaneous tissue, caused by Sporotrichum schenckii or beurmanni, 1898. Died June 30, 1920.

SCHENCK, Harry Paul, Am. otolaryngologist; b. Phila., Jan. 6, 1894; s. James Buchanan and Savilla Ettinger (Fries) S.; B.S., Haverford Coll., 1918; M.D., U. Pa., 1923; m. Edna Manvillier Leinbach, Dec. 1924. Staff U. Pa. Hosp., 1923-25, 38-—; curator Mutter Mus., Coll. Physicians of Phila., 1925-37; staff U. Pa., 1927-37, Evans Inst., 1932-59, head dept. otolaryngology U. Pa., 1938-59, prof. otolaryngology grad. sch. medicine, 1939-59, prof. emeritus, 1959-—; asso. otolaryngologist Lankenau Hosp., Phila., 1949-59; research consultant New York EE Infirmary, 1956-66. Consultant in otolaryngology VA, Trustee research fund Central Bureau of Research. Served as pvt. M.C., AEF, World War I; capt. M.C., USNR, World War II. Recipient Casselberry award Am. Laryngol. Assn., 1929; Newcomb award Am. Laryngological Assn., 1959. Diplomate Am. Bd. Otolaryngol. (bd. dirs.). Fellow Coll. Physicians of Phila., A.C.S. (board govs. 1956-66); mem. A.M.A., Am. (pres. 1957-58), Phila. (past pres.) laryngol. assns., Phila. Allergy Society (past president), New York Academy of Sciences, American Otol. Society, Am. Academy Allergy, Coll. Allergists, Am. Laryngol. Rhinol. and Otol. Assn., Am. Acad. Ophthalmology and Otolaryngology, Assn. Mil. Surgeons U. S., Soc. Med. Cons. to Armed Forces. Editor-in-chief: Cyclopedia of Diseases of the Ear, Nose and Throat. Contbr. profl. jours. Research and microscopic observations on allergic changes in living tissue; pathology of lymphocytes, plasma and mast cells; surg. management of osteomyelitis of frontal bone; devel. technic for estimation total precipitable protein in spinal fluid. Home: 1235 Wyngate Rd., Wynnewood, Pa. Office: 326 S. 19th St., Phila. 19103.*

SCHENCK VON GRAFENBURG, Johann, German physician; b. Freiburg im Breisgau, 1531; doctorate U. Tübingen (Germany), 1554; children include Johann Georg; began practice medicine, Freiburg, 1554; later worked in Strasbourg (now in France). Author: Observationum medicarum rariorum libri VII, in quibus nova abdita, admirabilia, monstruosaque exempla circa anatomen, aegritudinum causus, signa, eventus, curationes, per communes locos artificiose digesta proponuntur, 1584-97; De capite humano, 1584; De febrib —— morbis epidemics et con tagiosis, 1597, Made detailed studies of illness and disease, which he classified according to most probable cause. Died Freiburg im Breisgau, 1589.

SCHENDEL, Harold Eugene, Am. nutritional biochemist; b. Mankato, Minn., Oct. 1, 1928; s. William George and Emma Caroline (Schauer); B.A., N. Central Coll., Ill., 1950; M.S., U. Ill., 1951, Ph.D., 1954; m. Marilyn Womeldorff, Sept. 6, 1954; children—Mark, Robyn, Timothy, Gail. Postdoctoral fellow U. Ill., 1955-56, 60-62, Washington U., St. Louis, 1956; sr. research officer U. Capetown, S. Africa, 1956-60; asst. prof. nutrition U. Conn., Storrs, 1962; sr. prof. food and nutrition Winthrop Coll., Rock Hill, S.C., 1962-66; prof. nutrition Fla. State U., Tallahassee, 1966-—. Project leader, pres. So. Assn. Coll. Tchrs. Food and Nutrition, 1963-—. Mem. So. Nutrition Research Group (sec., S.C. rep. 1964-66), S.C. Dietetic Assn. (sec. 1965-66), A.A.A.S., Am. Inst. Nutrition, Am. Soc. Clin. Nutrition, Am Dietetic Assn., N.Y. Acad. Scis., S.C. Pub. Health Assn. Research, numerous publs. on handling of protein by body; malnutrition; lipids. Home: Route 8, Box 51, Tallahassee, Fla.

SCHENK, Erwin Karl, German geologist; b. Metz, Germany, Apr. 25, 1907; s. Karl Wilhelm and Klara (Dorrenberg) S.; student U. Cologne (Germany), 1928, U. Vienna (Austria), 1929; Dr.phil., U. Bonn (Germany), 1934, Dipl.-Geologist, 1934; m. Lotte Hoos, Mar. 24, 1937; children—Margret, Magdel, Frieder, Volker. Asst., Geol. Inst., U. Bonn., 1933-37; geologist Geologische Landesanstalt in Berlin, 1937-45; geologist Hess Landesamt fur Bodenforschung, Wiesbaden, Germany, 1945-60; dir. Geologische Forschungsstelle des Zweckverbandes Oberhessisohn Versorgungsbetriebetriebe, Hungen, Germany, 1960-—; faculty Justus Liebig U., Giessen, Germany, 1955-60; expert for hydrology UNESCO, 1954-—. Mem. Deutsche geologische Gesellschaft, Geologische Vereinigung, Deutsche Quartargeologische Vereinigung. Author: Mechanik der periglazialen Strukturböden; Die geologischen Erscheinungen der Subfusion des Basaltes; also numerous articles. Research on mechanics of mountain bldg., froststructures in soils, frost action and permafrost, geoelec. investigation of tectonic structures; interpretation of geomagnetic anomalies, and reversed magnetisation, volcanological research, hydrological investigations, laterite problems. Home: 6 Professorenweg, Giessen, Hessen. Office: 2 Niddaer Strasse, Hungen, Hessen, Germany.*

SCHENK, Worthington George, Jr., Am. surgeon; b. Buffalo, Feb. 10, 1922; s. Worthington George and Edna (Klein) S.; B.A., Williams Coll., 1942; M.D., Harvard, 1945; m. Jean L.K. Lyon, Mar. 9, 1945; children—Martha (Mrs. James Fike), Lura, Worthington George III, Elsa, Gregory, Molly, Andrew. With U. Buffalo, 1948-66, asso. prof. surgery, 1959-66; prof. surgery State U. N.Y. at Buffalo, 1966-—; dir. surgery E. J. Meyer Meml. Hosp., Buffalo, 1966-—. Diplomate Am. Bd. Surgery. Mem. A.M.A., Am. Surg. Assn., Am. Assn. Surgery Trauma, Soc. U. Surgeons, Soc. Vascular Surgery (treas.), Soc. Clin. Surgery (sec.), Internat. Cardiovascular Soc., A.C.S., Phi Beta Kappa. Editorial bd. Rev. Surgery. Research, numerous publs. on hemodynamics as applied to surg. problems and studies by biophys. methods, trauma and biochem. derangements of severely stressed patient. Home: 164 Doncaster Rd., Kenmore, N.Y. 14217. Office: 462 Grider St., Buffalo 14215.*

SCHENKER, Victor Joseph, neurochemist; b. Montreal, Que., Can., Nov. 25, 1908; s. Eugen and Pauline Anna (Schneider) S.; B.Sc., McGill U., Montreal, 1940, Ph.D. cum laude in exptl. Medicine, 1944; m. Anne Carol Malamud, May 4, 1934; 1 dau., Erika Anne. Came to U.S., 1948, naturalized, 1955. Research asso. McGill U., 1937-41, 44-51; research asso. Royal Victoria Hosp. U. Clinic, 1944-51, lectr., head nutritional research lab., attending staff, 1946-48, asst. prof. medicine, 1951-53; staff Worcester Found. for Exptl. Biology, Shrewsbury, Mass., 1951-53; faculty dept. psychiatry State U. N.Y. Downstate Med. Center, 1953-65, asso. prof., 1953-65, dir. biochem. research, 1958-65; asso. prof. biochemistry Albany (N.Y.) Coll. Medicine, Union U., 1965-—; chief biochemistry div. Research Labs. in Psychiatry and Aging, VA Hosp., Albany, 1965-—. Mem. N.Y. Acad. Scis., Soc. Biol. Psychiatry, Canadian Phsyiol. Soc., Am. Assn. U. Profs., A.A.A.S. Research, publs. on bioassay of adrenal cortical activity, protein metabolism in man in health and during convalescence from acute and chronic illness, biogenesis of adrenal cortical hormones, biochem. and endocrine factors in alcoholism and other neuropsychiat. disorders, nuclei acids and asso. factors as chem. substrates of memory and learning. Home: 7 Pepper Lane, Loudonville, N.Y. 12211. Office: VA Hosp., 151-H, Albany, N.Y. 12208.*

SCHER, Allen Myron, Am. physiologist; b. Boston, Apr. 17, 1921; s. Emmanuel and Celia (Richmond) S.; B.A., Yale, 1942, Ph.D., 1951; m. Maryonda Edmonstone, Aug. 30, 1952; children—Celia, John. Faculty dept. physiology and biophysics U. Wash. Sch. Medicine, Seattle, 1950-—, prof., 1962-—. Mem. computer research study sect. NIH, 1963-67, mem. cardiovascular study sect., 1967-—. AEC Predoctoral fellow, 1946-50. Mem. Am. Physiol. Soc., Biophys. Soc. Contbg. author: Physiology and Biophysics, 1965. Editor: Am. Jour. Electrocardiography, 1962-—. Research, publs. on pathway of elec. activity in heart, origin of electrocardiogram, control of heart stroke output, control system analysis of pressure control. Home: 8304 Avalon Dr., Mercer Island, Wash. 98040. Office: Dept. Physiology and Biophysics, U. Wash. Sch. Medicine, Seattle 98105.*

SCHERAGA, Harold Abraham, Am. phys. chemist; b. Bklyn., Oct. 18, 1921; s. Samuel and Etta (Goldberg) S.; B.S., Coll. City N.Y., 1941; A.M., Duke, 1942, Ph.D., 1946, Sc.D. (hon.), 1961; m. Miriam Kurnow, June 20, 1943; children—Judith Anne, Deborah Ruth, Daniel Michael. Fellow Harvard Med. Sch., 1946-47; faculty Cornell U., Ithaca, N.Y., 1947-—, prof., 1958-—, chmn. dept. chemistry, 1960-67, Todd prof., 1965-—. Vis. lectr. div. protein chemistry Wool Research Labs. C.S.I.R.O., Melbourne, Australia, 1959; co-chmn. Gordon Conf. on Proteins, 1963; mem. Commn. on Molecular Biophysics, Internat. Orgn. Pure and Applied Biophysics, 1965-—; mem. biochemistry tng. com. NIH, 1963-65, career devel. com., 1967-—. Recipient Eli Lilly award, 1957. Fulbright and Guggenheim fellows Carlsberg Lab., Copenhagen, Denmark, 1956-57, Weizmann Inst., Rehovoth, Israel, 1963. Mem. A.A.A.S., Nat. Acad. Sci., Am. Chem. Soc., Am. Soc. Biol. Chemists, Biophys. Soc., Faraday Soc., Am. Acad. Arts and Letters, Phi Beta Kappa, Sigma Xi, Phi Lambda Upsilon. Author: Protein Structure, 1961; also numerous articles. Co-editor: Molecular Biology, 1961-—. Research on phys. chem. studies of proteins, other macromolecules; structure of water. Home: 212 Homestead Terrace, Ithaca, N.Y. 14850.*

SCHERAGO, Morris, microbiologist; b. Rumania, Dec. 25, 1895; s. Israel and Bessie (Jacobs) S.; came to U.S., 1900, naturalized, 1920; B.S., cornell U., 1917, D.V.M., 1919; D.Sc., U. Ky., 1967; m. Jane Stone, Sept. 5, 1920; children—Emily (Mrs. Harvey L. Rubin), Earl James. Faculty, U. Ky., Lexington, 1919-—, prof., chmn. dept. microbiology, 1923-66; vis. prof. bacteriology Washington U. Sch. Medicine, St. Louis, Thailand Project, 1951-52. Cons. USPHS Hosp., Lexington, 1959-—. Recipient Alumni Assn. Faculty Research award, 1959; named Distinguished Prof. of Year Coll. Arts and Scis., 1950-51. Diplomate Am. Bd. Microbiology. Fellow Am. Acad. Microbiology, Am. Coll. Allergists, A.A.A.S., Am. Pub. Health Assn., Am. Soc. Clin. Pathologists, Am. Assn. Immunologists; mem. Am. Soc. Microbiology, Am. Soc. Profl. Biologists, Ohio Valley Allergy Soc., Am. Assn. U. Profs. (past chpt. pres.), Sigma Xi (past chpt. pres.). Asst. editor: Rev. Allergy and Applied Immunology, 1949-—; editorial bd. Annals of Allergy, 1947-65; editor allergy sect. Biol. Abstracts, 1959-—. Research, numerous publs. especially in vitro leukocyte hypersensitivity and its presence and absence in infectious diseases and recovery, and a test for diagnosis and treatment. Home: 255 Shady Lane, Lexington, Ky. 40503.*

SCHERBAUM, Otto Hermann, cell physiologist; b. Vienna, Austria, May 13, 1925; s. Hermann Joseph and Emilie (Bunes) S.; Ph.D., U. Vienna, 1950; m. Trude Schreier, Apr. 30, 1952; children—Claudia-Frances, Andrea-Victoria. Came to U.S., 1957, naturalized, 1962. Postdoctoral fellow Zoophysiology Lab., Copenhagen, Denmark, 1952-55; research fellow Cytophysiology Lab. and Fibiger Lab., Copenhagen, 1955-56; asst. research zoologist U. Cal. at Los Angeles, 1957-58, asst. prof., 1958-61, asso. prof. zoology, 1961-—. Fellow A.A.A.S.; mem. Soc. Protozoologists, Am. Soc. Zoologists, Am. Inst. Biol. Scis., Biophys. Soc. Gen. Physiology, Am. Phys. Soc., Am. Soc. Cell Biologists, Soc. Devel. and Growth, Internat. Soc. Cell Biologists. Contbr. numerous articles to tech. jours. Developed 1st methods for induction synchronous cell div.; research on nucleic acid and phosphorus metabolism div. synchronized protozoa. Home: 10507 Tennessee Av., Los Angeles 90064.*

SCHERER, Alexander Nicolaus von, see von Scherer, Alexander Nicolaus.

SCHERER, Eberhard, German radiobiologist; b. Coblenz, Germany, Oct. 5, 1918; s. Wilhelm and Henriette (Ebert) S.; student Univs. Berlin, Munich, Tubingen; M.D.; m. Felicitas Reisch, Oct. 12, 1947; children—Hans-Eberhard, Michael, Annette, Bettina. Successively became med. specialist in internal medicine, 1950, specialist in radiology, 1953, instr., 1955, prof. at large, 1961; became titular prof. radiology U. Munster, 1963; dir. Roentgen Inst., also Radiology Clinic, Essen, West Germany. Recipient Holthusen award, 1960. Mem. Vereiningung deutscher Strahlenschutzärzte Soc., German Röntgen Soc., Röntgen Soc. Rhine/Westphalia, Soc. Naturalists and Physicians. Editor: Strahlenpathologie der Zelle, 1963. Research and publs. on radiology and biology. Home: Heierbusche 64, Essen-Bredeney, West Germany. Office: Städtische Krankenmanstalten, Hufelandstrasse 55, Essen, West Germany.

SCHERER, William Franklin, Am. physician, microbiologist; b. Buffalo, Aug. 2, 1925; s. William F. and Helen (Seymour) S.; student U. Rochester, 1942-44, M.D., 1947; m. Janice Spicer, Aug. 10, 1946; children—Judith, John, Robert. Faculty, U. Minn., 1950-62, prof. microbiology, 1959-62; prof., chmn. dept. microbiology Cornell U. Med. Sch., 1962-—. Dir. com. on viral infections Armed Forces Epidemiology Bd., 1965-—; chmn. bd. sci. counselors Nat. Inst. Allergy and Infectious Diseases, 1966-—. NRC fellow, U. Minn., 1950-51, Markle scholar, 1953-55, 57-60. Recipient Theobald Smith award, 1959. Mem. Soc. for Exptl. Biology, Soc. Tropical Medicine, Am. Assn. Pathologists and Bacteriologists, Am. Assn. Immunologists, Am. Acad. Microbiologists, Am. Epidemiological Soc., Am. Soc. Cell Biology, Am. Soc. Microbiologists, Infectious Disease Soc. Research, publs. summer ecology of Japanese encephalitis virus in Japan; discovered poliovirus growth in Hela human cancer cells; developed animal cell cultures analogous to bacterial cell methods, method for freezing and storing living animal cells; discovered and characterized arthropod-borne viruses in Japan and Central Am. Home: 6 Sherman Av., Bronxville, N.Y. 10708. Office: 1300 York Av., N.Y.C. 10021.*

SCHERESCHEWSKY, Joseph Williams, physician; b. Peking, China, Mar. 6, 1873; s. Samuel I. J. and Susan Mary (Waring) S.; brought to U. S. in infancy; A.B., Harvard, 1895; M.D., Dartmouth Med. Sch., 1899; m. Bessie Berry Conklin, June 27, 1899; children—Dorothy Duncan, Mary Elizabeth, Henry Waring, Catherine Chapman, Helen Louise, John Forby, Benjamin Berry. Acting hosp. steward, U. S. Army, in Cuba, Spanish-Am. War, 1898; apptd. asst. surgeon USPHS, 1899, promoted through grades to med. dir., 1930, in charge Office of Field Investigations of Occupational Diseases, 1913-18; asst. surgeon gen. in charge of div. of sci. research, Bur. Pub. Health Service, 1918-22; later in charge investigations in cancer USPHS; in charge organizing Ga. Cancer Control Program, 1937. Recipient Distinguished Service Key, Am. Conf. of Phys. Therapy, 1936. Asso. editor Jour. Indsl. Hygiene. Specialist in indsl. hygiene and occupational diseases; founder Cancer Research Center, Harvard. Died July 9, 1940.

SCHERF, David, physician; b. Kimpolung, Austria, Oct. 18, 1899; s. Isaac and Regina (Kreindler) S.; M.D., U. Vienna, 1922; m. Gertrude H.B. Goetz, Apr. 11, 1930. Came to U.S., 1938, naturalized, 1944. Asst. prof. U. Vienna, 1935; faculty N.Y. Med. Coll., N.Y.C., 1938-—, prof. clin. medicine, 1956-—. Cons. physician Met. Hosp., N.Y.C., 1944-—; vis. physician Flower Fifth Av. Hosp., N.Y.C., 1944-—. Mem. Am. Coll. Cardiology (past v.p.), Am. Heart Assn.; hon. mem. French, Argentina, Swiss, Brasilian heart assns., Austrian, Argentinian Med. assns. Author: (with Boyd) Clinical

Cardiology, 1933, 57; (with Boyd) Clinical Electrocardiography, 1945; (with Schott) Extrasystoles, 1953; (with Cohen) The A-V Node, 1964; also numerous articles. Introduced electrocardiography exercise test into clin. medicine; research on mechanisms of irregularities of heart. Home: 10 E. 85th St., Office: 55 E. 86th St., N.Y.C. 10028.*

SCHERMERHORN, Richard Alonzo, Am. sociologist; b. Evanston, Ill., Oct. 18, 1903; s. William David and May (Hoffman) S.; B.A., Dakota Wesleyan U., 1924; M.A., Northwestern U., 1927; Ph. D., Yale, 1931; m. Helen Katherine Karban, Sept. 6, 1926. Asst. prof. Kan. Wesleyan Coll., 1932-33, Clark Coll. and Spelman Coll., 1933-38; asst. prof. Baldwin Wallace Coll., 1940-45, asso. prof., 1945-47; asso. prof. R.I. U., 1947-48; asso. prof. Case Western Res. U., Cleve., 1948-61, prof., 1962—; Fulbright prof. Lucknow U., India, 1959-60. Fellow Gen. Edn. Bd., 1938-39; Yale fellow, 1939-40. Fellow Am. Sociol. Assn.; mem. Indian Sociol. Soc., Soc. for Study Social Problems, Am. Assn. U. Profs., Phi Kappa Phi. Author: These Our People, 1949; Society and Power, 1961; Psychiatric Index for Interdisciplinary Research, 1964; also articles. Formulation of conceptual framework for research in comparative ethnic relations. Home: 2405 Overbrook Rd., Cleveland Heights, O. 44106. Office: Case Western Res. U., Cleve. 44106.*

SCHERR, Charles W., Am. physicist; b. Phila., Mar. 19, 1926; s. Edmund S. and Ruth Scherr; B.S., U. Pa., 1949; M.S., U. Chgo., 1951, Ph.D., 1954; m. Estelle Grace Roman, Apr. 6, 1953; 1 dau., Eve Susan. Research asso. U. Chgo., 1954-56; prof. physics U. Tex., Austin, 1956—. Mem. Am. Phys. Soc., Am. Inst. Physics. Author: (with others) Free-Electron Theory of Conjugated Molecules, A Source Book, 1964. Research on theoretical contbns. to basic understanding of the extra nuclear structure of atoms and molecules. Office: Dept. Physics, U. Tex., Austin, Tex. 78712.*

SCHERRER, Paul, Swiss physicist; b. St. Gallen, Switzerland, Feb. 3, 1890; s. Hermann and Ida (Zürcher) S.; ed. U. Göttingen (Germany), U. Paris, U. Konigsberg (now Kaliningrad, USSR); Dr.phil.; M.D., h.c., Zurich, Switzerland; D.Sc.h.c., U. Louvain (Belgium), U. Toulouse, (France); Dr.h.c., U. Louvain, U. Toulouse, U. Freiburg, U. Geneva, U. Zurich, U. St. Gall, U. Trondeim; m. Ina Sonderegger. Lectr., U. Gottingen; prof. physics, Zürich, 1916—, dir Physics Inst., 1920—. Pres., Swiss Atomic Energy Com. Recipient Mariel-Benoit Prize. Mem. Swiss, Am., Spanish physics socs., Philomatic Soc. Paris, Royal Inst. London (hon.), Swedish, Acad., Acad. Gottingen, Acad. Heidelberg. Author: Physics of Solid State X-rays; Nuclear Physics. Research in X-rays, nuclear physics, quantum theory, crystal structure, magnetism; developed (with Debye) process for application of X-rays in tech. Home: 8, Rislingstr., Zurich, Switzerland.

SCHERTENLEIB, Charles, economist, diplomat of Monaco; b. Paris, France, Sept. 28, 1905; s. Ernest and Lucie (Homery) S.; M.S. magna cum laude, U. Laval, 1944; Ph.D., U. Lausanne, 1944; LL.D., U. Paris, 1946, D.Sc., 1947; m. Lucie Meila, Nov. 11, 1936; children—Francis, Nadine. Came to U. S., 1956, naturalized, 1962. Cons., Internat. Inst. Applied Econs., 1930-36; prof. econs. Ecole Consulaire Mons, 1932-40; private-docent U. Lausanne, 1942-51; vis. prof. Loyola Coll., Montreal, Que., Can., 1953-55; vis. prof. Laval U., Quebec, Que., 1953-55; vis. prof. Sch. Fgn. Service, Georgetown U., 1956-58; asso. prof., 1959—; vis. prof. Howard U., Am. U., Washington, 1957—, Internat. Center, Washington, Inst. Econ. Devel. Bank Internat. Reconstrn. and Devel., 1963—; hon. consul of Monaco, Washington, 1964—. Internat. auditor Gulf Oil Corp., 1938-42, adminstr., cons. comml. and indsl. corps., 1942-52; cons. internat. edn. program AID, 1961—. Decorated chevalier de la Legion d'Honneur; chevalier de l'Ordre de la Couronne de Belgique; knight cross Order of St. Lazare, officer des Palmes Academiques, officer d'Academie de France, officer du Merite de Londres, Comdr. de l'Etoile Noire; comdr. of Order Honor and Merit of Haiti; comdr. du Merite Nat. Francais; comdr. de l'Ordre du Merite de la Cote d'Ivoire; Croix de lere Classe de Saint Jean de Latran (Vatican); recipient Gold medal Societe des Sauveteurs des Alpes Maritimes France; named laureate de l'Institut, French Acad., laureate of U. Paris Faculty of Law and Econs. Fellow Am. Soc. African Culture, African Studies Assn.; mem. Am. Econ. Assn., Am. Acad. Polit. and Social Sci., A.A.A.S., Am. Marketing Assn., Am. Assn. U. Profs., Canadian Inst. for Polit. and Social Studies, Assn. Suisse des Juristes (life), Internat. Assn. for Advancement Sci. Mgmt. (past v.p., 1944-52), Centro Studi e Seambi Culturali Internazionali of Palerme (hon.); fgn. corr. mem. Pontifical Tiberina Acad. Rome, Academie d'Alsace, Colmar, Acad. of Macon, Washington Acad. Scis.; hon. academician Academia Culturale Adriatica of Milan, Valdesia Alumni Assn., Kappa Alpha Phi. Author: Traite theorique et pratique de l'etude des marches, 1945; La Liberte de Commerce et de l'Industrie in droit suisse, 1954; (with others) Marketing, 1960. Research and publs. on comparative internat. growth and productivity, allocation of resources, econ. and social legislations, pub. administrn. orgn. and policies; pioneered work in managerial econs., indsl. mgmt., market structures. Home: 2614 Woodley Pl. N.W. Washington 20008; also 7 Crets de Champel, Geneva, Switzerland.*

SCHEUBEL, Johann, German mathematician; b. Kirchheim, Germany, 1494; prof. math. U. Tübingen (Germany). Author: De numeris et diversis rationibus, 1545; also articles. Editor: Elements (Euclid), 1558. Died Tübingen, 1570.

SCHEUBLE, Hugo Ignaz, Austrian physicist; b. Vienna, Sept. 6, 1881; s. Ignaz and Magdalena (Sura) S.; ed. U. Vienna, also Vienna Poly.; Ph.D. Asst., Vienna Poly., 1907-10, Prague (Czechoslovakia) Poly., 1911-12; asst. Leoben Sch. Mines, 1912-32, prof. electrophysics, electrotech. and applied geophysics, 1932-52, prof. emeritus, 1952—. Mem. Assn. Austrian Engrs. and Architects. Publns. on mechanics of deviation and divining rod. Address: Huettelbergstrasse 9, Vienna 14, Austria.

SCHEUCHZER, Johann Jakob, Swiss geologist, naturalist; b. Zurich, Switzerland, Aug. 2, 1672; s. Hans Jakob Scheuchzer; student, Altdorf and Utrecht; doctorate, 1702; children include Johannes, Hans Caspar, David. Librarian Art Cabinet of Zurich; named prof. math. Carolinum, 1710. Author: Specimen lithologiae Helveticae, 1702; Physica, 1703; Piscium vindiciae et querelae, 1708; Herbarium diluvianum, 1709; Naturhistorie des Schweizerlandes, 2 vols., 1716-18; Physica sacra Jobi, 1721; Homo diluvii testis (described skeleton of giant salamander which he took for a man, now in Haarlem mus.), 1726; Kupferbibel S. Physica Sacra, 4 vols., 1731-35. Founder paleontology in Switzerland; collected fossils; classified ammonites according to structure; researches encompassed mathematics, geodesy, geography, numismatics, philogy. Died Zurich, June 23, 1733.

SCHEUER, Paul Josef, chemist; b. Heilbronn, Germany, May 25, 1915; s. Albert and Emma (Neu) S.; came to U. S., 1938, naturalized, 1944; B.S., Northeastern U., 1943; M.A., Harvard, 1947; Ph.D., 1950; m. Alice Elizabeth Dash, Sept. 5, 1950; children—Elizabeth E., Deborah A. Mem. faculty U. Hawaii, Honolulu, 1950—, prof. chemistry, 1961—, chmn. dept., 1959-62. Mem. Am. Chem. Soc., Chem. Soc. London, A.A.A.S. Research and publs. in field of structural elucidation of organic compounds derived from natural sources particularly from Hawaiian plants and marine organisms. Home: 3271 Melemele Pl., Honolulu 96822.*

SCHEUERMANN, Holger Werfel, Danish orthopedist; b. Horsholm, Denmark, Feb. 12, 1877; practiced radiology, Copenhagen; described osteochondrosis of vertebrae (Scheuermann's kyphosis, or kyphosis dorsalis juvenilis), 1920, also a disorder characterized by necrosis of epiphyses of vertebrae (Scheuermann's disease), 1921. Died Copenhagen, Mar. 2, 1960.

SCHEURER-KESTNER, Auguste, French chemist, physicist; b. Mulhouse, France, Feb. 11, 1833; m. Kestner's dau.; dir. Sète Pyrotech. Co., during War of 1870; founder coop. soc. for workingmen, 1865; elected to Chamber Deps., 1871, to Senate, 1875; active in Dreyfus affair; encouraged Zola to write J'accuse. Recipient Gold medal Mulhouse Soc., 1868. Author: Principes élémentaires de la théorie chimique des types appliquée aux combinaisons organiques, 1863. Research on combustion, also on organic chemistry. Died Bagnères, France, 1899.

SCHEVING, Lawrence Einar, Am. biologist; b. Hensel, N.D., Oct. 20, 1920; s. Einar Larus and Mary E. (Brown) S.; B.S., DePaul U., 1949; M.S., Loyola U., Chgo., 1950, Ph.D., 1957; m. Virginia M. Krumdick, Aug. 6, 1949; children—Lawrence vertebrae (Scheuermann's kyphosis, or Kyphosis dor-Allen, Mary Kathleen, John Steven, Jennifer Lynne, Patricia (dec.). Faculty, Lewis Coll., Lockport, Ill. 1950-57; faculty dept. anatomy Chgo. Med. Sch., 1957-67, prof. anatomy, 1966-67; prof. anatomy La. State U., 1967—. chief profl. services 361st Med. Lab.,Chgo., 1960-67. Recipient Bd. Trustees Research award Chgo. Med. Sch., 1962; named Prof. of Year, Chgo. Med. Sch. Student Council, 1964. Mem. Am. Assn. Anatomists, Am. Soc. Zoologists, Internat. Soc. Biorhythm Research (dir.). Internat. Soc. Electro Myographic Kinesiologists, Pan Am. Assn. Anatomists, Sigma Xi. Research, publs. on biol. rhythms especially circadian rhythms, factors affecting growth of cells and tissues outside of body, function of some common muscles of man. Home: 2520 Ramsey Dr., New Orleans 70114.*

SCHIAPARELLI, Giovanni Virginio, Italian astronomer; b. Savigliano, Piedmont, Mar. 14, 1835; early edn. in Savigliano; studied engring. and math., Turin, 1850; degree, 1854; studied meteorology, Berlin, 1856-58; m. Maria Comotti, 1865; 2 sons, 3 daus. Trained in astronomy under Struve at Pulkova; became dir. Brera Obs., Milan, 1860; prof. geodesy Instituto techico superior di Milano, 1863-72. Mem. Royal Acad. Scis. Turin, Royal Astron. Soc., Internat. Geodetic Assn., Soc. Acad. Lincei, Royal Soc, 1896, French, Vienna acads. scis. Author: Obsservazioni astronomiche e fische sull'asse di rotazione e sulla topografia del pianeta Marte, 1878-1910; Die Astronomie im Alten Testment, 1904; also numerous articles. Contbd. immensely to knowledge of stars and planets by his calculations and observations; discoverer asteroid Hesperia, 1861; observed double stars, also markings (named them canals) on Mars; thought Venus and Mercury rotated on their axes simultaneously with revolutions around sun; demonstrated that meteor swarms move in cometary orbits; contbd. to study of ancient astronomy in Babylonia. Died Milan, July 4, 1910.

SCHIAVINATO, Giuseppe, Italian mineralogist; b. Padua, Dec. 10, 1915; s. Leopoldo and Melania (Marcolin) S.; Dr ès sc; m. Albina Rebula, Jan. 3, 1943. Prof. mineralogy U. Milan, also pres. Faculty Sci. Mem. Lombard Inst., Acad. Sci. and Letters, Pugliese Acad. Sci. Publns. on mineralogy, petrography, sci. of mines. Research on tertiary vulcanites of Veneto; mineragenesis and petrogenesis; intrusive massif of Mt. Adamello and metamorphites of Central Alps. Home: via. Baldissera 4. Office: via Botticelli 23, Milan, Italy.

SCHICHAU, Ferdinand, German engr.; b. Elbing, Germany, 1814; founder iron works and shipbldg. plant, Elbing, 1834; built 1st screw vessel in Germany, 1855, a Woolf steam engine, 1860, 1st compound ship engine in Germany for gunboat, 1878, 3-phase expansion engine for passenger steamship, 1883, torpedo boat with 3-phase expansion engine, 1887, 4-phase expansion engine for torpedo boats capable of 34 knots; established shipyard, Danzig, Poland. Died 1896.

SCHICK, Bela, physician; b. Boglar, Hungary, July 16, 1877; s. Jacob and Janka (Pichler) S.; M.D., Karl Franz U., Graz, Austria, 1900; hon. degree Albert Einstein Coll. Medicine, 1957; m. Catharine Fries, Dec. 3, 1925. Came to U. S., 1923, naturalized, 1929. Faculty, U. Vienna, 1902-23; dir. pediatric dept. Mt. Sinai Hosp., N.Y.C., 1923-42; clin. prof., lectr. diseases of children Columbia, 1936-43; vis. prof. pediatrics Albert Einstein Coll. Medicine, 1936-43; chief pediatric dept. Beth-El Hosp., Bklyn., 1950-62; cons. pediatrician Sea View, Willard Parker, Beth Israel hosps., N.Y. Infirmary for Women and Children. Recipient Gold medal N.Y. Acad. Medicine, 1938; Addingham Gold medal, Leeds, Eng., 1938; Gold medal Midwest Forum on Allergy, 1941; Semelvers medal Am. Hungarian Med. Soc., 1954; John Howland award Am. Pediatric Soc., 1954; John Brandeis Gold medal, 1956. Mem. Am. Soc. Pediatrics, Am. Acad. Pediatrics, Acad. Medicine; hon. mem. Harvey Soc., many others. Author: (with Clemens Pirquet) Serum Krankheit, 1905; (with Theodore Escherich) Scarlet Fever, 1912; Pirquetsche System der Ernahrung, 1918; (with William Rosenson) Child Care Today, 1933. Research, numerous publs. on allergic and immunological phenomena; devised Schick Test for determining susceptibility to diphtheria, 1913. Address: 1045 Park Av., N.Y.C. 10028.*

SCHIEMANN, Guenther, educator, chemist; b. Breslau, (now Wroclaw, Poland), Nov. 7, 1899; s. Robert and Elsbeth (Prager) S.; ed. univs. Breslau, Freiburg im Breisgau (Germany), Fedl. Tech. U. Zurich (Switzerland), 1919-25; Ph.D., U. Breslau, 1925; m. Hildegard Augustin, Dec. 30, 1930; children—Eberhard, Anneliese (Mrs. Lange), Waltraut (Mrs. Müller), Gottfried. Asst., chief asst. Tech. U. Hanover (Germany), 1926-35, docent organic chemistry, 1929-37, unscheduled prof., 1946—, prof., dir. Inst. Chem. Engring. Tech. U. Hanover, 1956—. indsl. chemist Cologne, Wiesbaden, Offenbach/Main (all Germany), 1935-50; prof., dir. Inst. Chem. Engring. U. Istanbul (Turkey), 1950-56. Mem. Soc. German Chemists, Leopoldina, Internat. Soc. Study Fats, DECHEMA. Author: Die Chemie der Natürlichen und Künstlichen Farbstoffe, 1935; Die Organischen Fluorverbindungen in Ihrer Bedentung für die Technik, 1951; (with Boy Cornils) Cyklische Organische Fluorverbindungen, 1967. Research on chem. engring. and organic chemistry; specialist on varnish raw materials, artificial resins, organic fluorine compounds, fluidization. Home: 14 Hohenzollernstrasse, Hanover, Germany.

SCHIFF, Harold Irvin, Canadian chemist; b. Kitchener, Ont., Can., June 24, 1923; s. Jack and Lena (Bierstock) S.; B.A. with honours in Chemistry and Physics, U. Toronto, 1945, M.A. in Chemistry, 1946, Ph.D. in Phys. Chemistry, 1948; m. Daphne Line, Dec. 30, 1948; children—Michael, Sherry. Postdoctoral fellow Nat. Research Council, Ottawa, Ont., 1948-50; asst. prof., asso. prof., prof. dept. chemistry McGill U., Montreal, Que., Can., 1950-63; prof., chmn. dept. chemistry York U., Toronto, Ont., 1963—, dean sci., 1966—. Cons., Def. Research Bd. Can., 1960—, Environmental Sci. Service Adminstrn., Boulder, Colo., 1963—. Fellow Chem. Inst. Can.; mem. Sigma Xi. Research and numerous publs. on transport phenomena in electrolytic solutions, chem. kinetics, mass spectrometry, chem. processes in upper atmosphere, chemiluminescence, reaction of gas phase atoms, ions and simple molecules in planetary atmospheres. Home: 60 Donwoods Dr., Toronto 12, Ont., Can.*

SCHIFF, Jerome A., Am. biologist; b. Bklyn., Feb. 20, 1931; s. Charles K. and Molly (Weinberg) S.; B.A. in Biology and Chemistry, Bklyn. Coll., 1952; Ph.D. in Botany, U. Pa., 1956. With biology dept. Brandeis U., Waltham, Mass., 1956—, prof., 1966—. Carnegie Found. fellow Stanford, 1962-63. Fellow Am. Acad. Arts and Scis.; mem. Soc. for Developmental Biology (sec. 1964-66), Am. Soc. Plant

Physiologists, Am. Soc. Protozoologists, Am. Soc. Biol. Chemists, Physiol. Soc. Am., Biophys. Soc., A.A.-A.S. Asst. editor Plant Physiology, 1964—. Research and publs. on physiology, biochemistry and devel. of algae, especially devel. of chlorophyll-containing structures of Euglena; biochemistry of Chlorella, especially its reduction of sulfate to sulfide. Home: 18 Bowdoin, Cambridge, Mass. Office: Biology Dept., Brandeis U., Waltham, Mass. 02154.*

SCHIFF, Leonard Issac, Am. physicist; b. Fall River, Mass., Mar. 29, 1915; s. Edward Ephraim and Matilda (Brodsky) S.; B.E., Ohio State U., 1933, M.Sc., 1934; Ph.D., Mass. Inst. Tech., 1937; m. Frances Margaret Ballard, Aug. 25, 1941; children—Ellen Margaret, Leonard Ballard. NRC fellow, research asso. U. Cal., Cal. Inst. Tech., 1937-40; instr. U. Pa., 1940-42, asst. prof., 1942-44, asso. prof., 1944-47, acting chmn. physics dept., 1942-45; research physicist NDRC, Columbia, 1941-45, U. Cal., 1944-45; mem. anti-submarine warfare operation research group, 1943-45; staff Los Alamos Sci. Lab., 1945-46; asso. prof. Stanford, 1947-48, prof., exec. head physics dept., 1948—; dir. Varian Assos., 1948-53; vis. prof. Ia. State Coll., 1952, U. Paris, 1956-57, Univs. Madras and Bombay (India), 1963; Guggenheim fellow, 1956-57; vis. com. physics dept. Mass. Inst. Tech., 1954-56; cons. editor McGraw Hill Book Co., 1954—; Stewart lectr. U. Mo., 1955; Phillips lectr. Haverford, 1963; Tektronix Found. lectr. Ore., 1964. Vice pres., dir. Varian Found., 1961—; mem. phys. scis. program com. Alfred P. Sloan Found., 1961—. Chmn. adv. com. physics div. Office Sci. Research, Air Research and Devel. Command USAF, 1955—; mem. sci. adv. group Office Aerospace Research, USAF, 1963—. Recipient Lamme medal Ohio State U., 1959. Mem. Internat. Conf. Theoretical Physics, Tokyo, also Kyoto, Japan, 1953. Fellow Am. Phys. Soc. (councillor 1953-57), A.A.A.S., Cal. Acad. Sci.; mem. Am. Assn. Physics Tchrs. (Oersted medal 1966); Fedn. Am. Scientists (council del.), Am. Assn. U. Profs., Nat. Acad. Scis., Sigma Xi, Tau Beta Pi. Author: Quantum Mechanics (rev. edit.), 1955; also sci. research and rev. articles. Asso. editor Rev. Sci. Instruments, 1943-45, Phys. Rev., 1945-47, 63-65, Physics Today, 1950-56, Rev. Modern Physics, 1951-54, Ann. Rev. Nuclear Sci., 1952-63, Jour. Math. Physics, 1960-62. Research on theoretical nuclear physics, elementary particle physics, relativity, gravitation. Home: 27 Berenda Way, Menlo Park, Cal. Office: Stanford U., Stanford, Cal.*

SCHIFF, Moritz, German physiologist, biologist; b. Frankfort/Main, Germany, 1823; ed. Heidelberg, Berlin, Göttingen (all Germany); M.D., 1844; dir. ornithology sect. Frankfort Zool. Mus.; prof. comparative anatomy, Berne, Switzerland, 1854-63, Florence, Italy, 1863-77, Geneva, Switzerland, 1877-96. Author: Gesammelte Beitrage zur Physiologie, 4 vols., 1894-96. Experimented with removal of thyroid gland in animals which proved fatal, 1859, but succeeded in saving the animals' lives by following removal of gland with its intraabdominal transplantation, 1884; experimented (with J.L. Reverdin) on prodn. of myxedema in animals, 1882. Died 1896.

SCHIFFER, Davide, Italian neuropathologist; b. Verzuolo, Piedmont, Italy, Jan. 28, 1928; s. Alessandro and Firmina (Boero) S.; student medicine U. Milan (Italy), 1946-49; Ph.D., U. Turin (Italy), 1950, M.D., 1952; m. Silvana Gaffuri, July 2, 1959; children—Cristina, Isabella. Asst., Clinic Nervous and Mental Diseases, U. Turin, 1952-53, chief lab. neuropathology and neurohistochemistry, 1957—, faculty neurosurg. pathology, 1961—; research staff Hirnforschungs Institut Neustadt (Germany), 1953. Mem. Italian Soc. Neurology (mem. directory sect. neuropathology). Contbg. author: Neuropathology. Numerous publs. on neuropathology and histochem. research on cerebral tumors, senescence, lesions of nervous tissue. Home: 49 C. so.M.D'Azeglio, Turin, Italy.*

SCHIFFER, John Paul, physicist; b. Budapest, Hungary, Nov. 22, 1930; s. Ernest and Elizabeth (Tornai) S.; came to U.S., 1947, naturalized, 1953; B.A., Oberlin Coll., 1951; M.Sc., Yale, 1952, Ph.D., 1954; m. Marianne Tsuk, June 28, 1960; 1 dau., Celia Anne. Research asso. Rice Inst., Houston, 1954-56; asst. physicist Argonne (Ill.) Nat. Lab., 1956-59, asso. physicist, 1959-64, sr. physicist, asso. div. dir., 1964—. Vis. asso. prof. Princeton, 1964. Guggenheim fellow, 1959-60. Fellow Am. Phys. Soc. Research, publs. on structure of atomic nucleus, gravitational red shift measurement using Mossbauer effect, expts. on search for fractionally charged particles.*

SCHIFFER, Menahem Max, mathematician; b. Berlin, Germany, Sept. 24, 1911; s. Chaim and Miriam (Alpern) S.; M.A., Hebrew U., 1934, Ph.D., 1938; m. Fanya Rabinovics, Dec. 3, 1937; 1 dau., Dinah B. Came to U. S., 1946, naturalized, 1955. Jr., sr. asst., lectr. Hebrew U., Jerusalem, 1934-46, prof., 1950-51; research asso. Harvard, 1946-49; vis. prof. Princeton, 1949-50; vis. prof. math. Stanford (Cal.) U., 1948, 50, prof., 1951—. Mem. Am. Math. Soc., Am. Acad. Arts and Scis. Author: (with Bergman) Kernel Functions and Elliptic Differential Equations in Mathematical Physics, 1953; (with Spencer) Functionals of Finite Riemann Surfaces, 1954; (with Adler and Bazin) Introduction to General Relativity, 1965. Editor: Archive for Rational Mechanics and Analysis, Jour. d'Analyse Mathematique. Research and numerous publs. on variational methods in theory of univalent functions, theory of fundamental solutions in partial differential equations by means of kernel function methods, theory of Riemann surfaces, problems of fluid dynamics. Home: 3748 Laguna St., Palo Alto, Cal. 94306.*

SCHILD, Albert, Am. mathematician; b. Hessdorf, Germany, Mar. 3, 1920; s. Josef and Sara (Wolf) S.; B.A., U. Toronto, 1946, M.A., U. Pa., 1948, Ph.D., 1950; m. Clara Shekter, Dec. 29, 1946; children—Hannah, Ruth, Rebecca, Eva, Judith, Beatrice, Batsheva. Came to U. S., 1946, naturalized, 1959. Asst. instr. math. dept. U. Pa., 1946-50; from instr. to asso. prof. math. Temple U., 1950-62, prof., 1962—, chmn. dept., 1963—. Lectr., NSF; cons. RCA, Cherry Hill, N.J., Sun Oil Co., Phila. Mem. Am. Math. Soc., Math. Assn. Am., Operations Research Soc. Research on properties of certain types of functions of complex variable, scheduling problems; introduced Schild functions. Am. Home: 4940 N. 9th St., Phila. 19141.*

SCHILD, Alfred physicist; b. Istanbul, Turkey, Sept. 7, 1921; s. Ignaz and Fanny (Mandel) S.; student U. Coll., U. London, 1938-40; B.A., U. Toronto, 1943, M.A., 1944, Ph.D., 1946; m. Winnifred Zara Beames, Aug. 7, 1943; children—Carol H.M., Frances K., John D. Came to U.S., 1946, naturalized, 1956. Faculty, Carnegie Inst. Tech., 1946-48, 49-56; Frank B. Jewett fellow Inst. for Advanced Study, Princeton, 1947-48; asst. prof. U. Toronto, 1948-49; adv. mathematician Westinghouse Research Labs., Pitts., 1956-57; faculty U. Tex., Austin, 1957—, Ashbel Smith prof., dir. Center for Relativity Theory, 1963—. Cons. to industry; sci. sec. Internat. Conf. on Relativistic Theories of Gravitation, Warsaw, Poland, 1962. Fellow Am. Phys. Soc.; mem. Am. Math. Soc., Am. Assn. Physics Tchrs., Math. Assn. Am. Author: (with J.L. Synge) Tensor Calculus, 1949; also numerous articles. Coeditor: Gravitational Collapse and Quasi-Stellar Sources, 1965. Translator (from German) The Mission of the Vega (sound play Friedrich Durrenmatt), 1962. Patentee in oil prodn. field. Research on gen. relativity; Einstein's theory of gravitation; math. applied to engring. problems, including elasticity theory and underground flow of oil. Home: 2712 Mt. Laurel Lane, Austin, Tex. 78703.*

SCHILD, Heinz Otto, pharmacologist; b. Fiume, May 18, 1906; s. Hermann and Thekla (Spiegel) S.; M.D., U. Munich (Germany), 1931; Ph.D., U. Edinburgh (Scotland), 1935; D.Sc., U. London, (Eng), 1950; m. Mireille Madeline Haquin, Aug. 2, 1938; children—Marion (Mrs. Ian Short), Renee, Barbara. Asst. pharmacology dept. U. Edinburgh, 1936; faculty dept. pharmacology U. Coll., London, Eng., 1937—, prof. pharmacology, head dept., 1961—; examiner univs. of London, Leeds, Liverpool (all Eng.), W. Africa, W.I., Makerere Coll., 1947-67. Mem. vis. team med. scientists WHO, S.E. Asia, 1952—, mem. com. on biol. standardization, 1963—. Fellow Royal Soc.; mem. Physiol. Soc., Brit. Pharmacological Soc., Immunology Soc., Biometric Soc., Soc. Exptl. Biology, Brit. Med. Assn. Author: (with A. Wilson) Applied Pharmacology, 1968; also numerous articles. Research on way drugs produce contraction and relaxation of smooth muscles, mechanism by which allergic reactions are produced. Home: Mole Ridge, St. Mary's Rd., Leatherhead, Surrey, Eng. Office: Dept. Pharmacology, U. Coll., Eng., London.*

SCHILDER, Franz Alfred, zoologist; b. Prag, Austria, Apr. 13, 1896; s. Franz and Marie (von Sterneck) S.; D.philosophy, U. Vienna, 1921; m. Maria Hertrich, Aug. 4, 1924; 1 dau., Franziska (dec. 1961). Asst., Entomol. Mus. Berlin (Germany), 1922-24, Inst. Genetics, Berlin, 1924; ofcl. Biologische Reichsanstalt Naumburg/Saale (Germany), 1925-47; hon. prof. U. Halle (Germany), 1945, prof. zoology, 1947-62. Hon. mem. malacological socs. London, Honolulu. Author: Anleitung zu biostatistischen Untersuchungen, 1951; Biotaxonomie, 1952; Die Bänderschnecken, 1953; Lehrbuch der Zoogeographie, 1956; also numerous articles. Research on taxonomy and evolution of living and fossil Cypraeacea, influence of heredity and environments on variations of European landsnail Cepaea, practical simplified methods in statis. research, taxonomy of several groups of beetles, statis. methods in zoogeography, rules in fixing type localities in animal species. Home: 19 Schleiermacherstrasse, Halle-Saale, DDR 402, East Germany.*

SCHILDER, Paul Ferdinand, psychiatrist; b. Vienna, Austria, Feb. 15, 1886; s. Ferdinand and Berta (Fürth) S.; M.D., U. Vienna, 1909, postgrad., 1909-10; Ph.D., U. Halle, 1911; m. Mitzi Moser, July 1919; m. 2d, Lauretta Bender, Nov. 1936; children—Michael, Peter. Asst. in psychiatry Univ. Hosps., Halle, 1910-12, Leipzig, 1912-14; physician in Austrian Army, 1914-18; 1st asst. in psychiatry Univ. Hosp., Vienna, 1918-28; extraordinary prof. U. Vienna, 1925; clin. dir. psychiatric div. Bellevue Hosp., N.Y., 1929—; research prof. psychiatry N.Y. U., 1929—. Author: Hypnosis (with O. Kauders), 1926; Introduction to a Psychoanalytic Psychiatry, 1928; Brain and Personality, 1931; The Image and Appearance of the Human Body, 1935; Psychotherapy, 1938; also several books pub. in Berlin and Vienna. Described encephalitis periaxialis diffusa (Schilder's disease), an inflammation of white matter of cerebrum, found mainly in children, 1912. Died Dec. 8, 1940.

SCHILDKNECHT, Hermann, German chemist; b. Fuerth, Bavaria, Aug. 2, 1922; s. Lorenz and Margarethe (Hitz) S.; student chemistry U. Erlangen (Germany), 1946-52, Masters Degree, 1952, Ph.D., 1953; m. Helga Mai, Aug. 25, 1953; 1 dau., Ulrike. Sci. asst. Inst. Organic Chemistry, Erlangen, 1952-53, pvt. asst., 1953, lab. asst., 1955-63, lectr. chemistry, 1959; asso. prof. U. Heidelberg (Germany), 1963, prof. organic chemistry, dir. Organic Chemistry Inst., 1964—. Author: Zonenschmelben Verlagchemie, 1964; Zonemelting, 1966; also articles. Research on plant and animal chemistry, chem. def. mechanism of arthorpodes, organic compounds in meteorites, zone melting of organic substances, origin of life; invented method of columncrystallization, method of electron pyrolysis. Home: 5 Wilckens-, Heidelberg, Germany.*

SCHILL, Göran, Swedish chemist; b. Nässjö, Sweden, July 21, 1918; s. Simon Johansson and Tekla Schill; S.; Apotekarexamen, Royal Inst. Pharmacy, 1943, Farmacie licentiatexamen, 1952, D.Pharmacy, 1965; m. Brita Lindholm, July 29, 1944; children—Jonas, Karin, Stefan. Analytical chemist Central Mil. Hosp., 1944-47; asst. prof. chemistry Royal Inst. Pharmacy, Stockholm, Sweden, 1947-58, asso. prof. analytical chemistry, 1958—. Author: (with B. Danielsson) Läkemedl, 1957; Analytisk farmacevtisk kemi, 1960; also articles. Research on quantitative determination of various substances important in pharmacy, conditions for extraction and determination of association complex-ese using physico-chem. methods. Home: 19C Poppelvägen, Sollentuna, Sweden. Office: 49 Kungstensgatan, Stockholm, Sweden.*

SCHILLER, Walter, physician; b. Vienna, Austria, Dec. 3, 1887; s. Friedrich and Emma (Friedmann) S.; grad. U. Vienna Med. Sch., 1912; m. Marie Papper, Feb. 25, 1923; children—Esther Marianne, Sue J. (Mrs. Howard Carlotn Udell). Lab. dir. Jewish Meml. Hosp., N.Y.C., 1937-38; dir. pathology Cook County Hosp., Chgo., 1938-43; staff Woman's and Children's Hosp., Chgo., 1943-52. Lectr. Royal Acad., Dublin, 1935, U. Manchester, Eng., U. Birmingham, Eng., 1936, various Am., Can. univs. Mem. Internat. Coll. Surgeons (certification of recognition 1958), Am. Assn. Pathologists and Bacteriologists (hon.), Austrian Soc. Gynecology and Obstetrics, Am. Assn. Cancer Research. Author: Pathologie und Klink der Granulosazelltumoren, 1934; also articles, chpts. in textbooks. Developed Schiller test for painting female cervix for early diagnosis of cancer; research on uterine cancer and ovarian tumors, classification of ovarian tumors; discovered an ovarian tumor (mesonephroma). Home: 1410 Lake St., Evanston, Ill. 60201. Died May 2, 1960.*

SCHILLING, John Albert, Am. surgeon; b. Kansas City, Mo., Nov. 5, 1917; s. Carl Fielding and Lottie Lee (Henderson) S.; A.B., Dartmouth, 1937; M.D., Harvard, 1941; m. Barbara Whipple, Feb. 13, 1943 (div. Aug. 1956); children—Christine, Katharine; m. 2d, Lucy West, June 8, 1957; children—Jolyon, John. Mem. faculty U. Rochester Sch. Medicine and Dentistry, 1944-56; prof., head dept. surgery U. Okla. Sch. Medicine, Oklahoma City, 1956—. Nat. cons. to surgeon gen. USAF, 1958—; mem. surgeon gen.'s com. U. S. Army, 1963—; cons. Nat. Cancer Inst., 1966—. Diplomate Am. Bd. Surgery. Fellow A.C.S.; mem. Am. Cancer Soc., Am. Assn. Cancer Research, Soc. Exptl. Biology and Medicine, Soc. U. Surgeons, Am. Surg. Assn., Soc. Exptl. Pathology, Assn. Am. Med. Colls., N.Y. Acad. Scis., Am. Fedn. Clin. Research, Am. Assn. Surgery of Trauma, Surg. Biology Club, Am. Physiol. Soc., Titulaire de la Societe Internationale de Chirurgie, Nat. Assn. Standard Med. Vocabulary, Soc. Med. Consultants Armed Forces, Pan-Pacific Surg. Assn., Soc. for Surgery of Alimentary Tract, Alpha Omega Alpha, Sigma Xi, others. Mem. editorial bd. Surgery, 1962—, Annals of Surgery, 1966—; adv. editorial bd. Am. Jour. of Surgery, 1958—. Research and publs. on biochemistry of wound healing, pulmonary and renal function, work capacity of man and animals, pathophysiology of intestinal disease. Home: 24 Oakwood Dr., Oklahoma City 73121. Office: 800 N.E. 13th St., Oklahoma City 73104.*

SCHILLING, Richard Selwyn Francis, English physician; b. Kessingland, Eng., Jan. 9, 1911; s. George and Florence Louise (Loweth) S.; student St. Thomas's Hosp., 1929-36, M.B., B.S., U. London, 1937, M.D., 1947, D.P.H. with distinction, D.I.H., 1947, D.Sc., 1964; m. Heather Maude Eilnore Norman, Aug 28, 1937; children—Christopher John, Elinore Marguerite, Erica Ruth. Obstetric house physician St. Thomas's Hosp., 1935; house physician Addenbrooke's Hosp., Cambridge Eng., 1936; asst. indsl. med. officer Imperial Chem. Industries (metals) Ltd., Birmingham, Eng., 1937, med. insp. factories, 1939-42; sec. indsl. health research bd. Med. Research Council, 1942-46; Nuffield fellow indsl. health 1946-47; reader occupational health U. Manchester (Eng.), 1947-56; dir. dept. occupational health London Sch. Hygiene and Tropical Medicine, U. London, 1956—, prof. dept., 1960—. Civil cons. in occupation health Royal Navy, 1966; civil cons. indsl. medicine RAF, 1967; cons. WHO, 1956. Fellow Royal Coll. Physicians (London); Royal Soc. Arts; mem. Soc. Occupational Medicine (past pres.), Brit. Occupational Hygiene Soc. (past pres.), Royal Soc. Medicine (past pres. occupational medicine sect.). Author: Modern Trends in Occupational Health, 1960; also articles. Devel. techniques for studying respiratory diseases in textile workers; field studies and identification of diseases in Gt. Britain, Netherlands, Egypt and U. S.; studies of occupational health hazards, including deep sea fishing. Home: 32 Inner Staithe Hartington Rd., London W.4., Gt. Britain.*

SCHILLING, Robert Frederick, Am. physician; b. Adell, Wis., Jan. 19, 1919; s. Edgar F. and Lillian (Bollard) S.; B.S., U. Wis., 1940; 1943; m. Mariam Alan Hansen, Feb. 2, 1946; children—Carla, Robert Frederick II, Fredericka, Richard, Anne. Faculty, U. Wis., Madison, 1951—, prof. medicine, 1962—, USPHS Research prof. medicine, 1963-64, chmn. dept. medicine, 1964—. Mem. Assn. Am. Physicians, Am. Soc. Clin. Investigation, Central Soc. for Clin. Research, Soc. for Exptl. Biology and Medicine, Phi Beta Kappa, Alpha Omega Alpha. Research, numerous publs. on absorption and utilization vitamin B12, mechanisms in causation anemia, gastrointestinal absorption nutrients. Home: 4153 Mandan Crescent, Madison, Wis. 53711.*

SCHILLING, Victor, hematologist; b. Torgau, Austria, Aug. 28, 1883; prof., Berlin. Author: Das Blutbild, 1912; Praktische Blutlehre, 1922. Developed a modification of Arneth's classification of neutrophils in which cells are divided into 4 groups (Schilling's hemogram), 1912. Died Rostock, Germany, May 30, 1960.

SCHIMITSCHEK, Erwin, botanist; b. Vysokopole, Mar. 31, 1898; s. Eduard and Rosa (Petzelbauer) S.; student Vienna Agr. Coll.; Dr. agrégé; m. Gertrud Schreckenthal, Dec. 27, 1933; children—Ingrid, Gerda. Agrégé, Vienna, Austria, 1930-45; became asso. prof., 1936; prof. Faculty Sylviculture Turkey, 1937-39; with Vienna Agr. Sch., 1939-45; dir. S.O.-Institut für Wald- und Holzforschung, 1942-45; joined Niederösterreichische Landesforstinspektion, 1947, dept. sylviculture Fed. Expt. Inst., Mariabrunn, 1951. Recipient Karl Escherich medal, 1963. Mem. German, Finnish (corr.) entomology socs. Soc. Forestry Sci. Finland. Author: Schlüsschadliche Käfer, 1937; Forstinsekten der Türkei und ihre Umwelt, 1944; Forstschutz Taschenbuch, 1949; Bericht über augetrossene Forstschaden und deren Bekämpfung in Niederösterreich in der Jahren, 1946-50. Die Bestimmung von Insektenschäden im Walde nach Schadensbild und Schädling, 1955. Research and numerous publs. on susceptibility of trees to attacks of primary and secondary insect pest; entomophagous; applied entomology, especially biol. and ecol. control. Home: Beethovenstrasse 8. Office: Forstzoologie Institüt, Schloss, Hannover-Münden, West Germany.

SCHIMPER, Andreas Franz Wilhelm, botanist; b. Strasbourg (now in France), May 12, 1856; s. Philipp Wilhelm Schimper; asso. prof., Bonn, Germany; prof. U. Basel (Switzerland); mem. extensive sci. expdns. Author: Pflanzengeographie (early may of plant distbn.), 1898. Pioneer in plant ecology; revealed stability of plant territories; introduced term chloroplast and distinguished it from chromatophore; studied devel. of starch grains. Died Basel, Sept. 9, 1901.

SCHIMPER, Karl Friedrich, German biologist, morphologist; b. Mannheim, Germany, Feb. 15, 1803; pvt. tchr. and researcher; pioneer in modern plant morphology; formulated theory of phyllotaxis. Died Schwetzingen, Germany, Dec. 21, 1867.

SCHIMPER, Philipp Wilhelm, geologist, paleobotanist; b. Dossenheim, Alsace, Jan. 12, 1808; became German citizen; at least 1 son, Andreas Franz Wilhelm; prof. geology, dir. Mus. of Strasbourg (now in France). Mem. French Acad. Scis., 1854. Author: Bryologia Europaea, 6 vols., 1836-55; Synopsis Muscarum Europearum, 1860; Traité de paléontologie végétale, 1869-74. Authority on mosses. Died Strasbourg, Mar. 20, 1880.

SCHINDEWOLF, Otto Heinrich, German geologist, paleontologist; b. Hanover, Germany, July 6, 1896; s. Heinrich and Helen (Wietfeld) S.; student U. Göttingen (Germany), 1914-18; Dr.Phil., U. Marburg (Germany), 1919; Dr.rer.nat., U. Bonn (Germany), 1966; m. Hedwig Scheel, Dec. 17, 1924; children—Eberhart, Ulrich. Faculty, U. Marburg, 1919-27; geologist, dir. Geol. Survey Berlin (Germany), 1927-47; prof. U. Berlin, 1947-48; prof. U. Tübingen (Germany), 1948-64, prof. emeritus, 1964—. Decorated Leopold von Buch-Plakette. Author numerous books, articles. Research on paleontology of corals and cephalopods, problems of evolution, biostratigraphy. Home: 17 Ob dem Viehweidle. Office: 10 Sigwartstrasses, Germany.*

SCHINDLER, Guenter Martin, mathematician; b. Ebersdorf, Germany, Sept. 15, 1928; s. August and Rosalie (Schmidt) S.; student U. Kiel (Germany), 1949-51, Vordiplom in Math., 1951; Hauptdiplom in Math., U. Goettingen (Germany), 1953; Dr.rer.nat., 1956; m. Helga M. Breitenstein, Feb. 28, 1957; children—Nicoline, Christoph. Came to U.S., 1957, naturalized, 1962. Staff, Kernreaktor G.m.b.H. Karlsruhe, Germany, 1956-57; sr. scientist USAF Missile Devel. Center, Holloman, N.M., 1957-58; asso. prof. math. U. N.M., 1957-58, U. Cal. at Santa Barbara, 1958-59; chief mathematician Advanced Tech. Corp., Goleta, Cal., 1959-61; project mgr. Gen. Electric Co., Santa Barbara, Cal., 1961-65; sr. tech. specialist N.Am. Aviation, Inc., Downey, Cal., 1965—. Cons. Astro-Research Corp., Santa Barbara, 1961-65, USN Marine Engring. Lab., Annapolis, Md., 1965. Mem. Am. Math. Soc., N.Y. Acad. Sci. Research, publs. on math. formulation and solution complex problems and interpretation results in terms phys. and tech. significance. Home: 624 in terms phys. and tech. significance. Home: 28026 Bechgate Dr., Palos Verdes Peninsula, Cal. 90274. Office: 12214 Lakewood Blvd., Downey, Cal. 90241.*

SCHINTLMEISTER, Josef Peter, Austrian nuclear physicist; b. Radstadt, Austria, June 16, 1908; s. Peter and Marie (Scharfetter) S.; Dr.phil., U. Vienna, 1932; m. Maria Renata Deinlein, Mar. 14, 1936; children—Wolfgang, Ingeborg; m. 2d, Alexandra Obolenskaja, June 2, 1955; children—Alexander, Josef. With Austrian Patent Office, Vienna, 1934-39; docent, asst. II. phys. inst., U. Vienna, 1939-45; prof., dir. inst. exptl. nuclear research Dresden (Germany) Inst. Tech., 1955-58; head Central Inst. Nuclear Research, German Acad. Scis. Berlin, Rossendorf/Dresden, 1958-64, dir., 1964—. Recipient Nat. prize, East Germany, 1964. Author: Die Elektronenröhre als physikalisches Messgerät, 1942; Tabellen der Atomkerne (with W. Runz), 1958-65; Der Isospin von Atomkernen, 1960; also numerous articles. Research on dE/dx particle discrimination, 1st multiparameter analysis, 1st precise measurements of long life alpha-emitters particle energies and applications of systematic of alpha decay for prediction of alpha emission, 1st separation of fission products by method of passing through thin foils, nuclear spectroscopy, nanosecond coincidences. Home: 3 Calberhastrasse, 8054 Dresden. Office: P.O. Box 19, Rossendorf, 8051 Dresden, East Germany.*

SCHINZ, Hans, Swiss botanist; b. Zurich, Switzerland, Dec. 6, 1858; prof., Zurich; studied southwest Africa. Author: Explorations dans le Sud-Ouest de l'Afrique, 1887; Deutsch-Südwestafrika, 1891; Conspectus florae Africae, 1895; (with others) Flora der Schweiz, 6 vols., 1900. Editor a Swiss flora, also proc. Swiss Natural Sci. Soc. Died 1941.

SCHINZ, Hans Rudolf, Swiss radiologist; b. Zurich, Switzerland, Dec. 13, 1891; s. Hans and Amalie (Frei) S.; student medicine U. Zurich; M.D., postgrad. Vienna, Stockholm, Paris; Dr. natural research honoris causa; m. Ginevra Ruesch. Dir. roentgen inst., radiotherapeutic clinic U. Zurich, 1919—; became privat docent radiology, Zurich, 1934; mem. Parliament, canton of Zurich. Mem. Am. Roentgen Ray Soc., Skandinavische Vereinigung für Radiologie Stockholm, Soc. medica chirurgica Bononiensis, Royal Soc. Medicine London, Deutsche Röntgenkunde, Am. Coll. Radiology, Soc. medica chirurgica Bononiensis. Author: Vom Krebs und seiner Bekampfung, 1949; Co-author: Lehrbuch der Röntgendiagnostik, 5th edit., 1950. Research in X-ray diagnostics, radiotherapy, radiation biology. Home: Kurhaustr. 78, Zurich, Switzerland.

SCHINZ, Rudolf Eduard, engr., inventor; b. Zurich, Switzerland, 1812; developed (with Hirn) exhaust gas preheater to re-use lost heat, 1843; devised spring pressure gauge; built suspension bridges, Marienburg, Prussia. Died 1855.

SCHIRREN, Carl Georg, German physician; b. Kiel, June 15, 1923; s. Carl Georg and Annelise (Reuter) S.; student univs. Kiel, Greifswald, Hamburg, Zurich, Munich; Dr.med.; m. Leonore Scharmer, 1950; 7 children. Faculty, U. Munich, 1954—, unscheduled prof., 1960—, dep. dir. dermatol. clinic. Corr., hon. mem. Danish, Italian, Mexican dermatol. socs. Co-editor: Dermatologisches Handbuch. Research, numerous publs. in dermatology: physiology, tumors, radiobiology, radiotherapy, autoagression diseases. Address: 5 Fiedlerstrasse, Munich-Solln, West Germany.*

SCHISHKIN, Boris Konstantinovich, Russian botanist; b. Apr. 19, 1886; grad. Tomsk U., 1911. Asst., Tomsk U., 1914-18, prof., 1925-30; became asso. Bot. Inst., USSR Acad. Scis., 1931, dir., 1938-49. Recipient State Prize, 1952. Corr. mem. USSR Acad. Scis. Editor series of works. Research and numerous publs. on flora of USSR; bot. studies in Siberia, Middle Asia, Calpathians, Caucasus. Office: V. L. Komarov Inst. Botany, USSR Acad. Scis., Ulitsa Popova 2, Leningrad, USSR.

SCHLÄFER, Hans Ludwig, German phys. chemist; b. Frankfort/Main, Germany, May 5, 1923; s. Ludwig and Ida (Röll) S.; Dipl.-Chem., U. Frankfort/Main, 1949, Dr.phil.nat., 1950. Asst., Inst. Phys. Chemistry, U. Frankfort/Main, 1952-55, faculty, 1955—, prof., 1965—. Recipient chemistry prize, Göttingen Acad. Sci., 1962. Author: Komplexbildung in Lösung, 1961; (with G. Glieman) Einführung in die Ligandenfeldtheorie, 1967; also numerous articles. Research on coordination chemistry: structure and bonding, absorption spectra and magnetic behavior, photochemistry and luminescence, reaction kinetics. Address: 5 Hansa-Allee, Frankfort/Main, West Germany.*

SCHLÄFLI, Ludwig, Swiss mathematician; b. Burgdorf, Switzerland, 1814; prof., Bern, Switzerland. Author: Theorie der vielfachen Kontinuität, 1901. Founder multidimensional geometry. Died 1895.

SCHLAGINTWEIT, Hermann, German geographer; b. Munich, May 13, 1826; engaged in phys. geog. research with brother Adolphe in Alps, 1846-53; on sci. mission with brothers Adolphe and Robert, to India, Deccan, N.W. provinces, summits and glaciers of Himalaya, Sikkim, Boutan, Assam, Turkestan, Karakoroum and Kun-Lun mountain chains, 1854-57. Author: Untersuchungen über die physikalische Geographie der Alpen, 2 vols., 1950-54; (with brother Robert) Results of a Scientific Mission to India and High Asia, 2 vols., 1861; Reisen in Indien und Hochasien, 4 vols., 1869-80. Died Munich, Jan. 19, 1882.

SCHLAGINTWEIT, Robert, German geographer; b. Munich, 1833; on mission to India with brothers Hermann and Adolphe, 1854; prof., Giessen, Germany, from 1864. Author: (with brother Hermann) Results of a Scientific Mission to India and High Asia, 1861; also publs. based on studies of Cal., 1871, of Mormons, 1874. Died Giessen, 1885.

SCHLAPP, Max Gustav, Am. neurologist; b. Ft. Madison, Ia., Nov. 4, 1869; s. George and Marie (Dupuis) S.; student Cornell U., 1891-93, also Bellevue Med. Sch. (New York U.); studied in Germany 6 yrs; M.D., U. Berlin, 1896. Specialist in mental and nervous diseases; asst. prof. neuropathology Cornell U., until 1914; prof. neuropathology Post-Grad. Med. Sch. Hosp., 1911—; apptd. by Gov. Glynn mem. N.Y. State Commn. to investigate Provision for Mentally Deficient, 1914; chmn. med. bd. Children's Cts. N.Y.C. Contbr. of numerous tech. papers pertaining to neurol. topics, causes of feeble-mindedness. Died Mar. 5, 1928.

SCHLATTER, Carl, Swiss surgeon; b. Zurich, Switzerland, Mar. 18, 1864; prof. surgery, Zurich, from 1899. Author: Die Veranderungen des Linmatwassers, durch den Abwasser der Stadt, 1890; Ohaffner Brief an meine Kollegen, 1910. Performed 1st successful complete stomach resection (gastrectomy), 1897; described (after R.B. Osgood) painful lesions of tibial tuberosity (Osgood-Schlatter disease), 1903; developed new techniques in accident surgery. Died 1934.

SCHLECHTWEG, Heinz, German didacticist; b. Neu-Isenburg, Germany, Nov. 15, 1905; s. Karl and Lina (Rind) S.; Dr., U. Frankfurt (Germany), 1929, tchr.'s examination, 1929, 31; Dr.phil.nat.habil., U. Göttingen (Germany), 1933; m. Erika Dürr, Feb. 27, 1943; 1 dau., Ursula. Theoretical physicist Inst. for Indsl. Research, 1934-46; asst. prof. U. Koeln, 1950-59; prof. didactics maths. div. Duisburg, Paedagogische Hochschule Ruhr, Kettwig, Germany, 1946—; staff Haus Der Technik, Essen, Germany, 1947—. Cons. geologist Gesellschaft Essen, 1963—; mem. Commn. internat. pour l'amélioration de l'enseignement des mathématiques, Brussels, Belgium, 1965—. Mem. Deutsche Physikalische Gesellschaft, Gesellschaft Deutscher Naturforscher und Ärzte, Gesellschaft für angewandte Mathematik und Mechanik, Verein Eisenhütenleutte. Author or co-author books; also numerous articles. Research on theory of metals and strength, didactics of primary schs. Address: 13 Brederbachstrasse, Kettwig, Germany.

SCHLEGEL, Hans Günter, German microbiologist; b. Leipzig, Germany, Oct. 24, 1924; s. Walther Johannes and Ilse (Hahn) S.; student U. Leipzig, 1946-48; Staatsexamen für das höhere Lehramt, U. Halle (Germany), 1949, doctor's degree, 1950, Habilitation für Mikorbiologie und Pflanzenphysiologie, 1954; m. Ingeborg Tegtmeyer, June 29, 1957; children—Peter, Dagmar, Uta-Susanne. Asst., Institut für Botanik, U. Halle, 1951; asst., sci. researcher Institut für Kulturpflanzenforschung, Deutschen Akademie der Wissenschaften, Berlin, 1952-56; fellow Max-Planck-Institut für Zellchemie, Munich, 1956; fellow dept. microbiology Western Res. U., Cleve., 1957-58; prof. microbiology, head Institut für Mikorbiologie, U. Göttingen (Germany), 1958—; head Institut für Mikrobiologie, Gesellschaft für Strahlenforschung, Neuherberg-München, Germany, 1966—. Mem. Akademie der Wissenschaften zu Göttingen. Author: Anreicherungskultur und Mutantenauslese, 1965; also numerous articles. Research on physiology and biochemistry of chemosynthetic and photosynthetic bacteria of soil and water, especially Knallgasbacteria utilizing molecular hydrogen as an energy- and carbon dioxide as a carbon-source; regulation mechanisms in autotrophic bacteria. Home: 2 Grünberger Strasse, 3406, Bovenden/Göttingen, Niedersachsen, W. Germany.*

SCHLEGEL, Richard, Am. physicist; b. Davenport, Ia., Aug. 29, 1913; s. Richard and Mayme (Hansen) S.; A.B., U. Chgo., 1935; M.A., U. Ia., 1936; postgrad. U. Colo.; Ph.D., U. Ill., 1943; m. Frances Stanley McKee, June 16, 1946; children—Thomas H., Catherine M. Lectr., Mus. Sci. and Industry, Chgo., 1938-40; asso. physicist Metall. Lab., U. Chgo., 1941-43; instr. physics Princeton, 1945-48; faculty Mich. State U., East Lansing, 1948—, prof., 1957—, acting head dept. physics, 1955-56; vis. prof. U. Cal., Berkeley, summer 1959. Vis. asso. Cavendish Lab., Cambridge, Eng., 1954-55; affiliate History and Philosophy of Sci. Group, Cambridge, Eng., 1961-62. Fellow Am. Phys. Soc. Author: Time and the Physical World, 1961; Completeness in Science, 1967. Research on Lorentz transformations, study of cosmology, phys. time concept, quantum theory and Zeno's paradoxes, philosophy of sci. Home: 660 Stoddard Av., East Lansing, Mich. 48823.*

SCHLEHUBER, Alva Marion, Am. plant breeder; b. Durham, Kan., Jan. 14, 1906; s. August and Matilda (Kreutziger) S.; B.S., Kan. State U., 1931; M.S., Wash. State U., 1933, postgrad., 1933-36, Ph.D., 1938; postgrad. Martin Luther U., Halle, Germany, 1936-37; m. Veronica Mary Horsch, June 12, 1940; children—Rita (Mrs. Michael McShea), Thomas George, Margaret Helen. Grammar sch. tchr., 1925-27; acting supt. Adams Br. Expt. Sta., 1936; asso. agronomist Mont. State Coll., Bozeman, 1938-41; asso. geneticist U.S. Sugar Plant Field Sta., Meridian, Miss., 1941-45; prof. agronomy in charge small grains investigations Okla. State U., Stillwater, 1945-66; crops specialist IRI Research Inst., N.Y.C., 1966—; Fulbright lectr., Brazil, 1965-67. Chmn., Nat. Wheat Improvement Com., 1960-66, Hard Red Winter Wheat Improvement Com., 1958-66; cons. on agronomic research, teaching and extension Ethiopia, 1965. Recipient award of merit Gamma Sigma Delta, 1966. Fellow Am. Soc. Agronomy, A.A.A.S.; mem. Crops Sci. Soc. Am., Genetics Soc. Am., Am. Genetic Assn., Brazilian Soc. Genetics. Research and numerous publs. on creation of varieties of wheat, oats and barley, mode of inheritance of disease and insect resistance; correlated inheritance between maturity and quality characteristics in wheat. Home: 369 Gen.Neto Pelotas, RS, Caixa P. 541. Office: Caixa Postal 541, Pelotas RS, Brazil.*

SCHLEICH, Carl Ludwig, German physician; b. Stettin, Germany, July 19, 1859; prof., Berlin. Author: Schmerzlose Operationen, 1894. Used subcutaneous injections of dilute solutions of cocaine as anesthesia, 1892; introduced infiltration anesthesia into medicine, 1894-95. Died Berlin, Mar. 7, 1922.

SCHLEICHER, August, German linguist; b. Meiningen, 1821; prof., Bonn, Germany, also Prague, Czechoslovakia, Jena, Germany. Author: Recherches de grammaire comparée, 1848-50; Manuel de Lithuanien, 1856-57; Abrégé de grammaire comparée des langues indo-germaniques (standard work), 1861; La Théorie de Darwin et la linguistique, 1863. Applied theories of Darwin to Indo-european linguistics. Died Jena, 1868.

SCHLEIDEN, Matthias Jakob, German botanist; b. Hamburg, Germany, Apr. 5, 1804; ed. Heidelburg, 1824-27; began career in law, then turned to botany; prof., Jena, Germany, Tartu (now Dorpat, Estonia), Dresden, Weisbaden, Frankfort, Germany. Author: Grundzüge der wissenschaftl. Botanik, 2 vols., 1842-43; Die Bedeutung der Juden fur den Erhaltung und Weiderbelebung der Wissenschaften in Ma., 1877. Made microscopic studies of plant tissue; elaborated cell theory for plants, 1838; recognized importance of cell nucleus and associated it with cell division; accepted Darwinian theory of evolution in spite of his assn. with nature philosopher sch. Died Frankfort, June 23, 1881.

SCHLENK, Hermann, biochemist, educator; b. Jena, Germany, July 28, 1914; s. Wilhelm and Mathilde (von Hacke) S.; dipl. chem. U. Berlin, 1936; Dr. rer. nat., U. Munich, Germany, 1939; m. Inge K. Schier, June 29, 1946; children—Thomas, Cornelia. Came to U.S., 1949, naturalized, 1954. Research chemist Bad. Anilin & Sodafabrik, Ludwigshafen, 1939-44; research asso. U. Munich, 1944-46; asst., lectr. U. Wurzburg, 1946-49; asso. prof. Tex. A. and M. Coll., 1949-52; with Hormel Inst. U. Minn., Austin, 1953—, prof. biochemistry, 1955—. Mem. Am. Chem. Soc., Am. Oil Chemists Soc., Am. Soc. Biol. Chemists, Sigma Xi. Contbr. numerous articles to profl. jours.; biochem. research on lipids. Office: 801 16th Av. N.E., Austin, Minn. 55912.

SCHLESINGER, Bernard, Brit. physician; b. Eng., Nov. 23, 1896; s. Richard Schlesinger; ed. Emmanuel Coll., Cambridge; M.A., M.D.; m. Winifred Regensburg; 2 sons, 2 daus, 1 dec. dau. Cons. physician N.W. Army and Central Command, India Command, 1939-45; cons. physician U. Coll. Hosp., Hosp. for Sick Children, Gt. Ormond St., Royal No. Hosp.; now ret.; cons. pediatrician to Army. Recipient Dawson Williams prize, 1961. Fellow Royal Coll. Physicians, Assn. Physicians, Royal Soc. Medicine (pres. pediatric sect. 1960-61); mem. Brit. Pediatric Assn. (pres. 1953-54). Author numerous books, articles. Research on rheumatism, children's diseases studied relation of throat infection to acute rheumatism in children, and showed hemolytic streptococcus is causative agt., 1930. Home: St. Mary Woodlands House, St. Mary Woodlands, Lambourn, Berkshire, Eng.

SCHLESINGER, Edward Bruce, Am. neurol. surgeon; b. Pitts., Sept. 6, 1913; s. Samuel B. and Sara Marie (Schlesinger) S.; B.A., U. Pa., 1934, M.D., 1938; m. Mary Eddy, Nov. 1941; children—Jane, Mary, Ralph, Prudence. Faculty, Columbia Coll. Phys. and Surg., N.Y.C., 1946—, prof. clin. neurol. surgery, 1964—; attending neurol. surgeon White Plains (N.Y.) Hosp., 1954-63, cons., 1963—. Diplomate Am. Bd. Neurosurgery. Fellow N.Y. Acad. Scis.; mem. Harvey Cushing Soc., Harvey Soc., A.A.A.S., Assn. Research in Nervous and Mental Disease, Neurosurg. Soc. Am., Am. Assn. Surgery of Trauma, Am. Rheumatism Soc., Am. Coll. Clin. Pharmacology and Chemotherapy, A.M.A., Eastern Assn. Electroencephalographers, Sigma Xi. Research, publs. on uses, effects of curare in neuromuscular disease, lesions of central nervous system, localization of brain tumors using radioactive tagged isotopes. Home: 67 Olyphant Av., Dobbs Ferry, N.Y. 10522. Office: 710 W. 168th St., N.Y.C. 10032.*

SCHLESINGER, Frank, Am. astronomer; b. N.Y., May 11, 1871; s. William Joseph and Mary (Wagner) S.; B.S., Coll. City of N.Y., 1890; M.A., Columbia, 1897, Ph.D., 1898; Sc.D., U. Pitts., 1920; Sc.D., Cambridge, 1925. In charge Internat. Latitude Obs., Ukiah, Cal., 1899-1903; astronomer Yerkes Obs., under auspices of Carnegie Instn., 1903-05; dir. Allegheny Obs. (U. Pitts.), 1905-20; dir. Yale U. Obs., 1920-41. Recipient Gold medal Royal Astron. Soc., 1927; Bruce Gold medal Astron. Soc. of Pacific, 1929; Townshend medal Coll. City N.Y., 1935. Fellow A.A.A.S. (past chmn. Sect. A), Am. Acad. Arts and Sciences; mem. Nat. Acad. Sciences, Am. Philos. Soc., Am. Astron. Soc. (past pres.), Internat. Astron. Union (past pres.), Phi Beta Kappa, Sigma Xi; hon. asso. Royal Astron. Soc. Can.; fgn. asso. Royal Astron. Soc. of London; hon. mem. Mexican Astron. Soc., Italian Soc. of Spectroscopists; corr. mem. French (Valz medal 1926), Swedish acads. scis., French Bureau de Longitudes. Collaborating editor Astrophys. Jour. Author catalog listing 4000 stellar distances, 1935; also monographs in scientific journals on reduction of phorographic plates, stellar parallaxes, variations of latitude, spectroscopic binaries, solar rotation, motion in line of sight; developed photographic method of determining distances of stars from earth. Died Lyme, Conn., July 10, 1943.

SCHLESINGER, Hermann, Austrian internist; b. Pressburg, Austria, June 2, 1866; prof., Vienna. Author: Die Krankheiten des höheren Lebensalters, 1914; Syphilis und innere Medizin, 3 vols., 1925-28; Klinikund Therapie der Alterskrankheiten, 1930. Died 1934.

SCHLEUSENER, Alfred Friedrich Karl Paul, German geophysicist; b. Staffelde, Germany, Mar. 1, 1898; s. Paul and Margarete (Beulter) S.; Dipl.Mining Engr., Tech. U., Berlin-Charlottenburg, 1925; Dr.-Ing., Tech. U., Breslau, Germany, 1934; m. Lore Eva Sophie Helene Schlickum, May 21, 1938; 1 son, Eckart Hermann Paul. With Seismos BmbH, Hannover, Germany, 1925-63, party chief sci. expdns. in Spain, Africa, U.S., Persia, Turkey, Italy, Poland, Iceland, 1925-63, dept. mgr., 1935-63; faculty Tech. U. Hannover (Germany), 1954—, prof. geophysics, 1958—. Mem. Deutsche Geophysikalische Gesellschaft (pres. 1965-66), Soc. Exploration Geophysicists, European Assn. Exploration Geophysicists. Author: (with O. Niemczck) Gravity Measurements across the Young Volcanic Zone of Iceland, 1943; also articles. Research on gravimetric problems, tables of gravity, gravity of central Europe; constructed 1st European transportable static gravity meter. Address: 17 Ludwig Bruns, 3 Hannover, Germany 3.*

SCHLIEMANN, Heinrich, archeologist; b. Neubukow, Mecklenburg-Schwerin, Germany, Jan. 6, 1822; m. Ekaterina Lishin Oct. 12, 1852; m. 2d, Sophia Engastromenos, Sept. 24, 1869; several children. Apprentice in grocery bus., Furstenberg, 1836-41; employed in Amsterdam, 1842-46; founded own export-import bus., 1847, traveled extensively; was in Cal. when it was admitted to union, 1850, thus automatically became U. S. citizen; operated bank for gold prospectors, Sacramento, 1851-52; spent several years in Russia, primarily in St. Petersburg, made fortune as mil. contractor during Crimean War; ret. from bus. to pursue archeol. interests, 1863, went to visit Homeric sites in Greece, 1868; later settled in Athens; continued excavations til his death. Author: Ithaka, der Peloponnes und Troja, 1869; Trojanische Altertümer, 1874; Troja und seine Ruinen, 1875; Mykenä, 1878; Ilios, 1881; Stadt und Land der Trojaner, Orchomenos, 1881; Troja. Ergebnisse meiner neuesten Ausgrabungen, 1884; Tiryns, 1886; an autobiography, 1892. Founder of modern archaeology of Greek Bronze Age; on basis of his faith in historicity of Homer's Iliad and his conviction of existence of Priam's Troy, theorized that Hissarlik, Turkey, (not Bunarbashi) was its site; dug at Hissarlik identifying seven strata, erroneously thought Troy II to be Priam's city because of treasure recovered from this level; excavated Mycenae and discovered grave circle within citadel, 1876; established at Mycenae evidence of culture described by Homer, refuting belief that it was myth; cleared tholos tomb at Orchomenus, 1880-81; cleared rock from Tiryns, revealing plan of Mycenaen palace, 1885; excavated at Ithaca, Alexandria, Cythera; tried to secure rights to site of Knossos, Crete. Died Naples, Italy, Dec. 26, 1890.

SCHLINGER, Evert Irving, Am. entomologist; b. Los Angeles, Apr. 17, 1928; s. William McKinnely and Julia (Gleason) S.; student U. Cal. at Davis, 1946-49, Ph.D., 1957; B.S., U. Cal. at Berkeley, 1950; m. Audrey Darlene Sirois, June 8, 1957; children—Mathew James, Jane Ann, Brian Thomas; m. Barbara Hellen Carll, Jan. 20, 1951 (div. June 1956); 1 son, Evert Irving. Profl. entomol. collector Assos. in Tropical Biogeography in Mexico, 1953-54, Cal. Acad. Scis., Peru, Ecuador, Colombia, 1954-55; staff U. Cal. at Riverside, 1956—, asso. entomologist, 1963—, prof. entomology, 1967—; leader NSF grants, 1963—. Guggenheim Found. fellow, 1966-67. Fellow A.A.-A.S., Cal. Acad. Scis.; mem. Entomol. Soc. Am., Ecol. Soc. Am., Soc. Systematic Zoology, Am. Pacific Coast, Kan. entomol. socs., So. Cal. Acad. Sci., Sigma Xi. Contbg. author, asst. editor: Biological Control of Insect Pests and Weeds, 1964. Editorial bd. U. Cal. Publs. in Entomology, 1966—; mem. editorial bd. Jour. Econ. Entomology, 1963-67, chmn., 1967—. Research and publs. on biology, systematics and host-relationships of internal spider parasites of the world; one of team which successfully controlled spotted alfalfa aphid in Cal., which developed machine for sampling insect populations; studies on internal parasites of aphids, seasonal distributional patterns, world wide dispersal and distbns. of insects. Home: 172 Knox Ct., Riverside, Cal. 92507.*

SCHLOEMANN, Ernst, physicist; b. Borgholzhausen, Germany, Dec. 13, 1926; s. H.W. and Auguste (Koch) S.; M.S., U. Goettingen (Germany), 1953, Ph.D., 1954; postgrad. (Fulbright scholar) Mass. Inst. Tech.; m. Gisela Mattiat, June 9, 1955; children—Susan C., Sonia G., Barbara I. Came to U.S., 1954, naturalized, 1965. Mem. solid state and molecular theory group Mass. Inst. Tech., 1954-55; mem. staff research div. Raytheon Co., Waltham, Mass., 1955-61, 62—, sci. fellow, 1962—; vis. asso. prof. Stanford, 1961-62; vis. prof. U. Hamburg (Germany), 1966. Fellow Am. Phys. Soc.; mem. Sigma Xi. Research, numerous publs. on theory of ferromagnetic materials and applications to microwave devices, theory heat conduction in dielectric solids. Home: 38 Brook Rd., Weston, Mass. 02193. Office: 28 Seyon St., Waltham, Mass. 02154.*

SCHLOERB, Paul Richard, Am. surgeon; b. Buffalo, Oct. 22, 1919; s. Herman George and Vera M. (Gross) S.; A.B., Harvard, 1941; M.D., U. Rochester, 1944; m. Louise Mary Grimmer, Feb. 25, 1949; children—Ronald G., Patricia J., Marilyn A., Dorothy E., Paul R. Faculty, Harvard Med. Sch., 1951, U. Rochester Sch. Medicine, 1952; faculty U. Kan. Sch. Medicine, 1952—, research prof. surgery, 1964—. Cons. in surgery Kansas City, Wichita VA hosps. Recipient Research Career Devel. award USPHS, 1962-67. AEC-NRC fellow, 1948-49. Mem. Soc. U. Surgeons, A.M.A., A.A.A.S., A.C.S., Central Surg. Assn., Am. Fedn. Clin. Research, Am. Heart Assn., Am. Physiol. Soc. Research on body water and mineral content, shock, kidney and liver failure, treatment before and after surg. operations. Home: 8024 Fontana St., Prairie Village, Kans. 66208. Office: U. Kan. Med. Center, Kansas City, Kan. 66103.*

SCHLOESING, Alphonse Théophile, French chemist; b. Paris, May 26, 1856; s. Jean-Jacques-Théophile Schloesing; ed. École polytechnique; prof. chemistry, dir. State Sch. Applied Mfg.; mem. French Acad. Scis., 1903, Acad. Agr. Author: Étude sur le dosage de la nicotine dans le jus de tabac, 1895; Étude sur l'acide phosphorique dissous par les eaux du sol, 1899. Research on fermentation, nitrification of manure, presence of azote and argon in blood, hygroscopic properties of tobacco and extraction of nicotine. Died Paris, July 9, 1903.

SCHLOESING, Jean-Jacques-Théophile, French chemist; b. Marseille, France, July 9, 1824; dir. state Sch. Applied Mfg.; became prof. chemistry Agronomic Inst., 1876; mem. Acad. Agr., French Acad. Scis., 1882. Conducted expts. in agrl. chemistry which contbd. to sci. base of agronomy; developed methods of analysing tobacco; inventor method of preparing sodium carbonate from ammoniac, 1855; gave definitive proof of role of bacteria in nitrification of organic azote, 1870. Died Feb. 8, 1919.

SCHLOFFER, Hermann, Austrian surgeon; b. Graz, Austria, May 13, 1868; prof., Innsbruck, Austria, also Prague, Czechoslovakia. Credited with 1st successful operation for pituitary tumor in acromegaly in man, 1906; began transnasal route in surgery of pituary ademona, 1907; concentrated on operations of intestinal tract; fgn. body granulation swellings named Schloffer tumors after him. Died 1937.

SCHLöMANN, Ernst, physicist; b. Borgholzhausen, Germany, Dec. 13, 1926; s. H. Wilhelm and Augusta (Koch) S.; M.S., U. Göttingen (Germany), 1953, Ph.D., 1954; m. Gisela Mattiat, June 14, 1955; children—Susan C., Sonia G., Barbara I. Mem. solid state and molecular theory group, Mass. Inst. Tech., 1954-55; with Raytheon Co., Waltham, Mass., 1955—, sci. fellow, 1964—; vis. asso. prof. Stanford, 1961-62; vis. prof. U. Hamburg (Germany), 1966. Fellow Am. Phys. Soc.; mem. Sigma Xi. Research, publs. on solid state physics, magnetic phenomena, ferromagnetic resonance, lattice dynamics, thermal conductivity of solids, microwave physics and tech., statis. mechanics. Patentee in field. Home: 38 Brook Rd., Weston, Mass. 02193. Office: 28 Seyon St., Waltham, Mass. 02154.

SCHLOSSMAN, Abraham, Am. physician; b. N.Y.C., May 20, 1918; s. Louis and Ida (Krasnow) S.; A.B., Wesleyan U., Conn., 1936; M.A., Columbia, 1937; M.D., N.Y. U., 1943; m. Carolyn Sondheimer, Apr. 3, 1952; 1 dau., Nina. Practice medicine specializing in ophthalmology, N.Y.C., 1947—; staff Manhattan Eye Ear and Throat Hosp., Beth Israel Hosp.; clin. asso. prof. ophthalmology State U. N.Y., Downstate Med. Center, 1955—. Co-recipient Schoenberg

Meml. award Nat. Soc. for Prevention Blindness, 1951. Diplomate Nat. Bd. Med. Examiners, Am. Bd. Ophthalmology. Fellow A.C.S.; mem. A.M.A., N.Y. State, N.Y. County med. socs., N.Y. Acad. Medicine, Am. Acad. Ophthalmology and Otolaryngology (award of merit 1957), Assn. for Research in Ophthalmology, N.Y. Soc. for Clin. Ophthalmology (past pres.), Am. Physicians Fellowship for Israel Med. Assn., Pan Am. Assn. Ophthalmology, Instituto Barraquer (Barcelona, Spain), Société Francaise d'Ophtalmologie, Internat. Assn. for Prevention of Blindness, Ophthalmological Soc. U.K., Société Belge d'Ophthalmologie, European Council for Strabismic Studies (corr.), Contact Lens Assn. Ophthalmologists (founder, 1st pres.), Internat. Contact Lens Council Ophthalmology (founder, 1st pres. 1966—), Oxford Ophthalmol. Congress. Author: (with F.H. Theodore) Occular Allergy, 1958; also numerous articles, chpts. in books. Research on ophthalmology especially glaucoma strabismus, contact lenses; co-discovered syndrome of Glaucomatocyclitic crises. Home: 240 E. 79th St., N.Y.C. 10021 Office: 667 Madison Av., N.Y.C. 10021.*

SCHLUMBERGER, Charles, French naturalist; b. Mulhouse, France, 1825; ed. École polytechnique; with Corps Marine Engrs.; investigated microscopic organisms; discovered dimorphism of foraminiferes, also determined law of their classification. Died Paris, 1905.

SCHLUTER, Robert Arvel, Am. physicist; b. Salt Lake City, Aug. 27, 1924; s. Arvel Riser and Florence (Leach) S.; B.S., U. Chgo., 1947, Ph.D. (AEC fellow, Eastman Kodak Co. fellow), 1954. Research asso. Enrico Fermi Inst. Nuclear Studies U. Chgo., 1954-55; asst. prof. physics Mass. Inst. Tech., 1955-60; asso. scientist Argonne (Ill.) Nat. Lab., 1960—; prof. physics Northwestern U., Evanston, Ill., 1961—. Mem. Am. Phys. Soc., Am. Civil Liberties Union, Am. Assn. U. Profs., Sigma Xi. Author: (with Enrico Fermi) Nuclear Physics, 1949; also numerous articles. Study of elementary particle and high energy physics; interaction of fundamental particles Home: 721 Foster St., Evanston, Ill. 60201. Office: Argonne Nat. Lab., Argonne, Ill. 60440.

SCHMÄHL, Dietrich Fritz Karl, German physician; b. Breslau, Germany, Sept. 30, 1925; s. Friedrich Karl and Ella Schmähl; Dr.med., Zum Heligen Geist, Breslau, 1954, Dozent, 1964; m. Christa Maurer, Oct. 8, 1965; 1 dau., Astrid. Cancer researcher Chirurg. U. Klinik in Cancer Research, Frieburg, Germany, 1950-61; dozent Path. Inst., U. Bonn, 1960-64; dir. Inst. Exptl. Cancer Devel. and Cancer Chemotherapy, German Cancer Research Center, Heidelberg, Germany, 1964—; prof. in ord. U. Heidelberg, 1964—. Mem. German Research Soc., Royal Soc. Medicine London (affiliate). Author: Entstehung Wachstum und Chemotherapie maligner Tumoren, 1963; also numerous articles. Research in carcinogenic mechanism of action of carcinogenic compounds, cancer chemotherapy, metastability of tumors, tumor-host-relationships. Home: Sandwingert 57a. Office: Berliner Strasse 27, Heidelberg, D 6900, Germany.*

SCHMALHAUSEN, Ivan Ivanovich, Russian zoologist; b. Apr. 23, 1884; s. I. F. Schmalhausen; grad. Kiev (USSR), U., 1907. Became prof. Voronezh, U., 1918, Kiev U., 1921, Moscow U., 1938-48; dir. Inst. Zoology and Biology, Ukrainian Acad. Scis., 1930-41; dir. Inst. Evolutionary Morphology (now Animal Morphology), USSR Acad. Scis., 1938-48; joined staff Zool. Inst., 1948. Fellow Acad. Zoology in Agra, German Acad. Leopoldina; academician Ukrainian Acad. Scis., USSR Acad. Scis. Author: The Organism as a Whole in Individual and Historical Development, 2d edit., 1942; Ways and Regularities of Evolutionary Process, 1939; Factors of Evolution (Theory of Stabilizing Selection), 1946; Problems of Darwinism, 1946; Basis of Comparative Anatomy of Vertebrate Animals, 4th edit., 1947; also articles. Research in comparative anatomy, evolutionary morphology, including regularities in growth of animals, characteristics of evolutionary processes, unpaired fins of fish, origin of limbs of land vertebrates, origin of land vertebrates; developed theory on animal growth, theory of stabilizing selection in evolution. Home: Ulitsa Chkalova 14/16. Office: Embryological Lab., Inst. Zoology, USSR Acad. Sci., Leninskii Prospekt, 33, Moscow, USSR.

SCHMEIDLER, Werner Johannes, German physicist; b. Berlin, Germany, June 7, 1890; s. Johannes and Elise (Heinemann) S.; Ph.D., U. Göttingen; Dr.engr-ing. honoris causa; m. Else Strobel, Apr. 19, 1919; children—Marie-Elisabeth, Hans-Otto, Erich-Arnold. Instr., Göttingen, also Kiel, Germany; prof. Breslau (now Wroclaw, Poland), Berlin tech. colls.; independent scientist; prof. Tech. U. Berlin, became prof. emeritus, 1958. Recipient Gauss medal Braunschweig Sci. Soc., 1958; named mem. honoris causa Berlin Tech. Coll., 1959. Mem. Gesellschaft fur angewandte Math. und Mechanik, German, Berlin math. socs., Wissenschaftliche Gesellschaft für Luft und Raumfahrt. Author: Integralgleichungen mit Anwendungen in Physik und Technik, 1955; Lineare Operatoren im Hilbertschen Raum, 1955. Research in integral equations, study of current. Address: Kreuzritterstrasse 3, 1 Berlin 28, West Germany.

SCHMEIL, Otto, German zoologist; b. Grosskugel, Germany, Feb. 3, 1860; rector, Madgeburg, Germany; ind. scholar, Heidelberg, Germany. Author: Deutschlands freilebende Süsswasser-Copepoden, 3 vols., 1892-96; Lehrbuch der Zoologie, 1899; Lehrbuch der Botanik, 2 vols., 1901. Reformed teaching of biology; his textbooks were widely used. Died Heidelberg, Feb. 3, 1943.

SCHMERLING, Louis, Am. chemist; b. Milw., Apr. 20, 1912; s. Max and Bertha (Frumin) S.; B.S., U. Wis., 1932; Ph.D., Northwestern U., 1935; m. Annette D. Frazin, July 25, 1937; children—Judith Miriam (Mrs. Jon Howard Kaas), Michael Allen. With Universal Oil Products Co., Des Plaines, Ill., 1935—, research asso., 1964—. Lectr. organic chemistry Northwestern U., 1941-42. Mem. nomenclature com., organic div. Am. Chem. Soc., 1946—; chmn. organic nomenclature com. NRC, 1955. Recipient Ipatieff prize Am. Chem. Soc. 1947, Precision Sci. Co. award in petroleum chemistry, 1951. Editor: (with B.T. Brooks, C.E. Board, S.S. Kurtz) The Chemistry of Petroleum Hydrocarbons, 1954-55; also many articles. Elucidated mechanism of formation of high-octane gasoline by catalytic alkylation of isobutane with olefins; discovered several catalyst and peroxide-induced reactions of hydrocarbons with chlorine-containing compounds; numerous patents in field petroleum tech., petrochemistry. Home: 183 Lawton Rd., Riverside, Ill. 60546. Office: 30 Algonquin Rd., Des Plaines, Ill. 60016.*

SCHMID, Dieter Otto, German veterinarian, microbiologist; b. Stuttgart, Germany, Apr. 27, 1925; s. Friedrich and Gretchen (Schramm) S.; student U. Tübingen (Germany), 1951-56; Degree in Vet. Medicine with approbation, U. Munich (Germany), 1956, Dr.med.vet., 1957, docent for microbiology and sci. infectious diseases, 1967; m. Dagmar Schlitt, May 20, 1958; children—Gisela, Ute. Research officer U. Munich Inst. Vet. Microbiology, 1957-59; research fellow dept. genetics U. Wis., Madison, 1959-60, U. Cal. at Davis Sch. Vet. Medicine, 1960; research officer Inst. Animal Blood Group and Resistance Research, Munich, 1960—. Mem. expert panel on blood group scientists FAO, 1963—. Mem. European Soc. Animal Blood Group Research, numerous publs. on bacteriology, immunology, serology, immunogenetics, blood groups in animals, exptl. chemotherapy and cancer, blood groups in cattle, horses, chickens, and sheep, hemoglobins in cattle and horses, serumprotein-polymorphism in cattle and horse, eggwhite-polymorphism in chicken; comparative blood group research. Home: 25 Reitmorstr., Munchen, Germany.*

SCHMID, Emil, Swiss botanist; b. Cannstadt, Feb. 18, 1891; s. Wilhelm and Marie (Koch) S.; Ph.D., U. Zürich (Switzerland); m. Margarete Gams, 1921; agrégé, 1936; children—Dieter, Walter, Barbara. Became titular prof., curator Bot. Mus. and Bot. Gardens, U. Zürich, 1944. Mem. Italian Acad. Forestry Sci., Svenska Växtgeografiska Sällskapet, Bavarian Bot. Soc. (corr.). Author: Die Relikförhrenwälder der Alpen; Vegetationskarte der Schweiz; Erläuterungen zur Vegetationskarte der Schweiz; Die Erfassung der Vegetationseinheiten mit theoristischen und epimorphologischen Analysen. Address: Pelikanstrasse, Zürich, Switzerland.

SCHMID, Gerhard Paul, German chemist; b. Stuttgart, Germany, Oct. 5, 1900; s. Hans and Emma (Bezner) S.; Dr.engring., Stuttgart Tech. Prof. agrégé, prof. at large Stuttgart Tech. Coll.; became prof. phys. chemistry U. Cologne (Germany), 1952. Mem. Wissenschaftliche Vereinigung für Ultraschallforschung; Gesellschaft Deutscher Chemiker; Deutsche Bunsen-Gesellschaft; Kolloid-Gesellschaft. Author: Chemische Anwendung von Ultraschall; Elektrochemie feinporiger Kapillarsysteme; Chemisches Glänzen von Metallen. Home: Kardinal Schultestrasse 30, Bensberg b. Cologne, West Germany.

SCHMID, Hellmut Heinrich; geodesist; b. Dresden, Germany, Sept. 12, 1914; s. Kurt Rudolf and Dora (Meissner) S.; B.S., Tech. U., Dresden, Germany, 1936, M.S., 1938, Dr.Engring., 1941; m. Ilse Hieby, Oct. 7, 1949; 1 dau., Monica Ilse. Came to U.S., 1945, naturalized, 1955. Chief trajectory measurement Electro-Mechanische Werke, Peenemünde, Germany, 1941-45; chief geodetic sect. ordnance sub-office-rocket, Ft. Bliss, Tex., 1945-50; supervisory research geodesist Army Ordnance, Ballistic Research Labs., Aberdeen Proving Ground, Md., 1950-62, cons., 1962—; sci. adviser to dir. U.S. Army Engrs. Geodesy Intelligence, Mapping Research and Devel. Agy., Ft. Belvoir, Va., 1962-63, cons. 1963—; sci. adviser U.S. Coast and Geodetic Survey, Rockville, Md., 1963-65; dir. Geodetic Research Lab., Inst. for Earth Scis., Environmental Sci. Services Adminstrn., Rockville, Md., 1965—. Recipient Robert H. Kent award U.S.A. Ballistic Research Labs., 1962; Colbert medal Soc. Am. Mil. Engrs., 1965; U.S. Dept Commerce Gold medal, 1966. Mem. Am. Geophys. Union, Am. Soc. Photogrammetry (Fairchild award 1958, Talbert Abram award 1966). Research, publs. on gen. analytical solution to problem of photgrammetry; gen. least sq. solution for hybrid measuring systems; design of STK stereocomparator; design of ballistic cameras. Home: Goshen Estates, Warfield Ct., Gaithersburg, Md. 20760. Office: Environmental Sci. Services Adminstrn., Washington Sci. Center, Bldg. 1, Rockville, Md. 20852.*

SCHMID, Josef, Austrian biochemist; b. Langua, Feb. 14, 1919; s. Alois and Therese (Schmied) S.; M.D., U. Vienna; m. Alice Hillelmanoach, Jan. 24, 1943; 1 son, Alexander. Physician in chief Wilhelmina Hosp., until 1945; asst., instr. 2d Med. Clinic, U. Vienna; dir. Rheumatism Inst., Deutsch Altenbourg; staff pvt. inst., Vienna. Mem. Internat. Med. Soc., Soc. Phys. Medicine, Order Physicians Vienna, European Hematology Soc. Home: Scheibengasse 13, Vienna 19, or Kitzbühl, Austria. Office: Walfischgasse 10, Vienna 1, Austria.

SCHMID, Rudi, physician; b. Glarus, Switzerland, May 2, 1922; s. Rudulf and Bertha (Schiesser) S.; M.D., U. Zurich (Switzerland), 1947; Ph.D., U. Minn., 1954; m. Sonja D. Wild, Sept. 17, 1949; children—Isabelle S., Peter R. Came to U.S., 1948, naturalized, 1954. Instr., U. Hosp., Mpls., 1952-54; spl. research fellow NIH dept. biochemistry Columbia Coll. Phys. and Surg., N.Y.C., 1954-55; sr. hematologist NIH, Bethesda, Md., 1955-57; asso. medicine Thorndike Meml. Lab., Harvard Med. Sch., Boston, 1957-59, asst. prof., 1959-62; prof. U. Chgo., 1962-66, U. Cal. Med. Sch., San Francisco, 1966—. Cons. NIH and Clin. Center, 1959—, U.S. Army Surgeon Gen., 1962—. Mem. Am. Soc. Biol. Chemists, Am. Soc. for Exptl. Pathology, Am. Soc. for Clin. Investigation, Am. Soc. Hematology, Am. Assn. for Study Liver Diseases (past pres.), A.C.P., Assn. Am. Physicians, Am. Gastroent. Assn., Central, Western socs. for clin. research, Western Assn. Physicians. Research, numerous publs. on hemoglobin, bilirubin, porphyria; developed methods for prodn. exptl. porphyria, 1st method for prodn. isotopically labeled bilirubin; demonstrated bile pigment formation from non-hemoglobin sources in liver, lack of muscle phosphorylase in McArdle's syndrome. Home: 211 Woodland Rd., Kentfield, Cal. 94904. Office: U. Cal. Med. Center, San Francisco 94122.*

SCHMID, Rudi, physician; b. Glarus, Switzerland, Dec. 26, 1926; s. Karl and Agathe (Tschumperlin) S.; Ph.D., in Organic Chemistry, U. Zurich, 1955; m. Edith Roth, Feb. 27, 1953; children—Michael, Markus, Beat. Research asso. organic chemistry dept. U. Zurich, 1952-55; research asso. biochemistry dept. U. Basle, 1955-56; head isotope lab. pharm. research div. CIBA Ltd., Basle, Switzerland, 1956—. Cons. on use radioisotopes in research and industries Swiss Soc. for Atomic Energy, 1960—. Mem. Swiss Biochem. Soc., Swiss Chem. Soc., Swiss Analytical Soc. Research, publs. on reaction mechanisms using radioisotopes, synthesis of radioactive labelled organic compounds, drug metabolism, pharmacocinetic, autoradiographic studies. Home: 36 Schonenbachstr., Reinach, BL, Switzerland. Office: CIBA Ltd., Basle BS, Switzerland.*

SCHMIDT, Anthony John, biologist: b. Winnipeg, Man., Can., May 11, 1927; s. Peter J. and Julia (Miller) S.; B.A., U. Wash., 1952, M.S., 1954; Ph.D., Princeton, 1957; m. Suzanne Eldredge, June 10, 1955; children—Peter West, Adrienne, Karl Frederick, Bradford Paul. Came to U.S., 1949, naturalized, 1966. Instr., Princeton, 1957-58; faculty dept. anatomy U. Ill. Coll. Medicine, Chgo., 1958—, prof., 1965—, mem. Grad. Coll., 1959—, dir. grad. tng. in developmental biology, 1961-67. Mem. Am. Soc. Zoologists, Am. Am. Soc. Anatomists, Soc. for Developmental Biology, Am. Soc. for Cell Biology, A.A.A.S., Sigma Xi. Author: The Cell Biology of Repair and Regeneration in Vertebrates, 1968. Research, publs. on metabolism in the chem. mediators of repair and regeneration processes in vertebrates, especially in adult urodele limb tissues. Home: 233 S. Ridgeland Av., Oak Park, Ill. 60302. Office: 1853 W. Polk St., Chgo. 60612.*

SCHMIDT, Carl Frederick, Am. physician; b. Lebanon, Pa., July 29, 1893; s. Jacob C. and Mary Ellen (Greth) S.; A.B., Lebanon Valley Coll., 1914, Sc.D., 1955; M.D., U. Pa., 1918, Sc.D., 1965; Med.Sc.D., Charles U., Prague, Czechoslovakia, 1963; m. Elizabeth Viola Gruber, June 16, 1920; children—Carl Frederic, Barbara (Mrs. R. P. DeLong). Faculty, U. Pa., Phila., 1918-19, 24—, prof. pharmacology, 1932-59, emeritus prof., 1959—, dir. dept., 1939-59; asso. pharmacology Peking (China) Union Med. Coll., 1922-24; research dir. aerospace med. research dept. U. S. Naval Air Devel. Center, Johnsville, Pa., 1962—. Mem. drug research bd. Nat. Acad. Scis., 1964—, chmn. subcom. on continuing med. edn., 1965—; mem. Pa. Bd. on Drugs, Cosmetics and Appliances, 1963—. Recipient Hachmeister award Georgetown U., 1951, Schmiedeberg Plaquette, Deutsche Pharmacologsches Gesellschaft, 1963. Mem. Am. Physiol. Soc., Am. Soc. for Pharmacology (past pres.), Assn. Am. Physicians, Aerospace Med. Assn., Pan Am. Med. Assn., Am. Heart Assn., Am. Acad. Arts and Scis., Nat. Acad. Scis., Internat. Union Physiol. Scis. (past pres. sect. on pharmacology), Sigma Xi, Alpha Omega Alpha. Author: (with K. K. Chen) Ephedrine and Related Substances, 1926; also numerous articles. Research on introduction of ephedrine; actions of drugs on respiration, kidney function, blood supply to organs; interrelations of chem. and nervous factors in control of respiration and circulation; aerospace physiology and pharmacology. Home: Pickwick Dr., Doylestown R.D. 3, Pa. 18901. Office: U. S. Naval Air Devel. Center, Johnsville, Warminster, Pa., 18974.*

SCHMIDT, Eduard Oskar, German zoologist; b. Torgau, 1823; prof., Jena, Germany, also Cracow, Poland, Graz, Austria, Strasbourg (now in France). Author: Eponges de la mer Adriatique, 1862-66; Descendance et Darwinisme, 1874; Les sciences naturelles et la philosophie de l'inconscient, 1878. Made studies on flatworms, sponges. Died Strasbourg, 1886.

SCHMIDT, Erich Friedrich, archeologist; b. Baden-Baden, Germany, Sept. 13, 1897; s. Erhard Friedrich and Frida (Loeffler) S.; student Friedrich Wilhelm U., Berlin, 1921-23; Ph.D., Columbia U., 1929; m. Mary Helen Warden, 1934 (dec.); m. 2d Lura Florence Strawn, 1943; children—Richard Roderick, Erika Lura. Came to U. S., 1923. Field dir. archeol. expdn. to Ariz. under auspices Am. Mus. Natural History, N.Y., 1924-26; co-dir. to Asia Minor under auspices Oriental Inst., 1927-29; field dir. to Mesopotamia and Persia under auspices of U. Mus., U. Pa., and Pa. Mus. Art, 1931-32; field dir. to Persia (Ray Expdn.) under auspices U. Mus., U. Pa., and Boston Mus. Fine Arts, 1934-36; field dir. Persepolis Expdn., Iran, 1935-39; organized Aerial Survey Expdn. in Iran (Mary-Helen Warden Found.); asso. prof. U. Chgo., 1954, prof., 1962. Mem. Am. Oriental Soc., Archaeol. Inst. Am. Author books relating to field; latest publ.; Flights Over Ancient Cities of Iran, 1940; Persepolis Vol. I, Structures—Reliefs—Inscriptions, 1953; Vol. II, Contents of the Treasury, 1957. Died Oct. 4, 1964.

SCHMIDT, Ernst Heinrich Wilhelm, German phys. engr.; b. Vögelsen, Germany, Feb. 11, 1892; s. Ernst Friedrich and Magdalena (Voss) S.; student Dresden, Munich Insts. Tech., U. Munich; Dipl. Ing. 1919, Dr. Ing., 1921; Dr. rer.nat.h.c., Aachen Inst. Tech., 1952; LL.D. U. Glasgow, 1961; m. Sophie Rothhammer, Apr. 12, 1921; children—Hertha, Waltraut (Mrs. Heinz Heier), Helmut, Reinhart. Prof., Danzig Inst. Tech., 1925-37; dir. Aircraft Research Inst., Braunschweig-Völkenrode (Germany), 1937-45; with Royal Aircraft Establishment, Farnborough (Eng.), 1946; prof. Braunschweig Inst. Tech., 1945-52; prof. thermodynamics Munich Inst. Tech., 1952——, rector, 1956-58. Recipient Leibniz prize, 1943; Gold Grashof medal, 1956; Arnold Eucken medal, 1958; Merit Cross, Fed. Republic Germany, 1961; Ludwig Prandtl ring, 1964; Bavarian order merit, 1964; Max Jakob meml. award, 1965; Dechema medal, 1967. Mem. Braunschweig Sci. Soc., (v.p. 1941-52), Carl-Cranz Soc. (pres. 1961——), Assn. German Engrs., Inst. Mech. Engring., Germany Phys. Soc., Soc. Applied Math. and Mechanics, German Assn. Refrigeration Tech. Author: Technische Thermodynamik, 10th edit., 1963; Thermodynamics, 1949, 1966; Wasserdampftafeln, 6th edit., 1963; also numerous articles. Editor: Zahlenwerte und Funktionen, 6th edit. (Landolt-Börnstein), 1955-66. Research on heat and mass transfer (Schmidt number), ciritical state of fluids, method of solving differential equations with finite differences, numerous inventions in field. Home: 8 Rümelin-Strasse, Munich, West Germany.*

SCHMIDT, Fred Henry, Am. physicist; b. Detroit, Sept. 12, 1915; s. Hugo W. and Maude (Schmidt) S.; B.S. Engring., U. Mich., 1937; M.A., U. Buffalo, 1940; Ph.D., U. Cal. at Berkeley, 1945; m. Margaret Cresswell, Sept. 5, 1939; children—Karla, Kurt. Engr., Am. Tel. & Tel. Co., Denver, 1937-39, physicist Manhattan Project, Berkeley, Oak Ridge and Los Alamos, 1942-45; physicist Radiation Lab., U. Cal., Berkeley, 1945-46; faculty U. Wash., Seattle, 1946——, prof., 1956——. Guggenheim fellow, Amsterdam, Geneva, 1956-57; NSF Postdoctoral fellow, Geneva, 1963-64. Fellow Am. Phys. Soc.; Mem. Sigma Xi. Contbr. articles to physics jours. Research, publs. on nuclear physics. Home: 7036 58th. Av. N.E., Seattle 98115.*

SCHMIDT, Gerhard Herbert, German zoologist, biochemist; b. Deutsch-Krone (W. Prussia), Feb. 20, 1928; s. Josef Bernhard and Martha (Hasenleder) S.; Dr. rer. nat., U. Münster, 1954; m. Ursula Stirnberg, May 16, 1959; children—Wolfgang, Annette, Sabine. Research fellow German Inst. Lipid Research, Münster, Germany, 1954-57; asst. Inst. Applied Zoology, U. Würzburg (Germany), 1957-64, docent, 1965——; asst. Inst. Organic Chemistry, U. Zurich (Switzerland), 1964-65. Mem. German Zoöl. Soc., German Soc. Lipid Research, Internat. Union for Studying Social Insects. Research and publs. on metamorphosis in holometabolic insects; caste differentiation in ants; biochemistry and physiology of pteridins and flavins in insects; ecology of orthoptera; temperature regulation mechanisms of insects. Address: 11 Ludwigstrasse, Würzburg, West Germany.*

SCHMIDT, Gernhard, optician; b. Neissaar Island, Estonia, Mar. 30, 1879; studied optics, Sweden; ground reflectors for telescopes; became asso. Hamburg Obs., Bergedorf, Germany, 1926. Devised corrector plate for eliminating distortion from astron. telescopes and cameras (Schmidt telescope or camera). Died Hamburg, Germany, Dec. 1, 1935.

SCHMIDT, Gustav, engr.; b. Vienna, Austria, 1826; prof. machine tech., Pzibram, then Polytechnikum, Prague, Czechoslovakia. Research on steam engines, condensation, steam boiler, circulation of compressed air in pipes, durability of hauling cable; formulated a theory of water turbines, 1862. Died 1883.

SCHMIDT, Hans, German physicist; b. Remscheid, Germany, July 14, 1920; s. Friedrich and Anna (Funke) S.; Dr.ès.sc., U. Iena, U. Bonn; m. Ilse Renkhoff, Sept. 27, 1943; children—Wilfried, Wolfram, Christoph. Asst., Bonn Inst. Theoretical Physics, 1942-43; physicist Siemens-Schuckert Co., Berlin, 1943-45; asst. Firme Sternwarte, Bonn, Germany, 1945-54; with Hoher List Obs., Bonn, 1954——; became instr. U. Bonn, 1951, prof. at large, 1958. Mem. German Phys. Soc., Astron. Soc. Author: Selen-Sperrschichtgleichrichter. Research on dense double stars, stellar statistics. Home: Office: Observatorium Hoher List, Daun/Eifel, West Germany.

SCHMIDT, Hermann, German geologist; b. Wuppertal, Germany, Nov. 3, 1892; s. Hermann and Wanda (Schmidt) S.; Ph.D., U. Gottingen; m. Helene, June 2, 1921; children—Felix, Gudrun. Became asst. Prussian Geology Office, Berlin, 1920; became curator, instr. U. Göttingen, 1922, prof., 1927, prof. emeritus, 1961. Recipient Hans Stille medal German Geol. Soc., 1954. Mem. Belgian Geol. Soc. (hon.), Belgian Paleontology Soc., London Geol. Soc. Author: Carbon. Coniatiten, 1924; Bionomie der Fossilen Meeresböden, 1934; Cornberger Fährten, 1959. Editor, Paläontologische Z, 1951-62. Address: Planckstrasse 6, Gottingen, W. Germany.

SCHMIDT, Ingeborg, physician; b. nr. Dorpat, Estonia, Dec. 26, 1899; d. Arthur F. Th. and Meta K. (Bergwolf) S.; student U. Dorpat, 1920-25; physicians license Estonia, 1926; M.D., U. Tübingen, Germany, 1927. Came to U.S., 1947, naturalized, 1955. Research asst. Physiol. Insts., Tübingen and Berlin, 1926-35; research asst. Aero-med. Research Inst., Berlin, 1936-45; scientist USAF Sch. Aviation Medicine, Randolph Field, Tex., 1947-52; research asst. Ophthal. Inst., Columbia, N.Y.C., 1952-53; prof. div. optometry Ind. U., Bloomington, 1954——. Mem. Aerospace Med. Assn., Am. Acad. Optometry, A.A.A.S., Am. Astronautical Soc., Am. Optometric Assn., German Soc. for Aerospace Medicine, German Opthal. Soc., Ind. Optometric Assn. (hon. mem.), Optical Soc. Am., Sigma Xi. Author: Bibliography of Aviation Medicine, 2 vols., 1938, 43; (with M. Richter) Grundriss der Farbenlehre der Gegenwart, 1940. Co-editor: Trendelenburg, Der Gesichtssinn, 1961. Research, numerous publs. on color and night vision, application of physiol. optics to astronautics and automobile driving; discovered a sign typical in color vision of female carriers of spl. type of color deficiency; designed apparatus for night vision, color vision testing. Office: Optometry Bldg., 800 E. Atwater St., Bloomington, Ind. 47401.*

SCHMIDT, J(akob) E(dward), medical and medicolegal lexicographer, physician; born at Riga, Latvia on June 16, 1906; the son of Michael E. and Rachel I. (Goldschmidt) S.; came to U. S., 1924, naturalized, 1929; grad. Balt. City Coll., 1929; Ph.G., U. Md., 1932, B.S., 1935, M.D., 1937; Litt.D., 1939. Engaged pvt. med. practice, Balt., 1940-53; resident, Charlestown, Ind., 1953——; lexicologist, neologist, 1945——; med. and medicolegal lexicographer, 1950——; author: What's the Good Word (Balt. Sun); Sharpen Your Tongue (American Mercury): The Medical Lexicographer (Modern Medicine); Medical Semantics (Medical Science); Underworld English (Police); Medical Vocabulary Builder (Trauma). Asso. medical editor, Trauma; Medical Dictionary Editor; compiler 50,000-word vocabulary test, 1956; cons. medical and medicolegal terminology also the consultant in medical lexicography, Mem. revision com. U. S. Pharmacopeia Xi; Am. Dialect Soc.; Am. Name Soc.; Am. Med. Writers' Assn. Internat, Soc. Gen. Semantics; A.M.A.; Med. and Chirurgical Faculty of Md.: Balt. City Med. Soc.; Rho Chi Hon. Pharm. Soc.; The Owl Club; chmn. Nat. Assn. on Standard Med. Vocabulary. Author: Reversicon—A Medical Word Finder, 1956; Medical Terms Defined for the Layman, 1958; Medical Discoveries—Who and When, 1959; Dictionary of Medical Slang, 1959; Narcotics—Lingo and Lore, 1960; Dictionary of Libido Terminology, 1961; A Baby Name Finder, 1961; Dr. Schmidt's Attorneys' Dictionary of Medicine, 1962; 1,000 Elegant Phrases, 1964; Attorneys' Dictionary of Medicine and Word Finder, 1965; New and Additional Medical Words for Lawyers, 1965; Controversial Medical Terms—with Recommendations, 1965; The Medical Lexicographer, 1966; Police Medical Dictionary, 1968; Practical Nurses' Medical Dictionary, 1968; Paramedical Dictionary, 1968. Invented iodine-pentoxide-shunt method and apparatus for detection of carbon monoxide in medicinal oxygen; effect of cesium and related metals on oxidation of organic matter in drinking water; the TV eye phenomenon. Home: Monroe at Park St. Office: 934 Monroe St., Charlestown, Ind. 47111.

SCHMIDT, Johann Adam, physician; b. Aub/Würzburg, Germany, 1759; ed. Würzburg; pupil of Joseph Barth; became asst. surgeon in war against Prussia, Prague, Czechoslovakia, 1778; with mil. surgeons sch., Vienna, after war. Lectr. diseases of eye, Vienna Sch. Ophthalmology, 1795-1809; founder clinic for poor with eye diseases; prof. Josefs Acad.; Research and publs. on tear ducts; introduced term iritis, 1801; 1st to detect rampant growth features; introduced formation of an artificial pupil through iridectomy; his theories confirmed with advent of pathology. Died Vienna, Feb. 18, 1807.

SCHMIDT, Johann Friedrich Julius, astronomer; b. Eutin, Germany, Oct. 26, 1825; with observatories at Altona, Bilk, Bonn (all Germany); became dir. nat. obs., Athens, 1858. Author: Resultate aus zehn Jährigenbeobachtungen über Sternschnupper, 1852; Das Zodiakallicht, 1856; Der Mond, 1856; Über Riller auf dem Mond, 1866. Editor map of moon, 1878. Studied comets, variable stars, zodiacal light; authority on selenography; 1st to observe changes in crater Linné and other areas of moon's surface. Died Athens, Feb. 7, 1884.

SCHMIDT, Johanna Gertrud Alice, German topographer; b. Leipzig, Germany, Aug. 7, 1909; d. Fritz and Alice (Fiedler) S.; student univs. Greifswald, Fribourg/Brisgau, Munich, Berlin; Ph.D. Reader in classical langs.; asst. for hist. geography U. Greifswald, also U. Berlin; prof. Inst. Ednl. Scis., Leipzig, Germany; dir. services Municipal Inst. Hist. and Cultural Research, Leipzig; instr. Diesterweg Coll., also W. Berlin Coll. Edn.; prof. Berlin Coll., dir. Institut für Kultur-und Heimatkunde. Author: Ethos; Mass und Harmonie; Heimat und Kultur. Editor: Kultur- und Heimatstudien. Numerous publs. on topography of Mediterranean countries, geog. history, European history. Home: Lupsteinerweg 54 a, 1 Berlin-Zehlendorf, 37, Germany. Office: Badensche Strasse 50-51, 1 Berlin-Schöneberg 62, W. Germany.

SCHMIDT, Johannes, German linguist; b. Prenzlau, 1843; prof. comparative grammar, Berlin; mem. Acad. Scis., Berlin. Author: Les liens de parente des langues indogermanique, 1872; Pour servire à l'histoire du vocalisme des langues indo-germanique, 1871-75; La forme du pluriel dans les neutres indogermanique, 1889; L'habitat primitif des Indo-Germains et le système numéral Européen, 1890; Critique de la théorie des sonnantes, 1895. Originator theory of linguistic oscillations in regard to continuity of grammar in dialects of same lang. family. Died Berlin, 1901.

SCHMIDT, Johannes, Danish biologist; b. Jägerspris, Denmark, 1877; dir. Carlsberg Physiol. Lab., Copenhagen; mem. French Acad. Scis., 1924. Author: Bacteria; Flora of Koh Chang, 1900. Research in oceanography and biology; discovered breeding ground of European eel on ocean bed southeast of Bermuda, 1904. Died Copenhagen, Feb. 22, 1933.

SCHMIDT, John Wesley, Am. plant geneticist and breeder; b. Moundridge, Kan., Mar. 13, 1917; s. John J. and Kathrina (Sperling) S.; student McPherson Coll., 1936-37, 40-41; A.B., Tabor Coll., 1947; M.Sc., Kan. State U., 1949; Ph.D., U. Neb., 1952; m. Olene Lucile Hall, June 23, 1943; children—Karen, Vicki, Wesley, Loren, Jerold. Faculty, Kan. State U., 1951-54, asso. prof., 1953-54; faculty U. Neb., Lincoln, 1954——, prof. plant genetics, 1962——. AEC-NRC fellow U. Neb., 1949-51. Fellow Am. Soc. Agronomy ; mem. Am. Genetic Assn.; Phi Kappa Phi, Gamma Sigma Delta, Sigma Xi. Research, publs. on inheritance and cytology of wheat, early devel. hybrid wheat; breeder wheat and oat varieties important in midwestern U.S. Home: 1130 N. 37th St., Lincoln, Neb. 68503.*

SCHMIDT, Leon Herbert, Am. pharmacologist; b. Huron, O., June 28, 1909; s. Herbert H. and Amelia (Baehr) S.; B.A., DePauw U., 1929; Ph.D., U. Cin., 1932; m. Ida Theresa Genther, July 28, 1931; children—Nancy Jeanne, Richard Allen. Faculty, U. Cin., 1938-63, research prof., 1950-63; prof. comparative pharmacology U. Cal. at Davis Sch. Vet. Medicine, San Francisco, 1963——, dir. Nat. Center For Primate Biology, 1963——; cons. NIH, NRC, Armed Forces Epidemiological Bd.; mem. adv. med. bd. Leonard Woods Meml.; mem. exec. com. VA Com. Chemotherapy Tb., expert com. WHO. Recipient, Trudeau medal, Nat. Tuberculosis Assn., 1967. Diplomate Am. Acad. Microbiology. Fellow A.A.A.S., Am. Soc. Tropical Medicine and Hygiene (Charles Francis Craig lectr. 1956), N.Y. Acad. Scis.; mem. Am. Assn. Cancer Research, Am. Assn. U. Profs., Am. Chem. Soc. (Eminent chemist 1956), Am. Soc. Pharmacology and Exptl. Therapeutics, Am. Soc. Biol. Chemists, Am. Soc. Microbiology, Am. Thoracic Soc., Soc. Exptl. Biology and Medicine, Sigma Xi. Editorial bd. Am. Rev. Tb., 1958-62; Jour. Nat. Malaria Soc., 1945-47; Jour. Pharmacology and Exptl. Therapeutics, 1944-50, Am. Jour. Tropical Medicine and Hygiene, 1958-66. Research, numerous publs. on metabolism of bile acids, therapy of acute bacterial infections, malaria Tb, cancer, metabolism of drugs, comparative biology of non-human primates. Home: 619 Oak Av., Davis, Cal. 95616. Office: Nat. Center Primate Biology, U. Cal., Davis, Cal. 95616.*

SCHMIDT, Leopold, Austrian ethnographer; b. Vienna, Mar. 15, 1912; s. Arnold and Rosa (Stadimann) S.; Ph.D., U. Vienna; m. Margarete, Dec. 22, 1938; children—Regine. Prof. regional ethnography U. Vienna; dir. Austrian Mus. Regional Ethnography. Recipient Wilhelm Heinrich Riehl prize German Ethnographical Soc., 1937, Cross of Merit, Fed. State of Burgenland, 1962. Mem. Vienna Ethnographic Soc. (pres.). Research and numerous publs. on ethnography. Home: Linke Wienzeile 36, Vienna 6, Austria. Office: Laudongasse 19, Vienna 8, Austria.

SCHMIDT, Ludwig Erich, pharmacologist, physiologist; b. Brasov, Rumania, May 6, 1921; s. Ludwig and Iha (Broser) S.; student U. Danzig; Dr.med., U. Freiburg im Breisgau, 1947; m. Margarethe Zimmermann, Nov. 14, 1959; children—Renna, Eva, Susanne, Christian. Asst., U. Freiburg im Breisgau, Germany, 1946-49, head physician, prof. inst. pharmacology, 1950-63; med. officer Med. Acad. German Armed Forces, Munich, 1963-64; sci. dir. Inst. Naval Medicine, Kiel, Germany, 1964—. Mem. German Soc. Pharmacology, German Soc. Sci. and Medicine, Freiburg Med. Soc. Research, numerous publs. on coronary circulation, circulation, respiration, laxatives, vegetative nervous system, supercooling and rewarming, kidneys, toxicology. Home: 66 Blitz-Strasse. Office: 12 Kopperpahler Allee, 23 Kiel, West Germany.*

SCHMIDT, Maarten, astronomer; b. Groningen, Holland, Dec. 28, 1929; s. Wilhelm and Antje (Haringhuizen) S.; Ph.D., U. Leiden (Holland), 1956; Sc.D., Yale, 1966; m. Cornelia Johanna Tom, Sept. 16, 1955; children—Elizabeth, Marijke, Anne. Sci. officer Leiden Obs., 1949-59; Carnegie Postdoctoral fellow Mt. Wilson Obs., 1956-58; asso. prof. Cal. Inst. Tech., Pasadena, 1959-64, prof. astronomy, 1964—. Named (with J. L. Greenstein) Cal. Scientist of Year, Cal. Mus. Sci. and Industry, 1964. Asso. Am. Astron. Soc. (Helen Warner prize 1964); mem. Internat. Astron. Union. Contbr. articles to tech jours. Determined spiral structure of our galaxy, redshifts for galaxies identified with radio sources; research on distbn. mass in our galaxy, effects of star formation on evolution star systems, spectra quasi-stellar radio sources. Office: Cal. Inst. Tech., Pasadena, Cal. 91109.*

SCHMIDT, Nathalie Joan, Am. microbiologist; b. Flagstaff, Ariz., Sept. 24, 1928; d. Joseph Francis and Gertrude (Hill) Schmidt; B.A., U. Ariz., 1950; M.S., Northwestern U., 1952, Ph.D., 1953. Bacteriologist, Evanston (Ill.) Hosp. Assn., 1953-54; research microbiologist Cal. Dept. Pub. Health Virus Lab., Berkeley, 1954—. Cons. NIH, 1963—. Fellow Am. Acad. Microbiology; mem. Am. Assn. Immunologists, Soc. Exptl. Biology and Medicine, N.Y. Acad. Scis., Am. Pub. Health Assn. Asso. editor, contbr. Diagnostic Procedures for Viral and Rickettsial Diseases, 1964. Contbr. numerous articles to profl. jours. Developed tests and reagts. for lab. diagnosis of poliomyelitis, varicella, herpes simplex, rubella and certain human enterovirus infections; characterization of antigenic components of enteroviruses and rubella virus; described new viruses causing human illness; characterization of viral inhibitors present in serum and other biol. materials. Home: 2008 Berkeley Way. Office: 2151 Berkeley Way, Berkeley, Cal. 94704.*

SCHMIDT, Otto, Russian mathematician, explorer; b. Moghilev, Russia, 1891. Became instr. math. U. Kiev, 1916; later prof. math. U. Moscow; chief No. Sea Route Adminstrn., 1932-39; became dep. to Supreme Soviet, 1937. Decorated Hero of Soviet Union, Order of Lenin, Order of Red Star. Mem. USSR Acad. Scis. (became v.p. 1939). Chief editor, Large Soviet Ency., 65 vols., 1926. Participated in Soviet-German expdn. to Pamirs which discovered largest glacier in world, 1928; leader several expdns. to Far North, including 1st air expdn. to N. Pole, 1937. Office: Physico-Math. Br., USSR Acad. Scis., Moscow, USSR.

SCHMIDT, Richard Penrose, Am. physician; b. Akron, O., July 27, 1921; s. Richard Homer and Frances Angeline (Speakman) S.; student Miami U., Oxford, O., 1939-40; B.S., Kent State U., 1953; M.D., U. Louisville, 1945; m. Betty Corinne Heminger, June 5, 1943; children—Victoria Frances, Richard Penrose. Faculty, U. Louisville, 1951-54, asst. prof. medicine, 1953-54; faculty U. Wash. Sch. Medicine, 1954-58, asst. prof. surgery, 1955-58; faculty U. Fla. Coll. Medicine, Gainesville, 1958—, prof. medicine, 1961-65, head dept. medicine, 1962-65, asso. dean, chief of staff, 1965—. Mem. adv. com. in psychiatry, neurology and psychology VA, 1965—. Diplomate Am. Bd. Psychiatry and Neurology. Mem. Am. Acad. Neurology (past trustee), Am. Epilepsy Soc. (past pres.), Am. chmn. program com. 1957—), So. (past pres.) electroencephalographic socs., Am. Neurol. Assn. (past mem. council), Alachua County Med. Soc. (past pres.), Am. Acad. Neurology (pres. 1967—); A.M.A., Assn. for Research in Nervous and Mental Disease, A.A.A.S., A.C.P., Sigma Xi, Phi Chi, Alpha Omega Alpha. Research, publs. on exptl. epilepsy, electrophysiology of epilepsy. Home: 2001 N.W. 31st Terrace, Gainesville, Fla. 32601.*

SCHMIDT, Wilhelm, ethnologist; b. Hörde, Germany, Feb. 16, 1868; ed. univs. Berlin, Vienna; entered Roman Catholic Soc. of Divine Word, 1890; prof. U. Vienna; founder, dir. anthrop. inst., Mödling; prof. U. Freiburg (Switzerland), from 1938. Author: Grundlinien einer Vergleichung der Religionen und Mythologien der austroneisischen Völker, 1910; Der Ursprung der Gottesidee, 2 vols., 1912-26; Handbuch der vergleichen Religionsgeschichte, 1930; The Origin and Growth of Religion, 1931; High Gods in North America, 1933; Handbuch der Methode der kulturhistorischen Ethnologie, 1937; Das Figentum auf den ältesten Stufen der Menschheit, 3 vols., 1937-42; Rassen und Völker in Vorgeschichte und Geschichte des Abendlandes, 3 vols., 1946-49. Made studies in comparative religion, langs., culture; specialized in langs. of S. Asia, Australia, Oceania. Died Freiburg, Feb. 10, 1954.

SCHMIDT-KOENIG, Klaus Otto, German zoologist; b. Heidelberg, Germany, Jan. 21, 1930; s. Otto and Marlies (Hummel) Schmidt; student U. Heidelberg, U. Munich (Germany), U. Kiel (Germany), U. Freiburg (Germany), Max Planck-Inst. fur Verhaltensphysiologie, Wilhelmshaven, Germany, 1955-57; Ph.D. summa cum laude U. Freiburg, 1958; m. Inka Rauch, Aug. 31, 1959; children—Ariane Karin, Martina Sabine. Staff, Max-Planck-Inst. fur Verhaltensphysiologie, 1958-63; staff dept. zoology U. Göttingen (Germany), 1963—, privatdozent, 1965—; research asso. dept. zoology Duke, Durham, N.C., 1959—. Mem. German Zool. Soc., German Ornithol. Soc., Sigma Xi. Contbg. author: Advances in the Study of Behavior, 1965. Research, publs. on population dynamic, ecology, migration in several passerine species, sun compass orienation, chronometry, navigation in homing pigeons, celestial orientation and navigation in passerines. Offices: I. Zoologisches Institut, Berlinerstr. 28, 3400 Göttingen, West Germany; also Dept. Zoology, Duke, Durham, N.C. 27706.*

SCHMIDT-NIELSEN, Bodil Mimi (Mrs. Knut Schmidt-Nielsen), physiologist; b. Copenhagen, Denmark, Nov. 3, 1918; d. August and Marie (Jorgensen) Krogh; D.D.C., U. Copenhagen, 1941, D.Odont., 1946, Ph.D., 1955; m. Knut Schmidt-Nielsen, Sept. 20, 1939 (div. Feb. 1966); children—Astrid (Mrs. John McHugh), Bent, Bodil. Came to U.S., 1946, naturalized, 1952. Faculty, Duke, 1952-64; prof. biology Western Res. U., Cleve., 1964—; trustee Mt. Desert Island Biol. Lab. Recipient Career award NIH, 1962. John Simon Guggenheim Meml. fellow, 1952-53; Bowditch lectr., 1958. Fellow N.Y. Acad. Scis., A.A.A.S.; mem. Am. Physiol. Soc., Soc. Exptl. Biology and Medicine. Research, numerous publs. on biochemistry of saliva, water metabolism of desert animals, comparative kidney physiology, comparatory physiology of excretory organs. Home: 13855 Superior Rd., Cleve. 44118. Office: 2080 Adelbert Rd., Cleve. 44106.*

SCHMIDT-NIELSEN, Knut, Am. physiologist; b. Norway, Sept. 24, 1915; s. Sigval and Signe Torborg (Sturzen-Becker) Schmidt-N.; Mag. Scient., U. Copenhagen, 1941, Dr. Phil., 1946. Came to U. S. 1946, naturalized, 1952. Research fellow Carlsberg Labs., Copenhagen, 1941-44, U. Copenhagen, 1944-46; research asso. zoology Swarthmore Coll., 1946-48; docent U. Oslo, 1947-49; research asso. physiology Stanford, 1948-49; asst. prof. Coll. Medicine, U. Cin., 1949-52; prof. physiology Duke, 1952—, James B. Duke prof. physiology, 1963—. Mem. panel environmental biology NSF, 1957-61; mem. nat. adv. bd. physiol. research lab. Scripps Instn. Oceanography, U. Cal. at San Diego, 1963—; Harvey Soc. lectr., 1962; Regents' lectr. U. Cal. at Davis, 1963; speaker numerous internat. confs. and congresses, 1955—. Guggenheim fellow, 1953-54; grantee Office Naval Research, 1952-54, 58-61, UNESCO, 1953-54, Office Surgeon Gen., 1953-55, Office Q.M.Gen., 1953-54, NIH, 1955-57, 63, 64-67, NSF, 1957-61, 59-60, 60-61, 61-63; recipient Research Career award USPHS, 1964. Fellow N.Y. Acad. Sci., A.A.A.S., Am. Acad. Arts and Scis.; mem. Nat. Acad. Scis., N.C. Acad. Sci. (Poteat award 1957), Am. Physiol. Soc., Am. Soc. Zoologists, Soc. Exptl. Biology. Author: Animal Physiology, The Physiology of Desert Animals: Physiological Problems of Heat and Water, 1964, also numerous articles. Sect. editor Am. Jour. Physiology, 1961-64, Jour. Applied Physiology, 1961-64; editorial bd. Jour. Cellular and Comparative Physiology, 1961—, Physiol. Zoology, 1959—; cons. editor Annals Arid Zone, 1962—; hon. editorial adv. bd. Comparative Biochemistry and Physiology, 1962-63. Research in comparative animal physiology, including respiration, blood chemistry, oxygen transport; water metabolism, renal function, salt excretion in marine birds and reptiles; temperature regulation and physiology of desert life; studies of camel, ostrich, other desert animals; Sci. expdns. to Sahara and other deserts. Office: Dept. Zoology, Duke Univ., Durham, N.C. 27706.*

SCHMIDT-ROHR, Ulrich, German physicist; b. Frankfort/Oder, Germany, May 25, 1926; s. Georg and Ruth (Rohr) S.-R.; student Berlin, Braunschweig insts. tech., 1943-47; M.S., U. Heidelberg, 1949, Ph.D., 1953; Fgn. student summer project fellow, Mass. Inst. Tech., 1954; m. Helma Wernery, Oct. 11, 1963; children—Volker, Ute, Klaus. Asst., Max Planck Inst., Heidelberg, 1954-60; sci. mem. Max Planck Inst. Nuclear Physics, Heidelberg, 1961-62, mem. directory, 1966—; prof. U. Heidelberg, 1961—; dir. Inst. Nuclear Physics, Nuclear Research Installation, Jülich, 1962-65. Mem. Max Planck Soc. Research, publs. on nuclear reactions, accelerator physics, beta decay. Home: 15 Im Hofert, 69 Heidelberg-Schlierbach. Office: Am Bierhelder Hof, 69 Heidelberg, West Germany.*

SCHMIDTKE, Hans Herbert, chemist; b. Rastenburg, Germany, July 9, 1929; s. Robert E. and Asta (Pietsch) S.; diploma Frankfurt (Germany) U., 1956, Dr. Degree, 1958; m. Helga G. Kühnel, Jan. 14, 1962; 1 dau., Jacqueline. Staff, Northwestern U., Evanston, Ill., 1959-60, Max-Planck Inst. for physics, Munich, Germany, 1960-61; chemist Cyanamid European Research Inst., Cologny, Geneva, Switzerland, 1961—; faculty Tech. High Sch., Zurich, 1966. Research, publs. on theory and spectra of coordination compounds basic quantum-mechanics, molecular orbital and ligand field theory. Home: 5, Chemin des Palettes, Gd. Lancy, Geneva 1212. Office: 91 Route de la Capite, Cologny, Geneva, 1223, Switzerland.*

SCHMIEDEBERG, Johann Ernst Oswald, pharmacologist; b. Laidsen, Kurland, Oct. 11, 1838; became prof., Dorpat (now Tartu, Estonia), 1871, Strasbourg (now in France), 1872. Author: Grundriss der Arzneimittellehre, 1883. Founder (with E. Klebs and Naunyn), editor (with Naunyn) Archives für experimentelle Pathologie und Pharmacologie (marked new era in exptl. pharmacology). First to prepare muscarin, 1869; made 1st studies of effects of poisons on heart of frog, circa 1870. Died Baden-Baden, Germany, July 12, 1921.

SCHMITT, Francis Otto, Am. biologist; b. St. Louis, Mo., Nov. 23, 1903; s. Otto Franz and Clara Elizabeth (Sniniger) S.; A.B., Washington U., St. Louis, 1924, Ph.D., 1927; Nat. Research Council fellow, Univ. of Calif., 1927-28; Univ. College, London, 1928, Kaiser Wilhelm Inst. Berlin-Dahlem, 1928-29; hon. D.Sc., Johns Hopkins University, 1950, Washington U., 1952, U. Chicago, 1957, Valparaiso U. 1959; M.D. honoris causa, U. Gothenburg, Sweden, 1964; LL.D. honoris causa, Wittenberg U., 1966; m. Barbara Hecker, June 18, 1927; children—David (dec.). Robert Hecker, Marion. Assistant prof. zoology, Washington U., 1929-34, asso. prof., 1934-38, prof., 1938-40, head dept., 1940; prof. biology, Mass. Inst. Tech., 1941, head dept., biology 1942-55, Inst. prof., 1955—, chmn. neuro-scis. research program, 1962—. Mem. Nat. Adv. Health Council, U. S. Pub. Health Service, 1959—; bd. sci. cons. Sloan-Kettering Inst. Cancer Research, 1963—. Trustee Mass. Gen. Hosp.; vis. com. Harvard Div. Sch., 1964-65. Recipient Alsop award, Am. Leather Chem. Assn., 1947; Lasker award, Am. Pub. Health Assn., 1956; T. Duckett Jones Meml. award Helen Hay Whitney Foundation, 1963. Fellow American Academy Arts and Sciences (council 1964-65), A.A.A.S., New York Academy Scis., mem. Soc. Gen. Physiologists, Histochem. Soc., Am. Phils. Soc. (council 1964-66), Am. Physiol. Soc. Am. Soc. Zoologists, Soc. Exptl. Biology and Medicine, Soc. Growth and Development (treas. 1945-46, pres. 1947), Am. Leather Chem. Assn., Biophys. Soc. (councilor), Electron Microscope Soc. Am. (dir. 1944-47, pres. 1949), Crystallographic Soc., Soc. Philomathique de Paris, Nat. Acad. Scis., Phi Beta Kappa, Sigma Xi, Editor: Fast Fundamental Transfer Processes in Aqueous Biomolecular Systems, 1960; Macromolecular Specificity and Biological Memory, 1962; Neurosciences—A Study Program, 1967. Research and numerous publications on biochem. nerve studies; established lipoprotein layered structure of nerve myelin through polarization-optical and x-ray diffraction studies (latered confirmed by electron microscopy); studied wound healing (led to investigations of chem. composition, molecular orgn., and biol. properties of collage molecule); discovered nature of soluble monomer, presence of peptide chains protruding from body of molecule, their role in fibrogenesis; axonal perfusion studies demonstrated molecular mechanism of ion gating and excitation. Home: 72 Byron Rd., Weston, Mass. 02193. Office: M.I.T., Cambridge, Mass. 02139.*

SCHMITT, Harold William, Am. physicist; b. Seguin, Tex., Aug. 11, 1928; s. Ben E. and Gertrude (Thiele) S.; student So. Meth. U., 1945-47; B.S., U. Tex., 1949, M.A., 1952, Ph.D., 1954; m. Jonell Britsch, May 4, 1952; children—Carol Emily, Anne Laine, Joy Diane. Research asst. Los Alamos Sci. Lab., 1952-54; physicist Oak Ridge Nat. Lab., 1954—, group leader physics of fission group, 1959—. Guest scientist Kernforschungszentrum, Karlsruhe, Germany, 1966-67; chmn. bd. dirs. ORTEC, Inc., Oak Ridge, 1960-64. Mem. Am. Phys. Soc., Sci. Research Soc. Am. Contbg. author: Fast Neutron Physics, 1963. Research, publs. on neutron-induced nuclear reactions, nuclear fission; reaction cross sects., sphere-transmission measurements of neutron capture cross sects., ionization and energy loss of fission fragments, devel. methods of energy calibration of solid state nuclear particle detectors for heavy ions and fission fragments; devel. and application of solid-state nuclear particle detector techniques in research on nuclear fission, exptl. and theoretical studies of nuclear fission. Home: 121 Canterbury Rd. Office: Physics Div., Oak Ridge Nat. Lab., P.O. Box X, Oak Ridge 37830.*

SCHMITT, Josef, chemist; b. Bamberg, Germany, Dec. 1, 1912; s. Karl and Marie (Heim) S.; grad. in engring. U. Munich (Germany), 1937, Docteur rer.nat. summa cum laude, 1939; m. Marcelle Lecocq, June 9, 1947; 1 son, Patrice. Asst., U. Lille (France), 1947-51; dir. chemistry Research Center, CLIN-BYLA, 1951—. Mem. Chem. Soc. France. Research, publs., patents on first synthesis of sterolic body,synthesis of squalene, polyenes and palmitic acid, in sterolic series, new medicine active on central nervous system. Home: 6 Sentier des Garennes, L'Hayles Roses, France. Office: Ets Clin-Byla 20, rue des Fossés-St-Jacques, Paris V, France.*

SCHMITT, Roman Augustine, Am. chemist; b. Johnsburg, Ill., Nov. 13, 1925; s. Joseph S. and Mary (Freund) S.; student Ill. Coll., 1946-48; M.S.,

U. Chgo., 1950, Ph.D., 1953; m. Jean Marie Vertovec, Dec. 28, 1954; children—Joseph, Mary, Peter, Katherine. Instr., U. Ill., Champaign, 1953-56; with Gen. Atomic, San Diego, 1956-66; faculty dept. chemistry Chemistry and Radiation Center, Ore. State U., Corvallis, 1966—. Faculty, U. San Diego, 1959. Mem. Am. Chem. Soc., Am. Geophys. Union, Geochem. Soc., Meteoretical Soc. Research, publs. on photodisintegraption and fission uranium and thorium nuclei, recoiling atoms in solids, various Szilard-Chalmers reactions, concentration trace elements determined in meteorites and terrestrial matter; application technique of instrumental neutron activiation to siliceous matter for multielement analysis. Home: 1830 Hawthorne Pl., Corvallis, Ore. 97330.*

SCHMITT, Waldo LaSalle, Am. zoologist; b. Washington, D.C., June 25, 1887; s. Ewald and Fanny Mathilde (Hesselbach) S.; B.S., George Washington U., 1913, Ph.D., 1922; M.A., U. of Calif., 1916; D Sc., University of Southern California, 1948; married Alvina Stumm, Nov. 19, 1914; children—Waldo Ernest (dec.), Barbara Ann (wife of Robert T. Lundy, United States Navy). Aid in econ. botany, U. S. Dept. Agr., 1907-10; scientific asst., U. S. Bur. Fisheries, 1910-13, naturalist, 1913-14; asst. curator, Div. Marine Invertebrates, U. S. Nat. Museum, 1915-20, curator, 1920-43, head curator of biology, 1943-47, zoölogy, 1947-57; research associate Smithsonian Instn., 1957—. Walter Rathbone traveling scholar, Smithsonian Instn., in S.A., 1925-27; at Tortugas Marie Lab., 1924-25, 1930-32; with Hancock Pacific Expdns., 1932-35; Smithsonian-Hartford Expdn. to West Indies, 1937; naturalist Pres. Cruise to Clipperton, Cocos, and Galapagos Islands, 1938; Hancock Atlantic Expdn., N. Coast South America-Netherlands West Indies, 1939; King Crab investigation for U. S. Fish and Wildlife Service in Alaska, 1940; on special detail to Latin America, 1941, 42, 43; study mus. Paris, Turin, 1954; member Smithsonian-Bredin Expdn., Belgian Congo, Sudan, Egypt, 1955, Caribbean, 1956, and 58, 59, Tahiti and Society Islands, 1957, also Quintana Roo, Yucatan, 1960, Bahamas, 1961, Palmer Peninsula survey USARP, Antarctica, 1962-63. Mem. bd. trustees Bears Bluff Lab., Wadmalaw Island, S.C., Internat. Oceanography Found.; bd. adv. Marine Lab. U. Miami; Serological Mus. Rutgers U., Inst. of Fisheries Research U. of N.C. Member USN Oceanographic Conf., 1925, chmn. conf. Importance and Needs of Systematics in Biology, NRC, 1953; del. Friday Harbor Symposium on Biology of Marine Boring and Fouling Organisms, 1957, Internat. Conf., Taxonomic Biochemistry, Physiology and Serology, U. Kansas, Lawrence, 1962. Fellow A.A.A.S., Cal. Academy Sciences; mem. Soc. Systematic Zoology (prin. founder 1948), Am. Soc. Limnology and Oceanography, Soc. Ichthyol. and Herptol., Am. Soc. Zool., Am. Mus. Assn., Am. Geophys. Union, Biol. Soc. Washington, Washington Acad. Sciences, Sigma Phi Epsilon, Beta Kappa Alpha, Sigma Xi; corr. mem. Zool. Soc. London, and hon. mem. other foreign scientific socs. Author: The Marine Decapod Crustacea of California, 1921; The Sun, and The Harvest of the Sea; Crustaceans; Applied Systematics: The Usefulness of Scientific Names of Animals and Plants. Research, publs. in biol. oceanography, marine invertebrate zoology; described number new species crustacea. Home: "Pinecrest," Highland Lane, Takoma Park, Md. Office: U. S. National Museum, Washington, D.C.*

SCHMITZ, Fredrick, German botanist; b. Sarrebruck, 1850; Author: Die Familiendiagramme der Rhoeadinen, 1878. Research on nuclei of thallophytes; made classic studies of chromatophores of algae. Died Greifswald, Germany, 1895.

SCHMITZ-DUMONT, Otto, inorganic chemist; b. Pretoria, S. Africa, Feb. 13, 1899; s. Winny and Margarethe (Fischer) S.-DuM.; Promotion, U. Bonn (Germany), 1923, Habilitation, 1926; m. Klara Hermann Dec. 26, 1921. Faculty, U. Bonn., 1926—, prof. inorganic chemistry Inst. Inorganic Chemistry, 1960—, dir. 1960—. Mem. Gesellschaft Deutscher Chemiker, Deutsche Bunsengesellschaft. Author: (with Rheinboldt) Chemische Unterrichtsversuche, 1962; also numerous articles. Research on constn. of polymer indoles, mechanism of polymerization of indole, polymerization of substituted diphenylethylenes, reactions in liquid ammonia, color and constn. of inorganic solids. Home: 14 Malteser Str., Bonn, Germany.*

SCHNATHORST, William Charles, Am. plant pathologist; b. Ft. Dodge, Ia., May 8, 1929; s. William T. and Elizabeth (Nelson) S.; B.S., U. Wyo., 1952, M.S., 1953; Ph.D., U. Cal., Davis, 1957; m. Rosemarie A. Meyer, Dec. 29, 1951; children—Diana Lynn, William John, Douglas Alan. Research asst. plant pathology U. Cal., Davis, 1954-56; plant pathologist Agrl. Research Service, U. S. Dept. Agr., Davis, 1956—; mem. steering com. Conf. on Soil Fungi, 1962—. Mem. Am. Phytopathol. Soc., Am. Chem. Soc., Am. Inst. Biol. Scis., Cal.-Wyo. Acad. Sci., Mycol. Soc. Am., Botan. Soc. Am., A.A.A.S., Am. Inst. Biol. Sci., N.Y. Acad. Sci., Cotton Disease Council, Sigma Xi, Phi Kappa Phi, Alpha Zeta. Studies, numerous publs. on biology and physiology of powdery mildew diseases; theory of inoculum and inoculum potential, host-parasite physiology of vascular plant disease fungi; survival of bacterial plant pathogens; single cell isolation and preservation of bacterial cultures; serology of host-parasite compatibility with plants and pathogens supporting a theory of host specificity being conditioned by sharing of common antigens between plants and pathogens; serol. structures and virulence in bacterial pathogens; serology and bacterial taxonomy. Home: 647 Hunt Way, Davis, Cal. 95616.*

SCHNECK, Jerome M., Am. physicians; b. N.Y.C., Jan. 2, 1920; s. Maurice and Rose (Weiss) S.; A.B., Cornell U., 1939; M.D., L.I. Coll. Medicine, 1943; m. Shirley R. Kaufman, July 24, 1943. Practice medicine, specializing in psychiatry, N.Y.C., 1947—; asso. vis. psychiatrist Kings County Hosp. Center, Bklyn., 1959—; faculty State U. N.Y. Downstate Med. Center, 1950—, clin. asso. prof. psychiatry, 1958—; supervising psychiatrist Community Guidance Services, 1955-60; vis. lectr. L.I. U., 1955, N.Y. Med. Coll., Met. Hosp., 1965; cons. Council on Mental Health A.M.A., 1956-58, NBC, 1962—, Morton Prince Clinic, 1965—. Recipient award of merit, 1955, Gold medal award, 1958, Best Book award, 1965, Bernard B. Raginsky award, 1966, all from Soc. Clin. and Exptl. Hypnosis. Diplomate Am. Bd. Psychiatry and Neurology, Am. Bd. Med. Hypnosis. Fellow Am. Med. Authors, Internat. Soc. Clin. and Exptl. Hypnosis, A.A.A.S., Am. Psychiat. Assn., Am. Psychol. Assn., Am. Med. Writers Assn., Am. Soc. Psychoanalytic Physicians, Am. Acad. Psychosomatic Medicine; mem. Soc. Apothecaries London, Inst. Practicing Psychotherapists, Inst. Research In Hypnosis (dir.), Am. Assn. History of Medicine, Assn. Advancement Psychotherapy, A.M.A., Am. Bd. Examiners Clin. Hypnosis, Caribbean Fedn. Mental Health. Author: Hypnosis in Modern Medicine, 1953; Studies in Scientific Hypnosis, 1954; A History of Psychiatry, 1960; The Principles and Practice of Hypnoanalysis, 1965; also numerous articles. Contbns., publs. studies on new methods and concepts in hypnosis; psychiatry and psychoanalysis; history of medicine and psychiatry. Home and office: 26 W. 9th St., N.Y.C. 10011.*

SCHNEIDER, Albert, Am. bacteriologist, inventor; b. Granville, Ill., Apr. 13, 1863; s. John and Elizabeth (Burcky) S.; M.D., Coll. Physicians and Surgeons Chgo., 1887; B.S., U. Ill., 1894; M.S. U. Minn., 1894; Ph.D., Columbia, 1897; m. Marie Louise Harrington, June 28, 1892; 1 dau., Cornelia Elizabeth. Instr. botany U. Minn., 1893; prof. pharmacognosy and bacteriology Northwestern U. Sch. Pharmacy, Chgo., 1897-1903; prof. pharmacognosy and bacteriology U. Cal., 1903-19, materia-medica and therapeutics, 1904-06; dir. expt. sta. Spreckels Sugar Co., 1906-07; pharmacognosist U. S. Dept. Agr., 1909-15; micro-analyst Cal. State Food and Drug Lab., 1915-19; prof. pharmacognosy Coll. Pharmacy of U. Neb., 1919-21 dean sch. pharmacy N. Pacific Coll., Portland Ore., 1922—. Mem. internat. jury awards Panama P.I. Expn., 1915. Author: Primary Microscopy and Biology, 1890; A Text-book of General Lichenology, 1897; Guide to the Study of Lichens, 1898; Microscopy and Micro-Technique, 1899; Hints on Drawing for Students of Biology, 1899; General Vegetable Pharmacography, 1900; The Limitations of Learning, and Other Science Papers, 1900; Powdered Vegetable Drugs, 1920; Bird and Nature Study Chart Manual, 1903; Medicinal Plants of California, 1909; Drug Plant Culture in California, 1912; Pharmaceutical Bacteriology, 1920; Bacteriological Methods in Food and Drugs Laboratories, 1915; The Microbiology and Microanalysis of Foods, 1920; Microscope in Detective Work (report., Office of Naval Intelligence), 1918; Laboratory Pharmacology and Toxicology, 1925. Translator: Westermaier's Compendium der Allgemeinen Botanik, 1896. Editor-in-chief Pacific Pharmacist, 1910-15. Contbr. many papers to scientific jours. Inventor of car ventilating system. Died Oct. 27, 1928.

SCHNEIDER, David M., Am. anthropologist; b. N.Y.C., Nov. 11, 1918; B.S., Cornell U., 1940, M.A., 1941; Ph.D., Harvard, 1949. Lectr., London Sch. Econs., 1949-51; asst. prof. Harvard, 1951-56; faculty U. Cal., Berkeley, 1956-60, prof., 1958-60; prof. U. Chgo., 1960—, chmn. dept., 1963-66. Cons. NIH, 1960-64. Fellow, Center Advanced Study Behavioral Sci., 1955-56. Fellow Am. Anthrop. Assn., Royal Anthrop. Inst. (Great Britain), Assn. Social Anthropologists. Author: (with Homans) Marriage, Authority and Final Causes, 1955; (with Gough) Martilineal Kinship, 1961; American Asso. editor, Am. Anthropologist, 1956-59. Contbr. to theory of culture and social structure by innovations in primitive and modern soc. Officer: Dept. Anthropology, U. Chgo., 1126 E. 59 St., Chgo. 60637.*

SCHNEIDER, Edward Christian, Am. biologist; b. Wapello, Ia., Aug. 21, 1874; s. John George and Augusta J. (Bauersfeld) S.; B.S., Tabor Coll., Ia., 1897; Ph.D., Yale, 1901; Sc.D., U. Denver, 1914; M.P. E., Internat. Y.M.C.A. Coll., 1923; Sc. D., Colo. Coll., 1932; m. Elsie M. Faurote, June 24, 1902; children—Edwin George, Marion Elsie (Mrs. R. E. Joyce). Instr. chemistry Tabor Coll., 1897-99, prof. biology and physiol. Chemistry, 1901-03; prof. biology Colo. Coll., 1903-07, head prof., 1907-19; Daniel Ayres prof. biology Wesleyan U., Conn., 1919-44. John Jeffries award for contrbts. to aeromedicine Inst. Aero. Scis., 1942. Fellow A.A.A.S., Am. Phys. Edn. Assn.; mem. Am. Physiol. Soc., Am. Soc. Biol. Chemists, Am. Pub. Health Assn., Soc. Exptl. Biology and Medicine, Soc. Am. Bacteriologists, Sigma Xi, Phi Beta Kappa. Author: Physiology of Muscular Activity, 1933, revised edition, 1939 revised (with P. K. Karpovick), 1948. Part Author: Report of Pike's Peak Expdn., 1911; Manual of the Med. Research Laboratory, Air Service, publishing chiefly studies of the influence of high altitudes and low oxygen on mankind, aviation physiology, and effects of phys. exercise and tng. Died Oct. 3, 1954.

SCHNEIDER, Edward Martin, Am. physician; b. Cleve., May 12, 1922; s. Solomon S. and Beatrice (Sicherman) S.; student Northwestern U., 1940-43; M.D., U. Cin., 1946; m. Jane Einstein, June 18, 1950; children—Douglas Allen, Robert Stewart. Asst. prof. medicine U. Okla., 1952-57; sr. physician Miners Meml. Hosp. Assn., 1957-61; chief medicine Cameron Meml. Hosp., Bryan, O., 1961-62; chief clin. research sect. Upjohn Co., Kalamazoo, 1962-67; dir. clin. pharmacology Lederle Labs., 1967—; teaching cons. Borgess, Bronson hosps. Diplomate Am. Bd. Internal Medicine. Fellow A.C.P., Am. Coll. Clin. Pharmacology and Chemotherapy, Internat. Acad. Law and Sci.; mem. Am. Assn. for Study Liver Disease, Mich. Heart Assn. (trustee), Sigma Xi. Research, publs. on human pharmacology and toxicology, evaluation of drug effectoon liver and gastrointestinal tract. Home: 368 Clover Lane, Wyckoff, N.J. 07481. Office: Middletown Rd., Pearl River, N.Y. 10965.*

SCHNEIDER, Erich Ernst, physicist; b. Berlin, Germany, June 1, 1911; s. Ernst and Hedhig (Schneider) Cohn; student Technische Hochschule Munich (Germany), 1929-31, U. Berlin, 1931-32; Dr.phil. nat., U. Freiburg (Germany), 1933; m. Barbara Elisabeth Schatzman, Oct. 6, 1938; children—Ursula Renate, Barbara Beate. Research staff Institut for Theoretical Physics, Copenhagen, Denmark, 1934; tech. work, Berlin, 1934-36; research staff photochem. processes of vision U. Coll., London, 1937-38; faculty U. Newcastle/Tyne, Eng., 1938—, reader solid state physics, 1963—; radar research Admiralty, 1943-45; staff Duke, Durham, N.C., 1953-54; reader solid state physics Newcastle div. U. Durham, 1960. Cons. for electronics mfrs., Newcastle area, 1958—. Fellow Inst. Physics, Phys. Soc.; mem. Am. Phys. Soc., Faraday Soc. Author: Nuclear Moments, 1958; also articles. Translator: Kernmomente (Kopfermann). Research on microwave transitions in hydrogen atoms, ultramicroscopy, 2d harmonic theory of defects; pioneered research in electron spin resonance and asso. instrumentation, including 1st use of superheterodyne detection in electron spin resonance spectrometer; discovered V-centres. Home: Lindisfarne, Stoneyhurst Rd., Newcastle/Tyne 3, Eng.*

SCHNEIDER, Hans, mathematician; b. Vienna, Austria, Jan. 24, 1927; s. Hugo and Bella Schneider; M.A., U. Edinburgh, 1948, Ph.D., 1952; m. Miriam Wieck; children—Barbara, Peter, Michael. Asst. lectr., lectr. Queen's U., Belfast, Ireland, 1952-59; vis. prof. Wash. State U., 1956-57; faculty dept. math. U. Wis., Madison, 1959—, prof., 1965—, chmn. math. dept., 1966-68. Mem. Am., Edinburgh, London math. socs., others. Author: (with G. P. Barker) Matrices and Linear Algebra, 1968. Research, articles on matrix theory, semi-groups. Home: 910 S. Midvale Blvd., Madison, Wis. 53711.*

SCHNEIDER, Howard Albert, Am. biologist; b. Milw., Dec. 25, 1912; s. George A. and Stella (Wendler) S.; B.S., U. Wis., 1934, M.S., 1936, Ph.D., 1938; m. Marie Gugler, Aug. 17, 1937; children—Susan, Cynthia. Fellow Wis. Alumni Research Found., 1934-36; Rockefeller Found. fellow, 1939-40; staff Rockefeller Inst. for Med. Research, 1940-65, asso. prof., 1957-65; mem. Inst. for Biomed. Research, Edn. and Research Found., A.M.A., 1965-67, acting director of the Institute, 1967-68. Fellow A.A.A.S. (mem. council 1956-58), N.Y. Acad. Sci.; mem. Am. Inst. Nutrition, Harvey Soc., Animal Care Panel, Am. Chem. Soc., Soc. for Exptl. Biology and Medicine, Am. Inst. Biol. Scis., Sigma Xi. (past chpt. pres.). Asso. editor Jour. Nutrition, 1957-61. Research, numerous publs. in biochemistry and nutrition, natural resistance to infections disease leading to discovery pacifarins which pacify infectious disease into innocuous and silent states. Home: 777 N. Michigan Av., Chgo. 60611. Office: 535 N. Dearborn St., Chgo. 60610.*

SCHNEIDER, Konrad Victor, German physician; b. Bitterfeld, Saxony, 1614; physician in ordinary, prof., Wittenberg. Author: Dissertatio de ossi cribriformi, 1655; De catarrhis, 1660. Described nasal mucous membrane (Schneider's, or Schneiderian, membrane), also demonstrated that nasal mucus originates there rather than in pituitary, 1660. Died 1680.

SCHNEIDER, Louis, sociologist; b. Vienna, Austria, Mar. 22, 1915; s. Gustave and Frieda (Salz) S.; came to U.S., 1921, naturalized, 1928; B.A., Coll. City N.Y., 1935; M.A., Columbia, 1938, Ph.D., 1947; m. Betty Sancier, May 24, 1947 (div. June 1954); children—David S., Dana A.; m. 2d. Josephine Ann Sundine, Jan. 3, 1956; 1 dau., Valerie S. Economist, WPB, OPA, Washington, 1943-47; asst. prof. sociology Colgate U., Hamilton, N.Y., 1947-49; asso. prof. Purdue U., Lafayette, Ind., 1949-54, prof., 1954-59; prof. U. Ill., Urbana, 1960-67, head dept. sociology, 1960-64; prof. U. Tex., Austin, 1967—. Center for Advanced Study in Behavioral Scis. fellow, Stanford, Cal., 1954-55. Author:

Freudian Psychology and Voblen's Social Theory, 1948; (with Ogle and Wiley) Power, Order and the Economy, 1954; (with Dornbusch) Popular Religion, 1958; also numerous publs. Editor: Problems of Economics and Sociology, 1963; The Scottish Moralists, 1967; editor, contbg. author Religion, Culture and Society, 1964. Home: 4210 Prickly Pear Dr., Austin, Tex. 78731.*

SCHNEIDER, Maurice Camille, French physician; b. Nancy, France, Mar. 10, 1933; s. Albert and Felice (Lubel) S.; M.D., U. Paris, 1960; m. Marie-Claude Leblond, May 7, 1966. Bacteriologist, hematologist Inst. Gustave Roussy, Villejuif, France, 1961; microbiologist Institut de Cancerologie et d'Immunogenetique, Villejuif, 1964, asst. microbiology, U. Paris, 1964—. Mem. N.Y. Acad. Scis., Internat. Soc. Hematology. Research, publs. on chemotherapy of leukemias and hematosarcomas, bone marrow transplantation, new antibiotics, treatment of infectious diseases. Home: 4 Roosevelt, Sceaux 92, France. Office: Institut G. Roussy, 16 bis Avenue P.V. Couturier, Villejuif 94, France.*

SCHNEIDER, Otto, geophysicist; b. Bohlitz, Germany, Jan. 20, 1912; s. Erwin Alexander and Martha (Dietze) S.; Dr.phil., U. Berlin (Germany), 1936; m. Elfriede Busse, May 18, 1937; 1 dau., Eva Marie. Adviser meteorology Askania-Werke AG, Berlin, 1936-39; staff Servicio Meteorológico Nacional, Buenos Aires, Argentina, 1939-58, head geophysics dept., 1947-58; prof. statistics Escuela Superior de Meteorologia, 1947-48, prof. geophysics, 1948-53; prof. geophysics U. Buenos Aires, 1953-58; chief sci. dept. Instituto Antartico Argentino, Buenos Aires, 1956—; prof. geomagnetism U. La Plata, 1968—; reporter Aurora for S.Am., 1956-65. Mem. expdns. to Antarctica, 1957-58, 61. Recipient prize Premio Academia Nacional de Ciencias de Buenos Aires, 1965. Mem. Internat. Assn. Geomagnetism and Aeronomy (exec. com. 1957-63, chmn. lunar com.), Internat. Assn. Meteorology and Atmospheric Physics, Sociedad Cientifica Argentina, Sociedad Argentina de Estudios Geograficos, Asociación Argentina de Geofisicos y Geodestas, (pres. 1967—), Am. Geophys. Union, Deutsche Gesellschaft für Polarforschung. Research, publs. on variability of, and solar effects on, lunar geomagnetic variations, determination of geomagnetic tides at Isla Ano Nuevo, 1st sci. work on Aurora Australis in S. Am.; determined location of auroral zone in Weddell Sea sector. Home: 3 de Febrero 1841, Buenos Aires. Office: Cerrito 1248, Buenos Aires, Republica Argentina.*

SCHNEIDER, Walter Carl, Am. biochemist; b. Cedarburg, Wis., Sept. 26, 1919; s. Carl F. and Martha (Neuenfeldt) S.; B.S., U. Wis., 1941, Ph.D., 1945; m. Edith May Janot, May 30, 1942; children—James C., Susan J., Walter Carl. Fellow, Jane Coffin Childs Fund, U. Wis., Madison, 1945-46, Rockefeller Inst. for Med. Research, 1946-47; instr. U. Wis., 1947-48; with lab. biology Nat. Cancer Inst., Bethesda, Md., 1948-61, head nucleic acids sect., lab. of biochemistry, 1961—, mem. splty. fellowship bd., 1964—. Mem. Am. Assn. Cancer Research, Am. Soc. Biol. Chemists, Am. Chem. Soc., Phi Beta Kappa, Phi Lambda Upsilon, Sigma Xi, Gamma Alpha. Asso. editor Jour. Nat. Cancer Inst., 1952-53. Research, numerous publs. on devel. methods for determination of nucleic acids in animal tissues and for isolation of subcellular particulates such as nuclei and mitochondria from tissues; research in occurrence of deoxyribose compounds other than DNA in normal and tumor tissue. Home: 10112 Ashwood Dr., Kensington, Md. 20795. Office: Nat. Cancer Inst., Bethesda, Md. 20014.*

SCHNEIDER, Wilhelm, German dermatologist; b. Braunfels/Wetzlar, Germany, May 22, 1910; s. Wilhelm and Lina (Schermuly) S.; student univs. Kiel, Marburg; Dr.med., U. Giessen, 1936; m. Marianne Schmidt, Aug. 15, 1947; children—Hans-Joachim, Renate. Faculty, U. Tübingen (Germany), 1943—, prof., 1950—, dir. dermatol. hosp. and out-patient dept., 1961—, dean medicine, 1966-67. Mem. German Soc. Phlebology (v.p.), German Soc. Angiology (pres.); hon. mem. Austrian Dermatol. Soc., French Soc. Phlebology; corr. mem. Argentine Dermatol. Soc. Author: (with others) 2 dermatology handbooks, textbook; also numerous articles in field. Research in indsl. medicine and preventive medicine, devel. new forms of treatment for skin disorders with emphasis on drugs for topical application, angiology, new concepts of pathogenesis and therapeutic approaches to various skin disorders. Address: 25 Liebermeisterstrasse, 74 Tübingen, West Germany.*

SCHNEIDERMAN, Howard Allen, Am. biologist; b. N.Y.C., Feb. 9, 1927; s. Louis and Anna (Center) S.; B.A., Swarthmore Coll., 1948; M.A., Harvard, 1949, Ph.D., 1952; m. Audrey MacLeod, Sept. 16, 1951; children—Anne Mercedes, John Howard. Faculty, Cornell U., 1952-61; prof., chmn. dept. biology, dir. Developmental Biology Center Western Res. U., Cleve., 1961—; instr. invertebrate zoology Marine Biol. Labs., Woods Hole, Mass., 1956-58 (trustee). Cons. Gen. Med. Sci. Inst. NIH, 1961—. Fellow A.A.A.S., N.Y. Acad. Scis.; mem. Soc. Exptl. Biology, Am. Soc. Zoologists, Entomol. Soc. Am., Radiation Research Soc., Soc. Exptl. Biology and Medicine, Lepidopterists Soc., Am. Soc. Naturalists, Am. Assn. U. Profs., Am. Soc. Cell Biology, Soc. Developmental Biology (pres. 1965-66), Internat. Inst. Embryology, Phi Beta Kappa, Sigma Xi. Editorial bd. Am. Zoologist, Gen. and Comparative Endocrinology, Am. Naturalist, Jour. Exptl. Zoology. Research, numerous publs. in biochemistry and endocrine control of insect growth and devel., radiation biology, plant physiology, gen. invertebrate zoology. Home: 2328 Lamberton Rd., Cleveland Heights, O. 44118. Office: 2127 Cornell Rd., Cleve. 44106.*

SCHNEIDERMAN, Marvin Arthur, Am. math. statistician; b. N.Y.C., Dec. 25, 1918; s. Alexander and Mollie (Simpkins) S.; B.S., Coll. City N.Y., 1939; postgrad. Harvard, Ohio State U., U. London (Eng.); M.A., Ph.D., Am. U.; m. Irene Wolfson, Dec. 20, 1941; children—Jo Harte, Sarah, Susan. With U. S. Census Bureau, 1940-43; statistician War Dept., 1946-48; math. statistician Nat Cancer Inst., NIH, Bethesda, Md., 1948—; professorial lectr. in med. statistics Georgetown U., 1960—. Mem. adv. bd. Consumers Union. Rockefeller Pub. Service fellow, 1959-60. Fellow Am. Statis. Assn., Royal Statis. Soc., A.A.A.S.; mem. Biometrics Soc., Am. Assn. Cancer Research. Editorial bd. Jour. Nat. Cancer Inst., Research, numerous publs. on math. and statis. techniques for evaluation of effectiveness med. treatments for screening and discovery new drugs and for elucidation biol. mechanisms, math. in devel. cell counting and sizing devices. Home: 6503 E. Halbert Rd., Bethesda 20034. Office: Nat. Cancer Inst., Bethesda, Md. 20014.*

SCHNEIERSON, S. Stanley, Am. microbiologist; b. N.Y.C., May 15, 1906; s. Isaac and Anna (Levine) S.; A.B., Cornell U., 1928; M.D., L.I. Coll. Medicine, 1932; m. Marianne Jacqueline Katz, July 21, 1950; 1 dau., Martine (Mrs. Jerome Kowal). Research asst. Mt. Sinai Hosp., N.Y.C., 1935-38, Pasteur Insts. of Paris, Brussels, 1938-39; asst. bacteriologist Mt. Sinai Hosp., 1939-41, asso. bacteriologist, 1946-55, dir. microbiology, 1955—, prof. microbiology Mt. Sinai Sch. Medicine, 1966—. Diplomate Am. Bd. Pathology, Am. Bd. Microbiology. Fellow Am. Acad. Microbiology, A.C.P., N.Y. Acad. Medicine, N.Y. Acad. Scis., A.A.A.S. Author: Atlas of Diagnostic Microbiology, 1965; also numerous articles. Research in microbiology and teaching. Home: 64 E. 86th St., N.Y.C. 10028. Office: 1 E. 100th St., N.Y.C. 10029.*

SCHNEIRLA, Theodore Christian, Am. animal psychologist; b. Bay City, Mich., July 23, 1902; s. Christian and Mary Emily (Badger) S.; B.Sc., U. Mich., 1924, M.Sc., 1925, Sc.D., 1928; m. Leone Margaretta Warner, June 12, 1926; children—Lois Janet (Mrs. Harold Adrian Drum), Donn Richard (dec.). Faculty dept. psychology N.Y. U., 1928-45, adj. prof., 1947—; asso. curator dept. animal behavior Am. Mus. Natural History, N.Y.C., 1942-47, curator, 1947—; adj. prof. dept. biology City U. N.Y., 1965—. NRC fellow, 1930-31; Guggenheim fellow, 1944-46. Fellow A.A.A.S., Am. Psychol. Assn., N.Y. Zool. Soc., Am. Acad. Arts and Scis.; mem. Am. Zool. Soc., Am. Soc. Naturalists, Internat. Union Study Social Insects (past pres. N.Am. sect.), N.Y. Entomol. Soc., Sigma Xi, Phi Sigma, Psi Chi, Gamma Alpha. Author: Learning and Orientation in Ants, 1929; (with Norman R.F. Maier) Principles of Animal Psychology, 1935; (with others) Recent Experiments in Psychology, 1938; also numerous articles. Research on maze problems with ants and rodents, nomadic and predatory behavior of Old and New World army ants in their colonies based on changing stimulative effects from successive broods, social devel. of kittens, theory accounting for approach and withdrawal behavior in terms of quantitative stimulative effects underlying developmental orgns. differing according to psychol. level. Home: 140 Cabrini Blvd., N.Y.C. 10033. Office: Am. Mus. Natural History, 79th and Central Park W., N.Y.C. 10024.*

SCHNELLER, George Otto, Am. inventor, mfr.; b. Nürnberg, Germany, Jan. 14, 1843; s. Henry and Elizabeth (Ruckert) S.; m. Clarissa Alling, May 1, 1873, 3 children. Came to U. S., 1860; cashier, accountant Osborne and Cheesman Co., brass manufactory, Ansonia, Conn.; returned to Germany, 1870, to U. S., 1872; obtained 4 patents on corset springs between 1872-73; bought spectacle factory, Shelton, Conn.; began mfg. brass corset eyelets; secured patents on die for making eyelets, eyelet machine, punch and die for eyelet machines, 1884; bought Osborne and Cheesman business, 1882, reorganized it as Ansonia Osborne and Cheesman Co., began manufacture of brass goods under his patent rights; made important contbns. to world corset industry by his inventions; founder, treas. Union Fabric Co., Ansonia; invented and patented hook and eye, bustle, machine for covering dress stays, button press, button-fastening device; active in building electric street ry. system between Derby, Ansonia and Shelton, Conn.; represented Ansonia in lower house of Conn. Legislature, 1891-93. Died Ansonia, Oct. 20, 1895.

SCHNEPF, Eberhard Friedrich Hermann, German botanist; b. Nürnberg, Germany, Apr. 4, 1931; s. Georg and Anne (Warnecke) S.; grad. U. München (Germany), 1952; Dr.rer.nat., U. Bonn (Germany), 1958; m. Rosemarie Langbein, Mar. 5, 1959; children—Christian, Ulrike. Staff, U. Marburg (Germany), 1959-64; faculty U. Göttingen, (Germany), 1964-66, wissenschaftlicher rat, prof., 1966; prof. cytology U. Heidelberg (Germany), 1966—. Research, publs. on structure and function of plant cell, ectodesmata, Golgi-apparatus, secretion processes, plastids, cell wall and compartmentation. Home: 3 am Neckardamm, 6801 Neckarhausen, Germany. Office: 15 Berliner Strasse, 69 Heidelberg, Germany.*

SCHNEYER, Leon Harold, Am. physiologist; b. N.Y.C., Mar. 20, 1919; s. Leon and Ida (Greenberg) S.; B.A. magna cum laude, Washington Sq. Coll., N.Y. U., 1945, D.D.S., 1943, Ph.D. in Physiology, 1949; m. Charlotte Alper, June 11, 1945. Instr. physiology N.Y. U. Coll. Dentistry, 1943-48; staff Montefiore Hosp., N.Y.C., 1948-52; faculty U. Ala. Med. Center, Birmingham, 1952—, prof. physiology, 1959—, prof. dentistry, 1958—. Mem. dental study sect. NIH, 1956-59; mem. Nat. Bd. Dental Examiners, 1959-61. Recipient Research Career award NIH, 1964—. Fellow N.Y. Acad. Scis.; mem. Am. Physol. Soc., Soc. for Exptl. Biology and Medicine, Am. Assn. U. Profs., Internat. Assn. for Dental Research, Sigma Xi. Editorial bd. Jour. Dental Research, 1958-60. Research, numerous publs. on secretory activity by salivary glands, methods collection saliva from individual glands in man, conditions in animals for secretion digestive enzyme, biophys. aspects secretory activity in salivary gland, energetics salt transport by glands, system for study glandular activity outside body. Home: 1618 A 29th Ct. S., Birmingham, Ala. 35209.*

SCHNITTER, Gerold, Swiss mech. engr.; b. Basel, Switzerland, Oct. 25, 1900; s. Hermann and Emilia (Jenny) S.; civil engring. diploma Fed. Poly. Zurich; m. Margarethe Behn von Eschenburg, June 15, 1925; children—Nicolas, Catherine, Susanne, Anne. Engr. pub. works various European countries; tech. dir. Conrad Zschokke Co., Geneva; became prof. hydraulics and found. works Fed. Poly., Zurich, 1952; named dir. Hydraulic Research Labs., European Packaging Fedn., 1953. Recipient awards, Toulouse, France, Swiss Soc. Engrs. and Architects. Mem. Nat. Swiss Com. Founds. (pres.), Swiss Union Rd. Research and publs. on hydroelectricity, earth mechanics, cement founds. Home: Erbstrasse 14, Küssnacht. Office: Gloriastrass 37-39, Zurich, Switzerland.

SCHNORE, Leo Francis, Jr., Am. sociologist; b. Elyria, O., Jan. 10, 1927; s. Leo Francis and Mildred (Kraus) S.; A.B., cum laude Miami U., 1954; m. Elinor Carole Schick, Sept. 9, 1950 (div. Mar. 1968); children—Carol Elizabeth, Barbara Lee. Instr., Brown U., 1954-56; asst. prof. Mich. State U., 1956-57, U. Cal. at Berkeley, 1957-59; faculty U. Wis., Madison, 1959—. prof., 1962—. Mem. Population Assn. Am. (dir. 1964—), Social Sci. Research Council (Aux. Research award 1962, dir. 1964—, treas. 1965—), Am. Sociol. Assn., Phi Beta Kappa. Author: The Urban Scene, 1965; also numerous articles. Editor: (with P.M. Hauser) The Study of Urbanization, 1965; (with Henry Fagin) Urban Research and Policy Planning, 1967; Social Science and the City, 1968. Discovered relationship between residential distbn. of social classes in Am. urban areas and size and age of urban areas, elite groups nr. center in small, new areas at periphery in large, old areas. Address: Dept. Sociology, U. Wis., Madison, Wis. 53706.*

SCHOCKEN, Klaus, physicist; b. Berlin, Germany, Apr. 24, 1905; s. Wilhelm and Sophie (Meyer) S.; Ph.D., U. Berlin, 1928. Came to U.S., 1936, naturalized, 1944. Physicist, Army Med. Research Lab., Fort Knox, Ky., 1948-52, St. Bonaventura U., Olean, N.Y., 1952-56, Chgo. Midway Lab., 1956-58, G.C. Marshall Space Flight Center, Huntsville, Ala., 1958—. Mem. Am. Phys. Soc., Math. Assn. Am. Research, publs. in application of field theory to thermophysics, application of fluid flow theory to blood circulation, application electromagnetic field theory to electrocardiogram. Home: 1903 Alexander Dr., Huntsville 35801. Office: G.C. Marshall Space Flight Center, Huntsville, Ala. 35812.*

SCHOECK, Helmut, sociologist; b. Graz, Austria, July 3, 1922; s. Stephan and Anna (Heigl) S.; student U. Munich, 1941-45; Ph.D., U. Tübingen, 1948; m. Margaret Weiler, June 24, 1947; children—Natalia, Raymond, Stephanie. Prof. social sci. Fairmont (W.Va.) State Coll., 1950-53; prof. sociology Emory U., 1954-65; prof. sociology, dir. Inst. Sociology, U. Mainz, Germany, 1965—. Fellow Am. Sociol. Assn., Am. Anthrop. Assn. Author: Nietzsches Philosophie, 1948; Soziologie, 1952; USA: Motive und Strukturen, 1958; Was heisst politisch unmoeglich, 1959; Scientism and Values, 1960; Relativism, 1961; Financing Medical Care, 1962; Soziologie und die Gesellschaften, 1964; Der Neid: Eine Theorie der Gesellschaft, 1966. Research on nature and limits of interdisciplinary discourse; cross-cultural study of health ins. systems; history of sociology; study of envy, using cross-cultural approach, as basis of social controls required for formation of lasting associations. Home: 30 Haideweg, 6200 Wiesbaden-Sonnenberg, Germany.*

SCHOELMERICH, Paul, German physician; b. Kasbach, Germany, June 27, 1916; s. Johannes and Katharina (Lichte) S.; student U. Bonn; M.D., U. Leipzig, 1941; m. Gerda Frowein, Oct. 5, 1946; children—Juergen, Axel, Uwe. Became asst., Univ. Hosp., Marburg/Lahn, Germany, 1946, asst. prof., 1952, asso. prof., 1958; dir. II. med. clinic, Univ. Hosp., Meinz, Germany, 1963——. Mem. German Assn. Internal Medicine, German Assn. Circulation Research, Rheumatology, Phys. Medicine. Author: (with Grill, Löhr, Scholtze) Beiträge zur Angiographie Chirurgischer Lungenerkrankungen, 1964; also numerous articles. Research on regulation of body temperature, theory of phonocardiography, electrocardiography; bacterial endocarditis, pericarditis, myocarditis. Home: 67 Weidmannstrasse. Office: 1 Langenbeckstrasse, Mainz, W. Germany.*

SCHOENBERG, Erich Karl Wilhelm, astronomer; b. Warsaw, Poland, Dec. 27, 1882; s. Gustav and Christine (Rabe) S.; student U. Dorpat, U. Kiel; Ph.D.; m. Hedwig Gudowius, May 5, 1953. Asst. (1906-12), observer (1912-18) U. Dorpat/Estonia Obs., collaborator Helsingfors (Finland) Obs., 1920-25; instr. U. Greifswald, Germany, 1925; prof., dir. Breslau Obs., 1926-45; prof., dir. Munich Obs., 1946-56. Recipient prize and medal Royal Russian Acad. Astronomy, 1908; named master U. Dorpat, 1913. Mem. Acad. Sci. Bavaria, Astron. Soc. Author: Theoretische Photometrie, 1929; Die aequatoriale Beschleunigung bei Jupiter, 1948; Die Absorption und die Dichte 31 Dunkelwocken, 1949; Eine hydrodynamische Theorie der aequatorischen Beschüleunigung der Sonne, 1958; Über die Veranderlichkeit des Sonnen Durchmessers. Address: Rimsting am Chiemsee, Am Assbaum 2, W. Germany.

SCHOENEBORN, Heinz, German mathematician; b. Remscheid, Oct. 24, 1921; s. Heinrich and Marie (Lemmer) S.; Dr.ès.sc., U. Bonn (Germany); m. Ursula Fuchs, Oct. 24, 1952. Instr. math. U. Bonn, 1949-53; prof. math. U. Aix-la-Chapelle, 1964——. Mem. Assn. German mathematicians. Contbr. articles to math. jours. Home: Nordstrasse 114, Bonn, Germany. Office: Templergraben 55, Aachen, W. Germany.

SCHOENHEIMER, Rudolf, biochemist; b. Berlin, Germany, May 10, 1898; Ph.D., U. Berlin; mem. faculty Coll. Phys. and Surg. (Columbia); introduced isotopic tracers in biochem. research, 1935; used fat molecules which contained deuterium to study body fat of animals, found that ingested fat is stored and stored fat used; used heavy isotope of nitrogen to study amino acids. Committed suicide, N.Y.C., Sept. 11, 1941.

SCHOENMACKERS, Jakob, German pathologist; b. Geldern, Germany, May 26, 1912; s. Hubert and Gertrud (Windbergs) S.; student U. Freiburg (Germany), 1931-32, U. Marburg, 1933, U. München (Germany), 1933-34; Med.Examen, U. Düsseldorf (Germany), 1936, Promotion, 1939, Habilitation, 1947; m. Ilse Schumacher, Apr. 25, 1949. Med. asst. Johannis-Hosp., Dortmund, 1936, Inst. Pathology, Gen. Hosp., Essen, Germany, 1937-47, Inst. Pathology, U. Düsseldorf, 1947-60; chief dept. pathology Gen. Hosp., Aachen, Germany, 1960-66; prof. pathology Tech. U., Aachen, 1966——, dean Med. Faculty, 1966——. Recipient Deutscher Rötgenpreis German Soc. Roentgenology, 1953, Arthur Weber-Preis for circulation research German Soc. Circulation Research, 1961. Fellow Concilii Scientiarum, Collegii Internationalis Angiologiae. Author: (with Vieten) Atlas postmortaler Angiogramme; (with Bargmann, Doerr) Herz d. Menschen, 1963; (with Ober, Rauen, Zander) Probleme der Krebsforschung, 1961; also numerous articles. Postmortem angiographical investigation of coronary arteries, valve ov heart, lung vessels, portal vein, integration of heart muscle, of elastic and non-elastic artheriosklerosis of coronary arteries. Home: 245 Ronheider Berg, Aachen, Germany.*

SCHOENTAL, Regina, cancer research worker; b. Dzialoszyce, Poland, June 12, 1906; d. Gershon and Dina (Eisenberg) Schoental; M.Chem., D.Phil., Jagiellonian U., Cracow, Poland, 1930; D.Sc., U. Glasgow (Scotland), 1950. Research, U. Oxford (Eng.) Sch. Pathology, 1938-46, dept. chemistry U. Glasgow, 1946-52, cancer research dept. Royal Beatson Meml. Hosp., 1952-54, dept. oncology Chgo. Med. Sch., 1952-53, toxicology research unit Med. Research Council, Carshalton, Surrey, Eng., 1955——. Fellow Chem. Soc. London, Biochem. Soc. Contbg. author: Clar's Polycyclic Hydrocarbons, 1964. Research, publs. on identification and synthesis of metabolites of polycyclic aromatic hydrocarbons, role of natural products present in plants and fungi in aetiology of human cancer; discovered carcinogenic actions of pyrrolizidine alkaloids, diazomethane, N-methyl-N-nitrosourethane. Home: 15 Birdhurst Ct., Woodcote Rd., Wallington, Surrey. Office: Med. Research Council Labs., Toxicology Research Unit, Carshalton, Surrey, Eng.*

SCHOETENSACK, Otto, German prehistorian, anthropologist; b. 1850; studied neolithic beds of middle Rhine; discovered lower jaw believed to come from extinct species of man (Heidelberg man), Mauer, nr. Heidelberg, 1907. Died 1912.

SCHOFFENIELS, Ernest Hubert, Belgian biochemist; b. Liège, Belgium, May 11, 1927; s. Jules René and Bertha-Christine (Schmetz) S.; M.D., U. Liège, 1953, Candidat in Biol. Scis., 1947, Agrégé Enseignement Supérieur, 1960; m. Marthe Gorin, Feb. 17, 1955; children—Michel, Francoise, Eric. Research asso. Columbia, N.Y.C., 1955-58, asst. prof., 1960-61; researcher FNRS, U. Liège, 1958-60, asso. prof. biochemistry, 1962——. Fulbright fellow U. S. Ednl. Found. in Belgium, 1955-58. Recipient T. Glüge prize, 1958; Jean de Meyer prize, 1959; Léon Fredericq prize, 1962; Alumni prize Belgian U. Found., 1963. Mem. Soc. Belge de Physiologie, Soc. Belge de Biochimie, Soc. Belge de Zoologie, Soc. Belge d'Entomologie, Centre Nat. de Biochimie et de Biologie Moleculaire. Author: (with M. Florkin) Biochimie et Biologie Moleculaire, 1966; Cellular Aspects of Membrane Permeability, 1967; also numerous articles. Cons. editor Archives Internationales de Physiologie et de Biochimie, 1960——, Life Scis., 1962——. Research on generation of bioelectric potentials in single isolated electroplax of electric eel, biochem. aspects of membrane permeability, relation between active transport of cations and nitrogen metabolism. Home: 282 rue de la Belle Jardinière, Angleur, Belgium. Office: Laboratoire de Biochimie, U. Liège, 17 place Delcour, Liège, Belgium.*

SCHOKALSKY, Julius Mikhailovitch, Russian geographer; b. St. Petersburg, circa Oct. 5, 1856; ed. Marine Acad.; navigated Baltic Sea; admitted to mil. maritime acad.; became dir. maritime meteorology sect., prin. physics obs.; dir. Navy Hygrographic Service; founder system of hydro-meteorol. stas. on Black Sea, 1907, nat. maritime obs., Sebastopol, 1909; prof. Soviet Naval Acad.; conducted oceanographic expdns., 1922-27; mem. French (1932), Soviet acads. scis.; pres. Soviet Geog. Soc. Author: Maritime Meteorology, 1882; Oceanography, 1917. Died Leningrad, Mar. 26, 1940.

SCHOKNECHT, Günter, German physicist; b. Berlin, Germany, Mar. 10, 1930; s. Wilhelm and Elisabeth (Wieczorek) S.; Diplom-Physiker, Freie U. Berlin, 1954; Dr.rer.nat., Technische U., Berlin, 1957. Asst., Fritz-Haber-Institut of Max-Planck-Gesellschaft, 1952-58; dir. phys. lab. Städt. Auguste-Viktoria Krankenhaus, Berlin, 1958—. Mem. Deutsche Physikalische Gesellschaft, Deutsche Röntgen Gesellschaft. Research, publs. on statistics in paracrystals, determination of electron-density-distbn. in crystals, application of computers in radiation dosimetry. Home: 5 Mülenstrasse, 1 Berlin 37. Office: 125 Rubensstrasse, 1 Berlin 41, Germany.*

SCHOLANDER, Per Fredrik, physiologist; b. Orebro, Sweden, Nov. 29, 1905; s. Torkel F. and Agnethe (Faye-Hansen) S.; M.D., U. Oslo, 1932, Ph.D., 1934; m. Susan Irving, June 20, 1951. Came to U. S., 1939, naturalized, 1945. Research fellow physiology U. Oslo, 1932-39; research asso. zoology Swarthmore Coll., 1939-43, research biologist, 1946-49; research fellow biol. chemistry Harvard, Med. Sch., 1949-51; physiologist Woods Hole Oceanographic Inst., 1952-55; prof. physiology, dir. Inst. Zoophysiology, U. Oslo, 1955-58; prof. physiology Scripps Instn. Oceanography, U. Cal., 1958——, dir. of Physiological Research Laboratory, 1963——. Served to maj. USAAF, 1943-46. Decorated Legion of Merit. Rockefeller fellow, 1939-41. Mem. Nat., Norwegian acads. scis., Am. Philos. Soc., Am. Acad. Arts and Scis., Arctic Inst. N. Am., Am. Physiol. Soc., Marine Biol. Lab. Assn., Am. Assn. Advancement Sci., Am. Soc. Plant Physiologists, Soc. Gen. Physiologists, Sigma Xi. Research and numerous publications on physiology of diving, climatic adaptations of arctic and tropical animals and plants, gas composition and secretion in swimbladder of fishes, composition of gas in glacier ice, osmosis mechanism; developed micro and ultra micro techniques for study respiration single marine eggs during cell divisions. Home: 8374 Paseo del Ocaso, La Jolla, Cal. 92037. Office: U. Cal. San Diego, La Jolla, Cal.*

SCHOLES, Samuel Ray, Jr., Am. chemist; b. Pitts., June 5, 1915; s. Samuel Ray and Lois E. (Boren) S.; B.S., Alfred U., 1937; Ph.D., Yale, 1940; m. Doris Emily Hann, Oct. 28, 1944; children—Susan, Jean Ann. Faculty dept. chemistry Alfred (N.Y.) U., 1940-41, 46—, prof., chmn. dept. chemistry, 1958—; faculty Tufts Coll., Medford, Mass., 1941-46. Mem. Am. Chem. Soc., Am. Assn. U. Profs., Sigma Xi. Determined dissociation constants for bicarbonate ion at temperatures from 0° to 55° C using reversible cells. Home 10 Park St., Alfred, N.Y. 14802.*

SCHOLS, Charles Mathieu, Dutch geodesist; b. Maastricht, Netherlands, Mar. 28, 1849; degree in civil engring. Poly. Sch., Delft, Netherlands; tchr. Royal Mil. Acad., Breda, Netherlands; became tchr. geodesy, Delft, 1878; supr. govt. geog. survey, 1886; mem. Royal Acad. Scis. (sec. royal surveying com. 1881); publs. on math. and geodesy. Died Delft, Mar. 28, 1897.

SCHOLTEN, Robert, geologist; b. Alkmaar, Netherlands, May 22, 1923; s. Jacques Hieronymus and Louise (Van Gennep) S.; Geol. Candidaat, U. Amsterdam, 1943; M.S., U. Mich., 1948, Ph.D., 1950; m. Marcia Alice Pettitt, Sept. 10, 1949; children—Peter Jacques, Sarah Lynn. Came to U.S., 1946, naturalized, 1953. Geologist, Sohio Petroleum Co., 1949-51; faculty Pa. State U., University Park, 1951—, prof., 1961—; Fulbright prof. geology U. Istanbul, 1958-59. Recipient Fulbright research award; Guggenheim fellow U. Milan, 1966. Author: (with J.P. Miller) Laboratory Studies in Geology, 1962. Home: 314 Arbor Way, State College, Pa. 16801. Office: Dept. Geology and Geophysics, Pa. State U., University Park, Pa. 16802.*

SCHOLTYSECK, Erich Oswald, German zoologist; b. Riegersdorf, Germany, Oct. 7, 1918; s. Andreas and Gertrud (Nemeller) S.; student U. Bonn (Germany), 1937-39, doctorate in zoology, 1952; m. Hildegard Mengeringhausen, Aug. 24, 1957; children—Joachim, Rolf, Bernward. With Inst. Zoology, U. Bonn, 1952——, prof. zoology, 1965——. Mem. Soc. Protozoologists, Deutsche Zoologische Gesellschaft, Deutsche Gesellschaft für Parasitologie. Research, publs. on devel., cytology, cytochemistry and electron microscopy of coccidia. Home: 7 Drachenfelstrasse, 53 Duisdorf/Bonn, Germany.*

SCHOLZ, Willibald Oscar, German psychiatrist; b. Greiz/Thuringe, Germany, Dec. 15, 1889; s. Rudolf Oscar and Marie Thérèse S.; student univs. Tubingen, Munich, Iena; M.D. honoris causa; m. Julie Wölfling, May 13, 1926. Asst., Psychol. and Neurol. Clinic, U. Tubingen, 1919-26; became prof. agrégé, 1925; physician in chief Leipzig Psychol. and Neurol. Clinic, 1926-31; became asso. prof., 1930; staff Inst. Psychol. Research, Kaiser Wilhelm Inst., Munich, 1931——; became dir. Brain Pathology Inst., 1936; named scholar Union Med. Coll., Peking, China, 1937; became dir., 1945; staff Duke U. Med. Sch., 1950; became dir. emeritus, cons. scientist, 1960. Mem. Deutsche Gesellschaft für Psychiatrie und Nervenheilkunde, Leopoldina Acad., Verein dt. Neuropathologen und Neuroanatomen (hon.), Am. Assn. Neuropathologists, French Soc. Neurology, Am. Psychiatry Assn. (corr.), Brit. Neuropath. Soc. Author: Über die Empfindlichkeit des Gehirns für Röntgen- und Radiumstrahlen, 1935; Myélite nécrotique (Foxi-Alajouanine)-Angiodysgenetische nekrotisierende Myelopathie, 1951; Die Krampfschädigungen des Gehirns, 1951. Editor: Archiv für Psychiatrie und Zeitschrift für die gesamte Neurologie; Acta Neuropathologica, Zentralblatt für die gesamte Neurologie und Psychiatrie; Handbuch der speziellen pathologischen Anatomie und Histologie, Teil Nervensystem, 1955-58. Address: Kraepelinstrasse 2, Munich, 23, W. Germany.

SCHOMAKER, Verner, Am. chemist; b. Nehawka, Neb., June 22, 1914; s. Edwin Henry and Anna (Heesch) S.; B.S., U. Neb., 1934, M.S., 1935, Ph.D., Cal. Inst. Tech., 1938; m. Judith Rooke, Sept. 9, 1944; children—David Rooke, Eric Alan, Peter Edwin. George Ellery Hale fellow Cal. Inst. Tech., 1938-40, sr. fellow in chem. research, 1940-45, asst. prof., 1945-46, asso. prof., 1946-50, prof. chemistry, 1950-58; with Union Carbide Research Inst., 1958-65, asst. dir., 1959-63, asso. dir., 1963-65; prof., chmn. dept. chemistry U. Wash., Seattle, 1965——. John Simon Guggenheim Meml. Found. fellow, 1947-48. Received Am. Chem. Soc. award in pure chemistry, 1950. Mem. Am. Chem. Soc., Am. Crystallographic Assn. (pres. 1961-62), A.A.A.S., N.Y. Acad. Scis., Sigma Xi, Phi Lambda Upsilon, Pi Mu Epsilon, Sigma Tau, Alpha Chi Sigma. Research and publs. on structure of gas molecules, crystal structures by X-ray diffraction, Fourier methods, developed visual method of interpreting electron diffraction photographs from gases; recognized phase-shift on scattering of electrons by heavy atoms. Home: 13224 42d Av. N.E., Seattle 98125.*

SCHOMBURGK, Richard, naturalist; b. Freiburg, Germany, 1811; explored (with brother Robert Hermann) Guiana, 1841-44; named dir. Bot. Garden, Adelaide, Australia, 1865. Author: Pub Reisen in Britisch Guiana, 3 vols., 1847-48; The Flora of South Australia, 1875; also papers. Died Adelaide, 1891.

SCHÖNBAUER, Leopold, Austrian surgeon; b. Thaya, Austria, Nov. 13, 1888; prof., Vienna. Author: (with Hoff) Hirnchirurgische Erfahungen, 1932; Das medizinische Wien, 1943; Lehrbuch der Chirurgie, 2 vols., 1849-50. Founder of neurosurgery in Austria; research on carcinoma, surgery of thyroid glands, history of medicine.

SCHÖNBEIN, Christian Friedrich, chemist; b. Metzingen, Swabia, Oct. 18, 1799; ed. Tübingen, Erlangen (both Germany); tchr. chemistry and physics, Keilhau, nr. Rudolstadt, 1823-25; visited Eng., Paris; became mem. faculty U. Basel (Switzerland), 1828, prof. physics and chemistry, 1835; mem. Grand Counsel, Basel; mem. French Acad. Scis. 1863. Author: L'action du feu sur l'oxygène 1837; Entstehung von Ozon bei der Elektrolyse von Schwefelsäure, 1840; Documents pour la chimie physique, 1844; De la production de l'ozone, 1844; Sur la combustion lente ou rapide des corps dans l'air atmospherique, 1845. Asst. to Johann Dingler with Polytechnisches Jour., 1820. Discovered ozone, 1840; studied hydrogen peroxide, passivity of metals; used nitrocellulose (guncotton) in firearms, 1846, also 1st to prepare collodion from it; rejected atomic theory and exact quantitative analysis. Died Sauersberg, nr. Baden-Baden, Germany, Aug. 29, 1868.

SCHÖNER, Johannes, German geographer; b. Karlstadt/Franconia, Germany, Jan. 16, 1477; Catholic priest, Bamberg, Germany; became Lutheran; prof. math. Coll. of Nuremberg (apptd. through influence of Melanchthon), from 1526. Author: Luculentissima quaedam terrae descriptio, 1515; Equatorii astronomici, 1522; De nuper sub Castiliae ac Portugaliae regibus repertis insulis, 1523; Opusculum geographicum, 1533; Regionamtus de triangulus omnimolus, 1533; Algorithmus demonstratus, 1834; Opusculum astrologicum, 1539; Sine Tables, 1541; Opera mathematica, 1561. Made 6 terrestrial globes still extant, including one signed and dated 1520; his globes of 1515 and 1520 are among earliest to designate America by name, also indicate a strait at south end of S.Am. Died Nuremberg, Germany, Jan. 16, 1547.

SCHÖNFELD, Eduard, German astronomer; b. Hildburgahausen, Germany, 1828; became dir. Mannheim (Germany) Planetarium, 1859; prof., astronomy, dir. planetarium, Bonn, from 1875; asst. to Argelander in preparing Durchmusterung, a catalog of stars in No. Hemisphere to 9-10th magnitude, later extended to 23 degrees, south declination. Author: Bonner Sternverzeichnis, viete sektion, 1899. Cataloged 133,659 stars from 2 to 23 degrees, south declination, also 489 nebulae. Died 1891.

SCHÖNFELD, Vilém, Czechoslovakian physician; b. B. Bystrica, Czechoslovakia, Sept. 6, 1914; s. Aron and Gabriela (Dubova) S.; M.D., Charles U., Prague, Czechoslovakia, 1939; Candidate Scis., 1960; m. Lucie Keppichová, May 31, 1944; children—Eva, Hana. Sr., Charles U., Polyclinic, 1950-51, asst. prof. hygiene faculty, 1960-64; mem. Research Inst. for Care of Mother and Child, Prague, 1951-60, head obstet. dept., 1964—. Research, publs. on pathogenesis, diagnosis and treatment of inflammatory diseases in gynecology, habitual abortions, menopause and menopausal troubles. Home: 52 Cernokostelecká, Prague 10. Office: 157 Nábr.K.Marxe, Prague 4, Czechoslovakia.*

SCHÖNLEIN, Johann Lukas, physician; b. Bamberg, Germany, Nov. 30, 1793; ed. Landshut, 1811-13; attended Julius Hosp., Würzburg, 1813; degree, 1816; apptd. instr., Würzburg, 1817, provisional chief of clinic, 1820, prof. spl. pathology and therapeutics, 1824-32 (dismissed for polit. reasons); went to med. faculty, Zurich, Switzerland; became head of Berlin clinic, 1840; physician in ordinary to kings Friedrich Wilhelm III and IV. Author: Allgemeine und spezielle Pathologie und Therapie, 1832; Klinische Vorträge, 1842. Physician of Enlightenment; turned German med. practice from idealist to mechanistic view of nature; founder exact natural sci. sch. based on knowledge of processes of organism, Berlin; introduced the new methods of percussion, auscultation, microscope and lab. into diagnosis; instituted postmortem examinations; laid (with Müller) found. of Berlin Sch. Medicine; wrote on typhoid crystals; credited with introducing term hemophilia, 1828; gave 1st description of rheumatoid peliose, 1837; discovered that favus is caused by a fungus (now called Achorion schoenleini or Trichophyton schoenleini), 1839. Died Jan. 23, 1864.

SCHOOF, Herbert Frederick, Am. entomologist; b. N.Y.C., Apr. 18, 1914; s. Charles and Elizabeth (Gaensley) S.; B.S. in Biology, N.C. State Coll., 1936, M.S. in Biology, 1938; Ph.D. in Entomology, U. Ill., 1940; m. Beulah Christine Weathers, Aug. 28, 1937; 1 son, Charles Vernon. Entomologist, N.C. Bd. Health, 1941; with USPHS, 1942—, entomologist charge research on fly biology and control, Atlanta, 1948-53, asst. chief Tech. Devel. Labs., Nat. Communicable Disease Center, Savannah, Ga., 1953—. Cons. on vector control problems to AID, WHO, Pan Am. Health Orgn., 1957—; tech. adviser Chatham County Mosquito Control Commn., 1963—. Mem. Am. Soc. Tropical Medicine and Hygiene, A.A.A.S., Am. Mosquito Control Assn., Sci. Research Soc. Am., Sigma Xi, Phi Kappa Phi, Phi Eta Sigma, Alpha Kappa Phi. Research, numerous publs. on insect resistance, biology and control of arthropods, anopheline and culicine biology and control. Home: 3 Pinewood Av., Parkersburg, Savannah 31406. Office: Tech. Devel. Labs., P. O. Box 769, Savannah, Ga. 31402.*

SCHOOLCRAFT, Henry Rowe, Am. explorer, ethnologist; b. Albany County, N.Y., Mar. 28, 1793; s. Lawrence and Margaret Anne Barbara (Rowe) S.; student Union Coll., Middlebury Coll.; LL.D., U. Geneva, 1846; m., 1823; m. 2d, Mary Howard, 1847. Visited mineral regions in So. Mo., Ark., 1817-18; agt. on N.W. frontier; supt. Indian affairs for Mich., 1836-41, negotiated several treaties with Chippewa Indians, including treaty which gave U. S. parts of Mich., 1836; a founder Hist. Soc., 1828, Algic Soc. of Detroit, 1832; projected an Indian ency. Author: A View of the Head Mines of Missouri, 1819; Narrative Journals of Travels through the Northwestern Regions of the United States . . . to the Sources of the Mississippi River, 1821; Narrative of an Expedition through the Upper Mississippi to Itasca Lake, the Actual Source of the Mississippi, 1834; Algic Researches (concerning Indian mental characteristics), 2 vols., 1839; Oneonta (describing Indian history), 1844-45; Notes on the Iroquois, 1847; Personal Memories of . . . Thirty Years with the Indian Tribes, 1851; Historical and Statistical Information Respecting the History, Condition, and Prospects of the Indian Tribes of the United States, 6 parts, 1851-57; Grammatical Construction of the Indian Languages. Died Washington, D.C., Dec. 10, 1864.

SCHOOLEY, Allen Heaten, Am. interdisciplinary scientist; b. Terril, Ia., Dec. 16, 1909; s. Alfred Heaten and Bertha Hope (Allen) S., B.S. in Elec. Engring., Ia. State U., 1931; M.S. in Elec. Engring., Purdue U., 1932, Dr.Engring., 1966; m. Meriel Mercedes Vestrem, Sept. 10, 1938; children—Robert Allen, Jean Suzanne, Mary Elena. Electronic engr. RCA, 1936-40; with U. S. Naval Research Lab., 1940—, supt. electronics div., 1954-56, asso. dir. research, 1957—; resident adviser to Brazilian Navy, Rio de Janeiro, 1956-57; research scientist NATO Anti-Submarine, Submarine Warfare Research Center, La Spezia, Italy, 1963-65; lectr. Geophys. Inst., U. Bergen (Norway), 1965. Mem. U. S. Navy Undersea Warfare Research and Devel. Planning Council, 1959-63. U. S. nat. com. Union Radio Sci. Internat., 1958—. Recipient Distinguished Civilian Service award USN, 1946, created knight Ordem do Merito Naval (Brazil), 1963. Fellow I.E.E.E. (Distinguished Service award Washington sect. 1963, Harry Diamond Meml. prize 1963; chmn. Washington sect. 1957-58), A.A.A.S., Washington Acad. Scis.; mem. Am. Geophys. Union, Sci. Research Soc. Am., Am. Inst. Physics, Washington Philos. Soc., Armed Forces Electronics and Communications Assn., Am. Assn. Physics Tchrs., U. S. Naval Inst., Sigma Xi, Theta Xi. Author articles interdisciplinary fields electronics, optics, oceanography, fluid dynamics, research mgmt. Patentee in field. First to develop miniature tubes; developed precision radar range measuring equipment; 1st to define dimensionless quantity (ratio facet size to flatness tolerance) and applied to radar sea surface interference; 1st to demonstrate inverted profile capillary water waves, vertical wake collapse of propelled submerged body in stratified. Home: 6113 Cloud Dr., Springfield, Va. 22150. Office: U. S. Naval Research Lab., Washington 20390.*

SCHOOLEY, John Campbell, Am. physiologist; b. Chgo., Apr. 24, 1928; s. Eldred and Vernetta (Campbell) S.; A.B., U. Cal., Berkeley, 1951, Ph.D., 1957; m. Caroline Penny Naus, Aug. 8, 1953; children—Diana, Karen, Peter. Specialist 3 (T), U.S. Army, Oak Ridge Inst. Nuclear Studies, 1954-56; research physiologist Donner Lab., U. Cal., Berkeley, 1957—. Mem. N.Y. Acad. Scis., A.A.A.S., Am. Physiol. Soc., Soc. Exptl. Biology and Medicine, Am. Soc. Anatomists, Transplantation Soc., Am. Assn. Anatomists. Research, publs. on formation of blood cells; devel. of an antibody against erythropoietin, which controls red cell formation. Home: 3036 Hillegass Av., Berkeley, Cal. 94705.*

SCHOONHOVEN, John James, Am. biologist; b. McIntyre, N.Y., July 3, 1864; s. George W. and Maria (Mead) S.; A.B., St. Francis Coll., 1893; A.M., Coll. St. Francis Xavier, 1894; postgrad. L.I. Coll. Hosp., 1888-90, New York U., 1900-01; m. Helen E. Butterfield, June 16, 1897; 1 son, George Otis. Staff lectr. N.Y., Newark, Jersey City bds. edn.; lectr., pres. dept. zoology, fellow Bklyn. Inst. Arts and Scis.; spl. lecturer, instr. in edn. N.Y. U., 1930—. Pres. Am. Assn. for Planting and Preservation of City Trees, 1910-12; del. 15th Internat. Congress on Hygiene and Demography; dir. Brooklyn Zoöl. Assn., 1915-25. Contbr. mags. and sci. publs. Specialist on animal parasites, entomogenous fungi, trypanosomiasis. Died June 1936.

SCHOPF, Johann David, physician, scientist; b. Wunsiedel, Germany, Mar. 8, 1752; studied medicine and natural sciences U. Erlangen (Germany), 1770-73, M.D., 1776. Served as surgeon in German regt. assigned to Brit. Army, N.Y., 1777-83; traveled throughout Eastern U. S. and Bahamas, 1783-84; pres. Ansbach Medicinal-Collegium, Prussia, 1795. Author: Beytrage zur Mineralogischen Kenntniss des Ostlichen Theils von Nordamerika und seiner Geburge (1st systematic work on Am. geology), 1787; Materia Medica Americana, Potissimum Regni Vegetabilis, 1787; Reise durch einige der mittlern and sudlichen vereinigten nordamerikanischen Staaten nach Ost-Florida und den Bahama-Inselen (his masterpiece), 2 vols., 1788; Historia Testiludinum, Iconibus Illustrata (Fasc. I-VI, Erlangen, 1792-1801; wrote papers on Am. climate and diseases, also 1st papers ever written on Am. ichthyology, Am. frogs and turtles. Died Sept. 10, 1800.

SCHOPPER, Herwig Franz, German physicist; b. Landskron, Germany, Feb. 28, 1924; s. Franz and Margarete (Stark) S.; Diploma in physics, U. Hamburg (Germany), 1949, Dr.rer.nat., 1951; m. Ingeborg Stieler, Apr. 14, 1949; children—Doris, Andreas. Sci. asst., Hamburg, 1951-53; oberassist., privatdozent U. Erlangen (Germany), 1953-57; asso. prof., dir. Inst. Nuclear Physics, U. Mainz (Germany), 1957-61; prof., dir. Inst. Exptl. Nuclear Physics U., also Nuclear Research Center, U. Karlsruhe (Germany), 1961—; vis. scientist Cavendish Lab., Cambridge, Eng., 1957; research asso. Cornell U., 1961; guest prof. CERN, Geneva, Switzerland, 1967. Recipient Physics prize Acad. Göttingen, 1957; Carus medal Acad. Leopoldina, 1959. Mem. Deutsche Phys. Gesellschaft, Am. Phys. Soc., Acad. Leopoldina. Author: Weak Interactions and Beta Decay, 1966; contbg. author: Teilchenbeschleuniger, 1962, Linear Accelerators, 1968; also numerous articles. Research on thin metallic layers, detection of parity violation in beta decay by beta-gamma circular polarization, elementary particle physics, nucleon structure by electron and neutron scattering, accelerator devel. Home: 6 Osteroder Strasse, 75 Karlsruhe, Germany.*

SCHORLEMMER, Carl, chemist; b. Darmstadt, Germany, Sept. 30, 1834; s. Johannes Schorlemmer; studied chemistry, Heidelberg, also Giessen, Germany; pupil of Bunsen and Kopp; LL.D., Glasgow, 1888; became lab. asst. Manchester (Eng.) Coll., 1861; lectr., 1873; named prof. organic chemistry Owens Coll., Manchester, 1874. Fellow Royal Soc., 1871, Author: Lehrbuch der Kohlenstoffrierbindungen, 1871; Rise and Development of Organic Chemistry, 1874; (with Sir Henry Roscow) Systematic Treatise on Chemistry, 3 vols., 1877-1911; also an incomplete history of chemistry, German and English transls. Established hypothesis that normal paraffins form a single rather double series (fundamental conception of modern organic chemistry); studied homologues of aliphatic series; showed relation of hydrocarbons in series to their substitution products; discovered gen. method for converting secondary into primary alcohol, 1866; made observations on valance of carbon. Died June 27, 1892.

SCHOTT, Charles Anthony, geodesist; b. Mannheim, Baden, Germany, Aug. 7, 1826; grad. Polytechnic Sch., Carlsruhe, C.E., 1847. Came to U. S., 1848. Asst., U. S. Coast and Geodetic Survey, asst., 1856—, dir. computing div. nearly 45 years; mem. govt. parties to observe total eclipse of sun, Aug. 1869 (Springfield, Ill.), Dec. 1870 (Cantania, Sicily); del. to Internat. Conf. on Terrestrial Magnetism, Bristol, Eng., 1898. Mem. Nat., Washington acads. scis. Contbr. many papers on hydrography, geodesy and particularly on terrestrial magnetism in reports of Survey, on tides, meteorology and physics of globe in Smithsonian Instn. publs., 1858-81. Died 1901.

SCHOTT, Gaspar, physicist; b. Königshofen, Germany, Feb. 5, 1608; ed. Rome; pupil of Kircher in Rome; became mem. Soc. Jesus, 1627; tchr. Jesuit coll., Palermo, Italy; returned to Germany, 1657; prof. math. and physics, Augsburg, until 1666. Author: Magia universalis naturae et artis, 1657-59; Cursus mathematicus; Mechanica hydraulico-pneumatica . . . , 1657; Thaumaturgus physicus . . . , 1657-59; Physica curiosa . . . , 1662; Technica curiosa . . . , 1664. Gave 1st published account of Guericke's expts. with air pump; conceived of fastening Napierean small slide rule on movable cylinders to permit logarithmic arithmetic operations. Died Augsburg, May 22, 1666.

SCHOTT, Wolfgang, German geologist; b. Hamburg, Germany, Feb. 1, 1905; s. Gerhard and Gertrud (Tietz) S.; student U. Göttingen, 1925-30, Dr.phil., 1930; 1st exam. Prueussische Geologische Landesanstalt, Berlin, 1931, 2d exam., 1936; Dr.phil.habil., U. Rostock/Mecklenburg (Germany), 1936; m. Lotte Grundig; 3 children. Staff, Preussische Geologische Landesanstalt, Reichsamt für Bodenforschung (now Budnesanstalt für Bodenforschung), Hannover, Germany, 1934—, now mng. dir.; hon. prof. U. Göttingen. Research, publs. on petroleum-geology, marine geology, palaeogeography. Office: P.O. Box 54, Bundesanstalt für Bodenforschung, 3 Hannover-Bucholz, Germany.*

SCHOTTEN, Carl Ludwig, chemist; b. 1853; prof. Physiol. Inst., Berlin, Germany; developed method (Schotten reaction) for introducing benzoyl group into compound; research on constitution of piperidine and acids of bile, also on ichthyol, source of hippuric acid in urine. Died 1910.

SCHOTTKY, Walter, physicist; b. Zurich, Switzerland, July 23, 1886; s. Friedrich and Henriette (Hammer) S.; doctorates in engring., tech. and natural research U. Berlin; m. Elisabeth Lintz, 1923; 3 sons. Corp. research, 1916-20; privat docent U. Würzburg (Germany), 1920-22; prof. U. Rostock (Germany), 1923-26. Recipient Hughes medal, 1936. Research in electron theory, electroacoustics, ion theory; invented screen-grid tube, 1915; discovered irregularity in thermion emission in vacuum tube (shot effect or Schottky effect).

SCHOTTMÜLLER, Hugo, German internist; b. Trebbin, Germany, 1867; prof., Hamburg, Germany. Author: Lietfaden fur den klinisch-bakteriologischen Kulturmethoden, 1924. Made basic studies in clin. bacteriology; introduced Pyramidon treatment of rheumatism of joints; established endocarditis lenta as ind. disease; isolated Streptococcus viridans (Schottmüller's bacillus), showed it causes bacterial endocarditis, 1910; isolated Streptothrix muris ratti, showed it causes rat-bite fever, circa 1914. Died Hamburg, May 19, 1936.

SCHOU, Axel Villiam, Danish geographer; b. Copenhagen, Denmark, Aug. 18, 1902; s. Francis and Petrea (Petersen) S.; Bachelor, U. Copenhagen, 1927, Ph.D., 1945; m. Laura Andersen, Oct. 17, 1927; children—Manfred S., Gunna S. (Mrs. Bertel Würtzen). Lectr. geography Rysensteen Coll., 1927-53; prof. geography U. Copenhagen, 1953—, head dept. geography, 1964—. Head ednl. wall map sect. Danish Geodetic Inst., 1946—; sec. Danish Nat. Com. for Geography. Decorated Knight 1st class Order of Danebrog. Mem. Internat. Geog. Union (chmn. commn. on coastal geomorphology), Royal Danish Geog. Soc. (sec.-gen.), Swedish Geog. Soc. (hon.), Societé Serbe de Géographie (hon.). Author: Det marine Forland, 1945; The Landscapes vol. I., Atlas of Denmark, 1949; Olie, 1961; Construction and Drawing of Block Diagrams, 1962; Geografi og velfaerdsstat, 1967; also numerous articles. Editor: atlases, wall-maps. Research on wind work indicating rules for orientation of shore line simplification (laws of Lewis and Schou), importance of marine foreland formation for shoreline devel. in various environments, including moraine archipelago (Denmark), tropical cliff and lagoon coasts (Brazil). Home: 112 Strandboulevarden, 2100, Copenhagen, Denmark.*

SCHOU, Jens Solver, Danish pharmacologist, physician; b. Copenhagen, Denmark, Oct. 10, 1929; s. Svend Aage and Ellen (Fjerdingstad) S.; M.D., U. Copenhagen, 1955, Ph.D. in Pharmacology, 1960; m. Annelise Baerentzen, Nov. 15, 1958; children—Charlotte, Henriette. Research asso. Inst. Pharmacology, U. Copenhagen, 1956-58, faculty, 1958—, prof. pharmacology, chmn. dept., 1967—; vis. asso. prof. U. Cal. at San Francisco, 1966. Chmn., Danish Pharmacopoea Commn., 1967—. Mem. Danish Acad. Tech. Scis., Danish Pharmacological Soc. (chmn. 1966—). Author: Subcutaneous Absorption of Drugs, 1959; Forordningslaere, 1966; also articles. Described self-depression in subcutaneous absorption of drugs; research on morphine, morphine-like drugs, capillary permeability, inflammation and antirheumatics, absorption mechanism from mucles. Home: 1 E Johannevej, Charlottenlund, Denmark. Office: 20 Juliane Mariesvej, Copenhagen, Denmark.*

SCHOU, Mogens Abelin, Danish physician; b. Copenhagen, Denmark, Nov. 24, 1918; s. Hans Jacob and Margrethe (Broddersen) S.; grad. Copenhagen U. Med. Sch., 1944; research fellow N.Y. State Psychiat. Inst., Columbia, 1949; Dr.med., Aarhus U., 1959; m. Agnete Henriette Jessen, June 23, 1943; children—Jette, Karen, Katrine, Bodil. Research asst. Inst. Cytophysiology, Copenhagen U., 1948-49; research asst. Inst. Physiology, Aarhus U., 1950-52, research asso. dept. psychiatry, 1952-56, reader in psychopharmacology, 1965—; research dir. neurochem. and psychopharmacol. research unit Inst. Psychiatry, Risskov, Denmark, 1956—. Panel mem. Internat. Brain Research Orgn., 1959—. Recipient Lundbeckfonden award, 1963. Mem. Danish Med. Assn., Danish Psychiat. Soc., Danish. Soc. Clin. Psychopharmacology, Collegium Internat. Neuro-Psychopharmacology, Internat. Soc. Neurochemistry, Scandinavian Soc. Biol. Psychiatry (pres). Author: Lithium i den psykiatriske terapi, 1959. Research, publs. on use of lithium in manic depressive psychoses; toxicology, pharmacology, metabolism of lithium. Home: 4 Syrenvej. Office: Inst. Psychiatry, Risskov, Denmark.*

SCHOUW, Joachim Frederik (or Schow), Danish botanist; b. Copenhagen, Denmark, 1789; ed. U. Copenhagen; apptd. prof. botany, Copenhagen, 1821; univ. rep. Danish Assembly, 1835, later pres. Author: Elements of a Universal Geography of Plants, 1822; Delineations of Nature, 1839; Earth Plants and Man; other works. Noted as tchr. of details of technique. Died 1852.

SCHRAER, Harald, Am. biologist, educator; b. Boston, June 10, 1920; s. Morris and Nettie (Belenko) S.; student Kan. State Coll., 1939-40; B.A., Syracuse U., 1948, M.A., 1949; postgrad. Harvard; Ph.D., Cornell U., 1954; m. Rosemary Schmidt, June 12, 1952; 1 son, David J. Research asso. Albert Einstein Med. Center, Phila., 1952-56; research asso. dept. physics Pa. State U., University Park, 1956-58, sr. research asso., 1958-61, asso. prof. biophys, 1961-67, prof., 1967—; lectr. Harvard Med. Sch., 1967-68. Vis. scientist NIH, 1961. Mem. Am. Physiol. Soc., Am. Soc. Cell Biology, Soc. for Exptl. Biology and Medicine, Biophys. Soc., Am. Zool. Soc., Fedn. Am. Scientists. Research in ultrastructure of calcium transporting cells and tissues, cellular control mechanisms of calcium transport, radiographic quantitation of bone mineral content in living structures. Home: Box 347, Boalsburg, Pa. 16827. Office: Life Sci. Bldg., Pa. State U., University Park, Pa. 16802.*

SCHRAMM, Gerhard Felix, biochemist; b. Yokohama, Japan, June 27, 1910; s. Conrad and Eva (Bubendey) S.; univs. Göttingen, Munich; Dr.phil., U. Göttingen, 1935; Dr.med.h.c., U. Giessen, 1964; m. Hildegard Schenkel, July 30, 1938; children—Gesa (Mrs. G. Stöffler), Hilke (Mrs. M. Stoffregen), Carsten, Henning. Head dept. Max Planck Inst. for Biochemistry (formerly Kaiser Wilhelm Inst.), Berlin, Tübingen, 1941-54; prof. U. Tübingen, 1953—, head biochemistry dept., dir. Max Planck Inst. for Virus Research, 1954—. Recipient Liesegang prize, 1957; Liebig Meml. medal Deutsche Chem. Gesellschaft, 1958; Lasker Found. prize, 1958, Jungius prize U. Hamburg, 1963. Mem. Max Planck Soc. (sci. bd. 1952—), Leopoldina (Carus medal 1957), numerous others; hon. mem. N.Y. Acad. Scis. Author: Biochemie der Viren, 1954. Research on steroid chemistry, structure of tobacco mosaic virus, isolation of infectious nucleid acids, chemistry of nucleic acids and proteins, chem. synthesis of polynucleotides. Address: 99 Hausserstrasse, Tübingen, West Germany.*

SCHRAMM, Wilbur (Lang), Am. communication scientist; b. Marietta, O., Aug. 5, 1907; s. Archibald and Louise (Lang) S.; A.B., Marietta, Coll., 1928, Litt.D., 1945; A.M., Harvard University, 1930; Ph.D., University of Iowa, 1932; m. Elizabeth Donaldson, August 5, 1934; children—Mary Barbara, Richard Michael. Began as newspaper reporter and desk editor; became correspondent of Associated Press, 1928; postdoctoral and National Research fellow, 1932-35; asst. prof. English, U. Ia., 1935-38, asso. prof., 1938-40, prof., 1940-43, dir. Sch. of Journalism, 1943-47; asst. to pres., dir. Inst. Communications Research, dir. Univ. Press and research prof., Univ. Ill., 1947-55, dean division of communications, 1950-55; professor communications, and journalism, also in Institute Communication Research Stanford U., 1955—, dir. Inst. Communication Research, 1957—; Janet M. Peck prof. internat. communication, 1962—; on leave as director ednl. services, O.W.I., Washington, D.C., 1942-43; cons. U. S. Dept. State, UNESCO, U. S. Information Agy., AID, U. S. Office Edn., U. S. Defense Dept., Ford Found., Nat. Educational Television, others; member of U. S. Defense Sci. Bd. 1956-61; Editor Am. Prefaces, 1936-41. Founded and directed the Writers Workshop, U. Ia., 1937-41. Chairman U. S. delegation to Asian Powers Conference on Communication, Bangkok, 1960. Founded Ill. Inst. of Communications Research, 1948. Received O. Henry prize for fiction in 1942; George Polk Meml. award for research, 1960. Mem. Am. Sociol. Assn., Am. Psychol. Assn., Am. Statis. Assn., Am. Assn. Pub. Opinion Research, A.A.A.S., Phi Beta Kappa. Author or co-author several books; latest publs.: Mass Communications, 1949; Process and Effects of Mass Communication, 1954; Four Theories of the Press, 1956; Responsibility in Mass Communication, 1957; One Day in the World's Press, 1959; Television in the Lives of Our Children, 1961; The People Look at Educational Television, 1963; Mass Media and National Development, 1964; The New Media, 1967; Communication and Change, 1967. Editorial bd. Pub. Opinion Quarterly, Journalism Quarterly. Contributions to the theory of human communication, especially to knowledge of the effects of the mass media, the use of modern communication in economic development and in education; and the storage, retrieval, and dissemination of scientific information. Home: 1518 Hamilton Av., Palo Alto, Cal. 94303.*

SCHRANK, Auline Raymond, Am. physiologist; b. Hamilton, Tex., Aug. 15, 1915; s. Herman and Emma Hulda (Schrank) S.; B.A., S.W. Tex. State Coll., San Marcos, 1937; Ph.D., U. Tex., 1942; m. Dorris Parke, Sept. 5, 1942; children—Kenton Parke, Karon. Mem. faculty U. Tex., 1939—, prof. zoology, 1959—, chmn. dept., 1963—. Mem. Biophys. Soc., Soc. Gen. Physiologists, Am., Scandinavian socs. plant physiologists, Tex. Acad. Scis., Sigma Xi, Alpha Chi. Research and publs. in bioelectric fields, growth, plant trophism, animal regeneration. Home: 2502 Tower Dr., Austin, Tex. 78703.*

SCHRAUZER, Gerhard Norbert, chemist; b. Franzenbad, Czechoslovakia, Mar. 26, 1932; s. Otto and Ludmilla (Mazanek) S.; B.Sc., U. Munich (Germany), 1952, M.Sc., 1956, Ph.D., 1963; m. Carol Ann Phipps, June 30, 1957; children—Richard-Daniel, Michael - Edgar, Bernice-Alexandra, Kenneth-Gerhard. Privatdozent, U. Munich, 1963-64; research supr. Shell Devel. Co., Emeryville, Cal., 1964-66—; prof. chemistry, U. Cal. at San Diego, La Jolla., 1966—; Mem. Am., German chem. socs. Contbg. author: Advances in Organometallic Chemistry 2, 1964. Research, publs. on complexes of transition metals, homogeneous catalysis, spectra of metal complexes, vitamin B12, related enzymatic reactions. Home: 7831 Boulevard Pl., La Jolla, Cal. 92032.*

SCHRECKER, Anthony Wolfgang, chemist; b. Vienna, Austria, Feb. 15, 1915; s. Paul and Leonie (Sobotka) S.; Doktorandum, U. Vienna, 1937, postgrad.; Licence-ès-sciences Sorbonne, Paris, France, 1940; Ph.D., U. Ill., 1948. Came to U.S., 1942, naturalized, 1948. Chemist, May Chem. Corp., Newark, 1943-44; tech. adviser French Supply Council, Washington, 1944-46; supervisory research chemist Nat. Cancer Inst., NIH, Bethesda, Md., 1948—; vis. investigator Scripss Clinic and Research Found., La Jolla, Cal. 1962-63, 67-68. Mem. Am. Assn. Cancer Research, Am. Soc. Biol. Chemists, Am. Chem. Soc., Washington Acad. Scis., Phi Beta Kappa, Sigma Xi. Research on structure and stereochemistry of lignans, mechanism action of antitumor agts., mechanism resistance to antileukemic drugs. Office: NIH, Bethesda, Md. 20014.*

SCHREIBER, Hans Karl Walter, German biophysicist; b. Brauna, Germany, June 9, 1902; s. Carl and Antonie (Heisig) S.; Dipl. Ing., Dresden Inst. Tech., 1926; Dr.Ing., Stuttgart Inst. Tech., 1929; m. Dora Bernard, Mar. 10, 1948; children—Gerhard Hans, Wolfgang Roland. Faculty, U. Berlin, 1934-64, prof., 1955-64, asst. Inst. Radiation Research, 1929-49, head prof., 1946-61; prof., dir. Inst. Radiation Research, Med. Acad., U. Kiel, Lübeck, Germany, 1964—. Mem. Comité International de la Lumiére (sec. gen. 1936-46), German Radiation Soc., German socs. for light research, phys. medicine, biophysics, Comité International de Photobiologie (hon.). Research, numerous publs. in plant and animal mutations caused by radiation, dosimetry of ultraviolet radiation, effects of and protection from radiation, ultrasonics, sensitivity of human eye to ultraviolet light. Died Apr. 3, 1968.

SCHREIBER, Klaus Johannes, German chemist; b. Lübeck, Germany, Jan. 25, 1927; s. Johannes and Wilma (Lorenz) S.; diploma in chemistry, U. Rostock (Germany), 1949, Dr.rer.nat., 1953; Habilitation, U. Jena (Germany), 1962; Prof, German Acad. Scis., Berlin, 1965; m. Dorothea Grobler, Aug. 2, 1952; 1 dau., Angelika. Chemist, Inst. for Plant Breeding Gross Lüsewitz, 1949-52; asst., chem. dept. U. Rostock, 1952-54; head chem. dept. Research Sta. Mühlhausen/Thuringia, German Acad. Agrl. Scis., Berlin, Germany, 1954-61; head dept. chem. physiology Inst. Cultivated Plant Research, German Acad. Scis., Gatersleben, Germany, 1961—; asso. prof. U. Halle, Germany, 1965—. Mem. Deutsche Akademie der Naturforscher Leopoldina, Chemische Gesellschaft in der Deutschen Demokratischen Republik, Gesellschaft Deutscher Naturforscher und Ärzte, Deutsche Gesellschaft für Arzneipflanzenforschung. Author: (with G. Schneider, G. Sembdner) Gibberelline, ihre Derivate und Abbauprodukte, 1966; also numerous articles. Editor: Chemie und Biochemie der Solanum-Alkaloide, 1961. Isolation, structural elucidation and synthesis of natural products; research on photochemistry, application of phys. methods. Home: 2 Leibnizweg. Office: 3 Corrensstrasse, Gatersleben, Germany/GDR.*

SCHREIBER, Raemer Edgar, Am. physicist, govt. ofcl.; b. McMinnville, Ore., Nov. 11, 1910; s. Michael and Bertha (Raemer) S.; B.A., Linfield Coll., 1931; M.A., U. Ore., 1932; Ph.D. in Physics (U. fellow), Purdue U., 1941, D.Sc., 1964; m. Marguerite Elizabeth Doak, Sept. 16, 1933; children—Paula C. (Mrs. David E. Hill), Sara B. Grad. asst. physics Ore. State U., 1932-35; grad. asst. physics Purdue U., 1935-37, instr., 1937-42; research asso. OSRD, Manhattan Dist. Project, 1942-43; with Los Alamos Sci. Lab., AEC, 1943—, design, constrn., operation 1st reactor using enriched U-235, 1943-45, mem. nuclear assembly team Trinity test bomb and Tinian combat operations, 1945, head nuclear assembly team Operation Crossroads, Bikini, 1946, head critical assembly group, 1946-47, alternate div. leader for atomic weapon engring., 1947-51, div. leader, 1951-55, div. leader for nuclear rocket devel., 1955-62, tech. asso. dir., 1962—. Cons. propulsion panel USAF Sci. Adv. Bd., 1958-60, mem. 1960-62; mem. research adv. com. for nuclear energy processes NASA, 1959-63. Fellow Am. Phys. Soc., Am. Rocket Soc., Am. Nuclear Soc. (exec. bd. 1962—, pres. 1967-68); mem. Sigma Xi, Sigma Pi Sigma. Home: R.F.D. 1, Pajarito Village, Espanola, N.M. 87532. Office: P.O. Box 1663, Los Alamos 87544.*

SCHREIBER, Vratislav, Czechoslovian endocrinologist; b. Prague, Czechoslovakia, June 29, 1924; s. Josef and Jarmila (Hrouzkova) S.; M.D., Charles U., 1950; m. Olga Chechova, Aug. 30, 1950; 1 son Michal. Research worker Lab. for Endocrinology and Metabolism, Prague, 1957-63, leading research worker, 1963—, vice dir., vice dean for research Med. Faculty, Charles U., 1966—. Mem. Collegium for Med. Scis., Czechoslovak Acad. Scis.; mem. pharm. cons. com. Ministry of Health. Recipient State prize, 1965; prizes Endocrine Soc., 1956, 62, Physiol. Soc., 1963, Sci. Council of Ministry of Health, 1964. Mem. Czechoslovak Endocrine Soc., Physiol. Soc. Author several books, including The Hypothalamo-Hypophysial System, 1963; also numerous articles. Studies of hypothalamic regulation of adenohypophyseal function, adenohypophyseal growth, metabolism and functions. Home: 39 Zitomirska, Prague 10, Czechoslovakia.*

SCHREIDER, Eugène, anthropologist, physiologist; b. Petersbourg, Russia, Mar. 21, 1901; s. Gregory and Flora (Henquine) S.; ed. U. Rome (Italy), Paris (France) U.; div.; 1 son, Charles-Gregory. Research staff Lab. Physiology, Conservatoire National des Arts et Metiers, Paris, 1932-39; lectr. comparative physiology human populations Inst. Ethnology, Paris U., 1939-40; dep. dir. Lab. Phys. Anthropology Sch. Advanced Studies, Paris, 1949-62, dir., 1962—; asso. dir. Lab. Human Biometry, Nat. Centre Sci. Research, Paris, 1952-61, dir., 1961—; prof. biometry Inst. Demography, Paris U., 1958—. Mem. Soc. Anthropology Paris (past pres.), Société Biotypologie (past gen. sec.), Société de Biométrie Humaine (gen. sec. 1966—). Author: Facteurs Physiologiques de la prédisposition aux accidents, 1935; Les types Humains, 3 vols., 1937; La Biométrie, 1960; La Biologie Humaine, 1964; also numerous articles. Research on phys., especially physiol., anthropology of Europe and Mexico, ecol. gradients and body-heat regulation in man, human variability, anat. and physiol. critical analysis of homeostasis concept, social stratification biol. characters, secular trends in modern populations, comparative biometry of vertebrates, evolutionary implications of biometrical data. Home: 38 bis rue du Fer à Moulin, Paris (5), France. Office: Labor de Biométrie 41, rue Gay-Lussac, Paris (5) France.*

SCHREINER, Werner Emil, Swiss physician; b. Zurich, Switzerland, May 8, 1921; s. Christian and Franziska (Gregoritsch) S.; m. Elizabeth Weber, Aug. 21, 1948; children—Bea Peter, Rachel Katherine, Elizabeth Regula. Research asso. Harvard Med. Sch., Boston, 1963-64; lectr. in obstetrics and gynecology U. Zurich, 1960—, head gynecol. and obstetrics policlinics, 1964—; sr. registrar in obstetrics and gynecology, 1964—. Mem. Assn. Mil. Surgeons U.S. (hon.), Swiss Soc. Gynecology and Obstetrics. Author: Fetus und Fruchtwasses, 1964; also articles. Research on placental metabolism, intrauterine oxygenation and metabolism of fetus examined by amniotic fluid. Home: Goldhaldenstr. 15, 8702 Zolliken, Switzerland 8702. Office: U. Zurich, Zurich 8006, Switzerland.

SCHREK, Robert, Am. exptl. pathologist; b. N.Y.C., Dec. 28, 1907; s. Herman and Helen (Goldshein) S.; B.A., N.Y. U., 1927; M.D., Vanderbilt U., 1931; m. Esther Weiss, Aug. 1, 1961. Research asst. in pathology Vanderbilt U., Nashville, 1931-36; resident pathologist Pondville State Cancer Hosp., Wrentham, Mass., 1936-37; asst. in pathology Yale Med. Sch., New Haven, 1937; asst. pathologist VA Hosp., Hines, Ill., 1938-41, chief tumor research lab., 1941—; asso. prof. pathology Northwestern U. Med. Sch., Chgo., 1962—. Cons. Argonne Nat. Lab., 1963—. Mem. Am. Assn. Exptl. Pathology, Am. Assn. Cancer Research, Leukemia Soc. (sec. Chgo. 1964—), Am. Assn. Pathology and Bacteriology, N.Y. Acad. Sci. Contbr. numerous articles to profl. jours. Mem. editorial bd. Am. Jour. Clin. Pathology, 1954—. Developed quantitative methods to study living blood cells from leukemic patients in lab., showed that adrenal cortex hormones and dimethyl sulfoxide are more toxic to leukemic than to normal blood cells, provided additional diagnostic criteria for some types of leukemia. Home: 436 Elm. Pl., Highland Park, Ill. 60035. Office: VA Hosp., Hines, Ill. 60141.*

SCHREMP, Edward Jay, Am. physicist; b. Newark, Aug. 20, 1912; s. Edward Joseph and Oleita (Hartman) S.; B.S., Mass. Inst. Tech., 1934, Ph.D., 1937; m. Magdalene Corbett, May 17, 1947; 1 dau., Magdalene Gwinnett. Faculty, Mass. Inst. Tech., 1937, Washington U., St. Louis, 1937-41, U. Cin., 1941-46; sci. liaison officer Office Asst. Naval Attaché for Research, Am. embassy, London, 1946-48; head, theory br. nucleonics div. U.S. Naval Research Lab., Washington, 1948-66, consultant in theoretical physics, 1966—. Mem. staff Radiation Lab., Mass. Inst. Tech., 1941-46; civilian OSRD, 1941-46. Fellow Am. Phys. Soc. Research in cosmic rays; contbns. to classical theory of dynamical systems, and theory of noise in amplifiers and in other linear networks; generalized Dirac equation; devel. theory predicting existence of two different kinds of neutrino and yielding a unified space-time model of elementary particles and resonances. Home: 226 S. Fairfax St., Alexandria, Va. 22314.*

SCHREUS, Hans, German dermatologist; b. Hückswagen, Germany, Sept. 10, 1892; s. Theodor and Maria (Dressen) S.; M.D.; m. Liesel Göhl, July 28, 1921; children—Gisela, Margret, Werner. Instr. dermatology Acad. Düsseldorf, Germany. Recipient Order of Merit, German Republic; Röntgen medal; Schaudinn-Hoffman medal. Research and numerous publs. on dermatology, radiology. Home: Wagnerstrasse 15, Düsseldorf, W. Germany.

SCHREYBER, Heinrich, see Grammateus, Heinrich.

SCHRIBAUX, (Pierre) Émile Laurent, French agronomist, botanist; b. Richebourg, Aug. 10, 1857; ed. Nat. Inst. Agronomy; became prof. Nat. Inst. Agronomy, 1890; founder ofcl. sta. for control and inspection of agrl. seed; mem. French Acad. Scis., 1934. Author: Botanique agricole, 1917. Died Paris, Oct. 29, 1951.

SCHRIEFFER, John Robert, Am. physicist; b. Oak Park, Ill., May 31, 1931; s. John Henry and Louise (Anderson) S.; B.S., Mass. Inst. Tech., 1953; M.S., U. Ill., 1954. Ph.D. in Physics, 1957; m. Anne Grete Thomsen, Dec. 26, 1960; children—Anne Bolette, Paul Karsten, Anne Regina. NSF postdoctoral fellow U. Birmingham (Eng.), Niels Bohr Inst., Copenhagen, Denmark, 1957-58; faculty U. Chgo., 1958-59, U. Ill., 1959-62; prof. physics U. Pa., Phila., 1962-64, Mary Amanda Wood prof. physics, 1964—. Guggenheim Found. fellow Niels Bohr Inst., 1966-67. Mem. Am. Phys. Soc., Sigma Xi. Author: Superconductivity, 1965. Research, publs. on solid state and low temperature physics especially superconductivity and magnetism. Home: 1100 Tower Lane E., Narberth, Pa. Office: Dept. Physics, U. Pa., Phila. 19104.*

SCHRIER, Stanley Leonard, Am. physician; b. Bronx, N.Y., Jan. 22, 1929; s. Harold and Nettie (Schwartz) S.; student N.Y.U., 1946-47; B.A. cum laude, U. Colo., 1949; M.D., Johns Hopkins, 1954; m. Peggy Pepper, June 7, 1953; children—Rachel D., Leslie E., David T. Sr. asst. surgeon USPHS, U. Chgo., Statesville Penitentiary, Army Med. Research Project, research asso., instr. dept. medicine U. Chgo., 1956-58; sr. asst. resident U. Chgo. Hosp., 1958-59; practice medicine, specializing in internal medicine, Palo Alto, Cal., 1959—; faculty Stanford, 1959—, asso. prof. medicine, 1963—. John and Mary R. Markle scholar, 1961. Diplomate Am. Bd. Internal Medicine. Mem. Am. Fedn. Clin. Research, N.Y. Acad. Scis., Internat., Am. socs. hematology, Phi Beta Kappa, Alpha Omega Alpha. Research projects include work on malaria, hemolytic anemia, red cell metabolism and transport; bacterial contamination of foods irrigated with polluted water. Home: 825 Ilima Ct. Office: 300 Pasteur Dr., Palo Alto, Cal.*

SCHRIRE, Velva, S. African physician; b. Kimberley, S. Africa, Oct. 24, 1916; s. Samand Sarah (Sendrowitz) S.; B.Sc. with triple honours, U. Cape Town (S. Africa), 1937, M.Sc. with honours in Physiology, 1938, Ph.D., 1940, M.B., Ch.B, 1941, M.D., 1963; m. Ruth Miriam Goldberg, Feb. 27, 1944; children—Robert Arthur, Barbara Jill, Stephen David, Jeremy Richard. Sr. asst. depts. medicine and pathology Groote Schuur Hosp., Cape Town, 1946-48, dir. cardiac clinic, 1951—; Nuffield Travelling fellow, London, 1949; sr. registrar Nat. Heart Hosp., London, 1949; practice medicine, specializing in cardiology, Cape Town, 1950—; sr. lectr. dept. medicine U. Cape Town, 1951—, dir. Council for sci. and indsl. research Cardiovascular-Pulonary Research Group, 1958—, prin. physician, cardiologist, 1953—, asso. prof. medicine, 1964—; sr. physician cardiologist Red Cross War Meml. Childrens' Hosp., Cape Town, 1959—. Mem. panel experts WHO, 1960—. Ella Lyman Cabot Travelling fellow, 1958; Oppenheimer Meml. Travelling fellow, 1961; fellow U. Cape Town, 1959, 64. Fellow Royal Coll. Physicians London, Royal Coll. Physicians Edinburgh, Am. Coll. Cardiology, N.Y. Acad. Scis.; mem. S. African Med. Assn. So. African (founder), Cape Town cardiac socs. Author: Clinical Cardiology, 1964; (with C. N. Barnard) Cardiac Surgery, 1967; also numerous articles. Research in heart diseases, phonocardiography, especially congenital heart disease; studies of effect of vasoactive substances on hemodiseases, cardiomyopathy, beriberi, cardiac aneurysm and aortic arteritis, incidence of ischaemic heart disease, pericarditis, tachycardias, mitral valve disease, cardiac surgery. Home: Julianne, Kilgetty Rd., Rondebosch, Cape Town. Office: Groote Schuur Hosp., Cape Town, S. Africa.*

SCHRODER, Harold Martin, psychologist; b. Tenterfield, Australia, Feb. 19, 1923; s. John Martin and Elizabeth (Heffernan) S.; B.A. with 1st class honours, Sydney (Australia) U., 1948; M.A., Ohio State U., 1951, Ph.D., 1953; m. Chloe Hayes, Dec. 13, 1952; children—Molly, Jeannie, Libby. Came to U.S., 1950, naturalized, 1958. Social Sci. Research fellow U. Cal. at Berkeley, 1952; faculty Princeton, 1954—, prof. psychology, 1960—. Author: Conceptual Systems and Personality Organizations, 1963; Human Information Processing, 1966; Information Processing a Perspective in Personality and Theory, 1967; also articles. Research on human information processing. Home: 66 Wittmer Ct., Princeton, N.J. 08540.*

SCHRÖDER, Klaus, metallurgist; b. Celle, Germany, Nov. 1, 1928; s. Fritz and Marta (Tönnies) S.; student Phillips U., Marburg, Germany, 1951; Dr. rer. nat., George-August U., Göttingen, Germany, 1954; m. Marianne Jahn, Apr. 11, 1957; children—Gordon K., Christoph R. Research officer Tribophysics Commonwealth Sci. and Indsl. Research Orgn., Melbourne, Australia, 1955-58; research asso., research asst. prof. U. Ill., Urbana, 1958-61; wissench. mitarbeiter U. Göttingen, 1961-62; asso. prof. metallurgy Syracuse (N.Y.) U., 1962—. Guest scientist Nat. Magnet Lab., Mass. Inst. Tech., Harvard, 1966—. Mem. Am. Phys. Soc., Am. Soc. for Metals, Sigma Tau. Research, publs. on phys. properties of alloys, specific heat, Hall and Seebeck effect, plastic properties of metals. Home 7226 Coventry Rd., East Syracuse, N.Y. 13057. Office: Dept. Metallurgy, Syracuse U., Syracuse, N.Y. 13210.*

SCHRÖDER, Wilhelm Georg, German chemist; b. Cologne, Germany, Apr. 25, 1900; s. Georg and Anna (Schilbe) S.; Dr.ès.sc., U. Bonn; m. Maria Schlüter, Dec. 24, 1926; children—Helga, Ute. Asst., Inorganic Chem. Inst., U. Bonn, 1921-23; asst. in chief, dept. head Inorganic Electrochem. Inst. Aix-la-Chapelle Tech., 1924-38; became instr. Stuttgart Tech. Coll., 1930, prof. at large, 1936, asst., 1939-45; prof. Clausthal Sch. Mines, 1948-57; collaborator Gmelin Inst. of Max Planck Soc.; collaborator sci., spl. editor, 1957—. Mem. Union Higher Edn. Research and publs. on heterogeneous equilibriums of salts, gaseous hydrates, oxidation of sulfuric acid and sulfites, reactions and transformation in solid state. Home: Niddastrasse 24, 638 Bad Hamburg v.d.H. Office: Varrentrappstrasse 40-42, 6000 Frankfort/Main, W. Germany.

SCHRÖDINGER, Erwin, Austrian physicist; b. Vienna, Austria, Aug. 12, 1887; s. Rudolph and (Bauer) S.; student U. Vienna, 1906-10; m. Annemarie Bertel, 1920. Asst., U. Vienna, from 1910; asst. to Max Wein, 1920; then extraordinary prof., U. Stuttgart (Germany); later prof. physics U. Zurich (Switzerland), for 6 years; prof. U. Berlin, 1927-33; fellow U. Oxford; with U. Graz (Austria), 1936-38; escaped to Italy, then U.S., followed by short stay at Princeton; then became dir. Sch. Theoretical Physics, Inst. Advanced Study, Dublin, Ireland, ret. 1955; returned to Vienna. Recipient (with P.A.M. Dirac) Nobel prize for physics, 1933. Author: Four Lectures on Wave Mechanics, 1928; Collected Papers on Wave Mechanics, 1928; Modern Atomic Theory, 1934; What is Life?, 1944; Statistical Thermodynamics, 1945. Research on specific heat of solids, statistical thermodynamics, atomic spectra; discovered Shrödinger wave equation, during early 1920's; showed that matrix mechanics can be replaced by wave mechanics, 1926; founder wave mechanics, which placed quantum theory on new basis, and on which he founded an atomic theory; contbd. to theory of color; investigated radium. Died Vienna, Jan. 4, 1961.

SCHROEDER, Gerhard, German chemist; b. Kassel, Germany, June 25, 1929; s. Johannes Franz and Margarete (Noeding) S.; Dr.rer.nat., Technische Hochschule, Karlsruhe, Germany, 1959; m. Helga Hess, Feb. 21, 1959. Postdoctoral fellow Yale, 1959-60; research asso. Union Carbide European Research Asso., Brussels, Belgium, 1961-64; privatdozent Technische Hochschule, Karlsruhe, 1964—. Author: Cyclooctatetraene, 1965; also articles. Isolation (with prof. Criegee) of 1st primary ozonide, a nickel complex of tetra-methyl-cyclobutadiene; isolation of 1st cis- and trans- ozonides; research on molecules with fluctuating bonds, chemistry of bullvalene and annulenes. Home: 7a Ellmendinger Str., 75) Karlsruhe-Durlach, Germany.*

SCHROEDER, Harry William, Am. plant pathologist; b. Linn, Kan., Aug. 31, 1916; s. Harry Phillip and Christine (Herchenroeder) S.; student Mankato State Coll., 1948-49; B.S., U. Minn., 1951, M.S., 1955, Ph.D., 1955; m. Maria de los Angeles Melendez de la Garza, Dec. 7, 1951; children—Christine Louise, Harry William II. Chief div. plant pathology Inst. Agrl. Investigations, Mexico City, Mexico, 1956-57; plant pathologist U.S. Dept. Agr., College Sta., Tex., 1957—, research plant pathologist, 1959—, investigations leader Market Quality Research div. Agrl. Research Service, 1965—; mem. grad. faculty Tex. A. and M. U., 1963—; cons. Ford Found. grants. Mem. Am. Phytopath. Soc., Am. Inst. Biol. Scis., A.A.A.S., Am. Soc. Microbiology, Sigma Xi. Research, numerous publs. on factors influencing microbiol. deterioration of rice during conditioning, handling, and storage; reported factors affecting devel. and accumulation of mycotoxins in peanuts, rice and cottonseed. Home: 1604 Jersey St., College Station 77840. Office: P.O. Box ED, Tex. A. and M. U., College Station, Tex. 77840.*

SCHROEDER, Hubert, German physicist; b. Landshut, Germany, July 14, 1913; s. Hugo and Marie (Finsterwald) S.; Dr.ès.sc., U. Munich; m. Isolde Certel, May 4, 1940; children—Meinhard, Regine. Began as physicist in optics industry, 1939; became collaborator Jenaer Glaswerk Schott und Gen., Landshut, also Mainz, Germany, 1950; became prof. agrégé U. Frankfort/Main, 1956, prof., 1962. Mem. Physikalische und opt. Gesellschaft. Research and publs. on physics of glass and magnetism, optics. Home: Nerotal 7, Wiesbaden, Germany. Office: Physics Inst. U., Robert Mayerstrasse 2, Frankfort/Main, W. Germany.

SCHROEDER, Hubert Ernst, dentist, electronmicroscopist; b. Königsberg (now Kaliningrad, USSR), Feb. 17, 1931; s. Ernst Richard and Ruth (Wollschläger) S.; Dr.med.dent., U. Frankfurt/Main (Germany), 1957; m. Jutta Schilling, Aug. 25, 1961; 1 dau., Alice Gabriele. Doctoral fellow U. Frankfurt/Main Inst. Pathology, 1955-57; with Dental Inst., U. Zürich (Switzerland), 1960—, sr. research fellow, 1963—; research fellow Inst. for Dental Research, N.Y. U. Coll. Dentistry, 1965; guest researcher U. Aarhus (Denmark), Royal Dental Coll., 1965-66. faculty operative dentistry and periodontology, 1964—. Mem. Biometric Soc., Internat. Assn. Dental Research, Deutsche Gesellschaft für Elektronenmikroskopie, Schweizerische Gesellschaft für Elektronenmikroskopie, Publs. on 1st detailed studies of mineralization of dental calculus; 1st experiments on calculus inhibition based on standardization of methods of epidemiological research in periodontology; description of ultrastructural details of oral epithelium, melanocytes and Langerhans' cells of oral epithelium in non-coloured people in a state of subfunction. Home: 498 Luegislandstr., 8051, Zürich, Switzerland.*

SCHROEDER, Manfred Robert, Am. physicist; b. Ahlen, Germany, July 12, 1926; s. Karl and Hertha (Kraemer) S.; D.Sc., U. Goettingen, 1954; m. Anny Menschik, Feb. 25, 1956; children—Marion, Julian, Alexander. Came to U.S., 1954, naturalized, 1963. With Bell Tel. Labs., Murray Hill, N.J., 1954—, dir. acoustics, speech and mechanics research, 1964—. Fellow Acoustical Soc. Am. (asso. editor Speech Communication), Audio Engring. Soc. (gov. 1960), German Phys. Soc., A.A.A.S. Research, numerous publs. on found. of statis. theory of sound propagation in rooms, psychoacoustics, speech processing devices for improved transmission, storage, intelligibility and security of speech signals; numerous patents in field. Home: 1004 Chimney Ridge Dr., Mountainside, N.J. 07092. Office: Bell Tel. Labs., Murray Hill, N.J. 07970.*

SCHROEDER, William Charles, Am. ichthyologist; b. Staten Island, N.Y., Jan. 10, 1895; s. William and Emma (Caffrey) S.; student George Washington U., 1922-23, Harvard, 1924-31; m. Adah Jensen, Aug. 27, 1916; children—W.H.E., Gloria M. (Mrs. William F. Gallagher). Bus. mgr. Woods Hole (Mass.) Oceanographic Inst., 1932-52, ichthyologist, 1952—, sr. scientist, 1964—; asso. curator fishes Harvard, 1936-60, hon. asso. in ichthyology, 1961—. Recipient Gold medal Geog. Soc. Cuba, 1940, citation for 25 years service Harvard, 1960, Gold pin for 30 years service Woods Hole Oceanographic Inst., 1961. Author: (with S. F. Hildebrand) Fishes of Chesapeake Bay, 1928; (with H. B. Bigelow) Fishes of the Western North Atlantic, 1948, 53, Fishes of the Gulf of

Maine, 1953; also numerous articles. Research, on marine fishes, especially sharks, skates and chimaeras, in the Western Atlantic Ocean, Can. to the W. Indies. Home: 167 Palmer Av., Falmouth, Mass. 02540. Office: Woods Hole Oceanographic Inst., Woods Hole, Mass. 02543.*

SCHROEDER VAN DER KOLK, Jacob Ludwig Conrad, Dutch pathologist; b. Leeuwarden, Netherlands, Mar. 14, 1797; M.D., U. Groningen (Netherlands), 1820; became mem. staff hosp., Amsterdam, 1821; prof. anatomy U. Utrecht (Netherlands), 1827-60; named insp. mental asylums, 1842; dir. asylum. of Utrecht, from 1827. Confirmed that cause of epilepsy is found in medulla, 1859. Died Utrecht, May 1, 1862.

SCHRÖTER, Carl, botanist; b. Esslingen, Germany, Dec. 19, 1855; prof., Zurich, Switzerland. Author: Die Flora der Eiszeit, 1882; Taschenflora des Alpenwanderers, 1894; Die Vegetation des Bodensees, 1902; (with others) Lebensgeschichte der Blütenplfanzen Mitteleuropas, from 1904; Das Pflanzenleben der Alpen, 1906. Made studies in plant geography. Died Zurich, Feb. 17, 1939.

SCHRÖTER, Johann Hieronymus, German astronomer; b. Erfurt, Germany, Aug. 30, 1745; ed. Göttingen, Germany; chief magistrate, Lilienthal, nr. Bremen, Germany; mem. French Acad. Scis., 1807. Author: Aphroditographische Fragmente zur genaueren Kenntnis des Planeten Venus, 1796; Selenotopographische Fragmente 2 vols., 1802; Kronographische Fragmente zur genaueren Kenntnis des Planeten Saturn, 1808. Made observations on phys. character of planets and moon, also telescopic observations of sun, from his pvt. obs. in Lilienthal. Died Erfurt, Aug. 29, 1816.

SCHRÖTTER, Anton Ritter von Kristelli, Austrian chemist; b. Olmütz, Austria, Nov. 26, 1802; prof. Graz, also Vienna, Austria. Author: Die Chemie in ihrem gegenwärtigen Zustande, 2 vols., 1847-49. Gave 1st description of red phosphorus, 1848; 1st to recommend hydrogen peroxide for bleaching hair. Died Vienna, Apr. 15, 1875.

SCHUBERT, Hermann, German mathematician; b. 1848; prof., Hamburg, Germany, 1876-1908. Author: Kälkul der abzählenden Geomtrie, 1879. First to show importance of principle of correspondence; extended enumerative geometry to n-dimensional space; stated preservation of number as fundamental principle of enumerative geometry; generalized principle of continuity to principle of conservation of number (from basis no longer held valid); worked on characteristics of system of curves. Died 1911.

SCHUBERT, Jack, Am. chemist; b. Chgo., Sept. 14, 1917; s. William and Lena (Adler) S.; B.S., U. Chgo., 1940, Ph.D., 1944; m. Mary A. Naeseth, Apr. 12, 1947; children—Ann N., Catherine N., Amy N. Research chemist Manhattan Project, U. Chgo. and Oak Ridge, 1942-47; asst. prof. U. Minn., Mpls., 1947-48; sr. chemist Argonne (Ill.) Nat. Lab., 1948-61; vis. prof. chemistry U. Buenos Aires, Argentina, cons. Argentine AEC, 1961-64; research prof. radiation chemistry U. Pitts. Grad. Sch. Pub. Health, 1965—. Sr. postdoctoral fellow NSF, 1956-57; exchange prof. Max Planck Inst. Biophysics, Germany, 1959-60. Fellow N.Y. Acad. Scis., A.A.A.S.; mem. Am. Chem. Soc., Am. Soc. Biol. Chemists. Author: (with R.E. Lapp) Radiation—What It Is and How It Affects You, 1957; Copper and Peroxides in Radiobiology and Medicine, 1964; also numerous articles. Co-editor: (with F.C. Nachod) Ion Exchange Technology, 1956. Pioneered devel. of ion exchange processes for separation of plutonium and fission products from irradiated uranium, theoretical and practical procedures for removal of radioelements from animals; antidote for beryllium poisoning, ion exchange and enzymatic methods for measurement of stability of complex ions; study of role of chelation and trace elements in biology and medicine.*

SCHUBERT, Leo, Am. chemist; b. N.Y.C., Mar. 1, 1916; s. Solomon and Katherine (Anker) S.; B.S., Coll. City N.Y., 1936; M.S., N.Y. U., 1943; Ph.D., U. Md., 1954; D.Sc. (hon.), St. Augustine Coll., 1964; m. Edith Broffman, Feb. 12, 1938; 1 dau., Leda Deirdre. With Nat. Bur. Standards, 1942-51; faculty Am. U., Washington, 1951—, prof., chmn. chemistry dept., 1957—, chmn. natural scis. and math. div., 1960—. Dir. insts. NSF, 1956—; mem. coms. Health Edn. and Welfare, 1965—. Fellow A.A.A.S.; mem. Chem. Soc. Washington (pres. 1964), Am. Chem. Soc. (editorial bd. Chemistry Jour. 1964—), Washington Acad. Scis. (pres. 1965), Sigma Xi, Phi Kappa Phi, others. Editorial bd. Sci. Tchr., Sci. Book Quar. Research, publs. on low temperature fluorescence, sci. edn. Office: 6813 Am. U., Washington 20016.*

SCHUBERT, Walter John, Am. chemist; b. Bklyn., May 14, 1925; s. William H. and Helen (Byzinski) S.; B.S., Fordham U., 1945, M.S., 1947, Ph.D., 1950. Asst. prof. John Carroll U., Cleve., 1950-51; asst. prof. St. John's U., Bklyn., 1951-53; Procter and Gamble fellow Fordham U., Bronx, N.Y., 1953-55, faculty, 1955—, prof. chemistry, 1962—, chmn. dept. chemistry Coll. Pharmacy, 1956—. Fordham U. Faculty fellow, 1959-60. Mem. Am. Chem. Soc., A.A.A.S., N.Y. Acad. Scis., Am. Soc. Biol. Chemists, Sigma Xi, Phi Lambda Upsilon, Rho Chi. Author: Lignin Biochemistry, 1965; also numerous articles. Research on chemistry lignin, chem. products resulting from decay lignin by wood-rotting fungi, proof intervention shikimic acid pathway in biogenesis lignin in growing trees; developed new method for isolation lignin from wood by application wood-destroying fungi. Home: 91 N. Henry St., Bklyn. 11222. Office: Fordham U. Coll. Pharmacy, Bronx, N.Y. 10458.*

SCHUCHHARDT, Karl, German archeologist, b. Hanover, Germany, Aug. 6, 1859; mus. dir., Hanover, also Berlin; helped excavate Pergamon. Author: Schliemanns Ausgrabungen, 1890; Der Goldfund vom Messingwerk bei Eberswalde, 1914; Alteuropa in seiner Kultur- und Stilenwicklung, 1919; Vorgeschichte von Deutschland, 1928; Die Burg im Wandel der Weltgeschichte, 1930. Died Aarolsen, Dec. 7, 1943.

SCHUCKBURG-EVELYN, Sir George Augustus William (6th baronet), Brit. mathematician; b. Eng., Aug. 23, 1751; s. Richard and Sarah (Hayward) S.; B.A., Balliol Coll., Oxford, Eng., 1772; m. Sarah Johanna Darker, July 3, 1782; m. 2d, Julia Annabella Evelyn, Oct. 6, 1785; 1 dau., Julia Evelyn Medley (Mrs. Charles Cecil Cope Jenkins). Mem. Parliament from Warwickshire, 1780-1804. Fellow Royal Soc., 1774, Soc. Antiquaries. Author: Observations made in Savoy to ascertain the Height of Mountains by the Barometer (results of investigations with William Roy), 1779; An account of the Equatoreal Instrument circa 1793; also papers. Reported to Royal Soc. results of expts. made to determine relation between English yard and an invariable standard, 1798; conducted expts. on capacity and weight. Died Aug. 11, 1804.

SCHUCKERT, Johann Siegmund, inventor; b. Nuremberg, Germany, 1846; worked with Edison, 1869; founder workshop, Nuremberg, built a dynamo, also Pilsen direct current arc lamp based on Krizik's patent, glass parabolic mirrors, completely valveless shaft and drain pumps; inventor rotary converter. Died 1895.

SCHUERCH, Conrad, Am. chemist; b. Boston, Aug. 2, 1918; s. Conrad, and Emma (Steinmuller) S.; B.S., Mass. Inst. Tech., 1940, Ph.D., 1947; m. Margaret Childs Pratt, June 26, 1948; children—Barbara Merle, Conrad, William Edward, Peter Henry. Sessional lectr., Hibbert Meml. fellow div. cellulose chemistry McGill U., Montreal, Que., Can., 1947-49; faculty State U. Coll. Forestry, Syracuse (N.Y.) U., 1949—, prof., chmn. dept. forest chemistry, 1956—. Guggenheim Meml. fellow, 1959-60. Mem. Am. Chem. Soc., Chem. Soc. (London, Eng.), A.A.A.S., T.A.P.P.I., Tech. Club Syracuse. Research, numerous publs. on chem. structure and phys. properties lignin, chem. reactions of wood, liquid ammonia plasticization wood, mechanism formation optically active and isotactic vinyl polymers; developed stereospecific chem. syntheses polysaccharides. Home: 125 Concord Pl., Syracuse, N.Y. 13210.*

SCHUESSLER, Karl Frederick, Am. sociologist; b. Quincy, Ill., Feb. 15, 1915; s. Hugo and Elsa (Westerbeck) S.; B.A., Evansville (Ind.) Coll., 1936; M.A., U. Chgo., 1939; Ph.D., Ind. U., 1947; m. Lucille Smith, June 27, 1946; children—Thomas Brian. Sociologist, Ill. State Prison, 1938; tchr. Highland Park (Ill.) High Sch., 1939-40; instr. Vanderbilt U., 1946-47; mem. faculty Ind. U., 1947—, prof., sociology, 1960—, chmn. dept., 1961—; research cons. Thammasat U., Bangkok, Thailand, 1963; vis. prof. U. Cal., Berkeley, 1965-66; visiting professor U. of Washington, 1967. Member adv. panel sociology NSF, 1962-63; mem. behavioral sci. study sect. NIH, 1964—. Mem. Am. Sociol. Assn., Am. Statis. Assn. Author: (with J. H. Mueller) Statistical Reasoning in Sociology, 1961; Social Research Methods; also numerous articles. Editor: (with Cohen and Lindesmith) The Sutherland Papers, 1956. Studies in the effectiveness of parole prediction, deterrent influence of the death penalty, personality of criminals, and variation in the crime rates of cities; methodological studies in attitude scaling and social prediction; research on the socialization of children, social basis of musical preferences, and the communality of different cultures. Home: 1820 E. Hunter St., Bloomington, Ind.*

SCHUETZ, Robert David, Am. chemist; b. Mpls., June 27, 1915; s. David Aloysius and Olive Margaret (Carver) S.; B.S., Northwestern U., 1937, M.S., 1939, Ph.D., 1947; m. Maryellen Murray, Nov. 28, 1942; children—Laura Catherine, Robert Edward. Research chemist, chem. engr. Richardson Co., 1940-46; research asso. Ohio State U., 1946-47; faculty Mich. State U., East Lansing, 1947—, prof. dept. chemistry, 1958-64, dir. labs., 1957-64, asst. dir. Inst. Biology and Medicine, 1964-66, asso. dir., 1966—. Mem. Am. Chem. Soc., N.Y. Acad. Scis., Sigma Xi, Alpha Chi Sigma. Author: Organic Chemistry; (with H. Hart) Laboratory Manual for Study Guide and Solutions Book for Organic Chemistry; also numerous articles. Patentee polymers, polymerization, medicinal chems.; research on reaction mechanisms and kinetics catalytic hydrogenations, carbohydrate chemistry, synthesis, reaction mechanism, stereochemistry organo sulfur compouids and sulfur heterocyclic compounds, biosynthesis naturally occuring organo sulfur compounds. Home: 626 Whitehills Dr., East Lansing, Mich. 48823.*

SCHUFFNER, Wilhelm August Paul, pathologist; b. Germany, 1867; introduced (with Herman Vervoort) oil of chenopodium to treat ancylostomiasis, 1900; described small round granules found in red blood cells of malarial patients as parasites develop (Schüffner's dots or granules), 1904; isolated (with A. Klarenbeek) Leptospira canicola from urine of dog, circa 1933; recorded (with C.M. Dhont, A. Klarenbeer, J. Voet) 1st case of Leptospira canicola in human being, 1934. Died 1949.

SCHUFLE, Joseph Albert, Am. chemist; b. Akron, O., Dec. 21, 1917; s. Albert Bernard and Daisy (Frick) S.; B.S., U. Akron, 1938, M.S., 1942; Ph.D., Western Res. U., 1948; m. Lois Carolyn Mytholar, May 31, 1942; children—Joseph Albert and Jean Ann (twins). Chemist, City of Akron, 1939-42; instr. Western Res. U., 1947; faculty N.M. Inst. Mining and Tech., Socorro, 1948-64, prof. chemistry, 1960-64; vis. prof. U. Coll., Dublin, Ireland, 1961-62; prof. chemistry, head dept. dir. Inst. Sci. Research, Highlands U., Las Vegas, N.M., 1964—. Mem. Am. Chem. Soc., A.A.A.S., Geochem. Soc., Soc. Chem. Industry, Inst. Chemistry (Ireland), N.M., Acad. Sci., History Sci. Soc., Sigma Xi. Co-editor: Water Improvement, 1962; contbg. author: Aridity and Man, 1963; also articles. Translator: Elective Attraction (T. Bergman), 1966. Research on phys. chemistry and geochemistry, structure of water, purification water, dating arid zone sediments, meteoritics, history sci.; patentee electrodialysis cell for water desalinization. Home: 1301 8th St., Las Vegas, N.M. 87701.*

SCHUH, Franz, Austrian surgeon; b. Vienna, 1804; prof., Vienna. Author: Über die erkenntnis der-pseudoplasmen, 1851; Pathologie und Therapie der pseudoplasmen, 1854. First to perform puncture of pericardium; performed 1st surgery with ether in Austria, 1847. Died Vienna, Dec. 23, 1865.

SCHUH, Frederik, Dutch mathematician; b. Amsterdam, Netherlands, Feb. 7, 1875; s. Frederik and Johann (Dinger) S.; student U. Amsterdam, U. Göttingen; Dr.ès.sci.; m. Maria Geertruida de Grijs, Aug. 2, 1902; children—Frederik, Karel Anton, Helena Dorothea Cleopatra. Prof. Apeldoorn, 1902-03, Sneek, 1903-07, Delft Tech. Sch., 1907-16, U. Groningen, 1961-45; actuary. Recipient Gold medal U. Amsterdam. Mem. Dutch Acad. Sci., Batavian Soc. Exptl. Philosophy. Research and publs. on algebra, differential and integral calculus. Author manual of mech. theory, manual of new geometry of triangles. Home: van Boetzelaerlaan 28, The Hague, Netherlands.

SCHUH, Friedrich Jacob Eduard, German geologist; b. Nurenberg, Germany, Dec. 8, 1889; s. Heinrich and Charlotte (Rhau) S.; ed. U. Fribourg; m. Margarethe Wienke, July 23, 1929; children—Lotte, Irmgard, Gretel. Asst., Rostock Geology Inst.; asso. prof., then prof., dir. Geology Inst., U. Munster, also geologist, became prof. emeritus, 1946. Mem. Geol. Union, German Geol. Soc., Paleontol. Soc. Contbr. articles to tech. jours. Introduced method of applied magnetic research to geol. prospecting, 1920; developed technique for using intensity-measurement devices in prospecting; originated theory of conformation of earth's crust and influence on climate of radioactivity on earth's crust. Address: Rudelsweiherstrasse 25, Erlangen, W. Germany.

SCHUHARDT, V(ernon) T(ruett), Am. microbiologist; b. San Antonio, Mar. 4, 1901; s. John W. and Kathryn B. (Sprague) S.; A.B., U. Tex., 1925; A.M., Rice Inst., 1930, Ph.D., 1940; m. Savannah Carter Girardey, Nov. 2, 1931. Sci. tchr. coach Hallettsville, (Tex.) High Sch., 1923-24, Del Rio (Tex.) High Sch., 1925-26; prof. biology, coach Schreiner Inst., Kerrville, Tex., 1926-28; tutor U. Tex., 1928-29; asst. Rice Inst., 1929-31; asst. dir. Tex State Dept. Health Lab., 1931-35; asst. prof. bacteriology U. Tex., 1935-37, asso. prof., 1937-42, prof. microbiology, 1942—; dir. Clayton Found. Brucellosis Research project, 1940-55. Fellow Am. Acad. Microbiology, Royal Soc. Tropical Medicine and Hygiene; mem. Soc. Gen. Microbiology, Am. Soc. Microbiology, A.A.A.S., Tex. Acad. Sci., Sigma Xi, Sigma Nu. Analysis of relapse phenomena (untreated and post-therapeutic) in spirochetal relapsing fever; established calloidal sulphur from amino acids as cause of peptone toxicity for Brucella species; codiscoverer of lysostaphin (bacterially produced lytic enzyme specific for genus staphylcoccus). Home: 1005 Gaston, Austin Tex. 78703.*

SCHUKNECHT, Harold Frederick, Am. physician; b. Chancellor, S.D., Feb. 10, 1917; s. J. G. and Dena (Weeldreyer) S.; student U. S.D., 1934-36; B.S., S.D. Sch. Med. Scis., 1938; M.D., U. Chgo., 1940; m. Anne Bodle, June 30, 1941; children—Judith, James. Asst. prof. otology U. Chgo., 1949-53; asso. surgeon Henry Ford Hosp., Detroit, 1953-61; chief otolaryngology Mass. Eye and Ear Infirmary, 1961—; Walter A. Lecompte professor of otology, professor laryngology, Harvard Med. Sch., 1961—; spl. research pathology ear and physiology hearing. Diplomate Am. Bd. Otolaryngology. Mem. A.M.A., Am. Acad. Ophthalmology and Otolaryngology, Am. Laryngological, Rhinological and Otological Soc., Acoustical Soc., Am., Am. Otol. Soc. Research and publs. on the ear and auditory disorders, especially deafness. Home: 263 Highland St., Weston, Mass. 02193. Office: 243 Charles St., Boston 02114.*

SCHÜLER, Hermann, German physicist; b. Posen, Germany, July 24, 1894; Dr.ès.sc.; Dr. honoris causa. Asst., U. Tübingen (Germany), 1920-24; collaborator Solar Physics Inst., Potsdam, Germany, also Kaiser Wilhelm Inst., Berlin, 1924-25; became instr. U. Berlin, 1926, prof. at large, 1932; hon. prof. physics U. Tübingen, 1948—; head spectroscopic research Max Planck Soc. Mem. Bunsen Soc., Faraday Soc., German Physics Soc. Research and publs. on atomic and molecular physics. Home: Flüthenweg 11. Office: Bunsenstrasse 10, Gottingen, W. Germany.

SCHULER, Max, German physicist; b. Zweibrücken, Germany, Feb. 5, 1882; s. Oscar and Bertha (Billhardt) S.; Dr.engring., Munich Tech. Sch.; Dr.engring. honoris causa Stuttgart Tech. Sch., 1962; m. Else Treitschke, Jan. 21, 1914; children—Wolfgang, Ilse. Dir. labs. Anschütz und Co., Kiel, Germany, 1908-14, dir., 1914-22; became prof. U. Göttingen (Germany), 1924, prof., 1928; named dir., prof. Inst. Applied Mechanics, U. Gottingen, 1934, prof. emeritus, 1946. Recipient Order of Merit of German Republic, 1962; gold bar Verein Deutscher Ingenieure, 1957. Mem. Gesellschaft für angewandt Mathematik und Mechanik, Deutsche Physikalische Gesellschaft, Verein Deutscher Ingenieure. Research and publs. on artificial horizons; Schuler's theorem on gyronavigation; discovered pendulum formula for fixed acceleration regarding earth; constrn. pendulum regulator for chronometry and barometry. Address: Rohnsweg 28, 34 Gottingen, W. Germany.

SCHULER, Robert Hugo, Am. chemist; b. Buffalo, Jan. 4, 1926; s. Robert H. and Mary (Mayer) S.; B.S., Canisius Coll., 1946; Ph.D., U. Notre Dame, 1949; m. Florence Forrest, June 18, 1952; children—Mary A., Margaret A., Carol A., Robert E., Thomas C. Asst. prof. chemistry Canisius Coll., Buffalo, 1949-52; asso. chemist Brookhaven Nat. Lab., Upton, N.Y., 1953-55, chemist, 1955-56; staff fellow, head radiation research labs. Mellon Inst., Pitts., 1956—, mem. adv. com. for research, 1962—. Recipient Centenary award U. Notre Dame, 1965. Mem. Am. Chem. Soc., Am. Phys. Soc., Radiation Research Soc., Faraday Soc., A.A.A.S. Research and many publs. on nature of effect of radiation on chem. systems. Home: Marlee Acres, Export, Pa. 15632. Office: 4400 5th Av., Pitts. 15213.*

SCHULL, William Jackson, Am. geneticist; b. Louisiana, Mo., Mar. 17, 1922; s. Eugene and Gertrude (Davenport) S.; B.S., Marquette U., 1946, M.S., 1947; Ph.D., Ohio State U., 1949; m. Victoria Novak, Sept. 7, 1946. Head genetics dept. Atomic Bomb Casualty Co., 1949-51; faculty U. Mich., Ann Arbor, 1951—, prof. human genetics, 1962—. Cons. Nat. Acad. Scis., NRC, WHO, NIH. Mem. A.A.A.S., Am. Assn. Phys. Anthropologists, Genetics Soc., Am. Soc. Human Genetics, Inst. Math. Statistics, Teratology Soc., Sigma Xi, Phi Sigma. Author: (with J.V. Neel) Human Heredity, 1954, Effects of A-Bombs, 1956, Effects of Inbreeding, 1965; (with F.W. Crowe, J.V. Neel) Neurofibromatosis, 1956; also many articles. Research on genetic effects of exposure to ionizing radiation, effects of inbreeding and relationship to genetic burden of a population, computer simulation of evolution in small human populations. Home: 2344 Yorkshire Rd., Ann Arbor, Mich. 48104.*

SCHULLER, Walter Harry, Am. chemist; b. Bayonne, N.J., Nov. 25, 1925; s. Harry Alvin and Catherine (Hellinger) S.; B.S. in Chemistry, U. Fla., 1946; Ph.D. in Organic Chemistry, Yale, 1951; m. Betty Jean Ogren, May 26, 1950; children—Walter Harry, Stephen Robert. Research fellow Cal. Inst. Tech., 1949-50; chemist Am. Cyanamid Co., Stamford, Conn., 1950-56, sr. research chemist, 1956; chief chemist Elkin Chem. Co., Miami, Fla., 1957-58; sr. chemist Naval Stores Lab., U.S. Dept. Agr., Olustee, Fla., 1958-60, prin. chemist, 1960—. Mem. Am. Chem. Soc. Research, publs. on enzymatic syntheses, polymerization diallyl monomers, amionic polymerization acrylonitrile, resin acid chemistry, structures and configurations, photosensitized oxidation, photchemistry dienes, color inhibition, paper chems., anti-static agts. Patentee in field. Home: 406 Oak Av., Lake City, Fla. 32055. Office: Naval Stores Lab., Olustee, Fla. 32072.*

SCHULMAN, Jerome Lewis, Am. child psychiatrist; b. N.Y.C., Nov. 15, 1925; s. Jacob and Pauline (Lewis) S.; B.A., U. Rochester, 1945; M.D., L.I., Coll. Medicine, 1949; m. Lenore Rosenberg, June 12, 1949; children—Ellen, Martha. Instr. psychiatry Johns Hopkins Hosp., Balt., 1955-57; faculty Northwestern U. Med. Sch., Chgo., 1957—, prof. pediatrics, prof. psychiatry and neurology, 1965—; dir. child devel., child guidance clinics, attending pediatrician, attending psychiatrist Children's Meml. Hosp., Chgo., 1957—. Diplomate Am. Bd. Pediatrics, Am. Bd. Psychiatry, Am. Bd. Child Psychiatry. Fellow Am. Acad. Pediatrics, Am. Psychiat. Assn.; mem. Am. Assn. on Mental Deficiency, Soc. for Biol. Psychiatry, Midwest Soc. for Pediatric Research, Am. Pediatric Soc. Author: (with Patricia M. Barger, J.C. Kaspar) The Therapeutic Dialogue—A Method for the Analysis of Verbal Interaction, 1964; (with Kaspar, F.M. Throne) Brain Damage and Behavior, 1965; (with D.T.A., Vernon, J.M. Foley, R.R. Sipowicz) Psychological Response of Children to Hospitalization and Illness, 1965; (with Kaspar, E.J. Barnes) Learning to Letter, 1965; The Management of Emotional Disorders in Pediatric Practice, 1966; also articles. Research on brain damage, psychol. effects of hospitalization, prevention of mental health disorders. Home: 414 Laurel Av., Wilmette, Ill. Office: 2300 Children's Plaza, Chgo. 60614.*

SCHULTE, Karl Ernst, German pharm. chemist; b. Deilinghofen, Westfalen, Germany, May 16, 1911; s. Karl and Auguste (Hellmann) S.; student U. Berlin 1934-39; Pharm. State Examination, U. München, 1936, Diploma-Chemist Examination, 1937, State Examination for Food Chemistry, 1938, Dr.Sci., 1939, Habilitation, 1941; Dr. (hon.) U. Lille (France), 1965; m. Maria Prollius, Sept. 7, 1942; children—Gisela, Ingrid. Asst., Inst. Pharmacy and Food Chemistry, München, 1939-42; privatdozent U. München, 1942-50, apl. prof., 1950-53, ordinarius for pharm. chemistry and food chemistry, 1959—; privatdozent Deutsche Forschungsanstalt für Lebensmittelchemie, 1942-50; extraordinarius for pharm. chemistry Freie U., Berlin, 1953-59; dir. Inst. Pharm. Chemistry, Münster, 1959; dean faculty maths. and sci., U. Münster, 1963-64. Mem. Soc. German Chemists, Soc. German Pharmacists, Soc. German Scientists and Physicians, Fédér. Internat. Pharmaceutique. Research and publs. on drug synthesis and drug analysis especially acetylenic Derivatives, food problems, isolation of medicinal plants and structure elucidation, isotope chemistry, compositions of pharm. preparations using isotopes. Home: 11, Fliednerstrasse, Münster, Westfalen 4400, Germany.*

SCHULTE-FROHLINDE, Dietrich, German chemist; b. München, Germany, Dec. 19, 1924; s. Heinrich and Emma (von Druffel) S.-F.; Diplom., U. Heidelberg (Germany), 1953, Dr., 1956; Priv.Doz.Dr., Technische Hochschule, Karlsruhe, Germany, 1963; m. Heidrun Andrä, Mar. 4, 1961; children—Heinrich, Verena. Asst., Max-Planck-Institut für medizinische Forschening, Heidelberg, 1956-59; leader radiation chem. group Inst. for Radiochemistry, Karlsruhe, 1959-65; head Inst. for Radiation Chemistry, Kernforschungszentrum, Karlsruhe, 1965—. Research and publs. in radiation chemistry of frozen aqueous solutions, cis-trans-isomerization, energy-transfer, free radical reactions, photochem. reactions. Home: 34 Kampmannstrasse, Karlsruhe-Grötzingen 7501, Germany. Office: Institut für Strahlenchemie, Kernsforschungszentrum, Karlsruhe, Germany.*

SCHULTEN, Hans, German historian; b. Elberfeld, Germany, May 27, 1870; prof., Erlangen, Germany; Author: Numantia, 4 vols., 1914-31; Geschichte von Numantia, 1933. Studied ancient history of Spain; discovered Celto-Iberian city Numantia. Died Erlangen, 1960.

SCHULTES, Johann, physician; b. Ulm/Donnau, Germany, 1595; ed. Padua, Italy; pupil of Fabricius ab Aquapendente and Adrian Vanden Speighel; doctorate, 1621. Practiced medicine, Padua, also Vienna, Austria; city physician, Ulm, until 1645. Author: Arman etarium chirurgicum (includes numerous case reports, catalog of surg. instruments of the time, illustrations of operative procedures and methods of bandaging and splinting), pub., 1653. Died Stuttgart, Germany, 1645.

SCHULTES, R. E., Am. botanist; b. Boston, Jan. 12, 1915; s. Otto R. and Maude B. (Bagley) S.; A.B., Harvard, 1937, A.M., 1938, Ph.D., 1941; M.H. (hon.), Universidad Nacional de Colombia (Bogotá), 1951; m. Dorothy Crawford McNeil, Mar. 27, 1959; children—R. E., Neil Parker and Alexandra Ames (twins). Plant explorer, rubber plants investigation Bur. Plant Industry, U. S. Dept. Agr., 1942-54; curator Harvard Bot. Mus., Cambridge, Mass., 1954—. Mem. Am. Acad. Arts and Scis., Linnean Soc. London, Colombian Acad. Scis., Soc. Econ. Botany, Am. Soc. Pharmacognosy Internat. Assn. Plant Taxonomy, New Eng. Bot. Club. Author: Native Orchids of Trinidad and Tobago, 1960; (with A. S. Pease) Generic Names of Orchids—Their Origin and Meaning, 1963. Editor: Botan. Mus. Leaflets, Harvard, 1958—; Econ. Botany, 1962—. Research, publs. chiefly in rubber producing, toxic, medicinal, narcotic plants of new world tropics. Home: 78 Larchmont Rd., Melrose, Mass. Office: Harvard Bot. Mus., Cambridge, Mass.*

SCHULTZ, Adolph H(ans), phys. anthropologist; b. Stuttgart, Germany (citizen of Zürich, Switzerland), Nov. 14, 1891; s. Julius and Sophie (Frick) S.; Ph.D., U. of Zürich, 1916; M.D. honoris causa, University of Bale, 1962; married Travis Bader, Dec. 10, 1924. Came to U. S., 1916, naturalized, 1934. Research asso. Carnegie Instn. of Wash., 1916-25, and 1937-38; asso. prof. of phys. anthropology, Johns Hopkins Med. Sch., 1925-51; professor anthropology and director anthropol. institute U. of Zurich, 1951-62, prof. emeritus, 1962—; mem. sci. expdn. to Asia, 1937. Fellow of the Zoological Society of London (foreign); member of the National Acad. Sciences, Am. Philos. Soc., Internat. Congress Anthropol. and Ethnol. Scis. (mem. permanent council, v.p. 1956), Soc. d'Anthropol. de Paris (hon.), N.Y. Acad. Sci. (hon.), Am. Assn. Phys. Anthropologists, Am. Anthrop. Assn., Am. Soc. Mammalogists, German Anthrop. Soc. (corr. mem.), Swiss Anthrop. Soc., Anthropol. Soc. Italy, Brit. Assn. Anatomists (hon.). Received Viking Fund Medal in Anthropology, 1948. Editor Handbook of Primatology; contbr. science articles to tech. jours., books. Established correlations and differentiations between development of man and other primates; demonstrated close similarity of man and apes early in life which diminishes through differing growth rate; noted certain human specializations as longest postnatal growth period and life span, latest beginning and ending of fertility. Home: 81 Dolderstr., Zurich, Switzerland.

SCHULTZ, Alfred Reginald, Am. geologist; b. Tomah, Wis., Mar. 26, 1876; s. John Fredrick and Ida M. (Kirst) S.; B.S., U. of Wis., 1900; Ph.D. (fellow), U. Chgo., 1905; m. Helene E. Burkhardt, Oct. 26, 1910; children—Irene Esther, Maxine Dorothy, John Burkhardt. Tchr. high sch., Wausau, Wis., 1900-02; resident hydrologist, Wis., 1903; geologist in charge Leith exploration party, Ont., Can., 1904; became connected with U. S. Geol. Survey, 1905, served as field asst., hydrologic aid, geologic aid, asst. geologist and geologist; mem. Coal Bd. (classification and valuation pub. coal lands), 1910; geologist Barbar Asphalt Co. and Bermuda Oil Co., Venezuela and Trinidad, S.Am., 1910-11; chmn. phosphate bd., metall. bd. land classification U. S. Geol. Survey, 1912-16, geologist in charge mineral div. land classification, 1916-18; geologist, oil div. U. S. Fuel Adminstrn., 1918; mgr. Burkhardt Milling & Electric Power Co., 1918-42; pres. Willow River Power Co., 1922-42; pres. Afton Power Co., 1921-42; pres. Hudson Hotel Co., 1929-42; dir. First Nat. Bank; cons. geologist. Mem. Geol. Soc. Am., A.A.A.S., Am. Forestry Assn., Sigma Xi. Writer numerous papers and govt. bulls. on stratigraphic and economic geology, gold, soda, oil, coal, leucite, potash, phosphate, water supplies. Died Sep. 30, 1943.

SCHULTZ, Gustav Theodor August Otto, German chemist; b. Finkenstein, Germany, Dec. 15, 1851; Ph.D., Königsberg (now Kaliningrad, USSR), 1874; prof. Tech. Inst. Munich, 1896-1928, also founder research and information center. Author: Farbstoff Tabellen; Die Chemie des Stein Kohdenteers, 1882. Contbd. to devel. of coal dye industry. Died Apr. 21, 1928.

SCHULTZ, Harry Pershing, Am. chemist; b. Racine, Wis., Mar. 9, 1918; s. Harry Carl and Mary (Olson) S.; B.S., U. Wis., 1942, Ph.D., 1946; m. Pearle Henriksen, Sept. 25, 1943; children—Stephanie, Tor, Alison. Research chemist NDRC, 1942-45, Merck & Co., Inc., Rahway, N.J., 1946-47; faculty U. Miami, Coral Gables, Fla., 1947—, prof. chemistry, 1952—. Mem. Am. Chem. Soc. (past chmn. Fla. sect.), Fla. Acad. Sci., Phi Beta Kappa, Sigma Xi, Pi Mu Epsilon, Phi Lambda Upsilon, Phi Delta Kappa. Author: (with Popp), Organic Chemical Preparations, 1964; also articles. Research on synthesis, reactions, physiol. activity, topology of organic chemistry. Home: 5835 S.W. 81st St., South Miami, Fla. 33143. Office: Chem. Dept., U. Miami, Coral Gables., Fla., 33124.*

SCHULTZ, Johann Heinrich, German psychologist; b. Göttingen, Germany, June 20, 1884; s. Hermann and J. (Gelzer-Sarasin) S.; student univs. Lausanne, Switzerland, Göttingen, Germany, Breslau (now Wroclaw, Poland); M.D. h.c. m. L. C. Woddislo, Nov. 19, 1944. Practice medicine, 1911—; physician in chief Lahmann Sanatorium, Dresden, Germany, 1920-24; neurologist, in pvt. practice, 1924—. Mem. Berlin Med. Soc., German Med. Hypnosis Soc. (pres.), Berlin Soc. Neurology. Author numerous books including: Des autogene Training, 1932; Seelische Krankenbehandlung, 1919; Grundfragen der Neurosenlehre, 1955; Handbuch der Neurosenlehr, 1958-62; also numerous articles in neurology, psychotherapy, med. psychology. Home: Bayernallee 17, 1 Berlin-Westend, W. Germany.

SCHULTZ, Julius, Am. biochemist; b. Rochester, N.Y., May 7, 1914; s. Benjamin and Ann (Duran) S.; student Cornell U., 1932-34; B.S., U. Mich., 1936, Ph.D. in Biol. Chemistry, 1940; m. Betty Jane Splane, Oct. 14, 1942. Research asso. U. Pa. Sch. Medicine, 1940-46; biochemist Jewish Hosp., Phila., 1947-49; prin. biochemist, asst. prof. research biochemistry Temple U., 1949-57; faculty Hahnemann Med. Coll., Phila., 1957-68, research prof. biochemistry, 1961-68; dir. Papanicolaou Cancer Research Inst., adj. prof. pharmacology U. Miami, 1968. Mem. Am. Chem. Soc. (sec. div. biochemistry), Am. Assn. Cancer Research, A.A.A.S., Am. Soc. Biol. Chemists, English Biochem. Soc. Research, numerous publs. on protection against phosgene poisoning, isolation and crystallization of prin. enzyme of leukemia tumor, structure, function and properties of enzyme of white blood cell, proteins of serum and their interrelationships. Home: 240 W. San Marino Dr., Miami Beach, Fla. 33136. Office: 1155 N.W. 14th St., Miami, Fla. 33136.*

SCHULTZ, Leonard Peter, Am. ichthyologist; b. Albion, Mich., Feb. 2, 1901; s. Charles A. and Della (Drumm) S.; A.B., Albion Coll., 1924, D.Sc., 1964; M.S., U. Mich., 1926; Ph.D., U. Wash., 1932; m. Dorothea C. Bowers, Aug. 20, 1927; children—James B., Marna C. (Mrs. George B. Parks), Gloria A. (Mrs. T. John Leppi). Faculty, U. Wash., 1928-36, asst. prof. fisheries, 1932-36; curator of fishes Smithsonian Instn., U.S. Nat. Mus., Washington, 1936-65, sr. zoologist, 1965—. Grad. council

George Washington U., 1949-—; dir. gen.'s panel experts on devel. fisheries FAO, 1963-—; also lectr., cons. Recipient Trophy award, Internat. Fedn. Aquarium Socs., 1958. Fellow A.A.A.S., Am. Inst. Fishery Research Biologists; mem. Am. Inst. Biol. Scis., Am. Soc. Ichtyologists and Herpetologists, Am. Soc. Limnology and Oceanography, Bermuda Biol. Sta., Nat. Geog. Soc., Soc. Systematic Zoology, Phi Beta Kappa, Sigma Xi, Phi Sigma, Delta Sigma Phi, others. Author: (with E. Stern) Ways of Fishes, 1948; (with H. Axelrod) Handbook of Tropical Fishes, 1955. Editorial bd. Fishes of the Western N. Atlantic, 1949-—, Funk and Wagnall's Universal Standard Ency., 1952-—. Research, publs. on classification, distbn. ecology of fishes. Home: 716 Scientists Cliffs, Port Republic, Md. 20676. Office: Smithsonian Instn., Washington 20560.*

SCHULTZ, Loris Henry, Am. animal physiologist; b. Mondovi, Wis., Feb. 9, 1919; s. Henry C. and Minnie (Switzenberg) S.; B.S., U. Wis., 1941, Ph.D. (Steenbock fellow), 1949; M.S. (Caleb Dorr fellow), U. Minn., 1942; m. Elmira Ann Baumann, May 28, 1949; children—Steven Karl, Mark Henry, David William. Asst. prof., asso. prof., prof. animal husbandry Cornell U., Ithaca, N.Y., 1949-57; prof. dairy sci. U. Wis., Madison, 1957-—. Mem. Am. Dairy Sci. Assn., Am. Soc. Animal Sci., Phi Eta Sigma, Alpha Zeta, Phi Kappa Phi. Research, publs. on metabolic disorders of ruminants and physiology of lactation, ketosis in dairy cows, developer use of propionates and propylene glycol for treatment, milk test for diagnosis and role of rumen acids in devel., milk composition, factors causing milk fat depression, effect of mastitis and mastitis treatment on milk composition. Home: 5055 Marathon Dr., Madison, Wis. 53705.*

SCHULTZ, Theodore John, Am. acoustical physicist; b. Jefferson City, Mo., Aug. 16, 1922; s. Otto Theodore and Juliaetta (Zeitz) S.; student Eastman Sch. Music, 1940-41, U. Mo., 1941-43, U. Tex., 1943-44; S.M. in Acoustics, Harvard, 1947, Ph.D. in Acoustics, 1954. Research physicist Naval Research Lab., 1947-48; instr., research fellow acoustics Harvard, 1948-55; asst. chief acoustics sect. Douglas Aircraft Co., Santa Monica, Cal., 1956-60; cons. in electronics and acoustics, 1960-66; prin. scientist Bolt Beranek & Newman, Inc., Cambridge, Mass., 1966-—. Fellow Acoustical Soc. Am.; mem. Am. Soc. for Testing and Materials, Internat. Standards Orgn., Tau Beta Pi. Research, publs. on acoustical instrumentation, archtl. acoustics, design of concert halls, acoustical test standards, electroacoustic transduction, undersea reverberation, noise control in aircraft and spacecraft. Patentee acoustic wattmeter, miniature condenser microphone. Home: 7 Rutland Sq., Boston 02118. Office: 50 Moulton St., Cambridge, Mass. 02138.*

SCHULTZ-HAUDT, Stig Deramm, Norwegian biochemist; b. Trysil, Norway, Dec. 2, 1920; s. Rolf D. and Signe (Klemetsdal) S.-H.; L.D.S., Sch. Dentistry, U. Oslo (Norway), 1949, Ph.D., 1958, postgrad.; M.S., U. Rochester, 1953, Ph.D., 1955. Research fellow U. Oslo Sch. Dentistry, 1955-58, faculty, 1958-—, asso. prof., 1960-—, chief histochem. lab. Inst. Path. Anatomy, 1963-—. Research, publs. on composition of connective tissue in health and inflammation. Home: 41 Stensgate, Oslo 4. Office: Inst. Pathological Anatomy, Rikshospitalet, Oslo, Norway.*

SCHULTZE, Max Johann Sigismund, German anatomist; b. Freiburg, Germany, Mar. 25, 1825; prof. Bonn, Germany, 1859-74. Author: On the Muscle Corpuscles; The Protoplasm of Rhizopods and Plant Cells, 1863. Showed that each muscle fiber develops from a single myoblast by division of its cell and nucleus, 1861, that protoplasm of all living cells is almost identical, 1863; described cells of olfactory mucous membrane (Schultze's cells), 1862; defined cell as lump of nucleated protoplasm, 1861, protoplasm as basis of life, 1863; studied life histories of various lower animals, including protozoa; demonstrated minute nerve endings in nose, ear and retina. Died Bonn, Jan. 16, 1874.

SCHULZ, George J., physicist; b. Brno, Czechoslovakia, Apr. 29, 1925; s. Felix and Elizabeth (Walerstein) S.; student U. Dr. E. Benes, Brno, 1945-47; B.S., Pa. State U, 1949, M.S., 1951; Ph.D., Mass. Inst. Tech., 1954; m. Rose Anfaenger, Sept. 20, 1949; children—Peter A., R. Stanley. Came to U.S., 1947, naturalized, 1957. Research asst. Pa. State U., 1949-51, Mass. Inst. Tech., 1951-54; adv. physicist Westinghouse Research Lab., Pitts., 1954-66; prof. applied sci. Yale, New Haven, 1966-—. Sec., Gaseous Electronics Conf., 1963. Fellow Am. Phys. Soc. (Davisson-Germer prize 1966). Research, publs. on electron collisions with atoms and molecules, mass spectrometry, vacuum techniques, electric discharges; discovered resonances in scattering of electrons on atoms and molecules, 1962. Home: Old Mill Rd., Woodbridge, Conn. 06525. Office: Mason Lab., Yale, New Haven 06520.*

SCHULZ, Karl Heinz, German physician; b. Molzen, Mar. 16, 1922; s. Friedrich and Dora (Heuer) S.; student univs. Göttingen, Leipzig, Hamburg; m. Hilde Kegel, Sept. 1, 1950; children—Heinz-Jürgen, Stefan. With allergies sect., dept. dermatology U. Hamburg, 1950-—, docent, 1960-66. sr. physician, 1961-—, prof., 1966-—. Mem. German, Hamburg dermatol. socs., German Soc. Allergy and Immunity Research, German Soc. Occupational Medicine. Author: Chemische Struktur und allergene Wirkung, 1962; also contbns. to handbooks, numerous articles in field. Research on hypersensitivity, desensitization and immunotolerance to simple chems., toxicology of chlorinated aromatic compounds, clin. dermatology. Address: 6 Beseler Strasse, Hamburg 52, West Germany.*

SCHULZ, Paul Herbert, German physicist; b. Rostock, Germany, Nov. 3, 1911; s. Heinrich and Marie (Mahrdt) S.; Dr.ès.sc., Ph.D., U. Rostock; m. Irene Kerp, July 19, 1938; children—Ingrid, Beata, Gabriele. Asst., U. Rostock, also U. Bonn; dir. labs. for study electric lights, Berlin; dir. Research Inst. Gaseous Body Physics, German Acad. Sci., Berlin; prof. U. Greifswald, also Karlsruhe (Germany) Tech. Sch. Mem. Internat. Lighting Commn., German Physics Soc., Tech. Lighting Soc. Research and publs. on spectroscopy, elec. discharges, elec. light sources; developed xenon lamp. Home: Bunsenstrasse 16. Office: Kaiserstrasse 12, Karlsruhe, W. Germany.

SCHULZ, Wilhelm Johann Gustav, geodesist; b. Berlin, Germany, Mar. 8, 1882; s. Wilhelm and Maria (Vogel) S.; Ph.D., U. Berlin; agrégé, Berlin; m. Eva Haack, Mar. 18, 1910; children—Hans, Wilhelm, Anamaria, Ilse-Eva. Head geodesy div. Cartographic Inst., Buenos Aires, Argentina; titular prof. univs. Buenos Aires, La Plata, Tucumán. Recipient Order of Merit Republic of Germany; Karl Ritter medal; Fuerra Aérea Argentina medal; gold medal and Expn. prize U. Gante, 1913. Mem. Deutscher Volksbund, Sociedad geografia (founder), Aero Club Argentina (hon.), Argentina Cartography Center, Nat. Geography Acad. Argentina. Author: Manual de Geodesia Superior; Manual de Gravimetría, Manual de Fotogrametría; Patagonia; Burmeiste, Viaje por los Etados del Plata, 1869-1943. Address: Rondean 338, San Isidoro-Beccar, F.C.N.G.B.M., Argentina.

SCHUMACHER, Berthold Walter, physicist; b. Karlsruhe, Germany, Apr. 15, 1921; s. Ernst Walter and Luise (Hofmann) S.; Diplom-Physicist, 1950, Dr.rer.-nat., 1953; m. Maja M. Gruen, Nov. 17, 1955; children—Reinhard, Thomas. Design and devel. engr. electronics and X-ray equipment Laboratorium Prof. Dr. Berthold, Wildbad, Germany, 1953-54; research fellow Ont. Research (Can.) Found., Toronto, 1954-58, dir. dept. physics, 1958-66; mgr. electron beam tech. Westinghouse Research Labs., Pitts., 1966-—. Mem. Canadian Assn. Physicist, Am. Phys. Soc., Inst. Physics and Phys. Soc. (Eng.), Verein Deutscher Ingenieure (Germany), German Phys. Soc. Research, publs. in electron beam physics, interaction electron beams with matter; inventions in electron beam machining and welding in air, analytical electron probes with specimen in air, electron-beam fluorescence probe for density, temperature and velocity measurements in rarefied gases, attenuation and single-scatter gauges for altitude and vacuum measurements, thermo-molecular vacuum pump, X-ray detector, X-ray spectrometer system. Home: 141 Kings Dale Rd., Pitts., 15221.*

SCHUMACHER, Clifford Rodney, Am. physicist; b. Waukegan, Ill., Dec. 19, 1929; s. Clifford Adolph and Delphine Marie (Barrett) S.; B.S., Wayne State U., 1951; Ph.D., Cornell U., 1962; m. Dorothy Elaine Kinley, June 17, 1950; children—Clifford Anthony, Kathleen Elaine, Carol Arlene. Mem. tech. staff Bell Telephone Labs., Murray Hill, N.J., 1954-57; mem. Inst. For Advanced Study, Princeton, N.J., 1961-62; NSF postdoctoral fellow Princeton, 1962-63; Enrico Fermi postdoctoral fellow U. Chgo., 1963-65; asso. prof. physics Pa. State U., University Park, 1965-—; research asso. Nuclear Studies Lab. Cornell U., summers 1962-63; vis. asst. physicist Brookhaven Nat. Lab., summer 1963. Mem. Am., Italian phys. socs. Contbr. chpts. to Crossed Field Microwave Devices, 1961. Research on theory of structure and interactions of elementary particles, analysis of high energy phenomena, quantum field theory, relativistic quantum mechanics. Home: 325 S. Garner St., State College, Pa. 16801. Office: Physics Dept., Pa. State U., University Park, Pa. 16802.*

SCHUMACHER, Heinrich Christian, astronomer; b. Bramstedt, Germany, Sept. 3, 1780; student 1st law, then astronomy; began as tchr. of law; became dir. Obs., Mannheim, Germany, 1813; named prof. astronomy, dir. Copenhagen (Denmark) Obs., 1815; apptd. to measure meridian arc between Skagen and Lawenburg and describe topography of Slesvig and Holstein by Danish govt., 1817; founder obs., Altona, Germany. Mem. French Acad. Scis. Fellow Royal Soc., 1821. Founder jour. Astronomische-Nachrichten. Author: Astron. Abhandlungen, 3 vols., 1823-25. Determined length of 2d pendulum; directed geodetic survey of Denmark. Died Altona, Dec. 28, 1850.

SCHUMANN, Hermann, German metallurgist; b. Dusseldorf, Germany, Sept. 12, 1924; s. Hermann and Emilie (Schlosser) S.; Dipl.Met., Bergakademie Freiberg (Germany), 1950, Dr. rer. nat., 1952, Dr.-Ing. habil, 1962; m. Ruth Bernhardt, June 12, 1953; children—Lutz, Dietrich. Sci. asst. Inst. fur Metallkunde und Materialprufung an der Bergakademie Freiburg, 1950-53; sci. staff Eisen-Forschungsinstitut Hennigsdorf, Berlin, Germany, 1953-60; prof. U. Rostock (Germany), 1960-—, dir. Inst. für Werkstofkunde, 1960-—. Author: Metallographie, 1955; also numerous articles. Research on hydrogen in steels, big steel forgings, steel for subzero temperatures, metastabile austenite, metallography; discovered gamma-epsilon alpha transformation in manganese steel, equivalence of alloying elements with pressure, cause of strengthening of Hadfield steel, forming of epsilon-Martensite in Austenite. Home: 16 Klement-Gottwald-Str., Rostock, Germany.

SCHUMANN, Theodor Eberhardt Werner, meteorologist; b. Carnarvon, S. Africa, Nov. 17, 1896; s. Adolf Wilhelm and Wilhelmine (Sterrenberg) S.; B.A. with honors, U. Stellenbosch, 1917, M.Sc., 1920, Ph.D., 1960; Ph.D. cum laude, U. Göttingen (Germany), 1924; Ph.D., U. S. Africa, 1959; m. Elizabeth van der Merwe, June 29, 1935; children—Werner Adolf, Ingrid Hechel, Hildegard Wilhelmine. Lectr., Rhodes Coll., Grahamstown, S.Africa, 1918-19, U. Stellenbosch, 1920-21; asst. Yale, 1924-25; research physicist Combustion Utilities Co., N.Y.C., 1926-30; research physics prof., Morgentown, W.Va., 1931; dir. Fuel Research Inst., Pretoria, S. Africa, 1931-33; dir. Weather Bur., Pretoria, 1933-57; mem. S. African Pub. Service Commn., 1957-58; dept. chmn. S. African Atomic Energy Bd., Pretoria, 1958-—. Recipient Alexander Buchan prize meteorology, 1943. Mem. S. African Acad. Arts and Sci. (Havenga prize for physics 1947, hon. life), Africa Inst. (mem. head com.). Author: The Abdication of the White Man, 1963; also articles. Research on theory of heat transfer, math. theory of flames, combusion of solid fuels, meteorology; introduced metric system for meteorol. observation in S. Africa. Home: 73 Anderson St. Office: Pvt. Bag 256, Pretoria, Republic of S. Africa.*

SCHUMANN, Viktor, German physicist; b. Markranstädt, Germany, Dec. 21, 1841; independent scholar, Leipzig, Germany. Research on ultraviolet spectrum, including discovery of new rays in outermost sect.; developed plate for photographing it. Died Leipzig, Sept. 1, 1913.

SCHUMB, Walter Cecil, Am. chemist; b. Boston, Sept. 10, 1892; s. Martin and Theresa (Laviska) S.; A.B., Harvard, 1914, A.M., 1916, Ph.D., 1918; m. Margaret Viola West, June 30, 1934 (dec.); m. 2d, Susan Alice Bridges, Oct. 29, 1955. Asst. gas chemist, Washington, 1918-19; asst. prof. chemistry Vassar Coll. Poughkeepsie, N.Y., 1919-20; faculty Mass. Inst. Tech., 1920-—, prof., 1934-—, now prof. emeritus. Mem. Am. Chem. Soc., Am. Acad. Arts and Scis. Research, numerous publs. on chemistry of fluorine and its compounds, silicon and its compounds. Home: 9 Garden St., Milton, Mass. 02186. Office: 77 Massachusetts Av., Cambridge, Mass.*

SCHÜMMELFEDER, Norbert, German chemist; b. Neuenkirchen, Nov. 9, 1916; s. Konrad and Hedwig (Eichhorn) S.; ed. univs. Munster, Greifswald and Berlin; M.D.; m. Ingeborg Hartlieb von Wallhor, Sept. 17, 1947; children—Sabine, Horst, Kurt. Sci. asst. Chem. Physiol. Inst., U. Greifswald, 1943-45; asst., then prin. asst. Pathol. Inst., U. Munster, 1945-54; asst., then curator Inst. Pathology, U. Bonn, 1954-62; dir. Pathology Inst., U. Cologne, 1962-—, prof. gen. pathology and anat. pathology, 1962-—. Mem. German Pathol. Soc., Soc. German Naturalists and Doctors, Histochem. Soc., N.Y. Acad. Sci., Germano-British Med. Union. Author: Probleme der Fermentcytochemie; Die experimentelle Strahlenschädigung des Zentralnervensystems; Die mikroskopische Diagnostik bösartiger Gechwülste; Über Fluoreszenmikroskopic. Home: Aachener Strasse 114, Junkersdorf, Kr. Cologne. Office: Pathologisches Inst., Univ. Cologne, Cologne-Lindenthal, W. Germany.

SCHUMPETER, Joseph Alois, economist; b. Triesch, Moravia (now Czechoslovakia), Feb. 8, 1883; s. Joseph Alois and Joan Marguerite (Gruener) S.; B.A., High Sch. (Theresianum), Vienna, Austria, 1901; Dr. in Law, U. of Vienna, 1906; Ph.D., Columbia U., 1913; Ph.D., Sofia, Bulgaria, 1939; m. Elizabeth Boody, Aug. 16, 1937. Came to U. S., 1932. Began practice of law, 1907; became lecturer U. of Vienna and prof. economics at U. of Gernowitz, 1909; prof. economics at U. of Graz, 1911-14; Austrian exchange prof. Columbia U., 1913; Austrian Minister of Finance, 1919-20; prof. economics, U. of Bonn, Germany, 1925-32, Harvard since 1932. Mem. Am. Econ. Assn., A.A.A.S., Am. Statis. Assn., Royal Econ. Soc., Econometric Soc. (pres. 1939-41). Author: Theory of Economic Development, 1911 (Eng. trans. 1934); Business Cycles, A Theoretical, Statistical and Historical Analysis of the Capitalist Process, 1939; Capitalism, Socialism and Democracy, 1942. Developer theory of bus. cycle in which entrepreneur is dynamic force; theory econ. devel. capitalism; thought influenced Am. sociology. Died Jan. 8, 1950.

SCHUR, Friedrich Heinrich, German mathematician; b. Maciejewo/Poznan, Poland, Jan. 27, 1856; prof., Leipzig, Germany, also Dorpat (now Tartu, Estonia), Breslau (now Wroclaw, Poland). Author: Lehrbuch der analytischen Geometrie, 1898; Grundlagen der Geometrie, 1909. Died Breslau, Mar. 18, 1932.

SCHUR, Wilhelm, German astronomer; b. Altona, Germany, Apr. 15, 1846; prof., dir. obs., Göttingen,

Germany. Author: Göttingen Sternkatalog für 1860, 1891; Bestimmung der Masse des Planeten Jupiter aus Heliometerbeobachtungen, 1882. Died Göttingen, July 1, 1901.

SCHÜRER-WALDHEIM, Fritz, Austrian surgeon; b. Wildalpen, June 9, 1896; s. Friedrich and Karoline (Marx) Schurer-W.; M.D., U. Vienna; m. Helga Adler, June 29, 1926; children—Helmut, Gerhart, Gerta, Erika. With Vienna Surg. Clinic, U. Graz Surgery Clinic, also 2d Surgery Clinic Vienna; doctor in chief surg. sect. Wilhelminenspital and Rudolfstiftung Hosp., Vienna; doctor in chief for Red Cross. Decorated Medal for Bravery, World War I and World War II; Austrian Order of Merit; recipient award Austrian Red Cross. Mem. Vienna Med. Soc., Soc. Austrian Surgeons, Vienna Sports Acad. (pres.). Publs. on surgery. Inventor anesthesia control device, also double glass bottle for blood transfusion. Address: Biberstrasse 3, Vienna 1, Austria.

SCHÜRMANN, Johann Wilhelm Walter, German bacteriologist; b. Lüdenscheid, May 14, 1880; s. Johann Heinrich and Helene (Crone) S.; ed. univs. Fribourg/Brisgau, Leipzig, Kiel, Giessen and Munich; M.D. Asst., Allgemeines Krankenhaus, Hagen; mil. ser-Mem. Vienna Med. Soc., Soc. Austrian Surgeons, Vienna Sports Acad. (pres.). Publs. on surgery. Inventor vice, Düsseldorf, 1908; asst. Inst. Exptl. Therapy and Clinic Infectious Diseases, Düsseldorf Acad. Medicine; doctor in chief Pathol. Inst., U. Marburg, 1909; chief sect. Research Inst. Infectious Diseases, also Swiss Serum and Vaccination Inst., Berne, 1910; qualified in hygiene and bacteriology U. Berne, 1912; prin. doctor, 1914-15; qualified at U. Halle, 1915, named prof. at large, 1918; at U. Giessen founder and dir. Pub. Health Service of Hamburg-Harburg and Wilhelmsburg, 1921; med. dir. Ruhr Miners Group, Bochum, 1922; hon. prof. U. Munster; dir. Reichsknappschat, Berlin, 1937; hygienist 2, A. K. Generalkommando, Poznan, 1941-42. Author: Mikrobiologie und Immunbiologie; Repetitorium der Hygiene, Bakteriologie und Serologie. Research on intravenous anesthetic and effects on animals; hygiene, bacteriology, serology, immunity research, chemotherapy, indsl. hygiene; occupational diseases of miners, silicosis and social medicine. Address: Kronenstrasse 18, Bochum, W. Germany.

SCHUSTER, Sir Arthur, math. physicist; b. Frankfort/Main, Germany, Sept. 12, 1851; s. Francis Joseph and Marie (Pfeiffer) S.; became Brit. citizen, 1875; ed. Owens Coll., Manchester, Eng.; Ph.D., Heidelberg, Germany, 1873; student of Kirchoff; several hon. degrees; m. Caroline Loveday, 1887; 1 son, 4 daus. Worked under W. Weber and E. Reicke at Göttingen, Germany, under Helmhotlz at Berlin, 1874; head Royal Soc. expdn. to observe solar eclipse at Siam, 1875; mem. expdns. to Colo., 1878, Egypt, 1882, W.I., 1886; mem. staff Cavendish Lab., Cambridge, 1876-81, worked with Clerk Maxwell, later Rayleigh; mem. staff Owens Coll., 1881-1907, as prof. applied math., then Langworthy prof. physics. A founder Internat. Research Council, 1919, sec. until 1928. Recipient Royal medal, 1893; Rumford medal, 1926; Copley medal, 1931. Fellow Royal Soc., 1879, sec., 1912-19, fgn. sec., 1920-24, v.p., 1919-24; mem. Brit. Assn. (pres. 1915), Internat. Assn. Acads. Author: An Introduction to the Theory of Optics, 1904; The Progress of Physics, 1911. Made (with Rayleigh) determination of ohm in absolute measure; conducted research in spectroscopy, electricity in gases, terrestrial magnetism, optics, math. theory of periodicity, calorimetry, radiometry, seismology; explained interference, absorption, Zeeman effect by means of electron; experimented on effect of optical lattices; made 1st photograph of spectrum of solar corona, 1882; 1st to demonstrate that electric current passes through a gas by means of gaseous ions, concluded that cathode rays are gaseous ions accelerated in strong field nr. cathode; 1st to show that ratio e/m may be obtained by deflecting cathode rays in magnetic field. Died Eng., Oct. 14, 1934.

SCHUSTER, Fritz, chem. engr.; b. Altenberg, Austria, May 7, 1901; s. Friedrich and Caroline (Krammer) S.; student Tech. U. Vienna (Austria), 1919-23; Techn.D., U. Graz (Austria), 1924; m. Grete Hösler, Sept. 7, 1927; children—Till, Ina. Asst., heal Lab. Recuperator A.B. Vienna, 1924-25; various positions in German industries, 1926-50; prof. U. for Mining and Metallurgy, Leoben, Austria, 1950-52; sci. dir. Inst. for Gas Technique, Essen, West Germany, 1952-65; prof. Tech. U. Aachen, West Ehrentafel, 1961; Austrian hon. cross for sci. and art 1st class, 1967. Hon. mem. Deutscher Verein von Gas und Wasserfachuauenrn, Osterrische Vereiniqung fur das Gas und Wasserfach. Author: Detoxification of Town Gas, 1935; Gas, Combustion, Heat (with G. Leggevie, I. Skunca), 1964; also numerous articles. Editor, Gaswärme, 1952-65, Hydrocarbon Gases, 1966-67. Research on detoxification of town gas, interchangeability of fuel gases, filling of spherical gasholders. Home: 8 Levering St., Essen, W. Germany 3-43. Office: 55, Templergraben, Aacher, West Germany 3-51.*

SCHÜTTE, Horst Robert, German biochemist; b. Hamburg, Germany, July 14, 1929; s. Walter K.H. and Matilde (Gieseken) S.; diplomexamen Martin Luther U. Halle (Germany), 1954, doktorexamen, 1957; m. Christa Dörrer, May 25, 1957; children—Wolfgang, Barbara, Andreas. Asst. organic chem. inst. Martin Luther U., Halle-Wittenberg, 1954-57, lectr. biochemistry, prof., 1966—; chief asst. acad. colleague Inst. Biochemistry Plants, German Acad. Scis., Halle 1958—, dir. isotope lab., 1960—. Author: Radioaktive Isotope in Organischer Chemie und Biochemie, 1966; also numerous articles. Biosynthesis of alkaloids especially biosynthesis of lupine alkaloids and nicotinic acids; use of isotopic technique in biochemistry; synthesis of isotopic labeled organic compounds. Home: 15, K.-Liebknecht Str., Halle/Saale. Office: 1, Weinbergweg, Halle-/Saale, Germany.*

SCHÜTTE, Karl Heinrich Christian, German astronomer; b. Brünsbuttel, Feb. 6, 1898; s. Carl and Dora (Bargob) S.; Ph.D., U. Breslau, 1922; m. Annemarie Hohle, June 18, 1926. Asst., Frankfort Obs., 1922-26, then Bayer Obs.; mem. Internat. Commn. Geodesy, Munich, 1926-40; instr. U. Munich, 1939; prof. astronomy U. Vienna, 1940-45; prof. U. Munich, 1951. Recipient Ernst Heinkel Astronautics prize, 1953. Fellow Royal Astron. Soc.; mem. Am. Inst. Aero. and Astronautics; corr. mem. Austrian Acad. Sci. Author: Die Bahnbestimmung aus dem Vektor der Bahngeschwindigkeit . . . Weltraumfahrt, 1953; Galaktozentrische Bahnelemente von Fixsternen, Sitzungsber. d. Österr. Akad. d. Wiss. Mathem.-naturw. K, 1952-58. Studies in theoretical astronomy; determination of orbits. Address: Schliessfach 110, Munich-Obermenzing, W. Germany.

SCHUTZ, (Johann) Wilhelm, German vet. pathologist; b. Berlin, 1839; pupil of Virchow; prof. Vet. U., Berlin, from 1876; discovered vaccination against piroplasmosis in cattle. Died 1920.

SCHÜTZENBERGER, Marcel P., French mathematician; b. Paris, Oct. 24, 1920; s. Pierre and Marie Louise (de Puis Aye) S.; M.D., Faculty Medicine, Paris, 1948, Ph.D., Faculty Sci., Paris, 1953; m. Hariati, June 1956. Staff Centre National de la Recherche Scientifique, 1956; research asso. Mass. Inst. Tech., 1956-57; prof. Faculty Sci., Poitiers, France, 1957-65; prof. Faculty Sci., Paris, 1965—; vis. prof. U. N.C., 1961-67, Harvard, 1962-63. Cons., WHO. Author: (with N. Chomsky) Algebraic Theory of Context-Free Language; also articles. Research in theory of algebra (monoids), application to problems in communication and computer scis. Home: 9 villa Poirier, Paris 15, France.*

SCHÜTZENBERGER, Paul, French chemist; b. Strasbourg, France, Dec. 23, 1829; s. Georges Frédéric Schützenberger; M.D., Strasbourg, 1855. Became préparateur to Prof. J. F. Persoz, Consevatoire des Arts et Métiers, 1853; became instr. chemistry Ecole superieure des sciences, Mulhausen, 1854, prof., 1865; asst. to Prof. A.V. Balard, Collège de France, prof. chemistry, 1876-97; directing prof. Municipal ecole de physique et de chemie, Paris, 1882-97. Mem. French Acad Scis., French Acad. Medicine. Author: Chimie appliquée à la physiologie animale à la pathologie et au diagnostic médical, 1864; Traité des matières colorantes, 1867; Leçons professées a la Société chimique, 1868-70; Sur le role de l'acide hypochloreux en chimie organique, 1871; Les fermentations, 1875; Traité de chimie générale, 1879-90. Research on physiol. chemistry, dyes, fermentation, proteins; isolated xanthopurpurine, purpine, pseudopurpurine, carminic acid; discovered carbonyl compounds of platinum, hyposulfurous acid; prepared cyanhydrines of glucose and levulose and hydrolyzed them to acids (led to method for proving constitution of sugars). Died Mézy, France, June 26, 1897.

SCHWAB, Georg Maria, German phys. chemist; b. Berlin, Feb. 3, 1899; s. Josef Berhard and Mary (Köglmayr) S.; Dr.phil., U. Berlin, 1923; Dr.h.c., Free U. Berlin, 1960, U. Paris, 1965, U. Liège, 1967; m. Elly Agallidis, Nov. 12, 1939; children—Maria, Andreas, Johanna. Asst., U. Berlin, 1923-25, U. Würzburg (Germany), 1925-27; faculty U. Munich (Germany), 1928-38, prof., 1933-38, conservator, 1928-38; chief research dept. Inst. N. Kanellopoulos, Piraeus, Greece, 1939-50; prof. phys. chemistry U. Munich, 1950—; prof. Tech. U. Athens (Greece), 1949. Decorated Officer Belgian Crown, St. George Cross of King of Greece. Mem. Internat. Union Physicists and Chemists (past pres. div. for phys. chemistry), German Chem. Soc. (pres. working group semiconductor chemistry 1963—, Liebig medal 1960), Fraunhafer Soc. Applied Research (pres. research council since 19——), Faraday Soc., Bunsen Soc., Swiss Chem. Soc. Author: Technology, 1927; Katalyse, 1931; Handbook of Catalysis, 1940; (with H. S. Taylor, R. Spence), Catalysis, 1937; also numerous articles. Research on pure ozone, active centers in catalysts, solid state chem. reactions, inorganic analysis by chromatography, electron exchange during catalytic reactions, mixed catalysts. Home: 9 St. Pauls Sq., Munich, Germany 8000.*

SCHWAB, John Joseph, Am. psychiatrist; b. Cumberland, Md., Feb. 10, 1923; s. Joseph L. and Eleanor (Cadden) S.; B.S., U. Ky., 1946; M.D., U. Louisville, 1946; M.S., in physiology (Med. fellow), U. Ill., 1949; postgrad. Duke, U. Fla.; m. Ruby Baxter Schwab, Apr. 4, 1945; 1 dau., Mary Eleanor. Internist, psychosomaticist Holzer Clinic, Gallipolis, O., 1954-59; Nat. Inst. Mental Mental Health Career tchr. U. Fla., Tallahasse, 1962-64, faculty, 1962-—, professor psychiatry and medicine, 1967—, dir. consultants liason program, 1964-67. chief Psychol. Consultation Service, Gainesville, Fla., 1961-64. Diplomate Nat. Bd. Med. Examiners. Mem. A.M.A., Fla. Med. Assn., Am. Psychiat. Assn., Psychosomatic Soc., Acad. Psychosomatic Medicine (exec. 1965—), Alpha Omega Alpha. Author: The Handbook for the Psychiatric Consultation, 1967; also articles. Asso. editor Psychosomatics, 1965—. Research on applicability of psychiat. concepts to gen. medicine, sociocultural aspects of mental illness; establishing guidelines for identification and mgmt. of med. patients whose illnesses are complicated by emotional distress. Home: 4500 Clear Lake Dr., Gainesville, Fla. 32601.*

SCHWABE, Heinrich Samuel, German astronomer; b. Dessau, Germany, Oct. 25, 1789; ed. Berlin; practiced pharmacy; began systematic daily survey of sun's surface, 1726, culminating in discovery of 10 year sun-spot cycle (later shown to be longer), 1843. Died Dessau, Apr. 11, 1875.

SCHWABE, Walter Wolfgang, plant physiologist; b. Berlin, Germany, June 1, 1920; s. Walter and Anne (Lagershausen) S.; B.Sc., Imperial Coll., U. London (Eng.), 1945, Ph.D., Diploma, 1949, D.Sc., 1957; m. Valerie Ellen Bramley, Nov. 15, 1958; children—Fiona, Anne Eileen, Ruth Karen Irene, John Walter Richard. Sci. officer Inst. Plant Physiology, Imperial Coll., Rothamsted Expt. Sta., 1946-58; prin. sci. officer unit plant morphogenesis and nutrition Agr. Research Council, Rothamsted and Wye Coll., 1958-65; prof. horticulture Wye Coll., U. London, Ashford, Kent, Eng., 1965—. Recipient Imperial Coll. Huxley Meml. medal and prize, 1958. Mem. Soc. for Exptl. Biology, Soc. for Devel. Biology, Inst. Biology Scandinavian Soc. for Plant Physiology. Research, publs. on effects of external environment growth and flowering in higher plants, partial analysis of underlying mechanisms. Home: Audlea, Bilting nr. Ashford. Office: Dept. Horticulture, Wye Coll., London U., Wye, Ashford, Kent, Eng.*

SCHWALBE, Gustav, German anatomist, anthropologist; b. Quedlingburg, Germany, Aug. 1, 1844; prof. in Leipzig, Germany, Jena, Germany, Königsberg (now Kalininberg, USSR), Strasbourg, France. Author: Lehrbuch der Neurologie, 1881; Lehrbuch der Anatomie der Sinnesorgane, 1887; Studien über den Pithecanthropus erectus, 1899; Der Neandertalerschädel, 1901; Die Vorgesch. des Menschen, 1904; (with V. E. Fischer) Anthropologie, 1923. Research on racial devel. of man, especially fossils; studied Neanderthal skull, skull remains from S. Am.; his result proved continuous devel. of homonoids. Died Strasbourg, Apr. 23, 1917.

SCHWAN, Herman Paul, biophysicist; b. Aachen, Germany, Aug. 7, 1915; s. Wilhelm and Meta (Pattberg) S.; student U. Goettingen, 1934-37; Ph.D., Frankfurt, 1940; Doctor habil. in physics and biophysics, 1946; m. Anne Marie DelBorello, June 15, 1949; children—Barbara, Margaret, Steven, Carol, Cathryn. Came to U. S., 1947, naturalized, 1952. Research scientist, prof. Kaiser Wilhelm Inst. Biophysics, 1937-47, asst. dir., 1945-47; research sci. U. S. Navy, 1947-50; prof. elec. engring., prof. elec. engring. in phys. medicine, asso. prof. phys. medicine U. Pa., 1950—, dir. electromed. div., 1952—, chmn. biomed. engring., 1962—; vis. prof., U. Cal. at Berkeley, 1956, University of Frankfurt (Germany), 1962; lectr. Johns Hopkins U., 1962. Fgn. sci. mem. Max Planek Soc. Adv. Research, Germany, 1962—; cons. NIH, 1962—; chmn. nat. and internat. meetings biomed. engring. and biophysics; 1959, 1960, 1961, 1965. Recipient Citizenship award, Phila., 1952; 1st prize Am. Inst. E.E., 1953; spl. award Phila. Inst. Elec. Engring. and Electronics, 1963. Fellow I.E.E.E. (chmn. and vice chmn. nat. profl. group biomed. engring. 1955, 62—; Morlock award 1967), A.A.A.S.; mem. Am. Standards Assn. (chmn. com. 1961-64), Am. Phys. Society, Biophysical Soc., Soc. for Cryobiology, Internat. Federation Medical, Electric and Biological Engineering, Sigma Xi. Co-author: Advances in Medical and Biological Physics, 1957; Therapeutic Heat, 1958; Physical Techniques in Medicine and Biology, 1963. Editor: Biol. Engring., 1968. Mem. editorial bd. Biophys. Zeitschrift, Jour. Med. Electronics Biol. Engring., Phys. Med. Biology; Transactions Biomed. Engring., Internat. Jour. Biomed. Engring., Biophysik; articles profl. jours. Developed special impedance techniques for biol. dielectric research; research on elec. and acoustical properties biol. material, capacitance and conductance changes of biol. membranes at very low frequencies; biophysics of microwave radiation (determination of physical parameters describing interaction of microwaves with biomatter). Home: 99 Kynlyn Rd., Radnor, Pa. 19088; also 59th St., Avalon, N.J. Office: Dept. Biomed. Engring., Moore Sch., U. Pa., Phila. 19104.*

SCHWANITZ, Franz, German botanist; b. Danzig, Aug. 6, 1907; s. August and Anna (Rybitzki) S.; Prof. Dr., U. Göttingen, 1927-31; m. Hanna Homann, May 15, 1938; children—Gesa, Hilke (Mrs. D. Winterhoff), Wiebke. Asst., Landwirtschaftl. Institut der Tech. Coll. Danzig, 1932-36; asst. Kaiser Wilhelm Institut fur Züchtungsforschung, Müncheberg, 1936-39; leader dept., 1939-49; dep. leader Max Planck Institut fur Bastfaserforschung, 1949-54; prof. botany U. Hamburg, 1954-60; dir. Inst. for Botany and Microbiology, Kernforschungsanlage, Jülich, 1961—. Hon.

prof. U. Münster. Author: Die Entstehung der Kulturpflanzen, 1957; Die Evolution der Kulturpflanzen, 1967. Research on polyploidy, radiation biology, analysis and diminution of radiation induced damages. Home: 2 Schlehdornweg, 517 Jülich. Office: Inst. fur Botanik und Mikrobiologie, Kernforschungsanlage, Jülich, Germany.*

SCHWANN, Theodor, German anatomist; b. Neuss, Germany, Dec. 7, 1810; studied under Johannes Müller; became prof. anatomy, Louvain, Belgium, 1848; became prof. physiology, Liége, Belgium, 1858. Fellow Royal Soc., 1879 (Copley medal 1845); mem. French Acad. Scis., 1879. Author: Mikroskopische Untersuchungen über die Übereinstimmung in der Struktur und dem Wachstum der Thiere und Pflanzen, 1839; Über das Wesen des Verdauungsprocesses, Archiv für Anatomische, Physiologische, und wissenschaftlichen Medicine, 1836. Recognized egg as cell, 1838; demonstrated all organisms were constructed of cells with individual limited life span; showed (with Jean Müller) that fermentation is produced by living organisms; described casing of nerve fibers (Schwann's envelope); studied artificial digestion, muscular contraction; discovered pepsin in gastric juice (1st enzyme to be prepared from animal tissue); his work laid foundations for modern histology. Died Cologne, Germany, Jan. 14, 1882.

SCHWARTE, Louis Harold, Am. microbiologist; b. Saratoga, N.Y., July 29, 1896; s. John C. and Virginia M. (Miller) S.; B.S., Cornell U., 1918, M.S., 1920; D.V.M., Ia. State U., 1928, Ph.D., 1934; m. Ella R. Murphy, June 2, 1934; 1 dau., Mary Kay. Pathologist, R.I. State U., 1928-30; faculty Ia. State U., Ames, 1930—, prof., 1945—. Cons. Ames lab. AEC, Nat. Animal Disease Lab. Mem. Osborn Research Club, Conf. Research Workers in Animal Diseases, Am. Ia. vet. med. assns., Ia., N.Y. acads. sci., Am. Pub. Health Assn., A.A.A.S., Internat. Acad. Pathology, U.S. Livestock San. Assn. Am. Assn. Avian Pathology, Sigma Xi, Gamma Sigma Delta, Phi Lambda Upsilon, Phi Zeta, Phi Kappa Phi. Research, publs. on virus infections, diagnosis, treatment and immunity, neurotropic infections and central nervous disturbances. Home: 1103 Roosevelt Av., Ames, Ia. 50010.*

SCHWARTING, Arthur Ernest, Am. pharmacognosist; b. Waubay, S.D., June 8, 1917; s. John E. and Johanna (Boelte) S.; B.S., S.D. State U., 1940; Ph.D., Ohio State U., 1943; m. Roberta L. Mitchell, June 14, 1941; children—J. Michael, Stephen A., Gerald A. Instr., U. Neb., 1943-45, asst. prof., 1945-49; faculty U. Conn., Storrs., 1949—, prof. pharmacognosy, 1953—. Recipient U. Conn. Alumni award for research and teaching excellence, 1966. Mem. Am. Pharm. Assn. (award for research achievement 1964), Am. Soc. Pharmacognosy, A.A.A.S., Am. Soc. Plant Physiologists. Author: (with V. E. Tyler, jr.) Experimental Pharmacogosy, 1955; also articles. Editor: Lloydia, 1961—. Research on therapeutic constituents in plants, fermentation prodn. of drugs, biosynthetic origin of drugs.*

SCHWARTZ, Abraham, Am. mathematician; b. N.Y.C., June 13, 1916; s. Nathan and Lena (Swoff) S.; B.S.S., Coll. City N.Y., 1936; M.S., Mass. Inst. Tech., 1937, Ph.D., 1939; m. Sylvia Paymer, June 16, 1940; children—Ina Barbara, Susan Doris, Margery Helen. Research asst. Inst. for Advanced Study, Princeton, N.J., 1939-41; instr. Pa. State Coll., State College, 1941-44; asst. prof., 1944-48; faculty Coll. City N.Y., 1948—, prof. math., 1966—, chmn. dept., 1964-66, dir. Sch. Gen. Studies, 1966—. Mem. Am. Math. Soc., Math. Assn. Am. (chmn. local sect. 1963-65), Am. Assn. U. Profs., Phi Beta Kappa, Sigma Xi. Author: Analytic Geometry and Calculus, 1960. Research, publs. in classical differential geometry. Home: 321 Windsor Rd., Englewood, N.J. 07631. Office: 139th St. at Convent Av., N.Y.C. 10031.*

SCHWARTZ, Irving Leon, Am. physiologist; b. Cedarhurst, N.Y., Dec. 25, 1918; s. Abraham and Rose (Doniger) S.; B.A., Columbia, 1939; M.D., N.Y. U., 1943; m. Felice T. Nierenberg, Jan. 12, 1946; children—Cornelia Ann, Albert Anthony, James Oliver. Fellow in physiology and medicine N.Y. U., 1947-50; fellow Rockefeller Inst., N.Y.C., 1950-52, asst., 1952-54, asso., 1954-58; sr. scientist Brookhaven Nat. Lab., Upton, N.Y., 1958-61; prof., chmn. dept. physiology U. Cin. Coll. Medicine, 1961-65; dean Grad. Sch., The Mt. Sinai Med. and Grad. Schs., N.Y.C., 1965—; pres. Life Scis. Found., Inc., N.Y.C., 1963—; research collaborator Brookhaven Nat. Lab., 1961—. NIH fellow, 1947-49; Dazian Found. fellow, 1949-50; Am. Heart Assn. fellow, 1951-52; Am. Physiol. Soc. Porter fellow, 1950-51. Mem. Am. Physiol. Soc., Biophys. Soc., Soc. Exptl. Biology and Medicine, Am. Soc. Clin. Investigation, Endocrine Soc., A.A.A.S., Harvey Soc., Sigma Xi. Research, publs. on methods for measuring distbn. body water and other small molecular constituents tissue, analysis secretion processes in excrine glands, gen. cellular membrane and transport phenomena, mechanism action peptide hormones posterior pituitary gland and insulin, hormonal influences on fat and carbohydrate metabolism, regulation hunger, appetite and satiety. Home: 10 E. End Av., N.Y.C. 10021.*

SCHWARTZ, Melvin, Am. physicist; b. N.Y.C., Nov. 2, 1932; s. Harry and Hannah (Shulman) S.; A.B., Columbia, 1953, Ph.D., 1958; m. Marilyn Fenster, Nov. 25, 1953; children—David, Diane, Betty Lynn. Asso. physicist Brookhaven Nat. Lab., Upton, N.Y., 1956-58; faculty Columbia, N.Y.C., 1958-66, prof., 1963-66; prof. Stanford, 1966—. Vice. pres., dir. Brevatome U.S.A., Inc. Fellow Am. Phys. Soc. (prize 1964). Research, publs. on high energy neutrinos demonstrating non-identity of muon and electronneutrinos; discovery E^0 hyperon, measurement of hyperon spins. Home: 2935 Pacific Av., San Francisco 94115. Office: Stanford U., Stanford, Cal. 94305.*

SCHWARTZ, Morton K., Am. biochemist; b. Wilkes-Barre, Pa., Oct. 22, 1925; s. Norman and Dorothy (Kanter) S.; B.A., Lehigh U., 1948; M.A., Boston U., 1949, Ph.D., 1952; m. June Epstein, Aug. 14, 1954; (dec.); children—Gary, Ronald. Chief biochemist L.I. Jewish Hosp., New Hyde Park, N.Y., 1956-57; asst. attending clin. biochemist dept. biochemistry Meml. Hosp for Cancer and Allied Diseases, 1957-58, asso. attending biochemist, 1958-67, attending biochemist, chmn. dept., 1967—; research fellow Sloan-Kettering Inst. for Cancer Research, N.Y.C., 1952-54, staff, 1955—, asso. mem., 1960—; faculty Cornell U. Med. Coll., Ithaca, N.Y., 1953—, asso. prof. biochemistry, 1962—; lectr. Hunter Coll., 1958-60, adj. prof., 1960-64. Mem. Am. Assn. for Cancer Research, Am. Assn. Clin. Chemists, Am. Chem. Soc., Am. Soc. Biol. Chemists, A.A.A.S., Sigma Xi. Research, numerous publs. on serum enzymes in patients with cancer, automation methods and studies kinetics enzymes. Home: 21 Chatham Pl., White Plains, N.Y. 10605. Office: 410 E. 68th St., N.Y.C. 10021.*

SCHWARTZ, Samuel, Am. research physician; b. Mpls., Apr. 13, 1916; s. Morris and Ida (Gorenberg) S.; B.S., U. Minn., 1938, M.D., 1943; m. Goldie Leader, June 17, 1937; children—Marcia (Mrs. Uri Kreisel), Ruth (Mrs. Jonathan Paradise), Jacky (Mrs. Arnold Turchick), David, Daniel, Michael, Joel, Jeremy, Rivka. Group leader Manhattan Project, U. Chgo., 1943-46; Commonwealth Fund fellow, Denmark, Sweden, 1946-48; faculty U. Minn., Mpls., 1948—, research prof. medicine, 1961—; vis. prof. in exptl. medicine and cancer research Hebrew U., Israel, 1961-62. Recipient Research Career award USPHS, 1963. Mem. Am. Soc. Biol. Chemists, Am. Soc. Clin. Investigation, Am. Assn. for Cancer Research. Research, numerous publs. in improved methods for analysis of blood and bile pigiments, studies of their formation and excretion in humans and exptl. animals, biol. effects of compounds derived from blood pigments, their possible use in treatment or diagnosis of cancer. Address: 2341 Texas Av., South Minneapolis, Minn. 55426.*

SCHWARTZ, Seymour I., Am. physician; b. N.Y.C., Jan. 22, 1928; s. Samuel and Martha (Paull) S.; B.A., U. Wis., 1947; postgrad. Syracuse U. Coll. Medicine; M.D., N.Y. U., 1950; m. Ruth Wainer, June 18, 1949; children—Richard, Kenneth, David. Faculty dept. surgery U. Rochester (N.Y.), 1957—, prof., 1967—; with Strong Meml. Hosp., Rochester, 1957—, sr. surgeon, 1967—, John and Mary R. Markle scholar in acad. medicine, 1960, dir. surg. research, 1962—. Mem. A.C.S., Am., Central surg. assns., Internat. Surg. Soc., Soc. U. Surgeons, Soc. Vascular Surgery, Internat. Cardiovascular Soc., Am. Assn. Thoracic Surgeons, Phi Beta Kappa, Sigma Xi, Alpha Omega Alpha. Author: Surgical Diseases of the Liver, 1964. Reseach, numerous publs. on blood coagulation, thrombosis, portal hypertension, essential hypertension; inventor carotid sinus nerve stimulator for reversal of hypertension, dead-space rebreathing tube for prevention of atelectasis. Home: 121 Southern Pkwy., Rochester 14618. Office: 260 Crittenden Blvd., Rochester, N.Y. 14620.*

SCHWARTZ, Steven Otto, physician; b. Kapolch, Hungary, July 6, 1911; s. Otto and Henrietta (Lemberger) S.; came to U. S., 1923, naturalized, 1929; B.S., Northwestern U., 1932, M.S., 1935, M.D., 1936. Research asso. hematology Beth Israel Hosp., Boston, 1937-38; Charlton Research fellow medicine Tufts Coll. Med. Sch., Boston, 1937-38; practice medicine specializing in hematology, Chgo., 1938—; dir. hematology dept., dir. Hektoen Inst. for Med. Research, Cook County Hosp., Chgo., 1939-68; asso. hematology Michael Reese Hosp., 1938-50; sr. attending physician Chgo. Wesley Meml. Hosp., 1955—; cons. hematologists Hines VA Hosp., 1956—; prof. internal medicine Cook County Postgrad. Sch., 1939-68; asst. prof. medicine U. Ill., 1942-47; prof. hematology Chgo. Med. Sch., 1947-55; faculty Northwestern U., 1955—, prof. medicine, 1959—. Diplomate Am. Bd. Internal Medicine. Fellow A.C.P., A.A.A.S.; mem. A.M.A., Am. Fedn. for Clin. Research, Chgo. Soc. Internal Medicine, Internat. (charter), European (corr.) socs. hematology, Acad. Internat. Medicine, Am. Assn. for Cancer Research, Am. Hematology Soc., Ill. (Gold medal 1955), Chgo. med. socs., Inst. Medicine, Chgo. Author: Hematology in Practice, 1961; also numerous articles. Research on hemolytic anemias, viral etiology of human leukemia. Home: 2185 Linden Av., Highland Park, Ill. 60035. Office: 720 N. Michigan Av., Chgo. 60611.*

SCHWARTZ, Wilhelm August Carl, German microbiologist; b. Radebeul-West, Saxony, Germany, Oct. 4, 1896; s. Carl Johannes and Antoinette (Saeger) S.; Dr.phil., U. München (Germany), 1922; Habilitation, Tech. U. Karlsruhe (Germany), 1928; m. Hanna Kraepelin, 1923 (div. 1930); children—Andreas Wera Stabel, Gunda Schwartz-Kraepelin; m. 2d, Edith Schulz 1935 (div. 1959); children—Arnold, Karin; m. 3d, Anna Adelheid Mueller, Aug. 14, 1959; children—Elke-Maria Dorothea. Asst., Botanisches Institut Marburg (Germany), 1922-24, Gärungsphysiologisches Institut Weihenstephan, 1924-26; staff Regierungsbotaniker, Landwirtschaftl. Versuchsanstalt Augustenberg, 1926-29; dir. Botanisch-Mikrobiologisches Institut, Tech. U. Karlsruhe, 1929-44; dir. Microbiol. Lab. Mahlum/Harz, 1944-55; dir. chair microbiology U. Greifswald (Germany), 1955-65; vis. asso. prof. Oceanographic Inst., Fla. State U., Tallahassee, 1966—. Mem. Deutsche Gesellschaft für Hygiene and Mikrobiologie, Am. Soc. Microbiologists, Geochem. Soc., Internat. Gesellschaft für Vitalstoff-Forschung. Author: (with Adelheid Schwartz) Allegemeine Mikrobiologie I/II, 1960, 61; also numerous articles. Editor: Zeitscher. f. allgem. Mikrobiol., vol. I, 1960-61. Research on gen. microbiology, ecology of microbes, symbiosis, geomicrobiology. Home: Hamburg-Blankense, Babendiekstr. 55, Germany. Office: Oceanographic Inst., Fla. State U., Tallahassee 32306.

SCHWARZ, Eugene Amandus, entomologist; b. Leignitz, Silesia, Apr. 2, 1844; s. Amandus and Luise (Harnwolf) S.; pupil of Gerhardt; came to U. S., 1872; tchr. Harvard; with U. S. Bur. Entomology, 1881-1926, custodian of Coleoptera, U. S. Nat. Mus., 1898-1926; founder (with H. G. Hubbard) Detroit Sci. Assn.; founder Entomol. Soc. Washington, hon. life pre. pres., 1916—. Contbr. numerous papers on Coleoptera and other biol. observations. Died Oct. 15, 1928.

SCHWARZ, Guenter, Am. physicist; b. Cologne, Germany, Nov. 26, 1913; s. Heinrich and Grete (Rosenbaum) S.; engring. diploma Tech. U. Berlin, 1938; Ph.D., Johns Hopkins, 1942; m. Inge Salomon, Aug. 9, 1942; children—Eva Harriet, Judith Ann. Came to U.S., 1938, naturalized, 1944. Faculty, U. Ill., 1942-46; research physicist Johns Hopkins, 1946-49; prof. physics Fla. State U., Tallahassee, 1949—, dir. Center for Research in Coll. Instrn. of Sci. and Math., 1966—. Vis. prof. Reed Coll., So. Meth. U., Andhra U., Waltair, India. Fellow A.A.A.S., Am. Phys. Soc. Research, publs. on fundamental properties of X-rays, magnitude of fundamental constants, teaching methods. Home: 701 Waverly Rd., Tallahassee, Fla. 32303.*

SCHWARZ, Helmut Julius, physicist; b. Wuppertal, Germany, Nov. 20, 1915; s. Gottlieb and Maria (Platte) S.; student U. Cologne (Germany), 1936-38; Ph.D. in Physics, U. Bonn (Germany), 1940; m. Ruth Else Mengel, Aug. 12, 1950; children—Yorck-Wedigo Michael, Arno. Faculty, U. Bonn, 1940; with German Air Force, 1940-42, tech. acad., 1942-44; mem. staff High Frequency Research, German Govt., 1944-45; dir. pvt. found. phys. sci., Wuppertal, 1945-49; faculty Centro Brasileiro de Pesquisas Fisicas, Rio de Janeiro, 1950-54, U. Rio de Janeiro, Brazil, 1951-57; also faculty Coll. Brazil Air Ministry and dir. Inst. Applied Physics, Centro Tecnico Aeronautica, S. Jose dos Campos, Brazil, 1955-57; dir. projects electronic physics RCA, 1957-60; sr. staff research scientist United Aircraft Corp., East Hartford, Conn., 1960-63; prof. physics Rensselaer Poly. Inst. Hartford Grad. Center, East Windsor Hill, Conn., 1963—; cons. various bus. firms. Fellow Am. Phys. Soc.; mem. German Phys. Soc., Am. Vacuum Soc., Internat. Orgn. Vacuum Sci. and Tech. (founding mem.), Electron Microscopy Soc. Am., I.E.E.E. (sr. mem.). Research and publs. on vacuum physics, especial ion pumping, charged particle optics, high-power-density electron beams; observation of non-linear reflection of electrons by laser beam. Home: 49 Carver Circle, Simsbury, Conn. 06070. Office: Rensselaer Poly. Inst. Hartford Grad. Center, East Windsor Hill, Conn. 06028.*

SCHWARZ, Hermann Amandus, German mathematician; b. Hermsdorf, Germany, Jan. 25, 1843; doctorate, Berlin, 1864; prof., Halle Germany, 1867-69, Zurich, Switzerland, 1869-75, Gottingen, 1875-92; prof. math., Berlin. Mem. French Acad. Scis. Collected papers pub. in 2 vols., 1890. Research in differential geometry, theory of functions, theory of minimal surfaces, elliptical functions, variable complexes and variation calculus; Schwarz inequality and Schwarz-Christoffel mapping in function of complex variable named after him. Died Berlin Germany, Nov. 30, 1921.

SCHWARZENBACH, Gerold, Swiss chemist; b. Horgen, Switzerland, Mar. 15, 1904; s. Johannes and Lina (Hotz) S.; diploma as Engring. Chemist, Swiss Fed. Inst. Tech., 1926, Dr.Sc.tech., 1928; m. Martha Tobler, Sept. 2, 1930 (dec. 1940); children—Ursula, Kurt, Dieter; m. 2d, Erica Schoch, Aug. 21, 1941; 1 dau., Annette. With U. Zürich, (Switzerland), 1931-55, prof., 1948-55; prof., dir. Inst. for Inorganic Chemistry, Swiss Fed. Inst. Tech., Zurich, 1955—; vis. prof. Ia. State Coll., Ames, 1951. Hon. mem. Danish Royal Acad. Sci., Am. Acad. Art and Sci., Deutsche Akademie der Naturforscher, Leopoldina. Recipient Marcel Benoist prize, 1964; Gold medal Talanta, 1964; Paul Karrer Medaille, U. Zurich, 1966. Author: Textbook of Inorganic Chemistry, 1941; Tables on Stability Constants of Metal Complexes, 1957; Die Komplexometrische Titration, 1955, 66; also numerous articles. Research on relationship between acidity and structure, color and structure, metal coordination chemistry; invented complexometric titration procedures. Home: 10 Schäppistrasse, 8006, Zürich, Switzerland.*

SCHWARZSCHILD, Karl, German astronomer; b. Frankfort/Main, Germany, 1873; prof., Göttingen, Germany; dir. astrophysics, Potsdam, Germany, 1909-16. Research on comets, preferential stellar motions (leading to his ellipsoidal hypothesis); studied equilibrium of stellar gas mass and developed theory of stellar atmosphere; used Einstein's equations to demonstrate that very massive bodies can gravitate themselves out of observable universe (Schwarzchild singularity), 1916. Died May 11, 1916.

SCHWARZSCHILD, Martin, astronomer; b. Potsdam, Germany, May 31, 1912; s. Karl and Else (Rosenbach) S.; Ph.D., U. Goettingen (Germany), 1935; D.Sc. (hon.), Swarthmore Coll., 1960; m. Barbara Cherry, Aug. 24, 1945. Came to U.S., 1937, naturalized, 1942. Research fellow U. Oslo (Norway), 1936-37, Harvard Coll. Obs., 1937-40; faculty Columbia, 1940-47; prof. astronomy Princeton, 1947——. Recipient Newcomb Cleveland prize A.A.A.S., 1957; Rittenhouse Astron. Soc. silver medal, 1966. Fellow Am. Acad. Arts and Scis.; mem. Am. (past councillor), Royal (Eddington medal 1963) astron. socs., Internat. Astron. Union (v.p. 1964——, Royal Astron. Soc. Can., Akademia der Naturforscher Leopoldina, Astron. Soc. Pacific (Bruce medal 1965), Nat. Acad. Scis. (Draper medal 1961), Royal Soc. Scis. Liege (corr.), Royal Netherlands Acad. Scis. and Letters (fgn.). Author: Structure and Evolution of the Stars, 1958; also numerous articles. Research on internal constn. stars and their evolution; astron. observations with telescopes carried by balloons into stratosphere. Home: 12 Ober Rd., Princeton, N.J. 08540.*

SCHWEBER, Silvan Samuel, Am. physicist; b. Strasbourg, France, Apr. 10, 1928; s. David and Dora (Edelman) S.; came to U.S., 1942, naturalized, 1951; B.S., Coll. City, N.Y., 1947; M.S., U. Pa., 1949; Ph.D., Princeton, 1952; m. Miriam Fields, June 14, 1965. Proctor fellow Princeton, 1950-51, instr., 1951-52; NSF fellow, Cornell U., 1952-54; sr. research physicist Carnegie Inst. Tech., 1954-55; faculty Brandeis U., Waltham, Mass., 1955——, prof. physics, 1961——, chmn. dept. physics, 1959-61, 64-65, chmn. Sch. Sci., 1963-64; vis. prof. Mass. Inst. Tech., 1961-62. Mem. Am. Phys. Soc., Math. Assn. Am., A.A.A.S., Phi Beta Kappa, Sigma Xi. Author: (with Bethe, de Hoffman) Filed Theory, 1955; Relativistic Quantum Field Theory, 1961. Research on interactions and structure elementary particles. Home: 22 Turning Mill Rd., Lexington, Mass. 02173. Office: Dept. Physics, Brandeis U., Waltham, Mass. 02154.*

SCHWEET, Richard S., Am. biochemist; b. N.Y.C., Oct. 6, 1918; s. Jacob and Betsey (Goldman) S.; B.S., Coll. City N.Y., 1938; M.S., Ia. State Coll., 1941, Ph.D., 1950; m. Shirley Woods, June 6, 1942; children—Lynn, Laurin, Richard. Fellow, Enzyme Inst., U. Wis., 1950-51; research fellow Cal. Inst. Tech., 1951-53, sr. research fellow, 1953-57; chief cardiac chemistry sect. City of Hope Med. Center, Duarte, Cal., 1957-60; prof. biochemistry U. Ky. Med. Sch., Lexington, 1960——, chmn. dept. cell biology, 1965——; established investigator Am. Heart Assn., 1956-59. Mem. fellowships rev. bd. USPHS, 1962-64, mem. physiol. chemistry study sect., 1965. Recipient Career Research award USPHS, 1960-65; mem. test com. Nat. Bd. Med. Examiners. Mem. Am. Soc. Biol. Chemists, Am. Chem. Soc., N.Y. Acad. Scis., Sigma Xi. Editorial bd. Archives Biochemistry and Biophysics, editor, 1965——; adv. bd. Biochemistry Genetics. Research, numerous publs. on pyruvate metabolism, amino acid metabolism, mechanisms protein synthesis. Died Apr. 3, 1967.

SCHWEIGERT, Bernard Sylvester, Am. biochemist; b. Alpha, N.D., Mar. 19, 1921; s. John S. and Barbara (Busche) S.; B.S., U. Wis., 1943, M.S., 1944, Ph.D., 1946; m. Alta Goede, Sept. 25, 1943; children—James Bernard, John Frederick. Research and teaching asst. U. Wis., 1942-46; asso. prof. biochemistry and nutrition Tex. A. and M. U., 1946-48; chief div. biochemistry and nutrition, asst. dir., dir. research and edn. Am. Meat Inst. Found., U. Chgo., 1948-60, asst. and asso. prof. biochemistry, 1948-60; prof., chmn. dept. food sci. Mich. State U., East Lansing, 1960——. Mem. food and nutrition bd., chmn. com. on marine protein resource devel. Nat. Acad. Scis., 1966——; chmn. sci. adv. com. on irradiation pasteurization of foods Am. Inst. Biol. Scis.-AEC, 1961——. Mem. Dairy Sci. Assn., Am. Inst. Nutrition, Inst. Food Tech., Am. Soc. Biol. Chemistry, Am. Chem. Soc., A.A.A.S., Am. Soc. Animal Sci. Research, numerous publs. on food sci., biochemistry and nutrition relating to nutritive value of foods, flavor changes in foods, biosynthesis of amino acids and nucleic acids, vitamin B12 and irradiation preservation of foods. Home: 1750 Woodside Dr., East Lansing, Mich. 48823.*

SCHWEIGGER, Johann Salomo Christoph, German physicist; b. Erlangen, Germany, Apr. 8, 1779; tchr. Erlangen; became successively prof. math. and physics Gymnasium of Bayreuth, 1803, Polytechnic Sch. Nuremberg, Germany, 1811, U. Erlangen, 1817, Halle, Germany, 1819. Founder, Jahrbuch für Chemie und physik, 1811; continued it under title Annales de chemie et de physique), 1811. Invented electrometer, using magnetic needle for measurement of electric force, 1808, string galvanometer, electromagnetic multiplier named after him. Died Halle, Sept. 6, 1857.

SCHWEIKART, Ferdinand Karl, German mathematician; b. Erbach, Germany, Feb. 28, 1780; became prof. law, Marburg, Germany, 1816, Giessen, Germany, 1809, Charkov, 1812, also Königsberg (now Kaliningrad, USSR), 1821-57. Author: Stackel und Angel; The Theorie of Parallel lines, 1895. Contributed to creation of non-Euclidean system; studied astral geometry. Died Aug. 17, 1857.

SCHWEINFURTH, George, explorer, naturalist; b. Riga, Latvia, 1836; collected bot. specimens in Egypt, Ethiopia, Khartoum, 1863; made expdn. to Equatorial Africa, 1868; explored Nile, areas of Niams-Niams and Mombouttous in ty. drained by Bahr el-Ghazal and its tributaries; discovered Ouelle River; founder Egyptian Inst., Cairo; dir. Museums of Capitol of Europe; numerous voyages to Egypt and nearby regions; explored Arabian desert, 1876-80, 84-85, Socotora, 1881, valley of Nile to Syout, at Assouan, 1882, Cyrenica, 1883, Arabia, 1888, Italian Erythaea, 1891-92. Author: Im Herzen von Afrika, 1874; Artes Africanae, 1876; Flora von Aegypten, 1887; Auf unbetretenen Wegen in Aegypten, 1922; Afrikanisches Skizzenburch, 1925. Died Berlin, Germany, 1925.

SCHWEINITZ, George Edmond de, see de Schweinitz, George Edmond.

SCHWEITZER, Albert, med. missionary, musicologist; b. Kaiserberg, Alsace, Jan. 14, 1875; s. Louis and Adele (Schillinger) S.; Ph.D., U. Strasbourg, 1899, Licentiate in Theology, 1900; M.D., 1913, later student advanced courses in theology, medicine, obstetrics and dentistry; studied tropical medicine Med. Faculty, Paris, and attended lectures Inst. of Tropical Hygiene, Hamburg; studied organ under Charles Marie Widor, Paris; hon. degrees: D.D., U. Zurich, 1920; Ph.D., U. Prague, 1926, Oxford, 1932; LL.D., St. Andrews, 1932, U. Chgo., 1949; D.D. and D.Mus., U. Edinburgh, 1932; m. Helen Bresslau, June 18, 1912; 1 dau., Rhena. Began as preacher, also deacon and curate Ch. of St. Nicholas, Strasbourg, 1899; acting prin. Theol. Coll., U. Strasbourg, 1901, prin., 1903, privatdozent in theology, 1902-12; founder hosp. at Lambaréné, Gabon, French Equatorial Africa, 1913, and since served as its missionary surgeon; visitor in U. S., making guest addresses at Goethe Festival, Aspen, Colo., 1949. Awarded Goethe Meml. prize, Frankfurt, 1928; recipient Nobel Peace prize 1952. Has toured in Europe as lectr., organist in ch. and concert; also made recs. of Bach organ music. One of founders of Paris Bach Soc., 1905, organist, 1905-11. As musicologist his critical edit. of Bach's organ works and his biography of Bach are regarded as authoritative. Author of autobiographical works and numerous books relating to theol. and mus. subjects, among which recently published in U. S.: J. S. Bach (translation of the German issue), 1935; Philosophy of Civilization (Part 2, Civilization and Ethics), 1947; On the Edge of the Primevela Forest, and More From the Primeval Forest (both in one vol.), 1948; Quest of the Historical Jesus, 1948; Psychiatric Study of Jesus, 1948; Goethe; four studies, 1940; Out of My Life and Thought, 1949. Died Lambaréne, Gabon, French Equatorial Africa, Sept. 4, 1964.

SCHWEITZER, George Keene, Am. chemist; b. Poplar Bluff, Mo., Dec. 5, 1924; s. Francis J. and Ruth E. (Keene) S.; B.A., Central Coll., Mo., 1945; M.S., U. Ill., 1946, Ph.D. in Chemistry, 1946; M.A. in Religion (NSF fellow), Columbia, 1960; Ph.D. in Philosophy, N.Y. U., 1964; Sc.D., Central Meth. Coll., Mo., 1964; m. Verna L. Pratt, June 4, 1948; children—Ruth, Deborah, Eric. Faculty, U. Tenn., Knoxville, 1948-59, prof. chemistry, 1960——; NSF fellow Columbia, 1959-60. Cons. to indsl. firms, govt. agys. Mem. Am. Chem. Soc., Am. Philos. Assn., History Sci. Soc., Acad. Religion, Phi Beta Kappa, Sigma Xi. Author: Radioactive Tracer Techniques, 1949; The Doctorate, 1965; also numerous articles. Research on solvent extraction metal chelates, volatile inorganic compounds, complex compound solution equilibria, history and philosophy of sci., sci. epistemology, sci. and religion. Home: 224 Golf Club Rd., Knoxville, Tenn. 37919.*

SCHWENKFELD, Kaspar, German zoologist, theologian; b. 1490; ed. Cologne and other universities. In service of Duke of Liegnitz, to 1529; as theologian sought to establish a via media between doctrines of Luther and Zwingli; lived in Strasburg, 1529-34; resided in south Germany; persecuted by Lutherans and expelled from Ulm, 1539; withdrew from Lutheran Church, 1543; traveled about to insure his personal safety. Adopted mysticism in his religious views; in zoology, categorized slugs and snakes as reptiles; described vertebrates by combining Greek roots and Latin qualifiers. Died Ulm, Dec. 10, 1561.

SCHWENDEMAN, J(oseph) R(aymond), Am. geographer; b. Waterford, Ohio, May 11, 1897; s. Francis John and Margaret (Tornes) S.; B.S., Ohio U., 1926; A.M., Clarke U., Worcester, Mass., 1927, Ph.D., 1941; student U. of Minn., 1927-28; m. Eithnea O'Donnell, Oct. 12, 1921; children—Elaine, Marion, Gerald, Joseph, Francis, Beth Ann. Part-time instr. U. Minn., 1927-28; head dept. geography, State Teachers Coll., Moorhead, Minn., 1928-44; prof. and head dept. geog., U. Ky., 1944-67; cons., distinguished prof. geography Eastern Kentucky University, Richmond, Ky., since 1967——; vis. lecturer summers, Columbia Coll., 1927, State Teachers College, Spearfish, S.D., 1928, Ohio U., 1941, St. John's U., Collegeville, Minn., 1943; worked on preparation of land use suitability maps, atlas and state base maps for state of Kentucky. Served as instr. U. S. Army Air Corps, State Teachers Coll., Moorehead, Minn., 1943, Recipient citation Am. Assn. Geographers, 1967. Fellow Nat. Council Geography Tchrs.; mem. Cath. Commn. on Intellectual and Cultural Affairs, Kentucky Economic Development Bd., Am. Men of Sci., Assn. Am. Geographers, Am. Assn. Univ. Profs., Ky. Acad. Sci. Author publications relating to field; compiler (with Irwin Sanders) Societies Around the World, 2 vols., 1948; rev. 1956; Geography of Kentucky, 1958, rev., 1963. Editor, pub. Directory of Coll. Geography of U. S., Vols. 1-18, 1949-67. Contbr. chpt. Lower Ohio Valley, Regional Geography of North American Mid-west, 1956; World Political Geography; revised geog. section Lincoln Library, 1952 edition. Holder patents and copyrights. Directed student travel groups in Europe, Canada, United States, Mexico; organized and conducted expdn. for study of weather on volcano Orizaba, Mexico, 1949. Inventor Geomatic World Map Projection, 1963. Home: 3512 Greentree Dr., Lexington, Ky.*

SCHWENDENER, Simon, botanist; b. Buchs, Switzerland, Feb. 10, 1829; prof. in univs. of Basel, Switzerland, Tübingen, Germany, Berlin, Germany. Mem. French Acad. Scis. Fellow Royal Soc., 1913. Author: Das mechanische Prinzip in anat. Bau der Monokotylen, 1874; Die mechanische Theorie der Blattstellung, 1878; Über Bau und Mechanik der Spaltöffnungen, 1881; Gesammelte botanische Abhandlung (collected works), 1898; Memoires, posthumous 1922. Proved histologically that lichens are composed of algae and fungi; developed mech. theory for devel. and arrangement of plant tissues. Died Berlin, May 27, 1919.

SCHWENTER, Daniel, German mathematician; b. Nuremberg, Germany, Jan. 31, 1585; became prof. Hebrew, U. Altdorf (Germany), 1608, prof. oriental langs., 1625, prof. math., 1628. First to use continued fractions for calculations, 1627. Author: Deliciae physico mathematica oder Mathematische und philosophische Erquickstunden, posthumous 1636. Died Altdorf, Jan. 19, 1636.

SCHWERDTFEGER, Werner, meteorologist; b. Köln, West Germany, July 12, 1909; s. Otto D. and Helene (Hueck) S.; student U. Freiburg, 1927-29; Dr.phil., U. Leipzig (Germany), 1931; Dr.phil. Habil., U. Königsberg (Germany), 1937; m. Marianne Noack, Mar. 11, 1933; children—Dietrich, Antje (Mrs. Paul Wickel), Wulf. Meteorologist, German Weather Service, 1931-48; faculty U. Königsberg, 1937-38, U. Vienna (Austria), 1942-43, U. München (Germany), 1947-48; sci. cons. Servicio Meteorológico Nacional, Argentina, 1948-56, Hydrographic Office Argentine Navy, 1957-61; faculty U. Buenos Aires (Argentina), 1957; vis. scientist U. Melbourne (Australia), 1958, at various U.S. instns., 1959-60; prof. meteorology U. Wis., Madison, 1962——. Mem. Am., German meteorol. socs., Am. Geophys. Union, Sociedad Cientifica Argentina, Assn. Am. U. Profs. Author: El Problema de la Previsión del Tiempo, 1952; also numerous articles. Research on meteorol. problems So. Hemisphere, particularly Antarctica. Home: 4801 Woodburn Dr., Madison, Wis. 53711.*

SCHWERT, George William, Am. biochemist; b. Denver, Jan. 27, 1919; s. George William and Agnes (Alford) S.; B.A. summa cum laude, Carleton Coll., 1940; Ph.D., U. Minn., 1943; m. Margaret Maxfield Houlton, June 26, 1943; children—George William III, Janet Margaret. Biochemist, Sharp & Dohme, Inc., 1943-44; faculty Duke Sch. Medicine, 1946-59, prof., 1957-59; prof., chmn. dept. biochemistry U. Ky. Coll. Medicine, Lexington, 1959——. Cons. USPHS, 1959-63; mem. biochemistry panel Wooldridge Com., 1964. Markle scholar in med. sci., 1949-54. Mem. Am. Soc. Biol. Chemists, Am. Chem. Soc., Biochem Soc. (Gt. Britain), A.A.A.S., Sigma Xi, Phi Beta Kappa, Phi Lambda Upsilon. Editorial bd. Jour. Biol. Chemistry, 1964-68. Contbr.. numerous articles to tech. jours. Home: 3316 Braemer Dr., Lexington, Ky. 40502.*

SCHWERTASSEK, Karl, Czechoslovakian chemist, fibre researcher; b. Schatzlar, Czechoslovakia, Dec. 20, 1903; s. Richard and Lina (Suchy) S.; Dipl. Ing., Deutsche Technische Hochschule, Brno, Czechoslovakia, 1927, Dr.Tech.Sc., 1932; m. Hella Gottermeier, Apr. 23, 1935 (dec. Mar. 1942); children—Karin (Mrs. Slabihoudek), Karl Kristian, Martina. Sci. collaborator Textile Research Inst., Liberec, Czechoslovakia, 1929-35, dir, 1935-48; sci. collaborator Central Textile Inst., Brno, 1948-54; head dept. Knitting Industries Research Inst., Brno, 1955——. Author: Defects in Viscose Fabrics, 1953; also articles; co-author: Handbuch der Werkstoffprüfung, 1960; also 4 books on textile testing. Founder, iodine sorptiometry (enabling study of structure of fibres; devel. number of methods on fibre research. Home: 54, Veveri, Brno. Office: 6, Vaclavska, Brno, Czechoslovakia.*

SCHWIDETZKY, Oscar, O. R., surg. instrument maker; b. Konitz, Polish Corridor, Dec. 31, 1874; s. August and Fredericke (Mohnke) S.; LL.D., Fairleigh Dickinson U., 1953; m. Anna Hasselhuhn, 1916. Came to U. S., 1900, naturalized, 1917. Apprentice to maker steel and surg. dental instruments, Berlin; worked with instrument makers, Berlin, Crefeld, also Aix-la-Chapelle; studied orthopedic work; importer surg. supplies, Germany, 1902-11; with Dr. Horace Greeley organized Greeley Lab., 1911-13; with Becton, Dickinson & Co., Rutherford, N.J., 1913-—; dir. South Bergen Savs. & Loan Assn., Wood Ridge, N.J. Hon. mem. Burge Tb Clinic, Phila. Mem. Assn. Med. Directors, Internat. Anesthesia Research Soc. (hon.), American Surgical Trade Assn. (hon.). Author articles, pamphlets on med. subjects. Originator numerous surg. instruments, including continuous caudal needle, syringes, Ace bandages, hypodermic needles for use in med. treatment, alleviation of pain. Died Oct. 9, 1963.

SCHWIETE, Hans Ernst, German chemist; b. Osnabrück, Germany, May 27, 1902; s. Friedrich-August and Katharina (Ippendorf) S.; Dipl.-Chem., U. Halle (Germany), 1923; grad. U. Frankfurt/Main (Germany), 1930, lectr., 1940; m. Elfriede Heinz, Dec. 17, 1938; 1 son, Jörg Wolfgang. Chemist, Fa. van Baerle & Co., Worms, also Dt. Betonalges, Frankfurt/Main, 1924-28; pvt. asst. U. Frankfurt/Main, also vol. asst. Buderus'sche Eisenwerke, Wetzlar, Germany, 1928-30; sci. co-worker, dept. head Kaiser-Wilhelm Inst. for Silicate Research, Berlin, 1930-37; dir. Fa. van Baerle & Co., Gernsheim, Germany, 1938; dir. Fa. E. Schwenk, Cement Works, 1941-48; expert for cement, plaster and concrete Württenberg-Baden Ministry Econs., 1948-—; prof. glass and ceramics, dir. Inst. for Gesteinshüttenkunde, Rhine Westph. Technische Hochschule, Aachen, Germany, 1951-—. Author: Die Verarbeitbarkeit von Zementen, 1939. Research and numerous publs. in phys. chemistry of silicates, binding agts. (cement, lime), ceramics, especially fire-proof glass and bldg. materials. Home: 253 Drimbornallee, Aachen, Bundesrepubl. Deutschland.*

SCHWIMMER, Sigmund, Am. biochemist; b. Cleve., Sept. 20, 1917; s. Solomon and Sarah (Bronsky) S.; student Ohio State U., 1935-36; B.S., George Washington U., 1940; M.S., Georgetown U., 1941, Ph.D., 1943; m. Sylvia Klein, Dec. 18, 1941; children—Susan Rose, Elaine June. With U.S. Dept. Agr., 1936-—, sr. biochemist, 1952-55, sr. research biochemist, 1955-63; prin. research biochemist, 1963-65, chief research biochemist, Albany, Cal., 1965-—; research asso. div. biology Cal. Inst. Tech., Pasadena, Sr. research fellow NSF, U. Copenhagen (Denmark), 1958-59. Recipient Superior Service award U.S. Dept. Agr., 1949, 63; Guggenheim fellow, 1958. Mem. Am. Soc. Biol. Chemists, Inst. Food Tech. Research, numerous publs. on enzymes plants, animals and microorganisms of importance in food sci., isolation, characterization, theory action, biogenesis, role in flavor, color, texture processed foods. Home: 1089 Keith Av., Berkeley, Cal. 94708. Office: 800 Buchanan St., Albany, Cal. 94710.*

SCHWINGER, Julian, Am. physicist; b. N.Y.C., Feb. 12, 1918; s. Benjamin and Belle (Rosenfeld) S.; A.B., Columbia, 1936, Ph.D., 1939; D.Sc. (hon.), Purdue University, 1961, Harvard, 1962, Columbia, 1966; m. Clarice Carrol, 1947. NRC fellow, 1939-40; research asso. U. Cal. at Berkeley, 1940-41; instr., then asst. prof. Purdue U., 1941-43; staff mem. Radiation Lab., Mass. Inst. Tech., 1943-46; staff Metall. Lab., U. Chgo., 1943; asso. prof. Harvard, 1945-47, prof., 1947-—, Higgins prof., 1966-—. Mem. bd. sponsors Bull. Atomic Sci.; J. W. Gibbs hon. lectr. Am. Math. Soc., 1960. Recipient C. L. Mayer nature of light award, 1949, univ. medal Columbia, 1951, 1st Einstein prize award, 1951; Nat. Medal of Sci. award for physics, 1964; co-recipient of Nobel prize in Physics, 1965. Mem. Nat. Acad. Scis., Am. Acad. Arts and Scis., Am. Phys. Soc., Civil Liberties Union. Editor: Quantum Electrodynamics, 1958. Formulated theory of electrodynamics, which represented a union of relativity and quantum theory and led to a better understanding of interaction between charged particles and an electromagnetic field, 1947. Home: 256 Slade St., Belmont, Mass. 02178.

SCHWYZER, Robert, Swiss chemist; b. Zürich, Switzerland, Dec. 8, 1920; s. Robert and Rosa (Schaetzle) S.; Dr.phil. II, U. Zürich, 1947; m. Rosa Nägeli, May 8, 1948; children—Arnold, Anna-Regula, Hans-Kaspar. Faculty, U. Zürich, 1951-63, prof., 1960-63; with CIBA Ltd., Basle, Switzerland, 1952-63, asst. mgr., 1959-63; research asso. U. Cal. at Berkeley, 1958; prof., head lab. molecular biology Swiss Fed. Inst. Tech., Zürich, 1963-—; vis. prof. dept. biochemistry U. Wash., Seattle, 1964. Recipient Alfred Werner award, 1957, Ruzicka award, 1959, Otto Nägeli award, 1964. Mem. Swiss Chem. Soc., Swiss Biochem. Soc., EMBO, A.A.A.S., N.Y. Acad. Scis. Research and numerous publs. on chem. synthesis of peptides, proteins and pharm. hormones. Home: Nägelistrasse 9, 8044 Zürich, Switzerland.*

SCIAMA, Dennis William, English cosmologist; b. Manchester, U.K., Nov. 18, 1926; s. Abraham and Nelly (Ades) S.; B.A., Trinity Coll., Cambridge, Eng., 1947, Ph.D., 1952; m. Lidia Dina, Nov. 26, 1959; children—Susan, Sonia. Fellow, Trinity Coll., Cambridge, 1952-56, faculty, 1961-—, lectr. dept. applied math. and theoretical physics, 1961-—, fellow Peterhouse Coll., 1964-—; mem. Inst. for Advanced Study, Princeton, N.J., 1954-55; Agassiz fellow Harvard, 1955-56; vis. prof. Cornell U., Ithaca, N.Y., 1960-61. Fellow Royal Astron, Soc.; mem. Phys. Soc. London, Cambridge Philos. Soc. Author: The Unity of the Universe, 1959; General Relativity, 1967; also articles. Research on relativity, cosmology and astrophysics. Home: 1 Sylvester Rd., Cambridge, Eng.*

SCITOVSKY, Tibor, economist; b. Budapest, Hungary, Nov. 3, 1910; s. Tibor de and Hanna (Hodosi) S.; doctor iuris, U. Budapest, 1932; postgrad. Trinity Coll., Cambridge (Eng.) U., Ecole Libre des Sciences Politiques, Paris, France; M.Sc. Econ., U. London (Eng.), 1938; m. Anne Marie Aickelin, Sept. 5, 1942 (div. 1967); 1 dau., Catherine. Came to U.S., 1939, naturalized, 1944. With Hungarian Gen. Creditbank, Budapest, 1933-34; research economist London and Cambridge Econ. Service, London, 1938-39; Leon Travelling fellow U. London, 1939-41; economist U.S. Dept. Commerce, Washington, 1946; faculty Stanford, 1946-57, prof., 1950-57; prof. econs. U. Cal. at Berkeley, 1957-—. With OECD, Paris France, 1967. Guggenheim fellow, 1949; Ford Faculty fellow, 1957. Fellow Royal Econ. Soc.; mem. Am. Econ. Assn., Am. Acad. Arts and Scis. Author: (with E.S. Shaw, L. Tarshis) Mobilizing Resources for War, 1951; Welfare and Competition, 1951; Economic Theory and Western European Integration, 1958; Papers on Welfare and Growth, 1964; also articles. Office: Dept. Econs., U. Cal. at Berkeley 94720.*

SCLATER, William Lutley, Brit. zoologist; b. Sept. 23, 1863; s. Philip Lutley and Jane Anne Eliza (Blair) S.; M.A., Keble Coll., Oxford, 1885; m. Charlotte Seymour Sclater, 1896. Asst. master Eton Coll., 1891-95; dir. S. African Mus., Cape Town, 1896-1906; pres. Brit. Ornithologists' Union, 1928-33; hon. sec. Royal Geog. Soc. Author (with F. F. Jackson) The Birds of Kenya and Uganda; other works on mammals and birds of S. Africa, birds of Colorado, Systema Avium Aethiopicarum, also papers. Editor Ibis, 1913-30, Zool. Record, 1921-38. Died July 1944.

SCMIDT, Nikolai Genrikhovich, Russian geophysicist; b. 1909; grad. Leningrad U., 1936. Tchr., 1938-40; engr. geophys. group All Union Research Geologist Inst., Leningrad, 1941-42; worked in No. Urals, 1942-47; dir. seismograph survey work KMA ty., 1948; chief engr. Kursk geol. expdn., 1952-58; chief geophysicist Glavgeologiia RSFSR, 1958-—. Recipient Lenin Prize, 1959. Author: A Short Review of Geographical Research on the Kursk Magnetic Anomaly, 1959. Contributed to discovery and survey of iron-ore deposits, Belgorod dist. of KMA. Address: USSR, Moskva, Glavnoe geologicheskoe upravlenie pri Ministerstve geologgi i okhrany nedr RSFSR.

SCOPOLI, Giovanni Antonio, Italian natural scientist; b. Cavalese, Italy, June 13, 1723; M.D., Innsbruck, Austria, 1743. Physician, Idria; became prof. mineralogy, Schemnitz, 1766; named prof. natural history, Pavia, Italy, 1777; 1st physician to Austrian miners of Tyrol. Author: Flora carriolica, 1772; Principia mineralogiae, 1772; Flora of Carniola, 1744; Entomologia Carnioliea; Deliciae florae et fauna insubriae; Tentamina physico-chymico-medica, 1761; Fundamenta chemiae, 1777. Genus Scopolia named after him by Linnaeus. Died Pavia, May 8, 1788.

SCORESBY, William, English mariner; b. Cropton, nr. Whitby, Eng., Oct. 5, 1789; s. William Scoresby; student chemistry, natural philosophy U. Edinburgh (Scotland), 1806-07, 09; grad. Queen's Coll., Cambridge (Eng.) U., 1825, B.D., 1834, D.D., 1839; m. Miss Lockwood, Sept. 1811; m.2d, 1828; m.3d, Sept. 1849; 2 sons. Whale fisherman (with father), from 1800, chief officer of Resolution, 1806, comdr., 1810, comdr. Esk, 1813-37, comdr. Fame, 1817, comdr. Baffin (supt. constrn.), 1820; ordained minister, curacy of Bessingay, Eng., from 1825; chaplain Mariner's Ch., Liverpool, Eng., from 1827; elected to incumbancy Bedford Chapel, Exeter, Eng., 1832; vicar Bradford, Eng., 1839-47. Monument erected to his memory Upton Ch. Fellow Royal Soc. Edinburgh, Royal Soc., 1824; mem. Wernerian Soc., French Acad. Scis., 1827. Author: Account of Arctic Regions, 2 vols., 1820; Journal of a Voyage to the Northern Whale Fishery and Discoveries on the East Coast of Greenland, 1823; Memorials of the Sea, 1833; Magnetical Investigations, 2 vols., 1839-52; Zoistic Magnetism, 1850; The Franklin Expedition, 1850; My Father . . . , 1851; Journal of a Voyage to Australia for Magnetical Research, 1859; others. Chief officer of 1st freely navigating ship to reach 81°30′ N. latitude, 1806; made survey of Balta Sound, Shetland Isles (also constructed original chart of it), 1807; extensive observations of natural phenomena of Arctic regions culminating in work considered to be one of founding documents of Arctic sci., 1820; made series of drawings of snowflakes as seen under microscope; collected many specimens of previously unknown plants; invented marine diver for recording deep-sea temperatures (established for 1st time that in Arctic seas temperatures at gt. depths are higher than nr. surface); influenced Royal Soc. to send expdns. to polar seas, 1818; research on magnetism; surveyed 400 miles of E. coast of Greenland, 1822; worked on improvements of conditions in factories. Died Torquay, Eng., Mar. 21, 1857.

SCOT, Reynold, see Scott, Reginald.

SCOTT, Alfred Witherspoon, Am. chemist; b. Macon, Ga., June 25, 1896; s. George Edward and Mamie Lee (Wing) S.; B.S., U. Ga., 1918; postgrad. U. Minn.; Ph.D., Princeton, 1921; m. Jane Sams, Aug. 8, 1923; children—Alfred Witherspoon, Richard Sams, George Edward. With CWS, 1918; faculty U. Ga., Athens, 1922-—, prof., head dept. chemistry, 1927-62, prof. emeritus, 1962, dir. div. phys. scis., 1932-62. Chmn. faculty adv. com. engring. expt. sta. Ga. Inst. Tech., 1938-41. Chmn. div. chemistry U. System of Ga. Div. Scis., 1935-62. Mem. Am. Chem. Soc. (councilor 1938, 42, chmn. Ga. sect. 1935), Ga. Acad. Scis. (pres. 1934), A.A.A.S., Am. Inst. Chemists, Am. Assn. U. Profs. Author: A Laboratory Manual of General Chemistry, 1925; A Laboratory Manual of General Chemistry and Semi Mocroqualitative Analysis, 1947; also articles. Research in preparation of unknown compounds of nitrogen, studies of rearrangement of their derivatives. Home: 238 Springdale St., Athens, Ga. 30601.*

SCOTT, Arnold Henry, Am. physicist; b. Sutherland, Ia., Aug. 20, 1899; s. Walter Stanley and Sylvia (Steele) S.; B.S., U. Miss., 1921, M.A., 1923; Ph.D., Johns Hopkins, 1928; m. Pinkie Kirkpatrick, June 16, 1925; 1 dau., Dorothy Gertrude (Mrs. Edward Ehredt Atkinson). Faculty, U. Miss., 1922-23; with Nat. Bur. Standards, Washington, 1924-—, chief dielectrics sect., 1963-65. Recipient Silver medal for meritorious service Dept. Commerce, 1961. Fellow A.A.A.S., Am. Phys. Soc., I.E.E.E., Washington Acad. Sci.; mem. N.Y. Acad. Sci., Am. Soc. for Testing and Materials (recipeint ward of merit 1958). Contbr. articles to tech. jours. on elec. properties rubber-sulphur compounds, measurement of multi-megohm resistance, precision dielectric measurements. Home: Mease Manor, Dunedin, Fla. Office: Nat. Bur. Standards, Washington 20234.*

SCOTT, Arthur Carroll, Am. surgeon; b. Gainesville, Tex., July 12, 1865; s. Rufus Franklin and Martha Helen (Moran) S.; M.D., Bellevue Hosp. Med. Coll., 1886; LL.D., Trinity U., Waxahachie, Tex., 1935; m. Maud Sherwood, Oct. 30, 1889; children—Arthur Carroll, Lucile (Mrs. Preston A. Childers), Helen (Mrs. Walker Saulsbury). Intern, resident surgeon Western Pa. Hosp., Pitts., 1886-88; practiced medicine, Gainesville, 1888-92; chief surgeon Gulf, Colo. & Santa Fe Hosp. Assn. and Ry. Co., 1892-—; organized (with Dr. R. R. White) Kings Daus. Hosp., Temple, Tex., supervised its devel., 1898-1904; sr. surgeon and pres. Scott & White Hosp., Temple, 1904-—. Chmn. Tex. com. Am. Soc. for Control of Cancer, 1913-25. Contbr. many papers and articles on med. subjects, especially cancer. Known for use of cautery in cancer removal, devel. of shadow test in diagnosis of breast cancer; pioneer in use of gas anesthesia; developed (with Raleigh and White) techniques for simultaneous operation on abdomen and pelvis by 2 surgeons. Died Oct. 27, 1940.

SCOTT, Arthur Ferdinand, Am. chemist; b. Coytesville, N.J., Aug. 14, 1898; s. George J. and Carrie (Kerwien) S.; B.S., Colby Coll., 1919, D.Sc. (hon.), 1964; M.A., Harvard, 1921, Ph.D., 1924; postgrad. U. Munich (Germany); m. Vera Prasilova, Mar. 19, 1925; children—Kytja (Mrs. Bruce R. Voeller), Nadja (Mrs. Robert L. Autrey), Dascha (Mrs. Lorimore R. Nicholl). Asst. prof. Reed Coll., Portland, Ore., 1923-26, prof. chemistry, 1937-—, dir. reactor project; asst. prof. chemistry Rice Inst., Houston, 1926-37; guest investigator Mass Inst. Tech., 1958-59. Head spl. projects in sci. edn. NSF, 1962-64; chmn. Gov.'s Nuclear Devel. Coordinating Com. Recipient Mfg. Chemists Assn. award, 1957. Mem. Am. Chem. Soc. (Sci. Apparatus Makers Assn. award 1960), Am. Nuclear Soc., Sigma Xi. Editor: Survey of Progress in Chemistry, vol. 1, 1963, vol. 2, 1964, vol. 3, 1966 vol. 4, 1968. Research, numerous publs. in atomic weight determinations and theory solutions, radiochemistry, especially application neutron activation analysis to iol. and med. problems. Home: 6608 S.E. Av., Portland, Ore. 97202.*

SCOTT, Donald Hyde, Am. horticulturist; b. Buxton, N.D., Dec. 3, 1911; s. David Brainard and Emma (Thompson) S.; B.S., N.D. State U., 1937; Ph.D., U. Md., 1949; m. Addie W. Larsen, Mar. 17, 1937; children—Sidney R., Lawrence B., Sharon L. With U.S. Dept. Agr., 1937-—, prin. horticulturist, Beltsville, Md., 1951-58, investigations leader, 1959-—. Mem. Am. Soc. Hort. Sci., Genetics Soc. Am., Am. Inst. Biol. Scis., Sigma Xi, Phi Kappa Phi. Research, numerous publs. on factors affecting winter hardiness of peach flower buds, inheritance of specific econ. and marker characters in tomato and cucurbits, effect of polyploidy derived by interspecific hybridization in strawberries on fertility or sterility of hybrids, new forms of blueberries by interspecific hybridization and polyploidy, new varieties of blueberries, blackberries, strawberries. Home: 4404 Howard Rd. Office: Plant Industry Sta., Beltsville, Md. 20705.*

SCOTT, Dukinfield Henry, English botanist; b. London, Nov. 28, 1854; s. George Gilbert and Caroline (Oldrid) S.; student Christ Ch., Oxford, 1872-76; 3 years tng. as engr.; studied under J. von Sachs, Würzburg, Germany, 1880-82; hon. degrees, univs. Manchester (Eng.), Aberdeen (Scotland); m. Henderina Victoria Klaassen, 1887. Became asst. to Daniel Oliver, Univ. Coll., London, 1882; placed in charge under

T. H. Huxley of bot. work at Normal Sch. Science, 1885; named dir. research Jodrell Lab., Kew, 1892; ret., 1906. Recipient Royal medal, 1906; Gold medal Linnean Soc., 1921; Darwin medal, 1926; Wollaston medal Geol. Soc., 1928. Fellow Royal Soc., 1894; pres. Linnean Soc., 1908-12, Royal Micros. Soc., 1904-06, S. Eastern Union Sci. Socs., 1909, botany sect. Brit. Assn., 1896, 1921, paleobot. sect. Internat. Bot. Congress, 1930; mem. French Acad. Scis. Author: Studies in Fossil Botany, 1900; Introduction to Structural Botany, 1894; The Present Position of Paleobotany, 1906; Extinct Plants and Problems of Evolution, 1924; also papers on plant evolution. Authority on fossil plants of carboniferous rock; collected more than 3000 slides of carboniferous plants; investigated (with W. C. Williamson) structure and affinities of fossil plants of paleozoic rocks. Died Basingstoke, Eng., Jan. 29, 1934.

SCOTT, Earl Bernard, Am. anatomist; b. Pitts., Mar. 25, 1908; s. Frank W. and Lydia A. (Miller) S.; B.S., U. Pitts., 1930, M.S., 1934, Ph.D., 1940; m. Miriam R. Kay, Nov. 15, 1942; 1 dau., Barbara Linda. Instr., U. Pitts., 1940-42; faculty U. S.D. Med. Sch., Vermillion, 1947—, prof. anatomy, 1955—, chmn. dept., 1964—. Mem. Am. Assn. Anatomists, Soc. for Exptl. Biology and Medicine, Electron Microscopy Soc. Am., Soc. for Cell Biology, A.A.A.S., Am. Soc. Zoologists, Gerontology Soc. Am., Sigma Xi. Research, publs. on path. changes of tissues in animals deprived of single essential dietary amino acids with both light and electron microscopy, electron microscope studies changes in tissues as a result of aging. Home: 403 Linden Av., Vermillion, S.D. 57069.*

SCOTT, Flora Murray, botanist; b. Craig, Scotland, Sept. 6, 1891; d. Robert and Mary (Jobson) Scott; M.A., St. Andrews U., Scotland, 1911, B.Sc., 1914; Ph.D., Stanford, 1925. Came to U.S., 1922, naturalized, 1935. Faculty, Constantinople Coll. for Women, 1921, Stanford, 1922-25; faculty U. Cal. at Los Angeles, 1925—, prof. botany, 1950-59, prof. emeritus, 1959. Mem. A.A.A.S., Bot. Soc. Am., N.Y. Acad. Scis., Electron Microscope Soc. Research, publs. on plant anatomy especially ultrastructure of cell wall. Home: 955 Stonehill Lane, Los Angeles 90049.*

SCOTT, Gayle, Am. geologist; b. Lamkin, Tex., Oct. 1, 1894; s. George M. and Ezora (Jennings) S.; B.S., Tex. Christian Coll., 1917, M.S., 1920; Ph.D., U. Grenoble, 1925; m. Mary Beth Waits, 1927; 1 dau. Tchr., Tex. Christian Coll., 1920-24, prof., 1926-48. Fellow Geol. Soc. Am.; mem. Geol. Soc. France, A.A.A.S., Am. Geophys. Union, Soc. Econ. Paleontologists and Mineralogists (pres. 1939). Author: The Geology of Johnson County, 1920; The Texas Cretaceous Section, 1933; also papers onerontic ammonites, geology of various locations. Editor jour. Am. Assn. Petroleum Geologists, 1944-47. Interpreted Woodbine as regressive phenomenon; correlated Woodbine with Cenomanian; studied correlation of Midway with Danion. Died 1948.

SCOTT, George Herbert, Brit. aero. pioneer; b. London, May 25, 1888; s. George Hall and Margaret (Wilkinson) S.; ed. Royal Naval Engring. Coll., Keyham; m. Jessie Campbell, 1919; 1 son, 3 daus. Gen. engring., Spain, 1908-14; joined Royal Naval Air Service, 1914; airship comdr.; flew 1st Brit. rigid airship, 1917; advanced to maj., 1918; joined tech. staff Royal Airship Works, 1920; became officer in charge flying, tng. and airship devel. Air Ministry, 1924; in charge opening airship communication in Brit. Empire. Devised system of mooring airships at head of mast; navigator of 1st rigid airship to make transatlantic flight, 1919; contributed to design R101 with unbraced traverse frame. Killed in crash of R101, France, 1930.

SCOTT, George Taylor, Am. physiologist; b. Troy, N.Y., Spet. 10, 1914; s. Robert Winfield and Helen Denison (Taylor) S.; B.S., Union Coll., 1938; A.M., Harvard, 1941, Ph.D., 1943; m. Elsie Mae Welling, Oct. 16, 1943; children—Helen Ann, Georganne Elsie. Austin teaching fellow Harvard, 1940-43; instr. zoology Oberlin Coll., 1943-45, asst. prof., 1945-49, asso. prof., 1949-52, prof., 1952—, chmn. dept. biology, 1956—; asso. scientist Woods Hole Oceanographic Inst., 1954-59. Trustee Bermuda Biol. Sta. for Research, 1955—, Marine Biol. Lab., 1956—. Fellow A.A.A.S., Ohio Acad. Sci., N.Y. Acad. Scis.; mem. Am. Soc. Zoologists, Soc. Gen. Physiologists, Am. Assn. U. Profs., Corp. Marine Biol. Lab., Sigma Xi, Theta Nu Epsilon, Gamma Alpha. Research and publs. on cellular ion active-transport in algae, control of color-change in fish, influence of tranquilizer and energizer drugs on pituitary gland secretion in frog, adrenergic blocking action of certain phenothiazine tranquilizer drugs. Home: 111 Forest St., Oberlin, O. 44074.*

SCOTT, Gordon Halter, Am. anatomist; b. Winfield, Kan., Apr. 10, 1901; s. John H. and Grace (Hatler) S.; A.B., Southwestern Coll., Kan., 1922, Sc.D., 1960; A.M., U. Minn., 1925, Ph.D., 1926; m. Luella Marguerite Smith, Dec. 25, 1923; children—Gordon H.S., John E.S. With Southwestern Coll., 1920-23, Johns Hopkins, 1922-23, U. Minn., 1923-26; asst. prof. Loyola U., Chgo., 1926-27; asst. Rockefeller Inst. for Med. Research, 1927-28; faculty Washington U., 1928-42, asso. prof. cytology, 1931-42; prof. anatomy, head dept. U. So. Cal., 1942-45; prof. anatomy Wayne State U., Detroit, 1945—, head dept. anatomy, 1945-50, dean, 1950-63, v. p. med. sch. devel., 1961—, acting dean, 1948-50. Mem. neurology program project com. NIH, 1961—. Recipient Outstanding Achievement award bd. regents U. Minn., 1958. Mem. Am. Assn. Anatomists, Am. Soc. Zoologists, A.A.A.S., Assn. Am. Med. Colls. Research, numerous publs. on histology of human growth changes, histochemistry, microincineration, spectography, aviation physiology, capillary circulation; developed electron microscope and electron microscopy. Home: 9 Fairwood St., Pleasant Ridge, Mich. 48069. Office: Mackenzie Hall, Wayne State U., Detroit 48202.*

SCOTT, Gordon Nickelsen, phys. chemist; b. Seacombe, Eng., Aug. 26, 1898; s. James and Laurey (Nickelsen) S.; A.B., U. Cal. at Berkeley, 1922, B.S. cum laude, 1922, Ph.D., 1926; m. Ethel Delilah Allen, Jan. 1, 1926; children—Maile Ruth (Mrs. Roger C. Olson), Lauredith (Mrs. Grisha Dotzenko), James Gordon, Sarah Allynne (dec.), Susan Angela (Mrs. Emil C. Beyer, III). Plant supt. James Wqkefield & Sons, Honolulu, 1926-27; research chemist Mid-Continent Petroleum Corp., Tulsa, 1927-28; research asso. Nat. Bur. Standards, Am. Petroleum Inst., Washington, 1928-34, pipe line technologist, Dallas N.Y.C., 1935-36; cons. in corrosion mitigation, Los Angeles, 1936—. Registered profl. engr., Cal. Mem. Nat. Assn. Corrosion Engrs. (Frank Newman Speller award 1955), Sigma Xi, Phi Lambda Upsilon, Alpha Chi Sigma. Research and publs. on corrosion especially elec. protection of buried structures, theory relating parameters of lognormal distbn. to factors in corrosion testing; relation between soil resistivity and corrosion. Home: 608 N. Cañon Dr., Beverly Hills, Cal. 90210. Office: 900 Wilshire Blvd., Los Angeles 90017.*

SCOTT, Gordon Ramsay, Brit. veterinarian; b. Arbroath, Scotland, July 6, 1923; s. William and Margaret (Jackson) S.; student Royal (Dick) Vet. Coll., 1941-46; B.Sc., U. Edinburgh, Scotland, 1946, Ph.D., 1959; M.S., U. Wis., 1953; m. Joan Henderson Walker, July 15, 1947; children—Ann, Andrew, Margaret. Asst. pvt. practice, Wigtown, 1946-47, Cranleigh, 1947-49; vet. research officer Vet. Research Lab., Kabete, Kenya, 1950-52; research fellow U. Wis., Madison, 1952-53; head virus sect. Vet. Research Lab., Kabete, 1953-56; tutor U. E. Africa, 1955-62; head virus div. E. African Vet. Research Orgn., Muguga, Kenya, 1956-62; lectr. tropical unit Faculty Vet. Sci., U. Edinburgh, 1963—. Lectr., Sirlba Coll., Kenya, 1950; cons., FAO, 1963-66. Recipient Beach award U. Wis., 1953. Mem. Royal Coll. Vet. Sci., Assn. Vet. Tchrs. in Research Works, Brit. Vet. Assn., Brit. Vet. Zool. Soc., Diagnostic Virology Group Scotland, Inst. Biology, N.Y. Acad. Scis., Path. Soc. Gt. Britain, Vet. History Soc. Author: Diagnosis of Rinderpest; also numerous articles. Devel. vaccines for rinderpest; studies on diagnosis, immunology and pathogenesis of rinderpest, African swine fever, bovine malignant catarrh, Newcastle disease, Rift Valley fever. Home: 22 Frogston Rd. W., Edinburgh, 10, Scotland. Office: Vet. Field Sta., Easter Bush, Roslin, Scotland.*

SCOTT, Harold George, Am. biologist; b. Williams, Ariz., Aug. 20, 1925; s. Milton Raymond and Lucile (Crosby) S.; student Mont. State Coll., 1943, U. Philippines, 1945-46, Temple U., 1946-49; B.S., U. N.M., 1950, M.S., 1953, Ph.D., 1957; m. Bettie Tabakin, Aug. 6, 1948; children—Jasmine, Lorelei, Rodger, Clifford, Curtis, Conrad, Dolores, Gloria. Entomologist, Med. Field Service Sch., Ft. Sam Houston, Tex., 1950-51, Kirtland AFB, N.M., 1951-55; scientist Communicable Disease Center USPHS, Atlanta, 1955—, chief vector-borne disease sect., 1964—. Recipient Phi Sigma Grad. award, 1955. Mem. Soc. Systematic Zoology, Entomol. Soc. Am., Research Soc. Am., Commd. Officers Assn. USPHS, Sigma Xi. Author: Epithalamion, 1955; (with K.S. Littig) Insecticidal Equipment, 1960, Moscas de importancia para la salud publica y su control, 1962, Flies of Public Health Importance and Their Control, 1962; Household and Stored-food Insects of Public Health Importance and Their Control, 1963; (with H.D. Pratt) Insecticide Formulation, 1964; (with M.R. Borom) Techniques of Public Health Entomology, 1964, Rodent-borne Disease Control Through Rodent Stoppage, 1965; (with others) Tularemia, 1964; (with C.J. Stojanovich) Illustrated Key to Mosquitoes of Vietnam, 1966, Illustrated Key to Anopheles Mosquitoes of Thailand, 1966, Illustrated Key to Anopheles Mosquitoes of Liberia, 1966; (with others) Clave Ilustrada Para Los Mosquitos Anofelinos de Venezuela, 1966, Pictorial Keys to Arthropods, Reptiles, Birds and Mammals of Public Health Significance, 1966, Clave Ilustrada Para Los Mosquitos Anofelinos de America Central y Panama, 1966; Plates of Vector-borne Diseases in Vietnam, 1966. Research, numerous publs. on ecol.-systematic analysis of collembola of N.M.; developed vector-borne disease tng. system for USPHS, instant tng. service for rapid response to disease emergencies, taxonomic system on mosquitoes of Vietnam; co-inventor of "SAD, SA/C Calculator" for use in pesticide formulations. Home: 8137 River Rd., Westwego, La. 70094. Office: USPHS Hosp., 210 State St., New Orleans 70094.*

SCOTT, Sir Henry Harold, English physician; b. London, Eng., Aug. 3, 1874; s. Douglas Lee Scott; ed. U. Coll., London U. (Milner research fellow sch. hygiene and tropical medicine), St. Bartholomew's and St. Thomas hosps., Eng.; M.D., diplomas pub. health, tropical medicine, Cambridge (Eng.) U.; m. Hariette Preston; 1 son (dec.); m. 2d., Eileen Anne Prichard, 1934. Resident med. officer Teignmouth, Dawlish, Newton Abbot Infirmary (all Eng.); house physician St. Thomas Hosp.; capt. Royal Army Med. Corps, South African (Boer) War; med. officer Hosp. Women and Children, Ft. Napier, Maritzburg, South Africa; pathologist Cambridge Hosp., Aldershot, Eng., World War I; bacteriologist Brit. Govt., Jamaica, B.W.I.; Hong Kong; lectr. bacteriology, med. jurisprudence, pub. health Victoria U.; lectr. tropical disease med. sch. Westminster Hosp., Eng.; dir. Bur. Hygiene and Tropical Diseases, 1935-42; cons. physician William Julien Courtauld Hosp., Eng.; bd. dirs. Chest Hosps. London. Examiner diplomas tropical med., tropical hygiene, parasitology (pub. health) and for Liverpool (Eng.) U. Fellow Royal Coll. Physicians (Fitz-Patrick lectr., London, examiner tropical medicine, surgery Conjoin bd.), Zool. Soc. (pathologist, London) Royal Soc. Edinburgh (Scotland); mem. Royal Soc. Tropical Medicine and Hygiene. Author: (with Maj. Gen. D. T. Richardson) British Red Cross Manual of Tropical Hygiene, 1946; A History of Tropical Medicine, 2 vols., 1939; Some Notable Epidemics, 1934; Post-Graduate Clinical Studies; (with Sir Andrew Balfour) Health Problems of the Empire; Tuberculosis in Man and Lower Animals. Editor proceedings 9th Internat. Congress Indsl. Medicine, 1948. Research on vomiting sickness, which he discovered in 1916 was caused by consumotion of fruit of ackee tree, and on central neuritis (Strachan-Scott syndrome), on pulmonary Tb. in man and animals, on actiology, pathology of fungus infection in birds and reptiles. Died Aug. 6, 1956.

SCOTT, Henry William, Jr., Am. surgeon; b. Graham, N.C., Aug. 22, 1916; s. Henry William and Lucie Claire (Turner) S.; A.B., U. N.C., 1937; M.D., Harvard, 1941; m. Mary Louisa Vanamee, Oct. 17, 1942; children—Henry William III, Mary Elizabeth, Virginia Wright, Patricia Vanamee. Faculty, Johns Hopkins, Balt., 1946-52, asso. prof. surgery, 1950-52; prof. surgery, chmn. dept. Vanderbilt U. Sch. Medicine, Nashville, 1952—, surgeon-in-chief U. Hosp., 1952—. Harvey Cushing fellow Children's Hosp., Boston, Peter Bent Brigham Hosp., Boston, 1945-46. Research, numerous publs. on surg. physiology and physio-pathology of gastrointestinal and cardiovascular systems. Home: 1050 Tyne Blvd., Nashville 37220.*

SCOTT, Hugh, English entomologist; b. Sept. 16, 1885; s. William Edward and Edith Truscott (Amos) S.; B.A., Trinity Coll., Cambridge (Eng.) U., 1906, M.A., 1910, Sc.D., 1919; m. Beatrice Emily Streatfeild, Jan. 15, 1913; 1 son, 2 daus. Asst. to David Sharp, Mus. Zoology, Cambridge U., 1906-08, curator in entomology, 1909-28; entomologist, hygiene dept. Royal Army Med. Coll., 1917-18; entomologist Dept. Agr., Baghdad, Iraq, 1928; asst. keeper entomology Brit. Mus., London, Eng., 1930-48. Prin. writer ofcl. handbook western Arabia and Red Sea, Naval Intelligence Div., Brit. Admiralty, 1942-46; mem. Entomol. Research Com., 1911-28. Leverhulme Research fellow, 1937. Fellow Royal Soc., 1941, Linnean Soc. (council 1933-36), Royal Entomol. Soc. (council 1924-27, v.p. 1926, editorial staff monthly mag. 1923). Author: In the High Yemen, 1942. Contbr. articles to sci. publs. Made several natural history expdns. to Seychelles Islands, 1908-09, to highlands Central Abyssinia, 1926-27, to western Aden Protectorate, Yemen, 1937-38, So. Ethiopia, 1948-49, No. Ethiopia (high Simien), 1952-53; shorter journeys to W.I., Basutoland, Kurdistan, several European countries. Died Nov. 1, 1960.

SCOTT, Irving Day, Am. geologist; b. Omaha, Sept. 16, 1877; s. William Henry and Sarah Emma Scott; A.B., Oberlin Coll., 1900; Pd.D., N.Y. State Coll. for Tchrs., 1901; Ph.D., U. Mich., 1912; m. Elizabeth Rogers, 1911; 4 children; m. 2d, Susan Cook. High sch. tchr., N.Y.; became asst. in mineralogy and geology U. Mich., 1907, instr, physiolgraphy, geology, 1907-13, asst. prof., 1913-20, asso. prof., 1920-30, prof., 1930-42, in charge physiography, summer camp, Mill Springs, Ky., 1924-35. Fellow Geol. Soc. Am.; mem. A.A.A.S., Mich. Acad. (pres. 1914-42). Research and publs. on fracture systems of land, ice push on lake shores, land lakes; studied karst region in Yugoslavia, glaciers of Switzerland, sand dunes in western Mich. Died 1955.

SCOTT, James Henderson, Irish dental anatomist; b. Dundalk, Ireland, Aug. 11, 1913; s. John and Nell (Henderson) S.; student Meth. Coll. Belfast, 1922-31; B.Sc., Queen's U., Belfast, 1939, M.B. B.Ch., 1942, M.D., 1946, D.Sc., 1957; F.F.D.R.C.S.I. (hon.), 1963; m. Olive Marron, Feb. 5, 1945; children—Ann, Geraldine, Mikael, Briged. Lectr., reader Queens U. Belfast, 1946-64, prof. dental anatomy, 1964—. Recipient Howard Mummery prize Brit. Dental Council, 1961. Mem. Anat. Soc. Gt. Britain and Ireland, Internat. Assn. Dental Research, European Orthodontic Soc. Author: (with N. B. B. Symons) Introduction to Dental Anatomy, 4th edit. 1964; (with A. D. Dixon) Anatomy for Students of Dentistry, 2d edit., 1966. Research, publs. on devel. and growth of human facial skeleton, devel. and eruption of teeth. Home: 3 Holyrood Malone Rd., Belfast, No. Ireland.*

SCOTT, Jesse Friend, Am. cytologist; b. Lexington, Ky., Oct. 5, 1917; s. Wellington F. and Sarah (Powell) S.; A.B. cum laude, Vanderbilt U., 1938, M.D., 1941; A.M., Harvard, 1963; m. Margaret V. Cox, Dec. 28, 1948; children—Elizabeth Friend, Edward Powell. Research asso. dept. biology Mass. Inst. Tech., 1948-55; faculty Harvard, 1948—, asso. prof. oncologic medicine, 1963—, chmn. bd. tutors biochem. scis., 1965-66; asso. biophysicist Mass. Gen. Hosp., 1952—, dir. NSF Undergrad. Research Participation Program. Mem. Am. Assn. Cancer Research, Am. Soc. Biol. Chemists, Am. Soc. Cell Biology, Biophys. Soc., Histochem. Soc., Soc. for Devel. and Growth, Phi Beta Kappa, Alpha Omega Alpha. Asso. editor: Jour. Histochemistry and Cytochemistry, 1960-67. Research, numerous publs. in biochemistry and biophysics of ribonuleic acids of normal and cancerous cells. Home: 14 Country Corners Rd., Wayland, Mass. 01778. Office: Huntington Labs., Mass. Gen. Hosp., Boston 02114.*

SCOTT, John Paul, Am. psychologist, zoologist; b. Kansas City, Mo., Dec. 17, 1909; s. John William and Vivian (Armstrong) S.; B.A., U. Wyo., 1930; B.A. (Rhodes scholar 1930-32), Lincoln Coll., Oxford (Eng.) U., 1932; Ph.D., U. Chgo., 1935; m. Sarah Fisher, June 18, 1933; children—Jean (Mrs. Eugene Franck), Vivian (Mrs. William B. Hixson), John, David. Assistant in zoology University of Chgo., 1932-35; faculty Wabash Coll., 1935-45, prof. zoology, 1942-45, chmn. dept., 1935-45; chmn. div. behavior studies Roscoe B. Jackson Meml. Lab., Bar Harbor, Me., 1945-58, trustee, 1946-49, sr. staff scientist, 1958-65; research prof., dir. grad. study in psychology Bowling Green (Ohio) University, 1965—; prof. biopsychology U. Chgo., 1958. Mem. Me. Bd. Psychol. Examiners, 1956-59. Recipient Jordan prize Ind. Acad. Sci., 1947; fellow Center Advanced Study in Behavioral Scis., 1963-64. Fellow Am. Psychol. Assn., N.Y. Zool. Soc., A.A.A.S.; mem. Me. Psychol. Assn. (pres. 1953-54), Am. Soc. Zoologists, Am. Eugenics Society (v.p. 1963—), Ecol. Society Am. (chairman sect. animal behavior and sociobiology 1957-58), Genetics Soc. Am., Phi Beta Kappa, Sigma Xi. Author: Animal Behavior, 1958; Aggression, 1958; (with John L. Fuller) Genetics and the Social Behavior of the Dog, 1965; also sci. articles, chpts. in books. Research on heredity and devel. social behavior in mammals; discovered critical period of primary socialization in the dog. Home: 918 Lyn Rd., Bowling Green, O. 43402. Office: Dept. Psychology, Bowling Green State U., Bowling Green, O. 43402.*

SCOTT, Milton Leonard, Am. nutritionist; b. Tempe, Ariz., Feb. 21, 1915; s. Squire Milton and Ione (Greenleaf) S.; A.B., U. Cal. at Berkeley, 1937; Ph.D., Cornell U., 1945; m. Dorothy Marie Jaeger, July 1, 1938; children—Grace Ione (Mrs. Alexander James Saroka), June Marie (Mrs. William J. Kopald). Chemist, Coop. G.L.F. Mills, Buffalo, 1937-42; faculty Cornell U., Ithaca, N.Y., 1946—, prof. poultry sci. dept., 1953—. Cons. to feed mfrs., feed stuff supplies, N.Y. State Conservation Dept. Recipient Am. Feed Mfrs. Assn. award, 1952; Nat. Turkey Fedn. Research award, 1958; Calcium Carbonate Co. Travel award, 1965. Mem. Am. Inst. Nutrition, Am. Soc Biol. Chemistry, Poultry Sci. Assn. (Borden award 1965), Am. Chem. Soc., A.A.A.S., Am. Soc. Animal Products, soc. for Exptl. Biology and Medicine, World's Poultry Sci. Assn., Sigma Xi, Phi Kappa Phi. Research, numerous publs. on exudative diathesis and muscular dystrophy in chicks and other animals, selenium as nutrient, interrelationship between selenium, vitamin E and sulfur amino acid metabolism, proteins, amino acids, metabolizable energy, vitamins, minerals, unidentified factor requirements. Home: 601 Hanshaw Rd., Ithaca, N.Y. 14850.*

SCOTT, Patrick, Scottish alchemist; b. Scotland, flourished, 1620. Followed James I to Eng. on his accession, 1618; occasionally acted as tutor to Charles I. Author: Calderwood's Recantation, or a Tripartite Discourse, 1622; The Tilage of Light, 1623; Vox Vera, 1625; Omnibus et singulis, 1619. Claimed to have made philosopher's stone, 1623. Died 1623.

SCOTT, Reginald (or Reynold Scot), English natural scientist; b. Kent, Eng., circa 1538; s. Richard and Mary (Whetemall) Scot; ed. Hart Hall, Oxford; collector of subsidies for lathe of Shepway, 1586, 87; capt. of untrained foot soldiers at country muster; mem. Parliament for New Romney, 1588-89. Author: Perfect Platform of a Hop-garden, (1st practical treatise on hop culture in Eng.), 1574; The Discourie of Witchcraft (attempt to prove that reason and religion both reject witchcraft and magic), 1584. Died 1599.

SCOTT, Robert E(dwin), Am. polit. scientist; b. Chgo., Apr. 27, 1923; s. Jules F. and Helen (Kasmar) S.; B.A., Northwestern U., 1945, M.A., 1946; Ph.D., U. Wis., 1949. Faculty, U. Ill., Urbana, 1949—, prof. polit. sci., 1961—; vis. prof. Yale, 1962; sr. staff mem. Brookings Instn., Washington, 1963-64. Cons. in El Salvador for Pub. Adminstrn. Service, Chgo., 1950, Protestant Episcopal Ch., Mexico, 1958, Central Am., 1957, Venezuela, 1966. Mem. Am. Polit. Sci. Assn., Midwest Conf. Polit. Scientists, Latin Am. Studies Assn. Author: Mexican Government in Transition, 1958, also articles. Research in politics of developing countries, especially Latin Am.; investigations in Mexico's polit. process, including its polit. culture; studies of Latin Am. party systems and polit. elites, Latin Am. univs. in polit. change. Home: 1505 Delmont Ct., Urbana, Ill. 61801.*

SCOTT, Robert Falcon, Brit. explorer; b. Outlands, Devonport, June 6, 1868; s. John Edward Scott; ed. Stubbington House, Fareham; D.Sc. (hon.), Cambridge U., Manchester U., 1905; m. Kathleen Bruce, 1908; 1 son. Entered Navy, 1882, advanced through ranks to capt., 1904; comdr. Nat. Antarctic Expdn., 1900-04 (on 1st voyage explored region around Ross Sea, founded the sea, discovered Edward VII peninsula, surveyed coast of Victoria land, reached new farthest South; 2d voyage was attempt to discover South Pole, reached it only to find that someone else had been there before them; comdr. Brit. Antarctic Expdn., 1910. Gold medallist Royal Geog. Soc., Royal Scottish Geog. Soc., Am., Swedish, Danish, Phila., Antwerp geog. socs. Author: The Voyage of the Discovery, 2 vols., 1905. Died Mar. 1912.

SCOTT, Robert Lane, Am. chemist; b. Santa Rosa, Cal., Mar. 20, 1922; s. Horace Albert and Maurine (Lane) S.; S.B., Harvard, 1942; M.A., Princeton, 1944, Ph.D., 1945; m. Elizabeth Sewall Hunter, May 27, 1944; children—Joanna Ingersoll, Jonathan Armat, David St. Clair, Janet Hamilton. Sci. staff Los Alamos Lab., 1945-46; Frank B. Jewett fellow U. Cal. at Berkeley, 1946-48, faculty U. Cal. at Los Angeles, 1948—, prof. chemistry, 1960—. John Simon Guggenheim fellow, 1955; NSF Sr. fellow, 1961-62. Mem. Am. Chem. Soc., Am. Phys. Soc., A.A.A.S., Faraday Soc., Sigma Xi. Author: (with J.H. Hildebrand) Solubility of Nonelectrolytes, 3d edit., 1950, rev., 1964, Regular Solutions, 1962; also articles. Research on statis. thermodynamics of liquids and solutions, solubility and phase equilibria, fluorocarbons and their solutions, high polymer solutions, sulfur solutions, heats and volume changes on mixing, critical phenomena, intermolecular forces and complex formation in solution, data processing with high speed electronic computers. Home: 3760 Grand View Blvd., Los Angeles 90066.*

SCOTT, Roland Boyd, Am. pediatrician; b. Houston, Apr. 18, 1909; s. Ernest John and Cordie (Clark) S.; B.S., Howard U., 1931, M.D., 1934; Gen. Edn. Bd. fellow U. Chgo., 1936-39; m. Sarah Rosetta Weaver, June 24, 1935; children—Roland Boyd, Venice Rosetta, Estelle Irene. Faculty, Howard U., Washington, 1937—, prof. pediatrics, 1952—; chief pediatrician Freedmen's Hosp., 1947—; staff D.C. Gen. Hosp., Washington Hosp. Center; cons. in pediatrics to NIH, hosps. Mem. com. Pub. Health Adv. Council, 1964—, U.S. Children's Bur., 1964—. Diplomate Nat. Bd. Med. Examiners, Am. Bd. Pediatrics. Mem A.M.A., Am. Pediatric Soc., Soc. Pediatric Research, Am. Acad. Allergy (v.p. 1966-67), Am. Acad. Pediatrics, Am. Fedn. Clin. Research, Nat. Med. Assn. (Distinguished Service medal 1966), A.A.A.S., Internat. Corr. Soc. Allergists, Assn. Ambulatory Pediatric Services, Am. Assn. U. Profs., Internat. Congress of Pediatrics, Am. Coll. Allergists, Kappa Pi, Beta Kappa Chi, Sigma Xi, Alpha Omega Alpha, others. Author: (with Althea D. Kessler) Sickle Cell Anemia and Your Child. Mem. editorial bd. Clin. Pediatrics, 1962—; cons. editor Medical Aspects of Human Sexuality. Research, publs. on sickle cell anemia, growth and devel. of infants and children, allergy in children. Home: 1723 Shepherd St. N.W., Washington 20011. Office: 114 Girard St. N.W., Washington 20009.*

SCOTT, Ronald Fraser, civil engr.; b. London, Eng., Apr. 9, 1929; s. Frederick Durham and Catherine (Fraser) S.; B.Sc., Glasgow U., Scotland, 1951; M.S., Mass. Inst. Tech., 1953, Sc.D., 1955; m. Pamela June Wilkinson, May 28, 1959; children—Grant Fraser, Craig Alistair, Roderick Jonathan. Teaching, research asst. Mass. Inst. Tech., 1951-55; soils engr. U.S. Army Engr. Corps, New Eng. div., Waltham, Mass., 1955-57; divisional soils engr. Racey, McCallum & Assos., Toronto, Ont., Can., 1957-58; faculty civil engring. Cal. Inst. Tech., 1958—, prof., 1967—. Cons. pvt. engring. firms, 1959—, Jet Propulsion Lab., 1960—, NASA, 1965—. Mem. Am. Geophys. Union, Am. Soc. C.E. Author: Principles of Soil Mechanics, 1963; also articles. Research on heat transfer in soils, including freezing and thawing, deformational behavior of soils, prin. investigator for soil testing device carried to moon on Surveyor Spacecraft 3, 4, 7; investigation of ocean floor soils. Home: 2752 N. Santa Anita Av., Altadena, Cal. 91001. Office: 1201 E. California St., Pasadena, Cal. 91109.*

SCOTT, Russell B(urton), Am. physicist; b. Ludlow, Ky., Apr. 17, 1902; s. Burton W. and Carrie May (Riggs) S.; B.S. cum laude, U. Ky., 1926, M.S., 1928; m. Leonora Downing, June 13, 1928; children—Marion Lee (Mrs. William F. Kenkel), Burton W. Instr. physics U. Ky., 1927-28; mem. staff Nat. Bur. Standards, 1928—, chief cyrogenic engring. lab., Boulder, Colo, 1953-62, acting dir. Boulder labs., 1962-63, mgr., 1963—. Mem. U. S. nat. com. Internat. Inst. Refrigeration, 1957-60, mem. ad hoc com. establish U. S. membership, 1956-57, mem. commn. I, 1959—. Fellow Am. Phys. Soc., A.A.A.S.; mem. Research Soc. Am. (pres. Boulder chpt. 1961), Phi Beta Kappa, Sigma Pi Sigma. Author: Cryogenic Engineering, 1959; also numerous articles. Am. editor, contbr.: Technology and Uses of Liquid Hydrogen, 1964. Am. editor Jour. Cyrogenics, 1960—. Research on properties of heavy hydrogen and its compounds, measurement of specific heats of solids, liquids, gases at low temperatures; improved helium liquefiers; worked on determination of properties of uranium and its compounds for devel. atomic bomb; also research on hydrogen bomb. Died Boulder, Colo., Sept. 24, 1967.

SCOTT, Walter Dill, Am. psychologist, mgmt. scientist; b. Cooksville, Ill., May 1, 1869; s. James Sterling and Henrietta (Sutton) S.; grad. Ill. State Normal U., 1891; A.B., Northwestern U., 1895; grad. McCormick Theol. Sem., 1898; Ph.D., University of Leipzig, 1900, (honorary), Northwestern University; LL.D., Cornell Coll., 1921; LL.D., U. of Southern Calif., 1932; m. Anna Marcy Miller, July 21, 1898; children—John Marcy, Sumner Walter. Asso. prof. psychology and edn., and dir. psychol. lab., Northwestern U., 1901-08, prof. psychology, 1908-20, pres., 1920-39, pres. emeritus, 1939—; dir. Bureau of Salesmanship Research, Carnegie Inst. Tech., 1916-17. Pres. The Scott Co., consultants and engrs. in industrial personnel, 1919-21. Trustee Wesley Memorial Hosp., Presbyn. Theol. Sem. of Chicago; mem. bd. trustees of Century of Progress, 1933, 34. Chmn. solid fuels advisory war council 1941-46; chmn. editorial board, American Peoples Ency., 1948. Mem. Am. Council on Edn., (chmn. 1927), Am. Psychol. Assn. (pres. 1918-19), Phi Beta Kappa, Sigma Xi. Author: Die Psychologie der Triebe, 1900; Theory of Advertising, 1903; Psychology of Public Speaking, 1907; Psychology of Advertising, 1908; Influencing Men in Business, 1911; Increasing Human Efficiency, 1911; Psychology of Advertising in Theory and Practice, 1921; Science and Common Sense in Working with Men, 1921; Personnel Management, 1941; The Life of Charles Deering, 1919; Biography of John Evans 1939; Life of John Evans; Life of Charles Deering. Joint Author: The Personnel System of the U. S. Army, Vol. I, History of the Personnel System, Vol. II, The Personnel Manual, 1919; Personnel Specifications, 10 vols., 1918-19; Dwellers by the Road, 1911; Aids in Selecting Salesmen, 1916; Stabilizing Business, 1923; Man and His Universe, 1929; Society Today, 1929; Life of Walter P. Murphy, 1947. Developed and supervised AUS personnel testing program; work influenced develop. personnel mgmt. in U. S. govt. and industry. Died Sept. 23, 1955.

SCOTT, William Berryman, Am. geologist; b. Cin., Feb. 12, 1858; s. William M. and Mary E. (Hodge) S.; A.B., Princeton, 1877; Ph.D., Heidelberg, 1880; LL.D., U. Pa., 1906; Sc.D., Harvard, 1909, Oxford U., 1912, Princeton, 1930; m. Alice A. Post, Dec. 15, 1883. Asst. in geology Princeton U., 1883, prof. geology and paleontology, 1884-1930. Mem. Nat. Acad. Sciences, Am. Philos. Soc. (pres. 1918-25), Geol. Soc. Am. (pres. 1924-25). Recipient E. K. Kane medal Geog. Soc. Phila.; Wollaston medal Geol. Soc., London, 1910; F. V. Hayden medal Acad. Nat. Sci., Phila., 1926; Elizabeth Clark Thompson Gold medal Nat. Acad. Scis., 1931; R. A. Penrose Gold medal Geol. Soc. Am., 1939; Daniel Girand Elliot gold medal, Nat. Acad. Sciences, 1944. Author: An Introduction to Geology, 1897, 3d edit., 1932; A History of Land Mammals in the Western Hemisphere, 1913, 2d Edit., 1937; The Mammalian Fauna of the White River Oligocens, 1935-41; Mammalian Fauna of the Duchesne River, 1945; The Theory of Evolution, 1917; Physiography, 1922; also about 60 monographs upon geol. and palaeontol. subjects. Editor and joint author of Reports Princeton University expdns. to Patagonia (9 vols.). Died Mar. 29, 1947.

SCOTT, William Taussig, Am. physicist; b. Yonkers, N.Y., Mar. 16, 1916; s. Carl Forse and Dorothea (Taussig) S.; student Haverford Coll., 1933-34; B.A., Swarthmore Coll., 1937; Ph.D., U. Mich., 1941; m. Helen Elizabeth Gabel, June 7, 1941 (div. Sept. 1961); children—Jennifer Smedley, Christopher Gephart, Stephanie Taussig, Melanie Hollister, Peter; m. 2d, Ann Howe Herbert, Sept. 29, 1961. Instr. physics Amherst Coll., 1941-44; instr. physics and math. Deep Springs Jr. Coll., 1944-45; faculty physics Smith Coll., 1945-61; prof., 1959-61; prof. U. Nev., Reno, 1961—. With Brookhaven Nat. Lab., summers 1947-53, U.S. Nat. Bur. Standards, summers 1954-56; summer faculty Harvard, 1957; research fellow in physics and Div. Sch., Yale, 1959-60, 62-63; with Nat. Center for Atmospheric Research, Bouled, Colo., summer 1964. Fellow Am. Phys. Soc; mem. Am. Assn. Physics Tchrs., Soc. for Social Responsibility in Sci. (past pres.), Phi Beta Kappa, Sigma Xi, Sigma Pi Sigma. Author: The Physics of Electricity and Magnetism, 1959, 66 (Spanish edit. 1962); also articles on cosmic ray showers, multiple scattering of particles, cloud physics to profl. jours. Home: 570 Cranleigh Dr., Reno, Nev. 89502.*

SCOTT, William Wallace, Am. physician b. Kansas City, Kan., Jan. 27, 1913; s. Rev. Virgil B. and Margaret Alena (Smith) S.; A.B., U. Mo., 1934; Ph.D., U. Chicago, 1938, M.D., 1939; m. Jessie Louise McGraw, June 18, 1936; 1 son, William Wallace, Intern in surgery U. Chgo. Clinics, 1940-41, research asso., resident surgery, 1941-45; instructor urology University Chicago, 1943-45, asst. prof. urology, 1945-46; prof. urology, The Johns Hopkins Sch. of Medicine since 1946; urologist in charge, the Johns Hopkins Hosp. since 1946. Mem. Medical-Chirurgical Faculty of Md. Mem. A.M.A., (gold medal, Group I, 1940), Am. Urol. Assn., Am. Assn. Genito-Urinary Surgeons, Am. Physiol. Soc., Societe Internationale d'Urologie, Clin. Soc. Genito-Urinary Surgeons, La Sociedad Mexicana De Urologica, So. Soc. Clin. Research, Am. Assn. Cancer Research, Balt. City Med. Soc., British Assn. Urol. Surgeons (corr.), Am. Soc. Study of Sterility, Venezuela Urol. Assn., Endocrine Soc., Halstead Soc., Peripatetic Club, Sigma Xi. Author: (with H. W. Jones) Hermaphroditism, Genital Anomalies, and Related Endocrine Disorders, 1958. Editor: Investigative Urology, 1963-—. Many contbns. dealing with the physiology of prostate gland, prostatic cancer and benign prostatic hyperplasia and endocrine subjects. Home: R.F.D. 1, Box 275, Freeland, Md. 21053. Office: Johns Hopkins Hosp., Balt. 21205.*

SCOTUS, Michael, see Michael Scotus.

SCOULER, John, Brit. physician, naturalist; b. Dec. 31, 1804; M.D., Glasgow, Scotland, 1827, LL.D., 1850; studied in Paris; surgeon, naturalist Hudson's Bay Co., on voyage to Columbia River, 1824-25; practiced medicine, Glasgow; prof. geology Andersonian U., 1829; became prof. of mineralogy, later of geology, zoology, botany Royal Dublin Soc., 1834. Scouleria (plant genus) and Scoulerite (mineral) named in his honor. Fellow Linnean Soc. Editor Cheek's Edinburgh Jour. Natural and Geog. Science; co-founder Glasgow Med. Jour. Contbr. papers on natural history and meteorology to jours. Died Nov. 13, 1871.

SCOVELL, Melville Amasa, Am. chemist; b. Belvidere, N.J., Feb. 26, 1855; s. Nathan and Hannah (Aller) S.; B.S, U. Ill., 1875, M.S., 1857, Ph.D..; m. Nannie D. Davis, Sept. 8, 1880. Instr. chemistry, 1875-76 U. Ill., asst. prof., 1876-80, prof. agrl. chemistry, 1880-84; mgr. sugar factories, 1884; spl. agt. U. S. Dept. Agr., 1884-85; dir. Ky. Agrl. Expt. Sta., 1885-1910; dir. Agrl. Expt. Sta. also dir. Agrl. Coll., State U., Ky., 1910-—. Mem. U. S. Food Standard Com.; in charge pure food control, fertilizer control, concentrated feed control and seed control work of State of Ky. Fellow A.A.A.S., Soc. Chem. Industry, London; mem. Am. Acad. Polit. and Social Science; pres. Assn. Ofcl. Agrl. Chemists, Am. Assn. Agrl. Colls. and Expt. Stas. Inventor (with Henry A. Weber) widely used process of clarifying cane sugars by superheating; modified Kjedahl method where nitrates are present. Died 1912.

SCOVILLE, William Beecher, Am. neurosurgeon; b Phila., Jan. 13, 1906; s. Samuel and Katharine (Trumbull) S.; B.A., Yale, 1928; M.D., U. Pa., 1932; m. Emily Barrett Learned, June 30, 1934; children—Barrett, Alison (Mrs. Derek Carter Pershouse), Peter Gallaudet, m. 2d, Helene Deniau, May 12, 1961; children—William Beecher, Sophie Deniau. Dir. dept. neurosurgery Hartford (Conn.) Hosp., 1939-—; asso. clin prof. neurosurgery Yale Med. Sch., New Haven. Mem. A.C.S., Am. Psychiat. Assn., Am. Assn. Neurol. Surgeons, Am. Acad. Neurol. Surgery, Am. Acad. Neurology, Soc. Biol. Psychiatry, Assn. for Research in Nervous and Mental Diseases, World Fedn. Neurosurg. Socs. (founding mem., editor Bull.), also mem. numerous fgn. neurosurg. socs. Research, numerous publs. on cervical disc techniques, cortical undercutting, memory studies in monkeys; developer neurosurg. operative techniques and instruments. Home: 27 High St., Farmington, Conn. 06032. Office: 85 Jefferson St., Hartford, Conn. 06103.*

SCOW, Robert O., Am. physiologist; b. Dos Cabezas, Ariz, Nov. 17, 1920; s. Oliver and Kate (Redden) S.; A.B., U. Cal. at Berkeley, 1943, M.A., 1944; M.D., U. Cal. Med. Sch., San Francisco, 1946; m. Millicent Thomas, Sept. 25, 1948; children—John Robert, Kate Marie, James Werner, Ann Bruce. Sr. asst. surgeon NIH, USPHS, Bethesda, Md., 1948-52, surgeon, 1952-56, sr. surgeon, 1956-—, med dir., chief sect. endocrinology, 1961-—; vis. prof. dept. exptl. medicine, cancer research Hebrew U. Hadassah Med. Sch., Jerusalem, Israel, 1965-66. Recipient Borden award for undergrad. research U. Cal. Med. Sch., 1946; John Simon Guggenheim Meml. Found., 1955. Mem. Am. Assn. Anatomists, Am. Physiol Soc., Endocrine Soc. (Schering scholar 1955), Sociedad Argentina de Endocrinologia y Metabolismo, Sociedad Argentina de Biologia, Phi Beta Kappa. Editorial bd. Am. Jour. Physiology, 1963-66, sect. editor Endocrinology and Metabolism, 1967-—. Research, numerous publs. on hormonal control of growth and fat metabolism, thyroid gland physiology, exptl. diabetes, ketosis of pregnancy, metabolism of adipose tissue and obesity, developed techniques for total pancreatectomy, perfusing isolated adipose tissue in lab. rat. Home: 11609 W. Hill Dr., Rockville, Md. 20852. Office: NIH, Bethesda, Md. 20014.*

SCOWEN, Eric Frank, English physician; b. Eng., Apr. 22, 1910; s. Frank Edward and Eleanor (Barnes) S.; M.D., St. Bartholomews Hosp. Med. Coll. U. London, 1934, D.Sc., 1962. Prof. medicine U. London; dir. med. professorial unit St. Bartholomew's Hosp. Mem. com. on safety of drugs Ministry of Health; chmn. Brit. Pharmacopoeia Commn.; mem. Poisons Bd. Home Office: vice-chmn. council Imperial Cancer Research Fund. Mem. Assn. Physicians, Endocrine Soc. Research in endocrinology and chemistry and genetics of inborn errors of metabolism. Home: 44 Lincoln's Inn Fields, London W.C.2. Office: St. Bartholomew's Hosp., London E.C. 1, Eng.*

SCRIBNER, Bourdon Francis, Am. chemist; b. Westernport, Md., Apr. 13, 1910; s. Bourdon Walter and Nellie (Mansfield) S.; B.S. in Chemistry, George Washington U., 1933, postgrad., 1933-36; M.S., U. Md., 1939; m. Sally Frances Mount, Feb. 23, 1963. Chemist spectroscopy sect. Nat. Bur. Standards, Washington, 1933-47, chief spectrochem. analysis sect., 1947-—. Recipient Meritorius award U. S. Dept. Commerce, 1948, award of merit Am. Soc. for Testing and Materials, 1958, Spectroscopy Soc. Pitts. award, 1959. Fellow Washington Acad. Scis., A.A.-A.S., Optical Soc. Am.; mem. Internat. Union Pure and Applied Chemistry, (mem. commn. on spectrochem. and other optical methods analysis 1961-—), Groupment pour L'Avancement Methodes Spectro graphique (hon., France), Am. Chem. Soc. Soc. for Applied Spectroscopy (medal award N.Y. sect. 1961), Sigma Xi. Research numerous publs. on atomic energy levels in spectra hafnium, lutetium, niobium, ytterbium; developed spectrochem. methods for analysis tin, uranium, bronze, cement, steels, carrier-distillation method for spectrographic analysis for trace elements, plasma jet excitation source. Home: 10618 Kenilworth Av., Bethesda, Md. 20014. Office: Nat. Bur. Standards, Washington 20234.*

SCRIBONIUS, Largus, Roman physician; flourished circa 1-50; student Rome at time of Tiberius. Accompanied Claudius on Brit. campaign, 43; his patron C. Julius Callistus obtained emperor's patronage for his writing; became personal physician to Claudius Caesar. Author: Compositiones medicamentorum (collections of prescriptions including description of opium extractions), circa 47. An empiricist whose work was used by Marcellus Empiricus and others.

SCRIMGER, Joseph Arnold, physicist; b. South Shields, Eng., Aug. 5, 1924; s John William and Winifred (Ryles) S.; B.S., King's Coll., U. Durham, Eng., 1949; M.Sc., U. Sask., Can., 1954, Ph.D., 1956; m. Stella Arkley, June 17, 1950; children—Paul, Joan Margaret, Elizabeth Anne, Malcolm George. Sci. asst. Ram-jet design Bristol (Eng.) Aeroplane Co., 1949-50; head devel. lab. Plessey Co., Eng., 1950-52; research scientist Pacific Naval Lab., Def. Research Bd. Can., Victoria, B.C., 1956-—, chmn. West Coast working party Oceanographic Services for Def. Can., 1965-—. Research, publs. in determination of amount of sodium in sodium layer of upper atmosphere, percentage of energy relative to continuum at bottom of solar Fraunhoffer D lines; illustrated effects of surface waves on normal mode sound propagation in shallow water ducts. Home: 1247 Rockcrest Av. Office: Def. Research Lab. Pacific, Victoria, B.C., Can.*

SCRIVEN, Michael, philosopher of sci.; b. Beaulieu, Eng., Mar. 28, 1928; s. Victor Reginald and Hilda (Grice) S.; B.A., U. Melbourne (Australia), 1948, M.A., 1950; D.Phil., Oxford (Eng.) U., 1956. Came to U.S, 1952, naturalized, 1965. Research office P.O. Labs., Melbourne, 1948, Commonwealth Sci. and Indsl. Research Orgn., Melbourne, 1949; instr., research asso. dept. philosophy., also Minn. Center for Philosophy Sci., U. Minn., 1952-56; asst. prof. Swarthmore Coll., 1956-60; prof. dept. history and philosophy sci. Ind U., Bloomington, 1960-—. Mem. adv bd. Behavioral Research Labs., Inc. Trinity Coll. scholar, Melbourne, 1947-48; Nuffield grantee Oxford U., 1951-52; research fellow Center for Advanced Studies in the Behavioral Scis., 1962. Mem. A.A.A.S., (council), Social Sci. Edn. Consortium (dir.). Author: (with A. Calvin) Psychology, 1961; (with others) Computers and Comprehension, 1964; Applied Logic: An Introduction to Scientific Reasoning, 1965; Primary Philosophy, 1966; also numerous articles. Improved prediction process for drift rate Australian fundamental standard time; research in parapsychology, analysis concepts sci explanation, prediction, classification, law, definition; refined exptl. design for placebo effect studies; designed psychotherapy outcome research. Home: Red Ridge, Rural Route 1, Bloomington, Ind. 47401.*

SCRIVNER, Lloyd Herbert, Am. veterinarian; b. Belleville, Kan., July 6, 1902; s. Herbert John and Ada Mary (Simmons) S.; D.V.M., Colo. State U., 1929; M.S., Cornell U., 1939; m. Gladys Myrtle Troutman, Dec. 26, 1929; children—Ronald Keith, Gary Wayne. Asst prof. U. Wyo., 1929-43; veterinarian Maplecrest Turkey Farms, Wellman, Ia., 1943-48; sta. veterinarian, head dept. vet. sci., U. Ida., Moscow, 1948-67; hon. prof. U. Guayaquil, Equador, 1954-56. Mem. Am., Ida. vet. med. assns., Poultry Sci. Assn., Sigma Xi, Phi Kappa Phi, Alpha Zeta, Alpha Psi Research, publs. on fowl paralysis, bacterial diseases of poultry; bacterial, viral diseases of cattle and sheep, genetic resistance to ovine parasitism studies. Home: 103 S. Polk St., Moscow, Ida. 83843.*

SCROPE, George Julius Paulett, Eng. geologist; b. London, Mar. 10, 1797; s. John Paulett and Charlotte (Jacob) Thomson; ed. Pembroke Coll., Oxford, Eng., St. John's Coll., Cambridge; B.A., 1821; m. Emma Phipps, 1821. Studied volcanic dists. in travels through Italy, Sicily, Germany; mem. Parliament from Stroud, Eng., 1833-68. Became fellow Royal Soc., 1826; mem. Geol. Soc. (became joint sec. 1825, Wollaston medal 1867). Author: Considerations on Volcanos, 1828; Geology of the Extinct Volcanoes of Central France, 1826; Principles of Political Economy, 1833; also articles. Attempted (with Sir Charles Lyell) to replace Wernerian theory of neptunism with uniformitarianism; witnessed eruption of Vesuvius, 1822. Died 1876.

SCUDAMORE, Sir Charles, Brit. physician; b. Wye, Eng., 1779; s. William and Elizabeth (Rolfe) S.; student Guy's, St. Thomas's hosps., London; M.D., Glasgow, Scotland, 1814; m. Georgina Johnson, 1811. Practiced medicine, London; apptd. physician to Prince Leopold of Saxe-Gotha, 1820; attending physician Duke of Northumberland (lord-lt.), Dublin, 1828; hon. mem. Trinity Coll., Dublin. Licentiate Royal Coll. Physicians. Fellow Royal Soc, 1824. Author: A Treatise on the Nature and Cure of Gout (early contbn. to study of distbn. of gouty changes throughout body, also contains 1st mention in English writing of frequent presence of circular chest in persons subject to gout, 1816, 17, 19, 23; A Chemical and Medical Report, 1820; An Analysis of the Tepid Springs of Buxton, 1820; An Essay on the Blood, 1824; on Colchicum, 1825; Observations on Le Laennec's Diagnosis, 1826; A Treatise on Rheumatism, 1827; Cases illustrating the Remedial Power of the Inhalation of Iodine and Conium in Tubercular Phthisis, 1830, 34; On Pulmonary Consumption, 1847. Died Aug. 4, 1849.

SCUDAMORE, Harold Hunter, Am. physician; b. Wayne City, Ill., Dec. 8, 1915; s. Faye Walter and Edith (Hunter) S.; B.S., Mont. State Coll., 1937; M.S., Northwestern U., 1940, Ph.D., 1942; B M., Northwestern U., 1945, M.D., 1946; m. Virginia Gordon Haskins, Dec. 26, 1942; children—James Allen, Gordon Hunter, Susan Edith, Walter Edwin. Fellow in medicine Mayo Found. for Education and Research, 1949-51; faculty Mayo Grad. Sch., Rochester, Minn., 1952-—, asso. prof. medicine, 1964-—. Cons. Mayo Clinic, St. Mary's Hosp, Methodist Hosp., Rochester, 1952-—. Fellow A.C.P.; mem. Minn. Med. Soc., A.M.A., Am. Gastroent. Assn., Am. Fedn. Clin Research, A.A.A.S., Am. Med. Writers Assn., Am., Minn., Internat. socs. internal medicine. Research, numerous publs. on crustacean physiology malabsorption syndromes, diseases of small intestine, colon and pancreas, metabolic balance studies, motility studies, radioisotopes in gastrointestinal diseases. Home: 610 12th Av. S.W. Office: Mayo Clinic, 200 1st St. S.W., Rochester, Minn. 55901.*

SCUDDER, Geoffrey George Edgar, zoologist, entomologist; b. Fawkham, Kent, Eng., Mar, 18, 1934; s. George T.A. and Eva L. (Chapman) S.; B.Sc., U. Wales (U.K.), 1955; D.Phil., Oxford (Eng.) U., 1958; m. Jacqueline Howard, Apr. 7, 1958; 1 dau., Nicola Claire. Faculty, U B.C., Vancouver, Can., 1958-—, asso. prof. zoology, 1964-—. NRC Can., Soviet Acad. Scis. exchange scientist, 1965. Royal Soc. and Nuffield Found. Commonwealth Bursary fellow Imperial Coll., London (Eng.) U., 1964-65. Mem. Royal Entomol. Soc. London, Soc. for Study Evolution, Canadian Zool Soc., Internat. Assn. Limnology, Systematic Assn. Gt. Britain. Research, numerous publs. on insect morphology, reinterpretation insect ovipositor, taxonomy and systematics Lygaeidae of world, fresh-water insects; revised and described new taxa. Office: Dept. Zoology, U. B.C. Vancouver 8, B.C., Can.*

SCUDDER, John Milton, Am. physician; b. Harrison, O., Sept. 8, 1829; s. John Scudder; grad. Eclectic Med. Inst., Cin., 1856; m. Jane Hannah, Sept. 8, 1849; m. 2d, Mary Hannah, after 1861; at least 4 children. Became prof. spl. and path. anatomy Eclectic Med. Inst., 1857, prof. diseases of women and children, 1858-60, dean, 1861, prof. pathology and principles and practice of medicine, 1860-67, prof. hygiene and phys. diagnosis, 1887-94. Author: Practical Treatise on the Diseases of Women, 1857; Materia Medica and Therapeutics, 1860; Eclectic Practise of Medicine, 1864; Domestic Medicine; or Home Book of Health, 20 editions, 1865; Principles of Medicine, 1867; Eclectic Practice in Diseases of Children, 1869; Familiar Treatise on Medicine, 1869; Specific Medication and Specific Medicines, 1870; Specific Diagnosis, 1874. Editor Eclectic Med. Jour, 1861-94, Jour. Health, The Eclectic (lit. jour.), 1870-71. Originator specific medication principle. Died Daytona, Fla., Feb. 17, 1894.

SCUDDER, Samuel Hubbard, Am. naturalist; b. Boston, Apr. 13, 1837; s. Charles and Sarah Lathrop (Coit) S.; A.B., Williams, 1857, A.M., 1860; B.S., Lawrence Sci. Sch., (Harvard), 1862; Sc.D., Williams, 1890; LL.D., Western U. Pa., 1890; m. Jeannie Blatchford, June 25, 1867. Assisted Louis Agassiz in Mus. Comparative Zoology, 1862-64; sec. Boston Soc. of Natural History, 1862-70, custodian, 1864-70, pres., 1880-87; asst. librarian Harvard, 1879-82; palaeontologist U. S. Geol. Survey, 1886-92. Fellow Am. Acad. Arts and Scis.; A.A.A.S. (gen. sec., 1875); mem. Nat. Acad. Scis., Am. Philos. Soc. Author: A Century of Orthoptera, 1879; Catalogue of the Scientific Serials of All Countries, 1879; Butterflies: Their Structure, Changes and Life Histories, 1881; Nomenclator Zoologicus, 1884; The Winnipeg Country, 1886; Butterflies of the Eastern United States and Canada, 1889; A Classed and Annotated Bibliography of Fossil Insects, 1890; The Fossil Insects of North America, 1890; Index to the Known Fossil Insects of the World, 1891; Tertiary Rhynchophorous Coleoptera of the United States, 1893; Brief Guide to the Common Butterflies, 1893; The Life of a Butterfly, 1893; Frail Children of the Air, 1895; Revision of the American Fossil Cockroaches, 1895; Guide to the Genera and Classification of N.A. Orthoptera, 1897; Revision of the Orthopteran Group, Melanopli, 1897; Everyday Butterflies, 1899; Catalogue of the Described Orthoptera of the United States and Canada, 1900; Adephagous and Clavicorn Coleoptera from the Tertiary Deposits at Florissant, Colo., 1900; Index to North Am. Orthoptera, Described in the 18th and 19th Centuries, 1901 (Boston Soc. Natural History). Died May 17, 1911.

SCURLOCK, Ralph Geoffrey, English physicist; b. Southampton, Eng., Aug. 21, 1931; s. Walter H. and Linda (James) S.; B.A. with honors in Physics, U. Oxford (Eng.), 1954, M.A., D.Phil., 1958; m. Maureen Mary Oliver, Aug. 6, 1956; children—Jonathan, Rogin, Timothy, Alexander. Research fellow Oxford (U., 1957-59; faculty U. Southampton, 1959—, sr. lectr., 1966—, course supr. postgrad. course in cryogenics and its applications, 1966—. Mem. Inst. Physics (asso.). Author: Low Temperature Behavior of Solids, 1966; also articles. Research on nuclear orientation at low temperatures, specific heats of ferromagnetic and antiferromagnetic materials in zero and high magnetic fields, broad nuclear magnetic resonances in ferromagnetic alloys using variable frequency, spin echo, NMR spectrometers. Home: 22 Brookvale Rd., Southampton, Eng.*

SCURO, Ludovico Antonio, Italian physician; b. Maglie, Italy, June 29, 1924; s. Corradino and Raffaella (Andretta) S.; M.D. with honors, U. Rome, 1947, internal medicine specilization diploma with honors, U. Rome, 1953; m. Anna Coletta, Apr. 19, 1952; children—Loredana Maria Rosaria, Alberto Corrado. Faculty, U. Rome, 1947-59, asso. prof. Inst. Med. Pathology, 1956-59; asso. prof. U. Cagliari, Sardenia, Italy, 1960-64, asst. dir.. Inst. Med. Pathology and Clin. Medicine, 1960-64; prof. Inst. Med. Pathology, U. Padua (Italy), 1964—, asst. dir., 1964—. Decorated commendator Italian Republic, 1965. Mem. Med. Acad. Rome, Tri-Veneto Soc. Infectious Disease (directive com.), Soc. Internal Medicine, Soc. Infectious Disease, Soc. Endocrinology, Soc. Allergy. Author: Side-Effects of Antibiotics: Secondary to Variations in Intestinal Microflora, 1955; Infectious Diseases, 1967; also numerous articles. Among 1st to study relationship between antibiotic therapy and immunity and exptl. med. surgery; research on relation between pantothenic acid and antibody prodn., ultra-structural alterations of hepatic cells and presence of virus-like bodies in liver cells of patients with infectious hepatitis; clin. and metabolic aspects of hemachromatosis in particular, regarding morphology of liver observed by electron microscope; standarization of desferrioxamine test in diagnosis of sideroses, 1966. Home: 17 A Aleardi, Padua, Italy.*

SCYLAX OF CARYANDA, Greek mathematician, geographer; b. Caryanda, nr. Halicarnasus, Caria; flourished circa 500 B.C. Sent by Darius to explore course of Indus River and sailed down it to sea, then westward through Indian Ocean to Red Sea. Author: Periphus of the Parts Beyond the Columns of Hercules. Made coastal survey around Mediterranean, Euxine, and other connecting seas.

SEABORG, Glenn Theodore, Am. chemist; b. Ishpeming, Mich., April 19, 1912; s. H. Theodore and Selma (Erickson) S.; A.B., U. Cal. at Los Angeles, 1934; Ph.D., U. Cal. at Berkeley, 1937; LL.D., U. Mich., 1958, U. Mass., 1963; D.Sc., U. Denver, 1951, Gustavus Adolphus College, 1954, Northwestern U., 1954, U. of Notre Dame, 1961, Ohio State U., 1961, Fla. State U., 1961, University Md., 1961, Temple U., 1962, Tulane U., 1962, Drexel Inst. Tech., 1962, Georgetown U., 1962, U. State of N.Y., 1962, Mundelein Coll., 1963, Trinity Coll., 1963, U. Detroit, 1965; L.H.D., No. Mich. Coll., 1962, Nebraska Wesleyan Univ., 1964; D.P.S., George Washington Univ., 1962; D.P.A., U. Puget Sound, 1963; m. Helen Griggs, June 6, 1942; children—Peter, Lynne, David, Stephen, John Eric, Dianne. Research chemistry, U. of Cal., Berkeley, 1937-39, instructor department chemistry, 1939-41, assistant prof., 1941-45, prof., 1945— (leave of absence 1942-46, also 1961—); director of nuclear chemical research, 1946-58, associate director radiation lab., 1954-61, chancellor Univ. of Cal. at Berkeley, 1958-61; chmn. of the Atomic Energy Commn. 1961—; sect. chief metall. lab. Univ. Chicago, 1942-46. Mem. gen. adv. com. AEC, 1946-50, mem. jt. commn. radioactivity Internat. Council Sci. Unions, 1946-56; mem. com. standards and units of radio-activity Nat. Research Council, 1947-51; mem. numerous sci, edn. advisory bds. for U. S. Office Edn., NSF, etc. Named 1 of Am.'s 10 outstanding young men Jr. C. of C., 1947; recipient Award in Pure Chemistry, Am. Chem. Soc., 1947; John Ericsson Gold Medal, Am. Soc. Swedish Engrs., 1948; Nobel Prize in Chemistry (with E. M. McMillan), 1951; John Scott award and Medal of City of Phila., 1953; Perkin medal Am. Sect., Soc. Chem. Industry, 1957; U. S. AEC Enrico Fermi award, 1959; Sci. and Engring. award Fedn. Engring. Socs., Drexel Inst. Tech., Phila., 1962; named Swedish Am. of the Year by Vasa Order of Am., 1962; Franklin medal Franklin Inst., 1963; Charles L. Parsons award, American Chemical Society, 1964, Gibbs medal Chicago section, 1966; Leif Erikson award, Leif Erikson Found., 1964, Washington award Western Society Engineers, 1965. Fellow American Phys. Society, Chem. Society, London (hon.), Royal Society Edinburgh (hon.), Am. Nuclear Soc., Cal. Acad. Scis., N.Y. Acad. Scis, A.A.A.S., Royal Soc. Arts (Eng.); mem. American Chem. Society, Am. Acad. Arts and Scis., Am. Philos. Soc., Royal Swedish Acad. Engring. Scis., Am. Inst. Chemists, Am. Nat. Acad. Scis., Am. Scandinavian Found., Phi Beta Kappa, Sigma Xi. Author: The Transuranium Elements, 1958; (with Joseph J. Katz) The Chemistry of the Actinide Elements, 1957; (with E. G. Valens) Elements of the Universe, 1958 (winner Thomas Alva Edison Found. award); Man-Made Transuranium Elements, 1963; (with D. M. Wilkes) Education and the Atom, 1964; (with E. K. Hyde and I. Perlman) Nuclear Properties of the Heavy Elements, 1964; also articles in professional journals. Member editorial adv. bd. Jour. Inorganic and Nuclear Chemistry, 1954—; adv. bd. Chem. and Engring. News, 1957-59; mem. editorial bd. Jour. Am. Chem. Soc., 1950-59; mem. hon. editorial adv. bd. Internat. Ency. Phys. Chemistry and Chem. Physics, 1957—; mem. panel Golden Picture Ency. for Children, 1957— mem. cons. and adv. bd. Funk and Wagnells Universal Standard Ency., 1957—. Research in nuclear chemistry and physics and artificial radioactivity and compilation data in this field. Co-discoverer in 1940 of element 94 (plutonium), in 1941 of nuclear energy source isotope Pu-239, and in 1941, fissionable isotope of U233, isotope of Np 237, and in 1944 of element 95 (americium) and element 96 (curium); co-discoverer in 1949 of element 97 (berkelium); co-discoverer in 1950 of element 98 (californium); in 1952, element number 99 (einsteinium), in 1953 of element number 100 (fermium), in 1955 of element number 101 (mendelevium), in 1958 of element number 102; originator actinide concept for placing heaviest elements in periodic system; co-discoverer numerous radioactive isotopes, including I-131, Fe-59, Mn-54 and Co-60. Home: 3825 Harrison St., N.W., Washington 20015. Office: AEC, Washington 20545.*

SEABURY, Paul, Am. political scientist; b. Hempstead, N.Y., May 6, 1923; s. Adam Alden and Maude (Harris) S.; B.A. magna cum laude, Swarthmore Coll., 1946; Ph.D. in Govt., Columbia, 1953; m. Marie-Anne Phelps, June 3, 1950; children—David Phelps, John Huntington. Lect., instr. Columbia, 1947-53; guest lectr. Free Univ., Berlin, Germany, 1951; mem. faculty U. Cal. at Berkeley, 1953-65, prof. polit. sci., 1963-65, asst. dean Coll. Letters and Sci., 1963-64, vice chmn. dept. polit. sci., 1964-65; prof. govt., provost U. Cal., Santa Cruz, 1966-67; prof. govt. U. Cal., Berkeley, 1967—, chmn. faculty Coll. Letters and Sci., 1967—; vis. research scholar Center Internat. Affairs, Harvard, 1965-66. Chmn. nat. exec. com. Americans for Democratic Action, 1961-62, nat. vice chmn, 1963-65; cons. State Dept., 1963—, mem. bd. fgn. scholarships, 1967—. Guggenheim fellow, 1961-62. Mem. Am. Polit. Sci. Assn., Phi Beta Kappa. Democrat. Episcopalian. Author: Wilhelmstrasse: A Study of German Diplomacy Under the Nazi Regime, 1954; Power, Freedom and Diplomacy (Bancroft prize 1964), 1963; The Balance of Power, 1965; The Rise and Decline of the Cold War, 1967. Research on purposes and characteristics of U. S. behavior in world politics. Home: 600 Alvarado Rd., Berkeley, Cal. 94720. Office: U. Cal., Berkeley, Cal. 94720.*

SEAGER, Lloyd D., Am. pharmacologist; b. Farina, Ill., July 26, 1902; s. Lely Daniel and Biance (Maxson) S.; B.A., Milton Coll., 1925; M.S., U. Ill., 1929; M.D., St. Louis U., 1934; m. Oma Pierce, Aug. 24, 1927; children—Irving Donald, Miram (Mrs. Harold Braswell, Jr.), Loretta (Mrs. James E. Durst). Asst. in physiology U. Ill., 1927-29; asst. in physiology, instr. pharamacology St. Louis U., 1929-39; faculty U. Tenn. Med. Sch., Memphis, 1939-44, Womans Med. Coll. Pa., Phila., 1944-49; prof., chmn. dept. physiology U. Ark. Med. Center, Little Rock, 1949-56, prof., chmn. dept. pharmacology, 1949—. Mem. Soc. Exptl. Biology and Medicine, Am. Soc. Pharm. and Exptl. Therapeutics, A.A.A.S. Contbr. to Cyclo. of Medicine Surgery and Specialities, 1948, 49. Research, publs. on chemotherapy of malaria and trypanosomiasis, use of oral penicillin in bacterial chemotherapy, nervous and chem. control of pulmonary circulation and mechanisms of pulmonary edema, influence on offspring of tranquilizer drugs given animals during pregnancy, toxic studies on digitalis. Home: 6925 Rockwood Rd., Little Rock 72207.*

SEAGRAVE, Frank Evans, Am. astronomer; b. Providence, Mar. 29, 1860; s. George Augustus and Mary Greene (Evans) S.; ed. sch. of Rev. Charles H. Wheeler, 1869-74, Harvard Coll. Obs., 1875-77; M.A., Brown U., 1911. Owner, dir. new obs., North Scituate, R.I. Made 3 total solar eclipse expdns. and observed visually and photographically the reversing layer, corona and prominence spectra 1878-1901; photographed transit of Venus, at Providence, Dec. 6, 1882; accomplished important work in relation to Halley's comet which appeared in 1909, 10; made numerous observations of comets, also computations and observations of variable stars of different types to determine their periods and forms of their light curves; studied orbits of comets, asteroids, the planet Pluto. Died July 15, 1934.

SEAL, Michael, physicist; b. Weston-super-mare, Eng., Apr. 15, 1930; s. Carl Cyril and Ina May (Hurford) S.; B.A. (Major open scholar) Corpus Christi Coll., Cambridge (Eng.) U., 1952, M.A., 1956, Ph.D., 1957, postgrad.; m. Cynthia Ida Austin Leach, Aug. 7, 1954; children—David James, Anne Caroline, Rosemary Elaine, Susan Elizabeth, Christopher Henry. Head diamond research Engelhard Industries, Inc., Newark, 1959-67, tech. coordinator, 1965-67; dir. research Amsterdam Diamond Test and Devel. Center, 1967—. Fellow Inst. Physics (London, Eng.); mem. Am. Chem. Soc., Mineral. Soc. Am., Nederlandse Natuurkundige Vereniging. Research, publs. on reflection electron microscopy, phys. and chem. properties diamonds, plastic deformation, wear of diamond abraded with diamond, variation in friction between diamonds, thermal graphitization, etching with oxidizing agts, growth structures and surface structures natural and synthetic diamonds, factors affecting performance and efficiency diamond grits used in abrasive wheels, catalysis and properties of precious metals, powder tech. Home: Guido Gezellestraat 5. Office: Amsterdam Diamond Test and Devel. Center, Sarphatikade 12, Amsterdam, The Netherlands.*

SEALANDER, John Arthur, Jr., Am. zoologist; b. Detroit Lakes, Minn., Dec. 9, 1917; s. John A. and Mary (Grimsgaard) S.; A.B., Luther Coll., 1940; M.S., Mich. State U., 1942; Ph.D., U. Ill., 1949; m. Lucile Rehm, Aug. 24, 1947; children—John Arthur III, Judith Anne, Sarah Kristine. Faculty dept. zoology U. Ark., Fayetteville, 1949—, prof., 1959—. USPHS spl. fellow Lab. Zoophysiology, U. Alaska, 1963-64. Mem. Am. Physiol. Soc., Am. Soc. Zoologists, Ecol. Soc. Am., Am. Soc. Mammalogists, Internat. Soc. Biometeorology, A.A.A.S., Am. Inst. Biol. Scis., Soc. Exptl. Biology and Medicine, S.W. Assn. Naturalists, Sigma Xi, Phi Sigma. Author: (with C.E. Hoffman) Laboratory Manual of Elementary Mammalian Anatomy, 1958; (with W.L. Evans, D.A. Becker) General Zoology Laboratory Manual, 1959. Research, articles on temperature tolerance in vertebrates, acclimation of small mammals to environmental temperature, seasonal changes in morphology and physiology of small mammals, taxonomy, natural history and ecology of mammals. Home: 1527 Markham Rd., Fayetteville, Ark. 72701.*

SEALE, Roy Quincy, Am. mathematician; b. Dallas, Jan. 9, 1898; s. William Quincy and Josie (McGlothlin) S.; A.B., So. Meth. U., 1919; M.A., Columbia U., 1920; Ph.D., Stanford, 1935; m. Georgie Hudspeth, Dec. 30, 1923; 1 son, George Quincy. Asst. engr. Southwestern Bell Telephone Co., 1920-21, engr., 1921-22; faculty So. Meth. U., Dallas, 1922—, prof. math., 1944-63, prof. emeritus, 1963—; prof. math. and physics, head dept. Del Mar Coll., Corpus Christie, Tex., 1935-42. Mem. Am. Math. Soc., Math. Assn. Am., Tex. Acad. Sci., A.A.A.S., Sigma Alpha Epsilon, Alpha Sigma Lambda, Kappa Mu Epsilon. Author: Some New Proofs of Minkowski's Theorems on Quadratic Forms, 1934; Aerial Navigation Sheets, 1941; also articles. Patentee in field. Inventor oil well testing and producing equipment, airplane engine control devices, automobile and truck safety equipment, manifold pressure control devices. Home: 3609 University Blvd., Dallas 75205.*

SEALY, Will Camp, Am. surgeon; b. Roberta, Ga., Nov. 6, 1912; s. Hugh K. and Parkie (Camp) S.; B.S., Emory U., 1933, M.D., 1936; m. Marian Sanford, May 17, 1941 (dec. Nov. 1959); children—William Leigh, Marjorie Susan, Neil Hugh, Brian Harris; m. 2d, Jageqeline Wombie, Apr. 30, 1965. Faculty, Duke Med. Center, Durham, N.C., 1946—, prof. surgery, 1955—, chief thoracic and cardiovascular surgery, 1950—. Mem. surg. study sect. NIH, 1962-65. Mem. Am. So. surg. assns., Am. Surgeons Assn., A.C.S., So. Thoracic Surg. Assn., Soc. U. Surgeons, Soc. Vascular Surgeons, Soc. Thoracic Surgeons, Am. Assn. Thoracic Surgery, Alpha Omega Alpha. Author: (with Postelwhaite) Disease of Esophagus, 1960; also numerous articles. Research on cardiovascular and pulmonary surgery and physiology, extracorporeal circulation, hypthermia, shock. Home: 2232 Cranford Rd., Durham, N.C. 27706.*

SEAMAN, Gerald Robert, Am. biologist; b. Pottsville, Pa., May 20, 1927; s. Aaron A. and Shirley (Pollak) S.; B.A., William Coll., 1945; M.S., Fordham U., 1947, Ph.D., 1949. Instr. biochemistry Creighton U. Med. Sch., 1949-50; faculty U. Tex. Med. Br., Galveston, 1950-63, asso. prof. physiology and microbiology, 1960-63; prof. biology Hunter Coll., N.Y.C., 1963-66; prof. Roosevelt U., Chgo., 1966—. James W. McLaughlin fellow infectious diseases, 1960. Mem. A.A.A.S., Soc. Protozoologists, Am. Soc. Microbiologists, Soc. Gen. Microbiology, Soc. for Exptl. Biology and Medicine. Author: Experiments in Microbial Physiology, 1962; also articles. Research on metabolism of free-living and parasitic protozoa.*

SEAMAN, William, Am. chemist; b. Providence, Nov. 27, 1901; s. Allen and Annie (Sugarman) S.; B.A., Brown U., 1922, M.S., 1922; Ph.D., Cornell

U., 1929; m. Sylvia Sybil Bernstein, Dec. 27, 1925; childen—Gideon, Jonathan. Staff, USN Dept., also U.S. Dept. Agr., Washington, N.Y.C., 1922-27; research chemist Standard Oil Co. N.J., Elizabeth, 1929-31, Internat. Paper Co., Glens Falls, N.Y., 1931-33; with Am. Cyanamid Co., Bound Brook, N.J., 1933-— group leader, 1952-56, research fellow, 1956-—. Fellow Am. Inst. Chemists; mem. Am. Chem. Soc. (chmn. N. Jersey sect. analytical group 1958), Phi Beta Kappa, Sigma Xi. Author: (with Frank M. Biffen) Modern Instruments in Chemical Analysis, 1956; also numerous articles, chpt. in book, encys. Research on methods analytical chemistry, organic chemistry, radiation chemistry, applications radioisotopes and radioactivity; patentee catalytic chemistry, sci. instruments. Home: 244 W. 74th St., N.Y.C. 10023.*

SEAMAN, William B., Am. physician; b. Chgo., Jan. 5, 1917; s. Benjamin and Dorothy Seaman; student U. Mich., 1934-37; M.D., Harvard, 1941; m. Veryl Swick, Feb. 26, 1944; children—Cheryl D., William David. Instr., Yale Sch. Medicine, 1948-49; faculty Washington U., St. Louis, 1949-56, prof. radiology, 1954-56; prof., chmn. dept. radiology Columbia Coll. Phys. and Surg., N.Y.C., 1956—; dir. radiology service Columbia Presbyn. Med. Center, N.Y.C. Mem. Radiol. Soc. N.Am., Am. Coll. Radiologists, Am. Roentgen Ray Soc., Assn. U. Radiologists (past pres.), N.Y. Roentgen Soc. (past pres.), N.Y. Gastroent. Soc. (past pres.). Research, numerous publs. on radiol. manifestation of disease of gastro-intestinal tract; clin. devel. high pressure oxygen radiotherapy. Home: 47 DePeyster St., Tenafly, N.J. Office: 622 W. 168th St., N.Y.C. 10032.*

SEARCY, Alan Winn, Am. chemist; b. Covina, Cal., Oct. 12, 1925; s. Claude Winn and Esther (Scofield) S.; A.B., Pomona Coll., 1946; Ph.D., U. Cal. at Berkeley, 1950; m. Gail Vaught, Oct. 30, 1945; children—Gay, William, Anne. Faculty, Purdue U., 1949-54, asst. prof. chemistry, 1950-54; faculty U. Cal., Berkeley, 1954-—, prof. materials sci., 1958-—, asso. div. head inorganic materials div. Lawrence Radiation Lab., 1961-64, asst. to chancellor, 1963-64, vicechancellor, 1964-67; Fulbright lectr. phys. chemistry Inst. Physics, Bariloche, Argentina, 1960. Cons. Gen. Motors Tech. Center, 1956-64, Union Carbide, 1956-—, Gen. Atomic 1957-—; mem. com. on high temperature chemistry NRC, 1961-—. Guggenheim fellow, 1967-68. Fellow Am. Ceramic Society ; member Am. Chem. Soc., A.A.A.S., Phi Beta Kappa, Sigma Xi. Contbr. numerous articles to sci jours. Measurements of vapor pressures and dissociation pressures for high temperature materials and studies of kinetics of high temperature sublimation and gas-solid reactions; derivation of chem. thermodynamic data from equilibrium pressure measurement; devel. a conceptual model for predicting bond angles in covalently bonded molecules and ions. Home: 55 Via Farallon, Orinda, Cal. 94563. Office: U. Cal., Berkeley, Cal.*

SEARES, Frederick Hanley, Am. astronomer; b. Cassopolis, Mich., ay 17, 1873; B.S., U. al., 1895; LL.D., 1930; postgrad., Paris, 1900-01; LL.D., U. Mo., 1934; m. 1896, 2d, 1942; 1 child. Fellow, U. Cal., 1895-96, 98-99, instr. astronomy, 1896-98; prof., dir. Lawa Obs. Mo., 1901-09; supt. computing div. and ednl. publs. Mt. Wilson Obs., Carnegie Instn., 1909-25, asst. dir., 1925-40, research asso., 1940-46; now ret. Mem. Internat. Astron. Union (chmn. com. on stellar photometry 1919-38), Nat. Acad. Scis., A.A.A.S. (became v.p. 1936), Am. Astron. Soc., Am. Philos. Soc., Astron. Soc. Pacific (became 2d v.p. 1897, 3d v.p., 1898, pres. 1929, Bruce medal 1940), Washington Acad. Scis., Royal Astron. Soc. (fgn. asso.). Collaborating editor Astrophys. Jour., 1927-34, 41-45, asso. editor, 1934-41. Research on theory of orbits, pertubations, practical astronomy, gen. magnetic field of sun, distbn. of stars. Home: 351 Palmetto Dr., Pasadena 2, Cal.

SEARLE, George Frederick Charles, English scientist; b. Oakington, Eng., Dec. 3, 1864; s. W. G. Searle. Ed. Peterhouse Cambridge (Eng.) U., 1888; m. Alice Mary Edwards, 1904. Asst. demonstrator Cavendish Lab., Eng., 1888-90, univ. demonstrator exptl. physics, 1890-1935, war time demonstrator, 1940-45; tech. asst. Royal Aircraft Establishment, Eng., 1917-19; lectr. exptl. physics U. Cambridge, 1900-35. Participant Pan-Anglican Congress, 1908. Author: Experimental Elasticity, 1908; Experimental Harmonic Motion, 1915; Experimental Optics, 1925; Experimental Physics, 1934; Research, publs. on electro-magnetism, phys. universe, measurement of magnetic hysteresis. Died Dec. 16, 1954.

SEARLE, George Mary, astronomer; b. London, Eng., June 27, 1839; s. Thomas and Anne (Noble) S.; came to U. S., 1840; A.B., Harvard, 1857, A.M., 1860; Ph.D. (hon.), Catholic U., 1896. Asst. Dudley Obs., Albany, 1858-59; with U. S. Coast Survey, 1859-62; asst. prof. U. S. Naval Acad., 1862-64; asst. Harvard, Obs., 1866-68; joined Paulists, 1868; ordained priest, 1871; became sci. tchr. Paulist Sem.; apptd. chief prof. math. Catholic U., Washington, 1895; superior gen. Paulist Fathers, 1904-09. Author: Elements of Geometry, 1877; Plain Facts for Fair Minds, 1895; How to Become a Catholic, 1906; The Truth About Christian Science, 1916. Discovered asteroid Pandora (while with Dudley obs.). Died July 7, 1918.

SEARLE, Shayle Robert, statistician; b. Wanganui, New Zealand, Apr. 26, 1928; s. Frank William and Hazel L. (Pearce) S.; B.A., Victoria U., Wellington, 1949, M.A., 1950; Dipl. Math. Stat., Cambridge U., Eng., 1953; Ph.D., Cornell U., 1958; m. Helen Mary Croshere, Oct. 11, 1958; children—Susan Powdrell, Heather May. Research statistician New ealand Dairy Bd., Wellington, 1953-62; sr. sci. officer Dept. Sci. Indsl. Research, Wellington, 1960-62; research Cornell U., Ithaca, N.Y., 1956-59, asst. prof. biol. statistics, 1962-65, asso. prof., 1965-—. Found. mem. New Zealand Computing and Data Processing Soc. Fellow Royal Statis. Soc.; mem. Biometric Soc., Am., New Zealand statis. assns., Am. Dairy Sci. Assn. Author: Matrix Algebra for the Biological Sciences, 1966; also numerous articles. Research on applications of math. statistics to dairy breeding and biology in gen. Home: 505 The Parkway, Ithaca, N.Y. 14850.*

SEARLS, Robert Louarn, Am. biochemist; b. Madison, Wis., Oct. 26, 1931; s. Ed. M. and Ann (Haughey) S.; B.S., U. Wis., 1953; Ph.D., U. Cal. at Berkeley, 1960; m. Ellen Donovan, June 10, 1961; children—Timothy, David. Research fellow Brandeis U., Waltham, Mass., 1960-63; asst. prof. U. Va., Charlottesville, 1963-—. Mem. Am. Soc. Zoology, Am. Inst. Biol. Sics., Am. Chem. Soc., Growth Soc. Research, publs. on flavoproteins and energy metabolism, differentiation and organogenesis. Home: Route 1, Earlysville, Va. 22936.*

SEARS, Ernest Robert, Am. geneticist; b. Bethel, Ore., Oct. 15, 1910; s. Jacob P. and Estella (McKee) S.; B.S., Ore. State Coll., 1932; M.A., Harvard, 1934, Ph.D., 1936; m. Caroline F. Eichorn, July 5, 1936; children—Michael, John, Barbara, Kathleen; m. 2d, Lotti Maria Steinitz, June 16 1950. Geneticist, U.S. Dept. Agr., Columbia, Mo., 1936—; faculty U. Mo., Columbia, 1936-—, prof., 1964-—. Recipient Hoblitzelle award Tex. Research Found., 1958. Mem. Agronomy Soc. (Stevenson award 1951), Nat. Acad. Scis., Am. Acad. Arts and Scis., Genetics Soc. Am., Genetics Soc. Can., Bot. Soc. Am., Am. Soc. Naturalists, Gamma Sigma Delta (award 1958), Sigma Xi. Research, publs. on origin, evolution, genetics, cytology and cytogenetics of wheat. Home: Route 1, Columbia, Mo. 65201.*

SEARS, Francis Weston, Am. physicist; b. Plymouth, Mass., Oct. 1, 1898; s. Walter Herbert and Ella (Blackmer) S.; S.B., Mass. Inst. Tech., 1921, S.M., 1924; Sc.D., Drexel Inst. Tech., 1960; m. Mildred Cornwall, Dec. 23, 1935. Faculty, Mass. Inst. Tech., Cambridge, 1921-55, prof. physics, 1941-55; Appleton prof. natural philosophy Dartmouth, Hanover, N.H., 1955-64, prof. emeritus, 1964-—. Mem. Am. Assn. Physics Tchrs. (pres. 1960, recipient Oersted medal 1960). Author: Principles of Physics, 1945; (with M.W. Zemansky) College Physics, 1947; (with Zemansky) University Physics, 1948; Thermodynamics, 1950 (books translated into 7 langs.). Home: Pine Tree Rd., Norwich, Vt. 05055.*

SEARS, Paul Bigelow, Am. ecologist; b. Bucyrus, O., Dec. 17, 1891; s. Rufus Victor and Sallie (Harris) S.; B.S., Ohio Wesleyan U., 1913; D.Sc., 1937; M.A., U. Neb., 1915, LL.D., 1957; Ph.D., U. Chgo., 1922; Litt.D., Marietta Coll., 1951; LL.D., U. Ark., 1957, Wayne State U., 1959; D.Sc., Oberlin Coll., 1958, Bowling Green State U. Instr. botany Ohio State University Columbus, 1915-19; faculty U. Neb., Lincoln, 1919-27, asso. prof., 1925-27; prof. botany, head dept. U. Okla., Norman, 1927-38; prof. botany, head dept. Oberlin (O.) Coll., 1938-50; chmn. conservation program Yale, New Haven, 1950-60. Mem. Nat. Sci. Bd., 1958-64; mem. plowshare adv. com. AEC, 1959-—. Recipient Book of Month award, 1936; Garden Club Am. medal, 1951; Louis Bromfield medal Friends of the Land, 1958; Richard Prentice Ettinger Program for Creative Writing medal, 1963. Guggenheim fellow, 1958. Mem. Ecol. Soc. Am. (past pres., Eminent Ecologist award 1965), Nat. Audubon Soc. (past chmn. bd.), Am. Soc. Naturalists (past pres.), A.A.A.S., Am. Acad. Arts and Scis., Phi Beta Kappa, Sigma Xi, Phi Sigma. Author: Deserts on the March, 1935; This is Our World, 1937; Who are These Americans? 1939; Life and Environment, 1939; This Useful World, 1941; Charles Darwin, 1950; Where There is Life, 1962; The Biology of the Living Landscape, 1964; The Living Landscape; also numerous articles. Editorial bd. Daedalus, 1957-—. Research on postglacial climatic sequences in N.Am.; analysis of cultural patterns in applied ecology. Home: 4 N. Forest Circle, New Haven 06515.*

SEARS, Robert Richardson, Am. psychologist; b. Palo Alto, Cal., Aug. 31, 1908; s. Jesse B. and Stella (Richardson) S.; A.B., Stanford, 1929; Ph.D., Yale, 1932; M.A. (hon.) Harvard, 1950; m. Pauline Snedden, June 25, 1932; children—David O'Keefe, Nancy Louise (Mrs. Jonathan Shedd Barker). Instr., U. Ill., 1932-36; faculty Yale, 1936-42, asso. prof., 1942; prof. child psychology, dir. Ia. Child Welfare Research Sta., State U. Ia., 1942-49; prof. edn. and child psychology, dir. Lab. Human Devel., Harvard, 1949-53; prof. psychology Stanford, 1953-—, head dept. psychology, 1953-61, dean humanities and scis., 1961-—. Mem. Social Sci. Research Council, 1947-53, NRC, 1947-50. Mem. Am. Psychol. Assn. (past pres.), Soc. for Research in Child Devel., Am. Philos. Soc., Soc. Exptl. Psychologists. Author: (with others) Frustration and Aggression, 1939; Objective Studies of Psychoanalytic Concepts, 1943; (with Eleanor Maccoby, Harry Levin), Patterns of Child Rearing, 1957; (with Lucy Rau, Richard Alpert) Identification and Child Rearing, 1965; also articles. Application of learning and action theories to psychodynamics and personality devel. in children; research on child rearing as sources of personality characteristics, effects early sucking experience on infant orality. Home: 110 Golden Oak Dr., Portola Valley, Cal. 94026. Office: Humanities and Scis., Stanford U., Stanford, Cal. 94305.*

SEARS, William Rees, Am. aerodynamicist; b. Mpls., Mar. 1, 1913; s. William Everett and Gertrude (Rees) S.; B.Aero. Engring., U. Minn., 1934; Ph.D., Cal. Inst. Tech., 1938; m. Mabel Jeanette Rhodes, Mar. 20, 1936; children—David William, Susan Carol. Faculty, Cal. Inst. Tech., 1937-41; chief aerodynamics Northrop Aircraft, Inc., 1941-46; dir. Cornell U. Grad. Sch. Aero. Engring., Ithaca, N.Y., 1946-63, dir. Center for Applied Math., 1963-67, John Laporte Given prof. engring., 1962-—. Named Distinguished Alumnus, U. Minn., 1950. Fellow Am. Inst. Aeros. and Astronautics, Am. Acad. Arts and Scis., Internat. Acad. Astronautics: mem. Soc. Indsl. and Applied Math. (trustee). Author: The Airplane and Its Components, 1940. Editor: High Speed Aerodynamics and Jet Propulsion, vol. VI, 1956. Research, publs. on theoretical analyses of wings in unsteady motion, flow around rotating blades, boundary layers of yawed cylinders, interference between moving blade rows, dynamics of electrically conducting fluids. Home: 126 Honness Lane, Ithaca, N.Y. 14850.*

SEASHORE, Carl Emil, psychologist; b. Mörlunda, Sweden, Jan. 28, 1866; s. Carl Gustaf and Emily Charlotta (Borg) S.; A.B., Gustavus Adolphus Coll., 1891, D.Litt., 1937; Ph.D., Yale, 1895; Sc.D., 1935; LL.D., Wittenberg Coll., 1927; Sc.D., U. Pitts., 1931; LL.D., U. So. Cal., 1935; L.H.D., Augustana Coll., 1939; Mus.D., Chgo. Mus. Coll., 1939; m. Mary Roberta Holmes, June 7, 1900; children—Robert Holmes, Carl Gustav, Marion Dubois, Sigfrid. Asst. in psychol. lab. Yale, 1895-97; asst. prof. philosophy, State U. Ia., 1897-1902, prof. psychology since 1902, head dept. psychology, 1905-37, dean Grad. Coll., 1908-36, dean emeritus since 1936, dean pro tempore, 1942-46. Chmn. div. anthropology and psychology NRC, 1921-22. Fellow A.A.A.S., Soc. Exptl. Psychologists, Brit. Psychol. Assn. (hon.); mem. Nat. Acad. Scis., Acoustical Soc. Am., Am. Psychol. Assn. (pres. 1911). Author: Elementary Experiments in Psychology, 1908; Psychology in Daily Life, 1913; The Psychology of Musical Talent, 1919; Introduction to Psychology, 1922; Learning and Living in College, 1927; Trends in Graduate Study, 1931; Psychol. of Music, 1938; A Preview to College and Life, 1938; The Junior College Movement, 1941; Why We Love Music, 1941; Pioneering in Psychology, 1942; Psychology and Life in Autobiography, 1949; also papers on work and fatigue, mental work, illusions, psychology of music, gifted students. Editor: Univ. of Iowa Studies in Psychology, vols. 2-12; Studies in the Psychology of Music. Originator, Seashore Measures of Musical Talents, 1919, revised, 1939; In Search of Beauty in Music, 1947. Inventor audiometer, sound perimeter, voice-tunoscope, spark chromoscope, psychergograph, multiple recorder, other instruments useful in psychology of hearing. Died Oct. 16, 1949.

SEASHORE, Stanley Emanuel, Am. psychologist; b. Wahoo, Neb., Sept. 4, 1915; s. August T. and Jennie (Rose) S.; B.A., U. Ia., 1937; M.A., U. Minn., 1939; Ph.D., U. Mich., 1954; m. Eva Danielson, Aug. 31, 1940; children—Karen Rose, Christine Sigrid. Head personnel research, asst. dir. personnel U.S. Steel Corp., Pitts., 1939-45; staff cons. A.T. Kearney & Co., Chgo., 1945-49; mem. staff Inst. for Social Research, U. Mich., Ann Arbor, 1950-—, asst. dir., 1963-—, prof. psychology U. Mich., 1955-—. Fulbright fellow, Norway, 1956-57; Guggenheim fellow, Eng., 1965-66. Fellow Am. Psychol. Assn. (rep. to council 1965-—); mem. Am. Sociol. Assn., Sigma Xi. Author: Group Cohesiveness in the Industrial Work Group, 1954; (with D. Bowers) Changing the Structure and Functioning of an Organization, 1963; (with A. Marrow, D. Bowers) Management by Participation, 1967; also articles. Contbns. to study of social-psychol. processes in formal orngs., including original theoretical contbns. relating to communication, leadership, organizational change, group and organizational decision processes, mgr. devel. Home: 2270 Manchester Rd., Ann Arbor, Mich. 48104.*

SEATON, Edward Cator, English physician; b. Rochester, Eng., 1815; s. naval surgeon; M.D., Edinburgh, Scotland, 1837; surgeon N. Aylesford Union, Rochester; began practice in London, 1841; named insp. Gen. Bd. Health, 1858; became med. officer to local govt. bd, 1876. Fellow Royal Coll. Physicians; mem. Epidemiological Soc. (charter). Author: Handbook of Vaccination, 1868; also articles. Drew up report on small-pox and vaccination which led to passage of compulsory vaccination by Parliament, 1853. Died Jan. 31, 1880.

SEATON, Michael John, English physicist; b. Bristol, Eng., Jan. 16, 1923; s. Arthur William Robert and Helen (Stone) S.; B.Sc., U. Coll., London (Eng.), 1948, Ph.D., 1951; m. Olive May Singleton, June 10, 1943 (dec. Feb. 1959); children—Jane, Richard John; m. 2d, Joy Clarice Balchin, Oct. 20, 1960; 1 son, Anthony Michael. Asst. chemist Brit. Indsl. Solvents, 1940-42; staff physics dept. U. Coll., 1950-—, prof., 1963-—; chargé de recherche Institut d'Astrophysique, Paris, France, 1954-55. Fellow U. Colorado, 1961 Fellow of the Royal Society, member Royal Astronomical Soc., Inst. Physics and Phys. Soc., Internat. Astron. Union. Research numerous publs. on atomic collision processes especially in astrophysics, phys. processes in earth's atmosphere, gaseous nebulae, solar corona, stellar atmospheres. Home: 50 Ravensbourne Park, London, S.E. 6, Eng.*

SEAURAT, Edme-Sébastien, French astronomer; b Paris, 1724; became asso. with dept. math. Ecole Militaire, 1753; succeeded Lalande as editor Connaissance des temps, 1772; mem. French Acad. Scis., 1763. Author: Essai de perspective à l'usage des artistes, 1750; Tables de Jupite, 1766. Described lunette of double image (diplantidienne), 1772. Died Paris, 1803.

SEAWARD, John, English civil engr.; b. Lambeth, London, Eng., Jan. 1786; began career as surveyor and architect for his father; dir. constrn. Vauxhall Bridge; mgr. lead mines, Wales; supt. constrn. Gordon's Dowson's, other Thames docks; made drawings for new London Bridge, 1823; founder Canal Ironworks, Millwall, 1824. Mem. Instn. Civil Engrs. Author: Observations on the Rebuilding of London Bridge, with an examination of the Arch of Equilibrium proposed by Dr. Hutton and an investigation of a new method for forming an arch of that description, 1824; Observations on the Advantages and Possibility of successfully employing Steam Power in navigation Ships between this country and the East Indies, 1829. Contbg. author: Steam Engines (Thomas Tredgold), 1850. Inventor tubular boilers used in navy, disconnecting cranks for paddle-wheel engines, telescopic funnel, self-acting nozzles for feed and for regulating saturation of water in marine boilers, double passages in cylinders (for both eduction and steam), cheese couplings for connecting and disconnecting screw propeller to and from engines. Died Mar. 26, 1858.

SEAWARD, Samuel, English engr.; b. Eng., 1800; became asso. with brother John in engring. work, 1826; asst. in formation of Diamond Steam Packet Co., 1829, built engines for boats which ran between Gravesend and London; brought out (with brother) direct-acting engines (Seaward's engines), 1836, also designed large wing-bridges, dredging machines, cranes, other dock apparatus, machinery for lead, sugar, saw mills; advocated auxiliary steam power for voyage to India. Fellow Royal Soc., 1841. Died May 11, 1842.

SEBASTIANI, Enzo, Italian chem. engr.; b. Terni, Italy, Mar. 6, 1928; s. Severino and Albertina (Dinelli) S.; Laurea in Chem. Engring., U. Rome (Italy), 1951, Libera docenza in Chem. Engring., Principles, 1966; m. Maria Druch, Aug. 7, 1954; children—Lilia, Paolo. Chem. engr., Polymer S.p.A. (Montecatini), Factory of Terni, 1951-63; faculty U. Rome, 1963-—. Author: (with A. R. Giona) Chemical Engineering Principles—Unit Processes; also articles. Research in vapor-liquid equilibria, thermodynamics of solutions, gas absorption with chem. reactions, very soluble gases. Home: Via Isonzo 9, Terni, Italy. Office: U. Rome, via Endossiana 18, Rome, Italy.*

SEBEK, Oldrich Karel, microbiologist; b. Prague, Czechoslovakia, July 3, 1919; s. Vaclav and Marie (Sediva) S.; D.Sc., Charles U., Prague, 1946; postgrad. Rutgers U.; m. Dawn Islea Birch, Apr. 5, 1958; 1 son, Karel William. Came to U.S., 1947, naturalized, 1955. Internat. fellow fermentation J.E. Seagram & Sons, Inc., Louisville, 1947-48; research scholar Rutgers U., 1948-49, Fordham U., 1949-50, Ohio State U., 1950-52; research asso. Upjohn Co., Kalamazoo, 1952-—, U. Cal., Berkeley, 1966-67. Mem. Am. Soc. for Microbiology (chmn. agrl. and indsl. div. 1965). Abstractor fgn. lit. Chem. Abstracts, 1950-—. Contbr. numerous articles to tech. jours., chpts. in books. Research on metabolic activity microorganisms, synthesis, transformation and function antibiotics, vitamins and their precursors, steroidal hormones, amino acids and pigments; prepared radioactive penicillin, potent hydrocortisone derivatives; devised and developed use of pigments in analytical microbiology; patentee in field. Home: 1002 Short Rd. Office: The Upjohn Co., Kalamazoo 49001.*

SEBERT, Hippolyte, French engineer; b. Verberie, France, Jan. 30, 1839; gen. of a brigade; dir. central lab. naval arty.; mem. French Acad. Scis., 1897. Research and publs. on ballistics; improved chronographic recording of ballistic phenomena, 1880; constructed device to record recoil action of gun barrels. Died Paris, Jan. 23, 1930.

SEBRELL, William Henry, Am. physician; b. Portsmouth, Va., Sept. 11, 1901; s. William Henry and Millicent (Overton) S.; M.D., U. Va., 1925; postgrad. George Washington U.; Sc.D., Alfred U., 1955, Rutgers U., 1956; m. Margaret Shirley Bruffey, June 16, 1926 (dec. 1945); children—Betty Clark (Mrs. William Dyer Liddle), William Henry; m. 2d, Eloise Hopkins Glover, May 13, 1950; 1 stepson, Page Glover. Commd. asst. surgeon USPHS, 1925, advanced through grades to asst. surg. gen., 1950; chief div. physiology NIH, 1943-48, dir. Exptl. Biology and Medicine Inst., 1947-50; dir. NIH, 1950-55; research cons., adminstr. instnl. research grants Am. Cancer Soc., 1955-57; dir. Inst. Nutrition Scis., Columbia, N.Y.C., 1957-—, R.R. Williams prof. pub. health nutrition, 1956-—; chmn. Williams Waterman Fund-Research Corp., N.Y.C., 1956-68. Mem. Food and Nutrition Bd., 1940-—; chmn. protein adv. group WHO-FAO-UNICEF, 1965-67; chmn. Com. on Recommended Dietary Allowances, 1965-—. Recipient Research medal So. Med. Assn., 1946; Army-Navy certificate appreciation, 1946; award distinction Grocery Mfrs. Am., 1950; Joseph Goldberger award in clin. nutrition A.M.A.; Conrad A. Elvehjem award Am. Inst. Nutrition, 1968. Fellow Am. Acad. Physicians, N.Y. Acad. Scis., N.Y. Acad. Medicine, Assn. Am. Physicians; mem. Nat. Vitamin Found. (past pres.), Am. Inst. Nutrition (past pres.; Mead Johnson award 1940), Am. Soc. Clin. Nutrition (past pres.), Med. Soc. Guatemala (hon.), Am. Chem. Soc., Am. Soc. Biol. Chemists. Author: (with Harris) The Vitamins, 3 vols., 1954; also numerous articles. Research on prodn. riboflavin deficiency dog, blood dyscrasias due to folic acid deficiency, differentiation liver necrosis and cirrhosis due to dietary deficiency, human malnutrition in world; pioneered descriptions human riboflavin deficiency adrenal lesion due to pantothenic acid deficiency. Home: 97 Aldershot Lane, Manhasset, N.Y. 11030. Office: 562 W. 168th St., N.Y.C. 10032.*

SECCHI, (Pietro) Angelo, Italian astronomer, astrophysicist; b. Reggio, Emilia, Italy, June 29, 1818; entered Society of Jesus, 1833. Lectr. in physics and math., Gregorian U., Rome, 1839, and at Jesuit college, Loreto, 1841; taught at Stonyhurst (Eng.) and Georgetown U. (U. S.), 1849; prof. astronomy and dir. observatory at Gregorian U., Rome, 1849 (allowed to remain in Italy when Jesuits expelled, 1870); principal founder, Assn. of Italian Spectroscopists; mem., French Acad. Scis., 1857. Author: Catalogo delle stelle di cui si è determinato lo spettro luminoso, 1867; Sugli spettri prismatici delle stelle fisse, 1868; Dell'unità delle forze fisiche, 1869; Le soleil, 1870; Le stelle, 1877. Best known for his work in stellar spectroscopy, made 1st systematic spectroscopic survey of heavens; pioneered in classifying stars by their spectral types (4 groups); studied sunspots, solar prominences; photographed solar corona during eclipse, 1860; invented heliospectroscope, star spectroscope, telespectroscope and meteorograph; also studied double stars, weather forecasting and terrestrial magnetism. Died Rome, Italy, Feb. 26, 1878.

SECHENOV (Setchenov), Ivan Michailovich, Russian physiologist; b. Teply Stan, USSR, Aug. 13, 1829; degree in physiology U. St. Petersburg (now Lenignrad, USSR), 1851; student under Helmholtz, DuBois-Reymond in Germany, 1856-63. Tchr. Mil. Acad., St. Petersburg, 1860-70, in Odessa, USSR, 1870-76; prof. physiology U. St. Petersburg, from 1876; chair physiology U. Moscow (USSR), 1891-1901. Author: Reflexes of the Brain, 1863; Who Must Investigate the Problems of Psychology and How, 1870. Pioneer physiology, neurology in Russia; studied breathing, central nervous system; work on reflexes influenced Pavlov; theorized that spinal reflexes are inhibited by action of cerebral cortex; localized reflex inhibitory centers in cord and oblongata (later known as Sechenov's centers). Died Moscow, Nov. 15, 1905.

SECHER, Ole, Danish physician; b. Copenhagen, Denmark, Mar. 6, 1918; s. Knud I. A. and Yrsa (Laursen) S.; grad. in medicine U. Copenhagen, 1937, D.Med.Sci., 1952; m. Lili Suell Kiersgaard, Dec. 16, 1945; children—Niels, Jorgen. Staff in pharmacology various hosps., Copenhagen, 1949-51; staff Pa. U., Children's hosps., Phila., 1951-52; chief anesthestist Danish Hosp. ship Jutlandia, Korea, 1952-53, U. Hosp. Copenhagen, 1953-—, Scandinavian Hosp., Seoul, Korea, 1958-59; prof. anesthesia U. Copenhagen, 1964-—. Instr., Anesthesiology Centre, Copenhagen, 1953-—; mem. traveling faculty WHO, Egypt, 1955, Bulgaria, 1966; rep. Danish Govt. to Pakistan, 1963. Mem. Scandinavian (pres. 1952-54, 64-66), Danish (pres. 1955-58, 65-—), socs. anesthesiology, European Congress Anesthesia (pres. Copenhagen 1966), Korean Med. Assn. (hon.), Assn. Anesthetists Gt. Britain and Ireland (asso.), Royal Soc. Medicine London, Am. Soc. Anesthetists. Author: From Ether inhalation to Anesthesiology, 1965; The Treatment of Thorasic Injuries, 1966; The Anesthesiologycal Treatment of Labour and Children, 1966; also numerous articles. Research on peripheral action of ether (neuromuscular blocking effect), clin. anesthesia, spl. children's anesthesia and anesthesia for ceasarian sect., thoracic anesthesia; constructured equipment for resuscitation of newborn, children's anesthesia. Home: 4, Kirkehl, Hellerup, Denmark. Office: Rigshospitalet, Copenhagen O, Denmark.*

SECRETAN, Marc Louis François, optical engr.; b. Lausanne, Switzerland, 1804; capt. engr. corps, Vaud, Switzerland; became prof. math. Acad. Lausanne, 1838; went to Paris, 1844; worked with optician Lerebours; built instruments for geodesics, electroplating, photography, a large equatorial telescope, also a telescope with silvered mirror as specified by Leon Foucault; supplied Paris obs. and other establishments. Died Paris, 1867.

SEDACER, Guillem (or Guillelmus Sedacerius, Sedacianus, Sedaciensis), Catalan alchemist, Carmelite. Author: Sedacina totius alchimiae, or Summa Sedacina (treatise on alchemy, includes recipes for transmutation, discusses alums, other salts, borax, oils, man, tortoise, cock), 1378; also (if he used the name Guillelmus Sadacious) De lapide philosophico, 4 vols. Died 1382 or 1383.

SEDDON, John Carl, Am. physicist; b. Schenectady, Dec. 18, 1911; s. Arthur W. and Winnie (Turpit) S.; B.S. in Elec. Engring., Union Coll., Schenectady, 1933, B.S. in Physics, 1934; M.A., U. Buffalo, 1936; postgrad. Cornell U.; m. Reba Mary Woodbridge, Dec. 10, 1949; children—Carol Lynne, Roy John. Instr., U. Cin., 1940-41; radio engr., physicist Naval Research Lab., Washington, 1941-58; cons., staff scientist Goddard Space Flight Center, Greenbelt, Md., 1958-—. Cons. space scis. bd. com. on ionospheric measurements, mem. working group on satellite ionospheric measurements Nat. Acad. Scis., 1958-59; mem. space scis. bd. com. on radio physics NASA, 1959-62. Recipient award in pure sci. Research Soc. Am., 1955. Fellow I.E.E.E.; mem. Internat. Sci Radio Union, Sigma Xi. Contbr. articles to tech. jours. Patentee in field. Developed new method making ionosphere measurements using rockets; demonstrated that ionosphere is a continuum; invented various electronic devices for use in communications and high frequency measurements. Home: 8720 Standish Rd., Alexandria, Va. 22308. Office: Goddard Space Flight Center, Greenbelt, Md., 20771.*

SEDGWICK, Adam, English geologist; b. Dent, Eng., Mar. 22, 1785; s. Richard and Margaret (Sturges) S.; B.A., Trinity Coll., Cambridge, 1808; LL.D., Cambridge U., 1866; D.C.L. (hon.), Oxford, 1860. Became Woodwardian prof. geology Cambridge, 1818; priest and cannon Cathedral of Norwich; became prebendary of Norwich, Eng., 1834. Fellow Geol. Soc. (became pres. 1830), Royal Soc., 1821 (Wollaston medal 1851, Copley medal 1863); mem. Brit. Assn. (became pres. 1833, pres. geol. sect. 1837, 45, 53, 60), London Soc. Geology (a founder), French Acad. Scis. Author: Discourse on the Studies of the University of Cambridge, 1833; also articles. Studies in geology Cornwall and Devon, red sandstone in No. part of Eng., geology of Lake Dist., also Wales, 1822; introduced (with Murchison) term, Devonian, for slates in Cornwall, W. Somerset, Devon. Died Cambridge, Jan. 27, 1873.

SEDGWICK, William Thompson, Am. biologist; b. West Hartford, Conn., Dec. 29, 1855; s. William and Anne Thompson (Barbour) S.; Ph.B., Sheffield Scientific Sch. (Yale), 1877; Ph.D., Johns Hopkins, 1881; Sc.D. (hon.), Yale, 1909; m. Mary Katrine Rice, Dec. 29, 1881. Instr. physiol. chemistry Sheffield Sci. Sch., 1878-79; fellow, asso. biology Johns Hopkins, 1879; successively asst. prof., asso. prof. and prof. biology, prof. biology and pub. health Mass. Inst. Tech. 1883-—. Biologist Mass. State Bd. Health, 1888-96; curator Lowell Inst., Boston, 1897-—; chmn. Pauper Instns.' Trustees, 1897-99, Instns.' Registrar, City of Boston 1899-1900; pres. bd. dirs. Sharon Sanatorium, 1902-—; mem. adv. bd. 1902-—. Mem. Dept. Health of Mass.; mem. Internat. Health Bd.; mem. Rockefeller Found.; chmn. Harvard Tech. Sch. Pub. Health. Pres. Soc. Am. Bacteriologists, 1900; fellow Am. Acad. Arts and Scis.; A.A.A.S. (v.p. 1905); Author: Principles of Sanitary Science and Public Health, 1902. Co-author: General Biology, 1886; The Human Mechanism, 1906; A Short History of Science, 1917. Asst. editor: Life and Letters of William Barton Rogers, 1896. Early advocate of pasteurization of milk; demonstrated use of chlorine to disinfect water and sewage; traced 2 typhoid epidemics to a polluted brook and contaminated milk supply. Died Jan. 26, 1921.

SÉDILLOT, Charles Emmanuel, French surgeon; b. Paris, Sept. 14, 1804; student intern in hosps.; began career in mil. medicine, 1825, became surgeon's aide; served in Polish campaign; named to faculty of Paris, 1835; apptd. head surgeon, prof. Val-de-Grace mil. hosp.; became prof. clin. surgery Faculty Strasbourg (France), 1841; also prof. Strasbourg mil. hosp.; promoted to rank maj. doctor of 1st class, 1850. Mem. French Acad. Scis., 1872, Acad. Medicine. Author: Manuel de médecine légale, 1830; De la plique polonaise, 1932; Relation de la campagne de Constantine de 1837, 1838; Traité de médecine opératoire, bandages et appareils, 1839; Recherches sur le cancer, 1846; De l'insensibilité produite par le chloroforme et par l'éther, 1848; De la l'infection purulente, or pyoémie, 1849; Nouvelles considerations sur l'emploi du chloroforme, 1850; Des règles de l'application du chloroforme aux operations chirurgicales, 1852. Applied 1st stomach fistula, named the operation gastrostomy, 1846; 1st to use word microbe, 1871. Died Sainte-Menehould, France, Jan. 29, 1883.

SEDLACEK, Jindrich, Czechoslovakian physiologists; b. Prague, Czechoslovakia, May 7, 1928; s. Karel and Anna (Merunkova) S.; M.D., Charles U., Prague, 1952, candidate med. scis., 1957; m. Milose Sedlackova Dec. 27, 1952; children—Lucie, Simona. With Inst. Physiology, Charles U., 1950-—, asst. prof.

physiology, 1961——, head lab. for Embryophysiology, 1962——. Recipient prize Czechoslovac Med. Soc. J.E. Purkynje, 1960. Mem. Czechoslovac Physiol. Soc., Czechoslovac Pediatric Soc. Author: Basis of Physiology of Childhood, 1963; Prenatal Development of Electric Properties of Brain Tissue, 1967; also numerous articles. Research on devel. central nervous system in prenatal period, electric properties of brain tissue, electric, membrane and biochem. and physiol. patterns of maturation in fetal brain tissue. Home: 13 Blanická, Prague 2. Office: Inst. Physiology, Albertov 5, Prague 2, Czechoslovakia.*

SEDLAK, Jirí, Czechoslovakian physician, bacteriologist; b. Rokyeany, Czechoslovakia, May 23, 1908; s. Joseph and Marie (Paliveorá) S.; student Charles U., Prague, 1933-39, M.D. 1946, Dr.med.scis., 1961; D.Sc., Czechoslovak Acad. Scis.; m. Kveta Hacová, June 21, 1941; children—Ivan, Milan. Staff, Inst. Epidemiology and Microbiology until 1954, Med. Faculty, Charles U., 1954-1961; prof. microbiology since 1961; later dir. reference lab. facultative pathogenic Enterobacteriaceae, chmn. Inst. Microbiology. Mem. Internat. Commn. Bacterial Nomenclature, Enterobacteriaceae subcom. Author: (with Risdee) Enterobacteriaceae-Infectionen, 1961; also numerous articles. Classification new serotypes of Enterobacteriaceae. Home: 15 Dobrovského, Prague 7, Czechoslovakia.*

SEDLIN, Elias David, Am. physician, orthopedic researcher; b. Bklyn., Jan. 21, 1932; s. Arnold Boris and Sonia Davidovna (Lipschitz) S.; B.S. in Biology, U. Ala., 1951; M.D., Tulane U., 1955; Medicinae Doktor, U. Göteborg (Sweden), 1966; m. Barbara Sue Zidell, July 9, 1960. Research asso., emergency room lectr., Henry Ford Hosp., Detroit, 1961-63, NIH fellow, 1963-64; jr. attending physician Detroit Receiving Hosp., 1962-63; spl. NIH fellow dept. orthopedic surgery Sahlgrenska Sjukhuset, Gothenburg, Sweden, 1964——; docent exptl. orthopedics U. Göteborg, 1965-——. Fulbright scholar, 1962; NSF Postdoctoral fellow, 1964; recipient P.D. McGehee award Mobile Gen. Hosp., 1956; Ludvic Hektoen gold medal A.M.A., 1963; Nicholas Andry award Assn. Bone and Joint Surgeons, 1964. Diplomate Am. Bd. Orthopedic Surgery. Mem. Orthopedic Research Soc., Phi Beta Kappa. Contbr. to Symposium on Bone Biodynamics, 1963, European Symposium on Calcified Tissues, 1965; also articles. Patentee stress X-rays of ankles. Research in metabolic bone diseases, soft tissue injuries of maj. joints, phys. properties of bone; developed microscopic index for diagnosis of osteoporosis, math. model for simulation of phys. properties of bone. Address: 2959 S.W. Bennington Dr., Portland 1, Ore.*

SEDMAN, Garrioch, Brit. psychiatrist; b. Barrowford, Lancashire, U.K., Apr. 2, 1933; s. William Douglas and Sarah (Wallbank) S.; M.B., Ch.B. (state scholar), U. Manchester (Eng.), 1957, diploma in psychol. medicine, 1962, M.D., 1964; m. Kathleen Poynton, Apr. 26, 1957; children—Kay, Ewen. House officer United Manchester Hosp., 1957; sr. officer Manchester Gen. Hosp., 1958; staff dept. psychiatry U. Manchester, 1959-64, sr. registrar, 1962-64; lectr. psychiatry U. Sheffield (Eng.), 1964——. Mem. Brit. Med. Assn., Royal Medico-Psychol. Assn. Research, publs. on hallucinations, mental states of unreality, effects of lysergic acid diethylamide, effects of brainwashing, psychol. experiences occuring in epilepsy, religious mystical experiences in psychiat. patients. Home: 303 Crimicar Lane, Sheffield 10, U.K.*

SEDOV, Leonid Ivanovich, Russian engr.; b. Rostov-on-Don, USSR, Nov. 14, 1907; s. Ivan Sedov; student Sch. Edn., North Caucusus U. (USSR), 1924-26; grad. Sch. Physics and Math., Moscow State U., 1930; Dr. Degree, Moscow U. 1938. Asst. prof., lectr., prof. Moscow Cergo Orjonikidze Aviation Inst., Moscow, 1930-35; chief engr., asso. chief lab. N.E. Jukovsky Aerohydrodynamic Inst., Moscow, 1930-47; prof. Lomonosov State U., Moscow, 1937——, dean Sch. Hydrodynamics, 1951; dean dept. theoretical mechanics V.V. Kuybishev Mil. Engring. Acad., Moscow, 1938-41; sr. sci. worker Math. Inst., USSR Acad. Scis., Moscow, 1945, head Math. Inst., 1953-——, head mech. dept., editor Mechanics mag. Inst. Sci. Information, 1952, co-editor Reports of Academy of Sciences mag., 1954, now co-editor Astranatica Acta. chairman Permanent Interdepartmental Com. on Coordination and Control Sci. Theoretical Endeavors in Field of Orgn. and Realization Interplanetary Travel, 1954-——; head Soviet delegation Internat. Astron. Congress, Copenhagen, 1955, Rome, 1956, Barcelona, 1957, Amsterdam, 1958, London, 1959, Stockholm, 1960, Washington, 1961, Varna, 1962; v.p. USSR nat. com. on theoretical and applied mechanics, 1956. Decorated Order Sign of Esteem, Order of Red Banner, Order of Lenin. Mem. USSR Acad. Scis., Internat. Astron. Fedn. (pres. 1959-61); hon. mem. Am. Acad. Arts and Sciences. Author of: Theory of Plane Flow of Liquids, 1939; Ricochets on Water, 1943; Some Irregular Movements of a Compressible Fluid, 1945; Distribution of Strong Shock Waves (Chapligin prize), 1946; Plane Problems of Hydrodynamics and Aerodynamics (States prize), 1950; Similarity and Dimensional Methods in Mechanics (States prize), 1951; On the Theory of the Flare-Up of Novae and Supernovae (Lomonosov prize), 1957; Theory of Constructing Mechanical Models of a Compact Fluid, 1960; Introduction to the Mechanics of Continuous Media, 1962. Research in ideal fluidity (including hydrofoil theory and gas dynamics), astrophysics, and inter-

planetary connections; important in devel. of Sputnik. Address: Mathematical Inst., Acad. of Scis. of USSR, Moscow, Russia.

SEE, Horace, Am. naval architect; b. Phila., July 17, 1835; s. Richard Colhoun and Margarita (Hilyard) S.; academic edn.; learned trade of machinist; m. Ruth Ross Maffet, Feb. 20, 1879. Supt., engr. William Cramp & Sons, Phila.; established in N.Y., 1889; cons. engr. Newport News Shipbldg. Co., other corps. Largely responsible for intro. of triple-expansion engines into vessels of U.S.N.; designed engines of cruisers Yorktown, Concord, Bennington, Philadelphia, Newark and Vesuvius; yachts, Atalanta, Corsair, Stranger, Peerless; steamer Monmouth; steamships, Mariposa, Queen of the Pacific, Caracas, Olivette; designed hull and machinery of steamships El Rio, El Valle, El Alba, Comus, Proteus, U. S. cruisers Yankee, Dixie, hosp. ship Solace; wrecking steamers Relief and Tasco, New York police launches, into which many of his inventions were introduced, such as hydropneumatic ash ejector, folding hatch cover; inventor cylindrical mandrel for face bearings, which makes it possible to produce with certainty a true bearing and journal, double furnace water tube boiler, pneumatic siphon fire hydrant, etc. Superintends the constn., performance and maintenance of vessels and machinery. Died 1909.

SEE, Thomas Jefferson Jackson, Am. mathematician, astronomer; b. near Montgomery City, Mo., Feb. 12, 1866; s. Noah and Mary Ann (Sailor) S.; A.B., L.B., S.B., U. of Mo., 1889; A.M., Ph.D., U. of Berlin, 1892; m. Frances Graves, June 18, 1907. In charge observatory of U. of Mo., 1887-89; volunteer observer Royal Observatory, Berlin, 1891; organized and had charge dept. of astronomy, and aided in organization of Yerkes Observatory, U. of Chicago, 1893-96; astronomer Lowell Observatory in charge survey of Southern heavens, 1896-98. Prof. mathematics, U. S. Navy, 1899; in charge 26-in. equatorial telescope, U. S. Naval Cbservatory, 1899-1902; prof. mathematics, U. S. Naval Acad., 1902-03, Naval Observatory, Mare Island, Calif., 1903-62. lecturer on sidereal astronomy, Lowell Inst., Boston, 1899. Commd. capt., 1913. Fellow or mem. numerous Am. and fgn. sci. socs. Author and editor of numerous books and papers including: Newton's Principia (1687) and Laplace's Mecanique Celeste (1790-1825): The Wave-Theory Outlined, 1938; Electrodynamic Waves, Currents, Magnets, 1938; Wave-Theory of Molecular Forces, 1939; Proofs of the Gravitational Waves, 1939; Nonsensical Theory of Expanding Universe Demolished, 1940; Several Profound Proofs of Waves in Nature Established, 1941; All the Disturbances of the Moon's Motion Fully Explained, 1942; Many New Tests of the Wave-Theory, 1943; Invariability of Sidereal Day (Royal Society Paper, Hill Lectures), 1949-50; Explosive Forces Investigated, also the Long Waves of Gravitation, 1946; Life of Laplace; Theory of Cosmical Magnetism, 1951-52, 53. Most important work appeared early in his career in astronomy; with 24 in. Clark refractor at Flagstaff, Ariz. and Mexico City examined about 200,000 fixed stars in zone between 15 and 65 degrees south inclination, which led to discovery of about 600 new double stars previously recognized by Sir John Herschel and other observers, 1896-98; when F. R. Moulton criticized his math. conclusions about multiple stars, he made an intemperate, unbalanced reply; after several other setbacks in his early career, he displayed a lack of balance in his claims and general thought; claimed his math. and astron. investigations proved, disproved, or extended several major astronomical theories; Newton was his hero; held theory of relativity to be highly immoral. Died Oakland, Calif., July 4, 1962.

SEEBECK, Louis Frederick, German physicist; b. Jena, Germany, 1805; s. Jean Thomas Seebeck; became gymansium tchr., Berlin, 1829; became asso. with Acad. War, Berlin, 1831; named dir. Tech. Inst., Dresden, Germany, 1843; prof. physics, Leipzig, Germany, from 1849. Author: Monographie sur l'angle de polarisation, 1830; Discours à la mémoire de A. Volta, 1846. Died Dresden, 1849.

SEEBECK, Thomas Johann, physicist; b. Reval, Estonia, Apr. 9, 1770; at least 1 son, Louis Frederick; became friend of Goethe in Jena, Germany, worked with him on color theory, effect of colored light; mem. Acad. Berlin, French Acad. Scis. 1825. Obtained 1st chem. combination of ammonia with mercuric oxide, 1808; discovered thermoelectricity, 1821; devised thermocouple; used thermoelement to measure temperature; built a polariscope; studied heat radiation, rotary effect of sugar solutions on plane-polarized light. Died Berlin, Dec. 10, 1831.

SEEBOHM, Henry, Brit. ornithologist; b. Horton Grange, Eng., July 12, 1832; s. Benjamin and Esther (Wheeler) S.; ed. Soc. Friends' Sch.; steel mfr., Sheffield, Eng.; made expdns. to collect and to study birds in Holland, Greece, Asia Minor, Scandinavia, Germany, Siberia, southern Europe, S. Africa. Fellow Royal Geog. Soc. (sec. 1890-95), Linnean Soc., Brit. Ornithologists' Union, Zool. Soc. Author: Siberia in Europe, 1880; Catalogue of Birds in the British Museum, 1881; Siberia in Asia, 1882; A History of British Birds and their Eggs, 1883-85; The Geographical Distribution of the Family Charadriidae, 1887; Classification of Birds, 1890, supplement, 1895; The Birds of

the Japanese Empire, 1890; Geographical Distribution of British Birds, 1893; Address to the Yorkshire Naturalists' Union, 1893; also papers. Obtained eggs of grey plover and of other rare bird species from valley of Lower Petchora, 1875; made valuable contbns. to ornithology during visit to valley of Genesei, 1877. Died Nov. 26, 1895.

SEEFELDER, Matthias, German chemist; b. Boos, Germany, Feb. 28, 1920; s. Mathias and Walburga (Fischer) S.; grad. chemist U. Munich (Germany), 1951, D. degree, 1951; m. Ingeborg Escher, May 17, 1952; children—Alexandra, Stephanie. Research chemist Badische Anilin- & Soda-Fabrik, Ludwigshafen/Rhein, Germany, 1951-57, group leader research, 1957-61, mng. dir. dyestuffs research lab., 1961-——. mem. Gesellschaft Deutscher Chemiker, Schweizerischer Verein der Chemiker-Coloristen, Verein der Textilchemiker und Coloristen. Research, publs. on new syntheses and reactions of organic compounds especially carbodiimides, pyridines, amide chlorides, carbamide chlorides, vinyl compounds and formamidines. Patentee in field. Home: 34, Freinsheimer Strasse. Office: 38 Carl-Bosch-Strasse, Ludwigshafen/Rhein, Germany.*

SEEGAL, Beatrice Carrier (Mrs. David Seegal), Am. immunologist; b. Santa Barbara, Cal., Jan. 29, 1898; d. Charles Frederick and Emily Hinckley (Baker) C.; A.B., U. Cal. at Berkeley, 1918, A.M, 1921; M.D., Johns Hopkins, 1924; m. David Seegal, July 8, 1925. Hooper Found. fellow, 1919-20; Am.-Scandinavian Found. fellow, Copenhagen, Denmark, 1921-22; NRC fellow pathology Harvard Med. Sch., 1924-26; faculty Columbia Coll. Phys. and Surg., 1927-——, prof. microbiology, 1959-66, prof. emeritus, 1966-——, acting exec. officer dept. microbiology, 1950-51. Staff OSRD, 1944. Mem. A.A.A.S., Soc. for Exptl. Biology and Medicine, Harvey Soc., Soc. for Exptl. Pathology, Assn. Pathologists and Bacteriologists, Assn. Immunologists. Research, publs. on physiologic and immunologic factors affecting microcirculation, immunopathology of exptl. and human renal. disease. Home: 39 Claremont Av., N.Y.C. 10027. Office: 630 W. 168th St., N.Y.C. 10032.*

SEEGER, Raymond John, Am. physicist; b. Elizabeth, N.J., Sept. 20, 1906; s. John and Nora Adelaide (Madden) S.; A.B., Rutgers U., 1926; Ph.D., Yale, 1929; D.Sc., Kent State U., 1958; D.Sc., U. Dubuque, 1960; m. Vivian Mary Millen, Sept. 4, 1929; children—Muriel Jane, John Mark. Sr. staff asso. (research) NSF, Washington, 1952-——; adjunct professor American Univ., 1956-——. Recipient Distinguished Service award USN, 1946. Mem. Am. Phys. Soc., Am. Assn. Physics Tchrs. (Distinguished Service citation 1956), History of Sci. Soc., A.A.A.S., Phi Beta Kappa, Sigma Xi, Sigma Pi Sigma, Omicron Delta Kappa. Author: Our Physical Heritage, 1937; Galileo Galilei, His Life and His Works, 1966. Editor: (with G. Temple) Research Frontiers of Fluid Dynamics, 1966; editorial cons. Philosophy of Sci. Jour., 1955-59; mem. adv. editorial bd. Physics Letters, 1962-——; asso. editor Jour. Applied Physics, 1949-51. Research on understanding electric breakdown of solids, interaction of shockwaves in fluids, critical analysis of nature and devel. of sci. thought, founds. of dynamics and quantum mechanics. Home: 4507 Wetherill Rd., Washington 20016. Office: NSF, 1800 G St., N.W., Washington 20050.*

SEEGERS, Walter Henry, Am. physiologist; b. Westgate, Ia., Jan. 4, 1910; s. William and Mary (Wente) S.; B.A., State U. Ia., 1931, M.S., 1932, Ph.D., 1934; D.Sc., Wartburg Coll., 1953; m. Lillian Entz, Dec. 31, 1935; 1 dau., Dorothy Margaret. Research asso., U. Ia., 1934-35, 37-42, Antioch Coll., 1936-37; research staff Parke, Davis and Co., 1942-45; prof. physiology U. Detroit, 1946-51; faculty Wayne State U., Detroit, 1945-——, prof. physiology, 1946-——, William D. Traitel prof. hematology, 1964-——, chmn. dept. physiology and pharmacology, 1948-——. Mem. Mayor's Com. Rehab. Narcotic Addicts, 1961. Recipient Research award Wayne chpt. Sigma Xi, 1957, Commonwealth Fund N.Y. award in support creative work, 1957-58. Fellow N.Y. Acad. Sci., Am. Coll. Clin. Pharmacology and Chemotherapy, A.A.A.S.; hon. mem. Chilean Hematology Soc., Turkish Med. Soc., Alpha Omega Alpha, Phi Beta Pi; mem. Am. Chem. Soc., Am. Soc. Biol. Chemists, Am., Canadian, Detroit (past pres.) physiol. socs., Am. Heart Assn., Soc. for Exptl. Biology and Medicine, Internat., Am. socs. hematology, Mich. Acad. Sci., Engring. Soc. Detroit, Harvey Soc. N.Y. (hon.), Societe Internationale Europeenne D'Hematologie (corr.), Sigma Xi, Phi Lambda Upsilon, Alpha Chi Sigma. Author: (with Elwood A. Sharp) Hemostatic Agents, 1948; (with John M. Dorsey) Living Consciously: The Science of Self, 1959; Prothrombin, 1961; Prothrombin in Enzymology, Thrombosis and Hemophilia, 1967; also numerous articles. Editor: Blood Clotting Enzymology, 1967; (with Shirley A. Johnson) Physiology of Hemostasis amd Thrombosis, 1967. Research on nutrition; introduced thrombin for hemostasis; enzyme isolation from blood; theory of blood coagulation mechanisms. Home: 406 Barclay Rd., Grosse Pointe Farms, Mich. 48236. Office: 1400 Chrysler Freeway, Detroit 48207.*

SEEGMILLER, Jarvis Edwin, Am. physician, biochemist; b. St. George, Utah, June 22, 1920; s. Edwin Dee and Eleanor (Jarvis) S.; student Dixie

Jr. Coll., 1938-40; A.B., U. Utah, 1942; M.D., U. Chgo., 1948; m. Roberta Eads, Nov. 1, 1950; children—Susan Dale, Robert, Lisa, Richard. Research asso. Thorndike Meml. Lab., Harvard Med. Sch., Boston City Hosp., 1951-52; vis. investigator Pub. Health Research Inst. N.Y.C., 1952-54; biochemist Nat. Inst. Arthritis and Metabolic Diseases, NIH, Bethesda, Md., 1949-51, clin. investigator, 1954-64, 65—, chief sect. on human biochem. genetics; vis. scientist U. Coll. Hosp., London, Eng., 1964-65. Research, numerous publs. on identification of intermediates in oxidative pathway of glucose metabolism, identification of enzyme defect in human hereditary diseases including alcaptunuria, histidinemia and x-linked uricaciduria causes of hyperuricemia in gout and mechanism of acute attack of gouty arthritis. Home: 10114 Parkwood Dr. Office. Clin. Center, NIH, Bethesda, Md. 20014.*

SEELE, Keith C(edric), Am. egyptologist, educator; b. Warsaw, Ind., Feb. 13, 1898; s. Henry D. and Ora Capitola (Dick) S.; B.A., Coll. of Wooster, 1922, L.H.D. (hon.), 1947; B.D., McCormick Theol. Sem., 1926; T. B. Blackstone fellow U. of Berlin, 1926-28; fellowship Inst. Internat. Edn., Berlin, 1926-27; further study, Berlin, 1930; Ph.D., U. of Chicago, 1938; m. Diederika A. H. Millard, June 29, 1929. Teacher Assiut Coll., 1922-23; mem. epigraphic survey expdn. Oriental Inst. U. of Chicago, Luxor and Sakkara, Egypt, 1929-36; instr. Egyptology, U. of Chicago, 1936-41, asst. prof., 1941-46. asso. prof., 1946-50, professor of Egyptology, 1950-64, professor emeritus, 1964—; director of Oriental Institute Egyptian Assuan High Dam Program, 1960—. Delegate to the twenty-third Congress Orientalists, Cambridge, 1954, Institut d' Egypte, Cairo, U.N. Conf. Conservation and Utilization Resources, Lake Success, N.Y., 1949. Mem. Am. Oriental Soc., Egypt Exploration Soc. (London), Societe Francaise d' Egyptologie (Paris), Deutsches Archäologisches Institut, Institut d'Egypte (Cairo corr. mem. 1949), Phi Beta Kappa. Author: Medinet Habu II: Later Hist. Records of Ramses III (with H. H. Nelson and J. A. Wilson), 1932; Medinet Habu III: The Calendar, the Slaughterhouse, and Minor Records of Ramses III (with H. H. Nelson, J. A. Wilson and S. Schott), 1934; Reliefs and Inscriptions at Karnak I. Ramses III's Temple within the Great Inclosure of Amon (with H. H. Nelson and S. Schott), 1936; II. Ramses III's Temple within the Great Inclosure of Amon and Ramses III's Temple in the Inclosure of Mut (with H. H. Nelson and S. Schott), 1936; The Mastaba of Mereruka (with P. Duell and C. F. Nims), 1938; Festival Scenes of Ramses III (with H. H. Nelson and others), 1940; When Egypt Ruled the East (with G. Steindorff) 1941, rev. 1957; Coregency of Ramses II with Seti I and the Date of the Great Hypostyle Hall at Karnak, 1940; Reliefs and Inscriptions at Karnak III; The Bubastite Portal (with S. Schott, G. R. Hughes and others), 1953; Medinet Habu V: The Temple Proper, Part I (with G. R. Hughes and others), 1957; The Tomb of Tjanefer at Thebes, published 1959; also articles in professional journals. Editor: Blackfeet and Buffalo, Memories of Life among the Indians, 1962; Jour. Near Eastern Studies, 1948—; Most Ancient Egypt, 1964. Excavated at Qustul, Nubia, 1962-64, discovering ancient Egyptian New Kingdom cemeteries. Home: 18531 Argyle Av., Homewood, Ill.

SEELEY, John Ronald, sociologist; b. London, Eng., Feb. 21, 1913; A.B. with honors, U. Chgo., 1942, also postgrad.; m. Margaret Mary DeRocher, June 19, 1943; children—Michael John, David Andrew, Ronald James, Peter Clair. Faculty, U. Chgo., 1945-46, U. Toronto, 1947-60, York U., 1960-63; with Canadian Mental Health Assn., 1946-47, Community Surveys, Inc., Indpls., 1953-56, Alcoholism Found. of Ont., 1957-60; prof. sociology Brandeis U., Waltham, Mass., 1963-66, chmn. dept., 1965-66; sociologist med. dept. Mass. Inst. Tech., 1963-66; dean program dir. Center Study Democratic Instns., Santa Barbara, Cal., 1966—. Cons. govt. agys., assns., 1954—. Fellow Center for Advanced Study in Behavioral Scis., Am. Sociol. Assn., A.A.A.S., Am. Assn. for Humanistic Psychology; mem. Am. Acad. Polit. and Social Sci., Canadian Polit. Sci. Assn., Canadian Psychol. Assn., Acad. Psychoanalysis, Internat. Assn. Social Psychiatry, Soc. for Study Social Problems, Phi Beta Kappa. Author: (with A.R. Sim, B. Loosley) Crestwood Heights, 1953; (with Junker, Jones) Community Chest, 1956; (with Keller) The Alcohol Language, 1958; The Americanization of the Unconscious, 1966; also numerous articles. Make 1st psychologically oriented study of community, study of community and its philanthropic activities, established relationship between price of alcohol and death rate from liver cirrhosis, attempt to restructure social theory in light of psychiat. knowledge and vice versa. Home: 8352 Vereda del Padre, Goleta, Cal. Office: Box 4068, Santa Barbara, Cal. 93103.*

SEELIGER, Heinz Paul Richard, German microbiologist; b. Bad, Warmbrunn, Silesia, Nov. 1, 1920; s. Wilhelm and Erna (Langer) S.; med. student U. Breslau, 1940; candidate medicine U. Leipzig (Germany), 1942-45; M.D., U. Bonn (Germany), 1950; m. Hildegard Barnitzke, Nov. 8, 1947; 1 dau., Dagmar. Staff, EUCOM Med. Lab., Heidelberg, Germany, 1946-49; faculty U. Bonn, 1950-65, asso. prof., 1960-65; prof., chmn. Inst. Hygiene and Microbiology, U. Würzburg, (Germany) 1965—. Cons., expert WHO, 1965—; permanent sec. med bacteriology Internat. Com. on Nomenclature of Bacteria, 1962—. WHO fellow, Denmark, 1955, U.S.A., Can., 1955, Togo, Western Africa, 1961. Recipient Pastour medal in silver, 1963. Mem. German, Am., English, Canadian, French microbiol. socs. Author: Listeriose, 1955; Listeriosis, 1961; Mykologische Serodiagnostik, 1958; also numerous articles. Research in enteric infections, food-borne diseases, listeriosis and mycoses, immunology of fungi, nomenclature and taxonomy of bacterial, paramed. edn., tropical hygiene. Home: 87 Würzburg, Armin-Knab-str. 12. Office: 87 Würzburg, Josef-Schneider Strasse 2, Germany.*

SEELIGER, Hugo, German astronomer; b. Bielitz-Biala, Germany, Sept. 23, 1849; ed. Heidelberg, also Leipzig, Germany; asst. Bonn Obs.; dir. obs., Gotha, Germany; prof. astronomy, dir. obs., Munich. Studied star distbn.; originator star and nebula theory for birth of nova. Died Munich, Dec. 2, 1924.

SEELY, Samuel, Am. engr. educator; b. N.Y.C., May 8, 1909; s. Abraham and Cecelia (Strulovici) S.; E.E. cum laude, Poly. Inst. Bklyn., 1931; M.S., Stevens Inst. Tech., 1932; Ph.D., Columbia, 1936; m. Helen Elizabeth Anderson, June 28, 1951; children—Kittredge, Martha Ann. Asst. prof. elec. engring. Coll. City N.Y., 1936-41; staff radiation lab. Mass. Inst. Tech., 1941-46; asso. prof. electronics Naval Postgrad. Sch., Annapolis, Md., 1946-47; prof. elec. engring. Syracuse (N.Y.) U., 1947-56, head dept., 1952-56; prof. elec. engring., head dept. Case Inst. Tech., Cleve., 1956-64; vis. prof. elec. engring. Johns Hopkins, 1965, City Coll., U. N.Y.., 1966. Fulbright lectr. Nat. Tech. U., Athens, Greece, 1950-60; head engring. sect. NSF, 1961-63. Fellow Am. Phys. Soc., I.E.E.E.; mem. Sigma Xi, Tau Beta Pi, Eta Kappa Nu. Author: (with J. Millman) Electronics, 1941; (with W.R. LePage) General Network Analysis, 1951; Electron Tube Circuits, 1950; Electrical Engineering, 1956; Radio Electronics, 1956; Introduction to Electromagnetic Fields, 1958; (with F. Reza) Modern Network Analysis, 1959; Electromechanical Energy Conversion, 1962; Dynamic Systems Analysis, 1964. Research in antenna theory, electronic instrumentation, radar, nuclear physics, solid state theory. Home: R.D. 1, Warpas Rd., Madison, Conn. 06443.*

SEEMAN, Melvin, Am. sociologist; b. Balt., Feb. 5, 1918; s. Morris and Sophie (Hirschman) S.; B.S., John Hopkins, 1944; Ph.D., Ohio State U., 1947; m. Alice R. Zerbola, June 30, 1944; children—Teresa Ellen, Paul Daniel. Tchr. pub. schs., Balt., 1937-44; faculty Ohio State U., Columbus, 1944-59; faculty U. Cal. at Los Angeles, 1959—, prof., 1961—. Recipient Helen L. DeRoy award for research in social problems, 1955. Mem. Am., Pacific (pres. 1965-66) sociol assns. Author: Social Status and Leadership, 1959; also articles. Editor: Sociometry, 1964-66. Research in clarifying alternative meanings of "alienation" and devel. of procedures for testing applicability of concept to problems in learning theory and mass soc. Home: 21532 Paseo Serra, Malibu, Cal. 90265. Office: U. Cal., Los Angeles 90024.*

SEEMANN, Berthold Carl, botanist; b. Hanover, Germany, Feb. 28, 1825; Ph.D., Göttingen, Germany; student botany, Kew, Eng., 1844-46; 1 dau. Naturalist, H.M.S. Herald in voyages to W. coast of Am. and Arctic seas, 1847-51; commissioned (with Col. Smythe) to report on Fiji Island, 1860; travelled to Venezuela and Nicaragua. Fellow Linnean Soc., Royal Geog. Soc. Genus, Seemannia named in his honor by Rogel. Author: The Botany of the Voyage, circa 1852, 57; Narrative of the Voyage, 1853. Described and collected plants during travels; introduced cannibal tomato, several species of palm and candle tree (Parmentiera cerifera) to cultivation. Died Javali, Oct. 19, 1871.

SEEMANN, Hugo Josef, German metallurgist; b. Stuttgart, Nov. 24, 1899; s. Jakob and Berta (Armbruster) S.; ed. Darmstadt Tech. Sch., also univs. Munich and Erlangen; Ph.D.; m. Dagmar Engelbrecht, July 2, 1938; 1 dau., Karin. Asst. scholar univs. Erlangen, Marburg, Greifswald and Munich; instr. Berlin Tech. Sch.; head metall. research dept. Dürener Metallwerke, Berlin; head dept. Metall. Research Inst., Saarbrücken; prof. physics of metals and metallurgy U. Saar, Saarbrucken; con. Metall mag. Mem. (hon.) Inst. del Hierro y del Acero, Madrid. Author articles on physics research and metals. Studies in ultrasonics, structure and properties of liquids; metallic conductivity; dampening effects; magneto-resistance; dispersion strengthening. Home: Hellwigstrasse 17, Saarbrücken 3. Office: Univ. des Saarlandes, Saarbrücken 15, W. Germany.

SEFSTRÖM, Nils Gabriel, Swedish chemist; mineralogist; b. Ilsbo, Sweden, June 2, 1787; med. degree, 1813; studied under Berzelius; practice medicine, 4 years; tchr. chemistry, Carlsberg, Sweden, 1813, Carolinum, Stockholm, Sweden; named dir. Sch. for Lords; Falun, Sweden; named prof. chemistry and sci. Caroline Institut Medicine and Surgery, 1820. Introduced sci. methods to iron industry; discovered element, vanadium (earlier reported by del Rio), 1831; developed theory of rock channels. Died Stockholm, Nov. 30, 1845.

SEGAL, Bernard Louis, physician; b. Montreal, Que., Can., Feb. 13, 1929; s. Irving and Fay (Shecter) S.; B.Ss., McGill U., Montreal, 1950, M.D., 1955; m. to Idajane Fischman, February, 17, 1963; one daughter, Jody Segal. Began as cardiology fellow Georgetown U., 1959; USPHS fellow St. George's Hosp., London, Eng., 1960-61; asso. prof. medicine Hahnemann Med. Coll. and Hosp., Phila., 1964—, asst. head cardiology sect., 1964—. Cons. VA hosps., Phila., 1964, Wilke-Barre, Pa., 1964. Recipient Research prize in cardiology, 1961. Diplomate Am. Bd. Internal Medicine, Am. Bd. Cardiovascular Diseases. Fellow A.C.P., Am. Coll. Cardiologist, Am. Coll. Chest Physicians. Author: (with William Likoff) Ausculations of the Heart, 1965; also articles. Editor: Theory and Practice of Auscultation, 1964; Engineering in the Practice of Medicine, 1966. Research in heart sounds and murmurs congenital heart disease, atrial and ventricular septal defects, coronary heart disease, pathology and arteriography, echocardiography, drugs in angina, microphone inside heart chambers, pericardial cysts, electrocardiolography.

SEGAL, Harry Louis, Am. physician; b. Phila., Jan. 13, 1900; s. Morris and Bertha (Adams) S.; B.S., Syracuse U., 1922, M.D., 1924; m. Evelyn Buff, Aug. 26, 1935. Practice medicine, specializing in internal medicine, gastroenterology, Rochester, N.Y., 1932—; faculty U. Rochester, Sch. Medicine, and Dentistry, 1932—, prof. medicine, 1960—, head gastrointestinal unit, 1959-65; physician-in-chief Genesee Hosp., 1956-67, coordinator med. edn. 1956-65; dir. tng. program in gastroenterology and cancer research NIH, 1959—, dir. gastrointestinal research project, 1946—. Recipient Rochester Acad. Medicine awards, 1938, 56, 59; Schindler award Am. Gastroscopic Soc., 1959. Diplomate Am. Bd. Internal Medicine, Subsplty. Bd. Gastroenterology. Fellow A.C.P., Nat. Am. Med. Assn.; mem. Am. Assn. U. Prof., Am. Heart Assn., Am. Soc. Gastrointestinal Endoscopy, Am. Gastroent. Soc., N.Y. Acad. Sci., A.A.A.S., Am. Psychosomatic Soc., Assn. Am. Med. Colls., Sigma Xi, Alpha Omega Alpha. Research, numerous publs. on methods of early diagnosis of gastric cancer, using photography, tubeless analysis, chem. studies of enzymes. Home: 30 Stoneham Rd., Rochester 14625. Office: 224 Alexander St., Rochester, N.Y. 14607.*

SEGAL, Irving Ezra, Am. mathematician; b. N.Y.C., Sept. 13, 1918; s. Aaron and Fannie (Weinstein) S.; A.B., Princeton, 1937; Ph.D., Yale, 1940; m. Osa Skotting, Feb. 15, 1955; children—William, Andrew, Karen. Instr., Harvard, 1941 Princeton, 1941-43; mem. Inst. for Advanced Study, Princeton, 1946-47; faculty U. Chgo., 1948-60, prof., 1957-60; prof. Mass. Inst. Tech., Cambridge, 1960—; vis. asso. prof. Columbia, 1953-54. Sr. NSF fellow U. Copenhagen, Denmark, 1958-59; exchange prof. Sorbonne, Paris, 1965. Guggenheim fellow, 1947, 51. Mem. Am. Math. Soc., Am. Phys. Soc., Am. Acad. Author: Mathematical Problems of Relativistic Physics, 1963; also many articles. Editor: (with W. T. Martin) Analysis in Function Space, 1964. Research in modern analysis and its applications to contemporary theoretical physics. Home: 57 Avon Hill St., Cambridge, Mass. 02140.*

SEGAL, S.; Dutch ecologist; b. Amsterdam, Holland, May 23, 1933; s. Meier and Anna (Scholte) S.; grad. U. Amsterdam, 1960; m. Lida Oldewelt, Jan. 27, 1958; 1 son, Joeske. Lectr. ecology Bot. Inst., U. Amsterdam. Mem. Brit. Ecol. Soc., Royal Netherlands Bot. Assn., Internat. Assn. Limnology, others. Author: Een vegetatiekundig ouderzoek van hogere waterplanten in Nederland; also numerous articles. Editor: Contactblad voor Oecologen. Research in ecology, especially of aquatic plants, vegetation of marshes, old walls, system of growth-forms aquatics, ecol. structure theories, succession theories, exptl. and cytological studies. Address: 103 Weesperzijde, Amsterdam, Netherlands.*

SEGALL, B., Am. physicist; b. N.Y.C., July 23, 1925; s. Jacob and Frances (Levine) S.; B.S., Bklyn. Coll., 1948; M.A., U. Ill., Ph.D., 1951; m. Annette Leisten, June 28, 1953; children—Paul, Lorraine, Jeffrey. Postdoctoral fellow U. Ill.; Oersted fellow U. Copenhagen (Denmark) at Inst. for Theoretical Physics, 1952-53; with Bendix Radio, Towson, Md., 1953; staff Radiation Lab., U. Cal. at Berkeley, 1953-54; with Gen. Electric Research and Devel. Center, Schenectady, 1954—. Fellow Am. Phys. Soc. Research, publs. on pion scattering from deuterium, nuclear structure and alpha decay in deformed nuclei, theoretical solid state physics especially in electronic structure of metals, semiconductors, electrical and optical properties of metals and semiconductors. Home: 2416 Brookshire Dr., Schenectady 12309. Office: Gen. Electric Research and Devel. Center, Schenectady 12301.*

SEGALOFF, Alert, Am. physician; b. West Haven, Conn., July 25, 1916; s. Samuel and Lena (Merrman) S.; B.S. with distinction, Sheffield Sci. Sch., Yale, 1937; M.S. in Anatomy, Wayne U., 1939, M.D., 1943; m. Ann Zaem, May 9, 1940; children—Ruth Tina, David Charles, Joel Paul. Faculty, Tulane U. Sch. Medicine, New Orleans, 1945—, prof. medicine, 1963—; mem. vis. staff Tulane unit Charity Hosp. La., 1945-65, cons. internal medicine, 1965—; mem. internal medicine staff Ochsner Clinic, New Orleans, 1945—; dir. endocrine research Alton Ochsner Med. Found.,

New Orleans, 1945—; attending physician Ochsner Found. Hosp., 1947—. Mem. or chmn. numerous coms., cons. Nat. Cancer Inst., NIH, 1956—; chmn. panel NRC-Nat. Acad. Scis., 1966—; vis. prof. oncology Cancer Research Inst., U. Cal. Med. Center, 1967. Recipient Distinguished Service citation Wayne State U. Coll. Medicine, 1966. Diplomate Am. Bd. Internal Medicine. Fellow A.C.P., A.M.A., Royal Soc. Medicine, Am. Coll. Clin Pharmacology and Chemotherapy; mem. A.A.A.S., Am. Assn. Cancer Research, Am. Soc. Clin. Oncology (dir. 1965—), Am. Fedn. Clin. Research, Am. Soc. Clin. Investigation, So. Soc. Clin. Research (pres. 1957-58), Am. Soc. Exptl. Pathology, Central Soc. Clin. Research, Endocrine Soc. (Ciba award 1951), Ewing Soc., Soc. Exptl. Biology and Medicine, So. Med. Soc., Sigma Xi, Alpha Omega Alpha. Contbr. numerous articles to med. jours., chpts. to med. textbooks. Editor: Steroids, 1966—; editorial adv. bd. Index. Chemicus. Research in advances in hormone therapy of breast cancer in animals and in man. Home: 231 Ridgewood Dr., Metairie, La. 70005. Office: 1520 Jefferson Hwy., New Orleans 70121.*

SEGERSTRALE, Sven Gustaf, Finnish zoologist; b. Borga, Aug. 9, 1899; s. Albert and Hanna (Forsterus) S.; Ph.D., U. Helsinki; m. Swendelin Synnöve, Aug. 4, 1929; children—Monica, Gunilla, Susanne. Asst., then head hydrobiology lab. Finnish Sci. Soc.; asst. biology labs. Inst. Marine Research; curator, head curator Zool. Museum, U. Helsinki; lectr. U. S. and Europe. Mem. Finnish Sci. Soc., Finnish Flora and Fauna Soc., Internat. Oceanography Found. Author: Det underbara livet; Krisen och svaret; also articles on marine fauna of Baltic Sea and its history. Research in biology of brackish water, Baltic sea fauna. Home: Advägen 5 A. Office: Norra Järnvägsgatan 13, Helsinki, Finland.

SEGNER, Johnann Andres (von), German naturalist, mathematician; b. Presbourg, 1704; tchr. physics and math., Jena, Göttingen and Halle, Germany. Fellow Royal Soc., 1738. Author: Elementa arithmetica et geometriae, 1730; De figurus superficierum fluidarum, 1751. Originated theory of capillarity; invented Segner's cycle; studied theory of spinning top; gave proof for Descartes' rule of signs for equations having only real roots. Died 1777.

SEGOND, Paul, French surgeon; b. Paris, 1851; named surgeon of hosps., prof. agrégé, 1883; became prof. operative medicine, 1905, prof. clin. surgery, 1909; mem. Acad. Medicine. Author: Les abces chauds de la prostate et le phlegmon periprostatique (thesis, commended by Acad. Scis. and Surg. Soc.), 1880. A founder of gynecol. teaching in Paris. Died Paris, 1914.

SEGRÉ, Corrado, Italian mathematician; b. Saluzzo, Italy, Aug. 20, 1860; prof. U. Turin, Italy. Studied n-dimensional space, curved geometric forms of 3 dimension in space of 4 dimensions, differential geometry of hyperspace. Died Turin, May 18, 1924.

SEGRE, Diego, immunologist; b. Milan, Italy, Feb. 3, 1922; s. Ulderico and Corinna (Cornaldi) S.; D.V.M., U. Milan, 1947; M.S., U. Neb., 1954; Ph.D., U. Wis., 1957; m. Mariangela Bertani, July 22, 1952; children—Carlo, Alberto. Came to U.S., 1952, naturalized, 1957. Asst. prof. U. Milan, 1947-51; research fellow French Ministry Edn., Pasteur Inst., Paris, 1949; Fulbright fellow Purdue U., Lafayette, Ind., 1951-52; asst. animal pathologist U. Neb., Lincoln, 1952-55; research asst. U. Wis., Madison, 1955-57, asst. prof., 1957-60; prof. U. Ill., Urbana, 1960—. Mem. Am. Assn. Immunologists, Soc. for Exptl. Biology and Medicine, A.A.A.S., Am. Vet. Med. Assn., Conf. Research Workers in Animal Diseases, U.S. Livestock San. Assn., Sigma Xi. Research, numerous publs. in exptl. support for regulation of antibody formation by natural antibody role of natural antibody in immunologic maturation of baby pigs; mechanism of action of adjuvants of immunity, termination of immunologic tolerance by antigen-antibody complexes. Home: 2010 Boudreau Dr., Urbana, Ill. 61801.*

SEGRÈ, Emilio, physicist; b. Tivoli, Italy, Feb. 1, 1905; s. Giuseppe and Amelia (Treves) S.; Ph.D., U. Rome, 1928; Dr. honoris cause, U. Palermo, Gustavus Adolphus Coll.; m. Elfriede Spiro, Feb. 2, 1936; children—Claudio, Amelia, Fausta. Came to U. S., 1938, naturalized, 1944. Fellow Rockefeller Found., 1930-31; asst. prof. U. Rome, 1932-36; dir. physics lab. U. Palermo, 1936-38; research asso. U. Cal. at Berkeley, 1938-43, prof. physics, 1945—; group leader Los Alamos Sci. Lab., 1943-46; hon. prof. San Marcos U., Lima, Peru; vis. prof. U. Ill., Purdue U. Guggenheim fellow, 1959. Decorated comdr. of merit Republic of Italy. Recipient Hofmann medal German Chem. Soc.; (with O. Chamberlain) Nobel Prize in Physics, 1959. Fellow Am. Phys. Soc., mem. Nat. Acad. Scis., Am. Philos. Soc., Heidelberg Akademie Wissenschaften, Acad. Scis. of Peru, Soc. for Progress of Sci., Uruguay, Soc. Italiani di fisica, Accad. Naz. Lincei, Italy. Author: Nuclei and Particles, 1964. Co-discoverer elements technetium, astatine, plutonium, slow neutrons, antiproton; research in nuclear, atomic and particle physics. Home: 36 Crest Rd., Lafayette, Cal. 94549. Office: Dept. Physics, U. Cal., Berkeley, Cal. 94720.*

SÉGUIER, Jean-François, French botanist; b. Nimes, France, Nov. 25, 1703; mem. expdns. of Maffei; mem. French Acad. Scis., Acad. Inscriptions, Acad. Belles-Lettres. Author: Bibliotheca botanica, 1740; Plantae veroneses, 1754. Specialist on plant life around Verona, Italy. Died Mimes, Sept. 1, 1784.

SEGUIN, Armand, French chemist; b. Paris, Mar. 1767; mem. French Acad. Scis. Author: Sur la manière de tanner les cuirs, 1796. Research (with Lavoisier) on physiol. chemistry, respiration, transpiration; stated that oxidation of carbon, hydrogen (supplied by food to produce heat) takes place in tubes of lungs; developed process of tanning leather in 3-week period. Died Paris, Jan. 24, 1835.

SEGUIN, Camille, French engr.; b. Annonay; developed (with brother Marc) idea of suspension bridges; worked on ry. from Lyons to St. Etienne, 1824, other ry. constrn., also port planning. Died Toulon, France, 1852.

SEGUIN, Edouard, psychiatrist; b. Clamency, France, Jan. 20, 1812; s. T. O. Seguin; ed. Coll. of Auxerre, Lycée St. Louis, Paris, France; M.D. (hon.), Univ. City N.Y. (now N.Y. U.), 1861; m. 2d, Elsie Mead, 1880; at least 1 son, Edward Constant. Opened sch. for idiots in France, 1839; came to U. S., circa 1850; an organizer sch. for defectives, Randall's Island, N.Y. Author: Traitement Moral, Hygiene et Education des Idiots, 1846; Idiocy and its Treatment by the Physiologic Method, 1866; New Facts and Remarks Concerning Idiocy, 1869; Family Thermometry, 1873; The Clinical Thermoscope and Uniformity of Means of Observations, 1875; Medical Thermometry and Human Temperature, 1876. Research on sci. method for treatment of insane. Died Mt. Vernon, N.Y., Oct. 28, 1880.

SEGUIN, Edward Constant, neurologist; b. Paris, France, 1843; s. Edouard Seguin; grad. Coll. Physicians and Surgeons, N.Y.C., 1764; m. Margaret Amidon, 3 children. Came to U. S., 1850. Served with M.C., U. S. Army, 1864-69, asst. surgeon, Little Rock, Ark., 1864-65, post surgeon in N.M., 1865-69; lectr. diseases of nervous system Coll. Physicians and Surgeons, 1868-73, clin. prof., 1873-87; founder, pres. Am. Neurol. Assn., N.Y. Neurol. Soc. Author: Opera Minora, 1884. Published paper on use of thermometer containing 1st temperature chart used in U. S., 1866; wrote papers on subcutaneous injection of quinine in malarial fevers in which he emphasized importance of a sterile hypodermic needle, 1867; studied mental and nervous diseases in Paris, 1869; contbd. to recognition of functional and organic nervous diseases; advocated use of drugs in mental and spinal diseases, developed treatment these diseases involving use of iodides. Died Feb. 19, 1898.

SEGUIN, Marc, engr., physicist; b. Annonay, Apr. 20, 1786; mem. French Acad. Scis., 1845. Author: On the Influence of Railways and the Art of Their Planning and Construction. Built 1st suspension bridge, 1823, 1st tunnel on line St. Etienne-Loire, 1823; invented bridge suspended by iron cables and constructed on Rhone between Tain and Tournon, 1824; built rail with rail head; invented tubular boiler and patented it, 1828; tested air propellor for helicopter, 1846, flaps for clack valves, 1864; proved theoretically existence of mech. heat equivalent; designed airplane with wings; brought about use of cast iron instead of iron rails and wood for iron crosspieces in rail tracks. Died Annonay, Feb. 24, 1875.

SEIBERT, Florence B(arbara), Am. biochemist; b. Easton, Pa., Oct. 6, 1897; d. George Peter and Barbara (Memmert) Seibert; A.B., Goucher Coll., 1918, LL.D., 1938; Ph.D., Yale U., 1923; Porter fellow, U. of Chicago, 1923-24, Sc.D., 1941; Sc.D., U. of Pa., 1945, Lafayette Coll., 1947; honorary Science Doctor, Woman's Med. Coll., 1950; Guggenheim fellow, U. of Uppsala, Sweden, 1937-38; unmarried. Chemist Hammersley Paper Mill, 1918-20; Van Meter fellow, Yale U., 1921-22; Porter fellow Am. Physiol. Soc., Yale U., 1922-23, U. of Chicago, 1923-24; instr. pathology and asst. Sprague Memorial Inst., Chicago, 1924-28, asst. prof. biochemistry and asso., 1928-32; asst. prof. biochemistry, Henry Phipps Inst., U. of Pa., 1932-37, asso. prof., 1937-55, prof., 1955-59; special cons. U. S. Public Health Service, 1958—, prof. biochemistry, emeritus, 1959—; dir. Cancer Research Lab. Mound Park Hosp. Found., St. Petersburg, Fla., 1964—; cons. VA, Bay Pines, 1963—. Experimental work on protein and distilled water pyrogens, 1922-24, on chemistry and immunology of tuberculosis under Com. on Med. Research of Nat. Tuberculosis Association, 1924-59; preparation National Standard for Tuberculins, 1939; International Standard for World Health Orgn., 1952; experimental work cancer, 1954-58, 64—. Awarded Ricketts prize, Chgo., 1924; Trudeau medal, Nat. Tuberculosis Assn., 1938; Garvan gold medal, Am. Chem. Soc., 1942; First Achievement award, Am. Assn., Univ. Women, 1943. National Achievement Award, 1944, Gimbel Philadelphia Award, 1946, John Scott Award, 1947; Distinguished Daughter of Pennsylvania, 1950; John Elliott award, Am. Assn. Blood Banks, 1962. Fel. A.A.A.S.; mem. Am. Assn. Blood Banks (hon.), Trudeau Soc. (hon.), Phi Beta Kappa, Sigma Xi. Has written many scientific articles and reviews. Pioneered in removal of pyrogens or fever-producing substances from parenteral solutions, blood used for transfusions; prepared tuberculin (reagent for skin testing for Tb.) standard; research on chemistry of tuberculin, immunology, isolation and elimination of bacteria in cancer. Home: 470 3d St. S. Office: 701 6th St. S., St. Petersburg, Fla. 33701.*

SEIBERT, Henri Cleret, zoologist; b. Caen, France, July 20, 1915; s. George K. and Louise (Cleret) S.; came to U. S., 1921, naturalized, 1934; B.S., Haverford Coll., 1937; M.S., Johns Hopkins, 1940; Ph.D., U. Ill., 1947; m. Alice White, June 16, 1941; children—Peter, Michael, Elizabeth. Fellow in zoology U. Ill., 1946-47; biologist Manhattan Project, 1943-46; faculty Ohio U., Athens, 1947—, prof., 1962—, chmn. dept. zoology, 1963—. Cons. Ohio Div. Wildlife, 1955— Mem. Am., Brit. ornithol. unions, Wilson, Cooper ornithol. socs., Ohio Acad. Sci., Ohio Herpetological Soc. Author: (with F.H. Krecker) Life Principles, 2 vols., 1953; also articles. Research in energy requirements of birds, vertebrate fauna of Southeastern Ohio. Home: Earich Rd., Athens, O., 45701.*

SEIBERT, R(ussell) J(acob), Am. botanist; b. Shiloh Valley, Ill., Aug. 14, 1914; s. Erwin W. and Helen A. (Renner) S.; A.B., Washington U., St. Louis, 1937, M.S., 1938, Ph.D., 1947; m. Isabelle L. Pring, Dec. 26, 1942; children—Michael, Donna, Lisa. With U. S. Dept. Agr., 1940-50, agt.-botanist, Haiti, Peru, 1940-46, botanist-geneticist, Costa Rica, 1947-49; dir. Los Angeles State and County Arboretum, Arcadia, Cal., 1950-55, Longwood Gardens, Kennett Square, Pa., 1955—; head dept. arboreta and bot. gardens Los Angeles County, 1952-55. Am. del. Internat. Soc. Hort. Sci., 1960—; chmn. Am. Hort. Council-U. S. A. hort. exhbn., 1960 Floriade., Rotterdam, Holland. Fellow A.A.A.S.; mem. Am. Hort. Soc. (pres. 1964-65), Am. Inst. Biol. Scis. Research on Bignoniaceae of Maya area, Hevea in Peru. Home: Longwood, Kennett Square. Office: Longwood Gardens, Kennett Square, Pa. 19348.*

SEIBOLD, Eugen, German geologist; b. Stuttgart, Germany, May 11, 1918; s. Josef and Marie (Geiger) S.; Abitur, Wilhelms-Oberschule Stuttgart, 1937; student U. Tübingen, U. Bonn (Germany), 1939-48; Dr.rer.nat., U. Tübingen, 1948; m. Ilse Usbeck, Apr. 18, 1952; 1 dau., Ursula, Asst. prof. Technische Hochschule Karlsruhe, 1952-54; faculty U. Tübingen, 1954-58; prof., dir. geol. dept. U. Kiel (Germany), 1958—. Author: The Sea, 1964; (with Brinkmann) Lehrbuch der Allgemeinen Geologie; also articles. Research on gen. geology especially sediments and marine geology. Home: 13 Moltkestrasse. Office: Geology Dept., U. Kiel, 23 Kiel, Germany.*

SEIBOLD, Herman Rudolph, Am. pathologist; b. Phila., Aug. 30, 1908; s. Frederick Leonard and Augusta (Meyer) S.; V.M.D., U. Pa., 1931; m. Clara Bond Taylor, Oct. 20, 1934; children—John Rudolph, Robert William, Jean Marie (Mrs. William Robert Brough), Edward Albert, James Richard. Research fellow Leukemia Research Fund, Henry Phipps Inst., U. Pa., Phila., 1931-32; with U.S. Dept. Agr., 1932-51, veterinarian, pathologist path. div., Washington, 1939-51, research veterinarian Plum Island Animal Disease Lab., Greenport, N.Y., 1956-67; head div. pathology Delta Regional Primate Center, Tulane, 1967— prof dept. pathology Auburn (Ala..) U. Vet. Sch., 1951-56. Mem. Am. Coll. Vet Pathologists (past pres.), Am. Vet. Med. Assn., N.Y. Acad. Scis., Internat. Acad. Pathology, Am. Soc. for Microbiology, Tissue Culture Assn. Contbg. author: Veterinary Necropsy Procedures, 1954. Editorial bd. Am. Jour. Vet. Research, 1962—. Research, publs. on effects of moldy feed in dogs, relation of Spirocerca lupi to malignant esophageal tumors in dogs, modification of foot-and-mouth disease virus by chronic residence in cell cultures. Home: 3746 Myrtle St., Slidell, La. 70458. Office: Three Rivers Rd., Covington, La. 70433.*

SEIDEN, Philip Edward, Am. physicist; b. Troy, N.Y., Dec. 25, 1934; s. Herman L. and Freida (Bress) S.; A.B., U. Chgo., 1954, B.S., 1955, M.S., 1956; postgrad. U. Cal. at Berkeley, 1957; Ph.D., Stanford, 1960; m. Lois L. Gotteiner, Sept. 12, 1954; children—Jeffrey A., Mark J. Scientist, Lockheed Missiles and Space div. Palo Alto, Cal., 1956-59; research asso. Stanford U. W. W. Hansen Labs. Physics, 1957-59; NSF Postdoctoral fellow U. Grenoble (France), 1960; staff mem., mgr. coop. phenomena group IBM Research Center, Yorktown Heights, N.Y., 1960—. Fellow Am. Phys. Soc., A.A.A.S., Sigma Xi. Research, publs. on solid state physics, especially magnetism, magnetic resonance, superconductivity, physics of metals, ferromagnetic resonance line width in ferrites; discovered field independent resonance mode in antiferromagnets; studies in superconductivity in simple metals. Home: 96 Stephen Dr., Tarrytown, N.Y. 10591. Office: P.O. Box 218 Yorktown Heights, N.Y. 10958.*

SEIDENBERG, Abraham, Am. mathematician; b. Washington, June 2, 1916; s. Harry and Fannie (Globus) S.; Ph.D., Johns Hopkins, 1943; m. Ebe S. Cagli, Sept. 11, 1939. Faculty, U. Cal., Berkeley, 1945—, prof. math., 1958—; vis. prof. U. Mexico, 1962, Harvard, 1963-64. Mem. adv. bd. Minnemath Coll. Geometry Project. Office Naval Research fellow Harvard, 1948-49; Guggenheim fellow, 1953-54. Mem. Am. Math. Soc., Am. Assn. Math. Author: Lectures in Projective Geometry, 1962. Studies of geometry via algebra. Home: 44 Florida Av., Berkeley, Cal. 94707.*

SEIFERT, Gerhard Johannes, German pathologist; b. Leipzig, Germany, Sept. 9, 1921; s. Johannes and Rosa (Starke) S.; M.D., U. Leipzig, 1947; m. Leonore Sallmann, Aug. 15, 1950; children—Andreas, Christoph. Faculty, Insts. Pathology, U. Leipzig, 1949-58, U. Munster, 1958-65; prof., chmn. Inst. Pathology, U. Hamburg, 1965—. Mem. Internat. Acad. Pathology, European, German socs. pathology, European Soc. Diabetology, English-German Med. Assn., others. Author: Pathologie kindl. Pankreas, 1956; (with J. Oehme) Pathologie and Klinik Cytomegalie, 1957; also numerous articles. Research on diabetes mellitus, pediatric pathology, pathology on salivary glands and pancreas, calcium metabolism and calciphylaxis. Address: 52 Martinistrasse, Hamburg 20, Germany.*

SEIFERT, Mathias Joseph, Am. physician, surgeon; b. Chgo., Mar. 2, 1866; s. Anthony V. and Margaret (Kannen) S.; ed. Catholic Normal Sch., St. Francis, Wis., 1885, Bryant & Stratton Bus. Coll., 1886, Chgo. Musical Coll., 1887; M.D., U. Ill., 1901; postgrad., clinics in 12 countries of Europe and U. S.; m. Mary C. Karst, Feb. 8, 1888; children—Earl, Myra, Marie. Tchr., choir dir., ch. organist, Chgo., 1855-1896; organizer and pres., Western Mus. Acad., 1888-96; intern and extern Marion Sims Hosp., 1899-1905; instr. gynecology Chgo. Polyclinic, 1900-05; asst. prof. physiology U. Ill. Coll. Med., 1900-05, instr. sr. medicine, 1901-05, adj. prof. operative surgery U. Ill. Med. Dept., 1904-09, prof. phys. diagnosis and anesthesiology, Dental Dept., 1904-09; prof., head dept. surgery Chicago Med. Sch., 1910-16; staff Alexian Bros. Hosp., 1901-06, St. Mary of Nazareth Hosp. since 1904; sr. surgeon and pres. exec. staff Columbus Hosp. since 1915. Fellow A.C.S.; mem. Ill., Chgo. med. socs., Miss. Valley Med. Editors Assn. (Ill. exec.). Author: Eccyesis, with Prolonged Lactation; Case Report, 1920; Synthesis of Medical Terminology, 1925; Gynecology for Nurses, 1925; Cardio-Vascular Health Maxims, 1927; Olympian Rules, 1928. Contbr. to (books) Obstetrics, Gynecology and Abdominal Surgery, 1920; International Clinics, 1920; also numerous articles to med. jours. Editorial staff Internat. Abstract of Surgery, Gynecology, and Obstetrics, 1913—. Made motion picture, A High Posterior Gastro-Enterostomy, which he exhibited from 1929. Radio speaker since 1926. Died Jan. 31, 1947.

SEIFERT, Richard, indsl. chemist; b. 1861; discovered reaction for prodn. of salicylic acid; introduced salicylic acid derivatives for rheumatism, neuralgia, also antiseptics for bladder and intestinal disorders; contbd. to comml. prodn. of saccharine, benzene antiseptics, therapeutics. Died 1919.

SEIFRIZ, William, Am. botanist; b. Washington, Aug. 1, 1888; s. Paul and Anna (Schmidt) S.; B.S., Johns Hopkins, 1918, Ph.D., 1920; postgrad, U. Geneva (Switzerland), Kings Coll., London, Eng., Kaiser Wilhelm Inst., Germany, 1920-22; m. Instr. botany U. Mich., 1923-25; prof. botany U. Pa., 1925-55. Mem. Bot. Soc. Am., Ecological Soc., Am. Chem. Soc., A.A.A.S., Phi Beta Kappa, Sigma Xi. Asso. editor Protoplasma, Biodynamica, and Jour. Colloid Science; writer of books and sci. articles on protoplasm, physiology, colloid chemistry and plant geography. Died July 13, 1955.

SEIGNETTE, Pierre, French apothecary; b. La Rochelle, France, Dec. 4, 1660; apothecary, La Rochelle; mem. French Acad. Scis.; 1st to prepare sodium potassium tartrate (Rochelle salt), 1672, introduced it as laxative, 1682. Died La Rochelle, Mar. 11, 1719.

SEIJI, Makato, Japanese dermatologist; b. Ka-Makura, Japan, Feb. 1, 1926; s. Susumu and Toka (Sato) S.; M.D., U. Tokyo, 1949, Ph.D. in Biochemistry, 1955; m. Yoshimi Murata, Dec. 9, 1953; children—Satoru, Hiroshi, Itaru. Asst. prof. Gumma U. Sch. Medicine, 1954-57; research asso. U. Ore. Med. Sch., 1957-58; staff Oxford (Eng.) U. Radcliff Infirmary, 1958-59; faculty Harvard Med. Sch., 1959-61; asso. prof. Juntendo U. Sch. Medicine, 1962-66; instr. Tokyo U. Sch. Medicine, 1962—; prof. dept. dermatology Toyko Med. and Dental U., 1966—. Recipient 2d Yamaji Sci. Promotion award, 1966; Golden award in exhibits Am. Acad. Dermatology and Syphilology, 1961. Mem. Japanese Dermatol. Assn., Japanese Biochem. Assn., Soc. for Investigative Dermatology, Royal Soc. Medicine, Am. Soc. for Cell Biology, N.Y. Acad. Sci. Author: Biochemistry of Skin, 1966; also articles. Electromicroscopic and biochem. research on mammalian melanocyte, melanin forming cell, mechanisms of melanogenesis; discovered melanosome concept. Home: 1-22-7 Okayama Meguro-ku, Tokyo, Japan.

SEIK-KOWA, Shinsuke, Japanese mathematician; b. Fujioka, Japan, 1642; founder sch. for math. Seki Sch. Author: Kai Fakudai no Ho, 1683; Hatsubi Sampo, 1674. Introduced Chinese algebraic methods to Japan; discovered functions, determinants; possibly developed yenri (form of integral calculus); solved 2 linear equations in 2 unknowns using determinants; studied magic squares and circles. Died 1708.

SEIKO, Sato, Japanese mathematician; flourished 1670; obtained value of 3.14 for pi; applied celestial element method to equations of 6th degree and lower. Author: Kongenki, 1666.

SEIPEL, John Howard, Am. neurologist; b. Pitts., Nov. 9, 1925; s. John Howard and Marie (Shaser) S.; B.S., Carnegie Inst. Tech., 1946, M.S., 1947; M.D., Harvard, 1954; Ph.D., Northwestern U., 1958; m. Janice Lois Duffney, July 4, 1959; children—Janice Marie, John, Tabitha, William. Research med. officer USPHS, Nat. Cancer Inst., Bethesda, Md., 1956-58; chief neurology lab. Georgetown Clin. Research Inst., FAA, Washington, 1961-66; cons. to Md. Psychiat. Research Inst., 1966—; attending physician in neurology Georgetown Hosp., D.C. Gen. Hosp., Fairfax Hosp., VA Hosp., 1961—. Mem. Am. Chem. Soc., A.A.A.A.S., Aerospace Med. Assn., N.Y. Acad. Sci., Am. Acad. Neurology (S. Weir Mitchell award 1966). Research, publs. in biomagnetics, biophys. field theory, passive elec. properties of biol. materials; discovered magnetic field arising from a biol. system, magnetic component of neuron impulse; reduced rheoencephalography to a clinically reliable method. Home: 5335 Summit Dr., Fairfax, Va. 22030.*

SEITELBERGER, Franz, Austrian physician; b. Vienna, Dec. 4, 1916; s. Franz and Berta (Heuritsch) S.; Dr.med., U. Vienna, 1940; m. Erika Weber, July 7, 1945; children—Edda, Linda, Rainald. Vis. scientist Max Planck Inst. Brain Research, Giessen (Germany), 1952-53; docent U. Vienna, 1954-58, prof. neurology, neuroanatomy, neuropathology, 1958-59, dir. neurol. inst., 1959—; vis. scientist Nat. Inst. Neurol. Diseases and Blindness, NIH, Bethesda, Md., 1960. Mem. World Fedn. Neurology (sec. research group neuropathology 1958—), Austrian Acad. Scis., Max Planck Soc. for Promotion Scis., Soc. Austrian Neurologists and Psychiatrists, Vienna Soc. Physicians, Assn. German Neuropathologists and Neuroanatomists, Am. Assn. Neuropathologists, French Soc. Neurology, British Neuropathol. Soc., others. Mgn. editor Acta Neuropathologica, 1961—. Contbr. numerous articles in field. Research in gen. neuropathology, especially degenerative changes of central nervous system; also on viral encephalitides and demyelinating diseases. Address: 97 Weimarerstrasse, Vienna 1190, Austria.*

SEITZ, Frederick, Am. physicist; b. San Francisco, Calif., July 4, 1911; s. Frederick and Emily Charlotte (Hofman) S.; A.B., Leland Stanford Jr. U., 1932; Ph.D., Princeton, 1934; Doctorate Hon. Causa University of Ghent, 1957; D.Sc., U. of Reading, 1960, Rensselaer Poly. Inst., 1961, U. Notre Dame, 1962, Marquette U., 1963, Carnegie Inst. Tech., 1963, Case Institute Technology, 1964, Princeton Univ., 1964, Northwestern U., 1965, University of Delaware, 1966; LL.D., U. Mich., 1965, Lehigh U., 1966; m. Elizabeth K. Marshall, May 18, 1935. Instr. in physics, U. of Rochester, 1935-36, asst. prof., 1936-37; research physicist, research labs., Gen. Electric Co., 1937-39; asst. prof., Randal Morgan Lab. of Physics, U. of Pa., 1939-41, asso. prof., 1941-42; prof. of physics and head of dept. Carnegie Inst. Tech., 1942-49; prof. physics University of Illinois, 1949-57, head dept., 1957-64, dir. control systems lab., 1951-52, dean Grad. Coll., v.p. research, 1964-65; pres. Nat. Acad. Scis., 1962-65, full-time pres., 1965—; dir. tng. program Clinton Labs., Oak Ridge, 1946-47. Chmn. Naval Research Adv. Committee, 1960-62; vice chmn. Def. Sci. Bd., 1961-62, chmn., 1964—. Chmn. governing bd. Am. Inst. Physics, 1954-59. Trustee Rockefeller Found. Recipient Franklin medal Franklin Inst. Phila., 1965. Fellow Am. Physical Soc. (pres. 1961); mem. Nat. Acad. Scis., Am. Acad. Arts and Sciences, Am. Soc. for Metals, Am. Inst. Mining, Metall. and Petroleum Engrs., Am. Philos. Soc. Author: Modern Theory of Solids, 1940; The Physics of Metals, 1943. Research and publications on theory of solids, nuclear physics. Home: 3025 Whitehaven St. N.W., Washington 20008. Office: 2101 Constitution Av. N.W., Washington 20418.*

SEITZ, Philip Franz Durham, Am. psychoanalyst; b. Evansville, Ind., Mar. 6, 1921; s. Charles L. and Arnia (Lloyd) S.; student Ind. U., 1938-41; M.D., U. Pa., 1944; div.; children—Charles L. III, Diane D., Philip Franz Durham, Jennifer K. Research fellow Hosp. U. Pa., Phila., 1948-49; dir. psychiat. research Ind. U. Med. Center, Indpls., 1949-55; staff Dept. Mental Health, 1965—. Diplomate, Am. Bd. Psychiatry. Fellow Am. Psychiat. Assn., Hofheimer award for research 1955; mem. Am. Psychoanalytic Assn., Chgo. Psychoanalytic Soc. (sec. 1964-66). Research, publs. on infantile experience and adult behavior in animals, substitution of symptoms by hypnosis. Home: 1819 N. Lincoln Park W., Chgo. 60614. Office: 1819 N. Lincoln Park W., Chgo. 60614; also 400 Jackson Av., Glencoe, Ill. 60022.*

SÉJOURNÉ, (Aignan-) Paul-Marie-Joseph, French engr.; b. Orléans, France, Dec. 21, 1851; insp. roads and bridges; prof. Sch. Civil Engring.; mem. French Acad. Scis. Author: Voûtes articulées; Voûtes inarticulées, 1913. Built Adolphe Bridge, Luxembourg, of reinforced concrete, 1904; began use of dry mortar to fill fissures in masonry. Died Jan. 14, 1939.

SEKIGUTI, Takeshi, Japanese climatologist; b. Nagano, Japan, Aug. 13, 1917; s. Makoto and Satsuki (Shima) S.; M.S., Tokyo U. Lit. and Sci., 1941; D.Sc., U. Tokyo, 1952; m. Toshiko Tanabe, May 7, 1944; children—Akiko, Koh. Staff, Far Eastern Inst. in Japan, 1941-43; climatologist Central Meteorol. Obs., Tokyo, 1943-52; vis. prof. Johns Hopkins Lab. Climatology Balt. 1952-53 prof. climatology Tokyo U. Edn., 1953—; vis prof U. Brazil, Rio de Janerio, 1960-61. Mem. Assn. Japan Geographers, Tokyo Geog. Soc., Japan, Am. meteorol. socs., Am. Geophys. Union, Assn. Am. Geographers, Internat. Biometrics Soc. Author: (with K. Takahashi) Our Japanese Weather, 1956; World Weather and Life, 1959; City Climatology; also articles. Editor: Tokyo Jour. Climatology, 1964—. Research on local and micro-scale climatology including city climate, heavy rain climatology and regional climatology of Japan and adjacent area; introduced regional geography of Far E. Home: 2-3-20 Naka-machi, Hoya, Tokyo, Japan.*

SEKIYA, Seikel, Japanese seismologist; b. Gifu Prefecture, Japan, 1854; ed. Kaisei Sch., Tokyo, Japan; went to Britain, 1876; apptd. to faculty Kobe (Japan) Normal Sch., 1879; became prof. sci. dept. Tokyo U., 1886, also 1st dir. Seismology Obs.; asso. with Tokyo Meteorol. Obs. Established (with Sir. James Ewing, John Milne) Seismology in Japan; studied earthquakes, volcanic activities. Died 1896.

SELA, Michael, biochemist; b. Tomaszow Maz., Poland, Mar. 6, 1924; s. Jakob and Rivka (Aleskowski) Salomonowicz; M.Sc., Hebrew U., Jerusalem, Israel, 1946, Ph.D., 1954; postgrad. Ecole de Chimie, U. Geneva, (Switzerland), 1947-48; m. Margalit Liebmann, June 20, 1948; children—Irit, Orlee. Staff Weizmann Inst. Sci., Rehovoth, Israel, 1950—, sr. scientist, 1957—, prof. immunology, 1963—, head dept. chem. immunology, 1963—; vis. scientist U. S. NIH, Bethesda, Md., 1956-57, 1960-61. Mem. panel on immunology WHO, 1963—, panel on molecular biology Internat. Cell Research Orgn., 1966—. Recipient Israel prize in natural scis. 1959. Mem. Israel Biochem. Soc. (pres. 1965—), Israel Chem. Soc., N.Y. Acad. Scis., Am. Chem. Soc., A.A.A.S., Brit. Biochem. Soc., Brit. Soc. Immunology, European Molecular Biology Orgn., Internat. Cell Research Orgn. Editor: New Perspectives in Biology, 1964. Editorial bd. Biochimica et Biophysica Acta, 1963—, Archives of Biochemistry and Biophysics, 1965—, Immunochemistry, 1964—. Research and numerous publs. on synthetic protein models, devel. synthetic antigens and their use for elucidation molecular basis of immunological phenomena; chem. structure and biol. function enzymes, antigens and antibodies. Address: The Weizmann Inst., Rehovoth, Israel.*

SELBERG, Atle, mathematician; b. Langesund, Norway, June 14, 1917; s. Ole Michael and Anna Kristina (Skeie) S.; Dr. Philos., U. Oslo, 1943; m. Hedvig Liebermann, Aug. 13, 1947; children—Ingrid Maria, Lars Atle. Came to U.S., 1947. Research fellow U. Oslo, 1942-47; mem. Inst. for Advanced Study, Princeton, N.J., 1947-48, permanent mem., 1949-51, prof., 1951—; asso. prof. Syracuse U., 1948-49. Recipient Fields medal and prize Internat. Congress Mathematicians, 1950. Mem. Am. Math. Soc., Norwegian Acad. Sci., Royal Danish Acad. Scis. and Letters, Am. Acad. Arts and Scis. Research and publs. on theory of numbers, analysis, discontinuous groups. Home: 7 Maxwell Lane. Office: Institute for Advanced Study, Princeton, N.J. 08540.*

SELBY, Augustine Dawson, Am. botanist; b. Athens County, O., Sept. 2, 1859; s. Warren and Emily (Garretson) S.; B.S., Ohio State U., 1893; postgrad. Washington U., Shaw Sch. Botany, 1899, Columbia, 1902-03; m. Libbie Glover, Dec. 15, 1883. Supt. schs., Huntington, W.Va., 1884-86; prin. high sch., Ironton, O., 1886-87, Garfield Sch., Columbus, O., 1887-89; tchr. botany Columbus High Sch., 1889-94; botanist and chemist Ohio Agrl. Expt. Sta., 1894-1902, chief dept. botany, 1902-23; prop. Selby Heights Fruit Farm, Sharpsburg, O. Sec. Columbus Hort. Soc., 1888-94. Author reports Ohio Agrl. Expt. Sta., Wooster, O.; also papers. Research on plant diseases and their remedies, plant breeding, comml. apple growing. Died May 7, 1924.

SELBY, John Prideaux, English naturalist; b. Alnwich, Eng., July 23, 1788; s. George and Margaret (Cook) S.; entered Univ. Coll., Oxford, Eng., 1806; M.A. (hon.), Durham, 1839; m. Lewis Tabitha, Dec. 17, 1810; 3 daus. High sheriff for Northumberland, 1823; expdn. (with Graham and Greville) for study of fauna and flora of Sutherlandshire, Scotland, 1833; 2d expdn. organized, 1834. Fellow Linnean Soc.; mem. Berwickshire Naturalists' Club. Author: Illustrations of British Ornithology (1st attempt to produce set of life-sized illustrations of Brit. birds), 1821-34; British Forest Trees, 1842; also papers. Founder (with William Jardine and G. Johnston) Mag. of Zoology and Botany, 1837, also joint editor. Died Mar. 27, 1867.

SELDEN, George Baldwin, Am. inventor; b. Clarkson, N.Y., 1846; s. Henry Rogers and Laura Anne (Baldwin) S.; student U. Rochester, 1861-64, Yale, 1865, Sheffield Sci. Sch., 1867; m. Clara Drake Woodruff, Dec. 14, 1871 (dec. 1903); m. 2d, Jean Shipley, Apr. 1909; 2 sons, 1 dau. Admitted to N.Y. bar, 1871; specialized in patent law. Developed lightweight, high speed, 3-cylinder gasoline compression engine run by hydrocarbon liquid fuels, 1887; designed a road locomotive (virtually an automobile). Died 1922.

SELDEN, Richard Thomas, Am. economist; b. Pontiac, Mich., Mar. 31, 1922; s. Arthur Willis and Florence L. (Seeley) S.; B.A., U. Chgo., 1948, Ph.D., 1954; M.A., Columbia U., 1949; m. Martha Mathiasen, Mar. 21, 1953; children—Phoebe Serena, Thomas Mathiasen. Instr. U. Mass., Amherst, 1949-50; faculty Vanderbilt U., Nashville, 1952-59. asso. prof., 1955-59; research asso. Nat. Bur. Econ. Research, N.Y.C., 1958-59, research staff, 1959—; asso. prof. banking Columbia U., 1959-63; economist 1st Nat. City Bank, N.Y.C., 1962-63; prof. econs. Cornell U., Ithaca, N.Y., 1963—. Mem. adv. council banking and financial research com. Am. Bankers Assn., 1964—; Cornell U. rep. on univs. Nat. Bur. Com. for Econ. Research, 1965—. John Simon Guggenheim Meml. Found. fellow 1965; Fulbright Advanced Research scholar, Brussels, Belgium, 1964-65. Mem. Am. Econ. Assn., Am. Finance Assn. Author: The Postwar Rise in the Velocity of Money, 1962; Trends and Cycles in the Commercial Paper Market, 1963; (with George R. Morrison) Time Deposit Growth and the Employment of Bank Funds, 1965; also articles. Research on demand for money, role financial intermediaries other than banks in affecting changes in monetary velocity. Home: 508 Cayuga Heights Rd., Ithaca, N.Y. 14850.*

SELDIN, Donald Wayne, Am. physician; b. N.Y.C., Oct. 24, 1920; s. Abraham Louis and Laura (Ueberal) S.; B.A., N.Y. U., 1940; M.D., Yale, 1943; m. Muriel Deborah Goldberg, Apr. 1, 1943; children—Leslie Lynn, Donald Craig, Donna Leigh. Instr. medicine Yale, New Haven, 1948-50, asst. prof., 1950-51; asso. prof. Southwestern Med. Sch., U. Tex., Dallas, 1951-52, prof., chmn. dept. internal medicine, 1952—; chief med. service Parkland Meml. Hosp., Dallas, 1952—. Mem. Gen. Research Center Com., NIH, USPHS, 1964—; mem. gen. med. study sect. USPHS, 1957-61; cons. to Surgeon Gen. Mem. So. (pres. 1964), Am. (pres. 1965—) socs. clin. investigation, Central Soc. Clin. Research (pres. 1963), Am. Fedn. Clin. Research, Assn. Profs. Medicine, Assn. Am. Med. Colls., Assn. Am. Physicians, A.C.P., Royal Soc. Medicine, Endocrine Soc. Am., Physiol. Soc. Mem. editorial bds. Jour. Lab. and Clin. Medicine, 1958-60, Am. Jour. Medicine, Medicine, Nephron. Research, many publs. in function of kidneys in normal and diseased states, factors governing maintenance of normal vol. and composition of body, mechanisms, concerned with exchange of fluids between blood and cells in normal and diseases states. Home: 11525 St. Michael Dr., Dallas 75230. Office: 5323 Harry Hines Blvd., Dallas 75235.*

SELEUCOS THE BABYLONIAN, astronomer; b. Seleucia/Tigris, flourished mid 2d century. Placed sun at center of earth's orbit before Copernicus' system propounded this; explained tides by resistance opposed by moon to diurnal rotation of atmosphere; discovered periodical inequalities in tides of Red Sea (which he connected with position of moon in zodiac).

SELF, John Teague, Am. parasitologist; b. Spur, Tex., Sept. 28, 1906; s. John N. and Mary (Kyzar) S.; A.B., Baylor U., 1931, A.M., 1932; Ph.D., U. Okla., 1936; m. Ida Adair Burleson, Sept. 14, 1937; children—Virginia Byrd (Mrs. James D. Brashear), Richard Burleson. Mem. faculty U. Okla., Norman, 1936—, prof. zoology, 1948—, chmn. dept., 1946-57. Sec.-treas. Rocky Mountain Biol. Lab., Crested Butte, Colo., 1951-57, trustee, 1959-65; exec. officer Sci. Edn. Improvement Program for Okla., 1960; asst. entomologist USPHS, 1945. Mem. Okla. Acad. Sci. (permanent sec. 1957—), A.A.A.S. (exec. council 1957), Am. Assn. Systematic Zoologists, Am. Soc. Zoologists, Am. Soc. Parasitologists, Am. Microscope Soc., Am. Soc. Exptl. Biology and Medicine, Am. Soc. Tropical Medicine and Hygiene. Research and publs. in field of parasitic diseases in fish and wildlife, function of migratory birds in spread of parasites, importance of animal parasites in human disease. Home: 1621 Rosemont Dr., Norman, Okla. 73069.*

SELIGMAN, Arnold Max, Am. surgeon; b. St. Johnsbury, Vt., Mar. 30, 1912; s. Maurice J. and Sylvia F. (Crestin) S.; B.A., Harvard, 1934; M.D., 1939; m. Bluem Appel, Mar. 3, 1940; children—Myron, Dale, Garry, Stanley. With Beth Israel Hosp., Boston, 1939-54, dir. tumor clinic, 1952-54; asst. prof. surgery Harvard, 1948-54; asso. prof. surgery Johns Hopkins, Balt., 1954-68, professor of surgery, since 1968—, also surgeon-in-chief Sinai Hosp., 1954—. Responsible investigator on research grants Nat. Cancer Inst., Bethesda, Md., 1946—, chmn. chemotherapy study sect., 1965-67. Diplomate Am. Bd. Surgery; Fellow A.C.S.; mem. Am. Surg. Assn., Am. Assn. Cancer Research (bd. 1958-61, 67—), A.M.A., Boylston Med. Soc., A.A.A.S., Am. Acad. Arts and Scis., Histochem. Soc. (pres. 1954), Am. Soc. Cell Biology, Soc. Exptl. Biology and Medicine, Phi Beta Kappa, Sigma Xi, Alpha Omega Alpha, others. Asso. editor Jour Histochemistry and Cytochemistry, 1950—, Jour. Biochem. and Biophys. Cytology, 1954-60, L'Histochemie, 1964—, Histochemie, 1967—. Research and publs. on synthesis of metlylcholonthrene and other carinogenic hydrocarbons, first induction of exptl. gliomas in mice, first clin. use of synthetic vitamin K, first successful use of peritoneal irrigation for treating uremia, preparation of radioiodoprotein and study in traumatic shock, devel. of many new histochem. methods for light and electron microscopy. Home: Kembrooke Farms, Park Heights Av., Owings Mills, Md. 21117. Office: Sinai Hosp., Belvedere and Greenspring Avs., Balt. 21215.*

SELIGMAN, Charles Gabriel, Brit. ethnologist; b. London, Eng., Dec. 24, 1873; s. Hermann and Olivia (Mendez da Costa) S.; M.B., St. Thomas's Hosp., 1896; m. Brenda Zara Salaman, 1905; 1 son. Became house physician to St. Thomas's Hosp., 1897; joined Cambridge Anthropol. Expdn. to Torres Strait and Borneo, 1898; returned to path. studies St. Thomas's Hosp., 1899, named supt. clin. lab., 1901; made expdn. to New Guinea, 1904; went to Ceylon, 1906; prof. ethnology U. London, 1913-34. Recipient Bristowe medal pathology, 1897, Rivers medal, 1925. Fellow Royal Soc., 1919; mem. Royal Anthropol. Inst. (pres. 1923-25). Author: The Melanesians of British New Guinea, 1910; The Veddas, 1911; The Pagan Tribes of Nilotic Sudan, 1932. Research on tropical disease and body abnormalities, Melanesians, Veddas, peoples of Sudan. Died Oxford, Eng., Sept. 19, 1940.

SELIGMAN, George Benham, Am. mathematician; b. Attica, N.Y., Apr. 30, 1927; s. George Frederick and Florence (Benham) S.; student Sampson Coll., 1946-47; B.A., U. Rochester, 1950; M.A., Yale, 1951, Ph.D., 1954; m. Irene Schwieder, July 31, 1959; children— Barbara Helen, Karen Alice. H.B. Fine inustr. math. Princeton (N.J.) U., 1954-56; mem. faculty Yale, New Haven, 12, 1916; s. Philip and Anna (Schloss) S.; B.S., U. Md., 1940; Sc.D., Johns Hopkins U., 1942; M.D., U. Utah, 1946; m. Harriet Tutelman, Jan. 7, 1949; children—Judith Toby, Ruth Louise, Daniel Albert. Research biochemist U.S. Dept. Agr., 1942-43; research fellow, instr. U. Pa., Phila., 1949-51; dir. div. biochemistry and Hosp. of U. Pa., Phila., 1953-58; dir. clin. labs. Yale-New Haven Hosp., 1958—; prof. medicine and pathology, chmn. sect. clin. pathology Yale U. Sch. Medicine, New Haven, 1959—. Fellow A.C.P., Coll. Am. Pathologists; mem. Am. Assn. Clin. Chemists (pres. 1961-62), Am. Chem. Soc., A.M.A., Am. Soc. Clin. Pathologists. Editor: Standard Methods in Clinical Chemistry, vols. 2, 3, 4, 1958-64. Research and many publs. in chemistry of uremia, instrumentation for clin. lab., application of computer techniques in clin. lab. Home: Bond Rd., Woodbridge, Conn. 06525.*

SELIGMAN, Henry, nuclear physicist; b. Frankfort/Main, Germany, Feb. 25, 1909; s. Milton and Marie (Gans) S.; ed. univs. Lausanne, Paris and Zurich; Ph.D.; m. Lesley Bradley, Dec. 31, 1941; children—Andrew, Paul. Expert atomic energy British Atomic Energy Project, 1941-43; with Canadian Nat. Center Atomic Research, 1944-46; dir. isotope sect. British Atomic Project, Harwell; research del. for isotopes Internat. Atomic Energy Agy., 1958—. Publs. on growing neutron absorption, measurement of quantitative effluence of Irish Sea, also isotopes in industry. Home: Peter Jordanstrasse 40, Vienna 19. Office: Internat. Atomic Energy Agy., Kärntnerring 11, Vienna 1, Austria.

SELIGMANN, Maxime Gerard, French immunologist; b. Paris, France, Mar. 14, 1927; s. Armand Pierre and Antoinette (Baer) S.; M.D., U. Paris, 1955; m. Francoise Brolliet, Mar. 11, 1953; children—Christophe, Virginie, Francois. Research asst. Pasteur Inst., Paris, 1953-60; asso. prof. U. Paris Sch. Medicine, 1961; head Lab. Immunochemistry Hosp. Saint-Louis, Paris, 1961—, head clin. dept., 1964—. Mem. Internat. Soc. Hematology, Brit. Soc. for Immunology, French Soc. Hosps. of Paris of Immunology, Hematology and Pediatrics. Research, numerous publs. on leucocyte antigens, immunology of systemic lupus, antibodies to desoxyribonucleic acid, human fibrinogen, myeloma globulins, macroglobulinaemia. Home: 6, Square du Trocadero, Paris 16e. Office: Lab. d'Immunochimie, Hosp. Saint-Louis, 2 Pl. Dr. Fournier, Paris 10 e, France.

SELIGSON, David, Am. physician; b. Phila., Aug. 1956—, prof. math., 1965—; Fulbright lectr. in math., Muenster, Germany, 1958-59. Mem. adv. com. on Fulbright lecturing and research grants to NRC Nat. Acad. Scis., 1963-65. Mem. Math. Assn. Am., Am. Math. Soc. Research and publs. in classification of certain abstractly defined algebraic systems and interpretation of their relationships to other questions in algebra, especially linear groups.*

SELIM, Prince, see Jahangir.

SELKURT, Ewald Erdman, physiologist; b. Edmonton, Alta., Can., Mar. 13, 1914; s. Ephraim and Amanda Olga (Stirle) S.; came to U. S., 1920, naturalized, 1930; B.A., U. Wis., 1937, M.A., 1939, Ph.D., 1941; m. Ruth Marion Gesley, June 21, 1941; children—Claire Elaine, Sylvia Ann. Instr., N.Y. U. Sch. Medicine, 1941-44; faculty Western Res. U. Sch. Medicine, Cleve., 1944-58, asso. prof. physiology, 1949-58, coordinator phase I med. curriculum, 1953-55; prof., chmn. dept. physiology Ind. U. Sch. Medicine, Indpls., 1958—. Mem. sub-com. on shock com. on med. scis. NRC, 1953-58. NSF fellow, Göttingen-Munich, Germany, 1964-65. Mem. Harvey Soc., Am. Physiol. Soc., Soc. for Exptl. Biology and Medicine, Am. Heart Assn., Phi Beta Kappa, Sigma Xi, Phi Sigma, Gamma Alpha. Author: Physiology, 1963; also numerous articles. Research on physiology renal blood flow and electrolyte handling, relationship glomerular filtration rate and sodium excretion, effect ischemia and hemorrhagic shock on kidney function, countercurrent mechanism and urinary concentration, physiology splanchnic circulation, physiology hemorrhagic shock; demonstrated autoregulation renal blood flow. Home: 3269 W. 42d St., Indpls. 46208.*

SELLA, Quintino, Italian crystallographer; b. Biella, Italy, July 27, 1827; prof. mineralogy Turin Engring. Sch.; mem. French Acad. Scis. Author: Théorie et pratique de la règle à calcul, French edit., 1862. Drew up accurate geol. map of Biella region. Died Beille, Mar. 14, 1884.

SELLER, John, English hydrographer; flourished Eng., 1658-98; made compasses and naut. instruments, also engaged in teaching and surveying; hydrographer to king for 3 reigns; pub. and sold maps, charts, geog. books, 1st English books of nav.; made observations of magnetic needle. Author: The English Pilot, 1671; The Sea Atlas, 1671; Pocket Book containing several choice Collections in Arithmetic, Geometry, Surveying, Dialling . . . , 1677; The Sea-Gunner, 1691.

SELLERS, Edward Alexander, Canadian pharmacologist; b. Winnipeg, Man., Can., Sept. 14, 1916; s. Henry Eugene and Irene (Maulson) S.; student Ridley Coll., 1930-33; M.D., U. Man., 1939; Ph.D., U. Toronto, 1947; m. Jean Moncrieff, Oct. 9, 1939; children—Edward, Hugh, Alexander. Faculty, U. Toronto (Ont., Can.), 1946—, prof. Banting and Best dept. med. research, 1948-50, prof., head dept. pharmacology, 1958-66, asso. dean Faculty Medicine, 1965—; chief supt. Def. Research Med. Labs., Def. Research Bd. Can., 1955-58. Chmn., Royal Canadian Navy Personnel Research Com., 1954-58. Mem. Am., Canadian physiol. socs., Am. Soc. Pharmacology and Exptl. Therapeutics, Pharmacol. Soc. Can., Nutrition Soc., Biochem. Soc. (U.K.). Contbr. numerous articles to tech. jours. Established role non-shivering thermogenesis as prin. metabolic adaption during prolonged exposure to cold, research on relationship endocrine glands, nutritional factors to cold adaptive process, feed-back control thyroid by pituitary. Home: 21 Valley View, Toronto 7, Ont., Can.*

SELLHEIM, Hugo, German gynecologist; b. Biblis/Worms, Germany, Dec. 28, 1821; prof., Freiburg, also Düsseldorf, Tübingen, Halle, Leipzig, Germany. Author: Topographische Atlas des weiblicher Beckens, 1900; Das Verhalten der Muskein des weiblicher Beckens, 1902; Die Beziehungen des Geburtskanals und des Geburtsobjecktes zur Geburtsmechanik, 1906; Über Geburtsvorgang und Geburtsleitung beim engen Becken, 1912. Elucidated passive movements of baby in passage through birth canal. Died Leipzig, Apr. 22, 1936.

SELLIE, William, Brit. physician; b. Lanark, Scotland, 1697; s. Archibald and Sarah (Kennedy) S.; M.D., U. Glasgow, 1745; studied midwifery, Paris, France; m. Eupham Borland, 1724. Began practice as surgeon and apothecary, Lanark, 1720; became mem. Faculty Physicians and Surgeons, Glasgow, 1733; settled in London, 1739; taught midwifery at his home, using a model of bones covered with leather, 1741. Author: A Treatise on the Theory and Practice of Midwifery, 1752; A Collection of Cases and Observations in Midwifery (superseded earlier description of mechanism of parturition and curves followed by infant during birth, importance of exact measurements of pelvis emphasized); A Set of Anatomical Tables with Explanations. Died Mar. 5, 1763.

SELLIN, Thorsten, sociologist; b. Ornskoldsvik, Sweden, Oct. 26, 1896; s. Jonas Theodore and Martha (Westman) S.; came to U.S., 1914, naturalized, 1922; A.B., Augustana Coll., 1915, LL.D., 1942; A.M., U. Pa., 1916, Ph.D., 1922; J.D., U. Uppsala (Sweden), 1957, U. Leiden (Netherlands), 1960; m. Amy Johanna Anderson, June 10, 1920; children—Theodore, David, Eric. High sch. tchr., Mpls., 1916-20; faculty U. Pa., Phila., 1921—, prof. sociology, 1930—; vis. prof. univs. Uppsala, Stockholm, Lund, 1946-47, Princeton, 1949, U. Cal. at Berkeley, 1958. Fulbright lectr., Cambridge (Eng.) U., 1959-60; cons. to fgn. govt. agys.; pres. Internat. Penal and Penitentiary Found., Bern, 1965—; pres. Internat. Criminology Congress The Hague (Netherlands), 1960. Decorated grand officer. Order N. Star (Sweden), Médaille Pénitentiaire (France); recipient award Augustana Coll., 1961, Ill. Acad. Criminology, 1958. Fellow Am. Sociol. Assn., A.A.A.S.; mem. Internat. Soc. Criminology (pres. 1956-67), Am. Acad. for Polit. and Social Sci. (editor Annals 1929—), Sociol. Research Assn., Social Sci. Research Council, Am. Soc. Criminology (award 1960), Internat. Soc. Social Def., Internat. Assn. Penal Law, Soc. d'Hist. du Droit, Am. Philos. Soc., Internat. Inst. Sociology, Phi Beta Kappa. Author: Crime in the Depression, 1937; Criminality of Youth, 1940; Culture Conflict and Crime, 1938; Pioneering in Penology, 1944; The Death Penalty, 1959; (with M.E. Wolfgang) The Measurement of Delinquency, 1964; also numerous articles. Research in penal history, criminology theory, death penalty; co-designer Sellin-Wolfgang index delinquency. Home: 4106 Locust St., Phila. 19104.*

SELMER, Ernst S., Norwegian mathematician; b. Oslo, Norway, Feb. 11, 1920; s. Ernst W. and Ella (Sejersted) S.; Cand. real, U. Oslo, 1945, Ph.D.,

1952; m. Lillemor Faanes, Sept. 21, 1945; 1 dau., Johanne-Sophie. Lectr. math. U. Oslo, 1946-57; prof. math. U. Bergen (Norway), 1957—, dean faculty math. and sci., 1965-68. Mem. Norwegian Council on Electronic Computers, 1961—. Mem. Norwegian, Am. math. socs., Norwegian Acad. Author: Differential and Integral Calculus, 1954. Editor: Nordisk Matematisk Tidskrift, 1952—. Research and publs. on electronic computers, number theory (indeterminate equations, linear recurrence), group theory (permutation groups), combinatorial problems (theory of arrays). Office: Dept. Math., U. Bergen, Norway.*

SELMI, Francesco, Italian chemist; b. Vignola, Italy, Apr. 7, 1817; prof., Modena and Bologna, Italy. Author: Enciclopedia di chimica scientific e industriale, 4 vols., 1870-76; Criteri per la ricerca dei alcaloidi vegetali in differenza delle ptomaine, 1880. Research on casein, cadaver poisons, alkaloids, collidal solutions using term pseudo-solutions. Died Bologna, Italy, Aug. 13, 1881.

SELOVE, Walter, Am. physicist; b. Chgo., Sept. 11, 1921; s. Abe and Rose (Feld) S.; S.B., U. Chgo., 1942, Ph.D., 1949; m. Fay Ajzenberg, Dec. 18, 1955. Asst. instr. U. Chgo., 1942-43; staff mem. Mass. Inst. Tech. Radiation Lab., 1943-45; physicist Argonne Nat. Lab., 1947-50; instr. Harvard, 1950-52, asst. prof., 1952-56; staff mem. U. Cal. Radiation Lab., Livermore, 1953-54; asso. prof. U. Pa., Phila., 1957-61, prof., 1961—. Cons. Joint Congl. Com. on Atomic Energy, Intermittantly 1957-62. NRC fellowship, 1946-47, NSF Sr. Postdoctoral fellowship, 1956-57. Fellow Am. Phys. Soc.; Fedn. Am. Scientists (vice chmn. 1958), Phi Beta Kappa, Sigma Xi. Research in nuclear physics and particle physics; discovery of the F meson, originator of "Fast neutron chopper" for neutron spectroscopy; inventor of devices for radar receivers.*

SELTZER, Albert Pincus, physician, surgeon; b. Myropolia, Rumania, Aug. 12, 1903; s. Pincus Judah and Ida (Sunshine) S.; came to U.S., 1905, naturalized, 1908; M.D., Temple U., 1928; M.Sc., in Medicine, 1943, Sc.D. in Medicine, 1944; LL.D., Shaw U., 1953; postgrad. at numerous schs. in U.S., Europe; m. Sylvia Superstein, Nov. 5, 1944; children—Benjamin Pincus, Marjorie Ann. Practice medicine specializing in ear, nose and throat and plastic surgery, Phila., 1933—; faculty Grad. Sch. Medicine, U. Pa., 1943—, asso. prof. otolaryngology, 1948—; dept. chief Phila. Gen., Mercy-Douglass, Community hosps., Albert Einstein, St. Luke's Children's med centers. Named Man of Year, V.F.W., 1959. Fellow Internat. Coll. Surgeons, A.C.S.; mem. A.M.A., Pa. State Med. Soc., Am. Acad. ophthalmology and Otolaryngology, Reading Eye, Ear, Nose and Throat Soc. (hon.), Am. Soc. Facial Plastic Surgery, Am. Otorhinologic Soc. for Plastic Surger, Nat. Med. Assn., Pan-Am. Assn. Otorhinolaryngology, Am. Med. Assn. Vienna, Am. Acad. Plastic Surgery for Head and Neck. Author: Plastic Surgery of the Nose, 1949; Diseases of the Ear, Nose, and Throate, 1950; Your Nasal Sinuses and Their Disorders, 1951; Ear, Nose and Throat for the General Practitioner, 1965; also numerous articles. . Developed numerous new instruments used in ear, nose, throat and plastic surgery. Home and office: 2104 Spruce St., Phila. 19103.*

SELTZER, Lawrence Howard, Am. economist; b. New York, N.Y., May 2, 1897; s. Herman and Jessie (Morrell) S.; A.B., U. of Mich., 1920, A.M., 1921, Ph.D., 1925; married Sarah Seltzer, August 29, 1927; children—Ruth Janet (wife of Dr. S. I. Harrison) (dec.), Ronald Anthony. Reader economics, U. of Mich., 1919-20, instr., 1920-21; instr. econs. and sociology Wayne State U., 1921-25, asst. prof., 1925-27, asso. prof., 1927-34, prof., 1934—, Franklin Meml. prof. of human relations, 1961-62, chmn. dept. econs., 1953-59; dep. chmn., exec. bd. Inst. Indsl. and Labor Relations U. Mich-Wayne State U., 1959-60, chmn., 1960-62; asso. prof. finance, sch. business adminstrn., U. of Mich., summer 1929; prof. econs., U. Cal., Berkeley, summer 1950; vis. prof. econs. University of Mich., 1959; Distinguished vis. prof. econs. Swarthmore Coll., 1966-67; vis. prof. econs. Wharton Sch. Finance and Commerce, U. Pa., 1967. Tech. adviser, Labor Adv. Bd., N.R.A., 1933; expert asst. to dep. gov. in charge finance, F.C.A., 1934; head economist and asst. dir. research and statistics, U. S. Treasury Dept., 1934-39, cons. expert, 1942-53; cons. expert Fed. Res. Bank of N.Y., summers 1940, 45; mem. directing com. study war financing, Nat. Bur. Econ. Research, 1941, chmn. directing com. study capital gain taxation, 1942, chmn. com. on fiscal research, 1950-60; consultant U.N., 1951. Awarded 1st prize Hart, Schaffner and Marx nat. prize essay competition in economics, 1927. Mem. Am. Finance Assn. (v.p. for pub. finance, 1947-48), Nat. Tax Assn., Tax Inst., Inc. (adv. council, 1947-49), Royal Econ. Soc., Am. Econ. Assn. (mem. bd. editors, 1945-48), Am. Statis. Assn., Phi Beta Kappa. Author or coauthor books relating to field; latest publs.: Capital Gains Taxation, 1946; Economic Theory in Review, 1950; The Nature and Tax Treatment of Capital Gains and Losses, 1951; Interest as a Source of Personal Income, 1956; The Personal Exemptions in the Income Tax, 1968. Contbr. to Ency. of Social Scis. Analyzed growth and financing of Am. automobile industry, theory and practice tax treatment capital gains and losses for income tax, role personal exemptions in U. S. income tax. Home: 19475 Stratford Rd., Detroit 48221. Office: Wayne State U., Detroit, Mich. 48202.*

SELVIDGE, Harner, Am. communications engr.; b. Columbia, Mo., Oct. 16, 1910; s. Robert Washington, and Ivy (Harner) S.; B.S., Mass. Inst. Tech., 1932, M.S. in Elec. Engring., 1933; S.M., Harvard, 1934, S.D. in Communications Engring., 1937; m. Eloise B. Campbell, Aug. 16, 1933; children—Robert C., Judith E., Margaret J. (Mrs. Robert H. Jubb), Ross S. Instr., Mass. Inst. Tech., 1934, Harvard, 1934-37; asso. prof. Kan. State U., 1938-41; sr. staff applied physics lab. Johns Hopkins, Silver Spring, Md., 1941-45; dir. spl. products devel. Bendix Corp., Detroit, 1945-60; exec. v.p. Meteorology Research, Inc., Altadena, Cal., 1960—, also dir.; dir. Cohu Electronics, San Diego, Cal. Recipient Naval Ordnance Devel. award, 1945, Warren Eaton Meml. award Soaring Soc. Am., 1961. Fellow I.E.E.E., Am. Inst. Aers. and Astronautics (asso.); mem. A.A.A.S., Am. Meteorol. Soc., Am. Phys. Soc., Am. Soc. for Engring. Edn., Air Pollution Control Assn., Soaring Soc. Am. (past pres., dir. 1957—), Nat. Aero. Assn. (dir. 1959-60). Contbg. author: American Soaring Handbook, 1963; also numerous articles. Pioneered research in ultra high frequency propagation and diffraction and high frequency transmission lines; research in proximity fuse and guided missiles. Patentee in field. Home: 3177 Maiden Lane, Altadena 91001. Office: 464 W. Woodbury Rd., Altadena, Cal. 91001.*

SELWOOD, Pierce Wilson, chemist; b. Vancouver, B.C., Can., Mar. 25, 1905; s. Harry Gordon and Jane (Gourlay) S.; B.A., U. B.C., 1927; M.A., U. Ill., 1928, Ph.D., 1931; m. Alice Taylor, Aug. 18, 1938; 1 son, Pierce Taylor. Postdoctorate research Princeton, 1931-35; faculty Northwestern U., Evanston, Ill., 1935-62, prof., 1948-62; prof. chemistry U. Cal. at Santa Barbara, 1962—. Mem. Am. Chemical Society. Author: Magneto chemistry, 1943; General Chemistry, 1950; Adsorption and Collective Paramagnetism, 1962; Chemical Principles, 1964; also numerous articles. Editor, Jour. Catalysis, 1962—. Research on applications magnetic methods to structural inorganic chemistry particularly catalytically active solids. Home: 1227 Viscaino Rd., Santa Barbara, Cal. 93103..*

SELWYN, Alfred Richard Cecil, Canadian geologist; b. Kilmington, Eng., July 28, 1824; s. Revi Townshend and Charlotte Sophia (Murray) S.; ed. privately; m. Matilda Charlotte, 1852 (dec. 1882); 3 sons, including Percy H., 1 dau. Asst. geologist Geol. Survey Gt. Britain; dir. Geol. Survey of Victoria, Australia, 1852-69, Geol. Survey Can., 1869-95. Mem. Royal Soc. Can. (became pres. 1896). Research on gold fields, mineral areas, water supply, distbn. goldbearing drifts on placer deposits; discovered certain tertiary strata produced from waste of older rocks had little or no gold, while older deposits were rich in gold. Died Oct. 18, 1902.

SELYE, Hans, physician; b. Vienna, Austria, Jan. 26, 1907; s. Dr. Hugo and Maria Felicitas (Langbank) S.; student Coll. of the Benedictine Fathers, Komarom, Czechoslovakia, 1916-24; med. student, U. Paris, France, 1925-26, U. Rome, Italy, 1926-27; M.D., German U. Prague, 1929, Ph.D., 1931; D.Sc., McGill U., 1942, Assumption U., Windsor, Catholic U. Chile, 1956; hon. degree U. San Carlos, 1959; Doctor Honoris Causa, Nat. U. Argentina, U. Montevideo; m. Gabrielle Grant, Feb. 17, 1949; children—Michel, Jean, Marie, André. Asst. in exptl. pathology, histological lab., German U., 1929-31; Rockefeller Research fellow, dept. biochem. hygiene, Johns Hopkins U., 1931; Rockefeller Research fellow, dept. biochemistry, McGill U., Montreal, 1932-33, lecturer in biochemistry, 1933-34, asst. prof. biochemistry, 1934-37, asst. prof. of histology, 1937-41, asso. prof. histology, 1941-45; prof. and dir. Institute of Experimental Medicine and Surgery, Universite de Montreal, since 1945. Expert consultant to Surgeon General, U. S. Army, 1947-57. Awarded Casgrain and Charbonneau prize (McGill U.) Gordon Wilson medal for 1948, award by Am. Clin. and Climatol. Assn.; Heberden medal, 1950; Medal of the Accademia Medico Fisica Fiorentina, 1950; Henderson gold medal, 1964. Fellow Royal Society (Canada), New York Acad. of Sciences, A.A.A.S. Found. mem., hon. mem., or corr. mem., fgn., nat. state and local profl. med. assns. and orgns., in both gen. and spl. fields. Author several specialized med. works; also Annual Reports on Stress, 1951-56; The Stress of Life, 1956; The Chemical Prevention of Cardiac Necroses, 1958, Calciphylaxis, 1962; From Dream to Discovery, 1964; The Mast Cells, 1965 plus 1,200 articles. Research on stress; general adaptation syndrome; hormonal conditioning; cardiac necroses; aging; steroid anesthesia; neurohumoral reflex of lactation; formalin-arthritis test; mechanical "tissue scaffoldings" for the topical induction of growth, metaplasia. Home: 659 Milton St. Office: 2900 Mount-Royal Blvd., Montreal, Can.*

SELZER, Arthur, physician; b. Lwow, Poland, July 3, 1911; s. Martin and Janina (Lam) S.; Med. Dipl., U. Lwow, 1935; M.D., U. Cracow, Poland, 1936; m. Jadwiga Winkler, July 31, 1936; children—Martin A., Peter M. Came to U.S., 1938, naturalized, 1943. Practice medicine, specializing in cardiology, San Francisco, 1941—; with Stanford U. Sch. Medicine, 1941—, clin. prof. medicine, 1957—; clin. prof. medicine U. Cal. Sch. Medicine, 1960—; dir. cardiopulmonary lab. Presbyn. Med. Center, San Francisco, 1960—; cons. cardiology VA Hosps. San Francisco, Martinez, Cal., Letterman Gen. Hosp., U.S. Naval Hosp., Oakland, Cal. Fellow Am. Heart Assn., A.C.P., Am. Coll. Cardiology; mem. Western Soc. Clin. Rsearch, Western Assn. Physicians. Author: The Heart: Its Function in Health and Disease, 1966. Research, numerous publs. in clin. cardiology; heart failure, congenital heart disease, digitalis, electrocardiography. Home: 5 Greenview Lane, Hillsborough, Cal. 94010. Office: Presbyn. Med. Center, 2351 Clay St., San Francisco 94115.*

SELZER, Melvin Lawrence, Am. psychiatrist; b. N.Y.C., Feb. 3, 1925; s. Louis and Anna (Greenberg) S.; B.S., Tulane U., 1949, M.D., 1952; m. Elizabeth J. Thompson, Sept. 22, 1957; children—Béla, Aaron, Noah. Asso. psychiatrist U. Mich. Health Service, 1957-59, faculty Med. Sch., Ann Arbor, 1959—, asso. prof. psychiatry, 1965—, chief adult inpatient service Neuropsychiat. Inst., 1962-63; practice psychiatry, Ann Arbor, 1957—. Examiner Am. Bd. Psychiatry and Neurology, 1962—; resource cons. Pres.'s Com. for Traffic Safety, 1964—; con. criminal code revision com. State Bar Mich., 1966—. Mem. Am. Psychiat. Assn., Assn. Am. Med. Colls., A.M.A., Phi Beta Kappa, Alpha Omega Alpha. Editor: Psychiatry for Lawyers Handbook, 1966. Research and publs. in psychotherapy, student mental health, renal function following electroshock, role psychopathology and alcoholism in traffic accidents. Home: 1201 Red Oak Rd., Ann Arbor, Mich. 48103.*

SEMAT, Henry, physicist; b. Poland, Aug. 15, 1900; s. Abraham and Eva (Levine) S.; came to U. S., 1905, naturalized, 1918; B.S., City Coll. N.Y., 1922; M.A., Columbia, 1923, Ph.D., 1934; m. Ray Koch, Sept. 1, 1923; children—Edith Joan (Mrs. Bernard A. Kemp), Barbara Ann. Mem. faculty City Coll., City Univ. N.Y., 1921—, prof. physics, 1951—, chmn. dept., 1959-62, exec. officer Doctoral Program in Physics, 1966-67; vis. summer prof. Columbia, 1951,56; spl. research X-ray spectroscopy. Recipient Townsend Harris award, City Coll., 1964. Mem. Am. Phys. Soc., Am. Assn. Physics Tchrs., A.A.A.S., Am. Assn. U. Profs. (pres. City Coll. N.Y. chpt., 1962-63), Phi Beta Kappa (president Gamma of N.Y. 1964—), Sigma Xi (pres. City Coll. N.Y. club 1958). Author: Physics in Modern World, 1949; Fundamentals of Physics, 4th edit., 1966; (with R. Katz) Physics, 1958; (with Harvey White) Atomic Age Physics, 1959; Introduction to Atomic and Nuclear Physics, 4th edit., 1962; (with R. Blumenthal) College Physics—A Programmed Aid, 1967. Research in X-ray spectroscopy; studied effect of chem. combination on structure x-ray absorption limits. Home: 42-20 Kissena Blvd., Flushing, N.Y. 11355. Office: City Coll., City U. N.Y., Convent Av. and 139th St. N.Y.C. 10031.*

SEMENÉKO, Nikolái Panteleimónovich, Soviet geologist; b. Nov. 16, 1905; grad. Dnepropetrovsk Mining Inst., 1927. Mem. staff Dnepropetrovsk Mining Inst., until 1941; became prof., 1937; chief sect. on geol. scis. Ukraine Acad. Sci; prof. Kiev U. Mem. Ukrainian Acad. Scis. (v.p. 1950—, acad. sec. 1948-50). Research on pre-Cambrian crystallographic massifs, Ukraine.

SEMENOV, Nikolai Nikolaevich, Russian chem. physicist; b. Saratov, Russia, Apr. 16, 1896; s. Nikolai Alex and Elena (Dmitrieva) S.; grad. Leningrad State U., 1917; hon. doctorate Oxford (Eng.) U., 1960, Bruxelles (Belgium) U., 1962, Milano (Italy) Poly. Inst., 1964, Carlovy U., Praha (Czechoslovakia), 1965, Budapest (Hungary) Tech. U., 1965, London (Eng.) U., 1965; m. Natalia Nikolaevna Burtseva, Sept. 15, 1924; children—Yurii Nikolaevich, Ludmila Nikolaevna (Mrs. Vitalii Iosifovich Goldanskii). Head, Lab. Electronic Phenomena, Inst. Physics and Tech., Leningrad, USSR, 1920-31; dir. Inst. Chem. Physics, Moscow, USSR, 1931—; prof. chem. physics Moscow State U., 1944—. Recipient Nobel prize in chemistry (with Cyril Hinshelwood), 1956; Order of Lenin. Fellow Royal Soc., 1958. Hon. fellow Royal Soc. Edinburgh; hon. mem. Indian Acad. Scis., Hungarian, N.Y., Rumanian acads. scis.; mem. Acad. Scis. USSR, Am. Nat. Acad. Scis. (fgn.), Czechoslovakian Acad. Scis. (fgn.), Naturalists Acad. Sci., Leopoldina. Author several books including: Chain Reactions, 1934; Some Problems of Chemical Kinetics and Reactivity, 1954, rev., 1958; also numerous articles. Developed theory of branched chain reactions and theory thermal combustion and explosion (valuable in jet and rocket engine design). Address: 2-b Vorobyevskoye shossé, Moscow, USSR.*

SEMENOV-TJANSANSKIJ, Petr Petrovic, Russian explorer; b. Urusovo, Russia, Nov. 14, 1827; made expdns. to Dzoungarie, and Tien-Shan Mountains, Central Asia, 1846-56; became dir. Empire Statis. Bur., 1863; dir. 1st Russian census. Mem. Russian Soc. Geography (became v.p. 1873). Editor geog. statis. lexicon of Russian empire. Author: Geograficeskstatisticeskij slovar' rossijskoj imperij, 5 vols., 1863-85. Translator: (with Grigoriev, Khanikov) The Asian (Ritter), 1856-60. Geog. and statis. research. Died St. Petersburg, Russia, Mar. 11, 1914.

SEMENZA, Giorgio, biochemist; b. Milan, Italy, June 23, 1928; s. Carlo and Clementina (Gerli) S.; M.D., U. Milan, 1951; m. Berit Andersson, May 31, 1958; children—Christina, Jan., André. Asst., Med. Clinic, U. Milan, 1951-55, privat dozent biochemistry, 1958-61; research fellow Physiol.-Chemisch. Institut, U. Zurich (Switzerland), 1954, asst., 1956-61, head asst., 1961-64, privat-dozent biochemistry, 1961-64, asso. prof., 1964—; research fellow Biochem. Inst., Uppsala (Sweden), 1955-56; vis. asso. prof. dept. biochemistry Chgo. Med. Sch., 1965. Recipient award (with others) Swiss Gastroent. Soc., 1963. Mem. several biochem. and sci. socs. Swiss and fgn. Research, publs. on path. changes in nucleic acids in leukemia; devel. chromatographic separation of polyelectrolytes, especially proteins; demonstrated proteinacious nature of an enzyme, cysteinyl-glycinase; studied (with Prader, Auricchio, Crane) intestinal sugar digestion and absorption, including enzymatic lack in genetically-transmitted or acquired disaccharide intolerances, mechanism of sugar absorption, kinetics of sodium activated transport systems. Home: 39 Langwattstrasse 8125 Zollikerberg, Switzerland. Office: 4 Zurichbergstrasse 8032 Zürich, Switzerland.*

SEMIROT, Pierre, French astronomer; b. Saint-Mandé, France, June 4, 1907; s. Henri and Marie (Couaillac) S.; Dr.ès sc., U. Bordeaux; m. Fernande Bouquet, July 25, 1933; children—Annie, Christiane, Monique. Tchr., 1926-31; asst. Bordeaux Obs., 1931, asst. astronomer, 1936; asst. astronomer, Paris, 1944; dir. Bordeaux Obs., 1947—; titular prof. chair astronomy U. Bordeaux, 1960—. Decorated Office Pub. Instrn., Order Legion of Honor. Author: Chocs imaginaires dans le problème des trois corps; Solutions périodiques dans le problème dans l'espace du mobile attiré par deux centres fixes et stabilité; Conditions de choc dans le problème des trois corps. Address: Observatoire, Floirac (Gironde), France.

SEMM, Kurt Karl Stephan, German gynecologist, obstetrician; b. Munich, Germany, Mar. 23, 1927; s. Karl S. and Margarete (Dillmaier) S.; Dr. degree U. Munich, 1950, Dr.med., 1964; m. Roswitha v. Morozwicz, May 19, 1927. Faculty, U. Munich, 1958—, prof. gynecology and obstetrics, 1964—, sr. physician II Frauenklinik, 1961—. Mem. German (1st sec.), Austrian, soc. for study fertility and sterility, European Sterility Congress Orgn. (gen. sec.), Internat. Fertility Assn. (nat. sec.), Soc. for Gynecology and Obstetrics of Brazil and Peru (corr.). Research, numerous publs. on gynecology and obstetrics, endocrinology and biochemistry; developed many instruments for diagnosis and treatment fertility and sterility, periteneoscopy, resuscitation of newborn. Home: 37 Melchiorstreet 8, Munich 71, W. Germany.*

SEMMELWEISS, Ignaz P., obstetrician; b. Budapest, Hungary, July 1, 1818; ed. U. Budapest; M.D., U. Vienna (Austria), 1844. Asst. prof. maternity dept. Vienna Gen. Hosp., 1846-49; head. obstetric physician St. Roch's Hosp., Budapest, from 1851, prof. obstetrics, from 1852. Author: Atiologie, Begriff und Prophylaxis des Kindbettfiebers, 1861, English edit., 1941. Suggested that physicians wash their hands in strong chems. to prevent their carrying childbed fever germs into maternity wards (this procedure greatly reduced death rate, methods not fully recognized until circa 1890. Died Vienna, Aug. 13, 1865.

SEMON, Sir Felix, laryngologist; b. Danzig, Germany, Dec. 8, 1849; s. Simon Joseph and Henrietta (Aschenheim) S.; student Heidelberger; M.D., Berlin, 1873, staatsexamen, 1874; student Vienna, Paris, London; m. Augusta Dorothea Redeker, 1879; 3 sons. Naturalized Brit. citizen, 1901. Became clin. asst. Hosp. for Diseases of Throat, London, 1875; physician in charge throat dept. St. Thomas's Hosp., 1882-97; became laryngologist to Nat. Hosp. for the Paralysed and Epileptic, 1888; named Royal Prussian prof., 1894; named hon. mem. staff Hosp. for Diseases of Throat, 1877; became physician extraordinary to Edward VII, 1901. Fellow Royal Coll. Physicians; mem. Laryngol. Soc. London (pres. 1894-96; Internat. Med. Assn. (named sec. laryngology sect. 1881). Recipient Order of Red Eagle, Kaiser Wilhelm II, 1888; Semon lectures in laryngology U. London, established in his honor, 1894. Author: Forschungen und Erfahrungen, 1880-1910, 1912. Research (with Sir Victor Horsley) on central motor innervation of larynx; developed law stating that in all progressive organic lesions of centres and trunks of motor laryngeal nerves the abductus of vocal cords succumb earlier than adductors (Semon's law); realized myxedema, cretinism and cachexia strumipriva are related and caused by loss of thyroid function, 1883. Died Great Missenden, Eng., Mar. 1, 1921.

SEMON, Waldo Lonsbury, Am. chemist; b. Demopolis, Ala., Sept. 10, 1898; s. Frank Emerson and Blanche (Lonsbury) S.; B.S. in Ch.E., U. of Wash., 1920, Ph.D., 1923; m. Marjorie Gunn, Sept. 10, 1920; children—Mary Blanche, Marjorie Elizabeth, Constance Anne. Began as civil engr., 1914, with U. S. Geol. Survey, 1917-18, Milwaukee R.R., 1918-20; asso. in chemistry. U. of Wash., 1920-23, instr., 1923-26, cons. chemist, 1920-26; research chemist B. F. Goodrich Co., Akron, O., since 1926, dir. synthetic research dept. since 1937; v.p. and dir. of research, Hydracarbon Chemical and Rubber Co., since 1940. Received Modern Pioneer award from Nat. Assn. of Mfrs., 1940; Cresson gold medal, Franklin Inst., 1964. Mem. Am. Chem. Soc., Am. Inst. Ch.E., Sigma Xi. Author: Organic Syntheses III (H. T. Clarke, editor), 1923; Organic Syntheses X (H. T. Clarke, editor), 1930; Chemistry and Technology of Rubber (C. C. Davis, editor), 1937; also articles for Indsl. and Engrng. Chemistry. Holds over 100 patents dealing with Koroseal, antioxidants, accelerators and compounding of rubber, synthetic rubber, etc. Home: 2990 Millboro Rd., Silver Lake, O. Office: B. F. Goodrich Co., Akron, O.*

SEMPLE, Ellen Churchill, Am. anthropo-geographer; b. Louisville, Ky., 1863; d. Alexander Bonner and Emerine (Price) S.; B.A., Vassar Coll., 1882, A.M., 1891; studied at U. of Leipzig, 1891-92, 1895; LL.D., U. of Ky., 1923; unmarried. Lectr. U. Chgo. 1906-21; Professor anthropo-geography, Clark Univ., Worcester, Mass., 1921-28. Pres. Assn. Am. Geographers, 1921. Gold medal, Am. Geog. Soc., 1914. Contributor scientific articles to Bulletin of Am. Geog. Soc., Journal of Geography, New York, and Geog. Journal of London. Author: American History and Its Geographic Conditions, 1903; The Influences of Geographic Environment, 1911; The Geography of the Mediterranean Region, 1931, Spl. field work on influence of geog. conditions on devel. soc.; pioneer in devel. of anthropogeography in U. S. Died May 8, 1932.

SEN, A. C., Indian biologist; b. Calcutta, India, July 21, 1906; s. Bireswar and Indumati Sen; B.Sc. with honors, M.Sc., 1st class, Calcutta U., 1921-28; Ph.D., Cambridge U., U.K., 1951; m. Renu, Apr. 18, 1935; 3 children. Sr. sci. asst. Locust Research Bur., India, 1931-35, Sugarcane Research, Bihar State, India, 1938-43; entomologist State of Bihar, 1947-60; prof. zoology-entomology State Agrl. Coll., Sabour, India, fellow Bihar U., 1947-56; dir. Agrl. Research Inst., Patna, 1957-60; prof. emeritus zoology dept. Ranchi U., Ranchi, Bihar, India, 1962—. Recipient certificate of merit Bihar Govt., 1952. Author: Store Grain Pests and Their control. Research, publs. on biology and control of cutworm pest, paddy pests and control, biol. control of plant pests, biology and control of plant parasitic nematodes. Home: Tagore Hill Rd., Morabadi, Ranchi-8, Bihar State, State, India. Office: zoology dept., Ranchi U., Ranchi, India.*

SEN, Hari Keshab, astrophysicist; b. Nagina, India, Feb. 9, 1905; s. Upendra Nath and Himangshu (Ray) S.; B.Sc., U. Coll. Sci., 1925, M.Sc., 1927, Ph.D., 1943; LL.B., U. Sch. Law, 1929; m. Kanika Mazumdar, Nov. 24, 1937; 1 dau., Manju Shree. Came to U. S., Sept. 1, 1947, naturalized, Oct. 13, 1953. Lectr. Indian univs. and colls., 1929-47; research fellow, lectr. astronomy Harvard Coll. Obs., Cambridge, Mass., 1947-51; physicist Central Radio Propagation Lab., Nat. Bur. Standards, Washington, also Boulder, Colo., 1951-54; sr. scientist Microwave Lab., Hughes Aircraft Co., Culver City, Cal., 1954-55, Air Force Cambridge Research Labs., Hanscom Field, Bedford, Mass., 1955—. Ripon prof. Indian Assn. Cultivation Sci., Jadavpur, India, 1956. Recipient Hill Meml. prize, 1942, Edn. Minister's Gold medal, 1949, research and devel. awards Space Edn. Found., 1958, 59. Mem. Internat. Astron. Union, Am. Astron. Soc., Am. Geophys. Union, I.E.E.E. (sr.), Sci. Research Soc. Am., A.A.A.S. Author: (with Donald H. Menzel, P1L. Bhatnagar) Stellar Interiors, 1963. Research on stellar constn., ionosphere, solar radio noise, solar and interplanetary plasma, radiation effects on shock wave structure, gravitational thermal tides in earth's atmosphere. Home: 29 Concord Av., Cambridge, Mass. 01731. Office: Air Force Cambridge Research Lab., Hanscom Field, Bedford, Mass. 01731.*

SEN, J., Indian botanist; b. Calcutta, India, Dec. 2, 1921; s. Mani Mohan and Prativa (Majumdar) S.; M.A., Calcutta U., 1944, Ph.D., 1955. Asst. lectr. botany Calcutta U., 1947-52, lectr., 1952-61; guest worker Riksmuseum, Stockholm, Sweden, 1956-57; research asso. Glasgow U., 1957-59; supt. Indian Bot. Garden, Calcutta, Howrah, India, 1961—; vis. lectr. agr., botany Calcutta U., 1961-66; vis. lectr. town and regional planning Bengal Engrng. Coll., Howrah, 1964-66. Fellow Linnean Soc. London, Geol. Soc. London and India; mem. Bot. Soc. Bengal. Research numerous publs. on nature of mineralization of plant tissues leading to formation of coal and petrifaction, morphology and distbn. of plant micro- and megafossils and distbn. pattern of some pleistocene fossils as compared to that of their living reps. constitute bulk of work; research in plant taxonomy, landscape designing; history of Indian concepts of organic evolution. Died Dec. 12, 1966.

SEN, Prafulla Kumar, Indian surgeon; b. Calcutta, India, Dec. 7, 1915; s. Pramatha Nath and Parijat Kumari (Gupta) S.; student Victoria Coll. Sci., U. Nagpur; M.B.B.S., U. Bombay, 1938, M.S. with distinction, 1940; m. Marie Teresa Barnes, Nov. 24, 1956. With King Edward VII Meml. Hosp. and Seth G.S. Med. Coll., U. Bombay (India), 1943-40, 51—, dir., prof. surgery, 1956—; vis. fellow Harrison dept. surg. research Hosp. of U. Pa., Phila., 1949-51. Cons. in thoracic and cardiovascular diseases; Recipient Fgn. Currency Credit award NIH, Bethesda, Md. 1965-—; Vishnevsky medal Acad. Med. Scis. USSR. Rockefeller Found. travel fellow; traveling prof. Rockefeller Found., 1962. Fellow Indian Acad. Med. Scis., Indian Acad. Scis., A.C.S.; mem. Am. Coll. Chest Physicians, Assn. Surgeons India (1st Sankaran prize), Soc. for Diseases Chest (v.p.), Soc. for Exptl. Med. Scis. (v.p.), Cardiol. Soc. India, Indian Council Med. Research (mem. cardiovascular diseases and hypertension expert group), Indian Council Med. Research (mem. emergency med. research com.), Internat. Surg. Soc., James IV Assn. Surgeons. Contbg. author: Textbook of Surgery, 1962. Research, numerous publs. in hpothermia; developed new methods for cooling organs, especially brain; described middle aortic syndrome; developed method of revascularisation of myocardium by multiple punctures through heart wall simulating primitive circulation of snake heart; developed isolated perfusion for treatment of hepatic failure and reported 1st clin. use with cadaveric human liver in cholaemia; studies in rheumatic heart disease, aneurysms of aorta, perforated peptic ulcers, burns and shock. Home: Denisandra-Pintoville, Gekhale Rd., Bombay 28, India. Office: K.E.M. Hosp., Bombay 12, India.*

SEN, S. P., Indian biologist; b. Calcutta, Apr. 6, 1927; s. Rabindra Nath and Ratnamala (Pal) S.; B.Sc. with Honors, Presidency Coll., Calcutta, 1946; M.Sc., Calcutta U., 1948, D.Phil., 1954, PRS, 1956; m. Mira Purkayastha, July 5, 1960; 1 dau., Nupur. J. C. Bose Research scholar Bose Research Inst., Calcutta, 1949-52; research fellow Purdue U., Lafayette, Ind., 1949-52; Mary and Richard Keating student Bose Inst., 1954-55, research fellow, 1955-61; hon. lectr. Calcutta U., 1961; reader botany, head dept. botany, Kalyani U., W. Bengal, India, 1961—. Recipient Mouat medal Calcutta U., 1955. Mem. Am., Scandinavian, Japanese, Indian (past v.p.) socs. plant physiologists, Bot. Soc. Bengal (editor Bull. 1966—), Soc. Biol. Chemists India, Indian Sci. Congress Assn. Contbg. author: Papier Chromatographie inder Botanik. Research and publs. on physiology of growth, reprodn. and aging, role of malic acid and oxalic acid in photosynthesis, genetic transformation of nitrogen fixing and antibiotic-producing microorganisms, yeasts and pseudomonads, occurrence of antibiotics in ferns and mosses, separation of plant growth substances, mechanism of action of antibiotics. Home: 68 Phear Lane, Calcutta-12, India. Office: Botany Dept., Kalyani U., W. Bengal, India.*

SEN, Sudhindra Nath, Indian chemist, microbiologist; b. Calcutta, India, Nov. 11, 1909; s. Manasa R. and Sushova (Majumdar) S.; B.Sc., Scottish Chs. Coll., Calcutta, 1931; M.Sc., U. Calcutta, 1933, D.Phil., 1950; m. Rama Roy, June 12, 1936; children—Krishna, Kishalaya, Swapna. Scientist dept. bacteriology and immunology Bengal Immunity Research Lab., Calcutta, 1935-48; bacteriologist Coll. Leather Tech., Calcutta, 1948-53; scientist leather microbiology Central Leather Research Instn., Council Sci. and Indsl. Research, Madras, I dia, 1953—; research fellow U. Leeds, Eng., 1951-52; faculty, research guide U. Madras, 1953—. Fellow Instn. Chemists India. Research and numerous publs. on composition of culture media and bacterial exo-toxin prodn., prodn. of microbial protease for use in leather industry, cause and prevention of coloration of salted hide, sterilization of hide and skin, microbial pigment formation. Home: 16 2d Main Rd., Kasturbangar Madras 20. Office: Central Leather Research Inst., Madras-20, India.*

SÉNAG, Jean Baptiste, French physician; b. Lombez, France, 1693; M.D.; 2 sons, including Gabriel Senac de Meilhan; named personal physician in ordinary to Louis XV, 1752. Mem. French Acad. Scis. Described leucocytes, pericarditis; introduced use of quinine for palpitations in heart disease, 1749; wrote commentary on Stahl's phlogiston theory. Died Paris, Dec. 20, 1770.

SENDERENS, Jean Baptiste, French chemist; b. Barbachen, France, 1856; prof. chemistry, dir. scis. Catholic Inst., Toulouse, France; mem. French Acad. Scis. Devel. (with Sabatier) catalytic hydrogenation of organic oils, 1899, used catalytic properties of reduced nickel to synthesize methane. Died Barbachen, Sept. 26, 1937.

SENDIVOGIUS, Michael, alchemist; b. nr. Cracow, Poland, 1556; s. Jacob and Catherine (Rogowska) Sedzimir; student univs. of Italy, Vienna, Leipzig, Altdorf, Germany. Made journeys to France, Spain, Eng., Sweden; rescued Scottish alchemist Alexander Seton from prison and after Seton's death married his widow; joined service of Rudolf II; became sec. to Sigismund II, Emperor of Poland, 1597; acted as liaison between Cracow and Prague, Czechoslovakia; imprisoned, 1599; visited Prague, 1604, France, 1605; imprisoned, Poland, 1605; (with Nicholas Wolski) placed in charge bldg. copper and iron foundries, Krzepice, Poland, Marburg, Germany, 1616, foundry, Silesia, Germany, 1620; made Italian journey, 1624; became councillor to Emperor Ferdinand II, 1626. Author: XII Tractatus de Lapide philosophorum, 1604. Novum Lumen Chymicum . . . , 1614; Tractatus de Sulphure; others. His works went through many editions and were translated into several languages. The Novum Lumen Chymicum (which may have been written by Seton) emphasizes the necessity of air for life and identifies the active part as a nitrous spirit which is needed for nutrition of plants. Died Aug. 1636.

SENDOV, Blagovest Hristov, Bulgarian mathematician; b. Assenovgrad, Bulgaria, Feb. 8, 1932; s Hristo Stoev and Marushka (Usheva) S.; dipl. math.

U. Sofia, (Bulgaria), 1956, doctor, 1963; m. Lilia Dimitrova Mindeva, Aug. 24, 1958; children—Marushka, Anna. Worker, Sofia, 1949-52; tchr. Boboshevo, Bulgaria, 1956-57; asst. U. Sofia, 1957-62, asst. prof. math., 1962—; chief dept. Math. Inst., Computing Centre, Bulgarian Acad. Scis., Sofia, 1962—. Mem. Union Scientists. Author: (with Peter Barnev) Computing Machines, 1966; also articles. Research on approximation theory in metrik of Hausdorff. Home: 17, Zelen sinur, Sofia 13. Office: 1 A. Ivanov, Sofia 26, Bulgaria.*

SENDROY, Julius, Jr., chemist; b. Zombor, Hungary, Sept. 26, 1900; s. Julius and Irene Kovacs (Birkas) S.; came to U.S., 1901, naturalized, 1913; B.S. cum laude, Coll. City of N.Y., 1923; M.A., Columbia, 1925, Ph.D., 1926; Sc.D., St. Bonaventure U., 1954; m. Jeannette Candee, June 21, 1932; children—Beatrice Marilyn (Mrs. Ray Williams), Peter Gyula. Mem. faculty Hosp. of Rockefeller Inst. for Med. Research, 1926-37, Loyola U. and Mercy Hosp., Chgo., 1937-48; chief chemist U.S. Naval Research Inst., Nat. Naval Med. Center, Bethesda, Md., 1948—, head div. chem., 1948-65, sci. adviser, 1965—; also cons. Naval Med. Sch., U.S. Army Chem Warfare Labs. Mem. NRC. Bd. dirs. Nat. Registry in Clin Chemistry. Recipient Van Slyke award N.Y. sect. Am. Assn. Clin. Chemists, 1962. Diplomate Am. Bd. Clin. Chemistry. Fellow N.Y. Acad. Scis., Am. Assn. Clin. Chemists (pres. 1964); mem. Am. Soc. Biol. Chemists, Am. Chem. Soc., Soc. Exptl. Biology and medicine (bd. editors 1941-48), Central Soc. Clin. Research, Aerospace Med. Assn., Am. Physiol. Soc., others. Mem. bd. editors Clin. Chemistry, 1961—, chmn., 1963-66. Research and publs. on phys. chemistry of blood gases and electrolytes; mineral, protein metabolism; analytical methods. Home: 6914 Selkirk Dr., Bethesda 20034. Office: Nat. Naval Med. Center, Bethesda, Md. 20014.*

SÉNEBIER, Jean, Swiss natural scientist; b. Geneva, Switzerland, May 6, 1742; studied privately in Paris libraries. Ordained minister, circa 1762; preached at Chancy several years; apptd. keeper pub. library, Geneva, 1773. Author: Essai sur l'art d'observer et de faire des expériences, 2 vols., 1775; Histoire littéraire de Genève, 3 vols., 1796; Catalogue raisonné des manuscrits conservés dans la Bibliotheque de Genève, 1779; Mémoires sur l'influence de la lumière solaire, 3 vols., 1782; Physiologie végétale, 5 vols., 1782-88; 1800; Essai sur l'art d'observer et de faire des expériences, 1802; Rapport de l'air atmosphérique avec les etres organisés, 1807. Demonstrated green plants change carbon dioxide to oxygen under influence of light, 1782; 1st to give connected view of entire process of vegetable nutrition in chem. terms; research in meteorology, physics, physiology. Died Geneva, July 22, 1809.

SENECA, Lucius Annaeus, Roman polit. philosopher, statesman; b. Cordova, Spain, circa 4 B.C.; s. Lucius Annaeus and Helvia (of Cordova) S.; ed. in Rome; m. Pompeia Paulina. Gained distinction as lawyer; named quaestor; banished to Corsica after romantic scandal, 41 A.D.; returned to Rome, 49 A.D., raised to praetorship; served as tutor to Nero; named consul, 57 A.D.; amassed tremendous wealth; became accomplice, then target of Nero's intrigues; Author: Epistolae morales ad Lucilium; Naturales Questiones; also many tragedies. Believed that polit. society through its degeneration had become separate from natural society which he viewed as fellowship of man; urged service to natural society; helped lay groundwork both for dualism of early Roman Catholic polit. thinkers and for back to nature idealism of later epochs; leading Stoic philosopher in Rome; in phys. scis. described earthquake at Campania, 63 A.D. and distinguished kinds of motion in quakes; saw connection between quakes and volcanoes; believed comets revolved in fixed orbits; wrote on geological, astronomical, physical and meteorological subjects which he explained from an atomistic viewpoint. Opened his veins and died in warm bath after being ordered to take his life, 65 A.D.

SENECAL, Jean Marcel, French pediatrician; b. Raincy, France, June 29, 1916; s. Marcel and Suzanne (Chaumeil) S.; ed. Paris Faculty of Medicine; m. Jeanne Demont-Vivot, Feb. 7, 1940; children—Claudine, Catherine, Christine. Intern Hosps. of Paris; prof. pediatry Kabul (Afghanistan) Faculty of Medicine, 1947-53, Dakar Faculty of Medicine, 1952-62, Rennes Faculty of Medicine, 1963-64; rep. Internat. Childhood Center for Africa, 1952-62; dir. Nat. Sch. Pub. Health, 1962—. Decorated Crder Palms Acad., Legion of Honor, Order Pub. Health, Order Social Merit (France); Nat. Order Republic of Senegal. Publs. on infant diseases, especially Tb and nutritional diseases. Address: 86, rue de Paris, Rennes (Ile-et-Vilaine France.

SENEFELDER, Aloys, inventor; b. Prague, Czechoslovakia, 1771; ed. Coll. of Munich, also Ingolstadt; children include Henry; actor; dir. state lithograph inst., Munich. Author: A History of Lithography, 1819; also several dramas. Inventor lithography, circa 1795, color lithography, 1826. Died 1834.

SENENSIS, Hugo, see Benzi, Ugo.

SENEVET, Georges Louis, French entomologist; b. Alger, Feb. 7, 1891; s. Edouard and Lucie (Bouthier) S.; Certificats Sci., U. Alger, 1922, Docteur en Médecine. Staff, Faculty Medicine, Paris, France, 1919-20; prof. Faculty Medicine, Alger, 1928-45; head lab. Pasteur Inst., Alger, lt. col. physician Army Lab., 1939-46; dir. Colonial Med. Inst., Alger, 1947—; head entomology lab. Antipaludian Service, Algeria, 1954-62; prof. parasitology and bacteriology Faculty Alger, 1962-66, hon. prof., 1966—. Decorated Legion of Honor, Médaille épidémiès. Officier Instruction publique, Médailles commemoratives. Mem. Société de Pathologie exotique Paris. Author: Anophèles France et Colonies, 1935; Ixodides de France, 1937; L. Andarelli Anophèles méditerrannée 1953; (with L. Andarelli) Culicides Mediteranée, 1956; Anopheles du globe, 1953. Research on mosquitos and ticks of France, N. Africa, Guyane, parasitic diseases. Home: 24 rue de Lourmel, Paris XV°. Office: Faculté de Médecine, Laboratoire de Parasitalogy, Rue de l'Ecole de Médecine, Paris 6, France.*

SENEZ, Jacques C., French microbiologist; b. Marseilles, France, Jan. 14, 1915; s. Charles and Juliette (Guigou) S.; M.D., Faculté des Médecine de Marseilles, 1941, Sci.D., 1955; m. Gillette (Gailleton, July 9, 1943; children—Catherine, Brigitte, Veronique. Dir. sci. Centre Nat. de la Recherche Sci., Marseille, dir. Laboratoire de Chimie Bactérienne. Mem. exec. com. Internat. Cell Research Orgn. Decorated chevalier la Legion d'Honneur. Mem. Am. Soc. Microbiology. Research, numerous publs. on energetics of bacterial growth, petroleum microbiology, gen. microbiology. Home: 96 reu St-Jacques, Marseille 6. Office: 31, Chmin J. Aiguier, Marseille 9, 13, France.*

SENFT, Alfred Walter, Am. physician; b. Windsor, Colo., Oct. 20, 1924; s. Hans and Elsie (Mueller) S. student Wartburg Coll., 1942, U. Zurich, 1946-47; B.A., U. Cal. at Berkeley, 1948; M.D., Harvard, 1952; Dr. Tropical Medicine and Hygiene, London Sch. Tropical Medicine, 1953; M.P.H., Harvard, 1956; m. Deborah Gates, Aug. 7, 1948; children—Stephen, Karen, Valerie, Physician, Lutheran Mission, Jaquaom Hosp., New Guinea, 1954; pvt. practice medicine, Falmouth, Mass., 1956-58, Woods Hole, Mass., 1958;66;ind. investigator Marine Biol. Lab., Woods Hole, 1956—; sch. physician Falmouth Schs., 1958—; cons. in parasitology to Merck Co., 1965; asso. prof. dept. med. scis. Brown U., Providence. Recipient A. Cressy Morrison award N.Y. Acad. Sci., 1965; Pub. Health award Chi Rho, 1956. Mem. Royal Soc. Tropical Medicine, Am. Soc. Parasitology, N.Y. Acad. Sci. Research and publs. in physiology and biochemistry of parasites; developed a chemically defined protein-free medium for schistosomes; obtained first EKG of whale. Address: 26 Halsey St., Providence, R.I.

SENGBUSCH, Reinhold von, see von Sengbusch, Reinhold.

SENGEL, Philippe, French biologist; b. Strasbourg, France, Oct. 9, 1928; s. Alfred and Marguerite (Liebrich) S.; student U. Strasbourg, 1947-55, College de France, U. Paris, 1955-58; m. Madeleine A. Kieny, Aug. 2, 1955; children—Valerie, Thomas. Under—dir. lab. exptl. embryology College de France, Paris, 1956-64; prof. zoology Faculty Scis., U. Grenoble, 1964—. Lalor Found. fellow Woods Hole, Mass., 1961; Sophie Fricke fellow Rockefeller Inst., N.Y.C., 1961-62. Recipient Godard prize Soc. Biology, Paris, 1962; Cocnacq-Jay Samaritaine prize Acad. Scis., Paris, 1965. Mem. French Soc. Zoology, Internat. Inst. Embryology, Am. Soc. Cell Biology, European Tissue Culture Club. Co-Author: The Epidermis, 1964. Research, numerous publs. on mechanisms of regeneration of pharynx in planaria, of differentiation of skin and cutaneous appendages in chick embryo, endocrinological studies on neural gland complex in ascidians. Home: 3 Blvd. Marechal Joffre, 38 Grenoble, France.*

SENGUERD, Wolferd, inventor; b. Utrecht, Netherlands, 1646; inventor air pump with double-bored stopcork, 1685, used it to produce both a vacuum and an excess pressure condition, 1697. Author: De tarantula (attempt to explain poisonous effects of tarantula by natural rather than occult reasons), 1668. Died 1724.

SENN, Milton J(ohn) E(dward), Am. physician; b. Milwaukee, Wis., Mar. 23, 1902; s. John and Louise (Rosenkranz) S.; B.S., U. of Wis., 1925, M.D., 1927; M.A. (hon.), Yale, 1948; m. Blanche Forsythe, Sept. 8, 1932; 1 dau., Corelyn F. Interne Columbia Hosp., Milwaukee, 1927-29, including resident in Children's Hosp., Milwaukee, 1927-28; fellow and instr. in pediatrics, Washington U., St. Louis, 1929-33; asso. in pediatrics Cornell U. Med. Coll., N.Y. City, 1933-39; Commonwealth Fund fellow in psychiatry, 1937-39; asst., asso. and later attending pediatrician, N.Y. Hosp., 1933-49; mem. faculty Cornell U. Med. Coll., 1939-49, prof. of pediatrics in psychiatry, 1948-49; dir. Child Study Center, Yale, 1948-66, also Sterling prof. of pediatrics and psychiatry, Yale, since 1948; pediatrician-in-chief Grace-New Haven (Conn.) Community Hosp., 1951-58. Member of board directors The Field Foundation. Member of American Pediatric Soc., Am. Acad. Pediatrics, Am. Orthopsychiatric Assn., Soc. Research Child Development, N.Y. Acad. Medicine. Contributor of numerous articles to professional journals. Interested in study of child rearing in various cultures; psychological development of infants and children; integration of psychiatry and pediatrics. Home: 972 Prospect St., Hamden, Conn., 06511. Office: 333 Cedar St., New Haven 06510.*

SENN, Nicholas, physician, surgeon; b. Buchs, Canton of St. Gall, Switzerland, Oct. 31, 1844; brought to U. S. by parents, 1853; settled at Ashford, Fond du Lac Co., Wis.; grad. Fond du Lac High School, 1864; taught school; grad. Chicago Med. Coll., 1868; Univ. of Munich, 1878; house physician, Cook Co. Hosp., 1868-69; practiced medicine, Fond du Lac, 1869-74; Milwaukee, 1874-93; was surgeon-gen. of Wis.; engaged in practice at Chicago; surgeon-gen. Ill. Nat. Guard; well known specialist in surgery; attending surgeon, Presbyn. Hosp., and surgeon-in-chief, St. Joseph's Hosp., Chicago, until 1898, prof. surgery, Coll. of Phys. and Surg., Chicago, 1884-87; prof. principles of surgery, 1887-90; from 1890 prof. practical and clinical surgery, Rush Med. Coll., also prof. surgery, Chicago Policlinic; professorial lecturer on mil. surgery, Chicago Univ.; del. Internat. Med. Congress, Berlin, 1890, Moscow, 1897, Madrid, 1903. Author: Four Months Among the Surgeons of Europe; Experimental Surgery; Intestinal Surgery; Surgical Bacteriology; Principles of Surgery; Pathology and Surgical Treatment of Tumors; Tuberculosis of Bones and Joints; Tuberculosis of the Genito-Urinary Organs; Syllabus of Practice of Surgery; Practical Surgery; Surgical Notes on the Spanish-American War; Medico-Surgical Aspects of the Spanish-American War; Practical Surgery; Nurse's Guide for the Operating Room, 1902; Around the World via Siberia; Our National Recreation Parks; Around the World via India—A Medical Tour. Noted for introduction of radiology in surgery. Died 1908.

SENNERT, Daniel, German physician; b. Breslau (now Wroclaw, Poland), Nov. 25, 1572; M.A., Wittenberg, 1598; studied at Leipzig, Jena and Frankfort on the Oder; M.D., Wittenberg, 1601; became prof., Wittenberg, Germany, 1602, also rector; physician to Elector of Saxony, Germany. Author: Institutionum medicinae . . . , 1611; Epitome naturalis scientiae, 1618; De chymicorum cum Aristotelicis et Galenicis Consensu ac Dirsensu, 1619; Practicae medicinae, 1635-52; Hypommemata physica, 1636; Opera omnia, 1641. Described scurvy, 1614, scarlet fever, 1619; 1st authenticated caesarian section on living woman; tried to compromise between Paracelsian and Aristotelian-Galenic medicine; attempted to explain chem. changes through minute corpuscles; opposed blind adherence to Paracelsus, but strongly supported the Paracelsian principles of matter; gave symptoms and antidotes for various poisons. Died Wittenberg, July 21, 1637.

SENNET, George Burritt, Am. ornithologist, mfr.; b. Sinclairville, N.Y., July 28, 1840; s. Pardon and Mary (Burritt) S.; m. Sarah Essex; 1 child. Mfr. oil-well machinery, Meadville, Pa., 1865-95, Youngstown, O., 1895-1900; mayor of Meadville, 1877-81; made ornithol. expdn. to Western Minn., 1867; expdns. to Rio Grande region of southern Tex., 1877, 78, 82; contbd. collection of birds to Am. Mus. Natural History, N.Y.C., 1883; discovered 10 new species of birds; 4 birds named in his honor. Author articles Notes on the Ornithology of the Lower Rio Grande, Texas, from Observations Made During the Season 1877 (Bull. U. S. Geol. and Geog. Survey of Territories, 1878), Descriptions of a New Species and Two New Subspecies of Birds from Texas (Auk), 1888. Died Mar. 18, 1900.

SEPPINGS, Sir Robert, English inventor; b. Norfolk, Eng., 1767; s. Robert and Lydia (Milligen) S.; D.C.L., Oxford U., 1836; m. Charlotte Milligen; children—John Milligen, Edward. Apprentice as working shipwright Plymouth stockyard, 1782; became master shipwright asst., 1800; master shipwright, Chatham, 1804; surveyor of navy, 1813-32. Recipient Copley medal, 1803. Fellow Royal Soc., 1814. Contbr. to Trans. Royal Soc. Inventor Seppings blocks for suspending vessels in dock, 1800, also system of diagonally bracing and trussing frame timbers of ships; recommended replacement of beakhead of ship by timbers run up sides forming a circular bow; 1807; introduced round stern; his improved methods of shipbuilding universally adopted. Died Sept. 25, 1840.

SEQUENZ, Heinrich Josef, Austrian elec. engr.; b. Vienna, Austria, Jan. 13, 1895; s. Heinrich and Katharina (Gall) S.; diploma in mech. engring., Tech. U. Vienna, 1923, Dipl.-Ing Elektrotechnik, 1924, Dr. techn., 1947; Dr.-Ing., Techn. Hochschule, Karlsruhe, Germany, 1935; Dr.phil., U. Vienna, 1938; Dr.-Ing.h.c., Technische Hochschule, Stuttgart, Germany, 1958; m. Ilse Wildt, July 2, 1950. Dozent, Technische Hochschule, Vienna, 1932-37, prof. electro-mech. engring., 1939-49, 54—, rector magnificius, 1942-45; staff ELIN A.G., 1949-54. Mem. Obersten Patent und Markensenates; mem. curatorium Austrian Research Council; v.p. Austrian nat. com. Internat. Elektrot. Kommission. Recipient Wilhelm-Exner medal Austrian Indsl. Assn., 1956; Gold medal of honor Austrian Assn. Engrs. and Architects, 1965. Mem. Austrian Acad. Scis. Vienna, German Acad. Scis. Berlin (editorial com. Technisches Zentralblatt). Author: Die Wicklungen elektrischer Maschinen, 4 vols., 1950-54; Elektrische Maschinen, 6th edit., 1952; also numerous articles. Editor, Elektrotechnik und Maschinenbau, 1938—; sci. dir. Elin-Zeitschrift, 1949—. Research on elec. engines, transformers, rectifiers, elec. installations, regulations, windings of elec. engines. Home: 118 Linke Wienzeile 1060, Vienna, Austria.*

SERAPION OF ALEXANDRIA, Greek physician; flourished circa 200-150 B.C.; author: Curationes; also treatise against med. sects. Founder empirical sch. medicine, rejected med. dogmatism and based his studies on observation and experiment, clin. cases, analogy.

SERAPION OF ANTIOCHEIA, mathematician, geographer; flourished 2d or 1st century B.C.; wrote on astrology; believed the sun 18 times larger than the earth.

SERBER, Robert, Am. physicist; b. Phila., Mar. 14, 1909; s. David and Rose (Frankel) S.; B.S., Lehigh U., 1930; Ph.D., U. Wis., 1934; m. Charlotte Leof, June 15, 1933. Nat. Research fellow U. Cal., 1934-36, research asso., 1936-38, prof. radiation lab., 1945-51; asst. prof. U. Ill., 1938-40, asso. prof., 1940-45; war work Metall. Lab., Chgo., 1942-43, Los Alamos (N.M.) Lab., 1943-45; prof. physics Columbia since 1951; cons. Brookhaven Nat. Lab. Mem. Solvay Conf., Brussels, Belgium, 1948, Atomic Bomb Group, Marianas, July-Sept. 1945; dir. phys. measurements Atomic Bomb Mission to Japan, Sept.-Oct. 1945. Mem. sci. policy adv. coms. Brookhaven Nat. Lab., Argonne Nat. Lab., Stanford Linear Accelerator Center, Nat. Accelerator Lab. Guggenheim fellow. Fellow Am. Phys. Soc.; mem. Nat. Acad. Scis., Sigma Xi. Contbr. sci. jours. Laid found. of orbit theory of high energy accelerators; contbns. in almost all aspects high energy physics, ranging from cosmic radiations to nuclear forces, meson theory and recent ultra-high energy scattering phenomena. Home: 450 Riverside Dr., N.Y.C. 10027.

SERENUS, mathematician from Antinoeia, Egypt; flourished circa 300-350. Author: De sectione cylindri (33 problems); De sectione coni (69 problems). Solved problem such that given an cone, a cylinder is found such that a section of both by the same plane give similar ellipses; discovered theorem which is found. of modern theory of harmonics.

SERGI, Giuseppe, Italian anthropologist; b. Messina, Italy, 1841; prof., Bologna and Rome, Italy. Author: Specie e varietà humane, 1900; Le origini umane, 1913. Founder Italian anthropology; studied origin and spread of Mediterranean race, origin of Aryans, distbn. of races; used photographs to catalogue skulls; helped clarify the problem of what is a useful measurement in phys. anthropology. Died 1936.

SERGIENKO, Semen Romanovich, Russian organic chemist; b. 1909; Mem. Turkmenia Acad. Scis. (v.p., exec. ed. News of Turkenia Acad. Scis.). Author numerous works including: Synthetic Rubber; An Outline of Chemistry and Petroleum Refining; High-Molecular Petroleum Compounds. Research on petrochemistry, polymerization mechanism of butadiene; developed petrochemistry of high-molecular petroleum compounds, 1951-59. Address: Turkmenia Acad. Scis., Ashkhabad, Turkmenia SSR, USSR.

SERGIEVSKY, Mikhail Vasilevich, Russian physiologist; b. 1898; grad. Med. Faculty, Kazan U., 1926; D.Med. Sci. Asst., lectr. dept. physiology Kazan Med. Inst., 1926-36; head chair physiology Kuybyshev Med. Inst., 1936——. Recipient Stalin prize. Mem. USSR Acad. Med. Scis. (corr.). Author over 150 works including: The Respiratory Center in Mammals and Regulation of Its Activity, 1950; The Cerebral Cortex and Regulation of Breathing, 1953; co-author, editor: Physiology and Pathology of Regulation of Respiration and Circulation, 1957; co-author: Summary of Research on the Physiology of Respiration in the Last Ten Years, 1961. Address: Kuybyshev Medical Institute, Chapaevskaya ulitsa 89, Kuybyshev, USSR.

SERIFF, Aaron Jay, Am. physicist; b. Eagle Pass, Tex., Apr. 26, 1924; s. Abraham Leon and Sarah (Dochen) S.; B.S. in Physics, U. Tex., 1944, M.A., 1946; Ph.D., Cal. Inst. Tech., 1951; m. Esther Kay Berman, Dec. 23, 1951; children—Jan Alison, Suzanne Catherine. Instr. math., U. Tex., 1946; physicist Shell Devel. Co., Houston, 1951-62, sr. research asso., 1962——. Mem. Soc. Exploration Geophysicists (mem. research com. 1962-65), Am. Phys. Soc., Am. Geophys. Union, A.A.A.S., Sigma Xi. Asst. editor Geophysics, 1964-65. Research in cosmic rays leading to discovery (with other) of new unstable particles V^0 and V ; exploration geophysics, particularly seismology. Home: 5003 Jason St., Houston 77035. Office: 3737 Bellaire Blvd., Houston 77025.*

SERIN, Bernard, Am. physicist; b. N.Y.C., May 22, 1922; s. Lewis and Rose (Kalvin) S.; B.S., Poly. Inst. Bklyn., 1941; M.A., U. Pa., 1944, Ph.D., 1946; m. Bernice Goldis, May 7, 1944; children—Judith Ann, Joan Alice. Faculty, Rutgers U., New Brunswick, N.J., 1947——, prof., 1958——; vis. research asso. prof. U. Ill., 1958-59. Fellow Am. Phys. Soc.; mem. Sigma Xi. Research, numerous publs. on super-conductivity; co-discoverer of isotope effect; studies of dilute alloys leading to concept of energy gap anistropy; specific heats: metals, solid neon and xenon; thermoelectric effects; semiconductors. Home: 32 Juniper Lane, Piscataway, N.J. 08854.*

SERMONTI, Giuseppe, Italian geneticist; b. Rome, Italy, Oct. 4, 1925; s. Alfonso and Letizia (Marchesano) S.; degree in Agrl. Scis., U. Pisa (Italy), 1947; degree in Biols. Scis., U. Rome, 1957; m. Isabella Spada, June 11, 1953; children—Andrea, Fabio, Valeria. Asst., Istituto di Genetica per la Cerealicultura, 1948-49, Istituto Superiore di Sanità, 1950-65 (both Rome); faculty U. Rome, 1958——, prof. genetics, 1965——, also tchr. biology of human races. Mem. Italian Genetic Soc., Italian Soc. for Molecular Biology. Editor in chief Annali dell' Istituto Superiore di Sanità, 1965——. Research, publs. on application of genetics to antibiotic producing microorganisms; discoverer (with G. Pontecorvo) sexual process in Penicillium chrysogenum; (with I. Spada-Sermonti) discovery of sexuality in Actinomycetes (Streptomyces); research on genetic system in Actinomycetes. Home: 12, Via O. Tommasini, 00162 Rome, Italy. Office: Istituto di Genetica, 22 via Archirafi, 90123 Palermo, Italy.*

SEROVY, George Kaspar, Am. mech. engr.; b. Cedar Rapids, Ia., Aug. 29, 1926; s. George and Helene (Kaspar) S.; B.S., Ia. State U., 1948, M.S., 1950, Ph.D., 1958; m. Janice Ann Radcliffe, Mar. 20, 1948; children—William Allan, Ann Elizabeth, Mary Lynn. Aero. research sci. NACA, Cleve., 1949-53; mem. faculty Ia. State U., Ames, 1953——, prof. mech. engring., 1960——. Cons. to indsl. orgns. on fluid mechanics of turbomachinery; project dir. NASA sponsored research on axial-flow pumps. Mem. Am. Soc. M.E. (chmn. compressor com., fluids engring. div. 1964-66), Soc. Automotive Engrs., Am. Soc. for Engring. Edn., Sigma Xi, Tau Beta Pi, Pi Tau Sigma. Research and publs. on fluid mechanics of internal flow, also methods for design and analysis of turbomachinery. Home: 3610 Oakland St., Ames, Ia. 50010.*

SERPE, J., Belgian physicist; b. Andrimont, Belgium, Aug. 1, 1914; s. J. and S. (Vandewiele) S.; Docteur en Sciences, U. Liège (Belgium), 1940, Agrégé de lá Enseignement supérieur, 1944; m. R. Vanbreuse, Apr. 22, 1939; children—Jacqueline (Mrs. A. Herion), Christiane, Claudine. With U. Liège (Belgium), 1938——, prof. physics, 1960——. Mem. commn. scientifique Institut Interuniversitaire des Sciences Nucléaires, 1959-62, pres., 1961, v.p. adminstrv. council, 1961. Recipient Prix Agathon de Potter pour la Physique; Chevalier de l'Ordre de Léopold; Prix des Amis de l'Université de Liège. Mem. Société Francaise de Physique, Société Royale des Sciences de Liège, Société Belge de Physique. Author: Les lois de conservation en Physique des particules élémentaires, 1959; also articles. Research on quantum field theory, including quantum electrodynamics, meson field; elementary particles, especially wave equations; showed Marjorana's theory for neutrino field is equivalent to Weyl's two-component theory. Home: 65 Rue Prof. Mahaim, Sclessin, Belgium. Office: 15, Avenue des Tilleuls, Liege, Belgium.*

SERPIERI, Alessandro, Italian astronomer; b. San Giovanni, Italy, Oct. 31, 1823; student Scolopians' Coll., Urbino, Italy, Ximenian Coll. Joined Order of Scolopians, 1838, ordained 1848; began teaching math. and astronomy Ximenian Coll. at age of 20; joined Urbino Coll., 1846, established obs., 1850, apptd. rector of coll., 1857, tchr. until 1884; rector Collegio della Badia Fiesolana. Research and publs. on earthquakes of 1873 and 75, electricity and physics, seismology; introduced concept of seismic radiant; observed shooting stars. Died Fiesole, Italy, Feb. 22, 1885.

SERRA, Gian Edoardo, Italian physician; b. Genoa, Italy, Dec. 1, 1928; s. Giacomo and Ildegarda (Cervetti) S.; M.D., U. Genoa, 1954, postgrad., 1958. House physician Ospedali Civili, Genoa, 1954-55, maternity asst., 1956; asst., div. obstetrics and gynecology Galliera Hosp., Genoa, 1958——; prof. Clinic Obstetrics and Gynecology, U. Rome, 1965——. Publs. on colposcopic diagnosis, lymphadenography, vaginocervical infections, treatment of tubal impermeability. Home: 14 via 20 settembre, Genoa, Italy.*

SERRE, Henri Marius Alphonse, French physician; b. Montpellier, Herault, France, Feb. 6, 1911; s. Alphonse and Marie (Gaillard) S.; M.D., Faculty Medicine of Montpellier, 1936, M.Sc., Faculty Sci. of Montpellier, 1937; m. Odette Grillon, July 12, 1934; children—Lucienne (Mrs. Jacques Debrus), Jean-Claude. Head dept. Med. Clinic Faculty Medicine of Montpellier, 1937-46, teaching staff gen. medicine, 1946-57, prof. clin. rheumatology, 1957——; resident physician Montpellier U. Med. Center, 1943-53, chief staff rheumatology, 1953——. Mem. French Nat. Coll. Rheumatology (pres. 1960), French League Against Rheumatism (pres. 1963), Soc. Paris Hosps., Internat. Soc. Internal Medicine. Research, numerous publs. on shock therapy in rheumatology, treatment of gout, primitive osteonecrosis of head of femur, brucellosis osteo-articular. Home: 6 Rue Boussairolles. Office: Clinique de Rhumatologie, 34 Montpellier, Herault, France.

SERRE, Jean-Pierre, French mathematician; b. Bages, France, Sept. 15, 1926; s. Jean and Diet Serre; B.A., Agrégé, Ph.D., Ecole Normale Superieure; m. Heulot Josiane, Aug. 10, 1948; 1 dau., Claudine. With Centre Nat. de la Recherche Sci., 1948-59; chmn. conferences Faculty of Sci., U. Nancy (France), 1954-56; prof. Coll. de France, Paris 1956——. Recipient Fields medal Internat. Congress Math., 1954. Fgn. mem. Am. Acad. Arts and Scis. Author: Algèbres de Die, 1955; Groupes Algebriques, 1959; Corps Locuax, 1962; Lie Groups, 1965. Research on homotopy groups, algebraic geometry, local fields, arithmetic. Home: 6 Av. Montespan, Paris 16. Office: Collège de France, Paris 5, France.*

SERRES, Antoine Étienne Reynaud Augustin, French physician, anatomist, embryologist; b. Clairac, France, Sept. 12, 1786; diploma in medicine, Paris, 1810; became head anat. work l'Amphitéatre central, 1814; named physician in chief Pitié, 1822; became prof. comparative anatomy Mus. Natural History, 1839. Mem. French Acad. Scis., 1823, French Acad. Medicine. Author publs. including: The Comparative Anatomy of the Brain in the Four Classes of Vertebrate Animals, 2 vols., 1824-26. Established laws of organogenesis; discovered devel. of animals and their organs goes from circumference toward center; studied physiology and pathology of brain and spinal cord; described gingival glands (epithelial masses found occasionally on gums of infants), 1817. Died Paris, Jan. 22, 1868.

SERRET, Joseph Alfred, French mathematician; b. Paris, Aug. 30, 1819; student Ecole polytechnique. Officer of arty.; became examiner for admissions Polytechnic Sch., 1848; named prof. suppleant higher algebra Sorbonne, Paris, 1849, suppleant astron. physics, 1856; named prof. celestial mechanics Coll. de France, 1861; became prof. differential and integral calculus Faculty Scis., 1863. Mem. French Acad. Scis., 1860. Author: Cours d'algèbre superieure, 1849; Traité de trigonométrie, 1850; Leçons sur les applications pratiques de la géométrie et de la trigonométrie, 1851; Traité d'arithmétique, 1852; Eléments de trigonométrie à l'usage des arpenteurs, 1853; Cours de calcul différentiel et intégral, 1867-68; Oeuvres de Lagrange, 1867-77. Studied theory of functions, group theory, differential equations. Died Paris, Mar. 2, 1885.

SERRIN, James, Am. mathematician; b. Chgo., Nov. 1, 1926; s. James Burton and Helen (Wingate) S.; B.A., Western Mich. Coll., 1947; M.A., Ind. U., 1949, Ph.D., 1951; m. Barbara West, Sept. 6, 1952; children—Martha Helen, Elizabeth Ruth, Janet Louise. Instr. math. Princeton (N.J.) U., 1951, Mass. Inst. Tech., Cambridge, 1952-54; faculty U. Minn., Mpls., 1954——, prof. math., 1958——, head sch. of math., 1964-65. Co-editor Archive for Rational Mechanics and Analysis. Research in fluid mechanics, partial differential equations, calculus of variations. Home: 4849 Aldrich Av. S., Minneapolis 55409.*

SERSALE, Riccardo, Italian chemist; b. Naples, Italy, Apr. 14, 1921; s. Luigi and Giulia (Attanasio) S.; Dr. degree in Chemistry, U. Naples, 1943. Faculty, Inst. Indsl. Chemistry, Faculty Engring., U. Naples, 1945-62, lectr., 1953-62; prof. materials tech. and applied chemistry, dir. Inst. Applied Chemistry, Faculty Engring., U. Bari (Italy), 1962-67; prof. applied chemistry Faculty Engring., U. Naples, 1967——. Mem. Società Nazionale di Scienze, Lettere ed Arti in Napoli, Soc. dei Naturalisti in Napoli. Research, publs. on chem. constn., structure, properties and tech. behaviour of inorganic materials such as silicates, hydraulic binders, blast-furnace slags, glasses, cast iron and steel, including corrosion phenomena. Home: 147 via Posillipo, 80123 Naples, Italy.*

SERTOLI, Enrico, Italian physiologist; b. Sondrio, Italy, June 6, 1842; prof. exptl. physiology, Milan, Italy; described elongated cells (Sertoli's cells) in seminiferous tubules of testes which support spermatids until they become spermatozoa, 1865. Died Jan. 28, 1910.

SERTÜRNER, Friedrich Wilhelm Adam, German chemist; b. Neuhaus, Germany, June 19, 1783; began as apothecary's asst., Paderborn, Germany, 1805; pharmacist, Hameln, Germany. Discovered morphine is main constituent of opium, 1804, and thus proved existence of organic bases containing nitrogen; originated term, alkaloid; discovered sulfurous acid, 1820. Died Hameln, Feb. 20, 1841.

SERULLAS, Georges Simon, French chemist; b. Poucin, France, Nov. 2, 1774; student pharmacy, Bourg, France, 1793; served in Corps Mil. Pharmacists, beginning 1793; became prof. Val-de-Grace, France, 1825; prof. chemistry Mus. Natural History. Publs. in chemistry; discovered iodoform, 1822, protochlorine of carbon, 1823, also iodine of cyanogen, and other iodine and bromine compounds. Died May 25, 1832.

SERVADIO, Emilio, Italian psychoanalyst; b. Genoa, Italy, Aug. 14, 1904; s. Cesare and Foustina (Finzi) S.; LL.D., Genoa U., 1926; m. Clara Valloscura, 1951; 1 dau. Became prof., fellow Andhra Research U., Madras Coll., India, 1938; pres. Psychoanalytic Center Rome, Italy, 1962——. Mem. Italian Soc. for Parapsychology (v.p. 1955-56), Italian Psychoanalytic Soc. (v.p. 1953-55), Parapsychoanalytic Assn. (charter), Internat. Psychoanalytic Assn., Italian Assn. for Psychol. Medicine and Psychotherapy, Italian

Soc. for Sci. Psychology, Soc. for Psychical Research (London), Internat. Com. for Study Methods in Parapsychology (chmn.). Author: La ricerca psichica, 1930; La psicoanalisi, 1953; Rôle des conflicts préodipiens, 1953; Il sogno, 1955. Contbg. author: Treccani Ency. Dictionary for Psychoanalysis and Para-Psychology. Contbg. author, editor: Enciclopedia Italiana. Translator psychoanalytic texts. Research on psychoanalytic approach to parapsychology. Address: Rome, 1 Via Annone, Italy.

SERVETUS, Michael, Spanish physician; b. Tudela, Spain, Sept. 29, 1511; student law, Toulouse, France, medicine, Lyons, France, Paris, theology and Hebrew, Louvain, Belgium; became student Med. Sch., Montpellier, France, 1540. Lectr. geometry and astrology; practice medicine, Avignon, France; later in Vienna, 1541-53; arrested and brought to trial before Inquisition, Vienne; escaped but captured at Geneva, Switzerland; imprisoned and burnt at stake as heretic by Calvin, 1553. Author: Syruporum universa ratio . . . , 1537; Christianismi restitutio, 1553. Discovered pulmonary circulation and blood purification by lungs; explained digestion as source of animal heat; one of 1st to try to introduce idea of connection between venous and arterial systems outside heart; opposed to idea of trinity, infant baptism. Died Geneva, Oct. 27, 1553.

SERVICE, Elman Rogers, Am. anthropologist; b. Tecumseh, Mich., May 18, 1915; s. Joseph Duaine and Ethel (Rogers) S.; A.B., U. Mich., 1941; Ph.D., Columbia, 1950; m. Helen Stevenson, Dec. 2, 1942. Faculty, Columbia, 1949-53; faculty U. Mich., Ann Arbor, 1953—, prof. anthropology, 1961—. Fellow Am. Anthrop. Assn.; mem. Phi Beta Kappa. Author: Tobati, 1954; Profile of Primitive Culture, 1958; (with M. D. Sahlins) Evolution and Culture, 1960; Profiles in Ethnology, 1963; The Hunters, 1966. Research and publs. on Indian socs. in Mexico and S.Am., theory of cultural evolutionism. Home: 401 Awixa Rd., Ann Arbor, Mich. 48104.*

SERVOIS, François Joseph, French mathematician; b. Mont-de-Laval/Doubs, France, July 19, 1767; prof., Chalons, also Metz, France; dir. arty. mus., Paris. Author: Solutions peu connues de différents problèmes de géométrie pratique, 1803. Contbd. to knowledge of infinitesimal calculus. Died Apr. 17, 1847.

SERY, Zdenek, Czechoslovakian surgeon; b. Vezky, Czechoslovakia, Sept. 23, 1917; s. Antonín and Marie (Será) S.; student Med. Faculty, U. Masaryk, Brno, Czechoslovakia, 1936-45, C.Sc., 1959; m. Daniela Hiklová, July 12, 1957. Faculty dept. surgery Palacky U. Med. Faculty, Olomouc, Czechoslovakia, 1945—, prof. surgery, 1967—. Author: Surgical Treatment of Achalasia of Esophagus, 1952; Pulmonary Resection in Children, 1962; Surgical Problems of Gastric Cancer, 1964; Indications and Results of Pulmonary Resection in Children, 1966. Research, numerous publs. on surg. pathology and treatment. Home: 6 I.P. Pavlova, Olomouc, Czechoslovakia.*

SESHADRI, Tiruvenkata Rajendra, Indian chemist; b. Kulitalai, South India, Feb. 3, 1900; s. Rajendra Tiruvengada and Namagiri Iyengar; M.A., Presidency Coll., Madras, India, 1924; Ph.D., U. Manchester (Eng.), 1929; D.Sc., Andhra U., Waltair, India, 1965, Banaras Hindu U., 1967; m. Kamala, 1924; children—Champaka, Kalyani, Chaya. Scientist, Agrl. Research Inst., Coimbatore, India, 1930-33; prof., head dept. chemistry Andhra U., Waltair, 1933-49; prof., head dept. chemistry U. Delhi (India), 1949-65, emeritus prof., 1965—. Mem. various coms. Indian Council Med. Research, Indian Council Agrl. Research, Indian AEC. Recipient Padma Bhushan, 1963. Fellow Royal Soc., 1960, Nat. Inst. Scis. India (pres.), Indian Acad. Scis., Royal Inst. Chemistry, Indian Chem. Soc., London Chem. Soc.; mem. German Acad. (Halle). Author: Chemistry of Vitamins and Hormones, 1952; also numerous articles. Research on chemistry of natural products, including vegetable coloring matters, drugs and insecticides, isolation of active prins., establishment of their constn. and confirmation by synthesis, biogenesis of these compounds and their synthesis. Home: 8 Cavalry Lines, Delhi 7, India.*

SESONSKE, Alexander, Am. nuclear, chem. engr.; b. Gloversville, N.Y., June 20, 1921; s. Abraham and Esther (Kreitzer) S.; B.Chem. Engring., Rensselaer Poly. Inst., 1942; M.S., U. Rochester, 1947; Ph.D., U. Del., 1950; m. Marjorie Ann Mach, Apr. 17, 1952; children—Michael Jan, Jana Louise. Engr., Chem. Constrn. Corp., N.Y.C., 1942; chem. engr. Manhattan project, 1943-45; chem. engr. Columbia-So. Chem. Corp., 1945-46; staff Los Alamos Sci. Lab., 1950-54, 60-61, cons., 1961-63; faculty Purdue U., Lafayette, Ind., 1954—, prof. nuclear and chem. engring., 1959—. Cons., Oak Ridge Nat. Lab., 1963—; chmn. reactor engring. div. rev. com. Argonne (Ill.) Nat. Lab., 1965-67. Mem. Am. Nuclear Soc., Am. Inst. Chem. Engrs., Am. Soc. Engring. Edn., Am. Chem. Soc., Sigma Xi, Omega Chi Upsilon. Author: (with Samuel Glasstone), Nuclear Reactor Engineering , 1963; also numerous articles. Research on liquid metal heat transfer and nuclear reactor engring. Home: 608 Elm Dr., West Lafayette, Ind. 47906.*

SESSLER, Gerhard Martin, physicist; b. Rosenfeld, Germany, Feb. 15, 1931; s. Martin and Else (Fischer) S.; student Freiburg U. (Germany), 1950-51; Vordiplom, Munich U., 1954; Diplom, Goettingen U., 1957, Dr. rer. nat., 1959; m. Renate Brigitte Schulz, Dec. 9, 1961; 1 dau., Cornelia Brigitte. Mem. tech. staff Bell Telephone Labs., Murray Hill, N.J., 1959—. Fellow Acoustical Soc. Am.; mem. Am. Phys. Soc., Deutsche Physikalische Gesellschaft. Contbr. chpt. to Physical Acoustics, 1967. Research on thermal relaxation processes in gases by means of sound, wave propagation in plasmas, in particular studies of ion acoustic waves, electroacoustic transducer research, room-acoustical investigations. Home: 10 Southgate Rd. Office: Bell Telephone Labs., Murray Hill, N.J. 07971.*

SESTAK, Bohdan, Czechoslovakian physicist; b. Bohostice, Czechoslovakia, Sept. 15, 1922; s. Bohdan and Karla (Bursikova) S.; student Tech. U., Prague, Czechoslovakia, 1945-49; Dr. in Exptl. Physics, Inst. Physics, Czechoslovak Acad. Sci., Prague, 1960; m. Miroslava Jedlickova, Sept. 2, 1950; children—Bohdan, Mojmir. Leader phys. lab. Poldi Steel Work, Kladno, Czechoslovakia, 1949-55; leader group for study mech. properties solids Inst. Physics, Czechoslovak Acad. Sci., 1956—; external tchr. Tech. U., Prague, 1965—. Mem. Czechoslovak Math.-Phys. Soc. Author: (with F. Kroupa, P. Kratochvil) Dislocations in Crystals, 1961. Research and publs. on preparation of iron-silicon alloys, single crystals from melt, preparation of high purity iron single crystals by recrystallization, revealing of dislocation in these crystals by etching, effect of temperature, deformation speed and of tension and compression on slip planes in silicon-iron alloys. Home: 29 Nad Manovkon. Office: 7 Vincina, Prague, Czechoslovaka.*

SESTINI, Benedict, mathematician, astronomer; b. Florence, Italy, Mar. 20, 1816; studied philosophy and theology Roman Coll. Entered Soc. of Jesus, Rome, Italy, 1836; asst. astronomer Roman Observatory until 1848; came to U.S., circa 1848; tchr. math. and natural scis. Georgetown U., Washington, 1848-69, continued research at univ. obs.; made studies of sun's surface, 1850-69; organized expdn. to Denver to observe total eclipse of sun, 1874, wrote account appearing in Am. Catholic Quar. Rev., 1878; planned and supervised constrn. Holy Trinity and St. Aloysius churches, Washington, also Jesuit Sem., Woodstock, Md., 1869-83; published Messenger of the Sacred Heart (had widest circulation of any Cath. mag. in U. S.), 1866-85; tchr. higher math. Gonzaga Coll., Washington; tchr. astronomy and geology Jesuit Sem., Woodstock, 1869-85. Author: Memoria Sopra; Colori delle Stelle del Catalogo di Baily Osservati, 1845; Memoria Seconda Intorno ai Colori delle Stelle (1st comprehensive study of its kind), 1847; A Treatise of Analytical Geometry, 1852; Astronomical Observations Made During the Year 1847 at the National Observatory, Washington, Vol. III, 1853; A Treatise on Algebra, 2 credits, 1855, 57; Elements of Geometry and Trigonometry, 1856; Manual of Geometrical and Infinitesimal Analysis, 1871; Theoretical Mechanics, 1873; Animal Physics, 1874; Principles of Cosmography, 1878. Died Frederick, Md., Jan. 12, 1890.

SETALA, Kai Martin Edvard, Finnish radiologist, pathologist; b. Pori, Finland, Sept. 13, 1913; s. Eino and Helmi (Snellman) S.; dr. medicine and surgery, U. Helsinki, 1941, specialist in X-ray diagnostics and therapy, 1946, in cancer, 1949; m. Inger Ekman, May 23, 1942; children—Christel, Eva, Anitra. Asst. in anatomy U. Helsinki, 1937-47, asst. in pathology, 1948-50, physician, dept. radiotherapy 1943-47, chief radiol. dept. U. Hosp., 1950-53, prof. pathology, 1953—, dep. physician in chief Radiologic Policlinic, U. Central Hosp., 1966—. Decorated Comdr. Order Finnish Lion, 1962. Mem. A.A.A.S., Coll. Orbis Radiology, Internat. Acad. Pathology, Royal Soc. medicine (London), Deutsche Gesellschaft Pathologie, others. Publs. on devel. of techniques for decorporation of radiostrontium; dose measurements in radiotherapy; clin., exptl. cancer research. Home: Uudenkaupungintie 8, Helsinki 35, Finland.*

SETCHELL, William Albert, Am. botanist; b. Norwich, Com., Apr. 15, 1864; s. George Case and Mary Ann (Davis) S.; A.B., Yale, 1887); A.M., Harvard, 1888, Ph.D., 1890; m. Clara Ball (Pearson) Caldwell, Dec. 15, 1920. Morgan fellow Harvard, 1887-88, asst. in biology, 1888-91, instr. biology Yale, 1894-95; prof. botany U. Cal., 1895-1934, emeritus;-instr. botany. Marine Biol. Lab., Woods Hole, Mass., 1890-95. Fellow Am. Acad. Arts and Scis., A.A.A.S., Cal. Acad. Scis., Torrey Bot. Club; mem. Nat. Washington acads. scis., Am. Philos. Soc., Bot. Soc. Am., Cal. Bot. Club, Soc. Biogégraphie, Soc. Linn. de Lyons, N.Y. Acad. Sci. (asso.), Bot. Soc. Japan (hon.); fgn. mem. Linnaean Soc., London, Kunglig Vetenskaps och Vitterhets Samhället i Göteborg. Author: Laboratory Practice for Beginners in Botany, 1897; (with N. L. Gardner) Algae of North America, 1903-24. Contbr. to bot. jours. Authority on algae of Pacific ocean. Died Apr. 5, 1943.

SETCHENOV, Ivan Michailovich, Russian neurologist; b. Simbirsk, Russia, 1829; credited with discovery that cerebrum can inhibit spinal reflexes (led to doctrine of synapeses of reflex arc), circa 1863. Died 1905.

SETH, Symeon, Byzantine physician; flourished 1071-78; translated various works from Arabic to Greek, including a dictionary on med. properties of foodstuffs (1st occidental work containing information on Hindu and Arabic drugs and spices, including camphor, musk, ambergis, hashish, julep, nutmeg, cloves. Wrote bot. dictionary, treatise on taste, smell, touch, also one on urine.

SETLOW, Richard Burton, Am. biophysicist; b. N.Y.C., Jan. 19, 1921; s. Charles M. and Elsie (Hurwitz) S.; A.B., Swarthmore Coll., 1941; Ph.D. in Physics, Yale, 1947; m. Jane Kellock, June 6, 1942; children—Peter, Michael, Katherine, Charles. Faculty, Yale, 1943-60, asso. prof. physics and biophysics, 1955-60; biophysicist Oak Ridge Nat. Lab., 1960—. Chmn. com. on photobiology Nat. Acad. Sci.-NRC, 1965—. Mem. Biophys. Soc. (mem. council 1965—), Radiation Research Soc., A.A.A.S. Author: Molecular Biophysics, 1962. Editorial bd. Radiation Research, 1963—, Photochemistry and Photobiology, 1963—. Research on effects ultraviolet on biol. systems and molecular error-correcting mechanisms. Home: 226 Outer Dr., Oak Ridge 37830. Office: Biology Div., Oak Ridge Nat. Lab., Oak Ridge 37830.*

SETON (or Setonius, Sitonius, Sedonius, Sidonius, Sutoneus, Suchtenius, Sydon, and the Cosmopolite), Alexander, alchemist: originally from Seton nr. Edinburgh, Scotland. Said to have changed lead into gold at Enkhuysen, Netherlands, 1602; gave demonstrations transmutation Amsterdam and Rotterdam (Netherlands), Germany, Switzerland, Italy, and Strasbourg, France. Died Cracow, Poland, Jan., 1604.

SETON, Ernest Thompson, (surname changed from Thompson to Seton), naturalist; b. South Shields, Eng., Aug. 14, 1860; s. Joseph L. and Alice (Snowden) Thompson; lived in backwoods of Can., 1866-70; on Western plains, 1882-87; ed. Toronto Collegiate Inst. and Royal Acad., London; m. Grace Gallatin, June 1, 1896; 1 dau., Ann (Mrs. Hamilton Chase); m. 2d, Julia M. Moss Butree, Jan. 22, 1935. Served as ofcl. naturalist govt., Manitoba; studied art in Paris, 1890-96; one of chief illustrators of Century Dictionary; delivered over 3,000 lectures. Recipient Silver medal Société d'Acclimatation, France, 1918; John Burroughs medal, 1926, Daniel Girard Elliot medal, 1928. Mem. Nat. Inst. Arts and Letters; pres. Seton Inst.; chief Woodcraft League of Am. chief scout Boy Scouts of Am., 1910-15; founder Woodcraft Indians, 1902. Author (and illustrator) numerous books including: Mammals of Manitoba, 1886; Birds of Manitoba, 1891; Art Anatomy of Animals, 1896; The Biography of a Grizzly, 1900; Pictures of Wild Animals, 1901; Natural History of the Ten Commandments, 1907; Life-histories of Northern Animals, 1909; The Arctic Prairies, 1911; Forester's Manual, 1911; Wild Animals at Home, 1913; Manual of Woodcraft Indians, 1915; Wild Animals' Ways, 1916; Lives of Game Animals, 1925-28; Mainly About Wolves, 1937; Great Historic Animals, 1937; Buffalo Wind, 1938; Trail of an Artist-Naturalist, 1940. Contbd. to knowledge of game animals; a founder modern sch. of animal fiction in which actual facts of animal life are given. Died Oct. 23, 1946.

SEVASTIKOGLOU, (Sevastik) John Anastase, orthopedic surgeon; b. Athens, Greece, Mar. 5, 1920; s. Anastase G. and Athina (Kalfopoulou) S.; grad. medicine U. Athens, 1946; M.D., Karolinska Inst., Stockholm, Sweden, 1958; m. Helen Valassaki, Mar. 5, 1944; children—Alf, Bob, Per. Research fellow Wenner-Grens Inst. Exptl. Biology, 1954-57; research asso. dept. orthopedic surgery U. Ill., Chgo., 1962-63; asso. prof. orthopedic surgery U. Uppsala (Sweden), 1959-67; prof., chmn. dept. orthopedic surgery U. Umea (Sweden), 1967—. Recipient Golden Cross, Greek Phoenix Order, 1959. Mem. Swedish Orthopedic Assn., Société Internat. de Chir. Orthopéd. et de Traumat., Société Internat. de Chirurgie. Author: Modern Treatment of Thromboembolic Diseases, 1951; The Early Stages of Osteogenesis in Tissue Culture, 1958; also articles. Research on bone biol. studies in embryonic osteogenesis, teratogenesis and osteoporosis. Home: 11 Luthagsesplanaden, Uppsala, Sweden.*

SEVCHENKO, Anton Nikiforovich, Russian physicist; b. 1903; grad. Belorussian U., 1931. With State Optical Inst., 1934-53; sect. head Inst. Physics and Math., Belorussian Acad. Sci., 1953-54, dir. Inst., 1954-57; rector Belorussian U., 1957—. Belorussian del. 3d gen. conf. IAEA, Vienna, Austria, 1959, 6th, 1962. Mem. Belorussian Acad. Scis. Research and publs. on properties of rare earths, photosynthesis, optical properties of uranyl compounds, complex organic molecules. Address: Belorussian University, Universitetsky gorodok, Minsk, Belorussian SSR, USSR.

SEVERI, Francesco, Italian mathematician; b. Arezzo, Italy, Apr. 13, 1879; s. Cosimo and Lucinia (Cambi) S.; ed. Inst. France; Hon.Dr., 1907. Prof., U. Parma (Italy), also U. Padua (Italy); rector U. Rome, 1923-25, also prof. differential calculus; became prof. higher math. Centro Italiano de Estudios Americanos, Rome, 1947; apptd. pres. Istituto di Alta Matematica, Rome, 1947. Named hon. prof. univs. of Padua, Ferraro, Toronto, Buenos Aires; recipient Gold medal, 1906; Guccia medal, 1908; Royal Prize in Math., Acad. Lincei, 1915. Mem. Royal Soc. Engring. Padua (dir.), Italian Acad., Italian Soc. Math.

and Physics (pres.), Acad. Lincei, insts. of Turin, Milan, Padua, Venice and Bologna, Nat. Assn. U. Profs., Italian Soc. Math. and Physics; fgn. mem. acads. of Halle, Liege, Barcelona, USSR. Author: Complementi di geometria proiettiva, 1906; Vorlesunger über algebraische Geometrie, 1921; Trattato di geometria algebrica, 1926; Geometria proiettiva, 1922; Elementi di geometri Firenze, 1926; Lezioni di analisi algebrica, 1933; Theory of Analytical Functions, 1936; Geometria piana e solida, 1943; Introduzione alla geometria algebrica, 1949. Research on algebraic geometry; independently reformulated concepts of enumerative geometry; originated proof for number of Picard integrals of 1st kind on surface equalling difference of genera. Home: Via L. Spallanzini 32, Rome, Italy.

SEVERIN, Sergei Evgenévich, biochemist; b. Dec. 21, 1901; grad. Moscow U., 1924. Staff, Physiol. Lab., Inst. Profl. Diseases, 1924-32; prof. Third Moscow Med. Inst., 1932-48; became prof. Moscow U., 1933, head biochemistry, 1939—; dir. Inst. Nutrition, USSR Acad. Scis., 1945-47, dir. Inst. Biol. and Med. Chemistry, 1948-49. Decorated Order of Lenin, Order of Red Banner of Labor. Mem. USSR Acad. Med. Scis., USSR Acad. Scis. (sec. dept. medico-biol. scis. 1949-57). Author: (with N. P. Meshkova) Practicum on the Biochemistry of Animals, 1950; (with P. P. Mitrofanov) Textbook of Physical and Colloidal Chemistry, 1941. Research on muscle tissue biochemistry, phosphorus derivatives of amino acids and peptides, appearance of carnosine and anserine in devel. of organism; carnoside and anserine effects on phosphorylation in skeletal muscles. Home: Novoslobodskaya, 57/65, Moscow, USSR.

SEVERINGHAUS, Aura Edward, Am. anatomist; b. Jeffersonville, Ind., May 5, 1894; s. Charles Edwin and Ida Henrietta (Mock) S.; student U. of Wis., 1912-15, A.B., Columbia, 1916, M.A., 1919, Ph.D., 1927, Sc.D., Doctor of Letters of Humanity; married to Sara M. Nay, July 9, 1923. Graduate asst. dept. of zoology, Columbia, 1916-17, 1919-20; asst. prof. of biology Peking (China) Union Medical Coll., Rockefeller Foundation, 1920-23, dean 1923-26; with Columbia U. since 1927, instr. in anatomy, Coll. of Physicians and Surgeons, 1927, asso., 1928, asst. prof., 1930, asso. prof., 1939, prof., asst. dean, 1942; asso. dean in medicine, sec. of faculty, 1945-62, professor emeritus, asso. dean emeritus, 1962—; dir. studies neurology and neurol. scis. Nat. Inst. Neurol. Disease and Blindness, 1962-68; dir. more Negroes for med. sch. program Macy Found., 1968—. Dir. N.Y. YMCA. Served with M.C., AUS, 1917-19. American del. Internat. Colloquium Les Hormones Sexuelles, Paris, 1937. Awarded Peoples Medal of Honored Merit with citation by Republic of China, 1943, for efforts in rehabilitation of med. libraries. Mem. Harvey Soc., Soc. Exptl.. Biology and Medicine, A.A.-A.S., Am. Assn. Anatomists, Marine Biol. Lab. (Woods Hole, Mass.), Assn. for Study of Endocrine Glands, Sigma Xi, Phi Beta Kappa. Awarded Order Brilliant Star (China). Co-author: Bailey's Text Book of Histology, 1944; Preparation for Medical Education in the Liberal Arts College, 1953; Preparation for Medical Education—A Restudy, pub. 1961. Contributor to the Pituitary Gland (The Association for Research in Nervous and Mental Diseases), 1938; Les Hormones Sexuelles (Singer Polingnac Foundation, Paris), 1939; Symposia on Quantitative Biology (Cold Spring Harbor Series), 1937; numerous sci. articles in sci. and med. jours. Research on cytology of endocrine glands; especially the thyroid and pituitary glands. Home: 375 W. 250th St., N.Y.C. 10471. Office: 277 Park Av., N.Y.C. 10017.*

SEVERINGHAUS, John Wendell, Am. physician; b. Madison, Wis., May 6, 1922; s. Elmer Louis and Grace (Colby) S.; B.S., Haverford Coll., 1943; postgrad. U. Wis., 1945-47; M.D., Columbia, 1949; m. Elinor Ford Peck, Aug. 28, 1948; children—Edwin Mark, Jean Colby, Anne Wendell, Jeffrey Peck. Staff anesthesiologist USPHS, NIH, 1953-58; faculty U. Cal. Med. Sch., San Francisco, 1958—, prof. anesthesia, 1965—, sr. staff Cardiovascular Research Inst., 1958—. Mem. anesthesia com. Nat. Acad. Scis.-NRC, 1954-68, mem. hyperbaric O2 com., 1964—. Mem. Am. Soc. Clin. Investigation, Am. Physiol. Soc., Assn. U. Anesthetists, Am. Soc. Anesthesiologists, Am. Fedn. Clin. Research. Mem. editorial bd. Jour. Applied Physiology, Am. Jour. Physiology, 1964—. Research and numerous publs. on respiratory physiology in hypothermia; defined alveolar dead space and related it to non-perfused pulmonary spaces; showed interaction of blood flow and brancho-constriction, role of cerebrospinal fluid in regulation of respiration and acclimatization to altitude; demonstrated pH regulation of spinal fluid, carotid body response to hypoxia to be absent in high altitude natives; measured uptake and solubility of various anesthetics; inventor carbon dioxide electrode, blood gas slide rule, rapid oxygen electrode, blood pH electrode. Home: Box 974, Allen Av., Ross, Cal. 94957. Office: University of Cal. Medical Center, San Francisco 94122.*

SEVERINUS, Marc Aurel, Italian anatomist; b. Tarsia, Italy, Nov. 2, 1580; M.D., U. Naples; prof. anatomy and surgery, Naples, Italy. Author: De recondita abscessuum natura libri VIII, 1632; Democritea, 1645; De efficaci medicina libri, III, 1646; Trimembris chirurgia, 1653; Zootomia Antiperipatias, hoc est, adversus Aristoteleos de respirone piscium diatriba; Phoca anatomicè specatus, 1661; Synopseos Chirurgicae libri vi, (extract of surg. works), 1664. Discussed how poisonous snake kills and antidotes for its venom, 1643, 51, elevated areas composed of lymph nodules in mucous membrane of small intestine, mainly in ileum, 1645; demonstrated analogous construction of anatomy of various animals; used tracheometry to treat diphtheria; described abscesses, granulatomas, buboes, neoplasms. Died Naples, Italy, July 16, 1656.

SEVERINUS, Petrus, Danish physician; b. Rypen, Jutland, 1542; practiced medicine, Flensburg, also Copenhagen, Denmark; canon of Roskild; prof. of poetry at Copenhagen, 1562; M.D., 1571, in France; physician to kings of Denmark for 30 years. Author: Idea medicinae philosophicae . . . , 1571. Systematized Paracelsus' medicine. Died Copenhagen, July, 1602.

SEVÉRNYI, Andréi Borísovich, Russian astronomer; b. May 11, 1913; grad. Moscow U., 1935. Joined Crimean Astrophys. Obs., USSR Acad. Scis., 1946, became dir., 1952. Recipient Stalin Prize, 1952. Corr. mem. USSR Acad. Scis. Author: The Stability and Oscillation of Gaseous Spheres and Stars, 1948; Solar Physics, 1957; Nuclear Processes in chromospheric Flares, 1963. Research and publs. on solar physics and astrophysics; solar chromospheric flares. Office: Crimean Astrophys. Obs., USSR Acad. Scis., Moscow, USSR.

SEVERSKY, Alexander P., de, aeronautical engr., airplane designer; b. Tiflis, Russia, June 7, 1894; s. Nicholas and Vera (Vasilieff) deS.; grad. Imperial Naval Acad. of Russia, 1914, post grad. student; Mil. Sch. Aeronautics, Russia; D.Sc. (hon.), Rollins College, Florida, 1944, Southern Illinois University, 1957; married Evelyn Olliphant, June 23, 1925. Came to U. S., 1918, naturalized, 1927. v. chmn. Russian Naval Aviation Mission to U. S.; asst. to Naval Attache, Russian Embassy, charge of aviation matters, 1918; aeronautical engr. and test pilot, U. S. govt., 1918; cons. engr. U. S. A. Service, 1921; pres., gen. mgr., dir. and founder Seversky Aero Corp., 1922-35; pres., gen. mgr., dir. and founder Seversky Aircraft Corp., 1931-39 (dir. successor firm, Republic Aviation Corp., 1939-40); cons. Chrysler Corp. 1933; founder Aviation Development Corp., 1940; founder and pres. Rotoflight Corp., 1940; founder and pres. Electronatom Corp., 1952; spl. consultant to Sec. of War to study and appraise role of air power in European and Pacific Theaters of Action, 1945; personal rep. Sec. of War at atom bomb tests, Bikini, 1946; ofcl. lectr. U. S. Air U., 1946—; cons. to chief staff U. S. Air Force, 1957—. Decorated Knight of St. George (Russia); Medal for Merit (U. S.), Officer of the French Legion of Honor (1948) and others. Awarded the Harmon trophy as outstanding airman, 1939; awarded Medal d'Officier d' Academie et de l'Instruction Publique (France), 1947; presented Harmon Trophy for 1940-45, by President Truman, 1947; awarded Cross of Lorraine (France), 1951; received Arts and Letters award, U.S.A.F., Association, 1951; Order of Daedalians, 1956. Licensed comml. pilot, profl. engr. Mem. bd. N.Y. State Aviation Edn. Commn., 1942. Fellow N.Y. Acad. Aeronaut. Scis.; mem. Air Force Hist. Found. (v.p., trustee); Nat. Aeronautic Assn., Soc. Am. Mil. Engrs., Aircraft Industries Assn., Am. Ordnance Assn., Am. Legion, Am. Rocket Soc., N.Y. State Soc. Profl. Engrs., Quiet Birdmen, Aviation Writers Association, Sportsman Pilot Association; hon. mem. Air Force Assn. of Republics of Argentina, Chile, and Uruguay. Author: Victory Through Air Power, 1942; Air Power—Key to Survival, 1950; America: Too Young to Die, 1961. Contributor McNaught Syndicate column on air power, 1943-44; United Press, 1941-43; King Features, 1950, 1952; articles to Reader's Digest, American Mercury, Cosmopolitan and others. Inventor many airplane devices including fully automatic bombsight; designed military aircraft including single-seater fighter plane, 1936, long range fighter escort, 1937, first air-cooled engine fighter with turbo-supercharger for high altitude combat, 1938. Has established many speed records including record for amphibians (230 miles per hr.), Detroit, 1935; transcontinental speed record, 1938. Work in development of air pollution control. Home: Asharoken Beach, Northport, L.I. Office: 30 Rockefeller Plaza, New York, N.Y. 10020.

SEVITT, Simon, pathologist; b. Dublin, Ireland, Oct. 17, 1914; s. Abram and Elizabeth (Armedir) S.; B.A., Trinity Coll., Dublin, 1937, M.Sc., 1938, M.B., 1939, M.D., 1943; D.P.H., U. Coll., Dublin, 1943; m. Betty Woolf, Mar. 2, 1941; children—Michael Andrew, Peter Anthony and Brian Maurice (twins). Research fellow Med. Research Council Ireland, Sch. Pathology, Trinity Coll., 1942-43; cons. pathology burns research and indsl. injuries unit Med. Research Council, 1947-62; dir., cons. pathologist Birmingham (Eng.) Accident Hosp., 1947—. External examiner pathology Irish Conjoint Bd., 1955—; dir. postgrad. courses in pathology trauma and burns Assn. Clin. Pathologists and Inst. Accident Surgery, 1956—; curator Mus. on Pathology of Injury, 1948—. Recipient Gold medal Natural Sci. Moderatorship, 1937; Hackett Research scholar, 1938-39; Adrian Stokes Travelling fellow in pathology, 1940-41. Fellow Royal Coll. Physicians in Ireland, Coll. Pathologists, Royal Irish Acad. Medicine; mem. Assn. Clin. Pathologists, Path. Soc. Gt. Britain and Ireland, Socialist Med. Assn., Assn. Sci. Workers. Author: Burns Pathology and Clinical Applications, 1957; (with Clarke, Badger) Modern Trends in Accident Surgery and Medicine, 1959; Fat Embolism, 1962; also numerous articles. Research on pathology and bacteriology of gastroenteritis of infants, burns and injury; discovered delayed permeability in burns inflammation; introduced prevention of thrombo-embolism by anticoagulant drugs in injured patients; designed blood-bank for use in hospitals. Home: 159 Green Rd., Birmingham 13, Eng. Office: Pathology Dept., Accident Hosp., Birmingham, Eng.*

SEWARD, Ralph Pray, Am. chemist; b. Chgo., Oct. 23, 1901; s. Ora Philander and Mary (Pray) S.; B.S., Kalamazoo Coll., 1922; M.A., Clark U., 1923; Ph.D., Brown U., 1925; m. Lucy Burnham, Apr. 28, 1928; children—Mary Anne (Mrs. Redding L. Crafts), Gordon Burnham, Marjorie Douglass. Prof. physics and math. Catawba Coll., 1925-27; chemist Aluminum Co. Am., 1927-28; prof. chemistry Brenau Coll., 1928-29; instr. chemistry Mass. Inst. Tech., 1929-34; faculty Pa. State U., University Park, 1934—, prof. chemistry, 1953—. Mem. Am. Chem. Soc., Sigma Xi, Phi Lambda Upsilon. Research and publs. in phys. and inorganic chemistry. Home: 720 Thomas St., State College, Pa. 16801.*

SEWELL, Duane Campbell, Am. physicist; b. Oakland, Cal., Aug. 15, 1918; s. Earl Ferris and Hazel (Campbell) S.; student Coll. of Pacific, 1938-40; B.A., U. Cal. at Berkeley, 1941; m. Ruth Elizabeth Lombardi, July 25, 1943; 1 son, Barre. Physicist, U. Cal. at Berkeley, 1941-52, Lawrence Radiation Lab., Livermore, Cal., 1952-59, asso. dir., 1959—. Sci. officer to gen. adv. com. U.S. AEC, 1963—; mem. Gov.'s Radiol. Def. Adv. Com., State of Cal., 1960—. Mem. Am. Phys. Soc. Research and publs. on uranium mass spectrography, high energy neutron cross sections, magnetic field measurements, separation processes for uranium isotopes, nuclear research and devel. Home: 4265 Drake Ct., Livermore, Cal. 94550.*

SEWELL, Walter Edwin, Am. mathematician; b. Newnan, Ga., Oct. 31, 1904; s. Thomas W. and Larookah (Drewry) S.; A.B., U. Ga., 1925, C.E., 1926, M.A., 1927; M.A., Harvard, 1932, Ph.D., 1936; m. Carolyne Willis Adams, Mar. 22, 1941; 1 son, Walter Adams. Instr. math. U. Ga., 1925-27, Harvard, 1934-36; asst. prof. Ga. Inst. Tech., 1936-40; commd. 2d lt. Army, Res., 1925, advanced to col., 1944, chief armed forces edn., 1945-49, prof. mil. sci. and tactics, State U. Ia., 1950-54, Mercer U., 1957-59, ret., 1959; research asso. in math. and logistics Army Research Office, Durham, N.C., 1960-62; asso. dir. spl. research in numerical analysis Duke U., Durham, 1963—, adjunct prof. math., 1965—. Mem. Am. Math. Soc., Phi Beta Kappa, Phi Kappa Phi. Author: Degree of Approximation by Polynomials in the Complex Domain, 1942; also many articles. Home: 2243 Cranford Rd., Durham, N.C. 27706.*

SEWELL, William, English veterinarian; b. Essex, Eng., 1780; diploma, 1799; became asst. to Edward Coleman, 1799; 2d prin. Vet. Coll., London; visited vet. establishments at Lyons and Paris, France, 1815, Vienna, Austria, Prague, Czechoslovakia, Berlin and Hanover, Germany, 1816; pres. Vet. Med. Soc., 1835-36, Royal Coll. Vet. Surgeons, 1852. Author report of visit to vet. schs. on continent, 1818, also several articles. Made supposed discovery of channel pervading medulla spinalis, 1803; rediscovered neurotomy, 1818. Died June 8, 1853.

SEWELL, William Hamilton, Am. sociologist; b. Perrinton, Mich., Nov. 27, 1909; s. William Hamilton and Lulu (Coller) S.; B.A., Mich. State U., 1933, M.A., 1934; Ph.D., U. Minn., 1939; m. Elizabeth Lucille Shogren, June 13, 1936; children—Mary E. (Mrs. Kenneth Knudson), William Hamilton III, Robert G. Instr. sociology U. Minn., 1934-37; faculty Okla. A. and M. Coll., 1937-44; faculty U. Wis., 1946—, prof. sociology, chmn. dept. sociology, 1957-62, Vilas research prof., 1964—, chancellor, 1967—. Chmn. behavioral scis. tng. com. NIH; adv. com., social sci. div. NSF; mem. NRC; mem. basic research com. U. S. Office Edn., 1964-66; cons. Govt. India, Ford Found., USDA, U. S. Dept. Health, Edn., Welfare. Fellow A.A.A.S., Center for Advanced Study in Behavioral Scis.; mem. Am. (v.p. 1962, chmn. social psychology sect. 1965), Rural (pres. 1955) sociol. assns., Sociol. Research Assn. (pres. 1953), Midwest (pres. 1953), Southwest (pres. 1941) sociol. socs., Am. Statist. Assn., Nat. Acad. Scis. Author: The Construction and Standardization of a Scale for Measurement of Socioeconomic Status of Farm Families, 1940; Research on Income and Levels of Living in the South, 1940; Farmers' Conceptions and Plans for Economic Security in Old Age, 1953; (with O. M. Davidsen) Scandinavian Students on an American University Campus, 1961; (with others) The Uses of Sociology, 1968; also numerous articles. Research on influence of social class on ednl. and occupational aspirations and achievement; influence of early social experiences on later personality; devel. of measures of social phenomena. Home: 6233 Countryside Lane, Madison, Wis. 53705.*

SEWELL, William Hawley, Am. cardiac surgeon; b. Tampa, Fla., Oct. 23, 1926; s. William H. and

Lucille (Sherman) S.; student Randolph Macon Coll., 1943; student Princeton, 1944-45; M.D., Yale, 1950-m. Karlene Wenzel, June 24, 1960; children—Susan, James, Sharon, Douglas. Clin. investigator VA, Oteen, N.C., 1960-63, staff cardiovascular surgeon, 1963-66; chief cardiac surgery, Guthrie Clinic, Sayre, Pa., 1966—. Recipient Thesis prize Yale, 1950; commended by surgeon gen. U. S. Navy, 1953. Fellow A.C.S., Am. Coll. Cardiology, Am. Coll. Chest Physicians, Internat. Coll. Cardiology; mem. Soc. Thoracic Surgeons, So. Thoracic Surg. Assn. Author: Surgery for Acquired Coronary Disease, 1966; also articles. Research on devel. effective operations for coronary heart diesese. Home: R.D. 2, W. Lockhart St. Office: Guthrie Clinic, Sayre, Pa. 18840.*

SEXTIUS, Quintus, philosopher; flourished circa 50 B.C.; founder philos. sch.; borrowed his ethical views from Stoicism, theory of soul as incorporeal entity from Plato, vegetarianism from Pythagoreans; probably identical to bot. and med. writer Sextius Niger. Probable author of Sententiae (translated from Greek into Latin by Rufinus).

SEXTUS EMPIRICUS, physician, philosopher; flourished end 2d-beginning 3d century; physician Empirical Sch.; traveled to Alexandria, Athens; leader sch. sceptics, circa 180-229. Author: Outlines of Pyrrhonism, book 1 (described sceptical basic terms, method of Pyrrhonism), book 2 (refutes dogmatic logic, theory of knowledge), book 3 (on physics, ethics); Against the Dogmatists, 5 books; Against the Schoolmasters (criticism of non-philosophical subjects in teaching, grammar, rhetoric, math., astronomy, music). Applied empirical methods to medicine; important in history of Greek philosophy.

SEXTUS PLACITUS PAPYRIENSIS, med. writer; flourished 5th century. Author: Liber de medicina ex animalibus (basis for later dung pharmacy which initiated rise of magic and decline of medicine).

SEYFERTH, Dietmar, chemist; b. Chemnitz, Germany, Jan. 11, 1929; s. Herbert C. and Elisabeth (Schuchardt) S.; came to U.S., 1933, naturalized, 1951; B.A. in Chemistry summa cum laude, U. Buffalo, 1951; M.A., Harvard, 1953, Ph.D., 1955; m. Helena A. McCoy, Aug. 25, 1956; children—Eric S., Karl D., Elisabeth M. Research chemist Dow Corning Corp., Midland, Mich., 1955-56; research staff Harvard, 1956-57; faculty, Mass. Inst. Tech., 1957—, prof. chemistry, 1965—. Fulbright scholar Technische Hochschule, Munich, Germany, 1954-55; Alfred P. Sloan Found. fellow, 1962-64; recipient Distinguished Alumnus award Coll. Arts and Scis., U. Buffalo, 1964. Mem. Am. Chem. Soc., Chem. Soc. (London), Gesellschaft Deutscher Chemiker, Royal Netherlands, Chem. Soc. Author: (with R.B. King) Annual Surveys of Organometallic Chemistry, Vol. 1, 1965, Vol. 2, 1966; also numerous articles. Regional editor for U.S. and Can., Jour. Organometallic Chemistry, 1963—; co-ordinating editor Organometallic Chemistry Revs., 1965—; editorial bd. Organometallic Synthesis, 1964—. Research on organometallics, especially organic derivatives of lithium, zinc, mercury, boron, silicon, germanium, tin and lead, certain aspects of organophosphorus chemistry. Office: Mass. Inst. Tech. Dept. Chemistry, Cambridge, Mass. 02139.*

SEYMOUR, Keith Morton, Am. chemist; b. Retreat, Wis., Nov. 5, 1903; s. Llewellen G.F. and Lovina Smith (Morton) S.; B.S., U. Wash., 1926, M.S., 1929, Ph.D., 1933; m. Mary Ann Hinckley, June 28, 1934; children—James Hinckley, Bruce Hanchett, Katherine Morton. Faculty Reed Coll., Portland, Ore., 1933-37, No. Mont. Coll., Havre, 1937-38, Carleton Coll., Northfield, Minn., 1938-42; chemist, research engr., tech. group supr. Stanolind Oil and Gas. Co., Tulsa, 1942-46; prof. chemistry, Butler U., Indpls., 1946—, head, dept., 1947—, John Hume Reade prof. chemistry, since 1968—. Mem. Am. Chem. Soc. (sec. Ind. sect. 1950-52), Chem. Soc. London, Sigma Xi, Phi Lambda Upsilon, Phi Kappa Phi. Author: Organic Chemistry, 1961; Introduction to Organic Chemistry, 1963. Home: 39 W. 46th St., Indpls. 46208.*

S'GRAVESANDE, Willem Jacob Storm van, see van s'Gravesande, Willem Jacob Storm.

S'GROOTEN, Christian, see Grooten, Christian s'.

SHACKLETON, Sir Ernest (Henry), Brit. explorer; b. Kilkee, Ireland, Feb. 15, 1874; s. Henry and Henrietta L. (Gavan) S.; ed. Dulwich Coll.; m. Emily Mary Dorman, 1904; 2 sons, 1 dau. 3d lt. Nat. Antarctic Exptn., 1901; comdr. Brit. Antarctic Exptn., 1907; on Antarctic Expdn., 1914; dir. Equipment and Transport Mobile Forces North Russia Winter Campaign, 1918-19; comdr. Brit. Oceanographical and Sub-Antarctic Expdn., 1921. Knighted, 1909. Author: The Heart of the Antarctic, 1909; The Diary of a Troopship, 1919. Sci. results of 1907-19 expdn. of great importance: Mt. Erebus ascended; south magnetic pole located; they reached to within 100 miles of South Pole. Died Jan. 5, 1922.

SHAFAREVICH, Igor Rostislavovich, Russian mathematician; b. June 3, 1923; grad. Moscow U., 1940; Joined staff Moscow U., 1944—, prof., 1953—; staff Math. Inst., USSR Acad. Scis. Recipient Lenin prize, 1959. Author: The Normability of Topological Fields, 1943; The General Law of Reciprocity, 1950; The Construction of Algebraic-Number Fields with a Given, Solvable Galois Group, 1954; The Merging Problem for Local Fields, 1959; coauthor: Homology Groups of Nilpotent Algebra, 1957. Mem. USSR Acad. Scis. (corr.). Research in algebra, theory of algebraic numbers; discovered gen. law of reciprocity and resolution of inverse problem of Galui for solvable groups. Office: Math. Dept., Moscow U., Moscow, USSR.

SHAFFER, Bernard William, Am. mech. engr.; b. N.Y.C., Aug. 7, 1924; s. Abraham and Eva (Ellinsky) S.; B.Mech. Engring., Coll. City N.Y., 1944; M.S., Case Inst. Tech., 1947; Ph.D. in Applied Math., Brown U., 1951; m. Feb. 23, 1947; children—Janet Ilene, Roberta Franceen. Aero. research scientist NACA (now NASA), Cleve., 1944-47; spl. lectr. applied mechanics Case Inst. Tech., 1946-47; research asso. grad. div. applied math. Brown U., 1947-50; faculty N.Y. U., 1950—, prof. mech. engring., 1958—, project dir. research div., 1951—. Cons. to pvt. cos., U. S. Army. Fellow Inst. Aeros. and Astronautics (asso.); mem. Am. Soc. M.E., Am. Soc. Engring. Edn., Sigma Xi, Tau Beta Pi, Pi Tau Sigma. Hon. editorial adv. bd. Internat. Jour. Mech. Scis. Research and publs. on theory of metal cutting, elastic-plastic analysis of wide curved bars and cylindrical shells; synthesis of mechanisms; stress analysis of solid propellant rocket assemblies; analysis of filament reinforced plastics. Home: 18 Bayside Dr., Great Neck, N.Y. 11023. Office: N.Y. U., University Heights, Bronx, N.Y. 10453.*

SHAFFER, Philip Anderson, Am. biochemist; b. Martinsburg, W.Va., Sept. 20, 1881; A.B., U. W.Va., 1900; Ph.D., Harvard, 1904; Sc.D. (hon.), U. Rochester, 1939, Washington U., St. Louis, 1953; m. 1904; 2 children. Asst. chemist McLean Hosp., Mass., 1900-03; asst. exptl. pathologist, med. coll. Cornell U., 1904-06, instr. path. chemistry, 1907-10; prof. biochem. Sch. Medicine, Washington U., 1910-46, Distinguished Service prof., 1946-52, emeritus prof., 1952—, dean 1915-19, 37-46; physiol. chemist Huntington Fund Cancer Research, 1907-10; head chem. sect. lab. Bellevue Hosp., N.Y.C., 1909-10. Mem. Am. Chem. Soc., Nat. Acad. Sci., Soc. Biol. Chemists (became sec. 1914, pres., 1923). Research in metabolism, diet, cystinuria, protein metabolism in typhoid, anti-ketogenesis, coupled reactions, oxidation-reduction, nature of kinetic barriers, semiquinones in catalysis, origin of acetone bodies, creatin and creatinin. Office: Washington U. Sch. Medicine, St. Louis 10.

SHAFRIR, Eleazar, biochemist; b. Cracow, Poland, Sept. 19, 1924; s. Markus and Rachel Shafrir; M.S., Hebrew U., Jerusalem, Israel, 1952, Ph.D., 1955; m. Aviva Rutenberg, June 8, 1950; children—Ehud Yehiel, Eviathar Samuel, Ora Rachel. Research asst. Hadassah U. Hosp., Jerusalem, 1951-55; sr. asst. Hebrew U. — Hadassah Med. Sch., 1955-57, lectr., 1960-64, dir. lab. clin. biochemistry, 1964-65, asso. prof. human biochemistry, 1965—. Mem. Israel Biochem. Soc., Soc. Exptl. Biology and Medicine U. S. A., A.A.A.S., Internat. Conf. Biochemistry Lipis. Author: (with others) Handbook of Physiology, 1965. Research, publs. on transport and storage of fat in mammalian body. Home: 57 Hehalutz, Jerusalem, Israel.*

SHAGASS, Charles, psychiatrist; b. Montreal, Que., Can., May 19, 1920; s. Morris and Pauline (Segal) S.; B.A., McGill U., 1940, M.D., C.M., 1949, diploma in psychiatry, 1953; M.S., U. Rochester, 1941; m. Clara Wallerstein, Nov. 1, 1942; children—Carla Louise, Kathryn Sharna, Thomas Alan. Came to U.S., 1958, naturalized, 1965. Faculty McGill U., 1952-58, U. Ia., 1958-66; research asso. psychology Allan Meml.. Inst. Psychiatry, Montreal, 1947-52, dir. electrophysiology lab., 1952-58; individual practice psychiatry, Montreal, 1953-58; mem. staff Royal Victoria Hosp.; prof. psychiatry Temple U., Phila., 1966—. Mem. Sigma Xi, Alpha Omega Alpha. Publs. on research on psychophysiology, psychopharmacology, and brain function in psychiat. disorders; introduction of concept of sedation threshold; studies of sensory evoked responses in mental disorders. Home: 1301 Spruce Lane, Wyncote, Pa. 19095. Office: Eastern Pa. Psychiat. Inst., 3300 Henry Av., Phila. 19129.

SHAGINYAN, Artashes Liparitovich, Russian mathematician; b. Leninakan, Armenia, 1906; grad. Yerevan U., 1930; postgrad. Leningrad U., until 1937; D.Physico-Math. Sci., 1945. With Yerevan U. 1929-44, prof., 1949—; dir. sect. math. and mechanics Armenian Acad. Scis., 1944-45, acad. sec. dept. physico-math. sci., 1950—, dir. Inst. Math. and Mechanics, 1955—. Mem. Armenian Acad. Scis. Research and publs. on theory of functions; analyzed problems of mean approximation per area, best approximations of weight in inorganic closed domains. Address: Yerevan University, ulitsa Abovyana 104, Yerevan, Armenian SSR, USSR.

SHAH, Swarupchand Mohanal, Indian mathematician; b. Deesa, India, Dec. 30, 1905; s. M.D. and Chunibehn Shah; B.A., Fergusson Coll., Bombay, India, 1927; M.A., U. London, 1930, Ph.D., 1942, D.Litt., 1951; m. Mafat Devi Varia, May 5, 1926; 1 son, Kanti Lal. Mem. faculty Muslim U., Aligarh, India, 1930-58, prof., chmn., 1953-58; vis. prof. U. Wis., Madison, 1958-59, Northwestern U., Evanston, Ill., 1959-60, U. Kan., Lawrence, 1960-66, University of Kentucky, Lexington, 1966—. Fellow Nat. Inst. Scis. of India, Indian Acad. Scis., Royal Soc. Edinburgh; mem. N.Y. Acad. Scis. Editor, Jour. of Indian Math. Soc., 1957-59; asso. editor Jour. Indian Math. Soc., 1959—, Proceeedings Am. Math. Soc., 1962—. Author: (with S.A.K. Rizvi) Plane Trigonometry, 1954; (with N.A. Khan) College Algebra, 1955; (with Ram Kumar and Govil) Hydrostatistics, 1957; also many articles. Research in theory of functions in particular entire and meromorphic functions, difference equations, theory of numbers, Fourier series. Home: 145 Virginia Av., Lexington, Ky. 40508.*

SHAHANI, Khem Motumal, dairy biochemist; b. Hyderabad Sind, India, Sept. 3, 1923; s. Motumal B. and Gujar (Hiranandani) S.; B.S., U. Bombay (India), 1943, Assoc. Ind. Dairy Res. Inst., 1947; Ph.D., U. Wis., 1950; m. Leona D'Acquisto, Aug. 28, 1954; children—Michael, Ray, Roger, Robbin. Came to U.S., 1947, naturalized, 1958. Instr., Agr. Coll., Sakrand, India, 1943-46; research asst. U. Ill., 1950-52; research asso. Ohio State U., 1953-57; faculty U. Neb., Lincoln, 1957—, prof. dairy biochemistry, 1961—; indsl. cons. Internat. Bus. Consultants, Bombay, 1952-53. Mem. Am. Dairy Sci. Assn. (Borden award 1964), Am. Soc. Microbiology, Inst. Food Technologists, Sigma Xi, Gamma Sigma Delta (Internat. award for distinguished service to agr. 1966), Gamma Alpha. Research and publs. on milk enzymes to determine their physiol. and technol. role and significance, lysozyme in bovine and human milk; isolated antibiotic from dairy culture. Home: 6610 Van Dorn St., Lincoln, Neb. 68506.*

SHAIN, Irving, Am. chemist; b. Seattle, Jan. 2, 1926; s. Samuel and Selma (Blockoff) S.; B.S. in Chemistry, U. Wash., 1949, Ph.D. in Chemistry, 1952; m. Mildred Ruth Udell, Aug. 31, 1947; children—Kathryn Ann, Steven Terry, John Robert, Paul Stuart. Faculty, U. Wis. Madison, 1952—, prof., 1961—, chmn. chemistry dept., 1967—. Mem. Am. Chem. Soc., Electrochem. Soc., Internat. Com. Electrochem. Thermodynamics and Kinetics. Mem. editorial adv. bd. Analytical Chemistry, 1961-64; asso. editor Monograph Series, Am. Chem. Soc., 1966-67. Research and publs. on rates and mechanisms of electrode reactions with applications in electroanalytical chemistry, theory and practice of stripping analysis (method for analysis of traces of metal ions in solution), stationary electrode polarography (including cyclic voltammetry) for electrode mechanism studies. Home: 5401 Russett Rd., Madison, Wis. 53711.*

SHAKERLEY, Jeremy, English astronomer, mathematician; b. possibly Lancashire, Eng.; flourished 1650; 2d observer of transit of Mercury at Surat, 1651; observed comet of 1652, also solar eclipse, circa 1652. Author: The Anatomy of Urania Practica, 1649; De Mercurue in sole vivendo; Tabulae Britannicae, the British Tables, 1653.

SHAKHOV, Féliks Nikoláevich, Russian geologist; b. Oct. 24, 1894; grad. Tomsk Technol. Inst., 1922. Instr., Tomsk Technol. Inst. 1922-35, became prof., 1935; joined West Siberian br. USSR Acad. Scis., 1944, staff Inst. Geology and Geophysics (Novosibirsk), 1957—. Recipient Order of Red Banner of Labor. Corr. mem. USSR Acad. Scis. Research and publs. in geology of rare and radioactive elements, ore deposits, formation of metals in various regions, methods of prospecting; survey work. Home: Academiya ul. 55, Apt. 6, Novosibirsk, 72, Siberia. Office: Institut Geology and Geophysics, Siberian Br., USSR Acad. Sci., Academgorolsk, Novosibirsk 72, Siberia.

SHALDON, Stanley, English physician; b. London, Nov. 8, 1931; s. Max and Sophie (Katz) S.; M.A., Middlesex Hosp. Med. Sch., 1953, M.B., B.Chir., 1955, M.D., 1961; m. Ann Josephine Faul, July 23, 1958; children—Christopher, Nicola. Registrar dept. medicine Hammersmith Hosp. and Postgrad. Med. Sch., 1959; registrar med. unit Royal Free Hosp., London, 1959-60, lectr. medicine, 1960-65, hon. cons., 1st asst. to renal unit, 1965-66; med. dir. Nat. Kidney Center, London, 1966—. Recipient Raymond Horton-Smith prize U. Cambridge, 1961. Fellow Royal Soc., Brit. Med. Assn.; mem. Med. Research Soc., Renal Assn., Am. Fedn. for Clin. Research, European Dialysis and Transplant Assn., Internat. Congress Nephrology. Research, numerous publs. on renal diseases, artificial organs; devel. of hosp.-based chronic haemodialysis program; tng. patients in use of artificial kidney machine in homes. Address: 5, Chessington Av., London N.3, Eng.*

SHALER, Amos Johnson, Am. metallurgist; b. London, Eng., July 8, 1917 (parents Am. citizens); s. Millard King and Mary Eleanor (Johnson) S.; B.S. in Physics, Mass. Inst. Tech., 1940, Sc.D. in Metallurgy, 1947; m. Patricia H. Bowman, Sept. 22, 1943; children—Louise, Cynthia Mary, James Lane. Research asst. New Consol. Gold Fields, Ltd., Johannesburg, S.Africa, 1940-42; devel. engr. C. H. Hirtzel & Co. (Pty.) Ltd., Johannesburg, 1942-43; tech. dir. Indsl. Rys. Equipment Co. (Pty.) Ltd., Johannesburg, 1943-47; faculty Mass. Inst. Tech., 1946-53; sci. liaison officer U. S. Office Naval Research, London, Eng., 1950-51; faculty Pa. State U., 1953-60; research asso. European Research Assos., S.A., Brussels, Belgium, 1959-60; v.p. tech. Electronics Devel., Inc., State College, Pa., 1960-63; full time self-employed cons., 1960-65; pres. Amos J. Shaler, Inc., State College, 1965—. Vice chmn. Acta Metallurgica, 1952-54; cons. Office Sci. and Tech., Exec. Offices of Pres., Washington, 1962—. Bd. dirs. Belgian-Am. Ednl. Found. Recipient Singer-Polignac Found. Bronze medal, 1939, Certificate of Merit U. S. Navy, 1963. Mem. Am. Inst. Mining, Metall. and Petroleum Engrs. (Rossiter Raymond Meml. award 1951), Am. Powder Metallurgy Inst. (plaque 1963), Am. Soc. Metals (Nat. Teaching award 1957), Iron and Steel Inst., Inst. Metals, Société Francaise de Métallurgie, Cum Laude Soc., Sigma Xi, Phi Kappa Phi, Phi Lambda Upsilon, Alpha Sigma Mu. Author: Solar Physics, Origins to 1875, 1941, Les Novae et les Naines Blanches, 1941; (with J. Wulff and H. F. Taylor) Metallurgy for Engineers, 1952. Translator: On Systems Formed by Points (M. A. Bravais), 1946. Patentee, research in mfr., consolidation, sintering and infiltration of metal powders and metal-powder products; advances in phys. metallurgy, methods of analysis of econs. of research and devel. Home and office: 705 W. Park Av., State College, Pa. 16801.*

SHALER, Nathaniel Southgate, Am. geologist; b. Newport, Ky., Feb. 20, 1841; grad. Lawrence Sci. Sch., Harvard, 1862, Sc.D., 1865; instr. zoology and geology Lawrence Sci. Sch., 1868-72; prof. palaeontology, 1868-87, then prof. geology Harvard; dir. Ky. geol. survey, 1873-80, devoting part of each year to that work; from 1884 geologist in charge Atlantic div. U. S. Geol. Survey; mem. Nat. Acad. Scis. Author: A First Book in Geology; Kentucky, a Pioneer Commonwealth; The Nature of Intellectual Property; The Story of Our Continent; The Interpretation of Nature (explanation of evolution by means of theory of reality of spiritual causes), 1893; Illustrations of the Earth's Surface; Sea and Land; The United States of America: a Study of the American Commonwealth; Fossil Brachiopods of the Ohio Valley; American Highways; Features of Coasts and Oceans; Domesticated Animals: their Relation to Man; The Individual: Study of Life and Death, 1900; The Neighbor, 1904; The Citizen, 1903. Died 1906.

SHALNIKOV, Aleksándr Iósifovich, Russian physicist; b. May 10, 1905; grad. Leningrad Poly. Inst., 1928. An organizer Inst. Physics, USSR Acad. Scis., 1935, mem. staff, 1935—; prof. Moscow U., 1938. Recipient Stalin Prize, several times, Order of Lenin. Author: Methods of Studying Organosols of Alkaline Metals, 1933, The Structure of Superconductors in Intermediary States, 1946; Surface Phenomena in Superconductors in an Intermediary State, 1946; (with A. Meshkovskii) Surface Phenomena in Superconductors in an Intermediary State, 1947; Level Indicator for Liquid Helium in Metal Containers, 1960. Research in low temperatures physics, colloids, thin metal films, superconductivity, especially in intermediates states. Home: Vorobévskoye Shosse 2, Moscow, USSR. Office: Leninski gory, Gosudarstvenny universitet, Moscow, USSR.

SHAMOV, Vladímir Nikoláevich, Russian surgeon; b. June 3, 1882; grad. Mil. Medicine Acad., 1908; Staff clinic Mil. Medicine Acad., until 1923; prof. Kharkov Med. Inst., 1923-39; staff Chief Mil. Medicine Adminstrn., Red. Army, 1941—; in charge chair Mil. Med. Acad., 1946—. Worked in clinics, U. S. A., Eng., 1913-14; prof. Mil. Medicine Acad., 1939-41, chief Fedorov Clinic, 1946—; dir. Polenov Neurosurg. Research Inst.; hon. sci. worker RSFSR, 1935—. Coeditor surg. sect. Large Medical Ency., 2d edit.; editorial council Problems of Neurosurgery; editorial bd. New Surg. Archives. Research and publs. on neurosurgery of sympathetic nervous system, blood transfusion, organ transplants.

SHAND, Samuel James, geologist; b. Oct. 29, 1882; s. James Shand; D.Sc., St. Andrews U.; Ph.D., U. Münster (Germany); m. Anna Davidson. Asst. in geology St. Andrews U., 1907-07; officer in charge geol. dept. Royal Scottish Mus., Edinburgh, 1907-11; prof. geology and mineralogy U. Stellenbosch (S. Africa), 1911-37; prof. geology Columbia U., N.Y.C., 1937-50. Recipient Draper medal Geol. Soc. S. Africa, 1927, Lyell medal. Geol. Soc. London, 1950. Fellow Royal Soc. Edinburgh, Geol. Soc. Author: Useful Aspects of Geology, 1925; Eruptive Rocks, 1927; The Study of Rocks, 1931; Earth-Lore, 1933; Rocks for Chemists, 1952; also papers. Died Apr. 19, 1957.

SHANDS, Alfred Rives, Jr., Am. orthopedic surgeon; b. Washington, Jan. 18, 1899; s. Aurelius Rives and Agnes Horner (Eppes) S.; B.A., U. Va., 1918, M.D., 1922; D.M.S. (hon.), U. Del., D.Sc. (hon.), Woman's Med. Coll.; m. Elizabeth Sheffer Prewitt, July 17, 1926; 1 son Alfred Rives III. Mem. faculty Johns Hopkins Sch. Medicine, 1926-27, George Washington U. Sch. Medicine, 1927-30, Duke, 1930-37; med. dir. Nemours Found., Wilmington, Del., 1937—; med. dir. Alfred I. duPont Inst., Wilmington, 1940—, surgeon-in-chief, 1940-62; vis. prof. U. Pa., 1941-66, emeritus, 1966; vis. prof. Albert Einstein Med. Coll., 1963—. Mem. coms. NRC; chmn. coordinating council Handicapped Child Del., Inc., 1963—; mem. nat. com. White House Conf. on Children and Youth, 1960. Adv. council Woodrow Wilson Rehab. Center Found.; bd. dirs. Assn. for Aid of Crippled Children, N.Y.C. Decorated Legion of Merit award; recipient Spl. award Fla. Fedn. Council on Exceptional children, 1964, Mancus Found. for Handicapped, 1963. Diplomate Am. Bd. Orthopaedic Surgery. Mem. Am. Acad. Orthopaedic Surgeons (a founder, v.p. 1939-40), Am. Rheumatism Assn. (a founder, sec.-treas. 1940-46), Del. Acad. Medicine (pres. 1957-59), Medical Academy of Delaware (president 1959), Latin-Am., Internat. socs. orthopaedic surgery and traumatology, Internat. Soc. Surgery, Orthopaedic Corr. Club (a founder), Am. Coll. Surgeons, So. Surg. Assn., Brit. Orthopaedic Assn. (hon.), Assn. Bone and Joint Surgeons (hon.), Alpha Omega Alpha, others. Author: Handbook of Orthopaedic Surgery, 1937, 7th edit., 1967; (with R.B. Raney) Orthopaedic and Surgical Histories of the U.S. Army Air Forces Medical Services in Wolrd War II, 1946; (with J. Vernon Luck) Primer on Prevention of Deformity, 1940. Editor, asso. editor various profl. jours., 1943—, asso. editor Mil. Medicine, 1960—, coeditor U.S. Army Orthopaedic History of World War II, 1946—. Research and publs. on rehab., problems of the handicapped child; adult orthopedic surgery. Home: 2401 Pennsylvania Av., Wilmington 19806. Office: Alfred I. duPont Inst., Rockland Rd., Wilmington, Del. 19899.*

SHANDS, Wayland Arthur, Am. entomologist; b. Darlington, S.C., Sept. 13, 1905; s. Arthur and Lillie (Wilson) S.; B.S., Clemson U., 1926; M.S., U. Minn., 1928; postgrad. U. Minn., 1929; m. Mary Lucretia Knox, Dec. 25, 1932; 1 son, Wayland Arthur. Staff, Fruit and Vegetable Insects Research Br., U. S. Dept. Agr., 1929—, sr. entomologist, 1940-62, investigations leader, research entomologist, 1962—; entomology research div. Agr. Research Service, U. S. Dept. Agr., 1953—; contact rep. U. Me., Orono, 1957—, prof. entomology Grad. Sch., 1966—. Fellow A.A.A.S.; mem. entomol. socs. Am. (chmn. Eastern br. 1965), Washington, Acadian Entomol. Soc., Can. (past v.p.), Potato Assn. Am., N.C. Acad. Sci., Statis. Assn. Am., Biometrics Soc., Sigma Xi, Phi Kappa Phi, Gamma Alpha. Research and publs. on ecology of beet leafhopper, biology and control of insects affecting flue-cured tobacco and of insects affecting potatoes; biol. control potato-infesting aphids, insect vectors and transmission of virus diseases affecting potatoes. Home: 17 Peters St., Orono, Me. 04473. Office: U. S. Dept. Agr. Bldg., U. Me., Orono, Me. 04473; also Aroostook Farm, Route 2, Presque Isle, Me. 04769.*

SHANE, Charles Donald, Am. astronomer; b. Auburn, Cal., Sept. 6, 1895; s. Charles Nelson and Annette (Futhey) S.; A.B., U. Cal. at Berkeley, 1915, Ph.D., 1920, LL.D., 1965; m. Ethel Haskett, Dec. 24, 1917 (dec. Jan. 1919); 1 son, Charles N.; m. 2d, Mary Lea Heger, Dec. 29, 1920; 1 son, William W. Instr., U. S. Shipping Bd., 1917-19; faculty U. Cal., Berkeley, 1920-45, prof. astronomy, 1936-45, asst. dir. sci. personnel Lawrence Radiation Lab., Berkeley, 1942-44, dir. personnel Los Alamos Lab., 1944-45, dir., astronomer Lick Obs., Mt. Hamilton, 1945-58, astronomer, 1958-63, astronomer emeritus, 1965—. Pres., Am. Assn. Univs. for Research in Astronomy, 1958-62; chmn. U. S. Nat. Com. of Internat. Astron. Union, 1966—. Mem. Am. Astron. Soc., Astron. Soc. Pacific, Royal Astron. Soc. (asso.), Nat. Acad. Scis., Am. Philos. Soc. Research, numerous publs. on stellar spectra, photometry absorption lines in solar spectrum, determination certain phys. constants from spectrum hydrogen, distbn. galaxies. Home: P.O. Box 582, Santa Cruz, Cal. 95061.*

SHANER, Ralph Faust, embryologist; b. Pottstown, Pa., Dec. 30, 1893; s. Ralph E. and Myra (Faust) S.; Ph.B., Lafayette Coll., 1915; Ph.D., Harvard, 1920; m. Jean Kimball, June 27, 1923; children—Dorothy (Mrs. Otto Juelich), Ralph K. Teaching fellow, instr. Harvard, 1915-21; faculty U. Alta., Can., 1921-59, prof. anatomy, 1926-59, prof. emeritus, 1959. Mem. Am. Assn. Anatomists, Canadian Assn. Anatomists, Royal Soc. Can. Research and publs. in comparative anatomy; normal devel.; congenital abnormalities of the heart. Home: 11608 Edinboro Rd., Edmonton, Alta., Can.*

SHANKLAND, Daniel Leslie, Am. physiologist; b. San Diego, Cal., June 18, 1924; s. Bert L. and Gussie (Williams) S.; student Cal. Inst. Tech., 1943, U. Colo., 1944, San Diego State Coll., 1946; B.S., Colo. State U., 1948; M.S., U. Ill., 1952, Ph.D., 1956; m. Colette Louise Savaiano, Oct. 22, 1955; children—Gregory Leslie, Christopher, Suzanne. Research rep. Stauffer Chem. Co., Omaha, 1956-57; faculty Purdue U., Lafayette, Ind., 1957—, prof. entomology, 1965—. Mem. Am. Soc. Zoologists, Entomol. Soc. Am., A.A.A.S., Sigma Xi. Research and publs. on properties and effects of secondary neurotoxins produced during poisoning of insects by DDT, involvement of peripheral nerves and spinal cord in prodn. of DDT poisoning syndrome in mammals, anatomy of nerves and muscles in insects, description and analysis of nerve network functions in insects, transducer characteristics of mechanoreceptors in insects. Home: 2819 Barlow St., West Lafayette, Ind. 47906.*

SHANKLIN, Douglas Radford, Am. pathologist; b. Camden, N.J., Nov. 25, 1930; s. John F. and Muriel (Morgan) S.; student Wilson Tchrs. Coll., 1949; A.B., Syracuse U., 1952; M.D., U. N.Y. in Syracuse, 1955; m. Virginia McClure, Apr. 7, 1956; children—Elizabeth, Leigh, Lois, John Carter, Eleanor. Instr. pathology Syracuse U., 1959, U. Fla., 1960-61; asst. prof. U. Fla., Gainesville, 1961-64, asso. prof. pathology and pediatrics, 1964-67; prof. obstetrics and pathology U. Chgo., also pathologist-in-chief Chgo. Lying-In Hosp., 1967—; research asst. Marine Biol. Lab., Woods Hole, Mass., summers 1952-54. Mem. Am. Assn. Pathologists and Bacteriologists, Internat. Acad. Pathology, Soc. Pediatric Research, So. Soc. Pediatric Research, So. Med. Assn., N. Fla. Path. Assn., Fla. Soc. Pathologists, Am. Soc. Exptl. Pathology, Phi Beta Kappa. Author: Diseases of Woman, Pregnancy, Child, 1964; also numerous articles. Developed animal models for hyaline membrane disease of premature infants, showed oxygen toxicity to be an important part of pathogenesis; added to descriptive knowledge of human placenta. Home: 2629 N.W. 12th Av., Gainesville, Fla. 32601.*

SHANKS, William, Brit. mathematician; b. 1812; calculated pi to 707 decimal places, 1873; computed e, modulus M, Euler's constant, numerous other constants. Died 1882.

SHANNON, Claude Elwood, Am. applied mathematician; b. Gaylord, Mich., Apr. 30, 1916; s. Claude Elwood and Mabel Catherine (Wolf) S.; B.S., U. Mich., 1936; M.S., Mass. Inst. Tech., 1940, Ph.D. in Mathematics, 1940; M.S. (hon.), Yale, 1954; D.Sc., U. Michigan, 1961, Princeton U., 1962; m. Mary Elizabeth Moore, Mar. 27, 1949; children—Robert, James, Andrew Moore, Margarita. Carnegie Instn., Washington, 1939, Bowles fellow, 1939-40; Nat. Research fellow Princeton, N.J., 1940; cons. Nat. Def. Research Com., 1941; research mathematician Bell Telephone Labs., 1941—; vis. prof. elec. communications, Mass. Inst. Tech., 1956, prof. communication scis. and math., 1957—; Donner professor sci., Mass. Inst. Tech., 1958—. Recipient Am. Inst. E.E. award, 1940, Morris Liebmann Memorial award, 1949; Stuart Ballantine Medal, Franklin Institute, Philadelphia, 1955; Research Corporation Award, 1956; fellow, Advanced Center Behavioral Scis., Stanford, Cal., 1957-58. Fellow Inst. Radio Engrs.; mem. American Acad. Arts and Sciences, Nat. Acad. Scis., Am. Math. Soc., Sigma Xi, Phi Kappa Phi. Author: Mathematical Theory of Communication, 1949; also papers on math. subjects. Research on Boolean algebra and switching currents, math. cryptography and computing machines; proposed math. theory of communication, 1948, important in fields involving problems of meaning, communication, and lang. Home: 5 Cambridge St., Winchester. Office: M.I.T., Cambridge, Mass.

SHANNON, Ira Lenwood, Am. dentist, clin. chemist; b. McGehee, Ark., Oct. 12, 1918; s. Ira I. and Lillie (Copher) S.; B.S., Pacific U., 1947; D.M.D., U. Ore., 1951; M.S. in Dentistry, U. Cal. at San Francisco, 1955; m. Hazel Scheff Shannon, Nov. 1, 1951; children—Michael, Stephen, Laurie, Cynthia. Chief exptl. dentistry USAF Sch. Aerospace Medicine, San Antonio, 1955-67; chief Spl. Lab. for Research in Oral Physiology, VA Hosp., Houston, 1967—. Fellow Am. Coll. Dentists; mem. Am. Dental Assn., Am. Assn. Pub. Health Dentists, Soc. for Exptl. Biology and Medicine, Research Soc. Am., Internat. Assn. Dental Research. Research and numerous publs. on stress and performance, preventive dentistry, systemic disease and oral health, diagnosis of human adrenal disease. Home: 112 Sequoia Dr., San Antonio, 78232. Office: Box 4188, USAF Sch. Aerospace Medicine, Brooks AFB, Tex. 78235.*

SHANNON, James Augustine, Am. physician, research adminstr.; b. Hollis, N.Y., Aug. 9, 1904; s. James A. and Anna (Margison) S.; A.B., Coll. Holy Cross, 1925, D.Sc., 1952; M.D., N.Y. U., 1929, Ph.D., 1935; LL.D., U. Notre Dame, 1957; D.Sc., Duke U., 1958, Providence Coll., 1958, Loyola U., Chgo., 1959, Cath. U. Am., 1960, W. Va. U., 1960, U. Md., 1965, N.Y. U., 1965, Jefferson Med. Coll., 1965; M.D., U. Louvain (Belgium), 1964, Karolinksa Inst., Stockholm, Sweden, 1964; D.H.L., Albert Einstein Coll. Medicine, 1962; m. Alice Waterhouse, June 24, 1933; children—Alice (Mrs. Peter Fuchs), James Anthony. Med. investigator, educator N.Y. U., 1931-46, dir. research service, Med. Div., 1941-46; dir. Squibb Inst. for Med. Research, 1946-49; asso. dir. research Nat. Heart Inst., 1949-52; asso. dir. NIH, Bethesda, Md., 1952-55, dir., 1955—. Mem. numerous panels, adv. coms.; mem. Pres.'s Sci. Adv. Com., 1959—; expert mem. adv. panel on malaria, WHO, 1956—; mem. adv. com. on med. research Pan Am. Health Orgn., 1962—; standing com. Fed. Council Sci. & Tech., 1959—. Recipient, Presidential medal for merit for World War II malarial studies, 1948, Mendel medal award, Villanova U., 1961, Pub. Welfare medal, Nat. Acad. Scis., 1962, Rockefeller Pub. Service award, 1964. Mem. Nat. Acad. Scis., Alpha Omega Alpha, Sigma Xi. Research on physiology of kidney, quantitative measurement of exogenous compounds in blood and urine, use of atabrine for malaria (World War II). Home: 12 N. Drive. Office: NIH, Bethesda, 20014.*

SHANNON, Lyle William, Am. sociologist; b. Storm Lake, Ia., Sept. 19, 1920; s. Bert Book and Amy (Sivits) S.; B.A., Cornell Coll., Mt. Vernon, Ia., 1942; M.A., U. Wash., 1947, Ph.D., 1951; m. Magdaline Eckes, Feb. 27, 1943; children—Mary

Louise, Robert William, John Thomas, Susan Michelle. Acting instr. U. Wash., 1952; vis. lectr. Wayne State U., 1956-57; faculty U. Wis., 1952-62, asso. prof., 1959-62; prof., chmn. dept. sociology and anthropology U. Ia., Iowa City, 1962—. Author: Underdeveloped Areas: A Book of Readings and Research, 1957; also articles. Developed scale differentiatating self-governing from non-self governing countries in terms of social and econ. devel. system for conducting surveys with indigenous interviewers and dirs.; research on juvenile delinquency, absorption of Mexican-Ams., and So. Negroes into indsl. community. Home: River Heights, Rural Route 1, Iowa City 52240.*

SHANOR, Leland, Am. botanist; b. Butler, Pa., July 21, 1914; s. Paul Leland and Marion (McCandless) S.; A.B., Maryville Coll., 1935; M.S., U. N.C., 1937, Ph.D., 1939; D.Sc., Ill. Wesleyan U., 1961; m. Mary Williams Ward, June 20, 1940; children—Charles Algernon, Paul Leland, II. Instr., Clemson Coll., 1939-40; faculty U. Ill., Urbana, 1940-56, prof., 1951-56; prof., head dept. biol. scis. Fla. State U., 1956-62; dean div. advanced studies Fla. Inst. Continuing U. Studies, 1962-65; prof., chmn. dept. botany U. Fla., Gainesville, 1965—. Sect. head div. sci. personnel and edn. NSF, 1964, dir. div. undergrad. edn. in sci., 1965. Pres., Highlands Biol. Sta., 1958-63, now trustee. John Simon Guggenheim Found. fellow, 1951-52; recipient Alumni citation Maryville Coll., 1965, Citation of Merit, Highlands Biol. Sta., 1964. Fellow A.A.A.S.; mem. Bot. Soc., Am., Mycol. Soc. Am., Assn. Southeastern Biologists, Nature Conservancy, Ill. Acad. Sci., Am. Inst. Biol. Scis., Sigma Xi, Phi Sigma. Research numerous publs. on culture, life cycles, and taxonomy entomogenous fungi, aquatic fungi, fungal plant parasites, Laboulbeniales. Home: 129 N.W. 30th St., Gainesville, Fla. 32601.*

SHANTZ, Edgar Moore, Am. plant physiologist; b. Rochester, N.Y., Oct. 20, 1916; s. Edgar and Grace (Moore) S.; Ph.D., U. Rochester, 1954; m. Margaret McLean, Apr. 12, 1941; children—Valerie McLean (Mrs. Randall K. Cole), Veronica Lee, Philip Edgar, Timothy Harold. With Eastman Kodak Research Labs., 1936-39, Distillation Products, Inc., 1939-45, Distillation Products Industries, 1946-49, Sloan-Kettering Inst., 1954-57; research chemist dept. botany Cornell U., Ithaca, N.Y., 1950-53, 58—, asso. prof. biology div., 1961-68; dir. natural product devel. Calbiochem Corp., 1968—. Mem. Am. Soc. Plant Physiologists, Am. Chem. Soc., A.A.A.S., Am. Inst. Biol. Scis., N.Y. Acad. Scis., Sigma Xi. Research and publs. on isolation of substances related to oil-soluble vitamins, isolation of vitamin A2, process for comml. synthesis of vitamin A, isolation of natural factors stimulating cell division in higher plants, growth of isolated plant cells and tissues in synthetic medium. Home: 809 Oakwood Pl., Pasadena, Cal. 91106. Office: 3625 Medford St., Los Angeles 90063.*

SHANTZ, Homer LeRoy, Am. botanist; b. Kent County, Mich., Jan. 24, 1876; s. Abraham K. and Mary E. (Ankney) S., B.Sc., Colo. Coll., Colorado Springs, Colo., 1901, Sc.D., 1926; Ph.D., U. of Nebraska, 1905; m. Lucia Moore Soper, Dec. 25, 1901; children—Homer LeRoy, Benjamin Soper. Inst. botany and zoölogy, Colorado Coll., 1901, 02; instr. botany, U. of Neb., 1903, 04, U. of Mo., 1905, 06; prof. botany and bacteriology, U. of La., 1907; spl. agt. and collaborator, alkali and drought resistant plant investigations, Bur. Plant Industry, U. S. Dept. Agr., summers, 1906, 07, expert, 1908, 09; plant physiologist same, 1910-20, sr. physiologist in charge physiol. and fermentation investigations, 1921-23, physiologist in charge plant geography in its relation to plant industry, 1924-26; prof. botany and head of dept., U. of Ill., 1926-28; pres. U. of Arizona, 1928-36; chief division of wildlife management, Forest Service, U. S. Dept. of Agriculture, 1936-1944; annuitant collaborator U. S. Dept. Agr., 1945—, prof. botany, U. Ariz., 1956, principal investigator Arizona African Expedition, 1956-57. Special lecturer on plant geography, Graduate Sch. Geography, Clark U., 1922-26. Mem. Ariz. State Bd. of Edn., State Bd. of Vocational Edn.; Ariz. State chmn. Rhodes Scholarship Com.; chmn. State Planning Bd. of Ariz. Fellow Am. Soc. Agronomy, Royal Soc. of Arts; mem. Phytog. Soc. Sweden, hon. pres., 7th Internat. Bot. Congress, Stockholm, 1950, also Paris, 1954. Mem. Botanical Soc., Washington, Assn. Am. Geographers, Ecol. Soc. Am., Soc. Plant Physiol. (Charles Reid Barnes life mem.), Wildlife Soc., Soc. pro Fauna et Flora Fennica, Internat. African Inst., Sigma Xi, Phi Beta Kappa. Contributor many articles to jours. and publs. of the U. S. Govt., dealing chiefly with natural vegetation and its value as an indicator of the agrl. capabilities of land, and the plant geography of N.A., S.A. and Africa; agrl. regions of Africa and management of wild animals on the national forests. Spl. detail to determine natural plant resources and crop producing possibilities of large portions of Africa and Latin America for use of Am. Commn. to Negotiate Peace, 1918-19; spl. detail as agrl. explorer with Smithsonian Instn. Africa Expdn., 1919-20; spl. detail as mem. Ednl. Commn. to East Africa under auspices of Phelps Stokes Fund and Internat. Ednl. Bd., 1924. Died June 23, 1958.

SHAPIRO, Alvin Philip, Am. physician; b. Nashville, Dec. 28, 1920; s. Samuel and Mollie (Levine) S.; A.B., Cornell, 1941; M.D., L.I. Coll. Medicine, 1944; m. Ruth Thomson, Aug., 1951; children—Debra, David. Faculty, U. Cin. Coll. Medicine, 1949-51, Southwestern Med. Sch. U. Tex., 1951-56; faculty U. Pitts. Sch. Medicine, 1956—, professor dept.. medicine, 1967—, chief psychophysiology lab., also dir. clin. pharmacology tng. program, 1960—, co-dir. hypertension-renal clinic Falk Clinic, 1956—; mem. staff VA, Presbyn., Univ. hosps., Pitts., 1960—; cons. staff Elizabeth Steel Magee Hosp., Pitts., 1962—. Diplomate Nat. Bd. Med. Examiners, Am. Bd. Internal Medicine. Fellow A.C.P.; mem. Am. Fedn. Clin. Research, Am. Psychosomatic Soc., A.A.A.S., A.M.A., Am. Heart Assn., Assn. Am. Med. Colls., N.Y. Acad. Scis., Am. Diabetes Assn., Am. Soc. Clin. Investigation. Soc. Exptl. Biology and Medicine, Am. Heart Assn., Alpha Omega Alpha. Author: (with S.O. Waife) Clinical Evaluation of New Drugs, 1959. Research and publs. on clin. pharmacology and psycho-physiology of hypertension; renal disease and pyelonephritis. Office: Dept. Medicine, U. Pitts. Sch. Medicine, Pitts., 15213.*

SHAPIRO, Anatole Morris, Am. physicist; b. Syracuse, N.Y., June 28, 1923; s. Louis H. and Mary (Shapiro) S.; B.A., U. Buffalo, 1944; Ph.D., Cornell U., 1952; m. Mildred Widgoff, June 7, 1945; children—Eve, Jonathan. Asso. physicist Brookhaven Nat. Lab., 1952-54; research fellow, lectr. Harvard, Cambridge, Mass., 1954-56, asst. prof., 1956-58; asso. prof. physics Brown U., Providence, 1958-63, prof., 1963—. Fellow Am. Phys. Soc. Research, publs. in nature of elementary particles. Home: 10 Watson Av., Barrington, R.I. 02806. Office: Dept. Physics, Brown U., Providence 02912.*

SHAPIRO, Arthur, Am. psychophysiologist; b. N.Y.C., June 24, 1910; s. Louis and Flora (Bersin) S.; student Columbia, 1927-30, M.D., 1934; m. Eva Posen, June 25, 1934; children—Judith Ellen (Mrs. Harry Baxter), Naomi Jo. Dir. R.M.K. Lab., Bklyn., 1934-37, research asso., 1938-40; practice internal medicine, 1940—; with Kings County Hosp., 1940—, internist in charge psychosomatic medicine, 1953-54, 60-61, med. cons., 1948—; staff St. John's Hosp.; faculty Sarah Lawrence Coll., 1943-46; faculty Downstate Med. Center dept. medicine State U. N.Y., 1953—, prof. psychiatry, 1964—; adj. prof. dept. elec. engring. Poly. Inst. Bklyn., 1964—; dir. psychophysiology lab. Downstate Med. Center, State U. N.Y., 1961—. Diplomate Am. Bd. Internal Medicine, Am. Bd. Clin. Hypnosis. Mem. Am. Psychosomatic Soc., Am. Heart Assn., N.Y. Acad. Scis., Soc. for Clin. and Exptl. Hypnosis, A.M.A., Acad. Psychotherapists, Assn. for Advancement Psychotherapy, Inst. for Research in Hypnosis (past founding pres.), Soc. for Psychophysiol. Research, Assn. for Psychophysiol. Study Sleep, Assn. for Advancement Med. Instrumentation, Ballistographic Research Soc., Instrument Soc. Am., Soc. for Gen. Systems Research. Research, publs. on microbial genetics, lipid metabolism, spinal anesthesia, blood nucleotides and treatment of multiple sclerosis, hypnosis, psychophysiology of dreaming, sleep; developed various instrumental methods for psychophysiol. research. Home: 225 Easter Pkwy., Bklyn. 11238. Office: 111 Woodruff Av., Bklyn. 11226.*

SHAPIRO, Ascher Herman, Am. mech. engr.; b. Bklyn., May 20, 1916; s. Bernard and Jennie (Kaplan) S.; student Coll. City N.Y., 1932-35; S.B., Mass. Inst. Tech., 1938, Sc.D., 1946; m. Sylvia Charm, Dec. 24, 1939 (div. 1959); children—Peter Mark, Martha Ann, Bennett Mary; m. 2d, Regina Julia Lee, June 4, 1961. Asst. mech. engring. Mass. Inst. Tech., 1938-40, instr., 1940-43, asst. prof., 1943-47, asso. prof., 1947-52, prof. mech. engring., 1952—, prof. charge Fluid Mechanics Div. Mech. Engring. Dept., 1954-62, Ford prof. engring., 1962—, chmn. faculty, 1964-65, head dept. mech. engring., 1965—; vis. prof. Applied Thermodynamics, U. Cambridge, 1955-56; delivered Akroyd Stuart Meml. Lectrs., Nottingham (Eng.) U., 1956; cons. United Aircraft Corp., M. W. Kellogg Co., Arthur D. Little, Inc., Hardie-Tynes Mfg. Co., Carbon & Carbide Chems. Corp., Oak Ridge, Rohm & Haas Co., Ultrasonic Corp., Jackson & Moreland, Engrs., Stone & Webster, Bendix Aviation, Oak Ridge Nat. Lab., others. Served on sub-coms. on turbines, internal flow, compressors and turbines NASA; chmn. Nat. Com. for Fluid Mechanics Films, 1960—; chmn. com. on ednl. films Commn. on Engring. Edn., 1962—. Editor Acad. Press, Inc., 1962—. Dir. lab. for devel. power plants for use in torpedoes Navy Dept., 1944-45; mem. Lexington Project to study and report on nuclear-powered flight to AEC, summer 1948; dir. Project Dynamo to study and report to AEC on tech. and econs. of nuclear power for civilian use, 1953; dir. Lamp Wick Study, Office Naval Research, 1955; mem. tech. adv. panel aeronautics Dept. Def.; cons. operations evaluation group Navy Dept.; mem. sci. adv. bd. USAF, 1964-67. Recipient Naval Ordnance Devel. award, 1945; joint certificate outstanding contbn. War and Navy depts., 1947; Richards Meml. award, Am. Soc. M.E., 1960, Worcester Reed Warner Medal, Am. Soc. M.E., 1965. Fellow Am. Acad. Arts and Scis., Am. Soc. M.E., Am. Inst. Aeros. and Astronautics; mem. Nat. Acad. Scis., Am. Soc. for Engring. Edn., A.A.A.S., Sigma Xi, Tau Beta Pi, Pi Tau Sigma. Author: The Dynamics and Thermodynamics of Compressible Fluid Flow, Vol. 1, 1953, Vol. 2, 1954; Shape and Flow, 1961; also ednl. films and articles tech. jours. Patentee on fluid metering equipment, combustion chambers, and propulsion apparatus and gas turbine auxiliary, magnetic disc, magnetic disc storage device. Mem. editorial board Jour. Applied Mechanics, 1955-56. Research in high speed flow of gases, propulsion engines, shock waves; devel. of improved form of vacuum pump; inventions in mixing apparatus, nuclear propulsion engine. Office: Mass. Inst. Tech., 77 Massachusetts Av., Cambridge, Mass.*

SHAPIRO, David, Am. psychologist; b. Bklyn., July 20, 1924; s. Benjamin and Sarah (Kramer) S.; student Bklyn. Coll., 1941-43; A.B., U. Ill., 1948; A.M., U. Mich., 1950, Ph.D., 1953; m. Shirley Jean Walrath, Dec. 31, 1951. Faculty, Harvard, 1953—, asst. prof. psychology, dept. psychiatry, 1966—; research psychologist Psychiat. Service Beth Israel Hosp., 1956-60; cons. NRC, Boston U. Recipient USPHS Research grant Nat. Inst. Mental Health, 1960-63, 64—, Research Career Devel. award, 1963, Fulbright scholar U. Paris, 1952-53. Fellow Am. Psychol. Assn.; mem. A.A.A.S., Soc. Psychophysiological Research, Phi Beta Kappa. Editor: (with P. H. Leiderman) Psychobiological Approaches to Social Behavior, 1964. Research, publs. chpts. in books on human group behavior; control of involuntary physiol. functions. Home: 165 Commonwealth Av., Boston 02116. Office: Mass. Mental Health Center, 74 Fenwood Rd., Boston 02115.*

SHAPIRO, Harry L(ionel), Am. anthropologist; b. Boston, Mass., Mar. 19, 1902; s. Jacob and Rose (Clemens) S.; A.B., Harvard, 1923, A.M., 1925, Ph.D., 1926; m. Janice Sandler, June 26, 1938 (dec. Apr. 1962); children—Thomas, Harriet, James. Tutor, Harvard U., 1925-26; asst. curator American Museum Natural History, 1926-31, associate curator, 1931-42, chairman of department of anthropology and curator since 1942; research professor Hawaii, 1930-35; prof., Columbia Univ., 1939—; associate Bishop Museum, Honolulu. Bd. directors Louise Wise Services, 1958—. Recipient Theodore Roosevelt distinguished service medal, 1964. Fellow National Academy of Sciences, American Academy of Arts and Sciences; honorary fellow Die Anthropologische Gesellschaft in Wien; mem. Nat. Acad. Scis., Am. Ethnol. Soc. (pres. 1942-43), Am. Assn. Physical Anthropologists (secretary 1935-39, v.p. 1941-42), Am. Anthrop. Association, (pres. 1948), A.A.A.S., National Research Council (chairman div. anthropology and psychology 1953-54), Am. Eugenics Soc. (pres. 1956-63), mem. Social Science Research Council. Author: Heritage of the Bounty, 1936; Migration and Environment, 1939; Aspects of Culture, 1956; The Jewish People, pub. 1960. Editor: Man Culture and Society, 1956. Contbr. sci. jours. Authority on phys. anthropology; studies of effects of inbreeding on populations, influence of environment on race and effect of world migrations. Home: 26 E. 91st St. 10028. Office: American Museum Natural History, N.Y.C. 10024.*

SHAPIRO, Irwin Ira, Am. physicist; b. N.Y.C., Oct. 10, 1929; s. Samuel and Esther (Feinberg) S.; A.B., Cornell U., 1950; M.A., Harvard, 1951, Ph.D., 1955; m. Marian Kaplun, Dec. 20, 1959; children—Steven, Nancy. Physicist, Lincoln Lab., Mass. Inst. Tech., Lexington, 1954-67, professor of geophysics and physics, 1967—. Faculty Boston U., 1956-57, Brandeis U., 1958-59. Recipient award Gravity Research Found., 1965. Mem. Am. Phys. Soc., Am. Astron. Soc., Am. Geophys. Union, Phi Beta Kappa, Sigma Xi. Author: Prediction of Ballistic Missile Trajectories from Radar Observations, 1958; also articles. Research on theories optical model nucleus, statis. mechanics, irreversible processes, statis. determination orbits ballistic bodies, satellite orbits, earth's dust belt, a fourth test Einstein's theory gen. relativity; determined astron. unit, axial rotation Venus from radar data; explained anomalous axial rotation planet Mercury; discovered resonance effects sunlight pressure on satellite orbits and predicted motion Echo balloon; developed theory motion West Ford dipoles. Home: 17 Lantern Lane, Office: Mass. Inst. Tech., Cambridge, Mass. 02139.*

SHAPIRO, Maurice Mandel, Am. physicist; b. Jerusalem, Palestine, Nov. 13, 1915 (parents Am. citizens); s. Asher and Miriam (Greenbaum) S.; B.S., U. Chgo., 1936, M.S., 1940, Ph.D., 1942; m. Inez J. Weinfield, Oct. 8, 1942 (dec. Oct. 1964); children—Joel Nevin, Ellen, Raquel Tamar. Asso. physicist Taylor Model Basin, U.S. Navy, 1942-44; group leader U. Cal. lab. at Los Alamos, 1944-46; sr. physicist Oak Ridge Nat. Lab., 1946-49; head cosmic ray br. U.S. Naval Research Lab., Washington, 1949-65, supt. nucleonics div., 1953-65, chief scientist Lab. for Cosmic Ray Physics, 1965—, chair of cosmic ray physics, since 1966—; instructor Chgo. City Colls., 1938-42; lectr. George Washington U., 1943-44, Nuclear Tng. Sch., 1947-48, ACF Industries, 1956-58, Fermi Internat. Sch. Physics, Varenna, Italy, 1962; vis. prof. Weizmann Inst. Sci., 1962-63; lectr. U. Md., 1949—. Mem. coms., panels U.S. Nat. Com. for IGY, 1957-62; mem. panel on emulsion experiments Nat. Acad. Scis. Space Sci. Bd., 1958-60; U.S. mem. steering com. Internat. Coop. Emulsion Flights, 1960-63. Guggenheim Found. fellow, Israel, 1962-63; recipient Manhatan Dist. award, 1945. Fellow Am. Phys. Soc., A.A.A.S., Washington Acad. Scis. (chmn. panel on awards in phys. scis. 1958); mem. Philos. Soc. Washington (pres. 1966), Italian Phys. Soc., Assn. Los Alamos Scis. (exec. com. 1946), Assn. Oak Ridge Engrs. and Scis. (chmn. 1948), Fedn. Am. Scis. (nat. council 1947, 55, 58-59), Phi Beta Kappa, Sigma Xi. A pioneer in application of nuclear emulsion techniques to cosmic radiation, heavy primary nuclei and helim component, abundance of Li, Be, B; isotopic composition of primary helium; stripped emulsion stacks for high-energy physics; helped establish existence of sigma hyperons; neutron and fission physics, hydrodynamics, piezoelectricity, underwater explosion phenomena; prin. investigator cosmic ray experiment Gemini Project. Home: 6511 Elgin Lane, Bethesda, Md. 20034. Office: Code 7020, U.S. Naval Research Lab., Washington 20390.*

SHAPIRO, Ralph, Am. meteorologist; b. Malden, Mass., Nov. 9, 1922; s. Samuel and Sadie (Aberblatt) S.; B.S., Mass. State Coll., 1943; M.S., Mass. Inst. Tech., 1948, Sc.D., 1950; m. Sylvia Ruth Olfson, Aug. 22, 1945; children—Susan Barbara, Gordon Paul, Nancy Ann. Staff meteorologist Lowell Obs., Flagstaff, Ariz., 1950-52; chief dynamics br. Air Force Cambridge Research Labs., Bedford, Mass., 1952—. Mem. Am. Meteorol. Soc., Royal Meteorol. Soc., Am. Geophys. Union, Research Soc. Am. Research, numerous publs. on dynamics large-scale atmospheric motions including variable solar emissions. Home: 30 Wayne Rd., Needham, Mass. 02194. Office: Cambridge Research Labs., Hanscom Field, Bedford, Mass. 02194.*

SHAPIRO, Victor Lenard, Am. mathematician; b. Chgo., Oct. 16, 1924; s. Joseph E. and Anna (Grossman) S.; B.S., U. Chgo., 1947, M.S., 1949, Ph.D., 1952; m. Florence Gilman, Mar. 21, 1948; children—Pamela, Laura, Charles, Arthur. Faculty, Rutgers U., 1952-54, 55-58, asso. prof., 1955-58; mem. Inst. for Advanced Study, 1954-55, 58-59; prof. Rutgers U., 1959-60; prof. U. Ore., 1960-64; prof. U. Cal. at Riverside, 1964—. NSF fellow, 1954-55. Mem. Am. Math. Soc., Math. Assn. Am., Sigma Xi. Author: Topics in Fourier and Geometric Analysis, 1961; also articles. Research on multiple trigonometric series, spherical harmonics, potential theory, geometric analysis, heat equation, group 2. Home: 3224 Celeste Dr., Riverside, Cal. 92507.*

SHAPLEY, Harlow, Am. astronomer; b. Nashville, Mo., Nov. 2, 1885; s. Willis Harlow and Sarah (Stowell) S.; A.B., U. of Mo., 1910, A.M., 1911, LL.D., 1927; Ph.D., Princeton, 1913, Sc.D., 1933; Sc.D., U. of Pittsburgh, 1931, U. of Pa., 1932, Harvard Univ., 1933, U. of Toronto, Canada, 1935; LL.D., Oglethorpe University, 1931; Sc.D., Brown U., 1933, New York U., 1946, U. of Copenhagen, 1946, U. of Delhi, 1947, U. of Hawaii, 1947; Dr. honoris causa, U. of Michocan, Mex.; Litt.D., Bates Coll., 1942; D. Honoris Causa, University of Mexico, 1951; D.Sc., U. Ireland, 1959, St. Lawrence U., 1963; married Martha Betz, April 15, 1914; children—Mildred (Mrs. Ralph Matthews), Willis, Alan, Lloyd, Carl. Astronomer the Mount Wilson Observatory, Cal., 1914-21; dir. obs. Harvard, 1921-52, Paine prof. astronomy, 1952-56, emeritus, 1956—; Wm. A. Neilsen research prof. Smith Coll., 1956-57; Phi Beta Kappa resident lecturer various American colls., 1957-58. Lecturer Lowell Institute, 1922; exchange lecturer Belgian univs., 1926; Halley lecturer, Oxford, 1928; Harry Todd lecturer for State of Mass., 1929; lecturer Jayne Foundation, Phila., 1930; Darwin lecturer, Royal Astron. Society, 1934; lecturer several fgn. universities, 1951—. Life mem. corp. Mass. Inst. Technology. Trustee Worcester Found. Exptl. Biology, pres., 1942-48; trustee of Science Service Incorporated. Fellow of American Academy Arts and Science (pres. 1939-44), A.A.A.S., (pres. 1947); mem. fgn., nat. and local scientific and profl. assns. and orgns.; hon. mem. Indian Acad. Scis., mem. French Acad. Scis., 1946, Nat. Acad. Scis., various other fgn. sci. assns., exec. officer of several. Trustee Institut Reserche Scientifique d'Afrique Centrale (Africa), Woods Hole Oceanographic Inst. Recipient numerous sci. awards, some of latest being: Pope Pius XI prize, 1941; Calcutta Science Soc. Medal, 1947; gold medal Indian Assn. for Cultivation Sci., 1947; Order of Aztec Eagle (Mexico), Crux de Honor (Puebla). Author: Starlight, 1926; A Source Book in Astronomy, 1930; Flights from Chaos, 1930; Star Clusters, 1943; Galaxies, 1943; The Inner Metagalaxy, 1957; Of Stars and Men, 1958; The View from a Distant Star, 1964. Important research on cosmogony, stellar and nebular photometry and spectroscopy; determined size and shape of Milky Way galaxy and position of solar system in it; developed pulsation theory of cepheid variables; devised technique for determining masses, radii, and other phys. properties of stars; established color-luminosity relation for bright stars in globular clusters; discovered first Sculptor-type galaxy. Address: Peterborough, N.H.

SHAPOSHNIKOV, Vladimir Nikolaevich, Russian microbiologist; b. Feb. 24, 1884; grad. Moscow U., 1910. Began work at Moscow U., after graduation; staff Chem. and Pharm. Inst., 1921-35; became dir. indsl. microbiology Inst. Microbiology, USSR Acad. Scis., 1935; prof. microbiology Moscow U., 1938—; organizer Lab. Indsl. Microbiology, Moscow U., 1935, also Lab. Antiobiotics; organizer 2 indsl. plants, 1935. Recipient Stalin Prize, 1949. Mem. USSR Acad. Scis. Author: (with others) New Developments in Pine Tapping, 1937; The Significance of Physiological Symptoms in the System of Microorganisms, 1942; Industrial Microbiology, 1948. Investigations in indsl. microbiology; classification of phases of fermentation; co-developer methods for prodn. lactic acids, citric acid, 1923; developed fermentation process for prodn. of acetone, butanol, 1929. Home: Leninskiye gory, sekt. K, Moscow, USSR. Office: Moscow U., Moscow, USSR.

SHAREFKIN, Jacob George, Am. chemist; b. N.Y.C., June 15, 1909; s. Joseph and Rose (Vernick) S.; m. Belle Drucker, June 15, 1937; children—Mark Frank, John Bard. Mem. faculty Bklyn. Coll. of City U. of N.Y., 1930—, prof. chemistry, 1965—. Fellow A.A.A.S., Am. Chem. Soc., Chem. Soc. of London. Author: Exercises in Organic Synthesis, 1948. Research and publs. in detection of functional groups in organic compounds, particularly olefins, organic iodine, boron compounds. Home: 1046 E. 28th St., Bklyn. 11210.*

SHARMA, Arun Kumar, Indian cytologist, cytochemist; b. Calcutta, India, Dec., 1924; s. C. C. and S. (Sen) S.; I.Sc., U. Calcutta, 1941, B.Sc., 1943, M.D., 45, D.Sc., 55; m. Archana Mookerjea, May 8, 1955. Mem. research staff U. Calcutta, 1944—, faculty, 1948—, reader, also in charge cytogenetics lab. Recipient medals and prizes for research and acad. distinctions. Mem. Bot. Soc. Bengal (v.p.), Indian Soc. Cytology and Genetics (treas.), Ecol. Soc. Am., Internat. Bur. Plant Taxonomists. Author: (with Archana Sharma) Chromosome Techniques—Theory and Practice, 1965. Chief editor: The Nucleus; mem. editorial bd. Indian Jour. Exptl. Biology. Research, publs. on new techniques for study of chromosome structure, concept. of origin of species in vegetatively reproducing plants through chromosome mosaics in somatic tissue, analysis of chromosomal basis of differentiation and chem. nature of differentiated nuclei, chem. nature of plant chromosomes and on basis of chromosome studies taxonomy of all monocotyledonous groups have been re-oriented. Home: 20/1/5/ Ballygunj Pl., Calcutta 19, India.*

SHARMA, Ganpati Parshad, Indian zoologist; b. Ambala, India, Dec. 11, 1917; s. Ram Chand and Bhagwanti Devi; B.Sc., Govt. Coll., Lahore (now Pakistan), 1938, B.Sc. with honors, 1940, M.Sc., 1941, Ph.D., 1945; Ph.D., Inst. Animal Genetics, Edinburgh, Scotland, 1947; m. Maitreyi Feb. 15, 1945; children—P. V., Trisha. Asst. research officer Indian Vet. Research Inst., Izatnagar, 1948-49; animal geneticist Govt. Live Stock Farm, Hissar, India, 1949-51; reader zoology Punjab U., 1951-59; prof., head dept. zoology, 1959—. Officer in charge scheme financed by WHO, Geneva, Switzerland, 1966. Recipient Omen prize Govt. Coll., Lahore, 1940. Fellow Zool. Soc., Royal Micros. Soc., Nat. Inst. Scis. India, Acad. Zoology Agra, Indian Acad. Zoology, Zool. Soc. India, Nat. Acad. Scis. Research, publs. on formation sperm in various groups of animals, chromosomes in relation to taxonomy and phylogenetic relationships of animals. Home: 6-I, Sector 16, Chandigargh, India.*

SHARMA, Narsingh Narain, zoologist; b. Jalaun, India, July 13, 1930; s. M. L. and F. K. Devi (Pipraiya) S.; B.S., Agra U. (India), 1952; M.S., U. Lucknow (India), 1954, Ph.D., 1958; Ph.D., U. Ga., 1962; m. Sudha Misra, July 6, 1964; 1 dau., Anita. Lectr. zoology Shia Degree Coll., Lucknow, 1958-60; research asst. U. Ga., 1960-62; adviser protozoology Yerkes Regional Primate Research Center, Emory U., Atlanta, 1962-63; officer scientists pool U. Lucknow, 1963-64; research asso. U. Mass., Amherst, 1964-66; microbiologist Chgo. Med. Sch., 1966—. Fellow Zool. Soc. London, Royal Micros. Soc. Eng.; mem. Soc. for Protozoologists, Sigma Xi. Research on morphology and systematics of intestinal protozoa of insects, particularly termites, immunological studies on poultry coccidia, cytochemistry and enzymology of protozoan parasites such as Trichomonas vaginalis and Balantidium coli, tissue cultures and their cytochemical response to infection with Trichomonas vaginalis; autoradiography of entamoebae. Home: 8800 Root St., Niles, Ill. 60648.*

SHARMAN, Geoffrey Bruce, Australian zoologist; b. Launceston, Australia, Jan. 13, 1925; s. Clifford Alfred and Jean (Ralston) S.; B.Sc., U. Tasmania, 1950; D.Sc., U. Western Australia, 1960; m. Barbara Veale, Feb. 23, 1952; children—John Andrew, Robert James. Lectr. zoology U. Adelaide, S. Australia, 1956-58, sr. lectr., 1959-61; prin. research scientist Commonwealth Sci. and Indsl. Research Orgn. div. Wildlife Research, Canberra, Australia, 1962-63, sr. prin. research scientist, 1964-65; prof. zoology Sch. Biol. Scis. U. New South Wales, 1966—. Fellow Zool. Soc.; mem. Australian and New Zealand Assn. for Advancement Sci., Australian Mammal Soc. (found. mem., 1st pres. 1961-63). Research, numerous publs. on chromosome numbers, sex-chromosome mechanisms, reproductive cycles and reproductive physiology of mammals, especially marsupials; discovered kangaroos exhibit embryonic diapause, prolonged storage of a totally quiescent embryo in uterus. Home: 3 Warrowa Pl., West Pymble, New South Wales, Australia. 2073.

SHARONOV, Vsévolod Vasílievich, Russian astronomer; b. Mar. 10, 1901; grad. Leningrad U., 1926. Mem. staff Chief Astron. Obs., USSR Acad. Sci., 1936-41; prof. Leningrad (USSR) U., 1944—, dir. U. Astron. Obs., 1951—. Author: The Visibility of Distant Objects and Flares, 1944; Mars, 1947; Measuring and Calculating the Visibility of Distance Objects, 1947; The Sun and its Observations, 2d edit., 1952; Photometric Studies of the Nature of Planets and Satellites, 1954; The Moon and Space Flights, 1957; Some Results of Observations of Mars During Opposition, 1958; Photometric and Colorimetric Comparison of the Surface of a Mars with Samples of Limonite and Rocks of Red Strata, 1960. Research and publs. on planets, atmospheric optics; devel. methods for absolute surface photometry of heavenly bodies; design of instruments for measurements of brightness of natural bodies and range of visibility. Office: A. USSR, Leningrad, Universitetskaya n. 7-9, Gosudarstvenny universitet.

SHARP, Aaron John, Am. botanist; b. Plain City, O., July 29, 1904; s. Prentice Daniel H. and Maude K. (Herriott) S.; A.B., Ohio Wesleyan U., 1927, D.Sc. (hon.), 1952; M.S., U. Okla., 1929; Ph.D., Ohio State U., 1938; m. Cora Evelyn Bunch, July 25, 1929; children—Rosa (Mrs. Odis Chambers), Katharine (Mrs. E. C. Clebsch), Mary Martha (Mrs. R. L. McFarland), Fred Prentice, Jennifer (Mrs. Peter Haskell), Faculty U. Tenn., Knoxville, 1929—, prof. botany, 1944-65, chmn. dept., 1951-61, curator herbarium, 1958—, distinguished prof., 1965—. Research asso. Tsali Inst. for Research on Eastern Cherokees, Cherokee, N.C., 1951-54; sect. sec. Inter-Am. Conf. on Conservation Renewable Natural Resources, Denver, 1948. Named Hon. mem. staff Hattori Bot. Lab., Japan, 1956—. John Simon Guggenheim Meml. Found. fellow, 1944-46, Gen. Edn. Bd. fellow, 1943, Cecil Billington lectr., 1947. Mem. A.A.A.S. (v.p., chmn. sect. 1963), Bot. Soc. Am. (treas. 1957-62, pres. 1965), Am. Soc. Plant Taxonomists (pres. 1961), Ecol. Soc. Am. (v.p. 1959), Assn. S.E. Biologists (v.p. 1955-56), Tenn. Acad. Sci. (v.p 1952, pres. 1953), So. Appalachian Bot. Club (pres. 1947-48), Am. Bryological Soc. (pres. 1935), Phi Beta Kappa, Sigma Xi, Phi Kappa Phi, Phi Epsilon Phi, Sigma Delta Pi. Author: (with C. C. Campbell, W. F. Hutson, H. L. Macon) Great Smoky Mountains Wildflowers, 1962; contbr. to Trees, Shrubs, and Woody Vines of the Great Smoky Mountains National Park (Arthur Stupka), 1964. Research on mosses, other plants of So. Appalachians and their place in hist. geology; relations of plants of N.Am. and East Asia; significance of botany in human culture and importance of plant conservation. Home: 1201 Tobler Rd., Knoxville, Tenn. 37919.*

SHARP, Abraham, English mathematician; b. Little Horton, Eng., 1651; s. John and Mary (Clarkson) S.; apprentice to William Shaw, mercer of York, then to a mcht. of Manchester, Eng.; became tchr. math., Liverpool, Eng.; employee of Flamsteed, Greenwich obs., 1676-90; engaged to make a mural arc (Flamsteed's 1st satisfactory instrument), 1688; tchr. math., London, 1690; became clerk King's shipyard, Portsmouth, 1691; ret. to make astron. instruments and models, Little Horton, 1694. Author: Geometry Improved, 1717. Noted for ability in graduating instruments; pub. 1st book on logarithms; also astron. maps. Died July 18, 1742.

SHARP, Benjamin, Am. zoologist; b. Germantown, Phila., Nov. 1, 1858; s. Benjamin and Hannah Ballinger (Leedom) S.; ed. Swarthmore Coll., 1871-76; M.D., U. Pa., 1879, Ph.D., 1880; Ph.D., U. Würzburg, 1883; postgrad. univs. Berlin, Leipzig, Zoöl. Sta., Naples, 1879-83; m. Virginia May Guild, Sept. 15, 1881. Prof. invertebrate zoology Acad. Natural Scis., Phila., 1883, made expdns., collecting in Caribbee Islands in winter, 1888-89; also corr. sec., 1890-1901, life mem.; prof. invertebrate zoology U. Pa., 1884-86; to H.I., collecting archaeol. and zool. specimens summer of 1893; also in Arctic, 1895, on U. S. revenue cutter Bear, in Alaska and Siberia. In charge zoology in 1st Arctic expdn., Lt. R. E. Peary, 1891 rep. in Gen. Court of Mass. for Nantucket, 1910-13. Vice pres. Nantucket Hist. Assn. Died Jan. 24, 1915.

SHARP, David Gordon, Am. biophysicist; b. Annandale, N.J., Nov. 15, 1909; s. William LaRue and Florence (Evans) S.; B.S., Rutgers U., 1932; M.A., Duke U., 1937, Ph.D., 1939; m. Marietta Jane Stiles, Apr. 15, 1939; children—Lora Louise (Mrs. F.W. Doolittle III), John David, Linda Lorraine, Leah Jane. Research engr. Westinghouse Elec. and Mfg. Co., 1932-36; virus research Duke U., Durham, N.C., 1939-57; prof. biophysics U. N.C., Chapel Hill, 1957—. Mem. NIH Tng. Com. on Biophysics, 1960—. Mem. Electron Microscope Soc. Am. (pres. 1960), Soc. Exptl. Biology and Medicine, Am. Soc. Immunologists, Am. Soc. Microbiologists. Research and many publs. in application of ideas and machines of physics to animal virology, especially electron microscopy of viruses; devised sedimentation method of counting viruses. Home: 307 Granville Rd., Chapel Hill, N.C. 27514.*

SHARP, Eugene Lester, Am. plant pathologist; b. Spokane, Wash., Nov. 19, 1926; s. Ernest L. and Mrytle (Day) Sharp; B.S. in Botany, U. Ida., 1949; M.S. in Plant Pathology Ia. State U., 1951; Ph.D., 1953; m. Mae O. Mehlenbacher, Feb. 19, 1954; children—Jane Elizabeth, Lorraine Mae, Jeffrey Eugene, Thomas Robert. Plant pathologist U. S. Army Chem. Corps Biol. Labs., Ft. Detrick, Md., 1953-55, research plant pathologist, 1955-57; faculty Mont. State U., Bozeman, 1957—, asso. prof. plant pathology, 1962—. Mem. adv. com. Am. type Culture Collection, 1962—. Fellow A.A.A.S.; mem. Am. Phytopath. Soc., Sigma Xi, Gamma Sigma Delta. Research and publs.

on techniques for preservation of fungus cultures, effects environment on infection process of plant parasitic fungi, genetic mechanisms of resistance of plants to cereal rust fungi and interactions of genes conditioning plant disease resistance with various environmental factors. Home: Route 1, Bozeman, Mont. 59715.*

SHARP, Lauriston, Am. anthropologist; b. Madison, Wis., Mar. 24, 1907; s. Frank Chapman and Bertha Staples (Pitman) S.; A.B., U. of Wis., 1929; certificate, U. of Vienna, 1931; A.M., Harvard, 1932, Ph.D., 1937; m. Ruth Burdick, Aug. 22, 1936; children—Alexander, Susannah. Asst. freshman dean, U. of Wis., 1929-30; mem. Logan Museum North African Expdn., 1930; mem. U. of Chicago Fox Indian Reservation study, 1932; asst. in anthropology, Harvard, 1932-33, 1936; fellow Australian Nat. Research Council, North Queensland Expdns., also lecturer in anthropology, U. of Sydney, 1933-35; field study in Java, Philippines, China, 1935; instr. in anthropology, Cornell U., 1936-39; asst. prof., 1939-43, asso. prof., 1943-47, prof. since 1947; instr. Chinese area, Army Specialized Training program, Cornell, 1943-44; acting asst. chief, div. of Southeast Asian Affairs, Dept. of State, 1945-46; lecturer, staff officers sch. for Asiatic studies, Yale, 1946; mem. Dept. of Asian Studies, Cornell, since 1946, dir. Cornell U. Studies in Culture and Applied Science 1947—, Thailand Project, 1947—; field study in Indochina, Malaya, Burma, India, 1948-49; chmn. dept. of sociology and anthropology, Cornell, 1949-56; dir. Cornell U. S. E. Asia Program, 1950-60; cons. Army War Coll., 1954-58; distinguished lecturer Anthropology Haverford College, 1955-56; chmn. acad. adv. com. on Thailand, U. S. AID, 1966—. Mem. UNESCO international panel for Humid Tropics Research, 1955—. Fulbright prof., Thailand, 1952-53; Guggenheim fellow, 1967-68, Fellow Royal Anthrop. Inst., Australian Inst. Aboriginal Studies, American Anthropological Association; mem. Nat. Research Council (mem. div. of anthropology and psychology, 1944-47, 1953-59; com. on Oceania, 1942-47, Pacific Sci. Conf., 1946-57, chmn. com. on Asia (1947-52, chmn. of committee South East Asia, 1956—), American Council of Learned Socs.-Social Science Research Council (mem. joint com. on So. Asia 1949-52, also on Asia, 1958—, Social Science Research Council), (member of the committee on area training 1949-51, com. on econ. development, 1949-50), Am. Inst. of Pacific Relations (com. on dependent territories study, 1945-46), Southeast Asia Inst. (dir. 1947-48), Soc. Applied Anthropology, Assn. Asian Studies (pres. 1961-62), Siam Soc., Asia Soc., Inc., Sigma Xi, Alpha Delta Phi. Author: Siamese Rice Village, 1953; Bibliography of Thailand, 1956; Handbook on Thailand, 1956; Some Principles of Cultural Change, pub. 1967. Co-author: Tribal Peoples of Chiengrai, 1964. Contbr. articles, reports, bibliographies and reviews to sci. jours. Discovered and described universal or cosmic totemism of Australian Aborigines; made, directed 1st studies Thai peasant cultures; (with L. M. Hanks) 1st ethnological studies Thailand hill tribes; 1st Am. program overseas studies in applied anthropology. Home: 880 Highland Rd., Ithaca, N.Y. 14850. Office: Cornell U., Ithaca, N.Y. 14850.*

SHARP, Robert Phillip, Am. geologist; b. Oxnard, Cal., June 24, 1911; s. Julian Heber and Alice (Darling) S.; B.S., Cal. Inst. Tech., 1934, M.S., 1935; Ph.D., Harvard, 1938; m. Jean Prescott Todd, Sept. 7, 1938; children—Kristin Todd, Bruce Todd. Mem. faculty U. Ill., 1938-43, U. Minn., 1946-47; prof. geology Cal. Inst. Tech., Pasadena, 1947—. Recipient Kirk Bryan award Geol. Soc. Am., 1964. Mem. Geol. Soc. Am. (councilor 1958-61), Am. Geophys. Union, Glaciological Soc. Research and many publs. in processes acting on surface of earth to produce land forms, including glaciers, wind, water, weathering. Home: 1410 E. Palm St., Altadena, Cal. 91001.*

SHARP, Samuel, surgeon; b. Jamaica, circa 1700; s. Henry Sharp; apprenticed to William Cheselden, 1724; diploma, 1732; student in France; became freeman Barber-Surgeons Co., 1731; surgeon to Guy's Hosp., London, 1733-57; traveled in Italy, 1765. Fellow Royal Soc., 1749; mem. Paris Royal Soc. Author: A Treatise on the Operations of Surgery, 1739; A Critical Enquiry into the Present State of Surgery, 1750. Supplied connecting link between methods of surgery represented by Cheselden and that represented by William Hunter; 1st to suggest the barrel of trephine be conical. Died Mar. 24, 1778.

SHARPE, Charles Bruce, elec. engr.; b. Windsor, Ont., Can., Apr. 8, 1926; s. Merton James and Anne (Meredith) S.; B.S., Northwestern U., 1946; B.S.E.E., U. Mich., 1947, Ph.D., 1953; S.M., Mass. Inst. Tech., 1949; m. Martha Jeanne Brunk, Apr. 24, 1954; children—Olivia Ann, Randall Bruce, David Alan, Thomas Merton, Donald Louis. Came to U.S., 1928, naturalized, 1947. Research asst. Mass. Inst. Tech., 1947-49; research asso. Aero. Research Center, U. Mich., Ann Arbor, 1949-50, Electronic Def. Group, 1951-53, staff cons. Cooley Electronics Lab., 1955-60, research engr. Inst. Sci. and Tech., 1961—; cons. Collins Radio Co., Cedar Rapids, Ia., 1960-61; faculty U. Mich., 1955—, prof., 1962—. Mem. I.E.E.E. (sr. mem., asso. editor Transactions on Circuit Theory, 1965-67), Sigma Xi, Tau Beta Pi, Eta Kappa Nu. Research and publs. in theory of linear antenna arrays and synthesis of nonuniform transmission lines; with D.S. Heim 1st investigated discontinuity problem in ferrite-filled waveguides; with C.G. Brockus devised technique for measuring microwave properties of ferroelectrics. Home: 2818 Glacier Way, Ann Arbor, Mich. 48105.*

SHARPE, Daniel, English geologist; b. Marylebone, Eng., Apr. 6, 1806; s. Sutton and Maria (Rogers) S.; ed. Mr. Cogan's sch., Walthamston; partner (with brother Samuel) in Portuguese mercantile bus.; lived in Portugal, 1835-38. Fellow Royal Soc., 1850, Linnean Soc., Zool. Soc.; mem. Geol. Soc. (treas. 1853, pres. 1856). Contbr. papers on geology of Portugal, Gt. Britain, Europe, to jours. Died May 31, 1856.

SHARPE, Francis Robert, mathematician; b. Warrington, Eng., Jan. 23, 1870; s. Alfred and Mary (Webb) S.; A.B., Cambridge U., 1892; Manchester U., 1900-01; Ph.D., Cornell U., 1907; m. Jeannette Welch, Sept. 1900; children—Elfreda J., Frances M., Edith J. Naturalized Am. citizen, 1910. Lectr. math. Queen's U., Kingston, Can., 1901-04; instr. math. Cornell U., 1905-10, asst. prof., 1910-19, prof., 1919-38, emeritus prof. since 1938. Mem. Am. Math. Soc., Sigma Xi. Contbr. papers on hydrodynamics and algebraic geometry. Died May 18, 1948.

SHARPE, Richard Bowdler, English ornithologist; b. London, Eng., Nov. 22, 1847; s. Thomas Bowdler S.; LL.D., U. Aberdeen (Scotland); m. Emily Burrows, 1867. First librarian Zool. Soc. London, 1867-72; sr. asst., dept. zoology Brit. Mus., 1872-95, asst. keeper vertebrata, from 1895. Pres. Internat. Ornithol. Congress, 1905. Recipient Gold medal for sci. Emperor Austria, 1891. Hon. fellow Zool. Soc. Author: Catalogue of the Birds in the British Museum, 27 vols., 1847-98; A Monograph of the Alcedinidae, 1868-71; A Monograph of the Hirundinidae, 1885-94; A Monograph of the Hirundinidae, 1885-94; Monograph of the Paradiseidae, 2 vols., 1891-98; Handbook to the Birds of Great Britain, 4 vols., 1894-97; Handlist of the genera and species of birds in the British Museum, 5 vols., 1899-1909; Catalogue of the Collection of Birds' Eggs in the British Museum, 5 vols., 1911-12. Editor: Naturalists' Library (Allen). Studies on kingfishers, swallows, birds of paradise. Died Dec. 25, 1909.

SHARPEY, William, English physiologist; b. Eng., Apr. 1, 1802; s. Henry and Mary (Balfour) s.; studied medicine and surgery, London, Paris; M.D., Edinburgh, Scotland, 1823, LL.D., 1859; practiced medicine, Arbroath, 1824-26; joint lectr. systematic anatomy, Edinburgh, from 1832; prof. anatomy and physiology Univ. Coll., London, 1836-74; examiner anatomy London U., from 1840. Fellow Royal Soc., 1839, mem. council, 1844, sec., 1853-72; mem. Gen. Med. Council, 1861-76. Contbr. article on echinodermata to Todd and Bowman's Cyclopaedia; contbr. to Baly's transl. of Physiology (Müller), 1837, 40. Editor: Elements of Anatomy (Jones Quain), 5th-8th edits. Described Sharpey's fibers (connective tissue which holds periosteum to bone), 1848. Died Apr. 11, 1880.

SHARPEY-SCHAFER, Sir. Edward Albert, English physiologist; b. Hornsey, London, Eng., June 2, 1850; s. James William Henry and Jessie (Brown) Schafer; grad. U. Coll., London (Sharpey scholar), 1871; M.D., univs. Berne (Switzerland), Groningen (Netherlands); D.Sc. in Medicine, U. Louvain (Belgium); Sc.D., univs. England, Ireland, Australia; LL.D., univs. Aberdeen, McGill, St. Andrews (Scotland); m. Maud Dixey, 1878; 2 sons, 2 daus.; m. 2d., Ethel Maude Roberts, 1900. Asst. prof. physiology, U. Coll., London, 1874-83, Jodrell prof., 1883-99; Fullerian prof. Royal Inst., Eng., 1878-81; prof. physiology U. Edinburgh (Scotland), 1899-1933, prof. emeritus from 1933. Mem., pres. Internat. Congress Physiologists, 1923. Fellow Royal Soc., 1878 (Royal medal, 1902, Copley medal, 1924); (hon.) Royal Coll. Physicians, Edinburgh (Baly medal, 1897); mem. Royal Soc. Edinburgh (pres. 1933), Brit. Assn. (gen. sec. 1895-1900, pres. 1912), Royal Life Saving Soc. (distinguished servicemedal, 1909, charter mem. Physiol. Soc. Author: A Course of Practical Histology, 1877; The Essentials of Histology, 1885; Experimental Physiology, 1912; The Endocrine Organs, 1916; History of the Physiological Society, 1927. Editor, author (with others) Advanced Text-book of Physiology; 2 vols., 1898-1900. Editor Quarterly Jour. Exptl. Physiology. Research, publs. on histology which he promoted as that part of physiology necessary for a proper understanding of functional activity; included research on wing structure insects, absorption of fat by intestinal villi of mammals, ablation of portions of cerebral cortex and effect of stimulation in the visual area of brain, the working of the ciliary muscle, effect of suprarenal gland extract upon arterioles and blood pressure, intervention and contraction of spleen, functions of the motor cortex, effect of chloroform upon various parts of body, pulmonary circulation and influence of vargus nerve on respiration and action of intercostal muscles; made discoveries regarding nerve functioning in jellyfish; demonstrated successful application of photography to record heartbeat of frog, 1884; demonstrated direct communication of canaliculi with blood capillaries in liver; developed new improved method artificial respiration, worked toward equal acceptance women in med. profession; credited (through extensive work on internatl secretions) with helping lay groundwork for field of endocrinology. Died North Berwick, Scotland, Mar. 29, 1935.

SHARPLESS, Stewart Lane, Am. astronomer; b. Milw., Mar. 29, 1926; s. Stewart Lane and Edith (Schiel) S.; Ph.B., U. Chgo., 1948, Ph.D., 1952; m. Roberta Tillman, Aug. 13, 1960. Carnegie fellow Mt. Wilson Obs., also Palomar Obs., 1952-53; astronomer U.S. Naval Obs., 1953-64, dir. astronomy and astrophysics div., 1963-64; prof. astronomy U. Rochester (N.Y.), also dir. C.E. Kennth Mees Obs., 1964—. Mem. Am. Astron. Soc., Internat. Astron. Union. Research and publs. on structure galaxies, classification stellar spectra, evolution groups of young stars. Home: West Hollow Rd., Naples, N.Y. 14512.*

SHARRARD, William John Wells, English orthopaedic surgeon; b. Lincoln, Eng., Nov. 8, 1921; s. William and Winifred Hannah (Wells) S.; M.B., Ch.B., Sheffield (Eng.) Med. Sch., 1944, M.D. with honors, 1954, Ch.M. with merit, 1966; m. Bessie Laura Petch, May 16, 1953; children—Sally Ann Margaret, John Christopher, Richard Michael, Mark Jonathan. Demonstrator anatomy U. Sheffield, 1945; cons. orthopaedic surgeon United Sheffield Hosps., 1955—; Hunterian prof. Royal Coll. Surgeons, Eng., 1956; vis. prof. to various med. schs. Gt. Britain, U. S., Can., 1964. Mem. adv. subcoms. on tetanus prophylaxis and neonatal surgery Ministry Health, London, 1964—. Fellow Brit. Orthopaedic Assn.; mem. Brit. Orthopaedic Research Soc. (founder, past pres.), Anat. Soc. Gt. Britain and Ireland, Spastics Soc. Gt. Britain (v.p. 1964—). Contbg. author: Recent Advances in Cerebral Palsy, 1958; Surgical Aspects of Medicine, 1959; Clinical Surgery—Orthopaedics, 1966. Research and publs. on prevention, correction, and treatment of paralysis and deformity in children; causes of poliomyelitis, cerebral palsy, spina bifida; prevention of tetanus after wounds; analysis of spinal cord and nerve root function in man. Editorial bd. Jour. Bone and Joint Surgery, 1964—, Jour. Devel. Medicine and Child Neurology, 1965—. Home: 140 Manchester Rd., Sheffield, 10, Yorkshire, Eng.*

SHARTLE, Carroll Leonard, Am. psychologist; b. Ruthven, Ia., June 26, 1903; s. Harry E. and Flora (Leonard) S.; A.B., State Coll. Ia., 1927; A.M., Columbia, 1932; Ph.D., Ohio State U., 1933; m. Doris Brown, Sept. 2, 1931; children—Alex B., Leonard H. With USES, 1935-39, Social Security Bd., 1939-42, War Manpower Commn., 1942-44; prof. psychology Ohio State U., Columbus, 1944—, dir. research div., 1964—, dir. research Human Resource Research Inst., U.S. Air Force, 1952-53; chief psychology and social scis. div. Office Sec. Def., 1961-64. Mem. med. adv. com. Social Security Adminstrn., 1955-60; mem. adv. group on human factors, sci. com. NATO, 1961-64. Fellow Am. Psychol. Assn. (past dir., treas.), A.A.A.S.; mem. Internat. Assn. Applied Psychology, Phi Beta Kappa (hon.). Author: (with others) Occupational Counseling Techniques, 1940; Occupational Information, 1946, rev., 1952, 59; Executive Performance and Leadership, 1956 (Spanish edit. 1960); also articles. Developer occupational classification methodology, methods for study leader and organizational behavior. Home: 6699 Olentangy River Rd., Worthington, O. 43085.*

SHATTOCK, Samuel George, English pathologist; b. London, Eng., Nov. 3, 1852; s. Samuel C. and Jane (Brown) Betty; M.D., U. Coll., London, 1874; m Emily Lucy Wood; 3 sons, 1 dau. Curator mus., U. Coll., London, elected fellow, 1910; tchr. pathology St. Thomas' Hosp., until 1924; curator path. mus. Royal Coll. Surgeons, 1897-1924. Fellow Royal Soc., 1917; mem. Path. Soc. London, Royal Soc. Medicine (editor Trans. and Proc.). Author: Thoughts on Religion, 1926. Research on morbid anatomy, secondary sex characteristics, virulence of bacteria, origin of cancer, healing of wounds in plants. Died Wimbledon, Eng., May 11, 1924.

SHATTUCK, Frederick Cheever, Am. physician; b. Boston, Nov. 1, 1847; s. George Cheyne and Anne Henrietta (Brune) S.; A.B., Harvard, 1868, A.M., 1872, M.D., 1873; Sc.D., 1912; LL.D., U. Cin., 1908; m. Elizabeth Perkins Lee, June 19, 1876; children—George Cheever, Henry Lee, Mrs. Elizabeth Perkins Bigelow, Mrs. Clara Lee Richardson. Practiced medicine, Boston, 1875—; clin. instr. auscultation and percussion, Harvard, 1879-84, instr. theory and practice of physic, 1884-88, Jackson prof. clin. medicine, 1888-1912, emeritus, 1912; cons. physician Mass. Gen. and various other hosps. Fellow Am. Acad. Arts and Scis.; mem. Mass. Hist. Soc. Early advocate of adequate feeding in typhoid fever; 1st successful drainage of pericardium accomplished under his direction, Boston. Died Jan. 11, 1929.

SHATTUCK, George Cheever, Am. physician; b. Boston, Oct. 12, 1897; s. Frederic Cheever and Elizabeth Perkins (Lee) S.; A.B., Harvard, 1901, M.D., 1905, A.M. (hon.), 1919; postgrad. U. Vienna (Austria), 1907-08; m. Virginia Grigby Chandler Peabody, July 9, 1932; stepchildren—Francis W. Peabody, Grigsby C. Peabody. Practice medicine specializing in internal medicine, Boston, 1908-15; faculty Harvard Med. Sch., 1908-15, 21—, clin. prof. tropical medicine, 1938-47, emeritus, 1947—; mem. A.R.C. San. Commn., Serbia, 1908-15; mem. League Red Cross Socs., Geneva, Switzerland, 1919-21. Cons. tropical diseases Mass. Gen. Hosp., 1928-

—, Boston City Hosp., 1941-66. Recipient Theobald Smith medal, 1954, Richard Pearson Strong medal Am. Soc. Tropical Medicine, 1962, Orden Nacional do Cruzeiro do Sul Official, Brazil, 1958, Orden Macional de Merito Commendador, Carlos J. Findlay, Cuba, 1950. Mem. Am. Soc. Tropical Medicine, Am. Acad. Tropical Medicine, Am. Acad. Arts and Scis., Royal Soc. Tropical Medicine and Hygiene. Author: Principles of Medical Treatment, 1926; The Peninsula of Yucatan, 1933; Medical Survey of the Republic of Guatemala, 1938; Diseases of the Tropics, 1950; also other books, numerous articles. Research on etiology elephantiasis; carified unrecognized vitamin deficiencies; showed alcoholic neuritis to be undistinguishable from beri-beri; value liver extract in sprue. Home: 450 Warren St., Brookline, Mass. 02146. Office: 25 Shattuck St., Boston 02115.*

SHAW, Alexis Eric, Australian pathologist; b. Brisbane, Australia, Feb. 2, 1917; s. John and Anna (Erzersky) S.; C.B.E., M.B., B.S., U. Queensland, 1941; postgrad. U. Sydney, 1942; m. Olwen Elizabeth Davies, May 17, 1944; children—Mary Anne, Alexis John, Bronwen, Margaret. First dir. Red Cross Blood Transfusion Service, Queensland, 1945, New Guinea and Papua, 1955-62; Colombo Plan expert on blood transfusion, Thailand, Burma, 1955; mem. Australian Nat. Blood Transfusion Com., 1945—; lectr. social and preventive medicine U. Queensland, 1954—, in surgery, 1963—. Fellow Internat. Soc. Haematology; mem. Coll. Pathologists of Australia, Australian, Asian, Pacific socs. haematology and blood transfusion. Co-author: Handbook of Sterilisation Procedures, 1953. Publs. on biochem. control of transfusions, exchange transfusions; significance of hypoproteinaemia in Beri-Beri, of maternal hypoproteinaemia in hydrops foetalis; designed one of first transfusion services housed in permanent quarters, 1946. Home: 589 Fairfield Rd., Yeronga, Brisbane. Office: 480 Queen St., Brisbane, Queensland, Australia.*

SHAW, Charles Frederick, Am. soil scientist; b. West Henrietta, N.Y., May 2, 1881; s. Frederick Franklin and Mary Anna (Tabolt) S.; B.S. in Agr. Cornell U., 1906; m. Helen Susannah Hosterman, June 19, 1909. Entered bur. soils U. S. Dept. of Agr., as sci. asst., 1906; instr. Pa. State Coll. 1907-09, asst. prof. agronomy, 1909-13; also soil scientist Bur. of Soils in charge soil survey of Pa.; prof. soil technology, U. Cal., 1913—, also in charge soil survey of Cal.; vis. prof. of soils, U. Nanking (China), 1930. Cons. engr. U. S. Reclamation Service, 1919-26, Nat. Commn. Irrigation, Mexico, 1926-28. Fellow Am. Soc. Agronomy, A.A.A.S., Am. Geog. Society. Author: Laboratory Guide in Soil Physics, revised edit., 1911; Soils of Pennsylvania, 1912; Key to Soils of California, 1928; Soils of China, 1931. Died Sept. 12, 1939.

SHAW, Charles Gardner, Am. plant pathologist; b. Springfield, Mass., Aug. 12, 1917; s. Walter Agustus and Hattie Mabel (Hendricks) S.; B.A., Ohio Wesleyan U., 1938; M.S., Pa. State Coll., 1940; Ph.D., U. Wis., 1947; m. Esther Anne Tennant, Aug. 17, 1940; children—Sharon Anne (Mrs. John Richard Taber), Charles Gardner III, Mark Tennant. Instr. bot. lab. Ohio Wesleyan U., 1937-38; lab. asst. Pa. State Coll., 1938-40; faculty U. Wis., 1940-47; faculty Wash. State U., Pullman, 1947—, prof., plant pathologist, 1957—, acting chmn. dept. plant pathology, 1960-61, chmn., 1961—, curator Mycol. Herbarium, 1947—. Mem. Wash. State Agrl. Pesticide Advisory Bd., 1966—. Mem. Am. Phytopath. Soc. (pres. Pacific div. 1966-67), Mycol. Soc. Am., Pacific Northwest Sci. Assn., A.A.A.S., British Mycol. Soc., Internat. Assn. Plant Taxonomists, Phi Beta Kappa, Sigma Xi, Phi Kappa Phi. Research and publs. on taxonomy of phytopathogenic fungi, especially downy mildews; foliar pathogens of forest trees, ecology and physiology of forest tree diseases; air pollution and flourine damage to plants. Home: 312 Howard St., Pullman, Wash. 99163.*

SHAW, Charles Henry, Am. physicist; b. Los Angeles, Jan. 26, 1908; s. William H. and Martha (Mimken) S.; A.B., U. Cal. at Los Angeles, 1930; Ph.D., Johns Hopkins U., 1933; m. Jane Carlisle Thomson, June 26, 1940; children—John William, Ann Carlisle. Nat. research fellow, Cornell U., Ithaca, N.Y., 1934-36; asst. prof. Johns Hopkins U., Balt., 1938-46, sr. physicist Allpied Physics Lab., Silver Spring, Md., 1943-46; asso. prof. physics, Ohio State U., Columbus, 1946-54, prof., 1954—. Fellow Am. Phys. Soc. Research on soft x-ray spectroscopy, structure of liquids by x-ray scattering. Died June 5, 1967.

SHAW, Dennis Frederick, English physicist; b. Teddington, Eng., Apr. 20, 1924; s. Albert and Lily (hill) S.; B.A., Christ Church, Oxford (Eng.) U., 1943, M.A., 1950, D.Phil., 1950; m. Joan Irene Chandler, June 25, 1949; children—Peter James, Margaret Denise, Katherine Joan, Deborah Mary. Jr. sci. officer Ministry Aircraft Prodn., London, 1944-46; staff Clarendon Lab., Oxford, 1946-64, sr. research officer, 1958-64; lectr. dept. nuclear physics Oxford U., 1964—, fellow Keble Coll., 1957—, tutor physics, 1957; vis. scientist CERN, 1961-62. Sci. adviser Home Office, London, 1962—, mem. sci. adv. council, 1965—. Fellow Phys. Soc. London, Zool. Soc. London; mem. Am. Phys. Soc. Author: Introduction to Electronics, 1962; also articles, chpt. in book. Research in nuclear physics; measured radius of helium nucleus with fast neutrons; designed and built deuterium bubble chamber; studied structure of deuteron using high energy protons; collaborated in design of hydrogen and helium bubble chambers; invented automatic detector for radioactive gases; research in sci. aids to crime detection. Home: 29 Davenant Rd., Oxford, Eng.*

SHAW, Eugene Wesley, Am. geologist; b. Delaware, O., July 29, 1881; s. William Bigelow and Irene (Gardner) S.; B.S., Ohio Wesleyan U., 1905, D.Sc., 1927; postgrad. U. Chgo., 1905-07; m. Abbie Potter Haylett, Oct. 18, 1907. With U.S. Geol. Survey most of time, 1907-21, in charge sub-sect. of sedimentation, 1919-20; cons. geologist, 1921-29; chief geologist Iraq Petroleum Co., London, 1929—. Made extensive investigations for oil corps. in various countries; in charge of natural gas valuation U. S. Treasury Dept., 1918-19; Mem. various profl. socs. Author: Coal, Oil and Gas of Foxburg Quadrangle, Pa. (U. S. Geological Survey), 1909; Mud Lumps at Mouth of the Mississippi River, 1912; Natural Gas of North Texas, 1916; Oil Fields of Allen County, Ky., 1919; also about 100 other publs. Explored and mapped 100,000 sq. miles along east base of Andes, S. Am.; engaged in estimation of underground reserves of oil and natural gas, appraisal of oil and gas properties, developed and prospective; studied behavior of streams involved in law suits and problems of river improvements, geologic history of mountain ranges, coasts and deltas of N. and S. Am., Europe, Asia, Africa; conducted studies of geophysics of great sedimentary basins. Died Oct. 7, 1935.

SHAW, George, English naturalist; b. Berton, Eng., Dec. 10, 1751; s. Timothy Shaw; ed. Magdalen Hall, Oxford, 1772; student medicine, Edinburgh, Scotland; M.B., M.D., Oxford, 1787. Ordained deacon, 1774; bot. lectr., Oxford; became asst. keeper natural history sect. Brit. Mus., 1791, keeper, 1807-13. Fellow Royal Soc., 1789, Linnean Soc. (co-founder, v.p.). Author: Speculum linnaeanium, 1790; Museum leverianum, 1792-96; Zoology of New Holland, 1794; Cimelia physica, 1796; General Zoology, 1800-12. Died July 22, 1813.

SHAW, James Headon, biochemist; b. Sharon, Ont., Can., Jan. 1, 1918; s. Merton and Myrtle (Foord) S.; came to U.S., 1939, naturalized, 1949; B.A., McMaster U., 1939; M.S., U. Wis., 1941, Ph.D., 1943, fellow, 1943-45; M.A. (hon.), Harvard, 1955; m. Vera Chapman, Dec. 18, 1943; children—Sandra Yvonne, James Stephen. Mem. faculty Harvard, Boston, 1945—, prof. nutrition Sch. Dental Medicine, 1965—. Mem. coms. NRC, 1953-63. Recipient Research Career award USPHS, 1965. Mem. Am. Inst. Nutrition, Internat. Assn. Dental Research, Soc. Exptl. Biology and Medicine, A.A.A.S., Am. Dental Assn., Sigma Xi, Omicron Kappa Upsilon. Research and publs. on nutrition and oral health; devel. of strains of lab. animals for study of dental caries and peridontal disease. Home: 10 Stiles Terrace, Newton Center, Mass. 02159. Office: 188 Longwood Av., Boston 02115.*

SHAW, John Harrison, physicist; b. Sheffield, Eng., Jan. 25, 1925; s. Herbert Ronald William and Ada (Harrison) S.; B.A., Cambridge (Eng.) U., 1946, M.A., 1949, Ph.D., 1952; m. Elizabeth Wroe, Sept. 10, 1949; children—Jennifer, Peter, Michael. Came to U.S., 1949. Faculty, Ohio State U., Columbus, 1949—, prof. physics, 1964—. Fellow Optical Soc. Am., Am. Meteorol. Soc., Royal Meteorol. Soc. Contbr. numerous articles to tech. jours. Determined amounts carbon monoxide and nitrous oxide present in unpolluted atmosphere; identificed absorption lines in atmospheric gases in infrared solar spectrum; measured absorption properties atmospheric gases in infrared.*

SHAW, Louis Agassiz, Am. physician; b. Sept. 25, 1886; prof. Harvard Sch. Pub. Health; co-inventor Drinker respirator for treatment of infantile paralysis and respiratory failure (John Scott medal 1931); authority on compressed-air illness (bends). Died Aug. 27, 1940.

SHAW, Margery Wayne Schlamp (Mrs. Charles Raymond Shaw), Am. geneticist; b. Evansville, Ind., Feb. 15, 1923; d. Arthur George and Louise (Meyer) Schlamp; A.B., U. Ala., 1945; M.A., Columbia, 1946; M.D., U. Mich., 1957; m. Charles Raymond Shaw, May 31, 1942; 1 dau., Barbara Rae. Instr., U. Alaska, 1951-53, vis. prof., 1955; faculty dept. human genetics U. Mich. Med. Sch., 1958-67; asso. prof. Grad. Sch. Biomed. Scis., U. Tex., Houston, 1967—. Recipient Billings Silver medal A.M.A., 1966. Mem. Am. Soc. Human Genetics, Genetics Soc. Am., Tissue Culture Assn., Phi Beta Kappa, Alpha Omega Alpha. Co-editor: Genetics and the Epidemiology of Chronic Diseases, 1963. Bd. editors Am. Jour. Human Genetics, 1962-68; asso. editor Cytogenetics, 1963—. Research and numerous publs. on structure and function of human chromosomes, effects of chems. on chromosomes, variation in chromosome morphology, localization of genes on specific chromosomes.

SHAW, Sir (William) Napier, English meterologist; b. Birmingham, Eng., Mar. 4, 1854; s. Charles Thomas and Kezia (Lawden) S.; B.A., Emmanuel Coll., Cambridge U., 1876, M.A., Sc.D.; postgrad. U. Berlin; LL.D., Aberdeen, Edinburgh; Sc.D., Athens, Dublin, Harvard, Manchester; m. Sarah Jane Dugdale Harland, 1885. Fellow, Emmanuel Coll., Cambridge U., 1877-1906, univ. lectr., 1887-99, Rede lectr., 1921, coll. senior tutor, 1890-99, asst. dir. Cavendish Lab., 1898; mem. Meteorol. Council, 1897-1905, sec. 1900-05; dir. Meterol. Office, 1905-20; reader in meteorology U. London, 1907-20; Halley lectr. Oxford U., 1918; prof. Royal Coll. Sci., 1920-24. Fellow Royal Soc. (Royal Medal, 1923), 1891; pres. Royal Meteorol. Soc., 1918-19, pres. math. and physics sect. Brit. Assn., 1908, ednl. sect., 1919, hon. fgn. mem. Am. Acad. Arts and Scis., Reale Accademia dei Lincei, Royal Swedish Acad., Norwegian Acad. Sci., Am., Russian geog. socs., Austrian, German meteorol. socs. Author: (with R. Glazebrook) Text-book of Practical Physics; (with R. G. K. Lempfert) The Life History of Surface Currents, 1906; Air Currents and the Laws of Ventilation, 1907; Forecasting Weather, 1911; The Weather Map and Meteorological Glossary, 1916; Manual of Meteorology, 4 vols., 1919-31; The Air and its Ways, 1923; (with J. S. Owens) The Smoke Problem of Great Cities, 1925; The Drama of Weather, 1933. Introduced Milibar and Tephigram in meteorology; studied air circulation, mechanics in atmosphere, upper atmosphere, air pressure. Died London, Mar. 23, 1945.

SHAW, Peter, English physician; b. Eng., 1694; s. Robert Shaw; M.D., Cambridge U., 1752; m. Frances Hyde; 1 dau., Elizabeth Warren. Practiced medicine, Scarborough, then London; physician extraordinary to George II, 1752; became physician in ordinary to George II, 1754, George III, 1760. Fellow Royal Soc., 1752, Royal Coll. Physicans. Author: The Dispensatory of the Royal College of Physicians, 1721; A Treatise of Incurable Diseases, 1723; Praelectiones Pharmaceuticae, 1723; The Juice of the Grape, or Wine preferable to Water, 1724; A New Practice of Physic, 1726; Three Essays in Artificial Philosophy, or Universal Chemistry, 1731; An Essay for introducing a Portable Laboratory, by means whereof all the Chemical Operations are commodiously performed for the purposes of Philosophy, Medicinal Metalurgy, and Family, 1731; Chemical Lectures read in London in 1731 and 1732, and at Scarborough in 1733, for the Improvement of Arts, Trades, and Natural Philosophy, 1734; An Inquiry into the Contents and Virtues of the Scarborough Spa, 1734; Examination of the Reasons for and against the Subscription for a Medicament for the Stone, 1738; Inquiries on the Nature of Miss Stephen's Medicaments, 1738; Essays for the Improvement of Arts, Manufactures, and Commerce, by means of Chemistry, 1761; Proposals for a Course of Chemical Experiments, with a view to Practical Philosophy, Arts, Trade, and Business, 1761. Editor works of Bacon and Boyle; pub. works of Stahl and Boerhaave, also several transls., adaptations. Died Mar. 15, 1763.

SHAW, Robert Harold, Am. climatologist; b. Madrid, Ia., June 26, 1919; s. Matthew McKnight and Lois (Stover) S.; B.S., Ia. State U., 1941, M.S., 1942, Ph.D., 1949; m. Adelaide Melba Urbutt, Jan. 9, 1945; children—Robert, Thomas, Mary Marjorie. Agt., Bur. Plant Industry, Ames, Ia., 1939; agr. research scientist, 1940; jr. agrl. statistician Agr. Marketing Service, Ames, 1941-42; with Ia. State U., Ames, 1946—, prof. climatology, 1957—. Cons., Presdl. Adv. Com. on Weather Control, 1955-56; mem. panel on natural resource sci. Commn. on Edn. in Agy. and Natural Resources, Agrl. Bd., NRC, 1965—. Fellow Am. Soc. Agronomy (past chmn. div. meterology), A.A.A.S. (past v.p. agr. div.); mem. Am. Meteorol Soc. (mem. com. on agrl. meteorology 1964—), Internat. Soc. Biometeorology, Phi Kappa Phi, Sigma Xi. Research and numerous publs. on sci. evaluation weather factors in crop prodn., agrl. climatology. Home: 1431 Duff Av., Ames, Ia. 50010.*

SHAW, Thomas, Am. inventor; b. Phila., May 5, 1838; s. James and Catherine (Snyder) S.; m. Matilda Garber, 1 dau. Apprenticed to machinist, Phila.; patented a gas meter (1st invention), 1858; patented a press mold for glass, gas stove, sewing machine, 1859; supt. Cyclops Machine Works, Phila., circa 1860; supt. Midvale Steel Works, 1867; produced many inventions, including a centrifugal shot making machine, a steam power hammer, spring-lock nut washer, 1867-70; established factory for mfg. his inventions, Phila., 1871; patented some 200 devices, including pressure gauges, pile drivers, hydraulic pumps, a device to detect and measure presence of noxious gases in mines (adopted by several European govts.), 1871-1901. Died Hammonton, N.J., Jan. 19, 1901.

SHAW, Trevor Ian, English, physiologist; b. York, Eng., Mar. 18, 1928; s. Donovan and Mona (Rawes) S.; B.A. (County Major and State Scholar), Cambridge (Eng.) U., 1949, M.A., 1954, Ph.D., 1954; m. Hannah Schmeltzer, Sept. 24, 1954; children—Susan Cordelia, Caroline Imogen. Mem. Spitzbergen Expdn., Cambridge U., 1953; mem. staff Marine Biol. Lab., Plymouth, Eng., 1955-66; prof. zoology Queen Mary Coll., London U., 1966—; vis. asso. prof. U. Cal. at Los Angeles, 1963; vis. prof., Duke, Durham, N.C., 1966. Recipient Scientific medal Zool. Soc., 1966. Mem. Physiol. Soc., Biophys. Soc., Marine Biol. Assn. U.K., Freshwater Biol. Assn. Research and publs. on ionic movements in red blood cells, iodine accumulation in brown sea weeds, active transport of sodium and potassium in squid nerve fibres, perfusion of giant axons, buoyancy studies on squid and gelatinous marine animals. Home: Oakapple House, Little Walden, Saffron, Walden, Essex, Eng. Office: Dept. Zoology, Queen Mary Coll., Mile End Rd., London E. 1, Eng.*

SHCHEGLOV, Vladimir Petrovich, Russian astronomer; b. 1904; grad. Moscow Surveying Inst., 1930; D.Physicotech. Sci. Dir., Tashkent Astron. Obs., 1941-—; prof. chair gen. mechanics Central Asian U., Tashkent, 1948-—. Mem. Uzbek Acad. Scis. (corr.). Research and publs. on practical astronomy and time service, history of astronomy. Address: Central Asian University, ulitsa Karla Marksa 32, Tashkent, Uzbeck SSR, USSR.

SHCHEGLYAEV, Andrei Vladimirovich, Russian heat engr.; b. Oct. 20, 1902; grad. Moscow Tech. Coll., 1926. Faculty, Moscow Tech. Coll., 1926-30; joined All-Union Heat Engring. Sci. Research Inst., 1924; instr. Moscow Inst. Energetics, 1930-48, became prof., 1948; prof. Moscow Tech. Coll., 1934-—. Recipient Stalin Prize, 1948, 52, Red Banner of Labor, 1962. Author: Testing of Steam Turbines, 1937; Regulating Steam Turbines, 1938; Some Problems of Exploiting Steam Turbines, 1947; Steam Turbines, 3d edit., 1955. Research on heat processes of steam turbines, including operation, testing, contributed to devel. new systems for control of turbines. Office: Moscow Inst. Energetics, Moscow, USSR.

SHCHELKIN, Kirill Ivanovich, Russian physicist; b. May 17, 1911; grad. Pedagogical Inst., Simferopol, 1932. After graduation joined staff Inst. Chem. Physics, USSR, Acad. Scis., 1932. Mem. USSR Acad. Sci. (corr.). Research and publs. on devel. gas dynamics in combustion, influence of turbulent flow in an initial mixture on flame acceleration, conditions for transfer of slow burning into detonation; originated theory of spin denotation. Office: Inst. Chem. Physics, USSR Acad. Scis., Vorobévskoye Shosse 2, Moscow, USSR.

SHCHERBAKOV, Dmitrii Ivanovich, Russian geologist; b. Jan. 13, 1893; grad. Simferopol U., 1922. Staff instns. of USSR Acad. Scis.; faculty Leningrad U., 1922; faculty Leningrad Poly. Inst., 1928-32; staff Inst. Geol. Scis., USSR Acad. Scis., 1939-54; became acad. sec. dept. geol. and geographic scis. USSR Acad. Scis., 1953, chmn. Antarctic Commn., 1961-—, became mem. radium expdns. beginning 1914; mem. Pamur expdn., 1928, Tadzhik-Pamir expdn. until 1936. Recipient Order of Lenin; prize World Peace Council, 1960. Mem. USSR Acad. Scis. (mem. presidum). Author: Genetic Types of Tin Deposits in Central Asia, 1936; The Raw Material Resources of Rare Metals in the USSR, 1938; High Temperature Ore Formations of the Central Caucasus, 1946; Prognostic Maps for Magmatogenic Ore Deposits, 1952; The Principles and Methods of Compiling Metallogenetic Maps, 1955. Research and publs. on geology and geochemistry of rare metals, radioactive elements. Home: n. Yakimanka, 3, Moscow, USSR. Office: Joint Antarctic Commn., USSR Acad. Scis., Leninskii Prospekt 14, Moscow, USSR.

SHCHERBAN, Aleksandr Nazarovich, Russian thermodynamicist; b. Dikanka, 1906; grad. Dnepropetrovsk Mining Inst., 1933; D. Tech. Sci. Dep. dir. Inst. Mining, Ukrainian Acad. Scis., 1946-53, chief learned sec. Presidium, 1953-57, v.p., 1957-62; chmn. State Com. for Coordinating Research, dep. chmn. Ukrainian Council Ministers, 1961 . Head Ukrainian delegation 2d Internat. Conf. on Peaceful Uses Atomic Energy, 1958. Mem. Ukrainian Acad. Scis. Author: The Principles of the Theory and Method of Thermal Calculations of Mine Methane, 1953; Air Conditioning in Mines, 1956. Research and numerous publs. on heat exchange in mine workings, mine ventilation, thermodynamics; developer equipment for automatic detection of methane in mines. Address: Soviet Ministry, ulitsa Kirova 12, Kiev, Ukrainian SSR, USSR.

SHCHUKIN, Aleksandr Nikolaevich, Russian radioengr.; b. July 22, 1900; grad. Leningrad Electrotech. Inst., 1927. Faculty, Leningrad Electrotech. Inst., 1929-41, Leningrad Mil.-Naval Acad., 1933-45; staff various research instns, including Leningrad Electrophys. Inst., Central Radio Lab. of Trust for Low Voltage Plants; maj.-gen. engring. Tech. Service. Mem. USSR Acad. Scis. Author: Propogation of Radiowaves, 1940; also articles. Research in propagation of short waves, short wave communication at great distances. Home: Alekseeskogo studgorodka 311 pr. 31, Moscow, USSR. Office: USSR Acad. Scis., Leninski, Prospekt 14, Moscow, USSR.

SHEAR, Murray J(acob), Am. biomed. scientist; b. Bklyn., Nov. 7, 1899; s. Victor J. and Henrietta (Robinson) S.; B.S. in Chemistry, City Coll. N.Y., 1920; M.A., Columbia, 1922, Ph.D. in Chemistry, 1925; m. Rose Roseman, Aug. 14, 1935; children—David Ben, Jonathan, Victory Henry. Chemist, Pease Labs., Inc., N.Y.C., 1922-23; asst. chemistry Columbia, 1923-25; research chemist, then adminstrv. officer pediatric research labs., Jewish Hosp., Bklyn., 1925-31; biochemist Office Cancer Investigation, USPHS, Harvard Med. Sch., 1931-39; biochemist Nat. Cancer Inst., NIH, Bethesda, Md., 1939-51, chief lab. chem. pharmacology, 1951-64, special adviser, 1964-—; instructor of pediatrics at Long Island College of Medicine, 1930-31; fellow Harvard Med. Sch., also Harvard U., 1931-39; cons. biochemistry Childrens Med. Center, Boston, 1948-61; chmn. bioassy panel, com. on growth, NRC, 1946-48; chmn. bd. civil service examiners NIH, 1947-51; chmn. chemotherapy com. Internat. Union Against Cancer, 1954-62, chairman of the finance com., 1958-62, mem. U. S. nat. com., 1961-64; mem. research commn. International Union Against Cancer, 1962-64, sec.-gen., 1964-66. Pres. Bethesda-Chevy Chase Jewish Community Group, 1942. Served with S.A.T.C., 1918. Mem. Am. Assn. Cancer Research (pres. 1960-61), Soc. Exptl. Biology and Medicine, Am. Soc. Biol. Chemists, Am. Soc. Pharmacology and Exptl. Therapeutics, Washington Acad. Medicine, Royal Soc. Medicine, Soc. Italiana di Cancerologia. Clubs: Harvard (Boston); Cosmos (Washington). Spl. research mechanism deposition of bone salts, genesis tumors with chemicals, chemotherapy and immunology of cancer. Home: 5203 Battery Lane, Office: Nat. Cancer Inst., Bethesda, Md. 20014.*

SHEAR, Theodore Leslie, Am. archeologist; b. New London, N.H., Aug. 11, 1880; s. Theodore R. and Mary Louise (Quackenbos) S.; A.B., N.Y. U., 1900, A.M., 1903; Ph.D., Johns Hopkins, 1904; studied Am. Sch. at Athens, 1904-05, U. Bonn, 1905-06; L.H.D., Trinity Coll., Hartford, Conn., 1934; m. Nora C. Jenkins, June 29, 1907 (dec. Feb. 1927); 1 dau., Chloe Louise; m. 2d, Josephine Platner, Feb. 12, 1931; 1 son, Theodore Leslie, Jr. Instr. Greek and Latin, Barnard Coll., N.Y.C., 1906-10; asso. in Greek, Columbia, 1911-23; lectr. on art and archeology Princeton U., 1921-27, prof. classical archeology since 1928; also curator of classical art in Museum of Hist. Art. Trustee Am. Sch. of Classical Studies, Athens, 1936-42; dir. excavation of Athenian Agora, 1930-42. Fellow Am. Acad. Arts and Scis., Am. Philos. Soc.; mem. Archeol. Inst. Am., Am. Philol. Assn., Am. Oriental Soc., Am. Numismatic Soc., Royal Soc. of Arts (London), Hellenic Soc. (London), Assn. des Etudes Grecques (Paris), Am. Geog. Soc., Phi Beta Kappa; hon. mem. Greek Archeol. Soc. (Athens). Conducted archeol. excavns. at Cnidus, 1911, Sardis, 1922, Corinth, 1925-31, Athens, 1931-40. Author: Influence of Plato on St. Basil, 1907; Sardis—Architectural Terracottas, 1925; Corinth—The Roman Villa, 1930; also numerous articles in archeol. periodicals. Died July 3, 1945.

SHEARD, Charles, Am. biophysicist; b. Dolgeville, N.Y., May 27, 1883; A.B., St. Lawrence, 1903, hon. Sc.D., 1930; A.M., Dartmouth, 1907; Ph.D. in Physics, Princeton, 1912; D.O.Sc., Los Angeles Coll. Optometry, 1953. Instr. pub. sch., N.Y., 1903-05; asst. physics dept. Dartmouth, 1905-07; instr. Ohio State U., 1907-09, asst. prof., 1909-14, prof., dir. applied optics, 1914-19; physiol. opticist, head div. ocular interests Am. Optical Co., 1919-24; dir. div. physics and biophysics research Mayo Clinic, 1919-24; prof. biophysics Mayo Found., Minn., 1924-49, emeritus dir., emeritus prof., 1949-—; mem. Mayo Aero. Med. Unit, 1941-45; prof. Rochester Jr. Coll., 1947-—; distinguished vis. lectr. Grad. Sch. Medicine, Tulane U., 1948-—; cons. Recipient Burdick award and medal, 1937, Nekon Achievement award 1951. Civilian bd. consultants Bur. Medicine and Surgury, U. S. Dept. Navy, 1947-53. Diplomate Am. Bd. Opticianry (pres. 1947-52, became sec. 1952). Fellow Am. Acad. Optometry, Am. Acad. Ophthalmology (hon.), Ophthal. Soc. U.K. (hon.); mem. A.A.A.S., Am. Optical Soc., Am. Physiol. Soc., Soc. Plant Physiologists, Soc. for Exptl. Biology and Medicine, Soc. Clin. Pathologists, Nat. Soc. for Prevention of Blindness, Assn. for Research in Ophthalmology, Optometry Assn. (hon.). Developed (with Arthur Hawley Sanford), photoelectric colorimeter for determining hemoglobin (Sheard-Sanford photelometer), 1930. Founder, Sheard Found. Edn. and Research in Vision, Grad. Sch., Ohio State U., 1943; developed tests in ocular refraction; studies in bioelectric potentials and currents, spectrophotometry and photoelectrometry in biology, dark adaptation, radiant energy effects in plant and animal tissues, energy exchanges between body and its environments, relation between accommodation and convergence. Died 1963.

SHEARER, Newton Henry, Jr., Am. chemist; b. Lynchburg, Va., Nov. 29, 1920; s. Newton Henry and Louise (Layne) S.; B.A. in Chemistry, Lynchburg Coll., 1941; M.S., U. Va., 1944, Ph.D. in Organic Chemistry, 1946; m. Betty Lou Johnson, May 14, 1955; children—Layne Elizabeth, Newton Henry III. With Tenn. Eastman Co. (Kingsport), 1946-61, Eastman Research A.G., Zurich, Switzerland, 1961-63, research asso. Tenn. Eastman Co., Kingsport, 1963-65, mgr. research labs. tech. information services, 1965-—. Mem. Am. Chem. Soc., Sigma Xi, Alpha Chi Sigma. Research, publs., patents on synthesis of antimalarial compounds, synthesis of vinyl monomers, vinyl polymers, organophosphorus chemistry catalysts for olefin polymerization, synthesis of polyesters and polyamides. Home: 2009 Westwind Dr., Kingsport 37660. Office: Research Labs., Tenn. Eastman Co., Kingsport, Tenn. 37662.*

SHEBANOV, Filipp Vasilevich, Russian phthisiologist; b. 1897; grad. Med. Faculty, 1st Moscow U., 1921; D.Med. Sci., 1944. Intern, dept. head, dep. dir., dir. Moscow Oblast Tb Research Inst., 1928-53; asst. Central Postgrad. Med. Inst., Moscow, 1936-41, lectr., 1941-48, prof., 1948-49, head chair Tb, 1949-52; head chair Tb 1st Moscow Med. Inst., 1952-—, also chmn. Tb Problems Commn. Former chmn. learned med. council, collegium mem. RSFSR Ministry Health; del. Internat. Tb Congress, Leipzig, 1951, Prague, 1953, Sofia, 1954, Paris, 1956, New Delhi, 1957. Decorated Order of Lenin. Mem. USSR Acad. Med. Scis. (corr.), Internat. Union against Tb (USSR rep. exec. com.), All-Union Soc. Phthisiologists (chmn.). Author: Tubercular Empyemas, 1946; Collapsotherapy of Pulmonary Tuberculosis, 1950; co-author, editor: Treatment of Tuberculosis Patients with PASA. Editor: Problems of Tb. Address: 1st Moscow Medical Institute, B. Pirogovskaya ulitsa 2-6, Moscow, USSR.

SHECUT, John Linnaeus Edward Whitridge, Am. physician; b. Beaufort, S.C., Dec. 4, 1770; s. Abraham and Marie (Barbary) S.; M.D., Coll. of Phila., 1791; m. Sarah Cannon, Jan. 26, 1792; m. 2d, Susanna Ballard, Feb. 7, 1805; 9 children. An early experimenter with use of electricity in treatment of yellow fever and crippled limbs, suggested that yellow fever was in part caused by lack of electricity in atmosphere; discouraged blood-letting and use of mercury as drug; organizer Antiquarian Soc. of Charleston (S.C.), 1813, incorporated as Literary and Philos. Soc. of S.C., 1814. Author: Flora Carolinaeenis (most thorough work on botany of S.C. then available), 1806. Died Charleston, June 1, 1836.

SHEDD, Solon, Am. geologist; b. Ill., May 25, 1860; s. Frank and Emily L. (Olin) S.; grad. Ore. State Normal Sch., Monmouth, 1889; A.B., Stanford U., 1896, A.M., 1907, Ph.D., 1910; m. Jeannette Wimberly, June 4, 1907. Tchr. natural scis. Ore. State Normal Sch., 1890-94; prof. geology and mineralogy Wash. State Coll., 1896-1925; asst. state geologist Wash. Geol. Survey, 1909-13, state geologist, 1921-25; curator Branner Meml. Geol. Library, Stanford U., since 1925. Acting asso. prof. geology Stanford U., summer and autumn 1921, summer 1922. Fellow Geol. Soc. Am.; mem. Am. Inst. Mining and Metall. Engrs., Seismol. Soc. Am., A.A.A.S., Am. Ceramic Soc. Author of reports on iron ores, building and ornamental stones, clays, cement materials, of Wash., bibliography of geology and mineral resources of Cal. Died Mar. 4, 1941.

SHEEHAN, J. Eastman, surgeon; b. Dublin, Ireland; s. Daniel Stanislaus and Catherine (Eastman) S.; M.D., Yale, 1908; studied at Oxford, London, Paris, Bern, Heidelberg, Berlin, Budapest, Vienna; m. Anastasia Dwyer, May 14, 1914; 1 dau., Marguerite Virginia. Prof. plastic reparative surgery N.Y. Polyclinic. Med. Sch. and Hosp.; surgeon to Morrisania and St. Clares Hosps., N.Y., lectr. to Internat. Clinic, Paris, 1922-—; guest lectr. univs. of Utrecht, Milan, Parma, Belgrade, Istanbul, Ankara, Rome, Paris, Dublin, Bucharest, Bruxelles, and Guy's Hosp., London; hon. surgeon, N.Y. police dept.; physician Nationalist Army, Spain; worked at Oxford, 1942; posted to burns center, Glasgow, 1943; worked under direction Royal Air Force Command, 1944. Apptd. observer at Parliamentary and Congressional groups, Bermuda, Washington, and Ottawa, Can., 1946. Fellow Royal Society of Medicine (London), A.C.S., N.Y. Acad. Medicine; mem. Am. Soc. of Oral and Plastic Surgeons (pres. 1935-36), French Soc. of Plastic and Reparative Surgeons (hon. pres.), French Surg. Soc., Société Française d'Oto-Rhino-Larynologie, Société Belge d'Oto-Rhino-Larynologie, Jornadas Medicas de Madrid (hon.), Les Journées Médicales de Bruxelle, Société Academique d'Historie Internationale (corr. editor), Bruxelles Médical (corr. editor), Royal Acad. Spain (corr. editor), Corr. editor: Bruxelles Med. Chirurgie Structive de Bruxelles La Crirurgia Plastica, Roma, Le Journal de la Chirurgie Plastique et Restauratrice, Paris. Author: Plastic Surgery of the Nose, 1925, rev., 1937; Plastic Surgery of the Orbit, 1927; Manual of Plastic Reparative Surgery, 1938; Surgery for War, 1944; General and Plastic Surgery, 1945. Pioneer in applying color to motion pictures of surgical operations; inventor method of rapid skin grafting with aid of tulle and glue. Died Jan. 8, 1951.

SHEEHAN, John Clark, Am. chemist, educator; b. Battle Creek, Mich., Sept. 23, 1915; s. Leo Clark and Florence B. (Green) S.; B.S., Battle Creek Coll., 1937; M.S., U. Mich., 1938, Ph.D., 1941; m. Marion M. Jennings, June 2, 1941; children—John Clark, David E., Elizabeth L. Research asso. U. Mich., 1941; sr. research chemist Merck & Co., Inc., 1941-46; asst. prof. chemistry Mass. Inst. Tech., 1946-49, asso. prof., 1949-52, prof. chemistry since 1952; cons. E. I. duPont de Nemours & Co., Bristol Labs., Arthur D. Little, Inc. Recipient Am. Chem. Soc. award in pure chemistry, 1951. Fellow Am. Acad. Arts and Scis., N.Y. Acad. Arts and Scis., Chem. Soc. (London, Eng.); mem. Am. Chem. Soc. (chmn. organic div. 1959), Nat. Academy Sciences, Sigma Xi. Mem. bd. editors Organic Synthesis, Jour. of Organic Chemistry, Archives of Biochemistry and Biophysics. Research in chemistry of penicillin, peptides, alkaloids, steroids, the synthesis of high explosives. Home: 10 Moon Hill Rd., Lexington 73, Mass. Office: Mass. Institute of Technology, Cambridge 39, Mass.*

SHEEHAN, Thomas John, Am. ornamental horticulturist; b. Bklyn., Apr. 13, 1924; s. Thomas and Emma (Terski) S.; A.B. in Botany, Dartmouth, 1948; M.S., Cornell U., 1950, Ph.D., 1952; m. Marion Elizabeth Ruff, June 23, 1950; children—Thomas Jerome, Peter Wilson, Marian Lynn. Asst. horticulturist Ga. Agr. Expt. Sta., 1952-54; staff U. Fla., Gainesville, 1954-—, ornamental horticulturist, 1967-—; vis. floriculturist U. Hawaii, 1962-63. Recipient Charles Botany prize Dartmouth, 1948. Mem. Fla. State Hort. Soc. (past v.p. ornamentals, dir. 1968-—, Silver medal

1958), Am. Soc. for Hort. Sci., Am. Hort. Soc., Am. Orchid Soc., Palm Soc., Fairchild Tropical Garden, Sigma Xi, Pi Apha Xi. Research, publs. on effects of photoperiod, temperature, nutrition and growth regulating chemicals on prodn. and keeping quality of floricultural crops, especially Orchidaceae. Home: 3823 S.W. 3d Av., Gainesville, Fla. 32601.*

SHEELY, Clyde Quitman, Am. chemist; b. Pelahatchie, Miss., July 12, 1904; s. Charlie Quitman and Ida (Marshall) S.; B.A. with Honors, Miss. Coll., 1925; M.S., La. State U., 1926; Ph.D., Ohio State U., 1930; m. Katherine T. Baskervill, Sept. 3, 1930; children—Clyde Quitman, William B., Robert A. Mem. faculty Miss. State U., State College, 1929—, now prof. chemistry. Exec. sec. Miss. Acad. Scis., 1940—. Fellow A.A.A.S.; mem. Am. Chem. Soc., Sigma Xi, Phi Kappa Phi, Gamma Alpha. Author: Freshman Chemistry Laboratory Manual, 1952; also articles. Research on vapor phase oxidation of hydrocarbons, aldehydes from naptha, zeolite softening of lime-treated water, relations of petroleum refining agts., also corrosion effects to naptha solutions of sulphur, thermo-allotropic modifications of sulfur. Home: Drawer CQ, State College, Miss. 39762.*

SHEEPSHANKS, Richard, English astronomer; b. Leeds, Eng., July 30, 1794; s. Joseph and Anne (Wilson) S.; fellow Trinity Coll., Cambridge, 1817, M.A., 1819; called to bar, 1825; took orders in Ch. of Eng., 1828; pursued sci. vocation; sci. adviser to Edward Troughton in south equatoreal case; mem. commns. on weights and measures, 1838, 43. Fellow Royal Soc., 1830; mem. Astron. Soc. (sec. from 1829). Author: (pamphlet) Letter in Reply to the Calumnies of Mr. Babbage, 1854; other pamphlets, 1845, also papers. Contbr. to Penny Cyclo. Determined longitudes of Antwerp and Brussels, 1838, Valentia, Kingstown, Liverpool, 1844; his reconstructed standard of length adopted, 1855; devised method of driving an equatoreal by clockwork. Died Aug. 4, 1855.

SHEETS, Herman Ernest, mech. engr.; b. Dresden, Germany, Dec. 24, 1908; M.E., U. Dresden, 1934; D.Tech. Scis., in Applied Mechanics, U. Prague (Czechoslovakia), 1936; m. Norma Sams, Oct. 17, 1942; six children. Chief engr. Chamberlain Research Corp., East Moline, Ill., 1939-42; dir. research St. Paul Engring. & Mfg. Co., 1942-44; project engr. Elliott Co., Jeannette, Pa., 1944-46; engring. mgr. Goodyear Aircraft Corp., Akron, O., 1946-53; chief research and devel. engr. Gen. Dynamics/Electric Boat, Groton, Conn., 1953-57, dir. research and devel., 1957-66, v.p. engring. and research, 1966—. Cited for work on Manhattan project, 1945. Fellow Nat. Acad. Engring., Am. Inst. Aeros. and Astronautics (asso.); mem. Am. Soc. M.E., Am. Nuclear Soc., A.A.A.S., Marine Tech. Soc., N.Y. Acad. Scis., Soc. Naval Architects and Marine Engrs., Pi Tau Sigma. Research and devel. power plants, applied mechanics and underwater tech. Home: 87 Neptune Dr., Mumford Cove, Groton 06340. Office: Gen. Dynamics Corp., Electric Boat Div., Groton, Conn. 06340.*

SHEETS, Raymond Franklin, Am. physician; b. Bentley, Ill., June 15, 1914; s. Raymond Franklin and Stella (McCallister) S.; A.B., Carthage Coll., 1936; M.S., U. Ill., 1940, M.D., 1940; m. Viola Miriam Smeining, Jan. 7, 1946; children—Peter Raymond, Michael Fredrick, Cordelia Ann. Faculty, U. Ia., Iowa City, 1949—, prof. internal medicine, 1959—, chmn. div. hematology dept. internal medicine, 1968—. Cons. VA Hosp., Iowa City, 1952—. Diplomate Am. Bd. Internal Medicine. Mem. A.M.A., A.C.P., Am. Soc. Hematology, Internat. Soc. Hematology, A.A.A.S., Soc. for Exptl. Biology and Medicine, N.Y. Acad. Medicine, U. Ia. Research Club, Central Soc. for Clin. Research. Contbg. author: The Heart and Lungs, 1959; Management of the Aged Surgical Patient, 1960; Modern Treatment, 1964. Research and publs. on mechanisms of red blood cell destruction in patients, kinetics of red cell destruction, hemolytic factor in pernicious anemia, sulfhydryl groups and reagents in red blood cells; erythroleukemia; May Hegglin anomoly; Vitamin B12. Home: 323 Koser Av., Iowa City 52240.*

SHEFER, David Grigorevich, Russian neuropathologist, neurosurgeon; b. 1898; grad. Med. Faculty, Saratov U., 1922; D.Med. Sci., 1936. Intern, Nervous Diseases Clinic, Astrakhan Med. Inst., 1922-25; asst. Clinic Nervous Diseases and Neurosurgery, Rostov-on-Don Med. Inst., 1925-37; prof., 1937—; head chair nervous diseases and neurosurgery Sverdlovsk Med. Inst., 1937—; sci. dir. Sverdlovsk Research Inst. Spa Treatment and Physiology. Mem. Sverdlovsk Soc. Neuropathologists, Psychiatrist and Neurosurgeons (chmn.), All-Russian (bd. mem.), All-Union (bd. mem.) socs. neuropathologists and psychiatrists, All-Union Soc. Neurosurgeons (bd. mem.). Author over 150 works, including X-Rays and the Central Nervous System, 1936; Diencephalic Syndromes, 1962; co-author: Diagnosis and Treatment of Injuries to the Peripheral Nerves, 1944. Mem. editorial council Problems on Neurosurgery, Korsakov Jour. Neuropathology and Psychiatry; co-editor Large Med. Ency., 2d edit. Address: Sverdlovsk Medical Institute, ulitsa Kommunarov 1, Sverdlovsk, RSFSR, USSR.

SHEFFER, Howard Eugene, Am. chemist; b. Schenectady, N.Y., Oct. 3, 1918; s. Roy Burgess and Florence (Drake) S.; B.S., Union Coll., 1939; M.S., Rensselaer Poly. Inst., 1940; Ph.D., Cornell U., 1943; m. Marjorie Reed, June 24, 1945; children—Dorothy, Roger, Alison, Peter, Nancy, Jane. With Union Carbide and Carbon Co., 1943-45; mem. faculty Union Coll., Schenectady, 1945—, prof. chemistry, 1960—. Cons. Schenectady Chem. Co., 1947—. Mem. Am. Chem. Soc., Sigma Xi, Phi Lambda Upsilon. Research and publs. in field of high temperature polyester wire enamels; patentee. Home: 14 Parkwood Dr., Burnt Hills, N.Y. 12027.*

SHEFFNER, Aaron Leonard, Am. biochemist, nutritionist; b. Chgo., Sept. 20, 1923; s. Nathan S. and Anna (Dolgin) S.; B.S., U. Chgo., 1943, M.S., 1946; Ph.D., U. Ill., 1951; m. Ruth Ellen Strauss, May 19, 1946; children—Paul W., Steven E., Susan I. Research asso., U. Chgo., 1946-49; asst. prof. So. Ill. U., 1951-54; charge biochemistry Nutrition br., tech. coordinator radiation preservation foods program Quartermaster Food and Container Inst., Chgo., 1954-57; biochemistry group leader nutritional, biochemical research Mead Johnson Research Center, Evansville, Ind., 1957-59, sect. leader, 1959-61, asst. dir. dept. nutritional biochemistry, 1961-62, dir., 1962-67, director of physiological chemistry, 1967—. Recipient Pres. award Mead Johnson & Co., 1961. Nutrition Found. fellow, 1949-50; USPHS Research fellow, 1950-51. Fellow N.Y. Acad. Sci., A.A.A.S.; mem. Am. Soc. Biol. Chemists, Am. Inst. Nutrition, Am. Fedn. Clin. Research, Am. Chem. Soc., Soc. Exptl. Biology and Medicine. Contbr. chpts. to Ency. of Chemistry, 1957, Internat. Ency. Pharmacology, 1967. Author: (with H. Spector) Radiation Preservation of Food, 1957; Newer Methods of Nutritional Biochemistry, 1966; also numerous articles. Research on devel. of a mucolytic used in standard therapy of pulmonary disease; PDR index for protein evaluation; tissue amino acids; enzyme synthesis in yeasts. Home: 1125 Glen Moor Ct., Evansville 47715. Office: 2404 Pennsylvania Av., Evansville, Ind. 47721.*

SHÉINDLIN, Aleksándr Efímovich, Russian thermophysicist; b. 1916; grad. Moscow Energetics Inst., 1937. Engr., Factory Constructor's Bur., 1937-39; faculty Moscow Energetics Inst., 1939-41, 45—, prof., 1955—. Recipient Lenin Prize, 1959. Research and publs. on thermo-phys. properties of water and water vapor under high parameters.

SHELDON, Eric, physicist; b. Pilsen, Bohemia, Oct. 24, 1930; s. Robert Bernard and Martha (Martin) S.; B.Sc., London (Eng.) U., 1951, B.Sc. in Physics, with honors, 1952, Ph.D., 1955; m. Sheila Harper, July 8, 1959; 1 son, Adrian. Lectr., research asso. Acton Tech. Coll., London, 1952-55; asso. physicist IBM Research Lab., Switzerland, 1957-59; research asso. E.T.H., Zurich, Switzerland, 1959-62, lectr., 1962-64, prof., 1964—; vis. prof., NSF fellow Va. U., 1968-69. Mem. Convocation London U., 1951—. Fellow Inst. Physics; asso. Royal Inst. Chemistry; mem. Chem. Soc. London, Royal Instn. Gt. Britain, A.A.A.S. Author: (with R. Szostak and P. Marmier) Kernphysik I, 1960, Kernphysik II, 1961; with U. Marmier) Physics of Nuclei and Particles, 1967; also articles. Research in cathodic electrode processes, theoretical nuclear physics especially elucidation of low-energy reaction mechanisms via angular distbn. and correlation studies of scattered and emergent particles, nuclear transfer mechanisms in heavy-ion reactions, applications of group theory.*

SHELDON, Huntington, physician, pathologist; b. N.Y.C., Jan. 14, 1930; s. Huntington D. and Magda (Merck) S.; B.A., McGill U., 1951; M.D., Johns Hopkins U., 1956; m. Suzanne Gross, June 11, 1955; children—Karen, Jennifer. Instr. pathology Johns Hopkins U., Balt., 1956-59, research fellow in orthopedic surgery, 1957-59; faculty McGill U., Montreal, Can., 1959—, prof. pathology, 1966—. Markle scholar in acad. medicine, 1959-64; sr. sci. cons. biol. scis. pavilion Montreal Worlds Fair, 1967. Mem. Am. Soc. Cell Biology, Soc. Exptl. Pathology, Internat. Acad. Pathology, Biophys. Soc. Author: (with S.M. Kurtz) Electron Microscopic Anatomy, 1964; also articles. Application histochemistry methods to preparations for electron microscopy, use of electron microscope to study effect of vitamin deficiencies on cellular structure, to exptl. pathology; use of autoradiographic and cell fractionation methods for study of adipose and other tissues. Home: 2862 Hill Park Circle, Montreal, Que., Can.*

SHELDON, John Lewis, Am. botanist, bacteriologist; b. Voluntown, Conn., Nov. 10, 1865; s. Samuel H. and Lucy A. (Lewis) S.; B.Sc., Ohio No. U., 1895, M.Sc., 1899; B.Sc., U. Neb., 1899, A.M., 1901, Ph.D., 1903; m. Clara Adams Fleming, Aug. 21, 1907; 1 son, Earl Fleming. Tchr. pub. schs., Conn., 1885-90, 1895-97; instr. Mt. Hermon (Mass.) Sch., 1892-94; instr. botany, prep. sch. U. Neb., 1898-99; acting head dept. biology Neb. State Normal Sch., 1899-1900; instr. botany U. Neb., 1901-03; prof. bacteriology W.Va. U., also bacteriologist, Agrl. Expt. Sta., 1903-07, prof. botany and bacteriology, 1907-13, botany, 1913-19. Fellow A.A.A.S., Bot. Soc. Am.; mem. Am. Phytopathol. Soc., Am. Genetic Assn., Sigma Xi, Phi Beta Kappa. Contbr. to bot. and agrl. publs.; investigator in plant pathology. Died Jan. 15, 1947.

SHELDON, Walter Herman, pathologist; b. Berlin, Germany, Feb. 21, 1911; s. Herman L. and Gertrude (Hepner) S.; student U. Heidelberg (Germany), 1929-31, U. Berlin (Germany), 1931-33; M.D., U. Catania (Italy), 1935; m. Marjorie Morton, June 16, 1940; children—Carol, John Walter, Barbara. Came to U.S., 1938, naturalized, 1942. Faculty Emory U. Sch. Medicine, Atlanta, 1943-60, prof. pathology, chmn. dept., 1949-60; prof. pathology Johns Hopkins Sch. Medicine, Balt., 1960—; chief pathologist Grady Meml. Hosp., Atlanta, 1943-54, Emory Hosp., Atlanta, Ga., 1949-60. Member American Association of Pathologists and Bacteriologists, Internat. Acad. Pathology, Am. Fedn. Clin. Research, Pan. Am. Med. Assn., So. Soc. Clin. Investigation, Soc. for Exptl. Biology and Medicine. Research, numerous publs. on inflammation, normal mechanism and interactions processes in inflammation. Home: 646 Charles St. Av., Towson, Md. 21204.

SHELDON, Warner Franklin, Am. physician; b. Milw., Aug. 2, 1910; s. C. Frank and Clara Belle (Warner) S.; B.S., St. Lawrence U., 1931; M.D., McGill U., Montreal, Que., Can., 1937; m. Margaret McKay, June 25, 1937; children—Margo (Mrs. R.T.A. Ross), Wendy (Mrs. James W. Wood), Lura (Mrs. William C. Clark), Frank W., Warner R. Faculty, U. Pa., Sch. Medicine, Phila., 1943—, prof. pathology, 1954-68, asso. chmn. dept., 1959-68; pathologist Luther Hosp., Eau Claire, Wis., 1968—. Chief Atomic Bomb Casualty Commn., Hiroshima, Japan, 1965-66; dir. Mt. Desert Island Biol. Lab., Salisbury Cove, Me., 1950-56, trustee, 1949—. Diplomate Nat. Bd. Med. Examiners, Am. Bd. Pathology. Mem. Am. Heart Assn., A.A.A.S., Path. Soc. Phila. (pres. 1957-58), Assn. Am. Med. Colls., Am. Assn. Pathologists and Bacteriologists. Research and publs. on cardiovascular pathology and physiology, effects radiation. Home: 351 Country Club Lane, Altoona, Wis. 54720. Office: Luther Hosp., 310 Chestnut St., Eau Claire, Wis. 54701.*

SHELDON, William Gulliver, Am. wildlife biologist; b. N.Y.C., Jan. 13, 1912; s. Charles and Louisa (Gulliver) S.; B.A., Yale, 1933; M.S., Cornell U., 1946, Ph.D., 1948; m. Louise Farwell Taylor, June 27, 1943; children—Charles, II, Margaret Soule. Wildlife biologist No. B.C., Can., 1934-35, W. China, Am. Mus. Natural History, Cornell U., 1945-48; leader Mass. coop. wildlife research unit U.S. Bur. Sport Fisheries and Wildlife, U. Mass., Amherst, 1948—. Recipient Harold Worthington Conservation award Western Mass., Hampshire County League Conservationists. Mem. Mass. Assn. Town Conservation Commns. (dir. 1965—), Eastern Bird Banding Assn. (councillor 1956-60), Wildlife Soc., Am. Soc. Mammalogists, Am. Com. for Internat. Wildlife Protection, Am. Ornithologists Union, Wilson Ornithology Soc., Am. Inst. Biol. Scis., Ecol. Soc., Sigma Xi. Author: The Book of the American Woodcock, 1967; also numerous articles. Research on life history and ecology of various vertebrates. Home: 874 N. Pleasant St., Amherst, Mass. 01002.*

SHELESNYAK, Moses Chiam, biodynamist; b. Chgo., June 6, 1909; s. Jonas and Fegel (Leavitt) S.; B.A. cum laude in Zoology, U.. Wis., 1930; postgrad. Alliance Francaise, Paris, 1929; Ph.D., Columbia U., 1933, postgrad. Coll. Phys. and Surg., 1933-35; postgrad. N.Y. Sch. Social Work, 1938-39; m. Roslyn Benjamin, Jan. 28, 1942; children—Betty Jane (Mrs. Franz Sondheimer), Henry Lawrence. Instr., Chgo. Med. Sch., 1935-36; lectr. human growth New Coll., N.Y., 1936-37; research asso. Mt. Sinai Hosp., N.Y.C., 1936-40, Beth Israel Hosp., N.Y.C., 1940-42; dean boys Hebrew Orphan Asylum, N.Y.C., 1938-40; head environmental, physiology and ecology br. U. S. Office Naval Research, 1946-49, acting head biophys. br., 1946-47; lectr. Johns Hopkins 1949-50; dir. Balt.-Washington Office, Arctic Inst. N.Am., 1949-50; sr. scientist Weizmann Inst. Sci., Rehovoth, Israel, 1950-57, faculty, 1957;—, prof., head dept. biodynamics, 1959—; vis. prof. College de France, 1960. Mem. expert adv. panel on human reprodn. WHO, 1964—; mem. hon. adv. panel human ecology of arid zones UNESCO, 1954—. Friedsam Research fellow Beth Israel Hosp., N.Y.C., 1940-42; Sir Simon Marks fellow, U.K., 1957-58; recipient Oliver Bird prize, 1958. Fellow A.A.A.S., Arctic Inst. N.Am., Eugenics Soc., Soc. for Research in Child Devel., Société Royale Bèlge Gynecologie et d'Obstetrique (hon. fgn. corr.), Societa Italiana per il Progresso della Zootecnicna; mem. Internat. Soc. for Research on Biology of Reprodn. (incorporator 1966), Aero. Med. Assn., Internat. Planned Parenthood Fedn. (research com.), Am. Physiol. Soc., Am. Polar Soc., Am. Soc. for Study Sterility (corr.), Arctic Circle, Assn. Advancement Sci. Israel (past mem. council). Biochem. Soc. Israel, Brit. Glaciological Soc., Ecol. Soc. Am., Am., Israel, Internat. (exec. council) endocrine socs., European Soc. for Drug Toxicity, Am., Israel chem. socs., Israel Fertility Assn. (v.p. 1965-67), Israel Soc. for Exptl. Biology and Medicine (past pres.), Physiol and Pharmacological Soc. Israel (past mem. exec. council), Soc. for Research in Child Devel., Interdisciplinary Brain Research Orgn. (neuroendocrinology panel), Sigma Xi. Research and numerous publs. in nidation, pregnancy, growth, devel., biol. bases for birth control, human ecology. Home: Beit No. 1, Neveh Weizmann. Office: Dept. Biodynamics, Weizmann Inst. Sci., Rehovoth, Israel.*

SHELFORD, Victor Ernest, Am. zoologist; b. Chemung, N.Y., Sept. 22, 1877; B.S., U. Chgo., 1903, Ph.D. in Zoology, 1907; m. 1907; 2 children. Asst., tutor zoology, W.Va., 1900-01; asst. U. Chgo., 1904-07, asso., 1907-09, instr., 1909-14; asst. prof. U. Ill., 1914-20, asso. prof., 1920-27, prof., 1927-46, emeritus prof., 1946—, in charge dept., 1939-41; biologist in charge research labs. Ill. Natural History Survey, 1914-29; in charge marine ecology Puget Sound Biol. Sta., 1914-30. Chmn. sci. adv. bd. Grassland Research Found., 1959; chmn. com. wildlife NRC, 1931-36, mem. com. ecology of grasslands, 1932-39. Fellow Entomol. Soc. Am.; mem. A.A.A.S., Am. Soc. Naturalists, Am. Soc. Zoologists, Am. Soc. Mammals, Ecol. Soc. Am. (past pres.), Wilson Ornithol. Soc., Bot. Soc. Am., Brit. Ecol. Soc., Natural History Soc. Mexico (corr.). Author: Animal Communities in Temperate America, 1913; Laboratory and Field Ecology, 1929; Naturalist's Guide to the Americas, 1925; (with F. E. Clements) Bio-Ecology, 1939. Research on life history, color pattern of tiger beetles; animal ecology, bioecology; marine communities; light effects and measurement; physiol. life histories; climate simulation apparatus.

SHELINE, Glenn Elmer, Am. radiologist; b. Flint, Mich., Mar. 31, 1918; s. Charles Earl and Onale (Newman) S.; B.S., U. Cal., Berkeley, 1939, Ph.D. in physiology, 1943; M.D., U. Cal., San Francisco, 1948; m. Patricia Brinton, June 14, 1948; 1 dau., Moira Lynn. Investigations plutonium chemistry Manhattan Dist. Engrs., 1942-45; NRC fellow, 1949-51; with dept. radiology U. Cal. Med. Sch., San Francisco, 1955—, prof., 1964—. Mem. San Francisco Med. Soc., Cal., Am. med. assns., Am. Thyroid Assn., Am. Radium Soc., Radiol. Soc. N.Am., Am. Roentgen Ray Soc., Soc. Nuclear Medicine. Research, publs. on chemistry of plutonium; use of radiation in treatment of malignancies, endocrine disorders; study of central nervous system tumors. Home: 2997 Mariposa Dr., Burlingame, Cal. 94010.*

SHELINE, Raymond Kay, Am. nuclear chemist, physicist; b. Port Clinton, O., Mar. 31, 1922; s. Raymond Kaiser and Rozena (Huard) S.; B.S. summa cum laude, Bethany Coll., 1943; Ph.D., U. Cal., Berkeley, 1949; m. Yvonne Faith Engwall, June 9, 1951; children—Yvette Ingrid, Raymond Kenneth, Jonathan Lee, Hans Eric, Rebecca Ruth, Martin Engwall, Christian Thomas. Faculty, Fla. State U., Tallahassee, 1951—, prof. chemistry, 1955—, prof. chemistry, physics, 1958—, chmn. Nuclear Sci. Program, 1959—; research prof. Inst. Theoretical Physics, Copenhagen, Denmark, 1955-58. Cons. Oak Ridge Nat. Lab., 1959—, Los Alamos Sci. Lab., 1961—, NSF, 1963—; mem. Nat. Acad. Sci. Post Doctoral-NSF fellowship com., 1960-63. Recipient Spl. citation for work on Atom Bomb Project Sec. Def., 1945, Silver Bowl from Niels Bohr for devel. nuclear chemistry lab. at Inst. Theoretical Physics, 1958, citation Am. Inst. Physics, 1963. Fulbright fellow Denmark, 1955-56; Guggenheim fellow Denmark, 1955-58, 64; Egyptian Nat. lectr. Ein Shams U., Cairo, 1956, Fulbright lectr., Europe, 1955-58. Fellow Am. Phys. Soc.; mem. Am. Chem. Soc., Sigma Xi. Research, numerous publs. in nuclear spectroscopy, models and structure; molecular spectroscopy, especially structure of metal carbonyls and their derivatives. Home: Route 4, Box 621, Tallahassee, Fla. 32301.*

SHELLABARGER, Claire J., Am. biologist; b. College Corner, Ohio, Oct. 23, 1924; s. Oliver Newton and Yvonne (Miller) S.; A.B., Miami U., 1948; M.A., Ind. U., 1949, Ph.D., 1952; m. Marilyn Edith Olsen, Dec. 27, 1948; children—Charles Martin, Nancy Ellen, Mary Anna. Asst. scientist Brookhaven Nat. Lab., 1952-57; USPHS spl. fellow to Nat. Inst. Med. Research, London, 1957-58; with Brookhaven Nat. Lab., 1958-60; prof. zoology, coordinator Kresge Isotope Lab., U. Mich., Ann Arbor, 1960—, chmn. radiation biology program, 1962—. Mem. Radiation Research Soc., Am. Physiol. Soc., Endocrine Soc., A.A.A.S., Sigma Xi. Publs. on studies of comparative aspects of thyroid hormone secretion; induction of tumors by radiation in animals. Home: 1514 Shadford Rd., Ann Arbor, Mich., 48104.*

SHELLENBERGER, John Alfred, Am. chemist; b. Moline, Ill., Jan. 8, 1900; s. Wilbur F. and Jenny (Johnstone) S.; B.S., U. Wash., 1928; M.S., Kan. State U., 1929; Ph.D, U. Minn., 1934; m. Annabel Frances Gangnath, June 3, 1939; children—Karen (Mrs. M. M. Stearns), Joan, Margo. Chemist, Fisher Flouring Mills Co., Seattle, 1923-28; asst. prof. agr. chemistry U. Ida., 1929-31; instr. biochemistry U. Minn., 1934-35; head products control Mennel Milling Co., Toledo, 1935-39; head biochem. div. Rohm & Haas Co., Phila., 1939-42; tech. adviser Corporacion para la Promocion del Intercambio, Buenos Aires, Argentina, 1942-44; head milling dept. Kan. State U., Manhattan, 1944-66, distinguished prof., 1966—. Recipient Outstanding Achievement gold medal U. Minn., 1965, 1st gold medal Assn. Operative Millers for outstanding contbns., 1963. Fellow A.A.A.S.; mem. Internat. Assn. Cereal Chemistry (pres. 1966—), Am. Assn. Cereal Chemists (past pres.), Am. Chem. Soc., Assn. Operative Millers, Am. Soc. Bakery Engrs., Inst. Food Technologists. Research and numerous publs. on micro methods of analysis, biochem. changes in grain during storage, cereal chemistry procedures, enzyme prodn., flour milling and baking tech. Home: 1715 Fairview, St., Manhattan, Kan. 66502.*

SHELLEY, Walter Brown, Am. physician; b. St. Paul, Feb. 6, 1917; s. Patrick Keary and Alfaretta (Brown) S.; B.S., U. Minn., 1940, Ph.D., 1941, M.D., 1943; m. to Marguerite Hilda Weber December 21, 1942; children—Peter B., Anne E. Practice medicine specializing in dermatology, Phila., 1950—; instr. Dartmouth, Med. Sch., 1949-50; faculty U. Pa. Med. Sch., 1950—, prof. dermatology, 1957—, chmn. dept. dermatology, 1965—; chief service in dermatology Phila. Gen. Hosp., 1965—. Recipient Spl. award Soc. Cosmetic Chemists, 1955. Fellow A.C.P.; mem. Am. Physiol. Soc., Am. Acad. Dermatology and Syphilology, A.M.A. (chmn. residency rev. com. for dermatology 1963—), Soc. for Investigative Dermatology (past pres.), Phila. Phys. Soc., Phila. Dermatol. Soc. (past pres.), Am. Dermatol. Assn., Coll. Physicians Phila., John Morgan Soc., Assn. Am. Physicians, Assn. Profs. Dermatology. Author: (with J.T. Crissey) Classics in Clinical Dermatology, 1953; (with others) Dermatology, 1956; (with H.J. Hurley, Jr.) The Human Apocrine Sweat Gland in Health and Disease, 1960; (with D.M. Pillsbury, A.M. Kligman) A Manual of Cutaneous Medicine, 1961; also numerous articles. Research on physiology of skin in health and disease especially sweating, itching and allergic states. Home: 505 County Line Rd., Radnor, Pa. 19088. Office: 3600 Spruce St., Phila. 19104.*

SHELLSHEAR, Joseph Lexden, anatomist; b. Sydney, Australia, July 31, 1885; s. Walter and Clara Mabel (Eddis) S.; grad. Sydney U., 1907; M.D., M.S.; m. Hildred Robertson. Sr. demonstrator, dept. anatomy Univ. Coll., London; prof. anatomy Hong Kong U., 1923-36; research prof. anatomy U. Sydney, 1937-48; radiologist. Author: The Anatomy of the Brain of the Chinese, Australian, Bushman and Prehistoric Man; also papers on anatomy of devel. of peripherial nervous system, blood supply of brain, prehistory of Hong Kong. Died Mar. 22, 1958.

SHELTON, E(berle) Kost, Am. internist, endocrinologist; b. Bloomfield, Ia., May 19, 1888; s. Eberle Kost and Katherine (Hayes) S.; student U. Denver, 1907-09; M.D., U. Colo., 1912, Sc.D., 1944; postgrad. various endocrine clinics, 1927-30; m. Margaret Norine, Feb. 25, 1914; 1 son, Paul Kingsley. Resident physician City and County Hosp., Denver, 1911-13; in gen. practice, 1913; internist, Denver and Antonito, Colo., 1914-26, Santa Barbara, Cal., 1930-40, Los Angeles, since 1938; asso. clin. prof. medicine, U. So. Cal., 1931-51; clin prof. medicine, U. Cal. at Los Angeles since 1951, sr. attending physician Harbor Gen. Hosp., since 1951; dir. endocrine clinic and mem. attending staff of hosp. Los Angeles Gen. Hosp.; mem. cons. staff St. Johns Hosp., Santa Monica, Cal. Diplomate Am. Bd. Internal Medicine. Fellow Am. Coll. Physicians; mem. Am. Med. Assn., Am. Therapeutic Soc., A.A.A.S., Assn. Study Internal Secretions, Los Angeles and Hollywood acads. medicine; Am. Soc. Research in Psychosomatic Problems. Author of pituitary chapter, vol. VIII, Tice's Practice of Medicine, and editor of endocrine volume, 1935. Author brochures on problems of internal medicine and endocrinology; contbr. to med. text books. Died Feb. 22, 1955.

SHEMIN, David, Am. biochemist; b. N.Y.C., Mar. 18, 1911; s. Louis and Mary (Bushkoff) S.; B.S., Coll. of City of N.Y., 1932; A.M., Columbia U., 1933, Ph.D., 1938; m. Mildred B. Sumpter, July 31, 1937 (dec. Apr. 1962); children—Louise P. (Mrs. Thomsa C. Homburger), Elizabeth; m. 2d, Charlotte J. Norton, Mar. 24, 1963. Mem. faculty Columbia U., N.Y.C., 1945-68, prof. biochemistry, 1953-68; prof. biochemistry Northwestern U., 1968—. Cons. to NSF, 1955-58, NIH, 1959—. Recipient Pasteur medal Pasteur Inst., Paris, 1955; Stevens award Columbia U., 1955. Mem. Nat. Acad. Sci., Am. Acad. Arts and Sci., Am. Soc. Biol. Chemists, Am. Chem. Soc., A.A.A.S., Soc. Bacteriology, Harvey Soc., Brit. Biochem. Soc., Swiss Biochem. Soc. (hon. mem.). Editor: Biochemical Preparations, vol. 5, 1957. Research and many publs. elucidating pathway by which cell synthesizes HEME, Vitamin B12 and related compounds; discovered origin of glycine; research on intermediary metabolism of amino acids. Home: 619 Library Pl., Evanston, Ill. 60201.*

SHEMYAKIN, Mikhail Mikhaylovich, Russian organic chemist; b. July 26, 1908; grad. Moscow U., 1930. With Research Inst. Organic Semi-Products and Dyes, Moscow Inst. Precision Chem. Tech., 1930-35, All-Union Inst. Exptl. Medicine, 1935-45, Moscow Textile Inst., 1937-59, Inst. Biol. and Med. Chemistry, USSR Acad. Med. Scis., 1945-47; prof., 1942—; asso. Inst. Organic Chemistry, USSR Acad. Scis., 1958—, dir. Inst. for Chemistry Natural Compounds, 1960—, acad. sec. dept. biochemistry, biophysics and chemistry of physiologically active compounds, 1964—, bur. mem. dept. chem. sci., until 1963, acad. sec., 1963-64. Mem. USSR Acad. Scis. Co-author: Chemistry of Antibiotic Substances, 1953; Oxidizing and Hydrolytic Conversions of Organic Compounds, 1957. Research and publs. on antibiotics, amino acids, vitamins, aldehydeacids, quinones, also other natural and bioactive agts.; developer methods to synthesize numerous types of organic compounds. Address: Institute for Chemistry of Natural Compounds, USSR Acad. Scis., 1-y Akademichesky pr. 18, Moscow, USSR.

SHEN, Chi-Neng, mech. engr.; b. Peiping, China, July 18, 1917; s. S.S. and H. (Huang) S.; came to U.S., 1950, naturalized, 1961; B.Engring., Nat. Tsing Hua U., Peiping, China, 1939; M.S., U. Minn., 1950, Ph.D., 1954; m. Ping-Wen Wu, Dec. 20, 1947; children—Harry H.L., Hilda H. Instr., U. Minn., Mpls., 1951-54; asst. prof. mech. engring., Dartmouth, Hanover, N.H., 1954-58; asso. prof. Rensselaer Poly. Inst., Troy, N.Y., 1958-60, prof., 1960—; vis prof. mech. engring. Mass. Inst. Tech., 1967-68. Arrangement chmn. Joint Automatic Control Conf., 1965. Mem. Am. Inst. Aeros. and Astronautics, Am. Soc. M.E., Am. Soc. Engring. Edn., Am. Nuclear Soc., Instrument Soc. Am., Sigma Xi. Research and publs. on nonlinear and optimum systems, synthesis and compensation, process dynamics, nuclear rocket control, astrodynamical guidance. Home: Hakes Rd., Eagle Mills, Troy, N.Y. 12180.*

SHĔN KUA, Chinese mathematician, astronomer; b. Ch'ien-T'ang, China, 1030; pres. Han-lin Coll. Author: (chem. work) Su-shen Tiang-fang; Meng-ch'i pi-t'an (Essays from the Torrent of Dreams). Measured areas and volumes; earliest Chinese summation of progression; earliest mention of magnetic needle; earliest description of moveable type in printing; prepared calendar, 1074.

SHÉNNIKOV, Aleksándr Petróvich, Russian botanist; b. Sept. 10, 1888; grad. St. Petersburg U., 1912. Mem. staff, Forestry Inst. (now Leningrad Forest Tech. Inst.), 1912-36; staff Petrograd U., Leningrad, 1919—; staff Botany Inst., USSR Acad. Sci., 1925—. Mem. Corr. mem. USSR Acad. Sci. Author: Theoretical Geobotanics During the Past Twenty Years, 1937; Meadow Vegetation in the USSR, 1938; Meadow Research, 1941; Plant Ecology, 1950; also articles. Research in theory of phytocenology, techniques of geobot. zoning, plants of USSR meadows, including classification. Office: Botanichesky Institut, AN SSR ul. Popova 2, Leningrad, USSR.

SHENTON, Walter Francis, Am. mathematician; b. Pottstown, Pa., Apr. 1, 1886; s. Robert Matlack and Helen Louise (Shick) S.; Sc.B., Dickinson Coll., 1907, A.M., 1909; Ph.D., Johns Hopkins, 1914; m. Mary Agnes Eslinger, Apr. 1, 1915; children—Helen Margaret, Robert Eslinger. Instr., Williamsport Dickinson Sem., 1907-10, Johns Hopkins, 1914-17; faculty U.S. Naval Acad., Annapolis, Md., 1917-25; prof. math., head math. dept. Am. U., Washington, 1925-57, prof. emeritus math., 1957—, marshal, 1930-50. Cons., NSF, 1958-62; adv. editor for math. Americana Ency., 1956-59. Mem. Math. Assn. Am. (charter, past chmn. Va., D.C., Md. sect.), Am. Math. Soc., Phi Beta Kappa, Omicron Delta Kappa, Phi Kappa Phi. Contbr. articles to tech. jours. Devised geometric methods for producing geometric sterograms from orthogonal projections. Home: 3605 Porter St., N.W., Washington 20016.*

SHEPARD, Charles Upham, Am. mineralogist; b. Little Compton, R.I., June 29, 1804; s. Mase and Deborah (Haskins) S.; grad. Amherst Coll., 1824, LL.D., 1857; M.D. (hon.), Dartmouth, 1836; m. Harriet Taylor, Sept. 23, 1831, 3 children. Lectr. botany Yale, 1830-31, lectr. natural history, 1833-47; in charge of Brewster Sci. Inst., New Haven, Conn., 1831-33; prof. chemistry S.C. Med. Coll., 1834-60, 65-69; lectr. natural history Amherst Coll., 1844-77; visited all known mineral localities east of Mississippi River; discovered phosphate of lime, 1865; his collection of meteorites largest in Am. by 1886; mem. Imperial Soc. of Natural Sci., St. Petersburg, Russia, Royal Soc. of Göttingen (Germany), Soc. Natural Sci., Vienna. Author: Treatise on Mineralogy (textbook), Part I, 1832, Part II, 1835. Contbr. papers to Am. Jour. Sci. and Arts. Died Charleston, S.C., May 1, 1886.

SHEPARD, Francis Parker, Am. geologist; b. Brookline, Mass., May 10, 1897; s. Thomas Hill and Edna (Parker) S.; B.A., Harvard, 1919; Ph.D., U. of Chicago, 1922; m. Elizabeth Buchner, June 12, 1920; children—Thomas Hill II, Anthony Lee. Began as instr., U. of Ill., 1922, prof. geology, 1939-46; research associate Scripps Instn. of Oceanography, La Jolla, California, 1942-45, principal geologist, 1945-48, professor of submarine geology since 1948; marine geologist, working on Navy projects for University of California, Div. War Research, and the Scripps Instn. since 1942; dir. Am. Petroleum Inst. project on sediments No. Gulf of Mexico, 1951-58. Mem. Geol. Soc. Am., Internat. Assn. Sedimentologists (pres. 1958-63, councilor, editor Sedimentology 1963—), Sigma Xi, Gamma Alpha, Alpha Sigma Phi. Author: Submarine Topography off Cal. Coast, 1942; Submarine Geology, 1948; Earth Beneath the Sea, 1959; (with R. F. Dill) Submarine Canyons and Other Sea Valleys, 1966. Editor: Recent Sediments Northwest Gulf of Mexico, 1960. Contbr. of science articles to geol. jours. Research on submarine canyons; study of sediment samples from numerous ocean floor environments to show continuous rise in sea rather than intermediate stage higher than present level. Home: 9090 LaJolla Shore Dr. Office: Scripps Instn., La Jolla, Cal. 92037. Office: Scripps Instn. of Oceanography, U. Cal., La Jolla, Cal. 92037.*

SHEPARD, Orson Cutler, Am. metallurgist; b. Del Rey, Cal., Dec. 9, 1902; s. Orson Ernest and Stella (Cutler) S.; A.B., Stanford U., 1925, Engr., 1928;

m. Grace Josephine Newland, June 22, 1927; children—Roger Newland, Cynthia Cutler. Research engr. Anaconda Copper Co., 1925, San Luis Mining Co., Mexico, 1925-27; mem. faculty Stanford (Cal.) U., 1928-—, prof. metallurgy, 1947-—, dir. div. mineral tech., 1952-57, head dept. metall engring., 1958-60, head dept. materials sci., 1960-67; vis. asso. prof. Mass. Inst. Tech., 1940-41. Engring. specialist Atomic Internat., 1957-58. Mem. Am. Soc. for Engring. Edn., Am. Inst. Mining and Metall. Engrs., Am. Soc. Metals, Sigma Xi, Tau Beta Pi. Author: (with W.F. Dietrich) Fire Assaying. Research in phys. metallurgy, corrosion, effect of environment on metals at elevated temperatures. Home: 600 Foothill Rd., Stanford, Cal. 94305.*

SHEPHARD, Roy Jesse, physiologist; b. London, Eng., May 8, 1929; s. Jesse and Esther Rose (Cummins) S.; B.Sc., London U., 1949, M.B., 1952, Ph.D., 1954, M.D., 1959; m. Muriel Neve Cullum, Aug. 18, 1956; children—Sarah Elizabeth, Rachel Judith. Practice medicine, specializing in physiology, London, 1952-56, Toronto, Ont., Can., 1964-—; research fellow cardiac dept. Guy's Hosp. U. London, 1952-54; med. officer RAF Inst. Aviation Medicine, 1954-56; Fulbright scholar, asst. prof. preventive medicine, applied physiology U. Cin., 1956-58; sr. sci. officer, prin. sci. officer Chem. Def. Exptl. Establishment, Porton Down, Wiltshire, Eng., 1958-64; prof. applied physiology Sch. Hygiene U. Toronto, 1964-68. Chmn. sci. com. Internat. Symposium Phys. Activity and Cardiovascular Health Ont. Heart Found., Ont. Med. Assn., Toronto, 1966-—. Mem. Am., U.K., Canadian physiol. socs., Med. Research Soc. U.K., Ergonomics Research Soc., Brit. Med. Assn., Canadian Assn. Sports Sci. (chmn. publs. com.), Canadian Assn. Health, Phys. Edn. and Recreation. Editor: Proceedings of Symposium on Phys. Activity and Cardiovascular Health, 1967. Research, numerous publs. on physiology of respiratory and cardiac systems, relationship between man's environment and his cardio-respiratory performance, lab. equipment for testing respiratory and cardiac function, air pollution and phys. fitness. Home: 42 Tollerton Av., Willowdale, Ont. Office: 150 College St., Toronto, Ont., Can.*

SHEPHERD, Dennis Granville, mech. engr.; b. Ilford, Eng., Oct. 6, 1912; s. George Granville and Lillian (Stubbins) S.; came to U.S., 1948, naturalized, 1951; B.S.E., U. Mich., 1934; m. Gertrude May Pitman, Sept. 1, 1939; children—Julian Granville, Joanna Ruth, Barbara Lynn. Engr., Crittall Mfg. Co., Eng., 1935-39, Power Jets Ltd., Eng., 1940-46; chief exptl. engr. A.V. Roe Can. Ltd., 1940-48; prof. mech. engring. Cornell U., Ithaca, N.Y., 1948-—, dir. Sibley Sch. Mech. Engring., 1965-—. Mem. Am. Soc. M.E., Combustion Inst., Sigma Xi, Phi Kappa Phi. Author: Introduction to the Gas Turbine, 1949, 60; Principles of Turbomachinery, 1956; Elements of Fluid Mechanics, 1965; also articles. Research on gas turbines, turbo machinery, combustion and heat transfer. Home: 142 N. Sunset Dr., Ithaca, N.Y. 14850.*

SHEPHERD, Ernest Stanley, Am. chemist; b. Remington, Ind., Mar. 30, 1879; s. William and Harriette Ellen (Lockwood) S.; student Ind. U., 1897-1900; A.B., Cornell, 1902; pvt. asst. to Prof. Bancroft, at Cornell, 1902; asst. in electrochemistry, Cornell, 1902; resigned to take up research on Carnegie grant, under Prof. Bancroft, 1902-04; phys. chemist Geophys. Lab., Washington, in charge research on lime-aluminasilica series of minerals, 1904-46. Member Am. Chem. Soc., Sigma Xi. Contbr. numerous papers on alloys, minerals and chemistry of volcanic phenomena to jours. Died Sept. 29, 1949.

SHEPHERD, John Thompson, physiologist; b. No. Ireland, May 21, 1919; s. William Frederick and Matilda (Thompson) S.; student Campbell Coll., Belfast, No. Ireland, 1932-37; M.B., B.Ch., Queen's U., Belfast, 1945, M.Chir., 1948, M.D., 1951, D.Sc., 1956; m. Helen Mary Johnston, July 28, 1945; children—Gillian Mary, Roger Frederick John. Extern surgeon Royal Victoria Hosp., 1946; lectr. physiology Queen's U., 1948-53, reader physiology, 1954-57; cons. physiology No. Ireland Hosps. Authority, 1953; asso. prof. physiology Mayo Found., cons. physician physiology sect. Mayo Clinic, 1957-62, prof. physiology Mayo Found., 1962-—, chairman of physiology section, 1966-—. Board dirs., chmn. research allocations com. Minn. Heart Assn., 1962. Brit. Med. Assn. scholar, 1949-50; Robert R. Leathen travelling scholar, 1953-54; Fulbright scholar, 1953-54; Anglo-French Med. exchange bursar, 1957. Mem. Soc. Exptl. Biology and Medicine, Am. Physiol. Soc., Am. Fedn. Clin. Research, Louis Rapkine Assn., Central Soc. Clin. Research, Am., Minn. heart assns., Physiol. Soc. Gt. Britain, Med. Research Soc. London, Sigma Xi. Author: Physiology of the Circulation in Human Limbs in Health and Disease, 1963. Editorial bd. Jour. Applied Physiology, Am. Jour. Physiology. Studies on regulation blood vessels in human limbs, mechanism of their control by nerves, chem. substances; comparative descriptions of heart and blood vessel adaptations to stress of exercise in animals, athletes, untrained subjects, patients with cardiovascular diseases. Home: 804 Fifth St. S.W. Office: Mayo Clinic, Rochester, Minn. 55901.*

SHEPPARD, Samuel Edward, research chemist; b. Hither Green, Kent, Eng., July 29, 1882; s. Samuel and Emily Mary (Taplin) S.; B.Sc., 1st class honors in chemistry, University Coll., London, 1903; D.Sc., 1906; 1851 Exhbn. scholar, 1907; studied Marburg U., Sorbonne, Paris and Cambridge, Eng.; m. Eveline Lucy Ground, Nov. 27, 1912; 1 son, Samuel Roger. Photographic research practice, Eng., 1910-11; chemist with Eastman Kodak Co., Rochester, N.Y., since 1912, asst. supt., in charge depts. of inorganic and physical chemistry of Research Lab., 1920, asst. dir. of research since 1923. Carried out chem. devel. of colloidal fuels with Submarine Def., 1917-19. Recipient Progress medal Royal Photog. Soc., 1928; Adelsköld medal Photog. Soc. of Stockholm, 1929; William H. Nichols medal, Am. Chem. Soc., 1930. Fellow Soc. Motion Picture Engrs., Chem. Soc., London, Photog. Soc. Am. (hon.), Royal Photographic Soc. (hon.); mem. Am. Chem. Soc., Am. Electrochem. Soc., American Standards Assn. Author: Photochemistry, 1914; Gelatin in Photography, Vol. I, 1923. Part Author: Investigations on the Theory of the Photographic Process, 1907; Silver Bromide Grain of Photographic Emulsions, 1921; Photography as a Scientific Implement, 1923. Contbr. numerous papers on chem. topics. Developed process of electrodeposition of rubber; sensitizing effect of sulphur on photographic emulsions; investigated development process, sensitivity, latent image, exposure. Died Rochester, N.Y., Sept. 29, 1948.

SHER, Samuel Alexis, Am. biologist; b. San Francisco, Apr. 24, 1923; s. Louis and Anna (Goldfein) S.; B.A., U. Cal. at Berkeley, 1948, Ph.D., 1952; m. Judith Korb, July 13, 1946; 1 son, Stephen Korb. Asst. nematologist U. Hawaii, 1952-53; sci. officer Sci. Service, Ottawa, Can., 1953; asst. nematologist dept. nematology U. Cal., Riverside, 1954-60, asso. prof., 1961-65, prof., 1965-—, vice-chmn. dept., 1956-63. Fulbright Research scholar Laboratoire des Nematodes, Antibes, France, 1963-64. Mem. Soc. Systematic Zoology, Soc. Nematologists, European Soc. Nematologists, Helminthological Soc. Wash., Sigma Xi. Author: Revision of the Hoplolaiminae Nematologica, 1966; also articles. Demonstrated proof of pathogenicity of a number of nematodes parasitizing ornamental plants; classification of soil and plant parasitic nematodes; description and illustration of soil and plant parasitic nematodes; morphology, phylogony, distbn. soil and plant parasitic nematodes. Home: 187 Nisbet Way, Riverside, Cal. 92507.*

SHERARD, James, Brit. botanist, physician; b. Eng., July 1, 1666; s. George and Mary S.; apprenticed to Charles Watts, apothecary, 1682; M.D., Oxford, 1731; m. Susanna Lockwood. Apothecary, London until 1720; became prof. botany, Oxford, 1728. Mem. Coll. Physicians. Fellow Royal Soc., 1706. Author: Hortus britanno-americanus (described 85 Am. trees which could be transplanted to Eng.), pub., 1767; Hortus Elthamensis, sive plantarum quas in Horto suo Elthami in Cantio collegit vir ornatissimus et praestantissimus . . . , 1732. Died Feb. 12, 1738.

SHERARD, William, English botanist; b. Bushby, Eng., Feb. 27, 1659; s. George and Mary Sherard; ed. Mcht. Taylor's Sch., St. John's Coll., Oxford; B.C.L., 1783; D.C.L., 1694; studied under Tournefort at Paris, under Paul Hermann at Leiden, Netherlands; made bot. expdns. to Geneva, Rome, Naples; consul for Turkish co. at Smyrna, 1702-16; made bot. and antiquarian expdns. in Asia Minor; founder chair of botany Oxford U. Plant in Linnean classification named in his honor. Fellow Royal Soc., 1718. Author: Schola Botanica, 1689; preface for Paradisus Batavus (Paul Hermann), 1698; also papers. Assisted Vaillant with his Botanicon Parisiense, Ray with last vol. of his Historia Plantarum; pub. catalogue of plants introduced at Paris by Tournefort. Brought John James Dillenius to Eng. Died Aug. 1728.

SHERBY, Oleg Dimitri, material scientist; b. Shanghai, China, Feb. 9, 1925; s. Dimitri Serge and Helen (Barashkova) S.; came to U.S., 1938, naturalized, 1944; B.S., U. Cal. at Berkeley, 1947, M.S., 1949, Ph.D., 1956; m. Ruth Juanita Slater, Sept. 4, 1949; children—Lawrence, Pamela Lynn, Stephen Gregory, Mark Paul. Teaching asst. U. Cal. at Berkeley, 1947-49, research metallurgist, 1949-56; postdoctoral fellow Sheffield U., Eng., 1956-57; sci. liason officer Office of Naval Research, London, Eng., 1957-58; faculty Stanford (Cal.) U., 1958-—, prof. materials sci., 1962-—. Cons. corps., govt. agys. Recipient Dudley award Am. Soc. Testing Materials, 1958. Mem. Am. Soc. Metals (chpt. pres. 1962), Am. Inst. Mining and Metall. Engrs., Sigma Xi, Tau Beta Pi, Theta Tau, Delta Chi. Research and many publs. on mech. behavior of solids and on atomic mobility at high temperatures. Home: 112 Emerson St., Palo Alto, Cal. 94301.*

SHERIF, Muzafer, social psychologist, author; b. Turkey, July 29, 1906; s. Sherif and Emine Basoglu; B.A., Ismir (Turkey) Internat. Coll., 1927; M.A., Istanbul (Turkey) U., 1929, Harvard, 1932; Ph.D., Columbia, 1935; m. Carolyn Wood, Dec. 29, 1945; children—Sue, Joan, Ann. Instr. Gazi Inst., 1937-38; from asst. to full prof. Ankara (Turkey) U., 1939-44; prof., research prof. U. Okla., 1949-—, dir. Inst. Group Relations, 1956-—. Vis. research prof. U. Tex., 1958-59; Ford vis. prof. U. Wash., 1960; distinguished vis. prof. Pa. State U., 1965. U. S. State Dept. fellow Princeton, 1945-47, research fellow Yale, 1947-49. Fellow Am. Psychol. Assn., Am. Sociol. Assn., Am. Ortho-Psychiat. Assn.; mem. Am. Assn. U. prof., research prof. U. Okla., 1949-66, dir. Inst. Group Relations, 1956-66; prof., dir. psychosocial studies program Pa. State U., University Park, 1966-—; Vis. research prof. U. Tex., 1958-59; Ford vis. prof. U. Wash., 1960; distinguished vis. prof. Pa. State U., 1965. U. S. State Dept. fellow Princeton, 1945-47, research fellow Yale, 1947-49. Fellow Am. Psychol. Assn., Am. Sociol. Assn., Am. Ortho-Psychiat. Assn.; mem. Am. Assn. U. Profs., Sigma Xi. Author: The Psychology of Social Norms, 1936; (with Hadley Cantril) The Psychology of Ego-Involvements, 1947; An Outline of Social Psychology, 1948; (with Carolyn W. Sherif) Groups in Harmony and Tension, 1953; Reference Groups. An Exploration into Conformity and Deviation of Adolescents, 1964; others. Studies, publs. on attitudes in group conflict, harmony, formation. Home: 507 Shannon Lane, State College, Pa. 16801. Office: Dept. Sociology, Pa. State U., University Park, Pa.*

SHERK, Kenneth Wayne, Am. chemist; b. Cottage Grove, Ore., Mar. 29, 1907; s. Alvin Lloyd and Florence (Bisbey) S.; A.B., Reed Coll., 1928; postgrad. Rice Inst., 1930-31; Ph.D., Cornell U., 1934; m. Dorothy Lucille Blacking, June 7, 1930; children—Larry Wayne, Lida Dee (Mrs. Martin Chipman), Betty Blacking (Mrs. Ernst Schöen-René), Kenneth Lloyd, Marjorie Joy (Mrs. Donal Dunphy), Jerome. Ray. Faculty, Smith Coll., Northhampton, Mass., 1935-—, prof. chemistry, 1951-—, dir. Grad. Study, 1959-—; asso. prof. U. Hawaii, Honolulu, 1949-50. Cons. to pvt. cos.; dir. NSF In-Service Inst., 1960-61, 63-64, 66-67. Mem. Am. Chem. Soc. (past chmn. Connecticut Valley sect.), A.A.A.S., New Eng. Assn. Chemistry Tchrs., Hawaiian Acad. Sci., Acacia, Sigma Xi. Research and publs. on migration aptitudes of various groups in intra-molecular rearrangements; complex ions of gold with ethylenediamine type ligands. Home: 25 N. Main St., Williamsburg, Mass. 01096.*

SHERMAN, Frederick G(eorge), Am. zoologist; b. McGregor, Mich., Apr. 16, 1915; s. William James and Mary Helen (McCullough) S.; B.S., U. Tulsa, 1938; Ph.D., Northwestern U., 1942; m. Barbara Corrine Tenney, Aug. 7, 1942; children—Susan, Martha, Sarah. Univ. fellow Sch. Medicine, Washington U., St. Louis, 1946; from instr. to prof. biology Brown U., 1946-60; prof. zoology, chmn. dept. Syracuse U., 1960-—; vis. biochemist Brookhaven Nat. Lab., Upton, N.Y., 1955-56. Fellww Picker Found., 1951-52; special research fellow USPHS, 1967-68. Fellow of the A.A.A.S.; mem. Am. Physiol. Soc., Sec. Gen. Physiologists (sec. 1957-59, councilor 1963-65); Soc. Study Growth and Devel., Am. Chem. Soc. Phi Beta Kappa, Sigma Xi. Research on effects of ionizing radiation on cellular metabolism; kinetics of cell populations; biochemical aspects of aging. Home: 106 DeWitt Rd., Syracuse, N.Y. 13214.*

SHERMAN, George Donald, Am. soil scientist; b. Ulen, Minn., June 4, 1904; s. George Chester and Ida (Otte) S.; B.S., U. Minn., 1933, M.S., 1937; Ph.D., Mich. State U., 1940; m. Minnie Jonasen, Aug. 26, 1944; 1 son, Keith Marlowe. Asst. chemist Ky. Agr. Expt. Sta., Lexington, 1941-42; chemist So. regional research utilization br., New Orleans, 1942-44; staff, faculty U. Hawaii, Honolulu, 1944-—, sr. soil scientist, sr. prof. soil sci., head dept., 1956-62, asso. dir. Hawaii Agr. Expt. Sta., 1962-—, dir. food processing lab., 1950-57. Program specialist Ford Found., U. Alexandria (Egypt), 1964-65. Cons. to pvt. cos. Fellow A.A.A.S., Am. Soc. Agronomy; mem. Soil Sci. Soc. Am., Internat. Soil Sci. Soc., Am. Chem. Soc., Sigma Xi. Research and numerous publs. on chem. and mineral. composition of tropical soils, mineral weathering rocks and soils, fertility and chemistry of soils of tropical regions, organic soils of temperate regions, devel. new tropical fruit products; identified processes of tropical soil formation; discovered bauxite deposits of Hawaiian Islands. Home: 828 Onaha, Honolulu 96816.*

SHERMAN, Henry Clapp, Am. chemist; b. Ash Grove, Va., Oct. 16, 1875; s. Franklin and Caroline (Alvord) S.; B.S., Md. Agrl. Coll., 1893; A.M., Columbia, 1896, Ph.D., 1897, D.Sc. (hon.), 1929; m. Cora Aldrich Bowen, Sept. 9, 1903 (dec.); children—Phoebe, Henry Alvord, William Bowen, Caroline Clapp Sherman Lanford. Asst. in chemistry Md. Agrl. Coll., 1893-95; fellow in chemistry Columbia, 1895-97, asst., 1897-98, lectr., 1899-1901, instr., 1901-05, adj. prof. analytical chemistry, 1905-07, prof. organic analysis, 1907-11, prof. food chemistry, 1911-24, Mitchill prof. chemistry, 1924-—, exec. officer, dept. of chemistry, 1919-39. Asst. in nutrition investigations U. S. Dept. Agr., 1898-99; research asso. Carnegie Instn., 1912-29, 33-—. Mem. com. on food and nutrition NRC, 1920-28, 40-—, chmn. subcom. on human nutrition, 1924-28. Chmn. com. on nutritional problems Am. Pub. Health Assn., 1919-33; pres. Am. Inst. of Nutrition, 1931-33, 1939-40; collaborator U. S. Nutrition Lab., 1940-—; chief Bur. of Human Nutrition, Dept. Agr., 1943-44. Medalist Am. Inst. of Chemists, 1933; Nichols medalist Am. Chem. Soc., 1934, Borden award Am. Inst. Nutrition; Franklin medalist and made hon. mem. Franklin Inst., 1946; Chandler medalist Columbia, 1949. Fellow A.A.-A.S.; mem. Am. Chem. Soc. (v.p. 1907-08), Am. Soc. Biol. Chemists (pres. 1926), Soc. Exptl. Biology and Medicine, Nat. Acad. Science; hon. mem. Harvey Soc.

Author: Methods of Organic Analysis, 1905, 12; Chemistry of Food and Nutrition, 1911, 8th rev. edit., 1952; Food Products, 4th edit., 1948; The Vitamins (with S. L. Smith), 1922, 2d edit., 1931; Food and Health, 1934 (rev., 1947); Essentials of Nutrition (with C. S. Lanford), 3d edit., 1951; Introduction to Foods and Nutrition (with C. S. Lanford), 1943; The Science of Nutrition, 1943; Foods; Their Values and Management, 1946; Calcium and Phosphorus, 1947; The Nutritional Improvement of Life, 1950. Noted for quantitative work on vitamins; introduced (with H. E. Munsell) Sherman-Munsell rate-growth unit of vitamin quantitative work on vitamins, mineral requirements of body, especially phosphorus, calcium; introduced (with H. E. Munsell) Sherman-Munsell rate-growth unit of vitamin A. Died Oct. 7, 1955.

SHERMAN, John Clinton, Am. geographer; b. Toronto, Ont., Can., May 3, 1916; s. Harold Clinton and Grace (Ubkes) S.; B.A., U. Mich., 1937; M.A., Clark U., 1942; Ph.D., U. Wash., 1947; m. Helen Jean Loyd, Mar. 15, 1941; children—Constance Jean (Mrs. Roger Newall), John Harold (dec.), Mary Helen, Barbara Lillian. From instr. to prof. U. Wash., Seattle, 1942-—, also chmn. dept. geography. Cartographer for An Introduction to Geography (Rhoads Murphey, 1961; Adv. co-editor: Oxford Regional Atlas of the United States and Canada, 1967; Research and publs. on cartography, including gen. communications function, design philosophy, spl. map devel. for blind and partially seeing. Home: 7424 55th Av. N.E., Seattle 98115.*

SHERMAN, Seymour, Am. mathematician; b. Apr. 30, 1917; s. Israel S. and Minnie S.; B.A., Cornell U., 1936, M.A., 1937, Ph.D., 1940; m. Rose Mary Dudley, Apr. 1964; children—Samuel, Deborah, Daniel, Sarah. Faculty, U. So. Cal., 1952-53; research asso. lab. applied math., statistics Stanford, 1954; vis. prof. Moore Sch. U. Pa., 1954-55, prof., 1955-60; prof. math. Wayne State U., 1960-64; prof. math. Ind. U., Bloomington, 1964-—. Reviewer for math. revs. Mem. Am. Math. Soc. Editor: Jour. Math. and Mechanics, 1964-—. Study of differential equations; numerical analysis; digital and analog computation; astrodynamics. Home: 1312 Nancy St., Bloomington, Ind. 47403.

SHERRINGTON, Sir Charles Scott, English neurophysiologist; b. London, Eng., Nov. 27, 1856; ed. Caius College, Cambridge (Eng.) U., M.D., 1885; studied in Berlin under Virchow and Koch; hon. degrees from Oxford, Paris, Manchester, Strasbourg, Louvain, Uppsala, Lyon, Budapest, Athens, London, Toronto, Harvard, Dublin, Edinburgh, Montreal, Liverpool, Brussels, Glasgow, Sheffield, Berne, Birmingham, Wales; m. Ethel Mary Wright, 1892; one son. Prof., physiology, U. Liverpool, 1895-1913; Fullerian prof. physiology, Royal Institution of Great Britian, 1914-17; Brown prof. pathology, London, 1914; Waynflete prof. physiology, Oxford U. 1917-36; Gifford Lectr., Edinburgh, 1936-38. Recipient (with Adrian) Nobel Prize in physiology and medicine, 1932. Fellow Royal Soc., 1893 (Royal medal, 1905; Copley medal, 1927; pres. 1920-25); mem. Brit. Assn. Advancement Sci. (pres. 1922); mem. French Acad. Scis., 1923; mem. Nat. Acad. Scis.; knighted, 1922; Order of Merit, 1924; Fellow, Royal College Physicians; Fellow, Royal College Surgeons, many other memberships, medals, awards, lectureships. Author: The Integrative Action of the Nervous System, 1906; Social Hygiene, 1913; Mammalian Physiology, 1916; Selected Writings, 1939; Man on His Nature, 1940; The Endeavour of Jean Fernel, 1946. Research in functioning of nervous system, e.g. reflexes and regeneration of nerve tissue; worked on decerebrate rigidity; studied reciprocal innervation; investigated function of neuron, function of synapse; introduced term and concept of integrative action of nervous system; introduced terms neuron, synapse, nociceptor, interoceptor, exteroceptor, teleceptor, proprioceptor, proprioception. Died Eastbourne, Sussex, Eng., Mar. 4, 1952.

SHERROD, Theodore Roosevelt, Am. pharmacologist; b. Town Creek, Ala., July 29, 1915; s. Woody and Martha (Reed) S.; A.B., Taladega Coll., 1938; M.S., U. Chgo., 1941; Ph.D., U. Ill., 1945, M.D., 1949; m. Jessie Maddox, Sept. 7, 1941; 1 son, Theodore Roosevelt. Faculty, U. Ill. Coll. Medicine, Chgo., 1945-—, prof. pharmacology, 1959-—. Cons. Dept. Pub. Aid; mem. Ill. Adv. Com. on Pharmacy. Mem. Inst. Medicine Chgo., Am. Soc. Pharmacology and Exptl. Therapeutics, Am. Soc. Clin. Pharmacology and Chemotherapy, Sigma Xi. Research, numerous publs. on drugs affecting heart, circulation, kidneys. Home: 500 E. 33d St., Chgo. 60616.*

SHERRY, Sol, Am. physician; b. N.Y.C., Dec. 8, 1916; s. Hyman and Ada (Greenman) S.; A.B., N.Y. U., 1935, M.D., 1939; m. Dorothy Sitzman, Aug. 7, 1946; children—Judith Anne, Richard Leslie. Fellow N.Y. U., 1939-40; faculty N.Y. U., 1946-51; dir. May Inst. for Med. Research, Cin., 1951-54; dir. medicine Jewish Hosp., St. Louis, 1954-58; prof. medicine Washington U., St. Louis, 1958-68, co-chmn. dept., 1964-68; chmn. dept. medicine Temple U., Phila., 1968-—. Chmn., Gen. Clin. Research Center com. NIH, 1961-63; mem. com. on thrombosis and hemorrhage and blood transfusions problems Nat. Acad. Scis.-NRC, 1962-66; chmn. Com. on Thrombolytic Agts., NIH, 1963-—; cons. Office of Surgeon Gen., U.S. Army Research and Devel. Comd., 1964-—; mem. sci. adv. bd. A.R.C., 1965-—. Recipient Career Research award NIH, 1962-67; Modern Medicine Achievement award Modern Medicine, 1963. Mem. A.M.A. (chmn. sect. on exptl. medicine and therapeutics 1963-—, mem. council on drugs 1964-—), Assn. Am. Physicians, Am. Soc. for Clin. Investigation, Central Soc. for Clin. Research, Am. Physiol. Soc. Research and numerous publs. in devel. of blood clot dissolving enzymes, in biochemistry and physiology of fibrinolysis; discovered action of thrombin and plasmin on synthetic substrates,, streptococcal desoxyribonuclease; elucidated significance of desoxyribonucleoprotein in purulent exudates; developed concept of enzymatic debridement and clin. use of enzymes for therapeutic purposes. Home: 408 Sprague Rd., Narberth, Pa. 19072. Office: 3400 N. Broad St., Phila. 19140.*

SHERWIN, Chalmers William, Am. physicist; b. Two Harbors, Minn., Nov. 27, 1916; s. Louis Blanchard and Georgia (Polhemus) Sherwin; B.S., Wheaton (Ill.) Coll., 1937; Ph.D., U. Chgo., 1940; m. Irene Ackerman, Sept. 18, 1937; children—Margaret Philippson, Priscilla, Catherine, Louise, Louis, Julia, Susan. Research asst. U. Chgo., 1940-41; mem. staff Radiation Lab., Mass. Inst. Tech., 1941-45; asso. Columbia, 1946; from asst. prof. physics to prof. U. Ill., 1946-60; chief scientist USAF, 1954-55; asso. dir. Coordinated Scis. Lab., U. Ill., 1959-60; v.p., gen. mgr. labs. div. Aerospace Corp., El Segundo, Cal., 1960-63, trustee, 1960; dep. dir. def. research and engring. for research and tech. Dept. Def., Washington, 1963-66; dep. asst. sec. commerce, sci. and tech., 1966-67; cons. Office Sci. and Tech., Exec. Office Pres., 1967-—; dir. tech. evaluation Gulf Gen. Atomic Corp., San Diego, 1968-—. Mem. sci. adv. bd. USAF, 1956-—. Recipient Presdl. Certificate of Merit, 1947; Sci. award Air Force Assn., 1956. Fellow Am. Phys. Soc., mem. Sigma Xi. Author: Introduction to Quantum Mechanics, 1959; Basic Concepts of Physics, 1961. Research in electronics, neutrino properties, spl. relativity, high resolution radar. Home: 7704 Whitefield Pl., La Jolla, Cal. 92037. Office: Gulf Gen. Atomic Corp., San Diego 92112.*

SHERWOOD, Thomas Kilgore, Am. chem. engr.; b. Columbus, O., July 25, 1903; s. Milton Worthington and Sadie (Tackaberry) S.; B.S., McGill U., 1923; D.Sc. (hon.), 1951; S.M., Mass. Inst. Tech., 1924, Sc.D., 1929; D.Eng. (hon.), Northeastern U., 1950; m. Virginia Howell, June 17, 1953; children—Thomas Kilgore, Richard M., Marcia. Mem. faculty Mass. Inst. Tech., Cambridge, 1930-—, prof. chem. engring., 1941-—, dean engring., 1946-52, Du Pont professor of chemical engineering. Sect., chief OSRD-NDRC, 1940-46. Recipient William H. Walker award Am. Inst. Chem. Engring., 1941; U.S. medal for merit, 1948; Founders award Am. Inst. Chem. Engring., 1963. Mem. Nat. Acad. Scis., Nat. Acad. Engring., Am. Acad. Arts and Scis. Home: Lowell Rd., Concord, Mass.*

SHETH, Uttamchand Khimchand, Indian pharmacologist; b. Lakhtar, India, Oct. 30, 1920; s. K. A. and Taralaxmi (Shah) S.; B.Sc., Bombay U., 1939, M.B., B.S., 1944, M.D., 1948; m. Sushila M. Kane, June 20, 1944; 1 son, Anil U. Asst. prof. pathology T.N. Med. Coll., Bombay, 1948-50, asst. prof. pharmacology, 1950-53, prof., 1953-55; prof., head dept. pharmacology Seth G.S. Med. Coll., Bombay, 1956-—. Mem. Bd. Goa Govt. for Promotion Med. Research in State Insts.; mem. bds. studies several Indian univs., several pharmacological research insts. Rockefeller fellow U. Utah, Salt Lake City, 1959-60. Fellow Coll. Physicians and Surgeons (mem. coll. council), Am. Coll. Clin. Pharmacology and Chemotherapy; mem. Assn. Physiologists and Pharmacologists India (gen. sec. 1964-65), Indian Assn. Advancement Med. Edn. (sec. pharmacology sub-com.), N.Y. Acad. Scis. Research, numerous publs. on diuretics, psychopharmacology, human pharmacology, med edn. Home: Pran. 7th Rd., Santaeruz (East), Bombay 55. Office: Seth G.S. Med. Coll., Bombay 12, India.*

SHETTLES, Landrum Brewer, Am. physician; b. Pontotoc, Miss., Nov. 21, 1909; s. Basil Manly and Sue (Mounce) S.; B.A., Miss. Coll., 1933, Sc.D., 1966; M.S. (fellow 1933-34), U. N.M., 1934; Ph.D., Johns Hopkins U., 1937, M.D., 1943; m. Priscilla Elinor Schmidt, Dec. 18, 1948; children—Susan Flora, Frances Louise, Lana Brewer, Landrum Brewer, David Ernest, Harold Manly and Alice Annmarie (twins). Instr. biology Miss. Coll., 1932-33; biologist U.S. Bur. Fisheries, 1934; instr. biology Johns Hopkins U., 1934-37, research fellow, 1937-38; research fellow Nat. Com. Mental Health, N.Y.C., 1938-43; attending obstetrician gynecologist, Columbia Presbyn. Med. Center, N.Y.C., 1951-—; Markle Found. scholar Columbia Coll. Phys. and Surg., 1951-56; asso. prof. clin. obstetrics and gynecology Columbia U., N.Y.C., 1951-—; prvt. practice, N.Y.C., 1951-—; Anglo-Am. lectr. Royal Coll. Obstetricians and Gynecologists, London, 1959. Research cons. Office Naval Research, Am. embassy, London, 1951-52; served to maj. M.C., Aus, 1944-46. Recipient Ortho medal and award Am. Soc. Study Fertility, 1960. Diplomate Pan Am. Med. Assn., Am. Bd. Obstetricians and Gynecologists. Fellow Royal Soc. Health, London, Am. Coll. Obstetricians and Gynecologists, A.C.S., A.A.A.S., World Med. Assn.; mem. Am. Soc. Zoologists, Am. Physiol. Soc., Soc. Exptl. Biology and Medicine, Soc. U. Gynecologists, Harvey Soc., Am., N.Y. State, N.Y. County med. socs., Phi Beta Kappa, Sigma Xi, Gamma Alpha. Author: Ovum Humanum, 1960; also numerous articles. Research on fisheries biology; physiology of human reprodn.; fertility and sterility; hemorragic disease of newborn infants; sperm biology; discovered and identified male and female producing sperms. Home: 622 W. 168th St., N.Y.C. 10032. Office: 180 Ft. Washington Av., N.Y.C. 10032.*

SHEVCHENKO, Fedor Iosifovich, Russian microbiologist; b. 1899; grad. Med. Faculty, Central Asian U., Tashkent, 1927; Cand. Med. Sci., 1936; D.Med. Sci., 1940. Med. technician, later head vaccine dept. Tashkent Health Bacteriology Inst., 1927-36; head chair microbiology Samarkand Med. Inst., 1939-—, dep dir. for sci. work and studies, 1952-58. Author: Medical Microbiology, 1961. Mem. editorial bd. Uzbekistan Med. Jour. Research and numerous articles on leishmaniasis, vaccine prodn., immunity and intestinal diseases in children. Address: Samarkand Medical Institute, Kommunisticheskaya ulitsa 35, Samarkand, Uzbekistan SSR, USSR.

SHEVYAKOV, Lev Dmitrievich, Russian mining engr.; b. Vetluga, Kostroma, Jan. 15, 1889; grad. Yekaterinoslav (now Dnepropetrovsk) Mining Inst., 1912; Dr.Tech. Sci., 1935. Asst., then lectr. Yekaterinoslav Mining Inst., 1913-20, prof., 1920-28; prof. Tomsk Technol. Inst., 1929-32, Sverdlovsk Mining Inst., 1932-44; dir. Geol. Mining Inst., also dep. chmn. Ural br. USSR Acad. Sci., 1939-44; prof., dir. chair for devel. blanket deposits Moscow Mining Inst., 1944-50; sci. dir. All-Union Coal Research Inst., 1944-46; chmn. council for study prodn. resources USSR Acad. Sci., 1946-49; mem. Council Sci. and Tech. Experts USSR Gosplan, 1943-46, chmn., 1946-58; dir. dept. for devel. mineral deposits Inst. Mining, USSR Acad. Scis., 1944-—, dep. acad. sec. dept. tech. sci., 1957-—. Chmn. Sci. Council on Problem of Kursk Magnetic Anomaly, 1952-—; mem. com. for Lenin prizes for sci. and tech. USSR Council Min, 1960-—; mem. com. to restore Donbas coal industry, 1920-22; cons. to various mining establishments. Recipient Stalin prize, 1942, Order of Red Banner of Labor, 1943, 45, Order of Lenin, 1948. Author: Collected Articles on Mining; Mining Water Drains; Mechanization of Mining Operations; Bracing of Open Pit Mines; Sinking of Open Pit Mines; Miner's Library; Working Mineral Deposits, 1928, 56; Synopsis of an Analytical Course in Mining Techniques, 1935; Working Blanket Deposits, 1936; The Theoretical Principles of Planning Coal Mines, 1950. Research and publs. on mine design, high prodn. and efficiency, local mining problems; devel. and improvement USSR Coal industry. Office: USSR Acad. Scis., Leninskii Prospekt, Moscow, USSR.

SHIBATA, Shoji, Japanese organic chemist; b. Tokyo, Japan, Oct. 23, 1915; s. Keita and Masako (Shogenji) S.; B.Pharm.Sci., U. Tokyo, 1938, D. Pharm.Sc., 1943; m. Michiko Nukiyama, Dec. 19, 1942; children—Takehiko, Tomohiko. Med. faculty Tokyo Imperial U., 1943-50, asst. prof., 1944-50; prof. chemistry U. Tokyo, 1950-—; vis. researcher London (Eng.) Sch. Hygiene and Tropical Medicine, 1953-54. Mem. Pharm. Soc. Japan (prize 1959), Pharmacognostical Soc. Japan, Chem. Soc. Japan, Chem. Soc. (London), Biochem. Soc. (London). Author: (with Y. Asahina) Chemistry of Lichen Substances, 1954; (with M. Yamazki) Biosynthesis of Natural Products, 1965; also numerous articles. Research on chemistry of lichen and fungal products and Chinese drug constituents, biosynthesis of lichen and fungal metabolites and higher plant products. Home: 4-10-2 Mejiro Toshima-ku, Tokyo, Japan.*

SHIBAYAMA, Gorosaku, Japanese physician; b. Tochigi Prefecture, Japan, 1871; grad. med. dept. Tokyo U., 1898; apptd. quarantine officer, 1900; successively ofcl. at Quarantine Bur., Contagious Disease Research Inst., Kanagawa Prefectural Office, Met. Police Bd. Discovered paratyphoid fever, 1905; studied Tb., bubonic plague, typhoid fever, beriberi. Died 1913.

SHIELDS, James, Brit. behavioral geneticist; b. Edinburgh, Scotland, Nov. 21, 1918; s. James and Julia (Macrae) S.; B.A., Merton Coll., Oxford U., 1945; Mental Health Certificate, London Sch. Econs., 1947; m. Elizabeth Joan Ede, Oct. 25, 1945; children—Julia, Mary. With genetics unit Inst. Psychiatry, Maudsley Hosp., U. London (Eng.), 1947-—, lectr., 1957-—. Mem. Genetical Soc., Eugenics Soc., Assn. Psychiat. Social Workers (asso.). Author: Monozygotic Twins, Brought Up Apart and Brought Up Together, 1962; also articles. Research on normal and psychiatrically disturbed twins including twins reared apart in an attempt to disentangle some effects of heredity and environment in intelligence, personality devel., neurosis and schizophrenia. Home: 51 Meadway, London N.W.11. Office: Genetics Unit, Maudsley Hospital, Denmark Hill, London S.E.5, Eng.*

SHIGEMATSU, Tsunenobu, Japanese chemist; b. Ehime, Japan, Dec. 28, 1916; s. Tasaburo and Misao S.; M.Sc., Kyoto (Japan) U., 1940, D.Sc.,

1952; m. Takako Shigematsu, Aug. 1, 1944; children—Toshihiko, Tatsuhiko. Faculty of science, Kyoto U., 1947—, prof. chemistry Inst. for Chem. Research, 1957—. Mem. Chem. Soc. Japan, Soc. for Japanese Analysts (award 1965), Atomic Energy Soc. Japan, Japan Soc. Materials Sci. Japan, Geochem. Soc. Japan, Soc. Sea Water Sci. Japan, Soc. Mass Spectroscopy Japan. Research and numerous publs. on radiochemistry, geochemistry and analytical chemistry; solvent extraction of metal chelates; trace elements in sea water, sea lives and sediments; spectrophotometric and fluormetric methods and activation analysis. Home: 9 Shimohananoki-ch., Koyama, Kita-ku, Kyoto, Japan.*

SHIGO, Alex Lloyd, Am. plant pathologist; b. Duquesne, Pa., May 8, 1930; s. Alex and Helen (Szilagyi) S.; B.S., Waynesburg Coll., 1956; M.S., W.Va. U., 1958, Ph.D., 1959; m. Marilyn Paul, May 22, 1954; children—Judy Ruth, Robert Paul. Prin. mycologist U.S.D.A. Forest Service, Laconia, N.H., 1959-65, Durham, N.H., 1965—; lectr. U. Me., 1964—; adj. prof. U. N.H., 1966—. Mem. Am. Phytopathol. Soc., Mycol. Soc. Am., Sigma Xi. Contbr. numerous articles in field to sci. jours. Research on parasitism by studying mycoparasites, fungi parasitic on other fungi; microelements, especially Mn affect on the degree of parasitism; research on diseases of northern hardwood trees; succession of organisms in discoloration and decay of wood. Home: R.F.D. 1, Denbow Rd. Office: Box 640, Dover Rd., Durham, N.H. 03824.*

SHIKHIEV, Ibrahim Abbas ohly, Russian organic chemist; b. Kirovabad, USSR, Mar. 22, 1904; s. Abbas Ismail and Sakina (Kodzhaeva) ohly; grad. Leningrad (USSR) State U., 1936, candidate chem. sci., 1940; m. Mamedova Zuleikha, Aug. 22, 1935; children—Mikhael, Mariam. Sci. worker Vitamin Inst., 1941-42, Azerbaijan br. USSR Acad. Scis., 1946-50, Inst. Organic Chemistry, USSR Acad. Scis., 1950-57; chief lab. elemento-organic compounds Inst. Petrochem. Processes, Azerbaijan br. USSR Acad. Scis., Baku, 1957—. Mem. Mendeleev Chem. Soc. Author several books. Research, publs. organic chemistry, elementoorganic derivatives acetylene (used in pharmacology, also as corrosion inhibitor in hydrochloric acid). Home: 31, 7-Hkrebtovaya, g.553a, r.21, Baku. Office: 30, Telnova, Baku-25, Azerbaijan, USSR.*

SHILLINGFORD, John Parsons, English physician; b. London, Eng., Apr. 15, 1914; s. Victor Sadler and Ethel (Parsons) S.; student London Hosp., 1937-42; M.D. magna cum laude, Harvard, 1943; M.D., U. London, 1943; m. Doris Margaret Franklin, Aug. 31, 1937; children—Michael John, Ann Elizabeth, James Hugh. First asst. med. unit London Hosp., 1947-50; dir. cardiovascular research unit Med. Research Council, Postgrad. Med. Sch. London, 1967—, prof. cardiology, 1966—. Fellow Royal Soc. Medicine, Assn. Physicians, Royal Coll. Physicians; mem. Med. Research Soc., Brit. Cardiac Soc. (sec. 1962—). Research and publs. on use of radioactive and other tracers in study of circulation, effect of coronary thrombosis on circulation. Home: 6 Hurlingham, Ct., London, S.W.6. Office: Royal Postgrad. Med. Sch., Ducane Rd., London W. 12, Eng.*

SHILOV, Yevgeniy Alekseevich, Russian chemist; b. 1893; grad. Moscow U., 1917. With Ivanovo Chem. Tech. Inst., 1919-47; asso. Inst. Organic Chemistry, Ukrainian Acad. Scis., 1947—; prof., 1936—; Del., Internat. Conf. on Uses Radioisotopes, Paris, 1957. Mem. Ukrainian Acad. Scis. Research and publs. on organic synthesis, kinetics and mechanism of organic reactions, theory of organic conversions. Address: Institute of Organic Chemistry, Ukrainian Acad. Scis., Vladimirskaya ulitsa 54, Kiev, Ukrainian SSR, USSR.

SHIMA, Etsuzo, Japanese seismologist; b. Kumamoto, Japan, Nov. 12, 1927; s. Tameo and Tsuchiko (Sakaida) S.; B.Sc., Geophys. Inst., U. Tokyo, 1951, Dr.Sc., 1962; m. Toshiko Motoki, May 28, 1953; 1 dau., Mari. Asst., Earthquake Research Inst., U. Tokyo, 1951-62, asso. prof., 1964—; postdoctoral fellow U. Wis., Madison, 1962-64. Mem. Seismol. Soc. Japan, Soc. Exploration Geophysicist Japan, Seismol. Soc. Am., Soc. Exploration Geophysicist. Research and publs. on modifications of seismic waves in superficial soil layers. Home: Chuo-5-21-12, Ota Ku Tokyo, Japan.*

SHIMAMOTO, Takio, Japanese physician; b. Mama Kochi-shi, Japan, Jan. 20, 1909; s. Shigenori and Tami S. Igaku-shi, Tokyo U. Med. Sch., 1929, Igaku-hakushi, 1938; m. Yasuko Hirosue, Apr. 30, 1936; children—Tatsuo, Nobuko, Chiyoko. Instr. dept. medicine Tokyo U. Med. Sch., 1933-47, asst. prof., 1947-52; vis. investigator dept. anatomy U. Cal. at Los Angeles, 1952-53; chmn. dept. physiology, prof. Med. Sch., Tokyo Med. and Dental U., 1953-59, prof., chmn. dept. medicine, 1959—, dir. Inst. for Cardiovascular Diseases, 1962—, dir. U. Hosp., 1966—. Pres., Japan Found. for Study Arteriosclerosis, 1965—. Mem. Physiol. Soc. Japan, Japanese Path. Soc., Japanese Circulation Soc., Japanese Soc. Internal. Medicine. Contbg. author: Method Achievements in Experimental Pathology, Vol. 1, 1966. Editorial bd. Jour. Atherosclerosis Research, 1962—. Research and numerous publs. on discovery of platelet clumping substance and pryridinol carbamate, venous bradykinin antagonist which is first drug capable of curing atheromatous lesions of athersclerosis, edmatous arterial reaction. Home: 13 Kitamachi Shinjukuku, Tokyo, Japan.*

SHIMBORI, Michiya, Japanese sociologist; b. Kobe, Japan, June 26, 1921; s. Jutaro and Hisashi (Araki) S.; A.B., Hiroshima Higher Normal Sch., 1943; M.A., Hiroshima U., 1945, Ph.D., 1956; m. Sumiko Kawase, Nov. 5, 1949; children—Rumiko Shimbori, Kumiko. Tchr. English, Hiroshima (Japan) Higher Normal Sch. for Girls, 1945; asst. prof. edn. Hiroshima Higher Normal Sch., 1946-52; asso. prof. ednl. sociology Hiroshima U., 1952—; vis. lectr. Ryukyu U., Okinawa, Japan, 1954; vis. prof. U. Chgo., 1959-60; researcher Ryukyu Islands, 1965. Recipient awards Japanese Ministry Edn., 1964, 66, Japanese Cabinet, 1965, 66. Mem. Am., Japanese sociol. assns., Comparative Edn. Soc., Japanese Soc. for Ednl. Sociology, Japanese Soc. for Study Edn. Author: The Problem of Love in Education, 1954; J. J. Rousseau, 1957; Academic Marketplace in Japan, 1965; Emile Durkheim, 1966; School Career, 1966; other; also numerous articles. Research on sociology edn. as pure sci., particularly theoretical founds. sociology edn. and sociol. analysis problems higher edn. Editor, Japanese Jour. Ednl. Sociology, 1953; adv. editor Indian Sociol. Bull., 1961, Internat. Rev. History and Polit. Sci., 1961. Home: 35 Kenei-jutaku, 11 Chome, Ujina, Hiroshima, Japan.*

SHIMER, Hervey Woodburn, Am. paleontologist; b. Martins Creek, Pa., Apr. 17, 1872; s. John C. and Maria (Engler) S.; A.B., Lafayette Coll., 1899, A.M., 1901; Ph.D., Columbia U., 1904; Sc.D. (hon.), Gettysburg Coll., 1916; m. Florence F. Henry, June 1, 1904; children—John A., Mary H. (Mrs. Charles R. Mangat-Rai). Tutor, Lafayette Coll., 1899-1901; asst. in paleontology Columbia U., 1901-03; mem. faculty Mass. Inst. Tech., Cambridge, 1903-42, prof. paleontology, 1920-42, prof. emeritus, 1942. Mem. Geol. Soc. Am., Paleontology Soc. Am., Boston Soc. Natural History, Washington Acad. Arts and Scis., Am. Acad. Arts and Scis., A.A.A.S., Sigma Xi. Author: An Introduction to the Study of Fossils, 1914, 33; North American Index Fossils, 1909; Evolution and Man, 1922; An Introduction to Earth History, 1925; (with R. R. Shrock) Index Fossils of North America, 1944. Died Dec. 13, 1965.

SHIMER, Porter William, Am. chemist, metallurgist; b. Shimerville, Pa., Mar. 13, 1857; s. Peter A. and Ellen (Werkheiser) S.; E.M., Lafayette Coll., 1878, Ph.D., 1899; m. Elizabeth Sandt, Oct. 12, 1880; children—Katharine (Mrs. J. Willard Paff), William Robert, Edward Bernard, Margaret (Mrs. Paul Hoffman). Propr. chem. and metall. lab., Easton, Pa., 1885—; pres. Shimer Chem. Co., 1924—; in analytical, cons. and investigation work; lectr. iron and steel Lafayette Coll., 1894-1902. Mem. Internat. Steel Standards Com. Recipient John Scott medal Franklin Inst., 1901, for invention of combustion crucible. Author of papers on new methods and apparatus used in analytical chemistry; application of chemistry to metall. and other problems. Inventor of a new process for case-hardening iron and steel, molten baths for steel treating, chaplet alloy used in iron founding; discovered titanium carbide. Died Dec. 7, 1938.

SHIMIZU, Sakae, Japanese physicist; b. Tokyo, Japan, July 18, 1915; s. Eijiro and Naka (Sakurai) S.; B.Sc., Kyoto (Japan) Imperial U., 1940; Ph.D., Kyoto U., 1950; m. Momoe Iwamura, Feb. 1, 1947; children—Toru, Masaru. Faculty, Kyoto Imperial U., 1943-46; faculty Kyoto U., 1946—, prof. physics, 1952—; lectr. physics Konan U., Kobe, Japan, 1953—. Mem. operation com. reactor sch. Japan Atomic Energy Research Inst., Tokai-mura, Ibaragiken, Japan, 1959—. Mem. Japan Radioisotope Assn. (dir. 1962—), Phys. Soc. Japan, Atomic Energy Soc. Japan, Japan Radiation Research Soc., Am. Phys. Soc., Am. Nuclear Soc., Health Physics Soc. (hon. editorial adv. bd. 1958—). Author: Measuring Instruments of Nuclear Physics, 1949; also articles. Research on exptl. nuclear physics, nuclear spectroscopy; gamma-ray scattering, exptl. finding of change of radioactive decay constant of Uranium 235 by chem. means; first exptl. evidence of radiationless annihilation of positrons. Home: 10 Higashihinokuchi-cho, Tanaka, Sakyo-ku, Kyoto, Japan.*

SHIMIZU, Takeo, Japanese radiologist; b. 1890; grad. sci. dept., Tokyo U., 1914; D.Sc.; student in Britain, 1921. Prof., Tokyo U., 1941-50; later mem. Sci. Research Inst. Recipient Crown Prince Marriage Meml. prize Imperial Acad., 1924. Remodeled Wilson's cloud chamber, allowing uninterrupted observation and photography of inflated steam.

SHIMKIN, Demitri Boris, anthropo-geographer; b. Omsk, Siberia, July 4, 1916; came to U. S., 1923; s. Boris Michael and Lydia (Serebrova) S.; A.B., U. Cal. at Berkeley, 1936, Ph.D, 1939; m. Edith Manning, Aug. 19, 1943; children—Alexander, Eleanor. Johnson scholar, univ. and research fellow U. Cal. at Berkeley, 1937-41; instr. Nat. War Coll., 1946-47, Inst. for Advanced Study, Princeton, 1947-48; research asso. Russian Research Center, Harvard, 1948-53; social sci. analyst, then sr. research specialist U. S. Bur. Census, Washington, 1953-60; prof. anthropology and geography, U. Ill., Urbana, 1960—; field work in Wyo., 1937-39, 66, Alaska, 1949, Ill., 1963, Miss., 1966; vis. prof. anthropology Harvard, 1964-65. Mem. NRC, 1964-67; mem. U. S. nat. com. Internat. Biol. Program, 1965—. Fellow A.A.A.S.; mem. Am. Anthropol. Assn., Assn. Am. Geographers, Phi Beta Kappa, Sigma Xi. Author: A Key to Soviet Power, 1953; (with others) Trends in Economic Growth, 1955; The Soviet Mineral-Fuels Industry, 1928-58, 1962. Research on industrialization and resource mgmt. in USSR and U. S., Uto-Aztecan ethnology and cultural history; instnl. and ecol. analyses of rural populations; devels. in human ecol. theory especially adaptive strategies. Home: 106 W. Pennsylvania Av., Urbana, Ill. 61801.*

SHIMKIN, Michael Boris, physician; b. Tomsk, Siberia, Oct. 7, 1912; s. Boris Michael and Lydia (Serebrova) S.; came to U.S., 1923, naturalized, 1928; A.B., U. Cal. at Berkeley, 1933; M.D., U. Cal. at San Francisco, 1937; m. Mary Louisa North, July 2, 1938; children—Peter Michael, Ann Mary, Philip North. Research fellow Nat. Cancer Inst., Harvard, 1938-39; with USPHS, 1939-63, med. dir., Washington, 1950-63; chief cancer biology Fels Research Inst., prof. medicine Temple U., Phila., 1963—, asst. v.p. for research, 1966—. Cons. med. research mission OSRD, USSR, 1943-44, UNRRA, 1944-45; adviser U.S. delegation to Internat. Health Congress, WHO, 1946; chmn. com. on lung cancer Am. Cancer Soc., 1955-57, on breast cancer, 1960; cons. Nat. Cancer Inst., Nat. Research Council, 1963—. Diplomate Am. Bd. Internal Medicine, Am. Bd. Preventive Medicine. Fellow A.C.P., Am. Coll. Preventive Medicine; mem. Am. Assn. Cancer Research, Am. Soc. Exptl. Pathology, Soc. Exptl. Biology and Medicine. Author: Science and Cancer, 1964; also numerous articles. Editor, Cancer Research, 1965—. Research on cancer, chem. carcinogenesis, breast and lung tumors in mice, clin. chemotherapy of cancer, epidemiology of cancer, internat. medicine especially of USSR. Office: Temple U. Health Scis. Center, Phila. 19140.*

SHIMODA, Koichi, Japanese physicist; b. Urawa, Japan, Oct. 5, 1920; s. Seishi and Kiyo (Takemasa) S.; B.Sc., Tokyo Imperial U., 1943; D.Sc., U. Tokyo, 1955; m. Kanako Fujiwara, Nov. 28, 1947; children—Sachiko, Etsuko, Junichi, Masataka. Mem. faculty U. Tokyo (Japan), 1948—, prof. physics, 1959—; chief microwave physics lab. Inst. Phys. and Chem. Research, Saitama, Japan, 1960—; research fellow radiation lab. Columbia, 1954-55; mem. staff Mass. Inst. Tech., 1962-63. Mem. Phys. Soc. Japan, Japan Soc. Applied Physics, Am. Phys. Soc., Optical Soc. Am. Author: Microwaves, 1955; Fundamentals of Electronics, 1958, rev. 1963. Research and publs. on microwave spectroscopy of atoms and molecules; atomic standard of frequency based on ammonia; theoretical and exptl. studies on molecular beam masers; devels. of novel methods of maser and laser spectroscopies. Home: 1-19-15, Kichijoji-minami, Musashino, Tokyo. Office: U. Tokyo, 7-3-1, Hongo, Bunkyo-ku, Tokyo, Japan.*

SHINAGAWA, Mutsuaki, Japanese chemist; b. Hiroshima-Ken, Japan, Nov. 3, 1913; s. Kinzo and Yoshiko (Ide) S.; B.Sc., Kyoto (Japan) Imperial U., 1937, D.Sc., 1945; m. Hiroko Ueda, Jan. 15, 1940; children—Taeko (Mrs. Takahiro Kumamaru), Ritsuko, Humiaki. Faculty, Kyoto Imperial U., 1937-51, asst. prof., 1945-51; prof. analytical chemistry Faculty Sci. Hiroshima (Japan) U., 1951-61; fellow M.D. Anderson Hosp. and Tumor Inst., U. Tex., Houston, 1955-56; prof. nuclear engring. chemistry Faculty Engring., Osaka (Japan) U., 1961—; faculty dept. chemistry Nat. Tsing Hua U., Taiwan, 1965-66; faculty Reactor Inst., Kyoto U., 1966-68. Mem. Japan Soc. for Analytic Chemistry (prize for analytical chemistry 1964), Chem. Soc. Japan, Atomic Energy Soc. Japan. Author: Analytical Method of Polarography, 1952; also numerous articles. Research on analytical chemistry using onium compounds as organic reagents especially from polarographic point of view, elucidation of mechanisms on Berdicka catalytic wave of hydrogen in polarography, hydrogen evolution using semi-conductor materials as electrolytic catalyser, focusing electrophoretic chromatography. Home: 56 Higashisonqda 3, Amagasaki-shi, Hyogo, Japan. Office: Yamadaue, Suita-shi, Osaka, Japan.*

SHINAGAWA, Shinryo, Japanese physician; b. Tokyo, June 22, 1923; s. Nobuji and Kiyo Shinagawa; grad. Tohuku U., 1946, M.D., 1952; m. Miho Kikawa, Oct. 31, 1949; children—Nobumichi, Yasuhiro, Keiko. Faculty, Hirosaki U. Sch. Medicine, 1951—, dir., prof. obstetrics and gynecology, 1958—. Chmn. obstetric hemorrhage com. Japanese Maternal Welfare Soc., 1966; adv. council World Assn. for Gynecol. Cancer Prevention, 1965. Author: Extraperitoneal Gynecologic Surgery, 1964; Obstetrics and Gynecology, 1967; also numerous articles. Editor: Hirosaki Med. Jour. Research on techniques in female pelvic surgery, blood coagulation and plasmin system. Home: 8-12 Miyuki-choh, Hirosaki, Japan.*

SHINOHARA, Kenichi, Japanese physicist; b. Kobe, Japan, Dec. 4, 1905; s. Sukeichi and Chika (Tanaka) S.; D.Sc., U. Tokyo, 1939; m. Ayako Uno, Mar. 31, 1933; children—Shoichi, Kunio. Research asst. Inst. Phys. and Chem. Research, Tokyo, 1929, fellow, 1940; prof. Kyushu U., Fukuoka, Japan, 1940-49; chief research fellow Inst. Phys. and Chem. Research, 1950-66; prof. physics Sci. and Engring. Lab., Waseda U.,

Tokyo, 1966—. Dir. Tokyo labs. Japanese Assn. for Radiation Research on Polymers, 1957-63. Research, numerous publs. on diffraction of electron beams by single crystals, properties of gamma rays and high energy electrons, irradiation of polymers by gamma rays and electrons. Home: 5-16-14 Mejiro, Toshima-ku, Tokyo, Japan.*

SHINOWARA, George Y(ukio), Am. biochemist; b. Seattle, Aug. 4, 1914; s. Joseph Giochero and Toyo Marie (Fukuda) S.; A.B., Wittenberg Coll., 1935; M.A., Ohio State U., 1937, Ph.D., 1938; D.Sc. (hon.), Wittenberg U.; 1961; m. Alice Marie Waterhouse, Aug. 8, 1941; 1 dau., Nancy Lee. Instr. pathology Ohio State U., 1938-42, asst. prof., 1942-48, asso. prof., 1948-52, prof., 1952-59; chief div. chem. pathology U. Hosp., 1938-59; prof. dept. pathology, N.Y. U. Sch. Medicine; dir. biochemistry Bellevue Hosp., 1959—. Mem. Harvey Soc., Am. Soc. Biol. Chemistry, A.A.A.S., Soc. for Sci. Investigation of Crime Ohio Acad. Sci., Soc. Exptl. Biology and Medicine, Sigma Xi. Author articles on mechanism human blood coagulation, isolation and characterization antihemophilic globulin, antithrombin from plasma, also lipoprotein from erythrocytes. Home: North Midland Av., Upper Nyack, N.Y. Office: 550 1st Av., N.Y.C. 16.*

SHIPLEY, Sir Arthur Everett, English zoologist; b. Walton/Thames, Eng., Mar. 10, 1861; s. Alexander and Amelia (Burge) S.; ed. St. Bartholomew's Hosp., Christ's Coll., Cambridge (Eng.) U.; M.A., Sc.D.; D.Sc. (hon.), Princeton; LL.D., U. Mich.; M.Sc. (hon.), Drexel Inst. Demonstrator comparative anatomy Cambridge U., 1885-94, fellow, 1887, lectr. advanced morphology of invertebrata, 1894-1908, apptd. sec. Museums and Lecture Rooms Syndicate, 1891, mem. council senate, 1896-1908, 17-19, reader zoology, from 1908, master Christ's Coll., from 1910, vice-chancellor univ., 1917-19. Fellow Royal Soc., 1904, Zool. Soc., Linnean Soc. (v.p.); mem. Research Defence Soc. (v.p.), Marine Biol. Assn. (chmn. council), Am. Assn. Econ. Entomologists (fgn.), Helminthological Soc. Washington (fgn.), other fgn. socs. Author: Zoology of the Invertebrata, 1893; Pearls and Parasites, 1908; J, a Memoir of John Willis Clark, 1913; The Minor Horrors of War; More Minor Horrors, 1916; (with A. Schuster) Studies in Insect Life, 1916; Britain's Heritage of Science, 1917; The Voyages of a Vice-Chancellor, 1919; Life, 1923; Cambridge Cameos, 1924; Islands. Joint editor, co-author: Cambridge Natural History; Grouse in Health and Disease. Studied parasitic worms; best known as author popular works zoology. Died Cambridge, Sept. 22, 1927.

SHIPLEY, Robert E., Am. physiologist; b. Dayton, O., July 24, 1912; s. Bernis Melvin and Grace (Trone) S.; B.S., Otterbein Coll., 1934; M.D., Western Res. U., 1938; m. Loretta Kardaszewski, Sept. 11, 1964. Head dept. physiology Eli Lilly & Co., Indpls., 1950—. Research and publs. on atherosclerosis, hypertension, kidney physiology. Home: 2615 Coldsprings Manor Dr., Indpls. 46222. Office: Lilly Lab. for Clin. Research, Marion County Gen. Hosp., Indpls.*

SHIPTON, Harold William, electronic engr.; b. Birmingham, Eng., Sept. 29, 1920; s. Albert W. and Emma (Dones) S.; A.M., Shrewsbury (Eng.) Tech. Coll., 1947; m. Janet Attlee, Nov. 15, 1947; 1 dau., Ann Helen. Came to U.S., 1957, naturalized, 1966. Electronic devel. engr. Burden Neurol. Inst., Bristol, Eng., 1946-57; asso. prof., head div. med. electronics U. Ia., Coll. Medicine, Iowa City, 1957—. Mem. Instrument Soc. Am. (dir. biomed. div. 1965—), Sigma Xi. Cons. editor Jour. Electroencephalography and Clin. Neurophysiology, 1957—. Research and publs. on application of electronic techniques and computer methods to medicine and biology. Home: 820 Woodside Dr., Iowa City 52240.*

SHIRANE, Gen, physicist; b. Nishinomiya, Japan, May 15, 1924; s. Jiro and Sawano (Nakamura) S.; B.Eng., U. Tokyo (Japan), 1947, Dr.Sci., 1953; m. Sakae Uchiyama, Apr. 9, 1950; children—Haruo, Tatsuo. Came to U.S., 1955, naturalized, 1962. Research asso. Tokyo Inst. Tech., 1947-52; research asso. Pa. State U., 1952-55, asst. prof., 1955-56; asso. physicist Brookhaven Nat. Lab., Upton, L.I., N.Y., 1956-57, physicist, 1963—; adv. physicist Westinghouse Research Lab., Pitts., 1957-63. Mem. Am. Phys. Soc., Phys. Soc. Japan. Author: (with F. Jona) Ferroelectric Crystals, 1962; also articles. Research on solid state physics especially ferroelectricity and magnetism. Home: 10 Livingston Rd., Bellport, N.Y. 11713. Office: Brookhaven Nat. Lab., Upton, N.Y. 11973.*

SHIRAS, George 3d, Am. biologist; b. Allegheny, Pa., Jan. 1, 1859; s. George Jr. and Lillie E. (Kennedy) S.; A.B., Cornell, 1881; LL.B., Yale Univ., 1883; Sc.D., Trinity Coll., 1918; m. Frances P. White, Oct. 31, 1885; children—Ellen Kennedy (Mrs. Frank J. Russell), George Peter. Admitted to bar State of Pa., 1883; asso. in practice with his father until 1892; mem. Shiras & Dickey, Pitts., until 1904. Mem. Pa. Ho. of Rep., 1889-90; elected to the 58th Congress (1903-05); writer since 1905, upon biol. subjects and legal questions connected with federal jurisprudence; promoter of legislation for protection of wild animals and birds; author of bills putting under fed. control migratory birds and migratory fish, the former becoming a law 1913. Recipient Gold medal Paris Expn., 1900, Grand prize at St. Louis World Fair, 1904, for photographs of wild animals. V.p. Am. Game Protective Assn. since 1912; mem. adv. bd. Migratory Bird Treaty Regulations, Dept. Agr., since 1914; trustee Nat. Geog. Society since 1908. Author: Hunting Wild Life with Camera and Flashlight, 2 vols., 1935, 2d edit., 1936. Inventor methods of photographing animals at night by flashlight; presented extensive lake shore frontage (now Shiras Park), City of Marquette, Mich.; donor Shiras Inst., incorporated 1938, for recreational and cultural benefits of Marquette and vicinity. Died Mar. 24, 1942.

SHIRAZI, Shah Fathullah, Indian inventor, scholar; b. circa 16th century; reformed mint and calendar at Ct. of Akbar; served on diplomatic missions for emperor Arhbar. Invented machine to clean 11 gun barrels simultaneously, wagon mill, multibarreled gun, multistoried traveling bath, portable cannon, water works system; in his calender abolished intercallary days and years and used duodenary cycles; it had 365 days, 5 hours, 45 minutes and 27 seconds in a year.

SHIREN, Norman Steven, Am. physicist; b. N.Y.C., Feb. 7, 1925; s. Samuel and Clara (Cinader) S.; B.S., Tufts U., 1945; postgrad. Columbia U., 1946-47; Ph.D., Stanford, 1956; m. Edith Y. Weisberg, Sept. 7, 1947; children—Leslie B., Jane H. Physicist, Hudson Lab., Columbia U., 1951-55, Gen. Electric Research Lab., 1955-61; research staff mem. IBM, T.J. Watson Research Center, Yorktown, N.Y., 1961-65, mgr. quantum physics, 1965—. Fellow Am. Phys. Soc.; mem. I.E.E.E. (asso.). Research and publs. on spin-phonon interactions, non-linear acoustics, parametric amplification, two quantum stimulated emission. Home: 219 Forest Dr., Mt. Kisco, N.Y. 10549. Office: P.O. Box 218 Yorktown Heights, N.Y. 10598.*

SHIRKEY, Harry Cameron, Am. pediatrician; b. Cin., July 2, 1916; s. Lewis Cameron and Pearl (Knight) S.; B.S. magna cum laude, U. Cin., 1939, M.D., 1945; m. Jo Ruth Bartholomew, Apr. 19, 1957; children—Wade, Kathryn, Jodi. Pharmacist, Children's Hosp., Cin., 1940-45, asst. dir. Out-Patient Dispensary, 1950-51; asso. prof. pharmacology Coll. Pharmacy, 1948-59, U. Cin., from instr. to asst. clin. prof. pediatrics Sch. Medicine, 1953-59; dir. Pediatric div. Cin. Gen. Hosp., 1954-59; prof. pediatrics, asso. prof. pharmacology, asso. prof. pediatrics Sch. Dentistry, dir. div. pharmacology, dept. pediatrics Med. Coll. Ala., U. Ala., Birmingham, also prof., chmn. dept. pharmacology Samford U. and dir., chief staff Children's Hosp., Birmingham; prof., chmn. dept. pediatrics, prof. pharmacology U. Hawaii, also med. dir. Kanikeolani Childrens Hosp., Honolulu, 1968—. Cons. pediatrician numerous hosps.; dir. Poison Control Center, Jefferson County, Ala.; vis. prof., lectr. various univs. Licentiate Am. Bd. Pediatrics. Fellow Am. Acad. Pediatrics, Am. Coll. Clin. Pharmacology and Chemotherapy, Am. Med. Writers' Assn.; mem. Am. Pediatric Soc., So. Soc. Pediatric Research, Am. Therapeutic Soc., N.Y. Acad. Scis., Am., So. med. assns., Am. Pharm. Assn., Am. Hosp. Assn., Am. Soc. Pharmacology and Exptl. Therapeutics, Drug Information Assn., Am. Assn. Poison Control Centers (pres. 1964-66), Sigma Xi, others. Editor: Pediatric Therapy, 3d edit. Contbr. Nelson's Textbook of Pediatrics, Ambulatory Pediatrics, others. Publs. on drugs, poisoning in children. Address: 226 N. Kuikani St., Honolulu 96816.*

SHIRKOV, Dmitrii Vasilievich, Russian theoretical physicist; b. 1928; grad. Moscow State U., 1949; Dr.Physico-Math. Scis., 1957. Staff, Joint Inst. Nuclear Research, 1956-60; joined Math. Inst., USSR Acad. Scis., Siberian Br., 1960. Corr. mem. USSR Acad. Scis. Research and publs. on theory of elementary particles, conductivity, dispersion relations, pertubation theory, pion-nucleon scattering at low energies, quantum field theory. Address: Math. Inst. Siberian Br., USSR Acad. Scis., Novosibirsk, Siberia.

SHIRLEY, David Allen, Am. chemist; b. Knoxville, Tenn., Sept. 15, 1918; s. John Fletcher and Tennie Marie (Beets) S.; B.S. in Chemistry, U. Tenn., 1939, M.S., 1940; Ph.D., Ia. State U., 1943, postdoctoral research fellow, 1943-44; m. Ruth Charlotte Wright, Aug. 30, 1941; children—Carolyn R., Allen C., Robert E., Elizabeth M. Research chemist E. I. duPont de Nemours & Co., Inc., 1944-47; asst., then asso. prof. Tulane U., 1947-53; mem. faculty U. Tenn., 1953—, prof. chemistry, 1955—, head dept., 1962—. Mem. Am. Chem. Soc. (chmn. E. Tenn. sect. 1959), Chem. Soc. (London, Eng.), Am. Assn. U. Profs., Tenn. Acad. Sci., Sigma Xi (pres. U. Tenn. chpt. 1960). Author: Preparation of Organic Intermediates, 1951; Organic Chemistry, 1964. Asso. editor: (Gilman) Organic Chemistry, 1964. Asso. editor: (Gilman Organic Chemistry—An Advanced Treatise, vols. III and IV, 1953; (Adams) Organic Reactions, Vol. VIII, 1954. Research and publs. on organic chemistry, including synthesis, organometallics, heterocyclics, natural products. Home: 1116 Montview Dr., Knoxville, Tenn. 37914.*

SHIRLEY, Ray Louis, Am. animal nutritionist; b. Martinsburg, W.Va., Dec. 11, 1912; s. Van I. and Bessie (Adams) S.; B.S., W.Va. U., 1937, M.S., 1939; Ph.D., Mich. State U., 1949; m. Helen Broadbent, Apr. 3, 1943; children—John, William, Susan, Henry, Bonnie. Faculty, Mich. State U., 1941-42, 47-49, Shepherd Coll., 1951-53; research chemist Hercules Powder Co. Research Center, 1942-47; faculty U. Fla., Gainesville, 1949-51, 53—, prof., 1949—, charge Animal Nutrition Lab., 1960—. Mem. Am. Soc. Animal Sci., Am. Inst. Nutrition, Soc. Exptl. Biology and Medicine (chmn. Southeastern sect.), Am. Chem. Soc., A.A.A.S., Sigma Xi (U. Fla. chpt. Faculty Research award 1960), Alpha Zeta, Gamma Sigma Delta, Phi Lambda Upsilon. Research and numerous publs. on effect of dietary and physiol. factors on deposition and excretion of radioactive tracer elements, activity of enzymes in tissues, growth and performance of animals. Home: 1523 N.W. 11th Rd., Gainesville, Fla. 32601.*

SHIROKOV, Yuri Michailovich, Russian physicist; b. Moscow, USSR, June 21, 1925; s. Michail F. and Elizabeth A. S.; grad. Moscow State U., 1948, Ph.D., 1951; m. Iren Leonova, Oct. 14, 1948; 1 dau., Elizabeth. Faculty Moscow State U., 1951-65; prof. physics Moscow Mining Inst., 1964—; research prof. Steklov Math. Inst., USSR Acad. Scis., Moscow, 1965—. UNESCO expert in India, 1964-65. Author: Lectures of the Foundations of the Relativistic Quantum Theory, 1964. Research, publs. on applications of Poincaré group to relativistic quantum theory, also other questions of quantum theory, nuclear theory. Office: 28 Vavilov, Moscow V-333, USSR.*

SHISHKIN, Boris Konstantinovich, Russian botanist; b. Apr. 19, 1886; grad. Tomsk U., 1911. Asst., Tomsk U., 1914-18, prof., 1935-30; staff Botany Inst., USSR Acad. Sci., 1931—, dir., 1938-49; head bot. research expdns., Siberia, Caucasus, Carpathians, Central Asia. Recipient Stalin Prize, 1952, Order of Lenin. Mem. USSR Acad. Sci. (corr.). Author: An Outline of Uryankhay Kray, 1914; Material on the Flora of Turkish Armenia, 1928; A Botanical and Geographic Outline of the Maritime Slope of the Ponti Range, 1930; The Vegetation of the Altai, 1937; Flora of the USSR, vols. 16-17, 1950-51. Editor: The Flora of the USSR; The Flora of Western Siberia (P. N. Krylov); The Flora of Leningrad Oblast. Office: AN SSSR, Leninsky prosp. 14, Moscow, USSR.

SHIVASTAVA, H. C., Indian horticulturist; b. Lucknow, India, Jan. 29, 1929; s. Narendra Bahadur and Sunder; M.Sc., Agra U., India, 1949, Ph.D., 1953; m. Sarla Srivastava, May 21, 1952. Mem. staff Central Food Technol. Research Inst., Mysore, India, 1953-64, head div. storage and preservation, 1959-64; sr. agr. mgr. Hindustan Lever Ltd., Ghaziabad, U.P., India, 1964—; tchr., guide postgrad. students various univs. Cons. fruit and vegetable industry on raw material. Recipient Gold medal for agrl. research, 1956; Ahmed Kidwai prize Govt. India, 1963. Author: Commercial Fruit and Vegetable Products; Commercial Storage of Fruit and Vegetable; Picking Handling and Transportation of Fruits and Vegetables. Research, publs. on standardization of conditions for refrigerated storage of fruits and vegetables and extended their storage life at their optimum conditions; improved methods of handling and transp. of perishables; standardized mushroom cultivation and submerged propagation for a new industry in India. Home: care Mr. Navendra Bahadur, Sit la Prasad Rd., Kundri Rakabgunj, Lucknow, U.P., India. Office: G.T. Road, Ghaziabad, U.P., India.*

SHIVITSKIS, Prantsishkus Kazimírovich, zoologist; b. Sept. 30, 1882; ed. Valparaiso U., U. S., 1911, U. Mo., 1917, U. Chgo., 1922. Prof. zoology U. Manila (P.I.), 1922-28; prof. anatomy and embryology Lithuanian U., Kaunas, 1929-40; prof. Vilnius U., 1940-48; sr. sci. worker, chief lab. Biology Inst., Lithuanian Acad. Scis., 1948—. Mem.. Lithuanian Acad. Scis. Author: Vistu belukscіu kiausiniu dejimas ir kaip jas nuo to apsaugoti, 1950; Parazitu apibudinimas, 1956. Research on sea and fresh water animals, embryology of spiders, molluscs, parasitology.

SHKAROFSKY, Issie Peter, Canadian physicist; b. Montreal, Que., Can., July 4, 1931; s. Frank and Sylvia (Alpert) S.; B.Sc. with first class honors in Physics and Math., McGill U., Montreal, 1952, M.Sc., 1953, Ph.D., 1957; m. Agnes Spira, Mar. 10, 1957; children—Marvin David, Sema, Lou Aaron. With RCA Victor Co., Ltd., Montreal, 1957—, sr. mem. sci. staff Plasma and Space Physics Research Lab. Mem. asso. com. on plasma physics NRC, Ottawa, Ont., Can., 1965—. Mem. Canadian Assn. Physicists, Am. Phys. Soc., Am. Geophys. Union. Author: (with T.W. Johnston, M.P. Bachynski) The Particle Kinetics of Plasmas, 1966; also articles. Research on radio propagation, microwave tubes, plasmas in lab. and in space. Home: 1957 Clinton Av., Montreal. Office: RCA Victor Research Labs., 1001 Lenoir St., Montreal 30, Que., Can.*

SHKLAR, Gerald, pathologist; b. Montreal, Que., Can., Dec. 2, 1924; s. Louis and Ann (Schleifstein) S.; B.Sc., 1947, D.D.S., McGill U., 1949; M.S., Tufts U., 1951; m. Judith Nisse, June 16, 1948; children—David, Michael, Ruth. With Tufts U., Boston, 1951—, prof. oral pathology, chmn. dept. oral pathology, 1962—, research prof. periodontology, 1961—, dir. cancer teaching program, 1952—, lectr. social dentistry, 1963—, acting asst. dean, 1962-63; lectr. oral pathology Forsythe Sch. for Dental Hygienist, 1953—; lectr. oral histopathology Harvard Univ.; vis. oral pathologists dept. dentistry Boston City Hosp., 1960—; cons. tng. grants

Nat. Cancer Inst., NIH, 1965—. Fellow Am. Coll. Dentists, Internat. Coll. Dentists, Am. Acad. Oral Pathology, A.A.A.S.; mem. Internat. Assn. Dental Research, New Eng. Soc. Pathologists, Mass. Soc. Pathologists, Am. Acad. Periodontology, Am. Cancer Soc., Am. Soc. Periodontology, Am. Acad. Dental Medicine. Author: (with P.L. McCarthy) Diseases of Oral Mucosa, 1965; also numerous articles. Research on pathology periodontal disease, surg. pathology bone, metabolism tumors, exptl. diabetes. Home: 33 Clinton Rd., Brookline, Mass. 02146.*

SHKLOVSKII, Iosif Samuilovich, Russian astrophysicist; b. July 1, 1916; grad. Moscow U., 1938; Dr. Physico-Math. Sci. Head dept. radio astronom Shternberg Astron. Inst., Moscow, 1944—; prof. Moscow U., 1938—. Author: The Solar Corona, 1951; The Nature of the Aurora Polaris' Radiance, 1952; The Origin of Cosmic Rays and Radio Astronomy, 1953; The Origin of the Crab Nebula's Radiance, 1953; Radio Astronomy, 2d edit., 1955; Cosmic Radio-Frequency Emission, 1956; The Nature of the Earth's Third Radiation Belt, 1960; On the Distant Planet of Venus, 1961. Divided radio-frequency emission of Galaxy into thermal and non-thermal types; research on aurora polaris, infra-red emission of night sky, origin of cosmic rays in envelopes of nova and super-nova stars; originated theory for ionization of solar corona. Office: USSR, Moskva, Leninski gory, Gosudarstvenny universitet.

SHLAPOBERSKY, Vasiliy Yakovlevich, Russian surgeon; b. 1901; D.Med. Sci., 1938. Asst. prosector, histologist Moscow Tb Inst., asso. Inst. Operative Surgery, 2d Moscow U., 1923-27; intern, asst. 2d Moscow Med. Inst., 1927-43, lectr., prof. Hosp. Surgery Clinic, 1943-52; head chair hosp. surgery Vilnius U., sr. asso. Clinic for Surgery Gastointestinal Tract, All-Union Inst. Exptl. Medicine, 1952-56; head dept. bone pathology Central Inst. Traumatology and Orthopedics, USSR Ministry Health, 1956—. Mem. Moscow (exec. sec.), All-Union (bd. mem., exec. sec.) socs. surgeons. Co-author: X-ray Treatment of Gunshot Injuries; author: Penicillin in Surgery, 3 edits.; Surgical Sepsis; Acute Purulent Peritonitis. numerous other works. Former sec., now mem. editorial council Surgery. Address: Central Institute of Traumatology and Orthopedics, USSR Ministry of Health, Teply p. 16, Moscow, USSR.

SHLYK, Alexander Arkadjevitch, Russian biologist; b. Minsk, USSR, Nov. 1, 1928; s. Arkadee and Maria Shlyk; student U. Bjelorussia, 1945-50; candidate Sci., BSSR Acad. Scis., 1954; D.Sc., Inst. Biochemistry, USSR Acad. Scis., 1963; m. Valeria S., Dec. 1951; two children. Staff, Inst. Biology BSSR Acad. Scis., Minsk, 1950-57, sr. sci. worker, 1956-57; head Biophysics and Isotopes Lab., BSSR Acad. Scis., Minsk, 1957—; faculty Bjelorussian U., Minsk, 1957—, prof. biophysics, 1965—. Mem. USSR Acad. Scis. (corr.). Author: Tagged Atom Method of Studying the Biosynthesis of Chlorophyll, 1956; Chlorophyll Metabolism in Green Plant, 1965; also numerous articles. Research on chlorophyll turnover and biosynthetic way in green plant; discovered metabolic heterogeneity of chlorophyll and dependence of state of molecule on its age; conception of chlorophyll b formation from new chlorophyll a molecules, reaction centres of chlorophyll metabolism and their role in regulation of 2 pigment systems of photosynthesis relationship. Home: 27 Academitcheskaya, Minsk, USSR.*

SHMALGAUSEN, Iván Ivánovich, Russian zoologist; b. Apr. 23, 1884; s. I. F. Shmalgausen; grad. Kiev U., 1907. Became prof. Voronezh U., 1918, Kiev U., 1921, Moscow U., 1938-48; dir. Inst. Zoology and Biology, Ukrainian Acad. Scis., 1930-41; dir. Inst. Evolutionary Morphology (now Severtsov Inst. Animal Morphology), USSR Acad. Sci., 1938-48, staff Zoology Inst., 1948—. Research on origin of ground vertebrates.

SHMIDT, Aleksándr, Russian biochemist; b. Mar. 18, 1892; M.D. Prof., 2d Leningrad Med. Inst., 1936-45, also dir. Vitamins Inst.; prof. Latvian U., 1945-51, Riga Med. Inst., 1953—; dir. sect. on metabolism and nutrition Inst. Exptl. Medicine, Latvian Acad. Scis., 1953—, acad. sec. sect. on biol. scis., 1946-52. Chmn., Learned Med Council, Latvian Ministry Health. Author: Ascorbic Acid, its Nature and Significance in the Living Body, 1941. Co-editor chemistry sect. Bolshaya meditsinskaya entisiklopediya, 2d edit. Research on biochemistry and prodn. of vitamins, protein-vitamin complexes, insulin. Address: USSR, Riga, Sovetsky b. 8, Meditsinksy institut.

SHNEIDEROV, Anatol James, cosmophysicist; b. Ekaterinburg, Russia, July 29, 1894; s. James G. and Alexandra (Petukhov) S.; C.E., Petrograd Mil. Engring. Sch., 1917, Mag. Mil. Eng., 1918; B. Elec. Engring., George Washington U., 1944; M.A., Tchrs. Coll. Columbia, 1948; postgrad. Johns Hopkins, 1945-46, Cath. U. Am., 1944-45, 55-58; m. Siren O. Martirosiantz, Sept. 30, 1930 (dec. Dec. 1958); 1 dau., Svetozara Anatol'evna (Mrs. Maxim D. Persidsky). Came to U.S., 1941, naturalized, 1950. Owner, Izida Assns., Harbin-Shanghai, China, 1922-41; managing dir. Shneider Process Co., Ltd., Hong Kong, Shanghai, Washington, 1941-45; chmn., pres. Polycultural Instn. Am., Inc., Washington, 1945—; geophysicist U.S. Geol. Survey, 1958-62; fellow European Center for Research on Gravitation, Rome, Italy, 1961, del. to U.S., 1962, prof., 1964; engr., geophysicist Arctic Bibliography Project, Arctic Inst. N.Am., Washington, 1962—. Mem. Am. Geophys. Union, Am. Phys. Soc., Philos. Soc. Washington, Fedn. Am. Scientists, A.A.A.S. Author: The Dreams I Dream and the Life I Lived, 1927; The Little Blue Book of Shanghai, published 1932. Research, publs. on radional field theory; pioneer in earth's expansion theory; astroseismology; author of expulsion planetary theory. Home: 1736 Columbia Rd., N.W., Washington 20009. Office: Arctic Inst. N.Am., Library of Congress, Annex C.R. 261, Washington 20540.*

SHOBEN, Edward Joseph, Jr., Am. psychologist; b. Oberlin, O., Oct. 3, 1918; s. Edward Joseph and Clara (Wharton) S.; A.B., U. So. Cal., 1939, M.A., 1945, Ph.D., 1947; m. Vera Anne Smith, Apr. 4, 1942; 1 son, Edward Joseph, III. High sch. tchr., 1939-41; counselor Vets Guidance Center, U. So. Cal., 1946, asst. in psychoednl. clinic, 1946-47; counselor, then dir. Student Counseling Office, State U. Ia., 1947-50, asst. prof. psychology, 1947-50; faculty Tchrs. Coll., Columbia U., 1950-65, prof. edn. and psychology, 1956-65, dir. clin. tng., 1962-64, coordinator coll. relations, 1964-65; prof. higher edn. and psychology, dir. Center for Research and Tng. in Higher Edn., U. Cin., 1965-66; dir. Commn. on Academic Affairs, Am. Council on Edn., 1966—. Commd. scientist USPHS, 1957—. Mem. numerous ednl. coms., bds. Bd. dirs. Washington Internships in Edn. Fellow Am. Psychol. Assn. (mem. council of reps.) past div. pres.), A.A.A.S., N.Y. Acad. Scis.; mem. Am. Assn. Humanistic Psychology (past pres.), Am. Coll. Personnel Assn. (past mem. exec. council), Am. Archaeol. Inst. Editor: Ednl. Record, 1966—; asso. editor Psychol. Record, 1965—. Author: (with L.F. Shaffer) The Psychology of Adjustment, 1956; Perspectives in Psychology, 1964; (with J.E. Fine, J.L. Malfetti) The Development of a Criterion for Driver Behavior, 1965; (with Ohmer Milton) Learning and the Professors; also numerous articles. Research in personality factors in academic learning, effects of coll. on personal devel. Office: Am. Council on Edn., 1785 Massachusetts Av. N.W., Washington 20036.*

SHOCK, Nathan Wetherill, Am. physiologist; b. Lafayette, Ind., Dec. 25, 1906; s. Joseph Henry and Blanche (Stults) S.; B.S., Purdue U., 1926, M.S., 1927, D.Sc., 1954; Ph.D., U. Chgo., 1930; m. Margaret Truman, Sept. 9, 1928; children—Joseph Baird, John Howard. Research Asso., U. Chgo., 1930-34; research asso. Inst. Child Welfare, asst. prof. physiology U. Cal., Berkeley, 1932-41; chief gerontology br. Nat. Heart Inst., also Balt. City Hosps., 1941—. Recipient Modern Medicine Achievement award in gerontology, 1960; Superior Service award Dept. Health, Edn., Welfare, 1965. Mem. Gerontological Soc. (pres. 1960), Am. Psychol. Assn. (pres. div. 20, 1952-53), Am. Physiol. Soc., Am. Heart Assn. (chmn. research com. 1963-64), Soc. Exptl. Biology and Medicine (chmn. md. sect. 1956), A.A.A.S. (chmn. sect. N, med. sci., 1959), Am. Geriatrics Soc. (Willard Thompson award 1965). Author: Trends in Gerontology, 1957, 63; Classified Bibliography Gerontology and Geriatrics, 1957. Editor, Conf. on Problems of Aging, Jour. Macy Found., 1950-54; editor-in-chief Jour. Gerontology, 1962—. Research, publs. on physiol. basis of aging. Home: 6505 Maplewood Rd., Balt. 21212. Office: Balt. City Hosp., Balt. 21224.*

SHOCKLEY, William Bradford, physicist; b. London, Eng., Feb. 13, 1910; s. William Hillman and May (Bradford) S.; B.S., Cal. Inst. Tech., 1932; Ph.D., Mass. Inst. Tech., 1936; Sc.D. (hon.), Rutgers University, 1956, U. Pa., 1955, Gustavus Adolphus Coll. (Minn.), 1964; m. Jean A. Bailey, 1933 (divorced 1955); children—Alison, William Alden, Richard Condit; married second, Emmy Lanning, 1955. Teaching fellow Mass. Institute Tech., 1932-36; mem. tech. staff Bell Telephone Labs., 1936-42, 45, became director transistor physics department, 1953; director Shockley Semiconductor Lab.; president Shockley Transistor Corp., 1958-60; director Shockley Transistor unit Clevite Transistor, 1960-63; lecturer Stanford U., 1958-63, Alexander M. Poniatoff prof. engring. sci., 1963—; deputy director and dir. research weapons systems evaluation group Department of Defense, 1955-58; expert cons. Office Sec. War, 1944-45; vis. lectr. Princeton, 1946; sci. advisor policy council Joint Research and Development Bd., 1947-49. Dir. research Antisubmarine Welfare Operations Research Group, U.S.N., 1942-44. Awarded Medal for Merit; co-winner (with John Bardeen and Walter H. Brattain) Nobel Prize in Physics, 1956; Wilhelm Exner medal Oesterreichischer Gewerbeverein of Austria, 1963; Holley medal Am. Soc. M.E., 1963. Fellow American Phys. Soc. (O. E. Buckley Prize), Am. Acad. Arts and Sci.; mem. Nat. Acad. Sci. (Comstock Prize), Institute Radio Engrs. (Morris Liebmann Prize), Sigma Xi, Tau Beta Pi. Author: Electrons and Holes in Semiconductors, 1950. Editor: Imperfections in Nearly Perfect Crystals, 1952. Inventor of junction transistor; research on energy bands of solids, ferromagnetic domains, plastic properties of metals, theory of grain boundaries, and order and disorder in alloys; 50 U. S. patents. Home: 23466 Corta Via, Los Altos, Cal. Office: Stanford U., Stanford, Cal. 94305.*

SHOCKMAN, Gerald David, Am. microbiologist; b. Mt. Clemens, Mich., Dec. 22, 1925; s. Solomon and Jennie (Madorsky) S.; B.S., Cornell U., 1947; Ph.D., Rutgers U., 1950; m. Arlyne Taub, June 2, 1949; children—Joel, Judith, Deborah. Research fellow Rutgers U., Brunswick, N.J., 1947-50; research asso. U. Pa., Phila., 1950-51, research fellow, 1951; research asso. Inst. for Cancer Research, Phila., 1952-60; asso. prof. microbiology Temple U. Sch. Medicine, Phila., 1960-66, prof., 1966—. British-Am. exchange fellow Oxford U., 1954-55. Fellow Am. Acad. Microbiology; mem. Am. Soc. Biol. Chemists, Soc. for Gen. Microbiology, Am. Soc. Microbiology (chmn. gen. div. 1966), Sigma Xi. Research on bacterial cell walls, membranes, protein synthesis, nutrition, growth and mode of action of antimicrobial agts. Home: 6401 N. 12th St., Phila. 19126.*

SHOEMAKER, David Powell, Am. phys. chemist; b. Kosskia, Ida., May 12, 1920; s. Roy Hopkins and Sarah (Anderson) S.; B.A., Reed Coll., 1942; Ph.D., Cal. Inst. Tech., 1947; m. Clara Brink, Aug. 5, 1955; 1 son, Robert B. John Simon Guggenheim Meml. Found. fellow, Inst. for Theoretical Physics, Copenhagen, Denmark, 1947-48; sr. research fellow chemistry Cal. Inst. Tech., 1948-51; faculty Mass. Inst. Tech., Cambridge, 1951—, prof. chemistry, 1960—. Cons., Humble Oil and Refining Co., 1957—. Mem., Am. Chem. Soc., Am. Phys. Soc., Am. Crystallographic Assn., Am. Acad. Arts and Scis., A.A.A.S., U.S.A. Nat. Com. for Crystallography (sec. treas. 1962-64), Internat. Union Crystallography (mem. crystallographic computing commn. 1960—, internat. tables commn. 1963—), Phi Beta Kappa, Sigma Xi, Phi Lambda Upsilon. Author: (with Carl W. Garland) Experiments in Physical Chemistry, 1962. Co-editor, Acta Crystallographica, 1964—. Research in crystal structures of metals, alloys, zeolites, organic compounds, crystal surfaces. Home: 50 Peacock Farm Rd., Lexington, Mass. 02173.*

SHOEMAKER, Frank Crawford, Am. physicist; b. Ogden, Utah, Mar. 26, 1922; s. Roy Hopkins and Sarah Parker (Anderson) S.; A.B., Whitman Coll., 1943; Ph.D., U. Wis., 1949; m. Ruth Elizabeth Nelson, July 11, 1944; children—Barbara Elaine, Mary Frances. Staff mem. Radiation Lab., Mass. Inst. Tech., 1943-45; instr. physics U. Wis., 1949-50; mem. faculty Princeton, 1950—, prof. physics, 1962—; asso. dir. Princeton-Pa. Accelerator, 1962—. Vis. scientist Rutherford High Energy Lab., Eng., 1965-66. Fellow Am. Phys. Soc.; mem. Am. Assn. Physics Tchrs., Phi Beta Kappa, Sigma Xi, Tau Kappa Epsilon. Co-author proposal for 3 billion electron volt Princeton-Pa. Accelerator, 1955. Home: 361 Walnut Lane, Princeton, N.J.*

SHOEMAKER, Robert Alan, Canadian mycologist; b. Toronto, Ont., Can., July 9, 1928; s. Stuart Alan and Verna (Winch) S.; B.S.A., U. Toronto, 1950, M.S.A., 1952; Ph.D., Cornell U., 1955; m. Eileen Patricia Constable, June 10, 1950; children—Christopher Alan, Frederick Robert, Leith Ellen, Jeffrey Carleton. With Plant Research Inst., Ottawa, Ont., 1955—, chief mycology sect., 1967—. Mem. Canadian Bot. Assn., Canadian Phytopath. Soc. Mycol. Soc. Am. Research and publs. on taxonomic studies of Pyrenomycetes including devel. and life-histories of asso. Fungi Imperfecti with spl. emphasis on Cochasso. Fungi Imperfecti with spl. emphasis on Cochliobolus, Pyrenophora, Broomella; description of new genera Bipolaris and Blogiascospora. Home: 93 Beaver Ridge, Ottawa 5. Office: Plant Research Inst., Central Exptl. Farm, Ottawa 3, Ont., Can.*

SHOEMAKER, William C., Am. physician; b. Chgo., Feb. 27, 1923; s. Frank O. and Frances (Morrison) S.; A.B., U. Cal. at Berkeley, 1943, M.D., 1946; m. Norma Johnson, Dec. 23, 1953; children—Robert, Stephan, Frank, Thomas. Research fellow surgery and biochemistry Harvard Med. Sch., 1956-60; prof. surgery Univ. of Ill., 1966—; dir. dept. surg. research Hektoen Inst., Chgo., 1964—; chief 3d surg. service Cook County Hosp., Chgo., 1964—. Recipient Research career award Nat. Heart Inst., NIH, 1962. Author: Shock: Chemistry, Physiology, and Therapy, 1966; also numerous articles. Research on hepatic physiology and metabolism, hemorrhagic, traumatic, and septic shock, electrolyte metabolism. Home: 4908 S. Kimbark Av., Chgo. 60615. Office: 1835 W. Harrison St., Chgo. 60612.*

SHOENBERG, David, physicist; b. St. Petersburg, Russia, Jan. 4, 1911; s. Isaac and Esther (Eisenstein) S.; B.A., Trinity Coll., Cambridge, Eng., 1932, Ph.D., 1935; m. Catherine Félicitée Fischmann, Mar. 14, 1940; children—Ann (Mrs. Yves La Rogue), Peter Jacques, Jane. Research staff Royal Soc. Mond Lab., Cambridge, 1932—, head lab., 1947—; faculty U. Cambridge, 1944—, reader physics, 1952—; UNESCO expert in low temperature physics Nat. Phys. Lab. India, New Delhi, 1953-54; Andrew W. Mellon vis. prof. U. Pitts., 1962; Gauss prof. U. Göttingen (Germany), 1964. Named mem. Order Brit. Empire, 1943; recipient Fritz London award for low temperature physics, Columbus, O., 1964. Fellow Royal Soc., 1953, Phys. Soc.; mem. Am. Phys. Soc. Author: Magnetism, 1949; Superconductivity, 1939, 52; also numerous articles. Research on measurement penetration depth of magnetic fields into superconductors; discovered de Haas-van Alphen effect in many metals and used to deduce shape and size of Fermi surfaces of many metals. Home: 2 Long Rd., Cambridge, Eng. Office: Royal Soc. Mond Lab., Free School Lane, Cambridge, Eng.*

SHOHNO, Naomi, Japanese physicist; b. Hiroshima, Japan, Nov. 19, 1925; d. Masami and Sakayo (Nogami) Shohno; M.Sc. in Physics, Kyushu U., 1947; D.Sc., Hiroshima U. Lit. and Sci., 1961; m. Chiseko Otani, Jan. 13, 1951; 1 son, Hiroyasu. Asst. physics, Kyushu U., 1947-50; faculty Hiroshima U., 1950—, prof. physics, 1961—. Mem. Phys. Soc. Japan. Research, publs. on compound model of elementary particles, atomic study on properties of liquid helium, effects of atom bomb in Hiroshima and Nagasaki, especially residual radiation effect. Home: Tosuien, Itsukaichi, Saiki-gun, Hiroshima, Japan.*

SHOLES, Christopher Latham, Am. inventor; b. Mooresbury, Pa., Feb. 14, 1819; s. Orrin Sholes; m. Mary Jane McKinney, Feb. 4, 1841; 10 children. Editor, Wis. Enquirer, 1840, Southport (later Kenosha, Wis.) Telegraph, 1841-45; postmaster Southport, 1845; mem. Wis. Senate, 2 terms, Wis. Assembly, 1 term; editor Milw. News, 1860, then Milw. Sentinel; collector Port of Milw., circa 1862; patentee (with Samuel Soule) paging machine, 1864; patentee improvement on numbering machine, 1867; patentee (with Glidden and Soule) lever typewriter, 1868; patentee improvements, 1871, sold patents to Remington Arms Co., 1873, became 1st typewriter used in industry. Died Milw., Feb. 17, 1890.

SHON, Frederick John, Am. nuclear engr.; b. Pleasantville, N.Y., July 24, 1926; s. Frederick and Lucy (Stelz) S.; B.S., Columbia, 1946; postgrad. Ohio State U., U. Cal. at Berkeley; m. Dorothy Theresa Patterson, June 8, 1946; 1 son, Robert Frederick. With Publicker Alcohol Co., Phila., 1946-47, Thermoid Co., Trenton, N.J., 1947-48, Mound Lab., Miamisburg, O., 1948-51; radiation chemist Atomics Internat., Canoga Park, Cal., 1951-52, cons., 1962-63; physicist, reactor operations supr. Lawrence Radiation Lab., Livermore, Cal., 1952-61; chief examining br., div. licensing and regulations U.S. AEC, Washington, 1961-62, chief reactor safety br., div. operational safety, 1963-67, asst. dir. for nuclear facilities, 1967—; cons. nuclear firms including Danish AEC, Risoe, Denmark, 1959; also lectr. nuclear engring. U. Cal. at Berkeley, 1955-63. Mem. Am. Nuclear Soc., Health Physics Soc., Tau Beta Pi. Research, publs. on nuclear reactor fuels, reactor operations, safety procedures. Home: 17401 Beauvoir Rd., Derwood, Md. 20855. Office: U.S. AEC, Washington 20545.*

SHOOK, Glenn Alfred, Am. physicist; b. Osgood, Ind., July 16, 1882; s. Alfred Smith and Olive (Gould) S.; student Moores Hill Coll.; A.B., U. Wis., 1907; Ph.D., U. Ill., 1914; m. Nellie Switzer, Nov. 15, 1911; 1 dau., Elizabeth Louise. Instr. physics Purdue U., 1907-11, U. Ill., 1911-14, U. Mich., 1914-15, Williams Coll., 1915-18; prof. physics Wheaton Coll., Norton, Mass., 1918-48, emeritus. Fellow Royal Soc. Arts, A.A.A.S.; mem. Am. Assn. Variable Star Observers, Am. Astron. Soc., Math. Soc. Am., Optical Soc. Am., Sigma Xi. Author: (with others) Practical Pyrometry, 1917; Mysticism, Sciences and Revelation (pub. Eng.), 1953. Pioneer worker in mobile color and applied optics. Died Aug. 26, 1954.

SHOOTER, Eric Manvers, biochemist; b. Mansfield, England, Apr. 18, 1924; s. Fred and Pattie (Johnson) S.; B.A., U. Cambridge, 1945, M.A., 1949, Ph.D., 1950; D.Sc., U. London, 1964; m. Elaine Staley Arnold, May 28, 1949; 1 dau., Annette Elizabeth. Postdoctoral fellow U. Wis., 1950; sr. scientist Brewing Industry Research Found., Nutfield, Eng., 1950-53; lectr. biochemistry University Coll., London, 1953-63; USPHS Internat. fellowship dept. biochemistry Stanford U. Sch. Medicine, Palo Alto, Cal., 1961-62, asso. prof. genetics, 1963-68, prof., 1968—. Mem. Am. Soc. Biol. Chemists, Chemical Soc., Biochem. Soc., Brit. Biophys. Soc. Contbr. numerous articles in field to sci. jours. Elucidation of the genetic control of the synthesis of hemoglobin and of relationship between presence of abnormal hemoglobins and disease; identification of the replicating subunit of deoxyribonucleic acid; molecular neurobiology. Home: 181 Gabarda Way, Menlo Park, Cal. 94025.*

SHOPPEE, Charles William, chemist; b. London, Eng., Feb. 24, 1904; s. J.W. and Elizabeth (Hawkswell) S.; Ph.D., U. London, 1921, D.Sc., 1930; D.Phil., U. Basle, 1941; m. Eileen Alice West, July 18, 1929; 1 dau., Adrienne. Asst. lectr. U. Leeds, 1929-34, lectr., 1935-39; Rockefeller research fellow U. Basle, 1939-45; reader in chemistry U. London, Royal Cancer Hosp., 1945-48; prof. chemistry U. Wales U. Coll. Swansea, 1948-56; prof. organic chemistry U. Sydney, Australia, 1956—. Vis. prof. U. Ga., Duke U. Fellow Royal Soc., 1956, Chem. Soc.; mem. Royal Inst. Chemistry, Royal Australian Chem. Inst. Research and numerous publs. on reaction mechanisms, steroids and sex hormones, conformational analysis. Home: 41 Klenthurst Rd., Stives, Sydney, Australia.*

SHOR, George G., Jr., Am. geophysicist; b. N.Y.C., June 8, 1923; s. George Gershon and Dorothy (Williston) S.; B.S., Cal. Inst. Tech., 1944, M.S., 1948, Ph.D., 1954; m. Elizabeth Louise Noble, June 11, 1950; children—Alexander Noble, Carolyn Elizabeth, Donald Williston. Computer, Seismic Explorations, Inc., Houston, 1948, party chief, 1949-50; asst. research geophysicist Scripps Instn. Oceanography, La Jolla, Cal., 1953-58, asso. research geophysicist, 1958-64, geophysicist, 1964—. Mem. Sigma Xi. Research in geol. structure of sea floor using seismic reflection and refraction methods. Home: 2655 Ellentown Rd., La Jolla 92037. Office: Box 109, La Jolla, Cal. 92038.*

SHORE, Ferdinand John, Am. physicist; b. Bklyn., Sept. 23, 1919; s. Ferdinand John and Magdalene (Schwarz) S.; B.S., Queens Coll., 1941; M.A., Conn. Wesleyan U., 1943; Ph.D., U. Ill., 1952; m. Paulina Barbara Pucko, May 25, 1946; children—Gregory, David, Carolyn, Pamela, Jonathan. Lab. asst. War Research NDRC Project Conn. Wesleyan U., 1941-45; research asst., U. Ill., 1946-51; asso. physicist Brookhaven Nat. Lab., 1952-60; with Queens Coll., Flushing, N.Y., 1960—, prof., 1964—; cons. physics dept. Brookhaven Nat. Lab., 1961—; mem. Nat. Council on Radiation Protection and Measurements, 1962—. Mem. Sigma Xi. Research on pyrotechnics, piezoelectricity, nuclear decay schemes, shielding, neutron cross sections, fission cross section, spins of neutron resonances, gamma-ray spectroscopy. Home: 77 Southern Blvd., East Patchogue, N.Y. 11772.*

SHORLAND, Francis Brian, chemist; b. Wellington, New Zealand, July 14, 1909; s. John Olive and Edith (Perry) S.; student Wellington Coll., 1921-26; M.Sc., Victoria Univ. Coll., 1932; Ph.D., U. Liverpool, 1937, D.Sc., 1950; m. Dorothy Myrtle Orr, Sept. 2, 1960; children—John Hunter, Alison Mary. Chemist, Agrl. Chem. Lab., Wellington, 1934-46; officer in charge, then dir. fats research div. D.S.I.R., Wellington, 1946-65, dir. food chemistry div., 1965—. Mem. nutrition research com., med. research council U. Otago, 1951—; a v.p. 1st Internat. Congress Food, Sci. and Tech., 1962; v.p. New Zealand Sci. Congress, 1965. Fellow or mem. Am. Oil Chemists Soc., Royal Soc. New Zealand (council 1964—), New Zealand Inst. Chemistry (pres. 1961), New Zealand Nutrition Soc. (sec. 1966—). Contbg. author: Annual Review Biochemistry, 1954; Chemistry Fats and Other Lipids, 1953; Comparative Biochemistry, 1962; Chemical Plant Taxonomy, 1963; also articles in periodicals. Research on relationship between fatty acid composition and classification of plants and animals, isolation and identification in ruminant and other fats of branched chain and eta-odd numbered fatty acids hitherto believed to be absent; elucidation of nature and formation of ruminant fats and trans-acids in particular. Home: 267 Karaka Bay Rd. Office: 109 Sydney St. W., Wellington, New Zealand.*

SHORT, James, Scottish optician; b. Edinburgh, Scotland, June 16, 1710; s. William Short; M.A., Edinburgh U.; math. tutor to duke of Cumberland, 1736; apptd. Gregorian for king of Spain, 1752; fellow Royal Soc., 1737. Contbr. to Philos. Trans. Royal Soc., 1736-63. Observed transits of Mercury, 1753, Venus, 1761, deduced authoritative solar parallax; determined difference of longitude between Greenwich and Paris on basis of observations of 4 transits of Mercury; 1st to give specula a true parabolic figure. Died June 14, 1768.

SHORT, James Franklin, Jr., Am. sociologist; b. Sangamon County, Ill., June 22, 1924; s. James Franklin and Ruth L. (Walbaum) S.; student Shurtleff Coll., Alton, Ill., 1942-43; B.A., Denison U., 1947; M.A., U. Chgo., 1949, Ph.D., 1951; m. Kelma E. Hegberg, Dec. 27, 1947; children—Susan Elizabeth, James Michael. Instr., Ill. Inst. Tech., 1950, Ind. U., South Bend extension, 1950-51; mem. faculty Wash. State U., Pullman, 1951—, prof. sociology, 1951—, dir. Sociol. Research Lab., 1962-65, dean Grad. Sch., 1964—; vis. asso. research prof. (Nat. Inst. Mental Health grantee) U. Chgo., 1959-62. Cons., Nat. Inst. Mental Health, 1961—, mem. behavioral scis. fellowship rev. panel, 1963-66; cons. NSF, 1960—, Ford Found., 1962-66. Faculty Research fellow Social Sci. Research Council, 1953-56. Mem. Am. (chmn. research com. 1963, mem. at large, council 1967—), Pacific (No. v.p. 1965-66, pres. 1967, adv. council 1968-69) sociol. assns. Author: (with A. F. Henry) Suicide and Homicide, 1954; (with F. L. Strodtbeck) Group Process and Gang Delinquency, 1965; Gang Delinquency and Delinquent Subcultures, 1968; also articles, chptls. in books. Asso. editor Social Problems, 1958-61, 67—, Am. Sociol. Rev., 1960-63, Am. Jour. Sociology, 1964—. Contbr. encys., yearbooks. Research on measurement and description of juvenile delinquent gangs and their relation to various social conditions; isolation of group processes accounting for delinquent character of some behavior episodes among gangs of adolescents. Home: 415 Dexter St., Pullman, Wash. 99163.*

SHORT, Roger Valentine, English vet. physiologist; b. Weybridge, Eng., July 31, 1930; s. Frank Arthur and Marian (Valentine) S.; B.V.Sc., U. Bristol, 1954; M.S., U. Wis., 1955; Ph.D., U. Cambridge, 1958; m. Mary Bowen Wilson, Apr. 19, 1958; children—Nicholas Robert Marcus, Fiona Mary, Clare Anna. Mem. ARC unit Reproductive Physiology and Biochemistry, U. Cambridge, 1956—, univ. lectr. dept. vet. clin. studies, 1961—; fellow Magdalene Coll. Mem. Endocrine Soc. (com. mgmt. 1967—), Soc. for Study Fertility, Zool. Soc. London. Contbr. numerous articles in field to sci. jours. Devel. methods for quantitative estimation of sex hormones in blood, comparative aspects of reprodn. in man, domestic animals, lab. animals and wild animals, including deer and elephants. Home: 92 New Rd., Haslingfield, Cambridge, Eng.*

SHORT, Sidney Howe, Am. inventor; b. Columbus, O., Oct. 8, 1858; s. John and Elizabeth (Cowen) S.; grad. Ohio State U., 1880; m. Mary Morrison, July 26, 1881, 4 children. Prof. physics and chemistry, v.p. U. Colo.; patented several electric traction inventions, 1880-85; joined U. S. Elec. Co., Denver, 1885, developed improved electric arc-lighting, new electric motor for streetcars; formed (with Charles Brush) Short Electric Ry. Co. for mfg. electric railroad equipment, Cleve., 1889, sold co. to Gen. Electric Co., 1892; joined Walker Mfg. Co., Cleve., 1893, made further improvements in electric traction equipment; obtained around 500 patents on elec. inventions, gained internat. reputation for knowledge in elec. ry. operation; went to Eng. to arrange for prodn. of inventions in electric ry. field, 1898; became tech. supt. English Electric Mfg. Co., Ltd., 1900. Died in Eng., Oct. 21, 1902.

SHORTHILL, Richard Warren, Am. physicist; b. Aberdeen, Wash., Dec. 28, 1928; s. William Warren and Elizabeth Ann (Boyle) S.; student Westminster Coll., Salt Lake City, 1947-48; B.A. in Physics, U. Utah, 1954, Ph.D., 1959; m. Ruth Louise Wood, Sept. 10, 1948 (dec. Sept. 1967); children—David Warren, Ann Louise; m. 2d, Ellen Lorrain Eggleston, June 19, 1968. TV cameraman sta. KSL-TV, 1949-50, TV audio engr., 1952-55; lab. instr. physics, then research asst. physics and teaching fellow U. Utah, 1955-59; research physicist, space physics, aero-space div. Boeing Co., 1960-62; research physicist, lunar physics, Boeing Sci. Research Lab., 1962—. Mem. Working Group Extra-Terrestrial Resources; discovered anomolous cooling lunar ray craters; spl. research thermal, photometric properties of lunar surface. Mem. Optical Soc. Am., Am. Astron. Soc., Am. Inst. Physics, Royal Astron. Soc. Can., Astron. Soc. Pacific, Sigma Xi. Research on thermal, photometric properties of lunar surface; discovered anomolous cooling Ray craters during eclipse; produced thermal, photometric maps, thermal images of moon through complete lunation. Home: 3315 58th St. S.W., Seattle 98116. Office: Boeing Sci. Research Labs., Seattle 98124.*

SHORTLEY, George, Am. physicist; b. Mpls., Mar. 3, 1910; s. George Hiram and Mabel (Johnson) S.; B.E.E., U. Minn., 1930; A.M., Princeton, 1930-33, Ph.D., 1933; postgrad. (NRC fellow) Harvard, also Mass. Inst. Tech., 1933-35; m. Irene Wilson, June 17, 1955. From instr. to prof. physics Ohio State U., 1935-42, 1946-49; div. chief Naval Ordnance Lab., 1942-46, Operations Research Office, Johns Hopkins, 1949-55; asso. dir. Borg-Warner Research Center, Des Plaines, Ill., 1955-57; v.p. Booz Allen Applied Research, Inc., Bethesda, Md., 1957—. Recipient Distinguished Civilian Service award Sec. Navy, 1946. Fellow Am. Phys. Soc.; mem. Am. Astron. Soc., Operations Research Soc. Am. (editor 1953-61, pres. 1965-66), Sigma Xi. Author: (with E. U. Condon) Theory of Atomic Spectra, 1935; (with Dudley Williams), Physics, 1950; Elements of Physics, 1953; Principles of College Physics, 1959. Research and publs. in quantum mechanics, astrophysics, applied math., numerical analysis, elasticity, mathematical biology, and atomic spectra. Home: 7200 Broxburn Dr., Bethesda 20034. Office: 4733 Bethesda Av., Bethesda, Md. 20014.*

SHOSTAKOVSKII, Mikhail Fedorovich, Russian organic chemist; b. 1905; grad. Irkutsk State U., 1929. Sr. sci. worker USSR Acad. Scis. Inst. Organic Chemistry, 1935-39; became chief Lab. Vinyl Compounds, 1939; named dir. Irkutsk Inst. Organic Chemistry Siberian br. USSR Acad. Scis., 1957. Recipient Stalin Prize. Corr. mem. USSR Acad. Scis. Research and publs. on vinyl esters, vinyl sulfides. Address: Institut Organic Chemistry, Siberian Br., USSR Acad. Scis., Irkutsk, Siberia.

SHOTTON, Frederick William, English geologist; b. Coventry, Eng., Oct. 8, 1906; s. Frederick John and Ada (Brookes) S.; B.A., Sidney Sussex Coll., Cambridge, Eng., 1927, M.A., 1930, Sc.D., 1947; m. Alice Louise Linnett, Sept. 16, 1930; children—Anne Elizabeth (Mrs. Donald Anthony Black), Margaret Alice (Mrs. David Abee). Lectr., U. Birmingham (Eng.), 1928-36, U. Cambridge, 1936-40; prof. geology, head dept. U. Sheffield (Eng.), 1945-49; prof. geology, head dept. U. Birmingham, 1949—, dean faculty sci. 1957-60, pro-vice chancellor, 1965—. Fellow Royal Soc., 1956, Geol. Soc. London (Prestwich medal, 1954) (past pres.); mem. Inst. Mining Engrs., Profl. Assn. Inst. Water Engrs., Soc. Belge de Géologie (hon.). Research, publs. on tectonic studies and stratigraphy of N. Eng. and English midlands, desert deposits of Permian and Trias, water supply on N. African deserts, mil. geology, succession and mode of origin of Pleistocene deposits, succession of insect faunas in Pleistocene. Home: 35 Park Av., Solihull, Warwickshire, Eng. Office: Dept. Geology, Univ., Birmingham 15, Eng.

SHOUDY, Loyal Amborse, Am. surgeon; b. Ellensburg, Wash., Sept. 23, 1880 s. John Alden and Mary Ellen (Stewart) S.; A.B., U. Wash., 1904; M.D., U.

Pa., 1909. Engaged in practice as surgeon, 1910—; intern German Hosp., Phila. (now Lankenau), 1910-13; physician in charge Mary Drexel Childrens Hosp., Phila., 1913-14; chief surgeon Bethlehem Steel Co., 1914-18, chief med. service, 1918-45, med. dir., 1945—; cons. surgeon St. Lukes Hosps., Bethlehem, Pa. Chmn. com. on indsl. health Am. Iron and Steel Inst.; dir. Nat. Safety Council; pres. Bethlehem area Boy Scouts Am.; mem. Health Adv. Council, U. S. C. of C.; mem. Adv. Group, N.A.M. Fellow A.C.S. (mem. fracture com.); mem. Am. Assn. Indsl. Physicians and Surgeons (pres.), A.M.A., Am. Pub. Health Assn., Conf. Bd. of Indsl. Physicians, Pa. Med. Soc. Contbr. numerous articles on traumatic and indsl. surgery, research on heat sickness and problems of indsl. hygiene. Died Aug. 30, 1950.

SHOUPP, William Earl, Am. physicist; b. Troy, O., Oct. 5, 1908; s. William Albert and Lydia Grace (Slack) S.; B.A., Miami U., Oxford, O., 1931, D.Sc., 1957; M.A., U. Ill., 1933, Ph.D., 1938; m. Kathryn Elizabeth Torbeck, June 13, 1932; 1 son, William Joseph. Grad. asst. instr. in physics, U. Ill., 1932-38; mgr. electronics and nuclear physics research dept. Westinghouse Electric Corp., Pitts., 1938-48, dir. research Bettis Atomic Power Lab., 1948-54, tech. dir. comml. atomic power, 1954-61, tech dir. Astronuclear Lab., 1961-62, v.p. research, 1962—. Mem. com. on biotech. and human research NASA, 1962-64; mem. divisional com. for engring, NSF, 1964-67; chmn. nuclear standards bd. U.S.A. Standards Inst. Named Man of Year in Sci., Pitts. Jr. C. of C., 1949; recipient Order of Merit, Westinghouse Electric Corp., 1953. Fellow Am. Phys. Soc., Am. Nuclear Soc. (pres. 1964-65, chmn. honors and awards com.), I.E.E.E.; mem. Am. Soc. M.E., Am. Inst. Aeros. and Astronautics, Nat. Acad. Engring. Research, patents, publs. on nuclear and reactor physics; devel. of nuclear devices; co-discoverer of photofission; research on radar and communication. Home: 343 Maple Av., Pitts. 15218. Office: Westinghouse Electric Co., Churchill Borough, Pitts. 15235.*

SHPIKITER, Vadim Olegovich, Russian biochemist; b. Krasnodar, USSR, Sept. 18, 1922; s. Oleg and Elena Shpikiter; grad. Rostov U., 1949, candidate scis., 1953, D.Sc., 1963; m. Galina A. Velikodvorskaya, 1956; children—Vera, Tatijana. Jr. research worker Inst. Biology and Med. Chemistry, Moscow, USSR, 1952-56, sr. research worker, 1956—; prof. biol. dept. Moscow U., 1964—. Mem. Biochem. Soc. USSR. Research, publs. on properties, structure and function of proteins especially soluble collagen, proteolytic enzymes, subunit structure of proteins; discovered alpha and beta components of soluble collagen. Home: V-415, Udaltsov St. 16, Moscow. Office: G-117, Pogodin St., 10, Moscow, USSR.*

SHRADER, Erwin Fairfax, Am. physicist; b. Pitts., Nov. 1, 1916; s. James Edmund and Ava (Ross) S.; A.B., Swarthmore Coll., 1937; Ph.D., Yale, 1940; m. Nancy Jane Siverd, Aug. 1, 1942; children—Stephen P., Ellen R., Ann E., Nancy Jane, Erwin Fairfax. Prof. physics Case Western Res. U., Cleve., 1940—; with div. war research Columbia, 1943-45; with U. S. Army C.E., Oak Ridge, 1945; with Clinton Lab., Oak Ridge, 1945-46; acting chief phys. and math. br., div. research AEC, 1955-57. Mem. Ohio Atomic Energy Adv. Com., 1958—. Fellow Am. Phys. Soc.; mem. Phi Beta Kappa, Sigma Xi. Study of photo nuclear reactions; interaction of high energy gamma rays with matter; time-of-flight neutron spectroscopy; nuclear detection devices. Home: 2042 Brunswick Rd., East Cleveland, O. 44112. Office: 10900 Euclid Av., Cleve. 44106.*

SHRAPNEL, Henry, English artillery expert; b. Eng., June 3, 1761; s. Zachariah and Lydia (Needham) S.; m. Esther Squires, May 5, 1810; 2 sons, including Henry Needham Scrope, 2 daus. Commd. successively arty. capt., 1795, brevet-maj., 1802, maj. in royal arty., 1803, regtl. lt.-col., 1804, col., 1813, maj.-gen., 1819, col., 1827, lt. gen., 1837; named 1st asst. insp. arty., 1804; worked at Royal Arsenal, Woolwich, Eng., for many years; named commandant Royal Artillery, 1827. Compiled range tables. Invented shrapnel shell, 1784, brass tangent slide, some fuses; improved small arms and ammunition; introduced parabolic chambers for mortars and howitzers. Died Southampton, Eng., Mar. 13, 1842.

SHREEVE, Herbert Edward, inventor, engr.; b. Cambridge, Eng., Aug. 1, 1873; s. William Sayer and Caroline (Ogden) S.; student elec. tech. City of London Guilds; m. Emily Loring Ames, May 7, 1904; a son, Herbert Prescott. Naturalized Am. citizen, 1917. With inspection dept. Am. Bell Telephone Co., 1895-98, engr., 1898-1908; with engring. dept. Western Electric Co., 1908-15; worked on 1st wireless telegraph tests between U. S. and France, Chgo. Radio Expts., beginning 1915; served with Signal Corps, U. S. Army, 1917-19; asst. to pres. Bell Telephone Labs., 1923-26; tech. rep. in Europe, Am. Tel. & Tel., also Bell Co., 1926-35. Named Pioneer Inventor, N.A.M. Patentee numerous tel. repeaters and transmission devices. Died 1942.

SHREEVE, Walton Wallace, Am. biochemist; b. Muncie, Ind., Aug. 2, 1921; s. Leonard Dale and Kathryn (Nichols) S.; B.A., DePauw U., 1943; M.D., Ind. U., 1944; Ph.D., Western Res. U., 1951; m. Phyllis Heidenreich, Dec. 30, 1945; children—Thomas, Daniel, James, Elizabeth. Research physician VA Hosp., Cleve., 1950-52; head radioisotope lab. U.S. Naval Hosp., Oakland, Cal., 1952-54; scientist Med. Research Center Brookhaven Nat. Lab., Upton, N.Y., 1954-64, sr. scientist, 1964—, dir. edn., 1961-66, chmn. radioisotopes com., 1963-65; clin. dir. Planned Parenthood chpt. Suffolk County, N.Y., 1965-66; guest prof. Karolinska Inst., Stockholm, Sweden, 1966-67. Mem. Am. Soc. Biol. Chemists, Am. Diabetes Assn., Endocrine Soc. Author: (with others) Principles of Nuclear Medicine, 1967. Research, numerous publs. on application of Carbon-14 and later tritium to study of intermediary carbohydrate metabolism in human subjects; first study of oxidation to CO2 and turnover of blood glucose in diabetics; gluconeogenesis; formation of plasma and liver lipids from 14C and 3H-labeled precursors in diabetic humans and mice. Home: 58 S. Country Rd., Bellport, N.Y. 11713. Office: Med. Research Center, Brookhaven Nat. Lab., Upton, N.Y. 11973.*

SHREVE, Forrest, Am. botanist; b. Easton, Md., July 1878; s. Henry and Helen Garrison (Coates) S.; A.B., Johns Hopkins, 1901, Ph.D., 1905; m. Edith Coffin Bellamy, June 1909; 1 dau., Margaret Bellamy. Asst. prof. botany Goucher Coll., Balt., 1906-08; mem. staff Div. Plant Biology, Carnegie Instn., Washington, 1908-43, in charge Desert Lab., Tucson, 1926-39; mng. editor The Plant World, 1911-19. Fellow A.A.A.S. (pres. southwest div., 1928-29); mem. Ecol. Soc. Am. (pres. 1922), Bot. Soc. America, Torrey Bot. Club, Cal. Bot. Soc., Western Soc. Naturalists, Assn. Am. Geographers (v.p. 1940), Assn. Pacific Coast Geographers (pres. 1942), Soc. Am. Foresters, Tucson Nat. Hist. Soc. (pres. 1932-33), Phi Beta Kappa, Sigma Xi (pres. Ariz. chpt. 1932-33). Author: Plant Life of Maryland (with others), 1910; A Montane Rain-Forest, 1914; Vegetation of a Desert Mountain Range, 1915; The Cactus and Its Home, 1931; also numerous papers in sci. jours. Joint author: Distribution of Vegetation in the United States, 1921. Editor: Naturalists Guide to the Americas, 1926. Died July 19, 1950.

SHRINER, Ralph Lloyd, Am. chemist; b. St. Louis, Mo., Oct. 9, 1899; s. George Bain and Edith Mercedes (Barnett) S.; B.S., Washington U., St. Louis, Mo., 1921; M.S., U. of Ill., 1923, Ph.D., 1925; m. Rachel Haynes, Aug. 17, 1929; 1 dau., Joan. Instr. chemistry, Washington U., 1921-22; asso. in research, N.Y. Expt. Station. 1925-27; asst. prof., U. of Ill., 1927-30, asso. prof., 1930-35, prof. organic chemistry, 1935-41; prof. chemistry and chmn. dept., Ind. U., 1941-46; prof. organic chemistry U. Ia., Iowa City, 1947-51, head dept. chemistry, 1951-64, prof. emeritus, 1964; vis. prof. chemistry of Southern Methodist University, 1963—; member of division Chemical and Chemical Tech., National Research Council, 1953-59; mem. math., phys., engring. div., Nat. Sci. Found, 1955-57. Mem. Am. Chem. Soc. (sec. organic div. 1935-40, councillor-at-large, 1943-46), A.A.A.S., Sigma Xi, Author or co-author books relating to field; also numerous sci. articles. Mem. editorial bd. Jour. Organic Chemistry, Organic Syntheses. Editor-in-chief: Chemical Reviews, since 1950. Research on synthesis and structure of organic compounds, including waxes, phytosterols, alkaloids; studies of anthocyanins and flavylium salts, lignin model compounds, synthetic drugs and stereoisomerism. Home: 2709 Hanover St., Dallas 75225.*

SHROCK, Robert Rakes, Am. paleontologist; b. Wawpecong, Ind., Aug. 27, 1904; s. Andrew and Stella (Glassburn) S.; A.B., Ind. U., 1925, A.M., 1926, Ph.D., 1928; m. Theodora Antoinette Weidman, Feb. 2, 1933; children—Wendolyn Theodora, Robert Ellsworth. Faculty, U. Wis., Madison, 1928-37, asst. prof., 1931-37; faculty Mass. Inst. Tech., Cambridge, 1937—, prof. paleontology and sedimentology, 1950—, head dept. geology, 1949-65; research asso. Harvard, 1950—. Mem. Geol. Soc. Am., Paleontol. Soc., Am. Assn. Petroleum Geologists, Soc. Econ. Paleontologists and Mineralogists, Nat. Assn. Geology Tchrs. Author: Invertebrate Paleontology, 1935; Index Fossils of North America, 1944; Principles of Invertebrate Paleontology, 1953; Sequence in Layered Rocks, 1948. Editor: International Series in the Earth Sciences, 1951-65. Research and publs. in use of fossils and sedimentary structures in determination of geology of past; discovered ancient Silurian coral reefs of Michigan Basin; described new graptolites, corals, and other Silurian fossils from Ind.; discovered deposits of aluminum ore in Haiti, Jamaica. Home: 18 Loring Rd., Lexington, Mass. 02173. Office: Mass. Inst. Tech., Cambridge, Mass. 02139.*

SHRUM, Gordon Merritt, Canadian physicist; b. Smithville, Ont., Can., Jan. 14, 1896; s. William Burton and Emma Jane (Merritt) S.; B.A., U. Toronto, 1920, M.A., 1921, Ph.D., 1923; D.Sc., U. B.C., 1961, McMaster U., 1963; m. Oenone Georgellen Baillie, May 20, 1930; children— Gordon Baillie, Laurna Jane (Mrs. Ian Strang). Physicist, Corning Glass Co., 1924-25; mem. faculty U. B.C., 1925-61, dir. grad. faculty, 1956-61; chmn. B.C. Hydro and Power Authority, Vancouver, 1961—; chancellor Simon Fraser U., Vancouver, 1963—. Dir., B.C. Research Council, 1952-61; mem. Nat. Research Council of Can., 1943-49, 50-56, Def. Research Bd., 1947-50, 51-54; chmn. B.C. Energy Bd., 1959—. Fellow Royal Soc. Can.; mem. Canadian Assn. Physicists. Research and publs. on liquefied helium; discovered origin of auroral green line, 1925. Home: 5941 Chancellor Blvd., Office: 970 Burrard St., Vancouver, B.C., Can.*

SHTERN, Lina Solomonovna, Russian physiologist; b. Aug. 26, 1878; b. Libava (now Liepaia, Latvia), Aug. 26, 1878; grad. U. Geneva (Switzerland); became prof. U. Geneva, 1917; prof. 2d Moscow Med. Inst. (formerly Med. Faculty 2d Moscow U.), 1925-49, dir. Physiology Inst., 1929-49; staff Inst. Biol. Physics, USSR Acad. Scis., 1954—. Recipient Stalin prize, 1943. Mem. USSR Acad. Med. Scis. Research on chem. basis of physiol. processes in animals and humans, tubercular meningitis, tetanus, encephalitis, antibiotics. Address: Inst. Biol. Physics, USSR Acad. Scis., Leninskii Prospekt, 33, Moscow, USSR.

SHTOKALO, Iosif Zakharovich, Russian mathematician; b. Skomorokhi (now Lvov Oblast), 1897; grad. Dnepropetrovsk U., 1931; D.Physico-Math. Sci., 1944. Instr., Kharkov Engring. Econ. Inst., Kharkov U., 1931-41; with Inst. Math., Ukrainian Acad. Scis., 1941-44; instr. Kiev U., 1944-46, prof., 1946-51, 56—. Mem. Ukrainian Acad. Scis. (chmn. Presidium Lvov br. 1949-56). Exec. editor A Ukrainian Mathematical Bibliography, 1963. Research and publs. on linear differential equations with variable, quasiperiodic and variable coefficients. Address: Kiev University, Vladimirskaya ulitsa 64, Kiev, Ukrainian SSR, USSR.

SHTRIKMAN, S., electronic engr.; b. Brisk, Poland, Oct. 21, 1930; s. Avraham and Esther (Kozlovsky) S.; B.Sc., Technion, Israel Inst. Tech., 1953, Diplomed Engr., Elec. Engr, 1954, D.Sc., 1958; m. Rachel Chodrovsky; children—Hadas, Ilan. With dept. electronics Weizmann Inst. Sci, Rehovoth, Israel, 1954—, sr. scientist, 1961-64, asso. prof., 1957—. Sr. vis. fellow dept. sci. and indsl. research, dept. math. Imperial Coll. Sci., London, 1960-62; sr. staff physicist Franklin Inst., Phila., 1964-65; vis. prof., NSF sr. fgn. fellow dept. physics U. Pa., Phila, 1964-65; cons. electronic industries, Haifa, Israel, 1965—. Mem. Am. Phys. Soc., I.E.E.E. Research, publs. on devel. of new theory of magnetization processes in ultrafine ferromagnetic particles which accounts for size dependence of coercive force; discovered new variational principles in elasticity which enable theoretical determination of electroproperties of heteroferous materials. Home: 1 Hasaron, Rehovoth, Israel.*

SHUBNIKOV, Aleksei Vasilévich, Russian crystallographer; b. Mar. 29, 1887; grad. Moscow U., 1912. Research staff, faculty People's U. of Shanyanvskii, Moscow, 1912-20; vis. prof. Urals Minng Inst., Sverdlovsk, 1920-25; with USSR Acad. Scis., 1925—, chief lab. crystallography, 1937-43, dir. Inst. Chrystallography, 1944-62; prof. cyrstallography Moscow U., 1953—. Recipient Stalin Prize, 1946, 50, Red Banner of Labor, 1962. Mem. USSR Acad. Scis., All-Union Mineral. Soc., Mineral. Soc. (hon.), French Mineral. Soc. (hon.). Author: How Crystals Grow, 1935; Quartz and its Application, 1940; (with E. E. Flint, G. B. Bokii) Fundamentals of Crystallography, 1940; Symmetry, 1940; Piezo-Electric Textures, 1946; Optical Crystallography, 1950; Symmetry and Anti-Symmetry of Finite Figures, 1951; (with others) Investigating Piezo-Electric Textures, 1955; Crystals in Science and Technology, 1956. Editor: (with N. N. Sheftal) Soveschanie po rostu kristallov, 1959; Growth of Crystals, vol. 3. Research on crystal growth; elec., optical, properties of crystals; crystal symmetry, including piezo-electric properties, methods of cutting, processing and polishing. Home: pl. Vosstaniya, 1, Moscow, USSR. Office: Inst. Crystallography, USSR Acad. Scis. Pyzhevskii Pereulok, 3, Moscow, USSR.

SHUIKIN, Nikolai Ivanovich, Russian organic chemist; b. Mar. 30, 1898; grad. Moscow U, 1927; With Moscow U., 1930-37, became prof, 1943; joined Inst. Organic Chemistry, USSR Acad. Scis. Mem. USSR Acad. Sci. (corr.). Research and publs. on preparation of catalysts for dehydrogenation of cyclanes and alkanes; hydrogenation and hydrogenolysis of furan nucleus. Home: 1-Aya Cheremushkinskaya, 3, Moscow. Office: N.D. Zelinskii Inst. Organic Chemistry, USSR Acad. Sci., Leninskii Prospekt, 31, Moscow, USSR.

SHULEIKIN, Vasilii Vladimirovich, Soviet geophysicist; b. Jan. 13, 1895; An organizer Black Sea Hydrophys. Sta., Crimea, 1929, Oceanographic Hydrophys. Lab., 1935, chair marine physics Moscow U., 1945, oceanographic sect. Moscow Hydrometeorol. Inst., 1930; mem. or leader various ocean and sea expdns. Recipient Semenov-Tian-Shanskii prize All Union Geog. Soc.; Stalin Prize, 1942; Order of Lenin, twice. Mem. USSR Acad. Sci. Author: Outline of Marine Physics, 1949; Marine Physics, 3d edit., 1953; The Theory of Sea Waves, 1956; More About Eddy Current in the Sea, 1960. Developed theory on thermal balance of sea (led to prediction of sub-surface thermal currents in Kara Sea), theory of sea waves, theory for thermal interaction between atmosphere, ocean, and continents; demonstrated origin of sea and lake color; devel. oceanographic instruments; worked out spectral curve equation on sea color. Office: AN SSSR, Leninsky prosp. 14, Moscow, USSR.

SHULL, Aaron Franklin, Am. biologist, geneticist; b. Miami County, O., Aug. 1, 1881; A.B., U. Mich., 1908; Ph.D. in Zoology, Columbia U., 1911; m. 1911; 2 children. Asst. zoology U. Mich., 1905-08, Columbia U., 1909-11; acting asst. prof. U. Mich., 1911, instr., 1911-12, asst. prof., 1912-14, asso. prof., 1914-21, prof., 1921-51, emeritus prof., 1951—. Fellow Entomol. Soc.; mem. A.A.A.S., Soc. Naturalists (sec. 1920-26, v.p. 1929, pres. 1934), Am Zool. Soc., Genetics Soc., Soc. for Study Evolution, Am. Eugenics Soc., Assn. Biol. Tchrs., Genetics Assn. Research on genetics of sex in insects, induced crossing over in Drosophilia, developmental physiology and genetics of aphids, genetics of lady beetles as related to evolution, life cycle of rotifiers.

SHULL, Clifford Glenwood, Am. physicist; b. Pitts., Sept. 23, 1915; s. David H. and Daisy (Bistline) S.; B.S., Carnegie Inst. Tech., 1937; Ph.D., N.Y. U., 1941; m. Martha-Nuel Summer, June 19, 1941; children—John G., Robert D., William F. Research physicist The Tex. Co., Beacon, N.Y., 1941-46; chief physicist Oak Ridge Nat. Lab., 1946-55; prof. physics Mass. Inst. Tech., Cambridge, 1955—. Chmn. vis. com. Brookhaven Nat. Lab., Upton, N.Y., 1961-62. Recipient Buckley prize Am. Phys. Soc., 1956. Fellow Am. Acad. Arts and Scis., Am. Phys. Soc. (chmn. solid state physics div. 1962-63); mem. A.A.A.S., Am. Crystallographic Assn., Research Soc. Am., Sigma Xi, Tau Beta Pi, Phi Kappa Phi. Research and publs. in solid state and neutron physics. Home: 4 Wingate Rd., Lexington, Mass. 02173.*

SHULL, George Harrison, Am. botanist; b. Clark County, O., Apr. 15, 1874; s. Harrison and Catharine (Ryman) S; B.S, Antioch Coll., 1901; LL.D., 1940, Ph.D., U. Chgo., 1904; Sc.D., Lawrence Coll., 1940, Ia. State Coll. Agr. and Mechanics Arts, Ames, 1942; m. Ella Amanda Hollar, July 8, 1906; 1 dau., Elizabeth Ellen; m. 2d, Mary J. Nicholl, Aug. 26, 1909; children—John Coulter, Georgia Mary, Frederick Whitney, David Macaulay, Barbara Weaver, Harrison. Bot. asst. U. S. Nat. Mus., 1902; bot. expert U. S. Bur. Plant Industry, 1902-04; asst. plant physiology U. Chgo., 1903-04; bot. investigator Sta. for Exptl. Evolution, Carnegie Instn. of Washington, Cold Spring Harbor, L.I., 1904-15; prof. botany and genetics Princeton, 1915-42, emeritus, 1942—; vis. lectr. in genetics Rutgers U., 1929-30, L. L. Kellogg meml. lectr., 1931; lectr. in Heterosis Conf., Ia. State Coll., 1950. Recipient Gold medal (for invention of hybrid corn) DeKalb Agrl. Assn., 1940; citation N.J. Bd. Agr., 1945; John Scott medal and premium, 1946; Marcellus Hartley medal Nat. Acad. Sci., 1949; elected to Hall of Fame Am. Mechanics Mag. Golden Jubilee, 1952. Fellow A.A.A.S.; mem. Deutsche Botanische Gesellschaft (corr.), Acad. Sci. (corr., Vienna), Gesellschaft fur Pflanzenzüchtung in Wien (hon.), John Torrey Club of Princeton (hon.), Deutsche Gesellschaft für Vererbungswissenschaft, Société Linnéenne de Lyon, Institut Internat. d'Anthropologie (Paris), Torrey Bot. Club (pres. 1947), Bot. Soc. Am., Am. Soc. Naturalists (v.p. 1911, pres. 1917), Ecol. Soc. Am., Am. Genetic Assn. (chmn. plant sect. 1912, advisory com. 1922—), Eugenics Research Assn. Eugenics Soc. Am., Genetics Soc. Am., Am. Geog. Soc., Am. Soc. Plant Physiology, Washington Acad. Sciences, Am. Philos. Soc., Sigma Xi. Lectr., contbr. papers on variation, heredity and plant-breeding. Founder and mng. editor Genetics (mag), 1916-25, asso. editor, 1925—; vice-pres. Genetics, Inc., 1940—; 1st editor genetics sect. Bot. Abstracts, 1918-22. Died Sept. 28, 1954; buried Santa Rosa, Cal.

SHULL, Harrison, Am. chemist; b. Princeton, N.J., Aug. 17, 1923; s. George Harrison and Mary Julia (Nicholl) S.; A.B., Princeton U., 1943; Ph.D., U. Cal. at Berkeley, 1948; m. Jeanne Johnson, Jan. 10, 1948 (div. 1962); children—James Robert, Kathy, George Harrison, Holly; m. 2d, Wilma Joyce Bentley, Aug. 17, 1962; children—Stanley Martin, Sarah Ellen. Asso. chemist U.S. Naval Research Lab., 1943-45; asst. prof. Ia. State U., 1949-54; mem. faculty Ind. U. Bloomington, 1955—, research prof., 1961—, dean Grad. Sch., 1966—; asst. dir. quantum chem. group Uppsala (Sweden) U., 1958-59; dir. Research Computer Center, Ind. U., 1959-63; vis. prof. U. Wash., 1960, U. Colo., 1963. Mem. Fulbright Selection Panel in Chemistry, 1959-67, chmn., 1963-67; mem. chemistry adv. panel NSF, 1964-67, cons., 1965—. NRC fellow, 1948-49, Guggenheim fellow, 1954-55. Fellow Am. Phys. Soc.; mem. Am. Chem. Soc., Assn. for Computing Machinery, A.A.A.S., Phi Beta Kappa, Sigma Xi, Phi Lambda Upsilon, Alpha Chi Sigma. Asso. editor Jour. Chem. Physics, 1952-54; editorial adv. bd. Spectrochimica Acta, 1957-63; editorial bd. Internat. Jour. Quantum Chemistry, 1966—. Research and numerous publs. on theoretical chemistry, nature of chem. bond, spectroscopy. Home: 3614 Longview St., Bloomington, Ind. 47401.*

SHULL J(ames) Marion, Am. botanist; b. Clark County, O., Jan. 23, 1872; s. Harrison and Catharine (Ryman) S.; ed. pub. schs., business coll., Valparaiso U., Art Students' League, N.Y.; m. Addie Virginia Moore, Dec. 20, 1906 (dec. Apr. 1937); children—Virginia Moore, Francis Marion; m. 2d, Mary Ethel Lerch, Apr. 1947. Student and instr. Antioch Coll., Yellow Springs, O., 1896-98; supr. music and drawing, Boise, Ida., 1898-99; tchr. pub. schs. Ohio, comml. artist, Memphis, Tenn., 1899-1906; in U. S. Post Office Dept 1906-07; dendrological artist U. S. Forest Service, Washington, 1907-09; bot. artist Bur. Plant Industry, 1909-25, asso. botanist, 1925-42; made over 1700 water color drawings and many in black and white for Dept. of Agr.; widely known as breeder of new varieties of iris and hemerocallis. Mem. A.A.A.S., Am. Hort. Soc., Bot. Soc. Washington, Am. Iris Soc. (Distinguished Service medal 1944, Silver medal), Nat. Carillon Assn. Author: Rainbow Fragments—A Garden Book of the Iris, 1931. Contbr. articles and illustrations to mags. Died Sept. 1, 1948.

SHULMAN, Yechiel, engr.; b. Tel-Aviv, Israel, Jan. 28, 1930; s. David and Rachel (Chonowsky) S.; S.B. in Aero. Engring., Mass. Inst. Tech., 1954, S.B. in Indsl. Mgmt., 1954, S.M. in Aero. Engring., 1954, Sc.D. in Aeros. and Astronautics, 1959; m. Ruth Danzig, June 29, 1950; children—Elinor, Deborah, Ron Eliezer, Orna Leah. Project engr. Aeroelastic and Structures Lab., Mass. Inst. Tech., 1954-59; asst. prof. Technol. Inst., Northwestern U., Evanston, Ill., 1959-62, asso. prof., 1962-67; research cons. Anocut Engring. Co., 1967—; cons. Am. Machine and Foundry Co., 1959-62, Gen. Am. Transp. Corp., 1962-67; mem. tech. staff Aerospace Corp., 1963. Fellow Am. Inst. Aeros. and Astronautics (asso., tech. com. on structural dynamics); mem. Am. Soc. M.E., A.A.A.S., Am. Soc. Engring. Edn. Research and publs. on aeroelastic instability of flight structures, dynamics of shell structures, orbital transfer and satellite rendezvous optimization, interaction of submerged structures with soils, coupled theory of dynamic thermo elasticity, control of electrochem. machining process. Home: 1248 Ash St., Winnetka, Ill. 60093. Office: 2375 Estes Av., Elk Grove Village, Ill. 60007.*

SHULTZ, George Pratt, Am. economist, educator; b. N.Y.C., Dec. 13, 1920; s. Birl E. and Margaret (Pratt) S.; B.A. cum laude, Princeton, 1942; Ph.D., Mass. Inst. Tech., 1949; m. Helena M. O'Brien, Feb. 16, 1946; children—Margaret Ann, Kathleen, Peter, Barbara, Alex. With Mass. Inst. Tech., 1946-57, asso. prof. indsl. relations, 1955-57; prof. indsl. relations U. Chgo., 1957—, dean Grad. Sch. Bus., 1962—. Mem. arbitration panels for labor mgmt. disputes in industries including elec. equipment, farm implements, textiles, chems., food products, metal fabricating, 1953—; sr. staff economist Pres.'s Council Econ. Advisers, 1955-56; cons. Office Sec. U. S. Dept. Labor, 1959-60, Pres.'s Adv. Com. on Labor-Mgmt. Policy; mem. steering com. Study of Collective Bargaining Basic Steel Industry, 1960; mem. Gov.'s Com. on Unemployment, Ill., 1961-62. Mem. research adv. bd. Com. for Econ. Devel., 1965—. Mem. Am. Econ. Assn., Indsl. Relations Research Assn., Nat. Acad. Arbitrators, Am. Assn. U. Profs. Author: (with C. A. Myers) The Dynamics of a Labor Market, 1951, Pressures on Wage Decisions, 1951; (with Thomas A. Whisler; Management Organization and the Computer, 1960; (with D. V. Brown) The Structure of Collective Bargaining, 1961. Contbr. articles in field to sci. jours. Home: 5731 S. Blackstone Av., Chgo. 60637.*

SHULUTKO, Lazar Ilich, Russian surgeon, orthopedist; b. 1897; grad. Med. Faculty, Odessa U., 1922; D.Med. Sci. Intern. Sevastopol 1st City Hosp., surg. clinic Kharkhov Tb Inst., chief physician, head surg. dept. Dukhovshchina Rayon Hosp., Smolensk Oblast, cons., chief physician Yevpatoriya orthopedic hosps., sr. asst. orthopedics and traumatology clinic Kazan Postgrad. Med. Inst., 1922-37; head chair orthopedics and traumatology Kazan Postgrad. Med. Inst., 1937—; dir. Kazan Research Inst. Crthopedics and Traumatology, 1945—. Mem. All-Union Soc. Traumatologists and Orthopedists (bd. mem.). Mem. editorial council Orthopedics, Traumatology and Prosthetics. Research and numerous publs. on osteomuscular pathology, osteosynthesis, osteoplastics, treatment of gunshot and other fractures, indsl. and agrl. traumatism. Address: Kazan Postgraduate Medical Institute, ulitsa Komleva 11, Kazan, RSFSR, USSR.

SHUMWAY, Bruce Wayne, Am. physicist; b. Shumway, Ariz., Dec. 16, 1918; s. Wallace Everett and Pearl (Denham) S.; B.S., U. Wash., 1944; student U. Cal. at Berkeley, 1946, 48-49, U. Paris, 1949-50; m. Michele Helene Tuchschmid, Apr. 6, 1950. Electronics engr. Nat. Cancer Inst. Lab. of Exptl. Oncology, San Francisco, 1947-48, biophysicist, 1950-53; nuclear physicist U. S. Naval Radiol. Def. Lab., San Francisco, 1953—. Mem. Am. Nuclear Soc., Health Physics Soc. Research and publs. in field of nuclear radiation shielding, dosimetry, biophysics. Home: 202 Hilton Av., South San Francisco, Cal. 94080. Office: U. S. Naval Radiol. Def. Lab., San Francisco 94135.*

SHUMWAY, Norman Edward, Am. heart surgeon; b. Kalamazoo, Feb. 9, 1923; s. Norman Edward and Laura (Vander Vliet) S.; M.D., Vanderbilt U., 1949; Ph.D., U. Minn., 1956; m. Mary Lou Stuurmans, June 21, 1951; children—Sara Jane, Norman Michael, Lisa Anne, Amy Martha. Faculty dept. surgery Stanford U. Sch. Medicine, Palo Alto, Cal., 1958—, prof., 1965—, chief div. cardiovascular surgery, 1964—. Mem. Am. Surg. Assn., Soc. U. Surgeons, Am. Assn. for Thoracic Surgeons, Soc. Thoracic Surgeons, Halsted Soc. Research and numerous publs. on transplantation of heart and heart valves. Home: 1291 Pitman St., Palo Alto, Cal.*

SHUMWAY, Waldo, Am. biologist; b. New Brunswick, N.J., May 8, 1891; s. Edgar Solomon and Florence (Snow) S.; A.B., Amherst Coll., 1911; A.M., Columbia, 1913, Ph.D., 1916; Mining Engr., Stevens Inst. Tech., 1954; m. Helen Davis, Nov. 20, 1920; 1 dau., Jean (Mrs. Peter Ferguson). Field worker Amherst Coll. Biol. Expdn. to Patagonia, 1911-12; asst. in zoology Columbia, 1914-15; asst. in biology Amherst, 1915-16, instr. in biology, 1916-17; asst. prof. biology Dartmouth, 1919-22; asso. prof. zoölogy U. Ill., 1922-29, prof., 1929-47; asst. dean Coll. Liberal Arts and Scis., 1926-31; dean Stevens Inst. Tech., 1947—, sec. bd. trustees 1947-55, provost, 1955—. Fellow A.A.A.S.; mem. Am. Soc. Zoölogists, Am. Soc. Engring. Edn., Am. Soc. Mil. Engrs., Am. Assn. Anatomists, Am. Soc. Naturalists, Am. Soc. Growth and Devel., Phi Beta Kappa, Sigma Xi. Author: Vertebrate Embryology, 1927; The Frog, a Laboratory Guide, 1928; Textbook of General Biology, 1931; Laboratory Manual for Vertebrate Embryology (with F. B. Adamstone), 1939. Contbr. to sci. and ednl. jours. Died Mar. 8, 1956; buried Arlington Nat. Cemetery.

SHUPE, James LeGrand, Am. veterinarian; b. North Ogden, Utah, Nov. 5, 1919; s. Parley Grant and Luella (Chadwick) S.; B.S., Utah State U., 1948; D.V.M., Cornell U., 1952; m. Faye Chambers, Sept. 3, 1957; children—James Grant, Dennis Chambers. Faculty, Utah State U., Logan, 1952-61, 66—, prof., 1959-61, 66—; research veterinarian Agrl. Research Service, U. S. Dept. Agr., Logan, 1961-66. Cons. Minister of Health, Ireland, 1963; cons. Govt. Switzerland, 1966. Recipient Honor Alumni award Utah State U. Mem. Am., Utah vet. med. assns., Internat. Acad. Pathology, Pan Am. Med. Assn., Soc. Toxicology. Author (with Frank Smith) Handbuck-Pharmakologie-Flourine, 1966; also numerous articles. Research on effects of flourides on livestock and wildlife; arthritis and degenerative arthrosis in cattle, polyarthritis in sheep; induced and hereditary congenital malformations in cattle. Home: 296 S. 250 East, Hyde Park, Utah 84318.*

SHURBET, Deskin Hunt, Jr., Am. seismologist; b. Lockney, Tex., Aug. 27, 1925; s. Deskin Hunt and Ethel (Ewing) S.; B.S., U. Tex., 1950, M.S., 1951; m. Larke Ann Harrington, Dec. 27, 1958; children—Pamela Lynn, Patricia Ann, Kari Larke. Dir., Bermuda Seismol. Obs., St. Georges, 1951-56; dir. Seismol. Obs., Tex. Technol. Coll., Lubbock, 1956—, prof. geoscis., 1956—. Cons. seismologist, 1956—. Fellow A.A.A.S.; mem. Seismol. Soc. Am., Am. Geophys. Union, Soc. Exploration Geophysicists, N.Y. Acad. Sci., Sigma Xi. Research, publs. on internal earth structure, earthquake seismology. Home: 5002 46th St., Lubbock, Tex. 79414.*

SHUSHANIYA, Platon Georgievich, Russian obstetrician, gynecologist; b. 1894; grad. Med. Faculty, Odessa U., 1924; D.Med. Sci., 1936. Head chair obstetrics and gynecology Faculty Mother and Child Welfare, Tbilisi Med. Inst., 1931-42, head chair obstetrics and gynecology Med. Faculty, 1942—. Chmn. obstetrics commn. Georgian Ministry Health, also mem. learned med. council. Decorated Order of Lenin. Mem. Georgian (lectr. 1931-36), All-Union (bd. mem.), Kiev (hon.), Kharkov (hon.) socs. obstetricians and gynecologists. Author numerous works, including: The Female Hormone Sexual Cycle and Disorders of It, 1936; Gynecology Textbook, 3 edits.; Gonorrhea in Women. Co-editor Obstetrics and Gynecology sect. Large Med. Ency., 2d edit.; mem. editorial council Obstetrics and Gynecology. Address: Tbilisi Medical Institute, ulitsa Melikishvili 16, Tbilisi, Georgian SSR, USSR.

SHUSTER, Carl Nathaniel, Jr., Am. marine biologist b. Randolph, Vt., Nov. 16, 1919; s. Carl Nathaniel and Edith (Gilman) S.; B.S., Rutgers U., 1942, M.S., 1948; Ph.D., N.Y. U., 1955; m. Helen Irwin, May 4, 1944; children—George Whitcomb, Kenneth Ashton, Chris Irwin, Carl Nathaniel, III, Forrest. Faculty, Rutgers U., 1948-55; dir. Marine Labs., U. Del., 1955-63; dir. N.E. Marine Health Sciences Labs., Pub. Health Service, U. S. Dept. Health, Edn. and Welfare, Narragansett, R.I., 1963—; adj. prof. of oceanography and zoology University of R.I., 1964—. Lank lecturer in the dept. biol. scis. U. Montreal (Que., Can.), 1962. Fellow A.A.A.S., N.Y. Acad. Scis.; mem. Am. Inst. Biol. Scis., Am. Soc. Limnology and Oceanography, Am. Soc. Zoology, Ecol. Soc. Am., Nat. Shellfisheries Assn., Del. Conservation Edn. Assn. (past pres.), Sigma Xi, Chi Psi. Research and publs. on estuarine ecology, horseshoe crabs especially natural history, morphology, and systematic relationships, ecol. role of leeches, biology mollusks; mem. Smithsonian-Bredin Caribbean expedition, 1958. Home: 51 Potter Rd., North Kingstown, R.I. 02852. Office: N.E. Marine Health Scis. Lab., USPHS, Narragansett, R.I. 02882.*

SHUSTER, Sam, English dermatologist; b. London, Eng., Aug. 24, 1927; s. Solomon and Anne (Feldman) S.; M.B., Coll. London, 1951, U. Ph.D., 1956, M.R.C.P., 1958; m. Rosemary Roberts Powell; children—David, Gabriel, Saskia. Research asst. dept. physiology U. Coll. London, 1952-55; med. registrar Brook Gen.'s Mile End Hosp., 1955-57 (all London); research fellow Royal Postgrad. Sch. Medicine Hammersmith Hosp., London, 1957-59; lectr. medicine Welsh Nat. Sch. Medicine, Cardif, 1959-61; sr. lectr. dermatology Inst. Dermatology, London, 1961-64; prof. dermatology U. New Castle Upon Tyne, 1964—. Recipient Parkes-Weber medal, 1967. Mem. Brit. Assn. Dermatology, Med. Research Soc., Skin Club, Brit. Soc. for Research in Aging. Research, numerous publs. in renal and pituitary adrenal physiology, physiology of skin and mechanism of certain skin diseases, effect of aging and hormones on skin, discovery that skin disease leads to severe dysfunction of internal organs including heart and bowel and of gen. metabolism and certain vitamins and minerals. Home: 2 Park Villas, New Catle-Upon-Tyne, Eng.

SHUTE, Evan Vere, Canadian physician; b. Lion's Head, Ont., Can., Oct. 21, 1905; s. Richard James and Jane (Treadgold) S.; B.A., U. Toronto, 1924, M.B., 1927; m. Marian Roberta Migger, Sept. 10, 1932; children—James, Roberta, Barry, Janet, Vere. Research asst. U. Western Ont., 1933-39; med. dir. Shute Inst., London, Ont., 1948—; practice medicine specializing in obstetrics, gynecology and vascular diseases, London, Ont., 1933—. Diplomate Am. Bd. Obstetrics and Gynecology. Fellow Royal Coll. Physicians and Surgeons; mem. Osler Soc. (past pres.), Am. Soc. Abdominal Surgery (Silver medal 1962), Am. Soc. for Study Sterility, Canadian Physiol. Soc., Brit. Soc. Endocrinologists, Canadian Soc. for Study Fertility (past pres.), Royal Soc. Medicine, (asso.), Am. Geriatric Soc. Author: Alpha Tocopheroi in Cardiovascular Disease, 1954; The Heart and Vitamina E, 1960; also many books of prose and verse; contbr. to anthologies; also numerous articles. Research on oestrogen treatment pre-eclampsia and eclampsia, Vitamin E use in cardiovascular renal diseases, burns, diabetes; morphine, congenital anaomalies, menorrhagia, menopause, prevention of prematurity, radiation menopause, hyperthyroidism, vulvovaginitis, retroversion, purpura, thrombosis. Home: Arva, Ont., Can. Office: 10 Grand Av., London, Ont., Can.*

SHUTTLEWORTH, Robert James, botanist, conchologist; b. Dawlish, Eng., Feb. 1810; s. James and Anna Maria (Roper) S.; ed. Geneva; studied medicine, Edinburgh, Scotland, 1830-32; Ph.D., (hon.), Basle, Switzerland; m. Susette, dau. of Count of Sury of Soleure, 1833; 1 son, Henry, 1 dau. Capt. Duke of Lancaster's regiment, 1833; settled in Berne and lived in Switzerland, from 1834; asst. sci. travellers; friend of Messner and Jean de Charpentier. Fellow Linnean Soc.; asso. Zool. Soc., Lyceum of N.Y.; original mem. Bot. Soc. Edinburgh. Author: Nouvelles Observations sur la Matière coloriante de la niege rouge, 1840; Notitiae malacologicae, 1856, Part II in German, 1878; also papers. Made one of the most important shell collections of his day; his herbarium (more than 150,000 specimens of flowering plants, 20,000 cryptograms) added to Brit. Mus. collection. Died Hyères, Apr. 19, 1874.

SHUYKIN, Nikolay Ivanovich, Russian organic chemist; b. 1898; grad. Moscow U., 1927. Instr., Moscow U., 1930-43, prof., 1943—; asso. Inst. Organic Chemistry, USSR Acad. Scis., 1937—. Recipient Stalin prize, 1946. Mem. USSR Acad. Scis. (corr.). Co-author: The Mechanism and Kinetics of Heterogeneous Catalysis in Organic Chemistry, 1955; author: Catalytic Synthesis of Nitrites, 1959. Research and publs. on furan nucleus hydrogenation and other catalytic contact conversions of furan compounds; developer methods to produce highly-active and stable catalysts for dehydrogenation of 5- and 6-membered cyclanes and alkanes. Address: Moscow University, Leninskie gory, Moscow, USSR.

SHVETSOV, Petr Filimonovich, Russian geologist; b. Jan. 27, 1910; grad. Moscow Geol. Survey Inst., 1935. Staff, Main Directorate No. Seaway, 1935-39; staff Inst. Permafrost, USSR Acad. Sci., 1939-59, became dep. dir., 1948, dir., 1956-59; mem. Kamchatka expdn., 1959. Chmn. commn. on problems of North, Council for Study Prodn. Resources, 1958—. Recipient Stalin prize, 1952. Corr. mem. USSR Acad. Scis. Author: Permafrost and Geological Engineering Conditions of the Anadyr' Region, 1938; (with V. P. Sedov) Gigantic Icing and Underground Waters of the Ridge of Tas-Khayatakh, 1941; The Earth's Frozen Strata, 1963. Home: ul. Chkalova 39/41. Cffice: V. A. Obruchev Inst., Permafrost, USSR Acad. Scis., Bolshoy Cherkasskiy Pereulok 2/10, Moscow, USSR.

SHY, George Milton, Am. physician; b. Trinidad, Colo., Sept. 30, 1919; s. James C. and Zella May (Henderson) S.; B.S., Ore. State Coll., 1940; M.D., U. Ore., 1943; M.R.C.P., Royal Coll. Physicians, 1947; M.S., McGill U., 1950; m. E. Doreen Shy, Jan. 21, 1945; children—Michael E., Kathleen E. NRC fellow Montreal Aero. Inst.; faculty U. Colo., Boulder, 1951-53; clin. dir. NIORB, NIH, Bethesda, Md., 1953-60; sci. dir., 1960-62; prof., chmn. dept. neurology U. Pa., Phila., 1962-67; prof., chmn. dept., dir. N.Y. Neurol. Inst., Columbia U., 1967. Mem. med. adv. bd., chmn. E. Pa. Multiple Sclerosis; cons. to Surgeon Gen., USPHS. Mem. Am. Neurol. Assn., Am. Assn. Neurologists, others. Research in neuromuscular disease, nuclear medicine, interim metabiology. Died Sept. 25, 1967.

SIBALD, Sir Robert, Scottish physician; b. Fifeshire, Soctland, Apr. 15, 1641; M.D., Leiden, Holland, 1661; apptd. geographer royal to Charles II; a founder, 1st prss. Coll. Physicians, U. Edinburgh (Scotland). Sibbaldia, genus of plants, named in his honor. Author numerous books on bot. history of Scotland, including Scotia Illustrata. Founder bot. garden, Edinburgh. Died 1712.

SIBATANI, Atuhiro, Japanese biologist; b. Osaka, Japan, Aug. 1, 1920; s. Zenjiro and Shizu Shibatani; B.Sc., Kyoto (Japan) Imperial U., 1946; D.Sc., Nagoya (Japan) U., 1953; D.med.Sc., Yamaguichi Med. Sch., Ube, Japan, 1960; m. Kiku Oshitani, May 26, 1966. Research staff Minophagen Pharm. Co., Tokyo, Japan, 1946-50; asst. Osaka U., 1950-53; faculty Yamaguchi Med. Sch., 1953-62, prof., 1955-62; prof. Hiroshima (Japan) U., 1962—; research asso. Rockefeller Inst., N.Y.C., 1960-61; sr. prin. research scientist Commonwealth Sci. and Indsl. Research Orgn., Sydney, Australia, 1966—. Mem. spl. com. Council Sci. and Tech., Japanese Govt., 1965-66. Mem. Japan Soc. Biochemistry, Japan Soc. Biophysics, Biochem. Soc. (Britain). Author: Nucleic Acids, 1953; Revolution in Biology, 1960; Looking Into Life, 1966; also numerous articles. Research on metabolic stability of deoxyribonucleic acid in animal cells and concentration of messenger ribonucleic acid by means of phenol treatment, sensitivity of nucleic acid synthesis to ultraviolet light. Office: Commonwealth Sci. and Indsl. Research Orgn., Delhi Rd., North Ryde, N.S.W., Australia; also Research Inst. for Nuclear Medicine and Biology, Hiroshima U., Kasumicho, Hiroshima, Japan.*

SIBBITT, Wilmer Lawrence, Am. mech. engr.; b. Greencastle, Ind., June 6, 1914; s. William Harrison and Arminta (Larch) S.; B.S. in Chem. Engring. with distinction, Purdue U., 1937, M.S., 1941, Ph.D., 1942; postgrad. U. Ill., 1938-39; m. Selma Doris Otte, Sept. 9, 1947; children—Sharon Martha, Wilmer Lawrence, Randy Robert, Sallye Lou, Tina Roberta, John Harrison. Research chemist E.I. du Pont de Nemours and Co., 1937-38; sr. chem. engr. Phillips Petroleum Co., 1945-46; sr. engr., group leader Monsanto Chem. Co., Clinton Nat. Lab., Carbide and Carbon Chems. Corp., Oak Ridge Nat. Lab., 1946-48; faculty Purdue U., 1941-45, 1948-55, prof., 1952-55; mem. staff Ramo-Wooldridge Corp., Los Angeles, 1955-56; staff nuclear propulsion div. Los Alamos Sci. Lab., 1957—; prof. mech. engring. Los Alamos Grad. Center, U. N.M., 1958—. Mem. Chem. Soc., Am. Soc. M.E., Am. Rocket Soc. Research and publs. on heat transmission, thermodynamics, fluid flow, materials, nuclear engring. Home: 939 Tewa Loop, Los Alamos 87544. Office: Los Alamos Sci. Lab., Box 1663, Los Alamos 87544.*

SIBLEY, William Arthur, Am. physicist; b. Fort Worth, Nov. 22, 1932; s. William Franklin and Sada (Rasor) S.; B.S., U. Okla., 1956, M.S., 1958, Ph.D., 1960; m. Joyce Elaine Gregory, Dec. 21, 1957; children—William Timothy, Lauren Shawn, Steven Marshall. Research physicist Kernforschungsanlage Julich and Inst. for Metal Research, Tech. U., Aachen, Germany, 1960-61; research physicist solid state div. Oak Ridge Nat. Lab., 1961—. Mem. Am. Phys. Soc. Research into effects of radiation, including gamma rays, neutrons, and high energy electrons, on optical and mech. properties of transparent solids; studies of color centers in alkali halides and II-VI compounds. Home: 108 Mohawk Rd., Oak Ridge 37830. Office: Solid State Div., Oak Ridge Nat. Lab., Oak Ridge 37830.*

SIBSON, Francis, anatomist; b. Cross Canonby, Eng., May 21, 1814; s. Francis and Jane S.; apprenticed to John Lizars, 1828; diploma Royal Coll. Surgeons Edinburgh, 1831; M.B., M.D., London, 1848; m. Sara Mary Ouvry, 1858. Resident surgeon, apothecary Nottingham Gen. Hosp., 1835-48; elected mem. senate U. London, 1865; curator Mus. at Coll. Physicians, became Gulstonian lectr., 1854, censor, 1874. Fellow London Coll. Physicians, Royal Soc., 1849. Author: Medical Anatomy; also numerous articles. Described extension of fascia which covered and strengthened dome of pleura; studied respiration. Died Geneva, Switzerland, Sept. 7, 1876.

SIBTHORP, John, English botanist; b. Oxford, Eng., Oct. 28, 1758; s. Humphry and Elizabeth (Gibbes); B.S., Oxford, 1777, M.A., 1780, M.B., 1783, M.D., 1784; student, Montpellier, France. Became Sherardian prof., Oxford, 1784; made bot. expdn. to Greece to determine plants named by Discorides; began another expdn. to Greece, 1794. Mem. French Acad. Scis. Fellow Royal Soc., 1788. Author: Flora Oxoniensis, 1794; Flora Graeca; Prodomus. Collected numerous plant species and some birds in Greece. Died 1796.

SICARD, Jean Athanase, French physician; b. Marseille, France, 1873; physician of hosps. Paris; prof. internal pathology Faculty of Paris. Studied cephalorachidian fluid, (with Wilris) sero-diagnosis of typhoid fever; introduced use of lipiodol as contrast medium in roentgenography, 1921, treatment of pain by alcoholization of nerves, (with J. Paraf, J. Lermoyez; injection of sodium salicylate to treat varicose veins; developed (with Forestier) use of opaque oil of iodine in radiography. Died Paris, 1929.

SICÉ, Jean, pharmacologist; b. Paris, France, Oct. 13, 1919; s. Adolphe and Madeline (Fouchereaux) S.; L.Sc., Faculty Scis. of Marseilles, 1945; D.Ph., Faculty of Medicine of Marseilles, 1943; children—Pierre, Genevieve, Monique. Came to U.S., 1948, naturalized, 1958. Instr. pharmacology Faculty of Medicine, Marseilles, 1945-48; instr. surgery U. Chgo., 1951-53; faculty Chgo. Med. Sch., 1953—, prof., 1966—; sr. research fellow USPHS, 1958-63. Recipient Perron award Nat. Acad. Medicine, Paris, 1945. Author: General Pharmacology, 1962. Research on metabolism and mechanism of action of ethers; also structures and activity. Home: 2020 Ogden Av., Chgo. 60612.*

SICHA, Milos Jirí, Czechoslovakian physicist; b. Praha, Czechoslovakia, Apr. 18, 1930; s. Frantisek and Vojteska (Preissig) S.; M.Sc., Charles U., Praha, 1956, Ph.D., 1960, Docent, 1965; m. Hana Sipova, Sept. 4, 1957. Mem. staff math. and physics, dept. electronics and vacuum physics Charles U., 1956-59, lectr. physics, 1959—, docent/asso. prof., 1967—; also head lab. plasmaphysics, 1963—. Mem. Assn. Czechoslovak Mathematicians and Physicicists. Author: (with others) Fundamental Concepts, Problems and Methods of Plasma Physics and Applications, 1961. Research and publs. on instabilities in low temperature plasma; using probe methods; devel. of original microwave method for measurement of local change of electron concentration in low temperature plasma. Home: 19/415 Severozapadnl V., Praha 4-Sporilov. Office: 5 Ke Karlovu, Praha 2, Czechoslovakia.*

SICHEL, Ferdinand, J.M., physiologist, biophysicist; b. Hamilton, Ont., Can., Oct. 22, 1906; s. Martin M. and M. (Morris) S.; B.Sc., McGill U., 1928; M.S., N.Y. U., 1930, Ph.D., 1934; m. Elsa M. Keil, June 10, 1937; 1 dau., Enid K. Came to U.S., 1928, naturalized, 1940. Demonstrator botany McGill U., 1927-28; asst. instr. N.Y. U., 1928-33, fellow, 1933-34; instr., U. Pa., 1934-35, Royal Soc. Can. fellow med. physics, 1935-36; instr. Howard U. Coll. Medicine, 1936-37; faculty U. Vt. Coll. Medicine, Burlington, 1937—, prof. physiology, chmn. dept. physiology, 1944-49, prof., chmn. dept. physiology and biophysics, 1949—; staff instr. physiology Marine Biol. Lab., Woods Hole, Mass., 1933-44. Cons.physiologist Mary Fletcher Hosp., DeGoesbriand Meml. Hosp., Burlington, 1953—. Fellow A.A.A.S.; mem. Am. Soc. Zoologists, Am. Physiol. Soc., Soc. Gen. Physiologists, Biophys. Soc., I.E.E.E. (sr.), N.Y. Acad. Scis., Vt. Heart Assn. (chmn. research com. 1963—), Vt. State Med. Soc. (hon.), Sigma Xi. Contbr. numerous articles to tech. jours. Research on phys. properties isolated skeletal muscle fibers, excitable and contractile systems in skeletal and cardiac muscle. Home: 35 Henderson Terrace, Burlington, Vt. 05401.*

SICK, Helmut, zoologist; b. Leipzig, Germany, Jan. 10, 1910; s. Paul and Luise (Sell) S.; Ph.D., U. Berlin (Germany), 1937; m. Marga Fehling, Aug. 17, 1938. With Kaiser Wilhelm Inst. Med. Research, Heidelberg, Germany, 1937-38; asst. curator Zool. Mus., U. Berlin, 1938-39; naturalist Cent. Brazil Found., Rio de Janeiro, 1946—; naturalist dept. birds Nat. Mus., Rio de Janeiro, 1959—. Mem. Am. Ornithol. Union, Wilson Ornithol. Club, Assn. Tropical Biology, Academia Brasileira de Ciências, Deutsche Ornithologen Gesellschaft, Deutsche Zoologen Gesellschaft, Deutsche Gesellschaft Naturforscher und Aerzte, Bayrische Ornith. Gesellschaft. Research and publs. on morphology, systematics, distbn., biology and physiology of birds, especially of Brazil. Home: 133 Rua Almirante Alexandrino, Rio de Janerio. Office: Museu Nacional, Quinta da Bôa Vista, Rio de Janerio, G.B., Brazil.*

SICUTERI, Federico, Italian physician; b. Florence Italy, Apr. 13, 1920; s. Pietro and Concetta (Berretti) S.; Degree in Medicine, Florence U., 1945, Specialist in Cardiovascular Diseases, 1954; m. Antonietta Baggiani, Sept. 15, 1949; children—Laura, Francesca. Dir. dept. medicine U. Hosp., Florence, 1955—; dir. Headache Centre, Florence, 1956—. Cons. heart diseases Nat. Heart Inst., 1955—. Recipient award Italian Med. Assn., 1951, Marzotto prize for med. investigations, 1955; H. Wolff award A.M.A., 1966. Mem. Royal Med. Soc. (London). Author: (with Greppi) L'emicrania, 1964; (with Back, Erdön) Hypotensive Peptides, 1966; also numerous articles. Discovered antimigraneous activity of antiserotoninic drugs, vasoconstrictive activity of new drug (indomethacin: indocin); developed new concept of symptoms of intracranial hemorrhages; studies in myocardial infection shock. Home: 17A J. Nardi, Florence. Office: Clinica Medica Università, Viale Morgagni, Florence, Italy.

SIDBURY, James Buren, Jr., Am. physician; b. Wilmington, N.C., Jan. 13, 1922; s. James Buren and Willie (Daniel) S.; B.S., Yale U., 1944; M.D., Columbia U., 1947; m. Alice Lucas Rayle, Aug. 29, 1953; children—Anne, Mary, Patricia, James, Robert. Instr. pediatrics Emory U., 1951-53; pediatrician USPHS Communicable Disease Center, Atlanta, 1951-53; pvt. practice medicine, specializing in pediatrics, Wilmington, 1953-54; with Johns Hopkins U., 1954-61, asst. prof. pediatrics, 1957-61; with Duke U. Med. Center, Durham, N.C., 1961—, dir. clin. research unit, 1961—, prof. pediatrics, 1965—. Diplomate Nat. Bd. Pediatrics; mem. Am. Pediatric Soc., Research Soc. Am., Am. Acad. Pediatrics, N.Y. Acad. Sci., Am. Chem. Soc., So. Soc. Pediatric Research, A.A.A.S., Md. Heart Assn., Am. Soc. Human Genetics, Endocrine Soc. Research and publs. on inherited disorders and metabolic diseases in children. Home: 4044 Nottaway St., Durham, N.C. 27707.*

SIDDAPPA, Gurunanjappa Siddappa, Indian biochemist and food technologist; b. Mysore, India, Jan. 11, 1906; s. Gurunanjappa and Gangamma Siddappa; M.A., Madras, 1930; Ph.D. (postdoctoral fellow), U.

Bristol, Eng., 1936; m. Sundaramma, June 1933; 9 children. Lectr. fruit preservation Agrl. Coll., Lyllapur, Punjab, India, 1937-38; officer-in-charge fruit canning lab., Quetta, Baluchistan, Pakistan, 1938-43; biochemist, govt. agrl. chemist, supt. malt factory Agrl. Coll. and Research Inst., Coimbatore, India, 1943-50; scientist Central Food Technol. Research Inst., Mysore, 1950-65, chmn. tng. program FAO and Nat., 1965-—. Recipient Keshalkar award for sci. contbn. to fruit preservation industry of India, 1963. Colombo Plan fellow, 1954-55. Fellow Royal Inst. Chemistry (London, Eng.); mem. Indian Horticulture Soc., Soc. Biol. Chemists India, Assn. Food Technologists India, others. Author: (with Girdharilal, G. L. Tandon) Preservation of Fruits and Vegetables, 1959. Research, publs, patentee in food sci. and tech. dealing with fundamental as well as applied aspects. Address: Central Food Technol. Research Inst., Mysore-2, India.*

SIDGWICK, Nevil Vincent, English chemist; b. Oxford, Eng., May 8, 1873; s. William Carr Sidgwick; ed. Christ Ch., Oxford, M.A.; Sc.D., Tübingen, Germany; D.Sc. (hon), Leeds, Eng.; LL.D., Liverpool, Eng. Baker lectr. chemistry Cornell U., 1931; prof. chemistry, 1933-45; del. Clarenden Press, 1922-48, Bodleian Library, 1931-48; fellow, tutor Lincoln Coll., Oxford; mem. adv. council Dept. Sci. and Indsl. Research, 1930-35; chmn. Chemistry Research Bd., 1932-35. Fellow Royal Soc., 1922 (Royal medal 1937); mem. Faraday Soc. (pres. 1932-34), Chem. Soc. (pres. 1935-37, Longstaff medal 1945), Am. Acad. Arts and Scis. (fgn.). Author: Organic Chemistry of Nitrogen, 1910; Electronic Theory of Valency, 1927; The Covalent Link in Chemistry, 1933; The Chemical Elements, 1950; also papers. Explained coordination in terms of electronic theory of valency, 1923. Died Mar. 15, 1952.

SIDHU, Surain Singh, physicist; b. Shamnagar, Amritsar, India, June 8, 1902; s. Sundar Singh and Balwant (Kaur) S.; came to U. S., 1920, naturalized, 1947; student U. Cal. at Berkeley, 1920-23; B.S., U. Pitts., 1925, M.S., 1926, Ph.D., 1937; m. Mary Elizabeth Homoney, June 8, 1929; children—Marion Lalita (Mrs. Forbes Linkhorn), Victor S., Ellen Diane. Grad. asst. physics U. Pitts., 1925-27, lectr. physics, asst. x-ray diffraction lab., 1933-37, dir. x-ray diffraction lab., 1938-50, faculty, 1942-50, prof., 1948-50; sr. physicist, group leader Argonne (Ill.) Nat. Lab., 1947-48, 50-—. Founder, Pitts. Ann. Diffraction Conf., 1942; mem. Ill. Bd. Radiation Physics, 1959-—. Fellow Am. Phys. Soc., Am. Coll. Radiology (asso.); mem. Am. Crystallographic Assn., Am. Assn. Physics Tchrs., A.A.A.S., Am. Inst. Physics, Sci. Research Soc. Am. (pres. Argonne chpt. 1958-59), Sigma Xi. Research numerous publs. on neutron and x-ray diffraction, chem. and magnetic structures, nuclear scattering amplitudes, solid state physics, radiography, radiology; co-discoverer negative nuclear scattering amplitude of titanium 48; developed nuclear null-matrices. Home: 1404 Thornwood Dr., Downers Grove, Ill. 60515. Office: 9700 S. Cass Av., Argonne, Ill. 60440.*

SIDIS, Boris, psychopathologist; b. Kiev, Russia, Oct. 12, 1867; came to U. S., 1887; A.B., Harvard, 1894, A.M., 1895, Ph.D., 1897, M.D., 1908. Asso. psychologist and psychopathologist Path. Inst., N.Y. State hosps., 1896-1901; dir. Psychopath. Hosp. and Psychopath. Lab. of N.Y. Infirmary for Women and Children, 1901; in practice at Boston. Med. dir. Sidis Psychotherapeutic Inst., Portsmouth N.H. Asso. editor Archives of Neurology and Psychopathology; asso. editor Jour. Abnormal Psychology. Author: Psychology of Suggestion; Multiple Personality (Sidis and Goodhart), 1905; An Experimental Study of Sleep, 1909; Philistine and Genius, 1911; The Psychology of Laughter, 1913; The Foundations of Normal and Abnormal Psychology, 1914; Symptomatology, Psychognosis and Diagnosis of Psychopathic Maladies, 1914; The Causation and Treatment of Psychopathic Diseases, 1916; The Source and Aim of Human Progress (A Study in Social Psychology and Social Pathology), 1919. Died Oct. 24, 1923.

SIDORENKO, Aleksàndr Vasilievich, Russian geologist; b. Oct. 19, 1917. Mem. staff Turkmen Affiliate USSR Acad. Scis., 1943-50; staff Kola Affiliate, 1950-—; chmn. Presidium, 1952-—. Corr. mem. USSR Acad. Scis. Author: The Main Characteristics of the Formation of Minerals in the Desert, 1956. Research and publs. on geomorphology, geologic structure of deserts; hypergenesis; mineral formation in desert climates; mineral deposits of Turkmen SSR. Address: Presidium of S.M. Kirov Kolski Br. of USSR, Kirovsk, Murmansk Oblast, USSR.

SIDRANSKY, Herschel, Am. pathologist; b. Pensacola, Fla., Oct. 17, 1925; s. Ely and Touba (Bear) S.; B.S., Tulane U., 1948, M.D., 1953, M.S., 1958; postgrad. U. Chgo.; m. Evelyn Lipsitz, Aug. 18, 1952; children—Ellen, David Ira. Vis. asst. pathology Charity Hosp. of La., New Orleans, 1954-58; instr. pathology Tulane U. Sch. Medicine, 1954-58; pathologist Lab. Pathology, Nat. Cancer Inst., NIH, 1958-61; prof. pathology U. Pitts. Sch. Medicine, 1961-—. Mem. Am. Assn. Pathologists and Bacteriologists, Am. Soc. Exptl. Pathology, Soc. for Exptl. Biology and Medicine, Am. Assn. for Cancer Research, Am. Inst. Nutrition, Internat. Acad. Pathologists, A.A.A.S., N.Y. Acad. Scis., Pitts. Pathology Soc., Sigma Xi. Research and publs. on exptl. nutritional deficiency diseases, exptl. fungus (aspergillus) infections, exptl. liver tumorigenesis. Home: 5075 Rosecrest Pl., Pitts. 15201. Office: 3550 Terrace St., Pitts. 15213; also Scaife Hall, U. Pitts. 15213.*

SIEBENS, Arthur A., Am. physiologist, physician; b. Atlanta, 1921; s. Arthur Robert and Irene (Westphal) S.; A.B., Oberlin Coll., 1943; M.D., Johns Hopkins, 1947; m. Barbara Siebens, June 7, 1947; children—Daniel, Arthur, Rebecca, Christopher, David, Janine. Faculty dept. physiology State U. N.Y., 1948-58; faculty dept. pediatrics and physiology U. Wis. Sch. Medicine, Madison, 1958-—, now prof., dir. Rehab. Center, 1959-—. Recipient Borden award. Mem. Am. Physiol. Soc., Am. Thoracic Soc., Congress Phys. Medicine and Rehab., Phi Beta Kappa, Alpha Omega Alpha. Research, publs. on representation of autonomic nervous system in brain, excitability of heart, mechanics of breathing, diffusion of gases in lungs, care of patients with spinal cord injury. Home: 205 Prospect Av., Madison, Wis. 53705.*

SIEBERG, August, German seismologist; b. Aachen, Germany, Dec. 23, 1874; prof., Jena, Germany. Author: Handbuch der Erdbebenkunde, 1904; Erdbebenkunde, 1923. Developed new methods in seismology. Died Jena, Nov. 18, 1945.

SIEBERT, Günther, German biochemist; b. Berlin, Germany, Jan. 28, 1920; s. Kurt and Magdalene (Schuchardt) S.; student medicine U. Freiburg, U. Munich, 1937-39; M.D., U. Berlin, 1945; m. Hannemarie Cram, Feb. 6, 1945; children—Karola, Konrad, Christa, Georg, Jürgen, Margret, Rudolf, Ruth, John-Walter, Hannemarie. Fulbright fellow N.Y. U., 1955-56; guest prof. Gondi Shapoor Med. Sch., Ahvaz, Iran, 1960, Baylor U., Houston, 1965; faculty Johannes Gutenberg U., Mainz, Germany, 1951-—, asso. prof. biochemistry dept. physiol. chemistry, 1957-—. Mem. Gesellschaft für Physiologische Chemie, Biochem. Soc., Am. Chem. Soc. Research and numerous publs. on isolation of nuclei from animal tissue cells for analytical, enzymic and metabolic studies of organelles, enzymic studies on fish tissues. Home: 61 Fort Gosenheim, Mainz 6500, Germany.*

SIEBOLD, see von Siebold.

SIEBURTH, John McNeill, bacteriologist; b. Calgary, Alta, Can., Sept. 1, 1927; s. Herman C. and Mary McNeill (Clark) S.; B.S.A., U. B.C., Vancouver, Can., 1949; M.S., Wash. State U., 1951; Ph.D., U. Minn., 1954; m. Janice Fae Boston, Sept. 2, 1950; children—Heather, Scott, Peggy, Leslie, Clark. Came to U.S., 1949, naturalized, 1956. Hormel fellow U. Minn., 1951-52, instr., 1952-53, research asso., 1953-55; asso. prof. vet. sci., Va. Poly. Inst., Blacksburg, 1955-60; research biol. oceanographer, U. R.I., Kingston, 1960-61, asso. prof. oceanography, 1961-66, prof., 1966-—. Bacteriologist for Am. Soc. for Microbiology, Argentine Navy in Antarctica, IGY, 1957-59. Fellow Royal Norwegian Council for Sci. and Indsl. Research, 1966-67. Mem. Am. Soc. for Microbiology, Am. Soc. Limnology and Oceanography, Am. Inst. Biol. Sci., Marine Tech. Soc. Research and numerous publs. on mechanism of antibiotic growth-promotion in poultry, nature of an enteritis in turkeys, a simplified method for detecting paratyphoid fever in chickens; traced cause bacteriological sterility in penguins, ecology cold requiring marine bacteria, role algal antibiotics in marine ecology. Home: 17 Locust Dr., Kingston, R.I. 02881.*

SIEDENTOPF, Henry Friedrich Wilhelm, German physicist; b. 1872; chief microscope div. Zeiss Works; prof. microscopy U. Jena (Germany); invented ultramicroscope, 1903. Died 1940.

SIEGBAHN, Manne Karl Georg, Swedish physicist; b. Örebro, Sweden, Dec. 3, 1886; s. Georg and Emma (Zetterberg) S.; D.Sc., (Hon.), U. Freiburg (Germany), 1931, U. Bucharest (Romania), 1942, U. Oslo (Norway), 1946, U. Paris, 1952. Prof. physics U. Lund (Sweden), 1914-23, U. Uppsala (Sweden), 1923-37; prof. Acad. Scis., Stockholm, Sweden, 1937-64, prof. emeritus, 1964-—; dir. Nobel Inst. Physics, 1939-64. Mem. Internat. Com. for Weights and Measures, 1939-—. Recipient Hughes medal Royal Soc., 1934, Rumford medal, 1940; Nobel prize in physics, 1924. Mem. Royal Soc. London, Royal Soc. Edinburgh, Academie des Sciences Paris, Russian Acad. Scis., Nat. Inst. India (hon.). Author: X-ray Spectroscopy, 1924; also articles. Research in X-ray spectroscopy, soft X-rays, extreme ultraviolet radiation. Address: Nobel Inst., 10540 Stockholm, 50, Sweden.*

SIEGEL, Armand, Am. physicist; b. N.Y.C., Oct. 10, 1914; s. Louis Alfred and Frances (Streitfeld) S.; A.B., N.Y. U., 1936; A.M., U. Pa., 1944; Ph.D., Mass. Inst. Tech., 1949; m. Mildred Helen Marks, May 23, 1943; children—Jonathan Marks, Andrew Francis, Jeffrey Nahum. Faculty, U. Pa., 1941-44; instr. Mass. Inst. Tech., 1944-45, research asso., 1952-53; faculty Boston U., 1950-—, prof. physics, 1960-—, chief investigator Air Force Office of Sci. Research grantee, 1955-—. Mem. Am. Phys. Soc. Research and publs. in fundamentals of quantum mechanics, stochastic processes in physics, kinetic theory. Home: 9 Tower Rd., Lexington, Mass. 02173.*

SIEGEL, Keeve Milton, Am. physicist; b. N.Y.C., Jan. 9, 1923; s. David P. and Rose (Jelin) S.; B.S. in Physics, Rensselaer Poly. Inst., 1948, M.S. in Physics, 1950; m. Ruth E. Boerker, June 22, 1951; children—Leigh Michael, David Alan. Head upper atmosphere group U. Mich., Ann Arbor, 1949-50, head theory and analysis group and dept., 1951-56, head Radiation Lab., 1957-61, prof. elec. engring., 1957-67. pres., chmn. bd. Conductron Corp., Ann Arbor, 1960-67; chmn. KMS Industries, Inc., 1967-—; dir. Gelman Instruments, Microwave Assos. Vis. prof. Oakland U., 1967-68. Mem. sci. adv. bd. U.S. Air Force, 1958-67; mem. adv. group radio expts. in space NASA, 1960-64; cons. U.S. Army, 1958-61. Fellow I.E.E.E. (asso. editor transactions profl. group on antennas and propagation), A.A.A.S., Am. Inst. Aero. and Astronautics (asso.); mem. Am. Inst. Physics, Am. Math. Soc. N.Y. Acad. Scis., Inst. Aero. Scis., Sigma Xi. Devel. practical methods for determination of radar cross sections of aircraft and missiles; determination of bounds on spl. functions of math. physics; inventor in fields of electromagnetic lenses, plasma lenses, radar and radar subsystems. Home: 1425 Hatcher Crescent, Ann Arbor 48103. Office: 220 E. Huron St., Ann Arbor, Mich. 48108.*

SIEGEL, Robert, Am. mech. engr.; b. Cleve., July 10, 1927; s. Morris and Mollie (Binder) S.; B.S., Case Inst. Tech., 1950, M.S., 1951; Sc.D., Mass. Inst. Tech., 1953; m. Elaine Jane Jaffe, July 19, 1951; children—Stephen D., Lawrence C. Research engr. Gen. Electric Co., Schenectady, 1953-55; research engr. NASA Lewis Research Center, Cleve., 1955-65, head analytical sect., 1965-—. Mem. Am. Soc. M.E., Sigma Xi, Tau Beta Pi. Research and publs. on heat transfer and fluid mechanics, heat transfer in tubes, transient free convection, transient solidification, boiling heat transfer in reduced gravity fields. Home: 3052 Warrington Rd., Shaker Heights, O. 44120. Office: NASA, 21000 Brookpark Rd., Cleve. 44135.*

SIEGEL, Sanford Marvin, biologist; b. Kansas City, Mo., Sept. 3, 1928; s. Samuel L. and Doris (Franklin) S.; student Mass. Inst. Tech., 1945; M.S., U. Chgo., 1950, Ph.D., 1953; m. Barbara Zenz, June 24, 1950; children—Stephanie, Andrea, Peter, David. Biochemist Army Chem. Corps, Ft. Detrick, Md., 1950-52; asst. prof. biology U. Tampa, 1954-55; research fellow Cal. Inst. Tech., 1953-54; asst. prof. biology U. Rochester, 1955-58; group leader phys. biochemistry Union Carbide Research Inst., Tarrytown, N.Y., 1958-67; prof. botany U. Hawaii, Honolulu, 1967-—. Guggenheim fellowship 1957-58. Mem. Am. Chem. Soc., Am. Soc. Plant Physiologists, Soc. Gen. Physiologists, Bot. Soc., Brit. Interplanetary Soc. Author: The Plant Cell Wall, 1962; also numerous articles Research on biochemistry of lignification, role of cell wall as a matrix; role of oxygen plant growth and development; oxygen toxicity and anoxia tolerance in plants; gen. and comparative physiology of environmental stress and extreme environments; cold heat, ammonia, salts, radiation on plants, animals, microorganisms; exobiology, paleobiology, environmental simulation, as exptl. subjects. Address: Dept. of Botany, U. Hawaii, Honolulu, 96822.*

SIEGERT, Arnold J(ohn) F(rederick), physicist; b. Dresden, Germany, Jan. 1, 1911; s. Frederick Ludwig and Lea (Rosenberg) S.; student Inst. Tech., Dresden, Germany, 1928-30, U. Vienna, 1929, U. Innsbruck, Austria, 1930; Ph.D., U. Leipzig, Germany, 1934; m. Anna B(eatrice) Brown, Feb. 12, 1944; 1 son, Allan. Came to U.S., 1936, naturalized, 1942. Lorentz Funds fellow, Leiden, Holland, 1934-36; grad. asst. Stanford, 1936-39; physicist, The Texas Co., Houston, 1939-42, Nat. Geophys. Co., Dallas, 1942, Stanolind Oil & Gas Co., Tulsa, 1942, Radiation Lab., Mass. Inst. Tech., 1942-45; asso. prof. physics Syracuse (N.Y.) U., 1946-47; prof. physics, Northwestern U., Evanston, Ill., 1947-—; mem. Inst. Advanced Study, Princeton, N.J., 1953-54; NSF sr. postdoctoral fellow, Instituut voor Theoretische Fysica, U. Amsterdam, 1962-63, Weizmann Inst. Sci., Rehovoth, Israel, 1963-64. Cons. Argonne Nat. Laboratory. Guggenheim fellow, 1953-54. Fellow Am. Phys. Soc., mem. Soc. Exploration Geophysicist, Sigma Xi. Contbg. author: Radiation Laboratory Tech. Series, 1947-53; Analysis in Function Space, 1964. Editorial bd., Jour. Math. Physics, 1961-63. Contbr. numerous articles to tech. jours. Patentee, geophys. exploration. Research in statis. mechanics and random processes. Home: 2347 Lake Av., Wilmette, Ill. 60091.*

SIEGFRIED, André, French economist; b. Le Havre, France, Apr. 21, 1875; Docteur ès Lettres; Dr. (hon.), U. Chgo.; prof. Collège de France, Paris, l'Ecole des Sciences Politiques, Institut d' Etudes Politiques, Paris. Mem. French Acad. Scis. Author: l'Angleterre d'aujourd'hui, 1924; Les Etats-Unis d'aujourd'hui, 1927; Aspects du vingtième siècle, 1955; Aufstieg zur Weltmacht, 1956; Tableau des partis en France; Le Canada, puissance internationale; La crise de l'Europe; l'Ame des Peuples; Tableau des Etats-Unis. Studied econ. and geog. bases for formation of society. Died Paris, Mar. 29, 1959.

SIEGLIN, Wilhelm, German geographer; b. Stuttgart, Germany, Apr. 19, 1855; prof. Leipzig, Berlin. Author: Karte der Entwicklung des Römischen Reiches, 1885; Die blonden Haare der idg. Völker des Alterums, 1935; also worked on hist. atlases of Droysen and Spruner. Died Munich, July 9, 1935.

SIEGMAN, Anthony Edward, Am. elec. engr.; b. Detroit, Nov. 23, 1931; s. Orra Leslie and Helen (Winnie) S.; A.B., summa cum laude, Harvard, 1952; M.S., U. Cal., Los Angeles, 1954; Ph.D., Stanford, 1957; m. Virginia Leigh Kelley, Mar. 31, 1956; children—Anne, Winn, Patrick. Faculty, Stanford (Cal.) U., 1954—, prof. 1964—; vis. prof. applied physics Harvard, 1965; cons. Sylvania Electronic Systems, Mountain View, Cal., Inst. Def. Analysis, Washington. Fellow I.E.E.E., Optical Soc. Am.; mem. Am. Phys. Soc., Am. Soc. Engring. Edn., Phi Beta Kappa, Sigma Xi. Author: Microwave Solid-State Masers, 1964. Editor: (with H. Heffner) McGraw-Hill Series in Physical and Quantum Electronics, 1964—. Research and numerous publs. on microwave electron devices, traveling-wave tubes, lasers, modulation and demodulation of light beams, application of lasers. Address: Box 5023, Stanford U., Stanford, Cal. 94305.*

SIEKER, Herbert Otto, Am. physician; b. Maplewood, Mo., Mar. 20, 1924; s. Otto H. and Lizzie (Moecklei) S.; student Miami U., Oxford, O., 1943-44; M.D., Washington U., St. Louis, 1948; m. Dorothy J. Linberg, June 14, 1948; 1 dau., Deborah Lynn. With Med. Center, Duke U., Durham, N.C., 1953—, prof. medicine, 1961—, asst. dean, 1964—. Mem. study sect. B. allergy and infectious disease NIH, 1965—. Mem. Assn. Am. Physicians, Am. Soc. for Clin. Investigations, Am. Fedn. for Clin. Research, A.C.P., Am. Clin. and Climatol. Assn., Am. Heart Assn. (past pres. N.C.), So. Soc. Clin. Investigation, Am. Trundeau Soc. (past pres. N.C.). Research and numerous publs. on heart and lung disease especially mechanisms controlling fluid volume of body, mechanisms for respiratory control, pathophysiology of cardiopulmonary disease especially emphysema, cellular response in tissue damage especially Tb, Identification of role of obesity in heart and lung disease. Home: 204 Forestwood Dr., Durham, N.C. 27707.*

SIEKERT, Robert George, Am. physician; b. Milw., July 23, 1924; s. Hugo Paul and Elisa (Kraus) S.; B.S., Northwestern U., 1945, M.D., 1948, M.S., 1947; m. Mary Jane Evans, Feb. 17, 1951; children—Robert George, John Eric-Spencer, Friedrich Anson Paul. Instr. anatomy U. Pa. Sch. Medicine, 1948-49; fellow neurology Mayo Grad. Sch. Medicine, Rochester, Minn., 1950-54, asso. prof. neurology, 1960—; cons. neurology Mayo Clinic, Rochester, 1954—. Diplomate Am. Bd. Psychiatry and Neurology. Fellow Am. Acad. Neurology, A.C.P.; mem. Am. Neurologic Assn. Research and numerous publs. on cerebrovascular disease in diagnosis, classification and treatment especially brief cerebral ischemic episodes and treatment with anticoagulant and surg. therapy. Home: 932 4th St. S.W., Rochester, Minn. 55901.*

SIEMENS, Frederick S., German engr., inventor; b. Mentzendorff, Germany, Dec. 8, 1826; s. Christian Ferdinand Siemens; worked in English br. of Siemens & Co. Invented regenerative smelting oven originally used for making glass products, 1856, gas recuperator which allowed prodn. of steel in open hearth and glass smelting in continuous fire, 1858. Died May 26, 1904.

SIEMENS, Sir William, metallurgist, electrician; b. Lenthe, Hanover, Germany, 1823; s. C. Ferdinand and Eleonore (Deichmann) S.; ed. Magdeburg, Göttingen (both Germany); student of Himly, Wöhler, Weber; D.C.L., Oxford U.; LL.D., Dublin, Glasgow univs.; m. Anne Gordon, 1859. Became Brit. citizen, 1859. Sold an elec. invention in Eng., 1843; engaged in glass making and other indsl. processes, Landore, 1869-88; London agt. of Siemens & Halske; established works at Charlton, 1866; laid Atlantic cable, also designed cable ship Faraday, 1874. Recipient medals at London, 1862, Paris, 1867, Bessemer medal, 1875; Howard prize, 1883; meml. window erected at Westminster Abbey, also elec. lab. at King's Coll. named in his honor. Fellow Royal Soc., 1862; mem. or fellow Inst. Civil Engs. (by spl. election), Brit. Assn. (pres. 1882), Soc. Telegraphy Engrs. (pres.), Soc. Mech. Engrs., Iron and Steel Inst. His collected works edited by E. F. Bamber, 1889. Introduced chronometric governor, anastatic printing, 1844; patentee regenerative steam engine and condenser, 1849; inventor water meter, 1851; Siemens brothers' regenerative furnace applied to melting and reheating of steel, 1857, later to glass making, other indsl. processes; announced (simultaneously with Charles Wheatstone and Cromwell Fleetwood Varley) principle of dynamo, 1867; inventor electric furnace, 1879, bathometer, electric thermometer; applied electric power to Portrush ry., 1883. Died Nov. 18, 1883.

SIENKO, Michell Joseph, Am. chemist; b. Bloomfield, N.J., May 15, 1923; s. Felix and Teofila (Kislova) S; A.B., Cornell U., 1943; Ph.D., U. Cal. at Berkeley, 1946; m. Carol Tanghe, Aug. 25, 1946; 1 dau., Tanya. Research asso. Stanford (Cal.) U., 1946-47; faculty Cornell U., Ithaca, N.Y., 1947—, prof. chemistry, 1958—; Fulbright lectr. U. Toulouse, France, 1956-57; vis. prof. Am. Coll. of Paris, 1963-64. Mem. Phi Beta Kappa, Sigma Xi. Author: Chemistry, 1957, 61, 66; Experimental Chemistry, 1958, 61; Physical Inorganic Chemistry, 1963; Stoichiometry, 1964; Equilibrium, 1964; Metal-Ammonia Solutions, 1964. Research in solid state and metal-ammonia. Home: 493 Ellis Hollow Creek Rd., Ithaca, N.Y. 14850.*

SIERPINSKI, Waclaw, Polish mathematician; b. Warsaw, Poland, Mar. 14, 1882; s. Constantin and Ludwika (Lapinska) S.; Ph.D., U. Cracow (Poland), 1906; hon. Dr., univs. of Lwow, Amsterdam, Tarru, Paris, Sofia, Bordeaux, Prague, Wroclaw, Lucknow; m. Anna Lesniewska, July 19, 1910; 1 son, Prof., U. Warsaw, 1919-60; lectr. U. Lwów (Poland), 1908-10. Mem. Polish Acad. Scis. and Letters, Inst. France, Warsaw Soc. Sci. and Letters (pres. 1931-52), Royal Dutch Acad. Sci., Czechoslovak Acad. Scis., Accademia dei Lincei, Polish Acad. Scis. (v.p. 1952-57, became mem. presidium 1957), Deutsche Akad. der Wissenschaften (corr.), N.Y. Acad. Scis. (hon. life), Math. Soc. (hon.), Czechoslovak Acad. Scis., Internat. Acad. Philosophy Scis. (v.p. 1962—). Author: Lecons sur les nombres transfinis, 1928; Hypothèse du continu, 1934; Introduction to General Topology, 1934; Les ensembles projectifs et analytiques, 1950; Algèbre des Ensembles, 1951; General Topology, 1952; On the Congruence of Sets and their Equivalence by Finite Decomposition, 1954; Cardinal and Ordinal Numbers, 1958; A Selection of Problems in the Theory of Numbers, 1964; Elementary Theory of Numbers, 1964. Research on logical founds. of math. and topology; leader modern Polish sch. math. Office: Konopczynskiego 5/7, m. 38, Warsaw, Poland.

SIEVER, Raymond, Am. geologist; b. Chgo., Sept. 14, 1923; s. Leo and Lillie (Katz) S.; B.S., U. Chgo., 1943, M.S., 1947, Ph.D., 1950; m. Doris Fisher, Mar. 31, 1945; children—Larry Joseph, Michael David. With Ill. Geol. Survey, 1943-56, geologist, 1952-56; NSF sr. postdoctoral fellow, research asso. Harvard U., Cambridge, Mass., 1956-57, mem. faculty, 1957—, prof., 1965—; asso. in geology Woods Hole Oceanographic Instn., 1957—; cons. oil cos., book pubs., govt. agys., ednl. testing. Fellow Geol. Soc. Am., Am. Acad. Arts and Scis. Author (with F.J. Pettijohn, P.E. Potter) Geology of Sandstones, 1966. Research, numerous publs. on conditions of sedimentation of cool measures sediments; phys. constn. of coal; geochemistry of formation of sediments; changes by which sediment is changed into rock; geochemistry of silica in rocks; marine geology of clay sediments. Home: 4 Madison St., Belmont, Mass. 02178.*

SIEVERS, Wilhelm, German geographer; b. Hamburg, Germany, Dec. 3, 1860; prof., Würzburg, Germany, also Giessen, Germany. Author: Venezuela, 1888; Allgemeine Landerkunde, from 1891. Editor multi-vol. geography of Leipzig (Germany). Died Giessen, June 11, 1921.

SIFOROV, Vladimir Ivanovich, Russian radio engr.; b. May 31, 1904; grad. Leningrad Electro-Tech. Inst., 1929. Faculty, Leningrad Electrotech. Inst., 1930-41, 46-53; staff Central Radio Lab., Leningrad, 1928-41; became prof. Leningrad Electro-Tech. Inst., 1938; joined Sci. Research Inst. Communication, 1953, Inst. Radiotechnics and Electronics, USSR Acad. Scis., 1955. Recipient Order Red Banner of Labor, 1964. Corr. mem. USSR Acad. Scis. Author: Receiving Devices, 1939; Resonance Amplifiers. Theory and Calculation, 1932; Band-Pass Amplifiers. Theory and Calculation, 1936; High Frequency Amplifiers. Theory and Calculation, 1939; Ultrashort-wave Pulse Signal Radio Receivers, 1947; Very High Frequency Radio Receivers, 1957; Radioelectronics in Space; Radio Engineering and Electronics; Lenin and Radioelectronics, 1960; coauthor: The Theory of Pulse Radio Communications, 1951. Research in high frequency and short wave radio engring.; developed theory and design of radio amplifiers and receivers. Home: Chistoprudniy bulv. 2. Office: Inst. Radiotechnics and Electronics, USSR Acad. Scis., Mokhovaya Ulitsa 11, K-9, Moscow, USSR.

SIFTON, Harold Boyd, Canadian botanist; b. Metcalfe, Ont., Can., June 16, 1889; s. William Sutton and Martha (Boyd) S.; B.A., U. Toronto, 1914, M.A., 1915, Ph.D., 1923; m. Jean Ethel Mitchell, June 25, 1919. Pub. sch. tchr., Ont., 1907, 1909, Sask., 1911; seed analyst Can. Dept. Agr., Ottawa, 1915-20; with U. Toronto, 1920—, prof., 1941-57, head dept. botany, 1952-57, prof. emeritus, 1957—. Fellow Royal Soc. Can., A.A.A.S., mem. Royal Canadian Inst., Bot. Soc. Am. Author (with R.B. Thomson): Poisonous Plants and Weed Seeds, 1922. Studies, publs. on comparative plant anatomy; autecology of plants, physiology of seed germination. Home: 198 Bessborough Dr., Toronto 17, Ont., Can.*

SIGAUD DE LA FOND, Joseph Aignan, French physicist; b. Bourges, France, Jan. 5, 1730; became prof. physics, Bourges, 1786, École Centrale, Paris, 1795; mem. French Acad. Scis. Author: Lecons de physique expérimentale, 2 vols., 1767. Dictionnaire de physique, 1781; De l'électricité médicinale, 1803. Began studies in physics, 1777; discovered (in expts. with Maquer) that combustion of hydrogen gas produces water; studied static electricity. Died Bourges, Jan. 26, 1810.

SIGEL, Mola Michael, microbiologist; b. Nieswiez, Poland, June 24, 1920; s. Zundel and Helen (Lubecka) S.; came to U.S., 1937, naturalized, 1941; B.A., U. Tex., 1941; Ph.D., Ohio State U., 1944; m. Mary Elizabeth Wynne, Dec. 22, 1941; children—Suzanne L.P., Vicki A.B., Rachel D.S., Valerie H.L., David E.B. Mem. faculty U. Pa., 1946-53; chief reference diagnosis, research unit Virus and Rickettsia sec. USPHS, Montgomery, Ala., 1953-55; faculty Sch. Medicine U. Miami (Fla.), 1955—, prof. microbiology, 1958—, prof. Inst. Marine Sci., 1961—; dir. virus labs. Variety Childrens Research Found., 1955-60, research dir., chmn. research staff, 1960—; spl. cons. WHO, Europe, 1956; research asso. Lerner Marine Lab., Bimini, Bahamas, 1963—; hon. cons. dept. microbiology U. West Indies, Kingston, Jamaica, 1964—. Diplomate Am. Bd. Microbiology. Fellow A.A.A.S., Am. Pub. Health Assn., N.Y. Acad. Scis.; mem. Am. Soc. Microbiology, Soc. Exptl. Biology and Medicine, Soc. Pediatric Research, Am. Assn. Immunologists, Am. Soc. Limnology and Oceanography, Am. Soc. Cell Biology, Am. Assn. Cancer Research, Tissue Culture Assn., Phi Beta Kappa, Sigma Xi. Author: Viruses, Cells and Hosts, 1965. Editor: Lymphrogranuloma Venereum, 1962. Contbr. numerous articles to profl. jours. Research, publs. on new variants of influenza virus; growth cycle of meningopneumonitis virus; immune response in marine fishes; virus induced neoplasia, its antigens and antibodies. Home: 7980 S.W. 58th St., Miami, Fla. 33143.*

SIGERIST, Henry Ernest, med. writer; b. Paris, 1891; s. Ernest Henry and Emma (Wiskemann) S.; ed. U. Coll., London, 1911, U. Munich (Germany), 1914; M.D., U. Zurich (Switzerland), 1917; D.honoris causa U. Madrid (Spain), 1935, D.Litt., U. Wittersrand, S. African, 1939; LL.D., Queen's U., Kingston, Ont., Can., 1941; D.Sc., U. London, 1953; m. Emma M. Escher, Sept. 14, 1916; children—Erica Elizabeth, Nora Beate (Mrs. Nora Beeson). Dir. Med. History Inst., Leipzig; came to U. S., 1931; prof., dir. Inst. History Medicine, Johns Hopkins, 1932-47; research asso. Yale, 1947-57; lived in Switzerland, beginning in 1947. Adviser to Govt. of Sask., Can., 1944, Govt. of India,1944. Recipient medals and numerous awards. Author: books, including: Man and Medicine, 1932; American Medicine, 1934; Socialized Medicine in Soviet Union, 1937; Civilization and Disease, 1943; The University at the Crossroads, 1946; Medicine and Health in the Soviet Union, 1947; Landmarks in the History of Hygiene, 1956; Medicine and Human Welfare; The Great Doctors; Letters of Jean De Carro, 1950; A History of Medicine, Vol. I, 1951. Editor books, reports, bulls., revs. Traveled through almost all European countries studying med. systems, histories and social systems. Died Pura Ticino, Switzerland, Mar. 17, 1957.

SIGHELE, Scipio, Italian psychologist; b. Brescia, Italy, June 24, 1868; studied under Lombroso; prof. domestic politics, Brussels, Belgium, 1899-1902. Author: La folla deliquente, 1891; La folla criminale, 1892; Delinquenza settaria, 1897. Publs. on social and criminal psychology, psychology of masses, collective suggestion. Died Florence, Italy, Oct. 21, 1913.

SIGLER, William Franklin, Am. wildlife scientist; b. Leroy, Ill., Feb. 17, 1909; s. John Adam and Bettie (Homan) S.; student Ill. Wesleyan U., 1927-28, Ill. State Normal U., 1937-39; B.S., Ia. State U., 1940, M.S., 1941, Ph.D., 1947; m. Margaret Brotherton, July 3, 1936; children—Elinor Jo, John. Technician, biologist U.S. Soil Conservation Service, 1935-37, 41-42; field adminstr. G.D. French Central Engring. Co., Davenport, Ia., 1940-41; with Ia. State U., 1942-47, research asso., 1945-47; with Utah State U., Logan, 1947—, prof., head dept. wildlife resources, 1950—. Cons. Surgeon Gen., NIH. Fellow A.A.A.S., Conf. Radiol. Health, Am. Inst. Fishery Research Biologists; mem. Am. Fisheries Assn., Am. Soc. Limnology and Oceanography, Ecol. Soc. Am., Wildlife Soc., Utah Acad. Sci., Sigma Xi. Author: Wildlife Law Enforcement, 1956; (with R.R. Miller) Fishes of Utah, 1963, The Collection and Interpretation of Fish Life History Data, 1952; also numerous articles. Research in fishery ecology, fish life histories; control of water pollution; aquatic toxicology. Home: 309 E. 2nd S., Logan, Utah 84321.*

SIGNAIGO, Frank Kerr, Am. chemist; b. Detroit, Dec. 1, 1907; s. Frank Edmund and Frances (Kerr) S.; B.S. in Chemistry, U. Mich., 1930; Ph.D., U. Wis., 1935; m. Virginia Estella Reynolds, Jan. 15, 1938; children—Ellen Virginia (Mrs. Dan. B. Brockman), John Francis, Therese Elizabeth (Mrs. Aubrey L. Raymond). Research chemist Gen. Motors Corp., Detroit, 1930-33, E.I. DuPont de Nemours, Wilmington, Del., 1935-45, research mgr., Buffalo, 1945-49, research dir. photo products dept., Wilmington, 1950—. Recipient Modern Pioneer award N.A.M., 1940; certificate merit OSRD, 1945. Mem. Am. Chem. Soc., A.A.A.S., Soc. Motion Picture and Television Engrs., Sigma Xi, Phi Lambda Upsilon, Alpha Chi Sigma. Research, publs., patentee in motor fuels, catalysis, organic sulfer compounds, synthetic films and fibers. Home: 1500 Brandywine Blvd., Wilmington 19809. Office: Nemours Bldg., Wilmington, Del. 19898.*

SIGNER, Rudolf, Swiss chemist; b. Herisau, Switzerland, Mar. 17, 1903; s. Jakob and Dora (Scherrer) S.; M.A., Swiss Fed. Inst. Tech., Zürich, 1928; m. Gret Meier, Apr. 30, 1928; children—Peter, Dieter, Martin, Emanuel, Anna, Rudolf. Teaching asst. U. Freiburg i. Br., 1926-35; prof. organic chemistry U. Berne (Switzerland), 1935—. Research, numerous publs. on flow birefringence of solutions of macromolecules, preparation of nucleic acid, counter current distbn. Home: 20 Bellevuestr., Gümligen, Switzerland. Office: Inst. Organic Chemistry, Freiestr. 3, Berne, Switzerland.*

SIGNORET, Victor Antoine, French entomologist; b. Paris, Apr. 6, 1816. Hon. fellow Entomol. Soc. London; mem. Entomol. Soc. France. Research and publs. on Coccidae, Am. Hemiptera, Tettigonides, Jassides, Cochenilles; tessellated palm, Mediterranean fig, aspidistra and Greedy scales named after him; his collection is said to be at Mus. Vienna. Died Paris, Apr. 3, 1889.

SIKORSKY, Igor Ivanovich, aero. engr.; b. Kiev, Russia, May 25, 1889; s. John S.; grad. Naval Coll., St. Petersburg, 1906; grad. Inst. of Technology, Kiev, 1908; M.Sc., Yale U., 1935; hon. degrees from Wesleyan and Lehigh U., Fla. So. Coll., R.I. State Coll., Northeastern U., U. Pa., U. Bridgeport, Conn., Yale; D.Sc., Colby College, 1955; D.Sc., Trinity Coll., 1965, Fairfield U., 1966; m. Elizabeth A. Semion, Jan. 27, 1924; 1 dau., 4 sons. Came to U. S., 1919, naturalized, 1928. Designed and built flying machines on own account, 1908-11; with Russo-Baltic Railroad Car Works, 1912-18, as head of the engineering department of its aviation factory; designed and built 75 large four-motored bombers used by Russian Army; went to France, 1918, and was commissioned by French govt. to build the Sikorsky plane for military use, but production cut short by the armistice; organized the Sikorsky Aero Engring. Corp., 1923, the Sikorsky Mfg. Corp., 1925, and in 1928, the Sikorsky Aviation Corp., United Aircraft Corporation; engring. mgr. Sikorsky Aircraft division until 1957, retired, but continues as adviser and consultant. Recipient: Potts Medal, Franklin Inst., 1933; hon. fellow Rochester Museum of Arts & Sciences, 1943; hon. fellow Am. Helicopter Soc., 1944, Gen. W. E. Mitchell award, 1944; Benjamin Franklin fellow Royal Society for Encouragement of Arts, Mfrs. and Commerce, London, 1960; First Fawcett Aviation Award, 1944; Warner Medal, Am. Soc. M.E., 1944; Hawks Memorial Trophy, 1947; Gold Medal, Fed. Aeronautique Internat., 1947; Presidential Cert. Merit, 1948; Silver Medal, Royal Aero. Soc., Eng., 1949; Alexander Klemin Award, Am. Helicopter Soc., 1950, Collier Trophy, 1950; Daniel Guggenheim Medal, 1951; Nat. Defense Transportation Award, 1952; Godfrey Lowell Cabot Award, N.E. Aero Club; one of 50 Americans chosen for Popular Mechanics Hall of Fame; John Scott medal City of Philadelphia, 1955; James Watt International gold medal (London, Eng.), 1955; United Aircraft Corp. established trophy in his honor, 1961; Engr. of Yr. Conn. Soc. Profl. Engrs., 1963; Chevalier, French Legion Honor; Grover E. Bell award Am. Helicopter Soc., 1960; Cross of Chevalier of the Legion of Honor, France, 1960; Elmer A. Sperry award, 1964; Modern Pioneers Creative Industry medal N.A.M., 1965; award of honor Wisdom Soc., 1966; Hall of Fame award Internat. Aerospace Hall of Fame, 1966; recipient of the National Medal of Science, 1967. Hon. fellow Royal Aero. Soc.; mem. Soc. Automotive-Engrs., Am. Soc. M.E., Nat. Aero. Assn., Aero, C. of C., Am. Helicopter Soc., Am. Inst. Aeros. and Astronautics, Am. Soc. French Legion of Honor, Early Birds Aviation, Profl. Engrs. State Conn., Royal Soc. Arts (Eng.). Author: Winged 'S,' 1938; Message of the Lord's Prayer, 1942; The Invisible Encounter, 1947. Built and flew 1st multimotored airplane, 1913; developed several types of planes, among them 1st successful long-range clippers, which pioneered transoceanic air service; developed flying boats; developed 1st successful helicopter, produced in Western Hemisphere, 1939; established world's record for sustained helicopter flight, 1941. Address: Sikorsky, Stratford, Conn.

SIKOV, Melvin Richard, Am. biologist; b. Detroit, July 8, 1928; s. Paul Merrill and Emma (Perlman) S.; B.S., Wayne U., 1951; Ph.D., U. Rochester, 1955; m. Shirley Dressler, June 1, 1952; children—Peter H., Stacy J., Thomas R.N. Asso. prof. radiobiology Wayne State U. Sch. Medicine, Detroit, 1955-65; sr. research scientist biology dept. Pacific N.W. Labs., Batelle Meml. Inst., Richland, Wash., 1965—. Asso. in radiology Detroit Receiving Hosp., 1955-65; cons. radiobiology VA Hosp., Dearborn, Mich., 1959-65, Detroit Meml. Hosp., 1963-65. Mem. Radiation Research Soc., Am. Soc. for Exptl. Pathology, Am. Assn. for Cancer Research, Soc. for Exptl. Biology and Medicine. Research, numerous publs. on effects of radiation of embryo, exptl. radiotherapy of animal tumors, biology of tumors and metastases.*

SILBER, Pierre, French chemist; b. Mulhouse, France, Sept. 17, 1923; s. Paul and Henriette (Golling) S.; Ingénieur chimiste Ecole Nationale Supérieure de Chimie, Paris, 1945; Docteur-ès-Sciences physiques, Faculté des Sciences de Paris, 1951; m. Suzanne Desjardin, Mar. 27, 1951; children—Martine, Joelle, Francoise, Pascal. Staff, Centre National de la Recherche Scientifique, Paris, 1946-52; faculty Faculté des Sciences de Montpellier, France, 1952-62, prof. chemistry, 1958-62; prof. Faculté des Sciences de Paris, 1962—. Mem. Société chimique de France, Société de Chimie-Physique. Contbg. author: Nouveau Traité de Chimie minérale, 1955. Sec. gen. Revue de Chimie Minrale. Research, publs. on synthesis, structure and reactivity relationships of new inorganic compounds. Home: 6, av. Constant Coquelin, 75, Paris 7, France.*

SILBER, Robert Howard, Am. biochemist; b. Hermann, Mo., Apr. 26, 1915; s. Victor Anton and Stella (Schuth) S.; A.B., Washington U., St. Louis, 1937, Ph.D., 1941; m. Ruth Whitney Bender, May 14, 1949; children—Lois (Mrs. Andrew Horner), V. Reed Littlefield, Lynn W. With Merck Inst., Rahway, N.J., 1941—, asso. dir. inst., 1958-66, sr. investigator drug metabolism, 1966—. Diplomate Am. Bd. Clin. Chemistry. Fellow Soc. Clin. Chemists, N.Y. Acad. Scis.; mem. Soc. Biol. Chemists, Endocrine Soc., Am. Chem. Soc., A.A.A.S. Editorial bd. Clin. Chemistry, 1966—. Contbr. numerous articles to tech. jours., chpts. to books. Research in endocrinology, clin. chemistry, nutrition; devel. methods for estimation adrenal function in man and animals. Home: 781 Hyslip Av., Westfield, N.J. 07090. Office: Merck Inst., Rahway, N.J. 07090.*

SILBERBERG, Ruth Katzenstein (Mrs. Martin Silberberg), pathologist; b. Kassel, Germany, Mar. 20, 1906; d. Ludwig and Kathe (Plaut) Katzenstein; student U. Freiburg, 1925, U. Berlin, 1926, U. Gottingen, 1927; M.D., U. Breslau (Germany), 1930; m. Martin Silberberg, Dec. 27, 1933. Came to U.S., 1937, naturalized, 1943. Staff, U. Breslau, 1934-35, 1932-35; research at Dalhousie U., Halifax, N.S., Can., 1934-36, faculty Washington U., St. Louis, 1937-41, faculty, 1944—, professor pathology, 1968—; faculty N.Y. U., 1941-44; sr. pathologists St. Louis City Hosp., 1948-59, Mo. Pacific Hosp., 1956-59. Mem. Am. Assn. Pathologists and Bacteriologists, Soc. Exptl. Pathologists, Soc. Exptl. Biology and Medicine, Am. Assn. Cancer Research, Am. Gerontol. Soc., A.A.A.S., Human Genetics Soc., Soc. Growth and Devel., Sigma Xi. Research and publs. on tissue growth, carcinogenesis, aging of skeleton, arthritis. Home: 18 S. Kingshighway, St. Louis 63108.*

SILBERG, Paul August, Am. physicist; b. N.Y.C., Oct. 2, 1922; s. August and Paula (Org) S.; B.S., Syracuse U., 1950, M.S., 1952, Ph.D., 1956; m. Julia Beatrice Hicks, Mar. 11, 1955; 1 son, Eric Rolfe. Research asso. U. Mich., Ypsilanti, 1955-56; devel. engr. Bell Aircraft, Inc., Buffalo, 1956-57; staff asst. Melpar, Inc., Arlington, Va., 1957-58, 1959-60; tech. staff Radiation, Inc., Silver Spring, Md., 1958-59; mgr. electromagnetic research dept. Raytheon Co., Wayland, Mass., 1960-65, tech. asst. to staff, Bedford, Mass., 1965-66; cons. sci. Northrop Nortronics, Norwood, Mass., 1966—. Mem. I.E.E.E., Am. Phys. Soc., Sigma Xi, Sigma Pi Sigma, Pi Mu Epsilon. Research, publs., patents on ball lightning, electromech. theory tornado, electromech. theory vorticity, pulse laser interaction with metals, non linear optical field. Home: 6 Richard Rd., Wayland, Mass. 01778. Office: Northrop Nortronics, 100 Morse St., Norwood, Mass.*

SILBERMANN, Jean Thiebaut, French physicist; b. Pont-d'Aspach, France, 1806; lab. asst. to Pouillet, Coll. Bourbon; joined Office of Bridges and Rds., 1829; lab. asst. in physics Faculty Scis.; lab. asst. Conservatory Arts and Crafts, until 1848, then curator collections. Built (with Favre) calorimeter for measurement of heat energy of gases; invented sympiezometer, cathetometer, focimeter, a dilatometer, heliostat, a pyrometer, diffraction bench; showed (before Jacoby) possibility of applying galvanoplastics to reprodn. of medals; studied speed of light and electricity. Died Paris, 1865.

SILBERRAD, Oswald John, English chemist; b. Buckhurst Hill, Eng., Apr. 2, 1878; s. Arthur Pouchin d'Artois and Lucy Clarissa (Savill) S.; Certificate 1st Pl., City and Guilds Tech. Coll., 1895-98; student Cambridge U.; Ph.D., (research scholar) Wurzburg (Germany) U., 1899; student Royal Instn., 1900; m. Lilian Glendora George, Feb. 27, 1922; 1 son. Founder, Silberrad Research Lab., 1906, also dir.; munitions cons. Brit. Govt., especially during World War I and II; founder armament research dept. Ministry of Supply, Royal Arsenal, Woolwich, Eng., 1910. Author: Treatise on Stability of Nitrocellulose, 1904; Treatise on Erosion of Bronze Propellers, 1909. Discovered explosive properties of tetryl, complete detonation of lyddite and dynamite, 1902-04, process for manufacture of millitic acid, 1907-08, cause of rapid deterioration of ships propellors, 1908-10, artificial retting of flax, 1910, procedure for making picric acid in large batches in iron nitrators, 1915, prodn. of dyes from T.N.T. residues, 1915, flameless and smokeless arty. powder; devel. plastic especially for aircraft. Died Loughton, Eng., June 17, 1960.

SILBERSTEIN, Ludwik, physicist; b. Warsaw, Poland, May 17, 1872; s. Samuel and Emily (Steinkalk; S.; studied at Cracow, Heidelberg and Berlin U., Ph.D. in Math. Physics, Berlin, 1894; m. Rose Eisenman, June 29, 1905; children—George Paul Hedwiga Renata, Hannah Emily. Naturalized U. S. citizen, 1935. Asst. in physics, Lemberg, 1895-97; lectr. math. physics U. Bologna (Italy), 1899-1904, U. Rome since 1904; math. physicist at research lab., Eastman Kodak Co., 1920-29; cons. math. physicist since 1930. Lectr. on relativity and gravitation. Cornell, 1920, Toronto U., U. Chgo., 1921. Mem. Am. Astron. Soc. Author: Vectorial Mechanics, 1913, 26; The Theory of Relativity, 1914, 24; Simplified Method of Tracing Rays Through Lenses, etc., 1918; Projective Vector Algebra, 1919; Elements of Electromagnetic Theory of Light, 1918; Elements of Vector Algebra, 1919; Theory of General Relativity and Gravitation, 1922; The Size of the Universe, 1930; Causality, 1933; also numerous papers on physics. Died Jan. 17, 1948.

SILLIMAN, Benjamin, Am. chemist; b. Trumbull, Conn., Aug. 8, 1779; s. Gold Selleck and Mary (Fish) S.; grad. Yale, 1796; M.D. (hon.), Bowdoin Coll., 1818; LL.D., Middlebury (Vt.) Coll., 1826; m. Harriet Trumbull, Sept. 17, 1809; m. 2d, Sarah (McClellan) Webb, Sept. 17, 1851; 9 children, including Benjamin. Admitted to Conn. bar, 1802; prof. chemistry, natural history Yale, 1802-53, gave 1st course of exptl. lectures ever given at Yale, 1804, largely responsible for Yale's acquisition of George Gibbs' mineral collection, began full course illustrated lectures in mineralogy, geology, 1813, instrumental in establishment Yale Med. Sch. (opened 1813), became prof. chemistry, induced Yale corp. to establish dept. philosophy and arts, 1847, became prof. emeritus, 1853; founder, propr., 1st editor Am. Jour. Science and Arts, 1818; delivered geol. lectures before Boston Soc. Natural History, 1835. Mineral sillimanite discovered by Bowen named in his honor. Mem. Am. Philos. Soc; 1st pres. Assn. Am. Geologists, 1840; original mem. Nat. Acad. Sciences, 1863. Author: Elements of Chemistry, 1830-31. Editor: Elements of Experimental Chemistry (William Henry), 1814. Improved deflagrator along lines of one made by Robert Hare, used it to study Voltaic current, early 1800's; introduced Priestly's soda water into U. S., 1806; observed (with colleague) meteorite fall, 1807. Died New Haven, Conn., Nov. 24, 1864.

SILLIMAN, Benjamin, Jr., Am. chemist; b. New Haven, Conn., Dec. 4, 1816; s. Benjamin and Harriet (Trumbull) S.; grad. Yale, 1837; m. Susan Huldah Forbes, May 14, 1840; 7 children. Became asso. editor Am. Jour. Sci. and Arts, 1838, editor, 1845-85; prof. practical chemistry Yale, 1846-53, founder Sch. Applied Chemistry in new Dept. of Philosophy and the Arts (later Sheffield Sci. Sch.), asso., 1847-69, prof. chemistry Yale Med. Sch. and Yale Coll., 1853-85; in charge chem. dept. World's Fair, N.Y.C., 1853. an original incorporator Nat. Acad. Scis., 1863; Author: First Principles of Chemistry, 1847; First Principles of Natural Philosophy, 1858; First Principles of Physics, 1859. Demonstrated that petroleum is essentially a mixture of hydrocarbons different in character from vegetable and animal oils and that it can be separated by distillation and simple means of purification into a series of distillates; identified what were to become major uses of petroleum for next 50 years, outlined principal methods of purifying those products. Died Haven, Jan. 14, 1885.

SILSBEE, Robert Herman, Am. physicist; b. Washington, Feb. 24, 1929; s. Francis Briggs and Clara (Gillis) S.; A.B., Harvard, 1950, A.M., 1951, Ph.D., 1956; m. Ann Livingston Loomis, Dec. 21, 1950; children—Douglas Wheeler, David Gillis, Peter Livingston. Staff mem. Oak Ridge Nat. Lab., 1956-57; faculty Cornell U., Ithaca, N.Y., 1957—, prof., 1965—. Mem. Am. Phys. Soc., Phi Beta Kappa, Sigma Xi. Research in properties of imperfections in solids through studies of electron spin resonance and optical absorption of these systems. Home: 117 Northview Rd., Ithaca, N.Y. 14850.*

SILVA, Paul Claude, Am. botanist; b. San Diego, Oct. 31, 1922; s. Roy Arthur and May (Henson) S.; B.A., U. So. Cal., 1946; M.A., Stanford, 1958; Ph.D. in Botany, U. Cal. at Berkeley, 1951. Faculty, U. Ill., Urbana, 1952-60, asso. prof., 1956-60; vis. asso. prof. U. Cal. at Berkeley, 1960-61, sr. herbarium botanist, 1961—. Recipient Darbaker award Bot. Soc. Am. for meritorious study algae, 1958. Mem. Internat. Phycological Soc. (past pres.), Phycological Soc. Am. (past pres.). Editor, Phycologia, 1961—. Research in structure, classification and distbn. of seaweed. Home: 1516 West View Dr., Berkeley, Cal. 94705.*

SILVATICUS, Jean-Baptiste (Sylvaticus, or Silvatico, Giambattista), Italian physician; b. 1550; student medicine, Pavia, Italy; med. prof., Pavia. Author: De secanda in putridis febribus salvatella, de que nostro in senandis venis modo cum antiquo comparato, 4 vols., 1583; Tractatus due de materia turgente et de aneurysmate, 4 vols., 1595; De iis qui morbum simulant deprehendensis (one of 1st works on feigning illness), 1595; Tractatus de compositione et usu theriacae, 8 vols., 1597; Galeni historiae medicinales enarratae, 1605; De unicornu lapide bezaar smaragdo et margaritis eorumque in febribus pestilentialibus usu, 1605; De anno climacterico, 1615. Died 1621.

SILVER, Arnold Herbert, Am. physicist; b. Bklyn., Sept. 27, 1931; s. Louis and Fannie (Sklar) S.; B.S. in Physics, Rensselaer Poly. Inst., 1952, M.S., 1954, Ph.D., 1958; m. Irene Mary Ten Eyck, May 24, 1952; children—Pamela Ann, Nancie Gail, Mark Edward, Lynn Alison, Susan Deborah. Research asst. Brown U., Providence, 1955-57; staff Sci. Lab., Ford Motor Co., Dearborn, Mich., 1957—, prin. research scientist asso., 1964-65, staff scientist, 1965—. Mem. Am. Phys. Soc., Sci. Research Soc. Am. Contbg. author: Modern Aspects of the Vitreous State, 1960. Research and publs. on application of nuclear resonance to study of structure of glass, structure of borides; participated in discovery and demonstration of quantum effects in macroscopic superconducting systems. Home: 30231 Wicklow Ct., Farmington, Mich. 48024. Office: Sci. Lab., Ford Motor Co., Dearborn, Mich. 48121.*

SILVER, Henry K., Am. pediatrician; b. Phila., Apr. 22, 1918; s. Samuel and Dora (Kreitzer) S.; B.A., U. Cal., 1938, M.D., 1942; m. Harriet Ashkenas, June 15, 1947; children—Stephen, Andrew. Instr., asst. prof. pediatrics U. Cal. Sch. Medicine, San Francisco, 1946-52; asso. prof., pediatrics Sch. Medicine, Yale, 1946-52; prof. U. Colo. Sch. Medicine, Denver, 1957—. Cons., Fitzsimons Gen. Hosp., U. S. Army and USAF, Denver. Rosenberg Found. fellow, 1945-47. Mem. Am. Acad. Pediatrics, Western Soc. for Pediatric Research (Ross award for edn. 1962), Am. Pediatrics Soc., Pediatric Research Soc., Rocky Mountain Pediatric Soc., Sigma Xi. Sr. author: Handbook of Pediatrics; Healthy Babies-Happy Parents, 1960; also articles. Research on provision of health care by allied profls., pediatric endocrinology. Home: 135 S. Ivy St., Denver 80222. Office: 4220 E. 9th Av., Denver 80220.*

SILVER, Samuel, Am. physicist; b. Phila., Feb. 25, 1915; s. Boris and Molly (Agrin) S.; B.A., Temple U., 1935, M.A., 1937, D.Sc., 1963; Ph.D., Mass. Inst. Tech., 1940; m. Marjorie Euster, Dec. 28, 1938; children—Daniel Ben, Deborah Ruth. Instr., U. Okla., 1941-42, asst. prof., 1942-43; staff mem. Radiation Lab., Mass. Inst. Tech., 1943-46; physicist Naval Research Lab., 1946-47; faculty U. Cal., Berkeley, 1947—, prof. engring. sci., 1952—, dir. Electronics Research Lab., 1957-60, dir. Space Scis. Lab., 1960—. Internat. chmn. Commn. VI, Internat. Sci. Radio Union, 1954-60, pres., 1966—; chmn. com. space radio research, 1963—; mem. adv. coms. govt., industry. Guggenheim fellow, 1953, 1960; recipient Distinguished Alumnus award, Temple U., 1964, fellow award, I.R.E., 1954. Fellow Am. Phys. Soc., I.E.E.E.; mem. Am. Geophys. Union, N.Y., Acad. Scis., Soc. Engring. Sci., A.A.A.S., Sigma Xi. Editor: Microwave Antenna Theory and Design, Vol. 12, 1948; Radio Waves and Circuits, 1963; (with K. Maeda) Space Radio Science, 1965. Research, publs. on applied electromagnetic theory, microwave optics; millimeter wave radioastronomy; atmospheric physics; devel. interdisciplinary program in space scis.*

SILVERMAN, Frederic Noah, Am. physician; b. Syracuse, N.Y., June 6, 1914; s. Max and Sophia (Silverman) S.; B.A., Syracuse U., 1935, M.D., 1939; m. Carolyn Rose Weber, Jan. 14, 1945. Dir. div. roentgenology Children's Hosp., Cin., 1947—, attending pediatrician, 1947—, asso. Research Found., 1949—; attending pediatrician Cin. Gen. Hosp., 1954—; mem. faculty Coll. Phys. & Surg., Columbia, 1946-47; faculty U. Cin., 1947—, prof. pediatrics, 1962—, prof. radiology, 1962—. Vis. prof. Antioch Coll., 1956—; asso. phys. growth Fels Research Inst., 1956—; cons. govt. agys. Bd. dirs. Children's Protective Service, Cin., 1962—. Diplomate Am. Bd. Pediatrics. Mem. Soc. Pediatric Research (v.p. 1959-60), Am. Acad. Pediatrics, A.A.A.S., Fedn. Am. Scientists, Assn. Am. Med. Colls., Am. Assn. Phys. Anthropoligists, Am. Pediatric Soc., Soc. Pediatric Radiology (pres. 1959-60), Am. Roentgen Ray Soc. (hon.), Sigma Xi, Phi Kappa Phi, Sigma Pi Sigma, Alpha Omega Alpha. Corr. editor Annales de Radiologie, 1958—; mem. editorial bds. Pediatrics, 1962—, Progress in Pediatric Radiology, 1964—. Contbr. sects. to pediatric and radiology textbooks. Research in application of x-ray diagnostic procedures in diseases of children. Home: 2825 Andrew Pl., Cin. 45209. Office: Children's Hosp., Cin. 45229.*

SILVERMAN, Paul Hyman, Am. zoologist; b. Mpls., Oct. 8, 1924; s. Adolph and Libbie (Idlekope) S.; student U. Minn., 1942-43, 46-47; B.S., Roosevelt U., 1949; M.S. in Biology, Northwestern U., 1951; Ph.D. in Parasitology, Sch. Tropical Medicine, U. Liverpool (Eng.), 1955; m. Nancy Josephs, May 20, 1945; children—Daniel Joseph, Claire. Research fellow Malaria Research Sta., Hebrew U., Rosh Pinna, Israel, 1951-53, dept. entomology and parasitology Sch. Tropical Medicine, U. Liverpool, 1953-56; sr. sci. officer dept. parasitology Moredun Inst., Edinburgh, Scotland, 1956-59; head dept. immunoparasitology Allen & Hansburys, Ltd., Ware, Eng., 1960-62; prof. zoology and vet. pathology and hygiene U. Ill., Urbana, 1963—, chmn., 1964-65, head dept. zoology, 1965—, sr. staff, mem. Center for Zoonoses Research, 1966—. Mem. Am. Assn. Immunologists, Am. Soc. Parasitologists, Brit. Soc. Immunologists, Royal, Am. socs. for tropical medicine and hygiene, Brit. Soc. for Parasitology, Sigma Xi. Editor: (with A.E.R. Taylor) Techniques in Parasitology, 1963; (with Ben Dawes) Advances in Parasitology, 1965; also numerous articles. Asst. editor Jour. Parasitology, 1966. Biochem. research on nutritional needs insecticide-resistance Levant housefly; elucidated epidemiology bovine cysiticercosis in Britain; research on mechanism resistance to metazoan parasitism; developed in vitro culture techniques for helminths and prodn. antigens for use in antiparasitic vaccines. Home: 1516 Waverly Dr., Champaign, Ill. 61820.*

SILVERMAN, William Aaron, Am. physician; b. Cleve., Oct. 23, 1917; s. Morris and Jenny (Berman) S.; B.A., U. Cal., 1939, M.D., 1941; m. Ruth G. Hirsch, June 8, 1945; children—Daniel, Jen, David. Practice medicine, specializing in pediatrics, N.Y.C., 1945—; faculty Coll. Phys. and Surg. Columbia, 1946—, now professor of pediatrics; mem. med. com. Planned Parenthood-World Population, 1965—; mem. pediatric adv. com. N.Y.C. Health Dept. Recipient E. Mead Johnson award for research in pediatrics, 1958. Career scientist Health Research Council N.Y.C. Mem. Am. Pediatric Soc., Soc. Pediatric Research, Harvey Soc., Am. Acad. Pediatrics. Author: Dunham's Premature Infants, 1961. Editor: (with A. Minkowski) Biologia Neonatorum, 1963. Mem. editorial bd. Pediatrics, 1964—. Research, numerous publs. on infantile cortical hyperostosis; discovered relationship between sulfixoxazole and neonatal kernicterus; investigation of influence of thermal environment on survival and well being of premature infant in first days of life. Home: 114 Alta Av., Yonkers, N.Y. 10705. Office: Babies Hosp., N.Y.C. 10032.*

SILVERS, Willys Kent, Am. zoologist; b. N.Y.C., Jan. 12, 1929; s. Lewis Julian and Miriam (Rosenzweig) S.; B.A., Johns Hopkins U., 1950; Ph.D., U. Chgo., 1954; m. Abigail Mae Adams, Sept. 29, 1956; children—Deborah, Willys. USPHS fellow Brown U., 1955-56, Jackson Lab., Bar Harbor, Me., 1956-57; asso. staff sci. Jackson Lab., 1957; asso. mem. Wistar Inst., Phila., 1958-65; asso. prof. med. genetics U. Pa. Sch. Medicine, Phila., 1965-67, professor of medical genetics, since 1967—, research asso. dept. dermatology, 1961—. Mem. Allergy and Immunology Study Sect., NIH, 1962—. Recipient Research Career Devel. award NIH, 1964. Mem. editorial bd. Transplantation, 1963—; asso. editor Jour. Exptl. Zoology, 1966—. Research and numerous publs. in field of Mammalian genetics with particular reference to coat-color determinants and immunogenetics, biology of tissue transplantation, biology of skin. Home: 210 Mill Creek Rd., Ardmore, Pa. 19003.*

SILVERSTEIN, Arthur Matthew, Am. immunologist; b. N.Y.C., Aug. 6, 1928; s. Sol D. and Beatrice (Pearl) S.; B.A., Ohio State U., 1948, M.S., 1951; Ph.D., Rensselaer Poly. Inst., 1954; m. Frances Swimmer, May 18, 1950; children—Lise, Mark William, Judith. Chief immunobiology br. Armed Forces Inst. Pathology, Washington, 1956-64; asso. prof. Med. Sch. Johns Hopkins, 1964-67, professor of ophthalmic immunology, 1967—; cons. USPHS, 1963—. Mem. A.A.A.S., Am. Assn. Immunologists, Brit. Soc. Immunology, Phi Beta Kappa. Author: (with A.E. Maumenee) Immunopathology of Uveitis, 1964. Research, numerous publs. on devel. of immunol. responses in mammalian fetus, technics for intra-uterine fetal surgery; pathogenesis of ocular disease, uveitis. Home: 2011 Skyline Rd., Ruxton, Md. 21204. Office: Wilmer Inst., Johns Hopkins Hosp., Balt. 21205.*

SILVESTRI, Francesco, Italian philosopher; b. Ferrara, Italy, 1474; studied theology and philosophy upon entering Dominican order, S. Maria of the Lambs monastery, Ferrara, circa 1488; at convent of S. Domenico, Mantua, Italy, 1498-1503, in Milan, Italy, 1503-07; master of students Studio, Bologna, Italy, 1507-16, then dir., became prior and regent, 1524; vicar gen. Lombard Congregation, 1518-20, Dominican order, from 1524. Reexamined Thomist lit.; made theoretical refutations of scholastic philosophers; infused newly rediscovered works of Aristotle and neo-Platonic writings of Florentine acad. into Intellectual works of his time. Died Sept. 19, 1528.

SILVETTE, Herbert, pharmacologist; b. McKee's Rocks, Pa., Dec. 23, 1907; B.S., M.S., U. Va., Ph.D., 1934; m. Brooks Johnson, 1931. Mem. faculty Med. Sch., U. Va., Charlottesville, 1928-47, asso. prof. pharmacology, 1946-47; vis. prof. pharmacology Meharry Med. Coll., 1949-52, U. Wash., 1953-54; faculty Med. Coll. Va., Richmond, 1954—. Guggenheim fellow, 1946-47. Mem. Am. Physiol. Soc., Am. Soc. Pharmacology and Exptl. Therapeutics, A.M.A. Author 10 novels. Contbr. articles to profl. jours. Address: "Low Gear", Stanardsville, Va. 22973.*

SIMANTON, William Aldrich, Am. entomologist; b. Fargo, N.D., Feb. 17, 1911; s. Frank L. and Gertrude (Aldrich) S.; B.S., Mich. State U., 1931; Ph.D., Ia State U., 1935, M.S., 1932; m. Rosemary Frost, July 7, 1935; children—William A., Donald F., Linda S. Chief entomologist Gulf Research and Devel. Co., Pitts., 1935-44; sr. technologist Shell Oil Co., San Francisco, 1945-48; tech. dir. Agrl. Chems. Co., Phoenix, 1949-50; prof., entomologist U. Fla. Citrus Expt. Sta., Lake Alfred, Fla., 1950—. Mem. Entomol. Soc. Am., Ecol. Soc. Am., Am. Inst. Biol. Scis. Research, publs., patents on insecticides and insect repellents; developed comprehensive insect survey procedures, specifications for superior tree spray oils; adapted ecol. data for computer analysis. Home: 405 W. Lake Summit Dr., Winter Haven, Fla. 33880. Office: Citrus Expt. Sta., Lake Alfred, Fla. 33850.*

SIMHA, Robert, phys. chemist; b. Vienna, Austria, Aug. 4, 1912; student Technol. Inst., Vienna, 1930-33; Ph.D., U. Vienna, 1935; m. Genevieve M. Cowling, June 7, 1941. Came to U.S., 1938, naturalized, 1944. Research asso. U. Vienna, 1935-38, Columbia, 1939-41; lectr. grad. div. Bklyn. Coll., 1940-42, Poly. Inst., Bklyn, 1941-42; asst. prof. Howard U., Washington, 1942-45; lectr. grad. sch. Nat. Bur. Standards, 1944-45, cons. research coordinator, 1945-51; prof. N.Y. U., 1951-58; prof. chemistry U. So. Cal., Los Angeles, 1958—. Indsl. research cons., 1940—. Recipient Lalor Found. award, 1940; award for meritorious service U. S. Dept. Commerce, 1949; award for superior accomplishments Nat. Bur. Standards, 1949. Fellow A.A.A.S., Am. Inst. Chemists, Am. Phys. Soc. (past vice chmn. div. high polymer physics), N.Y. Acad. Scis. (A. Cressy Morrison prize 1948), Washington Acad. Sci. (award for distinguished service 1946); mem. Am. Chem. Soc. (past chmn. div. polymer chemistry), Sigma Xi. Editorial bd. Jour. Colloid Sci., 1946—, Jour. Polymer Sci., 1966—. Research, numerous publs. on phys. properties, structure and reactions of macromolecules and high polymers (rubbery, plastic and materials). Home: 4564 Don Milagro Dr., Los Angeles 90008.*

SIMITCH, Tchedomir Pavle, Yugoslavian physician; b. Cumic, Yugoslavia, Nov. 28, 1896; s. Pavle Radoje and Perka (Durdevic) S.; D.M.S., Med. Faculty Strazbourg, 1924; m. Kosara Ristic, Oct. 14, 1924; children—Zorica (Mrs. Bogdan Plecas), Pavle. With Bacteriological Sta., Skopje, Yugoslavia, 1924-34; dir. Inst. for Hygiene, Skopje, 1934-36; faculty Vet. Faculty, Belgrade, Yugoslavia, 1936—, prof. parasitology, 1940—; chief Inst. Parasitology, Serbian Acad. Scis., 1949—; also lectr. Leader malariology course WHO, Belgrade, 1961-63. Recipient St. Sava and IVran, 1930, French Legion of Honour, 1932, Gold relief medal Belgrade U., 1966, others. Corr. mem. French Vet. Acad. Author: Malaria, 1948, 2d edit.; Protozoe books I, II, 1939-41; Helmints, 1948, (with Z. Petrovic) 2d edit., 1963; with (V. Zivkovic) Artropodes, 1958. Research, publs. on parasitic diseases. Home: 1 Kosovska, Belgrade, Yugoslavia.*

SIMMEL, Georg, German sociologist, philosopher; b. Berlin, Germany, Mar. 1, 1858; Ph.D., U. Berlin, 1881. Lectr. philosophy U. Berlin, 1885-1900, prof. extraordinary, 1900-14; prof. U. Strasbourg, 1914-18. Author: Einleitung in die Moralwissenschaft, 2 vols., 1892-93; Philosophie des Geldes, 1900; Schopenhauer und Neitzsche, 1907; Sociology, 1908; Kant, 1913; Goethe, 1913; Lebensanschauung, 1918. A leading figure in clarifying sociology's scope and making it a basic, precise social sci. in Germany; stressed study of forms of social interaction; held that change not stability was natural; analyzed patterns of authority and obedience; dealt with abstract sociol. theory rather than concrete detailed study. Died Strasbourg, Sept. 28, 1918.

SIMMONDS, Norman Willison, Brit. geneticist; b. Bedford, Eng., Dec. 15, 1922; s. William Henry and Ida (Willison) S.; B.A., Downing Coll., Cambridge (Eng.) U., 1943, Sc.D., 1966; AICTA, Imperial Coll. Tropical Agr., Trinidad, 1945; m. Christa Ebert, July 1, 1965. Staff Imperial Coll. Tropical Agr., Trinidad, 1945-59, sr. cytogeneticist banana research scheme, also Regional Research Centre, 1952-59; head dept. potato genetics John Innes Inst., Hertford, Eng., 1959-65; dir. Scottish Plant Breeding Sta., Edinburgh, 1965—. Cons. W.I. Sugar Cane Breeding Sta., Barbados, 1963—; hon. sr. lectr. botany Edinburgh U., 1966—. Fellow Linnean Soc., Inst. Biology. Author: Bananas, 1959; Evolution of the Bananas, 1962; also numerous articles. Research on cytogenetics and evolution of bananas and potatoes. Home: 9 McLaren Rd., Edinburgh 9. Office: Scottish Plant Breeding Sta., Pentlandfield, Roslin, Midl., Scotland.*

SIMMONS, Gustavus Lincoln, Am. physician; b. Hingham, Mass., 1832; M.D., Harvard, 1865; m. Celia Crocker, June 1, 1862; children—Gustavus, Samuel Ewer, Celia (Mrs. Dwight H. Miller), Carrie. Practiced medicine, Sacramento. Elected mem. Sacramento Bd. Edn., 1859; sch. supt.; mem. Sacramento Bd. Health. Mem. Sacramento Soc. for Med. Improvement (a founder 1868), Cal. Med. Assn. (pres. 1894-95). Pioneered blood vessel surgery by ligating common carotid artery (using silver wire for suture), 1864; repaired divided heel tendon; successfully treated miner with metal bar driven through his skull. Died Oct. 4, 1910.

SIMMONS, James Stevens, Am. physician; b. Newton, N.C., June 7, 1890; s. James Curtley and Angie Mary (Stevens) S.; B.S., Davidson Coll., 1911, Sc.D., 1937; postgrad. U. N.C., 1911-13; M.D., U. Pa., 1915; grad. Army Med. Sch., 1917; Ph.D., George Washington U. Med. Sch., 1934; Dr. P. H., Harvard, 1939; several hon. degrees; m. Blanche Scott, June 29, 1920; 1 dau., Frances Scott (Mrs. Frances Simmons McConnell). Resident and chief resident physician U. Pa. Hosp., 1915; bacteriologist William Pepper Lab., U. Pa., 1916. Served as 1st lt. M.R.C., 1916, 1st lt., M.C., U. S. Army, 1917, advancing through grades to brig. gen., 1943; retired from service, 1946; dean and prof. Harvard Sch. Pub. Health, Boston, Mass., 1946—. Served as chief of lab. services various U. S. Army hosps., comdg. officer various dept. labs., 1917-24; asst. dir. labs. Army Med. Sch., also chief bacteriol. dept. Army Med., Dental and Vet. Schs., 1924-28; pres. Army Med. Dept. Research Bd. Bur. Science, Manila, 1928-30; chief of dept. bacteriology Army Med. Sch., 1930-34, also dir. dept. preventive medicine, 1932-34; dir. of labs. Army Med. Center, 1932-34; pres. Army Med. Research Bd., Ancon, C.Z., 1934-35; asst. Corps Area surgeon, I.C.A., 1936-40; chief Preventive Medicine Service, Office of Surgeon Gen., U. S. Army, 1940-46, sr. cons., 1946—. Recipient Sternberg medal, 1910; Sedgwick Meml. medal, 1943; U. S. A. Typhus Commn. medal, 1943; Carlos J. Finlay medal, 1943, Walter Reed medal, 1944; D.S.M., 1945; Bruce medal, 1948; Charles V. Chapin medal, 1952; Gorgas award Assn. Mil. Surgeons U. S., 1952. Fellow A.A.A.S., A.M.A. (mem. bd. on preventive medicine); mem. Assn. Am. Physicians, Assn. Mil. Surgeons, Am. Society Tropical Medicine (pres. 1946), Am. Acad. Tropical Medicine (pres. 1946), Washington Acad. Sciences, Med. Assn. Isthmian Canal Zone (sec. 1935, pres. 1936), Nat. Malaria Com. (pres. 1942), Sigma Xi. Author of books and articles on exptl. bacteriology, preventive medicine and tropical medicine. Acting editor in chief Abstracts of Bacteriology, 1924-26; editor Med. Bacteriology sect. Biol. Abstracts, 1926-48; asst. editor Philippine Jour. Sci., 1929-30. Credited with conducting largest campaign in preventive medicine to that time, World War II; proved Aëdes albopictus is a vector of dengue, 1937. Died July 31, 1954.

SIMMONS, Norman Stanley, Am. biophysicist; b. N.Y.C., May 28, 1915; s. Irving Clay and Rhea (Dinitz) S; B.S., Coll. City N.Y, 1935; D.M.D., Harvard, 1939; Ph.D, U. Rochester, 1950; m. Alice Edith Beck, Sept. 5, 1937; children—Steven Mark, Peter Jay. Carnegie fellow in dental research U. Rochester Sch. Medicine and Dentistry, 1939-41, fellow dental research, 1947-50, sr. fellow, 1948-50; instr. periodontology Columbia, 1941-43; asso. research biochemist U. Cal. Los Angeles Atomic Energy Project, 1950-59, research biochemist, 1959—, prof. biophysics Center for Health Scis., 1963—, prof. oral medicine, 1963—; attending dentist Mt. Sinai Hosp., N.Y.C., 1946-47; cons. periodontology VA Hosps., Los Angeles, 1953-56; vis. scientist Childrens Cancer Research Found., Boston, 1958-60. Fellow in chemistry Harvard, 1955. Recipient USPHS, NIH Career Research award, 1963. Mem. Internat. Assn. Dental Research, Biophys. Soc., Am. Soc. Biol. Chemists, Internat. Assn. Pure and Applied Physics, Sigma Xi, Omicron Kappa Upsilon. Contbr. numerous articles to sci. jours. Discoverer lysozyme-mucoid interactions with implications to def. mechanisms of mucous membranes, conformation dependent cotton-effects and structure-function relationship in enzymes by optical rotatory dispersion and circular dichromism; 1st DNA isolation and characterization. Home: 409 Puerto Del Mar, Pacific Palisades, Cal. 90272. Office: 900 Veteran Av., Los Angeles 90024.*

SIMMONS, Ralph Oliver, Am. physicist; b. Kensington, Kan., Feb. 19, 1928; s. Fred Charles and Nellie (Douglass) S.; B.A., U. Kan., 1950; B.A. (Rhodes scholar), Oxford U., 1953; Ph.D., U. Ill., 1957; m. Janet Lee Lull, Aug. 31, 1952; children—Katherine Ann, Bradley Alan, Jill Christine, Joy Diane. Research asso. U. Ill., Urbana, 1957-59, faculty, physics, 1959—, asso. prof., 1961—. Sr. postdoctoral fellow NSF, 1964. Fellow Am. Phys. Soc.; mem. Phi Beta Kappa, Sigma Xi, Pi Mu Epsilon. Research on atomic defects in solids, thermodynamics of crystals, radiation damage. Home: 1005 Foothill Dr., Champaign, Ill.*

SIMON, Albert, Am. physicist; b. N.Y.C., Dec. 27, 1924; s. Emanuel D. and Sarah (Leitner) S.; B.S., Coll., City N.Y., 1947-Ph.D., U. Rochester, 1950; m. Harriet E. Rubinstein, Aug. 17, 1947; children—Richard, Janet, David. Physicist, Oak Ridge Nat. Lab., 1950-54, asso. dir. neutron physics div., 1954-61; head plasma physics div. Gen. Atomic Co., San Diego, 1961-66; prof. dept. mech. and aerospace scis. U. Rochester (N.Y.), 1966—. John Simon Guggenheim fellow, 1964-65. Fellow Am. Phys. Soc. (plasma physics exec. com. 1959-62, chmn. 1963-64); mem. Sigma Xi. Author: An Introduction to Thermonuclear Research, 1949; contbr. to Ency. Americana, 1964. Editor: Advances in Plasma Physics, 1967—. Theoretical research in plasma physics; research on reactor shielding, theory of nuclear reactions and polarization. Home: 263 Ashley Dr., Rochester, N.Y. 14620.*

SIMON, Alexander, Am. psychiatrist; b. N.Y.C., Oct. 13, 1906; M.D., Columbia, 1930; m. Olga Riedle, Oct. 26, 1934; 1 son, Francis. Asst. med. dir. Langley Porter Neuropsychiat. Inst., 1943-56, med. dir., 1956—; prof., chmn. dept. psychiatry U. Cal. Sch. Medicine, San Francisco, 1956—. Mem. com. on aging, Cal. Dept. Mental Hygiene, 1961—, chmn., 1964—; mem. Gov.'s Interdept. Com. on Problems of Aging, 1960-66. Fellow Am. Coll. Psychiatrists, Am. Gerontological Soc., Am. Geriatrics Soc.; mem. World (com. on tng. psychiatrists 1963—), Am. (chmn. com. on aging 1965—) psychiat. assns., A.M.A., Am. Psychopath. Assn., Am. Acad. Neurology, Soc. Biol. Psychiatry. Editor: Physiology of Emotions, 1961. Nat. Inst. Mental Health grantee, 1959—. Research, publs. on geriatric mental illness. Home: 2680 Jackson St., San Francisco 94115. Office: 401 Parnassus Av., San Francisco 94122.*

SIMON, Eric Jacob, Am. biochemist; b. Wiesbaden, Germany, June 2, 1924; s. Joseph and Paula (Meyer) S.; came to U. S., 1938, naturalized, 1945; B.S., Case Inst. Tech., 1944; M.S. in Organic Chemistry, U. Chgo., 1947, Ph.D., 1951; m. Irene Ronis, Aug. 9, 1947; children—Martin Allen, Faye Ruth, Lawrence David. Nat. Found. for Infantile Paralysis fellow Columbia, 1951-53; research asso. Cornell U. Med. Sch., 1953-58; vis. investigator dept. microbiology N.Y. U. Med. Sch., N.Y.C., 1958-59, faculty, 1959—, asso. prof. exptl. medicine, 1964—; lectr. chemistry Coll. City N.Y., 1952-59. Sec., Com. on Research on Narcotic Addition, 1963-65; career scientist Health Research Council N.Y.C., 1959—. Mem. Am. Soc. Biol. Chemists, A.A.A.S., Am. Chem. Soc., Am. Soc. for Cell Biology, Harvey Soc. Research on Synthesis s-succinyl coenzyme A, metabolism vitamin E, isolation and characterization vitamin E, effect morphine and related drugs on cell metabolism, inhibition RNA synthesis by levo phenol. Home: 375 Edgewood Av., Teaneck, N.J. 07666. Office: 550 1st Av., N.Y.C. 10016.*

SIMON, Eugène, French zoologist; b. Paris, Apr. 30, 1848; made expdns. throughout world in search of specimens; amassed a large spider collection; mem. French Acad. Scis., 1909; asso. mem. Paris Mus. Modern History. Author: Histoire naturelle des araignées, 1864-84; also a work on humming birds. Died Paris, Nov. 17, 1924.

SIMON, Sir Francis (Eugene), physicist; b. July 2, 1893; s. Ernst and Anna (Mendelssohn) S.; ed. univs. Munich, Goettingen, Berlin; Ph.D., Berlin; M.A., Oxford U.; m. Charlotte Muenchhausen, 1922; 2 daus. Became pvt. docent, Berin, 1924, asso. prof. physics, 1927; named prof., dir. lab. phys. chemistry, Breslau (now Wroclaw, Poland), 1931, resigned, 1933; vis. lectr. U. Cal. at Berkeley, 1932; researcher Clarendon Lab., Oxford, 1933, head, 1956, prof. thermodynamics U. Oxford, 1945-56, Lee prof. exptl. philosophy, 1956, also fellow Wadham Coll.; mem. Atomic Energy project, 1940-46; Guthrie lectr. Phys. Soc., 1956; Kelvin lectr. Instn. Elec. Engrs. Recipient Kamerlingh Onnes medal, 1950, Linde medal, 1952, Fellow Royal Soc. (Rumford medal 1948); fgn. hon. mem. Am. Acad. Arts and Scis. Author: The Neglect of Science, 1951; Atomic Energy, A Survey, 1954; also papers on physics of very low temperatures. Co-author: Low Temperature Physics, 1952. Died Oct. 31, 1956.

SIMON, Gustav S., German surgeon; b. Darmstadt, Germany, May 30, 1824; m.; 3 children; prof. surgery, Heidelberg, Germany; practice medicine, specializing in gynecology. Author monographs on treatment of vesicovaginal fistula, 1854, the excision of spleen, 1857, plastic surgery, 1868, surgery of kidneys, 1871-76; also numerous articles. First in Europe to perform nephrectomy, 1870; reintroduced splenectomy, 1857. Died 1876.

SIMON, Herbert Alexander, Am. behavioral scientist; b. Milw., June 15, 1916; s. Arthur and Edna (Merkel) S.; B.A., U. Chgo., 1936, Ph.D., 1943, LL.D., 1964; Sc.D., Yale, 1963, Case Inst. Tech., 1963; D.Phil., Lunds U., 1968; m. Dorothea Isobel Pye, Dec. 25, 1937; children—Katherine (Mrs. David L. Frank), Peter A., Barbara M. Research asst. U. Chgo., 1936-38; staff Internat. City Mgrs. Assn., Chgo., 1938-39, Bur. Pub. Adminstrn., U. Cal. at Berkeley, 1939-42, prof. polit. sci. Ill. Inst. Tech., Chgo., 1942-49; prof. adminstrn., computer sci., psychology, asso. dean Carnegie Inst. Tech., Pitts., 1949—. Dir. Social Sci. Research Council, 1958—, chmn. bd. dirs., 1961-65; chmn. div. behavioral scis. NRC, 1968—; mem. Pres.'s Sci. Adv. Com., 1968—. Recipient Adminstrs. award Am. Coll. Hosp. Adminstrs., 1958. Fellow Econometric Soc., Am. Psychol. Assn., Am. Sociol. Assn., A.A.A.S., Am. Assn. Arts and Scis.; mem. Nat. Acad. Scis. (com. on sci. and pub. policy), Am. Philos. Soc., Phi Beta Kappa, Sigma Xi. Author: Administrative Behavior, 1947, 57; Models of Man, 1957; (with J. G. March) Organizations, 1958; (with D. W. Smithburg, V. A. Thompson) Public Administration, 1950; also articles. Research on heuristic programming and computer simulation of human thinking. Home: 5818 Northumberland St., Pitts. 15217.*

SIMON, Italo, Italian pharmacologist; b. Sassari, Aug. 16, 1878; s. Vincenzo and Teresa (Solinas) S.; ed. medicine and surgery; m. Anna Mossa, Feb. 24, 1916; children—Antonio, Angelo, Angela. Prof. emeritus pharmacology U. Pisa; past prof. pharmacology univs. Calgiari, Sassari, Pavia and Padua. Recipient Gold medal Ministry Pub. Instrn. for Sci., Culture and Art. Mem. Adriatica Acad., Medica Acad. Rome; hon. mem. Ferrara Acad. Sci., Letters and Arts; corr. mem. Reale Acad. de Farmacia Inst. Spain. Author: Trattato di Farmacologia; Nozioni di Farmacologia veterinaria; Elementi di Farmacognosia; Farmacoterapia; Trattato di terapia; others. Address: via Monte delle Gioia 9, Rome, Italy.

SIMON, Sir John, English physician, surgeon; b. London, Oct. 10, 1816; s. Louis Michael Simon; apprentice of Joseph Henry Green, surgeon, St. Thomas; completed studies King's Coll.; several hon. degrees; m. Jane. Became 1st med. officer of health City of London, 1848; med. officer Gen. Bd. Health, 1855-58, Privy Council (under Pub. Health Act), 1858-71; surgeon, lectr., officer St. Thomas Hosp. Fellow Royal Soc., 1845. Author: English Sanitary Institutions, 1890; Personal Recollections, 1898, rev., 1903; also clin. surg. lectures. Contbr. article on inflammation (classic of its kind) to System of Surgery (Holmes). Engaged in glandular research; performed (before Edward Cook) operation of perineal puncture of urethra in cases of retention from stricture; ann. reports issued during his 8 years in office (reprinted in Pub. Health Reports, 2 vols., 1887) led to Pub. Health Act of 1875; reformed pub. health of London; began routine bldg. inspections; abolished cesspools; improved water supply and sewers; arranged procedures in cases of contagious disease; supported vaccination. Died July 23, 1904.

SIMON, Theodore, French psycholog'st; b. Dijon, France, July 10, 1873; studied medicine, Paris, wrote thesis, 1900. Attending psychiatrist Saint-Yon Hosp., 1905-20, med. dir., Perray-Vaucluse colony for retarded, nr. Paris, 1920-30; then med. dir. Henri-Rousselle hosp., retired 1936. Dir., Pedagogical Lab., Paris; prof. Teacher's Coll. of Seine. Recipient Prix Falret, 1907, Prix Baillarger, 1912. Author: (with Binet) Les enfants anormaux, 1907; l'Aliéné, l'asile, l'informier, 1911; Pédagogie expérimentale, 1924; editor; bulletin of societé Alfred Binet, 1912-60, Laboratorre et psychiatrie, 1937-39, l'Infirmier psychiatrique, 1953-58; a founder l'Année psychologique, 1895. Developed (with Binet) 1st test to measure intelligence (intelligence of subject is determined by comparison with intelligence of normal subjects of various ages, then mental age derived from test is divided by choronological age of subject, giving the I.Q. or intellectual quotient), for French Ministry of Education, used to detect mentally retarded children, 1905; devised scale of individual tests for children under 2 years of age, also P.V. group test, (developed at Perray-Vaucluse colony; important contbr. to the study of mental development of children and modern psychology. Died Sept. 4, 1961.

SIMONART, Paul, Belgian microbiologist; b. Louvain, Belgium, Apr. 16, 1907; s. Edgar Emire and Marie (de Booseré) S.; Ing.Chimiste Agricole, U. Louvain, 1930; Ph.D. in Biochemistry, U. London, 1933; m. Alberte Bosmans, Jan. 3, 1942; children—Luc Vincent, Paul Michel, Etienne, Marc, Francoise Marie. Fellow, Belgian Am. Ednl. Found., N.Y.C., Brussels, 1936; chargé de mission Govt. of Belgium, 1936-39; prof. U. Louvain, 1939—. Decorated comandeur l'Ordre de la Couronne; officier l'Ordre de Léopold; Lauréat de Travail. Mem. Soc. for Gen. Microbiology (London), Biochem. Soc. (London), Soc. Chimie Biol. (Paris), Am. Dairy Sci. Assn. Author: Introduction á la Microbiologie Générale, 1947; also numerous articles. Research on microbial chemistry and on microbiology; discovered (with A. E. Oxford, H. Raistrick) antibiotic, griseofulvin, used against mycotic infections in humans; invented process of bactofugation which is applied to improve bacteriological quality of foods, especially of dairy products. Home: 97 Haachtstraat, Veltem, Belgium. Office: 92 Kard, Mercierlaan, Heverlee-Louvain, Belgium.*

SIMONIS, Wilhelm Richard, German botanist; b. Neubrandenburg, Germany, June 25, 1909; s. Paul Ludwig and Mathilde (Ebbefeld) S.; Dr.Phil., U. Göttingen, 1935; Dr.habil, U. Tübingen, 1946; m. Berta Doring, Feb. 23, 1938; children—Barbara, Jürgen, Annette. Asst., Bot. Inst., Göttingen, 1935-39, Landw. Hochschule Hohenheim, 1939-42, U. Strassburg (France), 1942-46; staff U. Tübingen, 1946-49, privat dozent, 1946-49; dozent, dir. Bot. Inst., Technische Hochscule, Hannover, Germany, 1949-54; asso. prof. U. Hannover (Germany), 1954-58; prof. Bot. Inst., Bot. Gartens, U. Würzburg (Germany), 1958—, dir. Verwaltungsausschusses, 1964—. Mem. Deutsche Botanische Gesellschaft, Deutsche Gesellschaft für Lichtforschung. Author: (with K. Paech, W. Simonis) Pflanzenphysiologisches Praktikum, 1952; also numerous articles. Research, publs. on drought resistance of plants and photosynthesis, adaptation of photosynthesis in colored light, photophosphorylation and photosyntheis by higher plants and by algae, photosynthesis and ion uptake, effects of low doses of ionizing radiation on plants especially on plant physiology. Office: 64 Mittlerer Dallenbergweg, Würzburg, Deutschland.*

SIMON OF GENOA (Simon Januensis, Geniates a Cordo, or Simon de Cordo), Italian botanist, physician; flourished end of 13th century. Canon of Rouen, France; physician to Pope Nicholas IV; chaplain to Pope Boniface VIII. Compiled Synonyma medicinae seu Clauis Sanationis (standard work on medicine until 16th century), 1472; Liber servitoris liber XXVIII Bulchasin Benaberazerin translatus a Simone

Januensi interprete Abram judeo tortuosiensi, 1471; Liber Serapionis Aggregatus in medicinis simplicibus, 1473. Tried to identify medicinal plants mentioned by Greek and Arabic writers. Died Genoa, Italy, 1303.

SIMONS, Elwyn LaVerne, Am. geologist; b. Lawrence, Kan., July 14, 1930; s. Verne Franklin and Verna (Cuddeback) S.; A.B. in Biology, Rice U., 1953; M.A., Princeton, 1955, Ph.D. in Paleobiology, 1956; D.Phil. (Marshall scholar), Oxford U., 1959. Lectr. geology Princeton, 1958-59; asst. prof. zoology U. Pa., 1959-61; vis. asso. prof. geology, curator vertebrate paleontology Yale, 1960-61, faculty, 1961—, prof. paleontology, curator in charge div. vertebrate paleontology Peabody Mus., 1965—. Research asso. Am. Mus. Natural History. Richard C. Hunt Meml. fellow Wenner-Gren Found., 1965. Mem. Soc. Vertebrate Paleontology (co-editor Bull.), Inst. Human Paleontology, A.A.A.S., Assn. Phys. Anthropology, Geol. Soc. Am., Soc. for Study Human Biology, Soc. for Study Evolution, Am. Soc. Zoologists, Internat. Assn. Human Biologists, Sigma Xi. Research and publs. on anatomy, systematics of primates, evolutionary inter-relationships of fossil apes and their early human relatives. Address: Peabody Mus., New Haven 06520.*

SIMONS, Jakob Lennart, Finnish physicist; b. Vöra, Finland, Sept. 25, 1905; s. Mikael and Anna-Lisa Simons; M.S., U. Helsinki (Finland), 1928, Ph.D., 1932; m. Rut Gunhild Waselius, Aug. 1, 1937; children—Kai L., Tom M., Majlen. Faculty, Helsinki U., 1938—, prof. physics, 1941—, leader Van de Graaff accelerator lab., 1956—; researcher Inst. Theoretical Physics, Copenhagen, Denmark, 1939-40, Inst. for Advanced Study, Princeton, N.J., 1949-50. Rector, Vasa Summer U., 1965—. Named comdr. White Rose of Finland Order. Mem. Internat. Union Pure and Applied Physics (chmn. Finnish Nat. com. 1964—), Finnish (Homén prize 1963), Uppsala (fgn.), Gothenburg (fgn.) sci. socs., Phys. Soc. Finland, Am. Swedish Am. phys. socs., Acad. Tech. Finland, Finnish Chem. Soc. Author: Textbook of Physics, 1946; also numerous articles. Co-editor Nuclear Physics, 1955—. Research on polarization of Raman lines, structure of fluids, neutron-proton cross-sect., spectrum of fission fragments, theory of stability of positronium chloride, nuclear reactor and plasma physics, nuclear structure and nuclear spectroscopy. Home: Mörskomvägen 2, Helsinki 60, Finland. Office: U. Helsinki, Dept. Physics, Siltavuorenpenger 20, Helsinki 17, Finland.*

SIMONS, Joseph H., Am. chemist, chem. engr., educator; b. Chgo., May 10, 1897; s. David and Ester Simons; B.S. in Chem. Engring., U. Ill., 1919, M.S. in Chemistry and Math., 1922; Ph.D., U. Cal. at Berkeley, 1923; m. Eleanor Mae Whittaker, Aug. 22, 1936; children—Dorothy (Mrs. Eldon Lanning), Robert Whittaker. Head dept. chemistry and physics U. P.R., 1925; internat. fellow Cambridge (Eng.) U., 1929; asst. prof. Northwestern U., 1926-33; prof. chemistry Pa. State U., 1933-50; prof. chem. engring. and chemistry U. Fla., Gainesville, 1950—. Cons. dir. fluorine research Minn. Mining & Mfg. Co., 1944—. Mem. Am. Assn. U. Profs., Am. Inst. Chem. Engrs., A.A.A.S., Sigma Xi, Alpha Chi Sigma, Phi Lambda Upsilon, Sigma Pi Sigma. Editor: Fluorine Chemistry. Discovered fluorocarbons, hydrogen fluoride catalysis; invented processes for prodn.; developed techniques for low velocity ionic interactions. Home: 1122 S.W. 11th Ave., Gainesville, Fla. 32601.*

SIMONSEN, John Lionel, English chemist; b. Manchester, Eng., Jan. 22, 1884; s. Lional Michael and Anna (Bing) S.; B.S. with 1st class honors in chemistry, U. Manchester, 1903, D.Sc. (Schunck Research fellow), 1909; LL.D., St. Andrews (Scotland) U.; m. Jannet Dick Hendrie, Dec. 30, 1913. Asst. lectr., demonstrator U. Manchester, 1907-10; prof. chemistry Presidency Coll., Madras, India, 1910-17; chem. adviser Indian Munitions Bd., 1917-19; forest chemist Forest Research Inst., Dehra Dun, India, 1919-25; prof. organic chemistry Indian Inst. Sci., Bangalore, 1925-27; prof. chemistry U. Coll. N. Wales, Bangor, 1930-42; dir. research Colonial Products Research Council, Brit. Colonial Office, 1943-52. Mem. Agrl. Research Council, 1944-49; chm. bd. Pest Infestation Lab., 1949-52. Recipient Kaiser-i-Hind medal, 1921, Fritzche award Am. Chem. Soc., 1949. Fellow Royal Soc. (Davy medal 1950), 1932, Royal Inst. Chemistry, Royal Asiatic Soc. Bengal, Indian Assn. Cultivation Sci. (hon.); mem. Royal Soc. Arts, (v.p. 1949-54), Indian Sci. Congress (hon., pres. chemistry sect. 1917, hon. sec. 1914-26, pres. 1928), Chem. Soc. (v.p. 1940-43, 52-55, sec. 1945-49), Brit. Assn. (pres. sect. B. 1947). Author: The Terpenes, vols. I-V, 1930-32. Contbr. articles to sci. publs. Died Feb. 10, 1957.

SIMONSON, Ernst, Am. physician; b. Tiegenhof, Germany, June 26, 1898; s. Max and Kaethe (Paechter) S.; student U. Greifswald Med. Sch., 1918-24; m. Sophie Schemel, July 19, 1931; 1 son, Walter. Came to U. S., 1939, naturalized, 1945. Rockefeller fellow in pharmacology U. Greifswald, 1924-27; prof. U. Frankfurt (Germany), 1928-34; prof. emeritus, 1957—; prof. normal physiology 1st Med. Inst., Kharkov, Russia, 1931-37; head lab. of indsl. physiology Central Psychotech. Inst., Prague, Czechoslovakia, 1937-39; research asso. Mt. Sinai Hosp., Milw., 1939-44, asso. prof., 1944; prof. physiol. hygiene U. Minn., Mpls., 1957-66, emeritus, 1966—; dir. bio-electronic research Mt. Sinai Hosp., Mpls., 1966—. Mem. Am. Soc. Exptl. Biology, Am. Physiol. Soc., Gerontol. Soc., N.Y. Acad. Scis., Am. Heart Assn. (editorial bd. Jour.), Sociedad Peruana de Cardiologia (hon.), Sociedad Brasileria de Geratria (hon.), German Physiol. Soc. (corr.). Author: Differentiation Between Normal and Abnormal in Electrocardiography, 1961; Cerebral Ischemia, 1964; also numerous articles. Research in electrocardiography, vectorcardiography, applied physiology. Home: 5104 26th Ave. S., Mpls. 55417.*

SIMPLICIOS, Greek natural philosopher, Neoplatonist; b. Cilicia, Greece; flourished 6th century; disciple of Ammonius the Peripatetic and Damascius the Stoic; lived during period of Justinian; worked at Sch. of Athens, until 529; during persecution, took refuge with Chosroes, King of Persia (529-533) returned to Athens, 549; Author: Commentary upon the Enchiridion of Epictetus; Commentaries upon Aristotle; also commentary on Book I of Euclid; Commentary on Lunes of Hippocrates. Tried to unite Platonic and Stoic doctrines with the Peripatetic; explained stability of celestial bodies by excess of impetus over gravity; explained Antiphon's quadrature of circle.

SIMPSON, Charles Floyd, Am. veterinarian, pathologist; b. East Orange, N.J., Jan. 29, 1919; s. Charles F. and Josephine (Reuttner) S.; B.Sc., Rutgers U., 1940; D.V.M., Cornell U., 1944; M.Sc., Ohio State U., 1955; Ph.D., U. Minn., 1961; m. Lucy Virginia Allport, Apr. 27, 1946; children—Vicki A., Kim Charles. Practice vet. medicine, Summit, N.J., 1947-48; asso. pathologist, U. Fla., Gainesville, 1948-53, pathologist, 1955-58, 61—, prof. vet. sci., 1955—; acting chmn. dept. vet. sci., 1961; Am. Vet. Med. Assn. fellow Ohio State U., Columbus, 1954-55, Spl. NIH fellow, U. Minn., St. Paul., 1958-60. Mem. Am., Fla. vet. med. assns., Electron Microscope Soc. Am., Internat. Acad. Pathology, Soc. for Exptl. Biology and Medicine, Sigma Xi, Gamma Sigma Delta, Phi Zeta. Research, numerous publs. on electron microscope descriptions of bacteria and protozoa, cardiovascular system of birds, first E.M. descriptions of aortic lathyrism in turkeys and copper deficiency in chicks; produced dissecting aneurysms of turkeys using diethylstilbestrol. Home: 2225 N.W. 6th Pl., Gainesville, Fla. 32601.*

SIMPSON, Frank Edward, Am. dermatologist; b. Saco, Me., Sept. 7, 1869; s. Charles P. and Adelaide (Reade) S.; A.B., Bowdoin, 1890; M.D., Northwestern U. Med. Sch., 1896; postgrad., Paris, Berlin and Vienna; m. Beulah Lichty, Nov. 22, 1898 (dec.); m. 2d Beryl Lucile Kanagy, 1922; children—Frank Edward, Hugh Mills, William Langdon. Intern Cook County Hosp.; practiced at Chgo., since 1897; prof. skin and venereal diseases, Chgo. Policlinic, 1912-22; clin. prof. dermatology Northwestern U. Med. Sch. attending dermatologist Cook County, Wesley and Policlinic hosps. Pres. Am. Radium Soc.; mem. A.M.A., Chgo. Med. Soc., Chgo. Dermatol. Soc., Phi Beta Kappa. Author: Radium Therapy, 1922; Radium in Cancer and Other Diseases, 1926. Contbr. to Oxford Surgery, Lahrbuch der Strahlentherapie, and spl. articles on radium therapy and dermatology to fgn. and Am. books and jours. Died Dec. 13, 1948.

SIMPSON, Fred James, Canadian microbiologist; b. Regina, Sask., Can., June 8, 1922; s. Ralph James and Lillian (Anderson) S.; B.S., U. Alta., 1944, M.S., 1946; Ph.D., U. Wis., 1952; m. Margaret Christine Nelson, May 28, 1947; children—Christine Louise, Steven James, Leslie Colleen, Ralph Edwin, David Glen. Jr. research officer div. applied biology NRC, Ottawa, Can., 1946-48, asst. research officer Prairie Regional Lab., Sask., 1952-55; asso. research officer, 1957, sr. research officer, head physiology and biochemistry, bacteria sect., 1961—; vis. prof. microbiology U. Ill., 1964; lectr. biochemistry U. Sask., 1962, 64, 66. Sec. asso. com. on plant disease NRC, 1957-63, chmn. sub-com. on culture collections, asso. com. on taxonomy and culture collections. Mem. Can. Soc. Microbiologists, Am. Soc. Microbiology, Am. Soc. Biol. Chemists, Sigma Xi. Research, publs. in devel. of prodn. of 2, 3-butanediol by bacterial fermentation, prodn. of hydrolytic enzymes by micro-organisms to remove gummy pentosans as an aid in separating gluten and starch from wheat; determination of pathways by which sugars are fermented by Aerobacter aerogenes and pathway by which Aspergillus Flavus, other molds degrade flavonoids; discovered carbon monoxide is produced by Aspergillus Flavus from rutin and separation of enzymes involved. Office: Prairie Regional Lab., NRC, Saskatoon, Sask., Can.*

SIMPSON, Sir George Clarke, English meteorologist; b. Derby, Eng. Sept. 2, 1878; s. Arthur and Alice (Clarke) S.; B.Sc., Manchester (Eng.) U., 1900, D.Sc., 1906; D.Sc., Göttingen (Germany) U., 1904, Aberdeen (Scotland) U.; LL.D., hon: D.Sc., U., Sidney (Australia); m. Dorothy Jane Stephen, 1914; 4 children. Became lectr. meteorology Manchester U., 1905; imperial meteorologist Meteorol. Dept., India, 1906-20; physicist Brit. Antarctic Expdn., 1910-12; became dir. Meteorol. Office, London, 1920, 38. Fellow Royal Soc. London, 1915, Royal Meteorol. Soc. (pres. 1940-42); corr. mem. Akademie der Wissenschaft Vienna, Preussische Akademie der Wissenschaften; hon. mem. Internat. Meteorol. Soc. Research and numerous publs. on thunderstorms, water condensation, atmospheric electricity, radiation and climate; 1st determination of position of charges in thunderclouds; contributed to design of instruments for this determination. Address: 5 Mansfield House, Manor Fields, London, S.W. 15, Eng.

SIMPSON, George Gaylord, Am. vertebrate paleontologist; b. Chgo., June 16, 1902; s. Joseph Alexander and Helen Julia (Kinney) S.; student U. Colo., 1918-19, 20-22; Ph.B., Yale, 1923, Ph.D., 1926, Sc.D., 1946; Sc.D., Princeton, 1947, U. Durham, 1951, Oxford U., 1951, U. N.M., 1954, U. Chgo., 1959, Cambridge (Eng.), 1965, Kenyon Coll., 1967, U. Colo., 1968; LL.D., U. Glasgow, 1951; Docteur honoris causa, U. Paris, 1965; m. Lydia Pedroja, Feb. 2, 1923 (div. Apr. 1938); children—Helen Frances, Patricia (dec.), Joan, Elizabeth; m. 2d, Anne Roe, May 27, 1938. Marsha fellow, Peobody Mus., Yale, research on Mesozoic mammals, 1924-26; field asst. Am. Mus. Natural History, N.Y.C., 1924, asst. curator vertebrate paleontology, 1927, asso. curator, 1928-42, curator fossil mammals, 1942-59, chmn. dept. geology and paleontology, 1944-58; prof. vertebrate paleontology Columbia, 1945-59; Agassiz prof. vertebrate paleontology Mus. Comparative Zoology, Harvard, 1959—; prof. geology U. Ariz., Tucson, 1967—. Fellow NRC and Internat. Edn. in work on early fossil mammals, chiefly in British Mus., London, also other instns. Eng., France and Germany, 1926-27; expdns. to collect fossil animals include—No. Tex., Mont., N.M., Fla. and S.E. states, Argentina, Venezuela, Brazil. Fellow Am. Acad. Arts and Scis., Nat. Acad. Scis., Am. Philos. Soc., Geol. Soc. Am., Paleontol. Soc., A.A.A.S.; asesor honorario Museo de Ciencias Naturales (Caracas); mem. Soc. Vertebrate Paleontology (sec.-treas. 1940-41, pres. 1942), Soc. for Study Evolution (pres. 1946), Am. Soc. Mammalogists, Soc. Systematic Zoology (pres. 1962), Am. Soc. Zoologists (pres. 1962), Academia Nazionale dei Lincei (Italy), Academia de Ciencias (Argentina) (fgn. mem.), Sociedad Argentina de Estudios Geog. Gaea (hon. corr.). Zool. Soc. London (fgn. mem.), Royal Soc., 1915 (fgn. mem.), Academia de Ciencias (Venezuela, Brazil), Phi Beta Kappa, Sigma Xi. Author books relating to field, 1928—; latest publs.: The Meaning of Evolution, 1949; Horses, 1951; Life of the Past, 1953; The Major Features of Evolution, 1953; Evolution and Geography, 1953; Life, 1957, rev. (with W. S. Beck), 1965; Quantitative Zoology, 1960; Principles of Animal Taxonomy, 1961; This View of Life; The World of an Evolutionist, 1964; The Geography of Evolution, 1965; also articles. Recipient Lewis prize Am. Philos. Soc., 1942; Thompson medal Nat. Acad. Scis., 1943; Elliot medal, 1944, 65; Gaudry medal, Geol. Soc. France, 1947; Hayden medal, 1951; Penrose medal Geol. Soc. Am., 1952; André H. Dumont medal Geol. Soc. Belgium, 1953; Darwin-Wallace medal Linnean Soc., 1958; Darwin Plakette, Deutsche Akad. Naturforscher Leopoldina, 1959; Gold medal, Linnean Soc., 1962; Darwin medal Royal Soc., 1962; Nat. Medal of Sci., 1966; Verrill medal Yale, 1966. Collection, naming and description of early fossil mammals; evolution, zoogeography, and classification of living and extinct mammals; evolutionary theory; statis. methods in zoology; principles of biol. classification. Address: Mus. Comparative Zoology, Harvard Coll., Cambridge, Mass. 02138; also Dept. Geology, U. Ariz., Tucson 85721.*

SIMPSON, Howard Edwin, Am. geologist; b. Clarence, Ia., July 9, 1874; s. Hiram Garrison and Frances Abigail (Carter) S.; Ph.B., Cornell Coll., Ia., 1896, Sc.D., 1930; student U. Chgo., 3 summers; A.M., Harvard U., 1905; m. Carrie Esther Bonebrake, Dec. 30, 1903; children—Jessie Frances, Robert Bonebrake, Carolyn Cradock, Howard Edwin. Sci. tchr., later prin. high sch., Knoxville, Ia., 1897-1900; supt. schs., Columbus Jct., Ia., 1900-03; field and lab. asst. Harvard, 1903-04, asst. in physiography and meteorology, 194-05; instr. and asso. prof. geology Colby Coll., 1905-09; asst., prof. geology U. N.D., 1909-14, asso. prof., 1914-19, prof. geographic geology, 1919—, head dept. geology and geography and state geologist, 1933—. Vis. prof. U. Chgo., summer 1918, U. So. Cal., summer 1920; asst. U. S. Geol. Survey, various seasons; water geologist Can. Geol. Survey, summer 1929, Nat. Resources Bd., 1934; spl. meteorol. observer U. S. Weather Bur. Mem. N.D. Geog. Bd., N.D. Planning Bd. Fellow Geol. Soc. Am., Assn. Am. Geographers, A.A.A.S., Am. Meteorol. Soc., Am. Geog. Soc. Author or joint author: Geography of North Dakota; Underground Water Resources of Iowa; Ground Water Resources of N.D.; Conservation of Artesian Waters; A Method of Prospecting for Water. Died Jan. 31, 1938.

SIMPSON, Sir James Young, Scottish physician; b. Bathgate, Scotland, June 7, 1811; s. David and Mary (Jervie) S.; M.D., Edinburgh, 1832; D.C.L., Oxford U., 1866; m. Jessie Grindlay, 1839; 9 children, including Walter Grindlay. Became prof. midwifery, 1839; apptd. royal physician for Scotland, 1847. Recipient Monthyon prize French Acad. Scis., 1856; a maternity hosp. founded in his memory; bust erected in his honor at Westminster Abbey. Fellow or mem. Royal Med. Soc. Edinburgh (sr. pres. 1835), Acad. Medicine, Paris (fgn. asso.). Author: Obstetrics Memoirs and Contributions, 1855-56; Anaesthesia, 1871; Clinical Lectures on Diseases of Woman, 1872; Archaeological Essays, 1873; also papers. Introduced chloroform as anaesthetic, Nov. 1847; made important contbns. to obstetrics; a leading founder of gynecology; introduced uterine sound and sponge tent, thus aiding diagnosis and treatment hitherto impossible; anticipated discovery of Röntgen rays. Died May 6, 1870.

SIMPSON, Joanne Gerould, Am. meteorologist; b. Boston, Mar. 23, 1923; d. Russell and Virginia (Vaughan) Gerould; B.S., U. Chgo., 1943, M.S., 1945, Ph.D., 1949; m. Robert Homer Simpson, Jan. 6, 1965; children—David Malkus, Steven Malkus, Karen Malkus. Asst. prof. physics and meteorology Ill. Inst. Tech., 1947-51; meteorologist Woods Hole (Mass.) Oceanographic Inst., 1951-61; prof. meteorology U. Cal. at Los Angeles, 1961-65; chief exptl. meteorology br. Atmospheric Physics and Chemistry Lab., Environmental Sci. Services Administrn., U. S. Dept. Commerce, Washington, 1965——. Cons. U. S. Weather Bur., 1957-64, Esso. Research & Engring. Co., 1963-65, Rand Corp., 1964-65. Named Woman of Year, Los Angeles Times, 1964. Guggenheim Found. fellow, 1954. Mem. Am. (Meisinger award 1962), Royal meteorol. socs. Author: (with Herbert Riehl) Cloud Structure and Distributions over the Tropical Pacific Ocean, 1965; also numerous articles. Research on cumulus clouds and tropical hurricanes; use of aircraft as research tool; dir. Project Stormfury on hurricanes and cloud modification. Home: 780 Fairview Av., Annapolis, Md. 21403. Office: 8060 13th St., Silver Spring, Md. 20910.*

SIMPSON, John Alexander, Am. physicist; b. Portland, Ore., Nov. 3, 1916; s. John A. and Janet (Brand) S.; A.B., Reed Coll., 1940; M.S., N.Y. U., 1942, Ph.D., 1943; m. Elizabeth A. Hilts, Nov. 30, 1946; children—Mary Ann, John A. Research asso. OSRD, 1941-43; sci. group leader metall. lab. Manhattan (Atomic Bomb) Project, U. Chgo., 1943-46, faculty dept. physics and Enrico Fermi Inst. for Nuclear Studies, 1945——, prof., 1954——, Edward L. Ryerson Distinguished Service prof. physics, 1968——, established Lab. for Astrophysics and Space Research, 1964. Mem. spl. internat. com. for cosmic ray discipline IGY, 1954-60, U. S. Nat. Com., 1954——; pres. cosmic ray com. Internat. Union Pure and Applied Physics, 1963——; mem. astronomy missions bd. NASA, 1968——. Fellow, Center for Policy Study, 1966——. Fellow Am. Phys. Soc., Am. Geophys. Union; mem. Nat. Acad. Scis. (space sci. bd. 1958——), Internat. Acad. Astronautics, Am. Astron. Soc., Atomic Scientists Chgo. (chmn. 1945-46, mem. bd. Bull. 1945——), Phi Beta Kappa, Sigma Xi. Contbr. numerous articles to profl. jours. Discovered interplanetary origin of changes in cosmic ray intensity, interplanetary magnetic fields for modulation and storage cosmic ray particles; inventor neutron monitor pile for measurement interplanetary cosmic ray fluxes; research on solar flare particles, geomagnetically trapped radiation, acceleration charged particles in space; investigations on origin of cosmic radiation; inventor radioactivity measurement devices, expts. in earth satellites and deep space probes. Home: 5627 Blackstone Av., Chgo. 60637.*

SIMPSON, Maxwell, Irish chemist; b. Beach Hill, Ireland, Mar. 15, 1815; s. Thomas and (Browne) Simpson; B.A., Trinity Coll., Dublin, Ireland, 1837; M.B. in Chemistry, U. Coll., London, 1847; M.D. (hon.), Trinity, 1864, LL.D., 1878; D.Sc., Queen's U. Ireland, 1882; m. Mary Longhorne, 1845; 6 children. Lectr. chemistry Port Street and Peter Street Med. Sch., Dublin, 1847-51; in Marburg and Heidelberg, Germany, 3 years; lectr., Dublin, 1854-57; with Wurtz, Paris, 1857-59; set up lab. in home, 1860; prof. chemistry, Cork, Ireland, 1872-91; ret. and lived in London. Examiner, Indian Civil Service, Woolwich, Coopers Hill, Eng., also Queens U. Ireland. Fellow Royal U. Ireland. Fellow Royal Soc., 1862, Kings and Queens Coll. Physicians (hon., v.p. 1872-74). Research and publs. on devel. of method for determination of nitrogen in organic compounds difficult to burn; obtained for 1st time succinic and certain other di and tri-basic acids. Died Feb. 26, 1902.

SIMPSON, Oliver Cecil, Am. physicist; b. Wilderness, Mo., Dec. 6, 1909; s. William Peter and Eathel L. (Lowe) S.; B.S. with highest honors, U. Ill., 1930, Ph.D., 1934; m. Lorna Ellen Doege, Sept. 13, 1943; children—William Allen, Patricia Lynn. Research asst. Carnegie Inst. Tech., Pitts., 1934-39, asst. prof. physics, 1939-43, asso. prof., 1943-47; sr. scientist Metall. Lab., U. Chgo., 1944-47, sr. scientist Argonne (Ill.) Nat. Lab., 1947——, asso. dir. chem. div., 1947-59, dir. solid state sci. div., 1959——. Research in molecular physics, neutron scattering properties of ortho and para hydrogen, solid state physics and chemistry, high temperature thermochemistry. Home: 105 S. Prospect Av. Clarendon Hills, Ill. 60514. Office: 9700 S. Cass Av., Argonne, Ill.*

SIMPSON, Robert Homer, Am. meteorologist; b. Corpus Christi, Tex., Nov. 19, 1912; s. Clyde R. and Annie Laurie (Rainey) S.; B.S., Southwestern U., 1932; M.S., Emory U., 1935; Ph.D., U. Chgo., 1962; m. Mazie Houston, Dec. 23, 1935; children—Peggy Ann, Barbara Lynn; m. 2d, Joanne Gerould, Jan. 4, 1965. With Weather Bur., 1940——, dep. dir. research, 1960-64, dep. dir. nat. meteorol. services, Washington, 1964-65, asso. dir. Bur., 1965-68; dir. Nat. Hurricane Center, 1968——. Recipient Gold medal for exceptional service U. S. Dept. Commerce, 1962.

Mem. Am., Royal meteorol. socs., Am. Inst. Aeros, and Astronautics (mem. com. on meteorology 1961——), World Meteorol. Orgn. (mem. commn. on synoptic meteorology 1954-55), Am. Geophys. Union, Sigma Xi. Research, publs. on hurricane structure, processes for releasing energy in hurricanes, formulation 1st complete hypothesis for reducing hurricane winds by cloud seeding. Home: 11230 S.W. 173d Terrace, Miami, Fla. 33157. Office: P.O. Box 8286, U. Miami, Coral Gables, Fla. 33124.*

SIMPSON, Thomas, English mathematician; b. Market Bosworth, Eng., Aug. 20, 1710; self-educated; usher in sch.; became prof. math., Woolwich, Eng., 1743. Fellow Royal Soc., 1746. Author: A New Treatise on Fluxions, 1737; Treatise on the Nature and the Laws of Probability, 1740; Essays . . . on Mathematicks, 1740; Treatise on Algebra, 1745; Treatise on Geometry, 1747; Rectilinear and Spherical Trigonometry, 1748; Mathematical Exercises, 1752; Mixtures, 1757. Invented rule for finding area of figure given only a limited amount of data (named after him); introduced continuity into theory of math. probability, 1756; reintroduced (simultaneously with Euler) use of A,B,C, to designate angles and a,b,c, for the opposite sides of triangles; determined roots by reversion of series and infinite series; 1st to apply Newton-Raphson process to solution of transcendental equations; advocated use of arithmetic mean. Died Market Bosworth, May 14, 1761.

SIMS, Chester Thomas, Am. metallurgist; b. Winchester, Mass., Dec. 14, 1923; s. Ralph John and Elsie (Bartsch) S.; B.Sc. in Chem. Engring., Northeastern U., 1947; M.Sc. in Metall. Engring., Ohio State U., 1951; m. Sarah Grace Joseph, May 7, 1949; children—Laurie Anne, Lisa Joseph. Research engr. Battelle Meml. Inst., Columbus, O., 1947-50, asst. div. chief, nonferrous phys. metallurgy, 1952-58; project engr. Knolls Atomic Power Lab., Schenectady, 1958-60; mgr. high temperature materials, materials and processes lab. Gen. Elec. Co., Schenectady, 1960——. Cons. BMI/N.W. Labs., Hanford, Wash., 1964——; adj. lectr. Rensselaer Poly. Inst., 1963——. Mem. Am. Inst. Mining and Metall. Engrs. (pres. Hudson-Mohawk sect.), Am. Soc. for Metals, Am. Ordnance Assn. Developed processing techniques for metal rhenium and exploited its properties; studies in oxidation of columbium; co-author computer technique for predicting phase stability of nickel-base superalloys; developed nuclear fuel systems; patentee in new alloys and their applications. Home: 99 Midline Rd., Ballston Lake, N.Y. 12019. Office: care Gen. Elec. Co., Schenectady 12305.*

SIMS, Ethan Allen, Am. physician; b. Newport, R.I., Apr. 22, 1916; s. William Sowden and Anne (Hitchcock) S.; B.S., Harvard, 1938; M.D., Columbia U., 1942; m. Dorothea Foote Merriman, Aug. 5, 1939; children—Ethan Allen, Dorothea Foote, Nathaniel Merriman. J. Hudson Brown fellow chem. div. dept. medicine Yale Sch. Medicine, 1946-47, instr., 1947-50; with Coll. Medicine, U. Vt., Burlington, 1950——, now prof. medicine; attending physician Med. Hosp. Va., Burlington; program dir. tng. program Nat. Inst. Metabolic Diseases. Commonwealth fellow, 1964-65. Mem. Endocrine Soc., Am. Fedn. for Clin. Research, Am. Diabetes Assn., Sigma Xi. Research and publs. on electrolyte metabolism, renal function in pregnancy, aldosterone metabolism, carbohydrate metabolism. Home: 51 Old Farm Rd., South Burlington, Vt. 05401. Office: Dept. Medicine, Coll. Medicine, U. Vt., Burlington, Vt.*

SIMS, Henry John, Australian agrl. scientist; b. Shepparton, Victoria, Australia, Oct. 6, 1912; s. Samuel Thomas and Eliza (Nash) S.; B.Agr.Sc., Melbourne (Australia) U., 1933, M.Agr.Sc., 1940, B. Com., 1945; m. Ethel May Wakefield, Feb. 21, 1942. With Victoria Dept. Agr., 1934——, sr. cereal research officer, 1958-61, sr. agronomist in charge cereal exptl. work, 1961-66, sr. geneticist in charge plant breeding work, Melbourne, 1966——. Mem. Victorian Wheat Adv. Com., 1948——, Australian Barley Tech. Com., 1963——. Mem. Australian Inst. Agrl. Sci., Australian and New Zealand Assn. for Advancement Sci., Royal Australian Inst. Pub. Administrn., Australian Soc. Soil Sci. Research, numerous publs. on farming in cereal growing areas of Victoria, especially Mallee dist. including plant nutrition, soil fertility and devel. ley farming with medics; devel. wheat variety; grazing mgmt. of oats, sheep mgmt. Home: 21 Morwell Av., Watsonia, Victoria 3087. Office: Treasury Pl., Melbourne, Victoria 3002, Australia.*

SIMS, James Marion, Am. gynecologist; b. Lancaster County, S.C., Jan. 25, 1813; s. John and Mahala (Mackey) S.; grad. S.C. Coll., 1832; studied Charleston Med. Sch., 1833; grad. Jefferson Med. Coll., 1835, LL.D., 1881; m. Eliza Theresa Jones, Dec. 21, 1836; 9 children. Practiced medicine, Ala., 1835-50, performed unprecedented fistula operations with notable success, attracting wide attention; published history of vesicovaginal operations, on which he was recognized authority, 1852; founder Women's Hosp., N.Y.C., 1854; visitor to France and Eng. several times, 1861-82, consulting with gynecologists, lecturing, receiving honors in several nations; pres. A.M.A., 1876, Am. Gynecol. Soc., 1881. Author several works including: on Intra-Uterine Fibroid Tumors, 1874; Clinical Notes on Uterine Surgery, 1865; Treatise on Ovariotomy, 1873. Introduced Sims' gy-

necol. position in which patient lies on left side and chest with right thigh drawn up, circa 1854; 1st to use silver wire as suture material, 1858; introduced duckbill vaginal speculum (Sims' speculum), circa 1866; his work made possible surg. treatment of vesico vagina fistula. Died N.Y., Nov. 13, 1883.

SIMS, W(infield) Scott, Am. inventor; b. N.Y., Apr. 6, 1844; s. Lindsay D. and Catherine S.; grad. high sch., Newark; m. Lida Leek, June 11, 1867 (dec. 1888); 2d, Mrs. Josephine Courter French, June 24, 1891. Invented various devices in electro-magnets; constructed an electric motor for light work in 1872, weighing 45 pounds, and having a battery of 20 half-gallon Bunsen cells, to propel an open boat 16 feet long, with 6 persons on board, at 4 miles an hour; first to apply electricity for propulsion of torpedoes (patented 1882), his device of a torpedo being a submarine boat with a cylindrical hull of copper and with conical ends, furnished with screw propeller and a rudder, powered by electricity generated on shore or on shipboard, by means of which the torpedo is propelled, guided and exploded; devised a boat with a speed of 22 miles an hour and to carry a 500-lb. charge of dynamite; also invented the wireless dirigible torpedo, of which sold 5 to Japanese Govt., 1907; invented the Sims-Dudley Dynamite Gun, used by Cuban insurgents, and by the Rough Riders at battle of Santiago; worked on design of dynamite cruiser to carry 100 tons of high explosives, controlled by an operator on ship or shore; designed a dynamite gun for use with dirigible warships, an aeroplane dynamite gun. Died Jan. 7, 1918.

SIMSON, Sir Henry John Forbes, obstetrician; b. India, 1872; s. Robert and Amy (Inglis) S.; M.D.C.M., Edinburgh U., 1895; m. Lena Margaret Ashwell, 1908. Resident, Royal Infirmary, Edinburgh, Royal Maternity Hosp.; settled in London, 1900; dean post-grad. coll., surgeon Hosp. for Women, W. London Hosp.; examiner obstetrics and gynecology Conjoint Bd., also U. Leeds (Eng.); physician to Princess Royal and Duchess of York; cons. surgeon Hosp. for Women, Soho, Eng. Fellow Royal Coll. Surgeons Edinburgh, Royal Coll. Physicians. Contbr. numerous articles on obstetrics and gynecology to tech. jours. Died Sept. 1932.

SIMSON, Robert, Scottish mathematician; b. Kirktonhall, Scotland, Oct. 14, 1687; s. John and Agnes S.; M.A., Glasgow (Scotland) U., 1711; Hon. med. degree St. Andrews (Scotland) U., 1746. Regius prof. math. U. Glasgow, 1711-61. Author: Sectionum conicarum, 5 vols., 1735; Elements of Eculid, 1756; also articles. Translated Euclid's Elements, Loci plani of Appollonius. Restored Lieux plans (Plane locie) (by de Perge), 1748. Studied geometry of ancients. Died Glasgow, Oct. 1, 1768.

SINANOGLU, Oktay, chemist; b. Italy, Feb. 25, 1935 (parents Turkish citizens); s. Nuzhet Hasim and Ruveyde (Karacabey) S.; came to U. S., 1953; grad. Ankara (Turkey) Coll., 1953; B.S. in Chem. Engring. with highest honors, U. Cal. at Berkeley, 1956, Ph.D. in Phys. Chemistry (Univ. fellow 1958-59), 1959; M.S. in Chem. Engring. (Whitney fellow 1956-57), Mass. Inst. Tech., 1957; m. Paula Armbruster, Dec. 21, 1963; 1 son, K. Levni. Postdoctoral chemist AEC, Radiation Lab., Berkeley, 1959-60; mem. faculty Yale, 1962——, prof. chemistry, 1963——, fellow Trumbull Coll., 1961——; A. and R. Sloan vis. lectr. Harvard spring 1962; vis. prof. Middle East Tech. U., Ankara, summer 1962, now cons. professor; vis. scientist Center for Theoretical Studies, U. Miami, Coral Gables, Fla., 1965; dir. NATO International Summer Sch. of Quantum Chemistry, Istanbul, 1964. Contbg. author, editor: Modern Quantum Chemistry-Istanbul Lectures, 3 vols., 1965. Research on intermolecular forces, statis. mechanics gases, structure aqueous solutions, elementary particles structure and masses of mesons, quantum chemistry; developed theory of atoms and molecules dealing with many-body effects of electrons, theory on role of water and other solvents on shape, stability and photochemistry of biol. polymers like DNA. Mem. Am. Chem. Soc., Am. Phys. Soc., Am. Inst. Chem. Engrs. (award Western sect. 1955), Inst. Am. Chemists (award 1955-56), Phi Beta Kappa, Sigma Tau, Tau Beta Pi. Contbr. articles profl. jours., chpts. in books. Home: Berncliff Dr., Northford, Conn.*

SINCLAIR, Andrew, Brit. surgeon, naturalist; b. Paisley, Scotland; became asst. surgeon in navy, 1829; with H. M. S. Sulphur on surveying expdn. to S.Am. coast, 1834; collected plants in Mexico and C.Am., 1837, 38; apptd. surgeon to a convict ship, 1842; briefly asso. with Sir Joseph Hooker in New Zealand; pvt. sec. to Capt. Robert Fitzroy, 1843; colonial sec. of New Zealand, 1844-56. Honored by Hooker in name Sinclairia given to tropical Am. genus of Compositae (now merged in Liabum). Fellow Linnean Soc. Author: (contbns. to jours.) Remarks of Physalici pegalica, 1842; On the Vegetation of Auckland, 1851. Collected plants (described by Hooker in Arnott's Botany of Beechey's Voyage and Bentham's Botany of the Voyage of the Sulphur); his collection of zool. specimens presented to Brit. Mus., collection of plants to Sir W. J. Hooker. Died Mar. 26, 1861.

SINCLAIR, George, Scottish mathematician, experimenter; b. East Lothian, Scotland. Pedagogue, St. Andrews, Scotland, till 1654; master Glasgow (Scotland) U., from 1654, prof. philosophy, 1654-66, 89-91, prof. math., 1691-96; prof. math. Edinburgh (Scotland) U., from 1666; ind. expts.; studied extent, dip coal beds for mine owners, Lothian, 1672; supt. water pipe laying Edinburgh, 1673-74. Author: Tyrocinia Mathematica . . . , 1661; Ars nova et magna gravitatis et levitatis, 1669; The Hydrostaticks, or the Weight, Force and Pressure of Fluid Bodies made evident by Physical and Sensible Experiments, including within A Short History of Coal, 1672; Truth's Victory over Error, 1684; Satan's Invisible World Discovered, 1685; The Principles of Astronomy and Navigation, 1688. Worked to establish teaching of math. Scotland; one of 1st in Scotland to devote himself to study physics; used newly invented diving bell to explore undersea wreck, 1655; early research on barometer using it to establish altitudes, mine depths; suggested methods draining water from coal seams; supported belief in existence devils, spirits, witches, apparitions; studied magnetic needle; much work with barometer (studies weight, pressure, elasticity of air) duplication of Boyle's work; discussed fallacies in expts. purporting to prove possibility perpetual motion; commented on Linus' Funiculus, also Deusing's work on vacuum; gave account of comets, 1664-65; wrote on theory siphoning, hygroscope, chronoscope. Died, probably Glasgow, 1696.

SINCLAIR, George Morton, Am. metallurgist; b. Chgo., Mar. 9, 1922; s. James Polson and Helen (Stadelmaier) S.; B.S. in Metall. Engring., U. Ill., 1948, M.S., 1949; m. Ruth Winifred Nelson, June 5, 1943; children—Robert Bruce, Joan Ellen. With U. Ill., Urbana, 1947-52, 53—, prof. theoretical and applied mechanics, 1957—; research metallurgist Westinghouse Research Lab., 1952-53. Cons., Nat. Acad. Scis.-NRC. Mem. Am. Soc. M.E. (exec. com. metals engring. div.), Am. Soc. for Testing and Materials (mem. com. E-9), Am. Soc. for Metals, Ill. State Acad. Scis., Pi Kappa Alpha. Research and publs. on flow and fracture of metals, especially fatigue, mech. properties of engring. materials, relations of microstructure to mech. strength, heat treatment of metals, mech. testing apparatus. Home: 608 Harding Dr., Urbana, Ill. 61801.*

SINCLAIR, Hugh Macdonald, Brit. physician; b. Edinburgh, Scotland, Feb. 4, 1910; s. Hugh Montgomerie and Rosalie (Jackson) S.; B.A., Oxford (Eng.) U., 1932, M.A., B.M., 1936, D.M., 1939. Fellow, tutor Magdalen Coll., Oxford U., 1937—, v.p., 1956-57, dir. Oxford Nutrition Survey, 1941-47, dir. Lab. Human Nutrition, 1947-58, reader, 1951-58. Recipient U. S. medal of Freedom with silver palms; Dutch Order of Orange-Nassau. Fellow Royal Coll. Physicians, Chem. Soc., Inst. Biology; mem. Physiol. Soc. (hon. treas.), Biochem. Soc., Med. Research Soc., Nutrition Soc., Am. Inst. Nutrition, Am. Pub. Health Assn., Soc. for Exptl. Biology, Internat. Fedn. Medicine. Author: (with A. Robb-Smith) History of Anatomical Teaching in Oxford, 1950; (with R. McCarrison), 1953; The Work of Sir Robert McCarrison, 1953; (with Jelliffe) Tropical Nutrition and Dietetics, 1961; also numerous articles. Research on nutritional factors in causation of chronic degenerative diseases; showed a change to processed dietary fats was causing rapid increase in coronary heart disease, certain forms of cancer, duodenal ulcers. Home: Lady Place, Sutton Courtenay, Berks., Eng. Office: Magdalen Coll., Oxford, Eng.*

SINCLAIR, James Burton, Am. plant pathologist, botanist, biologist; b. Chgo., Dec. 21, 1927; s. James Lawrence and Helen M. (Thompson) S.; B.S., Lawrence U., 1951; Ph.D., U. Wis., 1955. Research asso. U. Wis., 1955-56; faculty La. State U., Baton Rouge, 1956—, prof. plant pathology, 1965—. Participant, Advanced Virology Seminar Inst., U. Md., 1963, Rhizoctonia Symposium, 1965. Mem. Am. Phytopath. Soc. (pres. So. div. 1965, editor Fungicide-Nematocide Data Book 1965—), Bot. Soc. Am., Mycol. Soc. Am., Cotton Disease Council (chmn. 1965-66), Internat. Orgn. Citrus Virologists, La. Acad. Scis., A.A.-A.S., Am. Inst. Biol. Scis., Sigma Xi, Gamma Sigma Delta. Research in control of cotton seedling diseases through seed and soil treatments with fungicides; studies on citrus fruit rots and virus diseases. Home: 6185 Esplanade, Baton Rouge, La. 70806.*

SINCLAIR, John G., Am. anatomist; b. Grand Haven, Mich., July 6, 1888; s. Peter and Ryntje (Van Westreinen) S.; student U. Ill., 1907-10; B.S., U. Chgo., 1911; M.S., U. N.D., 1926; Ph.D., U. Wis., 1928; m. Margaret Hancock, Sept. 11, 1919 (div. Jan. 1951); children—John G., Louise L. (Mrs. Robert Terry), Ruth (Mrs. John W. Placek); m. 2d, Katharine Springer, Dec. 22, 1959. Research asst. genetics Carnegie Inst. Desert Lab., Tucson, 1913-15; asso. zoology U. Chgo., 1915-17; asso. prof. U. N.D., 1920-26; instr. U. Wis., 1926-28; prof. anatomy U. Tex., Galveston, 1928-62, emeritus, 1962—. Fellow A.A.-A.S. (hon.); mem. Am. Assn. Anatomists, Tex. Acad. Sci. (hon. life, past pres.), Am. Genetics Soc., Sigma Xi. Author: (With John MacArthur) Comparative Vertebrate Anatomy Laboratory Text, 1917; Anatomy of the Fetal Pig, 1937; also articles, poetry. Research on quantitative relation of parathyroid gland structure and function in mother and fetus under diet control, developmental anomalies produced by drugs, placental weight and structure in relation to conditions of birth in man, structure of optic lens essential to accomodation mechanism, devel. of head structures peculiar to Cetacea (dolphins). Home: 4710n Woodrow St., Galveston, Tex. 77550.*

SINCLAIR, Rolf Malcolm, Am. physicist; b. N.Y.C., Aug. 15, 1929; B.S., Cal. Inst. Tech., 1949; M.A., Rice U., 1951, Ph.D., 1954; m. Margaret Lee Andrews, May 8, 1959; children—Elizabeth Ann, Andrew Caisley. Research physicist Westinghouse Research Labs., Pitts., 1953-56; research asst. Physics Inst., U. Hamburg (Germany), 1956-57, U. Paris (France), 1957-58; research staff Plasma Physics Lab., Princeton, 1958—; physicist Culham Lab., Abingdon, Berks., Eng., 1965-66. Mem. Am. Phys. Soc., Sigma Xi. Research, publs. in nuclear particles and nuclear structure, plasma physics and controlled thermonuclear fusion. Home: Port Mercer Rd., R.D. 3, Princeton, N.J. 08540.*

SINCLAIR, Walton Bunyan, Am. biochemist; b. Norwood, N.C., Nov. 15, 1900; s. William Thomas and Beatrice (Agnew) S.; B.S., Oglethorpe U., 1922; student U. Cal. at Berkeley, 1925; M.S., U. Minn., 1926, Ph.D., 1929; m. Louise Livingston, Oct. 2, 1936; 1 dau., Linda Louise (Mrs. James Robert Hawk). Teaching fellow biochemistry and pharmacology U. Cal. Med. Sch., Berkeley, 1925-26; instr. biochemistry U. Minn., 1926-28; research chemist biol. products and glandular derivatives Wilson & Co., Chgo., 1928-29; research chemist phys. and chem. properties cereal grains Nat. Biscuit Co., N.Y.C., 1930-32; research asso. biochemistry Citrus Expt. Sta., Riverside, Cal., 1932-39; asst. prof. U. Cal. at Riverside, 1939-44, asso. prof., 1944-48, prof., chmn. dept. biochemistry, 1948—. Mem. Am. Chem. Soc., Bot. Soc. Am., Am. Soc. Plant Physiology, Am. Soc. Hort. Sci., Sigma Xi. Author: (with E. T. Bartholomew) The Lemon, Its Composition, Physiology and Products, 1951. Author chpts., editor: The Orange, its Biochemistry and Physiology, 1960. Contbr. Ency. Brit., 1962. Research and numerous publs. on products and derivatives of animal endocrine glands, phys. and chem. properties of cereal grain, proteins in relation to baking quality, nutrition and enzyme action; biochem. changes in growth, devel. and maturation of citrus fruits. Home: 3649 Rosewood Pl., Riverside, Cal. 92506.*

SINCLAIR, Warren Keith, biophysicist; b. Dunedin, New Zealand, Mar. 9, 1924; s. Ernest W. and Jessie E. (Craig) S.; B.S., U. Otago, New Zealand, 1944, M.S., 1945; Ph.D., U. London (Eng.), 1950, postgrad.; m. Elizabeth J. Edwards, Mar. 19, 1948; children—Bruce W., Roslyn B. Came to U.S., 1954, naturalized, 1959. Radiol. physicist Dunedin Pub. Hosp., lectr. U. Otago, 1945-47; physicist Royal Cancer Hosp., London, 1947-54; extra-mural lectr. London U., 1951-54; head dept. physics M. D. Anderson Hosp., also prof. physics U. Tex., Houston, 1954-60; sr. biophysicist Argonne (Ill.) Nat. Lab., 1960—; prof. radiation biology U. Chgo., 1964—. Chmn., U. S. Nat. Com. for Med. Physics, 1963—, mem. Nat. Com. for Pure and Applied Biophysics, 1964—; mem. com. on radiology Nat. Acad. Scis.-NRC, 1964; mem. radiation study sect. NIH, 1966—; mem. Nat. Acad. Scis.-NRC com. Nat. Bur. Standards, 1964-66. Fellow Inst. Physics, Physics Soc.; mem. Hosp. Physicist Assn., Brit. Inst. Radiology (councillor 1953-54), Am. Assn. Physicists in Medicine (pres. 1960-61), Radiation Research Soc. (councillor 1964—), Soc. Nuclear Medicine, Radiol. Soc. Asst. editor Physics in Medicine and Biology, 1964—; editorial bd. Radiation Research, 1964—, Jour. Nuclear Medicine, 1960-66. Research in devel. of radioisotope techniques for use in radiation therapy and dosimetry, relative biol. effectiveness of high energy radiation beams and dosimetry, radiation effects of mammalian cells in culture, especially variation of X-ray sensitivity within generation cycle of synchronized mammalian cells. Home: 423 N. Madison St., Hinsdale, Ill. 60521. Office: Argonne Nat. Lab., Argonne, Ill. 60439.*

SINCLAIR, William John, Am. geologist; b. San Francisco, 1877; s. Samuel Fleming and Ellen (Milliken) S.; B.S., Cal. Coll., 1899; M.S., Ph.D., U. Cal., 1904; m. Della A. Coleman, Sept. 1, 1914. Joined as fellow, Princeton, 1904, successively became instr. geology, 1905, asst. prof., 1916, asso. prof., 1923, prof., 1930—; mem. various Am. Mus. expdns. to Bridger and Bighorn basics, Wyo.; asst. U. S. Geol. Survey. Mem. Geol. Soc. Am. Research and publs. on glacial deposits, ungulates, fossils; 1st demonstration of volcanic ash in Bridger beds, Wyo., using microscope; studied Tertiary mammals, reptiles, birds; discovered and described various extinct mammals from Cal. caves. Died 1935.

SINCLAR, George, see Sinclair, George.

SINELNIKOV, Kirill Dmitrievich, Russian physicist; b. May 29, 1901; grad. Crimean U., 1923. Prof., Kharkov U., 1936—; staff Physico-Tech. Inst., Leningrad, USSR, 1924-30; staff Physico-Tech. Inst. Ukraine Acad. Sci., 1930—, dir., 1944—. Hon. sci. worker Ukraine SSR, 1951—. Mem. Ukrainian Acad. Sci. Author: The Nature of Dielectric Losses, 1926; coauthor: Vacuum Metallurgy, 1957. Research and publs. in vacuum physics, high voltage engring., optics, photoelements, linear ion and electron accelerators, solid rectifiers. Address: USSR, Kiev, Ukraine SSR, Vladimirskaya 54, AN Ukraine SSR.

SING, Kenneth Stafford William, Eng. surface chemist; b. Bideford, Eng., Feb. 25, 1925; s. Reginald William and Edith (Popham) S.; student U. Coll., Exeter, Eng., 1942-48; B.Sc. first class degree Chemistry U. London, 1945, Ph.D., 1949; m. Ruby Corner, Mar. 25, 1950; children—Deborah Diane, Anne Millington, Margaret Claire, Jonathan Millington. Tech. officer Imperial Chem. Industries Ltd., Billingham, Eng., 1948-49; lectr. Royal Tech. Coll., Salford, Eng., 1949-55; head chemistry dept. Coll. Tech., Liverpool, Eng., 1956-65; prof., head chemistry dept. Brunel U., London, 1965—. Cons. to indsl. orgns.; mem. mineral processing research com. Warren Spring Lab., Ministry of Tech., 1966—. Fellow Royal Inst. Chemistry; mem. Chem. Soc., Soc. Chem. Industry. Author: (with S. J. Gregg) Adsorption, Surface Area and Porosity 1967; also articles. Research in applied surface chemistry of solids, dependence of surface properties of silica gels on conditions of preparation, surface properties of silica gels on conditions of preparation, of alumina, phys. adsorption of various vapors on solids having various types of porosity; attempted to elucidate mechanism of low temperature aging of alumina and to characterize porosity using gas adsorption. Home: Oak Grove, Coombe Lane, Hughenden Valley, High Wycombe, Buckinghamshire, Eng. Office: Dept. Chemistry, Brunel U., Woodlands Av., Acton, London, W.3, Eng.*

SINGER, Isaac Merrit, Am. inventor, machinist; b. Oswego, N.Y., Oct. 27, 1811; m. Catherine Maria Haley, 1835; m. 2d, Isabella Summerville, 1865. Patented rock-drilling machine, 1839, wood and metal carving machine, 1849; received 1st patent on sewing machine (superior because it could do continuous stitching), 1851; organized sewing machine mfg. firm I. M. Singer & Co., 1851, reached commanding position in sewing machine industry by 1854, brought about pooling of patents in industry; received 20 patents for improvements on his machine, including continuous wheel feed and yielding presser foot, 1851-63; developed 1st practical domestic sewing machine brought into wide use; withdrew from active connection with co., 1863. Died Torquay, Eng., July 23, 1875.

SINGER, Isadore Manual, Am. mathematician; b. Detroit, May 4, 1924; s. Simon and Freda (Rose) S.; B.S., U. Mich., 1944; M.S., U. Chgo., Ph.D., 1950; m. Sheila Ruff, Sept. 24, 1961; children—Steven, Eliot, Natasha. C. L. E. Moore instr. Mass. Inst. Tech., 1950-52, faculty, 1956—, prof. math., 1959—; asst. prof. U. Cal. at Los Angeles, 1952-54, Columbia, 1955; vis. mem. Inst. for Advanced Study, 1956. Sloan fellow, 1959-62. Mem. Am. Math. Soc. (editor Transactions 1960-66), Math. Assn. Am., Nat. Acad. Scis., Phi Beta Kappa, Sigma Xi. Research, publs. on differential geometry and functional analysis, especially topological invariants on a manofold expressible in analytic terms.*

SINGER, Jacob Jesse, physician; b. Leeds, Eng., July 12, 1885; s. Abraham and Rebecca (Silverblatt) S.; came to U. S., 1885; M.D., Washington U., 1904. Intern., 1904-06, specialized in Tb thereafter; med. dir. Los Angeles Sanitarium; asso. clin. prof. medicine Washington U., also U. So. Cal.; cons. Diseases of Chest, Cedars of Lebanon Hosp., Los Angeles, Barnes Hosp., Jewish Women's Hosp., Childrens' Hosp., U. S. Marines' Hosp. Mem. or fellow Am. Assn. Thoracic Surgeons, Am. Coll. Chest Physicians, A.M.A., Central Soc. Clin. Research. Author: (with Evarts Graham and Harry Ballen) Surgical Diseases of the Chest, 1935; also papers on surg. and med. treatment of pulmonary Tb. Pioneer in surg. treatment of pulmonary Tb, especially in use of artificial pneumothorax; successfully removed (with E. A. Graham) entire lung for carcinoma of bronchus, circa 1933. Died Apr. 13, 1954.

SINGER, James, Am. mathematician; b. N.Y.C., July 3, 1905; s. Mayer and Bessie (Cohen) S.; A.B., Cornell U., 1926; M.A., Princeton, 1928, Ph.D., 1931; m. Hortense Weinberg, June 26, 1932. children—Richard, John Alan. Instr. math. Princeton, 1927-34; faculty Bklyn. Coll., 1935—, prof. math., 1956—, chmn. dept., 1961—, sec. faculty, 1954-60, dir. NSF Math. Inst. at coll., 1961. Cons., Office Naval Research, 1950-54. Fellow A.A.A.S.; mem. Math. Assn. Am. (past chmn. met. N.Y. sect.), Am. Math. Soc., Indian Math. Soc., Fedn. Am. Scientists, Am. Assn. U. Profs., N.Y. Acad. Scis., Phi Beta Kappa, Sigma Xi, Phi Kappa Phi, Pi Mu Epsilon. Author: Elements of Numerical Analysis, 1965. Research and publs. on finite projective geometry and difference sets. Home: 3054 Bedford Av., Bklyn. 11210.*

SINGER, Jerome Ralph, Am. physicist; b. Cleve., Oct. 16, 1921; s. Sam R. and Rose (Malz) S.; B.S. cum laude, U. Ill., 1950; M.S., Northwestern U., 1952; Ph.D., U. Conn., 1956; m. Margaret B. Thaler, June 28, 1957; children—Sam Robert, Martha Rachel. Staff physicist U. S. Naval Ordnance Lab., 1956-57; chief physicist Nat. Sci. Labs., Washington, 1957-58; prof. physics U. Cal., Berkeley, 1958—; cons. Lockheed, MSD, Aeronutronics div. Philco Corp., Ampex Corp., Avco Corp., Inst. Def. Analyses. Fellow I.E.E.E. (guest editor 1st issue 1963); mem. Inst. Basic Research, Am. Brit. phys. socs., Optical Soc. Am., Sigma Xi. Author: Masers, 1960; Advances in Quantum Electronics, 1961; also articles. Research in lasers, nuclear and electron spin resonance, biophysics;

patentee NMR flow measurement system, frequency stblzn. systems, computer elements. Home: 17 El Camino Real, Berkeley, Cal. 94705.*

SINGER, Marcus (Joseph), Am. biologist, anatomist; b. Pitts., Aug. 28, 1914; s. Benjamin and Rachel (Gershenson) S.; B.S., U. Pitts., 1938; A.M., Harvard, 1940, Ph.D., 1942; m. Leah Horelick, June 8, 1938; children—Robert H., Jon F. Faculty, Harvard Med. Sch., 1942-51, Cornell U., 1951-61; H. W. Payne prof. anatomy, dir. dept., asso. dir. Developmental Biology Center, Western Res. U., Cleve., 1961—. Vis. prof. anatomy L.I. Coll. Medicine, 1950; vis. fellow Dutch Brain Inst., Amsterdam, 1959. Fellow Am. Acad. Arts and Scis., A.A.A.S.; mem. Am. Neurol. Assn. (asso.), Am. Assn. Anatomists, Am. Zool. Soc. (pres. developmental biology div. 1965), Assn. Research Nervous and Mental Disorders, Soc. Developmental Biology, Royal Soc. Medicine (Affiliate). Author: Dog Brain in Section, 1962; (with P. Yakovlev) Human Brain in Sagittal Section, 1954; also articles. Editor (with J. P. Schadé) vols. 13, 14 Progress in Brain Research; mng. editor Jour. Morphology, 1964—; asso. editor Jour. Exptl. Zoology, 1961-65. Research on nerve regeneration; neuroanatomy; exptl. morphology; regeneration; histochemistry. Home: 2905 Berkshire Rd., Cleveland Heights, O. 44118. Office: Dept. Anatomy, Sch. Medicine, Western Res. U., Cleve. 44106.

SINGER, Rolf, mycologist; b. Schliersee, Germany, June 23, 1906; s. Albert and Eva (Hennicke) S.; Ph.D., U. Vienna (Austria), 1931; Profesor honoris causa, Universidade de Recife (Brazil); m. Martha Kupfer, July 4, 1934; 1 dau., Amparo Heidi (Mrs. Roberto Eustaquio). Prof., Universidad Autónoma de Barcelona, Ayudante, 1934-35; sr. sci. expert Acad. Scis. USSR, Bot. Garden, Leningrad, 1935-40; asst. curator, acting dir. Farlow Herbarium, Harvard, 1941-48; prof., head botany dept. Universidad de Tucumán (Argentina), 1948-61; prof. Universidad de Buenos Aires (Argentina), 1961-68; vis. prof. U. Ill. Chgo. Circle, 1968—; curator, Field Museum, Chgo., 1968—; head chemistry dept. Neb. Wesleyan U., 1953-54, dir. research neuropsychiat. research program, 1957-58; sci. dir. Orgn. for Flora Neotropica, N.Y. Bot. Garden, 1965. Mem. Am. Mycol. Soc., Société Mycologique de France, Sociedad Argentina de Botánica, Sociedad Paleontológica Argentina. Author: The Boletineae of Florida, 1945; The Agaricales in Modern Taxonomy Lilloa, 1949, 62; Mushrooms and Truffles, 1961; Die Pilze Mitteleuropas V-VI, 1965; Monograph of the Genus Galerina (with A. H. Smith), 1964; also numerous articles. Research on classification, ecology evolution of fungi throughout world.*

SINGER, Ronald, anatomist; b. Cape Town, South Africa, Aug. 12, 1924; s. Solomon Charles and Sonia (Swirsky) S.; M.B., Ch.B., U. Capetown, 1947, D.Sc., 1962; m. Shirley Bernice Gersohn, Dec. 10, 1950; children—Hazel Lynn, Eric Gersohn, Sonia Esther, Charles Malcolm. Resident med. officer Kimberley Hosp., South Africa, 1947-48; govt. med. officer, supt. Gwelo Native Hosp., Rhodesia, 1948-49; with U. Cape Town, 1949-62, asso. prof., 1960-62; vis. prof. anatomy U. Ill., 1959-60; prof. anatomy U. Chgo. 1962—. Rotary Found. fellow Carnegie Instn. Washington Embryology Labs., Johns Hopkins, 1951-52. Fellow Royal Soc. So. Africa, Royal Anthrop. Inst.; mem. Am. Assn. Anatomists, Am. Assn. Phys. Anthropologists, Anat. Soc. Gt. Britain and Ireland, Soc. Study Human Biology, A.A.A.S., Sigma Xi. Research, publs. on human evolution in Africa and in genetics and biology of indigenous African populations, especially Hottentots and Bushmen; co-discoverer of Saldanha Skull. Home: 5822 Blackstone Av., Chgo. 60637.*

SINGER, S(iegfried) Fred, physicist; b. Vienna, Austria, Sept. 27, 1924; s. Joseph B. and Anne (Kelman) S.; B.E.E., Ohio State U., 1943; A.M., Princeton, 1944, Ph.D. in Physics, 1948. Came to U. S., 1940, naturalized, 1944. Instr. physics dept. Princeton, 1943-44; physicist, applied physics lab. Johns Hopkins, 1946-50; sci. liaison officer Office Naval Research, Am. Embassy, London, 1950-53; associate professor physics University of Maryland, 1953-59. prof., 1959-64, dir. Center of Atmospheric and Space physics, 1962-64; dean Sch. Environmental and Planetary Scis., U. Miami, 1964—; vis. research, Jet Prop. Lab., Cal. Inst. Tech., 1961-62; dir. Nat. Weather Satellite Center, 1962-64. Head sci. evaluation group Astronautics and Space Exploration Com., Ho. of Reps., 1958; mem. commn. IV, U. S. nat. com. Internat. Sci. Radio Union, 1954—; mem. tech. panels on rockets, cosmic rays U. S. nat. com. for IGY, 1957-58. Recipient gold medal for exceptional service U. S. Dept. Commerce, 1965; 1st Astronautics medal for sci. Brit. Interplanetary Soc., 1962. Fellow Am. Phys. Soc., Am. Inst. Aeros. and Astronautics, Am. Astronautical Soc.. (dir. 1960—), Brit. Interplanetary Soc., Royal Astron. Soc., Am. Geophys. Union; mem. Am. Meteorol. Soc., Pan Am. Med. Assn. (officer sect. on space medicine) Author: Geophysical Research with Artificial Earth Satellites, 1956. Editor: Progress in the Astronautical Scis. Research and publs. on origin cosmic radiations, aurorae and magnetic storms, meteorites, design research rockets and satellites. Home: 2451 Brickell Av., Miami, Fla. 33219. Office: Sch. Environmental and Planetary Scis., U. Miami, Coral Gables, Fla. 33124.*

SINGH, Baij Nath, Indian microbiologist; b. Banaras, India, July 31, 1914; s. Lal Bahadur and Subhadra (Devi) S.; Ph.D., U. Dublin (Ireland), 1938; Ph.D., U. London (Eng.), 1940, D.Sc., 1947. With Trinity Coll., Dublin, 1936-38; with Rothamsted Exptl. Sta., Eng., 1938-52, staff soil microbiology dept., 1940-52; head dept. microbiology Central Drug Research Inst., Lucknow, India, 1952-65, dep. dir. microbiology, 1965—. Mem. Indian Sci. Congress Assn. (pres. agrl. scis. sect. 1960), Soc. for Gen. Microbiology, Soil Sci. (Eng.), Soc. for Protozoology, Assn. Microbiologists. Editor: Indian Jour. Microbiology, 1961—. Research, publs. on selectivity in bacterial food, culture method for estimating numbers of amoebae in soil, classification of amoebae based on nuclear div., amino acids causing excystment of cysts of amoebae, cholesterol enhancing virulence of E. Histolytica, Acrasieae, Myxobacteria. Home: Vishwa nath Singh, Chetgang, Varanasi, India. Office: Central Drug Research Inst., Lucknow, India.*

SINGH, Harbhajan, Indian botanist; b. Pusa, India, Feb. 6, 1916; s. Arjan and Bakhtawar (Kaur) S.; B.Sc., Khalsa Coll., Punjab U., Amritsar, India, 1936; M.Sc., Agra (India) Coll., 1938; Asso. Indian Agr. Research Inst., 1940; m. Gurnitkaur, Mar. 15, 1940; children—Anup, Swaran (Mrs. Harjit Singh Sanga), Rajwant Kaur. Staff, Indian Agrl. Research Inst., New Delhi, India, 1940—, plant introduction officer, 1958-61, head div. plant introduction, 1961—. Mem. Indian Soc. Horticulture, Indian Soc. Plant Breeding and Genetics. Author: (monograph) Grain Amaranths, Buckwheat, Chenopods, 1961; Sweet Potato Cultivation, 1964; Leafy Vegetables, 1960; Farm Weeds, 1965; also numerous articles. Developed numerous varieties of crops and vegetables, including Pusa sawani, mosaic-resistant okra. Home: b-2 Library Rd., Indian Agrl. Research Inst., New Delhi, India.*

SINGH, Inderjit, Indian physiologist; b. Maymyo, Burma, Aug. 20, 1909; s. Meher and Sant (Kaur) S.; M.B.B.S., U. Rangoon, 1931; Ph.D., Cambridge (Eng.) U., 1936; m. Sunita I. Bhagat, May 9, 1940; children—Amarjit, Kunwarjit, Rita. Prof. physiology Dow Med. Coll., Karachi, India, 1944-48; prof., head dept. physiology Med. Coll. Agra., India, 1948—. Fellow Indian Acad. Scis., Nat. Inst. Scis., Indian Acad. Med. Scis. (founding). Research, numerous publs. on adminstrn. intravenous oxygen to non-breathing dogs, absence of all electrolytes in external media of tissues, vagus nerve in frog; demonstration of excitation of depolarised muscle, that relaxation in smooth muscle is active. Home: 32A, Lady Hardinge Rd., New Delhi-1, India. Office: Med. Coll., Agra, U.P., India.*

SINGH, Kunwar Suresh, Indian parasitologist; b. Barabanki, India, Aug. 15, 1924; s. Raghunath and Kamalini Singh; B.Sc., Lucknow U., 1943, M.Sc., 1945, Ph.D., 1952, D.Sc., 1961; postgrad. Liverpool Sch. Tropical Medicine, 1964-65; m. Prem Kumari, May 23, 1943; children—Manorama, Raghavendra, Kaushalendra. Prof. biology Colvin Taluqdar's Coll., Lucknow, India, 1945-46; asst. research officer I.C.-A.R., Lucknow U., 1946-51, asst. prof. zoology, 1951-58; research fellow U. Chgo., 1957; prof. parasitology Postgrad. Coll. Animal Scis. Indian Vet. Research Inst., Izatnagar, U.P., India, 1958—; guest researcher Liverpool Sch. Trop. Medicine, 1964-65. Research fellow Nuffield Found., 1964-65. Mem. Helminthological Soc. India, Am. Micros. Soc., Zool. Soc. India. Author: Introduction to Helminthology, 1968. Research, numerous publs. on morphology, taxonomy, biology, ecology, pathology and histochemistry of trematodes, cestodes and nematodes parasitic in animals, especially domesticated animals. Home: flat D-II, Ivri Campus, Izatnagar U.P., India.*

SINGH, Maharajah Jai, Indian astronomer, scholar; flourished circa 1730; studied astronomy under Ulugh Beg e Jamshed el Kashi. Ruled state of Jaipur, India; founder City of Jaipur, 1728; built Jantar Mantar, obs., New Delhi, India, 1724; also built observatories in Jaipur, Benares, Ujjain, Malthura, 1728-34. Translator, Ulugh Beg's catalogue of stars. Improved accuracy of astron. instruments; corrected errors of lunar tables of LaHire.

SINGLETON, Donald George, Australian physicist; b. Brisbane, Australia, Feb. 10, 1932; s. Thomas Edgar and Gladys (Stedman) S.; B.Sc. with 1st class honors, U. Queensland, 1955, M.Sc., 1956, Ph.D., 1962; m. Jessie Davina Mitchell, May 24, 1957; children—Helen Louise, Robert Jon, Cameron Bruce. Faculty U. Queensland, Brisbane, 1954-55, lectr. physics, 1956-63, sr. lectr., 1964; vis. scientist High Altitude Obs., U. Colo., Boulder, 1963; sr. research scientist Space Physics Wing, Weapons Research Establishment, Salisbury, Australia, 1965—. Mem. Inst. Physics London (asso.), Australian Inst. Physics (asso.), Am. Geophys. Union, A.A.A.S. Publs. on research on radio wave propagation in ionosphere; ground based, satellite and radio astron. propagation studies of ionization irregularities in F-layer of ionosphere. Home: 22 Berryman Dr., Modbury, South Australia. Office: Space Research Group, Weapons Research Establishment, Box 14244 G.P.O., Adelaide, South Australia, Australia.*

SINGWI, Kundan Singh, Indian physicist; b. 1919; M.Sc., D.Sc., U. Allahabad, India. Lectr., U. Allahabad, 1942-46; research fellow Nat. Inst. Scis. India, 1947-48; sr. lectr. U. Delhi (India), 1949; Brit. Council Research fellow U. Birmingham (Eng.), 1950-52; faculty physics U. Ill., 1952; reader Tata Inst. Fundamental Research, 1953—; head theoretical physics and applied math. div. Atomic Energy Establishment, India, 1955—. Research and publs. on theory of cold neutron scattering. Office: Atomic Energy Establishment, Apollo Pier, Rd., Bombay-1, India.

SINHA, Akhoury Purnendu Bhusan, Indian chemist; b. Buxar, India, Dec. 27, 1928; s. Skanda Narain and Surajbansi (Bakshi) S.; B.Sc., Sci. Coll., Patna U., 1948, M.Sc., 1954; Ph.D., Imperial Coll., London U., 1954, D.I.C., 1954; m. Snehlata Bakshi, Apr. 20, 1955; children—Ravi, Raj Kumar, Vijay Kumar. Asst. prof. Bihar U., Muzaffarpur, India, 1954; sci. officer Nat. Chem. Lab., Poona, India, 1955-63, asst. dir., head solid state material program, 1963—; counselor for Ph.D. students Poona U., 1957—, Banaras Hindu U., 1963—, Vikram U., 1965—, Karnatak U., 1966—. Recipient 5 U. Gold medals Patna U., 1948-50. Research, publs., patents on devel. new material with improved properties in fields of semiconductors and ferrites (resulting in establishment of new electroceramics industries in India), elec., magnetic and structural properties of transition metal compounds; discovered new phenomenon of dual negative resistance in Cadium-sulphide semiconductor. Home: B-4, NCL Colony Poona-8, India. Office: Nat. Chem. Lab., Poona-8, India.*

SINHA, Krityunjai Prasad, Indian physicist; b. Akhtiarpore, India, July 5, 1929; s. Maheshwari P. and Tara (Sahai) S.; M.Sc., Allahabad U., 1950; Ph.D., Poona U., 1956; Ph.D., Bristol (Eng.) U., 1959; m. Kanti Verma, May 22, 1955; children—Sarjai, Punita, Tripti, Jayant. Lectr., H. D. Jain Coll., Arrah, India, 1950-51; research staff phys. div. Nat. Chem. Lab., Poona, India, 1951-57; postdoctoral research fellow H. H. Wills Physics Lab., U. Bristol, 1957-59; sr. sci. officer phys. div. Nat. Chem. Lab., Poona, 1959-63, scientist, 1963—, group leader solid state and molecular physics group, 1964—; research guide physics and chemistry Poona U., Agra U., Bombay U., 1961—. Cons., Solid State Physics Lab., Nat. Phys. Labs., New Delhi, India, 1965—; vis. prof. Saha Inst. Nuclear Physics, Calcutta, India, 1965; staff Bell Telephone Labs., Murray Hill, N.J., 1968—. Fellow Indian Phys. Soc. Author: (monograph) Interactions in Magnetically Ordered Systems, 1965; also articles, chpt. in book. Research on atomistic aspects of phase transformations in solids, mechanisms of exchange interactions in magnetic crystals, relaxation processes determining interplay of excitations in lattice and spin systems, influence of coupling between electronic motion and lattice waves on transport properties of semiconductors, basic mechanism of superconductivity and effects of factors influencing their transition temperature; discovered spin effects on resistance minimum in semiconductors and its impact on thermoelectric devices. Home: Or. No. D-II/8 NCL Colony, Pashan Rd., Poona 8, India. Office: Bell Telephone Labs., Murray Hill, N.J.

SINNHUBER, Russell Otto, Am. food technologist; b. Detroit, Apr. 28, 1917; s. Fred and Rose (Bethke) S.; B.S., Mich. State Coll., 1939; M.S., Ore. State Coll., 1941; m. Dorothy Ann Collins, Aug. 22, 1942; children—John R., Carol A., Gary F. With Ore. Agrl. Expt. Sta., Astoria, 1943-57, asso. prof., 1950-57; faculty Ore. State U., Corvallis, 1957—, prof. dept. food sci., tech., 1963—. Recipient U. S. Dept. Interior Conservation Service award, 1966. Mem. Am. Chem. Soc., Am. Oil Chemists Soc., A.A.A.S., Inst. Food Technologists, Pacific Fisheries Technologists, Ore., N.Y. acads. sci., Sigma Xi, Phi Lambda Upsilon. Author: (with H. W. Schultz, E. A. Day) Symposium on Foods: Lipids and Their Oxidation, 1962. Research, numerous publs. on stability and quality of fats and oils fishery products, nutritional requirements of salmon and trout, devel. practical rations and feeds for fish, utilization of fishery by-products, irradiation preservation of fishery products, devel. new products, mold toxins and metabolites, cancer in fish, other animals. Home: 1420 Hillcrest Dr., Corvallis, Ore. 97330.*

SINNOTT, Edmund Ware, Am. botanist; b. Cambridge, Mass., Feb. 5, 1888; s. Charles Peter and Jessie Elvira (Smith) S.; A.B., Harvard, 1908, A.M., 1910, Ph.D., 1913; D.Sc., Northeastern U., 1948; Lehigh, 1950; m. Mabel H. Shaw, June 24, 1916; children—Edmund Ware, Mildred Shaw, Clara Richardson. Austin teaching fellow and asst. in botany, Harvard, 1908-10, 1911-12; Sheldon traveling fellow of Harvard, for bot. research in Australasia, 1910-11; instr. Harvard Forestry Sch. and Bussey instn., 1913-15; prof. botany and genetics, Conn. Agricultural Coll., 1915-28; prof. botany, Barnard Coll., 1928-39; prof. botany, Columbia, 1939-40; Sterling prof. of botany Yale, 1940-56, emeritus, chmn. dept. botany, 1940-50, director Sheffield Scientific School and chmn. Div. of Science, 1945-56, dean Graduate Sch., Yale, 1950-56. Member National Acad. of Sciences, American Philos. Society, American Acad. Arts and Sciences. Fellow A.A.A.S. (v.p. 1935, pres. 1948), Bot. Soc. of America (pres. 1937), Am. Society of Naturalists (pres. 1945), New England Botanical Club, Torrey Bot. Club (pres. 1931-34), Phi Beta Kappa, Sigma Xi. McNair Lecturer U. of N.C., 1949. Author: Botany Principles and Problems; (with L. C. Dunn and Th. Dobzhansky) Principles of Genetics;

Cell and Psyche; Two Roads to Truth, 1953; The Biology of the Spirit, 1955; Matter, Mind and Man, 1957; Plant Morphogenesis, 1960. Studies of plant morphology, histology, embryology, morphogenesis, and genetics, and interrelationships of these fields. Home: 459 Prospect St., New Haven.*

SINNOTT, John Alexander (Sinton), physician; b. Victoria, B.C., Can., Dec. 2, 1884; s. Walker Lyon and Isabella M. (Pringle) S.; M.D., Queens U., Belfast, Ireland, 1908; ed. Liverpool Sch. Tropical Medicine; m. Eadith Seymour Martin, Sept. 19, 1923; a dau., Eleanor Isabel Many. Staff pathology Queens; joined Indian Med. Service, 1911; med. officer to Bengal Lancers; began med. research supported by Indian Research Fund Assn., 1921; ret., 1938; sent to India, 1939; also in E. Africa, Middle E.; ret. again to No. Ireland, 1945. Cons. malariologist to War Office: pro-chancellor Queens U., Belfast, Ireland; high sheriff of County of Tyrone. Recipient Chalmers medal, Mary Kingsley medal, Manson medal. Mem. Royal Soc. London, A.M.A. Research and numerous publs. on sand flies and spread of disease, prevention of malaria and its treatment; pioneered use of plasmoquine with quinine or atebrin for malaria treatment. Died 1956.

SINSHEIMER, Robert Louis, Am. biophysicist; b. Washington, Feb. 5, 1920; s. Allen S. and Rose (Davidson) S.; B.S., Mass. Inst. Tech., 1941, M.S., 1942, Ph.D., 1948; m. Flora Joan Hirsch, Aug. 8, 1943; children—Lois June, Kathy Jean, Roger Allen. With Mass. Inst. Tech., 1942-49, research asso., 1948-49; faculty Ia. State Coll., 1949-57, prof. physics dept., 1955-57; prof. biophysics biology div. Cal. Inst. Tech., Pasadena, 1957—; mem. sci. adv. com. biology div. Argonne (Ill.) Nat. Lab., 1964—. Named Cal. Scientist of Year, 1968. Mem. Am. Acad. Arts and Scis., Am. Soc. Biol. Chemistry, Nat. Acad. Scis., Biophys. Soc., Optical Soc. Am., A.A.A.S., N.Y. Acad. Sci. Editor Jour. Molecular Biology, 1959-66, Ann. Rev. Biochemistry, 1965—. Research, publs. on structure and replication of DNA and RNA-containing viruses, nucleotide sequence in DNA, effect of ultraviolet radiation on nucleic acids and discovery of reversible hydration of pyrimidines; discovery of single-stranded viral DNA and of ring form viral DNA; mode of action of deoxyribonucleases and of RNA polymerase. Home: 1485 Chamberlain Rd., Pasadena, Cal. 91103.*

SION, Maurice, mathematician; b. Skopje, Yugoslavia, Oct. 17, 1928; s. Max and Sarah (Alalouf) S.; B.A., N.Y. U., 1947, M.S., 1948; Ph.D., U. Cal. at Berkeley, 1951; m. Emilie Grace Chisholm, Sept. 15, 1957; children—Crispin John, Sarah Anne, Dirk Andrew, Robin Robert. Mathematician, Nat. Bur. Standards, Washington, 1951-52; faculty U. Cal. at Berkeley, 1952-53, asst. prof., 1957-60; mem. Inst. for Advanced Study, Princeton, N.J., 1955-57, 62; faculty U. B.C., Vancouver, Can., 1960—, prof. math., 1964—; vis. asso. prof. U. Cal. at Berkeley, 1963, at Santa Barbara, 1965-66. Mem. Am. Math. Soc., Canadian Math. Congress, Phi Beta Kappa, Sigma Xi. Contbr. articles to tech. jours. Contbns. to devels. of theory of analytic sets, integration and differentiation on abstract and topological spaces. Home: 497 W. 6th Av., Vancouver 8, B.C., Can.*

SIPLE, Paul Allman, Am. explorer, geographer; b. Montpelier, O., Dec. 18, 1908; s. Clyde L. and Fannie Hope (Allman) Siple; B.S., Allegheny College, 1932, D.Sc. (honorary), 1942; Ph.D. (Geography), Clark University, 1930; D.Sc., U. Mass., 1958, Boston U., 1958, Clark U. (all honorary), 1958; LL.D., Gannon Coll., 1958; married Ruth I. Johannesmeyer, Dec. 1936; children—Ann Byrd, Jane Paulette, Mary Cathrin. Youngest mem. Admiral Byrd's Antarctic Expdn.; in charge of biol. and zoöl. work of expdn., bringing back specimens of penguins, seals for Am. Museum of Natural History, 1928-30, head of biological dept. Adm. Byrd's 2d expdn., 1933-35, and mem. Byrd's personal staff; in charge erecting and equipping the base in which Byrd lived alone 4½ mos. in 1934; leader Marie Byrd sledging party into newly discovered land; toured Europe, Asia Minor and N. Africa, 1932-33; geographer Div. Territories and Island Possessions, Dept. of Interior, assigned to U. S. Antarctic Expdn. as leader of West Base, Little America, 1939-41; geographer and tech. supervisor of supplies and equipment; on furlough, 1941, from U. S. Antarc. Expdn. and employed by the War Dept. as a civilian expert on design of cold climate clothing and equipment; head research and map projects for U. S. Antarctic Service, 1941-42; commd. in Army of U. S., Q.M. Corps, July 1942, discharged Aug. 1946. Mil. Geographer, sci. adviser Office Chief of Research and Development. Dept. Army Gen. Staff, 1946—; Sr. war dept. rep. Navy Antarctic Expdn. Highjump, 1946-47; dep. to Admiral Byrd, U. S. Antarctic Programs, sci. adviser Operation Deep Freeze I, 1955-56; sci. leader U. S. IGY Amundsen-Scott South Pole Sta., 1956-57. Mem. nat. council and camping com. of Boy Scouts of Am. Awarded Congl. medals, 1930, 37, 46; Heckel sci. prize, Hatfield award, 1931, Legion of Merit Award, 1946; exceptional civilian service award, Dept. Army, 1957; David Livingstone Centenary medal, Am. Geog. Soc., 1958; Hubbard medal, Nat. Geog. Soc., 1958; Distinguished Civilian Service award, Dept. Def., 1958; Patron's medal, Royal Geog. Soc. 1958; Hans Egede medal, Royal Danish Geographical Society, 1960. Fellow Arctic Institute of America, American Geographic Society; mem. Am. Polar Soc. (1st pres.), Assn. Am. Geographers (v.p. 1958, pres. 1959), Am. Geophys. Union, International Geophysical Year (U. S. com.), Clark University Geography Society, Phi Beta Kappa, Sigma Xi. Author: A Boy Scout with Byrd, 1931; Exploring at Home, 1932; Scout to Explorer, 1936; The Second Byrd Antarctic Expedition—Botany Report, 1938; Adaptations of the Explorer to the Climate of Antarctica, 1939; 90° South, 1959. Home: 131 N. Jackson St., Arlington 1, Va. Office: Army Research Office, OCRD, Pentagon, Washington 25.

SIRE, Georges Etienne, French physicist; b. Besancon, France, June 4, 1826; prof. mechanics, dir. Sch. Clock and Watch Making, Besancon. Mem. French Acad. Scis., 1891. Perfected technique for assay of gold and silver; solved numerous problems in chronology. Died Besancon, Sept. 12, 1906.

SIREK, Otakar Victor, physiologist; b. Bratislava, Czechoslovakia, Dec. 1, 1921; s. Otakar and Alice (Wagner) S.; M.D., Komenius U., Bratislava, 1946; M.A., U. Toronto, 1951, Ph.D., 1954; m. Anna Janek, July 27, 1946; children—Ann, Jan, Peter, Terese. Fellow dept. medicine Komenius U., 1946-47; research fellow Wenner-Gren Inst. Biol. Scis., Stockholm U., Sweden, 1947-50; faculty U. Toronto (Ont., Can.), 1950—, asso. prof. physiology, 1961—. Recipient Starr medal U. Toronto, 1958. Mem. Am. Diabetes Assn., Am. Physiol. Soc. Research, publs. in protein-anabolic effect of testosterone; growth hormone, insulin. Home: 93 Farnham Av., Toronto 7, Ont., Can.*

SIRI, William Emil, Am. physicist; b. Phila., Jan. 2, 1919; s. Emil Mark and Caroline (Schaedel) S.; student U. Chgo., 1937-43, U. Cal. at Berkeley, 1947-50; m. Margaret Jean Brandenburg, Dec. 3, 1949; children—Margaret Lynn, Ann Kathryn. Research engr. Baldwin-Lima-Hamilton Corp., 1943; physicist Radiation Lab., U. Cal. at Berkeley, 1943-45, physicist to Donner Lab., 1945—. Exec. v.p. Am. Mt. Everest Expdn., Inc. Mem. Am. Phys. Soc., Biophys. Soc., Am. Assn. Physicists in Medicine, Sigma Xi. Author: Nuclear Radiations and Isotopic Tracers, 1949. Field leader U. Cal. Peruvian Expdns., 1950-52; leader Cal. Himalayan Expdn., 1954; field leader Internat. Physiol. Expdn. to Antarctica, 1957; dept. leader Am. Mt. Everest Expdn., 1963. Research and publs. on gross body composition (total fat, water, protein), applications of radioisotopes to biology and medicine, high altitude physiology. Home: 1015 Leneve Pl., Richmond, Cal. 94532. Office: Donner Lab., U. Cal., Berkeley, Cal. 94270.*

SIRLIN, Julio Leo, biologist; b. Buenos Aires, Argentina, Dec. 18, 1926; s. Bernardo and Raquel (Herbstein) S.; D.Natural Sci., U. Buenos Aires, 1954. Demonstrator, Faculty Sci., U. Buenos Aires, 1949-52; research asso. Inst. Animal Genetics, U. Edinburgh (Scotland), 1955-59, lectr., staff, 1960-66; asso. prof. anatomy Cornell U. Med. Coll., N.Y.C., 1966—. Mem. Biochem. Soc., Internat. Inst. Embryology. Research, numerous publs. on cytochemistry of synthesis of nucleic acids and protein, biochemistry of nucleolar ribonucleic acid synthesis. Office: Dept. Anatomy, Cornell U. Med. Coll., 1300 York Av., N.Y.C. 10021.*

SIROTA, Nikolay Nikolaevich, Russian physicist; b. 1913; grad. Moscow Steel Inst., 1936, later postgrad.; D.Physico-Math. Sci. Instr., Moscow U., 1930-40, 45-55; asso. Inst. Gen. and Inorganic Chemistry, USSR Acad. Scis., 1941-54; instr. Moscow Inst. Non-Ferrous Metals and Gold, 1951—; prof., 1952—; asso. Physicotech. Inst., Belorussian Acad. Scis., 1957—; prof. Belorussian U. Author: The Thermodynamics of Intermetallic Compounds, 1944. Research and publs. on thermodynamics, kinetics of phase transition, structure and phys. properties of metals, physicochem. analysis of condensed media. Address: Belorussian University, Universitetsky gorodok, Minsk, Belorussian SSR, USSR.

SIROTININ, Nikolay Nikolaevich, Russian immunologist, pathophysiologist; b. 1896; grad. Med. Faculty, Saratov U., 1924; D.Med. and Biol. Sci., 1924. Prof., founder chair path. physiology Med. Faculty, Kazan U., prof. chair path. physiology Kazan Vet. Inst., head dept. path. physiology Kazan Vet. Research Inst., Tatar Inst. Exptl. Medicine, 1929-34; head dept. comparative pathology Inst. Exptl. Biology and Pathology, Ukrainian Peoples Commissariat Health, 1934-41; with Inst. Clin. Physiology, Ukrainian Acad. Scis., 1934-43, sr. asso. Inst. Microbiology, 1944-55, head dept. comparative physiology Inst. Physiology, 1946—; head dept. exptl. medicine, founder lab. Exptl. Medicine, Ukrainian Inst. Epidemiology and Microbiology, 1937-42; head Lab. Immunology and Path. Physiology Kiev Inst. Infectious Diseases, USSR Acad. Med. Scis., 1944-55, cons., 1955—; head chair path. physiology Kiev Med. Inst., 1955—. Presidium mem. learned med. council, mem. hosp. council Ukrainian Ministry health. Mem. Ukrainian Acad. Scis. (corr.), USSR Acad. Med. Scis., Kiev Soc. Pathologists (chmn. bur.), All-Union Soc. Pathophysiologists (bd. mem.), Ukrainian Soc. Microbiologists, Epidemiologists and Infectionists (bd. mem.). Author: Hypoergia and Its Importance in the Course of Infection, 1934; Life at Heights and Altitude Sickness, 1939; Evolution of Body Response, 1952; The Genesis and Course of Infection in the Light of Comparative Pathology, 1956; Problems of Allergy, 1961. Mem. editorial council Archives of Pathology; mem. editorial bd. Path. Physiology and Exptl. Therapy. Research and over 200 publs. on evolution of body response, effects of external stimuli, pathogenesis and prophylaxis of altitude sickness, oxygen starvation. Address: Kiev Medical Institute, b. Shevchenko 13, Kiev, Ukrainian SSR, USSR.

SIRSI, Madhav, Indian pharmacologist; b. Shimoga, Mysore, India, Oct. 24, 1912; s. Ramachandra Srinivasa and Sita (Bai) S.; M.B.B.S., U. Med. Coll. Mysore, 1935; L.T.M., Sch. Tropical Medicine, Calcutta, India, 1937; m. Suvarna N. Chouthoy, 1932; children—Suresh, Ramesh. Pathologist, Bowring, Lady Curzon and Isolation Hosps., Bangalore, Mysore, 1938-49; faculty Pharmacology Lab., Indian Inst. Sci., Bangalore, 1949—, asso. prof., 1965—. Adviser, Indsl. and Testing Labs., 1960—; mem. bd. med. research U. Mysore, 1964—; mem. Mysore State Ayurvedic Research Bd., 1963—. Mem. Indian Med. Assn., Indian Pharm. Assn., Indian Sci. Congress, numerous others. Research, numerous publs. on indigenous medicinal plants in clin. disorders, synthetic tuberculostats, drug resistance, antifungal antibiotics, chemotherapy and biochemistry of cancer. Home: 41/5 12th Block Kumarapark West, Bangalore 20. Office: Pharmacology Lab., Indian Inst. Sci., Bangalore, Mysore, India.*

SISAKIAN, Noraír Martirósovich, Soviet biochemist; b. Jan. 25, 1907; grad. Timiriazev Agrl. Acad., Moscow, 1932. Staff, Inst. Biochemistry, USSR Acad. Sci., 1939—, became mem., acad. sec. div. biol. scis., 1960. Mem. Armenian Acad. Scis. Recipient Stalin Prize, 1952, Bakh Prize, USSR Acad. Scis., 1949, Mechnikov Prize, 1950. Staff, Timiryazev Agr. Acad., Moscow, 1932-39. Author: The Biochemical Characteristics of Drought Resistance in Plants, 1940; The Biochemistry of Viticulture, 1947-57; The Enzyme Activity of Protoplasmic Structure, 1951; The Biochemistry of Metabolism, 1954; The Biochemistry of Plastids, 1954; The Chemism and Biochemical Functions of Plastics, 1956. Co-editor chemistry sect. Large Med. Ency. 2d edit. Research on influence of enzymes on metabolic processes; biol. effects of high radiation; fermentation processes. Office: USSR, Moskva, Leninsky prosp. 14, AN SSSR.

SISLER, Harry Hall, Am. chemist; b. Ironton, O., Mar. 13, 1917; s. Harry Chester and Minta (Hall) S.; B.S., Ohio State U., 1936; M.S., U. Ill., 1937, Ph.D., 1939; m. Helen Elizabeth Shaver, June 29, 1940; children—Elizabeth Ann (Mrs. Tom Rider), David Franklin, Raymond Keith, Susan Carolyn. Faculty, Chgo. City Colls., 1939-41, U. Kan., 1941-46; faculty Ohio State U., 1946-56, prof. chemistry, 1955-56; prof., head dept. chemistry U. Fla., Gainesville, 1956—, dir. div. phys. sci. and math., 1964—. Mem. chemistry adv. panel NSF, 1959-62; NSF lectr. to numerous colls., univs.; cons. indsl. firms. Arthur and Ruth Sloan vis. prof. Harvard, 1962-63. Mem. Am. Chem. Soc. (tour lectr., Outstanding Southeastern Chemist award Fla. sect. 1960), A.A.A.S., Sigma Xi, Alpha Chi Sigma, Phi Lambda Upsilon, Phi Delta Kappa, Gamma Sigma Epsilon, Phi Kappa Phi, Kappa Delta Pi. Author: General Chemistry-A Systematic Approach, 1949, 59; A Systematic Laboratory Course in General Chemistry, 1950, 61; Essentials of Chemistry, 1951, 59; Essentials of Experimental Chemistry, 1951, 59; Semi-Micro Qualitative Analysis, 1951, 66; College Chemistry-A Systematic Approach, 1953, 61; Chemistry in Non-Aqueous Solvents, 1961; Electron Structure, Properties and the Periodic Law, 1963; College Chemistry, 1967. Research, publs. on chemistry of chloramine and hydrazine derivatives, nitrogen-phosphorus derivatives, molecular addition compounds, aluminum alkyls. Home: 1025 N.W. 61st Terrace, Gainesville, Fla. 32601.*

SISMONDI, Jean Charles Leonard Simonde de, Swiss economist, historian; b. Geneva, May 9, 1773 of Italian aristocratic family; ed. in Geneva; m. Miss Allen, Apr. 1819. Became banker's clk., Lyons, France; took refuge with parents in Eng. during French Revolution; farmed in Pescia, Italy; joined circle of Madame de Stael at Coppel on Lake Geneva, Switzerland; lectured in Geneva; became sec. of chamber of commerce for then dept. of Leman; visited Paris, 1813; had interview with Napoleon I during Hundred Days. Author: Traité de la richesse commerciale, 1803; History of the Italian Republics in the Middle Ages, 1807-17; New Principles of Political Economy, 1819; Histoire des Français, 31 vols., 1821-44. Contended that scope of econs. should include human well-being; believed govt. should intervene to prevent dire poverty and concentration of wealth, laying groundwork for today's welfare econs.; analyzed technol. unemployment and recurrent unemployment crises; as historian gave weight to econ. influences. Died Geneva, June 25, 1842.

SISSON, Septimus, comparative anatomist; b. Gateshead, Eng., Oct. 2, 1865; s. George and Mary (Arnott) S.; came to U. S., 1882; S.B., U. Chgo., 1898; V.S., Ont. (Can.) Vet. Coll., 1891; postgrad. U. Berlin, 1905-06; D.V.Sc., U. Toronto, 1921; m. Katherine Oldham, Oct. 5, 1892. Prof. comparative anatomy Ohio State U., 1901—. Author: A Text-Book of Veterinary Anatomy, 1910; A Veterinary Dissection Guide, Part I, 1911; The Anatomy of the Domestic Animals, 1914. Translator: Ellenberger, Baum und Dittrich's Anatomie der Tiere für Künstler, 1906.

Main contbn. to anat. knowledge consists of 1st descriptions of natural form and topography of plastic organs and viscera of chief domestic animals as determined by fixation in situ by means of intravascular injection of formalin or other hardening fluid. Deceased.

SISSON, Wayne Andrew, Am. chemist; b. Shawsville, Va., July 21, 1903; s. James A. and Nora May (Jewell) S.; B.S., Roanoke Coll., 1925, M.S., 1928; postgrad. Cornell U.; Ph.D., U. Ill., 1931; m. Marian Benbow, June 18, 1931; children—James R., Ann Louise. Research asso. U. Ill., Urbana, 1931-33, 34-35; Tettile Found. fellow Leeds (Eng.) U., 1933-34; asso. cotton technologist U. S. Dept. Agr., Washington, 1935-36, collaborator, 1932-35, 50-54; research chemist Boyce Thompson Inst., Yonkers, N.Y., 1936-40; with Am. Viscose div. FMC Corp., Marcus Hook, Pa., 1940—, asst. to dir. research, 1960-64, mgr. Avicel research and devel., 1964—. Adj. prof. Bklyn. Poly. Inst., 1939-40. Recipient Alumni citation Roanoke Coll., 1958. Fellow A.A.A.S.; mem. Fiber Soc. (past pres.), Am. Chem. Soc. (past chmn., one of Ten Ablest Cellulose Chemists Chgo. sect. 1947, Anselme Payen award 1966), Textile Research Inst. (past chmn.), Rayon Producers Group (past chmn.), Sigma Xi, Alpha Chi Sigma, Gamma Alpha, Phi Lambda Upsilon, Epsilon Chi. Patentee in field. Research, publs. on cellulose, relation of structure to fiber properties of cotton and rayon, devel. crimped rayon and rayon tire cord. Home: 109 Quant Rd., Bowling Green, Media, Pa. 19063. Office: FMC Corp., Am. Viscose div., Marcus Hook, Pa. 19061.*

SITTER, Willem de, Dutch astronomer; b. Sneek, Holland, May 6, 1872; student under Kaptevyn, Groningen, Holland; became prof. theoretical astronomy U. Leiden (Holland), 1908; named dir. obs., Leiden, 1919. Author: The Expanding Universe, 1930; On the Motion and the Mutual Perturbations of the Material Particles in an Expanding Universe, 1933; The Astronomical Aspect of the Theory of Relativity, 1933. Computed size of universe as 200 million light years in radius with approximately 80,000 million galaxies; proposed the universe is an expanding a space-time continuum with motion and no matter; 1st application of theory of relativity to expansion and origin of universe; originated model of spherical universe in space time with finite radius; publ. on oribts of Jovian satellites. Died Leiden, Nov. 20, 1934.

SITTERLY, Charlotte Moore, Am. physicist; b. Ercildoun, Pa., Sept. 24, 1898; d. George Winfield and Elizabeth Palmer (Walton) Moore; A.B., Swarthmore Coll., 1920, D.Sc. (hon.), 1962; Ph.D. Astronomy, U. Cal. at Berkeley, 1931; m. Bancroft Walker Sitterly, May 30, 1937. Computer, Princeton Obs., 1920-25, 28-29; research asst., 1931-36, research asso., 1936-45; computer Mt. Wilson Obs., Pasadena, Cal., 1925-28; physicist Nat. Bur. Standards, Washington, 1945—. Recipient U. S. Dept. Commerce Meritorious Service award, 1951, Exceptional Service award, 1960; Fed. Woman's award, 1961; Annie Jump Cannon Centennial medal Wesley Coll., 1963. Fellow Optical Soc. Am., Am. Phys. Soc., Washington Acad. Sci.; mem. Royal (fgn. asso.), Am. (v.p. 1958-60, Annie Cannon prize 1937) astron. socs., A.A.A.S. (v.p. 1952), Astron. Soc. Pacific, Internat. Astron. Union (pres. commn. Fundamental Spectroscopic Data 1961-67), Triple Commn. Spectroscopy, Philos. Soc. Washington, Phi Beta Kappa, Sigma Xi. Author: Atomic Lines in the Sun-Spot Spectrum, 1932, A Multiplet Table of Astrophysical Interest, 1945; Atomic Energy Levels (3 vols.) 1949, 52, 58; An Ultraviolet Multiplet Table (5 sects.), 1950, 52, 62; (with H. N. Russell) The Masses of the Stars, 1940; (with H. D. Babcock) The Infrared Solar Spectrum, 1947; (with M. G. J. Minnaert, J. Houtgast) The Solar Spectrum 2935A to 8770 A, 1966; also numerous papers profl. jours. Research on analysis of atomic spectra; identification of lines in sunspot and solar spectra; atomic energy levels. Home: 3711 Brandywine St. N.W., Washington 20016. Office: Atomic Physic Div., Nat. Bur. Standards, Washington 20234.

SIVADJIAN, Joseph, pharmacologist; b. Constantinople, Turkey, Dec. 11, 1898; s. Garabet and Vasilik (Bezazian) S.; student Sch. Pharmacy, Sorbonne U., Paris, 1919, Docteur ès sciences, 1935, Dr. ès Lettres, 1938; m. Hripsimé Tulian, May 4, 1935; children—Jeanine (Mrs. Pierre Rouget), Annie Irene. Asst. chemotherapeutic lab. Pasteur Inst., Paris, France, 1926-46, head pharmacological lab., 1946—; conseiller scientifiquea la Société des Usines chimiques, Rhône-Poulenc, France, 1936—. Decorated Lauréat, Académie des Sciences. Mem. Société chimique de France, Société botanique de France, Société Thérapeutique, Société Physique. Author: Le Temps, Etudes Philosophiques, 1938; La Chimie des Vitamines et des Hormones, 1949; also numerous articles. Research on effects of drugs on conditioned reflexes; discovered hygrophotography and its applications. Home: 41 Rue Brancion. Office: 28 rue du Docteur Roux, Paris, France.*

SIZER, Irwin Whiting, Am. biochemist; b. Bridgewater, Mass., Apr. 4, 1910; s. Ralph W. E. and Annie (Jenkins) S.; A.B., Brown U., 1931; Ph.D., Rutgers U., 1935; m. Helen Whitcomb, June 30, 1935; 1 dau., Meredith Anne. Teaching asst. Rutgers U., 1931-35; faculty Mass. Inst. Tech., Cambridge, 1935—, prof. biochemistry, head dept. biology, 1955-67, dean Grad. Sch., 1967—. Cons. Johnson & Johnson Co., Ford Found., Commn. on Undergrad. Edn. in Biology. Fellow Am. Acad. Arts and Scis.; mem. Am. Physiol. Soc., Am. Chem. Soc., Am. Soc. Biol. Chemistry, Am. Soc. Zoology. Research in biochemistry of enzymes, proteins, amino acids, vitamins. Home: 52 Percy Rd., Lexington, Mass. 02173.*

SJOGREN, Hans Olof, Swedish microbiologist; b. Falun, Sweden, July 6, 1935; s. Erik Gustav and Ellen Laura (Lindquist) S.; D. Medicine, Karolinska Institutet, 1954, Docent, 1964; m. Barbro Elisabet Lindstrom, Apr. 29, 1961; children—Maria, Helena, Elisabet. Mem. staff Inst. for Tumor Biology, Karolinska Institutet, Stockholm, Sweden, 1958-67; docent in tumor biology, asso. prof. dept. med. microbiology, U. Lund (Sweden), 1967—. Research and publs. demonstrated for first time that polyoma virus-induced neoplasms have common specific antigens and showed that these specific antigens were different from the antigens of the virus particles; developed with I. Hellstrom invitro assays for humoral and cell-bound immunity to tumor specific antigens and demonstrated that specific antigens of viral tumors are in each group common for tumors of different animal species or classes. Home: 9 Markaskälsvägen, Lund, Sweden. Office: Inst. for Med. Microbiology, U. Lund, Lund, Sweden.*

SJÖGREN, Tage Anton Ultimus, Swedish surgeon; b. 1859; credited with 1st successful use of Xrays in cancer, 1899; described condition of dryness of mucuous membranes as result to deficient secretion of lacrimal, salivary and other glands (Sjögren's syndrome). Died 1939.

SJÖQVIST, Carl Olof, Swedish neurosurgeon; b. Stockholm, Sweden, Dec. 9, 1901; s. John and Maria (Von Hermanson) S.; Med.Kand., Karolinska Inst., 1923, Med.lic., 1928, Med.Dr., 1939; m. Inez Segardt, Nov. 2, 1929; 2 children. Jr. asso. neurosurgery Serafimerlassarettet, 1933-38; Rockefeller Found. fellow Yale, 1938-39; cons. neurosurgeon St. Eriks Hosp., 1940-43; cons. neurosurgeon Södersjukhuser Hosp., 1943-46; surgeon in chief, head dept. neurosurgery Sodersjukhuset, 1947—. Lectr. neurosurgery Karolinska Institutet, 1940-45. Recipient Lennmalm's prize Harvey Cushing Soc., Stockholm, 1941. Mem. Nordisk Neurokirurgisk Förening (became sec. 1946), Svenska Läkaresällskaper, Soc. Brit. Neurol. Surgeons. Author: Studies on Pain Conduction in the Trigminal Nerve, 1938; also articles. Divided tractus spinalis nervi trigemini to relieve pain caused by trigeminal neuralgia, 1937. Home: 24 Norr Mälarstrand. Office: Södersjukuset, Stockholm, Sweden.*

SJÖRS, Hugo, Swedish botanist; b. Stora Skedvi, Sweden, Aug. 1, 1915; s. Mattias and Lisa (Sundberg) S.; Fil. dr. Uppsala U., 1948; m. Gunnel Thelander, June 8, 1946; children—Kerstin, Anna, Gunnar. Faculty, Uppsala (Sweden) U., 1948-52, prof. ecol. botany, 1962—; asso. prof. Lund (Sweden) U., 1952-55, Sch. Forestry, Stockholm, Sweden, 1955-62; vis. lectr. Imperial Coll., London, 1961. Mem. state planning commns., 1960-62, 64-67; Hill Family vis. prof. U. Mpls., 1966. Mem. Swedish Phytogeog. Soc. (chmn.), Royal Swedish Acad. Sci. Author: Nordisk växtgeografi (Plant Geography of Scandinavia), 1956, 67; also articles. Editor, contbg. author: The Plant Cover of Sweden, 1965. Research on vegetation ecology, especially peat-forming vegetation; regional ecology of Boreal zone. Home: 7 Kabovägen, Uppsala, Sweden.*

SJOSTRAND, Fritiof Stig, biologist; b. Stockholm, Sweden, Nov. 5, 1912; s. Nils Johan and Dagmar (Hansen) S.; M.D., Karolinska Institutet, Stockholm, 1941, Ph.D., 1945; m. Märta Bruhn-Fähraeus, Mar. 24, 1941 (dec. June 1954); 1 son, Rutger; m. 2d, Ebba Gyllenkrok, Mar. 28, 1955; 1 son, Johan. Asst. prof. anatomy Karolinska Institutet, 1945-48, asso. prof., 1949-59, prof. histology, 1960-61; research asso. Mass. Inst. Tech., 1947-48; vis. prof. U. Cal. at Los Angeles, 1959, prof. zoology, 1960—. Recipient Jubilee award Swedish Med. Soc., 1959, Anders Retzius gold medal, 1967. Fellow Royal Micros. Soc. (hon., London, Eng.), Am. Acad. Arts and Scis.; mem. Electron Microscopy Soc. Am., Japanese Electron Microscopy Soc. (hon.). Author: Über die Eigenfluoreszenz Tierischer Gewebe Mit Besonderer Berücksichtigung der Säugertierniere, 1944; Electron Microscopy of Cells and Tissues, Vol. I, 1967; also numerous articles. Devel. technique for high resolution electron microscopy of cells, fluorescence microspectrography; invented ultramicrotome. Home: 1345 Casiano Rd., Los Angeles 90049.*

SJÖSTRAND, Nils Göran, Swedish physicist; b. Eksjö, Sweden, Mar. 28, 1925; s. Albert and Helga (Jeppson) S.; grad. Royal Inst. Tech., Stockholm, Sweden, 1949, Dr.Tech., 1959; m. Ingrid Strand, June 23, 1951; children—Bo Fritiof, Lena Gerda Elisabet. Scientist, AB Atomenergi, Stockholm, Sweden, 1949-60, cons., 1961—; prof. Chalmers U., Tech., Göteborg, Sweden, 1960—; staff Brookhaven Nat. Lab., Upton, N.Y., 1952-53, Atomic Energy Research Establishment, Harwell, Eng., 1957-58. Mem. expert mission, Turkey, 1965. Developed pulsed neutron source method for investigations in reactor physics and neutron physics. Home: 47, Skarsgatan, Göteborg S, Sweden.*

SKAE, David, Brit. physician; b. Edinburgh, Scotland, July 5, 1814; ed. Edinburgh U.; M.D. (hon.), U. St. Andress, 1842; m. Sarah Macpherson. Tchr., extra-academical med. sch., Edinburgh, 1836; surgeon Lock Hosp., from 1836; physician supt. Royal Edinburgh Asylum, 1846 Morningside, 1846-73; nominated Morisonian lectr. on insanity Royal Col. Coll. Physicians, Edinburgh, 1873. Fellow Royal Coll. Surgeons. Author: The Treatment of Dipsomaniacs, 1858; Legal Relations on Insanity, 1861, 67; Classification of the Various Forms of Insanity on a Rational and Practical Basis. Classified insanity according to underlying phys. condition of patient, thus bringing this aspect of mental illness to attention of psychiatrists; defined insanity as disease of the brain affecting the mind. Died Apr. 18, 1873.

SKAGGS, Lester S., Am. physicist; b. Trenton, Mo., Nov. 21, 1911; s. Frank B. and Etta (Clingingsmith) S.; student Trenton Jr. Coll., 1929-31; A.B., U. Mo., 1933, A.M., 1934; Ph.D., U. Chgo., 1939; m. Ruth Coffman, June 17, 1939; children—Margaret Leslie, John Baker, Mary Ann. Physicist, U. Cal., Los Alamos, 1944-45, Michael Reese Hosp., Chgo., 1945-49; asso. prof. dept. radiology U. Chgo., 1949-56, prof., 1956—. Research on devel. radiation sources for cancer therapy, extraction of electron beams from betatron. Home: 359 Osage St., Park Forest, Ill. 60466. Office: 950 E. 59th St., Chgo. 60637.*

SKALKA, Miloslav, Czechoslovakian biologist, radiobiologist; b. Bratislava, Czechoslovakia, Nov. 27, 1929; s. Antonin and Hilda (Kadlecova) S.; M.D., Purkyne U. Sch. Medicine, Brno, Czechoslovakia, 1953; C.Sc., Inst. Biophysics, Czechoslovak Acad. Sci., Brno, 1957; m. Otilie Hruskova, Feb. 11, 1953; children—Miloslava, Dagmar. Reader, Purkyne U. Sch. Medicine, 1952-53, postgrad. research fellow, 1953-56, asst. prof. path. physiology, 1964—; sr. research asso. Inst. Biophysics, Czechoslovak Acad. Sci., 1956—, head dept. cellular radiation damage, 1960—. Mem. European Soc. for Radiation Biology, Purkyne Soc. Czechoslovak Physicians, Czechoslovak Biol. Soc., Czechoslovak Biochem. Soc. Author: (with others) Methods in Radiobiology, 1966, Lectures on Biophysics, 1964; Pathophysiology of Radiation Damage, 1966; also numerous articles. Research on liver damage after total body irradiation or alkylating agts., damage by radiation and alkylating agts. to nucleoprotein complex of animal cell nucleus, to its stability and biol. function. Office: 135 Kralovopolska, Brno 12, Czechoslovakia.*

SKANDALAKIS, John Elias, surgeon; b. Molai, Sparta, Greece, Jan. 20, 1920; s. Elias N. and Vassiliki (Ritsos) S.; Grad. Med., U. Athens (Greece), 1946, Doct. Med., 1950; Ph.D., Emory U., 1962; m. Mimi M. Cutis, Oct. 15, 1950; children—Vickie, Elias, Demetrius. Came to U. S., 1951, naturalized, 1956. Dir. surg. edn. Piedmont Hosp., Atlanta, 1957—; prof. anatomy Emory U., Atlanta, 1956—; practice surgery, Atlanta, 1957—. Cons. Yerkes Regional Primate Research Center, 1965—. Recipient Meml. medal for nat. underground action Greek Govt., 1958; Golden award Assn. Abdominal Surgeons, 1962. Diplomate Greek Bd. Surgery. Fellow Southeastern Surg. Congress, A.C.S.; mem. A.M.A., Ga., So. med. assns., Fulton County Med. Soc., Greek, Ga. surg. socs. Author: (with others) Smooth Muscles Tumors of the Alimentary Tract, 1962; also articles. Research on gen. surgery, liver, smooth muscle tumors of gastrointestinal tract, surg. embryology. Home: 4107 Beechwood Dr. N.W., Atlanta 30327. Office: 2045 Peachtree St., N.E., Atlanta 30309.*

SKAPSKI, Adam Stanislaus, physicist-metallurgist; b. Cracow, Poland, May 21, 1902; s. Stanislas and Helena (Gostwicka) S.; diploma engring. Poly. Warsaw, 1922; M.Sc., Jagellonian U., 1924, Secondary Sch. Teaching certificate, 1925, Ph.D. Phys. Chemistry, also Ph.D. in Philosophy, 1927, degree of Docent, 1931; m. Mary Alice King; children—Irene (Mrs. J. Swiecicka), Barbara, Ellen Lenore. Came to U. S., 1946, naturalized, 1952. Metall. research Metallografiska, Inst., Stockholm, Sweden, 1933-34; prof. U. Mining and Metallurgy, dir. Metall. Research Inst., Cracow, 1934-39; imprisoned in Siberia for refusal to collaborate with USSR, 1939-42; sec. for edn. Polish govt.-in-exile, 1942-45; faculty U. Chgo., 1946-48; prof. U. Neb., 1949-53, U. Vt., 1953-60; tech. coop. adviser to AID in Nigeria, 1960-63; ednl. adviser Ford Found., Nigeria, 1963—, also spl. cons. Ministry Edn. West Nigeria, 1960—, North Nigeria, 1964—; mem. council Tech. Edn. No. Nigeria, 1965—. Mem. W. African Exam. Council. Decorated Golden Cross of Merit, Commandership Polonia Restituta (Poland); Order of Niger (Nigeria). others. Mem. Iron and Steel Inst. (London, Eng.), Am. Inst. Mining Metallurgy and Engring., Am. Soc. U. Profs., Am. Phys. Soc., others. Research, publs. on dilute and concentrated solutions, metall. equilibria, steel, free energy in metals and non-metals. Home: Main St., Milton, Vt. 05468. Office: Ford Found., 47 Marina, Lagos, Nigeria.*

SKAUEN, Donald Matthew, Am. pharmacist; b. Newton, Mass., May 14, 1916; s. Marcus and Mary (Duncan) S.; B.S., Mass. Coll. Pharmacy, 1938,

M.S., 1942; Ph.D., Purdue U., 1949; m. Rachel Margaret Burns, Oct. 25, 1942; children—Deborah Ellen (Mrs. Frederick Hinchliffe II), Bruce Jaye. Faculty, U. Conn., Storrs, 1948—, prof., 1960—. Mem. Am. Pharm. Assn., Radiol. Health Conf., Sigma Xi, Rho Chi, Phi Lambda Upsilon. Author: (with others) American Pharmacy, Husa's Pharmaceutical Dispensing, 1966. Research, numerous publs. on areas of ultrasonics, radiopharmaceutics and marine radiobiology. Home: 16 Storrs Heights Rd., Storrs, Conn. 06268.*

SKEHAN, James William, Am. geologist; b. Houlton, Me., Apr. 25, 1923; s. James William and Mary Effie (Coffey) S.; A.B. in Philosophy, Boston Coll., 1946, A.M., 1947; Ph.Licentiate in Philosophy, Weston Coll., 1947, S.T.B., 1954, S.T.L., 1955; A.M. in Geology, Harvard, 1951, Ph.D., 1953. Joined Soc. of Jesus, 1940, ordained priest Roman Cath. Ch., 1954; asst. prof. geology Boston Coll., 1951-61, associate professor of geology, 1962—, founder, 1958, since chairman dept. geology; dir. West Obs., 1956—; founder, dir. Geology-Astrogeology Research Center, 1963—. Dir. Nat. Sci. Found. undergrad. research participation program at Boston Coll., 1960-65. Am. geol. rep. to council A.A.A.S., 1963-65, Nat. Assn. Geology Tchrs. rep., 1965; participant Earth Sci. Curriculum Project Writing confs., 1964, 65. Fellow Geol. Soc. Am., Geol. Society of London; mem. Am. Fedn. Mineral. and Lapidaary Socs., Boston Geol. Soc. (pres. 1958-59), New Eng. Intercollegiate Geology Conference, National Association of Geology Teachers, American Society C.E., Am. Fedn. Mineral. Socs., Am. Inst. Mining, Metall. and Petroleum Engrs., Am. Soc. Photogrammetry, Assn. Geol. Study Deeper Zones of Earths Crust, European Assn. Exploration Geophysicists, Sigma Xi. Author: The Green Mountain Anticlinorium in the Vicinity of Wilmington and Woodford, Vermont, 1961; Geology of Basement Complex of Southeastern Nebraska, Northeastern Kansas and Vicinity, 1963; Geology of the North American Air Defense Combat Operations Center, Cheyenne Mountain and Peripheral Area . . . , 1965; Photography of the Sahara and the Kalahari from Tiros, 1965; also numerous papers. Research in the geology, especially geotectonics, of New Eng. area. Address: Dept. Geology, Boston Coll., Chestnut Hill, Mass. 02167.*

SKEMPTON, Alec Westley, English civil engr.; b. Northampton, Eng., June 4, 1914; s. Alec Westley and Beatrice (Payne) S.; B.Sc., Imperial Coll., U. London, 1935, M.Sc., 1936, D.Sc., 1949; m. Mary Nancy Wood, July 4, 1940; children—Judith (Mrs. Geoffrey Stevens), Katherine. Sci. officer Bldg. Research Sta., 1936-46; faculty Imperial Coll., U. London (Eng.), 1946—, prof. civil engring., 1957—. Cons. on civil engring. projects in Gt. Britain, Italy, Pakistan, Australia. Fellow Royal Soc., 1961, Geol. Soc.; mem. Internat. Soc. for Soil Mechanics (past pres.), Instn. Civil Engrs. Research, publs. on residual strength clays and particle orientation, pore water pressures, landslides, design earth dams, found. large structures; application sci. methods in engring. geology. Home: 16 The Boltons, London, Eng.*

SKEY, Frederic Carpenter, Brit. surgeon; b. Upton-on-Severn, Dec. 1, 1798; s. George Skey; ed. Edinburgh, Paris, St. Bartholomew's Hosp.; pupil of Abernathey; demonstrator anatomy St. Bartholomew's Hosp., circa 1826-31, became asst. surgeon, 1827, lectr. anatomy, 1843-65, surgeon, 1854-64; tchr. surgery Aldersgate St. sch. medicine, London; prof. human anatomy Coll. Surgeons, 1852; Hunterian orator, 1850. Fellow Royal Soc., 1837; pres. Royal Coll. Surgeons Eng., 1863, Med. and Chirurg. Soc., 1859. Author: Operative Surgery, 1851; Hysteria, 1867; also several pamphlets and letters. Died Aug. 15, 1872.

SKINNER, Aaron Nichols, Am. astronomer; b. Boston, Aug. 10, 1845; s. Benjamin Hill and Mercy (Burgess) S.; student Beloit Coll., 1867-68; spl. course in astronomy, U. Chgo., 1867-70; m. Sarah Elizabeth Gibbs, Feb. 9, 1874. Asst. Dearborn Obs., Chgo., 1867-70; asst. astronomer U. S. Naval Obs. Washington, 1870-98; prof. math. USN, 1898—; astronomer in charge of 9-inch Transit Circle, U. S. Naval Obs., 1893-1902; in charge 26-inch equatorial, 1902, 03; div. of equatorials, 1903—; In 1894-95 determined places of 8,824 stars in zone 14 degrees to 18 degrees south declination as a contbn. to the great Star Catalogue of Astronomische Gesellschaft, from 23° south declination to 80° north declination; discoverer 4 variable stars; active participant in all meridian circle work of Naval Obs., 1871— chief U. S. Naval Obs. expdn. to Sumatra to observe total solar eclipse of May 17, 1901; retired from active service with rank of comdr. U. S. N., 1907. Fellow A.A.A.S., Astronomische Gesellschaft, Astron. and Astrophys. Soc. Am. Author: Washington Zone Observations; Katalog der Astronomische Gesellschaft Zone 14° bis 18° (Leipzig); 1908. Died Aug. 14, 1919.

SKINNER, Brian John, geochemist; b. Wallaroo, South Australia, Dec. 15, 1928; s. Joshua Henry and Joyce (Prince) S.; came to U. S., 1951, naturalized, 1964; B.Sc. with honors, U. Adelaide (Australia), 1949; A.M., Harvard, 1952, Ph.D., 1955; m. Helen Catherine Wild, Oct. 9, 1954; children—Adrienne W., Stephanie W., Thalassa W. Geologist, Aberfoyle Tin Mines, 1949-50; lectr. mineralogy and crystallography U. Adelaide, 1955-58; research geochemist U.S. Geol. Survey, Washington, 1958-66, chief br. exptl. geochemistry and mineralogy, 1962-66; prof. Yale, 1966—, chmn. dept. geology and geophysics, 1967—. Dir. Econ. Geology Pub. Co. Fellow Geol. Soc. Am., Mineral. Soc. Am., Soc. Econ. Geologists; mem. Mineral. Soc. London, Mineral. Soc. Can., Geochem. Soc. Research, publs. on measurement thermal expansion in non-isotropic compounds, phase relating among naturally occurring metallic sulfides, polymorphism silica phases, geochemistry of ore deposits. Home: 193 E. Rock Rd., New Haven 06511.

SKINNER, Burrhus Frederic, Am. psychologist; b. Susquehanna, Pa., Mar. 20, 1904; s. William Arthur and Grace (Burrhus) S.; A.B., Hamilton Coll., 1926, Sc.D., 1951; M.A., Harvard, 1930, Ph.D., 1931; Sc.D., N.C. State Coll., 1960; Litt.D., Ripon Coll., 1961; Doctor of Science, University of Chgo., 1967; m. Yvonne Blue, Nov. 1, 1936; children—Julie, Deborah. Research fellow NRC, Harvard, 1931-33, jr. fellow Harvard Soc. Fellows, 1933-36; faculty U. Minn., 1936-45, asso. prof., 1939-45; war research Gen. Mills, Inc., 1942-43; faculty Ind. U., 1944-48, prof. psychology, chmn. dept., 1945-48; prof. psychology Harvard, 1948-57, Edgar Pierce prof., 1958—. Recipient Howard Crosby Warren medal, Soc. for Exptl. Psychologists, 1942, Distinguished Sci. Contbn. award, 1958. Mem. Am. Psychol. Assn., A.A.A.S., Nat. Acad. Scis., Am. Acad. Arts and Scis., Brit. Swedish psychol. socs., Am. Phil. Soc., Phi Beta Kappa, Sigma Xi. Author: Behavior of Organisms, 1938; Walden Two, 1948; Science and Human Behavior, 1953; Verbal Behavior, 1957; (with C. B. Ferster) Schedules of Reinforcement, 1957; Cumulative Record, 1959, 61; (with J. G. Holland) The Analysis of Behavior, 1961; The Technology of Teaching 1968. Developed techniques for exptl. analysis of behavior, particularly operant conditioning; analysis of schedules of reinforcement and their effects; application of exptl. analysis to verbal behavior; technological application of basic analysis to education (teaching machines and programmed instruction), psychotherapy, and design of cultures (Walden Two). Home: 11 Old Dee Rd., Cambridge, Mass. 02138. Office: Harvard U., Cambridge, Mass. 02138.*

SKINNER, Charles Gordon, Am. chemist; b. Dallas, Apr. 23, 1923; s. Charles Grady and Benona (Skiles) S.; B.S., N. Tex. State U., 1943, M.S., 1947; Ph.D., U. Tex., 1953; m. Lilly Ruth Brown, Apr. 15, 1944; children—Robert Gordon, Gary Wayne. Research chemist Celanese Corp. Am., Clarkwood, Tex., 1949-50; spl. instr. Del Mar Coll., Corpus Christi, Tex., 1949-50; research scientist U. Tex., Austin, 1951-53, Eli Lilly postdoctoral fellow, 1954-55; research scientist Clayton Found. Biochem. Inst., Austin, 1956-63; prof. chemistry N. Tex. State U., Denton, 1964—; asso. mem. grad. faculty U. Tex., 1961-64. Mem. Am. Soc. Biol. Chemists, Am. Chem. Soc., Am. Soc. Plant Physiologists, Alpha Chi Sigma, Sigma Xi, Phi Lambda Upsilon. Research, numerous publs. on synthesis and biol. studies of metabolite analogs, relationships between structure and physiol. activity in potential antimicrobial and anti tumor agts., isolation and identification of growth factors; patentee. Home: Route 1, Argyle, Tex. 76226. Office: Dept. Chemistry, N. Tex. State U., P.O. Box 5006, Denton, Tex. 76203.*

SKINNER, G(eorge) William, Am. anthropologist; b. Oakland, Cal., Feb. 14, 1925; s. John James and Eunice (Engle) S.; diploma in Chinese, U. Colo., 1946; B.A. in Chinese Studies, Cornell U., 1947, Ph.D. in Cultural Anthropology, 1954; m. Carol Bagger, Mar. 25, 1951; children—Geoffrey Crane, James Lauriston, Mark Williamson, Jeremy Burr. Asst. prof. sociology Columbia, 1958-60; asso. prof., full prof. anthropology Cornell U., Ithaca, N.Y., 1960-65; sr. specialist Inst. Advanced Projects, East-West Center, 1965-66; prof. anthropology Stanford, 1966—. Mem. Internat. Com. on Chinese Studies, 1963-64; dir. London-Cornell Project for Social Research, 1962-65; chmn. subcom. on Chinese soc. Social Sci. Research Council, 1963—; mem. Com. on Scholarly Liaison with Mainland China, 1966—; dir. Chinese Soc. Bibliography Project, 1963—. Fellow Am. Anthrop. Assn., Royal Anthrop. Inst.; mem. Assn. for Asian Studies (past dir.), Am. Acad. Polit. and Social Sci., A.A.A.S., Am. Ethnol. Soc., Siam Soc., Soc. for Applied Anthropology, Asia Soc., Am. Assn. U. Profs., Phi Beta Kappa, Sigma Xi. Author: Chinese Society in Thailand, 1957; Leadership and Power in the Chinese Community of Thailand, 1958; Marketing and Social Structure in Rural China, 1965; also articles. Editor: The Social Sciences and Thailand, 1956; Local, Ethnic, and National Loyalties in Village Indonesia, 1959. Research on assimilation of Chinese immigrant groups in Java, polit. culture of Indonesian Chinese, cycles of social control in Communist China, social structure in Chinese cities--traditional, modernizing and Communist. Home: 1200 Pine St., Palo Alto, Cal. 94301. Office: Dept. Anthropology, Stanford, Stanford, Cal. 94305.*

SKINNER, Morris Frederick, Am. geologist, stratigrapher, paleontologist; b. Springview, Neb., Sept. 14, 1906; s. Frederick Walter and Ezada (Phelps) S.; B.Sc., U. Neb., 1931; m. Shirley Marie White, Oct. 3, 1930; children—Barbara J. Lamb, Morris Frederick. With Frick Corp. (formerly Childs Frick), Pitts., 1927-66, asst. curator Frick Lab., 1956-66, asst. curator Frick Collection, Am. Mus. Natural History, N.Y.C., 1966—. Fellow Geol. Soc. Am., A.A.A.S.; mem. Soc. Vertebrate Paleontologists, N.Y., Neb. acads. sci., Sigma Xi. Systematic research on horses world, anatomy and evolution, stratigraphic sects. fossil localities; correleted fossil deposits in Neb., S.D., Wyo., Tex., N.D. Home: 34 W. 65th St., N.Y.C. 10023. Office: Am. Mus. Natural History, 77 Central Park W., N.Y.C. 10024.*

SKIPSKI, Vladimir P(avlovich), biochemist; b. Ugrojedy, Ukraine, Russia, Oct. 18, 1913; s. Pavel Georgievich and Anna (Novicka) S.; grad. Kiev State U., 1938, Inst. Exptl. Biology and Pathology, Kiev, 1941; Ph.D., U. So. Cal., 1956; m. Irina A. Lysloff, Sept. 5, 1959. Came to U. S., 1949, naturalized, 1956. Sr. research biochemist Inst. Exptl. Biology and Pathology, 1941-43; research asso. Kaviar Inst., Santa Barbara, Cal., 1949-51; research asso. U. So. Cal., Los Angeles, 1956; research asso. Sloan-Kettering Inst. for Cancer Research, N.Y.C., 1956-60, asso., 1960—; asst. prof. Sloan-Kettering div. Grad. Sch. Med. Scis., Cornell U., N.Y.C., 1961—. Swift & Co. fellow, 1951-55. Mem. Am. Soc. Biol. Chemists, Soc. Gen. Physiologists, Biochem. Soc. (Eng.), N.Y. Lipid Club, Am. Oil Chemists Soc., Sigma Xi. Am. Assn. for Cancer Research, Research on cancer and lipids, chromatography of lipids, lipid composition of serum lipoproteins and cell membranes. Home: 303 E. Prospect Av., Mount Vernon, N.Y. 10553. Office: 145 Boston Post Rd., Rye, N.Y. 10580.*

SKJELBREIA, Lars, civil engr.; b. Skedsmo, Norway, May 27, 1923; s. Anders and Karen (Hornslien) S.; came to U. S., 1945, naturalized, 1965; B.S. cum laude in Civil Engring. (scholar), Bucknell U., 1948; M.S. (fellow), U. Wash., 1949; Ph.D. cum laude (fellow), Cal. Inst. Tech., 1953; m. Evelyn George, June 25, 1950; children—Norman, Randi, Jimmy, Carol. Faculty, Bucknell U., Lewisburg, Pa., 1948; engr. Mont. Water Conservation Bd., Helena, 1949-50; design engr. Sandberg & Serell Corp., Pasadena, Cal., part-time 1951-54; group supr. Cal. Research Corp., La Habra, 1954-59; v.p. engring. Nat. Engring. Sci. Co., Pasadena, Cal., 1959-64; gen. mgr. Sci. Engring. Assos. div. Kaman Aircraft Corp., San Marino, Cal., 1964—. Mem. Am. Soc. C.E., Am. Soc. M.E., Am. Geophys. Union, Council on Wave Research, Engring. Found., Task Com. on Wave Forces, Sigma Xi, Pi Mu Epsilon. Author: Gravity Waves, Stokes' Third Approximation Tables of Functions, 1948. Research on devel. of higher order theories for calculating current flow in ocean waves and resulting forces exerted on fixed or floating objects in ocean. Home: 14845 E. Valeda Dr., La Mirada, Cal. 90638. Office: 2450 Mission St., San Marino, Cal. 91108.*

SKOBELTSYN, Dmitrii Vladimirovich, Russian physicist; b. Nov. 24, 1892; grad. Petersburg U., 1915. Staff, Poly. Inst., Leningrad, USSR, 1916-38; prof. Moscow U., 1940—; staff Physico Tech. Inst., USSR Acad. Scis., 1925-38, dir. Nuclear Physics Inst., 1946-61; dep. to USSR Supreme Soviet, 1954, 58, 62. Mem. Commn. on Fgn. Affairs Soviet Union; became chmn. Com. on Internat. Lenin Prize, 1950. Recipient Stalin Prize, 1951, S. I. Vavilov Gold Medal, 1952, Order of Lenin, 1962. Mem. USSR Acad. Scis. Research and publs. in nuclear physics, cosmic rays; began studies on interaction of substances with gamma rays from radium, 1923; studies in cosmic rays and showers, 1927-29; supported quantum character of Compton effect. Office: A. N. Lebedev Physics Inst., USSR Acad. Scis., Leninski Prospekt, 53, Moscow, USSR.

SKODA, Jan, Czechoslovakian biochemist; b. Nová Vcelnice, Czechoslovakia, Dec. 10, 1925; s. Jan and Anna (Nyvltová) S.; Ph.D., Inst. Chem. Tech., 1955; Dr.Sc., Czechoslovak Acad. Scis., 1964; m. Marie Polanková, Sept. 14, 1949; children—Eduard, Helena. Sr. research scientist Inst. Organic Chemistry and Biochemistry, Czechoslovak Acad. Sci., 1954—; docent Inst. Chem. Tech., Prague, Czechoslovakia, 1963—. Recipient Czechoslovak State prize for work in nucleic acids, 1961. Mem. Czechoslovak Biochem. Soc., Czechoslovak Microbiol. Soc. Co-author monographs. Research and numerous publs. on mechanism of action of antibiotics and antimetabolites; discovered (with others) 6-azauridine (used for control of psoriasis, clin. treatment choriocarinoma, others); contributed to discovery various antimetabolites and the elucidation of their mode of action. Home: 7 Nitranská Prague 3. Office: 2 Flemingovo, Prague 6, Czechoslovakia.*

SKODA, Joseph, Czechoslovakian physician; b. Pilsen, Dec. 10, 1805; med. degree, Vienna, 1831; asst. physician Vienna Gen. Hosp.; became parish doctor, 1839; given charge of ward for diseases of chest Gen. Hosp., 1840, became physician, 1841, prof. clin. medicine, 1846. Author: Abhandlung über Auskultation und Perkussion (modified and completed work of Laënnec), 1839. Reintroduced percussion and auscultation to Vienna, and improved terminology and clarified phys. causes of sounds. Died Vienna, June 13, 1881.

SKODA, Rastislav, Czechoslovakian vet. surgeon; b. Kromeríz, Czechoslovakia, Dec. 10, 1923; s. Ján and Vlasta (Kisnerová) S.; student High Sch. Vet. Medicine, Vienna, Austria, 1943; student U. Brno (Czechoslovakia), 1945-46, MVDr, 1949; student U. Alfort (France), 1946-47; C.Sc., Inst. Virology, Bratislava, Czechoslovakia, 1963; m. Bozena Famfuliková, Aug. 14, 1949; children—Jana and Pavel (twins), Vlasta. Practice vet. medicine, 1949-52; staff dept. virology Vet. Diagnostic Lab., 1953-59; staff Lab. for Exptl. Vet. Medicine, Czechoslovakian Acad. Agrl. Scis., Bratislava, 1959-62, staff Inst. Virology, 1963—. Recipient (with A. Zuffa, I. Brauner) State award of K. Gottwald, 1964. Research and publs. on attenuation of Aujeszky's disease virus in tissue cultures and preparation of live vaccine for swine. Home: 19 Devínska cesta. Office: 3/a pri Botanickej záhrade, Bratislava, Czechoslovakia.*

SKOOG, Douglas Arvid, Am. chemist; b. Willmar, Minn., May 4, 1918; s. Arvid C. and Hilma (Erickson) S.; B.S., Ore. State U., 1940; Ph.D., U. Ill., 1943; m. Judith M. Bone, Oct. 10, 1942; children—James A., Jon D. Research chemist Cal. Research Corp., Richmond, 1943-47; faculty Stanford, 1947—, prof. chemistry, 1962—, asso. exec. head dept., 1961—. Mem. Sigma Xi, Phi Kappa Phi, Phi Lambda Upsilon. Author: Fundamentals of Analytical Chemistry, 1963; Analytical Chemistry: An Introduction, 1965. Research, publs. in analytical and inorganic chemistry. Home: 719 Mayfield St., Stanford, Cal. 94305.*

SKOOG, Folke (Karl), plant physiologist; b. Fjäras, Sweden, July 15, 1908; s. Karl Gustav and Sigrid (Person) S.; B.S., Cal. Inst. Tech., 1932, Ph.D., 1936; Ph.D. (hon.), U. Lund (Sweden), 1956; m. Birgit Anna Lisa Bergner, January 31, 1947; 1 daughter, Karin. Came to U. S. 1925, became naturalized, 1935. Teaching asst., research fellow biology Cal. Inst. Tech., 1934-36; NRC fellow U. Cal., 1936-37, summer 1938; instr., tutor biology Harvard, 1937-41, research asso., 1941; asso., asso. prof. biology Johns Hopkins, 1941-44; chemist Q.M.C., also tech. rep., U. S. Army, ETC, 1944-46; asso. prof. botany U. Wis., 1947-49, prof., 1949—; vis. physiologist Pineapple Research Inst., U. Hawaii, 1938-39; asso. physiologist Nat. Insts. Health, USPHS, 1943; vis. lectr. Washington U., 1946, Lantbrukshogskolan, Ultuna, Sweden, 1952. Vice pres. physiol. sect. Internat. Bot. Congress, Paris, 1954, 64. Recip. Stephen Hales Award, Am. Soc. Plant Physiologists, 1954, certificate of merit Am. Bot. Soc., 1956. Mem. Nat. Acad. Scis. U. S., Bot. Soc. Am. (chmn. physiol. sect. 1954-55), Am. Soc. Plant Physiologists (v.p. 1952-53, pres., 1957-58), Scandinavian Soc. Plant Physiologists, Soc. Study Growth and Development, Am. Soc. Gen. Physiologists (v.p. 1956-57, pres. 1957-58), Am. Soc. Naturalists, Am. Society Biological Chemists, London Soc. Vis. Scientists, American Academy Arts and Sciences. Editor: Plant Growth Substances, 1951. Contbr. articles profl. jours. Patentee in field. Research on plant growth and development. Home: 2134 Chamberlain Av., Madison, Wis.

SKORYNA, Stanley Constantine, med. researcher, physician; b. Warsaw, Poland, Sept. 4, 1920; s. Constantine Gregory and Aleksandra Lydia (Fabian) S.; M.D., U. Vienna, 1943, Ph.D., 1962; M.S., McGill U., 1950; m. Halina Irene Grygowicz, Sept. 19, 1945; children—Christopher George Stanley. Faculty, McGill U., Montreal, Que., Can., 1957—, dir. Gastro-Intestinal Research Lab., asso. prof., 1959—; dir. Canadian Med. Expdn. to Easter Island, 1964-65; practice medicine, specializing in surgery. Medallist Royal Coll. Phys. and Surgs. Can., 1957; Outstanding Achievement award McGill U., 1966. Editor: Pathophysiology of Peptic Ulcer, 1963. Research, numerous publs. on exptl. prodn. of tumors systemic carcinogenesis, physiology of pylorus, pathophysiology of peptic ulcer, induction of tumors by radioactive strontium, effects of cortisone, intestinal binding of strontium by seaweed derivatives, mucolytic activity of amides, intestinal absorption of metal ions, natural gastric flora. Home: 14 Forden Av., Westmount, Que. Office: Donner Bldg. for Med. Research, McGill U., Montreal, Que., Can.*

SKRAUP, Zdenko Hans, chemist; b. 1850; became prof., Graz, Austria, 1886, Vienna, 1906; research on proteins, cocaine, morphine, maleic and fumaric acids, composition of quinine and cinchonine; synthesized quinoline and established its molecular structure; determined molecular weights of starch and cellulose; discovered acetyl derivatives of cellulose and formation of cellobioses. Died 1910.

SKRYABIN, Konstantin Ivanovich, Russian parasitologist; b. St. Petersburg (now Leningrad), Russia, Dec. 7, 1878; grad. Yurev Vet. Inst., Russia, 1905; Dr. Vet. Sci., 1934; M.D., 1938; Dr. Biol. Sci., 1943. Veterinarian Chimkent, Kazakhstan, 1905-07, Aulie-Ata (now Dzhambul), Kazakhstan, 1907-11; mem. univs. Koenigsberg (now Kaliningrad, Russia) and Neuchatel (Switzerland), also Alfort Vet. Sch., Paris, France, 1912-14; joined Central Lab. of Main Vet. Bd., Russia, 1912, asso., 1915-17, also head prof. vet. medicine and zoohygiene Stebutov Higher Agrl. Courses for Women, Russia; prof. parasitology Don Vet. Inst., Novocherkassk, USSR, 1917-20; head helminthology dept. All-Union (USSR) Inst. Exptl. Vet. Med., 1920-25; head helminthology dept. Central Tropical Inst., USSR, 1921-49; dir. All-Union Inst. Helminthology, 1931-33, founder lab. for study helminthiases in fur animals, 1926, founder phytohelminthology lab., 1933; head prof. parasitology Moscow (USSR) Vet. Inst., 1920-25, dean of Inst., 1922-23, prof. parasitology mil. vet. faculty, 1933-41; head prof. parasitology Leningrad Vet. Inst., 1925-27; head parasitology Moscow Zootech Inst. (formerly Moscow Zoovet. Inst.), 1927-48; prof. parasitology Moscow Vet. Acad. since 1943; mem. Lenin All-Union Acad. Agrl. Sci., 1935 (v.p. 1956-61), USSR Acad. Med. Sci., 1944, USSR Acad. Sci., 1939 (chmn. presidium Kirgiz br. 1943-52, founder study commn. on helminthofauna, 1922, reorganized as All-Union Helminthology Soc., biology dept, 1940); founder helminthology depts, tropical and sci. insts. republics USSR, 1923—. Named Hon. Sci. Worker, USSR, 1927, Kirgiz, 1945, Hero Socialist Labor, 1958; recipient Stalin prizes, 1941, 50, Lenin prize, 1957, Mechnikov gold medal USSR Acad. Sci., orders of Lenin, Red Banner Labor, Red Star. Author: (with R. S. Shul'ts) Helminthosis of Horned Cattle and Its Young Stock, 1937, Basis of Helminthology, 1940, Trychostrongyloidea of Man and Animals, 1954; Ascarides and Their Significance in Medicine and Veterinary Medicine, 1925; An Anatomical and Biological Outline of Tapeworms, 1927; Helminthology, 1927; Worm Invasions in Sheep and Their Significance in the Economics of Sheep-Breeding, 1931; Veterinary Parasitology and Invasive Disease in Domestic Animals, 1939; Trematodae in Animals and Man, vols. 1-12, 1947-56; (with others) Trematodology, Nematology, 1931, Filariasis of Animals and Man, 1948. Research on parasitic worms and methods of controlling helminthosis; directed 300 helminthological expdns. with Shul'ts; developed concept of transit hosts and additional reservoir; analyzed migrations and classified 200 varieties of bladder worms; introduced geo- and bio-helminthosis concepts. Address: Vsesoyuzhnaya Akademiya Selskokhozyaystvennykh Nauk, B. Kharitonyevskiy p. 21, Moscow, USSR.

SKUD, Bernard Einar, Am. marine biologist; b. Ironwood, Mich., Jan. 31, 1927; s. Ferdinand and Elma (Hendrickson) S.; B.S., U. Mich., 1949, M.S., 1950; postgrad. U. Wash.; m. Patricia Duffin, Aug. 20, 1950; children—Timothy, Ferd, Eric. Fishery research biologist U. S. Bur. Comml. Fisheries, Wash., Alaska, Me., Tex., 1950-61, Boothbay Harbor, Me., 1961—, lab. dir., 1961—. Mem. Internat. Commn. for N.W. Atlantic Fisheries, Internat. Passamaquoddy Fisheries Bd. Mem. Am. Soc. Ichthyologists and Herpetologists, Am. Soc. Zoologists, Am. Inst. Fishery Research Biologists, Am. Soc. Limnology and Oceanography, Am. Fisheries Soc. (asso. editor, book rev. editor). Research, and publs. on Pacific salmon, Alaska herring, Estuarines, Atlantic herring, behavior of marine organisms during solar eclipse, populations dynamics of marine organisms, lobsters. Home: P.O. Box 92, West Boothbay Harbor 04575. Office: U. S. Bur. Comml. Fisheries Biol. Lab., Boothbay Harbor, Me. 04575.*

SKULTETY, Francis Miles, Am. physician; b. Rochester, N.Y., June 6, 1922; s. Frank John and Hazel (von Kaitz) S.; B.S., U. Rochester, 1944, M.D., 1946; Ph.D., U. Ia., 1958; m. Constance Theresa Schmitt, Dec. 26, 1945; children—Miles Christian, John Scott, William Kent. Practice medicine, specializing in neurosurgery, Iowa City, 1952-66, Omaha, 1966—; faculty Coll. Medicine U. Ia., 1952-66, prof., 1963-66; Shackelford prof. neurosurgery, neuroanatomy Coll. Medicine U. Neb., Omaha, 1966—; com. mem. Nat. Inst. Neurol. Diseases and Blindness, 1966—; cons. neurosurgery Omaha VA Hosp. Fellow A.C.S.; mem. A.M.A., Harvey Cushing Soc., Am. Assn. Anatomists, Congress Neurol. Surgeons, Soc. Exptl. Biology and Medicine, Assn. For Research Nervous and Mental Diseases. Contbr. chpts. (with others) to Disability and the Law, 1962. Research on structure and function of midbrain and its relationship to other parts of brain, control of water and food intake by brain; discovered that exptl. damage to medial portions of midbrain results in a marked persistent increase in food intake and lesions in anterior hypothalamus result in a decrease in cats. Home: 1503 S. 83d St., Omaha 68124.*

SKYRING, Alan Paine, Australian physician; b. Cooktown, Australia, Sept. 30, 1926; s. Frank Reginald and Lucy (Paine) S.; M.B., Sydney (Australia) U., 1953; postgrad. Johns Hopkins Med. Sch., 1958-59; m. Shirley Ann Hatfield, Apr. 3, 1952; children—Keren Ann, Timothy Alan. Resident med. officer Royal Prince Alfred Hosp., Sydney, Australia, 1953-58, dir. A.W. Morrow dept. gastroenterology, 1959—; vis. fellow Johns Hopkins Hosp., 1958-59. Mem. Royal Australian Coll. Physicians (councillor 1965—), Australian Soc. for Med. Research (founder, past hon. sect.-treas., past pres.), Australian Gastroent. Soc., Australian Physiol. Soc. Editor: Modern Medicine of Australia, 1960—. Research, publs. on natural history of ulcerative colitis, gastric ulcer, disorders of absorption. Home: 133 Railway Pde Pennant Hills, N.S.W. Office: Royal Prince Alfred Hosp., Sydney, N.S.W., Australia.*

SLABAUGH, Wendell Hartman, Am. chemist; b. Nappanee, Ind., July 10, 1914; s. Floyd B. and Dinah (Hartman) S.; B.A., N. Central Coll., Naperville, Ill., 1936; M.S., N.D. State U., 1938; Ph.D., Wash. State U., 1950; m. Lois Bergeman, Aug. 16, 1939; children—Jane Louise, Peter Bergeman. Instr., Gogebie Jr. Coll., 1938-41, N.D. State U., 1941-43; fellow Mellon Inst., 1943; chemist Sherwin-Williams Co., 1944-49; instr. Wash. State U., 1947-50; asst. prof. Kan. State U., 1950-53; faculty Ore. State U., Corvallis, 1953—, prof., 1958—, asso. dean, 1961—. Cons. to industry in surface chemistry. Recipient spl. award in ordnance devel. USN, 1945. Mem. Am. Chem. Soc. (chmn. div. chem. edn. 1966), A.A.A.S., Am. Inst. Chemists, N.Y. Acad. Sci., Ore. Acad. Sci., Sigma Xi, Phi Kappa Phi, Phi Kappa Delta, Phi Lambda Upsilon. Author: College Physical Science, 1957, 65; General Chemistry, 1964, 66. Research in surface chemistry of clays. Home: 3435 Grant St., Corvallis, Ore. 97330.*

SLABY, Adolph Karl Heinrich, German physicist; b. Berlin, Apr. 18, 1849; prof., Berlin, 1883-1912; became dir. Electro-Tech. Sch., Berlin, 1884; Mem. Patent Ct., 1880-85. Author: Die Funkentelegraphie, 1897; Die Neuen Fortschritte auf dem Gebiete der Funkentelegraphie, 1901. Assisted Marconi and Arco in devel. wireless telegraph in Eng., then worked out system in Germany. Died Berlin, Apr. 6, 1913.

SLABY, O., Czechoslovakian embryologist, entomologist, b. Havlíchuv Brod, Czechoslovakia, May 26, 1913; s. Eduard and Barbora (Erazim) S.; MUDr., Med. Faculty, Charles U., Prague, Czechoslovakia, 1939; RNDr., Biol. Faculty, 1947, D.Sc., 1956; m. Roubalová Kveta, 1941; children—Ivan, Pvael, Otto. Faculty, Med. Faculty, Plzen, Czechoslovakia, prof. histology and embryology, dir. Inst. Histology and Embryology. Recipient diploma of honor and decoration Charles U. Mem. Internat. Embryological Inst. Utrecht, Zool. Soc. (mem. com.), Czechoslovak Soc. Entomologists. Research, numerous publs. on comparative embryology of vertebrates, including extremities, skull, organ of Jacobson, developmental anomalies; entomology, including lepidopterology, especially Zygaenidae and Parnassidae, zoogeography. Home: 13 K. Svetlé, Plzen, Czechoslovakia.*

SLACK, Glen Alfred, Am. physicist; b. Rochester, N.Y., Sept. 29, 1928; s. A. and Edith (Zohe) S.; B.S., Rensselaer Poly. Inst., 1950; Ph.D., Cornell U., 1956; m. Nancy Guttmann, Dec. 16, 1951; children—Margaret, David, Jonathan. Physicist, Gen. Electric Co., Schenectady, 1956—; vis. scholar Clarendon Lab. Oxford (Eng.) U., 1966—. John Simon Guggenheim Meml. fellow, 1966—. Fellow Am. Phys. Soc. Research, publs. on solid state physics and chemistry concerned with heat transport in solids at low and high temperatures, crystal growth, optical properties and magnetic properties of solids and sound wave propagation in crystals. Home: Box 225, Ridge Rd., Scotia, N.Y. 12302. Office: Gen. Electric Research Lab., P.O. Box 8, Schenectady 12301.*

SLACK, Lewis, Am. physicist; b. Phila., Apr. 15, 1924; s. Lewis and Martha (Fitzgerald) S.; S.B., Harvard, 1944; Ph.D., Washington U., St. Louis, 1950; m. Sarah Hunt Wyman, Dec. 29, 1948; children—Elizabeth Wyman, Susan Towne, Christopher Morgan. Physicist, Naval Research Lab., 1950-54; faculty George Washington U., 1954-62, prof., 1957-62, chmn. dept., 1957-61; asst. exec. sec. div. phys. scis. Nat. Acad. Scis.-NRC, Washington, 1962-67; Asso. dir. Am. Inst. Physics, N.Y.C., 1967—. Mem. A.A.A.S. (mem. coop. com. on teaching sci. and math. 1963—), Washington Acad. Sci., Philos. Soc. Washington, Am. Phys. Soc., Am. Assn. Physics Tchrs., Sigma Xi. Research, publs. on nuclear physics, angular correlation, beta and gamma ray spectroscopy. Office: 335 E. 45th St., N.Y.C. 10017.*

SLADE, Hutton Davison, microbiologist; b. London, Eng., Aug. 6, 1912; s. Francis and Florence (Davison) S.; came to U. S., 1913, naturalized, 1920; B.S., U. Md., 1935, M.S., 1936; Ph.D., Ia. State U., 1942; m. Eileen Fay Pryor, June 7, 1941; children—Richard Gary, Robert Bryan. Research biologist Wallerstein Co., N.Y.C., 1942-43; chief microbiology Rheumatic Fever Research Inst., Chgo., 1948-57; mem. faculty Northwestern U. Med. Sch., 1957—, prof. microbiology, 1959—. Cons. Naval Med. Research Unit 4, Great Lakes, Ill., 1958—, Office Naval Research, 1960—; mem. metabolic biology panel NSF, since 1966—; mem. grad. tng. grant com. NIH, 1967—. Mem. sch. bd. 37, Cook County, Ill., 1954-58. Served to maj., Med. Service Corps, AUS, 1943-46; col. Res. Established investigator Am. Heart Assn., 1956-61; recipient Research Career award NIH, 1962. Mem. A.A.A.S., Am. Acad. Microbiology, Am. Soc. Microbiology, Soc. Gen. Microgiology, Ill. Soc. Microbiology, Sigma Xi. Investigation of microbial physiology; cell structure, nutrition; enzymatic lysis; assimilation of carbon dioxide; nature of streptococci. Home: 3804 Lake Av., Wilmette, Ill. 60091. Office: 303 E. Chicago Av., Chgo. 60611.

SLADECEK, Vladimir, Czechoslovakian hydrobiologist; b. Horovice, Czechoslovakian, Jan. 17, 1924; s. Antonin and Anna (Horova) S.; student Faculty Sci., Charles U., Prague, Czechoslovakia, 1945-49; Rerum Naturalium Dr., Tech. U., Faculty Civil Engring., 1950, Candidate Tech. Scis., 1955; m. Alena Vinnikova, June 14, 1956; children—Jan, Petr, Helena. Staff, Hydraulic Research Inst., Prague, 1949-50; faculty dept. water tech. Inst. Chem. Tech., Prague,

1953—, lectr. hydrobiology, 1956—. Expert, WHO, Geneva, Switzerland, 1964—. Mem. Am. Soc. Limnology and Oceanography, Freshwater Biol. Assn. Author: (with Zelinka) Hydrobiology for Water Economy, 1964; also numerous articles, Editor: Internat. Assn. Limnology 1965—. Research on storage reservoirs, ponds, activated sludge, exptl. lagoon, sewage and indsl. wastes, ecol. studies of freshwater plankton; definition of new degrees; devel. system of water quality including saprobity, toxicity, and radioactivity. Home: 3 Havlovického, Praha Czechoslovakia.*

SLANSKY, Cyril Method, Am. chemist; b. Albuquerque, July 8, 1913; s. Joseph and Elizabeth (Kaba) S.; B.S., Coll. Ida., 1936; Ph.D., U. Cal. at Berkeley, 1940; m. Elvera Alice Tewell, June 7, 1939; children—Richard Cyril, Joanne Elizabeth, Marilyn Jeanne (Mrs. Robert Allen Hughes). Research chemist, chem. engr. Dow Chem. Co., Midland, Mich., 1940-44, research chemist, Pittsburg, Cal., 1944-47; group leader chem. research Gen. Electric Co., Hanford Works, Richland, Wash., 1947-52; head Works Lab., Am. Cyanamid, Idaho Falls, Ida., 1952-53; br. head chem. devel. Phillips Petroleum Co., Idaho Falls, 1953-66; tech. staff nuclear and chem. tech. Ida. Nuclear Corp., Idaho Falls, 1966—. Mem. Am. Chem. Soc., Am. Inst. Chem. Engrs., Am. Nuclear Soc. Contbg. author: Chemical Processing of Reactor Fuels, 1961. Research, publs. on chem. thermodynamics, prodn. magnesium from sea water. Patentee in magnesium tech.; recovery of uranium and plutonium from nuclear fuel, radioactive waste disposal. Home: 2815 Holly Pl. Office: P.O. Box 1845, Idaho Falls, Ida. 83401.*

SLARE, Frederick (or Slear), English physician, chemist; b. Northamptonshire, Eng., circa 1647; M.D., Oxford U., 1680; practiced medicine, London; fellow Royal Soc., 1680, mem. council 1682; fellow Royal Coll. Physicians, censor, 1692, 93, 1708, elector, 1708, also mem. council. Author: Experiments upon Oriental and Bezoar-Stones, 1715; also papers. Demonstrated expts. on spermatoza before Royal Soc., 1679; investigated phosphorus; demonstrated presence of salt in blood; repeated some of Boyle's expts. with ammoniacal copper solutions in which air was absorbed with accompanying change in color; showed that calculi are chemically unlike tartar, 1713; disproved miraculous virtues of animal calculi, 1717; defended inoculation. Died Sept. 12, 1727.

SLATER, George, English geologist; b. Sharow, Eng., 1874; ed. St. John's Coll., York, Imperial Coll. Sci., London; M.Sc., D.Sc., U. London; m. Anne Irwin, 1897; Asst. master, Hartwhistle, Northumberland, 1895-97, Ispwich, Eng., 1897-1918; demonstrator, asst. lectr. geology Imperial Coll. Sci. and Tech., 1918-39; glaciologist Oxford U. expdn. to Spitsbergen, 1921. Recipient Murchison fund Geol. Soc. London, 1928, Foulerton award Geologists' Assn., 1950. Research and publs. on structures in disturbed drift deposits. Died Jan. 27, 1956.

SLATER, John Clarke, Am. physicist; b. Oak Park, Ill., Dec. 22, 1900; s. John Rothwell and Katharine Southland (Chapin) S.; A.B., U. Rochester, 1920, Sc.D., 1964; A.M., Harvard, 1922, Ph.D., 1923, Sc.D., Ripon, 1946; studied Cambridge and Copenhagen, 1923-24; m. Helen Frankenfeld, 1926; children—Louise Chapin, John, Clarke; m. 2d Rose Mooney, 1954. Instr. physics Harvard, 1924-26, asst. prof., 1926-29, asso. prof., 1929-30; prof. physics in charge dept. Mass. Inst. Tech., 1930-51, inst. prof. physics, 1951-66; staff Radiation Lab., 1940-45; grad. research prof. physics and chemistry U. Fla., Gainesville, 1964—; fellow John Simon Guggenheim Meml. Found., Leipzig, 1929; mem. staff Inst. for Advanced Study, Princeton, N.J., 1937. Tech. staff Bell Telephone Labs., 1943-44; staff mem. Brookhaven Nat. Lab., 1951-52. Mem. Am. Phys. Soc., Am. Acad. Arts and Scis., Nat. Acad Scis., Am. Philos. Soc., Phi Beta Kappa, Delta Upsilon. Author: (with N. H. Frank) Introduction to Theoretical Physics, 1933; Introduction to Chemical Physics, 1939; Microwave Transmission, 1942; (with N. H. Frank), Mechanics, 1947, and Electromagnetism, 1947, Microwave Electronics, 1949; Quantum Theory of Matter, 1951; Modern Physics, 1955; Quantum Theory of Atomic Structure, vols. I and II, 1960; Quantum Theory of Molecules and Solids, Vol. I, 1963, Vol. II, 1965, Vo. III, 1966. Research and publs. on quantum theory of atoms, molecules and crystals, developed determinatal method for many electron problems, atomic orbitals for atoms, augmented plane wave method for energy bands in crystals. Home: 623 S.W. 27th St., Gainesville, Fla. 32601.*

SLATER, John Vernon, physiologist; b. Barrow-in-Furnace, Eng., Aug. 3, 1920; s. John Sheridan and Francis (Clow) S.; came to U.S., 1924, naturalized, 1935; A.B., Wayne U., 1946, M.S., 1947; Ph.D., U. Mich., 1951; m. Nina Florence Medlyn, Dec. 24, 1952; children—Arthur James, Kathleen Marilyn, Richard John, Barbara Jean. Asst. prof. U. Fla., Gainesville, 1952-55, U. Buffalo, 1955-57; asso. prof. U. Ariz., Tucson, 1958-61; asso. prof. U. Cal. at Berkeley, 1961—; asso. research biophysicist Donner Lab., 1963—. Cons. govt. agys. Mem. A.A.A.S., Am. Soc. Zoologists, Soc. Protozoology, Sigma Xi, Phi Sigma, Delta Omega. Research, publs. in ecology, comparative physiology, radiation biology, physiol., zoology. Home: 333 Peppertree Rd., Walnut Creek, Cal. 94597. Office: U. Cal., Berkeley, Cal. 94720.*

SLATER, Samuel, inventor; b. Belper, Eng., June 9, 1768; s. William and Elizabeth (Fox) S.; m. Hannah Wilkinson, Oct. 2, 1791; m. 2d, Esther Parkinson, Nov. 21, 1817; 9 children; apprentice in cotton bus., 1783; came to U. S. in response to search of R.I. legislature for experienced textile men, 1790; designed from memory Brit. machines for spinning cotton; built factory, Rehoboth, Mass., 1801; founder Village of Smithfield (later Slaterville), R.I., 1806; founder mills, Oxford (later Webster), Mass., 1812, began mfg. woolen cloth, 1814. A founder Cotton Industry. Died Webster, Apr. 21, 1835.

SLATER, Trevor Frank, English biochemist; b. London, Eng., Feb. 18, 1931; s. Samuel William Frank and Eliza Florence (Lock) S.; B.Sc. with honours in Chemistry, U. Coll., London, 1952, M.Sc., 1953, Ph.D., 1956, D.Sc., 1967; m. Hazel King, Mar. 23, 1961; children—Andrew Frank Guiscard, David John Trevor. Agrl. Research council grantee U. Coll., 1953-56, Beit Meml. Research fellow, 1956-59, Beit Meml. 4th Year fellow, 1959-60; lectr. U. Coll. Hosp. Med. Sch., London, 1960-64, sr. lectr., 1964—. Vacation cons. Ministry Agr., Fisheries and Food, 1966—. Fellow Royal Inst. Chemistry; mem. Biochem. Soc., Inst. Biology. Research, numerous publs. on biochem. disturbances in liver injury including necrosis, porphyrias, biliary tract interruptions, nucleotide changes in tissue damage. Home: 54 Bush Grove, Stanmore, Middlesex, Eng. Office: University Coll. Hosp. Med. Sch., London, Eng.*

SLÄTIS, Hilding Sören Eugen, Swedish physicist; b. Korsholm, Sweden, Feb. 12, 1902; s. Otto Wilhelm and Maria (Bengs) S.; ed. U. Helsingfors, also Abo Acad.; Dr.ès. sc.; m. Maja Svaetichin, Aug. 27, 1929; children—Otto, Anders. Prof., Abo Acad., 1946; now physicist Nobel Inst. Nuclear Physics, Stockholm. Decorated Order Polar Star. Mem. Swedish Nat. Physics Com. Author: Einfluss der Zuleitungen des Kondensators bei Messungen im Lechersystem, 1938; Optical Properties of Magnetic Lenses, 1945; (with Kai Siegbahn) Intermediate Image Beta-Ray Spectrometer, 1949; A Permanent Magnet Beta-Ray Spectrography, 1953; Intensity Determination of Photographically Recorded Conversion Lines II, 1962; others. Home: Kungl. Vetenskapsakademien, Stockholm 50. Office: Forskningsinstitutet för Atomfysik, Stockholm 50, Sweden.

SLATYER, Ralph Owen, Australian biologist; b. Melbourne, Australia, Apr. 16, 1929; s. Thomas Henry and Jean (Mackenzie) S.; B.Sc., U. Western Australia, 1951, M.Sc., 1954, D.Sc., 1959; m. June Helen Wade, May 16, 1953; children—Anthony James, Bethanne, Judith Jeanne. With div. land research Commonwealth Sci. and Indsl. Research Orgn., Canberra, Australia, 1951-67, chief research scientist, asso. chief div., 1966-67; prof. biology Research Sch. Biol. Scis., Australian Nat. U. Canberra, 1967—; research fellow Duke dept. botany, 1955-56, vis. prof., 1963-64. Liaison officer Australian Arid Zone Research, 1960-67. Recipient Edgeworth David medal, 1960. Mem. Am. Soc. Plant Physiology, Royal Meteorol. Soc., Australian Inst. Agrl. Sci., Australian Soc. Plant Physiology, Australian Soc. Soil Sci. Author: (with I. C. McIlroy) Practical Microclimatology, 1961; Plant-Water Relationships, 1967; also articles. Research on water utilization by plants, particularly phys. and physiol. mechanisms asso. with water transport through plants, effects of water shortage on plant prodn.*

SLAUCITAJS, Leonids, geophysicist; b. Jaunlaicene, Latvia, Apr. 10, 1899; s. Janis and Emilija (Berzonis) S.; Mag.math., U. Latvia, Riga, 1925, Habil.geoph., 1931, Dr.math., 1942; Dr.math., U. Stuttgart (Germany), 1948; D.Sc., Sydney (Australia) U., 1964; F.A., Conservatory of Music, Riga, 1933; m. Milda Hartmanis, Apr. 28, 1924; children—Tatjana (Mrs. Bracs), Andis A. Faculty, Latvian U., Riga, 1925-44, prof., dir. Inst. Geophysics and meteorology, 1940-44; prof., dean, rector Baltic U., Hamburg, Germany, 1946-48; prof., head dept. terrestrial magnetism Nat. U. La Plata (Argentina), 1949-68; Distinguished visitor U. Hawaii Inst. Geophysics, Honolulu, 1966-68. Head magnetic groups Argentine Antarctic Expdn., 1950-57; hon. asso. in applied math. U. Sydney. Recipient Three Stars Order, Latvia; Independency War medal and Commemorative medal, Latvia; also awards for papers. Fellow Australian-New Zealand Assn. for Advancement Sci., Deutsche Geophysir. Gessellschaft, N.Y. Acad. Scis. Author: Textbook for a University Course; Magnetismo terrestre, 1961; Oseanographie des Rigaischen Meersbusens, 1948; El conocimiento geomagnético de la Antartida Sudamericana, 1957; also numerous articles. Research on geomagnetic field in Baltic Sea and Latvian ter. which indicate its anomalistic character; Antarctic magnetic studies in anomalistic field and its variations; water layering investigations of Riga Gulf and Latvian lakes; global investigations on magnetic storms, daily variations and on magnetic secularvaration, including discovery of 50-year period. Home: 11274 Allendale Dr., Arvada, Colo. 80002.*

SLAUNWHITE, Wilson Roy, Jr., Am. biochemist; b. Waltham, Mass., Sept. 25, 1919; s. Wilson Roy and Grace (Smith) S.; B.S., Mass. Inst. Tech., 1941, M.S., 1942, Ph.D., 1948; m. Phyllis E. Perry, July 17, 1942; children—Wilson Roy III, William David, Peter Wayne, Michael Douglas. Research asso. Mass. Inst. Tech., 1942-45; with Naval Research Labs., Boston, 1945-46; research fellow Mass. Gen. Hosp., Boston, 1948-52; Damon Runyon fellow U. Utah, 1952-53; cancer research scientist Roswell Park Meml. Inst., Buffalo, 1953-67; research dir. Med. Found. Buffalo, 1967—; prof. biochemistry State U. N.Y. at Buffalo. Am. Cancer Soc. fellow, 1946-48. Mem. Am. Soc. Biol. Chemists, Am. Chem. Soc., Endocrine Soc., Soc. for Study Reprodn. Editor: Steroids, 1964—, Jour. Clin Endocrinology and Metabolism, 1968—. Research, publs. on biosynthesis of testosterone, metabolism, protein binding, microanalysis of steroids. Home: Pleasant Av., Lake View, N.Y. 14085. Office: 73 High St., Buffalo 14203.*

SLAVIK, Bohdan, Czechoslovakian botanist; b. Hradec Kralové, Czechoslovakia, Oct. 30, 1924; s. Cenek and Ruzena (Sklenarova) S.; Rer.nat.Dr., Charles U., Prague, Czechoslovakia, 1950; C.Sc., Inst. Biology, Czechoslovak Acad. Sci., Prague, 1954, Dr.-Sc., Inst. Exptl. Botany, 1967; m. Jirina Vesela, July 2, 1949; children—Jan, Dana. Sci. worker Inst. Biology Czechoslovak Acad. Sci., Prague, 1954-61, chief dept. plant physiology, 1958-61, chief Lab. Water Relations and Photosynthesis, Inst. Exptl. Botany, 1962—; spl. lectr. natural sci. faculty Charles U., 1960—; coordinator plant physiology research in Czechoslovakia, 1959—. Asso. mem. productivity process sectional com. Internat. Biol. Pragramme, 1964—. Editor, contbg. author: Methods in Studying Plant Water Relations, 1965; editor Water Stress in Plants, 1965; exec. editor Biologia Plantarum (jour. exptl. botany), 1959—. Research, publs. in water relations of plants, physiology and biophysics of water deficit in plants, distbn. water in plant tissues and inside cell; proved extrastomatal influence of hydration level on photosynthetic rate, influence of water deficit on stomatal transpiration rate. Home: 23 Karoliny Svetlé, Praha 1. Office: 2 Flemingovo, Praha 6, Czechoslovakia.*

SLAWSKY, Zaka Israel, Am. physicist; b. Bklyn., Apr. 2, 1910; s. Simon and Mollie (Brimberg) S.; B.S., Rensselaer Poly. Inst., 1933; M.S., Cal. Inst. Tech., 1935; Ph.D., U. Mich., 1938; m. Dorothy Altman, Jan. 19, 1945; children—Albert Altman, Richard Charles. Faculty, Bklyn. Coll., 1939-40, Union Coll., Schenectady, 1940-41; staff Naval Ordnance Lab., White Oak, Md., 1941—, chief physics research, 1960—; faculty U. Md., College Park, 1946—, prof., 1964—. Treas., Joint Bd. on Sci. Edn., Washington Acad., 1956—. Recipient Meritorious Civilian Service award USN, 1952, 58. Fellow Am. Phys. Soc., Washington Acad. Sci.; mem. N.Y. Acad. Scis., Washington Philos. Soc. Research, publs. on rotational fine structure from centrifugal distortion of symmetric molecules, effect of molecular structure on high speed gas dynamics, effect of electron sharing on relaxation times of molecules. Home: 9813 Belhaven Rd., Bethesda, Md. 20034. Office: White Oak, Silver Spring, Md. 20910.*

SLEAR, Frederick, see Slare, Frederick.

SLECHTA, Robert Frank, Am. biologist; b. N.Y.C., June 4, 1928; s. Frank C. and Helen (Pospisil) S.; A.B., Clark U., 1949, M.A., 1951; postgrad. Columbia, 1951-52; Ph.D., Boston U., 1955; m. Betty S. Youngren, May 16, 1953; 1 son, Marc William. Research asst. Worcester Found., Shrewsbury, Mass., 1952-53; biologist U.S. Army Med. Nutrition Lab., Denver, 1953-55; research asso., instr. Tufts U., Medford, Mass., 1955-58; faculty Boston U., 1958—, prof. biology, 1965—, asso. dean Graduate School, since 1967—. Mem. Boston ZZool. Soc. (exec. com., dir. 1964—), Am. Inst. Biol. Sci., A.A.A.S., Microcirculation Soc., Am. Soc. Zoologists, Sigma Xi. Author: (with M. Hawthorne and E. Blaustein) Laboratory Manual for General Biology, 1965; also articles. Research in limb regeneration in urodeles, starvation in prisoners of war, human factors in aircraft seating, effects of progestational compounds on reprodn. (early work on contraceptive pill), quantitative studies of blood flow in living microscopic vessels in mammals and amphibians. Home: 101 Wilson Rd., Bedford, Mass. 01730. Office: 2 Cummington St., Boston 02215.*

SLEESWYK, André Wegener, metallurgist; b. Bindjei, Indonesia, May 11, 1927; s. Fredrik Wegener and Cornelia H. (van Giffen) S.; student Naval Acad., 1946-48; ir. in Mech. Engring., Technische Hegeschool Delft (Netherlands), 1953; S.M., Mass. Inst. Tech., 1956; Dr. in Math. and Physics, U. Amsterdam (Netherlands), 1961; m. Jaaike van Bork; children—Constance Marjolyn, Frances Corine, Anneke, Marianne; m. 2d, Cornélie Christen, Mar. 10, 1960. Vis. fellow Mass. Inst. Tech., Cambridge, 1954-56; phys. engr. Koninklyke/Shell Laboratorium, Amsterdam, 1956-63; metallurgist Institut de Rederdres de la Sidérurgie Francaise (IRSID), St. Germain-en-Laye (S. & A.), France, 1963-64; prof. phys. engring. U. Gröningen (Netherlands), 1964—. Cons. to pvt. cos. Recipient Brandsma medal Netherlands Metall. Soc., 1962. Mem. Koninklyle Institut van Ingenieurs, Nederlandse Naturkundige Vereniging, Sigma Xi. Research, publs. on mech. engring., including displacement pumps, thermodynamics of liquification of gases, phys. metallurgy including plastic deformation of metals, formation of mech. twins, phase transitions between crystalline phases; devel. testing methods and equipment for mech. testing of materials. Home: 14 Lekstaat, Groningen, Netherlands.*

SLEISENGER, Marvin Herbert, Am. physician; b. Pitts., June 3, 1924; s. Albert and Celia Sleisenger; student Harvard, 1941-44, M.D., 1947; m. Lenore Ruth Cohen, June 27, 1948; 1 son, Thomas Paul. Practice medicine specializing in gastroenterology, N.Y.C., 1951—; with N.Y. Hosp., 1951—, chief div. gastroenterology, 1954—; asso. attending physician Bellevue Hosp., N.Y.C., 1957—; asso. in medicine Harvard Med. Sch., 1949-50; instr. medicine Tufts Coll. Med. Sch., 1949-50; with Cornell U. Med. Coll., 1951—, prof. medicine, 1965—. Cons. M.O.L. Project USAF, 1964, hosp. Rockefeller Inst., 1954—, Montrose VA Hosp., 1960—; mem. tng. grant com. USPHS, 1961—. Mem. A.A.A.S., Am. Fedn. Clin. Research (pres. Eastern sect. 1964—), N.Y. (sec. 1960-62), Am. gastroent. assns., Am. Soc. Clin. Investigation, N.Y. County Med. Soc., N.Y. Acad. Medicine, Cornell Med. Coll. Research Soc. (chmn. program com. 1959-61), Harvey Soc., A.C.P. Asso. editor Cecil-Loeb Textbook of Medicine; editor Gastroenterology, 1965. Research, publs. on mechanism of absorption, elucidation of protein excretion into GI tract, effect of gluten on adult celiac disease and relationship of immunological mechanisms to pernicious anemia and other diseases of GI tract. Home: 3755 Henry Hudson Pkwy., N.Y.C. 10063. Office: 525 E. 68th St., N.Y.C. 10021.*

SLEMMONS, David Burton, Am. geologist, geophysicist; b. Alameda, Cal., Dec. 31, 1922; s. Claude Hayes and Gladys (Hinton) S.; B.S. in Econ. Geology, U. Cal. at Berkeley, 1947, Ph.D., 1953; m. Ruth Marillyn Evans, Sept. 7, 1946; children—David Robert, Mary Anne. Faculty, U. Nev., Reno, 1951—, prof. geology, 1963—, dir. Seismographic Stas., 1952-65, chmn. geology-geography dept., 1966—. Cons. geologist, geophysicist to various mining cos., utilities, geol. engring. firms, govt. agys., 1954—; mem. earthquake investigation team UNESCO, 1965—. Recipient G. K. Gilbert award in seimic geology Carnegie Instn. Washington, 1962. Fellow Geol. Soc. Am.; mem. Seimol, Soc. Am., Am. Inst., Mining, Metallurgy and Petroleum Engring., Am. Geophys. Union, Sigma Xi, Phi Kappa Phi. Research, publs. on historic surface faulting in Basin-and-Range Province, volcanism and geology of Sierra Nev. and Nev., plagioclase feldspar optics, seismicity of Western U. S., ground control for remote sensing of environment. Home: 865 Ryan Lane, Reno 89503.*

SLEPIAN, Joseph, Am. research engr.; b. Boston, Feb. 11, 1891; s. Barnett and Annie (Bantick) S.; A.B., Harvard, 1911, A.M., 1912, Ph.D., 1913; postgrad. Sorbonne U., U. Göttingen, 1914; D.Engring., Case U., 1949; D.Sc., U. Leeds, Eng., 1955; m. Rose Myerson, Nov. 10, 1918; children—Robert Myer, David. Instr. math. Cornell U., 1915; cons. research engr. Westinghouse Research Labs., East Pittsburgh, Pa., 1916-38, asso. dir., 1938-56. Recipient John Scott medal Franklin Inst., 1932; Westinghouse Silver award, 1935. Fellow Am. Math. Soc., Am. Phys. Soc., Am. Inst. Elec. Engrs. (Lamme medal 1943), Edison medal 1948), I.R.E.; mem. Nat. Acad. Scis. Author: Conduction of Electricity in Gases, 1933. Contbr. papers to tech. jours. Patentee circuit interrupters, excitation of rotation machines, mech. rectifiers, de-ion circuit breakers, high voltage fuses, ignitron tubes. Home: 1115 Lancaster St., Pitts. 15218.*

SLETTEBAK, Arne, Am. astronomer; b. Danzig, Aug. 8, 1925; s. Nicolai and Valerie (Janczak) S.; came to U.S., 1927, naturalized, 1932; B.S., U. Chgo., 1945, Ph.D., 1949; m. Constance Loraine Pixler, Aug. 28, 1949; children—Marcia Diane, John Andrew. Asst., Yerkes Obs., U. Chgo., 1945-49; faculty Ohio State U., Columbus, 1949—, prof. astronomy, 1959—, chmn. dept., 1962—. Dir. Perkins Obs., Delaware, O., Ohio State and Ohio Wesleyan univs., 1959—. Mem. steering com. Earth Sci. Curriculum Project, 1965—. Fulbright fellow, Hamburg, Germany, 1955-56. Mem. Asso. Univs. for Research in Astronomy (dir. 1961—), Am. Astron. Soc. (mem. council 1964-67), Internat. Astron. Union. Research in stellar spectroscopy, stellar axial rotation and spectral classification. Home: 601 Seabury Dr., Worthington, O. 43085.*

SLICHTER, Charles Pence, Am. physicist; b. Ithaca, N.Y., Jan. 21, 1924; s. Sumner Huber and Ada (Pence) S.; A.B., Harvard, 1946, M.A., 1947, Ph.D., 1949; m. Gertrude Thayer Almy, Aug. 23, 1952; children—Sumner Pence, William Almy, Jacob Huber, Ann Thayer. Research asst. Underwater Explosives Research Lab., Woods Hole, Mass., 1943-46; faculty U. Ill., Urbana, 1949—, prof. physics, 1955—; Morris Loeb lectr. Harvard, 1961. Alfred P. Sloan fellow. Author: Principles of Magnetic Resonance; also articles. Research in magnetic resonance and solid state physics. Home: 319 Elmwood Rd., Champaign, Ill.*

SLICHTER, Louis Byrne, Am. geophysicist; b. Madison, Wis., May 19, 1896; s. Charles Sumner and Mary Louise (Byrne) S.; A.B., U. of Wis., 1917. Ph.D., 1922, Doctor of Science (honorary), 1967; m. to Martha Merry Buell. October 20, 1926; children—Susan Merry, Mary Louise (Mrs. Ward Whaling). Physicist with the Submarine Signal Corporation, Boston, Massachusetts, 1922-24; geophysical prospecting, Mason, Slichter & Hay, Madison, Wis., 1924-27, Mason, Slichter & Gauld, 1927-31; research asso., Calif. Inst. Tech., 1930-31; asso. prof. geophysics, Mass. Inst. Tech., 1931, prof. 1932-45; professor of geophysics, Univ. of Wis., 1946-47; dir. Inst. Geophysics, University of California, 1947-62, prof. geophysics, 1962, emeritus, 1963. Mem. div. 6. Office of Scientific Research and Devel., 1942-45. Recipient Presidential Certificate of Merit, 1948; Distinguished Service citation Univ. Wisconsin, 1957; Daniel C. Jackling award, Am. Institute Mining, Metall. and Petroleum Engrs., 1960; Alumnus of Yr. award Joint Wis. Alumni Assns. of So. Cal., 1963, William Bowie medal Am. Geophys. Union, 1966. Fellow American Physical Society; mem. American Academy of Arts and Sciences, Geological Society. Am. Mem. Nat. Acad. Sci., A.A.A.S., Am. Geophys. Union, Society Exploration Geophysicists (honorary life), American Sesmological Society Phi Beta Kappa, Sigma Xi. Contbr. technical articles. Research on earth physics; earth tides; seismology; electromagnetic prospecting methods. Home: 1446 Amalfi Dr., Pacific Palisades, Cal. Office: U. of Cal., Los Angeles 24.

SLIFKIN, Lawrence Myer, Am. physicist; b. Bluefield, W.Va., Sept. 29, 1925; s. Isaac Louis and Eva (Baden) S.; B.A., N.Y. U., 1947; M.S., Princeton, 1949, Ph.D., 1950; m. Miriam Kresses, July 4, 1948; children—Anne, Rebecca, Merle, Naomi. Research asso. U. Ill., Urbana, 1950-52, research asst. prof., 1952-54; asst. prof. U. Minn., Mpls., 1954-55; faculty U. N.C., Chapel Hill., 1955—, prof. physics, 1963—. Fellow Am. Phys. Soc. Research, publs. in imperfections in crystals. Home: 313 Burlage Dr., Chapel Hill, N.C. 27514.*

SLIJPER, Everhard Johannes, Dutch zoologist; b. Bolsward, Netherlands, Sept. 7, 1907; s. H.J. and C.G. (Kiel) S.; Dr.'s degree cum laude, Rijksuniversiteit Utrecht (Netherlands), 1936. Asst., prosector vet. anatomy U. Utrecht, 1935-48; mem. whaling expdn. in Antarctic waters, 1946-47; prof. vet. anatomy U. Indonesia, Bogor, 1949-51; prof. gen. zoology U. Amsterdam, 1951—, dir. Zool. Lab. Author: Die Cetaceen, 1936; Mens en Huisdier, 1944; De Vliegkunst in het Dierenrijk, 1952; Walvissen, 1958; Whales, 1962; Riesen des Meeres, 1962; De Geheimen van Reuzen en Dwergen, 1965; Riesen und Zwerge im Tierreich, 1967; also numerous articles. Research on functional anatomy of vertebrates; pure and applied research on whales and dolphins, evolution, behaviour of domestic animals. Office: Zoologisch Laboratorium der Universiteit van Amsterdam Plantage Doklaan 44, Amsterdam C., Netherlands.*

SLINGER, Stanley James, Canadian nutritionist; b. Guelph, Ont., Can., Nov. 20, 1914; s. John and Elizabeth (Robinson) S.; B.S.A., Ont. Agrl. Coll., 1937; M.S.A., U. Toronto, 1941; Ph.D., Cornell U., 1950; m. Mildred Victoria Simmons, Aug. 20, 1945; children—Anne Elizabeth, William Charles, John Thomas. Faculty, Ont. Agr. Coll., 1941-64, prof. charge nutrition poultry sci. dept., 1955-64; prof., head dept. nutrition U. Guelph, 1964—. Recipient Am. Feed Mfrs. award for poultry nutrition, 1956. Mem. Agrl. Inst. Can., Nutrition Soc. Can., Poultry Sci. Assn. (past dir.), World's Poultry Sci. Assn., Am. Inst. Biol. Sci. Research, numerous publs. on mode of action of antibiotics in nutrition, mineral, protein and vitamin requirements, measurement of available energy content of feedstuffs and factors affecting measurement of biologically available energy and nitrogen. Home: 10 Harcourt Dr., Guelph, Ont., Can.*

SLIPHER, Vesto Melvin, Am. astronomer; b. Clinton County, Ind., Nov. 11, 1875; s. David Clark and Hannah (App) S.; A.B., Ind. U., 1901, A.M., 1903, Ph.D., 1909, LL.D., 1929; hon. Sc.D., U. of Ariz., 1923, U. of Toronto, Canada, 1935; m. Emma Rosalie Munger, Jan. 1, 1904; children—Marcia Frances, David Clark. Astronomer, 1901-15, asst. dir., 1915-17, dir. since 1917, Lowell Obs., emeritus dir. since 1953; in charge Lowell solar eclipse expdn. to Syracuse, Kan., June 1918, and to Ensenada, Mexico, 1923. Awarded the Lalande prize and gold medal, Paris Acad. Sciences, 1919; Henry Draper gold medal of Nat. Acad. Sciences for discoveries in astron. physics, 1932; gold medal of Royal Astron. Soc., 1933, George Darwin lecturer, same society, 1933; awarded the Catherine Wolfe, Bruce gold medal by Astronomical Soc. of the Pacific, 1935. Mem. Nat. Acad. of Sciences, Am. Philos. Soc.; asso. Royal Astron. Soc. (London); fellow Am. Acad. Arts and Sciences, A.A.-A.S. (v.p. 1933); mem. Internat. Astron. Union, Am. Astron. Soc. (v.p. 1931), Société Astronomique de France, Phi Beta Kappa, Sigma Xi. Extensive investigations in astronomical spectroscopy; studies on the rotations and atmospheres of the planets; directed search that led to finding Lowell's trans-Neptunian planet—the new planet, Pluto. Discovered the rapid rotation and enormous space velocities of the nebulae, which furnished the observational basis for the expansion of the universe theory, that has grown out of Einstein's theory; high velocities of the star clusters; the cosmic radiations of the night sky; etc. Contritbuted numerous papers to astronom. publs. on the planets, nebulae, clusters, comets, stars and aurora. Address: Lowell Observatory, Flagstaff, Ariz.

SLOAN, David Harold, physicist; b. Lyman, Wash., Sept. 8, 1905; s. David Gardner and Edith (Jones) S.; B.S., Wash. State U., 1928, M.S., 1929; Ph.D. U. Cal. at Berkeley, 1941; m. Grace Cornog, June 1940 (div. Jan. 1941). Research asst. Gen. Electric Research Lab., Schenectady, 1928-30; research engr. Westinghouse Electric Co., East Pittsburgh, Pa., 1942-45; physicist Johns Hopkins Applied Physics Lab., Silver Spring, Md., 1945-46; faculty engring. U. Cal. at Berkeley, 1947-63, prof. emeritus, 1963—; physicist Physics Internat., San Leandro, Cal., 1963—. Mem. Am. Phys. Soc., I.E.E.E. Research, publs. on linear accelerator, megavolt X-ray, microwave power, supersonic ram jet, billion megawatt super power electron devices. Home: 2824 Forest Av., Berkeley, Cal. 94705. Office: 2700 Merced St., San Leandro, Cal. 94577.*

SLOANE, Sir Hans, English botanist, physician; b. Killyleagh, County Down, Ireland, Apr. 16, 1660; s. Alexander S.; student medicine, London, for four years; M.D., U. Orange (France), 1683; Dr.Physics, Oxford (Eng.) U., 1701; married in 1695; two daus. Physician to Duke of Albemarle (while there collected large number of subjects from animal kingdom, also over 800 species of plants), 1687-1689; physician Christ's Hosp., 1694-1730; attended Queen Anne, 1718-35; 1st physician to King George II, 1727. Fellow Royal Soc., (sec. 1693-1712, pres. 1727-41); mem. Royal Coll. Physicians (pres. 1719-35); fgn. mem. acads. sci. Paris, St. Petersburg, Madrid. Author: Natural History of Jamaica, 1707-23, also catalogue (in Latin) of Jamaican plants, 1696. Editor transactions Royal Soc. for twenty years. His museum and library of 50,000 volumes and 3560 manuscripts formed the nucleus of Brit. Mus.; established bot. garden at Chelsea, turned it over to state and left money for upkeep. Died Chelsea, Eng., Jan. 11, 1753.

SLOANE, T(homas) O'Conor, Am. inventor; b. N.Y., Nov. 24, 1851; s. Christian S. and Eliza M. (O'Conor) S.; A.B., St. Francis Xavier Coll., 1869, A.M., 1873, LL.D., 1912; E.M., Columbia, 1872, Ph.D., 1876; m. Isabel X. Mitchel, Sept. 18, 1877; 1 son, T(homas) O'Conor; m. 2d, Alice M. Eyre, Apr. 16, 1884; children—Charles O'Conor, John Eyre, Alice Mary. Prof. natural sciences Seton Hall Coll., S. Orange, N.J., 1888-89; sci. lectr., expert in many lawsuits about patents. Mem. adv. bd. N.Y. Elec. Sch. Mem. editorial staff Plumber and San. Engr., Sci. Am., Youth's Companion, Everyday Engring., Practical Electrics, and mng. editor The Experimenter; editor Amazing Stories. Author: Home Experiments in Science, 1888; Rubber Hand Stamps and the Manipulation of India Rubber, 1891; Arithmetic of Electricity, 1891; Electricity Simplified, 1891; Standard Electrical Dictionary, 1892; Electric Toy Making for Amateurs, 1892; How to Become a Successful Electrician, 1894; Liquid Air and the Liquefaction of Gases, 1899; The Electrician's Handy Book, 1905; Elementary Electrical Calculations, 1909; Motion Picture Projection, 1921; Rapid Arithmetic, 1922. Compiler: Facts Worth Knowing, 1890. Translator: Electric Light (Algave & Boulard), 1884; Jörgensen's Life of St. Francis of Assisi. Inventor self-recording photometer (1st instrument to record mechanically on index cards the illuminating power of gas); described new, accurate process for determining sulphur in illuminating gas, 1877. Died Aug. 7, 1940.

SLOCUM, Frederick, Am. astronomer; b. Fairhaven, Mass., Feb. 6, 1873; s. Frederick and Lydia Ann (Jones) S.; A.B., Brown U., 1895, A.M., 1896, Ph.D., 1898, Sc.D., 1938; m. Carrie E. Tripp, June 29, 1899. Began as instr. math. Brown U., 1895-1900, asst. prof. astronomy, 1900-09, acting dir. Ladd Obs., 1904-05; at Royal Astrophysical Observatory, Potsdam, Germany, 1908-09; lectr. N.Y. U., summer 1908; research asst., summer 1907; instr. in astrophysics Yerkes Obs., U. Chgo., 1909-11, asst., prof. astronomy, 1911-14; prof. astronomy and dir. Van Vleck Obs., Wesleyan U. Conn., 1914-18, since 1920; prof. nautical science Brown U., 1918-20. Research asso. Carnegie Inst., Washington, 1920; prof. astronomy Columbia U., summer 1923. Fellow Royal Astron. Soc., A.A.A.S.; mem. Am. Acad. Arts and Sciences, Astronomische Gesellschaft, Société Astronomique de France, Am. Astron. Soc. (v.p.), Internat. Astron. Union, NRC, Phi Beta Kappa, Sigma Xi. Author: Stellar Parallaxes From Photographs Made With the 20-inch Refractor of Van Vleck Observatory (with C. L. Stearns and B. W. Sitterly), 1938. Contbr. to publs. chiefly on observations on sun and determination of stellar distances. Died Dec. 4, 1944.

SLONIMSKI, Piotr P., biologist; b. Warsaw, Poland, Nov. 9, 1922; s. Piotr W. and Janina (Sobecka) S.; M.D., U. Cracow (Poland), 1947; Lic.Sc., U. Paris, 1951, D.Sc., 1952; m. Hanna Kulagowska, July 7, 1951; 1 dau. Agnès. With Nat. Center Sci. Research, France, 1947—, dir. research, 1962—, acting dir. lab. physiol. genetics, 1962—; asso. prof. genetics Sorbonne, Paris, 1960-65, prof., 1965—. Mem. French Nat. Council Sci. Research, 1960—; govt. cons. molecular biology, genetics, 1963—. Recipient Rapkine prize, 1952; 50th anniversary medal French Biochem. Soc., 1964. Mem. European Molecular Biology Orgn., Am. Soc. Biol. Chemists. Author: La Formation des Enzymes Respiratoires, 1953; also numerous articles. Established mitochondrial origin of inheritance of cellular respiration, chem. modifications of mitochondrial DNA causing cytoplasmic mutations; discovered respiratory adaptation and iso-cytochromes; regulation of synthesis of oxidation-reduction enzymes. Home: 204 G, Route de Chateaufort, GIF sur Yvette 91. Office: Lab. Genetics, Nat. Center for Sci. Research, GIF sur Yvette 91, France.*

SLOTH, Eric Niels, Am. chemist; b. Chgo., Apr. 23, 1922; s. Niels and Kathrine (Jensen) S.; B.S., U. Ill., 1949, M.S., 1950; postgrad. U. Cal. at Los Angeles, Ill. Inst. Tech.; m. Selma Christiansen Kildegaard, Sept. 13, 1947; children—Karen Ann, Susan Lynn, Lauri Jo, Eric Mark, Heidi Joy, Karl Peter. Asst., U. Ill., 1949-50; asst. chemistry U. Cal. at Los Angeles, 1950-53; staff Argonne (Ill.) Nat. Lab., 1954—, asso. chemist, 1956—. Mem. Am. Chem. Soc., Research Soc. Am., N.Y. Acad. Scis., Phi Lambda Upsilon, Alpha Chi Sigma. Research, publs. on surface ionization mass spectrometry. Home: 133 Springwood Dr., Naperville, Ill., 60540. Office: 9700 S. Cass Av., Argonne, Ill. 60439.*

SLOTTA, Karl Heinrich, chemist; b. Breslau, Germany, May 12, 1895; s. Karl and Marie (Simpig) S.; Ph.D., U. Breslau, 1923; Sc.D., U. Bonn, 1965; m. Maja Fraenkel, July 16, 1927; children—Sabine Crozier (Mrs. Ronald D. Crozier), Peter L. Came to U.S., 1956, naturalized, 1961. Faculty, U. Breslau, 1923-35; head chem. dept. Instituto Butantan, Sao Paulo, Brazil, 1935-38; sci. dir. Industria Pharmaceutica Endochimica, Sao Paulo, 1939-55; research prof. biochemistry U. Miami (Fla.) Sch. Medicine, 1956—. Mem. Deutsche Chemische Gesellschaft, Am. Chem. Soc., Am. Gerontol. Soc., Internat. Soc. Toxicology, Am. Soc. Exptl. Biology, Academia Brasileira de Ciencias. Author: Arzneistoff-Synthese, 1931; also numerous articles. Research in blood coagulation factors, thromboplastin, plasminogen; isolated and determined formula of progesterone, hormone of corpus luteum, 1934; isolated and crystallized Crotoxin, 1st pure active component of snake venom, 1938. Home: 5740 S.W. 52d Terrace, Miami, Fla. 33155.*

SLUNSKY, Rudolf, Czechoslovakian physician; b. Dobre Pole, Czechoslovakia, Nov. 24, 1926; s. Peter and Cecily (Heimel) S.; M.D, U T.G. Masaryk, Brno, Czechoslovakia, 1951; specialist obstetrics and Gynecology I and II degree, 1960, Cand. Med. Scis., 1963; m. Silvia Szilvasi, June 12, 1954. Staff, Regional Hosp., Kunz, Ostrava, Czechoslovakia, 1951, obstetrician, gynecologist, 1952-60, head outpatient clinic, gynecology and obstetrics, 1960-66, acting head dept. obstetrics, gynecology, 1966—; faculty Nurse Sch., Inst. Postgradual Medicine for Doctors in Prague, Czechoslovakia, 1966—; lectr. Socialist Acad., Ostrava, 1955—. Recipient award gynecol. sec. Purkynje's Soc., 1958, 65. Mem. Purkynje' Soc. Czechoslovak Physicians, Biol. Soc. Czechoslovakian Acad. Scis. Author: Die Blutgerinnungsstörungen in der Geburtshilfe, 1963; also numerous articles, brochures. Research, publs. on thromboplastic and fibrinolytic activity of amniotic fluid blood coagulation disorders in obstetrics. Home: 32 Syllabova. Office: 19 Syllabova, Ostrava 3, Czechoslovakia.

SLY, William Glenn, Am. chemist; b. Arcata, Cal., June 15, 1922; s. Altho Glenn and Esther (Sayles) S.; B.S., San Diego State Coll., 1951; Ph.D., Cal. Inst. Tech., 1955. Postdoctoral fellow Mass. Inst. Tech., 1957-58; faculty Harvey Mudd Coll., Claremont, Cal., 1959—, prof. chemistry, 1966—. Acad. guest prof. Eidg. Tech. Hoch., Zurich, 1965-66. Mem. Am. Chem. Soc., I.E.E.E., Am. Crystallographic Assn. Research in x-ray crystallography, molecular structure organic, metalorganic molecules. Home: 1036 N. College Av., Claremont, Cal. 91711.*

SLYE, Maud, Am. pathologist; b. Mpls., Feb. 8, 1879; d. James Alvin and Florence Alden (Wheeler) Slyde; A.B., Brown U., 1899, Sc.D., 1937; postgrad. U. Chgo., fellow, 1908-11. Prof. psychology and pedagogy R.I. State Normal Sch., 1899-1905; mem. staff Sprague Meml. Inst., Chgo., 1911-43; Instr. pathology U. Chgo., 1919-22, asst. prof., 1922-26, asso. prof., 1926-45, prof. emeritus, 1945-54. Research on mice many years to determine the nature of cancer, the relation of heredity to cancer, the laws governing malignancy and its localization, and age at which it will occur. Recipient Gold medal A.M.A., 1914; Rickets prize, 1915; Gold medal N.Am. Radiol. Soc., 1922. Mem. Assn. Cancer Research (v.p.), Chgo. Inst. Medicine, A.A.A.S., A.M.A., Am. Assn. Sci. Workers, N.Y., Ill. acads. sci.; hon. mem. Seattle Acad. of Surgery, So. Cal. Med. Soc., Phi Beta Kappa, Sigma Xi. Author 42 brochures on cancer; also Songs and Solaces (poems), 1934; I in the Wind (poems), 1936. Died Sept. 17, 1954.

SLYKHUIS, John Timothy, Canadian plant pathologist; b. Carlyle, Sask., Can., May 7, 1920; s. William and Emma (Hodgson) S.; B.S., U. Sask., 1942, M.S., 1943; Ph.D., U. Toronto, 1947; m. Ruth Enid Williams, July 6, 1946; children—Grace, Margaret, Dorothy, Timothy, Alan. Research officer agr. Sci. Service Lab., Lethbridge, Alta., 1952-57; head plant pathology unit sci. service Can. Agr., Ottawa, Ont., 1957-58, chief virology sect. Plant Research Inst., 1958-67, chief virology sect. Cell Biology Research Inst., 1967—; with Rothamsted Exptl. Sta. (Eng.), 1956-57. Mem. Agrl. Inst. Can., Am., Canadian phytopath. socs. Discovered several viruses of cereals and grasses including wheat striate mosaic in N.Am., Europe and their vectors, mite vector wheat streak mosaic virus, wheat spot mosaic and its vector, ryegrass mosaic in Eng., Hordeum mosaic in Can., undescribed soilborne viruses in Can. Home: 737 Broadview Av. Office: Cell Biology Research Inst., Research Br., Can. Agrl., Ottawa, Ont., Can.*

SMAILES, Arthur Eltringham, English geographer; b. Haltwhistle, Eng., Mar. 23, 1911; s. John Robert and Mary Elizabeth (Eltringham) S.; B.A. with 1st class honors, U. London (Eng.), 1931, M.A., 1933, D.Lit., 1965; m. Dorothy Forster, Aug. 14, 1937; 1 dau., Ruth Margaret. Faculty, U. Coll., U. London, 1928-53, head dept. geography Queen Mary Coll., 1953—, prof. geography U. London, 1955—. Recipient Royal Scottish Geog. Soc. Research medal, 1964. Fellow Royal Geog. Soc.; mem. Inst. British Geographers (founding, hon. sec. 1950-61, v.p. 1968—). Author: The Geography of Towns, 1953, rev., 1966; North England, 1960; also numerous articles. Research on urban and applied geography, regional geography. Home: 20 Marlborough Crescent, Sevenoaks, Kent, Eng. Office: Dept. Geography, Queen Mary Coll., Mile End Rd., London E.1., Eng.*

SMAKULA, Alexander, physicist; b. Dobrovody, Ukrainia, Sept. 9, 1900; s. Theodor and Maria (Juzwa) S.; Ph.D., U. Goettingen, 1927; m. Erika E. Bunde, Oct. 1, 1932; children—Ilse (Mrs. S. Armour), Elli (Mrs. C. Hand), Peter H., Fritz K. Came to U. S., 1946, naturalized, 1954. Research asso. U. Goettingen, 1927-30, Kaiser Wilhelm Inst., Heidelberg, Germany, 1930-34; head, research lab. Carl Zeiss, Jena (Germany), 1934-45; faculty Mass. Inst. Tech., Cambridge, Mass., 1951—, prof. physics, 1962—, dir. crystal physics lab., 1964—. Cons. U. S. Army, 1946-51. Fellow Optical Soc. Am.; mem. Ukrainian Engrng. Soc. (hon.), Am. Phys. Soc., Shevchenko Sci. Soc., Sigma Xi. Author: Single Crystals, 1962. Inventor nonreflective coating, 1935. Contbr. articles to German, Am. sci. jours. Home: 15 Orris St., Auburndale, Mass. 02166. Office: 77 Massachusetts Av., Cambridge, Mass. 02139.*

SMALE, Stephen, Am. mathematician; b. Flint, Mich., July 15, 1930; s. Laurence Albert and Helen (Morrow) S.; B.S., U. Mich., 1952, M.S., 1953, Ph.D., 1956; m. Clara Davis, Feb. 3, 1955; children—Nathan, Laura. Instr., U. Chgo., 1956-58; mem. Inst. Advanced Study, Princeton, N.J., 1958-60; asso. prof. U. Cal., Berkeley, 1960-61, prof., 1964—; prof. Columbia, 1961-64. Recipient Fields medal Internat. Union Mathematicians, 1966. Mem. Am. Math. Soc. (Veblen prize 1966). Research on differential topology, global analysis, especially qualitative ordinary differential equations; proved that one can turn a sphere inside out, proved Poincaré's conjecture for dimensions greater than four. Home: 69 Highgate Rd., Berkeley, Cal.

SMALES, Albert Arthur, English chemist; b. Wallsend-on-Tyne, Eng., May 4, 1916; s. Albert and Mary (Gowland) S.; B.Sc., U. London, 1941, D.Sc., 1965; m. Constance Bielby, July 19, 1941; children—Margaret, Valerie. Staff, Imperial Chem. Industries, Ltd., Billingham-on-Tees, Eng., 1932-47, Tenn. Eastman Corp., Oak Ridge, 1944-45; staff Atomic Energy Research Establishment, Harwell, Eng, 1947—, head analytical sci. div., 1966—. Decorated officer Order Brit. Empire. Fellow Royal Inst. Chemistry; mem. Chem. Soc., Soc. for Analytical Chemistry U.K. (pres. 1965-67), Geochem. Soc. Editor: (with L. R. Wager) Methods in Geochemistry, 1960; also numerous articles. Research on analytical chemistry and geochemistry including activation analysis and trace elements in meteorites, polarography, isotope dilution, radiochemistry. Home: 83 Bath St., Abingdon, Berkshire. Office: Bldg. 551 Atomic Energy Research Establishment, Harwell, Berkshire, Eng.*

SMALL, Albion Woodbury, Am. sociologist; b. Buckfield, Me., May 11, 1854; s. Rev. Albion K. P. and Thankful (Woodbury) S.; A.B., Colby, 1876, A.M., 1879; Newton Theol. Instn., 1876-79; univs. of Berlin and Leipzig, 1879-81; Ph.D., Johns Hopkins, 1889; (LL.D., Colby, 1900); m. Valeria von Massow, of Berlin, June 20, 1881. Prof. history and polit. economy, Colby Coll., 1881-88; reader in history, Johns Hopkins, 1888-89; pres. Colby Coll., 1889-92; prof. and head dept. sociology, 1892-1924, dean Grad. Sch. Arts and Lit., 1905—, U. of Chicago. Editor and founder, Am. Jour. of Sociology, 1895—. V.p. and Lit., 1905-1923, U. of Chicago. Editor Arts and Sciences, St. Louis Expn., 1904. Author: General Sociology, 1905; Adam Smith and Modern Sociology, 1907; The Cameralists, 1909; The Meaning of Social Science, 1910; Between Eras, from Capitalism to Democracy, 1913; Origins of Sociology, 1924. Contbrn. to establishment of limits of sociology and its acceptance as independent acad. field; important work in history of sociol. thought. Died Chicago, Ill., Mar. 25, 1926.

SMALL, James, English botanist; b. Brechin, England, 1889; s. William Small; ed. Pharm. Soc.'s Sch., Birkbeck Coll., London; B.Sc., London, 1913, M.Sc., 1916, D.Sc., 1919; m. Helen Pattison, 1917; 2 sons, 1 dau. Became lectr. botany Armstrong Coll., Newcastle, Eng., 1916; asst. lectr. botany Bedford Coll., 1916-20, Pharm. Soc.'s Sch., 1917-20; prof. botany Queen's U., Belfast, 1920-54, emeritus, 1954-55. Fellow Royal Soc. Edinburgh (Makdougall Brisbane prize for papers on quantitative evolution 1951), Royal Photog. Soc.; mem. Royal Irish Acad. Author: Origin and Development of Compositae; Application of Botany in Utilisation of Medicinal Plants; Textbook of Botany; What Botany Really Means; Hydrogen-ion Concentration in Plant Cells and Tissues; Geheimnisse der Botanik; El Secreto de la Vida de las Plantas; Pocket-Lens Plant Lore; Practical Botany; pH and Plants; Modern Aspects of pH, 1954; pH of Plant Cells, 1955; also papers. Died Nov. 28, 1955.

SMALL, John Kunkel, Am. botanist; b. Harrisburg, Pa., Jan. 31, 1869; s. George H. and Catharine K. S.; A.B., Franklin and Marshall Coll., 1892, Sc.D., 1912; Ph.D., Columbia, 1895; m. Elizabeth Wheeler, 1896; children—George Kunkel, Kathryn Wheeler, Elizabeth, John Wheeler. Curator, herbarium Columbia U., 1895-99; spl. agt. Ga. Geol. Survey, 1895; curator herbarium N.Y. Bot. Garden, 1898-1906, head curator, 1906-32, chief research asso. and curator, 1932—. Author: A Monograph of the North American Species of the Genus Polygonum, 1895; Flora of the Southeastern States, 1903, 2d edition, 1913; Flora of Miami, 1913; Flora of Lancaster County, 1913; Florida Trees, 1913; Flora of the Florida Keys, 1913; Shrubs of Florida, 1913; Ferns of Tropical Florida, 1918; Ferns of Royal Palm Hammock, 1918; From Eden to Sahara—Florida's Tragedy, 1929; Ferns of Florida, 1932; Manual of Southeastern Flora, 1932; Ferns of Vicinity of New York (illustrated), 1935. Interested in bot. exploration of southeastern U. S., interpretation and classification of its flora and its phytogeography, with special reference to its native palms, irises, flowering epiphytes, cacti, and ferns. Died Jan. 20, 1938.

SMALL, Walter Madison, Am. geologist; b. Cooperstown, Pa., July 5,, 1887; s. Madison M. and Ida (Alcorn) S.; A.B., Allegheny Coll., 1911, D.Sc., 1961; m. Berta Muehlbauer, June 17, 1933. Sci. tchr. Custer County High Sch., Miles City, Mont., 1911-13; field geologist ind. refining co., Mexico, 1914; field geologist Tex. Co., 1915; exploration mgr. Midcontinental Devel. Co., Tulsa, 1916; field geologist Carter Oil Co., Mountain States, 1917; chief geologist Midco, Mexico, 1918-19; cons., Tampico, Mexico, 1920-22; chief geologist Cia. del Agwi S.A., Mexico, 1923-27; field mgr. Oil Co. Angola, West Africa, 1928-30; geologist Eurogasco, Austria, 1931-36; field mgr., dir. Standard Oil Co., Egypt, 1936-38; chief geologist Romana Americana, 1938-41, cons., 1941-45; chief geologist Seaboard Dominicana, Dominican Republic, 1946-47; cons. geologist Husky Oil Co., Wyoming, 1948-49, Can., 1949-50, Ind., 1951-52, Colo., 1952-53; cons. Esso, 1955-56; cons., 1956—. Fellow A.A.A.S., Geol. Soc. Am., Am. Geol. Soc.; mem. No. Appalachian Geol. Soc., Am. Geophys. Union, Am. Assn. Petroleum Geologists, Am. Geol. Inst. Research, publs. on geol. and geophys. exploration, organizing fgn. personnel. Home: Route 1, Cooperstown, Pa. 16317.*

SMALLMAN, Raymond Edward, English phys. metallurgist; b. Wolverhampton, Eng., Aug. 4, 1929; s. David and Edith (French) S.; student U. Birmingham (Eng.), 1947-53; m. Joan Doreen Faulkner, Sept. 6, 1952; children—Lesley Ann, Robert Ian. With Atomic Energy Research Establishment, Harwell, Eng., 1953—, sr. sci. officer dept. phy. metallurgy and sci. materials, 1958—; faculty U. Birmingham, 1958—, prof. phys. metallurgy, 1964—; vis. prof. materials sci. Stanford, 1962. Fellow Inst. Metallurgists (chmn. metal sci. jour. publ. com.). Author: Modern Physical Metallurgy, 1962; (with K. H. G. Ashbee) Modern Metallography, 1966; also numerous articles. Research on deformation and defect structure of metals and alloys using mech. property measurements with X-ray diffraction techniques and thin film transmission electron microscopy; 1st observations of several crystal defects including dislocation loop, double dislocation loop, helical dislocation, climb source. Home: 59 Woodthorne Rd., Wolverhampton, Staffordshire, Eng. Office: U. Birmingham, Edgbaston, Birmingham 15, Eng.*

SMART, J. Samuel, Am. physicist; b. New Bloomfield, Mo., Aug. 31, 1919; s. Samuel Richard and Julia Ann (Glennen) S.; B.A., Westminster Coll., 1939; M.S. La. State U., 1941; Ph.D., U. Minn., 1948; m. Claudia Rordam, Mar. 3, 1942. Instr., AAC, Chanute Field, Ill., 1941-43, also Yale; physicist Bur. Ships, USN, Washington, 1943-46, U.S. Naval Ordnance Lab., Silver Spring, Md., 1948-55; sci. liasion officer Office Naval Research, London, 1955-57, physicist, Washington, 1958-60; vis. physicist Brookhaven Nat. Lab., Upton, N.Y., 1957-58; physicist IBM Research Center, Yorktown Heights, N.Y., 1960—. Recipient Distinguished Civilian Service award USN, 1951; Westminster Coll. Alumni Achievement award, 1964. Fellow Am. Phys. Soc. (sec.-treas. div. solid state physics 1960-63).

Contbr. chpts. to books, articles to tech. jours. Research on solid state physics; hydrology; origin of chem. elements; magnetism. Home: 71 Mt. Airy Rd., Croton-on-Hudson, N.Y. Office: IBM Research Center, Yorktown Heights, N.Y.

SMART, Robert Forte, Am. biologist; b. Tyro, Miss., Apr. 9, 1905; s. James Henry and Cornelia (Tucker) S.; B.A., Miss. Coll., 1927; M.A., Harvard, 1929, Ph.D., 1935; m. Eleanor Lucile Ferguson, Aug. 28, 1929; children—Robert Ferguson, Eleanor Tucker (Mrs. James McCormick Paxton). Asst. prof. biology Miss. Coll., 1928-29; Austin scholar Harvard, 1928-29, 32-33; faculty U. Richmond (Va.), 1929—, prof. biology, 1941—, chmn. div. scis., chmn. dept. biology, 1941-57, dean Richmond Coll., 1957-67, provost U., 1967—. Pres., Va. Sch. Bds. Assn., 1956-57. Fellow A.A.A.S.; mem. Bot. Soc. Am., Mycol. Soc. Am., Va. Acad. Sci. (pres. 1944-45), Phi Beta Kappa, Sigma Xi, Phi Kappa Sigma, Omicron Delta Kappa, Tau Kappa Alpha. Author: Workbook for General Biology, 1941; Studies in the Biology of Bacteria, 1957; also articles. Home: 7003 University Dr., Richmond, Va. 23229.*

SMEATON, John, English civil engr.; b. Austhorpe, Leeds, Eng., June 8, 1724; s. William and Mary (Stones) Sm.; ed. Leeds grammar sch.; m. Anne, June 8, 1756; 2 daus. Opened a shop in London, 1750; studied canal and harbor systems of Netherlands, 1754; built 3d Eddystone lighthouse, 1756-59; built arched bridges, Perth, Banff, Coldstream, also Forth and Clyde Canal; built Ramsgate harbor, 1774. Fellow Royal Soc., 1753; founder Smeatonian Club, 1771. Author: Experimental Enquiry concerning the Natural Powers of Wind and Water to Turn Mills (Gold medal), 1759; account of Eddystone lighthouse, 1792; also papers. Improved instruments used in astronomy and nav.; rediscovered hydraulic cement, 1756. Died Oct. 28, 1792.

SMEDS, Helmer, Finnish geographer; b. July 29, 1908; s. Werner and Sofia (Bjodklund) S.; Phil. mag., Helsinki (Finland) U., 1929, Ph.D., 1935; m. Fatiana Kulikoff, Nov. 23, 1945. Tchr., People's High Sch., 1932-33; tchr. Tchrs' Tng. Coll., 1936-37, secondary sch., 1937-41; prof. Helsinki Swedish Sch. Econs., 1942-50; prof. Helsinki U., 1950—; mem. Nat. Diet of Finland, 1945-58. Hon. fellow Geog. Soc. Paris, Geog. Soc. Lund. Author several books, articles. Pioneered historic-geog. research in Finland; human geog. research in Ethiopian highland; studies in population and settlement geography in Finland. Office: Dept. Geography, Helsinki U., Helsinki, Finland. Died Aug. 12, 1967.*

SMEKAL, Adolf, physicist; b. Sept. 12, 1895; s. Gustav and (Hauptmann) S.; ed. Tech. U., Vienna, U. Graz (Austria), U. Berlin; Phil.D., U. Graz (Austria), 1917; m. Gertrud Eschenburg, 1942. Lectr. physics U. and Tech. U., Vienna, 1920-21; asso. prof. U. Vienna, 1917; prof. theoretical physics U. Halle (Germany), 1928-45; prof. Tech. U., Darmstadt, Germany, 1946-49; prof. physics U. Graz, 1949—. Recipient Haitinger prize, 1923. Mem. Austrian Acad. Scis., Leopold Acad. Author: Allgemeinen Grundlagen der Quantenstatistik und Quantentheorie, 1926; Über Grenzflächenphysik, 1950; also numerous articles. Research in quantum theory, X-ray spectroscopy, crystal physics, physics of amorphous solids, thermodynamics; predicted Raman effect, athermic plasticity of solids. Office: U. Graz, Graz, Austria.

SMELIK, Pieter Gerard, Dutch endocrinologist; b. Zevenbergen, Holland, Oct. 18, 1928; s. Jan and Cornelia (Hart) S.; student U. Utrecht, 1946-49, U. Gröningen, 1949-54; m. Foekje van Dalen, June 29, 1955; children—Anneke, Jan, Willem, Pieter. Asst. research fellow in pharmacology U. Gröningen, 1955-63; asso. prof. pharmacology U. Utrecht, 1964—. Mem. Dutch Endocrine Soc., European Soc. Comparative Endocrinology. Author: Autonomic Nervous Involvement in Stress Induced ACTH Secretion, 1959; also articles. Research on central nervous regulation of hypophyseal ACTH secretions, mechanism of adrenocorticoid feed-back on ACTH secretion, role of nuerotransmitter in central nervous system on hypophysial functions. Home: 184 Deken Heinensk., Bunnik, Holland.*

SMELLIE, Robert Martin Stuart, Scottish biochemist; b. Rothesay, Scotland, Apr. 1, 1927; s. William Thomas and Jean (Craig) S.; B.Sc., U. St. Andrews (Scotland), 1947; Ph.D., U. Glasgow, (Scotland), 1952, D.Sc., 1963; m. Florence Mary Devlin Adams, June 19, 1954; children—William Stuart Adams, David Craig Shaw. With U. Glasgow, 1949-55, 56—, Cathcart prof. biochemistry, 1966—; research fellow N.Y. U., Coll. Medicine, 1955-56. Cons. Pfizer Ltd., Sandwich, Kent., 1964—. Fellow Royal Soc. Edinburgh, Inst. Biology (chairman of Scottish br. 1967—); mem. Biochem. Soc. (com. 1967—), Brit. Biophys. Soc., European Molecular Biology Orgn., Brit. Assn. for Cancer Research. Contbg. author: The Nucleic Acids, 1955; Progress in Nucleic Acid Research, 1963. Publs. on measurements of deoxyribonucleic acid content of cells; devel. quantitative method for determining nucleotide composition of ribonucleic acid; research on rate of synthesis of DNA and RNA in animal tissues and tumor cells, devel. enzymes of RNA synthesis in animal cells infected with RNA containing viruses; detection and characterization of enzymes responsible for synthesis of DNA and RNA in animal cells. Home: 39 Russell Dr., Bearsden, Glasgow, Scotland.*

SMELSER, George K., Am. anatomist; b. Anderson, Ind., Dec. 14, 1908; s. Walter B. and Katie (Keiser) S.; B.A., Earlham Coll., 1929; Ph.D., U. Chgo., 1932; m. Elizabeth H. Childs, Mar. 19, 1932; 1 dau., Elizabeth Ann (Mrs. Lewis F. Willey). Faculty, Columbia Coll. Phys. and Surgs., N.Y.C., 1934—, prof. anatomy assigned to ophthalmology, 1955—. Recipient Proctor medal, 1961. Mem. Assn. for Research Ophthalmology, Am. Acad. Ophthalmology, Am. Assn. Anatomists, Harvey Soc. Editor: Structure of the Eye, 1961. Research, numerous publs. in anatomy, embryology and physiology of eye. Home: 390 Lafayette Av., Westwood, N.J. 07675. Office: 630 W. 168th St., N.Y.C. 10032.*

SMELSER, Neil Joseph, Am. sociologist; b. Kahoka, Mo., July 22, 1930; s. Joseph and Susie Marie (Hess) S.; B.A., Harvard, 1952, Ph.D., 1958; B.A., Magdalen Coll., Oxford, Eng., 1954, M.A., 1959; m. Helen Thelma Margolis, June 10, 1954 (div. Mar. 1965); children—Eric Jonathan, Tina Rachel. Faculty, U. Cal. at Berkeley, 1958—, prof. sociology, 1962—, spl. asst. to chancellor for student polit. activities, 1965, asst. chancellor for ednl. devel., 1966—; research tng. candidate San Francisco Psychoanalytic Inst., 1962—. Mem. Social Sci. Research Council, 1961-65. Rhodes scholar, 1952-54. Mem. Am. (mem. council 1962—, exec. com. 1962—), Pacific sociol. assns. Author: (with T. Parsons) Economy and Society, 1956; Social Change in the Industrial Revolution, 1959; Theory of Collective Behavior, 1963; The Sociology of Economic Life, 1963. Editor: (with W.T. Smelser) Personality and Social Systems, 1963; Sociology: An Introduction, 1967. Research on relations between kinship and econ. instns. during periods of rapid econ. devel., methodological issues involved in comparing large-scale social systems. Home: 8 Mosswood Rd., Berkeley, Cal. 94704.*

SMERDON, Ernest Thomas, Am. agrl. engr.; b. Ritchey, Mo., Jan. 19, 1930; s. John Erle and Ada (Davidson) S.; B.S., U. Mo., 1951, M.S., 1956, Ph.D., 1959; m. Joanne A. Duck, June 9, 1951; children—Ernest Thomas, Katherine Anne, Gary Joe. Mem. faculty U. Mo., 1956-58; faculty Tex. A. and M. U., College Station, 1959—, prof. agrl. engring., 1962—, dir. Water Resources Inst., 1964—. Cons. to Coll. Agrl. Egring., Punjab Agrl. U., Ludhiana, India, 1965. Mem. Am. Soc. Agrl. Engrs. (tech. paper award 1961), Am. Soc. C.E., Am. Geophys. Union (vis. scientist 1965-66), Am. Soc. Engring. Edn., Sigma Xi, Tau Beta Pi, Gamma Sigma Delta, others. Publs., original research on erosion and sediment transport in relation to cohesive sediments; developed design procedures for surface irrigation systems concerning over land flow theory. Home: 703 Inwood St., Bryan, Tex. 77801. Office: Water Resources Inst., Tex. A. and M. U., College Station, Tex. 77843.*

SMETANA, Hans Frank, physician; b. Roemerstadt, Austro-Hungarian Empire, Apr. 5, 1894; s. Francis Joseph and Barbara (Sykora) S.; student U. Vienna (Austria) Sch. Engring., 1911-14, M.D., 1922; m. Florence Ennis Geraghty, Aug. 18, 1945; children—Roswitha Marie, Hans Rainer. Came to U.S., 1923, naturalized, 1936. Instr., Johns Hopkins, 1923-25; asso. Peking (China) Union Med. Coll., 1925-27; asst. pathologie Inst. Pathology U. Vienna, 1927-29; prof. pathology Sch. Tropical Medicine, San Juan, P.R., 1929-30; asst. prof. Columbia Coll. Phys. and Surg., N.Y.C., 1930-46; chmn. div. pathology, also dir. research Armed Forces Inst. Pathology, Washington, 1946-55, chmn. hepatic and pediatric pathology, 1958-63; vis. prof. Delhi U. Patel Chest Inst., All India Inst. Med. Sci., New Delhi, India, 1955-57, Chin Med. Bd. to Far Eastern Med. Instns., 1957-58; prof. pathology, pathologist-in-chief, dir. labs. Delta Regional Primate Research Center, Tulane U., Covington, La., New Orleans, 1963-66. Profl. lectr. George Washington U., 1948-55; cons. NIH, 1959-63. Decorated for meritorious civilian service, U.S. Dept. Army, 1963; recipient award for distinguished service Armed Forces Inst. Pathology, 1963; award for distinguished achievement Art and Sci. Club, Washington, 1955. Mem. Washington Soc. Pathologists (past pres.), 38th Parallel Med. Soc. (hon. mem. Korea), Korean Com. Zone Med. and Dental Assn. (hon.), Japanese-Am. Soc. Pathologists (hon.), Austro-Am. Med. Assn. (hon.), Philippines Soc. Pathologists (hon.), Soc. for Exptl. Biology and Medicine, Assn. Pathologists and Bacteriologists, Coll. Am. Pathologists, A.M.A., Internat. Acad. Pathology, N.Y. Acad. Medicine, New Orleans Acad. Pathology. Research, publs. on histopath. studies of degenerative, infectious, and toxic diseases of lung and liver, tumors lung and nervous system, Hodgkins disease. Home: Mockingbird Lane, Tchefuncta Club Estates, Covington, La. 70433.*

SMETANA, Karel, Czechoslovakian physician, morphologist; b. Prague, Czechoslovakia, Oct. 28, 1930; s. Karel and Marie (Mihulkova) S.; MUDr., Charles U., Prague, 1955, C.Sc., 1962, D.Sc., 1967; m. Vlasta Krouzkova, Oct. 23, 1953; 1 son, Karel. Faculty Charles U., 1951—, external lectr., 1962—; sci. officer lab. for Ultrastr. Research of Cells and Tissues, Czechoslovakian Acad. Sci., Prague, 1961—, head dept. blood morphology and cytology, 1964—; research fellow dept. pharmacology Baylor U. Coll. Medicine, Houston, 1962-63, vis. asso. prof., 1963, vis. prof., 1967; head lab. for electron microscopy Inst. for Exptl. Therapy, 1961-63; lectr. course cell biology UNESCO, 1964—. Recipient Hematology prize, 1962; co-recipient Biophys. prize, 1958. Mem. Czechoslovak Histochem. and Cytochem. Soc. (mem. com.), Nat. Com. for Electron Microscopy, Hematol. Soc., Anat. Soc. Research, publs. on denucleation of erythroblasts, fine morphology of nucleoli, substructre and cytochemistry of ribonucleoprotein particles in nucleoli, nuclei and cytoplasm, cytochemistry of nucleic acids, basic and acidic proteins in cell nuclei, nucleolar coefficient of mature lymphocytes in peripheral blood of patients with malignant diseases, ultrastructure of tumor and leukemic cells. Home: 2 Pod Sporilovem. Office: 4 Albertov, Prague, Czechoslovakia.*

SMETS, Georges Joseph, Belgian chemist; b. Louvain, Belgium, Aug. 11, 1915; s. Guillaume and Marguerite (van Marsenille) S.; D.Sc., U. Louvain, 1940; m. Elizabeth Stas, Oct. 11, 1941; children: Geneviève (Spaas), Marie-Rose, Martine, Guy, Brigitte. Research fellow, FNRS, 1941-43; asst. prof., 1944-48, full prof., 1948—, Faculty of Sci., U. Louvain; advanced fellow, Belgian Am. Education Found., 1949; Francqui prof., U. Liège, 1952-53; exchange prof. U. Aberdeen, Leeds, Birmingham, 1956; U. Paris, 1959, U. Strasbourg, 1961; U. Moscow, 1964. Awarded Stas prize, 1940, and Agathon de Potter prize, 1952, Royal Acad. of Begium. Mem., Société Chimique de Belgique; Am. Chem. Soc.; N.Y. Acad. Sci.; Koninklijke Vlaamse Academie voor Wetenschappen, letteren en Schone Kunsten van België (pres. 1962); Vlaamse Chemische Vereniging; Société Scientifique de Belgique; Soc. Chimie Industrielle de Paris; IUPAC; Conseil National de Chimie. Mem., ed. board, Journal Polymer Sci. (New York), 1946; ed. board, Die Makromolekulare Chemie, 1955. Author 140 published articles, papers. Home: 5 Beukenlaan, Heverlee, Belgium. Office: University of Louvain, Lab. of Macromolecular Chemistry, 96, Naamsestraat, Louvain, Belgium.*

SMID, Lucas Johannes, Dutch mathematician; b. Amsterdam, Holland, May 11, 1901; s. Lukas and Margo (Alberts) S.; Ph.D., Dr. ès sc., Amsterdam Communal U.; m. Jantje Vennema, July 28, 1932; children—Margo, Cornelius, Gerda. Prof. philosophy, 1924-36; actuary, 1936—; dir. ins. soc., 1957—. Mem. Philos. Soc., Soc. Actuaries, Actuaries Club. Author: Eine absolute Axomatik der Geometrie in einer begrenzten Ebene, 1935; Levensverzekeringswiskunde, 1949; also articles on circular geometry. Home: Troelstraweg 104, Leeuwarden. Office: Nieuwestad 9, Leeuwarden, The Netherlands.

SMIDT, Diedrich Trieno, German animal scientist; b. Breinermoor, Germany, July 14, 1931; s. Diedrich and Trientje (Strenge) S.; Dr. med. vet. Vet. Coll., Hannover, Germany, 1958; Dr.sc. agr., Göttingen (Germany) U., 1962; m. Heide von Geldern, July 5, 1958; children—Gernot, Wolfhart. Staff mem. Inst. Animal Breeding and Genetics, Göttingen U., 1961-66, Univ. tchr., 1966—. Mem. German Soc. Animal Sci., Soc. for Study Fertility, Soc. Lab. Animal Scis., N.Y. Acad. Scis. Author: Sexualpotenz und Fruchtbarkeitsvererbung beim Schwein, 1962; Die Schweinebesamung, 1965. Contbr. numerous articles to profl. jours. Editorial bd. Jour. Der Tierzüchter. Research on reproductive biology in farm animals, artificial insemination, estrous cycle control, embryo transfer, in vitro culture and fertilization of mammalian eggs, embryology in miniature pigs. Home: 7 Ludwig Beck Strasse, Göttingen, Germany.*

SMIGEL, Erwin O., Am. sociologist; b. N.Y.C., Nov. 3, 1917; s. Joseph O. and Ida (Sachs) S.; A.B., U. N.C., 1939, postgrad. student, 1939-40; M.A. (Penfield fellow), N.Y. U., 1942, Ph.D., 1949. From instr. to asso. prof. sociology Ind. U., 1948-57; sr. fellow law and behaviorial scis., U. Chgo. Law Sch., 1957-58; asso. prof. N.Y. U., 1958-59, prof. sociology, 1959—, chmn. dept. sociology Univ. Coll., 1958-62; chmn. dept. sociology Washington Square Coll., 1962-66, head all-university department sociology, 1966—. Served with USAAF, 1942-46. Mem. Am. Sociol. Assn., Soc. Study Social Problems, Indsl. Research Assn., N.Y. Acad. Sci., Am. Assn. U. Profs., Eastern Sociol. Soc. Co-author: Nursing Home Administration, 1962. Author: The Wall Street Lawyer: Professional Organization Man?, 1964; Editor: Work and Leisure, 1963. Contbr. to profl. jours. Editor: Social Problems, 1957-61, asso. editor, 1961-63; editorial bd. Estudioís de Sociologia, 1961-65. Study of occupations, organizations; added to Max Weber's concept of bureaucracy by discovery additional type ("professional bureaucracy"). Office: N.Y. U., N.Y.C. 10003.*

SMILES, Samuel, Brit. chemist; b. Belfast, Ireland, 1877; s. Samuel Smiles; ed. univs. London, Paris, Jena (Germany); D.Sc.; D.Sc. (hon.), Belfast; m. Minnie Patterson, 1920. Asst. prof. organic chemistry, prof. chemistry, fellow Univ. Coll., London; Daniell prof. chemistry, fellow King's Coll., London; prof. organic chemistry Armstrong Coll. Fellow Royal Soc., 1918; v.p. Chem. Soc. Author: Relations between Physical Properties and Chemical Constitution; also papers on organic compounds of sulphur. Died May 6, 1953.

SMILEY, Charles Hugh, Am. astronomer; b. Camden, Mo., Sept. 6, 1903; s. Herbert Leslie and Hattie (McCurry) S.; student U. Cal. at Los Angeles, 1920-22; A.B., U. Cal. at Berkeley, 1924, M.A., 1925, Ph.D., 1927; Sc.D., Monmouth Coll., 1961; m. Margaret Kendall Holbrook, June 12, 1928. Instr., U. Ill., 1927-29; Guggenheim fellow Royal Obs. Greenwich, 1929-30; faculty Brown U., Providence, 1930—, prof. astronomy, 1945—, chmn. dept. astronomy, dir. Ladd Obs., 1938—; leader solar eclipse expdns. to Peru, Brazil, Thailand, Pakistan, Can., U.S., also expdns. to observe atmospheric refraction at low angular altitudes. Recipient Franklin L. Burr prize Nat. Geog. Soc., 1949. Mem. Geog. Soc. Peru (hon.), Am. Astron. Soc., Royal (Eng.), Brit., Royal (Can.) astron. socs., Am. Assn. Variable Star Observers, Math. Assn. Am., Inst. Nav., Phi Beta Kappa, Sigma Xi, Pi Mu Epsilon. Research, numerous publs. on outer corona sun and zodiacal light nr. sun, atmospheric refraction at large zenith distances, hurricanes, Mayan astronomy. Home: 28 Montague St., Providence 02906.*

SMILEY, Malcolm Finlay, Am. mathematician; b. Monmouth, Ill., Dec. 15, 1912; s. Robert Rennsealar and Annie Laurie (Kerr) S.; B.S., U. Chgo., 1934, M.S., 1935, Ph.D., 1937; m. Dorothy Manning, Aug. 20, 1941. Mem. Inst. Advanced Study, 1937-38; instr. Lehigh U., 1938-40, asst. prof., 1940-42, asso. prof., 1946; vis. instr. U. Chgo., 1939, vis. asst. prof., 1941; instr. U. S. Naval Postgrad. Sch., 1942-46; asso. prof. Northwestern U., 1946-48; prof. State U. Ia., 1948-60; prof. U. Cal., Riverside, 1960—, chmn. dept. math. 1962-63. Fellow Fund Advancement of Edn., 1954-55. Mem. Am. Math. Soc., Math. Assn. Am., Circ. Mat di Palermo, Phi Beta Kappa, Sigma Xi. Author: Algebra of Matrices, 1965. Research on abstract algebra, including betweenness in lattice theory, measure theory in lattices, alternative rings, topological algebra and matrix theory. Home: 5748 Ives Pl., Riverside, Cal. 92506.*

SMILEY, Terah Leroy, Am. geochronologist; b. nr. Clay Center, Kan., Aug. 21, 1914; s. Terah Edward and Evangaline (Huls) S.; student U. Kan., 1934-36; B.A., U. Ariz., 1946, M.A., 1949; m. Winifred Whiting, June 10, 1946; children—John Terah, Maureen Kate, Kathlyn Elaine. Dir. geochronology labs. U. Ariz., Tucson, 1956—, acting dir. lab. tree-ring research, 1958-60. Mem. U.S. Com. on INQUA, Nat. Acad. Scis.-NRC; sec. Commn. on Dating the Pleistocene Internat. Union for Quaternary Research; hon. v.p. 2d Internat. Conf. on Palynology, Utrecht, 1966. Mem. Internat. Soc. Study Biometeorology, Internat. Union Quaternary Research, Geol. Soc. Am., A.A.A.S. (chmn. com. on arid lands), Am. Meteorol. Soc., Am. Geophys. Union, Am. Anthrop. Assn., Ecol. Soc. Am., Soc. Am. Archaeology, Tree-Ring Soc., N.Y., Ariz. acads. sci., Sigma Xi. Contbns. to creation and devel. of field of geochronology. Home: 2732 N. Gill Av., Tucson 85719.*

SMILLIE, Robert Maxwell, Australian biochemist; b. Sydney, Australia, July 9, 1933; s. Maxwell Oswald and Violet (Loch) S.; B.Sc. with 1st class honors, U. Sydney (Australia), 1954, M.Sc., 1955; Ph.D., Queens U., Kingston, Ont., Can., 1959; m. Barbara Anne Sparling, Nov. 8, 1958; children—Catherine Anne, Peter Kenneth, Michael Raymond. Asso. scientist Brookhaven Nat. Lab., Asso. Univs., Upton, L.I., N.Y., 1959-63; leader plant physiology unit, Commonwealth sci. and Indsl. Research Orgn., hon. asso. Sch. Biol. Scis., U. Sydney, 1963—. Mem. Am. Soc. Plant Physiologists, Am. Soc. Protozoologists, Australian Biochem. Soc., Australian Soc. Plant Physiologists, Royal Australian Chem. Inst. Research, publs. on isolation and properties of plant mitochondria, cellular regulation of plastid differentiation, physiology of aging of plant tissues, mechanism of photosynthetic electron transfer. Home: 59 Finlay, Warrawee, N.S.W., Australia. Office: Sch. Biol. Scis., U. Sydney, Sydney, N.S.W., Australia.*

SMIRK, Frederick Horace, physician; b. Accrington, Eng., Dec. 12, 1902; s. Thomas and Betsy (Cunliffe) S.; Baskill Math. scholar U. Manchester, 1919, M.B., C.L.B. with 1st Class Honors, 1925, M.D. (Gold medal), 1925; m. Aileen Winifrede Bamforth, Dec. 22, 1931; children—Adrian Thomas, Bernard Andrew, Pauline Marguerite, Cunliffe, Antony Hugh Cunliffe. Med. registrar Manchester (Eng.) Royal Infirmary, 1926-29; Dikenson Travelling scholar U. Vienna, 1930; Beit Meml. fellow, later asst. depts. pharmacology and medicine U. Coll., London, 1930-34; prof. pharmacology, physician postgrad. dept. Egyptian U., 1935-39; prof. medicine U. Otago, Dunedin, New Zealand, 1940-61, research prof. medicine, dir. Wellcome Med. Research Inst., 1962—; vis. prof. Brit. Postgrad. Med. Sch., London, 1949; McIlraith vis. prof. Royal Prince Alfred Hosp., Sydney, Australia, 1953. Chmn., Clin. Research Com., 1942-60; mem. expert com. on hypertension and ischaemic heart disease WHO. Fellow Royal Coll. Physicians London, Royal Australasian Coll. Physicians (past v.p.); mem. Med. Research Council New Zealand (mem. council 1944-60); mem. Internat. Soc. Cardiology (mem. council, mem. hypertension research sub-com.), Physiol. Soc., Med. Research Soc. Author: High Arterial Pressure, 1957; also numerous articles, chpts. in books. Research on normal circulation and hypertension, irregularities of heart action, including sudden death; selective breeding of rats with hereditary hypertension. Home: 68 Cannington Rd., Dunedin, New Zealand.

SMIRNOV, Nikolai Vasilevich, Russian mathematician; b. Oct. 17, 1900; grad. Moscow U., 1926. Prof., Lenin Moscow Pedagogical Inst., 1937-41; became asso. Inst. Math., USSR Acad. Sci., 1938; prof. Moscow City Pedagogical Inst., 1943—. Recipient Stalin prize, 1951. Corr. mem. USSR Acad. Scis. Author: Limited Laws of Distribution for Terms of Variational Series, 1949; Theory of Probability and Mathematical Statistics in Technics, 1955. Research in math. statistics, particularly nonparametric methods, probabilty. Home: 1-Aya Cheremushkinskaya 24 /1. Office: V.A. Steklov Inst. Math., USSR Acad. Scis., 1-y Akademicheski, Moscow, USSR.

SMIRNOV, Vasilii Ivanovich, Russian metallurgist; b. Feb. 11, 1899; grad. Leningrad Mining Inst., 1922; degree Dr. Tech. Scis., 1938. Engr., Katalinskii Copper Works, Urals, 1922-25; supr. reconstrn. of Karabashskii Copper Smelting Plant, 1925-27; chief metallurgist Urals Copper Trust, 1927-30, also dept. tech. dir.; became lectr. Ural Inst. Non-Ferrous Metals, 1930; prof. Ural Poly. Inst., 1933—. Sci. cons. Inst. Metallurgy and Ore-Dressing of Altai Mining and Metall. Inst., Acad. Sci. Kazakh SSR. Recipient Order Red Banner of Labor, twice. Mem. Kazakh SSR Acad. Sci., USSR Acad. Sci. (corr.). Author: Hydrometallurgy of Copper 1947; Metallurgy of Copper and Nickel, 1950; Reverberatory Smelting Theory and Practice, 1952; Shaft Furnace Smelting in Non-Ferrous Metallurgy, 1955; Pyrometallurgy of Copper; The Firing of Copper Ores and Concentrates; coauthor: Heat Content and Melting Point of Slags in Shaft-Furnace Smelting of Lead, 1959. Research on melting of nickel, copper ores and their concentrates. Office: Ural Poly. Inst., Kazakh SSR Acad. Scis., 1-uchebny Korpus, Sverdlovsk, USSR.

SMIRNOV, Vasilii Sergeevich, Russian metallurgist; b. 1915; grad. Ural Poly. Inst., 1937; Dr.Tech. Scis., 1948. Worked in industry, 1937-38, 41-42; asst. sr. lab. technician, 1938-41; staff Ural Poly. Inst., 1942-49; became chmn. dept. plastic treatment metals M. I. Kalinin Leningrad Poly. Inst., 1949, dep. dir., 1954-56, then named dir. Mem. USSR Acad. Scis. (corr.). Author: Transverse Metal Rolling, 1948; Calibration of Rollers by Coordinative Zones, 1953; Transverse Rolling and Machine Building, 1957; Longitudinal Periodic Rolling, 1962; Fundamentals in the Theory of Metal Rolling, 1962. Research in pressure treatment of metal, including transverse spiral and longitudinal periodic rolling, punching and pressing. Office: M. I. Kalin Leningrad Poly. Inst., Polytechnical ul. 3, Leningrad, K-64, USSR.

SMIRNOV, Vladimír Ivanovich, Russian mathematician; b. Leningrad, Russia, June 10, 1887; grad. St. Petersburg U., 1910; Dr.Physico-Math. Sci., 1936. Prof., Petersburg Inst. Engrs. Means Communication, 1912-30; intrr. Petrograd U., 1915-26; prof. Leningrad U., 1926—; staff Seismol. and Math. Insts. USSR Acad. Scis., 1929-35. Recipient Stalin prize, 1948, Order Lenin, twice. Mem. USSR Acad. Scis. Author: A Course of Advanced Mathematics, 5 vols., 1924-47, 4 vols., 1957; Conjugate Functions in Multi-Dimensional Euclidean Space, 1954. Research and publs. on theory of function of complex variable; developed (with S. L. Sobolev) method for obtaining solutions on propagation of waves in elastic media with plane boundaries; developed method for studying oscillations of elastic circle and sphere. Office: USSR Acad. Scis., Leninskii Prospekt, 14, Moscow, USSR.

SMIRNOV, Vladimir Ivanovich, Russian geologist, b. Jan. 1910; grad. Moscow Geol. Survey Inst., 1934. Instr., Moscow Geol. Inst., 1936-46; USSR dep. minister geology, 1946-51; simultaneously prof. Moscow Geol. Survey Inst., and Moscow Inst. Non-Ferrous Metals and Gold; named prof. Moscow U., 1951. Mem. Com. for Lenin Prizes for Sci. and Tech., 1960. Mem. USSR Acad. Scis. Author: The Laws Governing the Distribution of Ore Deposits and Prospecting Methods in the Talas Alatau, 1939; An Assessment of Mineral Raw Material Reserves, 1950; The Geological Principles of Ore Deposit Prospecting and Survey, 1957. Research on geology of ore deposits, prospecting and assaying. Office: Dept. Geology, Moscow U., Moscow, USSR.

SMIT, Jakob Van Rouendal, chemist; b. Stellenbosch, S. Africa, Jan. 12, 1931; s. Wynand Stephanus and Jacoba Maria (Van Rouendal) S.; B.Sc., Stellenbosch U., 1950, M.Sc. cum laude, 1954; D.Phil., Oxford (Eng.) U., 1956; m. Elsie Johanna Elizabeth Cronjé, Jan. 12, 1957; children—Wynand Jakobus, Thomas Frederik. Research officer, sr. research officer, head sect. inorganic and radiochemistry Council for Sci. and Indsl. Research, Pretoria, S. Africa, 1956-64; sr. lectr. dept. chemistry Stellenbosch U., 1964-65; editor publs. S. African Chem. Inst., Johannesburg, 1965-66; research mgr. Chem. Industries S. Africa, Johannesburg, 1966-67; sr. lectr., head dept. chemistry U. Coll. Western Cape, Bellville, Capetown, S. Africa, 1967—. Rhodes scholar Oxford U., 1953-56; Swiss Postdoctoral fellow, Zurich, 1959. Mem. S. African Chem. Inst., Chem. Soc. (London). Research, publs. and patents on activation analysis; pioneered usefulness of ion exchange properties of insoluble heteropolyacid salts showing their use for chromatographic separations. Home: 17 Van der Weshtuizen Av., Durbanville, S. Africa. Office: Chemistry Dept., Univ. Coll. Western Cape, P.O. Kasselvlei, S. Africa.*

SMITH, Adam, Scottish economist; b. Kirkcaldy, Scotland, June 5, 1723; s. Adam and Margaret (Douglas) S.; ed. U. Glasgow (Scotland), 1737-40, LL.D., 1762; also ed. Balliol Coll., Oxford (Eng.) U., 1740-46. Lectr. rhetoric, belles-lettres, Edinburgh (Scotland) U., from 1748; prof. logic U. Glasgow, 1751, chair moral philosophy, 1752-63, rector, from 1787; traveling companion Duke of Buccleuch, 1764-66, ret. Kirkcaldy, 1766-76, London, Eng., 1776-78; commr. customs Scotland, from 1778. Fellow Royal Soc., 1767, Royal Soc. Edinburgh. Author: Theory of Moral Sentiments, 1759; Inquiry into the Nature and Causes of the Wealth of Nations, 2 vols., 1886 (considered 1 of most influential books of 18th century); Essays on Philosophical Subjects, 1795. Founder classical sch. econs., precursor modern sci. polit. economy; proponent laissez-faire, developed theory div. labor; spokesman for rising industrialists and the colonies; advocated free trade between and within nations, competition and specialization; formulated doctrine that real value and market price can be equal only when there is an equilibirum between supply and demand; formulated theories of money and distbn.; emphasized individual self-interest as mainspring of benefit to soc.; opposed monopoly, mercantilism. Died Edinburgh, July 17, 1790.

SMITH, Alan B., Am. nuclear physicist; b. Chgo., Dec. 19, 1924; s. Hubert B. and Jeanette (Saunders) S.; B.A., Beloit Coll., 1949; M.S., Ind. U., 1950; Ph.D., 1952; m. Barbara Kelley, Dec. 26, 1949; children—Christopher Kelley, David Lindsey. Sr. physicist Argonne (Ill.) Nat Lab., now head applied nuclear physics sect. Reactor Physics div. chmn. nuclear cross sects. adv. group U. S. AEC. Fellow Am. Phys. Soc.; mem. Am. Nuclear Soc. (Hon. Achievement award), Phi Beta Kappa. Research, numerous publs. on fast neutron physics. Home: 645 63d St., Downers Grove, Ill. Office: Argonne Nat. Lab., Argonne, Ill.*

SMITH, Albert Charles, Am. biologist; b. Springfield, Mass., Apr. 5, 1906; s. Henry Joseph and Jeanette Rose (Machol) S.; A.B., Columbia, 1926, Ph.D., 1933; m. Nina Crönstrand, June 15, 1935; children—Katherine (Mrs. S. Robert Goldman), and Michael Alexis; m. 2d, Emma van Ginneken, Aug. 1, 1966. Engaged as assistant curator N.Y. Bot. Garden, 1928-31, asso. curator, 1931-40; curator herbarium Arnold Arboretum of Harvard U., 1940-48; curator div. phanerogams U. S. Nat. Museum, Smithsonian Inst., 1948-56; program dir. systematic biology Nat. Sci. Found., 1956-58; dir. Mus. of Natural History, Smithsonian Institution, 1958-62, asst. sec., 1962-63; prof. botany, dir. research U. Hawaii, Honolulu, 1963-65, Wilder professor of botany, since 1965—. Bot. expeditions, Colombia, Peru, Brazil, Brit. Guiana, Fiji, West Indies, 1926-56; del. Internat. Bot. Congresses, Amsterdam, 1935, Stockholm, 1950 (v.p. systematic sect.), Montreal, 1959, International Zoological Congress, London, 1958. Bishop Museum fellow Yale, 1933-34, Guggenheim fellow, 1946-47; trustee Hawaiian Bot. Gardens Found.; hon. asso. B. P. Bishop Museum, 1964—. Fellow Am. Acad. Arts and Scis., A.A.A.S.; mem. Am. Inst. Biol. Scis., Bot. Soc. Am., Am. Soc. Plant Taxonomists (pres. 1955; mem. council 1956-62, chmn. 1957-59), Am. Fern Soc. (asso. editor jour. 1952-60), Soc. to Study Evolution, New Eng. Bot. Club (corr. sec. 1944-47), Assn. Tropical Biology (pres. 1967-68), Internat. Assn. Plant Taxonomy (v.p. 1959-64), Biol. Soc. Washington (pres. 1962-64), Nat. Acad. Scis., Am. Soc. Naturalists, Nature Conservancy, Bot. Soc. Washington (pres. 1962); hon. mem. Soc. Cubana de Botanica, Fiji Soc. Editor Brittonia, 1935-40. Jour. Arnold Arboretum, 1941-48, Sargentia, 1942-48, editorial com. International Code Botanical Nomenclature, 1954-64. Author tech. articles. Research on phytogeography and taxonomy of flowering plants, especially those of southwest Pacific and tropical America. Office: U. Hawaii, Honolulu 96822.*

SMITH, Alec, biologist; b. Birmingham, Eng., Jan. 19, 1927; s. George Alexander and Ada (Shotton) S.; B.Sc., U. Birmingham, 1948, D.Sc., 1965; Ph.D., London U., 1950; m. Irene E. G. Forsdike, Feb. 6, 1954; children—Linda Catherine, Diana Mary. Entomologist, E. A. Filariasis Research Inst., Mwanza, Tanganyika, 1950-54; entomologist East African Inst. of Malaria and Vector-Borne Diseases, Amani, Tanganyika, 1954-59; entomologist Tropical Pesticides Research Inst., Arusha, Tanzania, 1959-64, dep. dir., 1965-66, dir., 1967—. Cons. WHO, 1963; mem. East African Forestry Research Coordinating Com., 1967, East African Agrl. Research Coordinating Com., 1966-67, East African Natural Resources Research Council, 1966-67. Contbr. numerous articles to sci. jours. Devel. entomol. technique for use in malaria control schemes and in assessment of new insecticides, incrimination of mosquitoes Anopheles gambiae and Anopheles funestus as vectors of bancroftial filariasis and their role in epidemiology of malaria in East Africa. Address: P. O. Box 3024, Arusha, Tanzania.*

SMITH, Alexander, chemist; b. Edinburgh, Scotland, Sept. 11, 1865; s. Alexander W. and Isabella (Carter) S.; B.Sc., U. Edinburgh, 1886, LL.D., 1919; Ph.D., U. München, 1889; m. Sara Bowles, Feb. 16, 1905. Asst. in chemistry U. Edinburgh, 1889-90; prof. chemistry and mineralogy Wabash Coll., 1890-94; asst. prof. chemistry U. Chgo., 1894-98, asso. prof., 1898-1903, prof. and dir. of gen. and phys. chemistry, 1903-11, dean Jr. Colls., 1900-11; prof. and head of dept. chemistry Columbia U., 1911-1921. Author: Lassar-Cohn Laboratory Manual of Organic Chemistry (translated), 1895; Laboratory Outline of General Chemistry, 1899 (transl. into German, Russian, Italian, Portuguese); The Teaching of Chemistry and Physics (with E. H. Hall), 1902; Introduction to General Inorganic Chemistry, 1906 (transl. into German, Russian, Italian, Portuguese); General Chemistry for Colleges, 1908; Text-book of Elementary Chemistry, 1914; Intermediate Chemistry, 1919. Research on forms of sulphur, vapor-pressure measurement at high pressure. Died Sept. 9, 1922.

SMITH, Alexander Goudy, Am. physicist; b. Clarksburg, W. Va., Aug. 12, 1919; s. Edgel Ohr and Helen (Reitz) S.; student Marshall Coll., 1938-39; S.B., Mass. Inst. Tech., 1943; Ph.D., Duke, 1948; m. Mary Elizabeth Ellsworth, Apr. 17, 1942; children—Alexander Goudy, Sally Jean. Physicist, Mass. Inst. Tech. Radiation Lab., 1942-46; instr. Duke, 1946-48; faculty U. Fla., Gainesville, 1948—, prof. physics, astronomy, 1956—, asst. dean Grad. Sch., 1961—, chmn. dept. astronomy, 1962—. Cons. USAF, 1954-64, Asso. U. for Research in Astronomy, 1964—; prin. investigator numerous research projects USAF, Army, Navy, NASA, NSF. Co-recipient Oak Ridge Inst. award for research, 1948. Fellow Am. Optical Soc., Am. Phys. Soc.; mem. Am. Geophys. Union, Internat. Astron. Union, Fla. Acad. Scis. (medal 1965), Astron. Soc. Pacific, Am. Assn. Physics Tchrs., Am. Assn. U. Prof., Am., Brit. astron. socs. Author: (with T.D. Carr) Radio Exploration of the Planetary System, 1964; Radio Exploration of the Sun, 1967; also numerous articles. Research on devel. tunable magnetrons for radar transmitters; pioneered devel. sci. microwave spectroscopy, application radio astronomy to study planets. Home: 1417 N.W. 17th St., Gainesville, Fla. 32601.*

SMITH, Anthony Joseph, Am. psychologist; b. Cin., Oct. 8, 1916; s. Anthony Joseph and Gladys (Bradley) S.; A.B., U. Cal. at Los Angeles, 1939, Ph.D., 1945; m. Barbara Josephine Garrison, June 16, 1941; children—Carol Leslie, Eric David Alan, Laurel Kimberly. Lectr., U. Minn., Mpls., 1944-45; instr. U. Ill., Urbana, 1945-46; faculty U. Kan., Lawrence, 1946—, prof., 1954—, chmn. dept. psychology, 1953-62, 66—. Cons. VA, 1953—. USPHS spl. research fellow Tavistock Clinic, London, Eng., 1962-63. Mem. Am., Kan. (pres. 1955), Midwestern, Southwestern psychol. assns., Sigma Xi. Research, publs. on phenomenon of consistency among interrelated concepts, methods by which introduction of new information induce series of attitude changes within a system of interrelated objects and concepts. Home: 1010 Sunset Dr., Lawrence, Kan. 66044.*

SMITH, Archibald, Brit. mathematician; b. Greenhead, Glasgow, Scotland, Aug. 10, 1813; s. James and Mary (Wilson) S.; ed. Glasgow U., LL.D., 1864; B.A., also elected fellow Trinity Coll., Cambridge, Eng., 1836, M.A., 1839; m. Susan Emma Parker, 1853; 6 sons, including James Parker, 2 daus. Became barrister Lincoln's Inn, 1841. Fellow Royal Soc., 1856, Gold medal, 1865. Author: (with Frederick John Owen Evans) Admiralty Manual for ascertaining and applying the Deviations of the Compass caused by the Iron in a Ship, 1862; Supplement to the Rules for ascertaining the Deviations of the Compass caused by the Ship's Iron, 1855; A Graphic Method of correcting the Deviations of a Ship's Compass, 1855. Editor: Journal of A Voyage to Australia (by William Scoresby, with additional material, including an exact formula for effect of iron of a ship on compass), 1859. Founder (with Duncan Farquharson Gregory) Cambridge Math. Jour., 1837. Deduced from Poisson's gen. equation practical formulas for correction of observations made on board ship, 1842-47; deduced convenient tabular forms from formulas, 1851; one of 1st mathematicians in Eng. to use symmetrical method in analytical geometry, in deducing algebraical equations for Fresnel's wave-surface. Died Dec. 26, 1872.

SMITH, Archibald William, physicist; b. Edmonton, Alta., Can., Jan. 6, 1930; s. Andrew and Ruth (Manual) S.; B.Sc., U. Alta., 1951, M.Sc., 1952; Ph.D., U. Toronto, 1955; m. Margaret Elaine Campbell, Sept. 23, 1953; children—Malcolm C., Elaine D., Cynthia R. Physicist, Def. Research Bd., Ottawa, Ont., Can., 1956-61, Thomas J. Watson Research Center IBM, Yorktown Heights, N.Y., 1962—. Mem. Am. Phys. Soc. Research, publs. on nuclear size effects in atomic spectra, anodic oxide films on aluminum, ferromagnetic resonance and domain structure in yttrium iron garnet, optical nonlinear effects, photon statistics in gas and injection lasers, diffraction of laser light by magnetoelastic waves in yttrium iron garnet. Home: 280 Macy Rd., Briarcliff Manor, N.Y. 10510. Office: P. O. Box 218, Yorktown Heights, N.Y. 10598.*

SMITH, Arthur Henry, Am. biochemist; b. Sandusky, O., Jan. 27, 1893; s. Norman Textor and Mary (Appell) S.; B.Sc., Ohio State U., 1915, M.S., 1916; Ph.D., Yale, 1920; m. Adeline C. Thomas, Apr. 8, 1922; 1 dau., Caroline Thomas (Mrs. William A. Schaub). Jr. chemist U.S. Bur. Mines, 1917; research asst. Rockefeller Inst. Med. Research, 1919; faculty Yale, 1920-37, asso. prof. physiol. chemistry, 1930-37; prof. physiol. chemistry Wayne State U., Detroit, 1937-63, prof. emeritus, 1963—. Fellow Am. Inst. Nutrition (Borden award 1954, past pres.), A.A.A.S.; mem. Am. Soc. Biol. Chemists, Am. Chem. Soc., Soc. for Exptl. Biology and Medicine, Sigma Xi, Phi Lambda Upsilon, Alpha Chi Sigma. Research, numerous publs. on nutrition, vitamins, minerals, metabolism, energy, inorganic-organic acids, growth, learning, nutrition. Home: 2240 Iroquois St., Detroit. 48214.*

SMITH, Bernard, Brit. geologist; b. 1881; s. Alfred and Henrietta M. (Bussey) Sm.; ed. Sidney Sussex Coll., Cambridge; M.A., 1907; Sc.D., 1924; m., 1912; 1 son. Demonstrator geology, 1904-06; geologist Geol. Survey, 1906; dist. geologist, 1920; asst. to dir. His Majesty's Geol. Survey, 1931-35; dir. geol. Geol. Survey Gt. Britain and Mus. Practical Geology, London, from 1935. Recipient Bigsby medal, 1927. Fellow Royal Soc., 1933, Geol. Soc. (award 1913). Author: Physical Geography for Schools; also papers and memoirs on stratigraphy, water supplies, mineral resources. Died Aug. 19, 1936.

SMITH, Byron Capleese, Am. ophthalmic plastic surgeon; b. Tonganoxie, Kan., Aug. 28, 1908; s. Fountain M. and Anna (Capleese) S.; B.S., U. Kan., 1929, M.D., 1931; m. Margurite E. Fox Tate, June 26, 1948; children—Constance (Mrs. J. Stanley Williams), Grady L., Lindsay B. Mem. staff N.Y. Eye and Ear Infirmary, N.Y.C., 1938—, cons. plastic surgeon, 1950—; staff Manhattan Eye, Ear and Throat Hosp., N.Y.C., 1946—, surgeon, 1950—, bd. dirs., 1951—, chmn. dept. ophthalmology, 1963—, dir. dept. ophthalmic plastic surgery, 1958—; sr. surgeon cons. Paek Med. Group, N.Y.C., 1963—; faculty Yale Sch. Medicine, 1935-37, Oxford U., Eng., 1945, N.Y. U., 1951-62; ophthalmic plastic surgeon N.Y. U. Med. Center, 1955-63. Ophthalmic cons. Kingsbridge VA Hosp., 1949-51; cons. Southampton, Littauer hosps., Monmouth Med. Center, Hackensack Hosp. Assn.; cons. U. S. Armed Forces, 1947—. Recipient Ignazzio Barraquer award, 1965; Am. Acad. Ocularists award, 1966; award U. Republic China, 1966. Diplomate Am. Bd. Ophthalmology; mem. A.M.A., Am. Ophthal. Soc., Soc. Med. Cons. World War II, Soc. Research Ophthalmologists, Pan Am. Ophthal. Assn., Oxford (Eng.) Ophthal. Congress, Soc. Rehab. Facially Disfigured, Phi Beta Pi. Contbr. to Atlas Eye Surgery, 1962; Plastic Surgery of the Eye and Adnexa, 1962. Research in fractures of orbit, deformities complicating orbital fractures, methods of mgmt. of sunken eye and double vision. Home: 200 E. 66th St., N.Y.C. 10021. Office: 722 Park Av., N.Y.C. 10021.*

SMITH, Carroll N., Am. entomologist; b. Menlo, Ia., Nov. 5, 1909; s. Ulysses G. and Alice (Butler) S.; A.B. with distinction, George Washington U., 1932, M.A., 1934, Ph.D., 1941; m. Charlotte S. Yochelson, Oct. 12, 1937; 1 dau., Alice Louise (Mrs. Stephen Landy). Biol. aid Entomology Research div. U.S. Dept. Agr., Washington, 1931-35, med. entomologist, Washington, 1936-37, Vineyard Haven, Mass., 1938-41, Savannah, Ga., 1942-46, Orlando, Fla., 1947-63, Gainesville, Fla., 1963—; courtesy prof. entomology U. Fla., 1963—. Mem. expt. panel on insecticides WHO, 1962—, expt. panel on tick-borne diseases FAO, 1966—; asso. mem. Commns. on Rickettsial Diseases and on Malaria, Armed Forces Epidemiol. Bd. Recipient U.S. Dept. Agr. Unit Superior Service award, 1958. Mem. Fla. Entomol. Soc. (award 1963), Fla. Anti-Mosquito Assn., Am. Mosquito Control Assn., Entomol. Soc. Am. (pres. 1964), Sigma Xi. Contbr. numerous articles to sci. jours. First successful use of personal repellents for ticks and chems. for area control of ticks; first field demonstration of utility of chemosterilants for insect control; contbns. to biology and control of ticks, house flies, mosquitoes, body lice, bed bugs, fleas, cockroaches, chiggers, and sand, black, stable fies. Home: 317 N.W. 32d St. Office: P. O. Box 1268, Gainesville, Fla. 32601.*

SMITH, Cedric Martin, Am. pharmacologist; b. Stillwater, Okla., Feb. 1, 1927; s. Otto Mitchel and Mary Catherine (Carr) S.; B.S., Okla. A. and M. Coll., 1949; B.S., U. Ill., 1950, M.S., 1953, M.D., 1953; m. Mary Ella Wylie, 1948; children—Cristine, Michael, Celia. Faculty dept. pharmacology U. Ill. Coll. Medicine, Chgo., 1954-66, prof. pharmacology, 1963-66; prof., chmn. dept. pharmacology Schs. Medicine and Dentistry, State U. N.Y. at Buffalo, 1966—. Cons. tech. staff Inst. For Def. Analyses, Arlington, Va., 1964-65; mem. NIH grants rev. study sect. for pharmacology; NIH Spl. fellow Physiologisches Institut der Universität Göttingen, Germany, 1961-62. Recipient Borden award for undergrad. research, 1953. Mem. Am. Soc. Pharmacology and Exptl. Therapeutics, A.A.A.S., N.Y. Acad. Scis., Am. Assn. U. Profs., Sigma Xi, Alpha Omega Alpha, Nu Sigma Nu. Editorial bd. Jour. Pharmacology and Exptl. Therapeutics, 1959-65. Research, numerous publs. on evaluation of muscle relaxant drugs; mechanisms of action of agents altering sensory nervous systems. Home: 670 LeBrun, Amherst, N.Y. 14226. Office: Capen Hall, State U. N.Y., Buffalo 14214.*

SMITH, Charles Roger, Am. vet. physiologist; b. Hartville, O., Mar. 31, 1918; s. Charles Roger and Ethel O. (Seeman) S.; student Ohio U., 1936-38, 40-41; D.V.M., Ohio State U., 1944, M.Sc., 1946, Ph.D., 1953; m. Genevieve Lorraine Taylor, Aug. 9, 1946; children—Ronald Roger, Debra Aileen, Eric William. Faculty, Ohio State U., Columbus, 1944—, prof. vet. physiology, 1957—, chmn. dept., 1954—; research asso. U. Minn., 1946, Purdue U., summers 1949-51; vis. scholar U. Wash., 1960. Consultant to Morris Animal Found., 1968—; mem. Central Ohio Heart Research Com., 1957-60, Fellowship Com., 1960-63. Mem. Am. Vet. Med. Assn. (research council 1959—, rep. div. med. scis. Nat. Acad. Scis. 1965—), A.A.A.S. (council 1957-65), Am. Physiol. Assn., Am. Heart Assn., Sigma Xi, Omega Tau Sigma, Phi Zeta. Reviewer, Am. Jour. Vet. Research, 1959—. Research, numerous publs. on comparative mammalian cardiology, cardiovascular sounds in horse, normal rhythms in equines, congenital heart disease in canine. Home: 4873 Chevy Chase Av., Columbus, O. 43221.*

SMITH, Clement Andrew, Am. physician; b. Ann Arbor, Mich., Nov. 19, 1901; s. Shirley Wheeler and Sara (Browne) S.; A.B., U. Mich., 1923, A.M., 1925, M.D., 1928; A.M. (hon.), Harvard, 1948; Sc.D. (hon.), Colby Coll., 1958; M.D. (hon.), U. Groningen, Netherlands, 1964; m. Margaret Beal Earhart, Feb. 6, 1926 (dec. Oct. 1960); children—Pamela, Margaret (Mrs. Eric Herz), Hilary Janet. Practice medicine, specializing in pediatrics, Boston, 1933-43; mem. staffs Chidren's Hosp., Boston, Boston Lying-in Hosp; cons. children's med. service Mass. Gen. Hosp., Boston, 1947—; faculty U. Mich., 1932, Wayne U., 1943-45; faculty Harvard Med. Sch., 1933-43, 45—, prof. pediatrics, 1963—. Chmn., Josiah Macy Confs. on Physiology of Prematurity, 1956-61; praelator in pediatrics Queens Coll., U. St. Andrews, 1957—. Recipient Ylppo medal U. Helsinki, 1957. Mem. Council on Food and Nutrition, A.M.A., Am. Acad. Pediatrics (Borden award 1962), Am. Pediatric Soc. (v.p. 1961-62), Corr. Soc. Pediatrics, Paris, Obstet. Soc. Boston. Author: The Physiology of the Newborn Infant, 1945. Editor: Pediatrics, 1962—; editorial bd. Excerpta Medica, 1946—, Biologia Neonatorum 1952—. Publs. on description of readjustment of normal and abnormal respiratory, circulatory, and metabolic processes following premature and term birth. Home: 37 Fayerweather St., Cambridge, Mass. 02138. Office: 221 Longwood Av., Boston 02115.*

SMITH, Clifford E(dward), Am. astronomer; b. Olmstead County, Minn., July 30, 1901; s. Frank J. and Jessie May (Phillips) S.; A.B., Carleton Coll., 1923; M.A., Swarthmore Coll., 1926; Ph.D., U. Cal. at Berkeley, 1936; m. Ruth Kinell, June 24, 1929; children—Helen (Mrs. Richard Stanley), Nelson Edward, Eric Thomas. Asst. math. and astronomy Swarthmore Coll., 1925-26; instr. Carleton Coll., 1926-27, Fresno (Cal.) State Coll., 1928-29, Mills Coll., 1920-31; asst. prof. Fresno State Coll., 1932-34; asst. computing Lick Obs., 1934 36; mem. faculty San Diego State Coll., 1937—, prof. astronomy, 1946—; cons. Convair Astronautics, 1953-60; project dir. geodetic observing program coop. U. S. Naval Obs., IGY, 1957-61. Fellow A.A.A.S.; mem. Am. Astron. Soc., Astron. Soc. Pacific, Sigma Xi, Sigma Pi Sigma. Research on star parallax, photog. position of stellar bodies, radial velocities, classification of spectral types and photog. determination of stellar magnitudes. Home: 8518 Chevy Chase Dr., La Mesa, Cal. 92041. Office: San Diego State Coll., San Diego 92115.*

SMITH, Cyril Stanley, Am. metallurgist; b. Birmingham, Eng., Oct. 4, 1903; s. Joseph Seymour and Frances (Norton) S.; B.S., U. Birmingham, 1924; D.Sc., Mass. Inst. Tech., 1926; D.Litt., Case Inst. Tech., 1965; m. Alice Kimball, Mar. 16, 1931; children—Stuart Marchant, Anne Kimball. Came to U.S., 1924, naturalized, 1940. Research metallurgist Am. Brass Co., Waterbury, Conn., 1927-42; asso. div. leader in charge metallurgy Los Alamos Lab., 1943-45; dir. Inst. For Study Metals U. Chgo., 1945-56, prof. metallurgy, 1946-61; Inst. prof., prof. history sci. and tech., prof. metallurgy Mass. Inst. Tech., Cambridge, 1961—. Mem. Pres.'s Sci. Adv. Com., 1959; mem. com. on sci. and pub. policy Nat. Acad. Scis., 1965—. Recipient Francis J. Clamer medal Franklin Inst., 1952; Am. Soc. Metals Gold medal, 1961; Mem. Am. Inst. Mining, Metall. Engrs. (Douglas medal 1963, past div. chmn.), Am. Phys. Soc. (past div. chmn.), Soc. For History Tech. (Leonardo da Vinci medal 1966, past pres.), Am. Philos. Soc., Internat. Acad. of History Scis. Author: A History of Metallurgy, 1960; Sources for the History of Steel, 1967. Editor, co-translator The Pirotechnica of Vannoccio Biringuccio, 1942, 66; Lazarus Ercker's Treatise on Ores and Assaying, 1951; On Divers Arts, the Treatise of Theophilus, 1963; editor: The Structure and Properties of Solid Surfaces, 1953, The Sorby Centennial Symposium on the History of Metallurgy, 1965. Research on microstructure of polycrystalline materials, plutonium metallurgy, effect of explosive shock on metals, metall. archaeology, hist. interaction of pure sci., tech. and art. Home: 31 Madison St., Cambridge, Mass. 02138.*

SMITH, David Eugene, Am. mathematician; b. Cortland, N.Y., Jan. 21, 1860; s. Abram P. and Mary E. (Bronson) S.; Ph.D., Syracuse U., 1881, Ph.M., 1884, Ph.D., 1887, LL.D., 1905; M. Pedagogics, Mich. State Normal Coll., 1898; student in Europe at various times, also 1907-08; D.Sc., Columbia, 1929; L.H.D., Yeshiva Coll., 1936; m. Fanny Taylor, Jan. 19, 1887; m. 2d, Eva May Luse, November 5, 1940. Practiced law, Cortland, 1881-84; tchr. math. State Normal Sch., Cortland, 1884-91; prof. math. Mich. State Normal Coll., 1891-98; prin. N.Y. State Normal Sch., Brockport, 1898-1901; prof. math. Teachers Coll. (Columbia), 1901-26 (emeritus). Librarian Am. Math. Soc., 1902-20, v.p., 1922, asso. editor Bull., 1902-20. Vice-pres. Internat. Commn. on the Teaching of Mathematics, 1908-20, pres., 1928-32, hon. pres. since 1932. Fellow Mediaeval Acad. Am., A.A.A.S.; mem. Math. Assn. Am. (pres. 1920-21), History and Science Soc. (pres. 1927), Am. Math. Soc., Deutsche Math. Verein, Phi Beta Kappa; hon. mem. Calcutta Math. Soc. Author: History of Modern Mathematics, 1896; Teaching of Elementary Mathematics, 1900; Rara Arithmetica, 1907; Teaching of Arithmetic, 1909, 13; Teaching of Geometry, 1911; Hindu-Arabic Numerals (with L. C. Karpinski), 1911; History of Japanese Mathematics, 1912; Union List of Mathematical Periodicals (with C. E. Seely), 1918; Number Stories of Long Ago, 1919; The Sumario Compendioso of Juan Diez, 1920; Our Indebtedness to Greece and Rome in Mathematics, 1922; Essentials of Geometry, 1923; Historical-Mathematical Paris, 1924; Mathematics Gothica, 1925; History of Mathematics (2 vols.), 1924, 25; Progress of Arithmetic in 25 years, 1924; Progress of Algebra in 25 years, 1925; Computing Jetons, 1924; Teaching of Junior High School Mathematics (with W. D. Reeve), 1927; Le Comput Manuel de Magister Anianus, 1928; Source Book in Mathematics, 1929; History of American Mathematics Before 1900 (with J. Ginsburg), 1934; The Rubáiyát of Omar Khayyám (metrical version), 1933; Poetry of Mathematics and Other Essays, 1934; Numbers and Numerals (with J. Ginsburg), 1937; The Wonderful Wonders of 1, 2, 3, 1937; also over 40 math. textbooks and many articles in various journals. Translator: Descartes's La Géométrie, 1925. Editor: A Portfolio of Portraits of Eminent Mathematicians, Part I, 1905, Part II, 1906, high school edit., 1907; DeMorgan's Budget of Paradoxes (2 vols.), 1915; Portraits of Eminent Mathematicians, Portfolio I, 1936, Portfolio II, 1937; Firdausi Celebration, 1936. Math. editor New Internat. Ency., 1902-16, Monroe's Cyclo. of Edn., 1911-13, New Practical Reference Library, 1912, Ency. Brit., 1927, Nat. Ency., 1933; asso. editor Am. Math. Monthly, 1916-—, Scripta Mathematica, 1932-—. Died July 29, 1944.

SMITH, David Macleish, Brit. mech. engr.; b. Elgin, Scotland, June 9, 1900; s. David T. and Mary (Pennycook) S.; B.Sc., Glasgow (Scotland) U., 1920, D.Sc., 1932; m. Doris Kendrick, Apr. 22, 1941. With Met. Vickers Elec. Co. Ltd. (later merged as Asso. Elec. Industries Ltd.), Trafford Park, Manchester, Eng., 1920-—, chief engr. mech. engring. devel., 1958-61, cons. mech. engr., 1961-—. Fellow Royal Soc., 1952. Royal Aero. Soc. Research, publs. on problems arising in turbine type machinery especially vibration problems and flow problems; devel. jet engines of axial flow type. Home: Flowermead, Winton Rd., Bowdon, Cheshire, Eng. Office: Turbine-Generator div. Asso. Elec. Industries Ltd., Trafford Park, Manchester 17, Eng.*

SMITH, David Tillerson, Am. physician; b. nr. Anderson, S.C., Oct. 1, 1898; S. William Whitaker and Florence (Sullivan) S.; A.B., Furman U., 1918, D.Lit., 1949; M.D., Johns Hopkins, 1922; m. Susan Gower, Sept. 12, 1922; 1 dau., Rosalind (Mrs. Robert Shields Abernathy). Faculty, Duke, Durham, N.C.,1930-—, James B. Duke, prof. microbiology, 1960-—, chmn. dept. preventive medicine, 1963-—; cons. VA Hosp., Durham, 1960-—, Tb VA hosps. Southeastern U.S., 1960-—. Recipient Trudeau medal for research on Tb, 1957; Medal for Distinguished Service, So. Tb Conf., 1958. Diplomate Am. Bd. Microbiology. Fellow Am. Acad. Microbiology; mem. Nat. Tb Assn. (dir. 1934-—, pres. 1950), Am. Soc. Pathologists and Bacteriologists, Soc. Am. Microbiologists, Am. Soc. Immunology, Am. Thoracic Soc., Am. Assn. Physicians. Author: Oral Spirochetes and Related Organisms in Fusospirochetal Diseases, 1932; Fungus Diseases of the Lungs, 1947; (with others) Manual of Clinical Mycology, 1944; (with others) Zinsser's Textbook of Bacteriology, 1948, 52, 57; (with others) Zinsser Microbiology, 1960, 1964. Research, publs. on diagnosis, treatment, prevention of chronic bacterial, mycotic, nutritional diseases; specifically, pellagra, brucellosis and Tb. Home: 3437 Dover Rd., Hope Valley, Durham, N.C. 27707.*

SMITH, Dietrich Conrad, Am. physiologist; b. Pekin, Ill., Mar. 21, 1901; s. Dietrich Conrad and Alma (Hippen) S.; A.B., U. Minn., 1923, A.M., 1924; Ph.D., Harvard, 1928; m. Margaret Odell Todd, Aug. 24, 1923; children—Dietrich Conrad 4th, Margaret Ann (Mrs. Thomas Buie Costen). NRC fellow Harvard, 1928-30, U. Munich (Germany), 1930-31; ind. investigator Naples (Italy) Zool. Sta., 1931-32; instr. U. Tenn., 1933-37; faculty U. Md., Balt., 1937-—, prof. physiology, 1949-65, asso. dean, 1955-65, prof. emeritus, 1965-—. Fellow A.A.A.S.; mem. Nat. (past dir.), Md. (sec. 1950-—) socs. for med. research, Am. Physiol. Soc., Endorcrine Soc., Am. Soc. Zoologists, Assn. Am. Med. Colls., Marine Biol. Lab., Bermuda Biol. Sta. Author: (with W.R. Amberson) Outline of Physiology, 1948; Euthanasia and Disposal Laboratory Animals in Methods of Animal Experimentation, 1965; also numerous articles. Research in pigmentary responses in vertebrates, venticular fibrillation in mammals, physiology thiamin deficiency, stress responses in mammals, physiology fish thyroid, electrolyte balance in mammals. Home: 216 Oak Forest Av., Balt. 21228.*

SMITH, Donald Morison, chemist; b. Montreal, Que., Can., Nov. 19, 1925; s. J. Thorold and Marjorie (Morison) S.; B.Sc. with honors in Chemistry, McGill U., 1946, Ph.D., 1956; M.S., U. Minn., 1949; m. Doris Markson, June 18, 1955; children—Alexander G.M., Eric J.M. With Carleton U., Ottawa, Ont., Can., 1949-51, Def. Research Chem. Labs., Ottawa, 1951-53, Harvard, 1955-56, Stanford Research Inst., 1956-58, U. B.C., Vancouver, 1958-59; research chemist Food and Drug Directorate, Tunney's Pasture, Ottawa, 1959-65; chief food standards, additives and regulations sect. food sci. and tech. br., nutrition div. FAO, Rome, Italy, 1965-—. Mem. Am. Chem. Soc., Chem. Soc. (London), A.A.A.S., Chem. Inst. Can., Soc. Chem. Industry, Can. Inst. Food Tech., Sigma Xi, Phi Lambda Upsilon. Research, publs. on thiophene chemistry, wood chemistry, map of urinary metabolites, food analysis, gas and paper chromatography as analytical tools to determine biochemically related constituents of natural material, methods for trace amounts of safrole, carcinogenic polynuclear aromatic hydrocarbons, aflatoxin, nitrates and nitrites, and uric acid as index of insect contamination, analysis of meat and meat products to establish standards based on protein content. Home: 40, Viale Africa. Office: Nutrition Div., FAO, Via delle Termedi Caracalta, Rome, Italy.*

SMITH, Donald Oscar, Am. physicist; b. Santa Fe, Feb. 9, 1925; s. Rollin Alanson and Fern (Green) S.; B.S., M.S., Mass. Inst. Tech., 1950, Ph.D. in Physics, 1955; m. Claire Vivian Wilson, June 3, 1951; children—Wendell Alanson, Rebecca Fern. Staff mem. Lincoln Lab., Mass. Inst. Tech., Lexington, 1955-—. Mem. Am. Phys. Soc. Research, publs. on magnetic films, methods to combine use lasers and magnetic films in computer memories; pioneered demonstration switching time in anaosecond region. Home: 16 Dewey Rd., Lexington, Mass.*

SMITH, Donald Ridgeway, Am. physician; b. Berkeley, Cal., May 23, 1909; s. Wilfred J. and Ethel R. (Davies) S.; A.B., U. Cal., 1931, M.D., San Francisco, 1935; m. Eleanor Wright, Sept. 21, 1931; children—Glenn David, Grant Ridgeway, Linda Eleanor. Practice medicine, specializing in urology, San Francisco, 1940-—; faculty U. Cal. Sch. Medicine San Francisco, 1940-—, chmn. div. urology, 1953-—, clin. prof., 1955-—. Fellow A.C.S.; mem. Am. Urol. Assn., A.M.A., Pacific Coast Surg. Assn., Am. Assn. Genito-Urinary Surgeons, Soc. Pediatric Urology, Am. Assn. Med. Colls. Author: General Urology, 1966. Research, numerous publs. on surgery and urology in pediatric urology, surg. correction of hypospadis, psychosomatic aspects of urology. Home: 1695 Funston Av., San Francisco 94122. Office: 384 Post St., San Francisco 94108.*

SMITH, Dorothy Gordon, microbiologist; b. Barbados, B.W.I., Mar. 5, 1918; d. Robert Garraway Smith and Alice (Nicholls) Smith King; came to U. S., naturalized, 1927; student Radcliffe Coll., 1935-38; B.A., Queen's U., Can., 1940; Ph.D., Rutgers U., 1947. With Canadian NRC, Kingston, Ont., 1940-41, meningococcal meningitis commn. Johns Hopkins, 1941-43; Merck Inst., Rahway, N.J., 1943-46; with U. S. Army, Ft. Detrick, Md., 1947-—, dep. chief Biol. Scis. Lab., 1966-67, chief Enviromental Biology office, 1967-—. Diplomate Am. Bd. Microbiology. Fellow Am. Acad. Microbiology, A.A.A.S.; mem. Sci. Research Soc. Am., Soc. Exptl. Biology and Medicine, Am. Soc. for Microbiology, N.Y. Acad. Scis., Sigma Xi. Research, publs. on Clostridia bacteria and effects of chemotherapeutic agts. on their infections, toxicity and properties of antibiotics; contbns. to discovery of streptomycin and chemistry of penicillin, research on all aspects of viral and rickettsial diseases of man and animals. Home: Brooklawn Apts. Office: Environmental Biology Office, U. S. Army Biol. Center, Ft. Detrick, Frederick, Md. 21701.

SMITH, Dudley Crofford, Am. physician; b. Lafayette Springs, Miss., Dec. 15, 1892; s. John General Marion and Carra (Powell) S.; B.S., U. Miss., 1914; M.D., U. Va., 1916; m. Lake Morrow, June 28, 1921; children—Dudley Crofford, Powell Morrow, Marjorie (Mrs. Hamilton Smithey). Intern, resident U. Va. Hosp.; instr. medicine U. Va., 1916-17, founded dept. dermatology and syphilology, 1924, prof., 1934-50; spl. cons. USPHS, since 1933; investigator studying and standardizing syphilis treatment with penicillin in collaboration with NRC and USPHS, 1943-1949. Diplomate Am. Bd. Dermatology and Syphilology (founder). Mem. A.M.A. (chmn. sect. dermatology and syphilology 1950), Am. Dermatol. Assn. (v.p. 1943-45), Am. Acad. Dermatology and Syphilology, Sigma Xi. Contbr. articles on epidemology of syphilis and dermatology to profl. jours. Died Aug. 30, 1950.

SMITH, Edgar Fahs, Am. chemist; b. York, Pa., May 23, 1856; s. Gibson and Susan (Fahs) S.; B.S., Pa. Coll., 1874; A.M., Ph.D., Göttingen, 1876; hon. degrees from various colls. and univs.; m. Margie A. Gruel, 1879. Instr. chemistry U. Pa., 1876-81; prof. chemistry Muhlenberg Coll., 1881-83, Wittenberg Coll., 1883-88; prof. chemistry U. Pa., 1888-1920, vice provost, 1899-1911, provost, 1911-20; pres. electoral course for Pa., 1925. Mem. Jury of Awards, Chgo. Expn., 1893; mem. U. S. Assay Commn., 1895, 1901-05; adviser in chemistry Carnegie Inst., 1902, research asso., 1915, 18; trustee Carnegie Found., 1914-20; pres. Wistar Inst., Phila., 1911-22. Recipient Elliott Cresson medal Franklin Inst., 1914, Chandler medal Columbia U., 1922. Mem. Nat. Acad. Scis., Am. Philos. Soc. (pres. 1902-06), Am. Chem. Soc. (pres. 1898, 1921, 22). Author (or editor): Classen's Quantitive Analysis, 1878; Clinical Analysis of Urine (with John Marshall), 1881; Richter's Inorganic Chemistry (5th edit.), 1900; Smith & Keller's Chemical Experimentation, 1902; Richter's Organic Chemistry, 1900; Smith's Electro-Chemical Analysis, 1911; Oettel's Practical Exercises in Electro-Chemistry, 1897; Oettel's Electro-Chemical Experiments, 1897; Elements of Chemistry, 1919; Shorter Course Chemical Experiments, 1913; Theories of Chemistry, 1913; Elements of Electrochemistry, 1913; also numerous books and pamphlets relating to history of chemistry in Am., and investigations in inorganic chemistry, including determination of atomic weights. Research on complex acids of tungsten, molybdenum, niobium; developed rotating anode, other electrolytic methods. Died May 3, 1928.

SMITH, Edward, English physician, med. writer; b. Heanor, Derbyshire, circa 1818; student Queens Coll., Birmingham; M.B., London U., 1841, M.D., 1843, B.A., LL.B., 1848. Lectr., demonstrator anatomy Charing Cross Hosp., 1853; asst. physician Brompton Hosp. for Consumption, 1861; cons. to govt. on poor-law and prison dietaries; apptd. med. officer of poor-law bd. Fellow Royal Soc., 1860, Royal Coll. Surgeons, Royal Coll. Physicians. Author: Structural and Systematic Botany, 1854; (with D. I. Ansted and others) Natural History of the Inanimate Creation, 8 vols., 1856; Experimental Inquiries into the Chemical and other Phenomena of Respiration, and their Modifications by various Physical Agencies, 1859; Cn the Action of Foods upon the Respiration during the Primary Processes of Digestion, 1859; Health and Disease, as influenced by the Daily, Seasonal, and other Cyclical Changes in the Human System, 1861; Consumption: its Early and Remediable Stages, 1862; Reports to Privy Council on the Dietary of Lancashire Operative, and of other Low-fed Populations, 1862-63; Practical Dietary for Families, Schools, and the Working Classes, 1864; How to get Fat, 1865; Foods, in Internat. Sci. Series, 1872; A Manual for Medical Officers of Health, 1873; A Handbook for Inspectors of Nuisances, 1873; Health: a Handbook for Households and Schools, 1874. Invented instrument to measure inspired air and to collect carbonic acid in expired air, 1859; as med. officer of poor-law bd. he placed poor-law dietaries on sci. practical basis; did work in reforming hygienically structural arrangements of workhouses and workhouse infirmaries. Died London, Nov. 16, 1874.

SMITH, Edward Bryon, pathologist; b. Petersburg, Ind., Apr. 24, 1912; s. Jacob Owen and Estella (Richardson) S.; student Hanover Coll., 1930-31; B.S., Ind. U., 1936, M.D., 1938; m. Edith M. Cash, Nov. 17, 1937; children—Michael E., Allen R., Philip A. Faculty, Washington U., St. Louis, 1940-42, asst. prof. pathology, 1949-51; asso. in pathology U. Pa., Phila., 1946-48; asso. prof. Baylor U. Med. Coll., Houston, 1948-49; prof. pathology, chmn. dept. Ind. U. Sch. Medicine, Indpls., 1951-62; chief hematologic pathology br. Armed Forces Inst. Pathology, Washington, 1962-65; clin. prof. pathology George Washington U., 1962-65; prof. pathology U. Mich., Ann Arbor, 1965-—. Sec.treas., trustee Am. Bd. Pathology, 1955-64, pres., 1965-—. Mem. Internat. Acad. Pathology (council 1952-58, pres. 1957), Am. Assn. Pathologists and Bacteriologists, Am. Soc. Clin. Pathology, Am. Soc. Exptl. Pathology, Coll. Am. Pathologists, Electron Microscope Soc. Am., N.Y. Acad. Sci. Author: (with P. Beamer, F. Vellios, D. Schulz) Principles of Human Pathology, 1959; also articles. Research on pathology of infectious mononucleosis, various tumors of lymphoid tissue, electron microscopic effects of contaminants and metabolites of phenacetin, phagocytosis in certain neoplasms. Home: 4940 N. Melrose St., Tampa, Fla.*

SMITH, Edward John, Am. physicist; b. Dravosburg, Pa., Sept. 21, 1927; s. Edward John and Margaret (Kerber) S.; B.A., U. Cal. at Los Angeles, 1951, M.A., 1952, Ph.D., 1959; m. Gloria Hawkins, Oct. 3, 1953; children—Michael, Claudia, Brian, Rachel. Research geophysicist Inst. Geophysics, U. Cal. at Los Angeles, 1955-59; mem. tech. staff TRW Systems, Redondo Beach, Cal., 1959-61; sr. scientist, research group supt. Jet Propulsion Lab., Pasadena, Cal., 1961-—. Mem. Am. Geophys. Union.

A.A.A.S. Publs. on measurement of magnetic fields in space using satellites and space probes, fluctuating magnet fields within ionosphere and lower magnetosphere, ring current and magnetic tail fields nr. Earth, magnetic fluctuations within Earth's magnetosheath, interplanetary magnetic fields, magnetic fields nr. Venus and Mars. Home: 2536 Boulder Rd., Altadena, Cal. 91001. Office: 4800 Oak Grove Dr., Pasadena, Cal. 91103.*

SMITH, Edward Staples Cousens, Am. geologist; b. Biddeford, Me., Aug. 23, 1894; s. James G.C. and Eva L. (Staples) S.; B.S., Bowdoin Coll., 1918; postgrad. Mass. Inst. Tech.; A.M., Harvard, 1920; m. Frances E. Shaver, Jan. 12, 1946. Faculty, Union Coll., Schenectady, 1923-60, prof. geology, 1931-60, prof. emeritus, 1960—; cons. geologist, 1940—. Fellow Geol. Soc. Am., Meteoritical Soc.; mem. Mineral. Soc. Am., Paleontol. Soc., Am. Geophys. Union, Sigma Xi, numerous others. Author: (with others) Applied Atomic Power, 1946. Research, numerous publs. in volcanic rocks and their composition, fossils in relation of land and sea and evolution, fluorescence of minerals that may have led to improvement of color tv screens. Home: 2244 N.W. 4th Pl. Office: P.O. Box 1154, Gainesville, Fla. 32601.*

SMITH, Edwin, Am. astronomer, geodesist; b. N.Y., Apr. 13, 1851; s. Edwin and Adelia O. (McIntyre) S.; ed. Coll. City of N.Y.; m. Lucy S. Black, Nov. 17, 1885. With U. S. Coast and Geod. Survey, 1870-95, 97—, asst., 1874; astronomer in charge party to observe transit of Venus at Chatham Islands, S. Pacific, 1874, at Auckland, New Zealand, 1882, determined force of gravity at Auckland, N.Z., Sydney, New South Wales, Singapore, Tokio, Japan, San Francisco and Washington, with the 3 Kater pendulums belonging to Royal Soc. Eng. and used in Gt. Indian Survey, in charge instrument div., 1879-94, during which time also carried on observations for variation of latitude at Rockville, Md., in cooperation with Internat. Geod. Assn.; with N.Y. State Land Survey, 1895-97; established Internat. Geod. Assn. Latitude Obs. at Gaithersburg, Md., 1899, made observations for variation of latitude to 1901; engaged on astron., magnetic, and geodetic work of survey, 1901-12. Wrote several papers pub. as appendices to Coast and Geodetic Service Reports. Died Dec. 2, 1912.

SMITH, Edwin Lee, Am. physiologist; b. Shelton, Neb., Aug. 12, 1907; s. George W. and Blanche (Lee) S.; Ph.G., U. Neb., 1929, B.Sc., 1935, M.Sc., 1938; Ph.D. (Sidney Walker III fellow), U. Chgo., 1941; m. Maxine Amelia Pierce, June 17, 1933; children—Suzanne Amelia (Mrs. Neal Carpenter), Edwin Lee. Instr., U. Ill. Med. Sch., 1941-43; asst. prof. physiology Med. Coll. Va., 1943-47; prof. physiology, chmn. dept. physiology, and pharmacology Dental Br., U. Tex., Houston, 1947—, lectr. physiology med. br., 1953—; vis. prof. physiology Baylor U. Coll. Medicine, 1951—. Mem. Am. Physiol. Soc., Soc. for Exptl. Biology and Medicine, Internat. Assn. for Dental Research, Sigma Xi. Research, numerous publs. on blood transfusions, blood pressure and blood vol. Home: 4322 Briarbend, Houston 77035.*

SMITH, E(lias) A(nthon) Cappelen, metall. engr.; b. Trondhjem, Norway, Nov. 6, 1873; s. Elias Anthon and Anna T. (Rovig) S.; grad. Tech. Coll., Trondhjem, Norway, 1893; m. Mary Ellen Condon (dec. 1927); m. 2d, Carmen Arlegui. Asst. chemist Armour & Co., Chgo., 1893-95; chemist Chgo. Copper Refining Co., 1895-96; supt. of electrolytic copper refinery Anaconda Copper Mining Co., Mont., 1896-1900; metall. engr. in charge metall. operations, Balt. Copper Smelting & Rolling Co., 1901-10; cons. metall. engr. Am. Smelting & Refining Co., N.Y., 1910-12; cons. metall. engr. Guggenheim Bros. of N.Y., Chile Exploration Co., Braden Copper Co., 1912-25; mem. Guggenheim Bros., since 1925; v.p. and dir. Peirce-Smith Converter Co. (holder of patent on his method of copper converting) since 1908; pres. and dir. Minerec Corp.; dir. Chilean Nitrate Sales Corp., Anglo-Chilean Nitrate Corp., Lautaro Nitrate Co., Ltd., Cia. Salitrera Anglo Chilena, Pacific Tin Corp. Recipient Gold medal Mining and Metall. Soc. Am. for distinguished service in art of hydrometallurgy, 1920. Mem. Mining and Metall. Soc. Am., Am. Inst. Mining and Metall. Engrs., Royal Norwegian Sci. Soc. Inventor of extraction method in use at Chuquicamata plant of Chile Exploration Co., Chile; originator of Guggenheim method of extracting nitrate from caliche. Died June 25, 1949.

SMITH, Elliott, Am. astronomer; b. Blue Earth County, Minn., Jan. 19, 1875; s. Frank Y. and Harriet Amanda (Cornish) S.; A.B., U. Minn., 1903; postgrad. (fellow), U. Cal., 1905-06; Ph.D., U. Cin., 1910; m. Louise Josephine Strautman, Nov. 28, 1908; children—Harriet Louise (Mrs. Paul Herget), Stephen E. Asst. in astronomy U. Minn., Licks Obs., 1903-05; asst. in obs. U. Cin., 1907-10, asst. prof. astronomy, 1910-20, asso., 1920-36, prof. since 1936, prof. and dir. of obs. since 1940. Mem. A.A.-A.S., Am. Astron. Soc., Astron. Soc. Pacific, Phi Beta Kappa, Sigma Xi. Author: Catalog of Proper Motion Stars (with others) 1916, 1930; A Catalog of 4,683 Stars Observed by Elliott Smith, 1922; The Luminosity of Meteors and Comets, 1940. Died Sep. 29, 1943.

SMITH, Emil L., Am. biochemist; b. N.Y.C., July 5, 1911; s. Abraham and Esther (Lubart) S.; B.S., Columbia, 1931, Ph.D., 1937; m. Esther Press, Mar. 29, 1934; children—J. Donald, Jeffrey B. Instr. Columbia, 1936-38; research asso. Rockefeller Inst., 1940-42; biochemist E. R. Squibb & Sons, 1942-46; asso. prof. U. Utah, 1946-50, prof. biochemistry, research prof. medicine, 1950-63, acting chmn., 1958-59; prof. biol. chemistry, chmn. dept. U. Cal. Med. Center, Los Angeles, 1963—. Mem. adv. com. USPHS, NRC, Office Naval Research, Am. Cancer Soc.; Internat. orgn. and programs com. Nat. Acad. Scis. Guggenheim fellow, 1938-40. Fellow N.Y. Acad. Sci., Am. Acad. Arts and Scis.; mem. Nat. Acad. Scis., Soc. Exptl. Biology and Medicine, Am. Chem. Soc., Am. Soc. Biol. Chemists, Am. Soc. Naturalists, Biochem. Soc. Gt. Britain, A.A.A.S. Author: (with others) Principles of Biochemistry, 3d edit., 1964. Editorial bd., Proc. Soc. Exptl. Biology and Medicine, Biochem. Preparations, Ann. Rev. Biochemistry, Physiol. Revs., Jour. Biol. Chemistry. Research and publs. on structure and function of enzymes and other proteins. Home: 10627 Le Conte Av., Los Angeles 90024.*

SMITH, Erminnie Adelle Platt, Am. geologist, ethnologist; b. Marcellus, N.Y., Apr. 26, 1836; d. Joseph Platt; grad. Troy (N.Y.) Female Sem., 1853; m. Simeon H. Smith; 4 children. Mem. staff Bur. Am. Ethnology of Smithsonian Instn., Washington, 1880, studied culture of Iroquois Indians, 1880-82, compiled Iroquois-English Dictionary; wrote Myths of the Iroquois, published by Bur. of Ethnology, 1883; 1st woman elected fellow N.Y. Acad. Scis.; mem. A.A.A.S.; reflected deep interest in geology and botany; founder, 1st pres. Aesthetic Soc. of N.J. Died Jersey City, June 9, 1886.

SMITH, Ernest Ketcham, Jr., Am. physicist; b. Peking, China, May 31, 1922 (parents Am. citizens); s. Ernest Ketcham and Grace (Goodrich) S.; B.A., Swarthmore Coll., 1944; M.S., Cornell U., 1951, Ph.D., 1956; m. Mary Louise Standish, June 23, 1950; children—Priscilla Farwell, Nancy Goodrich, Cynthia Cable. Mem. physics dept. N.Y. U., 1946; with engring. dept. MBS, N.Y.C., 1946-49; mem. elec. engring. dept. Cornell U., 1950-54; with Nat. Bur. Standards, Boulder, Colo., 1951-65, div. chief, 1960-65; dir. Aeronomy Lab. ESSA Boulder Labs., Boulder, 1965-67, dir. Inst. Telecommunication Sciences, 1967—; affiliate prof. atmospheric sci. Colo. State U., 1964—, chmn. Sporadic-E Propagation meeting, 1965, co-chmn. seminar on cause and structure of ionospheric sporadic-E, 1965; asso. Harvard Obs., 1966. Mem. A.A.A.S., Sci. Research Soc. Am., Am. Geophys. Union, Sigma Xi, numerous others. Author: Worldwide Occurrence of Sporadic E, 1957. Editor: (with S. Matsushita) Ionospheric Sporadic E, 1962. Research on ionosphere, refractive index of air. Home: 62 Wild Horse Circle, Pine Brook Hills, Boulder 80302. Office: ESSA Boulder Labs., Boulder Colo. 80302.*

SMITH, E(rnest) Lester, English biochemist; b. Teddington, Eng., Aug. 7, 1904; s. Lestern and Rose (Nettleton) S.; B.Sc., Chelsea Coll. Sci., 1925; M.Sc., U. London, 1926, D.Sc., 1933; m. Winifred Rose Fitch, Aug. 29, 1931. Staff, Glaxo Labs. Ltd., Greenford, Eng., 1926-64, sr. biochemist, 1952-64, cons., 1964—. Recipient Gold medal in therapeutics Worshipful Soc. Apothecaries, 1954; Hanbury Meml. medal Pharm. Soc., 1966. Fellow Royal Inst. Chemistry, Royal Soc., 1957; mem. Biochem. Soc., Hematological Soc. Author: Vitamin B12; also numerous articles. Research on phys. chemistry of soap solutions and saponification, prodn. methods for vitamins A and D, methods for extraction and crystallization of penicillin, chemistry of vitamin B12 and anti-B12 analogues; isolation of vitamin B12; preparation of radioactive penicillin and vitamin B12 for tracer studies. Home: Amberheath, Three Oaks Guestling, Hastings, Sussex, Eng. Office: Glaxo Research Ltd., Greenford, Middlesex, Eng.*

SMITH, Erwin F., Am. plant pathologist; b. Gilbert's Mills, N.Y., Jan. 21, 1854; s. R. K. and Louisa (Frink) S.; B.S., in biology, U. Mich., 1886, Sc.D., 1889, LL.D., 1922; Sc.D., U. Wis., 1914; m. Charlotte M. Buffett, Apr. 13, 1893 (dec. 1906); m. 2d, Ruth Annette Warren, Feb. 21, 1914. Expert pathologist U. S. Dept. of Agr., 1899—; later in charge lab. of plant pathology, Bur. of Plant Industry. Trustee Marine Biol. Lab., Woods Hole, Mass. (3 terms). Certificate of honor, A.M.A., 1913, for cancer in plants. Fellow Am. Acad. Arts and Scis., A.A.-A.S.; mem. Nat. Acad. Scis. (chmn. bot. sect. 3 yrs.) Am. Philos. Soc.; pres. Soc. Plant Morphology and Physiology, 1902, Soc. Am. Bacteriologists, 1906, Bot. Soc. Am., 1910, Am. Phytopathol. Soc., 1916. An incorporator of Nat. Carillon Assn. Author: Bacteria in Relation to Plant Diseases, Vol. I, 1905, Vol. II, 1911, Vol. III, 1914; For Her Friends and Mine (sonnets, issued pvtly.), 1915; Pasteur—the history of a Mind (transl. with Florence Hedges) 1920; An Introduction to Bacterial Diseases of Plants, 1920; various papers on general botany, mycology, sanitary science and bacteriology. Asso. editor Centralblatt für Bacteriologie, 25 vols.; contbr. to Standard Dictionary, 1st edit. Studied plant tumors; compared plant crown-gall to cancer in animals. Died Apr. 6, 1927.

SMITH, F. J., Irish math. physicist, mathematician; b. No. Ireland, Mar. 31, 1935; s. John and Mary (Nolan) S.; B.Sc. with 1st class honors, Queen's U., Belfast, Ireland, 1957, Ph.D., 1962; M.A., Cath. U. Am., 1959; m. Ann Maureen Dowling, June 21, 1965; children—Owen John Robert, Roty Maurice Michael. Faculty, Queen's U., Belfast, 1962-64, reader, 1966—; research asso. U. Md., 1964-66. Mem. Brit. Computer Soc. Publs. on devel. methods for computing transport properties of atomic gases (heat transfer, diffusion); quantal wave effects in atomic collision theory. Home: 181, Saintfield Rd., Belfast 8, N. Ireland.*

SMITH, Sir Francis Pettit, English inventor; b. Hythe, Eng., Feb. 9, 1808; s. Charles and Sarah (Pettit) S.; m. Anne Buck, 1830; 2 sons; m. 2d, Susannah Wallis, 1866; 2 sons. Engaged in farming; adviser to Brit. Admiralty, until 1850; became curator patent office mus., 1860. Recipient nat. testimonial with purse, 1857. Ass. Instn. Civil Engrs.; mem. Inst. Naval Architects, Royal Soc. Arts of Scotland, Am. Ins. Built model of screw-propelled boat, 1835, patented, 1836; built 6-ton boat Francis Smith with wooden screw, 1836, 1st screw steamship for Royal Navy, 1839, 1st screw warship (Rattler), 1841-43. Died South Kensington, Eng., Feb. 12, 1874.

SMITH, Frank, Am. zoologist; b. Winneconne, Wis., Feb. 18, 1857; s. Samuel Franklin and Aurelia (Shepard) S.; Ph.B., Hillsdale Coll., 1885, D.Sc., 1923; A.M., Harvard, 1893; m. Edith M. Fox, Sept. 8, 1887 (dec. 1888); 1 son, Donald Fisk; m. 2d, Isadora Stamats, July 12, 1890. Prof. chemistry and biology Hillsdale Coll., Mich., 1886-92; instr. biology Trinity Coll., Hartford, Conn., 1892-93, instr. zoology, 1893-96, asst. prof., 1896-1900, asso. prof., 1900-13, prof. zoology, 1913-26 (emeritus), U. Ill. Fellow A.A.A.S. Contbr. chiefly to morphology and taxonomy of land and fresh water annelids, of fresh water sponges, and on migration of birds. Died Feb. 3, 1942.

SMITH, Frank Ackroyd, Am. biochemist; b. Winnipeg, Man., Can., Feb. 14, 1919 (parents Am. citizens); s. Frank and Doris (Babcock) S.; B.A., Ohio State U., 1940, M.S., 1941, Ph.D., 1944; m. H. Jane McGuire, Apr. 15, 1944; children—Susan Jane, Deborah A. With Atomic Energy Project U. Rochester (N.Y.), 1944—, faculty Sch. Medicine and Dentistry, 1946—, asso. prof. radiation biology, 1960—. Mem. Am. Chem. Soc., Am. Soc. Pharmacology and Exptl. Therapeutics, Soc. Toxicology, Am. Indsl. Hygiene Assn., Am. Assn. Clin. Chemists, A.A.A.S., Am. Assn. U. Profs., Sigma Xi. Author: (with H.C. Hodge and P.S. Chen, Jr.) Biological Effects of Organic Fluorides, 1963; (with H.C. Hodge) Biological Properties of Inorganic Fluorides, 1965. Editor: Pharmocology of Fluorides, 1966. Contbr. chpts. to books, numerous articles to profl. jours. Research in toxicology of fluorides, metals of interest to atomic energy programs, effects, mechanisms of action in gen. toxicology. Home: 84 Raleigh St., Rochester, N.Y. 14620.*

SMITH, Frank Houston, Am. chemist; b. Cornelius, N.C., May 18, 1903; s. George Henderson and Louise (Bratton) S.; B.S., Davidson Coll., 1926; M.S., N.C. State Coll., 1931; m. Lois Ellington, Aug. 11, 1934; children—Frank Houston, David Richard. Jr. chemist N.C. Dept. Agr., 1927; asst. in animal nutrition N.C. State Coll., Raleigh, 1928-46, asso. prof. animal sci., 1947-62; prof. animal sci. N.C. State U., Raleigh, 1963—. Mem. Am. Chem. Soc., Am. Oil Chemists Soc., Am. Inst. Nutrition, Sigma Xi, Gamma Sigma Delta. Research, numerous publs. on methods for determination of gossypol content of cottonseed products and parts of cotton plant, for determining gossypol content of animal tissues; studies on genetic aspects of gossypol in leaves and flower buds of gossypium, isolation of gossypol from liver tissue of swine, effect of diet and iron on accululation of gossypol in livers of swine, effect of bound gossypol on nutritive quality of cottonseed protein, amino acid composition of cottonseed protein from glandless cotton. Home: 2506 Stafford Av., Raleigh, N.C. 27607.*

SMITH, Frederick George Walton, marine biologist, oceanographer; b. Bristol, Eng., Jan. 28, 1909; s. Frederick George and Nelly Maud (Belton) S.; B.S. (Royal scholar), Royal Coll. Sci., 1931, Ph.D., 1934; m. Florence May Walker, Jan. 21, 1939; children—Alexandra Walton. Came to U.S., 1940, naturalized, 1949. Commonwealth fellow, Princeton, 1934-36; biologist sponge fishery investigation Bahamas Govt., 1936-40; faculty U. Miami (Fla.), 1940—, prof. zoology, 1944-48, dir. Inst. Marine Sci., 1942—, prof. oceanography, 1948—; pres. Internat. Oceanographic Found., Miami, 1963-. Dir. Fla. Oyster Div., 1949-52; fisheries adviser Bahamas Govt., 1945—; tech. adviser govt. agys. Chmn., Fla. Commn. for Marine Scis. and Tech. Recipient Achievement medal Fla. Acad. Sci., 1942; Naval Ordnance Devel. award USN, 1945. Fellow Am. Geog. Soc.; mem. A.A.A.S., Am. Geophys. Union, Am. Soc. Limnology and Oceanography, Marine Biology Assn. U.K. Author: Atlantic Reef Corals, 1948; (with H. Chapin) The Ocean River, 1952, The Sun, The Sea and Tomorrow, 1954; also numerous articles. Editor: Sea Frontiers, 1954—; editorial bd. Bull. Marine Sci., 1951—. Research on embryology of marine invertebrates, comml. fisheries, tropical oceanography, biology of marine borers and fouling. Home: 4845 S.W. 78th St., Miami, Fla. 33143.*

SMITH, Frederick Viggers, psychologist; b. Hamilton, N.S.W., Australia, Jan. 24, 1912; s. Frederik Thomas and Agnes (Viggers) S.; B.A., U. Sydney (Australia), 1938, M.A., 1941; Ph.D., U. London (Eng.), 1948. Lectr., Tchrs. Coll., Sydney, 1938-46, Birbeck Coll., U. London (Eng.), 1946-48, U. Aberdeen (Scotland), 1948-50; prof., head dept. psychology U. Durham (Eng.), 1950—; vis. prof. Cornell U., Ithaca, N.Y., 1957. Fellow Brit. Psychol. Soc.; (past pres.); mem. Aristotelean Soc., Assn. for Study Animal Behavior. Author: Explanation of Human Behavior, 1951, rev., 1960; also articles. Research on explanation in psychology, evaluation of systems of psychology, instinct as factor in explanation of behaviour, instinctive behaviour and perceptual aspects of imprinting. Home: 8 Deyncourt, Lowes' Barns, Durham, Eng.*

SMITH, George, Brit. Assyriologist; b. Chelsea, London, Mar. 26, 1840; apprenticed to learn banknote engraving at age 14; later studied cuneiform inscriptions at Brit. Mus. Became asst. in Assyriology dept. Brit. Mus., 1867; earliest success was finding 2 inscriptions, one marking date of total eclipse of sun in month of Sivan in May 763 B.C., other referring to date of an invasion of Babylonia by Elamites in 2280 B.C.; achieved world-wide fame through his translation of Chaldaean account of flood given on Layard's tablets, 1872; went to Ninevah at expense of Daily Telegraph to find missing fragments of tablets of the Gilgmesh Epic recalling the Deluge; at that time found the missing fragments as well as tablets referring to succession and duration of Babylonian dynasties. Author: Annals of Assur-banipal, 1871; The Phonetic Values of Cuneiform Characters, 1871; Assyrian Discoveries, 1875; Ancient History from the Monuments: Assyria, 1875; The Chaldaean Account of Genesis, 1876; Ancient History from the Monuments: Babylonia, pub., 1877; The History of Sennacherib, pub., 1878. Died Aleppo, Syria, Aug. 19, 1876.

SMITH, George Edson Philip, Am. civil engr.; b. Lyndonville, Vt., Dec. 29, 1873; s. Franklin Horatio and Hattie Louisa (Powers) S.; B.S., U. Vt., 1897, C.E., 1898, D.Civil Engring., 1929; postgrad. U. Wis.; m. Maude North, Oct. 1, 1904; 1 son, George Edson Philip. Prof. civil engring. U. Ariz., Tucson, 1900-06, prof. irrigation engring., irrigation engr. Agrl. Expt. Sta., 1906-55, head dept. to 1944, part-time, 1944—; cons. engr., 1916-41. Recipient John C. Park Outstanding Civil Engr. award, 1966. Mem. Am. Soc. C.E., Am. Soc. Agrl. Engrs. Research, numerous publs. on measurement and importance of underflow beneath rivers, runoff from desert lands, fluctuations of groundwater level, groundwater law, physiography of Ariz. valleys, character and quantity of water supply in Oahu, Hawaii. Address: 1195 E. Speedway St., Tucson 85719.*

SMITH, George Elwood, Am. physicist; b. White Plains, N.Y., May 10, 1930; s. George F. and Lillian (Van Vorhees) S.; A.B., U. Pa., 1955; M.S., U. Chgo., 1956, Ph.D., 1959; m. E. Janet Carson, July 6, 1955; children—Leslie Ann, Lauren Lucille, Carson Eric. Mem. Solid state spectroscopy dept. Bell Telephone Labs., Murray Hill, N.J., 1959-64, head device concepts dept., 1964-68, head of electro optical device dept. 1968—. Fellow Am. Phys. Soc.; sr. mem. I.E.E.E. Editorial bd. Rev. Sci. Instruments, 1963-66; Solid State Electronics, 1966—. Research on electronic properties of semimetals, tech. direction of a solid state device group. Home; 65 Summit Rd. Office: Bell Telephone Labs., Murray Hill, N.J. 07971.*

SMITH, George Foster, Am. physicist; b. Franklin, Ind., May 9, 1922; s. John Earl and Ruth (Foster) S.; B.S. in Physics, Cal. Inst. Tech., 1944, M.S., 1948, Ph.D. (Standard Oil fellow 1950-51) magna cum laude, 1952; m. Jean Arthur Farnsworth, June 3, 1950; children—David Foster, Craig Farnsworth, Sharon Windsor. Research engr. Engring. Research Assos., St. Paul, 1946-48; teaching fellow, resident asso. Cal. Inst. Tech., 1947-52; physicist Hughes Research Labs., Malibu, Cal., 1952-57, co-head storage tube research dept., 1957-58, head exploratory studies dept., 1958-61, mgr. quantum physics dept., 1961-62, asso. dir., 1962—; v.p. Hughes Aircraft Co., 1965—, also mem. tech. bd.; adj. asso. prof. elec. engring. U. So. Cal., 1959—; vice chmn. tech. program com. Western Electronic Conv., 1964; mem. steering com. Hughes Sch.-Industry Sci. Program, 1956. Active Boy Scouts Am., YMCA. Served to lt. (j.g.) USNR, 1944-46. Fellow Am. Phys. Soc., Inst. Elec. and Electronic Engrs.; mem. Research Soc. Am., Sigma Xi, Tau Beta Pi (pres. Cal. Inst. Tech. chpt. 1957-58). Contbr. articles tech. publs. Patentee in field. Research in thermionic emission, secondary emission, direct-view storage tubes, lasers. Home: 6423 Riggs Pl., Los Angeles 90045. Office: 3011 Malibu Canyon Rd., Malibu, Cal. 90265.*

SMITH, George Francis Maurice, Canadian biologist; b. Toronto, Ont., Can., Jan. 1, 1912; s. Arthur and Lillian (Reesor) S.; B.A., U. Toronto, 1935, M.A., 1937, Ph.D., 1939; m. Elizabeth Gillespie. Jan 1, 1942; children—James, Allan, Bruce. Biologist, Biol. Bd. Can., St. Andrews, N.B., Can., 1934-42; research asso. NRC Can., Toronto, 1942-45; prof. zoology U. N.B., Fredericton, 1945-52; scientist Def. Research Bd. Can., Toronto, 1952-58; biologist Fisheries Research Bd. Can., Ottawa,Ont., 1958—. Decorated mem. Order Brit. Empire. Mem. A.A.A.S., Biometrics Soc., Am. Fisheries Soc. Research, publs. in marine biology, aviation medicine, statis. design of expts., population dynamics. Office: Fisheries Research Bd. Can., Ottawa, Ont., Can.*

SMITH, George Frederick, Am. chemist; b. Lucasville, O., July 29, 1891; s. Clarence Mitchell and Matilda (Pflaumer) S.; B.A., U. Mich., 1917, M.A., 1919, Ph.D., 1922; m. Mary Ellen Sweeney, Aug. 19, 1921; children—Ruby Mae, J.C. Clifton, Mary Eleanore. Faculty, U. Ill., Urbana, 1921—, prof. emeritus, 1957—; pres. G. Frederick Smith Chem. Co., Inc., Columbus, O., 1924—; pres. Aeration Processes, Inc., Columbus, 1934-63, chmn. bd. dirs., 1963—. Mem. Am. Chem. Soc. (Fisher award, Anachem award), Soc. Analytical Chemists Eng. Author: (with Harvey Diehl) Quantitave Analysis, 1952. Contbr. numerous articles to profl. jours. Founded first indsl. application of modern era of aerosol food products; mfr. perchloric acid, 1924-40 as its only Am. mfr. Home: 502 W. Main St., Urbana, Ill. 61801.*

SMITH, George Otis, Am. geologist; b. Hodgdon, Me., Feb. 22, 1871; s. Joseph O. and Emma (Mayo) S.; A.B., Colby Coll., 1893, A.M., 1896, LL.D., 1920; Ph.D., Johns Hopkins, 1896; Sc.D., Case Sch. Applied Sci., 1914, Colo. Sch. Mines, 1928; m. Grace M. Coburn, Nov. 18, 1896; children—Charles Coburn, Joseph Coburn, Mrs. Helen Coburn Fawcett, Elizabeth Coburn, Louise Coburn. Engaged in geol. work in Mich., Utah, Washington, and in N.E., 1893-1906; asst. geologist and geologist, 1896-1907, dir. U. S. Geol. Survey, 1907-30, except 1922-23 while mem. U. S. Coal Commn.; chmn. Fed. Power Commn., 1930-1933; dir. Central Me. Power Co. Fellow Geol. Soc. Am., A.A.A.S.; mem. Coal Mining Inst. Am. (hon.), Am. Inst. Mining and Metall. Engrs. (ex-pres.), Am. Forestry Assn., Wash. Acad. Scis., Mining and Metall. Soc. Am., Am. Assn. Petroleum Geologists, Nat. Geog. Soc. (trustee), Phi Beta Kappa. Author of reports on areal, economic, petrographic and physiographic geology in pubs. U. S. Geol. Survey, also papers and addresses on economics of mineral and power resources and adminstrn. of sci. work by govt.; editor and co-author of Strategy of Minerals, 1919. Died Jan. 10, 1944.

SMITH, G(eorge) Pedro, Am. chemist; b. Norfolk, Va., Oct. 26, 1923; s. George P. and Annie (Wilson) S.; B.S., U. Va., 1944, M.S., 1947, Ph.D., 1950; m. Helen Whitesell, July 6, 1945; children—Carol Ann, Beverly Lynn, Leslie Elane. Group leader Oak Ridge Nat. Lab., 1950—; faculty U. Tenn., Knoxville, 1952—, prof. chemistry, 1964—. Mem. Am. Chem. Soc., Am. Phys. Soc., Am. Soc. for Metals, A.A.A.S. Research on role crystal orientation in friction and cohesion between metal surfaces; corrosion of metals; devel. optical spectroscopy as tool to study mechanism of melting of crystals and structure of liquids at high temperatures. Home: 7925 Chesterfield Dr., Knoxville, Tenn. 37919. Office: Oak Ridge Nat. Lab., P.O. Box X, Oak Ridge 37831.*

SMITH, Gerard Edward, English botanist; b. Camberwell, Surrey, 1804; s. Henry Smith; student Merchant Taylors Sch., from 1814; B.A., St. John's Coll., Oxford, 1829. Vicar, St. Peter-the-Less, Chichester, 1835-36; rector of North Marden, Sussex, 1836-43; vicar of Cantley, 1944-46, of Osmaston-by-Ashbourne, 1854-71; perpetual curate of Ashton Hayes, 1849-53. Author: Remarks on Ophrys', 1828; A Catalogue of Rare or Remarkable Phanogamous Plants collected in South Kento, 1829; Are the Teachings of Modern Science Antagonistic to the Doctrine of an Infallible Bible?, 8 vols., 1863. First to describe statice occidentalis, 1831; also described filago apiculata, 1846. Died Ockbrook, Derby, Dec. 21, 1881.

SMITH, Sir Grafton Elliot, anatomist, anthropologist; b. Grafton, New S. Wales, Australia, Aug. 15, 1871; s. Stephen Sheldrick and Mary Jane (Evans) S.; M.B., U. Sydney (Australia), Ch.M., 1892, M.D., 1895; m. Kathleen Macredie, 1900; 3 sons. Research, Cambridge (Eng.) U., 1896-1900, also hon. fellow St. John's Coll., 1931; 1st prof. anatomy Govt. Med. Sch., Cairo, Egypt. from 1900; prof. anatomy U. Manchester (Eng.) U., 1909-19, U. Coll., London, Eng., 1919-32; Fullerian prof. physiology Royal Inst., from 1933. Recipient Huxley medal 1935, Prix Fauvelle, Anthrop. Soc. Paris, 1911. Fellow Royal Soc. (Royal medal 1912), 1907. Author: The Ancient Egyptians, 1911; The Migrations of Early Culture, 1915; Evolution of the Dragon, 1919; Elephants and Ethnologists, 1924; The Evolution of Man: Essays, 1924; Human History, 1930; Diffusion of Culture, 1933. Research on comparative cerebral morphology, comparative anatomy; with archaeol. survey Nubia; studied mummification in Egypt, Peking man, evolution of man and brain. Died Broadstairs, Eng., Jan. 1, 1937.

SMITH, Grant Gill, Am. chemist; b. Fielding, Utah, Sept. 25, 1921; s. Joseph H. and Bertha (Jensen) S.; B.A., U. Utah, 1943; PL.D., U. Minn.; m. Phyllis Cook, Dec. 30, 1946; children—Meredith Lynn, Kathleen, Vivienne, Geoffrey Gill, Randall Cook, Roger Todd. Research chemist Clinton Engr. Works, Tenn. Eastman Corp., Oak Ridge, 1944-46; insr. U. Minn., 1949; faculty Wash. State U., Pullman, 1949-61, asso. prof. chemistry, 1956-61; asso. prof. chemistry Utah State U., Logan, 1961-63, prof., 1963—; vis. prof. U. London Imperial Coll., Eng., 1957-58. Mem. Am. Chem. Soc. (past pres. Wash., Ida. and Salt Lake sects.), Chem. Soc. London, Utah Acad., Sigma Xi, Phi Lambda Upsilon, Alpha Chi Sigma. Research, publs. in field of phys. organic chemistry, relationships of structure to chem. reactivity and mechanism of organic reactions especially in vapor phase systems. Home: 805 River Heights Blvd., Logan, Utah 84321.*

SMITH, Grant Newey, Am. biochemist; b. Ogden, Utah, Feb. 13, 1918; s. Frederick and Maria (Newey) S.; A.S., Weber Coll., 1938; A.B. with honors, U. So. Cal., 1940; M.S., Utah State U., 1941; Ph.D., Cornell U., 1947; m. Mary Betsy Patterson, Dec. 20, 1947; children—Martha Louise, Diane Elizabeth, Carol Jean. Asst. plant physiologist Utah State Agrl. Coll., Logan, 1940-41; chemist N.Y. Coll. Agr., Ithaca, 1941-42; analytical chemist U. S. Dept. Interior, Ithaca, 1942; Rockefeller fellow Cornell U., 1942-44; sr. research biochemist Parke Davis & Co., Detroit, 1947-50; biochemist-microbiologist Gen. Electric Co., Washington, 1950-53; biochemist Dow Chem. Co., Midland, Mich., 1953-63, asso. research scientist, 1963—. Mem. A.A.A.S., Am. Chem. Soc., Am. Soc. Biol. Chemists, Biochem. Soc. Britain, Chem. Soc. Britain, Phi Beta Kappa, Sigma Xi, Phi Sigma. Contbr. chpts. to books, numerous articles to profl. jours. Developed new drugs and agrl. chems. by studying their mode of action in plants and animals, determination of their metabolization by plant or animal; determined safety for human use of Chloromycetin, Zoalene, Ronnel, Ruelene, Dalapon, Zectran, Dursban, related compounds. Home: 910 Love St. Office: P.O. Box 512, Midland, Mich. 48640.*

SMITH, Guy Donald, Am. soil scientist; b. Atlantic, Ia., June 20, 1907; s. Daniel Christian and Hope (Curtis) S.; B.S., U. Ill., 1930, Ph.D., 1940; M.A., U. Mo., 1934; D.Sc., U. Ghent, 1968; m. Jean Elizabeth Randall, July 5, 1934; children—Ann Hope (Mrs. William Spurgin), Guy Donald, Gail Elizabeth (Lisa Mockett), Randall Curtis, Arthur Randall. Research asst. U. Ill., 1930-33, asso., then asst. prof., 1935-42; with Resettlement Adminstrn., Champaign, Ill., 1934-35; sr. soil correlator U. S. Dept. Agr. Bur. Plant Industry Soils and Agrl. Engring., 1946-50, prin. soil correlator, 1951-52, dir. soil survey investigations Soil Conservation Service, 1952—; Francqui chair U. Ghent, Belgium, 1964-65. Recipient Soil Research award Am. Soc. Angronomy, 1964; Distinguished Service award Dept. Agr. Fellow Am. Soc. Agronomy; mem. Soil Sci. Soc. Am. (pres. 1959), Royal Flemish Acad. Fine Arts, Letters and Sci. Author: (with F. F. Riecken, R. W. Simonson) Understanding Iowa Soils; Soil Classification, A Comprehensive System, 1960. Developed principles of system for classifying soils as natural 3 dimensional bodies that support plants (widely used outside Communist countries for internat. comparisons of soils). Home: 6407 40th Av., University Park, Md. 20782. Office: South Agr. Bldg., 12th and Independence Av. S.W., Washington 20250.*

SMITH, Harlan Ingersoll, Am. anthropologist; b. East Saginaw, Mich., Feb. 17, 1872; s. Harlan Page and Alice Elvira (Ingersoll) S.; ed. U. Mich.; m. Helena E. Oakes, Nov. 25, 1897; children—Elizabeth Alice (Mrs. Allan T. Powell), Marjorie Oakes (Mrs. G. Douglas Mallory). Studied archeology of Saginaw Valley, before 1890; asst. Peabody Mus., Harvard, 1891; field asst., dept. anthropology Chgo. Expn., 1891-93; explored ancient mounds in Ohio, Ky., Wis., N.Y., Mich.; in 1891-93 had charge anthrop. collections in mus. U. Mich.; explored ancient garden beds nr. Kalamazoo, Mich., for Archaeol. Inst. Am., 1894; became connected with Am. Mus. Natural History, N.Y., 1895, mem. faculty, 1896-1914, asst. curator of archaeology, 1900-1910, asso. curator of anthropology, 1910-11, hon. curator of archaeology, 1912-14; archaeologist Geol. Survey Can., 1911-20; archaeologist Nat. Mus. Can. (formerly Victoria Meml. Mus.), 1920-37. Am. archaeologist Jesup N. Pacific expdn.; instr. evolution of industries Pratt Inst., Bklyn., 1906-07; lectr. N.Y. Bd. of Edn., 1898-1911. Archaeol. exploration in B.C., 17 seasons, 1897-1929, in N.S., 1914, other parts of Can., 1911-21. Fellow A.A.A.S. Author: Archaeology of Lytton, 1899; Archaeology of the Thompson River Region, 1900; Cairns of British Columbia and Washington, 1901; Shell Heaps of the Lower Fraser River, British Columbia, 1903; Archaeology of the Gulf of Georgia and Puget Sound, 1907; Archaeology of the Yakima Valley, Washington, 1910; The Prehistoric Ethnology of a Kentucky Site, 1910; An Album of Prehistoric Canadian Art, 1923; The Archaeology of Merigomish Harbour, Nova Scotia, 1929; also numerous papers on anthrop. and mus. subjects. Made motion pictures of Indians in British Columbia and Alberta, 1923-29. Died Jan. 28, 1940.

SMITH, Harlan James, Am. astronomer; b. Wheeling, W.Va., Aug. 25, 1924; s. Paul Elder and Anna Persis (McGregor) S.; A.B., Harvard, 1949, M.A., 1951, Ph.D., 1955; m. Joan Greene, Dec. 21, 1950; children—Nathaniel, Sarah (dec.), Julia, Theodore, Hannah. Research asst. astronomy, teaching fellow, research fellow Harvard, 1946-53; from instr. to asso. prof. astronomy Yale, 1953-63; prof. astronomy, chmn. dept., dir. McDonald Obs., U. Tex., 1963—. George R. Agassiz research fellow Harvard Obs., 1952-53. Mem. Am. (acting sec. 1961-62), Royal astron. socs., Am. Geophys. Union, A.A.A.S., Internat. Astron. Union, Internat. Sci. Radio Union, Sigma Xi. Co-editor Astron. Jour., 1960-63. Study of quasars; variable stars; planets; radio astronomy. Home: 2705 Pecos St., Austin, Tex.

SMITH, Harold Hill, Am. geneticist; b. Arlington, N.J., Apr. 24, 1910; s. Frederick Harold and Hilda (Burgess) S.; B.S., Rutgers U., 1931; A.M., Harvard, 1934, Ph.D., 1936; m. Mary Downing, Feb. 11, 1939; children—Frederick, Lucy (Mrs. Lawrence Keane), Hilda, Susan (Mrs. Peter Jurs). Faculty, Cornell U., 1946-56; sr. geneticist Brookhaven Nat. Lab., Upton, N.Y., 1956—; sr. scientist Internat. Atomic Energy Agy., 1958--59; vis. prof. U. Cal., Berkeley, U. Buenos Aires, 1966; Fulbright lectr. U. Amsterdam, 1953. Hon. research asso. U. Coll. London, 1966. Guggenheim fellow, 1952. Mem. Genetics Soc. Am., Am. Genetic Assn., Bot. Soc. Am., Radiation Research Soc., Soc. Devel. Biology, A.A.A.S., N.Y. Acad. Scis., Sigma Xi. Mem. editorial bd. Jour. Heredity, 1964; Mutation Research, 1964. Research, numerous publs. on quantitative inheritance, effect of extra chromosomes, inheritance of alkaloids and other characteristics of tobacco plant evolution, induction of mutations, radiation genetics, plant tumors, genetic control of devel., relative biol. efficiency of ionizing radiations. Home: Tower Hill Rd., Shoreham, N.Y. 11786. Office: Biology Dept., Brookhaven Nat. Lab., Upton, N.Y. 11973.*

SMITH, Harold Theodore Uhr, Am. geologist; b. Castle Shannon, Pa., July 4, 1908; s. William Henry and Martha (Uhr) S.; B.S., Wooster Coll., 1930; M.A., Harvard, 1933, Ph.D., 1936; m. Althea Waterman Page, June 8, 1935; children—Conrad, Myron, Roger. Faculty, U. Kan., 1935-43, asso. prof., 1946-56; geologist U. S. Geol. Survey, 1943-46; prof. geology, head dept. U. Mass., Amherst, 1956—. Cons. research and devel. bd. Def. Dept., 1952-53. Mem. Am. Soc. Photogrammetry, Geol. Soc. Am., Internat. Assn. for Quaternary Research, Assn. Am. Geographers, A.A.A.S., Am. Assn. Petroleum Geologists, Phi Beta Kappa, Sigma Xi. Author: Aerial Photographs and Their Applications, 1943; also numerous articles. Research on use aerial photography in geol. research, eolian and cryopedologic phenomena in interpretation of Quaternary geochronology and paleoclimatology, geomorphic phenomena in Antarctica. Home: Arnold Rd., Rural Route 2, Amherst, Mass. 01002.*

SMITH, Harold Wood, Am. elec. engr., educator; b. Brookfield, Mo., Feb. 8, 1923; s. William Bryan and Celia (Tarpening) S.; B.S., U. Tex., 1944, M.S., 1948, Ph.D., 1954; m. Patricia Yvonne Richardson, Oct. 17, 1942; 1 dau., Teresa Lynn. Faculty dept. elec. engring. U. Tex., Austin, 1946—, prof., mem. grad. faculty, 1959—, radio engr. Elec. Engring. Research Lab., 1946—, dir. Geomagnetics Lab., 1964—. Mem. I.E.E.E., Am. Geophys. Union, Sigma Xi, Tau Beta Pi, Eta Kappa Nu. Research, publs. on measurement and understanding of propagation of hydromagnetic waves resulting from interaction of solar wind and earth's magnetic field. Home: Route 7, Box 942B, Austin, Tex. 78703.*

SMITH, Harry L., Am. physician; b. Fairport, Mo., Sept. 3, 1887; s. Edward L. and Laura M. Smith; M.Sc., U. Minn., 1928; M.D., U. Ia., 1918; m. Mary Meredith, Dec. 31, 1939; 2 children. Staff, Mayo Clinic, Rochester, Minn., 1925—; prof. medicine Mayo Found., Rochester; also pvt. practice medicine. Mem. A.C.P., Central Soc. Clin. Research, A.M.A., Am. Heart Assn. Research in cardiology and internal medicine; introduced (with others in Mayo Clinic) cortisone for treatment rheumatic fever, 1949. Office: Mayo Clinic, Rochester, Minn.

SMITH, Harry Madison, Am. geneticist; b. Macomb, Ill., June 21, 1918; s. James Madison, Jr. and Grace (Smith) S.; B.S., U. Chgo., 1939, Ph.D., 1942; m. Mildred Rosetta Gutknecht, June 16, 1946. Faculty, U. Ark., 1946-48, U. Wyo., 1948-53; Fulbright research scholar U. Coll., Mandalay, Burma, 1951-52; lectr. Loyola U., Chgo., 1953; lectr. Middlebury Coll., 1954; asso. prof. Am. U., Beirut, Lebanon, 1954-58, research asso., dir. anthrop. blood grouping lab., 1961-65; NIH research fellow Columbia, 1958-60; U. Ibadan, Nigeria, 1960; asso. prof., prin. investigator NIH research grant Springfield (Mass.) Coll., 1965—. Sr. scientist res. USPHS; Nat. Security Seminar, Springfield, 1966. Fellow A.A.A.S.; mem. N.Y. Acad. Scis., Genetics Soc. Am., Am. Soc. Human Genetics, Am. Genetic Assn., Assn. Mil. Surgeons U. S., Am. Assn. U. Profs., Sigma Xi. Research, publs. on population genetics and ecology, biol. fitness and human evolution. Home: 340 Arcadia Blvd., Springfield 01118. Office: 263 Alden St., Springfield, Mass. 01109.*

SMITH, Harry Pratt, Am. pathologist; b. Johnson County, Ia., Feb. 18, 1895; s. Walter Z. and Estella M. Smith; A.B., U. Cal., 1916; George Williams Hooper Found. fellow U. Cal., 1917-19, M.S., 1918, M.D., 1921; unmarried. Asst. and instr. in pathology Johns Hopkins, 1921-23, nat. research fellow in chemistry, 1923-24; asst. and asso. prof. pathology U. Rochester, 1924-30; prof. pathology State U. Ia., Coll. Medicine, Iowa City, 1930-45; Delafield prof. pathology Coll. Phys. and Surg., Columbia U., 1945-60, head dept. until 1960, now Delafield prof. emeritus; librarian Am. Soc. Clin. Pathologists, 1960—. Mem. Am. Soc. Clin. Pathologists (past pres.; librarian), A.M.A., A.A.A.S., Am. Soc. Exptl. Pathol. (sec.-treas. 1940-46, pres. 1947-48), N.Y. Acad. Medicine, Harvey Soc., Coll. Am. Pathologists (bd. govs. 1947-1953), Am. Assn. Pathologists and Bacteriologists, Soc. Exptl. Biology and Medicine. Research in blood volume determination at high altitude; fate of intravenously injected dyes; bile salt metabolism; Vitamin K, blood clotting and bleeding tendency. Home: St. Clair Hotel, E. Ohio St. Office: 445 N. Lake Shore Dr., Chgo. 11.*

SMITH, Henry John Stephen, Brit. mathematician; b. Dublin, Ireland, Nov. 2, 1826; s. John and Mary (Murphy) S.; B.A., Balliol Coll., Oxford, 1850, M.A., 1855; student Sorbonne, Paris, France, also Collège de France, 1845-46. Named Savilian prof. geometry, Oxford, 1861. Fellow Royal Soc., 1861. Mem. various royal commns.; became chmn. mng. body Meteorol. Office, 1877. Author: Work of H.J.S. Smith, 1894. Research and publs. on number theory, linear indeterminate equations, ternary quadratic forms, elliptic functions. Died Oxford, Feb. 9, 1883.

SMITH, Herbert Huntington, Am. naturalist; b. Manlius, N.Y., Jan. 21, 1851; s. Charles and Julia Maria (Huntington) S.; studied Cornell U., 1868-72; m. Amelia Woolworth Smith, Oct. 5, 1880. Best known as collector of natural history specimens; traveled in Brazil, 1871, 73-77, 81-86, Mexico, 1889, W.I., 1890-95, Colombia, 1898-1901; in Mexico employed for Biologia Centrali-Americana, in W.I. for W. Indian Com. of Royal Soc. and Brit. Assn.; curator Carnegie Mus., Pitts., 1896-98, asso. with instn., 1902; curator Ala. Mus. of Natural History, 1910—. Author: Brazil—the Amazons and the Coast, 1880; De Rio de Janeiro á Cuyabá, 1886 (in Portuguese). Published His Majesty's Sloop Diamond Rock, over the pen-name. H. S. Huntington, 1904. Collaborator Century Dictionary, Century Cyclo. Names, Johnson's Cyclo. Collected at least 50,000 specimens, found in museums throughout world. Died Mar. 22, 1919.

SMITH, Hilton Albert, Am. chemist; b. Plymouth, N.Y., Sept. 4, 1908; s. Roy Leon and Ethel (Lewis) S.; A.B., Oberlin Coll., 1930; A.M., Harvard, 1932, Ph.D., 1934; m. Elizabeth Zorbaugh, June 28, 1933; children—Cynthia Jean (Mrs. William Vorih), Lewis Harvey, Judith Ellen (Mrs. James Busse), Roy Hilton, Charles Leeman. Postdoctoral research asst. Harvard, 1934-45; faculty Lehigh U., 1935-41; prof. chemistry U. Tenn., Knoxville, 1941—, dean Grad. Sch., coordinator research, 1961-66, v.p. for grad. studies and research, 1966—. Cons. to industry, govt. agys. Recipient Orins award for Outstanding Sci. Research in South; So. Chemist award, 1967. Mem. Am., London chem. socs., A.A.A.S., Am. Assn. U. Profs., Tenn. Acad. Scis., Tenn. Edn. Assn., Phi Beta Kappa, Phi Kappa Phi, Sigma Xi. Contbr. articles to profl. jours. Research on chem. kinetics; catalysis; photochemistry; phys. organic chemistry; isotope studies. Home: 5817 Toole Dr., Knoxville, Tenn. 37919.

SMITH, Hobart Muir, Am. herpetologist; b. Stanwood, Ia., Sept. 26, 1912; s. Charles Henry and Frances (Muir) S.; A.B., Kan. State U., 1932; M.S., U. Kan., 1933, Ph.D., 1936; m. Rozella Pearl Blood, Aug. 26, 1938; children—Bruce Dyfrig, Sally Frances. Instr. biology U. Rochester (N.Y.), 1941-45; asst. prof. zoology U. Kan., Lawrence, 1945; asst. prof. wildlife mgmt. Tex. A. and M. Coll., College Station, 1946; faculty U. Ill., Urbana, 1947—, prof. zoology, 1957—, curator herpetology Mus. Nat. History, 1947—. Author: Amphibians of Kansas, 1934; Mexican Lizards, Genus Sceloporus, 1939; Handbook of Lizards, 1945; Checklist and Key to Snakes of Mexico, 1945; Checklist and Key to Amphibians of Mexico, 1948; Handbook of Amphibians and Reptiles of Kansas, 1950; Checklist and key to Reptiles of Mexico, 1951; Evolution of Chordate Structure, 1960; Reptiles and Amphibians: A Guide to Familiar American Species, 1958; Poisonous Amphibians and Reptiles: Recognition and Bite Treatment, 1959; Herpetological Type-Specimens in the University of Illinois Museum of Natural History, 1964; Snakes as Pets, 1965; Laboratory Studies of Chordate Structure, 1966. Home: 601 E. Douglas St., St. Joseph, Ill. 61873. Office: Dept. Zoology, U. Ill., Urbana, Ill. 61801.*

SMITH, Hugh McCormick, Am. ichthyologist; b. Washington, Nov. 21, 1865; s. Thomas Croggon and Cornelia Frances (Hazard) S.; M.D., Georgetown U., 1888; postgrad. N.Y.; LL.D., Dickinson, 1908; m. Emma Hanford, Mar. 12, 1899. Entered U. S. Fish Commn. (now Bur. Fisheries), as asst., 1886, asst. in charge sci. inquiry, 1897-1903, dir. biol. lab., Woods Hole, Mass., 1901-02; co-spl. agt. in charge of fisheries, 10th Census; dep. commr. of fisheries, 1903—. Mem. med. faculty Georgetown U., 1888-1902, prof. normal histology, 1895-1902. Rep. internat. fishing congresses. Fellow A.A.A.S. Pub. numerous reports and papers on ichthyology, economic fisheries and pisciculture, in govt. reports and elsewhere; contbr. to tech. and popular periodicals. Editor Bur. Fisheries, 1904—. Died Sept. 28, 1941.

SMITH, Ian Maclean, physician; b. Glasgow, Scotland, May 21, 1922; s. Robert Kessen and Anna (Maclean) S.; M.B.Ch.B., Glasgow U., 1944, M.D., 1957; m. Jeanne Montgomery Smith, Dec. 23, 1948; children—Robert M., William M., Douglas M., Jan Scott M. Came to U. S., 1949, naturalized, 1957. Asst. prof. Rockefeller Inst., asst. physician R. I. Hosp., 1953-55; faculty, chief infectious disease div. U. Ia., also U. Hosps., Iowa City, 1955—, prof. internal medicine, 1965—; attending physician VA Hosp., Iowa City, 1961—; also cons. state agys. Recipient Fulbright award, 1949; Wyeth Research award, 1962. Nat. Cancer Inst. fellow, 1949-51. Fellow A.C.P.; mem. Royal Coll. Physicians and Surgeons, Coll. Pathologists, Glasgow Royal Medico-Chirurg. Soc., Path. Soc. Great Brit. and Ireland, Soc. Am. Microbiologists, A.M.A., Am. Thoracic Soc., Soc. Exptl. Biology and Medicine, Am. Soc. Tropical Medicine and Hygiene, Sigma Xi, others. Author: Staphylococcal Infections, 1958. Research, publs. on biochem. derangements resulting in death of infected host and their prevention, staphylococcal infections in man. Home: 1033 E. Washington St., Iowa City 52240.

SMITH, Sir James Edward, English botanist; b. Norwich, Eng., Dec. 2, 1759; s. James and Frances (Kinderley) S.; studied botany under John Hope, U. Edinburgh, 1781-83; studied botany, London, 1783, Europe, 1786; M.D., U. Leiden (Netherlands), 1786; m. Pleasance Reeve, 1796; lectr. on botany; founder, 1st pres. Linnean Soc., 1788. Fellow Royal Soc., 1785; mem. French Acad. Scis., Imperial Acad. Natural Curiosities, acads. Stockholm, Turin, Lisbon, N.Y., Phila., Uppsala. Author: English Botany, 36 vols., 1790-1819; Introduction to Physiological and Systematic Botany, 1807; English Flora, 6 vols., 1824-33; also numerous articles. Produced 1st complete botany of Eng. with Linnean system of classification; owned complete collection and library of Linneus. Died 1828.

SMITH, James Hammond, Am. physicist; b. Colorado Springs, Colo., Feb.2, 1925; s. James Hollingsworth Clemmer and Elizabeth (Hammond) S.; student Rutgers U., 1944; A.B., Stanford, 1946; Ph.D., Harvard, 1951; m. Jeanne Nelson, Sept. 9, 1950 (dec. Feb. 8, 1953); children—Nelson A., Jennifer E.; m. 2d, E. Jean Walker, Feb. 26, 1955; children—Rebecca R., Margaret A. Prof. physics U. Ill., Urbana, 1951—; vis. prof. Mass. Inst. Tech., 1962-63. Mem. phys. sci. study com. Ednl. Services, Inc., 1955-58. Guggenheim fellow, 1965-66. Fellow Am. Phys. Soc.; mem. Am. Assn. U. Profs., Am. Assn. Physics Tchrs., Alpha Kappa Lambda. Author: Introduction to Special Relativity, 1965. Research on photonnuclear reactions in light elements, weak interactions. Home: 1738 Westhaven Dr., Champaign, Ill. 61820. Office: Physics Dept. U. Ill., Urbana, Ill.*

SMITH, James John, Am. physiologist; b. St. Paul, Jan. 28, 1914; s. James Wendel and Catherine (Welsch) S.; M.D., St. Louis U., 1937; M.S., Northwestern U., 1940, Ph.D., 1946; m. Mary E. Schumacher, Mar. 17, 1946; children—Philip W., Lucy G., Paul R., Gregory K. Dean, asst. prof. physiology Loyola U. Sch. Medicine, Chgo., 1946-50; chief edn. div. VA Dept. Medicine and Surgery, Washington, 1950-52; vis. lectr. in pharmacology George Washington U. Sch. Medicine, Washington, 1950-52; prof., chmn. dept. physiology Marquette U. Sch. Medicine, Milw., 1952—. Jesse Horton Koessler fellow Inst. Medicine, Chgo., 1939-41; Fulbright Research Scholar, Inst. Physiology, U. Heidelberg, Germany, 1959-60. Mem. Soc. for Exptl. Biology, Am. Physiol. Soc., Reticulo-Endothelial Soc., Milw. Acad. Medicine, Sigma Xi. Research in physiology of shock, regulation of circulation. Home: 1423 N. 122d St., Milw. 53226.

SMITH, Janice Minerva, Am. nutritionist; b. Osco, Ill., Oct. 13, 1906; d. J. Heber and Minnie M. (Hadley) Smith; A.B., U. Ill., 1930, M.S. in Nutrition (Bronze Tablet scholar), 1932, Ph.D., in Biochemistry, 1937. Asst. prof. U. Ill., Urbana, 1930-36, prof. nutrition, 1944—, head dept. home econs., 1950—; asso. prof. Pa. State U., State College, 1937-43; nutritionist War Food Adminstrn., U.S. Dept. Agr., Washington, 1943-44. Mem. adv. council Pillsbury scholarship, 1962—; mem. Pres.'s Commn. on Food and Fiber, 1965—. Named Woman of Year, Altrusa, 1961—. Mem. Am. Dietetic Assn., Am. Inst. Nutrition, Am. Home Econs. Assn. (editorial bd. 1963—), Ill. State Nutrition, A.A.A.S., N.Y. Acad. Scis., Phi Beta Kappa, Phi Kappa Phi, Omicron Nu, Phi Upsilon Omicron, Iota Sigma Pi, Sigma Delta Epsilon, Phi Sigma, Sigma Xi. Contbg. author: Soybean and Soybean Products, 1951; also articles. Mem. rev. bd. Am. Jour. Clin. Nutrition, 1950-54; Research in spl. protein needs of adults, calcium needs of adolescents and preschool children, riboflavin and thiamine needs of pre-adolescents, nutritional status and dietary habits selected populations in Pa. and Ill., energy and dietary needs ageing. Home: 1112 S. Pine St., Champaign, Ill. 61820. Office: Bevier Hall, U. Ill., Urbana, Ill. 61801.*

SMITH, Jesse Graham, Jr., Am. physician, educator; b. Winston-Salem, N.C., Nov. 22, 1928; s. Jesse Graham and Pauline (Griffith) S.; B.S. in Medicine, Duke, 1947, M.D., 1950; m. Dorothy Jean Butler, Dec. 28, 1950; children—Jesse Graham III, Cynthia Lynn, Grant Butler. Instr., asst. prof.

dermatology U. Miami (Fla.) Sch. Medicine, 1957-60; asso. prof. dermatology Duke Med. Center, Durham, N.C., 1960-62, prof., 1962——. Mem. gen. medicine study sect. NIH, 1964——; Howard Fox Meml. lectr., 1966. Mem. A.M.A., Soc. Investigative Dermatology (dir.), Am. Acad. Dermatology (Clyde L. Cummer gold award 1963), Am. Dermatol. Assn., Am. Soc. Exptl. Pathology, Sigma Xi, Alpha Omega Alpha. Editorial bd. Arch. Dermatology, Jour. Investigative Dermatology. Research, numerous publs. on connective tissue in cutaneous disease, elastic tissue, acid glycoasaminoglycans, lyosomes in relation to cutaneous disease. Home 1118 Woodburn Rd., Durham, N.C. 27705.

SMITH, John Hammond, Am. civil engr., b. Wellsville, O., Oct. 14, 1867; s. John William and Almira (Hart) S.; E.E., Western U. Pa. (now U. Pitts.), 1898; postgrad. Cornell U., summer 1900; m. Anna D. Coleman, July 3, 1901 (dec. 1906); children—Anna Virginia, Lilian Isabella; m. 2d, Gertrude M. Smith, Jun 23, 1909 (dec. 1918); m. 3d, Helen C. Dalrymple, June 25, 1919 children—Helen Ilene, Evelyn Almira, Martha Louise. With Julian Kennedy, Riter-Conley Mfg. Co., Am. Bridge Co. and Concrete Products Co., Pitts., 1896-1910; supt. shops and instr. U. Pitts., 1898-1900, prof. drawing, 1900-09, prof. civ. engring., 1909——. Contbr. many papers on engring. and photo-sculpture. Inventor of new photo-sculpturing process and instruments for testing materials. Died 1932.

SMITH, John Lawrence, Am. chemist, physician; b. Charleston, S.C., Dec. 17, 1818; s. Benjamin Smith; student U. Va., 1835-37; M.D., Med. Coll. S.C., 1840; m. Sarah Julia Guthrie, June 24, 1952. Founder (with Dr. S. D. Sinkler) So. Jour. Medicine and Pharmacy (later became Charleston Med. Jour. and Review), 1846; adviser on cotton culture to Turkish govt., circa 1847, investigated their mineral resources and discovered emery and coal deposits, 1847-50 (useful in discovery of several emery deposits in U. S.); prof. chemistry U. Va., 1852; prof. med. chemistry, toxicology U. Louisville (Ky.), 1854-66; lectr. Smithsonian Inst.; his collection of meteoric stones (one of finest in Am.) sold to Harvard; pres. Louisville Gas Works; pres. A.A.A.S., 1872; mem. Nat. Acad. Scis. Author: Mineralogy and Chemistry: Original Researches, 1873. Inventor inverted microscope, 1850. Died Louisville, Oct. 12, 1883; buried Louisville.

SMITH, J(ohn) Warren, Am. meteorologist; b. Grafton, N.H., Sept. 21, 1863; s. John R. and Mary E. (Wadleigh) S.; B.S., N.H. Coll. Agr. and Mechanic Arts, 1888, M.S., 1900; Lawrence Sci. Sch. (Harvard), 1891-92; grad. Summer Sch. of Agr., Ohio State U., 1902; children—(by 1st marriage) Ruth Eaton, Russell Wellington; (by 2d marriage) Audrey. Began with U. S. Weather Bur., 1888; dir. N.E. sect., 1890-96, Mont. sect., 1896-97, Ohio sect., 1898-1900; dist. forecaster at St. Louis, 1909-10; prof. meteorology and dir. Ohio sect., 1910-15; also prof. meteorol. science Ohio State U., 1910-15; apptd. chief of div. agrl. meteorology Weather Bur., Washington, 1916. Pres. Ohio Acad. Science, 1914-15. Author: Agricultural Meterology, a Study in Weather and Crops (Rural Text Book Series), 1920. Died Jan. 21, 1940.

SMITH, Kathleen, Am. psychiatrist; b. Fayetteville, Ark., Oct. 9, 1922; d. Carl Anton and Mary Ann (Henderson) Smith; B.S., U. Ark., 1944; M.D., Washington U., St. Louis, 1949. Physician, Malcolm Bliss Psychiat. Hosp., St. Louis, 1953-62, dir. inpatient service Mental Health Center, 1957-60, dir. tng., 1960——, physician in psychiatry, 1962-64, clin. dir., 1964, supt., 1964——; physician in psychiatry St. Louis State Hosp., 1958-60; from instr. to asso. prof. psychiatry Washington U. Sch. Medicine, 1953——, vis. physician Unit I, 1956——, research asso. Social Sci. Inst., 1965——. Cons., St. Louis City Hosp. Nurses' Infirmary, 1955-57. USPHS research fellow in psychiatry Washington U., Malcolm Bliss Psychiat. Hosp., 1952-53. Diplomate Am. Bd. Neurology and Psychiatry. Mem. A.M.A., Am., Mid-Continent psychiat. assns., A.A.A.S., Pan Am. Med. Assn., Sigma Xi, others. Studies, publs. on organic aspects of schizophrenia, particularly on presence of taraxein in serum of schizophrenics and peculiarities in perspiration odor, also shock, drug treatment schizophrenia. Home: 1161 S. McKnight St., St. Louis 63117. Office: 1420 Grattan St., St. Louis 63104.*

SMITH, Kendric Charles, Am. biochemist; b. Oakwood, Ill., Oct. 13, 1926; s. Russell W. and Virginia (Mozley) Sm.; student Berea Coll., 1944-46; B.S. in Chemistry, Stanford, 1947; Ph.D. in Biochemistry, U. Cal. at Berkeley, 1952; m. Marion Edmonds, Feb. 5, 1955; children—Nancy Carol, Martha Ellen. USPHS fellow U. Cal. at Berkeley, 1952-54; research asst. radiology U. Cal. Sch. Medicine, San Francisco, 1954-56; research asso. radiology Stanford Sch. Medicine, Palo Alto, Cal., 1956-62, faculty, 1962——, asso. prof., 1965——, lectr. biophysics Stanford, 1957——. Mem. com. on photobiology Nat. Acad. Sci.-NRC, 1964——. Mem. Am. Soc. Biol. Chemistry, A.A.A.S., Am. Chem. Soc., Biophys. Soc., Radiation Research Soc., Photobiology Group (Eng.), No. Cal. Photobiology and Photochemistry Group (founder, pres. 1962-66), Sigma Xi. Contbg. author: Photphysiology, vol. II, 1964; also numerous articles. Exec. editor Photochemistry and Photobiology, 1966——. Research in photochem and radiation chem. effects on nucleic acid acids, biochem. mechanisms for repair of cellular radiation damage. Home: 4101 MacKay Dr., Palo Alto, Cal. 94306.*

SMITH, Lawrence Weld, Am. physician; b. Newton, Mass., June 20, 1895; s. William G. and Marion (Reynolds) S.; B.A., Harvard, 1917, M.D., 1920; m. Dorothy Matthews, Oct. 12, 1935; 1 dau., Shirley (Mrs. James T. Kilbreth, Jr.). Instr. pathology Harvard, 1920-21, asst. prof., 1923-28; prof. pathology U. Philippines, 1922-23; asst. prof. Cornell U., 1928-33, asso. prof. pathology, 1933-35; prof. pathology Temple U. Med. Sch., 1935-45; med. dir. Comml. Solvent Corp., N.Y.C., Terre Haute, Ind., 1945-52; cons. pathology, N.Y.C., 1953-59; pathologist Waldemar Med. Research Found., Woodbury, N.Y., 1955-65; dir. R.I.S.T. Clin. Lab., Huntington, N.Y., 1966——. Cons. to sec. war, World War II. Mem. Flemish Acad. Sci. (cons.) Soc. Exptl. Biology and Medicine, Soc. Exptl. Pathology, Am. Soc. Cancer Research, Am. Assn. Pathology and Bacteriology, Am. Soc. Clin. Pathology, Coll. Am. Pathologists, A.M.A., Am. Geriatrics Soc., N.Y. Acad. Medicine, N.Y. Acad. Sci., N.Y., Phila. pathology socs. Author: (with E.S. Gault) Essentials of Pathology, 1938; also numerous articles. Participant in exptl. devel. of cryotherapy. Home: 89 West Gate Dr., Huntington, N.Y. 11743. Office: 2036 New York Av., Huntington Station, N.Y. 11746.*

SMITH, Lawton Harcourt, Am. biologist; b. Poughkeepsie, N.Y., Nov. 15, 1924; s. Frank I. and Dorothy (Harcourt) S.; B.S., U. Conn., 1950; M.A., Syracuse U., 1952, Ph.D., 1954; m. Jeanette Parke, Oct. 6, 1946; 1 son, Lawton Bradley. Biologist biology div. Oak Ridge Nat. Lab., 1954——. NSF fellow, Holland, 1959-60. Mem. Am. Physiol. Soc., Radiation Research Soc., Soc. for Exptl Biology and Medicine. Home: 124 Dartmouth Circle. Office: Biology Div., Oak Ridge Nat. Lab., Oak Ridge 37830.

SMITH, Leonard Charles, Am. biochemist; b. Spokane, Wash., Jan. 31, 1921; s. Leonard Charles and Edith (McLellan) S.; A.B., Mont. State U., 1943; Ph.D., U. Ill., 1949; m. Mary Elaine Rush, Oct. 1, 1945; children—David E., Lynn Frederic, Steven Mark, Peter Douglas, Andrew Ian. Instr. biochemistry Northwestern U. Med. Sch., Chgo., 1949-56; asso. prof biochemistry U. S.D., Vermillion, 1956-61, prof., 1962-66; prof. chemistry Ind. State U., Terre Haute, 1966——. vis. lectr. Glasgow (U.K.) U., 1961-62. Research biochemist Hines (Ill.) VA Hosp., 1949-56. Mem. Am. Chem. Soc., Soc. Exptl. Biology and Medicine, Sigma Xi, Pi Mu Epsilon. Research, publs. on amino acid metabolism. Home: 422 Bluebird Dr., Terre Haute, Ind. 47803.*

SMITH, Leslie Garrett, physicist; b. Rotherham, Eng., Nov. 14, 1927; s. Leslie and Phyllis (Garrett) S.; B.A., Cambridge U., 1948, Ph.D., 1951; m. Cornelia Hicks, Dec. 17, 1955; children—Phyllis Eleanor, Jonathan Ellsworth. Came to U.S., 1951, naturalized, 1959. Chief lightning discharge studies group Air Force Cambridge Research Labs., Bedford, Mass., 1953-58, dir. atmospheric scis. lab. GCA Corp. Tech. div., 1958——. Recipient Darton prize Royal Meteorol. Soc., 1954. Fulbright travel scholar, 1951-53. Mem. Am. Geophys. Union, Am. Inst. Aeros. and Astronautics. Editor: Recent Advances in Atmospheric Electricity, 1958. Research in electric charge of raindrops, lightning discharges and other aspects of atmospheric electricity, ionosphere investigations using rockets, first direct observation of C-layer and first direct observation of ionospheric fine structure. Home: 20 Mayall Rd., Waltham, Mass. 02154. Office: Burlington Rd., Bedford, Mass. 01730.*

SMITH, Lloyd P(reston), Am. physicist; b. Reno, Nov. 6, 1903; s. Preston Brooks and Ida (Sauer) S.; B.S. in Elec. Engring., U. Nev., 1925; Ph.D., Cornell University, 1930; D.Sc., University of Nevada, 1961; married Florence S. Hunkin, Sept. 16, 1928; m. 2d, Elizabeth J. LeMat, Dec. 11. 1965. Research engr. General Electric Company, Schenectady, 1925-26; Coffin fellow physics Cornell U., 1926-27, instr. physics, 1927-30, asst. prof., 1932-36, prof., 1936-56, chmn. dept. physics, dir. dept. engring. physics, 1946-56, dir. Research Found., 1948-56, chmn. corp. com. Cornell U. Council, mem. council for Coll. Engring., 1956——; faculty mem. bd. trustees Cornell U., 1952-56; v.p. physics and applied scis. Stanford Research Inst., Menlo Park, Cal., 1965——. NRC fellow Cal. Inst. Tech., 1930-31; Internat. Research fellow U. Munich and Utrecht, 1931-32; lectr. Stanford, 1935; research physicist RCA Labs., Princeton, N.J., 1939, cons. war research, 1941-45, asso. research dir., 1945-46, cons., 1946-55, mem. research planning commn., 1952-55, mem. council Fund for Peaceful Atomic Development, 1956——; cons. atomic bomb research U. Cal., 1942; cons. Union Carbide Nuclear Co., formerly Carbide & Carbon Chemicals Corp., Oak Ridge, Tenn., 1947——, Brookhaven Nat. Labs., Upton, L.I., 1947-49, Haloid Co., 1952-53, Detroit-Edison Co., 1953-56; cons. AVCO Mfg. Corp., 1956, 1959, vice pres., dir., 1956-59, pres. research and advanced development div., 1956-58; reached dir. research laboratory Aeronutronic div. Ford Motor Co., 1959-63, v.p. research Philco Research Labs., 1963-65; own tech., research and devel. mgmt. bus., 1965——; director Research Corp. N.Y., Knox Glass Co. of New York; cons. Union Carbide Corporation, Douglas Aircraft, Inc., 1959. Mem. exec. com. Def. Sci. Bd.; advisory panel for physics Nat. Sci. Found., 1956-59; mem. ORNL Adv. Com. Thermonuclear Research, 1960——; member of adv. com. anti-submarine warfare Nat. Security Indsl. Assn. Pres., mem. bd. dirs. Orange County Philharmonic Soc. bd. trustees Children's Hosp. Orange County. Recipient certificate of merit USN, 1947. Mem. Am. Phys. Soc., Am. Assn. Physics Tchrs. (chmn. com. physics engring. edn. 1955-56), Am. Ordance Assn. (chairman of physics section 1959——), Am. Institute Physics, Am. Society Engineering Edn., N.Y. Acad. Scis., Sigma Xi, Phi Kappa Phi. Author: Mathematical Methods for Scientist and Engineers, 1953. Work includes discovery positive ion emission from high temperature metals, invention high intensity positive ion sources; contbn. to uranium isotope separation (Manhattan Project); invention magnitron microwave generator with spiral electron beam frequency modulation; built 1st mass spectrometer; built 1st linear accelerator for heavy ions at Cornell U. Home: 2315 Eastridge Av., Menlo Park 94025. Office: Stanford Research Inst., Menlo Park, Cal. 94025.*

SMITH, Louis Ezra, Jr., Am. physicist; b. Lexington, Ky.; s. Louis Ezra and Dorothy (Thompson) S.; A.B., San Diego State Coll., 1938; postgrad. U. So. Cal.; Ph.D., U. Wash., 1945; m. Dorothy Donna Sawyer, June 25, 1943; children—Randall Sawyer, Donna Lou, Joan Lura, Russell Louis. Instr., U. Wash., 1943-45; prof. physics San Diego State Coll., 1946——; physicist, phy. sci. study com. Ednl. Services, Inc., 1962-63. Mem. Am. Assn. Physics Tchrs. Research in cosmic rays; electricity and magnetism; nuclear physics; underwater sound. Home: 4580 73d St., LaMesa, Cal. 92041. Office: San Diego State Coll., San Diego 92115.*

SMITH, (Edith) Lucile, Am. biochemist; b. Jackson, Miss., Sept. 9, 1913; d. Harry Darley and Kathleen (Edmundson) S.; B.S., Newcomb Coll., 1935; M.S., Tulane U., 1937; Ph.D., U. Rochester, 1950. Faculty, Tulane U., 1935-47; research fellow U. Rochester, 1947-50; faculty U. Pa., 1950-58; vis. research fellow U. Cambridge (Eng.), 1954-55; faculty Dartmouth Med. Sch., Hanover, 1958——, prof. biochemistry, 1964——. Mem. Am. Soc. Biol. Chemists, Am. Chem. Soc., A.A.A.S., Sigma Xi. Contbg. author: The Bacteria, vol. II, 1961; Haematin Enzymes, 1965; Oxidases and Related Redox Systems, 1965; also articles. Research on purification mammalian cytochrome c oxidase, characterization kinetics and properties cytochrome c oxidase, respiratory chain enzymes bacteria, energy producing systems photosynthetic bacteria. Home: 19 S. Park St., Hanover, N.H. 03755.*

SMITH, M(ahlon) Brewster, Am. psychologist; b. Syracuse, N.Y., June 26, 1919; s. M. Ellwood and Blanche (Hinman) S.; student Reed Coll., 1935-38; B.A., Stanford, 1939, M.A., 1940; Ph.D., Harvard, 1947; m. Deborah Anderson, June 21, 1947; children—Joshua H., T. Daniel, Rebecca M., J. Torquil. Asst. prof. social psychology Harvard, Cambridge, Mass., 1947-49; prof., chmn. dept. psychology Vassar Coll., Poughkeepsie, N.Y., 1949-52; staff asso. Social Sci. Research Council, N.Y.C., 1952-56; prof., dir. grad. tng. in psychology N.Y. U., N.Y.C., 1956-59; prof. psychology U. Cal. at Berkeley, 1959-68, dir. Inst. Human Devel., 1965-68; prof., chmn. dept. physiology U. Chgo., 1968——. Vice pres. Joint Commn. on Mental Illness and Health, 1955-60; pres. Soc for Psychol. Study of Social Issues, 1958-59. Center for Advanced Study in Behavioral Scis. fellow, 1964-65. Fellow Am. Psychol. Assn. (dir. 1960-63); mem. Am. Sociol. Assn., Soc. for Research on Child Devel., A.A.A.S., Phi Beta Kappa, Sigma Xi. Author: (with S.A. Stouffer, et al) The American Soldier, vol. 2, Combat and Its Aftermath, 1949; (with J. Gillin, et al) For a Science of Social Man, 1954; (with J. S. Bruner, R. W. White) Opinions and Personality, 1956; also articles. Editor: Jour. Social Issues, 1951-55, Jour. Abnormal and Social Psychology, 1956-61. Co-authored classic formulation of effects of personality on attitudes and opinions; formulation and study of effectiveness and mental health; research on intergroup attitudes and relations among adolescents. Home: 4733 S. Woodlawn Av., Chgo. 60615.*

SMITH, Marion Russell, Am. entomologist; b. Pendleton, S.C., June 19, 1894; s. James Dawson and Lena (Russell) S.; B.S., Clemson Coll., 1915; M.S., Ohio State U., 1917; Ph.D., U. Ill., 1927; m. Myra Williamson Fant, Jan. 1, 1920; children—Marian Nevitt (Mrs. Eugene H. Stossel), David Hamilton. Asst. entomologist Clemson Coll., 1915-16, 18-19; entomologist N.C. Dept. Agr., 1919-20; instr. high sch., 1920-21; ant specialist Miss. State Plant Bd., State College, 1921-35, 36-37; asso. prof. Miss. State Coll., 1921-35, 36-37; entomologist U. S. Dept. Agr., 1917-18, 20, 31——, ant specialist, Washington, 1937-64, ret., 1964. Mem. Entomol. Soc. Am., Washington, Royal entomol. socs., Soc. Systematic Zoology, Internat. Union for Study Social Insects, Nat. Geog. Soc., Sigma Xi. Contbr. numerous articles to profl. jours. Authority on ants of U. S., their classification, biology, distbrn., econ. importance. Home: 519 N. Monroe St., Arlington, Va. 22201.*

SMITH, Maynard Edwin, Am. meteorologist; b. Cambridge, Mass., Oct. 22, 1919; s. Rufus Daniel and Georgia (Burr) S.; B.A., Princeton, 1941; M.S., N.Y. U., 1942; m. Doreen D. Dallam, Dec. 2, 1945; children—Kent M., Kathleen A. Supr. meteorology Am. Airlines, N.Y., 1945-48; leader meteorology group Brookhaven Nat. Lab., Upton, N.Y., 1948—. Cons. air pollution problems to numerous corps.; cons. wind load problems N.Y. World Trade Center, 1959—; mem. com. on atmospheric and indsl. hygiene NRC, 1941-45. Decorated Bronze Star. Mem. Am. Meteorol. Soc., Air Pollution Control Assn., N.Y. Acad. Sci. Research, publs. on atmospheric diffusion and deposition problems and devel. tracer techniques and instruments for such studies. Home: 103 Cedar Shore Dr., Massapequa, N.Y. 11758. Office: Brookhaven Nat. Lab., Upton, N.Y. 11973.*

SMITH, Mervyn Leslie, chemist; b. London, Oct. 3, 1906; s. Alfred and Lily (Holmes) S.; B.Sc. with honors in chemistry, U. Coll., London, 1929, Ph.D., 1931; diploma Imperial Coll., London, 1932; m. Ruby Violet Cole, Mar. 31, 1934; children—Tessa Ann Wendy (Mrs. Douglas Brian Smith), Brian Leslie, Alan Huw. Research chemist J. & E Sturge Ltd., Birmingham, U.K., 1932-39, R.O F. Wrexham, Wales, 1940-43, Philips Elec. Ltd., Material Research Lab., 1944-51; with Atomic Energy Research Establishment, Harwell, Eng., 1952-63, group leader, 1954-63; with Inst. Nuclear Sci., CENTO, Tehran, Iran, 1959-63, dir., 1961-63, sci. sec., 1967—. Fellow Royal Inst. Chemistry (Ramsay Meml. Research medal 1932); mem. Birmingham Micros. Soc. (past pres.), Soc. Chem. Industry, Chem. Soc., Royal Inst. Chemistry, Rheological Club, Materials Sci. Club. Author: (with Herdan) Small Particle Statistics, 1952; Enriched Isotopes, 1956; (with Koch) Science and Development, 1964; also numerous articles. Editor: Isotope Separators, 1958. Measurement of viscosity and abrasive characteristics of microscopic particles of different forms of calcium carbonate for toiletry; correclation of phys. properties of materials used in electronic components with their particle size and shape; improvement of procedures for electromagnetic separation of isotopes of all elements. Office: CENTO, Box 1828, Tehran, Iran.*

SMITH, Moncrieff Hynson, Jr., Am. psychologist; b. St. Louis, Sept. 21, 1917; s. Moncrieff Hynson and Nell (Galbraith) S.; A.B., U. Mo., 1940; Ph.D., Stanford, 1947; m. Mary M. Devine, June 12, 1960; children—Mason A., Virginia D., Parker G. (by former marriage); MacKenzie, Laurie. Instr., Harvard, Cambridge, Mass., 1947-49; prof. psychology U. Wash., Seattle, 1951—, chmn. joint Ph.D. tng. program psychology-physiology. Home: 6003 50th St. N.E., Seattle 98115.

SMITH, Nathan, Am. physician, surgeon; b. Rehoboth, Mass., Sept. 30, 1762; s. John and Elizabeth (Ide) Hills S.; M.B., Harvard, 1790, M.D. (hon.), 1811; M.D. (hon.), Dartmouth, 1801; m. Elizabeth Chase, Jan. 1791; m. 2d. Sarah Hall Chase, Sept. 1794; 10 children including Nathan Ryno. Practiced medicine Cornish, N.H., 1787-96; prompted establishment of professorship of medicine at Dartmouth, 1798, prof., 1798-1814; prof. Yale, 1813-29, through his personal efforts Conn. Legislature appropriated $20,000 to Yale Med. Sch. and development of bot. garden; pres. Vt. Med. Soc., 1811. Author: Practical Essay on Typhous Fever, 1824. Editor Am. Med. Rev., 1824-26. Performed successful ovariotomy, 1821; credited with performing 1st amputation of leg at knee joint, circa 1825; studied spotted fever, typhus. Died New Haven, Conn., Jan. 26, 1829.

SMITH, Nathan Ryno, Am. surgeon; b. Cornish, N.H., May 21, 1797; s. Nathan and Sarah Hall (Chase) S.; A.B., Yale, 1817, M.D., 1823; postgrad. U. Pa. Med. Sch., 1825, 26; m. Juliette Octavia Penniman, 1824; 8 children, including Alan Penniman. Established (with father) med. sch. at U. Vt., circa 1824; prof. anatomy, physiology, 1824-36; tchr. anatomy, mem. 1st faculty Jefferson Med. Coll., 1826-27; prof. anatomy U. Md., 1927, prof. surgery, 1929-38, 40-70; prof. surgery Transylvania U., Lexington, Ky., 1838-40. Editor (with father and others) Am. Med. Rev., circa 1825-26; founder Phila. Monthly Jour. Medicine and Surgery, 1827, editor, 1827-28; founder, editor Balt. Monthly Jour. Medicine and Surgery, 1830; published Surgical Anatomy of the Arteries, 1832. Pioneer in extirpation of thyroid gland; inventor instrument for lithotomy; constructed the anterior splint. Died Balt., July 3, 1877.

SMITH, Newbern, Am. physicist, elec. engr.; b. Phila., Jan. 21, 1909; s. Montgomery and Lillian (Newbern) S.; B.S., U. Pa., 1930; M.S., 1931, Ph.D., 1935; m. Mary Wynne de Mare, Oct. 17, 1936; m. 2d, Margaret H. Smith, Sept. 15, 1953; children—Newbern D., Sandra W., Anne J. Began as teaching asst. U. Pa., 1931, reasearch asst., 1933; asst. in physics Phila. Coll. Osteopathy, 1934-35; physicist Nat Bur. Standards, 1953-54; tech. head Interservice Radio Propagation Lab., 1942-46, chief, Central Radio Propagation Lab. 1948-54; research engr. U. Mich. Research Inst., 1954-59; prof. elec. engring. U. Mich., 1959—; part time tchrs., communications engring. and radio George Washington U., 1943-47. Recipient Harry Diamond award I.R.E., 1952. Fellow I.R.E. (bd. editors 1949-53); mem. Internat. Sci. Radio Union, Am. Standards Assn. Optical Soc. Am., Am. Geophys. Union, A.A.A.S., Washington Acad. Sci., Philos. Soc. Washington, Sigma Xi, Eta Kappa Nu, Tau Beta Pi, Pi Mu Epsilon, Sigma Alpha Epsilon. Mem. senate adv. com. on color TV, 1950. Tech. adviser to U. S. del. Internat. Radio Conf., Atlantic City, 1947, Provisional Frequency Bd., Geneva, Switzerland, 1948, Internat. Radio Consultative Com., Stockholm, Sweden, 1948; vice chmn. U. S. Del., Internat. Radio Consultative Com. Geneva, 1951. Research and numerous publs. on radio wave propagation, radio astronomy. Home: 1431 Greenview Dr., 48103.*

SMITH, Norman Obed, chemist; b. Winnipeg, Can., Jan. 23, 1914; s. Ernest and Ruth (Kilpatrick) S.; B.S. with honors U. Man. (Can.), 1935; M.S., 1936; Ph.D., N.Y.U., 1939; m. Anna Marie O'Connor, July 1, 1944; children—Richard Obed, Graham Michael, Stephen Housley. Came to U.S., 1950, naturalized, 1958. Faculty, U. Man., Winnipeg, 1939-50, asso. prof. chemistry, 1948-50; faculty Fordham U., N.Y.C., 1950—, prof. chemistry, 1965—. Fellow Chem. Inst. Can.; mem. Am. Chem. Soc., Geochem. Soc., Sigma Xi, Phi Lambda Upsilon. Research, publs. in field of heterogeneous equilibrium, solid solutions, clathrates. Home: 59 Monrovia Blvd., Tuckahoe, N.Y. 10707. Office: Dept. Chemistry, Fordham U., E. Fordham Rd., N.Y.C. 10458.*

SMITH, Paul Albert, Am. engr.; b. Morning Sun, Ia., Jan. 9, 1901; s. Jonas W. and Estella (McLellan) S.; B.S., U. Mich., 1924; m. Sylvia Juanita Ralston, July 9, 1923; children—Paul A., Kathryn Caroline (Mrs. Robert Gifford). Instr., U. Mich., 1923-24; served from ensign to rear adm. U. S. Coast and Geodetic Survey, 1924-54, asst. to dir., 194 -46; alternate, then U. S. rep. on council Internat. Civil Aviation Orgn., Montreal, Can., 1946-53; cons. to asst. sec. def. for research and devel., also to dir. Advanced Research Projects Agy., Dept. Def., Washington, 1953-59; engr. RAND Corp., Washington, 1959—. Mem. coms. Nat. Acad. Sci.-NRC.; sci. adv. bd. USAF, 1958. Recipient Exceptional Service awards Dept. Commerce, 1953, Dept. Air Force, 1964; named Distinguished Alumnae, U. Mich., 1953. Fellow Am. Soc. C.E., Geol. Soc. Am., A.A.A.S., Am. Geog. Soc., Wash. Acad. Sci., Am. Inst. Aeros. and Astronautics (asso.); mem. Sigma Xi, Tau Beta Pi. Research in submarine physiography, hydrography, cartography. Home: 4714 26th St., North Arlington, Va. 22207. Office: 1000 Connecticut Av. N.W., Washington 20006.*

SMITH, Paul Francis, Am. microbiologist; b. Brookville, Pa., Apr. 3, 1927; s. Leo F. and Josephine (Ferguson) S.; B.S., Pa. State U., 1949; M.S., U. Pa., 1950, Ph.D., 1951; m. Marie D. Rymshaw, July 1, 1951; children—Rebecca, Leigh Ann, Laurie, Graham. Research microbiologist Merck & Co., Inc., Rahway, N.J., 1951-52; faculty U. Pa. Sch. Medicine, Phila., 1952-61, asst. prof., 1956-61; prof., chmn. dept. microbiology U. S.D., Vermillion, 1961—. Cons. USN Med. Research Unit 4, 1965-66. Recipient Lederle Med. Faculty award, 1960. Diplomate Am. Bd. Microbiology. Fellow Am. Acad. Microbiology, Am. Pub. Health Assn., A.A.A.S.; mem. Am. Soc. for Microbiology, Am. Chem. Soc., Sigma Xi, Delta Chi. Mem. editorial bd. Jour. Bacteriology, 1964—. Author: (with C. Panos) The Mycoplasma, 1966; also numerous articles. Research on physiology of mycoplasma. Home: 11 N. Prentis St., Vermillion, S.D. 57069.*

SMITH, Peter Alan Somervail, chemist; b. Erskine Hill, Eng., Apr. 16, 1920; s. Harold B. and Eva (Barr) S.; came to U.S., 1923, naturalized, 1938; B.S. in Chemistry, U. Cal., Berkeley, 1941; Ph.D., U. Mich., 1944; m. Mary Charmine Walsh, Apr. 19, 1952; children—Kent Alan Kinloch, Leslie Charmaine. Research asso. penicillin project OSRD, Ann Arbor, Mich., 1944-45; faculty U. Mich., Ann Arbor, 1945—, prof. chemistry, 1958—. Cons. to industry, govt.; chmn. com. on nomenclature NRC, 1960—. Fulbright research scholar, New Zealand, 1951. Mem. Anthroposophical Soc., Royal Philatelic Soc. London, Phi Beta Kappa, Alpha Chi Sigma, Gamma Alpha, Phi Lambda Upsilon. Author: Chemistry of Open-chain Organic Nitrogen Compounds, 1965; contbg. author to vols. 3, 11 Organic Reactions, 1946-60, Molecular Rearrangements, 1963. Home: 811 Mt. Pleasant Av., Ann Arbor, Mich. 48103.*

SMITH, Philip E. L., Canadian anthropologist; b. Fortune, Nfld., Can., Aug. 12, 1927; s. George Frederick and Alice (Lake) S.; B.A., Acadia U., 1948; A.M., Harvard, 1957, Ph.D., 1962; postgrad. Bordeaux (France) U.; m. Fumiko Ikawa, 1959; 1 son, Douglas P.E. Faculty, U. Toronto (Ont., Can.), 1961—, asso. prof., anthropology. Dir. Canadian Expdn., Nubia, 1962-63. Fellow Am. Anthrop. Assn.; mem. Current Anthropology (asso.). Author: Le Solutréen en France, 1965; also articles. Research on synthesis Solutrean culture in France, slavage archaeology in Egypt, Iranian prehistory. Home; 99 Madison Av., Toronto 5, Ont., Can.*

SMITH, Philip Sidney, Am. geologist; b. Medford, Mass., July 28, 1877; s. Sidney Leroy and Katherine (Butler) S.; A.B., Harvard, 1899, A.M., 1900, Ph.D., 1904; m. Lenore Willis Kinney, Nov. 26, 1900; children—Sidney Butler, Katharine, Constance (Mrs. Robert F. Thurrell, Jr.). Geol. field worker, Mich., S.D.; instr. geology Harvard, Radcliffe Coll., 1900-06; joined Alaskan div. U. S. Geol. Survey, 1906, became adminstrv. geologist, acting dir., 1915, chief Alaskan geologist, 1925-45; made 2 exploratory expdns. in Alaska for USN, 1924, 26; connected with Arctic Inst. Fellow Geol. Soc. Am.; mem. Am., Nat. geog. socs., Washington Acad. Scis., A.A.A.S., Soc. Econ. Geologists, Inst. Mining and Metall. Engrs., Am. Polar Soc., Am. Geophys. Union, Soc. Profl. Geographers, other socs. Author: (with others) Aerial Geology of Alaska, 1939; also bulls., papers on Alaskan geology. Died May 10, 1949.

SMITH, Philip Wayne, Am. zoologist; b. Neoga, Ill., Dec. 2, 1921; s. Warde and Pauline (Brown) S.; student Eastern Ill. U., 1940-42, 46-47; B.S., U. Ill., 1948, Ph.D., 1953; m. Dorothy Dearnbarger, May 23, 1942; 1 dau., April (Mrs. John E. Hoffman). Asst. aquatic biologist Ridge Lake Field Lab., Charleston, Ill., 1942; with Ill. Natural History Survey, Urbana, Ill., 1947—, asso. taxonomist, 1955-63, taxonomist, 1963—; prof. zoology U. Ill., Urbana, 1965—. Mem. Herpetologists' League (pres. 1965-67), Am. Soc. Ichtyologists and Herpetologists (v.p. 1966, rep. to the NRC 1963). Author: The Amphibians and Reptiles of Illinois, 1961; also numerous articles. Research on systematics and biology fishes, amphibians, reptiles, biogeography. Home: 1407 Briarcliff Dr., Urbana 61801. Office: Ill. Natural History Survey, Urbana, Ill. 61801.*

SMITH, Ralph Grafton, pharmacologist; b. Woodstock, Ont., Can., Mar. 15, 1900; s. James and Sarah (Karn) S.; B.A., U. Toronto, 1921, M.A., 1922, M.D., 1925; Ph.D., U. Chgo., 1928; m. Barbara Emily Paton, Aug. 19, 1933; children—Alan M., Roger M., Robert G., Eric J. Came to U.S., 1926, naturalized, 1940. Faculty, U. Mich. Med. Sch., Ann Arbor, 1928-43, asso. prof. pharmacology, 1937-43; prof. Tulane U. Med. Sch., New Orleans, 1943-50; dir. div. new drugs FDA, Washington, 1950-66, asst. to dir. for Nat. Acad. Sci. liaison, 1966—. Recipient Henry Russell award U. Mich., 1934-35; Superior Service award Dept. Health Edn. and Welfare, 1958, Distinguished Service award, 1964. Mem. Am. Soc. Pharmacology and Exptl. Therapeutics, Soc. for Exptl. Biology and Medicine, Pan-Am. Med. Assn., Sigma Xi, Alpha Omega Alpha. Contbr. articles to tech. jours. Research on respiratory stimulants; cyanide poisoning and sulfur metabolism. Home: 1026 Noyes Dr., Silver Spring, Md. 20910. Office: Bur. Medicine, FDA, Washington 20204.

SMITH, Ray Fred, Am. entomologist; b. Los Angeles, Jan. 20, 1919; s. Ray M. and Elsie E. (Weisheit) S.; B.S., U. Cal. at Berkeley, 1940, M.S., 1941, Ph.D., 1946; m. Elizabeth J. McClure; children—Donald S., Thomas A., Katherine N. Tech., lab. asst. U. Cal. at Berkeley, 1941-45, asso. expt. sta., 1945-46, instr. entomology, jr. entomologist expt. sta., 1946-48, asst. prof., entomology, asst. entomologist expt. sta., 1948-54, asso. prof., asso. entomologist expt. sta., 1954-60, prof. entomologist expt. sta., 1960—, chmn. dept. entomology and parasitology, Berkeley-Davis, 1959-63, chmn. dept. entomology and parasitology, Berkeley, 1963—. Guggenheim fellow, 1950-51. Fellow A.A.A.S., Cal. Acad. Sci., mem. Entomol. Soc. Am., Brit. Ecol. Soc. Ecol. Soc. Am. Pacific Coast Entomol. Soc. Devel. integrated Pest control, ecology and bionomics of alfalfa insects; bionomics of non-social bees; biosystematics of Diabrotica beetles. Home; 3092 Hedaro Ct., Lafayette, Cal. Office: Giannini Hall., U. Cal., Berkeley, Cal. 94720.*

SMITH, R(aymond) Dale, Am. anatomist; b. Ambridge, Pa., July 1, 1914; s. George Joseph and Frances Elizabeth (Dale) S.; B.S., U. Pittsburgh, 1936, M.S., 1938, Ph.D., 1939; m. Violet Cecilia Trondle, June 21, 1938; children—Thomas, Joseph, Stephen, Paul, Raymond, Sheila, Mark. Instr., head premed. dept. Gonzaga U., 1939-40, prof., head dept. biology, 1944-46, Coll. Our Lady of the Elms, Chicapee, Mass. 1940-44; asst. prof. anatomy U. Md. Sch. Medicine, 1946-49, asso. prof. anatomy, 1949-50; prof., head dept. anatomy Creighton U. Sch. Medicine, Omaha, 1950—, acting assistant dean medical school, 1959-60, asst. dean, 1960-65. Fellow A.A.A.S.; mem. Am. Assn. Anatomists, Am. Assn. Phys. Anthropologists, Am. Assn. U. Profs., Am. Soc. Zoologists, Sigma Xi. Study of anatomy and physiology of muscles of mastication. Home: 11721 Shirley St., Omaha 68144.*

SMITH, Raymond James, Am. engr-geologist; b. Manchester, N.H., July 16, 1924; B.S., Cal. Inst. Tech., 1945, M.S., 1948; M.A., Princeton, 1950, Ph.D., 1951. Investigator for Princeton, Caribbean, 1948-54; prof. La. State U., 1954-58; civil engr., geologist USN Civil Engring. Lab., Port Hueneme, Cal., 1958—. Indsl. and municipal cons., 1948—. Mem. Geol. Soc. Am., Am. Soc. C.E., Am. Geophys. Union, Seismol. Soc. Am. Research, publs. on Caribbean Island arcs, engring. properties of marine sediments, soil mechanics. Home: 791 Via Ondulando, Ventura, Cal. 93003. Office: U.S. Naval Civil Engring. Lab., Port Hueneme, Cal. 93041.*

SMITH, Richard Thomas, Am. pediatrician, exptl. pathologist; b. Oklahoma City, Apr. 15, 1924; s. Harvey Taylor and Rachel (Grant) S.; student U. Tex., 1941-44; M.D., Tulane, 1950; m. Jean Whisenant,

Aug. 7, 1946; children—Mary Schell, Richard Thomas, Joseph Ryan, John Taylor, Claudia Jane. Intern pediatrics U. Minn. Hosps., 1950-51; resident pediatrics U. Minn., 1951-52, Helen Hay Whitney research fellow, 1952-55, asst. prof. pediatrics, 1955-57, research fellow NRC, 1952-53; sr. investigator Nat. Arthritis and Rheumatism Found., 1955-60; asso. prof. pediatrics U. Tex., Southwestern Med. Sch., 1957-58; prof. pediatrics, chmn. dept. U. Fla., 1958-67, prof., chmn. dept. pathology, 1967—; chief pediatrics U. Fla. Teaching Hospital and Clinics; consultant National Institutes of Health, also to surgeon gen. USPHS. Served as lt. (j.g.) USNR, 1944-46. TOYM award U. S. Jr. C. of C. 1958; E. M. Johnson Research prize, 1963. Mem. So. Soc. Clin. Research Soc. Pediatric Research, Am. Soc. Clin. Investigation Central Soc., Clin. Research, Am. Soc. Clin. Investigation Central Soc., Clin. Research Am. Assn. Immunologists Am. Soc. Exptl. Pathology, Soc. Exptl. Biology and Medicine, A.A.A.S, Am. Fedn. Clin. Research, Am. Pediatric Soc., Sigma Xi, Alpha Omega Alpha Contbr. research and clin. papers sci. periodicals. Investigation of clinical and general immunology, tumor biology, relationship between susceptibility to infection and age. Home: 1704 S.W. 8th Dr., Gainesville, Fla.*

SMITH, Robert, English mathematician; b. Lea, Eng., 1689; s. John and Hannah (Cotes) S.; B.A., Cambridge U., 1711, M.A., 1715, LL.D.; 1723, D.D., 1739; master of mechanics to George II; math. preceptor to Duke of Cumberland; Plumian prof. astronomy, Cambridge U., 1716-60, designed, had built telescope at Trinity obs., became master Trinity Coll., 1742, acting vice-chancellor, 1742-43, instituted Smith's prize for math. and sci. students. Author: A Compleat System of Opticks, 4 vols., 1738; Harmonics, 1749. Died 1768.

SMITH, Robert Angus, Brit. chemist; b. Glasgow, Scotland, Feb. 15, 1817; s. John and Janet (Thomson) S.; student Glasgow U.; Ph.D., Giessen, Germany, 1841; student of Liebig; LL.D., Glasgow, 1881, Edinburgh, 1882; became cons. chemist, Manchester, Eng., 1844; apptd. insp. of alkali works, 1863. Fellow Royal Soc., 1857. Author: Disinfectants and Disinfection, 1869; Air and Rain, the beginnings of a Chemical Climatology, 1872; also papers. Contbr. to Ures' Dictionary, also to Chem. News. Pioneer in chemistry of hygiene; studied organic impurities of air; proposed method (minimetric) of estimating carbonic acid in air, 1865. Died May 12, 1884.

SMITH, Robert Emrie, Am. physiologist; b. Ellensburg, Wash., Sept. 28, 1913; s. Emrie Beech and Bessie A. (Pippinger) S.; A.B., U. Cal. at Berkeley, 1934, M.A., 1938, Ph.D., 1946; m. Barbara Clark, May 21, 1938 (dec. Sept. 1966). Environmental physiologist Naval Med. Research Inst., Bethesda, Md., 1943-48; prin. physiologist NIH, Bethesda, Md., 1948-51; physiology faculty U. Cal. Med. Sch., Los Angeles, 1951-67, prof. physiology, 1962-67; prof., chmn. physiol. scis. Sch. Vet. Medicine, U. Cal., Medicine, 1967—. Organizing chmn. Internat. Symposia on Temperature Acclimation, Fedn. for Am. Soc. Exptl. Biology, 1959, 62, 65. Mem. Am. Physiol. Soc., A.A.A.S., others. Editor: Fedn. Proc., 1960, 63, 66. Contbr. articles to profl. jours. Office: Sch. Vet. Medicine, U. Cal., Davis; Cal. 95616.

SMITH, Robert Gillen, Am. educator; b. Dover, N.J., Oct. 16, 1913; s. John Wesley and Elizabeth (Gillen) S.; A.B., Drew U., 1936; M.A., Columbia, 1939, Ph.D., 1950; m. G. Lois Squier, Dec. 23, 1942; children—Robert L., Donald P. Faculty, Drew U., Madison, N.J., 1937-42, 44, 46—, prof. polit. sci., 1954—, dir. Inst. for Research on Govt., 1964—. Vis. prof. Grad. Sch. Pub. Adminstrn., N.Y. U., 1966-67; cons. city and county govts.; pres. N.J. Social Sci. Acad., 1965-67. Recipient plaque Coll-Fed. Agy., 1965, Alumni Achievement award in Arts Drew U., 1960. Mem. Am. Soc. Pub. Adminstrn., Am. Polit. Sci. Assn., Com. for Urban Research and Edn., Regional Plan Assn. Author: Public Authorities, Special Districts and Local Government, 1964. Contbg. editor Dictionary Polit. Sci., 1964. Extensive research on pub. authorities in U. S., Eng., Can.; differentiated between pub. authorities and other spl. govts.; work, particularly in N.Y.C., on coordinating spl. govts. with gen. purpose conventional govts. Home: 5 Wyndehurst Dr., Madison, N.J. 07940.*

SMITH, Robert Jay, Am. zoologist; b. Flint, Mich., July 26, 1922; s. Jay Perry and Alice (Green) S.; A.B., Alma Coll., 1947; M.S., U. Mich., 1949, Ph.D., 1953. Prof. zoology U. Dubuque (Ia.), 1953-55, Coll. William and Mary, 1955-56, U. Miami (Fla.), 1956-57; prof. U. Detroit, 1958—. NIH grantee, 1964-67. Mem. Am. Soc. Zoologists, Am. Soc. Parasitologists, Am. Micros. Soc., Sigma Xi. Research on Ancylid snails as carriers of trematode diseases in animals and man. Home: 17559 Pennington Dr., Detroit 48221.*

SMITH, Sir Robert Murdoch, Scottish archaeologist; b. Kilmarnock, Scotland, Aug. 18, 1835; s. Hugh and Jean (Murdoch) S.; ed. Glasgow (Scotland) U.; m. Eleanor Baker, 1869; 9 children. Comnd. Royal Engrs. Corps., 1855; participated expdn. Asia Minor, 1856-59; explored ancient city Cyrene with Lt. E. A. Porcher, 1860; constrn. Persian sect. telegraph line Eng.-India, 1863, dir. telegraph line at Teheran, from 1865; dir. Sci. and Art Mus., Edinburgh, Scotland, from 1885; dir. in chief Indo-European telegraph dept., from 1887. Mem. com. Scottish Nat. Portrait Gallery. Bd. dirs. Mfrs. of Scotland. Recipient Silver medal Soc. Arts Scotland. Discovered mausoleum at Halicarnassus, Asia Minor; findings of Greek sculpture and inscriptions now in Brit. Mus., London. Died Edinburgh, July 3, 1900.

SMITH, Robert Roland, Am. surgeon; b. Mansfield, O., June 28, 1912; s. Edward and Mary (Bowers) S.; B.S., Ashland (O.) Coll., 1933; M.D., Western Res. U., 1937; m. Melverda Marie Birch, June 18, 1938; children—Carolyn Marie, Robert Birch. Commd. officer USPHS, 1937-60; chief surgery Nat. Cancer Inst., Bethesda, Md., 1951-61; practice medicine, specializing in surgery, Atlanta, 1961—; asso. prof. surgery Emory U. Sch. Medicine, 1963—. Mem. James Ewing Soc., Soc. Head and Neck Surgeons. Research, numerous publs. on role of cancer cell contamination in operative wounds in relation to local recurrence and mitastases of cancer and means to prevent. Home: 2070 Chrysler Dr., Atlanta 30329. Office: Emory U. Clinic, Atlanta 30322.*

SMITH, Sedgwick Eugene, Am. animal nutritionist; b. Elkins, W. Va., Apr. 4, 1914; s. Charles J. and Anne (Lothes) S.; B.S., Pa. State U., 1935; Ph.D., Cornell U., 1939; m. Margaret Gainey, June 8, 1940; children—Edward, J. Smith, Eileen A., Mark F. Agt., U.S. Dept. Agr., 1939-41, animal physiologist, 1941-46; prof. dept. animal sci. Cornell U., Ithaca, N.Y., 1946—, acting head dept. animal husbandry, 1954, 59. Chmn. subcom. on rabbit nutrition NRC, 1954—. Mem. Am. Soc. Animal Sci., Am. Dairy Sci. Assn., Am. Inst. Nutrition. Author: Food for Life, 1952; Agriculture Chemistry, II, 1951; also numerous articles. Research on nutrition and physiology mineral elements, nutrition small animals and livestock. Home: 200 Roat St., Ithaca, N.Y. 14850.*

SMITH, Sidney Irving, Am. biologist; b. Norway, Me., Feb. 18, 1843; s. Elliot and Lavinia H. (Barton) S.; Ph.B., Sheffield Sci. Sch. (Yale), 1867; A.M., Yale, 1887; m. Eugenia P. Barber, June 29, 1882. Asst. in zoology Sheffield Sci. Sch., 1867-75, prof. comparative anatomy, 1875-1906, emeritus, 1906—. Had charge of deep water dredging in Lake Superior for U. S. Lake Survey, 1871; and for U. S. Coast Survey at St. George's banks, 1872; asso. with biol. work. U. S. Fish Commn. for many yrs. Mem. Nat. Acad. Scis. Contbr. of numerous papers, especially in marine zoology. Died May 17, 1926.

SMITH, Stanley Desmond, English physicist; b. Bristol, Eng., Mar. 3, 1931; s. Henry George Stanley and Sarah (Wear) S.; B.Sc., U. Bristol, 1952, D.Sc., 1966; Ph.D., U. Reading, 1956; m. Gillian Anne Parish, July 1, 1956; children—David Stanley, Nicola Jane. Sr. sci. officer Royal Aircraft Establishment, Farnborough, Eng., 1956-58; research asst. dept. meteorology Imperial Coll., 1958-59; research asst. U. Reading, (Eng.); 1959-61, lectr., 1961-65, reader in physics, 1965—; chmn., dir. Coating & Filter Design Ltd. Fellow Royal Meteorol. Soc. Author (with J. T. Houghton) Infra-Red Physics, 1966. Research, publs. on interference filters, design, double half wave type and experimentally in infra-red; magneto optics in semi-conductors particularly Faraday effect; lattice vibrations of impurity atoms; selective chopper radiometer for atmospheric temperature sounding from satellites. Home: 21, Simon's Lane, Wokingham, Berks, Eng. Office: J. J. Thomson Phys. Lab., U. Reading, Eng.*

SMITH, Stanley George, cytologist; b. Morden, Surrey, Eng., Feb. 6, 1909; s. Thomas Sidney and Ellen (Anthony) S.; B.S., McGill U., 1935, M.S., 1936, Ph.D., 1938; m. Margaret Jane Caldwell, June 22, 1939; children—Margaret (Mrs. Roger Omar Adams), Pamela (Mrs. Brian Hansen), Peter Stanley. Cytologist, Can. Dept. Agr., Sault Ste. Marie, Ont., 1945-60; head sect. cytology and genetics Can. Dept. Forestry, Sault Ste. Marie, 1960—. Recipient Maj. Hiram Mills Gold medal in biology, 1935. Royal Soc. Can. fellow in biology, 1937. Mem. Soc. For Study Evolution, Genetics Soc. Am., Genetics Soc. Can., Entomol. Soc. Can. Research, numerous publs. cytological attributes common to European spruce sawfly in Europe and Can., taxonomic problems mainly in field of forest entomology. Home: 112 Upton Rd. Office: 1195 Queen St. E., Sault Ste. Marie, Ont., Can.*

SMITH, Stephen, Am. surgeon; b. Onondaga County, N.Y., Feb. 19, 1823; s. Lewis and Chloe (Benson) S.; M.D., Coll. Phys. and Surg. (Columbia), 1850; A.M. (hon.), Brown, 1876; LL.D., U. of Rochester, 1891; m. Lucy E. Culver, June 1, 1858. Resident surgeon Bellevue Hosp., 1850-52, attending surgeon, 1854; prof. surgery Bellevue Hosp. Med. Coll., 1861-65, anatomy, 1865-74; prof. clin. surgery, med. dept. 1874; cons. surgeon Bellevue, St. Vincent and Columbus hosps. Investigated san. condition of N.Y., 1865, and reported to legislature; U. S. commr. to 9th Internat. San. Conv., Paris, 1894; founder and 1st pres. Am. Pub. Health Assn., Mem. city and nat. bds. of health, commr. in lunacy, and commr. State Bd. of Charities, N.Y. Author: Handbook of Surgical Operations; Principles of Operative Surgery; Doctor in Medicine; The City that Was; Who Is Insane?, 1916. Joint editor N.Y. Jour. Medicine, 1853-57, editor, 1857-60; editor N.Y. Med. Times, 1860-64. Died Aug. 26, 1922.

SMITH, Sterling Bishop, Am. chemist; b. New Haven, May 7, 1899; s. Herbert Hudson and Frederica (Bishop) S.; Ph.B., Yale, 1920, M.S., 1923; Ph.D., N.Y. U., 1927; m. Harriet Chamberlain, Aug. 15, 1928; children—Robert M., Edward C. Faculty, Trinity Coll., Hartford, Conn., 1923-65, prof. chemistry, 1952-65. Mem. Am. Chem. Soc., Am. Assn. U. Profs. Research, publs. on solubility relations in 3 and 4 component systems. Home: 35 Grandview Terrace, Wethersfield, Conn. 06109.*

SMITH, Theobald, Am. pathologist; b. Albany, N.Y., July 31, 1859; s. Philip and Theresa (Kexel) S.; Ph.B., Cornell, 1881; M.D., Albany Med. Coll., 1883; hon. degrees Harvard U. Chgo., Yale, Princeton, other univs.; m. Lillian H. Egleston, May 17, 1888; children—Dorothea Egleston, Lilian Hillyer (Mrs. Robert F. Foerster), Philip Hillyer. Dir. pathol. lab., Bur. Animal Industry, Dept. Agr., 1884-95; dir. pathol. lab. Mass. Bd. Health, 1895-1915. Prof. bacteriology Columbian (now George Washington) U., 1886-95; prof. comparative pathology Harvard, 1896-1915; dir. dept. of animal pathology Rockefeller Inst. for Med. Research, 1915-29; Harvard exchange prof. to Berlin, 1911-12; pres. Internat. Union Against Tb, 1926, Congress of Am. Physicians and Surgeons, 1928. Recipient Mary Kingsley, Kober, Flattery, Trudeau, Sedgwick, Holland Soc. and Gerhard medals; Copley medal of Royal Soc. of Gt. Britain. Fellow Am. Acad. Arts and Scis., Soc. Tropical Medicine and Hygiene (hon. London), Path. Soc. Gt. Britain and Ireland; mem. French Acad. Scis. Fellow Royal Soc., 1932. Contbr. numerous med. jours. on nature and causation of infectious and parasitic diseases. Discovered (with F. L. Kilborne) causative agt. of Tex. cattle fever, Pyrosoma bigeminum, 1893, proved disease is transmitted by cattle tick Boöphilus bovis; credited with 1st clear distinction between human and bovine tubercle bacilli, 1898; successfully used neutral toxin-anti-toxin mixtures for immunization against diphtheria, by 1909. Died Dec. 10, 1934.

SMITH, Thomas, English physicist; b. Leamington, Eng., Apr. 6, 1883; s. William Edward and Dorothy (Jameson) S.; B.A., Queens Coll., Cambridge (Eng.) U., 1905, M.A., 1919; m. Elsie Muriel Elligott, Mar. 29, 1913; children—Helen Muriel, Thomas Edward, Mary Elizabeth (Mrs. Bernard Lennox), Dorothy Jean (Mrs. Kenneth W. Rankin), John Hilary. Staff, Nat. Phys. Lab., Teddington, Eng., 1907-48, supt. optics, 1937-48. First pres. Internat. Commn. on Optics, 1947-49; chmn. Brit. Com. on Optics, 1947-55. Fellow Royal Soc., 1932; mem. Optical Soc. (pres. 1927-29), Phys. Soc. (pres. 1936-38), Inst. Physics (founding), Optical Soc. Am. (hon.). Research, numerous publs. in geometrical optics and methods of designing optical instruments. Home: Buschall, Prospect Rd., Heathfield, Sussex, Eng.*

SMITH, Warren DuPré, geologist; b. Leipzig, Germany, May 12, 1880; s. Charles Forster and Anna Leland (Du Pré) S.; brought to U. S. in infancy; B.S., U. Wis., 1902, Ph.D., 1908; M.A., Stanford, 1904; fellow U. Chgo., 1904-05; m. Phoebe Ellison, July 14, 1910; children—James Francis, Warren Ellison, Phoebe Hall. Field asst. Wis. Geol. and Natural History Survey, 1900-02; geologist U. S. Govt. Mining Bur., Manila, 1905-06, Div. of Mines, Bur. of Science, 1906-07; chief Div. of Mines, P.I., 1907-14; pres. Philippine Soc. Engrs., 1912; head dept. of geology U. Ore., 1914-17, ret. 1947; chief div. mines Bur. of Science, Philippine Govt., 1920, 21 (leave of absence from U. of Ore.); cons. geologist on Owyhee Dam. Ore., U. S. Bur. of Reclamation, 1927; ranger-naturalist Crater Lake Nat. Park, 1934, 35. U. S. del. Internat. Geol. Congress, Toronto, 1913; del. 1st Pan-Pacific Sci. Conf., Honolulu, 1920; mem. Governor's Spl. Mining Com., Ore., 1935; mem. Philippine com. Pacific Sci. Bd. Fellow Geol. Soc. Am. (pres. Cordilleran sect. 1925), Pacific Geog. Soc., Ore. Mining Congress (pres. 1934), Ore. Acad. Sci. (pres.-elect, 1948), Phi Beta Kappa, Sigma Xi. Geology and Mineral Resources of the Philippine Islands; Scenic Treasure House of Oregon, 1941; also articles and monographs on spl. phases of Philippine and Malayan geology and papers on Ore. and Pacific geology and geography. Editor: Physical and Economic Geography of Oregon, 1940. Died July 18, 1950.

SMITH, Warren LaVerne, Am. elec. engr.; b. Wayne, Neb., July 6, 1924; s. James Morrison and Mattie (Meng) S.; student Mich. Coll. Mining and Tech., 1942-43; B.S. in Elec. Engring., U. Wis., 1945, postgrad. in Physics; m. Frances Dowd, June 18, 1948; children—Marjory, Catherine, James, Gerald. With Bell Telephone Labs., Allentown, Pa., 1954—, supr. precision frequency control devices and monolithic crystal filters, 1962—. Sr. mem. I.E.E.E. Devel. precision crystal-controlled oscillators for use as frequency standards and master clocks, including devel. oscillator circuits and asso. equipment, monolithic crystal filters, and low-noise frequency sources for microwave generators. Home: 3046 Meadowbrook Circle. Office: 555 Union Blvd., Allentown, Pa. 18103.*

SMITH, William, English geologist, engr.; b. Churchill, Eng., Mar. 23, 1769; s. John and Anne Smith; LL.D., Trinity Coll., 1835; asst. to Edward

Webb, surveyor; hired by Somerset coal-field to survey a canal, 1793; made study trip to investigate constrn. and working of canals, 1794; employed on works Somerset Coal Canal, until 1799; lectr., 1824-28; became landsteward Hackness estate, 1828. Recipient Wollaston medal, 1831. Author: Strata identified by Organised Fossils, 1816; A Stratigraphical System of Organized Fossils, 1817; A Geological Map on a reduced scale, 1819; New Geological Atlas of England and Wales, 1819-24. Founder of Stratigraphical geology; projected great map of English strata; authority on drainage and irrigation; collected fossils which the Brit. Mus. acquired. Died Aug. 28, 1839.

SMITH, William Burton, Am. chemist; b. Muncie, Ind., Dec. 13, 1927; s. Merrill Mark and Felice Hoy (Richardson) S.; B.A., Kalamazoo Coll., 1949; Ph.D., Brown U., 1954; m. Marian Louise Roseborough, Aug. 9, 1954; children—Mark W., Frederick D., Mary F. Research asso. Fla. State U., 1953-54, U. Chgo., 1954-55; asst. prof., then asso. prof. Ohio U., 1955-61; R.A. Welch vis. prof. chemistry Tex. Christian U., 1960-61, prof. chemistry, chmn. dept., 1961-—. Research participant Oak Ridge Inst. Nuclear Studies, 1956-—. Mem. Am. Chem. Soc., Sigma Xi. Author: A modern Introduction to Organic Chemistry, 1961; also research articles. Research in organic reaction mechanisms, polymer chemistry, molecular structure, nuclear magnetic resonance parameters. Home: 3604 Wedghill Way, Fort Worth 76133.*

SMITH, William Norman, Am. mathematician; b. Glencoe, Ill., Jan. 14, 1914; s. Frederick Henry and Nella (Nichols) S.; B.S., Northwestern U., 1935, M.A., 1936; Ph.D., U. Wis., 1952; m. Ruth Rohde Saunders, May 29, 1943; children—Nancy Nella, Donald Quentin, William Frederick, Richard Vincent. Faculty, U. Wis., 1943-48, instr., 1946-48; faculty U. Wyo., Laramie, 1948-—, prof. math., 1960-—, dept. head, 1963-—. Mem. Math. Assn. Am. (sec.-treas. Rocky Mountain sect. 1964-67(, Am. Math. Soc., Phi Beta Kappa, Phi Eta Sigma. Author: (with L. Henkin, V. Varineau, M.J. Walsh) Retracing Elementary Mathematics, 1962; also articles. Research on differential equations, founds. of math. Home: 1815 Park Av., Laramie, Wyo. 82070.*

SMITH, William Sooy, Am. civil engr.; b. Tarlton, O., July 22, 1830; s. Sooy and Ann (Hedges) S.; A.B., Ohio U., 1849, later A.M.; grad. U. S. Mil. Acad., 1853; m. Elizabeth Haven; m. 2d, Josephine Hartwell, 1884; 1 son, Charles Sooysmith. Apptd. 2d lt. 3d Arty. U. S. Army; stationed in N.M.; resigned; went to Chicago, 1854, entered engring. service of I.C. R.R. Co.; soon apptd. asst. engr. to Col. Graham, U. S. engr. in charge of improvements of Lake Michigan harbors; conducted select sch. at Buffalo, 1855-57; practiced as civil engr., 1857-59; chief engr. of co. building iron bridge across Savannah River for Savannah & Charleston R.R. Co., 1860-61; served with U. S. Army 1861-1864; resumed practice as civil engr., Chgo.; work as engr. and contractor for U. S. Govt. and ry. cos. included reconstrn. of Waugoshanee Light House at western entrance of Straits of Mackinac; built 1st all-steel ry. bridge in world (Glasgow, Mo.), and sub-structures of 6 other bridges, by pneumatic process, which developed and greatly improved; introduced (with son) into U. S. freezing process for difficult subaqueous work, and sank 2 shafts through quick-sands and boulders, to depth of 100 feet, by this method; completely changed methods of constructing founds. for heavy bldgs. in Chgo., carrying the loads down to hard bottom, 50 feet or more, by means of piles cut off below water surface, or by sinking columns of concrete to hard bottom and resting bldgs. on them; aided in devel. of plans of high steel bldgs. in Chgo. and throughout world; leader in urging Govt. to create bd. to test Am. metals and mem. of that bd. during the 3 yrs. of its existence; inventor 1st pneumatic caisson ever built; designed new system of fireproof bldg. Died Mar. 4, 1916.

SMITH, William Thomas, Jr., Am. chemist; b. Wilmington, N.C., July 12, 1903; s. William Thomas and Emma (Underwood) S.; A.B., U. Va., 1925; M.A., Ohio State U., 1928, Ph.D., 1930; m. Elizabeth Avery Reagan, Mar. 31, 1934; 1 dau., Elizabeth Avery (Mrs. Sidney Carl Padgett). Faculty, U. Tenn., Knoxville, 1925-—, prof. chemistry, 1950-—. Cons. Oak Ridge Nat. Lab., 1950-—. Mem. Am. Chem. Soc., Tenn. Acad. Sci., Sigma Xi, Alpha Chi Sigma, Gamma Alpha. Author: (with J.H. Wood) Laboratory Manual For College Chemistry, 1966. Research, numerous publs. on chemistry of rhenium and technetium, metal-metal salt phase, thermodynamic properties of solutions. Home: 109 Fronda Lane, Knoxville, Tenn. 37920.*

SMITH, William Vick, Am. physicist; b. Dallas, Dec. 11, 1916; s. Charles Grover and Aurelia (Mayer) S.; A.B. summa cum laude, Harvard, 1937, M.A., 1939, Ph.D., 1941; m. Jeanette Luella Pedersen, Apr. 18, 1941; 1 dau., Jeanette Aurelia. Engr., Sylvania Elec. Products Corp., Salem, Mass., 1941-42; staff Mass. Inst. Tech. Radiation Lab., Cambridge, 1942-46; physicist Raytheon Mfg. Corp., Waltham, Mass., 1946; faculty Duke, 1946-51, U. Del., 1951-56; sr. staff mem. IBM, Poughkeepsie, N.Y., 1956-60, IBM Research Center, Yorktown Heights, N.Y., 1961-—. Exchange scientist IBM, Holland, 1960-61. Cons. to govt., industry; mem. sci. program com. 3d Internat. Quantum Electronics Symposium, 1963. Fellow A.A.A.S., Am. Phys. Soc. (sec.-treas. div. solid state physics); mem. Research and Engring. Soc. Am., Phi Beta Kappa, Sigma Pi Sigma, Pi Mu Epsilon. Author: (with W. Gordy, R.F. Trambarulo) Microwave Spectroscopy, 1953; (with Peter P. Sorokin) The Laser, 1966; contbr. to Microwave Magnetrons, vol. 6, 1948. Patentee in field. Research on elec., optical and microwave properties of gases and solids, including useful electronic properties. Home: 28 Garey Dr., Chappaqua, N.Y. 10514. Office: T.J. Watson Research Center, IBM, Yorktown Heights, N.Y. 10598.*

SMITH, Sir William Wright, Brit. botanist; b. Lochmaben, Scotland, Feb. 3, 1875; s. James T. Smith; ed. U. Edinburgh; M.A., D. ès Sc., LL.D.; m. Emma Wiedhofft; 3 daus. Lectr. botany U. Edinburgh, 1902-07, later Regius prof. botany; curator herbarium Royal Botanic Gardens, Calcutta, 1907-08; officiating dir. Bot. Survey of India, 1908; explored vegetation of N.W. Sikkim, Tibet-Nepalese frontier, 1908, Sikkim-Chumbi frontier, 1909; asst. keeper Royal Botanic Garden, Edinburgh, 1911-22, the Regius keeper. Recipient George Robert White medal Mass. Hort. Soc., 1952. Fellow Royal Soc., 1945, Linnean Soc., Royal Soc. Edinburgh (pres. 1944-49); mem. Royal Hort. Soc. (Victoria Medal of Honor 1925, Veitch Meml. medal 1930, hon. prof., 1938, v.p. 1944), Am. Acad. Arts and Scis. (hon.). Research and publs. on flora of India, Burma, China. Died Dec. 15, 1956.

SMITH, Wilson, English bacteriologist, virologist; b. Gt. Harwood, Lancashire, Eng., June 21, 1897; s. John and Nancy (Baron) S.; M.B., Ch.B., 1924, M.D., 1929, Manchester U.; m. Muriel Mary Nutt, July 2, 1927; 2 daus., Gillian Mary (Mrs. Richard John McClure), Lonice Muriel (Mrs. Colin Elliot Purdom). Staff mem., National Inst. of Med. Research, London, 1927-39; prof. bacteriology, U. Sheffield, 1939-46; prof. bacteriology, U. College Hosp. U. London, 1946-60; consultant, Microbiological Research Establishment, Ministry of Defense, 1960-64. Mem., Med. Research Council, 1961-65; governing body, Lista Inst., Royal Veterinary College, Foot and Mouth Disease Research Council. Fellow Royal Soc. (vice-pres., 1960, council, 1951 and 1960, Leeuwenhoek Lectr., 1957); Fellow Royal College Physicians; mem., N.Y. Acad. Scis., Pathological Soc. Great Britain and Ireland; Soc. General Microbiology. Author: (with others) Mechanisms of Virus Infection, 1963; approximately 130 published articles. Discovered (with others) virus of influenza, 1933; showed growth of influenza virus in fertile hen's eggs; research in poliomyelitis and development of safe vaccine. Died July 10, 1965.

SMITH-PETERSEN, Marius Nygaard, orthopedic surgeon; b. Grimstad, Norway, Nov. 14, 1886; s. Morten and Kaia (Ursin) S.; came to U. S., 1903, naturalized, 1924; student Chgo., 1906-07; B.S., U. Wis., 1910; M.D., Harvard, 1914; M.D. (hon.) U. Oslo; m. Hilda Dickinson, Sept. 1, 1917; children—Porter Cushing, Morten, Hilda Whitney. After internship engaged in practice of orthopedic surgery, Boston, since 1916; asst. instr. orthopedic surgery Harvard, 1920-30, instr., 1930-35, clin. prof. of orthopedic surgery, 1935-46; chief of orthopedic service Mass. Gen. Hosp., 1929-46. Hon. mem. Brit. and Canadian orthopaedic assns., Italian Soc. Orthopaedics and Traumatology, Royal Med. Soc. Edinburgh; mem. Internat. Soc. Surgery, Am. Acad. Orthopedic Surgeons, Am. Orthopedic Assn., Internat. Soc. Orthopedic Surgeons, A.C.S., A.M.A., N.E. Surg. Soc., Phi Beta Kappa. Introduced vitallium cup or cap in orthroplasty of hip, 1939. Died June 16, 1953.

SMITHBURN, Kenneth C., Am. physician; b. Noblesville, Ind., Oct. 19, 1904; s. Gustav and Mary (Cottingham) S.; B.S., Ind. U., 1926, M.D., 1928; m. Florence Elizabeth Bartley, June 17, 1928. Asst. in pathology and bacteriology Rockefeller Inst., N.Y.C., 1930-32, asso., 1932-38; mem. staff Rockefeller Found., N.Y.C., 1938-59, dir. Yellow Fever Research Inst., Uganda, 1946-48, leader Arbor Virus Research unit S. African Inst. Med. Research, Johannesburg, 1953-59. Mem. hemorrhagic fever commn., virus and rickettsial disease commn. U.S. Army Epidemiological Bds., 1952-53; mem. expert panel on yellow fever WHO, 1954-59, panel on virus diseases, 1959-64. Fellow N.Y. Acad. Sci., Royal Soc. Tropical Medicine and Hygiene; mem. Am. Soc. Tropical Medicine, Am. Assn. Immunologists. Research, publs. on meningitis, anemias, Tb, yellow fever, other viruse diseases; discoverer various new viruses. Home: 3339 W. 42d St., Indpls. 46208.*

SMITHLEY, William Royall, Jr., Am. chemist; b. Richmond, Va., Apr. 22, 1919; s. William Royall and Margaret (Logan) S.; B.S., Va. Mil. Inst., 1939; M.S., U. Va., 1941, Ph.D., 1949; m. Dorothy Belle Turner, Mar. 13, 1943; children—Lynn, Lucia, Susan, Sarah. Faculty dept. chemistry Birmingham-So. Coll., 1948-55, prof., 1951-55; group leader research and devel. dept. V-C Chem. Co. (merger with Socony Mobil Oil Co. 1963), Richmond, 1955-66; group leader Mobil Chem. Co., Richmond, 1966-—. Mem. Am. Chem. Soc., Va. Acad. Sci., Sigma Xi, Alpha Chi Sigma, Omicron Delta Kappa. Research on organic phosphorous chemistry, process studies, synthesis, process improvement, profit improvement. Home: 824 Westham Pkwy., Richmond 23229. Office: 401 E. Main St., Richmond, Va. 23208.*

SMITHSON, James, chemist, mineralogist; b. France, 1765; natural son of Hugh S. and Elizabeth Keate Macie; grad. Pembroke Coll., Oxford (Eng.) U., 1782, M.A., 1786. Became mem. Royal Soc., 1787; devoted life to study of chemistry and mineralogy, including studies of native minium (red lead), zeolite, aluminum; Smithsonite (carbonate of zinc) named for him; never visited U. S.; made his will, 1826; leaving his estate to a nephew, providing that should the nephew die childless, the estate would go to U. S. to found Smithsonian Instn., Washington (established by Congress 1846). Died Genoa, Italy, June 26, 1829.

SMITHSON, John Royston, Am. physicist; b. Norrisville, Md., Sept. 11, 1912; s. Lawrence and Nettie (Strawbridge) S.; B.S., Washington Coll., 1934; M.S., Ind. U., 1940; Ph.D., Cath. U. Am., 1955; m. Dorothy Leitch, Sept. 11, 1937; children—Martha Lynn, Mary Rae. Tchr. high schs., 1934-40; asst. prof. Ball State Coll., Muncie, Ind., 1940-43, 46-47; faculty U.S. Naval Acad., Annapolis, Md., 1947-—, prof., 1957-—. Research scientist, lectr. various univs. Mem. Acoustical Soc. Am., Am. Assn. Physics Tchrs., U.S. Naval Inst., Sigma Xi. Research, publs. in absorption of sound in electrolytic solutions, fluctuations of sound transmitted in open water. Home: 1309 Poplar St., Annapolis, Md. 21401.*

SMOLUCHOWSKI, Roman, physicist; b. Zakopane, Poland, Aug. 31, 1910; s. Marian and Sophia (Baraniecka) S.; M.A., U. Warsaw (Poland), 1933; Ph.D., U. Groningen (Holland), 1935; m. Louise Catherine Riggs, Feb. 3, 1951; children—Peter, Irene. Came to U. S., 1935, naturalized, 1946. Mem. Inst. for Advanced Study, Princeton, N.J., 1935-36, instr., research asso. physics dept., 1939-41, prof. solid state scis., head solid state and materials program, 1960-—; research asso., head physics sect. Inst. Metals, Warsaw, 1936-39; research physicist Gen. Electric Research Labs., Schenectady, 1941-46; asso. prof., staff Metals Research lab. Carnegie Inst. Tech., Pitts., 1946-50, prof. physics and metall. engring., 1950-56, prof. physics, 1956-60. Mem. solid state panel Research and Devel. bd. Dept. Def., 1949, sec., 1950-61; mem. tech. adv. bd. Aircraft Nuclear Propulsion, 1950; chmn. com. on magnetism Office Naval Research, 1952-56; chmn. com. on solids NRC, 1950-61, chmn. solid state scis. panel, 1961-67; chmn. advisory panel Nat., Bur. Standards, 1967-—. Fulbright prof. Sorbonne, Paris, France, 1955-56; lectr. U. Liege (Belgium), 1956, Internat, Sch. Solid State Physics, Varenna, Italy, 1957, Mol, Belgium, 1963; vis. prof. NRC, Brazil, 1958-59; adv. com. metallurgy Oak Ridge Nat. Lab., 1960-67; visiting professor, Paris, France, 1966-67. Fellow Am. Physical Soc. (chmn. div. solid state physics 1944-46), Am. Acad. Arts and Scis.; mem. Am. Soc. Metals, Am. Inst. Mining, Metall., and Petroleum Engrs., Am. Crystallographic Assn., Finnish Acad. Scis. (fgn.), Sigma Xi, Pi Mu Epsilon, Alpha Sigma Mu (hon.). Author: (with Mayer, Weyl) Phase Transformations in Solids, 1951; (with others) Imperfections in Nearly Perfect Crystals, 1952; (with others) Molecular Science and Molecular Engineering, 1959. Editor: (with N. Kurti) Monograph Series on Solid State, 1957. Research, publs. on theory of solids, radiation effects, magnetism, lattice defects. Home: 27 Westcott Rd., Princeton, N.J. 08540.*

SMORODINSKII, Ya A., Russian physicist; b. Vishera, USSR, Dec. 30, 1917; s. Abraham Ya and Rosa (Fitingoff) S.; D. Phys. Math. Scis., Leningrad (USSR) U., 1947; m. Esther Perskaya, Nov. 13, 1940; 1 dau., Nohami. Staff, Inst. Phys. Problems, USSR Acad. Scis., Moscow, 1939-56, Joint Inst. for Nuclear Research, Dubna, USSR, 1956-—. Secondary sch. lectr. elementary physics; (with L. Landau) lectrs. on nuclear physics, 1956. Recipient Order Lenin, State prize, 1951. Publs., investigation of scattering of particles, theory of nuclear structure, nuclear decay, group theory. Home: 4 Lesnaya, Dubna, USSR. Office: Head P.O. Box 79, Moscow, USSR.*

SMULLIN, Louis Dijour, Am. elec. engr., educator; b. Detroit, Feb. 5, 1916; s. Isaac M. and Ida (Dijour) S.; student Wayne U, 1932-34; B.S. in Elec. Engring., U. Mich., 1936; M.S., Mass. Inst. Tech., 1939; m. Ruth Frankel, June 14, 1939; children—Susan B., Frank M., Joseph I., David H. Engr., Ohio Brass, 1936-38, Farnsworth TV, Ft. Wayne, Ind., 1939-40, Scintilla div. Bendix Aviation Corp., Sidney, N.Y., 1940-41; dept. head Internat. Tel. & Tel. Labs., Nutley, N.J., 1946-47; staff Radiation Lab., Mass. Inst. Tech., Cambridge, 1941-46, staff Research Lab. Electronics, 1948-50, div. head Lincoln Labs., Lexington, 1950-55, prof. elec. engring., 1955-—, dept. head, 1966-—. Cons. to industry; mem. steering com. Kanpur-Indo-Am. Program, 1961-65, tech. adviser, 1965-66. Recipient Pres.'s certificate merit. Fellow I.E.E.E., Am. Acad. Arts and Scis., Am. Phys. Soc. Author: (with Montgomery) Microwave Duplexers; (with Haus) Noise in Microwave Devices. Research in radar, microwave tubes, plasma physics. Home: 76 Standish Rd., Watertown, Mass. Office: 77 Massachusetts Av., Cambridge, Mass.*

SMYTH, Charles Henry, Jr., Am. geologist; b. Oswego, N.Y., Mar. 31, 1866; s. Charles Henry and Alice (DeWolf) S.; Ph.B., Columbia, 1888, Ph.D., 1890; postgrad. U. Heidelberg, 1890-91; m. Ruth A. Phelps, July 30, 1891; children—Charles Phelps, Henry DeWolf. Prof. geology and mineralogy Hamilton Coll., 1891-1905; prof. geology Princeton, 1905-34, emeritus, 1934. Fellow Geol. Soc. Am., A.A.A.S. Contbr. many papers on pre-Cambrian geology, petrology, ore deposits. Died Apr. 4, 1937.

SMYTH, Charles Phelps, Am. chemist; b. Clinton, N.Y., Feb. 10, 1895; s. Charles Henry, Jr. and Ruth (Phelps) S.; grad. Lawrenceville Sch., 1912; A.B., Princeton, 1916, A.M., 1917; Ph.D., Harvard, 1921; m. Emily Ellen Vezin, Feb. 12, 1955. Asst. chemist U.S. Bur. Standards, 1917; faculty chemistry Princeton, 1920—, prof., 1938-58, David B. Jones prof. chemistry, 1958-63, prof. emeritus, 1963—. Chemist, Manhattan Project, 1943-45; expert cons. U.S. Army, Germany, 1945; cons. Office Naval Research, 1963—. Recipient certificate appreciation War Dept., certificate merit OSRD, 1945; certificate pub. service U.S. Department of State, 1958; Medal of Freedom. Fellow of Am. Phys. Soc.; member of the Dept. State, 1958. Fellow Am. Phys. Soc.; mem. Am. Chem. Soc. (Nichols medal 1954), Nat. Acad., Am. Philos. Soc., Faraday Soc., Groupment Ampere, Phi Beta Kappa, Sigma Xi. Author Dielectric Constant and Molecular Structure, 1931; Dielectric Behavior and Structure, 1955; also numerous articles. Editorial bd. Ann. Tables Phys. Constants, 1941-45; asso. editor Jour. Chem. Physics, 1933-36, 52-54; adv. editor Dielectrics, 1962-64. Research and publs. on determination of molecular dipole moments and structures; use of dielectric constants to study molecular rotation and phase transitions in solids, ultrahigh frequency measurements, electromotive forces, vapor pressure, heat capacities, molecular refraction, and infrared adsorption. Home: 245 Prospect Av., Princeton, N.J. 08540.*

SMYTH, Charles Piazzi, astronomer; b. Naples, Italy, Jan. 3, 1819; s. William Henry Smyth; ed. Royal Obs., Cape of Good Hope, 1835; LL.D., Edinburgh (Scotland) U.; m. Jessie Duncan, 1855. Apptd. astronomer royal for Scotland, profl. practical astronomy Edinburgh U., 1845-88. Visited Russian observatories, 1859. Recipient Makdougall-Brisbane and Keith prizes Royal Soc. Edinburgh. Fellow Royal Soc., 1857; mem. Royal Astron. Soc.; corr. mem. acads. Munich (Germany), Palermo (Italy). Author: Teneriffe, 1862; Three Cities in Russia, 2 vols., 1862; Life and Work at the Great Pyramid, 8 vols., 1867; Antiquity of Intellectual Man, 1868; On the Antiquity of Man, 1868; Madeira Spectroscopic, 1882; Our Inheritance in the Great Pyramid, 5th edit., 1890. Research, publs. on spectroscopy; promoted study of telluric absorption and brought rain-band into use for weather prediction; constructed map of solar spectrum, 1877-78; adopted end-on vacuum tubes for investigation of gaseous spectra; observed (with Alexander Herschel) harmonic character of carbonic-oxide spectrum; picked out 6 significant triplets in spectrum of oxygen; repeatedly measured citron-ray of aurora, 1871-72; observed spectrum of zodiacal light at Palermo, Apr. 1872; inferred subjection of earth's temperature to cycle identical with that of sunspots; executed large solar spectrographic chart; made photographic studies of cloud forms. Died Feb. 21, 1900.

SMYTH, David Henry, Brit. physiologist; b. Lisburn, N. Ireland, Feb. 9, 1908; s. Joseph and Mary Jane (Spence) S.; B.Sc., Queens U., Belfast, Ireland, 1929, M.B. B.Ch., B.A.O., 1932, M.Sc., 1934, M.D., 1935, Ph.D., 1940; m. Edith Mary Hoyle, July 27, 1942. House surgeon Royal Victoria Hosp., Belfast, 1932-33; demonstrator physiology, Belfast, 1933-36, Göttingen, Germany, 1936-37; lectr. U. Coll., London, 1936-46; prof. physiology U. Sheffield (Eng.), 1946—, pro vice chancellor, 1962-66. Cons. physiologist United Sheffield Hosps. Mem. Physiol. Soc., Biochem. Soc., Royal Soc. Medicine. Research, numerous publs. on absorption of nutrient substance by intestine. Home: The Swevic, Foolow, Derbyshire, Eng. Office: Dept. Physiology, Univ., Sheffield, Eng.*

SMYTH, Henry DeWolf, Am. physicist; b. Clinton, N.Y., May 1, 1898; s. Charles Henry, Jr., and Ruth Anna (Phelps) S.; Bachelor of Arts, Princeton, 1918, Master of Arts, 1920, Ph.D., 1921; Ph.D., Cambridge University, England, 1923; Dr. of Science, Drexel Inst., 1950. Case Inst. Tech., 1953; married Mary de Coningh, June 30, 1936. NRC fellow in Cambridge, Eng., 1921-23, Princeton U., 1923-24; instr. in physics, Princeton U., 1924-25, asst. prof., 1925-29, asso. prof., 1929-36, prof. 1936-66, Joseph Henry prof. physics, 1946-66, chmn. dept. of physics, 1935-50, chmn. of the university research bd., 1959-66; cons. to govt. agencies and industry; asso. editor Physical Review, 1927-30. Consultant on war research projects to Nat. Research Council and to Office of Scientific Research and Development 1940-45; Cons. Manhattan Dist. project (atomic bombs). U. S. Engrs., 1943-45; mem. U. S. Atomic Energy Commn. 1949-54; past chmn. Bd. Sci., Engring. Research, Princeton; U. S. rep. Internat. Atomic Energy Agy., 1961—. Trustee Asso. Univs., Ind. (Brookhaven Nat. Lab.), 1946-49; chmn. bd. trustees Univ. Research Assos., 1965—. Fellow Am. Phys. Society (member council 1940-44, v.p. 1956, pres. 1957), Am. Acad. Arts and Scis., member American Philosophical Society; mem. Phi Beta Kappa, Sigma Xi. Author: Matter, Motion and Electricity, 1939. Atomic Energy for Military Purposes (Official War Dept. Report on Atomic Bombs), 1945. Contbr. to Physical Review and other scientific jours. Mem. editorial bd. Princeton Univ. Press, 1946-49, 59-61. Research on atomic energy; ionization of gases; positive ray analysis; matter, motion, and electricity; molecular structure. Home: Lafayette Rd., W. Princeton, N.J. 08540.

SMYTH, James Carmichael, Brit. physician; b. Fifeshire, Scotland, 1741; s. Thomas and Margaret S. Carmichael; M.D., U. Edinburgh (Scotland), 1764; m. Mary Holyland, 1775; 8 sons, 2 daus. Physician, Middlesex (Eng.) Hosp., from 1768; physician-extraordinary to George III. Fellow Royal Soc., 1779, Royal Coll. Physicians. Author: An Account of the Experiments made on board the Union Hospital Ship to Determine the Effect of the Nitrous Acid in Destroying Contagion, 1796; A Description of the Jail Distemper, as it appeared among the Spanish Prisoners at Winchester in 1780, 1795; The Effect of the Nitrous Vapour in preventing and destroying Contagion, 1799; A Treatise on Hydrocephalus, 1814. Editor: Works of the late Dr. William Stark, 1788. First to use nitrous-acid gas for prevention of contagion in cases of fever. Died June 18, 1821.

SMYTH, Sir Warington Wilkinson, geologist, mineralogist; b. Naples, Italy, Aug. 26, 1817; s. William Henry and Annarella (Warington) S.; M.A., Trinity Coll., Cambridge (Eng.) U., 1844; student geology in Germany on Worts Found.; m. Antonia Story-Maskelyne, 1864; 2 sons. Mining geologist Geol. Survey, 1844; lectr. mining Sch. of Mines, 1951; insp. crown minerals, 1857; mineral surveyor to Duchy of Cornwall, 1852. Decorated fgn. orders S.S. Maurice and Lazare, Jesus Christ, S. Jago da Espada. Fellow Royal Soc., 1858, Geol. Soc., 1845 (hon. sec. 1856-66, pres. 1866-68, fgn. sec. 1873-90); mem. Royal Geol. Soc. Cornwall (pres. 1871-79. Author: Treatise on Coal Mining; A Year with the Turks; contbr. papers to Memoirs of the Geol. Survey, Quar. Jour. Geol. Soc., transactions Royal Geol. Soc. Cornwall. Possessed a knowledge of mineralogy and geology of Cornwall superior to anyone at the time. Died June 19, 1890.

SMYTH, William Henry, engineer; b. Birkenhead, Eng., May 16, 1855; s. Henry and Ann Jane (Finglass) S.; ed. Yorkshire Coll. of Tech., Leeds, followed by apprenticeship with Kitson & Co., Leeds; m. Helen Pauline Bradshaw, 1884. Draftsman, Asquith & Co., Leeds, 1874-76; came to U. S.; in gen. practice as cons. engr., 1879—. Fellow Royal Econ. Soc. London (life), Am. Geog. Soc.; mem. Am. Acad. Polit. and Social Sci. Author: Is the Inventive Faculty a Myth, 1895; Technocracy—National Industrial Management, 1917; Federation of Nations, 1922; The Story of the Stadium, 1923; Concerning Irascible Strong, 1926; Did Man and Woman Descend from Different Animals? A New Theory of the Origin of the Sexes, 1927; Coming Events, "Social Credit" Criticized, The Truth About Technocracy, 1933; National Master Code—Industrial Constitution for National Industrial Management, 1934; Women in Industry, 1934; Money and Currency, 1935; Problem of Crime, 1938; also many essays on economics and social science. Inventor of many machines and devices, including a drag-saw, 1879; many machines for making, soldering, testing and heading cans, 1889-1903; mech. movement, 1890; pneumatic apparatus, 1896; hydraulic and chain bucket dredger, 1898; air compressor valve, 1899; art of utilizing heat energy, 1900; internally fired engine, 1900; ore roasting furnace, mech. stoker, valve, etc., 1901; printing press, 1902; deep well pump, 1903; inventor system of raising water by direct explosion on its surface, later universally known as direct explosion pump; segmental cargo boat for war use, 1917; power drive-chain, 1920; roller-hinge for heavy-duty chain, 1921; track-layer chain, 1922; resilient-track tractor, 1922; high-speed tractor, 1922; track-layer track-assembly, 1923; two-point-support wheel-base track-layer, 1923; convertible tractor, 1924; combined track-layer and round-wheel tractor, 1925; friction-drive track-layer, 1928; military tractor, 1932; convertible tractor, 1932; new system of highspeed, non-stop, railroad transportation, 1933. Died Feb. 18, 1940.

SMYTHE, Hugh Heyne, Am. sociologist, govt. ofcl.; b. Pitts., Aug. 19, 1913; s. William Henry and Mary Elizabeth (Barnhardt) S.; A.B., Va. State Coll., 1936; M.A. (scholar 1936-37), Atlanta U., 1937; Ph.D., Northwestern U., 1945; grad. student Fisk U., 1938-39, U. Chgo., 1940, Columbia, 1950-51, Woodstock (Vt.) Sch. (grantee Am. Friends Service Com.), 1950; m. Mabel Hancock Murphy, July 26, 1939; 1 dau., Karen Pamela. Researcher, adminstrv. asst. Am. Youth Commn. of Am. Council Edn., 1937-38; research asst., instr. Fisk U., 1938-39; research asso. Atlanta U., 1942; asst. dir. research Negro Land Grant Coll. Coop. Social Studies Project, also prof. sociology Morris Brown U., 1944; prof. sociology Tenn. State Agrl. and Indsl. U., 1945-46; dep. dir. spl. research N.A.A.C.P., 1947-49; dir. research W. B. Graham & Assos., N.Y.C., 1949-50; vis. prof. sociology and anthropology Yamaguchi Nat. U., Japan, 1951-53; prof. sociology Coll. City N.Y. 1953—; ambassador to Syrian Arab Republic, 1965—; dir. research N.Y. State Senate Finance Com. evaluation project, State Commn. Against Discrimination, 1956; social affairs officer U. S. Mission to UN, 1961-62; Fulbright prof. Chulalongkorn U., also adviser NRC, Thailand, 1963-64; UN corr. Eastern World and Africa Trade and Devel. (London), 1964—; vis. prof., spl. lectr. govt., pvt. orgns., univs., 1950—; adviser, mem. delegations U. S., other internat. orgns. and meetings, 1957—. Harriet M. Strong Found. scholar, 1935-37; Rosenwald fellow, 1939-41; grantee Am. Council Learned Socs., 1940, Social Sci. Research Council, 1960; recipient award Brit. Colonial Office, 1948; Ford Found. fellow, 1952-58; named Distinguished Alumni, Va. State Coll., 1957. Fellow African Studies Assn., Soc. Applied Anthropology, Am. Anthrop. Assn.; mem. Eastern Sociol. Soc., Assn. Asian Studies, Japan Soc., Am. Assn. Tchrs. Chinese Lang. and Culture, Inst. Race Relations Gt. Britain, Siam Soc., Operations Crossroads Africa, Sociol. Soc. Japan, Am. Assn. U. Profs., Am. Com. on Africa, Com. on World Devel. and World Disarmament, Sigma Xi, Alpha Kappa Delta, Iota Sigma Lambda, Alpha Pi Zeta, Alpha Phi Alpha. Author: (with W. E. B. DuBois) Negro Land Grant Colleges Social Studies Project, 1944; (with M. M. Smythe) New Nigerian Elite, 1960; also pamphlets, articles, chpts. in books. Mem. editorial bd. Jour. Human Relations, 1954—, Africa Today, 1955—, Social. Abstracts, 1956—. Analysis of human relations, especially in minority peoples and new elite leadership class emerging in newly independent nations of Asia, Africa, Latin Am., and Pacific regions; implications of race and cultural relations, social class and polit. devels. in underdeveloped world. Home: 345 8th Av., N.Y.C. 10001. Office: Am. Embassy, Box D, APO N.Y.C. 09694.*

SMYTHE, Lloyd Earle, chemist; b. Suva, Fiji, June 15, 1922; s. Harold Earle and Gwenda (Bone) S.; B.Sc. with honors I, U. Sydney (Australia), 1943, M.Sc., 1946; Ph.D., U. Tasmania, 1953; m. Ann Jean Gilbert, Sept. 1, 1945; 1 dau., Lesley Ann. Lectr. chemistry U. Sydney, 1946-53; sr. lectr. inorganic chemistry U. Tasmania, 1953-55; prin. research officer Australian AEC, Sydney, 1955-60, sr. prin. research scientist, 1960-67, chief research scientist, 1967—. Corday Morgan Commonwealth fellow London Chem. Soc., 1955. Fellow Royal Australian Chem. Inst. (pres. N.S.W. br. 1967-68). Contbg. author: Progress in Nuclear Energy Series 9, vol. 3, 1963. Research, publs. on electrochemistry, reaction kinetics and analytical chemistry, chemistry aqueous nuclear reactor systems; forensic applications of neutron activation analysis. Home: 45 Newton St., Strathfield N.S.W. Office: Australian AEC Research Establishment, Lucas Heights, Sydney, N.S.W., Australia.

SMYTHE, William R(alph) Am. physicist; b. Canon City, Colo., July 5, 1893; s. William Rodman and Eva (DeCou) S.; A.B., Colo. Coll., 1916; A.M., Dartmouth, 1919; Ph.D., U. Chgo., 1921; m. Helen Flint Keith, Mar. 21, 1921; children—William Rodman, Sylvia (Mrs. David Woeller). Prof. physics U. Philippines, 1921-23; Nat. Research fellow Cal. Inst. Tech., Pasadena, 1923-26, research fellow, 1926-27, faculty, 1927—, prof., 1940-64, prof. emeritus, 1964—, head spl. ballastics sect. Caltech Rocket Project, 1942-46. Recipient Army-Navy certificate of Merit, 1946; Navy Bur. Ordnance Devel. award, 1946. Fellow Am. Phys. Soc., A.A.A.S.; mem. Phi Beta Kappa, Sigma Xi, Tau Beta Pi. Author: Static and Dynamic Electricity, 1937. Editorial bd. Jour. Applied Physics, 1953-55. First radio frequency mass spectrometer gave precise oxygen isotope ratio, isolation and identification of radioactive isotopes K40 and RB87. Home: 674 Manzanita Av., Sierra Madre, Cal. 91024.*

SNATZKE, Guenther, chemist; b. Hartberg, Austria, Oct. 26, 1928; s. Franz and Adele (Wendl) S.; Dr.phil., U. of Graz (Austria), 1953; Dozent, Bonn (Germany) U., 1966; m. Ingeborg Reinelt, Dec. 22, 1953. Fellow, Hamburg (Germany) U., 1953-57, asst., 1957-60; asst. Bonn U., 1960-61, oberassistent, 1961—. Cons. to pharm. and sci. instrument cos. Research, numerous publs. on structure elucidation of natural products, organic stereochemistry, application of phys. methods to organic chemistry. Home: 3 Baumschulallee, Bonn, Germany 53.*

SNELGROVE, Alfred Kitchener, Am. geologist; b. St. John's, Nfld., Can., Apr. 7, 1902; s. Gilbert and Eliza (Smith) S.; B.S., McGill U., 1927, M.S., 1928; Ph.D., Princeton, 1930; D.Sc., Meml. U., Nfld., 1964; m. Rachel M. Betts, June 6, 1936. Came to U.S., 1928, naturalized, 1943. Faculty, Mich. Technol. U., Houghton, 1940—, now prof., head dept. geology and geol. engring. Mem. Geol. Soc. Am., Soc. Econ. Geologists, Mineral Soc. Am., Am. Inst. Mining, Metall., Petroleum Engrs., Can. Inst. Mining and Metallurgy, others. Author: Mines and Mineral Resources of Newfoundland, 1938; Opportunities in Geology and Geological Engineering, 1960; Geohydrology of Indus River, West Pakistan, 1967. Discovered Alpine type chrome-bearing rock masses Western Nfld.; postulated structural control of courses of Indus River; contbd. to definition of geol. engring. field, use of ore guides in base-metal deposits of Central Eire, explanation of nodular chromite in West Pakistan. Home: 304 Vivian St., Houghton, Mich. 49931.*

SNELL, Arthur Hawley, physicist; b. Montreal, Que., Can., Mar. 10, 1909; s. John Ferguson and Evelyn (Morphy) S.; B.A., U. Toronto, 1930; M.Sc.,

McGill U., 1931, Ph.D., 1933; m. Ethelyn Towner, Aug. 20, 1941; children—Arthur Towner, Lynn Hawley. Came to U. S., 1938, naturalized, 1949. 1851 Exhbn. scholar U. Cal., Berkeley, 1934-37, research asso., 1937-38; research instr. U. Chgo., 1938-42, chief cyclotron sect. Metall. Lab., Manhattan dist., 1942-44; group leader Oak Ridge Nat. Lab., 1944-48, dir. physics div., 1948-57, asst. lab. dir., 1957—; dir. thermonuclear div., 1958—. Chmn. subcom. on instruments and techniques, com. on nuclear sci. Nat. Acad. Sci., 1954-61; participant Internat. Atoms for Peace Confs., 1955, 58. Recipient Gov.-Gen.'s medal for research, 1933. Fellow Am. Phys. Soc., Royal Soc. Arts; mem. Fed. Atomic Scientists. Asso. editor: Phys. Rev., 1958-61; editorial adv. bd. Jour. Nuclear Energy; editor Nuclear Instruments and Their Uses, 1962. Research in Stark effect in molecular hydrogen, induced radioactivity, nuclear isomers of bromine, identfication and assignment of delayed neutron activities, radioactive decay of neutron, neutrino recoil spectrometry, beta recoil spectrometry, charge spectrometry, controlled thermonuclear research. Home: Rt. 3, James Ferry Rd., Kingston, Tenn. 37763. Office: Oak Ridge Nat. Lab., P.O. Y, Oak Ridge 37831.*

SNELL, Esmond Emerson, Am. biochemist; b. Salt Lake City, Sept. 22, 1914; s. Heber Cyrus and Hedwig (Ludwig) S.; B.A., Brigham Young U., 1935; M.A., U. Wis., 1936, Ph.D., 1938; m. Mary Caroline Terrill, Mar. 17, 1941; children—Esmond Emerson, Richard T., Allan G., Margaret Ann. Research asso. U. Tex., 1939-41, asst. prof. chemistry, 1941-43, asso. prof., 1943-45; faculty U. Wis. 1945-53, prof., 1948-53; prof. chemistry, asso. dir. Biochem. Inst., U. Tex., 1951-56; faculty U. Cal., Berkeley, 1956—, prof. biochemistry, 1956—; chmn. dept. biochemistry, 1956-62, Berkeley. Recipient Eli Lilly award, 1945; Mead-Johnson & Co. B-Complex award, 1946, Osborne-Mendel award, 1951. Mem. Am. Soc. Biol. Chemists (pres. 1961-62), Am. Acad. Arts and Scis., Am. Soc. Microbiology, Nat. Acad. Scis. Am. Chem. Soc. Editor: Biochemical Preparations, vol. 3, 1953; editor-in-chief Ann. Rev. Biochemistry, vol. 32, 1963, vol. 33, 1965; co-editor Chemical and Biological Aspects of Pyridoxal Catalysis, 1963. Research, publs. on nutritional requirements of bacteria leading to devel. methods for determination of several vitamins and amino acids in foodstuffs; ind. discovery several vitamins with contbns. to their isolation and chem. identification; devel. theory explaining action of B6 in living organisms; clarified steps involved in converting excess vitamin B6 and pantothenic acid to their excretion products in certain bacteria. Home: 50 Arlmonte Dr., Berkeley, Cal. 94707.*

SNELL, Foster Dee, Am. chemist; b. Binghamton, N.Y., June 29, 1898; s. Dayton A. and Bertha Viola (Hickling) S.; B.S., Colgate University, Hamilton, N.Y., 1919, D.Sc. (honorary), 1963; M.A., Columbia U., 1922, Ph.D., 1923; m. Cornelia Tyler, June 18, 1921; 1 dau., Barbara Anne. Instr. in chemistry, Columbia U., 1919-20, Coll. of City of N.Y., 1920-23; in charge of tech. chemistry, Pratt Inst., Brooklyn, 1923-28; own business as cons. chemist, chem. engr. since 1923, incorporated as Foster D. Snell, Inc., Brooklyn, 1930, pres., 1930-61, chairman board, 1961-65, chmn. exec. com., 1965—; chmn. bd. Foster D. Snell, Research, Ins., Supplee Labs., Inc.; director Cargille Scientific, Inc. Mem. referee bd. Office of Production Research and Development, 1943-45; v.p. Fat and Oil Commn., Internat. Union of Pure and Applied Chemistry, 1947-51, pres. 1951-53. Mem. div. chemistry and chem. tech. NRC, 1949—. Alumni trustee of Columbia University, 1964—. Served with USN, 1918. Awarded gold medal, Soc. Chemistry Industry, Manchester, Eng., 1949; honor award N.Y. chpt. Am. Inst. Chem., 1952. Fellow Chem. Soc. (London), Royal Society of Arts (London), Am. Assn. for Advancement of Sci., Am. Inst. Chemists (pres. 1946-48), mem. Am. Chem. Soc. (councilor, 1925-28, 1936-42, council policy committee 1949-52), American Oil Chemists Society (vice pres. 1943-45; representative to NRC 1953-63). Am. Inst. Chem. Engrs., Assn. Cons. Chemists and Chem. Engrs. (pres. 1952-54, v.p. 1955-59), Am. Soc. for Testing Materials, Soc. of Chem. Industry (v.p. 1937-40, hon. sec. Am. Sect. 1926-39, vice chmn. 1940-42, chmn. 1942-44), Soc. de Chimie Industrielle, Alumni Assn. of Grad. Sch. of Columbia U. (pres. 1940-42), Phi Beta Kappa. Sigma Xi. Co-author: Colorimetric Methods of Analysis, 1923, second edition (two volumes), 1936-37, third edition (four volumes), 1948-54, volume IIA, 1959, volume IIIA, 1961; Chemicals of Commerce, 1939; Chemistry Made Easy, 4 volumes, 1943; Dictionary of Commercial Chemicals, 1962 (all with Dr. Cornelia T. Snell); Comml. Methods of Analysis (with Frank M. Biffen), 1944; also author or co-author numerous tech. articles. Holds over 50 patents. Specialist in chem. engring.; studied colorimetric analysis, indsl. alkalies, surfactants and syndets, bldg. materials and trade waste. Home: 2 Fifth Av. Office: 29 W. 15th St., N.Y.C. 11.

SNELL, Fred Manget, Am. biophysicist; b. Soochow, China, Nov. 11, 1921 (parents Am. citizens); s. John Abner and Grace (Birkett) S.; A.B., Maryville Coll., 1942; M.D., Harvard, 1945; Ph.D., Mass. Inst. Tech., 1952; m. Mary Larmour, Mar. 14, 1946; children—Susanna Jean, John Steven, Cynthia Anne. Research asso. dept. biology Mass. Inst. Tech., also Children's Cancer Research Inst.,

Cambridge, Mass., 1952-54; asso. biol. chemistry, Harvard Med. Sch., 1954-57, asst. prof. biol. chemistry, 1957-59; prof. biophysics, chmn. dept. biophysics State U. N.Y. at Buffalo, 1959-65, dean of graduate studies, since 1967——. Member of Biophys. Soc., Am. Chem. Soc., A.A.A.S., Am. Physiol. Soc., N.Y. Acad. Sci., Am. Inst. Biol. Sci. Am. Assn. U. Profs., Sigma Xi. Author: (with S. Shulman, R.P. Spencer, C. Moos) Biophysical Principles of Structure and Function, 1965; also numerous articles. Research on transport processes in biol. systems, fundamental theory transport processes, kinetic and thermodynamic; analysis and math. modelling biol. processes. Home: 4090 Elma Rd., Williamsville, N.Y. 14221. Office: Dept. Biophysics, Sherman Hall, State U. N.Y., Buffalo 14214.*

SNELL, James Laurie, Am. mathematician; b. Wheaton, Ill., Jan. 15, 1925; s. Roy Judson and Lucille (Ziegler) S.; B.A., U. Ill., 1947, M.A., 1948, Ph.D., 1951; m. Joan Perry, Dec. 30, 1952; children—John, Mary. Instr. math. Princeton, 1951-54; faculty Dartmouth, Hanover, N.H., 1954—, prof. math., 1962—, exchange prof. U. Paris, 1965-66. Mem. Am. Math. Soc., Math. Assn. Am. Author: (with John G. Kemeny, G.L. Thompson) Introduction to Finite Mathematics, 1957; (with John G. Kemeny, Hazelton Mirkil, G.L. Thompson) Finite Mathematical Structures, 1959; (with John G. Kemeny) Finite Markov Chains, 1960; (with J. Berger, B.P. Cohen, M. Zelditch) Types of Formalization in Small Group Research, 1962; (with John G. Kemeny) Mathematical Models in the Social Sciences, 1962; (with John G. Kemeny, A. Schleifer, G.L. Thompson) Finite Mathematics with Business Applications, 1962; (with John G. Kemeny, A. Knapp) Denumerable Markov Chains, 1965. Home: Main St., Norwich, Vt. 05055. Office: Dept. Math., Dartmouth, Hanover, N.H.*

SNELL, Richard Saxon, anatomist; b. Richmond, Surrey, Eng., May 3, 1925; s. Claude Saxon and Daisy (Wright) S.; L.R.C.P., M.R.C.S., 1948, M.B., B.S., 1949, Ph.D., 1955, M.D., King's Coll. U. London, 1961; m. Maureen Cashin, June 4, 1949; children—Georgina Sara, Nicola Ann, Melanie Jane, Richard Robin. House surgeon to Sir Cecil P. G. Wakeley, King's Coll. Hosp., 1948-49; lectr. anatomy King's Coll. U. London, 1949-59, lectr. in charge histology, 1953-59; lectr. anatomy U. Durham, Eng., 1959-63; faculty Yale, 1963-67, asso. professor anatomy, medicine, 1965-67; prof., chmn. dept. anatomy N.J. Coll. Medicine and Dentistry, 1967——. Mem. Anat. Soc. Great Britain, Am. Soc. Anatomists, Am. Acad. Dermatology. Research, numerous publs. on pigmentation of skin and its control, including description of histochem., electron micros. and endocrinological studies; histochem. investigations into distbn. of cholinesterase in periperal and central parts of nervous system. Home: 12 Powder Horn Dr., Morristown, N.J. 07961. Office: Dept. Anatomy, N.J. Coll. Medicine and Dentistry, Jersey City.*

SNELL, Willebrod van Roijen (Snellius), Dutch mathematician; b. Leiden, Holland, 1591; s. Rodolph Snell; became prof. math., Leiden, 1613. Author: Eratosthenes Batavus, 1617; Cyclometricus, de circuli dimensione, 1621; Concerning the Comet, which appeared in 1618; Tiphus Batavus, 1624; Doctrinae triangulorum canonicae, 1627; Hessian and Bohemian Observations; Libra astronomica philosophica. Used triangulation to measure meridian for 1st time; discovered refraction law, also properties of polar triangle in spherical trigonometry, prins. for determining length of arc of a meridian from measurement of any base line; attempted to measure size of earth by triangulation; discovered law stating that when a ray of light passes from one medium to another the sine of the angle of incidence divided by sine of angle of refraction is constant (refractive index being the constant). Died Leiden, Oct. 30, 1626.

SNELLEN, Herman, Dutch ophthalmologist; b. Zeijst, Holland, Feb. 19, 1834; M.D., Utrecht, Holland, 1858; became asst. Nederlandsh Gasthuis voor Ooglijders, Utrecht, 1858, first physician, 1862, dir., 1884-1903; became prof. ophthalmology U. Utrecht, 1858. Author: Opotoypi ad visum determinandum, 1862; also numerous books and articles on eye. Originated set of type of various sizes for testing central vision (Snellen's test), 1862. Died Utrecht, Jan. 18, 1908.

SNETSINGER, Robert John, Am. entomologist; b. Diamond Lake, Ill., Mar. 6, 1928; s. Clarence Joseph and Helen (Mills) S.; student No. Ill. U., 1946-47, U. Mo., 1947-48, So. Ill. U., 1951; B.S., U. Ill., 1952, M.S., 1953, Ph.D., 1960; m. Wendy Dawn Rigler, June 26, 1960; 1 dau., Laurel T. Research asst. Ill. Natural History Survey, Urbana, 1954-60; faculty Pa. State U., University 1960—, asso. prof. entomology, 1965—. Cons. entomologist, 1964——. Mem. Entomol. Soc. Am.; Canadian, Pa. (editor) entomol. socs., Internat. Commn. for Crop Protection, Sigma Xi, Phi Sigma. Editor: Procs. of 5th Internat. Mushroom Congress, 1962. Research, publs. in biology and control of arthropod pests of greenhouse crops, ornamental crops, and mushroom crops, also control of household and structural pests, including rodents. Home: Box 214B, R.D. 1, Bellefonte, Pa. 16823. Office: Pa. State U., University Park, Pa. 16802.*

SNIDER, Delbert Arthur, Am. economist; b. Fayetteville, O., Jan. 11, 1914; s. Chris C. and Florence (Berger) S.; A.B., U. Cin., 1936, M.A., 1937; Ph.D., U. Chgo., 1951; m. Helen Kuller, Sept. 4, 1939; children—Suzanne, Chris. Economist, chief rep. U.S. Treasury Dept., Washington, Hawaii, N. Africa, Eng., France, Belgium, 1940-43, 45-47; faculty Miami U., Oxford, O., 1947-48, 49—, prof. econs., 1953—. U.S. mem. Greek Currency Commn., Athens, 1948-49; cons. govt., 1965. Mem. Am. Midwest econs. assns., Am. Assn. U. Profs., Authors Guild. Author: Introduction to International Economics, 1954; Economics: Principles and Issues, 1962; Economic Myth and Reality, 1965; International Monetary Relations, 1966; also articles. Research on nature of conflict between internal and external econ. equilibrium, and appropriate policies to render internal and external stability consistent with each other. Home: 213 Oakhill Dr., Oxford, O. 45056.*

SNIDER, Luther Crocker, Am. geologist; b. Mt. Summit, Ind., Sept. 13, 1882; s. John and Lou (Leath) S.; student Rose Poly. Inst., 1903-04, U. Okla., 1910-11; A.B., Ind. U., 1908, A.M., 1909; Ph.D., U. Chicago, 1915; m. Ruth Gladys Marshall, Mar. 31, 1907; children—Hester Bernice, John Luther. Teacher common and high schs., 1901-03 and 1904-06; chemist, field geologist and asst. dir. Okla. Geol. Survey, 1909-15; field geologist Pierce Oil Corp., Tulsa, Okla., 1915-16, Cosden Oil & Gas Co., 1916-17; asst. chief and chief geologist Empire Gas & Fuel Co., Bartlesville, Okla., 1917-25; cons. geologist Henry L. Doherty & Co., 1925-35, Cities Service Co., 1935-40; prof. of geology, U. Tex., from 1941. Fellow A.A.A.S., Geol. Soc. America, Soc. Econ. Geology; mem. Am. Assn. of Petroleum Geologists (editor 1933-37; pres. 1940), Sigma Xi, Phi Beta Kappa (alumnus). Author: Petroleum and Natural Gas in Oklahoma, 1913; Oil and Gas in the Mid-Continent Fields, 1920; Earth History, 1932; also various bulls. of Okla. Geol. Survey, 1910-16. Contbr. to scientific jours. Field work and analysis on mineral resources of Okla. Died May 24, 1947; buried Mount Summit, Ind.

SNODGRASS, James M., Am. oceanographer; b. Marysville, O., May 3, 1908; s. William H. and Clara May (Hopkins) S.; B.A., Oberlin Coll., 1931; postgrad. U. Pa., Harvard; m. Eleanore Catherine Zwerner, July 23, 1936; children—William, Marian (Mrs. Clemens Deisenhammer). Asst. in psychology Oberlin Coll., 1932-34, 36-37; research asso. Free Hosp. for Women, Brookline, Mass., 1938-39; instr. Oberlin Coll., 1939-42; mem. tech. staff Div. War Research, 1942-46; chief engr. motion picture and sound div. Dayton Acme Co., 1946-48; mem. staff Scripps Instn. Oceanography, U. Cal. at San Diego, 1948—, head spl. devels. div. U. S. rep. to Intergovtl. Oceanographic Commn., 1962, chmn. working group on communications, 1962—; mem. Com. on Radio Frequency Requirements for Sci. Research, Ocean Engring. Panel of Com. on Oceanography, panel on gravitational effects Space Sci. Bd., Nat. Acad. Scis.-NRC. Named Man of Year, Nat. Telemetering Conf., 1966. Fellow Instrument Soc. Am.; mem. A.A.A.S., Acoustical Soc. Am., Physiol. Soc. Phila., Marine Tech. Soc., Am. Inst. Biol. Scis., I.E.E.E. (sr. mem.), Sigma Xi. Contbr. articles to profl. jours., chpts. to books. Research in design, devel. of oceanographic instruments; designed first electronic deep sea oceanographic instrument, ocean bottom sediment temperature gradient recorder, also radiance and irradiance meters, pressure equalized instrumentation. Home: 633 Gravilla St., La Jolla, Cal. 92037.*

SNOOK, H(omer) Clyde, Am. electrophysicist; b. Antwerp, O., Mar. 25, 1878; s. Wilson Hunt and Nancy Jane (Graves), S.; A.B., Ohio Wesleyan U., 1900, M.S., 1910, Sc.D., 1926; A.M., Allegheny Coll., 1902; postgrad. U. Pa., 1904-08; m. May Eusebia McKee, June 24, 1903. Prof. physics and chemistry High Sch., Ohio Soldiers and Sailors Orphans' Home, Xenia, O., 1900-01; asst. prof. chemistry Allegheny Coll., 1901-02; wireless telegraph expert Queen & Co., Phila., 1902-03; pres. Roentgen Mfg. Co., Phila., 1903-13, Snook-Roentgen Mfg. Co., Phila., 1913-16; v.p. Victor Electric Corp., Chgo., 1916-18; elec. engr. with Western Elec. Co., 1918-25, Bell Telephone Labs., 1925-27; cons. engr. since 1927. Chmn. noise elimination com. Nat. Safety Council, 1930. Recipient Edward Longstreth medal Franklin Inst., 1919; Gold medal Radiol. Soc. N.Am., 1923; Gold medal Am. Coll. Radiology, 1928, also hon. fellow. Fellow Am. Inst. E.E., Am. Phys. Soc.; mem. Am. Roentgen Ray Soc., Phila., Roentgen Soc., Phi Beta Kappa. Inventor X-ray transformer; numerous patented devels. in X-rays, radio, the communication art, metallurgy and optics. Died Sept. 22, 1942.

SNOOK, Theodore, Am. anatomist; b. Titusville, N.J., Apr. 14, 1907; s. Theodore S. and Carrie (Davis) S.; B.S., Rutgers U., 1929, M.S. 1930; Ph.D., Cornell U., 1933; m. Jane P. MacNamee, Sept. 2, 1933; 1 dau., Patricia J. (Mrs. William H. Fawcett). Faculty, Cornell U., 1930-34, Syracuse Med. Sch., 1934-46, Tulane Med. Sch., 1946-49, U. Pitts. Med. Sch., 1949-53; prof. anatomy U. N.D., Grand Forks, 1953—. Mem. Am. Assn. Anatomists, Biol. Photog. Assn., Am. Assn. U. Profs. (pres. U. N.D. chpt. 1963), Microcirculatory Soc., A.A.A.S., Reticuloendothelial Soc. Sigma Xi. Research in histology, histochemistry, circulation mammalian spleens and other lymphoid organs, 1934—. Home: 1623 Lewis Blvd., Grand Forks, N.D. 58201.*

SNOW, Baron (Charles Percy Snow), English physicist; born at Leicester, England, October 15, 1905; the son of William Edward and Ada (Robinson) Snow: B.Sc., Univ. Leicester (England), 1927, M.Sc., 1928; Ph.D., Christ's Coll., Cambridge U., 1930; LL.D., U. Leicester, 1955, U. Liverpool, 1960, Bklyn. Poly. Coll., 1962, St. Andrews U., 1962; D. Litt., Dartmouth, 1960, Bard Coll., 1962, Temple U., 1963, Syracuse U., 1963, U. Pitts., 1964; D.H.L. Kenyon Coll., 1961, Washington U., St. Louis, 1963, U. Mich., 1963; D. Philol. Scis., Rostov on Don, 1963; m. Pamela Hansford Johnson, July 14, 1950; 1 son, Philip Charles Hansford. Engaged as profl. scientist, 1928-40; fellow Christ's Coll., 1930-50, tutor, 1935-40, Rede lectr. Cambridge, 1959; with Brit. Civil Service Commn., 1945-60; dir. English Electric Co., Ltd., 1947-64; bd. directors London br. U. Chgo. Press. Parliamentary sec. to minister of tech., Eng., 1964—. Created Knight, 1957; named comdr. Order Brit. Empire; created Baron Snow of City of Leicester, 1964. Hon. mem. Am. Acad.-Nat. Inst. Arts and Letters; hon. fgn. mem. Am. Acad. Arts and Scis. Author: Death Under Sail, in 1932; New Lives for Old, 1933; The Search, 1934; Strangers and Brothers, 1940; The Light and The Dark, 1947; Time of Hope, 1949; The Masters, 1951; The New Men, 1954; Homecomings, 1956; The Conscience of the Rich, 1958; The Affair, 1969; Corridors of Power, 1964; Two Cultures and the Scientific Revolution (Rede lecture), 1959; Science and Government (Godkin lectures), 1960; View Over the Park, 1950. His works underline the cultural separation between sci. and the arts in the atomic age. Home: 199 Cromwell Rd., London S.W. 5. Office: English Electric House, Strand, London W. C. 2, Eng.

SNOW, Charles Ernest, Am. anthropologist, anatomist; b. Boulder, Colo., Apr. 11, 1910; s. Charles Frank and Georgia Etta (McNaughton) S.; A.B.,, U. Colo., 1932; A.M., Harvard, 1935, Ph.D., 1938; m. Katharine Meyer, Sept. 3, 1932 (div. Sept. 1952); children—Carolyn (Mrs. Richard Francis), Georgiana (Mrs. Norman Haaland), Marina (Mrs. Richard D. Horton), Kathleen McDonald; m. 2d, Katherine Burnett, Nov. 28, 1952. Faculty, U. Ky., Lexington, 1942—, prof. anatomy and phys. anthropology, 1965—. Cons. brs. of govt., museums. Recipient Civilian Meritorious Service award U.S. Army, 1948; Distinguished Prof. award Coll. Arts and Scis., 1952; 1st Distinguished Tchr. award U. Ky., 1960. Wenner-Gren Found. grantee, 1951-52, 55, 64-65; NSF grantee, 1964-65; others. Fellow Am. Anthrop. Assn.; mem. Am. Assn. Anatomists, Am. Assn. Phys. Anthropology, Am. Assn. U. Profs., A.A.A.S., Phi Beta Kappa, Sigma Xi, Omicron Delta Kappa, Lambda Chi Alpha. Research, publs. on skeletal remains of various Am. Indian mounds; methods of identification from skeletal remains. Died Oct. 5, 1967.

SNOW, Chester, Am. physicist; b. Salt Lake City, June 1, 1881; s. Willard and Dora (Pratt) S.; A.B. in Math. and Physics magna cum laude, Harvard, 1906; Ph.D. (fellow in Physics) U. Wis., 1914; m. May Maughan, Aug. 22, 1906; children—Chester Weston, Margaret (Mrs. Richard Ray Roberts), Robert Maughan. Prof. physics, head dept. Brigham Young U., Provo, Utah, 1906-11, prof. math, head dept., 1911-12; prof. math., head dept. U. Ida., Moscow, 1914-20; math. physicist Bur. Standards, Washington, 1920-50, Los Alamos, 1944-45; mathematician Diamond Ordnance Fuze Lab., Washington, 1950-59. Recipient Gold medal award in math. Dept. Commerce, Washington, 1949. Author: Hypergometric and Legendre Functions with Applications to Integral Equations of Potential Theory, 1952. Research, publs. on various applications of math. methods to theory of magnetism; participated in work essential to prodn. of atomic bomb. Home: 2520 Harrison Blvd., Ogden, Utah 84403.*

SNOW, Francis Huntington, Am. entomologist; b. Fitchburg, Mass., June 29, 1840; grad. Williams Coll., 1862, A.M., 1865, Ph.D., 1881; grad. Andover Theol. Sem., 1866; LL.D., Princeton, 1890; m. Jane Appleton Aiken, July 8, 1868. Mem. 1st faculty State U. Kan., 1866-70, as prof. math. and natural sci., prof. natural history, 1870-89, pres. faculties, 1889-90, chancellor, 1890-1901, prof. organic evolution, systematic entomology and meteorology, 1901-08. Kan. legislature appropriated $50,000 for Snow Hall of Natural History at the univ., 1885, to contain collections made by him; conducted 26 expdns. to Kan., Colo., N.Mex., Tex. and Ariz., making Kan. Univ. collection of 22,000 species of insects, one of the largest in U. S. Fellow A.A.A.S. Noted for work in ornithology, meteorology, systematic and economic entomology, especially in artificial application of fungus diseases to destruction of chinch bugs in the field. Died Sept. 20, 1908.

SNOW, John, English physician; b. York, Eng., Mar. 15, 1813; apprenticed to William Hardcastle, surgeon, Newcastle-on-Tyne, Eng.; student Hunterian Sch. Medicine, London, Westminister Hosp.; M.D., U. London, 1844. Practice medicine, London; became lectr. forensic medicine Aldersgate St. Sch. Medicine, 1849. Recipient prize Inst. France, 1849. Mem. Royal Coll. Surgeons, Soc. Apothecaries, Med. Soc. London (became pres. 1855). Author: On the Mode of Communication of Cholera, 1849; Chloroform and other Anaesthetics, 1858. Studied cholera outbreaks and showed they were related to water supplies; introduced use of ether as anesthetic for surgery to Eng., 1846-47. Died London, June 16, 1858.

SNOWDEN, Kenneth Ulmer, Australian metal physicist; b. Melbourne, Australia, Mar. 2, 1928; s. Harry and Doris (Ulmer) S.; B.Sc. in Physics, Melbourne U., 1951, Ph.D. in Metallurgy, 1959; m. Marilyn Margaret Manderson, Jan. 22, 1955; children—John, Richard, Andrea, Melinda. Sci. officer Commonwealth Dept. Works, Melbourne, 1953-53; research staff Broken Hill Asso. Smelters Pty. Ltd., attached to Baillieu Lab., metallurgy dept. Melbourne U., 1953-61, 63—; mem. metal physics group Franklin Inst., Phila., 1961-63. Mem. Inst. Physics (asso., London), Inst. Metals (London), Australian Inst. Metals, Am. Inst. Mining Engrs. Research, publs. on fatigue and creep in metals and alloys, effect of surface reactions and environmental conditions on these processes, plastic deformation and cyclic work hardening of metal single crystals; discovered discontinuities in air-pressure dependence of fatigue strength of lead and indium, relation between maximum shear stress and shape change of certain metal crystals under cyclic stress conditions. Office: Baillieu Lab., Metallurgy Dept., Melbourne U., Victoria, Australia.*

SNOWDON, John Colin, physicist; b. London, Eng., June 21, 1932; s. Cecil Fillis and Norah (Crowhurst) S.; B.Sc., U. London, 1952, B.Sc., 1953, D.I.C., 1956, Ph.D., 1956, D.Sc., 1965; m. Anne Joy Vickery, Sept. 8, 1956; children—Jane Louise, Mark Andrew. Research scientist Philco Corp., Phila., 1956-58; research asso. U. Mich. Willow Run Labs., Ann Arbor, 1958-59, asso. research physicist, 1959-60; vibration engr. Hawker-Siddeley Aviation Ltd., Kingston, Eng., 1960-61; Asst. prof. Ordnance Research Lab. Pa. State U., 1961-64, asso. prof., 1964-67, prof. engring. research, 1967—. Fellow Acoustical Soc. Am. (asso. editor Jour. 1966—), Brit. Inst. Physics; mem. Am. Soc. Engring. Edn., Brit. Acoustical Soc., American Society of Mechanical Engineers, also Sigma Xi. Author: Shock and Vibration in Damped Mechanical Systems, 1968; also articles. Theoretical treatment of vibration of damped mech. structures; develop. theories of vibration isolation with particular reference to properties of rubberlike materials; theories of shock isolation. Home 251 S. Osmond St.,, State College, Pa. 16801.*

SNYDER, Conway Wilson, Am. physicist; b. Kirksville, Mo., Jan. 24, 1918; s. Joseph Conway and Agnes (Wilson) S.; A.B., U. Redlands, 1939; M.S., State U. Ia., 1941; Ph.D., Cal. Inst. Tech., 1948; m. Marjorie Ada Frisius, Feb. 4, 1943; children—Donald Frisius, Sheryl Frances, Sylvia Gay. Acting head physics sect., Jet Propulsion Lab. Cal. Inst. Tech., 1956-58, staff scientist, 1960—. Mem. Am. Phys. Soc., Am. Geophys. Union, A.A.A.S., Sigma Xi. Research on interplanetary medium, especially using spacecraft; confirmation of solar-wind hypothesis of solar plasma. Home: 4409 Lowell Av., La Crescenta, Cal. 91214. Office: 4800 Oak Grove Dr., Pasadena, Cal. 91103.*

SNYDER, Donald DuWayne, Am. physicist; b. Benton Harbor, Mich., Apr. 11, 1928; s. Gilbert W. and Adah (Hoover) S.; B.A., Andrews U., 1948; Ph.D., Mich. State U., 1957; m. Elsie J. Sipchenko, May 3, 1951; children—Karen Lee, Keith Alan. Sr. research physicist Gen. Motors Research Lab., Warren, Mich., 1957-59; chmn. physics dept. Andrews U., Berrien Springs, Mich., 1959-67, chmn. honors program, 1964-67; asso. prof. Ind. U., South Bend, 1967—. Mem. Am. Phys. Soc., Am. Assn. Physics Tchrs., Am. Crystallographic Assn., Sigma Xi, Research, publs. in elucidation of effect of isotopic mass on phys. properties of metals, elucidation of mechinsms in photoconductors. Office: Physics Dept., Ind. U., South Bend, Ind. 46615.*

SNYDER, Harold Ray, Am. chemist; b. Mt. Carmel, Ill., May 20, 1910; s. Ira Martin and Sabra (Hobbs) S.; B.S., U. Ill., 1931; Ph.D., Cornell U., 1935; m. Mary Jane McIntosh, May 23, 1935; children—Jane, John Jerome, Mary Ann. Research chemist Solvay Process Co. (N.Y.), 1935-36; research asst. U. Ill., Urbana, 1936-37, faculty, 1937—, prof., 1945—, asso. head dept. chemistry and chem. engring., 1957-60, asso. dean for research Grad. Coll., research prof. organic chemistry, 1960. Mem. adv. bd. Organic Reactions, 1954—, Organic Syntheses, 1951—. Guggenheim fellow, 1939. Mem. Soc. Chem. Industry, A.A.A.S., Am. German chem. socs. Author: (with Reynold C. Fuson) Organic Chemistry, 1944; 60; (with Reynold C. Fuson, Charles Price, Ralph A. Connor) Brief Course in Organic Chemistry, 1941, 47, 2d edit. (with Reynold C. Fuson, Lyell C. Behr), 1959. Discovery new organic reactions and devel. methods for synthesis of organic compounds. Home: 708 W. Florida Av., Urbana, Ill. 61801.*

SNYDER, James Newton, Am. physicist; b. Akron, O., Feb. 17, 1923; s. Louis Emery and Mary (Sullivan) S.; B.S., Harvard, 1945; M.A., 1947, Ph.D., 1949; m. Betty Jane Cooper, July 28, 1944; 1 son, James Newton. Faculty physics dept. U. Ill., Urbana, 1949—, asso. prof. Digital Computer Lab., 1957-58, research prof., 1958-64, research prof., asso. head lab., 1964—; head computer div. Midwestern U. Research Corp., Madison, Wis., 1956-57. Mem. Am. Phys. Soc., A.A.A.S., Am. Soc. for Engring. Edn., Assn. for Computing Machinery, Phi Beta Kappa, Sigma Xi. Research, publs. on field theory, complex atomic processes, application of computer techniques to exptl. physics and air traffic control. Home: 210 Mumford Dr., Urbana, Ill. 61801.*

SNYDER, John Crayton, Am. physician; b. Salt Lake City, Utah, May 24, 1910; s. Crayton Chambers and Flora (Macdonald) S.; A.B., Stanford; M.D., Harvard, 1935, LL.D., 1964; m. Virginia Ferry, June 14, 1942; children—Virginia Townsend, John Macdonald, Gordon Mansfield. Mem. of the staff of the international health division of Rockefeller Foundation, 1940-46; member of U. S. A. Typhus Commission, 1942-46; professor and head of department of microbiol., Harvard Sch. Pub. Health, 1946—, dean faculty Pub. Health, 1954—, Henry Pickering Walcott professor, 1961—. Consultant of WHC and Armed Forces Epidemiology Board. Member of American Epidemiol. Society, Assn. Am. Physicians, Am. Public Health Assn., Am. Acad. Arts and Scis., N.Y. Acad. Sci. Research on epidemic typhus fever; discovery of therapeutic effect of para-aminobenzoic acid on typhus; research on trachoma; participation in development of vaccine against trachoma in Saudi Arabia. Office: 55 Shattuck St., Boston 02115.*

SNYDER, Laurence Hasbrouck, Am. geneticist; b. Kingston, N.Y., July 23, 1901; s. DeWitt C. and Gertrude (Wood) S.; B.S., Rutgers U., 1922; M.S., Harvard, 294, Sc.D., 1926; D.Sc., Rutgers U., 1947, Ohio State U., 1960; H.H.D., U. N.C., 1962; m. Guildborg M. Herland, Dec. 25, 1923; children—Clara Reed (Mrs. Stanley P. Converse), Margaret Neal (Mrs. Donald L. Petersen). Faculty, N.C. State U., 1924-30, Ohio State U., 1930-47; dean U. Okla. Grad. Coll., 1947-58; pres. U. Hawaii, Honolulu, 1958-63, pres. emeritus, 1963—. Pres., Human Relations Area Files, 1948-54. Mem. A.A.A.S. (past pres.), Am. Soc. Human Genetics (pst pres.), Genetics Soc. Am. (past pres.), Phi Sigma (nat. pres. 1954-58). Author: Blood Grouping in Clinical and Legal Medicine, 1929; Principles of Heredity, 1934; also numerous articles. Pioneered devel. statis. methods for study human heredity, application human genetics to medicine. Home: 2885 Oahu Av., Honolulu 96822.*

SNYDER, Lawrence Clement, Am. chemist; b. Ridley Park, Pa., Apr. 16, 1932; s. Lawrence C. and Barbara (Boyer) S.; B.S., in Chemistry, U. Cal. at Berkeley, 1953; M.S., Carnegie Inst. Tech., 1954, Ph.D., 1959; m. Anne Marie Svedi, Jan. 7, 1958; children—Lenore Anne, Evan Lawrence, Leland Thomas. Mem. tech. staff Bell Telephone Labs., Inc., Murray Hill, N.J., 1959—. Lectr. chemistry Columbia, N.Y.C., 1965—. Mem. Am. Phys. Soc., Am. Chem. Soc., A.A.A.S. Research, publs. in chem. physics, electronic structure of molecules and its relation to their chemistry and spectroscopy, molecular structure determination from NMR in liquid crystal solvents. Home: 213 Lawrence Dr., Berkeley Heights, N.J. 07922. Office: Bell Telephone Labs., Murray Hill, N.J. 07971.*

SNYDER, Nathan William, aerospace engr.; physicist; b. Montreal, Que., Can., Apr. 21, 1918; s. Joseph and Sarah (Catton) S.; came to U. S., 1923, naturalized, 1940; B.S., U. Cal. at Berkeley, 1941, M.S., 1944, Ph.D. in Mech. Engring. and Math., 1947, postgrad. in Physics; m. Rosalie Shaw, May 21, 1944; children—Christine Ellen, Lorraine Dale. Faculty, U. Cal. at Berkeley, 1942-58, chmn. dept., prof. nuclear engring., 1955-58; prin. scientist space tech. Inst. for Def. Analysis, Advanced Research Projects Agy., U. S. Dept. Def., 1958-61; v.p. research, engr. Royal Research Corp., Oakland, Cal., 1961-62; chief scientist Kaiser Aerospace & Electronics, Oakland, 1962-64; Neely prof. aerospace engring. Ga. Inst. Tech., Atlanta, 1964—; dir. Geonetics Corp., San Diego. Cons. to govt. agys.; mem. space tech. panel Pres.'s Sci. Adv. Com., 1964—; mem. adv. com. on isotope and radiation devel. AEC. Mem. Am. Rocket Soc. (past chmn. power source com.), Am. Inst. Aeros. and Astronautics (past mem. electric power sources com.), Am. Phys. Soc., I.E.E.E. (mem. new power sources com. 1959—), Am. Nuclear Soc., Acoustical Soc. Am., Am. Inst. Chem. Engrs., Marine Tech. Soc., Phi Beta Kappa, Sigma Xi, Tau Beta Pi. Author: Energy Conversion for Space Power, 1961; Space Power Systems, 1961; also numerous articles. Research in space tech., nuclear tech., energy conversion, rocket propulsion, heat transfers, fluid dynamics, emissivity of surfaces, weapons systems, acoustics. Home: 2577 Red Valley Rd. N.W., Atlanta 30305.*

SNYDER, Virgil, Am. mathematician; b. Dixon, Ia., Nov. 9, 1869; s. Ephraim and Elisa Jane (Randall) S.; B.Sc., Iowa State Coll., 1889; Cornell U., 1890-92; Ph.D., U. Göttingen, 1894, postgrad., 1899, 1903, Italy 3 yrs.; Heckscher fellow, Italy, 1921-22, 28-29; hon. doctorate U. Padua, 1922; m. Margarete Glesinger, Dec. 28, 1894; children—Herbert, Norman. Instr. math. Cornell U., 1895-1903; asst. prof., 1903-10, prof., 1910-38. vis. prof. math. Brown U., 1942-43, Rollins College, Winter Park, 1943-44. NRC, 1926-29. Fellow Am. Acad Arts and Scis.; mem. Am. Math. Soc. Bull. 1903-21; review editor since 1938; pres. 1927-28), Deutscher Mathematiker-Verein, Circola Matematico di Palermo, Sigma Xi. Author: Differential Calculus (with James McMahon), 1898; Differential and Integral Calculus, 1902; Elementary Text-book on the Calculus, 1912; Analytic Geometry of Space, 1913; Topics in Algebraic Geometry, 1928, Supplement, 1935. Editor: Plane Geometry, 1910; Solid Geometry, 1912; also semi-centennial publs. Am. Math. Soc., 1938. Died Jan. 4, 1950.

SOBCZYK, Andrew, Am. mathematician; b. Duluth, Minn., May 4, 1915; s. Stanley M. and Mary (Mazak) S.; student Duluth Jr. Coll., 1932-34; B.A., U. Minn., 1935, M.A., 1936; Ph.D. (J.S.K. fellow in Math.), Princeton, 1939; m. Aurellia Eugenia Wyrzykowski, Mar. 19, 1940; children—Josephine (Mrs. John Oliver Scott), Garret, Lawrence, Rowena, Stanley. Faculty, Princeton, 1938-39, Ore. State U., 1939-42, Boston U., 1946-51, U. Fla., 1956-60, U. Miami, Coral Gables, Fla., 1960-65; staff mem. Radiation Lab., Mass. Inst. Tech., 1942-46, U. Cal. Los Alamos Sci. Lab., 1951-56; S. Maner Martin prof. math. Clemson (S.C.) U., 1965—. Del., Internat. Congress Mathematicians, Harvard, 1950, Moscow, 1966, Prague, 1966; organizer Clemson Conf. on Projections, 1967. Mem. Math. Assn. Am., Am., Canadian, Belgian, Swiss math. socs. Author: Linear Projections, 1967; also numerous articles. Discoverer continuous linear transformations of spaces of infinitely dimensional vectors; analyzed and invented networks for stabilization of carrier-frequency feedback control systems, developer math. methods for determination of transient input compressions in bomb device. Home: 109 Hillcrest Av., Clemson, S.C. 29631.*

SOBEL, Irvin, Am. economist; b. Akron, O., June 13, 1917; s. Frank and Sara (Bogatin) S.; B.A., Ohio State U., 1939, M.A., 1946; Ph.D., U. Chgo., 1951; m. Peggy Strick, June 5, 1944; children—Michael Edward, Mark David, Frances Ellen. Instr., Ohio State U., 1946-47; Roosevelt Coll., 1947-48, Northwestern U., 1948-49; faculty Wash. U., 1949-67, prof., 1959-67; Ford faculty fellow, 1956-57; prof., chmn. dept. econs. Fla. State U., Tallahassee, 1967—; vis. prof. Hebrew U., 1956-57, U. Bologna, Italy. Fulbright scholar, lectr. U. Rome. Cons. Office Manpower Automation and Tng.; mem. adv. bd. to Sec. Labor on Europe. Mem. Am. Econ. Assn., Nat. Acad. Arbitrators, Indsl. Relation Research Assn., Am. Assn. U. Profs. Research publs. on functioning of labor markets, role of certain difficult to place groups in labor market such as older working, negroes; basic research on human resources in econ. devel. Home: 3524 Wesford St., Tallahassee 32301. Office: Dept. Econs., Fla. State U., Tallahassee.*

SOBELS, Frederik Hendrik, geneticist; b. Lunteren, Netherlands, Feb. 22, 1922; s. Frederik Henderik and Irmengard (Biermans) S.; Ph.D., U. Utrecht (Netherlands), 1962; m. Paula M. van Lynden, Mar. 5, 1949; children—Frits, Carla, Paulina, Godert, Marietta. Demonstrator dept. zoology U. Zurich (Switzerland), 1949; sr. research asst. dept. zoology U. Utrecht, 1950-52, sr. sci. officer dept. genetics, 1954-59; Brit. Council scholar Inst. Animal Genetics, 1953; prof. radiation genetics Faculty Medicine U. Leiden, 1959—; sci. dir. Inst. for Radiopathology and Radiation Protection, Leiden, 1963—. Mem. European Soc. for Radiobiology (pres. 1964-66), Netherlands Soc. for Radiation Biology (pres. 1964-67), Netherlands Soc. Genetics (pres. 1964—), Internat. Assn. Radiation Research (pres. 1966—). Editor: Repair from Genetic Radiation Damage and Differential Radiosensitivity in Germ Cells, 1963. Mng. editor Mutation Research, 1964—. Research, numerous publs. on discovery that in Drosophila the yield of X-ray induced mutations and chromosome aberrations can be modified by various post-radiation treatments; elucidation of role of oxygen in determining genetic radiosensitivity in different stages of sperm devel. Home: Voorstraat 25, Noordwijk/Binnen, Netherlands. Office: Dept. Radiation Genetics, 62 Wassenaarseweg, Leiden, Netherlands.*

SOBER, Herbert Alexander, Am. biochemist; b. N.Y.C., Feb. 24, 1918; s. Casper C. and Sarah (Victor) S.; B.S., Coll. City N.Y., 1938; M.S., U. Wis., 1940, Ph.D. (Indsl. fellow), 1942; m. Eva Katzenelbogen, Aug. 28, 1941; children—Lillian Sara, Barbara Jane. Toxicologist, Edgewood Arsenal, 1942-45; research asst. in gastroenterology Mt. Sinai Hosp., N.Y.C., 1945-47; with Nat. Cancer Inst., NIH, Bethesda, Md., 1947—, chief lab of biochemistry, 1959—. Dir., Found. for Advanced Edn. in Scis., 1960. Mem. Am. Chem. Soc. (councilor 1964), Soc. Biol. Chemists, Am. Cancer Soc., Biochm. Soc. Britain, Biophys. Soc. Editor: Handbook of Biochemistry, 1967; asso. editor Biochemistry, 1961; adv. bd. Analytical Biochemistry, 1966. Research, numerous publs. on nucleic acid and protein biochemistry, chromatographic separations of proteins and nucleic acids, cellulose ion-exchange absorbents, nucleoproteins, biochemistry of cancer. Home: 5200 W. Cedar Lane. Office: Lab. Biochemistry, Nat. Cancer Inst., NIH, Bethesda, Md. 20014.*

SOBERMAN, Robert Kenneth, Am. physicist; b. N.Y.C., Apr. 8, 1930; s. Julius and Helen (Mile) S.; B.S., City Coll. of N.Y., 1950; M.S., N.Y. U., 1952, Ph.D., 1956; m. Diana Helene Gross, June 6, 1954; children—Ellen Margo, June Ann. Research physicist Gen. Electric Vallecitos Atomic Lab., Pleasanton, Cal., 1955-57, group leader phys. aeronomy Space Scis. Lab., Phila., 1966—; sr. scientist Avco Research and Advanced Devel. Div., Wilmington, Mass., 1957-59; chief meteor physics br. Air Force Cambridge Research Labs., Bedford, Mass., 1959-66. Adj. asso. prof. Northeastern U., 1959-63; cons. European Space Research Orgn., 1963—, NASA, 1962—. Recipient Founders Day award N.Y. U., 1956. Mem. A.A.A.S., Am. Geophys. Union, Am. Astron. Soc., Sigma Xi. Research, publs. on noctilucent clouds and meteoric material through use of rockets and satellites; developed techniques for studying extra-terrestial and cloud particles by collection with high altitude rocket; developed technique for simulating meteors by accelerating pellets from a reentering rocket. Home: 1001 City Av., Phila. 19151. Office: Gen. Electric Space Scis. Lab., Box 8555, Phila. 19101.*

SOBERNHEIM, Johann, German physician; b. Bromberg, Germany, 1803; M.D., U. Königsberg (now Kaliningrad, USSR), 1828; practice medicine, Berlin. Author: Handbuch der pratischen Arzeneimittellehr, 1836; Deutschland's Heilquellen . . . 1836; Tabulae pharmalogicae, 1836. Introduced term myocarditis, circa 1837. Died 1846.

SOBOLEV, Sergéi Lvóvich, Russian mathematician, mech. engr.; b. Oct. 6, 1908; grad. Leningrad U., 1929, also Dr.Phys.-Math. Scis. Staff, Seismol. Inst., USSR Acad. Scis., 1929-32, asso. Math. Inst., 1932-58; prof. Moscow U., 1935-59; prof. Novosibirsk Inst., 1959—; became mem. presidium Siberian br. USSR Acad. Scis., 1961, also dir. Inst. Math. and Computation Center. Recipient Stalin Prize, 1941, Order of Lenin, Order of Red Banner of Labor. Mem. USSR Acad. Scis. Author: Some Uses of Functional Analysis in Mathematical Physics, 1950; Equations of Mathematical Physics, 3d edit., 1954; Research and publs. on n-dimensional space, elastic bodies, plane waves; developed theory of generalized functions, (with V. I. Smirnov) methodology for study of propagation of elastic waves from rectilinear boundaries. Office: USSR, Novosibirsk, RSFSR, Gosudarstvenny universitet.

SOBOLEV, Víktor, Víktorovich, Russian astronomer; b. Sept. 2, 1915; grad. Leningrad U., 1938. Staff, Leningrad U., 1941—, became prof., 1948. Corr. mem. USSR Acad. Scis. Author: Moving Star Envelopes, 1947; The Transfer of Radiant Energy in the Atmospheres of Stars and Planets, 1956; The Theory of the Evolution of Stars, 1960; coauthor: A Course in Astrophysics and Astral Astronomy, 1951; Theoretical Astrophysics, 1952. Research in astrophysics, especially radiation transfer, non-stationary stars, theories on spectrum lines, luminosity of moving media, non-stationary radiation field, atmospheres of planets, gas mists; determined relationship of early and late spectrum classes of giant stars. Office: USSR, Leningrad, Universitetskaya n. 7-9, Gosudarstvenny universitet.

SOBOLEV, Vladimir Stepanovich, Russian petrographer, mineralogist; b. Lugansk, USSR, May 30, 1908; grad. Leningrad (USSR) Mining Inst., 1930. Staff, All-Union Central Geol. Survey Research Inst., 1930-31; instr., Leningrad Mining Inst., 1931-39, prof., 1939-41; prof. Irkutsk U., 1941-45; prof. Lvov U., 1945—; dir. mineralogy dept. Leningrad Mining Inst., also Fedorov Inst., 1943-45; asso. Lvov Inst. Geology Minerals, Ukraine Acad. Scis., 1947—. Recipient Stalin Prize, 1949. Mem. USSR Acad. Scis., Ukraine Acad. Scis. (corr.). Author: The Petrology of Traprocks in the Siberian Plateau, 1936; An Introduction to the Mineralogy of Silicates, 1949; The Geology of the Diamond Deposits of Africa, Australia, Borneo and North America, 1951; Fedorov's Method, 1954. Research and publs. on iron deposits and traprocks of Siberian, petrography and mineralogy of Ukraine, alkaline rocks, crystalline schist and granitoids of So. Yakutia; predicted location of diamonds in Siberia using studies of fgn. diamond deposits; developed laws for crystallization of trappeau magna; related difference in ionic radii in isomorphic series to fusibility curves. Home: Inst. Geology of Minerals, Ulitsa Kopernika 15, Lvov, Ukrainian SSR, USSR.

SOBRERO, Ascanio, Italian chemist; b. Casale, Italy, Oct. 12, 1812; studied under Berzelius, also Liebig; became prof. chemistry Inst. Tech., U. Turin (Italy), 1847. Author: (with L.Ch.A. Barresuoil) Appendice a tous les traités d'analyse chemique, 1843; Manuale de chimica applicata alle arte, 3 vols., 1851-57. Discovered, observed and reported explosive powers in nitro-glycerine. Died Turin, May 26, 1888.

SOBRERO, Luigi Paolo, Italian physicist; b. Turin, Italy, Oct. 28, 1909; s. Francesco and Gabrielle (Rubatto) S.; Dr. ès sc. math. and physics. Prof. theoretical physics U. Rio de Janeiro; prof. Cagliari U.; now prof., dir. Mechanics Inst., U. Trieste. Author: Theorie der ebenen Elastizität, 1934; Elasticidade, 1942; Les Voutes minces, 1961; also numerous articles. Research on theoretical and applied mechanics, phys. math. problems, rheology. Address: viale R. Sanzio 36, Trieste, Italy.

SOCHAVA, Viktor Borisovich, Russian geobotanist, geographer; b. Jan. 20, 1905; grad. Leningrad (USSR) Agrl. Inst., 1924. Instr. Leningrad Agrl. Inst., 1924-26; with reindeer-breeding inst. All-Union Lenin Acad. Agrl. Sci., 1931-35; head reindeer-breeding dept. Arctic Inst., 1935-38; instr. Herzen Pedagol. Inst., Leningrad, 1939-50; instr. Leningrad U., 1938-50, prof., 1950—. Mem. USSR Acad. Scis. (corr., asso. botany inst. 1926-36, 43—, dir. inst. geography Siberian dept. 1960—). Research vegetation, landscapes USSR; developed ecol.-geog.-genetic classification system vegetation, principles classifying regions by geobot. and landscape characteristics, expdns. Far East, Siberia, Rumania, Czechoslovakia, Ural Caucasus, Carpathian mountains. Address: Institut Geografii Sibirskogo otdeleniya A.N. SSSR, Novosibirsk, Russian Soviet Federated Socialist Republic USSR.

SOCIN, August, Swiss surgeon; b. Vevey, Switzerland, 1837; prof., Basel, Switzerland; founder Basler Chirurgischen Klinik. Advocated antiseptic methods. Died Basel, 1899.

SOCRATES, Greek philosopher; b. Athens, circa 470 B.C.; s. Sophroniscus and Phaenarete; as youth studied gymnastics and music, later studied geometry and astronomy; m. Xanthippe; 2 sons. Influenced by Sophists; started career as sculptor; took part in 3 campaigns of war; Potidaea, 432-429 B.C.; Delium, 424; Amphipolis, 422; sat on senate of 500 in 406 B.C.; devoted life to educating Athenian youths, encouraging them to question all aspects of life (supposedly believed he had divine commission to teach); left no writings although accounts of his doctrines as well as his personal life and character recorded in works of Plato and Xenophon; his influence on Plato enormous; indicted as offender against public morality in 399 B.C.; his defense recorded in Plato's dialogue, the Apology; Socrates convicted and given death penalty; drank hemlock after refusing offers of his friends to help him escape in 399 B.C.

SOCZEWINSKI, Edward, Polish chemist; b. Lublin, Poland, Sept. 4, 1928; s. Adam and Magdalena (Golian) S.; Magister Chemistry, U. Lublin, 1952, D., 1960, Docent habilit., 1963. Pharm. faculty dept., inorganic chemistry Med. Acad., Lublin, 1952—, docent, 1964-65, head dept. inorganic chemistry, 1965—; sec. chromatographic analysis subcom. com. analytical chemistry Polish Acad. Scis., 1962—. Co-recipient State award II-d degree, 1964. Mem. Polish Chem. Soc., Polish Pharm. Soc., Trade Union Pub. Health Workers. Contbg. author: Chromatography, 1957. Research, publs. on theory partition chromatography, extraction and counter-current distbn. expecially parameters influencing zone migration rates and determining optimal condition for analysis, separation and purification of substances. Home: Ul. Junoszy 35/11, Lublin, Poland.*

SODDY, Frederick, Brit. physicist; b. Eastbourne, Eng., Sept. 2, 1877; s. Benjamin and Hannah (Green) S.; student Eastbourne Coll., 1893-94; B.A., Oxford. 1898; M.A., 1910; LL.D., Glasgow; m. Winifred Moller Beilby, Mar. 3, 1908 (dec. Aug. 1936). Demonstrator chemistry McGill U., Montreal, Can., 1900-02; established atomic disintegration of radioactive elements and existence of atomic energy (with Ernest Rutherford); proved spectroscopically the prodn. of helium from radium (with Sir W. Ramsay), 1903-04; London U. extension lectr., Western Australia, 1904; lectr. phys. chemistry, radioactivity, U. Glasgow, 1904-14; prof. chemistry U. Aberdeen, 1914-19. Oxford, 1919-36; ret. 1936. Recipient Nobel prize in chemistry, 1921, Cannizzaro prize by Acad. de Lincei, Rome, 1923. Author: The Interpretation of Radium, 1909; Matter and Energy, 1912; The Chemistry of Radio-Elements, (2 vols.) 1911-14; Science and Life, 1920; Interpretation of the Atom, 1932. Applied phys. laws of conservation to econs. and writer on new econs.; Wealth, Virtual Wealth and Debt, 1926; Story of Atomic Energy, 1949; Atomic Transmutation, 1953. Responsible for conception of isotopes and the displacement law of radioactive change which is at the root of nuclear physics. Died Sept. 22, 1956.

SODEMAN, William Anthony, Am. physician; b. Charleroi, Pa., June 13, 1906; s. William J. Carl and Anna (Dietz) S.; B.S., U. Mich., 1929, M.D., 1931; m. Mary Agnes Wagner, Apr. 21, 1928; children—William Anthony, Thomas M. Practice medicine, specializing in internal medicine, prof., head dept. tropical medicine Tulane U., New Orleans, 1941-52; specialist cardiology, Columbia, Mo., 1953-57, Phila., 1957-67, Rosemont, Pa., 1967—; prof., chmn. dept. internal medicine U. Mo. Sch. Medicine, 1953-57; Magee prof. medicine, head dept. medicine Jefferson Med. Coll., 1957-58, dean, prof. medicine, 1958-67, v.p. med. affairs, 1962-67, dean, prof. medicine emeritus, 1967—; sci. dir. Life Ins. Med. Research Fund, Rosemont, 1967—; attending physician emeritus Jefferson Med. Coll. Hosp., 1967—. Cons. cardiology Sch. Dist. Phila., 1959—; cons. internal medicine Lankenau Hosp., 1967—. Mem. bd. Nat. Bd. Med. Examiners, 1968—. Mem. A.M.A., Am. Soc. Clin. Investigation, A.C.P. (treas.), Am. Heart Assn., Am.

Pub. Health Assn., Am. Soc. Tropical Medicine, Central Soc. Clin. Research, Royal Soc. Tropical Medicine and Hygiene, Am. Fedn. Clin. Research, Am. Coll. Cardiology (trustee), Sigma Xi. Author: Pathologic Physiology, 1967 also numerous articles. Research in cardiology, clin. and instrumental methods in cardiology. Home: Rosemont Plaza, 1062 Lancaster Av. Office: 1030 E. Lancaster Av., Rosemont, Pa. 19010.*

SODERBERG, C(arl) Richard, mech. engr.; b. Ulvöhamm, Sweden, Feb. 3, 1895; s. Jonas Axel and Johanna Kristina (Nordquist) S.; naval architect Chalmers Inst. Tech., Göteborg, Sweden, 1919, Tech. Dr., 1951; fellow Am.-Scandanavian Found., 1919-20; S.B., Mass. Inst. Tech., 1920; D.Eng., Tufts U., 1958; came to U. S., 1919, naturalized, 1927; m. Sigrid Kristina Lofstedt, May 9, 1921; children—Carl Richard, Lars Olof, Barbro Kristina. Devel. engr. for large elec. machinery Westinghouse Electric & Mfg. Co., East Pittsburgh, Pa., 1922-28, chief turbine engr., 1931-38; in charge of turbo generators Swedish Gen. Electric Co., Västeras, Sweden, 1928-30; prof. mech. engring. Mass. Inst. Tech., 1938-59, Inst. prof., 1959-61, Inst. prof. emeritus, 1961—; head dept. mech. engring., 1947-54, dean engring. 1954-59. Decorated comdr. Royal Order N. Star. Recipient Capt. Joseph H. Linnard prize (with Ronald B. Smith) Soc. Navy Architects and Marine Engineers, 1944; Certificate of Appreciation, Army and Navy, 1948, John Ericsson Gold medal. Am. Soc. Swedish Engrs., 1952; Exceptional Service award USAF, 1955; Gustav de Laval medal, 1958; Am. Soc. M.E. Engrs. medal, 1960; New Eng. award Engring. Socs. New Eng., 1961. Fellow Am. Acad. Arts and Scis., Am. Inst. Aeros. and Astronautics, Nat. Acad. Scis.; mem. A.A.-A.S., Brit. Inst. M.E., Am. Soc. M.E. (hon.), Ingeniörsvetenskapsakademiens, Svenska Teknologföreningen, Franklin Inst., Soc. Naval Architects and Marine Engrs., Am. Soc. Swedish Engrs., Sigma Xi, Pi Tau Sigma, Tau Beta Pi. A designer and inventor of steam turbines and generators for land and marine applications. Author numerous monographs to Am. and European jours. on dynamics, virations, design of turbines and generators, gas turbines, engring. edn. Address: 6 Joy St., Boston 02108. Office: 77 Massachusetts Av., Cambridge, Mass. 02139.*

SÖDERBERG, Ulf Axel, Swedish neurophysiologist; b. Härnösand, Sweden, Mar. 1, 1928; s. Bror Axel Egron and Karin (Bergh) S.; Med.kand. Karolinska Institutet, Stockholm, Sweden, 1949, Med. dr., 1958; m. Birgitta Högstedt, Dec. 28, 1963; children—Monica, Johan, Helena. Asst. prof. Karolinska Institutet, 1958-62; asso. prof. psychiat. neurophysiology U. Uppsala (Sweden), 1962—; sci. staff Nobel Institute for Neurophysiology, Stockholm, 1949—. Author: Short-term Reactions in the Thyroid Gland, 1958; also numerous articles. Research in thermoregulation, energy metabolism, thyroid gland activity and brain blood flow. Home: 35 Drakskeppsvägen, Viggbyholm, Sweden. Office: Nobel Inst. for Neurophysiology, Stockholm 60, Sweden.*

SODERWALL, Arnold Larson, Am. biologist; b. Portland, Ore., Nov. 13, 1914; s. Axel Emanuel and Beda (Larson) E.; B.A., Linfield Coll., 1936; M.A., U. Ill., 1938; Ph.D., Brown U., 1941; m. Alice Blanche Southard, Nov. 1, 1938; children—Charles K., Kathryn Jean. Faculty, U. Ore., Eugene, 1941—, prof. biology, 1960—. AEC fellow Cornell U., 1965-66. Fellow A.A.A.S.; mem. Am. Assn. anatomists, Am. Soc. Zoologists, Fedn. Am. Socs. for Exptl. Biology. Research, numerous publs. on viabiity spermatozoa in female reproductive tract rat, guinea pig, effect aging on litter size and gestation length in golden hamster, beneficial action vitamin E on gestation in senescent female golden hamsters, changes in blood serum proteins in pregnant, senescent and X-irradiated female hamsters. Home: 2493 Harris St., Eugene, Ore. 97405.*

SODI-PALLARES, Demetrio, Mexican physician; b. Mexico City, Mexico, June 8, 1913; s. Demetrio Sodi and Carmen (Pallares VDA.) Guergue; ed. Mexican Nat. U.; m. Soledad de la Tijera, Jan. 3, 1941; children—Juan, Demetrio, Graciela, Marcela, Ana Alicia, Laura Maria. Faculty, Nat. U. Mexico Sch. Medicine, Mexico City, 1944—, prof. theory cardiovascular diseases, 1947-49, prof. postgrad. studies cardiology Inst. Cardiology, 1949—, chief prof. Cardiovascular Clinics, 1951—, bd. examiners, 1956—. Fellow A.C.P., Am. Coll. Cardiology (1st v.p.), Am. Coll. Chest physicians (hon.); mem. Argentinian (corr.) Mexican (past pres.), Brazilian, Chilean, Colombian, French (corr.), Peruvian, Venezuelan socs. cardiology, Soc. Dermatology Mexico, Nat. Acad. Medicine Mexico (pres.), Sao Paulo Acad. Medicine, Tex. Heart Assn., Interam. Soc. Cardiology (treas.-sec.), Academia de Medicina de Sevilla España, Sociedad Española de Cardiología (hon. pres.). Author: New Basis de Electrocardiography, 1956; Electrocardiografía y Vectocardiografía Deductivas, 1964; also numerous articles. Research on electrocardiography, polarizing treatment for cardiovascular ailments. Home: Bravo 30, San Jerónimo, Lidice, México, D.F. 20, Mexico.*

SOFFER, Alfred, Am. cardiologist; b. South Bend, Ind., May 5, 1922; s. Simon and Bessie (Rokach) S.; B.A., U. Wis., 1943; M.D., U. Wis., 1945; m. Isabel Weintraub, July 22, 1956; children—Jonathan, Joshua, Gil. Fellow physiology U. Rochester (N.Y.) Sch. Medicine and Dentistry, 1949; fellow clin. labs. Genesee Hosp., Rochester, 1950, fellow medicine, 1950-51, electrocardiographer, 1951-58; practice internal medicine, Rochester, 1951-58; chief cardiopulmonary labs. Rochester Gen. Hosp., 1958-62; asso. medicine Northwestern Med. Sch., Evanston, Ill., 1963-64; asso. prof. medicine Chgo. Med. Sch., 1965-66; chmn. sci. program Am. Coll. Cardiology, Chgo., 1966. Diplomate Am. Bd. Internal Medicine. Fellow A.C.P., Am. Coll. Chest Physicians, Am. Coll. Cardiology; mem. Am. Fedn. for Clin. Research. Author: Chelation Therapy, 1964; (with S.R. Elek, D. Scherf) Clinical Electrocardiograms, 1965; also articles. Research on diagnosis and treatment of digitalis poisoning, pharmacology of chelating agts. (to bind metals), clin. cardiology. Sr. editor Jour. A.M.A., 1962-67; editor-in-chief Diseases of the Chest, 1968—. Home: 325 Alexis Ct., Glenview, Ill. 60025. Office: 112 E. Chestnut St., Chgo. 60610.*

SOFFER, Louis Julius, physician; b. London, Eng., Dec. 29, 1904; s. Shamy and Jennie (Schuster) S.; came to U.S., 1905, naturalized, 1912; B.S., Johns Hopkins, 1932; M.D., L.I. Coll. Medicine, 1928; m. Helen Luber, Sept. 9, 1928; children—Richard L., Lucy J. (Mrs. Charles Blankstein). Faculty Johns Hopkins Med. Sch. and Hosp., Balt., 1930-35; dir. div. endocrinology attending physician Mt. Sinai Hosp., N.Y.C., 1950—; clin. prof. medicine State U. N.Y. Dowstate Coll. Medicine, N.Y.C., 1951—. Recipient Alumni Achievement medal State U. N.Y., 1957; Outstanding Alumnus award Bklyn. Jewish Hosp., 1962. Diplomate Am. Bd. Internal Medicine. Fellow A.C.P.; mem. N.Y. Acad. Medicine (chmn. com. med. edn. 1959-60, Harlow Brooks medal 1957), Am. Soc. for Clin. Investigation, Soc. Exptl. Biology and Medicine, Harvey Soc., Endocrine Soc., A.M.A., Sigma Xi. Author: Diseases of Adrenals, 1946, 48; Diseases of the Endocrine Glands, 1950, 56; (with Dorfman, Gabrilove) The Human Adrenal Gland, 1962; also numerous articles. Research in endocrinology with particular reference to adrenal physiology and disease; introduced treatment of collagen disease with glucogenic steroids; discovered Gonadotropin inhibiting substance which regulates action of luteinizing hormone on gonads. Home: 1185 Park Av. Office: 1175 Park Av., N.Y.C., 10028.*

SOGANDARES, Franklin, parasitologist; b. Ancon, Panama, C.Z., May 12, 1931; s. Anastasio and Blanca (Bernal) S.; B.S., Tulane U., 1954; M.S., U. Neb., 1955, Ph.D., 1958; m. Lucy Ann McAlister, June 12, 1960; children— Franklin McAlister, Maria, John. Marine parasitologist Fla. State Bd. Conservation Marine Lab., St. Petersburg, 1958-59; faculty Tulane U., New Orleans, 1959—, prof. biology, 1965—, coordinator for sci. planning, 1965—. Adv. panel for systematic biology, biomed. div. NSF; bd. sci. advisers, saltwater fisheries div. Fla. State Bd. Conservation. Mem. Am. Soc. Parasitologists, Conf. Biol. Editors, Soc. Wildlife Diseases, Am. Soc. Zoologists, Helminthological Soc. Washington, Am. Inst. Biol. Scis., A.A.A.S., N.Y. Acad. Sci., Sigma Xi. Editorial bd. Jour. Parasitology, 1964-68, Am. Midland Naturalist, 1966—. Research, publs. in field of host-parasite relationships with spl. interest in evolutionary biology of parasitism as reflected by digenetic trematodes. Home: 7926 Sycamore St., New Orleans 70118.*

SOHL, Norman Frederick, Am. paleontologist; b. Oak Park, Ill., July 14, 1924; s. Fred John and Florence (Wray) S.; B.S., U. Ill., 1949, M.S., 1951; Ph.D., U. Ill., 1954; m. Dorothy Martha Jansen, June 5, 1947; 1 son, Norman Frederick. Research asst. Ill. Geol. Survey, Urbana, 1949-50; instr. Bryn. Mawr Coll., 1952-53, U. Ill., Urbana, 1953-54; research geologist U.S. Geol. Survey, Washington, 1954—; asso. professorial lectr. George Washington U., 1963—; vis. prof. geology U. Kan., Lawrence, 1966. Mem. Inst. Malacology (trustee 1962—, v.p. 1966—), Soc. Systematic Zoology, Soc. for Study Evolution, Soc. Econ. Paleontology, Paleontol. Soc. Am., Paleontol. Soc. Washington (past pres.) Sigma Xi. Research, numerous publs. on classification, evolution, and systematics class Gastropoda; integration Mesozoic gastropods and oysters with stratigraphic and geog. occurence on global scale. Home: 7105 Vermillion Pl., Annadale, Va. 22003. Office: U.S. Nat. Mus., Washington 20242.*

SOHNCKE, Leonhard, German physicist; b. Halle/Saale, Germany, Feb. 22, 1842; prof., Karlsruhe, Jena and Munich, Germany. Author: Die Entwicklung einer Theorie der Kristallstruktur, 1879; Über Stürme und Sturmwarnungen, 1875. Studied crystal structure. Died Munich, Nov. 1, 1897.

SOILA, Anssi Pekka, Finnish radiologist; b. Hämeenlinna, Finland, Oct. 5, 1923; s. Paavo J. and Hilja (Salo) S.; Licentiate Medicine, U. Helsinki (Finland), 1951, M.D., 1958, Docent, 1960; m. Orvokki Vahala, May 29, 1948; children—Helena, Anssi, Kalevi. Docent roentgen diagnostics U. Turku, 1960-64; prof., chmn. U. Helsinki, 1964—; staff radiologist Pori, 1959-60, Kivelä Hosp., 1960-63, HUCH, 1963-64. James Picker Found. Research fellow., 1957-59. Mem. A.A.A.S., Internat. Coll. Angiology, Internat. Soc. Cybernetics. Author: Before During After, 1964; also numerous articles. Research on rheumatoid arthritis, technique of roentgen diagnostics, clin. roentgen diagnostics. Home: 8 B P.Laurintie, Helsinki 34, Finland.

SOIVA, Keljo Urpo, Finnish physician; b. Kanagasala, Finland, Sept. 26, 1920; s. Martti and Elli (Rautell) S.; med.lic., U. Helsinki (Finland), 1947, M.D., 1954; m. Outi Kalkkinen, Dec. 30, 1945; children—Auli, Martti. Gynecologist, obstetrician Womens' Clinic, U. Helsinki, 1948-52, Inst. Midwifery, Helsinki, 1951-54; asst. chief obstetrics and gynecology Women's Clinic, U. Turku (Finland), 1954-62, acting prof., 1963; chief obstetrician, gynecologist 1st dept. obstetrics and gynecology Central Hosp., Tampere, Finland, 1962—. Lectr., U. Helsinki, 1963—. Mem. Finnish Med. Assn., No. Assn. Obstetrics and Gynecology, Deutsche Gesellschaft fur Gynakologie, Internat. Coll. Surgeons. Author: Effects of Obstetric Factors and Oxytocic Drugs on Postpartum Hemorrhage, 1954; also numerous articles. Research in obstetrics and gynecology, chiefly postpartum hemorrhage, toxemia of late pregnancy and fetal prognosis. Address: Central Hosp., Tampere 10, Finland.*

SOKAL, Joseph Emanuel, research physician; b. Lwow, Poland, Apr. 11, 1917; s. Henry Bear and Ethel (Torten) S.; B.A., with honors, Columbia, 1936; M.D. cum laude Yale, 1940; m. Nancy Branan, Nov. 29, 1947; children—David C., Paul J. Jane Coffin Childs Meml. Fund fellow Yale Sch. Medicine, 1947-50, asst. prof. medicine, Markle scholar med. scis., 1950-55; chief cancer research internist Roswell Park Meml. Inst., Buffalo, N.Y., 1955—; faculty State U. N.Y. at Buffalo, 1955—, research prof. physiology, asso. research prof. medicine, 1962—; asst. physician Buffalo Gen. Hosp. Fellow A.C.P.; mem. Am. Physiol. Soc., A.A.A.S., N.Y. Acad. Scis., Soc. Exptl. Biology and Medicine, Endocrine Soc., Am. Assn. Cancer Research, Am. Fedn. Clin. Research, A.M.A., N.Y. State, Buffalo med. socs., Phi Beta Kappa, Sigma Xi, Alpha Omega Alpha. Research, publs. on Hodgkin's disease, treatment of leukemia, liver glycogen disease, function of glucagon, nodular goiter.*

SOKAL, Robert Reuven, biologist; b. Vienna, Austria, Jan. 13, 1926; s. Siegfried and Klara (Rathner) S.; came to U.S., 1947, naturalized, 1958; B.S., St. Johns U., Shanghai, China, 1947; Ph.D., U. Chgo., 1952; m. Julie Chen-chu Yang, Aug. 12, 1948; children—David J., Hannah J. Faculty, U. Kan., Lawrence, 1951—, prof. statis. biology, 1961—; Fulbright prof. Tel Aviv and Hebrew U., Israel, 1963-64. NIH career investigator, 1964—. Watkins scholar U. Ill., 1956; NSF fellow U. Coll., London, Eng., 1959-60. Am. Soc. Naturalists, Soc. Study Evolution, Genetics Society of Naturalists, Society for the Study of Evolution (vice president 1967), Genetics Soc., Am., Biometric Soc., Am. Statis. Assn., Ecol. Soc. Am., Japanese Soc. Population Ecology, Entomol. Soc. Am., Classification Soc., Kan. Entomol. Soc., Kan. Acad. Sci., Am. Inst. Biol. Scis., Am. Assn. U Prof., Sigma Xi. Author: (with P.H.A. Sneath) Principles of Numerical Taxonomy, 1963; also numerous articles. Co-founder prins. and meth- (with F.J. Rohlf) Biometry, 1968, Statistical Tables, 1968; also numerous articles. Co-founder prins. and methodology of numerical taxonomy, gen. systematic theory; population and ecol. genetics; research on application multivariate statis. methods and computer techniques in biology. Home: 933 Ohio St., Lawrence, Kan. 66044.*

SOKOLNIKOFF, Ivan Stephan, mathematician; b. Chernigov, Russia, Nov. 10, 1901; s. Stephan Ivan and Katherine (Triguboff) S.; B.S. in Elec. Engring., U. Ida., 1926; Ph.D. in Math., U. Wis., 1930; m. Ruth Lawyer, Dec. 23, 1947; 1 dau., Katherine Ann. Came to U. S., 1921, naturalized, 1927. Faculty, U. Wis., 1927-46, prof. math., 1941-46; prof. math. U. Cal. at Los Angeles, 1946—; commd. ofcl. investigator NDRC, 1941, chief tech. aide math. panel, 1944-45, cons., 1945-46. Recipient medal Classical Gymnasia, Anders, Russia, 1918; Presdl. Certificate of Merit. Wis. Alumni Research Found. grantee, 1945; Research Corp. N.Y. grantee, 1949, 50, 51; Guggenheim fellow, Cambridge, Eng., 1952-53, Swiss Fed. Inst. Techn., Zürich, 1959-60. Mem. Am. Math. Soc. (past mem. council), Math. Assn. Am. (past gov.), Circolo Math. di Palermo, Sigma Xi, Tau Beta Pi. Author numerous books including: Tensor Analysis, 1951; Mathematical Theory of Elasticity, 1946, rev., 1956; (with R. M. Redheffer) Mathematics of Physics and Modern Engineering, 1958, rev., 1966; Tensor Analysis, Theory and Application, 1964; also articles. Function-theoretic methods for solution of two dimensional problems in math. theory of elasticity and in mechanics of continuous media. Home: 17751 Tramonto Dr., Pacific Palisades, Cal. 90272. Office: U. Cal. at Los Angeles, 90024.*

SOKOLOFF, Boris Theodore, research physician; b. St. Petersburg, Russia, Nov. 13, 1895; s. Theodore and Marie (Verchovzev) S.; Ph.D., U. St. Petersburg, 1911; M.D., Med. Sch. Petrograd, 1916; m. Alice Hunt, Oct. 24, 1946; 1 son, Kiril. Research scientist specializing in field of cancer and artherosclerosis Pasteur Inst., Paris, 1928, Rockefeller Inst. Med. Research, 1929-30, U. Wash. Med. Sch., 1932-35, Columbia, 1935-39; dir. So. Bio-Research Inst., Cooke Meml. Cancer Lab. Fla. So. Coll., Lakeland, 1946—. Fellow Royal Soc. Medicine, Royal Soc. Arts, Am. Assn. Cancer Research; mem. Am. Chem. Soc., Am. Soc. Biol. Chemists, N.Y. Acad. Sci., Am. Soc. Biol. Editors, Soc. Econ. Botany. Author: Napoleon, Medical Biography, 1937; Cancer, 1941; The Civilized Diseases, 1944; Story of Penicillin, 1945; The Miracle Drugs, 1949; The Science and Purpose of Life, 1950; New Approaches, 1952; Biography of August Comte, 1961; also numerous articles. Mng. editor: Jour. Growth, 1962—. Discovery of anticancer antibiotics, Yucca alolfolia in Yucca flower, antiviral and anticancer antibiotics in fungi, Paecilomyces and Trichoderma genera, role of ascorbone, oxidized form of ascorbic acid, which is harmful to blood vessels. Home: Mt. Holly Rd., Katonah, N.Y. 10536. Office: So. Bio-Research Inst., Fla. So. Coll., Lakeland, Fla. 33802.*

SOKOLOFF, Louis, Am. physiologist; b. Phila., Oct. 14, 1921; s. Morris and Goldie (Levy) S.; A.B., U. Pa., 1943, M.D., 1946; m. Betty Jane Kaiser, Jan. 21, 1947; children—Kenny, Ann. Research fellow, instr., asso. in physiology and pharmacology U. Pa. Grad. Sch. Medicine, 1949-55; asso. chief sect. on cerebral metabolism Lab. Neurochemistry, Nat. Inst. Mental Health, NIH, Bethesda, Md., 1953-56, chief sect. on cerebral metabolism Lab. Clin. Sci., 1956—. Mem. Am. Physiol. Soc., Am. Soc. Biol. Chemists, Am. Neurol. Assn., A.A.A.S., Assn. for Research in Nervous and Mental Disease, N.Y. Acad. Scis., Biophys. Soc., Soc. Gen. Physiologists, Am. Chem. Soc., Phi Beta Kappa, Sigma Xi, Alpha Omega Alpha. Editor: (with others) Human Aging-A Biological and Behavioral Study, 1963; editorial bd. Jour. Neurochemistry. Research, numerous articles on physiology and pharmacology of cerebral circulation, methodology of measurement of cerebral circulation and metabolism in man and animals; contbns. to field mechanism of action of thyroid hormones and to biochemistry of brain. Home: 707 Hermleigh Rd., Silver Spring, Md. Office: Nat. Inst. Mental Health, Bethesda, Md. 20014.*

SOKOLOSKI, Walter Thomas, Am. microbiologist; b. Newark, Oct. 29, 1916; s. Bartholomew and Catherine (Konarski) S.; B.S., Ind. U., 1948; M.S., Purdue U., 1953, Ph.D., 1954; m. Marjorie Dean Mayer, Feb. 14, 1947; children— Thomas, Trese. With Upjohn Co., Kalamazoo, 1954—, head analytical microbiology group, 1956-60, head spl. problems in control, 1960—. Recipient Soc. Am. Bacteriologists award, 1953. Mem. Am. Soc. Microbiologists, Soc. Indsl. Microbiology, Registry Med. Technologists, Sigma Xi. Research numerous publs. on new antibiotics, separations of mixtures of antibiotics, methods for finding new antibiotics, assay devels. for antibiotics in body fluids and tissues, preservative efficacy in products, low temperature storage of microorganisms, automation of vitamin assays. Home: 3304 Cranbrook St. Office: 7000 Portage Rd., Kalamazoo 49001.*

SOKOLOVSKII, Vadim Vasilévich, Russian mech. engr.; b. Oct. 17, 1912; grad. Moscow Inst. Constrn. Engrs., 1933. With Math. Inst., 1936-39; instr. Inst. Mechanics, USSR Acad. Scis., 1939-40, asso., also prov., 1940—. Recipient Stalin prize, 1943, 52. Mem. Acad. Scis. USSR (corr.), Polish Acad. Scis. Author: Statics of Loose Media, 1942; The Theory of Plasticity, 2d edit., 1950; Diode Amplitutde Limiter; The Method of Reducing the Background Noise of Magnetic Drum Coatings in Automatic Devices, 1960. Developed theory of plane plastic tense condition, method for solution problems of plane terminal equilibrium of loose and cohesive media; proposed methods for analytical study of plasticity; solved problems of plane deformed conditions; studies in shells, statics of a loose medium. Home: B. Cheremushkinskaya 6/1. Office: Inst. Mechanics, USSR Acad. Scis., Leningradskii Prospekt, 7, Moscow, USSR.

SOLANDER, Daniel Charles, botanist, geologist; b. Norrland, Sweden, Feb. 12, 1736; studied under Linnaeus; probably received M.D., Upsala, Sweden; officer Brit. Mus., asst. librarian to catalogue natural history collections, 1763-68, became keeper natural history dept., 1773; accompanied Joseph Bank on Cook's voyage in Endeavor, 1768-71; visited (with Banks) Iceland, 1772. Island in Mergui Archipelago, also one S. on New Zealand, and a genus of Atropaceae named in his honor Fellow Royal Soc., 1764; mem. French Acad. Scis. Editor: Elementa Botanica (Linné), 1756; Natural History of Zoophytes (John Ellis), 1786. Introduced Linnean system to Eng. Died London, May 13, 1782.

SOLANO DE LUCQUES, Francisco, Spanish physician; b. Montilla, Spain, 1685; B.M., U. Granada (Spain), 1707; practice medicine, Illora, Spain, then Antequera, Spain, beginning 1717; hon. physician to Felipe V. Mem. Seville Acad. First to insist on use of pulse for prognosis and distinguished 3 types of pulse. Author: Origen morboso comn y universal, generante de los accidentes todos segun las doctrines del grande Hipócrates, 1718; Lapis Lydos Apollinis, 1722; Observaciones sobre el pulso. Died 1736.

SOLARINO, Giuseppe, Italian pathologist; b. Syracuse, Apr. 8, 1904; s. Giovanni and Concettina (Genovesi) S.; M.D.; m. Immacolata Vizioli, 1933; children—Giovanni, Concetta, Piervittorio, Nicola, Michele. Prof., U. Medicine; now dir. U. Bari Pathology Inst.; co-dir. Pathologica mag. Decorated knight Order Crown Italy, Crder Merit Italian Republic. Mem. Italian Pathology Soc., Internat. Microbiol. Soc., Bari Acad. Sci. (pres.); corr. mem. Rome Acad. Medicine. Author: Eritropenia digestiva, 2d edit., 1932; La plasmalipasi ematica, 1950; Vitamine e difese organiche, 1957; Sulla protezione chimica e biologica del danno da radiazioni jonizzanti, 1961; Gli antibiotici endogeni, 1961. Co-editor: Giornale Italian Patol. Sci. Home: via Sparano 35. Office: Ist. Pat. Gen. Univ. Bari, Palazzo, Ateneo, Bari, Italy.

SOLBRIG, Otto T(homas), botanist, biosystematist; b. Buenos Aires, Argentina, Dec. 21, 1930; s. Hans Joachim and Rose M. (Muggleworth) S.; came to U.S., 1955, naturalized, 1961; student Universidad Nacional La Plata (Argentina), 1950-54; Ph.D., U. Cal. at Berkeley, 1959; m. Roberta Mae Chittum, Aug. 4, 1956; children—Hans J., Heide F. Staff, Gray Herbarium, Harvard, 1961-66, asso. curator, 1963-66, lectr. dept. biology, 1964-66; asso. prof. botany U. Mich., Ann Arbor, 1966—. Recipient Cooley award Am. Soc. Plant Taxonomy, 1962. Mem. Internat. Orgn. Plant Biosystematics (sec. gen. 1964—), Bot. Soc. Am., A.A.A.S., Genetics Soc. Am., Internat. Assn. Plant Taxonomists, Phi Beta Kappa, Sigma Xi. Author: Evolution and Systematics, 1966; also numerous articles. Research in genetical, chromosomal and chem. relationships of wild plants. Home: 2475 Devonshire St., Ann Arbor, Mich. 48104.*

SOLEIL, Jean Baptiste François, French optician; b. Paris, 1798; invented photoelectric microscope, saccharimeter, gonimeter. Died Paris, 1878.

SOLE-LLENAS, J., Spanish physician; b. Barcelona, Spain, Nov. 10, 1923; s. Francisco and Maria (Llenas) S.; B.S.; Balmes Inst., 1942; M.D., U. Barcelona, 1950; m. Teresa Solanellas, Oct. 8, 1951; children—Juan, Jose Oriol, Teresa. Asst. prof. Sch. Medicine de Barcelona, 1958—; dir. radiology, neuroradiology Neurol. Inst. Barcelona, Spain, 1964—. Mem. Spanish Soc. Neuroradiology (pres. 1962—), Acad. Med. Sci. Barcelona, World Fedn. Neurology, Spanish Soc. Neurology, Spanish Soc. Radiology. Author: Neuroradiologia, 1962; (with A. Wackenheim) Diagnostico Radioneurologico, 1967. Research, numerous publs. on radiology of developmental abnormalities of sacral canal, spinal roots. Home: 34 Maestro Falla, Barcelona 17. Office: Neurol. Inst., Dept. Radiology, 8, Llull, Barcelona 5, Spain.*

SOLEM, G(eorge) Alan, Am. zoologist; b. Chgo., July 21, 1931; s. George Oliver and Lillian (Kinloch) S:; B.S. magna cum laude, Haverford Coll., 1952; M.A., U. Mich., 1954, Ph.D., 1956; m. Barbara Ellen King, May 14, 1960; 1 son, Anders Erik. Research asst. Mus. Zoology, U. Mich., Ann Arbor, 1952-55; asst. curator lower invertebrates Field Mus. Natural History, Chgo., 1957-59, curator lower invertebrates, 1959—; teacher of evolution, Northwestern U., evenings 1967—. Mem. A.A.A.S., Ecol. Soc. Am., Am. Malacological Union. Author: Systematics and Zoogeography of the Land and Freshwater Mollusca of the New Hebrides, 1959; The Neotropical Land Snail Genera Labyrinthus and Isomeria, 1966; also articles. Research on systematics, evolution, phylogeny of non-marine mollusks, particularly Pacific Island and Neotropical faunas, revision of camaenid land snail genera Labyrinthus, Isomeria, Amphidromus, families Helicarionidae and Endodontidae; 1st comprehensive zoogeographic analysis of Oriental, Australian and Pacific Island land snail faunas. Home: 53 Elizabeth Lane, Barrington, Ill. 60010. Office: Field Mus. Natural History, Lake Shore Dr. and Roosevelt Rd., Chgo. 60605.*

SOLE SABARIS, Luis, Spanish geologist; b. Gava, May 18, 1908; s. Felip and Maria Sabaris; Dr. ès sc., U. Barcelona; m. Concepcion Sugranes, July 24, 1940. Prof. Barcelona Lycée, 1932-40, U. Grenada, 1940-43, U. Barcelona, 1943—. Decorated Palms Académiques. Mem. Barcelona Acad. Sci., Council Sci. Research. Publs. on geology and morphology of Spain and especially the Pyrenees. Address: calle Mallorca 293, Barcelona, Spain.

SOLIMAN, Fouad Atalla, Egyptian physiologist; b. Cairo, Egypt., Sept. 12, 1923; s. Atalla Soliman and A. Awadalla (El-Afi) S.; B.Sc. with distinction, Cairo U., Faculty Vet. Medicine, 1947; M.S., Mich. State U., 1950, Ph.D., 1952; m. Amira Soliman Abdel Malek, May 21, 1953. Vet. insp. Ministry of Agr., 1947; demonstrator Faculty Vet. Medicine, Cairo U., 1948-56, lectr., 1956-62, asst. prof., 1962—. Author: Animal Physiology, 19—; also numerous articles. Research on thyroid function as related to reproductive stat of animals, hormonal changes during bone fracture healing; discovered that at stage of estrus thyroid gland is most active, also in humans at ovulation time and factors regulating such changes. Home: 13 El Gamaa, Giza, Egypt. Office: Faculty Vet. Medicine, Giza, Egypt.*

SOLINUS, Gaius Julius, Roman geologist; author: Polyhistor, seu de mirabilibus mundi (originally Collectanea rerum memorabilium), (geog. summary of known world including origins, history, customs of nations), circa 200. Described stone jet common in Britain (Tanatus), absence of snakes in Ireland; introduced term mare Mediterraneum.

SOLLAS, William Johnson, English geologist; b. Birmingham, Eng., May 30, 1849; s. William Henry and Emma (Wheatley) S.; student Royal Coll. Chemistry, Royal Sch. Mines; M.A., St. John's Coll., Cambridge; hon. degrees, Dublin, Bristol, Oslo, Adelaide; m. Helen Corin, 1874; 2 daus.; m. 2d, Amabel Nevill Jeffreys Moseley. Univ. extension lectr., 1873-79; curator Bristol (Eng.) Mus., also lectr. geology Univ. Coll., Bristol, 1879-80, prof. zoology and geology, 1880-83; prof. geology and mineralogy Trinity Coll., Dublin, 1883-93; petrologist Geol. Survey of Ireland, 1983-97; prof. geology and paleontology Oxford U., 1897-1936, also fellow Univ. Coll. Recipient Bigsby medal, 1893; Wollaston medal, 1907; Royal medal, 1914. Fellow Royal Soc., 1899, Imperial Coll. Sci and Tech. (hon.); pres. Geol. Soc., 1908-10. Author: The Age of the Earth, 1905; The Rocks of Cape Colville Peninsula, New Zealand, 2 vols., 1905; Ancient Hunters, 1911; Tetractinellida (24th vol., reports of Challenger expdn.); also papers. In charge Royal Soc. expdn. to Funafuti in S. Pacific to test theories of coral reef origin by boring, 1896; explored Paviland cave, other caves, river terrace sites in western Europe; authority on paleolithic man; revised classification of fossil sponeges, described new species; devised new method for determining specific gravity of minute particles; studied minute structure of crystals, origin of flints, fresh water fauna, petrology and glacial features of Ireland, igneous rocks of Carlingford, pleochroic haloes in biotite of Leinster granites; devised method of grinding fossils for purposes of study. Died Oct. 20, 1936.

SOLLE, Gerhard Karl Hugo, German geologist, paleontologist; b. Koblenz, Germany, Apr. 3, 1911; s. Richard and Hildegard (Nädelin) S.; student U. Graz (Austria); Dr.phil.nat., U. Frankfort/Main (Germany), 1937; m. Gisela Schwabe, May 8, 1959; children—Renate (Mrs. Joachim v. Köppen), Heike, Eva. Staff, U. Frankfort/Main, 1937-54, apl. prof. geology and paleontology, 1948-54; prof. Tech. U., Darmstadt, Germany, 1954—, dir. Geologisch-Paleontologisches Institut, 1954—. Mem. numerous geol. and paleontol. assns. in Germany, France, Eng., U.S., Australia. Editor: Rudolf Richter Festschrift, 1957. Research, publs. on stratigraphy and paleontology of the Devonian. Home: 1 Waldmühlenweg, 61 Darmstadt, Fed. Germany.*

SOLLNER, Karl, chemist; b. Vienna, Austria, Jan. 9, 1903; s. Anton Maria and Julie (Karplus) S.; Ph.D., U. Vienna, 1926; m. Herta Rosenberg, July 23, 1934; 1 dau., Barbara. Came to U.S., 1937, naturalized, 1943. With Kaiser Wilhelm-Inst. für physik Chemie., Berlin, Germany, 1927-33; privatdozent für chemie U. Berlin, 1933; research guest U. Coll., London, Eng., 1933-37; research chemist Cornell U., Ithaca, N.Y., 1937-38; faculty U. Minn. Med. Sch., Mpls., 1938-63, prof., 1947-63; prin. research analyst to head sect. on electrochemistry and colloid physics NIH, Bethesda, Md., 1947—. Fellow N.Y. Acad. Scis., Am. Inst. Chemists, A.A.A.S.; Washington Acad. Sci.; mem. Am. Chem. Soc., Electrochem. Soc., Soc. Gen. Physiologists, Sigma Xi. Research, numerous publs. on mechanisms of high intensity ultrasonics in colloidal systems, ultrasonic cavitation as cause of emulsification, dispersion of solids and fog formation, ultrasonic orientation, accumulation and coagulation, elucidation of mechanisms of electrostenolysis and anomalous osmosis, membranes of highest electrochemical activity, permselective and liquid ion exchanger membranes of highest ionic selectivities and high transmissivities, membrane electrodes. Home: 3714 Manor Rd., Chevy Chase, Md. 20015. Office: NIH, Bethesda, Md. 20014.*

SOLOFF, Louis Alexander, physician; b. Paris, France, Oct. 2, 1904; s. Abraham and Rebecca (Wagenfeld) S.; came to U.S., 1906, naturalized, 1930; B.A., U. Pa., 1926; M.D., U. Chgo., 1930; m. Mathilda Robin; 1 dau., Joann (Mrs. Robert Green). Dir. pathology St. Joseph's Hosp., Phila., 1933-45; dir. pathology Eagleville Sanitorium, Norristown, Pa., 1933-45; with Temple U. and Hosp., Phila., 1930—, chief div. cardiology, 1956—, prof. medicine, 1966—. Mem. A.C.P., A.M.A., Am. Heart Assn., Am. Fedn. for Clin. Research, Pa. Med. Soc., Am. Heart Assn. Council on Clin. Cardiology, Am. Assn. U. Profs., Assn. Univ. Cardiologists. Research, numerous publs. on cardiovascular disease. Home: 1901 Walnut St., Phila. 19103. Office: 3400 N. Broad St., Phila. 19140.*

SOLOMON, David Harris, Am. physician; b. Cambridge, Mass., Mar. 7, 1923; s. Frank and Rose (Roud) S.; A.B., Brown U., 1944; M.D., Harvard, 1946; m. Ronda L. Markson, June 23, 1946; children—Patricia Jean (Mrs. Richard E. Sinaiko), Nancy Ellen. Fellow in endrocinology New Eng. Center Hosp., Boston, 1951-52; faculty medicine U. Cal. Los Angeles Sch. Medicine, 1952—, prof., 1966—; chief med. service Harbor Gen. Hosp., Torrance, Cal., 1966—; attending physician VA Hosp., Los Angeles, 1952—. Cons. Fresno (Cal.) County Hosp.; cons. USPHS Metabolism Tng. Com., 1960-64; mem. adv.

com. Casa Loma Coll., Pacoima, Cal., 1967—. Mem. Am. Soc. Clin. Investigation, Am. Fedn. Clin. Research, Am. Physiol. Soc., Western Soc. Clin. Research (councillor 1963-66), Endocrine Soc., Am. Diabetes Assn., Am. Thyroid Assn., Soc. Exptl. Biology and Medicine, A.C.P. Los Angeles Soc. Internal Medicine (exec. council 1960-62), Am. Bd. Internal Medicine, Diabetes Assn. So. Cal., Los Angeles Endocrine Club, A.A.A.S., Western Assn. Physicians, Phi Beta Kappa, Sigma Xi, Alpha Omega Alpha. Research, numerous publs. on endocrinology including studies of thyroid physiology, actions of thyroid-stimulating hormone on thyroid gland, mechanism of action of antithyroid drugs, mechanisms for disposal of thyroid-stimulating hormone by body, significance of antibodies against thyroid-stimulating hormone, studies of adrenal physiology, pituitary physiology, disease, human hyperthyroidism (Graves' disease) have shown that this disease is probably autoimmune in origin; demonstrated that glomerulus was part of kidney which provided antigen which led to exptl. nephrotoxic nephritis in rats. Home: 317 McCarty Dr., Beverly Hills, Cal. 90212. Office: Harbor Gen. Hosp., 1000 W. Carson St., Torrance, Cal. 90509.*

SOLOMON, John Brian, immunologist; b. Cambridge, Eng., July 26, 1924; s. John Francis and Emily (Blake) S.; B.Sc. with honors in Chemistry, Nottingham (Eng.) U., 1950; Ph.D. in Biochemistry, London (Eng.) U., 1955; m. Kathleen Audley Pemberton, Mar. 31, 1948; children—Stephen Teague, Gale Rosalind. Biochemist, Glaxo Research Labs., Greenford, Middlesex, Eng., 1950-51; fellow Prof. J. Brachet's Labs., Burssels, Belgium, 1955; staff unit for chem. embryology Chester Beatty Research Inst., London, 1955-65; sr. research fellow microbiology dept. U. Adelaide (South Australia), 1965-67; research fellow Inst. Exptl. Biology and Genetics, Czechoslovak Acad. Sci., Prague, 1963. Mem. Brit. Soc. for Immunology, Internat. Inst. Embryology, Brit.. Soc. Developmental Biology, Australian Soc. Microbiology. Author: The Biochemistry of Animal Development, 1965; also articles. Research on metabolism of naphthalene, enzymes and function in embryos, immunity of young, active and passive immunity, transplantation reaction and induction of immunological tolerance in embryos, trophoblast in cancer problem. Address: Dept. Bacteriology, Med. Sch., Foresterhill, Aberdeen, Scotland.*

SOLOMON, Richard Lester, Am. psychologist; b. Boston, Mass., Oct. 2, 1918; s. Frank and Rose (Roud) S.; A.B., Brown U., 1940, M.Sc., 1942, Ph.D., 1947; M.A., Harvard, 1951; m. Sara-Grace Hahn, Feb. 14, 1944; children—Janet E., Elizabeth Grace. Instr., Brown U., 1946-47; faculty Harvard, 1947-60, prof., 1957-60; prof. exptl. psychology U. Pa., Phila., 1960—; with OSRD, 1944-45. Guggenheim fellow, 1957-58. Mem. Am., (pres. div. exptl. psychology 1965—), Nat. Acad. Scis., Eastern (past pres.) psychol. assns., Psychonomic Soc., Soc. Exptl. Psychologists. Research, publs. on conditioning and learning. Office: 3815 Walnut St., U. Pa., Phila. 19104.*

SOLON, Greek astronomer, mathematician; b. Salamis, Greece, circa 639 B.C.; Athenian parentage; desc. of Codrus; Became archon (highest magistrate), Athens, 594 B.C.; made voyage to Egypt, Cyprus, Lydia. Introduced leap month into Athenian calendar, 594 B.C.; studied astronomy; formulated code of laws, basis of laws of 12 tables, Rome. Died 559, B.C.

SOLOTOROVSKY, Morris, Am. bacteriologist; b. N.Y.C., Oct. 10, 1913; s. Samuel and Ida (Feinstein) S.; B.S., U. Va., 1934; M.S., N.Y. U., 1938; Ph.D., Columbia, 1946; m. Mary Louise Whitty, Aug. 27, 1945; children—Peter, Nina, Emilie, Julian. Research asst. Columbia Coll. Physicians, 1937-41, instr., 1942-43; exptl. plant operator Guggenheim Bros., N.Y.C., 1941-42; research asso. in bacteriology Merck Sharp & Dohme Labs., Rahway, N.J., 1946-58; prof. bacteriology Rutgers U., New Brunswick, N.J., 1958—. Fellow A.A.A.S., Am. Acad. Microbiology; mem. Am. Assn. Immunologists, Am. Pub. Health Assn., Am. Soc. Microbiology, Am. Thoracic Soc. Author: (with Hubert Lechevalier) Three Centuries of Microbiology, 1965; also numerous articles. Editor: Cell Culture in the Study of Bacterial Disease, 1965. Research in chemotherapy of Tb, devel. of various antibodies such as novobiocin and oxamycin, host-parasite interaction-applications of cell culture in study of mode of action of toxin and intracellular parasitism. Home: Heather Lane, Princeton, N.J. 08540.*

SOLTERER, Josef, economist; b. Vienna, Austria, Dec. 28, 1897; s. Carl B. and Maria (Widhalm) S.; B.S., So. Coll., Hattiesburg, Miss., 1927; A.M., Georgetown U., 1929, Ph.D., 1932, LL.D., 1957; D.Sc, U. Catolica de Chile, 1955; m. Hortense M. McClure, June 15, 1929 (dec. Oct. 1952); 1 son, Carl F.; m. 2d, A. Elizabeth Curran, June 14, 1955; 1 dau., Helen Maria. Came to U. S., 1924, naturalized, 1932. Navigation officer KPM, Java, Dutch East Indies, 1920-23; instr. fgn. lang. So. Coll., Hattiesburg, Miss., 1924-27; instr. to prof. econs. Georgetown U., 1929—, chmn. dept., 1940—. Cons. Credit Ceops., Haiti, 1945- cons., lectr. Econ. Orgn., Instituto de Estudios Superiories, Montery, Mexico, 1949, U. Catolica of Chile, 1954; Fulbright prof. univs. Innsbruck, Vienna, 1955. Commd. freg. lt., Austro-Hungarian Navy, 1917. Mem. Am., Cath. (pres. 1949-50) econ. assns., Econometric Soc., Soc. Indsl. and Applied Mathematics, Am. Statis. Assn. Author: Grundlagen der Pluralistischen Wirtschatt, 1964. Research on stable relations within an innovating economy, such as income distribution, relation to public policy. Home: 518 Meridian Av., Arlington, Va. 22213. Office: Georgetown U., Washington 20007.*

SOLTI, Francis, Hungarian cardiologist; b. Budapest, Hungary, Oct. 29, 1921; s. John Spilak and Elisabeth (Abel) S.; ed. U. Med. Sch. Budapest, 1941-47, candidate Med. Scis., 1958; m. Hilda Gosztonyi, May 12, 1959. Staff dept. medicine U. Med. Sch., Budapest., 1947—, leader cardiological lab., 1951—. Mem. Hungarian Cardiological Soc. Research, publs. on regulation of cerebral and extremital blood circulation; introduced new method for determination of cerebral and limb blood flow on human beings and animals. Home: 54-56. Ulloi, Budapest VIII. Office: 2/a Koranyi S. Budapest VIII, Hungary.*

SOLVAY, Ernest, Belgian chemist; b. Rebecq, Brussels, Belgium, Apr. 16, 1838; s. Alexandre Solvay; began as helper in gas factory, Brussels; Solvay and Co. was founded in Brussels, 1863. Founder, Institut Solvay, Brussels. Mem. French Acad. Scis. Invented process for prodn. of sodium carbonate (named after him), 1863. Died Brussels, May 26, 1922.

SOMBART, Werner, German economist, sociologist; b. Ermsleben-am-Harz, Germany, Jan. 19, 1863; s. Anton Ludwig Sombart; ed. univs. Riga and Berlin; Ph.D., LL.D. Sec-gen. Bremen (Germany) C. of C., 1888; asst. prof. U. Breslau, 1890-1906; prof. Berlin Coll. Commerce, 1906-17; prof. U. Berlin, 1917-31, emeritus, 1931-41. Author: Sozialismus und soziale Bewegung im 19. Jahrhundert, 1896; Das Lebenswerk von Karl Marx, 1900; Technik und Wirtschaft, 1901; Der moderne Kapitalismus, 3 vols., 1902-27; Die deutsche Volkswirtschaft im 19. Jahrhundert, 1903; Die Juden und das Wirtschaftsleben, 1911; Kreig und Kapitalismus, 1912; Luxus und Kapitalismus, 1913; Der Bourgeois, 1913; Der proletarische Sozialismus, 1924; Die drei nationalökonomien, 1930; Deutscher Sozialismus, 1934; Weltanschauung, Wissenschaft und Wirtschaft, 1938; Anthropologie, 1938. Research on western econ. history and origin of trend from capitalism to socialism; classified hist. devel. dialectically according to periods of culture; expanded Simmel's theory on importance of money economy in specializing and depersonalizing social relationships; influenced by Marxist econ. thought; late in life accepted authoritarian state and German Nat. Socialism. Died Berlin, May 18, 1941.

SOMERS, George Fredrick, Am. plant physiologist; b. Garland, Utah, July 9, 1914; s. George Fredrick and Elizabeth (Sorenson) S.; B.S., Utah State U., 1935; B.A. with honors, Oxford U., 1938, B.Sc., 1939; Ph.D., Cornell U., 1942; m. Beulah Rich Morgan, June 24, 1939; children—Ralph M., Steven J., Gary F. Faculty, Cornell U., 1944-51, asso. prof. biochemistry, 1949-51; asso. dir. Del. Agrl. Expt. Sta., 1951-59; faculty U. Del., Newark, 1951—, chmn. dept. biol. scis., 1959—, H. Fletcher Brown prof. biology, 1962—; vis. prof. U. Philippines, 1958-59. Union Pacific scholar, 1930; Rhodes scholar, 1936; Henry Strong Denison fellow, 1939. Fellow A.A.A.S.; mem. Am. Chem. Soc., American Institute of Biological Scis., American Assn. Plant Physiology, Bot. Soc. Am., Soc. Gen. Physiology, Sigma Xi, Phi Kappa Phi. Author: (with J.B. Sumner) Chemistry and Methods of Enzymes, 3d edit., 1953; Laboratory Experiments in Biological Chemistry, 2d edit., 1949; also articles. Co-editor biochem. sect. Chem. Abstracts, 1963—. Research on metabolism in higher plants as related to organic acids, carbohydrates and vitamins; influence of environmental factors on vitamin C content of vegetables; developed integrating sunlight recorder for field studies; viscoelasticity in plant tissues as related to cell wall properties. Home: 22 Minquil Dr., Newark, Del. 19711.*

SOMERVILLE, Mary Fairfax, mathematician, physicist; b. Manse of Jedburgh, Scotland, 1780; d. Sir William and Margaret (Charters) Fairfax; ed. boarding-sch., Musselburgh, Scotland; mainly self-educated; m. Samuel Greig, 1804; 1 son, Woronzow; m. 2d, William Somerville, 1812; children—Martha, Mary. Recipient Victoria Gold medal Geog. Soc., 1869, award Italian Royal Geog. Soc.; Somerville Hall, Mary Somerville scholarship for women in math. and Somerville Coll. at Oxford, named in her honor; her bust placed in great hall of Royal Soc.; presented with pension by Sir Robert Peel. Hon. mem. Royal Astron. Soc. Author: The Connection of the Physical Sciences, 1834; Physical Geography, 1848; Molecular and Microscopic Science, 1866; Translator: Mechanism of the Heavens (Laplace's), 1831; paper to Royal Soc., The Magnetic Properties of the Violet Rays of the Solar Spectrum. Died Naples, Italy, Nov. 29, 1872.

SOMMER, Otto Ambrose, German agronomist; b. Aschaffenburg, Germany, June 6, 1902; s. Heinrich and Pauline (Aulbach) S.; Dr.agronomy, Hohenheim Agrl. Sch.; m. Hanne Henze, Aug. 14, 1929; children—Hannelore, Heiner, Barbara, Alfred. Successively asst., prof. agronomy, adminstr. in cattle raising, instr. cattle raising; prof. U. Göttingen; with Hohenheim Sch. Agronomic Sci., State Agr. Sta.; dir. Institut für Konstitutionsforschung Brunswick-Völkenrode; prof. cattle raising Munich-Freising-Weihenstephan Tech. Sch. Author books on cattle raising. Research and numerous publs. on raising of, biology of, and selective breeding of cattle and hogs. Home: Hohenbacherstrasse 13, Freisingen-Vöttingen. Office: Weihenstephan Tierzucht-Institut, Freising, W. Germany.

SOMMERFELD, Arnold Johannes Wilhelm, German physicist; b. Königsburg (now Kaliningrad, USSR), Dec. 5, 1868; s. Franz and Cäcilie (Mathias) S.; Ph.D., Konigsberg U., 1891; m. Johanna Höpfner, Dec. 1897; 2 sons, 1 dau. Became asst. mineralogy Göttingen (Germany) U., 1891, instr., 1895; named prof. math. Mining Acad., Clausthal, Germany, 1897; named prof. mechanics Inst. Tech., Aachen, Germany, 1900; prof. theoretical physics Munich U., 1906-40; vis. prof. U. Wis., 1922-23, Cal. Inst. Tech., 1928, U. Mich., 1931. Recipient Planck medal, 1931, Lorentz medal, 1939, Oersted medal, 1948. Fellow Royal Soc., 1926; mem. Nat. Acad. Scis. (Washington), Am. Acad. Arts and Sci. acads. of Berlin, Calcutta, Göttingen, Madrid, Munich, Rome, and Vienna. Author: (with Felix Klain) Theory of the Gyroscope, 1897-10; Atombau und Spektrallinien, 1919; Wave- Mechanics, 1929; Vorlesungen über theoretische Physik, 1943-48; also articles on math. and physics. Research on wave character of X-rays; originated formula for structure of spectral lines and gen. quantum theory of spectral lines, 1916; developed theory of metallic electrons, 1927; studied wave spreading in wireless telegraphy, theory of gyroscope (with Kelin); application of quantum theory to Bohr atom model and spectroscopy. Died Munich, Apr. 26, 1951.

SOMMERS, Henry Stern, Am. phyicist; b. St. Paul, Apr. 21, 1914; s. Henry S. and Helen (James) S.; student Stanford, 1932-34; B.A., U. Minn., 1936; Ph.D., Harvard, 1941; m. Nancy Trenholm, July 12, 1938; children—Ann H. (Mrs. Jerol Harrington), Craig T., Heather (Mrs. Edward B. Sussman), Henry S. III. Instr. physics Harvard, 1941-42; staff mem. Radiation Lab., Mass. Inst. Tech., 1942-45; asst. prof. physics Rutgers U., 1946-49; staff mem. Los Alamos Sci. Lab., 1949-54; staff mem. RCA Labs., Princeton, N.J., 1954—. Mem. Princeton Study Center, 1962—. Recipient Naval Ordinance Devel. award, 1945. Guggenheim fellow, 1960-61. Fulbright fellow, 1959-60. Research in nuclear physics, low temperature physics, photoconductivity semiconductor devices, instrumentation. Patentee in field. Home: 207 Riverside Dr. Office: RCA Labs., Princeton, N.J. 08540.*

SOMMERS, Sheldon Charles, Am. pathologist; b. Indpls., July 7, 1916; s. Charles Birk and Leonore (Dickey) S.; B.S., Harvard, 1937, M.D., 1941; m. Edith Briggs, Nov. 9, 1943. Pathologist, Mass. Meml. Hosps., 1953-61, Scripps Meml. Hosp., La Jolla, Cal., 1961-63; asso. dir. labs., Delafield Hosp., New York City, 1963-67, director laboratories since 1967—; instr., asso., lectr. Harvard Med. Sch., 1950-61; asso. prof. pathology Boston U. Sch. Medicine, 1953-61; clin. prof. U. So. Cal. Sch. Medicine, 1962; asso. prof. pathology Columbia U. Coll. Phys. and Surg., 1963-65, prof., 1965—. Cons. VA hosps. Mem. New England Path. Soc. (past pres.), Bergen City Med. Soc., Am. Assn. Pathologists and Bacteriologists, Am. Soc. Clin. Pathologists, Coll. Am. Pathologists, Internat. Acad. Pathology, Histochem. Soc., Am. Soc. Exptl. Pathology, N.Y. Acad. Medicine, Soc. for Exptl. Biology and Medicine. Author: (with Langdon Parsons) Gynecology, 1962; also numerous articles. Editor: Pathology Annual, 1966—. Research on gen. and exptl. pathology, lesions and devel. regional ileitis, ulcerative colitis, stomach cancer, endometrial cancer in human, radiation reactions, cancer transplants, endocrine pathology, thyroid lesions, kidney changes with hypertension and pyelonephritis, renal juxtaglomerular cells and relation to hypertension, red cells and sodium, pituitary cytology, hypothalamic malformations, idiopathic pulmonary hemosiderosis, kidney changes in diabetes. Home: Cambridge Way, Alpine, N.J. Office: 99 Ft. Washington Av., N.Y.C. 10032.*

SOMMERVILLE, Duncan M'Laren Young, mathematician; b. Beawar, Rajputana, India, Nov. 24, 1879; s. James Sommerville; ed. Perth Acad.; M.A., D.Sc., St. Andrew's (Scotland) U.; m. Louisa Agnes Beveridge, 1912. Lectr. math. St. Andrew's U., 1902-14; prof. pure, applied math. Victoria U. Coll., Wellington, New Zealand, from 1915. Fellow New Zealand Inst. (Hector medal 1928), Royal Soc. Edinburgh, Royal Asiatic Soc.; mem. Edinburgh Math. Soc. (pres. 1911-12), Australasian Assn. Advancement Sci. (pres. Sect. A 1924). Author: Bibliography of Non-Euclidean Geometry, 1911; Elements of Non-Euclidean Geometry, 1914; Analytical Conics, 1924; An Introduction to the Geometry of n Dimensions, 1929; Analytical Geometry of Three Dimensions, 1933; memoirs in math. jours. Died Jan. 31, 1934.

SOMOGY, Stefano, statistician; b. Miskolc, Hungary, July 14, 1904; s. Adolfo and Giovanna (Leopold) S.; ed. Budapest, Berlin, U. Padua; Dr. ès soc.

sc.; m. Lydia Paroletti, July 31, 1932; children—Giovanni, Pietro. Became prof. demography, 1935, prof. san. statistics, 1955; head math. bur. Central Statistics Inst.; dir., later insp. gen. Econ. and Financial Statis. Services; demographic and san. statistics. Mem. Central Italian Biostatistics Assn. (pres.), Italian Medico-San. Statis. Soc. (v.p.), Italian Inst. Social Protection Studies, Italian Parapsychol. Soc. (pres.), Internat. Inst. Statistics, Internat. Union Demographic Studies. Author: Mutamenti demografici del Nord e Sud d'Italia e consequenze socioeconomiche, 1962; I bilanci demografici delle regioni italiane dal 1861 al 1961; Problemi sanitriali in aree sottosviluppata, 1963; Per un registro oncologico nazionale, 1963; Distribuzione territoriale dei morti per tumori, 1963. Home: Via Montesante 25, Rome. Office: via Marchese Ugo 57, Palermo, Italy; also Societa Italiana di statistische medico-santiari, Largo Corrado Ricci 44, Rome, Italy.

SONCK, Eric Carl, Finnish dermatologist; b. Wiborg, Nov. 10, 1905; s. Karl Joel and Siri Ingrid (Lubeck) S.; M.D., U. Turku (Finland); m. Karin Hertwig, Aug. 3, 1935; children—Christel, Klaus, Johan, Henrik. Specialist in dermatology and venerology; became instr. U. Turku, 1944, prof. dermatology and venerology, 1955—. Mem. Finnish Med. Soc., German Dermatology Assn. (hon.), Finnish Flora and Fauna Soc. Editor, Excerpta Medica. Research and publs. on mycology, histopathology of skin diseases, lymphogranuloma inguinale, venerology. Home: Linnankatu 19, Turku, Finland. Office: Keskussairaala, Turku, Finland.

SONDHAUSS, Karl Friedrich Julius, physicist; b. Breslau (now Wroclaw, Poland), 1815; dir. Royal Sch., Neisse. Studied tone vibrations; used ball filled with carbon dioxide to prove breaking of sound waves. Died 1886.

SONDHEIMER, E(rnst) H(elmut), math. physicist; b. Stuttgart, Germany, Sept. 8, 1923; s. Max and Ida (Oppenheimer) S.; M.A., U. Cambridge (Eng.), 1948, Ph.D., 1949; m. Janet Harrington Matthews, Aug. 18, 1950; children—Julian Philip, Judith Anne Sophia. Fellow Trinity Coll., Cambridge, 1948-52; lectr. math. Imperial Coll., U. London (Eng.), 1951-55, reader applied math. Queen Mary Coll., 1955-60, prof., head dept. math. Westfield Coll., 1960—; external examiner in math. physics U. Birmingham (Eng.), 1961-64, U. Edinburgh (Scotland), 1965—. Fellow Am. Phys. Soc., Phys. Soc. London; mem. London Math. Soc. (mem. council 1965—.). Research, publs. on electron theory of metals. Home: 51 Cholmeley Crescent, London N.6 Eng.*

SONDHI, Keshav C., biologist; b. Miani, West Punjab, India, July 27, 1930; s. Gurucharan Singh and Brij Rani (Chadha) S.; I.Sc., Holkar Coll., Indore, India, 1948; B.Sc., Maharaja's Coll., Jaipur, India, 1950; M.Sc., Birla Coll., Pilani, India, 1952; Ph.D., Agra (India) Coll., 1959, U. Coll., London, Eng., 1959; m. Gunthild Edeltraud Slawitzsch, Oct. 2, 1961. Scientist collaborator dept. zoology and comparative anatomy U. Coll., 1958-61; geneticist New Eng. Inst. for Med. Research, Ridgefield. Conn., 1961-62; vis. scientist gerontology br. NIH, Bethesda, Md., and scientist gerontology br. NIH, Bethesda, Md., and Balt. City Hosps., 1962-63; vis. asst. prof. dept. zoology Rutgers U., Newark, 1963-66, asso. prof., 1966—. Rutgers Research Coun. fellow, 1966-67. Fellow Acad. Zoology India (foundn., life), Zool. Soc. (London), Genetical Soc. (U.K.), Soc. for Exptl. Biology (U.K.), Genetics Soc. Am., Am. Soc. Zoologists, A.A.A.S. Research, publs. on internal orgn. tongue in reptiles, genetic control animal growth and form, physiol. genetics aging. Home: 39 A Cedar Grove Gardens, Cedar Grove, N.J. 07009. Office: 40 Rector St., Newark 07102.*

SONEA, Sorin, microbiologist; b. Cluj, Rumania, Mar. 14, 1920; s. Leonida Eusebiu and Cornelia (Chibulcuteanu) S.; M.D., U. Bucarest, 1944; diploma in Hygiene, U. Paris, 1949; m. Rodica Vlad, Feb. 2, 1946; children—Ioana, Peter, Michael, Alexander. Faculty, U. Montreal, Que., 1950—, head dept. microbiology and immunology, 1964—. Research, numerous publs. on mechanisms of virulence of some pathogenic bacteria and genetic and chem. factors influencing this virulence, fluorescence of bacteria. Home: 4282 Badgley St., Montreal, Que., Can.*

SONI, Narendar Nath, radiobiologist; b. Srinagar, India, Mar. 10, 1927; s. Ram Rakha Mal and Sarasvati Devi (Nanda) S.; B.D.S., Sir Currimby Ebraham Meml. Dental Coll., Bombay, India, 1949; M.S., U. Rochester, 1959; M.S.D., Ind. U., 1961; m. Hilda F. D'Cruz, Aug. 16, 1956; children—Anil M., Rajiv P. Research fellow Indian Council Med. Research, Bombay, 1950; clin. fellow Eastman Dental Dispensary, Rochester, N.Y., 1957-59; grad. asst. Ind. U. Sch. Dentistry, Indpls., 1960-61; asst. prof. pedodontia Howard U. Coll. Dentistry, Washington, 1962-65, asso. prof., 1965—. Mem. Internat. Assn. Dental Research, Am. Acad. Periodontology, Sigma Xi. Contbr. articles to sci. jours. Polarizing and microradiographic study of sound, carious and etched enamel, x-ray densitometric study of cementum, mitotic activity in human gingival epithelium, enamel decalicification in food-saliva mixture, microradiographic study of odontologic tissues in Cooley's anemia. Home: 1504 21st st. N.W., Washington 20036.*

SONKA, Jiri, Czechoslovakian physician; b. Prague, Czechoslovakia, Dec. 28, 1920; s. Vojtech and Frantiska (Faltusova) S.; M.D., U. Prague, 1947; Ph.D., Charles U., Prague, 1956, D.Sc., 1964; m. Libuse Neumannova, Dec. 28, 1954; children—Karel, Klara. Faculty, Charles U., 1953—, asso. prof. Lab. Endocrinology and Metabolism, 1964—, teaching staff 3d Med. Clin, Faculty Medicine, 1952—. Author: Pentosy-chemie, fysiologie a klinika, 1956; (with Dienstbier, Arient, others) Experimental Radiation Disease, 1966; (with Dobersky, Dolecek) Obesity, 1967; also numerous articles. Research on role of pentose cycle and its regulators in physiology and pathogenesis of diseases. Home: 26 Na Kodymce Prague 6. Office: Lab. Endocrinology and Metabolism, 1 Unemocnice, Prague 2, Czechoslovakia.*

SONNEBORN, Tracy Morton, Am. geneticist; b. Baltimore, Md., Oct. 19, 1905; s. Lee and Daisy (Bamberger) S.; A.B., Johns Hopkins, 1925, Ph.D., 1928, D.Sc., 1957; m. Ruth Meyers, June 6, 1929; children—Lee M., David R. Research asst. zoölogy, Johns Hopkins, 1930-31, research asso., 1931-33, asso., 1933-39; asso. prof. zoölogy, Ind. U., 1939-43, prof. 1943-53, distinguished service professor Indiana U., 1953—, acting chmn. div. of biological scis., 1963-64. Discoverer of sexes in ciliated protozoa, 1937, roles of genes, cytoplasm and environment, control cell heredity, 1943. Co-winner prize ($1000) Am. Assn. Advancement Sci. for outstanding research paper, 1946; Kimber Genetics award Nat. Academy Scis., 1959. Mem. Am. Soc. Naturalists (pres. 1949), American Society of Zoologists (president 1956), Genetics Society of America (pres. 1949), Society for Study Evolution (v.p. 1958), Nat. Acad. Sci. (1946); Am. Philos. Soc. (Phila.), Am. Acad. of Arts and Sci. (Boston), Am. Soc. Protozoology, Royal Society of London (foreign member), Internat., Am. socs. cell biology, Soc. Growth and Development, Am. Eugenics Soc. (dir. 1957), Am. Inst. Biol. Scis. (pres. 1960-61), Phi Beta Kappa, Sigma Xi; hon. member Facultad de Biologiá y Ciencias Médicas, Universidad de Chile, Sociadad de Biologiá de Santiago de Chile, Sociadad de Biologiá de Concepción (Chile) Sociadad Médica de Concepcion (Chile), French Society of Protozoology. Member of the editorial bd. Journal of Morphology, 1946-49, Genetics since 1947; Journal Experimental Zoology since 1948; Physiological Zoology, 1948— member editorial committee Annual Review of Microbiology, 1954-58. Contbr. of articles to mags. and chpts. to books on genetics. Editor: The Control of Human Heredity and Evolution, 1965. Discovered sex types in one-celled animals and extended gene theory to them; also principles of interrelations between genes, rest of cell, and environment in control of development and heredity; also several types inheritance not due to genes, chromosomes; effect of aging on induction of mutations; implications of genetics for control human heredity, evolution. Home: 1305 Maxwell Lane, Bloomington, Ind. 47401.*

SONNENSCHEIN, Ralph R(obert), Am. physiologist; b. Chgo., Aug. 14, 1923; s. Robert and Flora (Kieferstein) S.; student Swarthmore Coll., 1940-42, U. Chgo., 1942-43; B.S., Northwestern U., 1943, M.S., 1946, M.D., 1947; Ph.D., U. Ill., 1950; m. Patricia W. Niddrie, June 21, 1952; children—David, Lisa, Ann. Faculty, U. Cal. at Los Angeles, 1951—, prof. physiology, 1962—. Mem. Am. Physiol. Soc., Soc. for Exptl. Biology and Medicine, A.A.A.S. Research, numerous publs. on regulation intestinal secretion, pain mechanisms, action analgesic drugs, physiology and pharmacology skin, oxygen poisoning, circulation brain and skeletal muscle. Home 18212 Kingsport Dr., Malibu, Cal. 90265. Office: Dept. Physiology, U. Cal. Sch. Medicine, Los Angeles 90024.*

SONNENSCHEIN, Robert, Am. physician; b. Chgo., 1879; M.D., Rush Med. Coll., 1901; otolaryngologist Michael Reese Hosp., 1926-39; clin. prof. otolaryngology Rush Med. Coll., Post Grad. Med. Sch., Chgo., 1933-39. Pres. Chgo. Laryngol. and Otol. Inst.; dir. Jewish People's Inst.; mem. Adv. Med. Bd., 3d Ill. dist., World War I. Fellow A.C.S.; mem. Am. Acad. Ophthalmology and Otolaryngology, Am. Laryngol. Assn., Am. Laryngol., Rhino. and Otol. Soc., Am. Otol. Soc. Contbr.: The Nose, Throat and Ear and Their Diseases (Jackson and Coastes), 1929; A Text Book of Surgery (Christopher). Research and publs. on problems of testing hearing, 1929, surgery of ear, 1935. Died 1939.

SONNERAT, Pierre, French biologist; b. Lyons, France, Aug. 18, 1748; mem. Commeron's expdns. to Madagascar, the Seychelles, Manilla; mem. French Acad. Scis. Author: Voyage à la Nouvelle-Guinée, 1776; Le Palmier des Seychelles. Introduced cultivation of bread tree, coco, other useful plants into French colonies; gave 1st detailed descriptions of flora and fauna of New Guinea; contbd. to knowledge of mammals of Madagascar. Died Paris, Mar. 31, 1814.

SOOD, Baldev Sinah, Indian physicist; b. Lahore, Pakistan, Oct. 7, 1929; s. N. S. and Channan (Kaur) S.; S.L.C., Punjab U., 1946, B.Sc. with honors, 1951, M.Sc. with honors, 1952; Inter Sci., Delhi U., 1948; Ph.D., U. Birmingham (Eng.), 1959 m. Rita Singh, Jan. 21, 1961; children—Deepak, Anjali. Lectr. physics dept. Punjab U., 1953-56, reader physics, 1960-63; Govt. of India Research scholar Birmingham U., 1956-59, research fellow, 1959-60; reader Punjab U., Patiala, 1963-64, prof., head physics dept., 1964—. Fellow Indian Phys. Soc. Research and publs. on application of techniques of beta and gamma ray spectroscopy to investigation in low-energy nuclear physics and solid state physics. Home: 4-A University Campus, Patiala, India.*

SOOD, Gian Chand, Indian ophthalmologist; b. Ludhiana, India, Apr. 5, 1923; s. Ram Chand and Atma (Devi) S.; L.M.S.Pb., India Inst. Med. Sci., 1944, D.O., 1958, M.S. in Ophthalmology, 1962; m. Vimla Rani, Dec. 11, 1949; children—D. Manorma, S. Rajneesh, S. Rakesh, D. Amita, D. Vijay, Luxmi, D. Aruna, D. Punum. Ophthalmic surgeon Sant Parmanand Blind Relief Mission, Delhi, India, 1946-51, Blind Relief Work, Kathmandu, Nepal, India, 1951-57; staff W. Wales Gen. Hosp., Carmarthen, 1959-60; pool officer Council Sci. and Indsl. Research, 1960-63; faculty Maulana Azad Med. Coll., New Delhi, India, 1963—, reader ophthalmology, 1963—; dir. Nepal Blind Relief Mission, 1951-57. Recipient 3 Silver medals, Gold medal for blind relief work Birbunj Janta, 1948. Mem. All India Ophthalmic Soc., Ophthal. Soc. U.K. Research, publs. on relationship of phylogeny in heterogenous keratoplasty; described new procedure in squint surgery; invented muscle clamp for lengthening operation. Home: 174-176 Khyber Pass, Delhi. Office: M. Azad Med. Coll., New Dehli, Dehli, India.

SOOYSMITH, Charles, Am. civil engr.; b. Buffalo, N.Y., July 20, 1856; s. William Sooy and Elizabeth (Haven) Smith; C.E., Rensselaer Poly. Inst., 1876; postgrad. Polytechnicum, Dresden, and other places in Europe, 1876-78; m. Pauline Olmstead, Dec. 17, 1887. Asst. supt. maintenance dept. A.,T.&S.F. R.R., 1879-80; pres. Sooysmith & Co., contracting engrs., 1884-1900, builders of many important subaqueous engring. works. Introduced into U. S. so-called freezing process for excavating, and took out many patents covering its application to building of subaqueous tunnels; inaugurated pneumatic caisson method for foundations of large buildings; served as expert in connection with underground works, notably with Underground Rapid Transit R.R. in N.Y.; mem. Met. Sewerage Commn. of New York. Died June 1, 1916.

SOPWITH, Thomas, English mining engr.; b. Newcastle/Tyne, Jan. 3, 1803; s. Jacob and Isabella (Lowes) S.; m. Mary Dickenson, 1828; m. 2d, Jane Scott, 1831; m. 3d, Anne Potter, 1858; 1 dau., Ursula (Mrs. David Chadwick). Land surveyor, engr., Alston, Eng.; (with William South) made Northumbrian survey; instrumental in founding of mining records office, 1838; made mining survey County Clare, Eire; employed on devel. ry., Belgium, 1843. Fellow Royal Soc. Author: A Historical and Descriptive Account of All Saints Church in Newcastle-upon-Tyne, 1826; Treatise on Isometrical Drawing, 1834; Eight Views of Fountains Abbey . . . with Description, 1832; An Account of the Mining Districts of Alston Moor, Weardale and Teesdale, 1833; Description of Monocleid Writing Cabinets, 1841; An Account of the Museum of Economic Geology, 1843; The National Importance of Preserving Mining Records, 1844; Education: its Present State and Future Advancement, 1853; Notes of a Visit to Egypt, 1857; Notes of a Visit to France and Spain, 1865; Education in Village Schools, 1868; Three Weeks in Central Europe, 1869. Fellow Royal Soc., 1845. Directed attention of Brit. Assn., Royal Soc. to sci. importance of recording geol. features exposed in cuttings of rys. (Brit. Assn. made grant for this purpose, 1840). Died Jan. 16, 1879.

SORANOS OF EPHESUS, Greek physician; b. Ephesus, flourished 2d century; studied in Alexandria; practiced medicine, Alexandria and Rome; physician under Trajan and Hadrian, 98-138. Author of almost 20 books dealing with med. history, gynecology, med. terminology, hygiene of midwife and conception, infant care, pathology, also bibliography of Hippocrates, and work on bandages. Founder of obstetrics and gynecology; considered ablest gynecologist of classical times; restored Methodical sch. by adjusting its exaggerations to his time, while blending it with tradition; believed in gen. symptoms, but gave attention to individual factors; accurately observed course of illness and distinguished different forms of diseases; his work was dominate influence during middle ages, and his works were widely read in the West.

SORBY, Henry Clifton, English chemist, geologist; b. Woodbourne, Eng., May 10, 1826; s. Henry and Amelia Sorby; ed. privately. Pres. Firth Coll., Sheffield, Eng., 1882. A founder of Sheffield U. Fellow Royal Soc., 1957. Author: On the Microscopic Structure of Crystals, 1858; On the Microscopic Structure of Iron and Steel. Inventor method of microscopic examination of metals, 1963; devel. technique for preparing rock slides for microscope; student biology, archaeology, Egyptian hieroglyphics, geology, spectroscopy. Died Sheffield, Mar. 9, 1908.

SOREL, Georges, French social, polit. philosopher; b. Cherbourg, France, Nov. 2, 1847; ed. École Polytechnique. Civil engr., until 1892; later studied, wrote. Author: L'Ancienne et la Nouvelle Métaphysique, 1894; Procès de Socrate, 1889; Les Préoccupations métaphysiques des physiciens modernes, 1905; Réflexion sur la Violence, 1908; De l'utilité du pragmatisme, 1917; Plaidoyer pour Lénine, 1919. Theorist of syndicalism in France; rejected nation state, sought soc.

based on indsl. unionism; stressed need for revolutionary violence and myths of proletariat (proposed gen. strike as best revolutionary method); influenced archaeology, Egyptian hieroglyphics, geology, spectroscopy. Died Sheffield, Mar. 9, 1908.

SÖREMARK, Rune Sven Arnold, Swedish dentist, biophysicist; b. Götteborg, Sweden, Mar. 14, 1926; s. Arnold G. and Ester (Magnusson) S.; L.D.S., Karolinska Institutet, 1953, Dr. Odont., docent biophysics, 1960; m. Ulla Löfberg, Feb. 6, 1954; children—Heléne, Josefine, Patrik. Faculty, dept. prosthetics Faculty Odontology, Karolinska Institutet, 1955-63, asso. prof., 1962-63; prof. prosthetics Faculty Odontology, U. Umea (Sweden), 1963—; vis. prof. dentistry, acting chmn. depts. prosthetics and roentgenology Harvard Sch. Dental Medicine, Boston, 1964—. Recipient Swedish Dental Soc. award, 1961, Miller prize, 1964. Mem. Soc. Nuclear Medicine, Internat. Assn. Dental Research, Am. Dental Assn., Swedish Dental Soc., Scandinavian Soc. Odontology. Research, on devel. of some methods for studying microelements in biol. tissues, improvement of techniques of X-ray diagnosis especially quantitative roentgenography, accumulation of various microelements in mammalian body, prosthetics. Address: 188 Longwood Av., Boston 02115.*

SORENSEN, Henning, Danish geologist; b. Copenhagen, Denmark, Apr. 20, 1926; s. J. P. and Anna M. (Thomsen) S.; M.Sc., U. Copenhagen, 1951, Dr.phil., 1962; m. Helle A. Abildgaard, Apr. 1, 1961; children—Lise, Hans. Sci. asst. U. Oslo (Norway), 1952; with U. Copenhagen, 1953—, prof., 1962—; lectr. mineralogy Tech. U. Denmark, 1954-61. Mem. various geol. expdns. to Greenland, 1946-64; head Inst. Petrology, U. Copenhagen. Mem. Geol. Soc. Denmark (pres. 1967—). Author: Vor Jordklode, 1963; (with H. Wienberg Rasmussen, A. Berthelsön, J. Espersen) Geologi, 4th edit., 1968; also articles. Research on ultramafic rocks from Greenland and Norway, alkaline rocks from south Greenland and their rare minerals; dir. research project in S. Greenland on deposits of uranium, thorium, beryllium, niobium, rare earth minerals. Office: 5 Ostervoldgade, Copenhagen K 1350, Denmark.*

SORENSEN, Peder, (Severinus), Danish chemist, physician; b. Ribe, Denmark, circa 1540; student, Padua, Italy, 1566, Wittenburg, Germany, 1569; M.D., France, circa 1570. Became sgt.-surgeon to Ct. of Frederick II, 1571, later Christian IV. Author: Idea medicinae philosophicae. Expanded ideas of Paracelsus about invisible agts. of disease. Died 1602.

SÖRENSON, Sören Peer Lauritz, Danish biochemist; b. Havrebjerg, Denmark Jan. 9, 1868; Ph.D., Copenhagen, Denmark, 1899; dir. chemistry Carlsberg Lab.; prof., Copenhagen. Invented pH symbol for negative logarithm of hydrogen-ion concentration on Sörenson scale (Sorensen's symbol); pioneered research on hydrogen-ion concentration; research on amino-acids, proteins, enzymes, fermentation. Died Copenhagen, Feb. 1939.

SORGENFREI, Theodor, Danish geologist; b. Augustenborg, Denmark, Dec. 8, 1915; s. Theodor and Johanne (Dominicussen) S.; M.S., U. Copenhagen (Denmark), 1941, D.Sc., 1958; m. Inger Margrethe Petersen, Feb. 7, 1939; children—Lis (Mrs. Per Hellung Larsen), Steen, Henrik, Tim. Geologist, Geol. Survey Denmark, 1934-42, sect. geologist, 1942-54, state geologist, 1954-61; prof. tech. geology Tech. U. Denmark, Lyngby, 1961—; cons. geologist Danish Am. Prospecting Co., 1946-58, A.P. Moller Cos., 1962—; vis. prof. La. State U., 1965. Spl. lectr. U. London (Eng.), 1961, U. Stockholm (Sweden), 1965. Decorated knight Order of Danebrog (Denmark). Named chevalier Ordre des Palmes Académiques, France, 1964. Mem. Internat. Geol. Congress (gen. sec. XXI session in Nordic countries 1960), Internat. Union Geol. Scis. (sec. gen. 1961-64), Danish Acad. Tech. Scis., Royal Danish Acad. Scis. and Letters. Author: Molluscan Assemblages from the Marine Middle Miocen of South Jutland and their Environments, 1958; also articles. Research on regional and local geology strucutral geology, applied geology Denmark, biology extinct faunas Tertiary Europe. Home: 71 Kvaedevej, Virum, Denmark. Office: 100 Lundtoftevej, Lyngby, Denmark.*

SORIA-V, Jorge, geneticist; b. Patate, Ecuador, May 20, 1925; s. Jose S. and Celia (Vasco) S.; Ing. Agr., Central U., Quito, 1952; Magister Agriculturae, Inter Am. Inst.. Agrl. Scis., Turrialba, Costa Rica, 1954; Ph.D., Ind. U., 1958. Prof. genetics and plant breeding U. Guyaquil, Ecuador, 1954-55; head agronomy dept. Tropical Expt. Sta., Pichilingue, Ecuador, 1954-55; geneticist, plant breeder InterAm. Inst. Agrl. Scis., 1958—. Mem. Crop Scis. and Agronomy Soc. Am., Soc. for Study Evolution. Publs. on studies of inheritance of yield and its components, prodn. of high yielding biclonal cacao hybrids, use of numerical taxonomy to determine relationships in species of Solanum, sect. Morella. Address: InterAm. Inst. Agrl. Scis., Turrialba, Costa Rica.*

SORKIN, Ernst, Swiss immunologist; b. Rheinfelden, Basel, Switzerland, Feb. 5, 1920; s. Nikolaus and Lisa S.; Ph.D., U. Basel, 1946; m. Doris Brügger, 1946; 1 dau., Jeanette. Asst. U. Basel, 1946-52, privat dozent, 1950; staff Institut Biologie physicochimique, Paris, France, 1952; officer Tb Immunization Research Centre, WHO, Copenhagen, Denmark, 1953-61; dir. med. dept. research Inst., Davos, Switzerland, 1962—; staff Ford Hosp., Detroit, 1958. Mem. Swiss Soc. for Allergy and Immunology, (pres. 1967-68), Brit. Soc. Immunology, Swiss Chem. Soc., Swiss Soc. Natural Scis. Research, numerous articles on characterization of mycobacterial products, humoral and cellbound antibodies in immunity, mechanism of chemotaxis; discovered (with Boyden) cytophilic antibodies. Home: Davos. Office: Forschungsinstitute, Davos, Switzerland.*

SORM, Frantisek, Czechoslovakian chemist; b. Praha, Czechoslovakia, Feb. 28, 1913; s. Frantisek and Kamila (Durdilova) S.; D.Sc., Tech. U., Prague, Czechoslovakia, 1938; D.Sc., Université Libre Bruxelles, Brussels, Belgium, 1965; m. Zora Drapalova, July 4, 1940; children—Zora, Milan. Faculty, Tech. U., Prague, 1938-39, prof. chemistry, 1945-50; prof. Charles U., Prague, 1950—; with chem. industry, 1939-45; dir. Inst. Organic Chemistry and Biochemistry, Czechoslovak Acad. Scis., 1952—. Recipient Fritzsche medal Am. Chem. Soc., 1959, Stas medal Belgian Chem. Soc., 1962. Mem. Czechoslovak (pres. 1962—), USSR, Hungarian, German, Bulgarian, Roumanian, Polish, Leopoldina, Royal Danish acads. scis., Brit. Chem. Soc. (hon.). Author: (with L. Dolejs) Guaianolides and Germacranolides, 1965; also numerous articles. Research on alcaloid syntheses, discovery of new terpenoid structures (1st detection medium ring compounds in nature), synthesis of azulene; determination of protein structures (chymotrypsine, trypsine) and analysis of their similarities, synthesis of nucleic acid components and analogues as cancerostatics). Home: 9 Korejska, Prague 6, Czechoslovakia.*

SOROFF, Harry Solomon, thoracic surgeon; b. Sydney, N.S., Can., Feb. 2, 1926; s. Nathaniel and Fannie (Weinstein) S.; came to U. S., 1929, naturalized, 1948; student Temple U., 1942-44, M.D., 1948; m. Marilynn Wilson, Sept. 19, 1953; children—David, Daniel, Jonathan. Chief metabolic div. surg. research unit Brooke Army Med. Center, Ft. Sam Houston, San Antonio, 1953-56; fellow thoracic surgery Mt. Auburn Hosp., also Malden Hosp., 1957-60, research fellow surgery Harvard Med Sch., Boston, 1957-60; asst. in surgery Peter Bent Brigham Hosp., Boston, 1957-60, chief thoracic lab., 1960-61; faculty Tufts U. Sch. Medicine, Boston, 1961—, asso. prof. surgery, 1964—; asso. dir. Clin. study unit New Eng. Center Hosps., Boston, 1961—; established investigator Am. Heart Assn., 1961-66; staff New Eng. Center Hosp., Boston, 1964—. Jr. cons. in thoracic surgery Lemuel-Shattuck Hosp., Jamaica Plain, Mass., 1964—. Diplomate Am. Bd. Surgery, Nat. Bd. Med. Examiners, Bd. Thoracic Surgery. Mem. A.M.A. Research, publs. in surg. metabolism, organ transplantation, devices to assist circulation. Home: 2 Apple Hill Rd., Peabody, Mass. Office: 171 Harrison Av., Boston 02111.*

SOROKIN, Contantine, biologist; b. Tsaritsyn, Russia, Aug. 7, 1903; s. Alexis and Lydia (Arefiev) S.; student Crimea U., Simferopol, USSR, 1921-23; diploma Don Agrl. Inst., Novocherkassk, USSR, 1929; Candidate, Nauk. Acad. Agrl. Scis., Moscow, USSR, 1936; Ph.D., U. Tex., 1955; m. Tamara Amarsky, Feb. 28, 1932. Came to U. S., 1949, naturalized, 1955. Research asst. Don Agrl. Inst., 1926-29; dir. Seed Testing Sta., Stavropol, USSR, 1929-32; chmn. genetics plant breeding Kuban Inst., Krasnodar, USSR, 1932-39, lectr. plant physiology, 1940-42; research prof. plant physiology Agrl. Inst., Halbturn, Austria, 1943-45; research scientist II in algal physiology U. Tex., Austin, 1951-55; research asso. in cellular and algal physiology U. Md., College Park, 1955—. NIH Spl. Research fellow, 1956-57. Mem. N.Y. Acad. Scis., Soc. for Developmental Biology, Bot. Soc. Am., Am. Soc. Plant Physiologists, Scandinavian, Japanese socs. plant physiologists, Internat. Phycological Soc., Gerontological Soc., Sigma Xi. Research, publs. in cell physiology and devel., photosynthesis, cell secretion, developed concept of primary cellular aging. Home: 6805 Wells Pkwy., University Park, Md. 20782. Office: Dept. Botany, U. Md., College Park, Md. 20742.*

SOROKIN, Pitirim Alexandrovitch, sociologist; b. Village of Touria, Russia, Jan. 21, 1889; s. Alexander P. and Pelageia V. (Rimskych) S.; ed. Teachers Coll., Kostroma Province, Russia, 1903-06, evening sch., St. Petersburg (now Leningrad), Russia, 1907-09, Psycho-Neurol. Inst., St. Petersburg, 1909-10, U. of St. Petersburg, 1910-14; Magistrant of Criminal Law, 1915, Dr. of Sociology, 1922; hon. Ph.D., 1950; m. Helen Petrovna Baratynskaia, May 26, 1917; children—Peter, Sergei. Came to U. S., 1923, naturalized, 1930. Privat-dozent Psycho-Neurol. Institute, 1914-16; University of St. Petersburg, 1916-17, prof. sociology, 1919-22; prof. sociology, Agrl. Acad., 1919-22, U. of Minn., 1924-30; prof. sociology Harvard University, 1930-64, emeritus prof., 1964—; now dir. Research Center Creative Altruism. Co-editor New Ideas in Sociology, 1913-15; editor in chief Volia Naroda, newspaper, at Petrograd, 1917. Mem. exec. com. All-Russian Peasant Soviet, 1917; mem. Council of Russian Republic, 1917; sec. to prime minister. 1917; mem. Russian Constl. Assembly, 1918; pres. Internat. Congress Sociology, 1937. Mem. Internat. Soc. Comparative Study of Civilizations (pres. 1961-64), Am. Acad. Arts and Scis., American Sociological Association (president 1965—); hon. member Internat. Inst. Social Reform, Internat. Inst. Sociology, Czecho-Slovak Acad. Agr., German Sociological Soc., Belgian Royal Acad., Rumanian Royal Acad. Works include: Crime and Punishment, 1913; Sociology of Revolution, 1925; Social Mobility, 1927; Contemporary Sociological Theories, 1928; Social and Cultural Dynamics, 4 vol., 1937-41; Society, Culture and Personality, 1947; Altruistic Love, 1950; Social Philosophies of an Age of Crisis, 1950; Explorations in Altruistic Love and Behavior, 1950; S.O.S.: The Meaning of Our Crisis, 1951; The Ways and Power of Love, 1954; Forms and Techniques of Altruistic and Spiritual Growth, 1954; Fads and Foibles in Modern Sociology, pub. 1956; American Sex Revolution, 1957; Power and Morality, 1958; A Long Journey, 1963; The Basic Trends of Our Time, 1963; Sociological Theories of Today, 1966. Contbr. jours. Author classic study of social stratification; approached study of soc. historically finding ideational, idealistic and sensate cultures; stressed importance of integralist method. Died Feb. 10, 1968.

SORRE, Maximilien, French geographer; b. Rennes, France, 1880; student école normale de Saint-Cloud, France, 1899-1901; Doctor letters, 1913. Tchr. école normales of La Roche-sur-Yon, Perpignan, Montpellier (all France); prof. faculties letters of Bordeaux and Strasbourg, France; prof. Faculty Letters, Lille, France, became dean 1929; named rector Acad. Clermont-Ferrand, Acad. Aix-Marseille, 1934; with U. Montpellier, 1937-41; prof. Sorbonne, Paris, 1941-48. Author: Les Pyrénées, 1928; Les Fondaments de la géographie humaine, 4 vols., 1943-52; contbg. author: Géographie universelle. Established biol. and tech. founds. of human geography.

SORUM, Harald, Norwegian physicist; b. Modum, Norway, May 5, 1915; s. Martin and Ingeborg (Stende) S.; Sivil ingeior, Tech. U. Norway, 1941, Dr.techn., 1952; m. Karen Margrete Byberg, Mar. 10, 1922; children—Geir Inge, Harald Sigmund. With Inst. Physics, NTH, 1941-43, 1949-53, lectr. X-ray crystallography, 1956—; Brit. Council scholar Casvendish Lab., Cambridge, 1947-49; sr. research fellow Norwegian Def. Research Establishment, 1953-56; prof. physics Tech. U. Norway, 1966—. Mem. Chem. Soc., Phys. Soc., Royal Norwegian Soc. Sci. Research, publs. on X-ray spectroscopy, especially in range long wave lengths; determination of structure of crystals by aid of X-ray diffraction. Home: 15 Bugges vei, Trondheim, Norway.*

SOSA, Julio Maria, histologist; b. Mercedes, Uruguay, July 1, 1906; s. Eusebio C. and Amena (Soumastre) S.; D.Medicine and Surgery, U. of Republic, Montevideo, Uruguay, 1931; m. Haydée Maeta Savio, July 1, 1950; children—Julio Célmar, Julio César, Carlos Maria, Mariayde. Asst. div. histological research Neurol. Inst., Montevideo, 1927-43, head, 1943-58; asst. prof. histology Med. Sch., 1941; prof. microscopic anatomy and cytology Sch. Humanities and Scis., 1946-60, head dept., 1958-60, prof. histology Sch. Dentistry, 1954-60; prof. histology chmn. organizer Inst. Histology, U. Los Andes, Mérida, Venezuela, 1960—, dir. Electronmicroscopy Center, 1967—. Recipient Hippocratic medal U. Montpellier (France), 1948. Mem. Biol. Soc. Montevideo, Internat. Soc. Cell Biology, Anat. Soc. Gt. Britain and Ireland, Am. Assn. Anatomists, Electron Micros. Soc. Am., Linnean Soc. London, Gerontological Soc. U .S., Histochem. Soc. U. S., Uruguayan, Venezuelan assns. for advancement sci., A.A.A.S., N.Y. Acad. Scis., Deutsche Anatomische Gesellschaft, Assn. des Anatomistes (France). Research, publs. on structure and significance of Golgi apparatus, problems of normal and path. structure of nerve cells and nerve tissue, types of nerve cell div. Home: Apartado 149, Mérida, Venezuela.*

SOSIGENES, Greek astronomer, mathematician; commd. by Julius Caesar to reform Roman calendar, circa 46 B.C. Author: Revolving Spheres (only fragments remain); also 3 Commentationes concerned with astron. calculations. Mentioned idea that Mercury revolved around sun.

SOSIN, Abraham, Am. physicist; b. Chgo., June 23, 1925; s. Solomon and Sophie (Mason)) S.; B.S., U. Chgo., 1949, Ph.B., 1948; M.S., U. Ill., 1950, Ph.D., 1954; m. Dorothy Simon, Sept. 28, 1949; children—Howard B., Cynthia A. Faculty U. Fla., 1954-55; supr., group leader atomics internat. div. N. Am. Aviation, Thousand Oaks, Cal., 1955-65, physicist Sci. Center Div., 1965—; faculty U. Cal., Los Angeles, 1956-60, lecturing prof. engring., 1962-68; prof. physics and materials engring., 1968—. Mem. Am. Phys. Soc., Am. Acad. Arts and Scis., Research and Engring. Soc. Am., Sigma Xi. Investigations of defects in metal crystals, primarily by use of irradiation with high energy electrons; measurement and theory of displacement of atoms during irradiation; extensive study of diffusion of defects, interaction of defects, using kinetic methods. Home: 23436 Ladrillo St., Woodland Hills, Cal. 91364. Office: 1049 Camino dos Rios, Thousand Oaks, Cal. 91360.*

SOSKIN, Samuel, Am. physician; b. N.Y. City, Apr. 10, 1904; s. Elias and Bertha (Bergin) S.; M.D., U. Toronto (Ont., Can.), 1926, A.M., 1927, Ph.D., 1929; m. Palma Abraham, Aug. 31, 1935; children—Samuel Richard, Susan Jane. Research fellow, dept. physiology U. Toronto, 1926-27, demonstrator in physiology, 1927-29; dir. metabolic and endocrine research Michael Reese Hosp., Chicago; 1929-43, med. dir., 1943-50, dir. Med. Research Inst. and dean Postgrad. Sch., 1946-52; spl. cons. U.S.P.H.S., 1948-53; asst. prof. physiology University of Chicago, 1932-41, prof. and lectr. physiology 1941-51; asso. prof. medicine Northwestern U. Sch. Medicine, 1951-52; med. cons., Cedars of Lebanon Hosp., Los Angeles, 1952—; asso. clin. prof. medicine U. Cal. Los Angeles, 1953-60. President of the Greater Los Angeles Nutrition Council, 1960-62. Fellow American College of Physicians; mem. Am. Physiol. Society, Am. Inst. Nutrition, Assn. for Study Internal Secretions, Soc. for Exptl. Biology and Med., Am. Soc. for Clin. Investigation, Central Soc. Clin. Research, Inst. Med. Chgo., Chgo. Soc. Internal Med. (pres. 1945-46, council 1946-49), A.A.A.S., Sigma Xi. Mem. editorial bds. Jour. Clin. Endocrinology 1945-51, endocrinol. sect. Excerpta Medica, 1948—; editor-in-chief Metabolism, 1951-58, cons. editor, 1958—; Progress in Clinical Endocrinology, 1950; Carbohydrate Metabolism, 1946, rev. 1952. Research in fields of metabolism and endocrinology. Home: 748 S. Beverly Glen Blvd., Los Angeles 24. Office: 9735 Wilshire Blvd., Beverly Hills, Cal.*

SOSMAN, Merrill C(lary), Am. roentgenologist; b. Chillicothe, O., June 23, 1890; s. Francis Asbury and Mollie (Browning) S.; A.B., U. Wis., 1913; M.D., Johns Hopkins, 1917; postgrad. Mass. Gen. Hosp., 1921-22; M.A., Harvard, 1949; m. Arline Clark Adams, June 27, 1918; children—John Leland, Barbara Clark. Resident physician U. S. Soldiers Home Hosp., Washington, 1917; became roentgenologist in chief Peter Bent Brigham Hosp., Boston, 1922; cons. roentgenologist Childrens Hosp., Psychopathic Hosp., N. E. Peabody Home for Crippled Children (Boston), Cape Cod Hosp. (Hyannis); instr. in roentgenology Harvard Med. Sch., 1922-28, asst. prof., 1928-40, clin. prof., 1940-44, clin. prof. radiology, 1944-48, became prof. of radiology 1948; cons. radiology Mass. Gen. Hosp. Recipient Gold medal Radiol. Soc. of Am. Diplomate Am. Bd. Radiology. Fellow A.A.A.S.; mem. A.M.A., N.E. Roentgen Ray Soc. (pres.; George W. Holmes lectr. 1947), Radiol. Soc. N.A., Am. Roentgen Ray Soc. (pres. Caldwell lectr. 1947), Harvey Cushing Soc. (pres), Am. Coll. Radiology, Mexican Soc. Radiol. and Phys. Therapy (hon.), Venezuela Radiol. Soc., Am. Acad. Arts Scis., Sigma Xi. Contbr. of numerous articles on diagnosis and treatment of diseases or tumors by X-ray to sci. publs. Died Mar. 28, 1959.

SOSNOWSKI, Ryszard, Polish physicist; b. Bialowieza, Poland, Jan. 5, 1932; s. Franciszek and Wanda (Flanczewska) S.; M.Sc., Warsaw U., 1955, Ph.D., 1960; postgrad. Moscow U., 1955-57 m. Maria Lefeld, Aug. 13, 1957; 1 dau., Marzena. Asst., Inst. Physics, Warsaw (Poland) U., 1953-55, lectr., 1958-60, 62-65; research asso. Inst. Nuclear Research, Warsaw, 1955, group leader nuclear spectroscopy, 1958-60, group leader bubble chamber group, 1961—; research asso. Joint Inst. Nuclear Research, Dubna, Poland, 1957-58; IAEA fellow European Orgn. for Nuclear Research, Geneva, Switzerland, 1960-61. Mem. Polish. Phys. Soc. Research, publs. on new energy levels and transitions in nuclei of medium atomic numbers, beta-electron polarization in decays of different nuclei, high energy collisions with strange particle prodn. and collisions of high multiplicity. Home: 14 m 21 Wilcza. Office: 69 Hoza, Warsaw, Poland.*

SOSTRATOS, surgeon, zoologist; probably practiced in Alexandria, after 30 B.C.; author gynecol. and other med. works, also works on zoology; promoted surg. technique; ranks near Aristotle as a Greek zoologist.

SOTCHAV, Victor Borisovich, Russian geobotanist, geographer; b. June 20, 1905; grad. Leningrad Agrl. Inst., 1924. Staff, Leningrad Agr. Inst., until 1926; staff Bot. Inst., USSR Acad. Sci., 1926-36, 43—; staff Inst. Reindeer Breeding, Lenin All-Union Acad. Agrl. Scis., 1931-35; chmn. dept. reindeer breeding Arctic Inst., 1935-38; faculty A. I. Gertsen Pedagogical Inst., Leningrad, 1939-50; faculty Leningrad. U., 1938—, became prof., 1944; dir. Inst. Geography of Siberian and Far E., Irkutsk Siberian dept., USSR Acad. Scis., 1960; also head lab. geography and cartography vegetation V. L. Komarov Inst. Botany, USSR Acad. Scis., Leningrad. Recipient Order of Lenin; Pierre Fermat Silver medal 1960. Corr. mem. USSR Acad. Scis. Research and publs. on plants from various parts of USSR, including previously unexplored tys. USSR forest vegetation, theoretical and methodical aspects of vegetation mapping, problems of comprehensive mapping; organized food base for reindeer breeding. Office: Botanical Inst., USSR Acad. Scis. ul. Prof. Popova, 2, Leningrad 22, USSR.

SOTELO, Robert, Uruguayan cytologist, electronmicroscopist; b. Montevideo, Uruguay, Jan. 2, 1915; s. José Sotelo and O. Mercedes (Lotufo) Larriera; B.Medicine, U. Uruguay, 1933; m. Dora Sosa, Jan. 3, 1942; children—Dora Marta, José Roberto. With Instituto de Investigación de Ciencias Biológicas, Ministerio de Instrucción Pública y Previsión Social, 1937—, head lab. histoneurology, cytology and tissue culture, 1955-57, head dept. cell ultrastructure, 1956—; with Faculty Medicine, Montevideo, 1941-47, instr., 1944-47. Rockefeller Found. fellow, vis. investigator Rockefeller Inst. lab. cytology and electronmicroscopy, 1954-55; guest Rockefeller Inst., 1958. Mem. Sociedad de Biología de Monte video, Internat. Soc. for Cell Biology, Am. Soc. for Cell Biology. Research, publs. on normal innervation of gastrointestinal tract, electron microscopy of germinal cells and meiotic chromosomes. Home: 2238 Santiago Nievas. Office: 3318 Avda. Italia, Montevideo, Uruguay.*

SOUBEIRAN, Eugène, French pharmacologist; b. Paris, May 24, 1797; prof. Sch. Pharmacy, Paris. Author: Manuel de pharmacie, 1826; Nouvelle traité de pharmacie théorique et pratique, 2 vols., 1835-36; Fabrication des eaux minér artificielles, 1839. Discovered (independently of Liebig) chloroform, 1831. Died 1858.

SOUCEK, B., Yugoslavian engr.; b. Bjelovar, Yugoslavia, Apr. 25, 1930; s. Franjo and Marija (Lazic) S.; B.Sc., U. Zagreb, 1955, Ph.D., 1963; m. Erika Skuric, Aug. 8, 1959; 1 dau., Marina. Asst. researcher Ruder Boskovic Inst., Zagreb, 1955—, head data processing facility, 1966—; asso. prof. elec. engring. U. Zagreb, 1963—; vis. research asso. Brookhaven Nat. Lab., Upton, L.I., 1964-66. Recipient Nat. prizes Yugoslavia for achievements in digital electronics and measurement of stochastical, 1963, 64. Research, publs. on design of nuclear instruments, new effect in random triggering of monostable elec. systems, statis. distrbn., computer system for nuclear pulse spectroscopy with megachannel resolution, theory of filtering of random events. Home: 3 Vijenac. Office: 54 Bijenicka, Zagreb, Yugoslavia.*

SOUDEK, Dusan, Czechoslovakian biologist; b. Prague, Czechoslovakia, May 4, 1920; s. Stephan and Marie (Strouhalova) S.; M.D., Faculty Medicine, Brno, 1949, docent, 1964; m. Vera Pistecka, Aug. 28, 1947; children—Stephan, Dusan, Ivan. Asst., lectr. Faculty Medicine, Brno, 1945-53, 60-62, lectr. Faculty Pharmacy, Brno, 1953-60; head dept. genetics and cytology Research Inst. Pediatrics, Brno, 1963—. Mem. Internat. Soc. Cell Biology. Author: (with Necas) Outline of General Cytology, 1960; also articles. Research on cell biology, cell nucleus, human cytogenetics. Home: 10 Nezvalova. Office: 9 Cernopolni, Brno, Czechoslovakia.*

SOUDER, Wilmer, Am. physicist; b. Salem, Ind., 1884; s. Charles Losson and Elizabeth (Newlon) S.; A.B., Ind. U., 1910, A.M., 1911; Ph.D. magna cum laude, U. Chgo., 1916; certificate in law Columbus U., 1934; m. Lillie Martin, 1908; 1 dau., Martha Ella (Mrs. Walter Leser)). Asst. physicist Ind. U., 1908-11, U. Chgo., 1913-16; with Nat. Bur. Standards, Washington, 1911-61, div. chief, 1945-54, cons., 1928—. Recipient Meritorious Service award Dept. Commerce, 1951; Distinguished Alumni award Ind. U., 1954. Fellow Am. Coll. Dentists; mem. Internat. Assn. Dental Research (pres. 1940-41), D.C. Dental Soc. (hon.), Am. Dental Assn. (hon.), Am. Chem. Soc., Am. Phys. Soc., Internat. Assn. Chiefs Police, Internat. Assn. for Identification, A.A.A.S., Sigma Xi. Research, numerous monographs on applications of structural and chem. properties in selection and use of restorative dental materials, reliability and effectiveness of phys. evidence in detection of crimes. Address: 107 Ann Av., Landisville, Pa. 17538.*

SOUGUET, Jacques-Charles-Emile, French mathematician; b. Bessèges, France, 1871; mem. French Acad. Scis. Research and publs. on mechanics, thermodynamics, mechanics of fluids, explosives, theory of heat engines. Died Montpellier, France, 1943.

SOULAIRAC, André, French psychophysiologist, psychiatrist; b. Maubeuge, France, June 17, 1913; s. André and Mathilde (Sévin) Medicine Doctor, Faculte de Medecine de Paris, 1937; Sci. D., Faculte des Scis. de Paris, 1947; m. Marie-Louise, Monthillaud, Apr. 26, 1952; 1 son, Michel-Andre. Prof. psychophysiology Faculté Scis. Paris, 1949—; dir. research unity clin. psychophysiology French Inst. Med. Research, since 19——; médecin chief Psychiat. Hosp., Sainte-Anne, Paris, 1956—; mem. Nat. Center Sci. Research, 1956—. Decorated chevalier Légion of Honor. Fellow Assn. for Study Animal Behavior; mem., past pres. Soc. Endocrinology; mem. N.Y. Acad. Scis. Author: Les Lois de Lavie, 1958; also numerous articles. Editor: Annales d'Endocrinologie, 1950—, European Rev. Endocrinology, 1964—. Research on neuro-endocrinology and animal behavior, regulation and biol. roles of levels of vigilance, psychopharmacology, neurochemistry, enzymology of central nervous system. Home: 22 rue Pierre-Curie, Paris. Office: 9 Quai Saint-Bernard, Paris, France.*

SOULAVIE, Jean Louis Giraud, geologist, monk; b. 1752. Author: L'histoire naturelle de la France meridionale (publ. suspended because of attacks by clergy and prominent scientists such as Buffon). Founder of stratigraphical paleontology; precursor of theory of transformism; reached conclusions in attempts to determine chronological sequence of rocks which did not agree with Scriptures; discovered that lava flows along river beds; distinguished 5 kinds of sedimentary strata according to fossil content; stressed environmental factors in extinction of old and rise of new species. Died 1813.

SOULE, Arthur Bradley, Am. radiologist; b. St. Albans, Vt., Oct. 22, 1903; s. Arthur Bradley and Minnie (Miller) S.; A.B., U. Vt., 1925, M.D., 1928; m. June Yale Crouter, June 29, 1931; children—Caroline Yale, Arthur Bradley III. Practice medicine, specializing in radiology, Burlington, Vt.; chief radiol. service Mary Fletcher Hosp., 1933—; prof. radiology Coll. Medicine U. Vt., 1936—. Mem. Nat. Acad. Scis.-NRC Academic Radiology, USPHS Tng. Grants Com. Fellow Am. Coll. Radiology (chmn. com. on technologist tng., mem. residency rev. com. radiology); mem. Am. Roentgen Ray Soc., N.E. Cancer Soc. (pres.), Radiol. Soc. N. Am. Obersvations, publs. on use of radiology in dental and med. diagnosis; method of radiology. Home: Shelburne, Vt. 05482. Office: Mary Fletcher Hosp., Burlington, Vt.*

SOULE, Malcolm Herman, Am. bacteriologist; b. Allegany, N.Y., Dec. 5, 1896; s. Charles M. and Ida May (Ervin) S.; B.S., U. Mich., 1921, M.S., 1922, D.Sc., 1924; LL.D., St. Bonaventures Coll., 1928; m. Alma Dengler, Sept. 7, 1926; children—Mary Alma, Margaret Laura. Instr. analytical chemistry U. Mich., 1919-20; instr. bacteriology Sch. of Medicine 1923-25, asst.. prof., 1925-28, asso. prof., 1928-31, prof. since 1931, chmn., dept., and Hygienic Lab. since 1935; vis. prof., U. Chgo., 1931, Sch. Tropical Medicine, P.R., 1931; leprosy investigation, Leonard Wood Meml. P.I., 1933-34, chmn. Med. Adv. Bd., 1944; cons. to dir. div. health and sanitation. Coordinator of Inter-American Affairs since 1942. Del. of U. S. Govt. to various sci. and med. congresses adviser to dir. Leprosy Service and Malaria, Brazil, 1950; mem. med. ednl. mission to Japan, auspices SCAP, Unitarian Services Com., 1951; mem. Com. Internat. Affairs, NRC. Recipient Gold medal A.M.A., 1930. Fellow A.A.A.S. (mem. exec. com. since 1947); mem. Am. Assn. Pathologists and Bacteriologists (pres. 1947), Am. Acad. Tropical Medicine (council mem. since 1937), Am. Assn. Pathologists and Bacteriologists (council since 1940), Am. Assn. Immunologists, Am. Chem. Soc., Am. Micros. Assn., Am. Pub. Health Assn. (fellow), Am. Soc. Exptl. Pathology, Am. Soc. Tropical Medicine (v.p. 1941), Bot. Soc. Am. Internat. Leprosy Assn., Path. Soc. Great Britain and Ireland, Soc. Exptl. Biology and Medicine, Soc. Am. Bacteriologists, Sociedad Medico-Quirurgica del Guayas, Societe de Pathologie Exotique, Brazilian Leprosy Assn. Sigma Xi Mem. mem. editorial bds., Science; Am. Jour. Pathology; Am. Jour. Tropical Medicine. Contbr. scientific articles on microbic respiration, microbic dissociation, goitre, undulant fever, leprosy, poliomyelitis, relapsing fever, tropical medicine. Died Aug. 3, 1951.

SOURANDER, Bertil Johan Patrick, neuropathologist; b. Helsinki, Finland, Mar. 5, 1917; s. Bertil and Margit (Spare) S.; M.D., U. Helsinki (Finland), 1946; D.Sc., U. Gothenburg (Sweden), 1953; m. Brita Marianne Ljunggren, May 3, 1950; children—Patrick, Dag. Asst. prof. pathology U. Uppsala (Sweden, 1953-57; asso. prof. U. Gothenburg, 1957—. Cons. neuropathologist Lillhagen Mental Hosp., 1959—. Mem. nueropath. oommn. World Fedn. Neurology 1959—. Mem. Internat. Soc. Neuropathology, Brit., Scandinavian neuropath. socs. Research, publs. on structural and chem. changes in nervous system caused by infections, inherited metabolic diseases and irradiation, tissue culture studies on neuroretina. Home: 11 Norra Liden, Gothenburg, Sweden.*

SOURKES, Theodore Lionel, Canadian biochemist; b. Montreal, Que., Can., Feb. 21, 1919; s. Irving and Fannie (Golt) S.; B.Sc., McGill U., Montreal, 1939, M.Sc. magna cum laude, 1946; Ph.D., Cornell U., 1948; m. Shena Rosenblatt, Jan. 21, 1943; children—Barbara, Myra. Asst. prof. pharmacology Georgetown U. Med. Sch., Washington, 1948-50; research asso., dept. enzyme chemistry Merck Inst. Therapeutic Research, Rahway, N.J., 1950-53; sr. research biochemist Allan Meml. Inst, 1953-65, dir. Lab. Chem. Neurobiology, 1965—; from lectr. to prof. biochemistry dept. psychiatry McGill U, 1954—. Recipient Sr. Fellowship award Parkinson's Disease Found., N.Y.C., 1963. Mem. Canadian Biochem. Soc., Canadian Physiol. Soc., Pharmacological Soc. Can., Canadian Soc. Clin. Chemistry, Am. Soc. Biol. Chemists, Am. Soc. Pharmacology and Exptl. Therapeutics, N.Y. Acad. Scis., Internat. Brain Research Orgn., Canadian Soc. Study History and Philosophy Sci., Sigma Xi. Author: Biochemistry of Mental Disease, 1962; Nobel Prize Winners in Medicine and Physiology, 1901-1965, pub. 1967. Editorial bd. Canadian Jour. Biochemistry, Revue Canadienne de Biologie. Asso. editor: Methods in Medical Research, vol. 9, 1961. Research, publs. on drugs for treatment high blood pressure; 1st basic research on methyldopa; research functions of action substances in brain (elucidation of roles of dopamine, serotonin in nervous system), also biochem. deficiencies in Parkinson's disease. Home: 4645 Montclair Av., Montreal 28. Office: 1033 Pine Av. W., Montreal 2, Que., Can.*

SOUTH, Sir James, English astronomer; b. London, Eng., Oct. 1785; s. James South; LL.D., Cambridge (Eng.) U.; m. Charlotte Ellis, 1816. Equipped obs., Campden Hill, Kensington, Eng., 1826; astron. observer (with John Frederick William Herschel), London, (with Laplace), Paris, France, 1835. Fellow Royal Soc. (Copley medal 1826), 1821; mem. Astron. Soc. (co-founder, Gold medal 1826), St. Petersburg (USSR), Brussels (Belgium) acads. scis. Contbd. papers, pamphlets to sci. socs. Measured 458 compound stars, 160 of the new. Died Oct. 19, 1867.

SOUTHAM, Chester Milton, Am. physician; b. Salem, Mass., Oct. 4, 1919; s. Walter A. and Elizabeth (Furbish) S.; B.S., U. Ida., 1941, M.S., 1943; M.D., Columbia, 1947; m. Anna Lenore Skow, Sept. 24, 1939; children—Lawrence A., Lenore E., Arthur M. Med. fellow Meml. Hosp., N.Y.C., 1948-51, asst. attending physician, 1952-58, asso., 1959—; with Sloan-Kettering Inst., N.Y.C., 1948—, mem., 1963—; faculty Cornell U. Med. Coll., N.Y.C., 1951—, asso. prof. medicine, 1958—. Mem. sci. adv. com. Damon Runyon Meml. Fund, 1961—; panel on therapy Am. Cancer Soc., 1961-66; com. for research on tobacco and health Edn. and Research Found., A.M.A., 1965-66. Recipient certificate of achievement U.S. Army, 1955. Mem. Am. Assn. Cancer Research (director 1966—, president 1968—), American Assn. for Exptl. Pathology, Am. Assn. Immunologists, Am. Fedn. Clin. Research, Am. Soc. Tropical Medicine and Hygiene, Harvey Soc., James Ewing Soc., N.Y. Acad. Sci., Soc. Exptl. Biology and Medicine, Tissue Culture Assn., N.Y. County, N.Y. State med. socs., A.M.A., Am. Soc. Microbiology, Sigma Xi. Research, numerous publs. in cancer. including evaluation of chemotherapeutic drugs, effect of virus infections on tumor growth, cancerigenic viruses, combined effect of chems. and viruses in cancerigenesis, immunological status of cancer patients, host resistance to cancer, studies on cancer-specific antigens. Home: 50 Orchard Rd., Demarest, N.J. Office: 410 E. 68th St., N.Y.C. 10021.*

SOUTHERN, John Albert, Am. chemist; b. Sneedville, Tenn., Dec. 27, 1904; s. John M. and Pearl (Mitchell) S.; B.S., Furman U., 1927; M.S., Vanderbilt U., 1930; Ph.D., U. N.C., 1938; m. Ethel Dunn Carlisle, Sept. 5, 1936; children—Janet Carlisle (Mrs. D.T. Huskey), John Lawrence. Tchr. high sch., 1927-29; chemist Am. Cast Iron Pipe Co., 1931-33, TVA, 1933-34; asso. prof. Furman U., Greenville, 1934-47, prof. chemistry, 1958—, chmn. div. sci. and math., 1958-68; chmn. div. sci. and math. Cumberland U., 1947-51; prof. chemistry Howard Coll., Birmingham, Ala., 1952-58, chmn. dept., 1954-58. Mem. S.C. Acad. Sci. (sec. 1946-47), Am. Chem. Soc. (chmn. S.C. sect. 1946-47, Western Carolinas sect. 1961-62), A.A.A.S., Sigma Xi. Research, publs. on mono esters of dicarboxylic acids, metal salts of monoesters of dicarboxylic acids, qualitative analysis of zinc and magnesium; patentee in field. Home: 106 Duncan Chapel Rd., Greenville, S.C. 29609.*

SOUTHWELL, Byron Lester, Am. agrl. scientist; b. Manassas, Ga., Jan. 27, 1900; s. H. F. and Susie (Smith) S.; B.S., U. Ga., 1923, M.S., 1928; m. Harriet Evans, June 17, 1931. Head animal sci. dept. Ga. Coastal Plain Expt. Sta., Tifton, 1932—; chmn. sec. exec. com. Regional Breeding Projects. Recipient Distinguished Service award So. Pasture and Forage Crop Improvement Conf., 1965; Distinguished Service award Am. Soc. Animal Sci., 1967; named Ga.'s Man of Year in Agr., 1954. Fellow Am. Soc. Animal Sci. ; mem. Assn. So. Agrl. Workers, Am. Soc. Animal Prodn., Am. Soc. Range Mgmt., Am. Dairy Sci. Assn., Sigma Xi, Gamma Sigma Delta. Contbr. numerous articles to profl. jours. Assembled dairy, beef and swine herds for Coastal Plain Sta. and initiated sta's livestock research program with herd of polled herefords. Home: 1010 Hall Av. Office: Ga. Coastal Plain Expt. Sta., P.O. Box 748, Tifton, Ga. 31794.*

SOUTHWICK, Charles Henry, Am. zoologist; b. Wooster, O., Aug. 28, 1928; s. Arthur F. and Faye (Motz) S.; B.A., Coll. Wooster, 1949; M.S., U. Wis., 1951, Ph.D., 1953; m. Heather Milne Beck, July 12, 1952; children—Steven, Karen Leslie. NIH fellow, 1951-53; asst. prof. biology Hamilton Coll., 1953-54; NSF fellow Oxford (Eng.) U., 1954-55; faculty Ohio U., 1955-61; asso. prof. pathobiology Johns Hopkins University, Balt., 1961—, professor, since 1968—. Member of the primate adv. com. Nat. Acad. Sci.-NRC, 1963—. Recipient Fulbright Research award, India, 1959-60. Fellow A.A.A.S., Acad. Zoology; mem. Am. Soc. Zoologists, Ecol. Soc. Am., Am. Soc. Mammalogists. Editor: Primate Social Behavior, 1963. Research, numerous publs. on animal social behavior and population dynamics, influences animal social orgn. on demographic characteristis mammal populations; pioneered primate surveys in India. Home: 6724 Glenkirk Rd., Balt. 21212.*

SOUTHWICK, Harry Webb, Am. surgeon; b. Grand Rapids, Mich., Nov. 21, 1918; s. George Howard and Jessie (Webb) S.; B.S., Harvard, 1940, M.D., 1943; m. Lorraine Hinsdale, June 27, 1942; children—Harry, Sandra, Charles Howard, Gay. Individual practice surgery, Chgo., 1950—; attending surgeon U. Ill. Research and Ednl. Hosps., Presbyn.-St. Luke's Hosp., Chgo., 1963—; clin. prof. surgery U. Ill. Coll. Medicine, 1963. Mem. Ill. Gov's Adv. Com. Cancer Control, 1966—. Mem. Am. Cancer Soc. (pres. Chgo. unit 1964-67), also numerous other coms. and bds. asso. with cancer control and treatment. Author: (with W.H. Cole, S. Roberts, G.O. McDonald) Dissemination of Cancer, 1961; (with D.P. Slaughter and L.J. Humphrey) Surgery of the Breast, 1968; also articles. Home: 830 Heather Lane, Winnetka, Ill. 60093. Office: 1725 W. Harrison St., Chgo. 60612.*

SOUTHWICK, Philip Lee, Am. chemist; b. Lincoln, Neb., Nov. 15, 1916; s. Philip Orin and Dorothy (Harpham) S.; A.B., U. Neb., 1939, M.A, 1940; Ph.D., U. Ill., 1943; m. Helen Louise Cather, Sept. 1, 1942; 1 son, James Philip. Research chemist Merck & Co., Inc., Rahway, N.J., 1943-46; faculty Carnegie Inst. Tech., Pitts., 1946—. prof. chemistry, 1955—. Mem. exec. com. Samuel and Emma Winters Found., 1962—; mem. com. on awards in chemistry Fulbright Act, Nat. Acad. Scis.-NRC, 1960-63. Recipient Carnegie Teaching award, 1953. Fellow N.Y. Acad. Scis., A.A.A.S.; mem. Am. Chem. Soc., Pitts. Chemists Club, Phi Beta Kappa, Sigma Xi, Alpha Chi Sigma. Research, publs. on introduction of methods for synthesis of new organic compounds with physiol. activity. discovered agts. reducing blood pressure, affecting some functions of central nervous system or suppressing growth of certain disease organisms. Home: 36 Woodland Farms Rd., Pitts. 15238.*

SOUTHWORTH, George Clark, Am. physicist; b. Little Cooley, Pa., Aug. 24, 1890; s. Freedom and Mary (Fleek) S.; B.S., Grove City Coll., 1914, M.S., 1916, D.Sc., 1931; Ph.D., Yale, 1923; m. Lowene Smith, Aug. 14, 1913; children—Margaret Eleanor, (Mrs. Arthur G. Pulis), George Howard. Asso. physicist U. S. Bur. Standards, Washington, D.C., 1917-18; instr. and asst. prof. physics Yale, New Haven, 1918-23; radio research engr. Am. Tel. and Tel. Co., Bell Telephone Labs., Murray Hill, N.J., 1923-55, radio cons., 1955—. Recipient Morris Liebman prize Inst. Radio Engrs., 1938, Medal of Honor, 1962; Levy medal Franklin Inst., 1946, Ballantine medal, 1947. Fellow A.A.A.S., I.E.E.E., Am. Phys. Soc.; mem. Sigma Xi. Author: Principles and Applications of Waveguide Transmission, 1950; Forty Years of Radio Research, 1962. Original research leading to practical use of microwaves for radar, radio relay, radio astronomy; patentee in field. Contbr. numerous articles and research papers to sci. jours. Home: 19 Williams Rd., Chatham, N.J. 07928. Cffice: Bell Telephone Labs., Murray Hill, N.J.

SOUTHWORTH, Hamilton, Am. physician; b. N.Y.C., Apr. 7, 1907; s. Thomas Shepard and Jean (Hamilton) S.; B.A., Yale, 1929; M.D., Johns Hopkins, 1933; m. Katharine Robertson Jones, June 30, 1933; children—Hamilton Southworth, Jean (Mrs. John G. Leness), Thomas Shepard, Katharine (Mrs. R.A.K. Smith, Jr.). With Columbia Coll. Phys. and Surg., 1937—, clin. prof. medicine, 1957—; practice internal medicine, N.Y.C., 1937—; staff Presbyn. Hosp., N.Y.C.; vis. prof. medicine Am. U. Hosp., Beirut, Lebanon, 1962. Cons. in medicine Sta. Hosp., Mil. Acad., West Point, N.Y., 1947—. Diplomate Am. Bd. Internal Medicine (mem. examining bd. 1964—). Mem. A.C.P., N.Y. Acad. Medicine, Alpha Omega Alpha. Author: (with F.G.Hofmann) Columbia-Presbyterian Therapeutic Talks, vol. 1, 1963, vol. 2, 1964; also articles. Pioneered description acidosis produced by sulfanilamide, kidney stones produced by sulfapyridine; pioneered report syndrome red blood cell aplasia with tumor thymus gland. Home: 139 E. 79th St., Office 903 Park Av., N.Y.C. 10021.*

SOVA, Zdenek, veterinarian, physiologist; b. Trebíc, Czechoslovakia, Oct. 7, 1924; s. Vladimír and Aloisie (Hladka) S.; MV.Dr., Faculty Vet. Medicine, Brno, Czechoslovakia, 1950, Ph.D. in Vet. Medicine, 1961, D.Vet. Medicine, 1966; m. Magda Burilova, Mar. 25, 1953; 1 dau., Zdenek. Mil. vet. doctor, head dept. lab. diagnostics, 1950-55; head dept. internal vet. medicine U. Agr., Prague, Czechoslovakia, 1955-63, prof. physiology domestic animals, 1963—, vice dean, 1964—. Recipient Distinguished Service medal, 1955. Mem. Sci. Soc. J. E. Purkynje. Research and numerous publs. on diagnosis of liver diseases in domestic animals, especially in horse, leptospirosis in horse. Home: 4 Na ostrohu, Prague 6, Czechoslovakia.*

SOWERBY, James, English naturalist, artist; b. London, Eng, Mar. 21, 1757; s. John and Arabella S.; student flower painting Royal Acad., later botany; m. Anne Brettingham De Carle; children—James DeCarle, George Brettingham, Charles Edward. Illustrator plates for L'Herilier, French botanist, 1786-87, for Bot. Mag., 1787. Fellow Linnean Soc.; mem. Geol. Soc. Author: An Easy Introduction to Drawing Flowers According to Nature, 1788; English Botany, 36 vols., 1790-1814; Flora luxurians or The Florists Delight, 1791; Coloured Figures of English Fungi, 1797, 1815; British Mineralogy, 1803; The British Miscellany, 1804-06; The Mineral Conchology of Great Britain, 7 vols., 1812-46. Wrote, illustrated works on flowers, mineralogy, fossil shells; works contain over 3700 colored plates drawn by him (noted for care and fidelity of illustrations). Died London, Oct. 25, 1822.

SPACU, Gheorghe, Rumanian chemist; b. 1883; mem. Acad. Rumanian People's Republic. Author: (with R. Ripan) Sels complexes de magnésium, 3 vols., 1921-22; (with P. Spacu) Contribution à l'étude des sels complexes homogènes et hétérogenes en solution, 4 vols., 1930-32; Contribution à l'étude des périodures complexes, 2 vols., 1932. Research on ferramines, complex amines of manganese; drew up 125 widely used analytical methods for quantitative identification of metallic elements. Died 1955.

SPADOLINI, Igino, Italian physician; b. San Miniato, Italy, Nov. 26, 1887; s. Luigi and Enrichetta (Galli) S.; M.D., Faculty Medicine; m. Gina Vannucci, Aug. 14, 1912; children—Luigi, Enrico, Giorgio. Prof. physiology U. Florence, Italy. Recipient Silver medal for gt. merit in pub. health. Corr. mem. Lincei Acad. Research and numerous publs. on physiology. Home: via Pian dei Giullari 23, Florence, Italy.

SPAET, Theodore Hertzel, Am. physician; b. N.Y.C., June 24, 1920; s. Morris and Frances (Rosenberg) S.; B.A., U. Wis., 1942; M.D., N.Y. Med. Coll., 1945; m. Amy Elizabeth Abrams, Apr. 23, 1941; children—Linda Sue, Barbara Lynn. Practice medicine, specializing in internal medicine, N.Y.C., 1945—; instr., asst. prof. medicine Stanford Sch. Medicine, San Francisco, 1951-55; head dept. hematology Montefiore Hosp., N.Y.C., 1955—; asst. prof., asso. prof. pathology Columbia Coll. Phys. and Surg. 1955-66; prof. medicine Albert Einstein Coll. Medicine, 1966—; cons. Manhattan VA Hosp., 1960—, St. Lukes Hosp., 1966—. Diplomate Am. Bd. Internal Medicine. Fellow N.Y. Acad. Medicine; mem. Am. Soc Clin Investigation, Am., Physiol. Soc., Soc. Exptl. Biology and Medicine, Am., Internat. socs. hematology, Am. Fedn. Clin. Research, American Society for Clinical Investigation. Research, numerous publs. in normal and abnormal blood coagulation, physiology of hemostasis and devel. of thrombosis. Home: 23 Rectory Lane, Scarsdale, N.Y. 10583. Office: Montefiore Hosp., 111 E. 210th St., N.Y.C. 10467.*

SPAETH, George Link, Am. physician; b. Phila., Mar. 3, 1932; s. Edmund Benjamin and Lena (Link) S.; B.A. magna cum laude, Yale, 1954; M.D. cum laude, Harvard, 1959; m. Ann Ward, May 17, 1958; children—Kristin Lea, George Link, Eric Edmund. Intern, U. Mich. Hosp., 1959-60; resident Wills Eye Hosp., 1961-63, chief resident, 1963, sr. asst. surgeon, 1965-67, dir. glaucoma services, 1967—; clin. asso. NIH, 1963-65; practice medicine, specializing in ophthalmology, Phila., 1967—; served as lt. comdr. USPHS, 1963-65; asst. Grad. Hosp. U. Pa., 1965-66, asso., 1966—; instr. Med. Sch. U. Pa., 1965—; clin. instr. Temple U., 1966-67, asso. prof. ophthalmology, 1967—. Fellow Am. Acad. Ophthalmology and Otolaryngology, Phila. Coll. Physicians; mem. A.M.A., Assn. Research Ophthalmology, Phi Beta Kappa, Alpha Omega Alpha. Contbr. chpts. to Textbook of Dermatology, 1967; Corticosteroids and the Eye, 1967. Codiscoverer disease, homocystinuria, which is responsible for mental retardation and ocular abnormalities; noted diagnostic cataract in Glycolipid Lipidosis-Fabry; described advances in diagnosis and treatment of glaucoma; first cured human being with trichinosis, first treated homocystinuric patients with pyridoxine. Home: 15 Laughlin Lane, Phila. 19118. Office: 1930 Chestnut St., Phila. 19103.*

SPAHLINGER, Henry, Swiss bacteriologist; b. Geneva, Switzerland, Aug. 8, 1882; ed., Geneva; discovered serum treatment (named after him) for pulmonary Tb, 1912, and developed this treatment to destroy Tuberculous toxins, 1919.

SPAIN, D. M., Am. physician; b. Bklyn., July 25, 1913; s. Meyer and Clara (Rubin) S.; m. Ruth L. Borgeneht, Nov. 25, 1945; children—Julie, Amy, Lillian, Robert. Asst. prof. pathology Columbia, 1941-49, asso. prof., 1950-60; clin. prof. pathology State U. N.Y., 1961—; dir. dept. lab. research, Westchester County, N.Y., 1949-53; dir. pathology Brookdale Hosp. Center, Bklyn., 1953—. Mem. State Bd. Med. Examiners N.Y., 1964—. Fellow or mem. A.M.A., A.C.P., N.Y. Acad. Scis., Soc. Exptl. Biology and Medicine, other profl. socs. Author: Complications of Modern Medical Practice, 1965; Diagnosis and Treatment of Tumors of Chest, 1960; also articles. Editorial bd. Am. Rev. Respiratory Diseases, 1966—; asso. editor Diseases of Chest. Research in parthogenesis and devel., factors involved in coronary heart disease, emphysema, lung cancer. Home: 3 Tyleer Rd., Scarsdale, N.Y. Office: Brookdale Hosp., Brooklyn Plaza, Bklyn.*

SPAIN, Robert Jay, Am. physicist; b. N.Y.C., Dec. 4, 1937; s. Leo David and Mildred (Frisch) S.; B.S., M.S., Mass. Inst. Tech., 1959; D.Sc., Université de Paris a la Sorbonne, 1963; m. Simone Girou, Nov. 22, 1963; 1 dau., Bethsabee Laura. Physicist, IBM, Poughkeepsie, N.Y., 1957-58; physicist Lab. For Electronics, Inc., Boston, 1959—, now mgr. applied research dept. Mem. Sigma Xi, Eta Kappa Nu. Research on switching mechanisms, domain wall energies, motions and interactions in thin magnetic films, application of magnetic thin film phenomena to computer memory, logic and display devices. Home:1455 Commonwealth Av., Brighton, Mass. 02135. Office: 1075 Commonwealth Av., Boston 02215.*

SPALDING, John Frederick, Am. radiobiologist; b. Nashua, N.H., Apr. 8, 1920; s. Louis Lovejoy and Ada (Nutbrown) S.; B.S., U. N.H., 1948; M.S., Tex. A. and M., 1951, Ph.D., 1953; m. Patricia Lee Hutchins, Jan. 13, 1946; children—David John, Linda Lee. Research staff radiobiologist Atomic Warfare Directorate of Cambridge Research Center, Boston, 1952-53; sect. leader radiobiology sect. Health Research Group, Los Alamos Sci. Lab., 1953—. Mem. Radiation Research Soc., Soc. of Exptl. Biology and Medicine, Genetics Soc. Am., Sigma Xi. Research, publs. on radiation injury and recovery in mice, dogs, monkeys; radiation genetics including studies on 45 generations of irradiated mice pointing out genetic resilience of mammal. Home: 154 El Rayo St., Los Alamos 87544. Office: P.O. Box 1663, Los Alamos Sci. Lab., Los Alamos 87544.*

SPALDING, Lyman, Am. physician, surgeon; b. Cornish N.H., June 5, 1775; s. Dyer and Elizabeth (Parkhurst) S.; grad. at Charlestown, Mass., 1794; M.B., Harvard, 1797, M.D. (hon.); M.B., M.D. (hon.), Dartmouth; m. Elizabeth Coves, Oct. 9, 1802, 5 children. A founder Dartmouth Med. Sch., 1798, lectr. chemistry and materia medica, 1797-99; practiced medicine, Portsmouth, N.H., 1799-1812, also contract surgeon for U. S. Army troops in harbor; founded med. society which became Eastern Dist. br. of N.H. Med. Soc., 1802; lectured on chemistry and surgery at acad., Fairfield, N.Y., (became Coll. Physicians and Surgeons of Western Dist. N.Y., 1813), 1810-17, pres., 1813-17; practiced in N.Y.C., 1817-21. Author: Reflections on Fever, 1817. Studied yellow fever, vaccination, hydrophobia; founded U. S. Pharmacopoeia. Died Portsmouth, Oct. 21, 1821.

SPALLA, Celestino, Italian microbiologist; b. Rivanazzano, Italy, Sept. 22, 1927; s. Fulvio and Emilia (Mutti) S.; D.Agrl. Scis., U. Milan (Italy), 1953; m. Rosa Repossi, Nov. 14, 1954; 1 son, Massimo. Staff, Inst. Plant Pathology, U. Milan, 1951-54, faculty, 1965—, prof. mycology, 1965—; researcher, Research Labs., Farmitalia, Milan, Italy, 1954-59, mgr. research in indsl. microbiology, 1959—. Mem. Società Italiana di Microbiologia, Società Italiana di Scienze farmaceutiche. Research, publs. and patents in physiology, taxonomy, and genetics of microorganisms; sexual reprodn. of Mucorales; developed procedures to produce (by cultures of microorganisms), vitamin B12, ergot alkaloids, cortisol, antibiotics, anticancer substances and carotinoids. Home: 21 L. Soderini, Office: 35 dei Gracchi, Milan, Italy.*

SPALLANZANI, Lazzaro, Italian naturalist, biologist; b. Scandiano, Modena, Italy, Jan. 12, 1729; student Jessuit Coll., Reggio de Modena, 1744, U. Bologna, from 1747. Took holy orders; prof. logic, metaphyscis, Greek, U. Reggio, from 1754; prof., Modena, from 1760; held chair natural history U. Pavia (Italy), dir. Pavia Mus., from 1768. Fellow Royal Soc., 1768; mem. French Acad. Scis. Author: Saggio di osservazione microscopiche relative al sistema della generazione de' signori di Needham e Buffon, 1767; Prodromo di un'opera da imprimerse sopra le riproduzioni animale, 1768; De' fenomeni della circolazione, 1773; Dissertazioni di fisica animale e vegetabile, 1780; Viaggi alle due Sicilie ed in alcune parti dell' Apennino, 1792-97. Disproved Needham's theory of spontaneous generation of microorganisms; investigated circulation of blood, electricity of torpedo, breeding of eels; studied digestive action of saliva; described digestive action of gastric juices; studied sperm cells fertilizing ova; artificially inseminated dog; one of earliest observers to locate correctly respiratory center; investigated regeneration of legs in salamander, sensory perception bats; traveled in Mediterranean area collecting specimens for mus.; visited Naples while Vesuvius was erupting also studied volcanoes of Sicily, Lipari Islands. Died Pavia, Feb. 11, 1799.

SPANIER, Edwin Henry, Am. mathematician; b. Washington, Aug. 8, 1921; s. David H. and Anne (Goldman) S.; B.A., U. Minn., 1941; M.S., U. Mich., 1945, Ph.D., 1947; m. Toby R. Abramovich, June 14, 1942; children—Rita Vivian, Gail Nancy, Lawrence Michael. Faculty U. Chgo., 1948-59, prof. math., 1958-59; prof. U. Cal. at Berkeley, 1959—. Cons. System Devel. Corp., Santa Monica, Cal., 1960—, Nat. Security Agy., 1954-62, McGraw-Hill Book Co., N.Y.C., 1962—. Mem. Am. Math. Soc., Math. Assn. Am. Research, publs on algebraic topology and math. theory langs. Home: 1339 Contra Costa Dr., El Cerrito, Cal. 94530. Office: Dept. Math., U. Cal., Berkeley, Cal. 94720.*

SPANN, Othmar, Austrian economist; b. Altmannsdorf/Vienna, Austria, Oct. 1, 1878. Author: Die Haupttheorien in der Volkswirtschaftslehre, 1911; System der Gesellschaftslehre, 1914; Fundament der Volkswirtschaftslehre, 1918; Vom Geist der Volkswirtschaftslehre, 1919; Der wahre Staat, 1921; Irrungen des Marxismus, 1928; Hauptpunkte der universalistischen Staatsauffassung, 1929; Geschichtsphilosophie, 1932. Considered society an individual organism structured according to estates; opposed materialistic econ. theory with a complete spiritual theory of society of universalism; opponent of Marxism. Died Neustift/Burgenland, July 8, 1950.

SPAR, Jerome, Am. meteorologist; b. N.Y.C., Oct. 7, 1918; s. Nathan and Celia (Meltzer) S.; B.S., City Coll. of N.Y., 1940; M.S., N.Y. U., 1943; Ph.D., 1950; m. Frances Fernbach, Apr. 5, 1945; children—Susan Ellen, Richard Eric. Faculty, N.Y. U., N.Y.C., 1946—, prof. meteorology, 1950—, dir. meteorol. research U.S. Weather Bur., 1964-65. Mem. Am., Royal meteorol. socs., Am. Geophys. Union, Am. Assn. U. Profs., A.A.A.S. Author: Earth, Sea and Air, 1962, 65; The Way of the Weather, 1957, 62, 66; also articles. Research on theory of atmospheric tides, atmospheric energy transformations, small scale synoptic processes, numerical weather prediction, weather modification expts., radioactive fallout. Home: 18 Fieldmere Av., Glen Rock, N.J. 07452. Office: Dept. Meteorology, N.Y. U., University Heights, Bronx, N.Y.*

SPARGO, Benjamin H., Am. pathologist; b. Six Mile Run, Pa., Aug. 11, 1919; s. Benjamin H. and Lillian (Rankin) S.; B.S., U. Chgo., 1950, M.S., 1952, M.D., 1952; m. Barbara Scollard, Mar. 12, 1942; children—Janet, Patricia. Practice medicine, specializing in pathology, Chgo.; faculty U. Chgo., 1954—, prof. dept. pathology, 1964—. Mem. gen. med. research program-project com. Nat. Inst. Gen. Med. Scis., 1966—. Recipient USPHS Research Career award. Mem. Am. Soc. Exptl. Pathology, Am. Assn. Pathologists and Bacteriologists, Electron Microscope Soc. Am., Am. Soc. Clin. Pathologists, Internat. Acad. Pathology, Phi Beta Kappa, Sigma Xi, Alpha Omega Alpha. Research in patterns of renal ultrastructural reactions to injury; demonstration diagnostic lesions in serial renal biopsies taken at intervals throughout natural history various clin. syndromes has been basis for devel. exptl. models in animals to help evaluate significant sequential changes. Home: 5719 S. Kenwood Av. Office: 950 E. 59th St., Chgo. 60637.*

SPARRMANN, André, Swedish naturalist, botanist; b. Upland, Sweden, 1747; traveled to China and Cape of Good Hope; accompanied Cook as biologist on voyage around world; explored interior areas of Cape of Good Hope areas. Author: Voyage au cap de Bonne-Espérance, au cercle polaire austral et autour du monde, ainsi que dans le pays des Hottentots et des Cafres, 1787. Studied flora of China. Died 1820.

SPARROW, Arnold Hicks, radiobiologist; b. Saskatoon, Sask., Can., Oct. 2, 1914; s. Charles E. Saskatoon, Sask., Can., Oct. 2, 1914; s. Charles E. and Ella-May (Harrison) S.; B.Sc., U. Sask., 1935, M.Sc., 1938; Ph.D., McGill U., Montreal, Que., Can., 1941; m. Rhoda Cornish, Sept. 4, 1943; children—David, Alan, Paul, Carl. Came to U.S., 1941, naturalized, 1947. Instr., Harvard, 1945-47; radiobiologist Brookhaven Nat. Lab., Upton, L.I., N.Y., 1947—; vis. prof. Cornell U., 1956. Mem. div. biology and agr. NRC, 1962-67. Canadian NRC fellow McGill U., 1938-41; NRC fellow Harvard, 1941-42. Mem. Radiation Research Soc. (past mem. council), Genetics Soc. Am. (treas. 1966—, vice pres. 1967-68, pres. 1968-69), A.A.A.S. (past mem. council), Am. Soc. Cell Biology, Biophys. Soc., Bot. Soc. Am., Council Biology Editors, Soc. for Study Devel. and Growth, Sigma Xi. Author: (with John P. Binnington, Virginia Pond) Effects of Ionizing Radiations on Plants, 1958; also numerous articles. Editor-in-Chief Radiation Botany, 1960—. Research on radiobiology, structure of chromosomes, radiation cytology and genetics; factors determining specific radiosensitivity, effects of ionizing radiation on plants including tumor induction and space biology. Home: 11 Leisurely Lane, Bellport, N.Y. 11713. Office: Biology Dept., Brookhaven Nat. Lab., Upton, N.Y. 11973.*

SPARROW, Frederick Kroeber, Am. botanist; b. Washington, May 11, 1903; s. Frederick Kroeber and, Minnie (Tomlinson) S.; B.S., U. Mich., 1925; A.M., Harvard, 1926, Ph.D., 1929; m. Anna Gabler, Sept. 2, 1925; children—Frederick Tomlinson, George Burbank. Austin teaching fellow Harvard, 1926-29; asst. in botany Radcliffe Coll., 1927; asst. in plant ecology, Biol. Sta., Cold Spring Harbor, N.Y., 1926-31; instr., asst. prof., biol., Dartmouth, 1929-36; Nat. Research fellow in biol. scis. Cornell U., 1931-32, Cambridge U. (Eng.) and U. of Copenhagen (Denmark), 1932-33; research fellow Woods Hole (Mass.) Oceanographic Instn., 1934-36; research, Atkins Gardens and Research Lab., Harvard, Cienfuegos, Cuba, 1949, Botany Sch., Cambridge U., 1956; asst. prof. botany, U. Mich., 1936-44, asso. prof., 1944-49, prof. botany, 1949—; vis. prof. U. Hawaii, 1963, U. Cal. at Berkeley, 1966. Mem. com. on biology and agr. NRC, 1953-58; research asso. Mich. State Coll. Exptl. Sta., summers 1943-44. Collaborator div. cotton and other fibre crops and plant disease bur. plant industry, soils and agrl. engring., agr. research adminstrn. U. S. Dept. Agr. Fellow A.A.A.S.; mem. N.E. Bot. Club, Bot. Soc. Am., Mich. Acad. Scis., Arts and Letters (sec. 1943-47, editor 1947-52, pres. 1954-55); mem. Mycol. Soc. Am. (sec.-treas. 1945-48, v.p. 1948, pres. 1949; 7th Ann. lectr. 1958), Brit. Mycol. Soc., Washington Acad. Sci., Am. Soc. Limnology and Oceanography; asso. editor Mycologia, 1939-40, (asso.) Cambridge (Eng.) Philos. Soc., Sigma Xi. Author: Acquatic Phycomycetes, 1943, 2d edit., 1961; (with T. W. Johnson) Fungi of Oceans and Estuaries, 1961; Recipient Russell award for achievement in research and teaching U. Mich., 1944. Research and publs. on morphology, life history, fresh water and marine fungi; parasitic lower fungi. Home: 1922 Day St., Ann Arbor, Mich.*

SPATH, Leonard Frank, English geologist; b. Oct. 20, 1882; B.Sc., M.Sc., D.Sc., U. London (Eng.); m. Florence Elizabeth Sweet, 1915; 2 sons. Geol. explorations to Africa, also Am., before World War I; paleontologist geology dept. Brit. Mus. Natural History, from 1912; lectr. geology Birkbeck Coll., U. London, 1920-51. Recipient Lyell medalist Geol. Soc., 1945. Fellow Royal Soc., 1940, Geog. Soc.; mem. Royal Danish Acad. Scis. and Letters. Research, publs. on evolution, morphology and systematics of Fossil Cephalopoda. Died Mar. 2, 1957.

SPAULDING, Albert Clanton, Am. anthropologist; b. Choteau, Mont., Aug. 13, 1914; s. Thomas Claude and Willie (Clanton) S.; B.A., Mont. State U., 1935; A.M., U. Mich., 1937; Ph.D., Columbia, 1946; m. Charlotte L. Smith, 1931; children—Ronald R., Catherine E. Instr., then asst. prog. anthropology, also asst. curator Kan. U., 1946-47; mem. faculty U. Mich., 1947-61, prof. anthropology, 1960-61, curator archaeology, 1951-61; program dir. anthropology NSF, 1959-63; prof. anthropology, head dept. U. Ore., 1963-66; prof. anthropology U. Cal. at Santa Barbara, 1966—, dean Coll. Letters and Sci., 1967—. Fellow Am. Anthrop. Assn. (exec. bd. 1963—); mem. Soc. Am. Archaeology (sec. 1954-56, 1st v.p. 1958-59, pres. 1964-65). Research and publs. in Aleutian Islands and eastern U. S., archeol. theory and method, especially quantitative methods. Home: 1220 Dover Rd., Santa Barbara, Cal. 93103.*

SPAULDING, Earle Henry, Am. microbiologist; b. Rutland, Vt., Jan. 31, 1907; s. George W. and Jennie (Weaver) S.; A.B., Wesleyan U., 1929; Ph.D., Yale, 1936; m. Dorothy Wheeler, Aug. 12, 1933; children—Carolyn (Mrs. Alan C. Coman), Betty Jean (Mrs. Bruce Gladfelter), Richard K. Faculty, Temple U. Sch. Medicine, Phila., 1936—, prof. microbiology, chmn. dept., 1949—. Mem. Am. Bd. Microbiology, chmn., 1966—; com. bacteriology Nat. Bd. Med. Examiners, 1952-55. Fellow Am. Acad. Microbiology (bd. govs. 1956-60); mem. Am. Soc. Microbiology (councillor 1950-52), A.A.A.S., Am. Assn. Immunologists, A.M.A., N.Y. Acad. Sci., Reticuloendothelial Soc., Sigma Xi, Alpha Omega Alpha. Author: (with J.A. Kolmer, H. Robinson) Approved Laboratory Techniques, 1951; also chpts. in books, articles. Research on growth and characterization of anaerobic bacteria, effects of antibiotics on patients before bowel surgery. Home: 316 Evergreen Rd., Jenkintown, Pa. 19046. Office: Sch. Medicine, Temple U., Phila. 19140.*

SPEARMAN, Charles Edward, English psychologist; b. London, Eng., Sept. 10, 1863; s. Alexander Young and Louisa (Mainwaring) S.; ed. Leamington (Eng.) Coll., Würzburg, Göttingen (both Germany) univs.; Ph.D., U. Leipzig (Germany); LL.D., U. Wittenberg (Germany); m. Fanny Aikman, 1901; 1 son, 4 daus. Exptl. reader in psychology U. London, from 1907, Grote prof. mind and logic, from 1911, prof. psychology, 1928-31, emeritus, from 1931. Hon. psychol. adviser U. Chesterfield (Eng.), 1939-45. Fellow Royal Soc., 1924; mem. Kaiserlich Deutsche Akademie der Naturforscher (hon.), Brit. Psychol. Soc. (hon.), U. S. Nat. Acad. Sci. (fgn. asso.), Société Francaise de Psychologie (fgn. asso.), Deutsche Gesellschaft für Psychologie (hon.), Brit. Psychol. Soc. (pres. 1923-26), Brit. Assn. Advancement Sci. (pres. sect. J 1925); others. Author: Die Normaltäuschungen in der Lagewahrnehmung; An Economic Theory of Spatial Perception; The Method of Right and Wrong Cases without Gauss's Formulae; Correlations of Sums or Differences; The Theory of Two Factors; The Principles of Cognition, 1923; The Abilities of Man, 1927; Creative Mind, 1931; Psychology down the Ages, 2 vols., 1937; (with L. W. Jones) Human Ability, 1950. Formulated 3 qualitative, 5 quantitative gen. laws psychology; applied math. correlation to testing; laid found. of factor analysis in psychology, especially study intelligence. Died Sept. 17, 1945.

SPEARMAN, Richard Ian Campbell, English zoologist; b. London, Eng., Aug. 14, 1926; s. Alexander Young and Dorothy (Bower) S.; B.Sc. with spl. honors in Zoology, Birkbeck Coll., U. London, 1952, Ph.D., 1962. Asst. pest control research div. Ministry Agr., London, Eng., 1948-51; Rockefeller research asst. genetics dept. U. Coll., London, 1954-57; sci. staff Brit. Med Research Council dermatology dept. U. Coll. Hosp. Med. Sch., London, 1957-63, 1st asst. dermatology dept., 1963—. Fellow Linnean Soc. London, Royal Soc. Medicine; mem. Inst. Biology, Soc. for Exptl. Biology, Genetical Soc., Anat. Soc. Gt. Britain. Author (with A. Jarrett) Histochemistry of the Skin, Psoriasis, 1964; (with Jarrett, P. Riley) Dermatology, A Functional Introduction, 1966; also articles. Research on skin keratinization by histochem. methods from comparative evolutionary point view, alteration epidermal keratinization, genetics hair growth. Home: Oaks Bungalow, Oaks Av., London S.E. 19, Eng.*

SPECK, Marvin Luther, Am. microbiologist; b. Middletown, Md., Oct. 6, 1913; s. John Luther and Pearl (Wilhide) S.; B.S., U. Md., 1935, M.S., 1937; Ph.D., Cornell U., 1940; m. Jean Moler Critchlow, Sept. 11, 1940; children—Linda Jean, Martha Loraine, Susan Carol. Faculty, N.C. State U., Raleigh, 1947—, prof. food microbiology, 1951—, William Neal Reynolds prof., 1957—. Spl. cons. USPHS. Mem. Am. Acad. Microbiology, Am. Soc. Microbiology, Am. Dairy Sci. Assn. (Borden award in dairy mfg. 1959), Inst. Food Technologists, Sigma Xi, Alpha Zeta, Gamma Sigma Delta. Author: (with

R.N. Doetsch) Elementary Experiments in Dairy Bacteriology, 1955; (with others) Dairy Microbiology, 1957; numerous articles. Editorial bd. Jour. Dairy Sci., 1953——. Research on methods for more rapid and dependable growth of bacteria in mfr. cheeses; devel. of standards for high temperature pasteurization of milk products and for chem. sanitization of food equipment. Home: 3204 Churchill Rd., Raleigh, N.C. 27607.*

SPECTOR, Walter Graham, English pathologist; b. London, Eng., Dec. 20, 1924; student Queens' Coll., Cambridge (Eng.) U., 1942-44, B.A., 1945, M.A., 1949, M.B.B.ch., 1947; m. June Routley, 1957; children—Timothy David, Andrew John. Graham scholar in pathology U. Coll. Hosp. Med. Sch., 1949-51, Beit Meml. fellow., 1951-53, lectr., 1953-60, sr. lectr., 1960-62; prof. pathology U. London, St. Bartholomew's Hosp. Med. Coll., 1962——; cons. pathologist St. Bartholomew's Hosp., 1962——. Sec., Beit Meml. Fellowships for Med. Research, 1964——. Fellow Royal Coll. Physicians London; mem. Coll. Pathologists. Author: (with Roy Cameron) Chemistry of the Injured Cell, 1962; also numerous articles. Research on mechanism of inflammatory reaction especially role of histamine and other amines and polypeptides, analysis of cellular infiltration of inflamed tissues. Home: 99 Kingsley Way, London N.2. Office: Pathology Dept., St. Bartholomew's Hosp., London, Eng.*

SPEDDING, Frank Harold, Am. physicist; b. Hamilton, Ont., Can., Oct. 22, 1902 (father Am. citizen); s. Howard Leslie and Mary Ann Elizabeth (Marshall) S.; B.Chem. Engring., U. Mich., 1925, M.S., 1926, D.Sc., 1949; Ph.D., U. Cal. at Berkeley, 1929; LL.D., Drake U., 1946; D.Sc., Case Inst. Tech., 1956; m. Ethel Annie MacFarlane, June 21, 1931; 1 dau., Mary Ann Elizabeth (Mrs. Anthony J. Calciano). Faculty U. Cal., Berkeley, 1929-34, Cornell U., 1935-37; faculty Ia. State U., Ames, 1937——, prof. chemistry, 1941——, prof. physics, 1950——, prof. metallurgy, 1962——, dir. Inst. for Atomic Research, 1945——, dir. Ames Lab. AEC, 1947——. Guggenheim prof., Europe, 1935; mem. atomic bomb and Tolman coms., 1948——; tech. rep. AEC, Geneva Conf. on Peaceful Uses Atomic Energy, 1955; AEC, Dept. State rep. 5th World Power Conf., Vienna, 1956. Recipient James Douglas gold medal Am. Inst. Mining, Metall. and Petroleum Engrs., 1961; Distinguished Citizen award State of Ia., 1966. Fellow Am. Phys. Soc., A.A.A.S., mem. Am. Chem. Soc. (Langmuir award 1933, Ia. medal 1948, Nichols medal 1952), Nat. Acad. Scis., Faraday Soc., Am. Assn. U. Profs., Sigma Xi, Phi Lambda Upsilon (hon., Jubilee plaque for outstanding work in inorganic chemistry 1949), Phi Kappa Phi, Tau Beta Pi. Organized, directed chemistry div. Plutonium Project for atomic bomb research, Chgo., 1942-43; developed processes for prodn. of uranium, thorium metal, ion-exchange separations of rare-earth elements, and prodn. of rare-earth metals. Home: 520 Oliver Circle, Ames, Ia. 50010.*

SPEER, Vaughn C., Am. animal nutritionist; b. Milford, Ia., Apr. 5, 1924; s.Ira F. and Margie (Pierce) S.; B.S., Ia. State U., 1949, M.S., 1951, Ph.D., 1957; m. Mary E. Graf, Aug. 19, 1947; children—David M. James D., Elizabeth A., Diana L. With Ia. State U., Ames, 1949-51, 53——, prof. animal nutrition, 1966——; with product control dept. Ralston Purina Co., St. Louis, 1951-57, asso. animal husbandry, 1953-57. Mem. Am. Soc. Animal Scis., A.A.A.S. Research, numerous publs. on nutrient requirements of swine, mgmt. factors affection performance of swine. Home: 1428 Curtiss Av., Ames, Ia. 50010.*

SPEIDEL, Carl Caskey, Am. anatomist, biologist; b. Washington, D.C., Oct. 26, 1893; s. George and Emma (Caskey) S.; Ph.B., cum laude, Lafayette Coll., 1914; Ph.D., Princeton, 1918; Sc.D., Lafayette Coll., 1942; m. Margaret Knowles, June 17, 1920. Part-time asst. in biology, Princeton, 1914-16, Maule fellow in biology, 1916-17; instr. in zoölogy, U. Akron (O.), 1917-18; acting prof. biology, St. Lawrence U., 1919; with U. Va. Med. Sch.; 1920——, asso. prof., 1922-31, prof., 1931-64, emeritus prof., 1964——, chmn. Sch. Anatomy, 1949-59; vis. prof. biology. Randolph-Macon Woman's Coll., 1964-65, 66-67. Fellow A.A.A.S.; mem. Am. Assn. Anatomists (mem. exec. com. 1940-47), Am. Soc. Zoölogists, Am. Soc. Naturalists, Va. Acad. Sci., L.I. Biol. Assn., Phi Beta Kappa, Sigma Xi, Alpha Omega Alpha, Nu Sigma Nu. Apptd. mem. Examiners in Basic Sci. for Commonwealth of Va., 1944-56. Awarded First Barge Math. Prize, Lafayette Coll., 1912; Va. Acad. Sci. research prize, 1927; 30; Pres. and Visitors research prize, U. Va., 1930, 33, 34, 36; A.A.A.S. research prize, 1932, for notable contribution to science during 1931. Research and publs. on neurology, exptl. anatomy, radiation biology and cinephotomicrography; pioneering work in observing and recording by time-lapse movies microscopic changes in cells of living intact animals; studies of nerve, muscles and tissues in radiation-injured animals; observations in early devel. of sea urchins after irradiation of eggs and sperm. Address: 1873 Field Rd., Charlottesville, Va. 22903.*

SPEIRS, J. Murray, Canadian ornithologist; b. Toronto, Ont., Can., Apr. 7, 1909; s. Robert Miller and Jennie (McClure) S.; B.A., U. Toronto, 1931, M.A., 1938; Ph.D., U. Ill., 1946; m. Doris Louise Huestis, June 12, 1939. Demonstrator astronomy U. Toronto, 1932-33, demonstrator zoology, 1933-38, faculty, 1947——, bibliographer, instr., 1955——; with U. Ill., 1938-42, 1945-46; meteorologist Can. Dept. Transport and RAF, 1943-45. Sec., Great Lakes Research Com., 1953-55. Mem. Brit., Am. ornithologists unions, Wilson Ornithol. Soc., Fedn. Ont. Naturalists (editor 1953-61), Canadian Soc. Zoologists (sec. 1964-65). Editor: (with W. W. Judd) A Naturalist's Guide to Ontario, 1964. Research, publs. on migration of Am. robin, life history of Lincoln's sparrow, radar study of bird migration, nomenclature of channel catfish, Gt. Lakes bibliography. Home: 1815 Altona Rd., Pickering, Ont. Office: Dept. Zoology, U. Toronto, Toronto, Ont., Can.*

SPEISER, Andreas, Swiss mathematician; b. Basel, Switzerland, June 10, 1885; s. Paul and Elizabeth (Sarasin) S.; Ph.D., U. Göttingen, Germany; student U. Berlin; ed. U. London, 1909, Sorbonne, Paris, 1909-10; m. Emilie La Roche, Mar. 28, 1916; 4 children. Became Privatdocent, U. Strasbourg (France), 1911; named asst. prof. U. Zürich (Switzerland), 1917, prof., 1919-44; prof. math. U. Basel, 1944——. Dir. gen. Leonhardi Euleri opera omnia. Author: Gruppentheorie, 1923; Klassische Stücke der Mathematik, 1925; Die mathematische Denkideise, 1932; Ein Parmenideskommentae, 1936; Die geistige Arbeit, 1955. Published works of I. H. Lambert. Attempted to delineate significance of math. for sci. and art. Office: Sevogelstrasse 60, Basel, Switzerland.*

SPEKE, John Hanning, English explorer; b. Somerset, Eng., May 4, 1827; s. William and Georgina Elizabeth (Hanning) S.; ed. for army. Served in Punjab under 1st Viscount Gough, until 1854; explored Somaliland (under Richard Burton), from 1854, investigated Lake Nyassa, discovered Lake Tanganiyka, Lake Victoria Nyanza (latter independently of Burton), 1858. Recipient medal Royal Geog. Soc. Author: Journal of the Discovery of the Nile (gave information to Samuel White Baker which enabled him to discover Lake Albert Nyanza), 1863; What led to the Discovery of the Source of the Nile, 1864. Theory that Lake Victoria Nyanza was source reservoir of Nile later confirmed by him and Grant, summer 1862; (with Grant) 1st European to cross equatorial Africa. Died Bath, Eng., Sept. 18, 1864.

SPEMANN, Hans, German biologist, zoologist; b. Stuttgart, Germany, June 27, 1869; student medicine, physics, botany, zoology Heidelberg, Munich, Würzburg (all Germany) univs. Staff, Zool. Inst., Wurzburg, 1894-1908; prof. zoology, comparative anatomy Rostock, Germany, from 1908; dir. Kaiser Wilhelm Inst. Biology, Berlin-Dahlem, Germany, from 1914; prof. zoology, Freiburg-im-Breisgau, Germany, 1919-35. Recipient Nobel prize in medicine and physiology, 1935. Author: Embryonic Development and Induction, 1938; Experimentelle Beiträge zu einer Theorie der Entwicklung, 1936; Forschung und Leben, 1943. Extensive studies on embryo devel.; discovered directive function (organizer effect) certain tissues in embryo; demonstrated that morphogenesis of embryo, in main outline as well as in detail, is result of interactions between different regions of tissue; researches influential in final discredit of preformation, vitalism. Died Freiburg, Baden, Germany, Sept. 12, 1941.

SPENCE, Kenneth Wartinbee, Am. psychologist; b. Chicago, Ill., May 6, 1907; s. William James and Mary E. (Wartinbee) S.; B.A., McGill U., Montreal, Can., M.A., 1930; Ph.D., Yale, 1933; m. Isabel Ruth Temte, June 15, 1927 (div. 1960); children—Shirley, William James; m. 2d, Janet A. Taylor, December 26, 1960. Nat. Research Council fellow, Yale, 1933-34; research asst. in psychology, Yale, 1934-37; asst. prof. psychology, U. of Virginia, 1937-38; asso. prof. psychology, U. of Iowa, 1938-42, prof. and head dept. psychology, 1942-64; prof. dept. psychology U. Texas, Austin, 1964——; Silliman lectr., Yale, 1955. Recipient award Am. Psychol. Assn., 1956, Howard Crosby Warren medal, 1953. Mem. Am., Midwestern psychol. associations, National Academy of Science, Society for Experimental Psychologists, also the Sigma Xi. Author: Comparative Psychology Monographs, No. 40, 1932, No. 75, 1939; Behavior Theory and Conditioning, Yale Press, 1956. Co-author: Comparative Psychology, 1942; Handbook of General Experimental Psychology, 1951; Behavior Theory and Learning, 1960. Research on the nature of basic learning processes.

SPENCE, Leslie Percival, West Indian virologist; b. St. Vincent, W.I., Aug. 16, 1922; s. Louis Percival and Lillian (Clarke) S.; M.B., Ch.B., U. Bristol (Eng.), 1950; diploma tropical medicine and hygiene London Sch. Hygiene and Tropical Medicine, 1951, Dip. Bact., 1957; m. Phyllis Joan Haddaway, Apr. 18, 1953; children—Helen Joan, Michele Lilian. With Trinidad Govt. Med. Service, 1951-54; epidemiologist specialist pathologist, 1958-62; Trinidad Regional Virus Lab., 1954-58; faculty U. W.I., Port of Spain, 1962——, prof. virology, 1966——. Hon. cons. microbiology U. Hosp., Jamaica, W.I., 1962——. Rockefeller Found. fellow in virology, 1955. Fellow Am. Acad. Microbiology; mem. Coll. Pathologists (founding, Eng.), Royal Soc. Tropical Medicine and Hygiene (local sec. 1963——). Research, publs. on identification and determination of interrelationships among insect transmitted viruses in Eastern Caribbean and Guianas including description new virus types, viral infections of Trinidadian children especially respiratory and central nervous system virus infections. Home: 16-18 Jamaica Blvd., Port of Spain, Trinidad.*

SPENCE, Robert, English chemist; b. South Shields, Eng., Oct. 7, 1905; s. Robert and Rebecca (Robertson) S.; B.Sc. with 1st honors in Chemistry, Armstrong Coll., U. Durham (Eng.), 1926; Ph.D., Princeton, 1930, D.Sc., 1935; m. Kate Lockwood, July 25, 1936; children—Alan, Catherine, James. Lectr. phys. chemistry U. Leeds (Eng.), 1931-46; sci. adviser Hdqrs. RAF, Middle East, 1942-43, Hdqrs. MAAF, Caserta, 1943-44, Anglo Canadian Atomic Energy Lab., Montreal, Que., Can., 1945; head U.K. chemistry group Chalk River Lab., 1945-47, Atomic Energy Research Establishment, Harwell, Eng., 1947-68, head chemistry div., 1947-60, dep. dir., 1960-64, dir., 1964-68; master 3d coll., prof. applied chemistry U. Kent at Canterbury, Kent, Eng., 1968——. Companion of Bath. Commonwealth Fund fellow, 1928-31. Fellow Royal Soc., 1959, Chem. Soc., Royal Inst. Chemistry; mem. Faraday Soc. Translator: (with Hugh S. Taylor; Catalysis (G. M. Schwab), 1937. Research, publs. on chemistry of dinaphythylene compounds, 1926-28, kinetics of chain reactions, thermal and photochem., 1928-42; devel. anti-malaria spray techniques, smoke techniques; devel. plutonium separation processes; extraction of uranium from sea water. Address: Masters Lodge, 3d College, U. Kent at Canterbury, Kent, Eng.*

SPENCE, William, English entomologist; b. Hull, Eng., 1783. Research in field, London, Eng., from 1812. Fellow Royal Soc., 1834 (mem. council), Linnean Soc. (mem. council) Co-founder Entomol. Soc. London (pres. 1847) Author: (with William Kirby) Introduction to Entomology, 4 vols., 1815-26. Research, publs. on natural history. Died London, Jan. 6, 1860.

SPENCER, Donald Clayton, Am. mathematician; b. Boulder, Colo., Apr. 25, 1912; s. Frank Robert and Edith (Clayton) S.; B.A., U. Colo., 1934; B.Sc., Mass. Inst. Tech., 1936; Ph.D., U. Cambridge (Eng.), 1939, Sc.D., 1963; m. Mary Josephine Halley, July 25, 1936 (div. July 1950); children—Maredith (Mrs. Richard Arthur Church), Marianne; m. 2d, Natalie Robertson, July 7, 1951; 1 son, Donald Clayton. Instr., Mass. Inst. Tech., 1939-42; faculty Stanford, 1942-50, prof. math., 1946-50, 63——; asso. prof. Princeton, 1950-53, prof., 1953-63; mem. applied math. group NDRC, N.Y. U., 1944-45. Co-recipient Bocher prize Am. Math. Soc., 1948. Mem. Nat. Acad. Scis. Author: (with A. C. Schaeffer) Coefficient Regions for Schlicht Functions, 1950; (with M. Schiffer), Functionals of Finite Riemann Surfaces, 1954; (with H. K. Nickerson, N. E. Steinood) Advanced Calculus, 1959. Research on theory of deformation of complex structure of higher dimensional, compact complex analytic manifolds and its generalization to a systematic theory of deformation of structures on manifolds defined by transitive, continuous pseudogroups; developed procedure for defining a canonical resolution of the sheaf of germs of solutions of an arbitrary system of homogeneous linear partial differential equations; formulation of non-coercive boundary value problems. Home: 4123 Alpine Rd., Portola Valley, Cal. 94025. Office: Dept. Math., Stanford, Stanford, Cal. 94305.*

SPENCER, Edgar Winston, Am. geologist; b. Monticello, Ark., May 27, 1931; s. Terrel Ford and Allie Bell (Shelton) S.; student Vanderbilt U., 1949-50; B.S., Washington and Lee U., 1953; Ph.D., Columbia, 1957; m. Elizabeth Penn Humphris, Nov. 26, 1958; children—Elizabeth Shawn, Kristen Shannon. Lectr., Hunter Coll., N.Y.C., 1954-57; faculty Washington and Lee U., Lexington, Va., 1957——, head dept. geology, 1959——, prof., 1961——. Prin. investigator NSF research grant, 1959-61. NSF Sci. Faculty fellow, 1965-66. Fellow Geol. Soc. Am., A.A.A.S.; mem. Am. Assn. Petroleum Geologists, Nat. Assn. Geology Tchrs., Sigma Xi. Author: Basic Concepts of Physical Geology, 1962; Basic Concepts of Historical Geology, 1962; Geology, A Survey of Earth Science, 1965. Research in Precambrian history and structure of Middle Rocky Mountains, Paleozoic history and structure of Central and So. Appalachian Mountains, tectonics of Pacific Island Arc systems. Home: Poorhouse Mountain, Rockbridge County, Va. 24450. Office: Geology Dept., Washington and Lee U., Lexington, Va. 24450.*

SPENCER, Frank Cole, Am. surgeon; b. Haskell, Tex., Dec. 21, 1925; s. Frank Spencer and Lillian Josephine (Cole) S.; B.S. magna cum laude, N. Tex. State U., 1944; M.D. (Founder's Medal Scholarship award), Vanderbilt U., 1947; m. Corinne Ewell, Sept. 10, 1946; children—Elizabeth Kay, Patricia Jean, Frank Ewell. Faculty, Johns Hopkins Sch. Medicine, 1954-61, U. Ky. Sch. Medicine, 1961-65; prof., chmn. dept. surgery N.Y. U., N.Y.C., 1966——, dir. surg. divs. Bellevue Hosp., 1966——; dir. surgery U. Hosp., N.Y.C., 1966——; cons. in surgery Walter Reed Army Hosp., 1957-65, Nat. Heart Inst., NIH, Bethesda, Md., 1958-61, VA Hosp., Lexington, Ky., 1961-65. Recipient Legion Merit award USN, 1953; Markle scholar, 1956-61. Diplomate Am. Bd. Surgery, Bd. Thoracic Surgery. Mem. Am. Assn. Thoracic Surgery, A.C.S., Am. Heart Assn., A.M.A., Am. Surg. Assn., Internat. Cardiovascular Soc., Soc. Clin. Surgeons, Soc. Med. Consultants to Armed Services, Soc. U. Surgeons, Soc. Vascular Surgery. Research, numerous publs. on cardiac and vascular surgery. Home: 330 E. 33d St., N.Y.C. 10016. Office: 550 1st Av., N.Y.C. 10016.*

SPENCER, Frank Robert, Am. physician and surgeon; b. Burlington, Ia., June 12, 1879; s. Robert Spencer and Alice (Kendall) S.; A.B., U. Mich., 1900, M.D., 1902; m. Edith Clayton, Apr. 5, 1911; children—Donald Clayton, John Robert. Began as physician and surgeon, 1902; asst. Med. Faculty, U. Mich., 1902-04; mem. Med. Faculty, U. Colo., 1905—, prof. emeritus otolaryngology. Pres. Colo. Bd. Med. Examiners, 1924-26. Fellow A.C.S., Am. Otol. Soc., Am. Laryngol. Assn. (pres. 1947), Am. Laryngol., Rhinol. and Otol. Soc., Am. Acad. Ophthalmology and Otolaryngology (pres. 1941), Sect. of Laryngology, Otology and Rhinology of A.A. (chmn. of sect., 1928); mem. Am. Bd. Otolaryngology (charter), Colo. Otolaryngol. Soc. (pres.). Colo. State Med. Soc. (past pres.), Denver Clin. and Pathol. Soc. (asso.), Sigma Xi. Author of textbook on Laryngeal Tuberculosis; also author more than 75 articles in med. jours.; contbr. to textbook on nose, throat and ear and their diseases; also to Ency. of Medicine. Mem. editorial bd. Laryngoscope, St. Louis. Died 1957.

SPENCER, Herbert, English social philosopher; b. Derby, Eng., Apr. 27, 1820; son of William Geroge and Harriet (Holmes) S.; educated by father and uncle; unmarried. Engr., London & Birmingham R.R., 1837-46; an editor Economist, 1848-53; devoted fulltime to writing; 1853-1903; contrib. articles to Westminster Review. Author: Social Statics, 1850; Essays, 1851, 1863, 1874; Principles of Psychology, 1855; Synthetic Philosophy (10 vols., inc. Principles of Biology, of Sociology, of Ethics), 1860-96; Education, 1861; The Study of Sociology, 1873; The Nature and Reality of Religion, 1885; Various Fragments, 1897; Facts and Comments, 1902; Autobiography (2 vol.), 1904. attempted to synthesize scientific knowledge of his day, especially concept of evolution, and systematically apply it to all fields of human endeavor; advocate of polit. laissez-faire, believed that struggle for existence in polit. sense would lead to survival of fittest and best form of govt.; his philosophy was widely accepted in Am. Died Brighton, Eng., Dec. 8, 1903.

SPENCER, J. W. (Joseph William Winthrop), Canadian geologist; b. Dundas, Ont., Can., Mar. 26, 1851; s. Joseph and Eliza Eleanor (Coe) S.; B.A.Sc., McGill U., Montreal, 1874; A.M., Ph.D., Univ. of Göttingen, 1877; LL.D., from U. of Ala., 1913, from U. of Manitoba, 1919; m. Katherine Sinclair Thomson, 1896. Science master, Collegiate Inst., Hamilton, Ont., 1877-80; prof. geology and chemistry, King's Coll., 1880-82; prof. geology, U. of Mo., 1882-87; state geologist of Ga., 1888-93; geologist, W.I., 1894-1904; spl. commr. Geol. Survey Can., 1905-08. Fellow Geol. Soc. London, Geol. Soc. America, A.A.A.S. Author reports on Georgia, Reconstruction of the Antillean Continent, Evolution of the Falls of Niagara, History of the Great Lakes, Age of the Shores of Lake Ontario and the Modern St. Lawrence River, and many papers in scientific jours. relating to above subjects. First to show the Great Lake basins due to stream (not glacial) erosion, and to discover buried channels between them; also to describe and name their four great glacial ancestors (Lake Warren, Algonquin, Lundy, and Iroquois); by scientific measurements, was first to investigate physical changes and determine age of Niagara Falls (39,000+4,000 yrs.); demonstrated present stability of lake region, after measuring late maximum deformation and its direction; correlating these earth movements with anomalies of gravity and submarine canyons off the coast, demonstrated recent great changes of level of land and sea. "Founder of the scientific history of the Great Lakes; discoverer of the evolution of the Falls of Niagara; founder of the science of submarine valleys"; founder Mus. Historic Geology, U. of Manitoba. Died Oct. 9, 1921.

SPENCER, Jack Taif, Am. biologist; b. Mantua, O., Sept. 23, 1912; s. William B. and Margaret (Taif) S.; B.S., Kent State U., 1935; M.S., U. Wis., 1936; Ph.D., Ohio State U., 1939; m. M. Edith Woolley, Apr. 4, 1934; 1 son, George W. Agt., Bur. Plant Industry, U.S. Dept. Agr., 1936-39; asst. prof. Hiram Coll., 1939-40; asst. agronomist U. Ky., 1940-42, 45-49; research specialist Biol. and Chem. Warfare, Dept. Def., Washington, 1950-61; program dir., facilities and spl. programs NSF, Washington, 1961—. Cons. on constrn. and design of marine research facilities. Recipient Semicentennial award Kent State U., 1960. Mem. Am. Inst. Biol. Sci., A.A.A.S., Am. Soc. Botany, Am. Genetic Assn., Washington Acad. Sci., Archaeol. Inst. Am. Research, publs. in isolation of genetic factors influencing growth and devel. of root systems in plants, factors affecting interspecific and intergeneric hybridization in grasses, factors influencing seed devel in Poa pratensis and Dactylis glomerta. Home: 8302 Donnybrook Dr., Chevy Chase, Md. 20015. Office: NSF, 1800 G St., Washington 20550.*

SPENCER, Leonard James, English mineralogist; b. Worcester, Eng., July 7, 1870; s. James and Elizabeth (Bonser) S.; student Tech. Coll., Bradford, Eng., Royal Coll. Sci. Ireland, Dublin, 1886-89; B.A., Sidney Sussex Coll., Cambridge (Eng.) U., 1892, M.A., 1897; Sc.D., 1921; m. Edith Mary Close, 1899; 1 son, 2 daus. With His Majesty's Patent Office, from 1889; asst., mineral dept., keeper minerals Brit. Mus. Natural History, 1927-35; examiner, natural scis. tripos Cambridge U., 1899-1900, 31-32; faculty U. Durham (Eng.), 1907-08, 12-14. Referee for mineralogy vols. Internat. Catalogue Sci. Lit., 1901-04. Fellow Royal Soc., 1925, Mineral. Soc. Am. (hon. life, Roebling Gold medal 1940), Geol. Soc. Cornwall (Bolitho Gold medal 1931), Geol. Soc. (Murchison medal 1937), Chem. Soc., Royal Geog. Soc.; mem. German Mineral. Soc., Mineral. Soc. (pres. 1936-39), Royal Hort. Soc. Author: The World's Minerals, 1911; A Key to Precious Stones, 1936, 2d edit., 1946; Editor: Mineral. Mag., from 1900; Dictionary Lists of New Mineral Names, 1897-1946. Contbr. articles Ency. Britannica, 11th-14th edits., Dictionary Applied Chemistry (Thorpe). Helped introduce metric carat as legal unit of weight for precious stones, 1914; zinc phosphate and an iron carbide were named spencerite in his honor. Died Apr. 14, 1959.

SPENCER, Merrill Parker, Am. physician; b. Pawnee, Okla., Feb. 27, 1922; s. William Arthur and Charles Gertrude (McCabe). S.; student Okla. Bapt. U.; M.D., Baylor U., 1945; m. Jean Burdett, Aug. 18, 1964; children—Merrillyn Ann, Carla Lu, Gordon W., Joseph B. Faculty, Bowman-Gray Sch. Medicine, Wake Forest Coll., Winston-Salem, N.C., 1951-63, asso. prof. physiology and pharmacology, 1963; dir. Virginia Mason Research Center, Seattle, 1963—; established investigator Am. Heart Assn., 1959-61; vis. asso. in engring. Cal. Inst. Tech., 1962-63. Recipient USPHS Career Devel. award, 1961-63. Mem. Am. Physiol. Soc., Am. Heart Assn., N.W. Soc. for Clin. Research, I.E.E.E., Marine Tech. Soc., Alpha Omega Alpha. Research numerous publs. on hemo-dynamics heart circulation; devel. original sq. wave electro-magnetic flow meter for application to imtact arteries, transmural pressure guage for application to surgically exposes arteries, bubble detection during decompression.*

SPENCER, Richard Paul, Am. physician, biochemist; b. N.Y.C., June 7, 1929; s. David E. and Frances (Fried) S.; A.B., Dartmouth, 1951, postgrad. Med. Sch.; M.D., U. So. Cal., 1954; M.A. (NSF fellow, Helen Hay Whitney fellow), Harvard, 1958, Ph.D., 1961; m. Gwendolyn Enid Williams, Apr. 7, 1956; children—Carolyn Roberts, Jennifer Holt, Priscilla James. Faculty dept. biophysics U. Buffalo, 1961-63; chief radioisotope service VA Hosp., Buffalo, 1961-63; asso. prof. nuclear medicine Yale Sch. Medicine, New Haven, 1963—; chmn. radioisotope com. Yale-New Haven Hosp., 1963—. Mem. Am. Physiol. Soc., A.A.A.S., Soc. Nuclear Medicine, Biophys. Soc. Author: The Intestinal Tract, 1960; (with others) Biophysical Principles, 1965; also numerous articles. Research on intestinal metabolism and transport, quantitation of prenatal and postnatal growth, use of radioactive isotopes as tracers, determination organ dimensions and weight. Office: Yale-New Haven Hosp., New Haven 06504.*

SPENCER, Roy Clarence, Am. physicist; b. Pennellville, N.Y., Apr. 14, 1901; s. Clarence B. and Clarissa (Gillespie) S.; A.B., Cornell U., 1922; Ph.D., Columbia, 1932; m. M. Blanche Wheelwright, June 3, 1929; children—Barbara J. (Mrs. Dale R. Paape), Dana R. Fellow in astronomy Swarthmore Coll., 1922-23; research asst. x-rays Cornell U., 1923-24; radio tube devel. engr. Westinghouse Lamp Co., Bloomfield, N.J., 1926-27; instr. physics Columbia, 1927-31; instr. to asso. prof. physics U. Neb., 1931-41; staff mem. radiation lab. Mass. Inst. Tech., 1941-46; chief antenna lab. Air Force Cambridge Research Center, Bedford, Mass., 1946-55; sr. engring. specialist Sylvania Electric Products, Inc., Waltham, Mass., 1955-58; prin. staff scientist Martin Co., Balt., 1958-61; mgr. antenna research RCA, Moorestown, N.J., 1961-64, staff scientist, 1964—. Co-founder Interservice Antenna Group, 1954. Fellow A.A.A.S., Am. Phys. Soc., I.E.-E.E. (editorial bd. I.R.E. 1954-59); mem. Sci. Research Soc. Am. (1st pres. Air Force Cambridge Research Center Branch), Internat. Sci. Radio Union (del. internat. assemblies 1952, 54, 57, chmn. subcommn. on microwave optics 1952-57, mem. commn. 6), Sigma Xi. Pioneer in application of optics and Fourier methods to design of microwave antennas; research on x-rays with double crystal x-ray spectrometer, 1930-38, operational analysis instruments, 1931-41. Home: 102 Devon Rd., Cinnaminson, N.J. 08077. Office: RCA, Moorestown, N.J. 08057.

SPENCER, Sir Walter Baldwin, Brit. biologist, ethnographer; b. Stretford, Eng., June 23, 1860; s. Reuben and Martha (Circuit) S.; student Owens Coll., Manchester, Eng.; B.S., Exeter Coll., Oxford, Eng.; m. Mary Elizabeth Bowman, Jan. 1887; 1 dau. Became asst. to H. N. Moseley, 1885; named fellow Lincoln Coll., 1886, prof. emeritus, 1919; named hon. fellow Exeter Coll., 1906. Spl. commr. for Commonwealth Govt.; chief protector aborigines in No. Terr. Fellow Royal Soc., 1900. Author: (with F. J. Gillen) Native Tribes of Central Australia, 1889, Northern Tribes of Central Australia, 1904, The Arunta, a Study of a Stone Age People, 1927, Across Australia, 1912; Native Tribes of the Northern Territory of Australia, 1914; Wanderings in Wild Australia, 1928; Spencer's Last Journey, 1931, Ethnological research in continental Australia; discovered unknown tribes. Died Hoste Island off Tierra del Fuego, July 14, 1929.

SPENCER, William Franklin, Am. chemist; b. Carlinville, Ill., Mar. 4, 1923; s. Jesse H. and Mayme (Wohlert) S.; student Blackburn Coll., 1940-42; B.S., U. Ill., 1947, M.S., 1950, Ph.D., 1952; m. Marjorie Ann Hall, June 2, 1946; children—Barbara Annette, William Franklin II, Gary Alan. Asst. in soil physics U. Ill., 1948-49; asst. chemist U. Fla. Citrus Sta., Lake Alfred, 1951-54, asso. soil chemist, 1957-62; soil scientist U.S. Dept. Agr., Laramie, Wyo., 1954-55, Brawley, Cal., 1955-57, research soil scientist, Riverside, Cal., 1962—; cons. soils research Central U., Maracay, Venezuela, 1959. NRC AEC fellow U. Ill., 1949-51. Mem. Am. Soc. Agronomy, Soil Sci. Soc. Am., A.A.A.S., Clay Mineral Soc., Sigma Xi. Research, publs. field soil chemistry and plant nutrition with emphasis on clay minerals, cation exchange reactions, chemistry of soil phosphorus, fertilizer needs, and interactions between phosphates and other nutrient elements as related to growth and development of citrus and various field crops. Home: 2935 Arlington Av., Riverside, Cal. 92506. Office: U. Cal. Riverside, Cal. 92507.*

SPENCER-JONES, Sir Harold, English astronomer; b. London, Mar. 29, 1890; became dir. Cape Obs., 1923; dir. Greenwich Obs., 1933-55; mem. French Acad. Scis., Royal Acad. Eng., Royal Astron. Union (pres. 1937), Internat. Astron. Union (pres. 1945-48). Research on shifting of earth's poles, irregularities of earth's movement, calculation of masses of planets, value of solar parallax. Died June 1960.

SPENGLER, Oswald, German social philosopher; b. Blankenburg-am-Harz, Germany, May 29, 1880; ed. in math., philosophy, history, art Munich, Berlin (both Germany) univs.; Ph.D., U. Halle (Germany), 1904. Sch.-master, Hamburg, Germany, 1908-11; later writer, Munich. Author: Der Untergang des Abendlandes (presented philosophy of history based on premise that every civilization goes through cycle of devel., decline analogous to biol. life cycle), 1918-22; Preussentum und Sozialismus, 1920; Pessimismus, 1921; Politische Pflichten der deutschen Jugend, 1924; Neuban des deutschen Reiches, 1924; Der Mensch und die Technik, 1931; Jahre der Entscheidung, 1933; Politische Schriften, 1932; Reden und Aufsätze, 1937. Predicted downfall of Western civilization; supported German Nat. Socialism, but opposed racist doctrines. Died Munich, May 8, 1936.

SPERANKI, Georgil Nestorovich, Russian pediatrician; b. Feb. 20, 1873; grad. U. Moscow (USSR), 1898. Staff, U. Moscow, 1898-1909; asso. Central Sci. Research Inst. for Care of Mothers and Infants, 1922-48, dir., 1925-30; also pres. Central Inst. for Advancement Doctors; dir. dept. for newborn children Abrikosova Maternity Home, 1907-10; asst. Children's Clinic, Moscow Higher Women's Courses, 1915-18; head chair children's diseases Kuban U., Krasnodar, 1919-20; dir. Home for Child Care, Moscow, 1921-22; prof., then head chair child pediatrics Central Postgrad. Med. Inst., Moscow, 1934—; asso. Inst. Pediatrics, USSR Acad. Med. Sci., 1951—; founder Children's Consultation Clinic, Abrikosova Maternity Home, Moscow, 1907, 1st hosp. with polyclinic, consultation clinic, milk kitchen, 1910; founder chair children's diseases Kuban U., 1919, Home Child Care, 1921. Mem. learned med. council USSR Ministry Health; mem. qualification commn. Higher Certifying Commn. Recipient Order of Lenin (4). Mem. USSR Acad. Scis. (corr.), USSR Acad. Med. Scis. (chmn. commn. for awarding nominal prizes for best works on pediatrics 1959—), All-Union Soc. Pediatrics (hon., chmn. 1957), Moscow Soc. Pediatricians (hon. chmn.). Author: The Study of Septics in Children of an Early Age, 1947; Dysentery in Infants, 1952; Chronic Nutritive Disturbances in Young Children, 1953; The Mother's ABC; The Healthy Child; Mother and Child; Manual of Infantile Diseases. Chief editor Pediatriya, 7 vols., 1956; editor pediatrics sect. Large Med. Ency., 1st and 2d edit. Editor, Pediatriya, 1922—; founder Jour. for Study Infants, 1922; Research and publs. on disturbances in digestion, nutrition and gastric ailments in young children, pneumonia, grippe and septics in newborn, pathology of older children; tested use and dosage of drugs for various ages of children. Home: ul. Chaplygina 22, Moscow, USSR. Office: Acad. Med. Scis. USSR, Solyanka 14, Moscow, USSR.

SPERRY, Elmer Ambrose, Am. elec. engr.; b. Cortland, N.Y., Oct. 12, 1860; s. Stephen Decatur and Mary (Burst) S.; student State Normal and Training Sch., Cortland, N.Y., 1876-79; Cornell U., 1879; 80; E.D., Stevens, 1921, Lehigh, 1927; Sc.D.., Northwestern, 1925; m. Zula A., d. Edward Goodman, June 28, 1887; children—Mrs. Helen Marguerite Lea, Edward Goodman, Lawrence Burst (dec.), Elmer Ambrose. Founder, 1880, Sperry Electric Co., Chicago, mfrs. arc lamps, dynamos, etc.; founder Sperry Electric Ry. Co., mfrs. cars, Cleveland, O.; pres. Sperry Gyroscope Co., Brooklyn, 1910-26, chmn. bd., 1926-29; pres. Sperry Development Co., Inc.; mfg. own inventions, 1910. Member Naval Consulting Board, 1915— (chmn. coms. on aeronautics, mines and torpedoes, aids to navigation). Awards: First prize, Aero Club of France, 1914; Franklin medal, Philadelphia, 1914; grand prize for gyro-compass and gyroscopes, San Francisco Expn., 1915; Collier trophy, 1915, for drift set, 1916; Scientific Am. medal, Am. Mus. of Safety; also awarded John Fritz medal in 1927, Holley medal, 1927; Franklin Inst. medal, 1929; Am.

Iron and Steel Inst. medal, 1930. Decorated by Czar Nicholas of Russia for navigation equipment; decoated by Emperor of Japan with Order of Rising Sun, and Order of Sacred Treasure. Mem. Am. Inst. E.E. (a founder), Am. Electrochem. Soc. (a founder), Nat. Acad. Science, Nat. Research Council (chmn. div. engring. and industrial research, 1928-30); chmn. Am. Com. World Engring. Congress, Tokyo, 1929. Author numerous papers. Erected 350-foot electric beacon on Lake Michigan in 1883; also invented 1st electric chain mining machine, 1888; devised detinning and electrochem. processes and machinery for making fuse wires; designed electric automobile; held over 400 patents issued in U. S. and Europe; inventor of gyro-compass, aeroplane and ship stabilizers, highest intensity searchlight (1 1/2 billion candle power), compound internal combustion engine, fire control apparatus, gyro track recorder, transverse fissure detector, street lighting system, and numerous special devices. Died Brooklyn, N.Y., June 16, 1930.

SPERRY, Roger Wolcott, Am. psychobiologist; b. Hartford, Conn., Aug. 20, 1913; s. Frances Bushnell and Florence (Kraemer) S.; A.B., Oberlin Coll., 1935, A.M., 1937; Ph.D., U. Chgo., 1941; m. Norma Gay Deupree, Dec. 28, 1949; children—Glenn Tad, Janeth Hope. NRC fellow Harvard, 1941-42; research aso. Yerkes Labs., 1942-46; faculty U. Chgo., 1946-52, asso. prof., 1952-53; chief sect. on developmental neurology Labs. Anat. Scis., NIH, 1952-53; Hixon prof. psychobiology Cal. Inst. Tech., Pasadena, 1954—. Recipient Oberlin Coll. Distinguished Alumni citation, 1954. Fellow A.A.A.S., Am. Acad. Arts and Scis., Am. Psychol. Socs.; mem. Nat. Acad. Scis., Am. Assn. Anatomits, Am. Physiol. Soc., Psychonomic Soc., Am. Soc. Zoologists, Soc. for Study Devel. and Growth, Am. Soc. Naturalists, Internat. Brain Research Orgn. Research, numerous publs. on central nervous readjustment after nerve muscle, skin and eye transplantions, patterned growth of behavioral nerve nets in devel. and regeration, electric field theory and cerebral correlates of perception, split-brain approach to cerebral orgn., basis of conditioned response and mind-brain problems. Home: 1369 Boston St., Altadena, Cal. 91001. Office: 1201 E. California St., Pasadena, Cal. 91109.*

SPERTI, George Speri, Am. biophysicist; b. Covington Ky., Jan. 17, 1900; s. George and Caroline (Speri) S.; E.E., U. Cin., 1923; Sc.D., U. Dayton, 1934, Duquesne U., 1936, Bryant Coll., 1957. Baldwin fellow, research asst. U. Cin., 1923-24, research prof., dir., co-founder Basic Sci. Research Lab., 1925-35, fellow Grad. Sch., 1930-35; NRC fellow, 1929-31; Rockefeller Found. fellow, 1930-31; dir., pres., co-founder Institutum Divi Thomae, Cin., 1935—; dir. Franklin Corp.; tchr. Internat. U., Rome, 1958. Recipient appointment to Pontifical Acad. Scis., 1936; Cath. Action medal, Mendel medal, 1943; Christian Culture award, 1947; Star of Solidarity, Italian Republic, 1956; Internat. U. medal, 1958. Mem. A.A.A.S., Am. Phys. Soc., Am. Soc. Plant Physiologists, Am. Chem. Soc., Optical Soc. Am., Italian Soc. Physics, Ohio Acad. Sci., Sigma Xi, Eta Kappa Nu, Tau Beta Pi. Author: (with Herman Schneider) Quantum Theory in Biology, 1926; Correlated Investigations in the Basic Sciences, 1928. Editor-in-chief Studies of the Institutum Divi Thomae, 1937—. Contbr. numerous articles to profl. jours. Discovered some laws governing actions of radiations in biology; intercellular substances which regulate cellular metabolism; invented K-va polyphase power-factor meters; developed processes and products for food and drug industries, electronic and irradiation devises, gaseous discharge and illumination products, aviation instruments. Home and office: 1842 Madison Rd., Cin. 45206.*

SPETER, Max, historian of chemistry; b. Bistritz, Rumania, Apr. 1, 1883; s. Johann and Anna (Dollberg) S.; student Technische Hochschule, Budapest, Hungary, Hanover, Munich, Germany; Diploma in Engring., Technische Hochschule, Munich; Ph.D., U. Berlin, 1910. Asst. to Oskar von Miller, founder ISIS; staff Deutsches Mus.; worked for various chem. firms; asst. to Dr. R.S. Meyer, U. Berlin; operator tungsten lab., Chalottenberg, Germany, 1910-12. Author: The Chemical Elements, 1911; also numerous articles; contbg. author: Buch der grossen Chemiker, 1929-30. Discovered Boyle-Hanckknitz recipe for phosphorus, 1929; helped in solution of difficult hist. chem. questions. Died June 30, 1942.

SPETNER, Lee Mordecai, Am. physicist; b. St. Louis, Jan. 17, 1927; s. Abraham Isaac and Rose (Raskas) S.; B.S., Washington U., St. Louis, 1945; Ph.D., Mass. Inst. Tech., 1950; m. Julia Borvick, Oct. 22, 1950; children—Abba Isaac, Shlomo Abraham, Daniel Simeon, Sharon Pearl. Research asst. Mass. Inst. Tech., Cambridge, 1948-50, research asso., 1950-51; physicist, sr. staff Johns Hopkins Applied Physics Lab., Silver Spring, Md., 1951-58, prin. profl. staff, 1958—, lectr. evening coll., 1961—; lectr. Howard U., 1955-60. William S. Parsons fellow Johns Hopkins, 1962-63. Mem. Am. Phys. Soc., Internat. Sci. Radio Union, I.E.E.E., A.A.A.S., Sigma Xi. Showed how to make practical estimates of important statis. property of a random signal (power spectral density) from a finite sample; developed theory of radar reflections from ocean surface. Contbr. articles to profl. jours. Home: 7801 13th St. N.W., Washington 20012. Office: 8621 Georgia Av., Silver Spring, Md. 20910.*

SPETTOWA, Stanisawa Maria Janczyszyn, Polish physician; b. Lvov, Lemberg, Poland, Feb. 12, 1902; d. Julian and Julia (Amalowicz) Janczyszyn; M.D., U. Lvov, 1927; postgrad. U. Vienna (Austria), U. Stockholm (Sweden); m. Karol Spett, Apr. 21, 1930. Sr. asst. neurol. dept. U. Lvov, 1935-45; asso. prof. U. Warsaw (Poland), 1951-54; prof. neuroradiology Med. Acad. Cracow (Poland), 1954—; chief neuroradiol. dept. Med. Acad. Cracow, 1945—. Chief cons. neuroradiology for Poland, 1954—; head postgrad. studies in neuroradiology, 1950—. Recipient Golden Cross of Merit, 1954. Mem. Polish Neurol. Assn. (chmn. neuroradiol. sect. 1962—). Author: (with A. Kunicki) Choroby ukladu nerwowego, 1951; also articles. Angiographic studies of venous system and pathologic vascularisation in brain tumors; orgn. neuroradiology in Poland as autonomic br. radiology. Home: 3 Botaniczna. Office: 48 Kopernika, Cracow, Poland.*

SPEUSIPPOS, mathematician, natural philosopher; flourished 347-339 B.C.; s. Eurymedon and Potone (sister of Plato); asso. with Plato at Acad.; accompanied Plato on his last visit to Syracuse; head Acad., 347-339 B.C. Attempted to classify various animal and plant species by their resemblances to each other; emphasized differences and similarities rather than 1st principles; studies in math. theory, ethics.

SPICER, Brian Milton, Australian physicist; b. Melbourne, Australia, Oct. 7, 1928; s. William Milton and Mavis (Rees) S.; B.Sc., U. Melbourne, 1950, M.Sc., 1952, Ph.D., 1956, D.Sc., 1965; m. Lesley Patricia Parry, Feb. 21, 1953; children—Carol Mavis, David Brian, Trevor Leslie. Faculty, U. Melbourne, 1956—, prof. physics, 1965—, asso. dir. nuclear research, 1963-65. Vis. prof. physics U. Va., 1964, Ia. State U., 1964-65. Recipient David Syme Research medal, 1959. Fellow Australian Inst. Physics (chmn. Victorian br. 1963), Inst. Physics and Phys. Soc. (chmn. Victorian div. Australian br. 1962); mem. Am. Inst. Physics. Author: (with D. E. Caro, J. A. McDonell) Modern Physics, 1961 (pub. London 1962, U. S. under title An Introduction to Atomic and Nuclear Physics 1964); also articles. Research on interaction of x-radiation with atomic nuclei, theoretical and exptl. research on nuclear structure, structure of photonuclear giant resonance and its relation to shape of nucleus. Home: 3 Tovey St., North Balwyn, Victoria. Office: Sch. Physics, U. Melbourne, Parkville, Victoria, Australia.*

SPICER, Edward Holland, Am. anthropologist, b. Cheltenham, Pa., Nov. 29, 1906; s. Robert Barclay and Margaret (Jones) S.; student U. Del., 1925-27, Johns Hopkins U., 1927-28; B.A., U. Ariz., 1932, M.A., 1933; Ph.D., U. Chgo., 1939; m. Rosamond Brown, June 21, 1936; children—Robert Barclay, Margaret Pendelton, Lawson Allan. Instr. social anthropology Dillard U., 1938-39; instr. anthropology U. Ariz., Tucson, 1939-41, faculty, 1946—. prof., 1950—; vis. prof. Cornell U., 1947-50, U. Cal., Santa Barbara, 1958. John Simon Guggenheim Found. fellow Research in Mexico, 1941-42, 55-56; NSF Sr. postdoctoral fellow, 1963-64; social sci. analyst War Relocation Authority, Ariz., 1942-43, head community analysis sect., Washington, 1943-46.Recipient Southwest Library Assn. award, 1964. Mem. Am. Anthrop. Assn., A.A.A.S., Soc. Applied Anthropology, InterAm. Indian Inst., Internat. Congress Americanists. Author: Pascua, a Yaqui Village in Arizona, 1940; Potam, a Yaqui Village in Sonora, 1954; Cycles of Conquest, 1962. Editor: American Anthropologist, 1960-64. Research Yaqui Indian culture, culture history, acculturation among Southwest and Mexican Indians, theory acculturation. Home: 5344 E. Ft. Lowell Rd., Tuscon 85716.*

SPICER, Samuel Sherman, Jr., Am. pathologist; b. Denver, Aug. 12, 1914; s. Samuel Sherman and Eleanor (Kirk) S.; B.S., U. Colo., 1935, M.D., 1939; m. Gettrude E. McRae, Dec. 27, 1941; children—Kenneth, Eleanor, Samuel Sherman. Staff, NIH, Bethesda, Md., 1941-66, chief sect. biophys. histology Nat. Inst. Arthritis and Metabolic Diseases, 1960-66; prof. pathology Med. Coll. S.C., Charleston, 1966—. Mem. Am. Soc. for Cell Biology, Am. Soc. for Exptl. Pathology, A.M.A., Histochem. Soc., Internat. Acad. Pathology, Royal Micros. Soc. Contbg. author: Methods and Achievements in Experimental Pathology, vol. 2, 1967. Research, numerous publs. on nutrition including folic acid deficiency, indsl. hygiene including methemoglobinemia and erythocyte metabolism, chemistry of muscle proteins including action of trinucleotides on contractile muscle, histochemistry, cytochemistry, exptl. pathology; characterized various types of acid mucosubstances. Home: 546 N. Hobcaw Dr., Mt. Pleasant, S.C. 29464. Office: Med. Coll. S.C., Charleston, S.C. 29464.*

SPICER, William Edward, III, Am. physicist; b. Baton Rouge, Sept. 7, 1929; s. W.E., II and Kate Crystal (Watkins) S.; B.S., William and Mary Coll., 1949; B.S., Mass. Inst. Tech., 1951; M.A., U. Mo., 1953, Ph.D., 1955; m. Cynthia Stanley, June 12, 1951; children—William Edward, IV, Sally Ann. Tech. staff RCA Labs., Princeton, N.J., 1955-62; vis. scientist Lawrence Radiation Lab., U. Cal. at Livermore, 1961-62; faculty Stanford, 1962—, prof. elec. engring. and materials sci. depts., 1965—. Cons. to various cos., govt. agys.; mem. solid state scis. panel Nat. Acad. Sci.-NRC, 1965—, panel 222.00 physics div. Nat. Bur. Standards, 1966—. Recipient RCA Achievement award, 1957, 60. Fellow Am. Phys. Soc.; mem. Phi Beta Kappa, Maxwell Soc. Research, publs. on solid state physics and applications including theory of photoemission, optical properties of solids, electronic structure of metals, semiconductors and insulators. Home: 395 Cervantes Rd., Portola Valley, Cal. 94025. Office: Stanford U., McCullough Bldg., Stanford, Cal. 94305.*

SPIEGEL, Ernest Adolf, physician; b. Vienna, Austria, July 24, 1895; s. Ignaz and Elsie (Fuchs) S.; M.D., U. Vienna, 1918; M.D. (hon.), U. Zürich (Switzerland), 1965; m. Anna S. Adolf, Aug. 1, 1925. Came to U.S., 1930, naturalized, 1936. Docent anatomy and physiology nervous system U. Vienna, 1924-30; prof. exptl. and applied neurology, head dept. exptl. neurology Temple U. Med. Center, Phila., 1930-66, prof. emeritus, 1967—. Fellow A.A.A.S., Am. Electroencephalography Soc. (hon.); mem. Coll. Physicians Phila., Am. Acad. Neurology, Am. Neurol. Assn., Am. Physiological Society, Epilepsy League, Am. Therapeutic Soc., German Neurosurg. Soc. (Otto Foerster medal 1964, hon.), Vienna Med. Soc. (corr.), Vienna Psychiat. Neurol. Soc. (corr.). Author: Muscle Tonus, 1924; Centers of Autonomic Nervous System, 1926; Experimental Neurology, 1930; (with I. Sommer) Neurology of Eye and Ear, 1930; (with H.T. Wycis) Stereoencephalotomy, vol. I, 1952, vol. 2, 1962; also numerous articles. Chief editor: Confinia Neurologica, 27 vols. 1938; Progress in neurology and Psychiatry, 22 vols., 1945. Developed method introducing guided electrodes into human brain for treatment involuntary movements, intractable pain, emotional disorders, certain types convulsive disorders, brain tumors, study physiology of deep cerebral structures; research of part of internal ear related to equilibrium and its centers. Home: 6807 Lawnton Av., Phila. 19126.*

SPIEGEL, Melvin, Am. biologist; b. Bklyn., Dec. 10, 1925; s. Philip Edward and Sadie (Friedman) S.; B.S., U. Ill., 1948; Ph.D., U. Rochester, 1952; m. Evelyn Sclufer, Apr. 16, 1955; children—Judith Ellen, Rebecca Ann. Research fellow Cal. Inst. Tech., 1953-55; faculty Colby Coll., Waterville, Me., 1955-59; prof. biology Dartmouth, Hanover, N.H., 1959—. Mem. Marine Biology Lab. Corp., Woods Hole, Mass., 1956—, editorial bd. Biol. Bull., 1966—. Mem. cell biology study sect. NIH, 1966—. Fellow A.A.A.S.; mem. Soc. for Developmental Biology, Am. Inst. Biol. Sci., Am. Assn. U. Profs., Soc. Gen. Physiologists, Internat. Inst. Embryology, Am. Soc. Zoologists, Sigma Xi. Research, publs. in control mechanisms in enzyme and protein synthesis during devel., immunological reactions in cell adhesion. Home: 15 Barrymore Rd., Hanover, N.H. 03755.*

SPIEGEL-ADOLF, Mona, biochemist; b. Vienna, Austria, Feb. 23, 1893; d. Jacques and Hedwig (Spitzer) Adolf; M.D., U. Vienna, 1918, Docent, 1931; m. Ernest A. Spiegel, Aug. 1, 1925. Came to U.S., 1931, naturalized, 1936. Prof., head dept. colloid chemistry Temple U. Med. Sch., Phila., 1930-66. Fellow A.A.A.S.; mem. Am. Soc. Biol. Chemists, Biochem. Soc. London, Am. Optical Soc., Am. Chem. Soc., Am Crystallographic Assn. Author: The Globulins, 1930; (with G.C. Henry) X-Ray Diffraction in Biology and Medicine, 1947; also numerous articles. Research on proteins, lipids, effect of radiation; studies of Amaurotic fam. Idiocy disease. Home: 6807 Lawnton Av., Phila. 19126.*

SPIEKER, Edmund Maute, Am. geologist; b. Balt., Feb. 25, 1895; s. Edward Henry and Adelaide (Maute) S.; A.B., Johns Hopkins, 1916, Ph.D., 1921; m. Helen Frances Heard, Dec. 28, 1922; 1 son, Andrew Maute. With U. S. Geol. Survey, 1917-65, geologist, 1921-24, part-time 1924-65; with Imperial Oil, Ltd., 1919-40; faculty Ohio State U., Columbus, 1924—, prof. geology, 1932-52, research prof., 1952-65, research prof. emeritus, 1965—, chmn. dept. geology, 1944-52. Mem. panel on geology research and devel. bd. U. S. Dept. Def., 1950-57, chmn., 1953-57; cons. NSF, 1959-64. Fellow Geol. Soc. Am., A.A.A.S., Ohio Acad. Sci., Paleontol. Soc.; mem. Am. Assn. Petroleum Geologists (Distinguished lectr. 1950, 52), Soc. Econ. Paleontologists and Mineralogists, Geol. Soc. Washington, History of Sci. Soc., Am. Assn. U. Profs., Phi Beta Kappa, Sigma Xi. Author: Paleontology of the Zorritos Formation of the North Peruvian Oil Fields, 1922; The Wasatch Plateau Coal Field, 1931; The Transition Between the Colorado Plateaus and the Great Basin in Central and Eastern Utah, 1949; also articles. Research on prins. mountain bldg. chronology especially relating to central N.Am. Cordillera, geology coal fields and Cretaceous-Tertiary stratigraphy and history of central-eastern Utah, exploration of Rocky Mountain front, N.W. Can. Home: 4793 Olentangy Blvd., Columbus, O. 43214.*

SPIELMANN, Jacob Reinbold, chemist; b. Strasbourg, France, Mar. 31, 1722; M.D., 1748; studied under Pott, Marggraf, Henckel, Geoffroy. Practice medicine, Berlin; became prof. chemistry Strasbourg, 1749, prof. Greek and Latin poetry, 1754, prof. medicine, 1759; taught chemistry to Goethe. Fellow French Acad. Scis. Author: Institutiones chemiae praelectionibus academicis accomodatae, 1763; Institutiones materiae medicae, 1774; Delectus dissertationum medicarum argentoratensium, 1777-81; Kleine praktische medizinische und chemische schriften, 1786. Studies and publs. on soap, elastic mineral resin, tartar, phosphoric acid, acidum pingue, neutral salts, relation between solubility and temperature. Died Sept. 9, 1783.

SPIES, Joseph Reuben, Am. chemist; b. Madison, S.D., Nov. 5, 1904; s. Joseph M. and Helen (Dahlberg) S.; A.B., U. S.D., 1927; M.S., U. Md., 1931, Ph.D., 1934; m. Renice Numbers, Nov. 26, 1930; 1 son, Carl Joseph. With U. S. Dept. Agr., 1930—, research chemist allergens investigations dairy products lab. Eastern Utilization Research and Devel. div., Washington, 1966—. Co-recipient Hillebrand award Chem. Soc. Washington, 1950. Fellow A.A.A.S.; mem. Am. Chem. Soc., Soc. Exptl. Biology and Medicine, Am. Soc. Biol. Chemists, Am. Acad. Allergy, Phi Beta Kappa, Sigma Xi, Alpha Chi Sigma. Author: Cats and How I Photograph Them, 1958; The Compleat Cat, 1966; also numerous articles. Research in isolation, chemistry and toxicity on fish poisons, croton resin and crotooside, allergens of oil seeds, methodology of amino acid analysis, particularly the determination of tryptophan in proteins. Home: 507 N. Monroe St., Arlington, Va. 22201. Office: 12th and Independence Sts. S.W., Washington 20250.*

SPIES, Tom Douglas, Am. physician; b. Revenna, Tex., Sept. 21, 1902; s. John Earl and Mary (Love) S.; A.B., U. Tex., 1923; M.D., Harvard, 1927; Sc.D. (hon.), U. of South, 1944. Intern in pathology Peter Bent Brigham Hosp., Boston, 1928-29; 1st asst. to Dr. F. B. Mallory, Boston City Hosp., 1929-30; intern Lakeside Hosp., Cleve., 1930-31; teaching fellow Western Res. U., 1931-32, instr. in medicine, 1932-34, sr. instr. medicine, 1934-35; asst. prof. medicine U. Cin. Coll. Medicine, 1935-36, asso. prof. medicine, 1936-47; vis. prof. medicine U. Ala., 1941—; prof. nutrition and metabolism, chmn. dept. Northwestern U. Med. Sch., 1947—; dir. Nutrition Clinic, Hillman Hosp., Birmingham, Ala., 1936—. Apptd. to Food and Nutrition Bd., NRC, 1943; apptd. cons. to Sec. of War on Tropical Medicine Army Med. Sch., Washington, 1945. Recipient John Phillips Meml. award A.C.P., 1939; Scientific award Am. Pharmaceutical Mfrs. Assn., 1941; So. Med. Assn. Research medal, 1943; Distinguished Achievement award Modern Medicine mag., 1957; Distinguished Service award A.M.A., 1957; Oscar B. Hunter Meml. award in Therapeutics, Am. Therapeutic Soc., 1959. Diplomate Am. Bd. Internal Medicine. Fellow A.C.P., Royal Soc. Tropical Medicine and Hygiene (Eng.); mem. A.M.A., A.A.A.S., Am. Assn. Pathologists and Bacteriologists, Am. Inst. Nutrition, Am. Soc. Clin. Investigation, Am. Soc. Exptl. Pathology, Assn. Am. Physicians, Am. Soc. Tropical Medicine, Central Soc. Clin. Research, Research Club, Soc. Exptl. Biology and Medicine, So. Med. Assn., Sigma Xi, Phi Beta Kappa. Co-author (with Dr. R. R. Williams) Vitamin B1 and Its Use in Medicine, 1938; also numerous sci. articles to med. jours. Contbd. to knowledge to Tb, arteriosclerosis, deficiency diseases, diseases of metabolism; demonstrated (with Vileter, Koch, Caldwell) that folic acid is element in maturation of erythrocytes, 1945. Died Feb. 28, 1960.

SPIESS, Eliot Bruce, Am. geneticist; b. Boston, Oct. 13, 1921; s. George Nicholas and Rena (Bunce) S.; A.B. cum laude, Harvard, 1943, A.M., 1947, Ph.D., 1949; postgrad. Columbia; m. Luretta Davis, June 23, 1951; children—Arthur E., Bruce D. Faculty, Harvard, 1947-52; faculty U. Pitts., 1952-66; prof. biol. scis. U. Ill., Chgo., 1966—. U.S. AEC research grantee, 1956—. Mem. Genetics Soc. Am., Soc. Study Evolution (asso. editor), Am. Soc. Naturalists, A.A.A.S., Am. Inst. Biol. Scis., Am. Assn. U. Profs., Sigma Xi. Author: Papers on Animal Population Genetics, 1962; also numerous articles. Research in population genetics, particularly genetic potential of populations of Drosophila, dynamics of natural selection, genetics of behavior, origin of races and species. Home: 1153 Asbury Av., Winnetka, Ill. 60093. Office: Dept. Biol. Scis., U. Ill., Box 4348, Chgo. 60680.*

SPIESS, Fred Noel, Am. oceanographer; b. Oakland, Cal., Dec. 25, 1919; s. Fred Henry and Elva (Monck) S.; A.B., U. Cal. at Berkeley, 1941; M.S. in Communication Engring., Harvard, 1946; Ph.D., U. Cal. at Berkeley, 1951; m. Sarah Scott Whitton, July 25, 1942; children—Katherine, Mary E., John F., Helen S., Margaret J. Nuclear engr. Gen. Electric Co., Knolls Atomic Power Lab., Schenectady, 1951-52; with U. Cal. at San Diego, 1952—, research geophysicist, dir. Marin Phys. Lab., 1958-61, prof. oceanography Scripps Inst. Oceanography, 1961—, chmn. dept. oceanography, 1963-64, acting dir. Inst., 1961-63, dir., 1964-65, dir. Marine Phys. Lab., 1958—. Recipient Wetherill medal Franklin Inst., 1965. Fellow Acoustical Soc. Am.; mem. Am. Geophys. Union, Am. Phys. Soc., Phi Beta Kappa, Sigma Xi. Research, publs. on ways in which acoustic energy travels through sea, devel. of techniques for underwater detection, exploration and communication; codeveloper manned oceanographic research buoy. Home: 9450 La Jolla Shores Dr., La Jolla, Cal. 92037. Office: Marine Phys. Lab., San Diego 92152.*

SPIGELIUS, Adriaan van den, see van den Spigelius, Adriaan.

SPIKES, John Daniel, Am. biophysicist; coll. dean; b. Los Angeles, Dec. 14, 1918; s. John Lawrence and Gladys Alva (Murchison) S.; B.S., Cal. Inst. Tech., 1941, M.S., 1946, Ph.D., 1948; m. Anne Dorland, July 17, 1942; children—John N., Daniel A., Mary Anne. Mem. faculty U. Utah, 1948—, prof. molecular biology, 1955—, head dept., 1954-62, chmn. div. biol. scis., 1957-58, Distinguished Faculty lectr., 1962, dean Coll. Letters and Sci., 1964—. Cell physiologist AEC, 1958-60, cons., 1960—; mem. panels NSF, 1957—; cons. Smithsonian Instn. Mem. Am. Soc. Plant Physiologists, A.A.A.S., Biophys. Soc., Soc. Gen. Physiologists, Am. Chem. Soc., Am. Physiol. Soc., Radiation Research Soc., Sigma Xi. Research in photobiology and radiation biology, especially light absorption and energy-transfer processes in photosynthesis and dye-sensitized photo-oxidation reactions. Home: 1928 Hubbard Av., Salt Lake City 84108.*

SPILHAUS, Athelstan F(rederick), meteorologist, oceanographer; b. Cape Town, Union of S. Africa, Nov. 25, 1911; s. Karl Antonio and Nellie (Muir) S.; B.Sc., U. Cape Town, 1931, D.Sc., 1948; S.M., Mass. Inst. Tech., 1933; D.Sc., Coe Coll., 1961; m. Gail Griffin, Jan. 30, 1964; children by previous marriage—Athelstan F., Mary Muir, Eleanor, Margaret Ann, Karl Henry. Came to U. S., 1931, naturalized, 1946. Research asst. Mass. Inst. Tech., 1934-35; asst. dir. tech. services Union of S. Africa Defense Forces, Pretoria, 1935-36; research asst. Woods Hole Oceanographic Instn., Woods Hole, Mass., and Cambridge, Mass., 1936-37, investigator in phys. oceanography, 1938, phys. oceanographer, 1940—; asst. prof. meteorology New York U., 1937, asso. prof., 1937-42, prof., 1942, dir. research, 1946; meteorol. advisor to Union S. Africa Govt., 1947; dean Inst. Tech. U. Minn., 1949—. Mem. bd. trustees Aerospace Corp. (Los Angeles). United States Commissioner for Seattle World's Fair, 1961-62; chairman of national fisheries center and aquariam advisory board of United States Dept. Interior, until 1965. Mem. adv. coms. for armed forces; mem. nat. com. Internat. Geophys. year; mem. committee on oceanography, committee on polar research Nat. Acad. Scis.; mem. exec. bd. UNESCO, 1955-58. Trustee Woods Hole Oceanographic Instn., St. Paul Inst., Internat. Oceanographic Found., Pacific Sci. Center Found., Sci. Service, Inc. Decorated Legion of Merit, Exception Civilian Service medal USAF; recipient Patriotic Civilian Service award Dept. Army. Fellow of Royal Meteorol. Society, American Institute of Aeronautics and Astronautics, A.A.A.S. (dir.); mem. Marine Tech. Soc. (dir.), Society of Limnology and Oceanography, American Soc. Engring. Edn., Minn. Soc. Profl. of Engrs., Am. Meteorological Soc., Royal Society South Africa, American Geophys. Union, Tau Beta Pi, Sigma Xi, Iota Alpha. Contbr. Jour. of Marine Research, Jour. of Meteorology; author: Workbook of Meteorology (with James E. Miller), 1942; Weathercraft, 1951; Meteorgological Instruments (with W. E. K. Middleton), 1953; Satellite of the Sun, 1958; Our New Age, Turn to the Sea, 1959. Inventor Bathythermograph, 1938; research and devel. of meteorol. equipment, radar and radio upper wind finding, spherics, devel. meteorol. instruments for measurements from aircraft in flight. Home: Box 172, Mound, Minn. 55364. Office: U. Minn., Mpls. 55455.*

SPILLER, William Gibson, Am. neurologist; b. Balt., Sept. 13, 1863; s. Robert Miles and Anna Augusta (Maltby) S.; M.D., U. Pa., 1892, Sc.D., 1934; LL.D., Lafayette, 1934; m. Helen C. Newbold, Jan. 3, 1888; children—Helen Newbold (Mrs. Randolph G. Adams), Robert Ernest, William Raymond, Donald Percival. Asst. clin. prof. nervous diseases and asst. prof. neuropathology U. Pa., 1901-03, prof. neuropathology and asso. prof. neurology, 1903-15, prof. neurology, 1915-32 (emeritus); prof. nervous diseases, Phila. Polyclinic; clin. prof. Woman's Med. Coll. of Pa., 1902-25; hon. cons. neurologist Phila. Gen. Hosp. Fellow Coll. Physicians Phila.; mem. Am. Neurol. Assn. (pres. 1905); Phila. Neurol. Soc. (pres.). Extensive contbr. on neurology. Introduced (with Edward Martin) div. of anterolateral column of spinal cord to treat persistent pain of organic origin in lower part of body, 1912. Died Mar. 18, 1940.

SPILMAN, Edra Lavergene, Am. biochemist; b. Cambridge, O., Dec. 27, 1918; s. Frederick and Anna (Parker) S.; B.S., Ohio U., 1946; Ph.D., Western Res. U., 1953; m. Helen Wright, Aug. 21, 1959; children—Edra Shawn, Stanley David. Faculty, Western Res. U., Cleve., 1953-67; asst. prof biochemistry, 1955-67, lab. mgr. in lab. curriculum, 1953-63, dir. M-D Labs., 1964-67; prof., chmn. dept. lab. edn. Mt. Sinai Sch. Medicine, N.Y.C., 1967—. Mem. N.Y. Acad. Sci. Contbr. articles to profl. jours. Developed multi discipline labs. Home: 401 E. 89th St., N.Y.C. 10028.*

SPINA-FRANCA, Antonio, Brazilian physician; b. Jau, Brazil, Sept. 13, 1927; s. A. and Maria (Mattosinho) S.-F.; grad. in medicine, U. de Sao Paulo (Brazil), 1951, postgrad., 1952-53, M.D., 1960, Docente-Livre de Clínica Neurológica, 1963; m. Marilia Lange, July 1, 1954; children—Renato, Adriana, Fabio, Luciano. With Faculty Medicine, U. Sao Paulo, 1952-66, asst. prof., 1963-66; prof. neurology Faculdad de Ciëncias Médicas e Biológicas de Botucatu, Brazil, 1966—. Mem. Academia Brasileira de Neurologia, Academia de Medicina de Sao Paulo, Associacao Médica Brasileira, World Fedn. Neurology, Sociedad Argentina de Neurología. Research, publs. on clin. neurology, physiopathology of cerebrospinal fluid; electrophoretic investigation of cerebrospinal fluid proteins, especially cysticercosis of central nervous system. Home: 505 Rua Carlos Steinen, Sao Paulo, Sao Paulo, Brazil. Office: Dept. Neurology, Faculdade de Ciencias, Médicas e Biológicas, Botucatu, Sao Paulo, Brazil.*

SPINDLER, George Dearborn, Am. anthropologist; b. Stevens Point, Wis., Feb. 28, 1920; s. Frank Nicholas and Winifred (Hatch) S.; B.S., Wis. State Coll., 1940; M.A., U. Wis., 1948; Ph.D., U. Cal., Los Angeles, 1952; m. Louise Schaubel, May 29, 1942; 1 dau., Sue Carol. Tchr. pub. high schs. Wis., 1940-42; from research asso. to prof. Stanford, 1950—, prof. anthropology edn., 1960—, exec. head dept. anthropology, 1962-67; Burton lectr. Harvard, 1957. Center for Advanced Study in the Behavioral Scis. fellow, 1956-57; Wenner-Gren Found. for Anthropol. Research fellow, 1952-53. Fellow Am. Anthrop. Assn.; mem. Southwestern Anthrop. Assn. (pres. 1962-63), Sigma Xi. Author: Menomini Acculturation, 1955; (with L. Spindler, A. Beals) Culture in Process, 1966. Editor: Education and Anthropology, 1955; (with L. Spindler) Case Studies in Cultural Anthropology, 1960; Education and Culture, 1963; Am. Anthropologist, 1963-67; (with L. Spindler Methods in Cultural Anthropology, 1965, Case Studies in Education and Culture, 1966. Originator (with L. Spindler) of instrumental activities inventory, a technique for study of culture change; comparative studies of Am. Indian. Home: 4750 Alpine Rd., Portola Valley, Cal.*

SPINK, Wesley William, Am. physician; b. Duluth, Minn., Dec. 17, 1904; s. George C. W. and Caroline (Kuntz) S.; A.B., Carleton Coll., 1926, D.Sc., 1950; M.D., Harvard, 1932; m. Elizabeth Hamilton Hurd, Aug. 29, 1935; children—Helen (Mrs. Robert Dupoint, Jr.), William. Faculty, U. Minn. Med. Sch., Mpls., 1937—, prof. medicine, 1947—. Chmn. com. on brucellosis NRC, 1947—; mem. com. on shock, 1963—; mem. expert com. on brucellosis WHO, 1950—; mem. adv. bd. FDA, 1964—. Recipient Modern Medicine award Bd. Editors Modern Medicine, 1953; Chapin medal City of Providence, 1964. Hon. fellow N.Y. Acad. Scis., Royal Australasian Coll. Physicians; mem. A.C.P. (past pres.), Am. Soc. for Clin. Investigation (past pres.), Central Soc. for Clin. Research (past pres.), Minn. Med. Found. (past pres.), Am. Assn. Immunologists, Assn. Am. Physicians, Am. Clin. and Climatol. Assn., A.M.A., Am. Bd. Microbiology (charter). Author: Sulfanilamide and Related Compounds in General Practice, 1941; The Nature of Brucellosis, 1956; also numerous articles. Pioneered clin. research sulfonamides and antibiotics for therapy infectious diseases; research on brucellosis, introduction successful therapy for human brucellosis, septic shock, nature and mgmt. human shock. Home: 1916 E. River Terrace, Mpls. 55414.*

SPINKS, John William Tranter, chemist; b. Methwold, Eng., Jan. 1, 1908; s. John William and Sarah (Tranter) S.; B.Sc., U. London, 1928, Ph.D., 1930, D.Sc., 1957; LL.D., Carleton Coll., 1958; D.Sc., Assumption Coll., 1961; m. Mary Strelioff, June 5, 1939. Faculty, U. Sask., Saskatoon, Can., 1930—, prof., 1939—, head dept. chemistry, 1948—, dean grad. studies, 1949—, pres., 1959—. Mem. Can. Council, Sask Research Council, Am. Chem. Soc., Faraday Soc., Research Soc. Am., Research Soc. Can., Canadian Inst. Chemistry. Author: Atomic Spectra, 1937, Molecular Spectra, 1939; (with R.J. Woods) Radiation Chemistry, 1964. Research, publs. on photochemistry and chem. kinetics, use of radioactive isotopes in agr. and industry, free radicals; co-discovered nitroxyl perchlorate. Home: Pres.'s Residence, U. Sask., Saskatoon, Sask., Can.*

SPINOZA, (Baruch) Benedictus de, Dutch philosopher; b. Amsterdam, Netherlands, Nov. 24, 1632; s. Michael Espinoza; ed. under several rabbis, studying works of Ibn Ezra, Maimonides, other Jewish philosophers; student natural sci., Latin under Franz van den Ende. Excommunicated from synagogue for unorthodox views, left Amsterdam, 1656; lived in retreat of Collegiants, 1656-61; earned living as lens grinder; leader small philos. group; went to Rhijnsburg, (met Henry Oldenburg), 1661, to Voorburg, Netherlands, 1663, to the Hague, Netherlands; declined to become prof. philosophy Heidelburg (Germany) U., 1673. Author: Renati Des Cartes Principiorum Philosophiae, Pars I et II, More Geometrico Demonstratae per Bendictum de Spinoza, 1663; Tractatus Theological-Politicus, 1670; B.D.S. Opera Posthuma (including Ethica More Geometrico Demonstrata, Tractatus Politicus, Tractatus de Intellectus Emendatione, Correspondence, Hebrew Grammar), 1677; Korte Verhandeling van God, den Mensch, en Deszelfs Welstand, 1862. Philosophy exerted gt. influence on later thought; basic principle that all existence is embraced in God; focus on organic structure, devel. character as organic whole; postulated concept basic unity, rejecting doctrine separation of mind, body; 1st to criticize Bible in modern sense Biblical criticism; exponent dem. govt., separation ch. and state, individual freedom. Died The Hauge, Feb. 21, 1677.

SPINRAD, Bernard I., Am. physicist; b. N.Y.C., Apr. 16, 1924; s. Abraham and Rose (Sorrin) S.; B.S., Yale, 1942, M.S., 1944, Ph.D., 1945; m. Marion Eisen, June 29, 1951; children—Alexander A., Mark D., Jeremy P., Diana E. Sterling fellow, Yale, 1945-46; physicist Oak Ridge Nat. Lab., 1946-48; with Argonne (Ill.) Nat. Lab., 1949—, dir. reactor engring. div., 1957-64, sr. physicist reactor physics div., 1964—; vis. prof. nuclear engring. U. Ill., 1964; guest scientist Brookhaven Nat. Lab., Upton, N.Y., 1964. Cons. U.S. delegation Geneva Atomic Confs., 1955- 58; chmn. European-Am. Reactor Physics Com., 1961-63, mem., 1963—. Named Young Man of Year, Chgo. Jr. C. of C., 1956, Hinsdale(Ill.) C. of C., 1958. Fellow Am. Nuclear Soc. (dir. 1962-65, chmn. honors and awards com. 1963—), Am. Phys. Soc.; mem. A.A.A.S., Research Soc. Am., Am. Inst. Aeros. and Astronautics, Sigma Xi. Sr. editor Reactor Physics Constants. Patentee materials testing reactor, production reactors, teaching reactor; Research in devel. nuclear reactors for isotope prodn., research and education, codification reactor physics for improved design of new systems. Home: 845 Wellner Rd., Naperville, Ill. 60540. Office: 9700 S. Cass Av., Argonne, Ill. 60440.*

SPINRAD, Hyron, Am. astronomer; b. N.Y.C., Feb. 17, 1934; s. Emanuel B. and Ida (Silverman) S.; A.B., U. Cal. at Berkeley, 1955, M.A., 1959, Ph.D., 1961; m. Bette L. Abrams, Aug. 17, 1958; children—Michael, Robert. Investigation galaxies U. Cal. at Berkeley, 1960-61; planetary atmospheres work Jet Propulsion Lab., Pasadena, Cal., 1961-63; investigation atmospheres of coolest stars U. Cal. at Berkeley, 1964—; Lick Obs. fellow, 1959-61. Mem. Am. Astron. Soc., Astron. Soc. Pacific. Research on water vapors on Mars, molecular hydrogen on Jupiter, Saturn, Uranus and Neptune; temperature measurements on Venus atmosphere. Home: 19 Lance Ct., Moraga, Cal. 94556. Office: Dept. Astronomy, U. Cal., Berkeley, Cal.*

SPIRO, Herbert John, Am. polit. scientist; b. Hamburg, Germany, Sept. 7, 1924; s. Albert John and Marianne (Stiefel) S.; came to U. S., 1938, naturalized, 1944; student San Antonio Jr. Coll., 1942-43; A.B. summa cum laude, Harvard, 1949, M.A., 1950, Ph.D., 1953; m. Elizabeth Anna Petersen, June 7, 1958; children—Peter John, Alexander Charles. Adminstrv. asst. U. S. War Dept., Vienna, Austria, 1945-46; faculty Harvard, Cambridge, Mass., 1950-61, asst. prof., 1957-61; asso. prof. polit. sci. Amherst (Mass.) Coll., 1961-65; prof. polit. sci. U. Pa., Phila., 1965—; chmn. Asian and African Studies Program, Amherst-Smith-Mt. Holyoke colls., U. Mass., 1964-65; vis. asso. prof. U. Chgo., 1961, Stanford, 1963; Fulbright research prof. U. Coll. Rhodesia and Nyasaland, 1959-60; vis. prof. internat. affairs Woodrow Wilson Sch., Princeton, 1966. Lectr., mil., govt. schs., USIS; cons. Japanese Commn. on Revision Constrn., 1962, Brit. Commn. to Rev. Constn. of Federation of Rhodesia and Nyasaland, 1960. Guggenheim fellow, 1959; Social Sci. Research Council fellow, 1962; Rockefeller Found. fellow, 1958; Bowdoin prize and Sheldon Travelling fellow Harvard, 1953-54. Fellow Africian Studies Assn.; mem. Am. Polit. Sci. Assn., Am. Soc. for Polit. and Legal Philosophy, Peace Research Soc., Internat. Platform Assn., Phi Beta Kappa. Author: (with others) The Government of Postwar Germany, 1954; Politics of German Codetermination, 1958; (with others) Patterns of Government, 1958; 2d edit. rev., 1962; Government by Constitution, 1959; Politics in Africa, 1962; (with others) Five African States, 1963; World Politics: The Global System, 1966; (with others) Authority, Nomos I, 1958, Responsibility, Nomos III, 1960; Africa in a World of Change, 1967; (with others) Why Federations Fail, 1967; also articles. Editor: Africa: The Primacy of Politics, 1966. Research in theory of constn.-bldg., consensus-formation, devel. politics, methodological reintegration of splintered compartments of polit. sci. Home: 271 S. 3rd St., Phila. 19106.*

SPIRO, Karl, physiochemist; b. Berlin, Germany, June 24, 1867; prof., Strasbourg, France, also Basel, Switzerland. Author: Medizinische Kolloidlehre, 1932. Isolated (with August Stoll) ergotamine, 1921; discovered (with Otto Porges) gamma globulin, 1903; studied protoplasm, especially influence of hydrogen ion concentration, effect of osmotic pressure, role of buffer. Died Basel, Switzerland, Mar. 21, 1932.

SPIRO, Melford Elliot, Am. anthropologist; b. Cleve., Apr. 26, 1920; s. Wilbert I. and Sophie (Goodman) S.; B.A., U. Minn., 1941; Ph.D., Northwestern U., 1948; m. Audrey Goldman, May 26, 1950; children—Michael, Jonathan. Faculty, Washington U., St. Louis, 1948-52, U. Conn., Storrs, 1952-57, U. Wash., Seattle, 1957-64; prof. anthropology U. Chgo., 1964—. Mem. Am. Anthrop. Assn., Am. Ethnol. Soc. (pres.). Author: Kibbutz: Venture in Utopia, 1956; Children of the Kibbutz, 1958; Context and Meaning in Cultural Anthropology, 1965; Burmese Supernaturalim, 1967. Research, numerous publs. on relationships among personality, culture and soc., ethnology of Micronesia, Israel, Burma. Home: 5490 S. Shore Dr., Chgo. 60615.*

SPIRO, Robert Gunter, biochemist; b. Berlin, Germany, Jan. 5, 1929; s. Harry Leopold and Kate (Lowenstein) S.; came to U.S., 1940, naturalized, 1945; A.B. with honors, Columbia, 1951; M.D. cum laude, State U. N.Y. Coll. Medicine at Syracuse, 1955; m. Mary Jane Paisley, June 21, 1952; children—David, Mark. Research fellow Harvard Med. Sch., Boston, 1956-58, research asso., 1958-61, asso. dept. biol. chemistry, 1962-64, asst. prof., 1964—, asso. dir. Elliott P. Joslin Research Labs., 1961—. Established investigator Am. Heart Assn., 1961-66. Mem. Am. Soc. Biol. Chemists, Phi Beta Kappa, Alpha Omega Alpha. Research, publs. on metabolic sequence of events in exptl. diabetes mellitus, chem. structure of carbohydrate portion of several glycoproteins, basement membranes and collagens, biosynthesis of glycoproteins, enzymes and intermediates; developed techniques for study of carbohydrate of glycoproteins. Home: 33 Grove Hill Park, Newtonville, Mass. 02160. Office: 170 Pilgrim Rd., Boston 02215.*

SPITELLER, Gerhard, chemist; b. Vienna, Sept. 24, 1931; s. Otto and Erna (Fizia) S.; Dr., U. Innsbruck, 1956; m. Margot Friedmann, Oct. 7, 1960. Asst., U. Innsbruck, 1956-58, U. Vienna, 1958-64; asst., docent Mass. Inst. Tech., 1960-61; faculty U. Göttingen (Germany), 1965—, prof., 1966—. Author: Massenspektrometrische Strukturuntersuchung organischer on structure determinations of alkaloids, e.g. carpaine, Verbindungen, 1966; also numerous articles. Research fluorocarpamine, laburnamine, rindline; mass specta of steroids; temperature effect on mass spectra, stopping of mass spectrometric degradation processes at first stages. Address: 13 Ludwig Beckstrasse, 34 Göttingen, West Germany.*

SPITSYN, Víktor Ivánovich, Russian chemist; b. Apr. 25, 1902; grad. Moscow U., 1922. Faculty, Moscow U., until 1931, prof., 1942—, pro-rector, 1942-48; chief radio chemistry lab. Inst. Phys. Chemistry, USSR Acad. Scis., 1946—, inst. dir., 1953—; prof. Liebknecht Pedagogical Inst., Moscow, 1932-42. Recipient Order of Lenin, Order Red Banner of Labor. Author: The Chlorination of Oxides and their Compounds with Carbon, 1931; Soviet Chemistry Today, 1961; (with others) Techniques in the Use of Radioactive Indicators, 1955. Corr. mem. USSR Acad. Scis. Research and publs. in chemistry of rare elements, radiochemistry, inorganic chemistry, heat stability of oxyacid salts; demonstrated reversibility of oxide chlorination reactions at high temperatures; determined structure of aquapoly and heteroply compounds; discovered cause of oxide sublimation in chlorine or hydrogen chloride atmospheres. Address: USSR, Moskva, Leninskie gory, Gosudarstvenny universitet.

SPITZ, David, Am. polit. scientist; b. N.Y.C., Dec. 13, 1916; s. Geza and Irma (Lampel) S.; B.Social Sci., Coll. City N.Y., 1937; A.M., Columbia U., 1939, Ph.D., 1948; m. Ruth Sachere, Oct. 25, 1942; children—Deborah, Janet. Faculty, Ohio State U., Columbus, 1947—, prof. polit. sci., 1957—; vis. prof. Hunter Coll., 1954, Cornell U., 1958-59, Kenyon Coll., 1960-61, Johns Hopkins, Bologna (Italy) Center, 1962-63. Fund for Advancement Edn. fellow, 1951-52; Fund for Republic fellow, 1955; Rockefeller Found. fellow, 1955-56, 60; Fulbright fellow, 1962-63. Mem. Am. Polit. Sci. Assn, Midwest Conf. Polit. Scientists, Am. Soc. for Polit. and Legal Philosophy, Am. Assn. U. Profs. Author: Patterns of Anti-Democratic Thought, 1949, rev., 1965; Democracy and the Challenge of Power, 1958; The Liberal Idea of Freedom, 1964; Political Theory and Social Change, 1967; also numerous articles. Critical analysis and application to contemporary problems polit. concepts democracy, freedom, power, civil disobedience; def. liberalism against conservative attacks. Hime: 100 Webster Park, Columbus, O. 43214.*

SPITZER, Lyman, Jr., Am. astrophysicist; b. Toledo, June 26, 1914; s. Lyman and Blanche (Brumback) S.; grad. Phillips Andover Acad., 1931; B.A., Yale, 1935, D.Sc., 1958; postgrad. Cambridge (Eng.) U.; Ph.D., Princeton, 1938; D.Sc., Case Inst. Tech., 1960; LL.D., Toledo U., 1963; m. Doreen D. Canaday, June 29, 1940; children—Nicholas Canaday, Dionis Coffin, Sarah Lutetia, Lydia Strong. Instr. physics and astronomy Yale, 1939-42, asso. prof. astrophysics, 1946-47; scientist spl. studies group div. war research Columbia, 1942-44, dir. sonar analysis group, 1944-46; prof. astronomy Princeton, 1947-52, Charles A. Young prof. astronomy, 1952—, chmn. dept. astronomy, dir. obs., 1947—, dir. Project Matterhorn, 1953-61, chmn. exec. com. Plasma Physics Lab., 1961-66. Mem. Nat. Acad. Sci., Am. Acad. Arts and Scis. Author: Physics of Fully Ionized Gases, 1956, 2d edit., 1962. Editor: Physics of Sound in the Sea, 1956. Research on interstellar matter, cosmogony, stellar atmospheres, plasma physics; pioneered research on controlled thermonuclear fusion and in space astronomy. Home: 659 Lake Dr., Princeton, N.J. 08540.*

SPITZKA, Edward Anthony, Am. physician; b. New York, June 17, 1876; s. of Edward Charles and Catherine (Watzek) S.; ed. Coll. City of New York; M.D., Coll. Phys. and Surg. (Columbia), 1902; m. Alice Eberspacher, June 20, 1906. Demonstrator anatomy, 1904-06, prof. gen. anatomy, 1906-14, Jefferson Med. Coll.; dir. Daniel Baugh Inst. of Anatomy Phila., 1911-14; pvt. practice, nervous and mental diseases, New York, 1914—. Commd. capt., Med. R.C., June 1, 1917; lt. col. Med. R.C., Nov. 7, 1919. Certified by U. S. Civil Service Commn. as spl. expert, med. referee, 1921; medical referee, neuro-psychiatric sect., U. S. Veterans' Bur., Washington, Mar. 1-Aug. 25, 1921; chief med. rating sect., U. S. Veterans' Bur., New York Office, Aug. 26, 1921—. Mem. Commn. on Resuscitation from Electric Shock. Fellow A.A.A.S.; mem. Com. on 1st Aid Treatment of Surg., Med. and Gas Poisoning Cases. Editor 18th Am. edition of Gray's Anatomy. Performed autopsy and examined brain of Czolgosz, assassin of President McKinley, and attended many electrocutions, recording detailed observations upon electric death, anat. variations, etc., of criminals; studied brains of many eminent men and of various races. Died Sept. 5, 1922.

SPITZKA, Edward Charles, Am. neurologist; b. N.Y., Nov. 10, 1852; s. Charles Anthony and Johanna (Tag) S.; ed. Coll. City N.Y., 1870-73; M.D., Univ. Med. Coll. (New York U.), 1873; postgrad., Leipzig and Vienna, 1873-76; m. Catherine Watzek, June 30, 1875; 1 son, Edward Anthony S. In practice as specialist in internal diseases, particularly of nervous system; med. expert in cases of insanity or injury to brain or spinal cord; notably in the trial of Guiteau, assassin of President Garfield, where he testified to the prisoner's insanity; prof. of med. jurisprudence and neurology N.Y. Post-Grad. Med. Coll., 1885-87; cons. neurologist Sydenham Hosp. Vice pres. sect. neurology, 9th Internat. Med. Congress, 1887, chmn. Sect. of Somatology, Med. Congress, St. Louis Expn., 1904. Mem. Am. Neurol. Assn. (pres. 1890), N.Y. Neurol. Soc. (pres. 1883-84), Assn. Am. Anatomists. Author: Insanity: Its Classification, Diagnosis, and Treatment, 1883. Editor Am. Jour. Neurology, 1881-84. Discoverer inter-optic lobes of reptilian brain; classified mental disorders; anatomical studies of brain. Died Jan. 13, 1914.

SPITZNAGEL, John Keith, Am. immunologist, physician; b. Peoria, Ill., Apr. 11, 1923; s. Elmer Florian and Anna (Kolb) S.; B.A. (Honor scholar), Columbia, 1943; M.D., 1946; m. Anne Moulton Sirch, Feb. 2, 1947; children—John, Jean, Margaret, Elizabeth, Paul. Vis. investigator Rockefeller Inst. for Med. Research, 1952-53; chief medicine, chief infectious disease div. U.S. Army Hosp., Ft Bragg, N.C., 1953-57; professor bacteriology, immunology and medicine U. N.C. Sch. Medicine, Chapel Hill, 1957—. Sr. Research fellow, Research Career Devel. awardee USPHS, 1958-68. Fellow A.C.P.; mem. Am. Assn. Immunologists, Am. Soc. for Microbiology, Soc. for Exptl. Biology and Medicine, A.A.A.S., A.M.A., Sigma Xi. Research, publs. on basic proteins in subcellular granules white blood cells, their role as antibacterial agts. in resistance to infection and as intermediaries in acute inflammatory response, characterization normal antibodies, toxic action normal serum for radioactive bacteria, growth and toxicity of tubercle bacilli, immunity to typhoid, role kidney infection in hypertension. Home: Iris Lane, Chapel Hill, N.C. 27514.*

SPITZY, Hans Georg, Austrian chemist; b. Graz, Austria, Apr. 2, 1921; s. Josef and Georgine (Dobay) S.; Dr.chemistry; m. Hilde Minarik, Apr. 1, 1950; children—Hans Georg, Michaela. Became instr. Technische Hochschule, Graz, 1954, asso. prof. gen. chemistry, micro- and radiochemistry. Recipient Fritz Pregl prize Austrian Microchem. Soc., 1950; Fritz Pregl prize Austrian Acad. Sci., 1961. Mem. Commn. Microchem. Methods. Mem. Austrian Microchem. Soc., Internat. Union for Pure and Applied Chemistry. Research and publs. on analytical chemistry, especially trace analysis, micro- and radiochemistry; radio- and microanalytical methods for study of iodine metabolism (thyroid hormones). Home: Herrengasse 3, Graz, Austria.

SPIVAKOVSKII, Aleksandr Onisimovich, Russian transport engr.; b. Jan. 29, 1888; grad. Petrograd Poly. Inst., 1917. Tchr. Dnepropetrovsk Poly. Inst., 1919-33; staff Dreprope trovsk Mining Inst., 1921-23; staff Moscow Mining Inst., USSR Acad. Scis., 1949—, prof., 1933—. Recipient Stalin Prize, 1947. Corr. mem. USSR Acad. Scis. Author: Conveyer Units, 4 parts, 1933-35; Conveyers (Transport Machines of Continuous Action), 1941; Mining Transport, 1949; (with N. F. Rudenko) Lifting and Transport Machines, 1949; Cable Conveyers, 1951. Research in mining transp.; improved (with others) methods of transporting coal in Donbass mines, scraper conveyers. Home: Kutuzovskii Prospekt 27. Office: Moscow Mining Inst., USSR Acad. Scis., Moscow, USSR.

SPOCK, Benjamin (McLane), Am. physician; b. New Haven, Connecticut, May 2, 1903; son of Benjamin Ives and Mildred Louise (Stoughton) S.; B.A., Yale University, 1925; student Yale Med. Sch., 1925-27; M.D., Coll. Phys. and Surgeons, Columbia, 1929; m. Jane Davenport Cheney, June 25, 1927; children—Michael, John Cheney. Internship in medicine, Presbyn. Hosp., N.Y. City, 1929-31; pediatrics, N.Y. Nursery and Child's Hosp., 1931-32; psychiatry, N.Y. Hosp., 1932-33; practice of pediatrics, N.Y.C., 1933-44 and 1946-47; instr. pediatrics, Cornell Med. Coll., 1933-47; asst. attending pediatrician, N.Y. Hosp., 1933-47; cons. in psychiatry, Mayo Clinic, Rochester, Minn., asso. prof. psychiatry, Mayo Foundation, U. of Minn., 1947-51; professor of child development U. of Pitts. 1951-55, Western Res. U., 1955-67. Served as lt. comdr. M.C., USNR, 1944-46. Author: Baby and Child Care, 1946; A Baby's First Year (with J. Reinhart and W. Miller), 1954; Feeding Your Baby and Child (with M. Lowenberg), 1955; Dr. Spock talks with Mothers, 1961; Problems of Parents, 1962; Caring for Your Disabled Child (with M. Lerrigo), 1965; Dr. Spock on Vietnam (with M. Zimmerman), 1968. Research in preventive psychiatry; emotional development in children. Home: Lagoon Marina, Red Hook, St. Thomas, V.I. Office: 538 Madison Av., N.Y.C. 10022.*

SPOEHR, Herman Augustus, Am. chemist; b. Chicago, Ill., June 18, 1885; s. Charles A. and Frida (Baeuerlen) S.; S.B., U. of Chicago, 1906, Ph.D., 1909, D.Sc., 1929; studied U. of Berlin, 1907, U. of Paris, 1908; m. Florence Mann, Dec. 17, 1910; children—Alexander, Hortense. Swift fellow, U. of Chicago, 1909; asso. in chemistry, same, 1910; became staff mem. Lab. Plant Physiology, Carnegie Instn. Washington, 1910, asst. dir. Coastal Lab. of same, 1926-27, chmn. div. of plant biology, 1928-30; acting prof. chemistry, Stanford, summer 1924; director natural sciences Rockefeller Foundation, 1930-31; chmn. div. plant biology Carnegie Inst., 1932-47, chmn. emeritus, 1947-—. Consultant to Sec. of State, 1950-51. Mem. Am. Chem. Soc., Bot. Soc. of Am. A.A.A.S., Am. Academy of Arts and Sciences, Am. Philos. Soc., Am. Soc. Naturalists, Am. Soc. Plant Physiologists, hon. mem. Deutsche Botanische Gesellschaft, Linnean Soc., London. Author of "Photosynthesis," and of numerous papers on photosynthesis and chemistry of carbohydrate metabolism. Died June 21, 1954.

SPOERL, Edward Schurr, Am. biologist; b. Knowles, Wis., Apr. 3, 1918; s. Edward J. and Myrtle (Schnurr) S.; B.S., U. Wis., 1940, Ph.D., 1947; postgrad. Cornell U.; m. Barbara Gray,, Oct. 19, 1946; children—Patricia M., Robert E. Chief biochemistry sect. Mound Lab., Miamisburg, O., 1948-54; chief Cell Research br. Army Med. Research Lab., Ft. Knox, Ky., 1954-62, dir. biophysics div., 1962-—; lectr. U. Coll. U. Louisville, 1958-63. Mem. A.A.A.S., Biophys. Soc., Am. Soc. Microbiology, Am. Soc. Cell Biology, Radiation Research Soc., Bot. Soc., Am. Soc. Plant Physiology, Am. Chem. Soc., Sigma Xi. Research, publs. on aspects of effects of ionizing radiation and chems. on cell div. to discover specific biochem. changes with are produced. Home: 5113 Dawn Dr., Louisville 40216. Office: U.S. Army Med. Research Lab., Ft. Knox, Ky. 40121.*

SPON, Charles, French physician; b. Lyons, France, Dec. 25, 1609; student, Ulm, Germany; studied under Rodon and John Baptist Morin in Paris; doctorate, Montpellier, France, 1632; practiced medicine, Paris, 1627-31; became mem. Coll. Physic, Lyons, 1634; practiced medicine, Lyons, until 1684; named hon. physician to king, 1645. Author: Sibylla medica (prognostics of Hippocrates in verse), 1661. Assisted various authors in achieving publ. of their works. Died Feb. 21, 1684.

SPON, Jacob, archeologist; b. Lyons, France, 1647; tchr., med. coll., Lyons; collected inscriptions of Lyons; traveled to Rome to join numismatist Vaillant, Brit. botanist Wheller; visited Ionic Islands, Greece, Troy, Phrygia, Thrace, Constantinople, Asia Minor; returned to France with 2000 unedited inscriptions, 150 manuscripts, 600 medalions; fled to Switzerland upon revocation of edict of Nantes, 1685. Author: Recherches des antiquités et curiosités de la ville de Lyon, 1673; De l'origine des étrennes, 1674; Relation de l'état présent de la ville d'Athènes, ancienne capitale de la Grèce, 1874; Ignotorum atque obcurorum quorumdam deorum arae, 1674; Voyage d'Italie de Dalmatie, de Grèce et du Levant, 1678; Historie de la republique de Genève, depuis les premiers siècles de la foundation de la ville, tirée fidelment des manuscrits, 1680. Died Vevey, Switzerland, 1685.

SPOOR, William Arthur, Am. zoologist; b. N.Y.C., Dec. 14, 1908; s. William Harding and Amy (Toogood) S.; B.S., U. Wash., 1931; Ph.D., U. Wis., 1936; m. Iris Rumburg, Dec. 21, 1934; children—Jane Elizabeth, Richard Dean. Grad. asst. zoology, then teaching fellow U. Wis., 1931-36; mem. faculty U. Cin., 1936-—, prof. zoology, 1954-—, head dept., 1958-64; mem. summer staff F. T. Stone Lab., Ohio State U., 1948-62. Mem. aquatic life adv. com. Ohio River Valley Sanitation Commn., 1952-—; cons. aquatic physiology div. water supply and pollution control USPHS, 1960-—. Mem. A.A.A.S., Am. Soc. Zoologists, Am. Assn. U. Profs., Sigma Xi. Research and publs. on sexual dimorphism, diurnal activity, age and growth rate, respiratory metabolism, temperature, and environmental oxygen requirements of fish, polarographic measurement of dissolved oxygen, temperature acclimatization of crayfish. Home: 620 Evanswood Pl., Cin. 45220.*

SPÖRER, Gustav Friedrich Wilhelm, German astronomer; b. 1822; formulated Spörer's law of latitude variation of zones of sunspots; ascertained rotation period for various zones of sun, also position of sun's equator. Died 1895.

SPORN, Philip, elec. engr. b. Austria, Nov. 25, 1896; s. Isak and Rachel (Kolker) S.; brought to U. S. by parents, naturalized, 1907; E.E., Columbia U. Sch. Engring., 1917, post grad. work, 1917-18; D. Eng., Stevens Inst. Tech., 1947, Illinois Institute of Technology, 1953, Poly. Institute Brooklyn, 1955, Tri-State College, 1959; Docteur honoris causa, U. of Grenoble, (France), 1950; LL.D., Hanover Coll., 1953; L.H.D., Marshall Coll., 1956; D.Sc., Ohio State U., 1957, Ind. Technical College, 1958; D. Tech. Scis., Haifa Techneon (Israel) 1959; LL.D., Columbia University, 1966; m. Sadie Posner, Sept. 10, 1923; children—Deborah (Mrs. Andrew Gilbert), Arthur David, Michael Benjamin. Associated with Am. Electric Power Company, 1920-—, successively protection engr., communication engr., transmission and distbn. engr., chief elec. engr., chief engr., vice pres. charge engring. activities, v.p. and chief engr., 1920-45, exec. v.p. (Am. Electric Power Co. and Am. Electric Power Service Corp.), 1945, pres., chief executive officer, 1947-61, also president; dir. some 20 cos. in the Am. Electric Power System, 1947-61, chairman system devel. com., mem. exec. com., dir., 1961-—; pres., dir. Ohio Valley Electric Corp., Ind.-Ky. Electric Corp., 1952-—; pres., dir. Nuclear Power Group, Inc., 1955, v.p., dir. 1957-61, chmn. working com. East Central nuclear group, 1957-58, chmn. research and development com., 1958-—; Consultant WPB, 1944-45; consultant Oak Ridge Nuclear Power Project, Monsanto Chem. Co., 1947; mem. electric power com. NSRB, 1947-53; chmn. U. S. AEC ad hoc adv. com. on cooperation between electric power industry and AEC, 1949-51; mem. electric utility defense adv. council Defense Electric Power Adminstrn., 1950-52; vice chmn. U. S. AEC ad hoc adv. com. to evaluate Shippingport project, 1959; mem. AEC ad hoc adv. com. on reactor policies and programs, 1959. Lectr. Indsl. College of Armed Forces, 1948-57; mem. U. S. delegation Geneva Conf. for Peaceful Uses of Atomic Energy, 1955; chmn. U. S. nat. com. C.I.G.R.E.; mem. com. on dispersal and disposal radioactive wastes Nat. Acad. Scis.-NRC; council on indsl. atomic energy Nat. Indsl. Conf. Bd.; mem. U. S. State Dept. ad hoc adv. com. on U. S. Policy toward internat. atomic energy agy., 1962; chmn. Nat. Acad. Scis. adv. bd. on hardened electric power systems, 1963-—; chmn. exec. adv. com. FPC, Nat. Power Survey, 1962-65. Mem. vis. com. for nuclear engring. and reactor depts. Brookhaven Nat. Lab., 1953-57. Mem. adv. groups several colls. and univs. Recipient numerous awards and honors profl. assns., also Chevatier French Legion of Honor. Mem. numerous profl. assns., former dir. or officer several. Author and inventor in field elec. power. Research on generation, transmission, distribution, and utilization of electric energy and power, including hydropower, nuclear and steam power. Home: 320 E. 72 Street, N.Y.C. 10021. Office: 2 Broadway, N.Y.C. 10004.

SPOROS OF NICAEA, mathematician; b. circa 275. Possibly tchr. of Pappos. Author of work containing information relating to math. in early history, especially in relation to duplicating cube and squaring circle.

SPOTTISWOODE, William, English mathematician, physicist; b. London, Eng., Jan. 11, 1825; s. Andrew and Mary (Longman) S.; B.A., Oxford (Eng.) U., 1845; LL.D. (hon.), univs. Cambridge, Dublin and Edinburgh; D.C.L., Oxford U.; m. eldest dau. of William Urquhart Arbuthnot, 1861. Sr. univ. scholar, 1846, Johnson math. scholar, 1847; succeeded his father as Queen's printer, 1846; traveled to Eastern Russia, 1856, Croatia and Hungary; engaged in exptl. phys. sci. from 1870. Fellow Royal Soc. 1853 (treas. 1871, pres. 1878-83); mem. Brit. Assn. (pres. math. sect. 1865); corr. mem. French Acad. Scis., 1876. Author: Meditationes Analyticae, 1847; Elementary Theorems Relating to Determinants, 1851; A Tarantasse Journey Through Eastern Russia in the Autumn of 1856; The Polarisation of Light, 1874; Polarised Light, 1879; A Lecture on the Electrical Discharge, its Form and Functions, 1881. Wrote 1st elementary treatise on determinants; 1st to make wide use of symmetrical determination notation, research in physics on polarization and elec. discharge through rarefied gases. Died June 27, 1883.

SPRAGOS, John, economist; b. Athens, Greece, Nov. 4, 1926; s. Menelaus and Penelope (Voudouroglou) S.; M.A., U. Edinburgh, 1950; postgrad. U. Manchester; m. Mary Constance Whitwill, Sept. 26, 1956; children—Paul Byron, Helen Penelope. Research fellow U. Sheffield, 1953-57; faculty U. Coll., U. London (Eng.), 1957-—, prof. econs., 1965-—. Mem. Royal Econ Soc., Assn. U. Tchrs. Econs. (exec. com.), Assn. U. Tchrs. Author: The Decline of the Cinema: An Economist's Report, 1962. Mem. editorial bd. Rev. Econ. Studies, 1955-65. Research, publs. on theory and policy of forward exchange, internat. econs., theory and policy of fgn. exchange. Home: 72 Wildwood Rd., London N.W. 11, Eng.*

SPRAGUE, Frank Julian, Am. engineer, inventor; b. Milford, Conn., July 25, 1857; s. David Cummings and Frances Julia (King) S.; grad. U. S. Naval Acad., 1878; D.Eng., Stevens Inst., 1921; D.Sc., Columbia, 1922; LL.D., U. of Pa., 1924; m. Mary Keatinge, 1885; 1 son, Frank D'Esmonde; m. 2d, Harriet Chapman Jones, Oct. 11, 1899; children—Robert Chapman, Julian King, Frances Althea. Mem. of jury, Crystal Palace Expn., London, England, 1882, and had charge of tests of dynamo-electric machines, gas engines and electric lights, as reported to Navy Dept.; elec. studies and experiments at Stevens Inst., Brooklyn Navy Yard and U. S. Torpedo Sta., Newport; resigned, 1883, to devote attention to elec. work; asst. for a time to Thomas A. Edison; founded, 1884, Sprague Electric Ry. & Motor Co., pres. Sprague Development Corp., Sprague Safety Control & Signal Corp; cons. engr. Sprague, Westinghouse, Otis and Gen. Electric cos.; mem. Terminal Electrical Commn. N.Y.C. & H.R.R.R. As cons. engr. S.P. Co. made studies for electrification of Sierra Nevada sect. of that system; selected as mem. U. S. Naval Consulting Bd. by Am. Inst. E.E. and the Inventors' Guild, and engaged during World War I in development of fuses and air and depth bombs. Fellow and past pres. Am. Inst. E.E., New York Elec. Soc., Am. Inst. Consulting Engrs., Inventors' Guild. Awarded gold medal, Paris Expn., 1889, for elec. ry. development; Elliott Cresson medal, Franklin Inst., 1904, for multi-unit system; grand prize for "invention and development in electric rys.," St. Louis Expn., 1904; Edison gold medal, 1910; Franklin medal, 1921. Author of various scientific papers on electricity. Using his constant speed elec. motor, was first to engage in gen. mfr. and introduction of industrial elec. motors (endorsed by Edison Electric Light Co. 1885); developed pilot control of indsl. and other motors; pioneer in ry. electrification; equipped first modern trolley ry. in U. S. at Richmond, Va., 1887, later in Florence, Italy, Halle, Germany, and more than 100 rys. in 2 yrs.; developed A.C. induction smelting furnace, also high speed and house automatic elec. elevators and installed Central London equipment; invented method of operating two elevators on same rails in a common shaft, and acceleration control of car safeties; invented multiple-unit system of elec. train control, now in general use, and also system of regeneration used on mountain elec. rys. and on high speed elec. elevators; promoted high tension, direct current elec. ry. system; developed system of automatic signal and brake train control to enforce obedience to signals; etc.; engaged for yrs. in promoting underground rapid transit. Died Oct. 25, 1934.

SPRAGUE, James Mather, Am. anatomis;t b. Kansas City, Mo., Aug. 31, 1916; s. James P. and Lelia (Mather) S.; A.B., U. Kan., 1938, A.M., 1940; Ph.D., Harvard, 1942; m. Dolores Eberhart, Nov. 25, 1949; 1 son, James B. With Johns Hopkins U., 1942-50, asst. prof., 1946-50; faculty U. Pa., Phila., 1950-—, prof., 1958-—, mem. Inst. Neurol. Scis., 1956-—, asso. dir., 1958-61, chmn. Acad. Senate, 1962-63. Vis. investigator Northwestern U. Med. Sch., 1948, Rockefeller U., 1955, U. Cambridge (Eng.), 1956, U. Pisa, Italy, 1966. John Simon Guggenheim fellow, 1948-49. Recipient Lindbach Found. award. Mem. Am. Assn. Anatomists, Am. Physiol. Soc., Internat. Brain Research Orgn. Co-editor: Progress in Physiological Psychology, 1966. Research on orgn. of motor, visual, sleep and wakefulness centers of lower nervous system, moderation of behaviour by midbrain. Home: 631 Moreno Rd., Penn Valley Pa.*

SPRAGUE, Randall George, Am. physician; b. Chgo., Sept. 22, 1906; s. William Roger and Margaret (Hennessy) S.; B.S., Northwestern U., 1930, M.S., 1934, M.D., 1935; Ph.D., U. Minn., 1942; LL.D., U. Toronto, 1964; m. Anne Whitcomb, Apr. 8, 1939; children—Susan (Mrs. James K. Stroebel), Nancy (Mrs. Charles R. Hutchinson, Jr.), Linda (Mrs. Douglas E. Graham), William Randall. Practice medicine, specializing in internal medicine, Rochester, Minn., 1939-—; cons. physician sect. medicine Mayo Clinic, 1940-63, head sect. medicine, 1948-63, pres. voting staff, 1961-62, sr. cons. sect. medicine, 1963-—; prof. medicine Mayo Grad. Sch., 1953-—. Mem. metabolism and endocrinology study sect. NIH, 1947-51, mem. Nat. Adv. Dental Research Council, 1963-67. Recipient Victory medal Boston Diabetes Trust Fund, 1950; J. Howard Reber medal Phila. Metabolic Assn., 1953; award of merit Northwestern U. Alumni Assn., 1955; Centennial Merit award Northwestern U. Med. Sch., 1959. Fellow A.C.P.; mem. Am. Soc. Clin. Investigation, Assn. Am. Physicians, Endocrine Soc., A.A.A.S., Central Interurban Clin. Club (past pres.), Am. Diabetes Assn. (Banting medal 1954, Banting Meml. award 1965, (past pres.), Central Soc. Clin. Research (past pres.), Royal Soc. Medicine (hon.), Belgian Royal Acad. Medicine (corr.). Research, publs. on hyper and hypofunctioning lesions of adrenal cortex in man, pathologic physiology of diabetes mellitus, metabolic effects of adrenocortical hormones in man. Home: 1184 Plummer Circle. Office: Mayo Clinic, Rochester, Minn. 55901.*

SPRANGER, Eduard, German psychologist; b. Berlin-Grosslichterfelde, Germany, June 27, 1882; prof., Leipzig, Germany, also Berlin, Tübingen, Germany. Author: Grundlagen der Geschichtswissenschaft, 1905; Wilhelm von Humboldt und die Humanitätsidee, 1909; Lebensformen, 1914; Psychologie des Jugendalters 1924; Der deutsche Klassizismus und das Bildungsleben der Gegenwort, 1927; Goethes Weltanschauung, 1933; Pädagogische Perspektiven, 1951; Kulturfragen der Gegenwart, 1953. Exponent of psychology as an art; divided life forms into 6 ideal structural types according to dominant merit of person: theoretic, aesthetic, economic, social, polit., religious.

SPRATT, John Stricklin, Jr., Am. surgeon; b. San Angelo, Tex., Jan. 3, 1929; s. John Stricklin and Nannie (Lee) S.; student So. Meth. U., 1945-48; M.D., U. Tex., 1952; m. Beverly Winfiele, Dec. 27, 1951; children—John Arthur, Shelley Winfiele, Robert. Asst. resident surgery, Barnes Hosp., St. Louis, 1955-57, resident, Am. Cancer Soc. fellow in surgery, 1958-59, asst. surgeon, 1959-66; USPHS Cancer Research fellow in radiotherapy and surgery Mallinckrodt Inst. Radiology, St. Louis, 1957-58; faculty Washington U. Sch. Medicine, 1959-—, lectr., prof., surgery, 1965-—; faculty U. Mo. Sch. Medicine, 1961-—; advanced clin. fellow Am. Cancer Soc., 1960-63; dir. clin. research Cancer Research Center, Columbia, Mo., 1964-—, dir., 1965-—; chief surgeon Ellis Fischel State Cancer Hosp., Columbia, 1961-—, chief staff, 1964-—. Diplomate Am. Bd. Surgery. Fellow A.C.S. (pres. Central Mo. chpt. 1965-—); mem. A.A.A.S., Am. Assn. U. Profs., Soc. Head and Neck Surgeons, Central Surg. Assn., Biophys. Soc., Soc. U. Surgeons, Soc. Pelvic Surgeons, Sigma Xi, Alpha Omega Alpha. Mem. editorial bd. Cancer, 1964-—, Missouri Medicine, 1966-—. Research, publs. on natural history and treatment of cancer. Home: 105 S. Glenwood. Office: Ellis Fischel State Cancer Hosp., Columbia, Mo. 65201.*

SPRATT, Nelson Tracy, Jr., Am. zoologist; b. Atlanta Dec. 26, 1911; s. Nelson Tracy and Katherine (Guerard) S.; A.B., Emory U., 1935; Ph.D., U. Rochester, 1940; m. Jacqueline Rae Dahl, June 8, 1952; children—Nelson Tracy III, Jacquelita Joan, Gwendolyn Dahl, Melinda Vail, Jorgine Rae. Faculty, U. Minn., Mpls., 1949-—, prof., 1951-—, chmn. dept. zoology, 1959-66; vis. prof. Gustavus Adolphus Coll., 1956, U. P.R., 1966. Program dir. NSF, 1957-58. Mem. Am. Inst. Biol. Scis., Am. Soc. Zoologists, Am. Soc. Anatomists, Am. Soc. Naturalists. Author: Introduction to Cell Differentiation, 1965. Research, numerous publs. on growth in culture of chick embryos on synthetic nutrients, patterns of cell movements in early embryos, patterns of enzyme activity in young chick embryos. Home: 3208 Shorewood Dr., St. Paul 55112. Office: Dept. Zoology, U. Minn., Mpls. 55455.*

SPRAY, Robb Spalding, Am. bacteriologist; b. Omaha, Feb. 19, 1890; s. Charles Henry and Maggie (Augusta) S.; B.S., Purdue U., 1914; M.S., Pa. State U., 1918; Ph.D. (Logan fellow), U. Chgo., 1923; m. Sarah Cornelia Eighmy, Apr. 25, 1946; 1 dau., Jean Elizabeth (Mrs. Stanley Kieran Swanson). Faculty, U. W. Va. Sch. Medicine, 1921-66, prof., 1926-46, prof. emeritus, 1946-66; bacteriologist Cumberland (Wis.) Meml. Hosp., 1958-66; cons. Stella Cheese Co. Mem. Soc. Microbiology, Sigma Xi. Research, numerous publs. primarily in field of pulmonary disease of animals, human food poisoning epidemics. Home: 1270 Webb St., Cumberland, Wis. 54829. Died May 14, 1966.*

SPREITER, John Robert, Am. physicist; b. Oak Park, Minn., Oct. 23, 1921; s. Walter Floyd and Agda A. (Hokanson) S.; B. Aero. Engring., U. Minn., 1943; M.S., Stanford, 1947, Ph.D., 1954; postgrad. U. Grenoble (France) Les Houches Sch. Theoretical Physics, Enrico Fermi Internat. Sch. Physics, Italy; m. Brenda Owens, Aug. 7, 1953; children—Terry Anne S., Janet Lynne, Christine, Hilary. Aero engr. flight research sect. NACA, Ames Aero. Lab., Moffett Field, Cal., 1943-46, aero research scientist theoretical aerodynamics br., 1947-58, research scientist theoretical br. Ames Research Center, NASA, 1958-62, chief theoretical studies br., 1962-68; lectr. Stanford, 1950-58, prof. applied mechanics and aeros. and astronautics, 1968-—. Fellow Royal Astron. Soc., Am. Inst. Aeros. and Astronautics (chmn. com. on space and atmospheric physics 1967-—); mem. Internat. Sci. Radio Union (commn. mem. U.S. nat. com. 1963-—), Am. Geophys. Union, Am. Phys. Soc., Astron. Soc. Pacific, A.A.A.S., Internat. Assn. Geomagnetism and Aeronomy, Sigma Xi, Tau Beta Pi, Tau Omega. Research, numerous publs. on aerodynamics of wings and bodies, transonic flow theory, magnetosphere of earth, solar wind, electrostatic fields in ionosphere, magnetic moment of Venus, deformation of simple shell structure. Home: 11541 Herman Way, Los Altos, Cal. 94022. Office: Dept. Applied Mechanics, Stanford U., Stanford, Cal. 94305.*

SPRENBEL, Kurt Polykarp Joachim, German physician; b. Boldekow, Germany, Aug. 3, 1766; student theology, Greifwald; Ph.D., 1808; prof. medicine, later botany, Halle, Germany. Author: De medicina Ebraeorum, 1789; Versuch einer pragmatischen Geschichte der Arzneikunde, 5 vols., 1801-03; Geschichte der Chirurgie, 2 vols., 1805-19; Handbuch der Pathologie, 3 vols., 1806-11 Geschichte der Botanik, 2 vols., 1817. Handbuch der Pathologie, 1795-97; Instituationes mediciae, 1809-16; Demonstrated that most hermaphroditic flowers are not selfpollinating since stamens and ovules mature at different times, 1793, that insects are attracted to flowers by spl. scents and colors and those without them are usually wind pollinated. Died Halle, Germany, Mar. 25, 1833.

SPRENGEL, Christian Konrad, German botanist; b. Spandau, Germany, Sept. 22, 1750; rector of Spandau; gymnasium tchr., Berlin-Spandau. Author: Das endeckte Geheimnis im Bau und in der Befructung der Blumen, 1793. Showed role of insects and wind in fertilization of flowers, 1793. Died Berlin, Apr. 7, 1816.

SPRENGEL, Hermann Johann Philipp, chemist, physicist; b. Schillerslage, Germany, Aug. 29, 1834; ed. U. Göttingen, Germany, U. Heidelberg, Germany. Naturalized Brit. citizen. Worked at Oxford, then London. Fellow Royal Soc., 1878. Author: On the Vacuum, 1865. Inventor high vacuum pump (Sprengel pump), 1865; devel. U tube for measurement specific gravity and expansion of liquids; pioneered research in high explosives. Died Jan. 14, 1906, London.

SPRENT, John Frederick Adrian, parasitologist; b. Mill Hill, London, Eng., July 23, 1915; s. Frederick Pullar and Violet Agnes (Clay) S.; B.Sc., U. London, 1943, Ph.D., 1945, D.Sc., 1953; m. Muriel Florence Hines, Mar. 25, 1937; children—Jonathan, Anthony, Elizabeth. Vet. research officer No. Nigeria, 1942-44; research fellow Ministry Agr. Labs., Eng., 1945-46; research fellow U. Chgo., 1947-48; sr. research fellow Ont., (Can.) Research Found., 1948-52; with U. Queensland, Brisbane, Australia, 1952-—, prof. parasitology, 1956-—. Cons., mem. expert panel WHO, 1963-—. Recipient Henry Baldwin Ward award Am. Soc. Parasitologists, 1962. Fellow Royal Coll. Vet. Surgeons, Australian Acad. Sci.; mem. Australian Soc. for Parasitology (past pres.). Author: Parasitism—An Introduction to Parasitology and Immunology for Students of Biology, Veterinary Science and Medicine, 1963; also numerous articles. Research on life history of nematodes in man and domestic animals, immunology and pathology of nematode infections. Home: Livesay Rd., Moggill, Brisbane, Australia.*

SPRING, Erik Alfred, Finnish physicist; b. Helsinki, Finland, July 23, 1928; s. Erik Artur and Edit (Jakobsson) S.; Cand.Phil., U. Helsinki, 1959, Lic.Phil., 1962, Dr.Phil., 1963, Docent, 1964; m. Eila Elisabeth Haapoja, May 30, 1953; children—Christina, Marianne. Work study man Wärtsilä Corp., 1947-60; research asst., research fellow U. Helsinki, 1960-66; chief physicist U. Central Hosp., Radiotherapy Clinic, Helsinki, 1966-—. Research, publs. on nuclear decays and reactions especially angular correlation measurements of gamma-gamma cascades; developed new methods for measurements of internal pair formation. Home: Kontulankaari 3.G.163, Helsinki 94. Office: Radiotherapy Clinic, Haartmaninkatu 4, Helsinki 29, Finland.*

SPRINGER, Alfred, Am. chemist; b. Cin., Feb. 12, 1854; s. Lemuel and Antonie (Fries) S.; A.M., Ph.D., U. Heidelberg (Germany), 1872; Dr. Natural Sci., Ruperto Carola U., Heidelberg, 1931; m. Eda Elsas, Dec. 30, 1879; children—Elsa (Mrs. Christian Meyer), Alfred. Sole owner Alex, Fries & Bro. mfg. chemists, Cin., 1873, ret., 1936. Recipient John Scott Legacy premium and medal Franklin Inst., 1891. Fellow A.A.-A.S. (gen. sec., 1884; v.p., 1892); mem. Am. Chem. Soc.; corr. mem. Brit. Assn. for Advancement of Science. medal, Franklin Inst, 1891. Author: Glycholic Ether, 1879; Pentacholoramyl Formate, 1881; Reduction of Nitrates by Ferments; A Latent Characteristic of Aluminum, 1891; Increase of Segmental Vibrations, 1897. Contbr. papers on chem. and phys. subjects. Co-inventor, patentee torsion balance, also patentee aluminum soundboards for mus. instruments. Died Feb. 24, 1946.

SPRINGER, George, Am. mathematician; b. Cleve., Sept. 3, 1924; s. Jack H. and Helen (Finkel) S.; B.S., Case Inst. Tech., 1945; M.S., Brown U., 1946; Ph.D., Harvard, 1949; m. Annemarie Keiner, Mar. 26, 1950; children—Leonard, Claudia, Joel. Faculty, Mass. Inst. Tech., 1949-51, Northwestern U., 1951-54, U. Kan., 1955-64; prof. math. Ind. U., Bloomington, 1964-—. Math. analyst Computing Lab., Aberdeen Proving Ground, Md., 1953; vis. prof. U. Münster (Germany), 1954-55, U. Würzburg (Germany), 1961-62, Mackenzie U., Sao Paulo, Brazil, 1961. Mem. Am. Math. Soc., Math. Assn., A.A.A.S., Am. Assn. U. Profs., Sigma Xi. Author: Introduction to Rieman Surfaces, 1958. Contbr. articles on univalent functions, complex variables. Laplace's equation, quasi-conformal mapping to profl. jours.; research on complex analysis. Home: 2001 Southdowns Dr., Bloomington, Ind. 47401.*

SPRINGER, George Ferdinand, immunochemist; b. Berlin, Germany, Mar. 1, 1924; s. Ferdinand and Elizabet (Kalvin) S.; M.A., U. Heidelberg, 1947; M.D., U. Basel, Switzerland, 1951; m. Heather M. Bligh, Aug. 11, 1951; children—Martin, Elizabeth, Julia. Came to U. S., 1951; naturalized, 1962. Faculty, U. Pa. Sch. Medicine, 1951-62; prof. microbiology Northwestern U. Sch. Medicine, Chgo., 1963-—; dir. dept. immunochemistry research Evanston (Ill.) Hosp. Assn., 1963-—. Recipient Oehlecker prize German Soc. Blood Transfusion, 1966. Established investigator Am. Heart Assn., 1958-63. Woodward fellow, 1953-54. Fellow A.A.A.S.; mem. Soc. Microbiology, Assn. Immunologists, N.Y. Acad. Sci., Biochem. Soc. Gt. Britain, Internat. Soc. Blood Transfusion, German Soc. for Blood Transfusion, Am. Soc. Biol. Chemists. Editor: monograph series: Molecular Biology, Biochemistry and Biophysics, 1964-—, Transactions Polysaccharides in Biology I-V, (Josiah Macy, Jr. Found.), 1956-59; Adv. editor Transfusion 1964; asso. editor Klinische Wochenscrift, 1965-—. Contbr. to biochem. textbooks. Research in immunochemistry of blood-group active substances, carbohydrate chemistry, action of viruses on blood groups, infectious mononucleosis, phys. chemistry of antigen-antibody interactions, original contbns. to origin of isoagglutinins, chem. nature of human M and N blood-group antigens, endotoxin receptor of human red cells and blood-group active substances in microbes and higher plants, isolation of new anti-coagulant. Home: 20 Country Lane, Northfield, Ill. 60094. Office: 2650 Ridge Av., Evanston, Ill. 60201.*

SPRINGER, Stewart, Am. biologist, govt. ofcl.; b. Norwich, N.Y., June 5, 1906; s. Horace Stewart and Abigail (Fish) S.; student Butler U., 1924-27; A.B., George Washington U., 1963; m. Vergie Eleanor Fayard, Aug. 5, 1932; children—Diana (Mrs. Frank Murray), Philip, Michael, John, Christopher. Biol. collector, Fla., 1927-40; with Fla. Marine Products, Inc., 1940-42, Shark Industries div. Borden Co., 1942-50; investigator OSRD, 1942; cons. OSS, Naval Research Lab., 1943-44; with U.S. Fish and Wildlife Service, Pascagoula, Miss., 1950-55, chief br. exploratory fishing, Washington, 1956-62, research biologist, Stanford, Cal., 1963-—. Govt. adviser to delegation Internat. Conf. Safety Life at Sea, 1960. Mem. A.A.A.S., Am. Inst. Biol. Scis. (shark research panel), Am. Soc. Ichthyologists and Herpetologists, Soc. Systematic Zoology, Am. Fisheries Soc. Research, publs. on biology and taxonomy of sharks. Home: 509 N. Manchester St., Arlington, Va. 22203. Office: U.S. Nat. Mus., Washington 20560.*

SPRINGER, Victor Gruschka, Am. ichthyologist; b. Jacksonville, Fla., June 2, 1928; s. Leon and Rae (Dayan) S.; B.A., Emory U., 1948; M.S., U. Miami, 1954; Ph.D., U. Tex., 1957; m. Shirley B. Silverman, May 2, 1965. Ichthyologist marine lab. Fla. Bd. Conservation, St. Petersburg, 1957-61, sci. adviser, cons., 1964-—; asso. curator fishes div. fishes U.S. Nat. Mus., Washington, 1961-—. Mem. Am. Soc. Ichthyologists and Herpetologists (Frederick Stoye award 1957, treas. 1965-—), Soc. for Study Evolution, Am. Fisheries Soc., Soc. Systematic Zoology, A.A.A.S., Am. Inst. Biol. Scis., Conf. Biol. Editors, Sigma Xi. Editor: Proc. Biol. Soc. Washington, 1965-—. Studies on life history, classification, and distbn. marine shore fishes. Home: 8405 Ashwood Dr., Alexandria, Va. 22308. Office: Div. Fishes, U.S. Nat. Mus., Washington 20560.*

SPROULL, Robert Lamb, Am. physicist; b. Lacon, Ill., Aug. 16, 1918; John Steele and Chloe (Lamb) S.; student Deep Springs Coll., 1935-38; A.B., Cornell U., 1940, Ph.D., 1943; m. Mary Louise Knickerbocker, June 27, 1942; children—Robert Fletcher S., Nancy May S. Research physicist RCA LABS., Princeton, N.J., 1943-46; faculty Cornell U., Ithaca, N.Y., 1946-63, prof., 1956-63, dir. Lab. Atomic and Solid State Physics, 1959-60, Materials Sci. Center, 1960-63, v.p. for acad, affairs, 1965-—; dir. Advanced Research Projects Agy., U.S. Dept. Def., Washington, 1963-65. Prin. physicist Oak Ridge Nat. Lab., 1951; collaborateur sci. European Research Assos., Brussels, Belgium, 1958-59; chairman of the Defense Science Board, since 1968-—. Fellof Am. Phys. Soc.; mem. Telluride Assn. (pres. 1945-47), Author: Modern Physics, 1956. Editor: Jour. Applied Physics, 1954-57. Research on crystals. Home: 108 Northview Rd., Ithaca, N.Y. 14850.*

SPRUCE, Richard, Brit. botanist; b. Ganthorpe, Eng., 1817; Ph.D. (hon.), Imperial German Acad. Master, St. Peter's Sch., York, Eng. (began work on mosses there); collecting expdn. Pyrenees Mountains, 1846; issued sets of mosses and described them for Annals and Mag., 1849-50; S.Am. expdn. for William Jackson Hooker, George Bentham, others (Bentham received, named, distributed plants sent home), 1849; collected chinchona plants for India, 1859. Fellow Bot. Soc. Edinburgh (contbr. Transactions 1850), Royal Geog. Soc.; asso. Linnean Soc. Contbr. articles to bot. publs. Discovered many new plants in Amazon region (including new genera of Leguminosae, at least 200 species fungi), 1850; collected 250 species ferns in tarapoto at eastern foot of Andes (also brought home vocabularies of 21 Amazonian langs. and maps 3 previously unexplored rivers); flowering plants numbered 7,000 species; moss, Sprucea, liverwort, Sprucella, named for him. Died Dec. 28, 1893.

SPRUCH, Larry, Am. physicist; b. Bklyn., Jan. 1, 1923; s. Joseph and Gussie (Friedlander) S.; B.A., Bklyn. coll., 1943; Ph.D. (Tyndale fellow), U. Pa., 1948; m. Grace Marmor, Jan. 8, 1950. AEC fellow Mass. Inst. Tech., 1948-50; faculty N.Y. U., N.Y.C., 1950-—, prof. physics, 1961-—. Cons. Livermore (Cal.) Radiation Lab., 1960-—. NSF sr fellow Oxford U., U. Coll. London, Eng., Mem. Am. Phys. Soc., Am .Assn. U. Profs. Research, numerous publs. on nuclear physics, atomic physics, theory scattering. Home: 14 E. 8th St., N.Y.C. 10003.*

SPRUNT, Douglas Hamilton, Am. pathologist; b. Wilmington, N.C., Aug. 2, 1900; s. William H. and Bettie (Hamilton) S.; B.S., U. Va., 1922; M.D., Yale, 1927, M.S., 1929; D.Sc., Southwestern U., Memphis, 1967; m. Edith Charlescraft Lucas, Oct. 17, 1933; children—Alice Hamilton (Mrs. Walter Derryberry), Edith Lucas (Mrs. Edgar Toms, Jr.).

Practice medicine, specializing in pathology, Memphis, 1944—; prof. pathology, chmn. dept. U. Tenn., 1944—; chief of lab. City of Memphis Hosps., 1944—. Cons. Oak Ridge Inst. Nuclear Physics, Inc., 1945-55; mem. com. on growth NRC, 1947-50; mem. pathology sect. USPHS, 1955-60. Fellow A.C.P., Coll. Am. Pathologists (past bd. govs.); mem. Am. Soc. Exptl. Pathology (past pres.), Am. Soc. Clin. Pathologists, Am. Assn. Pathologists and Bacteriologists (past pres., sec.), Assn. Immunologists, Soc. Exptl. Biology and Medicine, A.M.A., A.A.A.S, So. Med. Assn., Internat. Acad. Pathology, Kappa Alpha, Nu Sigma Nu. Research, numerous publs. on viruses, toxins, endocrines, antibiotics and gerontology. Home: 191 E. Parkway S., Memphis 38104. Office: Inst. Pathology, U. Tenn., 858 Madison Av., Memphis 38103.*

SPUHLER, James Norman, Am. anthropologist; b. Tucumcari, N.M., Mar. 1, 1917; s. Frank Jacob and Hettie (Aylesworth) S.; B.A., U. N.M., 1940; M.A., Harvard, 1942, Ph.D., 1946; M.A., U. Oxford, 1962; m. Helen Margaret McKaig, Sept. 14, 1946; 1 son, Derek Drake. Instr., Ohio State U., Columbus, 1946-47, asst. prof., 1947-50; asso. biologist U. Mich., Ann Arbor, 1950-53, asso. prof., 1953-60, prof. anthropology and human genetics, 1960-68, chmn. dept., 1958-68; Leslie Spier prof. anthropology U. N.M., 1967—; vis. prof. U. Tex., 1965-66. Center for Advanced Studies in Behavioral Scis. fellow, 1955-56; dir. child health survey Atomic Bomb Casualty Commn., Japan, 1959; research fellow dept. human anatomy U. Oxford, 1962-63. Mem. Am. Anthrop. Assn., Am. Assn. Phys. Anthropologists, Am. Soc. Human Genetics, Royal Anthrop. Inst. Author: Natural Selection in Man, 1958; Evolution of Man's Capacity for Culture, 1958, 65; Genetic Diversity and Human Behavior, 1967. Research in human biology and human population genetics including human biology of Ramah Navaho Indians, assortative mating in human populations, inbreeding in Japanese children. Home: 8720 Rio Grande Blvd., Albuerque 87114.*

SPURR, Josiah Edward, Am. geologist; b. Gloucester, Mass., Oct. 1, 1870; s. Alfred and Cratia E. (Snow) S.; A.B., Harvard, 1893, A.M., 1894; m. Sophie C. Burchard, Jan. 18, 1899; children—Edward Burchard, John Constantine, William Alfred, Robert Anton, Stephen Hopkins. Mining engr., geologist to Sultan of Turkey, 1901-1902; geologist, U. S. Geol. Survey, 1902-06; chief geol. dept. Am. Smelting & Refining Co., Am. Smelters Securities Co., Guggenheim Exploration Co., 1906-08; Spurr & Cox (Inc.), cons. specialist in mining, 1908-11; v.p. charge mining Tonopah Mining Co. of Nev., 1911-17; mem. com. mineral imports U. S. Shipping and War Trade bds., 1917-18; exec. war minerals investigations, chief metal mining engr. Bur. Mines; chief engr. War Mineral Relief, 1918-19; editor Engring. and Mining Jour., 1919-27; prof. geology Rollins Coll., 1930-32. Mem. Mining and Metall. Soc. Am. (pres. 1921), Am. Inst. Mining and Metall. Engrs., Soc. Econ. Geologists (pres. 1923), Geol. Soc. Am., Am. Geog. Soc. Author: The Iron-Bearing Rocks of the Mesabi Range in Minnesota (Minn. Geol. and Nat. Hist. Survey), 1894; Through the Yukon Gold Diggings, 1900; Geology Applied to Mining, 1904; The Ore Magmas, 1923; Geology Applied to Selenology, 1944; Features of the Moon, 1945; Lunar Catastrophic History, 1947; The Shrunken Moon, 1949; also various monographs and reports on econ. geology, etc. Editor: Political and Commercial Geology, 1921. Mt. Spurr peak in Alaska named by U. S. Geol. Survey in honor of his explorations in Alaska, 1896, 98; gave exptl. evidence of age of Teritiary period as 45 to 60 million years. Died Jan. 12, 1950.

SPURR, Stephen Hopkins, Am. forestry scientist; b. Washington, Feb. 14, 1918; s. Josiah Edward and Sophie (Burchard) S.; B.S., U. Fla., 1938; M.F., Yale, 1940, Ph.D., 1950; m. Patricia Chapman Orton, Aug. 18, 1945; children—Daniel Orton, Jean Burchard. With Harvard, 1940-50, asst. prof., acting dir. Harvard Forest, 1943-45; asso. prof. U. Minn., 1950-52; prof. U. Mich., Ann Arbor, 1952—, asst. to v.p. acad. affairs, 1961-65, dean Sch. Nat. Resources, 1962-65, dean Horace H. Rackham Sch. Grad. Studies, 1964—. NSF Faculty fellow U. Cal., Berkeley, 1957-58; Fulbright Research fellow Australia, New Zealand, 1960. Mem. New Zealand Inst. Foresters, Soc. Am. Foresters, Ecol. Soc. Am. Author: Aerial Photographs in Forestry, 1948; Forest Inventory, 1952; Photogrammetry and Photo-Interpretation, 1962; Forest Ecology, 1964; also numerous articles. Study of physiology of growth in forest trees; application of forest ecology in silviculture; econ. devel. of natural resources; changes in forest composition. Home: 880 Colliston Rd., Ann Arbor, Mich. 48105.

SPURZHEIM, Johann Kaspar, phrenologist; b. nr. Trier, Germany, Dec. 31, 1776; studied medicine, Vienna, Austria. Student of Franz Joseph Gall (founder of what came to be phrenology), 1800-05, became Gall's asst., 1805; traveled from Vienna to France, Eng. and later Am. in effort to spread Gall's doctrines, 1813; lived and worked in U. S., circa 1813-32. Author: (with Gall) Anatomie et Physiologie du Système Nerveux en Genéral, et du Cerveau en Particulier, avec Observations sur la Possibilité de Reconnaitre Plusiers Dispositions Intellectuelles et Morales de l'Homme et des Animaux par la Configuration de leurs Tetes, 2 vols., 1810-19; The Physiognomical System of Gall and Spurzheim, 1815; Manuel du phrénologie, 1832. First to coin term phrenology; gave names to some mental faculties and parts of skull which are still in use today; believed that skull had 37 powers corresponding to 37 organs. Died Boston, Nov. 10, 1832.

SQUALERMO, Luigi (Anguillara), Italian botanist; b. Anguillara Sabazia, circa 1512; s. Francesco Squalermo; studied under Luca Ghini, Padua, Itlay; 10 children. First prefect Paduan Bot. Gardens, 1546-61; prepared collections of various bot. species for Duke of Savoy, 1561; worked for Duke Alphonse II of Ferrara, Italy. Author: Semplici, 1561. Studied medicinal effects of Italian plants; and improved methods for their use. Died Sept. 5, 1570.

SQUIER, Ephraim George, Am. archeologist; b. Bethlehem, N.Y., June 17, 1821; s. Joel and Catharine (Kilmer or Kulmer) S.; m. Miriam Florence Folline, 1858; m. 2d, Miriam F. F. Leslie, circa 1874. Founder, Poet's Magazine, Albany, N.Y., 1842, published only 2 issues; asso. N.Y. State Mechanic (organ for prison reform), 1842-43; editor Evening Jour. (Whig publ.), Hartford, Conn., 1844-45; publisher Scioto Gazette, Chillicothe, O., circa 1845; elk. Ohio Ho. of Reps., 1847, 48; interested in archaeology, examined native remains in N.Y., published Aboriginal Monuments of the State of New York (his chief work on subject) 1851; chargé d'affaires to C.Am., 1849-51; sec. Honduras Interoceanic Ry. Co., 1853; editor publishing firm Frank Leslie, supervised Frank Leslie's Pictorial History of the American Civil War, 2 vols., 1861-62; U. S. commr. to Peru, 1863-65; consul gen. of Honduras, N.Y.C., 1868. Author: The Ancient Monuments of the Mississippi Valley, 1848; Tropical Fibres, and Their Economic Extraction, 1861; Honduras, 1870; Is Cotton King?, 1861; Mound Builders of Ohio Valleys. Investigated pre-Columbian archeology of Mississippi Valley, N.Y., C.Am., Peru. Died Bklyn., Apr. 17, 1888.

SQUIER, George Owen, Am. electrical engr., inventor; b. Dryden, Mich., Mar. 21, 1865; s. Almon Justice and Emily (Gardner) S.; grad. U. S. Mil. Acad., 1887; fellow, Johns Hopkins, 1902-03 and 1903-04, Ph.D., 1903; hon. D.Sc. from Dartmouth College, 1922; unmarried. Signal officer volunteers, 1887-98; Signal Corps, U. S. Army, 1899; Maj. gen. 1917; Comd. U. S. Cable-ship Burnside, 1900-02, during laying of Philippine cable-telegraph system. U. S. mil. attaché at London, Eng., 1912; commd. lt. col. Signal Corps, Mar. 17, 1913; brig. gen., chief signal officer U. S. A., Feb. 14, 1917; maj. gen., Oct. 6, 1917; in charge of army air service, May 20, 1916-May 20, 1918. D.S.M. (U. S.); Knight Comdr. St. Michael and St. George (Great Britain); Commander Order of the Crown (Italy); Commander Legion of Honor (France). Elliott Cresson gold medal, Franklin medal. Researches: Electrochemical effects due to magnetization; the polarizing photochronograph; the sine wave systems of telegraphy and ocean cabling; the absorption of electro-magnetic waves by living vegetable organisms; tree telephony and telegraphy, multiplex telephony and telegraphy, over open circuit bare wire laid in the earth or sea. Inventor of the monophone for broadcasting over telephone wires and over power wires, also wired wireless, 1910; inventor of "Quickaid," a first aid kit for Army and Red Cross use. Died Mar. 24, 1934.

SQUIRES, Donald Fleming, Am. zoologist; b. Glen Cove, N.Y., Dec. 19, 1927; s. Charles and Freda (Fleming) S.; A.B., Cornell U., 1950, Ph.D., 1955; M.A., U. Kan., 1952; m. Jean Marie Buchanan, Mar. 29, 1951; children—Gregg Wallace, Lee Ann. Asst. curator invertebrate paleontology Am. Mus. Natural History, 1955-59, asso. curator, 1959-61; asso. curator div. marine invertebrates Smithsonian Instn., Washington, 1961-62, curator-in-charge, 1962-64, chmn. dept. invertebrate zoology, 1965-66, dep. dir. Mus. Natural History, 1966-68; dir. Marine Scis. Research Center, prof. earth and space scis. and biol. scis. State U. N.Y. at Stony Brook, 1968—. Fulbright Research fellow, New Zealand, 1959. Mem. Am. Soc. Limnology and Oceanography, Soc. Systematic Zoology, Geol. Soc. New Zealand, Am. Inst. Biol. Scis., A.A.A.S., Am. Assn. Museums, Geol. Soc. Washington, Biol. Soc. Washington, Paleontol. Soc. Washington, Sigma Xi. Research and publs. on systematics, ecology, biogeography and evolution of stony corals, especially those of deep sea; description and ecology of deep-water coral structures and coral reef-formation, especially of So. hemisphere. Home: Setauket, N.Y. 11785. Office: Marine Scis. Research Center, State U. N.Y., Stony Brook, N.Y. 11790.*

SQUIRES, Euan James, English physicist; b. Matlock, England, June 15, 1933; s. Wilfred James and Olive (Darnell) S.; B.Sc., U. Manchester, Eng., 1953, Ph.D., 1956; m. Phyllis Mary Burney, Aug. 6, 1954; children—Timothy James, Carolyn Mary. Jr. fellow Atomic Energy Research Establishment, Harwell, Eng., 1956-59; Imperial Chem. Industries fellow U. Cambridge, Eng., 1959-61; research fellow U. Cal., Berkeley, 1961-62; lectr. U. Edinburgh, 1963-64; prof. applied math. U. Durham, Eng., 1964—. Fellow Phys. Soc. London. Author: Complex Angular Momentum and Particle Physics, 1963; also numerous articles. Research on theoretical nuclear structure, nuclear scattering, elementary particles and their interactions. Home: 10 Hill Meadows, Shincliffe, Durham, Eng.*

SRB, Vladimir, Czechoslovakian biologist; b. Hradec Kralové, Czechoslovakia, Feb. 23, 1931; s. Ing. Frantisek and Marie (Svobodova) S; RNDr., Faculty Natural Sci., Brno, Czechoslovakia, 1956; C.Sc., J.E. Purkyne U., Brno, 1965; m. Eva Havelkova, July 28, 1956; 1 dau., Daniela. Lectr. dept. biology Mil. Med. Acad., Hradec Kralové, 1957-58; lectr. Faculty Pharmacy, Brno, 1958-59; sr. lectr. Med. Faculty, Charles U., Hradec Kralové, 1959—; external tchr. botany Faculty Pedagogy, Hradec Kralové, 1961-64. Mem. Czechoslovak Biol. Soc., Czechoslovak Histo- and Cytochemic Soc., Czechoslovak Botanic Soc., Nature Protection Soc. Nat. Mus. in Prague. Author numerous textbooks, articles. Research on hydro-activity in cereal and essential oil plants, mut. relations of plants, giant cells in plants, influence of X-ray irradiation of cell permeability, nucleus and chromosomes radiobiology of plants, karyotypology in vertebras. Home 881 Divisova, Hradec Kralové, Czechoslovakia.*

SREENIVASAN, Arunachala, Indian biochemist; b. Pamini, Tanjore, India, July 13, 1909; s. Arunachala and Krishnamma (Krishnasamy) Sastrial; B.A. with honors, Madras U., 1930, M.A., 1934, D.Sc., 1936; m. Sundar Venkataraman, July 12, 1937; children—Mrs. Indu S. Kumar, Shiva Kumar. Research biochemist Indian Inst. Sci., Bangalore, 1931-38; agrl. chemist Inst. Plant Industry, Indore, 1938-43; lectr., reader, prof. food tech. Bombay U., 1943-59; dep. dir. Central Food Technol. Research Inst., Mysore, 1959-64; head biochemistry and food tech. divs. Atomic Energy Establishment Trombay, Bombay, India, 1964—. Research asso. U. Wis., 1948, 52-53, Harvard, 1948-49; vis. scientist Mass. Inst. Tech., 1949; guest scientist Rockefeller U., 1958. Fellow Nat. Inst. Sci. India, Indian Acad. Sci.; mem. Soc. Biol. Chemists (pres.), Assn. Food Technologists India (exec. com.), Nutrition Soc. India. Research, numerous publs. on specific functions of folic acid and vitamin B12, aspects of cell metabolism on regulation of protein biosynthesis and maintenance of intracellular functions, effects of B vitamins, amino acids, caloric intake on utilization of dietary proteins, biochem. lesions in protein deficiency states and devel. and evaluation of processed protein-rich foods, radiation preservation of foods. Home: 6 Anand Bavan, Warden Rd., Bombay 26. Office: Atomic Research Center, Trombay, Bombay 74, India.*

SREERAMULU, T., Indian botanist; b. Avanigadda, A.P., India, Nov. 1, 1925; s. Shri T. and Smt T. (Mahalaksmi) Subrahmanyam; B.Sc., Andhra U., Waltair, India, 1945; M.Sc., Agra (India) U., 1947; Ph.D., London U., 1956, diploma Imperial Coll., 1956; m. Parvathavardhani, Aug. 15, 1951; 1 son, Prasad. Faculty sci. Andhra U., Waltair, 1947—, reader botany, 1959—. Fellow Linnean Soc., Indian Bot. Soc., Indian Phytopath. Soc. Research, publs. on spore dispersal and epidemiology of diseases of rice and sugarcane crops; described diurnal and seasonal rhythms in airborne spores of several Indian fungi. Home: U. Staff Quarters Waltair, Andhra, Pradesh, India.*

SRETENSKII, Leonid Nikolaevich, Russian mathematician; b. Feb. 27, 1902; grad. Moscow U., 1923; Dr. Physico-Math. Scis., 1936. Staff, Central Aero-Hydrodynamics Inst., 1931-41; became asso. Marine Hydrophys. Inst., USSR Acad. Scis., 1951; prof. Moscow U., 1934—. Corr. mem. USSR Acad. Scis. Author: The Theory of Wave Movements in Liquids, 1936; The Theory of Newtonian Potential, 1946; The Theory of Long Period High Tides, 1947; The Motion of a Goryachev-Chaplygin Gyroscope, 1933; The Spatial Problem of Steady Waves of Finite Amplitude, 1954; The Cauchy-Poisson Problem for Waves of Finite Amplitude, 1690. Research on equilibrium of rotating liquid, theory of liquid wave movements, movement of a heavy solid around a fixed point, integral equations and differential geometry, specific equations of math. physics. Office: Marine Hydrophysics Inst., USSR Acad. Scis., Adovaya Ulitisa 1, Lyublino, Moscow Oblast, USSR.

SRIDHARA, mathematician; b. India, flourished circa 991-1025. Author: Ganitasara (book on arithmetic); also lost work on algebra. Hindu method of solving quadratic equations attributed to him; work contains clear account of arithmetic operations and the zero; wrote on weights, measures.

SRINIVASAN, Ramachandran, Indian physicist; b. Nannilam, Madras, India, July 5, 1933; s. A. Ramachandran and R. Abirami; B.Sc. with honors, Madras Christian Coll., U. Madras (India), 1954, M.Sc., 1955, Ph.D., 1958. Faculty, U. Madras, 1958-61, 62—, prof. physics, 1964—. Commonwealth Research fellow U. Cambridge (U.K.), 1962; recipient Raman Research prize Madras U., 1963. Fellow Inst. Physics, Phys. Soc. London. Research, publs. on theoretical treatment and application of statis. methods in crystal structure analysis using X-ray diffraction data; devel. new methods of analysis in crystal structure determinations. Home: 21 4th Main Rd., Raja Annamalapuram, Madras-28, India.*

SRIVASTAVA, Anand Swarup, Indian entomologist, zoologist; b. Banda, India, Jan. 6, 1923; s. M. L. and M. L. Srivastava; B.Sc., Allahabad U., 1941, M.Sc., 1943, D.Phil., 1945; Ph.D. (Govt. of India State scholar), U. Wis., 1948; m. Krishna Kumari, May 20, 1943; children—Annapurna, Ranjana, Chandra Shekhar, Chandra Bhushan, Mira, Arun. Tech. officer Dte. P.P.Q., New Delhi, India, 1948; asst. systematic entomologist I.A.R.I., 1949; sr. sci. officer Ministry Def., 1949-54; entomologist to govt., also officer in charge protection, Kanpur, India, 1954—. Adviser, Union Pub. Service Com., New Delhi. Recipient award Sci. and Tech. Soc., 1964. Fellow Entomol. Soc. India (v.p., pres. Kanpur br.); mem. Sci. and Tech. Soc. (v.p.), Internat. Congress Entomologists. Author: Reptiles, 1945; also numerous articles. Developed new control measures against harmful pests with insecticides and synergists; studies on mode of action of DDT in insect cuticular wax; determination of free amino acids and enzymes-cholinesterase, insecticides of plant origin; devel. new technique of rodent control through use of chemosterilants resulting in their complete eradication. Home: 514/464 New Mumfordganj, Allahabad, U.P., India. Office: Entomologist to Govt., Kanpur, U.P., India.*

SRIVASTAVA, M. D. L., Indian cytogeneticist; b. Uttara Pradesh, India, July 1, 1905; s. Uma Pati and Lal (Chuni) S.; B.Sc., Allahabad U., 1931, M.Sc., 1933, D.Sc., 1937; m. Kanti Kumari Srivas, June 3, 1933; children—Ramesh, Maya Dhar (Mrs. Asha Srivastava), Kanchan (Mrs. V. D. N. Sahi), Sheela, Jyotsana. Faculty, Allahabad U., 1938—, prof. zoology, head zoology dept., 1955—. Recipient U.P. Edn. Minister's gold medal, 1948. Fellow Nat. Acad. Scis. India (past sectional pres, gen. sec. 1963—), Zool. Soc. India (mem. council). Author: An Introduction to the Comparative Anatomy of Vertebrates, 1952; (with U. S. Srivastava) A Textbook of Invertebrate Zoology, 1962; also numerous articles. Research on cytoplasmic inclusions in yolk formation in eggs, spermatogenesis and secretory phenomena of cells; structure and behavior of chromosomes in meiosis of different kinds of animals; lampbrush chromosomes of frog's egg; induction by heterogeneous inductors in frog eggs. Home: 4 D Beli Av., Allahabad, U.P., India.*

STAAB, Heinz A., German chemist; b. Darmstadt, Germany, Mar. 26, 1926; ed. univs. Marburg, Tubingen, Heidelberg; m. Ruth Müller; children—Doris, Volker. Became lectr. U. Heidelberg (W. Germany), 1956, instr., 1959, asso. prof., 1962, titular prof. organic chemistry, dir. Organic Chem. Inst., 1963. Mem. Assn. German Chemists, Bunsengesellschaft, Chem. Soc. London, Am. Chem. Soc. Author: Einführung in die Theoretische Organische Chemie, 1959. Research and publs. on organic compounds with unusual electronic structure, NMR and mass spectra; phys. organic chemistry; mechanisms of organic and biochem. reactions. Home: Schloss-Wolfsbrunnenweg 43, Heidelberg, W. Germany.

STAAR, Richard F(elix), Am. polit. scientist; b. Warsaw, Poland, Jan. 10, 1923 (parents Am. citizens); s. Alfred and Agnes (Gradalski) S.; A.B. with honors, Dickinson Coll., 1948; A.M., Yale, 1949; Ph.D., U. Mich., 1954; m. Jadwiga Maria Ochota, Mar. 28, 1949; children—Monica, Christina. Research specialist U.S. Dept. State, Washington, 1951-54; faculty Ark. State Coll., 1957-58; overseas lectr. U. Md., 1958-59; faculty Emory U., Atlanta, 1959—, prof. polit. sci., 1962—, chmn. dept., 1966—. Chester W. Nimitz prof. U.S. Naval War Coll., 1963-64; prof. fgn. affairs Nat. War Coll., 1967—. Recepient Gold Cross of Merit Polish Govt. in Exile, London 1964. Mem. Am., So. polit. sci. assns. Author: Poland, 1944-62; Sovietization of a Captive People, 1962; Communist Regimes of Eastern Europe: an Introduction, 1967; also articles. Editor: Aspects of Modern Communism, 1968. Documented process of sovietization imposed by USSR and its agents in Poland; surveyed polit., econ. and mil. relations throughout Eastern Europe. Home: 879 Clifton Rd. N.E., Atlanta 30307.*

STABLER, Howard Parker, Am. physicist; b. Bklyn., Oct. 26, 1903; s. Edward Lincoln and Elizabeth (Tubby) S.; S.B., Harvard, 1925, Ph.D., 1931; m. Margaret Van Alstyne, Apr. 5, 1932; children—Elizabeth, Robert Coleman, George Merrit. Faculty, Williams Coll., Williamstown, Mass., 1931—, prof. physics, 1946—; mem. staff radiation Lab., Mass. Inst. Tech., 1943-46; vis. prof. Cal. Inst. Tech., 1961-62. Mem. Am. Phys. Soc., I.E.E.E., Am. Assn. Physics Tchrs. (chmn. New Eng. 1960-61). Apparatus notes editor Am. Jour. Physics, 1960-62. Contbr. articles to profl. jours.; patentee. Developed reversible binary counter and shaft position indicator. Home: 186 Main St., Williamstown, Mass. 01267.*

STACKELBERG, Otto Magnus, archaeologist; b. Reval, 1787; Author: Der Apollotempel zu Bassae in Arcadien, 1826; Vues pittoresques et topographiques de la Grèces, 2 vols. 1829-38; Die Gräber der Griechen, 1837. Edited and illustrated publs. on travel and costumes. Discovered temple remains at Bassae and Hypogaien of Corneto. Died St. Petersburg, Russia, Apr. 8, 1837.

STACKHOUSE, John, English botanist; b. Trehane, Cornwall, Eng., 1742; s. William Stackhouse; ed. Exeter Coll., Oxford (Eng.) U.; m. Susanna Acton, Apr. 21, 1763; 4 sons, 3 daus. Fellow Exeter Coll., Oxford U., 1761-64, resigned, 1764; traveled abroad for several years; ret. to Bath, Eng., 1804. Fellow Linnean Soc. Author: Nereis Britannica (on sea wracks), 1795; Illustrationes Theophasti, 1811; Historia Plantarum, 1813-14. Erected Acton Castle, Perranuthnoe, for purpose of pursuing research on marine algae; studied seaweeds. Died Nov. 22, 1819.

STACY, Gardner Wesley, Am. chemist; b. Rochester, N.Y., Oct. 29, 1921; s. Gardner Wesley and May (Roberts) S.; B.S., U. Rochester, 1943; Ph.D., U. Ill., 1946; m. Mary Mullen, Apr. 1, 1968; children—Marcia Ann, Donald Gardner, Robert James, and Richard Neal. Research asst. antimalarial program OSRD, U. Ill., 1943-46; fellow Cornell U. Med. Coll., 1946-48; faculty Wash. State U., Pullman, 1948—, prof. chemistry, 1960—, faculty exec. com., 1961-64. Recipient Petroleum Research Fund Internat. award, Australia, New Zealand, 1963-64. Mem. Am. Chem. Soc. (councillor Wash.-Ida. sect. 1957—), Phi Beta Kappa, Alpha Chi Sigma, Phi Lambda Upsilon, Sigma Xi. Contbg. Author: Oxadiazines, Thiadiazines, and Their Benzo Derivatives, 1961; also articles. Research on synthesis of anti-malarials, corticosteroid analogs, anti-radiation agts. as prospective therapeutic compounds; tautomeric systems. Home: 200 N. State St., Pullman, Wash. 99163.*

STADELMAN, William Jacob, Am. food scientist; b. Vancouver, Wash., Aug. 8, 1917; s. William Henry and Eda (Huber) S.; B.S., Wash. State U., 1940; M.S., Pa. State U., 1942, Ph.D. (Grad. fellow), 1948; m. Margaret Jane Lloyd, Apr. 3, 1942; children—Ralph Lindsay, Paula Gardner. With Wash. State U., 1948-55, asso. prof., 1952-55; asso. prof. Purdue U., 1955-58, prof., 1958—. Dir. Research & Devel. Assos. Natick, Mass. Mem. sci. adv. bd. Refrigeration Research Found., 1967—; mem. research council Inst. Am. Poultry Industries, Chgo., 1957—. Recipient Poultry and Egg Nat. Bd. Christie award, 1955. Mem. Am. Chem. Soc., Inst. Food Technologists, Am. Pub. Health Assn., Poultry Sci. Assn., Poultry Edn. Assn., Sigma Xi. Contbg. author: Introduction to Livestock Production, 1962, 2d edit., 1966. Research and numerous publs. on quality evaluation and preservation of meat and poultry products; invented improved method for preservation of fresh quality characteristics of shell eggs; improving processing procedures so as to assure the consumer of tender, flavorful meat products. Home: 1429 N. Salisbury, West Lafayette, Ind. 47906. Office: Poultry Bldg., Purdue U., Lafayette, Ind. 47907.*

STADTMAN, Earl Reece, Am. biochemist; b. Carrizozo, N.M., Nov. 15, 1919; s. Walter William and Minnie Ethyl (Reece) S.; B.S., U. Cal. at Berkeley, 1942, Ph.D., 1949; m. Thressa Campbell, Oct. 17, 1943. With Alcan Hwy. survey Pub. Roads Adminstrn., 1942-43; research asst. U. Cal. at Berkeley, 1948-49, sr. lab. technician, 1949; AEC fellow Mass. Gen. Hosp., Boston, 1949-50; chemist lab. cellular physiology Nat. Heart Inst., 1950-58, chief enzyme sect., 1958-62, chief lab. biochemistry, 1962—; biochemist Max Planck Inst., Munich, Germany, also Pasteur Inst., Paris, France, 1959-60. Mem. adv. com. Oak Ridge Nat. Lab., 1963-66. Recipient biochem. study sect. research grants NIH, 1959-63; recipient Paul Lewis Lab. award in enzyme chemistry Am. Chem. Soc., 1952. Mem. Am. Chem. Soc. (exec. com. biol. div. 1959-60, 62-64, chmn. 1964-65), Am. Soc. Biol. Chemists, Am. Soc. Microbiology, Washington Acad. Scis. (award biol. chemistry 1957). Editor Jour. Biol. Chemistry, 1960-65; exec. editor Archives Biochemistry and Biophysics, 1960—. Research and publs on preservation dried fruits, bacterial metabolism and fat synthesis; cotactor requirements of bacteria. Home: Route 1, Derwood, Md. Office: National Heart Institute, Bethesda 14, Md.*

STAEBLER, George Russell, Am. forester; b. Ann Arbor, Mich., Mar. 19, 1917; s. Albert John and Ella (Goodell) S.; B.S.F., U. Mich., 1939, M.F., 1951; m. Stellajoe English, Nov. 30, 1943; children—Jo Ann, Gretchen, Rebecca. Reports editor WPA, Ohio Forest Survey, Columbus, Ohio, 1940-41; research forester TVA, Norris, Tenn., 1941-42; research forester Pacific N.W. Forest and Range Expt. Sta., Olympia, Wash., 1946-57; asst. prof. silviculture U. Wash., 1954-55; silviculturist Weyerhaeuser Co. Forestry Research Center, Centralia, Wash., 1957-66, dir. forestry research, 1966—. Mem. So. Am. Foresters. Research and numerous publs. in forestry, principally on thinning of Douglas fir; developed theoretical method of calculating schedules for thinning which permit evaluation of mgmt. alternatives and resulting financial implications.*

STAEHELIN, Rudolf, internist; b. Basel, Switzerland, Aug. 28, 1875; prof., Berlin, also Basel; coeditor of handbook for internal medicine. Wrote on metabolism, lung diseases, infectious diseases. Died Mar. 27, 1943.

STAFFORD, Helen Adele, Am. biologist; b. Phila., Oct. 9, 1922; d. Morton O. and Ethel (Scherer) Stafford; B.A., Wellesley Coll., 1944; M.A., Conn. Coll., 1948; Ph.D., U. Pa., 1951. Research asso. biochemistry U. Chgo., 1951-54, instr. botany, 1951-53; faculty dept. biology Reed Coll., Portland, Ore., 1954—, prof., 1965—. Guggenheim fellow Harvard, 1958-59; NSF sr. postdoctoral fellow U. Cal. at Los Angeles, 1963-64. Mem. Am. Soc. Plant Physiologists, Bot. Soc. Am., Soc. Biol. Chemists. Asso. editor Jour. Plant Physiology, 1964—. Research and publs. on plant biochemistry, carbohydrates, phenolic compounds. Home: 5426 S.E. 45th St., Portland, Ore. 97206.*

STAFFORD, Richard Anthony, Brit. physician; b. Cropredy, Eng., 1801; s. Egerton Stafford; apprenticed to Lawrence and Warner, physicians, St. Bartholomew's, London, 1820-24; student, Paris; m. twice. House surgeon St. Bartholomew's for Abernethy, 1923-24; began practice surgery, 1826; became sr. surgeon St. Marylebone Infirmary, 1831; surgeon-extraordinary to H.R.H. Duke of Cambridge; named Hunterian orator, 1851. Recipient Jacksonian prize for essay. Fellow Royal Coll. Surgeons (became mem. council 1848). Author: A Series of Observations on Strictures of the Urethra, 1828; Further Observations on Lancetted Stylettes, 1829; A Treatise on Injuries of the Spine, 1832; On Perforations of Strictures of the Urethra, 1834; An Essay on the Treatment of some Affections of the Prostrate Gland, 1840; On Treatment of Haemorroids, 1853. First description of sarcoma of prostate in a child of 5, 1839. Died Jan. 15, 1854.

STAGNER, Ross, Am. psychologist; b. Waco, Tex., June 15, 1909; s. William Arch and Della (Shumate) S.; B.A., Washington U., St. Louis, 1929; M.A., U. Wis., 1930, Ph.D., 1932; M.A., Dartmouth, 1947; m. Margaret Wieland, Dec. 14, 1928; children—Rhea (Mrs. B.C. Das), Martin William. Faculty, U. Akron (O.), 1935-39, Dartmouth, 1939-49; with Koppers Co., Pitts., 1943-45; prof. psychology and labor relations U. Ill., 1949-57; prof., chmn. dept. psychology Wayne State U., Detroit, 1957—. Bd. dirs. NRC. Recipient Pabst Postward Employment award, 1944. Social Sci. Research Council fellow, 1932-33; NSF fellow, 1962; Ford fellow, 1963-64; Fulbright prof., Italy, 1955-56, Eng., 1965-66. Mem. Am. Psychol. Assn. (past pres. div. personality, indsl division), A.A.A.S., Indsl. Relations Research Assn., Nat. Inst. Indsl. Psychology (London). Author: Psychology of Personality, 1937; (with T.F. Karwoski) Psychology, 1952; Psychology of Industrial Conflict, 1956; (with H. Rosen) Psychology of Union-Management Relations, 1965; Dimensions of Human Conflict, 1967; Psychological Aspects of International Conflict, 1967. Research and numerous publs. on personality measurement, attitude measurement, organizetional climate in industry, managerial decision-making. Home: 32 Norwich Rd., Pleasant Ridge, Mich. 48069. Office: Dept. Psychology, Wayne State U., Detroit 48202.*

STAHL, Georg Ernst, chemist; b. Ansbach, Bavaria, Oct. 21, 1660; studied med. at Jena under G. W. Wedel; M.D., 1683; married four times. Gave public lectures, 1683; lectured on chem. at Jena, 1684; court physician to Duke of Saxe-Weimar, Johann Ernst, 1687; 2d prof. of med., U. Halle, 1694-1716; physician to Frederick I of Prussia, 1716-34. Publ. journal, Observationes chymico-physico-medicae mensibus singulis bono cum Deo continuandae, 1697-98. Author: Fragmentorum Aetiologiae Physiologico-Chymicae ex Indagatione Senu-Rationali, 1683; Zymotechnia fundamentalis seu Fermentationis theoria generalis, Qua Nobilissimae hujus Artis, et Partis Chymiae, utilissimae . . . 1697; Observationes Chymico-Physico-Medicae curiosae, Mensibus singulis, 1697-98; Theoria medico vera, 1707; De vera Diversitate Corporis Mixti et Vivi et utriusque peculiarium conditionum atque proprietatum neccesaria discretione, demonstratio, 1707; Observationes Physico-Chymico-Medicae Curiosae, 1709; Dissertatio de Solutio Martis in puro alcali et Anatomia Sulphuris Communis, 1712; Opusculum Chymico-Physico-Medicum, 1715; Chymia rationalis et experimentalis, 1720; Fundamenta Chymico-Pharmaceutia Generalia, 1721; Billig Bedencken, Erinnerung und Erläuterung über D. J. Bechers Natur-Kündigung der Metallen, 1723; Fundamenta Chymiae, 1723; Elementa Chirurgiae Medicae ex mente, manu, methodoque Stahliana proflua jamque communis usus reddita, 1727; Fundamenta Pharmaciae Chymicae manu methodoque Stahliana posita, 1728; Materia Medica . . . , 1728; Experimenta, Observationes, Animadversiones, C C C Numero Chymicae et Physicae, 1731; Fundamenta Chymiae Dogmatico-Rationalis-Experimentalis . . . , 1732, 1747. Derived his chem. views from Johann Becher; renamed Becher's terra pinguis phlogiston; advanced phlogiston theory; believed in alchemy but warned against its frauds; opposed iatrochemical theory of acids and alkali; observed that acids have different strengths; propounded a view of fermentation which in some respects resembles that supported by Justus von Liebig, 150 years later; in medicine, he supported an animistic (vitalistic) system; gave account of lachrymal fistula; vaguely foreshadowed psychotherapy. Died Berlin, Prussia, May 14, 1734.

STAHL, S. Sigmund, peridontist; b. Berlin, Germany, June 16, 1925; s. Abraham L. and Rose (Kleinmann) S.; came to U. S., 1939, naturalized, 1944; student Bklyn. Coll., 1942-44; D.D.S., U. Minn., 1947; M.S., U. Ill., 1949; m. Phyllis R. Schloff, June 8, 1947; 1 dau., Jacquelyn Sue. Research asst. dept. histology U. Ill. Coll. Dentistry, 1947-49; with dept. periodontia and oral medicine N.Y. U. Coll. Dentistry, 1950—, prof., 1965—, exec. sec. Murry and Leonie

Guggenheim Inst. for Dental Research, 1965—; cons. periodontics VA Hosp., Bklyn., 1955—; attending oral surgery N.Y. U. Hosp., 1964—; vis. prof. Hebrew U., Jerusalem, 1963. Fellow Am. Coll. Dentists, Am. Med. Writers Assn; mem. Northeastern Soc. Periodontists (pres. 1965—), Am. Dental Assn., Am. Acad. Periodontology, A.A.A.S., N.Y. Acad. Sci., Internat. Assn. Dental Research, Am. Geriatrics Soc., Gerontol. Soc., Sigma Xi. Author: Textbook for Periodontia, 1950; Oral Diagnosis and Treatment, 1957, The Practice of Periodontia, 1960, Pharmatherapeutics of Oral Disease, 1964; also numerous articles. Editorial bd. Jour. Oral therapeutics and Pharmacology; Jour. Periodontal Research. Research on response of oral tissues to a variety of metabolic stressors, healing of gingival tissues, significance of cytologic smears in diagnosis of oral cancer. Home: 2621 Palisade Av., Bronx, N.Y. 10463. Office: 339 E. 25th St., N.Y.C. 10010.*

STAHMANN, Mark Arnold, Am. biochemist; b. Spanish Fork, Utah, Mar. 30, 1914; B.A., Brigham Young, 1936; postgrad. U. Wis., 1936-42, Ph.D. in Biochemistry, 1941; m. 1941; 2 children. Asst. chemistry Rockefeller Inst., 1942-44; research asso. organic chemistry Mass. Inst. Tech., 1944-45; faculty U. Wis., Madison, 1946—, prof. biochemistry, 1956—; researcher OSRD, 1944-46. Guggenheim fellow Pasteur Inst., Paris, 1955; Fulbright scholar, Nagoya, 1967. Mem. A.A.A.S., Am. Soc. Plant Physiologist, Am. Chem. Soc., Soc. for Exptl. Biology and Medicine, Phytopath. Soc., Am. Soc. Biol. Chemists. Research on plant diseases, plant proteins, synthetic polypetides, anticoagulants, polypeptidyl proteins, viruses and warfarin; synthesized (with others) dicumarol. Office: Dept. Biochemistry, U. Wis., Madison, Wis. 53706.

STAINBROOK, Edward, Am. psychiatrist; b. Meadville, Pa., Jan. 24, 1912; s. Charles Cochran and Harriet (Smith) S.; A.B., Allegheny Coll., Meadville, Pa., 1935; Ph.D., M.D., Duke, 1945; m. Elizabeth Selden, July 13, 1940; 1 dau., Judith. Faculty, Sch. Medicine, Yale, 1948-52, asso. prof., 1952, psychiatrist-in-charge Univ. hosp. and clinic, 1951-52; practice medicine, specializing in psychiatry, Syracuse, N.Y., 1952-56, Los Angeles, 1956—; prof., chmn. dept. psychiatry Coll. Medicine State U. N.Y., Syracuse, 1951-56; vis. lectr. dept. sociology and anthropology Syracuse U., 1953-56; prof., chmn. dept. psychiatry Sch. Medicine U. So. Cal., 1956—; attending psychiatrist U. Hosp., Meml. Hosp., Syracuse Psychopathic Hosp., 1952-56; acting chief psychiat. service VA Hosp., Syracuse, 1953-56; chief psychiatrist Los Angeles County Gen. Hosp., 1956—. Diplomate Nat. Bd. Med. Examiners, Am. Bd. Neurology and Psychiatry. Fellow A.C.P., A.C.S.; mem. Am. Psychiat. Assn., Group For Advancement Psychiatry, Am. Psychopath. Assn., A.M.A. Research and numerous publs. on psychol. effects of electroshock therapy, exptl. catatonic, Rorschach description of early schizophrenia, use of insulin psychiat. treatment, effects of induced convulsive reactions in lab. animals, med. edn., social psychiatry. Home: 1277 Parkview Av., Pasadena, Cal. 91103. Office: 1934 Hospital Pl., Los Angeles 90033.*

STAINBROOK, Merrill Addison, Am. paleontologist; b. Brandon, Ia., Feb. 27, 1897; B.A., Ia., 1921, M.S., 1922, Ph.D. in Paleontology, 1927; asst., Ia., 1922-25; instr. zoology U. Tenn., 1926-27; asst. prof. geology Tex. Tech. Coll., 1927-34, asso. prof., 1934-40, prof., 1940-46, 47-48. Fellow Paleontol. Soc., Assn. Petroleum Geologists, Geol. Soc. Am., Ia. Acad.; mem. Soc. Econ. Paleontology and Mineralogy. Research in invertebrate paleontology, stratigraphy; specialist in corals, also in brachiopods, especially those of Cedar Valley limestone of Ia. Died 1956.

STAINOV, Petko Stoyanov, Bulgarian jurist; b. Kazanlik, Bulgaria, May 19, 1890; s. Stoyan Petkov and Bonka (Shirova) S.; student Sch. Law, U. Paris (France), 1913; D.Laws, U. Paris, 1914; D.honoris causa; Warsaw U., 1966; m. Anna Madjarova Kamenova, July 13, 1919; children—Maria Ivan (Mrs. Secoulova), Michaila Petr (Mrs. Petrova). Editor newspaper Mir Sofia, 1915-19, 21-23, Slovo, Sofia, 1935-38; dir. press bur. Fgn. Office, Sofia, 1919-21; prof. U. Sofia (Bulgaria), 1923-65; minister transport, 1930-31; Bulgarian ambassador in Paris, 1934-35; minister fgn. affairs, Sofia, 1944-46; mem. Parliament, 1923—. Vice pres. Com. for Balkan Collaboration. Recipient Gt. Cross of Bulgaria; Polonia restituta, 1931; Civil merit 1st class, 1946; Bulgarian order People's Republic, 1965. Mem. Bulgarian Acad. Scis. Author: Administrative Law, 2 vols., 1936; Theory of the Administrative Act, 1951; Law of the Waters, 1957; The International Law Aspects of River Pollution Control, 1964. Determined guaranty for legality of adminstrv. activity for Bulgarian law, 1935-36, established methods for solving differences between states for utilization of waters of internat. rivers, studied internat. law aspects of river pollution control especially for Danube. Home: 36 Oborishte. Office: Inst. of Law, Benkovski St., 3, Sofia, Bulgaria.*

STAINTON, Henry Tibbats, English entomologist; b. London, Eng., Aug. 13, 1822; s. Henry Stainton; ed. King's Coll., London; m. Isabel Dunn, 1946. About 1840 turned his attention to entomology, especially micro-lepidaptera; established Entomologists Annual, 1855, Entomologists' Weekly Intelligencer, 1956; a founder Entomologists Monthly mag., 1864; instrumental in founding Zool. Record Assn. Fellow Royal Soc., 1867 (mem. council 1880-82, sec. 1861), Linnean Soc. (sec. 1869-74, v.p. 1883-85); mem. entomol. socs. London (sec. 1850-51, pres. 1881-82), France, Stettin, Italy; hon. mem. entomol. socs. Belgium, Switzerland. Author: An Attempt at a Systematic Catalogue of The British Tineidae and Pterophoridae, 1849; A Supplementary Catalogue of the British Tineidae and Pterophoridae, 1851; The Entomologists' Companion, 1852; Bibliotheca Stephensiana, 1883; Insecta Britannica Lepidoptera, Tineina, 1854; The Natural History of the Tineina, 1873; The Tineina of Southern Europe, 1869; also numerous papers. Research on tineidae; founder important sci. magazines. Died Dec. 2, 1892.

STAKER, William Paul, Am. physicist; b. Aberdeen, S.D., Apr. 9, 1919; s. Moses Roy and Anna (Fischer) S.; B.S. in Edn., Ill. State U., 1940; M.S., U. Ia., 1942; Ph.D., N.Y. U., 1950; m. Jane Hamlin, Dec. 27, 1949; children—Paul Howard, Susan Jane. Physicist, Burnside Lab., E.I. duPont de Nemours & Co., Inc., 1942-46; research asso. N.Y. U., 1946-50; asso. physicist Argonne Nat. Lab., 1950-52, group leader, 1952; with Engring. Research div. Standard Oil (Ind.), 1952-56; with Combustion Engring., Inc., Windsor, Conn., 1956—, mgr. reactor physics, 1959-60, mgr. engring. and physics Naval Reactors div., 1961-67, project manager nuclear power department, 1967—. Mem. Am. Nuclear Soc., Am. Phys. Soc., Sigma Xi. Research and publs. on interior ballistics and propellant applications, cosmic ray neutron density measurements, neutron and reactor physics, tracer studies. Patentee radioactive inspection instrument, radiation flowmeter. Home: 64 Glenbrook Rd., West Hartford, Conn. 06107. Office: Combustion Engring., Inc., Nuclear Power Dept., Windsor, Conn. 06095.*

STAKMAN, Elvin C(harles), Am. plant pathologist; b. Algoma, Wis., May 17, 1885; s. Frederick and Emelie (Eberhardt) S.; A.B., U. Minn., 1906, A.M., 1910, Ph. D., 1913; Dr. Nat. Sci. (hon.), U. Halle, 1938; D. Sc. (hon.), Yale, 1950, U. R.I., 1953, U. Minn., U. Wis., 1954, Cambridge, 1964; m. Louise Jensen, Sept. 6, 1917. Pub. sch. tchr., Minn., 1906-09; mem. faculty U. Minn., 1909-53, prof. plant pathology, 1918-53, head sect. plant pathology, 1913-40, head div. plant pathology and botany, 1940-53, now professor emeritus; pathologist charge barberry eradication campaign U. S. Dept. of Agr., 1918, pathologist, agt., 1919-55, collaborator, 1955—; Hitchcock prof. U. Cal., 1955; spl. consultant agr. Rockefeller Found., 1953—. Nat. Def. leader rubber expdn. to S. Am., 1940; sci. mission to Japan, SCAP, 1948; guest prof. U. Halle, 1930-31. Mem. div. biology, agr. Nat. Research Council, 1931-34, vice chmn., 1937-38, 47-48, mem. agrl. bd., 1950-58; mem. com. biology and medicine U. S. AEC, 1948-51, cons., 1954-59; member exec. com. Nat. Sci. Bd., 1950-54; mem. National Commn. for UNESCO, 1950-56; mem. U. S. delegation to UN Conf. Application Sci. and Tech. Benefit Less Developed Areas, Geneva, 1963; advisor conf., Florence, 1950, del. conf. Paris, 1951. Del. Pan Pacific Sci. Congress, 1923, Nat. Acad. Scis.; 6th Internat. Bot. Congress, 1935; pres. phytopath sect. 7th Internat. Bot. Congress, 1950, honorary vice pres. of 10th Congress, 1964; delegate 8th All-Pakistan Science Congress, 1956. Decorated Cruz de Boyacá (Colombia); recipient of the E. C. Hansen medal and prize in 1928; Centennial award from Michigan State Coll., 1955; Medalla del Mérito Agrónomico (Colombia), 1955; certificate of merit Botanical Society of America, 1956; Otto Appel medal, 1957. Fellow American Acad. Arts and Scis., Bot. Soc. Edinburgh (hon.), Am. Phytopathol. Soc. (pres. 1922; chmn. war emergency com. 1942-43); mem. Indian Phytopathol. Soc., N.Y. Acad. Scis., Am. Philos. Soc., Bot. Soc. Am., Am. Genetic Assn., A.A.A.S (pres. 1949), Brit. Am. mycological societies, Am. Soc. Naturalists, History Sci. Soc. Am. College Allergists (honorary), Norwegian Academy Sci. and Letters, National Academy Sci., Philos. Sci. Assn., Association Applied Biology (England, hon.), Karachi Bot. Society (honorary), Torrey Bot. Club, Wash. Acad. Sci., Kaiser Akad. Naturf. (Halle), Canadian (hon.), Japanese (hon.) phytopathol. socs., Swedish Royal Acad. Agr. (fgn. mem.), Phi Beta Kappa, Sigma Xi. Editor-in-chief Phytopathology, 1925-29; Am. editor Phytopathologische Zeitschrift since 1931; editorial com. Ann. Rev. Microbiology, 1946-54. Author (with J. G. Harrar) Principles of Plant Pathology, 1957; (with others) Campaigns Against Hunger, 1967; also bulls., articles prof. jours. Research on plant pathology; mycology; epidemiology of cereal rusts and smuts; aerobiology; plant disease resistance; genetics of plant pathogens; agricultural improvement. Home: 1411 Hythe St., St. Paul 55108. Office: U. Minn. Inst. of Agr., St. Paul 55101.

STAL, Carl, Swedish entomologist; b. Castle of Carlberg, Sweden, Mar. 21, 1833; passed medicophilos. exam, Uppsala, Sweden, 1858; studied anatomy and physiology, Stockholm; Ph.D., Jena, Germany; became asst. to C. H. Boheman in entomol. sect. Nat. Zool. Mus., Stockholm, 1859, prof. and supt. sect., 1867. Author: Hemiptera Africana, 1864-66; Enumeratio Hemipterorum, 1870-76; Recensio Orthapletorum VI-3, 1873-76. Leading Swedish hemipterist and orthopterist; engaged in fundamental, systematic and taxomonic work; named and described many Am. insects. Died June 13, 1878.

STALEY, Dean Oden, Am. meteorologist; b. Kennewick, Wash., Oct. 18, 1926; s. Oden Thurman and Georgia (Staley) S.; B.S., U. Wash., 1950, Ph.D., 1956; M.A., U. Cal., Los Angeles, 1951; m. Ethel Arlene Storlie, Nov. 29, 1963; children—Jane Susan, Martin Frank. Faculty, U. Wis., 1955-59; faculty dept. meteorology U. Ariz., Tucson, 1959—, prof., 1965—. Mem. Am. Meteorol. Soc., Am. Geophys. Union, Royal Meteorol. Soc., A.A.A.S. Developed theory of tropopause formation, theory of mass exchange between troposphere and stratosphere; clarified radiative cooling distbn. at inversions and the tropopause.*

STALKER, Harrison Dailey, Am. biologist; b. Detroit, July 3, 1915; s. John Nellis and Edith (Dailey) S.; B.A., Coll. Wooster, 1937; Ph.D., U. Rochester, 1941; m. Marion Louise Leffler, May 31, 1941; 1 son, George Harrison. Faculty dept. biology Washington U., St. Louis, 1942—, prof., 1956—; mem. genetics adv. panel NSF, 1960-63. Mem. Genetics Soc. Am., Am. Soc. Naturalists, Evolution Soc., Am. Soc. Human Genetics, Phi Beta Kappa, Sigma Xi. Studies and numerous publs. in mechanisms of evolution, chromosomal evolutionary changes in species of flies, evolution of parthenogenesis, devel. photog. comparison method for analysis of species differences in chromosome pattern. Home: 440 Bradford St., Webster Groves, Mo. 63119. Office: Biology Dept., Washington U., St. Louis 63130.*

STALLEY, Robert Delmer, Am. mathematician; b. Mpls., Oct. 25, 1924; s. Francis Charles and Florence (Goode) S.; B.S., Ore. State U., 1946, M.A., 1948; postgrad. Stanford, 1948-49; Ph.D., U. Ore., 1953; m. Dorothy Ann Jeffery, Aug. 27, 1950; children—Mark Frederick, Jeffery Alan, John Michael, Lorena Ellen. Faculty, U. Ariz., 1949-51, Ia. State U., 1953-55, Fresno State Coll., 1955-56; mathematician Sperry Rand Co., St. Paul, 1955, U.S. Naval Ordnance Test Sta., China Lake, Cal., 1956; prof. math. Ore. State U., Corvallis, 1956—. Mem. Am. Math. Soc., Math. Assn. Am., Sigma Xi. Contbr. articles on number theory and numerical analysis of profl. jours. Home: 1405 Forest Dr., Corvallis, Ore. 97330.*

STALNAKER, John Marshall, Am. psychologist; b. Duluth, Minn., Aug. 17, 1903; s. William Edgar and Sara (Tatham) S.; B.S. with honors, U. of Chicago, 1925; A.M. in Psychology, in 1928; LL.D., Purdue University, 1956; married Ruth Elizabeth Culp, July 29, 1933; children—John Culp, Robert Culp, Judith Culp. Teacher of rural sch. in Hardisty, Alberta, Can., 1922; teacher math. and science, Harvard Sch. for Boys, 1925-26; instr. psychology and spl. research asst. to the pres., 1926-30, asst. prof. edn. and psychology (on leave), 1930-31, Purdue U.; dir. attitude measurement, athletic survey, U. of Minn., 1930-31; examiner (instr.) bd. of examinations, U. of Chicago, 1931-36; asst. prof., 1936-37, asso. prof., 1937-44; prof., Princeton, 1944-45; research asso., Coll. Entrance Exam. Bd., 1936-37, cons. examiner, 1937-42; asso. sec., 1942-45; dir. Navy test research unit, 1942-45, contractor's tech. rep. for N.D.R.C. project N-106, 1942-45; dir. Army-Navy Coll. Qualifying Test, 1943-45; dean of students and prof. psychology, Stanford, 1945-49; prof. psychology, coordinator of psychol. scis. and services, Ill. Inst. Tech., 1949-51; consultant Fund for Advancement Edn., 1952-55, Nat. Sci. Found., 1952—; dir. studies for Assn. Am. Medical Colls., 1949-55; president National Merit Scholarship Corporation, 1955—. Trustee and director Pepsi-Cola scholarship bd. 1945-54. Mem. adv. com., Foreign Service Exam. Dept. of State, 1941-51, sci. adv. board to Chief of Staff, U. S. Air Force, 1950-53. Awarded Certificate Merit, Pres. U. S., 1948, Distinguished Civilian Service award Secretary Navy, 1946. Member Am. Psychol. Assn. Am. Statistical Assn., Psychometric Soc., Am. Edn. Research Assn., Nat. Edn. Asso., A.A.A.S., Phi Beta Kappa Assos., Sigma Xi. Contbn. to devel. of testing aptitude, intelligence and achievement. Home: 1875 Elm St., Winnetka, Ill. Office: 1580 Sherman Av., Evanston, Ill.

STAMLER, Jeremiah, Am. physician; b. N.Y.C., Oct. 27, 1919; s. George and Rose (Baras) S.; A.B., Columbia U., 1940; M.D., L.I. Coll. Medicine, 1943; m. Rose Steinberg, June 27, 1942; 1 son, Paul J. Fellow pathology L.I. Coll. Medicine, 1947; research fellow cardiovascular dept. Med. Research Inst., Michael Reese Hosp., Chgo., 1948, research asso., 1949-55, asst. dir. dept., 1955-58; established investigator Am. Heart Assn., 1952-58; dir. heart disease control program Chgo. Bd. Health, 1958—, dir. chronic disease control div., 1961-65; dir. div. adult health and aging, 1965—; asso. dept. medicine Northwestern U. Med. Sch., 1958-59, asst. prof. medicine, 1959-65, asso. prof., 1965—. Recipient Lasker award. Fellow Am. Pub. Health Assn., A.A.A.S.: mem. Am. Fedn. Clin. Research, Am. Heart Assn. (vice-chmn. exec. com. council arteriosclerosis), Am. Physiol. Soc., Am. Soc. Clin. Investigation, Am. Soc. Study Arteriosclerosis (past dir., past chmn. program com., past sec.-treas.), Am., Chgo. diabetes assns., Am. Psychosomatic Soc. (editorial bd. Psychosomatic Medicine), Am. Soc. Clin. Nutrition, Assn. Clin. Scientists, Middle States Pub.

Health Assn., Central Soc. Clin. Research, Chgo. Heart Assn. (chmn, epidemiology com.), Ill. Pub. Health Assn. (mem. exec. com.), Ill. Acad. Scis., Diabetes Assn. Greater Chgo. (dir.), Soc. Exptl. Biology and Medicine (sec. III. chpt.), Am. Inst. Nutrition, Chgo. Nutrition Assn., Chgo. Acad. Scis., Inst. Medicine Chgo., Phi Beta Kappa. Author: (with L. N. Katz): Experimental Atherosclerosis, 1953; (with A. Blakeslee) Your Heart Has Nine Lives-Nine Steps to Heart Health, 1963; Lectures on Preventing Cardiology, 1966. Editor: (with R. Stamler) The Epidemiology of Hypertensive Diseases—Proceedings of an International Working Conference. Western Hemisphere editor Jour. Atherosclerosis Research. Contbr. articles to profl. jours. Research and control of pub. health heart disease. Home: 1332 E. Madison Park, Chgo. 60615. Office: Chgo. Bd. Health, Chgo. Civic Center, Chgo. 60602.*

STAMM, Alfred J., Am. chemist; b. Los Angeles, Dec. 29, 1897; s. August Julius and Alice Elizabeth (Kottmeier) S.; B.S. in Chemistry, Cal., Inst. Tech., 1921; M.S. in Phys. Chemistry, U. Wis., 1923, Ph.D., 1925; m. Erdine R. Timberlake, Aug. 16, 1928; children—Virginia E. (Mrs. John C. Lemanczyk), Bonnie E. (Mrs. Paul Dougherty), Alfred John. Chemist, Gen. Petroleum Corp., 1921-22; asst. chemistry U. Wis., 1922-25; from asso. to prin. chemist to chief div. derived products, to subject matter specialist U. S. Forest Products Lab., Madison, Wis., 1925-59; mem. faculty N.C. State of U. N.C., Raleigh, 1959—, Reuben B. Robertson prof. wood chemistry, 1962—. Fellow Internat. Edn. Bd., U. Uppsala (Sweden), 1928-29; sr. Fulbright research fellow Commonwealth Sci. and Indsl. Research Orgn. Forest Products Lab., Melbourne, Australia, 1955-56; T.A.P.P.I. Honorarium lectr. Pacific N.W., 1961; Wood Sci. and Tech. Honorium lectr. at six univs., 1962-63. Mem. Am. Chem. Soc. (chmn. celloid div. 1934, cellulose div. 1953, Wis. sect. 1936), Forest Products Research Soc. (mem. bd. S.E. sect. 1963-65), Internat. Acad. Wood Sci. (hon.), Soc. Wood Sci. and Tech., Sigma Xi, Alpha Chi Sigma, Phi Lambda Upsilon, Tau Beta Pi, Xi Sigma Pi, Gamma Sigma Delta. Author: Wood and Cellulose Science, 1964; also articles, chpts. in books. Co-author: Chemical Processing of Wood, 1953. Research on surface chemistry and physics of cellulosic materials, wood-fluid relationships, swelling of wood and its chem. and phys. control resulting in devel. Several types of treated wood. Home: 3212 Rutherford Dr., Raleigh, N.C. 27609.*

STAMM, Martin, physician, surgeon; b. Thaygen, Switzerland, Nov. 16, 1847; s. August and Verona (Bernath) S.; student U. Pa. Coll. Medicine, 1868-70; M.D., U. Berne (Switzerland), 1872; m. Anna Marquerite Scheurer Walker, 1872; children—Till Edele (Mrs. George W. Haynes), Hans Eugene. Prof. operative and clin. surgery Coll. Physicians and Surgeons, Cleve., 1892-1901; established pvt. hosp., Fremont Fellow A.C.S.; mem. A.M.A., Am. Assn. Obstetricians and Gynecologists, Ohio Med. Soc. Contbr. to med. publs. First (with Nicholas Senn) to perform gastroenterostomy in U. S.; performed 1st hepaticostomy in world, 1888; 1st to perform Kocher's hernia operation, 1892; devised Stamm's gastrostomy 1894; performed early operation for resection of kidney; performed 1st Mickulicz operation for resection of large bowel, 1901; helped introduce Duhrrsen's vaginal caesarian sect. in U. S., 1903; devised Stamm's pole ligation for lessening danger in operations for exophthalmic goitre or Graves' disease, 1908. Died May 22, 1918.

STANBURY, John Bruton, Am. physician; b. Clinton, N.C., May 15, 1915; s. Walter Albert and Zula V. (Bruton) S.; B.A., Duke, 1935; M.D., Harvard, 1939; m. Jean Cook, Jan. 6, 1945; children—John, Martha J., Sarah K., David M., Pamala C. Research fellow pharmacology Harvard Med. Sch., 1947; chief, thyroid unit dept. medicine Mass. Gen. Hosp., also Harvard Med. Sch., 1949-66; prof. exptl. medicine Mass. Inst. Tech., 1966—. Mem. Am. Soc. Clin. Investigation, Assn. Am. Physicians, Am. Thyroid Assn., Soc. Human Genetics, Am. Acad. Arts and Scis. Author: (with Brownell, Riggs, Perinetti, Itioz and del Castillo) Endemic Goiter: The Adaptation of Man to Iodine Deficiency, 1954; (with Means and DeGroot) The Thyroid and Its Diseases, 1963. Editor (with Wyngaarden and Fredrickson) The Metabolic Basis of Inherited Disease, 1960, 2d edit., 1966. Research and publs. in pathophysiology of endemic goiter; pathophysiology of inherited thyroid disease and thyroid physiology. Home: 43 Circuit Rd., Chestnut Hill, Mass.*

STANDAERT, Frank George, Am. pharmacologist; b. Paterson, N.J., Nov. 12, 1929; s. George J. and Ethel (Miller) S.; A.B., Harvard, 1951; M.D., Cornell U., 1955; m. Joan F. Cairns, Feb. 7, 1959; children—David, Robert, Christopher. Fellow in pharmacology Cornell Med. Coll., 1956-57; faculty Med. Coll., Cornell U., N.Y.C. 1959-67, asso. prof. pharmacology, 1964—. Recipient USPHS Career Devel. award, 1960-67; Shering Found. prof., chmn. pharmacology Georgetown U. Schs. Medicine and Dentistry, 1967—. Recipient USPHS Career Devel. award, 1960-67. Mem. A.A.A.S., Am. Soc. for Pharmacology and Exptl. Therapeutics, A.M.A. (faculty affiliate), Harvey Soc., N.Y. State Soc. for Med. Research, Animal Care Panel. Research and publs. on means by which drugs affect ability of nerve to cause muscle contraction. Home: 8205 Stone Trail Dr., Bethesda, Md. 20034. Office: 3900 Reservoir Rd., Washington 20007.

STANDART, George Lenell, chem. engr.; b. Detroit, Jan. 29, 1921; s. Abram Charles and Lillian (Clarke) S.; B.S. in Applied Chemistry, Cal. Inst. Tech., 1946, postgrad., 1946-48; Dr.Sc., Czechoslovak Acad. Sci., 1964; m. Phoebe Neubauer, Apr. 5, 1945; children—Nancy S., Sally S. Staff, Inst. Chem. Tech., Prague, Czechoslovakia, 1948—, prof. dept. chem. engring., 1961—; staff Inst. Chem. Process Fundamentals, 1958-63, external sect. head. Indsl. cons., 1948—. Recipient Gottwald State prize in sci., 1959. Mem. Czechoslovak Chem. Soc. (chmn. sect. for chem. engring. 1955-62), Czechoslovak Sci.-Tech. Soc. Author: Chemical Engineering; also articles. Research on mass transfer, especially distillation tray efficiency, thermodynamics of irreversible processes on phase boundary, especially interphase heat and mass transfer. Home: 4 Hládkov, Prague 6, Czechoslovakia.*

STANDIL, Sidney, Can. physicist; b. Winnipeg, Man., Can., Oct. 19, 1926; s. David and Anne (Silverstein) S.; B.S., Queen's U., Kingston, 1948, M.S., 1949; Ph.D., U. Man., 1951; m. Adele Goldberg, June 14, 1950; children—Lynda Ruth, Alan Mark, Arthur Ian, Frederick Lloyd. Research asst. Atomic Energy Can. Ltd., Chalk River, Ont., Can., 1948-49; faculty U. Man., Winnipeg, 1949—, prof. physics, 1963—. Mem. Canadian Assn. Physicists, Am. Phys. Soc., Am. Geophys. Union. Research and publs. in devel. of scintillation counter and its applications in nuclear spectroscopy, also cosmic radiation. Home: 772 Campbell St., Winnipeg, Man., Can.*

STANG, Louis George, Jr., Am. chemist; b. Portland, Ore., Oct. 25, 1919; s. Louis George and Pearl (Graham) S.; B.A., Reed Coll., 1941; student Cal. Inst. Tech., 1941-42; m. Dorian Ruth Heintz, June 5, 1943; children—David James, Steven Cory, Mark Edward. Research asso. NDRC, Pasadena, Cal., Evanston, Ill., Tooele, Utah, 1942-43; research chemist Manhattan Project, Oak Ridge, Chgo., Dayton, 1943-45; research chemist Universal Oil Products, Riverside, Ill., 1945-47; head hot lab. div. Brookhaven Nat. Lab., Upton, N.Y., 1947—. Cons. radiochemistry and design of hot labs. Mem. Am. Nuclear Soc., Am. Chem. Soc., N.Y. Acad. Scis., A.A.A.S., Phi Beta Kappa, Sigma Xi. Author: Hot Laboratory Equipment, 1958. Editor: Nuclear Applications Jour. Am. Nuclear Soc., 1964—. Publs. on devel. of plant for processing of multi-curie amounts of radioactive chemicals and prodn. of short-lived radioisotopes. Home: 379 Greene Av., Sayville, N.Y. 11782. Office: 60 Rutherford Dr., Upton, N.Y. 11973.*

STANHOPE, Charles, English inventor; b. London, Aug. 3, 1753; studied under LeSage, Geneva, Switzerland; m. Hester Pitt, 1774; 4 daus.; also 3 sons from 2d marriage. Became mem. House of Commons, 1780, House of Lords, 1786. Fellow Royal Soc., 1772. Author: Principles of Electricity, containing Divers New Theories and Experiments, together with an Analysis of the Superior Advantage of High and Pointed Conductors, 1779. Invented 1st iron hand printing-press, cylindrical biconvex lens with ends of unequal curvature, calculating machines, method of stereotyping (used by Clarendon Press beginning 1805); explained return stroke phenomenon; studied tuning instruments with fixed tones, magnifiers, glass rod length L, fireproof constructions for bldgs., ship design. Patentee steam-propelled ship, 1790. Died Chevening, Eng., Dec. 15, 1816.

STANHOPE, George Edward, English archaeologist; b. Newbary, Eng., June 26, 1866; s. Henry Howard Molyneaux Herbert and Evelyn S.; student Trinity Coll., Cambridge, 1885-87; m. Almina Wambwell, 1895; Children—Henry George Alfred Marius Victor Francis, Evelyn. Began excavation in Thebes, 1903; worked with Howard Carter for 16 years; became 5th earl of Carnarvon, 1890. Author: Five Years Exploration at Thebes. Discovered tomb of king's son of Dynasty XVIII; funerary temple of Queen Hatshepsut; a tomb of Dynasty XII; tomb of King Tutankhamun, Dynasty XVIII, 1922 (all Egypt). Died Apr. 5, 1923.

STANIER, Roger Yate, microbiologist; b. Victoria, B.C., Can., Oct. 22, 1916; s. Francis Thursfield and Dorothy (Broadbent) S.; B.A., U. B.C., 1936; M.A., U. Cal. at Los Angeles, 1940; Ph.D., Stanford, 1942; m. Germaine Bazire, June 2, 1956; 1 dau., Jane Françoise. Came to U.S., 1946. Asst. prof. Ind. U., 1946; faculty U. Cal. at Berkeley, 1947—, prof. microbiology, 1951—. Recipient Eli Lilly award in bacteriology and immunology Soc. Am. Bacteriologist, 1950; Guggenheim fellow Institut Pasteur, Paris, 1951-52. Mem. Am. Acad. Arts and Scis., Am. Soc. Biol. Chemists, Am. Soc. Microbiology, Sigma Xi. Author: (with M. Dondoroff, E.A. Adelberg) The Microbial World, 1957; also numerous articles. Research on bacterial classification, degradation of aromatic compounds by bacteria, pigment synthesis and metabolism of photosynthetic bacteria, structure and orgn. of bacterial cell. Home: 200 Panoramic Way, Berkeley, Cal. 94704.*

STANINIMIR, Fempl, Yugoslavian mathematician; b. Zemun, Yugoslavia, July 26, 1903; s. Fempl Lukas and Ann (Banatovic) S.; D.Phil. with honors in math. Philos. Faculty Belgrade, Yugoslavia, 1956; m. Katharina Waldmann, Oct. 9, 1948. Research asst. Astro. Obs. Belgrade, 1927-30; tchr. secondary schs., Pancevo, Yugoslavia, 1930-33, Zemun, 1933-48; prof. Coll. for Secondary Sch. Tchrs., Belgrade, 1948-58; faculty Elec. Engring. Faculty Belgrade, 1958—, asso. prof., 1959—. Mem. Math. Inst. Belgrade. Author: Series, 1960; Elements Calculus of Variation, 1965. Research, publs. on elliptic integral and functions, non-analytic functions, brachistochrones in fields of different forces. Home: 11 Kozjacka, Belgrade, Yugoslavia.*

STANLEY, Edward, English surgeon; b. London, Eng., July 3, 1793; s. Edward and (Blizard) S.; ed. Merchant Taylors' Sch., 1802-08; apprentice to Thomas Ramsden, St. Bartholomew's Hosp. Lectr. anatomy St. Bartolomew's Hosp., 1926-48, surgeon, 1838-61; prof. human anatomy and physiology, from 1835; surgeon extraordinary to Queen Victoria, 1858. Hunterian Orator, 1939. Fellow Royal Soc., 1830; life mem. council Royal Coll. Surgeons, pres, 1848, 57, pres. Royal Med. and Chirurg. Soc. Author: Illustrations of the Effects of Diseases and Inquiry of the Bones, 1849; A Treatise on the Diseases of the Bones, 1849; A Manual of Practical Anatomy, 1818; An Account of the Mode of Preforming the Lateral Operation of Lithotomy, 1829; Hunterian Oration, 1839; His classical works represented for many years all that was known of pathology of bone disease. Died May 24, 1862.

STANLEY, George Mahon, Am. geologist; b. Detroit, Mar. 15, 1905; s. Louis Crandall and Jane (Mahon) S.; B.S. in Civil Engring., U. Mich., 1928, M.A. in Geology, 1930, Ph.D., 1932; m. Ellen Burden Stevenson, Aug. 14, 1934; children—Martha Alice, G(eorge) Patterson. Faculty, U. Mich., 1930-48, asso. prof., 1945-48; asso. geologist U.S. Geol. Survey, Iron River, Mich., 1945-46; faculty Fresno (Cal.) State Coll., 1948—, prof., 1954-67, chmn. dept. geology, 1955-65; mem. Labrador Crater Expdn., Nat. Geog. Soc., 1954, expdn. to ancient lakes Salton Sea Basin, NSF, 1963-65. Mem. Geol. Soc. Am., Mich. Acad. Sci. Author: Geology of the Cranbrook Area, 1936; Pre-historic Mackinac Island, 1945; also articles. Showed that uplifted beaches in northeastern Gt. Lakes slope S.W. beneath modern lakes, the Glacial Lake Algonquin was drained to very low levels several hundred feet below modern Lake Huron and fed from Lake Mich. basin by now submerged river channel as receding Pleistocene ice sheet evacuated Ottawa River Valley. Home: 545 E. Buckingham Way, Fresno, Cal. 93704.*

STANLEY, Sir Henry Morton, explorer; b. Denbigh, Wales, 1841; came to U. S., 1857; adopted by Mr. Stanley in New Orleans; m. Dorothy Tennant, July 12, 1890. Sent to Africa by N.Y. Herald, 1870, 74; found Livingstone; discovered course of the Congo; returned to Africa for King of Belgians, 1879-84; his work resulted in founding of Congo Free State, 1887-89; commanded Emin Relief Expdn.; mem. Brit. Parliament for N. Lambeth, as Liberal Unionist, 1895-1900. Author: How I Found Livingstone, 1872; My Kalulu, Prince, King and Slave, 1873; Coomassie and Magdala, 1874; Through the Dark Continent, 1878; The Congo and the Founding of Its Free State, 1885; In Darkest Africa, 1890; My Dark Companions, 1893; My Early Travels in America and Asia, 1894; Slavery and the Slave Trade, 1894; Through South Africa, 1898. Died 1904.

STANLEY, Neville Fenton, microbiologist; b. Port Moresby, Papua, New Guinea, Oct. 7, 1918; s. Evan Richard and Helen (Turner) S:; B.Sc., U. Adelaide, 1942, D.Sc. 1953; m. Muriel E.P. MacDonald, Dec. 13, 1941; children—Evan Richard, Fiona Juliet, Bryden Jane. Became research officer Inst. Epidemiology and Preventive Medicine, Prince Henry Hosp., Sydney, Australia, 1946, also dir. bacteriology; successively became dir. Inst. Epidemiology and Preventive Medicine, Sydney, 1954, Found. prof., chmn. dept. microbiology Sch. Medicine, U. Western Australia, 1956; dir. dept. microbiology Royal Perth Hosp.; dir. Virus Diagnostic Lab., Dept. Pub. Health for Western Australia. Chmn. state examining council Australian Inst. Med. Lab. Tech.; hon. cons. microbiologist to King Edward Meml. Hosp. for Women, Perth, 1961; dean faculty sci. U. Western Australia, 1962; Fulbright scholar, vis. prof. biology U. Notre Dame,1964 Mem. Australia and New Zealand Assn. for Advancement of Sci. (chmn. div. 1959), Australian Soc. Microbiology (chmn. W. Australia br.) Soc. Gen. Microbiology, Am. Soc. Microbiology, Soc. Exptl. Biology and Medicine, Royal Soc. Medicine London. Asst. editor Australian Jour. Exptl. Biology and Medicine, 1963—. Numerous publs. on isolation of new antibiotic, aspergillin; isolated and purified a bacterial lipid that stimulated monocytes and antibodies; showed that Coxsackie viruses cause aseptic meningitis; isolated and described new group of reoviruses; research on role of virus-induced autoimmune disease in leading to cancer. Home: 28 Keane St., Peppermint Grove, Western Australia, Australia.*

STANLEY, Paul Elwood, Am. elec. engr.; b. Huntington, Ind., Nov. 6, 1909; s. Noah E. and Bertha (Chalmers) S.; A.B., Manchester Coll., 1930; M.A., Ohio State U., 1933, Ph.D., 1937; m. Lucille

Klutz, Nov. 27, 1930; children—Mary Louise (Mrs. J.M. Johnson), Carol Ann (Mrs. Paul Conrad), Barbara Jean. Faculty, Purdue U., Lafayette, Ind., 1943—, interim head Sch. Aeros. and Astronautics Engring. Scis., 1963-65, prof., 1965—; cons. numerous indsl. firms. Fellow Am. Inst. Aeros. and Astronautics (asso.); mem. I.E.E.E., Am. Soc. Engring. Edn. Author: (with Joseph Liston) Creative Product Envolvement, 1964. Research in simulation of turbine engines by analog computers, hosp. elec. safety. Home: 100 Hideaway Lane, West Lafayette, Ind. 47906.*

STANLEY, Wendell M(eredith), Am. bio-chemist; b. Ridgeville, Ind., Aug. 16, 1904; s. James G. and Claire (Plessinger) S.; B.S., Earlham Coll., Richmond, Ind., 1926, hon. Sc.D., 1938; M.S. U. of Ill., 1927, Ph.D., 1929; Sc.D., 1959; Sc.D., Harvard, Yale, 1938, Princeton, 1947, U. Pitts., 1962, U. Pa., 1964; LL.D., U. Cal., 1946, Ind. U., 1951, Jewish Theol. Sem., 1953, Mills Coll., 1960; Dr. honoris causa, U. Paris, 1947; m. Marian Staples Jay, June 15, 1929; children—Wendell, Marjorie (Mrs. Robert J. Albo), Dorothy (Mrs. Roger Erickson), Janet E. Research asso. and instr. in chemistry, U. of Ill., 1929; Nat. Research fellow, Munich, Germany, 1930-31; with Rockefeller Inst. for Med. Research since 1931, mem., 1940-48; Hitchcock prof. U. Cal., 1940, chmn. dept. biochemistry, 1948-53; prof. biochemistry, dir. virus lab. Univ. Cal. at Berkeley, 1948—, professor, chmn. dept. virology, 1958-64, professor of molecular biology, 1964—. Vanuxem lectr. Princeton, 1942; Messenger lectr. Cornell, 1942; Silliman lectr. Yale, 1947. Trustee Mills Coll., 1951-58. Mem. expert adv. panel on virus diseases WHO, 1951—; nat. adv. cancer council USPHS, 1952-56; chairman section biochemistry Nat. Acad. Scis., 1955-58; dir.-at-large Am. Cancer Soc., 1955-61; mem. bd. scientific counselors National Cancer Inst., 1957-61, chmn. 1957-58. Recipient nat. award Am. Cancer Soc., 1963. Fellow N.Y. Acad. Scis.; mem. Am. Assn. Immunologists, Am. Philos. Soc., Am. Chem. Soc., Am. Phytopathological Society, Harvey Soc. (hon.), Am. Soc. Biol. Chemists (mem. council 1951-54), A.A.A.S., Sigma Xi. Awarded A.A.A.S. prize, 1936; Isaac Adler prize by Med. Sch. of Harvard, 1938; Rosenberger medal by U. of Chicago, 1938; John Scott medal, certificate, and preminum by the City of Phila., 1938; gold medal of Am. Inst. of the City of New York, 1941; Copernican Citation by the Copernican Quadricentennial Nat. Com., 1943; Nichols Medal of the N.Y. Sect. of American Chem. Society, 1946; (with Sumner and Northrop) Nobel Prize in Chemistry, 1946; Gibbs Medal of Chicago Sect. of Am. Chem. Soc., 1947; Franklin Medal of Franklin Inst., 1948; Presdl. Certificate of Merit, 1948; Modern Medicine award, 1958; Am. Cancer Soc. award, 1959. Contr. to scientific jours. Research on viruses; 1st to obtain isolated virus crystals, showing virus to be proteinaceous, 1935; isolated nucleic acid from crystallized virus, 1936; work on influenza, mutation and reprodn. of viruses. Address: Virus Lab., U. Cal, Berkeley, Cal. 94720.*

STANLEY, William, Am. elec. engr., inventor; b. Bklyn., Nov. 22, 1858; s. William and Elizabeth A. (Parsons) S.; ed. Williston Sem., Easthampton, Mass., also Yale; m. Lila C. Wetmore, Dec. 22, 1884. Research asst. to Hiram Stevens Maxim, later to Edward Weston; worked in own research lab., Englewood, N.J., 1883-85; chief engr. Westinghouse Electric Co., 1885-88, Stanley Elec. Mfg. Co., 1890-95, Stanley Instrument Co., 1898-1903. Inventor transformer, alternating-current system of long distance light and power transmission, 2-phase motors, generators, an alternating-current watt-hour meter with moving parts magnetically suspended; experimented on storage batteries, incandescent lamps. Died May 14, 1916.

STANNARD, James Newell, Am. radiation biologist; b. Owego, N.Y., Jan. 2, 1910; s. Jay Ellis and Miriam (Newell) S.; B.A., Oberlin Coll., 1931; postgrad. Yale, 1931-32; M.A. Harvard, 1934, Ph.D., 1945; m. Grace L. Kingley, Aug. 7, 1935; children—Susan Lou (Mrs. Joseph C. Stumpf). Instr. physiology U. Rochester, 1935-39; asst. prof. pharmacology Emory U., 1939-41; sr. pharmacologist, prin. physiologist NIH, USPHS, 1941-44, 46-47; faculty U. Rochester (N.Y.), 1948—, prof. radiation biology and biophysics, asso. dean for grad. studies, asso. prof. pharmacology Sch. Medicine and Denistry, asso. dir. edn. atomic energy project, 1959—. Cons. Nat. Center for Radiol. Health, 1968; mem. nat. adv. com. on radiation nat. council on radiation protection. USPHS. Mem. Am. Physiol. Soc., Radiation Research Soc. (chmn. com. on edn.), Health Physics Soc., chmn. com. on edn. and tng., pres. elect 1968-69), Biphysics Soc., Soc. Gen. Physiologist, Am. Soc. Pharmacology and Exptl. Therapeutics, Am. Indsl. Hygiene Assn., Editor: (with G. W. Casarett) Metabolism and Biological Effects of an Alpha Patticle Eitter, Polonium 210, 1964; also articles. Research on effect of certain chem. inhibitors on cellular respiration, biol. effects of alpha particle emitting radioisotopes, hazards of inhaled radioactive materials, effects of radiation on cell membranes. Home: 10 Tall Acres Dr., Pittsford, N.Y. 14534. Office: 260 Crittenden Blvd., Rochester, N.Y. 14620.*

STANNUS, Hugh Stannus, Brit. physician; b. June 18, 1877; s. Hugh Hutton Stannus; ed. St. Thomas's Hosp., U. London, Paris; M.D., Ph.D., U. London. Cons. physician French Hosp.; asso. staff Tropical Diseases Hosp.; med. adviser Bd. Inland Revenue; cons. Ministry Health, Ministry Pensions; examiner for diploma tropical medicine U. Liverpool (Eng.); Lumleian lectr., 1944. Recipient Gold medal U. London. Fellow Royal Coll. Physicians London, Royal Soc. Medicine, Royal Soc. Tropical Medicine and Hygiene, Soc. Genealogists, Nutrition Soc., Royal Anthrop. Inst. Gt. Britain. Author: Monograph on Wa-Yao. Sectional editor, Tropical Diseases Bull. Research and publs. on anthropology, tropical diseases. Died Feb. 27, 1957.

STANOJEVIC, Lazarus, educator, psychiatrist; b. Sombor (now Backa, Yugoslavia), 1883; s. Lazarus and Sofia (Klemen) S.; ed. Coll. Combor; grad. U. Budapest (Hungary), 1909; m. Vera Jovanovic. Mem. Mil. Applied Med. Sch., Vienna, Austria, 1909-10; physician mil. hosp., Innsbruck, Austria, 1910-11; capt. physician clinic U. Vienna, 1912-13; served at mil. hosp., Vienna, 1914-17; chief physician hosp. at Trnava, Czechoslovakia, 1918-19; dir. Insane Asylum nr. Zagreb, Yugoslavia, 1919-23; chief prof. psychiatry, psychology, neurology U. Belgrade (Yugoslavia), 1923—. Mem. German Assn. Psychology, German Assn. Neurologists and Natural Scientists (both Berlin), Serbian Med. Assns., Assn. Psychiatry and Neurology (Vienna). Author numerous works on psychiatry, neurology, exptl. psychology. Address: Obilicer Venac 19, Belgrade, Yugoslavia.

STANSEL, Valentin, astronomer; b. Olmütz, Czechoslovakia, 1621; joined Soc. of Jesus, 1637, ordained; tchr. Olmütz and Prague, Czechoslovakia; tchr. astronomy, Evora, Portugal; tchr. moral theology, Brazil; later superior San Salvador Sem., Bahia, Brazil. Contbr. numerous articles to tech. jours. Astron. observations, especially of comets. Died Bahia, Dec. 18, 1705.

STANSLY, Philip Gerald, Am. biochemist, microbiologist; b. N.Y.C., July 23, 1912; s. Avram and Bertha (Tuchman) S.; B.S., Cornell U., 1933; M.Sc., N.Y. U., 1936; Ph.D.., U. Minn., 1944; postdoctoral work U., Wis., 1952-54; m Marguerite R. Anzolut, July 15, 1938; children—Philip A., Pamela J. Sr. scientist Am. Cyanimid Co., Stamford Conn., 1944-52; sr. research scientist Mich., Cancer Found., 1954-67; asso. prof. microbiology Sch. Medicine, Wayne State U., Detroit, 1954-67; dir. dept. microbiology Mason Research Inst., Worcester, Mass., 1967-68; head microbiology dept. Biomed. Research Center, Litton Systems, Inc., Bethsda, Md., 1968—. Cons, Nat. Cancer Inst, 1967—. Fellow N.Y. Acad. Scis.; mem. Am. Soc. Biol. Chemistry, Am. Assn. Cancer Research, Soc. Exptl. Biology and Medicine, Sigma Xi. Contbr. articles to tech. jours. Discovered antibiotic polymyxin; research in causation and prevention of cancer; discovered virus causing reticulum cell sarcoma in mice.*

STANTON, Austin N., Am. electronic engr.; b. Cromwell, Ia., May 31, 1903; s. Jay Birney and Harriett (Laughlin) S.; B. Engring., State U. Ia., 1925, M.S. in Physics, 1927; m. Margaret Lovina Saveraid, June 12, 1926; children—Beverley (Mrs. Prince), Peggy (Mrs. Head). Chief engr. Automatic Refrigerator Corp., Chgo., 1934-36; v.p., chief geophysicist Geophys. Petroleum Surveys, Dallas, 1936-38; v.p. Tex. Geophys. Co., Dallas, 1938-40; pres. Petroleum Sci. Labs., Dallas, 1940-42; asso. prof. elec. engring. So. Meth. U., Dallas, 1942-45; pres. Varo Inc., Garland, Tex., 1946-58, 62—, chmn. bd., 1958—. Fellow Am. Astronautical Soc. (bd.), Brit. Interplanetary Soc.; mem. I.E.E.E., Am. Rocket Soc., Soc. Exploration Geophysicists, Am. Astron. Soc, Kappa Eta Kappa (founder 1923), Sigma Xi. Patentee, research, devel. electron devices, especially electrically driven tuning fork, static inverter, microcircuity, high-power microwaves, infrared viewing equipment, space tech., geophys. exploration. Home: 4240 Brian Creek Lane, Dallas 75214. Office: Garland Bank & Trust Bldg., P.O. Box 411, Garland, Tex. 75040.*

STANTON, Mearl Frederick, Am. pathologist; b. Staunton, Ill., Aug. 14, 1922; s. Fred W. and Bertha (Johnson) S.; student U. So. Ill., 1940-42, U. Wis., 1944; M.D., St. Louis, U., 1948; m. Margie Hartman, Apr. 7, 1951. Sr. instr. St. Louis U. Sch. Medicine, 1950-55, pathologist St. Mary's Group, 1953-55; pathologist Army Chem. Center, Edgewood, Md., 1955-56; pathologist Nat. Cancer Inst., NIH, Bethesda, Md., 1956—, sci editor jour. Nat. Cancer Inst., 1968. Research and publs. on host-parasite relationship in infectious diseases, pathology, etiology and mechanisms of carcinogenesis. Home: 4119 Stanford St., Chevy Chase, Md. 20015. Office: 9000 Rockville Pike, Bethesda, Md. 20014.*

STANTON, Ralph Gordon, Canadian mathematician; b. Lambeth, Ont., Can., Oct. 21, 1925; s. Gordon Wyman and Ida Maude (Robertson) S.; B.A., U. Western Ont., 1944; M.A., U. Toronto, 1945, Ph.D., 1948; certificado em Lingua Portuguesa, Rio de Janeiro, Brazil, 1946. Faculty, U. Toronto, (Ont.), 1949-57; prof. math., chmn. dept. U. Waterloo (Ont.), 1957-66, dean grad. studies, 1960-66; vis. prof. math. U. Man., 1966-67; professor York U., Toronto, Ont., 1967—; summer research officer NRC, 1946-53; sessional lectr. Carleton U., 1950; vis. prof. under Ford Found. grant U. Wis., 1961-62. Mem. Canadian Seminar on Inter-Am. Affairs, 1954; mem. Ont. Math. Commn., 1960-67. Govt. of Brazil fellow, 1945-46. Mem. Royal Statis. Soc., Biometric Soc., Am. Math. Soc., Math. Assn. Am., Canadian Math. Congress, Canadian Operational Research Soc., Am. Assn. U. Profs., Canadian Assn. U. Tchrs. Author: Numerical Methods for Science and Engineering, 1961; (with K.D. Fryer) Topics in Modern Mathematics, 1964, Algebra and Vector Geometry, 1965. Research in combinatorial analysis, Mathieu groups, math. biology, teaching of math. Home: Woodbridge, Ont. Office: York U., Toronto 12, Ont., Can.*

STAPF, Otto, botanist; b. Ischl, Austria, Mar. 23, 1857; s. Joseph Stapf; student botany under prof. Wiesner, Vienna, Austria; Ph.D.; m. Martha Beranek, 1892. Asst. to prof. Kerner von Marilaun, 1882-89; pvt. docent U. Vienna, 1887-91; asst. for India, Royal Bot. Gardens, Kew, 1891-99; principal asst. 1899-1908; keeper Herbarium and Library, 1909-22; traveled in Southwest, Central and North Persia, 1885. Recipient Victoria medal Honour in Horticulture, 1927, Veitch Meml. medal, 1932; Linnean medal, 1927. Fellow Royal Soc. 1908; mem. (corr.) Imperial Acad. Scis., Vienna, (hon.) Deutsche Botanische Gesellschaft; bot. sec. Linnean Soc., 1908-16. Author: Iconum Botanicarum Index Londinensis; also monographs of Ephedra Indian Aconites, Indian Oil-gasses. Research on Oriental, Indian, African, Malayan floras. Died Aug. 3, 1933.

STAPLE, Ezra, Am. biochemist; b. Phila., Sept. 3, 1917; s. Simon and Esther (Weimann) S.; B.S. with distinction, U. Pa., 1939, M.S., 1942, Ph.D., 1949. Research chemist OSRD, U. Pa., 1942; mem. sr. sci. staff SAM Labs., Columbia U., 1942-44; sr. research chemist, dept. head Carbide and Carbon Chems. Corp., Manhattan Dist., Oak Ridge, 1944-47, cons., 1949-51; faculty U. Pa. Sch. Medicine, Phila., 1951—, asso. prof. biochemistry, 1965—. DuPont fellow in chemistry, 1947-48; established investigator Am. Heart Assn., 1957-62; recipient Career Devel. award NIH, 1962—. Mem. Am. Chem. Soc., Am. Soc. Biol. Chemists, Chem. Soc. London, N.Y. Acad. Scis., Am. Oil Chemists Soc., A.A.A.S., Am. Heart Assn. (council on basic scis.), Sigma Xi. Contbr. numerous articles to profl. jours, chpts. to books. Research in catabolism of sterols and related compounds especially in mammals, study of first steps in conversion to steroid hormones and to bile acids. Home: 2450 N. 59th St., Phila. 19131.*

STAPP, Carl, German biologist; b. Biedenkopf, Germany, May 2, 1888; s. Carl and Caroline (Stephany) S.; student U. Munich, U. Marburg; dr. ès sc. honoris causa; Ph.D.; m. Charlotte Kunze, May 9, 1921; children—Irmgard, Renate, Wolfgang. Asst., Bot. Inst., U. Marburg, also Hygiene Inst., U. Frankfort; became collaborator Biologische Reichsanstalt, Berlin-Dahlem, Germany, 1920, later govt. counselor, mem. govt. cons.; govt. cons., dir. Bacteriology and Serology Inst., Biologische Bundesantalt, Brunswick, Germany; now ret. Recipient Otto Appel medal 1958. Mem. Nat. Leopoldina Acad. Author: Pflanzenpathogene Bakterien, 1958; Bacterial Plant Pathogens, 1961. Became editor Zentralblatt für Bakteriologie II, 1932, co-editor, 1952. Address: Magnitorwall 5, Braunschweig, W. Germany.

STARE, Fredrick John, Am. nutritionist, physician, educator; b. Columbus, Wis., Apr. 11, 1910; s. Frederick A. and Susan (Seidell) S.; B.S., U. Wis., 1931, M.S., 1932, Ph.D., 1934; M.D., U. Chgo., 1941; D.Sc., Suffolk U., 1963, Trinity Coll., Dublin, Ireland, 1964; m. Helen Elma Haxton, June 9, 1959; children—Fredrick A., David S., Mary S. NRC fellow Washington U., St. Louis, 1934-35; Rockefeller Found. fellow, Cambridge, Eng., 1935-36, Szeged, Hungary, 1936-37, Zurich, Switzerland, 1937; faculty Harvard, 1942—, prof. nutrition, 1946—, head dept., 1942. Nutrition cons. to numerous vol. and govt. health agys. Mem. A.M.A., Am. Chem. Soc., Am. Inst. Nutrition, Brit. Nutrition Soc., Am. Pub. Health Assn. Author: Eating for Good Health, 1964; also numerous articles. Editor: Nutrition Revs., 1942—. Research on food utilization, vein feeding, prevention heart disease. Home: 267 Cartwright Rd., Wellesley, Mass. 02121. Office: 665 Huntington Av., Boston 02115.*

STARIK, Iosif Evseevich, Russian chemist; b. Mar. 23, 1902; grad. Moscow U., 1924. Asso., Inst. Radium, 1924-46; became dep. dir. Radium Inst., USSR Acad. Scis., 1946, became chmn. commn. on absolute age geol. formations, 1961; prof. Leningrad U., 1946—. Recipient Order Lenin, 3 times. Corr. mem. USSR Acad. Scis. Author: The Colloidal Properties of Polonium, 1930-33; Radioactive Methods of Geological Dating, 1938; Radiochemical Analysis, 1936; The Form of Conditions for the Primary Migration of Radioelements in Nature, 1943; The Colloidal Properties of Polonium, 1956; Role of Secondary Processes in the Dating of Rocks by Radioactive Methods, 1956; The State of Microscopic Quantities of Radioelements in the Liquid and Solid States, 1957; Thorium Isotope Concentration in the Waters of the Black Sea, 1959; New Data on Determination of Uranium Content in Meteorites, 1960; (with Yu A. Barbanel) Investigation of Several Functions characterizing the State of Substances in Solution, 1962. Research on colloidal conditions of radioelements related to their absorption properties, geol dating by radioactive methods, conditions for migration of radioelements; devel. radiochem. analysis.

Address: Commission on Absolute Age of Geol. Formations, USSR Acad. Scis., Leninskii, Prospect 14, Moscow, USSR.

STARK, Guenther, German physician; b. Berlin, May 6, 1922; s. Alfred and Margarethe (Keup) S.; student univs. Berlin, Königsberg (now Kaliningrad, USSR); state exam., U. Marburg/Lahn, 1948; m. Eva Verres, July 24, 1953; children—Bettina, Sabine. Faculty, U. Mainz (Germany), 1958—, unscheduled prof., 1962—, head physician, dept. obstetrics and gynecology, 1959-65, head dept., 1965—. Recipient Carl Thomas award, 1960. Mem. German Assn. Obstetrics and Gynecology. Research and numerous articles on biochemistry of placenta; metabolism of electrolytes and aldosterone in normal and pathological pregnancy; postoperative changes of metabolism; problems of Rh factor. Address: 8 In den Gärten, 65 Mainz, W. Germany.*

STARK, Johannes, German physicist; b. Schickenhof, Germany, Apr. 15, 1874; student U. Munich, 1894-98. Asst., U. Munich, 1898-1900; privatdozent physics U. Göttingen (Germany), 1900-06; asso. prof. Tech. High Sch., Hanover, Germany, 1906-09; prof. Tech. High Sch., Aachen, Germany, 1909-17, U. Greifswald (Germany), 1917-20, U. Würzburg (Germany), 1920-33; became pres. Physikalisch-Technische Reichsanstalt, Charlottenburg, Germany, 1933. Recipient Nobel prize in physics, 1919. Studied radiation and atomic theory; discovered Doppler effect in canal rays spectra, 1905; 1st to split spectral lines with electric field (Stark effect), 1913. Died Traunstein, Germany, 1957.

STARK, John Thomas, Am. geologist; b. Jackson, Tenn., Sept. 1, 1888; s. John Thomas and Ella (Barton) S.; B.S., Northwestern U., 1920, M.S., 1921; Ph.D., U. Chgo., 1927; m. Terii Pahi, July 6, 1936; 1 dau., Teura (Mrs. Peter C. Morris). Mem. faculty Northwestern U., Evanston, Ill., 1927-52, prof. geology, 1936-52, chmn. dept., 1936-43; with U.S. Geol. Survey, 1952-63; ret., 1963. Recipient Meritorious Service award U.S. Geol. Survey, 1963; medal Royal Siamese Embassy in Washington, 1946. Mem. Geol. Soc. Am., Soc. Econ. Geologists, Am. Mineral. Soc., Am. Geophys. Union. Author: Geology of Borabora, Society Islands, 1938; (with others) Military Geology of Guam, Mariana Islands, 1963, Military Geology of Truk Islands, Caroline Islands, 1963; also articles. Research in structural geology and petrography; field research in Pre-Cambrian areas, Minn., Ont., Colo., islands of Guam, Truk, South Pacific. Home: 17 Ellis Dr., Jackson, Tenn. 38303.*

STARK, Ronald William, entomologist; b. Calgary, Alta., Can., Dec. 4, 1922; s. Robert Donaldson and Christina (Currie) S.; B.Sc. In Forestry, U. Toronto (Ont., Can.), 1948, M.A. in Zoology, 1951; Ph.D. in Forest Entomology, U. B.C., Vancouver, Can., 1958; m. Mary Laurita McMann, Sept. 2, 1944; children—Debra Jean, David Ronald. Came to U S., 1959, naturalized, 1966. Entomologist div. forest biology Can. Dept. Agr., Calgary, 1948-49; faculty, staff U. Cal. at Berkeley, 1959—, entomologist, prof., 1966—. Vice chmn. Cal. Forest Pest Action Council, 1964—. Named Distinguished Tchrs. award U. Cal. at Berkeley, 1960. Mem. Soc. Am. Foresters (sec. div. entomology and pathology 1965—), Entomol. Soc. Am., Entomol. Soc. Can., Ecol. Soc. Am., Internat. Soc. Biometeorology, Japanese Soc. Population Ecology, A.A.A.S., Am. Inst. Biol. Scis., Sigma Xi. Author: (with D. L. Wood and K. Graham) Forest Entomology Laboratory Manual, 1962; also numerous articles. Research on insect population fluctuations, methods for sampling to determine population size, destructiveness certain insects. Home: 111 Ardith Dr., Orinda, Cal. 94563. Office: U. Cal., Berkeley, Cal. 94720.*

STARK, Werner, sociologist; b. Marienbad, Czechoslovakia, Dec. 2, 1909; s. Adolf and Jenny (Schneider) S.; Diplomvolkswirt, U. Hamburg, 1931, Dr.rer.pol., 1934; student London Sch. Econs., 1930-31, U. Geneva, 1933; Dr. jur., U. Prague, 1936; M.A., U. Edinburgh, 1947; m. Kate Franck, Aug. 14, 1934. Faculty, Prague Sch. Polit. Sci., 1937-39, U. Cambridge (Eng.), 1941-42, U. Edinburgh, 1945-51, U. Manchester, 1951-63; guest prof. Purdue U., 1960-61; prof. Fordham U., N.Y.C., 1963—. Mem. Am., Am. Catholic sociol. assns. Author: Sozialpolitik, 1936; The Ideal Foundations of Economic Thought, 1943; The History of Economics in its Relation to Social Development, 1944; History of Economics (Italian translation), 1950, (Japanese translation), 1955, (German translation), 1960, (Spanish translation), 1961; America: Ideal and Reality. The United States of 1776 in Contemporary European Philosophy, 1947; Jeremy Bentham's Economic Writings, 3 vols., 1952-54; The Sociology of Knowledge, 1958, (German translation), 1960, (Japanese translation), 1961, (Italian translation), 1963, (Spanish translation), 1963; Social Theory and Christian Thought, 1959; Montesquieu: Pioneer of the Sociology of Knowledge, 1960; Ideal Foundations of Economic Thought (Japanese translation), 1961; The Fundamental Forms of Social Thought, 1962; Jeremy Bentham's Economic Writings (Spanish translation), 1965; The Sociology of Religion, 3 vols., 1966-67. Gen. editor: Rare Masterpieces of Philosophy and Science, 1950. Numerous publs. in field of sociology. Home: 5643 Mosholn Av., N.Y.C. 10471. Office: Fordham U., N.Y.C. 10458.*

STARKE, Kurt Walter Ernst, German chemist; b. Berlin, Nov. 11, 1911; s. Hugo and Emma (Brunzel) S.; Dr.phil., U. Berlin, 1937; Dr.phil. habil., U. Munich, 1945; m. Alexa Hartmann, Sept. 29, 1951; children—Rolf Peter, Hans Christoph. Research asst. Kaiser Wilhelm Inst. Chemistry, Berlin, 1937-41, U. Munich, 1941-44, Max Planck Inst. Med. Research, Heidelberg, 1944-48; research asso. McMaster U., Hamilton, Ontario Can., 1948-50; asst. prof. U. B.C., Vancouver Can., 1950-56; asso. prof. U. Ky., Lexington, 1956-59; prof., head inst. nuclear chemistry, U. Marburg (Germany), 1959—. Mem. Am. Chem. Soc., Soc. German Chemists. Contbr. numerous articles to profl. jours. Separations of radioactive nuclides; discovery of element 93; chem. effects of nuclear transformations; preparations of metal complexes; decontamination of radioactive solutions. Home: 10 Am Strauch, 3551 Wehrshausen bei Marburg. Office: Institut für Kernchemie, 12 Biegenstrasse, 355 Marburg, West Germany.*

STARKEY (or Storkey, Stirk), George, Brit. alchemist; b. Bermudas; s. George Stirk; M.A., Harvard, 1646. Apothecary, America, where he was said to have learned some secrets of transmutation from alchemist called Eirenaeus Philaletha; gave demonstrations of transmutations, sold quack medicines London, Eng., 1646-50. Author: Nature's Explication and Helmont's Vindication or a Short and Sure Way to a Long and Sound Life, 1657; Pyrotechny Asserted and Illustrated, 1658; The Admirable Efficacy of Oyl Which is Made of Sulphur-Vive, 1660; George Starkey's Pill Vindicated; A Brief Examination and Censure of Several Medicines, 1664; A Smart Scourge for a Silly, Saucy Fool, an Answer to Letter at the End of a Pamphlet of Lionell Lockyer, 1665; An Epistolar Discourse to the Author of Galeno-Pale, 1665; Liquor Alchahest or a Discourse of that Immortal Dissolvent of Paracelsus and Helmont, 1675. Editor Mirror of Alchymy. Claimed to be original maker of what was known as Richard Mathew's Pill. Died circa 1665.

STARKEY, Otis Paul, Am. geographer; b. Buffalo, Apr. 14, 1906; s. Frederick Robinson and Laura (Hirsch) S.; B.S., Columbia U., 1927, A.M., 1930, Ph.D., 1939; m. Evelyn Saxton Locke, June 25, 1960. Editorial asst. to Prof. J. Russell Smith, N.Y.C., 1928-31; faculty U. Pa., 1931-42, asst. prof., 1939-42; geographer Mil. Intelligence Service, Washington, 1942-45; dep. dir. planning div. Office Fgn. Liquidation Commr., U.S. State Dept., Washington, 1945-46; prof. geography Ind. U., Bloomington, 1946—, chmn. dept., 1946-56. Recipient Civilian Meritorious Service award U.S. War Dept., 1946. Mem. Am. Soc. Profl. Geographers (pres. 1947), Assn. Am. Geographers, Nat. Council for Geog. Edn., Ind. Acad. Scis., Ind. Acad. Social Sci., A.A.A.S., Am. Geog. Soc. Author: (with Lester E. Klimm) Introductory Economic Geography, 1937; (with William F. Christians) Exploring Our Industrial World, 1938; The Economic Geography of Barbados, 1939; Commercial Geography of the Eastern British Caribbean, 1961; also articles, abstracts, revs. Introduced econ. theory into econ. geography; studied the Lesser Antilles especially econ. geography and history of the Barbados; prepared planning studies on Euro-African terrain for World War II operations; used nat. accounting data in Anglo-Am. econ. geography. Home: 2602 N. Browncliff Lane, Bloomington, Ind. 47401.*

STARKS, William Joseph, Am. chemist; b. Wilkinson, Ind., Feb. 26, 1904; s. Charles E. and Daisy (McDaniel) S.; A.B., Ind. U., 1926, A.M. 1929, D.Sc., 1966; Ph.D., U. Ill., 1936; postgrad. Harvard; D.Sc., Mich. Technol. U., 1966; m. Meredith Pleasant, Dec. 31, 1930; children—Ruth (Mrs. James W. Foster), Katherine (Mrs. Richard L. Albrecht), Charles, John. With E. I. duPont de Nemours & Co., 1929-34; with Esso Research & Engring. Co., Linden, N.J., 1936-39, 40—, dir. chem. div., 1946, sci. adviser, 1958—; prin. chemist U. S. Dept. Agr., Peoria, Ill., 1939-40. Chmn. div. chemistry, chem. tech. NRC, 1952-55; chmn. U. S. delegation Internat. Union Pure Applied Chemistry, Stockholm, 1953. Recipient Distinguished Alumni Service award Ind. U., 1956; Charles Goodyear award, 1962; Perkin award, 1964; Priestly medal, 1965. Mem. Am. Chem. Soc. (pres.), Armed Forces Chem. Assn., Am. Inst. Chemists, (Gold medal 1954), Assn., Research Dirs., Soc. Chem. Industry, Sigma Xi, Alpha Chi Sigma, Phi Lambda Upsilon. Patentee in field. Research, publs in synthetic rubber, plastics, petrochems, petroleum product; pioneer Butyl synthetic rubber. Home: 704 Highland Av., Westfield, N.J. 07090. Office: Esso Research & Engring. Co., Linden, N.J. 07090.*

STARLEY, James, Brit. inventor; b. Albourne, Eng., Apr. 21, 1831; s. Daniel Starley; m. Jane Todd, Sept. 22, 1853; children—James, John Marshall, William. Became gardener to John Penn, 1846; joined Newton Wilson, London, 1855; Coventry Machinists' Co. formed for mfg. Starley's new sewing machine, 1857; owner of business for bicycles; partner (with Brothwick Smith) Smith, Starley & Co. Built bicycles and tricycles for general use; invented curved spring, small hind wheel, pivot centre steering, chain and chain wheels for rotary motion in tricycles, step for mounting, double throw crank, pinion steering gear. Died June 17, 1881.

STARLING, Ernest Henry, Brit. physiologist; b. Barnsbury Sq., London, Apr. 17, 1866; s. Matthew Henry and Ellen (Watkins) S.; student King's Coll., London; student Guy's Hosp., 1882-89; hon. degrees Trinity Coll., Dublin, Sheffield, Eng., Cambridge, Eng., Breslau (now Wroclaw, Poland), Strasbourg, France, Heidelberg, Germany; m. Florence Amelia Sieveking, 1891; 1 son, 3 daus. Became demonstrator physiology Guy's Hosp., 1889; Jodrell prof. physiology U. Coll., 1899-1923; named Foulerton research prof. Royal Soc., 1922. Recipient Baly medal Royal Coll. Physicians, 1907. Fellow Royal Soc., 1899 (Royal medal 1913). Author: Principles of Human Physiology, 1912. Studied (with R. Heidenhain) lymph formation, 1892; discovered functions of serum proteins, 1896; discovered (with Bayliss) secretin, 1902; 1st to use term, hormone, 1905; showed (with E.B. Vernay) that kidney tubules reabsorb water, 1925; studied intestinal movement, lymph secretion and body fluids; developed generalization that energy of contraction is function of length of muscle-fibers (Starling's law of heart). Died Kingston, Jamaica, May 2, 1927.

STAROS, Anthony, Am. mech. engr.; b. Freeport, N.Y., July 17, 1923; s. Theodore and Clara (Preziosi) S.; student Mass. Inst. Tech., 1941-43; B.S., Cornell U., 1944; M.S., Stanford, 1947; postgrad. Hofstra U., 1954-56; m. Carolyn Frank, Jan. 19, 1952; children—Paul Anthony, Philip Anthony, Christan Clare. Research engr. Franlin Inst. Lab., Phila., 1947-49; chief engr. U.S. VA Prosthetic Testing and Devel. Lab., N.Y.C., 1949-56, dir. VA Prosthetics Center, 1956—. Chmn. N.Am. subcom. Internat. Com. Prosthetics and Orthotics; mem. various coms. prosthetics-orthotics Nat. Acad. Scis.-NRC. Mem. Internat. Soc. For Rehab. Disabled. Contbg. author: Orthopedic Appliances Atlas Vol. II, 1960. Research and numerous publs. primarily in field of bioengineering prins., methods in prosthetics, orthotics. Home: 25 Blanchard Dr., Northport, N.Y. 11165. Office: 252 7th Av., N.Y.C. 10001.*

STARR, Chauncey, Am. physicist, nuclear mfg. co. exec.; b. Newark, Apr. 14, 1912; s. Rubin and Rose (Dropkin) S.; B.S., Rensselaer Poly. Inst., 1932, Ph.D. in Physics, 1935, D. Engring., 1964; m. Doris Evelyn Debel, Mar. 20, 1938; children—Ross M., Ariel E. Research fellow physics Harvard, 1935-37; research asso. Mass. Inst. Tech., 1938-41; research physicist D.W. Taylor Model Basin, USN Bur. Ships, 1941-42; staff U. Cal. Radiation Lab., 1942-43, Tenn. Eastman Corp., Oak Ridge, 1943-46, Clinton Labs., Oak Ridge, 1946; dir. atomic energy research dept. N.Am. Aviation, Inc., Downey, Cal., 1949-55, v.p., 1955-66, gen. mgr. Atomics Internat. Div., Canoga Park, Cal., 1955-60, pres., 1960-66; dean Coll. Engring., U. Cal., Los Angeles, 1955—. Mem. Cal. Adv. Council on Atomic Energy Devel. and Radiation Protection, 1960—; mem. com. on isotopes and radiation devel. AEC, 1964—; mem. sci. adv. bd. nuclear panel USAF, 1960-64. Fellow Am. Nuclear Soc. (pres. 1958-59), Am. Phys. Soc.; mem. Atomic Indsl. Forum (v.p. 1964), Am. Inst. Aero. and Astronautics (sr. mem.), N.A.M. (mem. nuclear energy com. 1956—), Am. Power Conf. (mem. industry com. 1954—), Nat. Acad. Engring., A.A.A.S., Soc. Nuclear Medicine, Sci. Research Soc. Am. (mem. bd. govs. 1960—, chmn. 1966-68), Sigma Xi, Eta Kappa Nu (eminent mem.) Author: (with R.W. Dickinson) Sodium Graphite Reactors, 1958. Contbr. articles to tech. jours. Directed devel. of world's 1st nuclear reactor to operate in space, of sodium-graphite and organic-cooled nuclear power reactors; 1st exptl. work on phenomena of paramagnetic dispersion at low temperature in crystalline compounds of iron group; built early cryogenic equipment for use in high magnetic fields; research in semi-conductor rectifiers, elec. and thermal properties of metals at very high temperatures. Home: 239 N. Cliffwood Av., Los Angeles 90049.*

STARR, Frederick, Am. anthropologist; b. Auburn, N.Y., Sept. 2, 1858; s. Frederick and Helen Strachan (Mills) S.; B.S., Lafayette Coll., 1882, M.S., Ph.D., 1885, Sc.D., 1907. Tchr. science Wyman Inst., 1882-83; prof. scis. State Normal Sch., Lock Haven, Pa., 1883-84; prof. biol. scis. Coe Coll., 1884-85; in charge of ethnology Am. Mus. Natural History, 1889-91; registrar Chautauqua U., 1888-89; asst. prof. anthropology U. Chgo., 1892-95, asso. prof., 1895-1923, and curator anthropol. sect. Walker Mus. Field work in ethnography and phys. anthropology, especially in Mexico; went to Japan, 1904, on behalf of St. Louis Expn., to secure a group of the Ainu, the aboriginal population of Japan, for which was awarded a grand prize; led expdn. into Congo Free State, 1905-06, investigating conditions there, visiting 28 different tribes; field study in P.I., 1908; in Japan, 1909-10, 17; in Korea, 1911, 13 and 1915-16; in Liberia, 1912; lectures on anthropology and his various travels and investigations. Recipient museums medal; Holland, 1900; various decorations. Author: American Indians, 1899; Indians of Southern Mexico, 1899; The Ainu Group at St. Louis, 1904; Readings from Modern Mexican Authors, 1904; The Truth About the Congo, 1907; In Indian Mexico, 1908; Filipino Riddles, 1909; Japanese Proverbs and Pictures, 1910; Congo Natives, 1912; Liberia, 1913; Korean Buddhism, 1918; Fujiyama, the Sacred Mountain of

Japan, 1924. Editor: The Anthropological Series; Central America (reader). Died Aug. 14, 1933.

STARR, Isaac, Am. med. scientist; b. Phila., Mar. 6, 1895; s. Isaac and Mary (Barclay) S.; B.S. magna cum laude, Princeton, 1916; M.D., U. Pa., 1920; m. Edith Nelson Page, Apr. 22, 1922; children—Vidal Davis (Mrs. Richard Clay), Isaac, Lynford Lardner, Harold Page. Faculty U. Pa., 1922—, Hartzell Research prof. therapeutics 1933-45, 48-61, dean chool of Medicine, 1945-48, emeritus prof. therapeutic research, 1961—; acting physian Hosp. U. Pa., 1925-33, asso., 1933-45, asst. chief clinic, 1945-62. Chmn. various coms. NRC; mem. cardiovascular study sect. NIH, 1945-50; cons. U.S. Naval Air Devel. Center, 1949-61. Recipient Lasker award Am. Heart Assn., 1957, Kober medal Assn. Am. Physicians, 1967. Mem. Am. Physiol. Soc., Am. Soc. Pharmacology and Exptl. Therapeutics, Soc. Clin. Investigation (past pres.), Assn. Am. Physicians, A.M.A. (past chmn. Council on Drugs), Am. Heart Assn., Am. Coll. Cardiology, Am. Coll. Clin. Pharmacology, Harvey Soc. (hon.). Author: Physiologic Therapy for Obstructive Vascular Disease, 1953; (with A. Noordergraaf) Ballistocardiography in Cardiovascular Research, 1967. Editorial bd. Am. Heart Jour., 1938-50, Circulation, 1950-55, 56-60. Research and numerous publs. in abnormalities of heart disease and effect of treatment upon them, renal physiology, peripheral vascular disease, clin. pharmacology, ballistocardiogram. Home: 505 Cresheim Valley Rd., Phila. 19118. Office: 851 Gates Meml. Pavilion, U. Hosp., Phila. 19104.*

STARR, Louis, Am. physician; b. Phila., Apr. 25, 1849; s. Isaac and Lydia (Ducoing) S.; A.B., Haverford Coll., 1868, LL.D., 1908; M.D., U. Pa., 1871; m. Mary Parrish, Sept. 16, 1882. Intern Episcopal Hosp., Phila., 1871-73, asst. physician, 1873-74, vis. physician, 1875-84; asst. physician Children's Hosp., 1874, vis. physician, 1879; physician So. Home for Destitute Children, 1874-78; out-patient physician Univ. Hosp., 1878-80; cons. pediatrist to Maternity Hosp., 1879—; Instr. physiology and therapeutics U. Pa., 1874-77, lectr. symptomatology, 1877-79, diseases of children, 1880-84, clin. prof. diseases of children, 1884-90. Fellow Coll. Physicians, Phila.; foundation mem. Pediatric Soc. Asst. editor Pepper's System of Medicine, 1885; Am. editor Goodhart's Diseases of Children, 1885, 89; editor An American Test Book of the Diseases of Children, 1895, 99; editor of dept. of Diseases of Children, In the Am. Year Book of Medicine and Surgery. Author: Disease of Digestive Organs in Infancy and Childhood, 1886, 91; Hygiene of the Nursery (7 edits., 1888-1906); Diets for Infants and Children in Health and in Disease, 1896; A Synopsis of the Physiological Action of Medicines, 1877, 80. Died Sept. 12, 1925.

STARR, Mortimer Paul, Am. bacteriologist; b. N.Y.C., Apr. 13, 1917; s. Morris and Fannie (Blank) S.; B.A., Bklyn. Coll., 1938; M.S., Cornell U., 1939, Ph.D., 1943; m. Phoebe Betty Butwenig, June 18, 1944; children—Nancy Sue, Janina Lynn, Pamela Jill. Faculty Bklyn. Coll., 1939-47; faculty U. Cal. at Davis, 1947—, prof. bacteriology, 1958—, spl. asst. to chancellor, research grants and contracts, 1963-67. Recipient Outstanding Alumnus award of honor Bklyn. Coll., 1947; NRC fellow, 1944-46; spl. fellow NIH, USPHS, 1953-54, 62; Guggenheim fellow, 1958, 68-69; medalla Bernard O'Higgins((Chile), 1968. Mem. Am. Soc. Microbiology, Am. Phytopath. Soc., Soc. Gen. Microbiology, Biochem. Soc., numerous others. Editor: Global Impacts of Applied Microbiology, 1964. Asso. editor: Phytopathology, 1953-56, Ann. Rev. Microbiology, 1958—. Publs. on biochemistry and classification of plant-disease bacteria; chemistry of bacterial pigments; enzymology; unusual bacteria; bacterial taxonomy; bacteriophage. Home: 751 Elmwood Dr., Davis, Cal. 95616.*

STARR, M(oses) Allen, neurologist; b. Bklyn., N.Y., May 16, 1854; s. Egbert and C. Augusta (Allen) S.; A.B., Princeton, 1876, A.M., 1879, Ph.D., 1884; M.D., Coll., Phys. and Surg. (Columbia), 1880; LL.D., Princeton, 1899; Sc.D., Columbia, 1904; m. Alice Dunning, June 7, 1898; 1 dau., Katharine Eunice. Lectr. diseases of mind and nervous system, Columbia, 1887-89, prof., 1889-1900, prof. neurology, 1903-15, emeritus, 1915-32. U. S. del. to Charcot Centennial, Paris, 1925. Fellow A.A.A.S., N.Y. Acad. sciences. Author: Familiar Forms of Nervous Diseases, 1893; Brain Surgery, 1895; Altas of Nerve Cells, 1897; Nervous Diseases, Organic and Functional, 1913. Investigated cerebral localization and brain tumors. Died Germany, Sept. 4, 1932.

STARR, Richard Cawthon, Am. phycologist; b. Greensboro, Ga., Aug. 24, 1924; s. Richard Neil and Ida (Cawthon) S.; B.S. in Secondary Edn., Ga. So. Coll., 1944; M.A., George Peabody Coll., 1947; Ph.D., Vanderbilt U., 1952; postgrad. (Fulbright scholar) Cambridge (Eng.) U., 1950-51. Faculty dept. botany Ind. U., Bloomington, 1952—, prof., 1960—, founder, dir. Culture Collection Algae, 1953—. Trustee Am. Type Culture Collection, 1961-66. Guggenheim fellow, 1959-60. Fellow A.A.A.S., Ind. Acad. Sci.; mem. Bot. Soc. Am. (sec. 1965—, Darbaker award for meritorious work in algae 1955), Phycological Soc. Am., Brit. Phycological Soc., Internat. Phycological Soc. (sec. 1964-66), Soc. Protozoologists. Mem. editorial bd. Jour. Phycology, Phycologia. Contbr. articles to profl. jours. Research on sexual process in various green algae, especially desmids and colonial green Flagellates. Home: 802 E. 3d St., Bloomington, Ind. 47401.*

STARZL, Thomas Earl, Am. physician; b. Le Mars, Ia., Mar. 11, 1926; s. Roman F. and Anna Laura (Fitzgerald) S.; B.A., Westminster Coll., 1947; M.A., Northwestern U., 1950, Ph.D., 1952, M.D., 1952; m. Barbara Brothers, Nov. 27, 1954; children—Timothy, Rebecca, Thomas. Faculty, Northwestern U. Med. Sch., 1958-61; faculty U. Colo. Med. Sch., Denver, 1962—, prof. surgery, 1964—; mem. staffs Colo. Gen., Denver VA hosps. Recipient Achievement award Lund U. and Malmo (Sweden) Surg. Soc., 1965, Most Outstanding Contbn. to Surg. Sci. award Prix de la Société Internationale de Chirurgie, 1965; Markle scholar, 1959-64. Mem. Am. Surg. Assn., Soc. U. Surgeons, Soc. Vascular Surgery, Johns Hopkins Med. and Surg. Soc., Central, Western surg. assns., Transplantation Soc. Author: Experience in Renal Transplantation, 1964. Research, publs. on mechanisms of subcortical brain function; physiology of heart block; carbohydrate metabolism; transplantation of liver and kidney. Home: 415 Krameria St., Denver 80220. Office: 1055 Clermont St., Denver 80220.*

STAS, Jean Servais, Belgian chemist; b. Aug. 21, 1813, Louvain, Belgium; prof. chemistry, Brussels, Belgium, for 30 years; worked with Dumas. Recipient spl. medal Belgium Royal Acad., 1891. Mem. French Acad. Scis. Fellow Royal Soc., 1879 (Davy medal 1895). Author: Nouvelles recherches sur les proportions chimiques, 1865; Recherches de statistique au sujet du chlorure et du bromure d'argent, 1872; Oeuvres completes de Stas, 3 vols., posthumous, 1895. Research on phlorozin, nicotine, (with Dumas) carbonic gas and atomic weights; results of his research disproved Prout's hypothesis; developed methods for determination of atomic weights and analysis, method for detecting vegetable alkaloids. Died Brussels, Dec. 13, 1891.

STASOV, Vladimir Vasilevich, Russian archeologist; b. 1824; librarian, St. Petersburg. Author: Proischozdenie russkich bylin, 1868; Slavjanskij i vostocnyj ornament po ruskopisjam ot IV do XIX v., 1887. Stated theory of oriental origin of Russian Byelorussians. Died St. Petersburg, Oct. 23, 1906.

STATE, David, surgeon; b. London, Ont., Can., Nov. 13, 1914; s. Louis and Sara (Rosenberg) S.; B.A., U. Western Ont., 1936, M.D., 1939; M.S., U. Minn., 1944, Ph.D., 1946; m. Avis Lorberbaum, Nov. 25, 1945; children—Norman, Claudia, Leslie, Roseanne, Matthew. Practice medicine, specializing in surgery, Los Angeles, 1953-58, Bronx, N.Y., 1958—; dir. surgery Cedars of Lebanon Hosp., 1953-58; prof. surgery Albert Einstein Coll. Medicine, 1958—, chmn. dept. surgery, 1959—; dir. surgery Albert Einstein Coll. Medicine Hosp., Bronx Municipal Hosp. Center, Lincoln Hosp. Diplomate Am. Bd. Surgery. Fellow A.C.S.; mem. Am. Surg. Assn., A.M.A., Soc. U. Surgeons, Halsted Soc., Am. Heart Assn., Am. Cancer Soc. (dir. N.Y. div.). Research, numerous publs. on gastric cancer, gastrointestinal tract and gastrointestinal surgery, surg. treatment of congenital megacolon (Hirschsprung's Disease), pyloric antrum, parotid tumor, acid inhibition and acid rebound, pulmonary artery pressure, studies of gastric antrum, homograft and allograft, homotransplantation in the G.I. tract. Home: 51 Paddington Rd., Scarsdale, N.Y. 10583. Office: Dept. Surgery, Albert Einstein Coll. Medicine, Bronx, N.Y. 10461.*

STATZ, Hermann, physicist; b. Herrenberg, Germany, Jan. 9, 1928; s. Franz and Lydia (Kohler) S.; M.S., Technische Hochschule, Stuttgart, Germany, 1949, Ph.D., 1951, postgrad. 1951-52; m. Ilse Dobler, Apr. 11, 1953; children—Eva Maria, Ingrid Elizabeth. Came to U.S., 1952, naturalized, 1958. Mem. staff Max Planck Inst. Metal Research, Stuttgart, 1949-51, Mass. Inst. Tech., 1952-53; asst. mgr. research div., mgr. quantum electronic and electrooptic dept. Raytheon Co., Waltham, Mass., 1953—. Fellow Am. Phys. Soc. Research and publs. on semiconductor surfaces, semiconductor devices, paramagnetic resonance, exchange interaction of magnetic ions, masers and lasers. Home: 10 Barney Hill Rd., Wayland, Mass. 01778. Office: 28 Seyon St., Waltham, Mass. 02154.*

STAUB, Hans, Swiss physicist; b. Wald (ZH), Switzerland, Jan. 20, 1908; s. Heinrich and Julia (Detiker) S.; Diplom, Fed. Inst. Tech., Zurich, 1931, Ph.D., 1935; Erika Weidemann, Mar. 7, 1935; children—Jeanette (Mrs. Olaf Leifson), Margrit, Marion. Came to U. S., 1937, naturalized 1943. Instr., Stanford, 1938-41, asst. prof., 1941-43; asso. prof., 1946-49, prof., 1949; group leader Los Alamos Labs., 1943-46; ordentlicher prof., dir. Physik Inst., U. Zurich, Switzerland, 1949—. Author: Kernspectroscopie; Magnetische Moment des Neutrons; Experimentelle Methoden der Kernphysik. Research in nuclear physics, especially nuclear spectroscopy. Home: 73 Drusbergstrasse, Zürich 8053. Office: 9 Schonberg Gasse, Zurich, Switzerland.*

STAUDINGER, Hermann, German chemist; b. Worms, Germany, March 23, 1881; s. Franz and Auguste (Wenck) S.; ed. univs. of Halle, Munich and Darmstadt; hon. degrees; Dr. Ing., U. Karlsruhe, 1950; Dr. Rer. Nat., U. Mainz, 1951; Dr. (C), U. Salamaca, 1954; Dr. Chem., U. Turin, 1954; Dr. Sci. Tech., Eidgenös Tech. Hochschule Zurich, 1955; Dr. b.c., Strassburg University, 1959; m. Magda Woir, 1928. Lecturer chemistry Strassburg University, 1907-08; prof. Karlsruhe Tech. High Sch., 1908-12; prof. Fed. Tech. High Sch., Zurich, 1912-26, U. Freiburg im Breisgau, 1926-51; prof. emeritus. Mem. Gottingen, Heidelberg, Halle, Munich acads. Scis., Frankfort, Zurich phys. socs., Royal Physiograph, Soc. Recipient Leblanc medal from the French Chemical Society, 1931, Cannizzarro prize, Rome, Italy, 1933, Nobel Prize in Chemistry, 1953, Nendrnck, 1961. Author: Die Ketene, 1912; Tabellen fur allgemeine und anorganische Chemie, 1935; Die hochmolekularen organischen Verbindungen, 1932; Organisch qualitative Analyse, 1939; Organische Kolloidchemie, 1941; Vom Aufstand der technischen Sklaven, 1947; Makromolekulare Chemie und Biologie, 1947. Research in macromolecular compounds; studies of polymerization basic to devel. of plastics and other synthetics; discovered ketene. Died Freiberg, Sept. 9, 1965.

STAUDINGER, Otto, lepidepterologist; b. Grosswüstenfelde, Germany, May 2, 1830. Author: Katalog der Lepidopterem des europäischen Faunengebietes, 1871; Exotische Schmetterlinge, 2 vols., 1888-92; Katalog der Lepidopteren des paläarktischen Faunengebietes, 1901. Studied palearctic and exotic butterflies. Died Lucerne, Switzerland, Oct. 13, 1900.

STAUFF, Joachim Hans Paul, German chemist; b. Berlin, Sept. 10, 1911; s. Paul Ernst and Claire (Labude) S.; student U. Göttingen (Germany), 1930 37, U. München, (Germany), 1932-33; Dr.phil., U. Berlin, 1936; m. Ilse Glasgow, Sept. 16, 1936. Asst., Kaiser Wilhelm Institut Physik. Chemie, Berlin-Dahlem, 1936-39; staff U. Frankfurt/Main (Germany), 1940-43, 53—, prof. chemistry, 1965—; chemist Farbwerke Hoechst, Frankfurt/Main, 1948-53; dir. Inst. Phys. Biochemistry and Colloidchemistry, U. Frankfurt/Main, since 1962—. Mem. D. Bunsen Gesellschaft, Gesellschaft Deutscher Chemiker, Am. Chem. Soc. Author: Kolloidchemie, 1960; also articles. Research on structure of soap solution, photochemistry of chlorination and sulfochlorination, heat denaturation of proteins, excited states of biol. systems, luminescence and chemiluminescence. Home: 14 Freiligrathstr., Bad Soden/Taunus, Germany. Office: 11 Robert Mayer Str., Frankfurt/Main, Germany.*

STAUFFACHER, Dietrich Werner, Swiss chemist; b. Zurich, Switzerland, Mar. 12, 1927; s. Heinrich and Rosina (Zwicky) S.; Dipl.sc.nat., Swiss Fed. Inst. Tech., 1951, Dr.sc.nat., 1953; m. Maya Spuehler, Apr. 30, 1955; children—Regula, Rudolf. Research chemist Sandoz AG Research Labs., Basle, 1953-61, sr. research chemist, 1961—. Mem. Swiss Chem. Soc. Research, publs. on isolation and structure elucidation of natural products, terpenes, alkaloids. Home: 10 Hinterlindenweg, 4153 Reinach Bl, Switzerland. Office: Research Labs. for Pharm. Chemistry, Sandoz AG, 4000 Basle, Switzerland.*

STAUFFER, Herbert Milton, Am. physician; b. Phila., Apr. 26, 1914; s. Milton and Anna (Hood) S.; A.B., Temple U., 1935, M.D., 1939, M.S., 1945; m. Joan Dunbar, July 19, 1941; 1 son, William Scott. Practice medicine, specializing in radiology, Phila., 1949 ; faculty Sch. Medicine, Temple U., 1949—, prof. radiology, 1952—, chmn. dept. radiology, 1957—. Mem. radiology tng. com. NIH, 1966—; cons. USPHS. Diplomate Am. Bd. Radiology. Mem. Assn. U. Radiologists, Radiol. Soc. N.Am., Am. Roentgen Ray Soc., A.M.A., Am. Coll. Radiology. Research and numerous publs. in radiologic image processing and its relation to perception of diagnostic information, radiologic techniques in cardiovascular dynamics; investigations in contrast media mechanisms and toxic effects. Home: 909 Hagy's Ford Rd., Narberth, Pa. 19072. Office: Radiology Dept., Temple U. Hosp., Phila. 19140.*

STAUFFER, Robert Burton, Jr., Am. polit. scientist; b. Philipsburg, Pa., May 3, 1920; s. Robert Burton and Ruth (Hons) S.; B.S. in Mus. Edn., State Coll., West Chester, Pa., 1942; M.A. in Polit. Sci., U. Okla., 1947; Ph.D., U. Minn., 1954; m. Joan Helen Hammond, Sept. 4, 1945; children—Robert H., James B., Ann M. Mem. faculty U. Hawaii, Honolulu, 1950—, prof., 1963—. Researcher, UN, 1957; field research, S.E. Asia, 1962; Fulbright prof. U. Philippines, 1963-64. Mem. Am., Western polit. sci. assns., Am. Assn. U. Profs., Assn. for Asian Studies, Soc. Applied Anthropology. Author: The Development of an Interest Group: The Philippine Medical Association, 1966; also articles. Editorial bd. Human Orgn., 1966—, Western Polit. Sci. Quar., 1967—. Research on measuring impact of aid programs on recipient nations; methods of measuring polit. change in new nations of world. Home: 4679 Kolohala St., Honolulu 96816.*

STAUSS, Henry Emanuel, Am. physicist; b. St. Louis, May 26, 1902; s. Emanuel and Louise (Kemper) S.; A.B., Washington U., St. Louis, 1923; M.A., U. Cal., Berkeley, 1925, Ph.D., 1927; m. Sylvia H. Albright, June 13, 1926; children—George H., Virginia L. (Mrs. Duff Tucker). Research physicist, metallurgist Baker & Co., Inc., Newark, 1930-42; indsl. specialist WPB, Washington, 1942-45; physicist Naval Research Lab., Washington, 1945-59; aerospace specialist NASA, Washington, 1959——. Mem. Am. Phys. Soc., Am. Inst. Mining, Metall. Engrs., Am. Soc. Metals, Brit. Inst. Metals, Research Soc. Am. Studies in reflection and refraction of X-rays, metallurgy of platinum group metals, metallurgy and solid state physics, space sci. Home: 8005 Washington Av., Alexandria, Va. 22308. Office: NASA, Washington 20546.*

STAVERMAN, Albert Jan, Dutch phys. chemist; b. Bussum, Netherlands, June 27, 1911; s. Werner Hendrik and Betsy (Keyzer) S.; student U. Groningen (Netherlands), 1929-36; Ph.D., U. Leiden (Netherlands), 1938; m. Catherina Pekelder, Dec. 23, 1938; children—Werner, Irene, Frederika. High sch. tchr., 1937-40; faculty U. Leiden, 1940-42, prof. phys. chemistry, 1958——; research worker Philips Labs., 1942-45; dept. head plastics Research Inst., TNO, 1947-54, dir. central lab., 1954——. Mem. Netherlands, Am. chem. socs., Faraday Soc., A.A.A.S. Contbg. author: Handbook of Physics, vol. XIII, 1960; Physics of Polymers, vol. IV, 1956. Research, numerous publs. on thermodynamics of polymers, membrane permeation, rheology, mech. properties of liquids and solids. Home: 6 Fruinlaan, Leiden, Netherlands.*

STAVITSKY, Abram Benjamin, Am. microbiologist; b. Newark, May 14, 1919; s. Nathan and Ida (Novak) S.; A.B., U. Mich., 1939, M.S., 1940; Ph.D., U. Minn., 1943; V.M.D., U. Pa., 1946; m. Ruth Bernice Okney, Dec. 6, 1942; children—Ellen Barbara, Gail Beth. Research fellow Cal. Inst. Tech., 1946-47; faculty Western Res. U., Sch. Medicine, Cleve., 1947——, prof. microbiology, 1962——. Mem. expert panel on immunology WHO, 1962——; mem. fellowship com. in microbiology USPHS, 1963-66. Fellow A.A.A.S.; mem. Am. Assn. Immunologists, Am. Soc. for Microbiology, Brit. Soc. for Immunology, Fedn. Am. Socs. Exptl. Biology (chmn. pub. information com. 1963-66). Editor, Jour. Cellular Physiology, 1965——. Research and numerous publs. on devel. methods for measuring antibodies, mechanisms whereby cells synthesize antibodies; discovered that many different kinds of substances caused prodn. of more than one kind of antibody during immunization of animals. Home: 14604 Onaway Rd., Shaker Heights, O. 44120. Office: 2109 Adelbert Rd., Cleve. 44106.*

STAVRAKY, George Woldemar, physiologist; b. Odessa, Russia, Feb. 19, 1905; s. Woldemar E. and Olga Stavraky; M.D., U. Odessa, 1926; M.D., C.M., McGill U., Montreal, Que., Can., 1932, M.Sc., 1930; m. Marie-Madeleine Gabrielle Lefebvre, July 4, 1934. Fellow Montreal Neurol. Inst., demonstrator physiology McGill U., 1932-35; faculty U. Western Ont., London, Can., 1936——, prof. physiology, 1949——. Cons. RCAF, 1950-60. Fellow Royal Soc. Can.; mem. Canadian Am. physiol. socs., Pharmacological Soc. Can., Canadian Neurol. Soc., Sigma Xi. Author: Supersensitivity Following Lesions of the Nervous System, 1961; also numerous articles. Research in chem. transmission nerve impulses in warm blooded organisms, secretory mechanisms gastro-intestinal tract, changes in threshold response neurones or effector cells after denervation. Home: 7 Thornton Av., London, Ont., Can.*

STEACIE, Edgar William Richard, Canadian phys. chemist; b. Westmount, Que., Can., Dec. 25, 1900; s. Richard and Alice Kate (McWood) S.; B.S., McGill U., 1923, M.Sc., 1924, Ph.D., 1926; postgrad. Frankfurt, Germany, 1934, Leipzig, Germany, 1935, King's Coll., London, 1935; D.Sc., McMaster U., 1946, N.B., Can., 1950, Laval, 1952, Toronto, Ont., Can., 1954, Man., Can., 1954; LL.D., Queen's U., 1952, Dalhousie, 1952, McGill U., 1953, Montreal U., D de l'U., 1956; m. Dorothy Catalina Day, 1925; children—John, Richard Brian, Diana Jeannette. NRC scholar 1925-26; Sterry Hunt fellow McGill U., 1926-28, lectr., 1928-30, asst. prof., 1930-37, asso. prof., 1937-39; dir. div. chemistry Nat. Research Council, 1939-52, v.p. sci., 1950-52, pres., 1952——; dep. dir. U.K.-Canada Atomic Energy Project, 1944-46. Mem. Def. Research Bd. Can., Atomic Energy Control Bd.; mem. sci. adv. commn. U. Ottawa. Recipient Profl. Inst. Pub. Service medal, 1949, Chem. Inst. Can. medal, 1953, Tory medal Royal Soc. Can., 1955. Mem. Am. Chem. Soc., Soc. Chem. Industry, Chem. Inst. Can. (pres. 1949-50), Royal Soc. Can. (pres. 1954-55), Internat. Union Pure and Applied Chemistry (v.p. 1951-53), Polish Chem. Soc. (hon.). Fellow Royal Soc. Author: (with F. M. G. Johnson) Viscosities of the Liquid Halogens, 1925; (with Otto Maass) Introduction to Physical Chemistry, 1926; Attempt to Determine the Osmotic Pressure of Very Dilute Solutions, 1929; Atomic and Free Radical Reactions, 1946; also numerous articles. Research in photochemistry, chem. kinetics, gas reactions; organized Canadian chemistry research for war and atomic energy. Died Aug. 28, 1962.

STEAD, Eugene Anson, Jr., Am. physician; b. Atlanta, Oct. 6, 1908; s. Eugene Anson and Emily (White) S.; B.S., Emory U., 1928, M.D., 1932; m. Evelyn Selby, June 15, 1940; children—Nancy W., Lucy S. (Mrs. Claude LaVarre), William W. Practice medicine, specializing in internal medicine, Durham, N.C., 1947——; prof. Emory U., 1942-46, dean Sch. Medicine, 1945-46; prof., chmn. dept. medicine Duke, 1947——. Mem. adv. council Nat. Heart Inst., Nat. Inst. Arthritis and Metabolic Diseases. Mem. Assn. Am. Physicians, Am. Soc. Clin. Investigation (pres.), Am. Heart Assn., Phi Beta Kappa, Alpha Omega Alpha. Contbr. numerous articles to profl. jours. Provided formulation for mechanisms governing cardiac failure in man. Home: 2122 Campus Dr., Durham. Office: Duke Hosp., Durham, N.C. 27706.*

STEARN, Allen Edwin, Am. chemist; b. Lafayette, Ind., Sept. 14, 1894; s. John Henry and Louise (Hagerty) S.; A.B., Stanford, 1915, A.M., 1916; M.S., U. Ill., 1917, Ph.D., 1919; m. Esther A. Wagner, Jan. 29, 1920; 1 dau., Enid W. (Mrs. John H. Bergstrom). Instr., U. Ill., 1919; asst. prof. W.Va. U., 1920; faculty U. Mo., 1921-24, 26——, prof. chemistry, 1938-62, prof. emeritus, 1962——, chmn. chemistry dept., 1938-58, acting dean Grad. Sch., summers 1933-55. Mem. Am. Chem. Soc., Am. Archeol. Soc., Am. Soc. Bacteriologists, Am. Pub. Health Assn., Phi Beta Kappa, Sigma Xi. Author: Elementary Outline of Analytic Chemistry, 1926; (with E. W. Stearn) Effect of Smallpox on Conquest of Amerindian, 1945, College Hygiene for Total Health, 1961, Physico-Chemical Behavior of Bacteria, 1928; also numerous articles Research on certain phases enzyme kinetics, some aspects physico-chem. behavior bacteria. Home: 535 N. Michigan St., Chgo. 60611. Office: 425 N. Michigan Av., Chgo. 60611.*

STEARN, Colin William, geologist; b. Bishops Startford, Eng., July 16, 1928; s. Clement Hodgson and Doris (Phillips) S.; B.Sc., McMaster U., 1949; M.S., Yale, 1950, Ph.D., 1952; m. Mary Joan Mackenzie, Aug. 22, 1953; children—Virginia, Patricia, Andrew. Asst. prof. dept. geol. scis. McGill U., Montreal, Que., Can., 1952-58, asso. prof., 1958-66, prof., 1966——, asst. dean faculty grad. studies, 1960-63. Mem. Geol. Soc. Am., Am. Assn. Petroleum Geologists, Geol. Assn. Can. Royal Soc. Can., Paleontol. Soc., Sigma Xi. Author: (with T. H. Clark) Geological Evolution of North America, 1960. Research and publs. on fossil coelenterates, chiefly an extinct group of reef-forming organisms called stromatoporoids. Home: 111 Biscayne St., Beaconsfield, Que. Office: McGill U., Montreal, Que., Can.*

STEARNER, Sigrid Phyllis, Am. radiobiologist; b. Chgo., Jan. 10, 1919; d. Carl Johan and Helga (Padderud) Stearner; student Wilson Jr. Coll., 1939-41; B.S., U. Chgo., 1942, M.S., 1943, Ph.D., 1946. Asso. biologist Argonne (Ill.) Nat. Lab., 1946——. Mem. N.Y. Acad. Scis., Am. Soc. Zoologists, Radiation Research Soc., Research Soc. Am., A.A.A.S. Research and publs. on relative contbn. of radiation dose and exposure time in acute radiation damage, effectiveness of early recovery processes, early radiation effects on microcirulation in chick and chick embryo, quantitative effects of ionizing radiations on growth and life span chicken. Home: 154 Juliet Ct., Clarendon Hills, Ill. 60514. Office: 9700 Cass Av., Argonne, Ill. 60440.*

STEARNS, Carl Leo, Am. astronomer; b. Westbrook, Me., Sept. 14, 1892; s. Albert Joseph and Cora (Weymouth) S.; B.A., Wesleyan U., Middletown, Conn., 1917; Ph.D., Yale, 1923; m. Mildred Parkhurst Booth, Aug. 9, 1923; children—Robert Leo, Elva Parkhurst (Mrs. George R. Creeger), Doris Elizabeth (Mrs. James B. Swain). Asst., Dudley Obs., Albany, N.Y., 1917-18; faculty Wesleyan U., Middletown, 1918-20, 25——, prof. astronomy, 1944-60, prof. emeritus, 1960——; research asst. Yale Obs., 1920-25; instr. Naval Flight Prep. Sch., 1943; vis. instr. Trinity Coll., Hartford, Conn., 1940-41. Recipient Donohoe Comet medal Astron. Soc. Pacific, 1927. Fellow Royal Astron. Soc. (Eng.); mem. A.A.A.S., Am. Astron. Soc., Internat. Astron. Union, Phi Beta Kappa, Sigma Xi. Author: (with F. Slocum, B.W. Sitterly) Stellar Parallaxes from Photographs made with 20-inch Refractor of the Van Vleck Observatory, 1938; also articles. Determined distances of stars by measuring displacement their photog. images produced by ann. motion of earth around sun, distance of sun from earth by measuring distance minor planet Eros. Home: 84 Home Av., Middletown, Conn. 06457.*

STEARNS, Genevieve, Am. chemist; b. Zumbrota, Minn., Dec. 24, 1892; d. Clayton Henry and Clara (Beierwalter) Stearns; B.S., Carleton Coll., 1912; M.S., U. Ill., 1920; Ph.D., U. Mich., 1927. Faculty, U. Ia., Iowa City, 1920——, research prof., 1954-58, prof. emeritus, 1958——; Fulbright prof. Ein Shams U., Cairo, U.A.R., 1960-61. Mem. U.S. team for WHO seminars in infant nutrition, Leiden, 1950. Recipient (with others) Borden award Home Econ. Assn., 1942; (with P.C. Jeans) award Am. Inst. Nutrition, 1946. Fellow Am. Inst. Nutrition; mem. Am. Soc. Biol. Chemists, Sigma Xi, Omicron Nu. Contbg. author: Infant Metabolism, 1956. Research and numerous publs. in nitrogen mineral and vitamin metabolism and requirements of infants, children in health and in disease, metabolism in diseases affecting bone. Home: 408 Myrtle St., Iowa City, Ia. 52240.*

STEARNS, Howard Oliver, Am. physicist; b. Haverhill, Mass., Sept. 20, 1891; s. William Dennett and Nettie Florence (Gould) S.; B.S., Dartmouth, 1915, M.S., 1917; postgrad. Johns Hopkins, 1918-19, Yale, 1924-25, Harvard, summers 1930-31; m. May Belle North, Dec. 25, 1924; 1 son. Howard Oliver. Faculty dept. physics Simmons Coll., 1917-18, 25-56, prof., 1930-56; asst. physicist Bur. Standards, Washington, 1918-19; physicist Mayo Clinic, Rochester, Minn., 1919-24, Yale, 1924-25. Physicist, chmn. Gravity Research Found., 1948——; chmn. Rainfall Utilization Found., 1965——. Mem. Am. Phys. Soc., Am. Geophys. Union, Am. Astronaut. Soc. Author: Elementary Medical Physics, 1947; Fundamentals of Physics and Applications, 1956. Contbr. articles to profl. jours. Research on elementary med. physics; radium; air speed meter testing. Address: 80 Prospect St., Wellesley Hills, Mass. 02181.

STEARNS, Martin, Am. physicist; b. Phila., Aug. 16, 1916; s. Celia (Birdman) S.; B.A., U. Cal. at Los Angeles, 1944, M.A., 1945; Ph.D., Cornell U., 1951; m. Mary Beth Gorman, Sept. 8, 1948; children—Daniel G., Richard G. Jr. sci. Manhattan Project, U. Cal. Radiation Lab., Los Alamos, 1945-46; research physicist Carnegie Inst. Tech., 1951-57; cons. Ramo-Wooldridge Corp., Los Angeles, 1955; staff sci. Gen. Atomic, La Jolla, Cal., 1957-60; chmn. dept. physics Wayne State U., 1960-62, dean Coll. Liberal Arts, 1962——. Bd. advisers Wayne State U. Press. Fellow Am. Phys. Soc.; mem. A.A.A.S., Am. Assn. Physics Tchrs., Fedn. Atomic Sci., Detroit Engring. Soc., Phi Beta Kappa, Sigma Xi. Contbr. tech. articles to profl. jours. Research on high energy radiation and pair prodn., photoproduction mesons, mesonic x-rays, plasma physics. Home: 19339 Starlane, Southfield, Mich. Office: Wayne State U. Coll. Liberal Arts, Detroit 48202.*

STEARNS, Richard Gordon, Am. geologist; b. Buffalo, Apr. 28, 1927; s. Howard G. and Grace (Harvey) S.; A.B., Vanderbilt U., 1948, M.S., 1949; Ph.D., Northwestern U., 1953; m. Jane Smallwood, July 1950; children—Jane, Bruce. Geologist, Standard Oil Co. Cal., San Francisco, 1951-53; asst. dir. Tenn. Div. Geology, Nashville, 1953-61; asst. prof. Vanderbilt U., Nashville, 1961-64, asso. prof., 1964——, chmn. geology dept., 1967——. Fellow Geol. Soc. Am.; mem. Am. Assn. Petroleum Geologists, Am. Inst. Profl. Geologists, Am. Geophys. Union. Author: The Cumberland Plateau Overthrust, 1954. Research in geol. history of Central South region with spl. emphasis on Tenn., ancient geography and earth movements.*

STEBBINS, George Ledyard, Am. plant geneticist; b. Lawrence, N.Y., Jan. 6, 1906; s. George Ledyard and Edith Alden (Candler) S.; A.B., Harvard, 1928, M.A., 1929, Ph.D., 1931; D.Sc., U. Paris (France), 1962; m. Margaret Chamberlaine, June 14, 1931 (div.); children—Edith Candler, Mrs. Nahas, Robert Lloyd, George Ledyard; m. 2d, Barbara Brumley, July 20, 1958. Instr., Colgate U., 1931-35; faculty U. Cal. at Berkeley, 1935-50, prof. genetics, 1947-50; prof. genetics U. Cal. at Davis, 1950——, chmn. genetics dept., 1957-63. Mem. Soc. for Study Evolution (past pres.), Bot. Soc. Am. (past pres.), Cal. Bot. Soc. (past pres.), Internat. Union Biol. Scis. (sec. gen. 1959-64), Nat. Acad. Scis., Am. Philos. Soc., Am. Acad. Arts and Scis., Cal. Acad. Scis., Genetics Soc. Am., Am. Inst. Biol. Sci., Am. Soc. Naturalists. Author: (with C.W. Young) The Human Organism and the World of Life, 1938; Variation and Evolution in Plants, 1950; Processes of Organic Evolution, 1966; also numerous articles. Synthesis of evidence regarding processes of evolution in plants; synthesis and establishment in nature of populations having genetic properties of new species; analysis of chromosomal variation and species relationships in plants, genetic and ecol. basis of plant distbn. in Cal.; analysis of gene action in devel. of plants. Home: 1009 Ovejas Av., Davis, Cal. 95616.*

STEBBINS, Joel, Am. astronomer; b. Omaha, Neb., July 30, 1878; s. Charles Sumner and Sara Ann (Stubbs) S.; B.S, U. of Neb., 1899, LL.D., 1940; student U. of Wis., 1900-01, hon. Sc.D., 1920; Lick Obs., U. Cal., Ph.D., 1903, LL.D., 1953; student U. Munich, 1912-13; Sc.D., U. Chicago, 1954; m. May Louise Prentiss, June 27, 1905; children—Robert P., Isabelle (Mrs. T. A. Dodge). Instructor in astronomy, University of Illinois, 1903-04, assistant professor, 1904-13, professor and dir. Obs., 1913-22; dir. Washburn Obs. and prof. astronomy, U. of Wis., 1922-48, emeritus, also research asso. Mt. Wilson Obs.; research asso. Lick Observatory since 1948. Recipient Royal Astronomical Soc. Gold Medal for 1950, Draper medal, Nat. Acad. Scis., Rumford medal, Royal Soc., Bruce medal, Astron. Soc. Pacific. Mem. National Academy Sciences, American Philosophical Society, Am. Acad. Arts and Sciences, A.A.A.S. (v.p.), Am. Astron. Soc. (sec., v.p., pres.); foreign asso. Royal Astron. Soc., Phi Beta Kappa, Sigma Xi. Made 1st elec. photometry of stars, interstellar space. Died Mar. 16, 1966.

STEBBINS, Robert Cyril, Am. herpetologist; b. Chico, Cal., Mar. 31, 1915; s. Cyril Adelbert and Louise (Beck) S.; B.A., U. Cal., Los Angeles, 1940, M.A., 1942, Ph.D., 1943; m. Anna-rose Cooper, June 8, 1941; children—John, Melinda (Mrs. Norman Broadhurst), Mary. With U. Cal., Los Angeles, 1943-45; with U. Cal., Berkeley, 1945——, prof. zoology, curator herpetology, 1958——. John Simon Guggenheim Memorial Found. fellow; NSF fellow. Mem. Soc. Ichthyologists and Herpetologists (v.p., 1950-51), Soc. Systematic Zoology (pres. Western div. 1954), Cooper Ornithol. Club, Herpetologists'

League, Western Soc. Naturalists, Biosystematists, Cal. Acad. Scis. Author: Amphibians of Western North America, 1951; Reptiles and Amphibians of Western North America, 1954; (with C.A. Stebbins) Birds of Yosemite, 1954; Reptiles and Amphibians of the San Francisco Bay Region, 1959; The Lives of Desert Animals in Joshua Tree National Monument, 1964; Field Guide to Western Reptiles and Amphibians, 1966; also motion pictures Nature Next Door, 1959; And No Room for Wildness, 1968. Herpetol. editor Copeia, Jour. Am. Soc. Ichthyologists and Herpetologists, 1955. Research, publs. on ecology, evolution of amphibians and reptiles; studies of reptile pineal mechanism, parietal eye. Home: 601 Plauteau Dr., Kensington, Cal. 94708. Office: Life Scis. Bldg., U. Cal., Berkeley, Cal. 94720.*

STECHKIN, Boris Sergrevich, Russian aero. engr.; b. July 24, 1891; grad. Moscow Higher Tech. Sch., 1918, Dr.Tech. Sci., 1953. A founder Central Aerodynamic Inst., 1918; an organizer Aero. Engrs. Acad., Moscow; instr. Moscow Higher Tech. Sch., 1918-29; became prof. Aero. Engrs. Acad., 1921; prof. Moscow Aviation Inst., 1933-37; named dir. engine lab. USSR Acad. Scis., 1954. Recipient Order of Lenin, Order of Red Banner of Labor, Order Red Star. Mem. USSR Acad. Scis. Author: Aircraft Engines, 1922; The Thermal Calculation of an Engine, 1927; A Theory of Jet Engines, 1929; Summary of Lectures on the Theory of Aircraft Turbocompressors, 1944; Theory of the Aviation Jet Engine, 1945. Research on theory and devel. of jet engines; studied flight and ground characteristics of aircraft engines; calculated heat balance and developed methods for constrn. of aircraft engines. Home: Leninskii Prospekt 13. Office: Laboratory of Motors, USSR Acad. Sci., Krasnoproletarskaya, Ulitsa 32, Moscow, USSR.

STECKLER, Robert, polymer chemist; b. Vienna, Austria, Nov. 27, 1914; s. Richard and Anna (Bondy) S.; Ph.D., U. Vienna, 1938; m. Grace Taylor, Oct. 7, 1949; 1 son, Richard L. Came to U. S., 1938, naturalized, 1944. Research asst. U. Graz (Austria), 1936-37; research mgr. resins and plastics div. Arco Co., Cleve., 1938-45; mgr. R. Steckler Labs., Cleve., 1945—; founder Synthetic Organics, Inc., Cleve., 1948—, exec. v.p., 1954—; pres. Ultra-Pane, Inc., Cleve., 1954-57; dir. Polycast Corp., Stamford, Conn. Fellow A.A.A.S. (mem. council 1963—), Am. Inst. Chemists; mem. Am. Chem. Soc., Soc. For Paint Tech., Soc. Plastics Engrs. (charter), Assn. Cons. Chemists and Chem. Engrs. (past pres.). Patentee in field. Research and devel. comml. processes for epoxy, vinyl, acrylic, phenolic, aniline, polyester, alkyd resins, bulk polymerization and casting, polymerization in solution, alkylation phenols, diallyl phthalate prepolymers and molding powders, polymeric plasticizers, sealants and insulating glass, reinforced plastics, laminates, water soluble polymers. Home: 21917 Halburton Rd., Beachwood, O. 44122. Office: 19220 Miles Av., Cleve. 44128.*

STEEB, Siegfried, German physicist; b. Stuttgart, May 28, 1931; s. Erich and Else (Uebel) S.; Diplom, Stuttgart Inst. Tech., 1956, Dr.rer.nat., 1958; m. Gisela Bregler, July 24, 1959; 1 dau., Karin. Chief asst. Max Planck Inst. for Refractory Metals, Stuttgart (Germany). Numerous publs. in field. Research on structure of liquid metals and alloys, solid niobium, tantalum and uranium, and their oxides, ternary uranium containing phases by X-ray and electron diffraction; x-ray microprobe analysis; mass spectroscopy. Address: 16 Achalmstreet, 7021, Oberaichen near Stuttgart, West Germany.*

STEEGMANN, Albert Theodore, Am. physician, neurologist; b. nr. Cole Camp, Mo., Sept. 4, 1902; s. Henry Edward and Charlotte (McGee) S.; B.S., U. Kan., 1926, M.D., 1928; postgrad. Cleve. City Hosp., Nat. Hosp., Queen's Sq., London, Eng., Max Planck Brain Research Inst., Munich, Germany; m. Frances Boten, Nov. 30, 1929; children—Carl Edward, Albert Theodore, Sara Catharine (Mrs. Spencer Dickson). Faculty Western Res. U. Med. Sch., Cleve., 1929-42; faculty U. Kan. Med. Sch., Kansas City, 1942—, prof. neurology, chief neurology sect., 1946—, dir. neuropathology lab., dept. internal medicine. Cons. neurology U. S. Vets. Facility, Kansas City, Mo., U. S. Fed. Prison, Leavenworth, Kan.; regional neurol. cons. U.P.R.R. Porter scholar, 1928. Diplomate Am. Bd. Neurology and Psychiatry. Fellow Am. Acad. Neurology; mem. Am. Neurol. Assn., Am. Assn. Neuropathologists, Alpha Omega Alpha. Author: Examination of the Nervous System, 1956, 2d edit., 1962. Studies, publs. on damage to central nervous system following oxygen deprivation (anoxia), also on alteration of circulation of cerebral cortex following microembolism, also on prevention of injury to brain in cardiac arrest by use of hypothermia. Home: 8314 Hemlock, Shawnee Mission, Kan. 66212. Office: Kan. U. Med. Center, Rainbow and 39, Kansas City, Kan. 66103.*

STEELE, John Dutton, Am. thoracic surgeon; b. Phila., Jan. 28, 1905; s. John Dutton and Edith (Williamson) S.; B.A., Williams Coll., 1926; M.D., U. Pa., 1932; M.S. in Surgery, U. Mich., 1937; m. Betsy Owen, July 2, 1936; children—Christopher, Polly (Mrs. W. Albert Munson), Wendy (Mrs. Harold Teasdale), Jenny (Mrs. Robert Allen). Practice thoracic surgery, Milw., 1938-55; chief surg. service Muirdale Sanatorium, Milw., 1938-55; asst. clin. prof. surgery Marquette U. Sch. Medicine, 1938-55; mem. faculty U. Cal. at Los Angeles Med. Sch., 1957—, clin. prof. surgery, 1961—; chief surg. service VA Hosp., San Fernando, Cal., 1955—; cons. thoracic surgeon Olive View Hosp., La Vina Sanatorium (both Los Angeles). Diplomate Am. Bd. Thoracic Surgery (founder mem.). Fellow A.C.S.; mem. Am. Assn. Thoracic Surgery (sr.), A.M.A., Am. Thoracic Soc. (v.p. 1946-47, sec. 1950-53, pres. 1954-55, councilor 1962-65), Nat. Tb Assn. (bd. dirs. 1950-55, v.p. 1953-54; Trudeau medal 1962), Wis. Anti-Tb Assn. (bd. dirs. 1944-55), Cal. Thoracic Soc., Soc. Thoracic Surgeons (founder mem.). Author: The Solitary Pulmonary Nodule, 1964. Editor: (with others) The Surgical Management of Pulmonary Tuberculosis, 1957; (with others) The Treatment of Mycotic and Parasitic Diseases of the Chest, 1964. Editor, John Alexander Monograph Series, 1957-65, Annals Thoracic Surgery, 1965—. Contbr. med. jours. Home: 19312 Romar St., Northridge, Cal. 91325. Office: VA Hosp., 13000 Sayre St., San Fernando, Cal. 91342.*

STEELE, John Murray, Am. physician; b. Newport, R.I., June 7, 1900; s. John Murray and Gertrude W. (Brooks) S.; A.B., Harvard, 1921; M.D., Johns Hopkins, 1925; m. Sylvia Moulton Ward, July 1, 1932; children—John M., Charles N., Lucy Ann. Dir. 3d N.Y. U. med. research div. Goldwater Meml. Hosp., N.Y.C., 1939—; prof. medicine N.Y. U. Sch. Medicine, N.Y.C., 1951—. Mem. adv. com. for med. research Sidney Hillman Health Center, 1947-51. Mem. Internat. Assn. Gerontology, Alpha Omega Alpha (hon.). Contbr. numerous articles to med. jours. Discovered that when fat is subtracted from body weight, oxygen consumption in female is not less than in male using body fat measurements. Home: 325 E. 72d St., N.Y.C. 10021. Office: Goldwater Meml. Hosp., Welfare Island, N.Y.C. 10017.*

STEELE, Robert, Am. biochemist; b. Pittsburgh, Kan., Dec. 1, 1915; s. Andrew Black and Eileen (French) S.; B.S. in Chemistry, U. Kan., 1936; Ph.D. in biochemistry, U. Chgo., 1940; m. Irene Grace Abrams, Sept. 8, 1963; 1 dau., Patricia Jean. Research asso. biochemistry Rutgers U., Rahway, N.J., 1941-42; instr. oncology U. Wis., Madison, 1946-47; asso. biochemistry Brookhaven Lab., Upton, N.Y., 1947-51, biochemist, 1951-56, sr. biochemist, 1966—; adj. prof. pharmacology, cons. dept. pharmacology N.Y. U. Med. Sch., 1960—. Mem. Am. Assn. Cancer Research, Am. Soc. Biol. Chemists, The Endocrine Soc., Am. Physiol. Soc. Research and numerous publs. on methods for conducting and interpreting isotope tracer experiments to measure turnover rates in blood of physiologically important compounds, influences of insulin, pituitary growth hormone, adrenal glucocorticoids on prodn. of blood glucose by liver and utilization blood glucose by tissues in mammals. Office: Biology Dept., Brookhaven Lab., Upton, N.Y.*

STEELL, Graham, English physician; b. Manchester, Eng., July 27, 1851; s. John Steell; ed. Edinburgh U.; M.D.; m. Agnes Dunlop M'Kie, 1886; 1 son. Prof. clin. and systematic medicine Victoria U. of Manchester; house physician Edinburgh Royal Infirmary, Fever House, Edinburgh Royal Infirmary; resident med. officer Stirling Royal Infirmary, Leeds Fever Hosp., Manchester Royal Infirmary; med. registrar London Fever Hosp.; cons. physician Manchester Royal Infirmary, Manchester and Salford Hosp. for Diseases of Skin, Christie Hosp. for Cancer. Fellow Royal Coll. Physicians. Author: The Sphygmograph in Clinical Medicine, 1899; The Physical Signs of Pulmonary Disease, 2d edit., 1900; Text-Book on Diseases of the Heart, 1906; Bradshaw Lecture, Royal College of Physicians, 1911; also articles. Described diastolic murmer in pulmonary valve insufficiency (Graham Steell murmur), 1888. Died Jan. 10, 1942.

STEELMAN, Sanford Lewis, Am. biol. chemist; b. Hickory, N.C., Oct. 11, 1922; s. John Avery and Blanche (Owenby) S.; B.S., Lenoir-Rhyne Coll., 1943; Ph.D., U. N.C., 1949; m. Margaret Elizabeth Abee, Jan. 10, 1945; children—Sanford Lewis, Brian L. Dir. biochem. research Armour Labs., Chgo., 1953-56, cons. biochemistry, 1956-58; asso. prof. Baylor U. Coll. Medicine, 1956-58, U. Tex., Houston, 1956-58; dir. endocrinology Merck Inst. for Therapeutic Research, Rahway, N.J., 1958—. Recipient Willard Sci. award, Lenoir-Ryne Coll., 1943. Mem. Am. Soc. Biol. Chemists, Endocrine Soc., Am. Chem. Soc., N.Y. Acad. Scis., Soc. for Exptl. Biology and Medicine, A.A.A.S., Biochem. Soc. (London, Eng.), Sigma Xi, Alpha Chi Sigma. Research and numerous publs. on natural products; isolation and biol. and physiochem. properties of protein and peptide hormones; physiology pharmacology and bioassay of steroidal hormones; vitamins; pharmacology of diuretics; pharmacology of antidiabetic agts.; fertility control compounds; domestic animal products. Home: 29 Brook Dr., Watchung, N.J. 07065. Office: Merck Inst. for Therapeutic Research, Rahway, N.J. 07065.*

STEEN, Frederick Henry, Am. mathematician; b. Bklyn., Nov. 26, 1907; s. Herman Claus Frederick and Frieda (Breuer) S.; A.B., Colgate U., 1929; M.A., Harvard, 1931, Ph.D., 1934; m. Marian Rodger, June 26, 1937; children—Martha, Robert, John, Rodger, Paul. Asst. instr., Colgate U., 1928-29; instr. Harvard, 1931-34; faculty Ga. Inst. Tech., 1934-42, asst. prof. math., 1936-42; faculty Allegheny Coll., Meadville, Pa., 1942—, prof. math., 1944—, chmn. sci. div., 1953-54, 62-64; asso. Danforth Found., 1950—. Mem. adv. panel NSF, 1959-60, 61, dir. insts., 1961-62. Mem. Am. Math. Soc., Math. Assn. Am. (past gov., past chmn. Allegheny Mountain sect.), Am. Assn. U. Profs., Phi Beta Kappa. Author: (with Ballou) Plane and Spherical, Trigonometry, 1943, 53, Analytic Geometry, 1943, 4th edit., 1966; Differential Equations, 1955; Mathematics of Finance, 1958; also articles. Research in motivating ideas and methodology of math. devel., learning processes in field. Home: R.D. 1, Meadville, Pa. 16335.*

STEENBOCK, Harry, Am. biochemist; b. Charlestown, Wis., Aug. 16, 1886; s. Henry and Christine (Oesau) S.; B.S., U. of Wis., 1908, M.S., 1910, Ph.D., 1916, Sc.D., 1938; grad. study Yale, 1912, U. of Berlin, 1913; Sc.D. (hon.), Lawrence Coll., 1947; m. Evelyn Carol Van Donk, March 6, 1948. Asst. in agricultural chemistry, U. Wis., 1908-10, Instr., 1910-16, asst. prof., 1916-17, asso. prof., 1917-20, prof., 1920-38, prof. biochemistry since 1938; fellow A.A.A.S.; mem. Am. Inst. of Nutrition, Am. Chem. Soc., Am. Soc. Biol. Chemists, Wis. Acad. Science, Royal German Acad. of Science (Halle), Sigma Xi, Phi Beta Kappa. Writer on human and animal nutrition with special attention to vitamins, mineral elements, and the effect of irradiation. Founder Wis. Alumni Research Found. Patentee ultra violet irradiation of food and synthesis of vitamin D. Died Dec. 25, 1967.

STEENROD, Norman Earl, Am. mathematician; b. Dayton, O., Apr. 22, 1910; s. Earl Lindsay and Sarah (Rutledge) S.; B.A., U. Mich., 1932; M.A., Harvard, 1934; Ph.D., Princeton, 1936; m. Carolyn Witter, Aug. 20, 1938; children—Katherine, Charles. Instr. mathematics Princeton, 1936-39; asst. prof. mathematics U. Chgo., 1939-42; asst. prof. mathematics U. Mich., 1942-45, asso. prof., 1945-47; asso. prof. mathematics Princeton, 1947-51, prof., 1951—. Staff operations research group Navy Dept., Washington, 1944-45. Guggenheim fellow, 1950-51. Mem. Nat. Acad. Sci., Am. Math. Soc., Am. Assn. U. Profs., Math. Assn. Am., Sigma Xi. Author: The Topology of Fibre Bundles, 1951; (with S. Eilenberg) The Foundations of Algebraic Topology, 1952. Editor Annals of Mathematics, 1950-62. Study of topology. Home: 129 Broadmead St., Princeton, N.J.

STEENSTRUP, Johann Japetus, zoologist; b. Vang, Norway, Mar. 8, 1813; student natural scis. and medicine; made exploratory voyages in Jutland, Ireland, Scotland, Norway, 1836-44; became lectr. botany and mineralogy Acad. Soroe, 1841; named prof. zoology U. Copenhagen (Denmark), 1845; named dir. Mus. Natural History, 1848. Mem. Acad. Scis. Copenhagen, French Acad. Scis. Fellow Royal Soc., 1863. Author: Sur la propagation et le développment des animaux à travers une série de générations alternantes, 1842; Recherches sur l'existance des hermaphrodites dans la nature, 1846. Discovered (independently of Chamisso) alternation of generations in heredity; publs. on hermaphroditism in nature, cephalopods. Died Copenhagen, June 20, 1897.

STEERE, William Campbell, Am. botanist, educator; b. Muskegon, Mich., Nov. 4, 1907; s. James Alabaster and Lois (Campbell) S.; B.S., U. Mich., 1929, M.A., 1931, Ph.D., 1932, D.Sc., 1962; postgrad. U. Pa.; D.Sc., U. Montreal, 1959; m. Dorothy Clara Osborne, June 14, 1927; children—Lois Dorothy (Mrs. Allan L. Beattie), Alice Helen (Mrs. Joseph H. Coulombe), William Campbell. Faculty dept. botany U. Mich., 1939-50, Stanford, 1950-58, dean grad. div., 1955-58; prof. biology Columbia, N.Y.C., 1958—; dir. N.Y. Bot. Garden, 1958—. Mem., past pres. Bot. Soc. Am., Am. Hort. Soc. (Liberty Hyde Bailey medal 1965), Am. Bryological Soc., Soc. Am. Naturalists, Torrey Bot. Club, Am. Soc. Plant Taxonomists, N.Y. State Assn. Museums, Cal. Bot. Soc. Author: Liverworts of Southern Michigan, 1940. Editor: Bryologist, 1938-54, Am. Jour. Botany, 1953-57. Research, numerous publs. on geog. distbn. plants especially mosses (field work in P.R., Yucatan, Ecuador, Colombia, Argentina, Alaska, Greenland, Iceland, arctic Can., Antarctics), chromosome numbers in mosses. Home: 122 Park Av., Bronxville, N.Y. 10708. Office: N.Y. Bot. Garden, Bronx, N.Y. 10458.*

STEERS, James Alfred, English geographer; b. Bedford, Eng., Aug. 8, 1899; s. James Alfred and Clara (Blott) S.; M.A., Cambridge (Eng.) U., 1923; m. Harriet Grace Wanklyn, July 10, 1942; children—James, Grace. Geography master Framlingham Coll., 1921-22; faculty Cambridge U., 1922-66, prof. geography, 1949-66; prof. emeritus, 1966—, fellow St. Catharine's Coll., 1925-66, emeritus fellow, 1966—, dean, 1925-46, tutor, 1933-46, pres., 1946-59; vis. prof. U. Cal. at Berkeley, 1959; vis. fellow, Canberra, Australia, 1967. Mem. Royal Geog. Soc. (past v.p., Victoria medal 1960), Brit. Assn. (pres. sect. E 1959), Instn. Brit. Geographers (past pres.), Geog. Assn. (past pres.), Nature Conservancy. Author: Introduction to the Study of Map Projections, 1926; The Unstable Earth, 1932; The Coastline of England and Wales, 1964; The Sea Coast, 1953; ScoltHead Island, 1960; also numerous publs. Editor: Field Studies in the British Isles, 1964; English edit. of

Processes of Coastal Developement (v.p. Zenkovich), 1967. Home: 3, Thornton Close, Girton, Cambridge, Eng.*

STEFAN, Josef, Austrian physicist; b. St. Peter, Austria, Mar. 24, 1835; Ph.D., U. Vienna, 1863; became prof. physics, U. Vienna, 1863, also dir. Phys. Inst.; tchr. of Boltzmann and Hasenöhr. Originated theory on diffusion of gases and studied heat conductivity of gases, 1872-75; originated law that total radiation from black body is proportional to 4th power of its absolute temperature (Stefan's or Stefan-Boltzmann law); studied kinetic theory of gases, hydrodynamics, theory of electricity. Died Vienna, Jan. 7, 1893.

STEFANESCU, Sabba, Rumanian geophysicist; b. Bucharest, Rumania, July 20, 1902; s. Sabba and Constanta (Negrea) S.; grad. Ecole Supérieure des Mines, Paris, France, 1923; D.Sc., U. Bucharest, 1945; m. Elena Ionescu Dolj, Jan. 24, 1937; 1 son, Ion-Sabba. Engr., Geol. Inst., Bucharest, 1927-50; prof. electromagnetic theory and elec. prospecting Inst. Geology and Mining Techniques, Bucharest, 1950-57, Inst. Oil, Gas and Geology, Bucharest, 1957—; dir. Centre Geophys. Research, 1961—. Mem. Geol. Com. Rumania, 1960—; mem. Council for Sci. Research, 1965—. Recipient State prize Laureate, 1952, Order of Labour, 1953, Star of Rumanian Popular Republic, 1964. Mem. Am. Geophys. Union, Carpatho-Balcanic Geol. Assn., Rumanian Acad. Author: Theoretical Studies on the Electrical Prospecting, 3 vols., 1929, 45, (with C. Schlumberger, M. Schlumberger), 1932; also articles. Established theory of electromagnetic field in media with plane parallel layers; created models of heterogenous media with continous variation of conductivity; found examples of open magnetic lines. Home: 8 Piatza Romana, Bucharest. Office: 64 Calea Grivitei, Bucharest, Rumania.*

STEFANINI, Mario, pathologist; b. Chieri, Italy, June 11, 1918; s. Eleuterio and Therese (Trivereau) S.; M.D., U. Rome, 1939; M.Sc., Marquette U., 1947; m. Elizabeth S. Just, Feb. 12, 1949; children—Marie Therese, Virginia Elizabeth. Came to U. S., 1946, naturalized, 1951. USPHS sr. research fellow, 1947-49; Damon Runyon sr. clin. research fellow, 1949-52; established investigator Am. Heart Assn., 1952-58; faculty Marquette U., 1946-49, Tufts U. Sch. Medicine, 1949-61; dir. research St. Elizabeth Hosp., Boston, 1954-61; hemopathologist St. Joseph Hosp., Chgo., 1963—; dir. labs. St. Elizabeth Hosp. Danville, Ill., 1966—. Cons. various hosps., Chgo. Recipient 1st awards Am. Assn. Blood Banks, 1953, 56; certificate of merit, Nat. Gastroenterologic Assn., 1948, A.M.A., 1953. Mem. A.M.A., Coll. Am. Pathologists, Am. Soc. Clin. Pathologists, Am. Soc. Clin. Investigation, Assn. Am. Immunologists, Am. Soc. Exptl. Pathology, Am. Physiologic Soc. Author: (with William Dameshek) Hemorrhagic Disorders, 1955. Editor: Progress in Clinical Pathology. Research and numerous publs. on relationship of morphology to platelet functions and fibrinolysis; developer blood platelets transfusion technics; discoverer aspergillin O. Home: 1104 Rossell Av., Oak Park, Ill. 60302. Office: 600 Sager Av., Danville, Ill. 61832.*

STEFANINI, Paride, Italian surgeon; b. Rome, Jan. 15, 1904; s. Arnaldo and Emilia (Tabasso) S.; M.D. in Surgery; m. Adriana Nelli, Dec. 28, 1932; children—Livia, Arnaldo, Mario, Alessandro. Dir., prof. Surg. Pathology Inst., U. Rome; practice surgery. Mem. Italian Gastroenterology Soc. (v.p.), Soc. Romana di chirurgica (pres.), Italian Surgery Soc. (v.p.). Contbr. numerous articles on surgery to tech. jours. Address: via Vincenzo Tiberio 24, Rome, Italy.

STEFANSSON, Vilhjalmur, sci. explorer, geographer, anthropologist; b. Arnes, Man., Can., Nov. 3, 1879; student anthropology Harvard; ed. also U. N.D.; B.A., U. Ia., 1903; m. Evelyn Baird, Apr. 1941. Archeol. researcher Iceland, 1904-05; made ethnol. survey of Central Arctic coasts of N.Am. for Am. Mus. Natural History, N.Y. and govt., Can., 1908-12; led expdn. to explore regions west of Parry Archipelago for govt. Can., 1913-16; adviser to U. S. govt. on defense conditions in Alaska, author Arctic reports and manuals for armed forces, World War II; adviser on northern operations Pan Am. Airways, from 1932. Arctic cons. northern studies Dartmouth, 1947-62. Recipient medals Geography Soc. Paris, Explorers Club, Geography Soc., Charles P. Daly medal Am. Geog. Soc., 1918; Arctic island named for him by govt. Can., 1952. Hon. fellow Am. Mus. Natural History, Italian Geog. Soc.; mem. Assn. Am. Geog. (medalist, v.p. 1945), Royal Geog. Soc. (medalist). Author: My Life with the Eskimo, 1913; Friendly Arctic, 1921; Northward Course of Empire, 1922; Hunters of the Great North, 1922; The Adventures of Wrangell Islands, 1925; The Three Voyages of Martin Frobisher, 1938; Unsolved Mysteries of the Arctic, 1938; Ultima Thule, 1940; Greenland, 1942; Not By Bread Alone, 1946; Great Adventures and Explorations, 1947. Discovered tribes hitherto unknown to whites, such as blond Eskimo of Victoria Land, Islands of Borden, Brock, Meighen, Loughheed, new land north of Prince Patrick Island, nr. 78th north latitude; located considerable parts of continental shelf north of Alaska and west of Banks Lands; demonstrated that adaptation to arctic life was comparatively simple. Died Hanover, N.H., Aug. 26, 1962.

STEFFENS, Hendrik, natural scientist; b. Stravanger, Norway, May 2, 1773; prof., Halle, Germany, also Breslau (now Wroclaw, Poland), Berlin. Author: Handbuch der Oryktognosie, 4 vols., 1811-24; Geschichten, Sagen und Märchen, 1823; Anthropologie, 2 vols., 1824. Attempted to unify all phenomena of life by means of history and nature. Died Berlin, Feb. 13, 1845.

STEGER, Eberhard, German chemist; b. Sebnitz, Saxony, Germany, Apr. 8, 1925; s. Walter and Johanna (Henke) S.; diploma Tech. Coll., Dresden, Germany, 1953, Dr.rer.nat., 1955, habil., 1959; m. Ursula Bergs, July 21, 1956; children—Wolfgang, Dorothée, Burkhard. Faculty, Tech. U., Dresden, 1960—, prof. spectroscopy, 1962—, dir. Inst. Spl. Analytical Chemistry, 1964—. Mem. commn. for spectroscopy German Acad. Scis. 1958—. Mem. German Bunsen Soc. for Phys. Chemistry, Chem. Soc. in GDR. Research and numerous publs. on infrared and Raman spectroscopy of inorganic compounds, phosophorous chemistry, chem. investigations on works of art. Home: 28 Reicker Str., Dresden, 8020, E. Germany.*

STEGGERDA, Frederic Russell, Am. physiologist; b. Holland, Mich., May 5, 1903; s. George John and Sena (Ter Vree) S.; A.B., Hope Coll., Holland, 1925; M.A., U. Minn., 1927, Ph.D., 1929; m. J. Marian Van Vessem, Sept. 5, 1930; children—Frederic John, Marina Britz, Janet M. Rowlings. Faculty, U. Ill., Urbana, 1933—, prof. physiology, 1947—. Dir. NSF Postdoctoral Research Participants, 1961—. Recipient research grants USAF, 1955, U.S. Dept. Agr., 1962, NIH 1958, Abbott Labs., 1953; Fulbright fellow, Hiroshima, Japan, 1961. Mem. A.A.A.S., Am. Physiol. Soc., Soc. For Exptl. Biology and Medicine, Aerospace Med. Assn., Gastroent. Soc. Author: Laboratory Manual of Physiology, 1938. Research and numerous publs. in gastroenterology in man and animals, circulation of turtle's heart; established daily calcium requirements in adult man. Home: 607 Nevada St., Urbana, Ill. 61801.*

STEHLI, Francis Greenough, Am. geologist; b. Upper Montclair, N.J., Oct. 16, 1924; s. Edgar and Emily (Greenough) S.; B.S., St. Lawrence U., 1949, M.S., 1950; Ph.D., Columbia U., 1953; m. Irene Comfort, June 19, 1948; children—Anne, Robert, John, Edgar. Asst. prof. invertebrate paleontology Cal. Inst. Tech., 1953-56; with research dept. Pan Am. Petroleum Corp., 1956-60, now cons.; prof. geology Western Res. U., Case Inst. Tech., Cleve., 1960—, chmn. dept. geology, 1961—. Mem. A.A.A.S., Geol. Soc. Am., Paleontol. Soc. Am., Geochem. Soc. Am., No. Ohio Geol. Soc. Research on Paleozoic and Mesozoic paleontology and stratigraphy in U. S., Mexico, Can., Madagascar, models for determining depositional water depts, reconstructing glacial age circulation pattern on oceans, test of assumed axial dipolar model of earth's magnetic field. Home: 16432 Stone Ridge Rd., Chagrin Falls, O. 44022. Office: Dept. of Geology, Case Western Res. U., Cleve. 44106.

STEIDTMANN, Waldo E(duard), Am. botanist; b. Prairie du Sac, Wis., Apr. 27, 1896; s. Charles Frederick and Bertha (Schoenberg) S.; rural tchrs. certificate Wis. State Tchrs. Coll., Whitewater, 1916; A.B., U. of Wis., 1923; M.S., U. Mich., 1929, Ph.D., 1935; m. Evelyn Katherine Dressel, Sept. 8, 1930; children—Sally Ann, James Richard. Country sch. teacher, Wis., 1916-17; instr. in botany Marquette U., Milw., 1923-32; instr. in biology Wis. State Teachers Coll., La Crosse, 1936; mem. faculty Bowling Green (O.) State U., 1936—, prof. of biology and chmn. biology dept., 1947—. Fellow Ohio Acad. Sci. (chmn. plant sci. sect., 1943-44); mem. A.A.A.S., Bot. Soc. Am., Sigma Xi. Research papers in paleobotany. Died June 21, 1955.

STEIGER, George, Am. chemist; b. Columbia, Pa., May 27, 1869; s. Benjamin F. and Martha L. (Young) S.; B.S., Columbian (now George Washington) U., 1890, M.S., 1892. Chemist, U. S. Geol. Survey 1892, chief chemist, 1916-30, chemist, 1930-39, ret., continued work through facilities of Geol. Survey. Fellow A.A.A.S.; mem. Am. Chem. Soc., Am. Inst. Mining Engrs., Geol. Soc. Washington, Mineral Soc., Wash. Acad. Scis. Contbr. various papers, mostly on original research on constitution of certain silicates and methods of chem. analysis. Died Apr. 18, 1944.

STEIGERT, Frederick Edward, Am. physicist; b. N.Y.C., Sept. 11, 1928; s. Karl Wilhelm and Margaret (Schuppert) S.; B.S., Union Coll., 1949; M.A., Ind. U., 1949, Ph.D., 1953; m. Lois Rae Copes, June 30, 1950; children—Frederick W., Richard E., Christine L., Heidi A. Instr. physics Yale, 1953-56, asst. prof., 1956-62; asso. prof. U. Conn., Storrs, 1962—. Mem. Am. Phys. Soc., Am. Assn. Physics Tchrs., Sigma Xi, Sigma Pi Sigma. Research and publs. primarily in field of nuclear scattering reactions. Home: Peck Rd., Bethany Conn. 06525. Office: Dept. Physics, U. Conn., Storrs, Conn. 06268.*

STEIGMAN, Joseph, Am. chemist; b. N.Y.C., May 10, 1913; s. Philip and Rose (Jawetz) S.; A.B., Columbia U., 1932, Ph.D., 1941; m. Margaret Patricia Robinson, Nov. 27, 1940. Research asso. dept. physics Columbia U., 1937-40; faculty Coll. City N.Y., 1939-42, Bklyn. Coll. Grad. Div., 1939-42; asso. dir. Tech. Research Labs., 1942-43; cons., 1943-47; faculty Poly. Inst. Bklyn., 1947—, prof. chemistry, 1961—. Fellow N.Y. Acad. Scis.; mem. Am. Chem. Soc. (nat. councillor 1963-65), Phi Beta Kappa, Sigma Xi, Phi Lambda Upsilon. Research and publs. on acid-base reactions in nonaqueous solvents, measurement ionbinding by proteins and electrolytes, effects of solvent orgn. on solubility, micelle formation and acidity. Home: 141 Joralemon St., Bklyn. 11201.*

STEIN, Calvert, neuropsychiatrist; b. Newcastle-on-Tyne, Eng., Apr. 6, 1903; s. Harry and Lily (Phillips) S.; came to U. S., 1912, naturalized, 1919; student Tufts Coll., 1921-24, M.D., 1928; LL.B., Northeastern U., 1938; postgrad. Yale Med. Sch., Columbia; m. Lucille I. Weinstein, Nov. 26, 1929; children—Elinor M. (Mrs. Aaron M. Leavitt), Mildred J. (Mrs. Harold Sobel). Practice gen. medicine, 1929-31, specializing in neuropsychiatry, Palmer, Mass., 1938-41, Springfield, Mass., 1946—; vis. lectr. Soc. Psychiatry and Analytical Psychiatry, Am. Internat. Coll., 1950-53, Orientat. Neurology and Psychiatry, Springfield Coll., Grad. Sch., 1948—, Columbia U. Med. Grad. Sch., 1964—. Cons. psychiatrist Smith Coll., 1956-57, Westover AFB, Springfield Hosp. Recipient 1st prize for psychiat. research New Eng. Psychiat. Assn., 1936. Fellow Am. Soc. Group Psychotherapy and Psychodrama (pres. 1964-66), Am. Psychiat. Assn. (life), Am. Soc. Clin. Hypnosis, (v.p.), Am. Acad. Psychosomatic Medicine; mem. Am. Acad. Neurology, Eastern Assn. Electroencephalographers, Eastern Psychiat. Research Assn. Author: Hidden Springs of Human Action, 1952; Psychotherapeutic Techniques in Non Psychiatric Specialities, 1967; also articles. Pioneered short-term psychotherapy, modified psychodrama, dynamic group therapy and coordinated hypnotherapy including clenched-fist displacement technique in treatment of convulsive disorders, reconditioning technique for compulsive suckers; research on heredity of epilepsy, mental competance and law, child guidance, marital and family counseling, psychosomatics, rehab. Home: 71 Meadowbrook Rd., Longmeadow, Mass. 01106. Office: 146 Chestnut St., Springfield, Mass. 01103.*

STEIN, Elias M., mathematician; b. Antwerp, Belgium, Jan. 13, 1931; s. Elkan and Hannah (Goldman) S.; came to U. S., 1941, naturalized, 1952; A.B., U. Chgo., 1951, M.S., 1953; Ph.D., 1955; student Columbia, 1951-52; m. Elly Intrator, Mar. 21, 1959; children—Jeremy, Karen Deborah. Instr., Mass. Inst. Tech., 1956-58; mem. faculty U. Chgo., 1958-62, asso. prof. math., 1961-62; mem. Inst. Advanced Study, Princeton, N.J., 1962-63; prof. math. Princeton, 1963—. Sloan Found. fellow, 1961-63; NSF sr. postdoctoral fellow, 1962-63. Mem. Am. Math. Soc. Contbr. profl. jours. Research in harmonic analysis. Home: 132 Dodds Lane, Princeton, N.J. 08540.*

STEIN, Eric Albert, Swiss biochemist; b. Geneva, Switzerland, Apr. 11, 1925; s. Leonard C. and Mella (Schnabel) S.; Ph.D. in Chemistry, U. Geneva, 1954; m. Frances van Scherpenzeel, June 28, 1955; children—Leonard, Caroline, Isabelle, Sylvie. Research asst. prof. dept. biochemistry U. Wash., Seattle, 1958-63, research asso. prof. dept. biochemistry, 1963-64; prof. U. Geneva, 1964—, head dept. spl. biochemistry, 1964—. Mem. Am. Soc. Biol. Chemists, Biochem. Soc. (Eng.). Research, publs. on structure of amylases of various origins, trace elements in biochem. systems, structure of metalloproteins, phys. biochemistry, enzymology. Home: La Trajurane, Collonge-Bellerive, 1245 Geneva, Switzerland.*

STEIN, Gabriel, chemist; b. Budapest, Hungary, May 24, 1920; s. Arthur and Zelma (Weisz) S.; M.Sc., Hebrew U., Jerusalem, Israel, 1946; Ph.D. (Cancer research fellow), King's Coll., U. Durham, (Eng.), 1950; m. Pauline Daphne Epstein, Aug. 2, 1950; children—Tamar Rachel, Asher Philip. Faculty, Hebrew U. Jerusalem, 1951—, prof. chemistry, 1956—. Sir Simon Marks fellow U. Cambridge (Eng.), 1960-61; NSF vis. prof. Boston U., 1965-66. Recipient Weizmann Sci. prize, 1960. Mem. Israel (chmn. 1959-60), Brit., Am. chem. socs., Council Profl. Orgns. Israel (chmn. 1961-65), Faraday Soc., Radiation Research Soc., N.Y. Acad. Scis., Research, publs. on mechanism of chem. and biol. actions of radiations, photochemistry of solutions, formation of solvated electron. Home: 15, Rehov Shmuel Hanagid, Jerusalem, Israel.*

STEIN, Lorenz, economist, lawyer; b. Barby, Germany, Nov. 15, 1815; prof., Kiel, Germany, Vienna, Austria, 1855-90. Author: Kommunismus und Sozialismus des heutigen Frankreich, 1842; Ideen zur Gesch. der Arbeit, 1849; Die Gesch. der sozialen Bewegung in Frankreich von 1789 bis auf unsere Tage, 3 vols., 1850; System der Staatswissenschaft, 2 vols., 1852-56; Die staatswiss. Theorie der Griechen vor Aristoteles und Platon und ihr Verhältnis zu dem Leben der Gesellschaft, 1853; Lehrbuch der Volkswirtsch., 1858, later published as Lehrbuch der Nationalökonomie; Die drei Fragen des Grundbesitzes, 1881; Verwaltungslehre, 8 vols., 1865-84. Helped to separate polit. sci. sociology from sociology and to establish it as a separate sci. Died Weidlingen/Vienna, Austria, Sept. 23, 1890.

STEIN, Sir Mark Aurel, archaeologist; b. Budapest, Hungary, Nov. 26, 1862; ed. Budapest,

Hungary, Dresden, Germany, U. Vienna, (Austria), Tübingen, Germany; hon. doctorates Oxford, Eng., Cambridge, Eng., Punjab U. Became prin. Oriental Coll., Lahore, India, also registrar Punjab U., 1888; joined Indian Edn. Service, 1889; became prin. Calcutta, Madrash, 1899; explored Central Asia, Western China, 1906-08; named supt. Indian Archael. Survey, 1910; explorations in Iran and Central Asia, 1913-16, also Iran, Iraq and Transjordan. Recipient Gold medal Royal Geog. Soc., 1909, Gold medal Royal Asiatic Soc., 1932, Huxley medal Royal Anthropol. Inst., 1934, Gold medal Soc. Antiquaries, 1935. Knighted, 1912. Author: Sanskrit Chronicle of Kings of Kashmir, 2 vols., 1900; Ancient Khotan, 2 vols., 1907; Ruins of Desert Cathary, 2 vols., 1912; The Thousand Buddhas, 1921; Serindia, 1921; Innermost Asia, 4 vols., 1929; On Alexander's Track to the Indus 1929; An Archaeological Tour in Gediosia, 1931; A Catalogue of Paintings Recovered from Tun Huang, 1931; Archaeological Reconnaisances in South East Iran, 1937; Old Routes of Western Iran, 1940; also treatises on Sanskrit lit. Archaeol. explorations for Indian Govt. in Chinese Turkestan, 1889; discovered site of Aornus. Died Kabul, Afghanistan, Oct. 28, 1943.

STEIN, Morris Isaac, Am. psychologist; b. N.Y.C., June 11, 1921; s. Samuel and Lena (Citron) S.; B.S., Coll. City N.Y., 1940, M.S., 1942; M.A., Harvard, 1943, Ph.D., 1949. Staff psychologist VA, N.Y.C., 1946-47, Boston, 1947-48; psychol. cons., Hines (Ill.) VA Hosp., 1948-51, 56-60, Chgo. Mental Hygiene Clinic, 1948-51, 56-60; faculty U. Chgo., 1948-60, asso. prof. psychology, 1954-60; prof. psychology N.Y. U., N.Y.C., 1960—, dir. Research Center for Human Relations, 1960-63, 66-67. Faculty, Wheaton Coll., 1947-48, Northwestern U. Med. Sch., 1956-60; mem. mental health study sect. B, NIH, 1962-66. Fellow Center for Advanced Study in Behavioral Scis., 1955-56. Mem. Am., Eastern, Midwestern psychol. assns., A.A.A.S., Soc. for Projective Techniques, Sigma Xi. Author: (with others) Assessment of Men, 1948; The Thematic Apperception Test, 1948, rev., 1955; (with G.G. Stern and B.S. Bloom) Methods in Personality Assessment, 1956; (with Shirley J. Heinze) Creativity and the Individual, 1960; Personality Measures in Admissions, 1963; Volunteers for Peace, 1966. Editor: Contemporary Psychotherapies, 1961. Cons. editor Jour. Abnormal Psychology, 1964—. Research and publs. on understanding and prediction of human behavior, effect of environment on a person's potentialities. Home: 7-13 Washington Sq. N., N.Y.C. 10003.*

STEIN, Robert, explorer; b. Rengersdorf, Prussia, Jan. 9, 1857; s. Joseph and Francisca (Kasper) S.; ed. at Glatz, 1873-75; came to U. S., M.D., Georgetown Coll., 1886. Came to U. S., 1875. Tchr.; entered U. S. Geol. Survey, 1885, became translator of German, French, Italian, Danish, Swedish, Russian, Dutch, Spanish, Portugese and other langs.; published plan for exploring Ellesmereland which attracted attention of Am. and European geographers, 1893; joined the 7th Peary expdn., 1897, landing with 3 Eskimo boatmen on western Greenland to explore and map coast afterward returning on the same vessel; took passage on the Diana, with 2 companions, to carry out his plan for exploring Ellesmereland, 1899, landed at Cape Sabine, spent 2 yrs. mainly in linguistic studies among the Eskimos, returned to U. S. in Oct. 1901. Contbr. on econ. and social subjects and on Arctic work in the Nat. Geog. Mag., on Arctic work in leading reviews. Advocated internat. conf. of experts in phonetics to devise uniform notation of sounds of speech, especially as uniform key to pronunciation in dictionaries. Died 1917.

STEIN, William Howard, Am. biochemist; b. N.Y.C., June 25, 1911; s. Fred M. and Beatrice (Borg) S.; B.S., Harvard, 1933; Ph.D., Columbia Coll. Phys. and Surg., 1938; m. Phoebe L. Hockstader, June 22, 1936; children—William Howard, David F., Robert J. Staff, Rockefeller Inst., N.Y.C., 1938—, mem., 1952—, prof. biochemistry, 1955—; vis. prof. U. Chgo., 1960, Harvard, 1963; Am.-Swiss Found. lectr., 1956; Harvey lectr., 1957; Philip Shaffer lectr., 1965. Chmn., U. S. Nat. Com. for Biochemistry, 1965—; sci. counselor Nat. Inst. Neurol. Disease and Blindness, 1961-65; med. adv. bd. Hebrew U., Hadassah Med. Sch., Jerusalem, Israel, 1957—. Fellow A.A.-A.S.; mem. Am. Soc. Biol. Chemists (chmn. editorial com. 1958-61), Nat. Acad. Scis., Am. Acad. Arts and Scis., Biochem. Soc. (London, Eng.), Am. Chem. Soc., Harvey Soc. Editorial bd. Jour. Biol. Chemistry, 1962-64, asso. editor, 1964—. Contbr. numerous articles to tech. jours. Devel. reagts. for amino acids, chromatographic methods permitting quantitative determination amino acids and peptides in protein hydrolysates and biol. fluids, automatic amino acid analyzer; derived structural formula ribonuclease; delineated active site enzyme; research on chromatography proteins, streptococcal proteinase, pepsin. Home: 168 E. 74th St., N.Y.C. 10021.*

STEINACH, Eugen, Austrian surgeon; b. Hohenems, Austria, Jan. 27, 1861; s. Simon Steinach; M.D., Vienna, 1886. Asst., Physiologische Institut, U. Innsbruck (Austria), 1886-89; became lectr., prin. asst. to Ewald Hering, German U., Prague, Czechoslovakia, 1890; asst. prof. U. Prague, 1895-1907, prof., 1907-18; named prof. U. Vienna, also dir. Biologisches Institut of Acad. Scis. Vienna, 1918. Author: Life and Sex, 1940; Verjüngung durch experimentelle Neubelebung der alternden Pubertätsaruse, 1920. Originated operation for treatment of impotency in which ductus degerens is occluded (Steinach's operation or method), 1920; tried to rejuvenate men and animals by grafting sexual glands from young animals; studied phys. and psychic sex characteristics; demonstrated that sex and reprodn. depend on internal secretion of sex glands; 1st to propose there is a spl. phys. organ designated as gland of puberty. Died Montreaux, Switzerland, May 13, 1944.

STEINBACH, Henry Burr, Am. biologist; b. Dexter, Mich, Oct. 7, 1905; s. Henry August and Mary (Laney) S.; A.B., U. Mich., 1928; A.M., Brown U., 1930; Ph.D., U. Pa., 1933; m. Eleanor Parsons, June 9, 1934; children—Alan Burr, Mary Parsons, Joseph Henry, James Burr. Demonstrator physiol. and bio-chemistry Brown U., 1928-30; instr. zoology, U. Pa., 1930-33, U. Minn., 1935-37, asst. prof., 1937-38; asst. prof. zoology Columbia, 1938-42; asso. prof. zoology, Washington U., 1942-46; prof., 1946-47; prof. zoology U. Minn., 1947-57; prof. and chmn. dept. zoology U. Chgo., 1957—. Asst. dir. National Sci. Found., 1952-53; chmn. div. biology and agr. NRC, 1958-62. Trustee Marine Biol. Lab., director of the Marine Biol. Lab., 1966—. NRC fellow U. Chgo. and U. Rochester med. schs., 1933-35, Guggenheim fellow, 1955-56. Fellow Am. Acad. Arts and Sciences (member bd. of directors 1964-66); member of American Society of Zoologists (pres. 1957), Soc. Gen. Physiologists (pres. 1953), Am. Physiol. Soc., Sigma Xi. Mem. editorial bd. Physiol. Zoology. Author papers on sci. subject. Research on bioelectric phenomena; injury potentials; enzyme systems of granular components; ontogenesis of enzyme systems in chicks; sodium potassium equilibrium of cells. Home: 5326 University Av., Chgo. 60615.

STEINBACH, Howard Lynne, Am. physician; b. Pitts., Sept. 17, 1918; s. Morris and Mary (Bortz) S.; B.A., U. Cal., Los Angeles, 1940; M.D., St. Louis U., 1943; m. Ilse Rosengarten, May 29, 1951; children—Lynn, Lisa. Practice medicine, specializing in radiology, San Francisco, 1945—; prof. radiology, vice chmn. department of radiology, Univ. of Cal. Med. Center, San Francisco, 1963—. Cons., VA Hosp., Letterman Gen. Hosp., San Francisco, Oak Knoll Naval Hosp., Oakland, Cal. Fellow Am. Coll. Radiology; mem. A.M.A., Am. Roentgen Ray Soc., Assn. U. Radiologists, Am. Rheumatism Assn., Radiol. Soc. N.Am., Alpha Omega Alpha. Research and numerous publs. in diagnostic radiology, metabolic, endocrine and systematic diseases. Home: 330 Robinwood Lane, Hillsborough, Cal. 94010. Office: U. Cal. Med. Center, San Francisco 94122.*

STEINBACH, Marc, Rumanian physician, biostatistician; b. Sibiu, Rumania, Aug. 16, 1920; s. Rubin and Louise (Goldrina) S.; M.D., Bucharest (Rumania) Sch. Medicine, 1946; m. Renée Berler, Apr. 24, 1955. Faculty, Bucharest Sch. Medicine, Colentina Hosp., 1948-52, asst. prof. internal medicine, 1951-52; chief div. ecology of atherosclerosis Inst. Internal Medicine, Rumanian Acad., 1957—, head div. biology and medicine Center Math. Statistics, 1964—. Mem. Internat. Soc. Internal Medicine, Internat. Med. Assn. for Study Living Conditions and Health, Biometric Soc. Author: Statistics in Biology and Medicine, 1961; also articles. Research on atherosclerosis, relationship between atherosclerosis and high blood pressure, environmental factors in atherosclerosis, role of biliary tract and terminal illeon in conversion of cholesterol into biliary acids. Home: 44 Pictor Iscovescu, Bucharest. Office: 19-21 Socs. Stefan cel Mare, Bucharest, Rumania.*

STEINBERG, Arthur Gerald, Am. geneticist; b. Port Chester, N.Y., Feb. 27, 1912; s. Bernard Aaron and Sarah (Kaplan) S.; B.S., Coll. City N.Y., 1933; M.A., Columbia U., 1934, Ph.D. (univ. fellow), 1941; m. Edith Wexler, Nov. 22, 1939; children—Arthur E., Jean E. (Mrs. Josef Shengili). Mem. faculty McGill U., 1940-44; faculty Antioch Coll., also staff Fels Research Inst., 1946-48; cons. div. Mayo Clinic, 1948-52; geneticist Children's Cancer Research Found., also research asso. Children's Hosp., Boston, 1952-56; prof. biology Western Res. U., Cleve., 1956—, asso. prof. human genetics dept. preventive medicine, 1960—; dir. heredity clinic Lakeside Hosp., Cleve., 1958—. Mem. adv. bd. Cystic Fibrosis Found. Cleve., Nat. Cystic Fibrosis Research Found., United Cerebral Palsy Research Found., Nat. Found. Neuromuscular Diseases; chmn. com. 3d Internat. Congress of Human Genetics, 1964-66; mem. expert com. genetics WHO. Fellow A.A.A.S., Ohio Acad. Sci. (v.p. sect. 1964, Australian Acad. Sci. (sr.); mem. Am. Soc. Hematology, Am. Soc. Human Genetics (pres. 1964), Am. Soc. Naturalists, Conf. Biol. Editors, Genetics Soc. Am., Biometrics Soc., Am. Assn. U. Profs., Am. Inst. Biol. Scis., Sigma Xi. Sr. editor vols. I-V, Progress in Med. Genetics; mem. internat. bd. editors Human Genetics Abstracts; cons. editor Transfusion, Vox Sanguinis. Research and publs. on transmission of various hereditary diseases in man; linkage studies in human chromosomes; genetic controls of evolution, population distbn. in the world. Home: 2196 Delaware Dr., Cleve. 44106.*

STEINBERG, Bernhard, Am. physician; b. N.Y.C., June 18, 1897; s. Murray and Sura (Sprinborg) S.; B.S., Fordham U., 1920; M.D., Boston U., 1922; m. Roberta Riman, Feb. 13, 1932; children—Bernard Evanbar, Michael Evanbar. Practice medicine, specializing in pathology (oncology, hematology), Toledo, 1927-64, Pomona, Cal., 1964—; dir. clin. labs., dir. Inst. Med. Research Toledo Hosp., 1929-64; lectr. forensic medicine U. Toledo Law Sch. and Toledo Police Acad., 1934-64; asso. clin. prof. research pathology Loma Linda (Cal.) U. Sch. Medicine, 1964—. Recipient award for method treatment peritoneal infections Soc. Clin. Pathologists, 1942; Cin. Proctological award for work on diagnosis of cancer, 1951; Hon. Phi Beta Phi award of year, 1952. Mem. Am., Internat. socs. hematology, Soc. Clin. Research, Exptl. Soc. Pathologists, Am. Assn. Pathologists and Bacteriologists, Am. Fedn. Biol. Socs., A.M.A., Soc. Exptl. Medicine and Biology. Author: Infections of the Peritonuem, 1944; It Was My Idea, 1951; also numerous articles. Research, publs. on treatment and prevention of peritonitis; described systemic form of nodular panniculitis and isolation of blood regulators controlling devel. and maturation of blood cells, identification of early cancer by RNA stain of cell cytoplasm, gradual cytoplasmic changes by electron microscopy in cytoplam of cancer developing cells, disease of lungs (Exfoliative Broncho-alveolar Disease), pathol. changes and developmental mechanism of allergic asthma, vascular transformation in pulmonary embolism and effect on survival; introduced quantitative measuring device for blood proteins separated by starch-gel electrophoresis; described methods for radiocurability and immunity devel. using model-rat lymphosarcoma; chemotherapy methods in malignant disease especially breast cancer. Home: P.O. No. 633, Claremont, Cal. 91711. Office: 142 Nemaha St., Pomona, Cal. 91766.*

STEINBERG, Daniel, Am. physician, biochemist; b. Windsor, Ont., Can., July 21, 1922 (parents Am. citizens); s. Maxwell Robert and Bess (Krupp) S.; B.S., Wayne U., 1941, M.D. summa cum laude, 1944; Ph.D., Harvard, 1951; m. Sara Murdock, Nov. 30, 1946; children—Jonathan Henry, Ann Ballard, David Ethan. Instr., Boston U. Sch. Medicine, 1947-48; vis. scientist Carlsberg Lab., Copenhagen, Denmark, 1952-53; research staff biochemistry and medicine Nat. Heart Inst., NIH, Bethesda, Md., 1953-68, chief Lab. Metabolism, 1962-68, chmn. sci. adv. com. on ednl. activities, NIH, 1955-60; prof. medicine and head div. metabolic disease U. Cal. San Diego, La Jolla, 1968—; pres. Found. for Advanced Edn. in Scis., Kensington, Md., 1959-62, 65—. Mem. adv. bd. Jour. Lipid Research, 1964—. Diplomate Nat. Bd. Med. Examiners. Mem. Am. Heart Assn. (exec. com., chmn. program com. 1965-67, chmn. council on arteriosclerosis 1967—), Soc. Biol. Chemists, Soc. for Clin. Investigation, A.A.A.S., Soc. for Exptl. Biology and Medicine, A.M.A. Editor-in-chief Jour. Lipid Research, 1961-64. Research and numerous publs. on mechanisms of protein biosynthesis including first demonstration of step-wise synthesis, mechanisms of fatty acid moblzn. from adipose tissues and its relations to blood lipid and liver lipid levels, identification of metabolic defects in Refsum's Syndrome, mechanisms by which dietary fat influences blood cholesterol levels; demonstrated feasibility lowering blood cholesterol levels by use inhibitors cholesterol biosynthesis. Home: 9551 La Jolla Farms Rd., La Jolla, Cal. 92037.*

STEINBERG, Ellis Philip, Am. nuclear chemist; b. Chgo., Mar. 26, 1920; s. Solomon and Sarah (Saphir) S.; B.S., U. Chgo., 1941, Ph.D., 1947; m. Esther Abraham, Dec. 16, 1944; children—Sheryl, David, Deborah. Jr. chemist U.S. War Dept. Elwood Ordnance Plant, Joliet, Ill., 1941-43; sr. chemist Argonne (Ill.) Nat. Lab., 1943—. Mem. subcom. on radiochemistry NRC, 1959—; chmn. Gordon Conf. on Nuclear Chemistry, 1961. Guggenheim fellow Bohr Inst., Copenhagen, Denmark, 1957-58. Mem. Am. Phys. Soc., Am. Chem. Soc., Research Engring. Soc. Am. Research and numerous publs. in mass and charge distbns. in low energy nuclear fission and spontaneous fission, characterization of decay properties of fission products, interactions of high energy protons with complex nuclei; devel. of absolute counting techniques. Home: 194 Westwood Dr., Park Forest, Ill. 60466. Office: 9700 S. Cass Av., Argonne, Ill. 60439.*

STEINBERG, John C., Am. physicist; b. Lakota, Ia., June 21, 1895; s. August Frederick and Anna (Salter) S.; B.S., Coe Coll., 1916, M.S., 1917; Ph.D., State U. Ia., 1922; m. Mary Isis Kinser, June 25, 1957; children—Valeda Rae, John C. Mem. tech. staff, sub-dept. head Bell Telephone Labs., Inc., Murray Hill, N.J., 1922-57; prof. marine sci. Inst. Marine Scis., U. Miami (Fla.), 1960—. Recipient Navy Ordnance Devel. award 1945, Army Navy Certificate of Appreciation, 1947, Alumni award of merit Coe Coll., 1956. Mem. Am. Phys. Soc., A.A.A.S., Acoustical Soc. Am. Research, publs. and patents in characteristics of speech, music and noise in relation to hearing and to requirements for transmission and reprodn., propagation of underwater sound in relation to environmental changes. Home: 8730 S.W. 48th St., Miami, Fla. 33165.*

STEINBERGER, Emil, Am. endocrinologist; b. Berlin, Germany, Dec. 20, 1928; s. Isaac and Itta (Schlanger) S.; came to U.S., 1948, naturalized, 1954; M.S., State U. Ia., 1955, M.D., 1955; m.

Anna Schneider, Dec. 24, 1950; children—Pauline Edith, Inette Emily. Research asst. dept. anatomy State U. Ia., 1950-55; endocrine fellow, Detroit Receiving Hosp., 1958-61; med. research officer Naval Med. Research Inst., Bethesda, Md., 1956-58; practice medicine, specializing in endocrinology, Phila., 1961—; asso. mem. div. endocrinology and reprodn. Albert Einstein Med Center, 1961-66, chmn., 1967—, chief sect. clin. endocrinology, 1965—, also mem. edn. com., co-dir. postgrad. course in endocrinology; mem. endocrine study sect. NIH, 1957-58. Mem. Soc. Am. Anatomists, Endocrine Soc., Nat. Acad. Scis., A.M.A., Sigma Xi. Research and numerous publs. on germinal epithelium of testes following various phys. and chem. insults, dynamics of spermatogenesis, physiology and endocrinology of testes in vitro, mechanisms concerned with pituitary gonadotropin prodn. and release, ovulation and spermatogenesis and therapy of disorders in this area; devel. of techniques for growth of gonadal tissue in vitro. Home: 1114 Rock Creek Dr., Wyncote, Pa. 19095. Office: Albert Einstein Med. Center, York and Tabor Rds., Phila. 19141.*

STEINER, Erich Ernst, botanist; b. Thun, Switzerland, Apr. 9, 1919; s. Gotthold and Emmy (Schmid) S.; B.S., U. Mich., 1940; postgrad., U. Va., 1940-42; Ph.D., Ind. U., 1950; m. Dorothy A. White, July 8, 1944; children—Kurt, Karl, Kim. Faculty, U. Mich., Ann Arbor, 1950—; prof., 1961—. Mem. Bot. Soc. Am., Genetics Soc. Am., Soc. Am. Naturalists, Soc. for Study Evolution, Sigma Xi. Author: (with Sussman, Wagner), Botany Laboratory Manual, 1957, rev., 1965; also articles. Discovered self-incompatibility in true-breeding hybrids of evening primrose; research on evolution in evening primroses. Home: 1309 Henry St., Ann Arbor, Mich. 48104.*

STEINER, Jakob, Swiss mathematician; b. Utzenstorf, Switzerland, Mar. 18, 1796; studied under Pestalozzi; student Heidelberg, Germany, Berlin, Germany. Became prof. geometry U. Berlin, 1834. Author: Systematische Entwicklung der Abhängigkeit geometrischer Gestalten voneinander (introduced "geometrical forms of synthetic geometry), 1832; Die geometrischen Constructionen ausgeführt mittelst der geraden Linie und eines festen Kreises, 1833; Vorlesungen über synthetische Geometrie, 1867; Allg. Theorie uber das Berühren und Schneiden der Kreise und der Kugeln, 1931. Mem. Prussian Acad. Scis., French Acad. Scis., 1854. Introduced principle of duality, 1832; studied principles of inversion (later called transformation by reciprocal radii), geometric theory of cubic surfaces, properties of a surface of 4th order. Died Bern, Switzerland, Apr. 1, 1863.

STEINER, Kurt, polit. scientist; b. Vienna, Austria, June 1, 1912; s. Jacob and Olga (Weil) S.; D.Jurisprudence, U. Vienna, 1935; Ph.D. in Polit. Sci., Stanford, 1955; m. Josepha, Aug. 26, 1939. Came to U. S., 1938, naturalized, 1944. Jud. asst., later atty., Vienna, 1935-38; successively instr., asst. dir., then dir. Berlitz Sch. Langs., Cleve. and Pitts., 1939-43; prosecutor, spl. asst. to chief counsel, prosecution sect. Internat. Mil. Tribunal Far East, Tokyo, Japan, 1948; pros. atty., legal sect. SCAP, Tokyo, 1948-49, legislative atty., later chief civil affairs and civil liberties br., legislation and justice div., legal sect., 1949-51; vis. research scholar Center Internat. Studies, Princeton, 1954-55; mem. faculty Stanford, 1955—, prof. polit. sci., 1962—, asso. exec. head dept., 1959-63, chmn. com. E. Asian Studies, 1964-65, dir. Stanford in Germany, summer 1961, dir. Stanford Center Japanese Studies, Tokyo, 1962. Ofcl. escort Japanese Legal Edn. Mission to U. S., 1950-51. Ford fellow, 1953-54. Mem. Am. Soc. Internat. Law (adv. group Japan 1962-64), Am. Polit. Sci. Assn., Assn. Asian Studies, Internat. House Japan, Am. Assn. U. Profs., Japanese Am. Soc. Legal Studies (councillor 1965—). Author: Local Government in Japan, 1965; also articles, chpts. in books. Studies of Japanese politics, especially on local level, Japanese law, especially family law, using a combined legal-sociol. approach; comparative politics. Home: 832 Sonoma Terrace, Stanford, Cal. 94305.*

STEINER, Maximilian, plant physiologist; b. Vienna, Apr. 29, 1904; s. Franz and Marianne (Brenner) S; Ph.D., U. Vienna; m. Hedwig Schumacher, Sept. 28, 1936; children—Waltraud, Maria, Katharina, Christine, Cäcilia. Lectr., asst. U. Vienna; asst. Stuttgart (Germany) Tech. Coll.; asst. U. Göttingen (Germany); named prof., Stuttgart, 1935, prof. at large, Göttingen, 1940; became asso. prof. Bonn, 1948, prof., 1953, also dir. Pharmacognostic Inst. Mem. Nat. Hist. Soc. Moselle, Naturalists' Union Rhineland and Westphalia (1st pres.). Author: Methoden der Fermentforschung; Moderne Methoden der Pflanzenphysiologie. Research and publs. on lichenology, plant ecology, chem. plant physiology, including nitrogen and fat metabolism, secondary plant substances. Home: Riesstrasse 22, Bonn, West Germany.

STEINER, Paul E(by), Am. pathologist; b. Columbus Grove, O., Oct. 9, 1902; s. Menno Simon and Clara (Eby) S.; A.B., Coll. City Detroit, 1926; M.S., Northwestern U., 1930, M.D., 1932; Ph.D., U. Chgo., 1933. Fellow in medicine NRC, 1932-33; faculty pathology U. Chgo., 1934-58, prof., 1948-58; prof. pathology U. Pa. Sch. Medicine, Phila., 1959—. Vice pres. Internat. Union Against Cancer, 1954-58; mem. Nat. Adv. Cancer Council, 1949-53; vis. scientist Nat. Cancer Inst., 1953; Thomas Dent Mutter lectr. Coll. Physicians Phila., 1960; Askanazy lectr. Soc. for Geog. Pathology, Milan, Italy, 1963. Recipient Howard Taylor Ricketts prize, 1934. Mem. Am. Assn. for Cancer Research (pres. 1951), Am. Assn. Pathologists and Bacteriologists, Am. Soc. for Exptl. Pathology, Société Belge de Médicine Tropicale. Author: Cancer: Race and Geography, 1954. Research and publs. on cancer and cancer producing substances, cirrhosis, anniotic fluid embolism, ulcerative colitis, lymphogranulomatosis, other diseases. Home: 2201 The Parkway, Phila. 19130.*

STEINER, Robert Emil, radiologist; b. Prague, Czechoslovakia, Jan. 2, 1918; s. Rudolf Max and Clary (Nordlinger) S.; student U. Vienna, 1935-38; M.B., Ch.B., B.A.O., U. Dublin, 1940, M.D., 1957; Diploma in Med. Radiology, U. London, 1945; m. Gertrud Konirsch, Mar. 17, 1945; children—Hilary Clare, Ann Elizabeth. Staff, Macclesfield Gen. Infirmary, 1941, Emergency Med. Service Hosp., Winwick, Eng., 1941-43, United Sheffield Hosps., 1943-50; dep. dir. X-ray diagnostic dept. Hammersmith Hosp., 1950-57, dir., 1957—; faculty Postgrad. Med. Sch. London, 1957—, prof. diagnostic radiology, 1960—. Fellow Faculty Radiologists, Royal Coll. Physicians, Am. Coll. Radiology; mem. Brit. Inst. Radiologists, Royal Soc. Medicine, Brit. Cardiac Soc. Author: (with R. Daley, J. F. Goodwin) Clinical Disorders of Pulmonary Circulation, 1960; also numerous articles. Editor: Brit. Jour. Radiology, 1961-65. Research on cardiovascular radiology and pulmonary circulation. Home: 12 Stonehill Rd., London S.W. 14, Eng.*

STEINER, Robert Frank, Am. biochemist; b. Manila, P.I., Sept. 29, 1926; s. Frank and Clara (Weems) S.; A.B., Princeton, 1947; Ph.D., Harvard, 1950; m. Ethel Mae Fisher, Nov. 3, 1956; children—Victoria, Laura. Jewett postdoctoral fellow Naval Med. Research Inst., 1950-51, staff biochemist, 1951—, head biol. macromolecules br., 1961-65, chief lab. phys. biochemistry, 1965—; lectr. Georgetown U., 1957-58, Howard U., 1958-61. Mem. Bd. Civil Service Examiners, Potomac River Naval Comd., 1957—. Recipient Superior Civilian Service award Navy Dept., 1966. Fellow Washington Acad. Sci.; mem. Soc. Biol. Chemists, Am. Chem. Soc., Biophys. Soc., N.Y., Md. acads. sci. Author: (with R. Beers) Polynucleotides, 1961; The Chemical Foundations of Molecular Biology, 1965; Molecules and Life, 1965; also many articles. Research on size, shape and structure of proteins and nucleic acids; developed methods for analyzing protein assns., fluorescence techniques for studying protein structure; observed organized structure in biosynthetic nucleic acids. Home: 8804 Tallyho Trail, Potomac, Md. 20854. Office: Naval Med. Research Inst., Bethesda, Md. 20014.*

STEINER, William Glenn, Am. psychologist; b. Peoria, Ill., Aug. 15, 1926; s. Glenn Victor and Verna (Johnson) S.; student N. Central Coll., 1944-45; A.B., Aurora Coll., 1948; postgrad. Northwestern U.; A.M., U. Ill., 1958, Ph.D., 1960; m. Anna Jean Ronchetti, Mar. 15, 1952; children—William Glenn, Brian Douglas, Renee. Clin. psychologist Kankakee (Ill.) State Hosp., 1959-60; med. research asso. Thudichum Psychiat. Research Lab., Galesburg (Ill.) State Research Hosp., 1960-64; research asso. Yale, 1964-65; asso. prof. U. Tenn., 1965-67; prof. psychology Bradley U., Peoria, Ill., 1967—; cons. VA, 1966—. Mem. examination com. Am. Assn. State Psychology Bds., 1966—. Mem. Am. Psychol. Assn., Am. Physiol. Soc., Am. Sociol. Assn., Psychonomic Soc., A.A.A.S., Sigma Xi. Research, publs. on physiol. correlates of disordered behaviour with research activities closely allied to exptl. neurologist and biol. psychiatrist; worked on problems relating to convulsant brain phenomena, physiology of tranquilizing drugs, site of action of psychotomimetic chems., immunoneurophysiol. attempts to destroy brain tissue; currently researching physiology of mental retardation and alcohol addiction. Office: Dept. Psychology, Bradley U., Peoria, Ill.

STEINETZ, Bernard George, Jr., Am. endocrinologist; b. Germantown, Pa., May 30, 1927; s. Bernard George and Hazel (Jefferds) S:; A.B., Princeton, 1950; Ph.D., Rutgers U., 1954; m. Jane Rutledge Nash, June 17, 1949; children—Carl Nash, Scott Jefferds, Ann Rutledge. Teaching asst. Rutgers U., 1951-54; sr. scientist dept. physiology Warner-Chilcott Labs., Morris Plains, N.J., 1954-63; sr. research asso. Warner-Lambert Research Inst., Morris Plains, 1963-67; head reproductive physiology CIBA Pharmaceutical Company, Summit, New Jersey, 1967—. Fellow N.Y. Acad. Scis.; mem. Endocrine Soc., Am. Physiol. Soc., Am. Soc. Zoologists, Reticuloendothelial Soc., Microcirculation Soc. Contbg. author: Recent Progress in Endocrinology of Reproduction, 1959; Methods in Hormone Research, 1962. Research and numerous publs. in important role of ovarian hormone, relaxin, in normal delivery process in mammals, studies showed that relaxin acts in concert with estrogenic and progestational hormones in regulating growth and biochem. and enzymatic processes in uterus, cervix and pubic symphysis. Home: 81 Lake Dr., Mountain Lakes, N.J. 07046. Office: 201 Tabor Rd., Morris Plains, N.J. 07950.*

STEINGRIMUR, Jonsson, Icelandic elec. engr.; b. Gaulverjabae, Iceland, June 18, 1890; s. Jon Steingrimsson and Sigridur Jönsdóttir; E.E. with honors, Tech. Coll. Denmark, 1917; m. Laura Margret Arnadottir, Oct. 13, 1918; children—Gudrun Sigridur, Sigridur Oloef, Thora, Jon, Arndis. With J. L. la Cour, Cons. Engr., 1917-20; insp. Reykjavik Muncipal Elec. Works, 1920-21, dir., 1921-61; dir. Sog River Power Devel., 1937-66. Cons. to Iceland municipalities; mem. numerous tech. coms. Mem. Iceland Elec. Engring. Soc., Civil Engring. Soc., Assn. Iceland Elec. Power Works, Iceland Light Tech. Soc., numerous others. Contbr. articles to sci. jours., profl. publs. Pioneer in problems of distbrn. elec. power in Iceland. Home: 73 Laufasveg. Office: Hafnarhus, Tryggvagata, Reykjavik, Iceland.*

STEINHARDT, Ralph Gustav, Jr., Am. chemist; b. Newark, Sept. 15, 1918; s. Ralph Gustav and Norma (Stein) S.; B.S. in Chemistry, Lehigh U., 1940, M.S., 1941, Ph.D., 1950; B.S. in Chem. Engring., Va. Poly. Inst., 1944; m. Mary Etzler Hawks, Sept. 11, 1946; children—Theresa Ruth, Sara Jean, Ralph Gustav, III. Fellow, Lehigh U., 1940-42, 46-50, research asso., 1950-53, asst. prof., 1953-54; staff Los Alamos Lab., 1944-46; asso. prof. Va. Poly. Inst., 1954-56; prof. Hollins (Va.) Coll., 1956—, chmn. dept., 1956-63, 66—, dir. undergrad. research program, 1959-61; vis. research prof. U. Wis., 1960; vis. lectr. chemistry Stanford, 1963-64. Cons. Aerospace Research Corp., 1965—. NSF Sci. Faculty fellow, 1958-60. Fellow A.A.A.S., Am. Inst. Chemists; mem. Am. Chem. Soc., Va. Acad. Sci. (chmn. chemistry sect. 1962-63), Sigma Xi, Phi Lambda Upsilon. Contbr. articles to tech. jours. Developed first analytical X-ray photoelectron spectrometer. Home: Faculty Lane 3, Hollins Coll., Hollins College, Va. 24020.*

STEINHART, John Shannon, Am. geophysicist; b. Chgo., June 3, 1929; s. John E. and Jessie (Shannon) S.; A.B., Harvard, 1951; Ph.D., U. Wis., 1960; m. Carol A. Elder, Dec. 20, 1958; 1 dau., Gail Shannon. Project leader Woods Hole (Mass.) Oceanographic Inst., 1956; research asso. U. Wis., 1957-60; fellow dept. terrestrial magnetism Carnegie Instn. Washington, 1960-61, staff mem., 1961—. Mem. sci. adv. bd. Air Force Office Sci. Research, 1962—; sci. adv. bd. NSF, 1964—. NSF predoctoral fellow, 1959-60, postdoctoral fellow, 1960-61. Fellow Royal Astron. Soc.; mem. Am. Geophys. Union, Seismol. Soc. Am. (chmn. Eastern sect. 1965-66), Soc. Exploration Geophysicists (scholar 1957-59), A.A.A.S., Sigma Xi. Author: (with R. P. Meyer) Explosion Studies of Continental Structure, 1961; also articles. Pioneer in study of structure of earth's crust from controlled explosions on continents, heat flowing out of earth, using a technique of measuring in bottom of deep lakes. Home: 9419 Overlea Dr., Rockville, Md. 20850. Office: 5241 Broadbranch Rd. N.W., Washington 20115.*

STEINHAUS, Edward A(rthur), Am. invertebrate pathologist; b. Max, N.D., Nov. 7, 1914; s. Arthur Alfred and Alice (Rhinehart) S.; B.S., N.D. State U., 1936, Sc.D., 1962; Ph.D., Ohio State U., 1939; m. Mabry Clark, June 14, 1940; children—Margaret Ann, Timothy Clark, Cynthia Alice. Asst. bacteriologist Rocky Mountain Lab., USPHS, 1940-41, asso. bacteriologist, 1942-44; faculty U. Cal., Berkeley, Irvine, 1944—, prof., dean biol. scis., Irvine, 1963—; also chmn. dept. insect pathology, Cons. Pacific Sci. Bd., 1951—, NRC, 1965—, USPHS, 1956-64, other orgns.; vis. prof. U. Wash., 1955, U. Wis., 1961; mem. Internat. Com. on Comparative Pathology, 1965—; Griswold lectr. Cornell U., 1959. Guggenheim fellow, 1960-61. Fellow A.A.A.S., Entomol. Soc. Am. (pres. 1962-63, Founders Meml. award 1959), Am. Acad. Microbiol., Soc. for Invertebrate Pathology (pres. 1967-68); mem. Nat. Acad. Scis., Soc. Parasitology, Soc. Sci. and Tech. (India, hon.), Council of Biology Editors, Soc. Exptl. Pathology, Soc. Tropical Medicine and Hygiene, Soc. Evolution, Internat. Congress Insect Pathology (exec. com. 1962-66), Soc. Gen. Microbiol., Microscop. Soc., Pub. Health Assn., Am. Soc. Microbiology, Am. Inst. Biol. Scis. (governing bd. 1966), Sigma Xi. Author: Insect Microbiology, 1946; Principles of Insect Pathology, 1949. Editor: Insect Pathology, An Advanced Treatise, 1963. Jour. Invertebrate Pathology, 1959—, Ann. Rev. Entomology, 1955-63; also past mem. editorial bds. numerous sci. jours. Publs. on diseases of insects and other invertebrates and use of this knowledge in control of pests; suppression of disease in beneficial invertebrates. Home: 2415 Blackthorn St., Newport Beach, Cal. 92660.*

STEINHAUS, John Edward, Am. physician; b. Omaha, Feb. 23, 1917; s. Emil F. and Pearl (Haynie) S.; B.A., U. Neb., 1940, M.A., 1941; M.D., U. Wis., 1945, Ph.D., 1950; m. Mila Jean Pinkerton, Feb. 21, 1943; children—Kathryn, Carolyn, Barbara, William, Elizabeth. Practice medicine, specializing in anesthesiology, Madison, Wis., 1951-58, Atlanta, 1958—; faculty U. Wis., 1951-58; faculty Emory U., 1958—, prof., chmn. anesthesiology, 1959—; chief anesthesiology service Grady Meml. Hosp., 1959—. Diplomate Am. Bd. Anesthesiologists. Mem. Am. (dir., ist v.p.), So. (past pres.) socs. anesthesiologists, A.M.A., A.A.A.S., Assn. U. Anesthesiologists, Soc. Pharm. Exptl. Therapeutics, Phi Beta Kappa, Sigma Psi, Alpha Omega Alpha. Studies and numerous publs. on mechanism of toxic reactions of drugs used in treatment of cardiac irregularities; pain and analgesic drugs used to treat pain. Home: 836 Castle Falls Dr., Atlanta 30329. Office: 80 Butler St., S.E., Atlanta 30303.*

STEINHAUSER, Ferdinand Ottomar, Austrian meteorologist; b. Schrattenthal, Austria, May 4, 1905; s. Ferdinand and Maria (Teimel) S.; Dr.Phil., U. Vienna, 1933; m. Margareta Krumpholtz, Feb. 10, 1935; 1 son, Peter. Faculty, U. Vienna, 1929—, prof. meteorology and geophysics, 1953—, dir. Zentralanstalt für Meteorologie 1953—. Permanent Austrian rep. WMO. Mem. Austrian, N.Y. acads. scis., Czechoslovak Meteorol. Soc. Author: Meteorologie des Sonnblicks, 1938; Klima und Bioklima von Wien, 3 vols., 1955-59; (with O. Eckel, F. Lauscher) Klimatographie von Osterreich, 2 vols., 1958-60. Research on evaluation of elastic deformation of earth's crust by local load, meteorol. and climatol. conditions of Alps, distbrn. of turbidity of air, variations of solar radiation, influence of atomic bomb trials on weather, bioclimatology. Home: 38 Hohe Warte, Vienna, Austria.*

STEINHÜBEL, Gejza, Czechoslovakian plant physiologist; b. Banská Bystrica, Czechoslovakia, Oct. 6, 1922; s. Gejza and Mária (Stollmannová) S.; RNDr., Komensky U., Bratislava, Czechoslovakia, 1949; m. Vlasta Zoulova, Feb. 6, 1954; 1 son, Gejza. Asst., Inst. Plant Physiology, Komensky U., 1946-51; scientist Slovak Acad. Scis., Bratislava, at Mlynany Arboretum Slovak Acad. Scis., Vieska nad Zit., Czechoslovakia, 1951—; faculty Agronomical U. in Nitra, 1951—, docent, 1964—. Mem. Czechoslovakian Bot. Soc., Czechoslovakian Soc. History Scis. and Technic. Author: Arboretum Mlynany in Past and Now, 1957; Problem of Shadow in Holly and Cherry-Laurel, 1961; Introduction to the Ecophysiology of Evergreens, 1967; also articles. Research on longevity of evergreen plant leaf, solid air-pollutants and their influence on physiol. processes in plant, inhibition of seed prodn. in tulip tree. Home and office: 175 Arboretum SAV, Vieska n. Z., Czechoslovakia.*

STEINMAN, David Bernard, Am. bridge engr.; b. N.Y.C., June 11, 1886; s. Louis Kelvin and Eva (Scollard) S.; B.S., summa cum laude, Coll. City of N.Y., 1906, D.Sc., 1947; C.E., Columbia, 1909, A.M., 1909, Ph.D., 1911, D.Sc., 1953; E.D., Manhattan Coll., also Rensselaer Poly. Inst., 1953, Mich. Coll. Mining and Tech., 1954, U. Mich., 1956; D.Sc., Ohio Northern U., Sequoia U., U. Ghent, Minerva Academy Advanced Studies (Italy), 1953, Haute Academie Latine Internationale (France), 1953; Bradley Univ., 1956; D.C.E., Univ. of Bologna (Italy), 1953; Dr. Higher Learning (Vidya Ratna), Shri Bhuvaneshwari Pith, Kathiawad, India, 1953; LL.D., Alfred U., 1953; m. Irene Hoffman, June 9, 1915; children—John Francis, Alberta, David. Engring. work until 1910; prof. civil engring. U. Ida., also practiced as cons., 1910-14; spl. asst. to Gustav Lindenthal on design and constrn. important bridges, 1914-17; prof. in charge civil and mech. engring. Coll. City of N.Y., 1917-20; in practice as cons., 1920—; designing or cons. engr. numerous notable bridges in U. S. and on four other continents, 1922-55, Kingston Bridge, 1952-56, Mackinac Bridge, 1953-75, also Sky-Ride and Observation Towers, Century of Progress Expn., Chgo., 1933, mil. bridges for U. S. govt., 1941-53; engaged in reconstrn. Bklyn. Bridge, 1948-54; v.p. Tioga-Nichols Bridge Co., Smithboro Bridge Co.; dir. Independence Bridge Co., Interboro Bridge Co., Richmond-Hopewell Bridge Co.; pres. Pan-Am. Pub. Works, Inc. Recipient many honors, awards, prizes, U. S. and abroad, among the latest: Chevalier Legion of Honor, Grand Cordon de l'Etoile du Bien et du Merite, Chevalier Ordre du Merite Scientifique, Croix d'Honneur Legion Franco-Belge, Knight Comdr. (with star) Order of Gold Cross of Mil. Chapter of Cyprus and Jerusalem, William Procter prize Sci. Research Soc. Am., Norman medal (second time), Croes medal, Rowland prize Am. Soc. C.E. Founder, pres. David B. Steinman Found. Vice pres. bd. trustees Ecole des Hautes Etudes, N.Y.C. Registered profl. engr. 20 states and fgn. countries. Fellow or mem. Am. and fgn. profl. and sci. socs. and assns., including Royal Soc. Arts, Soc. des Ingenieurs Professionels de France. Author: How Bridges Have Increased Man's Mobility, 1952; Famous Bridges of the World, 1953; Mackinac Straits Bridge, 1954; Suspension Bridges: The Aerodynamic Problem and Its Solution, 1954; Miracle Bridge at Mackinac, 1957; Bridges and Their Builders, (rev. edit.), 1957; numerous articles. Designed more than 400 bridges; co-developer new type of cable and stiffening; solved aerodynamic problems of suspension bridges. Died N.Y.C., Aug. 21, 1960.

STEINMANN, Bernhard Friedrich, Swiss physician; b. Berne, Switzerland, May 9, 1908; s. Fritz and Elisabeth (Mauerhofer) S.; student U. Geneva (Switzerland), 1927-30, U. Vienna (Austria), 1931; M.D., U. Berne, 1934; children (by previous marriage)—Ursula (Mrs. Feitknecht), Veronika (Mrs. Gonin), Matthias; m. 2d, Trudy Zeller, July 26, 1958; 1 dau., Barbara. Faculty, U. Berne, 1943—, prof., 1957—; chief asst. Med. Clinic Berne, 1943-46; practice medicine, Berne, 1946—; chief C.L. Loryhaus of Inselspital Berne, 1947—. Mem. Austrian Geriatric Soc. (hon.), German Gerontol. Soc. (corr.), Swiss Soc. Internists, Swiss Soc. Cardiology, Swiss Soc. Angiology, Swiss Soc. Gerontology, Swiss Commn. Rehab. Author: Herz beim Scharlach, 1945; (with P. Imhof) Behandlung der Hemiplegie, 1955; Pflege der Betagten und Chronischkranken, 1962; also numerous articles. Research on effect of high blood pressure on circulatory mechanism; rehab. in hemiplegia and paraplegia; epidemiology of apoplexy; studies of blood pressure using interarterial blood pressure measurements. Home: Ittigenstr. 6, 3063 Ittigen (BE) Switzerland. Office: C.L. Loryhaus-Inselspital, Berne, Switzerland.*

STEINMANN, Fritz, Swiss surgeon; b. Bern, Switzerland, 1872; became prof. surgery, Bern, 1899. Invented surg. nail or pin to insert in distal fragment of fractured bone as a hold for skeletal traction (Steinmann's pin or nail), 1907. Died 1932.

STEINMANN, Gustav, German geologist, prehistorian; b. Braunschweig, Germany, Apr. 9, 1856; prof., Jena and Bonn, Germany; made research expdns. to S. Am. Author: Einfuhrung in die Paläontologie, 1903; Die geologischen Grundlagen der Abstammungslehre, 1908; Die Eiszeit und der vorgeschichtl. Mensch, 1910; Geologie von Peru, 1929. Died Bonn., Oct. 7, 1929.

STEINMAURER, Rudolf, Austrian physicist; b. Wels/H.A., Germany, Mar. 11, 1903; s. Ignaz and Marie (Grief) S.; ed. U. Vienna, U. Graz; Ph.D.; agrégé exptl. physics, Innsbruck, Austria, 1935; m. Luise Mudrak, Dec. 28, 1935; children—Helmut, Erwin. Asst., U. Graz, U. Innsbruck; became asso. prof. U. Innsbruck, 1949, prof., dir. Physics Inst., 1955—. Mem. Internat. Assn. U. Profs. and Lectrs., Austrian Acad. Sci. (corr.), Austrian Physics Soc. Research and publs. on high energy physics, radiation. Home: Erzherzog-Eugenstrasse 15, Innsbruck, Austria.

STEINMAYER, Reinhard August, Am. geologist; b. La Salle, Ill., Feb. 14, 1892; s. Christian M. and Katherine (Feurer) S.; B.S., U. Ill. 1916; postgrad. U. Chgo., 1924; m. Dorothy Gibbs, Dec. 23, 1918; children—Reinhard August (dec.), Nadyne (Mrs. S. K. Manson), Katherine (Mrs. M. J. McLean). Engr., St. L.-S.F. Ry., 1916-17, ICC, 1917-18; geologist Roxana Petroleum Corp., 1918-19, Atlantic Oil Refining Co., 1919-20; appraiser C., B. & Q. R.R., C., R.I. & P. Ry., 1920-22; faculty Tulane U., New Orleans, 1922-57, prof., head dept. geology, 1922-57; cons. geologist, geol. expert witness, 1924—. Mem. Am. Assn. Petroleum Geologists, La. Engring. Soc., New Orleans Acad. Scis., Sigma Xi. Contbr. articles to profl. jours. Research on phases of sedimentation; bottom sediments of lakes; salt domes; engineering and petroleum geology. Address: 260 Homestead Av., Metairie, La. 70005.

STEINMETZ, Charles Henry, Am. physician; b. Logansport, Ind., Oct. 5, 1929; s. Henry G. and Amelia (Flaitz) S.; A.B., Ind. U., 1950, Ph.D., 1953; M.D., U. Cin., 1960; m. Phyllis Miller, Dec. 31, 1950; children—Marc A., Curtis G., Thomas M., Christopher A. Lab. asst., teaching fellow Ind. U., 1950-53; chief space biology USAF Aero Med. Field Lab., Holloman AFB, N.M., 1953-56; head ecology and controls sect. life scis. Aerojet Gen. Co., 1961-62; corporate dir. med. research and life scis. N.Am. Aviation, Inc., El Segundo, Cal., 1962-68; dir. health systems Litton Industries, Bethesda, Md., 1968—. Mem. Med. Research Assn. Cal. (dir., v.p. 1968), Aerospace Indsl. Life Scis. Assn. (pres. 1967-68), Phi Beta Kappa, Sigma Xi, Alpha Omega Alpha, Pi Kappa Epsilon. Research and publs. on function of thyroid gland, man in space, health systems. Office: 7300 Pearl St., Bethesda, Md. 20014.*

STEINMETZ, Charles Proteus, elec. engr., b. Breslau, Germany, Apr. 9, 1865; s. Carl Heinrich and Caroline (Neubert) S.; ed. Breslau, Berlin, Zürich, Switzerland; A.M., Harvard, 1902; Ph.D., Union U., N.Y., 1903. Came to U.S., 1889. Cons. engr. Gen. Electric Co., 1893-1923; prof. electrophysics Union Univ., 1902-23. Pres. Am. Inst. E.E., 1901-02, also Illuminating Engring., Soc., Nat. Assn. Corp. Schs. Author: Theory and Calculation of Alternating-Current Phenomena, 1897, 5th edition, 1916; Theoretical Elements of Electrical Engineering, 4th edit., 1915; Theory and Calculation of Transient Electric Phenomena and Oscillations, 1909, 3d edit., 1919; General lectures on Electrical Engineering, 5th edit., 1917; Radiation, Light and Illumination, 1909, 2d edit., 1911; Engineering Mathematics, 1910, 3d edit., 1917; Electric Discharges, Waves and Impulses, 1911; America and the New Epoch, 1916; Theory and Calculation of Electric Circuits, 1917; Theory and Calculation of Electrical Apparatus, 1917. Also math. papers and investigations and numerous papers on theoretical exptl. investigations in elec. engring. Established basic laws of magnetic hysteresis, 1891; 1st to use complex notation for alternating current problems, 1893; contbd. to electric machine design, electric traction, lightning protection, street lightning; patentee over 100 inventions; inventor improved generators, motors; research on theory and calculation of alternating current phenomena; devised lightning arresters for high power transmission lines. Died Oct. 26, 1923.

STEINMETZ, Sebald Rudolf, Dutch sociologist; b. Breda, Netherlands, Dec. 6, 1862; prof., Amsterdam, Netherlands. Author: Ethnologische Studien zur ersten Entwicklung der Strafe, 2 vols., 1894; Soziologie des Krieges, 1929; Ges. Kleinere Schriften, 3 vols., 1933-35. Made empirical sociol. studies with emphasis on ethnology. Died 1940.

STEINTHAL, Heymann, German linguist; b. Groebzig, Germany, 1823; studied Chinese lang. and lit. in Paris; became asso. prof. gen. linguistics, Berlin, 1863. Author: Précis de linguistique, 1850-71; L'origine du langage, 1851; Histoire de la linguistique chez les Grecs et les Romains, 1863; Les langues des Nègres Mandé, aux points de vue psychologique et phonétique, 1867. Follower of W. von Humboldt; emphasized metaphysics of language. Died Berlin, 1899.

STEJNEGER, Leonhard, naturalist; b. Bergen, Norway, Oct. 30, 1851; s. P. Stamer and Ingeborg C. (Hess) S.; grad. R. Frederic's U., Christiania, 1875 (Cand. jur.), Dr. Philosophy, hon. causa, 1930; m. Marie Reiners, Mar. 22, 1892; 1 dau., Inga. Came to U. S., 1881; on a natural history expdn. to Bering Island and Kamchatka, 1882-83, collecting for U. S. Nat. Museum, asst. curator of birds, 1884-89, curator reptiles since 1889, head curator of biology since 1911. Revisited Commander Islands, 1895, for Fish Commn. to study fur-seal question, 1896-97, as mem. U. S. Fur Seal Commn., and again in 1922 for Dept. of Commerce. Del. from Smithsonian Instn. to Zoöl, Congress 7 times, 1901-35, to Internat. Ornithologists Congress, 1905. Recipient Walker Grand prize Boston Soc. Natural History. Fellow Am. Ornithologists Union, A.A.A.S.; mem. Bergen Mus. (life), Nat. Acad. Scis., acads. scis. Christiania and Washington; fgn. mem. Zool. Soc., London, Ornithol. Soc. Bavaria, Acad. Natural Scis. Phila. Biol. Soc. Washington (pres. 1907, 08), Am. Soc. Ichthyol. and Herpetol. (pres.), Internat. Zool. Congress. Assn. Am. Geographers, Sigma Xi; hon. mem. other profl. socs. Author: Norsk Ornitologisk Ekskursjonsfauna, 1873; Norsk Mastozoologisk Ekskursjonsfauna, 1874; Results of Ornithological Explorations in the Commander Islands and in Kamchatka, 1885; Standard Natural History, Vol. IV, Birds (greater part), 1885; Report of the Rookeries of the Commander Islands, Season of 1897, 1897; The Asiatic Fur-Seal Islands and Fur-Seal Industry, 1898; The Relations of Norway and Sweden, 1900; The Herpetology of Porto Rico, 1904; The Herpetology of Japan and Adjacent Territory, 1907; The Origin of the So-called Atlantic Animals and Plants of Western Norway, 1907; Georg Wilhelm Steller, pioneer of Alaskan Natural History, 1936, also many monographs and contbns. on zool. subjects. Died Feb. 28, 1943.

STEKEL, Wilhem, Austrian psychiatrist; b. Bojan, Bucovinia, Mar. 18, 1868; studied under Krafft-Ebing, Vienna. Practiced medicine, specializing in nerve diseases, Vienna; asst. to Freud; founder jealousy clinic, 1935. Author: Dichtung und Neurose, 1909; Die Sprache des Traumes, 1910; Der Wille zum Schlaf, 1916; Störungendes Trieb- un Affektlebens, 10 vols., 1932; Der Seelenarzt, 1933; Fortschritte der Traumdeutung, 1935; Techniques of Analytical Psychotherapy, 1940. Psychoanalyzed over 10,000 people; studied techniques of psychoanalytic psychotherapy; emphasized role of therapist as teacher. Died London, Eng., June 21, 1940.

STELCK, C. R., Canadian geologist; b. Edmonton, Alta., Can., May 20, 1917; s. Robert Ferdinand and Florella (Stanbury) S.; B.Sc., U. Alta., 1937, M.Sc., 1941; Ph.D., Stanford, 1950; m. Frances Gertrude MacDowell, Apr. 24, 1945; children—David Richard, Brian Francis, Leland Bruce, John Warren. Well site geologist B.C. Dept. Lands, 1940-42; field geologist Imperial Oil Co., Alta., B.C., 1942-47; faculty U. Alta., Edmonton, 1947—, prof. geology. Registered profl. geologist, Alta. Fellow Royal Soc. Can.; mem. Geol. Soc. Am., Paleontol. Soc., Edmonton Geol. Soc., Geol. Assn. Can., others. Research, publs. on integration of Western Canadian stratigraphy. Home: 11739 91 Av., Edmonton, Alta., Can.*

STELLA, Erasmus, physician; b. Germany, circa 1490; ed. Leipzig, Germany, Bologna, Italy; apptd. town physician, Zwiekau (Stueler), Germany, 1501; later burgomeister, Zwickau. Author: Interpraetamenti gemmarum libellus, 1517. Grouped minerals according to color rather than alphabetically; described several minerals, including varieties and localities. Died 1521.

STELLER, George Wilhelm, zoologist; b. Windsheim, Germany, Mar. 10, 1709; s. Cantor and (Jacob) S.; student theology U. Willenberg, natural history U. Halle; m. circa 1738. Tchr. natural history, Halle, Germany; joined Berings expdn. at St. Petersburg; traveled across Russia collecting specimens, 3 years; sailed for Am. from Okhotsk to Alaska; explored various parts of N. Am.; returned to Russia. Mem. French Acad. Scis. Described Arctic fauna; accounts of expdn. with Bering in Asia and Alaska; described birds and animals, including Steller's sea cow (now extinct), Stellar's sea lion, fur seal colony; 1st to write on Alaskan natural history. Died Tjumen, Siberia, Nov. 12, 1746.

STELLMACHER, Karl Ludwig, mathematician; b. Brandenburg, Germany, Mar. 24, 1909; s. Peter E.W. and Marie (Stichert) S.; student U. Halle (Germany), 1927-28, U. Kiel (Germany), 1928; Ph.D., U. Göttingen (Germany), 1936; m. Ellinor Schrieber, Mar. 6, 1938; children—Inga, Irene.

Asst., Inst. for Applied Mechanics, Göttingen, 1936-48; docent mathematisches institut U. Göttingen, 1948-54, prof., 1954-56; prof. math. U. Md., 1956—. Research and publs. on math. found. of gen. relativity, stability theory, theory of hyperbolic differential operators. Home: 9828 Hedin Dr., Silver Spring, Md. 20903. Office: Dept. Math., U. Md., College Park, Md.*

STELLUTI, Francesco, Italian naturalist, poet; b. Fabriano, Italy, 1577; mem. Accademei de Lincei. Author: Il Parnasso, 1631; also 1st systematic observations with microscope to be published, 1625. Studied honey bees. Died after 1651.

STELSON, Paul Hugh, Am. physicist; b. Ames, Ia., Apr. 9, 1927; s. Hugh Eugene and Ada (Woolley) S.; B.S., Purdue U., 1947, M.S., 1948; Ph.D., Mass. Inst. Tech., 1950; m. Helen Campbell MacLachlan, Dec. 18, 1950; children—Hugh Carrier, James MacLachlan, Fred Woolley, Ben Howard. Research asso. Mass. Inst. Tech., Cambridge, 1950-52; physicist Oak Ridge Nat. Lab., 1952-62, sr. physicist, 1963—; vis. prof. physics Rice U., Houston, 1962-63. Research in application of Van de Graaff accelerators to study of properties of nuclei, especially inelastic scattering of neutrons by nuclei, excitation of nuclear energy levels by Coulomb excitation mechanism. Home: 886 W. Outer Dr., Office: Oak Ridge Nat. Lab., Oak Ridge.*

STELSON, Thomas E., Am. civil engr.; b. Iowa City, Aug. 24, 1928; s. Hugh Eugene and Ada (Woolley) S.; B.S., Carnegie Inst. Tech., 1949, M.S., 1950, D.Sc., 1952; m. Constance Anne Semon, Mar. 24, 1951; children—Kim Adair, Thomas Semon, Arthur Wesley, Rebecca Anne. Mem. faculty Carnegie-Mellon U., Pitts., 1952—, prof. civil engring., 1960-61, Alcoa prof. engring., 1961—, acting head dept. civil engring., 1957-59, head dept., 1959—; dir. projects Pa. Gov.'s Com. for Transp., 1967-68. Mem. Am. Soc. C.E., Pa. Soc. Profl. Engrs., Am. Soc. Engring. Edn., Am. Geophys. Union, Am. Concrete Inst., Internat. Assn. for Hydraulic Research, A.A.A.S., Sigma Xi, Tau Beta Pi, Phi Kappa Phi. Research and publs. in fluid mechanics, hydraulic transport, foam mechanics, concrete fatigue. Home: 133 Elmore Rd., Pitts. 15221.*

STEMBERA, Zdenek Karel, Czechoslovakian gynecologist; b. Pilsen, Czechoslovakia, June 23, 1920; s. Antonin and Anna (Horova) S.; M.D., Charles U., Prague, Czechoslovakia, 1948, D.Sc., 1966; m. Jarmila Librova, Nov. 18, 1944; children—Daniela, Renata. Physician dept. surgery Dist. Hosp., Chrudim, Czechoslovakia, 1948-49; asst. prof. Charles U. Med. Sch., 1950-51; asso. prof., head research unit Inst. for Care of Mother and Child, Prague, 1952—. Mem. gynecol. and obstet. sect. Czechoslovak Med. Soc. J.E. Purkyne. Author: The Hypoxie Fetus, 1967; also numerous articles. Proposed new method for early diagnosis of fetal hypoxia; research on metabolic relationship between mother and healthy or hypoxic fetus; measured blood flow in umbilical vessels immediately after delivery of healthy and hypoxic human newborn. Home: 4 Nam.Machka Prague 5, Czechoslovakia.*

STENDLER, Celia Burns, Am. psychologist; b. Naugatuck, Conn., May 18, 1911; d. Thomas and Mary (Hyland) Burns; student U. So. Conn., 1928-30, Yale Extension, 1930-33; B.S., Columbia U., 1937, M.A., 1940, Ph.D., 1947; m. Charles Stendler, Dec. 30, 1941; 1 dau., Faith. Instr., Columbia U., 1944-46; faculty U. Ill., Urbana, 1946—, now prof. edn., vis. prof. psychology U. Cal. at Berkeley, 1965-66, vis. research psychologist Sci. Curriculum Improvement Study 1966-67. Cons., U. City (Mo.) Sch. Comprehensive Sch. Project, Ford. Found. Mem. Am. Psychol. Assn., A.A.A.S., Soc. for Research in Child Devel. Author: Children of Brasstown, 1950; (with William E. Martin) Child Behavior and Development, 1953; Readings in Child Behavior and Development, 1964; Macmillan Science Series, Vols. 1-9, 1959, 62, 66; also articles. Research on socialization process in children, cognitive processes involved in learning sci. and math., implications of theory of Jean Piaget in teaching of sci. to children, curriculum devel. to foster logical thinking, presch. culturally-disadvantaged children. Home: 617 W. Hessel St., Champaign, Ill. Office: Coll. Edn., U. Ill., Urbana, Ill.*

STENHOUSE, John, Scottish chemist; b. Glasgow, Scotland, Oct. 21, 1809; s. William and Elizabeth (Currie) S.; ed. Glasgow U., Anderson's Coll., U. Giessen (Germany), 1837-39; LL.D., Aberdeen (Scotland) U., 1850. Lectr. chemistry St. Bartholomew's Hosp., London, 1951-57; assayer to mint, 1865-70. Fellow Royal Soc. (Royal medal 1871), 1848, Inst. Chemistry; mem. Chem. Soc. (co-founder 1841). Contbr. alone and with C. E. Groves, articles to sci. publs. Inventor charcoal air-filters, respirators; discovered betorcinol; patentee many ingenious, useful inventions in dyeing, waterproofing, sugar mfr., tanning. Died Dec. 31, 1880.

STENIJ, Sten Einar, Finnish mechanician; b. Helsinki, Finland, Sept. 30, 1900; s. Sten Edvard and Naëmi (Granfelt) S.; Magister Philosophy, U. Helsinki, 1923, licentiate, Ph.D., 1932; m. Helena Lindgren, Oct. 16, 1932; children—Eva (Mrs. Olli Heiskanen), Helmi (Mrs. Jaakko Salonen). Asst., Inst. Marine Research, Helsinki, 1924-31, chief dept., 1931-38; prof. mechanics Tech. U. Helsinki, 1938—, dean Faculty Gen. Sci., 1954-65, rector, 1965—. Mem. Societas Scientiarum Fennica (permanent sec. 1966—). Contbr. papers on oceanography, mechanics. Home: 28 Kulosaarentie. Office: Tecnical U., Otaniemi, Helsinki, Finland.*

STENO, Nicolaus, anatomist, geologist; b. Copenhagen, Denmark, Jan. 10, 1638; student Copenhagen U., 1656-60, U. Amsterdam (Netherlands); M.D. in absentia, U. Leiden (Netherlands), 1664. Prof. anatomy, Padua, Italy; house physician to Grand Duke Ferdinand II of Tuscany; anatomist Santa Maria Nouva Hosp., Florence, Italy; royal anatomist U. Denmark, from 1672; returned to Italy, 1674; ordained priest Roman Catholic Ch., 1675; apptd. titular bishop Titiopolis, 1677; apptd. vicar-apostolic to Hanover, Germany, 1677, to Hamburg, Germany, 1683-85; apptd. aux. bishop Numster, 1680. Mem. Accademia del Cimento. Author: Disputatio de glandulis ovis, 1660; Observationes anatomicae, 1662; De musculis et glandulis, 1664; Prodromus de solido intra solidum naturaliter contento, 1669; Epistola de propia conversione, 1677. Skilled anat. dissection; described lachrymal gland, upper and inferior lachrymal ducts; discovered excretory duct of parotid gland (ductus stenonianus); explained role of parotid gland in prodn. saliva; studied structure of muscles, sinews; recognized muscles composed of fibrils; applied principle of parallelogram of forces to muscles, also studied mechanics muscular contraction; investigated nature, function heart, showing it composed primarily of muscle; of reproductive organs, explaining function ovaries, of brain; pioneered in geology, paleontology, crystallography; discussed formation, displacement, destruction, successive nature stratified sedimentary rocks Tuscany; argued that sedimentary deposits originally horizontal; observed mountain ranges may originate by volcanic action, faulting, folding of crust, erosion of highlands; recognized organic origin fossils; measured angles crystal interfaces; explained crystal growth. Died Schwerin, Mecklenburg, Germany, Dec. 6, 1686.

STENRAM, Unne Nils Hakan, Swedish physician; b. Malmo, Sweden, Dec. 29, 1926; s. Nils Edvin and Ally (Jonasson) S.; M.B., U. Lund (Sweden) 1948, M.D., 1953, Ph.D., 1956; m. Ingrid Alexandra Rothstein, Mar. 15, 1952; children—Alexandra Elisabeth, Unna Margareta, Marie Thomasine, Maja Vedina Christianna. Staff, U. Lund, 1947-62, asst. prof. pathology, 1958-62; asso. prof. U. Uppsala (Sweden), 1962—. Research and publs. on histology, ultrastructure, and biochemistry of nucleic acid and protein synthesizing structures of cell, liver damage during usage of oral contraceptives, Pathology. Home: 1 Nackrosgatan, Uppsala, Sweden.*

STENSEN, Niels, see Steno, Nicolaus.

STENSIO, Erik Anderson, Swedish palaeozoologist; b. Parish of Döderhult, Sweden, Oct. 2, 1891; s. Johan Frederik Anderson and Ottilia (Eriandsson) S.; Ph.D., U. Upsala (Sweden), 1921; Dr.Hon. causa, Sorbonne, Paris, France, U. Copenhagen (Denmark), U. Oslo (Norway), Tübingen, Germany. Prof., dir. palaeozoology dept. Swedish Mus. Natural History, Stockholm, Sweden, 1923-33, 36-59; prof. hist. geology and palaeontology U. Upsala, 1934-35. Recipient numerous medals from learned socs. and acads. Mem. K.Vet. Akad. Stockholm, Royal Soc., 1946, Royal Soc. Edinburgh, Akad. Nauk Moscow, Am. Philos. Soc., Geol. Soc. Am. Publs. and anat. research on lower vertebrate fossils especially from Devonian period. Office: Palaeozoology Dept., Swedish Mus. Natural History, Stockholm, 50, Sweden.*

STENSTROM, K(arl) Wilhelm, physicist; b. Gothenborg, Sweden, Jan. 28, 1891; s. Anders Hugo and Aurora (Svenson) S.; Ph.D., U. Lund (Sweden), 1919; postgrad. Harvard; m. Annette Treble, June 1, 1922; 1 dau., Margaret Luxford (Mrs. A. MacDonnell Richards). Came to U. S., 1919, naturalized, 1939. Fellow Swedish Am. Found., physicist State Inst. for Malignant Disease, Buffalo, 1921-26; prof. biophysics and head radiation therapy U. Minn., Mpls., 1926-56. Decorated Order N. Star Sweden; recipient Am. Cancer Soc. award for Minn., 1958. Mem. Am. Phys. Soc., A.A.A.S., Radium Soc. Am., Am. Soc. Roentgenology, Am. Cancer Soc., N.Am. Radiol. Soc., Sigma Xi, Phi Rho Sigma. Author: Manual of Radiation Therapy, 1956; Rontgen Spektra, 1919; also numerous articles. Research on X-ray spectroscopy, especially refraction of X-rays, M-series; contbns. to field med. physics, especially radiation therapy. Home: 541 Key Royale Dr., Holmes Beach, Fla. 33510.*

STENT, Gunther Siegmund, Am. molecular biologist; b. Berlin, Germany, Mar. 28, 1924; s. George and Elizabeth (Karfunkelstein) S.; came to U.S., 1940, naturalized, 1945; B.S., U. Ill., 1945, Ph.D., 1948; m. Inga Loftsdottir, Oct. 27, 1951; 1 son, Stefan Loftur. Merck fellow Cal. Inst. Tech., 1948-50; Am. Cancer Soc. fellow U. Copenhagen, 1950-51; Am. Cancer Soc. fellow Inst. Pasteur, Paris, France, 1951-52; asst. research biochemist U. Cal., Berkeley, 1952-56, asso. prof. bacteriology, 1957-59, prof. molecular biology, 1959—; external mem. Max Planck Institut Molekulare Genetik, Berlin, Germany, 1967—; vis. prof. Collège de France, Paris, 1967. Mem. A.A.A.S., Genetics Soc. Am. Author: Molecular Biology of Bacterial Viruses, 1963; (with J. Cairns and J. Watson) Phage and Origins of Molecular Biology, 1966. Research and numerous publs. in analysis of replication and structure of genetic material by radioisotope decay inactivation; concomitant transcription and translation of genetic information. Home: 145 Purdue Av., Berkeley, Cal. 94708.*

STEPAN, Jan, clin. biochemist, rheumatologist; b. Kutna Hora, Bohemia, Dec. 20, 1913; s. Jaroslav Stepan and Jana (Rumlova) S.; Magister pharmaciae, Charles U., Prague, Czechoslovakia, 1935, doctor rerum naturalium, 1938, doctor medicinae, universae 1951, candidatus scientiarum medicarum, 1963; m. Iva Svorcikova, June 1, 1942; children—Jan, Sarka. Private docent Charles U., Hradec Kralové, Bohemia, 1950-51, prof. aggregé, dir. inst. med. biochemistry, Plzen Pilsen, 1951-60; docent clin. biochemistry in rheumatology Research Inst. for Rheumatic Diseases, Prague, 1961-66; vis. prof. Clin. and Inst. for Phys. Medicine and Balneology, U. Justus Liebig Giessen, Bad Nauheim, W. Germany, 1966-67. Mem. Czechoslovak Med. Soc. Author: Medical and Clinical Biochemistry, 1958; also numerous articles. Research on mineral metabolism with exptl. models, mucoid compounds in patients with articular diseases, abnormal metabolism in rheumatic and articular disease. Home: 120 Jeseniova, Prague 3. Office: Na Slupi 4, Prague 2, Czechoslovakia.*

STEPHAN, Edouard, French astronomer; b. Sainte-Pezenne, France, Aug. 31, 1837; prof. Marseille (France) Faculty Scis.; founder, dir. Marseille Obs.; mem. French Acad. Scis. Author: Étude sur la phenomène des étoiles filantes, 1871. Made early attempt to measure stellar distance by interferential method, Marseille, 1873. Died Marseille, Dec. 31, 1923.

STEPHAN, Heinz Hermann Karl, German neuroanatomist; b. Dessau, Germany, Feb. 7, 1924; s. Hermann Karl and Emma (Kursch) S.; Dr.rer.nat., U. Kiel (Germany), 1950; m. Ingrid Trede, 1948; children—Volker, Michael, Roland. Asst., Inst. for Brain Research, Neustadt, Germany, 1950-51, Inst. for Domestic Animals, Kiel, 1951-53, Max-Planck Inst. for Brain Research dept. neuroanatomy, Frankfurt, Germany, 1953-66, head Lab. Comparative Neuroanatomy, 1966—. Mem. World Fedn. Neurology (sec. commn. comparative neuroanatomy 1959—), Internat. Brain Research Orgn., Max-Planck Soc. Author: (with R. Hassler) Evolution of the Forebrain, 1966; also articles. Research on changes in brain weight and brain composition from wild to domestic animals, morphogenesis of brain, comparative anatomy of insectivore brains, brain-body size relationship, influence of adaptation to water life on brain, comparative anatomy of Septum telencephali, evolution of primate brain, comparative anatomy and quantitative comparison of limbic and olfactory systems. Home: 6078 Neu Isenburg, Rosenstr. 38, Germany. Office: 6 Frankfurt/Main-Niederrad Deutschordenstr. 46.*

STEPHANI, Ludolf, German archeologist; b. Beucha, nr. Leipzig, Germany, 1816; travelled in northern Greece, Asia Minor, southern Italy, Sicily; became prof. philology, Dorpat (now Tartu, Estonia), 1846; named curator classical antiquities, mem. Acad. Scis., St. Petersburg, 1850. Author: Voyage à travers quelques contrées de la Grèce du Nord, 1843; Hercule au repos, 1854; Antiquites du Bosphore Cimmérien, 1954; Le nimbe et l'auréole dans les oeuvres de l'art ancien, 1859; La collection de vases de l'Ermitage impérial, 1869; La collection d'antiques de Pavlovsk, 1872. Died Pavlovsk, 1887.

STEPHEN, Michael John, physicist; b. Johannesburg, S. Africa, Apr. 7, 1933; s. Henry and Theodora (de Kiewiet) S.; B.Sc., U. Witwatersrand, S. Africa, 1953; D.Phil., Oxford U., 1956; m. Johanna A. Pallotta, Aug. 13, 1966. Research asso. Columbia U., 1958-60; research fellow Oxford U., 1960-62; faculty Yale, New Haven, 1963—, asso. prof. physics, 1965—. Cons., Bell Telephone Labs., Inc., 1964—. Mem. Am. Phys. Soc. Research, publs. in fields of atomic physics, solid state physics, and super-conductivity. Office: Physics Dept., Yale, New Haven 06520.*

STEPHEN, William Irvine, Brit. chemist; b. Aberdeen, Scotland, Aug. 15, 1929; s. Henry George and Bonté (Irvine) S.; B.Sc. in Chemistry with 1st Class Honours, U. Aberdeen, 1950; Ph.D., U. Birmingham (Eng.), 1953; m. Kathleen Fraser Donald, Apr. 4, 1956; children—Gillian Kathryn and Susan Caroline (twins). Sci. staff Royal Naval Sci. Service, 1953-55; Imperial Chem. Industries Research fellow U. Birmingham, 1955-57, faculty, 1957—, sr. lectr., 1966—. Chmn. pesticides analysis adv. com. reagents panel Ministry Agr., 1963—. Fellow Royal Inst. Chemistry (chmn. Birmingham and Midlands sect. 1966—), Internat. Union Pure and Applied Chemistry (chmn. interdiv. com. on purity lab. chems.), Chem. Soc. London, Soc. for Analytical Chemistry London. Research and numerous publs. on classical analytical chemistry, exploitation of chem. reactions for quantitative analytical purposes and devel. new organic analytical reagents for inorganic analysis; discovered several reagents leading to improved analytical procedures.

Home: 85, Oakfield Rd., Selly Park, Birmingham, 29, U.K.*

STEPHEN, William Procuronoff, entomologist; b. St. Boniface, Man., Can., June 6, 1927; s. Stephen S. and Amalia (Hoppe) Procuronoff; B.S., U. Man., 1948; postgrad. Ia. State U., 1948-49; Ph.D., U. Kan., 1952; m. Dorris Jo Williams, June 8, 1952; children—Dana Ann, Jan Marie, Mary Beth, William Thaddeus. Came to U.S., 1953, naturalized, 1960. Asso. entomologist Can. Dept. Agr., Brandon, Man. and Ottawa, Ont., 1948-53; prof. entomology Ore. State U., Corvallis, 1953—, chmn. genetics com., 1957-58, exec. bd. Genetics Inst., 1965—. Recipient Basic Research award Ore. State U., 1960, OECD, award, 1962. Mem. A.A.A.S., Entomol. Soc. Am., Soc. Systematic Zoology, Soc. Study Evolution, Behavior Soc., Bee Research Assn. Research and numerous publs. in solitary and social bees, other than honey bees, for possible domestication as crop pollinators, overt behavior patterns among insects and their neurol. control, biochemistry of animals. Home: 1206 Fernwood Dr., Corvallis, Ore. 97330.*

STEPHENS, George, archaeologist; b. Liverpool, Eng., 1813; s. John and Rebecca Eliza (Rayner); student U. Coll., London; Ph.D., U. Uppsala (Sweden), 1877; m. Maria Bennett, 1834. Settled in Stockholm, Sweden, 1834; became naturalized Danish citizen, 1851; lectr. English lang. U. Copenhagen, prof. English and Anglosaxon, 1855-93. Fellow Soc. Antiquaries; mem. Soc. for Publ. of Ancient Swedish Texts (founder). Author: The Old Runic Monuments of Scandinavia and England, 1866-84; Pocket Dictionary of English and Swedish; also articles on religion, politics, grammar and origin of English lang. Interpreted inscriptions on ancient bldgs. of Eng. and Scandinavia and traced lang. in inscriptions to original source. Died 1895.

STEPHENS, John Lloyd, Am. archaeologist, traveller; b. Shrewsbury, N.J., Nov. 28, 1805; s. Benjamin and Clemence (Lloyd) S.; grad. Columbia, 1822; attended Litchfield (Conn.) Law Sch. Admitted to N.Y. State bar, 1825, retired from law practice to travel abroad, 1834; sent as Democrat on diplomatic mission to C.Am. by Van Buren, 1839; dir. Ocean Steam Navigation Co.; active supporter Hudson River R.R.; promoter Panama R.R., elected v.p. co., 1849; became pres., circa 1851; mem. N.Y. Constl. Conv., 1846. Author: Incidents of Travel in Egypt, Arabia, Petraea and the Holy Land, 2 vols., 1837; Incidents of Travel in Greece, Turkey, Russia and Poland, 2 vols., 1838; Incidents of Travel in Central America, Chiapas and Yucatan, 2 vols., 1841. Writings known for observations on ancient cultures; rediscovered ruins of Mayan city, Copán, 1839; stimulated Am. interest in study of pre-Columbian indian cultures. Died N.Y.C., Oct. 12, 1852.

STEPHENS, Robert Eugene, Am. physicist; b. Monessen, Pa., June 26, 1905; s. Robert Gilbert and Maude (Luce) S.; B.S. in Math., Washington and Jefferson Coll., 1926; M.S. in Physics, U. Pitts., 1933, Ph.D. in Physics, 1938; m. Mary Carr, Aug. 24, 1936; Physicist, Nat. Bur. Standards, Washington, 1935-36, 40-67, also cons. optics; with Mead Corp., Chillicothe, O., 1937-38; faculty U. Toledo, 1939-40. Mem. Optical Soc. Am., Washington Acad. Scis., Washington Philos. Soc. Asso. editor Jour. Optical Soc. Am., 1958-63. Research and publs. on systematic design procedures for Taylor-type anastigmatic lenses, viscosity of molten ceramics as function of temperature, measurement of index of refraction of optical substances for infrared, design of lenses and optical instruments; exptl. verification of superachromatism. Home: 4301 39th St., N.W., Washington 20016. Office: Nat. Bur. Standards, Washington 20016.*

STEPHENS, Stanley George, biologist; b. Dudley, Eng., Sept. 2, 1911; s. George and May (Selwood) S.; B.A., Cambridge (Eng.) U., 1933, M.A., 1936; Ph.D., Edinburgh (Scotland) U., 1941; m. Dorothy Louise Bolam, Sept. 2, 1938; children—David B., Michael G. Came to U.S., 1945, naturalized, 1951. Research asso. Carnegie Instn. Washington, L.I., N.Y., 1945-47; research prof. agronomy Tex. A. and M. U., 1947-49; prof. agronomy N.C. State U., Raleigh, 1949-52, William Neal Reynolds prof. genetics, 1952—. Mem. NSF Genetics Biology Panel, 1963-64; vis. prof. U. Hawaii, 1963; mem. Galapagos Internat. Sci. Project, 1964. Recipient Cotton Genetics award Joint Cotton Breeding Policy Com., 1951. Guggenheim fellow, 1959. Mem. Nat. Acad. Scis., Genetics Soc. Am., Soc. For Study Evolution, Am. Soc. Naturalists, Soc. Am. Archaeology, Soc. Tropical Biology, Gamma Sigma Delta, Phi Kappa Phi. Author: (with J.B. Hutchinson and R.A. Silow) The Evolution of Gossypium, 1947. Studies and numerous publs. on genetics and evolution of cottons, effects of natural and human means of dispersal on their present range of distbn., role of hybridization in race formation. Home: 3219 Darien Dr., Raleigh, N.C. 27607.*

STEPHENS, William Edwards, Am. physicist; b. St. Louis, May 29, 1912; s. Eugene and Marie P. (Gelwicks) S.; A.B., Washington U., St. Louis, 1932, M.S., 1934; Ph.D., Cal. Inst. Tech., 1938; m. Helen Elizabeth Burnite, Oct. 27, 1942; children—Richard Burnite, William Massie. Lectr., U. Pitts., 1938-39; instr. Stanford, 1940-41; faculty U. Pa., Phila., 1941—, prof. physics, 1948—, chmn. physics dept., 1963—; vis. prof. U. Zurich, Switzerland, 1957. Recipient Army-Navy certificate of merit for war research, 1947; Westinghouse research fellow, 1938-40. Mem. A.A.A.S., Am. Assn. U. Profs., Am. Phys. Soc., Phi Beta Kappa, Sigma Xi. Author: (with G.P. Harnwell) Atomic Physics, 1955. Editor: Nuclear Fission and Atomic Energy, 1949. Research on mass spectrometry; sector magnetic fields; photo fission; nuclear reactions; gamma ray induced photo nuclear effects. Home: 105 Rolling Rd., Phila. 19151.*

STEPHENSON, Charles Bruce, Am. astronomer; b. Little Rock, Feb. 9, 1929; s. Chauncey Alvaro and Ona (Richards) S.; student Little Rock Jr. Coll., 1946-47; B.S. in Math., U. Chgo., 1949, M.S. in Astronomy, 1951; Ph.D., in Astronomy, U. Cal. at Berkeley, 1955-58; m. Elizabeth Griffith Strong, June 21, 1952. With Army Map Service Lunar Occulation Propect, U.S. Army, 1953-55; faculty Case Inst. Tech., East Cleveland, O., 1958—, asso. prof. astronomy, 1964—. Mem. Internat. Astron. Union, Am. Astron. Soc., A.A.A.S. Research and publs. on spectral classification procedures for stars by objective prism method; discovered 1st white dwarf stars to be found by the objective prism method 2d preoutburst spectrum of a non-recurrent nova (exploding star) to be observed in 40 years; co-discoverer optical counterpart of the Scorpius X-Ray source. Home: 3362 Henderson Rd., Cleveland Heights, O. 44112. Office: Warner and Swasey Obs., East Cleveland, O. 44112.*

STEPHENSON, Charles V., Am. physicist; b. Centerville, Tenn., Oct. 1, 1924; s. Claude Bernarde and Carleen (Pettit) S.; B.A., Vanderbilt U., 1948, M.A., 1949, Ph.D., 1952; m. Luellen Hovey, Aug. 29, 1948; children—Charles Bruce, Frances Luellen, Gregory Brian. Research physicist Sandia Corp., Albuquerque, 1952-56, cons., 1956-58; asst. prof. physics Auburn (Ala.) U., 1956-58; head physics sect. So. Research Inst., Birmingham, Ala., 1958-62; asso. prof. elec. engring. Vanderbilt U., Nashville, 1962-65, prof., 1965—. Instr. U. Ala., University, 1958-62. Fellow Am. Phys. Soc., A.A.A.S.; mem. I.E.E.E. (sr.), Sigma Xi. Research and publs. on molecular spectroscopy, solid state physics, high polymer physics. Home: 871 Rodney Dr., Nashville 37205.*

STEPHENSON, Edward L., Am. nutritionist; b. Calhoun, Tenn., May 5, 1923; s. L. I. and Martha (Whiteside) S.; B.S.A., U. Tenn., 1946, M.S., 1947; Ph.D., Wash. State Coll., 1951; m. Mary Ellen Simpson, Mar. 15, 1947; children—Linda Jean, Kenneth Edward. Faculty, U. Ark., Fayetteville, 1949—, prof., 1958—, head dept. animal scis., 1964—. Named Ark. Poultryman of Year, 1963. Mem. Am. Inst. Nutrition, Animal Sci. Assn., Poultry Sci. Assn., Dairy Sci. Assn., Sigma Xi, Gamma Sigma Delta, Alpha Zeta. Research, numerous publs. on physiol. role Vitamin B12. Home: Route 8, Fayetteville, Ark. 72701.*

STEPHENSON, George, Brit. engr., inventor; b. Wylam, Eng., June 9, 1781; s. Robert and Mabel (Carr) S.; m. Frances Henderson, Nov. 28, 1802; m. 2d, Elizabeth Hindmarsh, Mar. 29, 1820; m. 3d, Elizabeth Gregory, Jan. 11, 1848; a son, Robert. With Dewley Colliery, 1794-1801; became brakesman, Black Callerton, 1801; engineman, Willington Ballast Hill, 1802; moved to Killingworth, 1804, Montrose, 1807, Killingworth, 1808; became engine-wright to colliery, 1812; laid ry. for Hetton Colliery, 1819-23; engr. Stockton & Darlington Ry., 1822-24; with Liverpool and Manchester Ry., 1824-30; chief engr. to Grand Junction Line, North Midland Line. Mem. Instn. Mech. Engrs. (founder, became pres. 1847), Brit. Assn. Invented (independently of Davy) miner's safety lamps, 1815; built locomotive, 1814; designed 1st steamblast locomotive, 1815; founder railroads in Eng. Died nr. Chesterfield, Eng., Aug. 12, 1848.

STEPHENSON, Hugh Edward, Jr., Am. physician; b. Columbia, Mo., June 1, 1922; s. Hugh Edward and Doris (Pryor) S.; A.B., B.S., U. Mo., 1943; M.D., Washington U., St. Louis, 1945; m. Sarah Norfleet Dickinson, Aug. 15, 1964; 1 son, Hugh Edward, III, Ann Dunlop. Faculty, U. Mo. Sch. Medicine, Columbia, 1953—, prof. surgery, 1956—, chmn. dept. surgery, 1956-60. Named one of Ten Young Men of Nation, Nat. Jr. C. of C., 1956; James IV Surg. Travelor, Gt. Britain, 1962. Diplomate Am. Bd. Surgery, Am. Bd. Thoracic Surgery. Mem. A.M.A., A.C.S., Vascular Surgery, Soc. Thoracic Surgeons, So. Thoracic Surg. Assn. Author: Cardiac Arrest and Resuscitation, 1958, 64; also articles. Research on cardio-vascular surgery. Home: 1004 Falcon, Dr., Columbia, Mo. 65201.*

STEPHENSON, John, inventor; b. County Armagn, Ireland, July 4, 1809; s. James and Grace (Stuart) S.; attended Wesleyan Sem.; m. Julia A. Tiemann, 1833; 3 children. Opened shop for repair of all kinds vehicles, 1831, conceived, built 1st omnibus or horse-car made in N.Y.C.; employed to build horse-drawn car for new N.Y. & Harlem R.R. (1st car for 1st street ry. in world), 1831; largest street-car builder in world, made horse cars, cable, electric, open cars; 1st patent granted, 1833; inventor U-shaped rail, double-ended streetcar; factory produced carriages and pontoons for U. S. Govt. during Civil War. Died New Rochelle, N.Y., July 31, 1893.

STEPHENSON, John, English zoologist; b. Padiham, Lancashire, Eng., 1871; B.Sc., U. London (Eng.), 1890, M.B., 1894, D.Sc., 1909; M.B., Ch.B., U. Manchester (Eng.), 1893; m. Gertrude Bayne, 1895. House physician Manchester Royal Infirmary, 1893-94, Royal Hosp. for Diseases of Chest, London, 1894; with Indian Med. Service, from 1895; in mil. employ, 1895-1900, with N.W. Frontier Expdn., 1897; civil surgeon, also on plague duty, 1900-06; prof. biology Govt. Coll., Lahore (now Pakistan), from 1906, prof. zoology, 1912, prin., 1912; vice chancellor Punjab (India) U., from 1913; lectr. zoology Edinburgh (Scotland) U., 1920-29. Fellow Zool. Soc., Linnean Soc., Royal Asiatic Soc., Royal Soc., 1930, Royal Coll. Surgeons. Author: The Hadiqatu-l-Haqiqat of the Hakim Sanai. Editor, translator zool. sect. Nuzhatu-l-Qulub. Contbr. monographs, articles on fauna to profl. publs. Research on fauna of Brit. India. Died Feb. 2, 1933.

STEPHENSON, Reginald Joseph, Am. physicist; b. Nottingham, Eng., Aug. 11, 1903; s. Alfred Ernest and Catherine (Granger) S.; B.Sc., U. London, 1924; M.Sc., U. Reading, Eng., 1927; Ph.D., U. Chgo., 1933; m. Helen Fiske Aldrich, Dec. 4, 1933; 1 son, John Aldrich. Came to U.S., 1929, naturalized, 1937; staff, Meml. Coll., St. John's, Nfld., Can., 1926-29; staff Worcester (Mass.) Poly Inst., 1929-31, U. Chgo., 1931-45; with Coll. Wooster (O.), 1945-59; William F. Harn prof., 1959—. Cons., Oak Ridge Inst. Nuclear Studies. Fellow, Am. Phys. Soc., Sigma Xi. Rotarian. Author: Exploring in Physics, 1937; (with A. W. Duff) Physics, 1938; (with H. B. Lemon and M. Ference Jr.) Experimental Analytical Physics, 1946; Mechanics and Properties of Matter, 1952; (with others) Selective Experiments in Physics, 1941. Research on X-rays, cosmic rays, history of science. Home: 827 N. Bever St., Wooster, O. 44691.*

STEPHENSON, Richard Montgomery, Am. chem. engr.; b. St. Louis, Sept. 10, 1917; s. Carl and Olive (Diall) S.; B.Chemistry, Cornell U., 1939, Chem.E., 1940, Ph.D., 1946; m. Lois Finney, June 14, 1966; children—Kathleen, John, Daniel. Research engr. Allied Chem., 1940-44; asst. prof. chem. engring. U. Minn., 1946-50; sr. design engr. Oak Ridge Nat. Lab., 1950-54; asso. prof. chem. engring. N.Y. U., 1954-57; prof. chem. engring. U. Conn., Storrs, 1958—. Cons. in chem. and nuclear engring.; Fulbright prof. Vienna Inst. Tech., 1957-58. Mem. Am. Chem. Soc., Am. Inst. Chem. Engring., Am. Assn. U. Profs., Am. Soc. Engring. Edn. Author: Introduction to Nuclear Engring., 1945; Introduction to Chemical Process Industries, 1966; also articles. Research in chem. and nuclear engring., chem. process industries, flux-trap nuclear reactor. Patentee in field. Home: 12 Hillyndale Rd., Storrs, Conn. 06268.*

STEPHENSON, Robert, English civil engr.; b. Willington Quay, Eng., Oct. 16, 1803; s. George and Frances (Henderson) S.; ed. Bruces Acad., Newcastle; Edinburgh U., 1822; D.C.L., Oxford, 1857; m. Frances Sanderson, June 17, 1829. Apprenticed to viewer of Killingworth Colliery, at age 16; employed in New Castle locomotive factory, 1823; supt. mines in Columbia, 1824-27; apptd. engr., 1826 took important part in constrn. of The Rocket, also in devising improvements, 1827-33; chief engr. to line connecting Birmingham with Manchester and Liverpool, from 1833, then lines between northern towns, 1836, Derby-Leeds Ry., 1837 (increased speed to 29 miles an hour by application of Gurney's steam-jet); tried to check ry. mania of 1844; overcame supporters of atmospheric rys., 1845; built high-level bridge, Newcastle, also Victoria Bridge, Berwick Menal tubular girder bridge (opened 1850) and Victoria Bridge, Montreal, 1859. M.P., Whilty, 1847-59. Recipient Gold medal for invention of system of tubular-plant ry. bridges French Exhibition, 1855. Fellow Royal Soc., 1849; pres. Instn. Civil Engrs. Contbr. article Iron Bridges to Encyclopaedia Britannica; also author numerous reports. Died Oct. 12, 1859.

STEPHENSON, Samuel Edward, Jr., Am. physician; b. Bristol, Tenn., May 16, 1926; s. Samuel Edward and Hazel B. (Walters) S.; B.S., U. S.C., 1946; M.D., Vanderbilt U., 1950; m. Dorthea Cole, June 12, 1950; children—Samuel Edward III, William Douglas, Dorthea Louise, Judith Maria. Practice medicine, specializing in surgery, Nashville, 1957—; faculty Sch. Medicine Vandebilt U., 1955—, asso. prof. surgery, 1960—; dir. S.R. Light Lab. Surg. Research, 1958-61; dir. pediatric surg. service, dir. intensive care div. Vanderbilt U. Hosp. Recipient certificate of merit So. Med. Assn., 1959, certificate of merit Southeastern Surg. Congress, 1962. Mem. A.A.A.S., Am., So. assns. thoracic surgery, A.C.S., Am. Fedn. Clin. Research, A.M.A., Am., So., Pan Pacific surg. assns., Assn. Am. Med. Colls., Internat. Cardiovascular Soc., Soc. Surgery Alimentary Tract, Soc. Vascular Surgery, Soc. U. Surgeons, So. Med. Assn., Sigma Xi. Research, numerous publs. on effect of coronary artery occlusion on cardiac arrhythmias, exptl. atherogenesis; devel. of electronic respirator control, synchronous cardiac pacemaker. Home: 4517 Sewanne Rd., Nashville 37220. Office: Vanderbilt U. Hosp., Dept. Surgery, Nashville 37220.*

STEPIEN, Lucjan Seweryn, Polish physician; b. Osiek, Poland, Dec. 18, 1912; s. Jan and Maria (Cukierska) S.; M.D., U. Warsaw, 1947, Veniam Legendi, 1951; m. Jadwiga Stepien, May 24, 1942; children—Wanda, Anna, Jadwiga. Dir. neurosurg. clinic Med. Acad., Lodz, Poland, 1948-56; dir. neurosurg. dept. Polish Acad. Scis., Warsaw, 1957-—; dir. neurosurg. clinic Med. Acad., Warsaw, 1960-—. Asso. dept. neurophysiology Nencki Inst. Exptl. Biology, Warsaw, 1949-—. Mem. Soc. Polish Neurosurgeons, World Fedn. Neurosurg. Socs. Research and numerous publs. on exptl. and clin. investigations on frontal lobes of brain, operative treatment of focal epilepsy, especially temporal lobe epilepsy, disturbances of higher nervous activity after focal lesions of brain, aphasia, studies on recent memory, denervation sinus caroticus in myasthenia gravis. Home: 71/4 Wilcza. Office: Neurosurgical Clinic, 6 Oczki, Warsaw, Poland.*

STEPP, Wilhelm Otto, German biochemist; b. Nuremberg, Germany, Oct. 10, 1882; s. Karl Ludwig and (Reuter) S.; ed. U. Munich; m. Margarete Krüger, 1913. Joined clinic U. Giessen (Germany), 1916; became prof., head Med. Polyclinic, Giessen, 1922; became Rockefeller fellow Inst. Hygiene and Pub. Health, Balt., 1924; prof., dir. Med. Clinic, U. Jena (Germany), 1924-26, U. Breslau (now Wroclaw, Poland), 1926-34, U. Munich, 1934-45; dir. Med. Clinic, U. Würzburg (Germany), 1947-48; ret., 1949. Author: (with Paul György) Avitaminosen und verwandte Krankheitszustände, 1926; (with I. Kühnau, H. Schroeder) Die Vitamine und ihre klinische Anwendung; ABC der Gesundheit, 1950; Das Bier, wie es der Arzt sieht, 1954; Allgem. Krantheitslehre für die Helfer des Arztes, 1956. Research and numerous publs. on internal medicine, including vitamins; discovered chemically pure fats without vitamins. Address: 29 Englschalkinger Str., Munich, 27.

STERIADE, Mircea, Rumanian neurophysiologist; b. Bucharest, Rumania, Aug. 20, 1924; s. Herman and Rachel S.; M.D., Faculty of Medicine, Bucharest, 1952; m. Stefana Kernbach, Apr. 21, 1949; 1 child, Donca. Asst. prof. anatomy Bucharest Acad., 1949-52, asst. prof. neurology, 1952, research fellow, 1951, sr. research fellow, 1955-65, head dept. for cortico-diencephalic relationship studies Inst. of Neurology, 1965-—. Mem. Neurol. Soc. Paris (hon. mem.), Physiol. Assn. France. Author: La Physiologie et la Physiopathologie du Cervelet, 1958. Research and numerous publs. on role of reticular formation in sensory transmission; potentiation in visual system; cerebellum, reticular formation, thalamus in genesis of epileptic seizures. Home: 33 13 Decembrie, Bucharest. Office: 42 Povernei, Bucharest, Rumania.*

STERLING, Guy, Am. civil engr.; b. Cleve., May 1, 1860; s. Theodore and Charlotte (Higgins) S.; student Kenyon Coll., 1877-78; C.E., Cornell U., 1887; spl. work in elec. engring., Lewis Inst., 1904; m. Harriot Brewer, Sept. 4, 1890. In wholesale and retail carpet bus. with Sterling & Co., Cleveland, 1878-83; asst. engr. maintenance of way, Denver, 1887-88, topog. engr. surveys in northeastern Cal., 1888-89, U.P. Ry.; topographer U. S. Geol. Survey, Clear Lake, Cal., also Snake River (Ida.) irrigation surveys, 1889; asst. engr. in charge constrn. Phyllis Canal, Ida., 1890; asst. chief engr. Sunnyside Canal, nr. Yakima, Wash., 1890-91, 92-94; chief engr. Cowiche and Wide Hollow irrigation dist. nr. Yakima, Wash., 1891-92; constrn. pvt. irrigation projects nr. N. Yakima, 1894-95; chief engr. Priest Rapids Project, on Columbia River north of Pasco, Wash., 1895; in practice in Utah and other states, 1895-—, principally hydraulic engring., examinations and reports for eastern investors, supervision of constrn., Chmn. com. on research and invention Utah Council Def. 1917-18. Inventor and patentee processes for extracting potash from mother liquor salts and from silicates, including wyomingite. Deceased.

STERN, Adolph John, Am. chemist; b. Nuremberg, Germany, Feb. 12, 1900; s. Hans and Katherine (Siegler) S.; diploma Ing., Tech. U. Munich., 1923, Dr.Ing., 1925, Dr. Habil., 1932; m. Margaret Scherbel, Mar. 4, 1938; 1 dau., Kathleen (Kathleen Sinibaldi). Came to U.S., 1938, naturalized, 1945. Faculty, Tech. U. Munich, 1925-37; research asso. Childrens Fund Mich., 1938; prof. chemistry Wagner Coll., S.I., N.Y., 1942-—, chmn. chemistry dept., 1950-52, dean Coll., 1952-66, spl. asst. to pres., 1966-—. Recipient Outstanding Alumni award Wagner Coll. Alumni, 1965. Fellow Am. Inst. Chemists; mem. A.A.A.S., Am. Chem. Soc. Author: (with Hans Fischer) Chlorophyll Die Chemie des Pyrrols, 1937. Contbr. numerous articles to profl. jours. Research and publs. primarily in field of Raman-effect, structure of organic compounds, photochemistry, absorption spectra, chemistry of porphyrines and chlorophyll, thermochemistry. Home: 16 Lloyd Ct., S.I., N.Y. 10310.*

STERN, Curt, zoologist, geneticist; b. Hamburg, Germany, Aug. 30, 1902; s. Barned S. and Anna (Liebrecht) S.; Ph.D., University of Berlin, Germany, 1923; D.Sc., MacGill, 1958; married Evelyn Sommerfield, Oct. 29, 1931; children—Hildegard, Holly Elisabeth, Barbara Ellen. Came to U. S., 1933, naturalized. Fellow Internat. Edn. Bd., 1924-26; privatdozent, U. of Berlin, 1928-33; fellow Rockefeller Foundation, 1932-33; research associate in zoology, U. of Rochester, 1933-35, asst. prof., 1935-37, asso. prof., 1937-41, prof., chmn., dept. zoology, 1941-47; chmn. div. biol. sciences, 1941-47, prof. exptl. zoology; prof. zoology, U. Cal., Berkeley, 1947-—, prof. genetics, 1958-—; vis. prof. Western Res. U., 1932; vis. lectr. Columbia U., 1944. Mem. adv. com. biology and medicine AEC, 1950-55. Fellow A.A.A.S.; mem. Am. Genetic Soc. (pres. 1950), Am. Soc. Human Genetics (president, 1957), Am. Philos. Soc., Nat. Acad. Scis. (recipient Kimber Genetics medal 1963), American Academy of Arts and Scis., Am. Soc. Zoologists (pres. 1962), Am. Soc. Naturalists, Soc. Growth and Devel., Sigma Xi. Author: Multiple Allelie, 1930; Faktorenkoppelung u. Faktorenaustausch, 1933; Principles of Human Genetics, 1949, 60. Editor. Genetics; Cytogenetics; Humangenetik. Contbr. articles on genetics to sci. pubs. Proved theory of chromosome crossing over in study of fruit flies, 1931; demonstrated crossing over in somatic as well as germ cells, 1936; proposed prepattern concept in developmental genetics, 1954. Home: 42 Kingston Rd., Kensington, Cal. 94707. Office: U. Cal., Berkeley, Cal. 94720.*

STERN, Frank, physicist; b. Koblenz, Germany, Sept. 15, 1928; s. Gustav and Louise (Beiersdorf) S.; came to U.S., 1936, naturalized, 1943; B.S., Union Coll., Schenectady, 1949; Ph.D., Princeton, 1965; m. Shayne Nemerson, July 9, 1955; children—David, Linda Ellen. Physicist, U.S. Naval Ordnance Lab., White Oak, Silver Spring, Md., 1953-62; research staff IBM Watson Research Center, Yorktown Heights, N.Y., 1962-—; faculty U. Md., College Park, 1955-62, prof. physics dept., part-time, 1959-62. Charles A. Coffin fellow, 1949-50; NSF fellow, 1952-53; recipient Meritorious Civilian Service award U.S. Naval Ordnance Lab., 1961. Fellow Am. Phys. Soc.; mem. Research Soc. Am., A.A.A.S., Phi Beta Kappa, Sigma Xi (asso.). Research and publs. on cohesive energy of iron, semiconductors, optical properties of solids, injection lasers. Home: 6 Robbins Rd., Pleasantville, N.Y. 10570. Office: IBM Watson Research Center, Yorktown Heights, N.Y. 10598.*

STERN, George Gordon, Am. psychologist; b. N.Y.C., Oct. 3, 1923; s. Fred and Dorothy (Gordon) S.; Ph.D., U. Chgo., 1949; m. Shirley Rosenthal, Nov. 13, 1942; children—Sally, Frederick, Patricia. Supr. research, examiner's office U. Chgo., 1949-53, lectr. psychology, 1951-53; prof. psychology Syracuse U., 1953-—, head evaluation Psychol. Research Center, 1953-—. Danforth lectr., 1964-65; mem. edn. panel President's Sci. Adv. Council, 1966-68. Fellow A.A.A.S., Am. Psychol. Assn., Soc. for Psychol. Study Social Issues; mem. Syracuse Psychol. Assn. (bd. dirs. 1956-57, 59-62, pres. 1957-58), Am. Ednl. Research Assn., Am. Sociol. Assn., Nat. Council on Measurement in Edn., Sigma Xi, Psi Chi. Author: (with M.I. Stein, B.S. Bloom) Methods in Personality Assessment, 1956; (with N. Haring and N. Cruickshank) Attitudes of Educators Towards Exceptional Children, 1958; also articles. Asso. editor Sociology Edn., 1963-—. Developed techniques for measuring psychol. characteristics of environment and their interaction with personality processes. Home: 421 Buffington Rd., DeWitt, N.Y. 13224.*

STERN, Herbert, biologist; b. Montreal, Que., Can., Dec. 22, 1918; s. Saul and Stella (Bloomstone) S.; B.Sc., McGill U., Montreal, 1940, Ph.D., 1946; m. Ruth Starrels, June 28, 1953; children—Rebecca Ann, David Benjamin, Jonathan Weil. Prof. botany U. Ill., Urbana, 1960-65; prof. biology U. Cal. at San Diego, 1965-—. Cons. NSF. Mem. Soc. for Developmental Biology (pres. 1964-65), Genetics Soc., Am. Soc. for Cell Biology, A.A.A.S., Am. Soc. Plant Physiologists. Author: (with D. L. Manney) Biology of Cells, 1965; also numerous articles. Editorial bd. Plant Physiology, Jour. Cell Biology. Research on chem. changes in dividing cells. Home: 5478 Soledad Rd., La Jolla, Cal. 92037.*

STERN, John Alexander, psychologist; b. Montabour, Germany, Jan. 4, 1925; s. Sol and Elsie (Berlin) S.; brought to U.S., 1935, naturalized, 1943; B.A., Hunter Coll., 1949; M.A., U. Ill., 1951, Ph.D., 1953; m. Carolyn Ingham, Oct. 22, 1953; children—Julie D., Nancy E., John C. Faculty, Washington U. Sch. Medicine, St. Louis, 1953-65, prof. psychology Washington U., 1966-—; dir. research Malcolm Bliss Mental Health Center, 1962-—. Mem. Am., Midwestern psychol. assns., Soc. Psychophysiol. Research (pres. 1966), Phi Beta Kappa. Research and publs. in psychophysiology and physiol. psychology. Home: 447 Belleview St., Webster Groves, Mo. 63119. Office: 1420 Grattan St., St. Louis 63104.*

STERN, Kurt, physician; b. Vienna, Austria, Apr. 3, 1909; s. Leopold and Elsa (Heller) S.; M.D., U. Vienna, 1933; m. Florence S. Sherman, May 28, 1939; children—Elsa L. (Mrs. Mark M. Slae), Josef Judah, David Michael. Came to U.S., 1938, naturalized, 1943. Practice medicine, specializing in pathology, Chgo., 1945-—; asst., asso. dir. Mt. Sinai Med. Research Found., 1948-60, dir. Mt. Sinai Blood Center, 1950-60; faculty Chgo. Med. Sch., 1949-60; prof. pathology, pathologist Research and Ednl. Hosp. Coll. Medicine U. Ill., 1960-—; cons. clin. pathology VA Hosp., Hines, Ill., 1960-—, West Side VA Hosp., Chgo., 1964-—. Recipient Parker Research award Chgo. Med. Sch., 1960. Fellow Coll. Am. Pathologists, Am. Soc. Clin. Pathologists, N.Y. Acad. Scis., A.A.A.S.; mem. Am. Assn. Blood Banks, Am. Assn. Cancer Research, Am. Assn. Immunologists, Am. Soc. Exptl. Pathology, Am. Soc. Human Genetics, Am. Assn. Pathologists and Bacteriologists, Internat. Acad. Pathology, Soc. Exptl. Biology and Medicine. Author: (with R. Willheim) Biochemistry of Malignant Tumors, 1943. Contbr. numerous articles to profl. jours. Studies, publs. on exptl. immunization of man to blood group factors, reticuloendothelial system; exptl. cancer research. Home: 6042 N. Lawndale Av., Chgo. 60645. Office: 1853 W. Polk St., Chgo. 60612.*

STERN, Kurt Guenter, biochemist; b. Tilsit, Germany, Sept. 19, 1904; s. John Kasper and Sonia (Goldberg) S.; Ph.D., Friedrich-Wilhelms U., 1930; m. Else E. Jacobi, Dec. 24, 1931; 1 son, Rudolph George. Came to U. S., 1935, naturalized, 1946. Carl Duisberg Found. fellow Rockefeller Inst., N.Y.C., 1930-31; sci. guest Courtauld Inst. Biochemistry, London, 1933-35; vis. lectr., Brown Coxe Research fellow Yale, 1935-38, research asst. prof., 1938-42; chief research chemist Overly Bio-chem. Research Found., N.Y.C., 1942-44; adj. prof. biochem. Poly. Inst., Bklyn., 1944-56. Cons. U. S. AEC Research Project, Montefiore Hosp., N.Y.C. Recipient Pasteur medal Soc. Chem. Biol. Paris, 1952. Fellow Am. Inst. Chemists, N.Y. Acad. Scis.; mem. Am. Chem. Soc., Am. Soc. Biol. Chemists, Am. Assn. Cancer Research, Harvey Soc. Author: General Enzyme Chemistry (with J. B. S. Haldane), 1932; Biological Oxidation (with C. Oppenheimer), 1939; Protoplasm, Ency. Britannica (with R. Chambers), 1948. Contbr. articles to reference books, and to publs. in biochemistry. Died Feb. 3, 1956.

STERN, Otto, physicist; b. Sohrau, Germany, Feb. 17, 1888; Ph.D., Breslau, 1912; LL.D., U. Cal., 1930. Private-docent, Tech. Hochschule, Zurich, 1913-14, Frankfurt, 1914-21; prof., Rostock, 1921-22, Hamburg, 1923-33; became research prof. physics dept. Carnegie Inst. Tech., 1933, now prof. emeritus. Recipient Nobel Prize in physics, 1943. Mem. Nat. Acad. Scis., A.A.A.S.; Philos. Soc.; Danish Royal Akad. Developed molecular-beam to study atoms and atom particles; researches confirmed magnetic moment in atoms and quantum theory; found magnetic moment of proton, 1933; work on kinetic theory.

STERN, Pavao, Yugoslavian pharmacologist; b. Varazdin, Yugoslavia, Mar. 17, 1913; s. Josip and Natalija (Schwarz) S.; M.D., Faculty Medicine, Zagreb, Yugoslavia, 1936; m. Bosiljka Srkoc, Nov. 19, 1945; 1 son, Milan. Head pharmacological lab. Pharm. Co., Factory Pliva, 1938-45; reader pharmacology faculty Medicine Zagreb, 1945-48; prof. pharmacology Faculty Medicine, U. Sarajevo (Yugoslavia), 1948-—, dir. Inst. Pharmacology, 1948-—, supr. postgrad. course, 1963-—. Recipient award for sci. work Republic of Bosna and Herzegovine, 1960. Mem. Acad. Scis. Bosna and Herzegovina, Brit. Royal Soc. Medicine, N.Y. Acad. Sci., Am. Acad. Allergy, French Soc. Allergy, Internat. Brain Research Orgn., German Pharmacological Soc. Numerous publs. on pioneer work on antihistaminics, action of substance P as a physiol. tranquilizer; studies on intention and resting tremors. Home: 6 Obala, Sarajevo, Yugoslavia.*

STERN, William, psychologist; b. Berlin, Germany, Apr. 29, 1871; pres., Breslau (now Wroclaw, Poland), Hamburg, Germany. Author: Zur Psychologie der Aussage, 1902; Die differentielle Psychologie und ihre methodischen Grundlagen, 1911; Person und Sache, 3 vols., 1906-24; Psychologie der frühen Kindheit, 1914; Die Intelligenz der Kinder und Jugendlichen, 1920; Allgemeine Psychologie auf personalistischer Grundlage, 1935. Studied psychology of youth, talent. Died Poughkeepsie, N.Y., Mar. 27, 1938.

STERN, William Louis, Am. botanist; b. Paterson, N.J., Sept. 10, 1926; s. Abram and Rose (Chrisman) S.; B.S., Rutgers U., 1950; M.S., U. Ill., 1951, Ph.D., 1954; m. Floraet Selma Tanis, Sept. 4, 1949; children—Susan Myra, Paul Elihu. Instr. wood anatomy Sch. Forestry Yale, 1953-55, asst. prof., 1955-60; curator div. plant anatomy Smithsonian Inst., 1960-64, chmn. dept. botany, 1964-67; prof. botany U. Md., College Park, 1967-—; expert FAO-UN, Philippines, 1963-64. Mem. Bot. Soc. Am., Internat. Assn. Wood Anatomists, Soc. Econ. Botany, Assn. For Tropical Biology, A.A.A.S., Am. Inst. Biol. Scis., Torrey Bot. Club, N.E. Bot. Soc., Soc. Tropical Foresters. Editor: Tropical Woods, 1953-60. Contbr. numerous articles to profl. jours. Research in relationships of poorly known plant families through use of comparative plant anatomy and allied bot. disciplines, anat. structure, plant phylogeny, systematic arrangement of wood collections, wood collecting for sci. purposes. Office: Dept. Botany, U. Md., College Park, Md. 20740.*

STERNBERG, Charles Hazellus, Am. naturalist; b. Middleburg, N.Y., June 15, 1850; s. Levi and Margaret Levering (Miller) S.; ed. Hartwick Sem. (N.Y.), Ia. Lutheran Coll. (Albion), Kan. State Agrl. Coll.; A.M., (hon.), Midland Coll., 1911; m. Anna Musgrove Reynolds, July 7, 1880; children—George Fryer, Charles Mottram, Maud, Levi. In charge of parties collecting fossils for Prof. E. D. Cope, 1876-79, 94, 96, 97; in charge for Prof. Agassiz, 1881, 82; for Prof. O. C. Marsh of Yale, 1884, and of expdns. for

Munich (Bavaria) Paleontol. Mus., 1892, 95, 1901, 02, 05; in charge collecting party for Geol. Survey of Can., and vertebrate paleontol. lab. at Victoria Meml. Mus., Ottawa, Can., head collector and preparator of vertebrate fossils Victoria Meml. Mus., Geol. Survey of Can. until 1916. conducted own lab., Lawrence, Kan., from 1917. Discovered 2 nearly complete skeletons 32 ft. long, Red Deer River, Alberta, 1916; also large skeleton Dimetrodon Permian of Tex., 1916, in Nat. Mus.; conducted expdn. to Permian of Tex. and Kan. Chalk, 1918, to Kan. Chalk and to Tex., in 1919; found four skeletons of Mosasaurs, 2 Pterodactyls, 3 of the great fish, a fine skeleton of Equus Scotti; explored the San Juan Basin, N.M., 1921-23; discovered new genera of ceratopsians and duckbilled dinosaurs and many turtles. Fellow A.A.A.S.; mem. Kan. Acad. Sci. (life), Soc. Am. Vertebrate Paleontologists. Author: The Life of a Fossil Hunter, 1909; Hunting Dinosaurs on Red River, 1916. Died July 20, 1943.

STERNBERG, George Miller, Am. bacteriologist; b. Hartwick Sem., Otsego County, N.Y., June 8, 1838; s. Levi and Margaret Levering (Miller) S.; M.D., Coll. Phys. and Surg. (Columbia), 1860; LL.D., U. Mich., 1894, Brown U., 1896; m. Martha L. Pattison, 1869. Apptd. asst. surgeon U. S. Army, 1861, ret. as brig. gen. surgeon gen., 1902; service began in army of Potomac, later in Dept. of Gulf; at end of Civil War in charge U. S. Gen. Hosp., Cleve.; served through cholera and yellow fever epidemics; had command of med. service in war with Spain, 1898; mem. and sec. Havana Yellow Fever Commn. Nat. Bd. Health, 1879. Author: Photo-Micrographs, and How to Make Them, 1883; Bacteria; Malaria and Malarial Diseases, 1884; Manual of Bacteriology, 1893; Text-Book of Bacteriology, 1895; Immunity, Protective Inoculations, and Serum-Therapy, 1897; Infection and immunity, 1903. Discovered (independently of Louis Pasteur) small diplococci in sputum of patients with lobar pneumonia, circa 1881; discovered pneumococci in saliva of healthy persons, also produced septicemia in rabbits by injection of human saliva, circa 1881; 1st to photograph Mycobacterium tuberculosis, circa 1893. Died Nov. 3, 1915.

STERNBERG, Hilgard O'Reilly, geographer; b. Rio de Janeiro, Brazil, July 5, 1917; s. Bruno Ludwig and Johanna Mary (O'Reilly Begg) S.; Bacharel Universidade do Brasil, 1940, Licenciado, 1941, Doutor, 1958; Ph.D., La. State U., 1956; Docteur Honoris Causa Université de Toulouse, 1965; m. Carolina da Silveira Lobo, July 28, 1942; children—Hilgard O'Reilly, Maria Inês, Ricardo, Leonel, Cristina. With U. Brasil, 1942-64, prof., 1944-64, head geography dept., 1958-64; prof. U. Cal., Berkeley, 1964—; vis. prof. Ind. U., 1959, U. Heidelberg, 1961, Stockholm Sch. Econs. 1961, U. Cal., Los Angeles, 1963, Columbia U., 1963-64; mem. advisory com. arid zone research UNESCO, 1955-57. Mem. Gesellschaft für Erdkunde zu Berlin, Societé de Géographie de Paris, Societé Serbe de Géographie, Royal Geograph. Soc., Deutsche Akademie der Naturforscher Leopoldina, Sociedade de Geografia de Lisboa, Assn. des Géographes Française, Suomen Maantieteellinen Seura, Academia Brasileira de Ciências, Associacao dos Geografos Brasilerios, Assn. Am. Geographers, Am. Geophys. Union. Author: A Agua e o Homen na Varzea do Careiro, 1956; A Terra e o Homem nos Trópicos, 1965. Research on hydrology of Amazon River; geomorphology of alluvial streams and flood plains; studies of erosion, floods and droughts; regional geography of Brazil; tropics; settlement geography. Home: 466 Michigan Av., Berkeley, Cal. 94707.*

STERNBERGER, Ludwig Amadeus, physician; b. Munich, Germany, May 26, 1921; s. Hugo and Emy (Seligstein) S.; B.A., Am. U. Beirut, 1941, M.D., 1945; m. Nancy Jeanne Hoy, Dec. 13, 1961. Came to U.S., 1948, naturalized, 1957. Practice medicine, specializing in allergy, Chgo., 1953-55, Balt., 1957—; faculty Sch. Medicine, Northwestern U., 1953-55; chief pathology br. Edgewood Arsenal, Md., 1957-67, acting chief physiology dept. Med. Research Lab., 1967—. research asso. surg. research Sinai Hosp., Balt., 1966—; faculty Sch. Medicine, John Hopkins, 1966—. Recipient U.S. Army Research and Devel. Achievement award, 1962, First prize CRDL Sci. Conf., 1963. Fellow Am. Acad. Allergy; mem. Am. Assn. Immunologists, Soc. Exptl. Biology and Medicine, Am. Heart Assn., Histochem. Soc. Editor: Symposium on Biochemical Electron Microscopy, 1966. Research and numerous publs. on electron immunocytochemistry, viruses and cell metabolism, molecular biology of mammalian cell. Office: Physiology Dept., Med. Research Lab., Edgewood Arsenal, Md. 21010.*

STERNE, Theodore Eugene, Am. physicist; b. N.Y.C., Nov. 23, 1907; s. Eugene Washington and Dora (Kohn) Stern; B.Sc., Princeton, 1928; Ph.D., Trinity Coll., Cambridge U., 1931; postgrad. (Nat. Research fellow) Harvard, Mass. Inst. Tech., 1931-33; M.A. (hon.), Harvard, 1956; m. Grace Isabel DeRoo, Aug. 5, 1932 (div. Nov. 1964); children—Theodore D., John R.; m. 2d, Lois Dorothy Cremins Isenberg, Nov. 28, 1964. Mem. staff Harvard, 1933-41, prof. astrophysics, 56-59; chief ballistician Ballistic Research Labs., Aberdeen Proving Ground, Md., 1946-56, chief spl. problems br., 1941-45, computing lab., 1945-47, 52-53, terminal ballistic lab., 1946-52, sci. adviser to dir., 1953-56; cons. Operations Research Office, Johns Hopkins, 1954-59, mem. staff, 1959-61; asso. dir. Astrophys. Obs., Smithsonian Instn., Cambridge, Mass., 1956-59; mem. staff Research Analysis Corp., McLean, Va., 1961-65; mem. staff Inst. for Def. Analyses, Arlington, Va., 1965—. Fellow Am. Acad. Arts & Scis., Am. Phys. Soc., Royal Astron. Soc.; mem. Am. Astron. Soc., Astron. Soc. Pacific, Operations Research Soc. Am., Am. Ordnance Assn., Philos. Soc. Washington, Phi Beta Kappa, Sigma Xi. Author: An Introduction to Celestial Mechanics, 1960. Inventor in mil. field. Research in statis. mechanics, binary and variable stars, astrophysics, statis. inference, ballistics, mil. operations, celestial mechanics, theory of combat. Home: 5502 Greystone St., Chevy Chase, Md. 20015. Office: 400 Army Navy Dr., Arlington, Va. 22202.*

STERNER, James Hervi, Am. physician; b. Bloomsburg, Pa., Nov. 14, 1904; s. Lloyd Parvin and Nora (Finney) S.; B.S., Pa. State U., 1928; M.D., Harvard, 1932; m. Frances Elkavich, Apr. 11, 1932; children—James, Susan, John. With Eastman Kodak Co., Rochester, N.Y., 1936—, asso. med. dir., 1948-51, med. dir., 1951—; faculty U. Rochester Sch. Medicine and Dentistry, 1940—, clin. prof. preventive medicine and community health, 1961—. Vice chmn. adv. com. for biology and medicine AEC, 1960-66; mem. nat. adv. com. on environmental health USPHS, 1964—; chmn. Nat. Air Conservation Commn., 1966—. Recipient Knudsen award Indsl. Med. Assn., 1957; 25th Albert David Kaiser medal Rochester Acad. Medicine, 1963. Mem. Am. Coll. Preventive Medicine (past pres.), Nat. Health Council (past pres.), Am. Acad. Occupational Medicine (award of honor 1959, past pres.) Am. Indsl. Hygiene Assn. (Cummings award 1955, past pres.), Council on Occupational Health (past chmn.), A.M.A. (chmn. council on environmental and pub. health 1963—), Phi Kappa Phi, Alpha Omega Alpha. Contbr. numerous articles to jours., chpts. to textbooks. Research on occupational and environmental health. Home: 195 Hollywood Av., Rochester, N.Y. 14618.*

STERNHEIMER, Rudolph Max, physicist; b. Saarbrucken, Germany, Apr. 26, 1926; s. John and Cornelia (Kahn) S.; came to U.S., 1941, naturalized, 1947; B.S., U. Chgo., 1943, M.S., 1946, Ph.D., 1949; m. Elizabeth Pieczur, May 8, 1952. Staff, Los Alamos Sci. Lab., 1949-51; asso. physicist Brookhaven Nat. Lab., Upton, N.Y., 1952-60, physicist, 1961-64, sr. physicist, 1965—. Fellow Am. Phys. Soc.; mem. Phi Beta Kappa. Contbg. author: Nuclear Physics, Vol. A, 1961, Vol. B, 1963. Research in atomic and nuclear physics, including theory of energy loss of charged particles, quadruple antishielding factors for ions and atomic states, polarization of nucleons in scattering by nuclei, mass relations for elementary particles and resonances. Home: 141 E. 55th St., N.Y.C. 10022. Office: Brookhaven Nat. Lab., Upton, N.Y. 11973.*

STETEFELDT, Carl August, metallurgist; b. Holzhausen, Gotha, Germany, Sept. 28, 1838; s. August Heinrich Christian and Friederika Christiane (Credner) S.; grad. U. Gottingen (Germany), 1862; m. Dec. 31, 1872. Came to U. S., 1863. Asst. to Charles A. Joy (prof. chemistry Columbia), N.Y.C.; asst. to cons. firm Aedlberg & Raymond, 1864; partner (with John H. Bialt) assay office and cons. bus., Austin, Nev., 1865; builder 1st lead blast furnace in dist. of Eureka (Nev.); designer Stetefeldt furnace (a metall. advance in processing sulphide ores containing gold and silver by chlorination process); v.p. Am. Inst. Mining Engrs., contbr. to inst.'s Trans. Author: The Lexivication of Silver-ores with Hyposulphite Solutions, 1888. Died Oakland, Cal., Mar. 17, 1896.

STETTEN, DeWitt, Jr., Am. biochemist; b. N.Y.C., May 31, 1909; s. DeWitt and Magdalen (Ernst) S.; B.A., Harvard, 1930; M.D., Columbia U., 1934, Ph.D., 1940; m. Marjorie A. Roloff, Feb. 7, 1941; children—Gail, Nancy, Mary, George D. Mem. faculty Columbia U., N.Y.C., 1940-47; asst. prof. biol. chemistry Harvard Med. Sch., Boston, 1947-48; dir. div. nutrition and physiology Pub. Health Research Inst. of N.Y.C., 1948-54; dir. intramural research Nat. Inst. Arthritis and Metabolic Diseases, NIH, Bethesda, Md., 1954-62; dean Rutgers U. Med. Sch., New Brunswick, N.J., 1962—. Mem. nat. sci. adv. com. Okla. Research Found., 1963-66, chmn., 1966; mem. div. med. scis. NRC, 1965—. Bd. dirs. Found. for Advanced Edn. in Scis. Recipient Alvarenga award Phila. Acad. Medicine, 1954; Banting medal Am. Diabetes Assn., 1957. Mem. Am. Soc. Biol. Chemists, Harvey Soc., Am. Chem. Soc., A.A.A.S. (v.p. 1962), Soc. Exptl. Biology and Medicine. Author: (with A. White, P. Handler and E. Smith) Principles of Biochemistry, 1952; also numerous articles. Mem. editorial bds. Jour. Biol. Chemistry, Jour. Internal Medicine, Sci., Jour. Chronic Diseases, Metabolism, Perspectives in Biology and Medicine. Research in intermediary metabolism of fats, carbohydrates and purines, elucidation of metabolic defects in certain diseases, especially diabetes, obesity, gout. Home: Highwood, Easton Av., Somerset, N.J. 08873. Office: Rutgers Med. Sch., New Brunswick, N.J. 08903.*

STETTEN, Marjorie Roloff, Am. biochemist; b. N.Y.C., July 13, 1915; d. George F. and Belle (Berel) Roloff; B.S., Rutgers U., 1937; Ph.D., Columbia U., 1944; m. DeWitt Stetten, Jr., Feb. 7, 1941; children—Gail, Nancy, Mary, George. With Columbia U. Coll. Phys. and Surg., 1940-47, research asso., 1942-47; research fellow Harvard Med. Sch., 1947-48; asso. div. nutrition, physiology Pub. Health Research Inst. City N.Y., 1948-54; biochemist Nat. Inst. Arthritis and Metabolic Diseases, NIH, Bethesda, Md., 1954-63; research prof. exptl. med. Rutgers Med. Sch., 1963—. Mem. Am. Soc. Biol. Chemistry, Marine Biol. Lab. Research and publs. on intermediary metabolism, amino acids and carbohydrate, including hydroxyproline glycogen, amino imidazole carboxamide; enzyme activities. Home: Highwood, Easton Av., Somerset, N.J. 08873. Office: Rutgers Med. Sch., New Brunswick, N.J. 08903.*

STETTER, Georg Karl Friedrich, Austrian physicist; b. Vienna, Austria, Dec. 23, 1895; s. Carl and Bertha (Schirer) S.; Dr.phil., U. Vienna, 1922; m. Marianne Kauscheder, Aug. 2, 1923; 1 dau., Elfriede. Asst., 2d Phys. Inst., U. Vienna, 1922-38, prof, 1938-45, head 1st Phys. Inst., 1953—. Recipient Schrödinger prize, 1966. Mem. Austrian Acad. Scis. (Haitinger prize 1926), Leopoldina. Research, numerous publs. on nuclear physics, aerosol physics. Address: 1 Littrowgasse, 1180 Vienna, Austria.*

STETTER, Hermann, German chemist; b. Bonn, Germany, May 16, 1917; s. Oskar and Clara (Schub) S.; student U. Bonn, U. Leipzig, 1936-40; Promotion, U. Bonn., 1947; m. Elisabeth Steinhaus, Mar. 12, 1944; children—Jörg, Eva. With U. Bonn., 1947-55, privatdozent, 1952-55; asso. prof. U. Munich (Germany), 1955-60; prof., dir. Inst. Organic Chemistry, U. Aachen (Germany), 1960—. Mem. Gesellschaft Deutscher Chemiker. Author: Enzymatische Analyse, 1951; also numerous articles. Research on syntheses and chemistry of long chain carboxylic acids, chemistry of adamantane and heterocycles of same structure, macrocyclic compounds. Home: 10 Am Pannhaus, Laurensberg, West Germany. Office: Institut für Organische Chemie der T.H., Aachen, West Germany.*

STEUNENBERG, Robert Keppel, Am. chemist; b. Caldwell, Ida., Sept. 18, 1924; s. Ancil Keppel and Lorraine (Brooks) S.; B.A., Coll. Ida., 1947; student Ida. State Coll., 1943-44; Ph.D., U. Wash., 1951; m. Jean Smylie, July 20, 1947. Chemist chem. engring. div. Argonne (Ill.) Nat. Lab., 1951—. Mem. Am. Chem. Soc., Am. Nuclear Soc., Research Engring. Soc. Am. Research, publs. on chemistry of fluorocarbons; methods of reprocessing nuclear reactor fuels using fluorides and pyrometals. Home: 60 Golden Larch Dr., Naperville, Ill. 60540. Office: 9700 S. Cass Av., Argonne, Ill. 60439.*

STEVELS, Johannes Marinus, Dutch chemist; b. Semarang, Indonesia, Apr. 1, 1913; s. Johannes Marinus and Maria J. M. (Pel), S.; D.Sc., U. Leiden (Netherlands), 1937; Schunck fellow U. Manchester (Eng.), 1937-38; m. Suzanna Bastiana Dronkers, Dec. 15, 1939; children—Josina M. C. (Mrs. Egbert Westphal), Albert L. N., Suzanna M. M. With Philips Research Labs., Eindhoven, Netherlands, 1938—, chief chemist, 1956—; prof. inorganic chemistry Eindhoven Tech. U., 1967—; prof. Glass Inst., 1966. Fellow Soc. Glass Tech.; mem. Inernat. Commn. on Glass (pres. 1966—), Nat. Com. Netherlands Glass Industry (sec. 1959—), Continental Sci. Union Glass (v.p. 1950—), German Glass Tech. Soc., Royal Netherlands Chem. Soc., Netherlands Ceramic Soc. Author: Polarisability and Cohesion Energy, 1937; Progress in the Theory of the Physical Properties of Glass, 1948; (with F. W. Klaazenbeek) Bibliography of Glass Literature, 1964; also numerous articles. Research on properties of vitreous systems: dielectric, optical properties, formation of color centers under radiation, spectra in relation to structure; influence of cooling rate on glass formation. Home: 32 Petrus Dondersstraat. Office: Research Labs., Philips' Gloeilampenfabriken, Eindhoven, Netherlands.*

STEVENS, Carl M(antle) II, Am. biochemist; b. Washington, Oct. 31, 1915; s. Neil Everett and Maude Minerva (Bradford) S.; B.A., Am. U., 1937; Ph.D. (fellow 1939-41), U. Ill., 1941; m. Mary Nasmith Corbett, Dec. 21, 1949; children—Ann Bradford, Nancy Gray, Rae Dickson, Alison Mary. Postdoctoral asst. Cornell U. Med. Coll., 1941-45; mem. faculty Wash. State U., Pullman, 1946—, prof. biochemistry; 1951—, chmn. dept. chemistry, 1961—; vis. lectr. U. Ill., 1957. Merck sr. fellow Cal. Inst. Tech. 1954-55. Mem. A.A.A.S., Am. Chem. Soc., Harvey Soc., Harvey Soc., Am. Soc. Biol. Chemists. Research on chemistry and biosynthetic pathway of pencillins; mechanism of action of mustard gas in producing genetic mutations; study of pathway of biosynthesis and metabolism of certain amino acids; biochem. genetics. Home: 1806 Creston Lane, Pullman, Wash. 99163.*

STEVENS, Charles Edward, Am. vet. physiologist; b. Mpls., June 5, 1927; s. Fred H. and Irene (Lamb) S.; B.S., U. Minn., 1951, D.V.M., 1955, M.S., 1955, Ph.D., 1958; m. Jacqueline H. Langlois, Sept. 10, 1952; children—Leslee Marie, Judith Mae, David Edward, Laura Ann. Research fellow Agr. Research Service, U. Minn., 1955-58, research asso., 1958-60, research physiologist, 1960-61; asso. prof. dept. physiology N.Y. State Vet. Coll., Ithaca, 1961-66, prof., 1966—. Mem. Am. Vet. Med. Assn., Am. Physiol. Soc., Am. Soc. Vet. Physiologists and Pharmacologists (pres. elect), Research Workers in Animal Diseases. Research and publs. in digestive and absorptive functions of ruminant forestomach, function and nervous reflex regulation of its motility and mechanisms by which inorganic ions and short-chain fatty acids are absorbed into the blood. Home: 1413 Ellis Hollow Rd., Ithaca, N.Y. 14850.*

STEVENS, Edwin Augustus, Am. inventor; b. Hoboken, N.J., July 28, 1795; s. John and Rachel (Cox) S.; m. Mary B. Picton, 1836; m. 2d, Martha Bayard Dod, 1854; 9 children, including Mary Picton Steven Barnett. Inventor, patentee (with brother Robert) a plow, 1821; took charge Union Stage Coach line between N.Y.C. and Phila., 1825, purchased it with his brothers, 1827; mgr. Camden and Amboy R.R. Transp. Co. (1st railroad in N.J.), 1830-65; inventor, patentee (with brother Robert) closed fire room system of forced draft, 1842, 1st used on Robert's steamboat N. America; permitted by Navy Dept. to build armored vessel designed by Robert, 1842; founder Stevens Inst. Tech. Hoboken; Stevens Battery begun, 1852, never finished. Died Paris, France, Aug. 7, 1868.

STEVENS, Frank Lincoln, Am. botanist; b. nr. Syracuse, N.Y., Apr. 1, 1871; s. Henry Benjamin and Helen C. (Lincoln) S.; B.L., Hobart Coll., 1891; B.S., Rutgers Coll., 1893, M.S., 1897; postgrad. Ohio State U., 1894-96; Ph.D. magna cum laude (fellow), U. Chgo., 1900; traveling fellow univs. Bonn and Halle, 1900-01; studied Naples Zool. Labs.; Sc.D., U. San Marcos, Lima, Peru, 1925; LL.D. U. Glasgow (Scotland), 1928; m. Adeline T. Chapman, June 16, 1897. Student asst. Rutgers Coll. and N.J. Agrl. Expt. Sta. 1891-93; tchr. science Racine Coll., 1893-94; tchr. chemistry and botany, high sch., Columbus, O., 1894-97; analyst Chgo. Drainage Canal Investigation 1899-1900; instr. biology, N.C. Coll. Agr. and Mech. Arts, 1901-02, prof. botany and vegetable pathology, 1902-12; for years biologist and head dept. of plant diseases N.C. Agrl. Expt. Sta.; dean Coll. Agr. and Mech. Arts, U. P.R., 1912-14; prof. plant pathology U. Ill., 1914-34; Bishop Museum fellow Yale Univ., 1921-22; prof. plant pathology U. Philippines, 1930-31. Lectr. at farmers' insts. Joint author: Agriculture for Beginners; The Hill Readers, 1906; Practical Arithmetic, 1909; Diseases of Economic Plants, 1910. Author: The Fungi That Cause Plant Disease; Plant Disease Fungi. Died Aug. 18, 1934.

STEVENS, George Thomas, Am. physician; b. Essex County, N.Y., July 25, 1832; s. Chauncey Coe and Lucinda (Hoadley) S.; ed. in N.Y. State; M.D., Castleton Med. Coll., 1857; Ph.D., Union Coll., 1877; m. Harriet W. Wadhams, 1861. Commd. surgeon 77th N.Y. Vols., 1861; prof., physiology and diseases of eye Union U., 1870-75; in practice N.Y. 1880—. Recipient highest prize from Royal Acad. of Medicine of Belgium for treatise on Functional Diseases of the Nervous System, 1883. Author: A Treatise on the Motor Apparatus of the Eyes, 1905; A Series of Studies of Nervous Diseases, 1911. Pub. early classification of motor anomalies of ocular muscles, 1886. Died Jan. 30, 1921.

STEVENS, Graeme Roy, New Zealand geologist; b. Lower Hutt, New Zealand, July 17, 1932; s. Alfred Laurie and Ida May (Bailey) S.; B.Sc., Victoria U., Wellington, New Zealand, 1953, M.Sc. with 1st class honors, 1955; Ph.D., U. Cambridge (Eng.), 1959; m. Diane Louise Ollivier, Oct. 20, 1962; children—Peter Michael, Robert Andrew. Jr. lectr. Victoria U., 1955; paleontologist New Zealand Geol. Survey, Lower Hutt, 1956—; Shell Postgrad. scholar Cambridge U., 1956-59; New Zealand Dept. Sci. and Indsl. Research, 1960. Mem. New Zealand Geol. Soc. (McKay Hammer award 1956), Royal Soc. New Zealand (Hamilton prize 1959), New Zealand Assn. Scientists, Paleontol. Assn. Author: New Zealand Geological Survey Palaeontological Bulletin, 36, 1965; also articles. Research on Quaternary period in New Zealand; stratigraphic and paleontol. studies of New Zealand and Indo-Pacific Jurassic and Cretaceous belemnites and Jurassic ammonites; studies of Jurassic and Cretaceous biogeography and climates. Home: 2 Christina Grove, Normandale, Lower Hutt. Office: New Zealand Geol. Survey, Dept. Sci. and Indsl. Research, Lower Hutt, New Zealand.*

STEVENS, John, Am. engr., inventor; b. N.Y.C., 1749; s. John and Elizabeth (Alexander) S.; grad. Columbia, 1768; m. Rachel Cox, Oct. 17, 1782; at least 7 children, including John Cox, Robert Livingston, Edwin Augustus, Mary, Harriet. Studied law, 1768-71, apptd. atty., N.Y.C., 1771; served from capt. to col., obtaining loans for Continental Army during Revolutionary War; loan commr. for Hunterdon County (N.J.); treas. N.J., 1776-79; surveyor gen. Eastern div. N.J., 1782-83; instrumental in framing 1st patent laws, 1790; became cons. engr. for Manhattan Co. (organized to furnish adequate water supply to N.Y.C.), circa 1880; became pres. Bergen Turnpike Co., 1802; received patent for multitubular boiler, 1803; his steamboat Little Juliana (operated by twin screw propellers) put into use on Hudson River, 1804; attempted to operate regular line of steamboats on Hudson between N.Y.C. and Albany and on other inland rivers, prevented by lawsuits; sent the Phoenix (1st sea-going steamboat in world) to Del., 1807, to Phila., 1809; built the Juliana, began regular ferry service, 1811; obtained 1st Am. railroad authorization from N.J. Assembly in 1815; authorized by Pa. Legislature to build Pa. R.R., 1823; designed, built exptl. locomotive on his estate in Hoboken, N.J. (1st Am.-made steam locomotive though never used for actual service), 1825; proposed a vehicular tunnel under the Hudson as well as an elevated railroad system for N.Y.C. Died Hoboken, Mar. 6, 1838.

STEVENS, Kingsley Morton, Am. physician; b. Lynchburg, Va., June 4, 1922; s. Arthur Kingsley and Susie (Krebbs) S.; B.S., Lynchburg Coll., 1943; M.D., Harvard, 1947; m. Virginia Dean Coffin, July 20, 1951; children—Anne, Heather, Lynn, Janet. AEC-NRC fellow Bowman-Gray Sch. Medicine, Winston-Salem, N.C., 1948-49, U. Chgo., 1949-50; research asso. surg. research unit Brooke Army Hosp., 1951-53; NIH fellow Hall Inst., Melbourne, Australia, 1953-54; biochemist Walter Reed Army Med. Sch., Washington, 1954-55; immunologist Merck Sharp and Dohme Research Labs., West Point, Pa., 1955-59; asst. prof. medicine U. Ky., Lexington, 1959-61, asso. prof., 1961—, asst. prof. microbiology, 1962—. Recipient Lederle Med. Faculty award, 1960-63; fellow Commonwealth Fund, 1965-66. Mem. Am. Assn. Immunologists, Soc. for Exptl. Biology and Medicine. Author: The Ecology and Etiology of Human Disease, 1967; also articles. Introduced concept that oxygen supply is limited and competition for supply by body cells and bacteria can cause either death of organism directly or by prodn. of cancer, arteriosclerosis, others. Home: 787 Hildeen Dr., Lexington, Ky. 40502.*

STEVENS, Neil Everett, Am. plant pathologist; b. Portland, Me., Apr. 6, 1887; s. Thomas Jefferson and Hattie (Mantle) S.; B.A., Bates Coll., 1908; Ph.D., Yale, 1911; m. Maude Bradford, Aug. 31, 1914; children—Russell Bradford, Carl Mantle II, Mary Christine, Inst. botany Kan. State Coll., 1911-12; pathologist Bur. Plant Industry, U. S. Dept. Agr., 1912-28, sr. pathologist 1928-36; adj. prof. George Washington U., 1931-36; prof. botany U. Ill., 1936-47, prof. plant pathology since 1947; sr. specialist Dept. Agr., summers 1937-44. Del., Internat. Botanical Congress, 1930, 35. Mem. Am. Phytopathol. Soc. (v.p. 1933; pres. 1934), Bot. Soc. Washington (sec. 1927; pres. 1931), A.A.A.S. (council 1947-50); v.p. and chmn. Sect. G, 1939), Mycol. Soc, Am. (council 1932; v.p. 1944), Bot. Soc. Am. (v.p. 1940, pres., 1946), NRC (mem. exec. com. dir. of biology and agr. 1944), Sigma Xi, Phi Beta Kappa. Advisory editor Rev. since 1935. Writer of government bulls. and tech. papers on diseases of plants and history of botany. Died June 26, 1949; buried Urbana, Ill.

STEVENS, Robert Livingston, Am. inventor; b. Hoboken, N.J., 1787; s. John and Rachel (Cox) L.; assisted father in building paddle steamer Phoenix, 1807, ferry boat Juliana (1st steam-ferry system), 1811; an organizer Camden & Amboy R.R. and Transp. Co., 1830, pres., chief engr.; began 1st steam ry. service in N.J.; engr. Mohawk and Hudson R.R. Leader in naval architecture; designed and built at least 20 steamboats and ferries; tested 1st steam ship with ship propellors (invented by father), 1904-05; made 1st trip in steam ship from N.Y. to Del., 1808; developed unified design for steam ships for waterways of U. S.; inventor, patentee (with brother Edwin) a plow, 1821, closed fire room system of forced draft (1st used on steamship N. America), 1842; built spl. boiler; inventor steam engine for belfry, also Vignole iron bands; introduced forced draft firing system under boilers, split paddle wheel, hog framing for boats, modern ferry slip; developed percussion shell for naval use which could be fired from cannons, during War of 1812; designed trail and hoor-headed spine, circa 1830; interested in use of armor on ships of war; designed and built yacht Maria (fastest ship of her day), 1850. Died Hoboken, Apr. 20, 1856.

STEVENS, S(tanley) S(mith), Am. psychophysicist; b. Ogden Utah, Nov. 4, 1906; s. Stanley S. and Adeline (Smith) S.; B.A., Stanford, U. 1931; Ph.D., Harvard, 1933; m. Maxine Leonard, Mar. 28, 1930 (dec. Oct. 1956) 1 son, Peter Smith m. 2d, Geraldine Stone, Apr. 11, 1963. Faculty, Harvard, Cambridge, Mass., 1933—, prof. psychology, 1946-62, prof. psychophysics, 1962—, dir. psycho-acoustic lab., 1944-62, dir. psychophysics lab., 1962—. Mem., chmn. various coms., NDRC, NRC, NIH, Nat. Acad. Scis. Recipient Howard Crosby Warren medal Soc. Exptl. Psychologists, 1943, Presdl. Certificate of Merit, 1948, Sci. award Am. Psychol. Assn., 1960 award Beltone Inst., 1966; NRC fellow, 1934-35. Fellow Acoustical Soc. Am. (exec. council 1946-49); mem. Nat. Acad. Scis., Am. Philos. Soc., Am. Acad. Arts & Scis., Soc. Exptl. Psychologists, Am., Eastern (pres. 1960-61) psychol. assns., Biophys. Soc., Am. Physiol. Soc., Psychonomic Soc. (gov. 1960-61), Am. Assn. U. Profs., Am. Inst. Physics, Optical Soc. Am., Philosophy Sci. Assn. (gov. 1957—), A.A.A.S. (v.p. sect. 1955), Sigma Xi, Phi Beta Kappa. Author: (with Hallowell Davis) Hearing: Its Psychology and Physiology, 1938; (with W.H. Sheldon and W.B. Tucker) The Varieties of Human Physique, 1940; (with W.H. Sheldon) The Varieties of Temperament, 1942; (with others) Hearing Aids: An Experimental Study of Design Objectives, 1947; (with F. Warshofsky) Sound and Hearing, 1965. Editor: Handbook of Experimental Psychology, 1951. Research, publs. on sensory processes, particularly hearing and vision; developed psychophys. law of sensation intensity. Home: 70 Francis Av., Cambridge, Mass. 02138.*

STEVENS, Thomas Stevens, Brit. chemist; b. Renfrew, Scotland, Aug. 8, 1900; s. John and Jane (Irving) S.; B.Sc., Glasgow (Scotland) U., 1921; D.Phil., Oxford U., 1925; m. Janet Wilson Forsyth, July 13, 1949. Asst. lectr. U. Glasgow, 1921-23, lectr., 25-47; faculty U. Sheffield (Eng.), 1947-66, prof. chemistry, 1963-66. Fellow Royal Soc., 1963, Royal Soc. Edinburgh; mem. Chem. Soc., Royal Inst. Chemistry, Soc. Chem. Industry. Contbg. author: Chemistry of Carbon Compounds, 1957-60; also articles. Research on electrophilic molecular rearrangements, alkaloid synthesis. Home: 313 Albert Dr., Glasgow S.1, Scotland.*

STEVENS, William Harmer, Canadian chemist; b. London, Ont., Can., Apr. 12, 1918; s. Alfred N. and Vallerie (Robison) S.; B.Sc., Queens U., Kingston, Ont., 1940, M.S., 1941; Ph.D., McMaster U., Hamilton, Ont., 1951; m. Mary E.T. Baker, May 9, 1942; children—Margaret (Mrs. Andrew W. Gemmell), Ann, Katharine, Shirley, Robert. Research officer dept. nat. def. Royal Mil. Coll., Kingston, 1942-45; lectr. Queen's U., 1945-47; staff devel. chemistry br. Atomic Energy of Can., Ltd., Chalk River, Ont., 1947—, sr. research officer, 1956, head devel. chemistry br., 1956—. Fellow Chem. Inst. Can.; mem. Chem. Soc. (London, Eng.). Research and publs. on chem. warfare, activated carbon adsorption, mustard gases, flame thrower fuels, carbon-14, infrared spectroscopy, fission gas diffusion in uranium dioxide, organic coolants for nuclear reactors. Patentee D20 analysis by infrared, electrolytic separation of hydrogen and deuterium. Home: 20 Le Caron St., Deep River, Ont. Office: Atomic Energy of Can., Ltd., Chalk River, Ont., Can.*

STEVENSON, Alan, Scottish civil engr.; b. Edinburgh, Scotland, 1807; s. Robert and Jean (Smith) S.; M.A. (Fellowes prize), Edinburgh U., 1826; LL.B. (hon.) U. Glasgow (Scotland). Succeeded father as engr. Scottish Lighthouse Commn., 1843; designed Skerryvore lighthouse tower, completed 1843. Recipient medals Russia, Prussia, Netherlands. Fellow Royal Soc. Edinburgh (council 1843-45), Instn. C.E.'s. Author: Account of the Skerryvore Lighthouse, 1848, expanded into Rudimentary Treatise, 1850. Contbr. articles Ency. Britannica. Introduced prismatic rings in constn. lighthouses; designed, carried out notable improvements on dioptric apparatus for lighthouses; built 10 lighthouses. Died Dec. 23, 1865.

STEVENSON, Alan Carruth, med. geneticist; b. Glasgow, Scotland, Jan. 27, 1909; s. Allan and Christina (Lawson) S.; B.Sc., Glasgow U., 1930, M.B., Ch.B., 1933, M.D., 1946; M.A., Oxford (Eng.) U., 1966; m. Annie Gordon Sheila Steven, Mar. 25, 1937; children—Robert, David, Aileen Gillian. Dep. med. officer health, Wakefield, Eng., 1936-39; reader pub. health London Sch. Hygiene and Tropical Medicine, 1946-48; prof. social and preventive medicine Queen's U., Belfast, N. Ireland, 1948-58; dir. population genetics research unit Med. Research Council, Oxford, 1958—. Mem. expert panel human genetics WHO, 1957—. Fellow Royal Coll. Physicians London; mem. Royal Soc. Medicine, Genetical Soc., Soc. for Study Fertility. Author: Recent Advances on Social Medicine, 1950; also articles. Research on frequencies of genetics traits and mutation rates of many genes in man; med. genetics. Home: The Haven, Newland St., Eynsham, Oxford. Office: Med. Research Council Propulation Genetics Research Unit, Old Rd. Headington, Oxford, Eng.*

STEVENSON, Arthur Francis Chesterfield, physicist; b. Sellinge, Eng., July 11, 1899; s. Alfred Leonard and Amy (Jenour) S.; B.A., Cambridge (Eng.) U., 1922, M.A., 1926, Ph.D., 1932. Came to U.S., 1953, naturalized, 1963. Faculty math. U. Toronto (Ont., Can.), 1922-50; asso. prof. Wayne State U., Detroit, 1953-56, prof. physics, 1956-66. Fellow Royal Soc. Can., Am. Phys. Soc.; mem. Am. Assn. Physics Tchrs. Research on quantum mechanics of atoms, theories of electromagnetic wave properties and light scattering. Home: 88 Cromwell Rd., London, S.W.7, Eng.*

STEVENSON, Charles Edward, Am. chemist; b. Mt. Vernon, N.Y., Feb. 26, 1913; s. Walter Samuel and Lula May (Helreigel) S.; B.S., Pa. State U., 1934, Ph.D., 1941; m. Eleanor Clara Conrad, Apr. 24, 1942; children—Elizabeth Jean, Alan Charles, David Robert. With Standard Oil Devel. Co., 1942-45, Diamond Glass Co., 1946-47, Phillips Petroleum Co., 1954-59; asso. dir. chem. engring. div. Argonne Nat. Lab., Idaho Falls, Ida., 1947-53, supr. exptl. breeder reactor-II fuel cycle facility, 1960—. Mem. Am. Inst. Chem. Engrs., Am. Chem. Soc. (chmn. nuclear chem. and tech. div. 1965), Am. Nuclear Soc. (standards com.), A.A.A.S., Sci. Research Soc. Am. Editor: Process Chemistry. Contbr. articles to profl. jours. Developer mechanism of oxidation hydrocarbon lubricating oils, synthesis carboxylic acids through Grignard reaction, solvent extraction processes for recovery nuclear fuels, methods for converting radioactive wastes to stable solids. Home: 1207 10th St., Idaho Falls 83401. Office: Argonne Nat. Lab., P.O. Box 2528, Idaho Falls, Ida. 83401.*

STEVENSON, David, Scottish civil engr.; b. Edinburgh, Scotland, 1815; s. Robert and Jean (Smith) S.; ed. Edinburgh U. Mng. partner Stevenson engring.

firm, works improvement rivers No. Eng., Scotland; engr. No. lighthouse bd., from 1853. Mem. Royal Scottish Soc. Arts (pres. 1869), Société des Ingénieurs Civils (Paris), Instr. C.E.'s (Council 1877-83 contbr. Proceedings); Royal Soc. Edinburgh (v.p. 1873-77). Author: The Application of Marine Surveying and Hydrometry to the Practice of Civil Engineering, 1842; Our Lighthouses, 1864; Reclamation and Protection of Agricultural Land, 1874. Contbr. articles to Ency. Britannica. Constructed numerous beacons, lighthouses (devised aseismatic arrangement for their constrn. Japan); introduced use paraffin lighthouses, 1870. Died July 17, 1886.

STEVENSON, George Salvadore, Am. physician; b. Phila., Oct. 5, 1892; s. George Edwards and Anna Ida (Musso) S.; Sc.B., Bucknell U., 1915, Sc.M., 1919, Sc.D., 1940; L.H.D., Monmouth Coll., 1965; m. Amy Llewellyn Patterson, Sept. 2, 1920; children—Anne Elizabeth, Amy Llewellyn (Mrs. Charles Farrington Bond), William Chandler. Asst., N.Y. Psychiat. Inst., 1920-22; instr. Cornell U. Med. Sch., 1920-22; research staff mental retardation Tng. Sch., Vineland, N.J., 1922-24; dir. Psychiat. Clinics, asst. prof. U. Minn., 1924-26; div. dir., med. dir. Nat. Assn. for Mental Health, N.Y.C., 1926-59. Mem. nat. mental health council U.S. Health, Edn. and Welfare Dept., 1947—. Recipient Lord and Taylor award Lord & Taylor Corp., 1953. Mem. Am. Orthopsychiat. Assn. (pres. 1934—), Am. Psychiat. Assn. (pres. 1949-50), World Fedn.. for Mental Health (pres. 1961-62), Assn. for Research in Child Devel. Author: (with Geddes Smith) Child Guidance Clinics, 1934; Mental Health Plan, 1954; (with Harry Mielt) Master Your Tensions, 1959; also numerous articles. Research on cerebrospinal fluid pressure, staining of Spirochaeta pallida, physiology of feeble minded, problems of coll. students. Home: 940 W. Front St., Red Bank, N.J. 07701.*

STEVENSON, Ian, psychiatrist; b. Montreal, Can., Oct. 31, 1918; s. John and Ruth (Preston) S.; student U. St. Andrews, Scotland, 1937-39; B.Sc., McGill U., 1940, M.D., 1943; m. Octavia Reynolds, Sept. 13, 1947. Came to U. S., 1945, naturalized, 1949. Intern, asst. resident Royal Victoria Hosp., Montreal, Can., 1944-45; intern, resident St. Joseph's Hosp., Phoenix, 1945-46; fellow internal medicine Alton Ochsner Med. Found., New Orleans, 1946-47; Commonwealth fellow medicine Cornell U. Med. Coll., 1947-49; asst. prof. psychiatry La. State U. Sch. Medicine, 1940-52, asso. prof., 1952-57; prof. psychiatry, chmn. dept. psychiatry and neurology U. Va. Sch., Medicine, 1957-67, Alumni prof. psychiatry, 1967—. Diplomate Am. Bd. Psychiatry and Neurology. Fellow Am. Psychiat. Assn.; mem. A.M.A., Med. Soc. Va.; Am. Psychosomatic Soc., A.A.A.S. Am. Soc. for Psychical Research. Author: Medical History-Taking, 1960; Twenty Cases Suggestive of Reincarnation, 1966. Work on exptl. psychoses, parapsychology, psychotherapy; studied relationship between agt. and percipient in extrasensory perception. Home: Wintergreen, Old Lynchburg Rd., Charlottesville, Va. 22901.*

STEVENSON, James, Am. ethnologist, explorer; b. Maysville, Ky., Dec. 24, 1840; m. Matilda Coxe Evans, Apr. 18, 1872. Spent several winters among Blackfoot and Sioux Indians; participated in survey of Yellowstone region, 1871, leader in making it a nat. park; in charge of exploration of Snake and Columbia rivers in Ida. and Wyo. territories, 1872, prepared maps of region; in survey trip of 1872 climbed Great Teton (1st white man known to have reached ancient Indian altar on its summit); served as col. with 13th N.Y. Volunteers, Union Army, 1861-65; engaged in research among Pueblo Indians and remains of their former settlements for Bur. Ethnology at its inception, 1879; outfitted, conducted expdns. investigating ancient ruins and the living Navaho, Zuni, Hopi, other Indian tribes; published 1st studies among the Navaho titled Ceremonial of Hasjelti Dailjis and Mythical Sand Painting of the Navajo Indians; his ornithol. collections in U. S. Nat. Mus., Smithsonian Instn. Died N.Y.C., July 25, 1888.

STEVENSON, John James, Am. geologist; b. N.Y.C., Oct. 10, 1841; s. Andrew and Ann Mary (Willson) S.; A.B., N.Y. U., 1863, A.M., 1866, Ph.D., 1867; LL.D., Princeton, 1893, Washington and Jefferson, 1902; m. Mary A. Macgowan, Apr. 13, 1865 (dec. 1871); m. 2d, Mary C. Ewing, Jan. 1, 1879. Prof. chemistry and natural history W.Va. U., 1869-71; prof. geology N.Y. U., 1871-82, chemistry and physics, 1882-89, geology and biology, 1889-94, geology, 1894-1909, emeritus prof., 1909. Aid on Ohio Geol. Survey, 1871-72, 74; geologist U. S. Geog. Survey, west of 100th Meridian 1873, 74, 78, 79, 2d Geol. Survey of Pa., 1875-78, 1881-82. Corr. or hon. mem. geol. socs. of Russia, Hungary, Belgium, Vienna, Edinburgh, Liverpool, Australasia, acads. of Halle, Dresden, Moscow, Padua, Palermo, Pisa, London; del. Internat. Geol. Cong., 1903 (v.p. for U. S.). Author: Geology of a Portion of Colorado, 1875; Report on Greene and Washington Districts, Pa., 1876; Report on Fayette and Westmoreland Districts, Pa., 1877-78; Geological Examinations in Southern Colorado and Northern New Mexico, 1881; Geology of Bedford and Fulton Counties, Pa., 1882. Died Aug. 10, 1924.

STEVENSON, Matilda Coxe, Am. ethnologist; b. San Augustine, Tex.; d. Alexander H. and Maria Matilda (Coxe) Evans; ed. Miss Anable's, Phila.; m. James Stevenson, Apr. 18, 1872 (dec. 1888). Explored (with husband) Rocky Mountain region, 13 years; with expdn. under his charge, from Bur. of Ethnology, to Zuñi, N.M., 1897; assisted in collecting archaic implements, ceramics and ceremonial objects for U. S. Nat. Mus.; on staff Bur. Am. Ethnology, Smithsonian Instn., 1889—; mem. jury for anthropology Chgo. Expn., 1893. Author: Zuñi and the Zuñians, 1881; Religious Life of the Zuñi Child, 1884; The Sia, Zuñi Scalp Ceremonials, 1890; Zuñi Ancestral Gods and Masks, 1898; The Zuñi Indians, Their Mythology, Esoteric Fraternities, and Ceremonies, 1903. Studied Zuñi mythology, philosophy and sociology and made extensive vocabulary; explored cave, cliff and mesa ruins of N. Mex.; visited all Pueblo tribes of N.Mex., Tusayan and Navajo of Ariz., Mission Indians of Cal.; received in secret orgns. of these peoples and studied their esoteric instns.; studied Taos and Tewa Indians, with spl. attention to philosophy, religion, symbolism and sociology, the Taos people, Zuñi medicinal and edible plants, preparation of cotton and wool for loom, 1904-10. Died June 24, 1915.

STEVENSON, Merlon Lynn, Am. physicist; b. Salt Lake City, Oct. 31, 1923; s. Merlon L. and Katie Lynn (Peterson) S.; student Weber Jr. Coll., 1941-43; A.B., U. Cal., Berkeley, 1948, Ph.D., 1953; m. Lois Griffin, July 16, 1948; children—Leslie Ann, Scott Griffin, Jeffery Lynn, Conrad Joseph, Cybele Marie. Physicist, Lawrence Radiation Lab., Berkeley, 1953—; faculty U. Cal., Berkeley, 1957—, prof. physics, 1964—. Mem. Am. Phys. Soc., Sigma Xi, Phi Beta Kappa. Contbr. articles to profl. jours. Research in exptl. determination of properties elementary particles. Home: 683 Vincente Av., Berkeley 94707. Office: Lawrence Radiation Lab., Berkeley, Cal. 94720.*

STEVENSON, Robert, Scottish civil engr.; b. Glasgow, Scotland, June 8, 1772; s. Alan and Jean (Lillie) S.; ed. Andersonian Inst., Glasgow, also U. Edinburgh; m. Jean Smith, 1796; children—Alan, David Thomas. Engr. to Scottish Lighthouse Bd.; designed, constructed 20 lighthouses, inventing intermittent and flashing lights; built Bell Rock Tower Lighthouse on novel design and with specially invented implements, 1807-12; designed many bridges including Hutchison Bridge; suggested modern rails; invented hydrophore; designed Eastern road approaches to Edinburgh; an originator Royal Conservatory Edinburgh. Fellow Royal Soc. Edinburgh, Geol. Soc. London, Astron. Soc. London, Antiguarian Soc. Edinburgh, Wernerian Soc. Edinburgh; mem. Inst. Civil Engrs. Author: Account of the Bell Rock Lighthouse, 1824; also articles in Edinburgh Ency., Ency. Britannica. Suggested modern rails; adopted malleable iron rails; invented intermittent and flashing lights for lighthouse, also hydrophore for procuring specimens of sea and river water. Died July 12, 1850.

STEVENSON, Robert Evans, Am. geologist; b. Des Moines, May 5, 1916; s. Caleb B. and Edythe Allyne (Astley) S.; B.S., U. Hawaii, 1939; M.S., State Coll. Wash., 1942; Ph.D., Lehigh U., 1950; m. Thelma Murray Morrison, Sept. 3, 1948; children—Roberta Jessamine, Sandra Edythe. Geologist, Wash. State Div. Geology, 1942-44; geologist Venezuelan Atlantic Refining Co., 1944-47; instr. Lehigh U., 1947-50; geologist S.D. Geol. Survey, 1950-51; faculty U. S.D., Vermillion, 1951—, now prof., chmn. dept. geology. Mem. Am. Assn. Petroleum Geologists, Geol. Soc. Am., A.A.A.S., Soc. Econ. Paleontologists and Mineralogists, Paleontol. Soc., Am. Geophys. Union, Internat. Sedimentological Assn. Research and publs. on distbn. of geologic units in S.D.; Cretaceous history of S.D., especially invertebrates. Home: 1225 Valley View, Vermillion, S.D. 57069.*

STEVENSON, Robert Everett, Am. oceanographer; b. Fullerton, Cal., Jan. 15, 1921; s. George and Zella (Hope) S.; B.A., U. Cal. at Los Angeles, 1946, M.A., 1948; Ph.D., U. So. Cal., 1954; m. Elizabeth C. Campbell, Nov. 26, 1963; children—Michael George, Robert Kurtz. Dir. inshore research U. So. Cal., Los Angeles, 1953-61; research scientist Tex. A. and M., Galveston, 1961-62, dir. marine lab., 1962-63; asso. prof. meteorology and geology Fla. State U., Tallahassee, 1963-65; asst. dir. biol. lab. Bur. Comml. Fisheries, Galveston, 1965—. Oceanographic cons. in marine pollution, oceanographic engring.; dir. NSF Teaching Inst., 1961-63; research scientist Office Naval Research, Eng., 1959. Fellow Geol. Soc. Am.; mem. Am. Meteorol. Soc., Am. Geophys. Union, Soc. Econ. Paleontologists and Minerologists. Author: Summer Climatic Environment of the Yorkshire Coast, England, 1961; also numerous articles. Defined environment and formation of marine marshes and deposits of estuaries, conditions of coastal marine waters which define local climate, evaluated effects of hurricanes on structure of surface ocean waters, application of space tech. to oceanography. Home: 1402 Bowie: Office: Bur. Comml. Fisheries, Galveston, Tex. 77550.*

STEVENSON, Stuart Shelton, Am. physician; b. Bridgeport, Conn., Nov. 11, 1914; s. Henry Cogswell and Marthena (Crump) S.; B.A., Yale, 1935, M.D. cum laude, 1939; M.P.H. with distinction Harvard, 1944. Instr. pediatrics Yale Sch. Medicine, 1942-43; asst. surgeon USPHS, 1943-46; staff mem. Health Commn. of IHD of Rockefeller Found., 1944-45; acting asst. pediatrician Bridgeport (Conn.) Hosp., 1946; pvt. practice pediatrics, Fairfield, Conn., 1945-46; asso. child health Harvard, 1946-47, asst. prof. child health, 1947-49; asst. physician Children's and Infants' Hosps., Boston, 1946-47, asso. physician, 1947-49; research prof. pediatrics U. Pitts., 1949-59, acting chmn. dept. pediatrics, 1958-59; asst. to med. dir. Children's Hosp. of Pitts., 1951-59, acting med. dir., acting chief of staff, 1958-59; prof. pediatrics chmn. dept. pediatrics Seton Hall Coll. Medicine, 1959-64; pediatrician-in-chief Jersey City Med. Center, 1959-64; chief of ward pediatrics Margaret Hague Maternity Hosp., N.J., 1959-64; attending pediatrician, dir. pediatrics St. Luke's Hosp. Center, N.Y.C., 1964—; attending pediatrician Columbia-Presbyn. Med. Center, N.Y.C., 1964—; clin. prof. pediatrics Columbia U. Coll. Phys. and Surg., 1964—. Spl. cons. Office Surgeon Gen., 1944-46. Rockefeller fellow, 1943-44. Diplomate Nat. Bd. Med. Examiners. Fellow Am. Bd. Pediatrics; mem. A.M.A., Med. Soc. County of N.Y., Med. Soc. State N.Y., N.Y. Acad. Medicine, N.Y. Clin. Soc., Am. Pub. Health Assn., Am. Pediatric Soc., Soc. Pediatric Research, Pediatric Travel Club. Editor, Am. Jour. Diseases of Children. Research and numerous publs. on nutrition neonatology, human growth and devel. Home: 2 Fifth Av., N.Y.C. 10011. Office: 421 W. 113th St., N.Y.C. 10025.*

STEVENSON, Thomas, Scottish engr., meteorologist; b. Edinburgh, Scotland, July 22, 1818; s. Robert and Jean (Smith) S.; ed. high sch.; m. Margaret Isabella Balfour; 1 son, Robert Louis. Joint-engr. bd. No. lighthouses, 1853-85; continued expts. of Alan Stevenson in lighthouse illumination. Mem. Royal Soc. Edinburgh (pres. 1885), Royal Scottish Soc. Arts (pres. 1859-60), Scottish Meteorol. Soc. (hon. sec. 1871), Instn. C.E.'s. Author: Design and Construction of Harbours, 1864. Research, publs. on lighthouse, harbor engring., lighthouse optics, meteorology; invented, perfected azimuthal condensing system lighthouse illumination; designed Stevenson screen for thermometers, 1864. Died May 8, 1887.

STEVER, Horton Guyford, Am. aero. engr.; b. Corning, N.Y., Oct. 24, 1916; s. Ralph Raymond and Alma (Mott) S.; A.B., Colgate U., 1938, D.Sc., 1958; Ph.D., Cal. Inst. Tech., 1941; m. Adelaide Louise Risley, June 29, 1946; children—Horton Guyford, Sara Newell, Margarette Risley, Roy Risley. Staff mem. radiation lab., instr. radar sch. Mass. Inst. Tech., Cambridge, asst. prof. aero. engring., 1946-51, exec. officer guided missiles program, 1946-48, asso. prof., 1951-55, prof., 1955-65, asso. dean engring., 1956-59, head mech. engring. dept., head naval architecture and marine engring. dept., 1961-65; pres. Carnegie-Mellon U., Pitts., 1965—. Dir. Koppers Co., United Aircraft Corp., Fisher Sci., System. Devel. Corp. Chmn. sci. adv. bd. USAF, 1961—; mem. research adv. com. on missile and space aerodynamics NASA, 1959—; mem. sci. adv. com. Com. on Sci. and Astronautics, U.S. Ho. of Reps., 1960—; mem. exec. com. Def. Sci. Bd., 1960—; cons. aero. industries; mem. adv. council dept. aero. engring. Princeton, 1963—. Recipient Scott gold medal Am. Ordnance Assn., 1960. Fellow Am. Inst. Aeros. and Astronautics (pres. 1961), Am. Acad. Arts and Scis., A.A.A.S., Am. Phys. Soc.; mem. Nat. Acad. Engring., Am. Soc. M.E., Am. Soc. for Engring. Edn., Phi Beta Kappa, Sigma Xi, Sigma Gamma Tau, Tau Beta Pi, Pi Tau Sigma. Contbr. articles to profl. jours. Research on hypersonic aerodynamics, nuclear propulsion aircraft; shock tubes; transonic aircraft; radar guided and ballistic missiles; cosmic rays; geiger counters. Home: 1045 Devon Rd., Pitts. 15213.

STEVIN, Simon, Dutch mathematician, physicist; b. Bruges (now in Belgium), 1548; m.; 2 children. Accountant in Antwerp; traveled in Prussia, Poland, Norway, and Denmark; became instr. for math. and sci. to Maurice of Orange; later served him as military and civil engr. and devised system of flooding lowlands against invading armies by manipulating sluices in dikes; dir. of so-called "waterstraet" (public works, roads, and waterways); quartermaster of Dutch army. Author: Tafeln van interest midtsgders de constructie der selver, 1582; Problematum geometricorum libri V, 1583; Dialectike ofte bewysconst, 1585; De Thiende, 1585 (contains 1st comprehensive system of decimal fractions and their practical applications); De Beghinselen der Weeghconst, 1586; De Beghinseln des Waterwichts, 1586; Weeghdaet, 1593; De Sterctenbouwing, 1594; De Havenvinding, 1599; Wiscontige Ghedachtenssen, 1605-08; Castrametatio, 1617; also numerous treatises; made 1st translation of Diophantos into a modern language. Leader of contemporary school of Dutch mathematicians; set forth theory of decimal fractions; urged their universal usage; declared universal introduction of decimal coinage, weights, and measures was only a matter of time; 1st to ennunciate basic theorems of statics on triangle of forces, 1586, and equilibrium of bodies on an inclined plane; discovered hydrostatic paradox that liquid exerts downward pressure on a surface independent of shape of its container and dependent only on height and area of surface; gave values of magnetic declination for 43 specific places on earth; made significant innovations in trigonometry, geography, fortification, and navigation; inventions include land carriage (which carried 26 passengers) propelled by wind sails. Died The Hague or Leyden, Holland, between Feb. 20 and Apr. 18, 1620.

STEWARD, Frederick Campion, biologist; b. London, Eng., June 16, 1904; s. Frederick W. and Mary (Daglish) S.; B.S. with first class honors in Chemistry, Leeds U., 1924, Ph.D. in Botany, 1926; D.Sc. in Botany, U. London, 1937; m. Ann T. Gordon, Sept. 7, 1929; 1 son, Frederick G. Came to U. S., 1945. Asst. lectr. botany Leeds U., 1929-34; reader in botany Birkbeck Coll., U. London, 1934-47; vis. prof. botany, chmn. dept. U. Rochester, N.Y., 1946-50; prof. botany Cornell U., 1950-65, Alexander prof. of biology and dir. Laboratory of Cell Physiology and Growth, 1965——. Dir. aircraft equipment Ministry Aircraft Prodn., London, 1940-45. Recipient Rockefeller fellowship, 1927-29, 33-34; Merit award Bot. Soc. Am., 1961; John Simon Guggenheim fellow, 1963; Stephen Hales award Am. Soc. Plant Physiologists, 1964. Fellow Am. Acad. Arts and Scis., Royal Soc., 1957. Author articles profl. jours. Research on mechanism salt adsorption in plants and its relation to plant nutrition; respiration and metabolism in osmotic work; nitrogen metabolism of plants, growth and morphogenesis. Home: 621 Highland Rd., Ithaca, N.Y.*

STEWARD, Julian H., Am. anthropologist; b. Washington, Jan. 31, 1902; s. Thomas G. and Grace (Garriott) S.; A.B., Cornell U., 1925; A.M., U. Cal., 1926, Ph.D., 1929; m. Jane Cannon, June 21, 1934; children—Gary Cannon, Michael. Instructor anthropology U. Mich., 1928-30; asso. prof. U. Utah, 1930-33; lectr. U. Cal., 1934; anthropologist Bur. Am. Ethnology, 1935-38, sr. anthropologist, 1938-43; dir. Inst. Social Anthropology, Smithsonian Instn., 1943-46; prof. Columbia, 1946-52; research prof. anthropology U. Ill., 1952——, acting head dept., 1959-60, mem. Center Advanced Studies, 1959——, dir. Studies Cross-Cultural Regularities; fellow Center Advanced Study Behavioral Scis., Stanford, 1960-61; dir. Kyoto (Japan) Am. Studies Seminar, 1956; Mem. Ethnohistorical Soc., Am. Assn. U. Profs., Andean Inst., Franciscan Hist. Soc. (hon. mem.), Nat. Acad. Sci., Am. Anthrop. Assn. (Viking medallist 1952), Soc. Am. Archaeology, A.A.A.S., Association Asian Studies, Soc Asiatic Studies, Academy de la Cultura Guarani Paraguay (hon.), Soc. Applied Anthropology, Am. Ethnol. Soc. Author: Theory of Culture Change, 1955, also papers and monographs, Co-author: People of Puerto Rico, 1956; Native People of South America, 1959. Editor: Handbook of South American Indians, vols. 1, 2, 1946, vols. 3, 4, 1948, vols. 5, 6, 1949; Contemporary change in Traditional Societies, 3 vols., 1967. Devel. methodology for ascertaining causes of cultural change in evolutionary transformations and for analysis of complex contemporary cultures and their changes. Home: Main St., Fithian, Ill. Office: Dept. Anthropology, Univ. Illinois, Urbana, Ill.*

STEWART, Alec Thompson, Canadian physicist; b. Windthorst, Sask., Can., June 18, 1925; s. Arthur and Nelly B. (Thompson) S.; B.S., Dalhousie U., Halifax, N.S., 1946, M.S., 1949; Ph.D., Cambridge U., 1952; m. Alta A. Kennedy, Aug. 4, 1960; children—Arthur James Kennedy, Hugh Donal. Asst., asso. research officer Atomic Energy Can., Ltd., Chalk River Labs., 1952-57; asso. prof. Dalhousie U., 1957-60; asso. prof. U. N.C., Chapel Hill, 1960-63, prof., 1963-67; head dept. physics Queen's U., Kingston, Ont., 1967——. Mem. Am. Inst. Physics, Am. Phys. Soc., Canadian Assn. Physicists. Author: (with L. O. Roellig) Position Annihilation, 1957; Perpetual Motion, Electrons and Atoms in Crystals, 1965. Research and numerous publs. on phonons in metals by neutron scattering, motions of electrons in many solids and liquids, chiefly metals, positrons and positronium in matter. Office: Physics Dept., Queen's U., Kingston, Ont., Can.*

STEWART, Alexander Patrick, English physician; b. Bolton, Eng., Aug. 28, 1813; s. Andrew and Margaret Stewart; ed. U. Glasgow (Scotland) Sch. Arts, M.D., 1838; postgrad. edn. Paris, France, Berlin, Germany. Practiced medicine London, Eng., from 1839; asst. physician Middlesex (Eng.) Hosp., 1850-55, physician, from 1855, also lectr. materia medica, medicine. Fellow Royal Coll. Physicians. Mem. Brit. Med. Assn. Author: Some Considerations on the Nature and Pathology of Typhus and Typhoid Fever Applied to the Solution of the Question of the Identity or Non-Identity of the Two Diseases, 1854; Sanitary Economics, 1849. Distinguished difference between typhus and typhoid in origin, cause, symptoms. Died London, July 17, 1883.

STEWART, Alfred Walter, Brit. chemist; s. William Stewart; ed. univs. Glasgow, Marburg, Univ. Coll., London; D.Sc.; m. Jessie Lily Coats, 1916; 1 dau. Carnegie research fellow, 1905-08; lectr. organic chemistry Queen's U., Belfast, Ireland, 1909-14, prof. chemistry, 1919-44; lectr. phys. chemistry and radioactivity Glasgow U., 1914-19; Author: Stereochemistry; Recent Advances in Organic Chemistry; Chemistry and its Borderland; Some Physico-Chemical Themes; Recent Advances in Physical and Inorganic Chemistry; also detective fiction (under pseudonym John Jervis Connington). Contbr. to sci. jours. Died July 1, 1947.

STEWART, Balfour, Scottish physicist, meteorologist; b. Edinburgh, Scotland, Nov. 1, 1828; s. William and Jane (Clouston) S.; ed. St. Andrews (Scotland) U., U. Edinburgh; m. Katherine Stevens, Sept. 8, 1863; 2 sons, 1 dau. Asst. to Prof. Forbes, Edinburgh U., from 1856; dir. Kew (Eng.) Obs., 1859-71; sec. to govt. meteorol. com., 1867-69; prof. natural philosophy Owens Coll., Manchester, Eng., 1870-87. Fellow Royal Soc. (Rumford medal (1868), 1862; mem. Soc. Phys. Research (pres. 1885-87), Phys. Soc. (pres.), Manchester Lit. and Philos. Soc. (pres. 1887). Author: Treatise on Heat, 1866; Lessons in Elementary Physics, 1870; The Conservation of Energy, 1872; (with Peter Guthrie Tait) The Unseen Universe, 1875; (with William Haldane) Lessons in . . . Practical Physics, 1885; Lessons in Practical Physics for Schools, 1888. Investigated sun spots; made calculations as to periodic irregularities in terrestrial, solar phenomena; research, publs. in radiant heat (helped lay found. spectrum analysis); demonstrated applicability of law of radiation to polarized rays of light, 1860; suggested variations in primary electric current in sun as cause of aurorae, magnetic storms, earth currents, 1860. Died Dec. 19, 1887.

STEWART, Bonnie Madison, Am. mathematician; b. Loveland, Colo., July 10, 1914; s. Magnus F. and Nannie (Lovett) S.; B.A. magna cum laude, U. Colo., 1936; Ph.M., U. Wis., 1937, Ph.D., 1940; m. Doris Odell, Sept. 7, 1940; children—Jan., Sidney. Faculty, Mich. State U., 1940-43, 44——, prof. math., 1953——. Mem. Math. Assn. Am. (gov. Mich. sect. 1956-59), Am. Math. Soc., Phi Beta Kappa, Sigma Xi, Pi Mu Epsilon. Author: Theory of Numbers, 1952, 2d edit., 1964; also articles. Research on number theory, matrix theory, graph theory, Euclidean geometry. Home: 4494 Wausau St., Okemos, Mich. 48864. Office: Dept. Math., Mich. State U., East Lansing, Mich. 48823.*

STEWART, David Denison, Am. physician; b. Phila., Oct. 10, 1858; s. Franklin and Amelia (Barron Jaques) S.; grad. Jefferson Med. Coll., 1879, followed by hosp. and lab. work; began practice in Phila. in 1885; physician St. Christopher's Hosp. for Children, St. Mary's Hosp., Episcopal Hosp.; chief med. clinic, lectr. diseases of nervous system, demonstrator clin. medicine, later lectr. medicine Jefferson Med. Coll.; prof. diseases stomach and intestines, Phila. Polyclinic. Published essays and research work on lead poisoning, diseases of nervous system, diseases of kidneys and on albumin testing, on treatment of aneurism by electrolysis through wire, on diseases of digestive system. Deceased.

STEWART, Donald Charles, Am. chemist; b. Salt Lake City, Dec. 15, 1912; s. John Caldwell and Nelle (Flandro) S.; A.B. in Chemistry, U. Cal. at Los Angeles, 1935; M.S. in Chemistry, U. So. Cal., 1940; B.S. in Chem. Engring., Va. Poly. Inst., 1944; Ph.D. in Biochemistry, U. Cal. at Berkeley, 1950; m. Dorothy W. Bockhop, Apr. 16, 1948; children—Katharine Ann, Deborah Jean. With Knudsen Creamery, 1935-42, Manhattan Project, 1944-46, U. Cal. Radiation Lab., Berkeley, 1946-50; chemist Argonne (Ill.) Nat. Lab., 1950-52, 54-59, asst. div. dir., 1952-54, asso. dir., 1959——. Observer, Bikini Bomb Tests, 1946; cons. atomic energy establishments in Japan, Korea, Taiwan, 1960, 65. Mem. Am. Chem. Soc., A.A.A.S., Am. Nuclear Soc., Sci. Research Soc. Am. Co-editor: Progress in Nuclear Energy, Series IX, Vols. 4-7, 1964——. Contbr. articles to profl. jours. Developer lab. scale separation processes involving rare earths, plutonium and transplutonium elements. Home: 4549 Lee St., Downers Grove, Ill. 60515. Office: 9700 S. Cass Av., Argonne, Ill. 60439.*

STEWART, George, Am. ecologist; b. Tooele, Utah, Nov. 7, 1888; s. William and Ellen (Speirs) S.; A.B., U. Utah, 1907; B.S., Utah Agrl. Coll., 1913; M.S., Cornell U., 1918; Ph.D. (Shevlin fellow), Minn., 1926; m. Wynona Barber, Sept. 5, 1918; 1 dau., Betty Ann. Instr. in agronomy Utah Agr. Coll., 1913-16, asst. prof., 1917-18, asso. prof., 1918-19, prof., 1919-30; became sr. ecologist, br. research U. S. Forest Service, 1930; supv. Utah State Agrl. Coll.-Fgn. Operations Adminstrn. Contract for Agr., Iran, 1954-56; prof. botany and agronomy Brigham Young U., 1956——. With U. S. Dept. Agr., 1951-53. Mem. Agrl. Commn. to Iran (7 year plan), 1949. Recipient Superior Service award, U. S. Dept. Agr.; Service award Utah Acad. Sci., 1951. Mem. Am. Soc. Agronomy, Am. Geog. Soc., Am. Genetic Assn., A.A.A.S., Soc. Am. Foresters, Am. Ecol. Soc., Sigma Xi. Author: Alfalfa Growing in U. S. and Canada, 1926. Co-Author: Principles of Agronomy, 1915; second edition of same completely revised, 1930; Development of Collective Enterprise, 1943. Co-editor: Western Agriculture, 1918. Asso. editor Jour. Forestry, 1939-46. Contbr. to profl. jours. Student of crops, soils, and irrigation of intermountain-region, and their social development and of important features in wheat genetics and new varieties, sugar-beet breeding; varieties of vegetation on desert, semi-desert, and foothill ranges, grazing on public lands, reseeding range lands and range conservation. Died Oct. 27, 1957.

STEWART, George Neil, physiologist; b. London, Can., Apr. 18, 1860; s. James Innes and Catherine (Sutherland) S.; A.M., U. Edinburgh, 1883, B.S., 1886, D.Sc., 1887, M.B., C.M., 1889, M.D., 1891, LL.D., 1920; D.P.H., U. Cambridge, 1890; m. Demonstrator physiology Owens Coll., Manchester, Eng., 1887-89; George Henry Lewes student U. Cambridge, 1889-93; examiner in physiology U. Aberdeen, 1891-94; instr. Harvard Med. Sch., 1893-94; prof. physiology and histology Western Res. U., 1894-1903: prof. exptl. medicine, 1907——; prof. physiology U. Chgo., 1903-07. Contbr. many original papers, 1887——. Inventor calorimetric method of measuring blood flow; studied epinephrine output of adrenal glands; established usefulness of adrenal cortex extracts. Died May 28, 1930.

STEWART, George Walter, Am. physicist; b. St. Louis, Feb. 22, 1876; s. Oliver Mills and Eleanor (Bell) S.; A.B., DePauw U., 1898, Sc.D., 1928; Ph.D., Cornell, 1901; Sc.D., U. Pitts., 1931; Sc.D., Kalamazoo Coll., 1949; m. Zella M. White, July 7, 1904. Asst. in physics Cornell, 1899-1901, instr., 1901-03; asst. prof. charge dept. physics U. N.D., 1903-04, prof. physics, 1904-09; prof. physics and head dept. U. Ia., 1909-46, prof. physics (retired), 1946; acting dean Graduate Coll., State U., 1921-22. Fellow A.A.A.S., Am. Acad. Arts and Scis., Ia. Acad. Sci., Am. Phys. Soc. (pres. 1949) Am. Acoustic Soc.; mem. Soc. for Promotion Engring. Edn., Am. Optical Soc., Nat. Acad. Scis., Am. Assn. Physics Tchrs. (Oersted medalist 1942), Phi Beta Kappa, Sigma Xi (pres. 1930-32). Author: Introductory Acoustics, 1933; Theoretical Acoustics (with R. B. Lindsay), 1930. Contbr. to current research in physics, upon radiation, archtl. acoustics, sound diffraction, liquid structure. Died Aug. 16, 1956.

STEWART, Harold Leroy, Am. physician; b. Houtzdale, Pa., Aug. 6, 1899; s. Alexander and Lillie (Cox) S.; student U. Pa., 1919-20, Dickinson Coll., 1921-22; M.D., Jefferson Med. Coll., 1926; grad. Army Med. Sch., Washington, 1929; research fellow, Jefferson Med. Coll., 1929-30, Harvard, 1937-39; Med. Sc. son Med. Coll., 1929-30, Harvard, 1937-39; Med. Sc.D., Jefferson Med. Coll., 1964; D. Medicine and Surgery (hon.), U. Perugia, 1965; m. Cecelia Eleanor Finn, Sept. 30, 1929; children—Robert Campbell, Janet Eileen. Instr. to asst. prof. pathology Jefferson Medical College, 1930-37; served as assistant pathologist Jefferson Med. Coll. Hosp., Phila. Gen. Hospital, 1929-37; pathologist Office Cancer Investigations, Harvard, USPHS, 1937-39; chief lab. pathology Nat. Cancer Inst., USPHS, Bethesda, Md., 1939——, chief path. anatomy dept. clinical center Nat. Insts. Health, 1954——; cons. Armed Forces Inst. Pathology; cons., mem. study groups WHO, 1957, member expert adv. panel cancer, 1957——. Mem. subcom. oncology NRC, 1947——, mem. com. on pathology, 1958——, com. cancer diagnosis and therapy, 1951-57, mem. U. S. A. Com. Internat. Council Socs. Pathology, 1957-62, chmn. subcom. on geog. pathology, 1960——; chairman of United States nat. com. International Union Against Cancer, 1953-59, U. S. del., 1951——, v.p. for U. S., 1962——. Member of adv. bd. Leonard Wood Meml., 1962. Diplomate Am. Bd. Pathology, Pan Am. Med. Assn. Mem. Am. Soc. Clinical Pathologists, Am. Geog. Soc., Am. Assn. Cancer Research (president 1958-59), American Society Study Growth, Am. Soc. Exptl. Pathology (president 1955), American Association Pathologists and Bacteriologists, Soc. Exptl. Biology and Medicine, Coll. Am. Pathologists, Md. (pres. 1950-51), Washington (sec.-treas. 1947-51) socs. pathologists, Internat. Acad. Pathology (pres. 1953-55), Internat. Union Against Cancer (exec. com. 1956——, v.p. 1962), Internat. Council Socs. Pathology (pres. 1962), Internat. Soc. Geog. Pathology, Colegio Anatomico Brasileiro, Societa Italiana di Cancerologia, Sociedad Columbiana de Patologia, Société Belge d' Anatomie Pathologique (hon.), Member ed. bd. Cancer Research, 1941-49, A.M.A. Archives of Pathology, 1957-62; editorial adviser Journal of National Cancer Inst., 1947-56. Research and publs. on exptl. induction of esophageal, gastric and intestinal carcinomas in mice, and study of their progress and histogenesis. Home: 119 S. Adams St., Rockville, Md. 20850. Office: National Cancer Institute, Bethesda, Md. 20014.

STEWART, Harris Bates, Jr., Am. oceanographer; b. Auburn, N.Y., Sept. 19, 1922; s. Harris B. and Mildred (Woodruff) S.; A.B., Princeton, 1948; M.S., 1956; m. Elise Bennett Cunningham, Feb. 21, 1959; 1 dau., Dorothy. Research asst. Scripps Instn. Oceanography, La Jolla, Cal., 1951-56; oceanographer U.S. Coast and Geodetic Survey, 1957-60, chief oceanographer, dept. asst. dir., 1960-65; dir. Essa, Inst. for Oceanography, Silver Spring, Md., 1965-67, dir. Atlantic oceanographic labs., 1967——. Chmn., survey panel Interagy. Com. on Oceanography, 1960-67; chmn. adv. bd. Nat. Oceanographic Data Center, 1965-66. Recipient Meritorious Service award U.S. Dept. Commerce, 1960, Exceptional Service award, 1965. Fellow A.A.A.S., Geol. Soc. Am., Washington Acad. Scis., Am. Geog. Soc.; mem. Am. Geophys. Union, N.Y. Acad. Scis. Author: The Global Sea, 1963; Deep Challenge, 1966; also articles. Research in marine geology and phys. oceanography Home: 737 N. Greenway Dr., Coral Gables, Fla. 33134. Office: 901 S. Miami Av., Miami, Fla. 33130.*

STEWART, Homer Joseph, Am. aeronautical engr.; b. Elba, Mich., Aug. 15, 1915; s. Earl Arthur and Alta Fern (Stanley) S.; student U. Dubuque (Ia.), 1932-33; B.Aero. Eng., U. Minn., 1936; Ph.D., Calif. Inst. Tech., 1940; m. Frieda Klassen, June 15, 1940;

children—Robert Joseph, Katherine Stanley, Barbara Ellen. Mem. faculty, Calif. Inst. Tech. since 1936, prof. of aeronautics since 1949, chief of research analysis sect., Jet Propulsion Lab., 1945-56, chief Liquid Propulsion Systems Div., 1956-58, special assistant to director, 1960-62, chief Advanced Technical Studies Office, 1963—. Dir. Cffice Program Planning and Evaluation, NASA, 1958-60. Member technical adv. bd. Aerojet-Gen. Corp., 1956—. Mem. tech. evaluation group of guided missile com., Research and Development Bd., 1948-50, chmn., 1951; mem. sci. advisory bd. USAF, 1949-56, 1959-64; sci. adv. com. Ballistics Research Lab., 1959—. Recipient Outstanding Achievement award, U. Minn., 1954. Mem. Inst. Aeros. Scis., Am. Meteorol. Soc., A.A.-A.S., Sigma Xi. Author: Kinematics and Dynamics of Fluid Flow, Sect. VI of Handbook of Meteorology, 1945. Contbr. articles on aeronautics and meteorology to tech. jours. Research on dynamic meteorology; theoretical aerodynamics; effect of shear instability on the transverse circulation in the atmosphere; supersonic and fluid flows; guided missiles. Home: 2393 Tanoble Dr., Altadena, Cal. Office: Cal. Inst. Tech., Pasadena, Cal.

STEWART, John D(unham), Am. surgeon; b. Monroe, N.C., Nov. 7, 1903; s. Henry Dixon and Ione (Wolfe) S.; B.A., U. Va., 1924; M.D., Harvard, 1928; m. Henrietta Seelye Rhees, Nov. 13, 1937; children—David Rhees, Harriet Seelye Rhees (Mrs. Leigh Callaway), Ione Wolfe. Prof. surgery U. Buffalo, 1941-65; v.p. State U. N.Y., Buffalo, 1963. Mem. surgery study sect. NIH, 1955-59; mem. adv. com. on surgery NRC, 1946-50. Mem. Danish Royal Soc. U. Surgeons (hon.), Am. Surg. Assn. (past pres.), Harvard Med. Alumni Assn. (pres. 1965—). Research and numerous publs. on wounds, shock, nutrition, liver function. Address: 2333 N.E. 30th Ct., Lighthouse Point, Fla. 33064.*

STEWART, John Morrow, Am. chemist; b. Greensboro, N.C., Oct. 31, 1924; s. David Henry and Mary (Morrow) S.; B.S., Davidson Coll., 1948; M.S., U. Ill., 1950, Ph.D., 1952; m. Joyce Loraine Clark, Sept. 3, 1949; children—Ellen, Susan, David. Instr., Davidson Coll., 1948-49; research asso. Rockefeller U., N.Y.C., 1952-57, faculty, 1957—, asso. prof., 1964-68; prof. biochemistry U. Colo. Sch. Medicine, 1968—. Mem. Am. Soc., Am. Soc. Biol Chemists, Phi Beta Kappa, Medicinal Chemistry, 1967—. Research on design, constrn. and applications of instrument for fully automatic synthesis of peptides; design and use of specific inhibitors to control diseases; relationship of structure to function of biologically active peptides, synthesis of amino acids and peptides, control of cancers. Home: 3690 E. Dartmouth St., Denver.*

STEWART, John Quincy, Am. physicist, generalist; b. Harrisburg, Pa., Sept. 10, 1894; s. John Quincy and Mary Caroline (Liebendorfer) S.; B.S., Princeton, 1915, Ph.D., 1919; m. Lillian V., d. John Howell Westcott, June 17, 1925; 1 son, John Westcott. Engr. dept. devel. and research, Am. Tel. & Tel. Co., N.Y., investigating speech and hearing, 1919-21; designed the first electrical voice; with dept. astronomy, Princeton, 1921-63, asso. prof. astron. physics, 1927-63; charter faculty Prescott (Ariz.) Coll., now prof. metaphysics of sci. Organized small party which successfully observed longest modern total solar eclipse, from S.S. Steelmaker in the Pacific, June 8, 1937; duration was more than seven minutes. Fellow Am. Phys. Soc., A.A.-A.S. (hon.), Am. Geog. Soc.; mem. Am. Astron. Soc., Am. Assn. Univ. Profs. (nat. 1st. v.p., 1940-41;, Phi Beta Kappa, Sigma Xi. Author: Astronomy (with H. N. Russell and R. S. Dugan), 1927, 38; Navigation (with N. L. Pierce), 1944; Coasts, Waves, and Weather, 1945. Research in physics astronomy, nav., meteorol., demography, phys. equations in which energy has leading role; organized studies in social physics. Home: Box 446, Sedona, Ariz. 86336. Office: Prescott Coll., Prescott, Ariz. 86301.*

STEWART, Matthew, Scottish mathematician; b. Rothesay in Bute, Scotland, 1717; s. Dugald and Janet (Bannatyne) S.; ed. Edinburgh (Scotland) U.; D.D., Glasgow (Scotland) U., 1756; m. Marjory Stewart; 1 son, Dugald. Minister of Roseneath, Scotland, 1745-47; prof. math. Edinburgh U., 1747-85, duties performed by son, Dugald, after 1772. Fellow Royal Soc., 1764. Author: General Theorems of Considerable Use in the Higher Parts of Mathematics, 1746 (reputation established by this book); Tracts Physical and Mathematical, 1761; Propositiones Geometricae More Veterum Demostratae, 1763. Applied geometrical demonstrations to astronomy. Died Jan. 23, 1785.

STEWART, Paul A(lva), Am. ecologist; b. Leetonia, O., June 24, 1909; s. William O. and Achsah (Burgett) S.; B.S., Ohio State U., 1952, M.Sc., 1953, Ph.D., 1957; m. Esther Mae Gephart, Nov. 26, 1947; children—David Enos, Seth Michael. Research fellow Ohio Coop. Wildlife Research Unit, Columbus, O., 1952-57; asst. to dir. div. fish and game Ind. Dept. Conservation, Indpls., 1958-59; biologist U.S. Fish and Wildlife Service, Laurel, Md., 1959-65; entomologist entomology research div. Agrl. Research Service, U.S. Dept. Agr., Oxford, N.C., 1965—. Life mem. Am. Soc. Mammlogists, Am. Ornithologists' Union, Cooper Ornithol. Soc., Wildlife Soc.; mem. Ecol. Soc. Am., Entomol. Soc. Am., Wilson Ornithol Soc., Inland Bird-Banding Assn., Gamma Sigma Delta. Research and numerous publs. on local and migratory movements, behavior in relation to weather, and population dynamics of birds; bird and mammal distbn.; measurement of numbers in wild animal populations; effect of a blacklight insect trapping program on ecosystem; biol. control of insects. Home: 203 Mooreland Dr., Oxford, N.C. 27565.*

STEWART, Ralph Randles, Am. botanist; b. West Hebron, N.Y., Apr. 15, 1890; s. Thomas Alvin and Mary (Randles) S.; A.B., Columbia U., 1911, A.M., 1915, Ph.D., 1916; D.Sc., U. Pumjab (India), 1953; LL.D., Alma Coll., 1963; m. Isabelle Caroline Darrow, Sept. 26, 1916; children—Jean Macmillan (Mrs. Bertrand Joel Andrews), Ellen Reid (Mrs. Stoddard Wilder Daniels). Faculty, Gordon Coll., Rawalpindi, West Pakistan, 1911-14, 16-33, prin., 1933-55, prof. botany, 1955-60, prin. emeritus, 1960—; research asso. U. Mich., Ann Arbor, 1960—; curator Oriental Herbarium, N.Y. Bot. Garden, 1942-43. Ednl. missionary United Presbyn. Ch., 1916-60. Decorated Kaisar i Hind Gold medal (Brit. Indian Govt.); Sitara i Imtiaz (Pakistan). Mem. Mich. Bot. Soc., Soc. Plant Taxonomists, Phi Beta Kappa, Sigma Xi. Author: (with Sultan Ahmad) Grasses of West Pakistan, 1958; also articles on grasses, ferns and flowering plants West Pakistan and Kashmir. Collected plants in many parts of Western Himalayas from Ganges Valley to Afghan line, mostly West Pakistan; discovered many new species. Home: 523 S. Forest Av., Ann Arbor, Mich. 48104.*

STEWART, Richard Cummins, Am. microbiologist; b. Reading, Pa., May 22, 1924; s. David Reid and Inez (Welch) S.; B.S., Albright Coll., 1949; Ph.D., U. Pa., 1953; m. Mary Ann Brady, Aug. 6, 1949; children—Catherine Ann, Joseph Patrick. Vitamin chemist Sharp & Dohme Co., Phila., 1949-50; head quality control Pitman-Moore Co., Indpls., 1953-55; head virology sect. Smith Kline & French Labs., Phila., 1955—. Mem. N.Y. Acad. Scis., Am. Soc. Microbiology, Soc. Indsl. Microbiology. Research, publs. on viral and bacterial chemotherapy, especially pathogenesis of viral infections in lab. animals. Home: 1001 Longview Rd., King of Prussia, Pa. 19406. Office: 709 Swedeland Rd., Swedeland, Pa. 19479.*

STEWART, Richard Donald, Am. physician, toxicologist; b. Lakeland, Fla., Dec. 26, 1926; s. LeRoy Hepburn and Zoa (Hachet) S.; student Bay City Jr. Coll., 1947-49; A.B. with distinction U. Mich., 1951, M.D., 1955, M.P.H., 1962; m. Mary Leeuw, June 14, 1952; children—Richard Scot, Gregory David. Indsl. physician Dow Chem. Co., 1956-59, dir. med. research lab., 1962-66; chmn. dept. environmental medicine, asso. prof. preventive medicine Marquette U. Sch. Medicine, Milw., 1966—. Recipient Authorship award Indsl. Medicine and Surgery's 1st Ann Awards Competition, 1963, Am. Indsl. Hygiene Assn. Jour., 1965. Mem. A.C.P. (asso.), A.M.A., Soc. Toxicology, Am. Fedn. for Clin. Research, Sci. Research Soc. Am., Am. Soc. for Artificial Internal Organs, Pan Am. Med. Assn. (mem. council sect. toxicology 1965—). Publs. on devel. of breath analysis techniques, hollow fiber artificial kidney, silastic catheter; research on exptl. human toxicology of chlorinated solvents. Home: 18525 Chevy Chase Dr., Brookfield, Wis. 53005. Office: Div. Preventive Medicine, 1725 W. Wisconsin Av., Milw.*

STEWART, Robert William, Can. physicist; b. Smoky Lake, Alta., Can., Aug. 21, 1923; s. Robert Edwards and May (Berry) S.; B.Sc., Queen's U., Kingston, Ont., Can., 1945, M.Sc., 1947; Ph.D., Cambridge (Eng.) U., 1952; m. Vera M. Brande, Aug. 28, 1948; children—Margaret Anne, Brian Keith, Philip Roy. Def. research sci. service officer Pacific Naval Lab., Esquimalt, B.C., Can., 1950-55; hon. asso. prof. Inst. Oceanography, U. B.C., Vancouver, 1955-60, prof. physics, 1961—; vis. prof. Dalhousie U., Halifax, N.S., Can., 1960-61, Harvard, 1964; distinguished vis. prof. Pa. State U., 1964; Commonwealth vis. prof Cambridge U., 1967—. Cons. Pacific Naval Lab., Can. Def. Research Bd., U.S. Naval Lab.; Canadian rep. Sci. Com. on Oceanic Research, 1963—. Author: (with J.S. Marshall, E.R. Pounder) Physics, 1967. Asso. editor Jour. Atmospheric Scis., 1965—, Jour. Marine Research, 1965—. Research and publs. on turbulent flows, underwater sound, interstellar gas dynamics, non-linear effects in waves, wave generation by wind, dynamic oceanography, air-sea interaction. Home: 4521 Langara Av., Vancouver 8, B.C., Can.*

STEWART, Ross, Canadian chemist; b. Vancouver, B.C., Can., Mar. 16, 1924; s. Colin and Jessie (Grant) S.; B.A., U. B.C., 1946, M.A., 1948; Ph.D., U. Wash., 1954; m. Greta Marie Morris, Sept. 7, 1946; children—Cameron Leigh, Ian Hampton. Faculty, Canadian Services Coll., Victoria, B.C., 1949-55; faculty U. B.C., Vancouver, 1955—, prof., 1962—. Fellow Chem. Inst. Can. Author: Oxidation Mechanisms, 1964; Investigation of Organic Reactions, 1966. Research and numerous publs. in mechanism of certain oxidation processes, strongly acidic and strongly basic solution, position of protonation of carboxylic acids, determination of acidity of feeble organic acids such as aromatic amines. Home: 4855 Paton St., Vancouver, B.C., Can.*

STEWART, Sarah Elizabeth, Am. physician; b. Tecatillan, Jalisco, Mexico, Aug. 16, 1906 (parents Am. citizens); d. Arthur J. and Coucha (Andrade) Stewart; B.S., N.M. State U., 1927; M.S., U. Mass., 1930; Ph.D., U. Chgo., 1939; M.D., Georgetown U., 1949; LL.D., New Mexico State University, 1962. Bacteriologist, Colo. Expt. Sta., 1930-33; research fellow U. Colo. Sch. Med., 1933-35; with USPHS, 1935—, med. dir., 1967—. Recipient Lenghi award Accademia Nazionale Dei Lincei, Rome, 1963, Lucy Worthan James award James Ewing Soc., 1964. Mem. Am. Assn. Cancer Research, Soc. Exptl. Biology and Medicine, Washington Acad. Sci., Med. Womens Assn., Alpha Omega Alpha, Sigma Xi. Author: (with Bernice Eddy, Mearl Stanton) Progress in Virus Research—Polyoma Virus, 1960; also articles. Contbg. author: Experimental Tumor Research, Vol. I; Advance in Virus Research, Vol. 7, 1960. Discovered polyoma virus—the 1st cancer inducing virus shown to grow in tissue culture, and to produce multiple kinds of cancer in mice, and cancer in other animals. Home: 9305 Kingsley Av. Office: Nat. Cancer Inst., Bethesda, Md. 20014.*

STEWART, Thomas Dale, Am. anthropologist; b. Delta, Pa., June 10, 1901; s. Thomas Dale and Susan (Price) S.; A.B., George Washington U., 1927; Franklin P. Mall Scholar in anatomy, M.D., Johns Hopkins Med. Sch., 1931; D.Sci. (hon.), Univ. of Cuzco, Peru, 1949; m. Julia C. Wright, Dec. 1, 1932 (deceased); one daughter, Cornelia E. (Mrs. Michael D. Gill); m. 2d, Rita Frame Dewey, July 12, 1952. Aid in div. physical anthropol., U. S. Nat. Museum, 1927-31, asst. curator, 1931-39, asso. curator, 1939-42, curator, 1942-61, head curator dept. of anthropology, 1961-62, director Museum Natural History, 1962-65, sr. research scientist, office anthropology, 1966—; visiting prof. anatomy Washington U. Sch. Med., St. Louis, 1943; vis. prof. physical anthropol., Escuela Nacional de Antropologia, Mexico, 1945; lectr. anatomy Med. Sch., George Washington U., 1958-67, Am. del. Internat. Congress Americanists, Mex., 1939; Am. del. Gen. Assembly Pan-Am. Inst. Geog. and History, Lima, Peru, 1941; Am. del. to Inter-Am. Conf. on Indian Life, Cuzco, Peru, 1949, La Paz Bolivia, 1954; mem. field trips under Smithsonian Instn. to Alaska, 1927, Mexico, 1939, 1945, Peru, 1941, 1949, Guatemala, 1947, 1949, Japan, 1954-55 Iraq, 1957, 60, 62. Recipient Viking Fund medal and award, 1953. Mem. Nat. Acad. Scis., Nat. Geog. Soc. (com. research and exploration), American Institute of Human Paleontology (president, 1955-62), Anthropol. Society of Washington (pres. 1944-46), Washington Acad. of Sci., Am. Orthopaedic Assn. (hon. mem.), Am. Assn. Phys. Anthropol. (sec.-treas. 1960-64, past pres.), Washington Acad. Med., Sigma Xi. Editor: American Journal Phys. Anthropol., 1943-48; contbg. editor Handbook Latin Am. Studies, 1938-60; adv. editor Clin. Orthopaedics, 1954—. Research on paleopathology; comparative human osteology; human identification. Home: 1191 Crest Lane, McLean, Va. Office: U. S. Nat. Mus., Smithsonian Instn., Washington 20560.

STEWART, Sir Thomas Grainger, Scottish physician; b. Edinburgh, Scotland, Sept. 23, 1837; s. Alexander and Agnes (Grainger) S.; M.D., U. Edinburgh, 1958; LL.D., Aberdeen U., 1897; M.D. (hon.), U. Dublin (Ireland), 1866, Royal U. Ireland, 1887; m. Josephine Dubois, 1863; m. 2d, Jessy Dingwall Fordyce Mcdonald, 1866; four sons, four daus. Studied at Prague, Berlin and Vienna; prof. practice of physics Fdinburgh U., 1876; physician in ordinary to Queen Victoria, 1882. Rep. Edinburgh U. at Berlin Congress on Tb. Fellow Royal Soc. Edinburgh; hon. fellow Royal Coll. Physicians Ireland; mem. Edinburgh Coll. Physicians (pres. 1889-91), Royal Med. Soc. (pres. while undergrad.). Author: A Practical Treatise on Bright's Disease of the Kidneys, 1869; On the Position and Prospects of Therapeutics, 1868; An Introduction to the Study of Diseases of the Nervous System, 1884; also numerous articles. Research on diseases of kidneys; one of 1st to draw attention to deep reflexes in neuritis under the title of Paralysis of the Hands and Feet from Disease of the Nerves; described condition known as multiple neuritis; induced Prof. Lister to perform operations on the brain for traumatic epilepsy. Died Feb. 3, 1900.

STEWART, W(ellington) (Buel), Am. pathologist; b. Chgo., June 18, 1920; s. George Ross and Helen Isabell (Skinner) S.; B.S., U. Notre Dame, 1942; M.D., U. Rochester, 1945; m. Helen M. Zimmerman, June 16, 1945; children—Douglas, Thomas, John. Rockefeller fellow pathology U. Rochester, 1948-49, Vet. Postgrad. fellow pathology, 1948-50; from asso. pathology to asso. prof. pathology Columbia, 1950-60; prof. pathology, chmn. dept. U. Ky. Coll. Medicine, 1960—. Mem. subcom. 13 Nat. Com. Radiation Protection and Measurement; chmn. Bd. Registry Med. Technologists, 1964—. Diplomate Nat. Bd. Med. Examiners, Am. Bd. Pathology (anat. and clin. pathology). Fellow Coll. Am. Pathologists; mem. Am. Soc. Exptl. Pathology, Am. Assn. Pathologists and Bacteriologists, Harvey Soc., Am. Soc. Exptl. Biology and Medicine, A.M.A., Ky. Soc. Pathologists (pres. 1966), Am. Soc. Clin. Pathologists, Sigma Xi, Alpha Omega Alpha. Contbr. articles sci. publs. Research on iron metabolism, erythrocytes, nutritional injury to liver. Home: 3034 Breckenwood Dr., Lexington, Ky.*

STEWART, William Sheldon, Am. botanist; b. San Diego, Nov. 14, 1914; s. Lee L. and Hazel (Sheldon) S.; B.A., U. Cal. at Los Angeles, 1936, M.A., 1937; Ph.D., Cal. Inst. Tech., 1939; m. Maria R. Markham, Sept. 10, 1940; children—Mary Lee, David Markham, Carol Ann. Plant physiologist U. S. Dept. Agr., 1939-45; asso. plant physiologist Citrus Exptl. Sta., U. Cal., Riverside, 1945-50, chmn. dept. horticulture, 1953-55; head dept. plant physiology Pineapple Research Inst., Honolulu, 1950-53; dir. Los Angeles County Dept. Arboreta and Botanic Gardens, Arcadia, Cal., 1955—; research asso. Cal. Inst. Tech., Los Angeles, 1955—. Mem. Bot. Soc. Am., Am. Soc. for Hort. Sci. Research, numerous publs. on application plant growth regulators in agr., particularly citrus. Home: 1945 Vista St., Sierra Madre, Cal. 91024. Office: 301 N. Baldwin Av., Arcadia, Cal. 91006.*

STEYERMARK, Julian Alfred, botanist; b. St. Louis, Jan. 27, 1909; s. Leo and Mamie (Isaacs) S.; A.B., Washington U., St. Louis, 1929, M.S., 1930, Ph.D., 1933; M.A., Harvard, 1931; m. Cora Shoop, Sept. 1, 1939. Staff, Herbarium, Chgo. Natural History Mus., 1937-58, curator, 1950-58; botanist Instituto Botanico, Ministerio de Agricultura y Cria, Caracas, Venezuela, 1959—; vis. prof. botany So. Ill. U., 1958, U. Mo., Columbia, 1958. Recipient Order of Quetzal, Guatemala Govt., 1961, Distinguished Service plaque Washington U., 1955; named Distinguished Botanist of Year, N.Y. Bot. Garden, 1965. Mem. Bot. Soc. Am., Internat. Assn. Plant Taxonomists, Fern Soc. Am., Torrey, New Eng. bot. clubs, Assn. Tropical Biology. Author: Flora of Missouri, 1963; Spring Flora of Missouri, 1940; (with P.C. Standley) Flora of Guatemala, 8 vols., 1946-59; also numerous articles; contbg. author: Flora of Venezuela, 4 vols., 1953-57. Discovered numerous new plant species in Guatemala, Ecuador, and Venezuela while on bot. explorations. Home: Residencia Magnolia, Calle 2, Los Palos Grandes, Caracas. Office: Instituto Botanico, Ministerio de Agricultura y Cria, Caracas, Venezuela.*

STEYSKAL, George Constance, Am. entomologist; b. Detroit, Mar. 30, 1909; s. Frank Joseph and Elizabeth (Norrick) S.; grad. high sch.; m. Beatrice Morgan, Oct. 13, 1933; 1 son, Neil M. Amateur entomologist to 1962; with Entomology Research div. U.S. Dept. Agr., Washington, 1962—. Mem. Entomol. Soc. Am., Entomol. Soc. Can., Soc. Systematic Zoology. Contbr. numerous articles to profl. jours. Research, publs. on taxonomy of Diptera (two-winged flies), including description of 188 new species. Home: 5622 Southwick St., Bethesda, Md. 20014. Office: 701 Lamont St., Washington 20560.*

STICKER, Georg, German hygienist, med. historian; b. Cologne, Germany, Apr. 18, 1860; prof. Giessen, Münster, Würzburg, Germany. Author: Die Pest, 2 vols., 1908-10; Geschichte der ansteckenden Geschlechtskrankheiten, 1931. Discovered flea could spread epidemics; publs. on tropical diseases, history of epidemics; described erythoma infectiosum (Sticker's disease), 1899. Died Zell, Germany, Aug. 28, 1960.

STICKLER, Roland, metallurgist; b. Grafenbach, Austria, May 4, 1931; s. Josef and Grete (Wittman) S.; B.S., Realgymnasium, Neunkirchen, Austria, 1949; M.S., Technische Hochschule Vienna, 1956, Ph.D., 1958; m. Erika Haubner, July 5, 1956; children—Eva, Christian. Came to U.S., 1958. Asst. prof. Institut f. Technologie Anorganischer Stoffe, Technische Hochschule Wien, Austria, 1956-58; mgr. phys. metallurgy Westinghouse Research Lab., Pitts., 1958—; lectr. Carnegie Inst. Tech., 1965-66. Mem. Am. Soc. Metals, Am. Inst. Metall. Engring., Am. Soc. Testing and Materials, Plansee Soc., Am. Crystallographic Assn. Research in microstructure of semiconductors, especially of imperfections in silicon; numerous publs. on electron microscopy, electron and X-ray diffraction, and electron beam microanalysis studies of properties of solids; studies on microstructure, crystalline perfection, composition of solids. Office: Westinghouse Research Lab., Pitts. 15235.*

STICKNEY, John Clifford, Am. physiologist; b. Vancouver, Wash., July 15, 1909; s. Henry James and Laura (Miller) S.; student U. Ore., 1929; B.S., Wheaton Coll., 1933; M.S., U. Wash., 1936; Ph.D., U. Minn., 1940. Faculty, Wheaton Coll., 1935-36, U. Minn., 1937-40; faculty W.Va. U. Sch. Medicine, Morgantown, 1940—, prof. physiology, 1957—. Mem. A.A.A.S., Am. Physiol. Soc., Soc. for Exptl. Biology and Medicine, Am. Inst. Biol. Scis., Sigma Xi. Author: (with Van Liere) Hypoxia, 1963; also numerous articles. Research on oxygen deficiency on mammals, adaptation to hypoxia. Home: 248 Wagner Rd., Morgantown, W.Va. 26505.*

STIEFEL, Eduard Ludwig, Swiss mathematician; b. Zurich, Switzerland, Apr. 21, 1909; s. Eduard and Paula (Schubert) S.; Dr. ès sc., Edigenössiche Technische Hochschule; m. Jeanette Beltrami, Oct. 28, 1939; children—Eduard, Eva. Prof. math., dir. Applied Math. Inst., Swiss Fed. Tech. Sch. Founder, Applied Math. Inst. Author books on descriptive geometry, numeric calculus; also articles. Recipient Silver medal Swiss Fed. Tech. Sch., U. Helsinki. Mem. Norwegian Acad. Sci. (hon.), Soc. Applied Math. and Mechanics, Swiss Math. Soc. Home: Drusbergstrasse 15, Zurich, Switzerland.

STIEGLITZ, Julius (Oscar), Am. chemist; b. Hoboken, N.J., May 26, 1867; s. Edward and Hedwig (Werner) S.; grad. Real-gymnasium, Karlsruhe, Germany, 1886; A.M., Ph.D., U. Berlin, 1889; D.Sc., Clark U., 1909; Ch.D., U. Pitts., 1916; m. Anna Stieffel, Aug. 27, 1891 (dec. 1933); m. 2d, Mary M. Rising, Aug. 30, 1934. With U. Chgo., 1892—, prof. chemistry, 1905-33, dir. analytical chemistry, 1909-15, dir. univ. labs., 1912-24, chmn. chemistry dept., 1915-33. Hitchcock lectr. U. Cal., 1909; Dohme lectr. Johns Hopkins, 1924; Fenton lectr. U. Buffalo, 1933. Mem. Internat. Commn. Annual Tables Constants, 1915-21; mem. div. of chemistry NRC, 1917-19, chmn. com. synthetic drugs, 1917-19; vice chmn. div. of chemistry, 1919-21; spl. expert Pub. Health Service, 1918—. Recipient Willard Gibbs medal, 1923. Fellow Am. Acad. Arts and Scis., Washington Acad. Scis., A.A.-A.S. (v.p.). Author: Elements of Qualitative Chemical Analysis, 2 vols., 1911-12. Studies and publs. on molecular rearrangements, ethers. Died Chgo., Jan. 10, 1937.

STIELTJES, Thomas Jean, French mathematician; b. Zwolle, Holland, 1856; naturalized French citizen; prof. U. Toulouse (France). Studied divergent and conditionally convergent series, theory of numbers, Riemann's function; defined integral; studied spherical harmonics. Died 1894.

STIERSTADT, Klaus Wilhelm Otto, German physicist; b. Göttingen, Germany, Oct. 28, 1930; s. Otto and Marie (Ohnesorge) S.; student U. Tübingen; Dipl.-Phys., U. Munich, 1955, Dr.rer.nat., 1956; m. Hildegard Nünemann, Aug. 2, 1957; children—Hanna, Helga. Asst., I. phys. inst., U. Munich (Germany), 1956-65, faculty, 1963—, sr. asst., 1965—; research scientist Nat. Center Sci. Research, Grenoble, France, 1965-66. Mem. German Phys. Soc. Author: Der Magnetische Barkhausen-Effekt, 1966; also numerous articles. Research on natural and artificial radioactivity of atmosphere, atmospheric aerosol particles, magnetism and magnetic materials. Address: 23 Ainmillerstrasse, 8 Munich, West Germany.*

STIFEL, Michael, German mathematician; b. Esslingen, 1487; ed. in monastery of native home; Augustinian monk converted to Protestantism by Luther's influence, 1523; became prof., Jena, 1959. Author: Arithmetica integra, 1544; Die deutsche arithmetimetica, 1545; Rechenbuch von der welschen und deutschen Practic, 1546. Introduced signs used in arithmetic today; predicted destruction of earth, Oct. 3, 1533, using astrological calculations; prepared table with numerical values of binomial coefficients for powers below the 18th; constructed arithmetical triangle; studied the beginnings of theory of exponents and logarithms. Died Jena, Apr. 19, 1567.

STIGLER, George Joseph, Am. economist; b. Renton, Wash., Jan. 17, 1911; s. Joseph and Elizabeth (Hungler) S.; B.B.A., U. Wash., 1931; M.B.A., Northwestern U., 1932; Ph.D., U. Chgo., 1938; m. Margaret L. Mack, Dec. 26, 1938; children—Stephen, David, Joseph. Asst. prof. Ia. State Coll., 1936-38; faculty U. Minn., 1938-46, prof., 1944-46; prof. Columbia, N.Y.C., 1946-58; prof. econs. U. Chgo., 1958—. Fellow Am. Statis. Assn.; mem. Am. Econ. Assn. (past pres.), Royal Econ. Soc. Author: Production and Distribution Theories, 1941; The Theory of Competitive Price, 1942; The Theory of Price, 1946, rev., 1952, 66; The Intellectual and the Market Place and Other Essays, 1963; Essays in the History of Economics, 1965; also articles, pamphlets. Studies in evolution of econs. as sci., sociology of sci., workings of econ. system, especially monopoly and econs. of information, methods and effects of public regulation of econ. activity. Home: 2621 Brassie Av., Flossmoor, Ill. 60422. Office: U. Chgo., Chgo. 60637.*

STILES, Charles Wardell, Am. zoologist; b. Spring Valley, N.Y., May 15, 1867; s. Samuel Martin and Elizabeth (White) S.; ed. Wesleyan U., Conn., 1885-86, Collège de France, 1886-87, U. Berlin, 1887-89, U. Leipzig, 1889-90, Trieste Zool. Sta., 1891, Pasteur Inst. and Collège de France, 1891; A.M., Ph.D., Leipzig, 1890; several hon. degrees; m. Virginia Baker, June 1897. Zoologist, Bur. Animal Industry, U. S. Dept. Agr. 1891-1902, cons., 1902-04; prof. of zoology, USPHS, 1902-30, asst. surgeon gen., 1919-30, med. dir. 1930-31; prof. medical zoology, Georgetown U., 1892-1906; spl. lectr. Army Med. Sch., 1894-1902, Johns Hopkins, 1897-1937, Navy Med. Sch., 1902—; hon. custodian helminthological collections U. S. Nat. Mus., 1893-1931; sec. advisory com. Smithsonian Table at Naples Zool. Sta., 1894—. U. S. Govt. del Internat. Zool. Congresses, Leyden, 1895, Cambridge, 1898, Berlin, 1901, Berne, 1904, Boston, 1907, Gratz, 1910, Monaco, 1913, Budapest, Hungary, 1927, Padua, 1930; sec., 1898-1936, Internat. Commn. on Zoöl. Nomenclature; sec., 1910-27, Internat. Commn. on Med. Zoology; detailed as agrl. and sci. attaché U. S. Embassy, Berlin, Germany, 1898-99; sci. sec. Rockefeller commn. for eradication of hookworm disease, 1909-14; asso. Smithsonian, 1931—; prof. zoology Rollins Coll. (winter faculty), 1931-38. Mem. many Am. and European scientific and medical societies; fgn. corr. Société de Biologie, France, Académie de Médecine, France; corr. mem. Zool. Soc., London. Author: Trichinosis in Germany, 1901; Index Catalogue of Medical and Veterinary Zoology from 1902 (continuing publ.); The Cattle Ticks (Ixodoidea) of the United States, 1902; Emergency Report on Surra, 1902; Report on Hookworm Disease (Uncinariasis) in the United States, 1903; Illustrated Keys to Trematode and Cestode Parasites of Man; Trematoda, 1908; Taxonomic Value of Stigmal Plates in Ixodoidea, 1910; Watsonius Watsoni, 1910; Cestoda, 1912; Nematoda, 1920; Studies on Intestinal Parasites (especially Amoebae) in Man, 1923; Key-Catalogue of the Protozoa Worms, Crustacea, Arachnoids, and Insects of Man, Primates, Chiroptera, Insectivora, Carnivora, 1925-32; Early History of the Hookworm Campaign in Our Own Southern U. S., 1939. Noted for discovery of prevalence of hookworm in southern U. S.; discovered Unicinaria Americana (Necator americanus), 1902. Died Jan. 24, 1941.

STILES, Karl Amos, Am. zoologist, geneticist; b. Banfield, Mich., Nov. 19, 1895; s. David C. and May (Kipp) S.; A.B., Albion Coll., 1927; M.S., U. Mich., 1931, Ph.D., 1935; D.Sc., Ferris State Coll., 1966; m. Nettie Rose Stockman, Aug. 31, 1928; 1 dau., Patricia Ann (Mrs. Norman E. Harris). Instr., Battle Creek (Mich.) Coll., 1927, U. Mich., Ann Arbor, 1931-34; prof., head biology dept. Coe Coll., Cedar Rapids, Ia., 1934-45, chmn. div. natural sci., 1938-45; faculty Mich. State Univ., East Lansing, 1946-61, prof. zoology, 1948-61, head dept. zoology, 1953-61. Mem. Biol. Stain Common., Evanston, Ill., 1934— editorial cons. Rinehart Co., 1946—. A.A.A.S. research grantee, 1938, 40, 43, 44; Collecting Net scholar Marine Biol. Sta., Woods Hole, Mass., 1929; Rockefeller Found. Gen. Edn. Bd. fellow, Columbia U., 1941-42. Fellow A.A.A.S., Population Reference Bur.; mem. Am. Soc. Zoologists, Am. Soc. Naturalists, Internat. Union Biol. Scis., Am. Inst. Biol. Scis., Ia. (past chmn. zoology sect.), Mich. (past chmn. zoology sect.) acads. sci., Am. Genetics Assn., Genetics Soc. Am., Am. Eugenics Soc., Am. Soc. Human Genetics, Am. Biol. Soc., Soc. for Study Evolution, Kelvin Soc., Physalia, Am. Inst. Biol. Scis. (mem. governing bd. 1960-64), Am. Soc. Zoologists, (mem. edn. com. 1959-65), Sigma Xi, ,Phi Kappa Phi, Beta Beta Beta (nat. hon. mem.), others. Author 20 books including: Handbook of Histology, 1940; (with R.W. Hegner) College Zoology, 1951; Laboratory Explorations in General Zoology, 1943; Laboratory Studies in General Zoology, 1964.; also numerous articles. Research on genetic defects, sci. method and teaching zoology, lab. procedures. Home: 4747 Woodcraft Rd., Okemos, Mich. 48864. Office: Natural Sci. Bldg., Mich. State U., East Lansing, Mich. 48823.*

STILES, Walter, English botanist; b. Hammersmith, London, Eng., Aug. 23, 1886; s. Walter and Elizabeth (Duny) S.; B.A., Emmanuel Coll., Cambridge, 1908, M.A., 1912, Sc.D., 1922; M.Sc., U. Birmingham, 1930; D.Sc., U. Nottingham, 1963; m. Edith Ethel May Harwood, July 7, 1920; children—Walter, Ruth Mary. Asst. lectr. botany U. Leeds, 1910-19; research worker food investigation bd. Dept. Sci. and Indsl. Research, London, 1918-20; prof. botany U. Coll., Reading, Eng., 1919-26, U. Reading, 1926-29; prof. botany U. Birmingham, 1929-51, dean faculty sci., 1932-35. Fellow Royal Soc., 1928 (past mem. council), Linnean Soc.; mem. Am. Soc. Plant Physiologists (corr.), Biochem. Soc., Brit. Ecol. Soc., Soc. for Exptl. Biology. Author: (with I. Jorgensen) Carbon Assimilation, 1917; The Preservation of Food by Freezing, 1921; Permeability, 1924; Photosynthesis, 1925; (with W. Leach), Respiration in Plants, 1932; An Introduction to the Principles of Plant Physiology, 1936; Trace Elements in Plants, 1946; also numerous articles. Research on water and salt relations of plant cells and tissues, respiration of seeds and seedling during germination period, respiration of storage tissues. Home: 21 Elsley Rd., Reading, Berkshire, Eng. Deceased.

STILL, Andrew Taylor, Am. osteopath; b. Jonesboro, Va., Aug. 6, 1828; s. Abraham and Martha Poage (Moore) S.; ed. log sch. house, Jonesboro, Va., Holston Coll., Newmarket, Tenn., and pvt. schs.; m. Mary M. Vaughn, 1837 (dec. 1859), m. 2d, Mary E. Turner, Nov. 15, 1860 (dec. 1910). Surgeon and maj. 21st Kan. Vols. in Civil War, founder osteopathy; practiced osteopathy, 1874—; pres. Am. Sch. Osteopathy, Kirksville, Mo., 1892—. Author: Autobiography of A. T. Still, 1897, rev., 1908; Philosophy of Osteopathy, 1899; Mechanical Principles, 1902; Practice and Research, 1910. Died Dec. 12, 1917.

STILL, Sir George Frederic, English pediatrician; b. Holloway, London, Eng., Feb. 27, 1868; s. George S. and Eliza (Andrew) S.; M.A., Caius Coll., Cambridge (Eng.) U.; M.D., 1896; ed. Guy's Hosp.; LL.D. (hon.) U. Edinburgh (Scotland), 1927. Murchison scholar Royal Coll. Physicians, 1894; Goulstonian lectr., 1902; Lumkeian lectr., 1918; Fitzpatrick lectr., 1928; Ingleby lectr. U. Birmingham (Eng.), 1927; physician-in-ordinary to Duke and Duchess of York, 1936; fellow, mem. council King's Coll., London; physician extraordinary to the King from 1937; cons. physician Hosp. Sick Children, Infant's Hosp., Dr. Bernardo's Homes and Soc. for Waifs and Strays, emeritus prof. diseases of children King's Coll. First pres. Brit. Pediatric Assn.; pres. Internat. Pediatric Congress, 1933; chmn. Nat. Assn. Prevention Infant Mortality, 1917-37. Recipient Dawson Williams Meml. prize for work for sick children, 1934. Fellow Royal Coll. Physicians, 1901; hon. fellow Royal Soc. Medicine, 1936, Soc. Medica Chirurgica Bologna (Italy); hon. mem. Am. Pediatric Soc., Canadian Soc. Study Diseases of Children; corr. mem. de la Société de Pédiatric, Paris, France. Author: Common Disorders and Diseases of Childhood, 1909; History of Pediatrics, 1931; Common Happenings in Childhood, 1938; Concerning Children and Other Things; (with Sir James Goodhart) A Textbook on Diseases of Children; also numerous articles. Described chronic rheumatoid arthritis in children, 1897; identified organism causing posterior basic meningitis; specialist on diseases afflicting children. Died Harnham Croft, Salisbury, Eng., June 28, 1941.

STILLÉ, Alfred, Am. physician; b. Phila., Oct. 30, 1813; grad. Univ. Pa., 1832, M.D., 1836, LL.D., 1876; LL.D., Pa. Coll., 1876. Resident physician Phila. Hosp., 1836; Pa. Hosp., 1840-41; vis. physician St. Joseph's Hosp., 1849-71; Phila. Hosp., 1865-72; prof. theory and practice of medicine Pa. Med. Coll., 1854-59, U. Pa., 1864-84. Author: Elements of General Pathology; The Unity of Medicine; Humboldt's Life and Character; War as an Element of Civilization; Othello and Desdemona; The National Dispensatory (with John M. Maish); Therapeutics and Materia Medica; Epidemic Meningitis; Epidemic or Malignant Cholera. One of 1st to distinguish typhus from typhoid fever. Died 1900.

STILLE, Bernd A., German biologist; b. Paderborn, Germany, Aug. 25, 1912; s. Bernhard and Maria (Henze) S.; Ph.D., U. Gottingen; m. Hildegard Leblanc, May 1, 1942; children—Elmar, Lothar, Reinhard, Berthold, Wolfgang. Joined Berlin Biol. Inst., 1937; dir. dept. Research Inst. on Refrigeration Foodstuffs, Karlsruhe, Germany, 1938-46; govt. counsel for Admiralty, 1942-45; became asst. U. Bonn, 1947, instr., 1951, then prof. Mem. German Hygiene and Microbiol. Assn. Research and publs. on Feulgen reaction in bacterial, microbiology of soil, and phytopathology. Home: Hausdorffstrasse 233, Bonn, West Germany.

STILLE, Hans, German geologist; b. Hannover, Germany, Oct. 8, 1876; prof., Hannover, also Leipzig, Göttigen, Berlin (all Germany). Author: Geologische Evolutionen und Revolutionen in der Erdrinde, 1913; Die Schrumpfung der Erde, 12 1922; Grundfragen der vergleichenden Tektonik, 1924; Einführung in den Bau Amerikas, 1940; Die Tendenz der Erdentwicklung, 1948. Research in geotectonics.

STILLER, Eric Thomas, chemist; b. Croyden, Gt. Britain, Jan. 8, 1907; s. Emil and Isa C. (Hastie) S.; B.Sc. with 1st class honors, U. Glasgow (Scotland), 1929; Ph.D., U. St. Andrews (Scotland), 1932; m. Suzanne Giesey, Jan. 10, 1948; 1 son, Eric Thomas. Came to U.S., 1939, naturalized, 1944. Commonwealth Fund fellow Rockefeller Inst., N.Y.C., 1932-34; hon. research fellow U. Manchester (Gt. Britain), 1934-35; research fellow Nat. Inst. for Med. Research, London, Eng., 1935-39; sr. research chemist. Merck & Co., Inc., Rahway, N.J., 1939-43; head dept. organic chemistry Wyeth Inst. for Biol. Research, 1943-47; dir. chem. devel. Squibb Inst. for Med. Research, New Brunswick, N.J., 1947-65, sr. research asso., 1965—. Cons., Biol. Inst., Oak Ridge Nat. Lab., 1946-47. Corecipient Mead Johnson award Am. Soc. Biol. Chemists, 1941. Mem. Am., Brit., Swiss chem. socs. Research, publs. and patents on carbohydrates, vitamin structures; devel. antibiotics, chemotherapeutics. Home: 1050 George St., New Brunswick, 08901. Office: E.R. Squibb & Sons, Inc., Georges Rd., New Brunswick, N.J. 08903.*

STILLING, Benedikt, German anatomist; b. Kirchhain, Germany, Feb. 22, 1810; surgeon in Kassel, Germany, also Vienna, Austria. Author: Untersuchungen über die Textur des Rückenmarks, 1842; Untersuchung über den Bau und die Verrichtung des Gehirns, 1840. Research on fine structure of brain; introduced microtome to histology; described solitary fasciculus of medulla oblongata (Stilling's bundle), 1843. Died Kassel, Jan. 28, 1879.

STILLINGER, Frank Henry, Jr., Am. chemist; b. Boston, Aug. 15, 1934; s. Frank Henry and Gertrude (Metcalf) S.; B.S., U. Rochester, 1955; Ph.D., Yale, 1958; m. Dorothea Anne Keller, Aug. 18, 1956; children—Constance Anne, Andrew Metcalf. NSF postdoctoral fellow Yale, 1958-59; mem. tech. staff Bell Telephone Labs., Murray Hill, N.J., 1959—. Fellow Am. Phys. Soc.; mem. N.Y. Acad. Sci. Editor: John Gamble Kirkwood Collected Works, 1965. Contbr. to Molton Salt Chemistry, 1964. Research in concentrated electrolytes and molten salts, phase transformations and critical phenomena, atomic and molecular quantum mechanics, surface properties of liquids, properties of anharmonic solids, structure and behavior of ordinary and heavy water. Home: 216 Noe Av., Chatham, N.J. 07928. Office: Bell Telephone Labs., Murray Hill, N.J. 07971.*

STILLINGFLEET, Benjamin, English naturalist; b Norfolk, Eng., 1702; s. Edward Stillingfleet; B.A., Trinity Coll., Cambridge, 1723; m. sister of Lord Barrington, 1746. Tutor to William Ashe-Windham, 14 years; traveled abroad, 5 years; apptd. surveyor of barracks in Savoy and guardroom at Tilt-yard, St. James's, and Kensington, 1760. Author: The Calendar of Flora, Swedish and English, 1755; Miscellaneous Tracts relating to Natural History, Husbandry and Physick, 1759; Select Works, 1811. Early defender of published Miscellaneous Tracts, Linnean system; 6 essays from Linnaeus's Amoenitates Academicae translated from Latin, included preface that was 1st fundamental treatise on principles of Linnaeus pub. in Eng.; helped establish Linnaean botany in Eng. Died Piccadilly, Eng., Dec. 15, 1771.

STILLMAN, John Maxson, Am. chemist; b. N.Y., Apr. 14, 1852; s. Jacob Davis Babcock and Caroline B. (Maxson) S.; Ph.B., U. Cal., 1874, Ph.D., 1885 LL.D., 1916; postgrad. in chemistry, Strasbourg (now in France), Würzburg, Germany, 1875-76; m. Emma Rodolph, June 1878. Asst. in chemistry U. Cal., 1873-75 instr. organic and gen. chemistry, 1876-82; chemist Boston and Am. Sugar Refining cos., 1882-91; prof. chemistry Leland Stanford Jr. U., 1891-1917, v.p., 1913-17, emeritus, 1917. Asst. ednl. dir. S.A.T.C., 1918. Author: Paracelsus, as Physician, Chemist and Reformer, 1920; The Story of Early Chemistry, 1924. Research on chemistry of vegetable substances, ammonia compounds of inorganic chlorides, molecular lowering of freezing point of napthymanun and diphenylanum. Died Dec. 13, 1923.

STIMMEL, Benjamin F., Am. endocrinologist, biochemist; b. Port Royal, Pa., Feb. 12, 1904; s. Benjamin F. and Margaretta (McMeen) S.; B.A., Bethany College, W. Va., 1926; Ph.D., U. Pittsburgh, Pa., 1934; m. Margaret Copland Scott, July 11, 1936; one dau., Margaret (Mrs. Leamond F. Lacey). Research fellow, dept. biochemical research, Cleveland Clinic, 1934-37; Macy Research Fellow, Cornell U. Med. College, 1937-40; research biochemist, dir. endocrine div., Rees-Stealy Clinic Research Found., San Diego, Calif., 1940-66, lectr. U. Calif., 1961; prof. San Diego State College, 1962. Mem., Am. Assn. Cancer Research; Am. Chem. Soc.; Assn. for Study Internal Secretions; Am. Soc. Biological Chemists; A.A.A.S. Author numerous published articles. Basic research, publs. in endocrinology, primarily in methodology, metabolism, and discovery of natural steroids in humans. Died Jan. 7, 1966.

STIMPSON, William, Am. naturalist, conchologist; b. Roxbury, Mass. Feb. 14, 1832; s. Herbert H. and Mary (Brewster) S.; attended Boston Latin Sch., 1848; studied under Louis Agassiz; M.D., Columbia, 1860; m. Annie Gordon, July 28, 1864, 3 children. First naturalist to employ deep sea dredging in work; apptd. to North Pacific Exploring Expdn., 1852-56, began classification of immense amount of data gathered during those years, with hdqrs. in Smithsonian Instn., Washington, 1856; results published in Smithsonian Miscellaneous Collections, 1907; became dir. Chgo. Acad. Scis., 1865, gathered collections and great manuscripts from naturalists all over world; became youngest mem. Nat. Acad. Sciences, 1868; never recovered from loss of bldg. and its treasures when Chgo. Acad. Sciences was destroyed by great Chgo. Fire of 1871. Author (in Latin): A Revision and Synonymy of the Mestaceous Mollusks of New England, 1851; Notes on North American Crustacea, 1859, Died Ilchester, Md., May 26, 1872.

STIMSON, Lewis Atterbury, Am. surgeon; b. Paterson, N.J., Aug. 1844; s. Henry C. and Julia M. (Atterbury) S.; A.B., Yale, 1863, LL.D., 1900; M.D., Bellevue Hosp. Med. Coll. (New York U.), 1874; traveled and studied abroad until 1873; m. Candace Wheeler, Nov. 1866; 1 son, Henry Lewis S. prof. physiology N.Y. U., 1883-85, anatomy, 1885-89, surgery, 1889-98; 1st prof. surgery Cornell U. Med. Coll., 1898— vis. surgeon Bellevue Hosp., N.Y. Hosp.; cons. Christ's Hosp., Jersey City. Author: Bacteria and Their Influence upon the Origin and Decelopment of Septic Complications of Wounds (wood prize essay 1875); Treatise on Fractures and Dislocations (classic work), 1883, 7 later edits.; Pasteur's Life and Work in Relation to the Advancement of Medical Science, 1893; Operative Surgery, 1900. Promoted tranverse incision in abdominal operation; 1st to show in treatment of old dislocation of elbow that bone formed on humerus as obstacle to reduction; (with Hodgen) brought together traction and suspension in fracture of femur; introduced molded plaster splint for treatment of fractures; introduced methods of reducing dislocations at hip and shoulder, 1888; contbd. to progress of orthopedics; performed ligation of ovarian and uterine arteries in sequence in their course in hysterectomy, 1889. Died Sept. 17, 1917.

STIMSON, Sister Miriam Michael, Am. chemist; b. Chgo., Dec. 24, 1913; d. Frank S. and Mary F. (Holland) Stimson; B.S., Siena Heights Coll., 1936; M.S., Institutum Divi Thomae, Cin., 1939, Ph.D., 1948. Faculty, Siena Heights Coll., Adrian, Mich., 1939—, prof. chemistry, chmn. div. natural sci., 1948—. Mem. Am. Phys. Soc., Am. Chem. Soc., Am. Soc. Cell Biology. Author: (with others) Introduction to Physical Science, 1966. Devel. of pressed potassium bromide disk method for handling organic solids for infrared work, extension of same for studying dimerization of pyrimidines under ultraviolent irradiation. Address: Siena Heights Coll., Adrian, Mich. 49221.*

STINCHCOMB, Thomas Glenn, Am. physicist; b. Tiffin, O., Sept. 12, 1922; s. George Alfred and Ruth (Brand) S.; B.S., Heidelberg Coll., 1944; M.S., U. Chgo., 1948, Ph.D., 1951; m. Maxine Orr Kohler, Nov. 22, 1945; children—James Alfred, William Jay, David Glenn, Dan Thomas. Asst. prof. Wash. State U., Pullman, 1951-54; prof., Heidelberg Coll., Tiffin, 1954-61; research physicist Ill. Inst. Tech. Research Inst., Chgo., 1961-68; prof., chmn. physics dept. DePaul U., Chgo., 1968—. Mem. Am. Phys. Soc., Am. Geophys. Union, Chgo. Physics Club, Sigma Xi. Research on space radiation, cosmic radiation, interaction of radiation in matter. Home: 1326 Rosalie St., Evanston, Ill. 60201. Office: 2323 N. Seminary Av., Chgo. 60614.*

STINE, Charles Milton Altland, Am. chemist; b. Norwich, Conn., Oct. 18, 1882; s. Milton Henry and Mary Jane (Altland) S.; A.B., Gettysburg Coll., 1901; B.S., 1903, A.M., 1904, M.S., 1905, Sc.D., 1926 ; Ph.D., Johns Hopkins, 1907; LL.D., Cumberland U, 1932, Temple U., 1941; Sc.D., U. Del., 1947; m. Martha E. Molly, Feb. 3, 1912; children—Mary Elizabeth (Mrs. F. Samuel Wilcox, Jr.), Barbara Ann (Mrs. J. Seth H. Cruice). Became prof. chemistry Md. Coll. for Women, 1904; fellow Johns Hopkins, 1906-07; joined staff E. I. du Pont de Nemours & Co. (Eastern Lab.), 1907, in charge organic chem. work, 1909-16; transferred to Wilmington office as head organic div., 1917, made asst. dir. chem. dept., 1919, chem. dir., 1924-30, v.p. and dir. since 1930, mem. exec. com. 1930-45, ret., 1945. Cons. to Chem. Warfare Service since 1942; mem. adv. com., dept. of chem. engring. Princeton U.; Acad. Natural Sciences Phila. Mem. Dirs. of Indsl. Industrial Research Assn., Am. Chem. Soc. (councillor; mem. com. to cooperate with C. W. S.), Am. Inst. Chem. Engrs. (councillor and pres. 1947), Franklin Inst. (life), Phi Beta Kappa. Hon. mem. Soc. Chem. Industry (Perkin medal 1950), Princeton Engring. Assn., chem. engring. socs. S. Africa, Australia. Developer numerous processes and products, many patented, in connection with high explosives, propellent powder, dyes, artificial leather, varnishes, paints; initiated research into break down of molecules into smaller molecules; his work led to advances in modern plastics, synthetic rubber, new plant hormones, vitamins, synthetic urea for fertilizer, Freon and other refrigerant fluids, motion picture films using synthetic camphor, med. chemicals such as sulfanilamide and sulfapyridine. Died May 28, 1954.

STINE, Wilbur Morris, Am. physicist; b. Tyrone, Pa., Nov. 3, 1863; s. John Sharp and Sarah (Riegel) S.; Ph.B., Dickinson Coll., 1886, Sc.M., 1889, Sc.D., 1893; postgrad. Ohio U., 1886-87, Ph.D., 1893; studied in Germany, 1889; m. Corinne Elizabeth Super, June 7, 1893; m. 2d, Grovina R. Boyer, Jan. 29, 1907. Instr., later asso. prof., prof. physics and engring. Ohio U., 1886-93; dir. elec. engring. Armour Inst. Tech., 1893-98; Williamson prof. engring. Swarthmore, 1898-1900; devoted attention to authorship, 1909. Mem. jury elec. awards Chicago Expn., 1893, Nat. Export Expn., Phila., 1899. Author: (in Science) Photometrical Measurements, 1900; The Contributions of Lenz to Electromagnetism, 1905, 21; The Discovery of the X or Ionic Ray, and Contributions to Its Physics, 1930, 31, and more than 125 papers, monographs and contributions to physics, technology and edn. Editor Elec. Engring., 1893-94. Began research in X-rays, 1891, obtained a true sciagraph, Feb. 14, 1892, earliest date on record; announced discovery of source of Roentgen rays, 1896, and first suggested remedial use of X-rays, especially for treatment of cancer, 1897. Died July 4, 1934.

STINSON, Benjamin Dale, Am. embryologist; b. Wilburton, Okla., Oct. 29, 1927; s. Lemuel Dale and Ruth Marie (Countiss) S.; B.S., U. Okla., 1950, M.S., 1953; Ph.D., Columbia, 1958; m. Katherine Barbara Krmpotich, Dec. 18, 1954. With Okla. Health Dept., 1950-51, U. Okla., 1951-53, Columbia, 1953-58; postdoctoral research fellow Princeton, 1958-60; faculty Cornell U. Med. Coll., N.Y.C., 1960—, asst. prof. anatomy, 1964—. Mem. Am. Soc. Zoologists, Soc. for Developmental Biology, Am. Soc. for Cell Biology, Am. Micros. Soc., Sigma Xi. Research, publs. on structural damage to chick embryo bone marrow produced by radioactive phosphorus treatment, protection by sulfhydryl chem. compounds of amputated limbs of amphibian larvae against x-ray effects; confirmed normal tissues transplanted into x-rayed limbs of adult salamanders cause formation of regenerate following amputation. Home: 1360 York Av., N.Y.C. 10021.*

STIPANICIC, Pedro Nicolas, Argentine geologist; b. Buenos Aires, Argentina, June 6, 1921; s. Nicolas and Sara (Vila) S.; D. Natural Scis. with Hon. Diploma, U. Buenos Aires, 1947, D.Natural Scis. in Petroleum Geology, 1947; m. Maria Ida Rosa Bonetti, July 28, 1948. Stratigrapher geologist Yacimentos Petroliferos Fiscales Argentina, 1948-55, asst. chief geology Central S. areas, 1955-56; with Argentina Nat. AEC, Buenos Aires, 1953—, chief geol. and mining dept., 1958-60, mgr. raw materials, 1960—; prof. hist. and Argentine regional geology Faculty Exact and Natural Scis., Buenos Aires U., 1955-59; prof. paleontology Nat. U. La Plata, 1960-63. Adviser, Mineralium Deposita, Berlin, Germany, 1965—; mem. Superior Council Geology Argentine Republic, 1966—. Mem. Internat. Union Geol. Scis. (Jurassic subcom.), Argentine Nat. Acad. Scis., Asociación Geológica Argentina, Centro

Argentina de Geólogos, Asociación Paleontológica Argentina, Sociedad Argentina de Estudios Geograficos, Instituto Argentino del Petróleo, Am. Soc. Econ. Geologists, Soc. for Geology Applied to Mineral Deposits (Berlin). Author: (with P. Groeber) Triassic, 1952; (with P. Groeber, A. Mingramm) Jurassic, 1952; (with A. Mingramm) Mineria, 1959; also articles. Research on Upper Paleozoic, Triassic and Jurassic geology in Argentina, diastrophic phases occurring in Argentina and Chile during Jurassic and Cretaceous; discovered several new fossil flora and fauna. Home: Aranguren 548, Buenos Aires. Office: Avda. Libertador 8250, Buenos Aires, Argentina.*

STIREWALT, Margaret Amelia (Mrs. David Richard Lincicome), Am. parasitologist; b. Hickory, N.C., Jan. 18, 1911; d. William Jacob and Mabel (Rhodes) Stirewalt; M.S. cum laude Randolph-Macon Woman's Coll., 1931; M.S., Columbia U., 1935; Ph.D., U. Va., 1938; m. David Richard Lincicome, Dec. 29, 1953. Tchr., Strasburg (Va.) High Sch., 1931-33; research asso. U. Va., Charlottesville, 1938-41; asst. prof. Flora Macdonald Coll., Red Springs, N.C., 1941-43; commd. ensign USN, 1943, advanced through grades to comdr., 1957, parasitologist Naval Med. Research Inst., Bethesda, Md., 1943—. Diplomate Am. Bd. Microbiologists. Fellow A.A.A.S.; mem. Washington Helminthological Soc. (past pres.), Washington Assn. Tropical Medicine (past v.p.), Am. Soc. Parasitologists, (program officer 1962-67), Am., Royal (London, Eng.) socs. tropical medicine and hygiene, Am. Inst. Biol. Scis., Sigma Xi. Contbr. articles to tech. jours. Described new species and genera flatworms; elaborated host-parasite relationships for blood flukes; developed apparatuses and methods for handling blood flukes; demonstrated invasive mechanisms for blood flukes, invasive fluke enzymes; proved resistance to hyperinvasion blood flukes. Home: 7118 Cedar Av., Takoma Park, Md. 20012. Office: Naval Med. Research Inst., Bethesda, Md. 20014.*

STIRLING, James (The Venetian), Brit. mathematician; b. Garden, Stirlingshire, Scotland, 1692; s. Archibald and Anna (Hamilton) S.; ed. Glasgow U., Balliol Coll., Oxford; m. Miss Watson; 1 child. Expelled from Oxford for corr. with Jacobites, 1715; studied in Venice, 10 years; came to London, circa 1725; mgr. Scots Mining Co., Leadhills, 1735. Fellow Royal Soc., 1726. Author: Lineae Tertii Ordines Newtonianae (intended to supplement Newton's Enumeratie Linearum Tertii Ordines, it gave 4 additional varieties to Newton's 72 forms of cubic curve), 8 vols., 1717; Methodus Differentialis sive Tractatus de Summatione et Interpolations Serierum Infinitatum, 1730; also papers to Royal Soc.: On the Figure of the Earth, and on the Variation of the Force of Gravity at its Surface, 1735; A Description of a Machine to Blow Fire by Fall of Water, 1745. Discovered trade secrets of Venetian glass-making; made first survey of the Clyde; made important contbns. to infinitesimal calculus and infinite series. Died Edinburgh, Dec. 5, 1770.

STIRLING, William, Brit. physiologist; b. Jan. 26, 1851; s. Francis and Isabella (Crawford) S.; M.D., D.Sc., U. Edinburgh; postgrad. Leipzig, Germany, Coll. de France, Paris; LL.D., Glasgow; m. Elisabeth Ferguson Crawford. Asst. to Regius prof. natural history and to prof. physiology U. Edinburgh; Regius prof., insts. medicine U. Aberdeen; prof. physiology and histology Victoria U., Manchester; Fullerian prof. physiology Royal Instn., London. Author textbooks on histology and physiology, various publs. on reflex functions of spinal cord, physiology of tetanus, red and pale muscle, structure of skin, cornea and muscles. Died Oct. 1, 1932.

STIX, Thomas Howard, Am. physicist; b. St. Louis, July 12, 1924; s. Ernest William and Erma (Kingsbacher) S.; B.S., Cal. Inst. Tech., 1948; Ph.D., Princeton, 1953; m. Hazel Rosa Sherwin, May 28, 1950; children—Susan Sherwin, Michael Sherwin. Mem. staff plasma physics lab. Princeton, 1953—, head exptl. div., 1961—, prof. astrophys, sci., 1962—. NSF sr. postdoctoral fellow physics Weizmann Inst. Sci., Rehovoth, Israel, 1960-61. Fellow Am. Phys. Soc. (chmn. div. plasma physics 1962-63); mem. Am. Assn. U. Profs., Sigma Xi, Tau Beta Pi. Author: The Theory of Plasma Waves, 1962. Adv. bd. McGraw-Hill Advanced Physics Monograph Series, 1963—. Research and publs. on primary cosmic radiation, excitation and absorption of plasma waves, plasma heating and confinement. Home: 231 Brookstone Dr., Princeton, N.J.*

STOCK, Alfred E., German chemist; b. Danzig (now Gdansk, Poland), July 16, 1876; studied under Emil Fischer, U. Berlin, for doctorate; also studied under Henri Moissan. Asst. to Emil Fischer; dir. Chem. Inst., Technische Hochschule, Karlsruhe, Germany; vis. lectr., Cornell U., 1932. Research on boron hydrides (compounds of boron and hydrogen), mercury poisoning, high-vacuum method of studying volatile substances, preparation and properties of beryllium. Died Karlsruhe, Aug. 12, 1946.

STOCK, C(harles) Chester, Am. biochemist; b. Terre Haute, Ind., May 19, 1910; s. Orion Louis and Jessie My (Blood) S.; B.S., Rose Poly. Inst., 1932, Sc.D. (hon.), 1954; Ph.D., Johns Hopkins, 1937; M.S., N.Y.U., 1941; m. Grace Elizabeth Knipmeyer, June 6, 1936. Instr. bacteriology N.Y.U., 1937-41; vis. investigator Rockefeller Inst. Hosp., 1941-42; tech. aide com. med. research OSRD, 1942-45, dep. chief div. 5, 1945-46, asst. to chmn. insect control com., 1945; tech. aide com. on treatment gas casualties Nat. Research Council, 1942-45, exec. sec. insect control com., 1945-46; chief div. chemotherapy Sloan-Kettering Inst., 1947—, asso. dir., 1957-60, sci. dir., 1960-61, vice pres., 1961—; dir. Walker Lab., Rye, N.Y., 1960—; prof. biochemistry Sloan-Kettering Div. Cornell Med. Coll., 1951—; mem. experimental therapeutics study sect. United States Public Health Service, 1949-54; screening panel, chmn. cancer chemotherapy Nat. Service Center, 1955-58, chmn. drug evaluation panel cancer chemotherapy, 1958-59; mem. chemotherapy review bd. Nat. Adv. Cancer Council, 1958-59; com. tumor nomenclature and statistics, chmn. panel nomenclature tumors exptl. animals Internat. Cancer Research Commn., 1952-54; mem. U. S. com. Internat. Union Against Cancer. Mem. sci. adv. bd. Roswell Park Meml. Inst. Research com. Music Research Found., 1951-55. Awarded Hemingway medal Rose Poly. Inst., 1932; recipient Army-Navy Certificate of Merit, 1948; Alfred P. Sloan award in cancer research, 1965. Fellow N.Y. Acad. Medicine (associate), New York Academy Sci. (vice president 1962-63); mem. of Am. Chem. Soc., Am. Assn. Cancer Research (dir. 1955), Harvey Soc., A.A.A.S., Am. Soc. Biol. Chemists, Soc. Exptl. Biology and Medicine (chairman of N.Y.C. sect. 1960-62), Societa Italiana Di Cancerologia (corr.), Sigma Xi, Tau Beta Pi, Alpha Chi Sigma, Blue Key. Editorial adv. bd. Cancer Research jour., 1951-58; mem. editorial bd. Jour. Medicinal and Pharm. Chemistry, 1958-64. Contbr. sci. jours., publs. Research on metal activation arginase, inactivation influenza, lymphatic choriomeningitis viruses by soaps; bactericidal action human serum; hypertensive materials in human blood; devel. agents useful in treatment human cancer. Home: One Gracie Terrace, N.Y.C. 28. Office: 410 E. 68th St., N.Y.C. 21; also 145 Boston Post Rd., Rye, N.Y.*

STOCKARD, Charles Rupert, Am. biologist, anatomist; b. Washington County, Miss., Feb. 27, 1879; s. Richard Rupert and Ella Hyde (Fowlkes) S.; B.Sc., Miss. Agrl. and Mech. Coll., 1899, M.S., 1901; Ph.D., Columbia, 1906; M.D., U. Würzburg, 1922; Sc.D., U. Cin., 1920; studied Carnegie Inst. Lab. for Tropical Biology, Dry Tortugas, Fla.; Naples Zool. Sta.; also visited chief zool. and anat. labs. of Europe; m. Mercedes Müller, Aug. 14, 1912; children—Marie Louise, Richard Robert. Comdt., and acting prof. mil. sci. and tactics Miss. Agrl. and Mech. Coll., 1898-1900, Jefferson Mil. Coll., 1900-03; asst. in zool. dept. Columbia, 1905, 06; asst. in embryology and histology Cornell Med. Coll., 1906-08, instr. comparative morphology, 1908-09, asst. prof. embryology and exptl. morphology, 1909-11, prof. anatomy, 1911—, also head dept.; investigator for Huntington Fund for Cancer Research, 1908—; lectr. colls. and univs. Fellow A.A.A.S. (v.p. 1933), N.Y. Zool. Soc., N.Y. Acad. Medicine. Author: Origin of Blood, 1915; Hormones and Structural Development, 1927; The Physical Basis of Personality, 1931. Mng. editor Am. Jour. Anatomy; editor Jour. Exptl. Zoology, Am. Anat. Memoirs. Research on origin of blood; developed (with George Nicholar Pana Papanicolaou) vaginal smear test for estrus, which helped reveal histologic changes occurring in vagina during menstrual cycle, circa 1917; studied chemicals and embryonic devel., regeneration, growth, oestrous cycle, cancer; worked on exptl. prodn. of cyclopean monsters, other monstrosities. Died Apr. 7, 1939.

STOCKBERGER, Warner W., Am. botanist, mgmt. scientist; b. Licking County, O., July 10, 1872; s. George Francis and Roena (Warner) S.; Ohio State U., summer courses in biology, 1900, 01; B.S., Denison Univ., 1902; Ph.D., George Washington Univ., 1907; D.Sc., Denison Univ., 1937; m. Maude N. Streeter, July 6, 1896; 1 dau., Lucile (Mrs. Earl W. Boyer). Teacher common schs., O., 3 yrs.; supt. schs., Hanover, O., 1895-97; student asst., Denison U., 1897-1900, instr. botany, 1901-03; entered service of U. S. Dept. Agr., July 1, 1903. Expert in histology, 1903-08, pharmacognocist, 1908-10, plant physiologist, 1910-13, in charge of drug, poisonous and oil plant investigations, July 1, 1913-40, personnel classification officer, 1923-25, dir. of personnel and business adminstrn., 1925-34, of personnel, 1934-38; special adviser to sec. of agriculture since 1938. Fellow A.A.A.S.; mem. Am. Pharm. Assn., Am. Soc. Pub. Adminstrn., Am. Polit. Science Assn., Civil Service Assembly of U. S. and Canada, Soc. for Personnel Adminstrn. (pres. 1937), Nat. Geneal. Society, Bot. Society Washington (pres. 1912-13), Am. Pharmaceutical Conv. (asst. sec., 1920-30), Am. Oil Chem. Society (hon.), Phi Beta Kappa, Sigma Xi, Editor Bulletin of Scientific Lab., Denison Univn., 1901-03; pres. Granville (Ohio) Bd. of Edn., 1901-03; service for Joint Congressional Commn. for Reclassification of Salaries, 1919-20. Author of various papers on medicinal plants and reports and articles in publs. of Dept. of Agr.; contbr. to scientific jours. and to The Book of Rural Life. Pioneer personnel mgmt. in U. S. Fed. Govt.; studied plant physiology and pharmacognosy, European methods hop prodn., utilization. Died May 27, 1944.

STÖCKER, Otto, German botanist; b. Freiburg im Breisgau, Germany, Dec. 17, 1888; s. Franz and Anna (Müller) S.; student U. Jena; Dr.phil.nat., U. Freiburg, 1922; m. Elisabeth Hager, Mar. 2, 1915; children—Burchard, Gerhild (Mrs. Klemp). Tchr. secondary sch., Bremerhaven, Germany, 1915-34; prof. botany, dir. Bot. Inst. and Garden, Darmstadt (Germany) Inst. Tech., 1934-58. Mem. German Bot. Soc. Author: Wasserhaushalt ägyptischer Wüsten-und Salzpflanzen, 1928; Exp. Ökologie der Pflanzen, 1929; Pflanzenphysiologische Übungen, 1942; Grundriss der Botanik, 1952; Handbuch der Pflanzenphysiologie, vol. 3, 1956; also numerous articles. Research on plant physiology, ecology, geography. Address: 10 Dachsbergweg, 61 Darmstadt, West Germany.*

STÖCKMANN, Fritz, German physicist; b. No. Germany, Oct. 19, 1918; s. Friedrich and Frieda (Heins) S.; student univs. Freiburg, Munich; Dr.rer.nat., U. Göttingen, 1942; m. Hildegard Brandt, Sept. 17, 1943; children—Hans-Jürgen, Ursula. Faculty, U. Göttingen (Germany), 1950-52; chief asst. Darmstadt (Germany) Inst. Tech., 1952-56, prof., 1956-59; vis. prof. Purdue U., Lafayette, Ind., 1954-55; prof., head inst. applied physics, Karlsruhe (Germany) Inst. Tech., 1959—. Mem. German Phys. Soc. Research, numerous publs. in solid state physics: photoconductivity and other related nonequilibrium properties of semiconductors and dielectrics. Address: 14 Stettiner Strasse, Karlsruhe, West Germany.*

STOCKMAYER, Walter Hugo, Am. chemist; b. Rutherford, N.J., Apr. 7, 1914; s. Hugo Paul and Dagmar (Bostroem) S.; S.B., Mass. Inst. Tech., 1935, Ph.D., 1940; B.Sc., Oxford (Eng.) U., 1937; M.A. (hon.), Dartmouth, 1961; m. Sylvia Bergen, Aug. 12, 1938; children—Ralph, Hugh. Instr. Mass. Inst. Tech., 1939-41, asst. prof., 1943-46, asso. prof., 1946-52, prof., 1952-61; instr. Columbia, 1941-43; prof. Dartmouth, 1961—. Cons. E. I. duPont de Nemours & Co., Wilmington, Del., 1945—; dir. Gordon Research Confs., 1963-66. Recipient Coll. Chemistry Tchr. award Mfg. Chemists' Assn., 1960. Fellow Am. Phys. Soc., Am. Acad. Arts and Scis.; mem. Am. Chem. Soc. (award in polymer chemistry 1966, chmn. div. polymer chems. 1968), Research, publs. on kinetics of polymerization, also sizes, shapes and motions of large molecules. Home: Box 361, Norwich, Vt. 05055. Office: Dartmouth Coll., Hanover, N.H. 03755.*

STOCKWELL, Edward Grant, Am. demographer; b. Newburyport, Mass., June 11, 1933; s. Frank Whitten and Margaret (Winters) S.; A.B., Harvard, 1955; M.A., U. Conn., 1957; Ph.D., Brown U., 1960; m. Janet Wiranis, Aug. 18, 1956; children—Edward Grant, Christopher, Susie. Population analyst U.S. Bur. Census, 1960-61; faculty U. Conn., Storrs, 1961—, prof. rural sociology, 1966—. Mem. Population Assn. Am., Am. Sociol. Assn., Rural Sociol. Soc., Internat. Union for Sci. Study Population, Am. Statis. Assn., Am. Acad. Polit. and Social Sci. Research and publs. on relation of population trends, changes to econ., social changes. Home: Little Lane, Storrs, Conn. 06268.*

STOCKWELL, John Nelson, Am. astron. mathematician; b. Northampton, Mass., Apr. 10, 1832; s. William and Clarissa (Whittemore) S.; ed. common schs., Brecksville, O.; A.M. (hon.) Western Res. U., 1862, Ph.D., 1876, m. Sarah Healy, Dec. 6, 1855. Fellow Am. Acad. Arts and Scis., A.A.A.S. Author: Memoir on the Secular Variations of the Planetary Orbits, in Smithsonian Contributions to Knowledge, 1872; Stock and Interest Tables, 1873; Theory of the Moon's Motion, 1881; Eclipse Cycles, 1901; Sheet Tax Tables, 1903; Theory of Planetary Perturbations, and the Cosmogony of Laplace, 1904; Ocean Tides, with elaborate tables for their computation, 1919; also papers. Died May 18, 1920.

STODDARD, George Edward, Am. animal nutritionist, educator; b. Boise, Ida., July 15, 1921; s. William Roy and Estella (Hansen) S.; B.S., U. Ida., 1943; M.S., U. Wis., 1948, Ph.D., 1949; m. Corinne Marie Pecora, Aug. 7, 1946; children—Karalee, Douglas G., Steven E., Gary W., Shauna M. Research asst. U. Wis., 1946-49; asst. prof. Ia. State U., 1949-52; asso. prof. dept. dairy sci. Utah State U., Logan, 1952-55, prof., 1955—, head dept., 1960—. Collaborator, Agrl. Research Service, U. S. Dept. Agr., 1961—; exec. com. Utah Am. Dairy Assn. and Utah Dairy Council, 1965—. Mem. Am. Dairy Sci. Assn., Am. Soc. Animal Sci., A.A.A.S., Utah Acad. Arts and Sci., Sigma Xi, Alpha Zeta. Research, numerous publs. on syndrome of chronic Bovine fluorosis, rumen physiology, radionucleotide uptake by cow and transfer to milk, forage quality and preservation, phosphorus nutrition, dairy cattle mgmt., carbohydrate digestion in calf intestine, pesticide secretion in milk. Home: 950 N. 4th E., Logan, Utah 84321.*

STODDARD, John Tappan, Am. chemist; b. Northampton, Mass., Oct. 20, 1852; s. William H. and Helen (Humphrey) S.; A.B., Amherst, 1874; student chemistry and physics, 1875-76; A.M., Ph.D., Göttingen, 1877; m. Mary Grover Leavitt, June 26, 1879; 1 son, William Leavitt. Asst. prin. Northampton High Sch., 1874-75; prof. physics and math. Smith Coll., 1878-81, Chemistry and physics, 1881-97, chemistry, 1897—. Author: Outline of Qualitative Analysis, 1883; Lecture Notes on General Chemistry, 2 vols., 1884, 85; Quantitative Experiments in General Chemistry, 1908; Introduction to General Chemistry, 1910;

The Science of Billiards, 1913; An Introduction to Organic Chemistry, 1914. Contbr. scientific articles and revs. to cyclos. and mags. Died Dec. 8, 1919.

STODDARD, Joshua C., Am. inventor; b. Pawlet, Vt., Aug. 26, 1814; s. Nathan Ashbel and Ruth (Judson) S.; m. Lucy Maria Hersey, Jan. 23, 1845; at least 2 children. Worked on father's farm for long period, interested primarily in bee culture, honey prodn.; inventor improvements and variants on horse-drawn hay rake, received 16 patents; his most famous invention was steam calliope, patented 1855; organized Am. Steam Music Co., Worcester, Mass., 1855, forced out of co. by 1860; many calliopes placed on side-wheelers, other river vessels of day; received no financial benefit from most of inventions; patented a fruit-paring machine, 1901. Died Springfield, Mass., Apr. 3, 1902.

STODOLA, Aurel, engr.; b. Liptó-Szent Miklós, Hungary, May 10, 1859; student Zurich, Switzerland, Berlin, Germany, Paris, France. Engr. for company mfg. steam engines; faculty U. Zurich, 1892-1929. Mem. French Acad. Scis., 1929. Research and publs. in math, physics and thermodynamics; contributed to devel. of steam and gas turbines. Died Dec. 25, 1942.

STODTMEISTER, Rudolf, German physician; b. Detmold, Germany, Jan. 19, 1908; s. Paul and Ella St. (Gädke) S.; student U. Tübingen (Germany), 1926-27, U. München (Germany), 1928; Dr.phil. U. Kiel (Germany), 1930, Dr.med., 1934; m. Hedwig Hartmann, Apr. 14, 1936; children—Walther, Richard, Gertrud, Martin, Herta. Asst., I. Med. Klinik, Charité, Berlin, 1934-36, Med. Klinik, Heidelberg, 1936-45; head physician Inn. Abt. Diakonisenkrankenhaus, Darmstadt, Germany, 1946-50; head physician div. internal medicine Städt. Krankenhaus, Pforzheim, Germany, 1950—; teaching staff Faculty Medicine, Heidelberg, 1938—; scientist collaborator Institut für Hämatologie, EURATOM, 1963—. Mem. Swiss Soc. Hematology (corr.). Author: (with Sandkuehler, Laur) Osteosklerose und Myelofibrose, 1953; (with T. M. Fliedner) Experimentelle und klinische Strahlenhämatologie, 1962; also numerous articles. Research on bone marrow in radiation hematology. Home: 55, Humboldstrasse, Office: Städt. Krankenanstalten, Pforzheim, Germany.*

STOECKEL, Walter, German gynecologist; b. Stobingen, Germany, Mar. 14, 1871; prof., Marburg, Kiel, Leipzig and Berlin, Germany; dir. univ. women's clinic Berlin Charité. Author: Lehrbuch der gynäkologischen Zystoskopie, 1910; Lehrbuch der gynäkologie, 1920; Lehrbuch der Geburtshilfe, 1920. Author textbooks. Editor: Handbuch der Gynäkologie. Founder gynecol. urology; studied cancer. Died Berlin, Feb. 13, 1961.

STOECKELER, Joseph Henry, Am. research forester; b. Höchst, Austria, June 27, 1908; s. Joseph Sebastian and Mathilde (Gerer) S.; came to U.S., 1913, naturalized, 1931; B.S., Ia. State Coll., 1930, M.S., 1931; Ph.D., U. Minn., 1956; postgrad. (NSF fellow) Royal Sch. Forestry, Stockholm, Sweden, 1959, U. Munich (Germany), 1960, U. Helsinki (Finland), 1960; m. Hazel Thorson Stoick, Dec. 27, 1947; 1 son, Joel Stoick. Research forester, La Crosse, Wis., 1931-32; head nursery and reforestation research Sandhills Expt. Sta., Towner, N.D., Rhinelander, Wis., 1933-43; forester-in-charge No. Lakes Forest Research Center, Rhinelander, 1946-55, sr. and prin. soil scientist watershed mgmt. div., 1956-65, prin. soil scientist forest mgmt., 1966—. Spl. lectr. Royal Sch. Forestry, also U. Munich, 1959-60. Mem. Soc. Am. Foresters (past mem. internat. relations com., past chmn. No. Wis. sect.), Soc. Foresters in Finland (corr.), Soil Sci. Soc. Am., Canadian Inst. Forestry, Ecol. Soc., Soil Conservation Soc. Am., Wildlife Soc., Minn. Acad. Sci., Sigma Xi, Gamma Sigma Delta. Author: (with G.W. Jones) Forest Nursery Practice in the Lake States, 1957; (with H.F. Arneman) Fertilizers in Forestry, 1960; Tree Planting in the Drylands of the Western United States for Reforestation, Afforestation, and Windbreaks, 1967; also numerous articles. Research in forest nursery prodn., silviculture no. hardwoods, soil-site relations in quaking aspen forests; established pine and other plantings in sandhills nr. Towner, N.D. Home: 2431 Como Av., St. Paul. 55108. Office: N. Central Forest Expt. Sta., St. Paul Campus, U. Minn., St. Paul 55101.*

STOECKHARDT, Jules Adolphe, German chemist; b. Roehrsdorf, Germany, 1809; student Berlin, Eng., France; worked in lab. of Struve, Dresden, Germany; became prof. natural scis. Blochmann Inst., 1838; prof. chemistry Chemnitz Sch. Industry, 1839-47; prof. agrl. chemistry Acad. Forestry and Agr., Tharandt. Author: Recherches sur la houille de Zwickau, 1840; Des couleurs et surtout des couleurs veneneuses, 1841; Chemie organique, 1846; l'Ecole de la chemie, 1846; le Petit livre du guano, 1856. Published (with Schober), Jour. d'economic rurale allemande, beginning 1840. Developed new methods for preparation of paints; popularized agrl. chemistry. Died Tharandt, 1886.

STOELTING, Vergil Kenneth, Am. anesthesiologist; b. Freelandville, Ind., Feb. 10, 1914; s. Andrew Philip and Ethel (Blanche) S.; B.S., Ind. U., 1936, M.D., 1936; postgrad. U. Wis., 1943-44, U. Ia., 1946-47; m. Bernice Blanche Markus, Sept. 5, 1936; children—Robert, Ann. Prof. anesthesiology Ind. U. Sch. Medicine, Indpls., also dir. dept. anesthesiology Ind. U. Hosp., 1947—; attending anesthesiologist Meth. Hosp., Community Hosp., 1956— (both Indpls.); also cons. hosps. Diplomate Am. Bd. Anesthesiology. Fellow A.M.A.; mem. U. Anesthesiologists, Acad. Anesthesiology, Am. Soc. Anesthesiology, Internat. Research Soc. Research and publs. on analgesic and anesthetic drugs; complications occurring during anesthesia. Home: 4706 Laurel Circle, Indpls. 46226.*

STÖFFLER, Johann, German astronomer, mathematician; b. Swabia, 1452; prof. math. astronomy and geography, Tübingen, Germany; tchr. of Melanchthon and Münster. Built astrolabe; predicted a deluge in 1524. Died 1531.

STOGDILL, Ralph Melvin, Am. psychologist, educator; b. Convoy, O., Aug. 29, 1904; s. John L. and Pearl (Foley) S.; student Ohio Wesleyan U., 1932-35; B.A., M.A., Ohio State U., 1930, Ph.D., 1934; m. Zoe Emily Leatherman, Mar. 19, 1928; 1 son, Robert Edmund. Psychologist, Bur. Juvenile Research, Columbus, O., 1934-42; asso. dir. Ohio Leadership Studies, Columbus, 1946-53; research social scientist Grad. Sch. Bus., Stanford, 1953-54; prof. Bur. Bus. Research, Ohio State U., Columbus, 1954—. Fellow Am. Psychol. Assn.; mem. Psychonomic Soc., Inst. Mgmt. Scis., Soc. Gen. Systems Research. Author: Individual Behavior and Group Achievement, 1959; Managers, Employees, Organizations, 1965; also articles. Research on attitudes, behavior prediction, leadership, group performance, orgn. Home: 3658 Olentangy Blvd., Columbus, O. 43214.*

STOHLER, Rudolf, zoologist; b. Basel, Switzerland, Dec. 5, 1901; s. Rudolf and Emma (Kiefer) S.; student U. Geneva, 1921; M.A., U. Basel, 1928, Ph.D., U. Cal., Berkeley (Internat. Exchange fellow), 1930; m. Genevieve J. Emerson, Sept. 16, 1929; children—Rudolf J., Genevieve Janet (Mrs. James Willett), Helen A. (Mrs. Howard Norskog), Constance D. (Mrs. Edmund Hake), Alice M. (Mrs. David Tiongco). Came to U.S., 1932, naturalized, 1937. Research, faculty U. Cal., Berkeley, 1932—, research zoologist, 1966—. Fellow Cal. Acad. Scis.; mem. Am. Malacological Union (past chmn. Pacific div.), Cal. Malacozoological socs., No. Cal. Malacozoological Club (past pres.), Conchological Club So. Cal., Swiss Zool. Soc., Sigma Xi. Editor: The Veliger, 1955—. Contbr. numerous articles to profl. jours. Discovered cause of mussel poisoning; described two new species of mollusks; invented techniques for preparing demonstration and study materials for student use. Home: 1584 Milvia St., Berkeley, Cal. 94709.*

STOICA, Emilian,, Rumanian physician, neurologist; b. Moreni, Ploiesti, Rumania, Sept. 7, 1929; s. Nicolae, and Margareta (Negulescu) S.; M.D., Faculty Medicine, Bucharest, Rumania, 1953; m. Maria Gilli, Mar. 31, 1951; 1 dau., Roxana. Asst. physiol. dept. Postgrad. Faculty, Medicine, Bucharest, 1954-58; prin. researcher Inst. Neurology, Acad. Socialist Republic Rumania, 1958—. Mem. Neurol. Soc. Author: (with V. Voiculescu) Starile comatoase, 1967; also articles. Research on quinidine treatment in epilepsy, relationship of cerebral vascular spasm to cold stress, disharmonious humoral syndrome at the onset of cerebral thrombosis, pathogenesis of cerebral thrombosis and hemorrhage. Home: 48 Popa Savu, Bucharest. Office: 42, Povernei, Bucharest, Rumania.*

STOICA, Ion, Rumanian physician; b. Fagaras, Rumania, Jan. 15, 1926; s. George and Itoafa (Sofia) S.; M.D., U. Bucharest, 1951; m. Cecilia Scurtu, July 30, 1954; 1 son, Liviu. Neurologist, Inst. Neurology, Bucharest, 1951-65, chief research sector, 1965—. Mem. Union Socs. Med. Scis. Author: (with A. Kreindler, E. Crighel) Infantile Epilepsy, 1961; Electro-Encephalography and Electro-Myography, 1963; also numerous articles. Research on epilepsy, applications of electroencephalography in human pathology. Home: 13 Decembrie, no. 13, raion 16 Februarie, Bucharest. Office: Inst. Neurology, Bucharest, Rumania.*

STOICHEFF, Boris Peter, physicist; b. Bitol, Yugoslavia, June 1, 1924; s. Peter and Vasilka (Tonna) S.; B.A.Sc., U. Toronto, 1947, M.A., 1948, Ph.D., 1950; m. Lillian Joan Ambridge, May 15, 1954; 1 son, Richard Peter. McKee-Gilchrist postdoctorate fellow U. Toronto, Ont., Can., 1950-51; postdoctorate fellow Canadian NRC, Ottawa, 1951-53, sr. research officer, 1954-64; vis. sci. Mass. Inst. Tech., Cambridge, 1963-64; prof. physics U. Toronto, 1964—. Fellow Royal Soc. Can., Optical Soc. Am.; mem. Am. Phys. Soc., Can. Assn. Physicists. Contbr. many articles to profl. jours. Developed techniques for high resolution Raman spectroscopy of gases, thus determined precise geometrical structures of many non-polar molecules including hydrogen, methane, ethylene, benzene; used laser radiation in various spectroscopic investigations including Brillouin and stimulated Raman scattering in liquids, two photon absorption in anthracene, observed stimulated Raman absorption and stimulated Brillouin scattering resulting in generation of intense hypersonic waves in solids. Home: 50 Himount Dr., Willowdale, Ont., Can.*

STOICHITA, Michaela Papilian, Rumanian physician; b. Cluj, Rumania, Sept. 21, 1921; d. Victor and Ecaterina (Iorga) Papilian; student Faculty Medicine, Cluj; m. Sandu Stoichita, July 31, 1946; children—Sandu, Victor, Ecaterina-Mihaela. Preparator, Med. Clinics, Faculty of Medicine, Cluj, 1945-49, asst. lectr., 1949-51; sci. researcher, chief physician Lab. Coagulation, Inst. Internal Medicine, Rumanian Acad., Bucharest, 1949—. Author: Coagulation and Fibrinolysis; also articles. Research on role of coagulation process in gen. econs. of organism, demonstrated existence of qualitative changes in fibrinogen in inflammatory and immunologic diseases, intravascular disseminated coagulation in pathogenesis of disease. Home: 29 Andrei Muresan. Office: 19-21 Sos.Stefan cel Mare, Bucharest, Rumania.*

STOILOV, Simion, Rumanian mathematician; b. 1887; prof. Bucharest (Rumania) Poly. Inst.; mem. Acad. Rumanian People's Republic. Author: Text-Book on Mathematical Analysis, 1940; The Theory of Functions of a Complex Variable, 2 vols., 1954-58; Cours sur les principes topologiques de la Théorie des fonctions analytiques, 1956; Oeuvres mathematiques, 1964. Founder contemporary Rumanian sch. of theory of functions; research on theory of equations with partial derivates in complex variables, also on theory of Riemann's surfaces. Died 1961.

STOKER, James J(ohnston), Am. mathematician, physicist; b. Dunbar, Pa., Mar. 2, 1905; s. James J. and Mary (Fischer) S.; B.S., Carnegie Inst. Tech., 1927, M.S., 1931; Dr. Math., Technische Hochscule, Zurich, Switzerland, 1936; m. Nancy X. Lynch, May 4, 1928; children—Nancy Jane (Mrs. Frank W. Peirce), James Johnston, Jerome L., Sue R. (Mrs. Jon Reilly). Engaged as instructor of mechanics Carneige Inst. Tech., 1928-31, asst. prof., 1931-37; asst. prof. N.Y.U., 1937-41, asso. prof., 1941-45, prof., 1945—, head all-univ. math. dept., 1958-66, dir. Courant Inst. Math. Scis., 1958-66. Dir. Service Bur. Corp. Research mathematician, applied mathematics panel NDRC, 1943-45. Fellow Am. Acad. Arts and Sciences; member of National Academy of Sciences, Am. Math. Soc. Math. Assn. American, Sigma Nu, Tau Beta Pi, Sigma Xi. Author: Nonlinear Vibrations, 1950; Water Waves (Heineman prize), 1957. Research in hydrodynamics; elasticity; vibration theory; and differential geometry. Office: 251 Mercer St., N.Y.C. 10012.*

STOKES, Adrian, bacteriologist, physician; b. Lausanne, Switzerland, Feb. 9, 1887; M.B., Trinity Coll., Dublin, Ireland. Officer, Royal Army Med. Corp, during European Wars; became prof. bacteriology and preventive medicine Trinity Coll., Dublin, 1919; named Sir. William Dunn prof. pathology London U., 1922; mem. Rockefeller Yellow Fever Commn., Lacos, until his death. Helped show that agt. of yellow fever is filterable virus and transmitted it (with J.H. Bauer, N.P. Hudson) to monkey Macaca rhesus, 1928; showed an epidemic of jaundice was caused by spirochete carried by rats; studied tetanus, typhoid. Died Sept. 19, 1927.

STOKES, Aldwyn Brockway, psychiatrist; b. Newport, Eng., Feb. 23, 1906; s. Reginald George Briant and Mary (Price) S.; B.A., Oxford U., Eng., 1927, M.A., 1934; M.B., Kings Coll., London, 1931, D.Ch., 1935, D.P.M., 1936; m. Margaret Fitzgerald, Apr. 29, 1935; children—Peter Brockway, Phillppa Margaret, Harriet Mary, David Aldwyn. Sr. med. registrar, tutor Kings Coll. Hosp., London, 1931-35; staff psychiatrist Maudsley Hosp., London, 1935-39, med. supt., 1945-47; Rockefeller traveling fellow, 1937-38; dep., then med. supt. Mill Hill Emergency Hosp., London, 1939-45; prof., head dept. psychiatry U. Toronto, Can., also psychiatrist-in-chief Toronto Psychiatric Hosp., 1947-66; psychiatrist-in-chief Clarke Inst. Psychiatry, Toronto, 1966—; cons. Toronto Gen. Hosp., Hosp. for Sick Children. Decorated Comdr. Order British Empire, 1947. Fellow Royal Coll. Physicians, Can., also London; mem. Toronto Med. Legal Soc. (pres. 1956), Ont. Neuropsychiatric Assn. (pres. 1956), Am. (v.p. 1963-64), Ont., Can. psychiatric assns., Royal Medico Psychol. Assn. Numerous publs. on longterm biol. and behavioral studies in periodic catatonia; research in ednl. and illness processes in different environmental settings. Home: 83 Woodlawn Av. W., Toronto 7. Office: 250 College St., Toronto, Ont., Can.*

STOKES, Sir Frederick Wilfrid Scott, English engr., inventor; b. Liverpool, Eng., Apr. 9, 1860; s. Scott Nasmyth and Emma Louisa Stokes; ed. St. Francis Xavier Coll., Liverpool, Kensington Catholic Sch., Catholic U. Coll., Kensington; m. Irene Ironsides. Apprentice civil engr. Gt. Western Ry. Co., 1878-81; designer bridges and other steelwork constrn. Hall and Barnsley Ry., 1881-85; with Ransomes and Rapier of Ipswich, 1885-1927, mgn. dir., 1897, chmn., 1907. Mem. Instn. Civil Engrs., Brit. Engrs. Assn. (pres. 1915-17). Patented rotary kilns for cement making; improved breakdown crane; invented shallow traverser for ry. carriage and wagon stock; superintended erection of sluices on Manchester ship canal, Assuan Dam, and Sennar Dam; designed famous Stokesgun (trench mortor). Died Ruthin, Eng., Feb. 7, 1927.

STOKES, Sir George Gabriel, Brit. mathematician, physicist; b. Skreen, County Sligo, Ireland, Aug. 13, 1819; S. Gabriel and Elizabeth (Haughton) S., ed. Bristol (Eng.) Coll., 1835, Pembroke Coll., Cambridge (Eng.) U. (sr. Wrangler, 1st Smith's prizeman 1841), 1837-41; m. Mary Robinson, 1857; 2 sons, 1 dau. Lucasian prof. math. Cambridge U., 1849-1903, mem. Parliament for univ., 1887-91, also fellow. Gifford lectr. Edinburgh (Scotland) U., 1891. Fellow Royal Soc., 1851 (Rumford medal 1852, sec. 1854-85, pres. 1885-90, Copley medal 1903), 1851; mem. Brit. Assn. Advancement Sci. (pres. 1869), French Acad. Scis. Author: On Light, 1887; Natural Theology, 1891; Mathematical and Physical Papers, 1880-1905. Discussed phenomena waves on water; created modern theory viscosity fluids (laid found. sci. hydrodynamics); developed math. theory motion viscous fluids (Stokes' law describes motion of small sphere in viscous fluid); discussed effect of wind on sound intensity, also how intensity is influenced by nature of gas in which sound is produced; studied aberration of light, Newton's rings, thick plates; examined principles of interference, polarization; investigated undulatory theory light; worked on concept of luminiferous ether; experimentally studied displacement of plane of polarization light by diffraction; established semi-convergent series used with Bessel function and other harmonic series; (from study Fourier series) elaborated complete and limited convergence infinite series; pioneer in spectrum analysis (sought to determine chem. composition of sun, stars from their spectra); showed fluorescence excited mainly by ultraviolet radiation, also explored ultraviolet spectrum; reported on double refraction; studied variation in gravity; considered founder sci. geodesy; suggested x-rays might be transverse electromagnetic waves traveling as innumerable solitary waves. Died Cambridge, Feb. 1, 1903.

STOKES, Jacob Leo, Am. microbiologist; b. Warsaw, Poland, Sept. 27, 1912; s. Louis and Rose (Kuperman) S.; brought to U.S., 1917, naturalized, 1922; B.S., Rutgers U., 1934; M.S., U. Ky., 1936; Ph.D., Rutgers U., 1939; m. Freida R. Robinson, Dec. 31, 1936; children—Anna Carol, Frances. Microbiologist Merck & Co., Inc., Rahway, N.J., 1939-47, head sect. on microbiol. metabolism, 1942-47; research asso. Hopkins Marine Sta., Stanford, 1948-50; asso. prof. bacteriology U. Ind., 1950-53; bacteriologist U.S. Dept. Agr., Albany, Cal., 1953-59; prof. bacteriology, chmn. dept. bacteriology, pub. health Wash. State U., Pullman, 1959—. Mem. Soc. Am. Bacteriologists, A.A.A.S., Sigma Xi. Research, numerous publs. on relation of algae to other microorganisms in nature; antibiotics; nutrition, physiology and biochemistry of microorganisms; iron bacteria, psychrophilic microorganisms. Home: 1908 Monroe St., Pullman, Wash. 99163.*

STOKES, Robert James, physicist; b. Devizes, Eng., Oct. 15, 1928; s. Ernest and Marie (Millin) S.; B.Sc., U. Bristol (Eng.), 1952; Ph.D., U. Birmingham (Eng.), 1955; m. Audrey Lovell Rogers, July 14, 1955; children—Neil Roger, Sandra Elizabeth. Research engr., asst. prof. U. Cal. at Berkeley, 1955-57; research scientist, staff scientist Honeywell Research Center, Hopkins, Minn., 1957—; vis. prof. Carnegie Inst. Tech., Pitts., 1965-66. Fulbright grantee, 1955. Mem. Brit. Inst. Metals, Am. Inst. Mining and Metall. Engrs., Am. Ceramic Soc. (Ross-Coffin Purdy award 1965). Research and publs. on strain hardening in metals; crystal defects in ceramics and their relation to plastic deformation, strengthening, and fracture; mech. properties of polycrystalline ceramics; discovered Cottrell-Stokes law, crack nucleation mechanisms; explained Joffe effect. Home: 4920 Kingsberry Lane, Minnetonka, Minn. 55343. Office: Washington Av. S., Hopkins, Minn. 55343.*

STOKES, Whitley, physician; b. Ireland, 1763; s. Gabriel Stokes; entered Trinity Coll., Dublin, Ireland, 1779, became scholar, 1781, B.Medicine, 1789, M.D., 1793; m. Mary Anne Picknell, 1782; 9 children. Prin. founder Bot. Gardens and Zool. Gardens; became fellow Trinity Coll., Dublin, 1787, suspended as tutor because of his polit. opinions for 3 years, 1798, became sr. fellow, 1805, lectr. natural history, 1816, Regius prof. medicine, 1830-43. First description of dermatitis gangrenosa infantum or ecthyma gangrenosum (also called ecthyma terebrans), 1807. Died 1845.

STOKES, William, Irish physician; b. Dublin, Ireland, Oct. 1, 1804; s. Whitley and Mary Anne (Picknell) S.; student chemistry and medicine, Glasgow; med. degree U. Edinburgh, 1825; M.D., U. Dublin, 1939; hon. degrees from univs. Edinburgh, 1861, Oxford U., 1863, Cambridge U., 1863; m. Mary Black, 1828; several children. Physician to Dublin Gen. Dispensary, 1925, to Meath Hosp., Dublin, 1825. to the Queen, 1861; Regius prof. medicine Dublin U., from 1843. Decorated Prussian Order Merit. Fellow Royal Soc., 1861; mem. Royal Med. Assn., Royal Irish Acad. (pres.). Author: An Introduction to the Use of the Stethoscope, 1825; Clinical Observations on the use of Opium; A Treatise on the Diagnosis and Treatment of Diseases of the Chest, 1837; The Diseases of the Heart and Aorta, 1854; Lectures on Fever, 1874; also many others; editor Dublin Quar. Jour. Med. Sci., from 1834. One of the founders of clin. teaching in Britain at Dublin Sch. Medicine; studied diseases of heart and lungs; one of first to state bleeding and purging harmful, opium beneficial for acute abdominal disease; published one of first treatises in Britain on use of stethoscope; credited with greatly raising standards of clin. medicine; said to be first to give full description of signs and symptoms of abdominal aneurysm; founded Pathol. Soc. of Dublin, 1838. Died Dublin, Ireland, Jan. 10, 1878.

STOKES, William Lee, Am. geologist; b. Hiawatha, Utah, Mar. 27, 1915; s. William Peace and Grace (Cox) S.; B.S., Brigham Young U., 1937, M.S., 1938; Ph.D., Princeton, 1941; m. Betty Asenath Curtis, Sept. 7, 1939; children—Betty Lee (Mrs. Kent C. Huff), Mary Susan, William Michael, Patricia Jane. Research asst. geology Princeton, 1941-42; staff U.S. Geol. Survey, 1942-47; faculty U. Utah, Salt Lake City, 1947—, prof., head dept. geology, 1954—, dir. Earth Sci. Mus., 1961—, dir. coop. dinosaur project, 1961—. Mem. A.A.A.S., Geol. Soc. Am., Soc. Vertebrate Paleontology, Am. Geophys. Union, Utah Geol. Soc. (past pres.), Soc. Econ. Paleontologists and Mineralogists, Utah Acad. Sci., Arts and Letters, Am. Assn. Petroleum Geologists. Author: (with D.J. Varnes) Glossary Selected Geologic Terms, 1955; Essentials of Earth History, 1960; (with S. Judson) Introduction to Geology, 1968; also numerous articles. Research on growth stages and habits dinosaurs, meaning of primary structures in sedimentary rocks, early Cretaceous rocks, paleogeography Western U.S., relation ore-deposits to faults in Gt. Basin; co-discoverer that most uranium deposits Western U.S. are localized by structures in sedimentary rocks and are most likely to occur with fossil material; discovered and opened quarry for Jurassic dinosaurs. Home: 1354 2d Av., Salt Lake City 84103.*

STOKINGER, Herbert Ellsworth, Am. toxicologist; b. Boston, June 19, 1909; s. William Herman and Charlotte (Greene) S.; A.B., Harvard, 1930; Ph.D., Columbia U., 1937; m. Helen Ackerman, June 10, 1950; 1 dau., Janet. Instr. chemistry City Coll. N.Y., 1932-39; research asso. Coll. Phys. and Surg., Columbia U., 1937-39; research asso. in bacteriology Sch. Medicine and Dentistry, Rochester U., 1939-43; chief indsl. hygiene sect. AEC, 1943-51; asst. prof. pharmacology and toxicology U. Rochester, 1945-48, asso. prof., 1948-51; chief toxicologist, div. occupational health USPHS, Cin., 1951—. Chmn. com. on toxicology USPHS Drinking Water Standards, 1958—; chmn. com. threshold limits for air Am. Conf. of Governmental Indsl. Hygienists, 1961—; com. on chem. toxicology Indsl. Hygiene Found.; liaison mem. air standards com. Am. Standards Assn. Recipient award of merit Am. Conf. Governmental Indsl. Hygienists, 1958, meritorious achievement award, 1965. Mem. Soc. Exptl. Biology, Soc. Bacteriologists, Chem. Soc., Soc. Pharmacology, Indsl. Hygiene Assn. (treas. 1961-64), Assn. Immunologists, Am. Pub. Health Assn. Editor: Beryllium—Its Industrial Hygiene Aspects, 1966. Contbr. chpts. to books, numerous articles in profl. jours. Demonstrated vanadium's capacity to reduce cholesterol and atherosclerosis, means by which toxic air pollutants cause lung changes; developed data leading to establishment of safe limits for air of worker and community environments, standards for drinking water in pub. supplies, tests for predicting worker hypersusceptible to chems., tests for early detection of incipient disease. Home: Dunham Hosp., Guerley Rd., Cin. 45238. Office: 1014 Broadway, Cin. 45202.*

STOKSTAD, E.L. Robert, Am. biochemist, nutritionist; b. China, Mar. 6, 1913 (parents Am. citizens); s. Christian and Elsie (Olson); B.S., U. Cal. at Berkeley, 1934, Ph.D., 1937; m. Edith Grandin, Sept. 8, 1934; children—Robert G., Paul A. Chemist, San Francisco Western Condensing Co., 1937-40; postdoctoral fellow, Lalor fellow Cal. Inst. Tech., Pasadena, 1940-41; with Lederle Labs., Am. Cyanamid Co., Pearl River, N.Y., 1941-62, dir. research biol. scis., agrl. div., 1959-61, research fellow Lederle Labs., 1961-62; prof. nutrition also biochemist Agrl. Expt. Sta., U. Cal. at Berkeley, 1962—. Mem. Am. Inst. Nutrition (Mead Johnson award for work on folic acid 1947), Poultry Sci. Assn. (Tom Newman medal 1951, Borden award for work in poultry husbandry 1952), Am. Chem. Soc., Am. Soc. Biol. Chemists, Soc. for Exptl. Biology and Medicine, Biochem. Soc. Eng., Am. Soc. for Animal Prodn., A.A.A.S. Contbr. numerous articles to tech. jours. Isolated vitamin K; research on water soluble vitamin studies, metabolic studies folic acid and vitamin B12, growth promoting properties for animals antibiotics. Home: 769 Arroyo Ct., Lafayette, Cal. Office: Agrl. Exptl. Sta., U. Cal., Berkeley, Cal. 94720.*

STOKVIS, Barend Joseph, Dutch physiologist; b. Amsterdam, Holland, 1834; M.D., Utrecht, Holland, 1856; prof. medicine U. Amsterdam; mem. Royal Acad. Medicine, Brussels, Acad. Medicine, Paris (corr.). Introduced term enterogenous cyanosis for a condition characterized by cyanosis, enteritis, 1902. Died Amsterdam, 1902.

STOLL, Arthur, Swiss indsl. chemist; b. Jan. 8, 1887; Ph.D., E.T.H., Zurich; student Kaiser Wilhelm Inst. for Chemistry, Berlin, Germany, Munich (Germany) U.; hon. Dr., Basle, Switzerland Berne, Switzerland, Sorbonne, Paris, Geneva, Switzerland, Zurich, Delt, Netherlands, Warsaw, Poland, Munich, Florence, Italy, Guatemala, Darmstadt, Würzburg, Toulouse, Turin, Strasbourg, Madrid univs. Became chief asst. Kaiser Wilhelm Inst., Berlin, 1912, Chem. Inst. Munich U., 1916; founder pharm. div. Sandoz Ltd., Basle, 1917, became mgr., 1923, vice chmn., 1934-63, chmn. bd., 1963—, pres. bd. mgmt., 1948-56; named Royal Bavarian prof., 1917; became asso. prof. U. Mexico, 1950. Recipient Benoist prize, Flückiger Gold medal, Chevreul medal, Pasteur medal, Paul Karrer Gold medal, Mem. Belgian Soc. Cardiology (hon.), Soc. de chimie industrielle Paris, Phys. and Med. Soc. Erlangen, Schweiz. Akademie der Medizinischen Wissenschaften, Schweiz Gesellschaft für analytische und angewandte Chemie, French, German, Brit., Spanish, Italian, Polish chem. socs., Consejo Superior de Investigaciones Cientificas Madrid (hon.), Acad. Halle, Acad. de Honor, Real Acad. de Farmacia Madrid, Vienna Biol. Assn., Am. Pharmacological Soc., Royal Soc. (fgn.), Vienna Biol. Assn. Author: Untersuchungen über Chlorophyll; Untersuchungen über die Assimilation der Kohlensäurer; Zusammenhänge zwischen der Chemie des Chlorophylls und seiner Funktion in der Photosynthese; Ein Gang durch biochem. Forschungsarbeiten; The Cardiac Glycosides; Quelques exemples illustrant la parenté entre les principes d'origine végétale et animale; Altes und Neues über Mutterkorn; Les alcaloides des l'ergot; Ueber Ergotamin; Alliin, the Specific Principle of Garlic; Ueber die Isomerie von Lysergsäure und Isolysergsäure; The Cardioactive Glycosides; Sennosides A and B, the Active Principles of Senna; Die spezifischen Wirkstoffe des Mutterkorns und ihre therapeutische Anwendung; Die Konstitution der Mutterkornalkaloide; The Cardiac Glycosides of Digitalis; Aus der Chemie der Naturstoffe. Research on chlorphyll, photosynthesis, cardiac glycosides, assimilation, ergot of rye; developed processing and use methods for natural substances. Address: Bildstöckliweg 11, Arlesheim (BL).

STOLL, Maximilian, physician; b. Swabia, 1742; practiced medicine, Vienna, Austria; added to prestige of Viennese clinic; enlarged ann. report. Author: Ratio Medindi, 3 vols., 1777-80; Aphorisms. Gave more detailed descriptions of lead colic, pulmonary Tb. Died 1788.

STOLL, Norman Rudolph, Am. parasitologist; b. North Tonawanda, N.Y., Sept. 4, 1892; s. Charles J. and Fredericka (Jenzen) S.; student Syracuse U., 1910-11; B.S., Mt. Union Coll., 1915, D.Sc., 1941; M.S., U. Mich., 1918; Sc.D., Johns Hopkins, 1923; m. Estella M. Scott, Aug. 23, 1919 (dec. 1947); children—Henry C., Margaret E. (Mrs. William A. Dawson), Louise M. (Mrs. Wm. S. Maddux); m. 2d, Helen Kennedy Stevens, Dec. 26, 1951. Faculty, Massillon (O.) High Sch., 1915-17, Detroit Central High Sch., Jr. Coll., 1919-21, Peking Union Med. Coll., 1923-24, Johns Hopkins, 1925-26; with Hookworm Commn., P.R., 1922, China, 1923-24, Panama, 1926, W.Africa, 1961; asso. mem. dept. animal pathology Rockefeller Inst. Med. Research, Princeton, N.J., 1927-50, N.Y.C., 1951; prof. emeritus Rockefeller U., 1963—. Mem. expert panel parasitic disease WHO, 1952—; mem. Internat. Commn. Zool. Nomenclature, 1944—, chmn. revision code, 1958-61; Theobald Smith lectr. N.Y. Soc. Tropical Medicine, 1962. Mem. Am. Soc. Parasitologists (past pres., editor Jour. Parasitology 1938-43), A.A.A.S., Am. Soc. Zoology, Am. Soc. Naturalists., Am. (Craig lectr. 1960), Royal socs. tropical medicine and hygiene, Royal Soc. Medicine, Sigma Xi. Author: This Wormy World, 1947. Research, numerous publs. on hookworms, related forms in man, domestic animals; pioneer axenic culture parasitic nematodes and entamoeba. Home: 256 Snowden Lane, Princeton, N.J. 08540. Office: Rockefeller U., N.Y.C. 10021.*

STOLL, Otto, Swiss natural scientist; b. Frauenfeld, Switzerland, Dec. 29, 1849; physician, Guatemala; later prof., mus. dir., Zurich. Founder, Ethnographic Soc., Zurich. Research on groups of peoples which were dying out, including Basques, Mayas, Ixil Indians. Died Zurich, Aug. 18, 1922.

STOLL, Robert Roth, Am. mathematician; b. Pitts., May 19, 1915; s. Harry Bachman and Elisabeth (Roth) S.; B.S., U. Pitts., 1936, M.S., 1937; Ph.D., Yale, 1942; m. Sarah Wherry Donaldson, Sept. 3, 1937; children—R. Kurt, Nancy D., David B. Instr. math. Rensselaer Poly. Inst., 1937-39; faculty Williams Coll., 1942-46, Lehigh U., 1946-52; prof. math. Oberlin (O.) Coll., 1952—, acting chmn. dept., 1953-54, 60-61, 66—. NRC fellow, 1945-46; Sci. Faculty fellow NSF, 1958-59, 67-68; Fulbright-Hayes Vis. prof. to Lebanon, 1964-65. Mem. Math. Assn. Am. (chmn. Ohio sect. 1954-55, bd. govs. 1961-64), Am. Math. Soc., Sigma Xi. Author: Linear Algebra and Matrix Theory, 1952; Sets, Logic and Axiomatic Theories, 1961; Set Theory and Logic, 1963; Linear Algebra, 1968; also articles. Research in algeraic theory of semigroups. Home: 290 Morgan St., Oberlin, O. 44074.*

STOLL, Wilhelm Friedrich, mathematician; b. Freiburg, Germany, Dec. 22, 1923; s. Heinrich and Doris (Eberle) S.; grad. U. Tübingen (Germany), 1949, Dr.rer.nat., 1953, Habilitation, 1954; m.

Marilyn J. Kremser, June 11, 1955; children—Robert, Dieter, Elisabeth, Rebecca. Asst., U. Tübingen, 1953-59, dozent, 1954-60; vis. lectr. U. Pa., 1954-55; staff Inst. for Advanced Study, 1957-59; apl. prof. U. Tübingen, 1960; prof. U. Notre Dame (Ind.), 1960—, head dept. math., 1966—. Mem. Am. Math. Soc., Math. Assn. Am., Deutsche Math. Ver., Hochschulverband, Sigma Xi. Research and publs. on complex analysis, complex manifolds and value distbrn. in several variables. Home: 711 White hall Dr., South Bend, Ind. 46615.*

STOLLER, Alan, psychiatrist; b. London, Eng., May 26, 1911; s. Maurice and Anne (Felman) S.; student U. London, 1929-35; M.R.C.S., L.R.C.P. (Eng.), 1935, D.P.M., 1938, F.A.N.Z.C.P., 1964; m. Joan Dorothy Leneveu, Aug. 6, 1948; children—Julian Alan, Melinda Ruth, Rodney Philip. Sr. med. officer, acting med. supr. Claremont Mental Hosp., West Australia, 1939-41; research asst. U. London, 1946-47; psychiatric cons. repatriation dept. Commonwealth of Australia Govt., 1947-53; chief clin. officer dept. mental health, Victoria, Australia, 1963—; dir. Mental Health Research Inst., Victoria, 1955—; sr. specialist Inst. Advanced Studies, East-West Center, Hawaii, 1966. Mem. bd. censor Australian and New Zealand Coll. Psychiatrists, 1963—; mem. bd. studies in psychol. medicine Melbourne U., 1956—; mem. mental health research fund com., 1956—; pres. World Fedn. for Mental Health, 1964-65; mem. expert adv. panel in mental health WHO, 1960—; pres. Marriage Guidance Council, Victoria, 1963—; v.p. Victorian Family Council, 1961—; mem. Youth Police Adv. Com., 1957—. Fellow Australia and New Zealand Coll. Psychiatrists, Am. Psychiatric Assn.; mem. Brit. E.E.G. Soc., Philippine Soc. Neurology and Psychiatry. Author: Mental Health Facilities and Needs of Australia, 1955; Growing Old, 1960. Editor: Family Today, 1962; New Faces, 1966. Publs. on research in biol., clin., social aspects of mental health especially problems involving mapping of community patterns of mental illness and mental retardation; hypothesis of virus-chromosome interaction as a cause of congenital anomalies. Home: 348 Dandenong Rd., East. St. Kilda, Victoria, Australia. Office: 300 Queen St., Melbourne C.1, Victoria, Australia.*

STOLPER, Wolfgang F(riedrich), economist; b. Vienna, Austria, May 13, 1912; s. Gustav and Paula (Deutsch) S.; came to U.S., 1934, naturalized, 1940; student U. Berlin (Germany), 1930-31, U. Bonn (Germany), 1931-33, U. Zurich (Switzerland), 1933-34; M.A., Harvard, 1935, Ph.D., 1938; m. Martha Vogeli, Aug. 11, 1938; children—Thomas, Matthew. Faculty, Harvard, 1936-41, Swarthmore Coll., 1941-49; faculty Mich. U., Ann Arbor, 1949—, prof. econs., 1954—, dir. Center for Research on Econ. Devel., 1963—; research asso. Nat. Bur. Econ. Research, 1946-52; staff Center for Internat. Studies, Mass. Inst. Tech., 1955, 56, 58, 59, Center Internat. Affairs, Harvard, 1962-63. Mem. U.S. Strategic Bombing Survey, 1945, Internat. Labor Office, 1946; chief econ. planning unit Fed. Ministry Econ. Devel., Lagos, Nigeria, 1960-62; head UN Econ. Mission, Malta, 1963. Cons., U.S. AID, also Internat. Bank for Reconstrn. and Devel., 1963—; Fullbright lectr., Germany, 1966. Guggenheim fellow, 1947. Mem. Am. Econ. Assn., Nigerian Econ. Soc. Author: Structure of the East German Economy, 1960; Germany Between East and West, 1960; Planning Without Facts, 1966; also articles. Research on effects of internat. trade on distbn. of income, interrelations between profitability of investment and the budget, balance of paydistbrn. in several variables. Home: 711 Whitehall ments and growth of underdeveloped countries.*

STOLTZ, Joseph Alexis, French physician; b. Andlau-au-Val, France, 1803; prof. obstetrics Strasbourg (France) U., last French dean before German occupation; prof., dean, Nancy, France, 1872-80; a leading French obstetrician. Died Andlau-au-Val, 1896.

STOLZY, Lewis Hal, Am. agronomist; b. Byron Center, Mich., Dec. 11, 1920; s. Hal C. and Bessie (Lilly) S.; B.S., Mich. State U., 1948, M.S., 1950, Ph.D., 1954; m. Ardyth Lee Marshall, June 18, 1947; children—Janice, Kevin, Lisa. Faculty, Mich. State U., 1950-54; faculty U. Cal., Riverside, 1954—, prof. soil physics dept. soils and plant nutrition, 1967—; with Mineral Nutrition Lab., Beltsville, Md., 1961. Fulbright Sr. Research scholar U. Adelaide (So. Australia), 1964-65. Mem. Am. Soc. Agronomy, Am. Soc. Hort. Sci., Am. Phytopathological Soc., Internat. Soc. Soil Sci., Sigma Xi, Phi Kappa Phi, Alpha Zeta. Research, numerous publs. on soil phys. conditions in relation to plant growth. Home: 5510 Fargo Rd., Riverside, Cal. 92506.*

STOMMEL, Henry Melson, Am. oceanographer; b. Wilmington, Del., Sept. 27, 1920; s. Walter H. and Marian (Melson) S.; B.S., Yale, 1942; M.A., Harvard, 1961; Ph.D. (hon.), U. Gotenberg, Sweden, 1964; m. Elizabeth H. Brown, Dec. 8, 1950; children—Matthew, Elijah, Abigail. With Woods Hole Oceanographic Instn., Woods Hole, Mass., 1944-60; prof. oceanography Harvard U., 1960-63; prof. oceanography Mass. Inst. Tech., Cambridge, 1963—. Recipient Sverdrup medal, Am. Meteorol. Soc.; recipient Albatross award. Mem. Nat. Acad. Scis. Author: The Gulf Stream, 1958, 1965; also numerous articles. Studies, publs. on theory of oceanic circulation; mixing processes in salt water; ocean surveys. Home: 766 Palmer Av., Falmouth, Mass. 02540.*

STONE, Benjamin Clemens, botanist; b. Shanghai, China, July 26, 1933; s. Benjamin Elizabeth (Crandall) S.; B.A. cum laude, Pomona Coll., 1954; Ph.D., U. Hawaii, 1960; postgrad. Washington U., St. Louis, Claremont Grad. Sch.; m. Michiko Muraoka, Aug. 7, 1965. Came to U. S., 1937, naturalized, 1954. Research asst. Smithsonian Instn., 1960; faculty Coll. Guam, Agana, 1961-65, prof., 1965, head biology dept., 1962-65; lectr. botany, curator U. Herbarium, U. Malaya, Kuala Lumpur, 1965—. Fellow Linnaen Soc. London, A.A.A.S.; mem. Bot. Soc. Am., Am. Soc. Plant Taxonomists, Malayan Soc. for Exptl. Biology, Micronesian Acad. Sci. (founder, past pres.). Research, publs. in plant taxonomy, geography, conservation, significance of natural vegetation to long-term economy of nations. Home: 33B Lorong Jambatan, Kuala Lumpur, Malaysia; also 5367 La Jolla Blvd., La Jolla, Cal. 92037.*

STONE, Calvin Perry, Am. psychologist; b. Portland, Ind., Feb. 28, 1892; s. Ezekiel and Emily (Brinkerhoff) S.; A.B., Valparaiso U., 1913; M.A., Ind. U., 1916, D.Sc., 1954; Ph.D., U. Minn., 1921; m. Minnie Ruth Kemper, June 30, 1917; children—James Herbert, Robert Kemper, Barbara Ruth. High sch. prin. and supt., 1910-14; teaching fellow U. Minn., 1916, 19-21, instr. psychology and histology, 1921-22; dir. research Psychol. Lab. of Ind. Reformatory, 1916-17; asst. prof. psychology Stanford, 1922-25, asso. prof, 1925-29, prof. 1929-54. Inst. Juvenile Research, 1928-29; research N.Y. Psychiat. Inst., 1945; Columbia U., summer 1945; U. summer 1947. Fellow A.A.A.S. (v.p. sect. I, 1938-39); mem. Western (pres. 1931-32), Am. (pres. 1941-42) psychol. assns., Am. Assn. on Mental Deficiency, Nat., Cal. acads. sci., Acad. Western Naturalists, Soc. Exptl. Biology, Sigma Xi. Contbr. of many articles on exptl. studies of instinct, sex behavior, learning, memory, and genetic psychology. Editor Jour. Comparative and Physiol. Psychology, 1947-50; Ann. Rev. Psychology, since 1948; editor Comparative Psychology, 1951. Died Dec. 28, 1954.

STONE, Edward James, English astronomer; b. London, Feb. 28, 1831; s. Edward Stone; student King's Coll., London; entered Cambridge, 1856, became scholar Queen's Coll., 1856, fellow, 1859-72, named hon. fellow, 1875, fifth wrangler, B.A., 1859, M.A., 1862; inc. M.A., Christ Church, Oxford, 1879; D.Sc., U. Padua (Italy), 1892. Became chief asst. to astronomer royal, Royal Obs., Greenwich, Eng., 1860-70; royal astronomer Cape of Good Hope, 1870-79; named Radcliffe observer, Oxford, 1878. Recipient Lalande prize French Acad. Scis., 1881. Fellow Royal Soc., 1868, Royal Astron. Soc. (pres. 1882-84). Author: Cape Catalogue (lists 12,441 star to 7th magnitude), 1880, later expanded it to include stars from 25° declination to equator; Tables for Facilitating Computation of Star-Constants, 1897. Observed reversal of Fraunhofer spectrum during eclipse of sun, 1874; deduced solar parallax 1st from Mars, then from transit of Venus, 1869. Died May 9, 1897.

STONE, Emerson Law, Am. physician; b. Waterford, N.Y., Apr. 7, 1895; s. Arthur J. and Agnes (Law) S.; A.B., Williams Coll., 1916; M.D., Johns Hopkins, 1920; Mus. B., Yale, 1930; m. Grace E. Kussmaul, July 15, 1926; children—Emerson Law II, John Arthur, Mary Diana. Resident house officer Johns Hopkins Hosp., 1920-21, resident in obstetrics, 1924-25; with New Haven Hosp., 1921-24; surgeon U. S. Lines, 1924; asso. prof. obstetrics and gynecology Yale U. Sch. Medicine, 1925-27, asso. clin. prof., 1927-44, clin. prof. since 1944; pvt. practice, New Haven, since 1927; cons. various regional hosps. since 1927. Diplomate Am. Bd. Obstetrics and Gynecology. Author: The New Born Infant, 1945. Musician. Died Jan. 10, 1953; buried Williamstown, Mass.

STONE, George Chester, Am. psychologist; b. Los Angeles, Feb. 21, 1924; s. Glenn Everett and Bessie (Jones) S.; student Cal. Inst. Tech., 1941-42; A.B. with highest honors U. Cal., Berkeley, 1948, M.A., 1951, Ph.D., 1954; m. Hannah Anita Zacharin, Nov. 14, 1946; children—Peter Franklin, Lisa Meredith, Sarah Ila, Laura Eleanor. Abraham Rosenberg research fellow U. Cal., Berkeley, 1951-52, teaching asst., 1952-53; research asso. Bur. Edn. Research U. Ill., 1953-55; sr. biol. investigator G. D. Searle & Co., Skokie, Ill., 1955-58; lectr. psychology U. Cal., Berkeley, 1960-62; faculty U. Cal. Sch. Medicine, San Francisco, 1959—, asso. prof. in residence, 1966—; research psychologist Cal. Dept. Mental Hygiene, Langley Porter Neuropsychiatric Inst., San Francisco, 1958—; cons. psychologist Stanford Research Inst., 1961; cons. Hine Labs., 1964—. Mem. Am., Western psychol. assns., A.A.A.S., Behavioral Pharmacology Soc., Soc. for Psychol. Study of Social Issues, Cal. Sch. Bd. Assn., Phi Beta Kappa, Sigma Xi. Studies and publs. on differences among animals in susceptibilities to certain psychotropic drugs; relation of differing susceptibilities to temperamental characteristics of animals; evidence against the existence of a single generalized central motive state, found in different effects of sound and hunger or thirst on behavior motivated by shock avoidance. Home: 801 Bolinas Rd., Fairfax, Cal. 94930. Office: Langley Porter Neuropsychiatric Inst., 401 Parnassus Av., San Francisco 94122.*

STONE, John Stone, Am. elec. engr.; b. Dover, Va., Sept. 24, 1869; s. Charles Pomeroy and Jeannie (Stone) S.; ed. Sch. of Mines (Columbia), 1886-88; Johns Hopkins, 1888-90; m. Sibyl Wilbur, Nov. 28, 1918. Experimentalist in laboratory, Am. Bell Te'ephone Co., 1890-99; gen. consulting elec. engr., 1899-1920; spl. lecturer on elec. oscillations, Mass. Inst. Tech., for a number of years; dir., v.p. and chief engr. from incorporation, 1902-08, pres. and chief engr. June 10, 1908-10, Stone Telegraph & Telephone Co. (mfg. and leasing wireless telegraph apparatus). Fellow Am. Acad. Arts and Sciences, A.A.A.S., Inst. of Radio Engrs. (v.p. 1913-14, pres. 1914-15, dir. 1912-18); organizer, vice chmn. Radio Engrs. Com. on National Defense; del. Internat. Elec. Congress, 1904, 2d Pan-Am. Scientific Congress, 1917; mem. advisory com. Am. Defense Society; pres. Soc. of Wireless Telegraph Engineers, 1906-09; member Franklin Institute, Alpha Delta Phi; asso. mem. Am. Inst. E.E. Awarded Edward Longstreth Medal for paper on "The Practical Aspects of the Propagation of High Frequency Waves Along Wires," by the Franklin Inst., 1913; medal of honor of Inst. of Radio Engrs. "for distinguished service in radio communication," 1923. Asso. Engr. at large dept. of development and research of Am. Tel. & Tel. Co., 1920-35. Patentee numerous inventions relating to improvements in telephony and telegraphy. Contbr. numerous papers on elec. subjects to sci. and tech. press. Spl. studies on elec. oscillation, radiation and electromagnetism; investigated Hertzian waves without wire conductors for possible telephone improvements. Died May 20, 1943.

STONE, L(awrence) Joseph, Am. psychologist; b. Washington, May 20, 1912; s. Nahum and Esther Bertha (Levinson) S.; A.B., Cornell U., 1933; A.M., Columbia, 1934, Ph.D., 1937; m. Beatrice Berlin, June 23, 1933 (dec. September 1962); children—Deborah (Mrs. Theodor Holm Nelson), Susannah (Mrs. Maurice Eldridge), Miriam (Mrs. Robert Leavitt). Instr. psychology Columbia, 1935-37; research asso. Sarah Lawrence Coll., 1937-39; instr. psychology Bklyn. Coll., 1938-39, Coll. City N.Y., 1940-42; research editor dept. child study Vassar Coll., 1939-41, asst. prof., 1941-47, asso. prof., 1947-49, prof., 1949-65, prof. psychology, 1965—, dir. film program, 1942—, acting chmn., 1953-55, chmn., 1955-61, chmn. of the dept. psychology, 1966—; dir. Vassar Film Series: Studies of Normal Personality Development; lectr. New Sch. Social Research, 1945-56; Fulbright Research grant, U. Oslo, Norway, 1957-58; Nat. Inst. Mental Health spl. fellowship, 1964-65; mem. staff Dutchess County Mental Health Clinic; past dir. Catharine Street Community Center, Poughkeepsie; producer films for Project Head Start, 1965-68, Dir. Com. for Certification of Psychologists N.Y. State, 1954-57, mem. adv. bd. psychology, Bd. regents U. State N.Y., 1958—. Mem. research career award com. Nat. Inst. Mental Health, 1962-66. Lieutenant to lieutenant commander, psychobiological unit, USPHS, 1944-46, sci. director USPHS Res. Nat. Inst. Mental Health spl. fellow, 1965. Certified psychologist N.Y. Diplomate Clin. Psychology, Am. Bd. Examiners Profl. Psychology, Fellow Am. Psychol. Assn., Am. Orthopsychiat. Assn., Soc. Projective Techniques (past dir.), Soc. Research Child Development (past bd. govs.); N.Y. State Psychol. Assn. (pres. 1952), World Fedn. Mental Health, Eastern Psychol. Assn., (pres. 1952), World Fedn. Mental Health, Eastern Psychol. Assn., Sigma Xi. Co-author: Childhood and Adolescence: A Psychology of the Growing Person (with J. Church), 1957, 2d edit., 1968. Cons. editor Random House. Contbr. to MacIver's The More Perfect Union, 1949. Contbr. articles profl. jours. Research on personal devel. in young children, established projective play techniques for study child, infant devel. Address: Vassar Coll., Poughkeepsie, N.Y. 12601.*

STONE, Leon Stansfield, Am. anatomist; b. Newton, N.J., Feb. 14, 1893; s. Thomas S. and Julia (Washer) S.; Ph.B., Lafayette Coll., 1916, D.Sc., Ph.D., Yale, 1921; m. Ruth Hoagland, Dec. 18, 1918; children—Mary Jane (Mrs. Carl Victor Hansen). Faculty, Yale Sch. Medicine, 1919-66, Bronson prof. comparative, dir. researches, grad. studies. Recipient Doyne Meml. medal Oxford U., 1947; cited by Lafayette Coll., 1960, 66. Mem. A.A.A.S., Soc. for Exptl. Biology and Medicine, Am. Assn. Anatomists, Oxford Ophthal. Congress, Sigma Xi. Research, numerous publs. on exptl. embryology, regeneration of retina and lens., devel. of vision in transplanted eyes of lower vertebrates, origin of cranial ganglia, dependence of sense organs on nerve supply. Home: 100 Bedford Av., Hamden, Conn. 06517. Office: 333 Cedar St., New Haven 06511.*

STONE, Livingston, Am. fish culturist; b. Cambridge, Mass., Oct. 21, 1835; s. Peter Robert Livingston and Lavina (Winship) S.; A.B., Harvard, 1857, A.M., 1860; grad. Meadville (Pa.) Theol. Sem., 1860; m. Rebecca Saulsbury Cushing, Apr. 8, 1875. Pastor Unitarian chs., Billerica, Mass., Detroit, Phila., 1860-64, Charlestown, N.H., 1864-68; established trout breeding sta., Charlestown, 1866 (2d attempt at practical fish culture in U. S.); apptd. dep. U. S. fish commr., 1872; established salmon hatching stas. for U. S. Govt. on McCloud River, Cal., 1872, 79, on Columbia River, for cannerymen, 1877; in early 70s transported car load of live Atlantic fish and fish eggs across continent and deposited them in Pacific Ocean (1st successful attempt of its kind); accompanied Govt. expdn. to Alaska to investigate salmon fisheries, as a result of which Afognac Island was designated as a reservation for salmon culture by President Harrison; in charge Cape Vincent (N.Y.) Sta., 1897-1906; ret., 1906. Recipient 2 diplomas at Internat. Expn., Berlin, 1880; diploma Internat. Expn., London, 1883; hon. Bronze medal Société d'Aclimatation, Paris. Author: Domesticated Trout, 1872. Died Dec. 24, 1912.

STONE, Marshall Harvey, Am. mathematician; b. New York, N.Y., Apr. 8, 1903; s. Harlan Fiske and Agnes (Harvey) S.; student Englewood (N.J.) pub. schs., 1908-19; A.B., Harvard U., 1922, A.M., 1924, Ph.D., 1926; grad. studies Université de Paris, 1924-25; ScD., Kenyon Coll., 1939; hon. Dr. Universidad de San Marcos, Lima, Peru, 1943; honorary Dr. Universidad de Buenos Aires, 1947; hon. Doctor, Athens, 1954); Sc.D., Amherst Coll., 1954, Colby Coll., 1959, U. of Mass., 1966; m. Emmy Portman, June 15, 1927 (div. July 1962); children—Doris Portman, Cynthia Harvey, Phoebe G.; m. 2d, Ravijojla Kostic neé Perendija, Aug. 8, 1962; 1 step-daughter, Svetlana Kostic. Part-time instr. Harvard U., 1922-23; instr. Columbia U., 1925-27; inst. Harvard U., 1927-28, asst. prof. mathematics, 1928-31; asso. prof. mathematics, Yale U., 1931-33; acting asso. prof. Stanford U., summer 1933; asso. prof. mathematics Harvard U., 1933-37; fellow Guggenheim Memorial Foundation, 1936-37; prof. mathematics, Harvard U., 1937-46, and chmn. of the dept. 1942; Walker Ames lecturer, U. of Wash., summer 1942; visiting lecturer, Facultad de Ingenieria, Universidad de Buenos Aires, 1943, Universidad do Brasil, Rio de Janeiro, 1947, Tata Inst., Bombay, 1949-50, Am. Math. Soc., 1951-52, College de France, 1953, Australian Math. Soc., 1959, Middle East Tech. U., Ankara, 1963, Pakistan Acad. Scis., 1964, U. Geneva, 1964; 1st Ramanujan vis. prof. Inst. Math. Scis., Madras, 1963; vis. prof. Research Inst. Math. Scis., Kyoto, Japan, 1965; vis. scientist C.E.R.N., Geneva 1966; hon. prof. U. Madurai (South India), 1967—; Fulbright fellow Australian National Univ., 1967; cons. to USN Dept. Bur. Ordnance and Office Vice Chief Naval Operations, 1942-43; Civil Service Employee, U. S. War Department, Office Chief of Staff, 1944-45, with overseas service in China-Burma-India and European theaters; Andrew MacLeish Distinguished Service Professor of Mathematics, U. Chgo., 1946-68, chmn. dept. mathematics, 1946-52, mem. com. on social thought, 1962-68; George David Birkhoff prof. math. U. Mass., Amherst, 1968—. Pres. Inter-Am. Committee Mathematical Edn., 1961—; vice chmn., div. mathematics National Research Council, 1951-52. Pres. International Math. Union 1952-54, Internat. Commn. on Math. Instrn., 1959-62, Inter Union Committee Teaching Science, Internat. Council Scientific Unions, 1962-65. Member of American Math. Society (president 1943-44, chmn. policy com. 1946-48), Nat. Acad. Scis., Am. Philos. Society; hon. mem. Union Matemàtica Argentine, Indian Math. Soc.; fgn. mem. Acad. Brasileira de Sciencias, Lund (Sweden) Physiographical Society. Author: Linear Transformations in Hilbert Space and Their Applications to Analysis, 1932. Contbr. research papers to domestic and foreign tech. jours. Made basic advances in application algebraic techniques in analysis, spectral Theory, geometry, topology and logic. Home: 260 Lincoln Av., Amherst, Mass. 01002.*

STONE, N. H., Am. physician; Mem. A.C.S., A.M.A., Alpha Omega Alpha, Phi Delta Epsilon. Research, publs. on causes of peptic ulcer, including hormones on gastric secretion and ulcer formation, treatment of thermal burns and complications, treatment of hand infections. Home: 2130 N. Lincoln Park W., Chgo. 60614. Office: 1825 W. Harrison St., Chgo. 60612.*

STONE, Ormond, Am. astronomer; b. Pekin, Ill., Jan. 11, 1847; s. Elijah and Sophia (Creighton) S.; student U. Chgo., 1866-70, A.M., 1875; m. Catharine Flagler, May 31, 1871 (dec. 1914); m. 2d, Mary Florence Brennan, June 9, 1915. Asst. U. S. Naval Obs., Washington, 1870-75, mem., sec. bd. visitors, 1901-03; dir. Cin. Obs., 1875-82; prof. astronomy U. Va., dir. Leander McCormick Obs., 1882-1912; ret. on Carnegie Found. In charge U. S. Naval Obs. eclipse expdn. to Colo., 1878, McCormick Obs. eclipse expdn. to S.C., 1900. Mem. A.A.A.S. (v.p. 1888); mem. Internat. Congress Arts and Sciences, St. Louis, 1904 (chmn. sect. astrometry). Founder, editor Annals of Math. Contbr. to match. and astron. jours. Made observations of double and variable stars, nebulae, satellites of Saturn. Died Jan. 17, 1933.

STONE, William Harold, Am. geneticist; b. Boston, Dec. 15, 1924; s. Robert and Rita (Scheinberg) S.; A.B., Brown U., 1948; M.S., U. Me., 1949; Ph.D., U., Wis., 1953; m. Trudi Stone, Apr. 25, 1965; children—Susan Joy, Debra M. Research asst. Jackson Meml. Lab., Bar Harbor, Me., 1947-48; faculty dept. genetics U. Wis., Madison, 1949—, prof., 1961—, prof. med. genetics, 1964—; NIH fellow Cal. Inst. Tech., 1960-61. Mem. panel blood group experts FAO, 1962—. Mem. Am. Inst. Biol. Scis., A.A.A.S., Am. Soc. Immunologists, Am. Genetics Assn., Am. Genetics Soc., Am. Soc. Human Genetics, Research Soc. Am., Internat. Soc. Transplant, Am. Soc. Animal Sci., Sigma Xi, Gamma Alpha. Author: Immunogenetics, 1967; also numerous articles. Research on immunology and genetics, transplantation tissues and organs, effects irradiation blood in twins, role immunologic phenomena in reproduction. Home: Box 393, Route 1, McFarland, Wis. 53558.*

STONE, Wilson Stuart, Am. geneticist; b. Junction, Tex., Oct. 6, 1907; s. Donald Stuart and Grace (Finney) S.; B.A., U. Tex., 1930, M.A., 1931, Ph.D., 1935; m. Julia Jean Lampman, Jan. 28, 1930; children—Charles Stuart, Laurie Jean Younglove, Michael (dec.), Faculty, U. Tex., Austin, 1932—, prof., dept. zoology, 1942—, chmn. dept. zoology, 1959-63, adviser to chancellor for grad. and research affairs, 1963-64, vice chancellor, 1964-66. Cons. genetics M.D. Anderson Hosp. and Tumor Inst., 1955—; mem. nat. adv. research resources council NIH, 1964—. Mem. Am. Soc. Naturalists (past sec.), Am. Soc. Zoologists, Genetics Soc. Am., A.A.A.S., Radiation Research Soc., Soc. for Study Evolution, Am. Soc. Human Genetics, Nat. Acad. Scis., Sigma Xi. Author: (with J. T. Patterson) Evolution in the Genus Drosophila, 1952; also numerous articles. Co-editor Genetics, 1957-62. Research in genetics and evolution including genic balance and sex determination, genetic control of differentiation, chromosomal abnormalities and their role in evolution, studies species relations and evolution in genus Drosophila, effect of radiation on genetic systems, effects radiation and other variables on island populations of Drosophila and protein variation in Drosophila. Home: 610 Bellevue Pl., Austin, Tex. 78705.*

STONE, Witmer, Am. naturalist; b. Phila., Sept. 22, 1866; s. Frederick D. and Anne E. (Witmer) S.; A.B., U. Pa., 1887, A.M., 1891, Sc.D., 1913; m. Lillie M. Lafferty, Aug. 1, 1904. Asst. curator Acad. Natural Scis., Phila., 1891-1908, curator, 1908-24, dir. mus., 1925-28, v.p., 1927, dir. emeritus, 1929-39. Fellow Am. Ornithologists' Union (pres. 1920-23), A.A.A.S.; mem. Internat. Com. on Zool. Nomenclature. Author: Birds of Eastern Pennsylvania and New Jersey, 1894; American Animals (joint author), 1902; Mammals of New Jersey, 1908; Birds of New Jersey, 1909; Flora of Southern New Jersey, 1912; Report on Birds of Yucatan and Southern Mexico; The Molting of Birds; Birds and Mammals of the McIlhenny Alaskan Expedition; The Phylogenetic Value of Color Characters in Birds; Birds of the Princeton Patagonian Expdn.; and a number of other papers in Proc. Acad. Natural Sciences, on birds, mammals, reptiles. Editor Auk, 1912-36. Died May 24, 1939.

STONEHOUSE, Bernard, biologist; b. Eng., May 1, 1926; s. Herbert and Kathleen (Smith) S.; B.Sc. with spl. honors, U. Coll., London, Eng., 1953; D.Phil., Merton Coll., Oxford, Eng., 1958, M.A., 1959; m. Sally Clacey, Sept. 17, 1955; children—Caroline, Felicity Ann, Paul. Meteorologist, pilot, later biologist Falkland Islands Dependencies Survey, 1946-50; research on penguins, seals, shore ecology, S. Georgia, 1953-55; leader Brit. Ornithologists' Union Centenary Expdn., Ascension Islands, 1957-59; faculty Oxford, 1958-60, Canterbury U., Christchurch, New Zealand, 1960—, reader zoology, 1964, head Antarctic Biology Unit, 1960-65. Recipient Polar medal, 1953. Author: Het bevroren continent, 1958; Wideawake Island, 1960; Penguins, 1968; Shorebirds of New Zealand, 1968; also children's natural history books on whales, gulls; also articles. Research on biology of penguins, including life history and ecology studies, synthesis explaining their biology and distbn. in terms of thermal balance, micro-ecology of Antarctica, biology of Antarctic animals in relation to climate; breeding cycles in polar and tropical seabirds; temperature regulation in deer and seals. Home: 849 Cashmere Rd., Christchurch 3, New Zealand.*

STONELEY, Robert, seismologist; b. London, Eng., May 14, 1894; s. Robert and Fanny (Bradley) S.; B.A., U. Cambridge, Eng., 1915, M.A., 1920, Sc.D., 1931; m. Dorothy Minn, Mar. 28, 1927; children—Robert, Anthony John Martin. Lectr., Sheffield U., 1920-23, U. Leeds, 1923-34; lectr. math. U. Cambridge, 1934-48, reader in theoretical geophysics, 1949-61, dir. studies in math. Pembroke Coll., 1935-61, fellow, 1943—; seismologist U. S. Coast and Geodetic Survey, 1961-63; prof. geophysics U. Pitts., 1964—; vis. prof. U. Cal., 1948, Am. U., 1955-56. Hon. research fellow Cal. Inst. Tech., 1956. Fellow Royal Soc., 1935, Am. Geophys. Union; mem. Internat. Assn. Seismology (pres. 1946-51), Royal Astron. Soc., London Math. Soc., Cambridge Philos. Soc., Am. Geophys. Union, Seismol.. Soc. Am. Research, numerous publs. on constn. of interior of earth, theory of propagation of earthquake waves, seismic sea-waves and earth tides. Office: Langley Hall, U. Pitts., Pitts. 15213.*

STONER, Dayton, Am. zoologist; b. North Liberty, Ia., Nov. 26, 1883; s. Marcus and Nancy (Koser) S.; A.B., U. Ia., 1907, M.S., 1909, Ph.D., 1919; m. Lillian Rebecca Christianson, Aug. 3, 1912. Began as asst. in mus. U. Ia., 1908-12, instr. in zoology, 1912-16, asso. in zoology, 1916-22, asst. prof., 1922-28; instr. ornithology and entomology U. Mich. Biol. Sta., summers 1919-20; temp. field asst. U. S. Bur. Entomology, winters 1928-31; field ornithologist Roosevelt Wild Life Forest Exptl. Sta., summers 1928-32; state zoologist N.Y. State Mus., Albany since 1932. Conducted ornithol., entomol. and mammalol. field work in Ia., Colo., Mich., Fiji Islands, New Zealand, W.I., Fla., N.Y., Vancouver Island. Fellow A.A.A.S., Ia. Acad. Sci.; mem. Am. Ornithol. Union, Am. Soc. Mammalogists, Wilson Ornithol. Club, Eastern Bird Band Assn., Wildlife Soc., Ia. Ornithol. Union, Sigma Xi. Author: Rodents of Iowa, 1918; Scutelleroidea of Iowa, 1920; Ornithology of Oneida Lake Region, 1932; Studies on The Bank Swallow, 1936; Wildlife Casualties on the Highways, 1936; Ten Years' Returns from Banded Bank Swallows, 1937; Temperature, Growth and Other Studies on the Eastern Phoebe, 1939; also other papers on birds, mammals, insects. Died May 8, 1944.

STONER, Edmund Clifton, English physicist; b. East Molesey, Surrey, Eng., Oct. 2, 1899; s. Arthur Hallett and Mary (Fleet) Stoner; B.A., Cambridge U., 1921, Ph.D., 1925, Sc.D., 1938; m. Jean Heather Crawford, Dec. 27, 1951. Faculty, U. Leeds (Eng.), 1924—, prof. theoretical physics, 1939-51, Cavendish prof., head dept. physics, 1951-63, prof. emeritus physics, 1963—; research fellow Emmanuel Coll., Cambridge U., 1928-31. Mem. sci. adv. com. Ministry of Supply, 1955-58; visitor wool industries research assn. Dept. Sci. and Indsl. Research, 1955-61, mem. postgrad. tng. awards com., also chmn. physics panel, 1957-62; Kelvin lectr. Instn. Elec. Engrs., 1944; Guthrie lectr. Phys. Soc., 1955. Fellow Royal Soc., 1937, Inst. Physics (v.p. 1957-60), Phys. Soc. Author: Magnetism and Atomic Structure and 1926; Magentism, 1930, 4th edit., 1948; Magnetism and Matter, 1934; also numerous articles. Research on distbn. atomic electrons, diamagnetic and paramagnetic properties of ionic compounds, densities in stars, calculations and application of Fermi-Dirac functions, coercivity of ferromagnetic materials, thermodynamics of magnetization. Home: 12 St. Chad's Dr., Leeds, 6, Yorkshire, Eng.*

STONER, Harry Berrington, English pathologist; b. Sheffield, Eng., Feb. 1, 1919; s. Harry John and Elizabeth (Sprigg) S.; B.Sc., U. Sheffield, 1939, M.B. Ch.B., 1942, M.D., 1946, M.C.Path., 1966. Research asst. in pathology U. Sheffield, 1946-48; Rockefeller Travelling fellow in medicine, 1948-49; research fellow Harvard, 1948-49; mem. sci. staff Med. Research Council, 1949—, with toxicology research unit, Carshalton, Surrey, Eng., 1953—. Cons. WHO, mem. WHO expert adv. panel on food additives, 1961—. Mem. Path. Soc., Physiol. Soc., Biochem. Soc., Coll. Pathology. Author: (with H. N. Green) Biological Actions of the Adenine Nucleotides, 1950; also numerous articles. Editor: (with C. J. Threlfall) The Biochemical Response to Injury, 1960. Research on biol. responses to phys. and chem. injuries. Home: 14 Sherwood Park Rd., Sutton, Surrey. Office: Med. Research Council Lab., Carshalton, Surrey, Eng.*

STONEY, George Johnstone, Irish math. physicist; b. Oakley Park, Ireland, Feb. 15, 1826; ed. B.A., Trinity Coll., Dublin, Ireland, 1848; M.A., 1852. D.Sc. (hon.) Queen's U., 1879; Sc.D. (hon.) Trinity Coll., Dublin, 1902. Asst., Parsonstown Obs., 1848-52; prof. natural philosophy, Queen's Coll., Galway, Ireland, 1852-57, sec., 1857-82; supt. civil service examinations in Ireland, until 1893; Fellow Royal Soc., 1861; mem. Royal Dublin Soc. (sec. for 20 years). Recipient Madden prize, 1852, 1st Boyle medal, 1899. Author: Memoirs on the Physical Constitution of the Sun and Stars. Introduced term electron, and calculated an approximate value for its charge, 1874; research and publs. on wave motion and optics, atomic structure and theory of spectra, kinetic theory of gases, planetary atmospheres, music and mus. echoes. Died London, July 5, 1911.

STOODLEY, Bartlett Hicks, Am. sociologist; b. Somerville, Mass., July 15, 1907; s. Harry Marr and Laura (Hicks) S.; A.B., Dartmouth, 1929; LL.B., Harvard, 1932, Ph.D., 1948; m. Helen Virginia Stark, Mar. 16, 1944; children—Bartlett Hicks, Ronald S. Practice law, Boston, 1932-42; faculty Wellesley (Mass.) Coll., 1947—, chmn. dept. sociology and anthropology, 1965—; Fulbright prof. U. Philippines, 1964-65. Recipient Am. Philos. Soc. award, 1962. Mem. Am. Sociol. Assn., Eastern Sociol. Soc. (past mem. exec. bd.), Assn. for Asian Studies. Author: The Concepts of Sigmund Freud, 1958; also articles. Editor: Society of Self, 1962. Research on relations between personality and society; showed that social motivation is a basic factor in Freudian theory. Home: 8 Service Dr., Wellesley, Mass. 02181.*

STOPES, Marie Carmichael, Brit. paleobotanist; b. Edinburgh, Scotland, 1880; s. Henry and C. C. S.; D.Sc., U. London (Eng.); Ph.D., U. Munich (Germany); m. Humphrey Verdon Roe, 1918 (dec. 1949); 2 sons. First woman apptd. to sci. staff U. Manchester, 1904; on sci. mission, Japan, 1907; with Imperial U., Tokyo, Japan, also explored country for fossils, for 1 1/2 years; founder (with H. V. Roe) Mother's Clinic Constructive Birth Control (1st birth control clinic in world), 1921; fellow, lectr. paleobotany U. Coll., London; lectr. U. Manchester. Fellow Linnean Soc., Geog. Soc., Royal Soc. Lit.; pres. Soc. Constructive Birth Control and Racial Progress. Author: The Study of Plant Life for Young People; Ancient Plants; A Journal from Japan; Botany, the Modern Study of Plants; Catalogue of the Cretaceous Plants in the British Museum, vols. 1, 2; The Lower Greensand (Aptian) Plants of Great Britain, 1915; Married Love, and Wise Parenthood, 1918; (with R. V. Wheeler)

The Constitution of Coal, 1918; Radiant Motherhood, 1920; A New Gospel to All Peoples, 1921; Contraception; its Theory, History and Practice, 1923; The First Five Thousand, 1925; Sex and the Young. The Human Body, 1926; Sex and Religion, 1929; Ten Thousand Birth Control Cases, 1930; Contraception, enlarged, rewritten, 1931; Roman Catholic Methods of Birth Control, 1933; Birth Control To-day, 1934; Marriage in My Time, 1935; Change of Life in Men and Women, 1936. Died Oct. 2, 1958.

STOPPANI, Andres Oscar M., Argentine biochemist; b. Buenos Aires, Argentina, Aug. 19, 1915; s. Oscar Carlos and Julia (Bahía) S.; M.D., U. Buenos Aires, 1941, D.Chemistry, 1945; Ph.D., U. Cambridge (Eng.), 1951; postgrad. U. Cal. at Berkeley, 1953. Faculty, U. Buenos Aires, 1936-45, prof. biochemistry Sch. Medicine, 1949—, dir. Inst. Biochemistry, 1953—; prof. U. La Plata Sch. Medicine, 1948-49; dir. cell metabolism Research lab. Nat. Atomic Energy Commn., 1957-59. Recipient Faculty award and Gold medal U. Buenos Aires, 1941. Fellow A.A.A.S.; mem. Nat. Council for Sci. and Tech. Research (dir. 1963—, exec. com. 1965—, Weismann prize 1963), N.Y. Acad. Scis., Soc. for Exptl. Biology and Medicine, Inst. Biology Gt. Britain, Argentine Assn. for Advancement Sci. (past pres.). Author: Physiological and Pharmacological Studies on Amphibian Melanophores, 1940; (with C. T. Rietti) Practical Methods in Biochemistry, 1962; (with V. Deulofeau, A. Marenzi) Textbook of Biochemistry, 1966; also numerous articles. Research on endocrine regulation of skin color in amphibia; metabolism and pharmacology of indoxylogenic compounds; pharmacology of respiratory multienzyme systems; carbon dioxide fixation in autotrophic and heteratrophic organisms; bacterial metabolism; action of steroid hormones on enzymes; biochemistry of animal venoms. Home: 2295 Viamonte, Buenos Aires, Argentina.*

STORCK, Hans Rudolf, Swiss physician; b. Zürich, Switzerland, Sept. 26, 1910; s. Emil and Gertrud (Weber) S.; student medicine, U. Geneva (Switzerland), U. Rostock (Germany), U. Paris; M.D., U. Zürich, 1938; m. Elinor Storck-Rettich, May 24, 1938; children—Dieter, Ulrich, Eva. With dermatol. Clinic, U. Zurich, 1938-39, Inst. Microbiology, 1941-42, dermatol. clinic, 1942-57, faculty, 1947—, prof. dermatology and venereology, dir. clinic, 1958—. Author: (with G. Miescher) Handbuch der Haut- und Geschlechtskrankheiten, 1962; also numerous articles. Clin. and exptl. research on eczema, allergy, drug-allergy, hemorrhagic diathesis, cutaneous infectious diseases, melanoma, biology of radiation. Home: 7 Mittelbergsteig, Zurich 8044, Switzerland.*

STORER, Francis Humphreys, Am. chemist; b. Boston, Mar. 27, 1832; s. David Humphreys and Abby Jane (Brewer) S.; student Lawrence Sci. Sch. (Harvard), 1850-51; S.B., Harvard, 1855, A.M., 1870; studied abroad, 1855-57; m. Catharine A. Eliot, June 21, 1871. Asst. to Prof. Cooke, 1851-53; chemist U. S. N. Pacific exploring expdn., 1853; practiced as chemist at Boston, 1857-65; prof. gen. and indsl. chemistry Mass. Inst. Tech., 1865-70; prof. agrl. chemistry Harvard, 1870-1907, dean Bussey Instn., 1871-1907. Fellow Am. Acad. Arts and Scis. Author: Dictionary of the Solubilities of Chemical Substances, 1864; Manual of Inorganic Chemistry, 1869; Manual of Qualitative Chemical Analysis, 1868 (both with Charles W. Eliot); Cyclopaedia of Quantitative Chemical Analysis, 1873; Agriculture in Some of Its Relations with Chemistry, 3 vols., 1897; Elementary Manual of Chemistry (with W. B. Lindsay), 1894; Manual of Qualitative Analysis (with W. B. Lindsay), 1899; Bulletin of the Bussey Institution; Alloys of Copper and Zinc; Manufacture of Paraffin Oil. Died July 30, 1914.

STOREY, Harold Haydon, plant pathologist; b. Manchester, Eng., June 10, 1894; s. Henry and Mary Louisa (Ford) S.; M.A., Ph.D., Cambridge U., 1928; m. Molly Everitt, Dec. 9, 1930; 1 dau., Mary Carol (Mrs. Philip Rees). With Dept. of Agr., Union of S. Africa, 1922-28; With East African Agr. and Forestry Research Orgn., 1928—, dep. dir., 1948-56, acting dir. 1956-57. Fellow Royal Soc., 1946. Decorated Companion of Order of St. Michael and St. George. Research and publs. on virus diseases of plants, activity of virus-transmitting insects. Address: Box 30148, Nairobi, Kenya.*

STORK, Gilbert (Josse), chemist; b. Brussels, Belgium, Dec. 31, 1921; s. Jacques and Simone (Weil) S.; B.S., U. Fla., 1942; Ph.D., U. Wis., 1945; D.Sc. (honorary), Lawrence Coll., 1961; married Winifred Stewart, June 9, 1944; children—Diana, Linda, Janet, Philip. Sr. research chemist Lakeside Labs., 1945-46; instr. chemistry Harvard, 1946-48, asst. prof., 1948-53; asso. prof. Columbia, 1953-55, prof., 1955—, Higgins prof. chemistry, 1967—; invited lectr. Swiss Am. Found. Sci. Exchange, 1959; Coover lectr., 1958; Folkers lectr., 1962; Bachmann lectr., 1962; spl. lectr. Internat. Union Pure and Applied Chemistry, Kyoto, 1963, London, 1968; Treat B. Johnson lecturer, Frank Burnet Dains lectr, 1964; Kharasch vis. professor U. of Chgo., 1966; consultant, chemistry panel NSF, 1958-61; adv. bd. Petroleum Research Fund, 1963-66, NIH, 1967—; cons. Am. Cyanamid, Syntex, S.A., Internat. Flavors and Fragrances; chmn. Gordon Steroid Conf., 1958-59; hon. adv. editor Tetrahedron; bd. editors Jour. Organic Chemistry, 1955-61; cons. editor Advanced Chemistry Series McGraw-Hill, 1961—. com. organic chemistry NRC, 1956-59, com. post-doctoral fellowships, 1959-62. Recipient Guggenheim fellow, 1959; Baekeland medal, 1961; Harrison Howe award, 1962; Edward Curtis Franklin meml. award Stanford, 1966; award for creative work in synthetic organic chemistry Am. Chem. Soc., 1967. Fellow Am. Acad. Arts and Scis., Nat. Acad. Scis.; mem. Am. Chem. Soc. (award in pure chemistry 1957, chmn. organic chemistry div. 1967), Chem. Soc. London, Swiss Chem. Soc. Devised new methods for construction complex molecules; total synthesis from elements complex natural products. Home: 153 Glenwood Av., Leonia, N.J. Office: Columbia U., N.Y.C. 10027.*

STORM, Robert MacLeod, zoologist; b. Calgary, Alta., Can., July 9, 1918; s. William Burns and Nellie (Haley) S.; B.Ed., No. Ill. State Tchrs. Coll., 1939; M.S., Ore. State Coll., 1941, Ph.D., 1948; m. Carol Grace Offner, Mar. 8, 1943; children—Robert H., Ellen C., David Bruce; m. 2d Marvene Patricia Christensen, Aug. 21, 1959; children—Susan Kay, Marjorie, Michael. Faculty zoology dept. Ore. State U., Corvallis, 1948—, now prof. zoology. Fellow Herpetologists League; mem. Am. Soc. Ichthyologists and Herpetologists, Am. Soc. Mammalogists, Sigma Xi. Research in distbn. and ecology of amphibians and reptiles of N.W., especially Ore. Home: 3800 Neer Av., Corvallis, Ore. 97330.*

STÖRMER, Fredrik Carl Mülertz, Norwegian mathematician, astronomer; b. Skien, Norway, Sept. 3, 1874; student Oslo (Norway) U. Prof. math. U. Oslo, 1903-46; organized network of stas. to photograph and measure height of aurora and spl. types of clouds in Norway. Mem. French Acad. Scis., 1947. Invented apparatus for photographing aurora borealis; studied integration of differential equations, composition of high atmosphere, motion of electric particles in earth's magnetic field, cosmic rays; discovered forbidden directions which lie in a cone (Störmer cone) when cosmic rays approach a magnetic dipole. Died Aug. 13, 1957.

STORMER, Leif, Norwegian paleontologist; b. oslo, Norway, July 1, 1905; s. Fredrik Carl and Ada (Clauson) S.; Dr.philos., Oslo U., 1931; m. Ingegerd Viborg Alten, May 23, 1932; children—Fredrik Carl, Erling, Inger. Curator, Paleontol. Mus., U. Oslo, 1940-46, prof. hist. geology, 1946—, dean Faculty Sci., 1957-59, A. Agassiz lectr. Harvard, 1960-64; pres. Internat. Commn. on Stratigraphy, 1948-52. Decorated knight Order of St. Olav; recipient Reusch medal, 1937; Bergen prize for advancement sci., 1945; Fridtjof Nansen award Acad. Sci., 1965. Mem. Internat. Paleontol. Union (past treas.), Geol. Soc. Am. (hon.), Paleontol. Soc. Am. (corr.), Geol. Soc. London, Acad. Sci. Copenhagen. Author: Jordens og livets historie, 1966; also articles. Research on fossil arthropods, including trilobites, horseshoe crabs, eurypterids and scorpions and their mutual relationships; stratigraphy of Paleozoic, especially Ordovician. Home: 2 Mogens Thorsens gt. Oslo, Norway.*

STORMONT, Clyde Junior, Am. geneticist; b. Viola, Wis., June 25, 1916; s. Clyde James and Lulu Elizabeth (Mathews) S.; B.A. in Zoology, U. Wis., 1938, Ph.D. in Genetics, 1947; m. Marguerite Butzen, Aug. 31, 1940; children—Bennie Lu, Michael Clyde, Robert Thomas, Charles James, Janet Jean. Faculty, U. Wis., Madison, 1946-50, asst. prof., 1947-50; faculty U. Cal. at Davis, 1950—, prof. genetics, 1959—, dir. serology lab., 1955—, acting dir. exptl. animal diseases research lab., acting chmn. dept. vet. microbiology, 1967—. Mem. Am. Assn. Immunologists, Am. Naturalists, Am. Soc. Human Genetics, Genetics Soc. Am., Soc. for Exptl. Biology and Medicine, Sigma Xi. Research and numerous publs. on genetic systems of blood groups in domestic cattle, domestic sheep, horses, Am. buffalo (bison); application of blood typing tests in forensic medicine. Home: 836 Douglass Av., Davis, Cal. 95616.*

STORMS, Lowell Hanson, Am. psychologist; b. Schenectady, Feb. 14, 1928; s. Charles Arba and Afton (Hanson) S.; B.A., U. Minn., 1950, M.S., 1951, Ph.D., 1956; m. Marilyn Isabel Hitchcock, June 16, 1955; children—Christopher Alan, Karen Afton, Bruce Lowell. Clin. psychologist Hastings (Minn.) State Hosp., 1954-56; Fulbright scholar U. London Inst. Psychiatry, 1956-57; psychologist, faculty U. Cal. at Los Angeles Neuropsychiat. Inst., 1957—, supervising psychologist, asso. prof., 1964—. Cons., V.A., 1964—; lectr. Cal. State Coll., 1965-66. Mem. A.A.A.S., Am., Western, Cal. psychol. assns., Phi Beta Kappa. Contbr. articles to profl. jours. Demonstrated sequences of mediators facilitating learning, situational factors affecting assn. strength, punishment effects and recovery related to intensity and duration of shock; conditions affecting schizophrenic disorganization. Home: 935 Wellesley Av., Los Angeles 90049.*

STORSTEIN, Ole, Norwegian physician; b. Bergen, Norway, July 16, 1909; s. Arne O. and Karoline (Stadheim) S.; M.D., Oslo U., 1936; Ph.D., U. Bergen, 1953; m. Ragnhild Eldbjorg Langva, Mar. 27, 1937; 1 dau., Liv (Mrs. Diethelm Spilker). Dist. physician, Skaanland, Norway, 1937-39; asso. prof. medicine U. Bergen, 1950-52; head physician Stokmarknes, Norway, 1952-54; head physician Bodo Hosp., Norway, 1954-59; asso. prof. cardiology Oslo U., 1959-63, prof., 1963—; research prof. Yale Sch. Medicine, 1957-58, Recipient Giornate Medici medal, 1963. Mem. Brit. Cardiac Soc. Author: The Effect of Pure Oxygen Breathing in Anoxemia, 1952. Mem. editorial bds. Am. Heart Jour., Acta Medica Scandinavica. Contbr. Research, numerous publs. on cardiopulmonary physiology, treatment of cardiac diseases. Home: 5 Noreveien St. Office: Rikshospitalet, Oslo, Norway.*

STORY-MASKELYNE, Mervyn Herbert Nevil, English, mineralogist; b. Wroughton, Eng., Sept. 3, 1823; B.A., Wadham Coll., Oxford, 1845, M.A., 1849, D.Sc. (hon.), Oxford, 1903. Became tchr. mineralogy and chemistry Oxford U., 1851, prof. mineralogy, 1856-95; keeper minerals Brit. Mus., 1857-80, rearranged and enlarged collections, also pub. catalogue, 1853, pub. guide, 1868; named hon. fellow Wadham Coll., Oxford, 1873; mem. Parliament, 1880-92. Recipient Wollaston medal, 1893. Fellow Royal Soc., 1870 (v.p. 1897-99), Geol. Soc. Author: Morphology of Crystals, 1895. Research on meteorites, diamond; introduced term solute to describe solid part of solution. Died Wroughton, May 20, 1911.

STORZ, Johannes, virologist; b. Hardt Rottweil, Germany, Apr. 29, 1931; s. Johannes and Therese Klausmann) S.; D.V.M., Vet. Coll., Hannover, Germany, 1957; Dr.med.vet., U. Munich, 1958; Ph.D., U. Cal. at Davis, 1961; m. Hannelore Roeber, Aug. 8, 1959; children—Gisela Therese, Johann Peter Konrad. Research asso. Fed. Research Inst. for Viral Diseases of Animals, Tübingen, Germany, 1957-58; lectr. U. Cal. at Davis, 1958-61; asst. prof., asso. prof. Utah State U., Logan, 1961-65; asso. prof. virology Colo. State U., Ft. Collins, 1965—. Mem. Am. Soc. for Microbiology, A.A.A.S., Am. Vet. Med. Assn., Conf. Research Workers in Animal Diseases, World Assn. for Buiatrics. Research, publs. on viral infections and diseases in animals, viral infections of pregnant animals and fetus, identification of cause of epizootic bovine abortion, demonstrator of significance of psittacosis agents as cause of polyarthritis in animals, indentification of eye diseases in animals caused by psittacosis agts., antigenic properties of these agents as related to their disease producing capacity, identification of viruses as causes of enteric disease in newborn animals. Home: 1300 Springfield Dr., Ft. Collins, Colo. 80521.*

STOTLAND, Ezra, Am. psychologist; b. N.Y.C., June 9, 1924; s. Isaac and Rose (Chaiken) S.; B.S. in Social Sci., City Coll. N.Y., 1948; M.A. in Psychology, U. Mich., 1949, Ph.D. in Social Psychology, 1953; m. Patricia H. Joyce, July 12, 1963; step-children—Barbara Hilyer; one daughter, Sheila Rose. Research asso. Research Center for Group Dynamics, U. Mich., 1953-56; research asso., lectr. U. Mich., 1956-57; faculty U. Wash., Seattle, 1957—, prof. psychology, 1965—. Cons., VA, 1959—. Fellow Am. Psychol. Assn., Soc. for Psychol. Study Social Issues. Author: (with A. Zander, A.R. Cohen) Role Relations in the Mental Health Professions, 1957; (with A. Kobler) The End of Hope, 1964, Life and Death of a Mental Hospital, 1965; also articles. Research on identification, perceived similarity between people, attitude theory and attitude change, social power, effects of social interaction, empathy, suicide, cognitive influences on anxiety, factors in viability orgns., social influence on mental patients, learning sense competence, self-esteem, birth order, authoritarian personality. Office: Dept. Physiology, U. Wash., Seattle.*

STOTZ, Elmer Henry, Am. biochemist; b. Boston, July 29, 1911; s. Herman Gustav and Mary (Hoch) S.; B.S., Mass. Inst. Tech., 1932; Ph.D., Harvard, 1936; m. Doris Mary White, Aug. 22, 1936; children—Cynthia, Gretchen, Lydia, Jonathan, Diana. Buhl fellow U. of Pittsburgh, 1936-37; research fellow U. of Chicago, 1937-38; dir. labs. McLean Hosp., Waverly, Mass., 1938-43; asso. in biochemistry Harvard Med. Sch., 1938-43; prof. agrl. biochemistry and head dept. food sci. and technology Expt. Sta., Cornell, 1943-47; prof. biochemistry and chmn. dept. U. of Rochester (N.Y.) Sch. of Medicine and Dentistry, 1947—; consultant in biochemistry. Trustee of Associated Universities, Incorporated. Fellow A.A.A.S.; mem. Am. Soc. Biol. Chemists (sec.), Am. Chem. Soc., Biol. Stain Commn. (treas., mem. bd. trustees), Internat. Union of Biochemistry, Sigma Xi. Research on enzymes involved in heart muscle oxidations; isolation, re-combination of cytochromos; processes involved in metabolism alcohol; evaluation of stains used in pathological examinations. Home: 840 East Av., Rochester, N.Y. 14607.*

STOUGHTON, Richard Baker, Am. physician; b. Duluth, Minn., July 4, 1923; s. Edward and Edna S.; B.S., U. Chgo., 1945, M.D., 1947; m. Gwendolen S. Schmidt, June 10, 1946; 1 son, Roland. Faculty, U. Chgo., 1950-56; asso. prof., dir. dermatology, dept. medicine Western Res. U., Cleve., 1957-67; practice medicine specializing in dermatology in Chgo., 1950-56, Cleve. 1956-67; head div. dermatology Scripps Clinic and Research Found., La Jolla, Cal., 1967—. Chmn. tng. grants com. dermatology Dept. Health Edn. Welfare, 1962-65; mem. commn. on cutaneous disease, Armed Forces Epidemiology Bd., 1962—; mem. med. research div. Nat. Acad. Sci. Chief editor Jour. Investigative Dermatology, 1967—.

Contbr. numerous articles on dermatology to med. jours. Home: Box 1264, Rancho Santafe, Cal. 92067. Office: 476 Prospect St., La Jolla, Cal. 92037.

STOUT, Arlow Burdette, Am. botanist; b. Jackson Center, O., Mar. 10, 1876; s. Hezekiah Milton and Harriet (Bond) S.; grad. State Normal Sch., Whitewater, Wis., 1903; A.B., U. Wis., 1909; Ph.D., Columbia, 1913; m. Zelda Judd Howe, June 22, 1909; children—Elizabeth Bond (Mrs. Herman Rausch), Arlow Burdette. Instr. botany U. Wis., 1909-11; dir. labs. N.Y. Bot. Garden, 1911-38, curator edn. and labs., 1938-47, emeritus. Recipient Roland medal Mass. Hort. Soc., William Herbert medal Am. Amarylis Soc.; Gold medal Hort. Soc. N.Y.; Bertram Farr award Am. Hemerocallis Soc.; Distinguished Service award N.Y. Bot. Gardens. Fellow A.A.A.S., N.Y. Acad. Scis.; mem. Soc. Am. Naturalists, Am. Hemerocallis Society, Baraboo Wis. Hist. Soc., Bot. Soc. Am. Torrey Bot. Club, Am. Soc. Hort. Sci., Am. Amaryllis Soc., Am. Genetic Assn., Genetics Soc. Am., Ohio State Hist. and Archaeol. Soc., Wis. Archaeol. Soc., Hort Soc. N.Y. (hon. life mem.), Pa., Royal hort. socs., Phi Beta Kappa, Sigma Xi. Spl. fields sci. research, genetics, plant breeding, cytology, sterilities in flowering plants. Died Oct. 12, 1957.

STOUT, Arthur Purdy, Am. physician; b. N.Y.C., Nov. 30, 1885; s. Joseph Suydam and Julia Frances (Purdy) S.; B.A., Yale, 1907; M.D., Columbia, 1912; m. Jean Stoddart, June 22, 1914; 1 dau., Julia Frances. Faculty Columbia, 1914—, prof. surgery, 1947-51, prof. pathology, 1950-54, prof. emeritus, 1954—; attending surg. pathologist Presbyn. Hosp., N.Y.C., 1914-51, cons., 1951—; pathologist Francis Delafield Hosp., N.Y.C., 1950-54, cons., 1954—. Consulting pathologist to hosps.; chmn. subcom. on oncology NRC, 1954-65. Recipient Am. Cancer Soc. medal, 1951; Am. Radium Soc. medal, 1952; Clement Cleveland medal N.Y.C. div. Am. Cancer Soc., 1953; James Ewing Soc. medal, 1957; Alumni medal Coll. Phys. and Surg., Columbia, 1964. Fellow Royal Soc. Arts, Royal Geog. Soc.; mem. Century Assn., N.Y. Cancer Soc. (past pres.), Halsted Soc. (past pres.), hon. mem. pathology socs. in Argentina, Mexico, France, Colombia, Belgium, Can. Author: Human Cancer, 1932; Atlas of Tumor Pathology, Fascicle on Tumors of the Peripheral Nervous System, 1949; Fascicle on Tumors of the Stomach, 1953; Fascicle on Tumors of the Soft Tissues, 1953; (with R. Lattes) Fascicle on Tumors of the Esophagus, 1957; (with P. B. Hudson) An Atlas of Prostatic Surgery, 1962. Research on behavior and recognition of soft tissue tumors, especially in children; developed (with M. R. Murray) methods of distinguishing malignant from benign tumors. Home: 157 E. 72d St., N.Y.C. 10021. Office: 630 W. 168th St., N.Y.C. 10032.*

STOUT, George Frederick, English psychologist; b. S. Shields, Eng., Jan. 6, 1860; s. George and Eliza (Frankland) S.; M.A., St. John's Coll., Cambridge (Eng.) U.; LL.D. (hon.), U. Aberdeen, 1899 St. Andrews U. 1937; D.Litt. (hon.), Durham U., 1923; m. Ella Ker, 1899; 1 son. Univ. lectr. moral scis. Cambridge U., 1894; fellow St. John's Coll., 1884 Anderson lectr. comparative psychology Aberdeen U., 1896-98; Wilde reader in mental philosophy U. Oxford, 1898; prof. logic and metaphysics St. Andrews U., 1903-36; retired, 1936. Author: Analytic Psychology, 1896; Manual of Psychology, 2 vols., 5th edit., 1938; Groundwork of Psychology, 1903; Studies in Philosophy and Psychology, 1930; Mind and Matter (Gifford lectures), 1931; God and Nature, 1952; also numerous articles in Mind and in proc. Aristotelian Soc. Studied nature of consciousness, relation between mind and matter; man's knowledge of material world. Died Sydney, Australia, Aug. 8, 1944.

STOUT, John Willard Jr., Am. phys. chemist; b. Seattle, Mar. 13, 1912; s. John Willard and Beatrice (Edson) S.; student Cal. Poly., 1925-30; B.S., U. Cal., Berkeley, B.S., 1933, Ph.D., 1937; m. Florence Louisa Parsons, Mar. 20, 1948; children—John Edward. Instr. chemistry, research asso. U. Cal., Berkeley, 1937-39, investigator NDRC, 1941-44; instructor chemistry Mass. Inst. Tech., 1939-41; group leader Manhattan Dist., Los Alamos, 1944-46; asso. prof. chemistry U. Chgo., 1946-54, prof., 1954—. Dir. Calorimetry Conf., 1953-55. Lalor fellow, 1938-39. Mem. Am. Phys. Soc., Am. Chem. Soc., A.A.A.S., Phi Beta Kappa, Sigma Xi, Tau Beta Pi, Chi Pi Sigma. Asso. editor; Phys. Rev., 1958-60; asso. editor Jour. Chem. Physics, 1954-56, editor, 1959—; editorial bd. Advances in Chem. Physics, 1964—. Research on magnetism, thermodynamics, electronic energy levels in crystals, cryogenics. Home: 5417 Greenwood Av., Chgo. 60615.*

STOUT, William Bushnell, Am. aero. engr.; b. Quincy, Ill., Mar. 16, 1880; s. James Frank and Mary L. (Bushnell) S.; ed. Hamline U., 1899-1900; U. Minn., 1901-02; m. Alma E. Raymond, June 16, 1906; 1 dau., Wilma Frances. Tech. and aviation editor Chgo. Tribune, 1912; joined staff Motor Age and Automobile; founded Aerial Age; moved to Detroit, 1914, as chief engr. Scirpps-Booth Co., advt. mgr., 1915, gen. sales mgr., 1916; apptd. chief engr. aircraft div. Packard Motor Car Co., 1917; during devel. of Liberty engine. Apptd. tech. adv. to Aircraft Bd., Washington,; built for bd. the 1st internally braced cantilever airplane in Am., veneer and wood constrn.; founded Stout Engring. Labs. and built 1st Am. comml. monoplane, known as Batwing, flown at Selfridge field, 1919; undertook contract, 1920, for 1st metal plane built in U. S., and all metal torpedo plane for U. S. Navy, flown by Eddie Stinson, Selfridge field, 1922; formed, 1922, Stout Metal Airplane Co. to build comml. metal planes; built air sedan, 3 seater cabin plane, all metal, and later Liberty engine eight passenger transport; sold Stout Metal Airplane Co. to Ford Motor Co., 1925, and served as v.p., gen. mgr. during devel. of Ford tri-motored transport plane from original single engine transport; founder Stout Air Services, 1926, operating 1st exclusive passenger airplanes in U. S., Detroit to Grand Rapids; on devel. of tri-motors, transferred line, Detroit to Cleve., 1927, and added the Detroit to Chgo. route; Stout Air Lines sold, 1929, to United Aircraft and Transport Co., N.Y.; Stout Engring. Labs. revived, 1929, for research and devel. in aeros.; built and developed all-metal Sky Car, new type airplane for pvt. owner use; designed and built under a contract with The Pullman Car and Mfg. Corp., a highspeed Railplane as basis of change in r.r. passenger work; developed new fibreglass automobile with engine in rear, at Graham-Paige Motors Corp., Willow Run, Mich. Commr. Mich. Aero. Commn. Dir. Nat. Aero. Assn. Past pres. Soc. Automotive Engrs., Inst. Aero. Scis.; mem. Detroit Aviation Soc. Died Mar. 20, 1956.

STOVER, Betsy Jones, Am. chemist; b. Salt Lake City, May 13, 1926; d. Richard Hugh and Bessie (Miers) Jones; A.B., U. Utah, 1947; Ph.D., U. Cal. at Berkeley, 1950; m. Clarence Nathan Stover, June 27, 1950; children—Susan, Steven Nathan. Faculty dept. chemistry U. Utah, Salt Lake City, 1950—, asso. research prof., 1958—, head chemistry, div. radiobiology dept. anatomy, 1950—. Former mem. subcom. 2 Nat. Com. on Radiation Protection. Mem. Am. Chem. Soc., Am. Phys. Soc., Radiation Research Soc., Phi Beta Kappa, Sigma Xi, Phi Kappa Phi, Alpha Lambda Delta. Author: (with Henry Eyring, Douglas Henderson, Edward Eyring) Statistical Mechanics and Dynamics, 1964; also articles. Application of chem. methods to biol. effects of internal irradiation. Home: 3400 Crestwood Dr., Salt Lake City 84109.*

STOW, Marcellus H(enry), Am. geologist; geology; b. Washington, May 19, 1902; s. James Warren and Lizzie R. (Miller) S.; A.B., Cornell U., 1926, A.M., 1927, Ph.D., 1931; m. Grace Wilhelmina Hammond, July 1, 1932. Asst. in geology Cornell U., 1924-26, instr., 1926-27, asst. prof., summers 1929-31; asst. prof. geology Washington and Lee U., 1927-34, asso. prof. and acting head dept. geol., 1934-37, Robinson prof. head of dept., 1937—, Thomas Ball prof., 1947—; on leave of absence as dept. dir. mining div. WPB, 1942-1946; chief mining br. Civilian Prodn. Adminstrn., 1945; with U. S. Geol. Survey, summers 1923, 24, 26; Lehigh Valley Coal Co., summer 1925; field geology and research Yellowstone-Bighorn Research Assn., summers 1933—; geol. cons. U. S. AEC, 1953—; geol. cons. to dir. Va. Dept. of Conservation and Devel. Fellow Geol. Soc. Am., A.A.A.S.; mem. Am. Assn. Petroleum geologists, Mineral Soc. Am., Am. Geophys. Union, Am. Inst. Metall. Engrs., Va., Washington acads. sci., Geochem. Soc., Yellowstone Bighorn Research Assn. (council 1935-44, pres. 1939-41, 44-45); mem. adv. bd. Va. Fisheries Lab., So. Assn. Sci. and Industry (exec. com.), NRC, So. Research Inst., Soc. Am. Mil. Engrs., Soc. Econ. Paleontology and Mineralogy, Sigma Xi, Phi Beta Kappa. Author: Mineral Resources and Mineral Industry of Virginia. Editor: The James River Basin—Past, Present and Future. Contbr. to geol. jours. Died Nov. 27, 1957.

STOWE, Emily Jennings, Canadian physician; b. South Norwich, Ont., Can., 1831; student Provicial Normal Sch., Toronto; grad. N.Y. Med. Coll. for Women, 1867; m. John Stowe; schooltchr.; admitted to Ont. Coll. Physicians and Surgeons, 1880; 1st woman officially permitted to practice medicine in Can.; a founder Toronto Woman's Lit. and Sci. Club, 1877; organized Dominion Woman Suffrage Assn., 1893, elected pres. Died 1903.

STOWELL, Elbridge Zebina, Am. aero. engr.; b. Arlington, Mass., Aug. 30, 1900; s. Elbridge W. and Nellie (Pangborn) S.; B.S., Tufts U., 1921; M.S., U. Neb., 1923; Ph.D., Am. U., 1927; m. Mary A. Gregory, Dec. 25, 1927. Physicist, U.S. Bur. Standards, 1923-27, Fed. Telegraph Co., Palo Alto, Cal., 1927-30; aero. scientist NACA, Langley Field, Va., 1932-59; inst. scientist S.W. Research Inst., San Antonio, 1959—; mem. Russian transl. panel Pergamon Press, 1960—. Named Inst. Scientist, Bd. govs. S.W, Research Inst., 1961. Mem. Phi Beta Kappa, Sigma Xi. Research and publs. on theory structural stability in elastic and plastic ranges of material, stress concentration in plastic range, behavior of metals at elevated temperatures, phenomenon of creep buckling, theory metal fatigue. Office: S.W. Research Inst., San Antonio 78228.*

STOWENS, Daniel, Am. pathologist; b. N.Y.C., Oct. 27, 1919; s. Oscar and Rose (Galkin) S.; A.B., Columbia U., 1940, M.D., 1943; m. Barbara J. Hagmann, Sept. 28, 1944; children—Daniel W., Christopher. Registrar, Am. Registry of Pediatric Surgery, 1954-58; pathologist, asso. prof., U. S.C., 1958-60; asso. prof. pathology U. Louisville, also dir. labs. Children's Hosp., Louisville, 1961-65; pathologist St. Lukes Memorial Hospital Center, Utica, New York, since 1966—. Recipient Gold medal Am. Soc. Clin. Pathology, 1958, Bronze medal, 1962; award for sci. paper Am. Coll. Gastroenterologists, 1959. Mem. A.M.A., Am. Assn. Pathologists, and Bacteriologists, Internat. Acad. Pathology, Am. Soc. Clin. Pathologists, Am. Acad. Pediatrics, Soc. for Pediatric Research, A.A.A.S., N.Y. Acad. Scis. Author: Pediatric Pathology, 1959; also numerous articles. Research on morbid anatomy childhood diseases particularly Wilms' tumors, neuroblastoma, leukemia, hyaline membrane disease, dermatoglyphics in genetics. Home: Fountain St., Clinton, N.Y. 13323. Office: St. Luke's Meml. Hosp. Center, P.O. Box 479, Utica, N.Y. 13503.*

STRABO, Greek geographer and historian; b. Amasia in Pontus, circa 63 B.C.; studied at Nysa under grammarian Aristodemos; studied geography under Tyrannion; philosophy under Xenarchos; also studied Aristotle, and knew Posidonios. Traveled extensively; went to Corinth, circa 29 B.C.; ascended Nile, circa 24 B.D., Upper Egypt, circa 25-24 B.C.; his early work apparently hist. in nature (sketches have been lost); his geography based on his own travels in addition to work of his predecessors including Homer, Eratosthenes, Polybios and Posidonios; inaccuracies are found for several reasons (he depended on Homer as factual source and disregarded work of Herodotos as well as Roman writers); his math. methods often faulty; his work on geography is 1st attempt to collect all geog. knowledge of time; he recognized Vesuvius as volcano (it erupted some 50 years after his death); discussed river action in connection with land formation; suggested for 1st time existence of unknown continents; his Geographica divided into 17 books (2 introductory, 8 on Europe, 6 on Asia and 1 on Africa); attempted to include math., phys., hist. and polit. geography; also wrote vast hist. work which is lost. Died circa 19 A.D.

STRACHEY, John, Eng. geologist; b. Eng., 1671; s. John Strackey; m. Elizabeth Elletson; m. 2d, Christiana Staveley. Became fellow Royal Soc., 1719. Author: Observations on the Different Strata of Earths and Minerals, 1727. First to suggest theory of stratification. Died June 11, 1743.

STRACK, Erich Martin Emil, German biochemist; b. Wollin (Pomerania), Germany, Nov. 20, 1897; s. Emil Gustav and Martha (Steffen) S.; student univs. Berlin, Würzburg; Dr.med., U. Greifswald, 1924, Diplom. Chem., 1927, Dr.phil., 1928; Dr.med.h.c., U. Leipzig, 1965, U. Rostock, 1967; m. Christa Gaul, May 9, 1962; 1 dau., Uta. Faculty, U. Leipzig, 1929—, prof., dir. Inst. Physiol. Chemistry, 1948-65, 1948-65, emeritus, 1962—; research Saxon Acad. and U. Leipzig, 1965—. Recipient Nat. prize, German Democratic Republic, 1960. Mem. Saxon Acad. Scis., Leopoldina, German Soc. Scientists and Physicians, German Physiol. Chem. Soc., German Soc. Nutrition, Biochem. Soc., Chem. Soc., Soc. Clin. Medicine. Research and numerous publs. in physiology, biochemistry, clin. medicine; bacterial dyes; animal bases; carnitin; waste water problems. Address: 19 Philipp-Rosenthalstrasse, 701 Leipzig, East Germany.*

STRADINS, Janis, Latvian chemist; b. Riga, Latvia, Dec. 10, 1933; s. Paul and Nina (Malysheva) S.; diploma in chemistry Latvian State U., 1956; Cand. Chem., Moscow State U., 1960; m. Laima Zutere, 1962; 1 son, Paul. Sci. collaborator Inst. for Forest Chemistry, Acad. Scis. of Latvia, Riga, 1956-57, Inst. Organic Synthesis, 1957-61, leader Lab. for Phys. Organic Chemistry, 1961—; lectr. history of chemistry Latvian State U., 1956—; leader dept. Paul Stradin Mus. History of Medicine, 1957-58. Mem. Commn. Electroanalytical Chemistry, Internat. Union Pure and Applied Chemistry, 1963—, USSR nat. rep., 1964—. Mem. Mendeleyev Soc. Chemists, Internat. Soc. History Medicine, Latvian Assn. Historians of Scis. and Tech. (v.p.), Author: Polarography of Organic Nitrocompounds (in Russian), 1961; Men, Experiments, Ideas (in Latvian), 1965; Theodor Grotthus (in Russian), 1966; Methods of Analysis in Nitrofuran Derivatives (in Russian) 1968. Editor: Paul Stradin's Selected Works, Vols. I-III (in Russian), 1963-65. Research, numerous publs. on polarography of organic compounds, mechanism of electroreduction, correlations between chem. structure and polarographic parameters, metabolism of nitrofuran remedies in human and animal body, electrochem. generation of nitrofuran free radicals and their ESR-spectra, history of chemistry especially in Baltics. Home: 19 Ventspils. Office: 21 Aizkraukles, Riga, Latvia, USSR.*

STRAFELDA, Frantisek, Czechoslovakian chemist; b. Suché Vrbné, Czechoslovakia, Apr. 22, 1925; s. Frantisek and Rozalie (Bláhová) S.; Ing., Inst. Chem. Tech., Prague, Czechoslovakia, 1950, Dr.cand.science, 1952, docent, 1960; postgrad. Tech. U. Dresden, Germany, 1956, Faculté des Sciences, Paris, 1964; m. Heda Machová, June 12, 1950; children—Petr, Helena. Faculty, Inst. Chem. Tech., Prague, 1946—, prof., 1966—, head dept. instrumental analysis, 1960—, head Faculty Inorganic Tech., 1960-64. Mem. adv. bds. several research and prodn. instns. Mem. Czechoslovak Chem. Soc. Research, publs., patents in quantitative interpretation of diffusion transport and relied electrode-phenomena on various stationary electrodes in flowing solutions and various moving electrodes in stationary solutions, electrochem. methods and instruments of automatic lab. and plant analysis. Home: 58/766 Nad Sárkou, Prague, Czechoslovakia.*

STRAHLER, Arthur Newell, geologist; b. Kolhapur, India, Feb. 20, 1918 (parents Am. citizens); s. Milton W. and Harriet (Brittan) S.; A.B., Coll. Wooster, 1938; A.M., Columbia, 1940, Ph.D. (Univ. fellow 1940-41), 1944; m. Margaret E. Wanless, Aug. 10, 1940; children—Alan H., Marjorie E. Mem. faculty Columbia, 1941—, prof. geomorphology, 1958—, chmn. dept. geology, 1959-62. Fellow Geol. Soc. Am, Am. Geog. Soc.; mem. Am. Geophys. Union, Phi Beta Kappa, Sigma Xi. Author: Physical Geography, rev. edit., 1960; The Earth Sciences, 1963; Introduction to Physical Geography, 1965. Research on quantitative statistical analysis slopes, drainage systems; terrain analysis. Home: 224 Warwick Lane, Leonia, N.J. Office: Dept. Geology, Columbia U., N.Y.C. 10027.*

STRAIN, Harold Henry, Am. chemist; b. San Francisco, Jan. 12, 1904; s. Everett R. and Mary I. (McCurdy) S.; A.B., Stanford U., 1923, A.M., 1924, Ph.D., 1927; m. Grace J. Talbott, Sept. 5, 1929; children—Bonnie Lee, Carol Jean (Mrs. Richard Lee Besore). Chemist, Libby McNeill and Libby, 1924; staff mem. Carnegie Instn. of Washington, coastal lab. Carmel, Cal., 1927-29, dept. plant biology Stanford, Cal., 1929-49; sr. chemist Argonne (Ill.) Nat. Lab., 1949—. Lectr. Ill. Inst. Tech., 1962; sr. research fellow Australian Acad. Sci., 1963. Recipient award in Chromatography and Electrophoresis Am. Chem. Soc., 1961, Midwest award St. Louis sect., 1964. Rockefeller Found. fellow, Copenhagen, 1937-38. Mem. Am. Chem. Soc., Sigma Xi, Phi Lambda Upsilon. Author: Leaf Xanthophylls, 1938; Chromatographic Adsorption Analysis, 1942; Chloroplast Pigments and Chromatographic Analysis, 1958. Publs. on invention and uses of chromatographic methods for resolution chem. mixtures; isolation of chloroplast pigments and determination of their properties related to photosynthesis and to evolution and classification of plants. Home: 5628 Plymouth Ct., Downers Grove, Ill. 60515. Office: 9700 S. Cass Av., Argonne, Ill. 60439.*

STRAIT, Louis Avrom, Am. biophysicist; b. Denver, Sept. 29, 1907; s. Abraham and Dora (Arbeitsman) S.; A.B., magna cum laude, U. Colo., 1930, M.A., 1932; Ph.D., U. Cal. at Berkeley, 1937; m. Elsa Barber, Oct. 13, 1934; 1 son, John Avrom. Faculty, U. Cal. at San Francisco, 1937-47, asso. prof. biophysics Coll. Pharmacy, dept. biophysics, 1945-47; faculty spectrographic lab. Coll. Pharmacy, U. Cal. Med. Center, San Francisco, 1947—, prof. biophysics, 1952—, cons. div. dermatology dept. medicine, 1950—. Cons. to hosps. Recipient Ebert award Am. Pharm. Assn., 1949. Mem. Am. Soc. Bioanalysts (trustee Herbert C. Johnston Meml. Fellowship fund), Am. Soc. for Testing and Materials (mem. com.), Am. Assn. Bioanalysts, A.A.A.S., Am. Chem. Soc., Am. Inst. Physics, Biophys. Soc., Coblentz Soc., Optical Soc. Am., Soc. for Applied Spectroscopy, N.Y. Acad. Scis., No. Cal. Soc. Spectroscopy. Research and publs. on theory and application of electronic, infrared and emission spectroscopy; correlation of molecular structure and chem. properties with spectra; application to biol. and med. problems including porphyria, cancer, convulsive states, spontaneous myohemoglobinuria. Home: 204 Grand View Av., San Francisco 94114.*

STRAITON, Archie Waugh, Am. elec. engr.; b. Arlington, Tex., Aug. 27, 1907; s. John and Jeannie (Waugh) S.; B.S., U. Tex., 1929, M.A., 1931, Ph.D., 1939; m. Esther McDonald, Dec. 28, 1932; children—Janelle (Mrs. Henry Holman), Carolyn (Mrs. Stanley Cameron). Staff engr. Bell Telephone Labs., 1929-30; faculty Tex. Coll. Arts and Industries, Kingsville, 1931-43; faculty U. Tex., Austin, 1943—, prof. elec. engring., 1947-63, Asbel Smith Prof., 1963—, dir. Elec. Engring. Research Lab., 1947—, chmn. dept. elec. engring., 1966—. Bd. dirs., v.p. Gulf Univs. Research Corp., 1964—; mem. council Univ. Corp. for Atmospheric Research, 1964—. Fellow I.E.E.E.; mem. Am. Soc. Engring. Edn., Am. Geophys. Union, Sigma Xi, Tau Beta Pi, Phi Kappa Phi, Eta Kappa Nu. Contbr. chpts. to books, numerous articles to profl. jours. Research in transmission of radio waves through atmosphere, devel. of millimeter wave spectrum. Home: 4212 Far West Blvd., Austin, Tex. 78731.*

STRAKHOV, Nikolai Mikhailovich, Russian geologist; b. Apr. 15, 1900; grad. Moscow U., 1928. Became lab. head Geol. Inst., USSR Acad. Scis., 1934; became mem. main editorial staff Bol'-shaya Sovetskaye Entsykl, 1953; asso., then dept. head Inst. Geology, USSR Acad. Scis., 1934-46; geologist So. Urals, 1928-34; prof. Moscow Geol. Survey Inst., 1930—; prof. Moscow U., 1955—. Recipient Lenin Prize, 1961, Stalin Prize, 1948, Red. Banner of Labor, 1960, Order of Lenin. Mem. USSR Acad. Scis. (editorial bd. Reports 1960—). Author: The Domanik Facies of the Southern Urals, 1939; Iron Ore Facies and their Analogies in the History of the Earth, 1947; Principles of Historical Geology, 1948; Calcareous Dolomite Facies in Modern and Ancient Reservoirs, 1951; Sedimentation in Modern Reservoirs, 1954; Theoretical Lithology and its Problems, 1957; Theoretical Principles of Lithogenesis. Research and publs. on diagensis in formation of sedentary rock, lime and iron ore; devel. comparative analysis in lithology.

STRALEY, H. W. III, Am. geologist, geophysicist; b. Princeton, W.Va., 1906; s. H. W. II and Rosa Lee (Walthall) S.; B.S. Concord Coll., 1925; postgrad. U. Chgo., 1928-33, U. N.C., 1934-38; m. Garnet Brammer, July 19, 1928; children—H. W. IV, William F. Faculty, Ga. Inst. Tech., Atlanta, 1949—, now prof.; mem. Tech. Mission to Cuba, 1943; del. to Internat. Geol. Congress, Copenhagen, 1960; cons. Lockheed Aircraft Co., Union Carbide Co. Pres. H. W. Straley; dir. Lilly Land Co. Mem. Am. Assn. Petroleum Geologists, Am. Geophys. Union, Am. Inst. Mining Engrs., Yorkshire (Eng.), Carolina geol. socs., Geol. Soc. Am., Soc. Exploration Geophysicists. Contbr. numerous articles to profl. jours. Research on folding, especially Appalachian, classification of folds, magnetic mapping of the Atlantic coastal plain; world mineral resources, tectonics. Home: 1635 W. Wesley Rd. N.W., Atlanta 30327.*

STRANATHAN, James Docking, Am. physicist; b. Kansas City, Mo., Nov. 2, 1898; s. Samuel W. and Bertha (Docking) S.; B.S., U. Kan., 1921, M.S., 1924; Ph.D., U. Chgo., 1928; m. Lena P. Monroe, Sept. 6, 1921; children—Ona Fern (Mrs. Eugene V. Nininger), Mary Maude (Mrs. Arthur Beebe). Mem. faculty U. Kan., 1920—, prof., 1934—, chmn. dept. physics, 1941-64, univ. rep. Argonne Nat. Lab., 1964. Cons. physicist, 1936—; asso. editor Am. Jour. Physics, 1954-56. Fellow Am. Phys. Soc., A.A.A.S.; mem. Am. Assn. Physics Tchrs., Kan. Acad. Sci. Author: The Particles of Modern Physics, 1942; also articles. Research on vapor, liquid and solid dielectrics, electric moments of molecules, electrets. Home: 1510 Crescent Rd., Lawrence, Kan. 66044.*

STRAND, Kaj Aage Gunnar, astronomer; b. Hellerup, Denmark, Feb. 27, 1907; s. Viggo Peter and Constance (Malmgren) S.; B.A., M.Sc., U. Copenhagen (Denmark), 1931, Ph.D., 1938; m. Emilie Rashevsky, June 10, 1949; children—Kristina Ragna, Constance Vibeke. Geodesist, Royal Geodetic Inst., Copenhagen, 1931-33; asst. to dir. U. Obs., U. Leiden, The Netherlands, 1933-38; mem. faculty Swarthmore (Pa.) Coll., 1938-46, U. Chgo., 1946-47, research asso., 1947—; prof. astronomy Northwestern U., also dir. Dearborn Obs., 1947-58; dir. astrometry and astrophysics div. U.S. Naval Obs., Washington, 1958-63, sci. dir., 1963—. Vis. prof. astronomy U. Copenhagen, 1954; mem. adv. bd. astronomy NSF, 1953-56, Office Naval Research, 1954-57. Fellow Am. Scandinavian Found., 1938-39, Danish Rask Oersted Found., 1939-40, Guggenheim Found., 1946. Mem. Am. Astron. Soc., Astron. Soc. Pacific, Internat. Astron. Union, Royal Danish Acad. Scis. and Letters, Sigma Xi. Research, publs. and developer high precision reflecting telescope for study of faint stars; discoverer of a series of dark companions in double star systems. Home: 3202 Rowland Pl. N.W., Washington 20008. Office: U.S. Naval Obs., Washington 20390.*

STRAND, Trygve, Norwegian geologist; b. Lillehammer, Norway, Jan. 16, 1903; s. Thor Jonsen and Ragnhild (Petersen) S.; cand.real., U. Oslo (Norway), 1928, dr.philos., 1934; m. Benedicte Gogstad, Mar. 15, 1930; children—Tor Gogstad, Benedicte (Mrs. Viggo Mohr). State geologist Geol. Survey Norway, 1936-56; prof. geology U. Oslo, 1956—. Mem. Norwegian Acad. Arts and Scis. Author: (with O. Külling) The Leandinavian Caledonides; also articles. Research in Cambro-Ordovician paleontology and stratigraphy, geology and structure of Norwegian Caledonides. Home: 10, Min. Ditleffi ver, Oslo 8, Norway.*

STRANDBERG, Malcolm Woodrow Pershing, Am. physicist, educator; b. Box Elder, Mont., Mar. 9, 1919; s. Malcolm and Ingeborg (Reistad) S.; S.B., Harvard, 1941; Ph.D., Mass. Inst. Tech., 1948; m. Harriet Bennett, Aug. 2, 1947; children—Josiah R.W., Susan A., Elisabeth G., Malcolm B. Research asso. Mass. Inst. Tech. Radiation Lab., 1941-42, 1943-45; faculty Mass. Inst. Tech., Cambridge, 1945—, prof. physics, 1960—; scientist RAF, 1942-43; tech. dir. Strand Labs., Inc., 1963. Mem. Nat. Acad. Scis. Panels Adv. to Nat. Bur. Standards, 1964. Fulbright lectr. U. Grenoble, France, 1961-62. Fellow Am. Phys. Soc., Am. Acad. Arts and Scis., A.A.A.S., Inst. Radio Engrs.; mem. Phi Beta Kappa, Sigma Xi. Author: Microwave Spectroscopy, 1954. Studies on microwave instrumentation; microwave spectroscopy and paramagnetic resonance; theory of noise of quantum mech. amplifiers; paramagnetic amplifiers. Home: 295 Harvard St., Cambridge, Mass. 02139.*

STRANDHAGEN, Adolf Gustaf, Am. engr.; b. Scranton, Pa., May 4, 1914; s. Daniel and Theresa (Lylick) S.; B.S., U. Mich., 1939, M.S., 1940, Ph.D., 1942; m. Lucile E. Perry, Aug. 22, 1940; children—Karen Perry, Gretchen Ann. Instr. Carnegie Inst. Tech., 1942-43, asst. prof., 1943-47; asso. prof. U. Notre Dame U., 1947-50, prof., head dept. engring. sci., 1950—. Mem. div. 2 Nat. Def. Research Council, 1943-45. Mem. A.S.M.E., Soc. Naval Architects and Marine Engrs., Sigma Xi, Iota Alpha. Author research articles. Made analytical determination of properties of maneuverability of surface ships; probability of lateral motion instability; pressure signatures of ships; three dimensional motion of towed submerged bodies. Home: 611 Edgewater Dr., South Bend. Office: University of Notre Dame, Notre Dame, Ind.*

STRANDJORD, Nels Magne, Am. radiologist; b. Grenora, N.D., Aug. 18, 1920; s. Selmer J. and Eunice (Langeland) S.; B.A., Luther Coll., 1942; M.D., U. Chgo., 1946; m. Margaret E. Fry, Sept. 10, 1944; children—David Christian, Sarah Eunice, Mark Charles, Daniel Theodore. Gen. practice medicine, Virginia, Minn., 1948-51; instr. radiology, 1958-59, asst. prof., 1959-61, asso. prof., 1961-65; prof., chmn. dept. radiology U. Kan. Kansas City, 1965—. Vis. prof. Nat Def. Med Center, Taipei, Taiwan, 1960-61; mem. physicians team Care-Medico and Dept. State, Algiers, 1962. Recipient James A. McClintock award for outstanding teaching U. Chgo., 1960; Picker scholar in radiol. research, 1959-62. Diplomate Am. Bd. Radiology. Mem. Am. Coll. Radiology, Assn. U. Radiologist, Radiol. Soc. N. Am., Internat. Paleopath. Assn., Chgo. Roentgen Soc., Sigma Xi. Contbr. profl. jours. Home: 2115 W. 61st Terrace, Shawnee Mission, Kan. 66208. Office: U. Kan. Med. Center, Rainbow Blvd. at 39th St., Kansas City, Kan. 66103.*

STRANDSKOV, Herluf Haldan, Am. human geneticist; b. Lindsey, Neb., Nov. 4, 1898; s. Nils Christian and Marie (Sorenson) S.; B.A., Ia. State Tchrs. Coll., 1923; M.S., U. Ill., 1926; Ph.D., U. Chgo., 1931; m. Florence Maltby, July 22, 1950; Faculty, U. Chgo., 1933-64; exchange prof. U. Frankfurt (Germany), 1953-54; vis. prof. U. R.I., Kingston, 1964—. Mem. Am. Soc. Human Genetics, Sigma Xi. Contbr. numerous articles to tech. jours., chpts. to books. Research on human inheritance. Home: Faculty Apts., Kingston, R.I.*

STRANDTMANN, Russell William, Am. acarologist; b. Maxwell, Tex., Apr. 9, 1910; s. Otto and Frieda (Schulz) S.; B.S., S.W. Tex. State Coll., 1935; M.S., Tex. A. and M. U., 1937; Ph.D., Ohio State U., 1944; m. Mary Ruth Chance, Sept. 5, 1936; children—Russell Lamar, Spurgeon Bruce. Prof., Tex. Technol. Coll., 1948-66; fellow entomology Bishop Mus., Honolulu, 1962-63, asst. dir. entomology, 1967—; vis. lectr. Inst. Acarology, summers 1953-57. Mem. Tex. Acad. Sci., Entomol. Soc. Am., Am. Parasitological Soc. Author: The Wasps of the Genus Philanthus North of Mexico, 1948; (with George W. Wharton) Manual of the Mesostigmatic Mites Parasitic on Vertebrates, 1958; also numerous articles. Described nesting habits of some ground nesting digger wasps; elucidated taxonomy of beekilling wasps Philanthus; described and illustrated several new species of mites.*

STRANSKI, Iwan N., physicist, chemist; b. Sofia, Bulgaria, Jan. 2, 1897; s. Nikola and Maria (Krohn) S.; student univs. Vienna, Sofia, Berlin; Dipl.-Chem., Dr.phil.; Dr.rer.nat.h.c., U. Breslau, 1940, Free U. Berlin, 1954; Dr.techn.h.c., Vienna Inst. Tech., Dr.-Ing. E.h., Aachen Inst. Tech., 1964; m. Martha Pötschke, 1926. Faculty, U. Sofia, 1926-44; mem. Kaiser Wilhelm Inst. Phys. Chemistry, Berlin, 1949—; prof., dir. Max Volmer Inst. Phys. Chemistry, Tech. U. Berlin, 1945—, rector, 1951-53, emeritus, 1963—; hon. prof. Free U. Berlin, 1949—; dep. dir. Fritz Haber Inst., Max Planck Soc., 1953—. Recipient Galvani medal, U. Bologna, 1938; Bulgarian Civil Service order, 1939; Polonia restituta, 1939; Service Cross, German Fed. Republic, 1964. Mem. Bavarian, Göttingen (corr.) acads. scis., German (Silver Hoffman medal 1939), Am. chem. socs., German Phys. Soc., German Mineral. Soc. (hon.); fgn. mem. Royal Swedish Soc. Sci. and Lit., Bulgarian Acad. Scis. (Kyrillus-Methodius prize 1940). Research, numerous publs. on growth of crystals, formation of crystal nuclei, evaporation and condensation processes, tribo luminescence. Home: 8 Faradayweg. Office: Fritz Haber Inst., 4-6 Faradayweg, 1 Berlin 33, West Germany.*

STRASBURG, Donald Wishart, Am. ichthyologist; b. Benton Harbor, Mich., Sept. 13, 1924; s. Herman Albert and Edith (Wishart) S.; B.S., U.S. Naval Acad., 1945; Ph.D., U. Hawaii, 1953; m. Donna Louise Derby, Jan. 31, 1952. Asst. to exec. v.p. Whirlpool Corp., St. Joseph, Mich., 1947-48; instr. Duke, 1953-55; ichthylogist Eniwetok Marine Biol. Lab., 1955; fishery research biologist Biol. Lab., U.S. Bur. Comml. Fisheries, Honolulu, 1955-67, chief research submarine program, 1963-67, spl. asst. to mgr. underwater tech. Electric Boat div., Groton, Conn., 1967—. Mem. Am. Ichthylogists and Herpetologists, Am. Soc. Zoologists, Marine Tch. Soc., Sigma Xi. Research, publs. on classification, ecology and biology of tropical marine fishes devel. submarines for research purposes. Home: Brewster Dr., Gales Ferry, Conn. 06335. Office: Electric Boat Div., Groton, Conn. 06340.*

STRASBURGER, Eduard Adolf, botanist; b. Warsaw, Poland, Feb. 1, 1844; ed. U. Paris (France), U. Bonn (Germany); Ph.D., U. Jena (Germany), 1866. Faculty U. Warsaw, 1868; prof. botany U. Jena, 1869-80; prof. U. Bonn. 1880-1912. Fellow Royal Soc., 1891. Author: Über Zellbildung und Zellteilung, 1875; Die Angiospermen und die Gymnospermen, 1879; (with others) Lehrbuch der Botanik, 1894; Chromosomenzahlen, Plasmastrukturen, Vererbungsträger und Reduktionsteilung, 1909. Fundamental contbns. bot. embryology; provided accurate descriptions of embryo sac in gymnosperms; studied life history of angiosperms; pioneer cytology; studied nuclear orgn., inheritance of sex differences in plants, plant cell behavior during mitosis; gave original description 3 consecutive stages mitosis (prophase, metaphase, anaphase); observed chromatic reduction (meiosis) in plants; investigated problems water transport in plants; showed forces moving sap upwards are phys., not physiol., also translocation of sap takes place in lumen of vessels. Died Bonn, May 18, 1912.

STRASSER, Hans, Swiss anatomist; b. Lavenen, Switzerland, May 20, 1852; prof. Freiburg im Breisgau, also Bern, Switzerland. Author: Lehrbuch der Muskel-und Gelenk-Mechanik, 4 vols., 1908-17; Einsteins Relativetätstheorie, eine Komödie der Irrungen, 1923. Application phys. laws to joint and muscle function, thus illuminating symptoms in paralysis; criticized Einstein's theory of relativity. Died Bern, Switzerland, Apr. 16, 1927.

STRASSL, Hans Ludwig, German astronomer; b. Rauischholzhausen, Germany, Jan. 10, 1907; s. Ludwig and Elisabeth (Oster) S.; ed. U. Marburg, U Göttingen; Ph.D.; m. Josefa Schlüter, July 27, 1940; Asst., Göttingen U. Obs.; collaborator Inst. for Aerodynamic Research, Göttingen; observer Astron. Inst., U. Bonn, Germany; dir. Astron. Inst., U. Münster (Germany), 1958——. Mem. Astron. Soc., Internat. Astron. Union. Research and publs. on dynamics of star systems, spectral astron. photometry, nomography in astronomy, radio astronomy. Home: Ochtrupweg 39, Munster-Westfalia, West Germany.

STRASSMANN, Fritz, German phys. chemist; b. Boppard, Germany, Feb. 22, 1902; s. Richard and Julie (Bernsmann) S.; Dr.engring., Hannover Inst. Tech.; m. Marie Heckter, July 20, 1937. Asst., Inst. for Phys. Chemistry, Hannover Inst. Tech.; head chem. dept. Kaiser Wilhelm Inst.; became prof. inorganic and nuclear chemistry, a dir. Chem. Inst., U. Mainz (Germany), 1946; named dir. chemistry dept. Max Planck Inst. for Chemistry, 1953. Mem. Kaiser Wilhelm Soc. Recipient (with O. Hahn and L. Meitner) Fermi awards, AEC, 1966. Research and publs. on nuclear chemistry, nuclear fission, (with Hahn) radioactive isotopes of uranium and thorium. Address: 40 Heideshelmerstr., Mainz-Gosenheim, Germany.

STRATFORD, William Samuel, Brit. astronomer; b. Eng., May 31, 1790; joined Navy, 1806, and went on bd. Pompée; participated in def. of Gaeta, reduction of Capri, passage of Dardanelles, destruction of Turkish squadron off Point Pesquies, bombardment of Copenhagen; served in North Sea, 1809-15; became lt., 1815; apptd. supt. Nautical Almanac, 1831. Recipient Silver medal Astron. Soc., 1827. Fellow Royal Soc., 1832; mem. Astron. Soc. (a founder, became 1st sec. 1820). Author: An Index to the Stars in the Catalogue of Royal Astronomical Society, 1831; On the Elements of the Orbit of Halley's Comet at its Appearance in the Years, 1935-36, 1835; Supplement to the Nautical Almanac of 1837, containing the Meridian Ephemeris of the Sun and Planets, 1836; Ephemeris of Encke's Comet 1838, 1838; Ephemeris of Encke's Comet 1839, 1838; Path of the Moon's Shadow over the Southern Part of France, the North of Italy and Part of Germany, during the total Eclipse of the Sun on 7 July 1842; Ephemeris of Faye's Comet, 1851. Compiled (with Francis Baily) catalogue of 2,881 fixed stars. Died Mar. 29, 1853.

STRATO (Straton of Lampsacus), Greek natural philosopher; b. Greece; flourished 330-270 B.C.; studied at Lyceum; traveled to Alexandria, Egypt; supposedly tutor to son of Ptolemy I; became 3d dir. Lyceum, succeeding Theophrastus, circa 288 B.C. First to propose intervention of deity was not necessary in nature; attempted to explain universe mechanistically; tried to reconcile Aristotelian and Democritan theories; described methods of forming vacuum; believed heavier bodies fell faster than lighter ones; understood law of lever.

STRATTON, Frederick John Marrian, English astronomer, astro-physicist; b. Birmingham, Eng., Oct. 16, 1881; s. Stephen Samuel and Mary Jane (Marrian) S.; ed. Mason U. Coll., Birmingham; B.A., Cambridge (Eng.) U., 1904, fellow, 1906 M.A., 1908; LL.D., Glasgow (Scotland) U.; Ph.D., U. Copenhagen (Denmark); hon. degrees. Lectr. math. Cambridge U. from 1911, dir. solar physics observatory, also prof. astro-physics, 1928-47; Halley lectr. Oxford (Eng.) U., from 1927. Dep. sci. adv. (Brit.) Army Council, 1948-50; gen. sec. Internat. Council Sci. Unions, 1937-52; sec. Internat. Astron. Union, 1925-48. Recipient Prix J. Janssen Société Astronomique de France; Companion of Distinguished Service Order 1917. Fellow Inst. Coimbra (Portugal). Fellow Royal Soc., 1947; mem. Royal Astron. Soc. (pres. 1933-35), Soc. Psychical Research (pres. 1935-55), Cambridge Philos. Soc. (pres. 1930-31), Brit. Assn. (gen. sec. 1930-35), (corr.) French Acad. Scis., Nat. Acad. Sci., Lima, Peru, Bureau des Longitudes, Paris, France, Royal Signals Instn. Japanese Order Sacred Treasure (hon.) Royal Soc. Can. Author: Astronomical Physics, 1925. Died Cambridge, Sept. 2, 1960.

STRATTON, George Malcolm, Am. psychologist; b. Oakland, Cal., Sept. 26, 1865; s. James T. and Cornelia A. (Smith) S.; A.B., U. Cal., 1888; A.M., Yale 1890; A.M., Ph.D., Leipzig, 1896; m. Alice Elenore Miller, May 17, 1894; children—Elenore (Mrs. Robert Fliess), Malcolm, Florence (Mrs. A. Reinke). Fellow in philosophy, 1891-93; instr. philosophy U. Cal., 1893-96, instr., asst. prof. and asso. prof. psychology, 1896-1904, dir. psychol. lab., 1899-1904, prof. psychology, 1908-35; mem. Institut für Experimentelle Psychologie, Leipzig, 1894-96; prof. exptl. psychology Johns Hopkins, 1904-08. Pres. Am. Psychology Assn., 1908; mem. NRC, 1921-24; chmn. anthrop. and psychol. div., 1925-26; mem. Nat. Acad. Scis., Nat. Inst. Psychology (hon.), Am. Inst. of Czechoslovakia (corr.). Author: Experimental Psychology and Its Bearing upon Culture, 1903; Psychology of the Religious Life, 1911; Theophrastus and the Greek Physiological Psychology Before Aristotle, 1917; Developing Mental Power, 1922; Anger, Its Religious and Moral Significance, 1923; Social Psychology of International Conduct, 1929; International Delusions, 1936; (with J. W. Buckham) George Holmes Howison, Philosopher and Teacher, 1934; Man, Creator or Destroyer, 1952; alos contbns. to various jours. on perception of change, eye movements, aesthetics of visual form, railway accidents and the color sense, race, nations, and internat. action. Died Oct., 1957.

STRATTON, Julius Adams, Am. physicist; b. Seattle, May 18, 1901; s. Julius A. and Laura (Adams) S.; student U. Wash., 1919-20; S.B., Mass. Inst. Tech., 1923, S.M., 1925; Sc.D., Eidgenossische Technische Hochshule, Zurich, Switzerland, 1927; D.Engring., N.Y. U., 1955; Sc.D., St. Francis Xavier Coll., 1957, Coll. William and Mary, 1964; LL.D., Northeastern U., 1957, Union Coll., 1958, Harvard, 1959, Brandeis U., 1959, Carlton Coll., 1960, U. Notre Dame, 1961, Johns Hopkins, 1962; L.H.D., Hebrew Union Coll., 1962, Oklahoma City U., 1963; m. Catherine Coffman, June 14, 1935; children—Catherine N., Ann Cary, Laura A. Mem. faculty Mass. Inst. Tech., Cambridge, 1924——, prof. physics, 1941-51, provost, 1949-56, v.p., 1951-56, chancellor, 1956-59, acting pres., 1957-59, pres., 1959——. Expert cons. Sec. War, 1942-46; mem. Nat. Sci. Bd.; mem. corp. Boston Mus. Sci. Recipient medal honor I.R.E., 1957; Faraday medal Instn. Elec. Engrs. London, 1961; Comdr. Order Boyaca, Columbia, 1964. Fellow Am. Acad. Arts and Scis., Am. Philos. Soc., Am. Phys. Soc., I.E.E.E., Nat. Acad. Scis. (v.p. 1961-65); mem. Sigma Xi, Tau Beta Pi, Zeta Psi, Eta Kappa Nu. Author: Electromagnetic Theory, 1941. Home: 111 Memorial Dr., Cambridge, Mass. 02142.*

STRATTON, Samuel Wesley, Am. physicist; b. Litchfield, Ill., July 18, 1861; s. Samuel and Mary B. (Webster) S.; B.S., U. Ill., 1884, D.Eng., 1903; D.Sc., Western U. Pa. (now U. Pitts.), 1903, Cambridge, 1909, Yale, 1919; LL.D., Harvard, 1923; Ph.D., Rensselaer Poly. Inst., 1924. Instr. math., asst. prof., prof. physics and elec. engring. U. Ill., 1885-92; successively asst. prof., asso. prof., prof. physics. U. Chgo., 1892-1901; drafted bill establishing Nat. Bur. of Standards, Washington, 1901, dir., 1901-23; pres. Mass. Inst. Tech., 1923——. Mem. Internat. Com. on Weights and Measures, Am. Inst. E.E. Nat. Acad. Scis., Nat. Adv. Com. for Aeronautics. Died Oct. 18, 1931.

STRAUB, Wolf Deter, Am. physicist; b. Boston, Apr. 27, 1927; s. Otto G.T. and Hilde (Lorenz) S.; B.S., Yale, 1950; M.S., U. Mich., 1952; m. Margrit A.E. Schmuziger, Mar. 11, 1961; children—Sibyl A., Dorothy A. Research div. Raytheon Co., Waltham, Mass., 1952-65; physicist Advanced Research br. NASA-ERC, Cambridge, Mass., 1965——. Mem. Am. Phys. Soc., Sigma Xi. Research in measurements of galvanomagnetic properties of semiconductors and semimetals as a function of temperature and magnetic fields. Home: 158 Barton Dr., Sudbury, Mass. 01776. Office: 575 Technology Sq., Cambridge, Mass. 02139.*

STRAUCH, Karl, physicist; b. Germany, Oct. 4, 1922; s. Georg and Garola (Bock) S.; came to U. S., 1939, naturalized, 1944; A.B. in Chemistry and Physics, U. Cal. at Berkeley, 1943, Ph.D. in Physics, 1950; m. Maria Gerson, June 10, 1951; children—Roger A., Hans D. Junior fellow of Soc. Fellows, Harvard, 1950-53; mem. faculty Harvard, 1953——, prof. physics, 1962——, director Cambridge Electron Accelerator, 1967——; spl. research high energy physics. Served with USNR, 1944-46. Mem. Am. Phys. Soc., Am. Assn. Physics Tchrs., Phi Beta Kappa, Sigma Xi. Contbr., profl. jours. Studies on nuclear structure with high energy protons; photoproduction of mesons and strange particles; devel. wide gap spark chamber. Home: 81 Pleasant St., Lexington, Mass. 02173. Office: Physics Dept., Harvard Univ., Cambridge, Mass. 02138.*

STRAUS, William L(evi), Jr., Am. phys. anthropologist, anatomist; b. Balt., Oct. 29, 1900; s. William Levi and Pauline (Gutman) S.; student Harvard, 1917-18; A.B., Johns Hopkins, 1920, Ph.D., 1926; m. Henrietta S. Hecht, Sept. 19, 1926 (dec. 1954); 1 dau., Pauline Gutman (Mrs. Edward Rauh); m. 2d, Bertha L. Nusbaum, June 15, 1955. Nat. Research Council fellow, 1926-27; instr. to asso. prof. anatomy Johns Hopkins, 1927-52, prof. phys. anthropology, 1952-57, prof. anatomy and physical anthropology, 1957-66, emeritus, 1966——, adminstrv. head dept. anatomy, 1947-48; Guggenheim fellow, 1937-38, vis. prof. anatomy Wayne U., 1950; mem. anatomy bd. State Md., 1947-52. Recipient Viking Fund Medal and award in phys. anthropology, 1952. Mem. adv. board Patuxent Instrn., Md., 1955-60. Fellow A.A.A.S. (editorial bd. 1953-64, v.p., chmn. anthropology sect. 1956), Am. Assn. Anatomist, Am. Anthropol. Assn; mem. Am. Inst. Human Paleontology, Am. Assn. Phys. Anthropologists (pres. 1953-55), Am. Soc. Mammalogists, Corp. Marine Biol. Lab. Woods Hole, Zool. Soc. London (corr.), Am. Assn. U. Profs., Am. Soc. Zoologists, Nat. Research Council (member of div. medical scis. 1960-64), Society of Vertebrate Paleontology, National Academy Scis., Phi Beta Kappa, Sigma Xi. Editor: (with others) Forerunners of Darwin: 1745-1859, 1959. Editor, contributor. The Anatomy of the Rhesus Monkey (with Carl G. Hartman), 1933; member editorial bd. Am. Jour. Anatomy, 1946-58, Human Biology, 1953——, Sci. and Sci. Monthly, 1953-64, Folia Primatologica, 1962——. Contbr. profl. jours. Research on anatomy of living and extinct primates; supporter of theory of origin of man from nonbrachiating generalized apes. Home: 7111 Park Heights Av., Balt. 21215.*

STRAUSS, Bernard S(amuel), Am. geneticist; b. N.Y.C., Apr. 18, 1927; s. Joseph and Kate (Silk) S.; B.S., Coll. City N.Y., 1947; Ph.D., Cal. Inst. Tech., 1950; m. Carol Maxine Dunham, Sept. 8, 1949; children—Paul Leonard, David Wilson, Leslie Joan. Faculty, U. Tex., 1950-52; faculty Syracuse U., 1952-60; faculty U. Chgo., 1960——, prof. microbiology, 1965——, chmn. interdepartmental com. genetics, 1962——; vis. Fulbright research prof. Osaka U., 1958-59. Mem. genetics tng. com. Nat. Inst. Gen. Med. Scis., 1962-66. Mem. Am. Soc. Biol. Chemists, Genetics Soc. Am., Am. Soc. Human Genetics, Am. Soc. Microbiology, Phi Beta Kappa, Sigma Xi. Author: An Outline of Chemical Genetics, 1960; also numerous articles. Research in field of bacterial mutagenesis; discovered new enzymes which degrade the genetic material (DNA); elucidation of one of the mechanisms by which bacterial viruses can carry bacterial genetic material from one organism to another (bacterial transduction). Home: 244 Indiana St., Park Forest, Ill. 60466.*

STRAUSS, Elliott William, Am. cell biologist, pathologist; b. Bklyn., Jan. 25, 1923; s. Joseph Maxwell and Sonia (Rapoport) S.; A.B., Columbia U., 1944; M.D., N.Y. U., 1949; m. Margaret Monica Crane, Dec. 9, 1951; children—William M., Monica M., Nicholas C. Head gastrointestinal service U.S. Naval Hosp., San Diego, 1954-55; asso. medicine Peter Bent Brigham Hosp., Boston, 1956-59; fellow Harvard Med. Sch., Boston, 1956-61, asso. pathology, 1961-65; asst. prof. pathology and medicine U. Colo. Sch. Medicine, Denver, 1965-67; asso. prof. med. sci. Brown Univ., 1967——. Research fel low Damon Runyon Fund for Cancer Research, 1958; USPHS fellow, 1959-60; recipient Research Career Devel. award USPHS, 1960-64. Mem. Am. Soc. for Cell Biology, Assn. Pathologists and Bacteriologists, Soc. Zoologists, A.A.A.S., Am. Soc. Exptl. Pathology. Contbr. articles to tech. jours. Demonstrated that absorption by isolated intestine can be studied by electron microscopy; research in lipid and vitamin B12 transport by microscopic and chem. methods, nontropical sprue in man, bacterial contamination intestinal diverticula of rat. Home: 58 Barnes St., Providence 02906.*

STRAUSS, Maurice Benjamin, physician; b. Bklyn., Mar. 5, 1904; s. henry M. and Ida (Igelheimer) S.; A.B., Amherst Coll., 1924; M.D., Johns Hopkins, 1928; m. Ruth Franc, Sept. 8, 1927; children—Peter Franc, Barbara Franc. Teaching staff Harvard Med. Sch., Boston, 1928-43; asso., 1946-52; staff Boston City Hosp., 1928——; chief med. service Boston VA Hosp., 1952——; prof. medicine Boston U. Sch. Medicine, 1952——. Lectr. Harvard Med. Sch., also Tufts U. Sch. Medicine, 1952——. Mem. A.C.P., Am. Fedn. for Clin. Research, Am. Soc. for Clin. Investigation, Assn. Am. Physicians, Boylston Med. Soc., Phi Beta Kappa, Sigma Xi, Alpha Omega Alpha. Author: (with L. G. Raisz) Clinical Management of Renal Failure, 1956; Body Water in Man, 1957; also numerous articles. Editor: Diseases of the Kidney, 1963; Familiar Medical Quotations, 1966. Research on cause and treatment of anemia, nutritional disorders of the nervous system, physiology of water and electrolytes in man, physiology and diseases kidney. Address: 150 S. Huntington Av., Boston 02130.*

STRAUSS, Maurice J., Am. dermatologist; b. New Haven, Conn., Jan. 3, 1893; s. Jacob and Theresia (Herrman) S.; A.B., Yale, 1914; M.D., Columbia, 1917; m. Carolyn Ullman, June 12, 1923; 1 son, John Steinert. Intern Bellevue Hosp., N.Y.C. 1917-18; pvt. practice of medicine, New Haven, 1919——; specialist in dermatology, 1929——; clin. asst. in dermatology and syphilology N.Y. Post-Grad. Hosp. and Med. Sch., 1917-29, asso. attending dermatologist to dispensary, 1929-32; asst. clin. prof. dermatology Yale U. Sch. of Medicine, 1932-36, asso. clin. prof., 1936-43, clin. prof. of dermatology, 1943-55, emeritus, 1955——; asst. attending physician New Haven Hosp. and Dispensary, 1932-37, attending physician, 1937——; dir. New Haven Venereal Disease Clinic, 1920-45; attending dermatologist Grace Hosp., 1928——. Hosp. of St. Raphael, 1931——; cons. dermatologist Laurel Heights Sanitorium Shelton, Conn.; cons. in dermatology and syphilology Norwich (Conn.) State Hosp., Newington (Conn.) V.A. Hosp., New Britain (Conn.) Hosp., and cons. in dermatology, Griffin Hosp., Derby, Conn. Diplomate Am. Bd. Dermatology and

Syphilology, 1933. Fellow in dermatology and syphilology of N.Y. Acad. of Medicne; mem. A.M.A., Soc. for Investigative Dermatology, Am. Acad. Dermatology and Syphilology, Am. Acad. Compensation Medicine, N.E. Dermatol. Soc., Atlantic Dermatologic Conf., Conn. State Med. Soc., Sigma Xi, Phi Beta Kappa. Died Feb. 2, 1958; buried New Haven.

STRAUSS, Ulrich Paul, chemist; b. Frankfurt, Germany, Jan. 10, 1920; s. Richard and Mariane (Seligmann) S.; A.B., Columbia U., 1941; Ph.D., Cornell U., 1944; m. Esther Lipetz, June 20, 1943 (dec. Sept. 1949); children—Dorothy, David; m. 2d, Elaine Greenbaum, Nov. 23, 1950; children—Elizabeth, Evelyn. Sterling fellow Yale, 1946-48; faculty Rutgers U., New Brunswick, N.J., 1948—, prof. phys. chemistry, 1960—, dir. Sch. Chemistry, 1965—. NSF Sr. fellow Nat. Center Sci. Research, Strasbourg, France, 1961-62. Fellow A.A.A.S., Am. Inst. Chemists, N.Y. Acad. Scis.; mem. Am. Chem. Soc. (chmn. phys. chemistry group N.J. sect. 1956, councillor, 1961—). Research and publs. on water soluble charged macromolecules; developed polysoaps.*

STRECKER, Edward Adam, Am. psychiatrist; b. Phila., Oct. 16, 1887; s. Adam and Mary (Weiler); prep. edn., St. Joseph's Coll., Phila., Sc.D., 1935; B.A., LaSalle Coll., Phila., 1907, M.A., 1911; Litt.D., 1938; M.D., Jefferson Med. Coll., 1911; LL.D., Franklin and Marshall Coll., Boston Coll.; L.H.D. St. Bonaventure U., 1956; m. Elizabeth Kyne Walsh, Jan. 1917. Resident physician St. Agnes Hosp., Phila., 1911-15; asst. physician Pa. Hosp. Dept. for Nervous and Mental Diseases, 1913-17, med. dir., 1917—, and dir. of clinic; staff neurologist, Pa., Phila. and Germantown hosps.; prof. of nervous and mental diseases Jefferson Med. Coll., 1925-31; prof. and head dept. psychiatry U. Pa., 1931-53, prof. psychiatry Schools of Medicine, clin. prof. psychiatry and mental hygiene, Yale U., 1926-32; chief of service and consultant Inst. for Mental Hygiene Pa. Hosp., Phila.; 17th Pasteur lecturer; cons. to govt. and pvt. instns. Thomas William Salmon Meml. lectr.; 1939; Pasteur, Menas Gregory Memorial, Bernard McGhie Meml. lectr.; 1946. Fellow A.C.P.; mem. Am. Neurol. Assn., Am. Psychiatric Assn. Author: Clinical Psychiatry, 1925; Clinical Neurology, 1927; Discovering Ourselves, 1931; Practical Examination of Personality and Behavior Disorders, 1936; Alcohol One Man's Meat, 1938; Beyond the Clinical Frontiers (Salmon lecture), 1940; Fundamentals of Psychiatry, 1942; Their Mothers' Sons, 1946; Their Mothers' Daughters (with V. T. Lathbury), 1956; also numerous articles and papers on nervous and mental disorders. Spl. researches in behavior disorder of children and normal and abnormal psychology of childhood. Died Jan. 2, 1959.

STRECKER, Herman, Am. lepidopterist; b. Phila., March 24, 1836; s. Ferdinand H. and Anna (Kern) S.; ed. pub. schs.; Ph.D., Franklin and Marshall Coll.; began granite and marble work when 12 yrs. old; designed Soldiers' Monument at Reading, Pa., numerous other works; spent all spare time in study of zoölogy, mineralogy, archaeology and botany; well known for works on butterflies and collection of over 370,000 of those insects. Author: Lepidoptera, Rhopaloceres and Heteroceres, Indigenous and Exotic, 1872-77; Butterflies and Moths of North America, 1878. Died 1901.

STREEMAN, John Robert, Am. phys. chemist; b. Ft. Worth, Apr. 12, 1930; s. Richard Edward and Bennie (Morrow) S.; B.S., Baylor U., 1951; M.A., U. Tex., 1953, Ph.D., 1955; m. Bonnie Greenwood, Aug. 4, 1951; children—Bonnie Jean, Stephen Austin. Aero. research scientist, NACA (now NASA), Cleve., 1955-56; sr. nuclear engr. Convair (now Gen. Dynamics), Ft. Worth, 1956-59; staff Los Alamos Sci. Lab., 1959—. Mem. A.A.A.S., Research and publs. in computer studies of nuclear reactor, passage of nuclear radiation through matter. Home: Box 414 Los Alamos 87544. Office: Group N-2, Los Alamos Sci. Lab., Los Alamos 87544.*

STREET, Herbert Edward, Brit. botanist; b. Newark, Eng., Dec. 17, 1912; s. Herbert and Lily Jane (Poulton) S.; B.Sc., U. Coll., Nottingham, 1938; Ph.D., Birkbeck Coll., London, 1943, D.Sc., 1954; m. Peggie May Sawyer, Sept. 13, 1940; children—Hilary Margaret, Rosemary Jane, Edward Wyman. Pharmacist North Middlesea Hosp., London, 1938-40; pharmacist Burroughs Wellcome Co. Ltd., Bexley, Kent, Eng., 1940-42; pharmacist Sch. Pharmacy U. Manchester, 1942-45; pharmacist Sch. Pharmacy U. Nottingham, 1945-47, lectr., 1947-49; sr. lectr. botany U. Manchester, 1949-54; prof. botany U. Wales, 1954-67; prof. botany, chmn. Sch. Biology U. Leicester, Eng., 1967—. Mem. Soc. Experimentive Biology U.K. Author: Plant Metabolism, 1963; Plant Tissue, Organ and Cell Culture in Cells and Tissues in Culture, Vol. 3, 1966; also numerous articles. Research in plant physiology by means of organ and cell cultures of plants, particularly carbohydrate, vitamin and mineral nutrition, hormonal control of growth and differentiation and exptl. studies of fine structure (electron microscopy). Home: 60 Shanklin Dr., Leicester, LEZ, 3QA, U.K.*

STREET, Jabez Curry, Am. physicist; b. Opelika, Ala., May 5, 1906; s. Jabez C. and Ann (Dunklin) S.; B.S. in Elec. Engring., Ala. Poly. Inst., 1927; M.A., U. Va., 1930, Ph.D., 1931; A.M. (hon.) Harvard, 1942; m. Leila Fripp Tison, July 1, 1939; children—Caroline Dunklin (Mrs. F. David Trickey), Curry Tison. Fellow Bartol Research Found., 1931-32; faculty Harvard, Cambridge, Mass., 1932—, prof. physics, 1947—, chmn. dept., 1956-60, asst. to dean for sci., 1966—; research asso., div. head Radiation Lab., Mass. Inst. Tech., 1940-45; chmn. physics vis. com. Brookhaven Nat. Lab., Upton, L.I., N.Y., 1954-55. Recipient Air Force Scroll for service on Sci. Adv. bd. USAF, 1956, U.S. certificate of merit for civilian war effort, 1948. Fellow Am. Phys. Soc.; mem. Nat. Acad. Sci., Am. Acad. Arts and Scis. Author: (with W.H. Furry, E.M. Purcell) Physics for Science and Engineering Students, 1952; also articles. Research on cosmic rays and high energy particles including measurements of intensity, absorption and nature of particles; discovered (with E.C. Stevenson) mu-meson; (with T.H. Johnson) devised first feed back voltage regulator; patentee radar modulators, Loran navigation system. Home: 56 Fletcher Rd., Belmont, Mass. 02178.*

STREETE, Thomas, Brit. astronomer; b. Cork, Ireland, flourished 1621-89; clk. Excise Office, London; tchr. math.; 1 of 6 chosen to resurvey after Gt. Fire, 1663-64; made astron. observations with Robert Hooke and Edmund Halley. Author: Astronomia Carolina, 1661. Devised new lunar tables; invented reflecting instrument, 1680.

STREETEN, David Henry Palmer, physician; b. Bloemfontein, S. Africa, Oct. 3, 1921; s. Reginald Craufurd and Olive Gladys (Palmer) S.; B.Sc., U. Orange Free State, 1941; M.B., B.Ch., U. Witwatersrand (S. Africa), 1946; D.Phil., Oxford (Eng.) U., 1951; m. Barbara Anne Wiard, Aug. 2, 1952; children—Robert Duncan, Elizabeth Anne, John Palmer. Came to U.S., 1951, naturalized, 1959. Faculty, U. Mich., Ann Arbor, 1953-60, asst. prof., 1955-60, investigator Howard Hughes Found., 1955-60; faculty State U. N.Y., Upstate Med. Center, Syracuse, 1960—, prof. medicine, 1964—. Rockefeller Traveling fellow in med. scis., 1951-52. Mem. Endocrine Soc., Am. Diabetes Assn., Central soc. for Clin. Research, Am. Fedn. for Clin. Research, A.M.A. Contbr. numerous articles to tech. jours., chpts. to books. Research on normal and abnormal intestinal peristalsis, effects adrenal cortical secretions, cortisol and aldosterone, sodium, potassium and water distbn. and excretion in health and in disease states, some aspects control secretion aldosterone, cortisol and vasopressin. Home: 334 Berkeley Dr., Syracuse, N.Y. 13210.*

STREETER, George Linius, Am. anatomist; b. Johnstown, N.Y., Jan. 12, 1873; s. George Austin and Hannah Green (Anthony) S.; A.B., Union Coll., New York, 1895, D.Sc., 1930; A.M., M.D., Columbia Univ., 1899; D.Sc., Trinity College, Dublin, 1928; LL.D., U. of Michigan, 1935; m. Julia Allen Smith, Apr. 9, 1910; children—Sarah Frances, George Allen, Mary Raymond. Asst. and instr. anatomy, Johns Hopkins 1902-06; asst. prof. anatomy, Wistar Inst. Anatomy, Phila., 1906-07; prof. anatomy and dir. anat. lab., U. Mich., 1907-14; research asso. Carnegie Instn. of Washington, 1914-18, dir. dept. of embryology, 1918-40, chmn. division animal biology, 1935-40, research asso. from 1940. Fellow Royal Society (Edinburgh). Mem. American Philos, Soc. of Philadelphia, American Soc. Naturalists, National Acad. Sciences, Am. Assn. Anatomists, Am. Soc. Zoologists, Inst. Internat. d'Embryologie; fgn. mem. Zoöl. Soc. London; hon. mem. Anat. Soc. Gr. Britain and Ireland. Died July 27, 1948.

STREHLER, Bernard Louis, Am. biochemist, biologist; b. Johnstown, Pa., Feb. 21, 1925; s. Bernard and Pauline (Steiner) S.; B.A. in Biology, Johns Hopkins, 1947, Ph.D. in Biology and Cell Physiology, 1950; m. Mary Theodore Penn, June 6, 1948; children—Bernard Louis, III, Jan A., Patricia A. Biochemist, Oak Ridge Nat. Lab., 1950-53; asst. prof. biochemistry U. Chgo., 1954-56; sect. chief gerontology br. NIH, 1956—; dir. research Aging Research Lab., VA Hosp., Balt., 1964—. Mem. Am. Soc. Biol. Chemistry, Growth Soc., Gerontological Soc., Sigma Xi. Author: Time, Cells and Aging, 1962; also numerous articles. Research on biochemistry firefly luminescence, amount and character age pigments, chemistry cell death, math. model age and mortality; developed assay techniques using firefly light; discovered plant luminescence and photosynthetic phosphorylation. Home: 41115 Westview Rd., Balt. 21218. Office: Aging Research Lab., VA Hosp., 3900 Loch Raven Blvd., Balt. 21218.*

STREHLOW, Roger Albert, Am. chemist; b. Milw., Nov. 25, 1925; s. Elmer John and Edna (Schmidt) S.; B.Sc., U. Wis., 1947, Ph.D., 1950; m. Ruby M. Miller, Aug. 21, 1948; children—John C., James E. Phys. chemist Ballistic Research Labs., Aberdeen Proving Ground, Md., 1950-58, chief physics br. Interior Ballistics Lab., 1958-60; vis. prof. aero. engring. U. Ill., Urbana, 1960-61, prof., 1961—. Fellow Am. Phys. Soc., Am. Inst. Aeros. and Astronautics (asso.); mem. Am. Chem. Soc., Combustion Inst., Sigma Xi, Alpha Chi Sigma. Research on combustion, detonations, properties of gases at high temperatures. Home: 505 S. Pine St., Champaign, Ill.*

STREIB, Gordon Franklin, Am. sociologist; b. Rochester, N.Y., July 7, 1918; s. Edward C. and Hattie (Molz) S.; B.A., N. Central Coll., Naperville, Ill.; M.S. Sociology, New Sch. for Social Research, 1947; Ph.D., Columbia, 1954; m. Ruth Boyer, Nov. 27, 1943; children—Marshall G., Carol I., Lawrence E., Nelson A. Research asst. Columbia U., 1947-49; lectr. Rutgers U., 1948-49; faculty Cornell U., Ithaca, N.Y., 1949—, prof., 1958—, chmn. dept. sociology, 1962—. Fulbright prof. Danish Nat. Inst. Social Research, Copenhagen, 1959-60. Social Sci. Research Council fellow, 1950-51. Fellow Am. Sociol. Soc., Gerontological Soc.; mem. Eastern Sociol. Soc., Soc. for Sci. Study of Religion, Soc. for Study of Social Problems, Nat. Council on Family Relations. Co-editor: Social Structure and the Family: Generational Relations, 1965. Research in application of survey methods in a semi-literate soc.; longitudinal research on impact of retirement; changes in family relations and structure in later stages of life cycle. Home: 112 N. Sunset Dr., Ithaca, N.Y. 14850.*

STREICHER, Hans-Joachim, German surgeon; b. Wolfratshausen, Germany, May 16, 1924; s. Ernst Theodor and Ottilie (Zürn) S.; student U. Berlin (Germany), 1943, U. Freiburg (Germany), 1943-44, U. Prague (Czechoslovakia), 1944-45; Staatsexamen U. Heidelberg (Germany), 1948, Dr. med., 1949; m. Irmgard Angelika Meier, June 19, 1950; children—Arne Christof, Uta Barbara, Ulrich Marko. Faculty, Chirurgische U.-Klinik Heidelberg, 1949-59, chief physician, 1957-59; first head physician Chirurgische U.-Klinik Marburg/Lahn, Germany, 1959—, prof. surgery, 1965—; lectr., apl. prof. surgery U. Heidelberg, 1959—, U. Marburg/Lahn, 1965—. Mem. Deutsche Gesellschaft für Chirurgie, Royal Soc. Medicine (London), Intersoc. Cytology Council, Société Internat. de Chirurgie. Author: Klinische Zytologie, 1953 (with Sandkühler) Milzchirurgie, 1961; Chirurgische Indikationen, 1967; also numerous articles. Research on clin. cytology, spleen surgery and pathophysiology, accident in childhood; bronchial and thorax injuries; exptl. investigation on revival time of heart; skin transplants; exptl. cirrhosis of liver; gastro-intestinal bleeding. Home: 10 Sonnhalde, 3551 Wehrshausen, Germany. Office: Chirurgische U.-Klinik, 355 Marburg/Lahn, Germany.*

STREIFF, E(nrico) B(ernard), ophthalmologist; b. Genoa, Italy, May 19, 1908; s. Josua Jakob and Gertrud (Pfister) S.; student medicine U. Genoa, 1927-33; m. Jacqueline Krafft, Mar. 14, 1957; 1 dau., Brigitte Muriel. Privat docent ophthalmology, Geneva, Switzerland, 1934-43; prof. ophthalmology U. Lausanne (Switzerland), 1944—; dir. Ophthalmol. Hosp., Lausanne, 1944—; head ophthalmic dept. Cantonal Hosp. Mem. Swiss, French, Italian, Greek, German, Belgian assns. ophthalmologists, Swiss Assn. Neurology, French Assn. Oto-neuro-ophthalmology. Author: (with M. Monnier) Der retinale Blutdruck im gesunden und kranken Organismus, 1946; also numerous articles. Research and publs. on marginal posterior dysplasia of cornea in irido-corneal malformations, mandibulofacial dysmorphosis and ocular alterations, traumatism of optica canals, ocular alterations in posterior cervical syndrome, role of uveal autoantibodies in pathogenesis of sympathetic ophthalmia, influence of hypersensitivity of sino-carotid artery and cardiodortic artery on ocular and brain circulation. Home: 2 Montbenon. Office: Ophthal. Hosp., Av. de France, 15, Lausanne, Vaud SZ, Switzerland.*

STREITWIESER, Andrew, Jr., Am. chemist; b. Buffalo, June 23, 1927; s. Andrew and Sophie (Morlock) S.; A.B., Columbia U., 1948, M.A., 1950, Ph.D., 1952; postgrad. (AEC fellow) Mass. Inst. Tech., 1951-52; m. Mary Ann Good, Aug. 19, 1950 (dec. May 1965); children—David Roy, Susan Ann. Faculty, U. Cal. at Berkeley, 1952—, prof. chemistry, 1963—. Cons. to industry, 1957—. Mem. Am. Chem. Soc. (Cal. sect. award 1964), Chem. Soc. London, A.A.A.S., Phi Beta Kappa, Sigma Xi. Author: Molecular Orbital Theory for Organic Chemists, 1961; Solvolytic Displacement Reactions, 1962; (with J. I. Brauman) Supplemental Tables of Molecular Orbital Calculations, 1965, (with C. A. Coulson) Dictionary of Electron Calculations, 1965; also numerous articles. Co-editor: Progress in Physical Organic Chemistry, 3 vols., 1963-65. Research on organic reaction mechanisms, optically active deuterium compounds, isotope effects, application molecular orbital theory to organic chemistry, effect chem. structure on hydrocarbon acidities. Office: Dept. chemistry, U. Cal., Berkeley, Cal. 94720.*

STRELETSKII, Nikolai Stanislavovich, Russian structural engr.; b. Sept. 14, 1885; grad. Petersburg Inst. Engrs. Communication Lines, 1911; postgrad. Charlottenberg Tech. High Sch., Berlin, Germany, 1911. Instr., Moscow Technol. Coll., 1915-18; prof., head bridge dept., 1918—; head chair bridges Higher Civil Engring. Sch., 1927-33, Kuybyshev Mil. Engring. Acad., 1933-36; prof., head steel structures Moscow Transp. Engring. Inst., 1933—; became staff cons. steel structures Desian Office, 1936. Dep. chmn. council on sci. constrn. USSR Acad. Bldg. and Architecture, 1947-49; chmn. com. on constrn. problems USSR Acad. Scis., 1949-53; dir. State Constrn. Inst. Recipient Order of Lenin, twice, Order of Red Bannor of Labor, twice. Mem. USSR Acad. Scis. (corr.), USSR Acad. Constrn. and Architecture. Author: Calculation Method for Strutless Girders with Parallel Booms and

Junction Loading, 1913; Historical Outline of the Development of Underwater Tunnel Construction, 1914; Draft Plans for the City and Rail Bridges Across the Oka River at Nizhniy Novgorod, 1915; Laws of Weight Changes in Metal Bridges, 1926; New Ideas and Opportunities in Industrial Metal Construction, 1934; Course on Bridges, 1931; Course on Metal Construction, Part 1-3, 1940-44; Basic Prerequisites for the Standardization of Bridge Installations in USSR Transportation, 1953; Material for A Course on Steel Structures, 1959. Research on bridges including design of bridges across Volga and Sunguri Rivers, Krymsky and Krasnokholmsky bridges; originated theories for calculation of structures; established sci. basis for transp. and indsl. structures standardization. Home: M. Levshinskii p. 14. Office: Acad. Constrn. and Architecture, USSR, Pushkinsaya Ulitsa 24, Moscow, USSR.

STRELKOV, Petr Geogievich, Russian physicist; b. 1899; grad. Leningrad Indsl. Inst., 1924. Research in Leningrad, 1923-26; sr. sci. worker then lab. supr. Inst. Phys. Problems, USSR Acad. Scis., 1936-56; dep. dir., lab. supr. All-Union Inst. Physico-Tech. and Radiotech. Measurements, Conn. on Measurements and Measuring Instruments, 1956-59; departmental chmn. Inst. Thermal Physics, Siberian br. USSR Acad. Scis., 1959. Recipient Stalin prize, 1943. Corr. mem. USSR Acad. Scis. Research and publs. on thermal and molecular processes. Home: Voroh'evskoye Shosse 2. Office: USSR Acad. Sci., Leninskii Prospekt 14, Moscow, USSR.

STRESEMANN, Erwin Theodor Friedrich, German zoologist; b. Dresden, Germany, Nov. 22, 1889; s. Richard and Marie (Dunkelbeck) S.; student U. Jena, 1908-09, U. Munich, 1909-10, Ph.D., 1920; student Freiburg U., 1912-14; m. Elisabeth Deninger, June 20, 1916 (div. July 1939); children—Rosemarie, Werner, Ernst; m. 2d Vesta Grote, Sept. 20, 1941. Zoologist 2d Freiburg Expdn. to Dutch East Indies, 1910-12; curator birds Zool. Mus. of Berlin U., 1921-61, prof., 1930—, dir. Zool. Mus. Berlin, 1957-59, chmn. Alex V. Humboldt Commn., 1960—. Mem. Academie Leopoldina, German Acad. Scis., German Ornithologists Soc. (pres. 1949-67). Author: Die Lautersheinungen In den Ambonischen Sprachen, 1918; Avifauna Macedonica, 1920; Aves, 1927; Die Entwicklung der Ornithologie, 1951; Die Mauser, 1966; Atlas der Verbreitung Palaearktischer Vogel, 1960—; also numerous articles. Linguistic studies, publs. in the Moluccan Islands; taxonomy, distbn. anatomy; physiology of birds; history of ornithology; genetics. Home: 28 Kamillenstrasse, Berlin 45, Germany. Office: 43 Invalidenstrasse, Berlin N. 4, Germany.*

STRICKER, Salomon, German pathologist; b. Waag Neustadtl, 1834; became prof., Vienna, Austria, 1868. Author: Vorlesungen über allgemeine und experimentelle Pathologie, 1877-83. Discovered that vasodilatation results from stimulation of posterior nerve roots, circa 1876. Died 1898.

STRICKLAND, Hugh Edwin, Brit. naturalist; b. Righton, Eng., Mar. 2, 1811; s. Henry Eustatius and Mary (Cartwright) S.; ed. Oriel Coll., Oxford, Eng., 1829, B.A., 1832, M.A., 1835; m. Catherine Dorcas, Mayle, July 23, 1845. Accompanied William John Hamilton in geol. tour, Asia Minor, Greece, Constantinople, Italy, Switzerland, 1835; visited N. of Scotland, 1837; Genus of brachipoda, also fossil plant named Stricklandia in his honor. Fellow Royal Soc., 1852; mem. Ashmolean Soc. (pres.). Author: Ornithological Synonyms, 1855; The Dodo, 1845; also numerous articles. Developed rules for zool. nomenclature, 1841. Died Sept. 13, 1853.

STRICKS, Walter, chemist; b. Vienna, Austria, Mar. 15, 1904; s. I.J. and Helen (Low) S.; Chem. Engr., Tech. U., Vienna, 1927, Ph.D., 1929; m. Helen Bickart, Dec. 29, 1951. Came to U.S., 1947, naturalized, 1952. Faculty, U. Minn., 1947-56; faculty Oglethorpe U., Atlanta, 1956-57; faculty Marquette U., Milw., 1957—, now prof. Fellow A.A.A.S.; mem. Am. Chem. Soc., Sigma Xi, Phi Lambda Upsilon. Research, numerous publs. in analytical, phys. chemistry, electrochemistry, spectrophotometry, plarographic, other phys.-chem. investigations of sulfhydryl and disulfide compounds of biochemical interest, raman spectra, kinetics. Home: 4951 N. Santa Monica Blvd., Milw. 53217.*

STRIEDER, John William, Am. thoracic surgeon; b. Boston, June 6, 1901; s. Joseph William and Elizabeth Merritt (Robinson) S.; S.B., Mass. Inst. Tech., 1922; M.D., Harvard, 1926; m. Helen Lucille Roberts, Aug. 17, 1935 (dec. Feb. 1961); children—Alison Tennant (Mrs. John S. Mayher), Helen Roberts (Mrs. Martin Linsky), Elizabeth Merritt; m. 2d, Denise Jouasset, Aug. 7, 1962. Practice medicine, specializing in thoracic surgery, Boston, 1927-30, 35—, Trudeau Sanatorium, N.Y., 1930-33, Ann Arbor, Mich., 1933-35; house surgeon Boston City Hosp., 1927-29, dir. thoracic surgery, 1946—; instr. thoracic surgery U. Mich. Hosp., 1933-35; instr. surgery Harvard Med. Sch., 1936-40; surgeon-in-chief thoracic surgery Univ. Hosp., Boston, Newton-Wellesley Hosp., Newton, Mass. Soldiers Home, Chelsea, Mass.; sr. surgeon New Eng. Baptist Hosp., New Eng. Deaconess Hosp.; surgeon-in-chief Boston Sanatorium, Mattapan, Mass.; prof. clin. surgery Sch. Medicine Boston U. Trustee Elizabeth Carleton House. Diplomate Am. Bd. Surgery, Am. Bd. Thoracic Surgery (founder's group; chmn.). Mem. A.M.A., A.C.S., Am. Assn. Thoracic Surgery, Soc. Thoracic Surgeons (exec. com.), Am. Cancer Soc., Am. Heart Assn., Am. Thoracic Soc., Nat. Tb. Assn., New Eng. (v.p.), Boston surg. socs., Sigma Xi, Lambda Chi Alpha. Mem. editorial bd. Alexander Monograph Series, 1956—. Contbr. articles profl. jours. First to litigate patent ductus arteriosus (congenital defect of human heart), 1938. Home: 143 Laurel Rd., Chestnut Hill, Mass. 02167. Office: 2000 Washington St., Newton Lower Falls, Mass. 02162.*

STRINGFELLOW, Henry Martyn, Am. horticulturist, b. Winchester, Va., Jan. 21, 1839; s. Horace and Harriet Louisa (Strother) S.; A.M., William and Mary Coll., Va., 1858; postgrad. Theol. Sem., Alexandria, Va., 1859-61; m. Alice Johnston, Dec. 15, 1863. Voted against secession, 1861; enlisted in C.S.A. as pvt., 1861; went to Tex. with Gen. Magruder, fall of 1862; pioneer in discovering value of gulf coast of Tex. for raising fruit and vegetables; planted 1st pear orchard on coast, 1882; planted 1st Satsuma oranges in Tex., 1884; in nursery business until 1895, later devoting attention to experiments in horticulture; moved to Fayetteville, Ark., 1909, and set out 1st English walnut grove of 500 trees in Ark. Author: The New Horticulture, 1896. Died June 17, 1912.

STRINGFELLOW, John, English inventor; b. 1799; industrialist, Chard, Eng.; asso. with W. S. Henson in Aerial Transit Co.; helped build Henson's airplane; built own airplane model with steam power, 1848; developed a compact, light but powerful model; flew about 40 meters in London. Died 1883.

STRINGHAM, Bronson, Am. mineralogist, educator; b. Salt Lake City, July 28, 1907; s. William and Lucy (Ferrin) S.; B.S., U. Utah, 1933; Ph.D., Columbia, 1941; m. Lucille Oblad, Sept. 25, 1941; children—Michael, Cynthia, Susan. Faculty dept. geology U. Utah, 1931-33, dept. mineralogy, 1937—, now prof., head dept. mineralogy, 1952—; faculty Columbia, 1933-36; geologist U. S. Geol. Survey, 1944-45. Fellow Geol. Soc. Am., Mineral Soc. Am.; mem. Soc. Econ. Geologists, Geochem. Soc., Am. Inst. Mining, Metall. and Petroleum Engrs., Clay Mineral Soc., Am. Assn. U. Profs., Sigma Xi, Phi Kappa Phi, Sigma Gamma Epsilon. Research, publs. on relationship intrusive porphyries to ore deposits, hydrothermal alteration and mineralization at Bingham copper mine. Home: 2169 Logan Av., Salt Lake City.*

STRINGHAM, (Washington) Irving, Am. mathematician; b. Yorkshire (later Delevan), N.Y., Dec. 10, 1847; s. Henry and Eliza (Tomlinson) S.; attended Washburn Coll., Topeka, Kan., 1866-73; grad. Harvard, 1877; in Paris, summer 1877; Ph.D., Johns Hopkins, 1880; in Europe, 1880-82, studying math. at Leipzig; in Spain, summer 1887; in Paris, academic year, 1899-1900; m. Martha Sherman Day, June 28, 1888. Prof. math. U. Cal., 1882-1909, dean, 1886-1909. Author: Uniplanar Algebra, 1893. Editor and author Am. edit. Charles Smith's Elementary Algebra, 1894. Studied regular solids in n-dimensional space, also geometry of n-dimensions along metrical line; gave pictures of projections on space of regular solids in n-dimensions. Died 1909.

STRODE, Thomas, English instrument maker; b. Somerset, Eng.; flourished 1642-97; s. Thomas Strode; student U. Coll., Oxford, 1642-44; student under Abraham Woodhead. Author: A Short Treatise of the Combinations, Elections, Permutatione and Composition of Quantities, 1678; A New and Easie Method to the Art of Dyalling, containing: (1) all Horizontal Dyals, all Upright Dyals, etc.) (2) the most Natural and Easie Way of describing the Curve-lines of the Sun's Declination on any Plane, 1688. Publs. on design of sundials, arithmetic, conic sections.

STROHMEYER, Friedrich, German chemist; b. Göttingen, Germany, Aug. 2, 1776; ed. Göttingen; studied under Vauquelin in Paris; became mem. faculty of U. Göttingen, 1802, prof. chemistry, 1810; insp. gen. of apothecaries in Hannover, Germany; mem. French Acad. Scis. Author: Grundriss der theoretischen Chemie, 2 vols., 1808. Noted for analysis of minerals; discovered cadmium, 1817; tchr. of Bunsen. Died Göttingen, Aug. 18, 1835.

STROMBIO, Gaetano, Italian clinician; b. Milan, Italy, 1753; dir. Milan Hosp. for Pellagra. Author: De Pellagra (most important work on pellagra of the period), 3 vols., 1786-89. First to show pellagra may occur without skin lesions, 1786. Died 1831.

STROMER, Martin, Swedish physicist; b. 1707; succeeded Celcius as prof. astronomy, Uppsala, Sweden; translated Euclid's Elements into Swedish; inverted Celcius' calibration for thermometer, 1750. Died 1770.

STROMEYER, Georg Friedrich Louis, German surgeon; b. Hannover, Germany, 1804; began practice medicine in Hannover, 1828; became asso. with surg. sch., 1829; founder orthopedic clinic, Hannover; became prof., Erlangen, Germany, 1838, Munich, 1841, Freiburg, 1842, Riel, 1848. Author: Erinnerungen eines Ärztes, 1875. Died 1876.

STRöMGREN, Bengt Georg Daniel, astronomer; b. Göteborg, Sweden, Jan. 21, 1908; s. Svante Elis and Hedvig (Lidforss) S.; M.S., Copenhagen U., 1927, Ph.D., 1929; m. Sigrid Caja Hartz, Mar. 31, 1931. Lectr., Copenhagen U., 1933, prof., 1938-40, dir. Observatory, 1940; asst. prof. U. of Chicago, 1936-37, asso. prof., 1937-38, vis. prof., 1946-47, prof., dir. Yerkes and McDonald Obs., Williams Bay, Wis., 1951-57; Sewell Avery distinguished service prof. U. Chgo., 1952; spl. lectr. astronomy London U., 1949; vis. prof. Cal. Inst. Tech., 1950; vis. prof. Princeton, 1950, mem. Inst. for Advanced study, 1957-67; occupant House of Honor, Copenhagen, 1967—; prof. astrophysics Copenhagen U., 1967—. Recipient Augustinus prize, 1950. Mem. Royal Danish Acad. Scis. and Letters, Danish Acad. Technical Sciences, American Academy of Arts and Sciences, Royal Astron. Soc. (associate), Societe Royale Des Sciences De Liege, American Astronomical Society (hon.; pres. 1965); Acad. Coimbra, Royal Swedish Acad. Scis., Physiographic Soc., Koninklijke Nederlandse Akademie van Wetenschappen, Internat. Astron. Unions (gen. sec. 1948), Internat. Council Sci. Unions (exec. com. 1948). Author: Laerebog i Astronomi (with Elis Stromgren), 1931; Lehrbuch der Astronomie (with Elis Stromgren), 1933. Editor: Handbuch der Experimentalphysik, Vol. 26 (Astrophysik) 1937. Developed theory of interstellar luminous clouds (called Strömgren spheres) as hydrogen ionized by high-temperature stars; research on internal make-up, pulsation and ionization of stars; spectral classification. Address: Observatoriet, Ostervoldgade 3, Copenhagen K, Denmark.*

STRöMGREN, Elis, astronomer; b. Hälsingborg, Sweden, May 31, 1870; prof., Copenhagen, Denmark. Author: Über den Ursprung der Kometen, 1914; Lehrbuch der Astronomie, 1931. Studied movement of double stars, comet paths. Died Copenhagen, Apr. 5, 1947.

STRONG, Dorothy Hussemann (Mrs. Frank Morgan Strong), Am. food scientist; b. Peoria, Ill., Feb. 15, 1908; d. Edward E. and Lillian (Sullivan) Hussemann; B.S., U. Ill., 1928, M.S., 1929, Ph.D., 1946; postgrad. U. Wis., Ore. State U.; m. Frank Morgan Strong, Dec. 21, 1957. Faculty U. Wis., Madison, 1930—, prof., 1949—, chmn. dept. foods and nutrition, 1955—, staff Food Research Inst., 1966—. Mem. Inst. Food Technologists, Am. Soc. for Microbiology, Am. Dietetic Assn., Am. Home Econs. Assn., N.Y. Acad. Scis., Royal Soc. Health (Brit.), Sigma Xi, Iota Sigma Pi, Omicron Nu, Sigma Delta Epsilon, Phi Upsilon Omicron. Research and publs. on food-borne illness, its control and bacteriological problems. Home: 625 Anthony Lane, Madison, Wis. 53711.*

STRONG, John (Donovan), Am. physicist; b. Riverdale, Kan., Jan. 15, 1905; s. Fred and Anna Laura (Bennett) S.; student Friends Univ., 1921-22; A.B., Kan. U., 1926; M.S., U. of Mich., 1928, Ph.D., 1930; D.Sc. (honorary), Southwestern at Memphis, 1962; m. Bethany June McLaughlin, Sept. 2, 1928; children—Patricia Ann, Virginia Marie. Nat. research fellow physics, Rockefeller Foundn., Calif. Inst. Tech., 1930-32, research fellow on 200-inch telescope project, Palomar Mtn., 1932-37; asst. prof. physics and astrophysics, Calif. Inst. Tech., 1937-42; research fellow Harvard, 1942-45 (on secret war research); prof. physics, Johns Hopkins, 1945-52, prof. exptl. physics, 1952—, dir. lab. of astrophysics and physical meteorology. Cons. Libbey-Owens Ford Corp. Mem., dir. Eppley Found. Research. Recipient Ives medal Optical Soc. Am., 1956. Fellow Am. Acad. Arts and Scis.; mem. Société Royal des Scis. de Liège (corr. member), Optical Society Am. (pres. 1959), Phi Beta Kappa, Sigma Xi. Author: Procedures in Experimental Physics, 1938; Concepts of Classical Optics, 1958. Work in exptl. physics, meteorology, evaporation in vacuum, optics, infrared spectroscopy, astrophys. observation; developed method of increasing speed of astron. instruments through reducing reflection; helped develop aluminum coating process for telescope mirrors. Home: 4033 Deepwood Rd., Balt. 18.

STRONG, Leonell C(larence), Am. mammalian geneticist; b. Renova, Pa., Jan. 19, 1894; s. Clarence A. and Ella (Meade) S.; B.S., Allegheny Coll., 1917, Sc.D., 1939; Ph.D., Columbia, 1922; M.D. (hon.), U. Perugia, Italy, 1957; m. Katherine Bittner, June 27, 1919; children—Leonell Clarence, Wilson W. Asso. prof. St. Stephen's Coll., 1921-25; research fellow Harvard Med. Sch., 1925-27; research asso. U. Mich., 1927-30, Roscoe B. Jackson Meml. Lab., 1930-33; research asso. cancer Yale Med. Sch., 1933-53; dir. biol. sta. Roswell Park Meml. Inst., Springville, N.Y., 1953-64; vis. fellow Salk Inst. Biol. Studies, San Diego, 1964-67; dir. Leonell C. Strong Research Found., 1968—. Belgium-Am. Found. and Anna Fuller Fund lectr., Europe, 1948; guest lectr. Internat. Gerontology Congress, St. Louis, 1951, London, 1954, Venice and Merano, Italy, 1957, Karolinska Inst., Stockholm, 1955, San Francisco, 2d Mammary Cancer Symposium, Perugia, 1957, Internat. Cancer Congress, Atlantic City, 1939, St. Louis 1947, London, 1958; research prof. U. Buffalo Grad. Sch., 1955-63. Mem. NRC com. gastric cancer Nat. Adv. Cancer Council. Mem. Royal Soc. Medicine (London), Genetics Soc. Am., Am. Assn. Cancer Research, Soc. Exptl. Biology and Medicine, Gerontological Soc., French Assn. Cancer Research (asso.), Sigma Xi, Phi Beta

Kappa. Author numerous articles on cancer research. Established 40 inbred strains of mice used in research throughout the world; genetic analysis of susceptibility to transplanted tumor, of tumors induced by chem. means; 1st germinal mutations by chem. means; determined that litter seriation (parental age) influences cancer susceptibility in mice; investigated possibility that aging process may influence morphological and physiol. characteristics of offspring. Home: 8533 Sugarman Dr., La Jolla, Cal. Office: 241 12th St., Del Mar, Cal. 92104.*

STRONG, Richard Pearson, Am. physician; b. Fortress Monroe, Va., Mar. 18, 1872; s. Richard P. and Marion B. (Smith) S.; Ph.B., Yale, 1893; M.D., Johns Hopkins, 1897; postgrad. U. Berlin, Institute für Infektionskrankheiten, Berlin, 1903; Sc.D. (hon.) Yale, 1914, Harvard, 1916; m. Grace Nichols, July 23, 1936. Resident house physician Johns Hopkins Hosp., 1897-98; 1st lt. asst. surgeon U. S. Army, 1898-1902; pres. bd. for investigation of tropical diseases in P.I., 1899-1901; established and directed Army Pathol. Lab.; dir. Govt. Biol. Lab., Manila, 1901-13 (resigned); resigned from army 1902; sent by govt. to Berlin, 1903, for scientific investigation; prof. tropical medicine Coll. Medicine and Surgery, Univ. of P.I., 1907-13; prof. tropical medicine Harvard, 1913-38, prof. emeritus since 1938. Del. to various med. congresses; mem. NRC. Recipient Theobald-Smith medal Am. Acad. Tropical Medicine, 1939, medal Am. Found. Tropical Medicine, 1944. Fellow Royal Soc. Tropical Medicine and Hygiene (hon., London), Am. Acad. Arts and Sciences, A.A.A.S. (v.p. sect. N, 1923-24); asso. mem. Société de Pathologie Exotique, Paris; mem. Assn. Am. Physicians (pres. 1925-26), A.M.A., Soc. Exptl. Biology, Acad. of Tropical Medicine (pres. 1936), Am. Soc. of Tropical Medicine (pres. 1914), Am. Assn. of Pathologists and Bacteriologists, Mass. Med. Soc., Boston Soc. Natural History, Am. Soc. Exptl. Pathology, Am. Soc. Parasitologists (pres. 1927), Soc. de Biologie (Paris), Soc. Belge de Med. Tropicale (honorary), Soc. Medico Chirurgica (Bologna). Author: (textbook) Diagnosis, Prevention and Treatment of Tropical Diseases (7th edit., 1944); reports on tropical diseases and expdns. to S. Am. and Africa. Editor med sect. Philippine Jour. Sci. circa 1910-13. Died July 4, 1948.

STRONG, Theodore, Am. mathematician; b. South Hadley, Mass., July 26, 1790; s. Joseph and Sophia (Woodbridge) S.; grad. Yale, 1812; m. Lucy Dix, Sept. 23, 1818; 2 sons, 5 daus. Tutor math. Hamilton Coll., Clinton, N.Y., 1812-16, prof. math. and natural sci., 1816-27; prof. math. Rutgers Coll., 1827-61, emeritus, 1861-69; fellow Am. Acad. Arts and Scis., 1832, Am. Philos. Soc., 1844; an incorporator Nat. Acad. Scis., 1863. Author: Treatise on Elementary and Higher Algebra, 1859; Treatise on Differential and Integral Calculus, 1869. Contbr.: Gill's Math. Miscellany; Silliman's Am. Jour. Sci.; Runkle's Math. Monthly. Began use of Liebnizian symbols and notations in calculus, 1820's; solved Cardan's irreducible case of cubic equations; gave geometrical demonstration of values of sines and cosines of sum and difference of 2 arcs. Died Feb. 1, 1869.

STRONG, William Walker, Am. physicist; b. Good Hope, Pa., May 16, 1883; s. William Harrison and Maria (Garretson) S.; B.S. with honors, Dickinson Coll., 1905; Ph.D., Johns Hopkins, 1908; m. Mary Alberta Kirk, June 17, 1916; children—Walker Albert, Margaret Kirk. Fellow by courtesy Johns Hopkins, 1908, asst., 1909-11; research asst. Carnegie Instn., Washington, 1908-11; fellow Mellon Inst., Pitts., 1911-13, also prof. elec. theory, U. Pitts., pres. Sci. Instrument & Elec. Machine Co., 1912; instr. Carnegie Inst., Pitts., 1914; physicist Research Corp., 1915, 19; cons. practice. Fellow A.A.A.S., Am. Phys. Soc.; mem. Phi Beta Kappa. Author: The Absorption Spectra of Solutions (Carnegie Instn.), 2 parts, 1910, 11; The New Science of Fundamental Physics, 1918; The New Philosophy of Modern Science, 1920; also vols. I, II, III, IV of collected papers from phys. and chem. jours.; Immortality in the Light of Modern Thought, 1923; Ourselves and Our Sciences, 1930. Developed fume mask for diphenylclorasin and other poisonous fumes; discovered effect of magnetic psychoanalysis, 1921; discovered Phoenician letters and words incised on stones in Pa., 1941; extended American Phoenician to 400 words. Died Oct. 25, 1955.

STROOBANT, Paul-Henri, Belgian astronomer; b. Ixelles, Belgium, Apr. 11, 1868; dir. Belgian obs.; prof. U. Brussels (Belgium); mem. French Acad. Scis. Author: La distribution des étoiles par rapport à la voie lactée, 1924. Research on small planets, also on Saturn, enlargement of sun's image at horizon, differential rotation of galaxy, distbn. of stars in Milky Way. Died St. Gilles, Belgium, July 15, 1936.

STROTHER, Charles Riddell, Am. psychologist; b. Yreka, Cal., Oct. 4, 1907; s. Charles Wesley and Nancy (Arnold) S.; B.A., U. Wash., 1929, M.A., 1932; Ph.D., U. Ia., 1935; m. June Voss, Aug. 26, 1935; children—Nancy (Mrs. Erich Luschei), Kathryn. Asst. prof. speech pathology U. Wash., 1935-39; asso. prof. psychology U. Ia., 1939-47; prof. U. Wash., Seattle, 1947—, prof. clin. psychology in medicine, 1947—, dir. Mental Retardation and Child Devel. Center, 1965—. Cons., U. S. Office Edn., 1965—; mem. research adv. coms. govt. agys., founds. Mem. Am. Psychl. Assn., Am. Assn. Mental Deficiency, A.A.-A.S., Phi Beta Kappa, Sigma Xi, Tau Kappa Alpha, Pi Sigma Alpha, Alpha Epsilon Delta. Author: Foundations of Speech, 1945; Psychology and Mental Health, 1957; also articles. Research on myokinetic factors in stuttering, effects of aging, central nervous system damage, electro-shock on cognitive functions; validities and reliabilities of various measures of intelligence. Home: 5012 N.E. 41st St., Seattle, 98105.*

STROTHER, Edward, English physician; b. Alnwick, Eng., 1675; s. Edward Strother; admitted pensioner Christ's Coll., Cambridge, 1695; M.D., U. Utrecht (Netherlands); practiced medicine, London; licentiate, mem. Royal Coll. Physicians. Author: Critical Essay on Fevers, 1716; Erodia—A Discoverie of Causes and Cures, 1718; Pharmocopia Practica, 1719; Smallpox, 1721; Practical Observations on Epidemic Fever, 1729. Translator: Materia Medica (Harman). Credited with introducing term puerperal fever, 1716. Died London, Apr. 14, 1737.

STROTZ, Robert Henry, Am. economist; b. Aurora, Ill., Sept. 26, 1922; s. John Marc and Olga (Koerfer) S.; student Duke, 1939-41; A.B., U. Chgo., 1942, Ph.D., 1951; m. Helen Berry, July 24, 1961; children—Vicki, Michael, Frances, Ellen, Ann. Faculty, Northwestern U., Evanston, Ill., 1947—, prof. econs., 1958—, dean Coll. Arts and Scis., 1966—. Mem. Am. Econ. Assn., Econometric Soc., Am. Statis. Assn. Mng. editor Econometrica, 1953—. Research and publs. on utility theory, welfare econs. and econometrics. Home: 2520 Sheridan Rd., Evanston, Ill. 60201.*

STROUD, William Daniel, Am. physician; b. Villa Nova, Pa., Nov. 20, 1891; s. Morris Wistar and Margaret P. (Rutter) S.; B.S., U. Pa., 1913, M.D., 1916; postgrad. U. Coll. Hosp., London, 1919-20, St. Andrew's Inst. Clin. Research, Scotland, 1920, L'Hopital de Pitiet, Paris, France, 1920; m. Agnes H. Shober, Sept. 19, 1923; children—William Daniel, Samuel Shober, Agnes Hutchinson, Margaret Rutter and Charlotte Wistar (twins). Engaged in practice of medicine at Phila. since 1921; chief of staff Children's Heart Hosp., Phila.; cons. cardiologist Pa. Hosp., Bryn Mawr Hosp., Abington Meml. Hosp., St. Christopher's Hosp., Phila., Norristown State Hosp.; prof. cardiology U. Pa. Grad. Sch. Medicine, asst. prof. clin. medicine U. Pa. Med Sch. Bd. dirs. Phila. Health Council and TB Com., Phila. Fellow A.M.A., Phila. Coll. Physicians; mem. Assn. Am. Physicians, Am. Clin. and Climatological Assn., Phila. Heart Assn., Am. Heart Assn., A.C.P. (Alfred Stengel Meml. award 1957), other med. socs. Contbr. articles to med. jours. Editor The Diagnosis and Treatment of Cardio-vascular Disease, 1957. Died Aug. 19, 1959; buried Bryn Mawr, Pa.

STRUBECKER, Karl, German mathematician; b. Hollenstein, Germany, July 8, 1904; s. Karl and Kathe (Wels) S.; ed. U. Vienna, Vienna Tech.; Ph.D.; m. Hildegard Salewsky, May 29, 1941; Became instr. Vienna Tech., 1931, asso. prof., 1938; named instr. U. Vienna, 1935, later asso. prof.; prof. U. Strasbourg (France), 1942; became prof. math. Karlsruhe (Germany) Tech. Sch., 1947. Corr. mem. Austrian Acad. Sci. Author: Einführung in die Höhere Mathematik, 1956; Differential geometrie; Kurventheorie der Ebene und des Raumes; Theorie der Flächenmetrik; Theorie der Flächenkrümmung; Vorlesungen über Darstellende Geometrie, 1958. Research on pure and applied math., including geometry. Home: Hansjakobstrasse 8, 75 Karlsruhe, West Germany.

STRUGHOLD, Hubertus, physician; b. Westtuennen, Westfalia, Germany, June 15, 1898; s. Ferdinand and Anna (Tillman) S.; Ph.D., U. Muenster, 1922; M.D., U. Wuerzburg, 1923. Came to U.S., 1947, naturalized, 1956. Research asst. Physiol. Inst., Wuerzburg, 1923-28; Rockefeller Found. fellow Western Res. U., U. Chgo., 1928-29; research asst., asso. prof. physiology, aviation medicine U. Wuerzburg, 1929-35; dir. Aeromed. Research Inst., Berlin, asso.. prof. physiology U. Berlin, 1935-45; prof. physiology, dir. Physiol. Inst. U. Heidelberg, 1946-47; with Air Force Sch. Aviation Medicine, Randolph AFB, Tex., 1947-59, chief dept. space medicine, 1949-57, advisor for research, 1959-62; chief scientist Aerospace Med. div. Brooks AFB, Tex., 1962—. Recipient John Jeffries award, 1958, Louis H. Bauer Founders award, 1965. Fellow Aerospace Med. Assn., Am. Inst. Aero. and Astronautics Aerospace Med. Assn.; mem. Am. Astronautic Soc., Internat. Acad. Astronautics, Internat. Acad. Aviation and Space Medicine, German Physiol. Soc. Author: The Green and Red Planet: A Physiological Study of the Possibility of Life on Mars, 1953; (with others) Principles of Aviation Medicine, 1939. Contbr. numerous articles in field to sci. jours. Contbns. to aviation medicine, space medicine, advancement of astronautics through med. research; creation of Hubertus Strughold award by Space Medicine br. Aerospace Med. Assn., 1961. Home: Menger Hotel, San Antonio, Tex. 78205. Office: Hdqrs. Aerospace Med. Div., Brooks AFB, Tex. 78235.*

STRUIK, Dirk Jan, mathematician; b. Rotterdam, Netherlands, Sept. 30, 1894; s. Hendrik Jan and Aartje (Schilperoort) S.; Ph.D., U. Leyden (Netherlands), 1922; m. Saly Ruth Ramler, July 14, 1923; children—Ruth Rebekka, Anne Nicolette (Mrs. Robert Macchi), Gwendolyn Jessica (Mrs. Roger Bray). Came to U.S., 1926, naturalized, 1934. Asst. mathematician Technische Hoogeschool, Delft, Netherlands, 1917-24; with U. Rome (Italy), 1924-25, U. Göttingen (Germany), 1925-26; faculty Mass. Inst. Tech., Cambridge, 1926—, prof. math., 1940-60, prof. emeritus, 1960—; guest prof. Nacional U. Mexico, 1934, 57, 60, 63, U. P.R., 1962, U. Utrecht (Netherlands), 1963-64, U. Costa Rica, 1965. Lobacevskii citation U. Kazan (USSR), 1928. Fellow Am. Acad. Arts and Scis.; corr. Royal Acad. Sci. Amsterdam; mem. Internat. Acad. History Sci. (Paris), Am. Math. Soc., Am. Math. Assn., Wiskundig Genootschap. Author: Grundzüge der Mehrdimensionalen Differentialgeometrie, 1922; (with J.A. Schouten) Einführung in die Neueren Methoden der Differentialgeometrie, 2 vols., 1935-38; Yankee Science in the Making, 1948; Concise History of Mathematics, 1948; Het Land von Stevin en Huygens, 1958; Economic and Philosophical Manuscripts of Karl Marx, 1964; Willem Gillisz Van Wissekerke, 1965; also numerous articles. Research on differential geometry, tensor calculus, history sci., hydrodynamics. Home: 52 Glendale Rd., Belmont, Mass. 02178.*

STRUM, Jacques Charles François, French mathematician; b. Geneva, Switzerland, 1803; tutor for Broglie family; became prof. spl. math. Rollin Coll., after July Revolution; named asst. master analysis Polytechnic Sci. sch., 1838, successor to Poisson, 1840. Recipient (with Colladon) Grand prize in math. for determining speed of sound in water, 1827. Mem. French Acad. Scis. Author: Cours de mecanique de l'Ecole Polytechnic, 1861; Cours d'analyse de l'Ecole polytechnic, 1857-63. Research on differential equations, optics, mechanics; originator theorem which permits determination of number of real roots of a numerical equation included between 2 numbers. Died Paris, 1855.

STRUMINSKII, Vladimir Vasil'evich, Russian mech. engr.; b. Apr. 29, 1914; grad. Moscow U., 1938. Became asso. Central Aero-Hydrodynamics Inst., 1941. Recipient Stalin prize, 1947, 48. Corr. mem. USSR Acad. Scis. Author: Wing Side-Slip in Viscous and Compressible Gas, 1946; Wing Side-Slip in a Viscous Liquid, 1946; The Theory of a Non-Stationary Boundary Layer, 1957; (with N. K. Lebed) A Method of Determining the Distribution of Circulation Along the Sweep of a Delta Wing, 1957; The Theory of Spatial Boundary Layer on a Side-Slip Wing, 1957. Research in aerodynamics, including devel. theories on a boundary layer on sliding wing, 1946, non-stationary boundary layer, 1948, 3 dimensional boundary layer for arbitrary surface, 1952. Office: Central Aero-Hydrodynamics Inst., USSR Acad. Scis., Moscow, USSR.

STRUNZ, Hugo, German mineralogist; b. Weiden, Germany, Feb. 24, 1910; s. Kuno and Bertha Strunz; Ph.D.; agrégé; Dr.ès.sc.; m. 1939; children—Rainer, Volker, Stephan. Author: Mineralogische Tabellen, Struktur und Klassifikaton der Mineralien, 1941-63; Mineralien und Lagerstätten in Ostbayern, 1952; Die Uranfunde in Bayern von 1804 bis 1962, 1962. Research and numerous publs. on crystallography, ore depositis, mineralogy, pegmatites, especially of Bavaria, S. Africa; discovered 20 new kinds of minerals, various kinds of isotypy and isomorphy. Address: Technische Universität, Berlin 12, West Germany.

STRUPP, Hans Hermann, psychologist; b. Frankfort/Main, Germany, Aug. 25, 1921; s. Josef and Anna (Metzger) S.; came to U. S., 1939, naturalized, 1945; student Coll. City N.Y., 1939-40; A.B., George Washington U., 1945, A.M., 1947, Ph.D., 1954; m. Lottie Metzger, Aug. 19, 1951; children—Karen Ruth, Barbara Jean, John Allen. Research psychologist USAF, 1949-54; supervisory research psychologist Adj. Gen. Office, U.S. Army, 1954-55; project dir. George Washington U., 1955-57; faculty U. N.C., Chapel Hill, 1957-66, prof. psychology, 1962-66; prof. Vanderbilt U., Nashville, 1966—. Cons. VA, 1961—; Lasker Meml. lectr., 1960. Recipient Helen D. Sargent Meml. Prize, Menninger Found., 1963. Diplomate in Clin. Psychology, Am. Bd. Examiners in Profl. Psychology. Fellow Am. Psychol. Assn.; mem. Sci. Asso.., Am. Acad. Psychoanalysis A.A.A.S., Psychologists Interested Advancement Psychotherapy (pres. 1965-66), Sigma Xi. Author: Psychotherapists in Action, 1960; also numerous articles. Editor: (with L. Luborsky) Research in Psychotherapy, Vol. 2, 1962. Research in psychoanalysis and psychotherapy, particularly therapist's contbn. to treatment process. Home: 5058 Villa Crest Dr., Nashville 37220.*

STRUTHERS, John, physician, anatomist; b. Brucefield, Eng., 1823; s. Alexander Struthers; ed. Edinburgh U.; m. Christina Alexander, 1857; lectr. anatomy, 1845-63; surgeon Edinburgh Royal Infirmary; prof. anatomy, Aberdeen, Scotland, 1863-89; mem. Gen. Med. Council, 1883-91; v.p. Royal Coll. Surgeons, Edinburgh, examiner in anatomy, from 1890. Author: Anatomical and Physiological Observations, 1854; Memoir on Anatomy of Humpback Whale, 1887; also papers. Introduced term Colles' fascia, 1854, for fascia described by Abraham Colles in 1811. Died 1899.

STRUTT, John William (3d Baron Rayleigh), English mathematician, physicist; b. Lanford Grove, nr. Maddon, Witham, Essex, Eng., Nov. 12, 1842; s. John

James and Elizabeth (Vicars) S.; studied math. Trinity Coll., Cambridge, 1861-65, B.A., 1865 (senior wrangler); fellow, Trinity Coll., 1870; m. Evelyn Balfour, 1871; 3 sons. Dir. Cavendish Lab., Cambridge U., 1879-84, univ. chancellor, 1908; prof. natural philosophy Royal Inst. Gt. Britain, 1887; helped found Nat. Phys. Lab., Teddington, Eng., 1900; pres. com. on aeronautics, 1909; sci. adviser Trinity House, 1896; mem. Privy Council, 1905. Recipient Rumford, Copley, Faraday medals; (with others) Nobel prize in physics, 1904. Fellow Royal Soc., 1873 (pres. 1905); pres. Internat. Congress of Electricians, 1908. Author: Treatise on the Theory of Sound, 1877; also papers. Research on light and color, dynamics of resonance and vibrations of gas and elastic solids, electricity, 1869-79; re-determined elec. units in absolute measurement, 1879-83; brought attention to memoir of J. J. Wateron which had anitcipated features of kinetic theory of gases in 1846; found that density of nitrogen from air is heavier than that from ammonia, 1893, discovered (with Ramsay) argon, 1894; investigated phys. optics and color vision; studied optical scattering, showing this why sky is blue; analyzed mathematically resolving powers of prisms, diffraction gratings. Died Witham, Eng., June 30, 1919.

STRUTT, Maximilian Julius Otto, electronics engr.; b. Soerakarta, Java, Oct. 2, 1903; s. Julius Otto and Hendrika (Heusser) S.; B.Sc.Eng., U. Munich, 1924; M.Sc.Eng., Tech. U., Delft, Netherlands, 1926, D. techn.Sc., 1927; m. Elfriede Schaefer, Oct. 18, 1932; 1 dau., Helga (Mrs. C. Villalaz). Teaching asst., patent engr. Tech. U. Delft, 1926-27; research fellow Philips Co., Eindhoven, Netherlands, 1926-46, electronics cons., 1946-48; prof., dir. dept. advanced elec. engring. Swiss Fed. Inst. Tech., Zurich, 1948—; indsl. electronics cons., 1948—; vis. prof. U. Cal. at Berkeley, 1961, 62, 63, 66-67. Mem. sr. sci. adv. com. NASA Project, 1963. Recipient C. F. Gauss medal, 1954. Dr.Eng. (hon.), U. Karlsruhe (Germany), 1950. Fellow I.E.E.E.; hon. mem. German and Japanese socs. Author numerous books, including: Skineffect en Temperatuurverdeeling in Electrische Geleiders, 1927; Lamé'sche-Mathieu'sche- und verwandte Funktionen in Physik und Technik, 1932; Moderne Mehrgitter-Elektronenröhren, Band 1, 1937; Ultra- and Extreme Short Wave Reception, 1947; Ferritas, 1950; Anleitung zur Vorlesung Höhere Elektrotechnik I und III, 1957; (with F. Vilbig) Forschritte der Hochfrequenztechnik, 1960; Vorlesung über Feldtheorie, 1964; Vorlesung über Lichttechik, 1965; Vorlesung über Festkörpertechnik, Transistoren, Elektronenröhren, 1965; Semiconductor Devices, Vol. 1 Semiconductors and Semiconductor Diodes, 1966; also numerous articles. Research and numerous patents loudspeaker systems and room acoustics, in multigrid electron tubes, UHF tubes, transistor and laser circuits, fast automatic spectrography. Home: 79 Kraehbuehl, Zürich 8044, Switzerland.*

STRUTT, William, English inventor; b. Eng., 1756; s. Jedediah and Elizabeth (Woollatt) S.; m. Barbara Evans; 1 son, Edward; 3 daus. Fellow Royal Soc., 1817 Developed method of ventilating and warming large bldgs. (used at Derbyshire Gen. Infirmary); invented Belper stove, 1806, form of self-acting spinning mule. Died Dec. 29, 1830.

STRUVE, Friedrich Georg Wilhelm von, astronomer; b. Altona, Germany, Apr. 15, 1793; ed. Dorpat (now Tartu, Estonia); at least 1 son, Otto Wilhelm; became observer Dorpat obs., 1813, dir., 1817; apptd. dir. Pulkovo obs., St. Petersburg, 1834. Mem. French Acad. Scis. Author: Catalogus novus stellarum duplicium, 1827; Stellarum duplicium mensurae micrometricae, 1937; Stellarum fixarum imprimis duplicium positiones mediae, 1852; Arc du méridien entre le Danube et la Mer Glaciale, 2 vols., 1857-60. Research on double stars; pioneer in measurement of stellar parallax (alpha Lyrae); made geodetic survey of Livonia, 1816-19; measured arc of meridian, 1822-27. Died St. Petersburg, Nov. 23. 1864.

STRUVE, Hermann von, German astronomer; b. Pulkovo/St. Petersburg, Russia, Oct. 3, 1854; prof., St. Petersburg, also Königsberg (now Kaliningrad, USSR), Berlin. founder obs., Neubabelsberg, Germany. Author: Beobachtungen der Staurntrabanten, 1898; Beobachtungen der Marstrabanten, 1898. Studied Saturn and Mars and their orbits. Died Herrenalb/Schwarzwald, Germany, Aug. 12, 1920.

STRUVE, Ludwig von, astronomer; b. 1858; s. Otto Wilhelm von Struve; became prof., dir. obs. U. Kharkov (Russia), 1894; apptd. prof., Simferopol, USSR, 1919. Research on determination of constant precession, also on proper motion of solar system. Died 1920.

STRUVE, Otto, astronomer; b. Kharkov, Russia, Aug. 12, 1897; s. Ludwig and Elisabeth (Grohmann) S.; student Michael Artillery Sch., Petrograd, Russia, 1916-17; Diploma of First Rank, U. of Kharkov, 1919; Ph.D., U. Chgo., 1923; hon. Sc.D., Case Sch. Applied Sci., 1939, U. Pa., 1956; Ph.D., (hon.), Copenhagen, 1946, U. Mexico, 1951; D.Sc. (hon.), Liège U., 1949; Wesleyan U., 1960; D. Phil., Kiel U., 1960; m. Mary Lanning, May 21, 1925. Came to America, 1921, naturalized, 1927. Asst. in astronomy, Yerkes Obs., 1921-23, instr., 1924-27, asst. prof., 1927-30, asso. prof., 1930-32, asst. dir., 1931-32, dir. 1932-47, chmn. and hon. dir., 1947-50; prof. astrophysics, U. of Chicago, 1932-47, Andrew MacLeish Distinguished Service prof., 1946-50; dir. McDonald Observatory of University of Texas, 1932-47, honoray director, 1947-50, chairman astronomy dept., 1947-49; professor of astrophysics, chmn. dept. dir. Leuschner Obs. U. Calif. at Berkeley 1950-59; director National Rario Astronomy Obs., 1959—; editor: The Astrophysical Journal, 1932-47; fellow International Education Board, Mt. Wilson Observatory, 1926; Guggenheim Foundation fellow, Cambridge (England) U., 1928. Trustee Associated Universities of N.Y., 1957-59, Associated Universities Incorporated of N.Y., 1959. Served as lt. in Imperial Russian Army, 1916-17; lt. White Russian Army, 1919-21. Fellow A.A.A.S.; mem. Nat. Acad. of Sciences, Am. Philos. Soc., Am. Astron. Soc. (pres. 1946), Astron. Soc. of Pacific (pres. 1951), Am. Phys. Soc., Wis., Cal. acads. sci., Am. Acad. Arts and Sci., Sigma Xi, Internat. Astron. Union (pres., 1952-55), Amsterdam, Stockholm and Oslo Acads., Uppsala Soc. Scis., Society Astr. de France, Astr. Gesellschaft; corr. mem. Société Royale des Sciences de Liège, Acad. Sci., Copenhagen, Haarlem (Holland) Soc. of Sciences; fgn. asso. mem. Royal Astron. Soc. (Eng.); fgn. mem. Royal Soc. London, Edinburgh; hon. mem. Royal Astr. Soc. Can. Decorated Chevalier, Comdr. Order of Crown (Belgium); Gold Medal, Royal Astr. Soc. (London), 1944; Bruce Gold Medal, San Francisco, 1948; Draper Gold Medal, Nat. Acad. Sci., 1950; Rittenhouse Medal, Phila., 1954; Janssengold medal Paris Acad. Sci., 1955; Bruce Blair award, 1956. Works include: Stellar Evolution, 1950; The Universe, 1962; (with Velta Zebergs) Astronomy of the 20th Century, 1962. Noted for interstellar studies; discovered interstellar hydrogen, 1938; 1st to show rapid rotation of hightemperature stars; work in stellar spectroscopy; developed theory of origin of the universe. Died Berkeley, Cal., Apr. 6, 1963.

STRUVE, Otto Wilhelm von, German astronomer; b. Dorpat (now Tartu, Estonia), May 7, 1819; s. Friedrich Georg Wilhelm von Struve; at least 1 son, Ludwig; dir. Pulkovo obs., St. Petersburg, 1862-89; mem. French Acad. Scis. Calculated movement of sun; discovered a satellite of Uranus, 1847, also about 500 new double stars; studied rings of Saturn; determined mass of Neptune; redetermined constant of precession. Died Karlsruhe, Germany, Apr. 16, 1905.

STRYCKERS, Joseph Maria Theodoor, Belgium weed scientist; b. Hasselt, Belgium, Oct. 14, 1921; s. Joseph Henri and Francisca (Berghs) S.; Landbouwkundig Ingenieur with distinction, Rijksfaculteit der Landbouwwetenschappen, Ghent, Belgium, 1946, Ingenieur voor Waters en Bossen with great distinction, 1947, Doctor in de Landbouwkundige Wetenschappen with greatest distinction, 1958; m. Alice Theelen, June 29, 1948; children—Reinhilde, Ingrid. With Rijksfaculteit der Landbouwwetenschappen, Ghent, 1947-—, docent, 1960-62, prof. (leerstoel voor plantenteelt), 1962-—, dir. center for weed research Centrum voor Onkruidonderzoek, 1964-—. Pres., European Weed Research Council, also pres. com. on herbicide evaluation. Decorated officer Order of Leopold II. Author: Onkruidbestrijding, 5th rev. edit. 1964; Flora en Gewassen in Oost-Vlaanderen, 1967. Research, publs on ecol. behavior of grassland plants, improvement of permanent pastures, herbicide evaluation research, weed research and application of herbicides in agronomic crops, hort. crops, others; research on weed control of non-cropped lands and on control of aquatic weeds. Home: 25, Europalaan, St.-Denijs-Westrem. Office: 235 Coupure Links, Ghent, Belgium.*

STRZALKOWSKI, Adam, Polish physicist; b. Tenczynek, Poland, Nov. 26, 1923; s. Karol and Natalia (Grychowska) S.; Magister, Jagellonian U., Crawcow, Poland, 1948, Doctor, 1960, Docent, 1963; m. Maria Bednarek, May 14, 1949; children—Wojciech, Wanda. Staff, Jagellonian U., 1945-61, asso. prof. Inst. Physics, 1963-—, head dept. nuclear physics and its applications, 1963—; research fellow Nuclear Physics Research Lab., U. Liverpool (Eng.), 1957-59; sci. officer Inst. Nuclear Physics, Cracow, 1955-—, head nuclear reaction lab., 1959—; vice dir. Center Nuclear Physics, Cracow, 1959-61. Decorated Polonia Restituta Cross. Mem. Polish Phys. Soc. Author: Introduction to Nuclear Physics, 1966; also articles. Research in astronomy, eclipsing variables, scattering of light in earth's atmosphere, solar radioastronomy, polarization of particles in nuclear reactions, problems of optical model of interaction of particles with nuclei, direct interaction nuclear reactions. Home: 10 Smoluchowskiego, Cracow, Poland.*

STRZELECKI, Sir Paul Edmund, explorer; b. nr. Poznan, Poland, 1796; ed. High Sch., Edinburgh, D.C.L., U. Oxford, 1860. Naturalized Brit. citizen, 1850. Fellow Royal Soc., 1853. Explored interior of Australia, beginning, 1839; discovered gold, Wellington Dist., Australia, named a commr. for distbn. Irish Famine Relief Fund, 1847-48; assisted in promotion emigration to Australia. Strzelecki range of hills, Victoria, Australia, Strzelecki Creek, S. Australia, also various species of plants and animals named in his honor. Author: Physical Description of New South Wales, 1845. Studies on flora, fauna, aborigines, geology and mineralogy of Australia. Died Oct. 6, 1873.

STUART, Herbert Arthur, physicist; b. Zurich, Switzerland, Mar. 27, 1899; s. William and Hildegard (Oeffinger) S.; ed. U. Göttingen; agrégé; m. Elisabeth, May 22, 1926; children—Renate, Karin, Eckhard. Became scholar U. Königsberg (now Kaliningrad, USSR), 1925, prof. at large, 1935; Rockefeller fellow U. Cal. at Berkeley, 1930-31; became prof. theoretical physics, dir. Physics Inst., Dresden Tech. Sch., 1936; became titular prof. phys.-chemistry U. Mainz (Germany), 1955. Research and publs. on molecular structure, free molecule structure, molecular structure and physics of high polymers. Home: Ft. Elisabeth 15, Mainz, West Germany.

STUART, John McDouall, explorer; b. Fifeshire, Scotland, Sept. 7, 1815; s. William Stuart; ed. privately Edinburgh; later student Mil. Acad., Edinburgh. Entered business, Scotland; emigrated to S. Australia, 1838; staff govt. survey; later pvt. surveyor; returned to Eng., 1864. Recipient prize for 1st colonist crossing country (however John McKinlay had preceded him); Gold medal Royal Geog. Soc.; Stuart's Creek named in his honor. Led several expdns. to Australian interior; lead 1st expdn. for discovery path across Australia, 1858. Died June 5, 1866.

STUBBLEFIELD, (Cyril) James, Brit. geologist; b. Cambridge, England, Sept. 6, 1901; s. James and Jane (Goodier) S.; student Chelsea Poly., 1918-21; B.Sc., Imperial Coll. Sci., London, 1923, Ph.D., 1925, D.Sc., 1942; D.Sc. (hon.), U. Southampton (Eng.), 1965; m. Emily Muriel Elizabeth Yakchee, June 11, 1932; children—Peter Jackson, Rodney George. Demonstrator geology Imperial Coll. Sci., 1923-28; staff Geol. Survey Gt. Britain and Mus. Practical Geology, 1928-66, asst. dir., 1953-60, dir., 1960-66; dir. Geol. Survey No. Ireland, 1960-66. Pres. 6th Internat. Congress Carboniferous Stratigraphy and Geology, 1967. Created Knight Bachelor, 1965. Fellow Royal Soc., 1944; mem. Geol. Soc. London (past pres.), Paleontographical Soc. (pres., past sec.), Brit. Assn. for Advancement Sci. (pres. sect. pres.). Co-editor: Handbook of Geology of Great Britain, 1929. Reviser: (Morley Davies) Introduction to Palaeontology, 1961. Research, publs. on stratigraphy and paleontology of older fossiliferous rocks chiefly those occurring in Gt. Britain, including those asso. with occurrence of coal, lead zinc minerals; studies on trilobites and graptolites. Home: 35 Kent Av., London W.13. Office: Imperial Coll. Sci., Prince Consort Rd., London S.W. 7, Eng.*

STUBBS, Morris Frank, Am. chemist; b. Sterling, Kan., May 25, 1898; s. Frank C. and Sarah (Morris) S.; A.B., Sterling Coll., 1921, D.Sc., 1960; M.S., U. Chgo., 1925, Ph.D., 1931; m. Lois K. MacCarthy, Dec. 26, 1923; 1 dau., Marilyn Joy. Sci. tchr. Elgin (Ill.) Jr. Coll. and Acad., 1921-23; head dept. phys. sci. Tenn. Wesleyan Coll., 1931-42; prof. chemistry, head dept. Carthage (Ill.) Coll., 1942-44, Tenn. Poly. Inst., Cookeville, 1944-46; prof., head dept. chemistry N.M. Inst. Mining and Tech., Socorro, 1946-63, dir. coll. div., 1962-63; prof. chemistry Tex. Technol. Coll., Lubbock, 1963-—. Recipient John Dustin Clark medal Central N.M. sect. Am. Chem. Soc., 1965. Fellow A.A.A.S. (past div. pres.); mem. Am. Chem. Soc. (past counselor, past chmn. sect.), Sigma Xi, Pi Kappa Delta. Author: (with W. Norton Jones) Laboratory Exercises in General Chemistry, 1954, Elementary Qualitative Analysis, 1956; Laboratory Exercises in General Chemistry, Including Qualitative Analysis, 1964; also articles. Research on chemistry indium; developed field tests for geochem. prospecting. Home: 3403 35th St., Lubbock, Tex. 79413.*

STUBEL, Alphonse, German geologist; b. Leipzig, Germany, 1835; ed. Leipzig, Heidelberg, Berlin; devoted career to study of volcanic phenomena. Author: (with Reiss) Excursion dans la region volcanique d'Egine, 1867, Histoire et description des éruptions de Santorin, 1868, La champ des morts d'Ancon, au Pérou, 1880-87, Voyages dans l'Amérique du Sud, 1890; (with Uhle) Les cites ruinées de Tiahuanaco dn dans le haut pays de l'ancien Pérou, 1892; Le massif volcanique de l'Equateur, 1897. Died Dresden, Germany, 1904.

STUCKENBERG, John Henry Wilburn, sociologist; b. Bramsche, Hanover, Germany, Jan. 6, 1835; s. Hermann R. and Anna Maria S.; A.M., Wittenberg Coll., O., 1860; studied divinity in same; divinity and philosophy in univs. of Halle, Göttingen, Berlin and Tübingen, Germany; D.D., Wooster Univ., 1875; LL.D., Pa. Coll., 1899; m. Mary Gingrich, Oct. 27, 1869. Pastor of chs. in Davenport, Ia., Erie, Pa., and Pittsburgh, Pa.; chaplain 145th Pa. vols., Civil war; with regt., Fredericksburg, Chancellorsville, Gettysburg. Prof. theol. dept., Wittenberg Coll., 1873-80; pastor Am. Ch., Berlin, Germany, 1881-94. Author: History of the Augsburg Confession, 1868; Christian Sociology, 1880; The Final Science, 1885; Introduction to the Study of Philosophy, 1888; The Age and the Church, 1893; Tendencies in German Thought, 1896; The Social Problem, 1897; Introduction to the Study of Sociology, 1898. Wrote several pamphlets and numerous articles on theol., ednl., philos. and sociol. subjects in Am. and English mags. and journals. Pioneer Am. sociologist; studied social ethics; first used term sociation to indicate forces that create a society as more than a mere collection of individuals. Died London, Eng., May 28, 1903.

STUCKEY, Jackson Henry, Am. physician; b. China, Tex., Jan. 12, 1916; s. Thomas W. and Ruth (Carter) S.; A.B., U. Tex., 1939; M.D., Yale, 1942; m. Charlotte Hathaway Vroom, 1962; 1 dau., Charlotte Elizabeth. Practice medicine, specializing in surgery, Bklyn., 1954—; prof. dept. surgery State U. N.Y., Downstate Med. Center, 1964—. Recipient Lederle Med. Faculty award, 1961-63. Career scientist Health Research Council City N.Y., 1963—. Diplomate Am. Bd. Surgery. Research, numerous publs. in cardiac surgery and cardiac physiology. Home: 160 Columbia Heights, Bklyn. 11201. Office: 450 Clarkson Av., Bklyn. 11203.*

STUDER, Alfred, Swiss pathologist; b. Liestal, Switzerland, Feb. 8, 1917; s. Emil and Sophie (Gerster) S.; state exam. Basel, U., 1942. Physician hosps., Geneva, Winterthur; faculty Path. Inst., U. Basel, 1953—, prof.; dep. dir., chief dept. exptl. medicine Hoffmann-La Roche & Co. Ltd., Basel. Mem. European Atherosclerosis Group, European Soc. for Study Drug Toxicity, Swiss, European hematological socs., Angiological Soc., Natural Sci. Soc., German Soc. Pathology, Free Assn. Swiss Pathologists, Swiss League Cancer Control. Author: Rheumatismus als Problem der experimentellen Medizin, 1959; Die Pathologie der Avitaminosen, 1962; Der Tierversuch in der Arterioskleroseforschung, 1963; also numerous articles. Research, publs. on blood coagulation, hematology, endocrinology, rheumatology, atherosclerosis, tuberculosis, vitaminology. Home: 11 Markgräflerstrasse. Office: 124 Grenzacherstrasse, Basel, Switzerland.*

STUDER, Bernhard, Swiss geologist; b. Büren, Switzerland, Aug. 21, 1794; ed. U. Göttingen, also in Freiburg, Berlin, Paris; became tchr. math. and physics Bern (Switzerland) Acad., 1816; prof. mineralogy and geology U. Bern, from 1834; pres. Commn. for Geol. Survey of Switzerland, from 1859. Recipient Wollaston medal Geol. Soc. London. Mem. French Acad. Scis., Societé helvetique des scis. naturelles, Geneva. Author: Geologie der westlichen-Schweizer-Alpen, 1834; Geologie der Schweiz, 2 vols., 1951-53; Geschichte der physichen Geographie der Schweiz, 1963; Index der Petrographie und Stratigraphie der Schweiz, 1872. Died Bern, May 2, 1887.

STUDIER, Martin Herman, Am. chemist; b. Leola, S.D., Nov. 10, 1917; s. Emil and Mathilda (Mueller) S.; B.A., Luther Coll., 1939; postgrad. Ia. State U., 1939-42; Ph.D., U. Chgo., 1947; m. Eleanor Anne Chrissinger, Dec. 31, 1944; children—James, Anne, John, Paul. With Metall. Lab., U. Chgo., 1943-47; staff Argonne (Ill.) Nat. Lab., 1947—, sr. chemist, group leader, 1948—. Mem. Am. Chem. Soc., Am. Phys. Soc., A.A.A.S., Research Soc. Am. Research, numerous publs. in nuclear chemistry of heavy elements, mass spectrometry with applications to noble gas compounds and organic matter in meteorites; co-discoverer einsteinium and fermium. Home: 4429 Downers Dr., Downers Grove, Ill. 60515. Office: 9700 S. Cass. Av., Argonne, Ill. 60440.*

STUDY, Eduard, German mathematician; b. Coburg, Germany, Mar. 23, 1862; prof. math., Bonn, also Greifswald, Germany. Author: Geometrie der Dynamen, 1903. Reformulated (independently of F. Severi) fundamental principle of enumerative geometry; studied (with J. L. Coolidge) straight lines in elliptic space; helped develop symbolic notation in theory of invariants; simplified method of differential operators and extended it to any quantic; used algebraic analysis instead of coordinate systems. Died Jan. 6, 1930.

STUECKELBERG VON BREIDENBACH ZU BREIDENSTEIN UND MELSBACH, Ernst Carl Gerlach (Baron Sovereign Holy Empire), physicist; b. Basel, Switzerland, Feb. 1, 1905; s. Alfred Stueckelberg and Alice von Breidenbach zu Breidenstein; student U. Munich; Ph.D., U. Basel, 1927; D.Sc.h.c., U. Neuchatel (Switzerland), 1962; Dr.phil.h.c., U. Bern, 1962; m. Blanche Morel, Sept. 17, 1931 (div. Aug. 1946); children—Alice Elisabeth May (Mrs. James Andre Favre), Bianca Maria (Mrs. Nicolas Plaoutine), Georg Heinrich; m. 2d, Charlotte Helena Lachat, Mar. 30, 1964. Research asso. Princeton, 1928-30, asst. prof., 1930-33; docent U. Zurich (Switzerland), 1933-35; faculty U. Geneva, 1935—, prof. theoretical physics, 1939—; faculty U. Lausanne, 1942—, prof., 1967—. With European Center Nuclear Research, Geneva, Switzerland. Fellow Am. Phys. Soc.; mem. Swiss Commn. Atomic Energy, Swiss Natural Sci. Soc., Soc. Physics and Natural History Geneva, Inst. Coimbra (Portugal); hon. mem. Phys. Soc. Rumania. Exptl. and theoretical work in molecular physics, relativistic quantum theory, theory of nuclear forces, gen. relativity and thermodynamics. Home: 20 Rue Henri-Mussard, Geneva, Switzerland; W. Germany. Office: Institute of Theoretical Physics, U. Geneva, 32 Bould. d'Yvoy, Geneva, Switzerland.*

STUETTGEN, Guenter, German physician; b. Düsseldorf, Germany, Jan. 23, 1919; s. Bernhard and Elisabeth (Semer) S.; student univs. of Freiburg, Marburg, Düsseldorf; M.D., U. Düsseldorf, 1943; m. Ruth Kanderske, June 10, 1944; children—Thomas, Ulrich. Resident physician, Jena, Düsseldorf, 1945. head physician Düsseldorf Dermatol. Hosp., 1953-65, became prof., 1957; head dept. dermatology U. Cologne, 1964-65; head physician Univ. Dermatol. Hosp., Frankfurt/Main, 1965—. Recipient Fed. Service Cross I, 1964. Mem. German, Brasilian, Italian dermatol. assns German Council Allergy. Author: Die normale und pathologische physiologie der Haut, 1965; also numerous articles. Research on biochemistry and pharmacology of skin, allergic dermatoses, burns, tropical dermatosis. Address: 14 Ludwig Rehnstrasse, Frankfurt /Main, West Germany.*

STUHLINGER, Ernst, physicist; b. Niederrimbach, Germany, Dec. 19, 1913; s. Ernst and Pauline (Werner) S.; Ph.D., U. Tuebingen, Germany, 1936; m. Irmgard Lotze, Aug. 1, 1950; children—Susanne, Tilman, Hans Christoph. Came to the United States in 1946, and became naturalized, 1955. Research cosmic rays, nuclear physics, 1934-41; asst. prof. Technische Hochschule, Berlin, Germany, 1936-41; guidance and control equipment Rocket Development Center, Peenemuende, Germany, 1943-45; with Guided Missile Development Office, Ft. Bliss, Tex., 1946-50; physicist Ordnance Missile Labs., Huntsville, Ala., 1950-56; physicist Army Ballistic Missile Agy., 1956-60; dir. Space Scis. Lab., George C. Marshall Space Flight Center, NASA, Huntsville, Ala., 1960—. Served from pfc. to cpl. German Army, 1941-43; Russian Campaign. Fellow Am. Astronautical Soc., Am. Rocket Society (dir.), Am. Inst. Aeros. and Astronautics (tech. dir.); mem. I.E.E.E., Internat. Acad. Astronautics, Rocket City Astron. Assn. (dir.), Deutsche Physikalische Gesellschaft, Deutsche Gesellschaft Fuer Raketen Technik und Raumfahrt (hon.), Hermann Oberth Gesellschaft (hon.), Assn. U. S. Army. Author: Ion Propulsion for Space Flight, 1964. Feasibility and design studies elec. propulsion systems for space ships, also artificial earth satellites. Home: 3106 Rowe Dr. Office: George C. Marshall Space Flight Center, NASA, Huntsville, Ala. 35801.*

STUHLMAN, Otto Jr., Am. physicist; b. Elberfeld, Germany, Nov. 12, 1884; s. Otto and Elizabeth (Henzman) S.; B.A., U. Cin., 1907; M.A., U. Ill., 1909; Ph.D. (Exptl. Sci. fellow), Princeton, 1911; m. Florence Lester, Aug. 22, 1962. Staff, Stevens Inst. Tech., 1912; instr., U. Pa., Phila., 1917; faculty U. Ia., 1918, U. W. Va., 1919; faculty U. N.C., Chapel Hill, 1920—, prof. physics, 1925-53, chmn. dept. physics, 1930, emeritus prof., 1953—. Recipient Potcat award N.C. Acad. Sci., 1947. Fellow Am. Physics Soc., A.A.A.S. Author: Introduction to Biophysics, 1943. Research on the photoelectric effect, spectra ultra violet, electron physics, biophysics, bioacoustics. Died Dec. 8, 1965.

STUIVER, Minze, geochemist; b. Vlagtwedde, The Netherlands, Oct. 25, 1929; s. Albert and Griet (Welles) S.; M.S. in Nuclear Physics, U. Groningen, 1953, Ph.D. in Biophysics, 1958; m. Anneke Hubbelmeyer, July 15, 1956; children—Ingrid, Yolande. Asst. prof. U. Groningen, The Netherlands, 1955-59; postdoctoral fellow Yale, New Haven, 1959-62, sr. research asso. biology and geology, 1962—, dir. Radiocarbon Lab., 1962—. Contbr. articles to profl. jours. Developer low level counting equipment for C14 age measurements, investigation relationship between solar activity and C14 variations in atmosphere during the past, Pleistocene climatic changes, carbon and sulfur metabolism of lakes; absolute sensitivity of olfactory receptors. Home: 53 Morse St., Hamden, Conn. 06517.*

STULBERG, Cyril Sidney, Am. microbiologist; b. Chgo., Apr. 11, 1919; s. Irwin and Mary (Fischel) S.; B.A., U. Minn., 1943, M.S., 1945, Ph.D., 1947; m. Elaine Helen Danzig, Aug. 20, 1944; children—Mary Beth, Peggy Ann, Mark Stephen. Instr., U. Minn., 1948-49; virologist Children's Fund Mich., Detroit, 1949-52; faculty Wayne State Coll. Medicine, 1953-61, asso. prof., 1958-61; sr. research asso. Child Research Center Mich., Detroit, 1953—; microbiologist Children's Hosp. Mich., Detroit, 1953—; prof. dept. microbiology Wayne State U. Sch. Medicine, Detroit, 1961—. Cons. hosps., govt. agys.; mem. Cell Culture Collection Com., 1960—; chmn. adv. com. on animal cell culture collection Am. Type Culture Collection, Rockville, Md., 1965—. Mem. Am. Soc. Microbiology, Am. Assn. Immunologists, Am. Assn. Cancer Research, Soc. for Exptl. Biology and Medicine, Tissue Culture Assn., Am. Acad. Microbiology, A.A.A.S., N.Y. Acad. Scis., Sigma Xi. Research, numerous publs. on human and animal virus infections, etiology and epidemiology infantile diarrheal disease; devel., characterization, and preservation animal and human tissue culture cells. Home: 23260 Berkley Av., Oak Park, Mich. 48237. Office: 660 Frederick St., Detroit 48202.*

STULBERG, Melvin Philip, Am. biochemist; b. Duluth, Minn., May 17, 1925; s. Irwin and Mary (Fischel) S.; B.S., U. Minn., 1949, M.S., 1955, Ph.D., 1957; m. Dorothy Elizabeth Bonnell, Sept. 8, 1955; children—Laurie Ann, Lisa Jean, Lynn Ellen. Research asso. Oak Ridge Nat. Lab., 1957-59, biochemist, 1959-61, 63—; biochemist U. S. AEC, Washington, 1961-63. Mem. Am. Soc. Biol. Chemists, Am. Chem. Soc., Sigma Xi. Research and publs. on mechanisms of protein biosynthesis including mechanism of amino acid activation, isolation of transfer RNA and their interactions, mechanism of enzyme action by use oxygen tracers. Home: 104 Woodridge Lane, Oak Ridge 37830. Office: Biology Div., Oak Ridge Nat. Lab., Oak Ridge 37831.*

STUMM, Erwin Charles, Am. paleontologist, geologist; b. Berkeley, Cal., Sept. 15, 1908; s. Ernest C. and Augusta (Eschle) S.; A.B., George Washington U., 1932, A.M., 1933; Ph.D., Princeton, 1936; m. Elizabeth Coon, Aug. 29, 1936; children—Virginia (Mrs. Roy H. Christensen, Diana, Ernest. Sci. aide U. S. Dept. Agr., Washington, 1926-30; staff U. S. Geol. Survey, 1930-33, 36-37; faculty Oberlin Coll., 1937-47, asst. prof., 1944-47; faculty U. Mich., Ann Arbor, 1947—, prof. geology, 1952—, curator Mus. Paleontology, 1947—. Fellow Geol. Soc. Am.; mem. Paleontol. Soc. (pres. 1966-67), Am. Assn. Petroleum Geologists, Soc. Econ. Paleontologists and Mineralogists, Mich. Geol. Soc., Mich. Acad. Sci., Arts, and Letters, Sigma Xi. Editor, Jour. Paleontology, 1957-63. Research and numerous publs. on Paleozoic paleontology and stratigraphy of Gt. Lakes region and Ohio Valley particularly fossil coral reefs. Home: 1610 Linwood Av., Ann Arbor, Mich. 48103.*

STUMPF, Carl, German psychologist, philosopher; b. Wiesenthied, Bavaria, Apr. 21, 1848; ed. U. Würzburg, 1865; student of Lotze, U. Göttingen; Dr. degree, Göttingen, 1868. Mem. faculty U. Würzburg, 1868-70; dozent U. Göttingen, 1870; prof. U. Würzburg, 1873; prof. philosophy U. Prague (Czechoslovakia), 1879; prof. U. Halle (Germany), 1884; lectr. U. Munich, 1889-94; prof. U. Berlin, 1894-97, rector, 1907-08; retired, 1921. Joint pres. Internat. Congress Psychology, Munich, 1896. Author: Über den psychologischen Ursprung der Raumvorstellung, 1873; Tonpsychologie, 2 vols., 1883-90; Über Leib und Seele, 1897; Beiträge zur Akustik und Musikwissenschaft, 1898; Erscheinungen und psychiashe Funktionen, 1907; Zur Einteilung der Wissenschaften, 1907; Philosophische Reden und Vorträge, 1910; Die Anfänge der Musik, 1911; Die Sprachlaute, 1926. Primarily philosopher, used psychology in its interest; influenced by Bretano; noted for pioneering work on psychology of music and tone; for expanding and bringing to great prominence psychol. lab. at Berlin; formulated theories on sensation and space perception; pioneered study child psychology; joint founder Berlin Verein fur Kinderpsychologie, 1900; founder archives for records primitive music, 1900; played important role in revision of concepts of psychophysics; formulated system of act psychology; forerunner of Gestalt psychology in work in exptl. phenomenology. Died Berlin, Dec. 25, 1936.

STUMPF, Paul Karl, Am. biochemist; b. N.Y.C., Feb. 23, 1919; s. Karl and Annette (Schreyer) S.; A.B. magna cum laude, Harvard, 1941; Ph.D., Columbia, 1945; m. Ruth R. Rodenbeck, June 13, 1947; children—Ann Carol, Kathryn Lee, Margaret Ruth, David Karl, Richard Frederic. Instr., U. Mich., 1946-48; faculty U. Cal. at Berkeley, 1948-58, prof. biochemistry, 1957-58; prof. biochemistry U. Cal. at Davis, 1958—, chmn. dept., 1965-68. Mem. physiol. chemistry study panel NIH, 1959-64; mem. adv. panel in metabolic biology NSF, 1965-68. NIH Sr. fellow, 1954-55; NSF Sr. fellow, 1961, 68; Guggenheim fellow, 1962, 69. Mem. Am. Soc. Biol. Chemists, Am. Soc. Plant Physiologists, Am. Soc. Oil Chemistry, Am. Chem. Soc., Biochem. Soc. (London, Eng.), NRC Author: (with J. B. Neilands) Outlines of Enzyme Chemistry, 1955; (with Eric E. Conn) Outlines of Biochemistry, 1963 also numerous articles. Exec. editor Archives of Biochem. Biophysiology, 1965—; editor Phytochemistry, 1965, Plant Physiology, 1967—, Biochem. Preparations. Research on lipid breakdown and synthesis in higher plants, photobiosynthesis lipids in chloroplasts, control mechanisms in lipid synthesis. Home: 764 Elmwood Dr., Davis, Cal. 95616.*

STUMPFF, Karl Johann Nikolaus, German meteorologist; b. Schleswig, Germany, May 17, 1895; s. Carl and Bertha (Nielsen) S.; ed. U. Kiel, U. Göttingen; Ph.D.; m. Elisabeth Grote, Jan. 5, 1924; children—Peter, Johanna Klaus-Hinrich, Ludwig. Asst., U. Breslau (now Wroclaw, Poland) Obs., 1925-34; observer U. Berlin Meteorol. Inst., 1934-42; dir. U. Graz (Austria) Obs., 1942-46; prof. U. Göttingen (West Germany), 1952—. Mem. Astron. Soc., Internat. Astron. Union. Author: Grundlagen und Methoden der Periodenforschung, 1937; Geographische Ortsbestimmungen, 1955; Astronomie, 1957; Himmelsmechanik, 1959; Wunder der Himmel, 1963. Address: Geismarlandstrasse 13, Göttingen, West Germany.

STUNKARD, Albert James, Am. physician; b. N.Y.C., Feb. 7, 1922; s. Horace Wesley and Frances (Klank) S.; B.S., Yale, 1943; M.D., Columbia, 1945. Research fellow psychiatry, 1951-52; research fellow medicine Columbia U. Service, Goldwater Meml. Hosp., N.Y.C., 1952-53; Commonwealth research fellow, then asst. prof. medicine Cornell U. Med. Coll., 1953-57; mem. faculty U. Pa., 1957—, prof. psychiatry, chmn. dept., 1962—. Mem. Am. Psychiat. Assn., Am. Psychosomatic Soc. Research and publs. on description of human obesity, its dependence on social factors, its behavioral pathology; disturbances in body image of obese persons. Home: 2415 Pine St., Phila. 19103. Office: Hosp. U. Pa., Phila. 19104.*

STUNKARD, Horace Wesley, Am. zoologist; b. Monmouth, Ia., Aug. 23, 1889; s. Hiram Wesley and Lula (Hopkins) S.; B.S. magna cum laude, Coe Coll., 1912, D.Sc., 1937; A.M., U. Ill., 1914, Ph.D., 1916; Sc.D., N.Y. U., 1954; m. Frances Grace Klank, June 12, 1920; children—Albert James, Eunice (Mrs. John

Ralph Latham). Faculty, N.Y. U., N.Y.C., 1916-—, prof., head dept. biology, 1925-54, prof. emeritus, 1954-—; research asso. Am. Mus. Natural History, N.Y.C., 1921-—. Lectr., U. B.C., Vancouver, Can., 1962; cons. U. S. Fish and Wildlife Service, 1951-62. Recipient Morrison prize N.Y. Acad. Sci., 1929; named Officier de l'Ordre du Merite pour la Recherche et l'Invention, Paris, France, 1964; Guggenheim Found. fellow, 1931-32; Oberlaender Trust fellow, 1938-39. Fellow Royal Soc. Tropical Medicine; mem. Am. Micros. Soc. (past pres.), Am. Soc. Parasitologists (past pres.), N.Y. Acad. Scis. (past pres.), N.Y. Soc. Tropical Medicine (past pres.), Soc. Systematic Zoology, (past Pres.), Am. Zool. Soc. (past v.p.), A.A.-A.S. (past v.p.). Contbr. numerous articles to tech. jours. Editor, Platyhelminthes and Mesozoa, 1926-—; editor Jour. Parasitology, 1944-54. Research on morphology, physiology, life-cycles, developmental and larval stages protozoan and helminthic parasites fishes, reptiles, birds and mammals and effects of these infections; origin, evolution and results of parasitic existence. Home: 5000 Waldo Av., N.Y.C. 10471. Office: Am. Mus. Natural History, N.Y.C. 10024.*

STURGEON, William, English inventor; b. Lancashire, Eng., May 22, 1783; s. John and Betsy (Adcock) S.; apprenticed to shoemaker trade; m. Mrs. Hilton; 3 children (all dec. in infancy); m. 2d, Mary Bromley, 1829; 1 child (dec. in infancy); 1 adopted dau. Ellen Coates (Mrs. Luke Brierley). Served in Army, 1802-20; opened shoemaker's shop, Woolwich, Eng.; apptd. lectr. sci. E. India Coll., Addiscombe, Eng., 1824; supt. Victoria Gallery Practical Sci., Manchester; became itinerant lectr., 1843; granted pension by Lord John Russell, 1849. Author: Experimental Researches, 1830; Scientific Researches (collected works), 1850; Twelve Elementary Lectures on Galvanism, 1843; also articles. Editor: Magnetical Advertisements (William Barlow), reissue 1843. Founder, Annals Electricity (1st elec. jour. in Eng.), 1836. Improved equipment for electromagnetic research; invented soft-iron electro-magnet; invented dynamo, 1823, electromagnetic rotary engine, 1832, electromagnetic coil machine, 1837; described method of amalgamating zinc plate of battery with film of mercury. Died Dec. 4, 1880.

STURHAN, Dieter, German zoologist; b. Meerbeck, Germany, Sept. 30, 1936; s. Eugen and Else (Glauert) S.; ed. U. Kiel, U. Munich; Dr.rer.nat., U. Erlangen, 1962; m. Hilke Gösselkeheld, May 5, 1964; 1 dau., Katja. Asst., Bayerische Landesanstalt für Pflanzenbau und Pflanzenschutz, Munich, 1958-60; nematologist Biologische Bundesanstalt, Institut für Hackfruchtkrankheiten und Nematodenforschung, Münster, 1962-—. Mem. Soc. European Nematologists, Deutsche Phytomedizinische Gesellschaft, Deutsche Zoologen Gesellschaft, Deutsche Ornithologen Gesellschaft. Author: Die Vogelwelt Schaumburg-Lippes, 1959; also articles. Research on avifaunistics and mammalogics, nematology, taxonomy, biol. races and genetics in plant and soil nematodes. Home: Rumphorstweg 29. Office: Toppheideweg 88, 44 Münster/Westf, Germany.*

STURKIE, Paul David, Am. physiologist; b. Hasse, Tex., Sept. 18, 1909; s. William D. and Lou M. (Luker) S.; B.S., Tex. A. and M. Coll., 1933, M.S., 1936; Ph.D., Cornell U., 1939; m. Betty Failmezger, Apr. 2, 1964; children— (by previous marriage) David P., Peggy; stepchildren—Robin, Stephanie, Victor. Asso. prof. poultry sci. Auburn (Ala.) U., 1939-44; mem. faculty Rutgers U., New Brunswick, N.J., 1944-—, prof. avian physiology, 1950-—, prof. physiology, 1963-—, chmn. dept. poultry sci., 1961-63, chmn. div. physiology, dept. animal sci., 1963-—. Recipient Poultry Sci. award Poultry Sci. Assn., 1947; Borden Co. award, 1956. Fellow A.A.A.S., Royal Soc. Edinburgh; mem. Am. Physiol. Soc., Poultry Sci. Assn., Am. Heart Assn., N.J. Acad. Medicine. Author: Avian Physiology, 1954, 65; also numerous articles. Research in heart and circulation of chickens, particularly blood pressure and relationship to health and disease, cardiac output, electro-cardiography, physiology of reproduction. Home: 103 Fern Rd., East Brunswick, N.J.*

STURM, Jacques Charles François, mathematician; b. Geneva, Switzerland, Sept. 29, 1803; ed. Coll. Geneva; tutor to son of Mme. de Staël; became prof. math. Rollin Coll., 1830, Poly. Sch., Paris, 1840; later prof. mechanics Faculty Scis., Paris, Fellow Royal Soc., 1840; mem. French Acad. Scis., 1836. Author: Mémoire sur la compressibilité des liquides, 1827; Cours d'analyse, 2 vols., 1857-59; Cours de mécanique, 1861; also treatise on theory of sight, 1845. Research on optics, mechanics, differential equations; originated Sturm's theorem for determining number and position of real roots of algebraic equation between given limits, 1829; measured (with Colladon) velocity of sound in water by means of bell submerged in Lake Geneva, 1826, also investigated compressability of liquids, 1834. Died Paris, Dec. 18, 1855.

STURM, John Christopher, German mathematician; b. Hippolstein, Bavaria, 1635; ed. U. Jena (Germany); at least 1 son, Leonhard Christoph; 1st minister, ch. in Germany, 5 years; prof. math. and natural philosophy, Altdorf, from 1669. Author: Collegium experimentale sive curiosum, in quo primaria hujus seculi inventa et experimenta physico-mathematica, an 1672, quibusdam naturae scrutatoribus spectanda exhibuit, et ad causas suas naturales demonstrativa methodo reduixit Johannes Christophorus Sturmius (mentions Leslie's differential thermometer); Mathesis Enucleata; Mathesis Juvenilis; also German transl. of Archimedes. Diffused knowledge and discoveries of 17th century by means of lectures and writings. Died Hippolstein, 1703.

STURMTHAL, Adolf Fox, economist; b. Vienna, Austria, Sept. 10, 1903; s. Leopold and Anna (Fuchs) S.; Dr.rerum Politicarum in Econs., U. Vienna, 1925; m. Hattie Ross, June 25, 1940; children—Jean Frances, Anne L.R., Suzanne M.L. Came to U.S., 1938, naturalized, 1943. Editor, Internat. Information, 1926-36, London corr., 1936-38; lectr. Am. U., 1939-40; faucity Bard Coll., Columbia U., 1940-55, prof., 1942-55; Philip Murray prof. Roosevelt U., Chgo., 1955-60; prof. labor and indsl. relations, U. Ill., Urbana, 1960-—; vis. prof. Cornell U., 1952-54, Grad. Sch. Bus., Columbia U., 1958-59, Yale, 1962-63; chief spl. reports sect. Fgn. Broadcasting Intelligence Service, 1944; lectr. U. Chgo., 1958-59. Cons. pvt. cos., govt. agys.; research asso. Center for Econ. Devel. and Cultural Change, U. Chgo., 1958-60; chmn. exec. com. Midwest Research Conf. on Econ. Devel., 1959-63. Mem. Am. Econ. Assn., Am. Polit. Sci. Assn., Indsl. Relations Research Assn. Author: Switzerland at the Crossroads, 1935; The Great Depression, 1938; A Survey of Literature on Postwar Reconstruction, 1944; Portrait der Amerikanischen Gewerkschaften, 1950; The Tragedy of European Labor, 1918-39, 1951; Unity and Diversity in European Labor, 1953; Contemporary Collective Bargaining, 1957; (with David Felix) U. S. Business and Labor in Latin America, 1960; Workers Councils, 1964; (with W. H. Franke) Current Manpower Problems, 1964; White Collar Trade Unions, 1966; also articles. Research on theory labor movement, analysis history European labor in terms distinction between pressure groups and polit. parties, relations between changes in structure labor force and devel. indsl. relations systems, methodology comparative internat. research in social scis. Home: 61 Greencroft Dr., Champaign, Ill. 61820.*

STURT, Charles, explorer; b. Bengal Presidency, Apr. 28, 1795; s. Thomas Lenox Napier and Jannette (Wilson) S.; ed. Astbury, Cheshire, Eng., Harrow; studied under Mr. Preston, nr. Cambridge, Eng.; m. Charlotte Christian Sheppey, 1834; 3 sons, including Napier George; 1 dau. Joined Army, 1813, entered Paris with his regt., 1815, participated in Whiteboy riots, Ireland; named mil. sec. to Sir Ralph Darling, gov. of New South Wales, Australia, 1827; leader various expdns. to interior of Australia; became asst. commr. lands, S. Australia, 1839; col. sec., 1849; returned to Eng., 1853. Recipient Founder's Gold medal Royal Geol. Soc., 1847. Fellow Royal Geog Soc., Linnean Soc. Author: Journals, 1833; Narrative of an Expedition into Central Australia, 1849. Surveyed and opened up largest river system of Australia and S. Australia. Died June 16, 1869.

STURTEVANT, Alfred Henry, Am. zoologist; b. Jacksonville, Ill., Nov. 21, 1891; s. Alfred H. and Harriet E. (Morse) S.; A.B., Columbia, 1912, Ph.D., 1914; Sc.D., Princeton, 1947, U. Pa., 1949, Yale, 1951; m. Phoebe Curtis Reed, Apr. 22, 1922; children—William C., Harriet M. (Mrs. Howard E. Shapiro), Alfred H. Employed as research assistant Carnegie Instn., Washington, 1915-28; prof. genetics, Cal. Inst. Tech., 1928-47, Thomas H. Morgan prof. biology, 1947-62, now emeritus; visiting professor Univ. Washington, 1960, University of Texas, 1962, Princeton University, 1963, University Wisconsin, 1964, U. Ore., 1965, U. Cal. at Santa Barbara, 1966; vis. Carnegie prof., Birmingham, 1932, Leeds and Durham, 1933; vis. lectr. Harvard, 1940. Recipient Kimber medal Nat. Academy Sci., 1957; John J. Carty medal Nat. Acad. Scis., 1965; Nat. Medal of Sci., 1968. Fellow A.A.A.S. (pres. Paciàc div. 1953-54); mem. Am. Soc. Zoologists (pres. 1934), Nat. Acad. Scis., Genetic Soc. Am. (pres. 1944), Am. Philos. Soc. Author: (with Morgan, Muller and Bridges) The Mechanism of Mendelian Heredity, 1915; (with G. W. Beadle) An Introduction to Genetics, 1939; A History of Genetics, 1965. Research on fruit flies; discovered gene position effect; developed methods of mapping chromosomes; proved crossover inhibitors in the fruit fly resulted from chromosome inversion; studied sex determination. Address: Cal. Inst. Tech., Pasadena, Cal. 91109.

STURTEVANT, Edward Lewis, Am. agrl. scientist; b. Boston, Jan. 23, 1842; s. Lewis W. and Mary (Leggett) S.; grad. Bowdoin Coll., 1863; grad. Harvard Med. Sch., 1866; m. Mary Mann, Mar. 9, 1864; m. 2d, Hattie Mann, Oct. 22, 1883; 5 children, including Grace. Commd. lt., Co. G, 24th Me. Volunteers, 1861, later capt.; with brother purchased and began devel. of Waushakum Farm, South Framingham, Mass., 1867; conducted numerous agrl. expts., particularly interested in physiology of milk and milk secretion (gained acceptance for his research); editor or co-editor Scientific Farmer, 1876-79; erected 1st lysimeter in Am. at Waushakum Farm; studied history of edible plants; 1st dir. N.Y. Agrl. Expt. Station at Geneva, 1882; leader movement for expt. stas. Author: (with brother Joseph) The Dairy Cow: A Monograph on the Ayrshire Breed, 1875; North American Ayrshire Register, 4 vols., 1875-80; Sturtevant's Notes on Edible Plants (edited by U. P. Hedrick), 1919. Died South Framingham, July 30, 1898.

STURTEVANT, Frank Milton, Jr., Am. biologist; b. Evanston, Ill., Mar. 8, 1927; s. Frank M. and Marguerite (Walsh) S.; B.A., cum laude, Lake Forest Coll., 1948; M.S., Northwestern U., 1950, Ph.D., 1951; m. Ruthann Patterson, Mar. 18, 1950; children—Barbara (dec.), Jill, Jan. Sr. investigator G. D. Searle & Co. Chgo., 1951-58; sr. pharmacologist Smith Kline French Labs., Phila., 1958-60; dir. sci. and regulatory affairs Mead Johnson & Co., Evansville, Ind., 1960-—. Fellow A.A.A.S.; mem. Soc. for Exptl. Biology and Medicine, Am. Soc. Pharmacology and Exptl. Therapeutics, N.Y. Acad. Scis., Drug Information Assn. Contbr. numerous articles to tech. jours. Pharm. research and devel. in hypertension, genetics, glucoregulation, endocrinology, central nervous system, pharmacokinetics, regulatory compliance and govtl. liaison. Home: 12410 Edgewater Dr., Evansville 47712. Office: Mead Johnson & Co., Evansville, Ind. 47721.*

STURTEVANT, Julian Munson, Am. chemist; b. Edgewater, N. J., Aug. 9, 1908; s. Edgar Howard and Bessie (Fitch) S.; A.B., Columbia, 1927; Ph.D., Yale, 1931; D.Sc., Ill. Coll., 1962; m. Elizabeth Caroline Reihl, June 8, 1929; children—Ann (Mrs. John W. Ormsby), Bradford. Faculty, Yale, New Haven, 1931-43, 46-—, prof. chemistry, 1952-—, chmn. dept., 1958-62, dir. Sterling Chem. Lab., 1950-58; staff mem. Radiation Lab., Mass. Inst. Tech., 1943-46. Cons., Socony Mobil Oil Co., 1946-—. Guggenheim fellow, Fulbright scholar U. Cambridge (Eng.), 1955-56; Fulbright scholar U. Adelaide (Australia), 1962-63. Mem. A.A.A.S., Am. Chem. Soc., Am. Fedn. Biol. Scientists, Fedn. Am. Scientists, Sigma Xi. Author: (with R. D. Coghill) Preparation and Identification of Organic Compounds, 1936; (with E. C. Pollard) Microwaves and Radar Electronics, 1948; also numerous articles. Research of heat effects on biochem. reactions, mechanism of action various enzymes; developer calorimetric equipment, apparatus for measuring rates of fast reactions in solution. Home: Indian Neck Point, Branford, Conn. 06405.*

STURTEVANT, William Curtis, Am. anthropologist; b. Morristown, N.J., July 26, 1926; s. Alfred H. and Phoebe (Reed) S.; B.A., U. Cal. Berkeley, 1949; Ph.D., Yale, 1955; m. Theda Maw, 1952; children—Kinthi D.M., Reed P.M., Alfred B.M. Instr., asst. curator anthropology Yale, 1954-56; with Smithsonian Instn. 1956-—, gen. anthropologist, curator Office Anthropology, 1965-—. Mem. Am. Anthrop. Assn., Linguistic Soc. Am., Soc. for Am. Archeology, Am. Ethnol. Soc., numerous others. Publs. on ethnography, linguistics and culture history, especially of N.Am. Indians, especially of the East. Home: 7009 Florida St., Chevy Chase, Md. 20015. Office: Office of Anthropology Smithsonian Instn., Washington 20560.*

STUTTE, Hermann, German child psychiatrist; b. Weidenau-Sieg, Westfalia, Germany, Aug. 1, 1909; s. Friedrich and Wilhelmine (Vitt) St.; student U. Freiburg, 1928-29, U. Bonn, 1929, U. Konigsberg, 1929-30, U. Paris (France), 1930-31, U. Frankfurt, 1931; M.D., U. Giessen, 1935; m. Marie-Luise Thraum, June 25, 1938; children— Klaus, Bernd. Faculty, U. Marburg (Germany), 1946-—, prof. child psychiatry, 1954-—, dir. Inst. Medico-Ednl. Youth Help, 1959-—; dir. U. Clinic of Child Psychiatry, Marburg, 1963-—; dir. Child Guidance Clinic, Marburg, 1959-—. Mem. French Child Psychiat. Assn. (hon.), German Child Psychiat. Assn., Union European de Pedopsychiatrie, German Assn. Psychiatrists and Neurologists, German Assn. Pediatrics, German Assn. Med. History. Author: Grenzen der Socialpädagogik, 1958; Psychiatr. und Gesellschaft, 1959; Kinder und Jugend Psychiatrie, 1960; also numerous articles. Editor: Jahrbuch für Jugendpsychiatrie, 6 vols., 1956-67; Monatsschrift fur Kriminologie und Straftrechtsreform, 1958-—. Research on psychoses, mental deficiencies, psychosomatic disorders in infants and children, juvenile delinquency, prevention and treatment of behavior disorders, heredodegenerative diseases in children. Home: von Harnackstrasse 20, 355 Marburg, Germany.*

STYCOS, J. Mayone, Am. demographer; b. Saugerties, N.Y., March 27, 1927; s. Steve and Clotilda (Mayone) S.; A.B., Princeton U., 1947; Ph.D., Columbia U., 1951; m. Maria Nowakowska, Nov. 25, 1964; children—Steve Andrew, Christie Mayone (from previous marriage), Marek. Bureau of Applied Social Research, Columbia U., 1948-50; project co-dir., U. Puerto Rico, 1950-53; post-grad. fellow, U. North Carolina, 1954-55; assoc. prof. sociology, St. Lawrence U., 1955-57; acting assoc. prof., 1957-60, assoc. prof 1960-63, prof. sociology, 1663-—, dir. Latin Am. Studies Program, 1962-66, dir. Internat. Population Program, 1962-—, chmn., dept. sociology, 1966-—, Cornell U. Consultant, Agency Internat. Development, 1962-64, NDEA Title VI Fellowship Committee, 1962-63; sr. consultant, Population Council, 1963-—; board mem., Latin Am Sci. Board, Nat. Acad. Scis., 1963-65; trustee, Population Reference Bureau, 1964-—; mem. various committees, NICHHD; mem. exec. committee, Internat. Planned Parenthood Federation, 1965-—; mem., Advisory Committee in Population and

Development, Organization of Am. States, 1968——. Mem. editorial board, Human Organization, 1962-64, Demography, 1965——. Author many articles in journals, reviews; articles translated and published in Spanish (some as books). Research on relation of culture and family structure to human fertility and migration. Home: Twin Glens Road, RD1, Ithaca, N.Y. 14850. Office: McGraw Hall, Cornell U., Ithaca, N.Y.*

STYRIKOVICH, Mikhail Adol'fovich, Russian heat engr.; b. Nov. 16, 1902; grad. Leningrad Technol. Inst., 1927. Staff, Leningrad Province Sci. Research Power Engring. Inst. (now Central Boiler Turbine Inst.), 1928-45; became asso. Inst. Energetics, USSR Acad. Scis., 1938, became asso. Moscow Inst. Energetics, 1939. Recipient Order of Red Banner of Labor, 1961. Corr. mem. USSR Acad. Scis. Author: Hydrodynamics and Heat Exchange in Steam Boilers and their Effect on Intraboiler Physicochemical Processes, 1951; Intraboiler Processes, 1954; Working Processes of Continously Operating Superhigh Pressure Coil Boilers, 1956; coauthor: A Course on Steam Boilers, Parts 1-2, 1934-39; Examination of Steam Content Distribution in the Boundary Fluidized Bed by the Beta-Ray Method, 1960. Research on thermal power, steam boiler processes; co-developer standards for heat and aerodynamic calculations for boiler units. Home: Leninskii Prospekt 13. Office: USSR Acad. Scis., Leninski Prospekt 14, Moscow, USSR.

SUAREZ, Francisco, Spanish natural philosopher; b. Granada, Spain, 1548; ed. Salamanca, Spain; became mem. Soc. Jesus, 1564; tchr., Segovia, Valladolid, Alcala, Salamanca, Rome; became prof. theology, Coimbra, Portugal, 1597. Author: Defensio catholique fidei contra anglicanae sectae errores, 1613. Attempted to disprove theory of divine right of kings. Died 1617.

SUAREZ-CAABRO, Jose Alfredo, biologist; b. Havana, Cuba, Nov. 25, 1913; s. Diego A. and Margarite (Caabro) S.; B.S., Inst. Havana, 1934; Chemist, U. Havana, 1939, Agric.Eng., 1940, Doct. Nat. Sci., 1945; m. María A. Sarmiento, Dec. 18, 1943; 1 son, Alfredo. Prof. biology Pre-U. Inst., Marianao, Havana, 1945-61; biologist Fisheries Research Center, BANFAIC, Havana, 1953-55; with Nat. Acad. Fis., Havana, 1954-55; prof. biology, dir. dept. marine scis. U. Villañova, Havana, 1956-61; research biologist Nat. Aquarium, Havana, 1961; dir. project fis. Research Inst., Bubano Inv. Tech., Havana, 1958-62; research asso. Inst. Marine Sci., UN, 1962-63; exptl. marine biologist UNESCO, 1963; Mem. Internat. Oceanographic Found. Author: (with F. F. Fernandez) Introduction to Agriculture, 1947; Fishing Gear and Methods, 1955; also articles. Research on Cuban and Caribbean Sea marine plankton, especially Chaetognatha, Gulf of Mexico marine plankton, shallow water and brackish water of Campeche Bank, life history of Cuban tuna and sardine-like fishes of Western part of Island. Home: 1820 S.W. 27th Av., Miami, Fla. 33145. Office: 63 Hamburgo, Mexico, D.F. 6, Mexico.*

SUBBOTIN, Mikhail Fédorovich, Russian astronomer; b. June 28, 1893; grad. U. Warsaw (Poland), 1914. Dir., Taskkent Obs., 1922-30; became dir. Inst. Theoretical Astronomy, USSR Acad. Scis., 1942; became prof. Leningrad U., 1930. Recipient Order of Lenin, twice. Corr. mem. USSR Acad. Scis. Author: A Course in Celestial Mechanics, 3 vols., 1937-49; Determining Special Points of the Analytical Function, 1916; The Astronomical and Geodetic Works of Gauss, 1956; Sur le Problème des Deux Corps de Masses Variables, 1936. Research and publs. in celestial mechanics, orbits of comets and planets including similarities in their motion. Address: Institute Theoretical Astronomy, Universitetskaya Naberezhnaya, 5, Leningrad, USSR.

SUBRAMANIAM, Manjeri Krishnier, Indian biologist; b. Calicut, South India, Apr. 20, 1909; s. Manjeri Appadurai Krishnier and Anantalakshmi Krishnier; B.A. with honors, Govt. Victoria Coll., 1928; M.A., Presidency Coll., Madras, India, 1931; M.Sc., D.Sc., U. Zool. Lab., Madras, 1932-35; m. Saraswathy Royan, Dec. 6, 1962. Staff, U. Zool. Research Lab., Madras, 1932-42; cytologist leprosy scheme Med. Coll., Hyderabad, India, 1943-44; with Indian Inst. Sci., Bangalore, 1944——, research scholar, 1944-45, cytologist Council Sci. and Indsl. Research Scheme, 1945-47, Imperial Chem. Industries research fellow, 1947-48, lectr. cytogenetics, 1948-53, asst. prof. cytogenetics, 1953——. Recipient Maharaja of Travancore Curzon prize, Madras U., 1937. Fellow Indian Acad. Scis. Research, publs. on methods of approach to study nucleus in living yeast cells, structural details of chromosomes. Home: 40, XVI Cross, Bangalore -3, Mysore, India.*

SUCHENWIRTH, Richard Matthias August, neurologist; b. Vienna, Austria, Nov. 1, 1927; s. Richard and Elisabeth (Kutsch) S.; M.D., U. Munich, 1950; m. Gertrud Meyer zu Hörste, July 12, 1955; children —Richard, Gertrud, Dietlinde, Roland. Sci. asst. U. Kiel, 1956-62; head physician Med. Acad. Lubeck, 1962-66; psychiat. and neurol. clinic U. Erlangen Germany; 1967——; lectr., 1965——. Mem. Internat. League Against Epilespy, Société Internationale pour Psychopathologie d'Expression. Author: Abbau der graphischen Leistung; 1967; co-author Begleitwirkungen und Misserfolge der psychiatrischen Pharmakotherapie, 1964; Problem der pharmakopsychiatrischen Langzeit und Kombinationsbehandlung, 1966; others. Research on sarcoidosis of nervous system, syndromes of low pressure of cerebrospinal fluid. Home: 60 1/2 Möhrendorferstr., D 852 Erlangen, Germany.*

SUCHER, Joseph, physicist; b. Vienna, Austria, Sept. 10, 1930; s. Max and Toby (Robinson) S.; B.S. summa cum laude, Bklyn. Coll., 1952; Ph.D., Columbia U., 1958; m. Dorothy Glassman, Aug. 6, 1952; children—Gabriel, Michael, Anatole, Anne. Faculty, U. Md., College Park, 1957——, prof. physics, 1964——. Columbia U. Higgins fellow, 1952; Boese fellow, 1954-56; NSF fellow, 1952-53, NSF sr. postdoctoral fellow, 1963-64; Guggenheim fellow 1968-69. Mem. Am. Phys. Soc., Phi Beta Kappa. Studies, publs. on theory of elementary particles and their interactions, theory of scattering in quantum mechanics, atomic physics. Office: Dept. Physics, U. Md., College Park, Md. 20742.*

SUCHET, Jacques Paul, French chemist, physicist; b. Paris, France, May 31, 1923; s. Charles and Odette (Menand) S.; Licencié ès Sciences, U. Paris, 1945, Docteur és Sciences, 1961; m. Jeannie Piroux, July 17, 1951. Engr., Lab. Physique du Métal, Grenoble, France, 1947-49; research engr. Cie Générale TSF, Paris, 1949-52, Sté An. Philips, Paris, 1952-55; group dir. Cie Saint-Gobain, Paris, 1956-60; staff Lab. Magnétisme et Physique du Solide CNRS, Bellevue/Meudon, France, 1960——. Mem. Societe Chim. de France, Societe Francaise Electroniciens et Radioelectriciens, Soc. Ecrivains Scientifiques de France. Author: Chimie Physique des Semiconducteurs, 1962; also numerous articles. Editor: Séminaires de Chimie de l'Etat Solide, 1968. Editorial adv. bd. Progress in Solid State Chemistry, 1965——. Research on ferrite materials including granular structure, disaccomodation; semiconducting compounds including prediction rules, crystallochem. model; interatomic bonds including ionicity, effective charge; transport phenomena in transition metal compounds, mixed valency mechanism in berthollides. Home: 71 av. du Général Leclerc, Gif-Sur-Yvette 91, France. Office: 1 pl. A. Briand, Bellevue/Meudon 92, France.*

SUCKLING, Eustace Edgar, biophysicist; b. Auckland, New Zealand, Aug. 4, 1915; s. Walter Edgar and Lois (Anthony) S.; M.S., Victoria Coll. (New Zealand), 1938; D.Elec. Engring., Bklyn. Poly. Inst., 1956; m. Joan Arawa Northey, May 25, 1946; children—Elizabeth Alice, David Maxwell. Instrumentation physicist New Zealand Govt. Med. Research Council at Med. Sch., Dunedin, 1946-48; instr. physiology L.I. Coll. Medicine, 1948-52; asst., asso. prof. dept. physiology Downstate Med. Center, Bklyn., 1952——, on leave to Makere U., Kampala, Uganda, 1967——. Mem. Am. Physiol. Soc., Harvey Soc., N.Y. Acad. Scis. (dir., sect. chmn.), Brit. Inst. Elec. Engrs. Author: (with others) Excitability of the Heart, 1955; Bioelectricity, 1962; The Living Battery, 1964; also numerous articles. Research in bioelectric phenomena in the heart and central nervous system, vibrations and lateral line organ of fish, sci. of visualizing by means of ultrasonic rays. Home: c/o Med. Sch., Makerere U., Kampala, Uganda. Office: Downstate Med. Center, State U. N.Y., 450 Clarkson Av., Bklyn. 11203.*

SUDA, Kanji, Japanese oceanographer, meteorologist; b. Gumma Prefecture, Japan 1892; grad. Tohoku U., 1921; tng. in Europe and Am. 1926-28. After graduation joined Marine Meteorol. Obs.; chief Fukuoka Meteorol. Obs.; became chief hydrographic dept. Maritime Control Bd., 1948. Mem. Investigation Com. on Radioactivity. Recipient Ministry Transp. award. Mem. Geophysics Scientists Assn., Ministry Edn. Land Survey Assn. Author: Marine Physics; Marine Science; The Ocean. Research on prevention of earthquakes, frost damage in Tohoku Dist., tidal currents.

SUDARSHAN, Ennackal Chandy George, physicist; b. Kottayam, India, Sept. 16, 1931; s. Ennackal Ipe and Achamma (Kaithail) Chandy; B.Sc., with honours, Madras Christian Coll., U. Madras, 1951, M.A., 1952; Ph.D., U. Rochester, 1958; postdoctoral fellow Harvard, 1957-59; m. Lalita Rau, Dec. 20, 1954; children—Pradip, Arvind, Ashok. Research asst. Tata Inst. Fundamental Research, Bombay, India, 1952-55; asst., asso. prof. U. Rochester, 1959-64; guest prof. U. Bern (Switzerland), 1963-64; prof. physics Syracus (N.Y.) U., 1964——; vis. prof. Brandeis U., summers 1959, 61, spring 1964, Inst. Math. Scis., Madras, India, 1962, 63, Delhi U., 1966. Fellow Am. Phys. Soc., Indian Acad. Scis. Author: (with R. E. Marshak) Elementary Particle Physics, 1961. Mem. editorial bd. Jour. Math. Physics, I.I.T. Jour. of Math. and Physics. Discovered (with Marshak) law of universal vecto-axial vector weak interaction; connection between classical and quantum theory of coherent light; proved (with T. F. Jordan and D. G. Currie) impossibility of relativistic action of a distance in classical mechanics; discovered master analytic representation of Lie algebra. Home: 2110 Euclid Av., Syracuse, N.Y. 13224.*

SUDDUTH, William Xavier, Am. physician; b. Springfield, Ill., Jan. 18, 1853; s. James McCreary and Amanda Elisabeth S.; Ph.B., Ill. Wesleyan U., 1873, A.M., 1889; D.D.S., Phila. Dental Coll., 1881; postgrad. Coll. Phys. and Surg. New York, 1883-84; M.D., Medico-Chirurg. Coll., Phila., 1885; postgrad univs. Berlin, Heidelberg, Vienna. Practiced dentistry, Bloomington, Ill., 1881-83; dir. Physiol. Lab., lectr. Medico-Chirurg. Coll., 1884-90; prof. pathology and oral surgery U. Minn., 1890-95; later in spl. practice, nervous diseases, Chgo.; from practice and engaged in mfg. alfalfa products, ranching and farming lectured in univ. extension courses; prof. morbid psychology and psycho-therapeutics and dir. psychophysical lab. Post-Grad. Med. Sch. Chgo.; well known as med. editor and writer. Died Mar. 7, 1915.

SUDHOFF, Karl, German physician; b. Frankfort/Main, Germany, 1853; practiced medicine; 1st dir. Inst. of History of Medicine, Leipzig, Germany. Author: Paracelsusforschungen, 1887-89; Arztliches aus griech Papyrusurbunelen, 1909; also writings on plague, medieval medicine, syphilis. Died 1938.

SUDO, Kingo, Japanese metallurgist; b. Kochi, Japan, May 28, 1918; s. Kintaro and Yukie (Kaida) S.; B.Eng., Tohoku U., 1941, Dr.Eng., 1955; postgrad. Imperial Coll., London U.; m. Teruko Yamamoto, Jan. 10, 1952; children—Yumiko, Hideyo, Atsuhiro. Prof., RIMDM, Tohoku U., Sendai, Japan, 1957-65, prof. nuclear engring., 1965——. Mem. Japan Inst. Metals, Atomic Energy Soc. Japan, Am. Inst. M.E. Research, publs. on smelting of sulphide ores, nuclear fuel metallurgy. Home: 12-20 Kawauchi Ohashidori, Sendai, Japan.*

SUESS, Eduard, geologist; b. London, Eng., Aug. 20, 1831. Asst. Imperial Mus., Vienna, Austria, from 1852; prof. geology U. Vienna, 1857-1901; rep. Diet, Lower Austria, 1869-96. Recipient Wollaston medal Geol. Soc. London, 1896. Fellow Royal Soc., 1894; mem. Austrian (pres. 1898-1911), French (fgn.) acads. scis. Author: Der Boden der Stadt Wien, 1862; Die Entstehung der Alpen, 1875; Das Antlitz der Erde, 3 vols., 1883-1909; Die Beziehungen der Erdrinde zu ausserirdischen Himmelskörpern, 1907; Erinnerungen, 1916. Publs. contained best geol. knowledge of time; specialist structural geology especially mountains; studied tertiary strata of basin of Vienna, former land link between Africa, Europe, also evolution of earth's surface. Died Vienna, Apr. 26, 1914.

SUESS, Hans Eduard, chemist; b. Vienna, Austria, Dec. 16, 1909; s. Franz Eduard and Olga (Frenzl) S.; Ph.D., U. Vienna, 1936; m. Ruth V. Teuteberg, Dec. 30, 1940; children—Beate, Stephen. Came to U. S., 1950, naturalized, 1955. Instr., U. Vienna, 1933-35; research asst. Fed. Tech. High Sch., Zurich, Switzerland, 1935-36; research asst. U. Hamburg, Germany, 1937-39, asst. prof., 1940-47, asso. prof., 1948-50; research fellow U. Chgo., 1950-51; phys. chemist U. S. Geol. Survey, Washington, 1951-55; research chemist Scripps Inst. Oceanography, La Jolla, Cal., 1955-58; prof. chemistry U. Cal., San Diego, 1958——. cons. Internat. Atomic Energy Agy., panel mem.; proposal reviewer NSF; adv. com. meteorite studies Ariz. State U.; adv. com. Dept. Mineral Scis., Smithsonian Instn. Guggenheim fellow. Mem. Nat. Acad. Scis., Heidelberg, Austrian acads. scis., Meteoritical Soc., A.A.A.S., Am. Geophys. Union, Fedn. Am. Sci. Research, numerous publs. on nuclear shell structure, radiocarbon dating, abundances of elements, geochemistry of cosmic ray-induced radioactivities, origin of meteorites, mixing rates of oceans. Home: 2680 Greentree Lane, La Jolla, Cal. 92037. Office: Dept. Chemistry, U. Cal.-San Diego, La Jolla, Cal. 92037.*

SUETONIUS, Gaius Tranquillus, Roman naturalist; b. circa 69-79; Author: De viris illustribus vitae duodecim caesarum, 120; Prata (on Roman antiquities and natural scis.). Died circa 160.

SUFRIN, Sidney Charles, Am. economist; b. N.Y.C., Mar. 4, 1910; s. Maurice N. and Sarah (Silverstein) S.; B.A., U. Pa., 1931, postgrad. U. Chgo., 1932-33; Ph.D., Ohio State U., 1940; m. Grace Romain DeJong, Nov. 23, 1937; children—Erica Marie (Mrs. Francis P. Kalibat), Jacoba Jetske, James Willard. Faculty, Syracuse (N.Y.) U., 1946——, now prof. econs. Cons. U.S. Govt., N.Y. State, pvt. industries in U.S., Europe, 1946——. Fellow A.A.A.S.; mem. Am. Econ. Assn., Indsl. Relations Research Assn., Regional Sci. Assn. Author: (with R. Sedgwick) Wage Policy in the Business Cycle, 1944, Labor Law, 1953, Labor Economics, 1955; (with C. Wolf) Capital Formation, 1959; Issues in Federal Aid to Education, 1962; Administration of National Defense Education Act, 1963; (with M. Buck) What Price Progress, 1964; Unions in Emerging Societies, 1964; Technical Assistance—Guide Lines and Theory, 1966; also numerous articles. Research on relations between gen. and social changes and specific changes in smaller parts of the soc., relations between econ. growth and union orgn., on aid programs and social devel. Home: 133 Humbert Av., Dewitt, N.Y. 13224.*

SUGAHARA, Tsutomu, Japanese biologist; b. Kyoto, Japan, Feb. 5, 1921; s. Kenji and Koto (Hirayama) S.; M.D., Kyoto U., 1944, D.Med., 1954; B.Sc., Osaka U., 1950; m. Akiko Fujita, Apr. 6, 1947; children—Kunio, Yumiko, Yoji. Dept. internal medicine and radiology Mie Prefectural U. Sch. Medicine, Tsu, Japan, 1950-56, asso. prof., 1955-56; asst. head dept. induced mutations Nat. Inst. Genetics, Misima, Japan, 1956-60; chief 2d Lab. div. radiation hazards Nat. Inst. Radiol. Scis., Chiba, Japan, 1960-61; prof. dept. exptl. radiology Kyoto U. Faculty Medicine, 1961——, chmn. dept. exptl. radiology Kyoto U., 1961——. Author: (with Y. Ueno) Fundamentals of Radiology, 1966; also numerous articles. Research on improvement of radiographic techniques and exptl. electrocardiography, determination radiation induced mutation rates in mice under chronic low dose gamma irradiations, radiosensitivity of cells and mammals and its modifying factors; demonstrated persistence of chromosome abberrations in peripheral blood of atomic bomb survivors; biol. implications of these aberrations in relation to delayed effects of radiation. Home: 14-46, Onoechyo, Zushioku, Yamashina, Higashiyama-ku, Kyoto, Japan.*

SUGAR, Hyman Saul, Am. ophthalmologist; b. Detroit, Sept. 7, 1912; s. Harry and Goldie (Gandleman) S.; A.B., U. Mich., 1932, M.D., 1935; m. Wilma Schiller, Sept. 26, 1941; children—Suzanne, Alan, Joel, David. Practice medicine, specializing in ophthalmology, Chgo., Detroit, 1940—; faculty Coll. Medicine Wayne State U., Detroit, 1947—, now clin. prof. ophthalmology; chief ophthalmology Sinai Hosp.; cons. ophthalmologist Detroit Meml. Hosp., Oakwood Hosp. Schoenberg lectr. N.Y.C., 1961. Diplomate Am. Bd. Ophthalmology. Mem. A.C.S., A.M.A., Am. Acad. Ophthalmology, Assn. For Research in Ophthalmology, Societe Francaise d'Ophthalmology, Ophthal. Soc. U.K., Sigma Xi, Alpha Omega Alpha, Phi Kappa Phi. Author: Extrinsic Eye Muscles, 1945, 7th edit., 1968; The Glaucomas, 1951; also numerous articles. Introduced concept and clin. entity of pigmentary glaucoma, original surg. techniques. Home: 25230 Southfield Rd., Southfield, Mich. 48075. Office: Fisher Bldg., Detroit 48202.*

SUGAR, Oscar, Am. physician; b. Washington, July 9, 1914; s. Nathan and Bertha (Miller) S.; A.B., Johns Hopkins, 1934; M.A., George Washington U., 1937, M.D., 1942; Ph.D., U. Chgo., 1940; m. Dorothy J. Cohn, Feb. 7, 1944; children—Lawrence Daniel, Ruth Louise, David Morris. Practice medicine, specializing in neurol. surgery, Chgo., 1948—; faculty Coll. Medicine U. Ill., 1948—, now prof. neurol. surgery. Mem. Soc. Neurol. Surgeons, Am. Assn. Neurol. Surgeons, Central Neurosurg. Soc., Am. Neurol. Assn., Central Neuropsychiatric Assn., A.M.A. Research, numerous publs. on growth of damaged peripheral nerve and spinal cord, connections of various parts of monkey brain; discovered buried motor area in the monkey; demonstration of blood vessels of human brain by injection of radio-opaque dye into the vertebral arteries, treatment of aneurysms. Home: 800 Edgewood Lane, Glenview, Ill. 60025. Office: 224 S. Michigan Av., Chgo. 60604.*

SUGARMAN, Nathan, Am. chemist; b. Chgo., Mar. 3, 1917; s. Barnett and Tessie (Fischer) S.; B.S., U. Chgo., 1937, Ph.D., 1941; m. Goldie Gertrude, Aug. 22, 1940; children—Tanya, Barry. With U. Chgo., 1942-45, 1946—, prof. chemistry, 1946—; with Los Alamos Scientific Lab. U. Cal., 1945-46. Mem. Am. Chem. Soc., Am. Phys. Soc., A.A.A.S., Fedn. Am. Scientists. Studies, publs. on high energy nuclear reactions, principally fission and spallation; recoil studies of products of nuclear reactions. Home: 1236 E. Madison Park, Chgo. 60615.*

SUGDEN, Samuel, English chemist; b. Leeds, Eng., 1892; s. Samuel Sugden; D.Sc., Royal Coll. Sci. London; m. Eleanor Dunlop, 1926. Research chemist Royal Arsenal, Woolwich, Eng., 1916-19; lectr. chemistry Birbeck Coll., 1919-28, reader in phys. chemistry, 1928-32, prof. phys. chemistry, 1932-37; univ. prof. chemistry U. Coll. London, from 1937. Fellow Royal Soc., 1934. Author: The Structure of Atoms, 1923; The Parachor and Valency, 1929; (with T. M. Lowry) A Class Book of Physical Chemistry, 1929. Research, publs. on molecule size and surface tension; furthered study of structure; introduced parachor. Died Oct. 20, 1950.

SUGDEN, Theodore Morris, Brit. physical chemist; b. Halifax, U.K., Dec. 31, 1919; s. Frederick Morris and Florence (Chadwick) S.; B.A., Cambridge U., 1942, M.A., 1945, Ph.D., 1950, Sc.D., 1961; m. Marian Cotton, Sept. 4, 1945; 1 son, Andrew Morris. Lectr. phys. chemistry Cambridge U., 1950-59, reader, 1959-63, fellow Queen's Coll., 1957-63; research dir. Shell Research Ltd., Thornton Research Center, Chester, Eng., 1964—; asso. prof. molecular scis. U. Warwick, 1965—. Fellow Royal Soc., 1963; mem. Faraday Soc., Chem. Soc. London, Combustion Inst. (medallist). Author: (with C.N. Kenney) Microwave Spectroscopy of Gases, 1965. Research, publs. on mechanism of reaction in high temperature flames, especially involving free radicals and ionic processes. Home: The Tithe Barn, Great Barrow, Chester. Office: Thornton Research Center, P.O. Box 1, Chester, Eng.*

SUGG, John Young, Am. microbiologist; b. McEwen, Tenn., Aug. 11, 1904; s. John A. and Pearl (Young) S.; student U. Tenn., 1922-24; A.B., Vanderbilt U., 1926, M.A., 1928, Ph.D., 1931; m. Marguerite Smith, Aug. 12, 1942. Instr., Vanderbilt U., Nashville, 1930-31; mem. faculty Cornell U. Med. Coll., N.Y.C., 1932—, prof. microbiology, 1964—. Mem. Soc. Am. Bacteriologists (chmn. program com. 1951-53), Am. Assn. Immunologists (sec.-treas. 1951-54), Am. Soc. Microbiology, Am. Acad. Microbiology, N.Y. Acad. Medicine, Harvey Soc., Soc. for Exptl. Biology and Medicine. Editor in chief Jour. of Immunology, 1954—. Research, numerous publs. on immunological relationships among pneumococci, hypersensitiveness to diptheria bacterial products, loss of immune substances from body, serological studies on sugar, biol. and immunol. properties of influenza viruses. Home: 365 Stewart Av., Garden City, N.Y.*

SUGIHARA, James Masanobu, Am. chemist; b. Las Animas, Colo., Aug. 6, 1918; s. William B. and Takeyo (Kubota) S.; A.A., Long Beach Jr. Coll., 1937; B.S., U. Cal. at Berkeley, 1939, postgrad., 1939-41; Ph.D., U. Utah, 1947; m. May Murakami, June 5, 1944; children—John, Michael. Faculty U. Utah, 1946-48, 49-64; research asso. Ohio State U., 1948-49; dean Coll. Chemistry and Physics, N.D. State U., Fargo, 1964—, prof. chemistry, 1964—. Cons. Sun Oil Co., 1964—. Mem. Am. Chem. Soc., Sigma Xi, Phi Kappa Phi. Author: Laboratory Exercises in Organic Chemistry, 1961; also articles. Research on carbohydrate chemistry, application carbohydrate derivatives in interpreting mechanisms of organic reactions, characterization metal compounds in crude oils, Gilsonite, related materials. Home: 1001 Southwood Dr., Fargo, N.D. 58102.*

SUGIHARA, Thomas T., Am. chemist; b. Las Animas, Colo., June 14, 1924; s. William and Takeyo (Kubota) S.; A.B., Kalamazoo Coll., 1945; S.M., U. Chgo., 1951, Ph.D., 1952; m. Fumi Anraku, Mar. 15, 1952; children—Sara, Edna. Research asso. Mass. Inst. Tech., 1952-53; faculty Clark U., Worcester, Mass., 1953-67, prof. chemistry, 1962-67, dept. chmn., 1963-66; prof., chmn. Tex. A. and M. U., 1967—; asso. scientist Woods Hole (Mass.) Oceanographic Instn., 1954—. Guggenheim fellow, 1961-62. Mem. Am. Chem. Soc., Am. Phys. Soc., A.A.A.S. Research, publs. on nuclear fission process, mechanisms by which nuclei interact with each other, oceanic circulation studies by means radioactive tracers. Home: 2300 Morningside Dr., Bryan, Tex. 77801. Office: Cyclotron Inst., Tex. A. and M. U., College Station, Tex. 77843.*

SUGIMOTO, Asao, Japanese physicist; b. Aomori Prefecture, Japan, 1911; grad. Tokoyo U., 1933; D.Sc., 1948. Became asst. Rigaku Research Inst. after graduation; later asst. researcher Sci. Research Inst. now chief researcher. Chmn. atomic furnace design subcom. Japan Sci. Devel. Council, 1955—. Translator: The Completion of the Atomic Bomb (H. O. Smythe). Research and publs. on beta ray, nucleus of magnetism; theoretical calculations for design and manufacture of atomic furnace.

SUGIMURA, Yukio, Japanese geochemist; b. Shizuoka, Japan, Dec. 30, 1931; s. Inokichi Nishio and Shizuko (Sugimura) S.; B.S., Tokyo (Japan) Met. U., 1954, M.S., 1957, Ph.D., 1961; m. Toshiko Kakinuma, Feb. 10, 1956; 1 dau., Hiroko. Research scientist Meteorol. Research Inst., Tokyo, 1960—; lectr. Tokyo U. Fisheries, 1964-65. Mem. com. Japanese Deep Sea Expdn., 1961—; participant Japanese Antarctic Research Expedition, 1966-67. Mem. Am. Geophys. Union, A.A.A.S., Oceanographic Soc. Japan, Japanese Chem. Soc., Japanese Geochem. Soc. Author: (with K. Noguchi) Planet Earth, 1960; (with others) Actinide Chemistry, 1962; also articles. Research on distbn. natural radioelements in ocean, determined rate of sedimentation of deep sea sediment. Home: 1-63 Sasazuka, Shibuya, Tokyo. Office: 4-35-8 Koenjikita, Suginami, Tokyo, Japan.*

SUGINOME, Harusada, Japanese chemist; b. Miyagi Prefecture, Japan, 1892; grad. Tohoku U., 1919; D.Sc., 1938. Asst. prof. Tohoku U., became prof. sci. dept., 1930; named dir. sci. dept. Hokkaido U., 1950, elected pres., 1954. Recipient Hattori Hokokai prize, 1925, prize Japan Chem. Soc., 1926. Author: Cyanide and Nitro Compound; also articles. Research in organic chemistry, including alkaloid of Acotinum Japonicum, carotinoid.

SUGITA, Gempaku (originally Yoku), Japanese physician; b. 1733; s. Hosen Sugita; studied surgery under Gentetsu Nichi, Chinese classics under Ryumon Miyase; also studied Dutch medicine. Author: Rangaku Kotohajime (intro. to Dutch science), other essays, also books on medicine. Translator: (from Dutch to Japanese, with Ryotaku and others) Kaitai Shinsho (provided Japanese physicians with knowledge of Western medicine), 1774. Died 1817.

SUGIURA, Kanematsu, chemist; b. Nagoya, Japan, June 5, 1892; s. Seisuke and Miyono (Aoki) S.; B.S., Bklyn. Poly. Inst., 1915; M.A., Columbia, 1917; D.Sc., Kyoto Imperial U., 1925; m. Zoe Marie Claeys, Oct. 20, 1923; 1 dau., Miyono Marie (Mrs. Franz Schmid). Asso., Sloan-Kettering Inst. Cancer Research, Rye, N.Y., 1947-58, mem., 1959-61, mem. emeritus, 1962—; abstractor chem. abstracts Biol. Bull. Am. Jour. Cancer Research, 1921-44. Recipient A. Leonard prize Am. Roentgen Ray Soc., 1925; Order Sacred Treasure 3d Class Emperor Hirohito Japan, 1960; citation, cup Japanese Govt., 1960; scroll of merit, plaques Tohoku Med. Assn. and U., Sendai, Japan, 1960; Honor certificate, Gold plaque Japan Med. Assn., 1965; certificates for cultural service Kanazawa U., 1966, Niigata U., 1966. Fellow A.A.A.S.; mem. Am. Assn. Cancer Research, Am. Chem. Soc., Soc. Exptl. Biology and Medicine, N.Y. Acad. Scis., Japanese Assn. Cancer Research (hon.). Research, numerous publs. on cancer, etiology, immunity, radiation and drug treatment, and nutrition; animal tumor storage for transplantation. Home: 18 Oak St., Harrison, N.Y. 10528. Office: 145 Boston Post Rd., Rye, N.Y. 10580.*

SUHL, Harry, Am. physicist; b. Leipzig, Germany, Oct. 18, 1922; s. Bernhard and Klara (Bergwerk) S.; came to U.S., 1948, naturalized, 1957; B.S., U. Wales, 1943; Ph.D., Oxford U., England, 1948; m. Rebecca Schnitzer, Aug. 19, 1949. Exptl. officer Admiralty Signal Establishment, Whitehall, Whitley, London, Eng., 1943-46; mem. tech. staff Bell Tel. Labs., Murray Hill, N.J., 1948-60; prof. physics U. Cal., San Diego, LaJolla, Cal., 1961—, chmn. dept. physics, 1965—. Fellow Am. Phys. Soc. Editor (with George Rado) Magnetism, 4 vols., 1963—. Studies, numerous publs. in theoretical solid state physics, its applications; name given to two phys. effects, concentration of holes and electrons in a semiconductor by a magnetic field, and indirect interaction between nuclear magnetic moments in ferromagnetic and antiferromagnetic materials; studies on resonance in magnetic materials; superconductivity and general theory of magnetism, particularly magnetic alloys; invented a type of low noise ferromagnetic amplifier, working on parametric principle. Home: 2301 Rue de Anne, La Jolla, Cal. 92037. Office: Physics Dept., U. Cal.-San Diego, La Jolla, Cal. 92037.*

SUHRMANN, Rudolf Johannes, German phys. chemist; b. Reichenberg (Silesia), Mar. 9, 1895; s. Karl Adolf and Wilhelmine (Auerbach) S.; Dr.rer.techn., Dresden Inst. Tech., 1921, Dr.rer.nat.h.c., 1960; m. Erna Stillmark, Aug. 29, 1925; children—Renate (Mrs. Schrader), Gerda (Mrs. Hahn), Ina (Mrs. Bertges). Faculty, Breslau (now Wroclaw, Poland) U. and Inst. Tech., 1925-46, prof., 1931-46, dir. phys.-chem. inst., 1933-46; prof., dir. phys.-chem. inst. Braunschweig (Germany) Inst. Tech., 1946-55, Hanover (Germany) Inst. Tech., 1955-64. Mem. Bunsen Soc., Chem. Soc., Phys. Soc., Braunschweig Sci. Soc., Leopoldina. Author: Lichtelektrische Zellen, 1932; Lichtelektrischer Effekt, 1958; Physikalisch-chemische Praktikumsaufgaben, 1928-64; also numerous articles in field. Research on photoelectric, thermionic, secondary emission in dependence on surface state; infrared, visible, ultraviolet absorption; elec. and optical behavior of thin metal films; adsorption phenomena on thin metal films, catalysis. Address: 12e Insterburger Strasse, Karlsruhe-Waldstadt, West Germany.*

SUIE, Ted, Jr., Am. microbiologist; b. Akron, O., June 20, 1923; s. Theodore and Anna (Pascu) S.; B.Sc., Akron U., 1947; M.Sc., Ohio State U., 1948, Ph.D., 1953. Faculty, Ohio State U., Coll. Medicine, Columbus, 1953—, asso. prof. ophthalmology, 1960—, dir. research, 1962—. Mem. Acad. Ophthalmology and Otolaryngology (instr. 1957—) Assn. Research in Ophthalmology, Sigma Xi, Phi Sigma Eta. Author: Microbiology of the Eye, 1958; also articles. Research in eye infections and immunology, pathogenesis and treatment involved in blinding diseases, mechanisms involved in immune reactions of eye. Home: 4500 Dublin Rd., Columbus, O. 43221.*

SUISETH, Richard, see Swineshead, Richard.

SUIT, Herman Day, Am. physician; b. Houston, Feb. 8, 1929; s. Lewis Avery and Bill (McVicker) S.; B.A., U. Houston, 1948; M.D., Baylor U., 1952, M.Sc., 1952; D.Phil., Oxford (Eng.) U., 1956; m. Joan Lucia Countryman, Nov. 11, 1960; Sr. asst. surgeon radiation br. Nat. Cancer Inst., NIH, Bethesda, Md., 1957-59; asst. radiotherapist M.D. Anderson Hosp. and Tumor Inst., U. Tex., Houston, 1959-63, chief sect. exptl. radiotherapy, 1962—, asso. radiotherapist, 1963—, asso. prof. radiotherapy, 1965—. Mem. space radiation study panel Nat. Acad. Sci., 1964—; mem. subcom. on radiobiology, com. on nuclear sci. Nat. Acad. Sci.-NRC, 1966. Research Career Devel. fellow USPHS, 1964—. Mem. Am. Coll. Radiology, American Therapeutic Radiologists, A.M.A., Am. Assn. for Cancer Research, Radiation Research Soc., Radiol. Soc. N.Am., Phi Kappa Phi, Alpha Omega Alpha. Research and publs. on radiation therapy administered under conditions of local tissue hypoxia, animal tumor systems to correlate radiation dose, tumor cure probability, tumor size, tumor growth rate, dose fractionation and immunological relationship between host and tumor, devel. radium applicators featuring after-loading. Home: 5516 Chenevert St., Houston 77004.*

SUITS, Chauncy Guy, Am. physicist; b. Oshkosh, Wis., Mar. 12, 1905; s. Chauncey Gibbs and Otillia (Berger) S.; B.A., U. Wis., 1927; D.Sc., Technische Hochschule, Zurich, Switzerland, 1929; D.Sc. (hon.) Union Coll., Hamilton Coll., Drexel Inst. Tech., Marquette U.; D. Eng. (hon.), Rensselaer Poly Inst.; m. Laura E. Struckmeyer, Oct. 28, 1931; children—James Carr, David Guy. With Gen. Electric Co., Schenectady, 1930-66, v.p., dir. research, 1945-66, ret., 1966; Mem. Nat. Acad. Scis., div. NDRC, 1942-46, OSRD, 1950; mem. sci. adv. councils to N.Y. state, U. S. Ho. of Reps. Recipient King's Medal (Great Britain); U. S. Medal for Merit; Procter Prize award Sci. Research Soc. Am., 1958; Distinguished Service award Am. Mgmt. Assn., 1959; Indsl. Research Inst. medal, 1962; Charles M. Schwab Meml. Lecture, Am. Iron and Steel Inst., 1963; Advancement of Research medal Am. Soc. Metals, 1966. Author: Suits: Speaking of Research, 1965. Research and publs. on non-linear electric circuits, discovery of many new types, devel. of analysis and practical application to indsl. control systems; high temperature and high pressure electric arc phenomena, devel. of velocity-of-sound methods of arc temperature measurement, devel. (with Poritsky) of heat transfer theory of high pressure arcs, studies of arcs in 1000 atmosphere pressure range, application of heat transfer concepts to circuit interruption tech. and practice. Address: Crosswinds, Pilot Knob, N.Y. 12844.*

SUKACHEV, Vladimir Nikolaevich, Russian botanist; b. June 7, 1880; grad. St. Petersburg (now Leningrad, USSR) Forestry Inst., 1902. Asst. botany St. Petersburg Forestry Inst., 1902-12; jr. botanist Bot. Mus., St. Petersburg Acad. Sci., 1912-18; prof. Geog. Inst., 1918-25; prof. Forestry Inst., later Forestry Tech. Acad., 1919-41; prof. Leningrad U., 1925-41; dir. Acclimatization Dept., 1924-26; dir. dept. geobotanics Main Bot. Gardens, USSR Acad. Scis., 1931-33, dir. Inst. Forestry, 1944—, dir. lab. sylviculture, 1959—; prof. Moscow Tech. Forestry Inst., 1944-48, Moscow U., 1948-51. Recipient Gold medal, 1901; V. V. Dokuchaev Grand Gold medal 1951; Geog. Soc. medal, 1912, 14, 29, 47; Order of Lenin (2). Mem. All-Union Bot. Soc. (founding, pres. 1946—), Moscow Naturalists Soc. (pres. 1955—), USSR Acad. Scis. Author: Swamps—Their Formation, Development and Properties, 1914; A Brief Guide to the Study of Forest Types, 1927; A History of Vegetation in the USSR During the Pleistocene Period, 1938. Research and numerous publs. on plants of various regions, cultivation of protective forests; spore-pollen analysis of glacial deposits; originated theory of swamp formation. Home: Leninskii Prospekt 13. Office: Lab. Forest Studies, USSR Acad. Sci., Moscow, USSR.

SULA, Ladislav, Czechoslovakian physician; b. Horní Hermanice, Czechoslovakia, Dec. 29, 1912; s. Antonin and Mathilde (Svestková) S.; D.degree, Masaryk U., Brno, Czechoslovakia, 1938; m. Jirina Kristufková, Mar. 27, 1956; children—Ladislav, Jan. Physician various Tb sanatoria, 1940-44; leading physician Tb bacteriol. lab. State Inst. Pub. Health, Prague, Czechoslovakia, 1944-52; chief microbiol.-epidemiological dept. Tb. Research Inst., Prague, 1952-57; chief Tb. Research Office, WHO, Copenhagen, Denmark, 1958-60, sr. med. officer Tb unit, Geneva, Switzerland, 1961-62, chief Tb Reference Lab., Prague, 1964. Recipient Govtl. Distinction for extraordinary service in fight against Tb, Pres. Czechoslovakian Socialist Republic, 1963. Mem. Med. Soc. J. E. Purkyne, Sociedad Venezolana de la Phtisiologia, Sociedad Venzolana de Microbiologia. Author: BCG Vaccination, 1955; Microbiology of Tuberculosis, 1965; also numerous articles. Devel. semi-synthetic lyophilized medium for growing Mycobacteria, concentrated freeze-dried medium, different tests for microbiol. differentiation of Mycobacteria. Home: 87 Ruská tr., Prague. Office: 48 Srobárova, Prague, Czechoslovakia.*

SULAMAA, Matti Veikko, Finnish pediatric surgeon; b. Pälkäne, Finland, May 30, 1910; s. Eemeli Bernhard and Hellin (Kyman) S.; M.D., U. Helsinki (Finland), 1935; m. Lea Kaarina Mannila, June 17, 1935; children—Simo, Jukka, Olli, Heikki. Surgeon-in-chief U. Children's Hosp., Helsinki, 1946—; lectr. surgery U. Helsinki, 1950—. Hon. dr. U. Tokio, 1963. Mem. Swedish Pediatric Soc. (hon.), Finnish Med. Assn. (chmn. council). Research, numerous publs. on gen. surgery, pediatric surgery, cardiovascular surgery, surg. rehab. of phocometric (thalidomide) babies. Home: 34 Topelius St. Office: 11 Stenback St., Helsinki, Finland.*

SULAYRES DE RENHAC, François-Louis-Joseph, French physician; b. Calhac, France, 1737; student Montpellier, France; B.M., 1765. Gave clear description of successive positions of fetus during descent and various presentations at birth; a founder theory of normal birth mechanism. Died 1772.

SULKIN, S(imon) Edward, Am. microbiologist; b. Boston, Oct. 21, 1908; s. Frank Samuel and Celia (Glazer) S.; B.S., U. R.I., 1930; Ph.D., Washington U., 1939; m. Lorraine Kahn Levy, Aug. 7, 1939; children—Sandra Lucille (Mrs. William Rex Wenneker), Daniel Ellis. Faculty, Washington U., 1939-43; dir. virus lab. St. Louis Health Div., 1940-43; faculty Southwestern Med. Coll., 1943-45; prof., chmn. dept. microbiology Southwestern Med. Sch. U. Tex., Dallas, 1945—; cons. USPHS, VA, Virus Commn. Armed Forces Epidemiological Bd. Recipient Health Service award Dallas Hosp. Council, 1950, citation Seoul Nat. U., 1962, Distinguished Alumnus award Washington U., 1965. Diplomate Am. Bd. Microbiology. Fellow Am. Pub. Health Assn., mem. A.A.A.S., Am. Acad. Microbiology, Royal Soc. Tropical Medicine and Hygiene; mem. N.Y. Acad. Scis., Soc. Exptl. Biology and Medicine, Am. Assn. Immunologists, Am. Soc. Microbiology, Am. Soc. Trop. Medicine. Contbr. chpts. to books, numerous articles to profl. jours. Research in toxinantitoxin reactions, bacteriophage, viruses, virus-bat interrelationship. Home: 3311 Chaparral Dr., Dallas 75234.*

SULLIVAN, James Francis, Am. physician; b. Peoria, Ill., Feb. 17, 1924; s. James Francis and Edna (Burkey) S.; B.S. Eureka Coll., 1949; M.D., St. Louis U., 1951; m. Jean Heighway, June 3, 1946; children—James, Mary, Patrick, Terrence, Margaret, Anne, Andrew. Faculty, St. Louis U., 1955-61, asst. prof. clin. medicine, 1959-61; faculty Creighton U., Omaha, 1961—, prof. medicine, 1964—, asst. chmn. dept. medicine, 1961—; chief medicine Omaha VA Hosp., 1966—. Fellow A.C.P.; mem. Central Soc. for Clin. Research, Am. Inst. Nutrition. Research and publs. on trace metal and lipid metabolism especially in alcoholic patients. Home: 5212 Webster St., Omaha 68132.*

SULLIVAN, John Daniel, Am. chemist, metallurgist; b. Columbia Falls, Mont., Feb. 4, 1900; s. Dennis D. and Mary Ann (O'Neill) S.; student U. Mont., 1917-19; B.S., U. Wash., 1921, M.S., 1922; grad. study in phys. chem., U. Cal., 1923-27, in mining and adminstrn, U. Ariz., 1927-28; m. Marguerite Cudahy Sicard, July 17, 1928. Fellow in chemistry U. Wash., 1921-22; analyst N.W. Expt. Sta., U. S. Bur. Mines, Seattle, 1922-23, asst. phys. chemist Pacific Expt. Sta., Berkeley, Cal., 1923-27, asso. metall. chemist S.W. Expt. Sta., Tucson, 1927-31; chief chemist Battelle Meml. Inst., Columbus, O., 1931-35, chief chemist, asst. dir. 1935-47, asst. dir., 1947-53, tech. dir., 1953-65, now cons. Mem. war metallurgy com. Nat. Acad. Sci., NRC, 1940-46; mem. div. chemistry and chem. tech. NRC, 1950-53, minerals and metals adv. bd. Recipient awards from War and Navy depts. in appreciation for service to OSRD, World War II; recipient first presdl. award Am. Ceramic Soc. Fellow Am. Ceramic Soc. (hon.; nat. treas., 1944-45, nat. v.p. 1946-47, nat. pres. 1947-48, Bleiniger award 1960), Soc. Glass Technology; mem. Am. Inst. Mining, Metall. and Petroleum Engrs. (chmn. extractive metallurgy div. 1948-50; dir. 1957-58, 59—; chmn. Rossiter Raymond award com., also chmn. Alfred Noble award com. 1958), Am. Soc. for Testing Materials (chmn. com. C-8 on refractories 1936-48; award of Merit 1952), Electrochem. Soc. (chmn. electrothermic div., 1930-36, mem. Acheson medal com., 1943-46, chmn. 1944), Brit. Ceramic Soc., Chem. Soc. London, Ohio Ceramic Industries Assn., Colo. Mining Assn., Soc. Chem. Industry Gt. Britain, Deutsche Keramische Gesallschaft, Faraday Soc., Geochem. Soc., Mining Assn. Mont., Sigma Xi, Keramos, Phi Lambda Upsilon. Contbr. numerous articles to tech. jours. Holder over 30 patents in ceramics and metallurgy. Home: 1521 W. Lewis Av., Phoenix 85007. Office: 2310 N. 15th Av., Phoenix 85007.*

SULLIVAN, Louis Wade, Am. physician; b. Atlanta, Nov. 3, 1933; s. Walter Wade and Lubirda (Priester) S.; B.S., Morehouse Coll., 1954; M.D., Boston U., 1958; m. Eva Williamson, Sept. 30, 1955; children—Paul, Shanta. Research fellow in pathology Mass. Gen. Hosp., Boston, 1960, in medicine, Harvard, 1961-63; instr. medicine Harvard Med. Sch., 1963-64; asst. prof. medicine N.J. Coll. Medicine, Jersey City, also asso. attending physician Jersey City Med. Center, 1964-66; asst. prof. medicine Boston U. Sch. Medicine, 1966—. Recipient Research Career Devel. award Nat. Inst. Arthritis and Metabolic Diseases, 1965. Mem. Soc. for Study of Blood (sec.-treas.), Am. Fedn. Clin. Research, Am. Hematology Soc., Soc. for Exptl. Biology and Medicine, Am. Soc. for Clin. Nutrition. Research and publs. on definition of minimal daily requirement for vitamin B12, effect of alcohol on blood cell prodn., metabolism of vitamin B12 and folic acid and their inter-relationships, studies of pernicious anemia in adults and children.*

SULLIVAN, Patrick Lee, Am. psychologist; b. Chgo., Oct. 18, 1919; s. Austin Edward and Anna (Cameron) S.; B.S., U. Ill., 1941; Ph.D., U. Cal. at Berkeley, 1950; m. Elizabeth Jane Wheaton, June 21, 1946; children—Patrick W., Robert E., David W. Faculty, Mich. State U., 1950-51, San Francisco State Coll., 1955-56; chief clin. psychologist VA, Oakland, Cal., 1951-55; cons. psychologist, asso., partner Glaser, Snowden & Assos., San Francisco, 1956-61; cons. psychologist, partner Snowden, Sullivan & Goodwin, San Francisco, 1961—. Mem. A.A.A.S., Cal., San Francisco, Western, Contra Costa County psychol. assns., Sigma Xi. Research, publs. on factors asso. with psychotherapeutic results, ethnocentrism, test devels. and applications, social and attitude influences on perception. Home: 3485 Springhill Rd., Lafayette, Cal. 94549. Office: Russ Bldg., San Francisco 94104.*

SULLIVAN, Robert Donald, Am. physician; b. Phila., Sept. 3, 1921; s. Guy P. and Hilda (Fulford) S.; M.D. cum laude, Syracuse U., 1948; m. Mary T. Pistovo, Aug. 31, 1957; children—Anthony, Leslie, Robert, Christopher, Laura Beth, Peter. Instr. medicine Cornell U. Coll. Medicine, 1955-61; chief oncology sect. N.Y. VA Hosp., 1955-61; v.p. Cancirco, N.Y.C., 1962—; dir. div. med. research Lahey Clinic, Boston, 1961—; staff Deaconess, Bapt., Hohneman, Brooks hosps. Mem. A.C.P., Am. Assn. for Cancer Research, Am. Fedn. for Clin. Research, Am. Soc. Oncologists (founding), Alpha Omega Alpha. Research, publs. on cancer, especially chemotherapy; developed (with D. Elton Watkins, Jr.) chronometric infusion pump for protracted ambulatory arterial infusion for use in liver cancer cases. Home: 79 Shornecliffe Rd., Newton, Mass. 02158. Office: Lahey Clinic, Boston 02215.*

SULLIVAN, Thomas Wesley, Am. nutritionist; b. Rover, Ark., Sept. 30, 1930; s. Tommy Hazel and Anna (Castleberry) S.; student Ark. Inst. Tech., 1947-49; B.S., Okla. State U., 1951; M.S., U. Ark., 1955; Ph.D., U. Wis., 1958; m. Thelma Hall, Dec. 25, 1955. Vet. agrl. instr. Alma (Ark.) Pub. Schs., 1951-52; research asst. U. Ark., 1954-55; teaching, research asst. U. Wis., 1955-58; faculty U. Neb., Lincoln, 1958—, prof. poultry sci., 1965—. Mem. Am. Inst. Nutrition, Poultry Sci. Assn., Soc. Exptl. Biology and Medicine, Sigma Xi, Phi Kappa Phi. Contbr. numerous articles in field to sci. jours. Established mineral requirements of turkeys and chickens; procedure for bioassay of feed phosphorus sources; evaluation of feed additives especially antibiotics and histomonostatic compounds. Home: 6941 Vine St., Lincoln, Neb. 68505.*

SULLIVAN, William Daniel, Am. biochemist; b. Boston, Nov. 18, 1918; s. William Patrick and Delia (Larkin) S.; A.B., Boston Coll., 1944, M.A., 1945; M.S., Fordham U., 1948; Ph.D., Cath. U. Am., 1959. Joined Soc. Jesus, 1938, ordained priest Roman Cath. Ch., 1951; tchr. Cranwell Prep. Sch., Lenox, Mass., 1945-46, Fairfield (Conn.) Prep. Sch., 1946-47, Cheverus High Sch., Portland, Me., 1952-53, Fairfield (Conn.) U., 1957-58; mem. faculty Boston Coll., 1958—, chmn. biology dept., 1958—, prof. biology, 1963—. Cons., WHIS-TV. Mem. Soc. Parasitology, Soc. Am. Bacteriology, Am. Soc. Zoologists, Soc. Protozoologists, N.Y. Acad. Scis., Am. Assn. U. Profs., A.A.A.S., Am. Polar Soc., Am. Micros. Soc., Am. Soc. Cell Biology, Am. Genetic Assn., Mass. Soc. Zoologists, Nat. Geog. Soc., N.E. Biology Soc., Am. Assn. Jesuit Scientists, Albertus Magnus Soc., Am. Internat. Biol. Soc., Sigma Xi (pres. Boston Coll. club 1961-63). Research and publs. on effects of radiation on microorganisms (tetrahymena pyriformis) during various phases of division, especially enzymatic activities. Address: Boston Coll., Chestnut Hill, Mass. 02167.*

SULLIVANT, William Starling, Am. botanist; b. Columbus, O., Jan. 15, 1803; s. Lucas and Sarah (Starling) S.; studied Ohio U.; grad. Yale, 1823; LL.D., Kenyon Coll., 1864; m. Jane Marshall, Apr. 7, 1824; m. 2d, Elisa Griscom Wheeler, Nov. 29, 1834; m. 3d, Caroline Sutton, Sept. 1, 1851; 13 children, including Thomas Starling. Compiler, A Catalogue of Plants, Native and Naturalized, in the Vicinity of Columbus, Ohio, 1840; contbr. 2 important sects. to 2d edit. Gray's Manual, 1856, republished separately as The Musci and Hepaticae of the United States East of the Mississippi River, 1856; greatest work, Icones Muscorum, 1864 (supplement 1874); distinguished as America's foremost bryologist; commemorated by genus Sullivantia which he discovered in Ohio; mem. Am. Acad. Arts and Scis., 1845. Died Columbus, Apr. 30, 1873.

SULLY, James, English psychologist, philosopher; b. 1842; Bridgwater, Mar. 3, 1842; s. J. W. Sully; ed. Ind. Coll., Taunton, Regent's Park Coll., London; M.A., U. London; postgrad. univs. Göttingen, Berlin (both Germany); m.; 1 son, 1 dau. Prof. philosophy Univ. Col., London; lectr. on edn. Univ. U. Cambridge, Coll. Preceptors. Author: Sensation and Intuition, 1874; Pessimism, 1877; Illusions, 1881; Outlines of Psychology, 1884; Teacher's Handbook of Psychology, 1886; The Human Mind, 1892; Studies of Childhood, 1895; Children's Ways, 1897; An Essay on Laughter, 1902; Italian Travel Sketches, 1912; My Life and Friends, 1918. Approached psychology in terms of psychol. acts; gave particular attention to questions of edn., social progress, art. Died Nov. 1, 1923.

SULMAN, Felix Gad, physician; b. Berlin, Germany, Mar. 30, 1907; s. Bernhard and Hedwig (Witkowsky) S.; D.V.M., Berlin U., 1930, M.D., 1933; m. Edith Grzebinasch, Jan. 5, 1934; children—Nourith, Irith. Asst., Research Inst. Hygiene and Immunology, U. Berlin, 1930-33; med. faculty Hebrew U., Jerusalem, 1934-54, lectr. dept. pharmacology Med. Sch., 1954, head dept. applied pharmacology, 1955; staff pub. health sect. Ministry Health, Israel, 1948-50. Cons. to govt. of Israel, pharm. industry. Hon. fellow Argentine Soc. Pharmacology and Therapeutics; mem. Soc. for Endocrinology (U.K.), Royal Soc. Medicine (U.K.), Endocrine Soc., Soc. for Exptl. Biology and Medicine, N.Y. Acad. Scis. Author: (with Zondek) Antigonadotropic Factor, 1942; (with others) Hormone and Psyche, 1960; also numerous articles. Research on sex hormones, antihormones, psychopharmacological drugs, serotonin. Home: 2 Abrabanel St., Jerusalem, Israel.*

SULZBERGER, Marion Baldur, Am. physician, dermatologist; b. N.Y. City, Mar. 12, 1895; s. Ferdinand and Stella (Ullmann) S.; student Harvard, 1912-13; studied medicine, France, Switzerland, Germany, 1920-29; received degree of basic med. science, U. of Geneva, Switzerland, 1921; M.D., U. of Zurich, Switzerland, 1926; m. Edna F. Lowenstein, 1915; 1 dau., Margaret L. (Mrs. Francis Dobo); m. 2d, Kathryn Mullen Conway, Oct. 23, 1933; m. 3d, Roberta Zechiel Merrill, Sept. 1958. Asst. clin. prof. dermatology and syphilology, Columbia U., 1935-46, asso. clin. prof., 1946-47; George Miller Mackee prof. dermatology and syphilology New York University-Bellevue Medical Center; prof. emeritus N.Y. U. Sch. Medicine; chmn. dept. dermatology and syphilology, Post-Grad. Med. Sch., N.Y. U., 1954-60, now prof. emeritus dermatology and syphilology; pres. Internat. League of Dermatologic Socs., 1957-62; mem. gen. medicine study sect. Nat. Insts. Health, Pub. Health Service, 1957-59; tech. dir. research U. S. Army Med. Research & Devel. Command, Office Surgeon Gen., 1961-64; tech. dir. research Letterman Gen. Hosp., San Francisco, 1964-67; sci. adviser U. S. Army Med. Research Unit, Presidio, 1968—. Awarded Legion of Merit (United States), 1949; Commander of the Cross of Anjouan, Legion of Honor (France), Fellow of New York Academy of Medicine, American Academy of Allergy, American Acad. Dermatology and Syphilology, A.C.P.; mem. American Dermatological Assn. (pres. 1959-60); mem. or hon. mem. fgn., nat. and state profl. socs. Editor: (with Dr. Fred Wise) Yearbook of Dermatol. and Syphilol., 1931-42, senior editor, 1943-55; editor numerous other tech. works. Author: Dermatologic Allergy, 1940; Dermatologic Therapy in General Practice (with Jack Wolf), 1940, 3d edit., 1948; (with Pillsbury, Livingood) Manual of Dermatology, 1942; (with others) Office Immunology, 1947; (with Wolf) Dermatology: Essentials of Diagnosis and Treatment, 1952; (with Herrmann) Clinical Significance of Disturbances in Sweating, 1954; Dermatology: Diagnosis and Treatment (with Wolf, Witten, Kopf), 2d edit., 1961; Vol. 6 of Traumatic Medicine and Surgery for the Attorney, 1962; Drugs of Choice—Dermatologic Medicaments, 1960, 61, 62; (with Witten) Drugs of Choice—Dermatologic Drugs, 1964. Research and publs. on effects and treatment of poisoning by chem. warfare and therapeutic agts.; debridement of burns; wound healing; various dermatitides. Home: 840 Powell St., San Francisco 94108. Office: Letterman Gen. Hosp., The Presidio, San Francisco 94129.*

SULZER, Johann Georg, Swiss physicist; b. Winterthur, Switzerland, 1720; s. Jean Georges Sulzer; ed. Winterthur, Zurich. Author: Allgemeine theorie der Schönen Künste. Made 1st observation of galvinism, noting taste produced by 2 different contiguous metals on tongue; conducted expts. on resistance experienced by shots through air and resistance in fluids, 1761; gave method of div. for thermometer scale. Died 1779.

SULZER-HIRZEL, Jean-Jacques, inventor; b. Winterthur, Switzerland, 1806; s. Jacques Sulzer; founder (with father and brother Salomon) iron works; built water tube boilers containing a furnace, 1858, steamships, 1860, hydraulic presses, horizontal steam engines, steam engines with valves, 1866, central heating systems, 1869, machine for producing condensed milk, 1874, Brandt drilling machines, 1876, automatic burners and Lind ice machine, 1877, also multi-celled high pressure centrifugal pump, mine drilling pumps, irrigation pumps, 1894, diesel engines, boilers, steam engines, high pressure compressors, from 1897. Died 1883.

SUMI, Franc, Yugoslavian geophysicist; b. Kranj, Yugoslavia, May 11, 1922; s. Franc and Nada (Pirc) S.; Mining Engr., Degree in Geophysics, U. Ljubljana (Yugoslavia), 1948; m. Sonja Dekleva, Sept. 9, 1949; children—Fran, Andrej. Geophysicist, Trepca Mines (Yugoslavia), 1948-51; head geophys. dept. Geol. Survey, Beograd, Yugoslavia, 1951-58, Ljubljana, Yugoslavia, 1958—; intermediate term geophysicist UN Devel. Programm, Kaduna, Nigeria, 1966—. Mem. Yugoslav Soc. Geologists, Yugoslav Soc. Mining Engrs. Author: Induced Polarization, 1966; also articles. Research on improvement and methods and interpretation techniques in applied geophysics, deduction of formula for direct determination of angle of dip of inclined geol. contact using resistivity method, deducation formulae for vertical gravimetrical sounding, devel. induced elec. polarization method and clarification of its causes in case of metallic and clay materials. Home: 16 Majde Vrhovnikova. Office: 33 Parmova, Geol. Survey, Ljubljana, Yugoslavia.*

SUMIKI, Yusuke, Japanese chemist; b. Niigata Prefecture, Feb. 10, 1901; Dr.Agr. Prof., Tokyo U. Mem. Japan Sci. Council. Author books including: Vegetable Hormones; Penicillin; Chemistry of Taste. Research on agrl. and organic chemistry, components of rice yeast; discovered new antibiotic for agrl. diseases, 1961.

SUMMERS-GILL, Robert George, Canadian physicist; b. Wadena, Sask, Can., Dec. 22, 1929; s. Herbert Reginald and Margaret (Robson) S.-G.; B.A., U. Sask., 1950, M.A., 1952; Ph.D., U. Cal. at Berkeley, 1956. Faculty, McMaster U., Hamilton, Ont., Can., 1956—, professor physics, 1966—. vis. scientist Lincoln. Lab., Mass. Inst. Tech., Lexington, 1960. Mem. Am. Phys. Soc., Optical Soc. Am., Phys. Soc. (London, Eng.), Canadian Assn. Physicist (registrar 1959-65), Sigma Xi. Research, publs. on photonuclear reactions in the light elements, scattering charged particles from beryllium, properties various radioactive nuclei deduced from atomic beam experiments. Home: 101 King St. E., Dundas, Ont. Office: Dept. physics, McMaster U., Hamilton, Ont., Can.*

SUMMERSKILL, William Hedley, physician; b. London, Eng., Jan. 8, 1926; s. William Hedley and Elaine (Gerard) S.; B.A., Oxford U., 1947, M.D., 1949, M.A., 1951, D.M., 1955; M.R.C.S., London U. Med. Sch., 1949, M.R.C.P., 1953; m. Elizabeth Anne Sheppard, Jan. 25, 1950 (dec. Oct. 1965); 1 son, William Storith Markham. Came to U. S., 1958, naturalized, 1965. Tutor London U. Med. Sch., 1953-55; research asso. Harvard Med. Sch., 1955-56; sr. registrar gastroenterology Central Middlesex Hosp., London, 1957-58; cons. medicine Mayo Clinic, Rochester, Minn., 1959—, head gastroenterology unit, 1962—; asst. prof. Mayo Grad. Sch. U. Minn., 1959-63, asso. prof., 1963—. Cons. NIH, 1966; mem. VA Research Program Evaluation Com., 1966. Recipient Evelyn Rothschild prize, 1944; Rockefeller Traveling fellow, 1955. Fellow A.C.P.; mem. Royal Coll. Physicians, Assn. Physicians of Gt. Britain and Ireland, Soc. Exptl. Biology and Medicine, Central Soc. for Clin. Research, Am. Gastroenterol. Assn., Am. Assn. for Study of Liver Disease, Am. Fedn. Clin. Research, Med. Research Soc., Med. Soc. London, Harvey Soc. (v.p. 1958), Royal Soc. Medicine, British Med. Assn., A.M.A., French Gastroenterology Assn., Sigma Xi. Research numerous publs. on functions of liver, biliary system and gastrointestinal tract in health and disease, particularly cause and treatment of liver failure, coma, ascites and jaundice. Home: 814 5th St. S.W. Office: 200 First St. S.W., Rochester, Minn. 55901.*

SUMMERSON, Charles Henry, Am. geologist; b. Catlettsburg, Ky., Nov. 15, 1914; s. John R. and Argolia (Wellman) S.; B.S., U. Ill., 1938, M.S., 1940, Ph.D., 1942; m. Harriet M. Rockwell, June 29, 1944; children—Henry C., Jane R., Philip C. Grad. asst. geology U. Ill., 1938-42; asst. aerial photo interpreter U.S. Air Force, 1943; asst. geologist U.S. Geol. Survey, 1943-45; asst. prof. geology Mo. Sch. Mines, 1946-47; faculty Ohio State U., Columbia, 1947—, now asso. prof. geology; asst. to dir. Ohio State U. Research Found., 1958-65. Cons. Ohio Geol. Survey, 1960—; staff asso. NSF, Washington, 1965-66. Mem. Am. Assn. Petroleum Geologists, Geol. Soc. Am., Paleontol. Soc., Soc. Econ. Paleontology and Mineralogy, Soc. Photographic Engrs., Internat. Assn. Sedimentology, Ohio Acad. Sci., A.A.A.S., Sigma Xi. Research and publs. on stratigraphy of Paleozoic rocks of Eastern N.Am., sedimentary petrology, carboniferous paleontology, Antarctic geology. Home: 76 Northmoor Pl., Columbus, Ohio 43214.*

SUMNER, Francis Bertody, Am. zoölogist; b. Pomfret, Conn., Aug. 1, 1874; s. Arthur and Mary Augusta (Upton) S.; B.S., U. of Minn., 1894; Ph.D., Columbia, 1901; m. Margaret Elizabeth Clark, Sept. 10, 1903; children—Florence Anne, Elizabeth Caroline, Herbert Clark. Tutor and instr. natural history Coll. City N.Y., 1890-1906; dir. biol. lab. U. S. Bur. Fisheries, Woods Hole, Mass., 1903-11; naturalist U. S. Bur. Fisheries steamer Albatross 1911-13; asst. prof. biology, Scripps Inst. for Bio. Research (later Inst. of Oceanography), U. Cal., 1913-19; asso. prof., 1919-26, prof. since 1926, acting dir., 1923-24; research asso. Carnegie Instn., Washington, 1927-30. Fellow A.A.A.S. (chmn. sect. F. 1938), Cal. Acad. Scis., San Diego Natural History Soc.; mem. Am. Soc. Zoölogists, Am. Soc. Naturalists, Western Soc. Naturalists (pres. 1921-22), Am. Soc. Mammalogists, Ecol. Soc. Am., Am. Genetic Assn., Am. Soc. Ichthyologists and Herpetologists, Nat. Acad. Scis., (corr.) Phila. Acad. Scis. Am. Philos. Soc., Soc. for Exptl. Biology and Medicine, Phi Beta Kappa, Sigma Xi. Contbr. papers on embryology and physiology of fishes, marine ecology, geographic variation, heredity and evolution. Made exptl. studies on nature and inheritance of adaptive variations, thus elucidating problems of organic evolution of new varieties in wild mammals and Mendelian inheritance of their characteristics. Died Sept. 6, 1945.

SUMNER, James Batcheller, Am. biochemist; b. Canton, Mass., Nov. 19, 1887; s. Charles and Elizabeth Rand (Kelly) S.; prep. edn., Roxbury (Mass.) Latin Sch., 1900-06; A.B., Harvard, 1910; A.M., 1913, Ph.D., 1914; grad. study U. of Brussels, 1921-22; m. Bertha Louise Ricketts, July 20, 1915 (divorced); children—Roberta Rand, Nathaniel (dec.), Prudence Avery, James Cosby Ricketts, Frederick Overton Burnley; m. 2d, Agnes Paulina Lundkirst, 1931 (div.); m. 3d, Mary Beyer, 1943; children—John Increase, Samuel B. (dec.). Acting prof. chemistry, Mt. Allison College, Sackville, N.B., Can., 1911; research asst. Worcester (Mass.) Poly. Inst., 1911-Jan. 1912; asst. prof. bio-chemistry, Cornell U., 1914-29, prof., 1929—, director of Lab. of Enzyme Chemistry, 1947-55; fellow Commn. for Relief in Belgian Ednl. Foundation, 1921-22; Guggenheim fellow, 1937. Awarded Scheele medal at Stockholm, Sweden, 1937; (with Stanley and Northrop) Nobel Prize in Chemistry, 1946. Mem. Am. Soc. Biol. Chemists, A.A.A.S., Soc. Exptl. Biology and Medicine, Nat. Acad. Science, Am. Acad. Arts and Sciences, Sigma Xi. Author: Textbook of Biological Chemistry, 1927; Laboratory Experiments in Biological Chemistry; Chemistry and Methods of Enzymes, 1943. Co-editor: The Enzymes, Chemistry and Mechanism of Action, 1950-52. Research on enzymes; 1st to isolate and crystallize an enzyme and show it to be protein, 1926; stimulated research in enzymes and viruses. Died Buffalo, N.Y., Aug. 12, 1955.

SUMNER, William Graham, Am. sociologist, economist; b. Paterson, N.J., Oct. 30, 1840; s. Thomas and Sarah (Graham) S.; grad. Yale, 1863; studied in Univs. of Göttingen, Germany, and Oxford, Eng.; LL.D., U. of E. Tenn.; m. Jeannie Whittemore Elliott, Apr. 17, 1871. Tutor Yale, 1866-1909; took orders in P.E. Church, and was asst. Calvary Ch., New York, and rector Ch. of the Redeemer, Morristown, N.J., until 1872; prof. polit. and social science, Yale, 1872—. Editor, Living Church, 1869-70; alderman, New Haven, 1873-76; pres., Am. Sociol. Soc., 1910. Author: A History of American Currency 1874; What Social Classes Owe to Each Other, 1882; Collected Essays in Political and Social Sciences, 1883; Economic Problems, 1884; Protectionism, 1885; Lives of Andrew Jackson; Alexander Hamilton and Robert Morris, 1891; The Financier and Finances of the Revolution, 1892; A History of Banking in the U. S., 1896; Folkways, 1907; The Forgotten Man and Other Essays (ed. A. G. Keller), 1918; Science of Society (4 vol., with Keller and Davie), 1927. Opposed any kind of govtl. regulation of free operation of economy; gave valuable analysis of folkways and mores, which reflects Social-Darwinism and free economy theories; developed theory that social customs were natural phenomena and could not be consciously developed. Died Englewood, N.J., Apr. 12, 1910.

SUN, James Ming Shan, mineralogist; b. Pingyuan, China, May 10, 1918; s. Hsi Wen and Hsiao Chieh (Ting) S.; B.S., Nat. Central U., 1940; M.S., U. Chgo., 1947; Ph.D., La. State U., 1950; m. Clare C. C. Yu, June 1, 1953; children—Linda F., Eugene F. Came to U. S., 1945, naturalized, 1962. Postdoctoral research fellow Columbia, N.Y.C., 1950-51; mineralogist N.M. Inst. Mining and Tech., Socorro, 1951-62; research specialist Jet Propulsion Lab., Cal. Inst. Tech., Pasadena, 1962-64; research physicist Air Force Weapons Lab., Kirkland AFB, N.M., 1964—. Fellow Mineral. Soc. Am.; mem. Am. Geophys. Union, Geochem. Soc., Am. Assn. Petroleum Geologists, A.A.A.S., Sigma Xi. Research publs. in exptl. mineralogy, x-ray crystallography and geochemistry, physics of high pressure, hypervelocity impact and cratering. Home: 7704 Sierra Azul St. N.E., Albuquerque 87110. Office: Air Force Weapons Lab., Kirkland AFB, N.M. 87117.*

SUN, Yun-Pei, insect toxicologist; b. Kaoyu, Kiangsu, China, June 20, 1910; s. Yun-Wu and Sao-Nan (Tso) S.; came to U. S., 1939, naturalized, 1960; B.S., Nat. Cheking U., China, 1932; M.S., U. Minn., 1941, Ph.D., 1943; m. Jung-Yi Tung, Jan. 15, 1938; children—James D., Louis J. Jr. chemist in charge insecticides lab. Nat. Agrl. Research Bur., China, 1935-39; China Found. Research fellow, 1939-41; research fellow Cornell U., 1944-48; research entomologist, dir. insecticide testing lab. Julius Hyman & Co., 1948-52; asst. mgr. in charge entomology and residue analysis labs. Shell Devel. Co., 1952-54, mgr. entomology dept., Denver, 1954-57, chief entomologist, Modesto, Cal., 1957—. Fellow A.A.A.S.; mem. Entomol. Soc. Am. (past sect. chmn.), Sigma Xi. Contbg. author: Advances in Pest Control Research, 1957; Methods of Testing Chemicals on Insects, vol. II, 1960; Analytical Methods for Pesticides, Plant Growth Regulators and Food Additives, vol. I, 1963, vol. II, 1964; Research, publs. on bioassay and evaluation of insecticides and their residues, synergism of insecticide combinations and method of evaluation, correlation of lab. and field toxicity data, correlation of chem. structure and insect toxicity. Home: 1918 La Villa Rose Ct., Modesto 95350. Office: P.O. Box 4248, Modesto, Cal. 95352.*

SUNDE, Milton Lester, Am. nutritionist; b. Volga, S.D., Jan. 7, 1921; s. Andrew C. and Clara (Mehl) S.; B.S., S.D. State Coll., 1947; M.S., U. Wis., 1949, Ph.D., 1950; m. Genevieve, Larson, Dec. 29, 1946; children—Roger, Scott, Robert. Faculty, U. Wis., Madison, 1947—, prof. poultry nutrition, 1957—; research scientist Rockefeller Found., Colombia, S. Am., 1960. Recipient Research award Am. Feed frs. Assn., 1961. Mem. Poultry Sci. Assn. (teaching award 1962, dir. 1961-64, v.p. 1965—), Inst. Nutrition, Soc. for Exptl. Biology and Medicine, World Poultry Sci. Assn., Sigma Xi, Alpha Zeta, Gamma Alpha. Research numerous publs. in poultry nutrition, amino acid requirement and metabolism, protein and energy needs, aspects calcium and magnesium metabolism. Home: 1111 Starlight Dr., Madison, Wis. 53711.*

SUNDERMAN, Duane Neuman, Am. chemist; b. Wadsworth, O., July 14, 1928; s. Richard B. and Carolyn (Neuman) S.; student Heidelberg Coll., 1945-47; A.B., U. Mich., 1949, M.S., 1954, Ph.D., in Chemistry, 1956; postgrad. Purdue U., 1949-50; m. Joan Catherine Huffman, Jan. 31, 1953; children—David Duane, Christine Joan, Richard Gail. Research chemist E.I. du Pont de Nemours, Argonne (Ill.) Nat. Lab., 1951-52, Savannah River Plant, 1952-54; prin. chemist Battelle Meml. Inst., Columbus, O., 1956, project leader, 1956-58, asst. chief, 1958-59, chief chem. physics div., 1959-65, asso. mgr. physics dept., 1965—. Mem. subcom. on radiochemistry NRC, 1961—. Phoenix fellow, 1955-56. Mem. Am. Nuclear Soc., Am. Chem. Soc., Am. Soc. for Testing and Materials, Sigma Xi, Phi Lambda Upsilon, Pi Kappa Delta. Research, numerous publs. on nuclear chemistry, nuclear reactor fuel devel., safety nuclear power reactors, fission products chemistry, radiochem. analysis, gamma ray spectrometry, radiation effects, applications radioisotopes, high temperature chemistry, chem. physics. Home: 2011 Pevensey Ct., Columbus 43221. Office: 505 King Av., Columbus, O. 43201.*

SUNDERMAN, Frederick William, Am. physician; b. Altoona, Pa., Oct. 23, 1898; s. William A. and Elizabeth (Lehr) S.; B.S., Gettysburg Coll., 1919, Sc.D., 1952; M.D., U. Pa., 1923, M.S., 1927, Ph.D., 1929; m. Clara Louise Baily, June 2, 1925; children—Frederick William, Joel Baily (dec.), Louise (dec.). Practice medicine, specializing in pathology, Phila., 1930—; dir. div. metabolic research Jefferson Med. Coll., 1951-65, clin. prof. medicine, 1951—; physician Jefferson Hosp., 1951—; dir. Inst. For Clin.

Scis., Inc., 1965—; cons., adviser numerous hosps., bus. firms, govt. agys., 1947—. Recipient Sci. Products awards Coll. Am. Pathologists, 1962, Distinguished Alumnus Honors award Gettysburg. Coll., 1963, medal of honor Armed Forces Inst. Pathology, 1964, Anniversary Goblet award Assn. Clin. Scientists, 1964. Ky. Col. Diplomate Am. Bd. Internal Medicine, Am. Bd. PathoMlogy (trustee), Am. Bd. Clin. Chemists. Fellow Am. Assn. Clin. Chemists, A.C.P., Am. Soc. Clin. Pathologists, Coll. Am. Pathologists (past trustee); mem. A.A.A.S., Am. Chem. Soc., Am. Inst. Biol. Scis., A.M.A., Am. Soc. Biol. Chemists, Am. Soc. Clin. Chemists, Am. Diabetes Assn., Assn. Clin. Scientists (dir. edn., past pres.), Endocrine Soc., Franklin Inst., Indsl. Med. Assn., Internat. Soc. Hermatology, Nat. Soc. Med. Research, Soc. Toxicology, Am. Coll. Clin. Pharmacology and Chemotherapy, Am. Indsl. Hygiene Assn., Pan Am. Med. Assn., Indian Soc. Endocrinology and Metabolism, Phi Beta Kappa, Sigma Xi, Alpha Omega Alpha. Author: (with F. D. Weidman) Xanthoma and Other Dyslipoidoses, 1941; (with F. Boerner) Normal Values in Clinical Medicine, 1949; (with others) Clinical Hemoglobinometry, 1953; (with F. W. Sunderman, Jr.) Lipids and the Steroid Hormones in Clinical Medicine, 1960, Measurements of Exocrine and Endocrine Functions of the Pancreas, 1961, Evaluation of Thyroid and Parathyroid Functions, 1963, Hemoglobin, its Precursors and Metabolites, 1964, The Serum Proteins and the dysproteinemias, 1964, Clinical Pathology of Serum Electrlytes, 1966; also numerous articles. Research on chem. alterations in body as result of disease. Home: 1833 DeLancey Pl. Office: 1930 Chestnut St., Phila. 19103.*

SUNDERMAN, Frederick William, Jr., Am. physician, pathologist; b. Phila., June 23, 1931; s. Frederick William and Clara Louise (Baily) S.; student Yale, 1948-49, Rice U., 1950-51; B.S., Emory U., 1952, postgrad. Sch. Medicine, 1951-53; M.D., Jefferson Med. Coll., 1955; m. Carolyn Lambeth Reynolds, Aug. 24, 1963. Fellow metabolic research Jefferson Med. Coll., 1956-58; instr., head chemistry div. U.S. Naval Med. Sch., Bethesda, Md., 1958-60; chief clin. chemistry div. NIH, Bethesda, 1960; instr., mem. div. metabolic research Jefferson Med. Coll., Phila., 1960-63; vis. investigator Cholera Research Centre, Indian Council Med. Research, Calcutta, India, 1963; asso. prof. pathology, dir. clin. labs. U. Fla. Coll. Medicine, Gainesville, 1964-67; prof. of pathology, 1967-68; prof., head dept. lab. lab. medicine U. Conn. Sch. Medicine, 1968—. Cons. to state agys., fed. agys., hosps. Bd. dirs. Inst. for Clin. Sci., Phila., Fellow Am. Assn. Clin. Chemists; mem. Assn. Clin. Scientists (past chmn. credentials com., pres. 1964-65), Internat, Acad. Pathology, Indsl. Med. Assn., Am. Assn. Pathologists and Bacteriologists, Soc. for Exptl. Biology and Medicine, A.C.P., Am. Fedn. for Clin. Research, Am. Assn. Cancer Research, Am. Soc. Clin. Pathologists, Am. Soc. for Exptl. Pathology, Coll. Am. Pathologists, Soc. Toxicology, A.M.A., Endocrine Soc., Phi Beta Kappa, Sigma Xi, Alpha Omega Alpha, Sigma Pi Sigma. Author: Lipids and Steroid Hormones in Clinical Medicine, 1960; Measurements of Exocrine and Endocrine Functions of the Pancreas, 1961; Evaluation of Thyroid and Parathyroid Functions, 1962; Hemoglobin its Precursors and Metabolites, 1963; Serum Proteins and the Dysproteinemias, 1964; Clinical Pathology of Serum Electrolytes, 1965; Clinical Pathology of Infancy, 1966; Laboratory Diagnosis of Liver Disease, 1967; Laboratory Diagnosis of Kidney Disease, 1968 (all with F.W. Sunderman); also numerous articles. Research in exptl. carcinogenesis, metabolism trace metals, fractionations of serum proteins, measurements of catecholamine metabolites. Home: Mountain Spring Rd., Farmington, Conn.*

SUNESON, Coit Alfred, Am. agronomist; b. Missoula, Mont., Jan. 3, 1903; s. Jonas F. and Selma (Swanson) S.; B.S., Mont. State U., 1928; M.S., Kan. State U., 1930; D.S., Montana State University, 1968; m. to Ann M. Nordquist, June 8, 1928; children—Ruth (Mrs. W.E. Nyquist), Alfred, Deanna (Mrs. Don C. Almeida), Lois (Mrs. R.F. Woodmansey), Paul. Research asso. Kan. State U., 1928-30; asso. U. Neb., 1930-36; asso. U. Cal. at Davis, 1936-68; research agronomist U.S. Dept. Agr., Davis, Cal., 1930-68. Fellow A.A.A.S., Am. Soc. Agronomy (Crops Sci. award 1966). Research and numerous publs. on plant breeding, population genetics; produced several comml. varieties; discovered genetic male sterility in barley; developed evolutionary method breeding. Home: East Lakeshore, Bigfork, Mont. 59911.*

SUN-TSU, Chinese mathematician; flourished 2d century. Author: Sun-Tsu Suan-Ching (arithmetical classic, includes explanation of sq. root and computation, information on Chinese metrology).

SUOMALAINEN, Esko, Finnish geneticist; b. Helsinki, Finland, June 11, 1910; s. Vihtori and Sanni (Lähde) S.; M.S. M.Sc., U. Helsinki, 1934, Ph.D., 1940; m. Valma Irja Tellervo Tynni, Sept. 12, 1943; 1 dau., Kaarina (Mrs. Eric Woirin Olavi). Faculty U. Helsinki, 1941—, prof. genetics, 1948—, head inst. genetics, 1949—. Vis. prof. U. Lund (Sweden), 1954. Chmn. Finnish Nat. Commn. for UNESCO, 1965. Mem. Finnish Acad. Sci. (sec. gen. 1964—), Acad. of Finland (sec. 1951-56), Societas Genetica Fennica (chmn. 1965—), Lepidopterists' Soc. (U. S., v.p. 1966), Societas Scientiarum Fenniae, Royal Physiographic Soc. (Lund), Deutsche Gesellschaft für Genetik. Author: Belträge zur Zytologie der parthenogenetischen Insekten I-II, 1940; Parthenogenesis in Animals, 1950; also papers. Research on parthenogenesis in animals, cytology of parthenogenesis, polyploidy. in animals, chromosomes and chromosomal evolution of Lepidoptera, chromosomes of different insect groups. Home: 18 Museokatu, Helsinki, Finland.*

SUPAN, Alexander, geographer; b. Innichen/Tyrol, Austria, Mar. 3, 1847; prof., U. Czernowitz, also at Gotha; dir. Perthes Geographic Inst., Gotha, 1884-1909; prof., Breslau (now Wroclaw, Poland). Author: Gründzuge der physischen Erdkunde, 1884; Oesterreich-Ungarn, 1889; Die territoriale Entwicklung der europaischen Kolonien, 1906; Leitlinien der allgemeinen politischen Geographie, 1918. Died Breslau, July 6, 1920.

SUPEK, Ivan, Yugoslavian theoretical physicist, philosopher, writer; b. Zagreb, Yugoslavia, Apr. 8, 1915; s. Rudolf and Marija (Sips) S.; D. Philosophy and Sci., 1940; m. Zdenka Tagliaferro, Aug. 3, 1945; children—Iris, Silva, Ivica. Prof. theoretical physics, history sci. U. Zagreb, 1945—; organizer Nuclear Research Inst., 1950—. Mem. Yugoslav Acad. Sci. and Art (chmn. Inst. Philosophy Sci. and Peace); Author: The Electrical Conductivity at Low Temperature, 1940; Great Piramide (play), 1959; On Atomic Volcanoes, 1959; Between War Lines (novel), 1962; Process of the Century (novel and play), 1962; On Atomic Island (play), 1962; Tale of the Modern Time (play), 1963; In Barracks (play), 1963; Theoretical Physics and Structure of Matter, 2 vols., 1963; Science and Philosophy, 1964. Editor; Priroda (Nature). Publs. on history sci. Research in quantum electrodynamics, analogue interactions between electron and phonon; noted predominance phonon interactions even at lowest temperatures; discovered differential equation of electric conductivity at low temperatures (replacing Bloch's integral equation), 1957; active resistance nuclear armament, 1944—. Home: 10 Rubeticeva. Office: 19 Marulicev trg, Zagreb, Yugoslavia.*

SUPEK, Zlatko, Yugoslavian pharmacologist; b. Tesanj, Yugoslavia, Nov. 8, 1914; s. Viktor and Marija (Vitkovic) S.; M.D., U. Zagreb (Yugoslavia), 1939; m. Ivana Gilic, May 22, 1943; children—Bojan, Mirna. Faculty, dept. pharmacology U. Zagreb, 1941—; prof., head dept., 1964—; chief Lab. for Exptl. Neuropathology of Radiation Injury, Inst. Rugjer Boskovic, 1953—. Recipient Rugjer Boskovic prize for sci. achievement, 1965. Mem. German Pharmacological Soc. (asso.). Research and publs. on role serotonin in genesis of radiation sickness and mechanism of action of some psychopharmacological agts., chem. protection against ionizing rediation, biol. indicators of severity of radiation sickness, internat. standards of drugs. Home: 4 Svibovac, Zagreb, Yugoslavia.*

SUPER, Donald Edwin, Am. psychologist; b. Honolulu, Hawaii, July 10, 1910; s. Paul and Margaret (Stump) S.; B.A., Oxford (Eng.) U., 1932, M.A., 1936; Ph.D., Columbia, 1940; m. Anne-Margaret Baker, Sept. 12, 1936; children—Robert Marion, Charles McAfee. Prof. psychology, edn. Tchrs. Coll., Columbia, N.Y.C., 1945—, dir. div. psychology, edn., 1965—; cons. govt. burs., colls., industry. Fulbright lectr. psychology U. Paris, France, 1958-59; Ford fellow, Poland, 1960; specialist U. S. Dept. State and Asia Found., Far East, 1961. Fellow Am. Psychol. Assn., A.A.A.S.; mem. Am. Personnel and Guidance Assn. (past pres.), Internat. Vocational Guidance Assn. (v.p. 1959—), Internat. Assn. Applied Psychology (dir.), Am. Ednl. Research Assn. Author: Appraising Vocational Fitness, 1949, 62; The Psychology of Careers, 1957; The Vocational Maturity of Ninth Grade Boys, 1960; La Psychologie des Interets, 1964, and others. Asso. editor, Jour. of Counseling Psychology, 1954-65; cons. editor, Jour. Ednl. Psychology, 1959-65. Studies, publs. on vocational devel. and occupational choice; determinants of careers, vocational maturity, aptitude and interest tests. Home: 124 Stonebridge Rd., Montclair, N.J. 07042.*

SUPNIEWSKI, Janusz Victor, Polish pharmacologist, chemist; b. Plock, Poland, Sept. 17, 1899; s. Michal and Antonina (Widulińska) S.; M.D., U. Warsaw (Poland), circa 1927; also Dr. Chemistry; studied under R. Adams at U. Ill., under Dixon, Pope at Cambridge, Danysz at Pasteur Inst.; m. J. H. Obrebska, July 9, 1936. Became asso. prof. pharmacology U. Warsaw, 1929; named prof. Jagiellonian U., Kraków, Poland, 1930; named dir. dept. pharmacology Polish Acad. Sci., 1953; imprisoned by Nazis in Sachsenhausen concentration camp, 1939; dean, organizer, Med. Faculty, also dept. pharmacology Jagellonian U., Kraków, 1945; organizer dept. pharmacology Polish Acad. Scis., 1953. Chmn. several sci. councils, especially in indsl. pharmacology. Recipient Gold Crosses of Merit, twice, Nat. Prize of 1st degree. Mem. Polish Acad. Sci. Author: Farmakologia; Receptura; Preparatyka nieorganiczna; also numerous articles. Synthesis of several metallo-organic chemotherapeutic compounds and methods of isolating sex hormones; synthesis of mescaline; research on biol. derivatives of plant and animal products, vitamins, carcinogenic compounds and analeptic drugs, sulfones, sulfonamides; methods of indsl. prodn. of various medicines, especially vitamins, vitamin C, antituberculosis, hydrazide of isoicotynic acid, antibiotics, chloromycetin. Died Kraków, June 3, 1964.

SUPPES, Patrick, Am. statistician, philosopher; b. Tulsa, Mar. 17, 1922; s. George Biddle and Ann (Costello) S.; B.S., U. Chgo., 1943; Ph.D. (Wendell T. Bush fellow), Columbia, 1950; m. Joan Farmer, Apr. 16, 1946; children—Patricia, Deborah, John Biddle. Instr. Stanford, 1950-52, asst. prof., 1952-55, asso. prof., 1955-59, prof. philosophy and statistics, 1959—, chmn. dept. philosophy, 1963—, asso. dean Sch. Humanities and Scis., 1958-61. Served to capt. USAAF, 1942-46. Murray. Butler silver medal Columbia U., 1965. Fellow Center for Advanced Study Behavioral Scis., 1955-56, Nat. Sci. Found., 1957-58. Mem. Math. Assn. Am., A.A.A.S.,Psychometric Soc., Am. Philos. Assn., Assn. Symbolic Logic, American Mathematical Society, American Psychological Association, Sigma Xi. Author: Introduction to Logic, 1957; Axiomatic Set Theory, 1960; (with Davison and Siegel) Decision Making, 1957; (with Richard C. Atkinson) Markov Learning Models for Multiperson Interactions, 1960; (with Shirley Hill) First Course in Mathematical Logic, 1964; also articles. Research on logic, set theory, decision making, learning models, philosophy of science, statis. bases of science. Home: 678 Mirada Av., Stanford, Cal.*

SURANYI, P., Hungarian physicist; b. Budapest, Hungary, Jan. 31, 1935; s. Joseph and Elizabeth (Szenes) S.; physicist's diploma, Roland Eötvös U., Budapest, 1958; candidate physics U. Dubna (USSR), 1964; m. Theresa Gàl, May 24, 1960. Sci. co-worker Central Research Inst. for Physics, Budapest, 1958-61, Joint Inst. for Nuclear Research, Dubna, 1961-65; sci. chief co-worker Central Research Inst. for Physics, Budapest, 1965—. Mem. Roland Eötvös Phys. Soc. Research, publs. on multiple meson prodn. in high energy nuclear reactions, high energy behaviour in several field theoretical models; breakdown of symmetries in elementary particles physics, mass differences of elementary particles. Home: 35 Lenin, Budapest VII. Office: Konkoly Thege, Budapest XII, Hungary.*

SURANYI-UNGER, Theodore Victor, economist; b. Budapest, Hungary, Feb. 4, 1898; s. John and Emma (Neugebauer) Suranyi-U.; Dr. iuris et rer. pol., U. Graz, Austria, 1919; Dr. phil., U. Budapest, 1920; Dr. oec. publ., Technology of Budapest; m. Dr. Nora baronesse de Braun, Dec. 20, 1926; children—Theodore, Nora. Came to U. S., 1928, naturalized, 1951. prof. econs., dean continental European univs., 1924-45; vis. prof. 1926-55, univs. of Vienna, Berlin, Paris, Munich, Göttingen, Hamburg, Tübingen, Breslau, Kiel, Mannheim, Muenster, Marburg, Tilburg, Sendai, Shanghai, Allahabad, Bombay; vis. prof. Am. univs., 1928-30, 33, 35, 37, 39; prof. econs., chmn. grad. econs. seminar Syracuse U., 1946—; research project, Göttingen, 1958—; dean faculty of economics and social sciences Göttingen University, Germany, 1964-65. Decorated Comdr. of Ordre de Mérite Agricole, Paris, 1939; awards Social Science Research Council, New York City, 1948, 57, Ford Found., 1958; Rockefeller Found. fellow, 1928-29, 32, Ford Found. fellow, 1956-57. Mem. Central N.Y. Econ. Assn. (pres. 1952). Author: Philosophy and Economics (2 vols.), 1923, 26; Twentieth Century Economics, 1931; Comparative Economic Systems, 1952, The Economic Growth of Southeast Europe, 1964. Contbr. articles prof. publs. Contributions to the fundamentals of the history of economics to economic philosophy, to a better understanding of comparative economic systems, to the economic analysis of the Southeast European countries and to the ways and possibilities of a reunion of the split world economy. Address: Dept. Econs., Syracuse U., Syracuse N.Y. 13210.*

SURAWICZ, Borys, physician; b. Moscow, Russia, Feb. 11, 1917; s. Josef and Mathilda (Soloweczyk) S.; M.D., Stafan Batory U., Wilno, Poland, 1939; m. Frida G. Van Klaveren, July 19, 1946; children—Christina M., Nina M., Tanya S., Serge J. Came to U.S., 1951, naturalized, 1956. Mem. staffs hosps. Germany, Norway, 1945-49; staff De Goesbriand Meml. Hosp., Burlington, Vt., 1951-53; Phila. Gen. Hosp., 1953-55; instr. cardiology U. Pa., Phila., 1954-55; instr. U. Vt., Burlington, 1955-57, asst. prof. clin. and exptl. medicine, 1957-62; chief div. cardiology U. Ky. Coll. Medicine, Lexington, 1962—, asso. prof. medicine, 1962-66, prof., 1966—. Cons. VA Hosp., Lexington, Ireland Army Hosp., Ft. Knox, Ky. Mem. A.M.A., A.C.P., Am. Heart Assn., Soc. U. Cardiologists, Am. Coll. Cardiology, Am. Physiol. Soc., Sigma Xi. Editor: (with E.D. Pellegrino) Sudden Cardiac Death, 1964. Research and numerous publs. on effect of electrolytes on electrocardiogram in man, correlation between electrocardiogram and transmembrane action potential in isolated rabbit hearts, role of electrolytes in genesis of cardiac arrhythmias. Home: 806 Overbrook Circle Dr., Lexington, Ky. 40506.*

SURDIN, Maurice, physicist; b. Melitopol, Russia, July 22, 1911; s. Mark and Fanny (Tarsis) S.; M.Sc. in Physics, U. Rennes and Paris, 1931; Elec. Engr., Ecole Supérieure d'Electricité de Paris (France), 1932; Dr. Degree in Physics, French U. Spl. Session,

London, Eng., 1942; m. Jeannine Madeleine Verhaeghe, Oct. 27, 1938; children—Yolande-Michèle (Mrs. Pierre Kerjan), Dina-Anne. Electronic engr. various cos., 1932-35; staff Laboratoire de Physique Experimentale, Collège de France, Paris, 1936-40; staff Free French Navy and Admiralty Signal Establishment, Gt. Britain, 1941-46; head Electronics dept. French AEC, 1946-64; dir. applied research European Space Orgn., Delft, Netherlands, 1964-68; professor Univ. of Bordeaux, 1968—. Chairman of com. in charge bldg. Saclay (France) Centre, 1949-52; vis. prof. physics U. N.C., Raleigh, 1960-61. Decorated Chevalier de la Légion d'Honneur; Officer des Palmes Académiques (France). Mem. Societe Francaise de Physique, Societe Francaise des Electroniciens et des Radioélectroniciens, Am. Phys. Soc. Research and publs. in electromagnetism particularly stochastic phenomena; discovered propagation of acoustic waves in plasmas, fluctuation of potential across a condenser having a polar dielectric; proposed theory of origin of cosmic radio waves; devel. stochastic electrodynamics. Home: 37 Blanchard, Fontenay-Aux-Roses, France. Office: Faculté des Sciences, U. de Bordeaux, 33 Talence, France.*

SURGENOR, Douglas MacNevin, Am. biochemist; b. Hartford, Conn., Apr. 7, 1918; s. William Hume and Flora E. (MacNevin) S.; A.B., Williams Coll., 1939; M.S., U. Mass., 1941; Ph.D., Mass. Inst. Tech., 1946; m. Lois Hutchinson, Nov. 2, 1946; children—Peter, Sara, Jonathan, Timothy, Stephen. Research chemist Mass. Inst. Tech., 1943-45; faculty Harvard, 1948-60, asst. prof., 1950-60, asso. mem. Lab. Phys. Chemistry related to Medicine and Pub. Health, 1950-54; sr. investigator Protein Found., Boston, 1956-60; prof., head dept. biochemistry U. Buffalo, 1960-62; dean Sch. Medicine, State U. N.Y. at Buffalo, 1962—, Provost of health sciences faculty, 1967—. Chmn. adv. bd. Erie County Lab., 1962—; dir. Western N.Y. Hosp. Rev. and Planning Council, 1962—, Internat. Com. for Nomenclature of Blood Clotting Factors. Mem. Am. Heart Assn. research com. 1961-67, chmn. 1964-66), Am. Chem. Soc., Am. Soc. Biol. Chemists, Am. Soc. Hematology, Biophys. Soc., Sigma Xi. Editor: (with Charles Bishop) The Red Blood Cell, A Comprehensive Treatise, 1964. Editorial bd. Circulation Research. Numerous publs. on devel. methods for fractionating proteins of human plasma including antihemophilic factors; research on biochem. enzyme system of blood coagulation.*

SURIAN, Joseph Donat, French physician, botanist; flourished late 17th century; physician, pharmacist, Marseille, France; author plant catelogue; made chem. analyses of plants to determine their med. value; noted for precision.

SURYANARAYANA, Dhanyamraju, Indian plant pathologist; b. Vijayanagaram, India, Sept. 20, 1922; s. D. and D. (Manikyamma) S.; B.Sc., P.R. Coll., Kakinada, India, 1941; M.Sc., Agra (India) U., 1943, Ph.D. in Plant Pathology, 1952; m. D. Manikyamma, Apr. 21, 1943; children—D. S. Rao, D. R. Ranjini, D. Venku, Lectr. botany Agra Coll., Agra U., 1943-52; asst. prof. mycology Central Coll. Agr., Delhi (India) U., 1952-56; plant pathologist, faculty Postgrad. Sch., Indian Agrl. Research Inst., New Delhi, 1956-63, organizer mycology sect. seed testing unit; prof., head dept. botany and plant pathology Punjab Agrl. U., Hissar, India, 1965—. Mem. Indian Phytopath. Soc. (past joint sec.), Indian Sci. Congress. Author: Botanical Hand Book Number I Algae and Fungi Thallophytes, 1952; also articles. Research on causes and control of plant diseases, including Green ear disease of pearl millet, downy mildew of corn, stack burn of rice, seed-borne diseases of certain Indian Vegetables. Home: 8/7 Punjab Agrl. U., Hissar, Punjab, India.*

SUSKIND, Raymond Robert, Am. physician; b. N.Y.C., Nov. 29, 1913; s. Alexander and Anna (Abramson) S.; B.A., Columbia, 1934; M.D., State U. N.Y., Bklyn., 1943; m. Ida Blanche Richardson, Dec. 27, 1944; children—Raymond Robert, Stephen Alexander. Instr. to asso. prof., dept. dermatology U. Cin. Coll. Medicine, 1948-62, asst. prof. to asso. prof. dept. preventive medicine and indsl. health, 1949-62; dir. dermatol. research program Kettering Lab., 1948-62; prof., head div. environmental medicine, prof. dermatology U. Ore. Med. Sch., Portland, 1962—. Cons. med. research and devel. to surgeon gen. U. S. Army, 1965—; mem. com. on cutaneous system NRC, Nat. Acad. Scis., 1958-65; cons. occupational health Pan Am Health Orgn. Fellow A.C.P.; mem. A.M.A., A.A.A.S., Am. Acad. Dermatology, Soc. for Investigative Dermatology, Assn. Am. Med. Colls., Soc. Exptl. Biology and Medicine, Am. Indsl. Hygiene Assn., Indsl. Med. Assn., N.Y. Acad. Scis., Ore. Dermatol. Soc., Sigma Xi. Contbr. numerous articles to profl. jours., chpts. to med. books. Measured physiol., structural and biochem. effects of water, environmental humidity, temperature, ultraviolet radiation and specific chems. on skin penetration, cellular reactions to irritants and allergic sensitizers, chem. components of sebaceous gland and its secretion, pathogenesis of chemically induced acne, skin protectants, sweat function defects in psoriasis. Home: 4424 S.W. Warrens Way, Portland, Ore. 97201. Office: 3181 S.W. Sam Jackson Park Rd., Portland, Ore. 97201.*

SUSRUTA, surgeon; b. India; emphasized importance of bones, muscles, ligaments, joints in anatomy; named and classified 1120 diseases; knew effects of 760 medicinal plants; recommended that bodies be allowed to decay in water to permit dissection (Hindus not permitted to cut bodies). Author: Susrutasamhita (med. ency. from last centuries of pre-Christian era, said to have come to Susruta from Dhanvantari, in fixed form by 7th century).

SUSSEX, Ian Mitchell, biologist; b. Auckland, New Zealand, May 4, 1927; s. Thomas Roy and Enid (Trevalyan) S.; B.S., Auckland U. Coll., 1947, M.S., 1949; Ph.D., U. Manchester, 1952. Faculty, Victoria U., Wellington, New Zealand, 1954-55, U. Pitts., 1955-60; asso. prof. Yale, New Haven, 1960—. Mem. Soc. Devel. Biology,A.A.A.S., Tissue Culture Assn., Am. Bot. Soc., Internat. Soc. Plant Morphologists, Soc. Exptl. Biology, Sigma Xi. Research in determination of leaf shape at the shoot apex, differentiation of cell types in tissue cultures, effects of nutrients and hormones on devel. and growth of plant cells. Home: 289 Park Rd., Hamden, Conn. 06514.*

SUSSEX, James Neil, Am. psychiatrist; b. Northcote, Minn., Oct. 2, 1917; s. Rollo Norton and Florence (Bartholomew) S.; A.B., U. Kan., 1939, M.D., 1942; m. Margaret Ann Garty, Apr. 25, 1943; children—Margaret Eileen, Mary Patricia, Barbara Lorraine, Teresa Virginia. Commd. lt., j.g. M.C., USN, 1942, advanced through grades to comdr., 1955, fellow in child psychiatry Phila. Child Guidance Clinic, U. Pa., 1949-51, asst. chief neuropsychiatry Nat. Naval Med. Center, Bethesda, Md., 1951-55; faculty U. Ala. Med. Center, Birmingham, 1955—, prof., chmn. dept. psychiatry, 1959—. So. rep. to adv. bd. Nat. Psychiat. Residency Selection Program, 1965—; planning dir. for mental retardation in Ala., 1964—; cons. Bur. Research, U. S. office edn., also Marshall Space Flight Center. Diplomate Am. Bd. Psychiatry and Neurology (dir. for child psychiatry 1966—). Mem. A.M.A., Am. Psychiat. Assn., Am. Orthopsychiat. Assn., Am. Acad. Child Psychiatry, Am. Coll. Psychiatrists, Am. Geriatric Soc., Am. Assn. Psychiat. Clinics for Children (council 1966—), Phi Beta Kappa. Research and publs. on clin. psychopharamcology, child psychiatry. Home: 1232 Graylynn Dr., Birmingham, Ala. 35216.*

SUSSMAN, Alfred Sheppard, Am. botanist; b. Portsmouth, Va., July 4, 1919; s. Morris and Celia (Rabinowitz) S.; B.S., U. Conn., 1941; A.M., Harvard, 1948, Ph.D., 1949; m. Selma Feinman, Nov. 28, 1948; children—Jean, Paul, Harold. Instr. microbiology Mass. Gen. Hosp., Boston, 1948-49; mem. faculty U. Mich., 1950—, prof. botany, 1961—, chmn. dept., 1963-68, asso. dean Coll. Lit., Sci. and Arts, 1968—. Cons., NSF panel devel. biology, 1963—; vis. scientist Brookhaven Nat. Labs., summers 1950, 51. NRC fellow, 1949-50; Lalor Found. summer fellow, 1956; NSF sr. fellow Cal. Inst. Tech., 1959-60; Atkins fellow Atkins Bot. Garden Soledad, Cuba, summer 1949. Mem. Am. Soc. Plant Physiologists (chmn. Midwest sect. 1961), Bot. Soc. Am. (chmn. microbiol. sect. 1958), Am. Inst. Biol. Scis. (steering com. biol. sci. curriculum study 1961-63, governing bd. 1968—), Soc. Study Growth and Devel., Am. Acad. Microbiology, Am. Soc. Microbiologists, Am. Soc. Cell. Biologists, Mycol. Soc. Am., Am. Assn. U. Profs., Sigma Xi, Phi Beta Kappa. Author: (with Steiner and Wagner) Botany Laboratory Manual, 2d edit., 1965; Biology Through Microbes, 1961; Microbes: Their Growth, Nutrition and Interaction, 1964; also numerous articles. Editor: (with G. C. Ainsworth) The Fungi, 3 vols., 1965, 66, 68; (with H. O. Halvorson) Spores: Their Dormancy and Germination, 1966. Mem. editorial bd. Am. Jour. Botany, 1960-66, Mycologia, 1963-67. Research on pathology of insects; comparison of spores and vegetative cells; proposed mechanism through which dormancy is imposed on mold spores; discovered and studied variant of a mold which grows rhythmically over 24 hours. Home: 1615 Harbal Dr., Ann Arbor, Mich. 48105.*

SUSSMAN, Marvin Bernard, Am. sociologist, educator; b. N.Y.C., Oct. 27, 1918; s. Louis M. and Gertrude (Clar) S.; B.A., N.Y.U., 1941; M.S., George Williams Coll., 1943; M.A., Yale, 1949, Ph.D., 1951; m. Ruth Annette Strahler, Aug. 15, 1942; children—Stuart Daniel, Martha Jean, Kenneth Brittan, Nancy Louise. Chmn. dept. sociology Union Coll., Schenectady, 1952-54; vis. asst. prof. U. Chgo., 1954-55; asso. prof. sociology Western Res. U., 1955-62, prof. sociology, chmn. dept. sociology, 1962—; consultant to the Social Security Administration, 1962—. Member of the research panel Vocational Rehab. Adminstrn., Welfare Adminstrn. F. Am. Sociol. Assn., Soc. Study Social Problems (pres. 1962-63), Soc. Applied Anthropol., Am. Pub. Health Assn.; mem. Am. Statis. Assn., Ohio Valley Sociol. Soc. (pres. 1962-63), Am. Anthrop. Assn., Nat. Ohio (pres. 1962-63) councils family relations, Nat. Rehab. Assn., Sociol. Research Assn., Am. Association U. Profs., Northeast Ohio Rehabilitation (pres. 1964—). Author: (with others) Social Class, Maternal Health and Child Care, 1957; (with R. Clyde White) Hough Area. Cleveland, Ohio: A Study of Social Life and Change, 1959; (with others) Tuberculosis and Rehabilitation; (with others) The Walking Patient: A Study in Ambulatory Care, 1966. Editor: Sourcebook in Marriage and the Family, rev. edit., 1968; Community Structure and Analysis, 1959; Sociology and Rehabilitation, 1966. Editor Jour. Marriage and The Family, 1963—; asso. editor Transaction, 1964—. Research on family and kinship structure in urban industrial societies. Home: 2837 E. Overlook Rd., Cleveland Heights, O. 44118. Office: 11027 Magnolia Dr., Cleve. 44106.*

SÜSSMILCH, Johann Peter, German statistician; b. Berlin, Sept. 3, 1707; minister, Berlin. Author: Die göttliche Ordnung in den Verhältnissen des menschlichen Geschlechtes, 1741. A founder of population statistics. Died Mar. 22, 1767.

SUSZ, Bernard Pierre, Swiss physicist; b. Geneva, Switzerland, Oct. 5, 1904; s. Paul and Caroline (Dessiex) S.; ed. U. Geneva, U. Berlin; Dr.ès.sci.; m. Anne Marie Martin, 1929; children—Jean-Philippe, Martin-Michel, Christian Paul. Prof. phys. chemistry, dean Faculty Sci., U. Geneva. Mem. Geneva Physics Soc. (past pres.), Central Swiss Soc. (past pres.). Research and publs. on infrared studies of donor-acceptor molecular compounds. Home: Combevalière, Cologny, Geneva, Switzerland.

SUTER, Emanuel, microbiologist; b. Basel, Switzerland, Feb. 7, 1918; s. Fritz and Clara (Vischer) S.; M.D., U. Basel, 1944; m. Joanne F. Otter, May 26, 1962; children—Bradley Jay, Brooke. Came to U.S., 1949, naturalized, 1954. Research fellow Sandoz, Inc., Basel, 1943-45; research fellow U. Basel, 1945-48; asst. Rockefeller Inst. Med. Research, N.Y.C., 1949-52; asso. in bacteriology and immunology Harvard U. Med. Sch., Boston, 1952-53, asst. prof., 1953-56; prof. microbiology, head dept. Coll. Medicine, U. Fla., Gainesville, 1956-65, dean Coll. Medicine, 1965—. Markle scholar, 1954-59. Mem. Assn. Immunologists, Soc. Am. Bacteriologists, Harvey Soc., N.Y. Acad. Scis., Soc. Exptl. Biology and Medicine, Schweiz Verein fuer Physiologie, Schweiz Microbiologische Gesellschaft, Sigma Zi, Alpha Omega Alpha. Research and numerous publs. on binding of drugs on plasma proteins such as cardiac glycosides and chemotherapeutic agts., physiology of macrophage and their role in immune response to infection, mode of action of bacterial endotoxins and their role in pathogenesis of disease. Home: 1238 N.W. 18th Terrace, Gainesville, Fla. 32601.*

SUTHERLAND, Edwin Hardin, Am. sociologist; b. Gibbon, Neb., Aug. 13, 1883; s. George and Lizzie (Pickett) S.; A.B, Grand Island (Neb.) Coll., 1904; Ph.D., U. of Chicago, 1913; m. Myrtle Crews, May 11, 1918; 1 dau., Betty Ann (Mrs. A. B. Sand). Teacher at Sioux Falls (S.D.) Coll., 1904-06, Grand Island Coll., 1909-11; prof. sociology, William Jewell Coll., Liberty, Mo., 1913-19; asst. prof. sociology, U. of Ill., 1919-25, asso. prof., 1925-26; prof. sociology, U. of Minn., 1926-29; prof. soiciology, U. of Chicago, 1930-35; prof. sociology, since 1935, head dept. sociology, Ind. Univ., 1935-49; visiting prof. sociology, U. of Kansas, 1918, Northwestern, 1922, U. of Washington, 1942, research associate in criminology, Bureau of Social Hygiene, New York, 1929-30; vis. prof. sociology San Diego State Coll., 1950. Pres. Indiana Univ. Inst. of Criminal Law and Criminology. Mem. Am. Sociol. Soc. (pres. 1939), Am. Prison Assn., Chicago Acad. Criminology (pres. 1932-34), Sociol. Research Assn. (pres. 1940-41). Author: Unemployment and Public Employment Agencies, 1913; Criminology, 1924; Principles of Criminology, 1947; An Ecological Survey of Crime and Delinquency in Bloomington, Indiana, 1937; The Professional Thief, 1937; White Collar Crime, 1949. Coauthor: Twenty Thousand Homeless Men, 1936; also chapters in Recent Social Trends, 1933, and Young's Social Attitudes, 1931. Author or editor various monographs and papers. Co-editor: Prisons of Today and Tomorrow, 1931. Maj. contbr. to study of criminology in U. S.; related factors leading to criminal behavior with community's problems of disorganization. Died Bloomington, Ind., Oct. 11, 1950.

SUTHERLAND, Gordon Brims Black McIvor, Brit. physicist; b. Watten, Caithness, Scotland, Apr. 8, 1907; s. Peter and Eliza Hope (Morrison) S.; M.A., St. Andrew's (Scotland) U., 1928, B.Sc., 1929, LL.D., 1958; Ph.D., Cambridge (Eng.) U., 1933, Sc.D., 1946; m. Gunborg Elisabeth Wahlström, Mar. 14, 1936; children—Ann Birgitta (Mrs. William Harris), Kerstin Elisabeth (Mrs. Thomas R. Stempel), Mary Seaton (Mrs. Richard McNutt). Fellow, Pembroke Coll., Cambridge U., 1936-49, asst. dir. research in colloid sci., 1944-47, reader spectroscopy, 1947-49, master Emmanuel Coll., 1964—; prof. physics U. Mich., 1949-56; dir. Nat. Phys. Lab., Teddington, Middlesex, Eng., 1956-64. Created knight, 1960. Fellow Royal Soc., 1949 (past v.p.), Inst. Physics and Phys. Soc. London (pres. 1964-66); mem. Internat. Union Applied Physics (v.p. 1963—). Author: Infra-red and Raran Spectra, 1935; also numerous articles. Research on molecular structure. Address: The Master's Lodge, Emmanuel Coll., Cambridge, Eng.*

SUTHERLAND, John MacKay, neurologist; b. Caithness, Scotland, Aug. 20, 1919; s. Donald Henry and Kathleen (Bremmer) S.; M.B., Ch.B., Glasgow (Scotland) U., 1943, M.D., 1950; m. Patricia Campbell, July 27, 1944; children—Gillian, Iain. House physician Western Infirmary, Glasgow, 1943-44, med. registrar, 1946-56; med. registrar Royal No. Infirmary, Inverness, Scotland, Royal Infirmary, Aberdeen, Scotland, 1946-56; faculty U. Queensland, Brisbane, Australia, 1956—, lectr. neurology, 1959—; sr. neurologist Royal Brisbane Hosp., Brisbane Children's Hosp., 1959—; hon. neurologist Spastic Centre, Brisbane, 1959—; referee neurologist Commonwealth of Australia, 1960—. Cons. neurologist specialist Commonwealth of Australia, 1960—. Cons. neurologist specialist list Royal Australian Air Force, 1960—. Fellow Royal Coll. Physicians Edinburgh, Royal Australasian Coll. Physicians; mem. Australian Med. Assn., Australian Assn. Neurologists. Author: (with J. H. Tyrer) Exercises in Neurological Diagnosis, 1967; also articles. Research on neurol. problems including epidemiology of multiple sclerosis in Scotland and Australia, clin. features of heredofamilial ataxias, mechanism of loss of consciousness asso. with coughing, drug therapy of epilepsy. Home: 162 Hart's Rd., Indooroopilly, Brisbane, Queensland, Australia.*

SUTOW, Wataru W., Am. physician; b. Guadalupe, Cal., Aug. 31, 1912; s. Yasaku and Yoshi (Sato) S.; A.B., Stanford U., 1939, student Sch. Medicine, 1940-43; 1940-43; M.D., U. Utah Sch. Medicine, 1945; m. Mary Hideo Korenaga, Sept. 1936; children—Ollie Ellen, Chiyono Jean, Edmund Keith. Head pediatric dept., dir. pediatric research Atomic Bomb Casualty Commn., Japan, 1947-50, 1953-54; fellow pediatrics Stanford U., 1950-51; asso. pediatrician U. Tex. M.D. Anderson Hosp. and Tumor Inst., Houston, 1954—. Research collaborator Brookhaven Lab., Upton, N.Y., 1957—; editorial bd. Yearbook of Cancer, 1956—; chmn. pediatric div. Southwest Cancer Chemotherapy Study Group, 1957—. USPHS Research Career award in pediatric oncology, 1963. Mem. Am. Acad. Pediatrics, Am. Fedn. Clin. Research, Soc. for Research in Child Devel., Am. Assn. Cancer Research, So. Soc. Pediatric Research. Contbr. numerous articles in field to sci. jours. Studied effects in children exposed to radiation to the atomic bombs in Hiroshima and Nagasaki, Japan and to fallout radiation in the Marshall Islands; studies on clin. cancer chemotherapy in children. Home: 3854 Palm St., Houston 77004. Office: 6723 Bertner St., Houston, Tex. 77025.*

SUTTER, Gilbert, French physicist; b. Strasbourg, France, June 29, 1931; s. Leon and Berthe (Sittler) S.; Licence ès-Sciences Physiques, U. Strasbourg, 1954, Doctorat ès Sciences Physiques, 1962; m. Genevieve Rozek, Aug. 20, 1957. Research physicist Centre Nat. de la Recherche Scientifique,-C.R.N., Strasbourg, 1956-62, 63-65; prof. U. Strasbourg, 1965—; Research asso. Columbia U., N.Y.C., 1962-63; prof. electronics Conservatoire Nat. des Arts et Metiers, Centre Associé de Mulhouse, France, 1966—. Recipient Bronze medal Centre Nat. de la Recherche Scientifique, 1962. Mem. Société Francaise de Physique. Research and publs. on electromagnetic transitions in nuclei, spectroscopy of nuclei, devel. nuclear electronics. Home: 15 Rue Jacques Peirotes, 67 Strasbourg, France. Office: C.R.N.-P.N., Rue du Loess, 67 Strasbourg-3, France.*

SUTTER, Jean Ulysse-Auguste, French geneticist; b. Anor, France, July 12, 1910; s. Jean Georges and Marthe (Lécoyer) S.; Docteur en Médecine, U. Paris, 1935, Docteur es Sciences, 1957; m. Monique Bazalgette, June 17, 1935; children—Martine (Mrs. Gosse), Francis, Michel. Practice medicine, Paris, 1935-58; staff Foundation Francaise pour l'Etude des Problémes Humains, Paris, 1942-45; chief dept. biol. demography Institut National d'Etudes Démographiques, Paris, 1945—; prof. population genetics l'Institut de Démographie, U. de Paris; researcher quantitative genetics Sorbonne, Paris. Decorated chevalier Légion d'Honeur; officier l'Ordre du Mérite. Mem. Société Francaise de Génétique (past pres.), Biometrics Soc. (past mem. council). Author: L'Eugénique, 1950; Recherches sur les effets de la consanguinité chez l'homme, 1957; Látteinte des incisives latérales supérieures, 1966; also numerous articles. Research on integration of human population genetics in demographical structures, including studies in consanguinity, endogamy, genealogy, isolates, mortality and fertility (differential), migrations, effects of birth control, incidence on pub. health by way of mutations and their genetic drift. Home: 3 Boulevard Voltaire, Paris XIe 75, Office: I.N.E.D., 23 Av. F. D. Roosevelt, Paris VIIIe 75, France.*

SUTTON, George Miksch, Am. ornithologist; b. Lincoln, Neb., May 16, 1898; s. Harry Trumbull and Lola Anna (Miksch) S.; B.S., Bethany Coll., 1923, Sc.D., 1952; Ph.D., Cornell U., 1932. Staff, Carnegie Mus., Pitts., 1920-24; state ornithologist State of Pa., Harrisburg, 1924-28; curator birds, instr., Cornell U., 1932-46; prof. zoology U. Okla., Norman, 1952-59, research prof. zoology, 1959—, curator birds, 1952—, ornithologist Biol. Survey, 1952—. Recipient Burroughs award for conservation writing John Burroughs Assn., 1962; named Conservation Tchr. of the Year, Nat. Wildlife Fedn., 1965. Mem. Wilson Ornithol.. Soc. (past pres.), Am. Ornithologists Union, Southwestern Assn. Naturalists, Arctic Inst. N.Am., Cooper Ornithol. Soc., Am. Polar Soc., Wildlife Soc., Sigma Xi, Phi Beta Kappa. Author: Introduction to Birds of Pennsylvania, 1928; Eskimo Year, 1933; Birds in the Wilderness, 1936; Mexican Birds, 1959; Iceland Summer, 1961; Oklahoma Birds, 1967; also numerous articles, chpts. in books. Illustrator for numerous books on birds. Research on bird distbn. in Am. and Eurasian arctic, Mexico, Okla.; Taxonomy of N.Am. birds. Participant expdns. to Labrador, Hudson Bay, Can. Arctic Archipelago, Iceland, Europe, Mexico. Home: 818 W. Brooks St., Norman, Okla. 73069.*

SUTTON, George Walter, Am. research mech. engr.; b. Bklyn., Aug. 3, 1927; s. Jack and Pauline (Aaron) S.; B.Mech. Engring., Cornell U., 1952; M.S., Cal. Inst. Tech., 1953, Ph.D. magna cum laude, 1955; m. Evelyn D. Kunnes, Dec. 25, 1952; children—James E., Charles S., Richard E., Stewart A. Research scientist Lockheed Missile Co., 1955; research engr. space sci. lab. Gen. Electric Co., 1956-61; vis. Ford prof. Mass. Inst. Tech., 1961-62; mgr. magnetohydrodynamic power generation space sci. lab. Gen. Electric Co., 1962-63; sci. adviser directorate devel. plans, Hdqrs. USAF, 1963-65; prin. research scientist AVCO Everett Research Lab., 1965—; lectr. magnetohydrodynamics U. Pa., 1960-63, Stanford, 1964; spl. research high temperature gas- and plasma dynamics. Recipient Arthur Flemming award for outstanding govt. service, 1965. Mem. Am. Inst. Aeros. and Astronautics (chmn. plasmadynamics tech. com., editor-in-chief jour.), Symposium Engring. Aspects Magnetohydrodynamics (pres.), Am. Soc. M.E., Am. Inst. Physics, A.A.A.S., Sigma Xi. Author: Proceedings 4th Symposium Engineering Aspects of Magnetohydrodynamics, 1964; Engineering Magnetohydrodynamics (with A. Sherman), 1965; Direct Energy Conversion, 1966—; also numerous articles. Measured instantaneous stresses due to cavitation in liquids; pioneered in devel. ablation heat protection for hypersonic re-entry (now used in all U. S. re-entry vehicles) and 1st measured heat of ablation; invented magnetically induced ionization for magnetohydrodynamic elec. power generators; predicted electron density fluctuations in hypersonic wakes. Home: 37 Winthrop Rd., Lexington, Mass. 02173. Office: 2385 Revere Beach Pkwy., Everett, Mass. 02149.*

SUTTON, Sir Graham, Brit. meteorologist; b. Cwmcarn, Mon, U.K., Feb. 4, 1903; s. Oliver and Rachel (Rhydderch) S.; B.Sc., U. Coll. Wales, Aberystwyth, 1923, D.Sc., 1949, LL.D., 1955; B.Sc., Jesus Coll., Oxford U., 1925; m. Doris Morgan, Apr. 2, 1931; children—Peter Morgan, Anthony Graham. Lectr., U. U. Coll. Wales, 1926-28 meteorologist Meterol. Office, U.K., 1928-41; supt. research Chem. Warfare, 1941-43; supt. tank armament research, 1943-45; chief supt. radar research and devel., 1945-47; Bashforth prof. math. physics Royal Mil Coll. Sci., 1947-53, dean, 1951-53; dir. gen. Meteorol. Office, U.K., 1953-65; chmn. Natural Environment Research Council, London, 1965—. Decorated Comdr. Order Brit. Empire; knighted. Fellow Royal Soc., 1949, Soc. Engrs. (hon.), Royal Meteorol. Soc. Author: Atmospheric Turbulence, 1949; Micrometeorology, 1953; Mastery of the Air, 1964; Mathematics in Action, 1954; Understanding Weather, 1965; Challenge of Atmosphere, 1962; also numerous articles. Research on popularization of aerodynamics and math. physics; helped establish math.-phys. base of study of micrometeorology. Home: 27 Woodend Dr., Sunninghill, Berks, U.K. Office: Natural Environment Research Council, Alhambra House, Charing Cross Rd., London, W.C.2. Eng.*

SUTTON, Harry Eldon, Am. geneticist; b. Cameron, Tex., Mar. 5, 1927; s. Grant Edwin and Myrtle (Fowler) S.; B.S. in Chemistry, U. Tex., 1948, M.A., 1949, Ph.D. in Biochemistry, 1953; m. Beverly Earlene Jewell, July 7, 1962; children—Susan Elaine, Caroline Virginia. With U. Mich., Ann Arbor, 1952-60, asst. prof. human genetics, 1956-60; faculty U. Tex., Austin, 1960—, prof. zoology, 1964—. Mem. adv. com. on personnel for research Am. Cancer Soc., 1961-64; mem. genetics study sect. NIH, 1963-67. Mem. Am. Soc. Human Genetics, Genetics Soc. Am., Am. Chem. Soc., A.A.A.S., Tex. Acad. Sci., Sigma Xi. Author: Genes, Enzymes, and Inherited Diseases, 1961; An Introduction to Human Genetics, 1965; also numerous articles. Editor: First Macy Conference on Genetics, 1960. Editor, Jour. Human Genetics, 1964—. Research on inheritance individuality metabolic patterns blood and urine, population studies inherited protein variants man, inheritance human haptoglobin types, chem. structure human transferrin variants, inheritance and structure human red cell acid phosphatase variants. Home: 1103 Gaston Av., Austin, Tex. 78703.*

SUTTON, Henry, instrument maker; flourished 1637-65; maker of precise instruments; engraver of scales, quadrants; drew diagrams for math. books; helped observe lunar eclipse, Aug. 8, 1663.

SUTTON, Richard Lightburn, Am. naturalist, dermatologist; b. Rock Port, Mo., July 9, 1878; s. John Grant and Virginia (Robertson) S.; student U. Mo., 1898-99; M.D., Univ. Med. Coll., Kansas City, Mo., 1901; postgrad. George Washington U., 1903-04, M.D.; U. S. Naval Med. Sch., 1904; postgrad. Johns Hopkins Med. Sch., in London, Hamburg, Berlin, Vienna and Paris; LL.D., U. Mo., 1922; Sc.D., Washburn College, Topeka, Kan., 1925; m. Lena Igel, Jan. 3, 1906; children—Richard Lightburn, Emma Louisa (Mrs. Lewis H. Moore). Asst. surgeon USN; prof. dermatology U. Kan.; spl. rep. dept. of natural history U. Mo., African expdn., 1923-24, expdns. for same dept. to Indo-China and India, 1925-26; headed African-Asiatic expdn., 1929-30, Arctic expdn., north of Spitsbergen, as spl. rep. dept. of natural history U. Kan., 1932; expdns. to New Zealand and Australia, investigating habits of swordfish, 1935, 36, 37, 38, 39, 40, Peru, 1941; with East Arctic Patrol, Northwest Territories, to Ellesmere Island, Somerset Island and Baffinland, summer of 1939. Trustee Kansas City Museum. Fellow Royal Geog. Soc. (life), Royal Soc., Edinburgh; mem. A.M.A. (chmn. dermatol. sect., 1913-14), Jackson County Med. Soc. (pres. 1913-14), Am. Dermatol. Assn., French Geog. Soc., Swedish, Brit. dermatol. socs. (corr.), Southwest Clin. Assn. (hon.). Author: An African Holiday, 1924; Tiger Trails in Southern Asia, 1926; The Long Trek, Around the World with Camera and Rifle, 1930; Diseases of the Skin (10th edit., with Richard L. Sutton, Jr.), 1939; An Introduction to Dermatology (4th edit., with Richard L. Sutton, Jr.), 1941; Synopsis of Diseases of Skin (with Richard L. Sutton, Jr.), 1942; An Arctic Safari, 1932; The Silver Kings of Aranas Pass and Other Stories, 1937; A Handbook of Dermatology (with R. L. Sutton, Jr.), 1949; sect. on The Mycoses in Tice's System, 1919, rewritten (with R. L. Sutton, Jr.), 1940-52; Collected Verse, 1946, also numerous sci. articles in Am. and fgn. jours. Contbr. to lay periodicals. Died May 18, 1952; buried Kansas City, Mo.

SUTTON, Thomas, English physician; b. Staffordshire, Eng., circa 1767; studied medicine, London, Edinburgh; M.D., Leiden, Netherlands, 1787; mil. physician; later settled in Greenwich, Eng.; cons. physician Kent dispensary. Licentiate Coll. Physicians. Author: Considerations regarding Pulmonary Consumption, 8 vols., 1799; Practical Account of a Remittent Fever frequently occurring among the Troops in this Climate, 8 vols., 1806; Tracts on Delerium Tremens, 1813; Letters to the Duke of York on Consumption, 1814; Credited with introducing term delirium tremens, also described it and distinguished it from other forms of delirium or frenzy, 1813; 1st modern Brit. physician to advocate bleeding and antiphlogistic treatment of fever. Died 1835.

SUZUKI, Keishin, Japanese astronomer; b. Tokyo, 1905; grad. Tokyo U., 1929. Successively tchr. Hiroshima First Middle Sch., part-time clk. Met. Mus. Sci., engr. Navy Fairway Dept.; became prof. Gakugei Coll. Tokyo, 1952. Recipient Transp. Minister's award, 1951. Mem. Japan Soc. Astronomy (councilor), Japan Geog. Edn. Research Assn. (v.p.). Author books including: Mathematical Calculation on Solar Eclipse; Principle of Astronomy; The Cosmos; Astronometric Map. Research and publs. on astron. phenomena.

SUZUKI, Masakuni, Japanese physiologist, gynecologist; b. Sendai, Japan, May 21, 1921; s. Masabumi and Akiyo (Miura) S.; M.D., Tokyo U., 1946; m. Teiko Kikuchi, Nov. 27, 1948; children—Coco, Yoshico. Faculty dept. obstetrics and gynecology Tohoku U. Sch. Medicine, Sendai, Japan, 1954-63, asso. prof., 1958-63; prof. dept. obstetrics and gynecology Niigata U. Sch. Medicine(Japan), 1963—; chief dept. obstetrics and gynecology Niigata U. Hosp., 1963—. Mem. Japanese Obstet. and Gynecol. Soc. (mem. bd.), Japanese Endocrinological Soc. (mem. bd.), Japanese Soc. Fertility and Sterility (mem. bd.). Author several books, numerous articles. Research on method determination of sex of fetus, induction of ovulation by plant extracts, treatment of choriocarcinoma by pelvic perfusion chemotherapy.*

SUZUKI, Sakaru, Japanese biochemist; b. Aichiken, Japan, Jan. 26, 1926; s. Kenshi and Aya (Sakakibara) S.; grad. Nagoya U., 1948, Ph.D., 1957; m. Kazuko Oida, Oct. 21, 1949; children—Yoshiro, Aiji. Postdoctoral fellow dept. pharmacology Washington U., St. Louis, 1957-59; faculty Nagoya (Japan) U., 1960—, prof. biochemistry, 1961—. Editor: Polysaccharide Biochemistry, 1968. Publs. on discovery of various sugar nucleotides from living cells; studies of their biol. function. Home: Wakamizu-Jutaku 1-14, Chikusa-cho, Nagoya, Japan.*

SUZUKI, Shigeo, Japanese chemist; b. Niigata, Japan, Jan. 31, 1920; s. Shigezo and Yori (Yachida) S.; grad. U. Tokyo, 1941, D.Agr., 1961; m. Mayuko Itoh, Aug. 8, 1946; children—Jun, Shinko. Sub-asst. U. Tokyo, 1941-48, lectr., 1962—; head potato and starch lab. Food Research Inst., Tokyo, 1948-62, head carbohydrate div., 1962—. Recipient award Technol. Soc. Starch, 1958, Ministry Agr. and Forestry, 1964. Mem. Am. Assn. Cereal Chemists, Inst. Food Technologists, Japan Agrl. Chem. Soc., Technol. Soc. Starch. Author: (with Z. Nikuni) Starch Handbook, 1961; (with Y. Sakurai) Sugar Handbook, 1964; (with T. Tamura) Starch for Corebinder, 1966; also numerous articles. Developed new enzymatic saccharification method for starch; discovered new methods of isomerization of dextrose into frucose, prodn. artificial rice. Patentee in field. Home: 2-26-8, Nagasaki, Toshima-Ku, Tokyo. Office: 2-Hamazonocho, Fukagawa, Koto-Ku, Tokyo, Japan.*

SUZUKI, Shin, Japanese chemist; b. Tokyo, Japan, Apr. 7, 1923; s. Shinkichi and Chiyoe Suzuki; grad. dept. applied chemistry U. Tokyo, 1945, completed grad. course, 1950; m. Akiko Kozuki, June 1953; children—Nobuo, Minoru. Faculty, U. Tokyo, 1950-64, asst. prof., 1959-64; prof. chemistry U. Chiba

(Japan), 1964—. Mem. Soc. Sci. Photography (award 1962), Printing Soc. Japan (award 1965), Electrochem. Soc., Chem. Soc., Soc. Photog. Sci. and Eng. (sect. editor 1962—). Author: Theory of Electrochemistry, 1950; Photochemistry, Photography, 1960; Measurements of Air Pollution, 1962; Air Pollution, 1966; Chemistry, 1960; also numerous articles. Developed theory of photsensitization of photog. emulsion and other material, unconventional photography; electrochem. studies of photochemistry, radiation chemistry, air pollution problems, atomic energy waste. Home: Miyanogimachi 1730, Chiba-shi, Chiba-ken, Japan.*

SUZUKI, Umetaro, Japanese agrl. chemist; b. Shizuoka Prefecture, Japan, 1872; grad. agr. dept. Tokyo U., 1896; protein studies in Germany and Switzerland, 1901; D. Agr.; became prof. Tokyo U.; mem. Imperial Acad. Discovered vitamin B; synthetized rice wine; developed new methods of mfg. salicylic acid and salvarsan. Died 1943.

SVANBORG, Alvar, Swedish physician; b. Umea, Sweden, Nov. 15, 1921; s. Arvid and Althea Svanborg; M.D., Karolinska Institutet, Stockholm, Sweden; m. Marianne Lindh, Nov. 12, 1948; children—Catharina, Elisabeth, Anna, Arvid. Faculty, U. Gothenburg (Sweden), 1955—, asso. prof. medicine, 1959—; head geriatric clinic II, Vasa Hosp., Gothenburg, 1966—. Research, numerous publs. on metabolic diseases. Home: 2, Splintvedsgatan. Office: Vasa Hosp., Gothenburg, Sweden.*

SVARDSON, Gunnar, Swedish zoologist; b. Stockholm, Sweden, Sept. 19, 1914; s. John F. and Eva (Larsson) S.; Fil.kand., Stockholm U., 1939, Fil.lic., 1941, Fil.dr., 1945; m. Britt-Marie Olsson, Sept. 30, 1941; children—Bjorn Gunnar, Asa Britta. Fishery biologist Fishery Bd. Sweden, 1942—, with Inst. Freshwater Research, Drottningholm, 1944—, dir. Inst., 1963—. Sci. adviser to Swedish Sportsmens Assn., 1959—, Anglers Assn., 1958—. Decorated Nordstjarneorden, riddartecken, 1964. Mem. Swedish Ornithol. Soc. (a founder, sec. 1946-52, head Ottenby Bird Sta. 1945-52). Author: (Svardson et Durango), 1951, Svenska Djur-Faglarna; Goda laxar och daliga, 1957; (with Nilsson) Fiskebiologi, 1965; also numerous articles. Research on bird migration, chromosomes and evolution of sibling species of salmonoid fish, fisheries biology. Home: Odmardsvagen 17, Bromma, Sweden.*

SVEDA, Michael, Am. chemist; b. West Ashford, Conn., Feb. 3, 1912; s. Michael and Dorothy (Druppa) S.; B.S., U. Toledo, 1934; Ph.D. (Eli Lilly Research fellow), U. Ill., 1939; m. Martha Augusta Gaeth, Aug. 23, 1936; children—Sally Anne, Michael Max, Marcia Lynne. Research chemist, Grasselli chems. dept. E. I. du Pont de Nemours & Co., Cleve., 1939-44, research supr., 1945-57, sales devel. supr. div., Wilmington, Del., product mgr., 1951-54, mgmt. cons., 1955-60; named project dir. for spl. projects in sci. edn. NSF, 1960; corporate asso. dir. research FMC Corp., N.Y.C., 1962-64; mgmt. counsel, 1964—. Named outstanding Alumnus U. Toledo, 1954; Paul Block scholar, U. Toledo. Mem. Am. Chem. Soc., A.A.A.S., Chem. Spltys. Mfrs. Assn., Am. Soc. for Testing and Materials. Publs. on discovery of sodium cyclohexyl sulfamate (an artificial sweetener); invented method for prodn. pyro-sulfyl chloride, thionyl chloride, vapor-phase sulfonation hydrocarbons with sulphur trioxide, method for separating two forms of DDT; 1st polymerization of silicon; supervised research on ludox colloidal silica and a form of solid silica used in lubricating grease; patentee (with Ludwig F. Audrieth) cycloaliphatic acids, 1842, Address: NSF, 1951 Constitution Av., Washington 25 and P.O. Box 211, Greenwich, Conn. 06830.

SVEDBERG, Theodor, Swedish chemist; b. Valbo, Sweden, Aug. 30, 1884; s. Elías and Augusta (Alstermark) S.; Ph.D., U. Uppsala, 1908; m. Andrea Andreen, 1909; m. 2d, Jan Frodi Dahlqvist, 1916; m. 3rd, Ingrid Blomquist Tauson, 1938; m. 4th, Margit Hellén Norbäck, 1948. Lectr. phys. chemistry U. Uppsala, 1907-12, prof. phys. chemistry, 1912—, dir. Inst. for Phys. Chemistry, 1949—; vis. prof. U. Wis., 1922-23. Mem. Nat. Acad. Scis. Recipient Nobel prize in chemistry, 1926. Author: Die Methoden zur Herstellung kolloider Lösungen anorganischer Stoffe, 1909; Die Existenz der Moleküle, 1912; Formation of Colloids, 1921; Colloid-Chemistry, 1923; other books; also numerous articles. Research in colloidal chemistry, determination of size of molecules, electrophoresis methods; devel. of ultracentrifuge for separation of colloidal particles from solution, 1923. Office: U. Uppsala, Uppsala, Sweden.

SVEHAG, Sven-Eric, Swedish immunologist; b. Karshamn, Sweden, Aug. 30, 1932; s. Edvard and Greta (Olsson) S.; D.V.M., Royal Vet. Sch., Stockholm, 1957; postgrad. Wash. State U.; Ph.D., Karolinska Inst. Sch. Medicine, 1965; m. Käth Nannborn, Sept. 21, 1957; children—Kristina, Anders. Asst., Royal Vet. Coll., Stockholm, 1955-57, Wash. State U., Pullman, 1958-60; postdoctoral fellow Pub. Health Research Inst. N.Y., 1960-64, Karolinska Inst. Sch. Medicine, 1964-65; docent Royal Vet. Coll., Stockholm, 1965-66; asso. prof. Nat. Bacteriol. Lab., Stockholm, 1966—. Hon. fellow Am. Scandinavian Found., 1958-64; research fellow Helen Hay Whitney Found., 1964-65. Mem. Swedish Assn. Microbiologists. Research, publs. on devel. sensitive systems for studies of antibodies produced against viruses in vivo and in vitro; demonstration maj. differences in formation of high and low molecular weight antibodies, ultrastructure of globulins, transplantation immunology.*

SVENSON, Erik, physicist; b. Riga, Latvia, Aug. 5, 1895; s. Woldemar and Helene (Reimers) S.; ed. U. Dorpat, U. Göttingen, U. Marburg; Ph.D.; m. Erne Hasselbaum, Oct. 31, 1941. From instr. to prof. math. Herder Inst., Riga, 1925-39; staff U. Posen, 1941-45, U. Heidelberg, 1949-50; instr. Ratisbonne Sch. Philosophy and Theology, also Bamberg Sch., 1950-59; became prof. emeritus U. Erlangen (Germany), 1959. Mem. German Union Mathematician, Applied and Mech. Math. Soc. Author: Molekular-statische Thermodynamik, 1928; Theorie gewisser Integraltypen, 1934; Theorie der Struktur der unendlichen periodischen Bruchenwicklungen, 1958. Home: Bissingerstrasse 35, 852 Erlangen, West Germany.

SVERDRUP, Erling, Norwegian math. statistician; b. Bergen, Norway, Feb. 23, 1917; s. Georg Johan and Gunvor (Gregusson) S.; actuarian U. Oslo, 1945, dr. philos., 1952. Statis. cons. Inst. Econs., U. Oslo, 1946-48, research asst. Inst. Math., 1948-52, prof., 1953—; vis. prof. Columbia, 1963-64. Rockefeller Found. fellow U. Cal. at Berkeley, U. Chgo., Columbia, 1949-50. Fellow Inst. Math. Statistics; mem. Royal Norwegian Acad. Sci., Internat. Statis. Inst. Author: Laws and Change Variations, 1964; also articles. Research on sampling distbns. in multivariate analysis; developed theorem on limit distbn. of continuous functions of random variables; minimax procedures in statis. inference and statis. prediction; discovered relationship between unbiased (similar) test procedures and conditional testing for gen. exponential classes of distbn., generalizing and reformulating Neymen-Pearson's theory; estimation and test procedures for forces of decrement and transfers in statis. follow-up studies. Home: 4 Olav Kyrres gate, Oslo 2, Norway.*

SVERDRUP, Harald Ulrik, Norwegian meteorologist and oceanographer; b. Sogndal, Norway, Nov. 15, 1888; s. Johan Edvard and Maria (Vollan) S.; A.B., U. of Oslo, 1911, A.M., 1914, Ph.D., 1917; LL.D. (honorary), University of California, 1946; m. Gudrun Bronn, June 8, 1928; 1 adopted dau., Anna Margrethe. Came to U. S., 1936. Asst. to Prof. V. Bjerknes in Oslo, 1911-12, in Leipzig, 1913-17; in charge scientific work on Maud Expdn. in Arctic, 1917-25; research asso., Carnegie Instn. of Washington, 1926, 1928-39; prof. meteorology, Geophysical Institute of Bergen, 1926-30; hon. professor Chr. Michelsen Inst., 1931-40; in charge scientific work on Wilkins-Ellsworth submarine Arctic expdn. on board Nautilus, 1931; prof. of oceanography, U. of Calif. and dir. Scripps Instn. of Oceanography, 1936-48; dir. North Polar Institute, Oslo, Norway, since 1948. Lieutenant in Reserve of Norway. Decorated Comdr. of St. Olav, Comdr. Dannebrog, Nordstjernen; D.S.M. (U.S.N.); Patrons medal, Royal Geog. Society; Bowie medal, American Geological Society; awarded Vega gold medal, Carl Ritter medal, Bruce memorial, Meteor, Agassiz medals. Member National Academy of Scis., Norwegian Academy of Sciences, American Academy of Arts and Sciences; honorary member Royal Meteorol. Soc., German Meteorol. Soc., Calif. Acad. Sciences, New York Academy Science, Am. Philosophical Soc.; hon. or corr. mem. 7 geog. socs. Author: Hos tundrafolket, 1938; also other books pub. in Norway; Oceanography for meteorologists, 1942; The Oceans (with M. W. Johnson, R. H. Fleming), 1942; (with W. H. Munk) Breakers and Surf, 1944. Editor and contributor scientific reports of Maud Expdn.; contbr. articles to sci. jours.; mem. bd. of editors Jour. of Marine Research. Author standard text on oceanography; developed theory of Pacific Ocean's salinity from Indian Ocean currents; helped explain equatorial countercurrent; co-developer of method of predicting surf and breakers. Died Oslo, Norway, August 21, 1957.

SVETOVIDOV, Anatoli Nikolaevich, Russian ichthyologist; b. Nov. 3, 1903; grad. Faculty Fisheries, Moscow Agrl. Acad. K. A. Timirgazen, 1925; Became asso. Inst. Zoology, USSR Acad. Scis., Leningrad, 1932; prof., 1928—; mem. bur. dept. biol. scis. USSR Acad. Scis., 1958—. Corr. mem. USSR Acad. Scis. Research and publs. on classification, morphology, geog. distbn. and origin of fish. Office: USSR, Leningrad, Univeritetskaya n. 1, Zoologichesky institut AN SSSR.

SVIEN, H. J., Am. neurologic surgeon; b. Dennison, Minn., Feb. 28, 1911; s. Olaus J. and Karen (Bestul) S.; B.A., St. Olaf Coll., 1931; postgrad. Mass. Inst. Tech., N.Y. U.; M.D., U. Minn., 1938; m. Nancy Gatch, June 8, 1946; children—Karen, Dagny, Hendrik T. Practice medicine, specializing in neurol. surgery, Rochester, Minn.; mem. sect. neurol. surgery Mayo Found., Mayo Clinic, 1948—; prof. Mayo Grad. Sch. Medicine U. Minn., 1966—. Diplomate Am. Bd. Surgery, Am. Bd. Neurologic Surgery. Mem. A.M.A., Harvey Cushing Soc., Neurosurgical Soc. Am., Congress Neurol. Surgeons, Soc. Neurologic Surgeons, Am. Acad. Neurol. Surgery, Pan Am. Med. Assn. Author: (with M. Y. Colby) Treatment for Chromophobe Adenomas, 1967; also numerous articles. Home: 827 8th St. S.W. Office: 200 1st St. S.W., Rochester, Minn. 55901.*

SVIRBELY, William J., Am. phys. chemist, educator; b. Duquesne, Pa., Mar. 2, 1908; s. Joseph and Rose (Dospoly) S.; B.S., Carnegie Inst. Tech., 1931, M.S., 1932, D.Sc., 1935; m. Dorothy M. Campbell, Nov. 19, 1941; children—Mary Christine, Jean Elizabeth. Asst. prof. Canisius Coll., 1935-37; faculty U. Md., 1937—, prof., 1946—. Mem. Am. Chem. Soc., Washington Philos. Soc., Sigma Xi, Alpha Chi Sigma, Tau Beta Pi, Phi Kappa Phi. Research, publs. on solution kinetics, thermodynamic studies of liquid metal systems, molecular structure, phase equilibria. Home: 9220 Limestone Pl., College Park, Md.*

SVIRIDENKO, Pavel Alekseievich, Russian zoologist; b. Putivl (now Sumy Oblast), Ukraine, Mar. 19, 1893; grad. Moscow (Russia) U., 1915. Asso., head zool. lab. entomol. bur., also head dept. plant protection, Transcaucasian Kray (USSR) Plant Protection Sta., 1916-30; dir. N. Caucasian Plant Protection Sta., Rostov-on-Don (USSR), 1923-30; lectr. U. Moscow, 1932-34, prof., 1934-41; mem. All-Union Assn. Struggle Against Damage Agrl. Plants, 1930-33; head entomology lab. All-Union Inst. Beet Cultivation, 1941-47 (both Moscow); corr. mem. Ukrainian Acad. Scis., 1945-48, mem., 1948 (head ecology dept. Inst. Zoology, Kiev, 1947-54, mem. Presidium, chmn. sect. biol. scis. from 1948). Author: The Appearance and Destruction of Mouse-Like Rodents, 1934; The Steppe Polecat and Its Agricultural Importance in Combating Rodents, 1935; Mouse-Like Rodents and the Protection of Crops, Stored Produce and Timber from Them, 1953; Storage of Food by Animals, 1957. Research on ecology of animals harmful to agr. and forestry. Address: Akedemiya Nauk Ukraine SSR, ul. Vladimirskaya 54, Kiev, Ukraine SSR, USSR.

SVORAD, Domin, Czechoslovakian physiologist; b. Hrusovany, Czechoslovakia, Oct. 10, 1920; s. Martin and Maria (Adamcik) S.; M.D., Bratislava U., 1946, Ph.D., 1950; C.Sc., Czechoslovak Acad. Sci., Prague, Czechoslovakia, 1955; Neuropsychiatrist State U. Bratislava (Czechoslovakia). 1946-51, asso. prof. psychiatry, 1961—; research physiologist Czechoslovak Acad. Sci., Prague, 1951-61; head Lab. Brain Physiology, Slovak Acad. Sci., Bratislava, 1961—, sci. sec. Inst. Normal and Path. Physiology, 1966—. Mem. Internat. Soc. Biometeorology, Czechoslovak Soc. for Physiology, Czechoslovak socs. for psychology, psychiatry, neurology, higher nervous activity. Author: Paraxysmal Inhibition (Experimental Analysis of Animal Hypnosis), 1956; also numerous articles. Regional editor, Physiology and Behavior, 1965—. Research on animal hypnosis (feigning death, tonic immobility, Totstellreflex), physiol. correlates of other manifestation of normal and path. behavior such as catatonia, exptl. neurosis, sleep, sleep deprivation. Home: 1, Sienkiewiczova, Bratislava, Czechoslovakia.*

SWADESH, Morris, linguistic anthropologist; b. Holyoke, Mass., Jan. 22, 1909; s. David and Clara (Talnoper) S.; Ph.B., U. Chgo., 1930, M.A., 1931; Ph.D., Yale, 1933; m. Evangelina Arana, Dec. 18, 1959; children—Deborah (Mrs. Robert Stucklen), Daisy, Joel Kalmen. Research asst. Yale, 1933-37; asst. prof. U. Wis., 1937-39; prof. Escuela Nacional de Antropologia e Historia, Mexico, 1939-42; asso. prof. Coll. City N.Y., 1948-50; prof. U. Nacional Autónoma de México, Mexico City, 1955—. Mem. Linguistic Soc. Am., Am. Anthropol. Assn., Linguistic Circle N.Y., Sociedad Mexicana de Antropologia, Academia de la Investigación Cientifica, W. African Langs. Soc. Research and numerous publs. on Am. Indian langs; devel. phonemics, method of dating past on basis of lang. characteristics. Died July 20, 1967.

SWAIN, George Fillmore, Am. civil engr.; b. San Francisco, Mar. 2, 1857; s. Robert Bunker and Clara Ann (Fillmore) S.; B.S., Mass. Inst. Tech., 1877; Royal Poly. Sch., Berlin, 1877-80; LL.D., N.Y. U., 1907, U. Cal., 1918; m. Katharine Kendrick Wheeler, July 7, 1891 (dec. 1901); 1 dau., Barbara; m. 2d, Mary Hayden Lord, Jan. 23, 1904 (died 1914); 1 dau., Clara; m. 3d, Mary Augusta Rand, Aug. 21, 1914; 1 step-dau., Alice Rand. Hydraulic expert 10th U. S. Census, 1880-84; Hayward prof. civil engring. Mass. Inst. Tech., 1887-1909; prof. civil engring. Harvard Engring. Sch., 1909-29. Cons. engineer Mass. R.R. Commn., 1887-1914; mem. Boston Transit Commn., 1894-1918, chmn. 1913-18; mem. many commns., and engr. for many structures and other works; made appraisal of assets and liabilities of N.Y., N.H.&H. R.R. for state commn., 1910; also valuations of N.Y. Central, Chicago elevated, Canadian railroads and other rys.; mem. delegation of Am. engrs. to France, 1918, and mem. Franco-Am. Engring. Commn., 1919. First recipient Lamme medal, 1928. Fellow Am. Acad. Arts and Sciences. Author: Notes on Hydraulics, 1885, 90; Conservation of Water by Storage, 1915; How to Study, 1917; The Young Man and Civil Engineering, 1922; Strength of Materials, 1924; Fundamental Properties of Materials, 1924; Stresses, Graphical Statics and Masonry, 1927; also articles on hydraulic and structural subjects. Died July 1, 1931.

SWAIN, Tony, English biochemist; b. Malton, U.K., Apr. 27, 1922; s. Charles and Isobel (Tench) S.; B.Sc., U. London, 1943, Ph.D., 1949; M.A., U. Cambridge, 1966; m. Wendy McCabe, July 26, 1947; children—Nicholas Jon, Jonathan Michael, Simon Charles Robert. Plant and food biochemist Low Temperature Research Sta., Cambridge, U.K., 1949-65;

sci. adviser Cabinet Office, London, 1965—. Sec. commn. on chem. plant taxonomy I.U.P.A.C., 1965—; mem. com. on wood chemistry I.U.F.R.O., 1964—. Mem. Bochem. Soc., Chem. Soc., Soc. Chem. Industry, Soc. Exptl. Biology, Phytochem. Soc. Editor: Chemical Plant Taxonomy, 1962; Comparative Phytochemistry, 1966; (with J. B. Pridham) Biosynthetic Pathways in Higher Plants, 1966. Exec. editor Phytochemistry jour., 1961—. Research and numerous publs. on application of micromethods to separation and elucidation of structure of flavonoid pigments, biosynthesis; taxonomic implications and physiology of flavonoids, structure and function of condensed tannins. Home: 58 Pymers Mead, Dulwich S.E.21, U.K. Office: Cabinet Office, Whitehall, London S.W.1., U.K.*

SWAINSON, William, naturalist; b. Liverpool, Eng., Oct. 8, 1789; m. dau. of John Parkes, 1825; 5 children; 3 daus. by 2d marriage. Joined service of Commissariat, Malta, 1807, also Sicily; returned to Eng., 1815; emigrated to New Zealand, 1837. Fellow Royal Soc., 1820, Linnean Soc. Author: Zoological Illustrations; contbr. 11 vols. to Cabinet Cyclopaedia, 1820-23; 3 vols. to Naturalist Library. Made collections of birds in Malta and Sicily; collection of birds, Brazil, 1816; adopted quinary system based on circular system of William Sharp Macleay. Died Dec. 7, 1855.

SWALLOW, George Clinton, Am. geologist; b. Buckfield, Me., Nov. 9, 1817; grad. Bowdoin Coll., 1843; A.M., M.D., Mo. Med. Coll. Lectr. on botany Bowdoin, 1843; prin. Hampden (Mo.) Acad., 1848; prof. geology and chemistry U. Mo., 1852, led movement for agrl. dept., established 1859, taught its 1st class, prof. geology and dean Agrl. Coll., 1870. State geologist, of Mo., 1852, of Kan., 1865; insp. miners, Mont. In 1858 announced discovery of permian rocks in Kan. (1st in Am.); helped build 1st silver furnace at Argenta, Mont., 1st silver mill at Philipsburg, Mont. Died 1900.

SWALLOW, John Crossley, English oceanographer, physicist; b. New Mill, Yorkshire, Eng., Oct. 11, 1923; s. Alfred and Elizabeth (Crossley) S.; M.A., U. Cambridge, 1947, Ph.D., 1954; m. Mary McKenzie Morgan, Aug., 1958; 1 stepdau., Lucy Morgan. With Admiralty Signal Establishment, Haslemere, Eng., 1943-47; staff Nat. Inst. Oceanography, Wormley, Surrey, Eng., 1954—, prin. sci. officer, 1958-63, sr. prin. sci. officer, 1963—, prin. scientist R.R.S. Discovery intermittently 1955—. Recipient Albatross award Am. Miscellaneous Soc., 1960; Henry B. Bigelow Gold medal Woods Hole Oceanographic Instn., 1962; Murchison grantee Royal Geog. Soc., 1965. Fellow Royal Soc. Research and publs. on deep circulation of ocean especially with neutrally buoyant floats. Home: Crossways, Witley, Surrey, Eng. Office: Nat. Inst. Oceanography, Wormley, Surrey, Eng.*

SWAMINATHAN, M., Indian nutritionist; b. Madras, India, June 1, 1912; s. M. S. Mahadevan and Meenakshi Swaminathan; B.A., Annamalai U., 1932, M.Sc., 1935; D.Sc., Madras U., 1940; m. Bhagyalakshmi, May 10, 1933; children—M. S. Venkatesan, M. S. Sundararajan, M. S. Lakshman. Asst. biochemist Nutrition Research Lab., Hyderabad, India, 1935-42; asst. prof. biochemistry and nutrition All India Inst. Hygiene, Calcutta, 1942-47; reader biochemistry Med. Coll., Jaipur, India, 1947-49; head dept. applied nutrition and dietetics Central Food Technol. Research Inst., Mysore, India, 1950—. Mem. expert com. on calcium requirements FAO/WHO, 1961; mem. expert com. on evaluation of protein quality Nat. Acad. Scis., U. S. A., 1963. Recipient Kidwai price ICAR, India, 1960. Fellow Nat. Inst. Scis. India. Author: (with R. K. Bhagavan) Our Food, 1966; also numerous articles in nutrition and dietetics. Developed method for assay of niacin in biol. materials; research on devel. processes for preparation infant food from buffalo milk and processed protein foods based on oilseed meals and legumes, amino acid and mut. supplementation of vegetable proteins, supplementary value of protein foods to diets of children. Home: 1109, Gita Rd., Mysore-4, Mysore State, India.*

SWAMMERDAM, Jan, Dutch naturalist; b. Amsterdam, Feb. 12, 1637; student univs. Leyden and Paris from 1661; M.D., U. Leyden, 1667. Never practiced medicine; eventually abandoned sci. for Bourignon sect, ca. 1673. Author: Tractatus physico-anatomicemedicus de respiratione usque pulmonum, 1667; Algemeine Verhandeling van floedeloose dierjens, 1669; Historia insectorum generalis, 1669; Miraculum naturae seu uteri mulieribus fabrica, 1672; Ephemeri vita, 1675; Biblia naturae, 1737-38. devised method of studying circulatory system by injecting wax and dyes; pioneer in use of microscope; investigated in detail insect microanatomy and habits; classified insects according to metamorphic devel., thus considered founder of modern entomology; observed red corpuscles for first time; discovered valves in lymph vessels and glands in amphibia (Swammeerdam glands); demonstrated that muscles change shape but not volume when they contract; a preformationist. Died Amsterdam, Feb. 15, 1680.

SWAN, Emery Frederick, Am. zoologist; b. Northfield, N.H., May 10, 1916; s. Frederick Mott and Margaret (Emery) S.; B.S., Bates Coll., 1938; Ph.D., U. Cal. at Berkeley, 1942; m. Lois Chamberlain, July 6, 1939; children—Olive (Mrs. Douglas H. MacGregor), Barbara Susan, Elizabeth Ann. Faculty State Tchrs. Coll., New Platz, N.Y., 1946-48, U. Wash., 1948-52; with Am. Brass Co., 1942-46; faculty U. N.H., Durham, 1952—, prof. zoology, 1964—. Asso. marine invertebrates Mus. Comparative Zoology, Harvard, 1953-57, 65—. Fellow A.A.A.S.; mem. Ecol. Soc. Am., Soc. Zoologists, Am. Soc. Limnology and Oceanography, Soc. Study Evolution, Soc. Systematic Zoology, Arctic Inst. N.Am., Phi Beta Kappa, Sigma Xi, Phi Sigma. Research, publs. on growth, regeneration, variation and classification in clams and echinoderms; identified tube-secreting glands in serpulid polychaete worms; effect substratum on growth of clams. Home: 3 Faculty Rd., Durham, N.H. 03824.*

SWAN, Harold Jeremy, Am. physician; b. Sligo, Ireland, June 1, 1922; s. Harold John and Marcella (Kelly) S.; student St. Vincent's Coll., Ireland, 1934-39; M.B., U. London, Eng., 1945, B.S., 1945, Ph.D., 1951; m. Pamela Winifred Skeet, June 3, 1946; children—Elizabeth, Carolyn, Jeremy, Eleanor, Geraldine, Catherine, Gordon. Came to U.S., 1951, naturalized, 1955. Cons. physician Mayo Clinic, Rochester, Minn., 1955-65; faculty U. Minn., 1956-65; dir. dept. cardiology Cedars-Sinai Med. Center, Los Angeles, 1965—; prof. medicine U. Cal. at Los Angeles, 1966—. Cons. Nat. Heart Inst., NIH, Bethesda, Md., 1964—; mem. research allocations com. Los Angeles County Heart Assn., 1965—. Recipient Walter Dixon Meml. award Brit. Med. Assn., 1950. Fellow Royal Coll. Physicians, A.C.P., American College of Cardiologists; member of the American and British physiol. socs., Am. Heart Assn., Central Soc. Clin. Research, Am. Soc. Clin. Investigation. Research, publs. on dynamics of diseased heart, congenital cardiac malformations of infants and children. Home: 1262 Coldwater Canyon Dr., Beverly Hills, Cal. 90210. Office: Cedars-Sinai Med. Center, 4833 Fountain Av., Los Angeles 90029.*

SWAN, Henry, Am. surgeon; b. Denver, May 27, 1913; s. Henry and Carla (Denison) S.; grad. Phillips Exeter Acad., 1931; B.A., Williams Coll., 1935, D.Sc., 1959; M.D., Harvard., 1939; m. Mary F. Wardwell, June 25, 1936 (div. Jan. 1964); children—Edith (Mrs. William Harrison), Henry, Helen (dec. 1963), Gretchen (Mrs. Frank Bering, Jr.); m. 2d, Geraldine Fairchild, Mar. 21, 1964. Faculty, U. Colo. Sch. Medicine, Denver, 1946—, prof. surgery, 1950—, head dept., 1951-61; prof. surgery Colo. State U., Fort Collins, 1963—; practice surgery, Denver, 1959—. Recipient Hoekten Gold medal for Original research A.M.A., 1958, Henry Christian award Harvard, 1939. Diplomate Am. Bd. Surgery (past mem.). Mem. Soc. for Vascular Surgery (past sec.), A.C.S. (past gov.), Am. Heart Assn. (past chmn. surg. research com.), A.A.A.S., Halsted Soc., Internat. Soc. Surgery (sci. council), Western, Central, Am. surg. assns., Am. Assn. Thoracic Surgery, Internat. Cardiovascular Soc., N.Y. Acad. Scis., Soc. Clin. Surgery, Soc. Cryobiology, Denver Clin. and Path. Soc., Denver Acad. Surgery. Research and numerous publs. on artificial bladder from colon, resection of aortic aneurysms, arterial homografts, auto-transplant of thyroid, open heart surgery using hypothermia, repair of pulmonary and aortic valves, atrial septal defects, bilateral posterior approach to adrenal glands, transplantation hypothermia, shock. Home and office: 6700 W. Lakeridge Rd., Denver 80227.*

SWAN, Sir Joseph Wilson, English physicist, chemist; b. Sunderland, Eng., Oct. 31, 1828; s. John and Isabella (Cameron) S.; m. Frances White, 1862 (dec. 1868); m. 2d, Hannah White, 1871; 4 sons, 3 daus. Apprenticed to firm of chemists and druggists, 1842; apprentice, later partner to John Mawson, chemist, Newcastle, Eng. Recipient Progress medal Royal Photog. Soc., 1902. Fellow Royal Soc. (Hughes medal 1904); pres. Inst. Elec. Engrs., 1898-99, Chem. Soc., 1900-01; 1st pres. Faraday Soc., 1903-04. Improved manufacture of collodion for photog. use; inventor dry plate for photography; introduced useful method of carbon printing; inventor incadescent electric lamp, 1860, miner's electric safety lamp; patentee bromide paper, 1879; 1st to produce practicable artificial silk; inventor cellular-surfaced leadplate storage battery. Died Warlingham, Eng., May 27, 1914.

SWAN, Kenneth Carl, Am. ophthalmologist; b. Kansas City, Mo., Jan. 1, 1912; s. Carl Edward and Blanche (Peters) S.; B.A., U. Ore., 1933, M.D., 1936; m. Virginia A. Grone, Feb. 5, 1937; children—Steven Karl, Kenneth Richard, Susan Virginia. Asst. prof. State U. Ia., 1940-44; prof. ophthalmology U. Ore. Med. Sch., 1945—, chmn. dept., 1945—. Cons. USPHS; mem. editorial bd. Investigative Ophthalmology, Archives of ophthalmology. Recipient Proctor Research medal, 1954, U. Ore. Med. Sch. Distinguished Service award, 1962. Mem. Ore. State, Am. med. assns., Am. Ophthalmol. Soc., Am. Acad. Ophthalmology and Otolaryngology, Assn. Research in Ophthalmology, A.A.A.S., Soc. Exptl. Medicine and Biology, Sigma Xi. Research, publs. in ocular pharmacology; physiology and pathology of binocular vision; surg. anatomy and techniques of the eye. Home: 4645 S.W. Fairview Blvd., Portland, Ore. 97221.*

SWAN, Lawrence Wesley, Am. biologist; b. Darjeeling, India, Mar. 9, 1922 (parents Am. citizens); s. Henry Marcus and Edna (Lundeen) S.; Ph.B., U. Wis., 1942; M.A., Stanford U., 1947, Ph.D., 1952; m. Ruth Chisholm Humphrey, Feb. 23, 1946; children—Rhonda Jane, Sharon Louise, Pamela Diane. Research officer Climatic Research Lab., Lawrence, Mass., 1943-46; instr. biology Stanford U., 1947-48; instr. biology Santa Clara U., 1952-53; with San Francisco State Coll., 1954—, prof. biology, 1964—; research asso. Cal. Acad. Scis., 1961—. Fellow Royal Geograph. Soc., Cal. Acad. Scis., mem. A.A.A.S., Am. Inst. Biol. Scis., Ecol. Soc. Am., Bombay Natural History Soc, Himalayan Club. Studies, publs. on high-altitude environments and kinds of life; various expeditions to Himalayas and Mexico. Home: 1032 Wilmington Way, Redwood City, Cal. 94062.*

SWANK, Roy Laver, Am. physician; b. Camas, Wash., Mar. 5, 1909; s. Wilmer and Hannah (Laver) S.; B.S., U. Wash., 1930; M.D., Northwestern U., 1934, Ph.D., 1935; m. Eulala F. Shively, Sept. 14, 1936; children—Robert Latimer, Susan Jane (Mrs. Joel Keizer), Stephen Wilmer (dec.). Commonwealth Fund fellow, Scandinavia, Montreal, 1939-41; instr. medicine Harvard Med. Sch., Peter Bent Brigham Hosp., Boston, 1941-48; asst. prof. neurology McGill U., Montreal, Can., 1948-54; prof. neurology Med. Sch., U. Ore., Portland, 1954—. Mem. Neurol. Study Sect., NIH, 1962-66. Mem. Am. Neurol. Assn., Am. Physiol. Soc., Sigma Xi. Author: (with Aagot Grimsgaard) Low Fat Diet, 1959; Biochemical Basis of Multiple Sclerosis, 1961; also numerous articles. Research in treatment of multiple sclerosis with low-fat diet; suspension stability of blood; aggregation of platelets by biol. agts.; screen filtration pressure method for determining aggregation of blood cells. Home: 4400 S.W. Schools Ferry Rd., Portland, Ore. 97225.*

SWANN, Howard G., Am. physiologist; b. Toronto, Ont., Can., Mar. 8, 1906 (parents Am. citizens); s. John Butler and Marguerite (Gray) S.; B.A., Harvard, 1928; M.S., U. Chgo., 1932, Ph.D., 1935; m. Ruth J. Baltzell, Jan. 5, 1933 (div. Mar. 1940); children—Cecila (Mrs. G.C. Taylor), Howard S.G.; m. 2d, Patti H. Fox, Aug. 4, 1940; children—William F., Elena, Hilary. Asst., instr. U. Chgo., 1935-41; faculty U. Tex. Med. Br., Galveston, 1942, 46—, prof. physiology, 1950—. Mem. Am. Physiol. Soc., Soc. For Exptl. Biology and Medicine, Sigma Xi. Contbr. numerous articles to tech. jours. Discovered subcortical nature olfactory learning, weak control by adenohypophysis of glomeruler layer adrenal cortex; research on process death, resuscitation; discovered high interstitial pressure and functional distention kidney, natural distension renal tubular lumina. Home: 2528 Av. O, Galveston, Tex. 77550.*

SWANN, Michael Meredith, Brit. biologist; b. Kent, Eng., Mar. 1, 1920; s. Meredith B. and Marjorie (Dykes) S.; M.A., Caius Coll., Cambridge U., 1945, Ph.D., 1948; m. Tess Gleadowe, Aug. 7, 1942; children—Meredith, Sylvia, Catriona, Peter. Demonstrator in zoology Cambridge U., 1946-52; prof. natural history Edinburgh (Scotland) U., 1952-65, dean Faculty of Sci., 1962-65, prin. and vice chancellor, 1965—. Mem. Med. Research Council, 1962-65, Council for Sci. Policy, 1965—. Fellow Royal Soc. 1962. Research, publs. in cell biology, particularly mechanisms of cell division, fertilization, effects of radiation. Home: Ormsacre, Barnton Av., Edinburgh 4, Scotland.*

SWANN, Sherlock, Jr., Am. chemist; s. Balt., Sept. 30, 1900; s. Sherlock and Edith R. (Deford) S.; B.S., Princeton, 1922; Ph.D., Johns Hopkins, 1926. Chemist, Columbia Gas Co., 1926-27; faculty U. Ill., Urbana, 1927—, research prof. chem. engring., 1941—; with Office Prodn. Research and Devel., 1944. Mem. Electrochem. Soc. (electroorganic editor Jour., past pres.), Am. Chem. Soc., A.A.A.S., Sigma Xi. Research, numerous publs. on electro-organic chemistry, influence cathode material on electrolytic reduction carbonyl, electrodeposition and catalysis. Home: 909 W. Oregon St., Urbana, Ill.*

SWANN, William Francis Gray, physicist; b. Ironbridge, Shropshire, Eng., Aug. 29, 1884; s. William Francis and Anne (Evans) S.; student Brighton (Eng.) Tech. Coll., 1900-03, Royal Coll. of Science (London), Univ. Coll., Kings Coll., City and Guilds of London Inst., 1903-07; B.Sc., London, 1905, D.Sc., 1910; asso. Royal Coll. of Science, 1906; hon. M.A., Yale, 1924; hon. D.Sc., Swarthmore Coll., 1929; hon. F.T.C.L., London, 1936; Litt.D. (hon.), Temple U., 1954; m. Sarah Frances Mabel Thompson, Aug. 14, 1909; (dec. 1954); children—William Francis, Charles Paul, Sylvia; m. 2d, Helene Laura Diedrichs, Dec. 23, 1955. Came to U. S. 1913. Chief Phys. div. Dept. Terrestrial Magnetism, Carnegie Instn. Washington, 1913-18; mem. faculties U. Minn., U. Chgo., Yale, 1918-27, dir. Sloane Lab., 1924-27, also chmn. advisory research com. Bartol Research Foundation of Franklin Inst., 1924-27, dir. same, 1927-59, dir. emeritus, 1959-62, sr. staff advisor Franklin Inst. Labs. for Research and Development, 1945-62. Fellow Imperial College of Science and Technology (London, Eng.), Phys. Soc., London, Am. Physical Soc. (v.p. 1929, 30; pres. 1931-33); mem., sometime officer numerous profl. assns. Mem. bd. dirs. Phila. Musical

Academy, chmn. 1951-58. Recipient Elliott Cresson Gold Medal, Franklin Inst., 1960. Author: The Architecture of the Universe, 1934; (with other) The Story of Human Error, 1936; Physics, 1941. Contbr. to study cosmic rays, atomic structure, relativity, and atmospheric electricity. Died 1962.

SWANSON, Arnold Arthur, Am. chemist; b. Rawlins, Wyo., Mar. 11, 1923; s. Arnold David and Gladys (Fern) S.; B.A., Duke U., 1948; M.A., Trinity U., 1959; Ph.D., Tex. A and M. U., 1961; m. Florence M. Peele, Dec. 27, 1949; children—Arnold Arthur II, Kathleen Rena, Kristina Rita, Annabeth Raye. Chemist Sch. Aerospace Medicine, San Antonio, Tex., 1950-59; head dept. chemistry USPHS, Southeast radiol. health labs., Montgomery, Ala., 1961-63; chief research lab. McKinney (Tex.) VA Hosp., 1963-65; chief research lab. VA Center, Temple, Tex., 1965-68; asso. prof. chemistry Med. Coll. S.C., 1968—; cons. Scott & White Meml. Hosp., Temple; cons. biochemistry; asst. prof. chemistry Baylor U. Coll. Dentistry; prof. Tex. Women's U. NIH fellowship, 1959-61. Fellow Am. Inst. Chemists, A.A.A.S.; mem. Am. Chem. Soc., Soc. Investigation Opthalmology, Sigma Xi. Research, publs. on radiation effects on eye tissue; lens proteolytic enzymes; cataract studies, chemistry of caries-lens. Home: 1417 Lenevar Dr., Charleston, S.C.*

SWANSON, Carl Pontius, Am. biologist; b. Rockport, Mass., June 24, 1911; s. Richard and Anna (Nordstrand) S.; B.S., Mass. State Coll., 1937; A.M., Harvard, 1939, Ph.D., 1941; D.Sc., U. Mass., 1957; m. Dorothy Jane Noggle, June 14, 1941; children—Michael Warren, Ann. Asst. prof. Mich. State Coll., East Lansing, 1941-43; asso. biologist NIH, Bethesda, Md., 1946; faculty Johns Hopkins U., Balt., 1946—, prof. biology, 1948-56, William D. Gill prof. in biology, 1956—, dean undergrad. studies, 1964—; vis. prof. U. Tex., 1958. Cons. biology div. Oak Ridge Nat. Lab., 1953—; pres. Comite Internationale de Photobiologie, 1964—; mem. study sects. NIH, NSF. Guggenheim fellow, 1954; Sheldon Travelling fellow, 1941. Fellow Am. Acad. Arts and Scis.; mem. Genetics Soc. Am., Am. Soc. for Cell Biology, Bot. Soc. Am., A.A.A.S., Am. Inst. Biol. Scis. Author: Cytology and Cytogenetics, 1957; The Cell, 1960, 64; also numerous articles. Biol. editor Prentice-Hall, Inc., Englewood Cliffs, N.J., 1958—. Research in structure and behavior chromosomes and cells particularly as influenced by X-rays, ultraviolet light and radiomimetic chems. Home: 1408 Carrollton Av., Ruxton, Md. 21204.*

SWANSON, Ernst Warner, Am. economist, econometrician; b. Ashtabula, O., June 26, 1902; s. Otto Efrim and Amanda Warner (Olson) S.; Ph.B. cum laude U. Chgo., 1930, Ph.D, 1940; postgrad. (Brookings' fellow) Brookings Instn., 1932-33; m. Mary Lorraine Cooney, Sept. 5, 1932; children—Eric, Karen (Mrs. Charles L. Hamilton). Asst. prof. Drake U., 1934-36; faculty Wash. State U., 1936-41, asso. prof., 1938-41; sr., prin. fiscal analyst Bur. Budget, Washington, 1941-44; asso. dir. research U. S. C. of C., Washington, 1944-48; prof. econs., dir. grad. program Emory U., 1948-56; interim cons., mil. scientist Lockheed Aircraft Co., Marietta, Ga., also Los Angeles, 1954-57; head econ. research Ga. Inst. Tech., 1956-60; prof. research econs., N.C. State U., Raleigh, 1960—. Interim cons. Ford Found., 1952-66; Mem. Pres.'s Regional Commn. on Appalachia, 1962-65. Named Best Tchr. Wash. State U., 1935-37. Mem. Am., So. (past pres.) econ. assns., Ednl. Found. Nuclear Sci. Author: (with Emerson Schmidt) Economic Stagnation or Progress, 1946; (with John Griffin) Education in the South, 1955; (with Cleon Harrell) Projections of Employment, 1960; also articles, chpts. in monographs. Editorial bds. S.E. Jour., 1951—, Jour. Pub. Law, 1957-61. Research on econ. analysis causes econ. growth and bus. cycles, derivation theory and its applications from theoretical and econometric analysis, restatement theory value under conditions econ. growth. Home: 2338 Champion Ct., Raleigh, N.C. 27606.*

SWANSON, Pearl Pauline, Am. nutritionist; b. Cokato, Minn., Sept. 13, 1895; d. Frank and Maria (Sigfridson) Swanson; B.S. in Chemistry, Carleton Coll., 1916; M.S. in Nutrition, U. Minn., 1924; Ph.D. in Physiol. Chemistry (Sterling fellow, also Alexander Broune Cox fellow), Yale, 1930. Instr. high sch., Fairbault, Minn., 1916-18; faculty Carleton Coll., 1920-22, Mont. State Coll., 1924-27; faculty Ia. State U., Ames, 1930-67, mem. grad. faculty, 1931-67, prof. nutrition, 1936-67, asst. dir. agrl. and home econs. expt. sta., 1944-61. Mem. nat. coms. U.S. Dept. Agr., 1948-58; mem. adv. com. North Central Home Econs. Research Adminstrs., 1955-61. Recipient Outstanding Achievement award U. Minn., 1951, Borden award, 1955, Outstanding Alumni Achievement Award Carleton Coll., 1966, others. Fellow N.Y. Acad. Scis.; mem. Am. Inst. Nutrition, Am. Dietetics Assn., Ia. Acad. Sci., Soc. Exptl. Biology and Medicine (emeritus), Phi Beta Kappa, Phi Kappa Phi, Sigma Xi, Phi Upsilon Omicron, Gamma Sigma Delta (Ia. Merit award 1961), Kappa Delta Gamma, others. Mem. editorial com. Ia. State U. Press, 1943-59, chmn., 1955-59; editorial com. Nutritional Status, U.S.A., 1950-55; asso. editor Jour. of Nutrition, 1949-53. Research and publs. on role of proteins, fats in nutrition; importance of diet in the gen. welfare of people. Home: Univ. Towers, 111 Lynn Av., Ames, Ia. 50010.*

SWANTON, John Reed, Am. ethnologist; b. Gardiner, Me., Feb. 19, 1873; s. Walter Scott and Mary Olivia (Worcester) S.; A.B., Harvard, 1896, A.M., 1897, Ph.D., 1900; postgrad. Columbia, 1898-1900; Ph.D., m. Alice Barnard, Dec. 16, 1903 (dec. Sept. 1926); children—Mary Alice, John Reed, Henry Allen. Ethnologist, Bur. Am. Ethnology, Washington, 1900-44. Chmn. U. S. De Soto Expdn. Commn., 1935—. Mem. Am. Anthrop. Association, Anthrop. Soc., Washington, A.A.A.S., Linguistic Soc. Am., Nat. Acad. Scis. Author: Contributions to the Ethnology of the Haida, 1905; Haida Texts and Myths, 1905; Haida Texts—Masset Dialect; Social Conditions, Beliefs, and Linguistic Relationship of the Tlingit Indians; Tlingit Myths and Texts; Indian Tribes of the Lower Mississsippi Valley and Adjacent Coast of the Gulf of Mexico; (with J. O. Dorsey) A Dictionary of the Biloxi and Of Languages; History of the Creek Indians and Their Neighbors; Social Organization and Social Usages of the Creek Indians; Religious Beliefs and Medical Practices of the Creek Indians; Social Conditions and Religious Beliefs of the Chickasaw Indians; Myths and Tales of the Southeastern Indians; (with A. S. Gatschet) A Dictionary of the Atakapa Language; Source Material for the Social and Ceremonial Life of the Choctaw Indians; Linguistic Material from the Tribes of Southern Texas and Northeastern Mexico; Source Material for the History and Ethnology of the Caddo Indians, Indians of the Southeastern U. S.; The Wineland Voyages; The Indian Tribes of North America, 1952. Died May 2, 1958.

SWARTOUT, John Arthur, Am. chemist; b. Madison, Wis., Mar. 17, 1916; s. John F. and Mildred (Brownell) S.; B.A., U. Buffalo, 1937; Ph.D., Northwestern U., 1940; m. Ellen McCance Rogers, Jan. 30, 1943; children—William M., Sue Ellen. Research chemist E.I. duPont, Waynesboro, Va., 1940-43, Metall. Lab., U. Chgo., 1943, Clinton Engring. works, 1943-44, Hanford Engring. Works, 1944-45; with Union Carbide Corp., Oak Ridge Nat. Lab., 1945-64, asst. research dir., 1950-55, dept. dir., 1950-55, dep. dir., 1955-64; asst. gen. mgr. for reactors U.S. AEC, Washington, 1965; gen. mgr. for research Union Carbide Corp., N.Y.C., 1966, dir. tech., 1966—. U.S. tech. adviser Atoms for Peace Conf., Geneva, Switzerland, 1955, 58, 64. Cited by U. Buffalo, 1957. Fellow Am. Nuclear Soc., A.A.A.S.; mem. Am. Chem. Soc., N.Y. Acad. Scis., Research Soc. Am., Indsl. Research Inst., Atomic Indsl. Forum. Research, publs. on abundance deuterium in natural waters, devel. chem. processes for separation plutonium and products fission, devel. nuclear reesarch and power reactors. Home: Route 2, Taylor Rd., Mount Kisco, N.Y. 10549. Office: 270 Park Av., N.Y.C. 10017.*

SWARTZ, Charles Dana, Am. physicist; b. Balt., July 25, 1915; s. Charles K. and Elizabeth (Howard) S.; A.B., Johns Hopkins, 1938, Ph.D., 1943; m. Katherine Hunt, Aug. 14, 1949; children—Timothy H., Douglas K., Christina H. Physicist, Manhattan Project, 1942-46; research asso. Lab. Nuclear Studies, Cornell U., 1946-48; faculty Johns Hopkins, 1948-56; faculty Union Coll., Schenectady, 1956—, prof. physics, 1962—. Fulbright prof. Ankara (Turkey) U., 1961-62. Mem. Am. Phys. Soc., Am. Assn. Physics Tchrs., Phi Beta Kappa, Sigma Xi. Research and publs. on low energy nuclear reactions, reaction in light elements. Home: 10 Crestwood Dr., R.D. 1, Ballston Lake, N.Y. 12019.*

SWARTZ, Charles Kephart, Am. geologist; b. Balt., Jan. 3, 1861; s. Joel and Adelia (Rosecrans) S.; A.B., Johns Hopkins, 1888, Ph.D., 1904; postgrad. U. Heidelberg, 1889; fellow, Clark U., 1889-90; B.D., Oberlin Theol. Sem., 1892; m. Elizabeth A. Howard, Dec. 12, 1892; children—Joel Howard, William Hamilton, Frank McKim, Howard Currier, Charles Dana. Instr. geology Johns Hopkins U., 1904-05, asso., 1905-06, asso. prof. geology and paleontology, 1907-10, collegiate prof. geology, 1910-31, emeritus since 1931. Fellow Geol. Soc. Am. (v.p. 1936), A.A.A.S.; pres. Paleontol. Soc., 1935. Contbd. to knowledge of Paleozoic geology, Silurian, Devonian, Carboniferous periods in Md. Died Nov. 28, 1949.

SWARTZ, Clifford Edward, Am. physicist; b. Niagara Falls, N.Y., Feb. 21, 1925; s. James Luke and Lena (Gee) S.; A.B., U. Rochester, 1945, M.S., 1946, Ph.D. (AEC fellow), 1951; m. Barbara Myers, June 22, 1946; children—Katherine, Paul, Christine, Cassandra, Erich, Tamara. Asso. physicist Brookhaven Nat. Lab., Upton, N.Y., 1951-62; asso. prof. physics State U. N.Y., Stony Brook, 1947-67, pro. fessor of physics, since 1967—. Member of the N.Y. State Edn. Dept. high sch. physics syllabus revision com., 1964-66; dir. quantitative sci. in grades project NSF, 1964, dir. conf. physics tchrs. from two year colls., 1965. Mem. Am. Phys. Soc., Am. Assn. Physics Tchrs., Phi Beta Kappa, Sigma Xi. Author: (with Lehrman) Foundations of Physics, 1965, Laboratory Experiments, 1965, Teachers' Guide, 1965; The Fundamental Particles, 1965; Microstructure of Matter, 1965; also articles. Research in instrumentation for high energy particle beam detection with accelerators, high energy gamma ray detection and analysis. Home: Box 1074 Wheeler Rd., Setauket, N.Y. 11785.*

SWARTZ, Joel Howard, Am. geophysicist; b. Bellevue, O., Nov. 10, 1893; s. Charles Kephart and Elizabeth (Howard) S.; A.B., Johns Hopkins, 1915, Ph.D., 1923; m. Virginia C.B. Markley, Dec. 20, 1920; children—Donald Markley, William Alan. Prof. chemistry, physics, and biology McKendree Coll., 1915-16; asst. geologist Md. Geol. Survey, 1916; fellow Johns Hopkins, 1919-20, asst., 1920-21, instr., 1921-23; geologist Compania Transcontinental de Petroleo, S.A., Tampico, Mexico, 1920; geologist Tenn. Geol. Survey, 1921-23, 25-28; faculty U. N.C., Chapel Hill, 1923-30; with U.S. Dept. Interior, 1930-63, geophysicist-in-charge Balt. field unit U.S. Geol. Survey, 1946-52, staff geophysicist, geophys. br., 1952-60, br. theoretical geophysics, 1960-63, emeritus, 1963—. Recipient Meritorious Service award U.S. Dept. Interior, 1942, 63, Certificate of achievement for Meritorious service AEC and Joint Task Force 1932, 1952. Fellow Geol. Soc. Am. (life) mem. Am. Geophys. Union (life), Soc. Exploration Geophysicists, Seismol. Soc. Am., Geol. Soc. Washington (life). Research and publs. on permafrost, volcanic activity, resistivity studies, seismic studies, geothermal investigations. Address: 226 Meadowvale Rd., Lutherville, Md. 21093.*

SWARTZ, Olof, Swedish botanist; b. Norrkoeping, Sweden, Sept. 21, 1760; student of Linnaeus; prof. natural history Carolina Inst., Stockholm, Sweden; made expdns. to S. Am. coast, also to India; mem. French Acad. Scis., Acad. Stockholm. Author: Nova genera et species plantarum, 1788; Flora Indiae, 1797. Discovered Swartzia, 2 other new genders of moss, 3 genders of ferns. Died Stockholm, Sept. 19, 1818.

SWARTZENDRUBER, Dale, Am. soil physicist; b. Parnell, Ia., July 6, 1925; s. Urie and Norma (Kinsinger) S.; student Goshen Coll., 1947-48; B.S., Ia. State U., 1950, M.S., 1952, Ph.D., 1954; m. Kathleen Jeanette Yoder, June 26, 1949; children—Karl Grant, Myra Mae, John Keith, David Mark. Instr., Goshen Coll., 1953-54; asst. soil sci. U. Cal. at Los Angeles, 1955-56; asso. prof. soil physics Purdue U., Lafayette, Ind., 1956-63, prof., 1963—. Mem. Soil Sci. Soc. Am. (asso. editor Proceedings 1965—, chairman div. 1966—), Am. Geophys. Union, A.A.A.S., Sigma Xi, Phi Kappa Phi, Gamma Sigma Delta. Research and publs. in physics and math. of water flow in soils and porous media. Home: 1953 Indian Trail Dr., West Lafayette, Ind. 47906.*

SWARUP, Sushiela Shyam, Indian hematologist, physician; b. Multan, West Pakistan, Nov. 7, 1925; d. Shyam and Bishan Devi (Menta) Swarup; M.B.-B.S., Punjab (India) U., 1948; D.Phil., U. Calcutta (India), 1961. Staff, Indian Council Med. Research, Calcutta, 1950—, sr. research officer, 1965—. Recipient Shakuntala Amir Shand prize, 1964. Mem. Indian Soc. Hematology (v.p. 1961—), Asian Pacific Soc. Hematology (councillor 1965—), Internat. Soc. Hematology, Indian Assn. Pathologists and Bacteriologists, Indian Med. Assn. Research and publs. on delineation of erythrocytic enzymes in thalassemia, hematological and biochem. characterization of Kyansanur Forest Disease. Home: 33 Dilkusa St., Calcutta 17. Office: Hematological Unit, Indian Council Med. Research, Sch. Tropical Medicine, Calcutta, West Bengal, India.*

SWASEY, Ambrose, Am. inventor; b. Exeter, N.H., Dec. 19, 1846; s. Nathaniel and Abigail Chesley (Peavey) S.; Dr. Eng., Case Sch. Applied Sci., 1905; Sc.D., Denison U., 1910, U. Pa., 1924, Brown U., 1931; LL.D, U. Cal., 1924, U. Rochester, 1925, U. N.H., 1930; m. Lavinia D. Marston, Oct. 24, 1871. In 1880 entered into partnership with W. R. Warner in firm of Warner & Swasey (inc. 1900), mfrs. of the 36-inch Lick telescope, 26-inch telescope of Naval Obs., Washington, and 40-inch Yerkes telescope, as well as exceptionally accurate dividing engine; 72-inch reflecting telescope for Canadian Govt., completed 1916; 60-inch, for Argentine Nat. Obs., completed 1922, 69-inch telescope for Ohio Wesleyan U., completed 1923; inventor Swasey range and position finder, adopted by U. S. Govt. Founder Engring. Found., 1914; provided astron. equipment, bldgs. at various univs. Recipient John Fritz Gold medal, 1924; Cleveland medal for public service, 1930; Franklin Gold medal, 1932; Am. Soc. M.E. medal, 1933; Washington award, 1935. Mem. Nat. Acad. Scis., Am. Soc. M.E. (pres.), Am. Soc. C.E. (hon.), Am. Inst. E.E. (hon.), Cleve. Engring. Soc., Instn. Mech. Engrs. Gt. Britain (hon.), Instn. Mining Engrs. Gt. Britain (hon.), Soc. Civil Engrs. France (hon.). Author: (papers) A New Process for Generating and Cutting the Teeth of Spur Wheels; Some Refinements of Mechanical Science. Inventor telescope mounting which made possible more accurate celestial photography; developed new gear cutting machine for generating and cutting teeth of spur gears at same time; devised epicyclodial milling machine for producing theoretical curves from which forgear teeth were made. Died June 15, 1937.

SWEAT, Max Leroy, Am. biochemist; b. Park City, Utah, July 8, 1914; s. Le Roy and Sarah (White) S.; B.S., U. Utah, 1938, Ph.D., 1949; M.S., Utah State U., 1941; m. Helen Allred, Jan. 11, 1951;

children—Teresa Anne, Thomas Scott, Kevin James, Linda Marie. Biochemist, NIH, Bethesda, Md., 1949-51; asst. prof. Western Res. U. Sch. Medicine, 1952-56; asst. research prof. U. Utah, Salt Lake City, 1951-52, asso. research prof. dept. obstetrics, gynecology, 1956—. Mem. Soc. Am. Biol. Chemists, Endocrine Soc., A.A.A.S., Sigma Xi. Research, publs. on enzymes, coenzymes of steroid hormone synthesis; first to show cortiol in human adrenal; developed fluorescence test for adrenal steroids; 1st to asso. steroid hormone metabolism with co-enzymes and cellular particles. Home: 2919 Oakridge Dr., Salt Lake City 84109.*

SWEET, John Edson, Am. mining engr., inventor; b. Pompey, N.Y., Oct. 21, 1832; s. Horace and Candance (Avery) S.; ed. dist. schs.; m. Caroline V. Hawthorne, Nov. 24, 1870 (dec. 1877); m. 2d, Irene A. Clark, May 9, 1889. Carpenter's apprentice, 1850; builder, architect in South, until 1861; attended London Expn., 1862; draughtsman, patent office Hazelt, Lake & Co.; with Patent Nut & Bolt Co., Birmingham, Eng.; asso. with Sweet, Barnes & Co., Syracuse, N.Y., 1864, became supt.; bridge builder, 1871-73; prof. practical mechanics Cornell U., 1873-79; pres. Straight-Line Engine Works, Syracuse, 1880—. Recipient John Fritz medal, 1914. Mem. Am. Soc. M.E. (a founder, pres. 1883-84), Engine Builders' Assn. U. S. (1st pres. 1889-1901), Syracuse Metal Trades Assn., John Fritz Medal Assn. Author: Things that are Usually Wrong. Inventor 1st straight line high speed engines built to rest on 3 points; built composing machine to form matrix for casting stereotype plates directly with use of moveable type, 1869, a measuring machine, incorporating several separate inventions, 1873, a traversing machine for working metal, 1886, a steam separator, 1893, a machine for boring chilled rolls. Died May 8, 1916.

SWEET, Walter Clarence, Am. invertebrate paleontologist; b. Denver, Oct. 17, 1927; s. Homer Gilbert and Marion (Paterson) S.; B.S. magna cum laude in Geology, Colo. Coll., 1950; M.S., U. Ia., 1952, Ph.D., 1954; m. Mona Louise Franken, June 1, 1957. Faculty, Ohio State U., Columbus, 1954—, prof. invertebrate paleontology 1966—; research fellow Oslo (Norway) U., 1956-57. Fellow Geol. Soc. Am.; mem. Geol. Soc. Norway (life), Paleontol. Soc., Internat. Paleontol. Union, Soc. Econ. Mineralogists and Paleontologists, Am. Assn. Petroleum Geologists, Nat. Assn. Geology Tchrs., Ohio Acad. Sci., A.A.A.S. Author: (with R. L. Bates) Geology: An Introduction, 1966; (with others), Treatise on Invertebrate Paleontology, Cephalopoda: Nautiloidea, 1964; also articles. Research on N.Am. and N. European Ordovician nautiloid cephalopods, Ordovician conodonts. Home: 3065 Shadywood Rd., Columbus, O. 43221.*

SWEET, William Herbert, Am. neurosurgeon; b. Kerriston, Wash., Feb. 13, 1910; s. Paul Williams and Daisy (Pool) S.; S.B., U. Wash., 1930; B.Sc., (Rhodes scholar), Oxford U. (Eng.), 1934, D.Sc., 1957; M.D., Harvard, 1936; m. Mary Elizabeth Rowland, Nov. 25, 1937; children—David Rowland, Gwendolyn, Paula Eleanor. Instr., Billings Hosp., Chgo., 1939-40; Commonwealth Fund fellow Harvard Med. Sch., 1940-41, faculty, 1945—, prof. surgery, 1965—; lectr. Med. Sch., Tufts Coll., 1947-51; staff Mass. Gen. Hosp., Boston, 1945—, chief neurosurg. service, 1961—. Mem. neurology B study sect. NIH, 1962—; mem. sci. and tech. adv. com. NASA, 1964—. Trustee Asso. Univers., Inc., Neuro-Research Found.; trustee, pres. Neuroscis. Research Found., Inc. Decorated His Majesty's medal for service in cause freedom, Eng., 1945. Fellow A.C.S., Am. Acad. Neurology, Am. Acad. Arts and Scis. (v.p. for biol. scis. 1964—); hon. mem. Soc. Brit. Neurol. Surgeons, Spanish-Portuguese Soc. Neurosurgeons, Assn. des Neurochirurgiens Suisses, Societe de Neuro-Chirurgie de Langue Francaise; mem. Soc. Neurol. Surgeons, Am. Neurol. Assn., American Surgical Association, American Academy of Neurol. Surgery, A.M.A., Am. Physiol. Soc., Electroencephalographic Soc., Halsted Soc., Harvey Cushing Soc., Mass. Med. Soc., New Eng. Neurosurg. Soc., Royal Soc. Medicine (Eng.), others. Author: (with J.C. White) Pain: Its Mechanisms and Neurosurgical Control, 1955; Pain and the Neurosurgeon, 1968; also numerous articles, and chpts. in books. Editor, Neurochirurgia, 1959—, Progress in Neurol. Surgery, 1963—. Research on mechanisms and neurosurg. control pain, use positron-emitting isotopes in diagnosis intracranial lesions, use slow neutrons and boron in brain tumors, methods study and treatment brain tumors, intracranial aneurysms, diabetic retinopathy. Home: 35 Chestnut Pl., Brookline, Mass. 02146. Office: Mass. Gen. Hosp., Boston 02114.*

SWEETMAN, Harvey Leroy, Am. ecologist, entomologist; b. Las Animas, Colo., Jan. 29, 1896; s. Philip Hay and Cora A. (Reynolds) S.; B.S., Colo. State U., 1923; M.S., Ia. State Coll., 1925; Ph.D., U. Mass., 1930; postgrad. U. Minn., 1925-27; m. Erma Bamesberger, June 12, 1965. Lab. asst. Colo. Expt. Sta., Grand Junction, 1923; jr. entomologist U.S. Dept. Agr. Bur. Entomology, Billings, Mont., 1923; research asst. Ia. Agr. Exptl. Sta., Ames, 1923-25, Minn. Agr. Expt. Sta., St. Paul, 1925-27; asst. research prof. Wyo. Agr. Expt. Sta., Laramie, 1927-29; faculty U. Mass., Amherst, 1930—, now prof. entomology; vis. prof. biology U. Mindanao (P.I.), 1966-67. Mem. A.A.A.S., Am. Inst. Biol. Scis., Ecol. Soc. Am., Entomol. Soc. Am., Royal Entomol. Soc. London, Animal Behavior and Sociology, Am. Soc. Limnology and Oceanography. Author: Biological Control of Insects, 1936; Principles of Biological Control, 1958; Identification of Structural Pests and Their Damage, 1965; also numerous articles. Research in animal ecology and basic and econ. entomology. Home: North Amherst, Mass 01059.*

SWEETSER, Frank Loel, Jr., Am. sociologist; b. Montclair, N.J., Feb. 5, 1913; s. Frank Loel and Lura (Parker) S.; A.B., Dartmouth, 1934; M.A., Columbia U., 1935, Ph.D., 1941; m. Barbara C. Brock, June 21, 1934 (dec. July 1959); children—Judith (Mrs. McFarland), Frank L.; m. 2d, D. Dorrian Apple, Nov. 23, 1960. Instr. sociology Ind. U., Bloomington, 1937-41; psychologist OSS, Washington, 1941-43; asst. dir. Community Analysis War Relocation Authority, Washington, 1943; asst. to dir. N.H.-Vt. Blue Cross/Blue Shield, Concord, N.H., 1946-48; faculty Boston U., 1948—, prof. sociology, 1959—; dir. area research Urban Renewal Demonstration Project, Boston, 1955-56. Cons. to state and city agys.; Fulbright lectr. urban sociology U. Helsinki (Finland), 1962-63. Mem. Am. Sociol. Assn., Eastern, Rural sociol. socs., Am. Statis. Assn., Westermarck Soc., Am. Scandinavian Found. Author: Social Ecology of Metropolitan Boston, 1950; Patterns of Change in Social Ecology of Metropolitan Boston, 1950-60; (with William C. Lring and Charles F. Ernst) Community Organization for Citizen Participation in Urban Renewal, 1957; Social Ecology of Metropolitan Boston, 1960; also articles. Clarification of theory of neighborhood and community groups and interaction through demonstration of continuing importance of neighboring activity in urban residential areas; codification of social orgn. for citizen urban renewal participation; developed theories of met. social and spatial structure through demonstration of universals and spl. factors underlying differentiation of neighborhoods in Am. and European metropolises. Home: Forest St., Norwell, Mass. 02061. Office: 232 Bay St. Rd., Boston 02215.*

SWENDENBORG, Emanual, Swedish mathematician, philosopher; b. Stockholm, Sweden, Jan. 29, 1688; s. Jasper Swendenborg; ed. U. Uppsala (Sweden), completed studies, 1710. Toured Europe, studied natural philosophy, from 1710; returned to Uppsala, studied natural sci., engring., from 1715; apptd. assessor-extraordinary Swedish Bd. Mines, 1716; lectr. economics House of Nobles; declined chair mathematics U. Uppsala (based on his belief that math. should not be limited to theory), 1724. Author: Prodromus principiorum rerum naturalium, 1721; Opera philosophica et Mineralia, 3 vols. (treats primary principles of universe, predicts nebular hypothesis and some modern ideas on atom), 1734; Prodromus philosophiae, 1734; also numerous others (his writings properly collected and examined for first time during late 19th century showed his ideas in almost every phase of science to be far ahead of his time). Began studies for sci. explanation of universe, 1721; sought to discover nature of soul and spirit by means of anat. studies, from 1734; first to form system of crystallography; made important physiol. discoveries; showed that motion of brain was synchronous with respiration, not heartbeat; made observations on functions of spinal cord, ductless glands (which concur with modern sci. views); moved from science to studies in theology and spiritualism; though he did not preach or attempt to form a sect, his followers (Swendenborgians) founded the New Church, based on his writings. Died London, Mar. 29, 1772.

SWENEY, Arthur Barclay, Am. psychologist; b. Champaign, Ill., Apr. 30, 1923; s. Merle Arthur and Edith (Sendenburgh) S.; B.S. in Elec. Engring., U. Ill., 1947, M.S.W., 1949; Ph.D., U. Houston, 1958; m. Martha Royce, June 7, 1944 (div. Apr. 1961); children—Ann, Dierdre, Michael; m. 2d, Raquel Zaldívar, Oct. 6, 1961; children—Rachel, Rebecca, Robert, Raymond. Dir., Rusk Settlement, 1949-54; lectr. U. Houston, 1954-58, Sch. Psychology, Houston, 1954-58; research asso. U. Ill., 1958-60, research asst. prof., 1960-62; asso. prof. Tex. Tech. Coll., Lubbock, Tex., 1962-66, prof., 1966—, dir. psychometric research bur., 1962—. Cons., Inst. for Personality and Ability Testing, 1959—, Big Spring State Hosp., 1963—, Girls Welfare Home, Albuquerque, 1964—. Mem. Psychometric Soc., Soc. Multivariate Exptl. Psychologists, Am. Psychol. Assn., S.W. Psychol. Assn. Author: (with Cottell, Radcliffe) The Nature and Measurement of Components of Motivation, 1963; also articles. Research in levels of motivation in humans; co-developer tests of motivation, behavioral test fear and anxiety, tests utilizing response set for measuring personality, def. mechanisms and conflict areas. Home: 4901 24th St., Lubbock, Tex.*

SWENSON, Clayton A., Am. physicist; b. Mpls., Nov. 11, 1923; s. Nels and Anna (Roth) S.; B.S., Harvard U., 1944; Ph.D., Oxford U., 1949; m. Heather M. F. Gell, Sept. 2, 1950; children—Anna, Paul, Wendy. Mem. staff Los Alamos Sci. Lab., 1944-46; instr. Harvard U., 1949-52; Div. Indsl. Cooperation staff mem. Mass. Inst. Tech., 1952-55; prof. physics, sr. physicist dept. physics AEC Ames Lab. Ia. State U., 1955—. Mem. Am. Phys. Soc., Am. Assn. Physics Tchrs., A.A.A.S., Phi Beta Kappa, Sigma Xi. Research, numerous publs. in solid state physics with emphasis on low temperature and high pressure and combinations of these; publs. on understanding of elementary solids (inert gases and alkali metals) at high pressures; devel. high pressure techniques for use at low temperatures. Home: 714 Lynn Av., Ames, Ia. 50010.*

SWENSON, George Warner, Jr., Am. electronics engr., radio astronomer; b. Mpls., Sept. 22, 1922; S. George Warner and Vernie (Larson) S.; B.S., Mich. Coll. Mining and Tech., 1944, E.E., 1950; S.M., Mass. Inst. Tech., 1948; Ph.D., U. Wis., 1951; m. Virginia Laura Savard, June 26, 1943; children—George Warner, III, Vernie Laura, Julie Loretta, Donna Joan. Asso. prof. elec. engring. Washington U., St. Louis, 1952-53; prof. U. Alaska, 1953-54; asso. prof. Mich. State U., 1954-56; faculty U. Ill., Urbana, 1956—, prof. elec. engring. and astronomy, 1958—. Cons. to govt. agys. and other sci. bodies. Fellow I.E.E.E.; mem. Am. Astron. Soc., Internat. Sci. Radio Union (mem. U.S. nat. com. 1965—), Sigma Xi, Eta Kappa Nu, Tau Beta Pi, Phi Kappa Phi. Author: Principles of Modern Acoustics, 1953; also articles. Research in radio astronomy, properties ionosphere, analog computation; designed radio telescopes. Home: 202 W. Vermont Av., Urbana, Ill. 61801.*

SWENSON, Knud George, Am. entomologist; b. Brookings, S.D., July 1, 1923; s. Andrew Peter and Helga (Jensen) S.; student Pomona Coll., 1943-44; B.S., S.D. State Coll., 1948; student Ia. State Coll., 1948-49; Ph.D., U. Cal., Berkeley, 1951; m. Vaudis Mary Andrus, Mar. 29, 1946; children—Elin Kay, Harry Charles. Asst. prof. N.Y. Agrl. Expt. Sta. Cornell U., 1951-53; asso. prof. Ore. State U., 1954-60, prof., 1960—; vis. scientist div. entomology Australian Commonwealth Sci. Indsl. Research Orgn., Canberra, Australia, 1960-61; agt. USDA, Corvallis, Ore., 1955-67; mem. editorial bd. Annals Entomol. Soc. Am., 1967—. Charles Atwood Kofoid Eugenics fellow U. Cal., 1949-50, Univ. fellow, 1950-51; Guggenheim fellow, 1960-61. Mem. Entomol. Soc. Am. (chmn. sect. C, 1966), Am. Phytopathol. Soc., A.A.A.S. Contbr. to Methods in Virology, Vol. I, 1967. Contbr. numerous articles to sci. jours. Studies on ecology of plant viruses and a formalization of concepts in this field; work on environmental and plant physiol. regulation of devel. of virus transmitting insects and mites; identification and transmission of viruses in legumes and fruit trees. Home: 12 Edgewood Way, Corvallis, Ore. 97330.*

SWENSON, Melvin John, Am. physiologist; b. Concordia, Kan., Jan. 14, 1917; s. John S. and Ida G. (Anderson) S.; D.V.M., Kan. State U., 1943; M.S., Ia. State U., 1947, Ph.D., 1950; m. Mildred Lucile Rockey, June 26, 1947; children—Myron, Pamela, Brita. Instr. La. State U., Baton Rouge, 1943; asst. prof. Ia. State U., Ames, 1949-50, prof., head dept. physiology and pharmacology Coll. Vet. Medicine, 1957—; faculty Kan. State U., 1950-56, asso. prof. physiology, 1953-56; prof. Colo. State U., 1956-57. Physiology rep. Nat. Bd. Vet. Med. Examiners, 1955—. Research fellow Am. Vet. Med. Assn., 1946-49. Fellow A.A.A.S.; mem. Am. Vet. Med. Assn. (research council 1951-57), A.A.A.S., Research Workers in Animal Diseases, Am. Soc. Vet. Physiology and Pharmacology, Ia. Vet. Med. Assn., Am. Assn. Vet. Nutrition, Soc. for Exptl. Biology and Medicine, Am. Physiol. Soc., Sigma Xi, Phi Kappa Phi, others. Research and publs on anemia in farm animals following birth, nutrition of sow during pregnancy; studies on injectable irondextran during pregnancy; studies on injectable irondextran in prevention of anemia in newborn swine. Home: 306 Westwood Dr., Ames, Ia. 50010.*

SWENSON, Wendell Monson, Am. psychologist; b. Marinette, Wis., Dec. 6, 1920; s. Emil and Sadie (Monson) S.; B.A., Gustavus Adolphus Coll., 1942; M.A., U. Minn., 1950, Ph.D., 1958; m. Constance Elizabeth Norman, Apr. 4, 1943; children—Wendell Norman, David Erik, Stephen Scott. Instr., dir. vets. affairs Gustavus Adolphus Coll., St. Peter, Minns., 1946-49; chief clin. psychologist St. Peter State Hosp., 1949-56; asso. prof., chmn. dept. psychology Gustavus Adolphus Coll., 1957-59; clin. psychologist Mayo Clinic, Rochester, 1959; faculty Mayo Grad. Sch. Medicine, U. Minn., 1959—, asso. prof. clin. psychology, 1965—; cons., clin. psychologist, Mankato, Minn., 1957-59. Mem. Am. Psychol. Assn., Gerontological Soc., Sigma Xi. Contbg. Author: Death and Identity, 1965; also articles. Research on psychol. changes in aging, perceptual changes in aging, psychol. reactions to phys. changes in aging, automated personality analysis. Home: 2067 Lenwood Dr., Rochester, Minn. 55902.*

SWETS, John Arthur, Am. psychologist; b. Grand Rapids, Mich., June 19, 1928; s. John A. and Sara (Heyns) S.; B.A., U. Mich., 1950, M.A., 1953, Ph.D., 1954; m. Maxine Ruth Crawford, July 16, 1949; children—Stephen A., Joel B. Instr. psychology U. Mich., Ann Arbor, 1954-56; asst. prof. Mass. Inst. Tech., Cambridge, 1956-60, asso. prof., 1960-63; sr. sci. Bolt Beranek and Newman, Inc., Cambridge, 1963-65, v.p., 1965—. Mem. Nat. Acad. Sci.-NRC Com. on Hearing and Acoustics, 1959-62. Fellow Acoustical Soc. Am. (chmn. com.

on psychol. and physiol. acoustics), A.A.A.S.; mem. Am. Psychol. Assn., Psychonomics Soc. Author: Signal Detection and Rocognition by Human Observers, 1964; (with D.M. Green) Signal Detection Theory and Psychonomics, 1966; also articles. Theoretical and exptl. investigation of sensory and decision processes in visual and auditory discrimination, measurement of effectiveness of information-retrieval systems, use of digital computers in psychol. research and instruction. Home: 8 Blueberry Lane, Lexington, Mass. 02173. Office: 50 Moulton St., Cambridge, Mass. 02138.*

SWIFT, Edgar James, Am. psychologist; b. Ravenna, O., July 24, 1860; s. Charles Edgar and Emily (Folger) S.; grad. Phillips Exeter Acad., 1882; A.B., Amherst, 1886; postgrad. univs. Leipzig and Berlin, 1889-92; Ph.D., Clark U., 1903; m. Claire Martha Coburn, Dec. 22, 1906. Sci. tchr. Lake Forest (Ill.) Acad., 1887-89; tchr. psychology Stevens Point Normal Sch., 1895-1900; prof. psychology and edn. Washington U., 1903-25, head dept. of psychol. summer session Peabody Coll. for Tchrs., Nashville, 1914, U. Wyo., 1915, U. Chgo., 1916, Kan. Agrl. Coll., 1917, U. of Calif., 1921; spl. lecturer applied psychology, Post Grad. Sch., U. S. Naval Acad., 1920-27, Naval War Coll., 1921-24. Author: Mind in the Making, 1908; Youth and the Race, 1912; Learning by Doing, 1914; Psychology and the Day's Work, 1918; Business Power Through Psychology, 1925; Psychology of Youth, 1927; How to Influence Men, 1927; The Psychology of Childhood, 1930; The Jungle of the Mind, 1931; also articles in sci. jours. Died Aug. 30, 1932.

SWIFT, Ernest Haywood, Am. analytical chemist; b. Chase City, Va., July 2, 1897; s. John Wesley and Anna (Williams) S.; student Randolph-Macon Coll., 1914-17, LL.D., 1960; B.S., U. Va., 1918; M.S., Cal. Inst. Tech., 1920, Ph.D., 1924; m. Elizabeth Flintoft Allen, Sept. 1, 1921; 1 dau., Mary Elizabeth (Mrs. Mortimer Kline). Faculty, Cal. Inst. Tech., Pasadena, 1919—, prof. chemistry, 1943—, chmn. div. chemistry and chem. engring., 1958-63, chmn. faculty, 1963-65. Mem. adv. panel Sci. Center, U. Va., 1965-67. Recipient Coll. Chem. Tchrs. award Mfg. Chemists Assn., 1963. Guggenheim fellow, 1957-58. Mem. Am. Chem. Soc. (Fisher award in analytical chemistry 55, Tolman award So. Cal. sect. 1962), Am. Acad. Arts and Scis., Sigma Xi. Author: Systematic Chemical Analysis, 1939; (with A.A. Noyes) Qualitative Chemical Analysis, 1942; Introductory Quantitative Analysis, 1950; (with W.P. Schaefer) Qualitative Elemental Analysis, 1962; also numerous articles. Research on rare element analysis, methods quantitative analysis, formal potentials of half-cells, distbn. metals between aqueous and non-aqueous solutions, coulometric and amperometric methods analysis, precipitation sulfides by thioacetamide. Home: 572 La Paz Dr., San Marino, Cal. 91108.*

SWIFT, Homer Fordyce, Am. physician; b. Paines Hollow, N.Y., May 5, 1881; s. Charles Fayette and Nancy Maria (Fordyce) S.; student Adrian Coll., 1898-1900; Ph.B., Western Res., 1902, postgrad., med. dept., 1902-04; M.D., U. and Bellevue Hosp. Med. Coll., 1906; D.Sc., N.Y. U., 1931; m. Emma Fordyce MacRae, Apr. 24, 1922; 1 step-dau., Alice MacRae (Mrs. Lester Kissel). Intern Presbyn. Hosp., N.Y., 1906-08; asst. in pathology and dermatology U. and Bellevue Hosp. Med. Coll., 1908-10; asst. physician, Rockefeller Hosp., 1910-12, physician, 1912-14; asso. prof. medicine Columbia, 1914-17, Cornell Med. Coll., 1917-19; asso. mem. Rockefeller Inst., 1919-22, mem., 1922-46, emeritus, 1946—; physician Hosp. of Rockefeller Inst. Med. Research, 1942-46, emeritus, 1946—. Kober lectr. Georgetown U. Med. Sch., 1949; spl. investigator OSRD study of streptococci 1942-45. Chmn. gen. adv. com. for cardiac program N.Y. State Dept. Health, 1941—; chmn. Am. Council Rheumatic Fever, 1945-46. Fellow A.A.A.S.; mem. A.M.A., N.Y. State Med. Soc., Assn. Am. Physicians, Am. Soc. Clin. Investigation (pres. 1928), N.Y. Acad. Medicine, Soc. Am. Bacteriologists, Am. Soc. Immunology, Harvey Soc. (pres. 1925-26). Collaborator: Trench Fever (report of commn. A.R.C. Research Com.), 1918. Contbr. to Forchheimer's Therapeusis of Internal Diseases, Practical Treatment (Musser and Kelly), Nelson's Loose-Leaf Medicine, Oxford Loose-Leaf Medicine, Text-Book of Medicine (Cecil); Bacterial and Mycotic Infections of Men (Dubos); also numerous articles in med. jours. Research on treatment of syphilis of central nervous system, also on rheumatic fever, streptococcus infections, trench fever; devised (with Arthur W. M. Ellis) Swift-Ellis treatment of gen. paresis or cerebrospinal syphilis by intradural injection of patient's blood serum taken after injection of arsphenamine, circa 1912. Died Sept. 24, 1953.

SWIFT, Jonathan Dean, Am. mathematician; b. Portland, Ore., Nov. 12, 1918; s. Mathew Dean and Marie (Serum) S.; B.A., U. Cal., Berkeley, 1939; Ph.D., Cal. Inst. Tech., 1947; m. Rosemary Moore, Mar. 15, 1947 (div. Sept. 1967). Faculty, U. Cal., Los Angeles, 1947—, prof. math., 1962—; mem. Inst. for Advanced Study, Princeton, N.J., 1954-55; mathematician Inst. for Def. Analyses, Princeton, 1961-62. Mem. Am. Math. Soc. (mem. council 1961—), Math. Assn. Am., Assn. for Symbolic Logic, A.A.A.S. Author (with L.J. Paige) Elements of Linear Algebra, 1961; also articles. Studies leading to solution of various problems relating to the structure of finite fields in number theory and algebra; shared in proof of uniqueness of plane of order 8, uniqueness of certain types of planes of order 9, other constructions of combinatorial geometry and projective geometry. Home: 1730 Glendon Av., Los Angeles 90024.*

SWIFT, Lewis, Am. astronomer; b. Clarkson, N.Y., Feb. 29, 1820; s. Gen. Lewis and Anna (Forbe) S.; ed. Clarkson Acad.; Ph.D. (hon.), U. Rochester, 1880; twice married. About 1854 took up study of astronomy, and made a 3-in. refractor; bought a 4 1/2-in. refractor with which he discovered comets; became dir. Warner Obs., Rochester, N.Y., 1882, subsequently of Lowe Obs. Recipient medals and prizes from leading Am. and fgn. socs. Fellow Royal Astron. Soc. Eng. Author: Simple Lessons in Astronomy, 1888. Discovered (with 16-refractor presented by people of Rochester) 900 nebulae there and over 300 at Echo Mountain, Cal., in both places a dozen comets, during the total solar eclipse, at Denver, 1878. discovered two intra-Mercurial planets; observed total eclipses of sun, 1869, 89. Died Jan. 5, 1913.

SWINARSKI, Antoni, Polish chemist; b. Torun, Poland, Dec. 20, 1910; s. Emil and Maria (Radonska) S.; ed. Institut de Chimie, U. Toulouse (France), 1933; Chem. Engr., U. Poznan (Poland), 1938, D.-Chemistry, 1948; m. Wanda Mieczkowska, Sept. 6, 1938; children—Maria (Mrs. Ryszard Grala), Wojciech, Jan, Magdalena. Engr., chem. works Poznanskie Zaklady Nawozów Fosforowych, Lubon, Poland, 1934-50; head dept. inorganic chemistry Copernicus U., Torun, 1951—, rector, 1962-65. Decorated Gold Order of Merit, Officer Cross of Polonia Restituta, Officer Cross of Légion d'Honneur. Mem. Societé Chimique de France. Author: Technology of Sulphuric Acid, 1951; also numerous articles. Intensified sulphuric acid prodn. process; studied effect of ligands on redox phenomena in coordination chemistry; application of activated carbon and natural iron oxides in water and gas purification. Home: 30/32 Sienkiewicza, Torun, Poland.*

SWINDLE, Percy Ford, Am. physiologist; b. Newtonia, Mo., Dec. 26, 1889; s. George and Mary (Ford) S.; A.B., B.S., U. Mo., 1911, A.M., 1912; Ph.D., U. Berlin, Germany, 1915; m. Ethel Wolcott, Aug. 14, 1917, (dec. Aug. 1942); children—Jeanne (Mrs. William A. Crawford), Roger W.; m. 2d Victoria Jakubiak, Dec. 26, 1955. Instr. exptl. psychology Ohio State U., 1916; instr. physiology, pharmacology Tufts Med. Sch., 1917-18; asst. prof. U. Mo., 1919-20; with Marquette U. Sch. Medicine, 1920—, prof., chmn. dept. Physiology, pharmacology, 1922-27, prof., chmn. dept. physiology, 1927-53, prof. physiology, 1953-57, prof. emeritus, 1957—. Mem. A.A.A.S., Am. Soc. Naturalists, Am. Soc. Mammalogists, Am. Soc. Eugenicists, Soc. Exptl. Biology and Medicine, Sigma Xi. Author: Quantum Reactions and Associations, 1922. Research, publs. on body fluid exchange mechanisms, vascular patterns of the body, pathology of intravascular red blood cell agglutination. Home: 2808 N. 40th St., Milw. 53210.*

SWINESHEAD, Richard, English natural philosopher; b. Glastonbury, Eng.; flourished 1337-48; known as the calculator; ed. Merton Coll., Oxford; became mem. Cistercian order, Swineshead, Eng. Author: Calculationes. Research on fundamental problems of mechanics and physics; introduced new abstractions; studied intention and remission of qualities, rarefraction, density, effect of speed of change on qualities, nature of force, resistance and reaction; anticipated graphical representation of functions; placed by Geralamo Cardono among 12 greatest thinkers of all time.

SWINGLE, Walter T(ennyson), Am. botanist, agriculturist; b. Canaan, Pa., Jan. 8, 1871; s. John Fletcher and Mary (Astley) S.; B.Sc., Kan. State Agrl. Coll., 1890, D.Sc., 1922; M.Sc., 1896; postgrad. U. Bonn, 1895-96; U. Leipzig, 1898; m. Lucie Romstaedt, June 8, 1901 (dec. 1910); m. 2d, Maude Kellerman, Oct. 2, 1915; children—John William, Stella (Mrs. Stanley F. Reed), Frank Anthony, Mary (Mrs. Francis L. Albert, Junior). Asst. botanist Kan. Agrl. Exptl. Sta., 1888; U. S. Dept. Agr., Bur. Plant Industry, 1891-41; collaborator since 1941; cons. tropical botany U. Miami, Coral Gables, Fla., since 1941; organizing large scale preparation of serial microtome sections herbarium and fresh material economic plants; investigated for Dept. Agr. the agr. and botany of France, Algeria, Morocco, Italy, Spain, Greece, The Balkans, Asia Minor, China, Japan, Philippines, Mexico. Recipient Meyer medal, 1948; Barbour medal, 1950. Fellow, life mem. A.A.A.S.; mem. Washington Acad. Scis. (original), Acad. Natural Sci. Phila., Am. Bot. Soc., Nat. Geog. Soc. (hon. life), Washington Acad. Medicine (original), others. Author: Our Agricultural Debt to China, 1945; (paper) The Botany of Citrus and Its Wild Relatives, (complete synopsis of orange sub-family), 1943; other papers. Introduced fig insect into Cal., 1899, thereby rendering possible culture of Smyrna type figs; 1st successful shipment standard varieties of date palms from Algeria into Cal. and Ariz., 1900, in charge establishment comml. culture of date palm, 1900-34; prop. exptl. date garden of Indio, Cal.; helped establish Egyptian cotton in Arizona; originated by hybridization in Fla., citranges, limequats, tangelos and other new citrus fruits; discovered neophyosis (rejuvenescence of old citrus varieties from nucellar buds seedlings, thereby eliminating all virus infections); 1st proved existence of centrosomes in plants; originated name and theory of metaxenia for direct effect of pollen on dates, 1928; introduced many new crop plants alkaloid-yielding species of Ephedra and high-yielding strains of tropical tung (abrasin); visited to advise regarding culture of cinchona, tung (abrasin), rubber and other tropical crops for export to U. S., 1939; conducted search for new rootstocks immune to Tristeza, fatal disease of Citrus; studied plants of China, supervised extensive translations from Chinese lit. on them; assisted the Librarian of Congress in building up largest collection of Chinese books outside Orient; wrote annual reports on same, 1915-35. Died Jan. 19, 1952.

SWINGLE, Wilbur Willis, Am. physiologist, zoologist; b. Warrensburg, Mo., Jan. 11, 1891; s. Jacob and Emma Lucy S.; A.B., A.M., U. Kan., 1916; Ph.D., Princeton, 1920; m. Emily Gerken, Nov. 2, 1916 (div.); m. 2d, Alice Sullivan, Apr. 1929; children—Stephen Grey, Philip Colin. Fellow in zoölogy U. Kan., 1916, instr., 1917-18; instr. zoölogy, Yale, 1920, asst. prof., 1921-26; prof., head dept. zoölogy State U. Ia., 1926-29; prof. biology Princeton, 1929—, Edwin Grant Conklin prof. biology, 1933-56, Henry Fairfield Osborn research prof., 1956-59, sr. research asso. biology, 1959-63, ret. 1959; head sect. adrenal physiology Bur. Research in Neurology and Psychiatry, N.J. Neuropsychiat. Inst., Princeton, N.J., 1963—. Recipient medal Endocrine Soc., 1959. Fellow N.Y. Acad. Scis., mem. Am. Soc. Zoölogists, Am. Assn. Anatomists, Am. Soc. Physiologists, Soc. Exptl. Biology and Medicine, Assn. for Study Internal Secretions (council, 1931-32), Am. Soc. Naturalists, Sigma Xi, Sigma Alpha Epsilon, Phi Chi. Research and numerous papers in relation of thyroid and pituitary glands to amphibian metamorphosis; extirpation and transplantation of parathyroid glands; relief of parathyroid tetany by various chem. compounds; preparation of active extracts of adrenal cortex; functional significance of adrenal cortex, cause of death from adrenal insufficiency. Home: 200 Jefferson Rd. Office: P.O. Box 1000, Princeton, N.J.*

SWINGS, Pol F., Belgian astrophysicist; b. Ransart, Belgium, Sept. 24, 1906; s. Jean and Marie Antoinette (Bils) S.; Ph.D., U. Liège (Belgium), 1927, Spl. D.Sc., 1931; Hon. Dr., U. Aix-Marseille (France), 1958, U. Bordeaux (France), 1963; m. Christiane Borgerhoff, July 30, 1932; 1 son, Jean Pierre. Asst. U. Liège, 1927-32, prof. astrophysics, 1932—; prof. U. Chgo., 1939-42. Named Comdr., Order Leopold (Belgium); recipient Francqui prize, 1947, Decennal prize in physics, 1958. Mem. Internat. Astron. Union (pres. 1964-67), Royal Acad. (Belgium), Paris Acad., Am. Acad. Arts and Scis., Am. Philos. Soc., Nat. Acad. Scis. (U. S.), Internat. Acad. Astronautics. Research and numerous publs. on cometary physics, stellar spectroscopy, shell stars, nebulae, lab. spectroscopy, molecular bands. Home: 23 Av. Léon Souguenet, Esneux, Belgium. Office: Institut D'Astrophysique, Sclessin, Belgium.*

SWINTON, William Elgin, palaeontologist; b. Kirkcaldy, Scotland, Sept. 30, 1900; s. William Wilson and Rachel (Cargill) S.; student Trinity Coll., Glenalmond, 1916-17; B.Sc., U. Glasgow, 1922, Ph.D., 1930. Demonstrator geology dept. Glasgow U., Scotland, 1922-24; with Brit. Mus., Dept. Palaeontology, London, Eng., 1924-61, prin. sci. officer, 1950-61, tng. officer, 1954-61; head life scis. div. Royal Ont. Mus., Toronto, Ont., Can., 1961-63, dir., 1963-66; prof. dept. zoology U. Toronto, 1962-67, prof. dept. geol. scis., 1963-66, Centennial prof., 1966—; trustee Centennial Centre for Sci. and Tech., Ont., 1964—. Strang-Steel scholar, U. Glasgow, 1922; recipient Geol. Soc. London Daniel Pidgeon award, 1928. Mem. Geol. Soc. Belgium, Royal Mus. for Central Africa, Museums Assn. (pres. U.K. 1958-60), Canadian (mem. council), Am. (mem. council 1964-67) museums assns. Author: Corridor of Life, 1948; Fossil Amphibians and Reptiles, 1954; Fossil Birds, 1958; Dinosaurs, 1962; Giants, Past and Present, 1966; also numerous articles. Research on the dinosaurs, the recreation of their environment and reactions to it; their diseases and general mode of life; fossil reptiles and birds. Home: 276 St. George St., Toronto 5, Ont., Can. Office: Massey Coll., U. Toronto, Toronto, Ont., Can.*

SWINTOSKY, Joseph Vincent, Am. research pharmacist; b. Kewaunee, Wis., Dec. 14, 1921; s. Joseph Frank and Mary (Paul) S.; B.S., U. Wis., 1942, Ph.D., 1948; m. Dorothy Ann Zevnik, June 13, 1953; children—Joseph Vincent, Denise Claire, Michael David, Steven John, John Anton, Paul Stuart. Instr. U. Wis. Sch. Pharmacy, 1946-49, asst. prof., 1949-53; sr. scientist Smith Kline & French Labs., Phila., 1953-54, unit head tabletting, 1954-57, group leader pharm. tech., 1957-59, prin. research pharmacist, 1959-64, head pharm. research sect., 1964—, dir. drug absorption program, 1957—; adj. prof. pharmacy Temple U., 1958—; vis. Scientist Am. Assn. Colls. Pharmacy, 1963—. Recipient Ebert award, 1958, Am. Pharm. Assn. Found. award, 1964. Fellow A.A.A.S. (vice chmn. 1960, chmn. pharmacy sect. 1960); mem. Am. Chem. Soc., Am. Pharm. Assn. (v.p. Phila. br. 1960-61, chmn. sci. sect. 1964-65), Acad. Pharm. (pres. elect 1965-66), Rho Chi (v.p. 1965). Editorial adv. staff Jour. Pharm. Scis., 1964-

—. Author numerous sci. research papers. Inventions include new drug compounds, layered tablets and coating techniques for prolonging therapeutic action of drugs; application of chemistry, math., and enzymology to problems of making new drugs, predicting and prolonging drug product shelf life, absorption and excretion of drugs, optimum frequency for drug admistrn. Home: Deep Creek Rd., Perkiomenville, Pa. 18074. Office: 1500 Spring Garden St., Phila. 19101.*

SWINYARD, Ewart Ainslie, Am. pharmacologist; b. Logan, Utah, Jan. 3, 1909; s. William and Bertha (Halverson) S.; B.S., Utah State U., 1932; B.S. in Pharmacy, Ida. State Coll., 1936; M.S., U. Minn., 1941; Ph.D. (univ. fellow, Winthrop Research fellow) U. Utah, 1947; m. Grace Chatterton Parkinson, June 19, 1934; children—Emma Lou (Mrs. Rodger David Kasteler), Richard Ewart. Faculty Ida. State Coll., 1936-47; faculty pharmacology U. Utah Coll. Medicine, Salt Lake City, 1945—, prof. pharmacology, 1967—, Distinguished Research prof., 1968-69. dir. research Coll. Pharmacy, 1947—, dir. Sch. Alcohol Studies, 1962—, head, dept. biopharm. scis., 1965—. Rennebohn lectr. U. Wis., 1960; George Beecher Kaufmann lectr. Ohio State U., 1963; vis. scientist Utah Acad. Sci., 1962—, Am. Assn. Colls. Pharmacy, 1962—. Recipient Research Achievement award in Pharmacodynamics Am. Pharm. Assn. Found., 1965. Faculty fellow Am. Coll. Apothecaries; mem. Am. Soc. for Pharmacology and Exptl. Therapeutics (editorial bd. 1965—), Am. Pharm. Assn. (chmn. physiol. testing com. 1955,57,59), Utah Pharm. Assn. (hon. life), Utah Acad. Sci., Arts, and Letters, Am. Coll. Neuropsychopharmacology, N.Y. Acad. Scis., Am. Assn. Colls. Pharmacy (sec.-treas. dist. 8, 1953—), Sigma Xi, Phi Kappa Phi, Phi Delta Chi, Rho Chi. Editor: Remington's Pharmaceutical Sciences, 11th-14th edits. Mem. editorial adv. bd. Jour. Pharm. Scis., 1953-60. Contbr. to Pharmacological Basis of Therapeutics, 1965—. Publs., contbns. to physiology of convulsive disorders; effects of stress and emotion on excitability of the brain; exptl. seizure patterns and modification by drugs. Home: 2055 Michigan Av., Salt Lake City 84108.*

SWITHINBANK, Charles Winthrop Molesworth, Brit. glaciologist; b. Pegu, Burma, Nov. 17, 1926; s. Bernard Winthrop and Dorothea (Molesworth) S.; B.A., Pembroke Coll., U. Oxford, 1949, M.A., 1953, Ph.D., 1955; m. Mary Stewart Fellows, Aug. 17, 1960; children—Anne Stewart, Carol, Kelvin. Asst. glaciologist Norwegian-Brit.-Swedish Antarctic Expedition, 1949-55; research fellow Scott Polar Research Inst., Cambridge, 1955-59; research asso. U. Mich., Ann Arbor, 1959-63; exchange sci. 9th Soviet Antarctic Expedition, 1963-65; glaciologist Scott Polar Research Inst., Cambridge, 1965—. Leader, U. Mich. Ross Ice Shelf Studies, U.S. Antarctic Research Program, 1959-62. Recipient Norwegian Medal of Honor, 1952; Brit. Polar medal, 1956; Swedish Retzius medal, 1966. Mem. A.A.A.S., Am. Geophys. Union, Am. Geog. Soc., Royal Geog. Soc., Glaciological Soc., Swedish Soc. Anthropology and Geography, Arctic Inst. of N.Am., Sigma Xi. Author: Norwegian-British-Swedish Antarctic Expedition 1949-52, Scientific Results, vol. 3, 1957; Ice Atlas of Arctic Canada, 1960; also articles. Research on morphology, regime, movement of polar ice sheets; measurements of ice movement on 7 of largest Antarctic valley glaciers; devel. techniques of oversnow travel; measurements of seasonal flow rates in Antarctic ice sheet; sea ice studies showing secular variability of ice distbrn. in arctic Can. Home: 7 Home End, Fulbourn, Cambridge, Eng. Office: Scott Polar Research Inst., Lensfield Rd., Cambridge, Eng.*

SWYER, Gerald Isaac MacDonald, English clin. endocrinologist; b. Bognor, Sussex, Eng., Nov. 17, 1917; s. Nathan and Nettie (Cohen) S.; B.A., U. Oxford, 1939, M.A., B.M., B.Ch., D.Phil., 1943, D.M., 1948; M.D., U. Cal. at Berkeley, 1943; m. Lynda Irene Nash, July 24, 1945; children—Rupert, Alexandra. Demonstrator anatomy Oxford U., 1940-41; house physician Radcliffe Infirmary, 1944; med. officer Southend Gen. Hosp., 1944; sci. staff Nat. Inst. for Med. Research, 1946-47; endocrinologist U. Coll. Hosp. Med. Sch., London, 1947—; cons. endocrinologist U. Coll. Hosp., 1951—. Mem. med. adv. council Family Planning Assn., 1958—; temporary sci. adviser WHO, 1965. Fellow Royal Coll. Physicians, Royal Soc. Medicine; mem. Soc. for Endocrinology, Soc. for Study Fertility, Internat. Fertility Assn. Author: Reproduction and Sex, 1954; (with S. C. Simpson, A. S., Mason) Major Endocrine Disorders, 1959; also numerous articles. Research on reproductive physiology, especially endocrine function, diagnosis and treatment of human infertility, control of fertility. Home: 34, Chester Close N., London, N.W. 1. Office: Univ. Coll. Hosp., London, W.C.1, Eng.*

SY, Michel, French organic chemist; b. Les-Ponts-de-Cé, France, Sept. 26, 1930; s. Emile and Geneviève (Blotière) S.; Docteur-ès Sciences, U. Paris, 1955; m. Pierrette Chantome, Apr. 3, 1954; children—Sophie, Valérie, Aurélie. With Centre National de la Recherche Scientifique, 1955-58, in charge research, 1957-58; became dir. lab. organic synthesis ecole Nationale Vétérinaire, Alfort, France, 1958, research master, 1962. Mem. French, Netherlands, Am. chem. socs., Chem. Soc. London, French Soc. Therapeutic Chemistry. Research, publs. on synthesis of organic compounds for biol. tests. Home: 17 rue Coysevox, Paris 18, France. Office: Laboratoire de Synthèse Organique, Ecole Nationale Vétérinaire, Maisons-Alfort (Val-de-Marne), France.*

SYDENHAM, Thomas, English physician; b. Wynford Eagle, Eng., Sept. 10, 1624; s. William and Mary (Jeffrey) S.; ed. Magdalen Coll., Oxford, B.M., 1648; M.D., Cambridge U., 1676; postgrad., Montpellier, France, 1660-61; m. Mary Gee; 3 sons, including William. Mem. All Souls Coll., 1648-55; began practice medicine, London, 1655; mem. parliamentary forces in English Civil War. Author: Methodus curandi febris, 1666; Tractatus de podagra et hydrope; Observationes medicae, 1676. Known as the English Hippocrates; friend of Robert Boyle and John Locke; emphasized clin. observation rather than theory; introduced laudanum or tincture of opium into medicine, also cooling system in treatment of smallpox, Peruvian bark in treatment of agues; gave classic descriptions of gout, fevers, hysteria, venereal disease; described scarlet fever and distinguished it from measles, 1675; credited with introducing term whooping cough, 1675; classified scarletina and chorea; made extensive studies on malaria; studied relation of seasons, years and ages to epidemics. Died London, Dec. 29, 1689.

SYLOW, Ludwig, Norwegian mathematician; b. 1832; prof., Christiania (now Oslo); pub. edit. of Abel's work, 1881; studied solvable groups; originated Sylow subgroups; developed fundamental theorem defining groups in accepted tech. sense, 1872. Died 1918.

SYLVATICUS, Matthaeus, Italian physician; b. Mantua, or Salerno, Italy; tchr., Salerno, by 1297; practiced medicine under patronage of Robert, king of Sicily, 1309-43. Author: Liber pandectarum medicine omnia medicine simplicia contiens quem ex omnibus antiquorum libris aggregavit (reference work on diseases and their cure, generally in alphabetical order, with plants named in Greek, Arabic and Latin, followed by descriptions of simples and their properties), 1317. Died circa 1342.

SYLVESTER, Bernard, astronomer. Author: De mundi universitate (also known as Cosmographia, based on Plato's Timaeus and neoplatonic writings, pantheistic in character), 1145-53; Experimentarius (adaptation of Arabic treatsie on geomoncy); Mathematicus. Emphasized influence of soil on plant and animal life.

SYLVESTER, Edward Sanford, Am. entomologist; b. N.Y.C., Feb. 29, 1920; s. Charles W. and Dolores (Worrell) Sylvester; B.S., Colo. State U., 1943; Ph.D., U. Cal. at Berkeley, 1947; m. Marian R. Uhl, Oct. 16, 1942; children—Kathryn L., Stephen S. Faculty, staff U. Cal. at Berkeley, 1947—, prof., entomologist, 1960—. Mem. Entomol. Soc. Am., Am. Phytopath. Soc., Am. Inst. Biol. Scis., Sigma Xi. Research, numerous publs. on analysis and measurement biotic and environmental factors responsible for transmission and spread plant viruses by insect vectors. Home: 366 Ocean View Av., Kensington, Cal. 94707.*

SYLVESTER, James Joseph, English mathematician; b. London, Eng., Sept. 3, 1814; s. Abraham Joseph; grad. St. John's Coll., Cambridge (Eng.) U., 1831 (barred from degree by Jewish faith, but awarded B.A. and M.A. after Test Act of 1872); B.A., M.A., U. Dublin (Ireland), 1841. Prof. natural philosophy U. Coll., London, 1837-41; prof. math. U. Va., 1841-42; returned to London, 1844, engaged in actuarial work, 1844-56; called to bar, Eng., 1850; prof. math. Royal Mil. Acad., Woolwich, Eng., 1855-70, Johns Hopkins U., Balt., 1876-83; Savilian prof. geometry Oxford (Eng.) U., 1883-97. Fellow Royal Soc., 1839; mem. French Acad. Scis. Author: The Collected Mathematical Papers of James Joseph Sylvester, 4 vols., 1904-12 (edited by H. F. Baker); The Laws of Verse, 1870; translations of Horace, German poets. First editor Am. Jour. Math., 1878-84. Leader in advancing math. research in U. S.; contbd. to algebra, number theory, theory of equations, pure and analytic geometry; founder (with Cayley) theory of algebraic invariants (essential to theory of relativity). Died Oxford, Mar. 15, 1897.

SYLVESTER II, see Gerbert of Aurillac.

SYLVESTER-BRADLEY, Peter Colley, Eng. geologist; b. Pinhoe, Devon, Eng., May 21, 1913; s. Charles Reginald and Juliet (Hawker) S.-B.; B.Sc., U. Reading, 1934; m. Joan Eveleen Mary Campbell, Aug. 10, 1945; children—Rowan, Roger, Rosemary, Benjamin. Lectr. in charge dept. geology Seale Hayne Agrl. Coll., Newton Abbot, Devon, 1936-39; faculty U. Sheffield (Eng.), 1946-59, sr. lectr. paleontology, 1956-59; Rose Morgan prof. U. Kan., 1955-56; F.W. Bennett prof. geology U. Leicester (Eng.), 1959—. Internat. commr. Zool. Nomenclature, 1953-58. Recipient Wollaston fund Geol. Soc., 1953, E.J. Garwood Fund. 1957. Mem. Yorkshire Geol. Soc. (past editor), Paleontol. Assn. (past treas.), Systematics Assn., (pres. 1964-67). Editor: The Species Concept in Palaeontology, 1956. Research and publs. in palaeontology, micro-palaeontology, stratigraphy, taxonomy, evolution and origin of life. Home: Noon's Close, Stoughton, Leicester LE2 2FJ. Office: The University, Leicester LE1, Eng.*

SYLVIUS, see Dubois, Jacques.

SYLVIUS DE LA BöE, Franciscus, chemist, physiologist; b. Hanau, Prussia, 1614; ed. various Dutch and German univs.; med. degree from Basel, Switzerland, 1637; postgrad., Paris, France, Holland. Practiced medicine in Hanau, 1637-39; lectr. on botany and anatomy, Leyden, Netherlands; practiced medicine at Amsterdam, Netherlands, from 1641; became prof. medicine, Leyden, 1658; built 1st univ. chem. lab. at Leyden. Author: Casus medicinales . . . , 1659; Disputationum medicarum . . . , 1663; Praxeos medicae idea nova, 1671; Methodus medendi, 1679; his collected Opera medica were printed 1st in 1671 and often thereafter until 1771. Leading advocate of Iatrochemical sch.; borrowed many ideas from Paracelsus, Descartes and Van Helmont; but discarded the archei and ferments of Van Helmont; rejected 4-humor theory of health; applied chemistry to medicine, in studies of body juices, pancreatic juice, saliva, bile; one of 1st to appreciate idea of chem. affinity; introduced ward instrn. into med. teaching; excelled as anatomist; disseminated Harvey's ideas in Leyden; discovered fissure in brain which separates anterior lobe of cerebrum from middle lobe (Aquaeductus Sylvii), 1641; one of 1st to indicate significance of nodules in lungs in pulmonary Tb. Died Leyden, Netherlands, Nov. 14, 1672.

SYME, James S., Scottish surgeon; b. Edinburgh, Nov. 7, 1799; s. John Syme; ed. U. Edinburgh; M.D., Bonn, Germany, 1869; D.C.L., Oxford U.; m. dau. of Robert Willis (dec. 1840); 2 daus., including Mrs. Lister; m. 2d, Jemima Burn, 1841; 1 son. Founder pvt. surg. hosp. where he introduced his system of clin. instrn., Edinburgh, 1829; apptd. Crown prof. clin. surgery Edinburgh U., 1833-48, 49—; named surgeon-in-ordinary to queen in Scotland, 1838; prof. clin. surgery Univ. Coll., London, 1848. Author: On Excision of Diseased Joints, 1831; On Diseases of the Rectum, 1838; On Structure of the Urethra and Fistula in Permeo, 1849; Excision of the Scapula, 1864. Introduced excision of elbow, also amputation of ankle joint (Syme's operation) into Britain; treated stricture of urethra by external division; studied inflammation, also function of periosteum in repair of bone; added to knowledge of diseases of rectum; left wounds open until blood ceased oozing (often attributed to Liston). Died Millbank, Scotland, June 26, 1870.

SYMEONIDIS, Alexander, Greek pathologist; b. Komotini, Greece, Aug. 29, 1909; s. Constantine and Formozi (Adamantini) S.; ed. univs Athens, Rome, Berlin; M.D.; m. Dora Tsatsou, Sept. 2, 1959; children—Constantine, George. Dir. cancer research dept. U. Berlin, 1936-38; named asst. prof. pathology U. Athens, 1941; dir. labs. Hellenic Cancer Inst., Evangelismo Hosp., 1939-47; became dir. geographic pathology dept. Nat. Cancer Inst., NIH, 1947; prof. gen. pathology and path. anatomy U. Salonika, 1956—; dir. Salonika Cancer Inst., 1957—. Vice pres. Greek Atomic Energy Commn. Research, publs on pathology of cancer. Home: Odos Emilius Riadis 1. Office: Panepistemion, Thessalonika, Greece.

SYMINGTON, Thomas, Scottish physician; b. Muirkirk, Ayrshire, Apr. 1, 1915; s. James and Margaret (Steven) S.; B.Sc., U. Glasgow, 1936, M.B., Ch.B., 1941, M.D., 1950, F.R.S.E., F.R.I.C., F.R.C.P., F.C.Path.; m. Esther Margaret Forsyth, Mar. 31, 1943; children—Robert, Alan, Esther. Asst. pathologist Glasgow Royal Infirmary, 1944-47; dep. asst. dir. pathology, R.A.M.C., Malaya, 1947-49; sr. lectr. U. Glasgow, 1949-54, St. Mungo-Notman prof. pathology, 1954—; vis. prof. Stanford, 1965-66. Cons. pathologist Western Regional Hosp. Bd.; mem. numerous sci. adv. coms. medicine, pathology, endocrinology; chmn. Testicular Tumour Panel. Mem. Path. Soc. Gt. Britain and Ireland, Soc. Endocrinology. Author: The Human Adrenal Cortex, 1962; also numerous articles. Research in endocrine pathology, especially adrenal gland—site of formation of hormones in human adrenal cortex, diseases of overactivity of human adrenal gland. Home: 1 Tavistock Dr., Glasgow, S.3. Office: U. Dept. Pathology, Royal Infirmary, Glasgow, Scotland.*

SYMINGTON, William, Brit. engr.; b. Leadhills, Oct. 1763; ed. Glasgow, Edinburgh univs. Patentee improved steam engine, 1787, system for paddle-wheel shaft (still used), 1801; developed 1st steam boat for practical use, Charlotte Dundas, 1802. Died Mar. 22, 1831.

SYMMERS, Douglas, Am. pathologist; b. Columbia, S.C., Sept. 17, 1879; s. George and Jessie (McKay) S.; student U. S.C., Univ. Tutorial Coll., London, Eng.; M.D., Jefferson Med. Coll., Phila., 1901. Instr. pathology Bellevue Med. Coll., N.Y., 1907, prof. pathology, 1911-18; instr. pathology, Cornell Med. Coll. 1908-11; asst. pathologist, N.Y. Hosp., 1908-

13; dir. labs. Bellevue and Allied Hosps., 1918-29; gen. dir. labs. Dept. of Hosps., N.Y.C., since 1929; cons. pathologist Englewood and Beekman St. hosps. Fellow N.Y. Acad. Medicine; mem. N.Y. Pathol. Soc. (pres. 2 terms), Assn. Am. Pathologists and Bacteriologists, Internat. Med. Museums. Editor Ziegler's General Pathology (5th Am. edit.). 1921. Contbr. papers on research work. Described giant follicular lymphadenopathy (Brill-Symmers disease) in papers pub. 1927, 38. Died Apr. 19, 1952.

SYMMONDS, Richard Earl, Am. surgeon; b. Greensberg, Mo., Mar. 19, 1922; s. Emmett E. and Gilda (Boone) S.; A.B., Central Coll., Fayette, Mo., 1943; M.D., Duke, 1946; M.S. in Gen. Surgery, U. Minn., 1953; m. Dorothy Jean Wall, Dec. 26, 1944; children—Richard Earl, Teresa Ellen, Jeffrey Boone. Practice medicine specializing in obstetrics, gynecology, and surgery, Rochester, Minn., 1953—; asso. prof. clin obstetrics and gynecology Mayo Grad. Sch. Medicine, U. Minn., Rochester, 1965—. Cons. St. Mary's Hosp., Methodist Hosp., Mayo Clinic (all Rochester). Fellow A.C.S.; mem. A.M.A., Coll. Am. Obstetricians and Gynecologists, Minn. Soc. Obstetricians and Gynecologists Soc. Pelvic Surgeons, Sigma Xi. Home: 13 Skyline Dr., Rochester, Minn. 55901.*

SYMONS, George James, English meteorologist; b. Pimlico, Eng., Aug. 6, 1838; s. Joseph and Georgina (Moon) S.; student Sch. Mines, Jermyn St., London; m. Elizabeth Luke, 1866; 1 child (dec. in infancy). Meteorol. reporter to registrar-gen., 1857-1900. Recipient Telford premium Instn. Civil Engrs., 1876, Albert medal Soc. Arts, 1897. Became fellow Royal Soc., 1878; mem. Royal Meteorol. Soc. (sec. 1878-79, 82-99, became pres., 1880, 1900), Scottish, Australasian meteorol. socs., Royal Bot. Soc. Issued 39 vols. statistics on rainfall, beginning 1860; founder monthly rain circular, 1863, became Monthly Meterorol. Mag., 1866. Author: Rain: How, When, Where, Why it is Measured, 1867; Pocket Altitude Tables, 1876; The Floating Island in Derwentwater, 1889; Merle's M.S. consideraciones temperici pro 7 annis, 1337-44, 1891; Theophrastus on Winds and Weather Signs, 1894; also essays, reports. Died Mar. 10, 1900.

SYNESIOS, neoplatonic philosopher; b. Cyrene; flourished circa 370-414; pupil of Hypatia; bishop of Ptolemais; produced early alchem. writings, also commentary on (pseudo) Democritos. Identity of bishop and alchemist has been questioned since 17th century, and some authorities date the alchemical writings as late as the 9th to 12th centuries.

SYNGE, John Lighton, Irish math. physicist; b. Dublin, Ireland, Mar. 23, 1897; s. Edward and Ellen (Price) S.; B.A., Trinity Coll., Dublin, 1919, M.A., 1922, Sc.D., 1926; m. Elizabeth Allen, 1918; children—Margaret, Cathleen, Isabel. Lectr. math., Trinity Coll., Dublin, 1920; asst. prof. math., Toronto, Ont., Can., 1920-25; fellow, prof. natural philosophy, Trinity Coll., Dublin, 1925-30; prof. applied math., Toronto, 1930-43, vis. lectr. Princeton, 1939; vis. prof., Brown Univ., 1941; ballistics mathematician U. S. Strategic Air Forces in Europe, 1944-45; chmn. dept. math. Ohio State U., 1943-46; head dept. math. Carnegie Inst. Tech., 1946-48; now with Dublin Inst. for Advanced Studies; vis. prof. Mass. Inst. Tech., summer 1947. Fellow Royal Soc. London, 1943, Royal Soc. Can.; mem. Royal Irish Acad. (pres. 1961-64), Am. Math. Soc., Math. Assn. Am., London Math. Soc. Author: Geometrical Optics, 1937; Principles of Mechanics (with B. A. Griffith), 1942. Editor: (with A. W. Conway) Mathematical Papers of Sir W. R. Hamilton, Vo. I, 1931; Tensor Calculus (with A. Schild), 1949; Science, Sense and Nonsense, 1951; Geometrical Mechanics and de Broglie Waves, 1954; Relativity: The Special Theory, 1956; The Relativistic Gas, 1957; Kandelman's Krim, 1957; The Hypercircle in Mathematical Physics, 1957; Relativity: The General Theory, 1960. Research and publs. on devel. of geometrical methods in classical mechanics and relativity; constructed (with others) relativistic models of gravitational fields; invented (with W. Prager) hypercircle for boundary value problems of electrostatics, elasticity and hydrodynamics; generalized Hamilton's ray-wave method with applications to de Broglie waves and water waves. Office: The Dublin Institute for Advanced Studies, 64 Merrion Sq., Dublin, Ireland.*

SYNGE, Richard Laurence Millington, English biochemist; b. Liverpool, Eng., Oct. 28, 1914; s. Laurence Millington and Katharine Charlotte (Swan) S.; student Winchester Coll., 1928-33; B.A., U. Cambridge, 1936, Ph.D., 1941; m. Ann Stephen, 1943; children—Jane, Elizabeth, Thomas Millington, (deceased), Matthew Millington, Patrick Millington, Alexander Millington, Charlotte, Mary. Biochemist Wool Industries Research Assn., Leeds, 1941-43, Lister Inst. Preventive Medicine, London, 1943-48; head protein chemistry Rowett Research Institute, Bucksburn, Aberdeen, 1948-67, on leave with the N.Z. Dept. Agr. at Ruakura Animal Research Sta., Hamilton, 1958-59; biochemist Food Research Inst., Norwich, 1967—. Vice pres. Brit. Peace Com. Received Nobel prize for chemistry (with A. J. P. Martin), 1952, for invention partition chromatography; John Price Wetherill medal Franklin Inst., Phila., 1959. Fellow Royal Soc., 1950, Royal Soc. Edinburgh, Royal Institute of Chemistry; member Biochemical Society, Society of General Microbiology, Nutrition Soc., Assn. Scientific Workers, Société de Chimie biologique, Am. Soc. of Biological Chemists (hon. mem.). Author sci. papers. Mem. editorial -bd. Biochem. Jour., 1949-55. Research on application of methods of phys. chemistry to isolation and analysis of proteins and related substances, with special reference to antibiotic peptides and higher plants. Home: 19 Meadow Rise Rd., Norwich, NOR 97 F. Office: Earlham Lab., Recreation Rd., Norwich, NOR 26 G, Eng.*

SYRKIN, Yakov Kovovich, Russian phys. chemist; b. Dec. 5, 1894; grad. Ivanovo-Vozresensk Poly. Inst., 1919. Prof., Lomonosov Inst. Fine Chem. Tech., 1931—; organizer, sci. chmn. dept. on molecular structure Karpov Physico-Chem. Prof. Ivanovo-Voznesensk Poly. Inst., 1919-25. Recipient Stalin Prize, 1943. Corr. mem. USSR Acad. Sci. Research and publs. in chem. thermodynamics, kinetics of reactions in solutions, Menshukin reactions in solution and in gas molecular structure, chem. bonds; 1st to use dipole moment in studying molecules in USSR. Office: Karpov Physico-Chem. Inst., USSR Acad. Sci., Obukha St., 10, Moscow, USSR.

SZABO, György, Hungarian physician; b. Pancsova, Hungary, Dec. 16, 1919; s. László and Trén (Ormos) S; student Med. Faculty, U. Beograd (Yugoslavia), 1938-41; M.D., U. Szeged (Hungary), 1944; C.Sc. Med., Hungarian Acad. Sci., 1952, D.Sc.Med., 1955; m. Zsuzsa Gyerkes, Nov. 30, 1946; 1 son, György; m. 2d, Zsuzsa Magyar, July 23, 1962. Physician, City Hosp., Keszthely, Hungary, 1944-45, Szt. István Hosp., Budapest, Hungary, 1945-46; staff dept. medicine U. Med. Sch., Budapest, 1946-61, asso. prof., 1958-61; chief research dept. Nat. Inst. Traumatology, Budapest, 1961—. Invited lectr. medicine U. Budapest, 1960—. Mem. Hungarian Physiol. Soc., Korányi Sándor Soc., Internat. Soc. Nephrology, Internat. Soc. for Lymphology. Author: (with I. Rusznyak, M. Foldi) Lymphatics and Lymph Circulation, 1960; (with M. Földi) Die Regulation der renalen Na und Wasserausscheidung, 1961; also numerous articles. Research on role of lymphatics in edema formation, mechanism of insufficiency of lymph circulation, effect of plasma vol. changes on renal function and circulation dynamics, mechanism of renal glucose excretion, steroid and pressor therapy in traumatic shock, mechanism of fat embolism. Home: 31. Rippl-Rónai ucca. Office: 17. Mező Imre ut, Budapest VIII, Hungary.*

SZABOLCS, Istvan, Hungarian soil scientist; b. Turkeve, Hungary, Feb. 23, 1924; s. Arpád and Iren (Lowinger) S.; Ph.D. in Chemistry, Debrecen U., 1948; postgrad. Timiryazev Acad., Moscow, USSR, 1949-53; C.Sc. in Agronomy, U. Moscow, 1953; D.Sc. in Soil Sci., U. Budapest (Hungary), 1959; m. Katalin Darab, July 4, 1955. Asst. prof. Debrecen U., 1947-49; aspirant fellow Timiryazev Acad., 1949-53; dep. dir. Inst. Irrigation and Soil Reclamation, Szarvas, 1953-54; dep. dir. Research Inst. Soil Sci. and Agrl. Chemistry, Hungarian Acad. Scis., Budapest, 1954-59, dir. Research Inst. Soil Sci. and Agrl. Chemistry, 1959—; lectr. prof. soil sci. Agrl. U., Gödöllö, Hungary, 1953-56, Budapest U., 1965—. Recipient Medal for Socialist Constrn., 1961; Tessedik Gold medal Hungarian Agronomic Soc., 1966. Mem. Hungarian Soil Sci. Soc. (chmn. 1966—), Internat. Soc. Soil Sci. (chmn. subcom. salt affected soils 1965—), Hungarian Acad. (mem. com. soil sci. 1953—), Hungarian Isotope Application Com. Author: Soils of Hortobágy, 1954; Effect of Drainage and Irrigation on Soil Formation Processes, 1961; Application of Isotopes, 1959; (with others) Methodbook on Large-Scale Gernetic Soil Mapping, 1966; also numerous articles. Research on salt affected saline and alkali soils including their mapping, physics, chemistry, improvement and utilization; large-scale soil mapping for agrl. purposes; soil survey methods: modern soil classification methods. Home: 36 Szilágyi E.-fasor, Budapest. Office: 15 Herman O.-út, Budapest, Hungary.*

SZAFRANSKI, Przemyslaw, Polish biochemist; b. Bialy Dwor, Poland, Feb. 25, 1925; s. Zygmunt and Emilia (Zdrojewska) S.; B.Sc. in Chemistry, Tech. High Sch., Gdansk, Poland, 1949; Ph.D, Inst. Biochemistry and Biophysics, 1958; Docent degree Warsaw (Poland) Med. Sch., 1960; m. Irena Weckowicz, July 20, 1949; children—Pawll. Staff organic chemistry div. Med. Sch., Gdansk, 1949-51, biochemistry dept. Warsaw Med. Sch., 1951-54; staff Inst. Biochemistry and Biophysics, Warsaw, 1954—, prof., 1967—, v.-dir. head dept. protein biosynthesis, 1967—. Staff Nat. Inst. for Med. Research, London, 1956-57, Rockefeller Inst., N.Y., 1960-61, Institut de Chimie Biologique, U. Strasbourg (France), 1964, Carnegie Instn. Washington dept. terrestrial magnetism, Washington, 1965. Mem. Polish Biochem. and Biophys. Com., Polish Biochem. Soc., Genetic Soc. Research, publs. on entose cycle in metabolism of mycobacterium, nucleotide-peptides in animal cells, coding of amino acids in silk biosynthesis, specificity of ribosomes in synthesis of protein, interaction between ribosomes, messenger ribonucleic acid and soluble ribonucleic acid. Home: 8, Ap. 22, Dlugosza. Office: 36, Rakowiecka, Warsaw, Poland.*

SZALAY, Sandor Alexander, Hungarian physicist; b. Nyiregyhaza, Hungary, Oct. 4, 1909; s. Sandor and soluble ribonucleid acid. Home: 8, Ap. 22, Dlu-1928, M.A., 1931, Ph.D., 1932, D.Sc., 1952; m. Eve Csongor, Sept. 25, 1948; children—Alexander, Andrew. Research scholar U. Leipzig, 1933-34, Tech. High Sch., Munich, 1934-35, Cavendish Lab., Cambridge, Eng., 1936; mem. faculty Kossuth U., Debrecen, Hungary 1935—, prof. exptl. physics, 1940—; head Inst. Nuclear Research, Hungarian Acad. Scis., Debrecen, 1954—. Recipient Nat. Kossuth award for merits in sci., 1952. Fellow Hungarian Acad. Sci.; mem. Geochem. Soc., numerous other sci. socs. Research and publs. on nuclear physics in low energy range, geochemistry of uranium; discovery of role of humic acids in accumulation of uranium and other cations in bioliths. Home: 6/A Poroszlay, Debrecen, Hungary.*

SZANTO, Jan, virologist; b. Budapest, Hungary, Oct. 21, 1925; s. Ladislav and Alzbeta (Kostalova) S.; M.D., Komensky U., 1951, Ph.D., 1955; m. Maria Durchankova, June 21, 1953; 1 son, Ladislav. With Inst. Virology, Czechoslovak Acad. Scis., Bratislava, 1951—, head dept. molecular biology of viruses, 1966—. Mem. Czechoslovak Biol. Soc., J. E. Purkyne Czechoslovak Med. Assn. Research, publs. on isolation of cell lines and strains able to multiply viruses, persistent infections of tissues cultures with different animal viruses, role of non-specific factors in influenza virus infection. Home: 13 Teplicka. Office: 1 Mlynska dolina, Bratislava, Czechoslovakia.*

SZARA, Stephen Istvan, physician, chemist; b. Hungary, Mar. 21, 1923; s. Janos and Maria (Katona) S.; D.Sc., in Chemistry, Petrus Pazmany U., Budapest, 1950; M.D., Med. U. Budapest, 1951; m. Madeline S. Gadanyi, Sept. 5, 1959. Asst. Prof. Med. U. Budapest, 1950-53; chief lab. biochemistry Central State Hosp. for Nervous and Mental Diseases, Budapest, 1953-56; vis. scientist Nat. Inst. Mental Health, NIH, Washington, 1957-62, chief sect. psychopharmacology NRC, 1962—. Mem. A.A.A.S., Am. Soc. for Pharmacology and Exptl. Therapeutics, Am. Coll. Neuropsychopharmacology. Author: (with Miomir Meszaros) Textbook of Organic Chemistry, 1955; also articles. Research on polysaccharide of fibrinogen and its role in blood clotting, hallucinogens, metabolism and hallucinogenic activity tryptamine derivatives. Home: 4201 Cathedral Av. N.W., Washington 20016. Office: William A. White Bldg., St. Elizabeths Hosp., Washington 20032.*

SZARSKI, Henryk, Polish zoologist; b. Cracow, Poland, Sept. 13, 1912; s. Adam and Anna (Gwiazdomorski) S.; M.Sc., Jagellonian U., Cracow, 1935, Ph.D., 1937, docent, 1946; m. Irena Warzycki, Sept. 1, 1937; children—Marcin, Maria. Asst. prof. dept. comparative anatomy Jagellonian U., Cracow, 1935-39, 45-48, prof., chmn. dept. comparative anatomy, 1967—; prof., head dept. gen. zoology Copernicus U., Torun, Poland, 1948-66, dean faculty, 1950-52, rector, 1956-59; vis. lectr. U. R.I., Kingston,, 1963. Decorated Polish Golden Cross of Merit, Officer of Order of Polonia Restituta; recipient several awards Ministry of Edn. Mem. Polish Acad. Sci. (corr.), Acad. Zoology (Agra, India), Polish Zoology Soc., Soc. for Study Evolution, Am. Soc. Ichthyologists and Herpetologists, Sigma Xi. Author: Plazy, 1938; Pstrag, 1950; (with W. Juszczyk) Plazy i gady krajowe, 1950; Pochodzenie Plazow, 1961; also numerous articles. Research on comparative anatomy and physiology of digestion in certain worms, amphibian phylogeny, anatomy, circulation. Home: Al. Slowackiego 15, m. 8, Krakow, Poland.*

SZASZ, Thomas Stephen, psychiatrist; b. Budapest, Hungary, Apr. 15, 1920; s. Julius and Lily (Wellisch) S.; came to U.S., 1938, naturalized, 1944; A.B. with honors in physics, U. Cin., 1941, M.D., 1944; m. Rosine Loshajian, Oct. 19, 1951; children—Margot Claire, Susan Marie. Research asst. Inst. for Psychoanalysis, Chgo., 1949-51, staff, 1951-56; prof. psychiatry State U. N.Y., Upstate Med. Center, Syracuse, 1956—. Mem. research adv. panel Inst. for Study Drug Addiction, 1964—. Recipient Stella Feiss Hofheimer award U. Cin., Coll. Medicine, 1944; hon. fellow Postgrad. Center for Mental Health, N.Y.C., 1962. Fellow Am. Psychiat. Assn.; mem. Am. Psychoanalytic Assn., Internat. Psychoanalytic Assn., Am. Psychosomatic Soc., A.A.A.S., N.Y. Acad. Scis. Author: Pain and Pleasure, 1957; The Myth of Mental Illness, 1961; Law, Liberty and Psychiatry, 1963; The Ethics of Psychoanalysis, 1965; Psychiatric Justice, 1965; also numerous articles. Research on sociology medicine and the physician-patient relationship, theory and practice psychoanalysis, ethical, polit. and social aspects psychiat. practice especially problems involuntary psychiat. hospitilization and treatment. Home: 116 Bradford Pkwy., Syracuse, N.Y. 13224.*

SZEGO, Clara Marian, Am. zoologist; b. Budapest, Hungary, Mar. 23, 1916; d. Paul S. and Helen (Elek) S.; came to U.S., 1921, naturalized, 1927; A.B., Hunter Coll., 1937; M.S., U. Minn., 1939, Ph.D., 1942; m. Sidney Roberts, Sept. 14, 1943. Instr. physiology U. Minn., 1942-43; Cancer Research Inst. fellow, 1943-44; research OSRD, Nat. Bur. Standards, 1944-45; research asso. Worcester Found. for Exptl. Biology, 1945-47; research instr. physiol. chemistry Yale U., 1947-48; with U. Cal. Los Angeles,1948—, prof. zoology, 1960—. Garvan fellow Am. Chem. Soc., 1937, Guggenheim fellow, 1956; recipient Ciba award, Endocrine Soc., 1953; named Woman of

Year in Sci., Los Angeles Times, 1957-58. Fellow A.A.A.S.; mem. Am. Physiol. Soc., Endocrine Soc., Soc. for Endocrinology (Eng.), Laurentian Hormone Conf., Soc. Exptl. Biology and Medicine, Phi Beta Kappa, Sigma Xi. Research, numerous publs. on steroid protein interactions and mechanisms of hormone action. Home: 1371 Marinette Rd., Pacific Palisades, Cal. 90272.*

SZEINBERG, Arieh, med. biochemist; b. Krzemieniec, Poland, Nov. 8, 1918; s. Zelman and Leah (Shapiro) S.; M.Sc., Hebrew U., Jerusalem, Israel, 1942, Ph.D., 1950; m. Bilha Beker, Nov. 6, 1945; children—Amir, Leora. Head lab. Govt. Hosp. Hayarkon, Tel-Aviv, Israel, 1944-53; head biochem. lab. Govt. Hosp. Tel-Hashomer (Israel), 1953-—; faculty Tel-Aviv U., 1956-—, asso. prof. chem. pathology Postgrad. Med. Sch., 1963-—, head dept. chem. pathology, 1964-—. Recipient Henrietta Szold award in med. research Tel-Aviv Municipality, 1959, Myer award in med. research Worker's Sick Fund, Israel, 1961. Mem. Internat. Biochem. Soc., Internat. Hematol. Soc., Internat. Genetic Soc. Research and numerous publs. on hereditary enzymatic deficiencies, phenylketonuria and other genetic causes of mental retardation, genetics of various Jewish ethnic communities. Home: 17 Lessing St., Tel-Aviv, Israel. Office: Govt. Hosp., Tel-Hashomer, Israel.*

SZEKERES, László, Hungarian chemist; b. Budapest, Hungary, June 10, 1911; s. Izsák Spitzer and Margit (Róna) S.; student U. Budapest, 1945-49; Ph.D., U. Sci., Szeged, Hungary, 1949; m. Hedvig Szekeres, Mar. 8, 1936; 1 son Gábor László. Staff, Chinoin Chem. and Pharm. Works, Budapest, 1932-45, Hungarian Chem. Works, Budapest, 1949; faculty dept. organic chemistry U. Scis., Szeged, 1950-52; faculty dept. chemistry U. Agrl. Scis., Budapest, 1952-61; prorector, 1957-60, head dept. chemistry, 1952-61; head dept. chemistry U. Vet. Medicine, Budapest, 1961-—. Recipient Order Work, 1958. Fellow Hungarian, Am., Helvetian chem. socs. Author: Textbook of General and Inorganic Chemistry for Students of Biology, 1966; Textbook of Organic Chemistry, 1967; also numerous articles. Research on analysis in pharm. chems., 1932-52, analytical chemistry, 1961-—; agrochem. and analytical chemistry research, 1952-61. Home: Budapest, Hungary VII., Peterdy u. 12.*

SZENDI, Blasius, Hungarian gynecologist; b. Tiszadob, Hungary, May 14, 1904; s. Ferenc and Julia (Szegedi) S.; M.D., U. Medicine, Debrecen, Hungary; 1930; m. Erika Karacsony. Asso. prof. obstet. and gynecol. unit U. Drebecen, 1930-41; staff obstet. and gynecol. unit County Hosp., Gyula, Hungary, 1950-67. Research and numerous publs. in obstetrics, gynecology, female urology, abdominal surgery; intrauterine life. Home: 28, b Arpád str., Gyula, Hungary.*

SZENTAGOTHAI, Janos, Hungarian anatomist; b. Budapest, Hungary, Oct. 31, 1912; s. Gustav and Margaret (Antal) Schimert; M.D., U. Budapest, 1936; m. Alice Biberauer, June 7, 1938; children—Cathrine, Claire, (Mrs. Miklos Réthelyi), Christine. Jr. asst. Budapest U. Med. Sch., 1936-37, sr. asst., 1938-40, prosector, 1941-45, prof. anatomy, head dept., 1963-—;prof. anatomy Pécs U. Med. Sch., 1946-63. Recipient Kossuth State award, 1950. Mem. Hungarian Acad. Sciences (co-editor Acta Biologica, 1955-—, Acta Morphologica 1951-—), Leopoldina Acad., Internat. Brain Research Orgn. (exec. council 1962-—). Author: Die Rolle der einzelnen Labyrinthrezeptoren bei der Orientation von Augen und Kopf in Raume, 1952; (with B. Flerkó, B. Mess, B. Halasz) Hypothalamic Control of the Anterior Pituitary, 1962; (with F.C. Eccles, M. Ito) The Cerebellum as a Neuronal Machine, published in 1968; also author numerous articles. Research on nerve junctions to elucidate circuitry and connectivity in central nervous system, spinal cord pathways and their connetions, structure of simplest reflex paths, pathways and mechanisms of labyrinthine eye-movement reflexes, visual system relay mechanisms, structural basis of nervous inhibition in gen., cerebellar pathways and neuronal machinery of cerebellar cortex, anat. bases of nervous control on endocrine functions. Home: 2-3 Magyar Jakobinnusok tere, Budapest XII, Hungary.*

SZENT-GYÖRGYI, Albert, biochemist; b. Budapest, Hungary, September 16, 1893; s. Nicholas and Josephine (Lenbossek) S.-G; M.D., U. of Budapest, 1917; Ph.D. Cambridge, Eng., 1927; m. Martha Borbiro, 1941; 1 dau., Cornelia (Mrs. Pollit). Came to U. S., 1947, naturalized, 1955. Professor medical chemistry, Szeged, 1931-45; professor biochemistry, U. of Budapest, 1945-47, director of research, Inst. Muscle Research. Marine Biol. Labs., 1947-—. Awarded Nobel prize in medicine, 1937, 1955; Lasker Award, Heart Association, 1954. Mem. Academy Sciences of Budapest (president) National Acad. Budapest (v.p.), Nat. Acad. Scis., Council of Edn. (chmn.). Author: Oxidation, Fermentation, Vitamins, Health and Disease 1939; Muscular Contraction, 1947; The Nature of Life, 1947; Contraction in Body and Heart Muscle, 1953; Bioenergetics, 1957; Submolecular Biology, 1960. Research on biol. combustion, cellular oxidation, muscle contraction, carbohydrate metabolism; isolated hexaronic acid (ascorbic acid, Vitamin C), 1928; later extracted Vitamin C from Hungarian paprika. Address: P.O. Box 187, Marine Biological Labs., Woods Hole, Mass. 02543.*

SZENT-GYORGYL, Andrew Gabriel, biologist; b. Budapest, Hungary, May 16, 1924; s. Imre and Elisabeth (Kronberger) S.; M.D., U. Budapest, 1947; M.A., Dartmouth Coll., 1963; m. Eva M. Szentkiralyi, Nov. 30, 1947; children—Christopher S., Kathryn A., David J. Came to U.S., 1948, naturalized, 1955. Mem. faculty U. Budapest, 1945-47; fellow Neurophysiol. Inst. of U. Kopenhagen, 1948; instr. physiology Marine Biol. Lab., Woods Hole, Mass., 1953-57; mem. Inst. for Muscle Research, Woods Hole, 1948-62; prof. biophysics Dartmouth Med. Sch., Hanover, N.H., 1962-66; prof. biology Brandeis U., Waltham, Mass., 1966-—. Established investigator Am. Heart Assn., 1955-62; Research Career award NIH, 1962-66. Fellow A.A.A.S.; mem. Nat. Acad. Scis., Am. Physiol. Soc., Am. Soc. Biol. Chemists, Biophys. Soc., Soc. Gen. Physiologists. Editor Archives of Biochemistry and Biophysics, 1960-—, Physiol. Revs., 1966-—. Research and numerous publs. on isolation and properties of muscle proteins, structure of fribrous proteins, structure of muscle, chem. events and structural changes in muscle contraction, conformation of fibrous proteins, subunit structure of myosin. Home: 9 Westgate Rd., Wellesley, Mass. 02181.*

SZILAGYI, Desiderius Emerick, physician; b. Nagykaroly, Hungary, June 20, 1910; s. Emmerich and Elizabeth (Ujlaky) S.; came to U.S., 1930, naturalized, 1931; M.D., U. Mich., 1935, M.S. in Surgery, 1940; m. Martha Evelyn Fowlkes, Mar. 4, 1950; children—Martha Anne, Elizabeth Christine. Mem. staff Henry Ford Hosp., Detroit, 1939-42, 45-—, chmn. dept. surgery, 1966-—. Med. dir. Ford Rubber Plantation, Para, Brazil, 1942-44. Mem. A.M.A., A.C.S., Am. Thyroid Assn., Internat. Soc. Surgery, Soc. Vascular Surgery, Am. Fedn. Clin. Research, Internat. Cardiovascular Soc. (past pres. N.Am. chpt.), Am., Western, Central (past pres.) surg. assns. Author: Reconstructive Surgical Treatment of Peripheral Arterial Occlusive Disease, 1965; also articles. Studies in vascular surgery; developer vascular substitute, plastic artery, and investigations in its exptl. and clin. behavior. Office: Henry Ford Hosp., Detroit 48202.*

SZILAGYI, Tibor, Hungarian physician; b. Debrecen, Hungary, Aug. 13, 1921; s. Mihaly and Emma (Gaal) S.; M.D., U. Debrecen, 1944; m. Hajnalka Hegyesi, Aug. 17, 1962. Faculty, U. Debrecen, 1942-—, prof. Inst. Pathophysiology, 1954-—. Named eminent worker U. Edn., 1963. Mem. Hungarian Physiol. Soc., Hungarian Microbiol. Soc., Hungarian Biol. Soc. Research, publs. on immunol. properties of myosin, actin, fibrinogen, fibrin, phosphorylase; inorganic ions and epinephrine, effects of hypothermia, Shwartzman phenomenon, diabetes and allergy. Home: 28 Darabos, Debrecen, Hungary.*

SZILARD, Leo, physicist; b. Budapest, Hungary, Feb. 11, 1898; s. Louis and Thekla (Vidor) S.; student engring., Budapest Inst. of Tech.; Dr.Phil., Univ. of Berlin, 1922; unmarried. Came to U. S., 1937, naturalized, 1943. Mem. teaching staff, U. of Berlin, 1925-32; research work in nuclear physics, St. Bartholomew's Hosp., London and Clarendon Lab., Oxford, Eng., 1935-38; worked on atomic energy, Columbia, 1930-42; with Metall. Lab., U. of Chicago, 1942-46, 1946-64. professor University of Chicago. Recipient Atoms for Peace award, 1959. Fellow Am. Phys. Soc. Mem. Staff Enrico Fermi Inst. Contbn. in devel. of atomic energy; with Walter Zinn proved possibility of self-sustaining nuclear-fission, 1939; with Enrico Fermi worked out amount, configuration and control of uranium and directed first nuclear chain reaction, 1942; research in molecular biology from 1946. Died La Jolla, Cal. May 30, 1964.

SZUTKA, Anton, chemist; b. Weldiz, Ukraine, Apr. 18, 1920; s. Nicholas and Maria (Tysiak) S.; student Technische Hochshule, Hannover, Germany, 1947-51; M.S., U. Pa., 1955, Ph.D., 1959; m. Nadia Dobrowolska, Aug. 1, 1957; 1 son, Nestor A. Came to U.S., 1951, naturalized, 1960. Chemist, Barret div. Allied Chem., Phila., 1951-54; research asso. U. Pa., Phila., 1954; research asso. Hahnemann Med. Coll., 1954-59, asst. prof., 1959-61; faculty U. Detroit, 1961-—, prof. chemistry, 1964-—, asst. chmn. dept. chemistry, 1966-—. Mem. Radiation Research Soc., Am. Chem. Soc., A.A.A.S., Assn. Analytic Chemists, Shevchenko Sci. Soc., Ukrainian Engrs. Soc. Am., Ukrainian Acad. Arts and Scis. Research and publs. on non-enzymatic synthesis of porphine-like substances from simple precursors under simulated primordial condition; measurements of rates of reactions of solvated electrons with various organic molecules. Home: 13323 Hart St., Huntington Woods, Mich. 48070. Office: 4001 W. McNichols Rd., Detroit 48221.*

SZYBALSKI, Waclaw, molecular geneticist; b. Lwow, Poland, Sept. 9, 1921; s. Stefan and Michalina (Rakowska) S.; Chem. Engr., Poly. Inst., Lwow, 1943; D.Sc., Poly. Inst., Danzig, 1949; m. Elizabeth Hunter, Feb. 5, 1955; children—Barbara A., Stefan H. Staff mem. Cold Spring Harbor (N.Y.) Biol. Labs., 1951-55; asso. prof. Inst. Microbiology Rutgers U., 1955-60; prof. McArdle Lab. U. Wis., Madison, 1960-—. Mem. Am. Soc. Biochemists, Am. Chem. Soc., Genetic Soc. Am., Am. Soc. Microbiologists, Soc. Gen. Microbiologists (London), A.A.A.S., Italian Soc. Exptl. Biology (hon.), Polish Med. Alliance (hon.). Research, numerous pubs. on elucidation of microbiological corrosion of water pipes, antibiotic resistance in bacteria; discovered radiosensitization of human cells by thymine analogs; devel. quantitative genetics of human cell lives; elucidation of mode of action of mitomycin and several other antibiotics; mechanism of initiation of RNA transcription. Home: 1124 Merrill Springs Rd., Madison 53705. Office: McArdle Lab., U. Wis., Madison, Wis. 53706.*

SZYMANSKI, Herman A., Am. chemist; b. Toledo, Sept. 24, 1924; s. Joseph John and Magdalene (Sczublewski) S.; B.Chem.Engring., U. Toledo, 1948; Ph.D., U. Notre Dame, 1952; m. Alice I. Rospond, Apr. 28, 1948; children—Diane, Herman, Paul, Joseph, Mark, John, Mary. Chemist, Am. Can Co., Maywood, Ill., 1951-52; chmn. dept. natural scis. Loyola U., Chgo., 1952-54; dir. office of research Canisius Coll., Buffalo, 1955-—, prof. chemistry, 1963-—, chmn. dept. chemistry, 1955-—. Recipient Tech. Socs. Council award, 1964. Mem. Am. Chem. Soc., Am. Inst. Chemists, Soc. Applied Spectroscopy (pres.-elect 1966-—). Author numerous books including IR —The Theory and Practice of Infrared Spectroscopy, 1966; also articles. Research on infrared and NMR spectroscopy, gas chromatography. Home: 286 Starin Av., Buffalo 14216.*

T

TABER, Richard, Am. zoologist; b. San Francisco, Nov. 22, 1920; s. Loren Bennett and Louise (Everett) T.; A.B. in Zoology, U. Cal. at Berkeley, 1942; M.S. in Wildlife Mgmt., U. Wis., 1949; Ph.D., U. Cal. at Berkeley, 1951; m. Barbara Fleming, Feb. 10, 1946 (div. Aug. 1967); children—Rebecca Louise, Douglass Fleming, Katherine Helen; m. Patricia Gannett, Apr. 11, 1968. Research biologist U. Cal. at Berkeley, 1948-55, acting asst. prof., 1955-56; faculty U. Mont., Missoula, 1956-68, prof. forestry, 1962-68, asso. dir. Mont. Forest and Conservation Expt. Sta., 1964-68; prof. forestry, asso. dir. Wash. Inst. Forest Product Research, U. Wash., Seattle, 1968-—; Am. specialist forest zoology U. S. Dept. State, 1960, 64; Recipient Fulbright Research award, W. Pakistan, 1963-64; Guggenheim fellow, 1964-65. Mem. Am. Soc. Mammalogists, Ecol. Soc. Am., Wildlife Soc., Am. Inst. Biol. Scis., A.A.A.S. Contbg. author: Techniques of Wildlife Investigation, 1960; also numerous articles. Research on biology wild birds and mammals, relations land-use to vertebrate ecology, wildlife conservation. Office: Coll. Forest Resources, U. Wash., Seattle.*

TABERNAEMONTANUS, Jacobus Theodorus, German botanist; b. Bergzabern, Bavaria, circa 1520; personal physician to Elector Johann Casimir, Heidelberg, Germany. Author: Neuw Kreuterbuch (illus. herbal, standard for 2 centuries), 2 vols., 1588-91; Died Heidelberg, 1590.

TABOR, David, English physicist; b. London, Eng., Oct. 23, 1913; s. Charles and Rebecca (Weinstein) T.; student Regent St. Poly. London, 1925-32; B.Sc., U. London, 1934; Ph.D., U. Cambridge (Eng.), 1939, Sc.D., 1956; m. Hannalene Stillschweig, Mar. 14, 1943; children—Daniel Charles, Michael. Research physicist div. tribophysics Commonwealth Sci. and Indsl. Research Orgn., Melbourne, Australia, 1940-45, acting dir., 1945-46; faculty Cambridge U., 1946-—, reader physics, 1964-—, dep. dir. subdept. for surface physics Cavendish Lab., 1946-—. Fellow Gonville and Caius Coll., Cambridge, 1957. Fellow Royal Soc., 1963. Author: Hardness of Metals, 1951; (with F. P. Bowden) Friction and Lubrications of Solids, Part I, 1950, rev., 1954, Part II, 1964. Research and publs. on basic mechanisms of friction, lubrication and adhesion, hardness of solids. Home: 8, Rutherford Rd., Cambridge, Eng.*

TACHENIUS (Tackenius, Tachen) Otto, German iatrochemist; b. Herford, Westphalia; M.D., Padua, 1652. Asst. to David Welman as pharmacist, Lemgo, then to physician Rotger Timpler; traveled to Warsaw, Kiel, Danzig (1640), Königsberg (1641), left for Italy (1644), then studied at Padua, settled at Venice. Author: Epistola de famoso liquore Alcahest Helmontii, 1652; Hippocrates chimicus, 1666; Antiquissimae Hippocraticae Medicinae Clavis Manuali experientia in Naturae fontibus elaborata, 1668; Tractatus De Morborum Principe . . . 1679. Described acid-alkali theory of Sylvius which he claimed to have found in writings of Hippocrates and Galen; had clear idea of composition of salts and insisted they were divided into 2 parts: acid and alkali; studied acids and alkalis in detail, noting their relative strengths and displacement powers; one of earliest chemists to discuss use of color indicators; gave first account of manufacture of soap; engaged in controversy with Johann Zwelfer over meaning of alkali. Reported to be still living in Venice, 1699 (date of death unknown).

TACHINNI, Pietro, Italian astronomer; b. Modena, Italy, 1838. Became dir. obs. of Spilamberto, Italy, 1859; joined obs. of Palermo, Italy, 1863; named dir. Roman Coll., Rome, 1879; joined mission to Indies for observation of transit of Venus, 1874. Author: Eclipses totales de soleil . . . , 1870; Le Passage de Vénus sur le soleil en 1874, 1875. Research in solar astronomy. Died Spilamberto, 1905.

TACQUET, André, Belgian mathematician; b. Antwerp, Belgium, 1612; prof., Antwerp, also Leuven, Belgium; mem. Soc. Jesus. Author: Cylindricorum et annularium libri IV, 1651; Éléments de géométrie plane et solide, 1654; Arithmétique théorique et practique, 1656. Died Oct. 22, 1660.

TADOKORO, Tetsutaro, Japanese chemist; b. Akita Prefecture, Japan, 1885; grad. Tohoku U., 1910; D.Agr., 1919, D.Sc., 1928; postgrad. in agrl. chemistry in Am., Eng., France, beginning 1918; Apptd. asst. prof. Tohoku U., 1911, later prof. agrl. dept., head sci. dept.; head sci. dept., prof. Hokkaido U., pres until 1947; pres Hokkaido U Arts and Scis., for short period, then resigned and became emeritus prof.; apptd. prof. Obihiro Livestock Coll., 1957. Author works including: Chemistry of Fibroid Protein; Multi-specific Living Chemistry; Oxygen Chemistry; Protein Chemistry. Research, publs., patents agr., physics, engring., including vitamin B dioxidizing enzyme, dehydrating enzyme, microbiol. protein, vegetable pulp, low and hgh temperature nutrition, vegetative hormone, saccarose-producing stimulant.

TAEUBER, Irene Barnes (Mrs. Conrad Taeuber), Am. demographer; b. Meadville, Mo., Dec. 25, 1906; d. N. C. and Lily (Keller) Barnes; A.B., U. Mo., 1927; M.A., Northwestern U., 1928; Ph.D., U. Minn., 1931; LL.D., Smith Coll., 1960; D.Sc., Western Coll. for Women, 1965; m. Conrad Taeuber, July 26, 1929; children—Richard Conrad, Karl Ernst. With Office Population Research, Princeton, 1936—, sr. research demographer, prof., 1962—. Demographic cons. to govt. agys., univs. Recipient U. Mo. award of merit, 1965, Regents award for distinguished achievement U. Minn., 1966. Fellow Am. Sociol. Assn., Am. Statis. Assn.; mem. Population Assn. Am. (past pres.), Internat. Population Union (past v.p.), UN World Population Conf. (v.p. 1966—). Author: General Censuses and Current Vital Statistics in the Americas, 1943; (with others) The Population of Europe and the Soviet Union, 1944, Public Health and Demography in the Far East, 1948; The Population of Tanganyika, 1949; The Population of Japan, 1958; (with C. Taeuber) The Changing Population of the United States, 1958; also numerous articles. Research on population growth, formation met. population, modernization Chinese in western Pacific including People's Republic of China. Home: 4222 Sheridan St., Hyattsville, Md. 20782. Office: Office of Population Research, Princeton U., 5 Ivy Lane, Princeton, N.J. 08540.*

TAFF, Joseph Alexander, Am. geologist; b. Ten Mile, Tenn., Nov. 20, 1862; s. Albert G. and Tirzah A. T.; student U. of Ark., 1886-88; B.S., U. of Tex., 1894; m. Mary M. Leverett, Dec. 24, 1891; children—Elizabeth Simonds, Charles Leverett, Mary Willis, Joseph Whitham, Rosa LeRoche. Mem. Ark. and Tex. geol. surveys, 1888-94; geologist on U. S. Geol. Survey, 1894-1909; geologists S.P. Co., 1909-25, cons. geologist, 1925-32; chief geologist, Associated Oil Co., 1921-29; cons. geologist, 1929-37; retired, 1937. Mem. Geol. Soc. American (hon.), Geol. Soc. Washington, A.A.A.S., Am. Inst. Mining Engrs., Seismol. Soc. America, Nat. Geog. Soc, Calif. Acad. Science, Am. Assn. of Petroleum Geologists (hon.), Sigma Xi. Author of papers, folios and bulletins in publs. of Tex., and U. S. geol. surveys and in jours. of tech. socs. on gen. and econ. geology of coal, oil, asphalt and cement resources in 11 states. Died, 1944.

TAGER, Joseph Michael, biochemist; b. Pietersburg, South Africa, June 13, 1925; s. Ellis Sellick and Hodda (Zway) T.; B.Sc. cum laude, U. Pretoria, 1947; Ph.D., U. Cal. at Los Angeles, 1952; Faculty U. Pretoria, South Africa, 1953-60, lectr., 1958-60; lectr., sr. lectr. U. Amsterdam, The Netherlands, 1961-66, reader in biochemistry, 1966—. Mem. South African (pres. sect. 1960), Amsterdam socs. for advancement sci., Soc. Physiologists, Biochemists and Pharmacologists South Africa (chmn. 1960), Biochem. Soc. U.K., Dutch Biochem. Soc., Sigma Xi, others. Author: (with others) Regulation of Metabolic Processes in Mitochondria, 1966; Mitochondrial Structure and Compartmentation, 1967; also articles. Research on mitochondrial metabolism, including respiration, oxidative phosphorylation, metabolism of amino acids, control mechanisms. Home: Leidsegracht 42, Amsterdam, The Netherlands.*

TAGER, Morris, microbiologist; b. Riga, Latvia, Dec. 25, 1909; s. Nathan B. and Masha (Tager) T.; Ph.B., Yale, 1931, M.D., 1936. Practice medicine, specializing in microbiology, Atlanta; prof., chmn. dept. microbiology Emory U., 1951—. Mem. Office Naval Research Microbiology Panel, 1959-65; mem. bacteriology, mycology study sect. NIH, 1962-64; bd. regents Nat. Library Medicine, 1964—; cons. Atlanta VA Hosp. Diplomate Nat. Bd. Med. Examiners. Fellow Am. Acad. Microbiology, Internat. Soc. Hematology; mem. Soc. Clin. Investigation, Am. Assn. Immunologists, Soc. For Am. Microbiology, A.A.A.S. Research, numerous publs. on mechanisms of blood clotting by bacterial products, such as staphylocoagulase med. mycology. Home: 4227 Carmain Dr. N.E., Atlanta 30305. Office: Woodruff Bldg., Atlanta 30322.*

TAGGART, John Victor, Am. physician; b. Brigham, Utah, Aug. 29, 1916; s. John Victor and Cecil (Townsend) T.; student U. Cal., Los Angeles, 1933-36; M.D., U. So. Cal., 1941; m. Theodora Janeway Lannon, July 10, 1959. Faculty, Columbia, N.Y.C., 1946—, prof. medicine, 1958—, prof., chmn. dept. physiology, 1962—; attending physician Presbyn. Hosp., 1958—. Recipient Gibbs prize N.Y. Acad. Medicine, 1952. Career investigator Am. Heart Assn., 1958-62. Mem. Am. Physiol. Soc., Assn. Am. Physicians, Am. Soc. Clin. Investigation, Sigma Xi, Alpha Omega Alpha. Research, numerous publs. in renal physiology, chemotheraphy of human malarias, enzyme chemistry, metabolic aspects of renal transport mechanisms. Home: 169 E. 78th St., N.Y.C. 10021. Office: 630 W. 168th St., N.Y.C. 10032.*

TAGLIACOZZI (or TALIACOTIUS), Gasparo, Italian surgeon, physician; b. Bologna, Italy, 1546; prof. anatomy, surgery, U. Bologna. Author: De Curtorum Chirurgia per Insitionem, 1597. Practised plastic surgery, specializing in repairing noses, ears, lips with skin grafted from arm (Taliacotian process) (methods little used until 1800's due to ch. opposition); wrote 1st treatise on rhinoplasty. Died Bologna, Nov. 7, 1599.

TAGLIAFERRI, Guido, Italian physicist; b. Rome, Italy, Jan. 27, 1920; s. Romeo and Beatrice (Corsi) T.; D.Physics, U. Pisa (Italy), 1941; Diploma in Physics, Scuola Normale Superiore, Pisa, 1942; m. Carla Silvia Pinto, Nov. 11, 1953; children—Federico, Beatrice. Asst. prof. U. Milan (Italy), 1945-53, 56-59, prof. radioactivity, 1961-63, prof. structure of matter, 1964—; research asso. Princeton, 1954-55; prof. exptl. physics U. Bari (Italy), 1960; dir. Milan sect. Nat. Inst. Nuclear Physics, 1964-67. Mem. Italian Phys. Soc. Research and publs. on penetrating component of cosmic rays; unstable particles produced in high energy nuclear reactions; instrumentation for nuclear research. Home: 2 Via Marsala, Milan, Italy.*

TAGUCHI, Kazumi, Japanese cardiovascular surgeon; b. Izu, Hyogo-ken, Japan, Sept. 26, 1925; s. Shigeichi and Yoshie T.; M.D., Okayama U., 1948, Ph.D., 1950; m. Aiko Kurihara, Dec. 1, 1954; children—Shinichi, Mami. Mem. adv. staff U. S. Acad. Sci., 1966; faculty Okayama U. Med. Sch., 1953-60; attending surgeon Tottori U. Hosp., 1960-61; faculty Hiroshima (Japan) U. Med. Sch., 1961—, asso. prof., head cardiovascular div., 1965—; dir. surg. dept. Hiroshima Citizens Hosp., 1961—. Hon. fellow U. Minn. Hosp., 1957. Fellow Hiroshima Med. Assn. (hon.); mem. Am. Coll. Chest Physicians, Japanese Soc. Surgery, Assn. for Thoracic Surgery, Internat. Coll. Surgeons. Author: Medical Treatment of Today, 1964; also numerous articles. Research on pulmonary hypertension and surg. mgmt., correction of tetralogy of Fallot, clin. application of superhemodilution technique in extracorporeal circulation, inventor prosthetic valve. Address: Dept. Surgery, Hiroshima U. Hosp., Kasumicho, Hiroshima, Japan.*

TAHMISIAN, Theodore Newton, biologist; b. Caesaria, Turkey, June 22, 1909; s. Vahram B. and Calliope (Serinides) T.; came to U. S., 1924, naturalized, 1926; B.A., Fresno State Coll., 1935; M.S., State U. Ia., 1939, Ph.D., 1942; m. Renie G. Sadoyan, Apr. 7, 1936; 1 son, Theodore Newton. Faculty Fresno State Coll., 1935-36; research asso. State U. Ia., 1942-43, U. Chgo., 1946; asso. scientist Argonne Nat. Lab. (Ill.), 1946-66, sr. scientist, 1966—. Fellow A.A.A.S., N.Y. Acad. Scis., Royal Micros. Soc. London; mem. Am Inst. Biol. Scis., Am. Sci. Affiliation, Am. Physiol. Soc., Am. Soc. Cell Biology, Am. Soc. Zoologists, Am. Micros. Soc., Electron Microscopy Soc. Am., Ill. Acad. Scis., Internat. Soc. Cell Biology, Mid-Western Soc. Electron Microscopists, Radiation Research Soc., Research Soc. Am., Soc. Exptl. Biology and Medicine, Soc. Gen. Physiologists, Sigma Xi. Contbr. to Mechanisms in Radiobiology, 1960. Research, publs. on responses of physiol. functions to irradiation; electron microscope studies of sperm, insect muscle. Home: 4843 W. 98th Pl., Oak Lawn, Ill. 60453. Office: 9800 S. Cass Av., Argonne, Ill. 60439.*

TAHORI, Alexander Shalom, entomologist; b. Memel/Klaipeda, Lithuania, Jan. 18, 1919; s. Heimann and Ella (Burrack) T.; student Hebrew U., Jerusalem, Israel, 1940-41, 46-47; M.Sc., U. Cal. at Berkeley, 1949, Ph.D., 1951; m. Eva Turk, Aug. 4, 1955; children—Michael, Dan. Head dept. entomology Army Med. Research Inst., Israel Def. Forces, 1951-57; head sect. insect toxicology Israel Inst. for Biol. Research, Ness-Ziona, 1957—; vis. prof. Hebrew U. Faculty Agr., Rehovot, Israel, 1957—. Mem. Entomol. Soc. Am., Israel Soc. Entomology, Sigma Xi. Author: Insect Toxicology, 1959; also articles. Editorial bd. Jour. Israel Soc. Entomology, 1966—. Research on action on synergists on insecticides, factors influencing resistance to insecticides and acaricides, insect biochemistry, compounds protecting plants from insect attack. Home: 50 Hasajit Ramat Hasharon, Israel. Office: Israel Inst. for Biol. Research, Ness-Ziona, Israel.*

TAI, Yuin Kuei, Chinese physicist; b. Foonghwa, Chekiang, China, Jan. 8, 1900; s. Shin Zu and Y. S. (Hu) T.; B.S., Kyoto (Japan) Imperial U., 1926; m. Wen-Lang Tien, Sept. 22, 1930; 5 children. Prof. physics Nat. Peping Normal U., 1926-28, Nat. Central U. Nanking, 1928-32, U. Nanking, 1932-45; dean studies, prof. physics Nat Taiwan U., Taipei, 1945-47, prof. physics, 1947—, chmn. dept., 1947-62, dir., 1962—. Mem. Acad. Sinica, Chinese Assn. for Advancement Natural Sci. (past pres.), Phys. Soc. Republic China, (past pres.). Author: General Physics for University, 1941; Nuclear Energy and its Uses in Peace, 1955; The World in Space, 1958; also articles. Research on neutron yields from 32 Mev protons on thick targets, origin of antiproton, total yield and angular distbn. of neutrons emitted from Zr-T target bombarded by deutron beam, neutron total cross sect. of arsenic at 14 Mev, angular distbn. protons from d (d.p) reaction, radiocarbon concentration in Taiwan wood. Home: 7 Chao-Chow St., Taipei, Taiwan.*

TAILLENS, Jean Pierre, Swiss physician; b. Lausanne, Switzerland, Aug. 1, 1910; s. Charles and Jeanne (Chapuis) T.; ed. Faculté de Médecine de Lausanne; m. Ramelet Marise, Dec. 1939, children—Pierre-Alain, Bernard, Nicole. Faculty otorhinolaryngology Faculté de Médecine de Lausanne, 1943—, prof., 1953—, dean Faculté de Médecine, 1954-56, also titular prof. otorhinolaryngology; in charge surg. service mil. hosp. Decorated by cities of Paris, Louvain, Madrid, Toulouse. Research and publs. on otorhinolaryngology. Home: Place Montbenon 2, Lausanne, Switzerland.*

TAINTER, Charles Sumner, Am. inventor; b. Watertown, Mass., Apr. 25, 1854; s. George and Abigail (Sanger) T.; pub. sch. edn.; m. Lila R. Munro, June 22, 1886; m. 2d, Laura Fontaine Onderdonk, Apr. 7, 1928. Inventor of the graphophone, also of the dictaphone; associate inventor of the radiophone, an instrument for transmitting sounds to a distance through the agency of light; member of the U. S. expedition sent to the South Pacific to observe transit of Venus, 1874. Awarded gold medal at Electrical Exhbn., Paris, 1881, for inventions in connection with the radiophone; Officer de l'Instruction Publique, France, for invention of graphophone, 1889; John Scott medal, 1900, by City of Phila., for the invention of the graphophone; gold medal, 1915, at Panama P.I. Expn. for work in connection development of the talking machine. Died Apr. 20, 1940.

TAINTER, Maurice (Lane), Am. physician; b. Carroll, Ia., May 10, 1899; s. Emory Baxter and Winnifred Alice (Lane) T.; A.B., Stanford U., 1921, A.M., 1924, M.D., 1925; D.Sc., Rensselaer Poly. Inst., 1951; m. Dorothy McGuinness, 1925; children—Shirley Anne (Mrs. Edwin O. Swenson), Cynthia Mae (Mrs. John C. Taylor, Jr.). Faculty Stanford, 1921-43; prof. applied physiology Albany Med. Coll., 1943-47; founding dir. Sterling-Winthrop Research Inst., 1946-60; dir. research Sterling Drug, Inc., 1943—, v.p., 1946—; vice chmn. Sterling Research Bd., 1960—. Mem. numerous adv. bds., coms. for univs., profl. orgns., govts.; mem. research adv. bd. Sch. Pharmacy Columbia, 1958—; mem. adv. council Grad. Research Center S.W., 1963—; com. on research N.A.M., 1960—; mem. N.Y. State Adv. Council for Advancement of Indsl. Research and Devel., 1960—, vice chmn., 1966—; med. adv. bd. Council for High Blood Pressure Research, 1947—. Bd. dirs. N.Y. State Found. for Sci. and Tech. Fellow N.Y. Acad. Scis. (past pres.), A.M.A., Am. Coll. Clin. Pharmacology and Chemotherapy; mem. Pharm. Mfrs. Assn., Am. Physiol. Soc., Am. Soc. Pharmacology and Exptl. Therapeutics, Soc. Toxicology, Am. Assn. Study Headache, Am. Soc. Anesthesiology (hon.), Soc. Exptl. Biology and Medicine, Nat. Soc. Med. Research, Am. Physicians Art Assn., numerous others, Sigma Xi, Alpha Omega Alpha, Tau Kappa Omega, Phi Lambda Upsilon. Editor-in-chief Stanford Med. Bull., 1940-43; mem. editorial bd. Archives Internationales de Pharmacodynamic et die Therapie, Clin. Medicine, Pharmacol. Revs., Toxicology and Applied Pharmacology. Studies, publs. in med. and dental pharmacology; therapeutics, adminstrn., history of medicine. Home: Kitchel Rd., Mt. Kisco, N.Y. 10549. Office: 90 Park Av., N.Y.C. 10016.*

TAISHI, Shotoku (Prince Shotoku), Japanese scholar; b. 573; s. Emperor Yomei and Anahobe-no (Hashito); studied Buddhism under Eji, a Korean priest; regent of Japan under Empress Guko, (593-628); prepared 17 article consitution, 604; opened diplomatic relations between Japan and China by sending an envoy to Emperor Yao of Sui Dynasty, 607; built Horyuji Temple and other Buddhist temples. Compiled 1st Japanese history book, 620. Improved calender, abacus, arithmetic. Died 621.

TAIT, James Francis, biophysicist; b. Stockton-on-Tees, Eng., Dec. 1, 1925; s. Herbert and Constance (Brotherton) T.; B.Sc. with 1st class honors in Physics, Leeds (Eng.) U., 1946, Ph.D. in Physics, 1948; Sylvia A. S. Wardropper, Sept. 1, 1956. Lectr. med. physics Middlesex Hosp. Med. Sch., London, 1948-55, sci. staff med. research council, 1955-58; sr. scientist Worcester Found. for Exptl. Biology, Shrewsbury, Mass., 1958—. Fellow Royal Soc., 1959; mem. Am. Acad. Arts and Scis., Endocrine Soc., Soc. for Endocrinology, N.Y. Acad. Scis., A.A.A.S., Biochem. Soc. (Eng.), Hosp. Physicist Assn. Editorial bd. Jour. Clin. Endocrinology and Metabolism. Contbr. numerous articles to jours., chpts. to books. Co-discoverer aldosterone; research on secretion, site of secretion, biosynthesis of aldosterone, dynamic theories steroid hormone transport and metabolism, measurement hormones in peripheral blood. Home: 5 High St., Southboro, Mass. 01772. Office: 222 Maple Av., Shrewsbury, Mass. 01545.*

TAIT, John Barclay, Scottish phys. oceanographer; b. Edinburgh, Scotland, June 7, 1900; s. Adam and Jane Russell (Barclay) T.; B.Sc., U. Edinburgh, 1922, Ph.D., 1932, D.Sc., 1939; Diploma in Applied Chemistry, Heriot-Watt Coll., Edinburgh, 1922; m. Euphemia Wallace Greig, June 16, 1927; children—John Barclay, Douglas Grieg, Elspeth Muriel (Mrs. John William Gadsby). Analytical chemist J. F. MacFarlan & Co., Edinburgh, 1923-25; with Fishery Bd. for Scotland, 1925-47, sr. naturalist, 1933-47; with Scottish Home Dept. fisheries div., Aberdeen, 1947-60, sr. prin. sci. officer, 1952-60; dep. chief sci. officer Scottish Dept. Agr. and Fisheries, Aberdeen, 1960-65; asso. Heriot-Watt Coll., 1922, fellow, 1951; Buckland Prof. 1938. Chmn. data exchange working group Intergovtl. Oceanographic Commn., 1964-66; leader Iceland-Faroe Expdn., organizer Internat. Council for Exploration Sea, 1960. Recipient Keith prize Royal Soc. Edinburgh, 1957-59. Fellow Royal Soc. Edinburgh (past mem. council), Royal Scottish Geog. Soc.; mem. Nat. Oceanographic Council, Royal Soc. London (mem. Brit. nat. com. for oceanic research 1959-—), Scottish Marine Biol. Assn. (mem. council 1948-—), Challenger Soc. for Promotion Oceanography, Internat. Assn. Phys. Oceanography. Author: Hydrography in Relation to Fisheries, 1952; Hydrography of the Faroe-Shetland Channel, 1927-52, 1957; also numerous articles. Prodn. of detailed surface current chart No. N. Sea and Faroe-Shetland Channel from analyses surface drift-bottle records; long-term studies water-masses Faroe-Shetland Channel transition region between oceanic and arctic waters evolving exact definition these water-masses by their salinities, long-term computations dynamics oceanic water-mass through Faroe-Shetland Channel. Home: Macbieknowe, West Linton, Peeblesshire, Scotland.*

TAIT, Peter Guthrie, Brit. mathematician, physicist; b. Dalkeith, Soctland, Apr. 28, 1831; s. John and Mary (Ronaldson) T; student Edinburgh (Scotland) U.; B.A, Cambridge (Eng.) U., 1852; LL.D., U. Glasgow (Scotland), 1901; Sc.D. (hon.), U. Ireland, 1875; m. Margaret Archer Porter, Oct. 13, 1857; children—John, Frederick, Harold, Wilfred, also 2 daus. Prof. math. Queen's Coll., Belfast, Ireland, 1854-60; prof. natural philosophy U. Edinburgh, 1860-1901; dir. Scottish Provident Instn. Recipient Gunning Victoria Jubilee prize, Royal medal Royal Soc., 1886. Fellow Royal Socs. Edinburgh (2 Keith prizes, sec. 1864); mem. several fgn. sci. acads. Author: (with William J. Steele) Dynamics of a Particle, 1856; (with W. Thomson), Sketch of Elementary Dynamics, 1863; Elements of Philosophy, 1863; Elementary Treatise on Quaternions, 1867; Natural Philosophy, (vol. I with W. Thomson), 1867; Thermo-Dynamics, 1868; (with Philip Kelland) Introduction to Quaternions, 1873; (with Balfour Stewart) The Unseen Universe, 1875; Recent Progress in Physical Science, 1876; Heat, 1884; Light, 1884; Properties of Matter, 1885; Dynamics, 1895; Scientific Papers, 2 vols., 1898-1900. Research in theory quaternions, ozone, kinetic theory gases, electricity, thermo-electricity, thermodynamics; devised 1st practical thermo-elec. diagram; studies on flight of golf ball, also effects of pressure on deep sea thermometer readings. Died Edinburgh, July 4, 1901.

TAIT, Robert Lawson, Scottish gynecologist; b. Edinburgh, Scotland, May 1, 1845; s. Archibald Campbell and Isabella Stewart (Lawson) T.; grad. Edinburgh (Scotland) U., 1866; LL.D., U. Albany; M.D., N.Y., St. Louis, Chgo.; m. Sybil Anne Stewart. House surgeon Wakefield (Eng.) Hosp., 1867-70; surgeon Birmingham (Eng.) Hosp. for Women, from 1871; prof. gynecology Queen's Coll., Birmingham, from 1887. Prof. anatomy Royal Soc. Artists, Birmingham Sch. Design. Recipient Hastings Gold medal Brit. Med. Assn. Founder Brit. Gynecol. Soc. Author: Diseases of Women, 1877; Abdominal Surgery, 1898; essays on operations performed. Perfected and established several operations for abdominal diseases, operations for removal of ovaries, 1872; introduced hepatotomy, 1880; 1st to successfully operate for ruptured tubal pregnancy, 1883. Died June 13, 1899.

TAJIMA, Eizo, Japanese radiation physicist; b. Otamura Saitama, Japan, Apr. 28, 1913; s. Senzou and Tome (Bondow) T.; Ph.D., Tokyo U. Edn., 1935; m. Namiko Akita, Jan. 5, 1942; children—Yusuke, Kensuke. Research fellow phys. and chem. inst. Nishina Lab., 1935-52; asso. research fellow U. Chgo., 1949-50; prof. Rikkyo U., Tokyo, Japan, 1952-—, dean univ., 1967-—. Alternate sci. com. Atomic radiation UN, 1956-—; mem. Radiation Council of Govt., 1956-—; mem. spl. com. radiation Atomic Energy Commn., 1955-—. Recipient award for publ. Mainich-Shinbun, 1949. Author: Bushitzu-no-Kyukyoku, 1948; also articles, chpt. in book. Research on particle acceleration by cyclotron, nuclear physics, radioactive fallout and radiation protection, neutron in space; constructed 23 inch and 60 inch cyclotrons. Home: 12-12 I-chome, Sasuke, Kamakura, Kanagawa, Japan.*

TAKABAYASHI, Takehiko, Japanese physicist; b. Hyogo Prefecture, Japan, 1919; grad. Tokyo U., 1941. Successively asst., lectr., asst. prof. Nagoya U.; staff Daiichi Arsenal Lab. Author: History of Science of Heat; History of Quantum Dynamics. Research on colloid and high polymers.

TAKACS, Lajos Ferenc, mathematician; b. Maglod, Hungary, Aug. 21, 1924; s. Lajos and Maria (Leiti) T.; Ph.D., Tech. U., Budapest, 1948; Dr. Math. Sci. Hungarian Acad. Scis., 1957; m. Dalma S. Horvath, Apr. 9, 1959; children—Judith, Susan. Came to U. S., 1959, naturalized, 1965. Mathematician Tungsram Research Inst., Budapest, Hungary, 1945-55; mathematician Research Inst. Math., Hungarian Acad. Scis., Budapest, 1950-58; asso. prof. Eotvos U., Budapest, 1953-58; asso. prof. Columbia U., 1959-66; prof. math. Case Western Res. U., Cleve., 1966-—. Fellow Inst. Math. Statistics, Royal Statis. Soc.; mem. Am. Math. Soc., Math. Assn. Am., Am. Statis. Assn. Author: Stochastic Processes, 1960; Introduction to the Theory of Queues, 1962; Combinatorial Methods in the Theory of Stochastic Processes, 1967. Research, numerous publs. in the theory and applications of probability and stochastic processes. Home: 2410 Newbury Dr., Cleveland Heights, O. 44118.*

TAKAGI, Atsushi, Japanese bacteriologist; b. Hiroshima, Japan, Jan. 15, 1918; s. Banji and Tsune (Ishii) T.; M.D., Kyushu U., 1941, Dr.Med.Sci., 1946; m. Iseko Moriwaki, Feb. 25, 1945; children—Kenichi, Eiko. Prof. bacteriology Yongo Med. Coll., 1945-51; prof. bacteriology Tottori U. Sch. Medicine, 1951-—. Mem. Japan Bacteriological Soc., Japanese Assn. Infectious Diseases, Japanese Soc. Electron Microscopy, Japan Mycological Soc. Author: (with others) Toda's Textbook of Bacteriology. Research, publs. on measurement of isoelectric point of bacteria by buffered dye solution; discover influenza C virus in Japan; isolation of causative agt. of Izumi feber in Japan; device of agglutination reaction in Tb with sensitized kaolin particles; demonstration of fine structure of spores of Clostridium group of bacteria with ultrathin sectioning technique, fine structure, function intracytoplasmic membrane system of bacteria, fungi. Home: 9-1 Uchi-machi. Office: 86 Nishi-cho, Yonago, Tottori Pref., Japan.*

TAKAGI, Kentaro, Japanese physiologist; b. Fukuoka, Japan, Mar. 17, 1910; s. Shigeru and Toku (Makino) T.; H.B. Kyusyu U., 1934, M.D., 1939; m. Setsu Komiya, May 6, 1936; children—Keiko (Mrs. Hiroji Oota), Masamichi, Masatake, Zyunko (Mrs. Sukehiko Koga). Asso. prof. Niigata Med. U., Asahi-machi-dori, Niigata-City, Japan, 1939-45, prof., 1945-56; prof. Nagoya (Japan) U. Sch. Medicine, 1956-—; research fellow Mil. Army, 1944-45. Recipient Sci. Bounty of Asahi Press and Chunichi Press. Mem. Japanese Soc. Physiology, others. Author: Introduction to Physiology, 1932; Physiology, 1937; also numerous articles. Research, publs. on central mechanism of respiration, regulation of respiration, regulation of body temperature, somato-vegetave reflex, especially due to application of pressure onto the skin, med. electronics, peripheral circulation, biometeology. Home: No. 1, 6-Chome Syogetsu-cho, Mizuho-ku, Nagoya, Japan. Office: Dept. Physiology, Nagoya U. Sch. Medicine, Tsuruma-cho, Syowa-ku, Nagoya, Japan.*

TAKAHASHI, Hachiro, Japanese physicist; b. Maebashi, Japan, Oct. 18, 1919; s. Tzunezo and Shino (Yamada) T.; B.Sc., Tohoku U., Sendai, Japan, 1951, D.Sc., 1961; m. Asako Sato, Mar. 11, 1956; children—Mizue, Yuri. Teaching asst., research asst. dept. physics Iwate U., Morioka, Japan, 1951-57, lectr., 1957-58, asst. prof., 1958-64, prof., 1964-—; vis. researcher Cosmic Ray Lab., Inst. Phys. and Chem. Research, Tokyo, 1960-61; vis. prof. dept. geophysics U. Western Ont., London, Can., 1965-66. Mem. Am. Geophys. Union, A.A.A.S., Phys. Soc. Japan, Japan Soc. Applied Physics. Contbr. articles to sci. jours. Devel. of cosmic ray physics, space physics. Home: 39-22 Tatemukai-cho, Morioka. Office: Ueda, Morioka, Iwate-ken, Japan.*

TAKAHASHI, Shinji, Japanese radiologist; b. Nihonmatsu, Japan, Jan. 28, 1912; s. Jonosuke and Saki (Uchiike) T.; M.D., Tohoku U. Sch. Medicine, 1938, M.D., Ph.D., 1943; m. Utako Watanabe, Apr. 16, 1941; children—Masaki, Hiroko. Asst. dr. Tohoku U. Dept. Radiology, Sendai, Japan, 1938-42, lectr., 1942-45; prof., chief dept. radiology Hirosaki (Japan) U., 1945-54; prof., chief dept. radiology Nagoya (Japan) U., 1954-—; cons. Aichi Cancer Center, Nagoya. Contbr. numerous articles to sci. jours. Pioneer field axial transverse tomography, macroradiography in high magnification, conformation radiotherapy, radiol protection. Home: 1-47 Yamanakacho, Showaku. Office: Hosp., 65, Tsurumaicho, Showaku, Nagoya, Japan.*

TAKAHASHI, Yasushi, physicist; b. Osaka, Japan, Dec. 12, 1924; s. Momochi and Chise (Miyata) T.; M.Sc., in Physics, Nagoya (Japan) U., 1951, D.Sc., 1954; m. Elizabeth McLoughlin, Sept. 15, 1959; children—Atsushi Mark, Hiroshi Jerome. Research asso. U. Rochester, 1953-54; postdoctoral fellow NRC, Ottawa, Ont., Can., 1954-55; research asso. State U. Ia., 1955-57; research scholar Dublin (Ireland) Inst. for Advanced Studies, 1957-58, asst. prof., 1958-60, prof. physics, 1960-—. Mem. Royal Irish Acad. Research, numerous publs. on gen. method relativistic field quantization, identities between charged particles propagator and electromagnetic vertex, Hamiltonian method with linear supplementary condition. Home: 12 Mt. Carmel Av., Roebuck, Dublin, Ireland.*

TAKAMINE, Jokichi, chemist, pharmacologist; b. Takaoka, Japan, Nov. 3, 1854; s. Dr. Scichi and Yuki T.; grad, in chem. engring., Engring. Coll., Imperial U. of Tokyo, 1879; Japanese Govt. student U. of Glasgow and Andersonian U., Glasgow, 3 yrs., 1879-81; (Dr. Chem. Engring., Imperial U. of Japan, 1899, Dr. Pharm., 1906); m. Caroline Hitch, 1885. Head chemist Imperial Dept. Agr. and Commerce, Tokyo, 1881-84; Imperial Japanese commr. to Cotton Centennial Expn., New Orleans, 1884-85; organized and erected 1st superphosphate works at Tokyo, 1887; came to America, 1890, and applied new process of conversion and fermentation to practical use resulting in production of diastatic enzyme ("Takadiastase"), now largely used as a starch digestant; established research lab. in New York and originated a process for isolating the active principle of the suprarenal glands, the product being known as ("adrenalin"); cons. chemist. Decorated by Emperor of Japan with 4th Order of the Rising Sun, 1915; apptd. by Emperor mem. Royal Acad. Sciences, Japan, 1913. Author: The blood-pressure-raising principle of the suparenal glands, 1901. Died New York, N.Y. July 22, 1922.

TAKARO, Timothy, surgeon; b. Budapest, Hungary, Aug. 30, 1920; s. Geza and Irene (Francsek) T.; came to U. S., 1922, naturalized, 1943; B.A., Dartmouth, 1941; M.D., N.Y. U., 1943; m. Marilyn Anne Keen, Apr. 3, 1949. Practice medicine, specializing in surgery, Oteen, N.C., 1957-—; dir. surgery Wanless Tb. Sanatorium in India, Presbyn. Bd. Fgn. Missions, 1954-57; dir. research, edn. VA Hosp., Oteen, 1957-60, chief cardiovascular surg. sect., 1960-62, chief surg. service, 1962-—; U. S.-USSR Exchange Research fellow, 1962. Recipient certificate of merit for sci. exhbn. A.M.A., 1966. Fellow A.C.S., Am. Coll. Chest Physicians; mem. So. Thoracic Surg. Assn., Am. Assn. Thoracic Surgery, Internat. Cardiovascular Soc., Soc. Thoracic Surgeons. Contbr. chpt. to Vascular Surgery, 1965. Mem. editorial bd. Annals of Thoracic Surgery, 1965-66. Contbr. numerous articles to profl. jours. Research on surg. stapling devices, devel. of magnification techniques in angiography, esophagopleural fistulas, pleuropulmonary amebiasis, fungus diseases of the lungs, pulmonary blastomas. Home: 12 Westchester Dr., Asheville, N.C. 28803. Office: VA Hosp., Oteen, N.C. 28805.*

TAKASU, Tsurusaburo, Japanese mathematician; b. Yamaguchi Prefecture, Japan, May 29, 1890; grad. Kyoto U., 1916. Joined faculty Seventh Higher Sch., Kagoshima, after graduation; became instr. Tohoku U., 1920, prof., 1934, emeritus prof., 1950; prof. Yokohama Municipal U., 1952-—. Mem. Internat. Mathematicians Soc. Author: Advanced Solid Geometry; Advanced Differential and Infintisimal Calculus; An Outline of Analytical Geometry. Research and numerous publs. on math., especially geometry and theory of functions.

TAKAYANAGI, Kazuo, Japanese physicist; b. Tomakomai, Japan, Nov. 28, 1926; s. Takeo and Kiwa (Ogawa) T.; grad. physics Faculty Sci., U. Tokyo (Japan), 1948, postgrad., 1948-50, D.Sc., 1955; m. Kazuko Nagasawa, Apr. 11, 1958; children—Masao, Hiroko. Faculty, Saitama U., Urawa, Saitama, Japan, 1950-66, prof., 1965-66; prof. Inst. Space and Aero. Sci., U. Tokyo, 1966-—; vis. fellow Joint Inst. Lab. Astrophysics, Boulder, Colo., 1963-64. Brit. Council scholar U. Coll., London, Eng., 1955-57. Mem. Phys. Soc. Japan, Soc. Terrestrial Magnetism and Electricity Japan. Research and publs. on theory of atomic collisions especially vibrational and rotational transitions in molecular encounters, electron collisions with simple molecules, atomic processes in interstellar matter and upper atmosphere of earth. Home: 1-3-10 Oji-honcho, Kita-ku, Tokyo, Japan.*

TAKEBE, Hikojiro Kenko, Japanese mathematician; b. Yedo, Tokyo, June, 1664; s. Takebe Chokuko; pupil of Seki Kowa. Shogunate samurai, 1703; ofcl. in dept. of ceremonies; drew map of Japan, 1719' astron. adviser to Yoshimune the 8th Tokagawa Shogun, until 1733. Author: Kenki Sampo, 1683; Hatsubi Sampo Yendan Genkai, 1685; (editor) Chu Shi-chien's Suan-Heiw Chi-Meng, 1690; Fukyu Tetsujutsu, 1722; Taisei Sankyo, 1766; Ynri Tetsujutsu. Attributed with discovery of Yenri, a type of integral calculus; used infinite series in study of circle and value of calculated pi to 41 decimal points, 1722. Died July 20, 1739.

TAKENOUCHI, Katsu, Japanese dermatologist, physician; b. Kochi City, Japan, Mar. 31, 1905; s. Kintaro and Shun (Miyabe) T.; M.D., Chiba (Japan) U., 1931, Ph.D. Dermatology, 1939; m. Kana Nagaoka, Nov. 14, 1936; children—Kazu Kashima, Jun Takahashi, Yasu, Tama, Haru. Faculty, Chiba U. Sch. Medicine, 1933-—, prof. dermatology, chmn. dept. dermatology, 1952-—. Recipient Minami prize for dermatology 1953, prize Japanese Vitamin Soc., 1964. Mem. Internat. Coll. Surgeons, Dermatol. Soc. India, Soc. Internat. Dermatologica tropica, Japanese Dermatol. Assn. (trustee). Author: Textbook of Dermatology, 1946, 68; Thiamine and Riboflavin Metabolism of the Skin, 1965; Treatment of Syphilis, 3 vols., 1963; Diagnostical Table of Dermatology, 1947; also numerous articles. Research on salt free diet for treatment of skin tuberculosis, fat free diet therapy for psoriasis, influences of vitamin A, B, D, and C to infection, systematic studies as allergic skin disease on lupus erythematosus, Libman-Sachs syndrome and erythema multiforme with muco-cutaneo-ocular syndrome, treatment and epidemiological survey of syphilis. Home: 22-13 1-chome Midori-cho, Chiba-city, Japan.*

TAKESHITA, Kenji, Japanese physicist; b. Kumamoto, Japan, Apr. 21, 1926; s. Daishiro Takeshita; grad. dept. physics Faculty Sci., Kyushu U., 1951, Dr.Sci. Medicine, 1961; m. Keiko, May 19, 1950; children—Gen. Hitoshi. With Kyushu U., 1953-65, asst. prof. Faculty Medicine, 1964-65; prof. Research Inst. for Nuclear Medicine and biology dept. radiation biology Hiroshima (Japan) U., 1965—. Lectr., Tottori U. Faculty Medicine, 1965—; adviser dept. radiology Atomic Bomb Casualty Commn., Hiroshima. Mem. Nippon Societas Radiologica (councilor), Japan Radiation Research Soc. Research, publs. on dosimetry of high energy radiation, including X-rays, electrons, 14 MEV neutrons; studies in elucidation of biol. action mechanism of radiations with different let, lethal effect in mice exposed to massive doses of gamma ray; survey exposure due to radioactive fallout and med. use X-rays. Home: RCU 13-238, Ushita-Machi, Hiroshima, Japan.*

TAKEUCHI, Tadashi, Japanese pathologist; b. Tokyo, Japan, Aug. 19, 1914; s. Kumpei and Chyoko (Narabayashi) T.; grad. Sch. Medicine U. Tokyo, 1941, higher degree medicine, 1952; m. Misako Harada, May 13, 1945; children—Totaro, Lina. Asst. dept. pathology U. Tokyo, 1941-44; prof. Matsumoto Med. Sch. 1944-48; pathologist 2d Tokyo Nat. Hosp., 1948-56; prof. pathology Nihon U. Sch. Medicine, Tokyo, 1956—; supervising pathology div. Nat. Inst. Radiol. Scis., Chiba, 1960—; vis. prof. U. Chgo., 1963. Mem. Japanese Soc. Nephrology (mem. bd. mgrs.), Japan Soc. Pathology and Clin. Pathology (mem. adv. com.), Japan Radiol. Scis. Soc. Author: A Modern Textbook of Pathology, 1963; Surgical Pathology, 1963; also numerous articles. Research on pathology of encephalitis Jap B.; introduction of surg. pathology in Japan; new radiol. method in blood circulation of kidney and other organs. Home: 1-36, Mabashi, Suginami-ku, Tokyo, Japan.*

TAKEYA, Kenji, Japanese med. microbiologist; b. Moji-shi, Japan, Mar. 29, 1922; s. Shoyu and Kinu (Tsuda) T.; M.D., Kyushu U., Sch. Medicine, 1944, D.M.S., 1949; m. Yasuko, Aug. 26, 1944; children—Shunichi, Makiko. Research asso. Kyushu U. Sch. Medicine, Fukuoka, Japan, 1946—, faculty, 1950—, prof., chmn. dept. bacteriology, 1963—, mem. staff research insts. for chest diseases, 1955—, for cancer, 1966—. Rockefeller Found. fellow U. Pa., U. Cal. at Los Angeles, 1955-56. Mem. expert adv. panel for cholera WHO, 1965—. Recipient Goto's prize Kyushu U., 1944. Mem. Japan Soc. Electron Microscope (Seto prize 1962), Japan Soc. Bacteriology, Am. Soc. Microbiology, Japan Soc. Tb, others. Author: (with Koike) Textbook of pathogen microbiology, 1963; (with others) Electron Microscopy in Biology and Medical Fields, 1964; others. Research, publs. on fine structures of bacteria and bacteriophages in relation to their functions, mycobacteriophages and virio-phages from a med. stand point; studies on thymus and immunity. Home: 17 - 7, Kusagae 2, Fukuoka, Japan.*

TAKIMOTO, Kiyoshi, Japanese geologist; b. Yamaguchi-ken, Japan, Oct. 12, 1905; s. Natsuziro and Fude T.; B.Sc., U. Kyoto (Japan), 1935, D.Sc. in Geology, 1943; m. Hanako, Oct. 2, 1936; children—Kiyohiko, Masahiko, Kazuko, Yoshihiko. Research fellow U. Kyushu, Fukuoka-shi, Japan, 1935-39, asst. prof. 1939-48; prof. Faculty Engring., U. Kyoto, 1948—; asso. lectr. Osaka City U., 1960—. Recipient Acad. award Japan Geol. Soc., 1944. Mem. Japan Mining and Geol. Soc., Japan Geol. Soc. Author: Mineral Deposits, 1953; Coal, 1955; Minerals in Metal Mining, 1955; also articles. Exploitation of tin ore deposits, wolfram ore deposits, copper ore deposits, non-metallic ore deposits. Home: Izumidencho 61, Yoshida, Sakyo-ku, Kyoto-shi, Japan. Office: Dept. Mineral Sci. and Tech., Faculty of Engring., Kyoto U., Yoshida, Sakyo-ku, Japan.*

TAKINO, Masuichi, Japanese physician; b. Santocho, Hyago Prefecture, Japan, Mar. 14, 1905; s. Kichitaro and Yukiko (Toda) T.; M.D., Kyoto U., 1929; m. Yoshiko Aoyagi, Nov. 17, 1931; children—Susumu, Toru, Hiroko (Mrs. Mitzuaki Fujimoto), Nobuko. Lectr. Med. Sch., Kyoto (Japan) U., 1939-40; adviser Dainippon Zoki Inst. for Med. Research, 1941-46, dir., 1947-61; dir. Nippon Zoki Inst. for Constl. Diseases, Higashiku, Osaka, Japan, 1962—; mng. staff Japanese Soc. Allergology, 1952—. Fellow Am. Coll. Chest Physicians; mem. Japanese Soc. Endocrinology and Cardiology (counselor 1941—), Internat. Soc. Biometeorology (mem. subcom. 1962—), Japanese Soc. Internal Medicine, Japanese Soc. Autonomic Nervous System (counselor 1956—). Author: A New Direction in Asthma Treatment, 1952; Allergy and Asthma, 1957; also numerous articles. Research on autonomic nervous system and its application to treatment bronchial asthma, surg. removal of carotid body and carotid sinus nerve in asthmatics; discovered antiallergic and rheumatic factors. Home: 25-5 Hoenzaka jutaku, Higashiku. Office: Heiwa Bldg., 2-10 Hiranomachi, Higashiku, Osaka, Japan.*

TALA, Pekka Mikael, Finnish physician; b. Lahti, Finland, June 28, 1919; s. Mikko and Hilja (Syvänen) T.; M.D., U. Helsinki (Finland), 1947, D.Medicine and Surgery, 1952; m. Tyyne Elisabeth Anttila, Sept. 23, 1945; children—Tarja Tuulikki, Timo Pekka, Olli Tapio. Asst. surgeon I. and II surg. clinics U. Helsinki, 1950-56; sr. asst. surgeon Clinic for Thoracic Surgery, U. Central Hosp., Helsinki, 1956-65, asso. prof. Clinic for Thoracic Surgery, 1965—, vacant prof., 1966—; faculty U. Helsinki, 1958—; surgeon Muurolan, Oulainen, Kiajava Tb sanatoriums, Mehiläinen Pvt. Hosp., Duodecim Oy, Finland. Fgn. mem. Nordisk Kirurgisk Förening, Nordisk Thoraxkirurgisk Förening, Am. Coll. Chest Physicians Societé Européene de Chirurgie Cardiovasculaire, Deutsche Gesellschaft für Chirurgie, Sociéte Internationale de Chirurgie. Research and numerous publs. on surgery of thoracic and cardiovascular diseases. Abstract writers Excerpta Medica Section Cardiovascular, 1957—, Annales Chirurgiae et Gyneacologiae Fenniae, 1959—; editorial sec. Annales Chirurgiae et Gynaecologiae Fenniae, 1966—. Home: Dosentintie 7 C 18, Helsinki 33. Office: Hosp., Haartmaninkatu 4, Helsinki 29, Finland.*

TALALAY, Paul, physician; b. Berlin, Germany, Mar. 31, 1923; s. Joseph Anton and Sophie (Brosterman) T.; came to U. S., 1940, naturalized, 1946; S.B., Mass. Inst. Tech., 1944; postgrad. U. Chgo. Sch. Medicine, 1944-46; M.D., Yale, 1948; m. Pamela Judith Samuels, Jan. 11, 1953; children—Antony Benjamin Charles, Susan Julie, Rachel Tanya. Research asso., asst. prof. surgery U. Chgo., 1950-51, asst. prof. biochemistry, 1955-57, asso. prof., then prof., 1957-63, asst. prof. Ben May Lab. Cancer Research, 1951-57, asso. prof., then prof., 1957-63; John Jacob Abel prof., dir. dept. pharmacology and exptl. therapeutics Johns Hopkins Sch. Medicine, 1963—. Sr. asst. surgeon USPHS, 1951-53. Mem. vis. com. dept. biology Mass. Inst. Tech. Scholar cancer research Am. Cancer Soc., 1954-58, recipient Career Professorship, 1958. Mem. A.A.A.S. (Theobald Smith award med. scis. 1957), Am. Soc. Biol. Chemists, Am. Soc. Clin. Investigation, Biochem. Soc., Am. Chem. Soc., Sigma Xi, Alpha Omega Alpha. Hon. editorial adv. bd. Biochem. Pharmacology, 1963-68; editorial bd. Jour. Biol. Chemistry, 1961-66. Research in steroid metabolism; devel. specific enzymatic methods for microestimations of steroids. Home: 400 Overhill Rd., Balt. 21210.*

TALAMON, Charles, French physician, bacteriologist; b. 1850. Discovered, demonstrated pathogenic role pneumococcus (differentiating if from Friedlander's bacillus penumoniae), 1883; distinguished rubella from rubeola, 1890; described complications of appendicitis, 1892. Died 1929.

TALAS, Miloslav, Czechoslovakian gynecologist, chemist; b. Retechov, Czechoslovakia, Jan. 13, 1927; s. Alois and Marie (Krajcová) T.; MUDr, Masaryk U., Brno, Czechoslovakia, 1950, RNDr, 1952; C.Sc., Med. Faculty, Charles U., Prague, 1966; m. Jana Kourová, Dec. 5, 1953; children—Blanka, Jitka. Research asst. dept. pharmacology Palacky U., Olomouc, Czechoslovakia, 1950-54, research asst., head biochem. lab. dept. obstetrics and gynecology, 1954—. Research fellow Population Council Physiology of Reprodn., Harbor Gen. Hosp., Torrance, Cal., 1967—. Research, publs. on estrogens, androgens and gonadotropins in urine in normal menstrual cycle and in Stein-Leventhal syndrome. Home: 2, Leninova, Olomouc, Czechoslovakia.*

TALBOT, Arthur Newell, Am. engineer; b. Cortland, Ill., Oct. 21, 1857; s. Charles A. and Harriet (Newell) T.; B.S., U. of Ill., 1881, C.E., 1885, LL.D., 1931; Sc.D., U. of Pa., 1915; D.Eng., U. of Mich., 1916; m. Virginia Mann Hammet, June 7, 1886 (died Dec. 4, 1919); children—Kenneth Hammet, Mrs. Mildred Virginia Gilkey, Mrs. Rachel Harriet Westergaard, Mrs. Dorothy Newell Goodell. Engring. work on rys., roads, bridges, bldgs. and municipal pub. wks. since 1881; asst. prof. engring. and mathematics, U. of Ill., 1885-90, prof. municipal and sanitary engring. and in charge theoretical and applied mechanics, 1890-1926, prof. emeritus; made many investigations on properties of steel, brick, concrete, reinforced concrete, etc., and on bldgs. in connection with U. of Ill. Engring. Exptl. Sta. and as consultant, also investigations of water purification, sewage treatment and hydraulic questions; dir. investigation conducted by Am. Soc. C.E. and Am. Ry. Engring. Assn. to determine stresses in railroad track since 1914. Mem. Am. Soc. C.E. (pres. 1918; hon. mem. 1925), Soc. for Promotion Engring. Edn. (pres. 1910-11), Am. Soc. Testing Materials (pres. 1913-14; hon. mem. 1923), Am. Water Works Assn. (hon. mem. 1930), Western Soc. Engrs. (Washington award, 1924; hon. mem. 1927), Am. Public Works Assn., Am. Ry. Engring. Assn. (dir. 1915-18, 1928-31; hon. mem. 1933), Am. Soc. Mech. Engrs., New Eng. Water Works Assn., A.A.A.S. (v.p. 1928), Am. Concrete Inst. (Turner medal, 1928; hon. mem. 1932), Instn. Civ. Engrs. (London); hon. mem. Instn. Structural Engrs. (London); mem. bd. visitors U. S. Naval Acad., 1918-21; etc. Awarded Henderson medal, Franklin Institute, 1931; Lamme medal, Soc. for Promotion Engring. Edn., 1932; John Fritz medal of the Founder Societies, 1937. Clubs: University, Engineers (Chicago). Author: The Railway Transition Spiral, 1901, 27; also many tech. articles and Experiment Sta. bulls. on engring. researches. Died Apr. 3, 1942.

TALBOT, Mary, Am. zoologist; b. Columbus, O., Nov. 30, 1903; d. Frank Thresher and Paulena (Schmitz) Talbot; B.S., Denison U., 1925; M.A., Ohio State U., 1927; Ph.D., U. Chgo., 1934. Instr., U. Omaha, 1927-28, Stephens Coll., Columbia, Mo., 1928-30, Mundelein Coll., Chgo., 1935-36; chmn. biology dept., prof. zoology Lindenwood Coll., St. Charles, Mo., 1936—. Mem. Ecol. Soc. Am., De l'Union International pour l'Étude des Insectes Sociaux, Assn. for Study Animal Behavior, Kan. Entomol. Soc., Am. Inst. Biol. Scis. Research and publs. on ants including population, nest structure, ecol. distbn. and selection of habitat, fluctuations in daily above ground activities, slave-making raids, behavior during flights. Home: 315 N. 6th St., St. Charles, Mo. 63301.*

TALBOT, William Henry Fox, English chemist, physicist, archeologist; b. Lacock Abbey, Wiltshire, Eng., Feb. 11, 1800; s. William Davenport and Elizabeth Theresa (Fox-Strangways) T.; ed. Harrow; B.A., Trinity Coll., Cambridge (Eng.) U., 1821, M.A., 1825; m. Constance Mundy, Dec. 20, 1832. Sci. research, also studied lit., archeology; M.P. for Chippenham, 1833-34. Recipient Great Gold medal Paris Exhbn., 1867. Fellow Royal Soc. (Royal medal 1838, Rumford medal 1842), 1831; mem. Royal Astron. Soc. Author: The Pencil of Nature, 6 parts, 1844-46 (contains 1st photographic illustrations produced without being partially drawn); Legendary Tales, in Verse and Prose, 1830; Hermes or Classical and Antiquarian Researches, 2 vols., 1838-39; The Antiquity of the Book of Genesis, 1839; English Etymologies, 1847; Assyrian Texts Translated, 1856; appendix to History and Handbook of Photography (G. Tissandier), 2d edit., English trans., 1878. Contbr. articles to sci. jours. Invented photogenic drawing (independently of Daguerre) which consisted of producing print on paper from photographic negative, 1839 (method, Talbotype, foundation for modern photography), patented, 1841; discovered method for taking instantaneous photographs, 1851, photographic engraving, 1852; research on spectrum; in mathematics, developed Fagnani's theorem; one of 1st to decipher cuneiform tablets of Nineveh. Died Lacock Abbey, Sept. 17, 1877.

TALCOTT, Andrew, Am. astronomer, engr.; b. Glastonbury, Conn., Apr. 20, 1797; s. George and Abigail (Goodrich) T.; grad, U. S. Mil. Acad., 1818; m. Catherine Thompson, Apr. 1826; m. 2d, Harriet Randolph Hackley, Apr. 11, 1832; 11 children. Brevetted 2d lt. U. S. Army, 1818; 1st lt. Engr. Corps, 1820, capt., 1830; engr., aide-de-camp to Gen. Henry Atkinson in establishment of posts on Upper Missouri and Yellowstone rivers, 1820-21; chief engr. Ft. Delaware, Del., 1824-25; supt. constrn. canal through Dismal Swamp in Va., 1826-28; supervised constrn. Ft. Monroe, Ft. Calhoun, Hampton Roads, Va., 1828-34; astronomer for determining boundaries between Ohio and Mich., 1828-35; chief engr. in charge of Western div. N.Y. & Erie R.R., 1836-37; supt. improvement of delta of Mississippi River, 1837-39; chief engr. Richmond & Danville R.R., 1848-55; astronomer, surveyor for making the northern boundary of Ia., 1852-53; chief engr. Ohio & Miss. R.R. from Cincinnati and St. Louis, 1856-57; located, constructed railroad from Vera Cruz to Mexico City, 1857-60, 61-67; mgr. Sonora Exploring & Mining Co., chief engr. State of Va., 1860-61; elected chief engr. Va., 1861; devised method of determining terrestrial latitudes through the observation of stars near the zenith; mem. Am. Philos. Soc. Died Richmond, Va., Apr. 22, 1883.

TALIACOTIUS, Gasparo, see Tagliacozzi, Gasparo.

TALIAFERRO, Lucy Graves (Mrs. William Hay Taliaferro), Am. microbiologist, immunologist; b. Cleve., July 25, 1895; d. Herbert Cornelius and Clara (Walter) G.; A.B., Goucher Coll., 1917; Sc.D., Johns Hopkins, 1925; m. William Hay Taliaferro, June 6, 1919. Staff, U. Chgo., 1925-60, vol. research asso. in microbiology, prof., 1959-60; vol. research asso. Argonne (Ill.) Nat. Lab., 1960—. Mem. Am. Soc. Immunologists. Author: (with W. H. Taliaferro, B. N. Jaroslow) Radiation and Immune Mechanisms, 1964; also articles. Research on malaria in man and other primates, serological and cellular aspects immunity to protozoan and worm parasites, dynamics antibody formation, effects ionizing radiation on immune mechanisms. Home: 5803 S. Kensington Av., La Grange, Ill. 60525. Office: Argonne Nat. Lab., 9700 S. Cass Av., Argonne, Ill. 60439.*

TALIAFERRO, William Hay, Am. microbiologist, immunologist; b. Portsmouth, Va., Feb. 10, 1895; s. William Hay and Mary Watkins (Leigh) T.; B.S., U. Va., 1915; Ph.D., Johns Hopkins, 1918; Sc.D., U. N.C., 1946; LL.D., Temple U., 1949; m. Lucy Engel Graves, June 6, 1919. Faculty, Johns Hopkins, 1919-24; faculty U. Chgo., 1924-60, Eliakim H. Moore Distinguished Service prof. microbiology, 1939-60, E. H. Moore Distinguished Service prof. emeritus, 1960—, chmn. dept. microbiology, 1932-60, asso. dean biol. scis., 1931-35, dean, 1935-44; sr. immunologist Argonne (Ill.) Nat. Lab., 1960—. Recipient Chalmers medal Royal Soc. Tropical Medicine and Hygiene, 1935, Mary Kingsley medal Liverpool Sch. Tropical Medicine, Eng., 1949, Condecoracion al Merito, Bernardo O'Higgins de primera Clase, Govt. of Chile, 1960. Mem. Royal Soc. Tropical Medicine and Hygiene (hon., Eng.), Nat. Acad. Scis., Am. Philos. Soc., Am. Soc. Parasitologists, Am. Soc. Tropical Medicine and Hygiene, Am. Soc. Immunologists, Am. Soc. Microbiology. Author: (with R. W. Hegner) Human Protozoology, 1924; The Immunology of Parasitic Infections, 1929; (with L. G. Taliaferro, B. N. Jaroslow) Radiation and Immune Mechanisms, 1964; also numerous articles. Editor: Medicine and the War, 1944.

Research on cytology of eye spots in lower organisms and their function in orientation to light; description immunity mechanisms, both humoral and cellular, which protect man and animals from infection by larger parasites; biochemistry of antibody formation, action ionizing radiations on antibody formation. Home: 5803 S. Kensington Av., La Grange, Ill. 60525. Office: Argonne Nat. Lab., 9700 S. Cass Av., Argonne, Ill. 60439.*

TALMAGE, David Wilson, Am. immunologist; b. Kwangju, Korea, Sept. 15, 1919 (parents Am. citizens); s. John Van Neste and Eliza (Emerson) T.; B.S., Davidson Coll., 1941; M.D., Washington U., 1944; m. LaVeryn Hunicke, June 23, 1944; children—Janet, Marilyn, David, Mark, Carol. Practice medicine, specializing in allergy, Chgo., 1952-59, Denver, 1959—; faculty U. Chgo., 1952-59; prof. medicine U. Colo., 1959—, prof. microbiology, 1960—, chmn. dept. microbiology, 1963-66, asso. dean faculty, 1966—. Recipient Distinguished Service award U. Chgo. Alumni, 1967. Markle scholar, 1956-60. Mem. Am. Assn. Immunologists, Am. Acad. Allergy (past pres.), Phi Beta Kappa, Alpha Omega Alpha. Author: (with John Cann) The Chemistry of Immunity, 1961; also numerous articles. Editor, Jour. of Allergy, 1963-67. Editor: (with Max Samter) Immunological Diseases, 1965. Research in mechanism of synthesis and action of antibodies involved in immunity, allergy and transplant rejection; proposed selective theory of antibody prodn., new theory of antibody specificity; devised new methods of antibody purification and measurement. Home: 280 S. Hudson St., Denver 80222. Office: 4200 E. 9th Av., Denver 80220.*

TALMAGE, Roy VanNeste, Am. physiologist; b. Moppo, Korea, Feb. 9, 1917 (parents U. S. citizens); s. John VanNeste and Eliza (Emerson) T.; A.B., Maryville (Tenn.) Coll., 1938; M.A., U. Richmond, 1940; Ph.D., Harvard, 1947; m. Helena Schlichter, June 6, 1942; children—Helena Louise, John VanNeste, Charles Emerson. Mem. faculty Rice U., Houston, 1947—, prof. biology, 1954—, chmn. dept., 1956-64, master Wiess Coll., 1957—. Dir. research grants AEC, NIH and NSF; cons. VA Hosp., Houston. Trustee, treas. Houston Museum NIH spl. fellow U. Leiden (Holland) Med. Sch., 1964. Mem. Am. Soc. Zoologists, Am. Soc. Physiologists, Endocrine Soc., A.A.-A.S., Soc. Exptl. Biology and Medicine, Orthopedic Research Soc. Author: The Armadillo (with Buchanan), 1954; The Parathyroids (with Greep), 1961; (with others) The Parathyroid Glands; Ultrastructure, Secretion, and Function, 1965, also numerous articles. Editorial bd. publs. Soc. Exptl. Biology and Medicine, Gen. and Comparative Endocrinology, Clin. Orthopaedics. Studies and publs. on effects of hormones on animals in various conditions, calcium content of bone. Home: Wiess House, Rice Univ., Houston 77001.*

TALMI, Igal, physicist; b. Kiev, Ukraine, Jan. 31, 1925; s. Moshe and Lea (Weinstein) T.; M.Sc., Hebrew U. Jerusalem, 1947; Dr.Sc.Nat., Swiss Fed. Inst. Tech., 1952; m. Chana Kivelewitz, Aug. 16, 1949; children—Yoav P., Tamar. Vis. fellow Princeton, 1952-54, vis. asso. prof., 1956-57, vis. prof., 1961-62, 66-67; faculty Weizmann Inst. Sci., Rehovoth, Israel, 1954—, prof. physics, 1958—. Mem. Israel Atomic Energy Commn., 1965—. Recipient Weizmann prize, 1963; Israel prize in exact scis., 1965. Mem. Israel Acad. Scis. and Humanities. Author: (with A. de Shalit) Nuclear Shell Theory, 1963; also numerous articles. Research in physics of nuclear structure; use of shell model to calculate nuclear energy levels with two-body effective interactions between nucleons. Home: Neve Weizmann, Rehovoth, Israel.*

TALMUD, David Lvovich, Russian phys. chemist; b. Oct. 24, 1900; grad. Odessa Chem. Inst., 1923. Faculty, Odessa U., 1923-25; asso. Leningrad Inst. Chem. Physics, 1930-34; asst. Inst. Biochemistry, USSR Acad. Scis., 1934, also mem. I. P. Pavlov Inst. Physiology, Leningrad. Recipient Stalin Prize, 1943, Order of Lenin. Mem. USSR Acad. Scis., Inst. Biochemistry. Author: Structure of Albumin, 1940; (with S. E. Bresler) Surface Phenomena, 1934; Research and publs. in phys. chemistry of surface layers, colloidal chemistry, protein structure, indsl. problems. Home: Leninskii Prospekt 13, Moscow, USSR. Office: A.N. Bakh Inst. Biochemistry, USSR Acad. Scis., Leninskii Prospekt, 33, Moscow, USSR.

TALSO, Peter Jacob, Am. physician; b. Ishpeming, Mich., Sept. 22, 1921; s. Jacob and Jennie (Mattson) T.; student No. Mich. Coll. Edn., Marquette, 1939-41; A.B., Wayne U., 1943, M.D., 1945; m. Evelyn M. Lucynski, Dec. 18, 1943; children—Jennifer, Cassandra, Kathryn Ann, Peter. Practice medicine, specializing in internal medicine, Chgo., 1950—; mem. staffs Mercy Hosp., St. Joseph's Hosp., Cook County Hosp. (all Chgo.); faculty U. Chgo., 1950-52; faculty Stritch Sch. Medicine, Loyola U., Chgo., 1952—, prof., chmn. dept. medicine, 1963—; cons. Hines (Ill.) VA Hosp. Mem. Gov.'s Adv. Council on Mental Retardation, Ill., 1964—; mem. med. adv. bd. Kidney Found. Ill., 1966—. Diplomate Am. Bd. Internal Medicine. Fellow A.C.P., Am. Coll. Cardiology, Inst. Medicine Chgo., Am. Coll. Clin. Pharmacology and Chemotherapy; mem. Central Soc. for Clin. Research, Ill., Chgo. socs. internal medicine, A.M.A., Am. Fedn. for Clin. Research, Am. Soc. Nuclear Medicine, Am. Soc. Human Genetics, N.Y. Acad. Sci., A.A.A.S., Chgo. Heart Assn. (bd. govs. 1955—), Ill. Psychiat. Tng. and Research Authority, Midwest Com. on Drug Investigation. Publs. on studies in electrolyte and water metabolism; hypertensive disease. Home: 10359 S. Longwood Dr., Chgo. 60643. Office: 1400 S. 1st Av., Hines, Ill. 60141.*

TALWANI, Manik, geophysicist; b. Patiala, India, Aug. 22, 1933; s. Birsain and Sarawati (Khosla) T.; B.S., Delhi (India) U., 1951, M.S., 1953; Ph.D., Columbia U., 1959; m. Anni Fitler, Apr. 3, 1958; children—Rajeev, Indira, Sanjay. Research scientist Lamont Geol. Obs., Palisades, N.Y., 1959—; asso. prof. Columbia U., N.Y.C., 1966—. Recipient 1st Krishnan medal Indian Geophys. Union, 1965, James B. Macelwane award Am. Geophys. Union, 1967. Fellow Geol. Soc. Am., Royal Astron. Soc.; mem. European, Am. socs. exploration geophysicists, Seismol. Soc. Am., Am. Geophys. Union, Sigma Xi. Editorial bd. Marine Geology, 1964—; asso. editor Jour. Geophys. Research, 1966—. Research and publs. on application of marine gravimetry to studies of earth's crust and upper mantle, gravity measurements at sea; developed instruments to eliminate error in these measurements, computing techniques to aid in interpretation. Home: 293A Brookway Av., Valley Cottage, N.Y. 10989. Office: Lamont Geol. Obs., Palisades, N.Y. 10964.*

TALWAR, Gursaran Parshad, Indian biochemist; b. Hissar, India, Oct. 2, 1926; s. Dina Nath and Sumitra (Nair) T.; B.Sc. with honors, Govt. Coll., Lahore, India, 1946; M.Sc., Punjab U., 1948; Docteur es Sciences, Sorbonne, Paris, France, 1953; m. Raksha Mehta, Nov. 18, 1961; children—Pratap, Priyae. French Govt. scholar Inst. Pasteur, Paris, 1950-52, fellow Nat. Inst. Hygiene, 1952-54; Alexander Von Humboldt fellow, W. Germany, 1954-55; asso. prof. biochemistry All-India Inst. Med. Scis., New Delhi, India, 1956-65, prof., head dept. biochemistry, 1965—; vis. prof., guest investigator Population Council Labs., Rockefeller U., 1964, 66. Mem. neurochemistry panel UNESCO, 1963—. Mem. Biochem. Soc. London, Soc. Biol. Chemists India, Assn. Physiologists and Pharmacologists India, Internat. Brain Research Orgn. Contbg. author: Macromoleucytes and Behaviour, 1966. Research and publs. on mechanism of action of growth hormone and estradiol, problems related to brain function. Home: All-India Inst. Med. Scis., C-11/17, Ansari Nagar, New Delhi, India.*

TAMIYA, Hiroshi, Japanese biochemist; b. Osaka, Japan, Jan. 5, 1903; s. Koreharu and Masayo (Kohno) T.; M.Sc., Imperial U. Tokyo, 1926, D.Sc., 1932; D.Sc., U. Tubingen, 1963; m. Nobuko Seida, July 7, 1928; children—Takako (Mrs. Hisashi Horie), Hajime. Faculty, U. Tokyo (Japan), 1933-63, prof. plant physiology, 1943-63, dir. Inst. for Applied Microbiology, 1955-63; dir. Tokugawa Inst. for Biol. Research, 1946—. Recipient prize and medal Japanese Assn. Sci., 1963, Acad. prize Acad. Japan, 1965. Mem. Deutsche Akademie der Naturforscher (hon.), Nat. Acad. Scis. (U. S.), Deutsche Botanische Gesellschaft (hon.), Bot. Soc. Am. (hon.) Author: Le bilan materiel et l'energetique des syntheses biologiques, 1935; also numerous articles. Analyses of material and energetic balance using Aspergillus; research on mechanism of photosynthesis, life cycle of alga using new technique of synchronous culture. Home: I-363 Shimo-ochiai, Shinjuku-ku, Tokyo. Office: Tokugawa Biol. Inst., Toshima-ku, Mejiro, Tokyo, Japan.*

TAMM, Christoph, Swiss chemist; b. Basel, Switzerland, Mar. 13, 1923; s. Walter Albert and Anna (Socin) T.; student U. Geneva, 1942; Ph.D., U. Basel, Switzerland, 1948; m. Lisa Hotz, Sept. 1, 1953; children—Lukas Vaspar, Micheal Andreas, Bettina Irene, Noëmi Christine. Post-doctoral fellow Squibb Inst. Med. Research, New Brunswick, N.J., 1949-50; post-doctoral fellow Columbia U., 1950-52; research asso. U. Basel, 1952-55, asst. prof., 1955; group leader Pharm. Research dept. Sandoz A.G., Basel, 1957-61; asso. prof. dept. chemistry, U. Basel, 1961-66, full prof. organic chemistry, 1966—, dean faculty scis., 1968—, cons. Sandoz A.G., 1961—. Mem. Swiss Chem. Soc., Am. Chem. Soc., Swiss Soc. Natural Scis. Studies, numerous publs. on structure of cardiac glykosides, steroids, synthesis of naturally-occurring sugars, microbiol. transformation of natural products; structure and biosynthesis of antibiotics and other natural products; nucleic acids of nucleotides. Home: 28 Meierweg, Riehen, Basel-Stadt, Switzerland. Office: Inst. fur Organische Chemie, 19 St. Johannsring, Basel, Switzerland.*

TAMM, Igor, Russian physicist; b. Vladivostok, USSR, July 8, 1895; s. Eugen and Olga (Davydova) T.; grad. Moscow State U., 1918, Dr. Physics and Math. Scis., USSR, 1933; m. Natalie Shuiskaia, Sept. 16, 1917; children—Irene, Eugen. Tchr. physics, univs. and tech. instns., Simpheropal and Odessa, 1919-22; with U. Moscow, 1924—, prof., 1927—, chair theoretical physics, 1930-37; head theoretical dept. Lebedev's Physical Inst., Acad. Scis. of USSR, Moscow, 1934—. Recipient (with P. Cerenkov, I. Frank), Nobel prize in physics, 1958; Stalin Prize, 1946. Mem. Am. Acad. Arts and Scis. (foreign member), Acad. Scis. USSR, Acad. Scis. of Poland. Author: The Principles of Electricity Theory, 1929; Über die Wechselwirkung der freien Elektronen mit der Strahlung nach der Diraschen Theorie des Elektrons und nach Quantenenlektrodynamik, 1930; Über eine mögliche Art der Elektronenbindung an Kristalloberflächen, 1932; Exchange Forces Between Neutrons and Protons and Fermi's Theory, 1934; (with S. Altshuler) Magnetic moment of the neutron, 1934; Radiation Emitted by Uniformly Moving Electrons, 1939; Relativistic Interaction of Elementary Particles, 1945; (with others) The Relativistic Theory of Neucleon Interaction, 1953; The Theory of Magnetic Thermonuclear Reactions, 1958. Research in quantum theory, controlled thermonuclear reactions; co-discoverer and interpreter of Cerenkov's radiation; with Shubin developed theory of photoeffect in metals, 1931; with Sakharov, suggested use of electric discharge in plasma, 1950. Address: Lebedev's Physical Institute, Academy of Sciences of USSR, Moscow B-134, Russia.

TAMMANN, Gustav, chemist, physicist; b. Jamburg, May 28, 1861. Prof., U. Tartu (Dorpat), U. Göttingen. Author: Lehrbuch der Metallkunde, 1914; Die Aggregatzustände, 1922; Lehrbuch der heterogenen Gleichgewichte, 1924. Research in phys. chemistry. Died Göttingen, Germany, Dec. 17, 1938.

TAMMI, Olli Eino, Finnish mathematician; b. Helsinki, Finland, Dec. 24, 1924; s. Eino W. and Aili (Julku) T.; M.A., Helsinki U., 1949, Ph.D., 1952; m. Annikki Tolsa, July 28, 1951; children—Pekka O., Eeva L., Heikki J. With Helsinki U., 1950-60, asso. prof., 1964—; prof. Oulu (Finland) U., 1960-61; asso. prof., Inst. Techn. Helsinki, 1961-64. Research, publs. on maximalization using step-function method. Home: 20 A 9 Ohjaajantie, Helsinki 40, Finland.*

TAMRES, Milton, chemist; b. Warsaw, Poland, Mar. 12, 1922; s. Morris and Lillian (Solberg) T.; came to U. S., 1929, naturalized; B.A. in Sci., Bklyn., 1943; Ph.D., Northwestern U., 1949; m. Françoise R. P. M. Lucot, Aug. 16, 1960; 1 dau., Louise R. Control chemist Celanese Corp. Am., 1943-44; faculty U. Ill., 1948-53, asst. prof., 1951-53; faculty U. Mich., Ann Arbor, 1953—, prof. chemistry, 1963—; vis. scientist NSF program, 1961—. Mem. Adv. Council on Coll. Chemistry, 1962-66. Guggenheim fellow, 1959-60; Petroleum Research Fund Internat. fellow, 1966-67. Mem. Am. Chem. Soc., A.A.A.S., Alpha Chi Sigma, Phi Lambda Upsilon (past nat. sec., past nat. editor, nat. pres. 1966—). Contbr. articles to tech. jours. Research in molecular assn., acid-base theory, effect ring size on properties cyclic compounds. Home: 1307 Brooks St., Ann Arbor, Mich. 48103.*

TANAKA, Kuniyoshi, Japanese chemist; b. Kyoto City, Japan, Nov. 21, 1913; s. Kunigoro and Kiyo (Nagano) T.; grad. Kyoto U., 1941, D.Ph., 1950; m. Jun-ko Amatsutsu, Feb. 27, 1943; children—Kunihiro, Sawa-ko, Yo-ko. Researcher, Chem. Research Labs. research and devel. div. Takeda Chem. Industries Ltd., Osaka, Japan, 1942-53, researcher-in-chief, 1953-60, head organic chemistry dept., 1960—. Mem. Pharm. Soc. Japan (prize 1950), Japanese Biochem. Soc., Japanese Cancer Assn., Chem. Soc. Japan. Contbr. articles to sci. jours. Discovery of method and theory of steric version of ephedrine, especially from physiologically inactive isomer of ephedrine to active one; established preparative methods of nucleotides and nucleotide coenzymes and synthesized new compounds which showed inhibitory activity upon nucleic acid metabolism and cancerostatic activity. Home: 5-87, Hon-machi, Toyonaka, Japan. Office: 4-56, Nishino-cho, Juso, Higashiyodogawa-ku, Osaka, Japan.*

TANAKA, Nobuo, Japanese microbiologist; b. Shizuoka-ken, Japan, Jan. 5, 1926; s. Manzo and Fusako (Aoki) T.; M.D., U. Tokyo, 1948; m. Teruko Tamada, Oct. 9, 1951; children—Yoko, Nobuyuki. Asst., Inst. for Infectious Diseases, U. Tokyo (Japan), 1950-53, asst. prof. Inst. Applied Microbiology, 1955-64, prof., 1964—; research fellow dept. bacteriology Harvard Med. Sch., 1953-55; researcher Inst. Microbial Chemistry, Tokyo, 1963—. Research and numerous publs. on antibiotics and their mechanism of action and biosynthesis; discovered several new antibiotics. Home: 1-7-12 Komazawa, Setagaya-ku, Tokyo, Japan.*

TANAKA, Nobuyuki, Japanese chemist; b. Tokyo, Japan, Mar. 8, 1920; s. Keijiro and Miki (Kurata) T.; grad. Tokyo Imperial U., 1941, postgrad., 1943-46, D.Sc., 1948; m. Hisako Idei, Nov. 17, 1944; 1 dau., Reiko. Asso. prof. Inst. Radiation Chemistry, Tokyo U., 1946-49, Inst. Sci. and Tech., 1949-55; research fellow dept. chemistry U. Minn., Mpls., 1952-54; prof. dept. chemistry Tohoku U., Sendai, Japan, 1955—. Recipient Am. Chem. Soc.-UNESCO award, 1952, Chem. Soc. award, 1965. Mem. Internat. Union Pure and Applied Chemistry (titular mem. analytical chemistry div. 1957—), Chem. Soc. Japan (mem. bd. 1949-52), Am. Chem. Soc., Japan Soc. for Analytical Chemists. Contbg. author: Electrodeposition, Treatise on Analytical Chemistry, Vol. 4, 1963; also numerous articles. Research on theory of electrochem. reactions of inorganic and organic compounds especially introduction of concept of chem. kinetics to theory of polarography; structure, equilibria and reactions of metal complexes in solutions, in solid state and at electrodes. Home: 58, Kitago-junin-machi, Sendai, Miyagi-ken, Japan.*

TANANAEV, Ivan Vladimirivich, Russian chemist; b. June 4, 1904; grad. Kiev Poly. Inst., 1925. Staff, Kiev Poly. Inst., 1925-34, became chief of lab., 1939; named chief of dept. Inst. Gen. and Inorganic Chemistry, USSR Acad. Scis., 1949, dept. dir., 1948-54. Exchange staff Nat. Bur. Standards, U. S. A., 1960. Recipient Order of Lenin. Mem. USSR Acad. Scis. Author: Physico-chemical Analysis in Analytical Chemistry, 1950; Kurnakov's Method of Physicochemical

Analysis in Analytical Practice, 1939. Research in analytical and inorganic chemistry, compounds of rare elements, fluorides, ferro-cyanides of various metals; application of physico-chem. methods to analytical chemistry. Home: 1-Aya Cheremushkinskaya 3, Moscow. Office: N.S. Kurnakov Inst. Gen. Chemistry, USSR Acad. Scis., Leninskii Prospekt 31, Moscow, USSR.

TANASE, Iona, Rumanian chemist; by Mares, Rumania, Apr. 24, 1925; s. Dobre P. and Maria Tanase; grad. Faculty Chemistry, Bucharest; m. Mogos Gheorghe, Nov. 28, 1958; 1 dau., Mogos Tiberius. Staff, dept. chemotherapy Acad. Rumania, Bucharest, 1952-—, head biochemistry labs., 1966-—. Mem. Union of Socs. Med. Scis. Author: Chromatographic Technique, 1967; also numerous articles. Introduced new chromatographic techniques for isolation and quantitative analysis of amino acids in biol. products; metabolic alterations occurring in amino acids, nucleic acids and purine and pyrimidine bases coincident with aging. Home: 1 Garibaldi. Office: 8 Bd.Dr.P. Groza, Bucharest, Rumania.*

TANCKE (Tancius), Joachim, German anatomist, surgeon; b. Perleberg, Prussia, Dec. 9, 1537. Prof. anatomy, surgery, Leipzig, Germany. Author: Alchimistisch Waiten Waitzenbäumlein, 1605; Sullinta . . . Brevis Artis Chemiae Instruction, 1605; Metallurgia . . . , 1728. Contbd. verses to Haligraphia (Tholde), preface to Triumph Antimonii, 1604, preface to collection med. prescriptions (attributed to Fidejustus Reinneccerus); pub. works attributed to Basil Valentine (in Promptuarium Alchemiae), 1614. Died Leipzig, Nov. 27, 1609.

TANDON, Ram Narayan, Indian plant pathologist; b. Shikohabad, India, Nov. 27, 1903; s. Nihal Chand and Matabdei (Matabdei) T.; B.Sc., Allahabad U., 1925, M.Sc., 1927; Ph.D., London U., 1939; Diploma, Imperial Coll., London, 1939; m. Sarawati Devi, Jan. 1923; children—Bishan N., Prakash N., Anand N., Gopal N. Faculty botany dept. U. Allahabad, 1927-65, prof. botany, plant pathology, head dept., 1959-65. Fellow Nat. Inst. Scis. (mem. council), Indian Acad. Scis.; mem. Nat. Acad. Scis. (v.p. 1966-68), Indian Sci. Congress Assn. (pres. botany sect. 1967-—), Indian Phytopath. Soc. (past pres.), Internat. Soc. Plant Morphologists (founding). Author: Supplement to the List of Indian Fungi, 1957-62; Physiology of Fungi, 1961; also numerous articles. Research on taxonomy and physiology of fungi, leaf spot diseases, post harvest diseases of fruits and vegetables, physiology of fungi; described several new fungi from India; devel. methods for control of some post harvest diseases. Home: 9 Edmonstone Rd., Allahabad, U.P., India.*

TANDON, S. L., Indian botanist; b. Nowshera (now Pakistan), Aug. 11, 1922; s. P. L. and Karam (Devi) T.; M.Sc. in Agrl. Botany, Banares Hindu U., 1945; Ph.D., Wash. State U., 1952; m. Pushpa, 1948; children—Nalini, Naresh, Mukesh. Reader botany, U. Delhi (India), 1958-—; asso. Indian Agrl. Research Inst. Mem. Indian Soc. Genetics and Plant Breeding, Hort. Soc. India. Research and publs. on colchicine induced polyploidy in relation of floriculture, polyploidy in relation to evolution in some angiosperms. Home: 8 D, Mauric Nager, Delhi-7, India.*

TANFORD, Charles, phys. biochemist; b. Halle, Germany, Dec. 29, 1921; s. Max and Charlotte (Eisenbruch) T. Came to U. S., 1939, naturalized, 1947; B.A. with honors in Chemistry, N.Y. U., 1943; M.A., Princeton, 1944, Ph.D., 1947; m. Lucia Lander Brown, Apr. 3, 1948; children—Victoria, James Alexander, Sarah Lander. Research chemist Oak Ridge Nat. Lab., 1944-45; postdoctoral fellow Harvard, 1947-49; faculty State U. Ia., 1949-60; prof. phys. biochemistry Duke, Durham, N.C., 1960-—. Cons., USPHS, 1959-63. Guggenheim fellow Yale, 1956-57. Recipient Research Career award USPHS, 1962. Mem. Am. Soc. Biol. Chemists, Am. Chem. Soc., A.A.A.S. Author: Physical Chemistry of Macromolecules, 1961; also numerous articles. Mem. editorial bd. Jour. Am. Chem. Soc., 1958-—, Jour. Biol. Chemistry, 1962-—. Research on properties of proteins, antibody protein, hemoglobin. Address: Duke U. Med. Center, Durham, N.C. 27706.*

TANG, Anthony Matthew, economist; b. Shanghai, China, May 6, 1924; s. Shu-ping and Yueh-pao (Hsu) T.; came to U. S., 1945, naturalized, 1953; B.B.A., Loyola U., New Orleans, 1949; Ph.D., Vanderbilt U., 1955; m. Jane Pentecost, Nov. 27, 1946; children—Matthew, Jeremiah, John, Lyda Anne, Arthur, Patricia. Mem. faculty Vanderbilt U., Nashville, 1955-—, prof. econs. and bus. adminstrn., 1963-—, chairman dept., 1968-—; vis. lectr. Osaka U., Japan, 1959-60; vis. prof. U. Cal. at Berkeley, 1963-64; vis. prof. econs., chmn. dept. Chinese U. Hong Kong, 1966-—. Cons. FOA, 1954. Mem. Am. Econ. Assn. (adv. com. to U. S. Census 1960-63, mem. policy bd. for Econs. Insts., 1963-—), Am. Farm Econs. Association, Society for International Development, Econometric Soc., So. Econ. Assn. (editor Jour. 1963-—). Author: Economic Development of the Southern Piedmont, 1958; also articles, chpts. in books. Research in testing explanations of regional agrl. poverty, measurement of productivity growth and contbns. of edn., research, devel. and diffusion to growth, devel. of procedure to evaluate policy and performance of agr. of Soviet-type economies. Home: 6720 Currywood Dr., Nashville 37205.*

TANG, Peter Sheng-Hao, polit. scientist; b. Hofei, Anhwei, China, Apr. 11, 1919; s. Ching Chow and Yun-kwei (Kuo) T.; A.B., Nat. Chengchih U., 1942; A.M., Columbia, 1947, Ph.D., 1952; m. Mary Mannien Cheng, Sept. 19, 1957; children—Catherine Teh-Yun, Grace Teh-Hua, Peter Teh-Li. Came to U. S., 1945, naturalized, 1955. Attaché Chinese embassy, Moscow, 1942-45; research asso. U. So. Cal., 1952-53, U. Wash., 1955-56; sr. research analyst Georgetown U., Washington, 1957-59, prof. adj. internat. relations, 1957-62; vis. prof. history Ind. U., 1962; prof. polit. sci. Boston Coll., 1962-—. Exec. dir. Research Inst. Sino-Soviet Bloc, Washington and Chestnut Hill, Mass., 1959-—, also trustee; cons. to univs. Am. Council Learned Socs. and Social Sci. Research Council grantee, 1959, 60, 62, Am. Philos. Soc. grantee, 1959. Mem. Am. Polit. Sci. Assn., Am. Hist. Assn., Am. Assn. Advancement of Slavic Studies, Assn. for Asian Studies, Am. Acad. Polit. and Social Sci., Pi Sigma Alpha. Author: Communist China Today, 2 vols., 1957-58, rev., 1961; Russian and Soviet Policy in Manchuria and Outer Mongolia 1911-31, 1959; Communist China as a Developmental Model for Underdeveloped Countries, 1960; The Training of Party Cadres in Communist China, 1961; Russian Expansion Into the Maritime Province, 1962; (with Joan Maloney) The Chinese Communist Impact on Cuba, 1962; The 22d Congress of the CPSU and Moscow-Tirana-Peking Relations, 1962; The Chinese Communist Struggle against Modern Revisionism, 1964. Analyzed and interpreted parallel expansionist policies of the Tsarist and Soviet govts. in Manchuria, Outer Mongolia, and Sinkiang prior to Communist takeover on the Chinese mainland; polit., econ., social, mil., cultural systems in Communist China; cause-effect relations of Moscow-Peking disputes; emphasized importance of ideological struggle in the Sino-Soviet rift over territorial, leadership, and econ. issues. Address: Dept. Polit. Sci., Boston Coll., Chestnut Hill, Mass. 02167.*

TANGEN, Roald, Norwegian physicist; b. Modium, Norway, Feb. 8, 1912; s. Sören and Anna (Hervig) T.; civil engr. Tech. U. Norway, 1936, dr.techn., 1948; m. Mally Falchenberg Möller, May 27, 1939; children—Dag, Svein. With dept. physics Tech. U. Trondheim, Norway, 1937-42, prof. physics, 1948-52; with U. Oslo (Norway), 1942-48, prof., 1952-—, dean Sci. Faculty, 1963-—. Mem. Norsk Fysisk Selskap, Det norske videnskapsakademi i Oslo, Det kongelige norske videnskabers selskaz Trondheim. Research, publs. on nuclear reactions, especially proton induced reactions in light elements. Home: 97C Bestumvegen, Oslo, Norway.*

TANIDA, Hiroshi, Japanese chemist; b. Osaka, Japan, Nov. 26, 1924; s. Saburo and Fusako (Umino) T.; M.S., Kyoto (Japan) U., 1947, Ph.D., 1959; postgrad. Tokyo (Japan) U., Brown U., Mich. State U.; m. Reiko Kitagawa, Mar. 8, 1951; children—Tomoko, Kazuko, Kyoko. Research chemist Shionogi Research Lab., Osaka, 1947-—. Mem. Chem. Soc. Japan, Pharm. Soc. Japan, Am. Chem. Soc., Chem. Inst. Can. Author: New Aromatics and Their Reactions, 1966; also articles. Syntheses and reactions of bridged benzocyclenes (a new type of aromatic compound); studies in mechanisms of carbonium ion rearrangements and elimination reactions, some reactions of pyridine- and quinoline-1-oxides. Home: 1-5-29, Mandai Abeno-ku, Osaka. Office: Shionogi Research Lab., Fukushi-ma-ku, Osaka, Japan.*

TANIEWSKI, Joseph, Polish physician; b. Radom, Poland, Mar. 4, 1899; s. Michael and Adeline (Adler) T.; M.D., Med. Faculty, Warsaw, U., 1925; m. Pauline Temerson, Nov. 12, 1927; children—Marius, Hedvige (Mrs. Andreas Szymczewski). Staff, Municipal Hosp., Warsaw, Poland, 1925-39, Mil. Hosp., Poznan, Poland, 1945-48; dir. otolaryngol. clinic Med. Acad., Szczecin, Poland, 1948-—. Recipient several Polish medals and awards for profl. and sci. activity. Mem. Otolaryng. Soc. in Szczecin (head 1948-—). Author: Audiology, 1951; Differential Diagnosis of Ear, Nose and Throat Diseases, 1960; also numerous articles. Co-editor, Polish Otolaryngology, 1948-—. Research in physiology and pathology of organ of hearing and equilibrium; diagnosis and treatment of diseases of upper air passages. Home: 39 u.Wyspianskiego, Szczecin, Poland.*

TANIEWSKI, Marian, Polish chemist; b. Radom, Poland, Feb. 5, 1930; s. Michal and Krystyna (Morawska) T.; Inz., Politechnika Slaska, 1952, Mgr. inz., 1954, Dr. nauk techn., 1959; Doc. habilit, 1962; m. Barbara Milowka, July 17, 1954; 1 dau., Maria. Staff, Inst. Chem. Synthesis, 1954-55; head lab. Inst. Heavy Organic Synthesis, 1955-61; aspirant, sr. asst. Politechnika Slaska, Gliwice, Poland, 1955-59, faculty, 1959-—, docent etat. (asst. prof.), 1962-—, head lab., 1962-—, dep. vice-chancellor, 1964-—. Postdoctoral scholar Oxford (Eng.) U., 1960-61; recipient award Minister of Higher edn. for spl. sci. achievements, 1964. Fellow Chem. Soc. London; mem. Polish Chem. Soc. Research and publs. on chemistry of hydrocarbons, theoretical and indsl. aspects of olefins prodn.; new methods of pyrolysis of gaseous and liquid hydrocarbons; discovered new catalysts for catalytic dehydrogenation of alkylbenzenes and paraffin hydrocarbons; kinetics and mechanism of thermal decomposition of hydrocarbons (olefins and paraffins); new catalytic transformation of olefins. Home: 9/2 Barlickiego. Office: 19 Strzody; Gliwice, Poland.*

TANIKAWA, Yasutaka, Japanese physicist; b. Hyogo Prefecture, Japan, 1916; grad. Osaka U., 1939; D.Sc. Became asst. instr. Osaka U. after graduation; instr., later asst. prof. Nagoya U.; prof. Kobe U. Research in quantum dynamics, elementary particle theory, reciprocal reaction of neuter meson and elementary particles, theory of reciprocal action of elementary particles; originated theory of beta collapse.

TANK, Franz, Swiss physicist; b. Zurich, Switzerland, Mar. 6, 1890; s. Theodor and Louise T.; m. Louise Wegmann, July 28, 1925; Dr.Phil., 1916; Dr.ès Sc. Tech. h.c., 1954; spl. student City and Guild Coll., London; 1 son, Theodor. With gymnasium, Zurich, 1902-08; with Swiss Fed. Inst. Tech., Zurich, 1908-12, prof. physics, 1922-33, prof. high frequency engring. and physics, 1934-60; rector, 1943-47; asst., lectr. physics Zurich U., 1912-22. Laurea in Honorem, Torino, Italy, 1960. Fellow I.E.E.E.; hon. mem. Swiss Assn. Elec. Engrs. Research and publs. on spectroscopy, electronics, TV, dielectric losses, electronics, high frequency engring. Home: 174 Frohburgst., Zurich, Switzerland.*

TANNENWALD, Peter Ernest, physicist; b. Kiel, Germany, Mar. 30, 1926; s. Bruno and Alma (Mendel) T.; came to U. S., 1940, naturalized, 1945; A.B. in Physics, U. Cal. at Berkeley, 1947, Ph.D., 1952. Staff, U. Cal. Radiation Lab., Berkeley, 1950-52; staff Lincoln Lab., Mass. Inst. Tech., Lexington, 1952-—, asso. head solid state div., 1965-—. Fellow Am. Phys. Soc.; member of Phi Beta Kappa, Sigma Xi. Editor: (with P. L. Kelley, B. Lax) Physics of Quantum Electronics, 1966. Research and publs in magnetic resonance and other microwave aspects solid state physics and quantum electronics, especially magnetic resonance in ferrites and magnetic films, masers, millimeter waves, microwave acoustics, stimulated Raman and Brillouin laser spectroscopy. Home: 1800 Massachusetts Av., Cambridge, Mass. 02140.*

TANNER, James Taylor, Am. animal ecologist; b. Homer, N.Y., Mar. 6, 1914; s. Clifford John and Emma (Taylor) T.; B.S., Cornell U., 1935, M.S., 1936, Ph.D., 1940; m. Nancy B. Sheedy, Aug. 15, 1941; children—David B., Betsy K., Jane T. Faculty, U. Tenn., Knoxville, 1947-—, prof. zoology, 1963-—; cons. Oak Ridge Nat. Lab., 1963-—. Mem. A.A.A.S., Am. Ornithol. Union, Ecol. Soc. Am., Internat. Soc. Biometeorology, Wilson Ornithol. Soc. Author: The Ivory-billed Woodpecker, 1942; General Zoology Laboratory Guide, 1950; also numerous articles. Editor: The Migrant, 1947-55. Research on history and status of rare species of birds, gen. ornithology and effects of climate on breeding of birds, characteristics of animal populations. Home: Route 10, Knoxville, Tenn. 37920.*

TANNER, William Francis, Jr., Am. geologist; b. Milledgeville, Ga., Feb. 4, 1917; s. William Francis and Robbie (Carter) T.; B.A., Baylor U., 1937; M.A., Tex. Tech. Coll., 1939; Ph.D., Okla. U., 1953; m. Julia Katherine Rigby, July 17, 1938; children—William Francis III, Bruce, Julianne. Asst. prof. Okla. Baptist U., Shawnee, 1946-51; spl. instr. Okla. U., Norman, 1951-54; asso. prof. geology Fla. State U., Tallahassee, 1954-66, prof., 1966-—. Vis. scientist La. Tech. Coll., Lamar Tech. Coll., N.W. La. Coll., 1965; instl. rep. Gulf U. Research Corp., 1965; chmn. Tertiary Sea Level Symposium, 1967. Fellow A.A.A.S., Geol. Soc. Am. (nat. program chmn. 1964); mem. Soc. Econ. Paleontologists and Mineralogists (chmn. grain size com. 1963-—), Southeastern Geol. Soc. (v.p. 1965), Am. Meteorol. Soc. (asso. editor Jour. Applied Meteorology 1966-—), Am. Assn. Petroleum Geologists, Am. Geophys. Union, Seismol. Soc. Am., Internat. Assn. Sedimentology, Internat. Assn. Hydrologic Research, Geologische Vereinigung, Am. Soc. Photogram., Geochem. Soc., Soc. Rheology, Soc. Exploration Geophysicists. Author: Geology of Seminole County, Okla., 1956 also numerous articles. Editor, Coastal Research Notes, 1963-—. Research on theory sediment transport, use sediment characteristics to determine ancient environments deposition, detailed field work in many western and so. states, theory mountain-making, theory ice ages, use models in geol. experimentation, statis. methods, improved methods teaching geology, reconstrn. ancient climates, beach erosion and maintenance. Home: 2004 High Rd., Tallahassee, Fla. 32303.*

TANNER, Wilson Pennell, Am. psychologist; b. N.Y.C., Dec. 23, 1912; s. Wilson Pennell and Marie (Rayman) T.; student Kenyon Coll., 1931-32; A.B., Wesleyan U., 1937; M.A., U. Fla., 1949; Ph.D., U. Mich., 1960; m. Helen Frances Hornbeck, Nov. 22, 1940; children—Frances Marie, Margaret Helen, Wilson Pennell, III, Robert Paul. Mem. electronic def. group dept. elec. engring. Cooley Lab., U. Mich., Ann Arbor, 1952-62, lectr., 1960-—, prof. psychology, 1963-—, dir. Sensory Intelligence Lab., 1962-—. Mem. adv. com. Directorate Information Scis., Air Force Office Sci. Research, 1962-63; mem. sci. adv. com. Center for Sensory Aids Evaluation and Devel., 1965-—; mem. Armed Forces-NRC Com. on Vision, 1961-65. Fellow Acoustical Soc. Am.; mem. Psychonomics Soc. (charter), Optical Soc. Am., Sigma Xi. Editor: (with Samuel Alexander, Robert E. Machol)

System Engineering Handbook, 1965. Research and publs. on sensory behavior of human beings, systems engring; introduced statis. decision theory as theoretical framework for human sensory behavior. Home: 1319 Brooklyn Av., Ann Arbor, Mich. 48104.*

TANNERY, Jules, French mathematician; b. Mantes, France, Mar. 24, 1848; brother of Paul Tannery; ed. l'École normale supérieure, became its vice prin., 1886, also prof., dir. sci. studies; mem. French Acad. Scis., 1907. Author: Introduction à la theorie des fonctions d'une variable, 1886; Éléments de la theorie des fonctions elliptiques, 1893-98; Leçons d'arithmétique, 1896; Science et philosophie, 1912. Died Paris, Nov. 11, 1910.

TANRET, Charles, French chemist; b. Joinville-le-Pont, France, 1847; pharmicist, Troyes, France, then Paris, from 1880; author thesis on albumin, 1872; described new reagent iodomercurate of potassium (reagent of Tanret); discovered various alcoloides and glucosides, including ergotinine, ergosterine, fongisterine, ergothioneine, pelletierine, valdivine, piceine; studied amyloses, glucose, starches; discovered (with Maquenne) racemic inositol. Died Paris, 1917.

TANSETTER VON THANNAU, Georg, Austrian astronomer; b. 1480. Prof. astronomy, Vienna. Prepared edit. of Witelo's work on perspective (with Apianus), pub. in Nuremberg, 1533, 51. Died 1530.

TANTTILA, Walter Hjalmer, Am. physicist; b. Sax, Minn., Nov. 21, 1922; s. Hermann and Rose (Koskeniska) T.; B.Chem. Engring., U. Minn., 1948, M.A., 1950; Ph.D., U. Wash., 1955; m. Katherine Patricia Turner, May 10, 1964; children—Patricia Rose, Harvey Maurice, Margaret Ellen, Ann Elizabeth. Instr., Minn. State Tchrs. Coll., Winona, 1950-51; research physicist Mpls.-Honeywell, 1951-52; asst. prof. Mich. State U., East Lansing, 1955-58; faculty U. Colo., Boulder, 1958—, prof. physics, 1963—. Cons. to pvt. cos., govt. agys. Recipient Sebastian-Kerrer award U. Wash., 1954. Mem. Am. Phys. Soc., Tau Beta Pi, Phi Lambda Upsilon. Contbr. articles to tech. jours. Research in magnetic resonance, low temperature ultrasonic excitation nuclear magnetic spin levels in solids, optical resonance on impurities in solids; co-discoverer ultrasonic excitation nuclear spin levels. Home: 3502 22d St., Boulder, Colo. 80302.*

TAQUINI, Alberto C., Argentinian physician; b. Buenos Aires, Argentina, Dec. 6, 1905; M.D., Med. Sch., U. Buenos Aires, 1929; m. Haydée Lucia Azumendi, July 4, 1933; children—Alberto C., Graziela Taquini de Blaquier, Carlos M. Chief research U. Buenos Aires, 1937, chief medicine, 1940-42, chmn., dir. Centro de Investigaciones Cardiologica, 1942—, chmn. postgrad. course cardiology, 1948-52, prof. medicine, 1952—, chmn. physiology, 1962—; research fellow Harvard, 1938-39, vis. research prof. U. Recife, 1952, U. Lima (Peru), 1959, U. Mich., 1960, U. San Pablo, 1961, U. Pa., 1962. Lectr. U. Cal. at Los Angeles, 1948, U. Toronto, U. Kingston (Ont.), 1962. Recipient award Faculty Medicine, 1936, 56, Nat. Com. Culture Periods, 1936-38, 42-44, 54-56. Mem. panel experts in cardiology WHO, 1959—. Mem. Argentine Soc. Cardiology, Argentine Soc. Physiology, Argentine Soc. Clin. Investigation, Argentine Soc. Rheology, Argentine Soc. Biology; hon. mem. socs. cardiology of Chile, Peru, Mexico, Venezuela, Centro Am., France. Author: Exploración del Corazón por Via Esofágica, 1936; Hipertensión Arterial Nefrógena 1943; El Corazón Pulmonar, 1954; also numerous articles. Research on mechanisms of adaptation to chronic anoxia, physiopathogenesis of arterial hypertension, renin angiotensin and neural mechanisms. Home: 1390 Montevideo, Buenos Aires, Argentina.*

TARBELL, Dean Stanley, Am. chemist; b. Hancock, N.H., Oct. 19, 1913; s. Sanford McClellan and Ethel (Millikan) T.; A.B., Harvard, 1934, A.M., 1935, Ph.D., 1937; m. Ann Tracy, Aug. 15, 1941; children—William Sanford, Linda Tracy, Theodore Dean. Postdoctoral fellow U. Ill., 1937-38; faculty U. Rochester (N.Y.), 1938-67, prof., 1948-61, Charles Frederick Houghton prof. chemistry, 1961-67, chmn. chemistry dept., 1964-67; Distinguished prof. chemistry Vanderbilt U., Nashville, 1967—. Cons., USPHS, 1955—. Guggenheim fellow Oxford U., 1946-47, Stanford, 1961-62. Mem. Am. Chem. Soc., Chem. Soc. (London), Nat. Acad. Scis., Am. Acad. Arts and Scis. Research and numerous publs. on chemistry of antibiotics, sulfur compounds, mechanism of reactions. Home: 6033 Sherwood Dr., Nashville 37215.*

TARDIEU, Auguste Ambroise, French physician; b. Paris, 1818; brother of Eugène-Amedée Tardieu; became doctor of hosps., 1850; named asst. prof., then prof. legal medicine, 1861; dean Sch. Medicine, from 1864; mem. Acad. Medicine. Author: Relation médicolégale de l'assassinat de la comtesse de Goerlitz, 1850; Dictionnaire d'hygiène publique et de salubrité, 1852-44; Etude médico-légale sur l'attentat aux moeurs, 1858; Etude médico-légale sur l'avortement, 1864; Etude sur l'infanticide, 1868; Etude médico-légal sur la pendaison, la strangulation et la suffocation, 1870. Died Paris, 1879.

TARDY DE MONTRAVEL, Louis François, French geographer; b. Vincesses, France, 1811; marine officer; accompanied Dumont d'Urville, 1837-40; explored Amazon Basin and adjacent areas of Brazil and Guiana, from 1840; established (with others) 1st French colony at New Caledonia, 1854; gov. Guiana, until 1864. Author: Instructions sur les côtés de la Guyane; Instructions sur la côté septentrionale du Brésil et le fleuve des Amazones, 1847; Instructions sur la Nouvelle-Calédonie, les mers du Japon et la mer d'Okhotsk, 1857. Died 1864.

TARJAN, Armen Charles, Am. phytonematologist; b. Cambridge, Mass., Dec. 10, 1920; s. Charles Harry ad Marie (Zakarian) T.; B.S., Rutgers U., 1947; M.S., U. Md., 1949, Ph.D., 1951; m. Mary Ann Coury, May 4, 1945; children—Susan Andrea, Alan Charles. Asst. nematologist U. S. Dept. Agr., Beltsville, Md., 1950-51; asst. prof. U. R.I., Kingston, 1951-55; faculty U. Fla. Citrus Expt. Sta., Lake Alfred, 1955—, prof., 1958—. Mem. Soc. Nematologists, Helminthological Soc. Washington, European Soc. Nematologists. Author: Check List of Plant and Soil Nematodes, 1960, and Supplement, 1967; also numerous articles. Research on systematics and control plant parasitic nematodes. Home: 631 W. Lake Elbert Dr., Winter Haven, Fla. 33880. Office: Citrus Expt. Sta., Lake Alfred, Fla. 33850.*

TARKIEWICZ, Stanislaw, Polish vet. physician; b. Sokal, Poland, Feb. 4, 1920; s. Jan and Stanislawa (Zelechowska) T.; Vet. physician U. Maria Curie-Skodowska, Lublin, Poland, 1948, D.V.M., 1951, habilitate, 1960; m. Irena Halina Marecka, Nov. 11, 1946; 1 son, Stanislaw Zbigniew. Faculty, High Agrl. Sch. (formerly Maria Curie-Skodowska) Lublin, 1948-61, curator, 1961-1964, sci. worker dept. vet. medicine, head dept. obstetrics and pathology reprodn. Vet. Faculty, 1965—. Mem. Polish Soc. Vet. Sci., Sci. Soc. in Lublin. Research and publs. on peritonel fluid of cattle and pathogenesis of reticulitis traumatica, influence of udder insufflation on glandular tissue and biochem. indices of blood serum, mastitis in cattle. Home: 2/41 R.D.M. Lublin, Poland.*

TARNIER, (Étienne) Stéphane, French surgeon; b. Aiserey, France, 1828; prof. obstet. clinic Faculty Medicine; pres. Acad. Medicine; leader of sch. of modern obstetrics; pioneer in application of ideas of Pasteur and Lister to obstetrics; inventor traction forceps. Died Paris, 1897.

TARNOWSKI, George Serge, physician, chemist; b. St. Petersburg, Russia, Aug. 29, 1911; s. Serge V. and Olga (Yurkevich) T.; M.D., U. Belgrade (Yugoslavia), 1935; m. Irene N. Muravyov, Sept. 6, 1942; children—George, Natalie. Came to U. S., 1949, naturalized, 1963. Med. bacteriologist, Europe, 1937-46; med. officer UNRRA, Germany, 1947-49; with Sharp & Dohme Co., 1949-51; exptl. cancer chemotherapy research Sloan-Kettering Inst., N.Y.C., 1951—. Mem. Am. Assn. for Cancer Research, Soc. Gen. Physiologists, Am. Chem. Soc. (div. med. chemistry). Contbr. articles to profl. jours. Screening chemicals for antitumor activity, mechanisms of action of antitumor drugs.*

TARR, Hugh Lewis Aubrey, biochemist; b. Clevedon, Eng., Nov. 17, 1905; s. Gilbert and Florence (Coates) T.; B.S.A., U. B.C., Vancouver, Can., 1926, M.S.A., 1927; Ph.D., McGill U., Montreal, Que., Can., 1928-31; Ph.D., Cambridge (Eng.) U., 1934; m. Patricia Garry, Feb. 1, 1956; children—Robin, Michael John, Nicholas Garry. Demonstrator microbiology U. B.C., 1927-29, McGill U., 1929-31; bacteriologist Rothamsted Exptl. Sta., Eng., 1934-38; staff Vancouver Lab., Fisheries Research Bd. Can., 1938—, dir., 1956—. Recipient Gold medal in pure and applied sci. Profl. Inst. Pub. Service Can., 1957. Mem. Fedn. Socs. for Exptl. Biology, Biochem. Soc., Canadian Biochem. Soc., Inst. Food Technologists, Royal Soc. Can. Asso. editor Jour. Food Sci. and Food Tech., 1952—, Fishing News Internat., 1957—. Research and numerous publs. on use of refrigerated sea water in fish transp., use of antibiotics in flesh food preservation: developed fat stabilizing agts. (antioxidants) for protecting flavor, odor, or nutritive properties of frozen fish products. Home: 5696 Eagle Harbour Rd., West Vancouver, B.C. Office: 6640 N.W. Marine Dr., Vancouver, 8, B.C., Can.*

TARR, Ralph Stockman, Am. geologist; b. Gloucester, Mass., Jan. 15, 1864; s. Silas Stockman and Abigal (Saunders) T.; S.B., Lawrence Sci. Sch. (Harvard), 1891; m. Kate Story, Mar. 28, 1892. Asst. U. S. Fish Commn. and Smithsonian Instn., 1882-83; asst. geologist, Tex. Geol. Survey, 1889; asst., U. S. Geol. Survey, 1888 and 1891; asst. in geology, Harvard, 1890-91; asst. prof. geology, 1892-97, prof. dynamic geology and physical geography, 1897-1906, prof. physical geography, 1906—, Cornell U.; spl. field asst. U. S. Geol. Survey, 1903-06. Author: Economic Geology of United States, 1893; Elementary Physical Geography, 1895; Elementary Geology, 1897; First Book of Physical Geography, 1897; Tarr and McMurry Geographies, 1900, 1902; Physical Geography of New York State, 1902; New Physical Geography, 1904. Authority on glaciers; contributed to knowledge of glacial erosion, nature of ablation moraine, and effect of earthquakes on glaciers; interested in geology of Finger Lake region in upper New York state Died Mar. 21, 1912.

TARSKI, Alfred, mathematician, logician; b. Warsaw, Poland, Jan. 14, 1902; s. Ignace and Rose (Prussak) T.; Ph.D., U. Warsaw, 1924; m. Maria Witkowski, June 23, 1929; children—Jan A., Eva Kristina (Mrs. Andrzej Ehrenfeucht). Faculty, U. Cal., Berkeley, 1942—, prof., 1946—, faculty research lectr., 1962-63; vis. prof. Nat. U. Mexico, summer 1957; research prof. Miller Inst. Basic Research in Sci., 1958-60. Recipient Fulbright award, 1954; Rockefeller Found. fellow, 1935; Guggenheim Found. fellow, 1941-42, 55-56; Colloquium lectr. Am. Math. Soc., 1952; Shearman Meml. lectr. London U. Coll., 1950, 65. Mem. Assn. Symbolic Logic (past pres.), Am., Dutch math. socs., Royal Netherlands Acad. Scis. and Letters, Nat. Acad. Scis., Brit. Acad. Author: The Concept of Truth in the Languages of Deductive Sciences, 1933; (with others) Geometry, 1935; Introduction to Logic and to the Methodology of Deductive Sciences, 1936; (with B. Jónsson) Direct Decompositions of Finite Algebraic Systems, 1947; A Decision Method For Elementary Algebra and Geometry, 1948; Cardinal Algebras, 1949; (with others) Undecidable Theories, 1953; Logic, Semantics, Metamathematics, 1956; Ordinal Algebras, 1956; also numerous articles. Study of general algebra; set theory; measure theory; math. logic and meta-mathematics. Home: 462 Michigan Av., Berkeley, Cal. 94707.*

TARTAGLIA, Niccolo (or Nicola Fontana), Italian mathematician; b. Brescia, Italy, circa 1500; prof., Verona, also Vicenza, Brescia, Venice. Author: Della nuova scienza, 1537; Quesiti ed invenzioni diverse, 1546; General trattato de'numeri e misure, 3 vols., 1556-60. Placed ballistics on sci. basis; credited with discovery of solution of cubic equation, 1541, pub. by G. Cardano as his own, 1545; claimed to have invented gunner's quadrant; studied fortifications, diving bell, raising of sunken ships. Died Venice, Dec. 13, 1557.

TARTAR, Vance, Am. zoologist; b. Corvallis, Ore., Sept. 15, 1911; s. Herman Vance and Stella (Parsons) T.; B.S., U. Wash., 1933, M.S., 1934; Ph.D., Yale, 1938; m. Emogean Eva Saunders, June 19, 1950; children—Helen Sarah, Karl Nicholas, Wanda Gale. Biologist, Wash. Dept. Fisheries, Gig Harbor, 1943-49; research prof. zoology U. Wash., Seattle, 1951—. Mem. Soc. for Developmental Biology, Soc. Protozoologists, Phi Beta Kappa, Sigma Xi. Author: The Biology of Stentor, 1961. Asso. editor Jour. Protozoology, 1960—. Discovered and developed method for cell grafting and nuclear transplantation in ciliate protozoa. Home: R.F.D., Nahcotta, Wash. 98637.*

TARUFFI, Cesare, Italian pathologist; b. Bologna, Italy, Mar. 27, 1821; s. Gaetano and Aurelia (Bevilacqua) T.; degree in surgery, 1842, in medicine, 1844; became mem. staff Hosp. for Cure of Beggars, Bologna, 1844; surgeon Venetian campaign, 1848; attended wounded in Rome, 1849; became prof. anatomy and patholgoy, 1859, prof. emeritus, 1893; lectr. various fgn. and Italian sci. socs. Mem. Acad. Sci. of Bologna Inst. (pres.), Med. Surg. Soc. Bologna (pres. 1897-98), Acad. Medicine Turin, Lombardian Inst. Sci. Author: Storia della Teratologia, 8 vols., 1881-94. Cofounder Italian Bibliography of Med. Sci. Died July 7, 1902.

TARVER, Harold, biochemist; b. Wigan, Eng., June 7, 1908; s. George G. and Margaret (Hayhurst) T.; B.S., U. Alta., Can., 1932, M.S., 1935; Ph.D., U. Cal., 1939; children—Frank R., Terrance. With U. Cal. at Berkeley and San Francisco, 1936—, prof. biochemistry, acting chmn. dept. biochemistry Sch. Medicine, 1963-67. Mem. A.A.A.S., Am. Soc. Biol. Chemists, Biochem. Soc. (Eng.), Soc. for Exptl. Biology and Medicine, N.Y. Acad. Scis., Am. Chem. Soc. Research, numerous publs. in protein, amino acid, fat metabolism, radioisotope labelling techniques.*

TASHIRO, Shiro, biochemist; b. Kagoshima, Japan, Feb. 12, 1883; s. Shirobe and A. (Tashiro) T.; came to U. S., 1901; B.S., U. of Chicago, 1909, Ph.D., 1912; Dr. Med. Science, Kyoto Imperial U., Japan, 1923; m. Shizuka Kawasaki, of Honolulu, Nov. 9, 1915; children—Kiyoshi, Kazuo, Mitsu. Asst. in biochemistry, U. of Chicago, 1909, fellow in physiol. chemistry, 1910-12, asst. in same, 1912-13, asso., 1913-14, instr. in same, 1914-18, asst. prof., 1918; asso. prof. biochemistry, U. of Cincinnati, 1919-21, prof. since 1921; asst. dir. biochem. service, Cincinnati Gen. Hosp. Fellow A.A.A.S.; mem. Am. Soc. Biol. Chemistry, Am. Physiol. Soc., Am Chem. Soc., Soc. de chim. Biology (France), Ohio Acad. Science (ex-v.p.), Phi Beta Kappa, Sigma Xi. Awarded Crown Prince Memorial prize, Imperial Acad., Japan. Author: A Chemical Sign of Life, 1917. Editor: (with N. C. Foot) Kawamura's Studies on Tsutsugamushi Disease, 1926. Contbr. numerous articles to scientific jours. Invented biometer to measure quantities of carbon dioxide given off by small amounts of tissue. Home: 257 Loraine Av., Cincinnati, O.

TASSILO, Antoine, Austrian gynecologist; b. Vienna, Oct. 25, 1895; s. Rudolf and Helen (Ullmann) T.; M.D., U. Vienna; postgrad. U. Clinic Surgery and Gynecology; agrégé, 1936; m. Lore Trappen, 1921; a child, Duglore. Dir. gynecol. sect. Vienna-Laniz Hosp., 1937-40; dir. Gynecol. Clinic, U. Innsbruck (Austria), 1940-43; dir. U. Clinic Gynecology, Vienna, 1943—. Recipient Silver decoration of Austrian Order of Merit. Mem. Austrian Acad. Sci., Leopoldina Soc. Halle. Contbr. numerous publs. to tech. jours. Home: Wickenburgstrasse 26, Vienna 8. Office: I. Frauenklinik, Spitalgasse 23, Vienna 1, Austria.

TATARINOV, Pavel Mikhailovich, Russian geologist; b. Nov. 6, 1895; grad. Leningrad Mining Inst., 1925. Staff Geol. Com. (All Union Sci. Research Geol. Inst.), 1924-49, 54; instr. Leningrad Mining Inst., 1930-40, became prof., 1940. Corr. mem. USSR Acad. Scis. Author: Conditions for the Formation of Ore and Non-Metallic Mineral Deposits, 1955; Mining Mechanics, 1960. Coauthor: A Course on Mineral Deposits, 1946; A Course on Non-metallic Mineral Deposits, parts 1-2, 1934-35. Research and publs. on ore deposits of iron and non-metallic indsl. minerals, particularly in Urals. Office: Leningrad Mining Inst., Leningrad, USSR.

TATARINOV, Yuri Semyonovitch, Russian physician; b. Astrakhan, USSR, Sept. 2, 1928; s. S. A. and A. F. (Tatarinova) T.; grad. Astrakhan Med. Inst., 1952, postgrad., 1952-55; M.D., 1965; m. Elvira K. Emelyantachick, Sept. 25, 1951; 1 dau., Nina. Asst. physiology Volgograd Med. Inst., 1955-57; reader biochemistry Astrakhan Med. Inst., 1957-59, chief chair biochemistry, 1959—, also rector. Mem. All-Union Soc. Biochemistry. Author numerous articles. Discovered two new antigenic components with electrophoretic mobility of beta globulins, also embryo-specific alpha globulin in blood serum of persons with hepatoma (basis for working out a new immunochem. method in diagnosing this disease); demonstrated group characteristic of the fetal sera according to embryo-specific globulins. Home: 23/20 Lenin. Office: 20 Metchnokov, Astrakhan, USSR.*

TATE, Alexander Norman, Brit. chemist; b. Wells, Eng., Feb. 24, 1837; s. James and Emma (Norman) T.; studied chemistry in lab. of James Sheridan Musfratt, Liverpool, Eng.; m. Elizabeth Millicent Faulkes, 1860; 2 daus. Joined lab. Messrs. J. Hutchinson & Co., 1860; began analytical and cons. practice Liverpool, 1863; built and managed oil-refining works, Isle of Man, also Flintshire until 1860; purchased practice, Hackins Hey, Eng.; editor mag. Research, 1888-90. Mem. Soc. Chem. Industry (charter, pres., v.p.). Author: Petroleum and its Products, 1863; also articles. Analysis of oils and fats; research in chem. geology, tech. chemistry. Died Orton, Gt. Britain, July 22, 1892.

TATE, Thomas Turner, Brit. engr., mathematician; b. Alnwick, Eng., Feb. 28, 1807; s. Ralph and (Turner) T.; studied under an architect, Edinburgh, Scotland; m. twice. Became lectr. chemistry York Med. Sch., 1835; became master math. and sci. dept. Battersea Tng. Coll., 1840, Kneller Coll., 1849. Fellow Royal Astron. Soc. Author: Principles of Geometry, Mensuration, Trigonometry, Land Surveying and Levelling, 1848; Mathematics for Working, 1848. Philosophy of Education, 1854; also articles. Invented Tate double-piston air pump; studies in math. and exptl. scis. Died Liverpool, Eng., Feb. 18, 1888.

TATSUMOTO, Mitsunobu, geochemist; b. Junsho, Japan (now Korea), Mar. 19, 1923; s. Zenshiro and Sai (Kihara) T.; B.S., Tokyo Bunrika U., 1948; D.Sc., 1957; m. Kimiko Okubo, Mar. 17, 1948; children—Kuniyasu, Mariko. Research asst., lectr. Tokyo Kyoiku U., 1948-57; research fellow Scripps Instn.-Oceanography U. Cal., La Jolla, 1957-58; research fellow dept. oceanography Tex. A. and M. Coll., 1959; research fellow div. geol. scis. Cal. Inst. Tech., 1959-62; chemist U. S. Geol. Survey, Denver, 1962—; research fellow Welch Found., 1958. Mem. Geochem. Soc., Am. Geophys. Union. Research in using lead isotopes as tracers, problems of source rocks of volcanic rocks, differentiation history of source rocks, mode of generation of magma and effect of crustal contamination; minor element distbn. in deep sea deposits. Office: U. S. Geol. Survey, Fed. Center, Denver 80225.*

TATUM, E(dward) L(awrie), Am. biochemist, geneticist; b. Boulder, Colo., Dec. 14, 1909; s. Arthur Lawrie and Mabel (Webb) T.; A.B., U. of Wis., 1931, M.S., 1932, Doctor of Philosophy, 1934; married June Alton, 1934; children—Margaret Carol, Barbara Ann; married second, Viola Kantor, 1956. General Edn. Bd. fellow, U. of Utrecht, Holland, 1936-37; research asso. in biology, Stanford, 1937-41, asst. prof. biol., 1942-45; asso. prof. botany, Yale, 1945-46, prof. microbiol., 1946-48; prof. biology Stanford, 1948-56, prof. biochemistry, head dept., 1956-57; mem., prof. Rockefeller Inst., 1957—. Recipient (with G. W. Beadle and J. Lederberg) Nobel Prize for physiology and medicine, 1958. Mem. A.A.A.S., Am. Chemical Society, American Philosophical Soc., American Society of Biological Chemists, Bot. Soc. Am., Genetics Soc., Nat. Acad. Sci., Nat. Sci. Bd. Mem. editorial bds. profl. jours.; co-editor, Frontiers of Biology. Codeveloper of "one gene-one enzyme" concept demonstrating that biochem. processes are regulated by genes; research in genetic mutations and with mold, Neurospora. Home: 450 E. 63d St., N.Y.C. 10021. Office: Rockefeller U., 66th St. and York Av., N.Y.C. 10021.

TAUB, Abraham Haskell, Am. applied mathematician; b. Chgo., Feb. 1, 1911; s. Haskel Joseph and Mary (Sherman) T.; B.S. honors in Math. and Physics, U. Chgo., 1931; Ph.D. in Math. Physics Princeton, 1935; m. Cecilia Vaslow, Dec. 26, 1933; children—Mara, Nadine, Haskell. Asst., Princeton, 1934-35; mem. Inst. for Advanced Study, 1935-36, 40-41, 59, 62-63; Instr. math. U. Wash., Seattle, 1936-38, asst. prof., 1938-42, asso. prof., 1942-46, prof., 1946-48; research prof. applied math. U. Ill., Urbana, 1948-64, head digital computer lab., 1961-64; prof. math. U. Cal., Berkeley, 1964—, dir. Computer Center, 1964-68; theoretical physicist div. 2 Nat. Def. Research Conf., Princeton, 1942-45. Mem. applied math adv. com. Nat. Bur. Standards, 1949-54, chmn. adv. panel applied math div., 1951-60; chmn. tng. and com. research mathematician NRC, 1952-54; mem. rev. com. applied math div., cons. Argonne Nat. Lab., Lemont, Ill., 1960—; cons. Brookhaven Nat. Lab., Upton, N.Y., 1957—, Los Alamos Sci. Lab., 1953-67, Lawrence Radiation Lab, Livermore, Cal., 1961—; chmn. math. and computer research adv. com. AEC, 1962-65. Recipient president's certificate of merit, 1948; Guggenheim fellow, 1948, 54. Fellow Am. Phys. Soc.; mem. A.A.A.S., Am. Math. Soc., Math. Assn. Am., Sigma Xi, Phi Beta Kappa. Editor: Collected Works of John Von Neumann, 1962. Editor, Ill. Jour. Math., 1957-62, Communications in Math. Physics, 1965—. Research on relativity, shock tubes and interaction of shock waves; determined classes of vacuum solutions of Einstein field equations; formulated generalization of Rankine-Hugonoit equations in general relativistic hydrodynamics; participated in design and constrn. of high speed stored programmed digital computers. Home: 242 Yale Av., Berkeley, Cal. 94708.*

TAUB, James Monroe, Am. metall. engr.; b. Cleve., July 26, 1918; s. Morris and Rhea (Altfeld) T.; B.S. in metall. engring. Case Inst. Tech., 1940, M.S., 1943; m. Selma E. Pearlman, Dec. 11, 1943; children—Janet B., Judith A. Metallurgist, Republic Steel Co., Cleve., 1940-42; research asst. Case Inst. Tech., 1942-44; group leader materials tech. group Los Alamos Sci. Lab. of U. Cal., Los Alamos 1944—. Recipient Ernest O. Lawrence award AEC, 1964. Mem. Am. Soc. Metals. Devel. fabrication techniques for uranium and other materials; devel. of refractory materials and fabrication techniques for high temperature application. Home: 1479 46th St., Los Alamos 87544. Office: Box 1663, Los Alamos 87544.*

TAUBE, Henry, chemist; b. Neudorf, Saskatchewan, Can., Nov. 30, 1915; s. Samuel and Albertina (Tiledetski) T.; B.S., U. Saskatchewan, 1935, M.S., 1937; Ph.D., Univ. Cal., 1940; married Mary Alice Wesche, Nov. 27, 1952; children—Linda, Marianna, Heinrich, Karl. Came to U. S., 1937, naturalized, 1942. Instr. U. Cal., 1940-41; instr., asst. prof. Cornell U., 1941-46; asst. prof. U. Chgo., 1946-48, asso. prof., 1948-52, prof., 1952-62, chmn. dept. chemistry, 1955-59; prof. chemistry Stanford U., 1962—; Brotherton lectr., vis. prof. U. Leeds, 1955; Arthur D. Little vis. prof. Massachusetts Inst. Technology, 1960; Baker lectr. Cornell U., 1965. Guggenheim fellow, 1949, 55; recip. award for nuclear applications in chemistry Am. Chem. Soc., 1955; Harrison Howe award, 1961; also Chandler medal Columbia University, 1964. Mem. Am. Acad. Arts and Scis., Nat. Acad. Scis., Am. Chem. Soc., Phi Beta Kappa, Sigma Xi. Work on inorganic reactions in solution, complex ions; (with J. P. Hunt) gave 1st proof of solvated cation in solution formula; correlated electronic structure with inorganic complexes substitution rate; research on oxidation reduction, electron transport. Home: 229 Grove Dr., Portola Valley, Cal. Office: Dept. Chemistry, Stanford U., Stanford, Cal. 94305.

TAUBE, Meczyslaw, Polish chemist; b. Warsaw, Poland, Aug. 23, 1918; s. Michal and Maria (Lelczuk) T.; M.S., Polytech. Inst., Warsaw, 1948; Ph.D., Polish Acad. Sci., 1960, Docent of radiochemistry, 1963; m. Zofia Wolowicz, Apr. 13, 1946; children—Michal, Piotr. Staff research and devel. dept. Govt. Planning Commn., 1949-56; head Transuranium Lab., Inst. Nuclear Research, Warsaw, 1956—; head radiochem. dept. Warsaw, U., 1961—. Mem. Polish Council Atomic Energy, 1960—. Recipient Grunwald Cross. Mem. Polish Chem. Soc., Am. Nuclear Soc. Author: Plutonium, 1958; Transuranium Elements, 1961; Hydrogen—the Carrier of Life, 1965; also articles. Coeditor, Nukleonika, 1957—. Research on molten plutonium chlorides as fuel for nuclear reactors with fast neutrons, synthesis of amino acids from primitive substrates; developed new concept of salt-boiling nuclear reactors. Home: 5 Pl. Konstytucji, Warsaw. Office: 16 Dorodna, Warsaw-Zeran, Poland.*

TAUBER, Henrik, Danish geochemist; b. Nyköbing, Denmark, Oct. 10, 1921; s. Frederik and Minna (Christensen) T.; M.Sc. in Chem. Engring., Tech. U. Copenhagen (Denmark), 1945; m. Anita Knötel, Apr. 5, 1950; children—Astrid, Ulrik, Vibeke. Chemist, H. Lundbeck & Co., Copenhagen, 1945, Industrias Votorantin S.A., Sao Paulo, Brazil, 1946-47; chem. engr. F. L. Smith & Co., Copenhagen, 1947-51; research asst. Carbon-14 Dating Lab., U. Copenhagen, 1951-57, faculty radioisotope techniques, 1957—; head C-14 Dating Lab., Nat. Mus., Copenhagen, 1957—. Sec. commn. on absolute age of Quarternary deposits INQUA, 1960—. Mem. Geol. Soc. Denmark, Geophys. Soc. Denmark, Archael. Soc. Denmark, Soc. Applied Nuclear Physics. Author: Differential Pollen Dispersion and the Interpretation of Pollen Diagrams, 1965; also articles. Research on devel. of C-14 dating method, time dependent and geographic variations in atmospheric C-14 activities, C-14 dating of Late and Postglacial horizons, pollen zone borders in Denmark and Western Europe; studies of pollen dispersion and pollen filtration with applications in pollen analysis and interpretations of pollen diagrams. Home: 223 Skodsborgvej. Office: Nat. Mus., 10 Nyvestergade, Copenhagen, Denmark.*

TAUSSIG, Frank William, Am. economist; b. St. Louis, Mo., Dec. 28, 1859; s. William and Adele (Wuerpel) T.; A.B., Harvard, 1879, Ph.D. and A.M., 1883, LL.B., 1886, Litt.D., 1916; Litt.D., Brown U., 1914, from U. of Cambridge, England, 1933; LL.D., Northwestern University, 1920, and from U. of Michigan, 1927; hon. Ph.D., U. of Bonn, Germany, 1928; m. Edith Thomas Guild, June 29, 1888; children—William Guild, Mrs. Mary Guild Henderson, Catharine C., Helen B.; m. 2d, Laura Fisher, 1918. Instr. polit. economy, 1882-86, asst. prof., 1886-92, prof., 1892-1901, Henry Lee prof., 1901-35 (emeritus), Harvard U.; chmn. U. S. Tariff Commn., 1917-19. Fellow Am. Acad. of Arts and Sciences. Author: Tariff History of United States, 1888; Silver Situation in United States, 1892; Wages and Capital, 1896; Principles of Economics, 1911, 4th edit., 1939; Some Phases of the Tariff Question, 1915; Inventors and Money-Makers, 1915; Free Trade, the Tariff and Reciprocity, 1919; International Trade, 1927; Social Origins of American Business Leaders, 1932. Editor of Quarterly Journal of Economics, 1896-1937. Authority on questions of international commerce, especially U. S. tariff; developed import-export theory and wage-fund theory; interested in applying economics to public policy; co-founder of School of Business Administration at Harvard. Died Nov. 11, 1940.

TAUSSIG, Helen Brooke, Am. physician; b. Cambridge, Mass., May 24, 1898; d. Frank William and Edith (Guild) Taussig; student Radcliffe Coll., 1917-19; A.B., U. Cal., 1921; research and study Johns Hopkins Sch. Medicine, 1922-24, M.D., 1927. Fellow heart sta. Johns Hopkins Hosp., 1927-28, intern pediatrics, 1928-30; charge childrens heart clinic Harriet Lane Home, Balt., 1930-63; mem. faculty Johns Hopkins Sch. Medicine, 1930-63, prof. pediatrics, 1959-63, prof. emeritus, 1963—. Named chevalier Legion of Honor (France); co-recipient Passano award, 1948; recipient Hon. medal Am. Coll. Chest Physicians, 1953, Antonio Feltrinelli prize, Rome, Italy, 1954, Lasker award Am. Pub. Health Assn., 1954; co-recipient Gardiner award, 1959; recipient First Thomas River Meml. Research Fellowship award Nat. Found., 1963, Medal of Freedom, 1964, Johns Phillips Meml. award A.C.P., 1966. Founders award Radcliffe Coll., 1966. Mem. Am. Heart Assn. (pres. 1965-66; Gold Heart award 1963). Assn. Am. Physicians, Am. Acad. Arts and Scis., Phi Beta Kappa. Author: Congenital Malformations of the Heart, 1947, 2d edit., 2 vols., 1960; also numerous articles to prof. journals. Specialist in congenital malformations of the heart; co-developer of "blue baby" operation on cyanotic children, 1944; 1st physician to alert U. S. to dangers of thalidomide, 1962; 1st to demonstrate that changes in heart and lungs could be diagnosed by x-ray and fluoroscope. Home: 1100 Hollins Lane, Balt.

TAUSSKY, Olga (Mrs. John Todd), mathematician; b. Olomouc, Czechoslovakia, Aug. 30, 1906; d. Julius David and Ida (Pollach) Taussky; Ph.D., U. Vienna, 1930; attended U. Zurich, 1929; postgrad. Bryn Mawr Coll., 1934-35; M.A. (hon.), Cambridge (Eng.) U., 1937; m. John Todd, Sept. 29, 1938. Came to U. S., 1947, naturalized, 1953. Sci. officer Ministry Aircraft Prodn., London, Eng., 1943-46; faculty U. London, 1937-44; cons. math. Nat. Bur. Standards, Washington, 1948-57; research asso. math. Cal. Inst. Tech., Pasadena, 1957—. Recipient Sr. Research award Dept. Sci. and Indsl. Research, London, 1947; fellow Girton Coll., Cambridge, 1934-38, 39-40; Fulbright prof. U. Vienna, 1965. Mem. Am. Math. Soc., Math Assn. Am., London Math. Soc., Edinburgh Math. Soc. Co-editor: Hilbert's Collected Papers in Number Theory, 1932; Applied Mathematics Series, Vol. 29, 1952, Vol. 39, 1954; editor Jour. Linear Algebra. Contbr. numerous articles to profl. jours., also chpts. to books. Research in number theory, algebra, matrix theory, topological algebra. Home: 1025 Sierra Bonita Lane, Pasadena, Cal. 91106.*

TAVEL, Ernst, Swiss surgeon, bacteriologist; b. Bretonnières/Waadt, Switzerland, Feb. 2, 1858; ed. Lausanne, Strasbourg, Berlin, Bern, Paris; worked under Koch, 1885; asst. to Keeher (surgeon); prof. bacteriology, Berne, Switzerland; practiced medicine, specializing in surgery, Berne. Author: (with Th. Kocher) Vorlesungen über chirurgischen Infektionskrankheiten, 1895; Chirurgische Infektion und deren Prophylaxe, 1905. Developed antistreptoccus serum; emphasized prevention of infection of surg. wounds, antisepsis and reduction of spread of bacteria. Died Oct. 6, 1912.

TAVERAS, Juan Manuel, physician; b. Moca, Dominican Republic, Sept. 27, 1919; s. Marcos M. and Ana L. (Rodriguez) T.; B.S., Normal Sch. Santiago (Dominican Republic), 1937; M.S., U. Santo Domingo (Dominican Republic), 1943; postgrad. U. Pa. Sch. Medicine, 1944-45; M.D., U. Pa., 1949; m. Berenice Helen McGonigle, June 12, 1947; children—Angela (Mrs. Perry Maynard), Louise Helen, Jeffrey Lawrence. Came to U. S., 1944, naturalized, 1950. Instr., U. Santo Domingo, 1943-44; asst. instr. Grad. Sch. Medicine, U. Pa., 1947-48; faculty Coll. Phys. and Surg., Columbia U., N.Y.C., 1950-65, prof. radiology, 1959-65; dir. radiology Neurol. Inst., N.Y.C., 1952-65; staff Presbyn. Hosp., N.Y.C., 1950-65; prof., chmn. dept. radiology, dir. Mallinckrodt Inst. Radiology, Washington U. Sch. Medicine, St. Louis, 1965—; radiologist-in-chief Barnes and Allied Hosps., Washington U. Med. Center, St. Louis, 1965—. Cons. to hosps. Diplomate Pan-Am. Med. Assn. Fellow Am. Coll. Radiology; hon. mem. Phila. Roentgen Ray Soc., Radiol. Soc. Venezuela, Tex. Radiol. Soc., Radiol. Assn. Central Am. and Panama; mem. A.M.A., Am. Neurol. Assn., N.Y. Roentgen Soc., Am. Roentgen Ray Soc., Radiol. Soc. N.Am., N.Y. County, St. Louis med. socs., Inter-Am. Coll. Radiology, World Fedn. Neurology, Am. Soc. Neuroradiology (past pres.), Harvey Cushing Soc. (asso.), N.Y. Acad. Scis., Assn. U. Radiologists, Mo. State Med. Assn., Mo. Radiol. Soc., St. Louis Soc. Neurol. Scis., Greater St. Louis Soc. Radiologists. Author (with Ross Golden), Roentgenology of the Abdomen 1961; (with Ernest H. Wood) Diagnostic Neuroradiology, 1964; also articles. Research on radiologic diagnosis of vascular and neoplastic diseases of central nervous system. Home: 17 Carrswold, St. Louis 63105.*

TAVERNIER, John Francis, French radiologist; b. Algiers, Algeria, Mar. 6, 1928; s. George and Marcelle (Girard) T.; B.Math.; m. Françoise Marty, Dec. 22, 1962. Asst. radiology, U. Bordeaux, 1958-66, prof., 1966——. Mem. French socs. Radiology, Med. and Biol. Physics. Research, numerous publs. on radiocinematography of stomach and kidney, selective catheterizations of subhepatic and renal veins, splenoportography, neuroradiology in ophthalmology. Home: 20 Cours de Verdun. Office: Saint-André Hospital, Bordeaux 33, France.*

TAVERNIER, René Jordaan Ferdinand, Belgian soil scientist; b. Nevele, Belgium, Aug. 26, 1914; s. Charles and Maria (Goethals) T.; Lie. Sci. Geology, U. Ghent (Belgium), 1935, Dr.Sc., 1941; m. Cecilia Vereecken, Dec. 21, 1940; children—Dirk, Stefaan, Karel. Faculty, U. Ghent, 1935——, prof. geomorphology and soil sci., 1948——, dir. Internat. Tng. Centre for Postgrad. Soil Scientists, 1962——; dir. Soil Survey Center Belgium, 1947——. Mem. Internat. Soc. Soil Sci. (past pres.), Acad. Belgium (comdr. de l'ordre de Leopold), Société belge de Pédologie, Société belge de géologie, Soil Sci. Soc. Am. Author: (with G. V. Jacks, D. H. Boalch) Multilingual Vocabulary of Soil Science, 1960; also numerous articles. Research on stratigraphy of Pleistocene and Gholacene deposits of Belgium; classification of soils of Belgium; soil maps of Belgium; study genesis of soils especially of temperate regions. Home: 97 Ryssenbergstreet, Ghent, Belgium.*

TAVOLGA, William, Am. animal behaviorist; b. N.Y.C., Feb. 9, 1922; B.S., Coll. City N.Y., 1943; M.S., N.Y. U., 1946, Ph.D., 1949; m. Margaret Cordsen, Apr. 12, 1946. Asso. prof. dept. biology Coll. City N.Y., 1946——; research asso. dept. animal behavior Am. Mus. Natural History, N.Y.C., 1952——. Mem. A.A.A.S., Animal Behavior Soc., Am. Soc. Ichthyologists and Herpetologists, Soc. for Devel. and Growth, Am. Assn. U. Profs. Author: (with W. E. Lanyon) Animal Sounds and Communication, 1961; Marine Bio-Acoustics, 1964; also numerous articles. Research on embryology of fishes, reproductive behavior fishes, hearing and sound prodn. in fishes, marine animals sounds. Home: 60 Durie Av., Closter, N.J. 07624. Office: Dept. Biology, Coll. City N.Y., 139th St. and Convent Av., N.Y.C. 10031.*

TAWDE, Nanasaheb Ramji, Indian physicist; b. Salel, Malwan, India, Jan. 15, 1898; s. Ramji Raghoji and Laxmibai (Sawant) T.; B.A. with honors, St. Xavier's Coll., Bombay (India) U., 1919, B.Sc., 1922, M.Sc., 1928; Ph.D., U. London (Eng.), 1933; m. Shanta N. Bagwe, May 24, 1924; children—Shishir, Ashok, Vijay, Ajit, Vijaya (Mrs. Bhaskar Suryavanshi). Demonstrator physics Karnatak Coll., Dharwar, India, 1922-24; lectr. physics Inst. Sci., Bombay, 1924-46, prof. physics, 1946-53; prof. physics Karnatak U., Dharwar, 1953-64, acting vice chancellor, 1962; vice chancellor Marathwada U., Aurangabad, India, 1964——; vis. scientist U. Chgo., also U. Cal. at Berkeley, India Wheat Loan Exchange Program, 1956-57. Expert adviser to govt. and statuatory bodies, 1936——; mem. council Nat. Inst. Scis. India, 1953——. Fellow Nat. Inst. Scis., Indian Acad. Scis., Inst. Physics London, Indian Phys. Soc. (pres. 1963——); mem. Indian Assn. for Cultivation Sci. (mem. council 1958——), Indian Sci. Congress (past pres. physics), Instn. Nuclear Engrs. Author series: Jeevan Vidnyan (with B. S. Patil), 1954. Research and numerous publs. on gross intensity distbns. in band spectra of diatomic molecules and spectral temperature determination, vibrational transition probabilities and electronic transition moment variation, quantum analysis of band-spectra of diatomic molecules, combusion spectroscopy of several hydrocarbon radicals in flame sources, polarization of fluorescence in some dystuffs. Home: Vice Chancellor's Lodge, Marathwada U., Aurangabad (Dn.), Maharashtra State, India.*

TAWNEY, Richard Henry, Brit. sociologist, economist; b. Calcutta, India, 1880; s. C. H. and Constance C. (Fox) T.; grad. (fellow 1918-21), Balliol Coll., Oxford (Eng.) U.; hon. degrees from univs. Oxford, Manchester, Birmingham, Sheffield, London, Chgo., Melbourne and Paris; m. Annette J. Beveridge, 1909. Asst., Glasgow (Scotland) U., 1906-08; tchr. tutorial classes com. Oxford U., 1908-14; mem. exec. Workers Ednl. Assn., 1905-47, pres., 1928-44; dir. Ratan Tata Found., U. London (Eng.), 1913-14; mem. cons. com. Bd. Edn., 1912-31; mem. Coal Industry Commn., 1919, Chain Trade Bd., 1919-22, Cotton Trade Conciliation Com., 1936-39; adviser Brit. embassy, Washington, 1941-42; mem. Univ. Grants Com., 1943-48; prof. econ. history U. London, 1931-49, prof. emeritus, 1949——. Hon. fellow Balliol Coll., 1938, Peterhouse, Cambridge (Eng.) U., 1946. Fellow Brit. Acad., 1935; mem. Am. Philos. Soc. Author: The Agrarian Problem in the Sixteenth Century; English Economic History, Select Documents (with Bland and Brown), 1914; Tudor Economic Documents (with E. Power); The Acquisitive Society, 1921; Education, The Socialist Theory; Thomas Wilson, A Discourse of Usury; Religion and the Rise of Capitalism, 1926; Equality, 1931; Land and Labour in China, 1932; The Attack and Other Papers, 1953; Business and Politics Under James I, 1958. Critic of modern capitalistic society; demanded soc. compensate individual for contbn. to common good; thought capitalism interested only in pvt. wealth rather than service to soc. Address: 21 Mecklenburg Sq., London W.C.1, Eng.

TAY, Warren, English physician; b. Eng., 1843; ed. London Hosp. Became asst. surgeon, ophthalmologist London Hosp., 1869, ret. as cons. surgeon, 1902; specialized in ophthalmology, dermatology and pediatrics combined with gen. surgery; surgeon to Northeastern Hosp. for Children, Hosp. for Diseases of Skin, Blackfriars, Royal London Ophthalmic Hosp. Mem. Royal Coll. Surgeons. Translator, editor: On Diseases of the Skins (Ferdinand Hebra), 1866-80. Research and publs. on skin diseases, tumors, syphilis, eyes; described degenerative condition of choroid observed in amaurotic familial idiocy, a recessive trait found predominently in Polish Jewish families, characterized by yellow spots around macula (Tay's choroiditis), 1881. Died 1927.

TAYLOR, Abraham, physicist; b. Manchester, Eng., Aug. 20, 1911; B.Sc. first class honours, Manchester U., 1933, M.Sc., 1934, Ph.D., 1936, D.Sc., 1967; m. Renée Manch, Apr. 2, 1941; children—Victor Lawrence, Sara Corinna. Came to U. S., 1952, naturalized, 1951. Head physics sect. Mond Nickel Co., 1947-52; scientist Horizons Inc., Cleve., 1952-54; adv. physicist Westinghouse Research Labs., Pitts., 1954——. Lectr. X-ray crystallography Carnegie Coll. Tech., Pitts., 1963——. Darbishire Research Fellow, 1936. Fellow Inst. Physics (Gt. Britain). Author: Introduction to X-Ray Metallography, 1945; X-Ray Metallography, 1961; Crystallographic Data on Metals and Alloys, 1963. Research and publs. on X-ray methodology, atomic structure of alloys and high pressure physics. Home: 2415 Beechwood Blvd., Pitts. 15217. Office: Westinghouse Research Labs., Pitts. 15235.*

TAYLOR, Alastair MacDonald, Canadian polit. scientist, geographer; b. Vancouver, B.C., Can., Mar. 12, 1915; s. James and Bertha (Redman) T.; B.A., U. So. Cal., 1937, M.A., 1939; postgrad. Columbia U., 1941-42; D.Phil., U. Oxford (Eng.), 1955; m. Mary E. Clements, July 17, 1944; children—Angus, Graeme, Duncan. With Nat. Film Bd. Can., 1942-44; with UNRRA, Washington, 1944-46; dept. pub. information UN Secretariat, N.Y.C., 1946-52; vis. prof. geography U. Edinburgh (Scotland), 1959-60; prof. geography and polit. studies Queen's U., Kingston, Ont., Can., 1960——. Mem. Canadian Assn. Geographers, Soc. for History Tech., Am. Hist. Assn., Canadian Inst. Internat. Affairs, Am. Acad. Polit. and Social Sci. Author: Indonesian Independence and the United Nations, 1960; (with T. W. Wallbank) Civilization—Past and Present, 1942; For Canada—Both Swords and Ploughshares, 1963; also articles. Analysis impact technology upon spatial patterns and societal instns. particularly evolution conceptualizing process. Home: Cartwright's Point, Kingston, Ont., Can.*

TAYLOR, Albert Hoyt, Am. physicist and radio engr.; b. Chgo., Jan. 1, 1879; s. Albert H. and Harriet (Getschell) T.; B.S., Northwestern U., 1902; Ph.D., Goettingen U., Germany, 1909; D.Sc. (hon.), U. N.D., 1953; m. Sarah E. Hickman, Aug. 9, 1911; children—Albert H., Barbara M., Harriet, Margaret A. Instr., later asst. prof. physics U. Wis., 1903-08; prof., head physics dept. U. N.D., 1909-17; commd. lt. USNR, 1917, advanced through grades to comdr., 1922, supt. radio div. naval research lab., 1923-45, pioneer research in development radar, ret. 1948. Awarded medal of honor by Inst. Radio Egrs., 1941, John Scot medal and premium, 1942, medal for merit by U. S. Pres., 1944. Fellow Inst. Radio Engrs. (past pres.), Am. Phys. Soc., A.A.A.S.; Am. inst. E.E.; mem. Naval Inst., Sigma Xi. Instrumental in development of radar; research on electrical measurements, radio and radar wave propagation, radio administration and electronics. Died December 11, 1961.

TAYLOR, Albert Lee, Am. nematologist; b. Florence, Colo., Nov. 30, 1901; s. James Thomas and Edith (Reed) T.; student U. Colo., 1919-22; B.S., George Washington U., 1934; m. Josephine Smith, Aug. 20, 1928; 1 son, James Daniel. Nematologist, U. S. Dept. Agr., Tifton, Ga., 1935-46, Beltsville, Md., 1949-56, Orlando, Fla., 1966——, leader nematology investigations, Beltsville, 1956-64; with Shell Chem. Corp., 1946-49; nematologist FAO, 1964-65. Mem. Soc. Nematologists, Soc. European Nematologists, Am. Phytopath. Soc. (councilor 1961), Helminthological Soc. Washington (pres. 1955). Research and numerous articles on plant-parasitic nematodes injurious to crop and ornamental plants, including devel. methods for control. Office: 2120 Camden Rd., Orlando, Fla. 32803.*

TAYLOR, Alfred Swaine, Brit. chemist; b. Northfleet, Eng., Dec. 11, 1806; s. Thomas and Susan Mary (Badger) T.; ed. Hosps. of Guy and St. Thomas; student Paris; M.D. (hon.), U. St. Andrews, Scotland, 1852; m. Caroline Cancellor, 1834; 1 dau., Edith (Mrs. F. J. Methold). Prof. med. jurisprudence Guy's Hosp., 1831-77, became joint lectr. on chemistry, 1832, sole lectr., 1851-70. Recipient Swiney prize Soc. Arts, 1859. Fellow Royal Soc., 1845, Royal Coll. Physicians; mem. Royal Coll. Surgeons. Author: A Manual of Medical Jurisprudence, 1844; The Principles and Practice of Medical Jurisprudence, 1865; On the Art of Photogenic Drawing, 1840; A Thermometrical Table on the Scales of Fahrenheit, Centrigrade and Reaumur . . . , 1845; On the Temperature of the Earth and Sea in Reference to the Theory of Central Heat, 1846; Poisons in Relation to Medical Jurisprudence and Medicine, 1848. Editor, London Med. Gazette, 1844-51. Editor: Elements of Physics (Arnott), 1876; (with G. O. Rees) Elements of Materia Medica (Pereira), 3d. and 4th edit., 1849. Presented 1st lectures on med. jurisprudence in Eng.; writings codified then existing legal precedents, chem. and anat. data on med. jurisprudence and poisons. Died London, May 27, 1880.

TAYLOR, Amos Elias, Am. economist; b. Glenville, Pa., July 4, 1893; s. William Franklin and Emeline (Albright) T.; A.B., Gettysburg Coll., 1915; M.A., U. of Chicago, 1920; Ph.D., U. of Pa., 1924; m. Leah Tipton, Nov. 12, 1921; 1 son, Amos Elias (deceased). Instr. in science, East Stroudsburg, Pa., High Sch., 1915-16; prof. Latin and English, Pa. Mil. Coll., 1916-17; instr. in economics, U. of Pa., 1920-23, asst. prof. finance, 1923-29; asso. prof. finance, Northwestern U., 1929-30; with U. S. Dept. Commerce, 1930-47, asst. chief finance div., 1931-39, chief, 1939-42, chief div. research and statistics, 1942-43, dir. bureau Fgn. and Domestic Commerce, 1943-47; exec. com. on Econ. Fgn. Policy, 1944-45; com. on Reciprocity Information, 1944-45; dir. Office of Bus. Econ., 1946-47; dir. dept. economic and social affairs, Orgn. of Am. States, 1947-58; exec. sec. Inter-Am. Econ. and Social Council 1947-58; adj. professor economics, American University Graduate Sch., 1932-59, prof. econs., 1959-64. Pvt. AEF, 1918-19; mem. Sorbonne Detachment, Feb.-June 1919. Asso. mem. com. statistical experts, League of Nations, 1938-46; mem. bd. trustees Gettysburg Coll., 1932-59. Mem. Nat. Acad. Econs. and Polit. Sci. (chmn. bd.), Am. Econ. Assn. (v.p., 1945), A.A.A.S., Royal Econ. Soc., Acad. Polit. Sci., Phi Beta Kappa. Author: Balance of International Payments of the United States, 1931-40; Foreign Investments in the U. S., 1937; Economic and Social Problems of the Americas, 1960. Contbr. articles to econ. jours. Work in development of techniques and statistical methods for determining the dollar receipts and payments, in a given period, from exports and imports of merchandise, services, and capital items. Home: Parkton, Md. Office: Am. U., Washington 6.*

TAYLOR, Angus Ellis, Am. mathematician; b. Craig, Colo., Oct. 13, 1911; s. David and Elizabeth (Ellis) T.; S.B., Harvard, 1933; Ph.D., Cal. Inst. Tech., 1936; m. Mary Kathleen Lapham, July 15, 1936; children—Gordon O., Kenneth L., Kathleen M. Faculty, Cal. Inst. Tech., 1936-37; NRC fellow Princeton, 1937-38; faculty U. Cal. at Los Angeles, 1938——, prof. math., 1947——, chmn. dept. math., 1958-64, v.p. acad. affairs, 1965——. Fulbright research fellow Johannes Gutenberg U., Mainz, Germany, 1955. Mem. Am. Math. Soc., Math. Assn. Am. Author: (with G. E. F. Sherwood) Calculus, 1942, rev., 1946, 54, Elementary Differential Equations, 1943; Advanced Calculus, 1955; Introduction to Functional Analysis, 1958; Calculus with Analytic Geometry, 1959; General Theory of Functions and Integration, 1965. Research and numerous publs. on theory of real and complex functions, analysis in abstract spaces and modern functional analysis, spectral theory of linear operators. Home: 82 Norwood Av., Berkeley, Cal. 94707. Office: University Hall, U. Cal., Berkeley, Cal. 94720.*

TAYLOR, Antony James William, psychologist; b. London, Eng., Aug. 14, 1926; s. Benjamin John and Olive (Watts) T.; Cert.Soc.Sci., London U., 1949; M.A., New Zealand U., 1955, Ph.D., 1965; m. Monica Jean Megroz, July 8, 1950; children—Martin, Matthew, Stephen. Probation officer, Essex, U.K., 1949-51; with Justice Dept., New Zealand, 1952-55, psychotherapist, part-time, 1964——; prison psychologist, 1956-61; lectr. Victoria U. Wellington (New Zealand), 1961-63, student counsellor, 1964——. Hon. cons. Samaritan Telephone Service, 1965——, Catholic Social Services, 1965——; selection adviser New Zealand Vol. Service Abroad, 1963——. Mem. Brit. Psychol. Soc. (fellow, mem. council New Zealand br. 1958——), Australian and New Zealand Student Health Assn., N. Z. Assn. for Mental Health (pres.), N.Z. Psychol. Soc. (found. chmn. div. clin. psychologists), Assn. Int. Cssn. Jurist. Research and publs. on diagnosis of criminal psychopathy, techniques of group psychotherapy, guilt patterns, clin. studies transvestites, aspects of student counselling, tatooing. Home: 17 Fairview Cresent, Wellington, New Zealand.*

TAYLOR, Brook, English mathematician; b. Edmonton, Middlesex, Aug. 18, 1685; s. John and Olivia (Tempest) T.; ed. St. John's Coll., Cambridge, LL.B., 1709, LL.D., 1714; m. 1st, Miss Brydges, 1721 (dec. 1723); m. 2d, Sabetta Sawbridge, 1729; 1 dau., Elizabeth Taylor Young. Pursued interests in law, philosophy, music, painting, also religious studies after 1715; on commn. which ruled on the claims of both Newton and Leibnitz to have discovered calculus; fellow Royal Soc., 1712 (1st sec., 1714-18). Author: Methodus Incrementorum Directa et Inversa, 1715; Linear Perspective, 1715; New Principles of Linear Perspective, 1719; Contemplatio Philosophica, 1793; also contbr. articles to Philos. Transactions. Research in math. physics; discovered Taylor's theorem on functions; solved problem of center of oscillation, 1708; developed solution to problem connected with Kepler's 2d law of planetary motion, 1712; wrote first treatise on calculus of finite differences, 1715; discovered formula for transversal string oscillation, also method of computing logarithms; made studies on fluid surface areas, adhesion, and trajectory. Died London, Dec. 29, 1731.

TAYLOR, Calvin W., Am. psychologist; b. Salt Lake City, May 23, 1915; s. Elliot Campbell and Mary (Walker) T.; B.A., U. Utah, 1938, M.A., 1939; Ph.D., U. Chgo., 1946; m. Dorothy Cope, June 18, 1943; children—Craig Elliot, Stephen Cope, Nancy Elizabeth. Jr. occupational analyst, statistician USES, FSA, Washington, 1941-42; faculty U. Utah, Salt Lake City, 1946-52, 54-64, 65—, now prof.; prin. investigator for various projects supported by founds. and agys.; dir. research Office Sci. Personnel, NRC/Nat. Acad. Sci., Washington, 1952-54; dir. personnel measurement research dept. U. S. Navy, San Diego, 1964-65. Cons. to govt. agys., numerous sch. dists.; mem. coms. NIH, 1957-61, 64—. Mem. A.A.A.S., Am., Rocky Mountain, Utah psychol. assns., Psychometric Soc., Psychonomic Soc., Phi Beta Kappa, Sigma Xi, Phi Kappa Phi. Author: Creativity: Progress and Potential, 1964; Widening Horizons in Creativity, 1964; also numerous articles, monographs. Editor: (with F. Barron) Scientific Creativity: Its Recognition and Development, 1963; (with F. Williams) Instructional Media and Creativity, 1966. Editorial bd. Jour. Creative Behavior. Initiated research in creativity and communication abilities; studied career performances of scientists, physicians, nurses, tchrs.; developed ednl. theory based on behavioral sci. findings; helped initiate archtl. psychology. Home: 2594 E. 3900 S. Salt Lake City 84117.*

TAYLOR, Charles Fayette, Am. orthopedic surgeon; b. Williston, Vt., Apr. 25, 1827; s. Brimage and Miriam (Taplin) T.; attended lectures N.Y. Med. Coll., 1855; M.D., U. Vt., 1856; m. Mary Salina Skinner, Mar. 7, 1854, 4 children including Henry Ling. Began practice of medicine, N.Y.C., 1857; became interested in phys. therapy, invented braces and orthopedic devices to treat bone and joint lesions, notably Taylor splint for treatment of spinal diseases and long extension hip splint; devised cure for Pott's disease (an infection of spinal vertebrae), 1857-63. Author: Mechanical Treatment of Angular Curvature or Pott's Disease of the Spine, 1863; Mechanical Treatment of Diseases of the Hip, 1873. Died Los Angeles, Jan. 25, 1899.

TAYLOR, Charles Vincent, Am. zoologist; b. Whitesville, Mo., Feb. 8, 1895; s. Isaac Newton and Christina (Bashor) T.; A.B., Mt. Morris (Ill.) Coll., 1911; M.A., U. of Calif., 1915, Ph.D., 1917; m. Lola Lucille Felder, May 6, 1921; children—Jeanne Lucile, Elouise Christine, Lola Lenore, Isaac Newton. Prin. high sch., Valley City, N.D., 1911-14; teaching fellow, U. of Calif., 1914-17, instr. zoology, 1917-18; Johnston scholar, Johns Hopkins U., 1918-20; asst. prof. zoology, U. of Calif, 1920-23, U. of Mich., 1923-24, U. of Calif., 1924-25; research asso. Carnegie Inst., Tortugas Lab., 1924, 26; acting asst. prof., Hopkins Marine Sta., summers, 1922, 23, asso. dir. since 1925; asso. prof. biology, Stanford U., 1925-26, prof., 1926-30; prof. zoology, U. of Chicago, 1925-26, prof., 1926-30; prof. zoology U. of Chicago, 1930-31; Herztein prof. biology, head dept. zoology, Stanford Univ., 1931-34; Herztein prof. biology since 1931, dean Sch. of Biol. Sciences, since 1934. Mem. Nat. Acad. of Sciences, A.A.A.S., Am. Soc. Zoologists, Soc. Exptl. Biology and Medicine, Nat. Geog. Soc., Am. Naturalists Soc., Western Soc. Naturalists (former pres.), Pacific Oceanographic Soc., Sigma Xi. Edited Contributions to Marine Biology, 1930; The Cell and Protoplasm, 1940. Research contributions on living cells: function of fibrillar systems, role of micronucleus, development of egg fragments; on protoplasm: sol-gel reversibility; cataphoresis of ultramicroscopic inclusions. Died Feb. 22, 1946.

TAYLOR, David Watson, Am. engineer; b. Louisa Co., Va., Mar. 4, 1864; s. Henry and Mary Minor (Watson) T.; Randolph-Macon Coll., Va., 1877-81; grad. U. S. Naval Acad., 1885, head of class and excelled highest record ever made there up to that time; sent to Greenwich in 1885, received highest honors of Royal Coll., 1888, again making the highest record of any student there up to that time; hon. D.Engring., Stevens Inst., Hoboken, N.J.; 1907, D.Sc., George Washington U., 1915; LL.D., Randolph-Macon Coll., 1922, U. of Glasgow, Scotland, 1924; m. Imogene Maury Morris, Oct. 26, 1892; children—Dorothy Watson, May Coleman, David Watson, Imogene Morris. Capt. U.S.N., Mar. 4, 1901; promoted to rank of rear admiral, 1917. Awarded gold medal by British Instn. Naval Architects, for best original paper on Ship-Shaped Stream Forms (first American so honored). In 1899 constructed (and had charge of) first experimental tank ever built in U. S. Retained by British Govt. as expert in suit growing out of Hawke-Olympic collision, 1911. Chief constructor U.S.N. and chief of Bur. of Constrn. and Repair, Navy Dept., 1914-22; retired, Jan. 15, 1923; awarded D.S.M. (U. S.); Comdr. Legion of Honor (France). Vice chmn. Nat. Advisory Com. for Aeronautics; mem. Soc. Naval Architects and Marine Engrs. (pres. 1925-27), (British) Instn. of Naval Architects (hon. v.p. 1931). John Fritz Medalist, 1931; gold medalist (British) North East Coast Instn. of Engrs. and Shipbuilders, 1931. Awarded 1st David Watson Taylor Gold Medal (established in his honor) by Soc. Naval Architects and Marine Engrs., 1936; new U. S. David Watson Taylor Model Basin named in his honor, 1937. Author: Resistance of Ships and Screw Propulsion, 1893; Speed and Power of Ships, 1910. Devised standards for calculating displacement and stability of ships. Died July 28, 1940.

TAYLOR, Donald Wayne, Am. psychologist; b. Boulder, Colo., May 9, 1919; s. Ralph O. and Mamie K. (Lauer) T.; A.B., Baker U., 1939, awarded Doctor of Laws (honorary), 1963; M.A., U. Kan., 1940; M.A., Harvard, 1942, Ph.D., 1943; M.A. (hon.), Yale, 1955; m. Ruth I. Spence, Sept. 1, 1940; children—Laird E., Patricia A., Barbara K. (dec.), and Roderic West. Special research associate at Radio Research Laboratory, Harvard University, 1943-45, also part-time instr. psychology, 1943-44; civilian specialist radar countermeasures, Am. embassy, London, Eng., 1944; asst. prof. psychology Stanford, 1945-48, asso. prof., 1948-55; prof. personnel administrn. Yale, 1955-60, prof. psychology, fellow Pierson Coll., 1956—, chmn. dept. industrial administrn., 1963-66, chmn. dept. psychology, 1966—, dir. div. social scis., 1966—; research fellow in cognitive Harvard U., 1961-62. Mem. div. anthropology and psychology Nat. Research Council, 1949-52; chmn. electronics tng. working group, Research and Development Bd., Office Sec. Def., 1951-52; dir. study undergrad. edn. Stanford, 1954-55. Rhodes scholar, 1939. Fellow Am. Psychol. Assn. (member publications bd. 1960-66, policy and planning bd. 1965—), Am. Assn. Advancement Sci.; mem. Phi Beta Kappa, Sigma Xi (national lectr., 1960-61), Psychonomic Society. Co-editor: Dorsey Series in Psychology, 1959—. Asso. editor Ann. Rev. Psychology, 1949-53. Author numerous articles. Cons. editor Irwin Dorsey series Behavioral Science in Business, 1960-63. Psychological studies of thinking, including studies of problem sclving, decision making and creative thinking; explored influence on problem solving of group participation, non-intellectual differences, differences between men and women; conducted field studies of variables related to creative thinking among engineers and physical scientists. Home: 1486 Ridge Rd., North Haven, Conn. Office: Yale U., 333 Cedar St., New Haven.*

TAYLOR, Edward Curtis, Am. chemist; b. Springfield, Mass., Aug. 3, 1923; s. Edward Curtis and Louise (Anderson) T.; student Hamilton Coll., 1942-44; A.B., Cornell U., 1946, Ph.D., 1949; m. Virginia Dion Crouse, June 29, 1946; children—Edward Newton, Susan Raines. Faculty, U. Ill., 1951-54; faculty Princeton, 1954—, prof. chemistry, 1964—, A. Barton Hepburn prof. organic chemistry, 1966—; vis. prof. Technische Hochschule, Stuttgart, Germany, 1960; vis. lectr. Weizmann Inst., 1960; mem. chem. adv. com. Office Sci. Research, USAF, 1962—; cons. indsl. firms. Recipient Research awards Smith, Kline & French Found., 1955, Hoffmann-LaRoche Found., 1964, 65. Fellow N.Y. Acad. Scis.; mem. Am., German chem. socs., Chem. Soc. (London), Phi Beta Kappa, Sigma Xi, Phi Kappa Phi. Editor: (with R. A. Raphael, H. Wynberg) Advances in Organic Chemistry, Vol. I-V, 1961-65; (with W. Pfleiderer) Pteridine Chemistry, 1964; (with A. Weissberger) Chemistry of Heterocyclic Compounds, Physical Chemistry of Heterocyclic Compounds, 1968. Research and numerous publs. in pteridine chemistry, new methods of synthesis in organic chemistry, synthesis of natural products, chemistry of heterocyclic compounds, medicinal chemistry; synthesized active principle of marijuana, 1966. Home: 288 Western Way, Princeton, N.J. 08540.*

TAYLOR, Ellison Hall, Am. chemist; b. Kalamazoo, Sept. 6, 1913; s. Laurence and Etta (Hall) T.; B. Chem., Cornell U., 1935; Ph.D., Princeton, 1938; m. Ruth Young, June 22, 1938; children—Laurence R., William E. Instr. chemistry U. Utah, 1938-40; instr. chem. engring. Cornell U., 1940-42; research staff div. war research Columbia U., 1942-45; with Clinton Labs. (now Oak Ridge Nat. Lab.), 1945—, dir. chemistry div., 1954—. Mem. Am. Chem. Soc., Am. Phys. Soc., A.A.A.S. Studies, publs. on application of molecular beams to chemical reactions; use of ionizing radiation in study of catalysis. Home: 143 Orchard Lane. Office: Oak Ridge Nat. Labs., Oak Ridge 37830.*

TAYLOR, Frank Bursley, Am. geologist; b. Ft. Wayne, Ind., Nov. 23, 1860; s. Robert Stewart and Fanny (Wright) T.; student Harvard, 1882-86; m. Minnette Amelia Ketchum, Apr. 24, 1899. Became spl. asst. U. S. Geol. Survey, 1900, later field asst. Research and publs. on history of Gt. Lakes region, glacial lake history; proposed scheme of continental drift and mountain bldg. with tidal forces; mapped moraines and shorelines in Ohio, Ind., Mass., Mich., N.Y., Can. Died Ft. Wayne, Ind., June 12, 1938.

TAYLOR, Frederick Winslow, Am. management scientist, engineer; b. Germantown, Phila., Pa., Mar. 20, 1856; s. Franklin and Emily (Winslow) T.; prepared for Harvard at Phillips Exeter Acad., 1874; left because of impaired eyesight; M.E., Stevens Inst. Tech., 1883; (Sc.D., U. of Pa., 1906; LL.D., Hobart Coll., 1912); m. Louise M. Spooner, May 3, 1884. Entered employ Midvale Steel Co., Phila., 1878, and was successively gang boss, asst. foreman, foreman of machine shop, master mechanic, chief draftsman, and chief engr. to 1889; left 1889 to begin work of organizing management of mfg. establishments of various kinds, in shop, office, accounting and sales dept.; has organized many kinds, including the Bethlehem Steel Co., Cramp's Shipbuilding Co., Midvale Steel Co. Inventor of Taylor-White process of treating modern high-speed tools, for which received personal gold medal from Paris Expn., 1900, and Elliot Cresson gold medal of Franklin Inst. Has received about 100 patents for various inventions. Author: Concrete, Plain and Reinforced (with S. E. Thompson), 1905; Art of Cutting Metals, 1906; Concrete Costs (with S. E. Thompson), 1911; The Principles of Scientific Management, 1911; Shop Management, 1911. Father of scientific management; developed philosophy of management stressing common purpose and common method accepted by both labor and management. Died Philadelphia, Pa., Mar. 21, 1915.

TAYLOR, Sir Geoffrey Ingram, Brit. meteorologist, physicist; b. London, Eng., Mar. 7, 1886; s. Edward Ingram and Margaret (Boole) T.; grad. U. Coll., Cambridge, 1908; became fellow Trinity Coll., Cambridge, 1910; D.Sc. (hon.) U. Bristol (Eng.), 1959; Hon. Dr., U. Paris, 1961; Oxford U., Cambridge, London, Birmingham, Liverpool, Edinburgh, Vancouver, Aachen, Oslo. Meteorologist, Scotia expdn. to N. Atlantic, 1913; exptl. aeronautics and meteorology, World War I; Yarrow Research prof. Royal Soc.; Mem. adv. council Ministry Supply, 1939-45; mem. Civil Def. Research Com., 1939. Recipient Royal medal, 1933; Copley medal, 1944; U. S. medal for merit, 1947; Gold medal Royal Aero. Soc., 1954; Exner medal; De Morgan medal London Math. Soc., 1956; Österreichischer Gewerbeverein, 1954; award Internat. Panetti prize and medal Acad. delle Scienze di Torinto, 1958; Timoshenko medal Am. Soc. M.E., 1958; Kelvin medal Inst. Civil Engrs., 1959; Médaille Trasenster, U. Liège (Belgium), 1962; Franklin medal Franklin Inst., 1962; Platinum medal Inst. Metals, 1964; James Watt Internat. Gold medal, 1965. Fellow Royal Soc., 1919, Indian Acad. Sci. (hon.); corr. mem. French Acad. Scis., 1946, nat. acads. of sci. at Washington, Amsterdam, Oslo, Swedish Royal Acad. Sci., Acad. dei Lincei, Royal Soc. Edinburgh, Royal Aero. Soc., Mem. group creating 1st nuclear explosion, Los Alamos, 1944-45; research in aeros., meteorology, engring., math.

TAYLOR, Harden Franklin, Am. chemist, ichthyologist; b. West Plains, Mo., July 15, 1890; s. Robert Marion and Nettie Clay (Laffoon) T.; A.B., Trinity Coll. (now Duke Univ.), Durham, N.C., 1913 (Hon.) D.Sc., 1936; m. Ella Wolstenholme, July 3, 1919 (dec.); m. second, Eloise Aley, Sept. 10, 1963. Asst. to prof. of biology Duke Univ., 1911-13; high sch. teacher, Tarboro, N.C., 1913-14, principal, 1914-15; scientific asst. Marine Biological Lab., U. S. Bur. of Fisheries, summers, 1911-14, scientific asst. div. of scientific inquiry, Washington, D.C., 1915-18, chief technologist, 1918-22, chief of div. of fishery industries, 1922-23; dir. of research Atlantic Coast Fisheries Co., 1923-25, vice pres. for scientific research, 1925-30, dir., 1927-44, managment com., 1929-30, pres. 1930-44; scientific consultant, 1944; cons. and member exec. com. Inst. of Fisheries Research, U. N.C., 1947-60; sci. director N.C. Shrimp Survey, 1947-50; cons. Oceanographic Inst., Fla. State University, 1951. Delegate World Engring. Congress, Tokyo, 1929, Internat. Congress of Refrigeration, The Hague, Netherlands, 1926; mem. Pacific Science Congress (com. on fish technology); fishery advisory com. United States Department Commerce, 1936-38 (chmn. subcom. on research, 1939); mem. Internat. Institut du Froid, Commission III, London meeting, 1938; mem. adv. council Refrigeration Research Foundation; chmn. fisheries sec. div. of biology and agriculture, Nat. Research Council; mem. ad hoc com. on fisheries Food & Agrl. Orgn. of UN, 1945; exec. dir. Survey of Marine Fisheries of N.C., U. N.C.; director Nortex Oil & Gas Corp. Recipient of Wildlife Society award, 1952. Fellow A.A.A.S., American Institute Chemists, N.Y. Acad. Sciences (trustee, treasurer president 1946-48); member American Chemical Society, American Inst. of N.Y.C. (trustee, pres. 1951), Princeton Engineering Assn., Am. Society of Limnology & Oceanography, Newcomen Society, Phi Beta Kappa Alumni Assn. N.Y. (pres. 1965—) Alpha Chi Sigma, Phi Beta Kappa, Omicron Delta Kappa, Sigma Upsilon. Contbr. scientific articles to government publs. and other jours., and reviews. Holder U. S. and fgn. patents in field. Studies of technology of fisheries, preservation of fish nets, high vitamin oils, and determination of Vitamin A. Died 1966.

TAYLOR, Harry William, Canadian physicist; b. nr. Sturgeon Valley, Sask., Can., Sept. 28, 1925; s. William and Gladys (Evans) T.; B.Sc., U. Man., 1951, M.Sc., 1952, Ph.D., 1954; m. Wanda Jason, June 18, 1949; children—Allison Leslie, Karen Elizabeth. Lectr. U. Man., Winnipeg, Can., 1952-53; fellow NRC, Ottawa, Ont., Can., 1954-55; faculty Queen's U., Kingston,

Ont., 1955-61, asst. prof., 1957-61; asso. prof. U. Alta., Edmonton, Can., 1961-65; asso. prof. physics U. Toronto (Ont.), 1965—. Mem. Am. Phys. Soc., Canadian Assn. Physicists, Am. Astron. Soc., N.Y. Acad. Scis., Royal Astron. Soc. Can. Research and publs. in nuclear structure through gamma-ray spectroscopy; observation of angular correlations of cascade gamma rays. Home: 3525 Grand Forks Rd., Cooksville, Ont. Office: Physics Dept., U. Toronto, Toronto 5, Ont., Can.*

TAYLOR, Howard Canning, Jr., Am. physician; b. N.Y.C., Feb. 17, 1900; s. Howard Canning and Alice C. (Gibbs) T.; Ph.B., Yale Sheffield Sci. Sch., 1920; M.D., Columbia College Phys. and Surgs., 1924; Doctor of Science, N.Y. U., 1955; m. Caroline Colgate, Sept. 1923; children—Barbara (Mrs. Donald Schoen), Caroline Alice (Mrs. Lincoln Day), Howard C. III. Asso. prof. obstetrics-gynecology N.Y. U., 1935-39, 1940-43, prof., chmn. dept., 1943-46; attending gynecol. 1936-46, cons. gynecol., 1946—, Roosevelt Hosp., N.Y.C., attending gynecol. Meml. Hosp., 1943-46. Prof. gynecology, U. Pa., 1939-40; prof. obstetrics-gynecol., chmn. dept., Columbia Coll. Phys. and Surgs., 1946-65; vis. gynecologist and obstetrician in charge Bellevue Hosp., 1943-46; director dept. obstetrics and gynecology Presbyterian Hospital, 1946-65; dir. gynecology Francis Delafield Hosp., 1949-65; dir. Internat. Inst. Study Human Reproduction, Columbia U., 1965—, prof. emeritus obstetrics and gynecology. Honorary life time director American Cancer Society, Incorporated, (president 1954-55). Diplomate Am. Bd. Obstetrics-Gynecology. Fellow A.C.S. (regent 1951-60); Am. Coll. Obstetricians and Gynecologists (president 1966-67), N.Y. Acad. of Medicine, N.Y. Acad. Sci., Royal Coll. Obstetricians and Gynecologists London (hon.); mem. Internat. Fedn. Gynecologists and Obstetricians (pres. 1961-64), A.M.A., Am. Assn. Obstetricians and Gynecologists, Am. Gynecol. Soc. (pres. 1957-58), Am. Radium Soc., Am. Assn. Cancer Research, N.Y. Obstet. Soc., N.Y. Gynecol. Soc. (pres. 1962-63); hon. mem. Central Assn. Obstetricians and Gynecologists, S. Atlantic Soc. Obstetricians and Gynecologists; fgn. hon. mem. Obstet. and Gynec. Soc. Sweden, Sociedad Peruana de Obstetricia y Ginecología, Obstet. Soc. Edinburgh, Obstet. and Gynec. Soc. Jugoslavia, German Gynec. Soc., Italian Obstet. and Gynec. Soc., Japanese Obstet. and Gynec. Soc., Obstet. and Gynec. Soc. Brazil. Editor in chief: Am. Jour. Obstetrics-Gynecology. Contbr. articles to med. jours. Contrib. to study of physiology of pregnancy; cancer of female reproductive organs. Home: 200 E. 66th St., N.Y.C. 10021. Office: 630 W. 168th St., N.Y.C. 10032.*

TAYLOR, Hugh Stott, chemist; b. St. Helens, Lancashire, Eng., Feb. 6, 1890; s. James and Helen (Stott) T.; B.Sc., Liverpool U., Eng., 1909, M.Sc., 1910, D.Sc., 1941; m. Elizabeth Sawyer, June 12, 1919 (dec. July 1958); children—Joan Mary (Mrs. W. W. Ashley) Elizabeth Sylvia (Mrs. M. F. Healy, Jr.). Instr., Princeton, 1914, prof. chemistry, 1922-58, dean grad. sch., 1945-58; pres. Woodrow Wilson Nat. Fellowship Found., 1958—; vis. prof. chemistry, U. Cal., Berkeley, 1930, U. Manchester, Eng., 1932, U. Louvain, Belgium, 1937; Decorated Knight Comdr. Brit. Empire (Eng.); Knight Comdr. St. Gregory (Papal); C. Leopold II (Belgium); recipient Nichols medal, 1928, Mendel medal, 1933, Longstaff medal, 1941, Franklin medal, 1951, William Procter award Research Soc. Am., 1964, Priestley medal. Fellow Royal Soc., 1932; mem. Chem. Soc. (hon. mem.), Faraday Soc. (pres. 1952-53), Pax Romana (pres. 1952-55), Acad. Lincei (fgn. mem.), Belgian Acad., Sigma Xi. Author numerous texts on phys. chemistry fuel prodn., indsl. hydrogen, catalysis. Editor: Treatise on Physical Chemistry, 1924. Research in catalysis and chemical kinetics. Home: 191 Library Pl. Office: 32 Nassau St., Princeton, N.J. 08540.*

TAYLOR, James Haward, English geologist; b. Esher, Surrey, Eng., Feb. 24, 1909; s. James and Lilian (Haward) T.; student Clifton Coll., 1923-26; B.Sc., U. London, 1931, Ph.D., 1936; A.M., Harvard, 1934. Henry fellow. Harvard, 1933-34; prin. geologist Geol. Survey Great Britain, 1935-49; prof. geology, head dept. U. London King's Coll., 1949—; cons. ore supply Iron & Steel Bd., 1954-67; mem. natural environment NRC, 1965-68. Fellow Royal Soc., 1960; mem. Internat. Assn. Sedimentologists (pres. 1963-67), Instn. Mining and Metallurgy (v.p. 1962-65), Mineral. Soc. (pres. 1963-65), Geol. Soc. (mem. council 1950-54). Author: Petrology of the Northampton Sand Ironstone Formation, 1949; The Northampton Sand Ironstone, 1951. Research, numerous publs. on origin, distbn. and assessment of ore deposits, especially those of sedimentary origin; processes of sedimentation. Home: 18 Kingsdowne Rd., Surbiton, Surrey, Eng. Office: Geology Dept., King's Coll., Strand, London, W.C.2, Eng.*

TAYLOR, James Herbert, Am. geneticist; b. Corsicana, Tex., Jan. 14, 1916; s. Charles A. and Delia May (McCain) T.; B.S. in Biology, Southeastern State Coll., 1939; M.S., U. Okla., 1941; Ph.D., U. Va., 1944; m. Shirley C. Hoover, May 1, 1946; children—Lynn Sue, Lucy Delia, Michael Wesley. Asst. prof. plant sci. U. Okla., 1946-47; asso. prof. botany U. Tenn., 1947-52; research cons. biol. div. Oak Ridge Nat. Lab., 1949-51; faculty Columbia U., 1951-64, prof. cell biology, 1958-64; prof. biol. scis. Fla. State U., Tallahassee, 1964—. Mem. genetics research panel NSF, 1963-66. Guggenheim fellow, 1958-59. Mem. Biophys. Soc., Am. Soc. Cell Biologists, Genetics Soc. Am., Bot. Soc. Am., Am. Inst. Biol. Scis. Author: Molecular Genetics, Part I, 1963, II, 1967. Selected Papers on Molecular Genetics, 1965; also numerous articles. Editorial bd. Jour. Cell Biology, 1963-66, Genetics, 1967—; regional editor The Nucleus, 1960—, Chromosoma, 1966—. Research on cell biology, reprodn. of chromosomes with particular reference to molecular mechanism deoxyribonucleic acid synthesis with use of radioisotopes, autoradiography. Home: 1414 Hilltop Dr., Tallahassee 32303.*

TAYLOR, John, Brit. mining engr.; b. Norwich, Eng., Aug. 22, 1779; s. John and Susannah (Cook) T.; ed. Dr. Houghton's Sch., Warwick, Eng. Became mgr. Wheal Friendship Mine, Tanistock, 1798; became chem. mfr., Stratford, Eng., 1812; founder consol. Mines, Gwennap, 1819; mineral agt. to Duke of Devonshire, also to commrs. of Greenwich Hosp.; a founder, treas. U. Coll., London. Fellow Geol. Soc. (treas. 1816-44), Royal Soc.; mem. Brit. Assn. (a founder, treas. 1832-61). Author: Statements Concerning the Profits of Mining in England; also articles. Introduced pulverization with grinders to Cornwall mines, 1808. Died London, Apr. 5, 1863.

TAYLOR, John Fuller, Am. biochemist; b. Jamestown, N.Y., June 10, 1912; s. Walter Alexander and Lucy Eva (Fuller) T.; B.A., Cornell U., 1933; Ph.D. (Francis P. Garvan fellow 1933-37), Johns Hopkins, 1937; m. Portia A. Hopper, June 22, 1935; children—Walter Fuller and Herbert Hopper (twins). NRC fellow med. scis. Harvard Med. Sch., 1937-39, Austin teaching fellow biol. chemistry, tutor biochem. scis., 1939-41; biochemist Lederle Labs., 1941-43; asst. prof. biol. chemistry Washington U. Med. Sch., St. Louis, 1943-52; prof. biochemistry, chmn. dept. U. Louisville Med. and Dental Sch., 1952—; vis. prof. U. Oslo (Norway), 1955. Commonwealth Fund fellow, 1961-62. Mem. Soc. Exptl. Biology and Medicine (pres. Ohio Valley sect. 1961-63, bd. editors Proceedings 1966—), Biochem. Soc. Eng., Am. Soc. Biol. Chemists, Am. Chem. Soc., Biophys. Soc., Phi Beta Kappa, Sigma Xi. Contbr. articles profl. jours. Contrib. in physical chemistry of proteins and polysaccharides; sulfhydryl proteins; enzyme proteins; hemoglobin; hemochromogens; quinones; oxidation-reduction potentials; ultracentrifugation. Home: 2 Normandie Village, Louisville 5.

TAYLOR, John Hall, Am. psychologist; b. Greenfield, Mass., Feb. 18, 1922; s. Kenyon Yale and Miriam (Pierce) T.; A.B., Wesleyan U., Middletown, Conn., 1947; M.A., U. Mich., 1950, Ph.D., 1952; m. Carrie Lenore Yoffe, Jan. 17, 1948; children—Deborah Helen, Hali Sarah, David Yoffe, Caleb Hall. Tech. aide Armed Forces-NRC Com. on Vision, Ann Arbor, Mich., 1948-54, mem. exec. com., Washington, 1964—; research asso. psychology U. Mich., 1952-54; research psychologist U. Cal. at San Diego, 1954—. Cons. N.R.C. Mem. Optical Soc. Am., Am., Western psychol. assns., Optical Soc. San Diego, Sigma Xi. Research, publs. on undersea and aerospace environmental factors, peripheral vision. Home: 1229 Via Barranca, La Jolla, Cal. 92037. Office: Visibility Lab., U. Cal., San Diego 92152.*

TAYLOR, John Keenan, Am. chemist; b. Mt. Rainier, Md., Aug. 14, 1912; s. Frank Stuart and Ida Mabel (Markward) T.; B.S., George Washington U., 1934; M.S., U. Md., 1938, Ph.D., 1941; m. Helen Isabelle Addis, Aug. 27, 1938; children—Helen Addis (Mrs. Norman Evans Prince, Jr.), Barbara Jean (Mrs. Stephen Cole Davis.). Chemist, Nat. Bur. Standards, Washington, 1929—, chief analysis and purification sect., 1961—; adj. prof. chemistry dept. Am. U., Washington, 1955—. Recipient U. S. Dept. Commerce Silver medal, 1960. Mem. Am. Chem. Soc., Electrochem. Soc., Washington Acad. Scis. (pres. 1966—), Chem. Soc. Washington (past chmn.). Author: (with P. J. Knipling, F. Smith) Project Ideas for Young Scientists, 1960; also numerous articles. Research on definitive measurements of refractive index water, devel. of methods of isotope separation based on differences in electrolytic migration, electro-analytical chemistry, devel. of sensitive polarographic methods for determination of trace and minor constituents, devel. of ultraprecise coulometric methods analysis. Home: 12816 Tern Dr., Route 3, Gaithersburg, Md. 20760. Office: Nat. Bur. Standards, Washington 20234.*

TAYLOR, Julius David, Am. pharmacologist; b. Erie, Pa., Dec. 18, 1913; s. James Daggett and Rella (Landis) T.; B.S., U. Pitts., 1936; Ph.D., U. Rochester, 1940; m. Frances E. Shearer, Dec. 30, 1938; 1 son, Jay David; m. 2d, Dorothy Wiederhold, Oct. 8, 1951. Research biochemist Eaton Labs., Inc., 1943-47; chem. pharmacologist Abbott Labs., North Chicago, Ill., 1948-60, head dept. pharmacology, 1960-64, drug evaluation specialist, 1965—; lectr. Med. Sch. Northwestern U., 1959-62; prof. pharmacology Chgo. Med. Sch., 1962—. Mem. A.A.A.S., Soc. Exptl. Biology and Medicine, Pharmacol. Soc., Am. Chem. Soc., Drug Information Assn., Soc. Gen. Systems Research, Sigma Xi. Research and numerous publs. in lipid chemistry, isolation of vitamins A and E, biochemistry of wound healing, drug enzymology, drug metabolism, radioactive tracers, drug kinetics and analogue computer, toxicology. Home: 905 Baldwin Av., Waukegan, Ill. 60085. Office: Abbott Labs., North Chicago, Ill. 60064.*

TAYLOR, Lauriston S(ale), Am. physicist; b. Bklyn., June 1, 1902; s. Charles and Nancy Bell (Sale) T.; student Stevens Inst. Tech., 1920-22; A.B., Cornell, 1926; D.Sc., U. Pa., 1960, St. Procopius Coll., 1965; postgrad. Columbia, 1930, Cornell, 1927, 30, 31; m. Azulah Frances Walker, Dec. 28, 1925; children—Lauriston Sale, Nelson Walker. With Bell Telephone Labs., N.Y.C., 1922; joined Nat. Bur. Standards, 1927, chief X-ray sect., 1940-49, chief proving grounds sect., 1941-42, asst. chief ordnance devel. div., 1942-43, asst. chief optics div., later atomic physics div., 1948-51, chief Radiation Phys. Lab., 1949-51, chief atomic and radiation physics div., 1951-60, chief radiation physics div., 1960-62, asso. dir., 1962-65; spl. asst. to pres. Nat. Acad. Scis., Washington, 1965—. Sec. Internat. Commn. Radiol. Units, 1934-50, chmn., 1953—; mem. Internat. Commn. Radiol. Protection, 1928—, sec., 1937-50; chmn. Nat. Com. on Radiation Protection, 1929-64; pres. Nat. Council on Radiation Protection and Measurements, 1964—; chief, biophysics br., div. biology and medicine A.E.C., 1949. Decorated Bronze Star medal, Medal of Freedom; recipient Exceptional Service medal, Dept. Commerce; Sylvanus Thompson medal, Brit. Inst. Radiology; Janeway Medal, Am. Roentgen Ray Soc., 1954; gold medal, Radiol. Soc. of N.Am., 1954, gold medal Am. Coll. of Radiology, 1965; Diplomate Am. Bd. Health Physics, Am. Bd. Radiology. Fellow Am. Phys. Soc.; mem. Washington Acad. Sci., Radiation Research Soc., Am. Radium Soc., Radiol. Soc. N.Am., Washington Acad. Medicine, Deutsche Rontgen-Gesellschaft (hon.), Health Physics Soc. (pres. 1959). Author sects. 16 sci. books; also numerous sci. articles. Devel. guarded field standard ionization chamber; studies in radiation measurement and protection; ionization of liquids; variabletime oscilators. Home: 7407 Denton Rd., Bethesda, Md. 20014. Office: Nat. Academy of Sciences, 2101 Constitution Av. N.W., Washington 20418.*

TAYLOR, Lewis Walter, Am. geneticist; b. East Troy, Wis., Oct. 23, 1900; s. Walter Andrew and Elvina (Hassold) T.; M.S., Kan. State U., 1925; B.S., U. Wis., 1922, Ph.D., 1931; m. Ethel Davis Putnam, Dec. 22, 1923; children—Martha Ann (Mrs. Robert Bradley Bartter). Asst. U. Ky., 1927-28; Faculty, U. Cal. at Berkeley, 1928—, prof. poultry husbandry, 1943-68, professor emeritus, since 1968—, chairman of the dept. poultry husbandry, 1933-51. Fellow A.A.A.S., Poultry Sci. Assn. (past pres., Teaching award 1956); mem. Cal., N.Y. acads. sci., Genetics Soc. Am., Am. Soc. Naturalists, Am. Genetics Soc., World's Poultry Sci. Assn., Teratology Soc., Soc. for Study Evolution, Cooper Ornithol. Soc., Am. Orntho1. Union. Editor: Fertility and Hatchability of Chicken and Turkey Eggs, 1949. Research and numerous publs. on genetic, nutritional and physical factors in avian reproduction, hereditary disease resistance, lethal genes and developmental defects. Home: 904 Keeler Av., Berkeley, Cal. 94708.*

TAYLOR, Michael Waistell, Brit. physician, antiquary; b. Portobello, Scotland, Jan. 29, 1824; s. Michael Taylor; M.D., Edinburgh (Scotland) U., 1843; student surgery, Paris, France, 1844; m. Mary Rayner, 1858; 3 sons, 1 dau. Asst. to John Hutton Balfour, Edinburgh U.; practiced medicine, Penrith, Eng., from 1845. Fellow Soc. Antiquaries London, Scottish Soc. Antiquaries; mem. Epidemiological Soc., Cumberland and Westmoreland Antiquarian and Archeol. Soc., Hunterian Med. Soc. (co-founder, pres.), Brit. Med. Assn. (co founder, pres. border counties br.), Royal Archeol. Inst. (council). Author: Old Manorial Halls of Cumberland and Westmoreland, 1892; med. treatises, also important paper on fungoid nature diphtheria, 1881. Discovered that scarlet fever could be caused by contamination milk, 1858; discovered traces Celtic occupation of Ullswater, starfish cairns of moor Divock, prehistoric remains at Clifton, Eng., Croglin moulds for casting spear heads in bronze. Died London, Eng., Nov. 24, 1892.

TAYLOR, M(ilton) Wight, Am. biochemist; b. Chatham, Mass., Jan. 12, 1904; s. Herman and Bessie (Smith) T.; B.S., U. Mass., 1925; M.S., Ia. State U., 1927, Ph.D., 1930; m. Sadie Josephine Perley, Sept. 17, 1927. Faculty, Rutgers U., New Brunswick, N.J., 1930—, prof., chmn. dept. agrl. biochemistry, 1954-62, prof. dept. biochemistry and microbiology, 1962—. Recipient Lindbach award for distinguished teaching Rutgers U., 1966. Mem. Am. Chem. Soc., Am. Inst. Nutrition, Poultry Sci. Assn. Research and numerous publs. in vitamins A and D, assay methods, stability and requirement for poultry, methods for preservation and analysis of forages, factors affecting intestinal absorption of amino acids. Home: 9 Dewey Dr., New Brunswick, N.J. 08901.*

TAYLOR, Moddie Daniel, Am. chemist; b. Nymph, Ala., Mar. 3, 1912; s. Herbert L. and Hattie (Oliver) T.; B.S., Lincoln U., 1935; S.M., U. Chgo., 1938, Ph.D., 1943; m. Vivian E. Taylor; 1 son, Herbert M. Faculty, Lincoln U., 1935-41, 45-48; asso. chemist Manhattan Project, 1943-45; prof. chemistry Howard U., Washington, 1948—. Vis. prof. Prairie View (Tex.) Coll., 1960-61. Recipient Outstanding Chemistry Tchr. award Mfg. Chemists Assn., 1960. Mem. Am. Chem. Soc., A.A.A.S., Washington Acad. Sci. Author: First Principles of Chemistry, 1960. Research and publs. on thermodynamic acid-base reactions and hydrogen bonding, chemistry of rare earth metal compounds. Home: 4560 Argyle Terrace N.W., Washington 20011.*

TAYLOR, Monica, protozoologist; b. St. Helens, Eng., Nov. 1, 1877; d. Joseph and Agnes (Picton) Taylor; student Mt. Pleasant Coll. Edn., 1898-1900; B.Sc. with honors, London (Eng.) U., 1910; D.Sc., U. Glasgow (Scotland), 1917, LL.D., 1953. Joined Order Notre Dame, 1900; lectr. sci. Coll. Education, Glasgow, 1901-46; vis. staff Carnoy Inst. Louvain, Belgium, Catholic U. Am., Trinity Coll. Recipient Neill medal Royal Soc. Edinburgh, 1958. Mem. Royal Philos. Soc. Glasgow (v.p. 1954-57), Brit. Assn. Sci. (v.p. 1952—). Author: (with Waddington) Principles of Biology, 1935; Sir Bertram Windle, A Memoir, 1932; also articles. Described devel. of Symbrarchus, chromosomes complex of Culex pipiens; research on amoeba, polyploidy and its evolution; developed methods of lab. growth material for protozoa. Home: 74 Victoria Crescent Rd., Glasgow, Scotland.*

TAYLOR, Norman, botanist; b. Hereford, England, May 18, 1883; s. James Durham and Mary Ann (Preece) T.; student Cornell U., 1901-02; D.Sc. (honorary), Washington College, 1958. Came to U. S., 1889, naturalized, 1896. Asst. curator N.Y. Botanical Garden, Bronx Park, 1905-11; curator of plants Brooklyn Botanic Garden, 1911-29; botany, horticulture and forestry editor, Webster's New Internat. Dictionary, 2d edition; editor of The Garden Dictionary, Houghton, Mifflin Co., 1933-36; director of Cinchona Products Inst., Inc., 1937-50; advisor Cinchona Instituut, Amsterdam, 1951-53; plant exploring expdns. to Bahamas, Haiti, Cuba, Santo Domingo, Puerto Rico, Yucatan, Guatemala, Bolivia, Peru. Ecuador and Brazil. Awarded Massachusetts Horticultural Society gold medal, 1936, Distinguished Service Award, N.Y. Bot. Garden, 1961, Liberty Hyde Bailey gold medal Am. Hort. Soc., 1963. Fellow New York Academy of Sciences, A.A.A.S.; associate fellow N.Y. Acad. Medicine; mem. Massachusetts Horticultural Soc., Torrey Botanical Club. Author: Botany, Science of Plant Life, 1924; Guide to the Wild Flowers, 1928; Cinebona in Java, 1945; Taylor's Encyclopedia of Gardening, 1948; Flight From Reality, 1949; The Permanent Garden, and Color in the Garden, 1953; Fragrance in the Garden, Herbs in the Garden, 1954; Fruits in the Garden, The Everblooming Garden, 1955; Wild Flower Gardening, 1956; Guide to Garden Flowers, 1957; Taylor's Garden Guide, 1957; The Ageless Relicts, 1962; 1001 Questions Answered About Flowers, 1963; Narcotics: Nature's Dangerous Gifts, 1963. Contbr. mags., Ency. Brit. and Book of Knowledge. Authority on flora and ecology of Long Island, on tropical American plants, Cinchona plants, and drug plants. Home: Elmwood Princess Anne, Md. Address: 20 W 10th St., N.Y.C. 11.

TAYLOR, Philip, civil engr.; b. Norwich, Eng., 1786; s. John and Susannah (Cook) T.; ed. Dr. Houghton's Sch., Norwich; studied surgery; m. Sarah Fitch, 1813; 8 children. Became druggist; partner (with his brother John) chem. works, London, from 1815; apptd. dir. Thames Tunnel Co.; with Brit. Iron Co.; worked in Paris, 1828; lived nr. Genoa, 1847-52, then returned to France. Contbr. articles to Quarterly Jour. Sci., 1819, Philos. Magazine, 1822. Invented method of lighting buildings by oil-gas, patented 1824, applied high-pressure steam in evaporating processes, patented 1816; horizontal steam engine, patented 1824; founded engrng. works, France, 1828; developed hot-blast process in iron mfg., patented 1832. Died St. Marquerite, France, July 1, 1870.

TAYLOR, R(aymond) Dean, Am. phys. chemist; b. Okemah, Okla., Aug. 18, 1928; s. H. Ray and Vera (Bay) T.; B.S., Kan. State Coll., 1950; Ph.D., Rice U., 1954; m. Janis Dexter, June 24, 1961; children—Scott Edward, Kay Suzanne. Staff, Los Alamos Sci. Lab., 1954—. Humble Oil Co. fellow, 1952-54. Mem. Am. Phys. Soc., Sigma Xi. Research and numerous publs. in low temperature physics especially liquid helium, solid state physics incorporating Mossbauer effect. Home: 210 Navajo Rd., Los Alamos 87544. Office: Los Alamos Sci. Lab., Los Alamos 87544.*

TAYLOR, Richard Moreland, Am. physician; b. Owensboro, Ky., Oct. 10, 1887; s. Edwin Pendleton and Betty (Moreland) T.; student U. Ky., 1904-05; M.D., U. Mich., 1910; Dr.Pub. Health, Johns Hopkins, 1926; m. Mary Ellen Stevick, Mar. 12, 1926; children—Mary Moreland, Elizabeth Dare (Mrs. Themistocles G. Michos), Suzanne France (Mrs. Henry M. Dater). Asst. bacteriologist U. and Bellevue Hosp. Med. Coll., 1910-11; instr. N.Y. Post-Grad. Sch. Medicine and Hosp., 1911-12, pathologist, 1912, faculty, 1913-19, prof., 1916-19; med. dir. A.R.C., Poland, 1920-21; mem. staff internat. health div. Rockefeller Found., 1922-52; dir. Virus Lab., Naval Med. Research Unit Number 3, Cairo, Egypt, 1951-55; lectr. dept. epidemiology Yale U. Med. Sch., 1956-59; lectr. epidemiology Sch. Pub. Health, U. Cal. at Berkeley, 1960—. Chmn. subcom. for exchange information, editor Catalogue on Arthropod-borne Viruses of World, 1960—. Decorated Legion of Honor (France). Mem. Am. Soc. Immunologists, Am. Soc. Tropical Medicine and Hygiene, A.A.A.S., Am. Pub. Health Assn. Author: (with G. K. Strode) Yellow Fever, 1951; also numerous articles. Research in epidemiology brucellosis, virus diseases, influenza, arthropod-borne viruses. Home: 4139 Los Arabis Dr., Lafayette, Cal. 94549. Office: Warren Hall, U. Cal., Berkeley, Cal. 94720.*

TAYLOR, Robert Cooper, Am. phys. chemist; b. Colorado Springs, Colo., May 5, 1917; s. Clarence Egbert and Marjorie C. (Cooper) T.; A.B., Kalamazoo Coll., 1941; Ph.D., Brown U., 1947; m. Evelyn Letitia Seeley, Dec. 27, 1942; children—David Robert, Donald Cooper. Research chemist Manhattan Project, 1942-45; faculty Brown U., 1947-49; faculty U. Mich., Ann Arbor, 1949—, prof. chemistry, 1963—. Mem. Am. Chem. Soc., Am. Phys. Soc., A.A.A.S., Sigma Xi. Asst. editor Inorganic Chemistry, 1964. Research and publs. on molecular structure and spectra of inorganic compounds, boron hydrides, boron hydride derivatives. Home: 850 Heatherway, Ann Arbor, Mich. 48104.*

TAYLOR, Robert William, dermatologist; b. London, Eng., Aug. 11, 1842; ed. Coll. Phys. and Surg., Columbia U., 1868. Prof. diseases of skin Womens Med. Coll., N.Y.C., also U. Vt.; surgeon venereal dept. Charity Hosp., also Bellevue Hosp. Dispensary. Author: A Clinical Atlas of Venereal and Skin Diseases, 1889; also articles. Mem. Am. Dermatol. Assn. (v.p.). Research in syphilis, expecially as related to skin diseases; 1st to describe diffused idiopathic atrophy of skin, 1875 (Taylor's diseases). Died 1908.

TAYLOR, Samuel Gale, III, Am. physician; b. Elmhurst, Ill., Sept. 2, 1904; s. Samuel Gale, Jr. and Anna (Mead) T.; B.A., Yale, 1927; M.D., U. Chgo., 1932; m. Eleanor Roberts, June 1, 1938; children—Constance (Mrs. William Blackwell), John W., Samuel Gale, IV. Practice medicine, specializing in internal medicine, Chgo., 1937—; asst. prof., then prof. medicine U. Ill., 1941—. Cons. cancer control program USPHS, 1958-64. Mem. A.M.A., A.C.P., A.C.S. (mem. cancer com.), Am. Coll. Radiology (mem. cancer soc.), James Ewing Soc., Am. Soc. Cancer Research, Am. Cancer Soc. (past dir.), Central Soc. Clin. Research, Am. Soc. Clin. Oncology, Eastern Cooperative Oncology Group, Am. Radium Soc. Research and numerous publs. in endocrinology, hematology and cancer. Home: 1386 N. Green Bay St., Lake Forest, Ill. 60045. Office: 1753 W. Congress St., Chgo. 60612.*

TAYLOR, Theodore Brewster, Am. nuclear physicist; b. Mexico City, Mexico, July 11, 1925; s. Walter Clyde and Barbara (Howland) T.; grad. Phillips Exeter Acad., 1942; B.S., Cal. Inst. Tech., 1945; Ph.D., Cornell U., 1954; m. Caro Dwight Arnim, June 13, 1948; children—Clare E., Katherine W., Christopher H., Robert P., Jeffrey J. Theoretical physicist U. Cal. Radiation Lab., Berkeley, 1946-49, Los Alamos Sci. Lab., 1949-56; sr. research adviser Gen. Atomic div. Gen. Dynamics Corp., San Diego, 1956-64, dep. dir. (sci.) Def. Atomic Support Agy., Dept. Def., 1964—. Cons. Los Alamos Sci. Lab., 1956-64, Aerospace Corp., 1960-61, Air Force Sci. Adv. Bd., 1955-58. Chmn. Los Alamos Study Group, Air Force Space Study Com., 1961; mem. panel on outer space U. S. Arms Control Disarmament Agy., 1961. Served to ensign USNR, 1942-46. Recipient Ernest Orlando Lawrence award, 1965. Mem. A.A.A.S., Am. Geophys. Union, Am. Phys. Soc., Am. Nuclear Soc. Work on the control and devel. nuclear power; co-developer of safe TRIGA reactor; improved nuclear weapons; studied use of nuclear energy to propel spacecraft; nuclear explosives and effects of nuclear explosives. Home: 7604 Glennon Dr., Bethesda, Md. Office: Hdqrs. DASA, The Pentagon, Washington.

TAYLOR, Thomas Glanville, Brit. astronomer; b. Devonshire, Eng., Nov. 22, 1804; s. Thomas Taylor; entered Royal Obs., Greenwich, Eng., 1820; m. Eliza Eley; 3 sons. Placed in charge transit instrument Royal Obs., Greenwich, 1822; became asst. to Sir Edward Sabine in pendulum expts., 1829; became dir. Madras Obs., 1830. Fellow Royal Soc., 1842, Royal Astron. Soc. Author: Madras General Catalogue, of Star, 1844; also observations made at Madras Obs. in 5 vols., 1831-39. made calculations for Stephen Groombridge's star catalogue. Determined longitude of Madras; his observations were important for studies on motions of So. stars. Died Southampton, Eng., May 4, 1848.

TAYLOR, T(homas) Ivan, Am. chemist; b. Farwest, Utah, Sept. 11, 1909; s. Thomas G. and Harriet Mae (Hill) T.; B.S., U. Ida., 1931, M.S., 1933; fellow, Columbia, 1936-37, Ph.D., 1938; m. Frances Rachel Nichols, Sept. 22, 1939; children—Mary, Arthur. Instr. in chemistry U. Ida., 1931-35; teaching asst. Columbia, 1935-38; instr. in chemistry U. Minn., 1938-40, asst. prof., 1940-42; chemist Nat. Bur. of Standards, Washington, 1942-44; chemist Research Control Group, Manhattan Dist., Washington, Oak Ridge, Tenn. and N.Y.C., 1944-45; prof. chemistry U. Ia., 1945; asso. prof. chemistry Columbia, 1945-48, prof., 1949—. Chairman Gordon Conf. on Physics and Chemistry of Isotopes, 1956; chmn. adv. com. chemistry div. Nat. Bur. Standards, 1961-64; mem. com. chemistry department of Brookhaven National Laboratory, 1965. Member Am. Chem. Society (councilor 1950-54, chmn. N.Y. sect. 1957-58), Am. Nuclear Society, American Physical Society, A.A.A.S., Society for Applied Spectroscopy, Sigma Xi. Author: Experiments in General Chemistry (with J. L. Maynard), 1944; Semimicro Qualitative Analysis (with H. H. Barber), 1942. Mem. editorial bd. Nucleonics, 1948-49. Author research reports sci. jours. Studies of separation and use isotopes in chem. research; reaction kinetics; catalysis, instrumental analysis; neutron interactions. Home: 140 Lakeview Av., Leonia, N.J.*

TAYLOR, Thomas Mayne Cunninghame, botanist; b. Pretoria, South Africa, July 7, 1904; s. Lionel E. and Hilda (Stirling) T.; B.A., U. B.C. (Can.), Vancouver, 1926; M.S., U. Wis., 1927; Ph.D., U. Toronto (Ont.), Can.), 1930; m. Barbara D. Howell, Sept. 10, 1928; children—Charles P. S., Jean D. (Mrs. Job Kuijt), Suzanne L. (Mrs. Charles M. Johnson), Phyllis M. (Mrs. William W. Aylsworth). Faculty, U. Toronto, 1928-45, asst. prof. botany, 1935-45; prof. U. B.C., 1946—, head dept. biology and botany, 1954-64. Gager fellow Bklyn. Botanic Garden, 1953. Mem. Bot. Soc. Am., Am. Fern Soc., Am. Soc. Plant Taxonomists, Cal. Bot. Soc., Canadian Bot. Assn. Author: Ferns and their Allies of British Columbia, 1964; Lilies of British Columbia, 1966; also articles. Research on occurrence and distbn. of ferns and their allies of Pacific N.W. and flowering plants of B.C. Home: Millstream Rd., Rural Route 6, Victoria, B.C. Office: Dept. Botany, U. B.C., Vancouver, B.C., Can.*

TAYLOR, Walter Willard, Am. anthropologist; b. Chgo., Oct. 17, 1913; s. Walter Willard and Marjorie (Wells) T.; grad. Hotchkiss Sch., 1931 A.B., 1935, Yale fellow Lab. Anthropology, 1936; student U. N.M., 1937; Ph.D. (Hemenway fellow), Harvard, 1943; m. Lyda Averill Paz, Sept. 6, 1937 (dec. May 1960); children—Peter Wells, Ann Averill, Gordon McAuliffe; m. 2d, Nancy Thompson Bergh, Nov. 24, 1962; children—Karin, Christina. Instr. Ariz. State Coll., 1937-38; asst. instr. Harvard, 1940, 42; instr. Tex. U., 1942-43, Prisoner of War Schs., 1944-45; lectr. Friends Internat. Seminars, 1948-53; vis. asst. prof. U. Wash., 1949, vis. prof., 1953; vis. prof. Escuela Nacional de Ante. Hist., Mexico, 1955-58, U. Merida, Mexico, 1956; prof. Mexico City Coll., 1956-57; professor of anthropology So. Illinois University, 1958-63, chmn. dept., 1963—. Collaborator anthropology United States National Museum, 1940—; director No. Mex. Archeol. Fund, Smithsonian Instn., 1940-49, 57—, dir. Southwest Archeol. Fund, 1949-57; pres. Indian Arts Fund, Santa Fe, 1948; mem. com. on archeol. identifications NRC, 1955-57, dir. program cultural ecology, 1959—; v.p. Centro de Investigaciones Anthropologicas de Mexico, 1955-56; pres. Found. for Anthrop. Research in Latin Am., 1956-57. Pres. Santa Fe Little Theatre, 1950; mem. bd. Players, Inc., Mexico City, 1957-58. Served with USMC, World War II; capt. Res. Decorated Bronze Star medal with citation, Purple Heart; Rockefeller Found. fellow humanities, 1946; Guggenheim fellow, 1950-51. Fellow A.A.A.S., Am. Anthrop. Assn.; mem. Pecos Southwestern Conf. (program chmn. 1958, 61), Am. Assn. U. Profs. Soc. Am. Archeol., Delta Kappa Epsilon. Research on theory and methods of archeology; pioneer work in archeology of northern Mexico. Home: 1005 San Acacio, Santa Fe.*

TAYLOR, William Randolph, Am. botanist; b. Phila., Dec. 21, 1895; s. William Long and Caroline (Sower) T.; B.S., U. Pa., 1916, M.S., 1917, Ph.D., 1920; m. Jean Falconer Grant, Dec. 18, 1926; children—William Randolph, James Keith. Prof. botany, curator U. Mich., Ann Arbor, 1930—; research asso. Acad. Natural Scis. (Phila.), Farlow Herbarium of Harvard. Trustee Marine Biol. Lab., Woods Hole, Mass., 1939—, Bermuda Biol. Sta., 1953—; mem. council, adv. bd. Charles Darwin Sta., Galapagos Islands. Recipient Retzius medal K. Fysiografiska Sallskapet, Lund, Sweden, 1948. Russel lectr. U. Mich., 1963. Fellow Royal Soc., 1934; corr. mem. Academie des Scis., Inst. Française, Soc. Venezolana Sci. Nat.; fgn. mem. Linnean Soc. London, Konikl. Acad. Wentensch., Litt. and Schone Kunsten Bruxelles, Belgium; mem. Am. Acad. Arts and Sci. Author: Marine Algae of Florida, 1928; Marine Algae of the Northeastern Coast of North America, 1937; Pacific Marine Algae of the Allan Hancock Expeditions to the Galapagos Islands, 1945; Plants of Bikini and other Northern Marshall Islands, 1950; Marine Algae of the Eastern Tropical and Subtropical Coasts of the Americas, 1960; also numerous articles. Studies on abnormal plant anatomy, alpine freshwater algae of Nfld., marine algae of Fla. and New Eng. coast, Caribbean and S.Am., Western Pacific, Indonesia, Philippines and Indian Ocean. Home: 785 Arlington Blvd., Ann Arbor, Mich. 48104.*

TCHEBYCHEV, Pafnutii Lvovich, Russian mathematician; b. Borovsk, USSR, May 26, 1821; ed. by pvt. tutor; student Moscow (USSR) U.; master pure math., 1846; doctorate, 1849. Prof. geometry U. St. Petersburg (now Leningrad, USSR); founder Russian Sch. Math. Recipient Silver medal for essay on theory of equation. Mem. French (corr., fgn.), St. Petersburg acads. sci., Royal Soc. (corr.), Berlin Acad. (corr.). Devised straight line motion; research on probabilities, quadratic forms, theory of integrals, theory of numbers, polynomials, prime numbers; work on problem of obtaining rectilinear motion; invented various math. machines. Died St. Petersburg, Dec. 8, 1894.

TCHEN, Tche Tsing, biochemist; b. Peiping, China, Oct. 1, 1924; s. Tze Fan and T.R. (Tsao) Cheng; Chem.E., Aurora U., Shanghai, China, 1948; Ph.D. in Biochemistry, U. Chgo., 1954; m. Ina Lin, Dec. 10, 1960; children—Terence, Vincent. Came to U. S., 1950, naturalized, 1964. Research asst., asso. U. Chgo., 1950-55; research fellow Harvard, 1955-58; faculty Wayne State U., Detroit, 1958—, prof. biochemistry, 1962—. Vis. prof. biochemistry U. Cal. at Berkeley, 1961, Harvard, 1965; cons. USPHS,

1960-65. Am. Cancer Soc. scholar, 1956-58. Mem. Am. Chem. Soc., Soc. Am. Biol. Chemists, Am. Soc. Microbiologists. Research and publs. on mechanism of action of enzymatic reactions, sterol biogenesis and metabolism, biochem. basis of differentiation of pigment cell system, control of genetic expression. Home: 19110 Midway, Southfield, Mich. 48076. Office, Wayne State U., Detroit 48202.*

TCHLENOV, Yehiel (Tschlenow, Jehiel), physician; b. Kremenchug, Lithuania, Oct. 31, 1863; ed. U. Moscow, 1888; practiced otolaryngology, Moscow; joined polit. Zionist movement, 1897; convened Helsingfors Conf., 1906; became leader Russian Zionist Movement, 1910; moved to Berlin, Germany, then to Copenhagen, Denmark 1915, later to London. Research and publs. on hemorrhagic diarrhea, movement of leukocytes, leukopenia, cerebral complication in otitis media; responsible for establishment spl. schs. for tng. of deaf mutes by Russian Govt. Died London, 1918.

TEACHER, John Hammond, Scottish embryologist, physician; b. Old Kilpatrick, Dumbartonshire, Scotland, Oct. 25, 1869; s. William Teacher; ed. Glasgow; M.A., U. Glasgow, 1888, M.B., C.M., 1893; M.D., 1904; m. Rachel Mitchell Hutcheson, 1904. Became curator Hunterian Mus., 1894; apptd. asst. prof. physiology U. Glasgow, 1899, of pathology, 1903, lectr. path. histology, 1908, St. Mungo prof. pathology, 1911; pathologist Glasgow Royal Infirmary, from 1909. Recipient Bella Houston Gold medal for M.D. thesis. Fellow Royal Coll. Physicians and Surgeons Glasgow, 1900. Author: Manual of Obstectrical and Gynaecological Pathology (edited by Alice J. Marshall, 1935); also several works on devel. of human embryos. Research on embryonic devel.; demonstrated (with Bryce) human embryo of 13-14 days (earliest known ovum developmental period in 1908); also studies on diseases associated with pregnancy, especially chorion epithelioma and symmetrical cortical necrosis of kidneys. Died Nov. 21, 1930.

TEAL, John Jerome, Am. human ecologist, animal domesticator; b. N.Y.C., Feb. 7, 1921; s. John Jerome and Isabelle (O'Sullivan) T.; B.S., Harvard, 1944; M.A., Yale, 1946; m. Penelope Holden, May 6, 1950; children—Pamela G., Ptarmigan P., John A., Lansing H. Arctic research, field work, 1946-50; Carnegie Found. Travelling fellow Svalbard, Norway, No. Scandinavia; 1st sr. fellow Arctic Inst. N.Am., research asso. McGill U., 1951-52; pres. Inst. No. Agrl. Research, Huntington Center, Vt., 1954—; asso. prof. geography and anthropology U. Vt., 1957-59; prof. human ecology U. Alaska, College, 1964—. Cons., dir. Canadian Reindeer Industry, MacKenzie Delta, Alaska, 1960—; Lithgow Osborne lectr., 1966-67. Mem. Conn. Acad. Arts and Scis., Am. Soc. Mammalogists, Am. Anthrop. Assn., A.A.A.S., Soc. for Am. Archaeology, Norsk Polarklubb. Author: Gift of Dominion, 1966; also numerous articles. Domesticated musk ox for use in no. agr.; research on use organic resources through domestication and genetic expts., origins agr. Home: Tunturi, Huntington Center, Vt. 05462. Office: U. Alaska, College, Alaska 99735.*

TEAL, John Moline, Am. biologist; b. Omaha, Nov. 9, 1929; s. Clarence W. and Valentine (Moline) T.; A.B., Harvard, 1951, Ph.D., 1955; m. Mildred Mann, Dec. 30, 1950; children—Eric Peter, Tanya Christine, Asst. prof. U. Ga., 1955-59; asst. prof. dept. zoology, Inst. Oceanography, Dalhousie U., 1959-61; research asso. Woods Hole (Mass.) Oceanographic Inst., 1961-65, asso. scientist, 1965—. Mem. A.A.A.S., Ecol. Soc. Am., Am. Ornithol. Union, Am Soc. Limnology and Oceanography. Author: (with Mildred Teal) Portrait of an Island, 1964. Studies of function in natural ecosystems, productivity and orgn. of salt marshes, metabolic rates of animals, effects of pressure in deep sea, heat balance of marine animals and plants, body temperature in warm-bodied fish. Home: Box 273, North Falmouth, Mass. 02556. Office: Woods Hole Oceanographic Inst., Woods Hole, Mass. 02543.*

TEALL, Sir Jethro Justinian Harris, Brit. geologist; b. Northleach, Eng., Jan. 5, 1849; s. Jethro and Mary (Hathaway) T.; B.A., Cambridge, 1873, M.A., 1876, also D.Sc.; D.Sc., U. Dublin, Oxford U.; LL.D., St. Andrews; m. Harriet Cowen, 1879; 2 sons. Lectr., St. John's Coll., Cambridge; staff Gt. Britain Geol. Survey Team, 1888, dir., 1901-14. Mem. Royal Commn. on Coal Supplies, 1901-05. Recipient Sedgwick prize, 1874; Bigsby medal, 1889; Wollaston medal, 1905; Delesse prize, 1907. Fellow Royal Soc., 1890, v.p. 1900-01; fellow Geol. Soc. (sec. 1893-97, pres. 1900-02). Author: British Petrography, 1888; (with J. R. Dakyns) Plutonic Rocks of Garagal Hill, 1892; (with J. Horne) Borolanite, 1893; also articles. One of 1st to study tectonics, natural history and genesis of several metamorphic minerals; studies on Brit. igneous rocks, Lewisian gneisses, torridon sandstone, Post-Cambrian igneous rocks. Died Dulwich, Eng., July 2, 1924.

TEAR, James DeGraff, Am. physicist; b. Warren, Ill., June 8, 1892; s. Henry C. and Mary (DeGraff) T.; B.S., Beloit Coll., 1916; Ph.D., Yale, 1922; m. Lillian Ward Grant, Sept. 1, 1922; 1 son, Daniel Grant. Staff, Nela Research Lab., Cleve., 1920-25, Gen. Elec. Research Lab., 1925-33; with Ford Instrument Co. div. of Sperry Rand Corp. N.Y.C., 1933-57, research dir., 1937-57. Fellow A.A.A.S., Am. Phys. Soc. Research on short electric waves, infrared, radiation pressure, aircraft instruments, automatic control of aircraft, hydraulic servomechanisms, gunfire control, gyroscopic stabilization. Home: 10 Bancroft Lane, Great Neck, N.Y. 11024.*

TECCE, Giorgio, Italian molecular biologist; b. Naples, Italy, Apr. 13, 1923; s. Raffaele and Ines (Venier) T.; Degree in Chemistry, U. Rome (Italy), 1946; m. Vanda Mcrello, July 28, 1951; children—Francesca, Marco. Faculty, U. Rome, 1950—, prof. chemistry fermentations, 1951—, dir. Inst. Gen. Physiology, 1958-64; dir. Center Nucleic Acids, Nat. Research Council, 1966—. Mem. commn. for biol. scis. curriculum study Italian Ministry Pub. Edn., 1964—; mem. commn. biophysics C.N.R., 1966—. Mem. European Molecular Biology Orgn., Soc. for Biophysics and Molecular Biology, Nat. Assn. Asst. Profs. (past pres.). Research, publs. on induction of pentose metabolism enzymes in microorganisms, physiology of thermophylic bacteria and molecular properties of their cell constituents, errors in translation of genetic code. Home: 91 Viale Ippocrate, Rome, Italy.*

TECLU, Nicolae, Rumanian chemist; b. 1839; mem. Rumanian Acad. Author: Zur Kennzeichnung der Flamme, 1899; Studien Behelfe für den Unterricht in der allgemeinen und technischen Chemie, 1905. Research on combustion and gas flame, properties and analysis of mineral substances of paper and ligneous fibers, also of oils and pigments for paints. Died 1916.

TEDESCHI, David Henry, Am. neuropharmacologist; b. Newark, Feb. 20, 1930; s. Edward David and Carmela (Beltrani) T.; B.S., Rutgers U., 1952; Ph.D., U. Utah, 1955; m. Irene Laura Mankiewicz, May 31, 1952; children—Paul David, Douglas James, Linda Diane. Research fellow in pharmacology U. Utah, 1952-55; sr. pharmacologist Smith Kline & French Labs., Phila., 1955-57, area dir., 1957-60, group leader, 1961-62, sect. head for neuropharmacology, 1962-67, asso. dir. pharmacology, 1967—. Mem. Acad. Pharm. Sci. (vice chmn. sect. on pharmacology and biochemistry). Research, numerous publs. and patents on mechanism of action of tranquilizers and antidepressants; developed several test procedures for assessing activity of tranquilizers, antidepressants, anorectics, muscle relaxants, antitussives and cardiovascular agts. in exptl. animals; developed exptl. clin. and marketable tranquilizers, antidepressants, analgesics, muscle relaxants, antiemetics, anti-appetite agts. Home: 509 Gatwood Rd., Cherry Hill, N.J. 08034. Office: 1500 Spring Garden St., Phila. 08034.*

TEDESCHI, Ralph Earl, Am. pharmacologist; b. Newark, Nov. 20, 1927; s. Edward David and Iris (Beltrani) T.; B.S. in Pharmacy, Rutgers U., 1951; Ph.D. in Pharmacology, Med. Coll. Va., 1954; m. Eleanor Snagg, June 10, 1951; children—Ralph Edward, Ronald Edmund. Researcher dept. pharmacology Oxford (Eng.) U., 1954-55; research asso. dept. medicine Jefferson Med. Coll., 1955-56; head pharmacology sect. structure-function-natural products dept. Smith, Kline & French Labs., Phila., 1956-67, associate director of pharmacology, since 1967—. Mem. Am. Soc. for Pharmacology and Exptl. Therapeutics, Soc for Exptl. Biology and Medicine, A.A.A.S., N.Y. Acad. Scis. Research and publs. on neurol. and cardiovascular areas, tranquillizers, analgetics, antidepressants, muscle relaxants, antihypertensive drugs. Patentee antidepressant drug tranylcypramine. Home: 1366 Paddock Way, Cherry Hill, N.J. 08034. Office: 1530 Spring Garden St., Phila. 19101.*

TEEVAN, Richard Collier, Am. psychologist; b. Shelton, Conn., June 12, 1919; s. Daniel Joseph and Elizabeth (Hallowell) T.; B.A., Wesleyan U., Middletown, Conn., 1951; M.A., U. Mich., 1952, Ph.D., 1955; m. Virginia Agnes Stehle, July 28, 1945; children—Jan, Kim, Clay, Allison. Faculty, U. Mich., 1955-57, Smith Coll., 1957-60; faculty Bucknell U., Lewisburg, Pa., 1960—, prof. psychology, 1964—, chmn. dept., 1966—. Mem. Am., Eastern, Pa. (pres. acad. div.) psychol. assns., Phi Beta Kappa, Sigma Xi. Author: (with Robert Birney) Instinct, 1961, Reinforcement, 1961, Color Vision, 1961, Measuring Human Motivation, 1962, Theories of Motivation in Learning, 1964, Theories of Motivation in Social and Personality Psychology, 1964, Fear of Failure, 1967, Readings in Introductory Psychology, 1966; (with Barry Smith) Motivation, 1967. Research and publs. on fear of failure, achievement motivation. Home: R.D.1, Lewisburg, Pa. 17837.*

TEFFÉ, Antonio Luiz von Hoonholtz, Baron, Brazilian geographer; b. Rio de Janeiro, Brazil, May 9, 1837; ed. Marine Academy, Rio de Janeiro. Explored Brazilian coast around St. Catherine Island; in charge of marking boundaries of Brazil and Peru, 1871; promoted to rear-admiral, Brazilian navy. Corr. mem., French Acad. Scis., 1889. Author of 1st treatise on hydrography ever written in Portuguese. Head of Brazilian mission at St. Thomas, Antilles, that observed transit of Venus; founded 1st geographic soc. in Rio and organized Brazilian hydrographic service. Died Petropolis, Brazil, Feb. 7, 1931.

TEGETMEIER, William Bernhard, English naturalist; b. Colnbrook, Eng., 1816; s. G. C. Tegetmeier; ed. U. Coll., London, Eng.; 1 son, 2 daus. Lectr. Govt. Tng. Coll.; Davis lectr. Zool. Soc.; writer, editor The Field mag., for 50 years. Recipient medals U. London, Soc. Apothecaries. Author: The Poultry Book, 1870; The Homing Pigeon, 1872; Pigeons, 1879; Table and Market Poultry, 1895; Natural History of the Cranes, 1881; Pallas's Sand Grouse, 1888; Pheasants, 5th edit., 1911; Horses, Zebras and Mule Breeding, 1895; The House Sparrow, 1899. Contbd. (with Sutherland) to Ency. Britannica. (With Darwin) conducted study of variation in animals; showed primary circular form of bee's cell; demonstrated pigeon's homing ability; (with Walter Gilbey) helped increase poultry industry; (with Miss E. Crmerod) research on sparrow's ruinous effects. Died Nov. 19, 1912.

TEICHERT, Curt, geologist; b. Königsberg, Germany, May 8, 1905; s. Richard and Louise (Zander) T.; student U. Munich (Germany), 1923-25, U. Freiburg (Germany), 1925, U. Königsberg (Germany), 1925-28; Ph.D., Albertus U., 1928; D.Sc., U. Western Australia, Perth, 1944; m. Gertrud Margarete Kaufmann, Dec. 28, 1928. Came to U. S., 1952, naturalized, 1959. Rockefeller fellow, 1930; privatdozent Techn. U., Berlin, 1931-35; research paleontologist U. Copenhagen (Denmark), 1933-37; research lectr. U. Western Australia, 1937-46; asst. chief govt. geologist Mines Dept., Melbourne Australia, 1946-47; sr. lectr. U. Melbourne, 1947-53; prof. geology N.M. Sch. Mines, 1953; geologist U. S. Geol. Survey, 1954-64, chief Petroleum Geology Lab., 1954-58, research geologist, 1958-61; AID adviser Geol. Survey, Pakistan, 1961-64; Distinguished prof. geology, dir. Paleontol. Inst., U. Kan., Lawrence, 1964—, geologist Danish N.E. Greenland Expdn., 1931-32; guest prof. U. Bonn (Germany), U. Gottingen, U. Freiburg, 1958, U. Tex., 1960. Cons. to oil cos., 1940-53, Australian Bur. Mineral Resources, 1948-52. Recipient David Syme prize sci. research U. Melbourne, 1948. Mem. Internat. Geol. Congress (past sec. internat. commn. Gondwana system), soc. géol. France (fgn. corr.), Paleontol. Soc. (past corr.). Author: Ordovician and Silurian Faunas from Arctic Canada, 1937; (with Clarke, Prider) Elements of Geology, 1944, Elementary Practical Geology, 1946; (with others) Treatise on Invertebrate Paleontology, Part K, 1964; also numerous articles. Research on older paleozoic faunas of No. Europe and Arctic, arctic geology, stratigraphy and paleontology of maj. sedimentary basins Australia, fossil Cephalopoda, modern and ancient coral reefs, gen. stratigraphy of Pakistan, nature of maj. evolutionary breaks, Devonian rocks and paleogeography of central Ariz. Home: 14 Westwood Rd., Lawrence, Kan. 66044.*

TEICHMANN, Bodo Alfred, German chemist; b. Leipzig, Germany, Mar. 9, 1932; s. Hugo Alfred and Elisabeth (Reichenbach) T.; Dipl.Chem., U. Leipzig, 1956, Dr.rer.nat., 1959; Dr.rer.nat.habil., Humboldt U., Berlin, Germany, 1965; m. Annelies Margit Hacker, May 16, 1959; 1 dau., Elke Margit. Staff, Inst. for Organic Chemistry, U. Leipzig, 1956-60, chief asst., lectr., 1958-60; chief asst. Inst. Cancer Research German Acad. Scis., Berlin, 1960—. Mem. Chem. Soc. E. Germany, German Acad. Scis. (mem. sect. cancerogenesis and basis of cancer therapy 1964—). Research and numerous publs. on insecticides on basis of chloral and brominated aliphatic dicarboxylic acids, cellular defense test of chemotherapeutics; discovered way to restore power of defense of lymphoid tissues after X-irradiation by humoral X-ray sensitive factor of thymus. Home: 2 Akademie-Neubau, 115 Berlin. Office: 70 Lindenberger Weg, 1115 Berlin, East Germany.*

TEICHMANN, Theodor, physicist; b. Königsberg, Germany, Sept. 16, 1923; s. Markus and Goldi (Rottenberg) T.; B.Sc. in Elec. Engring., U. Cape Town, 1943, M.Sc. in Math. (Croll scholar), 1945; M.A. in Physics, Princeton, 1947, Ph.D. in Physics, 1949; m. Selma Caroline Swift, Oct. 24, 1953; children—Valerie Susan, Greta Naomi. Came to U. S., 1946, naturalized, 1955. Lectr. elec. engring. U. Cape Town, 1944-46; research asso. Princeton, 1950-52; research physicist Hughes Aircraft Co., 1952-55; head systems lab., mgr. nuclear physics dept., asst. to dir. research, cons. scientist satellite systems, Lockeed Missiles & Space Co., 1955-60; mem. research staff Gen. Atomic div. Gen. Dynamics Corp., San Diego, 1960-68; prin. scientist Heliodyne Corporation, 1968—. Mem. Am. Phys. Soc., Am. Math. Soc., Am. Nuclear Soc., Am. Astronaut. Soc., Soc. Indsl. and Applied Math., Am. Ordnance Assn. Contbr. articles to profl. jours., also numerous tech. book revs. Research on resonance theory of nuclear reactions and analogy with electromagnetic circuits theory, application math. methods to problems neutron and radiation transport, space systems. Home: 1645 Copa de Oro Dr., La Jolla, Cal. 92037. Office: 11689 Sorrento Valley Rd., San Diego 92121.*

TEICHMEYER, Hermann Friedrich, German physicist, chemist; b. Minden, Germany, Apr. 30, 1685; M.D., Jena, 1705. Became prof. exptl. physics, 1717; prof. medicine extraordinary, 1719; lectr. chemistry, 1720-43; became prof. botany, surgery, anatomy, Jena, Germany, 1727; physician to the Dukes of Saxe-Weimar and Saxe-Eisenach. Mem. Berlin Acad. Scis., Academia Naturae Curiosorum. Author: Elementa philosophiae naturalis experimentalis, 1717; Institutiones chemiae dogmaticae . . . , 1729; other. Claimed to have described cobalt sympathetic ink before Hellot. Died Jena, Feb. 5, 1744.

TEILHARD DE CHARDIN, Pierre, paleontologist, geologist; b. Orcines (Puy-de-Dome), May 1, 1881; s. Emmanuel T. de C. D.Sc., Sorbonne, 1922. Ordained Jesuit priest, 1912; taught sci., Cairo, Egypt, Inst.

Human Paleontology, Mus. Natural History; prof. geology Catholic Inst., Paris, 1922-28; adviser to Geol. Service of China, 1929-45; dir. Laboratoire de Géologie Appliquée à l'Homme, Ecole des Hautes Etudes, from 1938; dir. research Centre National des Recherches Scientifiques, 1947; asso. Wenner-Gren Found., N.Y.C.; taught Aurora U., Peking; made several geol. and paleontol. expdns. to China. Mem. French Acad., 1947. Author: Le phénomène humain, 1955; Le milieu divin, 1964; Letters from a Traveller, 1962; Future of Man, 1964; Making of a Mind, 1965; Hymn of the Universe, 1965, also numerous articles. One of discoverers of Peking Man, 1928; known for his work on Cenozoic geology and on history of early man in eastern Asia; did preliminary research on terrain of Channel Islands, Egypt and Eng.; studied prehistoric phys. geography; published studies establishing idea of genetic structure of humanity; propounded philosophy (pub. posthumously because supressed by his Religious superiors during his lifetime) fusing science and religion and analyzing past, present and future of mankind. Died N.Y.C., Apr. 10, 1955.

TEIR, Harald Gustav, Finnish pathologist; b. Lappfjärd, Finland, Oct. 27, 1914; s. Frans Viktor and Hilja (Hannelius) T.; M.D., Helsinki (Finland) U., 1946, M.D.Sce., 1946; m. Brita Gunvor Johanna Granlund, Nov. 22, 1932; children—Hendrik, Gustav, Margareta, Christian, Eva. Staff, U. Helsinki, 1944——, prof. pathology, 1955——, dean Faculty Medicine, 1963——, head II dept. pathology, 1955——. Mem. Finnish State Med. Research Council, 1961——, vice chmn., 1961-66; cons. pathologist to Finnish hosps. Decorated Comdr. Finnish Order White Rose. Mem. or hon. mem. many Finnish and Scandinavian med. and odontological socs. Research and numerous pubs. on cellular growth control from biol. point of view, regeneration phenomenon, pathology of various organs., kinetics of blood and tissue granulocytes. Home: 5.A. Rakuunantie, Helsinki 33, Finland. Office: 3 Hartmanninkatu, Helsinki 29, Finland.*

TEISSEIRE, Paul, French chemist; b. Nice, France, Nov. 6, 1922; s. Honoré and Caroline (Pellegrin) T.; Ingénieur chimiste, École de Chimie, Marseille, France, 1944; m. Francine Monquet, Aug. 1, 1943; children—Richard, Arlette. Prof., 1944-45; research engr. Ste Roure-Bertrand Dupont, 1945-47; dir. research services, 1947——. Mem. French Soc. Phys. Chemistry, Chem. Soc. France, Am. Chem. Soc. Research and publs. on extraction of volatile oils, and the nature of their structure, including palmarosa, lanaudin, weld, geranium, laurel, galbanum; chromatography of gases; pure synthesis, including terpenes, sesquiterpenes, monocycles; stereochem. studies of seneterpenes. Home: 14 Avenue Pierre Semard. Office: 27 Avenue Pierre Leinard, Grasse, France 061.*

TEISSERENC DE BORT, Leon Philippe, French meteorologist; b. Paris, France, Nov. 5, 1855. Chief meterologist Central Meteorol. Bur., Paris, 1892-96; Founder Meteorol. Obs., Trappes; mem. French Acad. Scis., 1910; Société d'Agriculture. Author: Les Bases de la Meteorologis Dynamique, 2 vols., 1898-1907. Discovered of Stratosphere (isothermal layer); pioneered use of unmanned highflying instrumented balloons; first to systematically investigate atmospheric heights. Died Cannes, Jan. 2, 1913.

TEITEL, Alfred Bernard, pharmacologist; b. Dorohoi, Rumania, Feb. 2, 1900; s. Bernard H. and Sofia (Waldman) T.; student Faculty Medicine, Bucharest, 1917-23; docent pathophysiology, 1931; m. Margaret L. Broder, June 24, 1926; children—Paul, Theodor. Mem. Faculty Medicine, Bucharest, 1920——, asst. III med. clinic, 1920-47, prof. pharmacology, 1947-——. Mem. Physiol. Soc. Bucharest, Soc. Française de Therapie et de Pharmacodynamie, Internat. Soc. Biochem. Pharmacology, N.Y. Acad. Scis. Research, numerous publs. on morpho-physiol. properties of red blood cells, circulation of red blood cells in spleen, cholinergic activity of hyosciamine, double action of acetylcholine on nerves and muscles, correlations between acetylcholine and potassium, muscular action of triiodothyronine. Home: Bul. Hristo Botev 7, Bucharest. Office: Faculty of Medicine, Bul. Dr.P. Groza 8, Bucharest, Rumania.*

TEITEL, Robert Jerrell, Am. materials engr.; b. Indpls., Aug. 4, 1922; s. Harry S. and Ida (Bloom) T.; B.S. in Metall. Engring., Purdue U., 1944; Sc.D. in Metall. Engring., Mass. Inst. Tech., 1948; m. Faith E. Frankstein, Aug. 23, 1952; children—Sheri Lynn, Steven Lee, Bruce David. Devel. engr. in atom bomb project Houdialle-Hershey Corp., 1944; research staff Armour Research Lab., Chgo., 1944-45; research asso. Manhattan project, dept. metall. engring. Mass. Inst. Tech., 1945-48; scientist nuclear reactor sci. and engring. dept. Brookhaven Nat. Lab., Upton, L.I., N.Y., 1948-55; scientist, project mgr. nuclear and basic research dept. Dow Chem. Co., Midland, Mich., 1955-60; mgr., alkali metal research Rocketdyne div. NAR Corp., Canoga Park, Cal., 1960; scientist nuclear and advanced missile materials research Missile and Space div. McDonald Douglas Corp., Santa Monica, Cal., 1961——. Mem. Am. Nuclear Soc. (chmn. membership com. 1966——), Am. Soc. for Metals, Am. Inst. Aeros. and Astronautics, A.A.A.S. Research, publs. on basic thermodynamics of materials, nuclear reactor materials, nuclear reactor design, nuclear fuel processing, advanced nuclear power for space application, advanced materials research for missiles. Patentee in field. Home: 17159 Citronia Av., Northridge, Cal. 91324. Office: 3000 Ocean Park. Blvd., Santa Monica, Cal.

TEITELBAUM, Philip, Am. psychologist; b. Bklyn., Oct. 9, 1928; s. Bernard and Betty (Schechter) T.; B.S., Coll. City N.Y., 1950; M.A., Johns Hopkins, 1952, Ph.D. in Psychology, 1954. Faculty, Harvard, 1954-59; faculty U. Pa., Phila., 1959——, prof. psychology, 1963——. Cons. editor Jour. Comparative and Physiol. Psychology. Mem. Am. Psychol. Assn., A.A.A.S., Am. Physiol. Soc., Soc. Exptl. Psychologists, Phi Beta Kappa, Sigma Xi. Author: Fundamental Principles of Physiological Psychology, 1966. Research and publs. on hypothalamus, recovery of hypothalamic function after damage, role of taste and motivation in regulation of food intake.*

TELESIO, Bernardino, Italian philosopher, natural scientist; b. of noble parentage at Cosenza, near Naples, Italy, 1509; studied in Milan (under his uncle Antonio), Rome, and Padua, 1518-35; spent about nine years in a monastery; resigned to his brother the Archbishopric of Cosenza offered to him by Pope Pius IV; lived and lectured at Naples, 1545-52, 1563-87; lived at Cosenza and founded the Academia Cosentia. Author: De natura rerum juxta propria principia, 1565 (the complete nine books, 1586); De Somno; De his quae in aere fiunt; De Mari; De Cometis et Circulo Lacteo; De usu respirationis; also essays on physical science. Inaugurated empiricist reaction against authority of abstract reason as represented by scholastic Aristotelianism; opened way to naturalism; emphasized importance of sci. knowledge based upon experience and experiment; regarded matter as a positive reality that has no need to look outside for its sufficient explanation; sought to explain all forms of life, great and small, from the two opposing fundamental forces (dry-warm and moist-cold); derived his views from early Greek naturalistic philosophy; in his psychology and ethics had no clear qualitative distinction between the physiological and spiritual; set example of independent investigation (which Tommaso Campanella and others soon followed); Telesio's pupil Francesco Patrizi pointed out defects in his system. Died Cosenza, 1588.

TELFORD, Ira Rockwood, Am. anatomist; b. Lincoln, Ida., May 6, 1907; s. John Witt and Martha Starr (Rockwood) T.; student Ricks Coll., 1924-26; A.B., U. Utah, 1931, A.M., 1933; student U. Cal., at Berkeley, 1937-40; Ph.D., George Washington U., 1942; m. Thelma Challis Shrives, June 13, 1933; children—Ira Ralph, John Larry, Kent Matthews, Martha Ann. High sch. adminstr., 1933-37; instr., asst. prof. anatomy George Washington U., 1940-47, prof., exec. officer dept. anatomy, 1953——; prof., chmn. dept. anatomy U. Tex. dental br., 1947-53; cons. anatomy U. Hosp., 1953——. Fulbright scholar, Gt. Britain, 1960. Fellow A.A.A.S. (council 1958——); mem. Am. Assn. Anatomists, Soc. Exptl. Biology and Medicine, Am. Acad. Neurology, Washington, Tex. acads. sci., Internat. Soc. Dental Research, A.M.A. (asso.), Sigma Xi (chpt. pres. 1958-59), Author: Synopsis of Gross Anatomy (with John B. Christensen), 1966. Contbr. articles biol., med. jours. Vitamin E deficiency studies on muscle and nervous tissue; role of anoxia and hyperoxia in fetal development; muscular dystrophy in animals. Home: 3424 Garrison St. N.W., Washington 8. Office: 1335 H St. N.W., Washington 5.*

TELFORD, Thomas, Brit. engr.; b. Dumfriesshire, Scotland, Aug. 9, 1757; student chemistry. Mason, Dumfriesshire; studied, sketched old architecture, Edinburgh, Scotland, 1780-82; supervised erection several bldgs., London, from 1782; surveyor pub. works, Shropshire, Eng.; apptd. sole agt., engr., architect Ellesmere Canal, 1793. Fellow Royal Soc., 1827, Royal Soc. Edinburgh; mem. Inst. C.E.'s (founder soc. from which it evolved 1818, pres.). Contbr. archtl., navigational articles to Edinburgh Ency. Designed several archtl. works which showed gt. originality; built iron bridge over Severn River, 1795-98, Aqueduct Bridge over Dee Valley, Caledonian ship-canal, suspension bridge over Straits of Menai, 1819-25; designed Gotha Canal, 1808-10; improved rds. in Eng., 1814; completed Birmingham-Liverpool canal system, 1825, bridges at Tewkesbury, Eng., 1826, Gloucester, Eng., 1828, Dean Bridge, Edinburgh 1831, Clyde Bridge Glasgow, Scotland 1833-35 improved several harbors, including Dover, Eng., Aberdeen, Dundee (both Scotland); built St. Catherine Docks, London. Died Westminster, Eng., Sept. 2, 1834.

TELKES, Maria, phys. chemist; b. Budapest, Hungary, Dec. 12, 1900; d. Aladar and Maria (Laban) Telkes; Ph.D., U. Budapest, 1924. Came to U. S., 1925, naturalized, 1937. Research asst. U. Budapest, 1923-24; biophysicist Cleve. Clinic, 1925-37; research engr. Westinghouse Research Labs., 1937-39; research asso. Mass. Inst. Tech., 1939-53; project dir. Coll. Engring., N.Y. U., 1953-58; research dir. solar energy lab. Princeton div. Curtiss-Wright Co., 1958-60; dir. research and devel. Cryo-Therm Co., 1961-64; head solar energy applications lab. MELPAR, Inc., subsidiary Westinghouse Air-Brake Co., 1965——; specializes in solar energy research; developed solar stills for life rafts, thermoelectric materials and solar thermoelectric generators, designed solar heating equipment for house. Mem. Am. Chem. Soc., Soc. Women Engrs. (hon. life; recipient 1st award 1952), Solar Energy Soc. (dir., mem. publ. bd.), Hellenic Soc. Solar Energy (hon.). Author numerous articles profl. jours., chpts. in books. Research on solar energy, solar heating; phase-change thermal control in terrestrial and space applications; invented means for storage of solar heat; conversion of solar radiation into electric power; constructed solar stills to convert salt water into pure water. Home: 2301 E St. N.W., Washington 20037. Office: MELPAR, Inc., 7700 Arlington Blvd., Falls Church, Va. 22046.*

TELLER, Edward, physicist; b. Budapest, Hungary, Jan. 15, 1908; s. Max and Ilona (Deutch) T.; student Inst. of Tech., Karlsruhe, Germany, 1926-28; Ph.D., U. of Leipzig (Germany), 1930; D.Sc. (hon.), Yale, 1954, U. Alaska, 1959, Fordham U., 1960, George Washington U., 1960, U. So. Cal., 1960, St. Louis U., 1960, Rochester Inst. Tech., 1962; LL.D., Boston Coll., 1961, Seattle U., U. Cin., 1962; LL.D., U. Pitts. 1963; D.Sc., U. Detroit, 1964, Clemson U., 1966; L.H.D. Mt. Mary Coll., 1964; m. Augusta Harkanyi, Feb. 26, 1934; children—Paul, Susan Wendy. Naturalized, 1941. Research association, Gottingen, Germany, 1931-33; Rockefeller fellow, Copenhagen, Denmark, 1934; lecturer U. of London, 1935; prof. of physics George Washington U., Washington, 1935-41, Inst. for Nuclear Studies, U. Chicago, 1945; prof. physics, U. Chgo., 1946-49, 1951-52; professor physics, University California, 1953-60; professor at large, 1960——, chmn. dept. applied sci. Davis and Livermore, 1963-66; asst. dir. Los Alamos Sci. Lab., N.M., 1949-51; cons. Livermore br. U. Cal. Radiation Laboratory, 1952, staff mem., 1953; asso. director Lawrence Radiation Laboratory, University of Cal., 1954-——, dir., Livermore, 1958-60; concerned with planning and prediction function atomic bomb and hydrogen bomb, Manhattan Dist. of Columbia U., Metall. and Laboratory of Argonne Nat. Lab., U. Chicago, and Los Alamos, New Mexico, 1941-51, Radiation Lab., Livermore, Cal., 1952——. Mem. sci. adv. bd. USAF. Recipient Joseph Priestley Meml. award Dickinson Coll., 1957; Albert Einstein award, 1958; Research Institute of America Living History Award, 1960, Thomas E. White and Enrico Fermi awards, 1962. Fellow of American Nuclear Soc., Am. Phys. Soc.; mem. Am. Acad. Arts and Scis., Am. Ordnance Assn., Nat. Acad. Scis. Author: (with Francis Owen Rice); The Structure of Matter, 1948; (with A. L. Latter) Our Nuclear Future, 1958; The Legacy of Hiroshima, 1962; The Reluctant Revolutionary, 1964; (with W. K. Talley, G. H. Higgins, G. W. Johnson) The Constructive Uses of Nuclear Explosives, 1968. Contributions in chem., molecular, nuclear physics and in quantum theory; one of 1st to study thermonuclear reactions; research in practical application of thermonuclear principles in devel. thermonuclear weapons; spectroscopy of polyatomic molecules; theory of atomic nucleus; active in devel. Sherwood Project (controlled thermonuclear program), Project Plowshare (peaceful uses nuclear explosives). Home: 1573 Hawthorne Terrace, Berkeley, Cal. 94708.*

TELLIER, Charles Albert Abel, French engr., inventor; b. Amiens, France, 1828; called Father of Refrigeration; invented refrigeration apparatus and methods; introduced use of methylic ether, 1863, also trimethaline, compression refrigerating machines; introduced refrigerating in transatlantic trade; studied use of tar for roads, uses of compressed air; inventor engine run by ammonia, 1867. Author: L'ammoniaque dans l'industrie; Le Froid appliqué à la biere; La conservation de la viande par le froide; L'impôt unique et ses conséquences. Died Paris, 1913.

TEMKIN, Aaron, Am. physicist; b. Morristown, N.J., Aug. 15, 1929; s. Maxwell M. and Miriam (Feigelstein) T.; B.S., Rutgers U., 1951; Ph.D., Mass. Inst. Tech., 1956; m. Miriam Hannah Schachter, Aug. 10, 1958; children—Philip Herman, Jean Michelle. Fulbright fellow Heidelberg (Germany) U., 1956-57; physicist U. S. Naval Research Lab., Washington, 1957-58, Nat. Bur. Standards, Washington, 1958-60; aerospace technologist Goddard Space Flight Center, NASA, Greenbelt, Md., 1960——. Mem. Internat. Adr. Com. on Confs. on Physics Electronic Collisions. Am. Phys. Soc., Phi Beta Kappa, Sigma Xi, Delta Phi Alpha. Editor: Autoionization, 1966. Contbr. articles to profl. jours., also chpt. to Methods of Computational Physics, Vol. II. Research on electron scattering from atoms, method of polarized orbitals, nonadiabatic theory, decomposition of Schrödinger equation, theories of molecular structure to calculate electron molecule scattering. Home: 10822 Margate Rd., Silver Spring, Md. 20901. Office: Goddard Space Flight Center, NASA, Greenbelt, Md. 20771.*

TEMMINCK, Conrad Jacobus, Dutch naturalist; b. Amsterdam, Netherlands, Mar. 31, 1778; dir. Acad. Scis., Harlaam, Netherlands; dir. Mus. Natural History of Netherlands; mem. French Acad. Scis., 1852. Author: Historie naturelle générale des pigeons et des gallinacés, 1813-15; Manuel d'ornithologie, 1820-39; (with Laugier) Recueil de planches coloriées d'oiseaux, 1920; Monographie de mammalogie, 1835-41; Natuur Kundige Verhandelinge, 1839-44. Died Leiden, Netherlands, Jan. 30, 1858.

TEMPEL, Ernst Wilhelm Leberecht, astronomer; b. Cunnersdorf, Saxony (Germany), Dec. 4, 1821. Studied lithography; then became interested in lit, Copenhagen; practiced lithography, Venice, Italy, from 1850, also studied astronomy; became asst. astronomer to Benjamin Valz, Marseilles, France, 1860; returned to Italy; with Milan Obs., from 1871; apptd.

dir. Arcetri Obs., nr. Florence. Discovered a comet and a nebula, 1859, also 6 planetoids, almost 60 nebulae, and 20 comets (including the famous comet of 1866), 1856-77. Died Arcetri, Italy, Mar. 16, 1889.

TEMPERLEY, Harold Neville Vazeille, English math. physicist; b. Cambridge, Eng., Mar. 4, 1915; s. Harold W.V. and Gladys (Bradford) T.; B.A. with 1st class honors in Physics, Kings Coll., Cambridge U., 1937, M.A., 1941, Sc.D., 1958; m. Geraldine H. la C. Bartrop, Feb. 6, 1940; children—Virginia M.H., Julian H.V., Humphrey P.N. With Admiralty, 1939-45, sr. sci. officer, 1944-45; fellow Kings Coll., Cambridge, 1941-54, lectr. physics, 1948-53; sr. prin. scientist Aldermaston, 1955-65; prof., head dept. applied math. U. Coll. Swansea (U.K.), 1965—. Recipient Smiths prize for math., Cambridge U., 1939; Leverhulme Research award, 1945; Smithson Research fellow Royal Soc., 1947. Fellow Inst. Physics, Cambridge Philos. Soc., Inst. Math. and Applications; mem. Faraday Soc. Author: Properties of Matter, 1953; Changes of State, 1956; A Scientist Who Believes in God, 1961; also numerous articles. Research on theory of underwater explosions and damage to structures and effects on human beings, relations between molecular interactions and properties of bulk matter, liquids and changes of state. Home: Holt End House, Ashford Hill, Newbury, Berks., Eng. Office: Dept. applied Math., U. Coll., Swansea, U.K.*

TEMPLE, George Frederick James, English mathematician; b. London, Eng., Sept. 2, 1901; s. James and Frances (Compton) T.; B.Sc., London U., 1922, Ph.D., 1924, D.Sc., 1932; M.A., Oxford U., 1953; D.Sc., Dublin U., 1962; D.Sc., Louvain U., 1966; m. Dorothy Lydia Carson, Sept. 2, 1930. Mem. faculty London U., 1922-53, prof. math. King's Coll., 1932-53; Sedleian prof. natural philosophy Oxford U., 1953—. Chmn., Aero. Research Council, 1961-64. Decorated Comdr. Brit. Empire, 1955. Fellow Royal Soc., 1943; mem. Phys. Soc., London Math. Soc., Royal Aero. Soc. Author: An Introduction to Quantum Theory, 1931; Rayleigh's Principle, 1933; General Principles of Quantum Theory, 1934; An Introduction to Fluid Dynamics, 1958; Cartesian Tensors, 1960; also articles. Editor: (with R. J. Seeger) Research Frontiers in Fluid Dynamics, 1965. Office: Math. Inst., U. Oxford, 24-29 St. Giles, Oxford, Eng.*

TEMPLEMAN, Peter, English physician; b. Dorchester, Eng., Mar. 17, 1711; s. Peter and Mary (Haynes) T.; B.A., Trinity Coll., Cambridge (Eng.) U., 1731; M.D., U. Leiden (Netherlands), 1737. Curator of reading room Brit. Mus., 1758-60. Mem. French Acad. Scis. (corr.), 1762, Economical Soc. Bern (corr.), Soc. Arts, Manufactures and Commerce (sec. 1760). Author: Curios Remarks and Observations in Physics, Anatomy, Chirurgery, Chemistry, Botany and Medicine . . . , 2 vols., 1753-54; Practical Observations on the culture of Lucerne, Turnips, Burnet, Timothy Grass and Fowl Meadow Grass, 1766. Editor: Select Cases and Consultations in Physic (Woodward), 1757. Translator: (from Danish) Travels in Egypt and Nubia (Norden), 2 vols., 1756-57. Studied uterine tumors, heart polyps; encouraged communication of med. knowledge. Died London, Aug. 23, 1769.

TEMPLEMAN, Wilfred, Canadian marine biologist; b. Bonavista, Nfld., Can., Feb. 22, 1908; s. Charles and Sarah (Fisher) T.; student Meml. U. Coll., St. Johns, Nfld., 1927-28; B.Sc., Dalhousie U., Halifax, N.S., Can., 1930; M.A., U. Toronto (Ont., Can.), 1931, Ph.D., 1933; m. Eileen Eliza McGrath, Sept. 7, 1937; children—Margaret Elizabeth (Mrs. Ronald Nowe), Barbara Eileen (Mrs. James Hartling), Sheila Joan, Sandra Louise. Lectr., McGill U., 1933-36; asso. prof. Meml. U. Coll., St. John's, 1936-43, prof., 1943-44; dir. Nfld. Govt. Lab., 1944-49; dir. Fisheries Research Bd. Can., Biol. Sta., St. John's, 1949—. Chmn. standing com. on research and statistics Internat. Commn. for N.W. Atlantic Fisheries, 1964-67. Decorated Order Brit. Empire. Fellow Royal Soc. Can.; mem. Marine Biol. Assn. U.K., Canadian Soc. Zoologists. Research and numerous publs. on life histories of lobster, cod, haddock, redfish, Am. plaice, spiny dogfish, sharks, skates, capelin, and others. Home: 12 Darling St. Office: Fisheries Research Bd. Can., Water St. E., St. John's, Nfld., Can.*

TEMPLETON, John, Irish naturalist; b. Belfast, Ireland, 1766; s. James and Mary Eleanor (Legg) T.; ed. in pvt. schs.; m. Katherine Johnston, 1799; a son, Robert. Set up exptl. garden on his farm, 1793. Recipient prize Royal Irish Acad., 1795. Mem. Linnean Soc. (asso.), Belfast Soc. for Promoting Knowledge, Belfast Natural History Soc. (1st hon. mem.). Research and publs. on birds, mosses, liverworts; collected lichens; added Rosa hibernica, Orobanche rubra to list of Irish flora. Died Cranmore, Ireland, Dec. 15, 1825.

TENDOLKAR, Gangadhar Sadashiv, Indian metall. engr.; b. Malvan, India, Mar. 1, 1921; s. Sadashiv Govind and Gouri (Parulekar) T.; B.Sc., Bombay (India) U., 1940, B.Sc. (Tech.), 1942; Ph.D., Sheffield (Eng.) U., 1947; m. Kashi Vishnu Gosavi, May 17, 1949; children—Anil Gangadhar, Nayana Ganghar. Lectr. fuel tech. dept. chem. tech. Bombay U., 1947-53; prin. sci. officer Bhabha Atomic Research Centre (formerly Atomic Energy Establishment), Bombay, India, 1953-62; prof., head dept. metall. engring. Indian Inst. Tech., Bombay, 1962-64, sr. prof., 1963—. Mem. Indian Inst. Metals (asso. editor Trans. 1966—). Editorial adv. bd. Internat. Jour. Powder Metallurgy, 1966—. Research and publs. on powder metallurgy including deposition of metal powders, sintering of thorium and thoria; electrodeposition of uranium metal, activated sintering. Home: 386 King's Circle, Bombay. Office: Indian Inst. Tech., Bombay, Maharashtra, India.*

TENG, Lee C., physicist; b. Peiping, China, Sept. 5, 1926; s. Tsuy-Ying and Chien-Min (Ho) T.; came to U. S., 1947, naturalized, 1962; B.S., Fu-Jen U., Peiping, 1946; M.S., U. Chgo., 1948, Ph.D., 1951; m. Nancy L. S. Huang, Sept. 21, 1961; 1 son, N. H. Teng. Asst. prof. physics U. Minn., Mpls., 1951-53; asso. prof. physics U. Wichita (Kan.), 1953-55; asst. physicist Argonne (Ill.) Nat. Lab., 1955-56, asso. physicist, 1956-61, sr. physicist, 1961—, leader theory group Particle Accelerator div., 1956-62, dir. Particle Accelerator div., 1962-67; head accelerator theory sect. Nat. Accelerator Lab., Oak Brook, Ill., 1967—. Dir, U. S.-Korea and U. S.-China Sister Lab Arrangements, 1964—. Vis. prof. Tsing Hua U., Hsing-chu, Taiwan, 1959; professorial lectr. U. Chgo., 1963—. Fellow Am. Phys. Soc; mem. Am. Math. Soc., A.A.A.S., Academia Sinica, Am. Assn. U. Profs., Inst. for Aero-Space Scis., Sigma Xi. Research and numerous publs. in high energy and accelerator physics; designed zero gradient synchrotron, 200/400 BeV synchrotron. Patentee particle accelerator designs. Home: 503 N. Bruner St., Hinsdale, Ill. 60521. Office: Nat. Accelerator Lab., 1301 W. 22d St., Oak Brook, Ill. 60521.*

TENNANT, Smithson, chemist; b. Selby, Yorkshire, Eng., Nov. 30, 1761; s. Caivert and Mary (Daunt) T.; ed. Edinburgh (Scotland) U., 1781-82, Christ's Coll., Cambridge (Eng.), U., 1782-86; M.B., Emmanuel Coll., Cambridge U., 1788, M.D., 1796. Early career devoted to travel, farming, sci. research; lectr. mineralogy, 1812; prof. chemistry Cambridge U., from 1813; lectr. 1814. Fellow Royal Soc. (Copley medal 1804), 1785. Contbr. papers to Philos. Trans. Discovered osmium, iridium, 1804; provided analytical proof of composition of fixed air (synthetically proved by Lavoisier 1791); proved diamond consists of pure carbon, 1797; demonstrated harmful effect magnesia and its carbonate have on vegetation, 1799; invented bleaching with chlorine alkali. Died Boulogne, France, Feb. 22, 1815.

TENNENT, David Hilt, Am. biologist; b. Janesville, Wis., May 28, 1873; s. Thomas and Mary (Hilt) T.; B.S., Olivet (Mich.) Coll., 1900; Ph.D., Johns Hopkins, 1904; m. Esther Margaret Maddux, Apr. 8, 1909; 1 son, David Maddux. Acting prof. biology and physics, Randolph-Macon Coll., Va., 1903; lectr. biology, 1904-05, asso., 1905-06, asso. prof., 1906-12, prof., 1912-41, Bryn Mawr Coll. Mem. A.A.A.S., Am. Soc. Naturalists, Am. Soc. Zoölogists (pres. 1916), Nat. Acad. Scis., Phila. Acad. Nat. Scis. In charge of instruction in dept. embryology, Marine Biol. Lab., Woods Hole, Mass., 1920-22; vis. prof. biology, Kelo Univ., Tokyo, 1930-31. Investigations in marine biology at Marine Biol. Lab., Woods Hole, Mass., Brooklyn Inst. Arts and Scis., Cold Spring Harbor, L.I., Bur. Fisheries Lab., Beaufort, N.C., Hopkins Marine Sta., Pacific Grove, Calif., Dept. Marine Biology, Carnegie Inst. Washington, at Dry Tortugas, Fla., Jamaica, Naples Sta., Torres Strait, Australia, Japan. Studied photodynamic effects of vital dyes. Died Jan. 14, 1941.

TENNEY, S(tephen) Marsh, Am. physiologist; b. Bloomington, Ill., Oct. 22, 1922; s. Harry Houser and Caroline (Marsh) T.; A.B., Dartmouth College, 1943; M.D., Cornell U., 1946; m. Carolyn Cartwright, Oct. 18, 1947; children—Joyce B., Karen M., Stephen M. Instr. medicine U. Rochester Sch. Medicine, 1951-54, instr. physiology, 1953-54, asst. prof. physiology and medicine, 1954-56, asso. prof., 1956; prof. physiology Dartmouth Medical School, 1956—, dean, 1960-62, acting dean, 1966, dir. med. scis., 1957-59, also chmn. dept. physiology, Markle scholar in med. sci., 1954-59. Chmn. physiology study sect. NIH, 1962—; exec. com. NRC; physiology panel NIH study, Office Sci. and Tech. Mem. Am. Physiol. Soc., Am. Soc. Clin. Investigation, N.Y. Acad. Scis., Gerontol. Soc., Am. Heart Assn., Assn. Am. Med. Colls., Sigma Xi. Author sci. articles. Editorial bd. Circulation Research, Am. Jour. Physiology. Jour. Applied Physiology, Respiration Physiology. Contribution in cardiorespiratory physiology; high altitude physiology; regulatory biology. Home: 18 Rope Ferry Rd., Hanover, N.H. 03755.*

TENON, Jacques René, French surgeon; b. Sipaux, France, Feb. 21, 1724; became surgeon la Salpetrière, Paris, 1748; named prof. Coll. Surgery, 1757. Mem. French Acad. Scis., Soc. Agr. Author: Memoire sur les hôpitaux de Paris, 1788. Originated plan for reform in French hosp. system; described fibrous capsule covering posterior two-thirds of eyeball and serving as synovial sac, (Tenon's capsule), 1806, lymph space between eyeball and fibrous capsule of eye (Tenon's space), 1806. Died Paris, Jan. 15, 1816.

TENORE, Michelo, Italian botanist; b. Naples, Italy, 1781; prof. U. Naples; founder bot. garden, Naples. Author: Flora neapolitana, 1811-38; also numerous other bot. works. Died Naples, 1861.

TENTORI, Leonardo, Italian biochemist; b. Rome, Italy, Oct. 28, 1917; s. Rosario and Matilde (Capristo) T.; Degree in Medicine, U. Rome, 1941; m. Angela Bevacqua, Oct. 15, 1945; 1 son, Lucio. Asst., Istituto Superiore di Sanità, Rome, 1947-55, researcher, 1955-60, 1st researcher, 1960—; libero docente biochemistry U. Rome, 1955—. Mem. Biochem. Soc., Internat. Soc. Hematology. Research and numerous publs. on biochemistry of hematopoiesis, metabolism of some amino acids in vivo in rat, relationship between structure and function of hemoglobins of vertebrates and invertebrates. Home: 14 Via Monterotondo, Rome. Office: Istituto Superiore di Sanità, Viale Regina Elena 299, Rome, Italy.*

TENTORI, Tullio, Italian anthropologist; b. Naples, Italy, Apr. 11, 1920; s. Egidio and Ida (Merola) T.; ed. U. Rome; Dr.ès letters; m. Silvana Zanetti, May 12, 1947. Dir. Nat. Mus. Art and Popular Traditions; instr. indigenous Am. civilization and cultural anthropology U. Rome; prof. cultural anthropology Internat. U. Social Scis., Rome, also Faculty Sociology, U. Trent. Mem. Etnofonico (pres.), Italian Centre for Cultural Anthropology (pres.), Italian Soc. Social Sci. (exec. com.), Soc. for Rural Sociology. Research and numerous publs. on cultural anthropology, Am. indigenous civilization. Home: Largo E. Bertolotti 10, Rome. Office: Museo Tradizioni Popolari, piazza Marconi 8, Rome, Italy.

TEODORESCU, Emanoil, Rumanian botanist; b. 1866; mem. Rumanian Acad. Author: Température mortelle pour quelques disstases d'origine animale et vegetale, 1914; Complete Works, 3 vols., 1919; La volubilité à l'obscurité, 1925; Materiaux pour la flore algologique de la Roumanie. Research on algology, plant physiology, influence of monochromatic light on plant growth and devel.; discovered ficoerythrine in Cyanoficeae. Died 1949.

TEODORESCU, Leonid Barbu, Rumanian physician; b. Petruseni, Rumania, Sept. 3, 1918; s. Barbu D. and Maria (Timus) T.; grad. Faculty Medicine, Iassy, Rumania, 1944, O.R.L. specialist physician, 1948, D.medicine, 1959; m. Octavia Ieremia, July 2, 1947; 1 dau., Horia. Staff clinics Iassy, 1942—, primary physician, head Clinica O.R.L., 1960—; faculty Inst. Medicine, Iassy, 1946—, reader, head dept., 1965—. Mem. Union Socs. Med. Scis. Rumania, Balkan Med. Union. Co-author: Rino-sinusites Treatment, 1958, Otorhinolaryngology, 1951; author Manual Otorhinolaryngology, 1966; also articles. Clin. and paraclinical research in audiology; physiopathology of deafness, including occupational deafness, and its prophylaxis and cure; research on precancerous conditions and cancer; contbns. to surg. technique for tumors and malformations of ear, nose, and throat. Home: 30.N. Balcescu, Iasi, Rumania. Office: Clinica O.R.L., Spit. M.C.F., 1 Flamura Rosie Iasi, Rumania.*

TERADA, Kikuo, Japanese chemist; b. Kanazawa C., Japan, Apr. 14, 1927; s. Sadao Kimiko Terada; M.Sc., Nagoya (Japan) U., 1955, D.Sc., 1959; m. Itsuko Okubo, Apr. 14, 1957; children—Tsugako, Sachiko, Makiko. Asst. water research lab. Nagoya U., 1958-60; asst. prof. dept. chemistry Faculty Sci., Kanazawa (Japan) U., 1960—. Mem. Chem. Soc. Japan, Japan Soc. for Analytical Chemistry, Geochem. Soc. Japan. Author: Comprehensive Treatise on Inorganic Chemistry, III, Halogens, 1958; (with others) Handbook of Experimental Chemistry, 1958; also articles. Research on geochemistry of iodine in atmosphere and hydrosphere, separation and determination of technetium and rhenium and platinum metals by solvent extraction. Home: 3-10-7 Nomachi, Kanazawa C., Japan.*

TERANISHI, Roy, Am. chemist; b. Stockton, Cal., Aug. 1, 1922; s. Harukichi and Natsu (Kawamura) T.; B.S., U. Cal. at Berkeley, 1950; Ph.D., Ore. State U., 1954; m. Yasuko Moriya, Dec. 16, 1944; 1 son, Allan Y. Research chemist Western Utilization Research and Devel. Dir., U. S. Dept. Agr., Albany, Cal., 1954—, research chemist, 1954—. Mem. Am. Chem. Soc., Sigma Xi. Research and publs. on gas chromatographic techniques in study aroma chemistry, programmed temperature of capillary columns for analysis of complex mixtures, dual hydrogen flame detectors for direct vapor analyses from food products, combined capillary column gas chromatography with mass spectrometry for identification of very small quantities, high resolution capillary and packed columns for gas chromatographic separations for analytical purposes. Home: 1645 Cedar St., Berkeley, Cal. 94703. Office: 800 Buchanan St., Albany, Cal. 94710.*

TERAYAMA, Hiroshi, Japanese chemist; b. Okayama-Ken, Japan, Sept. 16, 1922; s. Bunji and Kino (Kuroda) T.; B.Sc., Tokyo Imperial U., 1944, D.Sc., 1948; m. Emi Sameshima, Apr. 1, 1947; children—Mitsuhiro, Masahiro, Emily Lisa. Chief, Enzyme Lab., Nat. Inst. Hygiene, Tokyo, Japan, 1948-50; asso. prof. dept. chemistry U. Tokyo, 1951—, research adviser for grad. students, 1957—, chief Lab. for Cytological Biochemistry, 1958—; post-doctoral fellow, research asso. dept. chemistry and chem. engring. U. Ill., 1952-56; lectr. Yamanashi U., Tokyo Ednl. U., 1964—. Mem. Japanese Cancer Soc., Japanese Biochem. Soc., Japanese Chem. Soc., Japanese Bio-phys. Soc., Japanese Soc. Analytical Chemistry, Japanese Soc. Chem. Cytology. Author: Fundamentals of Biophysical Chemistry, 1964; Fundamental Biochemistry, 1966. Publs. on studies of interactions between polyanions and polycations; properties of polyelectrolytes in solution; co-enzymes; azo-dyes; biosynthesis of nucleic acids and proteins; invention of colloid titration method. Home: 155 Tamagawa-Seta-machi, Setagaya-Ku, Tokyo, Japan.*

TERENIN, Aleksandr Nikolaevich, Russian phys. chemist; b. May 6, 1896; grad. Petrograd (Leningrad) U., 1921. Prof., Leningrad U., 1932—, founder, dir. catalysis lab.; dir. photochemistry lab. USSR Acad. Scis.; dep. dir. State Optics Inst.; mem. bur. dept. chemistry USSR Acad. Scis., 1960—. Recipient Order of Lenin; Order of Red Banner; Order Red Star; Stalin Prize, 1953; S. I. Vavilor Prize. Mem. USSR Acad. Scis. Author: Introduction to Spectroscopy, 1933; Photochemistry of Dyes and Related Organic Compounds, 1947; The Photochemistry of Salts, 1934; Photochemical Processes in Aromatic Compounds, 1944. Research and publs. on photosynthesis; phys. chem. processes occurring under influence of light, including devel. optical methods for their study; 1st to obtain infra-red spectra of gases under high pressure. Office: Inst. Physics, University, Leningrad B-164, USSR.

TERENTEV, Aleksandr Petrovich, Russian organic chemist; b. Jan. 20, 1891; grad. Moscow U., 1913. Instr., Moscow U., 1913-34, prof., 1934—. Recipient Stalin Prize, 1948, Order of Lenin, Order of Red Banner of Labor. Corr. mem. USSR Acad. Scis. Author: Sulfonation of Acidophobic Compounds, 1947; Chelate Compounds as Contact Insecticides, 1960; Detecting Carbon Impurities in Silico-Cupric Alloys, 1960. Research and publs. on organic functional analysis, synthesis of sulfonic compounds, heterocyclic compounds, stereochemistry, nomenclature of organic compounds. Home: Leninskii gory, sekt L, Moscow. Office: Chemistry Dept., Moscow U., Moscow, USSR.

TERENTIUS, Jean, natural scientist; b. Constantinople, 1580; missionary; mem. Acad. Lincei. Author: L'abrégé des plantes exotiques. Brought information about Chinese plants to the West. Died China, 1630.

TERJESEN, Sven G., Norwegian chem. engr.; b. Cslo, Norway, Sept. 18, 1914; s. Sven A. and Nathalie (Ditlefsen) T.; Chem. Tech., Tech. U. Norway, 1937; diploma chem. engring., Univ. Coll., London, 1939; hon. doctorate Aabo Acad., Finland; M. Brigitte Rading, May 22, 1948; children—Einarandreas, Marianne. Sci. asst. phys. chemistry Tech. U. Norway, Trondheim, 1937-38, prof. chem. engring., 1949-65; researcher Imperial Chem. Industries Ltd., Manchester, Eng., 1940-47; devel. work with Norsk Spraengstofindushi, Gullaug, Norway, 1947-49; research mgr. Norsk Hydro, Hevoga, Norway, 1965—. Mem. bd. Norwegian Def. Research Establishment, EUROCHEMIE, Mol, Belgium, Norwegian Inst. Atomic Energy. Mem. Norwegian Acad. Tech. (pres. 1958-64), Norwegian Chem. Soc. (v.p. 1964—), Inst. Chem. Engrs. (asso. London), Engring. Soc., Royal Norwegian Acad. Scis. Research on bacterial filters for antibiotics, reactions between solids and liquids, phenomena asso. with movement of liquid drops in liquid media, including mass transfer and resistance to movment. Home: Solhageveien 7, Heroya. Office: Research Dept., Norsk Hydro, Heroya, Porsgrunn, Norway.*

TERMAN, Frederick Emmons, Am. electronics engr.; b. English, Ind., June 7, 1900; s. Lewis Madison and Anna Belle (Minton) T.; A.B. Chem. Engring., Stanford, 1920, E.E., 1922; Sc.D. in Elec. Engring., Mass. Inst. Tech., 1924; D.S. (hon.), Harvard, 1945; Sc.D., U. B.C., 1950, Syracuse U., 1955; m. Sibyl Walcutt, Mar. 22, 1928; children—Frederick Walcutt, Terence Christopher, Lewis Madison. Instr. to prof. elec. engring., Stanford U., 1925-37, exec. head elec. engring. dept., 1937-45, dean engring., 1945-58, provost, 1955-65, v.p., 1959-65, emeritus, 1965—; pres. SMU Found. for Sci. and Engring., 1965—; dir. Harvard Radio Research Lab., 1942-45. Mem. Divs. 14 and 15 of Nat. Def. Research Com., 1942-45; mem. Army Adv. Com. on contractual and adminstrv. procedures for research and devel., 1948; mem. spl. tech. adv. group and TAPEC Com. Dept. of Def., 1953-56, research and devel. adv. com. Signal Corp, 1954-62; mem. Bd. Fgn. Scholarships, Dept. State, 1960-65; adv. council Army Electronics Proving Ground, 1954-57; mem. Naval Res. Adv. Com., 1956-64, chmn., 1957-58; dir. Hewlett-Packard Co., Watkins-Johnson Co., Stanford Bank, Granger Assos.; trustee Inst. Def. Analysis. Decorated by Brit. govt., 1946, U. S. Medal for Merit, 1948; Medal of Honor, I.R.E., 1950, Founders Award 1962. Fellow I.R.E. (pres. 1941), Am. Inst. E.E. (edn. medal 1956); mem. Nat. Acad. Sci. (chmn. engring. sect. 1953-55; mem. council 1956-59), Nat. Acad. Engring. (founding), NRC, Am. Phys. Soc., Am. Soc. Engring. Edn. (v.p. 1949-51, Lamme Medal 1964; hon. mem.), A.A.A.S., Audio Energy Society (hon.), Am. Philos. Phi Beta Kappa, Sigma Xi (exec. com.), Author several books 1927—; latest: Radio Engineers' Handbook, 1943; Electronic Measurements (with J. M. Pettit), 1952; Electronic and Radio Engineering, 1955. Contbr. tech. mags. Work on transmission of electricity long distances, circuits and tubes, resonant transmission lines, elec. instrumentation; organized and directed anti-radar research, 1942. Home: 445 El Escarpado, Stanford 94305. Office: McCullough Bldg., Stanford U., Stanford, Cal. 94305.*

TERMAN, Lewis Madison, Am. psychologist; b. Johnson County, Ind., Jan. 15, 1877; s. James William and Martha Parthenia (Cutsinger) T.; A.B., Central Normal Coll., Danville, 1898; A.B., Ind. U., 1902, A.M., 1903, LL.D., 1929; fellow in psychology and edn., Clark U., 1903-05, Ph.D., 1905; LL.D., U. Cal., 1945, U. So. Cal., 1949; Sc.D., U. Pa., 1946; m. Anna Belle Minton, Sept. 18, 1899; children—Frederick Emmons, Helen Clare. Prin. high schs., Smiths Valley, Ind., 1898-1901, San Bernardino, Cal., 1905-06; prof. psychology and pedagogy, State Normal Sch., Los Angeles, 1906-10; asst. prof., edn., Stanford, 1910-12, asso. prof., 1912-16, prof., 1916—, exec. head dept. psychology, 1922-42, prof. emeritus, 1942—. Mem. com. Psychol. Exam. Recruits, Com. on Classification of Personnel, U. S. Army, 1918-19; served as maj. in div. of psychology, Surgeon General's Office, Washington. Fellow A.A.A.S., Brit. Psychol. Soc., Ednl. Inst. Scotland (hon.); mem. Am. Psychol. Assn. (pres. 1923), N.E.A., Am. Sch. Hygiene Assn. (pres. 1917), Nat. Soc. Study Edn., Nat. Acad. Scis., Sigma Xi. Author: The Teacher's Health, 1913; The Hygiene of the School Child, 1914; (with Dr. E. B. Hoag) Health Work in the Schools, 1914; The Measurement of Intelligence, 1916; The Stanford Revision of the Binet-Simon Intelligence Scale, 1916; The Intelligence of School Children, 1919; The Terman Group Test, 1920; (with T. L. Kelley and G. M. Ruch) The Stanford Achievement Test, 1923; (with others) Genetic Studies of Genius, Vol. I, 1925, Vol. II (with Catharine M. Cox), 1926, Vol. III (with Barbara Burks and Dortha Jensen), 1930; Children's Reading (with Margaret Lima), 1925; Sex and Personality (with Catharine Cox Miles), 1936; Measuring Intelligence (with Maud A. Merrill), 1937; Marital Happiness, 1938; The Terman-McNemar Test of Mental Ability (with Q. McNemar), 1942; The Gifted Child Grows Up (with Melita Oden), 1947; The Gifted Group at Mid-Life (with Melita Oden), 1959. Editor The Measurement and Adjustment Series; asso. editor Brit. Jour. Ednl. Psychology; Jour. Genetic Psychology; Genetic Psychology Monographs. Assisted in revision of intelligence tests for U. S. Army; developed Terman Group Test (1920). Died Dec. 21, 1956.

TERMIER, Henri François Emile, French geologist; b. Lyons, France, Dec. 13, 1897; s. Joseph and Marguerite (Armand) T.; ed. Faculty Sci., Grenoble, also Paris; Dr.ès.Sci.; m. Geneviève Delpey, Nov. 11, 1942; a son, Michel Jean Gabriel. Asst. instr. Montpellier (France) Faculty Sci., 1923-25; geologist Morroco Mine Service, 1925-40; head geol. services, Morocco, 1940-45; prof. geology Algiers Faculty Sci., 1945-55; prof. stratigraphic paleontology Sorbonne, 1955-61, chmn. gen. geology, 1961—. Laureate A French Acad. Scis. 1936, 45, Geol. Soc. France, 1937; recipient Prix du Maroc, 1940, Internat. Spendiarov prize, 1952. Mem. French Geol. Soc., French Soc. Mineralogy and Crystallography, Deutsche geologie Gesellschaft, Deutsche geologie Verein, Belgian Geol. Soc., Geochem. Soc., Paleontology Soc., Soc. Econ. Paleontology and Mineralogy, French Assn. Sci. Writers. Author: Eutes sur le Maroc central et le Moyen Atlas septentrional; Paléontologie marocaine; Traité de géologie, 1952-61; Traité de stratigraphie; Erosion et sédimentation, 1960; also numerous articles. Home: 131 rue V.-Cousin, Sorbonne, Paris, France.

TERMIER, Pierre-Marie, French geologist; b. Lyons, France, July 3, 1859; ed. École polytechnique, École des Mines. Became engr. Corps des Mines, Nice, France, 1883; later insp. gen. mines; named prof. Ecole des mines de Saint-Etienne, France, 1885; became prof. mineralogy École des Mines, Paris, 1894, prof. geology, 1911; apptd. dir. Service de la carte géologique de France, 1911; head (with M. Bertrand, W. Kilian) French Sch. for Study Peripheral Tangential Movements in Alps. Mem. French Acad. Scis. (became v.p. 1930). Author: Etude sur la constitution géologique de la Vanoise, 1891; Le Massif des Grandes-Rousses, 1894; Les Montagnes entre Briancon et Vallouise, 1903; Les Nappes des Alpes orientales et la synthèses des Alpes, 1903; Notes sur la tectonique du Bassin du Gard, 1919; Nouvelles observations geologiques sur la Corse orientale, 1928; A la gloire de la terre, 1922; La Joie de connaitre, 1926; La Vocation de savant, 1929. Research in geology of mountains between Briancon and Vallouise, Gard Basin, Eastern alps, Eastern Corsica. Died Grenoble, France, Oct. 23, 1930.

TERNOVSKII, Vasilii Nikolaevich, Russian anatomist; b. Aug. 6, 1888; grad. natural sci. dept. Physico-Math. Scis. Faculty, Moscow (USSR) U., 1912, grad. Med. Faculty, 1915; M.D., 1922. Asst. Moscow U., 1918-24; staff Kazan Med. Inst., 1924-44; prof., from 1935; chair normal anatomy 2d Moscow Med. Inst., 1944—; specialist human, comparative anatomy USSR Acad. Med. Scis., from 1944, dir. Med. Bur., presidium, 1945-50. Mem. edn. com. for studying publ. of classics of Soviet and fgn. medicine, learned internat. council USSR Ministry Health. Mem. Soviet Nat. Assn. Natural Sci., English Historians, All-Union and Moscow Soc. Anatomists, Histologists and Embryologists (bd.), Internat. Soc. Med. Historians. Author: Anatomy of the Vegetative Nervous System, 1922; The Vegetative Nervous System and its Pathology, 1925; Morphology of the Ganglions of the Plexus Coeliacus, 1934; (with others) Morphology of the Vegetative Nervous System, 1953. Editor: (with others) History of Medicine sect. Large Med. Ency., 2d edit.; Anatomy of Man (F. Merkel), 6 vols.; Atlas of Human Anatomy (Frey). Research, publs. history of anatomy, medicine; supervised transl. into Russian De humani corporis fabrica (Vesalius), annotated, pub. 1950-54, also Canons of Medical Science (Ibn-Sin), vols. I, II, 1954-56. Home: Novoslobodskaya 57-65, Moscow. Office: 2d Moscow Med. Inst., Malaya Pirogovskaya 1, Moscow, USSR.

TER-POGOSSIAN, Michel, physicist; b. Berlin, Germany, Apr. 21, 1925; s. Michel and Anna (Sovratoff) T.-P.; B.A., U. Paris, (France), 1943; postgrad. Inst. Radium, Paris, 1945-46; M.S., Washington U., St. Louis, 1948, Ph.D., 1950; m. Ann Scott Dodson, Mar. 3, 1966. Came to U. S., 1947, naturalized, 1954. Faculty dept. radiology Washington U., Sch. Medicine, St. Louis, 1950—, prof. radiation physics, 1961-64, prof. biophysics in physiology, 1964—. Mem. radiation study sect. NIH, 1965—; mem. Internat. Com. Radiol. Units. Fellow Am. Physics Soc.; mem. Am. Nuclear Soc., Am. Radium Soc. Contbg. author: Planning Guide for Radiologic Installations, 1953; Planning Guide for Radiologic Installations, 1966; Technological Needs for Reduction of Patient Dosage from Diagnostic Radiology, 1963; Gynecology—Obstetric Guide, 1963; The Reduction of Patient Dose by Diagnostic Radiologic Instrumentation, 1964; Progress in Atomic Medicine, 1956. Research and publs. on biophysics and med. physics including applications of ionizing radiations; inventor scintillation detectors, therapeutic radium applicators, X-ray intensifying screens, new type of isotope camera, laser excited X-ray tube. Home: 7314 Maryland St., St. Louis 63130.*

TERPSTRA, Pieter, Dutch crystallographer; b. Barrddeel, Netherlands, Oct. 14, 1886; s. Johannes and Janke (Miedema) T.; D.Sc. cum laude, Groningen U., 1916; m. Jacoba Gesiena Kalverkamp, July 14, 1914; 1 son (dec.), 1 dau. Asst. lectr. crystallography Groningen U., 1924-48, prof., 1948-56. Author: A Thousand and One Questions on Crystallographic Problems, 1952; Introduction to the Space Groups, 1955; Crystallometry, 1965, also articles. Research on fundamentals of geometrical crystallography, developed new crystalgoniometer adapted to very small crystals, method for crystallographic calculation suited for use with modern calculating machines. Address: Ya Emmaningel, Groningen, Netherlands.*

TERQUEM, Alfred, French physicist; b. Metz, France, 1831; tchr. physics, lycée, Metz, also faculty scis., Strasbourg, then at Lille, France; mem. French Acad. Scis., 1886. Author: Capillarité, 1881; La science romaine k l'époque d'Auguste, étude historique d'après Vitruve, 1885. Studied timber of tones produced by discontinuous impacts, 1870. Died Lille, 1887.

TERQUEM, Olry, French mathematician; b. Metz, France, 1782; ed. Ecole polytechnique; asst. lectr. Ecole polytechnique; prof. math., lycée, then at Ecole d'artillerie de Mayence; became librarian Dépôt d'artillerie de St.-Thomas-d'Aquin, Paris, 1804. Editor Nouvelles annales de mathématiques. Died Paris, 1862.

TERRELL, Nelson James, Jr., Am. physicist; b. Houston, Aug. 15, 1923; s. Nelson James and Gladys (Stevens) T.; B.A., Rice U., 1944, M.A., 1947, Ph.D. (AEC fellow), 1950; m. Elizabeth Anne Pearson, June 9, 1945; children—Anne Pearson, Barbara Pearson, Jean Stevens. Asst. prof. Western Res. U., Cleve., 1950-51; staff mem. Los Alamos Sci. Lab., U. Cal., Los Alamos, N.M., 1951—. Fellow Am. Phys. Soc.; mem. A.A.A.S., Am. Astron. Soc., Phi Beta Kappa, Sigma Xi. Research and publs. on gamma-ray spectrometry, nuclear cross sects., fission neutron numbers, fission theory, theory of relativity, astrophysics. Home: 85 Obsidian Loop. Office: U. Cal., Los Alamos Sci. Lab., P.O. Box 1663, Los Alamos 87544.*

TERRIER, Louis-Félix, French surgeon; b. Paris, 1837; became intern, prosector, 1870, agrégé, 1872, surgeon of hosp., 1873; named prof., Faculty Medicine, 1892, Surg. Clinic, 1900; chief of services Laennec, also Salpetrière. Mem. Acad. Medicine. Author: De l'oesophagotomie externe, 1870; Des anévrismes cirsoïdes, 1872; Eléments de pathologie chirurgicale générale, 1885; La suture intestinale, 1898; also papers on gall stone removal, treatment of hernias, appendicitis, osteomyelitis. Placed patients with contagious and noncontagious diseases in different hosp. wards; one of chief proponents of ovariotomy. Died Paris, 1908.

TERRILLON, Octave, French surgeon; b. Oigny-sur-Seine, France, 1844; intern, 1868; prosector, doctor, 1873; surgeon, 1876; agrégé, 1878; tchr., Lourcine, also Salpetrière, Paris. Author: (with Monod) Traité des maladies du testicule et de ses annexes, 1889; (with Chaput) Asepsie et antisepsie chirurgicale, 1892. Research on disorders of male and female genital organs, abdominal surgery, cerebral localizations; promoted aseptic surgery. Died Paris, 1895.

TERRIS, Milton, Am. physician; b. N.Y.C., Apr. 22, 1915; s. Harry and Gussie (Dokshitski) T.; A.B., Columbia, 1935; M.D., N.Y. U., 1939; M.P.H. Johns Hopkins, 1944; m. Rema Lapouse, Nov. 23, 1941; children—Andrew David, Eugene Charles. Practice medicine, specializing in preventive medicine, Buffalo,

1951-58, N.Y.C., 1960—; asst. dean postgrad. edn. Sch. Medicine, U. Buffalo, 1951-58, asso., 1952-54, asst. prof., 1954-55, asso. prof. preventive medicine, 1955-58; prof. epidemiology Sch. Medicine Tulane U., 1958-60; head chronic disease unit dept. epidemiology Pub. Health Research Inst., N.Y.C., 1960-64; prof. preventive medicine N.Y. Med. Coll., 1964—. Fellow N.Y. Acad. Medicine; mem. Am. Pub. Health Assn. (pres.), Assn. Tchrs. Preventive Medicine (past pres.), Am. Epidemiological Soc., Phi Beta Kappa. Author: Goldberger on Pellagra, 1964. Research on chronic disease epidemiology, cancer, sarcoidosis, cirrhosis of liver, prematurity. Home: 6 Overdale Rd., Rye, N.Y. 10580. Office: N.Y. Med. Coll., Fifth Av. and 106th St., N.Y.C. 10029.*

TERRY, Luther Leonidas, Am. physician; b. Red Level, Ala., Sept. 15, 1911; s. James Edward and Lulu Maria (Durham) T.; B.S., Birmingham-So. Coll., 1931, D.Sc. (hon., 1961; M.D., Tulane U., 1935, D.Sc., (hon.), 1964; D.Sc. (hon.), Jefferson Med. Coll., 1962; LL.D. U. Alaska, 1964; D.M.S. (hon.), Woman's Med. Coll. Pa., 1964, others; m. Beryl Janet Reynolds; children—Janet Reynolds, Luther Leonidas, Michael D. Chief gen. med. and exptl. therapeutics Nat. Heart Inst., 1950-58, asst. dir., 1958-61; asst. prof. medicine Johns Hopkins Sch. Medicine, 1953-61; surgeon gen. USPHS, 1961-65; v.p. for med. affairs U. Pa., 1965—. Recipient George B. Clendening Meml. award D.C. Dental Soc., 1965, Robert D. Bruce award A.C.P., 1965, Distinguished Service medal USPHS, 1965. Diplomate Am. Bd. Internal Medicine. Fellow Am. Coll. Chest Physicians (hon.), Royal Soc. Health (hon. Gt. Britain); mem. Am. Hosp. Assn. (hon.), D.C. Trudeau Soc. (pres.), Am. Heart Assn., A.C.P. (gov.), A.M.A. (Ho. of Dels.) Am. Pub. Health Assn., Pub. Health Service Clin. Soc. (pres.) Nat. Resuscitation Soc. (adv. bd.), Nat. Bd. Med. Examiners, Assn. Am. Physicians. Studies, publs. on pneumococcal meningitis and pneumonitis; treatment of acute respiratory infections and bacterial endocarditis; malignant carcinoid; health communications to public. Office: U. Pa., 121 College Hall, Phila. 19104.*

TERRY, Michael, geographer; b. Newcastle-on-Tyne, Eng., May 3, 1899; s. Arthur Michael and Katherine (Neagle) T.; ed. King Edward's Coll., Birmingham, U. Durham; m. Ursula Livingstone-Learmonth, 1950. Head 14 expdns. to center of Australia, 1919—. Mem. Royal Geog. Soc. Australasia, Royal Exchange Sydney, Australian Soc. Authors, Pathfinders Assn. New S. Wales, Royal Geog. Soc. (life). Author: Across Unknown Australia; Through a Land of Promise; Hidden Wealth and Hiding People; Untold Miles; Sand and Sun; Bulldozer; also articles. Studies in cartography, anthropology, meteorology, minerals. Address: care of Rural Bank, Martin Pl., Sydney, New S. Wales, Australia.

TERRY, Robert James, Am. biologist; b. Crockett, Tex., May 1, 1922; s. James Arthur and Clara (Jackson) T.; B.S., Tex. So. U., 1946; M.S., Atlanta U., 1949; Ph.D., U. Ia., 1954; m. Mercedes Gwendolyn Riley, Oct. 2, 1948; 1 dau., Willa Michelle. Faculty Tex. So. U., Houston, 1948—, prof., head dept. biology, 1954—, dir. NSF undergrad. research program, 1962-64, dir. NSF Inst. for Tchrs., 1959-65; vis. scientist Tex. Acad. Sci., 1960-65. Cons. in sci. Madras (India) U.; 1965. NSF grantee, 1961—. Mem. Am. Soc. Zoologists, Tex. Acad. Sci., Sigma Xi, Beta Beta Beta, Beta Kappa Chi, Beta Kappa Chi (nat. pres. 1964—). Research publs. on regenerative capacities different regions nervous system, developmental relationship between peripheral organs and specific region of nervous system. Home: 3324 Rosedale St., Houston 77004.*

TERRY, Theodore Lasater, Am. ophthalmologist; b. Ennis, Tex., Feb. 19, 1899; s. John Stone and Lucy (Lasater) T.; ed. So. Methodist U.; M.D., U. Tex., 1922; m. Helen Pio, 1935. Asso. prof. U. Tex., 1925-26; became instr. ophthalmology Harvard, 1931; pathologist, asst. surgeon Mass. Eye and Ear Infirmary; became asst. surgeon Mass. Gen. Hosp., Boston, 1931. Pres., a founder Found. for Vision; asst. sec. Am. Bd. Cphthalmology. Diplomate Nat. Bd. Med. Examiners. Fellow A.M.A.; mem. Assn. for Research in Ophthalmology (trustee), A.A.A.S., Am. Ophthalmol. Soc., Acad. Ophthalmology and Otolaryngology. First description of retrolenthal fibroplasia (Terry's Syndrome), 1942. Died Sept. 28, 1946.

TERWILLIGER, Kent Melville, Am. physicist; b. San Jose, Cal., June 17, 1924; s. Ivan Melville and Clara (Kennedy) T.; B.S. in Physics, Cal. Inst. Tech., 1949; Ph.D., U. Cal. at Berkeley, 1952; m. Doris Lucile Heisig, Dec. 19, 1951; children—Steven K., Paul M., Thomas C., John R. Faculty, U. Mich., Ann Arbor, 1952—, prof. physics, 1965—. Chmn. Argonne Accelerator Users Group, 1962-64. Guggenheim fellow CERN, Geneva, Switzerland, 1964-65. Mem. Am. Phys. Soc. Research and publs. on photonuclear reactions, constrn. of particle accelerators, high energy physics and high energy physics instrumentation.*

TERZAGHI, Karl, engineer; b. Prague, Austria-Hungary, Oct. 2, 1883; s. Anton and Amalia (Eberle) T.; M.E., Technische Hochschule, Graz, Austria, 1904, D.Tech., 1912; Doctor honoris causa, U. Istanbul, U. Mexico, Eidgen. Poly. Zurich; Eng.D. (hon.), Lehigh U.; Sc.D., Trinity Coll., Dublin, Ireland; E.D., Technische Universität, W. Berlin, Germany; Tekniske Hogskole, Trondheim; Tech. Hochschule, Graz; Ohio State U. m. Ruth Allen Doggett, June, 1930; children—Eric Anthony, Margaret. Came to U. S., 1912, naturalized, 1943. Junior engr., Adolf Baron Pittel, Vienna, 1905-06; designing engr. and supt. construction in Austria, Croatia, Russia, U. S., 1906-14; prof. found. engring, École impériale d'Ingénieurs, Constantinople, Turkey, 1916-18, Am. Robert Coll., Constantinople, 1918-25; prof. soil mechanics and foundation engring., Mass. Inst. Tech., and cons. in N. and Central America, 1925-29; prof. Technische Hochschule, Vienna and cons. in Europe, Asia and Africa, 1929-38; hon. pres. Internat. Conf. Soil Mechanics and Found. Engring., Cambridge, Mass., 1957—; guest lectr. in U. S., 1936; cons. in U. S., 1939—; prof. of the practice of civil engring., Harvard U., 1938-56; emeritus, 1956—; vis. lectr. research cons., U. Ill. Consulting work U. S. South Am., India, England, France, North & Central Africa; pres. Internat. Conf. Soil Mech. Found. Eng., Cambridge, Mass., 1936, Rotterdam, 1948, Zurich, 1953. Reed. Norman Medal, Am. Soc. Civil Engrs., 1930, 1942, 1947; FitzGerald Medal, Clemens Herschel Prize, Boston Soc. Civil Engrs., 1926, 1928 and 1943; Brown Medal, Franklin Soc., Phila.; James Forrest lectr., Inst. Civil Engrs., London, 1939. Mem. Am. Soc. C.E. (hon.), Boston Soc. C.E. (hon.), Inst. Civil Engrs. (London), Am. Acad. Arts and Scis., Sigma Xi. Author: Erdbaumechanik, 1925; Ingenieurgeologie, 1929; Theorie der Setzung von Tonschichten, 1936; Theoretical Soil Mechanics, 1943; Soil Mechanics in Engineering Practice, 1948; From Theory to Practice in Soil Mechanics, 1960. Specialist in study, application of tech. geology; work in soil mechanics, soft ground tunneling, dam and bldg. founds. Died Winchester, Mass., Oct. 25, 1963.

TERZANO, Guillermo, Argentinian physician; b. Buenos Aires, Argentina, Oct. 2, 1908; s. Humberto and María Luisa (Bencini) T.; M.D., U. Buenos Aires Med. Coll., 1933. With Hosp. Alvear, 1934-37, Hosp. Rawson, 1938-41, Hosp. Rivadavia, 1941—; mem. staff Inst. Physiology U. Buenos Aires, 1941-44; Commonwealth Fund fellow Cornell U. Med. Coll, N.Y.C., 1946-47; Am. Cancer Soc. fellow, 1948-49; dir. Centro de Citologia Fundacion Williams, Little Co. of Mary Hosp., Buenos Aires, 1955—; asst. dir. Papanicolaou Cancer Research Inst., Miami, Fla., 1961-62; dir. cytology research lab. Hosp. Italiano, U. Buenos Aires, 1963—. Recipient Wien award, 1957. Fellow Internat. Acad. Cytology; mem. Am., Brazilian, Latin Am. Argentine socs. cytology. Author: (with Jose Maria Mezzadra) Colpocitologia, 1955. Research, numerous publs. on early diagnosis of cancer. Home: 2151 Bustamante St. Office: 2952 San Martin de Tours, Buenos Aires, Argentina.*

TERZIN, Alexander L., microbiologist; b. St. Endre, Hungary, Sept. 16, 1911; s. Lazar L. and Ada (Stana) T.; M.S., Med. Faculty, Zagreb, Yugoslavia, 1939; dipl. bacteriol. Med. Faculty, Belgrade, Yugoslavia, 1943; postgrad. (research fellow) Harvard Med. Sch., 1946-48; m. Radmila Oljacha, Feb. 11, 1952. Asst. lectr. Med. faculty, Belgrade, 1940-45; head dept. bacteriology Inst. Hygiene, Belgrade, 1945-50, head dept. virology, 1950-52; prof. microbiology Med. Faculty, Sarajevo, Yugoslavia, 1953-64; dir. Inst. Microbiology, prof. microbiology Med. Faculty, Novi Sad, Yugoslavia, 1964—. Mem. expert adv. panel on virus disease WHO, 1951—, dir. Yugoslav Region Influenza Center, 1951-65. Decorated Yugoslav Decorations of Labour, 3d order, 1949, 2d order, 1959, 1st order, 1965. Mem. Serbian Acad. Sci. and Arts (corr.), Am. Assn. Immunologists, Brit. Soc. Gen. Microbiology. Research and numerous publs. in sero-epidemiology viral infections, live vaccine against bacillary dysentery, bioelec. activity of chick embryos in viro; developed lab. procedures and reagents for diagnosing microbial diseases, method for detection and in vivo measurement of minor illness in mice. Home: 21 Fruskogorski put. Office: Medicinski Fakultet, Novi Sad, Yugoslavia.*

TERZUOLO, Carlo Agostino, physiologist; b. Acqui, Italy, Sept. 2, 1925; s. Domenico and Matilde (Vola) T.; M.D., U. Torino (Italy), 1949; m. Margherita Ertallini, Aug. 9, 1954; 1 son, Eric Robert. Came to U. S., 1954, naturalized, 1960. Research asst. Italian Nat. Research Council, 1947-49; research fellow Italian Nat. Research Council, U. Pisa (Italy), 1950-51; asst. prof. Universite Libre de Bruxelles, Brussels, Belgium, 1951-53; research asst. U. Cal. at Los Angeles, 1954-56, Multiple Sclerosis Soc. fellow, 1956-57, research asso., 1957-59; Hill prof. physiology U. Minn., Mpls., 1959—. Mem. Am. Physiol. Soc., Internat. Brain Research Orgn., Italian Neurophysiol. Soc. Contbr. numerous articles to tech. jours., books. Research in physiology nerve cells and synoptic transmission, sensory receptors physiology, central nervous system integrative actions. Home: 4134 Coffman Lane, Mpls. 55456.*

TESCHNER, Jacob, Am. surgeon, physician; b. N.Y.C., Aug. 7, 1858; s. Abraham and Rose (Popper T.; ed. Coll. City N.Y., Coll. Pharmacy, Coll. Phys. and Surgeons at Columbia U., N.Y. U.; m. Goldine Strauss, Jan. 28, 1886; children—Helen (Mrs. Tas), Ruth. Practice gen. surgery, 1881-93, orthopedic surgery, 1893-1915; gen. cons. from 1915. Mem. A.M.A., N.Y. State Med. Soc., Acad. Medicine, Met. Med. Soc., Physicians Mut. Aid Assn., Harlem Med. Ass., German Med. Soc. Contbr. articles to med. jours. Invented methods for treating rotary lateral curvature, flat foot, and paralytic conditions without using braces; used exercise to treat cardiac insufficiency. Died Oct. 23, 1927.

TESLA, Nikola, electrician, inventor; b. Smiljan, Lika (border of Austria-Hungary), July 9, 1856; s. of Greek clergyman and orator, and of Georgina Mandic, who was an inventor, as was her father; ed. 1 yr. at elementary sch., 4 yrs. at Lower Realschule, Gospic, Lika, and 3 yrs. at Higher Realschule, Carlstadt, Croatia, graduating 1873; student 4 yrs. at Polytechnic Sch., Gratz, in mathematics, physics and mechanics; afterward 2 yrs. in philos. studies at U. Prague, Bohemia; hon. M.A., Yale, 1894; LL.D., Columbia, 1894; D.Sc., Vienna Polytechnic. Began practical career at Budapest, Hungary, 1881, where made 1st electrical invention—a telephone repeater—and conceived idea of his rotating magnetic field; later engaged in various branches of engring. and manufacture. Since 1884 resident of U. S., becoming naturalized citizen. Inventor and discoverer: System of arc lighting, 1886; Tesla motor and system of alternating current power transmission, 1888; system of elec. conversion and distribution by oscillatory discharges, 1889; generators of high frequency currents, and effects of these, 1890; Tesla coil, or transformer, 1891; system of wireless transmission of intelligence, 1893; mech. oscillators and generators of elec. oscillations, 1894-95; researches and discoveries in radiations, material streams and emanations, 1896-98; high-potential magnifying transmitter, 1897; system of transmission of power without wires, 1897-1905; Tesla's steam and gasturbine and pump, etc. Chiefly engaged, since 1903, in development of system of telegraphy and telephony, and designing plant for transmission of power without wires, to be erected at Niagara. Died N.Y.C., Jan. 7, 1943.

TESSIER, Alexandre-Henri, French agronomist; b. Anderville, France, Oct. 16, 1741; dir. Establissement rural de Rambouillet; mem. French Acad. Scis., 1783, Acad. Medicine, Soc. Agr. Author: Dictionnaire d'agriculture et d'economie rurale, 1787-1816. Collaborator: Historie de l'introduction et la propagation du merinos en France, 1838. Founder Jour. d'agriculture à l'usage des habitants de la campagne, 1791. Research on acclimation of merino sheep in France; made comparative studies of various French and fgn. wheats. Died Paris, Dec. 11, 1837.

TESSIER, Fernand Victor Louis, French geologist; b. Marseilles, France, Aug. 11, 1920; s. Maximilien A. C. and Berthe M.-L. (Vidal) T.; Licence ès Scis., Marseille, 1942; Ingénieur-Géologue, Nancy, France, 1943; Doctorate ès Sciences Naturelles, Marseille, 1950; m. Marguerite A. J. Quenard, Aug. 14, 1943; children—Olivier M. H., Christian J. M., Marc F. L., Alain A. C. Successively asst. geologist, geologist, prin. geologist Geol. Survey French W. Africa, 1944-53; with U. Dakar, 1949-63, prof., 1957-63, became hon. prof., 1963; prof. U. Nice, 1963-66; prof. U. Marseilles, 1966—. Decorated Chevalier del'Ordre des Palmes Académiques, Chevalier de l'Ordre National du Sénégal. Mem. Société Géologique de France, Union Francaise des Géologues. Author: Contributions à la stratigraphie et à la paléontologie de la partie ouest du Sénégal (Crétacé et Tertiaire), 1952; also articles. Research on geology of Cretaceous and Tertiary in Senegal, Ivory Coast, Niger; described new fossils; discovered Maestrichtian and Paleocene stages in Senegal; studies in stratigraphy of Lower Paleozoic in Mali and Mauritany; lateritic rocks in Senegal; stratigraphy, structures, action of termites; paleoclimatology; sedimentary petrology and applied geology; discovered (with R. Dars, J. Sougy) large hercynian overthrusts in Senegal and Mauritany. Home: 3 rue de Bir Hakeim, Marseilles 13, France.*

TESSMAN, Irwin, Am. molecular biologist; b. Bklyn., Nov. 24, 1929; s. Morris and Anna (Chodack) T.; A.B., Cornell U., 1950; Ph.D., Yale, 1954; m. Ethel Stolzenberg, June 19, 1949; 1 son, Adam A. NSF Fellow dept. physics, Cornell U., 1954-55; Am. Cancer Soc. fellow, 1955-57, dept. biology Mass. Inst. Tech., 1957-58, research asso., 1958-59; faculty Purdue U., West Lafayette, Ind., 1959—, prof. biology, 1962—. Sterling fellow Yale, 1953; recipient Gravity Research Found. prize, 1953. Mem. A.A.-A.S., Am. Biophys. Soc., Genetics Soc. Am., Phi Beta Kappa. Asso. editor: Growth in Living Systems, 1961. Contbr. articles to tech. jours. Asso. editor Virology, 1959-63, 66—, editor, 1963-66. Research on reproductive mechanism of bacterial viruses, structure of hereditary material, transmission of genetic information from parent to progeny, mechanism of genetic recombination and orgn. genes on viral chromosomes. Office: Dept. Biol. Scis., Purdue U., West Lafayette, Ind. 47907.*

TESTA, Emilio, chemist; b. Chiasso, Switzerland, Aug. 1, 1925; s. Mario and Clotilde (Torriani) T.; Ph.D., U. Zurich, 1950, postgrad. research fellow; m. Emilia Mazzola, May 26, 1951; children—Silvano, Daniele, Donata. With Lepetit S.P.A. Research Labs., Milan, Italy, 1953—, dir. organic synthesis dept., 1959—, dir. pharm. dept., 1966—. Lectr. indsl. organic chemistry U. Milan, 1959—. Mem. Italian Chem. Soc., Italian Soc. Pharm. Sci. Research, numerous publs. on fructose metabolism of liver, synthesis of new heterocyclic compounds, synthesis of compounds active in central nervous system, chemistry of steroids. Home: San Simone, Vacallo, Ticino, Switzerland. Office: 38 Via Durando, Milan, Italy.*

TESTELLI, Mario Romano, physician; b. Ferrara, Italy, Mar. 19, 1928; s. Dante and Pia (Mattarelli) T.; M.D., U. Parma, 1953; m. Luz Maria Vaquero, Apr. 20, 1961; children—Maria Pia, Paolo, Sandro. Research asst. dept. electrocardiography Inst. of Cardiology, Mexico City, 1959-61, research asso., 1961-—; prof. physiology Nat. U., Mexico City, 1965-—. USPHS grantee, 1962-65. Mem. Am. Heart Assn., Am. Coll. Cardiology, Am. Coll. Chest Physicians, Cardiology Soc. Mexico. Contbr. numerous articles to med. jours., chpts. to books. Research on intracardiac phenomena in dogs and man by heart catheterization. Address: Instituto N. de Cardiologia, Av. Cuauhtemoc 300, Mexico City, Mexico.*

TESTER, Albert Lewis, zoologist; b. Toronto, Ont., Can., Nov. 27, 1908; s. Jesse ond Fanny (Kinge) T.; B.A., U. Toronto, 1931, M.A., 1932, Ph.D., 1936; m. Laura May Arnold, Apr. 21, 1934; children—Murray Arnold, Loretta May (Mrs. Edward W. Fink). Came to U. S., 1948, naturalized, 1954. From sci. asst. to sr. biologist Fisheries Research Bd. Can., Nanaimo, B.C., Can., 1931-48; faculty U. Hawaii, Honolulu, 1948-55, sr. prof. zoology, 58-—; dir. pacific oceanic fishery investigation U. S. Bur. Comml. Fisheries, Honolulu, 1955-57, chief div. biol. research, 1957-58. Chmn. fisheries div. 9th Pacific Sci. Congress, Bangkok, Thailand, 1957; chmn. div. biol. scis. 10th Pacific Sci. Congress, Honolulu, 1960. Mem. A.A.A.S., Am. Soc. Zoologists, Am. Soc. Limnology and Oceanography, Am. Fisheries Soc., Animal Behavior Soc., Am. Inst. Biol. Scis., Sigma Xi. Research and numerous publs. on fluctuations in abundance of Pacific herring, life history tropical baitfish, response of tuna to stimuli, factors affecting behavior of sharks, sensory perception in sharks. Home: 1208 The Palms, 431 Nahua St., Honolulu 96815.*

TESTUT, Jean-Léo, French anatomist; b. St.-Avit-Senieur, France, 1849; chief of anat. work, agrégé Faculty Medicine, Bordeaux, France; prof. anatomy Faculty of Lille (France), also at Lyons, France, 33 years. Author: Les anomalies musculaires chez l'homme expliquées par l'anatomie comparée, 1884; Traité d'anatomie descriptive, 1889; (with Jacob) Traité d'anatomie topographique. Died Bordeaux, 1925.

TETENYI, Peter, Hungarian chemotaxonomist; b. Budapest, Hungary, June 8, 1924; s. Imre and Iren (Meitin) T.; candidate agrl. scis. U. Agrl. Scis., 1955, d.biol. scis., 1964; m. Magdalena Erdösi, July 11, 1955; children—Peter, Tamás. Investigator, Inst. Genetics, 1949; 1st asst. U. Agrl. Scis., Budapest, 1950-56, dir. research Inst. Research Medicinal Plants, 1957-—. Mem. Fédération Internat. Pharmaceutic, Société Botanique de France, Hungarian Biol. Soc., Hungarian Pharm. Soc. Author: (with A. Bálint, F. Eördögh) Biology, 1949; (with A. Haraszthy, T. Hortobágyi) Botany, 1950. Editor: Infraspecific Chemical Taxa of Medicinal Plants, 1968; also articles. Introduced peanut to Hungary; verified existence of infraspecific chem. taxa at belladonna, poppy, mint, mustard; irradiation as breeding method of poppy, fennel, nightshade, for creation saprophytic strains of ergot; interspecific hybridization of popples; breeding new varieties of poppy, mustard, yarrow, tarragon. Home: 9 Népstadion, Budapest XIV. Office: 40 rue Daniel, Budapest, Hungary.*

TETER, Jerzy, Polish physician; b. Radomsk, Poland, Apr. 18, 1912; s. Ludwik and Janina (Pawlowska) T.; Med. Degree, U. Warsaw (Poland), 1938, postgrad., 1941-44; postgrad. Med. Acad., Lódz, Poland, 1948-49, Inst. Oncology, Warsaw, 1948-49. Sr. asst. dept. gynecology Med. Acad., Warsaw, 1950-51, dept. pathology Inst. Oncology, Warsaw, 1952-56; asso. prof. Med. Acad., Warsaw, 1955-—, head central oncological out-patient dept. for women, 1955-—, head endocrine unit, 1957-64, head dept. clin. endocrinology, 1964-—. Mem. sci. assembly Inst. Oncology in Poland, 1958-—. Recipient Gold Cross of Merits, 1957, Cross of Polonia Restituta for pub. service in orgn. early cancer detection in Warsaw, 1959. Mem. Polish Endocrine Soc. (pres. 1960-—), Internat. Soc. Endocrinology (exec. com. 1964-—), Internat. Acad. Cytology. Author: Hormonal Disorders in Women, 1958; Hormonal Disorders in Men, 1961; (with Tarlowska) Cancer of the Corpus Uteri, 1965; also articles. Editor, Polish Endocrinology, 1963-—. Research on neoplastic changes in gonadal dysgenesis, germ cell tumors, cytohistological patterns of early cervical cancer, sexual immaturity. Home: 8 Krakowskie Przedmiescie str., Warsaw, Poland.*

TETI, Mario, Italian microbiologist; b. Naples, Italy, Apr. 28, 1914; s. Giuseppe and Vincenzina (Niccoli) T.; M.D.; m. Bianca Raucci, Oct. 22, 1942; children—Diana, Maria, Giuseppe. Dir. Med. Microbiol. Inst., U. Messina (Italy); dir., instr. microbiology Catholic U. Rome. Mem. Sicilian Microbiol. Soc. (pres.). Contbr. numerous articles to tech. jours. Home: piazza Iuvara 3 bis 464. Office: Istituto di Microbiologia medica, Università, via p. Castelli, Messina, Italy.

TEUBER, Hans Joachim, German chemist; b. Berlin, Oct. 26, 1918; s. Paul and Johanna (Teige) T.; M.D., U. Heidelberg, 1946, Dr.rer.nat., 1949. Brit. Council fellow U. Oxford, 1952-53; faculty U. Heidelberg, 1953; exchange prof. U. Chgo., 1954; faculty Frankfurt U., 1954-—, prof. chemistry, 1960-—. Mem. Deutsche und Schweizerische Chemische Gesellschaft, Chem. Soc. London. Research, numerous publs. on structure and synthesis of natural products, new oxidation methods, synthesis of pharmacol. compounds, model reactions of biochem. pathways. Home: 9 Heideweg, Oberhoechstadt in Taunus, Germany.*

TEUBER, Hans-Lukas, physiol. psychologist; b. Berlin, Germany, Aug. 7, 1916; s. Eugen and Rose (Knopf) T.; baccalaureat, Coll. Francais, Berlin, 1934; student U. Basel (Switzerland), 1935-39; Ph.D., Harvard, 1947; m. Marianne Liepe, Aug. 1, 1941; children—Andreas Wolfgang, Christopher Lawrence. Came to U. S., 1941, naturalized, 1944. Asst. psychology Harvard, 1941-42; from research asso. to prof. psychiatry and neurology N.Y. U. Coll. Medicine, also dir. psychophysiol. lab. N.Y. U. Bellevue Med. Center, 1947-61; prof., chmn. psychology dept., also dir. psychophysiol. lab. Mass. Inst. Tech., 1961-—. Past mem. biol. and behavioral sci. panels Office Sci. Research, USAF; mem. bioscis. subcom. NASA; research adv. bd. United Cerebral Palsy Assn.; mem. Neurology A study sect. NIH; area cons. VA; mem. Internat. Brain Research Orgn., WHO. Served with USNR, 1944-46. Recipient of Karl Spencer Lashley award, 1966. Mem. Am. Acad. Arts and Scis., Am. Neurol. Assn., Am. Acad. Neurology, Assn. Research Nervous and Mental Diseases, Soc. Exptl. Psychologists, Am. Psychological Association, New York Acad. of Science, Eastern Psychological Assn., A.A.A.S., Sigma Xi. Author: Visual Field Defects after Penetrating Missile Wounds of the Brain, 1960; (with J. Semmes, S. Weinstein and L. Ghent) Somatosensory Changes after Penetrating Brain Wounds in Man, 1960; also articles, chpts. in books. Editor Internat. Jour., Neuropsychology, 1962-—, Jour. Comparative and Physiol. Psychology, 1956-—, Jour. Nervous and Mental Diseases, and Jour. Psychiat. Research, 1961-64; co-editor Exptl. Brain Research Jour., 1965-—. Elucidation of effects of brain injury in man; theories of perception and voluntary movement. Office: Psychology Dept., Bldg. E10-012, Mass. Inst. Tech., Cambridge, Mass. 02139.*

TEXTER, E. Clinton, Jr., Am. physician; b. Detroit, June 12, 1923; s. Elmer Clinton and Helen (Rotchford) T.; A.B., Mich. State U., 1943; M.D., Wayne State U., 1946; postgrad. N.Y. U. Med. Sch., 1948-49, Northwestern U., 1959-60; m. Jane Starke Curtis, Feb. 19, 1949; children—Phyllis Cardew, Patricia Ann, Catherine Jane. Instr., Duke Sch. Medicine, 1951-53; asso. in medicine Northwestern U. Med. Sch., 1953-56, asst. prof., 1956-61, asso. prof. medicine, 1961-—, dir. tng. programs in gastroenterology, 1956-63; practice medicine specializing in gastroenterology, Durham, N.C., 1951-53, Chgo., 1953-—; staff Passavant Meml. Hosp., VA Research Hosp., Cook County Hosp. (all Chgo.). Cons., U. S. Naval Hosp., Great Lakes, Ill., 1963-—, Skokie Valley Community Hosp., 1964-—; mem. Hektoen Inst. for Med. Research, 1966-—. Clarence F. G. Brown fellow Inst. Medicine, Chgo., 1953-56. Diplomate Am. Bd. Internal Medicine and Gastroenterology, Pan-Am. Med. Assn. Fellow A.C.P., Am. Med. Writers' Assn. (dir. 1965-—, mem. exec. com. 1965-66) mem. A.M.A., Am. Assn. for Study Liver Diseases, Am. Gastroent. Assn., Am. Soc. for Gastrointestinal Endoscopy, Am. Physiol. Soc., Am. Fedn. for Clin. Research, Gastroenterology Research Group (founder, chmn. 1959-60), Central Clin. Research Club, Central Soc. for Clin. Research, Chgo. Soc. for Internal Medicine, Sigma Xi, Alpha Omega Alpha, Theta Alpha Phi, Nu Sigma Nu, Delta Chi. Author: Peptic Ulcer, Diagnosis and Treatment, 1955; also numerous articles, book chpts. Editorial bd. Ill. Med. Jour., 1965-—; The Internist, 1966-—; co-editor Am. Jour. Digestive Diseases, 1956-65, editor, 1966-—. Research on gastrointestinal and cardiovascular physiology, smooth-muscle function of gastrointestinal tract and peripheral blood vessels, gastric secretions. Home: 201 E. Walton Pl., Chgo. 60611. Office: 700 N. Michigan Av., Chgo. 60611.*

THADDAEUS FLORENTINUS, see Alderotti, Taddeo.

THADDEUS, Patrick, Am. physicist; b. Wilmington, Del., June 6, 1932; s. Victor and Elizabeth (Ross) T.; B.Sc., U. Del., 1953; M.A., Oxford (Eng.), 1955; Ph.D., Columbia U., 1960; m. Janice Farrar, Apr. 6, 1963; children—Eva, Michael. Research asso. Radiation Lab., Columbia U., 1960-61; fellow Goddard Inst. for Space Studies, N.Y.C., 1961-64, staff, 1964-—; adj. asso. prof. physics Columbia U., 1967-—, State U. N.Y., Stony Brook, 1966-—. Fulbright fellow, 1953-55. Mem. Am. Phys. Soc., Am. Astron. Soc., Am. Geophys. Union, Am. Meteorol. Soc., N.Y. Acad. Scis. Research and publs. on high resolution studies of molecular hyperfine structure with beam masers; level crossing and double resonance studies of stable and radioactive atoms; also research on microwave spectroscopy of gases at high pressure, radioastron. investigation of planets, measurement of cosmic microwave background, radio studies of earth from spacecraft. Home: 606 W. 116th St., N.Y.C. 10027. Office: 2880 Broadway, N.Y.C. 10025.*

THAER, Albert, German agronomist; b. Zelle, Germany, May 14, 1752; became prof. rural economy, Berlin, 1810; Founder agrl. instn., Zelle, Agrl. Inst., Moeglin, Germany. Mem. Société nationale d'agriculture de France (fgn.), French Acad. Scis., 1820. Author: Description des nouveaux instruments d'agriculture les plus utiles, 1803. Died Moeglin, Oct. 26, 1828.

THAKER, Kumar Ambalal, Indian chemist; b. Borsad, Gujarat, India, Oct. 4, 1924; s. Ambalal Motilal and Kashiben A. (Bhatt) T.; B.Sc. with honours Bombay (India) U., 1944, M.Sc., 1947; M.Sc., Sayajee Jubilee Technol. Inst., 1947; Ph.D., London (Eng.) U., 1956; m. Kanubhai, Feb. 10, 1946; children—Pinakin, Medhavini, Pragna, Ponal. Research fellow Sayajee Jubilee Technol. Inst., Baroda, India, 1944-47; lectr. chemistry V.P. Sci. Coll., 1947-54; prof., head dept. V.P. Sci. Coll., Vallabhvidyanagar, India, 1956-59; reader organic chemistry Gujarat U., Ahmedabad, India, 1959-—. Fellow Chem. Soc. London, Indian Chem. Soc.; mem. Royal Inst. Chemistry (asso.), Indian Sci. Congress. Author: General Knowledge, 1955; also articles. Research on organic chemistry particularly stereochemistry; discovered different methods of optical resolution of racemates; use of optically active compounds for medicinal properties or study reaction mechanisms. Address: 92, Shardanagar, Ahmedabad 7 (Gugarat), India or Borsad Dist., Kaira, India.*

THAL, Alan Philip, Am. surgeon; b. Cape Town, South Africa, July 15, 1925; s. Alex M. and Bess (Clouts) T.; M.B., Ch.B., U. Cape Town, 1949; Ph.D., U., Minn., 1956; m. Felicia S. (Jacobs), Dec. 14, 1949; children—Alyson, John, Douglas. Came to U. S., 1950, naturalized, 1962. Faculty U. Minn., 1956-61, asso. prof. surgery, 1959-61; prof. Wayne State Sch. Medicine, 1961-66; prof. surgery U. Kan. Med. Center, 1966-—; vis. prof. Coll. Phys. and Surg., Columbia U., 1966. Established investigator Am. Heart Assn., 1956-61; mem. com. on shock Nat Acad. Scis.-NRC, 1962-—, mem. com. on trauma, 1962-—. Author: Shock, 1965; also numerous articles. Research on mode action bacterial exotoxins, pancreatities, shock, cardiac surgery, esophagogastric reconstrn. Home: 4101 W. 99th St., Shawnee Mission, Kan. Office: Surgery Dept., U. Kan. Med. Center, Kansas City, Kan. 66103.*

THALER, Heribert Leopold, Austrian physician; b. Badgastein, Austria, Nov. 4, 1918; s. Alois and Margarethe (Kreuter) T.; student U. Innsbruck, 1936-39; M.D., U. Vienna, 1941; m. Margarethe Kautzky, Mar. 23, 1949; children—Erik, Arnulf, Reinhart, Norbert, Michael, Georg. Physician 1st med. dept. Med. Sch., U. Vienna, 1947-67, docent internal medicine, 1962-—; head 4th med. dept. Wilhelminen Hosp., Vienna, 1967-—; research fellow Harvard Med. Sch., 1953-54. Mem. Vienna Soc. Physicians, Austrian, German socs. internal medicine, German Soc. Digestive and Metabolic Diseases, Internat. Assn. for Study of Liver, Soc. Research Vegetative Nervous System. Research and numerous publs. on clin. pathology of liver, including bodies in viral hepatitis, histogenesis of posthepatitic cirrhosis, histogenesis of alcoholic cirrhosis. Home: 7 Sebastianplatz, HA-1030 Vienna. Office: 37 Montleartstrasse, 11H71 Vienna, Austria.*

THALER, R(aphael) M(orton), Am. physicist; b. Bklyn., May 19, 1925; s. Herman T. and Ida (Breslin) T.; student Cornell U., 1942-46; A.B., N.Y. U., 1947; M.S., Brown U., 1949, Ph.D., 1950; m. Ruth Cornfield, June 1, 1952; children— Jon Frederick, Leslie Howard, William Jordon, Roberta Lynn. Faculty, Yale, 1950-53; staff Los Alamos Sci. Lab., 1953-60; research asso. fellow Inst. for Advanced Study, Mass. Inst. Tech., 1957-58; prof. Case Inst. Tech., Cleve., 1960-67; prof., vice chmn. dept. physics Case Western Res. U., Cleve., 1967-—. Cons., Argonne (Ill.) Nat. Lab., 1962-—, Lewis Research Lab., NASA, 1960-—. Fellow Am. Phys. Soc. Author: (with L. S. Rodberg, R. M. Thaler) Introduction to the Quantum Theory of Scattering, 1967; also articles on theoretical nuclear physics. Research on nuclear forces, interactions of nucleons with nuclei, theory of scattering, theoretical nuclear and elementary particle physics. Home: 3144 Huntington Rd., Shaker Heights, O. 44120. Office: 10900 Euclid Av., Cleve. 44106.*

THALER, William John, Am. physicist; b. Baltimore, Md., Dec. 4, 1925; s. Thomas Joseph and Catherine Loretta (Russanowski) T.; B.S., Loyola Coll., Balt., 1947; M.S., Cath. U. Am., 1949, Ph.D., 1952; m. Barbara Jane Jarnagin, June 16, 1951; children—Mark, Paul, Alice, Gregory, Geoffrey. Physicist., Baird Assos., Inc., Cambridge, Mass., 1947; research asst. Cath. U. Am., 1947-51; physicist acoustics br. Office Naval Research, 1951-52, field projects br., 1952-55, acting head field projects br., 1955, physicist, head field projects br., 1955-62; prof. physics Georgetown U., 1960-—. Mem. Am. Physics Soc., Acoustical Soc. Am., Optical Soc. Am., Washington Acad. Sci., Am. Geophys. Union, Am. Assn. U. Profs., Sigma Xi. Contbr. articles profl. publs. Work in ultrasonic research on relaxation in gases; ultrasonic effects on biological matter; shock waves in liquid, solid and gas media; work with laser. Home: 1621 Tilton Dr., Silver Spring, Md. Office: Dept. Physics, Georgetown U., Washington 7.

THALES OF MILETOS, Greek mathematician, astronomer; b. circa 636 B.C., probably at Miletos of Greek parents; traveled to Egypt, possibly Babylonia;

a pre-Socratic philosopher; founder of Ionian school of philosophy. Four treatises ascribed to him in antiquity: Nautical Astronomy; On Beginnings; On the Solstice; On the Equinox; (none is extant). Held that everything in nature composed of one basic stuff (water), and that earth a flat disk floating on water; efforts generally held to be 1st recorded attempt to find naturalistic instead of mythological interpretation of nature; some accounts say he introduced geometry to Greece, invented logical proof in geometry, proved theorems (that circle bisected by its diameter, that base angles of isosceles triangle are equal, that 2 straight lines interesecting each other produce equal and opposite angles, that angle inscribed in semicircle is right angle); also said sometimes to have been good astronomer (predicted solar eclipse of May 28, 585 B.C., determined sun's course from solstice to solstice, estimated size of sun and moon); also credited by some with discovery of static electricity; usually placed first on lists of 7 wise men of ancient Greece. Died circa 546 B.C.

THALGOTT, Fred William, Am. physicist; b. Zeigler, Ill., July 27, 1915; s. J. G. and Hattie (Rogers) T.; B.Ed., U. So. Ill., 1937; M.S. in Math. U. Ill., 1941, B.S. in Mech. Engring., 1947; m. Joan Knudson, Nov. 22, 1951; children—Susan, Kathleen, Fred William, Judith, David, James. Tchr., Ill. high sch., 1937-42; asso. mech. engr. power pile div. Oak Ridge Nat. Lab., 1947-48; staff Argonne Nat. Lab., 1948-—, sr. physicist, 1955, asso. dir. Ida. div., 1956-—. Mem. reactor physics planning group AEC, 1956-57; mem. U. S. fast reactor team, U.K. and Euroatom, 1956, 60, 62, USSR, 1964. Fellow Am. Nuclear Soc. (charter); mem. Am. Phys. Soc., Am. Math. Soc. Research and publs. on reactor physics especially fast reactor critical exptl. research including survey studies on submarine nuclear propulsion through pressurized water reactor, boiling water reactors, fast reactions. Home: 2810 W. Morningside, Idaho Falls 83401. Office: P.O. Box 2528, Idaho Falls, Ida. 83401.*

THALMANNER, O., Austrian pediatrician; b. Klagenfurt, May 14, 1922; s. Hermann and Stefanie (Kollmann) T.; Dr.med., U. Vienna, 1947; m. Liselotte Herrmann, Aug. 25, 1962; children—Florian, Georg. Physician, children's clinic U. Vienna, 1947-—, docent, 1958-62, asso. prof., 1962-—. Recipient Cardinal Quitzer prize, 1962. Mem. German Soc. Pediatrics (Moro prize, 1958), Vienna Soc. Physicians; corr. mem. Catalonian Soc. Pediatrics. Author: Die Toxoplasmose bei Mensch und Tier, 1957; Pränatale Erkrankungen des Menschen, 1967; also numerous articles. Research in pediatrics, toxoplasmosis, prenatal diseases. Address: 23 Feldgasse, Vienna 8, Austria.

THAMER, Burton John, chemist; b. Kitchener, Ont., Can., June 22, 1921; s. Owen and Stella (Worden) T.; brought to U. S., 1923, naturalized, 1929; B.S., U. Cal. at Berkeley, 1943; Ph.D., Ia. State Coll., 1950. Research asst. Manhattan Project, 1943-44, Clinton Labs., Oak Ridge, 1944; analyst Hanford Engr. Works, Richland, Wash., 1944-45; research asst. chemistry Inst. for Atomic Research, Ames, Ia., 1946-50; staff mem. Los Alamos Sci. Lab., 1950-—. Member of American Chemical Society, American Nuclear Soc., A.A.A.S., Sigma Xi, Phi Lambda Upsilon. Research and publs. on devel. and containment liquid nuclear fuels and blankets, complex-ion equilibria, determination of overlapping acid constants. Office: P.O. Box 1663, Los Alamos 87544.*

THATCHER, Everett Whiting, Am. physicist; b. Jefferson, O., Jan. 24, 1904; s. Winthrop Foster and Edith (Whiting) T.; A.B., Oberlin Coll., 1926, A.M., 1927; postgrad. Purdue U., 1926-27, U. Neb., 1929; Ph.D., U. Mich., 1931; m. Tennie Marie Klotz, June 23, 1928; children—Lucy Whiting (Mrs. James Roger Woodward), James Winthrop. Asst. in physics Purdue U., 1926-27; instr. physics U. Neb., 1927-29; instr., research asst. U. Mich., 1929-31; asst. prof. physics Union Coll., Schenectady, 1931-43, asso. prof., 1943-46; ground instr., coordinator CAA Pilot Tng. Program, 1940-43; research asso. NDRC, Princeton, 1943; sci. liaison officer, OSRD, Washington, London, 1944-45; dep. tech. dir. Operation Crossroads, Bikini, 1946; head research div. USN Electronics Lab., San Diego, 1946-62, prin. scientist, tech. planning and cons. staff, 1962-—. Recipient War Service certificate of merit OSRD, 1945; Commendation, Sec. Navy, 1947. Fellow Am. Phys. Soc.; mem. I.E.E.E. (sr. mem.), Amateur Radio Orgn. (v.p. San Diego council 1958, pres. 1959), Phi Beta Kappa, Sigma Xi (charter pres. San Diego 1948). Research and publs. on propagation and fading of radio waves, multiple space charge, elec. fluctuation in conductors, microwave reflection from Arctic sea ice, elec. properties of mono-molecular films; studied shot effect in electron tubes, discovery and measurement of space charge suppression. Home: 3803 Liggett Dr., San Diego 92106. Office: U.S. Navy Electronics Lab., San Diego 92152.*

THAUER, Rudolf, German physiologist; b. Frankfort on the Main, Germany, Sept. 24, 1906; s. Rudolf and Agnes (Knoblauch) T.; Dr.med., U. Frankfort, Berlin, Germany, 1932; m. Lotte Kalberlah, June 30, 1934; children—Sabine (Mrs. Kurt Greeff), Rudolf Kurt, Hans Jörg, Christoph. Faculty, U. Frankfort, 1934-43, prof. dept. physiology, 1939-43; dir. dept. physiology Acad. Danzig, Germany, 1943-45; physiologist Aero Med. Equipment Lab., Naval Air Material Center, Phila., 1947-51; dir. Kerckhoff Inst., Max-Planck-Gesellschaft, Bad Nauheim, also head dept. physiology U. Giessen (Germany), 1951-—. Mem. Sci. Adv. Council, 1964-—, Hessischer Forschungsrat, 1959-—. Recipient Senkenberg prize U. Frankfort, 1943. Mem. German, Am. physiol. socs. Editor, Pflügers Archiv für Physiologie, 1954-—, Verhandlungen der Deutschen Gesellschaft für Kreislaufforschung, 1952-—. Research in temperature regulation, blood circulation, physiology of central nervous system; discovery of peripheral mechanisms of temperature regulation. Home: 1 Parkstrasse, 6350 Bad Nauheim, Hessen, Germany.*

THAXTER, Roland, Am. botanist; b. Newton, Mass., Aug. 28, 1858; s. Levi L. and Celia (Laighton) T.; A.B., Harvard, 1882, Ph.D., and A.M., 1888; m. Mabel Gray Freeman, June 8, 1887; children—Charles Eliot, Katharine, Elizabeth, Edmund Lincoln. Asst. in biology, Harvard, 1886-88; mycologist, Conn. Agrl. Expt. Sta., 1888-1891; asst. prof. cryptogamic botany, 1891-1901, prof., 1901-19, prof. emeritus, 1919-—, Harvard. Fellow Am. Acad. Arts and Sciences, French Acad. Scis., 1925; Am. Philos. Soc., Nat. Acad. Scis., Phi Beta Kappa, A.A.A.S. Contbr. scientific publs. Work on devel., structure and relationship major groups fungi; wrote important monograph on Laboulbeniaceae. Died Cambridge, Mass., April 22, 1932.

THAYER, Sidney Allen, Am. biochemist; b. Milw., July 2, 1902; s. Alfred Merrill and Harriet (Osmond) T.; B.S., Beloit Coll., 1925; Ph.D., St. Louis U., 1930; m. Mary Helen Clark, May 25, 1929; children—Joan, Sidney Allen, Helen, Robert Joseph, Boggs. Faculty, St. Louis, 1930-—, prof. biochemistry, 1952-—. Mem. Am. Soc. Biochemists, Soc. Endocrinology, Am. Chem. Soc., Soc. for Exptl. Biology and Medicine, Sigma Xi, (past chmn.). Co-author: Vitamins and Hormones, 1946; also numerous articles. Research on isolation of female sex hormones, bioassay, metabolism of female sex hormones, isolation and characterization of vitamins K1 and K2, hormone effect on pyrmidine biosynthesis. Home: 32 Hill Dr., Kirkwood, Mo. 63122. Office: 1402 S. Grand St., St. Louis 63104.*

THAYER, Thomas Prence, Am. geologist; b. Scarsdale, N.Y., May 22, 1907; s. James Warren and Stella (Moore) T.; B.A., U. Ore., 1929; M.A., Northwestern U., 1931; Ph.D., Cal. Inst. Tech., 1934; m. Harriet Marjorie Clark, Nov. 27, 1931; children—Thomas Clark, Carolyn. Instr., U. Nev., 1934-35; with U. S. Engr. Dept., 1935-37; with U. S. Geol. Survey, 1937-—, research geologist, Washington, 1964-—. Cons., Liberia Mining Co., 1949-50. Decorated Order of Star of Africa (Liberia). Mem. Geol. Soc. Am., Soc. Econ. Geologists, Am. Geophys. Union, A.A.A.S. Research and publs. on chromite in U. S. and Cuba, iron ore in Liberia; found evidence that chromite-bearing alpine-type peridotites were formed in the mantle with asso. gabbroic rocks by fractional crystallization and re-emplaced into the crust as crystal mushes, that soda-rich dioritic rocks are part of magmatic sequence; prin. designer magnifying single-prism stereoscope for field use. Home: 4818 Essex Av., Chevy Chase, Md. 20015. Office: Resources Research Br., U. S. Geol. Survey, Washington 20242.*

THAYER, William Sydney, Am. physician; b. Milton, Mass., June 23, 1864; s. James B. and Sophia B. (Ripley) T.; A.B., Harvard, 1885 (Phi Beta Kappa; pres. 1929), M.D., 1889; LL.D., Washington Coll., Chestertown, Md., 1907, Edinburgh U., 1927, McGill U., 1929; hon. Dr. U. Paris, 1928; Sc.D. from U. Chgo.; m. Susan Chisolm Read, Sept. 3, 1901. Vis. phys. Johns Hopkins Hosp.; prof. emeritus of medicine, Johns Hopkins U. (Phi Beta Kappa). Mem. Bd. Overseers, Harvard, two terms; mem. Bd. Trustees, Carnegie Inst. of Washington, 1929. Fellow Am. Acad. Arts and Scis.; mem. numerous Am. and fgn. socs. Awarded distinction badge, Red Cross of Russia, 1918; D.S.M. (U. S.), 1919; Comdr. Legion of Honor, France, 1928; Bright medalist, Guy's Hosp., London, 1927. Author: Lectures on the Malarial Fevers, 1897; (with Dr. Hewetson) The Malarial Fevers of Baltimore (Johns Hopkins Hosp. Reports), 1895; Studies on Bacterial Endocarditis (pub. by same), 1925; America—1917, and other Verse, 1926. Research on circulatory system; described 3d heart sound, heart murmurs and block, arteriosclerosis, angina pectoris, bacterial endocarditis; studied blood in malaria, leukemia, thyphoid fever. Died Washington, Dec. 10, 1932.

THEILER, Max, virologist, physician; b. Pretoria, South Africa, Jan. 30, 1899; s. Arnold and Emma (Jegge) T.; student U. Capetown, 1917-18; L.R.C.P., Mem. Royal Coll. Surgeons, St. Thomas' Hosp., 1922; D.T.M. and H., London Sch. Trop. Med., 1922; m. Lillian Graham, Feb. 18, 1928; 1 dau., Elizabeth. Came to U. S., 1922. Asst., instr. dept. tropical medicine, med. sch. Harvard, 1922-30; staff mem. Rockefeller Found., N.Y.C., since 1930, dir. div. medicine and pub. health labs. since 1951; prof. Yale U. Sch. Med., 1964-—. Awarded Chalmers' medal, 1939; Flattery medal, 1945; Lasker award, 1949; Nobel prize in physiology and medicine, 1951. Mem. Am. Soc. Tropical Medicine, Harvey Soc., Royal Soc. Tropical Medicine and Hygiene. Author chpts. med. books; also numerous sci. articles. Developed 17D vaccine against yellow fever in humans; research on arthropod-borne virus, monkey encephalomyelitis; viruses isolated in the tropics. Address: Yale, 1303 A Yale Sta., New Haven 06520.

THEIMER, Otto Helmut, physicist; b. Vienna, Austria, Feb. 22, 1918; s. Hans and Josephine (Schanda) T.; B.S., U. Vienna, 1940; M.S., Tech. U. Munich, 1943, Ph.D., 1945; m. Rose Blattner, Nov. 19, 1945; children—Marvin M., Eileen I. Came to U. S., 1955, naturalized, 1961. Asso. prof. U. Okla., 1955-59; research prof. N.M. State U., Las Cruces, 1959-—; vis. staff mem. Los Alamos Sci. Lab., 1965-—; vis. scientist Institut fur Plasma Physics, Munich, 1965-66. Fellow Am. Phys. Soc., A.A.A.S.; mem. Am. Assn. Physics Tchrs., Sigma Xi. Research and numerous publs.. in structure of matter, properties of solids and plasmas by means of light scattering, high temperature fusion plasmas. Home: 2000 Turrentine Dr., Las Cruces, N.M. 88001.*

THEKAEKARA, Matthew Pothen, physicist; b. Changanachevvy, India, Mar. 21, 1914; s. Pothen Chacko and Mary (Kannam) T.; A.B., Madras U., 1937, M.S., 1939; Ph.D., Johns Hopkins U., 1956. Came to U. S., 1952, naturalized, 1962. Prof., chmn. physics Loyola Coll., Madras U., 1948-52; research asst., instr. Johns Hopkins U., Balt., 1952-57; faculty Georgetown U., Washington, 1957-64, asso. prof., 1960-64; aerospace technologist for solar studies Goddard Space Flight Center, NASA, 1964-—. Dir. summer conf. Coll. Profs. Astro-geophysics, 1960-62; dir. Inservice Inst., 1959-63; vis. prof. Pace Coll., N.Y., 1965. Recipient Jubilee Gold medal, French Acad. medal, 1938. Mem. Am. Phys. Soc., Optical Soc. Am. (pres. nat. capitol sect. 1963), A.A.A.S., Am. Assn. Physics Tchrs., Sigma Xi, Phi Beta Kappa. Author: Practical Physics, 2 vols., 1942, 44; Solar Constant, 1965; also numerous articles. Editor: (with others) Symposium on Recent Advances in Astrogeophysics, 3 vols., 1962-64. Research in spectra of xenon and other rare gases, precision measurement of wavelengths for absolute standard of length, automatic techniques in spectroscopy and interferometry, study of sources for solar simulation, techniques for measurement of radiant energy, mercury pool as precision tool for alignment of instruments for Orbiting Solar Obs. Home: 1527 Potomac Av., Balt. 21227. Office: NASA Goddard Space Flight Center, Greenbelt, Md. 20771.*

THELANDER, Hulda Evelin, Am. pediatrician; b. Little Falls, Minn., Jan. 11, 1896; d. John August and Ida (Olson) Thelander; M.A., U. Minn., 1924, M.D., 1925. Practice medicine specializing in pediatrics, China, 1925-26, San Francisco, 1926-—; chief dept. pediatrics Childrens Hosp., San Francisco, 1951-61; clin. prof. pediatrics U. Cal. at San Francisco, 1949-63, also Stanford Med. Schs. Recipient Muscular Dystrophy award Muscular Dystrophy Assn., 1958, Humanitarian award Variety Club, Internat., 1964. Diplomate Nat. Bd. Pediatrics. Mem. Am. Acad. Pediatrics, Am. Pediatrics Soc., Am. Acad. Neurology, Am. Acad. Cerebral Palsy, Am. Assn. Mental Deficiency, World Med. Assn., Sigma Xi, Alpha Omega Alpha. Research and numerous publs. on pediatrics, child devel., contagious diseases and chronic handicaps including mental retardation. Home: 40 Seafirth Pl., Tiburon, Cal. 94920. Office: 3641 California St., San Francisco 94118.*

THEMISON OF LAODICEA, physician; flourished in Rome under Augustus, 31 B.C.-14 A.D.; pupil of Asclepiades; founder methodical sch. of medicine; 1st to study chronic diseases methodically; accepted Asclepiades' theory of corpuscles; introduced concept of communia (excessive or too little excretion which hinders free movement of atoms), also new drugs, use of leeches into medicine; precursor of Thessalos of Tralles.

THEMO, Judaei, German physicist; natural philosopher; b. Westphalia, Germany; flourished 14th century; grad. Sorbonne, Paris, France, 1349; disciple of Albert of Saxony. Lived in Münster, Westphalia; went to Paris; proctor English nation, 1353, 55, 56, treas., 1357-61; German emissary to Pope Innocent VI at Avignon, France, 1359-60. Author commentary on Aristotle's Meteorologica (discusses gravity, densities of four elements: fire, air, water, earth, influences of sun and moon on tides, Roger Bacon's theory of vision). Several of his ideas can be found in Leonardo da Vinci's treatise Dal moto e misura dell'acqua.

THÉNARD, Louis-Jacques, French chemist; b. La Louptiere, Aube, France; b. May 4, 1777; studied chemistry under Vauquelin, Paris; 1 son, Arnold-Paul-Edmond. Asst. in Vauquelin's lab.; asst. prof. École Polytechnique, 1798; prof. chemistry Coll. France, 1802; dean Faculté Des Sciences, Paris, 1821; chancellor U. Paris, 1832; mem. Chamber of Deps., 1828-32. Fellow Royal Soc., 1824; mem. French Acad. Scis. (v.p., 1822, pres., 1823); Académie de Médecine. Author: Traité Élementaire de Chimie (standard chemistry text for 25 years), 1813-16. Discovered hydrogen peroxide, 1818, also cobalt ultramarine compound used to stain porcelain blue (Thenard's Blue), discovered (with Gay-Lussac) Boron and process for preparing sodium and potassium, and proved hydrogen and oxygen is present in caustic soda and potash; research on composition of metallic oxides, chemistry of cobalt compounds. Died Paris, June 21, 1857.

THEOBALD, Samuel, Am. ophthalmologist, otologist; b. Baltimore, Nov. 12, 1846; s. Elisha Warfield and Sarah Frances (Smith) T.; M.D., U. Md., 1867; studied ophthalmology under Arlt and Jaeger, in Vienna, and at Royal London Ophthalmic Hosp. and otology under Politzer, in Vienna, 1870-71; m. Caroline Dexter DeWolf, Apr., 1867. Began practice in Baltimore, 1867; clin. prof. ophthalmology and otology, 1894-1912, clin. prof. ophthalmology, 1912-25, prof. emeritus of ophthalmology, Jan. 1, 1925—, Johns Hopkins; ophthalmic surgeon, Johns Hopkins Hosp., 1889-1925, vis. ophthalmologist; ophthalmic surgeon, Baltimore Eye, Ear and Throat Charity Hosp., 1882—; cons. ophthalmic and aural surgeon, S. Baltimore Gen. Hosp. Pres., Am. Ophthal. Soc., 1910. Author: Prevalent Diseases of the Eye, 1906. Invented lachrymal probes, 1877; introduced boric acid in treatment diseases of eye, 1880; studied clin. use cocain as anesthetic. Died Dec. 20, 1930.

THEOCARIS, Pericles Stavros, Greek mech. engr.; b. Athens, Greece, Sept. 24, 1921; s. Stavros P. and Zoe (Fussekis) T.; grad. Mech. and Elec. Engr., Athens Nat. Tech. U., 1948; D.Applied Sci., U. Brussels (Belgium), 1952; D.Phys.Sc., Sorbonne, Paris, France, 1953. Asst. applied mechanics Athens Nat. Tech. U., 1948-51, prof., dir. Lab. Testing Materials, 1960—; vis. research asso. Mass. Inst. Tech., 1951; vis. research fellow Ill. Inst. Tech., Chgo., 1957-58; vis. asst. prof. Brown U., Providence, 1958-59; vis. prof. Pa. State U., 1963-64. Decorated Golden Cross of Bravery, Gold Crosses Royal Orders of George I and Phoenix. Mem. Am. Soc. M.E., Soc. Exptl. Stress Analysis, Internat. Assn. Bridge Structural Elements, Sigma Xi. Author: Moiré Fringes in Strain Analysis, 1967; Elements of Experimental Stress Analysis, 1967; also numerous articles. Research on theory of elasticity, exptl. stress analysis, plasticity, rheology, applications of interferometric methods in strain analysis. Home: 3A Stavropoulou St., Athens (814) Greece.*

THEODORICUS TEUTONICUS DE VRIBERG, see Dietrich of Freiberg.

THEODOROS OF CYRENE, mathematician; b. circa 460 B.C.; pupil of Protagoras; tchr. of Plato and Theaetetos; lectured on incommensurable line segments; classified all line segments which produce commensurable squares; studied problems of squares and sq. roots, also of proportionality; showed that sq. roots of 3 and other non-sq. numbers to 17 are irrational; demanded rigor in math. proofs. Died 369 B.C.

THEODOROU, Nicos, Greek physicist; b. Eretria, June 25, 1899; s. Panagiotis and Mata (Magou) T.; student U. Athens, U. Munich; Dr.ès.sci.; m. Veta Paspali, Apr. 18, 1954; Prof., dir. high sch.; prof. indsl. tech. Piraeus Tech. Sch. Mem. Assn. Greek Physicists (pres.), Athens Acad. Sci. Author: Kompendium der modernen Physik, 1935; Viscosität wässriger Elektrolytlösungen, 1938; Kosmische Strahlung, 1948; Radio Indikatoren; Physik Vorlesungen, 1962. Home: Odos Psychari 3, Athens, Greece. Office: Odos Karaoli 40, Piraeus, Greece.

THEODORSON, George A., Am. sociologist; b. N.Y.C., Oct. 19, 1924; s. Achilles George and Anna (Debos) T.; B.A., Cornell U., 1950, M.A., 1951, Ph.D., 1954; postgrad. U. Chgo. Family Study Center, 1953-54; m. Lucille Ann, Sept. 6, 1950; 1 dau., Carol Jean Ann. Faculty, U. Buffalo, 1954-56; faculty Pa. State U., University Park, 1956—, prof. sociology, 1966—, dir. grad. studies sociology and anthropology, 1957-58, 1965—. Fulbright prof. U. Rangoon, 1958-59, U. Vienna, 1962-63. Mem. Am. Sociol. Assn., Eastern Sociol. Soc., Burma Research Soc. (life). Editor, contbr. Studies in Human Ecology, 1961. Author: (with A. Theodorson) Modern Dictionary of Sociology, 1968. Contributor of articles and book revs. to profl. publs. Research on urbanization and industrialization of Western and non-Western socs., family studies, devel. social cohesion and role of hostility in social adjustment, human ecology. Home: 259 E. Waring Av., State College, Pa. 16801. Office: Dept. Sociology and Anthropology, Pa. State University, University Park, Pa. 16802.*

THEODOSIO OF BITHYNIA, Greek mathematician, astronomer; b. Bithynia, flourished circa 100 B.C.; author: Spherics; On Habitations; On Days and Nights; provided math. basis for geography; compiled early discoveries on spherical geometry; divised astron. tables for various sections of earth; invented a universal sun-dial.

THEOPHRASTOS, Greek botanist, philosopher; b. Eresos, Lesbos, circa 372 B.C; s. Melantas; student of Plato and Aristotle; Aristotle's successor in conducting the Lyceum, and as leader of Peripatetics, also continued work in Aristotelian biology. Author: The Characters; Enquiry into Plants; Aetiology of Plants; From the Metaphysics; also treatises. Described over 500 plants; considered founder of botany as systematic study; known for series of caricatures depicting various human ethical types; wrote on numerous subjects including mineralogy, meteorology, physiology, physics, philosophy. Died Athens, circa 287 B.C.

THEORELL, Axel Hugo Teodor, Swedish biochemist; b. Linköping, Sweden, July 6, 1903; s. Ture and Armida (Bill) T.; M.D., Caroline Inst., Stockholm, 1930; Dr. (hon.), U. Sorbonne, Paris, U. Brazil, U. Pa., Université Libre de Bruxelles; Dr. (honorary), University of Louvain, Belgium; m. Margit Alenius, 1931; 3 sons. Lectr. physiol. chemistry Caroline Inst., Stockholm, 1930-32; asst. prof. med. chemistry U. Uppsala, 1932-36; with Prof. Otto Warburg, Kaiser Wilhelm Institut für Zellphysiologie, Berlin-Dahlem, 1933-35; prof., dir. dept. biochemistry Nobel Med. Inst., 1937—. Chairman Stockholm Symphony Soc.; chmn. board directors Wenner-Gren Society, Wenner-Gren Center Foundation, Stockholm. Recipient of Nobel prize in physiology and medicine, 1955. Fellow Royal Soc., 1959; mem. Swedish Acad. Scis., Swedish Acad. Engring. Scis., Royal Danish Acad. Sci. and Letters, Norwegian Acad. Sci., Am. Acad. Arts and Scis., l'Accademia Nazionale del XL of Rome, Swedish Soc. of Physicians and Surgeons (pres. 1957-58; hon. mem.), Swedish Soc. Med. Research, Swedish Chemists' Assn. Sci. Council Swedish Bd. Health, Swedish Acad. Music, Nat Acad. Scis., Am. Philos. Soc., Polska Akademia Nauk Warszawa. Research on oxidizing enzymes of cell respiration; 1st to crystallize myoglobin, 1932; purified, crystallized and reversibly split "yellow" enzyme into protein and coenzyme parts, 1934; work on alcohol dehydrogenases. Home: Sveavägen 166, H, Stockholm Va, Sweden. Office: Medicinska Nobelinstitutet, Stockholm 60, Sweden.

THESLEFF, Stephen Wilhelm, pharmacologist; b. Helsinki, Finland, Jan. 6, 1924; s. Wilhelm A. and Mary af (Schulten) T.; M.D., Caroline Inst., Stockholm, Sweden, 1950, Ph.D., 1952; m. Ulla Ericson, Apr. 7, 1951; children—Peter, Jan. Asst. prof. pharmacology Caroline Inst., 1952; Rockefeller Research fellow U. Ill., 1952-53; asso. prof. pharmacology Sch. Dentistry, Malmö, Sweden, 1953-61; vis. prof. dept. zoology Duke, Durham, N.C., 1961-63; prof. pharmacology U. Lund (Sweden), 1963—. Fellow N.Y. Acad. Sci.; mem. Internat. Brain Research Orgn., Am. Soc. Pharmacology and Exptl. Therapeutics, Physiol. Soc. (Britain), Am. Acad. Neurology. Research and numerous publs. in neuromuscular transmission and trophic influence of innervation. Home: 4 Spoleg., Lund, Sweden.*

THESSALOS OF TRALLES, Greek physician; came from Lydia, flourished during time of Nero and Trajan practiced medicine in Rome; belonged to Methodist sch. medicine; completed writings of Themison; developed new methods of treatment; introduced clinical instruction; helped simplify med. theory; distinguished between acute and chronic illnesses.

THÉVENOT, Melchisédech, French physician; b. Paris, circa 1620; sent to mission to Rome, 1652; apptd. librarian Royal Library, 1684; a founder French Acad. Scis. Author: Account of Many Curious Voyages hitherto unpublished, 2 vols., 1672; Bibliotheca Thevenotiana (catelogue of collection of Royal Library). Described spirit-level (air- or bubble-level) method of directly measuring altitude, 1661, thus permitting constrn. of compact leveling instruments and clinometers; noted for knowledge of Oriental langs. Died Issy, nr. Paris, Oct. 29, 1692.

THIBAUD, Jean, French physicist; b. Lyons, France, 1901; ed. Lyons Faculty of Scis.; became engr. Ecole supérieure d'électricité, Paris, also asst. to Maruice de Broglie, 1921; named prof. Lyons Faculty of Scis., 1935; tchr. Sorbonne, 1940; founder, dir. Inst. Am Atomic Physics, Lyons. Author: La spectrographie des rayons X, 1925; Vie et transmutations des atomes, 1942; Energie Atomique et univers, 1945; Puissance de l'atome. Research on reactions of disintegration and transmutation; realized (at same time as Holweck) relationship between x-rays and ultraviolet radiation; studied annihilation of matter and prodn. of photons by bombarding platinum with positive electrons, 1933. Died Lyons, 1960.

THIBAUT, J. W., Am. psychologist; b. Marion, O., Apr. 30, 1917; s. Ralph G. and Marie (Walter) T.; AB., U. N.C., 1939; Ph.D., Mass. Inst. Tech., 1949; m. Ann Elliott Hommann, Jan. 5, 1944; children—Constantia de Massot, Charles Hommann. Asst. prof. Boston U., 1949-51; lectr., research asso. Harvard, 1951-53; faculty U. N.C., Chapel Hill, 1953—, prof. psychology, 1955—, chmn., 1960-66. Cons. NSF, 1957-60, Nat. Inst. Mental Health, 1961-63. Fellow Center for Advanced Study in the Behavioral Scis., Palo Alto, Cal., 1956-57; fellow Sorbonne, Paris, France, 1963-64. Fellow Am. Psychol. Assn., Soc. for Psychol. Study Social Issues. Author: (with H. H. Kelley) The Social Psychology of Groups, 1959; also articles, chpts. in books, monographs. Editor: Jour. Exptl. Social Psychology, 1965—. Research on bargaining and negotiation behavior using lab. expts. cross-nat. comparisons of bargaining behavior. Home: 1004 Highland Woods, Chapel Hill, N.C. 27514.*

THIEBLOT, Louis Philippe Félix, French biologist; b. Paris, July 12, 1909; s. Victor Octave and Georgette (Paillion) T.; M.D., Faculty of Medicine, Clermont, 1948; m. Gabrielle Thieblot, July 6, 1936; 1 son, Philippe. Practice of medicine, Ivry, France, 1938-49; prof. physiology Faculty of Medicine, Clermont, 1955—, chief of dept. biology, 1963—. Dir., Regional Inst. Sports. Decorated Palmes Academques. Mem. Endocrinol. Soc., Physiol. Soc., others. Author: Pineal Gland, 1955; also numerous articles. Research on physiology of pineal body, isolation of antigonadotropic factor. Home: 2 Rue de Salins, Clermont, Ferrand 63, France.*

THIELE, Johannes, German chemist; b. 1865. Prof. in Munich and Strasbourg. Research on organic chemistry; discovered semicarbazide, 1894; research on guanidine derivatives, 1892, tetrazole derivatives, 1895, nitramide, 1895, unsaturated lactones, 1899-1902; discovered derivatives of fulvene, 1900; advanced theory of partial valencies to explain addition reactions to double bonds and constitution of benzene. Died 1918.

THIELE, Otto Wolfgang, German biochemist; b. Cologne, Germany, Jan. 25, 1917; s. Otto and Mathilde (Kneip) T.; Dr.med., U. Cologne, 1940; student univs. Giessen; Dipl.-Chem., U. Göttingen, 1961; m. Christa Jäger, July 29, 1952; children—Lothar W., Jutta C. Asst., U. Cologne, 1949-52, U. Giessen, 1952-55; asst. Physiol. Chemistry Inst., U. Göttingen, 1955-58, faculty, 1958—, unscheduled prof., 1964—. Mem. Soc. Biol. Chemists, Soc. German Chemists, Soc. Clin. Chemists. Research and numerous articles on biochemistry of lipids, particularly plasmalogens and bacterial lipids. Address: 13 Ludwig-Beck-Strasse, 34 Göttingen, West Germany.*

THIELER, Sir Arnold, veterinarian; b. Frick, Switzerland, Mar. 26, 1867; ed. Aarau, Zurich, Switzerland; D.Med.Vet., Dr.Phil., Berne; D.Sc. (hon.), Syracuse; hon. D.Vet.Sci., U. S. Africa; m. Emma Jegge, 1893; 2 sons, 2 daus. Went to S. Africa, 1891; govt. vet. surgeon to So. African Republic, 1896-1900; govt. vet. bacteriologist, Transvaal, 1900-10; dir. vet. research Union S. Africa, 1910-18; dir. lamsierte research Union S. Africa, 1918-20, dir. vet. edn. and research, 1920-27; prof. tropical vet. medicine, Transvaal U. Coll., 1920-27; dean Faculty Vet. Sci., S. Africa, 1923-27; ret., 1927. Recipient grant and medal S. African Assn. Advancement Sci., Scott medal S. African Biol. Soc., Medaille d'or Société de pathologie exotique, Gold medal Royal Agrl. Soc. Eng., Budapest Gold medal for research in vet. sci. Fellow Royal Soc. S. Africa (hon.), Royal Soc. Medicine (hon.), Royal Soc. Tropical Medicine (hon.), Am. Soc. Tropical Medicine (hon.); mem French Acad. Scis. (corr. 1930). Research and publs. on tropical diseases of domesticated animals. Died London, July 24, 1936.

THIERET, John William, Am. botanist; b. Chgo., Aug. 1, 1926; s. Hans John and Lorena (Groves) T.; B.S., Utah State U., 1950, M.S., 1951; Ph.D., U. Chgo., 1953; m. Mildred Ann Wolf, Mar. 13, 1950; children—Robert Gene, Nancy Lee, Richard Lane, Jeffrey Grant. Curator econ. botany Field Mus. Natural History, Chgo., 1953-62; asso. prof. biology U. Southwestern La., Lafayette, 1962-67, prof. biology, 1967—. Adviser botany Ency. Britannica, 1959—. Mem. Soc. Econ. Botany. Research and numerous publs. on flora and vegetation of subarctic Gt. Plains of N.Am., flora of La., taxonomy of Gramineae and Scrophulariaceae. Home: 107 Brighton Dr., Lafayette, La. 70501.*

THIERSCH, Karl, German surgeon; b. Munich, Apr. 20, 1822; prof. in Erlangen and Leipzig; author: Der Epitheliakrebs, 1865; introduced epidermal grafting in plastic surgery; research on histogenesis of cancer; recognized Virchow's error in believing in connective tissue origin of cancer; discovered epithelial origin of cancer. Died Leipzig, Apr. 28, 1895.

THIERY, Michel, Belgian physician; b. Ghent, Belgium, Nov. 14, 1924; s. Michel and Augusta (de Taeye) T.; B.S. maxima cum laude, U. Ghent, 1945, M.D. maxima cum laude, 1949; m. Huguette Thiery, Aug. 28, 1957; 1 dau., Dominique. Faculty, U. Ghent, 1954-63, lector obstetrics/gynecology, 1959-63; prof. obstetrics, chmn. dept. obstetrics Ghent State U., 1963—, dir. obstet. clinic. Mem. Société royale belge de Gynécologie et d'Obstétrique (pres. 1965—). Author: Het experimentele carcinoma colli uteri, 1962; also numerous articles. Research on localization of enzymes in female genitalia and tumors, morphology and behavior of experimental cancer in neck of womb of lab. animals, blood chemistry of fetus, effects of abnormal labor on fetus. Home: Aan de Bocht 6, Ghent, Belgium.*

THIESSEN, Garrett William, Am. chemist; b. Stanwood, Ia., Jan. 9, 1902; s. William F. and Susan (Steiner) T.; A.B., Cornell Coll., Mt. Vernon, Ia., 1924; M.S., U. Ia., 1925, Ph.D., 1927; m. Isabel Davidson, July 10, 1940. Faculty, Geneva Coll., 1927-30; faculty Monmouth (Ill.) Coll., 1930—, head dept. chemistry, 1952-63, Pressly prof., 1954—. Recipient medal Am. Assn. Mfg. Chemists, 1957. Fellow A.A.A.S.; mem. Am. Chem. Soc. (chmn. Ill.-Ia. sect.), Electrochem. Soc. (chmn. electro-organic div.), Ill. Acad. Sci. (pres.), Phi Beta Kappa, Sigma Xi, Phi Lambda Upsilon. Author: (with Babor) How To Solve Problems in Physical Chemistry, 1944. Research and publs. on Kolbe electrolysis, conductivity of organic systems, small-scale identifications. Home: 408 N. 10th St., Monmouth, Ill. 61462.*

THIMANN, Kenneth Vivian, plant physiologist; b. Ashford, Kent, Eng., Aug. 5, 1904; s. Israel Phoebus and Muriel (Harding) T.; B.Sc., Imperial Coll., London U. (Eng.), 1924, A.R.C.S., 1924, D.I.C., 1925,

Ph.D., 1928; A.M. (hon.) Harvard, 1939; D.Sc., U. Basel (Switzerland), 1960; Doc.Hon., U. Clermont-Ferrand, France, 1961; m. Ann Mary Bateman, Mar. 20, 1929; children—Vivianne (Mrs. Jacob Nachmias), Karen (Mrs. James H. Romer), Linda (Mrs. John Thompson Dewing, Jr.). Came to U. S., 1930, naturalized, 1941. Instr., Kings Coll. for Women, London, 1928-30, Cal. Inst. Tech., 1930-35; lectr. Harvard, 1935-36, faculty, 1936-65, prof. biology, 1946-60, Higgins prof. biology, 1960-65; prof. biology, provost Crown Coll., U. Cal. at Santa Cruz, 1965—. Recipient Frank Hopkins prize, Imperial Coll., 1924. Mem. Soc. Gen. Physiologists (past pres.), Am. Soc. Plant Physiologists (past pres., Stephen Hales prize 1936, Barnes hon. life mem.), Am. Soc. Naturalists (past pres.), Bot. Soc. Am. (past pres.), Soc. Developmental Biology (past pres.), Am. Inst. Biol. Scis. (past pres.), Nat. Acad. Scis. (councillor), Am. Acad. Arts and Sci., Philos. Soc., Am. Soc. Biol. Chemists, Biochem. Soc. (Gt. Britain); hon. fgn. mem. Nat. Acad. Lincei (Rome, Italy), Akademie Leopoldina (Halle, Germany), Bot. Soc. Japan, Netherlands Bot. Soc., Rumanian Nat. Acad. Sci. Author: Phytohormones (With F. W. Went), 1938; The Life of Bacteria, 1955, 2d edit., 1963; also numerous articles. Co-discoverer plant growth hormone Auxin and its functions in plants; research on growth, plant biochemistry and effects of light and gravity in higher and lower plants and on formation of anthocyanins. Home: 36 Pasatiempo Dr., Santa Cruz, Cal. 95060.*

THIRRING, Walter Eduard, Austrian physicist; b. Vienna, Austria, Apr. 29, 1927; s. Hans and Antonia (Krisch) T.; Ph.D., U. Vienna, 1949; m. Helga Georgiades, Dec. 22, 1952; children—Klaus, Peter. Scholar, Dublin Inst. Advanced Studies, 1949-50; fellow Glasgow (Scotland), U., 1950; fellow M.P.I., Gottingen, Germany, 1950-51; UNESCO fellow E.T.H., Zurich, Switzerland, 1951-52; asst. U. Bern (Switzerland), 1952-53, docent, 1954-56, prof., 1958-59; mem. Princeton Inst. Advanced Studies, 1953-54; vis. prof. U. Wash., Seattle, 1957-58, Mass. Inst. Tech., Cambridge, 1956-57; dir. Inst. for Theoretical Physics, prof. U. Vienna, 1959—. Mem. Austrian Acad. Sci. Author: Einführung in die Quantenelektrodynamik, 1954; Principles of Quantum Electrodynamics, 1958; (with E. M. Henley) Elementary Quantum Field Theory, 1962; also articles. Research on theory of elementary particles, theory of gravitation, solid state physics. Home: Wien 19, Daringerg, 12a/5, Osterreich.*

THIRY, Simon Gabriel, physician; b. Sophia, Bulgaria, Apr. 21, 1923; s. Joseph and Clemence (Germeau) T.; Diploma in Humanities, Royal Athenaeum Seraing, 1941; grad. with honors in natural scis. and med. studies U. Liège (Belgium), 1944, Docteur en Médecine with honors in surgery and childbirth, 1948; Diploma of Fgn. Asst., U. Paris, 1950; m. Lucy Nicolas, June 28, 1950; children—Daniel, Sabine. With dept. surgery Liege Mil. Hosp., 1948-49; staff Clinic Nervous Disorders, la Salpétrière Hosp., Paris, 1949-50; scholarship candidate Nat. Found. Sci. Research, 1950-52; asst. surg. clinic U. Liege, 1952-58, head clinic neurosurgery, 1958-60; head dept. neurosurgery St. Joseph's Clinic, Liège, 1960—. Research fellow dept. neurosurgery Mass. Gen. Hosp., Boston, Harvard Med. Sch., also Labs. for Atomic Research, Brookhaven, L.I., N.Y., 1956-57. Research and numerous publs. on effects of intermittent light stimulation in electroencephalography, cerebral thrombophlebitis, epilepsy, meningitis, hydrocephalus, tonsillectomy; developed push-pull detection method of monopolar recording in electroencephalography. Home: 590 route du Coudroy, Liège. Office: 35 boulevard d'Avray, Liège, Belgium.*

THISELTON-DYER, William Turner, Brit. botanist; b. Westminster, Eng., July 28, 1843; s. William George and Catherine Jane (Firminger) T.-D.; ed. King's Coll., London, Eng., 1861-63, Christ Ch., Oxford (Eng.) U., 1863-65; B.Sc., M.A., LL.D., Sc.D., D.Sc., Ph.D.; m. Harriet Ann Hooker, 1877; 1 son, 1 dau. Prof. natural history Royal Agrl. Coll., Cirencester, Eng., from 1868; prof. botany Royal Coll. Sci., Dublin, Ireland, 1870-72, Royal Hort. Soc., South Kensington and Chiswick, Eng., 1872-75; asst. dir. Kew (Eng.) Gardens, 1875-85, dir., 1885-1905, also founder Bull., 1887; bot. adviser to sec. of State for Colonies, 1902-06. Demonstrator for T. H. Huxley, Royal Sch. Mines, South Kensington, 1872; royal commr. Melbourne Centennial Exhbn., 1888, Paris Internat. Exhbn., 1900, St. Louis Exhbn., 1904. Fellow Royal Soc. (v.p. 1896-97); mem. Brit. Assn. (pres. sect. D. at Bath 1888, Ipswich new bot. sect. K 1895). Author: (with Henry Trimen) Flora of Middlesex, 1869. Editor: How Crops are Grown (S. W. Johnson), 1869; Text-Book of Botany (Julius von Sachs, translated by A. W. Bennett), 1875; Flora Carpensis (William Henry Harvey), 1896-1925; Icones Plantarum, 1896-1906; Flora of Tropical Africa, 1897-1913; Bot. Mag., 1905-06. Contbd. revised vocabulary for Greek plant names Greek-English Lexicon (Liddell and Scott), 9th edit., also bot. chpts. for A Companion to Greek Studies, 1905, A Companion to Latin Studies, 1910. Helped to develop cocoa and rubber cultivation in Ceylon; research, publs. on cycads. Died Witcombe, Eng., Dec. 23, 1928.

THODAY, John Marion, English geneticist; b. Chinley, Eng., Aug. 31, 1916; s. David and Gladys (Sykes) T.; B.Sc., U. Coll. N. Wales, 1939; Ph.D., Cambridge (Eng.) U., 1948, Sc.D., 1965; m. Doris Rich, July 4, 1950; children—Antonia, Jonathan. Cytologist, Brit. Empire Cancer Campaign, 1946-47; faculty Sheffield U., 1947-59, head new dept. genetics, 1954-59; Arthur Balfour prof. genetics Cambridge U., 1959—. Fellow Royal Soc., 1965; mem. Genetical Soc., Eugenics Soc., Soc. for Exptl. Biology. Research, publs. on biol. effects of X-rays, neutrons and alpha-particles; discovered oxygen effect in radiobiology; studies of genetic effects of artificial selection, especially on genetic diversity of population; identification of individual genes concerned with graded characters. Home: 7 Clarkson Rd., Cambridge, Eng.*

THODE, Henry George, Canadian chemist; b. Dundurn, Sask., Can., Sept. 10, 1910; s. Charles Herman and Zelma Ann (Jacoby) T.; B.S., U. Sask., 1930, M.S., 1932, LL.D., 1958; Ph.D., U. Chgo., 1934; D.Sc., U. Toronto, 1955, U. B.C., 1960, Acadia U., 1960, Laval U., 1963, Royal Mil. Coll. Can., 1964, McGill University, 1966, Queen's University, 1967; m. to Sadie Alicia Patrick, February 5, 1935; children—John Charles, Henry Patrick, Richard Lee. Faculty, McMaster U., Hamilton, Ont., 1939—, pres., vice-chancellor, 1961—. Mem. NRC Can., Def. Research Bd. Can., 1955-61. Dir., exec. com. Atomic Energy Can. Ltd.; bd. govs. Ont. Research Found.; trustee Western N.Y. Nuclear Research Centre, State U. N.Y., Buffalo. Recipient Chem. Inst. Can. medal for distinguished achievement in chemistry, 1957, Tory medal Royal Soc. Can. for achievement in sci. research, 1959; named companion Order of Canada, 1967. Fellow Royal Soc. Can. (past pres.), Royal Soc., 1954; mem. Chem. Inst. Can. (past pres.). Research on mass spectrometry; isotope chemistry and applications to medicine; geochemistry of sulphur isotopes. Home: President's House, McMaster U., Hamilton, Ont., Can.*

THODORIC BORGOGNONI (or Teodorico Borgognoni, Theodoricus Cerviensis), Italian physician; b. Lucca, 1205; pupil of Hugh Borgognoni; Entered Dominican order; penitentiary to Bishop of Bitonto, 1266-98; became Bishop of Cervia, 1266-98; author treatises: De sublimatione arsenici, de aluminibus et salibus; Practica equorum; Cyrurgia (written before middle of 13th century). Improved his father's method of treating wounds, especially in use of wine to cleanse wound, also made improvements in use of spongia soperifera to bring about unconsciousness; studied use of mercury salts for skin diseases. Died Bologna, Dec. 24, 1298.

THOINOT, Léon-Henry, French physician, hygienist; b. Paris, 1858; became intern, 1882, doctor, 1886, prof. legal medicine, 1906; mem. Acad. Medicine. Author: Précis de microbie medicale et vétérinaire; Cours d'hygiène, 1889. Studied problems of hygiene, sanitation, disinfection, epidemic diseases. Died Paris, 1915.

THÖLDE, Johannes, German chemist; b. Frankenhausen, Germany; councillor, salt boiler of Frankenhausen. Author: Bericht der abschewlichen Kranckheit, 1599; (under pretext of transl. from a fictitious Basilius Valentinus) Haliographia . . . (includes descriptions of salt works, also oldest post-classical description and illustration of hydrometer), 1603; Cursus triumphalis antimonii (on use of antimony in fevers), 1604; others.

THOLOZAN, Joseph-Désiré, physician; b. Diego-Garcia, Ile Maurice, Oct. 8, 1820; ed. Paris; with army med. corps; became chief physician, 1862; prof., Val-de-Grace, France; physician to Shah of Persia; mem. French Acad. Scis., 1874, Acad. Medicine. Research on plague and cholera; paved way for ethnol. work of Dieulafoy. Died Teheran, Persia, July 31, 1897.

THOM, Herbert C(onrad) S(chlueter), Am. meteorologist; b. St. Paul, May 21, 1910; s. Robert and Margarette (Schlueter) T.; B.S., George Washington U., 1937, M.S., 1945; m. Marcella Daigle, Dec. 2, 1961. Lab. asst. Nat. Bur. Standards, 1930-34; hydraulic engr. Soil Conservation Service, 1935-38; hydrologic engr. U. S. Weather Bur., Washington, 1938-41, statistician, 1941-42, chief statistics div., 1942-44, chief Ia. Weather Service, 1944-48, climatol. specialist, 1948-51, chief climatologist, 1952-66, sr. research fellow, 1966—; prof. statistics Ia. State Coll., 1944-48; prof. climatology U. Md., 1949-51; vis. prof. statistics Cornell U. 1952; chief climatologist President's Adv. Com. Weather Control, 1955-57; Lockwood lectr. Conn. Agr. Expt. Sta., 1957; NATO Seminar lectr, Copenhagen, 1963-64, Ireland, 1965. U. S. mem. climatol. com. World Meteorol. Orgn., 1952, chmn. working groups statis. methods and applied climatology, 1960-65, chief U. S. delegation climatol. com. meeting, Washington, 1953, del., Washington, 1957, London, 1960, Stockholm, 1965. Recipient Dept. Commerce Silver medal, 1959, Gold medal, 1965, Applied Meteorology award Am. Meteorol. Soc., 1963. Mem Am. Meteorol. Soc., Am. Statis. Assn., Inst. Math. Statistics, Am. Geophys. Union, A.A.A.S., Sigma Xi. Pioneer in application of modern statis. analysis to climatology. Home: 14310 Bauer Dr., Rockville, Md. 20853. Office: Environmental Science Services Administration, Rockville, Md. 20852.*

THOM, William Taylor, Jr., Am. geologist; b. Roanoke, Va., June 9, 1891; s. William Taylor and Elizabeth Porter (Miller) T.; B.S. in Engring., Washington and Lee U., 1913, D.Sc., 1936; Ph.D. in Geology, Johns Hopkins, 1917; m. Rachel Trimble Hoopes, Nov. 4, 1916; children—William Taylor III, Judith Preston (Mrs. David J. Phelps), Elizabeth Pearson (Mrs. Robert Allen Bolster). Mem. exploratory staff U. S. Geol. Survey, Washington, 1912-17, 19-27; faculty Princeton, 1927-56, Blair prof. geology, 1937-56, Blair prof. emeritus, 1956—. Chmn. com. on studies in petroleum geology NRC, 1931-35. Recipient John Fleming medal Am. Inst. Geonomy and Natural Resources, 1957. Fellow World Acad. Art and Sci. (trustee Am. div. 1966—), Geol. Soc. Am.; mem. Am. Inst. Geonomy and Natural Resources Inc. (pres. 1961—), Am. Assn. Petroleum Geologists (hon.), Am. Inst. Mining, Metall. & Petroleum Engrs. (hon. sr.), Am. Geophys. Union (sr.), Am. Geol. Inst. (past dir.), Yellowstone-Bighorn Research Assn. (past pres.), Seismol. Soc., Phi Beta Kappa. Author: Petroleum & Coal—The Keys to the Future, 1929; The Goal of Democracy, 1940; Science and the Future of Mankind, 1961; also articles, bulls. Research on magnitude and human importance of world's mineral fuel resources; forms, dimensions and patterns of arrangement, and orders of magnitude of deformational features of earth's crust with which ore and mineral deposits are causally related; impacts on nat. and cultural devel. induced by irregular and unequal geog. distbn. of mineral resource deposits. Home: 434 Greenwood Av., Trenton, N.J. 08609. Office: Guyot Hall, Princeton, N.J. 08540.*

THOMA, Roy Elliott, Jr., Am. chemist; b. San Antonio, May 12, 1922; s. Roy E. and Michal (Miller) T.; B.A., U. Tex., 1943, M.A., 1948; postgrad. U. Colo., 1949, U. Tenn., 1952-53; m. Nancy Sour, July 4, 1953; children—Roy Elliott, III, Alvin Louis, Asso. prof. San Houston State Coll., Huntsville, Tex., 1948-51; asst. prof. Tex. Technol. Coll., Lubbock, Tex., 1951-52; chemist Oak Ridge Nat. Lab., 1952—. Fellow Am. Ceramic Soc. (editorial com., 1962—, chmn. 1964-66); mem. Am. Chem. Soc., Am. Nuclear Soc., Research Soc. Am. Contbg. author: Ency. Sci. and Tech.; Advances of Science and Technology of Rare Earths, Volume 2, 1966; Handbook of X-Rays, published 1968. Research and publications on application of fluoride systems to nuclear reactor tech.; developed theory of formation of complex fluoride compounds; advances of crystal chemistry of rare-earths. Patentee nuclear reactor fuels, and coolants, reprocessing tech. Home: 119 Underwood Rd., Oak Ridge 37830. Office: P.O. Box X, Oak Ridge Nat. Lab., Oak Ridge 37830.*

THOMALSKE, Günther Reinhold Ernst, German neurosurgeon; b. Hohkirch bei Görlitz, Germany, June 8, 1925; s. Paul Melchior and Eleonore (Pletz) T.; student U. Berlin (Germany), 1943, U. Freiburg (Germany), 1943-44, U. Prague (Czechoslovakia), 1944-45; M.D., U. Marburg (Germany), 1949; m. Anne This, July 11, 1958; children—Christine, Catherine. Pvt. asst. surgeon neurosurg. dept. Pasteur Hosp., Colmar, France, 1952-58; 1st asst. surgeon neurosurg. dept. Clinic for Neurol. Diseases, U. Frankfort (Germany), 1958-61, head surgeon, 1961—. Mem. German, French neurosurg. socs., German EEG Soc., French Soc. EEG and Neurophysiology. Research and publs. on neurophysiology of anesthetic agts., neurosurg. treatment of epilepsy, traumatology, neuroradiology, anesthesia and reanimation problems, psychopharmacology, cranioplastic surgery. Home: 12 Tulpenstrasse, 6078, Neu-Isenburg, Germany. Office: Schleusenweg, Neurosurg. Clinic, 6000 Frankfort/Main, Germany.*

THOMAS, Alan Francis, chemist; b. London, Eng., July 5, 1928; s. Harold and Elizzbeth (Simm) T.; B.Sc., Cxford U., 1952, M.A., D.Phil., 1953; children—Christine Anne, Steven Francis. Postdoctoral fellow NRC Can., Ottawa, Ont., 1954; research staff CIBA AG, Basel, Switzerland, 1955-59; lectr. Leeds (Eng.) U., 1960; research staff Firmenich et Cie., Geneva, Switzerland, 1960—; vis. prof. U. Ga., Athens, 1967. Mem. Chem. Soc. (London), Soc. Chem. Industry (London), Swiss Chem. Soc. Research, publs. on structural analysis of natural products, especially alkaloids, perfumes and flavors; mass spectrometry, especially small organic molecules. Home: 16 ch. de l'Esplanade Vernier-Geneve 1214. Office: Firmenich et Cie, Genève 1200, Switzerland.*

THOMAS, Caroline Bedell, Am. physician; b. Ithaca, N.Y., Nov. 29, 1904; d. Frederick and Mary (Crehore) Bedell; A.B., Smith Coll., 1925; D.Sc. (hon.), 1955; M.D., Johns Hopkins, 1930; m. Henry M. Thomas, Jr., June 23, 1934; children—Henry M., III, Eleanor Carey, Mary Whitall. Faculty, Johns Hopkins, 1935—, asso. prof. sch. medicine, 1952—, lectr. dept. epidemiology Sch. Hygiene and Pub. Health, 1960—; vis. physician Johns Hopkins Hosp., 1945—; practice medicine specializing in cardiovascular problems, Balt., 1947—. Cons. Surgeon Gen. of Army, 1944-46; cons. Smith Coll. Health Com., 1959—, bd. counselors, 1962-66. Recipient James D. Bruce Meml. award in preventive medicine A.C.P., 1957, Elizabeth Blackwell citation N.Y. Infirmary, 1958. Diplomate Am. Bd. Internal Medicine. Fellow A.C.P., Assn. Am. Physicians, Am. Heart Assn., Council for High Blood Pressure Research (chmn. 1961-

62), Council on Epidemiology; mem. A.M.A., Am. Physiol. Soc., Am. Soc. Clin. Investigation, Phi Beta Kappa, Sigma Xi, Alpha Omega Alpha. Author: (with D. C. Ross and E. S. Freed) An Index of Rorschach Responses, 1964, An Index to the Group Rorschach Test, 1965. An Atlas of Figure Drawings, 1966. Publs. on pioneer research in prevention of rheumatic fever; exptl. hypertension; study of the precursors of hypertension and coronary artery disease; preventive medicine. Home: 314 Overhill Rd., Balt. 21210. Office: 725 N. Wolfe St., Balt. 21205.*

THOMAS, Charles Allen, Am. chemist; b. Scott County, Ky., Feb. 15, 1900; s. Charles Allen and Frances (Carrick) T.; A.B., Transylvania Coll., 1920, D.Sc., 1933; M.S., Mass. Inst. Tech., 1924; LL.D., Hobart Coll., 1950; D.Sc., Washington U., 1947, Kenyon Coll., 1952, Princeton University, 1952, Ohio Wesleyan University, 1953, Brown University, 1956; D.Sc., Bklyn. Polytech, 1957; D.Sc., U. Ala., 1958; LL.D., Lehigh U., 1960; D.Sc., St. Louis U., 1965, Simpson Coll., 1967; Dr. Engring., U. Mo., 1965; m. Margaret Stoddard Talbott, Sept. 25, 1926; children—Charles, Margaret T. Thomas, Frances (Mrs. T. R. Martin), Katherine (Mrs. Stephen O'Neil). Research chemist General Motors Research Corporation, 1923-24, Ethyl Gasoline Corporation, 1924-25; president Thomas & Hochwalt Laboratories, 1926-36; vice president, tech. dir., mem. exec. com. Monsanto Co., 1945, exec. v.p., 1947-51, pres., 1951-60, chmn. bd., 1960-65, now dir., chmn. of finance committee; director Met. Life Insurance Company, Rand Corp., Southwestern Bell Telephone Co., First Nat. Bank in St. Louis, St. Louis Union Trust Co. Mem. business council Dept. Commerce; dep. chief Nat. Def. Research Committee, 1942-43, sect. mem. since 1943; mem. Manhattan project, in charge Clinton Labs. Oak Ridge, Tenn.; mem. sci. panel U. S. Rep. to U.N. AEC; apptd. chmn. sci. manpower adv. com. NSRB, 1950, mem. sci. adv. com. ODM, 1951; cons. Nat. Security Council, 1953. Chmn. bd. trustees Washngton University; curator Transylvania College; member of the advisory bd. St. Louis council Boy Scouts of Am. Awarded Medal for Merit, by Pres. Truman, 1946; Gold medal, Am. Inst. Chemists, 1948; Mo. Honor award for Distinguished Service in Engring., 1952; Perkin medal, American section Society Chemical Industry, 1953; Palladium medal American section Society de Chimic Industrielle, 1963. Fellow A.A.A.S., Am. Acad. Arts and Scis.; mem. Nat. Acad. Engring., American Chem. Soc., director, member bd. editors 1937-38, pres. 1948, chmn. bd. dirs., 1950-53; recipient Priestley Medal 1955), Corp. Massachusetts Institute of Technology, National Academy Science, Am. Philos. Soc., American Institute Chem. Engrs., Am. Inst. Chemists, Chem. Soc. London, Soc. Chem. Industries, Phi Beta Kappa, Sigma Xi. Research on rocket propellants; synthetic resins; development of tetraethyl lead; bromine from sea water; synthetic styrene and rubber; effect of alkali metals on combustion. Holds many United States patents. Home: 609 South Warson Road, St. Louis 24. Office: Monsanto Co., 800 N. Lindbergh Blvd., St. Louis 66.

THOMAS, Cyrus, Am. ethnologist, archeologist; b. Kingsport, Tenn., July 27, 1825; s. Stephen and Maria (Rogan) T.; ed. village schs. and acad.; admitted to bar, 1851, and practiced until 1865; m. Dorothy Logan, June 13, 1853; m. 2d, Miss L. V. Davis, Apr. 20, 1865; 6 children, County clerk Jackson Co., Ill., 1850-53; minister Evang. Luth. Ch., 1865-69; asst. on U. S. Geol. and Geog. surveys of Territories, under Ferdinand V. Hayden, 1869-73; prof. natural sciences, Southern Ill. Normal U., 1873-75; state entomologist of Ill., 1874-76; mem. U. S. Entomol. Commn., 1876-77; archeologist U. S. Bur. of Ethnology, 1882—. Author: Synopsis of the Acrididae of North America; Noxious and Beneficial Insects of Illinois (5 vols. reports as state entomologist); Aid to the Study of Maya Codices; The Cherokees and Shawnees in Pre-Columbian Times; Mound Explorations of Bureau of American Ethnology; Prehistoric Works East of the Rocky Mountains; Introduction to American Archaeology, 1898; Numeral Systems of the Mexican and Central American Tribes; The Mayan Calendar Systems; Indians of North America in Historic Times, 1903; (with William J. McGhee) Prehistoric North America, 1905; Languages of Mexico and Central America. Excavations proved that mound builders were Am. indians; pioneer study Maya culture. Died Washington, June 26, 1910.

THOMAS, Dan Anderson, Am. physicist; b. Ooltewah, Tenn., Oct. 1, 1922; s. Daniel Bryson and Blanche (Sylar) T.; B.S., U. Chattanooga, 1945; Ph.D., Vanderbilt U., 1952; m. Margaret Elizabeth Glaze, Mar. 19, 1944; children—Roger Nelson, Rebecca Lynn. Faculty dept. physics U. South, 1949-51, Rollins Coll., Winter Park, Fla., 1952-63; research physicist U. S. Naval Ordnance Lab., White Oak, Md., 1951-52; prof. physics Jacksonville (Fla.) U., 1963—, dean faculty, 1963-67, vice president, dean faculties, 1967—. Cons., U. S. Navy Underwater Sound Reference Lab., 1953-63. Fellow A.A.A.S. (council); mem. Acoustical Soc. Am., Am. Assn. Physics Tchrs., Am. Phys. Soc., Fla. Acad. Scis. (past pres.), Am. Assn. U. Profs., Am. Conf. Acad. Deans, Conf. Acad. Deans So. States, So. Conf. Deans Faculties and Acad. Vice Presidents, Blue Key, Sigma Xi. Research and publs. on beta-ray spectroscopy, electron capture, underwater acoustics, flexure waves in plates, theoretical mechanics. Home: 3272 University Blvd., Jacksonville, Fla. 32211.*

THOMAS, Dorothy Swaine, Am. sociologist, demographer; b. Baltimore, Md., Oct. 24, 1899; d. John Knight and Sarah (Swaine) Thomas; A.B., Columbia U., 1922; Ph.D., U. of London, 1924; m. William I. Thomas, 1935 (dec.). Research asst. Fed. Reserve Bank, 1924; fellow, Social Science Research Council, 1925-26; research, Laura Spellman Rockefeller Found., 1926; research asso. and asst. prof. Teachers Coll., Columbia, 1927-30; research asst. and asso. prof., Yale U., 1931-35, dir. of research in social statistics, 1935-39; staff mem. Carnegie Corp. Study of Negro in Am., 1939-40; Lectr. in sociology, U. of Calif., 1940-41, prof. of rural sociology, 1941-48; research prof. sociology U. Pa., 1948—, co-dir. study of population redistbn. and economic growth, 1952—, co-director of Population Studies Center; vis. professor Social Sci. Inst., U. of Stockholm, Sweden, 1933, 35, 36; cons. N.Y. State Research Found. for Mental Hygiene. Spl. analyst Fed. Emergency Relief Adminstrn., 1935-36; cons. Nat. Resources Com., 1936-37; chmn. com. on migration differentials, dir. Social Sci. Research Council; demographer Tech. Assistance Administration, UN, Bombay, India, 1957; mem. tech. adv. com. on 1960 census, U. S. Bureau Census; mem. Census Adv. Com. on Population statistics; com. on agrl. sci. Department of Agriculture; chmn., Pacific Coast Regional Committee of Social Science Research Council, 1942-46; dir. U. of California. Evacuation and Resettlement Study, 1942-48; cons. demography UN Tech. Assistance Bombay and Cairo, 1963-64; vis. prof. U. Gothenburg (Sweden), 1964; cons. Central Statistical Bd., Sweden, 1965. Hutchinson Research Medallist, London Sch. Economics, 1924. Fellow Am. Statis. Assn.; v.p., 1946, mem. American Sociological Society (pres. 1952), American Philosophical Society (councillor 1966-67), Population Assn. Am. (pres. 1958-59), International Union Sci. Study Population, Phila. Art Alliance. Author: Social Aspects of the Business Cycle, 1925; Child in America (with W. I. Thomas), 1928; Some New Techniques for Studying Social Behavior, 1929; Observational Studies of Social Behavior, 1933; Research Memorandum on Migration Differentials, 1938; Social and Economic Aspects of Swedish Population Movements, 1941; The Spoilage (with R. S. Nishimoto), 1946; The Salvage (with C. Kikuchi and J. Sakoda), 1952; Population Redistribution and Economic Growth, United States, 1870-1950 (with others), Vol. I, 1957, Vol. II, 1960, Vol. III, 1964. Studies on interrelations between economic and social developments and demographic phenomena, especially migration. Home: 118 S. Van Pelt St., Phila. 3.*

THOMAS, E(dward) D(onnall), Am. physician; b. Mart, Tex., Mar. 15, 1920; s. Edward E. and Angie (Hill) T.; B.A., U. Tex., 1941, M.A., 1943; M.D., Harvard, 1946; m. Dorothy Martin, Dec. 20, 1942; children—Edward Donnall, Jeffery A., Elaine. NRC fellow medicine dept. biology Mass. Inst. Tech., 1950-51; instr. medicine Harvard Med. Sch., Boston, also hematologist Peter Bent Brigham Hosp., Boston, 1953-55; research asso. Cancer Research Found. Children Med. Center, 1953-55; physician in chief Mary Imogene Bassett Hosp., also asso. clin. prof. medicine Coll. Phys. and Surg., Columbia U., 1955-63; prof. U. Wash. Sch. Medicine, Seattle, 1963—. Diplomate Am. Bd. Internal Medicine. Mem. Am. Soc. for Clin. Investigation, Assn. Am. Physicians, Am. Soc. Hematology, Am. Fedn. for Clin. Research, Internat. Soc. Hematology, N.Y. Acad. Scis., Western Assn. Physicians. Research and numerous publs. on hematology, marrow biochemistry and irradiation biology including whole body irradiation, transplantation of bone marrow and other organs, extracorporeal irradiation and cross circulation. Home: 1920 92d Av. N.E., Bellevue, Wash. 98004. Office: USPHS Hosp., Seattle 98144.*

THOMAS, Eva Maria Balling, physicist; b. Vienna, Austria, Aug. 30, 1923; d. Franz Karl and Maria (Vort) Balling; brought to U. S., 1939, naturalized, 1944; B.S., Good Counsel Coll., 1943; M.S., Fordham U., 1944; postgrad. Mass. Inst. Tech., 1950—; m. Abdelnour S. Thomas, Mar. 26, 1951; children—Robert, David, Paul, Simon, Mary, Joan. Faculty, St. Elizabeth Coll., 1944-46, Champlain Coll., 1946-47, Newton Coll., 1947-48, Wellesley Coll., 1948-50; research asst. on cosmic ray and ionospheric studies Dr. Victor F. Hess, 1943-45; research asso. isotopic studies Mass. Inst. Tech., 1949-51; physicist A. S. Thomas, Inc., Westwood, Mass., 1955—. Cons. field numerically controlled machine tools, 1963—. Mem. Am. Rocket Soc., Numerical Control Soc., Am. Meteorol. Soc. Research and numerous publs. on dielectric materials, devel. electro-magnetic absorber materials. Developer 4 Pi counter for detection iodine in thyroid gland. Home: 51 Church St., West Roxbury, Mass. 02132. Office: 355 Providence Hwy., Westwood, Mass. 02091.*

THOMAS, Frank Lincoln, Am. entomologist; b. Waltham, Mass., Oct. 16, 1887; s. Charles Albert and Harriet (Dewey) T.; B.Sc., U. Mass., 1910, Ph.D., 1914; m. Mabel Clare Randall Wrenn, Apr. 25, 1917; children—George M. Wrenn, Mabel Claire (Mrs. J. Fred Smith, Jr.), Frank Lincoln, Edith (Mrs. Ashford M. Groves), Helen (Mrs. John V. Perry, Jr.). Asst. entomologist Ala. Expt. Sta., 1915-18, asso. entomologist, 1920-23; faculty Ala. Poly. Inst., 1915-18, acting head dept. zoology and entomology, 1924; extension entomologist Ala. Agr. Extension Service, 1918-19; chief div. entomology, state entomologist Tex. Agr. Expt. Sta., 1924-46; prof. A. and M. Coll. Tex., College Station, 1946-54; dir. gen. cotton insect control com. Nat. Soc. Agr. Peru, Lima, 1954-57. Cons. vet. supply depot Am. Marketing Inst., 1962—. Mem. Entomol. Soc. Am., Apiary Insps. Am., Tex. Beekeepers Assn., Peruvian Entomol. Soc. (hon.), Phi Kappa Phi, Gamma Sigma Delta. Research and publs. on life history and control Mexican bean beetle E. of Rocky Mountains; pioneered research in airplane dusting for cotton insect control. Home: 1309 Walton Dr., College Station, Tex. 77840.*

THOMAS, Garth Johnson, Am. psychologist; b. Pittsburg, Kan., Sept. 8, 1916; s. Leslie Homer and Opal (Johnson) T.; A.B., Kan. State Tchrs. Coll., 1938; M.A., U. Kan., 1941; Ph.D., Harvard, 1948; m. Mary Mona Gee, Sept. 21, 1945; children—Gregory Allen, Barbara Elizabeth. Faculty, U. Chgo., 1948-54; faculty Coll. Medicine, Neuropsychiat. Research Inst., U. Ill., Chgo., 1954-66, research prof. dept. physiology, and biophysics, 1957-66; professor center for Brain Research, U. Rochester (N.Y.), 1966—; vis. prof. psychology Emory U., Atlanta, 1953. Mem. Am. Physiol. Soc., Psychonomics Soc., A.A.A.S., Am. Assn. U. Profs. Research and publs. on neurology of behavior by analysis of learning, emotion, instincts in animals after discrete exptl. lesions in brain. Home: 186 Buckland Av., Rochester, N.Y. 14618.*

THOMAS, Harold Allen, Am. civil engr.; b. Terre Haute, Ind., Aug. 14, 1913; s. Harold A. and Katherine (Sass) T.; B.S. in Civil Engring., Carnegie Inst. Tech., 1935; S.D., Harvard, 1938; m. Gertrude A. Green, July 2, 1935; children—Harold Allen, III, Stephen C., Calvin R. Faculty, Harvard, Cambridge, Mass., 1939—, Gordon McKay prof. civil and san. engring., 1956—. Cons. to govt. agys., Ford Found. Recipient (with R. S. Archibald) Herring award Am. Soc. C.E., 1952. Mem. Am. Soc. C.E., Boston Soc. Civil Engrs., Am. Acad. Arts and Scis., Am. Geophys. Union, Water Pollution Control Fedn. Author: (with others) Design of Water Resource Systems, 1962; also numerous articles. Developed new methods operations research for combining engring. design econ. analysis, and govtl. planning comprehensive water-resource devel. schemes. Home: 108 Oakland Av., Arlington, Mass. 02174.*

THOMAS, Harold Edgar, Am. hydrologist; b. Chgo., June 16, 1906; s. John David and Jeannette (Ritter) T.; B.S., U. Chgo., 1926, Ph.D., 1947; m. Ruth Melissa Newton, Oct. 9, 1937; children—Sheila Rae (Mrs. Rollie Gilliam), Sharon Lee (Mrs. Will Cross), Jon David, Patricia Jean, Evan Scott, Marianne. Inst. geology Ohio State U., 1927-29; geologist U. S. Geol. Survey, 1929-42; lt. col. C.E., U. S. Army, 1942-45; ground-water geologist U. S. Geol. Survey, Salt Lake City, 1946-49; research sci. Conservation Found., N.Y.C., 1949-50; hydrologist U. S. Geol. Survey, Menlo Park, Cal., 1950—. Tech. adviser AID, Tunisia, 1960-62; cons. Kan. Finance Com., 1954; cons. govts. of Panama, P.R., V.I., Nigeria, Kuwait; mem. Africa Sci. Bd. of Nat. Acad. Scis., 1963—. Mem. Am. Geophys. Union, Geochem. Soc., Soc. Econ. Geologists, Internat. Assn. Sci. Hydrologists, Am. Water Resources Assn. Author: Conservation of Ground Water, 1949; Drought in Southwestern United States, 1963; Geology and Ground Water of Cedar City and Parowan Valleys, Utah, 1946; also numerous articles. Developed interrelations of water resources used by man, particularly ground water and surface water, based on concept of hydrologic cycle; research on conflicts of hydrologic concepts with various cultural and legal traditions, effects of droughts on various water resources. Home: 1339 Portola Rd., Woodside, Cal. 94061. Office: 345 Middlefield Rd., Menlo Park, Cal. 94025.*

THOMAS, Hugh Owen, Brit. physician, surgeon; b. Anglesey, Wales, 1833; ed. U. Coll., London, Edinburgh U., also in Paris. Surg. practice, Liverpool, from 1857; author: Diseases of the Hip, Knee and Ankle Joints; Contributions to Surgery and Medicine, 8 parts, 1883-90. Pioneer in orthopedic surgery; developed several splints, including Thomas posterior splint for fracture of femur, and Thomas knee splint. Died Liverpool, 1891.

THOMAS, Jacob Earl, Am. physiologist; b. Steilacoom, Wash., Jan. 31, 1891; s. John Calvin and Nettie (Wyckoff) T.; student U. Wash., 1911-12; B.S., M.D., St. Louis U., 1918, M.S., 1924; Sc.D., Jefferson Med. Coll., 1960; m. May Johnson, Nov. 28, 1917 (dec. July 1957); children—Jacob Earl II, Marjorie Ellen (Mrs. John Fredrick Larkin); m. 2d Grace Neal Webster, June 15, 1958. Instr. physiology St. Louis U. Sch. Medicine, 1916-18, asst. prof., 1918-20, 21-27; asso. prof. physiology W.Va. U. Sch. Medicine, 1920-21; prof. physiology, chmn. dept. Jefferson Med. Coll. of Phila., 1927-56; prof. physiology, chmn. dept. Loma Linda U., 1956-64, prof. physiology, biophysics, 1964—. Recipient Certificate of Appreciation for contributions to advancement of med. research, Med. Research Assn. Cal., 1962. Mem. A.M.A. (chmn. sect. on pathology and physiology 1954), Am. Physiol. Soc., Am. Soc. for Pharmacology and Exptl. Therapeutics, Am. Gastroent. Assn., Soc. for Exptl. Biology and Medicine, Med. Research Assn. Cal., Phila. Coll. Physicians. Author: The External Secretion of the Pancreas, 1950; contbr. author: Ency. Britannica, 1961. Contbr. numerous articles to sci. jours. Introduced the concept of negative feedback control of gastric emptying (that emptying of the stomach after a meal is regulated and controlled by inhibitory reflexes and hormones initiated in duodenum by contact of gastric efflux with the duodenal mucosa); introduced several new methods in gastro-intestinal research, the most important being the devel. practical methods of preparing and maintaining tubulated gastric and duodenal fistulas; the duodenal fistula is now widely used in pancreatic research to provide access to entrance of pancreatic duct into the duodenum. Home: 30319 Sunset Dr., Redlands, Cal. 92373. Office: Loma Linda U., Loma Linda, Cal. 92354.*

THOMAS, Jan Bartholomeus, Dutch biophysicist; b. Bandung, Netherlands East Indies May 11, 1907; s. Emil and Christina (Snellen) T.; ed. Breda State U., Utrecht, Netherlands, 1928-39; m. Maria J. W. van der Staal, Apr. 15, 1950; 1 dau., Christina Anna (Mrs. U. P. van der Wal). Staff, U. Utrecht, 1947—, leader biophys. research group, 1947—, prof. biophysics, 1962—; fellow Treub Fonds, Bot. Gardens, Buitenzorg, Java, 1939-40, co-worker, 1940-46; vis. prof. botany U. Ill., Urbana, 1959-60. Mem. Royal Dutch Bot. Soc., Soc. for Biochemistry, Soc. for Cell Biology, Physics Soc., Sigma Xi, Phi Kappa Phi. Author: Primary Photoprocesses in Biology, 1965; also numerous articles. Research on structure and function photosynthetic apparatus.*

THOMAS, John Lawrence, Am. sociologist; b. Berlin, Wis., Apr. 9, 1910; s. John Lawrence and Anna (Masick) T.; A.B., St. Louis U., 1933; M.A., U. Montreal, 1936, 37; S.T.L., St. Mary's Coll., 1944; Ph.D., U. Chgo., 1949. Faculty dept. sociology St. Louis U., 1949-65, prof., 1964-65; research asso. Inst. Social Order, St. Louis, 1949-65, Cambridge (Mass.) Center for Social Studies, 1965—. Guggenheim fellow, 1952-53. Fellow Am. Assn. Marriage Counselors (affiliate); mem. Population Assn. Am.; Am., Am. Catholic (past pres., Ann. Research award 1957) sociol. socs., Religious Research Assos. Author: The American Catholic Family, 1956; Marriage and Rhythm, 1958; Catholic Viewpoint on Marriage and the Family, 1958; Religion and the American People, 1960. Research and numerous publs. on background, characteristics and changes in Am. Catholic family, analyzed its spl. problems as distinctive religious minority with regard to ethnic differences, mixed marriages and marital breakdown, devel. of Christian views of sex. Home: 42 Kirkland St. Office: 40 Kirkland St., Cambridge, Mass. 02138.*

THOMAS, John William, Am. nutritionist; b. Spanish Fork, Utah, Mar. 25, 1918; s. John B. and Kate (Tolhurst) T.; B.S., Utah State U., 1940; postgrad. U. Wis., 1940-41; Ph.D., Cornell U., 1946; m. Carolyn M. Palmer, Jan. 6, 1945; children—Linda, John, Barbara, Christopher. Research asso. for NDRC, Northwestern U. also Carnegie Inst. Tech., 1942-45; nutritionist, biochemist Dairy Cattle research br. Agrl. Research Adminstrn., U. S. Dept. Agr., Beltsville, Md., 1942-60; prof. dairy nutrition Mich. State U., East Lansing, 1960—. Recipient Am. Food Mfgrs. Assn. award, 1953, U. S. Dept. Agr. Superior Service award, 1959. Mem. Am. Chem. Soc., Am. Inst. Nutrition, Am. Dairy Sci. Assn., Am. Soc. Animal Sci., Sigma Xi, Phi Kappa Phi. Fellow A.A.A.S. Research on function of nutrients in body, amount of nutrients required for various levels productivity, relationship of animal performance with chem. constituents. Home: 316 John Rd., East Lansing, Mich. 48823.*

THOMAS, Joseph Miller, Am. mathematician; b. Ridley Park, Pa., Jan. 16, 1898; s. Joseph Miller and Sallie Alexa (Tyler) T.; A.B., U. Pa., 1918, Ph.D., 1923; m. Hilda Bell Howes, Jan. 22, 1938; children—Abby Pamela, Peter Chew. Faculty, U. Pa., 1919-24, 27-30; faculty Duke, Durham, N.C., 1930-65, prof. math., 1935-65; NRC fellow math. Princeton, 1924-26, Harvard, 1926, U. Paris (France), 1926-27. Mem. Am. Math. Soc. (past v.p.), Société Mathématique de France, Phi Beta Kappa, Sigma Xi. Founder, mng. editor Duke Math. Jour., 1935-44. Author: Differential Systems, 1937; Theory of Equations, 1938; Elementary Mathematics in Artillery Fire, 1942; Systems and Roots, 1962; also articles. Research on solution of systems of algebraic and differential equations with applications to geometry. Home: 2215 Cranford Rd., Durham, N.C. 27706.*

THOMAS, Lewis, Am. physician; b. Flushing, N.Y., Nov. 25, 1913; s. Joseph S. and Grace (Peck) T.; B.S., Princeton, 1933; M.D. cum laude, Harvard, 1937; m. Beryl Dawson, Jan. 1, 1941; children—Abigail (Mrs. Lucian Waddell), Judith, Eliza. Tileny Meml. Research fellow, asst. medicine Thorndike Meml. Lab., 1941-42; asst. prof. pediatrics Johns Hopkins Med. Sch., also pediatrician, dir. bacteriology lab. Harriet Lane Home for Invalid Children, Johns Hopkins Hosp., 1946-48; asso. prof. medicine, dir. div. infectious disease Tulane U. Sch. Medicine, New Orleans, 1948-50, prof., 1950; prof. pediatrics and internal medicine Am. Leagion heart research U. Minn. Med. Sch., Mpls., 1950-54; chmn. dept. pathology, prof. N.Y. U. Sch. Medicine, 1954-58, prof., chmn. dept. medicine, 1958-66; dir. 3d, 4th med. divs. Bellevue Hosp., N.Y.C., 1958-66; dir. medicine U. Hosp., N.Y.C., 1959-66; dean N.Y. U. Sch. Medicine, N.Y.C., 1966—. Cons. to govt. agys., pvt. hosps.; mem. Nat. Adv. Child Health and Human Devel. Council, 1964—; mem. Bd. Health, N.Y.C., 1956—; mem. narcotics adv. com. N.Y.C. Dept. Health, 1960—; mem. President's Sci. Adv. Com., 1967—. Fellow Am. Acad. Arts and Sci., N.Y. Acad. Scis.; mem. Assn. Am. Physicians, Am. Pediatric Soc., Am. Soc. Clin. Investigation, Am. Assn. Immunologists, Soc. Exptl. Biology and Medicine, Am. Soc. Exptl. Pathologists, Practitioners Soc., Harvey Soc., Am. Acad. Microbiology, Interurban Clin. Club, Royal Soc. Medicine, Societe Francaise d'Allergie. Research and numerous publs. on immunology. Home: 1 Washington Mews, N.Y.C. 10003.*

THOMAS, Llewellyn Hilleth, physicist; b. London, Eng., Oct. 21, 1903; s. Charles James and Winifred (Lewis) T.; B.A., U. Cambridge (Eng.), 1924, Ph.D., 1927, M.A., 1928, D.Sc., 1965; m. Naomi Estelle Frech, Sept. 27, 1933; children—James Rhys, Ann Rhonwen (Mrs. Howard Jay Viele), Margaret Olwen (Mrs. Peter John de Angelis). Mem: sr. staff Watson Sci. Computing Lab., N.Y.C., 1946—; prof. physics Columbia U., 1950—; fellow IBM, 1963-67; cons. Los Alamos Sci. Lab.; mem. sci. adv. com. Aberdeen Proving Ground. Mem. Nat. Acad. Sci., A.A.A.S., Am. Phys. Soc., Am. Assn. U. Profs., Am. Assn. for Computing Machinery. Research and numerous publs. in passage of electrified particles through matter, electron distbn. in atoms and molecules, use of computing machines in sci. problems. Home: 223 Glenwood Av., Leonia, N.J. 07605. Office: 612 W. 115th St., N.Y.C. 10025.*

THOMAS, Richard Sanborn, Am. biophysicist; b. Madison, Wis., June 14, 1927; s. Lyell Jay and Ethel (Sanborn) T.; B.A., Oberlin Coll., 1949; Ph.D. (NSF fellow), U. Cal. at Berkeley, 1955; m. Valborg Emilia Nedergaard Aegidius, Nov. 16, 1957; children—Eric, Karen. Postdoctoral fellow cytochem. dept. Carlsberg Lab., Copenhagen, Denmark, 1955-57; asst. research biophysicist virus lab. U. Cal. at Berkeley, 1958-60; research physicist Western Regional Research Lab., U. S. Dept. Agr., Albany, Cal., 1960—. Am. Cancer Soc. Jr. Postdoctoral fellow, 1955-56; NSF fellow, 1956-57; NIH Sr. Postdoctoral fellow dept. gen. botany Swiss Fed. Inst. Tech., Zurich, 1967-68. Mem. Electron Microscopy Soc. Am., Am. Soc. for Cell Biology, Biophys. Soc., A.A.A.S., No. Cal. Soc. for Electron Microscopy, Am. Soc. for Microbiology, Sigma Xi. Research and publs. on isolation and chem. analysis of subcellular components of liver (microsomes and ribosomes); isolation and chem. identification of intracellular crystals in amoebae; chem. and electron-microscopic cytochem. analysis of virus particle structure; electron microscopic techniques for intracelluar mineral localization; electron-microscopic study of bacterial spores, keratin, microfibrillar proteins. Home: 10 Windsor Av., Berkeley, Cal. 94708. Office: Western Regional Research Lab., Albany, Cal. 94710.*

THOMAS, Robert Glenn, Am. radiobiologist; b. Watertown, N.Y., Oct. 9, 1926; s. Glenn R. and Mildred (VanHorn) T.; B.S., St. Lawrence U., 1949; Ph.D., U. Rochester, 1955; m. Randi Lie, Sept. 24, 1965; children—Carol, Glenn, Paula. Faculty, U. Rochester, 1950-61; head dept. radiobiology Lovelace Found. for Med. Edn. and Research, Albuquerque, 1961—; cons. SNAP Hazards Evaluation Program, SLAM Hazzards Evaluation Program Operation Plumbbob, 1957; asst. project dir. BOOT Test Gen. Electric, Ida. Operations Office, 1958. Mem. Radiation Research Soc., Health Physics Soc., Am. Indsl. Hygiene Assn., A.A.-A.S., Sigma Pi Sigma, Pi Mu Epsilon. Research and numerous publs. in gen. field of evaluation of biol. hazards from ionizing radiation, biol. retention and excretion patterns of various radioactive materials. Home: 9116 Bellehaven St. N.E., Albuquerque 87112. Office: 5200 Gibson St. S.E., Albuquerque 87108.*

THOMAS, Sidney Gilchrist, English metallurgist, inventor; b. London, Apr. 16, 1850; s. William and Melicent (Gilchrist) T.; ed. Dulwich Coll.; studied applied chemistry and metallurgy independently. Asst. master sch. in Essex, 1867; clerk Marlborough Street Police Ct., 1867, Thames Ct., 1868-79. Recipient Bessemer Gold medal, 1883. Mem. council Iron and Steel Inst., 1882. Author: Elimination of Phosphorus in the Bessemer Converter, 1879; also articles to Iron jour. Discovered practical method for dephosphorizing iron in Bessemer converter, 1875; attempted to develop new typewriter. Died Paris, Feb. 1, 1885.

THOMAS, Theodore Gaillard, Am. physician; b. Edisto Island, nr. Charleston, S.C., Nov. 21, 1831; s. Rev. Edward T. and Jane Marshall (Gaillard) T.; Grad. Med. Coll. of Charleston, 1852; resident physician Bellevue and Ward's Island hosps., New York, and Rotunda Hosp., Dublin, Ireland; m. Mary Guillard; m. 2d Mary Willard, 1862; four sons, one dau. Prof. obstetrics and gynecology, Coll. Phys. & Surg., New York; cons. surgeon to several New York hosps. and mem. many med. socs. in U. S. and abroad. Autor: Diseases of Women, 1868, translated into French, German, Italian, Spanish and Chinese; assisted in preparing A Century of American Medicine, 1876; large contbr. to med. journals. Pioneer in gynecology; originated laparoelytrotomy and other surgical methods; first to remove ovarian tumor by cutting vagina; among first to distinguish cervix and uterus as separate organs; used incubator, 1867. Died Thomasville, Ga., Feb. 28, 1903.

THOMAS, Tracy Yerkes, Am. mathematician; b. Alton, Ill., Jan. 8, 1899; s. Tracy Reeve and Blanche Ailene (Yerkes) T.; A.B., Rice Inst., Houston, Tex., 1921; M.A., Princeton, 1922, Ph.D., 1923; m. Virginia Rowland, June 28, 1928; 1 son, Tracy Alexander. Nat. research fellow in physics, U. of Chicago, 1923-24, mathematics, Zürich, Switzerland, 1924-25, Harvard and Princeton, 1925-26; asst. prof. mathematics, Princeton, 1926-31, asso. prof., 1931-38; prof. mathematics, U. of Calif., 1938-44; prof. math. and chmn. dept., Indiana U., 1944-54, head grad. inst. for applied mathematics, 1950-54, dir. grad. inst. mathematics and mechanics, 1954-56, distinguished service prof. mathematics, 1956—; faculty research lecturer U. Cal., Los Angeles, 1943; vis. prof. U. Cal., San Diego, at La Jolla, 1962-63; vis. prof. engring. U. Cal. at Los Angeles, 1965-66, 67-68. Mem. S.A.T.C., Rice Inst., 1918. Vice pres. Am. Math. Soc., 1940-42; elected Nat. Acad. Scis., 1941. Fellow Royal Astronomical Soc., Indiana Academy of Science; member Am. Math. Soc., Math. Assn. of Am., A.A.A.S., Am. Assn. U. Profs., Sigma Xi. Author: The Elementary Theory of Tensors, 1931; The Differential Invariants of Generalized Spaces, 1934; Concepts from Tensor Analysis and Differential Geometry, 1961, 2d edit., 1965; Plastic Flow and Fracture in Solids, 1961. Work on theories of relativity, plasticity, shock waves, tensors and differential geometry; extended theory of conditions for discontinuities over moving surfaces. Address: Mathematics Dept., Ind. Univ., Bloomington, Ind. 47401.*

THOMAS, Walter Ivan, Am. agronomist; b. Elwood, Neb., Mar. 27, 1919; s. Percy E. and Ethel (Major) T.; B.S., Ia. State U., 1949, M.S., 1953, Ph.D., 1955; m. Margaret Ann Thompson, Feb. 15, 1941; 1 dau., Linda Margaret. Clk., Soil Conservation Service, Civilian Conservation Corps, Broken Bow, Neb., 1937-39, U. S. Civil Service, Washington, 1940-41; grad. asst. Ia. State U., 1949-50, instr., then asst. prof., 1953-59; mem. faculty Pa. State U., 1959—, prof. agronomy, 1963—, head dept., 1964—. Fellow A.A.A.S.; mem. Am. Soc. Agronomy, Am. Inst. Biol. Sci., Sigma Xi, Gamma Sigma Delta, Alpha Gamma Rho, Phi Mu Alpha. Sci. contbns. to knowledge of genetic control of differential accumulation of chem. elements. Home: 523 Sunset Rd., State College, Pa. 16801. Office: Tyson Bldg., Pennsylvania State Univ., University Park, Pa. 16802.*

THOMAS, William Hewitt, Am. biologist; b. Riverside, Cal., Dec. 25, 1926; s. Cleo Elson and Abigail (Roblee) T.; B.A., Pomona Coll., 1949; M.A., U. Md., 1952, Ph.D., 1954; m. Sara Sussmann, Dec. 29, 1956; 1 dau., Ann Muir. Lab. technician U. Cal. at Riverside, 1949-50; staff Scripps Inst. Oceanography, U. Cal. at La Jolla, 1954—, asst. research biologist, 1956-64, asso. research biologist, 1964—. Mem. Bot. Soc. Am., Am. Soc. Plant Physiologists, A.A.A.S., Am. Soc. Limnologists and Oceanographers, Phycological Soc. Am., Internat. Phycological Soc., Sigma Xi, Phi Kappa Phi. Research and publs. on evaluation of radioactive tracer method for measuring microscopic plant photosynthesis in sea; showed the Mississippi River flow increases productivity of part of Gulf of Mexico, that fertility of Eastern tropical Pacific Ocean is controlled by availability of nitrogen. Home: 5355 Pacifica Dr., San Diego 92109. Office: U. Cal. at LaJolla, Cal. 92037.*

THOMAS, William Isaac, Am. sociologist; b. Russell County, Va., Aug. 13, 1863; s. Thaddeus Peter and Sarah (Price) T.; A.B., U. of Tenn., 1884; student Berlin, and Göttingen, 1888-89; Ph.D., U. of Chicago, 1896; m. Harriet Park, June 6, 1888; 5 children including William Alexander and Edward Brown; m. 2d, Dorothy Swaine Feb. 7, 1935. Instr. English and modern languages, U. of Tenn., 1884-88; prof. English, Oberlin Coll., 1889-94, prof. sociology, 1894-95; instr. sociology. U. of Chicago, 1895-96, asst. prof., 1896-1900, assoc. prof., 1900-10, prof., 1910-18; lecturer New School for Social Research, 1923-28, Harvard, 1936-37; in charge Helen Culver fund for race psychology, 1908-18, research on Jewish Culture and behavior since 1918. Member Am. Sociol. Soc. (pres. 1927), Am. Acad. of Arts and Sciences. Author: Sex and Society, 1907; Source Book for Social Origins, 1909; Standpoint and Questionnaire for Race Psychology, 1912; Suggestions of Modern Science Concerning Education, 1914; The Polish Peasant in Europe and America (with F. Znaniecki), 1918-21; The Unadjusted Girl, 1923; The Unconscious (with others), 1927; The Child in America (with D. S. Thomas), 1928; Primitive Behavior, 1936. Pioneered in social psychology; used approach of ethnology to study customs and social patterns of civilized peoples; (with Znaniecki) 1st to use personal data as sources for sociol. investigation; developed concept of social disorgn. as perspective from which to study social phenomena; developed 4 categories of goals of human wishes (stability, response, new experiences, recognition). Died Berkeley, Cal., Dec. 5, 1947.

THOMAS OF CANTIMPRE (or Thomas of Brabant, Thomas Brabantinus, De Cantiprato, Cantimpratewsis), English encyclopaedist; b. Brabant, circa 1204; studied in Léige; lived in Chantimpré; entered Dominican Order, 1232; went to Paris, 1238; became sub-prior, Louvain, 1246. Author: De natura rerum, circa 1228; Bonum Universale de Apibus. Wrote sci. encyclopedia containing discussions on human body, the soul, races of man, zool. and bot. studies, astrology, and the elements. Died Louvain, circa 1271.

THOMPSON, A(lexander) Ralph, chem. engr.; b. Toronto, Ont., Can., Sept. 11, 1914; s. Allen and Mabel (Gilmour) T.; B.A.Sc., U. Toronto, 1936; Ph.D. in Chem. Engring., U. Pa., 1945; m. Roberta

Evelyn Diehl, July 24, 1937; children—Michael, Nancy. Came to U. S., 1940, naturalized, 1952. Chem. engr. Canadian Industries, Ltd., Beloeil, Que., New Toronto, Ont., 1936-40; faculty U. Pa., 1940-52; prof., chmn. dept. chem. engring. U. R.I., Kingston, 1952—. Dir. R.I. Water Resources Center, 1966—; cons., devel. engr. Sun Oil Co., 1945-52; cons. Monsanto Chem. Co. (Mass.), 1955-57. Mem. Am. Inst. Chemical Engineers (chmn. R.I. sect. 1957-58), Am. Chem. Soc., American Society for Engineering Education, A.A.A.S., Am. Assn. U. Profs., Sigma Xi (pres. U. R.I. chpt. 1962-63), Tau Beta Pi, Phi Lambda Upsilon, Phi Kappa Phi (treas. U. R.I. chpt. 1965—). Research, publs. on greater understanding of effect of molecular structure on refraction and disperson of binary solutions. Home: Summit Av., Tower Hill Heights, Wakefield, R.I. 02879. Office: Dept. Chem. Engring., U. R.I., Kingston, R.I. 02881.*

THOMPSON, Benjamin (Count Rumford), physicist; b. Woburn, Mass., Mar. 26, 1753; s. Benjamin and Ruth (Simonds) T.; attended grammar sch., apprenticed as storekeeper and importer, Salem, Mass., 1770; attended lectures Harvard; m. Sarah Walker Rolfe, Nov. 1772; 1 dau., Sarah; m. 2d, Marie Anne Pierrette Paulze Lavoisier, Oct. 24, 1805 (separated 1809). Taught sch., Rumford (now Concord), N.H., 1771; decided on mil. career, became commd. maj. 2d Provincial Regt., N.H., 1773; went to Eng. as loyalist refugee, 1776; apptd. sec. Province of Ga., 1776; under-sec. of state Am. Dept., 1780-81; apptd. lt. col. Brit. Army for Sevice in Am., Charleston, S.C., 1781; commdr. Queen's Rangers, Brit. Legion, King's Dragoons, Long Island, N.Y., 1782; returned to Eng., 1783; apptd. col. by King George III; col. and aide-de-camp to Charles Theodore of Salzbach, Elector of Bavaria and Palatine and performed services as initiating army reforms, philanthropic work for poor, 1784-98; maj. gen., head War Dept. Bavaria, from 1788; supr. constrn. of Eng. garden, Munich, Germany, 1789; introduced steam engine into Palatinate at Mannheim, 1791; prevented (at request of regency) French and Austrian armies from entering neutral city of Munich, 1796; apptd. head police dept., Bavaria, 1796; retuned to London, 1798, retired from Bavarian Service, 1798; made proposals, 1799, which resulted in incorporation of Royal Instn., London, 1800; planned orgn. of Bavarian Acad. Arts and Scis., Munich, 1801; travelled to Paris, 1801. Fellow Royal Soc., 1779 (Copley medal, 1792, first Rumford medal, 1802); mem. French Acad. Scis., 1802, Munich, Berlin, Mannheim acads. Best known for his cannon expts. showing heat to be a mode of motion, thereby disproving prevalent notion of heat as fluid material substance (caloric); tried to calculate mech. equivalent of heat; performed expts. to show there is no change of weight accompanying heating or cooling of bodies (thus fluid caloric, if it existed, must be weightless); conducted expts. to determine the most advantageous constrn. of firearms, also explosive force of gunpowder, and velocities of bullets; pioneer in establishing workshop and soup kitchen as efficient and humane means of helping poor; investigated warmth provided by natural and artificial clothing; discovered remedy for smoky chimneys; improved constrn. of fireplaces, chimneys, and cooking appliances; pioneer of central steam and hot water heating, and efficient prodn. of heat and its controlled conveyance for heating of large halls and rooms; made expts. to determine thermal conductivity of liquids; invented shadow photometer, water compensation calorimeter, and passage thermometer; improved Argand lamp; endowed Rumford professorship, Harvard, also Rumford medal of Royal Soc. and Am. Acad. Arts and Scis., 1796. Died Auteuil, nr. Paris, France, Aug. 21, 1814.

THOMPSON, Browder Julian, Am. engr.; b. Roanoke, La., Aug. 14, 1904; B.S. in Elec. Engring., U. Wash., 1925; also postgrad. work. With research lab. Gen. Electric Co., Schenectady, 1926-31, in charge research and devel. on radio receiving tube problems; head research sect. research and engring. dept. Radiotron div. RCA, Harrison, N.J., 1931-42; asso. dir. gen. research RCA Labs., Princeton, N.J., 1942-44. Became civilian cons. War Dept., 1944. Recipient Morris Liebmann Meml. prize I.R.E., 1936. Research and devel. of radio and TV tubes; developed mech. and elec. design for operation of tubes at ultra-high frequency; studies on radio-electron tube theory and design, amplifiers; extended useful radio-frequency range. Died Italy, July 4, 1944.

THOMPSON, David Grosh, Am. geologist; b. Lockland, O., May 12, 1888; s. David de Camp and Ribia Louise (Grosh) T.; ed. Northwestern U., 1911; master's degree U. Ill., 1913; postgrad. U. Chgo., Johns Hopkins; m. Frances E. Goodrich, July 11, 1922; children—David Goodrich, Clifford Francis. Mem. geol. surveys of Ill., Wis., Md.; instr. Lehigh U., Goucher Coll., Johns Hopkins; staff ground water div. U. S. Geol. Surveys, 1917-43. Recipient Goodel prize Am. Water Works Assn. Fellow A.A.A.S., Geol. Soc. Am., Am. Geophys. Union, Soc. Econ. Geologists, Washington Acad. Scis., Am. Water Works Assn. Research and publs. on ground-water hydrology of Coastal Plain, problems on legal regulation of ground-water supplies; worked on water supplies for mil. and naval establishments, war industries; water problems of Mojave Desert region; pioneered use of automatic water-stage recorders; developed rating curve of Artesian wells, pumping method for study of salt water problems. Died Washington, Feb. 19, 1943.

THOMPSON, Edgar Tristram, Am. sociologist; b. Little Rock, S.C., Sept. 13, 1900; s. John Sanders and Annie (Smith) T.; A.B., U. S.C., 1922; M.A., U. Mo., 1924; Ph.D., U. Chgo., 1932; m. Alma Louise Macy, June 15, 1929; 1 dau., Alma Lee (Mrs. Richard Schaffer). History tchr. Plant City (Fla.) High Sch., 1922-23; instr. rural sociology U. N.C., 1924-26; adj. prof. sociology U. Tex., 1927-28, Earlham Coll., 1928-29; instr. YMCA Coll., Chgo., 1929-30; asst. prof. sociology U. Wash., 1930-31; research prof. sociology U. Hawaii, 1932-35; faculty Duke, Durham, N.C., 1935—, prof. sociology, 1946—, acting chmn. dept. sociology, 1963-64, chmn. Center for So. Studies in Social Scis. and Humanities, 1966—. Hugh le May fellow Rhodes U., Grahamstown, Africa, 1956. Mem. Am., Soc. (past pres.) Sociol. socs., African Studies Assn. Author: Race Relations and the Race Problem, 1939; Race and Region, 1949; (with Everett C. Hughes) Race: Individual and Collective Behavior, 1958; also articles. Editor: Perspectives on the South, 1967. Research on comparative race relations, sociology of South. Home: 138 Pinecrest Rd., Durham, N.C. 27706.*

THOMPSON, Edward Herbert, Am. archeologist; b. Worcester County, Mass., Sept. 28, 1856; s. Josiah A. and Mary E. (Thayer) T.; grad. Worcester Tech. Inst., 1879. Am. consul to Yucatan, 1885-1909; devoted many yrs. to research and exploration of ruins of the Maya civilization; spent 14 mos. collecting material, under auspices of the Peabody Museum, for exhibition at the World's Fair, Chicago, 1893. Author: Children of the Cave, 1929; People of the Serpent, 1932; various reports. Discovered the long sought "Hidden City" buried in the interior of Yucatan; purchased plantations of Chichen and tracts adjoining, rebuilt the plantation house, later center of a colony of scientists, and instituted many modern improvements; uncovered the "Maya Venus," the mausoleum of the high priest, the Temple of the Painted Columns, and the ancient city, Old Chichen, also the "Chichen Tablet," and successfully explored the "Sacred Well" of Chichen Itza, recovering many objects of great archaeol. interest; became writer and lecturer. Died Plainfield, N.J., May 11, 1935.

THOMPSON, George Albert, Am. geophysicist; b. Swissvale, Pa., June 5, 1919; s. George Albert and Maude (Harkness) T.; B.S., Pa. State U., 1941; S.M., Mass. Inst. Tech., 1942; Ph.D., Stanford, 1949; m. Anita Kimmell, July 20, 1944; children—Albert J., Dan A., David C. Geologist, U. S. Geol. Survey, 1942-44; faculty Stanford (Cal.), 1946—, prof. geophysics, 1960—. Mem. earth scis. panel NSF, 1963-66; mem. geophysics working group NASA, 1965—. Recipient G. K. Gilbert award in seismic geology, Carnegie Instn. Washington, 1964. NSF postdoctoral fellow, 1956-57; Guggenheim fellow, 1963-64. Mem. Seismol. Soc. Am., Am. Geophys. Union, Geol. Soc. Am., A.A.A.S. Research and publs. on geology and geophysics of mountain structures to determine processes and origin of earth movements and volcanism. Home: 421 Adobe Pl., Palo Alto, Cal. 94306. Office: Geophysics Dept., Stanford, Cal. 94305.*

THOMPSON, George G., Am. psychologist; b. Bucklin, Kans., Mar. 17, 1914; s. Otie K. and Georgia (Beach) T.; B.A., Ft. Hays (Kans.) State Coll., 1937, M.S., 1938; Ph.D., State U. Ia., 1941; m. Evelyn W. Schuller, Oct. 25, 1940; 1 son, Kenrick Steven. Prof. psychology So. Ill. U., Carbondale, 1941-42, Syracuse U., 1942-59; spl. research asso. Psycho-Acoustic Lab., Harvard, 1943-45; prof. psychology Ohio State U., Columbus, 1959—. Mem. rev. com. research grants div. Nat. Insts. Mental Health, 1965—. Fellow Am. Psychol. Assn., Soc. Research Child Devel., Am. Ednl. Research Assn., A.A.A.S.; mem. Sigma Xi. Author: Child Psychology, 1952; (with R. G. Kuhlen) Psychological Studies of Human Development, 1952; (with E. F. Gardner and F. J. DiVesta) Educational Psychology, 1959; (with E. F. Gardner) Social Relations and Morale in Small Groups, 1957. Cons. editor Jour. Genetic Psychology, 1964—, also monographs and psychol. reports. Developer and pub. (with E. F. Gardner) Syracuse Scales of Social Relations, 1959; conducted and pub. (with W. J. Meyer) study of differential treatment and influence of female tchrs. in interactions with elementary sch. boys and girls.*

THOMPSON, Gerald Luther, Am. mathematician; b. Rolfe, Ia., Nov. 25, 1923; s. Luther and Sylva C. (Larson) T.; B.S. in Elec. Engring., Ia. State U., 1944; S.M. in Math., Mass. Inst. Tech., 1948; Ph.D. in Math U. Mich., 1953; m. Dorothea Vivian Mosley, Aug. 25, 1954; children—Allison Mosley, Emily Ann, Abigail Elizabeth. Faculty, Mass. Inst. Tech., 1946-48, U. Mich., 1948-51, Princeton, 1951-53, Dartmouth, 1953-58, Ohio Wesleyan U., 1958-59; faculty Carnegie Mellon U., Pitts., 1959—, prof. math. and indsl. adminstrn., 1963—. Cons., IBM, 1958-60, Econ. Research Inst., Princeton, 1956—, Bethlehem Steel Corp., 1964—. McKinsey & Co., 1965—. Ford Found. faculty research fellow, 1963-64. Mem. Am. Math. Soc., Math. Assn. Am., Inst. Mgmt. Scis., Operations Research Soc., Econometric Soc., Soc. for Indsl. and Applied Math. Author: (with others) Introduction to Finite Mathematics, 1957, Finite Mathematical Structures, 1959, Calculus of Functions of One Argument, 1960, Finite Mathematics with Business Applications, 1962. Editor: (with J. F. Muth) Industrial Scheduling, 1963. Research and publs. on theory of games, signaling strategies, von Neumann model of expanding economy, non-linear prodn. functions, linear programming, chance constrained programming, mixed integer programming, problem extraneous constraints and redundant variables, critical path analysis and linear programming from risk viewpoint, problems from heuristic programming viewpoint, complex project scheduling. Home: 15 Wedgewood Lane, Pitts. 15215.*

THOMPSON, Sir Harold Warris, English phys. chemist; b. Wombwell, Eng., Feb. 15, 1908; s. William and Charlotte Emily (Warris) T.; B.A., Oxford (Eng.) U., 1929, D.Sc., 1954; D.Phil., Berlin (Germany) U., 1930; m. Grace Penelope Stradling, June 27, 1938; children—Richard S., Alison S. Faculty, Oxford (Eng.) U., 1930—, prof. chemistry, 1964—, fellow, tutor St. John's Coll., 1930-64. Decorated comdr. Order Brit. Empire. Fellow Royal Soc., 1946 (past mem. council, fgn. sec. 1965—, Davy medal 1965); mem. Internat. Council Sci. Unions (pres. 1963-66), Faraday Soc. (past mem. council), Chem. Soc. (past mem. council), Royal Inst. Author: A Course in Chemical Spectroscopy, 1938; also numerous articles. Research on chem. spectroscopy, especially infrared range. Home: 33 Linton Rd., Oxford, Eng.*

THOMPSON, Sir Harry Stephen Mueysey, Brit. agriculturist; b. Yorkshire, Eng., Aug. 11, 1809; s. Richard John and Mary (Meysey) T.; ed. Trinity Coll., Cambridge; studied entomology under Charles Darwin; grad. with math. honors, 1832; m. Elizabeth Ann Croft, Aug. 26, 1843; 5 sons, including Henry Meysey Meysey-Thompson; 5 daus. Chmn., Northeastern Ry. Co.; Liberal mem. Parliament for Whitby, 1859-65; dep. lt. of Yorkshire; justice of peace; high sheriff. Mem. Royal Agrl. Soc. (a founder, chmn. jour. com.), Yorkshire Agrl. Soc. (became pres. 1862), Research and publs. on ability of soil to absorb and assimilate ammonia, 1845; discovered value of covered fold yards. Died Yorkshire, Eng., May 17, 1874.

THOMPSON, Sir Henry, Brit. surgeon; b. Framingham, Eng., Aug. 6, 1820; s. Henry and Susannah (Medley) T.; T.; M.B., U. Coll., London, 1851; studied surgery of urinary organs under Jean Civiale, Paris; m. Kate Fanny Loder, Dec. 16, 1861; 1 son, Henry Francis Herbert; 2 daus. Became house surgeon U. Coll. Hosp., 1850, asst. surgeon, 1853, surgeon, 1863; named prof. clin. surgery, 1866; became cons. surgeon, emeritus prof. clin. surgery, 1874; named prof. pathology and surgery Royal Coll. Surgeon, 1884; surgeon extraordinary to Leopold I, Belgium. Recipient Jacksonian prize, 1852, 60. Fellow Royal Coll. Surgeons; mem. Cremation Soc. Eng. (became pres. 1874). Author: The Pathology and Treatment of Stricture of the Urethra; Diseases of the Prostate, 6th edit., 1886; Practical Lithotrity and Lithotomy, 1863; Clinical Lectures on Diseases of the Urinary Organs, 8th edit., 1888; On Tumours of the Bladder, 1884; On Suprapubic Lithotomy, 1885; Preventive Treatment of Calculous Disease, 1888; Modern Cremation its History and Practice, 4th edit., 1901; On Food and Feeding, 1901; Diet in Relation to Age and Activity, 18th edit., 1901; The Motor Car, 1902; The Unknown God?, 1902, rev., 1903; (novels) Charley Kingston's Aunt, 1885, All But, 1886. Publs. on cremation; improved Civiale's technique for removing kidney stones by crushing; developed two-glass urine test for gonorrhea, 1860; attempted reform of death certification and registration. Died London, Apr. 18, 1904.

THOMPSON, Homer Armstrong, archeologist; b. Devlin, Ont., Can., Sept. 7, 1906; s. William James and Gertrude (Armstrong) T.; B.A., U. B.C., 1925, M.A., 1927, LL.D., 1949; Ph.D., U. Mich., 1929, LL.D., 1957; LL.D., Dartmouth, 1957, Univ. Toronto, 1961, Lyon, 1963, Athens, 1963, Freiburg i. Br., 1966; fellow Am. School of Classical Studies, Athens, 1929-33; m. Dorothy Burr, August 15, 1934; children—Hope, Hilary, Pamela. Instr. classics, U. of B.C., 1925-27; asst. prof. classical archaeology U. of Toronto, 1933-41, asso. prof., 1941-46, professor and head of department of art and archaeology, 1946-47; prof. Classical archaeology, Inst. for Advanced Study, Princeton, N.J., since 1947; asst. dir. and curator of the classical collection, Royal Ont. Mus., Toronto, 1933-47; field dir. Agora Excavations, Am. Sch. of Classical Studies, Athens, 1945-67; George Eastman vis. prof. Oxford U. 1959-60; Geddes-Harrower prof. Greek art and archaeology Aberdeen U., 1965. Served as lt., Royal Canadian Naval Vol. Res., 1942-45; on loan to the Royal Navy, 1943-45; officer in charge of Naval Intelligence in the Adriatic, 1943-45. Decorated Commander Order of the Phoenix (Greece). Fellow Royal Society Canada (retired); corr. fellow Brit. Academy; hon. mem. German Archaeol. Inst., Soc. Promotion Hellenic Studies, Greek Archaeol. Soc., Royal Soc. Arts and Scis., Göteborg, Heidelberg Academy of Sciences (hon.); mem. American Philosophical Soc., Am. Acad. Arts and Scis., Archaeol. Inst. Am. (v.p. 1938-46), Am. Numismatic Soc., Soc. Archtl. Historians. Contbr. studies on Athenian topography and classical archaeology to Hesperia, Am. Jour. of Archaeology, etc. As dir. of excavations of the Am. zone of Athenian Agora, reconstructed Stoa of Attalus, determined plan of Agrippa's Odeum from only few remaining blocks; these excavations known as models in archaeol. recording and reporting. Address: Cherry Valley Rd., Princeton, N.J. 08540.*

THOMPSON, James Burleigh, Jr., Am. geologist; b. Calais, Me., Nov. 20, 1921; s. James Burleigh and Edith (Peabody) T.; A.B., Dartmouth, 1942; Ph.D., Mass. Inst., Tech., 1950; m. Eleanora Mairs, Aug. 3, 1957; 1 son, Michael A. Instr. geology Dartmouth, 1942; research asst. Mass. Inst. Tech., 1946-47, instr., 1947-49; instr. petrology Harvard, 1949-50, asst. prof., 1950-55, asso. prof. mineralogy, 1955-60, prof., 1960-—. Guggenheim fellow 1963. Recipient A.L. Day medal Geol. Soc. Am., 1964. Fellow Mineral. Society America, Geological Society America, Am. Acad. Arts and Scis.; mem. A.A.A.S., Am. Geophys. Union, Sigma Xi. Research on geology of New England; metamorphic petrology. Home: 20 Richmond Rd., Belmont, Mass. Office: Harvard U., Cambridge 38, Mass.

THOMPSON, John Fanning, Am. plant physiologist; b. Ithaca, N.Y., May 24, 1919; s. Homer C. and Clara (Smith) T.; A.B., Oberlin Coll., 1940; Ph.D., Cornell U., 1944; m. Regina R. Machata, Oct. 9, 1943; children—Mark S., Ellen R., James R., William H., Edward J. Faculty, Cornell U., Ithaca, N.Y., 1944-45, 50-—, asso. prof. botany, 1955-—; research asso. dept. botany U. Chgo., 1946-47; research asso. dept. botany U. Rochester, 1947-49, asst. prof., 1949-50; plant physiologist Plant Soil and Nutrition Lab., U. S. Dept. Agr., Ithaca, 1952-—. NIH Postdoctoral fellow, 1947-49; NSF Sr. Postdoctoral fellow, 1959-60. Mem. Am. Chem. Soc. (exec. com. Cornell chpt. 1955-56), Am. Soc. Plant Physiologists, Am. Soc. Biol. Chemistry, Sigma Xi. Asst. editor Plant Physiology, 1960-61, asso. editor, 1966-—. Publs. on devel. methods of measuring amino acids; discovered several amino acids and depeptides; elucidated pathways of synthesis and degradation of amino acids and peptides. Home: 105 Texas Lane, Ithaca, N.Y. 14850.*

THOMPSON, John Richard, Am. marine zoologist; b. Holland, Mich., May 19, 1928; s. Oscar Edward and Verda (Rice) T.; B.A., Albion Coll., 1953, M.A., Duke, 1956, Ph.D., 1963; m. Mary Elizabeth Hamm, June 10, 1950; children—Richard Jonathan, Jeffrey James. With Bur. Comml. Fisheries, U. S. Fish and Wildlife Service, 1956-—, asst. base dir. Exploratory Fishing and Gear Research Base, Pascagoula, Miss., 1963-—. Lectr. to coll. groups, 1965-—. Mem. A.A.A.S., Am. Soc. Zoologists, Soc. Systematic Zoologists, Sigma Xi. Contbr. articles to profl. jours. Marine explorations, faunal survey in Western N. Atlantic from Cape Hatteras to Brazil, Gulf of Mexico, Caribbean Sea from 0-2000 fathoms. Home: 808 Woodmont Rd. Office: 239 Frederic St., Pascagoula, Miss. 39567.*

THOMPSON, John Vaughan, zoologist; b. Eng., Nov. 19, 1779. Asst. surgeon at taking of Demerara and Berbice, in wat against Dutch, 1803; became surgeon, 1803; returned to Eng., 1809; became dist. med. insp., Cork, Ireland, 1816; dep. insp. gen., 1830; became office of health in charge convict med. dept., Sydney, Australia, 1835. Author: A Memoir on Pentacrinus Europaeus, a Recent species discovered in the Cove of Cork, 1823; Zoological Researches, 1828. Research and numerous publs. on marine zoology; described new pouched rat from Jamaica, 1809; studies on extinct Mascarene birds in Madagascar and Mauritius, 1812-16; discoveries on crustacea of Ireland; 1st explanation of spawning of land crabs in sea, 1815, and 1st description of changes during their devel., 1828; described life history of feather star, 1823, barnacle, 1830; demonstrated barnacles are a member of Crustacea not Mollusca; discovered Polyzoa, 1830. Died Sydney, Jan. 21, 1847.

THOMPSON, Joseph Osgood, Am. physicist; b. Weymouth, Mass., July 29, 1863; s. Samuel and Mary Ann (Eaton) T.; student Thayer Acad., S. Braintree, Mass., 1878-79; B.A., Amherst Coll., 1884; Ph.D., U. of Strassburg, Germany, 1891; courtesy fellow, Yale, 1920; m. Lulu Lester Burbank, May 22, 1912; children—Rebecca Burbank, Samuel Mountfort. Teacher of science, Park Coll., Parkville, Mo., 1884-86; asst. in physics, Amherst Coll., 1886-87, instr. in mathematics, 1887-89; instr. in physics, Haverford (Pa.) Coll., 1891-94; asso. prof. physics, Amherst Coll., 1894-1918, prof., 1918-28, emeritus prof., 1928-—. Fellow A.A.A.S., Am. Physical Soc.; mem. Phi Beta Kappa. Established law of elastic lengthening in metals. Died Dec. 12, 1953.

THOMPSON, Laura (Mrs. Sam Duker), Am. anthropologist; b. Honolulu, Jan. 23, 1905; d. William and Maud (Balch) Thompson; B.A., Mills Coll., 1927; postgrad. Radcliffe Coll., 1928; Ph.D., U. Cal. at Berkeley, 1933; m. Sam Duker, June 7, 1963. Asst. ethnologist Bishop Mus., Honolulu, 1929-34; cons. Govt. of Guam, 1938-40; social scientist Ter. of Hawaii, 1940-41; coordinator personality research U. S. Office of Indian Affairs, 1941-47; cons. Inst. Ethnic Affairs, Washington, 1947-54; prof. anthropology Coll. City N.Y., 1954-56; faculty U. N.C., 1957, N.C. State Coll., Raleigh, 1959, Pa. State Coll., 1960, U. So. Ill., 1961, San Francisco State Coll., 1962; lectr. Bklyn. Coll., City U. N.Y., 1964. Mem. Policy bd. advisers Nat. Indian Inst., 1948; cons. Hutterite socialization project Pa. State U., 1962-65. Fellow A.A.A.S., N.Y. Acad. Scis., Am. Anthropol. Assn., Soc. for Applied Anthropology (founder), Phi Beta Kappa. Author: Fijian Frontier, 1940; The Hopi Way (with Alice Joseph), 1944; Guam and its People, 1947; Culture in Crisis, 1950; Personality and Government, 1951; Toward a Science of Mankind, 1961; also numerous articles. Research on hist. and edn. anthropology in Guam, ednl. anthropology in Hawaii, ednl. and psychiat. anthropology among Hopi, Papago, Sioux, Navaho, and Zuni Indians, cultural anthropology and values research among Icelanders and Lower Saxons of W. Germany, values research, Hutterites; ethnographic field work in Lau, Fiji. Address: 3215 Av. H, Bklyn. 11210.*

THOMPSON, Lloyd James, Am. physician; b. Princeton, Mo., Jan. 23, 1895; s. George R. and Martha (Foster) T.; A.B., U. Mo., 1917; M.D., Washington U., St. Louis, 1919; m. Dorothy A. Cannon, June 20, 1930; children—Nancy M. (Mrs. John Knight Tipton), James L. Faculty, Harvard Med. Sch., 1920-22, Yale Med. Sch., 1926-46, Wake Forest Coll. Med. Sch., 1946-56; clin. prof. psychiatry U. N.C., Chapel Hill, 1960-65. Cons. psychiatry to surgeon gen. Dept. Army, 1946-—, VA Hosp., Salisbury, N.C., 1954-60. Fellow Am. Psychiat. Assn. (life), Am. Orthopsychiat. Assn. (life), Am. Acad. Child Psychiatry; mem. Assn. Research Nervous Mental Disease, Group for Advancement Psychiatry, Phi Beta Kappa, Alpha Omega Alpha. Author: Mental Health Study of North Carolina, 1937; Reading Disability: Developmental Dyslexia, 1966. Contbr. articles to med. jours. Research in neurosyphilis and presenile psychoses, student mental hygiene and community psychiatry, paranatal mental health, lang. disturbances in brain-damaged children, developmental dyslexia. Address: Kings Mill Rd., Chapel Hill, N.C. 27514.*

THOMPSON, M(arcus) L(uther), Am. geologist; b. Liberty, Miss., July 12, 1906; s. Thomas Luther and Bessie (Lusk) T.; B.S., Miss. State Coll., 1930; M.S., State U. Ia., 1933, Ph.D., 1934; m. Velma Hagemeier, July 7, 1933; children—Thomas Luther, Jon Edward. Instr., Miss. State Coll., 1930-31; research asso. State U. Ia., 1934-36; asst. prof. N.M. Sch. Mines, 1939-40, asso. prof., 1940-42; asst. prof. U. Kan., 1943-44, asso. prof. 1944-46; asso. prof. U. Wis., 1946-48, prof. geology, 1948-54; prof. and chmn. dept. geology U. Kan., 1954-57; principal geologist and head of the geological group Illinois State Geological survey, 1957-67, principal research geologist, 1967-—; vis. Fulbright professor Kyushu U., Japan, 1960-61; paleontologist Shell Oil Co., Houston, 1937-38, geologist Phillips Petroleum Co., Lafayette, La., 1938-39, N.M. Bur. Mines, Socorro, 1939-42, Kan. Geol. Survey, Lawrence, 1942-46. Geol. expdn., So. Mexico and Guatemala, 1940, B.C., Yukon and Alaska, 1950. Del. 20th Internat. Geol. Congress, Mexico, 1956. Fellow Geol. Soc. Am.; mem. Paleontol. Soc., Am. Assn. Petroleum Geologists, Soc. Econ. Paleontologists and Mineralogists, Sigma Xi. Editor: Jour. of Paleontology, 1957-62. Research articles in sci. pubs. Research on fusulinid foraminifera; upper Paleozoic stratigraphic studies. Home: 303 W. Vermont St., Urbana. Office: Ill. State Geol. Survey, Natural Resources Bldg., Urbana, Ill. 61801.*

THOMPSON, Mary Harris, Am. physician, surgeon; b. Ft. Ann, N.Y., Apr. 15, 1829; d. John Harris and Calista (Corbin) T.; M.D., New Eng. Female Med. Coll., Boston, 1863; grad. Chgo. Med. Coll., 1870. Staff, Elizabeth Blackwell's Infirmary for Women and Children, N.Y.C.; began practice medicine, Chgo., 1863; joined staff U. S. San. Commn., 1863; founder Chgo. Hosp. for Women and Children (later Mary Thompson Hosp. Chgo.), 1865; founder Women's Med. Coll., 1868; attending physician, then head physician and surgeon Chgo. Hosp. for Women and Children, also lectr., clin. prof. gynecology. Mem. Chgo. Med. Soc., Internat. Med. Soc., A.M.A. (chmn. div. on children's diseases). First woman physician in Chgo.; pioneered opening of med. profession to women in Ill.; introduced various techniques for abdominal and pelvic operations. Died Chgo. May 23, 1895.

THOMPSON, (James) Maurice, Am. naturalist, geologist; b. Fairfield, Ind., Sept. 9, 1844, reared in Ga.; ed. by pvt. tutors on plantation in Ga. for civil engr.; carefully trained in Greek, Latin and French; m. Alice Lee, 1868; 1 son, 2 daus. served in Confed. army; after war chief engr. of a railroad in Ind.; later practiced law, Crawfordsville, Ind.; mem. Ind. legislature, 1878; del. to Dem. Nat. Conv., St. Louis, 1888; State geologist of Ind., 1885-89. For years a literary editor on staff of N.Y. Independent. Author: Poems: A Tallahassee Girl; Stories of the Cherokee Hills; Ethics of Literary Art; Toxophilus in Arcadia; His Second Campaign; At Love's Extremes; A Fortnight of Folly; The Ocala Boy; King of Honey Island; Hoosier Mosaics; The Witchery of Archery; Songs of Fair Weather; Byways and Bird Notes; Sylvan Secrets; The Story of Louisiana; Lincoln's Grave (poem); Explored, 1867, Lake Okeechobee, listing its birds, animals and plants; made ornithol. explorations of Okefenokee Swamp, Terre aux Boeufs and islands So. La., wilds of No. Mich., hill country Ala., Miss. and Ga. Died Crawfordsville, Ind., Feb. 15, 1901.

THOMPSON, Milton John, Am. aero. engr.; b. Grand Rapids, Mich., July 28, 1904; s. Schuyler Deane and Jennie (Albertson) T.; B.S. in Aero. Engring., U. Mich., 1925, M.S., 1926; Sc.D., Polytech. Inst. Warsaw (Poland), 1930; m. Helen Barbara Frank, Aug. 22, 1931; children—Richard Deane, Barbara Jean (Mrs. Ewing K. Evans). Mem. faculty U. Mich., 1926-28, 30-41; prof. aero. engring. U. Tex., Austin, 1941-—; v.p., cons. engr. Haneman Assos., Inc., Richland Hills, Tex., 1960-—. Fellow Am. Soc. M.E., A.A.A.S., mem. Am. Inst. Aeros. and Astronautics, Am. Soc. Engring. Edn., Am. Astron. Soc., Sigma Xi, Phi Kappa Phi, Tau Beta Pi, Sigma Gamma Tau. Author: (with R. A. Dodge) Fluid Mechanics, 1937. Research and publs. in airfoil theory, boundary layer skin friction and heat transfer, mechanics of flight. Home: 205 McConnell Dr., Austin, Tex. 78746.*

THOMPSON, Paul Everett, Am. biologist; b. Swainsboro, Ga., May 7, 1911; s. William Benjamin and Mary (Kitchens) T.; B.S., Ga. So. Coll., 1931; M.Ed., Duke, 1936; M.S., U. Ga., 1938; Ph.D., U. Chgo., 1943; m. Ruth Marguerite Ivey, May 29, 1933; 1 dau., Virginia (Mrs. William Talbert McEwan, Jr.). Faculty, Ga. Soc. Coll., Statesboro, 1939-41, Tulane U., New Orleans, 1944-45; dir. microbiology Mary I. Bassett Hosp., Cooperstown, N.Y., 1945-46; research parasitologist Parke, Davis & Co., Ann Arbor, Mich., 1946-52, lab. dir. parasitology, 1952-63, dir. parasitology 1963-—. Mem. expert panel on parasitic diseases WHO, 1963-—. Recipient Howard Taylor Ricketts prize U. Chgo., 1943. Mem. Am. Soc. Parasitology (v.p. 1959), Am. Soc. Tropical Medicine and Hygiene, Soc. Protozoologists. Contbr. to profl. textbooks. Research, publs. on biology of malarial parasites in lizards, including discovery of a new pattern of life cycle devel. in the vertebrate, response to chemotherapeutic agts., immunology, and description of 4 new species of plasmodia; research on anti-parasitic drugs culminating in drugs such as pyrvinium pamoate, biallylamicol, paromomycin, amopyroquine, cycloguanil pamoate.; devel. new methods for studying antiamebic drugs, anthelmintics, and antimalarial drugs. Home: 2411 Londonderry Rd., Ann Arbor 48104. Office: 2800 Plymouth Rd., Ann Arbor, Mich. 48106.*

THOMPSON, Raymond Harris, Am. anthropologist; b. Portland, Me., May 10, 1924; s. Raymond and Eloise (MacIntyre) T.; B.S., Tufts U., 1947; A.M., Harvard, 1950, Ph.D., 1955; m. Molly Coit Kendall, Sept. 9, 1948; children—Margaret Kelsey, Mary Frances. Fellow div. hist. research Carnegie Instn. Washington, 1949-51; asst. prof. anthropology, curator mus. anthropology U. Ky., 1952-56; faculty U. Ariz., Tucson, 1956-—, prof. anthropology, head dept., dir. Ariz. State Mus., 1964-—. Mem. adv. panel program in anthropology NSF, 1963-65, grad. fellowship panel, 1964-66. Fellow A.A.A.S., Am. Anthrop. Assn.; mem. Soc. Am. Archaeology (editor 1958-62, exec. com. 1962-65), Am. Ethnol. Soc., Am. Soc. Ethnohistory, Ariz. Acad. Sci., Ariz. Archaeol. and Hist. Soc., Seminario de Cultural Maya, Soc. Mexicana de Antropologia. Author: Modern Yucatecan Maya Pottery, 1958; also articles. Editor: Migrations in New World Culture History, 1958. Home: 6130 Cerrada el Ocote, Tucson, Ariz. 85718.*

THOMPSON, Reginald Campbell, English archeologist; b. London, Eng., Aug. 21, 1876; s. Reginald Edward and Anne Isabella (De Morgan) T.; ed. Caius Coll., Cambridge; M.A., D. Litt.; m. Barbara Broderick, 1911; 2 sons, 1 dau. Asst., Egyptian and Assyrian dept. Brit. Mus., 1899-1905; accompanied L. W. King to nr. East to obtain more accurate version of rock inscription of Darius at Bisitun, 1904; with Sudan Survey Dept., 1906; asst. prof. semitic languages U. Chg., 1907-09; dir. expdns. Brit. Mus. at Nineveh, 1904, 27, 29, 30, 31, Carchemish, 1911, Abu Shahrain, 1918; also expdns. for Byzantine Fund, at Wadi Sarga, 1913. Fellow, Merton Coll., Oxford U., 1923, Shillito reader in Assyriology, sub-warden, 1933-35. Fellow Brit. Acad., Soc. Antiquaries. Author: Reports of the Magicians and Astrologers of Nineveh and Babylon, 2 vols., 1900; The Devels and Evil Spirits of Babylonia, 2 vols., 1903-04; Late Babylonian Letters, 1906; (with L. W. King) The Inscription of Darius the Great at Behistun, 1907; Semitic Magic, 1908; A Pilgrim's Scrip, 1915; A Small Handbook to the History and Antiquities of Mesopotamia, 1918; Assyrian Medical Texts, 1923; The Assyrian Herbal, 1924; and On the Chemistry of the Ancient Assyrians, 1925; A Catalogue of Late Babylonian Tablets,1927; The Epic of Gilgamish (Translation), 1928; (with R. W. Hutchinson) A Century of Exploration at Nineveh, 1929; The Prisms of Esarhaddon and of Ashurbanipal, 1931; A Dictionary of Assyrian Chemistry and Geology, 1936; 500 Plates of Cuneiform Texts from Babylonian Tablets, 1900-06; The Seatonian Prize Poem, Ignatius, 1933; Digger's Fancy, 1938; contbr. to Cambridge Ancient History, Hastings' Dictionary of Religion and Ethics; also articles to jours. Editor Iraq. Excavated mound of Nineveh; revealed site of last capital of Assyrian Empire; regarded as leading authority on Assyrian natural sci. Died Moulsford, May 23, 1941.

THOMPSON, Richard Frederick, Am. physiol. psychologist; b. Portland, Ore., Sept. 6, 1930; s. Frederick A. and Margaret (Marr) T.; B.A., Reed Coll., 1952; M.S., U. Wis., 1953, Ph.D., 1956; m. Judith K. Pedersen, May 22, 1960; children—Kathryn M., Elizabeth K. Faculty U. Ore. Med. Sch., Portland, 1959-67, prof. dept. med. psychology, 1965-67; prof. dept. psychology U. Cal. at Irvine, 1967-—. Recipient Commonwealth Fund award, 1966. Mem. Am., Western (past program chmn.) psychol. assns., Am. Physiol. Soc., Psychonomic Soc., Sigma Xi. Author: Foundations of Psysiological Psychology, 1967; also numerous articles. Research in gen. field of brain function and behavior, role of cerebral cortex, analysis of neural processes underlying learning. Home: 3515 Seabreeze Lane, Corona del Mar, Cal. 92625. Office: Dept. Psychobiology, U. Cal., Irvine, Cal. 92664.*

THOMPSON, Robert Walder, Am. physicist; b. Minn., Dec. 28, 1919; s. Clarence Rudolph and Ruth (Taylor) T.; B.S., U. Minn., 1941; postgrad. Princeton, 1941-43; Ph.D., Mass. Inst. Tech., 1948; m. Wanda Lee Detmer, Dec. 18, 1948; children—Eric, Niels, Karen, Alfred Otto, Anders, Ingrid. Research physicist Princeton, 1941-43; staff Los Alamos Sci. Lab., 1943-45; research asso. Mass. Inst. Tech., 1945-48; asst. prof. physics Ind. U., Bloomington, 1948-53, asso. prof., 1953-57, prof., 1957-59; prof. U. Chgo., 1959—. Fellow Am. Phys. Soc. Research on cosmic radiation, elementary particle physics, mass spectroscopy. Home: 4940 Woodlawn Av., Chgo. 60615.*

THOMPSON, Roy Charles, Am. biochemist; b. Kansas City, Mo., June 19, 1920; s. Roy Charles and Mabel (Gieschen) T.; B.A., U. Tex., 1940, M.A., 1942, Ph.D., 1944; m. Eva Pellock, Mar. 30, 1946; children—Paul, Karl, Lee Ann, Megan. With U. Tex., Austin, 1940-44, asst. prof. chemistry, 1947-50; research chemist Manhattan Dist., U. S. Army Engrs., Plutonium Project, U. Chgo., 1944-46, Radiation Lab., U. Cal. at Berkeley, 1946-47; radiobiologist, mgr. metabolism operation biology sect. Gen. Electric Co., Richland, Wash., 1950-65; research asso. biology dept. Battelle Meml. Inst., Richland, 1965—. Mem. com. on permissible internal dose Nat. Council on Radiation Protection, 1961—. Mem. Am. Chem. Soc., Radiation Research Soc., Health Physics Soc. Research and publs. on analysis, isolation and characterization of bacterial growth factors, separation and characterization of radionuclides, deposition, distbn. and retention of various radionuclides, evaluation of hazard from internally deposited radionuclides. Home: 74 Whitten St. Office: Battelle-N.W., Richland, Wash. 99352.*

THOMPSON, Silvanus Phillips, Brit. physicist; b. York, Eng., June 19, 1851; s. Silvanus and Bridget (Tatham) T.; B.A., London U., 1869, B.Sc., 1875, D.Sc., 1878; hon. M.D., LL.D.; m. Jane Henderson, 1881; 4 daus. Sci. master, York, 1870-75; prof. exptl. physics U. Coll., Bristol, Eng., 1876-85; prin., prof. physics City and Guilds Tech. Coll., Finsbury, Eng., 1885-1916; prof. applied physics U. London. Fellow Royal Soc., 1891; mem. Phys. Soc. (became pres. 1901), Instn. Elec. Engrs. (became pres. 1899), Illuminating Engr. Soc. (pres.), Optical Conf. (became pres. 1912), Röntgen Soc. (pres.), Optical Soc. (became pres. 1905). Author: Elementary Lessons in Electricity and Magnetism, 1881; The Life of Philipp Reis, 1883; Dynamo-Electric Machinery, 1884; The Electro-Magnet and Electromagnetic Mechanisms, 1891; Light Visible and Invisible, 1897; Magnetic Mechanisms, 1891; The Life of Michael Faraday, 1898; Design of Dynamos, 1903; The Life of Lord Kelvin, 1910; Calculus Made Easy, 1910; Translator: De Magnete (Gilbert). Pioneered devel. applied electricity; studied in optics. Died Hampstead, Eng., June 12, 1916.

THOMPSON, Stanley Gerald, Am. nuclear chemist; b. Los Angeles, Mar. 9, 1912; s. Stanley and Bessie (Sims) T.; A.B., U. Cal. at Los Angeles, 1934; Ph.D., U. Cal. at Berkeley, 1948; m. Alice Isobel Smith, Nov. 27, 1938; children—Ruth Ann (Mrs. Kenneth Lincoln), Joyce Ellen. Chemist Standard Oil Co. Cal., 1934-42; chemist Metall. Lab., U. Chgo., 1942-44, 45-46, DuPont Co., Hanford, Wash., 1944-45; staff Lawrence Radiation Lab., U. Cal. at Berkeley, 1946—, lectr. phys. chemistry, 1958-59. Cons., Cal. Research Corp., 1946-49. Guggenheim fellow Nobel Inst. for Physics, Stockholm, Sweden, 1955, Copenhagen, Denmark, 65; recipient award for nuclear applications in chemistry, 1965. Mem. Am. Chem. Soc., Am. Phys. Soc. Contbr. numerous articles to tech. jours. Research in nuclear chemistry, alpha decay of heavy isotopes, spontaneous fission; discovery and early devel. bismuth phosphate separations process for plutonium; co-discoverer elements berkelium, 1949, californium, 1950, einsteinium, fermium, mendelevium.*

THOMPSON, Theophilus, English physician; b. Islington, Eng., Sept. 20, 1807; s. Nathaniel Thompson; ed. St. Bartholomew's Hosp., London; M.D., Edinburgh, 1830; studied in Paris; m. dau. of Nathaniel Watkin. Became physician Northern Dispensary, London; lectr. Grosvenor Place Sch. Medicine; physician Marlborough Street Consumption Hosp., from 1847. Fellow Royal Soc., 1846; pres. Med. and Harveian socs. Author: On the Improvement of Medicine, 1838; History of the Epidemics of Influenza in Great Britain from 1510 to 1837, 1852; Clinical Lectures on Pulmonary Consumption, 1854; Lettsomian Lectures on Pulmonary Consumption; also articles to jours. First to introduce cod-liver oil, and other useful medicines such as bismuth, zinc oxide into Eng.; suggested use of nomenclature of phys. signs in lung affections. Died Aug. 11, 1860.

THOMPSON, Theos Jardin, Am. scientist, nuclear engr.; b. Lincoln, Neb., Aug. 30, 1918; s. Theos Jefferson and Mabel E. (Dow) T.; A.B., U. Neb., 1941, M.A., 1942, Doctor of Science (honorary), 1964; Ph.D. in Nuclear Physics, University of Cal. at Berkeley, 1952; m. Dorothy Sibley, Feb. 14, 1947; children—Jeff, Edward, Robert, Elizabeth. Physicist U. Cal. Radiation Lab., 1948-52, lectr. physics, Berkeley, 1949-52; staff Los Alamos Sci. Lab., 1952-55, reactor design and constrn.; asso. prof. nuclear engring. Mass. Inst. Tech., 1955-58, prof., 1958—, dir. nuclear reactor for design, constrn. and operation; cons. Mem. AEC adv. com. reactor safeguards, 1959-66, chmn., 1960; head, co-editor AEC writing project Tech. of Nuclear Reactor Safety. Recipient E. O. Lawrence Meml. award, Atomic Energy Commn., 1964; Guggenheim Found. fellow, 1963-64. Served from 2d lt. to maj., C.W.S., AUS, 1942-46. Fellow American Nuclear Soc. (director); mem. Am. Academy of Arts and Sciences, American Phys. Soc., Phi Beta Kappa, Sigma Xi. Research in neutron physics; ionization of fundamental particles; design and construction of nuclear reactors. Home: 14 Everett St., Winchester, Mass. Office: 138 Albany St., Cambridge 39, Mass.

THOMPSON, Warren, Am. geologist; b. Palo Alto, Cal., Oct. 14, 1898; s. Frank E. and Clara (Gussefeld) T.; B.A., U. Colo., 1922; Ph.D., Stanford, 1935; m. Jane Patterson, Sept. 7, 1923; children—Peter Mark, Thomas Luman, Robert Wayne. Faculty, U. Colo., Boulder, 1926—, prof. geology, 1941-67, professor emeritus, since 1967—, head of the department of geology, 1949-60; geologist, cons. various oil cos., 1926—. Recipient Stearns award for teaching U. Colo., 1962. Mem. Am. Assn. Petroleum geologists (Distinguished lectr. 1956), Geol. Soc. Am., Rocky Mountain Assn. Geologists (hon.), Soc. Econ. Paleontologists and Mineralogists, Sigma Psi. Research and publs. on description and interpretation of ancient sediments particularly beaches and other littoral forms, modern beaches. Home: 985 Gilbert St., Boulder, Colo. 80302.*

THOMPSON, Wayne Edwin, Am. sociologist; b. Hunter, Kan., Mar. 3, 1927; s. Carlin T. and Zella (Van Leewen) T.; B.A. cum laude in Econs. and Polit. Sci., U. Colo., 1949, M.A., 1951; Ph.D. in Sociology, Cornell U., 1956; m. Ruth C. Cortis-Stanford, Sept. 27, 1953; 1 son, David C. Field dir. study occupational retirement dept. sociology and anthropology Cornell U., Ithaca, N.Y., 1952-53, research asso., 1956-58, co-dir. study occupational retirement, 1956-58, faculty, 1957-68, prof. sociology, 1964-68, sr. researcher studies of community issues, 1958-68, asst. dir. Social Sci. Research Center, 1958-61, dir., 1961-66; prof., head dept. sociology and anthropology Univ. of North Carolina at Greensboro, 1968—. Fulbright lectr. Sch. Social Scis., Tampere, Finland, 1964-65. Mem. Am. Sociol. Assn., Eastern Sociol. Soc., Am. Assn. U. Profs., Phi Beta Kappa, Phi Kappa Phi. Research and publs. on social gerontology especially on relationship of retirement to health, relationship of pre-retirement factors to adjustment in retirement, family in later years; also studies in community politics especially the negativist role of the politically alienated in local referenda. Address: U. N.C., Greensboro, N.C.*

THOMPSON, William Rae, Am. math. statistician; b. Bklyn., July 29, 1896; s. Robert Frederick and Maria Elizabeth (Lilliman) T.; A.B., Columbia U., 1923; Ph.D., Yale, 1930; m. Mary Elspeth Dalton, Nov. 21, 1942. Asst. lab. biophys. research Meml. Hosp., N.Y.C., 1922-24; research asst. pathology Sch. Medicine, Yale, New Haven, 1924-36; sr. biochemist N.Y. State Dept. Health, Albany, 1936-65, sr. research sci. in biometrics, div. labs. and research, 1965—. Recipient Alfred E. Smith award Capital dist. chpt. Am. Pub. Adminstrn. Assn., 1952; achievement award Albany chpt. Am. Statis. Assn., 1963. Fellow Royal Soc., 1933, Inst. Math. Statistics, Am. Statis. Assn., A.A.A.S.; mem. Am. Math. Soc., Am. Soc. Biol. Chemists, Biometric Soc., Econometric Soc., Soc. Indsl. and Applied Math., N.Y. Acad. Scis. Author monographs, also numerous articles. Research in methodology of sci. statis. inference with assumption economy; inventor volumetric calibration apparatus. Home: 1 Darooch Rd., Delmar, N.Y. 12054. Office: Div. Labs. and Research, N.Y. State Dept. Health, New Scotland Av., Albany 12201.*

THOMPSON, William Robert, psychologist; b. Toulon, France, July 10, 1924; s. William Robin and Mary (Carmody) T.; B.A., U. Toronto (Ont., Can.), 1945, M.A., 1947; Ph.D., U. Chgo., 1951; m. Mary T. Forde, Apr. 11, 1953; children—Judith Clare, William Robert. Research asso. McGill U., 1951-54; lectr. Queens U., Kingston, Ont., Can., 1954-56, prof., head dept. psychology, 1966—; faculty Wesleyan U., Middletown, Conn., 1956-66, prof., 1961-66; prof., head dept. psychology Queens U., Kingston, Ont., 1966—. Guggenheim fellow, 1959-60; fellow Center for Advanced Studies in Behavioral Scis., Stanford, Cal., 1963-64. Mem. Am., Canadian, Eastern psychol. assns., Psychonomic Soc., A.A.A.S., Sigma Xi. Author: (with J. L. Fuller) Behavior Genetics, 1960; also articles. Research on influence of genotype on behavior, effects of early experience in animals, evolution and behavior. Home: 158 Fairway Hills, Kingston, Ont., Can.*

THOMSEN, Christian Jurgensen, Danish archaeologist; b. Copenhagen, Denmark, Dec. 29, 1788; s. a merchant. Managed father's business until 1840; 1st curator Danish National Museum, 1816-65; sec. of commission to collect and preserve antiquities, 1816; dir. royal cabinet of medals, 1842. Author: Treatise on Northern Antiquities, 1831; A Guide to Northern Antiquities, 1836. Known as developer of Stone, Bronze and Iron Age categories of prehistorical man based on index of predominant type of tools and artifacts used in particular epoch. Died Copenhagen, May 21, 1865.

THOMSEN, Hans Peter Jörgen Julius, Danish chemist; b. Copenhagen, Denmark, Feb. 16, 1826. Tchr. chemistry Copenhagen Tech. High Sch., 1847-56, Mil. High Sch., 1856-66; prof. chemistry U. Copenhagen, 1866-91; dir. Copenhagen Poly., 1883-92; mem. Copenhagen Municipal Council, 35 years. Recipient Davy medal, 1883. Fgn. mem. Royal Soc., 1902. Author: Thermochemische Untersuchungen, 4 vols., 1882-86. Founder, Jour. Chemistry and Physics, editor, 1862-78. Invented process for prodn. soda from cryolite, 1853, polarization battery used in Danish telegraphic service; predicted existence of group of inactive gases, 1895; confirmed Kirchhoff's equation and Guildberg and Waage's theory of mass action, 1867; numerous calorimetric measurements; originated term, acidity, to refer to acid's tendency to unite with base; 1st tables of relative strength of acids; determined atomic weights of oxygen and aluminum. Died Copenhagen, Feb. 13, 1909.

THOMSEN, Wilhelm, German orthopedist; b. Friedrichstadt, Germany, Mar. 31, 1901; s. Wilhelm and Margarethe (Mahrt) T.; student univs. Kiel, Munich, Gottingen; M.D.; m. Ingeborg von Meyer, Sept. 19, 1931; children—Knut, Dagmar, Astrid, Jörg. Sports and med. tchr. Odenwaldschule; local med. and sports editor Bad Orb Children's Sch.; enetered orthopedic medicine with Prof. Lange, 1927; individual practice orthopedic medicine, Munich; staff U. Orthopedic Clinic; worked under Prof. Ludloff, U. Frankfort Orthopedic Clinic, from 1928; worked under Prof. Hohmann, 1930-45; head physician U. Clinic, 1934—; became prof., surgeon-in-chief Bad Homburg, Germany, 1941; med. controller Red Cross Hosp., Frankfort/Main, West Germany. Recipient Order of Merit Republic of Germany, 1963). Mem. Deutsche Lebensrettungsgesellschaft (pres. 1953—), Akademische Segler-Verein. Research and numerous publs. on orthopedy, feet and shoes, mechanism of articulation, prothesis equipment. Home: Feldstrasse 75, Bad Homburg, West Germany. Office: Königwarter Strasse 10, Frankfort/Main, West Germany.

THOMSON, Allen, Brit. biologist; b. Edinburgh, Scotland, Apr. 2, 1809; s. John and Margaret (Millar) T.; student Edinburgh, Paris; M.D., Edinburgh, 1830, LL.D., 1871; LL.D., Glasgow, Scotland, 1877; D.C.L. (hon.), U. Oxford, 1882; m. Ninian Jane Hill; 1 son, John Millar. Visited schs. and museums of Germany, Holland, Italy; lectr. physiology and microscopic anatomy; prof. anatomy Marischal Coll., Aberdeen, Scotland, 1839-41; prof. physiology Edinburgh, Scotland, 1842-48; prof. anatomy and physiology, Glasgow, 1848-77. Fellow Royal Soc., 1848, became councillor, 1877, a v.p., 1878; fellow Royal Coll. Surgeons Edinburgh, Royal Soc. Edinburgh; mem. Royal Med. Soc. Edinburgh (became pres. 1830), Philos. Soc. (pres.), Medico-Chirurg. Soc. (pres.), Sci. Lectures Assn. Glasgow (pres.), Brit. Assn. (became pres. 1876), Brit. Med. Assn. (1st pres. local chpt.). Editor: Life (Cullen), 2d vol. Research and publs. in embryology, devel. of organs (especially circulatory and genito-urinary systems), physiol. optics. Died London, Mar. 21, 1884.

THOMSON, Anthony Todd, Brit. physician; b. Edinburgh, Scotland, Jan. 7, 1778; s. Alexander Thomson; student U. Edinburgh, 1795-97; m. Christina Maxwell, 1801; 1 son, 2 daus.; m. Katherine Byerley, 1820; 3 sons, including Henry William (Byerley); 5 daus. Became practice medicine, London, 1800; founder Chelsea Dispensary, 1812; became prof. materia medica and therapeutics London U., 1828, prof. med. jurisprudence, 1832; also physician to dispensary of U. Coll. Fellow Royal Coll. Physicians; mem. Speculative Soc., Royal Med. Soc., Royal Coll. Surgeons. Author: Conspectus Pharmacopoeiae, 1918; The London Dispensatory, 1811; Lectures on the Elements of Botany, 1822; Elements of Materia Medica and Therapeutics, 1832. An editor: London Med. Repository, vols. I-VIII, 1814-17; editor: Practical Synopsis of Cutaneous Diseases (Bateman), 1829; The Season, 1847. Translator: The Philosophy of Magic, Prodigies and Apparent Miracles (A.J. Eusèbe Baconnière Salverte), 2 vols., 1846. Research on alkaloids and iodides, diagnosis and treatment of skin diseases, botany. Died Ealing, Eng., July 3, 1849.

THOMSON, Arthur, Brit. anatomist; b. Edinburgh, Scotland, Mar. 21, 1858; s. John and Mary (Arthur) T.; M.B., Edinburgh (Scotland) U., 1880, hon. degree, 1915; hon. degrees, Durham, Eng., Oxford, Eng., 1933; m. Mary Walker MacBeth, 1888; 2 daus. Demonstrator anatomy to Sir William Turner; became univ. lectr. in human anatomy Oxford, 1885, asso. prof. human anatomy, 1893, 1st Dr. Lee's prof., 1919-33, rep. of Oxford U. on Gen. Med. Council, 1904-29, assisted in devel. Med. Sch.; named prof. anatomy Royal Acad., 1900. Fellow Royal Coll. Physicians; mem. Oxford Med. Grad. Club (pres.), Anthrop. Soc. Paris, Anat. Soc. Gt. Britain and Ireland (pres.), Anat. Soc. (pres.), Internat. Med. Congress. Author: Handbook of Anatomy for Art Students, 1896; (with David Randall-MacIver) The Ancient Races of the Thebald, 1905; Anatomy of the Human Eye, 1912. Research on early devel. of human embryo, ophthalmology, anthropology, squatting facets on knee and ankle bone. Died Oxford, Feb. 7, 1935.

THOMSON, Sir Charles Wyville, Scottish naturalist; b. Bonsyde, Scotland, Mar. 5, 1830; s. Andrew Thomson; ed. Edinburgh (Scotland) U.; LL.D., Aberdeen, Scotland, 1853, Queen's U., Ireland, 1860, Dublin, Ireland, 1878; D.Sc., Queen's U., 1871; Ph.D., Jena, Germany; m. Jane Ramage Dawson, 1853; 1 son, Frank Wyville. Successively became lectr. botany King's Coll., Aberdeen, 1850, prof. Marischal Coll., 1851, prof. natural history Queen's Coll., Cork, Ireland, 1853; named prof. mineralogy and geology Queen's Coll., Belfast, Ireland, 1854, prof. natural sci., 1860; named prof. botany Royal Coll. Sci., Dublin, 1868; prof. natural history U. Edinburgh, 1870; dir. sci. staff Challenger Expdn., 1872-76. Recipient Royal medal, 1876. Fellow Royal Soc., 1869, Royal Soc. Edinburgh, Linnean Soc., Geol. Soc., Zool. Soc.; mem. Royal Irish Acad. Author: Depths of the Sea, 1873; Voyage of the Challenger in the Atlantic, 2 vols., 1877; also articles. Organized sounding and dredging expdns. to N. of Scotland, 1868, Mediterranean, 1870; proved animal life is abundant down to 650 fathoms; discovered deep-sea temperatures are not constant, 3-dimensional phenomenon of ocean, studies on echinoderms. Died Bonsyde, Scotland, Mar. 10, 1882.

THOMSON, Elihu, electrician; b. Manchester, Eng., Mar. 29, 1853; s. Daniel and Mary A. (Rhodes) T.; A.B., Central High Sch., Phila., 1870, A.M., 1875; hon. A.M., Yale, 1890; Ph.D., Tufts, 1894; Sc.D., Harvard, 1909; LL.D., U. of Pa., 1924; D.Sc., Victoria Univ., Manchester, England, 1924; m. Mary L. Peck, May 1, 1884 (died 1916); children—Stuart (dec.), Roland D., Malcolm, Donald T.; m. 2d, Clarissa Hovey, 1923. Prof. chemistry and mechanics, Central High Sch., Phila., 1870-80; 1880—; electrician for Thomson-Houston and General Electric cos., which operate under his inventions, more than 700 patents having been obtained; dir. Thomson Lab. of Gen. Electric Co., Lynn, Mass. Pres. Internat. Elec. Congress and chamber of official delegates thereto, St. Louis, 1904; pres. Internat. Electrotech. Commn., 1908-11. Fellow Am. Acad. Arts and Sciences (v.p.); mem. Nat. Acad. Sciences. Awarded Grand Prix in Paris, 1889 and 1900, for elec. inventions; decorated, 1889, by French Govt., Chevalier and Officer Legion of Honor, for elec. research and inventions; grand prize, St. Louis, 1904, for elec. work; Rumford medal, 1902; awarded 1st Edison medal, Am. Inst. E.E., 1910; Elliott Cresson medal, John Fritz medal, and Hughes medal of Royal Soc., London, 1916; Kelvin medal, 1924; the Franklin medal, by the Franklin Inst., 1925; Faraday medal, by Instn. of Elec. Engrs., Gt. Britain, 1927; also twice awarded the John Scott Legacy medal and premium, by City of Phila., and medals, Trans. Miss. Expn., Omaha, and Columbian Expn., Chicago; also Grashof Medal awarded by the Verein Deutscher Ingenieure of Germany, Mar. 29, 1935. Trustee and pres. Peabody Mus., Salem, Mass.; mem. Corp. Mass. Inst. Tech. and its acting pres., 1920-22; v.p. Am. Philos. Soc. Mem. Nat. Research Council. Contbr. to tech. jours. Invented high frequency generator, röntgen stereoscope, electric welding which bears his name; many additional inventions in electric lighting, power, etc. Died Mar. 13, 1937.

THOMSON, George, English physician; b. circa 1620; M.D., Leyden, 1648; m. Abigail Nettleshipp, Nov. 2, 1667; m. 2d, Martha Bathurst, Oct. 31, 1672. A Royalist, served in English civil war under Prince Maurice; went to Leyden after overthrow of Royalists; lived in London during plague of 1665, made spl. study of symptoms. Author: Loimologia: a Consolatory Advice, and some brief Observations concerning the present pest, 1665; Galeno-pale, or a chymical Trial of the Galenists, that their Dross in Physick may be discovered, 1665; A Gag for Johnson . . ., 1665; The Pest anatomised, 1666; The true Way of preserving the Blood, 1670; A check given to the insolent garrulity of H. Stubbe, 1671; Epilogismi chymici Observationes necnon Remedia Hermetica Longa in Arte Hiatrica exercitatione constabilita, 1673; The direct Method of Curing Chymically, 1675. Made detailed studies of plague; wrote against practices of excessive bleeding and purging, also method of trying to cure diseases by contraries. Died after 1684 (or possibly before 1680).

THOMSON, Sir George Paget, English physicist; b. Cambridge, Eng., May 3, 1892; s. Joseph John and Rose Elisabeth (Paget) T.; ed. Trinity Coll., Cambridge, 1910-14; D.Sc., U. Lisbon (Portugal), 1935, LL.D., Aberdeen, Scotland, 1948; Sc.D., U. Dublin (Ireland), 1948, U. Sheffield (Eng.), 1956, U. Wales, 1960, U. Reading (Eng.), 1960, Westminster Coll., 1964; L.H.D., Ursinus Coll., 1963; m. Kathleen Adam Smith, Sept. 18, 1924; children—John Adam, (Lilian) Clare (Mrs. Jan de V Graaff), David Paget, (Caroline) Rose Buchanan (Mrs. Colin Bell). Fellow, lectr. Corpus Christi Coll., Cambridge, 1914, 19-22; prof. U. Aberdeen (Scotland), 1922-30; prof. physics Imperial Coll. Sci., 1930-52; master Corpus Christi Coll., Cambridge, 1952-62. Chmn., 1st Brit. Com. on Atomic Energy, 1940-41; hon. cons. Atomic Energy Research Establishment, 1952—. Knighted, 1943. Recipient Nobel prize for physics (with C. J. Davisson), 1937; Faraday medal Instn. Elec. Engrs., 1960. Fellow Royal Soc., 1930 (Hughes medal 1939, Royal medal 1949); mem. Instn. Physics (past pres.), Brit. Assn. for Advancement Sci. (past pres.). Author: Applied Aerodynamics, 1919; The Atom, 1930; (with J. J. Thomson) Conduction of Electricity Through Gases, Vol. 1, 1928, Vol. 2, 1933; Wave Mechanics of the Free Electron, 1930; (with W. Cochrane) Theory and Practice of Electron Diffraction, 1939; The Foreseeable Future, 1955; The Inspiration of Science, 1961; J. J. Thomson and the Cavendish Laboratory, 1964; also numerous articles. Independently but simultaneously with Davisson discovered exptl. confirmation of wave properties of electrons; research on aerodynamics; dir. (with Chadwick) Brit. wartime research on atomic energy; research on nuclear power from fusion. Home: Little Howe, Mt. Pleasant, Cambridge, Eng.*

THOMSON, James, Brit. engr.; b. Belfast, Ireland, Feb. 16, 1822; s. James Thomson; M.A., Glasgow (Scotland), 1839, LL.D., 1870; LL.D., Dublin, 1878; D.Sc., Queen's U., Ireland, 1875; student Horsley Ironworks, Tipton, Eng., Fairbairn & Co's Works; m. Elizabeth Hancock, 1853; 1 son, 2 daus. Became civil engr., Belfast, 1851; named resident engr. to water commrs., 1853; prof. civil engring. Queen's Coll., 1857-73; prof. civil engring., Glasgow, Scotland, 1873-89. Fellow Royal Soc., 1877. Contbr. numerous articles to tech. jours. Invented Vortex water-wheel, 1850; developed various improvements in water wheels; studies on whirling fluids which lead to improvement of blowing fans, river currents, grand currents of atmospheric circulation, continuity of gaseous and liquid states of matter. Invented centrifugal pump; designed jet-pump for drainage of low lands. Died May 8, 1892.

THOMSON, John, Scotch physician, surgeon; b. Paisley, Scotland, Mar. 15, 1765; s. Joseph and Mary (Millar) T.; apprenticed to Dr. White of Paisley, 1785-88; student U. Glasgow, 1788-89, Edinburgh, 1789-90; M.D., John Hunters Sch. Medicine, 1792; student U. Edinburgh, King's Coll. Aberdeen, 1808; m. Margaret Crawford Gordon, 1793; 3 children, including William Thomson; m. 2d, Margaret Millar, 1806; a dau.; a son, Allen. Became asst. apothecary Royal Infirmary, Edinburgh, Soctland, 1790, house surgeon, 1791-92, surgeon, 1793-98; tchr. pvt. classes for gentlemen connected with Parliament House, 1799-1800; began teaching surgery Royal Infirmary, 1800; apptd. prof. surgery Coll. Surgeons Edinburgh, 1805; became prof. mil. surgery U. Edinburgh, 1808; founder chair gen. pathology; a founder Edinburgh New Town Dispensary, 1815; became prof. gen. pathology, Edinburgh, 1832. Licentiate, fellow Royal Coll. Physicians Edinburgh; mem. Med. Soc. (became pres. 1791). Author: The Elements of Chemistry and Natural History, 1798; Observations on Lithotomy with a New Manner of Cutting for Stone, 1808; Lectures in Inflammation: A View of the General Doctrines of Medical Surgery, 1813; also autobiography, 1832. Editor: The Works of William Cullen, M.D., 1827. Research on treatment war wounds, effectiveness smallpox vaccination. Died Edinburgh, Oct. 11, 1846.

THOMSON, Sir John Arthur, Scotch naturalist, biologist; b. East Lothian, Scotland, July 8, 1861; student univs. Edinburgh, Jena, Berlin; M.A., LL.D., Edinburgh; LL.D., McGill U., U. Cal., Aberdeen; m. 1889; 3 sons, 1 dau. Lectr. biology and zoology Sch. Medicine, Edinburgh; Regius prof. natural history Aberdeen U., 1899-1930; Gifford lectr., St. Andrews, Scotland, 1915; Terry lectr., Yale, 1924; Morse lectr. Union Sem., N.Y.C., 1924. Author: The Study of Animal Life, 1892, rev., 1917; Outlines of Zoology, 8th edit. 1929; The Natural History of the Year; The Science of Life; Progress of Science in the Nineteenth Century, 1904; Herbert Spencer, 1906; Heredity, 1908, rev. 1926; The Bible of Nature, 1909; Darwinism and Human Life, 1910, rev. 1916; The Biology of the Season, 1911; (with Patrick Geddes) Evolution, 1911; Introduction to Science, 1911; (with Patrick Geddes) Sex, 1914; The Wonder of Life, 1914; Secrets of Animal Life, 1919; The System of Animate Nature, 1920; Nature All the Year Round, 1921; The Control of Life, 1921; The Haunts of Life, 1922; What is Man?, 1923; The Biology of Birds, 1928; Science, Old and New, 1924; (with Patrick Geddes) Biology, 1924; Science and Religion, 1925; Concerning Evolution, 1925; The Gospel of Evolution 1925; The New Natural History, 1925-26; Towards Health, 1927; Modern Science, 1929; (with Patrick Geddes) Life: Outlines of General History, 1932; Scientific Riddles, 1932; Research on Alcyonarians (soft coral); attempted to reconcile science and religion. Died Limpsfield, Eng., Feb. 12, 1933.

THOMSON, Joseph, Brit. explorer; b. Dumfriesshire, Scotland, Feb. 14, 1858; s. William and Agnes (Brown) T.; ed. Edinburgh (Scotland) U. Geologist with Alexander Keith Johnston's expdn. to Central Africa, 1878-80, leader (on death of Johnston), 1879; leader expdn. for opening route between east coast and northern shore of Victoria Nyanza, 1882-83 (traversed country of Masaii, visited lake, reached Rabai); leader expdn. for Nat. African Co. to Sekoto, also made valuable treaties, 1885, expdn. from Quilimane to Kwa Chepo, 1890-91; explored Atlas Mountains, Morocco, 1888. Recipient Gold medal Royal Geog. Soc., 1885. Author: To the Central African Lakes and Back, 1881; Through Masai Land, 1885; Travels in the Atlas and Southern Morocco, 1889; Mungo Park and the Niger, 1890. Made extensive additions to geol. map Africa; serious contbns. to zoology, botany; several newly described Central African bot. species named for him. Died London, Eng., Aug. 2, 1895.

THOMSON, Sir Joseph John, English physicist; b. Chetham Hill, nr. Manchester, Eng., Dec. 18, 1856; s. Joseph James and Emma (Swindells) T.; ed. Owens Coll., Manchester; Trinity Coll., Cambridge, B.A., 1880 (2d wrangler in math. tripos and 2d Smith's prizeman); Fellow, 1880; hon. degrees from Oxford, Dublin, London, Victoria, Columbia, Cambridge, Durham, Birmingham, Göttingen, Leeds, Oslo, Paris, Edinburgh, Reading, Princeton, Glasgow, Johns Hopkins, Aberdeen, Athens, Cracow, Pennsylvania. Lectr. math., Trinity College, 1892; Univ. lecturer, Cambridge U., 1893; Cavendish prof. experimental physics, Cambridge U., 1894-1919; prof. natural philosophy, Royal Institute of Great Britain, 1905-18; during WWI, advisor to various govt. depts. and mem. Board of Inventions and Research; Master of Trinity Coll., Cambridge U., from 1918. Fellow, Royal Soc., 1884 (Royal medal, 1894; Hughes medal, 1902; Copley medal, 1914; pres. 1916-20); mem. Brit. Assn. Advancement Sci. (pres. 1909); Cambridge Philos. Soc. (pres.); fellow or mem. Royal Soc., Edinburgh, Royal Soc., Turin, Royal Irish Acad., Royal Soc., Uppsala, Am. Acad. Arts and Scis., Am. Philos. Soc.; French Acad. Scis. Recipient Nobel Prize for Physics, 1906; knighted 1908; Order of Merit, 1912; Hodgkins Medal, Smithsonian Inst., 1902; Franklin Medal, 1923; Scott Medal, 1923; Mascart Medal, 1927; Dalton Medal, 1931; Faraday Medal, 1938. Author: A Treatise on the Motion of Vortex Rings, 1884; Applications of Dynamics to Physics and Chemistry, 1886; Recent Researches in Electricity and Magnetism, 1892; Elements of Mathematical Theory of Electricity and Magnetism, 1895; Discharge of Electricity through Gases, 1897; Conduction of Electricity through Gases, 1903; The Structure of Light; The Corpuscular Theory of Matter, 1907; Rays of Positive Electricity, 1913; The Electron and Chemistry, 1923; Recollections and Reflections, 1936. Working with highly evacuated tubes, studied electric conduction in gases; deflected cathode rays in electric as well as magnetic fields; demonstrated conclusively that cathode rays negatively charged particles; measured ratio of charge to mass of cathode ray particles; found mass to be circa 1837 times smaller than mass of hydrogen atom, thereby opening up study of sub-atomic particles; these particles were called electrons by Lorentz; hence Thomson considered discoverer of electron, 1897; suggested plum pudding model of atomic structure (like raisins in pudding, electrons imbedded in sphere of positively charged electricity in such a way as to neutralize positive charge); investigated positive rays (positive ion beams); his pupil, Aston, established existence of isotopes in further research along this line. Died Cambridge, Eng., Aug. 30, 1940.

THOMSON, Junius Richard, Am. biologist; b. Rutherfordton, N.C., Feb. 24, 1927; s. Richard Lewis and Mattie Lee (Withrow) T.; B.S., Emory U., 1948, M.S., 1949; m. Margie Ellen Klein, Oct. 16, 1948; children—Karen, Stephen. Asst. bacteriologist Communicable Disease Center, Atlanta, 1949-50; faculty U. Chattanooga, 1950-51; asst. scientist Oak Ridge Nat. Lab., 1951-52; scientist So. Research Inst., Birmingham, Ala., 1952-64; sr. biologist Midwest Research Inst., Kansas City, Mo., 1964-66; sci. information analyst R. J. Reynolds Tobacco Co., Winston, Salem, N.C., 1966—. Free-lance writer on bionics, biomed. electronics, biodetection, 1962—. Fellow A.A.A.S., Ala. Acad. Sci.; mem. Am. Soc. Zoologists, Am. Soc. Profl. Biologists, Sci. Research Soc. Am., Mo. Acad. Sci., Sigma Xi. Contbr. articles to profl. jours. Research on cancer chemotherapy, screening drugs for anticancer and antileukemic activity, mechanism of action of drugs, structure-activity relationships, tissue transplantation, combination chemotherapy, drug resistance, synthetic diets. Home: 2308 Bitting Rd., Winston-Salem 27104. Office: Research Dept., R. J. Reynolds Tobacco Co., Winston-Salem, N.C. 27102.*

THOMSON, Robb Milton, Am. physicist; b. El Paso, Tex., Feb. 4, 1925; s. Rollins and Annie (MacMorrough) T.; student Tex. Western Coll., 1942; student U. Chgo., 1942-44, M.S., 1950; Ph.D., U. Syracuse, 1953; m. Alice Clendenin, May 8, 1948; children—Eric R., Judith E. From research asso. to prof. U. Ill., Urbana, 1953-68, prof. depts. physics and metallurgy, 1961-68; prof., head dept. materials scis. State U. N.Y., Stony Brook, 1968—; dir. materials scis. Advanced Research Projects Agy., Washington, 1965-68. Author: (with C. Wert) Physics of Solids, 1964. Research on imperfections in solids. Home: 5 Brewster Ct., Setauket, N.Y. Office: State U. N.Y., Stony Brook, N.Y.*

THOMSON, Robert Dundas, English physician; b. Eccles, Eng., Sept. 21, 1810; s. James and Elizabeth (Skene) T.; ed. Edinburgh (Scotland) U.; M.D., C.M., Glasgow (Scotland) U., 1831; studied under Liebig, Giessen, Germany; m. Miss Thomson. Asst. surgeon on voyage to India and China, E. India Co.; began practice medicine, London, 1835; asst. to Thomas Thomson, Glasgow; became chem. lectr. St. Thomas's Hosp., London, 1852; named med. officer health Marylebone, London, 1856. Fellow Royal Soc., 1854, Royal Coll. Physicians; mem. Brit. Meteorol. Soc. (pres.), Met. Assn. Med. Officers Health (pres.). Author: Records of General Science, 1835; British Annual and Epitome of the Progress of Science, 1837; Digestion:

The Influence of Alcoholic Fluids in that Function, and on the Value of Health and Life, 1841; Experimental Research on the Food of Animals and the Fattening of Cattle, with Remarks on the Food of Man, 1846; School Chemistry, or Practical Rudiments of the Science, 1848; Cyclopaedia of Chemistry, Mineralogy, and Physiology, 1854; Report to Government on the Waters of London during Cholera, 1854; The British Empire, 1856; Annual Report on the Health of the Parish of St. Marylebone, 1857. Research on blood compositions, food composition related to systems of animals, sanitation; water analysis. Died Richmond, Eng., Aug. 17, 1864.

THOMSON, Robert William, Scottish engr.; b. Stonehaven, Scotland, 1822; sent to Charleston (U. S.) to be educated as mcht., returned home shortly and began self-edn. Encouraged by Faraday and employed by Robert Stephenson; ry. engr., 1844; made survey for proposed line in eastern counties of Eng.; went to Java as agt. for engring. firm, 1852. Author: (article) On the Formation of Coal and on the changes produced in the composition of the strata by the solvent action of water slowly penetrating through the Earth's crust during long periods of geological time. Received patent for india rubber tires, 1845, for fountain pen, 1849; devised new machinery for mfg. sugar in Java, 1852; designed 1st portable steam crane, hydraulic docks, 1860, traction engine, 1867. Died Edinburgh, Mar. 8, 1873.

THOMSON, Thomas, Scottish chemist; b. Crieff, Scotland, Apr. 12, 1773; s. John and Elizabeth (Ewan) T.; student St. Andrews, 1787-90; M.D., Edinburgh, 1799; m. Agnes Colquhon, 1816; children—Thomas, Mrs. Robert Dundas Thomson. Tutor in family of Mr. Kerr, Blackshields, 1790; lectr. chemistry Edinburgh, 1800-11; opened 1st chem. lab. for practical instrn. in U.K., Edinburgh; lectr. chemistry U. Glasgow, 1817, regius prof., 1818. Fellow Royal Soc., mem. Philos. Soc. Glasgow (pres. 1834). Author: System of Chemistry, 1802; Elements of Chemistry, 1810; An Attempt to establish the First Principles of Chemistry by Experiment, 2 vols., 1825; History of Chemistry, 2 vols., 1830-31; An Outline of the Sciences of Heat and Electricity, 1830; Chemistry of Inorganic Bodies, 1831; Outlines of Mineralogy, 1836; Chemistry of Animal Bodies, 1843; Brewing and Distillation, 1849; also over 200 papers in Trans. of Royal Soc., Records of Sci.; contbr. to Ency. Brit. Editor: Anns. of Philosophy, 1813-21. Made known and supported Dalton's atomic theory, 1807 (1st detailed publ. on theory); determined several atomic weights (his results favorable to Prout's hypothesis), 1825; observed multiple proportions in oxalates, 1808; discovered sulphur chloride, 1803, chromyl chloride, 1827. Died July 2, 1852.

THOMSON, Sir William, 1st Baron Kelvin of Largs, Brit. physicist; b. College Square East, Belfast, Ireland, June 26, 1824; s. Prof. James and Margaret (Gardiner) T.; ed. by his father, who was prof. math. in Belfast and later U. Glasgow; attended U. Glasgow, 1834-40; studied at Peterhouse, U. Cambridge, 1941-45 (2d wrangler in math. tripos and 1st Smith's prizeman, 1845); Fellow, 1845-52; 1873-1907; visited Faraday's lab. at Royal Institution, London; attended U. Paris, 1845; numerous hon. degrees; m. Margaret Crum, 1852 (d. 1870); m. 2d Frances Anna Blandy. Prof. natural philos., U. Glasgow, 1846-99 (resigned 1899); refused 3 times Cavendish professorships of physics at Cambridge U.; chancellor, U. Glasgow, 1904. Mem., admiralty cttee. on designs of ships of war, 1871, 1904-05; mem. Brit. Assn. Advancement Sci. (pres. 1871); foundation mem., Soc. of Telegraph Engrs. (vice-pres. 1871, pres. 1874); privy councilor, 1902; knighted, 1866; raised to peerage, created baron, 1892. Fellow, Royal Soc., 1851 (Copley medal 1883, pres. 1890-94); mem. Royal Soc. Edinburgh (pres.); original mem., Order Merit, 1902; grand officer Legion of Honor; Prussian Order Pour le Mérite; mem. or fellow numerous other sci. socs. and acads. Author: (with Tait) Treatise on Natural Philosophy, 1867-74; Electrostatics and Magnetism, 1874; Mathematical and Physical Papers, 5 vols., 1882-1811; Popular Lectures and Addresses, 3 vols., 1889-94; The Baltimore Lectures on Molecular Dynamics and the Wave Theory of Light, 1904; and circa 700 sci. papers. Created, with very inadequate resources, 1st physics lab. for teaching purposes in Great Britain; famous for researches in electrodynamics, thermodynamics, thermoelectricity, and properties of matter; investigated transformation of heat into work; proposed absolute (later called Kelvin's) scale of temperature, 1848; experimented on heat developed by compression of air; verified lowering by pressure of melting point of ice; formulated 1st and 2d laws of thermodynamics; enunciated doctrines of available energy and dissipation of energy; estimated age of earth at circa 100 million years (led to controversy with Huxley); suggested process of refrigeration (sudden expansion of compressed cooled air); laid foundation for theory of electric oscillations, which formed basis for wireless telegraphy; invented mirror galvanometer; played leading role in laying of 1st transatlantic submarine cable, 1858, and in subsequent cable layings; patented siphon recorder and curb-transmitter; studied atmospheric electricity; invented water-dropping collector; improved electrometers; urged adoption of metric system and centimeter-gram-second absolute system; investigated mathematical theory of magnetism; introduced terms permeability, susceptibility; studied secular cooling of earth, changes of form during rotation of elastic spherical shells; contributed to theory of elasticity; advocated vortex theory of atoms, 1867; studied gyrostatic problems; reformed mariner's compass, 1873-78; devised apparatus for taking flying soundings, 1877; studied tides; invented tide-predicting machine; invented 1st harmonic analyzer; held numerous other patents; famous for dictum: "I can never satisfy myself until I can make a mechanical model of a thing." Died Netherhall, nr. Largs, Scotland, Dec. 17, 1907.

THOREAU, Henry David, Am. naturalist; social theorist; b. Concord, Mass., July 12, 1817; s. John and Cynthia (Dunbar) T.; grad. Harvard, 1837. Opened pvt. sch. with brother John, 1838; gave 1st of nearly annual lectures to Concord Lyceum, 1838; lived at home of Ralph Waldo Emerson, 1841-43, 47-49; edited The Dial during Emerson's absence, 1843; tutor in home of William Emerson on S.I., N.Y., 1843-44; resided at Walden Pond, arrested for failure to pay poll tax, 1845; wrote journals which formed basis of Walden; lived in family home in Concord, 1849-62; published Walden (his only organized work), 1854; Thoreau was part of extremely close-knit family and participated in pencil mfg. conducted in family home; his influence on later Am. literature has been ascribed to his pithy and forceful style which employed vivid imagery; mem. Transcendentalist Club (including Amos Bronson Alcott, J. F. Clarke, Margaret Fuller, F. H. Hodge). Author: A Week on the Concord and Merrimac Rivers (notable for literary criticism and nature studies), 1849; from his journals describing trips to Me., Cape Cod, and Can., 1848-53, there were published posthumously: Excursions, 1863; The Maine Woods, 1864; Cape Cod, 1865; A Yankee in Canada, 1866; Letters to Various Persons, edited by R. W. Emerson, 1865; Early Spring in Massachusetts, 1881, Summer, 1884, Winter, 1888, Autumn, 1892, selections from journals edited by H. G. O. Blake; Poems of Nature, edited by Salt and Sanborn, 1895; Walden edition of Writings of Henry David Thoreau, 20 vols., 1906, contains all journals except Apr. 1843-July 1845. His life at Walden Pond was an experiment designed to prove that a meaningful existence was possible apart from industrialism and materialism of New Eng.; set his own conscience above laws of govt. thus incurring charge of anarchism; philosophy of his essay on civil disobedience (classic statement of individualism in conflict with State) present in Walden; descriptions of nature contained in this and other works established him as naturalist (though not as a scientist); though not a reformer, deeply concerned with problem of slavery, spoke vigorously in defense of John Brown, 1859; distrusted group action and relied upon own definitions of morality; believed in human perfectibility; writing skill made him foremost interpreter Am. flora and fauna. Died Concord, Mass., May 6, 1862.

THOREK, Max, surgeon; b. Hungary, Mar. 10, 1880; s. Isaac and Sarah (Mahler) T.; prep. edn., Budapest, Hungary; came to U. S., 1900; M.D., Rush Med. Coll. (U. of Chicago); 1904; hon. LL.D., Lincoln Memorial U., Sc.D., Wesleyan Coll.; Honoris Causa, U. Istanbul, Turkey, 1954; married Fannie Unger, April 16, 1905; 1 son, Philip (M.D.). Practiced at Chgo., 1904—; asst. in gynecology to late Prof. Henry Banga, for 5 yrs.; prof. clin. surgery, Loyola U., 3 yrs.; chief surgeon to Am. Hosp., Chicago; prof. of clinical surgery, Cook County Grad. Sch. of Medicine; attending surgeon Cook County Hosp., cons. surgeon Municipal Tuberculosis Sanitarium; Cushing lecturer McGill U., Montreal, 1954. Fellow A.M.A., Internat. College Surgeons (life; internat. founder and permanent sec. general; editor in chief of jour.), Nat. Acad. Medicine (Colombia, S.A.), Internat. Coll. Anesthetists, Royal Soc. of Arts, Royal Photographic Soc. of Gt. Britain, Am. Coll. Gastroenterology; hon. fellow Royal Surg. Society (Sofia, Bulgaria), Internat. Society of Gastroenterologists; hon. mem. Belgian Society Gastroenterology; corr. fellow Peruvian Surg. Society; mem. Orden del Sol, Republic of Peru; member American Eugenics Society, Internat. Anesthetic Research Soc., Am. Soc. French Legion of Honor, corr. academician Nat. Acad. Sciences of Mexico; corr. mem. Soc. des Chirurgiens (Paris), Soc. Scientifique Française de Chirurgie Reparatrice, Russian Endocrinol. Assn. (Moscow), Soc. of Neurology and Endocrinology (Jassy, Roumania), Sociedad das Sciences Medicas (Lisbon), Brazilian Coll. Surgeons; hon. corr. mem. Egyptian Med. Assn. (Cairo, Egypt); Collaborator Rassegna Internazionale di Clinica Terapia, Milano, Italy; U. S. del. Internat. Congress Hepatic Insufficiencies, Vichy, France, 1927. Decorated Knight of Legion of Honor (France); Knight Order of Crown of Italy; others. Author: The Human Testis and Its Diseases, 1924; Surgical Errors and Safeguards, 1931, 4th editions, 1943; Modern Surgical Technic (3 vols.), 1941; Plastic Surgery of the Breast Abdominal Wall, 1942; A Surgeon's World, 1944. Contbn. to devel. surg. methods in several fields of medicine. Died Jan. 25, 1960.

THORÉN, Lars Olof, Swedish surgeon; b. Gothenburg, Sweden, Nov. 18, 1921; s. Lars Einar and Anna (Thorn) T.; med.kand., U. Uppsala (Sweden), 1943, med.lic., 1949, med.dr., 1958; m. Ingrid Richter, Aug. 10, 1946; children—Ingrid Birgitta, Lars Gunnar, Karin Maria, Anna Gunilla. Staff surg. clinic U. Uppsala, 1949-50, 53-58, asso. prof., 1958-63, prof. surgery, 1965—; staff surg. clinic Central Hosp., Orebro, Sweden, 1950-53; head surg. clinic Region Hosp., Örebro, 1964. Author: Bile Peritonitis, 1958; Västskebalans, 1960; also numerous articles. Research on postoperative pulmonary ventilation and complications, treatment of some fractures of hand, mammary cancer, pathophysiology of bile peritonitis, shock and body fluid imbalance. Home: 1 Bruksvägen, Uppsala, Sweden.*

THORN, George Widmer, Am. physician; b. Buffalo, N.Y., Jan. 15, 1906; s. George Widmer and Fanny (Widmer) T.; M.D., U. Buffalo, 1929; M.A., Harvard, 1942; D.Sc., Temple U., 1951; M.D., Catholic U. Louvain (Belgium), 1960; m. Doris Weston, June 30, 1931; 1 son, Weston Widmer. Practice medicine, specializing in internal medicine, Boston, 1942—; physician-in-chief Peter Bent Brigham Hosp., 1942—; Hersey prof. theory and practice physic Med. Sch. Harvard, 1942—, Samual A. Levine prof. medicine, 1967—; dir. research Howard Hughes Med. Inst. (Miami), 1956—. Recipient Oscar B. Hunter Meml. award Am. Therapeutic Soc., 1967. First Lilly lectr. Royal Coll. Physicians, London, 1966. Mem. Soc. Exptl. Biology and Medicine, Am. Soc. Clin. Investigation, Endocrine Soc., Am. Clin. and Climatological Assn., Sigma Xi. Author: (with others) Principles of Internal Medicine, 1950. Research and numerous publs. in endocrinology and metabolism; salt and water metabolism; discovered extract of adrenal glands; first to show complete adrenalectomy could be performed safely in man; developed ACTH for use in treatment numerous diseases in man. Home: 983 Memorial Dr., Cambridge, Mass. 02138. Office: 721 Huntington Av., Boston 02115.*

THORN, Niels Anker, Danish physiologist; b. Horsens, Denmark, Aug. 1, 1924; s. Niels Johan and Dagny (Nielsen) T.; M.D., U. Copenhagen (Denmark), 1951, D.Med. Scis., 1960; m. Ingrid Christiansen, Mar. 7, 1954; children—Peter, Soren. Research asso. Rockefeller Inst., N.Y.C., 1953-56; research asso. Inst. Med. Physiology U. Copenhagen, 1957-62, asso. prof. physiology, 1962—. Cons. in physiology Royal Dental Sch., Copenhagen, 1960. Mem. Harvey Soc., Soc. for Exptl. Biology and Medicine, Brit. Biochem. Soc. Contbg. author: Handbuch d. Exper. Pharmakologie—Ergänzungswerk XIII, 1960; Mineral Metabolism, An Advanced Treatise, 1960. Research and publs. on mechanism of secretion and action of antidiuretic hormone in normal and path. states. Home: 49 Egernvej, Copenhagen, F., Denmark.*

THORNBERRY, Halbert Houston, Am. virologist; b. Corydon, Ky., Dec. 28, 1902; s. Thomas Lynn and Gertie (Walker) T.; B.S., U. Ky., 1925, M.S., 1926; postgrad. U. Minn., 1926-28, Ph.D., 1934; postgrad. U. Ill., 1928-31, Columbia U., 1931-32; m. Kathryn Winder, June 18, 1946; 1 dau., Martha Lynn. Staff, Rockefeller Inst., Princeton, N.J., 1931-34; faculty U. Cal. at Riverside, 1934-35, U. Ky., Lexington, 1936; staff U. S. Dept. Agr., Washington, 1937; faculty U. Ill., Urbana, 1938—, prof., 1955—. Mem. Am. Chem. Soc., Am. Phytopath. Soc., Am. Soc. for Microbiologists, A.A.A.S. Author: Index of Plant Virus Diseases, 1966; also numerous articles. Research on factors influencing filtration and ultrafiltration of plant viruses, infection of plants by viruses, chem. inactivation of plant viruses, purification of plant viruses; description of new bacterial pathogens of plants; isolation of wilt-inducing protein from root-rot fungus; discovered bacteria overwintering in invisible infections in peach twigs. Patentee synthetic medium for prodn. streptomycin. Home: 1602 S. Hillcrest St., Urbana, Ill. 61801.*

THORNBURY, William David, Am. geologist; b. English, Ind., Apr. 23, 1900; s. Frank Anderson and Belle (Flanary) T.; A.B., Ind. U., 1925, Ph.D., 1936; A.M., U. Colo., 1928; m. Doris Bonita Groan, July 21, 1928; 1 son, David William. Tchr., Alton (Ind.) High Sch., 1919-20, Marengo (Ind.) High Sch., 1920-22, Cripple Creek (Colo.) High Sch., 1925-26; with Ind. U., 1928—, prof., 1955—. Mem. Ind. Acad. Sci., A.A.A.S., Geol. Soc. Am., Nat. Assn. Geology Tchrs. Author: Principles of Geomorphology, 1954; Regional Geomorphology of United States, 1965. Research and publs. in field glacial geology, including knowledge regarding glacial history of Indiana. Home: 926 S. Highland St., Bloomington, Ind. 47401.*

THORNDIKE, Alan Moulton, Am. physicist; b. Peekskill, N.Y., June 27, 1918; s. Edward L. and Elizabeth (Moulton) T.; B.A., Wesleyan U., 1939; A.M., Columbia U., 1940; Ph.D. in Chem. Physics, Harvard, 1947; m. Mary Louise Van Dyke, Sept. 7, 1942; children—Jean (Mrs. Todd C. Gould), Alan S., Charlotte A., Karl E., Margaret E. Staff, OSRD, 1941-46; physicist Brookhaven Nat. Lab., Upton, N.Y., 1947-57, 1959—; vis. prof. physics Johns Hopkins U., 1958-59. Fellow Am. Phys. Soc., Phi Beta Kappa. Author: Mesons, 1952; (with D. H. Frisch) Elementary Particles, 1964. Research in high energy physics; bubble chamber and cloud chamber observations of high energy particles; devel. analysis methods. Home: 72 S. Howell's Point Rd., Bellport, N.Y. 11713. Office: Brookhaven Nat. Lab., Upton N.Y. 11973.*

THORNDIKE, Edward Lee, Am. psychologist; b. Williamsburg, Mass., Aug. 31, 1874; s. Edward R. and Abby B. (Ladd) T.; A.B., Wesleyan U., Conn., 1895; A.B., Harvard, 1896, A.M., 1897, LL.D., 1933; Ph.D., Columbia U., 1898, Sc.D., 1929; Sc.D., Wesleyan and LL.D., Iowa, 1923; Sc.D., U. of Chicago, 1932; LL.D., Edinburgh, 1936; Sc.D., Athens, 1937; m. Elizabeth Moulton, Aug. 29, 1900; children—Elizabeth Frances, Virginia (dec.), Edward Moulton. Robert Ladd, Alan Moulton. Instr. edn. and teaching, Western Reserve U., 1898-99; instr. genetic psychology, Teachers Coll., Columbia, 1899-1901, adj. prof. ednl. psychology, 1901-04, prof., 1904-40, emeritus prof. since 1941; Wm. James lecturer, Harvard University, 1942-43. Fellow A.A.A.S. (president, 1934), N.Y. Acad. Sciences; mem. Nat. Acad. Science, Am. Philos. Soc., Am. Phychol. Assn., Am. Acad. of Arts and Sciences; hon. mem. Brit. Psychol. Assn. Clubs: Century (New York); Cosmos (Washington). Author: Educational Psychology, 1903; Mental and Social Measurements, 1904; Elements of Psychology, 1905; Principles of Teaching, 1905; Animal Intelligence, 1911; The Original Nature of Man, 1913; The Psychology of Learning, 1914; Psychology of Arithmetic, 1922; Psychology of Algebra, 1923; The Measurement of Intelligence, 1926; Fundamentals of Learning, 1932; Your City, 1939; Human Nature and the Social Order, 1940; Man and His Works, 1943; also various monographs and articles on psychological and educational subjects. Pioneer in characteristically Am. psychology; interested in intelligence and learning; studied measurement and differences in intelligence; developed theory that punishment is less effective in learning than repetition and reward. Died Montrose, N.Y., Aug. 9, 1949.

THORNE, Sir Richard Thorne, Brit. physician; b. Leamington, Eng., Oct. 13, 1841; s. Thomas Henry Thorne; student St. Bartholomew's Hosp., London; M.B., London, 1866; LL.D., hon. D.Sc.; m. Martha Rylands, 1866; 3 sons; 1 dau. Became physician to Hosp. for Diseases of Chest, London, 1868; named insp. med. dept. Privy Council, 1871; named prin. med. officer to local govt. bd., 1892; named permanent lectr. on pub. health St. Bartholomew's London, 1891; named Crown nominee on gen. med. council, 1895. Became mem. Royal Commn. on Tb, 1896, Royal Commn. on Sewage Disposal, 1898. Licentiate Royal Coll. Physicians. Fellow Royal Soc., 1892, Royal Coll. Physicians; mem. Royal Coll. Surgeons. Knighted, 1897. Author: On the Use and Influence of Hospitals for Infectious Diseases, 1882; On the Progress of Preventive Medicine during the Victorian Era, 1887; On the Natural History of Prevention of Diphtheria, 1891; The Administrative Control of Tuberculosis, 1898. Leader in pub. health; 1st to prove that typhoid fever is water-borne disease; contributed to establishment of hospitals for isolation of infectious disease. Died Dec. 18, 1899.

THORNTON, Henry Gerard, English microbiologist; b. London, Eng., Jan. 22, 1892; s. Francis Hugh and Adelaide Ethel (Burchell) T.; B.A. with 1st class honors, New Coll., Oxford U., 1914; D.Sc., U. London, 1935; m. Gerda Norregaard, Apr. 23, 1924; 1 son, Peter Kai. With Rothamsted Exptl. Sta., Harpenden, Eng., 1919-57, head bacteriology dept., 1920-41, soil microbiology dept., 1941-57. Mem. arid zone adv. com. UNESCO, 1951-54. Created knight, 1960. Fellow Royal Soc., 1941 (fgn. sec. 1955-60, v.p. 1959-60), Inst. Biology (pres. 1960-62), Royal Geog. Soc., Linnean Soc.; mem. Soc. Gen. Microbiology, Soc. Exptl. Biology, Genetics Soc., Eugenics Soc. Research, publs. on soil micropopulation, particularly nitrogen fixing bacteria in leguminous plants. Home: 3 Romeland Cottage, St. Albans, Eng.*

THORNTON, Robert John, Brit. botanist, physician; b. London, Eng., circa 1768; s. Bonnell and Sylvia (Brathwaite) T.; M.B., Trinity Coll., Cambridge, 1793; student Guy's Hosp., London; M.D., St. Andrews, Scotland, 1805; 1 son, 1 dau. Began practice medicine, London, 1797; physician Marylebone Dispensary; lectr. med. botany United Hosps. of Guy and St. Thomas. Licentiate Royal Coll. Physicians. Author: The Philosophy of Medicine being Medical Extracts, 1796; New Illustrations of the Sexual System of Linnaeus, 1797-1807; Facts decisive in Favour of the Cow Pock, 1802; Lectures on Botany, 1804-05; Plates of the Heart Illustrative of the Circulation, 1804; Vaccinae Vindiciae or a Vindication of the Cow Pock, 1806; Practical Botany, 1808; Botanical Extracts or Philosphy of Botany, 2 vols., 1810; Alpha Botanica, 1810; A New Family Herbal, 1810; A Grammar of Botany, 1811; The British Flora, 5 vols., 1812; Elements of Botany, 2 vols., 1812; Outline of Botany, 1812; Juvenile Botany, 1818. Introduced digitalis for scarlet fever. Died Jan. 21, 1837.

THORNTON, R(obert) L(yster), physicist; b. Wootton, Bedfordshire, Eng., Nov. 29, 1908; s. Dudley L. and Katharine (Foster) T.; B.Sc. McGill U., 1930, Ph.D., 1933. m. Mary Elizabeth Edie, Aug. 29, 1938; children—Katharine E., Denis L., Margaret A. Came to U. S., 1933, naturalized 1940. Instr. physics U. Mich., Ann Arbor, 1936-38; asso. prof. Washington U., St. Louis, 1939-45; asst. dir. process improvement div. Tenn. Eastman Corp., Oak Ridge, 1943-45; prof. U. Cal., Berkeley, 1945——, asso. dir. Lawrence Radiation Lab., 1959——. Study of nuclear physics; accelerator design and construction. Home: 522 Cragmont Av., Berkeley, Cal. 94708.*

THORNYCROFT, Sir John Issac, English naval architect, engr.; b. Rome, Italy, Feb. 1, 1843; s. Thomas and Mary Thornycroft; ed. Glasgow U.; studied shipbuilding, Govan on the Clyde; m. Blanche Charles, 1870; 2 sons, 5 daus. Built torpedo boats for Eng. and other European nations; founded shipbuilding works at Chiswick, 1866. Fellow Royal Soc., 1893; mem. Inst. Naval Architects (v.p.). Knighted, 1902. Contbr. articles to Transactions of Brit. Assn. Built first torpedo boat with locomotive steam boiler, 1875; invented turbine propeller for shallow-draft ships, coastal torpedo speedboat called Scooter, during World War I; his improvements in Naval architecture and marine engring. contbd. to devel. of high-speed ships. Died Benbridge, Eng., June 23, 1928.

THORODDSEN, Thorvaldur, geologist, geographer; b. Flatey, Breidi Fjord, West Iceland, June 6, 1855; ed. U. Copenhagen (Denmark); schoolmaster, Iceland, from 1880. Author: History of Icelandic Volcanoes, 1882; History of Icelandic Geography, 1892-1902; Earthquakes in Iceland, 1899; Geological Map of Iceland, 1901; Die Geschichte des isländischen Volkes, 1925. Made geol. explorations which included expdns. to study volcanic structure and history of interior of Iceland, 1876-98. Died Copenhagen, Sept. 28, 1921.

THORP, James, Am. soil scientist; b. Westtown, Pa., Jan. 12, 1896; s. George Smedley and Anne (Palmer) T.; B.S., Earlham Coll., 1921, Sc.D. (hon.), 1951; m. Eleanor Mahalah Ballard, June 19, 1923. With Soil Survey, U. S. Dept. Agr., 1921-52, prin. soil sci., 1942-52; head dept. geology and soil sci. Earlham Coll., Richmond, Ind., 1952-61, emeritus, 1961——. Cons. in China, Australia, Kenya; cons. Stanley W. Hayes Research Found., 1967——. Mem. Geol. Soc. Am., Soil Sci. Soc. Am., Internat. Soc. Soil Sci., Am. Soc. Agronomy, A.A.A.S., Assn. Am. Geographers, Internat. Assn. Quaternary Research (hon. v.p. congress 1965), Brit. Soc. Soil Sci., Sigma Xi. Author: Geography of the Soils of China, 1936; Field Study of Soils of Australia, 1956; also articles. Research on soil morphology, genesis and classification with emphasis on soils in relation to geology and geomorphology. Address: 606 S. W.A. St., Richmond, Ind. 47374.*

THORP, John, Am. machinist, inventor; b. Rehoboth, Mass., 1784; s. Reubin and Hannah (Bucklin) T.; m. Eliza A. Williams, Aug. 18, 1817. Received 1st patent for hand water loom, 1812, renewed, 1843, 2d patent for power loom, 1816; received 3 patents for improvements in spinning and twisting cotton (called "ring spinning"; basic method of continuous spinning still used), 1828; patent for netting machine, 1828; received 4 patents including narrow fabric loom (possibly 1st gang loom operated by power), 1829; established as machine builder, Providence, R.I., later North Wrenthan, Mass., 1830's. Died Nov. 15, 1848.

THORPE, Sir Jocelyn Field, English chemist; b. Dec. 1, 1872; s. William George Thorpe; ed. Worthing Coll., King's Coll., London, Eng., Royal Coll. Sci., London; Ph.D., U. Heidelberg (Germany), 1895; M.Sc., U. Manchester (Eng.), 1905; m. Lilian Briggs, 1902. Asst. lectr. organic chemistry. U. Manchester, from 1896, lectr., from 1899, sr. lectr. chemistry, bio-chemistry, 1908-10; prof. organic chemistry Imperial Coll., U. London, 1914-38, emeritus, 1938-40. Mem. adv. council Dept. Sci. and Indsl. Research, 1916-22, chem. defence com. War Office, mine safety orgns., dye-stuffs devel. com. Bd. Trade; pres. Indian Chem. Services Com., 1919-20. Research fellow Owens Coll. Fellow Royal Soc. (Sorby research fellow 1909-11, Davy medal 1922, council 1923 25), 1908, Inst. Chemists (pres. 1933-36), Chem. Soc. (Longstaff medal 1921, pres. 1928-31). Knighted, 1939. Author: The Synthetic Dyestuffs, 1905, 7th edit., 1933; Vat Colours, 1923; A Students' Manual of Organic Chemical Analysis, 1925. Editor: (with others) Thorpe's Dictionary of Applied Chemistry, 1937, supplementary vols., 1933-34. Research, publs. in applied chemistry, especially organic chemistry. Died June 10, 1940.

THORPE, Louis Peter, Am. psychologist; b. Battle Creek, Mich., May 15, 1893; s. Christian Adolph and Mary (Anderson) T.; A.B., Andrews U., 1925; A.M., Northwestern U., 1929, Ph.D., 1931; m. Alice Clair Kegebein, Feb. 4, 1918; 1 dau., Norma Mae (Mrs. Robert Clarence Tucker). Faculty, U. So. Cal., Los Angeles, 1937——, prof. edn., psychology, 1945——, vice chmn. acad. senate, 1958-59; faculty summer sessions Northwestern, 1932, 35, U. Cal., Berkeley, 1946, 47, 48, Johns Hopkins, 1955, U. Hawaii, 1957. Diplomate Am. Bd. Examiners Profl. Psychology. Mem. Am. Psychol. Assn., Nat. Soc. For Study Edn., Internat. Fedn. Mental Health, Internat. Council Psychologists; N.Y. Acad. Scis. Author: Psychological Foundations of Personality, 1938; Personality and Life, 1941; Personality and Youth, 1949; (with Allen M. Schmuller) Contemporary Theories of Learning, 1954, Personality: An Interdisciplinary Approach, 1958; The Psychology of Mental Health, 1960; (with Barney Katz and Robert T. Lewis) The Psychology of Abnormal Behavior, 1961; Child Psychology and Development, 1962. Basic research, numerous publs. on devel. of SRA Achievement Series for sch. children. Home: 969 Hilgard Av., Los Angeles 90024.*

THORPE, Sir Thomas Edward, English chemist; b. Barnes Green, Eng., Dec. 8, 1845; s. George and Mary (Wilde) T.; ed. chemistry dept. Owens Coll., Manchester, Eng., 1863-67; Ph.D., U. Heidelberg, 1869; D.Sc., LL.D., U. Bonn; m. Caroline Emma Watts, 1870. Staff lab. of F.A. Kekulé, Bonn, Germany, 1869-70; prof. chemistry Andersonian Coll., Glasgow, Scotland, 1870-74; with Yorkshire Coll. Sci., Leeds, Eng.; prof. chemistry Royal Coll. Sci., South Kensington, Eng., 1885-94, 1909-12; dir. Govt. Chemistry lab., 1894-1909; mem. 4 eclipse expdns.; magnetic survey (with Sir Arthur Rucker) of Brit. Isles, 1884-88. Recipient Longstaff medal Chem. Soc., 1881. Fellow Royal Soc., 1876, recipient Royal medal, 1889, fgn. sec., 1899-1903; mem. Soc. Chem. Industry (became pres. 1895), Chem. Soc. (pres. 1899-1901), Brit. Assn. (became pres. 1921). Author: Essays in Historical Chemistry, 1894; Joseph Priestley, 1906; Dictionary of Applied Chemistry, 3 vols., 1893, 3d edit., 7 vols., 1921; Chemical Problems, 1870; Qualitative Analysis, 1873; Inorganic Chemistry, 2 vols., 1874; Quantitative Analysis, 1874; Life of Sir Henry Roscoe, 1916. Supervised research on problems of detection of arsenic in beer, elimination of lead from pottery glazes, white phosphorus from matches; pioneered (with Roscoe) research on vanadium, phosphorus fluorides and oxides; studies on relation of molecular weights to specific gravities; determined atomic weights of various elements. Died Salcombe, South Devon, Eng., Feb. 23, 1925.

THORSELL, Walborg Susanna, Swedish chemist; b. Stockholm, Sweden, Feb. 12, 1919; d. Ernst Gottfrid and Allida (Danielsson) Thorsell; Fil.mag., U. Stockholm, 1947, Fill.lic., 1958. With W. Becker, Stockholm, 1939; asst. Swedish Inst. for Testing of Material, 1940; faculty U. Stockholm, 1940-49; faculty Vet.Coll., Stockholm, 1949——, asso. prof. chemistry, 1962——. Mem. Swedish Chemists Assn., Am. Chem. Soc., European Soc. Biochem. Pharmacology, Swedish Biochemists Assn. Author: Qualitative and Quantitative Inorganic Analysis, 1950; Physiological and Pathological Chemistry, 1952; also articles. Research on chem. nature and action of some enzymes, differences between some microorganisms, some metazoan parasites and their hosts. Home: 45 Karlskronavägen, Stockholm, Sweden.*

THORSON, G. A., Danish marine biologist; b. Copenhagen, Denmark, Dec. 31, 1906; s. Charles and Astrid (Gjertsen) T.; Mag. scient., Copenhagen (Denmark) U., 1930, Dr. phil., 1936; m. Ellen Johanne Jundrun Joergensen, Dec. 23, 1940; children—Ole, Bodil (Mrs. Soeren Larsen). Zoologist, Danish E.-Greenland Expdns., 1931-33; marine biol. researcher, Persian Gulf, 1937, Canary Islands, 1947, S. Cal., 1949, W. Africa, 1951, Fla., 1956, Thailand, 1966; faculty Copenhagen U., 1946——, prof., 1957——, dir. Elsinore (Denmark) Marine Biol. Lab., 1958——. Founder marine biol. labs., Island of Ven, Sweden, 1936, Elsinore, 1941; pres. Nordic Council for Marine Biology, 1960; mem. Danish Council Oceanography, 1959; mem. adv. bd. marine inst., Miami, Paris, Naples; lectr., Europe, U.S., Asia. Mem. Royal Danish Acad. Scis.; corr. or hon. mem. sci. socs. in Finland, France, Italy, Spain, Sweden, U.K., U.S. Research and publs. on reproductive and larval ecology of marine bottom invertebrates, on animal communities of sea-bottom, on balance between predators and prey in the sea. Home: 28, Nationernes Allé. Office: Marine Biol. Lab., Elsinore, Denmark 3000.*

THORSON, Thomas Bertel, Am. zoologist, b. Rowe, Ill., Jan. 12, 1917; s. Thomas B. and Hertha (Fylpaa) T.; student Waldorf Jr. Coll., 1934-36; B.A., St. Olaf Coll., 1938; M.S., U. Wash., 1941, Ph.D., 1952; m. Margaret L. Overgaard, Dec. 31, 1941; children—Sharon M., Joel T. Faculty, Yakima Jr. Coll., 1946-48, San Francisco State Coll., 1952-54, S.D. State U., 1954-56; faculty U. Neb., Lincoln, 1948-50, 56——, prof. zoology, 1961——, chairman department zoology, 1967——. Fellow A.A.A.S.; mem. Am. Inst. Biol. Scis., Ecol. Soc. Am., Am. Soc. Zoologists, Am. Soc. Ichthyologists and Herpetologists, Am. Fisheries Soc., Sigma Xi, Phi Sigma. Research and publs. on water economy of amphibians, rate of water loss and gain through skin, tolerance to desiccation, patterns of body water partitioning in cold-blooded vertebrates, osmoregulation of freshwater sharks and sawfish of Lake Nicaragua. Home: 2527 S. 54th St., Lincoln, Neb. 68506.*

THOSAR, Baji Vinayak, Indian physicist; b. Khamgaon, India, Apr. 3, 1913; s. Vinayak Naravan and Savitri (Sathaye) T.; M.Sc., Coll. Sci., Nagpur, India, 1935; Ph.D., Birmingham (Eng.) U., 1949; m. Leela Samel, Feb. 1, 1941; children—Sunila (Mrs. R. Thergaonkar), Ravindra, Aruna, Satish. Asst. prof. Coll. Sci., Nagpur, 1937-51; faculty Tata Inst. Fundamental Research, Bombay, India, 1951——, prof. physics, 1963——, dean Physics Faculty, 1963——. Fellow Indian Acad. Scis.; mem. Indian Phys. Soc. (v.p. 1967——). Research and publs. on crystal luminescence, nuclear spectroscopy, internal conversion processes in nuclei, positron effects in solids, polarization of emission lines of ruby, exptl. verification of internal conversion coefficients, positron annihilation in irradiated solid polymers. Home: 13, Jenkins House, Henry Rd., Bombay 1, India.*

THRALL, Robert McDowell, Am. mathematician; b. Toledo, Sept. 23, 1914; s. Charles Haven and Gertrude (Gerking) T.; B.A., Ill. Coll., 1935, Sc.D., 1960; M.A., U. Ill., 1935, Ph.D., 1937; m. Natalie Hunter, Sept. 3, 1936; children—Charles Alexander, James Hunter, Mary Emily. Faculty, U. Mich., Ann Arbor, 1937-40, 42——, prof. math., 1955——, prof. indsl.

engring. in operations analysis, 1956—, head operation research dept. Research Inst., 1957-60; with Inst. for Advanced Study, Princeton, 1940-42. Cons., RAND Corp., 1951—, Weapons System Evaluation Group, 1956—; Phillips visitor Haverford Coll., 1957-58. Recipient Henry Russell award U. Mich., 1948, Distinguished Faculty Achievement award 1965. Mem. A.A.A.S., Am., Indian, Swiss, London, French math. socs., Canadian Math. Congress, Econometric Soc., Inst. Math. Statistics, Math. Assn. Am., Nat. Council Tchrs. Math., Operational Research Soc., Operations Research Soc. Am., Psychometric Soc., Soc. for Indsl. and Applied Math., Inst. Mgmt. Scis., Mich. Acad. Sci., Arts and Letters, Sigma Xi. Author: (with E. B. Miller) College Algebra, 1950; (with L. Tornheim) Vector Spaces and Matrices, 1957. Research and publs. on algebra, classical groups and representation theory, lattice and game theory, math. models in operations research and behavioral scis. Home: 953 Spring St., Ann Arbor, Mich. 48103.*

THREEFOOT, Sam Abraham, Am. physician; b. Meridian, Miss, Apr. 10, 1921; s. Sam A. and Ruth (Lilienthal) T.; B.S., Tulane U., 1943, M.D., 1945; m. Virginia Rush, Feb. 6, 1954; children—Ginny Ruth, Tracyann, Shelley Ann. Faculty, Tulane U., New Orleans, 1948—, prof., 1963—; staff Touro Infirmary, New Orleans, 1953-56, 60—, sr. dept. medicine, 1963—, dir. research, 1953—; staff Charity Hosp. New Orleans, 1950-57, sr. vis. physician, 1963—. Diplomate Am. Bd. Internal Medicine. Fellow A.C.P., Am. Coll. Cardiology (gov. for La. 1967—); mem. Am. (mem. at large central com. medicine, programs 1965-68, v.p. elect 1969—, dir. 1966—), La. (chmn. research com. 1960-66, pres. 1966-67) heart assns., So. Soc. for Clin. Investigation (pres. 1966-67), Am. Fedn. for Clin. Research, Soc. for Exptl. Biology and Medicine, Central Soc. for Clin. Research, N.Y. Acad. Scis., Soc. Nuclear Medicine, Internat. Soc. Lymphology (exec. bd. 1966—, chmn. editorial bd. Lymphology 1966—). Research, publs. on fluid and electrolyte metabolism; function of lymphatics and capillaries and effects of drugs and pathology. Home: 519 Burdette St., New Orleans 70118.*

THRON, Wolfgang Joseph, mathematician; b. Ribnitz, Germany, Aug. 17, 1918; s. Ludwig G. and Annemarie (Joseph) T.; came to U. S., 1936, naturalized, 1944; A.B., Princeton, 1939; postgrad. State U. Wash., 1939-40; M.A., Rice U., 1942, Ph.D., 1943; m. Ann Lukach, June 7, 1953; children—Jonathan, Penelope, Peter, Karin, Rajinder. Faculty, Harvard, 1943-44, Washington U., St. Louis, 1946-54; faculty U. Colo., Boulder, 1954—, prof. math., 1957—. Vis. prof. Free U., Berlin, Germany, 1951; vis prof., Fulbright lectr. Panjab U., 1962-63. Mem. Am. Math. Soc., Math. Assn. Am., Phi Beta Kappa, Sigma Xi. Author: Introduction to the Theory of Functions of a Complex Variable, 1953; Topological Structures, 1966; also articles. Research in analytic theory of continued fractions, complex analysis, gen. topology. Home: 430 Christmas Tree Dr., Boulder, Colo. 80302.*

THROW, Francis Edward, Am. physicist; b. Ottumwa, Ia., Oct. 4, 1912; s. Frank Wesley and Amy (Smith) T.; B.A., Park Coll., 1933; M.S., U. Mich., 1936, Ph.D., 1940; m. Volena Viola Lochner, June 24, 1938; children—Carol Elizabeth, Jeannette Ethel, Edward Robert. Faculty, Poly. Inst. P.R., 1940-41, Pa. State Coll., 1941-42, State U. Ia., 1943-44; prof., head physics dept. Cornell Coll., Mt. Vernon, Ia., 1944-52, Wabash Coll., 1952-56; asst. dir. physics div. Argonne (Ill.) Nat. Lab., 1956—. Mem. Am. Phys. Soc., Am. Assn. Physics Tchrs., Sigma Pi Sigma. Editor: Nuclear Physics with Reactor Neutrons, 1963; Nuclear Spectroscopy with Direct Reactions, 1964. Contbr. articles to profl. jours. Home: 719 S. Gables Blvd., Wheaton, Ill. 60187. Office: Argonne Nat. Lab., Argonne, Ill. 60439.*

THUNBERG, Karl Peter, Swedish botanist, naturalist; b. Jönköping, Sweden, Nov. 11, 1743; ed. Uppsala (Sweden) U.; pupil of Linnaeus; M.D., 1770; became surgeon Dutch E. India Co., 1770; sailed as botanist to Cape of Good Hope., 1772-75; physician on ship bound for Japan, 1775, collected plants there for 3 years; returned to Sweden, 1779; became acquainted with Sir Joseph Banks in Eng.; apptd. demonstrator botany Uppsala U., 1781, succeeded Linnaeus as prof. botany, 1784. Fellow Royal Soc., 1788; mem. French Soc. Agr., French Acad. Scis. (corr.); hon. mem. many sci. socs. Author: Flora japonica, 1784; Resa uti Europa, Africa, Asia, 4 vols., 1788-93; Icones plantarum japonicarum, 1794-1805; Prodromus planatarum, 1800; Flora capensis, 1807-12; also numerous memoirs. Pioneer among Europeans in study of Japanese natural history; 2d only to Siebold as authority on Japan; eponym of tropical plant genus Thunbergia. Died nr. Uppsala, Aug. 8, 1828.

THURBER, George, Am. botanist, horticulturist; b. Providence, R.I., Spet. 2, 1821; s. Jacob and Alice Ann (Martin) T.; attended Union Classical and Engring. Sch., Providence; M.D., N.Y. Med Coll., 1859. Partner in a pharmacy, stimulated his interest in chemistry; lectured on chemistry with Franklin Soc., Providence; botanist, q.m., commissary on survey of boundary between U. S. and Mexico, 1850; with U. S. Assay Office, N.Y.C., 1853-56; lectr. botany Coll. of Pharmacy, N.Y. 1856-61, 65-66; lectr. Cooper Union; prof. botany and horticulture Mich. State Agl. Coll. (later Mich. State Coll.), 1859-63; editor Agriculturist, N.Y.C., 1863-85; pres. Torrey Bot. Club, 1873-80; life mem. Am. Pomol. Soc.; became corr. mem. Royal Hort. Soc. of London, 1886. Author: American Weeds and Useful Plants (a revision of William Darlington's Agricultural Botany, published 1847), 1859. Specialized in study of Am. grasses; early Am. exponent of agrl. botany. Died Passaic, N.J., Apr. 2, 1890.

THUREAU, P., French physician; b. Pouques les Eaux, France, Aug. 14, 1922; s. Maurice Antonin and Françoise (Bourdillon) T.; student Ecole Normale Supérieure, 1948-50, Doctorat ès Sciences Physiques, 1955; m. Jacqueline Paturot, June 20, 1945; children—Francoise, Catherine, Michéle. Prof. physics Ecole Nationale Supérieure de Mécanique et d'Aerotechnique, Poitiers, France., 1950-59; faculty Faulcté des Sciences, Caen, France., 1959—, prof., 1961—; dir. Ecole Nationale Supérieure d'Electronique et d'Electromécanique, Caen, 1967—. Mem. French Soc. Physics. Author: Electronique de Base, 1959; also articles. Research on kinetics of crystalline luminescences, luminescence of injection in metal structures; measurement of surface temperatures by photoluminescence. Home: 123, rue Basse, Caen (14) France.*

THURET, Gustave-Adolphe, French botanist; b. Paris, May 23, 1817; founder bot. garden (named after him), Antibes; corr. mem. French Acad. Scis., 1857; studied mushrooms and algae; identified zoospores and antheridies; (with Bornet) successfully impregnated Floridae, made other discoveries in sexual reproduction in seaweed. Died Nice, France, May 11, 1875.

THÜRING, Bruno Jakob, German astronomer; b. Warmensteinach-Bayreuth, Germany, Sept. 7, 1905; s. Hans and Anna (Troppmann) T.; Ph.D., U. Munich; Dr.ès.sc.; m. Marga Sulzer, Feb. 27, 1936. Sci. asst. State Obs., Munich, Germany, 1928-33; with Breslau (now Wroclaw, Poland) U. Obs., 1933-34; observer Heidelberg (Germany) Obs., 1934-35; with State Obs., Munich, 1935-40; instr. astronomy U. Munich, 1937-40; prof. astronomy U. Vienna, also dir. U. Obs. Vienna, 1940-45. Author: Theorie der Gravitation, 1950; Theorie Der Bewegungen der Planeten des Jupitergruppes; Die Sterne, 1952; Methodologische Untersuchungen zur Kosmologie, 1954; Methoden der Programmierung elektronischer Rechenanlagen, 1957-61. Research on cosmology, astronomy, especially celestial mechanics; theory of gravitation, precession, digital computer programming. Address: Marstallstrasse 22, Karlsruhe-Durlach, West Germany.

THURLBECK, William Michael, physician; b. Johannesburg, S. Africa, Sept. 7, 1929; s. William and Alice (Mears) T.; B.S., U. Cape Town, 1950, M.B., Ch.B., 1953; m. Elizabeth Anne Tippett, June 10, 1955; children—Sarah Margaret, David William, Alison Mary. Asst. resident, chief resident pathology Mass. Gen. Hosp., 1955-58, research fellow, 1960-61; research asst. Brompton Hosp., London, 1959; practice medicine, specializing in pathology, Montreal, Que., Can., 1961—; faculty McGill U., 1961—, prof. pathology, 1966—. Examiner in pathology Royal Coll. Physicians and Surgeons Can. Mem. Am. Assn. Pathologists and Barteriologists, Internat. Acad. Pathology, Path. Soc. Gt. Britain and Ireland. Contbg. author: Pulmonary Function in Disease, 1964; Comparative Arteriosclerosis, 1965. Publs. on research in pathology, natural history and prevalence studies of bronchitis and emphysema. Home: 602 Victoria Av., Westmount 6, Que. Office: 3775 University St., Montreal 2, Que., Can.*

THURMAN, Ernestine Hogan, Am. entomologist; b. Atkins, Ark., Mar. 7, 1920; d. Walton Cloud and Clara LaVanche (Lewis) Hogan; B.S., Coll. Ozarks, 1944; postgrad. Tulane U., 1944; Ph.D., U. Md., 1958; m. Deed C. Thurman, Jr., Apr. 18, 1948 (dec. Apr. 1953); children—Phyllis L., Deed C., III; m. 2d, J. Clyde Swartzwelder, June 20, 1964. Joined USPHS, 1945, commd. lt., 1951, advanced through ranks to Scientist Dir., capt., 1963; malaria control tng. adviser Chigmai, Thailand, 1951-53; entomologist, spl. cons. USOM/Thailand, Smithsonian Instn., Washington, 1953-54, exec. sec. div. research grants NIH, Bethesda, Md., 1954-64; research adminstr. collaborative studies Nat. Heart Inst., La. State U. Sch. Medicine, New Orleans, 1964—; vis. specialist Smithsonian Instn., 1953-58; faculty mil. entomology Nat. Naval Med. Center Grad. Sch., 1962. Named Profl. Woman of Year, Washington Bus. and Profl. Women's Clubs, 1960; Alumni Achievement award Coll. Ozarks, 1958; Nat. Def. Service medal, 1964. Fellow Am. Pub. Health Assn., Royal Soc. Health (London, Eng.), Acad. Zoology India, Washington Acad. Sci., A.A.A.S.: mem. N.Y. Acad. Scis., Research Soc. Am., Sigma Xi, Pi Kappa Delta, Phi Delta Gamma, Sigma Delta Epsilon. Author: The Calicidae of Northern Thailand, 1959; also articles. Research on mosquitoes No. Thailand in connection malaria eradication program, cardiovascular disease. Home: 30 Versailles Blvd., New Orleans 70125.*

THURN, Peter, German radiologist; b. Marmagen, Germany, June 15, 1920; s. Alois and Christine (Milz) T.; student medicine U. Münster, 1940—, U. Bonn, 1941—, U. Munich, 1944-45; state exam, 1946; m. Christa Gercken, July 31, 1965. Staff med. clinic U. Bonn (Germany), 1946-50, faculty, 1955—, prof., 1960-65, full prof., 1965—, dir. radiology, 1966—; dir. Radiol. Clinic, Aachen, Germany, 1963-65. Recipient Internat. Schleussner prize for radiology, 1956. Fellow concilii Scientarium Collegium Internationale Angiologiae. Author: Haemodynamik des Herzen im Röntgenbild, 1955; Lehrbuch der röntgenologischen Differential-diagnostik, 1958 (with Teschendorf); (with Cocchi) Einführung in die Rontgendiagnostik, 1959; also numerous articles. Research on roentgenology of heart; angiocardiography; selective levocardiography, coronary and renal angiography, roentgenography of liver, spleen, bladder, cerebral; lymphangioadenography; radiotherapy of bronchial carcinoma, leukemia, lymphogranulomatosis. Home: 26 auf dem Steinchen, Bonn, Germany.*

THURNEYSSER ZUM THURN (or Thurneisser), Leonhardt, alchemist; b. Basel, Switzerland, Aug. 6, 1530. Administrator Tyrolean mines; physician to elector of Brandenburg, Germany, from 1571; left Berlin, Germany, 1580; practiced alchemy, Rome, Italy, Cologne, Germany. Author: Archidoxa, Darin der Recht War Lauff der Planeten . . . , 1569; Quinta Essentia, 1570; Onomasticon Polyplosson, Multa pro Medicis et Chymicis Continens, 1574; Historia s. Descriptio Plantarum, 1578; Ermeneia, D.I. Onomasticon und Interpretatio, 1573; Onomasticon Terminorum Paracelsi, 1574; Metallchymia vel Magna Alchymia, 1583. Reorganized alum and saltpetre works; introduced improvements in glass mfg.; aided devel. Brandenburg chem. industry; wrote on preparation of sulfur, salts, mercury, metals, milk sugar; made crude form water analysis. Died Cologne, 1596.

THURNWALD, Richard, sociologist; b. Vienna, Austria, Sept. 18, 1869; prof. U. Halle, Germany, U. Berlin. Author: Ethnopsychologische Studien an Südsee-Völkern, 1913; Primitive Psychologie, 1922; Die menschliche Gesellschaft in ihren ethno-soziologischen Grundlagen, 5 vols., 1931-35; Lehrbuch für Völkerkunde, 1939; Des Menschengeistes Erwachen, Wachsen und Irren, 1951. Research in psychology and culture of primitive peoples; ethnological and sociol. studies in Solomon Islands, E. Africa, New Guinea. Died Berlin, Germany, Jan. 19, 1954.

THURSTONE, Louis Leon, Am. psychologist; b. Chicago, Ill., May 29, 1887; s. Conrad and Sophie (Stroth) T.; M.E., Cornell U., 1912; Ph.D., U. of Chicago, 1917; Ph.D. (hon.), U. Gothenburg, Sweden, 1954; m. Thelma Gwinn, July 17, 1924; children—Robert Leon, Conrad Gwinn, Fredrick Louis. Prof. of psychology, Carnegie Inst. Tech., 1915-23, University of Chicago since 1924; psychologist Inst. for Govt. Research, Washington, D.C., 1923-24, Charles F. Grey distinguished service prof., 1938; now research prof. and dir. psychometric lab. U. of N.C.; vis. prof. of Frankfurt, Germany, 1948; vis. prof. University of Stolkholm, Sweden, 1954. Author of trade tests for occupational classification in U. S. Army, World War I; mem. com. on classification of mil. personnel, Adj. Gen's Office, World War II. Editor of intelligence tests of Am. Council on Edn. Recipient Centennial Award Northwestern U., 1951. Fellow A.A.A.S.; hon. fellow British Psychol. Society, Swedish Psychol. Soc.; mem. Am. Philosophical Society, Am. Psychological Assn. (member council; pres. 1932-33), Soc. Promotion Engring. Edn. (council), Nat. Acad. Sciences, Academy of Arts and Sciences, Sigma Xi. Author: The Learning Curve Equation, 1918; The Nature of Intelligence, 1924; Fundamentals of Statistics, 1924; The Measurement of Attitude, 1929; The Vectors of Mind, 1935; Primary Mental Abilities, 1938; (with Thelma Gwinn Thurstone) Factorial Studies of Intelligence 1941; A Factorial Study of Perception, 1944; Multiple Factor Analysis, 1947. Known as developer of paper-and-pencil tests of intelligence, attitude. Died Sept. 29, 1955.

TICE, Linwood Franklin, Am. pharm. chemist, coll. dean; b. Salem, N.J., Feb. 17, 1909; s. Walter C. and Bessie (Waddington) T.; Ph.G., Phila. Coll. Pharmacy and Sci., 1929, B.Sc., 1933, M.Sc., 1935; D.Sc., St. Louis Coll. Pharmacy and Allied Scis., 1954; m. Marjorie Purnell, Aug. 4, 1929; children—Gregory, David P. Research fellow Iodine Edn. Bur., 1929-30, William R. Warner & Co., 1931-35, Edible Gelatin Mfrs.' Research Soc. Am., 1935-38; prof. pharmacy Baylor U., 1930-31; faculty Phila. Coll. Pharmacy and Sci., 1938—, prof. pharmacy, 1940—, dir. dept. pharmacy, 1941—, asst. dean, 1941-56, asso. dean, 1956-59, dean pharmacy, 1959-63, dean, 1963—. Cons., Gelatin Mfrs. Inst. Am., 1936-58, Merck, Sharp & Dohme, 1958—. Recipient Research Achievement award in advancement pharmacy Am. Pharm. Assn. Found., 1963, Alumni award Phila. Coll. Pharmacy and Sci., 1966. Fellow Am. Inst. Chemists, A.A.A.S., Am. Coll. Apothecaries; mem. Am. Assn. Colls. Pharmacy (past pres.), Am. Pharm. Assn. (pres. 1966-67, Pharm. Achievement award Phila. chpt. 1961), Am. Council on Pharm. Edn. (dir. 1960-66), U. S. Pharmacopeia (dir. 1960—), Am. Chem. Soc., N.Y. Acad. Scis., Franklin Inst., Am. Soc. Hosp. Pharmacists, Pa. Pharm. Assn., Am. Inst. History of Pharmacy, Rho Chi, Kappa Psi. Editor, Am. Jour. Pharmacy, 1946—; tech. editor Pharmacy Internat., also El Farmaceutico, 1941-59; asso. editor Remington's Practice of Pharmacy, 1948-63. Research and numerous publs. on devel. materials used in pharms.; devel. modern dosage forms. Home: 322 Morrison Av., Salem,

N.J. 08079. Office: Phila. Coll. Pharmacy and Sci., 43d St. and Kingsessing Av., Phila. 19104.*

TICHY, Vladmir, Czechoslovakian biologist; b. Boskovice, Czechoslovakia, May 10, 1923; s. Vladimir and Bozena (Stara) T.; RNDr., U. Brno, 1951; C.Sc., Charles U., Prague, 1957; m. Bozena Hrazdirova, Aug. 16, 1950; 1 child, Libuse. Asst., Inst. Plant Physiology U. Brno, 1950-56, sci. worker, 1956—, asso. prof., 1962—, sub dean Faculty Natural Scis., 1963-65. Mem. Czechoslovakian Bot. Soc., Czechoslovakian Sci. Soc. for Mycology. Research and numerous publs. on lichen and fungal physiology, humus formation, biol. activities of humus substances. Home: 48, Chudobova, Brno 15, Czechoslovakia. Office: 2, Kotlárská Brno, Czechoslovakia.*

TIDMAN, Derek Albert, physicist; b. London, Eng., Oct. 18, 1930; s. Albert Horace and Florance (Oscar) T.; B.Sc., Imperial Coll. Sci., London (Eng.) U., 1953, Ph.D., 1956; postgrad. London U. external student Sydney (Australia) U., 1953-56; m. Pauline Dunavan Harrell, Apr. 25, 1959; children—Katherine Fleming, Mark Harrell. Came to U. S., 1957, naturalized, 1963. Research asso. Sydney U., 1956-57; asst. prof. U. Chgo., 1957-60; faculty U. Md., College Park, 1960—, research prof. physics, 1964—. Cons., NASA, Goddard Space Flight Center, 1961—, Space Scis., Inc., Natick, Mass., 1961-62. Named Distinguished Young Scientist, Md. Acad. Scis., 1965. Mem. Am. Phys. Soc., Am. Geophys. Union, Sigma Xi. Author: (with D. C. Montgomery) Plasma Kinetic Theory, 1964; also numerous articles. Research in plasma physics and astrophysics, theoretical plasma models for solar radio emission, acceleration of energetic particles in turbulent plasmas. Home: 10714 Gatewood Av., Silver Springs, Md. 20903. Office: U. Md., College Park, Md. 20740.*

TIDWELL, Eugene Delbert, Am. spectroscopist, physicist; b. Lehi, Utah, Sept. 5, 1926; s. Delbert Cleon and Vernett (Webb) T.; B.S. in Math., Brigham Young U., 1951; postgrad. Md. U., 1954, 57, 60; m. Leah Mae Call, Dec. 17, 1948; children—Michael, Thomas, Anthony, Julia, Christopher, David, Bruce. Asst. gas chemist Kennecott Copper Corp., McGill, Nev., 1944-45, 47, 51-52; physicist Nat. Bur. Standards, Washington, 1952-63; engr. ARO, Inc., Arnold Air Force Sta., Tenn., 1963—. Asst. physics instr. Brigham Young U., 1950-51; math. instr. U. Tenn., 1963-64. Mem. A.A.A.S., Optical Soc. Am., Soc. for Applied Spectroscopy, Coblentz Soc. Contbg. author: Tables of Wave Numbers, 1961; also articles. Discovered molecular forces and rotational constants for many light gases by high resolution infrared spectra; first to reach high resolution 0.025 cm-1 with a spectrometer in nr. infrared region spectrum; perfected photog. methods measuring rocket nozzle ablation.*

TIDWELL, Herbert Collier, Am. biochemist; b. Mexia, Tex., Sept. 30, 1897; s. John Minor and Eva (Collier) T.; A.B., M.A., Baylor U., 1919; Ph.D., Johns Hopkins, 1930; m. Lutye Cogdell Suttle, Nov. 16, 1919. Faculty, Southwestern Med. Coll., U. Tex., Dallas, 1943—, prof., 1949-68 chmn. dept., 1949-63, prof. emeritus, 1968—. Mem. Am. Chem. Soc., Soc. Exptl. Biology, Soc. Biol. Chemists, Am. Inst. Nutrition, Am. Oil Chemists Soc. Research and numerous publs. in esterification, intravenous fat, nutrition, fat and carbohydrate metabolism, methionine excretion, ketosis, lipid absorption, transport and metabolism. Home: 4228 Stanhope St., Dallas 75205. Office: 5323 Harry Hines Blvd., Dallas 75235.*

TIDY, Charles Meymott, Brit. chemist, physician; b. Hackney, Eng., Feb. 2, 1843; s. William Callender and Charlotte (Meymott) T.; C.M., M.B., Aberdeen, Scotland, 1866; m. Violet Fordham Dobell, 1875; 1 son, 1 dau. Became prof. chemistry London Hosp., 1876; reader med. jurisprudence Inns of Court; pub. analyst, dep. med. officer health for London; med. officer health for Islington, also ofcl. analyst to Home Office. Licenciate Soc. Apothecaries. Mem. Royal Coll. Surgeons. Author: A Handy Book of Forensic Medicine and Toxicology, 1877; A Handbook of Modern Chemistry, 1878; Legal Medicine, 2 vols., 1882-83; The Story of A Tinder Box, 1889; also numerous articles. Invented method for analysis of water, 1879; modified Forchammer's process for determination of quantity of organic matter in water (Tidy's process); studies in legal medicine, pub. health, sanitation. Died London, Mar. 15, 1892.

TIEDEMANN, Friedrich, German physiologist, anatomist; b. Cassel, Germany, Aug. 23, 1781; s. Dietrich Tiedemann; student of Cuvier and Lamarck; prof., Landshut, Germany, then Heidelberg, Germany, from 1816. Fellow Royal Soc., 1832; corr. mem. French Acad. Scis. Author: Zoologie, 3 vols., 1808-14; Anatomie des Fischherzens, 1809; Anatomie und Bildungsgeschichte des Gehirns im Foetus, 1816; Die Verdauung nach Versuchen, 2 vols., 1826-27. Research on digestion, formation of human brain; ascertained internal capacity of skull by filling it with millet seed, which he then weighed, 1836. Died Munich, Jan. 22, 1861.

TIEMANN, Johann Karl Ferdinand, German chemist; b. Rubeland, Germany, June 10, 1848; studied chemistry and pharmacy under Otto and I. Knapp, in lab. of W. Hoffmann, Berlin; became asst. prof., Berlin, 1871, pvt. docent, 1878, prof. chemistry, 1882. Author: Handbuch der Untersuchung und Beurteilung der Wässer, 1889; Hygiene des Wassers, 1915. Research on nitriles, camphors, terpenes; synthesized vanillin; discovered ionone. Died Berlin, Nov. 14, 1899.

TIETZE, Christopher, demographer; b. Vienna, Austria, Dec. 11, 1908; s. Hans and Erica (Conrat) T.; M.D., U. Vienna, 1932; m. Sarah Lewit, Mar. 12, 1951. Came to U. S., 1938, naturalized, 1944. Staff mental hygiene study Johns Hopkins, Balt., 1938-43; research asso. Nat. Com. on Maternal Health, Balt., 1943-44, dir. research, N.Y.C., 1958-66; chief population and labor staff U. S. Dept. State, Washington, 1949-57; asso. dir. bio-med. div. Population Council, N.Y.C., 1967—. Lectr. obstetrics, gynecology Columbia U., 1959—; mem. adv. com. obstetrics, gynecology FDA, Washington, 1965—; mem. sci. groups WHO, 1965—. Recipient Commendable Service award Dept. State, 1954, Isidor Rubin award, 1956. Mem. Population, Am. Fertility Soc., Am. Sociol. Assn., Am. Pub. Health Assn. Author: The Condom as a Contraceptive, 1960; Surgical Sterilization of Men and Women, 1962; Bibliography of Fertility Control: 1950-1965, 65. Research and numerous publs. in statis. methodology for evaluation of human fertility and its control; developed Coop. Statis. Program for evaluation of intra-uterine devices.*

TIEWS, Klaus Friedrich Wilhelm, German fisheries biologist; b. Stettin, Germany, Jan. 17, 1928; s. Paul and Toni (Polumsky) T.; student U. Hamburg (Germany), 1947-51; Dr.rer.nat., U. Kiel (Germany), 1953; m. Ute Boysen, Apr. 17, 1954. Sci. asst. Inst. für Meereskunde, Kiel, 1953; fisheries biologist Inst. für Küsten-und Binnenfischerei, Bundesforschungsanstalt für Fischrei, Hamburg, 1953-68, professor and director, since 1968—; marine fisheries biologist FAO, Manila, Philippines, 1956-58; chief German Fisheries Mission to Thailand, German Bilateral Aid, 1961-66. Fisheries adviser Govt. of Ceylon, 1963, Govt. of Malaysia, 1966; sec. German Commn. for Exploration of Sea. Mem. Internat. Council for Exploration of Sea. Research and numerous publs. on biology and population dynamics of common shrimp, population dynamics of bluefin tuna, exploration of demersal fish resources in Gulf of Thailand and their econ. utilization, biology of herring, cod. Home: 30 Trenknerweg 2, Hamburg 52. Office: 9 Palmaille 2, Hamburg 50, Germany.*

TIFFENEAU, Marc-Émile-Pierre-Adolphe, French chemist, pharmacologist; b. Mouy, France, Nov. 7, 1873; pharmacist's apprentice; student Sch. Pharmacy; intern; became pharmacist of hosps., 1904; prof. Sorbonne, then Faculté de Médecine of Paris. Mem. French Acad. Scis., 1939, French Acad. Medicine. Research on chemistry of isomerizations and molecular shifts; made pharmacological studies of poisons, hypnotic drugs; showed action of volatile bromides is centered in brain. Died Paris, May 20, 1945.

TIGERTT, William David, Am. physician; b. Wilmer, Tex., May 22, 1915; s. Carl D. and Mary Lou (Strain) T.; M.D., Baylor U., 1937, A.B., 1938; m. Mary Helen Braack, July 1, 1938; children—Susan, David. Fellow, instr. Baylor U., 1938-40; commd. 1st lt. U. S. Army, 1940, advanced through grades to col., 1956; asst. comdt. Army Med. Service Grad. Sch., 1949-54; dir. spl. operations div. Walter Reed Army Inst. Research, Washington, 1956-61, dir., comdt., 1963—; comdg. officer U. S. Army Med. Unit, Ft. Detrick, Md., 1956-61; asso. prof. Med. Sch. Medicine, Balt., 1958—. Cons., Surgeon Gen., U. S. Army, 1960—, WHO, 1961. Diplomate Am. Bd. Pathology. Fellow A.C.P.; mem. Assn. Immunologists, A.M.A. Research and publs. in pathology of infectious disease and immunology, relationship of size infectious dose to time of disease onset, methods prophylaxis by vaccines, chemotherapy, methods of treatment. Home: 6924 15th St. N.W. Office: Walter Reed Army Inst. Research, Walter Reed Army Med. Center, Washington 20012.*

TIKHONOV, Andrei Nikolaevich, Russian mathematician, geophysicist; b. Gizhatsk, Smolensk Oblast, Russia, Oct. 30, 1906; grad. Moscow U., 1927; Dr. Physical-Math. Scis. Instr., Moscow U., 1927-36, prof., 1936—; also staff Inst. Terrestrial Physics. Recipient Order of Lenin. Corr. Mem. USSR Acad. Scis. Author: (with A. A. Samarskii) Equations of Mathematical Physics, 1953; On Homogeneous Canonical Difference Systems, 1960; The Stability Coefficients of Difference Systems, 1960. Research and publs. in theoretical topology, math. physics and geophysics; introduced concept of electromagnetic sounding of geol. layers, concept of product of topological spaces. Home: Leninskii Prospekt 13, Moscow, USSR.

TIKKANEN, Matti Haakon August, Finnish metallurgist; b. Turku, Finland, Nov. 28, 1915; s. Peter and Aurora (Tuominen) T.; Dipl.Eng. in Chem. Engring., Abo (Finland) Akademi, 1938; postgrad. Chalmers Inst. Tech., 1946; D.Sc., Inst. Tech., Helsinki, 1949; m. Satu Sisko Lehtonen, Aug. 23, 1941; children—Matti Juhani, Lena Marjatta. Supt., metallurgist State Airplane Works, Finland, 1940-46; chief research and devel. powder metallurgy Husquarna Gun Factories, Sweden, 1947-48; dir. metallurgy dept. State Inst. Tech. Research, 1949-63; prof. metallurgy Inst. Tech., Otaniemi, Finland, 1949—. Mem. Permanent Council in Metallic Corrosion, 1962—; cons. powder metallurgy, France, Switzerland, Italy, 1949, Sweden, 1949—, U. S. A., 1963; cons. in corrosion tech., U. S. A., 1963—. Mem. European Fedn. Corrosion, Finnish Chem. Soc. (past pres.), Am. Inst. Mining, Metall. and Petroleum Engrs., Am. Chem. Soc., A.A.A.S., Ingeniörsvetenskapsakademin Sweden (fgn.). Author: (with others) Corrosion and its Prevention; (with R. Asanti) Metals Technology; also articles. Research in sintering theory, powder metallurgy and ceramics, structure of metal carbides, solid state chemistry, defect structure versus reactivity, electrochemistry, passivity theory, corrosion theory, sintering of hard metal, vacuum melting of silver alloys; recovered vanadium from magnetite leaching of ilmenite; tech. inventions. Home: 1.N. Takojantie Tapiola, Finland. Office: Inst. Tech., Otaniemi, Finland.*

TILDEN, Evelyn Butler, Am. microbiologist; b. Lawrence, Mass., Mar. 28, 1891; d. Howard Benjamin and Harriette (Butler) Tilden; A.B., Brown U., 1913; M.S., Columbia U., 1926, Ph.D., 1929. Asst. bacteriology and pathology Rockefeller Inst. for Med. Research, 1928-31; asst. prof. Colo. State Coll., 1931-32; research asso. dept. research bacteriology Northwestern U. Med. Sch., 1932-37, asso. prof. Dental Sch., 1942-48, prof., 1948-54, chmn. dept. microbiology, 1942-54; bacteriologist NIH, Bethesda, Md., 1937-42; curator labs. Animal Hosp., Chgo. Zool. Park, Brookfield, Ill., 1954-63, emeritus, 1963—. Mem. Am. Soc. for Microbiology, Sigma Xi, Sigma Delta Epsilon, Iota Sigma Pi. Author: Outline of Bacteriology, 1948. Research and publs. on inclusion conjunctivitis, protozoa, bacterial prodn. of rare carbohydrates, oral bacteria, animal infections, fungal endotoxins. Home: 2117 N. Hudson Av., Chgo. 60614. Office: Animal Hosp., Chgo. Zool. Park, Brookfield, Ill. 60513.*

TILDEN, Sir William Augustus, Brit. chemist; b. London, Eng., Aug. 15, 1842; s. Augustus Tilden; ed. Royal Coll. Chemistry; D.Sc., London; Sc.D., U. Dublin (Ireland), Victoria; LL.D., U. Birmingham (Eng.). m. Charlotte Bush, 1869 (dec. 1905); m. 2d, Julia Mary Ramie, 1907; 1 son. Demonstrator Pharm. Soc., 1864-72; sci. master Clifton (Eng.) Coll., 1872-80; prof. chemistry Mason Coll, Birmingham, Eng., 1880-94, Royal Coll. Sci., London, 1894-1909, dean, from 1905; emeritus prof. Imperial Coll. Sci. and Tech., from 1909. Fellow Royal Soc. (Davy medal 1908), 1880; mem. Inst. Chemistry (pres. 1891-94), Chem. Soc. (treas. 1899-1903). Author: Introduction to Chemical Philosophy, 1876; Practical Chemistry, 1880; Hints on Teaching Chemistry, 1895; A Manual of Chemistry, 1896; A Short History of the Progress of Scientific Chemistry, 1899; The Elements, 1910; Chemical Discovery and Invention in the Twentieth Century, 1917; Life of Sir William Ramsay, 1918; Famous Chemists, 1921. Contbr. sci. papers to publs. First to make isoprene artificially (led to synthetic prodn. rubber, 1892); studied alkaloids brucine, strychnine, caffein, also terpenes; determined composition of NOCL. Died Dec. 11, 1926.

TILFORD, Shelby Grant, Am. physiol. chemist; b. nr. Clarkson, Ky., Jan. 11, 1937; s. Ernest and Dolores (McGrew) T.; student Ky. Wesleyan Coll., 1954-55; B.S., Western Ky. U., 1958; Ph.D., Vanderbilt U., 1962; m. Jacqueline E. Fulkerson, Jan. 24, 1956; children—Michael Grant, Michelle Renee. Research asso. U. S. Naval Research Lab., Washington, 1961-63; spectroscopy cons., 1963—. Mem. Research Soc. Am., Philos. Soc. Washington, Sigma Xi. Research and publs. on energy levels of molecules and atoms and how these energy levels interact, atmospheric and astrophys. phenomena in terms of such interactions. Home: 8805 Church Field Lane, Laurel, Md. 20810. Office: Code 7140, Overlook Av., Washington 20390.*

TILGHMAN, Richard Albert, Am. chemist; b. Phila., May 24, 1824; s. Benjamin and Anna Maria (McMurtrie) T.; B.A., U. Pa., 1841; studied chemistry under James C. Booth, Phila.; m. Susan Price Toland, 1860, 5 children. Wrote and delivered paper On the Decomposing Power of Water at High Temperatures (1st systematic study of hydration) before Am. Philos. Soc., 1847; developed process of hydrolysis for extracting acid from animal fat; produced caustic soda through hydrolysis; developed processes for mfg. potassium dichromate, paper pulp, and for producing gas from coal; sold discoveries to various industries in U. S. and Eng.; developed sandblast process for forming articles made of hard, brittle materials; dir. George Richards & Co., Ltd., mfrs. machine tools, also Tilghman Sand Blast Co., nr. Manchester, Eng. Died Mar. 24, 1899.

TILHO, Jean Auguste Marie, French scientist, explorer; b. Domme, France, May 1, 1875; grad. Saint-Cyr, 1895. Officer, French Colonial Army, from 1895, served to gen. Recipient Gold medals, geog. socs. London, Paris, Antwerp. Mem. French Acad. Sci., 1932, Acadèmie des Sciences Coloniales, Bureau des Longitudes. Conducted sci. explorations in Africa; established new frontier of Chad, 1907-08; explored Lake Chad and surrounding regions, also mountainous regions Tibesti, Erdi, Ennedi. Died Paris, France, 1956.

TILLER, Frank Monterey, Am. chem. engr.; b. Louisville, Feb. 26, 1917; s. Frank M. and Nellie (Lawson) T.; B. Chem. Engring., U. Louisville, 1937; M.S., University of Cin., 1939, Ph.D., 1946; Doutor Honoris Causa, University of Brazil, 1962, Universidade do Estado de Guanabara, 1966; m. Ann W. Quiggins, Dec. 20, 1941; children—Faith Lee, Richard Bertrand. Chem. engr. C. M. Hall Lamp Co., Detroit, 1940; instr. chem. engring. U. Cin., 1940-42; asso. prof. chem. engring. Vanderbilt U., 1942-52; dean engring. Lamar State Coll. Tech., 1951-55; dean engring., prof. chem. and elec. engring. U. Houston, 1955-63, M. D. Anderson prof. chem. engring., 1963-—, dir. internat. affairs, 1963-67, dir. Center for Study Higher Edn. in Latin Am., 1967-—; lecturer Instituto de Oleos, Rio de Janeiro, 1952, Universidade do Brasil, 1962-63; honorary professor Universidad Central, Quito, Ecuador, Universidad Guayaquil, 1958, Pontificia Universidade Catolica Rio de Janeiro, 1962; Fulbright grantee in Ecuador, 1958; Agy. Internat. Devel. coordinator, Guayaquil, Ecuador, 1960-65, Rio de Janeiro, 1963-—. Dir. Gupton-Jones Coll. Mortuary Sci., 1945-51. Recipient Streng award U. Louisville, 1937; pub. award, 1950, presentation award, 1952, 62, Am. Inst. Chem. Engrs. Mem. Am. Inst. Chem. Engrs., Am. Soc. Engring. Edn., Nat. Soc. Profl. Engrs., Filtration Soc. (London), Tex. Filtration Soc. (pres.), Sigma Xi, Sigma Tau, Phi Lambda Upsilon, Theta Chi Delta. Author: Vector Algebra, 1958; Introduction to Vector Analysis, 1963; also tech. articles. Developed applications calculus of finite differences in stagewise operations of chem. engring.; developed theory of variable flow rate in filter cakes; active in establishing grad. schs. engring. in Latin Am. Address: U. Houston, Houston 77004.*

TILLETT, William S(mith), Am. physician; b. Charlotte, N.C., July 10, 1892; s. Charles Walter and Carolyn (Patterson) T.; A.B., U. of N.C., 1913, D.Sc., 1942; M.D., Johns Hopkins, 1917; D.Sc. (hon.), U. Chgo., 1952, D.Sc. (hon.) Northwestern University, 1959; married Dorothy Stockbridge, September 8, 1928; 1 dau., Louise E. (Mrs. Douglas MacAgy). Asst. resident, resident physician, asso. Hosp. of Rockefeller Inst., 1922-30; asso. prof. of medicine, Johns Hopkins Med. Sch., 1930-37; prof. bacteriology, N.Y.U. Coll. Medicine, 1937-38, prof. medicine, 1938-58, emeritus, 1958-—; dir. research trainee grant USPHS, 1958-—. Dir. 3d med. div. and mem. exec. com. (sec. since 1941), Bellevue Hosp., N.Y. City, since 1938. Consultant to sec. of war, epidemic diseases, 1941-46. Lasker award, Am. Pub. Health Assn., 1949; Borden Award Am. Assn. Med. Colls., 1952. Mem. Nat. Acad. Scis., American Association Physicians (president 1957-58), Society for Clinical Investigation (pres. 1936-37), Harvey Soc. (pres. 1957-58) New York Academy of Medicine (chmn. sect. on medicine, 1944-45), N.Y. Acad. Sciences, Am. Soc. of Bacteriology, Am. Assn. of Immunology (editorial bd. jours.), Soc. Exptl. Biology and Medicine (editorial bd. jour.), Interurban Clin. Club (pres. 1946-47), A.A.A.S., Am. Med. Assn.; Sigma Xi. Contbr. articles on exptl. research and clin. studies to tech. jours., including Jour. Exptl. Medicine, Jour. Clin. Investigation, Bull. Johns Hopkins Hosp. Research in microbiological fields of experimental and clinical infections, especially pneumococcal and streptococcal. Home: 235 E. 22d St. Office: 550 1st Av., N.Y.C.*

TILLEY, Cecil Edgar, petrologist; b. Adelaide, S. Australia, May 14, 1894; s. John Edward and Jane (Nicholas) T.; student Adelaide U., 1913; B.Sc., U. Sydney (Australia), 1916; Ph.D., Cambridge (Eng.) U., 1923; m. Doris Marshall, June 21, 1928; 1 dau., Anne. Faculty, U. Cambridge, 1923-61, prof. mineralogy and petrology, 1931-61, fellow Emmanuel Coll., 1931-—, vice master, 1952-58; research asso. Carnegie Instn. Washington, 1956-—. Recipient Bigsby medal, 1937, Wollaston medal, 1960, Roebling medal, 1954. Fellow Royal Soc., 1938 (Royal medal 1967), Royal Soc. Edinburgh (hon.), Geol. Soc. Am. (hon.); mem. Internat. Mineral. Assn. (pres. 1964-—), Royal Swedish Acad. Scis. (fgn.), Nat. Acad. Scis. (U. S.) (fgn. asso.), Am. Acad. Arts and Scis. (fgn. hon.), Mineral. Soc. (past pres.), Geol. Soc. (past pres.), Nat. Acad. Scis. Research and numerous publs. on regional and exptl. petrology., igneous and metamorphic rocks. Home: 30 Tenison Av., Cambridge, Eng.*

TILLO, Aleksei Andreevich, Russian geographer; b. Kiev, Russia, Nov. 1839; gen. Russian army; corr. mem. French Acad. Scis., 1892; produced atlases of isobars of Russia and Asia; research in terrestrial magnetism, geodesy, meteorology, hypsometry. Died St. Petersburg, Jan. 11, 1900.

TILLY, Joseph Marie de, see de Tilly, Joseph Marie.

TILTON, George Robert, Am. geochemist; b. Danville, Ill., June 3, 1923; s. Edgar Josiah and Caroline (Burkmeyer) T.; student Blackburn Coll., 1941-43; B.S., U. Ill., 1947; Ph.D., U. Chgo., 1951; m. Elizabeth Jane Foster, Feb. 8, 1948; children—Linda Ruth, Helen Elizabeth, Elaine Lee, David Foster, John Robert. Staff dept. terrestrial magnetism and geophys. lab., Carnegie Instn. Washington, 1951-65; prof. geochemistry U. Cal. at Santa Barbara, 1965-—; gastdozent U. Bern (Switzerland), 1960-61. Fellow Am. Geophys. Union; mem. A.A.A.S., Geochem. Soc., Sigma Xi. Asso. editor Jour. Geophys. Research, 1963-—. Research and publs. on methods of determining ages of rocks, age of earth, heat prodn. by rocks in earth due to radioactivity. Home: 3425 Madrona Dr., Santa Barbara, Cal. 93105.*

TIMASHEFF, Serge Nicholas, chemist; b. Paris, France, Apr. 7, 1926; s. Nicholas S. and Tatiana (Rouzsky) T.; B.S., Fordham U., 1946, M.S., 1947, Ph.D., 1951; m. to Marina J. Gorbunoff, 1953; 1 dau., Marina S. Came to U. S., 1939, naturalized, 1944. Instr., Fordham U., 1947-50; postdoctoral fellow Cal. Inst. Tech., 1951; research fellow Yale, 1951-55; head phys. chem. investigation Eastern Regional Research Lab., U. S. Dept. Agr., Wyndmoor, Pa., 1955-66, head Pioneering Research Lab. Phys. Biochemistry, 1966-—; prof. dept. biochemistry Brandeis U., 1966-—; adj. prof. Drexel Inst. Tech., Phila., 1962-63. Recipient Student medal Am. Inst. Chemistry, 1946, Arthur S. Flemming award Jr. C. of C. Washington, 1964, Distinguished Service award U. S. Dept. Agr., 1965; sr. postdoctoral research fellow NSF, 1959-61. Mem. Am. Chem. Soc. (nat. award in chemistry of milk 1963, Phila. sect award for creative research, 1966), Am. Soc. Biol. Chemistry, Biophys. Soc., A.A.A.S., Sigma Xi, Alpha Chi Sigma. Editor: (monograph series) Biological Macromolecules, 1966-—; editorial bd. Archives Biochemistry and Biophysics, 1966-—. Research on solution structure of proteins, changes in protein conformations and thermodynamics protein interactions in solution, structure in solution of nucleic acids, methods of high polymer investigations. Office: Grad. Dept. Biochemistry, Brandeis U., Waltham, Mass. 02154.*

TIMBY, Theodore Ruggles, Am. inventor; b. Dover, Dutchess County, N.Y., Apr. 5, 1819; s. George W. and Sarah Johnson T., of Pittsfield, Mass.; ed. pub. schs.; hon. degress: A.M., Madison U., 1867; S.D., U. of Troy, O., 1882; LL.D., Iowa Wesleyan U., 1891; m. Charlotte M. Ware, 1844. Author: Bridging the Skies, 1883; Beyond, 1886; Stellar Worlds, 1896; Lighted Lore for Gentle Folk, 1902. Invented floating dry-dock, 1836; invented, 1841, the revolving turret used on the Monitor, and battleships; invented practical method of raising sunken vessels, 1841, the Am. turbine waterwheel or motor, 1844, and in 1857, the first commercially portable 33-inch mercurial barometer. From 1861 to 1891, invented and patented, at home and abroad, a system of coast defenses known as follows: The sighting and firing of heavy guns by electricity; the tower and shield; the cordon across the channel; the planetary and subterranean systems, with 15 other modifications of the turret system; invented and patented improvements in pneumatic and hydraulic power; also a new prin. in Turbine waterwheels or motors. Died Bklyn., Nov. 9, 1909.

TIMIRJAZEV, Kliment Arkadevic, Russian botanist; b. St. Petersburg, Russia, June 3, 1843; studied under Kirchoff, Helmholtz and Bunsen. Prof. in Moscow. Author: Socinenija (collected works), 1937-40. Darwinist; research in photosynthesis; co-founder Russian plant physiology. Died Moscow, Apr. 28, 1920.

TIMM, Gisela, German ophthalmologist, ophthalmopathologist; b. Kolberg, Pommern, Germany, Sept. 21, 1924; s. Erich and Käthe (Sabin) T.; M.D., U. Greifswald (Germany), 1955; Resident Eye Clinic, U. Leipzig (Germany), 1963-—, head physician, 1964-—, conductor dept. pathology and histology, ophthalmologist dept. neurosurgery, 1963-—. Mem. Deutsche Ophthalmologische Gesellschaft, Gesellschaft für Morphologie, Deutsche Gesellschaft der Naturforscher und Ärzte, Deutsche Gesellschaft für Allergie. Contbg. author: Handbuch der Pathol. Anatomie; also articles. Research on histopathology of eye in oxalosis, in leucoencephalitis, tumorous states. Office: 14 Liebigstr., Leipzig, Germany.*

TIMMONS, Francis Leonard, Am. agronomist; b. Little River, Kan., Feb. 12, 1905; s. Elmer Cyrus and May (Thurstin) T.; student McPherson Coll., 1922-24; B.S., Kan. State Coll., 1928, M.S., 1932; Ph.D., Wyo. U., 1963; m. Bessie Lavina Smith, Dec. 25, 1927; children—Shirley (Mrs. Thomas A. Goetz), Myrna Franklin, Irma Means, Kay Diane. With Kan. State Coll., Manhattan, 1928-35, extension agronomist, 1935; with Bur. Plant Industry, U. S. Dept. Agr., 1935-57, sr. agronomist, regional coordinator weed investigations, Logan, Utah, 1948-54, Laramie, Wyo., 1954-57; investigations leader weed control aquatic and noncrop areas in U. S. Agr. Research Service, Laramie, 1957-—. Recipient Outstanding Performance award U. S. Dept. Agr., 1960, Superior Service award, 1963. Fellow Am. Soc. Agronomy (chmn. program div. 6 1955, 65; mem. Weed Soc. Am. (hon., exec. com. 1965-—), Ecol. Soc. Am., Am. Inst. Biol. Scis., Western Weed Control Conf. (past chmn. research com.), Sigma Xi. Contbr. numerous articles to tech. jours. Pioneered devel. cultivation, cropping and chem. methods for control field bindweed, and ann. weeds in winter wheat; developed methods for controlling weeds in irrigated crops and on irrigation systems, ann. weeds in corn and onions, dodder in alfalfa seed crops, ann. weeds in legume seedlings, aquatic and bank weeds in irrigation and drain canals, phreatophytes on irrigation systems and flood plains; research on ecology and physiology Carex lanuginosa and nebvaskiensis. Home: 900 S. 13th St. Office: P.O. Box 3354 U. Sta., Laramie, Wyo. 82070.*

TIMOCHARIS, astronomer; flourished in Alexandria, 1st century B.C. Among 1st to find and record positions of chief stars; measured distances of chief stars from fixed points in skies; perhaps author of 1st real star catalogue; observed planets, sun; observations drawn upon by Ptolemy, Hipparchos.

TIMOFEEV, Pyotr Vasilevich, Russian elec. engr.; b. June 25, 1902; grad. Moscow U., 1925. Became asso. All Union Electro-Tech. Inst., 1928; faculty Moscow U., Moscow Inst. Energetics. Recipient Stalin Prize, 1946, 51. Corr. mem. USSR Acad. Scis. Author: (with V. V. Sorokina) The Shape of a Field for Electrostatic Lenses, 1948; Electron Emission from Complex Surfaces, 1957; Vacuum Beta-Emission and its Uses, 1960. Research and publs. on electronic optics, electron emission, photo-effect; design of photocells, electronic multipliers, transmitting tarbes. Home: Fil'skoe sh.5. Office: USSR Acad. Scis., Leninskii Prospekt 14, Moscow, USSR.

TIMOFEIEVSKII, Aleksandr Dmitrievich, Russian pathologist; b. Moscow, Russia, Feb. 20, 1887; student Paris (France) U., 1904-06; grad. Tomsk (Russia) U., 1912 M.D., 1919; m. S. V. Bonevolenskaya. Successively lab. asst., prosector, lectr., prof., head chair pathology Tomsk U., 1912-34; head exptl. dept. X-Ray Oncological Inst., Kharkov, Russia, 1934-41; with Biginikets Inst. Clin. Physiology, Ukranian Acad. Sci., Kiev, 1941-54; organizer, head dept. etiology and pathogenesis of tumors Inst. Exptl. Pathology and Theory of Cancer, USSR Acad. Med. Sci., 1954-—. Recipient Stalin prize, 1948; decorated Order of Lenin; named Honored Sci. Worker of Ukranian SSR, 1947. Mem. USSR Acad. Med. Sci., 1945; corr. mem. Ukranian Acad. Sci., 1939. Author: The Lymphocyte, Monocyle and Myeloblast of Normal and Leukemic Blood in Explantates, 1947; Current Theories on the Origin of Tumors, 1954; Explantation of Human Tumors, 1947; co-author: Models and Methods in Experimental Oncology, 1960. Co-editor Med. Jour. Ukranian Acad. Sci.; mem. editorial bd. jour. Problems of Oncology. Specialist in pathol. research with tissue culture on hematology, oncology, Tb; studied differentiation of cells in extended tumor cultures; studied (with wife) devel. leukemia cells, tubercular tumors; induced cancer in connective tissue. Address: Solyanka 14, Inst. eksperimentalnoy patologii i terapii raka Akademii medisinskikh nauk, Moscow, USSR.

TIMOSHENKO, Stephen, mechanical engr.; b. near Kiev, Russia, Dec. 23, 1878; s. Prokop and Jozefina (Sarnavskaja) T.; grad. Inst. of Engrs. of Ways of Communication, Russia, 1901; D.Sc., Lehigh U., 1936; D.Eng., Michigan University, 1938; D.Eng., l'École Polytechnique, Zürich, 1948, Technische Hochschule, Munchen, 1948, University of Zagreb, 1956; LL.D., University of Glasgow, 1951; married Alexander Archangelskaja, March 3, 1902; children—Anna (Mrs. F. Hetzelt), Gregor, Marina (Mrs. J. N. Goodier). Came to United States, 1922, naturalized, 1927. Instructor Inst. of Engr. of Ways of Communication, 1902-03; asst. prof. Polytechnical Inst., St. Petersburg, Russia, 1903-06; prof. Polytechnical Inst., Kiev, 1906-11, Electrotechnical and Polytechnical Inst., St. Petersburg, 1912-17, Polytechnical Inst., Zagreb, Yugoslavia, 1920-22; research engr. Westinghouse Electric and Mfg. Co., Pittsburgh, 1923-27; prof. of engring. mechanics, U. of Mich., 1927-36; prof. of theoretical and applied mechanics Stanford, 1936-44, emeritus. Corr. mem. Acad. Sciences (Russia), Acad. Tech. Sciences since 1939. Mem. American Philos. Soc., 1939; mem. Nat. Acad. Sciences, 1940. Foreign mem. Royal Soc. of London, since 1944, Accademia Nationale dei Lincei, Rome, since 1948. Mem. Greek Orthodox Ch. Author: Strength of Materials, 1930; Elements of Strength of Materials, 1935; Theory of Elastic Stability, 1936; Engineering Mechanics, 1937; Theory of Plates and Shells, 1940; Theory of Structures, 1945; (with D. H. Young) Advanced Dynamics, 1948; History of Strength of Materials, 1953; The Collected Papers of Stephen P. Timoshenko, 1953. Influential in field theoretical and applied mechanics; theory of elasticity; vibration in engineering. Home: 56 Wuppertal-Elberfeld, Böcklin Str. 35, Germany. Office: Dept. of Engr. Mech. Stanford U., Stanford, Cal. 94305.*

TIMS, Eugene Chapel, Am. plant pathologist; b. Learned, Miss., July 19, 1894; s. William Francis and Julia (Hanna) T.; B.S., Miss. A. and M. Coll., 1919; Ph.D., U. Wis., 1924; m. Maud Virginia Bingham, July 7, 1920; 1 son, Eugene Francis. Faculty, La. State U., 1920-—, prof. emeritus plant pathology, 1965-—. Fellow A.A.A.S.; mem. Am. Phytopath. Soc. Research and numerous publs. in control of diseases of sugar cane, figs, onions, shallots, tomatoes, cabbage. Home: 3155 E. Lake Shore Dr., Baton Rouge, La. 70808.*

TINBERGEN, Nikolaas, zoologist; b. The Hague, Netherlands, Apr. 15, 1907; s. Dirk Cornelis and Jeannette (Van Eek) T.; D.phil., Leiden (Netherlands) U., 1932; m. Elisabeth Amelie Rutten, Apr. 14, 1932; children—Jacob, Catharina (Mrs. August Loman), Dirk, Jannetje, Gerardina. Faculty, Leiden U., 1933-49, prof. exptl. zoology, 1947-49; faculty Oxford (Eng.) U., 1949-—, reader animal behavior, 1962-66, prof., 1966-—. Fellow Royal Soc., 1962; mem. Am. Acad. Arts and Scis. (fgn.), Akademie van Wetenschappen, Am. Mus. Natural History, Max Planck Gesellschaft. Author numerous books including: The Study of Instinct, 1951; Social Behaviour in Animals, 1953; Curious Naturalist, 1956; Animal Behavior, 1965; also articles. Research on social behavior of animals and responses to complex stimuli. Home: 88 Lonsdale Rd., Oxford, Gt. Britain.*

TINDAL, John Stuart, English endocrinologist; b. Birkenhead, Eng., Mar. 29, 1931; s. Reginald James and Phyllis (Pickering-Jones) T.; B.Sc. with honors, Liverpool (Eng.) U., 1951, Ph.D., 1954; m. Philippa

Margaret Kerridge, July 23, 1955; children—Sarah Jane, Miles John, Andrew James. With physiology dept. Nat. Inst. for Research in Dairying, Shinfield, Eng., 1953—, prin. sci. officer, 1966—; vis. staff dept. anatomy and Brain Research Inst., U. Cal. at Los Angeles, 1961-62. Sir Henry Wellcome Travelling fellow in medicine, 1961-62. Mem. Soc. for Endocrinology, Physiol. Soc., Soc. Exptl. Biology. Research and publs. on endocrine and neuroendocrine control of lactation and reprodn. Home: 68 Oatlands Rd., Shinfield. Office: Nat. Inst. for Research in Dairying, Shinfield, Reading, Berks., Eng.*

TING, Er Yi, physician; b. Shantung, China, June 3, 1919; s. Yu Chuang and Mary (Chang) T.; M.D., Nat. Def. Med. Coll., Shanghai, 1948; m. Theresa Wang, Nov. 8, 1958; children—Selene, Sandra, Selwyn. Instr. medicine Nat. Def. Med. Center, Taipei, Formosa, 1952-54, State U. N.Y., 1957-59, Albert Einstein Coll. Medicine, 1959-63; postdoctoral fellow U. Buffalo, 1956-57; asst. prof. medicine N.Y. Med. Coll., N.Y.C., 1963—; dir. pulmonary function research lab. Met. Hosp. Center, N.Y.C., 1963—; mem. staff Bronx Municipal Hosp. Center, 1959-63, King's County Hosp., 1957-59, Flower and Fifth Av. Hosp., 1963—, Beekman-Downtown Hosp., 1964—. Mem. Am. Physiol. Soc., Am. Coll. Chest Physicians, Am. Fedn. Clin. Research, Am. Thoracic Soc., A.M.A., Chinese Med. Assn., N.Y. Trudeau Soc., Internat. Union Physiol. Scis. Med. editor, hygiene sect. The China Tribune, 1964—. Research and numerous publs. on respiratory physiology, especially mechanics of breathing of lung, airway and chest wall under physiol. and diseased conditions. Home: 451 West End. Av., N.Y.C. 10024. Office: 50 Bayard St., N.Y.C. 10013.*

TING, Lu, aerodynamicist; b. Kiangsi, China, Apr. 18, 1925; s. Kao Y. and Y.F. (Yao) T.; came to U. S., 1947, naturalized, 1956; B.S. in Mech. Engring., Chiao-Tung U., Shanghai, China, 1946; M.S., Mass. Inst. Tech., 1948; M.S. in Aero. Engring., Harvard, 1949; Sc.D. in Aeros., N.Y. U., 1951; m. Alice Y.D. Pan, June 2, 1951; children—Luke, Diana, Mary. Spl. design engr. Foster Wheeler Corp., N.Y.C., 1952-55; faculty Poly. Inst. Bklyn., 1955-64, research prof., 1960-64; prof. aeros. and astronautics N.Y. U., 1964-68, professor of mathematics, since 1968—. Cons. to Gen. Applied Sci. Lab., Inc., Merrick, N.Y., 1960-67. Mem. Am. Inst. Aeros. and Astronautics, Am. Phys. Soc., Soc. for Indsl. and Applied Math., Sigma Xi. Research and publs. on blast wave, supersonic wing-body interference, viscous mixing problems, space trajectories, optimum techniques.*

TING, Yu Chen, geneticist; b. Honan, China, Oct. 3, 1920; s. Chin-yung and Yi-ying (Wang) T.; B.S., Nat. Honan U., 1944; M.S., U. Ky., 1949; M.S. in Agr., Cornell U., 1952; Ph.D., La. State U., 1954; m. Jovina Yueh-hung Chen, June 25, 1960; children—Andrew, Claire. Came to U. S., 1948, naturalized, 1959. Fellow Harvard, 1954-62; asst. prof. Boston Coll., Chestnut Hill, Mass., 1962-64, asso. prof. cytogenetics, 1964—; research collaborator Brookhaven Nat. Lab., Upton, L.I., N.Y., 1964—. Mem. Genetics Soc. Am., Am. Genetic Assn., Bot. Soc. Am., A.A.A.S., New Eng. Botanic Club, Sigma Xi. Author: Chromosomes of Maize-Teosinte Hybrids, 1964; also numerous articles. Determined chromosome number and induction flowering sweet potato plant; research on chromosomes maize, teosinte and Tripsacum, chromosome inversions in maize and teosinte, evidence teosinte introgression in maize. Home: 230 Bonad Rd., Brookline, Mass. 02167. Office: Dept. Biology, Boston Coll., Chestnut Hill, Mass. 02167.*

TINKHAM, Michael, Am. physicist; b. nr. Ripon, Wis., Feb. 23, 1928; s. Clayton Harold and LaVerna (Krause) T.; A.B., Ripon Coll., 1951; M.S., Mass. Inst. Tech., 1951, Ph.D., 1954; m. Mary Stephanie Merin, June 24, 1961. NSF fellow Clarendon Lab., Oxford (Eng.) U., 1954-55; research physicist U. Cal. at Berkeley, 1955-57, faculty, 1956-66, prof. physics, 1961-66; Gordon McKay prof. applied physics, prof. physics Harvard, 1966—. Fulbright lectr., France, 1961; cons. to pvt. cos.; mem. adv. panel for physics NSF, 1963-66. Guggenheim fellow, 1963-64. Fellow Am. Phys. Soc. (chmn. div. solid state physics 1966-67); mem. Sigma Xi, Phi Beta Kappa. Author: Group Theory and Quantum Mechanics, 1964; Superconductivity, 1965; contbg. author: Modern Physics for Engineers, 1961; Low Temperature Physics, 1962; also numerous articles. Research on infrared spectroscopy solids; pioneered demonstration energy gap in superconductors, field-free resonances in magnetic crystals, temperature-dependent soft mode in ferroelectric crystals; research in magnetic field effects in superconductors. Home: 98 Rutledge Rd., Belmont, Mass. 02178. Office: Harvard, Cambridge, Mass. 02138.*

TINKLE, Donald Ward, Am. zoologist; b. Dallas, Dec. 3, 1930; s. Maurice Ward and Rubye (Still) T.; B.S., So. Meth. U., 1952; M.S., Tulane U., 1955, Ph.D., 1956; m. Marjorie Anne White, Feb. 24, 1951; children—Donna Lynn, Randall Troy, Steven Lance, Melanie Ann. Asst. prof. W. Tex. State U., Canyon, 1956-57; faculty Tex. Technol. Coll., Lubbock, 1958-65, prof. biology, 1963-65; prof. zoology, curator Mus. Zoology, U. Mich., Ann Arbor, 1965—. Fellow Herpetologists League, A.A.A.S.; mem. Am. Soc. Ichthyologists and Herpetologists (bd. govs. 1961—), Southwestern Assn. Naturalists (editor 1958-63, pres. 1965-66). Research and publs. in reproductive cycles in reptiles, reptile populations, taxonomy and geog. distbn. of various reptile species. Home: 3527 Weber Rd., Saline, Mich. 48176. Office: Mus. Zoology, U. Mich., Ann Arbor, Mich. 48104.*

TINKLE, William John, Am. geneticist; b. nr. Marion, Ind., Nov. 20, 1892; s. John William and Mary (Miller) T.; A.B., Manchester Coll., 1916; postgrad. Bethany Bibl. Sem., 1919-20; M.A., Ohio State U., 1927, Ph.D., 1932; m. Lula Rench, Aug. 1, 1916; children—Helen (Mrs. Guy Tinley), David. Faculty, Taylor U., 1933-39, 44-48, head sci. dept., 1944-48; faculty biology dept. Anderson Coll., 1956-58. Mem. A.A.A.S., Am. Genetic Assn., Creation Research Soc. (sec.). Author: Fundamentals of Zoology, 1939; Modern Science and Christian Faith, 1948; Heredity: A study in science and the Bible, 1967; also numerous articles. Research showed that nomadism (an unreasonable desire to wander) and several types of deafness are inherited. Home: 118 W. South St., Eaton, Ind. 47338.*

TINNILÄ, Aulis Tapio, Finnish biologist; b. Nastola, Finland, Aug. 4, 1931; s. Toivo and Elma (Herrasmanni) T.; B.Sc., U. Helsinki (Finland), 1956, M.S., 1957; m. Pirkko Irmeli Kaipainen, Apr. 15, 1956; children—Markku, Sinikka, Liisa. Staff dept. pest investigation and plant protection service Agrl. Research Centre, Tikkurila, Finland, 1954-60, Huhtamäki-koncern, Turku, Finland, 1960-62; staff Bang & Co., Helsinki, 1962—, asst. dir., 1966—; faculty U. Helsinki, 1956-59. Mem. Finnish Aerosol Assn. (dir. 1964—), Plant Protection Soc., Agr. Sci. Soc., Scandinavian Agr. Sci. Soc. Writer books, booklets and numerous articles. Research on biology of insect pest in clover, nematodes in clover, damage and control of leafhoppers in oats, biology of thimothee calmidge and its control, phytotoxicity of aerosol formulations, biol. effect of Pyrethrum and other insecticides. Home: 1, Punatuvantie, Helsinki 68. Office: Box 10079, Helsinki 10, Finland.*

TINOZZI, Francesco Paolo, Italian surgeon; b. Naples, Italy, Aug. 31, 1894; s. Stefano and Emilia (Perretti) T.; ed. U. Naples; M.D.; m. Giovanna Rozzera, Dec. 9, 1939; children—Emilia, Stefano. Asst. surgery clinic U. Naples, 1919-34; prof. surg. pathology, U. Naples, U. Bologna, 1934-38; prof. surg. pathology, 1939-50; prof. U. Pavia (Italy), 1950—. Decorated Order of Merit Italian Republic, Order Crown of Italy, Order of Sts. Maurice and Lazarus; recipient Gold medal for Teaching, Culture and Arts. Mem. Italian Surg. Soc., Consiglio nazionale delle ricerche, Internat. Coll. Surgeons, Deutsche Zentralkomitee zur Erforschung und Bekämpfung der Kresbkrankheiten, Nat. Sci. Soc. Research and numerous publs. on pathology, clin. and diagnostic surgery, semeiology. Home: viale della Libertà 75, Pavia, Italy.

TIPSON, Robert Stuart, chemist; b. Derbyshire, Eng., Nov. 23, 1906; s. Herbert James and Mary (Stuart) T.; B.S., U. Birmingham (Eng.), 1926, B.S., 1927, Ph.D., 1932, D.Sc., 1945; m. Constance Margaret Goodwin, Mar. 12, 1932. Came to U. S., 1930, naturalized, 1940. Research chemist Rockefeller Inst., N.Y.C., 1930-39, Mellon Inst., Pitts., 1939-57; chemist Nat. Bur. Standards, Washington, 1957—. Mem. com. biol. chemistry Nat. Acad. Scis.-NRC, 1962—. Recipient Outstanding Performance rating Nat. Bur. Standards, 1965. Fellow London Chem. Soc. (life), A.A.A.S.; mem. Am. Chem. Soc., Am. Soc. Biol. Chemists, Sigma Xi, Alpha Chi Sigma. Editor: Specifications and Criteria for Biochemical Compounds, 1967; (with W. W. Zorbach) Synthetic Procedures in Nucleic Acid Chemistry, 1968. Asso. editor Advances in Carbohydrate Chemistry, 1954—; U. S. editor Carbohydrate Research, 1965—. Research and numerous publs. on carbohydrates, cinchona alkaloids, antipneumoccal drugs, antimalarial drugs, antiviral and anticancer agts., infrared spectroscopy; determined structure of levan; discovered the trans rule; made first synthesis of a natural nucleotide. Home: 10303 Parkwood Dr., Kensington, Md. 20795. Office: Nat. Bur. Standards, Washington 20234.*

TIPTON, Samuel Ridley, Am. physiologist; b. Sylvester, Ga., Oct. 6, 1907; s. James Harrison and Rosalie (Mangham) T.; A.B., Mercer U., 1928; Ph.D., Duke, 1933; m. Isabel Hanson, Dec. 18, 1934; children—Jennifer, Samuel Ridley, Hiram, Joseph. Faculty, U. Rochester Med. Sch., 1934-35, Ohio State U. Med. Sch., 1935-41, Wayne State U. Med. Coll., 1941-43, U. Ala. Med. Coll., 1943-47; prof. zoology U. Tenn., Knoxville, 1947-67, head dept. zoology and entomology, 1967—. Cons. biology div. Oak Ridge Nat. Lab., 1948—. Mem. Am. Physiol. Soc., Am. Soc. Zoologists, Soc. Gen. Physiologists. Author: (with Helen Ward) Laboratory Manual for Elementary Physiology, 1955. Research and publs. on action of thyroid hormones on enzymes in mammalian tissues concerned with energy release in cell, effects of hormones on formation of aggregates of cytoplasmic particles concerned with synthesis of specific proteins, effects of x-ray on particles aggregates. Home: 4617 Wye Way Dr., Knoxville, Tenn. 37920.*

TISCHER, Frederick Joseph, applied physicist; b. Plan, Austria, Mar. 14, 1913; s. Frederick J. and Wilhelmine (Jirkowsky) T.; M.S.E., U. Prague, Czechoslovakia, 1936, Ph.D., 1938; postgrad. U. Berlin, Germany, 1938, Princeton, 1962; m. Alma Schoeller, Dec. 23, 1942. Came to U. S., 1954, naturalized, 1961. Research asso., supr. Telefunken, Berlin, Germany, 1938-42; owner Microwave Research Lab., Aigen, Austria, 1942-47; research supr., lectr. Royal Inst. Tech., Stockholm, Sweden, 1947-54; br. chief electronic guidance lab. Ordnance Missile Labs., Huntsville, Ala., 1954-56; mem. faculty Ohio State U., 1956-62, U. Ala., Huntsville, 1962-64; prof. elec. engring. N.C. State U., Raleigh, 1965—. Guest lectr. Inst. Tech., Helsinki, Finland, 1952, 65. Cons. govt. and industry. Fellow I.E.E.E. (chmn. Huntsville sect. 1964-65). Author: Microwave Measurements, 1958; Basic Theory of Space Communications, 1965; also articles. Contbns. to microwave measurements, guided waves, plasma diagnostics, blackout time at re-entry, receiving type antennas; inventor traveling-wave resonator, H-guide, groove guide. Home: 2300 Avent Ferry Rd., Raleigh, N.C. 27606.*

TISELIUS, Arne (Wilhelm Kaurin), Swedish biochemist; b. Stockholm, Sweden, Aug. 10, 1902; s. Hans J. and Rosa (Kaurin) R.; Dr. phil., U. Upsala, 1931; student Princeton, 1934-35; Dr. honoris causa, univs. of Paris, Bologna, Glasgow, Madrid and Cambridge, Caroline Inst., Stockholm, U. Oxford, Oslo U., U. Lyons, U. Cal., Berkeley, Gustavus Adolphus College; m. Ingrid Margareta Dalen, Nov. 26, 1930; children—Eva, Per. Research asst. in phys. chemistry U. Upsala, 1925, became asst. prof., 1930, prof. biochemistry, 1938—. Mem. nat. sci. research com. Atomic Energy Research Com., Com. for Reformation of the Univs., Med. Research Council of Sweden, 1944-47; president Swedish Natural Science Research Council, 1946-50; vice president Nobel Foundation, 1947-60, pres., 1960-64, chmn. chemistry com.; mem. sci. adv. council Swedish Govt.; pres. Internat. Union Pure and Applied Chemistry, 1951-55. Recipient Nobel Prize in chemistry, 1948. Fellow Royal Soc., 1957; hon. fellow Royal Inst. Chemistry London; hon. mem. French Chem. Soc., Swedish Society Physicians, Harvey Soc. N.Y., N.Y. Acad. Scis., N.Y. Acad. Medicine, Royal Inst. Great Britain, Chem. Soc. London, Internat. Assn. Allergists Zurich, Real Sociedad Española de Fisica y Quimica Madrid, Soc. Sci. Helsingfors, Consejo Super. de Investigaciones Scientificas Madrid, Am. Acad. Arts and Sci., Royal Dutch, Swiss chem. socs., National Acad. Scis. India; fgn. mem. Soc. Chem. Industry London, American Philosophical Society; corresponding member Society Philomat. Paris, Academie des Ciencias de Lisboa, Acad. des Sciences, Paris; mem. Pontificia Sci. Acad. of Vatican, Royal Swedish Acad. Sci., Nobel Com. Chemistry, Royal Acad. Engring. Scis., Royal Soc. Sci. Upsala, Royal Soc. Scis. and Letters Gothenburg, Nat. Acad. Scis. Washington, Royal Danish Sci. Soc., Copenhagen and Accademia Nazionale De quaranta Roma, Polish Acad. Sci., Warsaw, Rumanian Acad. Scis. (hon.). Developed methods for separation, purification biochem. substances, particularly electrophoresis and chromatography; isolated the virus of murine paralysis and developed synthetic blood plasma. Home: Thunbergsvägen 22, Upsala. Office: Inst. of Biochemistry, Upsala U., Upsala, Sweden.

TISHLER, Max, Am. chemist; b. Boston, Oct. 30, 1906; s. Samuel and Anna (Gray) T.; B.S., Tufts Coll., 1928; M.S., Harvard, 1933, Ph.D., 1934; D.Sc., Tufts U., 1956, Bucknell U., 1962; m. Elizabeth M. Verveer, June 17, 1934; children—Peter Verveer, Carl Lewis. Teaching asst. Tufts Coll., 1924; Austin teaching fellow Harvard, 1930-34, research asso., 1934-36, instr. chemistry, 1936-37; research chemist Merck & Co., Inc., 1937-41, sect. head, process development, 1941-44, dir. developmental research, 1944-51, became asso. dir. research and devel., 1951, dir., 1962—, now pres. Merck Sharp & Dohme Research Lab. div.; pres. Merck Inst. Therapeutic Research; Rennebohm lectr. sch. pharmacy U. Wis., 1963. Mem. Nat. Def. Research Com. Mem. vis. com. dept. chemistry Harvard U.; trustee Tufts U.; asso. trustee sci. U. Pa.; adv. council Newark Coll. Engring., 1959; adminstrv. bd. Tufts New Eng. Med. Center; adv. bd. Coll. Engring., N.Y. U.; board trustees Union Junior Coll., 1964—. Recipient Merck & Co., Inc., Board of Dirs. Sci. award (resulting in establishment Max Tishler Vis. Lectureship at Harvard, and Max Tishler Scholarship (ann.) at Tufts Coll.), 1951; medalist, Industrial Research Institute, 1961, Society Chem. Industry, 1963; Julius W. Sturmer Meml. Lecture award Phila. Coll. Pharmacy, 1964; Chemistry Lectr. Royal Swedish Acad. of Engring. Sciences, 1964. Registered pharmacist, Massachusetts. Fellow N.Y. Acad. Sci., Chem. Soc. London, Am. Inst. Chemists, Soc. Chem. Industry (exec. bd. 1963—, vice chmn. 1965), A.A.A.S.; mem. Am. Acad. Arts and Sciences, Indsl. Research Inst. (dir.), Nat. Acad. Scis., Am. Chem. Soc. (chmn. div. organic chemistry), Swiss Chem. Soc., Harvard Assn. Chemists, Phi Beta Kappa, Sigma Xi, Pi Lambda Phi (big Pi award 1951). Author: Chemistry of Organic Compounds (with J. B. Conant), 1937; Streptomycin (with S. A. Waksman), 1949; sci. articles. Mem. bd. editors Organic Syntheses, 1953-61; editor Organic Syntheses, 1960-61. Headed group which first synthesized hydrocortisone and developed commercial syntheses for vitamin B2, K1, pantothenic acid; developed prodn. processes for penicillin, streptomycin, cortisone; discovered sulfaquinoxaline (1st effective drug against coccidosis); isolated (with S. A. Waksman) 1st actinomycin in crystalline form. Home: 857 Knollwood Terrace, Westfield, N.J. 07090. Office: Merck & Co., Inc., Rahway, N.J. 07065.*

TISON, Fernand, French biologist, pneumologist; b. Amiens, France, Mar. 16, 1918; s. Gustave and Germaine (Duthilloeul) T.; Doctor in medicine, Faculty Medicine, Lille, France, 1943, Diplome de phtisiologie, 1944; postgrad. Institut Pasteur, Paris, Institut Fournier, Paris, 1943; m. Regine Thillier, July 20, 1943. Departmental dir. French Red Cross, Lille, 1943; dir. sanatorium Praz-Coutant, 1945-56; chief Surg. Thoracic Center, Altitude, France, 1954-56; chief Central Lab., 1945-56; chief Central Lab., Med. Social Office for Commerce and Industry, Roubaix-Tourcoing, 1956—; chief lab. Pasteur Inst., Lille, 1958—. Recipient Laureate Nat. Medicine Acad., 1953-61; Bronze medal Nat. Med. Acad., 1964. Mem. Soc. Tuberculose, Soc. Microbiology, Pathologie respiratoire, Pathologie Thoracique du Nord, Soc. Med. Passy (past v.p.). Author: (with J. Audrin) Recherche isolement et étude du bacille tuberculeux, 1956; also numerous articles, chpt. in book. Research on diagnostic and hygiene of mycobacterioses, taxonomy of mycobacteria including description of new varieties; discovered culture process of mycobacteria, identification tests. Home: 25 rue Albert Samain-MARCQ-en-Baroeul (Nord) France.*

TISSERAND, François Félix, French astronomer; b. Nuits-Saint-Georges-Côte-D'Or, France, Jan. 13, 1845; D.Sc., École Normale Supérieure, Paris, 1868. Became asst. astronomer Paris Obs., 1866; observed solar eclipse, 1868; became dir. obs., Toulouse, France, 1873; mem. expdns. for observation Venus transits, Japan, 1874, Santo Domingo, 1882; became prof. celestial mechanics, Sorbonne, Paris, 1883; named dir. Paris Obs., 1892; named prof. Faculté de Toulouse, 1873. Mem. French Acad. Scis., 1878. Author: Observation des taches du soleil à Toulouse en 1874 et 1875; Traité de mécanique céleste, 4 vols., 1889-96; Leçons sur la determination des orbites, 1899; Electrodynamique de Weber, 1872; Sur les étoiles filantes, 1873; also numerous articles. Founder, Bull. astronomique, 1884, editor until his death. Research on capture of periodic comets. Died Paris, Oct. 20, 1896.

TISSERAUD, Louis Eugène, French agronomist; b. Flavigny-sur-Moselle, France, May 26, 1830;; ed. Institut agronomique de Versailles. Head various study missions in several countries; became dir. agrl. stas. under civil service, 1858, insp. gen. under minister of agr.; became head Institut agronomique, 1878. Mem. French Acad. Sci. Author: Etudes économiques sure le Holstein le Slesvig et le Danemark, 1865; Considerations générales sur l'agriculture, 1867; la Vegetation dans les hautes latitudes, 1875; Statistique agricole en France, 1894; l'Agriculture danoise, 1907. Reorganized sci. research at Institut agronomique; assisted in growth of agrl. teaching. Died Paris, Oct. 31, 1925.

TISSOT, Simon-André, Swiss physician; b. Grancy, Switzerland, 1728; practised medicine, Lausanne, Switzerland; taught at U. Pavia (Italy), for 3 years, then returned to Lausanne. Author: L'Inoculation justifiée, 1754; L'Onanisme, 1760; Avis au peuple sur sa santé, 1761; Traité de l'epilepsie, 1772 ;Traité des neres et de leurs maladies, 1782. Early advocate of inoculation. Died Lausanne, 1797.

TISZA, Laszlo, physicist; b. Budapest, Hungary, July 7, 1907; s. Bela and Camilla (Herzog) T.; student U. Gottingen (Germany), 1928-30, U. Leipzig (Germany), 1930; Ph.D., U. Budapest, 1932. Came to U. S., 1941, naturalized, 1946. Research asso. Ukrainian Physico-tech. Inst., Kharkov, USSR, 1935-37, College de France, Paris, 1937-40; faculty Mass. Inst. Tech., Cambridge, 1941—, prof. physics, 1960—. John Simon Guggenheim Meml. fellow 1962-63. Vis. prof. U. Paris, 1962-63. Fellow Am. Phys. Soc., Am. Acad. Arts and Scis.; mem. Am. Assn. Physics Tchrs., A.A.A.S. Author: Generalized Thermodynamics, 1966; also articles. Research on theory of liquid helium (initiated two-fluid concept, predicted existence of second sound; conceptual problems of thermodynamics, quantum mechanics. Address: Mass. Inst. Tech., 77 Massachusetts Av., Cambridge, Mass. 02139.*

TITCHENER, Edward Bradford, psychologist; b. Chichester, Eng., Jan. 11, 1867; s. John and Alice Field (Habin) T.; B.A., Brasenose Coll., Oxford U., 1890; Ph.D., Leipzig, 1892; M.A., Oxford, 1895; D.Sc., Oxford, 1906; LL.D., U. of Wis., 1904; D.Litt., Clark, 1909; D.Sc., Harvard, 1909; m. Sophie Kellogg Bedlow, June 19, 1894; children—Margaret Seymour, John Bradford, Frances Haliburton, Alice McLellan. Sr. scholar of Brasenose Coll. (in classics and philosophy) and sr. Hulmean exhibitioner, 1885-89; research student in physiology, Oxford, 1889-90; extension lecturer in biology, Oxford, 1892; asst. prof. psychology, 1892-95, Sage prof., 1895-1910; prof. in charge of music, 1896-98, Sage prof. psychology Grad. Sch., 1910—, Cornell U. Lecturer Columbia U., 1907-08, U. of Ill., 1909, Lowell Inst., 1911. Am. editor of Mind (a quarterly review of philosophy and psychology), 1894-1920; asso. editor, 1895-1921, editor, 1921-25, Am. Jour. Psychology. Fellow Zoöl. Soc. London, A.A.A.S., Royal Soc. Medicine; mem. Aristotelian Soc. London, Am. Philos. Soc.; mem. internat. com. 3d, 4th and 5th congresses of Psychology, v.p. 6th Congress; foreign mem. Polish Acad. Arts and Science. Author: An Outline of Psychology, 1896; A Primer of Psychology, 1898; Experimental Psychology, 2 vols., 1901-05; Elementary Psychology of Feeling and Attention, 1908; Lectures on the Experimental Psychology of the Thought-Processes, 1909; Textbook of Psychology, 1910; A Beginner's Psychology, 1915; Systematic Psychology, 1929. Translator of several psychol. works. Advocate of structural psychology stressing experimental research and controlled introspection. Died Ithaca, N.Y., Aug. 3, 1927.

TITEICA, Gheorghe, Rumanian mathematician; b. 1873; mem. Rumanian Acad., several fgn. sci. acads. and socs. Author: Sur les congruences cycliques et sur les systemes triplement conjugués, 1899; Sur certaines courbes gauches, 1911; Géométrie differentielle projective des réseaux, 1924; Introduction à la géométrie differentielle projective des courbes, 1931; Affine Geometry of Curves in Space, 1937. A founder of affine différential geometry of surfaces; developed new theories for S surfaces and for a class of oblique curves; introduced a class of surfaces and curves named after him. Died 1939.

TITEICA, Serban, Rumanian physicist; b. 1908; prof. Bucharest U.; mem. Acad. Socialist Republic Rumania. Author: Resistance Variation in a Magnetic Field, 1934; Theory of the Positron, 1940; Second Principle of Thermodynamics and Statistical Mechanics, 1953. Research on transport phenomenons; 1st to take into account quantification of electron's movement in magnetic field.

TITHERIDGE, John E(dward), New Zealand physicist; b. Auckland, New Zealand, June 12, 1932; s. Leslie Edward and Clarice (Barnes) T.; B.Sc., U. New Zealand, 1953, M.Sc., 1955, Dip. Hons, 1956; Ph.D., U. Cambridge (Eng.), 1960. Sr. research fellow U. Auckland (New Zealand), 1960—. Mem. Am. Geophys. Union. Research and publs. on ionosphere including new methods for analysis of ionosonde records, original investigations of distbn. of electrons in ionosphere from measurements on satellite signals. Home: 59 Maioro St., Auckland, New Zealand.*

TITIEV, Mischa, anthropologist; b. Kremenchug, Russia, Sept. 11, 1901; s. Jacob and Rose (Ornstein) T.; came to U. S., 1905, naturalized, 1912; A.B., Harvard, 1923, A.M., 1924, Ph.D., 1935; m. Estelle Berman, Aug. 12, 1935; 1 son, Robert Jay. Faculty, U. Mich., Ann Arbor, 1936—, prof. anthropology, 1951—. Social Sci. Research Council grantee, 1951; Fulbright fellow, Australia, 1954; NSF grantee, 1966. Fellow Am. Anthrop. Assn.; mem. Am. Ethnol. Assn., Am. Soc. Phys. Anthropologists, A.A.A.S., Sigma Xi, Phi Kappa Phi. Author: The Science of Man, 1954, rev., 1963; Introduction to Cultural Anthropology, 1959. Research on Hopi and Araucanian Indian cultures, study of culture change among Hopi Indians. Home: 910 Heather Way, Ann Arbor, Mich. 48104.*

TITLEY, Spencer R., Am. geologist; b. Denver, Sept. 27, 1928; s. Luther Rowe and Ruth (Spencer) T.; B.S. in Geol. Engring., Colo. Sch. Mines, 1951; Ph.D. in Geology, U. Ariz., 1958; m. Clara H. Ruxton, May 26, 1951; children—Ronald Alan, Jane Ellen, Jennifer Elizabeth. Geologist, N.J. Zinc Co., Colo., N.M., Ariz., 1951, mining geologist, 1953-55, 58-60; prof. geology U. Ariz., 1960— Co-editor: Geology of the Porphyry Copper Deposits, Southwestern North America, 1966. Studies and publs. of age related radiation damage properties of crystals; porphyry copper deposits; lunar stratigraphy and lunar processes; problems of mineral deposit; discovery dating of geol. events and history and origin of moon. Home: 6920 Taos Pl., Tucson 85715.*

TITTERTON, E(rnest) W(illiam), nuclear physicist; b. Tamworth, Eng., Mar. 4, 1916; s. William Alfred and Elizabeth (Smith) T.; B.Sc., U. Birmingham (Eng.), 1935, B.Sc. with honors, 1936, M.Sc., 1938, Dip.Ed., 1939, Ph.D., 1941; m. Peggy Eileen Johnson, Sept. 19, 1942; children—Elizabeth Jennifer, Andrew Brian, Ashley Clare. Research officer H.M. Admiralty, 1939-43; mem. Brit. Atomic Bomb Mission to Los Alamos, 1943-47; sr. mem. timing group Alamagordo trial of 1st atomic bomb, 1945; adviser on instrumentation Bikini atomic weapons test, 1946; head electronic div., Los Alamos, 1946-47; dir. research group Atomic Energy Research Establishment, Harwell, U.K., 1947-51; prof. nuclear physics Australian Nat. U., Canberra, 1950—, dean Research Sch. Phys. Scis., 1965—. Chmn., Atomic Weapons Test Safety Com., 1956—; mem. sci. adv. com. Australian AEC, 1956-64; mem. adv. com. Nat. Radiation, 1957—; mem. Def. Research and Devel. Policy Com., 1958—. Fellow Royal Australian Acad. Sci., Royal Soc. Arts and Sci., Am. Phys. Soc. Author: Facing the Atomic Future, 1955; also numerous articles. Research in low energy nuclear physics, structure of light nuclei, ternary fission and reaction mechanisms, pulse circuitry; initiated and maintained fallout monitoring service in continental Australia. Home: 8 Somers Crescent, Forrest, Canberra, ACT, Australia.*

TITTLE, Charles William, Am. physicist; b. Bonham, Tex., Nov. 11, 1917; s. Walter Lee and Minnie Olive (Farrow) T.; B.S., N. Tex. State U., 1939, M.S., 1940; Ph.D., (Gulf Oil Corp. fellow), Mass. Inst. Tech., 1948; m. Fredrieka Ellen Roberson, Feb. 3, 1943; children—Charles Eric, Paul Allen, Glen Michael, Carol Anne, Mark Edward, Carl Kent, Alan Lee. Faculty, N. Tex. State U., 1940-51, prof. physics, 1950-51; head nuclear physics sect. Gulf Research and Devel. Co., 1951-55; dir. Western div., asso. tech. dir. Tracerlab, Inc., Waltham, Mass., 1955-57; prof. physics and engring. So. Methodist U., Dallas, 1957—, chmn. dept. physics, 1965—. Cons. physicist to indsl. firms., 1948—. Mem. Am. Nuclear Soc. (dir. 1964-67), Am. Phys. Soc., Sigma Xi, Sigma Pi Sigma, Sigma Tau, Pi Tau Sigma. Author: Geophysics, 1960; also articles. Research on neutron diffusion, theory of neutron well logging, phenomena of neutron detection, new methods solving math. problems in physics and engring. especially boundary value problems; patentee methods and apparatus of nuclear well logging. Home: 3710 Shenandoah Av., Dallas 75205.*

TITUS, Harry Waltner, Am. nutritionist; b. Laramie, Wyo., Mar. 30, 1896; s. Edward De Villow and Emma (Waltner) T.; A.B., U. Wyo., 1918, A.M., 1925; Ph.D., George Washington U., 1931; m. Annie Laurie Greene, June 20, 1923; 1 son, Harry Edwin. Asst. state chemist for Wyo., 1918-19; head dept. chemistry Okla. Sch. Mines and Metallurgy, 1920-21; nutrition chemist N.M. State Coll., 1921-24, nutritionist, 1924-26; with Bur. Animal Industry, U. S. Dept. Agr., 1926-43, sr. biochemist, 1937-43; head feed spltys. div. War Food Adminstrn., Washington, 1943-44; dir. research, tech. counselor Limestone Products Corp. Am., Newton, N.J., 1944-64; ind. cons. in animal nutrition, 1964—. Fellow Washington Acad. Scis., Poultry Sci. Assn., A.A.A.S.; mem. Am. Chem. Soc., Am. Soc. Biol. Chemists, Am. Inst. Nutrition, Soc. Exptl. Biology and Medicine, Washington Acad. Scis., Poultry Sci. Assn., Am. Soc. Animal Sci., Am. Dairy Sci. Assn., Animal Nutrition Research Council. Author: The Scientific Feeding of Chickens, 1941, 49, 55, 61; Some Applications of the Equation of the Curve of Diminishing Increment in Animal Nutrition, 1959; also numerous profl., popular articles. Research on nutrition of beef cattle and poultry, nutritive value of spl. feedstuffs, digestibility of feedstuffs by chickens, metabolizable energy value of feedstuffs for chickens, requirements of chickens for minerals, role of minor elements in nutrition. Address: 7 Lakeview Av., Andover, N.J. 07821.*

TIZARD, Sir Henry Thomas, English physicist; b. Gillingham, Kent, Eng., Aug. 23, 1885; s. Thomas Henry and Mary (Churchward) T.; M.A., Magdalen Coll., Oxford (Eng.) U., 1908; student Berlin U., 1908-09; D.C.L. (hon.), U. Durham; LL.D. (hon.), univs. Edinburgh and Scotland; m. Kathleen Eleanor Wilson, 1915; 3 sons. Fellow Oriel Coll., Oxford U., 1911-20; reader thermodynamics Oxford U., also prin. asst. sec. Dept. Sci. and Indsl. Research, 1921-27, permanent sec., 1927-29; rector Imperial Coll. Sci. and Tech., London, Eng., 1929-42; chmn. Aero. Research Com., 1933-43; mem. Air Council, also Aircraft Supply Council, 1941-43; pres. Magdalen Coll., 1942-46. Devel. commnr., 1943-45; chmn. Brit. Adv. Council on Sci. Policy, also Brit. Def. Research Policy Com. Trustee Brit. Museum, 1937—. Hon. fellow Magdalen Coll., Oriel Coll.; recipient Am. Medal of Merit; Gold medal Franklin Inst., Phila.; decorated knight comdr. Order Bath, 1937. Fellow Royal Soc., 1926, Inst. Physics, Royal Aero. Soc., Chem. Soc. Arts (Albert medal 1944). Author numerous articles in field, 1910—. Joint editor Sci. of Petroleum, Aero. specialist; worked on devel. radar; with David Randall Pye did research on adiabatic compression gases; influenced devel. internal combustion engines. Died Fareham, Hampshire, Oct. 9, 1959.

TJIO, Joe-Hin, biologist; b. Java, Indonesia, Feb. 11, 1919; s. Eng Bok and Kow Nio (Thung) T.; came to U. S., 1959, naturalized, 1966; Ph.D., U. Colo., 1960; m. Inga Bildsfell, Oct. 4, 1948; 1 son, Yu Hin. Head dept. cytogenetics Estación Experimental de Aula Dei, Zaragoza, Spain, 1948-59; research biologist NIH, Bethesda, Md., 1959—; research asso. Inst. Genetics, U. Lund (Sweden), 1950—. Recipient J. P. Kennedy, Jr. Found. award, 1962. Mem. Am. Soc. Human Genetics, Am. Genetic Assn., Genetics Soc. Am., Am. Soc. Naturalists, Am. Inst. Biol. Scis. Editorial bd. Cytologia, 1956—, Cytogenetics, 1962—, Caryologia, 1966—. Research and publs. on plant, animal and human cytogenetics. Home: 120 Center Dr. Office: Nat. Insts. Health, Bethesda, Md. 20014.*

TOALDO, Giuseppe, Italian meteorologist; b. Pianezza, Italy, 1719; ed. Padua Seminary; became archpriest, Montegalda, 1754; prof. astronomy, Padua, from 1762; founder obs. in Chateau Ezzelino, 1767; Fellow Royal Soc., 1777. Editor works of Galileo, 1744. Author: Della maniera di difendere gli edificii dal fumine, 1772; also several works on meteorology and math. Research on atmospheric electricity and gnomonics; conducted several meteorol. investigations; credited with discovering periodicity of meteorol. cycles. Died Padua, 1798.

TOBIAN, Louis, Jr., Am. physician; b. Dallas, Jan. 26, 1920; s. Louis and Isabel (Franklin) T.; B.A., U. Tex., 1940; M.D., Harvard, 1943; m. Frances Williams, Oct. 18, 1951; 1 dau., Anne. Practice medicine specializing in internal medicine, Mpls., 1954—; faculty U. Minn., 1954—, prof. medicine, 1964—. Investigator Am. Heart Assn., 1951-56; mem. med. adv. bd. Council For High Blood Pressure Research, 1959. Diplomate Am. Bd. Internal Medicine. Fellow N.Y. Acad. Scis., A.C.P.; mem. Am. Soc. Clin. Investigation, Am. Physiol. Soc., Central Soc. Clin. Research, Am.

Soc. For Study Arteriosclerosis, Soc. Exptl. Biology and Medicine, Assn. Am. Physicians. Research, numerous publs. on hypertension and function of renal juxtaglomerular apparatus. Home: 1437 E. River Rd., Mpls. 55414.*

TOBIAS, Charles William, chem. engr.; b. Budapest, Hungary, Nov. 2, 1920; s. Karoly and Elizabeth (Milko) T.; dipl. chem. engring., U. Tech. Scis. Budapest, 1942, Ph.D., 1946; m. Marcia Rous, Sept. 10, 1950; children—Carla, Eric, Anthony. Came to U. S., 1947, naturalized, 1952. Research, devel. engr. United Incandescent & Elec. Co. Ltd., Ujpest, Hungary, 1942-47; instr. phys. chemistry U. Tech. Scis., 1945-46; faculty U. Cal., Berkeley, 1947—, prof. chem. engring., 1960—, chmn. dept., 1967—, prin. investigator Lawrence Radiation Lab., 1954—; asso. research prof., Miller Inst. Basic Sci., 1958-59. Fellow A.A.A.S.; mem. Am. Chem. Soc., Am. Inst. Chem. Engrs., Electrochem. Soc. (asso. editor Jour. 1955—), Internat. Com. for thermodynamics and Electrochem. kinetics, Sigma Xi. Editor: (with Paul Delahaye) Advances in Electrochemistry and Electrochemical Engineering, 1961. Research, numerous publs. on field mass and charge transport in electrochem. systems, design of electrolytic processes, batteries and fuel cells electrolysis in non-aqueous solvents. Home: 524 Moraga Way, Orinda, Cal. 94563. Office: Dept. Chem. Engring. U. Cal., Berkeley, Cal. 94720.*

TOBIAS, Cornelius Anthony, biophysicist; b. Budapest, Hungary, May 28, 1918; s. Charles and Elizabeth (Milko) T.; student Jozsef Müegyetem, Budapest, Hungary, 1935-39; M.A., in Physics U. Cal., Berkeley, 1940, Ph.D., 1942; m. Ida Lanning, Dec. 28, 1943; children—Eve Helen (Mrs. Joseph Lippold), Martin Howard. Came to U. S., 1939, naturalized, 1948. Faculty, U. Cal., Berkeley, 1947—, prof., 1955—, chairman for med. physics, 1967—, staff Lawrence Radiation Lab., 1946—. Staff, NRC. Recipient Kauser-Eigel prize U. Budapest, 1936; Ernest Lawrence Meml. award AEC, 1963. Hungarian-Am. Exchange fellow, 1939-40; Guggenheim fellow, 1956-57. Fellow Am. Phys. Soc.; mem. Internat. Acad. Astronautics, Radiation Research Soc. (past pres.), Biophys. Soc. (past mem. council), Soc. Nuclear Medicine, Med. Physics Soc., A.A.A.S., Am. Inst. Biol. Sci., N.Y. Acad. Sci., Internat. Commn. on Radiation Biophysics (v.p. 1966—).*

TOBIAS, Stephen Albert, mech. engr.; b. Vienna, July 10, 1920; s. Béla and Zelma (Pisker) T.; ed. in Hungary, Gt. Britain; M.A., Ph.D., D.Sc., engring. diploma; m. Stephanie, Oct. 9, 1945; children—Martin, Andrew. Designer machine tools, 1943-47; research scholar Imperial Chem. Industries, 1951-55; asst. dir. research U. Cambridge, 1955-59; prof. mech. tech. U. Birmingham (Eng.), 1959—. Mem. Instn. Mech. Engrs., Instn. Prodn. Engrs. Research and publs. on instability of machine tools, linear and non-linear vibrations; designed machine tools, high energy rate forming machines. Author: Schwingungen an Werkzeugmaschinen, 1961. Home: 28 Selly Wick Rd., Selly Park, Birmingham 29, Gt. Britain.

TOBIE, John Edwin, Am. biologist; b. Collison, Ill., Dec. 26, 1911; s. Edwin Lester and Carrie (Mansfield) T.; A.B., U. Ill., 1935; M.S., La. State U., 1936; Ph.D., Tulane U., 1940; m. Eleanor Marietta Johnson, May 30, 1947. Research fellow Tulane U., 1940-41, instr., 1941-43; chief depts. parasitology 19th and 406th med. gen. labs. U. S. Army, 1943-46; parasitologist Lab. Tropical Diseases, NIH, Bethesda, Md., 1943-57, biologist lab. immunology, 1957-61, acting chief lab. immunology, 1961-63, chief lab. germ free animal research Nat. Inst. Allergy and Infectious diseases, 1963—. Recipient Superior Service award U. S. Dept. Health, Edn., and Welfare, 1963. Fellow A.A.A.S.; mem. Am. Assn. Immunologists, Am. Soc. Tropical Medicine and Hygiene, N.Y. Acad. Scis., Sigma Xi. Publs. on devel. fluorescent method for detection malarial antibodies, for following antibody prodn. in humans with malaria; research on formation and quantitation of immune globulins in malaria, localization by fluorescence of tetracycline antibiotics in bone; discovered flotation method for concentration of protozoan cysts and worm eggs. Home: 5902 Wilson Lane, Bethesda 20034. Office: 9000 Rockville Pike, Bethesda, Md. 20014.*

TOBIS, Jerome Sanford, Am. physician; b. Syracuse, N.Y., July 23, 1915; s. David George and Anna (Feinberg) T.; B.S., Coll. City N.Y., 1936; M.D., Chgo. Med. Sch., 1943; m. Hazel Weisbard, Sept. 18, 1938; children—David, Heather, Jonathan. Practice medicine, specializing in phys. medicine, rehab., N.Y.C., 1948—; prof., dir. dept. phys. medicine, rehab. N.Y. Med. Coll. Flower and Fifth Av. Hosp., 1948-61; dir. vis. physician Met. Bird S. Coler Hosp., 1948-61; chief div. rehab. medicine Montefiore Hosp. and Med. Center, 1961—; prof. dept. rehab. medicine Albert Einstein Coll. Medicine, 1964—. Chmn., subcom. on med. care Mayor's Adv. Com. on Aging, 1957. Named Physician of Year, N.Y. Gov's. Com. to Employ Handicapped, 1957. Fellow Am. Acad. Cerebral Palsy, N.Y. Acad. Scis.; mem. Am. Acad. Phys. Medicine and Rehab., A.C.P., N.Y. Acad. Medicine, A.M.A. Author: Evaluation and Management of the Brain-Damaged Patient, 1959-60. Research, numerous publs. on programs of clin. care for chronically ill, cerebral palsy, aphasia, hemiplegia, amputees. Home: 229 Chestnut Rd., Manhasset, N.Y. 11030. Office: 111 E. 210th St., N.Y.C. 10467.*

TOBY, Jackson, Am. criminologist, sociologist; b. N.Y.C., Sept. 10, 1925; s. Phineas and Anna (Weissman) T.; B.A., Bklyn. Coll., 1946; M.A. in Econs., Harvard, 1947, M.A. in Sociology, 1949, Ph.D. in Sociology, 1950; m. Marcia Lifshitz, Aug. 1, 1952; children—Alan Steven, Gail Afriat. Research asso. Lab. Social Relations, Harvard, 1950-51; mem. faculty Rutgers U., 1951—, prof. sociology, chmn. dept., 1961—; cons. Youth Devel. Program, Ford Found., 1959-63; Mem. Am. Sociol. Assn., Eastern Sociol. Soc., Am. Assn. Pub. Opinion Research, Soc. Study Social Problems, Soc. Social Sci. Study Religion, Am. Assn. U. Profs., A.C.L.U., Ams. Dem. Action, Author: (with H. C. Bredemeier) Social Problems in America, 1960; Contemporary Society, 1964; also numerous articles. Bd. editors Trans-Action, 1963—. Made internat. surveys of extent and causes of adolescent delinquency in U. S., Sweden, Japan, other countries. Home: 17 Harrison Av., Highland Park, N.J. 08904. Office: Dept. Sociology, Rutgers U., New Brunswick, N.J. 08903.*

TOCH, Maximilian, Am. chemist; b. New York, N.Y., July 17, 1864; s. Moses and Caroline (Levy) T.; spl. course in chemistry, New York U., 1882, LL.B., 1886; post grad. special course, Columbia U., 1896; Chem. E., Cooper Union; D.Sc., Peking Univ., 1924; m. Hermine E. Levy, Oct. 14, 1891; children—Elaine, Constance, Alma, Maxine. Lecturer on organic chemistry, Columbia, 1905-06; municipal lecturer on paint, Coll. City of New York, 1909; adj. prof. industrial chemistry, Cooper Union, 1919-24; hon. prof. chem. engring. and industrial chemistry, U. of Peking and Nat. Inst. Technology, China, 1924; prof. chemistry of artistic painting. Nat. Acad. Design, New York, 1924-36; pres. and chief chemist Toch Bros., Inc.; chmn. bd. Standard Varnish Works, mfrs. paints, varnishes, enamels and chemicals. Fellow A.A.A.S., Micros. Soc. of New York, Royal Photographic Soc., Chem. Soc. of London, Am. Inst. Chemists (pres.); mem. Am. Chem. Soc., Soc. Chem. Industry, Am. Inst. Chemical Engineers, Hon. member American Institute of Chemists. Society of American Magicians. In charge of camouflage, United States, World War I. Author: Materials for Permanent Painting, 1911; How to Paint Permanent Pictures, 1921; Chemistry and Technology of Paints, 3d edit., 1925; Protection and Decoration of Concrete, 1930; Paint, Painting and Restoration, 1931, 2d edition, 1945. Specialist in chemistry of paints; originator of Toch system of camouflage; worked on permanence of paints. Died May 26, 1946.

TOCHER, Don, Am. seismologist; b. Hollister, Cal., May 19, 1926; s. William John and Helen (Stone) T.; student Harvard, 1942-43; A.B., U. Cal. at Berkeley, 1945, M.A., 1952, Ph.D., 1955; m. Carol Jane Wooldridge, Jan. 12, 1963; 1 son, Paul William Van Auker. Research fellow in seismology Harvard, 1955; research seismologist U. Cal. at Berkeley, 1956-64, research asso., 1964—; dir. Earthquake Mechanism Lab., U. S. Inst. for Earth Scis., San Francisco, 1965—. Cons. seismologist, 1955—. Recipient Grove Karl Gilbert award in seismic geology Carnegie Inst. of Washington, 1964. Fellow Royal Astron. Soc., Geol. Soc. Am.; mem. Seismol. Soc. Am. (editor Bull. 1957-61), Am. Geophys. Union, Soc. Exploration Geophysicists, Am. Assn. Petroleum Geologists, A.A.A.S., Sigma Xi. Contbr. articles. to profl. jours. Research on seismicity of Cal., earthquake mechanism and regional tectonics, fault breakage and creep, stress-induced seismic anisotropy in rocks. Home: 2740 Derby St., Berkeley, Cal. 94705. Office: 390 Main St., San Francisco 94105.*

TOCHOWICZ, Leon, Polish physician; b. Igolomia, Poland, July 18, 1897; s. Leon and Dorota (Nawrocka) T.; M.D., Jagellonian U., Cracow, 1926; m. Alina Teodora Kpcinska, Sept. 24, 1946. With Clinic of Internal Diseases, Jagellonian U., 1927-39, 45—, head of clinic, 1947—, dean faculty medicine, 1954-57, prof., 1955—. Cons., chief care council Polish Red Cross Citizens Aid Com., 1940-45; v.p. Acad. Medicine, Cracow, 1954-57, pres., 1957-65. Recipient medal of 10th Anniversary of Peoples Poland, 1958; named comdr. Order of Polonia Restituta, 1958; Order of Standard of Labor, 1964; J. Purkyne medal Assn. of Physicians of Prague, 1965. Mem. Polish Acad. Sci., Soc. Specialists in Internal Health, Am. Cardiol. Soc. Research, numerous publs. on cardiovascular diseases. Died Cracow, Poland, July 29, 1965.

TOCQUEVILLE, Alexis Henri Maurice Clérel de, French political theorist; b. Verneuil, France, July 29, 1805; s. Herve Louis Clérel, Comte det.; studied law; married an English woman. Asst. magistrate, Versailles, 1827; sent to Am. by French Govt. to study penitentiary system, 1831; elected to French Chamber of Deputies, 1839 and 1848; Minister of foreign affairs, 1849. Mem. Acad. of Moral and Political Sci., 1838. Author works including: Du systeme pénitentiaire aux États-Unis et de son application en France, 1832; De la démocratie en Amérique (2 vol.) 1835; L'Ancien Régime et la Révolution, 1856. Wrote classic study of popular democracy; political thought reflects 19th century liberalism and perceptive observation. Died Cannes, France, Apr. 16, 1859.

TODA, Tadao (pen name as poet, Kahanshi), Japanese bacteriologist; b. Gumma Prefecture, 1899; degree, 1928. Became prof. Manchuria Med. Coll., 1931; apptd. prof. Kyushu U., 1936, also dean med. dept. Mem. Japan Sci. Council. Author works including: A New Bacteriology by Toda; Tuberculosis and B.C.G. Research on manufacture of B.C.G. vaccine, chemotherapy of Tb, inoculation of humans with tuberculin and B.C.G. vaccine, leprosy, cancer, dengue fever.

TODD, Alexander Robertus, Baron of Trumpington, Brit. chemist; b. Glasgow, Scotland, October 2, 1907; son of Alexander and Jane (Lowrei) T.; D.Phil., U. Oxford, 1933; Dr. Phil. nat., U. Frankfurt, 1931; D.Sc., U. Glasgow, 1937, LL.D., 1949; M.A., University Cambridge, 1944; Dr. rer. nat., U. Frankfurt, 1931; D.Sc., U. Glasgow, 1937, LL.D., 1949; M.A., University Cambridge, 1944; Dr. rer. nat., U. of Kiel, 1950; D.Sc., London, 1958; D.Sc., Madrid, 1959, Exeter, 1960, Leicester, 1960, Melbourne, 1960, Aligarh, 1960, Wales, 1961, Yale, 1961, Sheffield, 1961, Strasbourg, 1962, Liverpool, 1963, Harvard Univ., 1963, Adelaide, 1965, also Strathclyde, 1965; LL.D. Edinburgh U., 1962; m. Alison Dale, Jan. 30, 1937; children—Alexander Henry, Helen Jean, Hilary Alison. Prof. chemistry U. Manchester, 1938-44; prof. organic chemistry University Cambridge, 1944—, master Christ's College, 1963—; visiting professor California Institute of Technology, 1938, U. Chgo., 1948, U. Sydney, 1950, Mass. Inst. Tech., 1954; Nieuwland lectr. Notre Dame U., 1948; Hitchcock professor University of California, 1959; vis. professor Rockefeller Institute, 1961-64. Chmn. adv. council on Sci. policy to Brit. Govt., 1952—; chmn. Royal Commn. on Medical Education, 1965—. Mng. trustee Nuffild Foundation, 1950-64. Recipient Lavoisier medal, French Chem. Soc., 1948; Davy medal Royal Society, 1949; Nobel prize in chemistry, 1957; Cannizarro medal Italian Chem. Soc., 1958; Stas medal Belgian Chem. Soc., 1962; Paul Karrer medal U. Zurich, 1962; Longstaff medal Chem. Soc. London, 1963. Created Knight Bachelor, 1954; created Baron, 1962. Fellow Royal Soc., Chem. Soc. (pres. 1960-62); hon. mem. French, Belgian, German, Spanish chem. socs.; mem. Deutsche Akad. Naturforscher Leopoldina, Halle, American Philosophical Society; foreign member National Academy Science U. S. A., Austrian Acad. Sci., Royal Australian Acad. Sci. Research, publs. on establishment of structure of nucleic acids; nucleotides; nucleotidic coenzymes; vitamins B1, B12 and E; phosphorylation and mechanisms of biological reactions involving phosphates; biol. pigments; numerous contbns. to field of biol. organic chemistry demonstrated that the hemp plant could be used for prodn. of narcotics. Author articles profl. jours. Address: Master's Lodge, Christ's Coll., Cambridge, Eng.

TODD, David, Am. astronomer; b. Lake Ridge, N.Y., Mar. 19, 1855; s. Sereno Edwards and Rhoda (Peck) T.; A.B., Amherst, 1875, A.M., 1878; Ph.D., Washington and Jefferson Coll., 1888; m. Mabel Loomis, Mar. 5, 1879; 1 dau., Millicent (wife of Dr. W. V. Bingham). Asst. U. S. Transit of Venus Commn., 1875-78; chief U. S. Naval Observatory eclipse parties in Tex., 1878; chief asst. on U. S. Nautical Almanac, 1878-81; prof. astronomy and nav. and dir. of obs., Amherst Coll., 1881-1920; prof. emeritus (Carnegie Foundation;, 1920. Astronomer in charge Lick Obs. observations, transit of Venus, 1882; prof. astronomy and higher mathematics, Smith Coll., 1882-87; astronomer in charge Am. eclipse expdn. to Japan, 1887; chief U. S. scientific expdn. to W. Africa, 1889-90; chief Amherst eclipse expdn. to Japan, 1896, to Tripoli, Barbary, 1900, to Dutch E. Indies, 1901, Tripoli, 1905, Russia, 1914, Florida, 1918, South America, 1919, chief of the Lowell Mars expedition to the Andes, 1907. Fellow A.A.A.S., Astron. Soc. America, Royal Soc. Arts, London. Imperial Saki cup from Mikado of Japan for services to Japanese education. Author: A New Astronomy, 1897; Stars and Telescopes, 1899; Népszerü Csilla gászat (Popular Astronomy in Hungarian, Budapest), 1901; Lessons in Astronomy, 1902; Astronomy To-Day, 1924; also articles in mags. and revs. Editor: Columbian Knowledge Series (3 vols.), 1893-95. Designed and erected new observatories at Smith Coll., Northampton, 1886-87, and at Amherst Col., 1903-05; made first photograph of solar corona from airplane, 1925; designed automatic camera for photographing eclipses. Died Madison Heights, Va., June 1, 1939.

TODD, David Keith, Am. hydrologist; b. Lafayette, Ind., Dec. 30, 1923; s. Marion W. and Evangeline (Klinkel) T.; B.S. in Civil Engring., Purdue U., 1948; M.S. in Meteorology, N.Y. U., 1949; Ph.D., in Civil Engring., U. Cal. at Berkeley, 1953; m. Caroline Lark, June 15, 1948; children—Stuart K., Brian W. Hydraulic engr. U. S. Bur. Reclamation, Denver, 1948-50; faculty dept. civil engring. U. Cal. at Berkeley, 1950—. Cons. to UN, govt. and state agys. NSF fellow, 1957-58, 64-65. Mem. Am. Geophys. Union (pres. sect. hydrology 1964—), Univs. Council on Hydrology (chmn. 1962-64), Am. Soc. C.E. (Research prize 1960), Am. Meteorol. Soc., A.A.A.S., Am. Water Works Assn. Author: Ground Water Hydrology, 1959; also numerous articles. Research on frequency rainfall and floods, ground water flow theory, tracers in ground water, sea water intrusion of ground water zones, recharge of ground water artificially, mgmt. of ground water basins, application of nuclear energy to water resources. Home: 2938 Avalon Av., Berkeley, Cal. 94705.*

TODD, Eli, Am. physician; b. New Haven, Conn., July 22, 1769; s. Michael and Mary (Rowe) T.; grad. Yale, 1787; studied medicine; m. Rhoda Hill, Aug. 9, 1796; m. 2d, Catherine Hill, Nov. 1828. Treated epidemic of "spotted fever", Farmington, Conn., 1808; practiced medicine in Hartford, Conn., 1820; mem. Conn. Med. Soc., v.p., 1823, pres., 1827-28; investigated conditions in insane asylums and became interested in treatment of mentally ill, 1812; a founder Soc. for Relief of Insane, 1822; a founder Conn. Retreat for Insane, Hartford, 1824, 1st supt., 1824-33, used trained personnel, recognized alcoholism as a mental disease. Died Hartford, Nov. 17, 1833.

TODD, John, mathematician; b. Carnacally, Northern Ireland, May 16, 1911; s. William Robert and Catherine (Stewart) T.; student Meth. Coll., Belfast, Ireland, 1922-28; B.S., Queen's U., Belfast, 1931; postgrad. St. John's Coll., Cambridge, Eng.; m. Olga Taussky, Sept. 29, 1938. Came to U. S., 1947, naturalized, 1953. Faculty, Queen's U., 1933-37, King's Coll., U. London, 1937-47; with U. S. Nat. Bur. Standards, 1947-57; prof. math. Cal. Inst. Tech., Pasadena, 1957——. Fulbright prof., Vienna, 1965. Mem. Am. Math. Soc., Soc. Indsl. and Applied Math., Assn. for Computing Machinery, Math. Assn. Am. Author: Tables of Arc Tangents, 1951; Survey of Numerical Analysis, 1962; Constructive Theory of Functions, 1963; also numerous articles. Editor: Mathematical Tables and Other Aids to Computation, 1950-57; editor-in-chief Numerische Mathematik, 1960——; asso. editor Jour. Approximation Theory, 1968——, Aequationes Mathematical, 1968——. Research on application various techniques of math to proper exploitation of automatic computing equipment. Home: 1625 Sierra Bonita Lane, Pasadena, Cal. 91106.*

TODD, John Arthur, English mathematician; b. Liverpool, Eng., Aug. 23, 1908; s. John Arthur and Agnes (Perfect) T.; B.A., Trinity Coll., Cambridge U., 1928, Ph.D., 1932. Asst. lectr. math. U. Manchester, 1931-37; lectr. math. Cambridge U., 1937-60, reader in geometry, 1960——, fellow, dir. studies in math. Downing Coll., 1958——. Fellow Royal Soc., 1948; mem. London Math. Soc. (sec. 1951-67, pres. 1967——). Author: Projective and Analytical Geometry, 1947. Research, publs. on algebraic geometry, group theory, algebraic invariants. Home: Downing Coll., Cambridge, Eng.*

TODD, John L., physician; b. Victoria, B.C., Can., Sept. 10, 1876; B.A., McGill U., Montreal, Que., Can., 1898, M.D., 1900; D.Sc. (hon.), Liverpool, Eng., 1909; m. Marjory Clouston, 1912; 3 daus. House surgeon Royal Victoria Hosp., Montreal, 1900-01; asst. lectr. Liverpool Sch. Tropical Medicine, 1901-05, also mem. expdns. to Senegal, Gambia, Congo Free State, 1902-05, 11; dir. Runcorn Research Lab., Liverpool Sch., 1905-07; asso. prof. parasitology McGill U., 1907-26. Recipient Mary Kingsley medal, Liverpool, 1910. Licentiate Royal Coll. Physicians London. Mem. Royal Coll. Surgeons Eng., Assn. Pathologists and Bacteriologists, Soc. for Exptl. Biology, Soc. Tropical Medicine, others. Research on tropical medicine, especially Trypanosomiasis; demonstrated (with J. E. Dutton) that Spirillum duttoni (Borrelia duttonii) causes relapsing fever in monkey, 1905. Died Aug. 26, 1949.

TODD, Robert Bentley, Brit. physician; b. Dublin, Ireland, Apr. 9, 1809; s. Charles Hawkes and Eliza (Bentley) T.; B.A., Trinity Coll., Dublin, 1829; M.A., Pembroke Coll., Oxford, 1832, B.M., 1833; D.M., Oxford, 1836; 1 son, James Henthord, 3 daus. Lectr., London; prof. physiology King's Coll., London, 1836-53; Gulstonian lectr., 1839; Lumelian lectr., 1849; examiner London U., 1839-40; also pvt. practice medicine; a founder King's Coll. Hosp., London, 1840, St. John's House Instn. for Nurses. Fellow Royal Soc., 1838, mem. council 1838-39; fellow Royal Coll. Physicians (censor 1839-40), Royal Coll. Surgeons; mem. Brit. Assn. (mem. subcom. 1836-37). Author: Physiological Anatomy and Physiology of Man, 1843; Practical Remarks on Gout, Rheumatic Fever and Chronic Rheumatism of the Joints, 1843; Description and Physiological Anatomy of Brain, Spinal Cord and Ganglions, 1845; Lumelian Lectures on the Pathology and Treatment of Delirium and Coma, 1850. Editor: (with Grant) The Cyclopaedia of Anatomy and Physiology, 1835-59; Gulstonian Lectures on the Physiology of the Stomach, 1839. Administered alcoholic stimulants for treatment of fever. Died Jan. 30, 1860.

TÖDT, Fritz, German chem. engr.; b. Hohenwestedt, Germany, Jan. 14, 1897; s. Peter and Thekla (Ziese) T.; student Charlottenburg Tech. U., Berlin; engring. degree; agrégé in edn., 1932; m. Charlotte Strohbach, June 10, 1938; children—Gisela, Fritz-Peter. Dir. dept. Berlin Indsl. Inst., 1924-44; named prof., 1942; head dept. Max-Planck Inst. Berlin-Dahlem, 1945-50; head profl. group Fed. Inst. Examination Materials, Berlin-Dahlem, West Berlin, 1951——. Mem. Dechema, German Chem. Assn., German Bunsen Soc. Author books, numerous articles. Research in sugar tech., electrochemistry, oxygen measurements, metal corrosion. Home: Dreilindenstrasse 40, 1000 Berlin 39, West Germany. Office: Bundesanstalt für Materialprüfung, U.D. Eichen 86, 1000 Berlin 45, West Germany.

TOEI, Kyoji, chemist; b. Hong Kong, Apr. 8, 1920; s. Kanzaburo, and Yoshi (Ueda) T.; grad. Faculty Sci., Kyoto (Japan) U., 1943, Ph.D., 1959; m. Noriko Aoki, Mar. 30, 1952; children—Jun-ichi, Keiji, Yôzo. Prof., Sixth Higher Sch., 1946-49; faculty Okayama (Japan) U., 1949——, prof. chemistry, 1961——. Mem. Chem. Soc. Japan, Japan Soc. for Analytical Chemistry. Research, publs. on syntheses of organic reagents for various cations and anions and their analytical applications, determination of acid dissociation constants and stability constants spectrophotmetrically and potentiometrically. Home: 4-10-39 Minamikata, Okayama, Japan.*

TOEPLER, August, physicist; b. Brühl/Bonn, Germany, 1836; became prof. physics Rigo Polytechnikum, 1865, U. Graz (Austria), 1869, Technische Hochschule Dresden (Germany), 1876. Inventor vacuum pump without taps which could create vacuum of 100 millionth atmosphere; inventor negative-streak device for making visible density differences of gases, 1864, improved it, 1868; inventor (with Boltzmann) optical method for analysis of oscillation phases of air column, 1870; research in acoustics. Died 1912.

TOFT, Robert Jens, Am. endocrinologist; b. Beloit, Wis., Mar. 2, 1933; s. George Peter and Ruth (Wood) T.; B.A., Beloit Coll., 1955; M.A., Rice Inst., 1957, Ph.D., 1960; m. Helen Suzanne Wadsworth, June 17, 1956; children—Thomas Andrew, Erik Kristian. Faculty, Bowdoin Coll., 1960-63, asst. prof., 1962-63; asst. physiologist Argonne (Ill.) Nat. Lab., 1963-64, cons. div. biol. and med. research, 1965-68; asst. prof. prof., 1966——; asso. program dir. Nat. Sci. Found., 1968-69. John Morse Meml. Found. scholar, 1951-55. Mem. Am. Assn. U. Profs., A.A.A.S., Am. Soc. Zoologists, Am. Inst. Biol. Scis., Ia. Acad. Sci., Sigma Xi, Beta Beta Beta. Contbg. author: The Parathyroids, 1961; Clinical Orthopaedics, 1960; also articles. Research on relationship parathyroid glands to growth and destruction bone, quantitative assessment parathyroid function as revealed by assay utilizing spl. bone cells, effects excess hormone remodeling and degenerative changes in bone, disbn. and retention radioactive isotopes iodine in thyroid glands beagle dogs, effects thyroid damage on distbn. radioactive cesium in beagles. Home: 509 2d Av. S., Ia., 52314.*

TOIVONEN, Sulo Ilmari, Finnish embryologist; b. Somero, Finland, Jan. 12, 1909; s. Vihtori and Kustaava (Hiltunen) T.; Ph.D., M.D.h.c., U. Helsinki, 1940, M.D., 1966; m. Aini Tellervo Andersson, Aug. 22, 1937; children—Auli (Mrs. Kari Rahkamo), Ilkka, Timo, Kirsti. Faculty, U. Helsinki, 1931——, prof. exptl. zoology, 1952——. Decorated Comdr. Finnish Lion. Mem. Soc. Zoologists (treas. 1945——), Internat. Inst. Embryology, Finnish Acad. Sci. Author: (with Lauri Saxén) Primary Embryonic Induction, 1962; also numerous articles. Research on differentiation of vertebrate embryo as influenced by chem. factors producing regional segregation. Home: 17 B 28, Mäntytie, Helsinki 27, Finland.*

TOKUNAGA, Tohru, Japanese microbiologist; b. Tokyo, Japan, Sept. 15, 1927; s. Shintaro and Kijuko (Yoshioka) T.; M.D., Kyushu U., 1952; M.D., Grad. Sch. Microbiology, 1956; M. Haruko Oka, Apr. 20, 1958; children—Shin Tokunaga, Yukiko Margaret, Ken. Research asso. dept. bacteriology Med. Sch., Kyushu U., Fukuoka, Japan, 1956-59; research asso. NIH, Tokyo, 1959-63, chief research officer, 1965——; postdoctoral fellow dept. med. microbiology and immunology Med. Center, U. Cal. at Los Angeles, 1963-64. Mem. Japan Bacteriological Soc., Soc. Japanese Virologists, Japanese Soc. for Tb. Research, numerous publs. on mycobacteriophage, phage typing for Mycobacteria, persistent infection RNA containing phage in E. Coli; discovered infectious DNA extracted from mycobacteriophage; research on interactions between infectious DNA and physiologically competent Mycobacteria. Home: 2-443, Horinouchi, Suginami-ku. Office: NIH, Kamiosaki, Shinagawa-ku, Tokyo, Japan.*

TOLAND, Hugh Huger, Am. surgeon; b. Guilder's Creek, S.C., Apr. 16, 1806; s. John and Mary (Boyd) T.; grad. (1st in class) Transylvania U., Lexington, Ky., 1828; m. Mary Goodwin, 1833; m. 2d, Mary Avery, 1844; m. 3d, Mrs. Mary B. (Morrison) Gridley, 1860; 3 children. Practiced in S.C., 1828-30; studied in Paris, 1830-33; began practice in Columbia, S.C., 1833; went to Cal., 1852; chief surgeon Marine Hosp., San Francisco, 1853; staff mem. county hosp.; founder Toland Med. Coll., San Francisco, 1864, pres., prof. surgery, 1864-80, gave all facilities of sch. to U. Cal., 1873; wrote 71 articles, many published in Pacific Medical and Surgical Jour.; wrote textbook on surgery. Performed successful operations relief clubfoot and strabimus (using lithotomy forceps); studied bone regeneration; popularized Antyllus' method preventing hemorrhage; specialized lithotimies; noted for work on bladder stones, plastic surgery. Died San Francisco, Feb. 27, 1880.

TOLANSKY, Samuel, English physicist; b. Newcastle upon Tyne, Eng., Nov. 17, 1907; s. Barnett and Rose (Chait) T.; Ph.D., U. Newcastle, 1931; postgrad. U. Berlin (Germany); Ph.D., diploma, U. London (Eng.), 1934; D.Sc., U. Manchester (Eng.), 1940; m. Ottilie Pinkcasovich, Apr. 7, 1935; children—Ann (Mrs. S. Saunders), Jonathan. Lectr., U. Manchester, 1934-46; prof. physics, head dept. Royal Holloway Coll., London U., 1946——. Mem. council Nat. Phys. Lab. Eng., 1957-62. Recipient Boys medal in physics, 1948; Silver medal Royal Soc. Arts., 1960. Fellow Royal Soc., 1952; mem. Inst. Physics and Phys. Soc., Royal Soc. Arts (past mem. council), Royal Astron. Soc. Author: High Resolution Spectroscopy, 1948; Surface Microtopography, 1960; History and Use of Diamond, 1962; Introduction to Atomic Physics (5th edit.), 1963; Optical Illusions, 1964; also numerous articles. Research, publs. on nuclear physics especially hyperfine structure and line spectra, multiple-beam interferometry, studies of surface structure and properties of diamonds. Home: 16 Sheen Common Dr., Richmond, Surrey. Office: Royal Holloway Coll., U. London, Englefield Green, Surrey, Eng.*

TOLA-PASQUEL, José, Peruvian mathematician; b. Lima, Peru, Feb. 12, 1914; s. Luis and Julia (Pasquel) T.; C.E., Cath. U., Lima, 1938; Ph.D., Nat. U. Engring., Lima, 1941; m. Paula Nosiglia, Mar. 2, 1945. Prof., Cath. U., Lima, 1938-49, pro-rector, 1965——; prof. U. San Marcos, Lima, 1938——; prof. Nat. U. Engring., 1946——, dir. Inst. Pure and Applied Math, 1962——. Decorated knight comdr. Peru Laurels of Learning. Mem. Peruvian, Am. math. socs., Acad. Ciencias del Peru (pres. 1965——). Author: Energy of Deformation, 1963; Linear Algebra, 1967; also articles. Research on topology and algebra, theory of elastic structures. Home: 374 Cerdena, Lima, Peru. Office: P. O. Box 4153, Lima, Peru.*

TOLBERT, Bert Mills, Am. biochemist; b. Twin Falls, Ida., Jan. 15, 1921; s. Ed and Helen (Mills) T.; student Ida. State U., 1938-40; B.S., U. Cal. at Berkeley, 1942, Ph.D., 1945; postgrad. Fed. Inst. Tech., Zurich, Switzerland, 1952-53; m. Anne Grace Zweifler, July 20, 1959; children—Elizabeth Dawn, Margaret Anne, Caroline Joan, Sarah Helen. Chemist, Lawrence Radiation Lab., Berkeley, 1944-57; faculty U. Colo., Boulder, 1957——, prof., 1961——; vis. prof. IAEA, Buenos Aires, Argentina, 1961-62. Biophysicist, U. S. AEC, Washington, 1967-68; cons. pvt. cos., govt. agys. Fellow A.A.A.S.; mem. Am. Chem. Soc., Am. Soc. Biol. Chemists, Radiation Research Soc., Soc. for Exptl. Biology and Medicine. Author: (with others) Isotopic Carbon, 1948; also numerous articles. Research on phys. organic chemistry, including use of isotopes in chemistry and biology, radiation chemistry, radiation effects in protein, intermediary metabolism, metabolism of ascorbic acid, nutritional role, instrumentation in radioactivity. Home: 444 Kalmia Av., Boulder, Colo. 80302.*

TOLEDO, Paulo Saraiva, Brazilian physicist; b. Brazil, May 27, 1920; s. Arthur Barros and Olinda (Saraiva) de Toledo; Mech. Elec. Engr., U. Sao Paulo (Brazil), 1944, postgrad. Faculty Scis.; m. Geralda Pereira, Mar. 1, 1946; children—Selizette, Rogerio, Marcio. Faculty, U. Sao Paulo, 1945-56, prof. theoretical physics dept. physics Faculty Scis., 1964——; head reactor physics div. Inst. Atomic Energy, Sao Paulo, 1956-64. Cons. Nat. Nuclear Energy Commn. Brazil, 1961-64. Bd. dirs. Centre Electronic Computation, U. Sao Paulo. Mem. Am. Nuclear Soc. Research, publs. on theoretical nuclear physics in field of nuclear polarization of neutrons and optical model; reactor physics, reactor operation and supervision, nuclear power plant econs., neutron transport theory. Home: 2595 R. Oscar Freire, Sao Paulo, Sao Paulo, Brazil.*

TOLL, John Sampson, Am. physicist; b. Denver, Oct. 25, 1923; s. Oliver Wolcott and Merle d'Aubigne (Sampson) T.; B.S. with highest honors, Yale, 1944; A.M., Princeton, 1948, Ph.D., 1952. Mng. editor, acting chmn. Yale Sci. Mag., 1943-44; with Princeton, 1946-49, Proctor fellow, 1948-49; Friends of Elem. Particle Theory Research grant for study in France, 1950; theoretical physicist Los Alamos Sci. Lab., 1950-51; staff mem., asso. dir. Project Matterhorn, Forrestal Research Center, Princeton, 1951-53; prof., chmn. physics and astronomy U. Md., 1953——, chmn. com. faculty research, 1956-59. Physics cons. to editorial staff Nat. Sci. Tchrs. Assn., 1957-61; U. S. del., head sci. secretariat Internat. Conf. on High Energy Physics, 1960; mem.-at-large U. S. Nat. Com. for Internat. Union of Pure and Applied Physics, 1960-63; chmn. research adv. com. on electrophysics for NASA, 1962-64; mem. Gov.'s Sci. Resources Adv. Bd., State Md., 1963-64. Recipient Benjamin Barge prize in math. Yale, 1943, George Beckwith medal for Proficiency in Astronomy, 1944. Mem. Am. Assn. Physics Tchrs., Am. Assn. U. Profs., Washington Acad. Scis., Philos. Soc. Washington, Assn. Higher Edn., Nat. Sci. Tchrs. Assn., Yale Engring. Assn., Phi Beta Kappa, Sigma Xi, Phi Kappa Phi, Sigma Pi Sigma. Contbr. articles to sci. jours. Research on elementary particle theory; scattering. Home: 2 Newlands St., Chevy Chase, Md. 20015.*

TOLLENS, Bernhard Christian Gottfried, German chemist; b. Hamburg, Germany, July 30, 1841; student of Wöhler; prof. agrl. chemistry, Göttingen, Germany. Author: Kurzes Handbuch der Kohlenhydrate, 2 vols., 1888-95. Research on hydrolysis of starch by sulfuric acid, polarization of sugar, relation of sugars to formaldehyde; synthesized (with Fittig) toluene 1864; ascertained molecular weights of arabinose, xylose, raffinose; devised Tollens test for aldehyde, also aldehyde lamp for san. work. Died Göttingen, Jan. 31, 1918.

TOLLES, Walter Edwin, Am. biophysicist; b. Moline, Ill., Feb. 1, 1916; s. Walter Edwin and Aileen (Sessions) T.; B.S., Antioch Coll., 1939; M.S., U. Minn., 1941; m. Gudrun Kercher, June 11, 1939; 1 dau., Mary Wesley. Asst., Kettering Found., Antioch Coll., 1937-39; teaching asst. U. Minn., 1939-42; physicist div. war research Airborne Instruments Lab., Columbia, 1942-45; supr. Airborne Instruments Lab., Inc., 1945-54, head dept. med. and biol. physics, 1954—. Bd. dirs. Inst. Oceanography and Marine Biology, 1959. Fellow N.Y. Acad. Scis. (sci. council 1965—), I.E.E.E. (chmn. profl. group bio-med. electronics 1959), A.A.A.S.; mem. Biophys. Soc., Am. Phys. Soc., Am. Soc. Limnology and Oceanography, Am. Soc. Cytology. Research on magnetic techniques in undersea warfare, spl. purpose radio receiving and transmitting systems, electronic countermeasures, high speed micro-scanning systems; measurement of morphological and chem. properties of microscopic objects, data handling systems, physiol. monitoring systems, clin. instrumentation, diagnostic computer methods. Home: Cove Neck Rd., Oyster Bay, N.Y. 11771. Office: Airborne Instruments Lab., Inc., Deer Park, N.Y. 11729.*

TOLMAN, Carl, geologist; b. Lacombe, Alberta, Can., May 7, 1897; s. John Albert and Emma (Jordan) Tolman; B.A., U. B.C., 1924; M.S., Yale U., 1925, Ph.D., 1927; D.Sc. (honorary), U. Missouri; m. Susan Irene Robertson, Dec. 28, 1927; children—Lexie (Mrs. Andrew W. McCourt), Joan (Mrs. Robert Mayer). Came to U. S., 1924, naturalized, 1943. Field asst. Geol. Survey Can., 1921-26, geologist, 1927-30, 1935-36; asst. prof. geology Washington U., 1927-38, asso. prof. geology, 1939-44, prof. geology and geol. engring., 1945-65, professor emeritus, 1965—, chairman department, 1945-54, dean graduate school arts and scis., 1946-54, acting vice-chancellor, dean of faculties, 1953-54, vice chancellor, dean faculties, 1954-61, chancellor, 1961-63; science attache Tokyo (Japan) Embassy, 1963-65; cons. U. S. State Dept., 1965-66; project mgr. Inst. Applied Geology, UN Devel. Program, Manila, 1966—; assistant prof. geology University Colorado, summer 1928; vis. prof. econ. geology U. Philippines, 1966—. Mineral specialist Foreign Economic Administration, 1943-45, consultant geologist since 1927; geologist Bur. Mines and Geology, Mo., 1931-32, 1940; exploration geologist Kayak Syndicate, Labrador, 1933; geologist Bur. Mines, Quebec, 1938-41, U. S. Geol. Survey, 1942, St. Louis Smelting and Refining div. Nat. Lead Co., since 1946. Fellow Geol. Soc. Am., Mineral. Soc. Am., A.A.A.S. (vice president, chairman section E 1955); member Soc. Econ. Geologists, Am. Inst. Mining and Metall. Engrs., Am. Soc. Engring. Edn., Nat. Council Research, Adminstrs., N.E.A., Mo. Soc. Profl. Engrs., American Assn. Petroleum Geologists, Sigma Xi. Author tech. articles in profl. jours. Contribution to economic and mining geology; Pre-Cambrian geology. Address: Dept. Earth Scis., Washington U., St. Louis 63130.*

TOLMAN, Edward Chace, Am. psychologist; b. West Newton, Mass., Apr. 14, 1886; s. James P. and Mary (Chace) T.; S.B., Mass. Inst. Tech., 1911; A.M., Harvard, 1912, Ph.D., 1915, Sc.D., Yale U., 1951 McGill U., 1954; m. Kathleen Drew, August 30, 1915; children—Deborah (Mrs. James G. Whitney), Mary (Mrs. Thomas John Kent), Edward. Began as instr. psychology, Northwestern U., 1915; instr. psychology, U. Calif., 1918-20, asst. prof., 1920-23, asso. prof., 1923-28, prof., 1928-54, prof. emeritus, 1954—. Fellow A.A.A.S.; mem. Am. Psychol. Assn. (mem. council 1931-34; pres. 1937; bd. dirs. 1945-47), Nat. Acad. Scis., Am. Philos. Soc., Western Psychological Assn., Soc. for Psychological Study of Social Issues (council 1938-40; chmn. 1940), Sigma Xi, Phi Beta Kappa. Author: Purposive Behavior in Animals and Men, 1932. Co-author: Comparative Psychology, 1934; Drives Toward War, 1942; Collected Papers, 1951. Contbr. sci. articles to periodicals. Purposive psychologist; believed behavior directed by willingness to continue toward a goal through trial and error; considered drives as basic to motivation; experimented with rats; helped reestablish theorizing in methods of psychology through his concept of intervening variable. Died 1959.

TOLMAN, Richard Chace, Am. physicist; b. West Newton, Mass., Mar. 4, 1881; s. James Pike and Mary (Chace) T.; S.B., Mass. Inst. Tech., 1903, Ph.D., 1910; Sc.D., Princeton U., 1942; studied at Charlottenburg and Crefeld, Germany, 1903-04; m. Ruth Sherman, Aug. 5, 1924. Instr. in theoretical chemistry, 1907-09, research asso. in physical chemistry, 1909-10, M.I.T.; instr. physical chemistry, U. Mich., 1910-11; asst. prof. U. Cin., 1911-12, U. Calif., 1912-16; prof. physical chemistry, U. Ill., 1916-18; chief dispersoid sect. Chem. Warfare Service, rank maj., 1918; asso. dir., 1919-20, and dir., 1920-22; Fixed Nitrogen Research Lab., War Dept.; prof. physical chemistry and math. physics and dean of the Grad. Sch., Calif. Inst. Tech., since 1922. Vice chmn. Nat. Defense Research Com., 1940. Sci. adv. to U. S. rep., U. N. Atomic Energy Comn., 1946. Fellow Am. Acad. Arts and Scis., A.A.A.S.; mem. Am. Chem. Soc., Nat. Acad. Scis., Am. Philos. Soc., Am. Phys. Soc., Wash. Acad. Scis. Author: The Theory of the Relativity of Motion, 1917; Statistical Mechanics with Applications to Physics and Chemistry, 1927; Relativity, Thermodynamics and Cosmology, 1934; The Principles of Statistical Mechanics, 1938; Investigations on theory of colloids, theory of relativity, theory of similitude, mass of the electron, nature of the fundamental quantities of physics, partition of energy, behavior of smokes, electric discharge in gases, reactions of nitrogen compounds, rate of chem. reaction, specific heat and entropy of gases, quantum theory, statistical mechanics, relativistic thermodynamics, cosmology, etc. Died Pasadena, Cal., Sept. 5, 1948.

TOLSTOY, Ivan, geophysicist; b. Baden-Baden, Germany, Mar. 30, 1923; s. Andre and Mary (Schouvaloff) T.; Licence ès Sciences, Sorbonne U., Paris, France, 1945; M.A., Columbia, 1947, Ph.D., 1950; m. Marie L. Simon, Jan. 7, 1947 (div. Jan. 1961); 1 dau., Alexandra; m. 2d, Margie M. Lugthart, Aug. 12, 1964; 1 dau., Eline. Came to U. S., 1945, naturalized, 1948. Asst., Lamont Geol. Obs., Columbia U., 1947-51, sr. research scientist Hudson Labs., Columbia, 1953-64, sr. research asso., asst. dir., 1964—; sr. research engr., Stanolind Oil & Gas Co., Tulsa, 1951-53; faculty U. Tulsa, 1952-53. Cons. Gen. Electric Corp., 1960-61, Schlumberger Well Surveys, 1963-64. Fellow Acoustical Soc. Am.; mem. Am. Inst. Physics, Am. Geophys. Union, Fedn. Am. Scientists, A.A.A.S., Sigma Xi. Author: (with C. S. Clay) Ocean Acoustics, 1966; also articles. Research in seismology, ocean bottom studies, theory wave propagation in oceans and atmosphere, underwater sound. Home: N. Broadway, Nyack, N.Y. Office: Hudson Labs., Columbia, Dobbs Ferry, N.Y.

TOMAN, Walter, psychologist; b. Vienna, Austria, Mar. 15, 1920; s. Karl and Paula (Hradil) T.; Diploma Psychology, U. Vienna, 1944, Ph.D., 1944, venia legendi, 1951; m. Eleonore Grüner, July 19, 1951; children—Christina E., Adrienne V. Instr., lectr., U. Vienna, 1945-51; lectr. Harvard, 1951-54; faculty Brandeis U., 1954—, prof. psychology, 1962—; prof. U. Erlangen-Nürnberg (Germany), 1962—; dir. Inst. Advanced Studies, Vienna, 1966-67. Mem. Am. Psychol. Assn., Deutsche Gesellschaft für Psychologie, Internat. Psychoanalytical Assn. Author: Einführung in die moderne Psychologie, 1951; Dynamik der Motive, 1954; Psychoanalytic Theory of Motivation, 1960; Family Constellation, 1962; also articles. Introduction of quantitative aspects to clin. psychology, quantification of social motivation, psychol. environments, especially family constellations. Home: 85 Tobey Rd., Belmont, Mass. 02178; also 15 Hedenusstrasse, Erlangen 852, Germany. Office: 1 Bismarckstrasse, Erlangen 852, Germany.*

TOMASHEFSKI, Joseph Francis, Am. physician; b. Plymouth, Pa., Dec. 30, 1922; s. Alexander and Mary (Rawa) T.; student Temple U., 1940-43; M.D., Hahnemann Med. Coll., 1947; m. Marguerite M. Kochuba, May 21, 1949; children—Joseph, Paul, Ann. Practice medicine specializing in preventive medicine, pulmonary problems, Columbus, O., 1953-57; chief research, dir. cardio-pulmonary labs., dir. inhalation therapy dept. Ohio Tb Hosp., Columbus, 1953-57; asst. prof. medicine Ohio State U. Coll. Medicine, Columbus, 1953—, asso. prof. preventive medicine and physiology, 1959—; staff U. Hosp., Columbus; fellow mech. engring. dept. Battelle Meml. Inst., Columbus, 1964—, med. dir., med. research adviser, 1957—, resident med. cons., 1963—. Cons. hosps., govt. agys. Diplomate Am. Bd. Preventive Medicine. Fellow Am. Coll. Chest Physicians (mem. com. on occupational disease of chest 1960—, past pres. Ohio); mem. Am. Fedn. for Clin. Research, Am., Central Ohio (mem. com. on cardiac work evaluation 1967—) heart assns., A.M.A. (past chmn. pulmonary function exhibit com.), Am. Physiol. Soc., Am., Ohio thoracic socs., Nat. Tb Assn., Columbus Acad. Medicine, Ohio State Med. Assn., Central Research Soc., Columbus Soc. Internal Medicine, Aerospace Med. Assn. Research and publs. on pulmonary function testing, treatment pulmonary diseases; devel. inhalation therapy techniques and devices, environmental problems. Home: 2071 Ellington Rd., Columbus 43221. Office: 505 King Av., Columbus O. 43210.*

TOMBAUGH, Clyde William, Am. astronomer; b. Streator, Ill., Feb. 4, 1906; s. Muron D. and Adella Pearl (Chritton) T.; A.B., U. Kan., 1936, M.A., 1939; D.Sc., Ariz. State Coll., Flagstaff, 1960; m. Patricia Irene Edson, June 7, 1934; children—Annette Roberta, Alden Clyde. Became asst., Lowell Observatory, Flagstaff, Ariz., 1929, asst. astronomer, 1938; instr. sci., Ariz. State Coll., Flagstaff (Navy V-12), 1943-45; vis. asst. prof. astronomy, U. Calif. at L.A., 1945-46; astronomer, Aberdeen Ballistics Labs. Annex, White Sands Proving Grounds, Las Cruces, N.M., since Aug. 1946, chief optical measurement sect., Ballistics Labs., 1948, chief research and evaluation br. planning dept. Flight Determination Division, 1948-53, chief investigator of search for natural satellites project, 1953-58, Planetary Astrophysical Research, 1958—; research asso. prof. astron. N.M. State U., 1955-59, prof., 1965—, planetary astrophysics research Research Center, 1959—; expdn. extension satellite research project, Quito, Ecuador, 1956-58. Recd. Jackson-Guilt Medal and Gift, Royal Astron. Soc. Eng., 1931; Edward Emery Slosson scholar in sci., 4 yrs., U. Kan. Fellow Soc. for Rsrch. on Meteorites, American Rocket Society; member of American Astronomical Soc., Internat. Astron. Union, Astron. Soc. Pacific, Sigma Xi. Discovered planet Pluto, 1930; 1 globular star cluster, 1932; 6 galactic star clusters, variable stars, asteroids, clusters of nebulae; extensive search for distant planets, natural earth satellites, studies in apparent distbn. extragalactic nebulae; geol. studies surfaces of Mars and moon; prodn. telescope mirrors. Home: 3302 West St. Mesilla Park, N.M. Office: N.M. State U. Research Center, University Park, N.M.*

TOMES, Sir John, English dental surgeon; b. Weston/Avon, Eng., Mar. 21, 1815; s. John and Sarah (Baylies) T.; ed. med. schs. King's Coll., Middlesex (Eng.) Hosp.; m. Jane Sibley, Feb. 15, 1844; 1 son, Charles Sissmore. House surgeon Middlesex Hosp., 1839-40, lectr., 1845; practiced dentistry, from 1840. Co-founder Dental Hosp., 1858. Recipient Gold medal Soc. Arts, 1845. Fellow Royal Soc., 1850, Royal Coll. Surgeons (hon.); mem. Ondontological Soc. (co-founder, pres.). Knighted, 1886. Author: A Course of Lectures on Dental Physiology and Surgery, 1848; A System of Dental Surgery, 1859. Contbd. papers Philos. Trans., 1849-56. Patented machine for copying irregular curved surfaces in ivory, 1845; invented dental forceps; studied histology of bone, teeth; administered ether for dental operations, 1847; induced Royal Coll. Surgeons to grant license in dental surgery, 1858; secured passing Dentists Act (professionalized dentistry), 1878; proved existence of protoplasmic processes from odontoblasts (called Tome's fibrils). Died July 29, 1895.

TOMIN, Mikhail Petrovich, Russian lichenologist; b. July 24, 1883; grad. Moscow Agrl. Inst., 1913. Staff, Voronezh Agrl. Inst., until 1929; faculty botany Archangel Forest Tech., Inst., Orenburg, Minsk, USSR; chief sect. on flora and herbarium Biology Inst., Byelorussian Acad. Sci. Mem. Byelorussian Acad. Sci. Research in lichens, especially of forest regions and saline semi-arid areas; described numerous new species, subspecies and varieties of lichens; four species of lichens are named after him.

TOMITA, Masao, Japanese chemist; b. Tokyo, Japan, Aug. 15, 1903; s. Harufusa and Mayuko (Yoshimachi) T.; B.S., Pharm. Inst., Faculty Medicine, Tokyo Imperial U., 1926; Ph.D., U. Tokyo, 1933; m. Michiko Yamamoto, May 4, 1928; children—Fujiko (Mrs. Toshiro Morii), Keiko (Mrs. Yasuji Ohoto), Tetsuo Tomita. Research asso. Itsuu Chem. Lab., Tokyo, 1926-39; prof. chemistry Kyoto (Japan) U., 1939-67, prof. emeritus, 1967—, dean Faculty Pharm. Scis., 1960-64; dean Kyoto Coll. Pharmacy, 1967—. Mem. Japan Sci. Council, 1957-60. Recipient Japan Acad. prize, 1960. Mem. Pharm. Soc. Japan, Chem. Soc. Japan. Author: Modern Course in Organic Chemistry, 1958; also numerous articles. Research on alkaloidal composition of plants of Menispermaceae, use of metallic sodium in liquid ammonia which effects cleavage of diphenyl ether bridge; discovered new method for structure studies of bisbenzylisoquinoline alkaloids; established method of isolating quarternary alkaloid contents in plants. Home: 30-7, Katsura-Hitsujisarumachi, Ukyo-ku, Kyoto. Office: Kyoto Coll. Pharmacy, Misasagi Yamashina, Higashiyama-ku, Kyoto, Japan.*

TOMIZUKA, Carl Tatsuo, physicist; b. Tokyo, Japan, May 24, 1923; s. Tatsuzo and Natsu (Tamaru) T.; B.S., U. Tokyo, 1945; M.S., U. Ill., 1954; Ph.D. in Physics, 1954; m. Karyl Dawn Sellmyer, Dec. 22, 1956; children—Faye Mariko, Frank Morito. Came to U. S., 1950, naturalized, 1959. Research asso. physics U. Ill., 1954-55, research asst. prof. physics and elec. engring., 1955-56; asst. prof. physics Inst. for Study Metals, U. Chgo., 1956-60; prof. U. Ariz., Tucson, 1960—. Mem. Am. Phys. Soc., Phys. Soc. Japan. Author: (with R. M. Emrick) Physics of Solids at High Pressures, 1965; also articles. Research on devel. of exptl. procedure to measure rate of atom movement inside solids; studies on nature of a vacant site created within solids; devel. reliable method to study atom movement in solids at high pressure and at high temperatures. Home: 5003 E. Cooper St., Tucson 85711.*

TOMKINS, Gordon Mayer, Am. biochemist; b. Chgo., June 4, 1926; s. Howard and Jean (Gordon) T.; A.B., U. Cal. at Los Angeles, 1945, Ph.D., at Berkeley, 1953; M.D., Harvard, 1949; m. Millicent Hanson, July 19, 1951; children—Leslie, Tanya. Chief sect. matabolic enzymes NIH, Nat. Inst. Arthritis and Metabolic Diseases, Bethesda, Md., 1955—, chief lab. molecular biology, 1963—. Recipient award in biology Washington Acad. Sci., 1966. Mem. Am. Soc. Biol. Chemists, Am. Chem. Soc. Research, numerous publs. in molecular mechanisms of biol. regulation, particularly in mammalian organisms. Home: 3804 Shepherd St., Chevy Chase, Md. 20005. Office: NIH, Bethesda, Md. 20074.*

TOMLINSON, Charles, Brit. natural scientist; b. North London, Eng., Nov. 27, 1808; s. Charles Tomlinson; ed. London Mechanics Inst.; m. Sarah Windsor. Kept (with brother) day sch., Salisbury, Eng.; lectr. exptl. sci. King's Coll. Sch., London; Dante lectr. U. Coll., London, 1878-80. Fellow Chem. Soc., Royal Soc., 1872; mem. Phys. Soc. (a founder), Brit. Assn. for Advancement Sci. (mem. council 1864). Author: The Student's Manual of Natural Philosophy, 1838; Introduction to the Study of Natural Philosophy, 1848; Pneumatics for the Use of Beginners, 1848; Rudimentary Mechanics, 1849; Experimental Essays, 1863; Illustrations of Science, 1867. Editor: Cyclopaedia of Useful Arts, 1852-54. Translator: Inferno (Dante), 1877. also numerous articles. Discoveries in surafce tension of liquids; studies on rotation of camphor on surface of water. Died Highgate, Eng., Feb. 15, 1897.

TOMMASI, Donato, Italian chemist, physicist; b. Naples, Italy, 1848; D.Sc. Author: Theoretical and Practical Treatise on Electrochemistry, 1889; Treatise on Electric Batteries and Accumulators, 1890; also several memoirs. Research in electrochemistry, thermochemistry, also electricity (batteries, accumulators, lighting, telegraphy); prin. work related to chem. action of batteries, thermal constants, nascent state of bodies; discovered thermal constants and expressed their gen. law. Died Rome, Italy, 1944.

TOMMASI, Michel Joseph-Marie, French pathologist; b. Marseilles, France, Nov. 7, 1928; s. Luc Georges and Suzanne (Granier) T.; M.D., U. Lyons (France), 1957; m. Denise Favre, Feb. 14, 1953; children—Christine, Laurence, Geneviève. Chief, Lab. Pathology, U. Lyons, 1957-61, prof. agrégé pathology, 1961—; biologist Hosps. Lyons, 1963—. Mem. Société Anatomique de Paris, Société Française de Neurologie, Société Europèene de Pathologie, Club Neuropathologique Français. Author: Les Encéphalopathies des Alcooliques, 1957; (with Garde, Aimard) Les Complications neurologiques des cancers viscéraux, 1958; also numerous articles. Research on pathology of brain of alcoholics, prolonged comatose states following brain trauma, neuropathology of anoxia. Home: 183 rue Cuvier, Lyon 6 (69), France.*

TOMMILLA, Eero Akseli, Finnish chemist; b. Merikarvia, Finland, Mar. 5, 1900; s. Johan Arvid and Lydia (Gran) T.; cand.phil., U. Helsinki, 1925, dr.phil., 1932; postgrad. U. Oxford, 1938; m. Salli Anna Margaretha Niininen, July 13, 1929; children—Juhani, Päiviö, Eeva Marjatta. Lectr., Inst. Edn., Jyväskylä, Finland, 1930-44; prof. phys. chemistry U. Helsinki, 1944-67. Mem. Finnish Chem. Soc., Finnish Acad. Scis. and Letters, Societas Scientiarum Fennica, Chem. Soc. (London), Faraday Soc., Deutsche Bunsen-Gesellschaft, Société de Chimie Physique. Author: Fysikaalinen kemia, 1950; also numerous articles. Research on anodic oxidation of organic compounds, kinetics of reactions in solution, solvent effects, chem. thermodynamics, history of chemistry. Home: 9 Vironkatu, Helsinki, Finland.*

TOMONAGA, Sin-itiro, Japanese physicist; b. Tokyo, Japan, Mar. 31, 1906; degree in atomic physics Kyoto U., 1929; postgrad. Leipzig U., Germany, 1937-39; m. Ryoko Sekiguchi, Oct. 27, 1940; children—Shigeko, Atsushi, Makoto. Research asst. Kyoto U., 1929-32; staff Sci. Research Inst., Tokyo, 1932-37; became lectr. Hokkaido U., 1937; lectr. Tokyo U. Edn., 1940, prof., 1941—; pres. Tokyo U. Edn., 1956-62; dir. Inst. Optical Research, 1963—. Mem. Japan Sci. Council, 1949—, pres., 1963—. Recipient Japan Acad. prize, 1948; Lomonosov medal, USSR, 1964; Nobel prize in physics (with R. P. Feynman and J. S. Schwinger), 1965. Mem. Nat. Acad. Sci. (fgn. asso.); Japan Acad., Royal Swedish Acad.; Deutsche Akademie der Naturforscher Leopoldina. Independently developed theory of quantum electrodynamics, 1941-43; research in quantum dynamics, electromagnetics.*

TOMPKINS, Charles Brown, Am. mathematician; b. Jasper, Fla., Aug. 30, 1912; s. Raymond Dean and Elizabeth (Massey) T.; B.S., U. Md., 1932, M.S., 1933; Ph.D., U. Mich., 1935; m. Mary Lewis, May 25, 1937; children—Brooke, Sarah Dean, Peter Lewis, William Edward. Faculty, U. Wis., Madison, 1941-42; with Engring. Research Assos., Inc., St. Paul, 1946-49; logistics researcher George Washington, U., Washington, 1949-52; mathematician, dir. Inst. for Numerical Analysis, Nat. Bur. Standards, Los Angeles, 1952-54; prof. math. U. Cal. at Los Angeles, 1954—, dir. computing facility, 1961-66. Mem. Am. Math. Soc. Contbg. author: High Speed Computing Devices, 1950. Research, numerous publs. in calculus of variations in large numerical analysis, devel. electronic computers; patentee magnetic drum for digital computer storage. Home: 29035 Cliffside Dr., Malibu, Cal. 90265. Office: 405 Hilgard Av., Los Angeles 90024.*

TOMPKINS, Edward Raymond, Am. chemist; b. Winterset, Ia., Apr. 7, 1908; s. Ernest R. and Harriett (Haney) T.; student Simpson Coll., 1927-29; A.B., Greeley State Coll., 1931; M.A., Ph.D., U. Cal. at Berkeley, 1942; m. Barbara Harris, July 14, 1939; children—Kenneth, Robert. Research chemist Ill. Inst. Tech. Research Inst., Chgo., 1942-43; group leader Oak Ridge Nat. Lab., 1943-48; dir. research Microchem. Spltys. Co., Berkeley, Cal., 1948-50; program leader Lawrence Radiation Lab., Berkeley, 1950-51; asso. tech. dir. Naval Radio. Def. Lab., San Francisco, 1951—. Mem. Am. Chem. Soc., Sigma Xi. Research, publs. on uranium fission products; first clean separation method for rare earths; handling of radioactive sources. Home: 621 Chestnut St., San Carlos, Cal. 94070. Office: Naval Radiol. Defense Lab., San Francisco 94135.*

TOMPKINS, Frederick Clifford, English chemist; b. Yeovil, Eng., Aug. 29, 1910; s. Frederick Wallir and Ethel Maud (Culliford) T.; student Bristol (Eng.) U.; B.Sc., King's Coll., London, Ph.D., D.Sc.; m. Catherine Livingstone McDougall, June 4, 1937; 1 dau., Josephine Charlotte Rozanne (Mrs. Simon Grant Harris). Asst. lectr., King's Coll., 1934-37; sr. lectr. Natal U., 1937-46; faculty Imperial Coll. Sci. and Tech., London, 1946—, prof. chemistry, 1946—. Fellow Royal Soc., 1955. Research, numerous publs. in surface chemistry, catalysis, phys. adsorption, solid state chemistry, decomposition of solids, kinetics of cations in solution. Home: 25 Clarence Gate Gardens, London, N.W.1, Eng.*

TOMPKINS, Harvey John, Am. physician; b. Chgo., June 14, 1906; s. Harvey John and Josephine (Burke) T.; B.S., Loyola U., Chgo., 1927, M.D., 1932; postgrad. U. Wis., N.Y. U.; m. Margaret Catherine Ruddy, Feb. 13, 1934; children—Harvey John, Patricia Marie. Commd. 1st lt. U. S. Army, 1934, advanced through grades to col., 1946; career physician U. S. VA, 1935-55; moblzn. designee Surgeon Gen.'s Office, 1948-66; ret., 1966; practice medicine, specializing in psychiatry, 1935—; chmn. dept. psychiatry St. Vincents Hosp. and Med. Center N.Y., 1955—; coordinator psychiat. activities Catholic Archdiocese N.Y., 1955—; chmn. N.Y.C. Community Mental Health Bd., 1958—; clin. prof. psychiatry Georgetown Med. Sch., 1947-55, Sch. Medicine N.Y. U., 1955—. chmn. study com. neurology, psychiatry Health Research Council City N.Y., 1958—; chmn. bd. trustees N.Y. Sch. Psychiatry. Diplomate Am. Bd. Psychiatry and Neurology (dir.). Fellow A.C.P.; mem. Am. (past pres.), World (asso. sec. gen.) psychiat. assns., N.Y. Acad. Medicine, A.M.A., Group For Advancement Psychiatry, Nat. Assn. Mental Health (chmn. profl. adv. com., research com.), Royal Medico-Psychol. Assn., Acad. Religion and Mental Health (past pres.). Contbr. to Psychiatric Unit in A General Hospital, 1965; Psychiatry in the Mid-Sixties, 1966; also numerous articles. Developed psychiat. program for vets. following World War II; promoted orgn. and function psychiat. treatment services in local community gen. hosps.; developed mental health program for states and met. areas, original work in use of films for pub. mental health edn. and tng., shorter forms treatment techniques for psychiatrically ill. Home: 157 Chapel Rd., Manhasset, N.Y. 11030. Office: 144 W. 12th St., N.Y.C. 10011.*

TOMPKINS, Paul Carter, Am. chemist; b. Walla Walla, Wash., Apr. 11, 1914; s. Richard Jackson and Daisy T. (Wilson) T.; A.B., Whitman Coll., 1936; postgrad. U. Chgo.; Ph.D., U. Cal. at Berkeley, 1941; m. Edythalena Anderson, Feb. 15, 1942; children—James Andrew, Judy Ann. Prin. chemist, Oak Ridge Nat. Lab., 1947-49; staff adviser to sci. dir. Naval Radiol. Def. Lab., San Francisco, 1949-50, asso. sci. dir., 1950-51, sci. dir., 1951-60; chief research br. div. radiol. health USPHS, Washington, 1960-61; dep. dir. office radiation studies AEC, 1961-62; exec. dir. Fed. Radiation Council, Washington, 1962—. Recipient Sec. Navy Distinguished Civilian Service award, 1960; Sec. Def. Distinguished Civilian Service award, 1960. Mem. Health Physics Soc., Radiation Research Soc., Am. Nuclear Soc., Am. Chem. Soc. Research, publs. on preparation of pure, high specific activity samples of radionuclides; devel. asso. equipment, spl. procedures, and radiation protection procedures to achieve adequate control of radioactive contamination. Home: 6808 Millwood Rd., Bethesda, Md. 20034. Office: Fed. Radiation Council, Washington 20449.*

TONG, Lee Karl Jan, chemist; b. Canton, China, Jan. 12, 1913; s. Young Kew and Ho-Jen (Ho) T.; came to U. S., 1920, naturalized, 1946; B.S., U. Cal. at Berkeley, 1938, Ph.D., 1942; m. Nancy Lim, Mar. 19, 1944; children—Vivian Jean, Lillian. Teaching fellow NDRC research staff U. Cal. at Berkeley, 1942-44; with Eastman Kodak Co., Rochester, N.Y., 1944—, research asso., 1953-66, sr. research asso., 1966—. Mem. Am. Chem. Soc., Sigma Xi. Research, publs. on reaction kinetics, polymerization reactions, mechanism of dye formation in color photography. Home: 571 Thomas Av., Rochester 14617. Office: Research Labs., Kodak Park, Rochester, N.Y. 14650.*

TONI, see de Toni, Giovanni Battista.

TONNA, Edgar Anthony, cellular biologist; b. Malta, May 10, 1928; s. John and Philomena (Doublet) T.; B.S., St. John's U., 1951; M.S., N.Y. U., 1953, Ph.D., 1956; m. Patricia Anita Kennedy, Aug. 18, 1951; children—Edgar, Paul, Patricia, John. Research biochemist Hosp. for Spl. Surgery, N.Y.C., 1953-56; asso. prof. biology L.I. U. Grad. Sch. Arts and Scis., 1956-62; head dept. histo-cytochem. research Hosp. for Spl. Surgery, N.Y.C., 1956-59, head histo-cytochem. research sect., head histo-pathology service div. exptl. pathology, 1959-67, research collaborator, 1967—; research collaborator Brookhaven Nat. Lab. Med. Research Center, 1957-59; cons. radiobiology N.Y. U. Coll. Dentistry and Inst. for Dental Research, 1965—, prof. histology, dir. Lab. for Cellular Research, 1967—. Recipient Ralph Morton Meml. certificate in sci.; Gregor Mendel's medal. Mem. Am. Assn. Anatomists, Orthopaedic Research Soc., Soc. Exptl. Biology and Medicine, Histochem. Soc., Gerontol. Soc., Royal Micros. Soc., Internat. Assn. Dental Research, Research Soc. Am., Soc. Cranio-Facial Biology, Royal Soc. Medicine. Research, numerous publs. on chemistry and physiology of cells of skeletal system and teeth during growth, aging and repair. Home: 9 Geiger Pl., Huntington, L.I., N.Y. 11746. Office: Inst. for Dental Research, N.Y. U. Coll. Dentistry, 339 E. 25th St., N.Y.C. 10010.*

TONNELAT, Marie Antoinette Baudot, French physicist; b. Charolles, France, Mar. 5, 1912; d. Louis and Alix (Menetriex) T.; Licence és Sciences, Paris; Licence ès Lettres, Doctorat és Sciences, Paris; m. Jacques Tonnelat, Nov. 10, 1936; children—Jacques, Anne-Silvie, Francoise. Dir. research Centre National de la Recherche Scientifique; titular prof. Faculté des Sciences, Paris; faculty history sci. Faculté des Lettres, Paris; vis. prof. U. Berne, Switzerland, U. Lisbon, Portugal, U. Rome, U. Moscow. Recipient various prizes French Acad. Scis., Acad. Moral and Polit. Scis. Author: Les Principes de la théorie électromagnetique et de la Relativité générale, 1960; Les Théories unitaires de la luminère de la gravitation, 1966; Louis de Broglie, 1967; also numerous articles. Research on wave mechanics of photon, gen. relativity, gen. solutions of basic theories of Einstein and Schrodinger; exptl. work deduced from gen. relativity, including calculations of frequencies on satellites, Euclidian theories of gravitation; philos. investigations on light and gravitation; history of 17th, 18th and contemporary sci. Home: 11 rue Monticelli, Paris, France.*

TöNNIES, Ferdinand, German sociologist; b. Eiderstedt, Schleswig-Holstein, 1855; s. prominent agrl. family; student classical philology and philosophy at Univs. Jena, Leipzig, Bonn, Berlin; Ph.D., U. Tubingen, 1877. From lectr. to prof. U. Kiel, 1887-1933 (dismissed by Nazis). Mem. German (a founder, long-time pres.), Am. (hon.) sociol. socs. Author: Gemeinschaft und Gesellschaft, 1887; Thomas Hobbes, Leben und Lehre, 1896; Die Sitte, 1909; Der englische Staat und der deutsche Staat, 1917; Marx, Leben und Lehre, 1921; Kritik der offentlichen Meinung, 1922; Soziologische Studien und Kritiken, 3 vols., 1925-29; Einfuhrung in die Soziologie, 1931; Inledning till Soziologien, 1932; Geist der Neuzeit, 1935, also numerous papers. Classified sociol. disciplines as pure or theoretical, applied and empirical; traced evolution of society from Gemeinschaft (community) to Gesellschaft (assn.) conceptually; analyzed social entities in relation to rational social will, norms and values; saw modern polit. state as new associational order that erodes traditional community patterns. Died 1936.

TONOLLI, Vittorio, Italian biologist; b. Milan, Italy, Sept. 16, 1913; s. Alessandro and Orestina (Vigano) T.; M.D., U. Milano, 1939, Libera docenza, 1951; m. Pirocchi Livia, Oct. 26, 1950. Research asst. Istituto Italiano di Idrobiologia, Verbania Pallanza, Italy, 1944—, dir., 1951—; asso. prof. limnology Milano U., 1951—. Author: Introduzione allo studio della Limnologia, 1964; also articles. Research on gen. limnology, population dynamics in lakes zooplankton. Home: 19 Miralago, Verbania-Suna, Italy. Office: Istituto Italiano di Idrobiologia, Verbania-Pallanza, Italy.*

TOOKE, Thomas, Brit. economist; b. Cronstadt, Eng., Feb. 29, 1774; s. William Tooke; m. Priscilla Combe, 1802; 3 sons. Began in business at the age of 15; later became partner firms Stephen, Thornton & Co., also Astell, Tooke & Thornton, London; ret. from his bus., 1836; gov. Royal Exchange Assurance Corp., 1840-52; chmn. St. Katharine's Dock Co. mem. Polit. Economy Club (a founder 1821). Mem. Factories Inquiry Commn., 1833. Tooke professorship econ. sci. and statistics at King's Coll., London founded in his honor. Fellow Royal Soc., 1821, French Acad. Scis., 1853; Author: Thoughts and Details on the High and Low Prices of the Last 30 Years, 1823; Considerations on the State of the Currency, 1826; Letter to Lord Grenville, 1829; History of Prices, 6 vols., 1838-57; Enquiry into the Currency Principle, 1844. An early supporter of free-trade movement; opposed bank restriction, resumption in raising or depressing gen. prices, currency theory. Died London, Feb. 26, 1858.

TOOL, Arthur Quincy, Am. physicist; b. Monroe, Ia., Dec. 5, 1877; s. Quinn Hammond and Selina (Oldham) T.; B.S., Grinnell Coll., 1904; M.A., U. Neb., 1906; Ph.D., U. Berlin, 1914; m. Eunice Elhel Woody, Nov. 24, 1915; 1 son, Arthur Quincy. Asst. prof. U. Neb., 1914-18; physicist Nat. Bur. Standards, 1918-48. Fellow Am. Phys. Soc., Am. Ceramic Soc., Soc. Glass Tech. (Eng.); mem. Washington Acad. Sci. Contbr. numerous articles on research concerning phys. properties of glass to tech. jours. Home: 7325 Carroll Av., Takoma Park, Md. 20112.*

TOOTH, Howard Henry, English physician; b. Eng., Apr. 22, 1856; s. Frederick Tooth; C.B., St. John's Coll., Cambridge, 1918, M.A., M.D., M.D., (hon.), Malta U.; m. Mary Beatrice Price, 1881; 1 dau.; m. 2d, Helen Katharine Chilver, 1908; 1 son, 2 daus. Cons. physician to St. Bartholomew's, Nat. Hosp. for Paralyzed and Epileptics, Met. Hosp.; examiner in medicine, Cambridge, Durham U.; censor Royal Coll. Physicians; cons. physician to troups in Malta, Italy. Fellow Royal Coll. Physicians. Contbr. articles to sci. and med. jours. Described perineal form of progressive muscular atrophy (Charcot-Marie-Tooth disease), 1886. Died May 13, 1925.

TOPINARD, Paul, French physician, anthropologist; b. l'Isle-Adam, France, 1830; studied in U. S.; became doctor, Paris; practiced medicine, Paris, until 1871; curator collections of Anthrop. Soc., 1872-80, succeeded Broca as sec.-gen., 1880; asst. dir. lab. anthropology École des Hautes Études; became prof. Ecole d'Anthropologie, 1876. Author: L'anthropologie, 1876; Éléments d'anthropologie générale, 1885; L'homme dans la nature, 1891; Les dernières étapes de la généalogie de l'homme. Described hypothetical line between glabella and mental point (Topinard's line), also angle formed at anterior nasal spine by lines from auricular point and glabella (Topinard's angle), 1876. Died Paris, 1911.

TOPSELL, Edward, Brit. naturalist; grad. Christ's Coll., Cambridge, 1587; B.A., circa 1591-92; later M.A.; m. Mary Seaton, Aug. 12, 1612. Ordained and inducted to rectory of East Hoathly, 1596; lived in Datchworth, Eng., 1598-1601; perpetual curate St. Botolph, Aldersgate, Eng., 1604-38; vicar of Syresham, Northants, Eng., 1602-08; vicar, Mayfield, Eng., 1605-06, East Grinstead, Eng., 1610-16; became chaplain of Hartfield, 1610. Author: The Reward of Religion . . . , 1596; The Householder or Perfect Man, 1610; The Historie of Foure-Footed Beastes, 1607; The Historie of Serpants, 1608; also contbr. to Time's Lamentation, 1599. Described zool. traditions and beliefs of his time. Died 1638.

TORCHIO, Menico, Italian biologist; b. Turin, Italy, Nov. 12, 1932; s. Severino and Annetta (Pironetti) T.; Laurea lode in Scienze Naturali, U. Turin (Italy), 1957; m. Cecilia Roggero, Feb. 1, 1961. Asst. prof. anthropology U. Turin (Italy), 1958-59; curator zoology Museo Civico Storia Nat. Milano, 1960-64, vice dir., 1964—; dir. Acquario Civico di Milano, 1963—. Mem. Internat. Commn. for Sci. Exploration of Mediterranean, 1964—. Recipient Encomium from Mayor of Milan, 1964. Mem. Società Italiana di Storia Naturale (bd. dirs.), Società Malacologica Italiana (bd. dirs.), Consiglio Nazionale Richerche. Author: Biologia marina, 1964; La vita nel mare, 1967; Gli animali d'acquario, 1968; also numerous articles. Research on systematic zoology and zoogeography, parasitology, history of biology. Home: 32 Martinengo, Office: 2 Gadio, Milan, Italy.*

TOREK, Franz, surgeon; b. Breslau, Germany (now Wroclaw, Poland), Apr. 14, 1861; s. Albert and Anna (Wiesner) T.; came with parents to U. S., 1872; A.B., Coll. City N.Y., 1880, A.M., 1887; M.D., Coll. Phys. and Surg., Columbia, 1887; m. Minnie Volkening, Apr. 29, 1896; children—Gretchen (Mrs. Edwin Stein), Paul. Tchr. pub. sch., N.Y., 1880-84; began practice at N.Y.C., 1887; surgeon N.Y. Skin and Cancer Hosp. (now Stuyvesant Sq. Hosp.), 1889-1928, cons. surgeon, 1928—; surgeon German Hosp. of New York (now Lenox Hill Hosp.), 1904-26, cons. surgeon, 1926—; instr. in surgery, N.Y. Post-Grad. Med. Sch., 1891-99, adj. prof. surgery, 1899-1915. Contbr. chpt. on surgery of the esophagus to Nelson's Surgery; chapter on thoracic surgery, Johnson's Operative Therapeusis; article on esophagectomy in Cyclo. of Medicine. Specialist in thoracic surgery, especially of esophagus; work in surg. resection in cases of carcinoma. Died Montclair, N.J., Sept. 19, 1938.

TORELL, Otto Martin, Swedish geologist; b. Varberg, Sweden, June 5, 1828; mem. Nordenskjöld's expdns. to Spitsbergen, 1858, 61; became prof. zoology and geology U. Lund (Sweden), 1866; chief Swedish Geol. Survey, 1871-97. Author: Bidrag till Spitzbergens molluskfauna, 1859. Research on invertebrate fauna, phys. changes of pleistocene and recent times, glaciology of Scandinavia, Iceland, Switzerland; demonstrated that drift deposits of northern Europe are mainly glacial in origin. Died Stockholm, Sept. 11, 1900.

TORFIMUK, Andrei Alekseevich, Russian geologist; b. Aug. 16, 1911; grad. Kazan U., 1933. Worked in oil industry after grad.; joined All-Union Oil-Gas Sci. Research Inst., 1953, dep. dir., 1953-55, dir., 1955-57; became dir. Inst. Geology and Geophysics, Siberian br. USSR Acad. Scis., 1957, became chmn. commn. for conservation nature, 1961. Named Hero of Socialist Labor, 1944, Stalin Prize 1st Degree, 1946, 50, Order of Lenin, twice. Corr. mem. USSR Acad. Scis. Author: The Oil Deposits of the Paleozoic Era in Bashkir, 1950; Conditions in the Formation of Oil Deposits of the Ural-Volga Oil Bearing Territory, 1955; The Petroleum and Gas Content of the Siberian Plateau, 1960. Research and publs. in tectonics and deposits of Volga-Ural oil bearing area; developed method for prospecting Ishimbaevo deposits; assisted in dividing Volga-Ural ty. into tectonic dists.; devel. method of flooding oil fields and increasing oil prodn.; devel. and introduction of contour flooding. Office: Novosibirsk 72, Akademgorodok, Siberia.

TORII, Tetsuya, chemist; b. Takao-city, Taiwan, May 14, 1918; s. Nobuhei and Masako (Suzuki) T.; grad. dept. chemistry U. Tokyo, 1943, D.Sc., 1956; m. Noriko Morita, Dec. 19, 1954; children—Tohru, Mari, Nobuya. With Mitsubishi Kasei Co. Ltd., 1943-52; lectr. Kanagawa U., 1952-55; faculty Chiba U., 1955-63, prof., 1962-63; prof. Chiba Inst. Tech., Narashino, 1963—; exec. sec. Japan Polar Research Assn., Tokyo, 1964—. Sec. nat. Antarctic com. Sci. Council Japan, 1961—, sec. nat. join com. for SCAR-SCOR, 1962—; prof. Tezukayama U., Nara, Japan, 1964—. Recipient Silver cup Prime Minister of Japan, 1962. Mem. Chem. Soc. Japan, Am. Geophys. Union, Limnological Soc., Explorer Club. Research, publs. on new colorimetric determination of cobalt and iron with orthonitrosoresorcin-monomethylether, chem. studies on nutrient matter and uranium in Antarctic Ocean, geochem. work at Dry Valley in Antarctica; discovered antarcticite in Antarctica. Home: 290, 2-chome, Nishiokubo, Shinjuku, Tokyo. Office: Japan Polar Research Assn., Shoko-kaikan 3-4-2, Kasumigaseki, Chiyoda-ku, Tokyo, Japan.*

TORKAR, Karl, Austrian chemist; b. Kematen, Austria, Apr. 20, 1920; s. Johann and Maria (Koss) T.; Dipl.Ing., Tech. U., Graz, Austria, 1949, Dr.techn., 1951; m. Helene Hybner, Oct. 8, 1946; 1 son, Klaus. Hochschulasst., Inst. Inorganic and Phys. Chemistry, 1948-62; prof., dir. Inst. Phys. Chemistry, Tech. U., Graz, 1962—, dean Faculty Sci., 1963-65. Recipient Rudolf Wegscheider award Austrian Acad. Sci., Vienna, Austria, 1957. Mem. Verein Österreichischer Chemiker, Gesellschaft Deutscher Chemiker, Am. Phys. Soc. Research, numerous publs. on influence of impurities and defects on reactions between and with solids. Home: 9, Mandellstrasse, Graz, A-8010, Austria.

TORNQUIST, Leo Waldemar, Finnish statistician; b. Jeppo, Finland, Feb. 14, 1911; s. Anders and Anna (Finskas) T.; M.S., Abo (Finland) akademi, 1934; Ph.D., Abo Akademi, 1937; m. Maj-Lis Rafaela Lindberg, July 25, 1935; children—Maj-Britt Anita, Sofia (Mrs. Michael Rosenbaum), Nils Arthur, Anna, Elisabeth. Tchr. statistics Abo akademi, 1937-38; traffic insp. State Rys., 1938-43; faculty U. Helsinki (Finland), 1943—, prof. statistics, 1950—; vis. prof. Cowles Commn., U. Chgo., 1952-53. UN service in Djakarta, Indonesia, 1960-61. Decorated Order Finnish Lion. Fellow Econometric Soc.; mem. Internat. Statis. Inst., Finnish Soc. Scis., Am. Statis. Assn., Biometric Soc., Union Internationale pour l'Etude Scientifique de la Population, Skandinaviska Aktuariesällskapet. Author: Kriterien fur die reellen algebraischen Zahlen, aritmetische Ketten und diophantische Approximationen, 1937; The Theory of Replicated Systematic Cluster Sampling with Random Start, 1963; Minimaxing and Optimal Programming, 1967. Research on theory of decision making, econometric research on devel. Finnish post and telegraph adminstrn. Home: 10 Unionsgatan, Helsinki 13, Finland.*

TÖRÖ, Imre, Hungarian physician; b. Debrecen, Hungary, Sept. 28, 1900; s. Imre and Berta (Bányász) T.; M.D., U. Debrecan, 1926; m. Lily Böhm, Sept. 19, 1932; 1 son, Imre. Faculty, U. Debrecan, 1936—, prof. anatomy and biology, 1947—, dean Med. Faculty, since 1948—, head dept. histology, embryology, Budapest, 1950—, rector magn., 1961-64; head morphology dept. Exptl. Research Inst., Hungarian Acad. Sci., sec. med. dept., 1951-55, sec. biol. dept. Acad., 1955-59. Pres., Nat. Bd. Popular Edn.; v.p. Hungarian com. UNESCO, 1965—. Recipient Kossuth prize, 1954; Hufeland gold medal, 1964; Janus Comenius medal. Scholar Inst. Biology, Berlin-Dahlen, Germany, 1929; Rockefeller fellow Columbia U. Med. Center, Woods Hole (Mass.) Marine Biol. Lab., 1947. Mem. Hungarian, Yugoslavian acads. sci., Hungarian Biol. Soc. (pres.), European Tissue Culture Club, Assn. des Anatomists, Internat. Soc. for Cell Biology, Soc. Mexicana de Anat., Am. Tissue Culture Assn. Author: Basic Lines of Man's Development, 1936; Development of Man, 1942; Histology, 1948; Tissue Culture in Research Methods of Experimental Medicine, vol. IV, 1958; (with others) General Biology, 1958; (with others) Man's Normal and Pathological Development, 1964; also numerous articles. Research on lens regeneration, effect of embryonic heart-muscle extract; reimplantation of cultivated tissue and orgn.; cytological def. mechanism, thymico-lymphatic organ. Home: II. Herman Otto ut 12, Budapest, Hungary.*

TOROPOV, Nikita Aleksandrovich, Russian phys. chemist; b. June 28, 1908; grad. Leningrad Poly. Inst., 1930; Dr.Tech. Scis. Staff, Lensovet Leningrad Technol. Inst., 1930-41, 44-53, became prof., 1940; staff Giprocement Inst., 1941-44; became dir. Inst. Silicate Chemistry, USSR Acad. Scis., 1953. Recipient Stalin Prize, 1952. Mem. USSR Acad. Constrn. and Architecture, USSR Acad. Scis. (corr.). Author: (with V. F. Zhuravlev) Physical and Colloidal Chemistry of Silicates, 1941; (with L. N. Bulak) A Course in Mineralogy and Petrography with Fundamentals of Geology, 1953; (with K. S. Evstrop'ev) The Chemistry of Silicon and the Physical Chemistry of Silicates, 2d edit., 1956; Chemistry of Cements, 1956. Research and publs. on mineralogy of silicates; physical chemistry of silicate systems, semi-conductors, and ferrite materials. Office: Inst. Chemistry of Silicates, USSR Acad. Scis., Makarova 2, Leningrad V-164, USSR.

TORRE, Giovanni Maria della, see della Torre, Giovanni Maria.

TORREALBA, José Francisco, Venezuelan physician; b. Santa Maria de Ipire, Venezuela, June 16, 1896; s. Tereso and Ana Maria (Gonzalez) T.; M.D., Central U. Venezuela, 1923; m. Rosa Tovar, Jan. 2, 1932; children—Ana Isabel (Mrs. Elias Lopez), Pedro Aquilino, Jose Witremundo, Ana Rosa (Mrs. Tulio Colmenares), Jose Francisco, Ana Teresa (Mrs. Carlos Ron), Sara del Pilar (Mrs. Alexis Ramos), Rafael Tereso, Jesus Rafael, Jose Ramon, Jose Nicolas, Ana Benigna (Mrs. Humberto Diaz). Dir., Mental Asylum, Caracas, 1924-27; practice medicine, Zaraza, 1928-41; attending physician Gen. Penitentiary of Venezuela, 1943-46; dir. Chagas Investigation Center, San Juan de los Morros, 1942—. Recipient Vargas prize, 1938; Brault prize Paris Acad. Medicine, 1942; Order del Libertador Grado Comendador, 1949; Bernardo Guzman Blanco prize, 1963. Author: Investigaciones sobre Enfermedad de Chagas en Zaraza, 4 vols., 1933-53; Investigaciones sobre Enfermedad de Chagas en San Juan de los Morros y otras Notas, 3 vols., 1954-62; also numerous articles. Research on Chagas disease, Herrick's illness, Kala-azar, Bilharziosis; discovered two tripanosomas; biotherapy of anticancerous plants particularly pilon and silk cotton. Home: 1 Avenida Sucre, San Juan Morros. Office: Avenida Los Banos, San Juan Morros, Guarico, Venezuela.*

TORRENS, Robert, Brit. polit. economist; b. Ireland, 1780; s. Robert and Elizabeth (Bristow) T.; m. Charity Chute; 1 son, Robert Richard. Served to col. Royal Marines, 1837, defended Isle of Anholt against Dutch, 1811, col. Spanish legion in Peninsular War; M.P. from Ashburton dist., 1831-35. Lake Torrens, S. Australia, also River Torrens named for him. Fellow Royal Soc., 1818; mem. Polit. Economy Club (co-founder). Author: An Essay on Money and Paper Currency, 1812; An Essay on the External Corn Trade, 1815; A Comparative Estimate of the Effects which a Continuance and a Removal of the Restriction of Cash Payments are respectively calculated to produce . . . , 1819; An Essay on the Production of Wealth, 1821; Letters on Commercial Policy, 1833; On Wages and Combinations, 1834; On the Colonisation of South Australia, 1835; An Enquiry into the Practical Working of the Proposed Arrangements for the Renewal of the Charter of the Bank of England and the Regulation of the Currency . . . , 1844; The Budget, or a Commercial and Colonial Policy, 1844; Self-Supporting Colonisation, 1847; The Principles and Practical Operation of Sir Robert Peel's Act of 1844 Explained and Defended, 1848; Tracts on Finance and Trade, 1852. Editor: Traveller; Globe. Contbr. pamphlets, letters. One of 1st economists to attribute prodn. wealth to joint operation of land and capital; influenced Peel, Ricardo; 1st stated law of diminishing returns; demonstrated how territorial div. of labor increases indsl. productivity. Died London, Eng., May 27, 1864.

TORRESANI, Jean Joseph, French cardiologist; b. Marseilles, France, Dec. 15, 1929; s. Louis and Rose (Dupre) T.; Baccalaureat, Lycée Thiers, Marseilles, 1947; Doctor in Medicine, Faculté de Médicine, Marseilles, 1959; m. Janine Cecile Cayet, Apr. 2, 1960; children—Jean-Louis Alain, Bruno Simon, François Bernard. Chief cardiology clinic Faculté de Médecine Marseilles, 1960-62; asst. J. Cantini Cardiovascular Center Marseilles, 1962-66; master of research Nat. Inst. Health and Med. Research, Marseilles, 1966—. Recipient Nat. Medicine Acad. award, 1960. Mem. French Cardiologic Soc. (award 1960), French Soc. Study Resuscitation and Artificial Organs. Author: Epicardial Leads in Man, 1959; (with R. Courbier) Circulatory Arrest, 1964. Research, publs. on cardiac resuscitation and artificial hearts. Home: 127 Jaubert. France. Office: Exptl. Cardiology Lab., J. Cantini Cardiovascular Center, 60 R. Garros Av. 13, Marseille 9°, France.*

TORRES QUEVEDO, Leonardo, Spanish engr., mathematician; b. Santa Cruz, Spain, Dec. 28, 1852; engr. of bridges and roads, Spain; corr. mem. French Acad. Sci., 1920, fgn. mem., 1927; credited with invention of some algebraic calculating machines, devices for remote control by Hertzian waves; built Niagra transporter bridge, also a dirigible. Died Madrid, Dec. 18, 1936.

TORREY, Harry Beal, Am. biologist; b. Boston, Mass., May 22, 1873; s. James Morrell and Elizabeth Jane (White) T.; B.S., U. Cal. at Berkeley, 1895, M.S., 1898; Ph.D., Columbia U., 1903; M.D., Cornell U., 1927; m. Grace Harbison Crabbe, July 17, 1902; children—Elizabeth Harbison (Mrs. John Graves Andrews). Faculty, U. Cal. at Berkeley, 1898-1900, 1901-12, asso. prof., 1908-12; prof. biology Reed Coll., 1912-20; prof. zoology, prof. exptl. biology U. Ore. Med. Sch., 1920-25, dir. edn. in med. sci., 1920-25; cons. Am. Soc. Hygiene, 1926-28; prof. hygiene and phys. edn., dir. student health service Stanford, 1928-38, prof. biology, 1933-38, prof. emeritus, 1938—; dir. Children's Hosp., Oakland, Cal., 1938-42; practice medicine specializing in internal medicine, Berkeley, Cal., 1938—. Fellow A.A.A.S., A.M.A.; mem. Soc. for Exptl. Biology and Medicine, Western Soc. Naturalists (past pres.), Phi Beta Kappa, Sigma Xi, Phi Delta Theta. Research, numerous publs. on devel. and behavior lower organisms, physiology endocrines and hormones as factors in differentiation. Home: 2801 Stuart St., Berkeley, Cal. 94705.*

TORREY, Henry Cutler, Am. physicist; b. Yonkers, N.Y., Apr. 4, 1911; s. John Cutler and Mabel (Kelso) T.; B.Sc., U. Vt., 1932; A.M., Columbia, 1933, Ph.D., 1937; m. Helen Post Hubert, Sept. 11, 1937; children—John Cutler, Merle Hubert. Instr. physics Princeton, 1937; instr. Pa. State U., State College, 1937-41, asst. prof., 1941-42; staff mem. Radiation Lab., Mass. Inst. Tech., 1942-46, chief crystal rectifier sect., 1944-46; asso. prof. Rutgers U., New Brunswick, N.J., 1946-50, prof., 1950—, chmn. physics dept., 1960-64; dean Grad. Sch., dir. research council, 1965—. Cons. Cal. Research Corp., Standard Oil Co. Cal., 1952—. Recipient certificate of Appreciation, War and Navy Dept., 1946; Guggenheim fellow, Faculté des Sciences de Paris, 1964-65. Fellow Am. Phys. Soc.; mem. Sigma Xi, Sigma Phi. Author: (with C. A. Whitmer) Crystal Rectifiers, 1948; also numerous articles. Editorial bd., Rev. Sci. Instruments, 1962-64. Collaborated (with E. M. Purcell and R. V. Pound) in first observation of nuclear magnetic resonance in bulk matter. Home: 32 Harrison Av., Highland Park, N.J.*

TORREY, John, Am. botanist, chemist; b. N.Y.C., Aug. 15, 1796; s. Capt. William and Margaret (Nichols) T.; M.D., Coll. Physicians and Surgeons, N.Y.C., 1818; A.M. (hon.), Yale, 1823; LL.D. (hon.), Amherst Coll., 1845; m. Eliza Shaw, Apr. 20, 1824, at least 4 children. Catalogued plants growing near N.Y.C., 1817; gave spl. attention to plants of Northeastern U. S.; as result of govt. sponsored expedition (1820), reported on plants collected by David Bates Douglass near source of Mississippi River, 1820; apptd. prof. chemistry, mineralogy and geology U. S. Mil. Acad., 1824-27; prof. chemistry Coll. Physicians and Surgeons, 1827-55, prof. emeritus until 1873; prof. chemistry and natural history Coll. N.J. (now Princeton), 1830-54; worked with Asa Gray on Flora of North America, 1838-43; apptd. N.Y. State botanist, 1836; wrote reports of exploring expdns. of Frémont, Marcy and others, circa 1836-58; elected fgn. mem. Linnean Soc. of London, 1839; mem. Am. Acad. Arts and Scis., 1841; U. S. assayer, 1853-73. Author: Flora of Northern and Middle Sections of the United States, 1823; A Compendium of the Flora of the Northern and Middle States, 1826; Flora of the State of New York, 2 vols., 1843. Contbd. research on Cyperaceae; built one of most valuable bot. libraries and herbariums in U. S.; Torrey's Peak (Colo.) named for him; plants named in his honor include Torreya Taxifolia, Torreya Californica, Torreya Nucifera, Torreya Grandis. Died N.Y.C., Mar. 10, 1873.

TORREY, Theodore Willett, Am. zoologist; b. Woodbine, Ia., Jan. 15, 1907; s. Henry Lee and Pansy (Willett) T.; A.B., U. Denver, 1927; M.A., Harvard, 1929, Ph.D., 1932; m. Marcella Loge, Aug. 14, 1938. Faculty, Ind. U., Bloomington, 1932-—, prof., 1947-—, chmn. dept. zoology, 1948-66. Mem. A.A.A.S., Am. Inst. Biol. Sci., Am. Soc. Zoologists, Soc. Exptl. Biology and Medicine, Sigma Xi. Author: Morphogenesis of the Vertebrates, 1962, 67; also numerous articles. Research in degeneration and regeneration of nerves and taste organs, embryology of sense organs, comparative embryology of vertebrate excretory organs, embryology of mammalian ovaries and testes, history and philosophy of sci. Home: 421 Clover Lane, Bloomington, Ind. 47401.*

TORRICELLI, Evangelista, Italian mathematician; b. Faenza, Italy, Oct. 15, 1608; studied sci. under Benedetto Castelli at Collegio di Sapienza, Rome, 1627; became sec. to Galileo, 1641-42 (3 months) and his successor as mathematician and philosopher to Grand Duke of Florence. Author: Trattato del moto, circa 1640; Opera geometrica, 1644. Suggested that air has weight and that we live in a sea of air, which continually exerts a pressure on us and on things in it; to measure this atmospheric pressure, invented mercury barometer, 1643; also produced vacuum (or nr. vacuum) in his barometers, a blow to dictum that nature abhors a vacuum; improved telescope; invented primitive microscope; investigated theory of projectiles; proved bodies falling from equal heights on differently inclined planes travel with same speed; showed two connected bodies cannot move spontaneously unless their common center of gravity descends; demonstrated that flow of liquid through an orifice proportional to square root of height of liquid (Torricelli's theorem); in mathematics, expert manipulator of curved indivisibles; found volume of acute hyperbolic solid; squared hyperbolas; engaged in controversy with Roberval on priority of solution of problem on properties of cycloid; rediscovered logarithmic spiral (he called it geometric spiral) and established its properties; invented method of drawing tangents. Died Florence, Italy, Oct. 25, 1647.

TORRIGIANO DE'TORRIGIANI, Pietro, b. nr. Florence, Italy, circa 1275; studied under Taddeo Alderotti in Bologna, Italy; studied and taught in Paris, France, 1306-11; became Carthusian monk. Author: Plusquam commentum. Proposed that fevers are not produced by putrid humors, that sensations are located in brain, that one set of nerves controls movements and sensations.

TORTI, Francesco, Italian physician, pharmacologist; b. Modena, Italy, Dec. 1658; M.D., U. Bologna, 1678; became prof. medicine U. Modena, 1681. Fellow Royal Soc., 1717. Author: Therapeuticae specialis ad febres quasdam perniciosas, inopinalo ac repente, lethales, una vero china-china pecual ri methodo ministrata, 1709; Responsiones iatro-aploogeticae . . . , 1715. Credited with introducing term malaria (bad air), 1690 (recorded 1718), also with introducing use of cinchona to treat the disease in Italy, 1712. Died Mar. 1741.

TORTORA, Mario, Italian physician; b. Casamarciano, Naples, Italy, Dec. 23, 1909; s. Salvatore and Carmela (Barone) T.; M.D., U. Naples, 1934. Faculty, U. Naples, 1943-55, U. Sassari, 1956-60; prof. obstetrics and gynecology U. Ferrara (Italy), 1960-—. Expert, WHO; organizer, dir. 1st Italian Center for Early Diagnosis of Gynecol. Cancer. Fulbright fellow, 1950, 53. Mem. Italian Cytological Soc. (pres.), Ferrara Acad. Scis. (v.p.), Am. Cytology Council, Anesthesia Research Council, Internat. Acad. Cytology, Brazilian Soc. Obstetrics and Gynecology, World Assn. for Gynecol. Cancer Prevention (European pres.). Research, numerous publs. on immunohematology, exptl. erythroblastosis in animals, cancer cytology, cytogenetics. Home: Casamarciano, Naples, Italy. Office: Dept. Obstetrics and Gynecology, U. Ferrara, Ferrara, Italy.

TORY, Henry Marshall, Canadian physicist; b. Guysboro, N.S., Can., Jan. 11, 1864; s. Robert Kirk and Anorah (Ferguson) T. B.A., McGill U., Montreal, Ont., Can., 1890, M.A., 1896, D.Sc., 1903, LL.D., 1908; numerous hon. degrees; m. Annie Gertrude Frost, June 30, 1893. Lectr. math. McGill U., 1893-1902, asso. prof., from 1903, gov.'s fellow, 1905-08, non-resident fellow 1908-15; pres. U. Alta. (Can.), 1908-28; dir. Nat. Research Labs., Ottawa, Ont. Pres. Canadian NRC; mem. exec. com. Brit. Empire Univs. Bur., 1912-26, numerous sci. commns. Fellow Royal Soc. Can., Royal Hist. Soc., Royal Empire Soc. Author: Manual of Laboratory Physics. Contbr. articles on pyrometry, edn. to sci. jours. Had large part in planning, equipping Nat. Research Labs., Ottawa. Died 1947.

TOSBERG, William August, prosthetist, orthotist; b. Bielefeld, Germany, Aug. 13, 1905; s. Hermann F. and Martha (Kronsbein) T.; student Engring. Coll., Bielefeld, 1924-27; m. Lina Bierbaum, July 8, 1930. Came to U. S., 1927, naturalized, 1938. Research staff U. S. VA, N.Y.C., 1945-52; tech. dir. prosthetic services Inst. Rehab. Medicine, N.Y. U. Med. Center, 1952-—; instr. rehab. medicine N.Y. U. Sch. Medicine, 1956-—. Prosthetic cons. UN, Internat. Soc. for Rehab. Disabled, City N.Y., 1952-—; mem. adv. com. Die Rehabilitation, Germany, 1961-—. Fellow Soc. Orthotists and Prosthetists; mem. Internat. Soc. for Rehab. of Disabled (U S. com. award 1963), Am. Orthotic and Prosthetic Assn. Author: Upper and Lower Extremity Prostheses, 1962; also articles. Developed new types of prostheses including physiol. knee; designed and manufactured first prosthesis for hemicorporectomy; participated in establishment of internat. prosthetic standards. Home: Box 112, Big Indian, N.Y. 12410. Office: 400 E. 34th St., N.Y.C. 10016.*

TOSCANELLI, Paolo dal Pozza, Italian physician, map-maker; b. Florence, Italy, 1397; ed. Padua, Italy, M.D., 1424. Became consevator Library Florence; author treatises: Prospettiva; Meteorologia Agricola; translator Ptolemy's geography. Studied math. and astronomy; made map of Atlantic Ocean, which placed Asia 3000 miles west of Europe and influenced Columbus to make westward voyage to China; made observations on orbits of comets of 1443, 1449-50, 1457, 1472, also Halley's comet, 1456; built gnomon on Santa Maria Del Fiore cathedral, in order to establish altitudes of solsitices, 1468. Died Florence, May 15, 1482.

TOSTESON, Daniel Charles, Am. physiologist, educator; b. Wauwatosa, Wis., Feb. 5, 1925; s. Alexis H. and Dilys (Bodycombe) T.; M.D., Harvard, 1949; m. Penelope Kinsley, Dec. 17, 1949; children—Heather, Tor, Zoe. Fellow dept. physiology Harvard Med. Sch., 1947-48; intern, asst. resident in medicine Presbyn. Hosp., N.Y.C., 1950-51; research fellow Med. dept. Brookhaven Nat. Lab., N.Y.C., 1951-53, Lab. Kidney and Electrolyte Metabolism, NIH, 1953-55, Dept. Biol. Isotope Research, Copenhagen, Denmark, 1955-56, Physiol. Lab., Cambridge, Eng., 1956-57, Lab. Kidney and Electrolyte Metabolism, Nat. Heart Inst., 1957; asso. prof. dept. physiology, Washington U. Sch. Medicine, St. Louis, 1958-61; prof., chmn. dept. physiology Duke Sch. Medicine, Durham, N.C., 1961-—. Cons. sci. rev. com. NIH, 1964; mem. molecular biology panel NSF, 1961-64, adv. com. Nat. Kidney Disease Found.; com. on blood and transfusions NRC, 1963. Mem. Am. Physiol. Soc., Soc. Gen. Physiologists, Biophys. Soc., Boylston Soc., Alpha Omega Alpha. Research, publs. on cellular functions and molecular mechanism biol. transport processes. Home: Route 1, Box 209AA, Piney Mountain Rd., Durham, N.C. 27706.*

TOTO, Patrick Daniel, Am. oral pathologist; b. Niles, O., Jan. 6, 1921; s. Vincent J. anl Esmeralda (Mazza) T.; B.S., Kent State U., 1943; D.D.S., Ohio State U., 1948, M.S., 1950; m. Eleanor P. Mitrikeff, Sept. 1, 1944; children—James Robert, Michael George, Robert Daniel. Faculty, Loyola U. Sch. Dentistry, Chgo., 1950-—, prof. oral pathology, coordinator research, 1957-—; cons. Hines VA Hosp., 1964-66. Diplomate Am. Bd. Oral Pathology, Am. Bd. Oral Medicine. Mem. Am. Dental Assn., A.A.A.S., Am. Acad. Oral Pathologists, Am. Acad. Dental Medicine, Internat. Assn. Dental Research, Sigma Xi. Research, numerous publs. on precursor cells to fgn. body giant cells, striated muscle repair, osteoclasts and plasma cells, bone loss in periodontal disease, carious teeth pathogenesis. Home: 433 Gillett St., Waukegan, Ill. 60085. Office: 1757 W. Harrison St., Chgo. 60612.*

TOTTLE, Charles Ronald, English metallurgist; b. Sheffield, Eng., Sept. 2, 1920; s. George and Minerva (Baker) T.; B.Metallurgy with 1st class honours, U. Sheffield, 1941, M.Metallurgy, 1949; M.Sc. (hon.), U. Manchester (Eng.), 1962; m. Eileen Pauline Geoghegan, June 10, 1944; children—Elizabeth Lauraine, Patrick Alan. Metallurgist, English Electric Co. Ltd., Rugby, 1941-45; lectr. King's Coll., Newcastle on Tyne, U. Durham (Eng.), 1945-50; prin. scientist metall. devel. lab. Atomic Energy Div., Ministry of Supply, Springfields Works, Preston, Eng., 1950-51; research mgr., dep. head Culcheth Research and Devel. Labs., U.K. Atomic Energy Authority, 1951-55; head labs., dep. dir., Dounreay, 1956-59; prof., head dept. metallurgy U. Manchester, 1959-67; prof., head Sch. Materials Sci., Bath (Eng.) U. Tech., 1967-—; resident research asso. Argonne (Ill.) Nat. Lab., 1964-65. Tech. cons. United Gas Industries Ltd., 1964-—; cons. U.K. Atomic Energy Authority, 1959-—. Fellow Instn. Metallurgists (v.p. 1967-—, joint editor text-book series, 1961-—), Inst. Physics, mem. Inst. Metals, Inst. Welding, Brit. Nuclear Energy Soc. Author: Science of Engineering Materials, 1965; also articles. Research on plastic flow in cast iron, nucleation of cast metals by coating on mould face, residual stress in cast and welded components, phys. properties of uranium compounds, metal-oxide cermets, super plasticity in alloys and metal compounds; devel. materials for nuclear power reactors, niobium metal fabrications; patentee fast reactor fuel elements. Home: Third Acre, Church St., Hilperton, Trowbridge, Wiltshire, U.K. Office: U. Bath, Claverton Down, Bath, Somerset, U.K.*

TOUÉRY, Pierre-Fleurus, French pharmacologist; b. Solomiac, France, 1802; ed. Montpellier; asst. to Dr. Balard; discovered use of animal black as gen. antidote, also as bleaching and extracting agt. of plant bitters; discovered Artemisia (genus of composite flowering plants from which santonin is derived). Died Solomiac, France, 1883.

TOULMIN, Lyman Dorgan, Am. geologist; b. Mobile, Ala., July 4, 1904; s. Lyman D. and Bessie (Stein) T.; B.A., U. Ala., 1926, M.S., 1934; Ph.D., Princeton, 1940; m. Geraldine Hulsart, Aug. 23, 1945; Faculty, Tex. A. and M. Coll., 1938-42, asst. prof., 1941-42; geologist Geol. Survey Ala., Tuscaloosa, 1942-45; asso. prof. Birmingham So. Coll. (Ala.), 1945-48; faculty Fla. State U., Tallahassee, 1948-—, prof. geology, 1951-—; vis. prof. La. State U., Baton Rouge, 1953-54. Mem. Geol. Soc. Am. (past sec. S.E. sect. 1957-59), Soc. Econ. Paleontologists and Mineralogists (past councilor, past pres. Gulf Coast sect.), Am. Assn. Petroleum Geologists, Paleontol. Soc., Paleontol. Research Instn., Sigma Xi. Contbr. articles to tech. jours. Described fossils from beds early Eocene age, formations Paleocene and Eocene age in Ala.; research on location Paleocene-Eocene boundary in eastern Gulf Coast region, Paleocene and Eocene guide fossils eastern Gulf Coast region. Home: 2006 E. Forest Dr., Tallahassee 32303.*

TOULOUSE, Edouard, French physician, physiologist; b. Marseilles, France, 1865; asst. physician Seine asylum; chief physician Villejuif asylum; chief physician, founder exptl. psychology lab. Sainte-Anne asylum, Paris; founder League of Mental Hygiene, Assn. Sexual Studies. Author: E. Zola, 1896; Les conflits intersexuels et sociaux, 1904; Comment former un esprit, 1908; H. Poncaré, 1910; Comment se conduire dans la vie, 1910; La question sexuelle et la femme, 1918; La vie nouvelle, 1919. Made med.-psychol. studies of men of superior intelligence with nervous diseases. Died Paris, 1947.

TOUNTAS, Constantin John, Greek physician; b. Athens, Greece, Apr. 14, 1917; s. John Michael and Filomila (Stavridou) T.; grad. U. Athens Med. Sch., 1940, Ph.D., 1942; m. Mary Mihalopoulou, Aug. 15, 1947; children—John, Ileana. Instr. Surg. Clinic, Laikon, U. Hosp., Athens, 1947-55; head dept. surgery Piraeus Gen. Hosp. Queen Frederica, 1955-66; prof. surgery Aristotelian U., Thessaloniki, Greece, 1961-—; head A dept. surgery AHEPA U. Hosp., Thessaloniki, 1961-—; pro-tempore prof. Med. Sch. U. Washington (D.C.), 1963. Decorated Greek Cross of Phoenix. Mem. Internat. Coll. Surgeons, Internat., French, Greek surg. assns., Internat. Cardio-Vascular Soc., Am. Coll. Chest Physicians (mem. com. on pulmonary surgery 1965-—), Med. Soc. Athens, Med. Soc. Thessaloniki, Turkish Med. Assn. (hon.). Author: (with others) Textbook on Thoracic and Cardiovascular Surgery, 1961; also numerous articles. Research on cardiac and thoracic surgery, portal hypertension, devel. method for transesophageal ligation of bleeding varices, transplantation of organs. Home and office: 13 Vas. Konstantinou St., Thessaloniki, Greece.*

TOUR, Charles Cagniard de la, see Cagniard de la Tour, Charles.

TOURAINE, Alain, French sociologist; b. Hermanville, France, Aug. 3, 1925; s. Albert and Odette (Cleret) T.; student Ecole Normale Superieure, 1945-50; Agrégé; U. Paris, 1950; m. Adriana Arenas Pizarro, Feb. 9, 1957; children—Marisol, Philippe. Research fellow Center for Sociol. Studies, 1950-58; dir. lab. indsl. sociology Sch. for Higher Studies, Sorbonne, Paris, 1958-—; prof. Faculty Letters, Paris-Nanterre, 1966-—; head Documentation Center for Latin Am., 1962-—. Mem. Nat. Council for Sci. Research, 1959-—. Mem. French Sociol. Soc., French Psychol. Soc., French Soc. for Polit. Sci. Author: L'évolution du travail ouvrier, 1955; Sociologie de l'action, 1965; La Conscience ouvrière, 1966; also articles. Research on indsl. sociology, social stratification and mobility, social aspects of econ. devel., sociol. theory. Home: 2,

Allée du Cèdre, Chatenay, Malabry-92, France. Office: 10, Rue Monsieur-le-Prince, Paris VI°-75, France.*

TOURAINE, Roger, French physician; b. Mouhers, France, May 9, 1921; s. Didier and Jeanne T.; student Faculté de Médecine, Lyons, France, 1939-44; m. Lucienne Alix, Apr. 22, 1948; children—Pascale, Bernard. Physician hosps. of Lyons; head pneumophtisiology dept. Hôpital Jules Courmont, 1960—; aggregate prof., lectr. Faculté de Médecine, Lyons. Mem. French Soc. Tuberculosis, French Soc. Respiratory Pathology, French Soc. Allergy. Author: Le Poumon Opaque Tuberculeux, 1951; (with Galy, Brune) Les Tumeurs de Médiastin, 1963; also numerous articles. Research in Tb, respiratory diseases, especially allergic respiratory diseases; discovered Ambrosia (ragweed) in vicinity of Lyons and studied its role in a hay-fever of region. Home: 6 Bis, Avenue Debrousse-69, Lyon 5, France. Office: Hôpital Jules Courmont, 69 Lyon, Pierre Bénite, France.*

TOURKY, Ahmed Riad, Egyptian chemist; b. Tanta, Egypt, Mar. 16, 1902; s. Ahmed Abulfutuh and Hassiba (El Khalifa) T.; Ph.D., U. Munich (Germany), 1928; Hon. Doctorate U. Algiers, 1964; m. Helen Aumüller, Feb. 4, 1930; 1 son, Adel. Faculty, U. Cairo, 1928-53, prof. phys. and inorganic chemistry, 1948-53, now part-time, dean higher studies, 1964—; dir. Nat. Research Center, 1953-64, head electrochemistry unit; minister sci. research, 1964-65; pres. Supreme Council Sci. Research, 1965—; prof. Ain-Shams U., part-time. Decorated Order Republic 1st class, State prize Merit, 1964. Mem. Egyptian Chem. Soc. (chmn. 1954—), Acad. Scis. USSR (fgn.), German Acad. Scis. (corr.), Internat. Union Pure and Applied Chemistry (mem. bur. 1961-63). Translator English and German textbooks. Research, numerous publs. on electrochemistry, electroanalytical chemistry including electrodic behavior of metals and alloys, abnormal mobility of hydrogen ion, dielectric properties of various inorganic compounds, causes of variation of color in certain oxides, hydrometallurgy of local ore deposits. Home: 15 Abdel Wahid El-Wakil, Heliopolis, Cairo. Office: 101 Kasrel Aini, Cairo, Egypt.*

TOURNEFORT, Joseph Pitton de, French botanist; b. Aix, France, June 5, 1656; ed. Jesuit Sem., Aix, until 1677; student medicine U. Montpellier (France), U. Barcelona (Spain); M.D., 1698. Prof. botany Jardin des Plantes, Paris, France, from 1688; prof. medicine Collège de France, from 1702. Mem. French Acad. Sci. Author: Herbal; Relation d'un Voyage du Levant, 1717; Eléments de Botanique, 3 vols., 1694, rev. edit., 1700; Historie des Plantes qui Naissent aux Environs de Paris, 1698; Traité de Matière Médicale. Pioneer systematic botany; 1st to group plants into genera; devised classification system which used form of corolla as determining factor; sci. expdns. to Pyrenees, Asia Minor, Greece (collected 1300 plant species); genus Tournefortia named for him by Linnaeus. Died Paris, Dec. 28, 1708.

TOURTELLOTTE, Wallace William, Am. neurologist; b. Great Falls, Mont., Sept. 13, 1924; s. Nathaniel Mills and Frances (Charlton) T.; Ph.B., U. Chgo., 1945, B.S. in Anatomy, 1945, Ph.D. Biochem. Pharmacology, 1948, M.D., 1951; m. Jean Ester Toncray, Feb. 14, 1953; children—Wallace William, George Mills, James Millard, Warren Gerard. Instr., U. Chgo., 1958-61; faculty U. Mich., Ann Arbor, 1957—, prof. neurology, 1966—, dir. neurology research lab., 1957 , dir. multiple sclerosis clinic, 1957—, lab. dir. bioengring. program, 1966—; vis. asso. prof. neurology and pharmacology Washington U., St. Louis, 1963-64; staff VA Hosp., Ann Arbor, Wayne County Gen. Hosp., Eloise, Mich., chief neurology, 1959—. Founding mem. com. to evaluate drug testing in patients with multiple sclerosis USPHS, 1960—. Diplomate Am. Bd. Neurology and Psychiatry. Fellow Am. Acad. Neurology (S. Weir Mitchell Neurology Research award 1959); mem. Am. Neurol. Assn., Clin. Soc. Neurologists, Am. Assn. Neuropathology (asso.), Central Soc. for Neurol. Research (pres. 1965-66), Assn. for Research in Nervous and Mental Disease, World Fedn. Neurology (founding mem. internat. commn. for correlation neurology and neurochemistry 1959—), Internat. Soc. Neurochemistry (founding mem. com. on clin. problems 1965—), World Assn. Neurol. Commns. (found., mem. neurochemistry commn. 1965—), Am. Soc. for Pharmacology and Exptl. Therapeutics, A.A.A.S., A.M.A., Pan Am. Med. Assn., Sigma Xi. Author: (with A. F. Haerer, G. L. Heller, J. E. Somers) Post-Lumbar Puncture Headaches, 1964; also numerous articles, chpts. in books. Research on gamma globulin and nervous system barriers in patients with multiple sclerosis; standardized ultramichrochem. procedures for quantitation of lipids in cerebrospinal fluid; developed techniques for quantitating neurol. function for patients with multiple sclerosis. Home: 2264 Manchester Rd., Ann Arbor, Mich. 48104.*

TOUSEY, Richard, Am. physicist; b. Somerville, Mass., May 18, 1908; s. Coleman and Adella Richards (Hill) T.; A.B., Tufts U., 1928, Sc.D. (hon.), 1961; A.M., Harvard, 1929, Ph.D., 1933; m. Ruth Lowe, June 29, 1932; 1 dau., Joanna. Instr. physics Harvard, 1933-36, tutor div. phys. scis., 1934-36; research instr. Tufts U., 1936-41; physicist U. S. Naval Research Lab. optics div., 1941-58, head instrument sect., 1942-45, head micron waves br., 1945-58, head rocket spectroscopy br., atmosphere and astrophysics div., 1958—. Mem. com. vision Armed Forces-NRC, 1944—; line spectra of elements com. NRC, 1960—; mem. Rocket and Satellite Research Panel, 1958—; cons. astronomy subcom., space sci. steering com. NASA, 1960-62; mem. com. Aeronomy Internat. Union Geodesy and Geophysics, 1958; U. S. nat. com. Internat. Commn. Optics, 1960; mem. sci. steering com. Project Vanguard, 1956-58. Bayard Cutting fellow Harvard, 1932-33, 35-36; recipient Meritorious Civlian Service award U. S. Navy, 1945; E. O. Hulburt award Naval Research Labs., 1958; Progress medal photog. Soc. Am., 1959; Prix Ancel Soc. Francaise de Photographie, 1962; Henry Draper medal Nat. Acad. Scis., 1963; Navy award, 1963; Eddington medal Royal Astron. Soc., 1964. Fellow Am. Phys. Soc., Optical Soc. Am. (dir. 1953-57); Frederic Ives medal 1960); mem. International Academy of Astronautics, National, Washington academies scis., Am. Astron. Society (v.p. 1964—), Soc. Applied Spectroscopy, A.A.A.S., Am. Geophys. Union, Philos. Soc. Washington, Sci. Research Society America, International Astronomical Union, Nutall Ornithol. Club, Audubon Naturalists Soc., Phi Beta Kappa, Contbr. articles in field to sci. jours. and books. Contbn. in solar ultraviolet spectroscopy; 1st to send spectrograph on rocket to record ultraviolet spectrum of the sun, 1946; work on atmospheric and physiol. optics, vacuum ultraviolet. Home: 7725 Cxon Hill Rd. S.E., Washington 21, Office: U. S. Naval Research Lab., Washington 25.

TOUSTER, Oscar, Am biochemist; b. N.Y.C., July 3, 1921; s. Mayer Herman and Henrietta (Silberstein) T.; B.S., Coll. City N.Y., 1941; M.A., Oberlin Coll., 1942; Ph.D., U. Ill., 1947; m. Eva Katherine Beach, Aug. 24, 1944; 1 dau., Alison Blair. Chemist, Atlas Powder Co., Tamaqua, Pa., 1942-43; research biochemist Abbott Labs., North Chicago, Ill., 1944-45; faculty Vanderbilt U., Nashville, 1947—, prof., 1958—, chmn. dept. molecular biology, 1963—; investigator Howard Hughes Med. Inst., Vanderbilt and Oxford univs., 1957-60; vis. lectr. chemistry U. Ill., 1959; vis. prof. biochemistry Cornell U., 1963. Mem. USPHS biochemistry tng. com., 1961-66, med. scientist tng. com., 1966—. Guggenheim fellow Oxford U., 1957-58. Mem. Am. Soc. Biol. Chemists, Biochem. Soc., A.A.A.S. (Theobald Smith award in med. scis. 1956), Am. Chem. Soc., Am. Assn. U. Profs., Sigma Xi, Phi Beta Kappa. Translator, revisor: Hollmann's Nonglycolytic Pathways of Metabolism of Glucose, 1964. Editorial bd. Jour. Biol. Chemistry, 1964—; asso. editor Organic Reactions, vol. 7, 1952. Contbg. author The Chemistry of Penicillin, 1948. Research, publs. on new products of bacterial origin; investigated aspects of carbohydrate metabolism in animals; elucidated biochem. basis of essential pentosuria, a genetic metabolic abnormality in man, by finding liver enzymes involved in pentose metabolism. Home: 215 La Vista Dr., Nashville, 37215.*

TOWER, Beauchamp, Brit. engr., inventor; b. 1845; apprenticed to Elswich Works, Newcastle, Eng., 1861-66. Built iron steamers at Tyne Ironworks, 1866; asst. to William Froude, Admiralty Exptl. Works, Torquay, 1869; made series of exptl. torpedoes for Sir William Armstrong, 1874; asst. to Lord Rayleigh at Terling, Essex, on hydraulic investigations, 1875; made expts. for com. on friction Inst. Mech. Engrs., 1882-91. Author numerous reports. Research on friction in ry. journal bearings instrumental in devel. of hydrodynamic lubrication theory, 1883; Died Brentwood, Essex, 1904.

TOWER, Donald Bayley, Am. neurochemist; b. Orange, N.J., Dec. 11, 1919; s. Walter Sheldon and Edith (Jones) T.; A.B., Harvard, 1941, M.D., 1944; M.S., McGill U., 1948, Ph.D., 1951; m. Arline Belle Croft, Aug. 5, 1947; 1 dau., Deborah Alden. Chief sect. clin. neurochemistry Nat. Inst. Neurol. Diseases and Blindness NIH, Bethesda, Md., 1953-61, chief lab. neurochemistry, 1961—; med. dir. USPHS, 1958—; asso. prof. neurology Georgetown Med. Sch., 1954—. John and Mary R. Markle scholar in med. scis., 1951-53. Mem. Am. Acad. Neurology, A.A.A.S., Am. Chem. Soc., Am. Soc. Biol. Chemists, Am., Canadian neurol. assns., Canadian Physiol. Assn., Internat. Soc. Neurochemistry, Internat. Brain Research Orgn. Author: Neurochemistry of Epilepsy, 1960; (with R. O. Brady) Neurochemistry of Nucleotides and Amino Acids, 1960; (with J. P. Schade) Structure and Function of Cerebral Cortex, 1960; (with S. A. Luse and H. Grundfest) Properties of Membranes and Diseases of Nervous System, 1962; also numerous articles. Research in biochemistry of nervous system, mechanisms causing epilepsy and applications to exptl. therapy, history of neurochemistry. Home: 7105 Brennon Lane, Chevy Chase, Md. 20015. Office: Nat. Inst. Neurol. Diseases and Blindness NIH, Bethesda, Md. 20014.*

TOWNES, Charles Hard, Am. physicist; b. Greenville, S.C., July 28, 1915; s. Henry Keith and Ellen Sumter (Hard) T.; B.A., B.S., Furman U., 1935, D.Litt., 1960; M.A., Duke, 1937; Ph.D., Cal. Inst. Tech., 1939; also numerous hon. degrees; m. Frances H. Brown, May 4, 1941; children—Linda Lewis, Ellen Screven, Carla Keith, Holly Robinson. Mem. tech. staff Bell Telephone Lab., 1939-48; asso. prof. physics Columbia, 1948-50, prof. physics, 1950-61, exec. officer physics dept. 1952-55; Fulbright and Guggenheim fellow, 1955-56; provost and prof. physics, Mass. Institute Technology, 1961—, v.p., dir. research Inst. for Def. Analyses, Washington, 1959-61; vis. prof. U. Paris, 1955-56, U. Tokyo, 1956; mem. editorial bd. Rev. Sci. Instruments, 1949-52; nat. lectr. Sigma Xi, 1950; asso. editor Phys. Rev., 1951-54. Recipient Page One award for Sci., 1958. Research Corp. Ann. award, 1959; Nobel prize for physics (with N. G. Basov and A. M. Prokhorov), 1964; others. Fellow Am. Phys. Soc. (mem. council 1959—, v.p. elect 1965—), Optical Soc. Am., I.E.E.E.; mem. Am. Phil Soc., Astron. Soc., Biophys. Soc., Instrument Soc. Am., Am. Acad. Arts and zscis., Nat. Acad. Scis.. Société Francaise de Physique (mem. council 1956-58), Phys. Soc. Japan. Author: (with A. L. Schawlon) Microwave Spectroscopy, 1955. Contbr. articles sci. publs. Patentee electronics, including masers and optical and infrared masers. Study of radio astronomy; nuclear and molecular structure; quantum electronics; optics; microwave spectroscopy. Home: 5 Follen St., Cambridge, Mass.*

TOWNES, Henry, Am. entomologist; b. Greensville, S.C., Jan. 20, 1913; s. Henry K. and Ellen (Hard) T.; B.S., Furman U., 1932, B.A., 1933; Ph.D., Cornell U., 1937; m. Marjorie Chapman, Oct. 9, 1937; children—David Keith, Jean Whilden. Instr., Syracuse U., 1937-38, Cornell U., 1938-40; insect taxonomist U. S. Nat. Mus., U. S. Dept. Agr., 1941-49; econ. survey of Micronesia, 1946; research asso. N.C. State Coll., 1949-52, prof. entomology, 1951-56; entomol. adviser to Philippine Govt., 1952-54; research asso. U. Mich., Ann Arbor, 1956—. Am. Philos. Soc. grantee 1947-48. Fellow Entomol. Soc. Am.; mem. Am. Entomol. Inst. (trustee, dir. 1961—), Dutch (corr.), Am., Can., Washington, Mich. (pres. 1965) entomol. socs. Author several books including: (with others) Ichneumon-flies of America North of Mexico, 1960; (with M. Townes, V. K. Gupta) A Catalogue and Reclassification of the Indo-Australian Ichneumonidae, 1961; (with M. Townes) Ichneumon-flies of America North of Mexico: 3 Subfamily Gelinae, Tribe Mesostenini, 1962; (with V. K. Gupta) Ichneumon-flies of America North of Mexico: 4 Subfamily Gelinae, Tribe Hemigasterini, 1962; (with S. Momoi, M. Townes) A Catalogue and Reclassification of the Eastern Palearctic Ichneumonidae, 1965; also numerous articles. Improved classification of N.Am. spider wasps and gnats, classification, nomenclature and bibliography of ichneumon-flies of world; introduced more efficient and safer control practices for insect pest of tobacco, improved control methods in Philippines for rats and insect pests of rice and corn; improved design of Malaise trap for insect. Address: 5950 Warren Rd., Ann Arbor, Mich. 48105.*

TOWNSEND, Charles H(enry) T(yler), Am. biologist, physicist; b. Oberlin, O., Dec. 5, 1863; s. Nathan Haskin and Helen Jeannette (Tyler) T.; Columbian (now George Washington) U. Sch. Medicine, 1887-91; B.S., George Washington U., 1908, Ph.D., 1914; m. Caroline W. Hess, Sept. 10, 1889 (dec. 1901); children—Karl Hess, Leland, Helen Tyler; m. 2d, Margaret C. Dyer, June 1, 1908; children—Charles Henry Tyler, Edward Dyer, Nathaniel Ostend, Mary Louise. Asst. entomologist, U. S. Dept. Agr., 1888-91; prof. entomology, zoölogy, and physiology, N.M. Agrl. Coll., and entomologist, Expt. Sta., 1891-93; curator Museum, Inst. Jamaica, 1893-94; field agt. div. entomology, U. S. Dept. Agr., 1894-98; again with N.M. Agrl. Coll. Expt. Sta., 1898-99; prof. biol., etc., Batangas Provincial Sch., P.I., 1904-06; expert, Gipsy Moth Lab., Bur. Entomology, U. S. Dept. Agr., 1907-09; govt. entomologist and dir. entom. stations, Peru, 1909-14; entom. asst., Bur. Entomology, U. S. Dept. Agr., 1914-19; hon. custodian muscoid diptera, U. S. Nat. Museum, 1914-25; chief entomologist, State of Sao Paulo, Brazil, 1919-22; ant expert in Brazil for Am. Cyanamid Co., 1923; dir. Cotton Plagues Lab., Piura, Peru, 1923-24; cotton plagues expert, Chamber Commerce and Agr., Iquitos, Peru, 1925; chief Inst. Parasitologia Agricola, Lima, Peru, 1926; chief entomologist Estacion Experimental Agricola S.N.A., Lima, 1927-29; head firm Charles Townsend & Filhos, Sao Paulo, since 1929; cons. entomologist Cla. Ford Industrial do Brasil, Rio Tapajós, Pará, since 1932. Author: Manual of Myiology (12 parts); also about 1000 titles on muscoid flies, cotton plagues, med. entomology, biogeography, ecology and physics. Pioneer work on American cotton weevils; discovered mode of transmission of disease, verruga, in Peruvian Andes; first analysed insect environments; demonstrated Cephenemyia as the swiftest organism; established about 1000 valid muscoid genera; explained gravity; recorded exact atomic weights; determined exact velocity of light; defined cosmic units of length, time and mass; explained moon's origin and earth's axial inclination; set Pleistocene duration and man in America at two million years. Died Mar. 17, 1944.

TOWNSEND, John Chalmer, Am. psychologist; b. Finleyville, Pa., Mar. 28, 1920; s. Melvin William and Rose Lillian (Stradley) T.; student Washington and Jefferson Coll., 1939-41; B.S., U. Pitts., 1946, M.S., 1947, Ph.D., 1948; m. Eunice Lucille Sampson, Aug. 13, 1921; children—Joyce Carol, John Chalmer, Joann Cynthia. Research asst. Western State Psychiat. Inst. and Clinic, Pitts., 1947-48; lectr. psychology U. Pitts., 1946-48; faculty W.Va. U., Morgantown, 1948-53, asso. prof., 1952-53; supervisory physiol. and exptl. psychologist USAF Air Research and Devel. Command, 1953-57; chief scientist Interceptor Pilot Research Lab., Tyndall AFB, Fla., 1956-57; prof. psychology Catholic U. Am., Washington, 1957—. Cons. to pvt. cos., govt. agys. Mem. Am., Eastern, D.C. psychol. assns., Sigma Xi, Psi Chi. Author: Introduction to the Experimental Method, 1953; also articles. Research in electro-shock therapy, tng. and proficiency measurements, decision making process; devel. tests of decision making ability, new parameters of measurement and statis. determination of biomed. measurements; invented electronic voice key for measuring verbal reaction time. Home: 9839 Cherry Tree Lane, Silver Spring, Md. 20901. Office: Dept. Psychology, Catholic U., Washington 20017.*

TOWNSEND, John Kirk, Am. ornithologist; b. Phila., Aug. 10, 1809; s. Charles and Priscilla (Kirk) T.; m. Charlotte Holmes, 1 child. Joined overland expdn. to Ore., 1835, to H.I., 1835; surgeon Ft. Vancouver, 1835-36; assembled valuable collection of birds and mammals; new birds from Ore. country described by him in Jour. of Acad. of Natural Scis. of Phila.; conceived idea of preparing work on birds of U. S., pub. one part of Ornithology of the United States of North America, 1840; Townsend's Bunting named for him; birds from his collection painted for last vol. of Audubon's Birds of America, 1844, his mammals described and painted by Audubon and John Bachman in Viviparous Quadrupeds of North America; secured and mounted birds for Nat. Inst., Washington, 1842; studied dentistry, Phila., 1845; life mem. Acad. Natural Sciences of Phila. Author: Narrative to a Journey Across the Rocky Mountains to the Columbia River, 1839. Died Washington, Feb. 6, 1851.

TOWNSEND, Joseph, English geologist; b. London, Eng., Apr. 4, 1739; s. Chauncy and Bridget (Phipps) T.; ed. Clare Hall, Cambridge (Eng.) U., B.A., 1762, M.A., 1765; student medicine, Edinburgh, Scotland; m. Joyce Namkivell, 1773 (dec. 1785); children—Thomas, Charles, James, Henry, Charlotte, Sophia; m. 2d, Lydia Hammond Clarke. Traveled in Ireland, 1769, France, Holland, Flanders, 1770, later Spain, Switzerland; apptd. chaplain to Duke of Atholl, also rector of Pewsey, Wiltshire, Eng. Author: The Physician's vade mecum, 1781; Journey through Spain, 1791; A Guide to Health, 1795-96; The Character of Moses Established for Veracity as an Historian, recording Events from the Creation to the Deluge, 1812-15; Etymological Researches, 1824; also others. Studied geology of Cornwall; had good knowledge of geology and mineralogy; criticized Hutton's uniformitarianism; firmly convinced the Mosaic account literally correct; collected minerals and fossils. Died Pewsey, Wiltshire, Nov. 9, 1816.

TOWSON, John Thomas, English sci. writer; b. Devonport, Eng., Apr. 8, 1804; s. John Gay and Elizabeth (Thomas) T.; ed. Stoke Classical Sch.; m. Margaret Braddon, Nov. 19, 1840. Began as chronometer and watchmaker; sci. examiner masters and mates, Liverpool, 1850-73; then became chief examiner in compasses; apptd. to prepare manual on compass deviation in iron ships, by Bd. Trade, 1863. Author: Tables for Facilitating the Practice of Great Circle Sailing, 1849; Tables for the Reduction of Exmeridian Altitudes, 1849; Practical Information on the Deviation of the Compass for the Use of Masters and Mates of Iron Ships; also articles. Research on daguerroytpe, 1839, deviation of compasses in iron ships; devel. new photog. materials; determined that Gt. Circle is shortest route across Atlantic; showed luminious and chem. rays do not focus at same distance from object. Died Liverpool, Jan. 3, 1881.

TOYAMA, Yoshiyuki, Japanese chemist; b. Tokyo, Japan, Sept. 6, 1896; grad., Tokyo U., 1920; doctorate. Mem. staff Tokyo Indsl. Research Inst., 1920-40; became prof. Nagoya U., 1940. Recipient Imperial prize Japan Acad., 1950. Research on aquatic animal oil, hardening of oil (established found. for margarine industry); elucidated structure of fish and whale oil.

TOYNBEE, Arnold, English philosopher, economist; b. London, Eng., Aug. 23, 1852; s. Joseph and Harriet (Holmes) T.; ed. King's Coll., London; Pembroke Coll., Oxford (Eng.) U., 1873, pass degree in Literae Humaniores, Balliol Coll., 1878; m. Charlotte Atwood, June, 1879. Tutor in charge of studies of men preparing for Indian Civil Service, Balliol Coll., Oxford U., from 1878; lectr. on current indsl. problems, Bradford, Newcastle, Bolton, Leicester, and London; active worker in charity orgn., co-operation, ch. reform. Toynbee Hall in Whitechapel named in his memory. Author: The Industrial Revolution, 1884. Applied historical method to econs.; practical reformer, who supported efforts by the people to raise standard of living and felt that the state had duty to assist these efforts. Died Wimbledon, Eng., Mar. 9, 1883.

TOYNBEE, Joseph, English surgeon; b. Heckington, Eng., Dec. 30, 1815; s. George Toynbee; apprenticed to Wade of Westminster Dispensary; studied anatomy under G. D. Dermott, Little Windmill St. Sch. Medicine; attended practice St. George's Hosp., U. Coll. Hosp.; m. Harriet Holmes, Aug. 1846; 9 children, including Arnold. Admitted to Coll. Surgeons Eng., 1838; became asst. to Sir R. Owen, Coll. Surgeons, Lincoln's Inn Fields, 1838; later became surgeon St. James' and St. George's dispensary; aural surgeon, lectr. on diseases of ear St. Mary's Hosp., 1852-64; aural surgeon Earlswood Asylum for Idiots; cons. aural surgeon Asylum for Deaf and Dumb. Fellow Royal Soc., 1842, Royal Coll. Surgeons; mem. Quekett Micros. Soc. Author: Hints on the Formation of Local Museums, 1863; Wimbledon Museum Notes; Diseases of the Ear, their Nature, Diagnosis and Treatment, 1860; On the Use of Artificial Membrana Tympani in Cases of Deafness, 1853; A Descriptive Catalogue of Preparations Illustrative of the Diseases of the Ear in the Museum of Joseph Toynbee, 1857. Demonstrated non-vascularity of articular cartilage and other tissues, 1842; description of corneal corpuscles (Toynbee's corpuscles), 1841. Died July 7, 1866.

TRACY, Martha, Am. physician; b. Plainfield, N.J., Apr. 10, 1876; d. Jeremiah Evarts and Marth Sherman (Green) Tracy; A.B., Bryn Mawr Coll., 1898; M.D., Woman's Med. Coll. Pa., 1904; Dr.P.H., U. Pa., 1917. With research dept. exptl. pathology Cornell U. Med. Sch., N.Y.C., 1904-07; asst., meningitis commn. N.Y. Bd. Health, 1905; asst. (under Huntington Fund for Cancer Research) Dept. Exptl. Pathology, N.Y.C., 1907-19; prof. physiol. chemistry Woman's Med. Coll. Pa., 1913-21, dean coll., 1918-40, prof. hygiene, 1921-23, prof. preventive medicine, 1923-31. Mem. Phila. Bd. Health, 1936-40; Asst. dir. pub. health Phila., 1940. Fellow Am. Coll. Physicians, A.M.A.; mem. Am. Med. Women's Assn. (past pres.), Am. Assn. U. Women, Nat. Assn. Deans of Women. Research, publs. on cancer and meningitis; work in pub. health. Died 1942.

TRACZYK, Wladyslaw Zygmunt, Polish physiologist; b. Warsaw, Poland, June 27, 1928; s. Zygmunt Teodor and Wladyslawa (Dzikiewicz) T.; M.D., Sch. Medicine Warsaw, 1951; Ph.D., First Sch. Medicine, Moscow, 1955; m. Zdzislawa Andrysik, Dec. 7, 1954; 1 son, Zdzislaw. Asst. to head Lab. Physiology, Polish Acad. Scis., Warsaw, 1956-62; asso. prof. physiology Sch. Medicine, Warsaw, 1963; prof., chmn. dept. physiology Sch. Medicine, Lodz, Poland, 1963—. Mem. com. physiol. scis. Polish Acad. Scis., 1966—. Mem. Royal Soc. Medicine London, Polish Physiol. Soc. (dir.). Author: Modern View on the Hypothalamus, 1966; Physiological Mechanism of Motivation and Emotion, 1967; also articles. Research on acetylcholine metabolism in caudate nucleus in relation to muscle tonus and brain elec. activity, properties of hypothalamic centers in motivation, control of hypophysis. Home: 2a Lwowska, Warsaw 10. Office: 3 Lindleya, Lodz 1, Poland.*

TRADESCANT, John, Dutch botanist; b. 1567; 1 son, John. Gardener to Charles I of Eng., from 1629. Genus Tradescantia named for him and his son. Collected plants of Asia and Mediterranean. Died 1637.

TRADESCANT, John, botanist; b. 1608; s. John Tradescant. With sci. expdn. to Va. Genus Tradescantia named for him and his father. Author: Museum Tradescantium, 1656. Increased and catalogued father's collection of plants. Died 1662.

TRAGER, William, Am. biologist; b. Newark, Mar. 20, 1910; s. Leon and Anna (Emilfork) T.; B.S., Rutgers U., 1930, Sc.D., 1965; M.A., Harvard, 1931, Ph.D., 1933; m. Ida Sosnow, June 16, 1935; children—Leslie, Carolyn, Lillian. Asso. mem. Rockefeller Inst., 1950-55, asso. prof., 1955-64; prof. Rockefeller U., N.Y.C., 1964—; vis. prof. Fla. State U., 1962, U. P.R., 1963, U. Mexico, 1965. Mem. study sect. Nat. Inst. Allergy and Infectious Diseases, 1954-58, 66—, mem. tng. grant com., 1960-64; guest investigator W. African Inst. Trypanosomiasis Research, Vom, No. Nigeria, 1958-59; mem. malaria commn. Armed Forces Epidemiological Bd., 1965—; pres. Am. Found. for Tropical Medicine, 1966—. Fellow A.A.A.S., N.Y. Acad. Scis., Entomol. Soc. Am.; mem. Am. Soc. Zoologists, Am. Soc. Parasitologists, Am. Soc. Exptl. Pathology, Soc. Exptl. Biology and Medicine, Soc. Protozoologists (past pres.), Am. Soc. Tropical Medicine and Hygiene (past v.p.), Liberian Inst. for Tropical Medicine (pres. 1966—). Editor: Jour. Protozoology, 1954-65; editorial bd. Jour. Parasitology, 1947-50. Numerous publs. on first successful cultivation of insect tissue in vitro, for silkworm virus propagation; mosquito larvae cultures; physiology, biochemistry of protozoan parasites. Home: 89 Lee Rd., Scarsdale, N.Y. 10583. Office: Rockefeller U., 66th St. and York Av., N.Y.C. 10021.*

TRAGUS, see Bock, Hieronymus.

TRAINOR, Lynne E. H., Canadian physicist; b. Chamberlain, Sask., Can., Dec. 4, 1921; s. Anthony R. and Aline (Souply) T.; B.A., U. Sask., 1946, M.A., 1947; postgrad. Columbia, 1947-48; Ph.D., U. Minn., 1951; m. Helen Marie Dixon, June 28, 1947; children—Patricia Ann, Laurel Jean, Charles R. W. NRC Can. postdoctoral fellow, 1952-53; asst. prof. Queen's U., Kingston, Ont., Can., 1953-55; vis. prof. U. B.C., Vancouver, Can., 1955-56; asso. prof. U. Alta., Edmonton, Can., 1956-63, prof., 1963; prof. physics U. Toronto (Ont.), 1963—, exec. sec. Theoretical Physics Inst., 1961-63. Mem. Canadian Assn. Physicists (sec. 1966—), Am. Inst. Physics, Am. Phys. Soc., Royal Astron. Soc. Can. Editor: (with others) Physics in Canada: Survey and Outlook, 1967. Research, publs. on theory of electron scattering on atoms, nuclear structure and nuclear spectroscopy, composite models of elementary particles; discovered new selection rules for nuclear electromagnetic transitions (isotopic spin selection rules), (with D. F. Goble) new correlation length in Bose-Einstein gas models of liquid helium. Home: 37 Farmcote Rd., Don Mills, Ont., Can. Office: U. Toronto, Toronto, Ont., Can.*

TRAMEZZANI, Juan Humberto, Argentine neuroendocrinologist, anatomist; b. Buenos Aires, Argentina, Oct. 12, 1928; s. Felix José and Rosa (Piovano) T.; M.D., U. Buenos Aires, 1952; m. María Estela Corrales, Nov. 5, 1960; children—Maria Laura, Juan Pablo. Research asst. Instituto Biología y Medicina Exptl., 1952-54; instr. dept. histology U. Sao Paulo Sch. Medicine, 1954-56; fellow Rockefeller Found. dept. anatomy U. Cal. at Los Angeles Sch. Medicine, 1956-58; faculty dept. histology U. Buenos Aires Sch. Medicine, 1958-59; dir. Instituto de Neurobiología, Buenos Aires, 1960—. Recipeint Gold mdal Colegio del Salvador, 1945; Diploma de Honor, U. Buenos Aires, 1956, prize Sch. Medicine, 1958, Prize Cámara Jr., 1966. Mem. Sociedad Argentina de Biologia, Asociación Latinoamericana de Ciencias Fisiológicas, Asociación Latinoamericana de Investigaión en Reproducción Humana Alirh, Consejo Argentino de Estudio sobre la Reproducíón, Asociación Argentina de Ciencias Naturales. Research, publs. on structure and function of hypothalmus and neurosecretory phenomena, including ultrastructural studies on neurohypophysis identification of adrenalin and noradrenalin storing cells with description of their ultrastructural features, including new technique for their identification; cellular types in carotid body. Home: 1578 Callao, Buenos Aires. Office: 2490 Obligado, Buenos Aires, Argentina.*

TRANSUE, William Reagle, Am. mathematician; b. Pen Argyl, Pa., Nov. 30, 1914; s. Ray Frank and Emily (Reagle) T.; B.S., Lafayette Coll., 1935; M.A., Lehigh U., 1939; Ph.D., 1941; m. Monique Serpette, May 27, 1936; children—William, Jacques, John. Asst., Inst. for Advanced Study, 1942-43, 48-49; chief ballistic research sect. Office Chief Ordnance, Washington, 1943-45; faculty Kenyon Coll., Gambier, O., 1945-66, prof. math., 1948-66; prof. math. State U. N.Y. at Binghamton, 1966—. Fulbright Research grantee U. Pavia, Italy, 1951-52; NSF faculty fellow, Paris, France, 1960-61. Mem. Am. Math. Soc., Math. Assn. Am., Société Mathematique de France. Research, publs. on properties Frechet variation and applications in analysis, theory subharmonic functions, lattice theory, theory integration. Home: 28 E. Hamton Rd., Binghamton, N.Y. 13903.*

TRAPEZNIKOV, Vadim Aleksandrovich, Russian engr.; b. Nov. 28, 1905; grad. Moscow Technol. Inst., 1928; Dr.Tech. Sci. Engr., All-Union Electro-Tech. Inst., 1928-33; faculty Moscow Inst. Energetics, 1930-41, became prof., 1939; joined Inst. Automation and Telemechanics, USSR Acad. Scis., 1941, apptd. dir., 1951. Recipient Stalin Prize, 1951, Order of Red Banner of Labor. Mem. USSR Acad. Scis., Soviet Union for Automatic Control (became chmn. nat. com. 1961). Author: Basis of Planning Series of Asynchronous Machines, 1937; (with others) Automatic Control of Linear Dimensions of Products, 1947. Directed devel. systems of automatic control; designed electronic modeling units; proposed calculating transverse field electric machines, techniques for econ. analysis; constrn. of elec. machinery.

TRAPMANN, Heinz, German pharmacist; b. Eschweiler, Germany, Nov. 17, 1926; s. Heinrich and Johanna (Dederichs) T.; Dr.nat.sci., U. Munich, 1954; m. Charlotte Ohneberg, Oct. 30, 1954; children—Heinrich Alexander, Martina Johanna. Faculty U. Munich, 1955—, sci. chief asst. Inst. Pharmacy and Food Chemistry, 1960—, lectr. pharmacy, 1959—. Teaching officer German Red Cross, 1962—. Mem. German Pharm. Soc. Research, publs. on splitting bonds in different compounds by metal-ions of rare earths, thorium, and zirkonium; dephosphorrylation of nucleotides by metal-ions and by ultra-violet irradiation. Home: 2 Ridlerstrasse, Munich 12, Germany.*

TRASK, James Dowling, Am. pediatrician; b. Astoria, L.I., N.Y., Aug. 21, 1890; s. James Dowling and Julia Norton (Hartshorne) T.; Ph.B., Yale, 1913; M.D., Cornell U., 1917; m. Phyllis Hayden Randall, June 4, 1921; 1 dau., Phyllis Randall. Became asso. attending pediatrician New Haven Hosp. and Dispensary, 1927; instr. medicine Yale, 1921-25, asst. prof., 1925-27, asso. prof. pediatrics, 1927—. Mem. Am. Pediatric Soc., Soc. Am. Bacteriologists, Soc. Am. Immunologists, Am. Soc. Clin. Investigation, Harvey Soc. Author: Scarlet Fever, 1928; Pneumonia and Bronchitis, 1937; also articles. Research in infantile paralysis, childhood pneumonia, capillary bronchitis, diagnostic value of nasal swab cultures; established (with F. Blake) viral etiology of measles; studies (with Yale) on hemolytic streptococci and scarlet fever. Died May 24, 1942.

TRAUB, Hamilton Paul, Am. physiologist, lineagicist; b. Crozier, Ia., June 18, 1890; s. Lorenz and Elizabeth (Graf) T.; student Harvard, 1914-15; B.A., U. Minn., 1923, M.S., 1925, Ph.D., U. Wis., 1941. Plant scientist, 1924-30; physiologist to prin. physiologist U. S. Dept. Agr., Orlando, Fla., Beltsville, Md., 1930-52; editor Plant Life, 1945—. Recipient Stout medal Am. Hemerocallis Soc., 1964. Fellow A.A.A.S.; mem. Soc. Study Evolution, Am. Soc. Plant Taxonomists, Internat. Assn. Taxonomy, Am. Plant Life Soc., Sigma Xi, Gamma Alpha, Alpha Zeta. Author: The Amaryllis Manual, 1958; The Phyla of Organisms, 1962; The Genera of Amaryllidaceae,

1963; Lineagics, 1964. Founder-editor: National Horticultural Magazine, 1922-24; Herbertia, 1934-48. Publs. on integration of lineagics, comml. use of auxins in fruit storage, devel. desirable flavors in expressed citrus juices, colchine-induced polyploidy. Home: 5804 Camino de la Costa, La Jolla, Cal. 92037.*

TRAUB, Robert, Am. med. entomologist; b. N.Y.C., Oct. 26, 1916; s. D. S. and Jeanette (Moses) T.; B.S. cum laude, Coll. City N.Y., 1938; M.S. in Med. Entomology, Cornell U., 1939; Ph.D. in Med. Entomology, U. Ill., 1947; m. Renée Gluck, Aug. 18, 1939; children—Jeanette Roberta (Mrs. Jerry A. Fisher), Roger Dennis. Army entomologist, malariologist, India, 1942-44; mem. Army Typhus Commn. India, Burma, 1944-45; chief dept. entomology Walter Reed Army Med. Center, Washington, 1946-55; comdg. officer, field dir. med. research units, Malaya, Borneo, Korea, 1955-59; col. Med. Sci. Corps., Med. Research and Devel. Command, Washington, 1959-62; ret., 1962; research prof. microbiology U. Md. Med. Sch., Balt., 1962—. Hon. research asso. Smithsonian Instn., Field Mus., Bishop Mus.; mem. commns. Armed Forces Epidemiological Bd.; cons. to govt. agys. Recipient U. S. Typhus Commn. medal; commended by His Majesty's Prin. Sec. of State for Fgn. Affairs concerning med. research in Malaya and Borneo, 1949, 52. Mem. Washington Acad. Scis., Phi Beta Kappa, Sigma Xi. Research, numerous publs. on 1st demonstration of scrub typhus infection in Himalayan alpine and subarctic terrain and semi-deserts and mountain-deserts of central Asia jungle cycle, from variety of rodents and chiggers in various Asian areas; prophylaxis and immunization of scrub typhus; isolation of mosquito- and tick-borne viruses; descriptions of many new genera and species of fleas and trombiculid mites; prevention and treatment of certain infectious diseases. Home: 5702 Bradley Blvd., Bethesda, Md. 20014. Office: 660 W. Redwood St., Balt. 21201.*

TRAUBE, Isidor, phys. chemist; b. Hildesheim, Germany, Mar. 31, 1860; Ph.D., U. Berlin, 1882. Became prof. Technische Hochschule, Berlin, 1900; left Germany, 1939; with Edinburgh (Scotland) until 1943. Research on molecular and atomic covolumes, osmosis, surface tension, colloids, critical temperature, theory of catalysis (led to theory of drug action); phys.-chem. studies of milk, urine, gastric juice, blood; designer viscometer, capillarimeter; originated Traube's rule relating surface tension of capillary active organic compounds to number of CH2 groups attached. Died Edinburgh, Scotland, Oct. 27, 1943.

TRAUBE, Ludwig, German physician, pathologist; b. Ratibor, Silesia, Jan. 12, 1818; studied under J. E. Purkinje and Johannes Müller; became prof. Charité, Berlin, 1853, Friedrich-Wilhelm Inst., 1857, U. Berlin, 1872. Author: Gesammelte Beiträge zur experimentellen Pathologie, 1871-78. Regarded as founder of exptl. pathology in Germany; used animal expts. to solve clinical problems; investigated pulmonary disorders caused by sect. of vagus nerve; credited with being first to give clear description of bigeminal pulse; introduced temperature curves, 1850; investigated relation of heart and kidney disorders; believed that often cases of cardiac hypertrophy are asso. with or due to contracted kidney in which there is marked loss of capillaries (at the time this theory received much attention), 1865; studied aspects of lead poisoning, suffocation, crisis and critical days, pathology of fever, effects of digitalis and other drugs. Died Berlin, Apr. 11, 1876.

TRAUBE, Moritz, German chemist; b. Ratibor, Germany, Feb. 12, 1826; pvt. researcher, Berlin; produced artificial semipermeable membranes (led to measurement of osmotic pressure), Breslau, Germany (now Wroclaw, Poland), 1867; studied fermentation, sugars, plant respiration, protoplasm, and muscles and oxidation. Died Berlin, June 28, 1894.

TRAVER, Jay R., Am. entomologist; b. Willoughby, Ohio, Aug. 3, 1894; s. Jay R. and Mabel (Dodd) T.; B.A., Cornell U., 1918, M.A., 1919, Ph.D., 1931. Acting head biology dept. Shorter Coll., 1923-24; faculty Women's Coll. U. N.C., 1924-30; research in entomology Cornell U., 1932-36; faculty U. Mass., Amherst, 1938—, prof. zoology, 1960-62, emeritus prof., 1962—. Mem. A.A.A.S., Am. Inst. Biol. Sci., Entomol. Soc. Am., Entomol. Soc. Washington, Soc. Limnology and Oceanography, Sigma Xi, Sigma Delta Epsilon. Author (with J. G. Needham, Yin-Chi Hsu) Biology of Mayflies, 1935; also articles. Research on food and breeding habits of black-nosed dace; the mite dermatophagoides as a human parasite; taxonomy of N.Am., neotropical mayflies; emphasis on accurate assn. of nymphal and adult stages, notes on habitats and habits of both stages. Home: 19 Moorland St., Amherst, Mass. 01002.*

TRAVERS, Alexandre, French chemist; b. Parentignat, France, 1883; ed. École Normale Supérieure; prof. indsl. chemistry U. Nancy (France); founder Ecole Nationale Supérieure of chem. industries. Author: Lecons de chimie; Notions modernes sur l'atome et la valence. Theoretical research on oxyredution and phosphoric acids; applied research on spl. steels, cements, metal corrosion, glucinium preparation. Died 1949.

TRAVERS, Benjamin, surgeon; b. Apr. 1783; s. Joseph and (Spilsbury) T.; pupil resident House of Astley Paston Cooper, 1800-06; student Edinburgh; m. Sarah Morgan, 1809; m. 2d, Millet, 1813; m. 3d, Stevens, 1831; several children. Became demonstrator anatomy Guy's Hosp., 1807-08; apptd. surgeon E. India Co. warehouses and brigade, 1809; surgeon to London Infirmary for Diseases of Eye, 1810-18; elected surgeon St. Thomas's Hosp., 1815, joint lectr. until 1819, and from 1834; began pvt. practice, 1816; surgeon extraordinary to Queen Victoria; surgeon in ordinary to Prince Consort, also sergeant-surgeon. Fellow Royal Soc., 1813; mem. Hunterian Soc. (became pres. 1827), Royal Med. and Chirurg. Soc. (became pres. 1827), Royal Coll. Surgeons Eng. (mem. council 1830, chmn. bd. midwifery examiners 1855, v.p. 1845-46, 54-55, pres. 1847-56). Author: An Inquiry into the Process of Nature in Repairing Injuries of the Intestines, 1812; A Synopsis of the Diseases of the Eye and their Treatment, 1820; An Inquiry Concerning . . . Constitutional Irritation, 1826; A Further Inquiry, 1835; The Physiology of Inflammation and the Healing Process, 1844. First hosp. surgeon in Eng. specializing in eye surgery; studies in pathology. Died London, Mar. 6, 1858.

TRAVERS, Morris William, English chemist; b. London, Eng., Jan. 24, 1872; s. William and Annie (Pocock) T.; ed. Univ. Coll., London, also U. Nancy (France); Sc.D., univs. London and Bristol. Jr. lectr. Univ. Coll., 1894, asst. prof. chemistry, 1896-1903; prof. Univ. Coll., Bristol, 1904-06; dir. Indian Inst. Sci., Bangalore, 1906-14; sci. dir. Duroglass Ltd., 1914-18; cons. practice, also dir. concerns engaged in constructing chem. engring. plants, 1937-39; prof. applied chemistry U. Bristol, 1927-37, prof. emeritus, 1937—; engaged as cons., 1927-37; cons. practice, research work U. Bristol, 1937-39; cons. explosives mfg. div. Ministry Supply, World War II; cons. Imperial Smelting Corp. Ltd., 1945—. Fellow Univ. Coll. Fellow Royal Soc., 1904; mem. Chem. Soc., Soc. Chem. Industry, Faraday Soc. (founder, past pres.), Soc. Glass Technology (founder, past pres.), Inst. Fuel (founder; Melchett medal). Author: The Experimental Study of Gases, 1900; The Discovery of the Rare Gases, 1928; Life of Sir William Ramsey, 1956. Discoverer (with Sir William Ramsey) inert gases, neon, krypton and xenon, 1898; expert on glass technology; conducted researches on low temperature. Died Aug. 25, 1961.

TRAVIS, Dorothy Frances, Am. biologist; b. Atlanta, Dec. 3, 1920; d. James E. Travis and Mary Z. (Huie) Travis Anderson; B.S. in Zoology, George Washington U., 1945; A.M. (Thomas Dana fellow), Radcliffe Coll., 1950, Ph.D. in Biology (Atomic Energy fellow), Radciffe-Harvard, 1951. Atomic Energy fellow Bermuda Biol. Sta., 1951-53; instr. U. N.H., 1953-55, asst. prof., 1955-58; research fellow Harvard, 1958-60, Med. Sch., 1960-62, research asso. dept. orthopedic surgery and Mass. Gen. Hosp., 1962-65, asst. biologist, 1965—, faculty asso., 1968—. Mem. A.A.A.S., Am. Assn. U. Profs., Am. Zool. Soc., N.Y. Acad. Scis., Electron Microscopy Soc. Am., Sigma Xi, Phi Beta Kappa, others. Research, publs. on structure, orgn. and relationship between inorganic crystals and organic matrix of mineralized tissues among invertebrates. Home: 247 Washington St., Winchester, Mass. 01890. Office: Orthopedic Research Labs., Mass. Gen. Hospital, Boston 02114.*

TRAYNHAM, James G(ibson), Am. chemist; b. Broxton, Ga., Aug. 5, 1925; s. James G. and Eddie Louise (Greer) T.; student S. Ga. Coll., 1942-43; B.S. in Chemistry, U. N.C., 1946; Ph.D., Northwestern U., 1950; m. Margaret A. Egert, Dec. 22, 1948; children—David F., Peter C. Instr., Northwestern U., 1949-50; asst. prof. Denison U., 1950-53; faculty La. State U., Baton Rouge, 1953—, prof. chemistry, 1963—, chairman of the department of chemistry, 1968—; postdoctoral research fellow Ohio State U., 1951-53. Petroleum Research Fund-Am. Chem. Soc. Type D award Eidg. Technische Hochschule, Zurich, Switzerland, 1959-60; Charles E. Coates award Baton Rouge sects. Am. Chem. Soc., Am. Inst. Chem. Engrs., 1965. Mem. Am. Chem. Soc. (past chmn. Baton Rouge sect.), Am. Assn. U.Profs., La. Acad. Scis., Sigma Xi, Phi Beta Kappa, Phi Lambda Upsilon. Author: Organic Nomenclature: A Programmed Introduction, 1966; also articles, book revs. Research on mechanisms of reactions of alkenes, quantitative measures of effects of substituent groups, mechanisms of carbonium ion reactions; discovered transannular hydrogen atom rearrangement in medium ring compounds. Home: 573 Magnolia Woods Dr., Baton Rouge 70808.*

TREADWELL, Carleton Raymond, Am. biochemist; b. Calhoun County, Mich., Dec. 28, 1911; s. Gilbert B. and Mina (Eyre) T.; A.B., Battle Creek Coll., 1934; M.S., U. Mich., 1935, Ph.D., 1939; m. Gloria E. Bennett, Sept. 3, 1941; children—Gloria Ann, Carla Maxine. Asst. biochemistry U. Mich., 1935-39; instr. biochemistry Baylor U., 1939-41, asst. prof., 1942-43; asso. prof. Southwestern Med. Coll., 1943-45; faculty George Washington U., 1945—, prof. biochemistry, head dept., 1959—. Cons. med. research VA Center, Martinsburg, W.Va. 1954—; cons. biochemistry to chief med. officer VA, Washington, 1957-59; mem. physiol. chemistry study sect. Nat. Insts. Health, 1959-63, member heart program project com., 1963—; chairman of Gordon Research Conf. Lipid Metabolism, 1959-60; mem. frontiers of sci. com. Joint Sci. Bd., 1959-60. Fellow A.A.A.S.; mem. Am. Soc. Biol. Chemists, Am. Inst. Nutrition, Soc. Exptl. Biology and Medicine (council D.C. 1956-58, nat. council 1955), Am. Assn. U. Profs., N.Y. (co-chmn. conf. atherosclerosis and hormones 1958), Washington (chmn. biol. awards com. 1959-61) acads. sci., Washington Acad. Medicine, Am. Chem. Soc., Sigma Xi. Club: Cosmos. Author numerous papers. Editorial bd. Jour. Nutrition, 1955-58; Arch. Biochem. Biophys., 1966—. Research on chemistry metabolism of fats and sterols. Home: 6901 Alpine Dr., Annandale, Va. 22003. Office: 1335 H St. N.W., Washington 20005.*

TREANOR, Charles Edward, Am. physicist; b. Buffalo, Oct. 22, 1924; s. William Michael and Margaret (Powers) T.; B.A., U. Minn., 1947; Ph.D., U. Buffalo, 1956; m. Ruth Ziegelmaier, Jan. 28, 1950; children—Timothy C., John J., Peter A., Melissa A., Michael P. Faculty, U. Buffalo, 1948-53; research staff physics and aerodynamics Cornell Aero. Lab., Buffalo, 1953-—, asst. head aerodynamics research dept., 1966—; vis. prof. aeros. and astronautics Stanford, 1966. Mem. Am. Phys. Soc., Combustion Inst., Sigma Xi. Research, publs. in high temperature gases, optical radiation from heated gas, chem. reaction rates and chem. composition of high speed gas flows. Home: 140 Segsbury Dr., Williamsville, N.Y. 14221. Office: 4455 Genesee St., Buffalo 14221.*

TRÉCUL, Auguste Adolphe Lucien, French botanist; b. Mondoubleau, France, Jan. 8, 1818. Mem. French Acad. Scis., 1866. Author: Observations sur le fruit du prismatocarpus, 1842. Research on secretions, adventives, gen. morphology, organogenesis. Died Paris, Oct. 15, 1896.

TREFALL, Harald, Norwegian physicist; b. Oslo, Norway, Nov. 10, 1925; student U. Bergen; Ph.D., U. Oslo, 1961. Research fellow U. Bergen, 1951-55, faculty, 1955—, prof. physics, 1964—. Mem. Norwegian Phys. Soc. (dir. 1963—), Am. Geophys. Union. Research and publs. on cosmic-ray time-variations especially meteorol. effects; time-variations and morphology of auroral-zone electron precipitation events, using high-altitude balloon recordings of asso. x-ray Bremsstrahlung. Home: Professorvegen 16, Bergen-Minde, Norway.*

TREFOUEL, Jacques, French physician; b. Le Raincy, France, Nov. 9, 1897; s. Eugène and Jeanne (Pottiez) T.; Docteur ès-sciences physiques, Licencié-ès-sciences (5 certificates); Dr. Honoris causa, 10 times; m. Thérèse Tréfouel Boyer, June 1, 1921. With Institut Pasteur, 1920—, asst., lab. chief, head, dir. Service Therapeutical Chemistry, 1940-65, hon. dir., 1965—. Mem. French Acad. Scis. (past pres.), French Acad. Medicine (pres. 1967—), Chem. Soc. France (hon. pres.), Superior Council Pub. Hygiene France (pres. sect. serums and vaccines), Soc. Microbiology French Lang. (past pres.), Am. Chem. Soc., Assn. Am. Physicians, Acad. Scis. and Letters Finaland (fgn.), Royal Acad. Medicine Belgium (hon. fgn.). Author: (with Thérèse Tréfouel, Father Palfrey) Syntheses organique; also numerous articles. Research in local anesthetics, arsenic, antipaludians, sulfamides, sulfones and antimonies (trypanocide properties of urea). Home: 207, rue de Vaugirard, Paris XV. Office: Institut Pasteur, 28, rue du Docteur Roux, Paris XV, France.*

TREGER, Abraham, biologist; b. Brest Litovsk, Poland, Jan. 12, 1909; s. Simon and Esther (Steingard) T.; M.S., U. Warsaw, 1937; m. Judith Schechner, May 17, 1949; children—Jacob, Estelle. Came to U. S., 1951, naturalized, 1957. Head microbiology lab., Pruzana, Poland, 1939-41; head bacteriology dept. Sanatorium Gauting, Munich, Germany, 1945-51; research asst. Tufts U. Med. Sch., Boston, 1952-57; research asso. Bio-Research Inst., Cambridge, Mass., 1957—. Exptl. biologist Bio-Research Consultants, Cambridge, 1957—. Mem. A.A.A.S., New Eng. Animal Care Panel. Contbr. articles to profl. jours. Co-discoverer new technique for collection endometrial secretions in rodents leading to discovery of plasminogen activator in mouse uterine fluid and ophthalmic hemorrhagic phenomenon; co-discoverer new class of synthetic auxins in exptl. tumor chemotherapy. Home: 21 Pope Hill Rd., Milton, Mass. 02186. Office: 9 Commercial Av., Cambridge, Mass. 02141.*

TREHERNE, John Edwin, English zoologist; b. Swindon, Wiltshire, Eng., May 15, 1929; s. Arnold Edwin Wilson and Marion (Spiller) T.; B.Sc., Bristol (Eng.) U., 1947, Ph.D., 1953; M.A., Cambridge (Eng.) U., 1966; m. June Vivienne Freeman, June 4, 1955; children—Jonathan Mark, Rebecca Sian. Mem. unit insect physiology dept. zoology U. Cambridge, 1955-68, fellow Downing Coll., 1966—, Univ. lectr. in zoology, 1968—; sec. Co. Biologists Ltd., 1963—. Recipient Sci. medal Zool. Soc. London, 1967. Mem. Soc. for Exptl. Biology (mem. council 1966—), Royal Entomol. Soc. London (council 1966—, v.p. 1967). Author: The Neurochemistry of Arthropods, 1966; also articles. Editor: The Physiology of the Insect Central Nervous System, 1965; Advances in Insect Physiology, 1963—, U. Revs. in Biology, 1962—. Research on insect physiology especially of nervous system, physiol. orgn. of insect central nervous tissues, chem. environment of nerve cells in complex nervous structures. Home: Park End, Swaffham, Bulbeck, Cambridge, Eng.*

TREIMAN, Sam Bard, Am. physicist, educator; b. Chgo., May 27, 1925; s. Abraham and Sarah (Bard) T.; student Northwestern U., 1942-44; S.B., U. Chgo., 1949, S.M., 1950, Ph.D., 1952; m. Joan Little, Dec. 27, 1952; children—Rebecca, Katherine, Thomas. Faculty dept. physics Princeton, 1952—, prof., 1963—. Cons. to AEC, Dept. Def. Mem. Am. Phys. Soc. Author: (with M. Grossjean) Formal Scattering Theory, 1960. Research, numerous publs. on subnuclear particles, beta-radioactivity, quantum field theory. Home: 60 McCosh Circle, Princeton, N.J.*

TREITZ, Wenzel, pathologist, anatomist; b. Hovermic, Bohemia, 1819; became physician, Prague, 1846; worked in anatomy under Hyrtle; later asst. of Olaudy and Engel; named prof. pathology and anatomy U. Cracow (Poland), 1852; became head Pathology/Anatomy Inst., prosector Gen. Hosp., Prague, Czechoslovakia, 1855. Research and publs. on description of anat. structures and conditions, including Treitz's Hernia, 1857, Treitz's fossa, 1857, Treitz's ligament, 1857, Treitz's muscle, 1857. Died Prague, Czechoslovakia, Aug. 27, 1872.

TRÉLAT, Emile, French surgeon; b. Paris, 1821; s. Ulysse Trélat; became prof. external pathology Faculty Medicine, 1872, prof. clin. surgery, 1890; mem. Acad. Medicine. Author: Sur la nécrose par le phosphore, 1854; Sur l'oesophagotomie interne, 1870; Sur la necrose phosphatée, 1874; Leçons de clinique chirurgicale, 1877. Died Paris, 1907.

TRÉLAT, Ulysse, French surgeon; b. Montargis, France, 1795; became mil. surgeon, 1813, doctor, 1821; named asst. physician Salpetrière, Paris, 1837; apptd. commissaire gen. for departments of Allier, Puy-de-Dôme, Cruese, Haute-Vienne; France, col., cav. Nat. Guard; constitutional rep. from Puy-de-Dôme; minister pub. works, 1848; mem. Paris municipal council, 1871-74. Author: Recherches historiques sur la folie, 1839; Des causes de la folie, 1856; La folie lucide, 1861. Died Menton, France, 1879.

TRELEASE, William, Am. botanist; b. Mt. Vernon, N.Y., Feb. 22, 1857; s. Samuel R. and Mary (Gandall) T.; B.S., Cornell, 1880; Sc.D., Harvard, 1884; LL.D., U. of Wis., 1902, U. of Mo., 1903, Washington U., 1907; m. Julia M. Johnson, July 19, 1882; children—Frank Johnson, Marjorie (dec.), Sam Farlow, Sidney Briggs, William. In charge Summer Sch., Botany, Harvard, 1883-84; lecturer botany, Johns Hopkins, 1884; instr. botany, U. of Wis., 1881-83, prof., 1883-85; prof. botany, Washington U., 1885-1913; dir. Mo. Bot. Garden, 1889-1912; prof. botany, U. of Ill., 1913-26, emeritus since 1926. Mem. Ill. State Board Natural Resources and Conservation since 1917. Chmn. Am. board editors Botanisches Centralblatt, 1900-21. Fellow Am. Acad. Arts and Sciences (1892), A.A.A.S.; mem. Nat. Acad. Sciences (1902), Am. Philos. Soc. (1903), etc., directeur (pres.) Académie Internationale de Géographie Botanique, 1836; 1st pres. Bot. Soc. America, 1894-95 and 1918; pres. Am. Soc. Naturalists, 1903, Cambridge Entomol. Club, 1889, Engelmann Bot. Club, 1898-99 (hon. pres. since 1900); sec. Wis. Hort. Soc., 1882-85, Acad. Science St. Louis, 1896 (president 1909-11). Mem. of Sigma Xi, Phi Beta Kappa. Edited (with Asa Gray) Botanical Works of the late George Eagelmann. Translated Poulsen's Botanical Micro-Chemistry and Salomonsen's Bacteriological Technology. Author: Agave in the West Indies, 1913; The Genus Phoradondron, 1916; Plant Materials of Decorative Gardening, 1917, 21, 26, 30; Winter Botany, 1918, 25, 30; The American Oaks, 1925; also many papers and reports on botany and entomology. Commemorated in many plant names and in Mount Trelease (12,500 ft. high), at head of Clear Creek, Colo., at first ascent to Loveland Pass. Noted for studies of genus Agave. Died Jan. 1, 1945.

TREMBLAY, Marc-Adélard, Canadian anthropologist; b. Les Eboulements, Que., Can., Apr. 24, 1922; s. William and Lauretta Tremblay; A.B., Montréal (Que., Can.), 1944, L.S.A., 1948; M.A., Laval U., Ste Foy, Que., 1950; Ph.D., Cornell U., 1954; m. Jaqueline Cyr, Dec. 27, 1949; children—Genevieve, Lorraine, Marc, Colette, Dominique, Suzanne. Research asso. dept. sociology and anthropology Cornell U., 1953-56; faculty Laval U., 1956—, anthropology, 1963—. Fellow Am. Anthrop. Assn., Am. Sociol. Assn., mem. Soc. for Applied Anthropology, Canadian Assn. U. Tchrs., Agrl. Econ. Research Council Can. (chmn. 1965—), Assn. des Sociologues de langue français, Canadian Sociol. and Anthropol. Assn. (chmn. 1965—), Union Internationale des anthropol. (dir. 1966—). Author: (with others) People of Coye and Woodlot, 1960; (with Gérald Fortin) Les Comportements économiques de la famille salariée du Québec, 1964; Les Fondements sociaux de la maturation chez l'enfant, 1965; also numerous articles. Pioneered research in culture and mental health; study on Acadian culture of Maritimes Can., econ. behavior of French Canadian family. Home: 835 Nouvelle-Orléans, Ste. Foy, Que. 10, Can.*

TREMBLEY, Abraham, naturalist; b. Geneva, Switzerland, Sept. 3, 1710; s. Jean and Anne (Lullin) T.; ed. Geneva; math. degree; tutor to children of Count Bentrick, La Haye, Holland, also to young duke of Richemond; became asst. dir. library, Geneva, 1757. Fellow Royal Soc., 1943. Author: Mémoires pour servir à l'histoire d'un genre de polypes d'eau douce, 1744. Research on transmission of paternal characters in corn; discovered regenerating powers of hydra, 1740. Died Geneva, May 12, 1784.

TRENDELENBURG, Friedrich, German physician, surgeon; b. Berlin, Germany, May 24, 1844; s. Adolf Trendelenburg; student, Edinburgh, Glasgow, Scotland; M.D., Berlin, 1866; M.D. (hon.), U. Aberdeen. Asst. to Langenbeck, Berlin, 1868-74; became med. dir. surg. regulations Berlin State Hosp., 1874; named prof. surgery, Rostock, Germany, 1875, Bonn, 1882; became prof., Leipzig, 1895, also dir. surgery U. Clinic; head physician, State Hosp., St. Jacob, Germany, Introduced various surg. procedures; Trendelenburg's cannula, Trendelenburg's operation for treatment of Tb, Trendelenburg's operation for excision of varicose veins, 1890, Trendelenburg's gait caused by paralysis of gluteal muscles, Trendelenburg's test of incompetency of venous valves, Trendelenburg's position for surg. patients, 1890, all named after him; described surg. treatment of hydronephrosis, 1886; 1st attempt (unsuccessful) to excise pulmonary embolus, 1908; used endotracheal anesthesia by 1869. Died Berlin, Dec. 15, 1924.

TRENDELENBURG, Ullrich Georg, pharmacologist; b. Rostock, Germany, Dec. 31, 1922; s. Paul Georg and Vroni (Wilcken) T.; M.D., Göttingen U. (Germany), 1952; Ph.D., Oxford U., 1956; M. Christine Elisabeth Teschemacher, May 30, 1953; 1 dau., Marie Charlotte. Came to U. S., 1957, naturalized, 1965. Asst. prof. dept. pharmacology Mainz U. (Germany), 1956-57; faculty dept. pharmacology Med. Sch. Harvard, Boston, 1957—, asso. prof., 1963—, acting head dept. pharmacology, 1966—. Mem. Am. Soc. Pharmacology and Exptl. Therapeutics, German, Brit. pharmacological socs., Brit. Physiol. Soc. Contbr. numerous articles to profl. jours. Study of normal function of nerves which innervate structures which are not under our voluntary control, drugs acting on these nerves so as to obtain better understanding of why drugs may be helpful under certain conditions. Home: 30 St. Paul St., Brookline, Mass. 02146. Office: Dept. Pharmacology, Harvard Med. Sch., Boston 02115.

TRENNER, Nelson Richard, Am. phys. chemist; b. N.Y.C., Dec. 2, 1905; s. George L. and Ida R. Trenner; B.S., N.Y. U., 1929, M.S. (Inman fellow), 1930, Ph.D. (U. fellow), 1932; postgrad. (Nat. Research fellow), Princeton; m. Kathryn T. Farrell, June 24, 1939; children—Idamae, Kathryn, Georganna, Robert, Nelson Richard. Mem. staff Merck, Sharp & Dohme Research Labs. div. Merck & Co., Inc., Rahway, N.J., 1937—, now head biophysics sect. Recipient certificate of distinction award for contbns. to sci. N.Y. U. Merck Fgn. fellow Biochem. Inst., Upsala, Sweden, 1952-53. Mem. Sigma Xi, Tau Beta Pi, Phi Lambda Upsilon. Research, numerous publs. on gaseous reaction kinetics, catalysis with deuterium tracer techniques and reaction kinetics of molecular and atomic deuterium, application of isotope dilution prin. to analytical problems concerned with organic substances, phys. chem. prins. to structure elucidation of various biol. chems., phase rule prins. to problems chem. homogeneity, nuclear magnetic resonance spectroscopy to chem. and biophys. problems; use of displacement absorption chromatography in separation of adrenocortical steroids. Home: 656 Shadowlawn Dr., Westfield, N.J. 07090. Office: Merck Sharp & Dohme Research Labs., Lufberry St., Rahway, N.J. 07065.*

TREON, Joseph Frederick, Am. toxicologist; b. Aurora, Ind., Apr. 12, 1908; s. James F. and Margaret C. (Coleman) T.; A.B. with distinction (Dearborn County Ind. scholar), Ind. U., 1930; M.A., U. Cin., 1932, Ph.D., 1935; m. Gertrude S. Beaman, Aug. 4, 1935; 1 dau., Andrea Louise. Mem. faculty Kettering Lab., U. Cin., 1936-57; mgr. toxicology and biochemistry sect. Atlas Chem. Industries, Inc., Wilmington, Del., 1957—; cons. Manhattan project U. Rochester, 1943-45, Gen. Electric, Lockland, O., 1946-50. Trustee Indsl. Hygiene Found. Am. Mem. Am. Chem. Soc., Am. Indsl. Hygiene Assn., Soc. Toxicology, Am. Soc. Pharmacology and Exptl. Therapeutics, Research Soc. Am., Sigma Xi, Phi Beta Kappa, Phi Lambda Upsilon, Alpha Chi Sigma, others. Research, publs., chpts. in books; pioneered in the devel. of toxicological methods in inhalation studies at ambient conditions, at reduced pressures and of thermal decomposition products and dietary feeding studies to ascertain the safe levels of exposure of workers to indsl. chemicals and safe levels of food additives to the human diet; physiologic and metabolic studies on solvents, pesticides, food additives, polyols, surfactants, and drugs; co-discover of isosorbide oral osmotic diuretic. Home: 1600 Tudor Place, Wilmington 19803. Office: Atlas Chem. Industries, Inc., Wilmington, Del. 19899.*

TRESSAN, Count Louis Elisabeth de la Vergne de, see de Tressan, Count Louis Elisabeth de la Vergne.

TRESSLER, Willis Lattanner, Am. oceanographer; b. Madison, Wis., Sept. 30, 1903; s. Albert Willis and Arlouine (Dart) T.; student Colo. Coll., 1920-21, Occidental Coll., 1921-22; B.A., U. Wis., 1926, M.A., 1928, Ph.D., 1930; m. Eleanor Coombs, Nov. 5, 1927; children—Sara (Mrs. Hans Winkler), Sheila (Mrs. Richard E. Mellinger). Faculty, U. Buffalo, 1930-40, U. Md., 1940-46; with OSS, 1944-45, CIA, 1946-50; oceanographer U. S. Naval Oceanographic Office, Washington, 1950-65, dir. edn. and tng., 1962-65. Oceanographer, Can.-U. S. expedition to Beaufort Sea, 1954, USN Antarctic expedition, 1954-55, Operation Deep Freeze, 1956, 57, U. S. Antarctic Research Program, McMurdo, 1959-61; sta. sci. leader Wilkes Internat. Geophys. Year Sta., Antarctica, 1958-59. Mem. A.A.A.S., Am. Soc. Zoologists, Internat. Limnology Soc., Ecol. Soc. Am., Am. Micros. Soc., Am. Phycological Soc., Am. Soc. Taxonomy, Antarctic Soc. New Zealand, Antarctican Soc., Washington Acad. Sci., Sigma Xi. Research, numerous publs. on limnology of Wis., N.Y., Philippine lakes, plankton of Chesapeake Bay, ecology and taxonomy of marine and fresh water ostracoda, Arctic and Antarctic oceanography. Address: Star Route, Granby, Colo. 80446.*

TRETHEWIE, E. R., Australian biologist; b. Launceston, Tasmania, Australia, Apr. 15, 1913; s. Arthur John and Emily (Perkins) T.; M.B., B.S., Melbourne U., 1935, M.D., 1938, M.R.A.C.P., 1939, D.Sc., 1945; M.D. Adelaide U., 1944; m. May 25, 1940 (div.); children—Dougals, Robert, Rosemary, Rodney. Faculty, Melbourne U., 1941-44, 1950—, reader physiology, 1958—; staff Royal Melbourne Hosp., 1940-44, 1950—, asst. physician and demonstrator medicine, 1958—; prof. U. Adelaide, dir. Inst. Med. and Vet. Sch., 1944-50; lectr. various countries. Mem. Australian Research Council, 1946-54. Recipient Embley prize, 1938. Found. mem. World Soc. Toxinology; mem. Australian Med. Assn., Royal Australian Coll. Physicians. Author: Simplified Electrocariography, 1953; ABC Atlas of Electrocardiography, 1966. Home: 27 City View Rd., Melbourne, Victoria, Australia.*

TRETYAKOV, Yuri Dmitrij, Russian chemist; b. Rostov on the Don, USSR, Oct. 4, 1931; ed. Chem. Faculty, Rostov U.; D. Chemistry, Moscow U., 1965; m. Ludmila Ovsenjeva, Oct. 20, 1961; 1 son, Alexej. Research fellow Chem. Faculty, Moscow U., 1958-64, head lab. chemistry of defect states in solids, 1966—; research fellow Max-Planck Inst. for Phys. Chemistry, Göttingen, Germany, 1964-66. Recipient 1st Lomonossov prize Moscow U. Mem. Mendeleyev Soc. Chemistry. Author: (with others) Thermodynamics of Ferrites, 1967; also numerous articles. Research in chemistry and thermodynamics of solids in defect states; investigation of disorder in magnetic oxydes; new methods of synthesis of ferrites with desired properties. Home: Niclucka-Maclaj St. 52. Office: Chem. Faculty, Moscow U., USSR.*

TREUB, Melchior, Dutch botanist; b. Voorschoten, Netherlands, Dec. 26, 1851; ed. U. Leiden (Netherlands); doctorate in natural scis., 1873; asst., bot. inst., Leiden; dir. state bot. garden, Buitenzorg, Java; became dir. dept. agr., Dutch E. Indies, 1885. Fellow Royal Soc., 1899; fgn. mem. acads. Amsterdam, Paris, Berlin, Munich, Copenhagen. Author: Sur l'aigrette des composées, 1872; also memoires. Research in plant physiology. Died St.-Raphael, France, Oct. 3, 1910.

TREVAN, John William, English physiologist; b. Bodmin, Cornwall, Eng., July 23, 1887; s. J. W. S. Trevan; M.B., B.S., B.Sc., St. Bartholomew's Hosp., London, Eng.; m. Ida Kathleen Keys (dec. 1937); 3 sons, 2 daus.; m. 2d, Margaret Smith; 1 son, 2 daus. Demonstrator physiology, casualty physician St. Bartholomew's Hosp., 1914-20; pharmacologist Wellcome Physiol. Research Labs., Beckenham, Eng., 1920-40, dir., 1940-53. Cons. Wellcome Found. Ltd.; hon. lectr. U. Coll. London; mem. bd. studies pharmacology U. London. Fellow Royal Soc., 1946, Royal Coll. Physicians; mem. Royal Soc. Medicine (pres. therapeutics sect. 1937-38). Research, publs. on physiology, pharmacology. Died Oct. 13, 1956.

TREVES, Sir Frederick, Brit. surgeon; b. Dorchester, Eng., Feb. 15, 1853; s. William and Jane (Inight) T.; student Merchant Taylor's Sch., 1964-71, London, Hosp. Med. Sch., 1871-75; numerous hon. degrees; m. Anne Elizabeth Mason, 1877; 2 daus. House surgeon, London Hosp., 1875-76, asst. surgeon, 1879, surgeon, 1884, in charge practice teaching anatomy, 1881-84, lectr. anatomy, 1884-93; tchr. operative surgery, 1893-94, lectr., 1893-97; resident med. officer Royal Nat. Hosp. for Scrofula, Margate, Eng., 1876; began pvt. practice, Derbyshire, Eng., 1877; Hunterian prof. anatomy, Wilson prof. pathology Royal Coll. Surgeons, 1881-86; examiner surgery, Cambridge, 1891-96; cons. surgeon to forces in field, 1899; pres. Hdqrs. Med. Bd. War Office, World War I; rector U. Aberdeen (Scotland), 1905-08; became surgeon extraordinary to Queen Victoria, 1900; named sgt.-surgeon to Edward VII, 1902, George V, 1910. A founder, 1st chmn. exec. com. Brit. Red Cross. Recipient Jacksonian prize Royal Coll. Surgeons, 1883; fellow Royal Coll. Surgeons Eng. Licentiate Apothecaries Hall. Knighted, 1901. Author: Surgical Applied Anatomy, 1883; A Manual of Operative Surgery, 1891; The Students Handbook of Surgical Operations, 1892; A System of Surgery, 2 vols., 1895; Scrofula and its Gland Diseases, 1882; Intestinal Obstruction, its Varieties with their Pathology, Diagnosis and Treatment, 1899, 1902; The Anatomy of the Intestinal Canal and Peritoneum, 1885; Tale of a Field Hospital, 1900; The Other Side of the Lantern, 1903; Highways and By-ways of Dorset, 1906; The Cradle of the Deep, 1908; Uganda for a Holiday, 1910; The Land that is Desolate, 1912; The Country of the Ring and the Book, 1913; The Riviera of the Corniche Road, 1921; The Lake of Geneva, 1922; The Elephant Man and Other Reminiscences, 1923; Physical Education; Surgical Applied Anatomy; also articles. Editor: A Manual of Surgery, 3 vols., 1886. Research in scrofula, Tb, intestinal obstruction; devel. operative treatment for appendicitis; described fold of periteneum asso. with appendix (bloodless fold of Treves), 1885. Died Vevey, Switzerland, Dec. 7, 1923.

TREVES, Samuel Blain, Am. geologist; b. Detroit, Sept. 11, 1925; s. Samuel and Stella (Stork) T.; B.S., Mich. Technol. U., 1951; M.S., U. Ida., 1953; Ph.D., Ohio State U., 1959; m. Jane Patricia Mitoray, Nov. 24, 1960; 1 son, John Samuel. Geologist Ford Motor Co., 1951, Ida. Bur. Mines and Geology, 1952, Otago Catchment Bd., New Zealand, 1953-54; faculty U. Neb., Lincoln, 1958—, prof. geology, 1966—, chmn. dept., 1964. Fellow Geol. Soc. Am.; mem. Soc. Econ. Mineralogists and Paleontologists, Royal Soc. New Zealand, Mineral. Assn. Can., Inst. Polar Studies, Nat. Assn. Geology Tchrs., Am. Polar Soc., A.A.A.S., Sigma Xi, Tau Beta Pi, Sigma Gamma Epsilon. Research, publs. on geology of igneous and metamorphic rocks of Ida., New Zealand, Mich., Antarctica, Greenland, emphasis on origin of Precambrian granitic complexes and origin granitic rocks. Home: 1716 B St., Lincoln, Neb. 68502.*

TREVIRANUS, Gottfried Reinhold, German naturalist; b. Bremen, Germany, Feb. 4, 1776; s. Joachim Johann Jakob Treviranus; student medicine and math., Göttingen, 1793-96; doctorate, 1796. Prof. math. and medicine Bremen Lyceum, 1797; practice medicine, Bremen. Author: Biologie oder die Philosophie der Lebenden Natur, 6 vols., 1802-22; Erscheinugen und Gesetze des Organischen Lebens, 2 vols., 1831-33; also numerous articles. Founder of histology; used microscope extensively; attempted to unite natural scis. into biology; originated term botany; histological and anat. studies of vertebrates; developed concept of descent with modification by phys. influences which foreshadowed evolutionary theory; attempted to find genealogical relationships of organized species. Died Bremen, Feb. 16, 1837.

TREVIRANUS, Ludolf Christian, German naturalist; b. Bremen, Germany, Sept. 18, 1779; s. Joachim Johann Jakob Treviranus; M.D., Jena, Germany, 1801; m. Began practice medicine, Bremen, 1801; became 3d prof. medicine, Bremen Lyceum, 1807; became prof. natural history, Rostock, Germany, 1812, prof. botany, dir. Bot. Gardens, Breslau (now Wroclaw, Poland), 1816; named prof., Bonn, Germany, 1830. Recipient prize Göttingen Soc. Sci., 1806. Mem. French Acad. Scis., 1835. Author: Physiologie der Gewächse, 2 vols., 1835-38; Die Anwendung des Holzschnittes zur Biolichen Darstellung von Pflanzen, 1855. Discovered intercellular spaces; 1st description of effects of chem. agts. on plant life; studies in internal structure of plants. Died Bonn, May 6, 1864.

TREVITHICK, Richard, English engr., inventor; b. Illogan, Cornwall, Eng., Apr. 13, 1771; s. Richard and Anne (Teague) T.; ed. Camborne Sch.; m. Jane Harvey, Nov. 7, 1797; children—Richard, John Harvey, Francis, Frederick, Henry, Anne, Elizabeth. Engr. Ding Dong Mine, Penzance, Eng., from 1797, Pen-y Darran Ironworks, Merthyr Tydvil, Wales, from 1803; supt. silver mines on Cerro de Pasco, nr. Lima, Peru, 1817-27. Developed double-acting water pressure engine (thus perfecting vacuum engine), 1782; built 1st steam carriage to carry passengers (Puffing Devil or Captain Dick's Puffer) (proved practicability of high pressure engine), 1801; proved that smooth wheels on smooth rail produced enough traction to pull trains when he built 1st steam rail locomotive, 1804; built simpler locomotive for steam circus (circular ry.), 1808; improved steam dredger, 1803-07; erected 1st Cornish engine (type of high pressure steam engine), 1812, used several in Lima mines, 1814; invented 1st mech. rock-boring machine, 1813; improved steamboat propulsion, marine boilers, recoil gun carriage, heating apparatus for apartments, use of superheated steam. Died Dartford, Kent, Eng., Sept. 22, 1832.

TREW (or Treu), Abdias, German astronomer, mathematician; b. Anspach, Germany, July 29, 1597; ed. Tng. Coll. of Monastery of Heilsbronn; M.A., Wittenberg, Germany, 1621; m. twice; 22 children. Apptd. rector City Sch., Heilsbronn, Germany, 1625, also tchr., 10 years; named prof. math Nuremberg U. of Altdorf, Germany, 1636, also prof. physics, 1650. Author: Janitur lycae musici, 1635; Manuale geometriae practicae, 1636; A Short Mathematical Treatise on Fortification, 1640; Geodaesia universalis, 1641; Directorium mathematicum . . . , Compendium fortificationis, 1641; Astrologia medica nucleus astrologiae correctae, 1651; Ingenieur-Stab, 1649; A Practical Compendium of all Mathematical Knowledge, 1657; Methodus genethliaca . . . , 1663; Lehrbuch der sphärischen Astronomie Grundliche Calenerkunst in 2 Teilen verfasst, 1666. Studies in fortification, geometry, calender, astronomy; defended Aristotle's Physics, opposed Copernicus; reported on comets appearing in his lifetime; assisted in adoption of Gregorian calender. Died Altdorf, Apr. 12, 1669.

TREW (or Treu), Christoph Jacob, German botanist, physician; b. Lauf, Germany, Apr. 16, 1695; began study medicine, Altdorf, Germany, 1711; M.D., 1716; studied anatomy and botany under Jussieu, Paris. Went to Leyden, Netherlands, 1718; began practice medicine, Lauf, 1719, Nurenberg, Germany, 1720; physician in ordinary to Margrave of Ansbach, 1736. Hon. mem. London, Berlin acads. sci., Bot. Soc. Florence; mem. Royal German Acad. Sci. (dir.). Author: (with Vogel) Plantae selectae, 10 parts, 1750-60; Hortus Nitidissimus, 3 parts, 1750; Plantae rariores, 1763; Catalogus auctorum, 1750; Catalogus operum botanicorum, 1755. Co-editor, Commercium literarium physico technicum medicum, 1730, editor, 1734-45. Worked on collection of plants from Asia minor of Leonh-Rauwolff: publs. in anatomy, surgery, botany; 1st to distinguish between gymnosperms and angiosperms. Died Nurenberg, July 18, 1769.

TRIBUS, Myron, Am. aero. engr.; b. San Francisco, Oct. 30, 1921; s. Edward and Marie D. (Kramer) T.; B.S. in Chemistry, U. Cal. at Berkeley, 1942, Ph.D. in Engring., at Los Angeles, 1949; D.Sc., Rockford (Ill.) Coll., 1966; m. Sue Davis, Aug. 30, 1945; children—Louann, Kamala. From instr. to prof. engring. U. Cal. at Los Angeles, 1946-60; cons. heat transfer Gen. Elec. Co., 1950; dir. aircraft icing research U. Mich., 1951-54; dean engring. Thayer Sch. Engring., Dartmouth, 1961—; cons. engr., 1950—. Dir., Carpenter Steel Co., Reading, Pa., 1967—. Adviser NATO, 1953; chmn. Heat Transfer and Fluid Mechanics Inst., 1954. Recipient Thurman H. Bane award Inst. Aero. Scis., 1945, Wright Bros. medal Soc. Automotive Engrs., 1945, Alfred Noble prize Engring. Socs., 1952. Registered profl. engr., Cal., N.H. Mem. Am. Soc. M.E., Inst. Aero. Space Scis., Am. Soc. Engring. Edn. Author: Thermostatics and Thermodynamics, 1961; also numerous articles. Study of thermodynamics; heat transfer. Home: 7 Willow Spring Lane, Hanover, N.H. 03755.*

TRICAVELLI, Victor, Italian physician; b. Venice, Italy, 1496; ed. Padua, Bologna; revolutionized med. teaching at U. Padua by introducing study of Hippocrates. Died Venice, 1568.

TRICOMI, Francesco Giacomo, Italian mathematician; b. Naples, Italy, May 5, 1897; s. Arturo and Corinna (Di Lustro) T.; Dr.math., 1918; m. Susanna Fomm, 1931. Asst., instr. U. Padua, U. Rome, 1921-24; became prof. U. Florence, U. Turin, 1925; head mission Cal. Inst. Tech., Pasadena, 1948-51. Recipient Gold medal of merit in edn., 1957; gold medal Nat. Acad. XL, 1956; Feltrinelli prize dei Lincei Acad. Mem. dei Lincei, Acad. Bologna (corr.), Bayerische Akedemie der Wissenschaft. Author numerous publs. and articles. Equation of transsonic aerodynamics named after him. Home: corso Tassoni 34, Turin. Office: Università, Turin, Italy.

TRIESNECKER, Francis a Paula, Austrian astronomer; b. Kirchberg, Austria, Apr. 2, 1745; ed. Vienna, Tyrnau; became mem. Soc. Jesus, circa 1761; tchr. Jesuit colls.; after suppression of Soc., went to Graz, Austria, where he was ordained; asst. dir. Vienna obs., dir., 1792-1817. Author: Novae motuum lunarium tabulae, 1802; also treatises contbd. to astron. jours. Editor Ephemerides (in which he pub. his Tabulae Mercurii, Martis, Veneris, Solares, and observations of heavenly bodies 1787-1806). Measured triangulation of Galicia and Lower Austria. Died Vienna, Jan. 29, 1817.

TRILLING, Leon, areo. engr.; b. Bialystok, Poland, July 15, 1924; s. Oswald and Regina (Zakhejm) T.; came to U. S., 1940, naturalized, 1946; B.S., Cal. Inst. Tech., 1944, M.S., 1946, Ph.D., 1948; m. Edna Yuval, Feb. 17, 1946; children—Alex R., Roger S. Research fellow Cal. Inst. Tech., 1948-50; Fulbright scholar U. Paris, 1950-51, vis. prof., 1963-64; faculty Mass. Inst. Tech., Cambridge, 1951—, prof. aeros. and astronautics, 1962—; cons. McGraw-Hill, Raytheon. Mem. adv. com. Air Force Office Sci. Research; mem. Engring. Edn. Mission to Soviet Union, 1958. Pres. Met. Com. Ednl. Opportunity. Guggenheim fellow, 1963-64. Asso. fellow Am. Inst. Aeros. and Astronautics; mem. A.A.A.S. Studies in aerodynamics; Gas surface interactions and dynamics; kinetic theory of gases. Home: 89 Mason Terrace, Brookline, Mass. 02146. Office: Mass. Inst. Tech., Cambridge, Mass. 02139.*

TRIMMER, John Dezendorf, Am. physicist; b. Washington, Sept. 19, 1907; s. Daniel K. and Louise (Dezendorf) T.; A.B., Elizabethtown Coll., 1926, Sc.D. (hon.), 1953; M.S. in Physics and Math., Pa. State U., 1933; Ph.D. in Physics, U. Mich., 1936; m. Mildred L. Ebersole, Dec. 31, 1930; children—Daniel Ross, Maud Alice. High sch. tchr., 1927-32; tchr., research aero. engring. dept. Mass. Inst. Tech., 1937-41, underwater sound research, 1941-43; engaged in electro-magnetic isotope separation, Oak Ridge, 1943-46; prof. physics U. Tenn., 1946-57, also cons. automatic control nuclear reactors and wind tunnels; prof. physics, head dept. U. Massachusetts, 1957-63, professor of physics, 1963-66; prof., head dept. physics Washington Coll., 1966—. Fellow A.A.A.S.; mem. Am. Phys. Soc., Acoustical Soc. Am., N.Y. Acad. Scis. Author: Response of Physical Systems, 1950; also articles. Research in acoustics, instrumentation, cybernetics, physics of ionized fluids; applied principles of instrumentation and automatic control to wind tunnels and to nuclear reactors. Home: 116 Cedar St., Chestertown, Md. 21620.*

TRIMMER, Joshua, English geologist; b. North Cray, Eng., July 11, 1795; s. Joshua Kirkey Trimmer; studied under William Davison, curate of New Brentford. Sent to manage father's copper mine in N. Wales, 1814; in charge farm, Middlesex, Eng.; became overseer slate quarries nr. Bangor and Carnarvon, Wales, 1825; joined Geol. Survey Eng., 1840. Fellow Geol. Soc. Author: Practical Geology and Mineralogy, 1841; also articles. Discovered sands with marine fossils of existing species 1,350 ft. above sea level; work in classification of superficial deposits of earth's crust; drainage, planting, other aspects of agr. Died London, Sept. 16, 1857.

TRINKAUS, John Philip, Am. cell and developmental biologist; b. Rockville Centre, L.I., N.Y., May 23, 1918; s. Charles Edward and Fransiska (Krueger) T.; B.A., Wesleyan U., Conn., 1940; M.A., Columbia, 1941; Ph.D., Johns Hopkins, 1948; m. Madeleine Francine Marguerite Bazin, Oct. 6, 1963; children—Gregor, Tanya, Erik. Faculty, Yale, 1948—, prof. cell and developmental biology, 1964—; fellow Branford Coll., 1953—, master, 1966—, dir. grad. studies in biology, 1965-66; staff embryology Marine Biol. Lab., Woods Hole, Mass., 1953-57. Panelist, Commn. on Undergrad. Edn. in Biol. Scis., 1965—; cons. biology Thomas Y. Crowell Book Co., 1965—. John Simon Guggenheim Meml. fellow Coll. France, Paris, 1959-60; cited by Wesleyan U., 1960. Mem. Am. Soc. for Cell Biology, Am. Soc. Naturalists, Am. Soc. Zoologists, Internat. Inst. Embryology, Soc. for Developmental Biology, Tissue Culture Assn. Author: Cell Movements and Morphogenesis, 1968; also articles. Asso. editor Jour. Exptl. Zoology, 1964—. Research on cell adhesion, cell movement in multicellular systems; cellular differentiation and maintenance of differentiation. Home: 80 High St., New Haven.*

TRITHEMIUS, Johann (or Tritheim, Trittenhemius), German alchemist; b. Trittenheim, near Trier, 1462; s. Johann Heidenberg; went to Trier, then Heidelberg to study. Joined Benedictine order; became abbot, 1483, at Heidelberg; abbot at Würzburg, 1506. Author: Trithemii de Sponheim, Abts zu Kreutzburg, Güldenes Kleinod, oder: Schatzkästlein, 1482; Steganographia. Wrote commentaries, sermons, epistles and histories. Died Würzburg, 1519.

TRLIFAJ, Ladislav, Czechoslovakian physicist; b. Nove Zamky, Czechoslovakia, June 8, 1925; s. Josef and Ruzena (Mrazova) T.; R.N.Dr., Charles U., 1950; candidate phys. and math. sci. Leningrad State U., 1953; D.Sc., Tech. U. Prague, 1963; m. Jarmila Pekarkova, Aug. 8, 1951; children—Jan, Zora. Phys. research Skoda Works, Prague, 1949-50; head theoretical dept. Phys. Inst., Czechoslovakian Acad. Scis., Prague, 1954-55, head theoretical dept. Nuclear Physics Inst., 1955-58, head reactor physics and techniques div., 1958-61, head nuclear physics div., Rez, 1964—; physicist Niels Bohr Inst., Copenhagen, 1962-63; lectr. J. A. Komensky U., Bratislava, 1958; asso. prof. Tech. U. Prague, 1961—. Mem. sci. council Joint Inst. for Nuclear Research, Dubna, 1965. Research, publs. on theoretical and applied problems in meson field theory, neutron transport theory, reactor theory, theory of atomic nuclei. Home: 79/279 Cimicka, Prague. Office: Nuclear Research Inst., Rez, Czechoslovakia.*

TRNAVSKY, Karel, Czechoslovakian pharmacologist, rheumatologist; b. Brno, Czechoslovakia, Dec. 6, 1930; s. Karel and Anna (Stastná) T.; M.D., U. Palacky, Olomouc, Czechoslovakia, 1955, Ph.D., 1961; m. Zdena Grmelová, Feb. 27, 1955. Research fellow dept. medicine Med. Sch., Olomouc, 1953-55; research asst. Research Inst. for Rheumatic Diseases, Piestany, 1955-61, sr. researcher, 1961-62, head sect. on exptl. therapy, 1962—; lectr. Postgrad. Sch. Medicine, Bratislava, Czechoslovakia. Mem. European Soc. Biochem. Pharmacology. Author: (with I. Skidmore) Anti-inflammatory Drugs, 1967; also numerous articles. Research on pharmacology of anti-rheumatic drugs including their influence on biochemistry of connective tissue and description of new properties. Home: 923 Kúpelny ostrov, Piestany. Office: Research Inst. for Rheumatic Diseases, Piestany, Czechoslovakia.*

TROELTSCH, Ernst, German sociologist; b. Augsburg, Germany, 1865; s. physician. Studied theology at Erlangen and Göttingen (grad. 1891). Lectr. in Göttingen, 1891; prof. theology in Bonn, 1892-94; prof. U. Heidelberg for 21 years until apptd. prof. philosophy at U. Berlin, 1915-23; elected to Prussian Diet, 1919; acted as parliamentary undersec. to Prussian minister edn. Author: Protestantism and Progress, 1906; Gesammelte Schriften (of which vol. I was translated as The Social Teaching of the Christian Churches). Made landmark contbn. to sociology of religion; analyzed how society and religion developed independently of one another and interrelatedly. Died Berlin, Feb. 1, 1923.

TROISIER, Emile, French physician; b. Sevigny, Ardennes, France, 1844. Hosp. staff physician, agrégé U. Paris. Mem. Acad. Medicine. Described enlargement of lymph node asso. with abdominal tumors, 1886 (known as Virchow-Troisier node or gland); research on typhoidal meningitis, cancer of lymphatic passages. Died Paris, 1919.

TROLL, Carl Theodor, German geographer; b. Gabersee, Bavaria, Dec. 24, 1899; s. Theodor and Elizabeth (Hufnagel) T.; Dr.phil., U. Munich (Germany), 1921, postgrad.; Dr.sc.h.c., U. Louvain (France), 1963; Dr.phil.h.c., U. Vienna (Austria), 1965; m. Elisabeth Kürschner, Sept. 15, 1930; children—Irene (Mrs. Kristen Rohlfs), Georg, Maria Theresa (Mrs. Dieter Strauss), Christian, Valentin, Liselotte (Mrs. Ludwig Bley), Margarete (Mrs. Peter Strauss), Richard, Franz. Dozent, U. Munich, 1925-30; faculty U. Berlin (Germany), 1930-38, prof., 1936-38; prof., dir. U. Bonn (Germany) Geog. Inst., 1938-66, prof. emeritus, 1966—. Recipient Gold medal Halle-Leopoldina, 1938, Berlin, 1959, Nuremberg, 1959; Vega medal, Sweden, 1953; Victoria medal, London, 1962; also numerous silver medals. Mem. Internat. Geog. Union (past pres. v.p. 1956-60, 64—). Author numerous books including: Pleistocene Geology and Climate, 1944; Die Tropischen Gebirge, 1959; Ecology of Landscape and High Mountain Research, 1966; Luftbildforschung, 1966; Die Entwicklungsländer, 1966; also numerous articles. Explorations in high mountain regions of S. Am., Africa, and Himalayas; comparative studies of climates, vegetation, frost soils, and pleistocenegeomorphology of world in three dimensions; interpretation of air photographs; founds. of landscape ecology, periglacial geology. Home: 55 Rheinbacherstr., Bonn 53, Germany.*

TRÖLTSCH, Baron Anton Friedrich von, see von Tröltsch, Baron Anton Friedrich.

TROMMSDORFF, Johann Bartholomaus, German chemist; b. Erfurt, May 8, 1770; trained as apothecary under Buchholz. Pub. his Journal de Pharmacie, 1794-1834; prof. chemistry and physics at Erfurt, 1795; founder Pharm. Inst., 1795. Mem. several acads. Author: Systematische Handbuch der Pharmazie, 1792; Chemische Receptirkunst oder Taschenbuch für practische Aerzte welche bey dem Verordnen der Arzneyen Fehler in chemischer und pharmacevtischer Hinsicht vermeiden wollen, 1797; Systematische Handbuch der geschichtlichen Chemie, 8 vols., 1805-07. Helped improve sci. position of pharmacy; used electrolysis to obtain ammonium amalgam. Died Mar. 8, 1837.

TRONCHIN, Théodore, physician; b. Geneva, Switzerland, May 24, 1709; pupil of Boerhaave at U. Leiden (Netherlands); doctorate, 1730; practiced medicine, Amsterdam, 1730-50, Geneva, from 1750; named hon. prof. medicine, Geneva, 1750; named physician to Duke of Orleans, 1766; presided over Coll. Physicians, Amsterdam; mem. Royal Acad. Physicians, Paris; fgn. asso. French Acad. Scis., 1778. Pioneer in innoculation and vaccination against smallpox with smallpox bacteria. Died Paris, Nov. 30, 1781.

TROOST, Gerard, mineralogist; b. Bois-le-Duc, Holland, Mar. 15, 1776; s. Everhard Joseph and Anna Cornelia (van Haeck) T.; M.D., U. Leyden; Master in Pharmacy, U. Amsterdam, 1801; pupil of Rene Haüy in mineralogy, crystallography; m. Margaret Tage, Jan. 14, 1811; m. 2d, Mrs. O'Reilly; 2 children. Collected minerals for Cabinet of King of Holland, 1807-09; apptd. mem. Dutch scientific commn. to Java, 1809; corr. mem. Museum of Natural History of France, Paris; arrived in Phila., 1810; established a pharm. and chem. lab., Phila., 1812-17; prof. mineralogy Phila. Museum, 1821; prof. pharm. and gen. chemistry Phila. Coll. Pharmacy, 1821-22; prof. mineralogy, chemistry U. Nashville (Tenn.), 1828-50; state geologist Tenn., 1831-50; his meteorite collection now at Yale; mem. Am. Philos. Soc., Geol. Soc. Pa., Amassed important geology and natural history collection, including items from archeology of Am. mound Indians; important research on fossil crinoids in Tenn. pub. 1909. Died Nashville, Aug. 14, 1850.

TROOST, Louis Joseph, French chemist; b. Paris, Oct. 17, 1825; student École normal, 1848-51; agrégé de l'Université, D.Sc., 1857; tchr. in provinces; prof. chemistry Lycée Bonaparte, Paris; prof. chemistry Sorbonne, 1874-1900. Mem. French Acad. Scis. (v.p. 1904, pres. 1905). Author: Recherches sur le lithium et ses composés, 1857; Un laboratoire de chimie au XVIIIe siècle, 1866; Précis de chimie, 40 edits., 1912. Research on metal alloys formed with hydrogen by combination or decomposition, especially iron, nickel cobalt, manganese; discovered that hexagonal zinc sulphide gives out rays that pass through black paper; studied porosity of metals at high temperatures, also solubility of gases in metals, effects of silicon and manganese on comml. irons, synthesis of minerals and dissociation of compounds at high temperatures, properties of zirconium and lithium; produced zirconium by reducing its halides with aluminum, iron or sodium; eponym of troostite (solid solution of carbon in iron); did considerable work with Deville. Died Paris, Sept. 30, 1911.

TROSHIN, Afanasii Semenovich, Russian cytologist; b. Alza, Mordvinian USSR, 1912; grad. Leningrad State U., 1936; became postgrad. Physiol. Inst. of U., 1940. Asso., All-Union Inst. Exptl. Medicine, 1940-41, 46-50; staff Inst. Oncology, USSR Acad. Med. Scis., 1950-51, Inst. Zoology, 1951-57, became lab. supr. cell physiology, Inst. Cytology, 1957, dir., 1958. Recipient Order Red Star. Corr. mem. USSR Acad. Scis. Author: Cell Permeability, 1956; The Creative Career of Dimitriy Nikolaevich Nasonov, 1959; Principle Tasks of Cytology, 1961; Micrurgy, its Potentials and Tasks, 1962; Das Problem der Zellpermeabilitat, 1958. Became chief editor Cytology. Research on cell permeability, bioelectric phenomena, radioactive calcium, and phosphorus in insects and fish, water content of protoplasm. Office: Inst. Cytology, USSR Acad. Scis., Prospekt Maklina 32, Leningrad F-121, USSR.

TROSMAN, Harry, psychoanalyst; b. Toronto, Ont., Can., Dec. 9, 1924; s. Samuel and Esther (Sherman) T.; M.D., U. Toronto, 1948; postgrad. Chgo. Inst. for Psychoanalysis; m. Marjorie Susan Goldman, June 22, 1952; children—Elizabeth, Michael, David. Came to U. S., 1948, naturalized, 1953. Faculty, U. Chgo., 1952—, asso. prof., 1959—; dir. psychiatry clinic U. Chgo., 1956—; faculty Chgo. Inst. for Psychoanalysis, 1965—. Cons. Chgo. Police Dept., 1960—, Ill. State Psychiat. Inst., 1962—. Recipient Franz Alexander award Chgo. Inst. for Psychoanalysis, 1965. Mem. Am. Psychiat. Assn., Ill. Psychiat. Soc., Am. Internat. psychoanalytic assns., Chgo. Psychoanalytic Soc. Author: The Student Physician as Psychotherapist, 1962; also articles. Research on dreams, application of psychoanalytic knowledge and theory to history, lit. and aesthetics.*

TROTMAN-DICKENSON, Aubrey Fiennes, Brit. chemist; b. Wilmslow, Cheshire, Eng., Feb. 12, 1926; s. Edward Newton and Violet (Nicoll) T.-D.; B.A., Balliol Coll., Oxford (Eng.) U., 1947, B.Sc., 1948; Ph.D., U. Manchester (Eng.), 1952; D.Sc., Edinburgh (Scotland) U., 1958; m. Danusia Hewell, July 11, 1953; children—Casimir, Beatrice, Dominic. Postdoctoral fellow NRC, Ottawa, Ont., Can., 1948-50; asst. lectr. Manchester U., 1950-53; exptl. officer E. I. DuPont de Nemours & Co., 1953-54; lectr. Edinburgh U., 1954-60; prof. chemistry U. Coll. Wales, Aberystwyth, 1960—, Author: Gas Kinetics, 1955; Free Radicals, 1959; (with G. D. Parfitt) Chemical Kinetics and Surface Chemistry, 1965; also numerous articles. Research on kinetics of free radical reactions in gas phase, kinetics of unimolecular reactions and reactions of chemically activated molecules. Home: 1 Lon Hendre, Aberystwyth, Cards, U.K.*

TROTSKY, Leon (original name, Lev Davidovich Bronstein), Russian political theorist; b. near Elisavetgrad (now Kirovograd), Ukraine, 1879; s. David Bronstein. Joined Social Democratic movement, 1896; arrested and exiled to Siberia, 1898; escaped to London, 1902, where he became associated with Lenin, Plekhanov and other Marxian socialists; wrote for Iskra, Lenin's paper; mem. 2nd Congress, Menshevik until 1904; returned to Russia, 1905; exiled and escaped to Vienna, where he became editor of Pravda; went to Zurich, then Germany, 1914, and France, 1915; in Paris, edited Nashe Slovo, expelled from France for denunciations of war, 1916; to New York, 1917, editor of Novy Mir; sailed for Russia when revolution broke out, March 1917, taken off ship at Halifax by Brit. and interned in Canada, later allowed to proceed to Petrograd; founded Vperiod; arrested by Kerenski's provisional govt., soon released; joined Bolsheviks at 6th Congress and took leading part in organizing and leading October Revolution; chmn. Military Revolutionary Committee; mem. Central Exec. Committee, Communist Party, 1917-27; People's Commissar for Fgn. Affairs, 1817-18; leading delegate to conference at Brest-Litovsk, created and organized Red Army, 1918-25; pres. Military Revolutionary Sov. of Rep. and People's Commissar for Transport, 1920-21; mem. Exec. Committee, Communist Internat.; expelled from Politburo and Communist Party for opposition to Stalin and official party policy, 1927; exiled to Alma-Ata, Kazakstan, 1927; deported, 1929; lived abroad; assassinated in Mexico City. Author: Defense of Terrorism, 1920; Lenin, 1924; Literature and Revolution, 1925; My Life, 1930; History of the Russian Revolution, 3 vols., 1932; The Revolution Betrayed, 1937; The Stalin School of Falsification, 1937. Co-father with Lenin of Russian Oct. 1917 Revolution and principal leader in founding of U.S.S.R.; believed in internat. "permanent revolution" as opposed to Stalin's nationalism; urged reliance upon peasantry as well as industrial worker in growth of socialism; advocated govt. based upon worker representation as opposed to Stalinist bureaucracy. Died Mexico City, Mexico, Aug. 21, 1940.

TROTTER, James, chemist; b. Dumfries, Scotland, July 15, 1933; s. James and Annie (Burns) T.; B.Sc., U. Glasgow, 1954, Ph.D., 1957, D.Sc., 1963; m. Ann White Ramsay, Sept. 14, 1957; children—Martin James, David Mark. Asst. lectr. U. Glasgow, 1954-57; NRC fellow, Ottawa, Can., 1957-59; Imperial Chem. Industries research fellow U. Glasgow, 1959-60; faculty U. B.C., Vancouver, Can., 1960—, prof., 1965—. Meldola medallist Royal Inst. Chemistry, 1962; Alfred P. Sloan fellow, 1965—. Fellow Royal Inst. Chemistry, Chem. Inst. Can. (chmn. Vancouver sect. 1965-66); mem. Chem. Soc. (London), Chem. Inst. Can., Am. Crystallographic Assn. Co-editor: Structure Reports, 1962. Research, publs. on determination of structures of crystals and molecules. Home: 4035 W. 27th Av., Vancouver 8, B.C., Can.*

TROTTER, Mildred, Am. anatomist; b. Monaca, Pa., Feb. 3, 1899; d. James Robert and Jennie Bruce (Zimmerly) Trotter; A.B., Mount Holyoke College, 1920; M.S. Washington U., 1921, Ph.D., 1924; Nat. Research Council fellow, Oxford U., 1925-26; D.Sc. (honorary), Western College for Women, 1956, Mount Holyoke Coll., South Hadley, Mass., 1960. Member faculty, Washington U., St. Louis, since 1920, prof. gross anatomy, 1946-58, prof. anatomy, 1958-67, prof. emeritus, lectr., 1967—; vis. prof. anatomy Makerere U. College, Kampala, Uganda, E. Africa, 1963; cons. Rockefeller Found., 1963; special cons. USPHS, 1943-45; anthropologist, Schofield Barracks, Hawaii, U. S. Department of Army, 1948-49, Fort McKinley, The Philippines, 1951. Recipient Viking Fund medal phys. anthropology, 1956. Mem. Am. Assn. Anatomists, Am. Assn. Phys. Anthrop. (pres. 1955-57), Anatomical Society of Great Britain and Ireland, American Anthrop. Assn., Mo. State Anat. Bd. (president 1957-67), Anat. Board of St. Louis (pres. 1941-48, 49-67,), Phi Beta Kappa, Sigma Xi. Contbr. chpts. to books and articles to med. and biol. jours. Contbn. to quantification human anatomical variation; studied variation in distbn. and quantity of hair; interracial variation between length of bones and total stature; race, sex and age variation effects of skeletal weight. Home: 18 S. Kingshighway Blvd., St. Louis 63108. Office: 4580 Scott Av., St. Louis 63110.*

TROTTER, Wilfred Batten Lewis, English surgeon, physiologist; b. Coleford, Gloucestershire, Eng., Nov 3, 1872; s. Howard Birt and Frances (Lewis) T.; grad. Univ. Coll., London, Eng., 1899; M.S., M.D., Univ. Colls. Hosps.; LL.D. (hon.), U. Edinburgh (Scotland); D.Sc. (hon.), U. Liverpool (Eng.); m. Elizabeth May Jones, 1910. Surg. registrar Univ. Coll. Hosp., 1901-04, surgeon, 1906-39, prof., dir. surg. unit from 1935; demonstrator anatomy Univ. Coll., 1904-06; asst. surgeon E. London Hosp. Children, 1906; hon. surgeon to the King, 1929-32, sergeant surgeon, 1932-39; Hunterian prof., 1913. Mem. Med. Research Council, 1929-33. Fellow Royal Soc. (v.p.), 1931, Royal Coll. Surgeons (mem. council, v.p.; Hunterian orator 1922). Author: Instincts of the Herd in Peace and War, 1916; Collected Papers of Wilfred Trotter, 1941; also papers on surgery. First important expositions of herd psychology, 1908-09; essays influenced social psychology; studied human cutaneous sensation; work on advanced brain, spinal cord and malignant tumor surgery; defined structure of afferent nervous system. Died Blackmoor, Hampshire, Eng., Nov. 25, 1939.

TROTTI, Leopoldo, Italian physicist; b. Perugia, Italy, Aug. 2, 1906; s. Paolo and Cesira (Bartoccini) T.; Liber Docente in Comparative Anatomy, 1949, Liber Docente in Oceanography, 1952; m. Lina Galleani, 1948; 1 son, Paolo. Norsk Polar Inst. grantee, Oslo, Norway, 1953-54; prof. oceanography U. Genoa (Italy), 1954-62; dir. Istituto Sperimentale Talassografico, Trieste, Italy, 1963—; research prof. Scripps Inst. Oceanography, La Jolla, Cal., 1955-56; vis. prof. oceanography U. Cape Town (S. Africa), 1959-61; lectr. oceanography U. Trieste, 1964—. Mem. Internat. Commn. for Exploration of Sea, 1963—; mem. Coordinating Com. for S. Africa, Research, Pretoria, 1960-65. Mem. Consiglio Nazionale delle Ricerche Rome, Royal Soc. S. Africa, Academia Scienze e Lettere Ligure, Com. Internat. for Exploration of Mediterranean. Research, publs. on marine biology, phys. oceanography, submarine geology. Home and office: 2 via Gessi, Trieste, Italy.*

TROTULA (Trotula di Ruggiero), physician; b. circa 1050; m. Joannes Platearius the Elder. Woman physician and med. tchr., Salerno, Italy; possibly author (or work may have been named after her): De mulierum passionibus et eorum cura (also called The Trotula), first book on obstetrics written in Western Europe by Christian author, deals with conduct of labor, care of newborn, feeding of infants, treatment of prolapse of uterus and uterine polyps, also treats of epilepsy and disease of teeth and gums; work was highly regarded and widely read for almost 500 years after author's death.

TROUESSART, Edouard-Louis, French naturalist; b. Angers, France, 1842; M.D.; dir. Angers mus.; prof. Mus. Natural History. Author: Historie naturelle de France: Mammifères, 1885; Catalogue des mammifères vivants et fossiles, fascicule 4, carnivores, 1888; Les micorbes, les ferments et les moisissures, 2d edit., 1890; (with P. Megnin) Les sarcoptides plumicoles or analgésinés, Ire partie, les ptérolichés; Geographie zoologique, 1890. Died Paris, 1927.

TROUG, Emil, Am. agriculturist; b. Independence, Wis., Mar. 6, 1884; B.S., U. Wis., 1909, M.S., 1912; married 1925; three children. Mem. faculty U. Wis., 1909—, prof. soils, 1921-52, head dept., 1939-52, emeritus prof., 1952—. Fellow Am. Soc. Agronomy (pres. 1938); hon. mem. Internat. Soc. Soil Sci.; mem. Am. Chem. Soc., Soil Sci. Soc. Am. (pres. 1954). Author numerous papers on soil analysis, fertilizing. Cons. editor Soils Sci. Specialist in study soil; work on soil acidity, phosphorus, potassium, colloids; methods soil analysis and fertilizing; prosphate availability; ceramic clay processing; ability of plants to feed. Address: 1108 Grant St., Madison, Wis.

TROUGHTON, Edward, English instrument maker; b. Corney, Eng., Oct. 1753; s. Francis Troughton. Apprenticed to brother, 1770; became partner until brothers death; then independent work until 1826; became partner with William Simons, 1826. Recipient Copley medal, 1809; Gold medal King of Denmark, 1830. Fellow Royal Soc, 1810; mem. Royal Astron. Soc. (charter). Contbg. author: Edinburgh Cyclopedia (Brewster). Publs. on descriptions of his instruments; built double-framed sextant, 1788, Brit. reflecting circle, 1796, 1st modern transit circle, 1806, dip sector, beam compass, hydrostatic balance; improved marine and mountain barometers, portable universal dial and pyrometer, snuff-box sextant, compensated mercurial pendulum, marine top; built equipment which Baily used to restore standard yard; built instruments for Am. Coast Survey, 1815; developed new method for graduating arcs of circles, 1778. Died Jan. 12, 1835.

TROUSSEAU, Armand, French physician; b. Tours, France, Oct. 14, 1801; student Coll. Lyons (France); Agrégé, Faculté de Paris, 1826; studied under Bretonneau, Tours, France. Prof. rhetoric, Chateauroux, France; tutor Coll. Blois; became doctor hosps., 1830; became prof. therapeutics, 1839; named head chair of med. clinic Hôtel-Dieu, Paris, 1852. Mem. French Acad. Medicine. Author: Du tubage de la glotte, 1859; Cliniques médicales de l'Hôtel-Dieu de Paris, 1861; (with Pidoux) Traité de thérapeutique. Showed compression of upper arm in tetany causes carpal spasm (Trousseau's phenomenon or sign), 1861; 1st description of hemochromatosis (Trousseau's disease), 1865; discovered meningeal streak (streak produced by drawing nail across skin) in various nervous or cerebral diseases. Died Paris, June 27, 1867.

TROUT, Edrie Dale, Am. radiation physicist; b. Franklin, Ind., Nov. 3, 1901; s. Claud and Nancy (Heck) T.; B.S., Franklin Coll., 1922, D.Sc. (hon.), 1951; m. Thomasina Watson, June 19, 1946. Tchr. high schs., Charleston, Ill., 1922-24, Atwood, Ill., 1924-28; with x-ray dept. Gen. Elec. Co., Chgo., Milw., 1928-62, cons. radiation physicist; prof. radiol. physics Ore. State U., Corvallis, 1962——. Jerman lectr. Soc. Radiologic Technologists, 1959. Hon. fellow Am. Coll. Radiology; mem. Am. Phys. Soc., Am. Roentgen-Ray Soc., Health Physics Soc., Sigma Xi. Investigations into x-ray protection, med., dental and indsl. applications of x-ray. Home: 3630 Chintimini Dr., Corvallis, Ore. 97330.*

TROWBRIDGE, Arthur Carleton, Am. geologist; b. Glasgow, Mo., Mar. 4, 1885; s. Samuel Hoyt and Julia (Goodhue) T.; B.S., U. Chgo., 1907, Ph.D., 1911; L.H.D., Augustana Coll., 1963; m. Susan Estelle Bussey, Aug. 29, 1911; children—Charles Lambert, Carolyn Frances (Mrs. Leslie E. Edwards). Prof. geology U. Ia., Iowa City, 1911-52, prof. emeritus, 1952——, head dept., 1934-52; geologist Turkish Petroleum Co. Ltd., Iraq, 1925-26; cons. geologist Gulf Research & Devel. Co., Houston, 1952-55. Chmn. com. on sedimentation NRC, 1931-34. Recipient Neil A. Miner award Nat. Assn. Geology Tchrs., 1960. Mem. A.A.A.S., Am. Assn. Petroleum Geologists (hon.), Am. Geol Inst. (editor Glossary on Geology and Related Scis. 1955-56), Geol. Soc. Ia. (1st Hon. Membership award 1960), Geol. Soc. Am., Ia. Acad. Scis., Am. Geophys. Union, Am. Assn. State Geologists, Soc. Econ. Paleontologists and Mineralogists. Research, numerous publs. on Ia. geology, glacial geology, The Driftless Area, Gulf Coast geology; a founder sedimentology. Home. 325 Morningside Dr., Iowa City 52240.*

TROWBRIDGE, Augustus, Am. physicist; b. New York, Jan. 2, 1870; s. George Alfred and Cornelia Polhemus (Robertson) T.; Phillips Acad., 1886-87; Columbia U., 1890-93, D.Sc., 1929; A.M., Ph.D., U. of Berlin, 1898; m. Sarah Esther Fulton, Sept. 20, 1893. Instr. physics, U. of Mich., 1898-1900; asst. prof. physics, 1900-03, prof. 1903-06, U. of Wis.; later prof. physics, Princeton; dean of Grad. Sch. Princeton, 1928——, Mem. Internat. Congress Applied Chemistry, Berlin, 1903; sec. Physics Sect., Internat. Congress Arts and Sciences, St. Louis, 1904. Trustee Princeton U. Press. Officer Legion of Honor (France); Knight of Order of St. Olav (Norway). Member National Acad. Sciences (chmn. div. physics, 1921); chmn. div. physical sciences and mem. research fellowship bd., Nat. Research Council, 1920-21, and mem. research fellowship bd. in physics and chemistry, 1920-25; European dir. for Science Internat. Edn. Bd.; fellow A.A.A.S. Research radiations, theory of coherer; developed devices for locating and finding range arty. batteries. Died Taormina, Sicily, Mar. 14, 1934.

TROWBRIDGE, John, Am. physicist; b. Boston, Aug. 5, 1843; s. John Howe and Adeline T.; S.B., Lawrence Scientific Sch. (Harvard), 1865, S.D., 1873; m. Mary Louise Thayer, 1875. Tutor Harvard, 1866-69; asst. prof. physics, Mass. Inst. Tech., 1869-70; asst. prof. physics, 1870-80, prof., 1880-88, Rumford prof. applied science, 1888——, dir. Jefferson Physical Lab., 1884——, Harvard. Asso. editor Am. Jour. of Science. Mem. Internat. Congress of Electricians, Paris, 1883; del. U. S. Congress of Electricians, Phila., 1884. Mem. Nat. Acad. Sciences, Am. Acad. Arts and Sciences (pres.). Author: The New Physics, 1884; The Electrical Boy, 1891; Three Boys on an Electrical Boat, 1894; What Is Electricity?, 1896; The Resolute Mr. Pansy, 1897; Philip's Experiments in Electrical Science, 1898. Research on conduction electricity through gases, spectrum analysis; designed elec. dynamometer; worked in thermal physics, wireless communication. Died Cambridge, Mass., Feb. 18, 1923.

TROWBRIDGE, William Pettit, Am. engineer, scientist, educator; b. Troy, N.Y., May 25, 1828; s. Stephen Van Rensselaer and Elizabeth (Conkling) T.; grad. U. S. Mil. Acad., 1848; A.M. (hon.), Rochester U., 1856, Yale, 1870; Ph.D. (hon.), Princeton, 1879; LL.D. (hon.), Trinity Coll., 1883, U. Mich., 1887; m. Lucy Parkman, Apr. 21, 1857, 6 surviving children. Asst. prof. chemistry U. S. Mil. Acad., 1847-48; commd. 2d lt. Corps Topog. Engrs., U. S. Army, 1848; served on Atlantic coastal survey, 1849; commd. 1st lt., 1854; prof. mathematics U. Mich., 1856-57; asst. supr. Coast Survey; selected to install self-registering instrument of permanent magnetic observatory established at Key West, 1860; executed hydrographic survey of Narragansett Bay, 1860, established navy yard; in charge of army engr. agy. for supplying materials for fortifications and for constructing engring. equipage for armies in field, N.Y.C., 1861; superintending engr. of constrn. at fort, Willets Point, N.Y., 1861; v.p., gen. mgr. Novelty Iron Works, N.Y.C., 1865-71; prof. dynamic engring. Yale, 1871-77; prof. engring. Columbia, 1877-92; mem. New Haven (Conn.) Board Harbor Commrs.; councilor N.Y. Acad. Sciences, 1878-84, v.p., 1885-89; prominent mem. A.A.A.S.; mem. Nat. Acad. Sciences. Designed instruments for gauging ocean depths and recording depths by soundings; designed cantilever bridge. Died New Haven, Aug. 12, 1892.

TRUANT, Joseph Paul, microbiologist; b. San Martino, Italy, Aug. 10, 1923; s. Liberale and Adele (Bearzatti) T.; B.Sc., U. Toronto (Ont., Can.), 1945; M.Sc., U. Western Ont., London, 1952, Ph.D., 1954; m. Flora Rina Lenardon July 3, 1948; children—Linda, Allan, Anita. Lectr., asso. prof. U. Windsor, 1945-51; fellow microbiology U. Western Ont., 1951-54; dir. bacteriology and mycology Henry Ford Hosp., Detroit, 1954-66; dir. microbiology Providence Hosp., Southfield, Mich., 1966——. Mem. Am. Soc. Microbiologists (chmn. clin. sec. 1963-66), Am. Pub. Health Assn., Canadian Assn. Microbiology. Co-editor: Textbook of Clinical Microbiology, 1967——. Research, publs. on synergistic activity using colistin sulfate and sulfonamides; described technique identifying acid fast bacilli by fluorescence, electronic procedure for detecting significant bacteriuria in humans. Home: 3617 Bradford Dr. W., Birmingham, Mich. 48010, Office: 16001 9 Mile Rd., Southfield, Mich. 48075.*

TRUAX, Robert Collins, Am. aero. engr.; b. Tolliston, Ind., Sept. 3, 1917; s. Darwin Hoskins and Alida (Gleason) T.; student U. Cal. at Berkeley, 1934-35; B.M.E., U. S. Naval Acad., 1939, B.S. in Aero. Engring., 1952; M.S., Ia. State U., 1953; m. Rosalind Heath Schroeder, July 19, 1941 (div. Oct. 1964); children—Ann Heath (Mrs. Albert C. Fleming), Kathleen Rosalind (Mrs. Steven Johnson), Steven Robert, Gary Hale; m. 2d, Sally Ann Sabins, Oct. 11, 1964; one son, Scott Alan Truax. Commissioned ensign U. S. Navy, 1939, advanced through grades to capt., 1957, in charge pioneer rocket project Engring. Expt. Sta., Annapolis, Md., 1941-45, dir. Propulsion Lab. Naval Missile Center, Pt. Mugu, 1945-46, head rocket sect. Bur. Aeros., 1946-49, head surface missiles, 1953-54, formulated intermediate ballistic missile program Western Devel. Div., USAF, 1955, organized, dir. Air Force Space Program, 1956-58; ret., 1959; dir. advanced devels. Aerojet-Gen. Corp., Sacramento, 1959-66, Inst. Def. Analyses, 1966-67, TRW Systems, 1967——. Recipient Legion of Merit for contbns. of Polaris concept. Mem. Am. Rocket Soc. (Goddard award 1951, dir. 1947-55, v.p. 1956, pres. 1957), Am. Inst. Aeros. and Astronautics, A.A.A.S., U. S. Naval Inst., Sigma Xi. Contbr. chpts. to Sounding Rockets, 1953, Jet Propulsion, 1955, Handbook of Astronautics, 1960, Ency. Sci. and Tech., 1961. Research, publs. on rocket performance, propellants and design; invented pumping, pressurizing and throttling systems; 1st proposed Polaris concept of sub-surface launched missile. Home: 26190 Edgemont Dr., Highland, Cal. 92346. Office: TRW Systems Group, 600 E. Mill St., San Bernardino, Cal. 92408.*

TRUCE, William Everett, chemist; b. Chgo., Sept. 30, 1917; s. Stanley C. and Frances (Novak) T.; B.S., U. Ill., 1939; Ph.D., Northwestern U., 1943; m. Eloise Joyce McBroom, June 16, 1940; children—Nancy Jane, Roger William. Faculty, Purdue U., Lafayette, Ind., 1946——, prof. chemistry, 1956——, asst. dean Grad. Sch., 1963——. Mem. numerous univ. dept. and profl. coms.; chmn. various profl. meetings. Guggenheim fellow Oxford U., 1957. Mem. Am. Chem. Soc., Phi Beta Kappa, Sigma Xi. Contbr. numerous articles to profl. jours., chpts. to books. Developed rule of trans-nucleophilic addition; research in new methods of synthesis, devel. new kinds of compounds and reactions. Home: 156 Creighton Rd., West Lafayette, Ind. 47902.*

TRUCHET, Jean, French physicist; b. Lyons, France, July 13, 1657; joined Carmelite Order. Mem. French Acad. Scis., 1699. Contributed to constrn. of works bringing water to Versailles gardens; dir. bldg. of Orléans (France) canal; various inventions, including machines for minting, bleaching linen, carpenter's devil. Died Paris, Feb. 5, 1729.

TRUDEAU, Edward Livingston, Am. physician; b. New York, Oct. 5, 1848; s. Dr. James and Cephise (Berger) T.; M.D., Coll. Phys. and Surg. (Columbia), 1871; hon. M.S., Columbia, 1899; LL.D., McGill U., Can., 1904; LL.D., U. of Pa., 1913; m. Charlotte G. Beare, June 29, 1871. Began practice in New York, 1872, but ill health forced him to go to the Adirondack Mountains, where became resident; founded, 1884, Adirondack Cottage Sanitarium for treatment of incipient consumption in working men and women, first of its kind in America; founded, 1894, Saranac Lab. for the study of tuberculosis, first research lab. for the purpose in America. Mem. Nat. Assn. Study and Prevention Tb (pres. 1904), Assn. Am. Physicians (pres. 1905), Congress Am. Physicians and Surgeons (pres. 1910). Pioneer student Tb in U. S.; lab. research on cure for Tb; 1st expts. on immunity in U. S.; expert early clin. diagnosis Tb. Died Saranac Lake, N.Y., Nov. 15, 1915.

TRUE, Alfred Charles, Am. agriculturist; b. Middletown, Conn., June 5, 1853; s. Rev. Charles Kittredge and Elizabeth Bassett (Hyde) T.; A.B., Wesleyan U., Conn., 1873, A.M., 1876, hon. Sc.D., 1906; Harvard, 1882-84; Ph.D., Erskine Coll., S.C., 1886; m. Emma Fortune, Nov. 23, 1875; children—Elizabeth Fortune (Mrs. James Herbert Twamley), Henry Hyde. Prin. high sch., Essex, N.Y., 1873-74; instr. State Normal Sch., Westfield, Mass., 1875-82; instr. Wesleyan U., 1884-88; editor, 1889-90, vice-dir., 1891-92, dir. 1893-1915, Office of Expt. Stas., Dept. of Agr. Editor-in-chief "Experiment Station Record" and "Experiment Station Work"; dir. states relations service, Dept. of Agr., 1915-23; counselor to sec. of agr., on states relations, July 1, 1923——. Dean Grad. Sch. of Agr., of Ohio State U., 1902, U. of Ill., 1906, Cornell U., 1908, Iowa State Coll., 1910, Mich. Agrl. Coll., 1912, U. of Mo., 1914, Mass. Agrl. Coll., 1916. Agrl. editor New Internat. Ency. and Year-Book. Fellow A.A.A.S. Author of monographs on agrl. expt. stas. in U. S. and agrl. edn. Advisory mem. of exec. com. Am. Farm Bur. Fedn. Trustee Am. Univ. Supervised expenditures of agrl. expt. stas. in U. S.; agrl. studies in Alaska, P.R., H.I. and Guam; studied food and nutrition of man, irrigation and drainage problems; specialist agrl. edn. Died Apr. 23, 1929.

TRUE, Rodney Howard, Am. botanist, physiologist; b. Greenfield, Wis., Oct. 14, 1866; s. John M. and Mary Annie (Beede) T.; B.S., U. of Wis., 1890, univ. fellow in botany, 1890-92, M.S., 1892; student botany under Pfeffer, at Liepzig, 1893-95, Ph.D., 1895; m. Katharine McAssey, July 1, 1896 (dec.); 1 son, Rodney Philip; m. 2d, Martha A. Griffith, Dec. 22, 1927. Taught common schs. in Wis. 2 yrs.; prin. Wis. Acad., Madison, 1892-93; instr. pharmacognosy, 1895-96, asst. prof. pharmacognosy, 1896-99, U. of Wis.; lectured at Harvard, winter, 1899-1900, and asst. Radcliffe Coll.; lecturer in botany, Harvard, 1900-01; plant physiologist, U. S. Dept. Agr., 1901-20, in charge physiol. investigations; prof. botany and dir. Botanic Garden, U. of Pa., 1920-37, emeritus prof. of botany, 1937-40; dir. Morris Arboretum, U. of Pa., 1933-40. Mem. gen. com. for revision of 9th U. S. Pharmacopoeia; adv. council of Allegheny Forest Expt. Sta., 1934——, chmn., 1936——. Fellow A.A.A.S. (sec. com. one hundred on scientific research, 1925, mem. council, 1926). Contbr. of papers on original research to Annals of Botany, Botanisches Centralblatt, and other scientific jours. and govt. bulls. Expert in plant physiology; studied growth and functions of mineral foods in plants. Died Apr. 8, 1940.

TRUEBLOOD, Emily Walcott Emmart (Mrs. Charles Kingsley Trueblood), Am. cytologist; b. Balt., Aug. 8, 1898; d. William Wirt and Hattie (Frist) Emmart; A.B., Goucher Coll., 1922; M.A., Johns Hopkins, 1924, Ph.D., 1930; m. Charles Kingsley Trueblood, June 16, 1949. Faculty, Western Md. Coll., 1924-27, Johns Hopkins, 1931-36; entomologist U. S. Dept. Agr., Mexico City, Mexico, 1930-31; cytologist NIH, Bethesda, Md., 1936——. Mem. Histochem. Soc., Am. Soc. Pharmacology and Exptl. Therapeutics, Am. Inst. Biol. Scis., Am. Arthritis and Rheumatism Soc., Soc. Am. Bacteriologists, Soc. Exptl. Biology and Medicine, A.A.A.S., Washington Acad. Sci. Author: The Badiamus Manuscript, 1940. Research, numerous publs. on carcinogenesis of liver tumors, chemotherapy of Tb, localization of antigenic substances by means of fluorescent antibodies, immunochem. studies on enzymes and hormones, localization of prolactin in rats, cats, rabbits, mice, fish. Home: 7100 Armat Dr., Bethesda 20034. Office: NIH, Bethesda, Md. 20014.*

TRUEBLOOD, Kenneth Nyitray, Am. chemist; b. Dobbs Ferry, N.Y., Apr. 24, 1920; s. Howard Moffitt and Louise (Nyitray) T.; A.B., Harvard, 1941; Ph.D., Cal. Inst. Tech., 1947. Postdoctoral fellow chemistry dept. Cal. Inst. Tech., Pasadena, 1947-49; faculty U. Cal. at Los Angeles, 1949——, prof. chemistry, 1960——, chmn. dept. chemistry, 1965——; vis. prof. chemistry, U. Ibadan (Nigeria), 1964-65. Mem.

U. S. A. Nat. Com. on Crystallography, 1960-65, vice chmn., 1963-65; exchange visitor to USSR, 1965. Recipient Fulbright award Oxford (Eng.) U., 1956-57; U. Cal. at Los Angeles Distinguished Teaching award, 1961. Mem. Am. Chem. Soc., Am. Crystallographic Assn., Phi Beta Kappa, Sigma Xi, Alpha Chi Sigma. Research, publs. on theory of chromatography, chem. crystallography including high speed computer applications; analysis of molecule motion in crystals. Home: 1645 Roscomare Rd., Los Angeles 90024.*

TRUELL, Rohn, Am. physicist; b. Washington, Apr. 6, 1913; s. Karl O. and Anna M. (Rohn) T.; B.S. in Engring. Physics, Lehigh U., 1935; postgrad. Columbia; Ph.D., Cornell U., 1941; m. Marjory Ann Schminck, Sept. 12, 1942; children—Ann Rohn, Marcia Lee. With RCA, 1935-38, RCA Labs., Princeton, 1941-44; staff Stromberg Carlson, 1944-46; faculty Brown U., Providence, 1946—, prof. physics, 1951—, chmn. div. applied math., chmn. phys. scis. council, 1963—, dir. metals research lab., 1950—, men. athletic adv. council, 1964—, space scis. council, 1963—. Guest scientist Brookhaven Nat. Lab., 1956—. Guggenheim fellow, 1959-60. Fellow Am. Phys. Soc.; mem. I.E.E.E. (sr.), Am. Math. Soc. Research, publs. on interactions related to ultrasonic waves and phys. properties of crystalline solids. Died, Jan. 10, 1968.

TRUELOVE, Sidney Charles, English physician; b. London, Eng., Feb. 24, 1913; s. Albert Alick and Ellen (Cooper) T.; B.A., King's Coll. Hosp., London, 1934, M.A., 1938, M.B., B.Chir., 1938; M.D., U. Cambridge (Eng.), 1946; m. Joan Savill, Aug. 15, 1939; children—Anne Philippa (Mrs. David Buckley), Richard Savill. Med. registrar Norfolk and Norwich Hosp., 1946-47; physician Nuffield dept. clin. medicine U. Oxford (Eng.), 1947—, dir. clin. studies, 1949-54. Author: (with P. C. Reynell) Diseases of the Digestive System, 1963; (with K. Lunsden) Radiology of the Digestive System, 1965; also numerous articles. Research on diseases of digestive system, especially large intestine, aetiology and treatment of ulcerative colitis. Home: 38 Woodstock Rd., Oxford, Eng. Office: Nuffield Dept. Medicine, Radcliffe Infirmary, Oxford, Eng.*

TRUEMAN, Sir Arthur Elijah, English geologist, paleontologist; b. Nottingham, Eng., Apr. 26, 1894; s. Elijah and Thirza (Cottee) T.; grad. U. Coll., Nottingham, Eng, 1914; M.Sc., U. London, 1916, D.Sc., 1918; LL.D., Rhodes, Glasgow, Wales, Leeds; m. Florence Kate Offler, 1920; 1 son, E. R. Lectr. geology U. Coll., Cardiff, Wales, 1917-20; head dept. geology and geography, prof. geology U. Coll., Swansea, Wales, 1920-33; Channing Wills prof. geology U. Bristol (Eng.), 1933-37; prof. Glasgow (Scotland) U., 1937-46. Chmn., Geol. Survey Bd., 1943-54. Recipient Bigsby medal Geol. Soc., 1939; Wollaston medal 1955. Fellow Royal Soc., 1942; mem. Geol. Soc. (pres. 1945-47), Brit. Assn. (became pres. sect. C 1948). Author: The Scenery of England and Wales, 1938; An Introduction to Geology, 1938. Editor: The Coalfields of Great Britain, 1954; This Strange World, 1940. Research and numerous publs. on geology and paleontology, including Brit. coalfields, Liassic rocks and fossils, amonites; paleontol. correlations of coal measures of seams within one coal field; correlation of divisions or zones of one coal field with another. Died Jan. 5, 1956.

TRUEX, Raymond Carl, Am. anatomist; b. Norfolk, Neb., Dec. 11, 1911; s. Robert Edwin and Anna O. (Eminger) T.; A.B., Neb. Wesleyan U., 1934; M.S., St. Louis U., 1936; Ph.D., U. Minn., 1939; m. Elizabeth Ann Doner, Dec. 21, 1938; children—Raymond Carl, Mary Elizabeth. Faculty, Columbia U., N.Y.C., 1938-48, asso. prof., 1945-48; 1942-61, prof., chmn. dept. anatomy Hahnemann Med. Coll., Phila., 1948-61; Career Award prof. anatomy Temple U. Sch. Medicine, Phila., 1961—. Cons. neurology study sect. NIH, 1962-66. Re-cipient Career award USPHS, 1961; Hektoen Research award A.M.A., 1952; Research award Chgo. Dental Soc., 1943. Mem. Am. Assn. Anatomists, Am. Acad. Neurology, A.A.A.S., A.M.A., Am. Vet. Med. Assn. (hon. life mem.), Gerontological Soc., Harvey Soc., Sigma Xi. Author: Human Cross Section Anatomy, 1942; (with C. E. Kellner) Atlas of Head and Neck, 1948; (with M. B. Carpenter) Human Neuroanatomy, 1963; also articles. Research on brain structure and innervation of the heart, circulation and conduction systems of heart. Home: 219 Avon Rd., Narbeth, Pa.*

TRUHAUT, René, French toxicologist; b. Pouzauges, France, May 23, 1909; s. Jules Alexis and Sidonie (Lucas) T.; D.Pharmacy, Faculty Pharmacie and Scis., Sorbonne, U. Paris, 1947, D.Scis., 1952. Lab. chief Cancer Inst. Villejuif, Cancer Inst. Gustave Roussy, 1930-41, head chem. research, 1941-48; asso. prof. Faculty Pharmacy, Paris, 1948-55, titular prof., 1955-60, titular prof. toxicology and indsl. hygiene, 1960—. Expert for food additives and pesticides WHO, 1955—; expert for toxic substances in industry ILO. Labor orgn., 1959—. Decorated Légion d'Honneur. Mem. Internat. Union Pure and Applied Chemistry, Eurotox (gen. sec., 1961—), Internat. Union Contra Cancrum (past chmn. com. on causative factors cancer). Research, numerous publs. on chem. carcinogenesis, chemotherapy of cancer, toxicology of thallium, fluorine, arsenic, lead, mercury, cadmium, indsl. solvents, food additives, pesticides; contributed to establishment of methodology for toxicological evaluation of these chemicals for the purpose of establishing acceptable daily intakes and tolerances; biochem. orientation of toxicological research. Home: Ville-Evrad, 2, rue Jean-Jaurès, 93-Neuilly S/Marne, France. Office: 4 Avenue de l'Observatorie, Paris, France.*

TRUM, Bernard Francis, Am. radiation biologist; b. Natick, Mass., Dec. 10, 1909; s. Richard J. and Anna (McGee) T.; A.B., Boston Coll., 1931; D.V.M., Cornell U., 1935; m. Mary Margaret Maroney, Aug. 20, 1936; children—B. Michael, John M., Margaret, Elizabeth. prof. zootechnia U. Mayor de San Simon (Bolivia), 1949-50, U. Tenn., 1950-56; veterinarian biology and medicine U. S. AEC, 1956-58; lectr. Mass. Inst. Tech., 1959-61; lectr. on vet. medicine Harvard Med. Sch., 1958—, dir. Animal Research Center, 1958—; dir. New Eng. Regional Primate Research Center, Southborough, Mass., 1962—. Mem. Nat. Council on Radiation Protection and Measurement, 1947—; mem. adv. council div. biology and agr. Nat. Acad. Scis.-NRC, 1959—. Mem. Animal Care Panel (pres. 1965), Radiation Research Soc., Soc. for Exptl. Biology and Medicine, N.Y. Acad. Scis., Am. Vet. Med. Assn., Albertus Magnus Guild. Research, numerous publs. on species specific variations in acute and chronic response to total body radiation, reprodn. physiology in equines, zootechnics pertaining to reprodn. domestic animals and simian primates in captivity. Home: 247 Washington St., Sherborn, Mass. Office: Harvard Med. Sch., 25 Shattuck St., Boston 02115.*

TRUMPY, Bjorn, Norwegian physicist; b. Bergen, Norway, July 6, 1900; s. George and Hilde (Jahnsen) T.; Dipl.engr., Tech. U., Trondhjem, Norway, 1922, Dr.techn., 1927; postgrad. U. Göttingen (Germany), 1931-33; m. Esther Hvoslef, June 27, 1927; children—George, Karen Marie (Mrs. Helge Martens), Bjorn. Asso. prof. Tech. U., Trondhjem, 1932-35; prof. physics Bergen (Norway) Mus., 1935-48; prof. physics Bergen U., 1948—. Chmn. bd. Norwegian Inst. Atomic Energy, 1951-55, mem. bd., 1951—; chmn. bd. Michelsens Inst., 1951—; mem. Dutch-Norwegian Joint Commn. for Atomic Energy, 1957—; chmn. Norwegian delegation to CERN, Geneva, Switzerland, 1958—, Norwegian del. to council, mem. com., 1958—. Decorated Comdr. St. Olav Order Norway, Comdr. Oraje Nassau Order (Netherlands). Mem. Am. Phys. Soc., Royal Norwegian Acad., Acad. Scis. in Bergen, Acad. Scis. in Tronhjem. Author: Research and numerous publs. on intensity and width of spectral lines, Raman effect and constitution of molecules, earth magnetism, cosmic rays, nuclear physics, high energy physics; studies on solar produced rays in the outer room with rockets and satellites. Home: 2a Storhaugen, Bergen. Norway. Office: Inst. Physics, Bergen, Norway.*

TRUSCOTT, Basil Lionel, Am. neurologist; b. Chambers, Neb., Aug. 4, 1916; s. Basil Reginald and Annie (Fryer) T.; B.A., Drew U., 1939; M.A., Syracuse U., 1940, M.S., 1942, Ph.D., 1943; M.D., Yale, 1950; m. Elizabeth Jane Worrill, June 19, 1948; children—Elizabeth Hood, Thomas Reginald. Instr., Georgetown U. Med. Sch., Washington, 1943-45, Yale Sch. Medicine, 1947-49; asst. prof. U. N.C. Med. Sch., 1951-54; faculty Albany (N.Y.) Med. Coll., 1960—, prof. neurology, 1962—; chief neurology sect. Albany VA Hosp., 1960—. Cons. neurology U. S. Army in Europe, 1958-60. Fellow Am. Acad. Neurology, Am. Assn. Anatomists, Alpha Omega Alpha, Sigma Xi. Research, publs. on interaction between pituitary and sex hormones, human learning and memory, factors producing strokes in humans. Home: 22 Wisconsin Av., Delmar, N.Y. 12054. Office: VA Hosp., Albany, N.Y. 12208.*

TRYON, George Washington, Am. conchologist; b. Phila., May 20, 1838; s. Edward K. and Adeline (Savidt) T.; Manufactured and sold firearms and hunting equipment; mem. Acad. Natural Scis. of Phila., 1859, largely responsible for erection of new bldg., an organizer conchological sect., 1866, donated his private collection over 10,000 species, curator, 1869-76, conservator conchological sect., 1875-88; wrote paper On the Mollusca of Harper's Ferry, Virginia, 1861, also more than 70 other papers on land, freshwater and marine mollusks; edited, published Am. Jour. Conchology, 1865-72; wrote Manual of Conchology, Structural and Systematic, with Illustrations of the Species (chief work), 1st vol., 1879, 3 vols., 1888, continued by Dr. Henry A. Pilsbry; wrote comic opera Amy Cassonet or the Elopement, 1875; went to Europe, 1874, 77; wrote account of earlier trip, The Amateur Abroad, 1875; published Structural and Systematic Conchology, 3 vols., 1882-84. Died Feb. 5, 1888.

TRZEBIATOWSKI, Wladimir, Polish chemist; b. Grodzisk, Poland, Feb. 25, 1906; s. Casimir and Vanda (Grossmann) T.; Chem-Eng., Politechnika, Lvov, Poland, 1929; D.Sc., Politechnika Lvov, 1931; m. Buguslawa, Jezowska, Sept. 5, 1935. Staff, Inst. Inorganic Chemistry, Inst. Tech. Politechnika, Lvov, 1929-38, asst. prof., 1935-38; prof. inorganic chemistry U. Lvov, 1938-45; prof. inorganic chemistry Tech. U., Wroclaw, Poland, 1945—; dir. Inst. Low Temperature and Structure Research, Polish Acad. Scis., Wroclaw, 1966—. Recipient Cross of Merit, 1950; Commdrs. Cross of Order Polonia Restituta, 1954; State Sci. Prize, 1955. Mem. Polish Acad. Scis., Polish, Am. chem. socs., Chem. Soc. (London). Author: Chemia nieorganiczna, 1966; Lehrbuch anorg. Chemie, 1967; (with K. Lukaszewicz) Zarys rentg. analizy strukturalnej, 1960; also numerous papers. Research on inorganic structural chemistry, magnetochemistry of uranium compounds. Home: 20, ul. Gierymskich, Wroclaw 12, Poland.*

TSAI LUN, Chinese inventor; b. Kueiyang, Kweichow, China, circa 50; eunuch at Later Han Ct.; invented paper making from such things as tree bark, hemp, and rags, 105. Died circa 118.

TSCHERMAK VON SEYSENEGG, Armin, Austrian physiologist; b. Vienna, Austria, Sept. 21, 1870; s. Gustav Tschermak von Seysenegg. Prof. physiology U. Vienna, also German U., Prague, Czechoslovakia, from 1913. Author: Allgemeine Physiologie, 2 vols., 1916-24. Research on nervous system, heredity. Died 1952.

TSCHERMAK VON SEYSENEGG, Erich, Austrian botanist; b. Vienna, Austria, Nov. 15, 1871; s. Gustav Tschermak von Seysenegg. Prof., Vienna Agrl. Sch. Mem. French Acad. Scis. (corr.), 1937, Agrl. Acad. Research on genetics (with others) rediscovered Mendel's laws of heredity, 1900. Died 1962.

TSCHERMAK VON SEYSENEGG, Gustav, mineralogist; b. Littau, Moravia, Apr. 19, 1836; children—Erich, Armin. Prof. mineralogy, petrography U. Vienna (Austria); curatory mineral. collection, Vienna. Corr. mem. French Acad. Scis., 1897. Author: Die Feldspatgruppe, 1865; Die Glimmergruppe, 1877-78; Mikrospkopische-Beschaffenheit der Meteoriten, 1883; Lehrbuch der Mineralogie, 1883. Research in petrography, crystallography, paleontology, meteorites; mineral. study of Austria; (with J. F. C. Hessel) proved that plagioclase felspars series can be considered as an isomorphous series made up of albite and anorthite combined in all proportions. Died Vienna, May 4, 1927.

TSCHESCHE, Rudolf, German chemist; b. Liegnitz, Germany, May 11, 1905; s. Max and Martha (Schade) T.; Ph.D., U. Breslau, 1927; Dr.phil.habil., U. Göttingen, 1935; m. Annemarie Hirsche, Aug. 4, 1934; children—Harald T., Helga T., Heidi Wolz, Hildegard T. Vis. research worker K.W. Inst. Biochemistry, Berlin, 1937-40; head dept. chemotherapy Schering AG., Berlin; 1940-45; faculty U. Hamburg, 1947-60, prof., 1951-60; prof. chemistry, dir. U. Bonn, 1960—. Recipient C. Duisberg Meml. prize, 1936. Mem. Chem. Soc. Switzerland, Verein Deutscher Chemiker, Am. Chem. Soc., Gesellschaft Deutscher Naturforscher und Ärzte. Research, numerous publs. on steroids, triterpenes, pteridines, alkaloids, other natural occurring compounds, biochem. pathways. Home: Herzogsfreudenweg 22, 5301 Röttgen, Germany.*

TSCHIRCH, Alexander, German pharmacologist, botanist; b. Guben, Germany, Oct. 17, 1856; named prof. pharmacology, Bern, Switzerland, 1890. Author: Grundlagen der Pharmakognosie, 1885; Angewandte Pflanzenanatomie, 1889; Anatomische Atlas der Pharmakognosie und Nahrungsmittelkunde, 2 vols., 1893-1900; Die Harze, 2 vols., 1899; Handbuch der Pharmakognosie, 6 vols., 1908-26. Died 1939.

TSCHIRGI, Robert Donald, Am. physiologist, univ. ofcl.; b. Sheridan, Wyo., Oct. 9, 1924; s. Frank Horace and Kathleen (German) T.; S.B., U. Chgo., 1945, S.M., 1947, Ph.D., 1949, M.D., 1950. Asst. prof. U. Chgo., 1950-53; faculty U. Cal. at Los Angeles Sch. Medicine, 1953-66, prof. depts. physiology and anatomy, 1959-66, acad. asst. to pres., 1960-63, univ. dean planning, 1964-66; prof. dept. neuroscis. U. Cal. at San Diego, 1966—, vice chancellor acad. affairs, 1966—. Project dir. med. edn. study U. Hawaii, Honolulu, 1963-64; cons. to pvt. cos., govt. agys., Greek Govt. Mem. Am. Physiol. Soc., Biophys. Soc., Internat. Brain Research Orgn., Sigma Xi. Research, numerous publs. in neurophysiology on carotid chemoreceptor control of respiration, metabolism of brain, exchange of solutes between blood and brain, slowly changing elec. potentials in brain; developed phylogenetic theory of perception of space and time. Home: 151 12th St., Del Mar, Cal. 92014. Office: U. Cal. San Diego, P.O. Box 109, La Jolla, Cal. 92037.*

TSELIKOV, Aleksandr Ivanovich, Russian mech. engr.; b. Apr. 20, 1904; grad. Moscow Tech. Coll., 1928; Dr.Tech.Scis. Began teaching, 1935; joined Central Construction Bur. Metall. and Mech. Engring., 1945; became prof. Moscow Higher Tech. Sch., 1949—. Recipient Stalin Prize, 1947, 48, 51, Order of Red Banner of Labor. Corr. mem. USSR Acad. Scis. Author: The Calculation and Design of Rolling Equipment, 1938; Rolling Mills, 1946; Rolling Mill Mechanisms, 1946; Progressive Pressure-Working Processes at Mass Machine-Building Plants, 1955. Research on theory of rolling and design of rolling equipment; developed method for calculating rolling mills; directed constrn. of mechanized rolling mills. Home: B. Afanse'evskii p. 3. Office: Central Constrn. Bur. Metall. and Mech. Engring., Moscow, USSR.

TSESEVICH, Vladimir Platonovich, Russian astrophysicist; b. Oct. 11, 1907; grad. Leningrad U., 1927. A organizer, dir. Stalinabad Obs., 1933-37; staff Astronomy Inst., USSR Acad. Scis. (now Inst. Theoretical Astronomy), 1937-42; prof., dir. obs. Odessa U., 1945—. Author: Methods of Studying Variable Stars, 1948; Studies of Eclipsed Variable Stars, 1953-54; What and How to Observe in the Sky, 1955; The Processing of Radar Observations of Meteor Echoes, 1960. Research and publs. on variable stars, double stars, changes in radiance; developed tables of elements of eclipsed stars. Office: USSR, Odessa, Ukrainian SSR, ul. Petra Velikogo 2, Gosudarstvenny universitet.

TSIOLKOVSKY, Konstantin Eduardovich, Russian physicist; b. Izhevsk, Russia, Sept. 17, 1857; became instr. Borovosk Sch., 1882, Kaluga Sch., 1892. Author: The Investigation of Outer Space by Means of Reaction Apparatus, 1903. Research and publs. on theory of rocketry; built large balloon out of metal; multistage rocket, 1929; worked out kinetic theory of gases independently of Maxwell, 1881; 1st to suggest possibility of space station. Died Kaluga, Russia, Sept. 19, 1935.

TSITSIN, Nikolai Vasilevich, Russian botanist; b. Dec. 18, 1898; grad. Saratov Inst. Agr., 1927; hon. dr., Jena (Germany) U. Joined All-Union Inst. Grain Farming for S.E., 1927; staff Omsk Zonal Expt. Sta. (later Siberian Research Inst. Grain Farming), 1932-36, dir., 1936-38; v.p. All-Union Lenin Acad. Agrl. Sci., 1938-48; chmn. State Commn. for Testing Varieties of Grain Crops, Olaceae and Grasses, 1939-48; dir. Research Inst. Grain Farming in Non-Black Earth Areas of the USSR, 1940-49; dir. All Union Agrl. Exhbn., Moscow, 1938-49, 54-57; dir. Main Bot. Gardens, USSR Acad. Scis., 1945—. Recipient Order of Lenin, (4), Stalin Prize, 1943; Silver medal and Scroll of Honor, World Peace Council. Mem. USSR Acad. Scis., All-Union Lenin Acad. Agrl. Sci., Rumanian Acad. Scis. (hon.), Czechoslovakian Acad. Agr. (hon.), German Acad. Agrl. Sci. (hon.). dep. USSR Supreme Soviet, 1937, 50, 54. Author: The Problem of Winter and Perennial Wheats, 1935; The Outcome of Crossing Wheat with Couch Grass, 1937; Research and numerous publs. in hybridization; crossed ordinary and treelike tomatoes; originated new perennial wheat, stable hybrid of winter branching wheat. Office: Main Bot. Garden, USSR Acad. Sci., Ostankino, USSR.

TSO, Tien Chioh, phytochemist; b. Hupeh, China, July 25, 1917; s. Ya Fu and Suhwa (Wang) T.; B.S., Nanking U., China, 1941, M.S., 1944; Ph.D., Pa. State U., 1950; postgrad. Oak Ridge Nuclear Studies; m. Margaret Lu, Aug. 28, 1949; children—Elizabeth, Paul. Came to U. S., 1947, naturalized, 1961. Supt. exptl. farm Ministry Social Affairs, China, 1944-46; exec. sec. Tobacco Improvement Bur., 1946-47; research chemist Gen. Cigar Research Lab., 1950-51; with U. S. Dept. Agr., 1952, 59—, prin. plant physiologist crops research div. Agrl. Research Service, Beltsville, Md., 1964-66, leader tobacco quality investigations, tobacco and sugar crops research br., 1966—. Cons. Taiwan (China) Tobacco Research Inst., 1958—. Fellow A.A.A.S.; mem. Am. Chem. Soc., Am. Soc. Plant Physiologists, Phytochem. Soc. N.Am., Tobacco Chemists Research Conf. (symposium chmn. 1965), Tobacco Workers Conf., Sigma Xi. Contbg. author: Annual Review of Plant Physiology, vol. 9, 1958. Research, publs. on establishment of loci of alkaloid formation, biosynthetic pathway, interconversion and fate of alkaloids in tobacco plants, chem. composition as affected by macro and micro elements, health-related factors including mycotoxins and phenolics, established source of radioelements and their elimination in tobacco; developed fatty esters and fatty alcohols as sucker control and plant pruning agts. Home: 4306 Yates Rd. Office: South Bldg., Plant Industry Sta., Beltsville, Md. 20705.*

TSOU, Kwan Chung, chemist; b. Shanghai, China, Apr. 5, 1922; came to U. S., 1947, naturalized, 1954; B.S., Central U. Chunking, 1944, M.S., 1948; Ph.D., U. Neb., 1950; m. Teresa Lee, June 8, 1949; children—Walter, Stephen, Jennifer. Research asso. Harvard, 1950-54; devel. mgr., dir. research Monomer-Polymer Lab., Dajac Lab., 1954-59; lab. dir. Central Research Lab., Borden Chem. Co., Phila., 1960-63; asso. prof. chemistry U. Pa. Sch. Medicine, Phila., 1963—. Mem. Am. Chem. Soc., Chem. Soc. London, Histochem. Soc., Am. Assn. Cancer Research, N.Y. Acad. Scis., Internat. Assn. Dental Research. Fellow A.A.A.S. Research, numerous publs. on cancer chemotherapy, organic synthesis, enzyme histochemistry, polymer chemistry. Home: 370 Heathcliffe Rd., Huntingdon Valley, Pa. 19006. Office: U. Pa. Hosp., 3400 Spruce St., Phila. 19104.*

TSUBAKI, Tadao, Japanese physician; b. Tokyo, Japan, Mar. 16, 1921; s. Nobutsugu and Ito (Fukui) T.; M.D., U. Tokyo, 1945, Dr. Med. Sci., 1958; m. Hisako Kanehara, Feb. 4, 1950; children—Hidemi Tsubaki, Megumi Tsubaki. Asst. dept. medicine U. Tokyo Sch. Medicine, 1945-56, asst. prof. Inst. Brain Research, 1957-65; prof. neurology Brain Research Inst., Niigata (Japan) U., 1965—. Mem. Japanese Soc. Neurology (trustee 1965—). Author: (with Sano, Goto) Clinical Neurology, 1966; also numerous articles. Research on epidemiological, clin. and path. studies on multiple sclerosis, cerebrovascular, neuromuscular and other degenerative diseases. Home: 31 2-Chome, Hamaura cho, Niigata, Japan.*

TSUBOI, Chuji, Japanese geophysicist; b. Tokyo, Japan, Sept. 9, 1902; s. Shogoro and Nao (Mitsukuri) T.; B.S., U. Tokyo, 1926, D.Sc., 1934; m. Masako Shimazono, Feb. 9, 1929; children—Yosiko (Mrs. Akira Ikushima), Teiichi, Atsuko (Mrs. Kanau Kawashima). With U. Tokyo, 1926—, prof. geophysics, 1943-63, emeritus prof., 1963—; sr. specialist Nat. Diet Library, Tokyo, 1963—; vis. prof. Cal. Inst. Tech., 1955-56, Ohio State U., 1959. Recipient Asahi Newspaper Cultural prize, 1957. Mem. Seismol. Soc. Japan, Geodetic Soc. Japan, Japan Acad. (Acad. prize 1952), Deutsche Acad. der Naturforscher. Author: Gravity, 1935; Earthquakes, 1941; Theory of Vibration, 1942; Constitution of the Earth, 1961; Geophysics, 1966; also numerous articles. Research on deformation of earth's crust, distbn. gravity anomalies, mode of earthquake occurrence, devel. ship-borne gravity meter, gravity and earthquakes. Home: 4-3-13 Bunkyo-ku, Tokyo, Japan.*

TSUCHIYA, Henry Mitsumasa, Am. bioengineer; b. Seattle, Dec. 9, 1914; s. Seishiro and Masa (Kanagaki) T.; B.S., U. Wash., 1936, M.S., 1938; Ph.D., U. Minn., 1942; m. Miyo Kitagawa, July 28, 1941; children—Marilyn (Mrs. Jon Lauglo), Arthur Kitagawa. Bacteriologist, No. Regional Lab., U. S. Dept. Agr., 1947-56; asso. prof. U. Minn., Mpls., 1956-63, prof., 1963—. Cons. NSF, 1962-64; NSF vis. scientist, vis. prof. U. Tokyo, 1966. Recipient Superior Service award U. S. Dept. Agr., 1953, Distinguished Service Team award, 1955. Mem. Am. Soc. Microbiology, Am. Chem. Soc., A.A.A.S., Sigma Xi. Author: (with C. E. Skinner, C. W. Emmons) Molds, Yeasts, Actinomycetes, 1947. Research, numerous publs. on dynamics of microbial populations, enzymatic polymerization and depolymerization of polysaccharides, chemotherapy, disposal of indsl. wastes. Home: 201 Burntside Dr., Mpls. 55422.*

TSUCHIYA, Kenzaburo, Japanese physician; b. Numazu, Japan, Sept. 3, 1921; s. Kunita and Ikuyo (Inagaki) T.; M.D., Keio U., 1944, D.Med. Sci., 1950; M.P.H., U. Cal. at Berkeley, 1959; m. Haruko Koda, Jan. 11, 1944; 1 son, Takeaki. Staff, Keio U. Sch. Medicine, Tokyo, 1944—, prof. preventive medicine and pub. health, 1967—. Mem. Japan Soc. Indsl. Medicine, Japanese Soc. for Hygiene (editor Jour. 1961-64), Japan Pub. Health Assn. (editor Jour. 1955-58). Author: (with Juko Kubota) Differential Diagnosis of Occupational Diseases, 1964; also articles. Research on toxicology of indsl. chems., epidemiology of cancer especially in relation to occupation. Home: 2-10-15 Nakaochiai, Shinjuku-ku, Tokyo, Japan.*

TSU CH'UNG-CHIH, Chinese mathematician, mechanic; b. Fan-Yang, China, 430. Author: Chui-Shu. Determined value of pi as between 3.1415926 and 3.1415927 (accurate to 6 decimal places); designed new calendar which was not adopted, 463; invented (or revived) south pointing vehicle; constructed motor boat. Died 501.

TSUJI, Ichiro, Japanese urologist; b. Okayama, Japan, Feb. 6, 1919; s. Shoshiro and Tsune (Kawate) T.; M.D., Tokyo U., 1949; m. Eiko Yuge, Apr. 10, 1947; children—Naoko, Masakazu. Instr. dept. urology Tokyo (Japan), 1950-52; prof. Hokkaido (Japan) U. Sch. Medicine, 1952—. Mem. Société Internationale D'Urologie, Internat. Soc. Nephrology, Japanese Urol. Assn. Author: Clinical Pediatric Urology, 1962; Urachus and its Diseases, 1949; Muscles of the Kidney, 1953; also numerous articles. Research on regeneration and reconstrn. of urinary tract, obstruction and neurogenic bladder dysfunction, clin. pathology of bladder cancer, anatomy of kidney muscles. Home: North 16, West 7, Sapporo, Hokkaido, Japan.*

TSUJI, Jiro, Japanese applied physicist; b. Tokyo, Japan, 1896; grad. Tokyo U., 1923; D.Eng., 1930. Became chief research worker Phys. and Chem. Research Inst., 1937, councilor, dir., v.p., 1945-46; also dir., pres. motion picture co., pres. Gauge Mfg. Co. Recipient Imperial Acad. prize. Mem. Applied Physics. Soc., Precision Apparatus Soc. (dir.), Japan Standard Soc., Japan Weights and Measures Inst. Author: Experimental Technology of Elasticity; Polarimeter; Single Colored Lamp; Revolving Mirror; Lecture on Science. Research in phonographic fringe system in expt. on elasticity of light; discovered spontaneous combustion in coal mines; invented explosive gas detector for coal mines.

TSUJI, Jiro, Japanese chemist; b. Shigaken, Japan, May 11, 1927; s. Tokumatsu and Tsuma Tsuji; B.S., Kyoto U., 1951; M.S. (Fulbright scholar) Baylor U., 1956-57; Ph.D., Columbia, 1960; m. Yoshiko Takahashi, Oct. 27, 1952; children—Atsuko, Takashi, Hiroko. Researcher, Nippon Shinyaku Co., Kyoto, Japan, 1951-55, 60-62; research asso. Basic Research Lab., Toyo Rayon Co., Kamakura, Japan, 1962—. Mem. Japan Chem. Soc. Contbr. articles to profl. jours. Discovered several new and useful reactions using palladium and rhodium catalysts especially reactions of carbon monoxide with olefins, acetylenes, amines. Home: 820 Tsu. Office: 1111 Tebiro, Kamakura, Kanagawa, Japan.*

TSUJI, Shusuke, Japanese physician; b. Kyoto, Japan, Mar. 8, 1912; s. Iwao and Yuka (Taga) T.; grad. Kyoto U. Sch. Medicine, 1935, M.D., 1939; m. Atsuko Takeoka, May 4, 1959; children—Hisako, Hiroko. Faculty, Kyoto U. Tb Research Inst., 1942—, prof. 2d dept. medicine, 1952—. Mem. Japanese Assn. Tb, Japanese Assn. Allergy, Japanese Assn. Internal Medicine. Author: Non-Surgical Therapy of Tuberculous Cavity, 1952; also articles. Devised extrapleural plombage for treatment Tb cavity, diagnosis and therapy of non-Tb lung diseases, immunology of sarcoidosis, mechanism of Tb immunity, transfer factor of delayed sensitivity in animals; discovered antituberculous agt. of peptide nature in human urine. Home: 9-5 Tanaka Harunacho, Sakyo-ku, Kyoto, Japan.*

TSUKAMOTO, Kempo, Japanese radiologist; b. Tokyo, Japan, Sept. 16, 1904; s. Michito and Hama (Ogawa) T.; grad. U. Tokyo Faculty Medicine, 1931, D.Med. Sci., 1947; m. Teiko Kobayashi, Oct. 26, 1928; 1 dau., Rurko (Mrs. Tetsuya Tsukamoto). Asst. dept. internal medicine Hosp. U. Tokyo, 1931-34; asst. physician dept. radiology Hosp. Cancer Inst., Tokyo, 1934-46, radiologist in chief, 1946-58; dir. Nat. Inst. Radiol. Scis., Chiba, Japan, 1958-67; dir. hosp. Nat. Cancer Center, Tokyo, 1967—. Mem. Radiation Council Japanese Govt., 1958—; chmn. spl. com. on radioactivity survey Japanese Atomic Energy Commn., 1960. Fellow Am. Coll. Radiology (hon.); mem. Societas Radiologiae Medicae Italiace (hon.), Internat. Congress of Radiology (exec. com. 1959—), Japan Soc. Radiation Research (pres. 1960). Author: Fall-out and Its Countermeasures, 1962; New Data on Atomic Bomb Survivors in Hiroshima and Nagafor cancer, effects of radiation on man and his environment including problems of radiation protection. Office: Nat. Cancer Center, 1-1, 5 chome, Tsukiji, saki, 1965; also articles. Research on radiotherapy Tokyo, Japan.

TSUKAMURA, Michio, Japanese bacteriologist; b. Matsuyama, Japan, July 4, 1923; s. Yoshio and Kaoru (Uwagawa) T.; M.D., Nagoya (Japan) U., 1946, D. Med. Sci., 1955; m. Junko Shigemi, May 5, 1949. Med. staff Obuso Nat. Sanatorium (now Chubu Central Hosp.), Aichi-Pref., Japan, 1947—, chief labs., 1954-56, chief internal medicine dept., 1956-66, chief labs., chief 1st dept. internal medicine, 1966—. Lectr. bacteriology Nagoya U., 1965—. Mem. Japanese Soc. Bacteriology, Japanese Soc. Tb, Japanese Soc. Genetics, Internat. Union against Tb (corr. mem. bacteriological com. 1964—). Research, numerous publs. on drug resistance of mycobacteria, radiation biology, taxonomy of mycobacteria and identification system for mycobacteria. Address: Nat. Sanatorium, Chubu Central Hosp., Obu, Aichi-Pref., Japan.*

TSUNEWAKI, K., Japanese geneticist; b. Ono, Japan, Nov. 26, 1930; s. Suntaro and Toyo (Shimidu) T.; B.S., Kyoto (Japan) U., 1953, M.S., 1955; Ph.D., Kan. State U., 1958; m. Sumiko Shimidu, May 5, 1957; 1 son, Hiroshi. Postdoctorate fellow NRC Can., 1957-59; researcher Nat. Inst. Genetics, Japan, 1959-65; faculty Kyoto U., 1965—, prof., 1966—. Mem. Genetics Soc. Japan, Japanese Soc. Breeding, Genetics Soc. Am. Publs. on comparative gene analysis of common wheat and its relatives, including origin and differentiation of cultivated wheat species using maj. genes as tracer. Home: 228-60, Iwakura Nakacho, Sakyo-ku, Kyoto, Japan.*

TSUTSUI, Minoru, chemist; b. Wakayama City, Japan, Mar. 31, 1918; s. Juntaro and Tazu (Hirata) T.; B.A., Gifu U., 1938; M.S., Tokyo (Japan) Bunriak U., 1941; M.S., Yale, 1953, Ph.D., 1954; D.Sc., Nagoya (Japan) U., 1960; m. Ethel Ashworth, Mar. 3d, 1956; children—William Minoru. Asst. prof. Tokyo Gakugei U., 1950-53; vis. research fellow Sloan-Kettering Inst., 1954-56; research chemist Monsanto Chem. Co., 1957-60; research scientist N.Y. U., N.Y.C., 1960—; project dir., 1962—, lectr., 1964, asso. prof. chemistry, 1965—. Cons. Union Carbide Co., 1963—, Research Inst., Glidden Co., 1963—. Recipient Cressy Morrison award in natural scis. N.Y. Acad. Scis., 1960. Mem. N.Y. Acad. Scis. (v.p. 1965—), Am. Chem. Soc., Chem. Soc. London, Chem. Soc. W. Germany, Chem. Soc. Japan. Author: (with S. Ishikawa) General Chemistry, 1947, Organic Chemistry, 1947; also numerous articles. Research on organotransition metal chemistry, arene metal pi-complexes, reprodn. Hein's polyphenylchromium compounds and structure elucidation; discovered new chem. bond. Home: 2400 Sedgwick Av., Bronx, N.Y. 10068.*

TSWETT (or Tsvett), Mikhail Semenovich, botanist; b. Asti, Italy, May 19, 1872; s. Simeon and Maria (Dorozza) T.; studied in Lausanne, Geneva; magister, U. Kazan (Russia), 1901. Russian subject; settled in Warsaw; asst., U. Warsaw, 1901, instr., 1902, prof. botany and agromony, 1907; prof. of botany and microbiology, Inst. of Technology, Warsaw, 1908; in Moscow, 1915; prof. and dir. botanical garden, Estonia, 1917. Recipient Lady Davy prize, 1895; prize of St. Petersburg Acad. Scis., 1910. Author: The Chromophylles in the Vegetable and Animal Kingdom; five published articles. Research on plant pigments, chromatography, chlorophyll; made 1st chromatographic analysis, 1906; developed percolation technique of separating plant pigments by extracting leaves with petrol and percolating solution through calcium carbonate. Died Voronezh, U.S.S.R., 1919.

TSYTOVICH, Nikolai Aleksandrovich, Russian geophysicist; b. May 13, 1900; grad. Leningrad Inst. Civil Engrs., 1927. Began teaching at various instns., 1930; became prof. Moscow Engring. Structural Inst.,

1951; chmn. presidium Yakut br. USSR Acad. Scis., 1947-53, joined staff Inst. Permafrost, 1943, dep. dir., 1948-53, asso., 1953—; prof. Moscow Civil Engring. Inst., 1951—. Recipient Stalin Prize, 1950. Mem. Acad. Construction and Architecture USSR, USSR Aero. Sci. (corr.). Author: (with M. I. Sumgin) Fundamentals in the Mechanics of Frozen Grounds, 1937; Estimation of Foundation Depressions, 1941; Ground Mechanics, 3d edit., 1951; also articles. Research on frozen ground mechanics, geocryology. Office: V.A. Obrachev Inst. Permafrost, Bol'shoy Cherkasskii Pereulok, 2/10, Moscow, USSR.

TUCK, Leo Dallas, Am. chemist; b. San Francisco, Oct. 12, 1916; s. Leo Clyde and Sarah (Black) T.; A.B., U. Cal., Berkeley, 1939, Ph.D., 1948; m. Grace Mildred Gardner, Aug. 15, 1953; children—Richard Dallas, David Michael. Faculty, Sch. Pharmacy U. Cal., San Francisco, 1948—, prof. chemistry, pharm. chemistry, 1963—. Fellow A.A.A.S.; mem. Am. Chem. Soc., Am. Phys. Soc., Am. Pharm. Assn., Sigma Xi, Rho Chi. Research on thermodynamic theory of non-isothermal processes, devel. of volatile compounds of boron and uranium; studies electron spin resonance spectroscopy, nuclear magnetic resonance, properties of electrolytes in non-aqueous solvents. Home: 833 Eucalyptus Av., Novato, Cal. 94947. Office: U. Cal. Med. Center, San Francisco 94122.*

TUCKER, Howard Gregory, Am. mathematician, educator; b. Lawrence, Kan., Oct. 3, 1922; s. Louis Harold and Jeannette (Garbarsky) T.; A.B., U. Cal. at Berkeley, 1948, M.A., 1949, Ph.D., 1955; m. Elizabeth Everett Lyser, July 9, 1946; children—Deborah, William, Paula, Alice. Faculty, Rutgers U., 1952-53, U. Ore., 1955-56; faculty U. Cal. at Riverside, 1956-68, University of California at Irvine, 1968—, asso. prof. math., 1962—. Mem. Inst. for Advanced Study, Princeton, N.J., 1963-64. Mem. Am. Math. Soc., Math. Assn. Am., Inst. Math. Statistics. Author: An Introduction to Probability and Mathematical Statistics, 1962; A Graduate Course in Probability, 1967; also articles. Research on probability and math. statistics, Lebesgue properties of infinitely divisible distbns., stochastic processes with independent increments estimation problems. Home: 18215 Meadowsweet Way, Irvine, Cal.*

TUCKER, Josiah, Brit. economist, polit. theorist; b. Carmarthenshire, Eng., 1712; B.A., St. John's Coll., Oxford, 1736, M.A., 1739, D.D., 1755; m. Woodward; m. 2d, Mrs. Crowe. Became curate St. Stephens Ch., Bristol, Eng., 1737; named rector All Saints Ch., Bristol, 1739; minor canonry in cathedral; domestic chaplain to Bishop Butler; became rector St. Stephens, 1749; became 3d prebendal stall, Bristol, 1756; named dean of Gloucester, 1758; resigned position as rector, Bristol, 1790. Author: Brief History of the Principles of Methodism, 1742; The Elements of Commerce and Theory of Taxes, 1755; Instructions for Travellers, 1757; Letters to the Rev. Dr. Kippis, 1773; Four Tracts together with Two Sermons on Political and Commercial Subjects, 1775, 3d edit. with 5th tract, 1775; Treatise concerning Civil Government, 1781; Union or Separation, 1799; also other publs. Anticipated arguments of Adam Smith against monopolies; opposed war for the sake of trade, also war against Am. colonies. Died Nov. 4, 1799.

TUCKER, Robert Charles, Am. polit. scientist; b. Kansas City, Mo., May 29, 1918; s. Charles and Adele (Steinfels) T.; student U. Mich., 1935-37; A.B., Harvard, 1939, M.A., 1941, Ph.D., 1958; m. Eugenia Pestretsova, Aug. 21, 1946; 1 dau., Elizabeth Adele. Attache embassy, Moscow, USSR, also editor embassy's joint press reading service, 1944-53; asso. prof., then prof. govt. Ind. U., 1958-62; vis. prof. Soviet studies Johns Hopkins Sch. Advanced Internat. Studies, 1962; prof. politics, dir. program Russian studies Princeton, 1962—. Served with OSS, 1942-44. Fellow Center Advanced Study Behavioral Scis., 1964-65; Guggenheim fellow, 1968-69. Mem. Am. Polit. Sci. Assn. (chmn. conf. Soviet and Communist studies 1963-64; Pi Sigma Alpha award 1966), Am. Philos. Assn., Am. Soc. Polit. and Legal Philosophy, Am. Assn. Advancement Slavic Studies (bd. 1963-64). Author: Philosophy and Myth in Karl Marx, 1961; The Soviet Political Mind, 1963. Editor: (with S. F. Cohen) The Great Purge Trial, 1965. Contributions to the comparative politics of modern authoritarian systems, with special reference to the concept of "movement-regimes"; analysis of Soviet politics, especially in Stalin and post-Stalin periods; development, with others, of a "conflict model" of the Soviet system; advancement of scholarship in Marxism, particularly the clarification of the Marx-Hegel relation and the relationship between original and mature Marxism of Marx. Home: 44 Hartley Av., Princeton, N.J. 08540.*

TUCKER, William Boose, Am. physician; b. Peitaiho, North China, Aug. 17, 1905 (parents Am. citizens); s. Francis Fisher and Emma Jane (Boose) T.; A.B., Oberlin Coll., 1929; 4-year certificate U. Chgo. Sch. Medicine, 1933, M.D., 1934; m. Sara Julia Jones, Oct. 29, 1932; children—William Kirkby, Sara W. Instr., Bennington (Vt.) Coll., 1934-35; faculty Gen. Coll., U. Minn., 1935-36, 46-54, prof. medicine, 1951-54; faculty U. Chgo., 1939-46, asst. prof. medicine, 1943-46; prof. medicine Duke Sch. Medicine, Durham, N.C., 1954-56; dir. Tb service VA Central Office, Washington, 1956-59, dir. pulmonary disease service, 1959-61, dir. med. service, 1961—; chief pulmonary disease service VA Hosp., Durham, 1954-55, chief med. service, 1955-56. Mem. nat. adv. heart council NIH, 1958—. Recipient Distinguished Service award U. Chgo. Med. Alumni Assn., 1951. Mem. Nat. Tb Assn. (dir. 1962—, Trudeau medal 1966), A.M.A., A.C.P., A.A.A.S., Am. Clin. and Climatol. Assn., Inst. Medicine Chgo., Am. Thoracic Soc. (pres. 1960-66), Am. Assn. Phys. Anthropology. Author: (with W. H. Sheldon, S. S. Stevens) The Varieties of Human Physique, 1940; also numerous articles, chpts. in book. Research on devel. classification of body build and its correlation with health and disease, effects of artificial pneumothorax in treatment pulmonary Tb, coordination controlled trials in efficacy of chemotherapy of Tb. Home: 106 Grafton St., Chevy Chase, Md. 20015. Office: 810 Vermont Av. N.W., Washington 20420.*

TUCKERMAN, Edward, Am. botanist; b. Boston, Dec. 7, 1817; s. Edward and Sophia (May) T.; B.A., Union Coll., 1837, M.A., 1843; grad. Harvard Law Sch., 1839, B.A., Harvard, 1847, grad. Harvard Div. Sch., 1852; m. Sarah Eliza Sigourney Cushing, May 17, 1854. Curator coll. museum Union Coll., 1842-43; lectr. in history Amherst (Mass.) Coll., 1854-58, prof. botany, 1858-86; contbd. articles to N.Y. Churchman on biog., hist. and theol. topics. mem. Nat. Acad. Scis., 1868. Author: Enumeratio Methodica Caricum Quarundam, issued privately, 1843; Enumeration of North American Lichens, 1845, also supplement, "Synopsis of the Lichens of New England, the Other Northern States, and British America," Lichens of California, Oregon, and the Rocky Mountains, 1866; General Lichenum: An Arrangement of North American Lichens, 1872; Catalogue of Plants Growing without Cultivation within Thirty Miles of Amherst College, issued privately, 1875; A Synopsis of North American Lichens, Part I, 1882. Authority in field of Am. lichenology; described new genus of flowering plants, Oakesia, 1842; first to explore mountains of New Eng. for lichens. Died Amherst, Mar. 15, 1886.

TUCKEY, Stewart Lawrence, Am. dairy technologist; b. Browns Valley, Minn., Aug. 24, 1905; s. William Ernest and Olive (Hanson) T.; B.S. in Agr., U. Ill., 1928, M.S., 1930, Ph.D. in Dairy Mfg., 1937; m. Frances Catherine Griswold, June 6, 1936; 1 dau., Frances. Staff, U. Ill., Urbana, 1928—, prof. dairy tech., 1959—. Cons., adviser to dairy industry, 1946—; staff Netherlands Inst. for Dairy Research, 1967. Recipient Borden Co. award in dairy mfg., 1939. Mem. A.A.A.S., Am. Chem. Soc., Am. Dairy Sci. Assn., Central Ill. Dairy Tech. Soc. (sec. 1957—), Sigma Xi, Phi Kappa Phi. Author: (with P. H. Tracy) Laboratory Exercises in Dairy Technology, 1939; (with D. E. Emmons) Cultured Products, 1967; also articles. Research on flavor compounds in cheddar cheese, microbial rennet from Bacillus cereus replacing calf rennet extract in cheese, acidity in cottage cheese. Home: 919 W. Charles St., Champaign, Ill. 61820. Office: 101 Dairy Mfg. Bldg., Urbana, Ill. 61801.*

TUDDENHAM, Read D(uncan), Am. psychologist, educator; b. Salt Lake City, Sept. 4, 1915; s. John C. and Helen (Underwood) T.; A.B., U. Utah, 1935; Ph.D., U. Cal. at Berkeley, 1941; m. Eileen Whelan, May 30, 1943; children—William John, Helen Anne. With U. Cal. Inst. Child Welfare, 1941-43, Adj. Gen.'s Office, War Dept., 1944-45; CBS, 1945-46; faculty U. Cal. at Berkeley, 1946—, prof. psychology, 1959—. Cons. VA, 1949—, Letterman Army Hosp., 1955—, Cal. Dept. Mental Hygiene, 1958—. USPHS spl. fellow U. Geneva, 1961-62. Diplomate clin. psychology Am. Bd. Examiners in Profl. Psychology. Fellow Am. Psychol. Assn., Soc. Research Child Devel.; mem. Cal., Western psychol. assns., Phi Beta Kappa, Sigma Xi, Phi Kappa Phi. Research, publs. on factors determining susceptibility to social pressure, longitudinal studies of phys. growth, intelligence and personality, cognitive devel. in childhood.*

TUDOROVSKII, Aleksandr Ilarionovich, Russian physician; b. Aug. 24, 1875; grad. Petersberg U., 1897. Faculty, Petersberg Poly. Inst., 1902-19, Petersberg, U., Leningrad, 1919-29; became head 1st Russian Calculating Bur. on Calculation of Optical Systems, 1916; joined State Optical Inst., 1918. Recipient Stalin Prize, 1942, 46. Corr. mem. USSR Acad. Sci. Author: Electricity and Magnetism, Part 1-2, 1933; Theory of Optical Devices, 1-2, 2d edit., 1948-52. Research and publs. in geometric optics, optics techniques, electromagnetic phenomena, calculation and devel. of photog. lenses; organized optics calculations. Office: USSR Acad. Scis., Leninskii Prospekt, 14, Moscow, USSR.

TÜDOS, Ferenc, Hungarian chemist; b. Szuhakalló, Hungary, Apr. 1, 1931; s. Ferenc and Maria (Tronka) T.; grad. U. Szeged (Hungary), 1953; Candidate's degree Techn. Inst. Leningrad (USSR), 1956; Ph.D., U. Budapest (Hungary), 1962; Acad. D. Chem. Studies, U. Leningrad, 1964; m. Helga Feuer, Aug. 24, 1956; children—Vera, Ann, Eve. Staff, Central Research Inst. for Chemistry, Hungarian Acad. Sci., Budapest, 1957—, head research group, 1957-60, head polymerization kinetics dept., 1960—, head and mem. various coms. Decorated Order Labor. Mem. Hungarian Chem. Soc. Author: Discussion of the Kinetics of Radical Polymerization on the Basis of the Hypothesis of Hot Radicals, 1966; also numerous articles. Hon. editorial adv. bd. European Polymer Jour., 1965. Research on kinetics of radical polymerization, inhibition involving practical aspects, polymerization in solid state, fundamental aspects of reaction kinetics, radical chemistry (preparation and investigation by ESR of some stable free radicals). Home: 26 Muraközi, Budapest, II. Office: 57/69 Pusztaszeri, Budapest, II, Hungary.*

TUFFIER, Théodore, French surgeon; b. Bellème, France, 1857; ed. Paris; surgeon at his own clinic, also Hôpital de la Pitié, Hôpital Beaujon, Paris; became prof. agrége Paris Faculty Medicine, 1889; organizer med. div. French Army, World War I, also during campaign, Morocco, 1920; prof. agrégé surgery U. Paris. Hon. mem. surg. socs. of Belgium, Brazil, Greece, Portugal, Japan; Fellow Royal Coll. Surgeons (hon.); mem. French Acad. Medicine. Research and publs. on surgery of kidneys, surg. treatment of Tb, by excising apex of lung, 1897; one of 1st to perform lung and heart surgery; 1st successful operation for chronic valvular heart disease, 1914; advocated use of spinal anesthesia. Died Paris, 1929.

TUFT, Louis, Am. physician; b. Phila., Sept. 14, 1898; s. Harry and Edith (Kofsky) T.; M.D., U. Pa., 1930; m. Carlyn J. Manasses, July 1, 1930; children—Janet Garvin, Betsy Ann, Harry M. Chief of allergy clinic Temple U. Hosp., Phila., 1931-64, clin. prof. medicine, 1955-64, prof. emeritus, 1964—; dir. Pa. State Health Labs., 1937-39. Mem. Argentine (hon.), Cuban (hon.) allergy soc., A.M.A., Phila. Pathol. Soc., Am. Assn. Immunologists, Am. Acad. Allergy, Phila. Allergy Soc., Phila. Coll. Physicians, A.C.P., Phila. Med. Soc., A.A.A.S. Author: Clinical Allergy, 1937, 49; (with J. A. Kolmer) Clinical Immunology, Biotherapy and Chemotherapy, 1941; also numerous articles. Home: 165 Moreland Rd., Bethayres, Pa. 19006. Office: 1530 Locust St., Phila. 19102.*

TUKE, Daniel Hack, Brit. physician; b. York, Eng., Apr. 19, 1827; s. Samuel and Priscilla (Hack) T.; student St. Bartholomew's Hosp., London, 1850; M.D., U. Heidelberg, 1853; LL.D., Glasgow, Scotland, 1883; m. Esther Maria Stickney, Aug. 10, 1853; 1 son, H. S. Staff, York Retreat, 1847-49; vis. physician Retreat, also York Dispensary, until 1859; cons. physician mental diseases, London, 1875-95; examiner mental philosophy U. London; gov. Bethlehem Royal Hosp.; lectr. mental diseases Charing Cross Hosp. Recipient prize Assn. for Improving Condition of Insane, 1854. Mem. Royal Coll. Surgeons, After-Care Assn. (a founder), Medicopsychol. Assn. (became pres. 1881). Author: (with Bucknill) A Manual of Psychological Medicine, 1858; Illustrations of the Influence of the Mind on the Body, 1872; Insanity in Ancient and Modern Life with chapters on Prevention, 1878; History of the Insane in the British Isles, 1882; Sleep-walking and Hypnotism, 1884; A Book on the Insane in the U. S. and Canada, 1885; Past and Present Provision for the Insane Poor in Yorkshire, 1889; Prichard and Symonds in especial relation to Mental Disease with a Chapter on Moral Insanity, 1891; Dictionary of Psychological Medicine, 1892. Early studies in insanity; reformed treatment of insane, especially in Can. Died Mar. 5, 1895.

TUKEY, Harold Bradford, Am. horticulturist; b. Berwyn, Ill., Sept. 30, 1896; s. James Bradford and Armenia (Mehrhof) T.; B.S., U. Ill., 1918, M.S., 1920; Ph.D., 1932; D.H.C. (hon.), U. Hanover (Germany), 1957; m. Margaret Davenport, Nov. 23, 1918 (dec. 1930); children—Loren Davenport, Lois (Mrs. William Degrove Baker, Jr.), Ronald Bradford; m. 2d, Ruth Schweigert, Nov. 23, 1932; children—Harold Bradford, Jr., Ann. Staff, N.Y. State Agrl. Expt. Sta., Geneva, 1920-45, chief in research, 1927-45; prof. Cornell U., Ithaca, N.Y., 1927-45; prof., head dept. horticulture Mich. State U., East Lansing, 1945-64, prof. emeritus, 1964—. U. S. tech. adviser Internat. Conf. on Atomic Energy, 1955; del. Internat. Hort. Congress, London, 1952, Scheveningen, 1955, Brussels, 1962, U. S. A., 1966; pres. XVII Internat. Hort. Congress, 1966. Recipient Jackson Dawson medal Mass. Hort. Soc., 1948; N.J. Colman award Am. Assn. Nurseryman, 1956; cited by Am. Hort. Council, 1957. Fellow A.A.A.S., Royal Hort. Soc. (v.p. 1964—), Am. Soc. for Hort. Sci. (past sec.-treas., pres., past editor); mem. Am. Inst. Biol. Scis. (organizing bd. 1946-47), Am. Pomological Soc. (Marshall P. Wilder medal 1956, past pres.), Internat. Soc. Hort. Sci. (pres. 1962-66), Bot. Soc. Am., Am. Soc. Plant Physiology, Soc. Nat. d'Hort. de France (hon.). (with others) The Pears of New York, 1921; The Pear and Its Culture, 1929; Plant Regulators in Agriculture, 1954; Dwarfed Fruit Trees, 1964; also numerous articles. Hort. editor Rural-New Yorker, 1920-64; asso. editor Am. Fruit Grower, 1947—. Research in clonal dwarfing rootstocks for fruit trees, plant propagation, stock-scion relations, use radioactive tracers in hort. research, uptake and loss nutrients by plant leaves and other above-ground parts; pioneered culture excized plant embryos, use plant regulators as herbicides; developmental morphology apple, peach, cherry. Home: The Maples, Woodland, Mich. 48897. Office: Dept. Horticulture, Mich. State U., East Lansing, Mich. 48823.

TUKEY, John Wilder, Am. statistician; b. New Bedford, Mass., June 16, 1915; s. Ralph H. and Adah M. (Tasker) T.; Sc.B., Brown U., 1936, Sc.M., 1937, awarded Doctor of Science degree, in 1965; A.M., Princeton Univ., 1938, Ph.D., 1939; Sc.D. (hon.), Case Institute of Technology, 1962; m. Elizabeth L. Rapp,

July 19, 1950. Instr. mathematics, Princeton, 1939-41, asst. prof., 1941-48, asso. prof., 1948-50, prof. since 1950; research asso. fire control research office, Princeton, 1941-45; mem. tech. staff Bell Labs., 1945—, asst. dir. research communication principles, 1958-62, asso. exec. dir. research Bell Telephone Labs., 1961—. Mem. Pres.'s Science Adv. Com., 1960-63. Guggenheim Fellow, 1949-50; Center Advanced Study in Behavioral Scis. fellow, 1957-58, NRC, 1951-60. Fellow A.A.A.S., Am. Statis. Assn., American Soc. Quality Control, Inst. Math. Statis., N.Y. Acad. Sci., Operations Research Soc. Am., Royal Statistical Soc.; mem. Am. Math. Soc., Assn. Computing Machinery, Biometric Soc., Math. Assn. Am., Nat. Academy of Scis., Am. Philos. Soc., Am. Acad. Arts and Scis., Internat. Statis. Inst., Sigma Xi. Author: Denumerability and Convergence in Topology, 1940; Statistical Problems of the Kinsey Report (with Cochran & Mosteller), 1954; (with R. B. Blackman) The Measurement of Power Spectra from the Point of View of Communications Engineering, 1958. Contbr. articles in scientific and tech. publs. Study of mathematical, theoretical, and applied statistics; point set topology; military analysis; fire control equipment. Home: 115 Arreton Rd., Princeton, N.J.

TULASNE, Edmond (Louis-René), French botanist; b. Azay-le-Rideau, France, Sept. 12, 1815; student law, Potiers, France; mem. Auguste de Saint-Hilaire's expdn. to Brazil; became asst. naturalist Paris Mus. Natural History, 1842, ret., 1872. Mem. French Acad. Scis., 1854. Author: Selecta fungorum carpologia, 3 vols., 1857-65. Founder modern mycology; research and publs. on mushrooms, leguminous plants, lichens, embryology, exotic flora; discovered polymorphism in fungi; several species of fungi named in his honor. Died Hyères, Var, Dec. 22, 1885.

TULECKE, Walter R., Am. botanist; b. Detroit, Feb. 10, 1924; s. Walter M. and Bertha (Ruehmkorf) T.; B.A., U. Mich., 1946, M.S., 1950, Ph.D. in botany, 1953; m. Hazel Batchelor, Mar. 31, 1946; children—Peg, Kari, Heidi, Kim. Asst. prof. botany Ariz. State Co., 1953-55; research asso. Bklyn. Bot. Gardens, 1955-57; research botanist Charles Pfizer & Co., 1957-59; asso. plant physiologist Boyce Thompson Inst., 1959-67; asso. prof. biology Antioch Co., Yellow Springs, O., 1967—. Cons. India Sci. Program, NSF, 1967. Mem. Am. Inst. Biol. Scis., A.A.A.S., Am. Soc. Plant Physiologist, Bot. Soc. Research, publs. in tissue culture and morphogenesis of plants; 1st culture of tissue from pollen; continuous liquid cultures of tissues from higher plants; growth of haploid male and female tissues in vitro. Home: 903 Xenia Av., Yellow Springs, O. 45387.*

TULLIS, James Lyman, Am. physician; b. Newark, O., June 22, 1914; s. Don Delano and Agnes (Luther) T.; student Rollins Coll., 1932-35; M.D., Duke, 1940; m. Marjorie White, Sept. 11, 1937; children—Virginia (Mrs. David Latham), Ann (Mrs. Paul Hunter III), James L., Susan. Research fellow in biochemistry Harvard Med. Sch., 1945-48, research asso., 1954-64, asst. clin. prof. medicine, 1965—, dir. blood characterization and preservation lab., 1953-55; asst. in medicine Peter Bent Brigham Hosp., Boston, 1946-50, asso. in medicine, 1955-58, sr. asso. in medicine, 1958—; attending physician W. Roxbury VA Hosp., Boston, 1948-64; attending physician hematologist New Eng. Deaconess Hosp., Boston, 1949—, chief Hematology and Chemotherapy Clinic, 1957—, chmn. dept. medicine, 1964—; attending physician L. I. Hosp., Boston, 1950-58; cons. Cambridge City Hosp., 1954-60, Monadnock Gen. Hosp., Peterborough, N.H., 1950—. Cons. to sec. def., Research and Devel. Bd., div. Med. Scis., Panel on Mil. and Field Medicine, 1952-54; alternating mem. Civilian Health and Med. Adv. Council Dept. Defense, 1954-58, mem. 1958—; sr. investigator Protein Found., 1956—, dir. Cytology Labs., 1960—. Recipient Glycerol Producers Research award, 1957; Silver medal Pasteur Inst., Paris, 1962. Diplomate Am. Bd. Internal Medicine, Nat. Bd. Med. Examiners. Fellow N.Y. Acad. Scis., mem. A.C.P., Mass. Med. Soc., Internat. Soc. Hematology (Katsunuma award 1960, sec.-gen. Western Hemisphere 1958—), Am. Soc. Hematology (pres. 1957-59); mem. Am. Heart Assn., European Soc. Hematology (hon.), VI Internat. Congress Hematology (v.p. and treas. 1955-56), Soc. Internat. de Transfusion Sanguine, A.M.A. (Hoekten medal 1959). Editor: Blood Cells and Plasma Proteins, 1953. Contbr. numerous articles to sci. jours. Research involving devel. methods for separation and preservation of cellular components of blood; isolation of coagulation factors of blood for use in treatment of human disease. Home: 188 Franklin St., Newton, Mass. 02158. Office: 110 Francis St., Boston 02215.*

TULLOCH, George Sherlock, Am. entomologist; b. Bridgewater, Mass., Aug. 3, 1906; s. Douglas James and Irene S. (Settle) T.; B.S., U. Mass., 1928; M.S., Harvard, 1929, Ph.D., 1931; m. Dorothy Gorton Gooch, Sept. 17, 1931; children—George Sherlock, James Douglas. Asst. entomologist State of Mass., 1930; entomologist Fairbanks Exploration Co. (Alaska), 1931; faculty Bklyn. Coll., 1932-65, prof., 1952-65, chmn. dept. biology, 1962-64; faculty Columbia Sch. Pub. Health, 1965; vis. scientist Arctic Aeromed. Lab., 1965-67; vis. scientist Sch. Aerospace Medicine, USAF, Brooks AFB, Tex., 1967—. Asso. entomologist for U. S. Govt., Mayaguez, P.R., 1934-35; entomologist Rockefeller Found., Rio de Janeiro, Brazil, 1940-41; staff U. Queensland (Australia), 1961. Fellow A.A.A.S., Entomol. Soc. Am.; mem. Bklyn. Entomol. Soc. (past pres.), Cambridge Entomol. Club, Am. Soc. Parasitologists, Am. Assn. U. Profs., Assn. for Study Social Insects, Sigma Xi. Author: Introduction to Animal Parasitology, 1963; also articles. Research on life history of mosquitoes of Mass., P.R., Alaska; recognized gynergates in ants (forms intermediate between queens and workers), worker ants with wings in genus Diacama (later named diacammathatigynes by William Morton Wheeler); co-discoverer 9 plus 1 structure of flagellar and ciliary structure as revealed by electron microscopy. Home: 4919 Pecan Grove, San Antonio 78222. Office: Sch. Aerospace Medicine, Brooks AFB, Tex. 78235.*

TULLY, John Patrick, Canadian oceanographer; b. Brandon, Man., Can., Nov. 29, 1906; s. John and Agnes (Mott) T.; B.Sc., U. Man., 1931; Ph.D., U. Wash., 1947; m. Ethel Lorraine Hamilton, Sept. 17, 1938; children—Jean Agnes (Mrs. Howard C. Prout), Anne Lorraine, James H. With Fisheries Research Bd. Can., 1931—, oceanographer in charge Pacific group, 1945-66, cons., Ottawa, 1966—. Sec., Can. Com. on Oceanography, 1966—. Decorated Order Brit. Empire, 1945; Coronation medal, 1954; medaille commemorative, Alberto Ier de Monaco et la Mer, 1967. Fellow Royal Soc. Can.; mem. Am. Geophys. Union, A.A.A.S., Am. Soc. Limnology and Oceanography, Sigma Xi. Research, numerous publs. on estuarine circulation and flushing, prediction of pollution, hydraulic models of sea ways, structure and circulation of N.E. Pacific, mechanics of heating and cooling in sea, under-water acoustics, airborne radiation thermometry. Home: 518-790 Springland Dr. Office: Sir Charles Tupper Bldg., Ottawa 8, Ont., Can.*

TULP, Nicolas, Dutch physician; b. Amsterdam, Holland, Oct. 11, 1593; ed. Leiden, Holland; M.D. Mem. senate, also burgomaster; prof., Amsterdam. Author: Observationes medicae (includes anat. study of orangutan), 1671. Described ileocecal valve. Died Sept. 12, 1674.

TUMANOV, Ivan Ivanovich, Russian plant physiologist; b. June 30, 1894; grad. Kiev Agrl. Inst., 1923. Staff, All-Union Inst. Horticulture, Leningrad, 1925-42; joined Inst. Plant Physiology, USSR Acad. Scis., 1940, became prof., 1947. Corr. mem. USSR Acad. Scis. Author: Physiological Basis of Frost Resistance of Cultured Plants, 1940; Main Achievements of Soviet Science in the Study of Frost Resistance of Plants, 1951; The Cultivation of Plants on Gravel for Research Purposes, 1960. Developed lab. test of drought and frost resistance of plants; research on rotting and fertility of agrl. crops, adaptations to winter conditions. Home: Sokol'nicheskaya slob. 14/18. Office: K. A. Timiryazev Inst. Plant Physiology, USSR Acad. Scis., Leninskii Prospekt, 33, Moscow, USSR.

TUMMERS, Jozef Hendrik, Dutch physicist; b. Sittard, Netherlands, Oct. 29, 1879; s. Jan Willem and M. (Alofs) T.; student univs. Leiden, Utrecht, Göttingen; Dr.ès.sc., Ph.D.; m. Hubertina Maria Verhagen, Apr. 23, 1919; 1 child, Arnoldus Maria Franciscus. Prof., Venlo, Netherlands, 1914-41, Tilburg, Netherlands, 1917-57; prof. philosophy sci., 1926-46; prof. natural philosophy Nimengen U., 1932-45; now ret. Mem. Thymgenootschap, Wiskundig Genootschap, Vereniging van de Wijsbeg der exakte vakken. Author: Die Spezielle Relativitätstheorie und die Logik, 1929; Evolutie der Wetenschap, Opbouw der Meetkunde, Meetkunde en Ervaring, Verantwoording Parallelenpostulaat, 1941-50; Une Transformation quadratique nouvelle, 1946; Construction des courbes algébriques d'une classe spéciale, 1959; Le Cercle d'Euler, nouveaux Théorèmes, 1960; La Physique théorique et le Métaphysique, 1961; contbr. theorems to De Opgaven van het Wiskeundig Genootschap, 1933-63. Address: S. Rochusstrasse 42, Steyl (Tegelen), Netherlands.

TUMULTY, Philip Anthony, Am. physician; b. Jersey City, N.J., Nov. 4, 1912; s. Joseph Patrick and Mary Alicia (Byrne) T.; A.B. cum laude, Georgetown U., 1935; M.D., Johns Hopkins, 1940; m. Claire Cotter, Jan. 24, 1942; children—Claire, Philip, Kathy, Mary, Alicia. Instr. medicine Johns Hopkins Hosp., 1946-49, asst. dir. med. clinics, 1948-50, dir. med. clinics, 1950-53, physician charge med.-surg. group clinic, 1948-53, chmn. pvt. med. service, 1955—, physician charge pvt. patient clinic, 1956—; asst. prof. medicine Johns Hopkins Sch. Medicine, 1949-51, asso. prof., 1951-53, 55-63, prof. medicine, 1963—; prof., dir. dept. medicine St. Louis U. Sch. Medicine, 1953-54; physician-in-chief pro tempore dept. medicine R.I. Hosp., 1957; cons. physician Balt. City Hosp.; spl. cons. medicine NIH; cons. medicine USPHS, VA Hosp., Balt., Walter Reed Gen. Hosp. Clin. fellow A.C.P., 1947-48. Diplomate Am. Bd. Internal Medicine. Mem. Am. Clin. and Climatol. Assn., Assn. Am. Physicians, Phi Beta Kappa. Contrib. to study of collagen diseases, bacterial endocarditis. Home: 5001 St. Albans Way, Balt. 21212. Office: 601 N. Broadway, Balt. 21205.

TUNELL, George, Am. geochemist; b. Chgo., Apr. 4, 1900; s. George Gerard and Caroline (Baum) T.; S.B. in Mining, Harvard, 1922, Ph.D. in Geology, 1930; m. Charlotte Ruth Philips, July 20, 1931. Petrologist, Geophys. Lab., Carnegie Instn., Washington, 1941-45; acting asso. prof. mineralogy and metalliferous geology Cal. Inst. Tech., 1946-47; asso. prof., prof. geology U. Cal. at Los Angeles, 1947-62; prof. geology U. Cal. at Riverside, 1962—. Mem. Mineral. Soc. Am. (past pres.), Geol. Soc. Am. (past v.p.), Geol. Soc. Washington (past pres.), Geochem. Soc. (past pres.), Soc. Econ. Geologists, Am. Crystallographic Assn., Mineral. Soc. Gt. Britain and Ireland, Societe Mineralogique de France, Washington Acad. Scis. (past v.p.), Sigma Xi, Tau Beta Pi. Author: Relations Between Intensive Thermodynamic Quantities and Their First Derivatives in a Binary System of One Phase, 1960; (with J. Murdoch) Introduction to Crystallography, 1964; (with D. V. Higgs) Angular Relations of Lines and Planes—Basic Graphical and Numerical Methods for the Solution of Geologic and Crystallographic Problems, 1967. Research and numerous articles on oxidation and enrichment of porphyry copper deposits in southwestern U. S., paragenetic relations of gold-silver telluride minerals at Cripple Creek, Colo., atomic structures of gold-silver telluride minerals, calaverite, krennerite, sylvanite, paragenetic and geochem. relations of mercury and antimony ore minerals in western U. S. and Mexico. Home: 3514 Lou Ella Lane, Riverside, Cal. 92507.*

TUNEVALL, (Thure) Gösta, Swedish physician, microbiologist; b. Karlstad, Sweden, Apr. 2, 1916; s. Thure Pettersson and Elisabet Andersson; med.lic., Carolinean Inst., Stockholm, 1943, med.dr., 1952; m. Gunnel Petterson, 1943; children—Göran, Magnus. Asst., State Bacteriological Lab., 1945-47; with Central Bacteriological Lab., Stockholm, 1947—, chief bacteriologist, 1950-56, dep. dir., 1956—; asst. prof. Carolinean Inst., Stockholm, 1953—. Cons. Hosp. Infectious Diseases, Stockholm, 1953—; Swedish del. World Med. Assn., 1951-63; mem. WHO expert panel on Pub. Health Lab. Methods, 1959—; sec. gen. permanent com. microbiol. documentation IAMS; sec. several state coms. on med. care and edn. Mem. Swedish Med. Assn. (v.p. 1961-63), Swedish Central Orgn. Academics (v.p. 1963—), Swedish Med. Soc. Author: Professional Activities of Swedish Doctors, a statis. study, 1957; also numerous articles. Editor: Abstracts and Symposia publs., VII Internat. Congress Microbiologists, 1958. Research on H influenzae, antibiotics, hypothermia, etiology of acute respiratory diseases. Home: 4 Karlaplan. Office: Box 177, Stockholm, Sweden.*

TUNIS, Marvin, Am. biochemist; b. N.Y.C., Apr. 18, 1925; s. Joseph and Edith (Flaschner) T.; A.B., Hunter Coll., 1950; Ph.D., U. Ill., 1954; m. Gilda Greenberg, Aug. 23, 1952; children—Karen Debra, Richard Barry, Susan Beth, Adam Michael. Research asso. U. Ill., 1954-56; USPHS fellow Columbia Coll. Phys. and Surg., 1955-56; sr. cancer research scientist Roswell Park Meml. Inst., Buffalo, 1957-68, State U. College, Buffalo, N.Y., 1968—. Mem. Am. Soc. Biol. Chemists, Am. Assn. for Cancer Research, Am. Chem. Soc., Sigma Xi, Phi Beta Kappa, Phi Lambda Upsilon. Research, publs. on enzymology, nucleotide synthesis by enzymes, enzyme inhibition, chem. studies with glyoproteins, tumor transplantation, biol. properties several agglutinins found in kidney bean extract. Home: 34 Clark Rd., Kenmore, N.Y. 14223. Office: 1300 Elmwood Av., Buffalo 14222.*

TUPOLEV, Andrei Nikolaevich, Russian aero. engr.; b. Pustomazovo, Oct. 29, 1888; grad. Moscow Higher Tech. Sch., 1918; Dr. Tech. Sci. An organizer Central Aerodynamic Inst., dir., 1918-35; lt. gen. Engr. Tech. Service; Recipient Stalin Prize, 1957; Lenin Prize; Order of Lenin (8). Mem. USSR Acad. Scis. Designed 1st wind tunnel, 1st all metal plane in USSR, glider, hydroplane, 2-seat airplane; dir. constrn. single-seat ANT-1, 1922, also numerous jet aircraft for commercial and military use; studies on aerodynamic calculations, material strength. Office: USSR Acad. Scis., Leninskii Prospekt, 14, Moscow, USSR.

TUPPER, Charles John, Am. physician; b. Miami, Ariz., Mar. 7, 1920; s. Charles Ralph and Grace (Alexander) T.; B.A. in Zoology, San Diego State Coll., 1943; M.D., U. Neb., 1948; m. Mary Hewes, Aug. 4, 1942; children—Mary Elizabeth, Charles John. Practice medicine, specializing internal medicine, Ann Arbor, 1954-66; research asst. Inst. Indsl. Health, U. Mich., 1951-52, research asso., 1954-56, instr. internal medicine, 1954-56, asst. prof., 1956-59, asso. prof., 1959-66, sec. Med. Sch., 1957-66, asst. dean, 1959-61, asso. dean, 1961-66; prof. medicine, dean U. Cal. Med. Sch., Davis, 1966—; mem. USPHS study group on periodic exams. Diplomate Am. Bd. Internal Medicine. Fellow A.C.P.; mem. A.M.A., Internat., Am., socs. internal medicine, Am. Coll. Health Assn., Assn. Am. Med. Colls., Am. Assn. Automotive Medicine, Alpha Omega Alpha, Nu Sigma Nu. Research in ox cell hemolysin test for infectious mononucleosis. Home: 3408 Lakeview Dr., El Macero, Davis, Ca. 95616.*

TUPPY, Hans, Austrian biochemist; b. Vienna, Austria, July 22, 1924; s. Karl and Emma (Grossmann) T.; Ph.D., U. Vienna, 1948; postgrad. U. Cambridge (Eng.), 1949-50, Copenhagen, Denmark, 1950-51, Stockholm, Sweden, 1954; m. Gertrude Veitsmeier, June 29, 1964; children—Eva, Christine. Staff, Inst. Organic Chemistry, U. Vienna, 1951—, prof., head Inst. Biochemistry, 1958—; vis. prof. Faculty Medicine, Montevideo, Uruguay, 1959, Yale, 1960. Recipient Wegscheider award Austrian Acad. Scis., 1955.

Mem. Austrian Acad. Scis., German Acad. Scis. Leopoldina. Research, publs. on structure and function of proteins and peptides, proteolytic enzymes, nucleic acids, carbohydrates. Home: 1A Kreindlgasse, Vienna, A-1190, Austria.*

TURANO, Luigi, Italian physician; b. Croton, Italy, July 26, 1899; s. Carlo and Henretia (Macry) T.; M.D., 1923; m. Maria Bonacome; 3 children. Asst., U. Rome Inst. Radiology, 1923-43; prof. radiology U. Florence, 1943-55, Med. Faculty Rome, 1955—; dir. Inst. Radiology, Radiol. Clinic, U. Rome. Recipient Gold medal Cultural Sch. Arts. Mem. Internat. Soc. Radiologists (pres.). Research, numerous publs. on physiology of esophagus, histograms of bone, primary hyperthyroid maladies, use of betetron in treatment malignant tumors, cobalt therapy. Home: 35 Via G. Ceracchi, Rome, Italy.*

TURCHINI, Jean Pascal, French histologist; b. Bastia-Corse, France, Oct. 22, 1928; s. André and Catherine (Zuccarelli) T.; Doctorat Medicine, Montpellier U., 1952, Doctorat Sci., 1962; m. Germaine de Zerbi, Aug. 13, 1959; children—Marc-André, Joseph. Agregation concours Histologie, 1958; maitre confs. agrégé Sch. Medicine, Limoges, France, 1958; maitre confs. agrégé Faculty Medicine, Clemont-Ferrand, France, 1959—, prof. histology, 1964—. Research, numerous publs. on histochemistry and overall infrastructures of organs in the newborn. Home: 47 Bd A. Briand, Clermont-Ferrand, 63, France.*

TURCK, Fenton Benedict, Am. physician; b. Milwaukee, Aug. 25, 1857; s. J. Byron and Sarah A. (Ashby) T.; prep. edn. Markham Acad., Milwaukee; M.D., Chicago Med. Coll. (Med. Dept. Northwestern U.), 1891; m. Avis L. Paine, June 10, 1897; children—Katherine Paine, Fenton Benedict. House surgeon, Alexian Bros. Hosp., Chicago, 1891-92; prof. internal medicine, Post-Grad. Med. Sch., 1893; lecturer Jefferson Med. Coll., Phila., 1896, Coll. Phys. and Surg., Chicago, 1901-02, U. of Rome, Italy, 1906. Admitted Feb. 21, 1913, to practice in N.Y. without examination by Bd. of Regents Univ. State of N.Y., "because of having attained a position of eminence and authority in his profession"; dir. Research Lab. Turck Foundation. Capt. Reserve M.C., U. S. A. Fellow New York Acad. Medicine. Del. Internat. Med. Congress 5 times between 1894 and 1913. Author: Experimental Studies in Biology. Devised instruments, including gyromele for exploration and scientific research in the alimentary tract, 1893; original research on gastritis, peptic ulcer, traumatic shock, etc.; studies in immunity on the shock phenomena and related living processes; investigations on cytost-anticytost reaction in cell division, regeneration and metabolism in plants and animals. Died Nov. 16, 1932.

TURGOT, Anne-Robert-Jacques (Baron de l'Aulne), French economist; b. Paris, May 10, 1727; s. Michel Etienne Turgot; ed. Collège Louis Le-Grand, Collège du Plessis, Séminaire de Saint-Sulpice, U. Paris. Dep. prosecutor gen., 1752, then counsellor to Parliament; master of petitions, 1753; intendant of Limoges, 1761-74; minister of Navy, 1774; comptroller gen. of finance, 1774-76; built new roads, reformed interest rates and taxation system, founded schs., established charity bureaus; while minister of finance guided by principles of no bankruptcy, no increase in taxation, no borrowing; reorganized postal and trans. service, restored free movement of grain among provinces; favored recall of the Parliaments, abolished some feudal privileges with introduction of his six edicts, 1776; reduced annual expenditures and interest on loans to restore credit; advised against French support for Americans against British; predicted it would ruin French finances; his work, for most part, undone by successor, Jacques Necker. Author: Questions Importantes sur le Commerce, 1755; Eloge de Gournay, 1759; Lettres sur la Tolérance, 1753-54; Reflexions sur la formation et la distribution des richnesses, 1766; Mémoire sur la surcharge, 1766; Mémoire sur les prets à intéret, 1769; Les six edits, 1776. Prin. founder of sci. polit. economy; Louis XVL's finance minister; removed from office because of opposition to his reforms. Died Mar. 18, 1781.

TURING, Alan Mathison, English mathematician; b. London, Eng., June 23, 1912; s. Julius Mathison and Ethel Sara (Stoney) T.; student King's Coll., Cambridge (Eng.) U., from 1931, Princeton, 1936-38; studied with Alonzo Church. With Brit. Fgn. Office, 1939-45, Nat. Phys. Lab., 1945-48; reader in math. Manchester (Eng.) U., 1948-54. Fellow of Royal Soc., 1951. Author: Programmers Handbook for the Manchester Electronic Computer, 1950. Contbr. articles to profl. jours. Involved in design, constrn., use electronic computers, 1945-54; worked (with F. C. Williams, T. Kilburn) on constrn. computer. Died June 7, 1954.

TURK, Herman, Am. sociologist; b. N.Y.C., May 29, 1924; s. Leo Franz and Lotta (Jaenke) T.; B.S. in Elec. Engring., U. Neb., 1947; M.A. in Sociology, Columbia, 1952; Ph.D. in Sociology with distinction, Am. U., 1959; m. Theresa Guminski, Dec. 7, 1959; children—Gregory, Norman. Research asst. Columbia Bur. Applied Social Research, 1950-52; staff Human Resources Research Office, George Washington U., 1952-56, research scientist, 1954-56; sociol. research contractor, Washington, 1956-57; social sci. analyst Nat. Inst. Mental Health Lab. for Socio-Environmental Studies, NIH, Bethesda, Md., 1957-58; research asso., asst. prof. sociology Duke, 1958-63; asso. prof. sociology, dir. Bur. Sociol. Research, U. Neb., 1963-66; prof. sociology, dir. Lab. for Organizational Research, U. So. Cal., Los Angeles, 1966—; adj. investigator U. S. Dept. Labor, 1966—. Cons. to univs., govt. agys. Fellow Am. Sociol. Assn.; mem. Am. Assn. for Pub. Opinion Research (past regional officer), So. (past rep. and sect. chmn.), Midwest, Eastern sociol. socs., Pacific Sociol. Assn., A.A.A.S., Am. Assn. U. Profs., Alpha Kappa Delta (exec. com. 1964—). Author: (with Thelma Ingles) Clinic Nursing: Explorations in Role Innovation, 1963; (with others) Social Aspects of Aging, 1966; also articles. Editor (with Richard L. Simpson), contbr. Institutions and Social Exchange, 1968; editor-in-chief Sociological Inquiry, 1964—. Demonstrated positive consequences of diversity in social values for social integration, conditions of overlap between affective and work relations in groups, consequences of authority relations for social intercourse, contrasting requirements placed on self and others; developed theory of representation between groups. Office: Dept. Sociology and Anthropology, U. So. Cal., Los Angeles 90007.*

TÜRKER, Kazim Rüstü, Turkish pharmacologist; b. Erzurum, Turkey, May 1, 1928; s. Mehmet and Fatma (Tayfur) T.; M.D., U. Istanbul (Turkey), 1952; m. Iffet Egitman, Jan. 3, 1957; 1 dau., Fatma Alev. With U. Istanbul Faculty Medicine, 1952-56, 58-62, docent pharmacology, 1961-62; docent pharmacology U. Ankara (Turkey) Faculty Medicine, 1962-64; research fellow Cleve. Clinic Found., 1964-66; prof. pharmacology U. Ankara Faculty Medicine, 1966—. Mem. A.A.A.S., Turkish Pharmacological Soc., Turkish Med. Assn. Author: (with S. Kaymakcalan) Deneysel Farmakoloji, 1964; also articles. Research on antibradykinin effect of noramidopyrine, effect of bradykinin on intestinal motility and its relation to catecholamines, specificity of distbn. of adrenergic receptors in cat intestine and tracheal muscle, neuro-muscular blocking effect of pronethalol, antiarrhythmic action of angiotensin on digital intoxication, action of angiotensin on sodium efflux and influx of uterus and vascular smooth muscle, mechanism of potentiation of biogen polypeptides on vascular smooth muscle. Home: 12/8 66 Sokak, Bahcelievier, Ankara, Turkey.*

TURKEVICH, Anthony, Am. nuclear chemist; b. N.Y.C., July 23, 1916; s. Leonid Jerome and Anna (Chervinsky) T.; A.B., Dartmouth, 1937; Ph.D., Princeton, 1940; m. Ireene Podlesak, Sept. 20, 1948; children—Leonid, Daria. Research asso. dept. physics U. Chgo., 1940-41, faculty Enrico Fermi Inst. for Nuclear Studies and chemistry dept.; with Manhattan Project, Columbia, 1942-43, U. Chgo., 1943-45, Los Alamos, 1945-46. Cons. AEC Labs. Recipient E. O. Lawrence award AEC, 1962. Mem. Nat. Acad. Scis., Am. Chem. Soc., Am. Phys. Soc., A.A.A.S. Research on chemical composition of the moon; chemical composition and radioactivity of meteorites; reactions of energetic particles with complex nuclei. Home: 1445 E. 56th St., Chgo. 60637.

TURKEVICH, Nicholas Joseph, Ukranian chemist; b. Ponykwa, Ukraine, USSR, Oct. 18, 1912; s. Michael and Melanie (Hrabovenska) T.; Dr.Tech. Sci., Poly. Inst. at Lvov, 1939, Dr. Pharmacy, 1955; m. Romana Kazanowska, Feb. 14, 1936; children—Andrew, George. Faculty, Poly. Inst. at Lvov, 1935—; prof. pharm. chemistry Med. Inst. at Lvov, 1945—. Author: (with G. Karpenko) The Antagonism of Drugs and Their Incompatabilities, 1958; The Pharmaceutical Chemistry, 1960; Chemistry of New Hypotensive Drugs, 1961; Nomenclature of Drugs, 1962; also numerous articles. Research, patents in chemistry of thiazolidones and complex compounds of bismuth. Home: 8 Chasanskaya, Lvov, Ukraine, USSR.*

TURMAIR, see Aventinus, Johannes.

TURNBULL, David, Am. phys. chemist; b. Elmira, Ill., Feb. 18, 1915; s. David and Luzetta Agnes (Murray) T.; B.S., Monmouth (Ill.) Coll., 1936, Sc.D. (hon.), 1958; Ph.D., U. Ill., 1939; A.M. (hon.), Harvard, 1962; m. Carol May Cornell, Aug. 3, 1946; children—Lowell D., Murray M., Joyce M. Tchr. research Case Sch. Applied Sci., 1939-46; scientist Gen. Elec. Co. Research Lab., 1946-62, mgr. chem. metallurgy sect., 1950-58; adj. prof. metallurgy Rensselaer Poly. Inst., 1954-62; Gordon McKay prof. applied physics Harvard, 1962—. Chmn. Gordon Conf. Physics and Chemistry Metals, 1952, internat. confs. Crystal Growth, 1958, Chem. Physics of Non-metallic Crystals, 1961, Office Naval Research panel study growth and morphology crystals, 1959-60. Fellow Am. Phys. Soc, N.Y. Acad. Scis., Am. Acad. Arts and Scis.; mem. Nat. Acad. Scis., Am. Soc. Metals (chmn. seminar com. 1954), Am. Inst. Mining and Metall. Engrs. (lectr. Inst. Metals div. 1961), Am. Chem. Soc. Editor: (with Frederick Seitz) Solid State Physics, 14 vols., 1955—; (with Doremus and Roberts) Growth and Perfection of Crystals, 1958. Asso. editor Jour. Chem. Physics, 1961-63; editorial adv. bd. Jour. Physics and Chemistry Solids, 1955—. Contributed to the theory for mechanism of phase changes, especially solidification of liquids and structural changes in solids; demonstrated importance of impurity particles in initiating solidification of ordinary liquid metals; contributed to measurements and theory of atom mobility in solids and liquids. Home: 77 Summer St., Weston, Mass. 02193. Office: Applied Physics Dept., Harvard Univ., Cambridge, Mass. 02138.*

TURNER, Billie Lee, Am. biologist; b. Yoakum, Tex., Feb. 2, 1925; s. James Madison and Julia Irene (Harper) T.; B.S., Sul Ross State Coll., 1949; M.S., So. Meth. U., 1950; Ph.D., Wash. State U., 1953; m. Virginia Ruth Mathis, Sept. 27, 1944, children—Billie Lee II, Matt Warnock. Faculty dept. botany U. Tex., Austin, 1953—, prof., 1959—, dir. Herbarium, 1959-65. Asso. investigator U. Ariz., 1956-57; vis. prof. U. Liverpool (Eng.), 1965-66. Recipient award N.Y. Bot. Garden, 1965. NSF sr. postdoctoral fellow, 1965-66. Mem. Bot. Soc. Am. (past sec.), Internat. Assn. Plant Taxonomists, A.A.A.S., Am. Inst. Biol. Scis., Soc. Study Evolution, Am. Assn. Plant Taxonomists, Southwestern Naturalists (gov., past pres.), Phi Beta Kappa, Sigma Xi. Author: Legumes of Texas, 1958; (with H. L. Shantz) Ecological Study of Africa, 1959; (with R. E. Alston) Biochemical Systematics, 1963. Research, numerous publs. on flora and vegetation of N.Am., Africa, Australia, leguminosae and compositae, chromosone studies of legumes and compositae, biochem. systematics, plant classification. Home: 4411 Ramsey Av., Austin, Tex. 78756.*

TURNER, C. Donnell, Am. biologist, endocrinologist; b. Curryville, Mo., Nov. 30, 1903; s. Jesse H. and Jennie (Flatford) T.; A.B., Westminster Coll., 1926; M.A., U. Mo., 1930, Ph.D., 1936. Instr., Hardin Coll., 1927-28; asst. prof. U. Ga., 1930-35; faculty Northwestern U., 1936-47; prof. biology Utica Coll., Syracuse U., 1948-49; vis. prof. zoology U. Rangoon (Burma), 1949; vis. prof. biology various univs. Japan, 1950-60; prof. biol. scis. Duquesne U., Pitts., 1960—. Mem. A.A.A.S., Am. Soc. Zoologists, Endocrine Soc. (U. S. A.), Endocrine Soc. (Japan), Am. Assn. Anatomists, Pa., N.Y. acads. scis., Internat. Fertility Assn., Limnology Soc. Am., Soc. for Exptl. Biology and Medicine, Am. Assn. U. Profs., Acacia, Sigma Xi, Gamma Alpha. Author: General Endocrinology, 1947, rev., 1955, 60, 66; also numerous articles. Research on birth defects from abnormal concentrations of hormones in blood stream mother during pregnancy. Office: Dept. Biology, Duquesne U., Pitts. 15219.*

TURNER, Charles Wesley, Am. endocrinologist; b. Chgo., Jan. 20, 1897; s. Charles Wesley and Florence (Wakeman) T.; B.S. in Agr., U. Wis., 1919, Ph.D., 1927; A.M., U. Mo., 1921; m. Katherine Conley, Dec. 2, 1961; children (by previous marriage)—Charles W. III, Marilyn Ann (Mrs. Robert Dahl), Barbara (Mrs. Lane Bauer). Faculty, U. Mo., Columbia, 1919—, prof. dairy husbandry, 1937-67, prof. emeritus, 1967—. Recipient Borden award in dairy prodn., 1940. Fulbright fellow, New Zealand, 1951-52; Japan Zootechnic Soc. fellow, 1964. Fellow A.A.A.S.; mem. Am. Dairy Sci. Assn. (dir. 1949-52), Am. Soc. Animal Prodn., Am. Poultry Sci. Assn., Am. Assn. Anatomists, Endocrine Soc. Author: Comparative Anatomy of Mammary Glands, 1939; The Anatomy of the Udder of Cattle and Domestic Animals, 1952; also articles. Research on hormones stimulating mammary gland growth and intensity of milk secretion; developed method for determining thyroxine secretion in domestic and lab. animals; devel. method for synthesis of thyroprotein. Home: 717 Hilltop Dr., Columbia, Mo. 65202.*

TURNER, Clarence Lester, Am. zoologist; b. Beaver, O., May 19, 1890; s. Marion Bartlett and Ida (Benner) T.; A.B., Ohio Wesleyan U., 1912, A.M., 1914; Ph.D., U. Wis., 1918; m. Inez Irene Kissner, Sept. 3, 1919; children—James Edmiston, Dorothy Jean (Mrs. John David Reed). Prof. zoology Wooster (O.) Coll., 1918-19, Beloit (Wis.) Coll., 1920-27; prof. zoology Northwestern U., Evanston, Ill., 1927-55, chmn. dept. zoology, 1937-41, prof. emeritus, 1955—; investigator Ohio Biol. Survey, 1918-20, Wis. State Bd. Health, 1924-26. Mem. Am. Soc. Zoologists, Am. Soc. Naturalists, Am. Soc. Ichthyologists and Herpetologists, Chgo. Acad. Scis. (past mem. bd. govs.) Research and numerous publs. on adaptive structures, embryology and endocrine control in viviparous fishes, seasonal reproductive cycles in other fishes. Home: 2518 Central St., Evanston, Ill. 60201.*

TURNER, Clarence Marshall, Am. physicist; b. Economy, Ind., Aug. 14, 1911; s. Ira Otis and Bertha (Marshall) T.; A.B., DePauw U., 1937; Ph.D., U. Wis., 1943; m. Jessie Gilchrest Brown, June 13, 1942; children—Edward Charles, Virginia Louise. Physicist, OSRD isotope separation project Princeton, 1942-43, Manhattan Project, Los Alamos, 1943-45, U. Cal. Radiation Lab., 1945-49, Brookhaven Nat. Lab., Upton, L.I., N.Y., 1949—. Mem. Am. Phys. Soc., Am. Vacuum Soc., Phi Beta Kappa. Research, publs. on electrostatic accelerator, ion source devel. for isotope separation, atomic bomb devel., Be7 branching ratio, electron and ionization loading electrostatic accelerators, reentrant electrode acceleration tube, high intensity pulsed ion sources. Home: 8 Hawkins Rd., Stony Brook, N.Y. 11790. Office: 17 Cornell St., Upton, N.Y. 11973.*

TURNER, Daniel, English physician; b. London, Eng., 1667. Mem. Barber-Surgeons Co., until 1711; surgeon. Apologia Chyrurgica, a Vindication of the Nobel Art of Chyrurgery, 1695; A Remarkable Case in Surgery, 1709; De Morris Cutaneis, a Treatise of Diseases incident to the Skin, 1714, 4th edit., 1731; Syphilis, 1717; The Art of Surgery, 2 vols., 1721, 6th edit., 1741; Discourse Concerning Fevers, 1727, 3d edit., 1731; A Discourse On Gleets, 1729; The Drop and Pill of Mr. Ward Considered, 1735. Aphrodisiacus, 1736; The Ancient Physicians Legacy, impartially surveyed, 1733. Composed cerate of calamine, wax, olive oil (Turner's cerate), 1700; med. works had little lasting value except to contain records of cases. Died London, Mar. 13, 1740/41.

TURNER, Edward, chemist; b. Jamaica, B.W.I., 1798; M.D., U. Edinburgh (Scotland), 1819; postgrad. (under Stromeyer) chemistry, mineralogy U. Göttingen (Germany), 1819-21. Lectr. chemistry U. Edinburgh, 1824-28; 1st to hold chair chemistry U. Coll., London, Eng., 1828-37. Fellow Royal Soc., 1831, Royal Soc. Edinburgh. Author: Introduction to the Study of the Laws of Chemical Combination and the Atomic Theory, 1825; Elements of Chemistry, 8 edits., 1827. Research, publs. on constn. many minerals, salts, especially ores, oxides of manganese; most important work on atomic weights of elements; concluded that all atomic weights are not simple multiples of that of hydrogen. Died Hampstead, Eng., Feb. 13, 1837.

TURNER, Edward Felix, Jr., Am. physicist; b. Newport News, Va., Apr. 21, 1920; s. Edward Felix and Marguerite (Fox) T.; B.S. in Physics, B.A. in Math., Washington and Lee U., 1950; M.S. in Physics, Mass. Inst. Tech., 1952; Ph.D. in Physics, U. Va., 1954; m. Pauline Swartz, Sept. 23, 1945; children—Andrea Lee, Elaine Felice, Paula Virginia. Faculty, George Washington U., 1954-57; faculty Washington and Lee U., Lexington, Va., 1957—, prof., 1959—, head physics dept., 1961—. Cons. Ford Found., 1964—, U. S. Office Edn., 1965—, other govt. labs. Mem. Am. Inst. Physics, Am. Assn. Physics Tchrs., Washington Philos. Soc., Va. Acad. Sci., Sigma Xi, Sigma Pi Sigma, Phi Beta Kappa. Theoretical explanation Germanium diode as switch of microwave power, developer techniques for measuring high resistances, devel. ultracentrifuge to study thin metal films. Home: 23 Sellers Av., Lexington, Va. 24450.*

TURNER, Francis John, geologist; b. Auckland, New Zealand, Apr. 10, 1904; s. Joseph Hurst and Gertrude (Reid) T.; B.Sc., Auckland U., 1924, M.Sc., 1926, D.Sc., 1933, D.Sc. (hon.), 1965; m. Esme Rena Bentham, Aug. 29, 1930; 1 dau., Gillian Bentham (Mrs. James Duncan McKercher). Came to U. S., 1946, naturalized, 1953. Lectr. geology U. Otago (New Zealand), 1926-46; faculty U. Cal. at Berkeley, 1946—, prof., 1949—, chmn. dept. geology, 1954-59. OAS tech. adviser to Brazil, 1962; vis. fellow Australian Nat. U., 1965. Guggenheim fellow, 1951, 60; Fulbright fellow, 1956. Fellow Royal Soc. New Zealand (Hector medal 1951), Geol. Soc. Am., Mineral. Soc. Am.; mem. Nat. Acad. Scis., Geol. Soc. Edinburgh (fgn.), Academia delle Scienze di Bologna (fgn.). Author: (with J. Verhoogen) Igneous and Metamorphic Petrology, 1951; (with Williams and Gilbert) Petrography, 1955; (with L. Weiss) Structural Analysis of Metamorphic Tectonites, 1963; Evolution of Metamorphic Rocks, 1948; Metamorphic Petrology, 1968; also numerous articles. Research on igneous and metamorphic rocks in New Zealand and Cal., nature and origin of metamorphic rocks, flow of crystals of minerals in solid state at high temperatures and pressures. Home: 2525 Hill Ct., Berkeley, Cal. 94708.*

TURNER, Sir George, English bacteriologist; b. Eng., 1836; ed. Cambridge U., Guy's Hosp., Montpellier; M.B. Became civil servant, med. officer of health, Cape Colony, 1895, Transvaal, 1900-08, med. supt. Pretoria Leper Asylum (all Union of South Africa). Fellow Royal Soc., 1865; mem. Royal Coll. Surgeons, licentiate Royal Coll. Physicians. Research in leprosy; developed serum used against rinderpest cattle plague, South Africa, 1896-1901. Died Mar. 12, 1915.

TURNER, Herbert Hall, English astronomer; b. Leeds, Yorkshire, Eng., Aug. 13, 1861; s. John and Isabella (Hall) T.; grad. Trinity Coll., Cambridge (Eng.) U., 1882, D.Sc., Fellow; D.Sc., Leeds, Strasbourg, Sydney, Wales univs.; D.C.L., Durham U.; fellow New Coll., Oxford U., 1893; m. Agnes Margaret Whyte, 1899; 1 dau. Chief asst. Royal Obs., Greenwich, 1884-93; Savilian prof. astronomy Oxford U., 1893-1930; pres. Royal Astron. Soc., 1903-04; gen. Sec. Brit. Assn., 1913-22; pres. seismology sect. Internat. Geophysical Union, Rome, 1922. Recipient Bruce medal Astron. Soc., 1927. Fellow Royal Soc., 1896; corr. mem. French Acad. Sci., 1908. Author: Modern Astronomy; Astronomical Discovery; The Great Star Map; A Voyage in Space. Worked in coordinating use of photography in astronomy; research in seismology; supported adoption of daylight saving time. Died Stockholm, Aug. 20, 1930.

TURNER, Louis Alexander, Am. physicist; b. Cleve., Jan. 1, 1898; s. Elmer Canfield and Lucy (Mason) T.; A.B., M.A., Cornell U., 1920; Ph.D., Princeton, 1923; m. Margaret Mather, July 20, 1931; children—Almon Richard, Elizabeth Chase. From asst. prof. to prof. Princeton, 1925-46, lectr. in physics, adviser to chmn. U. research bd., 1963—; with Radiation Lab., OSRD, Mass. Inst. Tech., Cambridge, 1940-46; prof, head physics dept. State U. Ia., 1946-50; dir. physics div. Argonne (Ill.) Nat. Lab., 1950-58, dep. dir. lab., 1958-63. Mem. Ill. Bd. Higher Edn., 1962-63. Recipient Presdl. certificate, 1946. Mem. Delta Upsilon. Research in atomic physics, spectroscopy, nuclear physics and thermodynamics, radar devel. Home: 96 Mason Dr., Princeton, N.J. 08540.*

TURNER, Malcolm Elijah, Jr., Am. biometrician; b. Atlanta, May 27, 1929; s. Malcolm Elijah and Margaret (Parker) T.; student Emory U., 1947-48; B.A., Duke, 1952; M.Exptl. Statistics, N.C. State U., 1955, Ph.D., 1959; m. Ann Clay Bowers, Sept. 16, 1948; children—Malcolm Elijah, IV, Allison Ann, Clay Shumate. Research asso. U. Cin., 1955, asst. prof., 1955-58; asst. statistician N.C. State U., Raleigh, 1957-58; asso. prof. Med. Coll., Va., Richmond, 1958-63, chmn. div. biometry, 1959-63; prof., chmn. dept. statistics and biometry Emory U., Atlanta, 1963—; analytical statistician Communicable Disease Center, USPHS, Atlanta, 1953; instr. Yale, summer 1966. Cons., Lockheed Ga. Co., 1964—. Fellow Am. Statis. Assn. (hon.), mem. A.A.A.S.; mem. Biometric Soc. (mng. editor Biometrics 1962—), Inst. Math. Statistics, Math. Assn. Am., Soc. for Indsl. and Applied Maths., A.M.A. (affiliate), Sigma Xi, Phi Kappa Phi. Research, publs. on theory and applications nonlinear regression methods in biosci. Home: 5220 Vermack Rd., Dunwoody, Ga. 30043. Office: Dept. Statistics and Biometry, Emory U., Atlanta 30322.*

TURNER, M(anson) Don, Am. physiologist; b. Pleasanton, Tex., Nov. 15, 1928; s. M. S. and D. M. (Fitzgerald) T.; B.S. in Biology, M.S., Baylor U., 1951; Ph.D. in Physiology, U. Tenn., 1955; m. Anne Jeter Cole, Apr. 17, 1953; children—John Linthicum, Deborah Ann, James Manson. Faculty. U. Miss. Sch. Medicine, Jackson, 1955—, asso. prof. surgery, asst. prof. physiology and biochemistry, 1961—; physiologist VA Hosp., Jackson, 1958—. Cons. Oak Ridge Inst. Nuclear Studies, 1962-65. Mem. A.A.A.S., Am. Chem. Soc. (past sec. Miss. sect.), Am. Physiol. Soc., Sigma Xi. Research, numerous publs. on use elec. anesthesia in clin. areas; developed practical infrared method for determination deuterium in biol. fluids, new methods for study blood flow, high resolution current gradient electrophoresis instrument; determined sensitivity protein synthetic process to hypoxia in intact animal. Home: 5847 Orchard Dr., Jackson, Miss. 39211.*

TURNER, Ralph Harold, Am. psychologist; b. Bowling Green, O., June 21, 1916; s. Virgil Erret and Mildred (Kurrley) T.; B.A., Ohio Wesleyan U., 1938, M.A., 1940; Ph.D., Ohio State U., 1947; m. Louise E. Ault, June 26, 1941. NRC fellow, 1941-42; research fellow Upjohn Inst. for Community Research, 1946-47; faculty dept. psychology Oberlin (O.) Coll., 1947—, prof., 1958—, chmn. dept. psychology, 1959—. Mem. Am. (pres. div. 2 1964-65), Midwestern, Ohio psychol. assns., A.A.A.S., Phi Beta Kappa, Sigma Xi. Author: (with T. M. Newcomb, P. E. Converse) Social Psychology, 1965; also articles. Research on responses of organisms to severe conflict, motives and attitudes of adult humans in critical choice circumstances, self concept as related to gen. behavior and adjustment, teaching methods as applied to instrn. in psychology classes. Home: 198 Shipherd Circle, Oberlin, O. 44074.*

TURNER, Ralph Herbert, Am. sociologist; b. Effingham, Ill., Dec. 15, 1919; s. Herbert Turner and Hilda Pearl (Bohn) T.; B.A., U. So. Cal., 1941, M.A., 1942; postgrad. U. Wis., 1942-43; Ph.D., U. Chgo., 1948; m. Christine Elizabeth Hanks, Nov. 2, 1943; children—Lowell Ralph, Cheryl Christine. Research asso. Am. Council Race Relations, 1947-48; mem. faculty U. Cal. at Los Angeles, 1948—, prof. sociology and anthropology, 1959—, chmn. dept. sociology, 1963—. Mem. behavioral scis. study sect. NIH, 1961-66, chmn., 1963-64; dir.-at-large Social Sci. Research Council, 1965-66. Faculty Research fellow Social Sci. Research Council, 1953-56; Sr. Fulbright scholar, U.K., 1956-57; Guggenheim fellow, 1964-65. Mem. Am. (council 1959-64, chmn. social psychology sect. 1960-61, pres. 1968), Pacific (pres. 1957) sociol. assns., Soc. Study Social Problems, (exec. com. 1962-63), Am. Council Family Relations. Author: (with L. Killian) Collective Behavior, 1957; The Social Context of Ambition, 1964; Robert Park on Social Control and Collective Behavior, 1967. Editor Sociometry, 1962-64, editorial cons., 1959-62; adv. editor Am. Jour. Sociology, 1954-56, Sociology and Social Research, 1961—; editorial staff Am. Sociol. Rev., 1955-56; asso. editor Social Problems, 1959-62, 67—. Measured relative employment status of Negroes in U. S.; extended theory of social roles and reference group behavior; studies in psycho-sociological characteristics of upward social mobility and ambition; formulated emergent norm theory for study of crowds and social movements. Home: 1126 Chautauqua Blvd., Pacific Palisades, Cal. 90272. Office: 405 Hilgard Av., Los Angeles 90024.*

TURNER, Richard Baldwin, Am. chemist; b. Mpls., Oct. 7, 1916; s. Hubert Michael and Jessie (Baldwin) T.; A.B., Harvard, 1938, M.A., 1940, Ph.D., 1942; m. Halina Deschko, June 14, 1952; children—Richard Baldwin, Tamara Elizabeth, William Reynolds. Research asso. Mayo Found., 1942-45, Mass. Inst. Tech., 1945-46; research fellow Harvard, 1946-51; faculty Rice U., Houston, 1951—, prof. chemistry, 1956—, chmn. dept. chemistry, 1960-63. Mem. Am., Swiss chem. socs., Chem. Soc. (London, Eng.), Nat., N.Y. acad. scis. Research, numerous publs. on chemistry natural products including steroids, terpenes, alkaloids, total stereospecific synthesis, heats catalytic hydrogenation. Home: 6219 Chevy Chase St., Houston 77027.*

TURNER, Thomas, Brit. metallurgist; b. Birmingham, Eng., 1861; s. H. Turner; ed. Royal Sch. Mines, London; M.Sc.; m. Christian Smith, 1887; 2 son, 2 daus. Became demonstrator chemistry Mason Coll., 1883, lectr. metallurgy, 1887; dir. tech. instrn. to Staffordshire County Council, 1894-1902; prof. metallurgy U. Birmingham (Eng.) 1902-26, then prof. emeritus, also life gov.; examiner metallurgy univs. London, Edinburgh, Glasgow, Manchester, Birmingham, Wales; hon. fellow Imperial Coll. Mem. adv. council Imperial Inst. Recipient De la Beche medal; Bessemer Gold medal; Seaman Gold medal Am. Foundrymen's Assn.; Fox Gold medal Inst. Brit. Foundrymen. Fellow Royal Inst. Chemistry; mem. Inst. Metals (pres.), Brit. Cast Iron Assn. (mem. council), Brit. Non-Ferrous Research Assn. (mem. council). Author: Metallurgy of Iron, 1895; Lectures on Iron Founding, 2d edit. 1910; Practical Metallurgy, 2d edit., 1919; also numerous articles. Research on effect of silicon and other elements on iron and steel (applications used in iron foundries worldwide); studies on structure, hardness, volatility of metals. Died Jan. 31, 1951.

TURNER, Thomas Bourne, Am. physician; b. Prince Frederick, Md., Jan. 28, 1902; s. George Dorsey and Virginia (Lyles) T.; B.S., St. John's Coll., Annapolis, 1921; M.D., U. Md., 1925, Sc.D., 1966; m. Anne Parran Somervell, Oct. 22, 1927 (dec. Feb. 1960); m. 2d, Lorna M. Levy, Sept. 16, 1961. Practice medicine, specializing in microbiology, Balt., 1927—; faculty Med. Sch. Johns Hopkins, 1927—, prof. microbiology, 1939—, dean med. faculty, 1957—; physician Moore Clinic, 1961—. Com. chmn. Nat. Found. Fellow Am. Pub. Health Assn.; mem. Assn. Am. Med. Colls. (pres.), Assn. Am. Physicians, Am. Social Hygiene Assn., Am. Soc. Clin. Investigation, Am. Venereal Disease Assn. (past pres.), A.M.A. Author: Biology of the Treponematoses, 1957; Fundamentals of Medical Education, 1963. Research, numerous publs. on immunology, diagnosis, characterization, control, treatment of syphillis, particularly in mil. personnel. Home: 1426 Park Av., Balt. 21217.*

TURNER, Victor Witter, anthropologist; b. Glasgow, Scotland, May 28, 1920; s. Norman and Violet (Witter) T.; B.A. with honors, U. Coll. London (Eng.), 1949; Ph.D., U. Manchester (Eng.), 1955; m. Edith Lucy Brocklesby Davis, Jan. 30, 1943; children—Frederick, Robert, Irene Helen, Alexander Lewis Charles, Rory Peter Benedict. Research asst. U. Manchester (Eng.), 1949-50; research officer Rhodes-Livingstone Inst. for Sociol. Research, Lusaka, No. Rhodesia, 1950-54; faculty U. Manchester, 1955-63, Simon Research fellow, 1956-58, sr. lectr., 1961-63; prof. anthropology Cornell U., Ithaca, N.Y., 1963—. Center for Advanced Study in Behavioral Scis. fellow, Palo Alto, Cal., 1961-62; Soc. for Humanities fellow Cornell U., 1967-68. Mem. Am. Anthrop. Assn., Assn. Social Anthropologists Gt. Britain, Royal Anthrop. Inst., Internat. African Inst. Author: Schism and Continuity in an African Society, 1957; Ndembu Divination, 1961; Chihamba, the White Spirit, 1962; (with M. Gluckman, D. Forde, M. Fortes) Essays in the Ritual of Social Relations, 1962; The Forest of Symbols, 1967; also articles. Research on intra-group conflict in preindsl. soc. using extended-case method; analysis of ritual symbols showing different levels of meaning, relationship between ritual and social processes in both repetitive and changing socs. Home: 100 Winthrop Dr., Ithaca, N.Y. 14850.*

TURNER, William, Brit. botanist; b. Morpeth, Eng., circa 1510; s. William Turner; B.A., Pembroke Hall, Cambridge, 1529-30, M.A., 1533, postgrad. until 1540; postgrad., Oxford; studied botany under Ghini, Bologna, Italy, 1542; M.D., Bologna or Ferrara, Italy; M.D., Basel, Switzerland, 1543, Cologne, Germany, 1544; m. Jane Auder; children—Peter, Winifred (Mrs. Parker), Elizabeth (Mrs. Whitehead). Collected plants in Rhine country, Holland, East Frieseland; physician to Earl of Emden, Frieseland; returned to Eng. on accession of Edward VI; chaplain, physician to Duke of Somerset; mem. Commons; became prebend, York, Eng., 1550, dean, Wells, 1550; named lectr., Isleworth, 1551; ordained priest, 1552; deprived of office of dean, 1553; lived abroad during Mary's reign; returned to Eng., 1559; again named dean of Wells, 1560; suspended for nonconformity, 1564. Author: Unio Dissidentium, 1538; Libellus de re herbaria novus (1st essay on sci. botany in Eng.), 1538; The huntynge and fyndynge out of the Romishe Fox . . . hyd among the Bysshoppes of Englande, 1543; Historia de naturis herbarium; Avium praecipuarum . . . , 1544; The rescuynge of the Romishe Fox . . . deuised by steven gardiner, 1545; The names of herbes in Greke, Latin, Englishe, Duche, and Frenche, 1548; A newe Dialogue . . . examination of the Messe, 1548; Dialogus de avibus et edrum nominibus (1st modern ornithological work); A Preservative or Triacle agaynst the poyson of Pelagius, 1551; A newe Herball wherein are conteyned the names of Herbes, 1551; The huntyng of the Romyshe Wolfe, circa 1554; The booke of Merchants newly made by the lord Plantapole, before

1555; The Spiritual Nosegay; A new Booke of Spirituall Physick for dyverse diseases of the Nobilitie and Gentlemen of Englande, 1555; The seconde parte of W. T.'s Herball; Hereunto is joined a book of the bath of Baeth, 1562; A New Boke of the natures and properties of all Wines commonlye used here in England; 1568; The first and seconde partes of the Herbal . . . , 1568. First English Sci. botanist; introduced lucern (horned clover to Eng.); named goatsbeard and Hawkweed. Died London, July 7, 1568.

TURNER, Sir William, Brit. anatomist; b. Lancaster, Eng., Jan. 7, 1832; s. William and Margaret (Aloren) T.; apprenticed to Dr. C. Johnston; grad. St. Bartholomew's Hosp., 1853; M.B., U. London, 1857; LL.D., Glasgow, St. Andrews, Aberdeen, Montreal, Western U. Pa.; D.C.L. (hon.), Oxford, Durham, Toronto; D.Sc. (hon.), Dublin, Cambridge; m. Agnes Logan, 1863; 3 sons, 2 daus. Became sr. demonstrator dept. anatomy U. Edinburgh, Scotland, 1854, prof. anatomy 1867, dean Faculty Medicine, 1878-81, prin., 1903-16. Mem. Royal Commn. to Consider Problem of Fixing Requirements for Profl. Practice, 1881; rep. of univs. Edinburgh and Aberdeen on Gen. Med. Council, 1873-83, rep. U. Edinburgh, 1886-1905. Fellow Royal Soc., 1877, Royal Soc. Edinburgh, Royal Coll. Physicians, Royal Coll. Physicians Edinburgh; mem. Royal Irish Acad. (hon.), Brit. Assn. for Advancement Sci. (became pres. 1900). Knighted, 1886. Editor: Lectures on Surgical Pathology (Paget); Atlas and Handbook on Human Anatomy and Physiology, 1857. A founder, editor Jour. Anatomy and Physiology. Research and numerous publs. on anthropology, anatomy, physiology, especially placentation of mammals, craniology of man; comparative anatomy of sea mammals. Died Edinburgh, Feb. 15, 1916.

TURPIN, Eugene, French chemist, inventor; b. Paris, France, 1848; Recipient Montyon prize, 1877. Author: Comment on a vendu la mélinite. Developed harmless dyes; studied explosives; made picric acid usable; invented melinite, 1877; pantentee process for mfg. shells; also explosives: panclastites, pyrodialites, chlorated powders. Died Pontoise, 1927.

TURPIN, Pierre-Jean-François, French botanist; b. Vire, France, Mar. 11, 1775; army druggist at Saint-Dominque; botanist Jardin des Plantes; mem. French Acad. Scis., 1833. Author: Flores Parisienses, 1818; Essai d'une iconographie des végétaux, 1820; also editor of bot. work with plates. Researches in plant physiology. Died Paris, France, May 1, 1840.

TURPIN, Raymond Alexandre, French genticist; b. Pontoise, France, Nov. 5, 1895; s. Charles Emile and Marguerite (Crunel) T.; M.D., Paris, 1925; m. Simone Henriette Gaillochet, Apr. 29, 1931; children—Bernard, Jean-Claude, Jacques, Marie-Hélène (Mme. Couturier), Gérard, Béatrice. Chef de service Hérold, Bretonneau, St.-Louis, Trousseau, Enfants Malades hosps.; agrégé, 1934; became prof. Paris U., 1947; founder, dir. Institut de Progénèse, 1959—; named prof. Clinique medicale des enfants, 1956. Mem. Nat. Acad. Medicine, French Acad. Scis. Commdr. of Fr. Legion of Honor. Author: Tétanie, 1925; Hérédité des prédispositions morbides, 1951; (with J. Lejeune) Progénèse, 1955, Chromosomes humains, 1965; also articles. Research on problems of human genetics, including sex determination, ionizing radiations, mongolism; discovered 1st examples of human chromosomal aberrations in mongolism (numerical), polydysspondyly (structural), heterocaryotic monozygotic twins; discovered electromyographic syndrome of human botany; developed (with Weill-Halle) 1st prevention against Tb in human beings by BCG. Home: 94, av. Victor Hugo, Paris. Office: Hopital des Enfants Malades, 149, rue de Sevres, Paris, France.*

TURQUET DE MAYERNE, Sir Theodore, naturalist, physician; b. Mayerne, nr. Geneva, Switzerland, Sept. 28, 1573; s. Louis and Louise (le Maçon) T. de M.; ed. U. Heidelberg; M.B., U. Montpellier, 1596, M.D., 1597; m. Marguerite de Boetslaër; 2 children; m. 2d, Elizabeth Joachimi; 2 sons, 3 daus., including Elizabeth de Caumont. Royal dist. physician, Paris, 1600; lectr. on medicine, openly used and defended chem. remedies; condemned unanimously by Coll. Physicians, U. Paris, 1603, recommended he be deprived of his office; went to Eng., presented to king who apptd. him physician to queen, 1606; incorporated M.D. at Oxford, 1606; returned to Paris, until 1611; returned to London, apptd. 1st physician to King James I, acquired large practice; physician to Charles I, Charles II. Fellow Royal Coll. Physicians. Author: Apologia in qua videre est inviolatis Hippocratis et Galeni legibus remedia chymice preparata, tuto usurpari posse, 1603; Pictoria Sculptoria et quae subalternarum artium, 1620-46; Prophylactica pro Principibus in regia Sti Jacobi habitantibus, 1644; Pharmacopoea, 1703. Made pharm. expts. with useful results; brought calomel into use; 1st prepared mercurial lotion called blackwash; expts. on pigments and enamels, discovered purple color used in carnation tints in enamel painting; invented washable varnished memorandum tablet; described hydrogen and probably observed inflammability of hydrogen before Boyle; wrote valuable account of typhoid fever, 1612. Died Chelsea, Eng., Mar. 22, 1655.

TURRELL, Franklin Marion, Am. plant physiologist; b. Pitts., Mar. 21, 1905; s. Marion Charles and Sarah (Waddell) T.; B.E.E., Eastern Ill. State U., 1929; M.S., U. Ia., 1932, Ph.D., 1935; m. Mary Margaret Eastabrooks, June 4, 1929; 1 son, Vincent Eastabrooks. With U. Ia., 1931-35, research asst., 1932-35; instr. plant physiology U. Cin., 1935-36; with U. Cal., Riverside, 1936—, prof. biochemistry, 1961—, radiol. adviser, 1949-54, dir. atomic energy commn. project, 1950-55. Fellow A.A.A.S.; mem. Synapsis Club (pres.), Am. Meteorol. Soc. (v.p.), Am. Soc. Plant Physiology, Bot. Soc. Am., Am. Chem. Soc. (sect. pres.), Am. Hort. Soc., Math. Assn. Am., N.Y. Acad. Sci., Scandinavian Soc. Plant Physiology, Internat. Soc. Bioclimatology and Biometeology, Sigma Xi. Author: Tables of Surfaces and Volumes of Spheres and of Prolate and Oblate Spheriods and Spheroidal Coefficients, 1946. Publs. on devel. of methods, elucidation of laws of plant growth; role of water in higher plant growth; effects of pesticides on plants. Home: 3574 Bandini Av., Riverside, Cal. 92506, and 33-851 Golden Lantern, Dana Point, Cal. 92629.*

TURRITTIN, Hugh Lonsdale, Am. mathematician; b. Rice, Minn., Apr. 24, 1906; s. Albert Hugh and Bessie (Bouck) T.; B.S. in Civil Engring., U. Minn., 1927; M.S. in Math., U. Wis., 1932, Ph.D., 1933; postgrad. U. Rome (Italy); m. Adele Fritz, June 18, 1932; children—Anton Hugh, Rachel Mary (Mrs. Paul Louis René Schmitt), Ute Porsche (adopted). Faculty, Tex. Western U., 1934-38, asst. prof., 1936-38; faculty U. Minn., Mpls., 1939—, prof. math., 1952—, acting head math. dept. Inst. Tech., 1958-59, dir. undergrad. studies in math., 1964—. Fulbright math. prof. U. Innsbruck (Austria), 1955-56. Mem. Am. Math. Soc., Am. Math. Monthly. Research, publs. on solutions ordinary differential and difference equations. Home: 4046 Beard Av., S., Mpls. 55410.*

TUTHILL, Sir George Leman, English physician; b. Halesworth, Eng., Feb. 16, 1772; s. John and Sarah (Jermyn) T.; B.A., Cambridge, 1794, M.A., 1809, Licence Adractcandub, 1812, M.D., 1816; m. Maria Smith, circa 1794; a dau., Laura Maria (Mrs. Bowett). Prisoner in France, circa 1796-1809; became Gulstonian lectr., 1818; physician to Westminster, Bridewell, Bethlehem hosps.; became mem. Com. for preparation Pharmacopdeia Londinensis, 1824; apptd. Harveian orator, 1835. Fellow Royal Soc., 1810, Coll. physicians. Author: Vindiciae medicae, 1834. Made (with William George Maton) various reforms in Royal Coll. Physicians. Died Apr. 7, 1835, London.

TUTHILL, Leonard Dale, Am. entomologist; b. Baxter Springs, Kan., Jan. 24, 1911; s. Leonard Sylvester and Myrtle (Hodgkins) T.; B.A., U. Kan., 1929, M.A., 1930; Ph.D., Ia. State Coll., 1941; m. Shirley Bragg Penruddocke, May 11, 1941; children—Rosalind Lenore, Jonathan Dale, Edmund Leonard, Maile Letitia. Instr., Ia. State Coll., 1937-45; entomologist Ida. Pest Control Commn., asst. entomologist Ida. Expt. Sta., 1945-46; asso. prof., asst. extension entomologist U. Ida., 1946-47; faculty U. Hawaii, Honolulu, 1947—, prof. entomology, 1949-67; provost Leeward Oahu Community Coll., 1967—, acting dean Grad. Sch., 1963-64, acting dean Internat. Coll., 1961, acting dean Faculties, 1959-60, acting dean Grad. Sch., dir. research, 1957-58, chmn. dept. zoology, 1954-57. Fulbright lectr. U. Cairo (Egypt), 1965-66. Fulbright Research fellow to New Zealand, 1950-51, Peru, 1958-59. Fellow A.A.A.S., Entomol. Soc. Am.; mem. Hawaiian Acad. Sci. (pres. 1962-63), Hawaiian, Kan. entomol. socs., Hawaiian Bot. Soc. (pres. 1957), Sociedad Entomol. del Peru, Société Entomol. d'Égypte (hon.), Sigma Xi (pres. local chpt. 1952-53). Research, publs. on discovery and description various new kinds of insects of the family Psyllidae or jumping plant lice. Home: 2610 Manoa Rd., Honolulu 96822.*

TUTTLE, Charles Wesley, Am. astronomer; b. Newfield, Mass., Nov. 1, 1829; s. Moses and Mary (Merrow) T.; attended Harvard Law Sch., 1854, A.M. (hon.), 1854; Ph.D. (hon.), Dartmouth, 1880; m. Mary Louisa Park, Jan. 31, 1872. Asst., Harvard Observatory, 1850; participated in eclipse expdn. to summit of Mt. Washington, 1854; admitted to Mass. bar, 1856; practiced, Boston, 1856; admitted to practice in U. S. circuit cts., 1858, U. S. Supreme Ct., 1861; took testimony for use before Ct. of Ala. Claims, 1874; made hist. and antiquarian studies of Me. and N.H. Author: Captain John Mason (edited by J. W. Dean), published posthumously, 1887; also numerous articles published in New Eng. Hist. and Geneal. Register, Proceedings of Mass. Hist. Soc., Notes and Queries. Made important contbn. to astronomy by explaining Saturn's "dusky" ring, 1850; discovered a comet, 1853; computed cometary orbits and ephemirides. Died Boston, July 17, 1881.

TUTTLE, Orville Frank, Am. geochemist; b. Olean, N.Y., June 25, 1916; s. Orvel Delano and Lucy (Holmes) T.; B.S., Pa. State U., 1939, M.S., 1940; Ph.D., Mass. Inst. Tech., 1948; m. Dawn Hardes, Nov. 21, 1941; children—Jean Lynn, Anne Laura. Research asso. Mass. Inst. Tech., 1942-43; phys. chemist Geophys. Lab., Washington, 1943-45, Naval Research Lab., 1945-47; petrologist Geophys. Lab., Washington, 1945-52; prof. geochemistry, chmn. div. earth scis. Pa. State U., 1953-59, dean Coll. Mineral Industries, prof. geochemistry, 1959-60, prof. geochemistry, 1960-65; prof. geochemistry Stanford, 1965—. Fellow Geol. Soc. Am., Mineral Soc. Am. (mem. council 1960-63, award 1951); mem. Nat. Acad. Scis., Mineral. Soc. London, Am. Geophys. Union, Geochem. Soc. (mem. council 1960-63), Geol. Soc. London (fgn.), Sigma Xi. Contbr. numerous articles to tech. jours. Invented pressure vessel for high pressure studies rocks and minerals; discovered new modification feldspar and demonstrated its presence in volcanic rocks; prepared synthetic granite in lab.; demonstrated that certain granites crystallized from magnas at reasonably shallow depths in earth's crust. Home: 3409 Greer Rd., Palo Alto, Cal. 94303. Office: Dept. Geology, Stanford, Stanford, Cal. 94305.*

TUVE, Merle Antony, Am. research physicist; b. Canton, S.D., June 27, 1901; s. Anthony G. and Ida Marie (Larsen) T.; prep. edn. Augustana Acad., Canton, S.D., 1915-18; B.S. in elec. engring., U. Minn., 1922, A.M., 1923; Ph.D., Johns Hopkins, 1926; D.Sc., Case, Kenyon, Williams, Johns Hopkins, U. Alaska; LL.D., Augustana College, Carleton; m. Winifred Gray Whitman (M.D.), Oct. 27, 1927; children—Trygve Whitman, Lucy Winifred. Teaching fellow U. Minn., 1922-23; instr. in physics, Princeton U., 1923-24, Johns Hopkins, 1924-26; staff mem., dept. of terrestrial magnetism, Carnegie Inst., 1926-46, dir. since 1946; on leave, 1940-46, war work (proximity fuse, etc.). Chmn. Sect. T. OSRD, 1940-45; dir. Applied Physics Lab., Johns Hopkins (Navy) 1942-46; U. S. exec. com. Nat. Com. for IGY, 1954—; mem. Pres.'s Sci. Adv. Com.-Internat. Science Panel. Received A.A.-A.S. prize (with L. R. Hafstad and O. Dahl), 1931, Presidential Medal of Merit, 1946; Research Corp. Award 1947; comdr. Order of Brit. Empire, 1948; John Scott Award, 1948; Comstock Prize, Nat. Acad. Sci., 1949; Howard N. Potts medal Franklin Inst., 1950; Achievement Medal, University Minn., 1950, Barnard medal, Columbia U., 1955, Medal Condor de los Andes (Bolivia), Bowie medal Am. Geophys. Union, 1963. Fellow Am. Physics Society, Institute Radio Engineers, A.A.A.S., Am. Acad. Arts and Sci.; mem. Am. Philos Soc., Nat. Acad. Scis., Philos. Soc. Wash., Wash. Acad. Scis., Sigma Xi, Phi Beta Kappa. Contbr. to Physical Review and other scientific jours. on nuclear physics, geophysics and biophysics. Editor Journal Geophysical Research, 1949-58. Co-developer radio-pulse transmission techniques; co-producer beta and gamma rays, high velocity protons; conducted seismic measurements of earth's crust; studied geophysics of Andean altiplano; worked on telescope image tubes, radio astronomic study hydrogen clouds, nuclear physics. Home: 135 Hesketh St., Chevy Chase, Md. Office: 5241 Broad Branch Rd., N.W., Washington 15, D.C.

TWEDDELL, Ralph Hart, engr., inventor; b. South Shields, Eng., May 25, 1843; s. Marshall Tweddell; ed. Cheltenham Coll.; m. Hannah Mary Grey, 1875. Apprentice, R.&W. Hawthorn of New Castle-on-Tyne engrs., 1861. Recipient Telford medal and premium, Gold medal Soc. Arts, Bessemer premium, 1890. Mem. Instn. Civil Engrs., Instn. Mech. Engrs. Inventor portable machinery which effected great changes in shipbldg. and constrn. of bridges and boilers; designed portable riveting machine, 1871. Patentee portable hydraulic apparatus, 1865, stationary hydraulic machine, 1866. Died Kent, Eng., Sept. 3, 1895.

TWEEDELL, Kenyon S., Am. zoologist; b. Sterling, Ill., Mar. 28, 1924; s. Henry Kirkham and Agnes (Wiederin) T.; student DePauw U., 1943-44; B.S., U. Ill., 1947, M.S., 1949, Ph.D., 1953; m. Joan Ellen Werber, Feb. 4, 1956; children—Eric P., Karen M., Kristin L., Lisa A., Brigit. Research asso. Control Systems Lab., U. Ill., 1951-54; faculty U. Me., 1954-58; faculty U. Notre Dame (Ind.), 1958—, professor dept. biology, 1968—, staff mem. Radiation Lab., 1960—. Corp. mem. Marine Biol. Lab., Woods Hole, Mass., 1956—. Mem. Am. Soc. Zoologists, Soc. for Developmental Biology, A.A.A.S., Ind. Acad. Sci., Sigma Xi. Research, publs. on twinning, egg devel., regeneration, tumor etiology, cell chemistry, adaptation kidney cancer of amphibia, irradiation and embryonic devel. Home: 210 E. Bartlett St., South Bend, Ind. 46601. Office: Dept. Biology, U. Notre Dame, Notre Dame, Ind. 46556.*

TWENHOFEL, William Henry, Am. geologist; b. Covington, Ky., Apr. 16, 1875; s. Ernst August Herman Julius and Helena (Steuwer) T.; B.A., Nat. Normal U., Lebanon, O., 1904; B.A., Yale, 1908, M.A., 1910, Ph.D., 1912; m. Virgie Mae Stephens, Sept. 10, 1899; children—Lillian Helena, Helen Vivian, William Stephens. Tchr. village and country schs., Ky., 1896-1902; tchr. sci. and mathematics, E. Tex. Normal Coll., Commerce, Tex., 1904-07; asst. and asso. prof. geology, U. Kan., 1910-16; state geologist of Kan., 1915-16; asso. prof. geology, 1916-21, U. Wis., prof., 1921—, chmn. dept., 1940—, ret., 1945. Prof., summer session, Stanford U., 1930. Chmn. Com. on Sedimentation, Nat. Research Council, 1923-31, chmn. Div. Geology and Geography, 1931-34, chmn. com. on Paleocology, 1934-37, geol. work in Kan., St. Lawrence, Baltic regions, Upper Miss. Valley. Mem. Geol. Soc. Am., Paleontol. Soc. Am. (pres. 1930), A.A.A.S., Wis. Acad. Sci., Am. Assn. Petroleum Geologists, Soc. Econ. Paleontologists and Mineralogists (pres. 1935); chmn. research com., 1938—, Sigma Xi, Phi Beta Kappa. Author: Treatise on Sedimentation, 1926; Invertebrate Paleontology, 1935; Principles of Sedimentation, 1939; Methods of Study of Sediments, 1941. Contbr. on sedimentation, stratigraphy and paleontology. Died Jan. 4, 1957.

TWITCHELL, Amos, Am. surgeon; b. Dublin, N.H., Apr. 11, 1781; s. Samuel and Alice (Willson) T.; A.B., Dartmouth, 1802, A.M., B.M., 1805, M.D., 1811; studied medicine under Nathan Smith; m. Elizabeth Goodhue, 1815. Practiced medicine, Keene, N.H., 1810-1850; overseer, Dartmouth, 1816; pres. N.H. Med. Soc., 1829-30; mem. A.M.A., Coll. Physicians Phila., Nat. Instn. Promotion Sci., 1841. Leading surgeon in Northern New Eng.; one of 1st in U. S. to perform extensive amputations for malignant disease, operations for stones in bladder and ovarian tumors, tracheotomy, trephining of long bones for suppuration; performed 1st tying off of carotid artery in U. S., 1807. Died Keene, N.H., May 26, 1850.

TWITTY, Victor Chandler, Am. biologist; b. nr. Shoals, Ind., Nov. 5, 1901; s. John McMahon and Emma (Chandler) T.; B.S., Butler Coll., 1925; Ph.D., Yale, 1929; m. Florence Mae Eveleth, Aug. 3, 1934; children—John, Anne (Mrs. Cecil Cutting). Instr., Yale, 1929-31; NRC fellow Kaiser Wilhelm Institut, Berlin, Germany, 1931-32; faculty Stanford, 1932-—, prof., 1936-—, Herzstein prof. biology, 1963-—, exec. head dept. biol. scis., 1949-63. Mem. Am. Soc. Zoologists (past pres.), Soc. for Study Devel. (past pres.), Western Soc. Naturalists (past pres.), Internat. Soc. Cell Biology, Internat. Inst. Embryology, Nat. Acad. Scis., Am. Acad. Arts and Scis. Research, publs. on microsurg. study embryos and larvae amphibians; discovered new species Californian salamanders, used for research in embryological studies, evolution, genetics and behavior. Home: 714 Alvarado Row, Stanford, Cal. 94305.*

TWORT, Frederick William, Brit. bacteriologist; b. Camberley, Eng., Oct. 22, 1877; s. William Henry and Elizabeth Crampton (Webster) T.; ed. St. Thomas's Hosp. Med. Sch.; m. Dorothy Nony Banister, 1919; 1 son, 3 daus. Asst. supt. clin. lab. St. Thomas' Hosp., 1901-02; asst. bacteriologist London Hosp., 1902-09; prof. bacteriology Faculties Medicine and Sci., U. London; became supt. Brown Inst., 1909. Recipient Roger's prize U. London, 1908. Licentiate Royal Coll. Physicians. Fellow Royal Soc., 1929; mem. Royal Coll. Surgeons. Research on Johne's Disease; described 1st vitamin (K); cultured leprosy bacillus, 1910; discovered bacteriophage (virus which destroys certain types of bacteria), 1915. Died Camberley, Eng., Mar. 20, 1950.

TYGSTRUP, Niels, Danish physician; b. Vejle, Denmark, May 27, 1926; s. Svend and Thora (Tork) T.; grad. U. Copenhagen (Denmark), 1952, M.D., 1966; m. Inge Sort, May 27, 1950; children—Pernille, Frederik. Registrar med. dept. B, also cardiologic Lab., Rigshospitalet, U. Copenhagen, 1957-61, faculty, 1963-—, lectr. internal medicine, 1966-—; sr. registrar med. dept. B, Bisperbjerg Hosp., Copenhagen, 1962-65. Mem. N.Y. Acad. Scis., Scandinavian Assn. Internal Medicine (sec. 1965-—), Soc. for Theoretic and Applied Therapy (mem. bd. 1966-—), European Assn. for Study Liver (mem. selection com. 1966-—), European Soc. Clin. Investigation. Author: Liver Function Determinations by an Intravenous Galactose Test, 1966; also numerous articles. Research on methods for determination of liver blood flow and functional liver mass by galactose, effects of alcohol on liver, demonstrated serum cholesterol lowering effect of paraaminosalicylic acid in hypercholesterolemic man, description of spl. type of intermittent jaundice. Home: 26, Berlings Bakke, Charlottenlund, Denmark. Office: Med. Dept. A, Rigshospitalet, Copenhagen, Denmark.*

TYLER, Albert, Am. biologist; b. Bklyn., June 26, 1906; s. Manuel H. and Lena (Miller) T.; A.B., Columbia, 1927, A.M., 1928; Ph.D., Cal. Inst. Tech., 1929; m. Betty Spivack, Apr. 19, 1926; children—James S., Steven R. Faculty, Cal. Inst. Tech., Pasadena, 1928-—, prof. biology, 1950-—; vis. prof. zoology U. Cal., Los Angeles, 1962. Cons. NSF, Washington, 1956-—; Marineland of Pacific, 1959-—, VA, Washington, 1966-—; Philips visitor Haverford Coll., 1960; del. 6th Internat. Conf. Planned Parenthood, New Delhi, India, 1959, 7th Internat. Conf., Singapore, 1963; mem. sci. group on biology human reproduction WHO, 1963; mem. various coms. Nat. Acad. Sci., Nat. Inst. Child Health and Human Devel. Fellow A.A.A.S., Institut Internat. d'Embryologie; mem. Am. Soc. Zoologists, Am. Assn. Immunologists, Am. (past pres.), Western (past pres.) socs. naturalists, Soc. Developmental Biology, Soc. Cell Biology, Soc. Gen. Physiologists (past pres.), Soc. Exptl. Biology and Medicine, Sigma Xi. Contbr. numerous articles to profl. jours. Formulation of immunological concepts of fertilization and devel., of origin and pathogenesis of tumors; discovered role of masked messenger ribonucleic acids in control of protein synthesis; analysis of energetics of devel., changes in respiration, elec. potential, other physico-chem. properties. Office: Cal. Inst. Tech., Pasadena, Cal. 91109.*

TYLER, David B(ernard), Am. physiologist, pharmacologist; b. N.Y.C., Jan. 13, 1905; s. Manuel Harry and Lena (Miller) T.; A.B., U. So. Cal., 1933, Ph.D., Sch. of Medicine, 1937; m. Matilda Fischer, June 16, 1929. Lab. instr. U. So. Cal. Sch. Medicine, 1933-37, lectr., 1937-47; research asso. and Hixon Fund fellow Cal. Inst. Tech., 1940-46; asst. chief physiologist Army Chem. Center, 1947; staff dept. embryology Carnegie Inst., Washington, 1947-50; prof., head dept. pharmacology U. P.R. Sch. Medicine and Dentistry, 1950-61; program director, member panel regulatory biology NSF, 1961-63, head physiol. sci. sect., 1964-—; vis. prof. Med. Coll., Trivandrum, India, 1959-60; fgn. service officer Tech. Coop. Mission, ICA, New Delhi, India, 1959-60. Investigator, OSRD, 1942-46, mem. panel on motion sickness, 1943-46. Mem. Am. Physiol. Soc., Soc. Exptl. Biology and Medicine, Am. Soc. Pharmacology and Exptl. Therapeutics, A.A.A.S., Am. Assn. U. Profs., Am. Acad. Neurology, Am. Assn. Med. Colls. Author sci. articles. Research in brain metabolism; mechanisms producing brain damage during insulin shock; biochem. basis for functional changes in brain during development; effect prolonged wakefulness on brain function; experimental psychoses; battle fatigue; motion sickness. Home: 8717 Jones Mill Rd., Chevy Chase, Md. 20015. Office: Nat. Sci. Found., Washington 20550.*

TYLER, Frank Hill, Am. physician; b. Villisca, Ia., Jan. 5, 1916; s. Royal F. and Fausta (Hill) T.; A.B., Willamette U., 1938; M.D., Johns Hopkins, 1942; m. Alida Woolley, Mar. 19, 1962; children—Karen J. (Mrs. James A. Stewart), Royal H., F. Peter. Practice medicine, specializing in metabolism, Salt Lake City, 1947-—; faculty U. Utah, 1947-—, prof. medicine, 1959-—. Mem. Am. Soc. Clin. Investigation, Assn. Am. Physicians, Endocrine Soc., Am. Clin. and Climatol. Soc. Research, numerous publs. on classification and clin. manifestation of muscle disease, particularly muscular dystrophy; physiol. regulation of cortisol secretion and adrenal disease. Home: 1130 Vine St., Salt Lake City 84121. Office: 50 N. Medical Dr., Salt Lake City 84112.*

TYLER, John Edwards, Am. physicist; b. Boston, Nov. 11, 1911; s. Lucius Spaulding and Florence Eaton (Edwards) T.; B.S., Mass. Inst. Tech., 1940; m. Frances Christine Blackwood, Aug. 10, 1940; children—Jean Christine (Mrs. Donald Walton), Elizabeth Edwards (Mrs. Thomas Coleman), Florence Edwards, Marguerite Larnard. Research asso. Mass. Inst. Tech., 1940-41; research physicist Nat. Research Corp., 1941-44, Interchem. Corp., N.Y.C., 1944-52; with Scripps Inst. Oceanography, U. Cal., La Jolla, 1952-—, research physicist, 1959-—. Mem. various coms. UNESCO, Nat. Acad. Scis., Fellow Optical Soc. Am., A.A.A.S.; mem. Internat. Assn. Phys. Oceanography, Sigma Xi. Contbg. author: The Sea, 1962; McGraw Hill Ency. Sci., 1960; also numerous articles. Inventions in optics and electro-optical devices; research in optics, optical instrumentation, spectroscopy, optical oceanography. Home: 7740 E. Roseland Dr., La Jolla, Cal. 92037.*

TYLER, Max Ezra, Am. microbiologist; b. Groveland, N.Y., June 1, 1916; s. Leon C. and Bessie (Dorn) T.; B.S., Cornell U., 1938; M.Sc., Ohio State U., 1940, Ph.D., 1948; m. Charlotte Dean Moelchert, May 3, 1945; children—Gail Ann, Scott Dean. Research bacteriologist, br. chief Biol. Labs., Ft. Detrick, Md., 1943-53; prof., chmn. dept. bacteriology U. Fla., Gainesville, 1953-—. Cons. U. S. Army Biol. Labs., 1953-56. Recipient Certificate of Appreciation Fla. Blue Key, 1963. Fellow Am. Pub. Health Assn., A.A.A.S.; mem. Am. Soc. for Microbiology (past pres. S.E. br.), Sigma Xi, Gamma Sigma Delta. Research, publs. on physiology, nutrition, anatomy of marine and estuarine bacteria, nutrition of lactic bacteria and Brucella, bacterial aerosols, mass culture pathogenic bacteria. Home: 1800 N.W. 10th Av., Gainesville, Fla. 32601.*

TYLER, Varro Eugene, Am. pharmacognosist; b. Auburn, Neb., Dec. 19, 1926; s. Varro Eugene and Venus (Leamer) T.; B.S., U. Neb., 1949; postgrad. Yale; M.S., U. Conn., 1951, Ph.D., 1953; m. Virginia May Demel, Aug. 20, 1947; children—Jeanne, David. Asso. prof., chmn. dept. pharmacognosy U. Neb., Coll. Pharmacy, 1953-57; asso. prof., chmn. pharmacognosy, dir. drug plant gardens U. Wash. Coll. Pharmacy, 1957-61, prof. pharmacognosy, 1961-66; research fellow Institut für Biochemie der Pflanzen, Halle/Saale, Germany, 1963-64; dean Purdue U. Sch. Pharmacy and Pharmacal Scis., Lafayette, Ind., 1966-—. Recipient Lehn & Fink Products Corp., medal 1949. Eli Lilly Research fellow, 1950-51; Gustavus A. Pfeiffer Meml. Research fellow, 1963-64. Fellow A.A.A.S.; mem. Am. Assn. Colls. Pharmacy (exec. com. 1966-—), Acad. Pharm. Scis. (chmn. elect sect. pharmacognosy and natural products 1966), Am. Soc. Pharmacognosy (past pres., past mem. exec. com.), Soc. for Econ. Botany, Deutsche Gesellschaft für Arzneipflanzenforschung, Am. (Found. Research Achievement award 1966), Ind. pharm. assns., Am. Soc. Plant Physiologists, Rho Chi, Sigma Xi, Kappa Psi (Scholarship key 1949), Phi Gamma Delta. Author: (with E. P. Claus) Pharmacognosy, 1965; (with A. E. Schwarting) Experimental Pharmacognosy, 1955; also numerous articles. Editorial adv. bd. Jour. Pharm. Scis., 1959-65, Lloydia, 1961-—, Planta Medica, 1963-—. Research on biosynthesis, physiology and distbn. of secondary plant prins. especially alkaloids, medicinal and toxic constituents of higher fungi especially ergot and mushrooms, sources of naturally occurring psychotropic drugs, application of chemotaxonomy to mushrooms. Home: 640 Ridgewood Dr., West Lafayette, Ind. 47906. Office: Purdue U. Sch. Pharmacy and Pharmacal Scis., Lafayette, Ind. 47907.*

TYLOR, Sir Edward Burnett, English anthropologist; b. London, Eng., Oct. 2, 1832; s. Joseph and Harriet (Skipper) R.; ed. Grove House Sch., Tottenham; D.C.L., Oxford U., 1875; m. Anna Fox, 1858. Keeper, U. Mus., Oxford, 1875; reader anthropology Oxford U., 1884, first prof. anthropology, 1896, hon. fellow Balliol Coll., 1903; Gifford lectr. Aberdeen U., 1888. Fellow Royal Soc., 1871; mem. Anthrop. Soc. (pres. 1891). Author: Anahuac, Mexico and the Mexicans, Ancient and Modern, 1859; Researches into Early History of Mankind, 1865; Primitive Culture, 1871; Anthropology, 1881. Known for studies on mythology, magic and mentality of primitive peoples, also for derivation of primitive religion from animism; as Darwinian, studied evolution of man, body and soul, and all activities of man, as language, religion, law, morality, art; used psychol. methods; applied statistical method to devel. of instns.; laid great stress on anthrop. facts; coined several anthrop. terms, as cross-cousin, local exogamy, teknonymy. Died Wellington, Somerset, Jan. 2, 1917.

TYLOR, Joseph John, Brit. engr., Egyptologist; b. Stoke, Newington, Eng., Feb. 1, 1851; s. Alfred and Isabella (Harris) T.; ed. U. London (Eng.), 1868, Poly. Sch., Stuttgart, Germany, 1868-70; m. Marion Young, Sept. 15, 1887; children—Alfred, George Cunnyngham. With Bowling Ironworks, Yorkshire, Eng., 1870-72; partner J. Tylor & Sons, 1872-84, sr. partner, from 1884; turned to Egyptology, from 1891; expts. with pictorial reprodn. ancient sculptures, paintings. Mem. Instn. (C.E.'s (asso.). Author: Wall Drawings and Monuments of El Kab (series of monographs individually titled: The Tomb of Pakeri, 1895; The Tomb of Sebeknekht, 1896; The Temple of Amenketep III, 1898; The Tomb of Renni, 1900). Patentee many successful inventions, particularly in relation to hydraulic meters; developed photographic technique to reproduce and clarify ancient Egyptian wall paintings. Died Alpes-Maritimes, France, Apr. 5, 1901.

TYNDALL, John, Brit. physicist; b. Leighlin Bridge, county Carlow, Ireland, Aug. 2, 1820; s. John and Sarah (Macassey) T.; attended local National School; mechanics' institute, 1842; U. Marburg, Hesse-Cassel, 1848-50, Ph.D., 1850; studied in Marburg and Berlin, 1851; hon. M.D., U. Tübingen (for his researches on bacteria), 1877; hon. degrees from Cambridge, 1865, Edinburgh, 1866, Oxford, 1873, Trinity College, Dublin, 1886, Columbia, 1887; m. Louisa Hamilton, Feb. 29, 1876. Civil asst., ordinance survey of Ireland, 1839-42; employed on English survey, 1842-44; employed as railway engr., 1844-47; tchr. mathematics and surveying, Queenwood College, Hampshire, 1847-48, 1851-53; prof. natural philos., Royal Institution of Great Britain (where he was friend and colleague of Faraday), 1854-87; succeeded Faraday as superintendent, 1867-87; hon. prof., from 1887; visited Switzerland frequently (as scientist and mountaineer); sci. advisor, Trinity House and Board of Trade, 1866-83; lectured in U. S., 1872-73. Fellow Royal Soc., 1852 (mem. council, 1856; Bakerian lectr., 1861, 1864, 1881; Rumford medal, 1864); mem. Brit. Assn. Advancement Sci. (pres. 1874). Author: The Glaciers of the Alps, 1860; Mountaineering, 1862; Heat Considered as a Mode of Motion, 1863; Cn Sound, 1867; Faraday as a Discoverer, 1868; Researches on Diamagnetism and Magne-crystallic Action, 1870; Notes on Light, 1870; Notes on Electrical Phenomena and Theories, 1870; Fragments of Science for Unscientific People, 1871; Hows of Exercise in the Alps, 1871; Contributions to Molecular Physics in the Domain of Radiant Heat, 1872; The Forms of Water in Clouds and Rivers, Ice and Glaciers, 1872; Six Lectures on Light, 1873; Lessons in Electricity, 1876; Essays on the Floating Matter of the Air in relation to Putrefaction and Infection, 1881; New Fragments, 1892; about 145 sci. papers. Studied diamagnetism and influence of crystalline structure and mechanical pressure upon manifestations of magnetic force; demonstrated conclusively polarity of bismuth and other diamagnetic bodies, 1850-56; explained glacier motion and structure, 1857-59; studied absorption and radiation of heat by gases and vapors; showed high absorptive and radiative power of clear aqueous vapor (of great meteorological significance); demonstrated dispersion of light beam by suspended particles in colloids and gases (Tyndall effect); showed blue of sky and polarization of sky-light due to excessively fine particles floating in our atmosphere; demonstrated optically pure air (moteless or germ-free) incapable of producing putrefaction in infusions; thereby giving final death blow to theory of spontaneous generation; introduced process of discontinuous heating, which successfully destroyed most resistant bacteria; discovered non-homogeneity of atmosphere affects sound; demonstrated that atmosphere exercises a selective and done in connection with lighthouse and siren work); champion of Mayer's claims to priority in conservation of energy; famed for lucidity of expression, for making difficult things clear, and for diffusion of sci. knowledge. Died Hind Head, Surrey, England, Dec. 4, 1893.

TYREE, Sheppard Young, Jr., Am. chemist; b. Richmond, Va., July 4, 1920; s. Sheppard Young and Rosa Dove (Burton) T.; B.S., Mass. Inst. Tech., 1942, Ph.D., 1946; m. Barbara Doris Jones, June 5, 1943; children—Susan Scott, Pamela Fore, Peter Burton, Sally Blair, Rebecca Young. Instr., Mass. Inst. Tech., Cambridge, 1943-46; faculty U. N.C., Chapel Hill, 1946-66, prof., 1958-66; prof. chemistry Coll. William and Mary, Williamsburg, Va., 1966-—. Research chemist Oak Ridge Nat. Lab., summer 1949; sci. officer Office Naval Research, Washington, 1954-55, liaison scientist, London, 1965-66. Fellow A.A.A.S.; mem. Am. Assn. U. Profs., Am. Chem. Soc. (Herty medal 1964), Faraday Soc., Chem. Soc., Sigma Xi. Author: (with Kerro Knox) Inorganic Chemistry, 1960; also articles. Editor: Inorganic Synthesis, vol. IX, 1967. Research in properties of inorganic substances dissolved in water, synthesis and phys. properties of metal halides and halide complexes of metals. Home: 110 Argall Town Lane, Williamsburg, Va. 23185.*

TYRRELL, David Arthur John, English virologist, physician; b. Ashford, Middlesex, Eng., June 19, 1925; s. Sidney Charles and Agnes (Blewett) T.; M.B., Ch.B. with honors, Sheffield U., 1948, M.D. with distinction, 1953; m. Betty Moyra Wylie, Apr. 15, 1950; children—Frances, Susan, Stephen. House physician, jr. med. registrar, Sheffield, Eng., 1948-50; research registrar Sheffield United Hosps., 1950-51; asst. physician Rockefeller Inst. Hosp., N.Y.C., 1951-54; mem. sci. staff Med. Research Council, 1954—, Virus Research Lab., Sheffield, 1954-57, common cold research unit, Salisbury, Eng., 1957—. Hon. lectr. medicine U. Sheffield, 1957—; hon. cons. virologist Salisbury Gen. Hosp., 1964—. Fellow Royal Coll. Physicians; mem. Coll. Pathologists, Path. Soc., Soc. Gen. Microbiology. Author: Common Colds and Related Diseases, 1965; also numerous articles. Research on growth and inhibition of influenza viruses and the viral role in causation of pneumonia and milder respiratory diseases, diseases of polioviruses and related organisms; a contbr. to discovery of rhinoviruses, a prin. cause of common colds in man. Address: Common Cold Research Unit, Salisbury, Wiltshire, Eng.*

TYRRELL, Joseph Burr, Canadian geologist, explorer, mining engr.; b. Weston, Ont., Can., Nov. 1, 1858; s. William and Elizabeth (Burr) T.; B.A., U. Toronto, 1880, M.A., 1889, LL.D., 1930; B.Sc., Victoria U., 1889; LL.D., Queen's U., 1940; m. Mary Edith Carey, Feb. 1894; children—Mrs. Dalton, George, Thomas. Staff, Canadian Geol. Survey, 1881-98; explored (with G. M. Dawson) Rocky Mountains, 1883; explored country N. of Calgary between Bow and Saskatchewan rivers, 1884-87, N.W. Man., 1887-89, Lake Winnipeg area, 1890-91, Lake Athabaska area, 1892-93, and crossed Barren Lands; recrossed Barren Lands, to Beach W. coast of Hudson Bay, 1894; then traveled to Lake Winnipeg; explored unknown country N.W. of Lake Winnipeg, 1896; became mining engr. Dawson, Yukon Ter., 1898, then Toronto, Ont., 1906-25. Recipient Medallion, Am. Inst. Mining Engrs., 1904, Murchison medal Geol. Soc., London, 1918; Wollaston Palladium medal, 1947; Daly Gold medal Am. Geog. Soc., 1930; Flavelle Gold medal Royal Soc. Can., 1933; Back award Royal Geog. Soc. London. Fellow Royal Soc. Can., Geol. Soc. London (sr.); mem. Nat. Geog. Soc., Geol. Soc. Am., Am. Inst. Mining, and Metall. Engrs. (Legion of Honor), Soc. Econ. Geologists, Inst. Mining and Metall. Engrs. (Eng.), Canadian Inst. Mining and Metall. Engrs. (hon.), Assn. Profl. Engrs. Ont., Royal Scottish Geog. Soc. (corr.), Arctic Inst. N.Am. (hon.), Royal Canadian Inst. (pres. 1910-13). Author: David Thompson, Explorer, 1910; also numerous articles. Surveyed unexplored region S.E. of Lake Athabaska, 1892-93, Barren Land, 1893, unexplored country between Churchill and Lake Winnipeg, 1894; discovered large areas of pre-Cambrian rocks rich in ore, copper, N.W. of Lake Winnipeg. Died 1957.

TYSON, Edward, English physician, anatomist; b. Clevedon, Somerset, Eng., 1650; s. Edward Tyson; B.A., Magdalen Coll., Oxford (Eng.) U., 1670, M.A., 1673; M.D., Cambridge (Eng.) U.; 1 son, Richard Tyson. Physician, Bridewell and Bethlehem Hosp., lectr. anatomy Barber-Surgeons, London, Eng., until 1699. Fellow Royal Soc., 1679, Royal Coll. Physicians (censor 1694). Author: Phocaena, or the Anatomy of a Porpess, 1680; The Anatomy of the Rattlesnake, 1683; Carigueya seu Marsupiale Americanum, 1698; Orangoutang, Sive Homo Sylvestris, 1699; 2d edit., 1751; A Philosophical Essay Concerning the Rhymes of the Ancients, 1699. Translator: Ephemeri Vita (Summerdam). Contbd. monographs Philos. Trans., also to Acta Medica (Bartholinus), to Historica Piscium (Willughby), 1686, to Natural History of Oxfordshire (Plot). First Englishman to pub. monographs on anatomy of particular animals; credited with introducing concept of missing link in anthropology; declared that pigmies, cynocephali, satyrs and sphinges of ancients were merely apes; demonstrated that hydatid or lumbricus hydropicus is animal and not mere morbid growth; gave more complete report on dissections of several animals than any previous. Died Aug. 1, 1708.

TYTELL, Alfred A., Am. microbiol. chemist; b. N.Y.C., Aug. 1, 1915; s. Boris H. and Rebecca (Korets) T.; B.S., Tufts U., 1936; Ph.D., Mass. Inst. Tech., 1940; m. Alice M. Griner, Aug. 6, 1945; children—Elnora, Holly. Faculty, Coll. Medicine, Cin., 1940-52, asso. prof., 1950-52; research asso. in antibiotics Merck Inst. for Therapeutic Research, West Point, Pa., 1952-58, dir. cell biology, virus and cell biology research div., 1958—. Mem. Am. Soc. Biol. Chemists, Am. Assn. Immunologists, Soc. for Exptl. Biology and Medicine, A.A.A.S., Am. Soc. Cell Biology, Tissue Culture Assn., N.Y. Acad. Sci. Research, numerous publs. in microbiology of bacterial toxins and toxoids, immunology of toxoids, biochemistry and nutrition of microorganisms and mammalian cells, microbial enzymes, chemistry and microbiology of antibiotics, virus chemistry, virus biology, virus vaccine immunology. Home: 113 Church Rd., Lansdale, Pa. 19446. Office: Merck Sharp & Dohme Research Labs., Merck & Co., Inc., West Point, Pa. 19486.*

TZENOV, Angelov Ivan, Bulgarian mathematician; b. Wratza, Bulgaria, Jan. 2, 1883; s. Tzeno Angelov and Katja (Kostowa) T.; mathematician, physicist U. Kliment Ochridski, Sofia, Bulgaria, 1908; m. Raina Stefanova Ganewa, Jan. 14, 1914; children—Stefan, Tzeno, Weselin, Anastasia (Mrs. Atanas Hubenow), Katja (Mrs. Ivan Arschinkow). Faculty, Kliment Ochridski, 1908—, prof. math., 1919—, deacon Phys. and Math. Faculty, 1925-30. Recipient Dimitrows award I, Kiril I. Methodyi, Narodna Republika Bulgaria I. Author: Analytical Mechanics, vols. I, III; High Mathematics; Vector Calculus; also numerous articles. Mem. Bulgarian Acad. Sci., Bulgarian Phys. and Math. Soc. (past dir.). Established new form of equations of movement of Holonom Systems (Equations of Tzenow) using kinetic energy; proved applicability of principles of analytical dynamics for nonholonom systems using equations of Tzenow; applied new equations for nonholonom systems with nonholonom bounds; introduced new gen. equation of dynamics (ful differential of a function K of generalized accelerations. Home: 10 Avicena, Sofia, Bulgaria. Office: Bulgarian Acad. Scis. 7noemvri str. 1.*

U

UBBELOHDE, Alfred Rene John Paul, physicist; b. Dec. 14, 1907; s. F. C. and Angele (Verspreeuwen) U.; grad. Oxford U., 1931. Sr. research fellow Oxford U., 1931-35; unmarried; Dewar research fellow Royal Instn., 1935-40, Ministry Supply, 1940-45; prof. chemistry Queen's U., Belfast, also dean faculty, 1945-54; prof. thermodynamics Imperial Coll., London, Eng., 1954—, head dept. chem. engring. and chem. tech., 1961—. Decorated C.B.E., 1962. Fellow Royal Soc., 1951; Inst. Physics, Royal Inst. Chemists; mem. Chem. Soc. (Liversidge lectr. 1960 Soc. Chem. Industry (C. Tennant lectr. 1961), Faraday Soc. (pres. 1963-64, v.p. 1964—). Author: Time and Thermodynamics, 1947; Introduction to Thermodynamical Principles, 1952; Man and Energy, 1954; Graphite and Its Crystal Compounds, 1960; Melting and Crystal Structure, 1965; also numerous articles. Research on gas kinetics; solid-solid and solid-melt phase transformations; electronic properties of graphite and its lower aromatic homologues. Address: Flat 3, 177 Queen's Gate, S.W. 7, England.

UCHUPI, Elazar, Am. geologist; b. N.Y.C., Oct. 31, 1928; s. Alfonso and Carmen (Urbizu) U.; B.S., Coll. City N.Y., 1952; M.S., U. So. Cal., 1954, Ph.D., 1962. Research asst. U. So. Cal., 1956-61; asso. scientist Woods Hole (Mass.) Oceanographic Instn., 1962—. Mem. Am. Assn. Petroleum Geologists, Geol. Soc. Am., Soc. Econ. Paleontologists and Mineralogists, Am. Soc. Limnology and Oceanography, Am. Geophys. Union, Sigma Xi. Research in topography, sediments and structure of continental margin off western, Eastern and Gulf Coast of U. S. Office: Woods Hole Oceanographic Instn., Woods Hole, Mass. 02543.*

UDENFRIEND, Sidney, Am. biochemist; b. N.Y.C., Apr. 5, 1918; s. Max and Esther (Tabak) U.; B.S., Coll. City N.Y., 1939; M.S., N.Y. U., 1942, Ph.D., 1948; m. Shirley F. Reidel, June 20, 1943; children—Alice Rosalind, Elliot. Lab. asst. N.Y.C. Dept. Health, 1940-42; research and teaching staffs N.Y. U. Med. Sch., N.Y.C., 1942-48; biochemist Lab. Chem. Pharmacology, Nat. Heart Inst., NIH, Bethesda, Md., 1950-53, head, sect. on cellular pharmacology, 1953-56, chief Lab. Clin. Biochemistry, 1956-68; dir. Roche Inst. Molecular Biology, Nutley, N.J., 1968—. Lectr., George Washington U., 1962—. Recipient Superior Service award Dept. Health, Edn. and Welfare, 1965, Distinguished Service award, 1966. Mem. Am. Soc. Biol. Chemists, Am. Soc. Pharmacology and Exptl. Therapeutics, Am. Assn. Clin. Chemists (Van Slyke award N.Y. met. sect. 1967), Am. Chem. Soc., Sigma Xi. Author: Fluorescence Assay in Biology and Medicine, 1962; also articles. Research in metabolism of aromatic amino acids, biosynthesis and metabolism of epinephrine and serotonin. Address: Roche Inst. Molecular Biology, Nutley, N.J.*

UDUPA, K. N., Indian surgeon; b. Katil, S. Knara Dist., India, July 7, 1920; s. K. Thammaya and Gangamma Udupa; A.M.S., Banaras Hindu U., 1943; M.S., U. Mich., 1948; m. Leela Udupa, June 30, 1947; 1 dau., Anjali. Civil surgeon Mandi Dist., Himachal Pradesh, India, 1949-54; research fellow surgery Harvard, asst. in surgery Peter Bent Brigham Hosp., Boston, 1954-55; civil surgeon, surg. specialist Snowdon Hosp., Simla, India, 1955-58; prof. surgery, prin. Coll. Med. Scis., Banaras Hindu U., Varanasi, India, 1959—, chief surgeon, supt. Sir Sunder Lal Hosp., 1959—. Mem. Assn. Surgeons India. Author: Principles of General Surgery, 1961; also numerous articles. Research on regeneration and healing of wounds and fractures, phytogenic anabolic steroid from indigenous herb. Address: Coll. Med. Scis., Banaras Hindu U., Varansi, India.*

UEDA, Hideo, Japanese physician; b. Zenkichi and Chiyo (Aoki) U.; M.D., U. Tokyo, 1935, D.Med. Sci., 1944; m. Takako Nagamitsu, June 1, 1936; children—Michiko Nagasawa (Mrs. Toshihiko Nagasawa), Kumiko Sugishita (Mrs. Yasuro Sugishita), Reiko, Jun. Lectr., U. Tokyo, 1948-50; prof. medicine Tokyo Jikeikai Med. Coll., 1950-58; prof. 2d dept. internal medicine U. Tokyo, 1958—. Cons. Japanese Inst. Cardiovascular Diseases, 1962—; vice chmn. cardiovascular com. Japanese Sci. Council, 1966—; mem. adv. panel on cardiovascular diseases WHO, 1960—. Mem. Japanese Soc. Nuclear Medicine (pres. 1962—), Japanese Soc. Internal Medicine (dir. 1957—), Japanese Circulation Soc., Japanese Soc. Gastroenterology. Author: Cardiology, 1966; (with E. Kimura, R. Kashida) Clinical Electroscardiography, 1965; (with G. Kaito, T. Sakamoto) Clinical Phonocardiography, 1963; also numerous articles. Discovered blood pressure regulating area in right subclavian artery of dog and rabbit, method for measurement of hepatic arterial blood flow and portal blood flow using dye and radioisotopes, method for detection of intra- and extra-hepatic shunt blood flow in hepatosplenic diseases, blood pressure regulating function of thalamus and human kidney, spl. type of anomalous atrioventricular conduction. Home: 3-70 Ogikubo, Suginamiku, Tokyo, Japan.*

UEDA, Takeo, Japanese pharm. chemist; b. Kyoto, Japan, Nov. 17, 1908; s. Kametaro and Komako (Koga) U.; grad. Tokyo (Japan) U. Sch. Medicine, 1932, Ph.D., 1938; m. Ariko Ohta, Apr. 3, 1933; children—Hiromi (Mrs. Toshihiko Kagawa), Mitsuko Kondo (Mrs. Denbei Kondo), Tadashi, Atsushi. Faculty, Kyoto U., 1939-45, prof., 1945; prof. Faculty Medicine, Keio U., 1945-63; prof. pharm. chemistry Kitasato U., Tokyo, 1963—, dean Coll. Pharm. Scis., 1964—. Mem. Pharm. Soc. Japan (prize pharm. scis.), Pharmacological Soc. Japan, Virological Soc. Japan, Soc. Chemotherapy Japan, Soc. Infectious Diseases Japan. Author: Ricron Kagaku, 1935; also numerous articles. Research on chemotherapeutic drugs against viruses, bacteria, protozoans, anti-inflammatory drugs, analgesic drugs, anti-neoplastic diseases drugs, pharm. analysis. Home: 1066 Izumi, Ktatama-gun, Tokyo, Japan.*

UEDA, Y., Japanese chemist; b. Osaka, Japan, June 9, 1912; s. Shoji and Shizuko Ueda; M.Pharm. Sci., Tokyo (Japan) U., 1944, D.Pharm. Sci., 1955; m. Reiko Iwasaki, Feb. 2, 1949. Asst., Inst. Pharm. Scis., Tokyo U., 1944-52; asso. prof. Kyushu U., Fukuoka, Japan, 1952—, prof. Inst. Pharm. Sci., 1966—. Vis. scientist NIH, Bethesda, Md., 1959-61. Mem. Pharm. Soc. Japan, Japan Soc. for Analytical Chemistry. Author: Instrumental Analysis, 1964; also articles. Research on syntheses of 4-aminomethylphenylsulfonamide, sulfur-containing steroid derivative, tea seed saponin, its aglycone and relation between hemolytic activities and film-expanding power of saponins; spectroscopic studies on sulfon, sulfonamide, tetralin, elucidations of several organic color reaction mechanisms.*

UEHLING, Edwin Albrecht, Am. physicist; b. Lowell, Wis., Jan. 27, 1901; s. Albert Edward and Sophia (Albrecht) U.; A.B., U. Wis., 1925; M.A., U. Mich., 1930, Ph.D., 1932; m. Margaret Ruth Ransom, June 21, 1928. Postdoctoral fellow U. Leipzig (Germany), 1933; NRC fellow U. Cal. at Berkeley, Cal. Inst. Tech., Pasadena, 1934-36; faculty U. Wash., Seattle, 1936-41, 45—, prof. physics, 1947—, acting head dept., 1954-55. With NDRC, Mass. Inst. Tech., Boston, 1941-43; operations analyst OSRD with vice chief naval operations, Washington, 1943-45; spl. assignment operations research analysis, U. S. Naval Forces, Far E., 1950; vis. prof. Eidgenössische Technische Hochschule, Zurich, Switzerland, 1966. Guggenheim fellow Harvard, Inst. for Theoretical Physics, Utrecht, Holland, 1955-56; recipient Pres.'s certificate of merit, 1947. Mem. Am. Inst. Physics, A.A.A.S., Am. Assn. U. Profs., Sigma Xi. Contbg. author: Annual Rev. Nuclear Sci., Vol. 4, 1954. Research, publs. on transport phenomena in gases, electron-positron theory and polarization of vacuum, penetration of charged particles in matter; derived certain vacuum polarization terms; contbd. to a better understanding of hydrogen bond and its connection with ferroelectric phase transition. Home: 5045 N.E. 70th St., Seattle 98115.*

UEMURA, Yasutada, Japanese physicist; b. Tokyo, Japan, Apr. 18, 1921; s. Kogoro and Yoshiko (Tomii) U.; grad. dept. physics U. Tokyo, 1944, D.Sc., 1958; m. Haruko Sakatani, Oct. 23, 1951; children—Yasuyoshi, Yasutomo. Asst. prof. Chuo U., Tokyo, 1949-53; research staff Toshiba Electric Co., Tokyo, 1953-56; faculty U. Tokyo, 1956—, prof. physics, 1963—. Mem. Phys. Soc. Japan, Soc. Applied Physics Japan, Am. Phys. Soc. Author: (with Makoto Kikuchi) Semiconductors—Physics and Applications, 1959; also articles. Research on spl. field theoretical solid state physics especially semiconductors and color centers in alkalihalides. Home: 191 Seijo-machi Setagaya-ku, Tokyo, Japan.*

UFFEN, Robert James, Canadian geophysicist; b. Toronto, Ont., Can., Sept. 21, 1923; s. James Frederick and Elsie May (Harris) U.; B.A., U. Toronto,

1949, M.A., 1950; Ph.D., U. Western Ont., 1952; D.Sc., Queen's U., 1967; m. Mary Ruth Paterson, May 3, 1949; children—Joanne Grace, Robert Ross. Faculty, U. Western Ont., 1951-66, prin. U. Coll., 1961-65, dean Coll. Sci., 1965-66; vice chmn. Def. Research Bd., Ottawa, Ont., 1966, chmn., 1967—. Mem. NRC, 1963-66; Canadian del. Internat. Union Geologic Sci., New Delhi, 1964; chief del. for Can., Internat. Union Geodesy and Geophysics, 1967; mem. council regents Colls. Applied Arts and Tech. Province Ont., 1966—. Research Council Ont. scholar, 1950, 51; research fellow Inst. Geophysics U. Cal., Los Angeles, 1953. Fellow Royal Soc. Can., Geol. Soc. Am.; mem. Am. Inst. Mining, Metall. and Petroleum Engrs., Canadian Inst. Mining and Metallurgy, A.A.A.S., Soc. Exploration Geophysicists, Am. Geophys. Union, Canadian Assn. Physicists. Editor: Earth and Planetary Science Letters, 1965—; editorial bd. Tectonophysics, 1964—. Co-discoverer Lac Allard ilmenite deposits; publs. on temperature of earth's interior, origin of volcanoes and influence of evolution of earth's interior on origin and evolution of geomagnetic field and evolution of life. Home: 50 Lyttleton Gardens, Rockcliffe Park, Ottawa 7. Office: 125 Elgin St., Ottawa 4, Ont., Can.*

UFFORD, Charles Wilbur, Am. physicist; b. Cambridge, Mass., Feb. 15, 1900; s. Frank Parker and Bertha (Tierney) U.; B.A., Haverford Coll., 1920; postgrad. Cambridge (Eng.) U.; B.S., Mass. Inst. Tech., 1922; M.A., Harvard, 1924, Ph.D., 1928; m. Beatrice Gaylord Wistar, Sept. 20, 1930; children—Charles Wilbur, Beatrice Gaylord (Mrs. Henry Zenzie). With testing dept. Gen. Electric Co., 1922-23; instr. Princeton, 1928-33; faculty Allegheny Coll., 1933-45, prof. physics, 1939-45; asso. prof. engring. Haverford Coll., 1945-47; asso. prof. physics U. Pa., Phila., 1947-49, prof., 1949—, chmn. dept., 1953-55, 56-63, chmn. div. phys. scis., 1955-56; physicist div. war research U. Cal., 1944-46. Fellow Am. Phys. Soc.; mem. Am. Assn. Physics Tchrs., Phi Beta Kappa, Sigma Xi. Author: (with H. D. Smyth) Matter, Motion and Electricity, 1937. Research, publs. on structure of atoms, nuclei and solids. Home: 730 Panmure Rd., Haverford, Pa. 19041. Office: David Rittenhouse Lab., U. Pa., Phila. 19104.*

UGALDO, Guido, Italian mathematician; b. Ubaldo, 1540; student of Frédéric Commandin; engaged in study at Chateau de Monte-Barroccio throughout life. Author (in Latin): Th ory of Universal Planispheres. Conceived of principle of virtual velocities. Died 1607.

UGI, Ivar, organic chemist; b. Arensburg, Estonia, Sept. 5, 1930; s. Theodor and Ellen (Behm) U.; D.Sc., U. Munich, 1954; m. Helga G. Dudzinski, Mar. 30, 1960; children—Ian. Faculty, U. Munich, 1959-62, U. Cologne, 1967—; with Bayer A. G. Leverkusen, 1962—, research dir., 1966—, chmn. basic research com. Recipient Acad. award Gottingen Acad. Sci., 1964. Mem. Am. Chem. Soc., German Chem. Soc. Research, numerous publs. on pentazole chemistry, syntheses of ioscyanides, isocyanide reactions; multicomponent condensations, stereoselective syntheses, gen. theory of chirality. Home: 8 Am Mittelberg. Office: Farbenfabriken Bayer AG., Leverkusen, Germany.*

UGOLINO DA MONTECATINI, Italian physician; b. Montecatini, circa 1345; s. Giovanni de' Caccini; M.D., Bologna, 1367. Town physician in Pistoia, later in Pescia; taught medicine in Pisa for 25 years, then in Florence, 1393-95; practiced and/or taught medicine in Pisa, Lucca, Pesaro, Perugia, returned to Florence. Author: De balneorum Italiae (Etruriae) proprietatibus ac virtutibus, written at beginning of 15th century. Cne of earliest balneologists, wrote one of earliest balneological treatises; also wrote consilium for Averardo de' Medici, commentary on 16th fen of 3d book of Qanun of Ibn Sina, treatise on plague. Died Florence, Oct. 10, 1425.

UHLARIK, Sándor Kálmán, physician; b. Bratislava, Czechoslovakia, Jan. 14, 1934; s. Agoston and Iren (Polak) U.; M.D., Med. Sch., U. Pecs (Hungary), 1959; m. Annegret Schreiber, Sept. 30, 1966. Asst. dept. anatomy U. Pecs, 1959-60; asst. dept. obstetrics Med. Sch., U. Szeged (Hungary), 1961-66; Alexander von Humboldt Found., Dusseldorf, West Germany, 1964-65; asst. dept. obstetrics and gynecology U. Marburg, West Germany, 1966—. Recipient Hungarian Republik Stipendium, Rakosi award, 1954-58. Research and publs. on pituitary ovarian and pituitary-hypothalamic feedback regulation between pituitary gland and hypothalamus for gonadotropins; demonstration that placental tissue transplants which undergo fatty necrosis in intact animals survive in hypophysectomized animals. Home: 7 Berliner-Str., Marburg, West Germany.*

UHLENBECK, George Eugene, physicist; b. Batavia, Java, Dutch East Indies, Dec. 6, 1900; s. Eugenius Marius and Annie Marie Constance Julie (Beeger) U.; student preliminary and high schs., The Hague; Ph.D., U. of Leyden, 1927; Sc.D. (honorary), Notre Dame University, 1953, Case Institute of Technology, 1960; married Else Renée Ophorst, Aug. 23, 1927. Came to U. S., 1927. Asst. in theoretical physics, U. of Leyden, 1925-27; instr. physics, U. of Mich., 1927-28, asst. prof., 1928-30, asso. prof., 1930-35, prof. theoretical physics, 1939-59, formerly Henry S. Carhart professor; professor theoretical physics, Utrecht, Holland, 1935-39; War research (radar), 1943-45, radiation Lab., Mass. Inst. Tech., Cambridge, Mass.; prof. physics, mem. Rockefeller Inst., 1961—. Fellow Am. Physical Soc., (pres. 1959), Dutch Physical Soc.; mem. Am. Philos. Soc., Nat. Acad. Scis. Research in nuclear physics, quantum mechanics, kinetic theory, statis. mechanics, atomic structure; work on fine structure of emissions spectra led to devel. (with Goudsmit) of electron spin hypothesis in classic model of atom, 1925. Office: Rockefeller Inst., N.Y.C. 21.

UHLENBRUCK, Gerhard, German biochemist; b. Cologne, Germany, June 17, 1929; s. Paul and Ruth (Padberg) U.; Dr.med. s.c.l., U. Cologne, 1955; m. Kathrin Kruchen, Nov. 25, 1960; children—Georg, Katja, Ruth. Sci. asst. Inst. for Physiol. Chemistry, 1955-60, 61-63; scholar Lister Inst., London, 1960-61; head dept. biochemistry Max-Planck-Inst. for Brain Research, Cologne-Lindenthal, Germany; lectr. immunochemistry U. Cologne, 1964—. Mem. Brit. Soc. for Immunology, German Soc. for Physiol. Chemistry, Internat. Soc. Blood Transfusion. Author: (with O. Prokop) Lehrbuch der menschlichen Blut-und Serumgruppen, 1963, 66; also numerous articles. Contbg. editor Vox sanguinis, 1965—. Research on neuraminic acid-containing glycoproteins and glycolipids, mucoid structure of red cell surface, biochemistry of blood group substances, use and action of enzymes in blood group serology, immunochemistry of heterophile antigens, red cell aggregation and pan-agglutination; new proposals for a nomenclature in blood-group serology, molluscagglutinins. Address: 308 Gleueler Strasse, Cologne-Lindenthal, West Germany.*

UHLENHUTH, Paul, German bacteriologist, hygienist; b. Hannover, Germany, Jan. 7, 1870; s. Karl U.; ed. Kaiser Wilhelm Akademie; M.D., 1894/5. Asst. of R. Koch, Institute of Infant Illnesses, Berlin, 1897-99; Prof. U. Griefswald, 1903; privat docent, hygiene and bacteriology, U. Griefswald, 1905; dir. bacteriological section, Reichsgesundheits-Amt., 1906; prof. U. Strasburg, Freiburg, Marburg. Author: (with R. Kraus) Handbuch der mikrobischen Technik, 1922; others; numerous articles. Research on typhoid, paratyphoid, Tb, influenza, syphilis, chemotherapy; discovered technique of differentiating animal and human blood; developed arsenical treatment of syphilis; developed serums against hog cholera and foot-and-mouth disease. Died Freiburg-im-Breisgau, Germany, Dec. 13, 1957.

UHLER, Philip Reese, Am. entomologist, geologist; b. Baltimore, Md., June 3, 1835; s. George Washington and Anne Maria (Reese) U.; ed. at D. Jones' Latin School and under pvt. tutors; LL.D., New York U., 1900; m. Sophia Werdebaugh, 1869 (died 1883); m. 2d, Pearl Daniels, Apr. 29, 1886. Spent nearly 3 yrs. at Harvard as librarian and asst. to Prof. Louis Agassiz, in his mus. of comparative zoology; explored parts of Island of Haiti for him; became connected with Peabody Library, Baltimore, 1862, later librarian, devising new methods adopted in its catalogue. Pres. Md. Acad. Sciences; asso. in natural sciences, Johns Hopkins U.; fellow A.A.A.S. Author: Descriotions of a Few Species of Coleoptera Supposed to be New, 1861; Synopsis of the Newroptery of North America; numerous papers. Authority on entomology, described many new forms; studied geology and archaeology. Died Oct. 21, 1913.

UHLIG, Herbert Henry, Am. electrochemist; b. Haledon, N.J., Mar. 3, 1907; s. Henry C. and Ida (Schuster) U.; Sc.B., Brown U., 1929; Ph.D., Mass. Inst. Tech., 1932; m. Greta Dorothy Johnson, Mar. 8, 1941; children—Karin Ida (Mrs. James Young), Maida Ruth, Kristin Greta. Research chemist Rockefeller Inst. Med. Research, N.Y.C., 1932-33; asst. chief chemist Lever Bros., Cambridge, Mass., 1934-36; research asso. Gen. Electric Co., Schenectady, 1940-46; faculty Mass. Inst. Tech., Cambridge, 1936-40, 46—, prof., 1953—. Mem. Meteoritical Soc., Am. Acad. Arts and Scis., Electrochem. Soc. (Palladium medal 1961), Am. Chem. Soc., Am. Soc. Metals, Nat. Assn. Corrosion Engrs. (Willis R. Whitney award 1951), Deutsche Bunsen Gesellschaft, Sigma Xi. Author: Corrosion and Corrosion Control, 1963; also many articles. Cons. editor McGraw-Hill Yearbook on Sci. and Tech., 1964—; adv. editor Corrosion Sci., 1961—; editor Corrosion Handbook, 1948. Research on corrosion and oxidation of metals, nature and source of corrosion resistance exhibited by stainless steels and other corrosion resistant metals and alloys, effect of chem. environments on fracture of metals, metall. evidence regarding origin of meteorites, survey of ann. cost of corrosion to U. S. Office: Mass. Inst. Tech., Cambridge, Mass. 02139.*

UHR, Leonard Merrick, Am. psychologist; b. Phila., Jan. 26, 1927; s. Saul Irving and Elizabeth (Fishtein) U.; B.A., Princeton, 1949; M.A., Johns Hopkins, 1951; M.A., U. Mich., 1953, Ph.D. 1957; m. Elizabeth Stern, July 1, 1949; children—Frank, Steven. With U. Mich., 1951-65, research psychologist, asso. prof. psychology, 1957-65, coordinator psychol. scis. Mental Health Research Inst., 1964-65; prof. computer scis. dept. U. Wis., Madison, 1965—. Cons. System Devel. Corp., Santa Monica, Cal., 1960—. Mem. Am. Psychol. Assn., Assn. Computing Machinery, I.E.E.E. Author: (with J. Pollard, E. Stern) Drugs and Phantasy, 1966; also numerous articles. Editor: Drugs and Behavior, 1960; Pattern Recognition, 1966. Research on behavioral psychopharmacology; dynamic computer models of cognitive and perceptual processes. Home: 211 Lathrop St., Madison, Wis. 53705.

ULAM, Stanislaw Marcin, mathematician; b. Lwow, Poland, Apr. 13, 1909; s. Jozef and Anna (Auerbach) U.; Dr.Sci., U. Lwow, 1933; m. Francoise Aron, Aug. 19, 1941; 1 dau., Claire Anne (Mrs. Don Grusin). Staff, Princeton Inst. for Advanced Study, 1936; jr. fellow, lectr. Harvard, 1936-40; prof. U. Wis., 1941-43; research adviser Los Alamos Sci. Lab., 1943—; prof. U. Colo., Boulder, 1965—; vis. prof. Harvard, 1951, Mass. Inst. Tech., 1956, U. Cal., La Jolla, 1962. Mem. Am. Acad. Arts and Scis., Nat. Acad. Scis., Am. Math. Soc., Am. Phys. Soc. Author: Collection of Mathematical Problems, 1960; Problems of Modern Mathematics, 1964; also numerous articles. Research in set theory and founds. maths., topology, group theory, probability theory, math. physics, Monte Carlo method, thermonuclear reactions, nuclear propulsion for space vehicles, applications math. to biol. scis., electronic computers. Office: Los Alamos Sci. Lab., Los Alamos 87544; also Dept. Math., U. Colo., Boulder, Colo. 80304.*

ULEHLA, Ivan, Czechoslovakian physicist; b. Skalica, Czechoslovakia, Oct. 17, 1921; s. Miloslav and Anna (Tilschova) U.; RN Dr., Charles U., Prague, Czechoslovakia, 1949, Ph.D., 1956; m. Ludmila Ulehlova, Oct. 31, 1942; children—Ivan, Josef, Katerina; m. 2d, Libuse Pouchla, Sept. 29, 1966. Asst., Charles U., 1949-54, Faculty math., physics, 1967—; sr. scientist Inst. Nuclear Research, Prague, 1954-60; prof. theoretical physics Faculty Tech. and Nuclear Physics, Prague, 1960-67. Sci. sec. Internat. Conf. on Peaceful Uses of Atomic Energy, UN, 1955, spl. asst. to gen. sec., 1958; mem. bd. govs. Internat. Atomic Energy Assn., Vienna, 1962-63; vice dir. Joint Inst. for Nuclear Research, Dubna, USSR, 1963-67. Author: From Physics to Philosophy, 1962; Optical Model of Atomic Nucleus, 1964; also other books, articles. Studies on equations for elementary particles, reactor physics, theory of atomic nuclei, physics and philosophy. Home: 14 Vhvluhu 14, Prague 4, Czechoslovakia.*

ULFELDER, Howard, physician; b. Mexico D.F., Mexico, Aug. 15, 1911; s. Sidney and Ethel May (Housel) U.; grad. Phillips-Exeter Acad., 1928; A.B. cum laude, Harvard, 1932, M.D., 1936; m. Ethel Louise Huse, Aug. 18, 1932; children—Howard, John Carleton, William Harrison, Thomas Huse. Chief gynecol. service, Mass. Gen. Hosp., 1955—, vis. surgeon, 1955—; mem. faculty Harvard Med. Sch., 1940—, clin. prof. gynaecology, 1955-62, Joe V. Meigs prof. gynaecology, 1962—; chief staff Vincent Meml. Hosp., Boston, 1955—; gynecologist Pondville State Hosp., Norfolk, Mass., 1946—; cons. gynaecologist Royal Prince Alfred Hosp., Sydney, Australia, Boston Lying-In Hosp., Mass. Eye and Ear Infirmary; McIlrath vis. prof. Royal Prince Alfred Hosp., 1961. Dir. Westgate Corp., Va. Bd. dirs. Am. Cancer Soc., United South End Settlements Boston. Diplomate Am. Bd. Gynecology, Am. Bd. Surgery, Fellow A.M.A., A.C.S. (bd. govs. 1959-62); hon. fellow N.J., Pitts., obstet. and gynecol. socs.; mem. Pan Pacific Surg. Assn., Eastern, New Eng. surg. socs., Am. Surg. Assn., Soc. Univ. Surgeons, Soc. Pelvic Surgeons (founder, pres. 1958-59), A.A.A.S., Halsted Soc., New Eng. Cancer Soc., Am. Coll. Obstetricians and Gynecologists, Internat. Soc. Surgery, Am. Soc. Study Sterility, N.Y. Acad. Scis., Aesculapian Soc. (pres. 1961), Am. Gynecol. Soc., Am. Assn. Obstetricians and Gynecologists, Am. Gynecol. Club. Contbr. profl. papers, chpts. textbooks. Co-author: An Atlas of Pelvic Operations, 1953. Research on endometriosis; carcinoma of the cervix; pituitary gonad tropin and ovary steroid metabolism; biol. mechanics of pelvic support. Home: 15 Sheffield Rd., Winchester, Mass. 01890. Office: 32 Fruit St., Boston 02114.

ULLBERG, Sven Gustav, Swedish pharmacologist; b. Stenstorp, Sweden, Dec. 23, 1920; s. Gustav Fredrik and Karin (Ahlenius) U.; V.M.D., Royal Vet. Coll., Stockholm, Sweden, 1949, Ph.D., in Pharmacology, 1954; m. Lillemor Nohrman, Apr. 6, 1947; children—Mats, Mans, Charlotte, Sophie. Asst., Royal Vet. Coll., Stockholm, 1950-54, faculty, 1955—, asso. prof. dept. pharmacology, 1964—. Cons., Swedish Def. Research Inst., 1959—. Mem. Swedish Physiol. Soc., Swedish Radiobiol. Soc., A.A.A.S. Research and numerous publs. on drug distbn. and metabolism studies, mode of action of drugs, antibiotics, vitamins, hormones, biogenic amines, methods for micro-autoradiography using water soluble substances; devel. technique for whole-body autoradiography; constrn. automatic microtome. Home: Klubbvägen 7, Danderyd, Sweden. Office: Dept. Pharmacology, Royal Vet. Coll., Stockholm 50. Sweden.*

ULLERY, John Calvin, Am. physician; b. Bradford, Ohio, June 29, 1907; s. William George and Armina (Warner) U.; A.B., Ohio State U., 1928; M.D., Jefferson Med. Coll., Philadelphia, Pa., 1932; m. Margaret Brown, Nov. 10, 1932. Intern, Pennsylvania Hosp., Philadelphia, 1932-34; resident obstetrician and gynecologist, Phila. Lying-In Hosp., 1934-36; in med. practice, Philadelphia, and Upper Darby, Pa., since 1935; became asst. prof. obstetrics Jefferson

Med. College, 1950; became assistant professor obstetrics and gynecology, professor, chairman dept. obstetrics and gynecology College of Medicine, Ohio State University, 1954—, chief of division obstetrics and gynecology at University Hospital, 1954—; obstetrician and gynecologist, chief of service, Phila. Lying-In Hosp.; chief of service, obstetrics and gynecol., Philadelphia Gen. Hosp., Del. Co. Hosp., Drexel Hill, Pa.; acting chief obstetrician and gynecol., Del. Co. Hosp., Landsdowne, Pa.; cons. obstetrician and gynecologist Mt. Carmel, White Cross, St. Ann's hosps.; pres. Ohio State Health Center, 1960—; nat. cons. in obstetrics and gynecology to Surgeon Gen., USHF, 1962—; area cons. VA, Dayton, O. and Ohio and Wright-Patterson AFB Hosp., O.; cons. cancer control program USPHS; cons. obstetrics and gynecology VA, Columbus Med. Area. Diplomate Am. Bd. Obstetrics and Gynecology. Fellow A.C.S., Coll. Physicians Phila., Am. Assn. Obstetricians and Gynecologists, Internat. Coll. Surgs. Am. Gynecol. Soc., Am. Coll. Obstetrician and Gynecologists (sec. 1956-59), Intersoc. Cytology Council, A.M.A., Pa. Med. Soc., American Bd. Obstetrics and Gynecology Sphinx Society, Sigma Xi. Author textbooks: Stress Incontinence in the Female, 1955; Obstetric Mechanisms, pub. 1957; Textbook of Obstetrics, pub. 1966; also manual on female cancer, articles in med. jours., (film) Continuous Spinal Analgesia in Caesarean Section, 1946. Asso. editor: Am. Jour. Surgery, 1957—. Research in obstetrics and gynecology. Home: 3139 Leeds Rd., Upper Arlington, Columbus 21. Office: Univ. Hosp., Columbus 10, O.

ULLMAN, Albert Daniel, Am. sociologist; b. Boston, Apr. 6, 1918; s. Adolph and Mary (Klebanoff) U.; A.B., Yale, 1940; Ed.M., Harvard, 1941, Ph.D., 1950; m. Althea L. Hayes, June 17, 1941; children—Stephen H., Michael A., Peter W., Richard A., Mary-Leslie. With Tufts U., Medford, Mass., 1946—, prof. sociology, 1958—, chmn. dept. sociology, 1953-64, acting dean Coll. Liberal Arts, 1966-67, dean, 1967—. Cons. to state and fed. agys. Mem. Am. Sociol. Assn., Am. Psychol. Assn., Soc. for Study Social Problems, Sigma Xi. Author: To Know the Difference, 1960; Sociocultural Foundations of Personality, 1965; also articles. Research on relationship of drinking customs to alcoholism, theory of personality in its devel. in a sociocultural context. Home: 35 Hillside Terrace, Belmont, Mass. 02178. Office: Tufts U., Medford, Mass. 02155.*

ULLMAN, Edward L., Am. geographer; b. Chgo., July 24, 1912; s. B. L. and Mary Louise (Bates) U.; B.S., U. Chgo., 1934, Ph.D., 1941; A.M., Harvard, 1938; m. Doris E. Jackson, Sept. 5, 1942 (div. (1959); children—Edward J., Anne B.; m. 2d, Joan Connelly, June 14, 1967. Asst. geography Harvard, 1934-35, 37-38, asst. prof. regional planning, 1946-49, asso. prof., 1949-51; instr. econ. geography Wash. State Coll., 1935-37, Ind. U., 1941-42; chief transport sect. OSS, 1942-43; dir. joint intelligence study pub. bd. Joint Chiefs of Staff, 1943-46; spl. estimates staff Dept. State, 1946; transportation economist U. S. Maritime Commn., 1946; prof. geography U. Wash., 1951—, asso. dean Grad. Sch., 1962-65; Fullbright research prof. U. Rome, Italy, 1956-57; vis. prof. econ. geography, dir. Meramec Basin Research project Washington U., 1959-61; vis. prof., fall 1962; vis. prof. Salzburg Seminar in Am. Studies, 1965; pres. Washington Center for Met. Studies, 1965-67. Transport cons. Stanford Research Inst. for Internat. Coop. Adminstrn. and Phillippine Govt., Manila. 1956; cons. European Productivity Agency, Orgn. European Economic Cooperation, 1957; cons. on transp. Asian Devel. Bank, 1968. Served as lt. USNR, 1943-46. Western region econ. analysis committee of Social Sci. Research Council, 1951-56. Mem. Assn. Am. Geographers (citation award 1958, council 1949), New England Geog. Conf. (pres. 1948-49), Nat. Research Council (com. on geography Office Naval Research, com. econ., urban geography), 1949-53, mem. at large behavioral sci. div. 1968—), Am. Geog. Soc., Regional Sci. Association (council 1958-60, pres. 1960-61). Am. Assn. U. Professors, Italian Geography Soc. (corr.) Sigma Xi. Author: Mobile: Industrial Seaport and Trade Center, 1943; American Commodity Flow, 1957; The Meramec Basin, published in 1962. Advisory editor of the Annals Assn. Am. Geographers 1951-57. Contbr. articles sci., profl. jours. Introduced central place theory to Am., 1941; originated "amenities" as a factor in regional growth, 1954; pioneered spatial interaction and commodity flow approach, 1949-56; devised analog-interaction model, 1957, and applied (with Donald J. Volk) to prediction of recreation attendance and benefits, 1962; invented Minimum Requirements method and applied (with Michael Dacey) to determination of urban economic base, 1959. Home: 1111 McGilvra Blvd. E., Seattle 98102.*

ULLMAN, Joseph Leonard, Am. mathematician; b. Buffalo, Jan. 30, 1923; s. David and Ida (Bir) U.; B.A., U. Buffalo, 1942; Ph.D., Stanford, 1949; m. Barbara Eloise Whalley, Aug. 25, 1961; children—Katharine Susan, Sara Elizabeth. Faculty, U. Mich., Ann Arbor, 1949—, prof., 1966—. Mem. Am. Math. Soc., Math. Assn. Am., Phi Beta Kappa, Sigma Xi. Research in math. analysis, developing methods for solving problems of approximation theory using complex variable theory and potential theory. Home: 308 Wilton St., Ann Arbor, Mich. 48103.*

ULLMAN, Montague, Am. psychiatrist; b. N.Y.C., Sept. 9, 1916; s. William and Nettie (Eisler) U.; B.S., Coll. City N.Y., 1934; M.D., N.Y. U. Coll. Medicine, 1938; m. Janet Simon, 1941; children—Susan, William, Lucy. Faculty, N.Y. U. Coll. Medicine, 1950-61, asst. clin. prof. psychiatry, 1959-61; mem. psychoanalytic faculty, tng. analyst N.Y. Med. Coll., 1950-62, asst. clin. prof., 1956-62; faculty State U. N.Y., Downstate Med. Center, 1961—, prof., 1963—; asso. dept. pediatrics div. pediatric psychiatry Jewish Hosp. Bklyn., 1955-57; asst. attending psychiatrist Flower and Fifth Av. Hosp., N.Y.C., 1956-62; dir. Clinic for Retarded Children, Putnam County, 1957-61; dir. dept. psychiatry Maimonides Hosp. Bklyn., 1961—; dir. Maimonides Community Mental Health Center, 1967—. Diplomate Nat. Bd. Examiners, Am. Bd. Neurology and Psychiatry. Fellow A.A.A.S. (past mem. council), Am. Psychiat. Assn., N.Y. Acad. Medicine, Acad. Psychoanalysis (charter); Mem. Am. Acad. Neurology, A.M.A., Am. Soc. for Psychical Research (trustee), Assn. for Research in Nervous and Mental Disease, Assn. for Psychophysiol. Study Sleep, N.Y. Acad. Scis., Parapsychol. Assn. (pres. 1966—), Soc. for Gen. Systems Research, Soc. for Psychophysiol. Research, Soc. Biol. Psychiatry, Soc. Med. Psychoanalysts (past pres.), Am. Assn. U. Profs., Alpha Omega Alpha. Author: Behavioral Changes in Patients Following Strokes, 1962; also articles. Research on techniques for integration of psychiat. concepts into basic collegiate nursing curriculum, directional alterations in life history stroke patients especially on operation of denial mechanisms and anosognosia, creativity and dreams. Home: 55 Orlando Av., Ardsley, N.Y. 10502. Office: 4802 10th Av., Bklyn. 11219.*

ULLMANN, Leonard Paul, Am. psychologist; b. N.Y.C., May 28, 1930; s. Siegfried and Irma (Lichtenstadter) U.; A.B. magna cum laude with honors in psychology, Lafayette Coll., 1951; Ph.D., Stanford, 1955, A.M., 1953; m. Rina Kalb, June 5, 1951; children—Jeremy Michael, Nancy Leigh. Coordinator, VA Psychiat. Evaluation Project, 1956-64; faculty U. Ill., Urbana, 1963—, prof. psychology, 1966—. Cons. to hosps., state agys. Fellow Am. Psychol. Assn.; mem. A.A.A.S., Soc. for Research in Child Devel. Author: (with L. Krasner) Case Studies in Behavior Modification, 1965; Research in Behavior Modification, 1965; Institution and Outcome: A Comparative Study of Psychiatric Hospitals, 1967; also numerous articles. Research on application learning concepts to investigation and treatment hospitalized psychiat. patients at individual and instl. level. Home: 403 Buena Vista Dr., Champaign, Ill. 61820. Office: Psychology Dept., U. Ill., Urbana, Ill. 61801.*

ULLMO, Jean, French sci. philosopher; b. Paris, France, June 3, 1906; s. Edouard and Pauline (Dreyfus) U.; ed. Ecole Polytechnique, 1924-26; m. Mlle. Richard, Nov. 12, 1924; children—Yves, Bernard. From asst. to examiner Ecole Polytechnique, Paris, 1934-59. Prof. econ. scis. Ecole Nationale d'Administration, 1946. Mem. Math. Soc. France, Philos. Soc. France, Econometric Soc., others. Author: La brise de la Physique Quantique, 1953; La Pensée Scientifique moderne, 1958. Research, publs. on devel. quantum mechanics, 1928-34; introduced Keynesian econs. in France, 1936; contbd. to extension, devel. French econ. thought after World War II, also to philosophy of sci., role of groups in theory of knowledge. Home: 118 Malesherbes Blvd., Paris, France.*

ULLSTRUP, Arnold John, Am. plant pathologist; b. Milw., Jan. 24, 1907; s. Otto E. and Judith (Van Loghem) U.; B.S., U. Wis., 1931, Ph.D., 1934; m. Sara L. Hoopes, June 19, 1934; children—Karen (Mrs. D. M. Newman), Arne Peter. Plant pathologist U. S. Dept. Agr., Purdue U., Lafayette, Ind., 1938—. Mem. Am. Phytopath. Soc., Mycol. Soc. Am., Torrey Bot. Club, Sigma Xi, Alpha Zeta. Research, numerous publs. on pathology of maize, inheritance of resistance to disease, life histories of pathogens of maize, description of new diseases of maize and inciting agts.; devel. of disease resistant inbred lines and hybrids of maize. Home: 1610 Ravinia Rd., West Lafayette, Ind. 47906.*

ULMER, Melville Jack, Am. economist; b. N.Y.C., May 17, 1911; s. Saul Kingsley and Lillian Ulmer; B.S., N.Y. U., 1937, M.A., 1938; Ph.D., Columbia, 1948; m. Naomi Zinkin, June 1, 1937; children—Melville Paul, Stephanie Marie. Journalist, N.Y. Am., 1932-37; chief gen. price sect. U. S. Bur. Labor Statistics, 1940-45; asst. chief reports div. Smaller War Plants Corp., 1945-46; chief financial orgn. sect. office Bus. Econs., U. S. Dept. Commerce, 1946-48, editor Survey Current Bus., 1948-50; faculty econs. Am. U., 1943-60, prof., 1952-60, chmn. dept., 1954-60; prof. econs. U. Md., College Park, 1960—; vis. prof. Mexico City Coll., 1950, Netherlands Sch. Econs., 1958-59, 65-66. Research asso. Nat. Bur. Econ. Research, 1951-60; econ. adviser for various govt. burs. Wilton Park fellow, Sussex, Eng., 1966. Recipient Meritorious Service award U. S. Govt., 1946; Sr. Fulbright awards U. S. Govt., 1958, 65; Univ. medal Free U. Brussels, 1959. Fellow A.A.A.S.; mem. Am. Econ. Assn., Econometric Soc., Pi Gamma Mu. Author: (with J. M. Blair) A History of Prices During World War II, 1945; (with C. Wright Mills) Small Business and Civic Welfare, 1946; The Economic Theory of Cost of Living Index Numbers, 1949; Trends and Cycles in Capital Formation by U. S. Railroads, 1860-1950, 1957; Economics: Theory and Practice, 1959, 65; Capital in Transportation, Communications and Public Utilities: Its Formation and Financing, 1960. Research, publs. on devel. of measures and patterns of growth and long cycles in transp. and pub. utility investment, measures of actual and intended bus. expenditures on plant and equipment; theoretical found. for measuring changes in cost of living; history of prices during World War II; clear and rigorous exposition of prins. of econs. Home: 10401 River Rd., Potomac, Md. 20854. Office: U. Md., College Park, Md. 20742.*

ULRICH, Edward Oscar, Am. paleontologist; b. Cincinnati, O., Feb. 1, 1857; s. Charles and Julia (Schnell) U.; studied German Wallace Coll., Berea, O.; hon. A.M., 1886, D.Sc., 1892; studied Ohio Med. Coll.; m. Albertine Zuest, June 29, 1886; m. 2d, Lydia Sennhauser, June 20, 1933. Curator geology, Cincinnati Soc. Natural History, 1877-81; paleontologist to geol. surveys of Ill., Minn. and Ohio, 1885-96; geologist U. S. Geol. Survey since 1897. Asso. in paleontology, U. S. Nat. Museum, 1914-32, retired; asso. editor Am. Geologist for 10 years. Mem. Nat. Acad. Sciences, Washington Acad. Sciences; fellow Geol. Soc. America, Paleontol. Soc.; foreign mem. Geol. Society London; corr. member Geol. Society of Sweden, Senkenberg, Naturforsch. Gesells.; corr. Phila. Acad. Science, 1932. Awarded Mary Clark Thompson medal, 1930. Penrose medal, 1932. Author: American Paleozoic Bryozoa, 1884 (Cincinnati Soc. Natural History); American Paleozoic Sponges and Paleozoic Bryozoa (Vol. VIII. Ill. Geol. Survey), 1890; monographs in Vol. III Geol. Survey of Minn., on the Lower Silurian Bryozoa, Lamellibranchiata, Ostracoda and Gastropoda of Minn. (2 parts), 1893-97; Geology of the Lead, Zinc and Fluor Spar District of Western Ky., 1804; Revision of Paleozoic Systems, 1911; The Ordovician-Silurian Boundary, 1914; Formations of the Chester Series, 1917; Correlation by Displacements of the Strandline, 1916; Major Causes of Land and Sea Oscillations, 1922; Silurian Formations of the Appalachian Region, 1923; Monograph of Silurian Ostracoda and New Classification of Paleozoic Ostracoda, 1923; Formations and Breaks between Paleozoic Systems in Wisconsin, 1924; Classification of the Conodonta, 1925; Relative Values of Criteria Used in Defining Paleozoic Systems, 1927; Monograph of the Telephidae, 1929; Monograph Dikellocephalidae, Parts 1 and 2, 1932. Contbns. to textbooks and journals. Research in paleontology and stratigraphy, especially Paleozoic era. Died Feb. 22, 1944.

ULRICH, Henri, chemist; b. Rheinsberg, Germany, May 4, 1925; s. Hermann and Ella (Seeling) U.; diplom chemiker U. Berlin (Germany), 1952, Dr.rer.nat., 1954; m. Franziska Schimitzek, June 2, 1954; children—Stefan, Tomas, Barbara, Bertram. Came to U. S., 1955, naturalized, 1966. Instr., U. Berlin, 1953-54; research asso. Ohio State U. Research Found., 1955-56; research chemist Olin Mathieson Chem. Corp., Columbus, O., 1956-59; group leader organic research Carwin Co., North Haven, Conn., 1959-62; head organic research Upjohn Co., Carwin Research Lab., North Haven, 1962-65, mgr. chem. research, 1965—. Mem. Am. Chem. Soc., A.A.A.S., Gesellschaft Deutscher Chemiker. Author: Cycloaddition Reactions of Heterocumulenes, 1967; also articles. Research on fluorine chemistry, agrl. chems., photog. developers, polyurethanes, chem. intermediates, chemistry reactive double bond systems. Patentee photog. developers, bldg. blocks for polymers. Home: Surrey Dr., Northford, Conn. 06472. Office: 410 Sackett Point Rd., North Haven, Conn. 06473.*

ULRICH, Roger Ellwood, Am. psychologist; b. Peoria, Ill., Aug. 30, 1931; s. Ralph C. and Della Irene Ulrich; B.S., N. Central Coll., 1953; M.A., Bradley U., 1956; Ph.D., So. Ill. U., 1961; postgrad. Ind. U., 1962; m. Carole Lou McNish, Aug. 28, 1959; children—Thomas Alan, Traci Ellen, Kristan Sue. Research asst. Behavior Research Lab., Anna (Ill.) State Hosp., 1959-61; asso. prof. psychology Ill. State U., 1962-63; asso. prof., chmn. dept. psychology Ill. Wesleyan U., 1961-65; prof., head dept. psychology Western Mich. U., Kalamazoo, 1965-67, research prof., 1967—, dir. Behavior Research Lab., 1965—. Mem. Am., Midwest psychol. assns., A.A.A.S., Psychonomic Soc. Author: Lectures in Psychology, 1963; (with others) The Control of Human Behavior, 1966; also articles. Adv. editor Jour. Exptl. Analysis of Behavior, 1966—. Research on aggression, animal and human learning, application of principles of behavioral analysis to social settings. Home: 122 Sydelle St., Kalamazoo, Mich. 49001.*

ULSTROM, Robert A., Am. physician; b. Mpls., Feb. 23, 1923; s. Alger Roy and Catherine (Dougall) U.; B.S., U. Minn., 1944, M.D., 1946; m. Mary Janet McGrath, June 7, 1946; children—Jane Elizabeth, Susan Jean, Cynthia Louise. Faculty dept. pediatrics U. Minn., Mpls., 1950-64, prof., 1961-64; asst. prof. U. Cal. Med. Sch., Los Angeles, 1953-56, prof., chmn. dept. pediatrics, 1964-66; asso. dean Coll. Med. Scis., prof. pediatrics U. Minn., Mpls., 1967—. Mem. NRC, 1961-64; gen. med. scis. study sect. NIH, 1963-68. Markle scholar in med. scis., 1954-59. Mem. A.A.A.S., Am. Pediatric Soc., Am., Central, Western socs. for clin. research, Endocrine Soc., Soc.

for Exptl. Biology and Medicine, Phi Rho Sigma. Publs., biochem. research on metabolism, endocrinology of newborn infant and related clin. problems. Home: 4616 Sunset Ridge, Mpls. 55416.*

ULUGH-BEG, astronomer, mathematician; b. Soltaniyeh, Persia, 1394; s. Shah Rokh; 1 son, Abd al-Latif. Grandson of Timur (Tamerlane); became gov. of Transoxiana as prince-regent; ruled Samarkand, from 1408; built obs. at Samarkand, begun 1428; succeeded his father to Timurud throne, 1447, then murdered at his son's instigation, 1449. Compiled first original star catalogue since Ptolemy's; made tables of plants, sun and moon; redetermined position of 992 fixed stars with unusual precision. Died Samarkand, Oct. 27, 1449.

UMBACH, Wilhelm, German physician; b. Frankfort/Main, Germany, Aug., 21, 1915; s. Wilhelm and Charlotte (Wust) U.; ed. U. Frankfurt/Main, U. Munschen, U. Jena; approbation, Promotion, 1945; Habilitation, 1955; m. Hilde Stein, June 7, 1947; children—Wolf-Ulrich, Claudia Freux. Faculty, U. Freiburg (Germany), 1955-—, prof. neurosurgery, 1960-—; dir. Neurosurg. Clinic, Siegen/Westfalen, Germany, 1966-—. Fellow Internat. Soc. Surgery, neurosurgery, neurology, angiology, electroencephalography, neurovegetative research. Author: Gesichtsneuralgien, 1960; Phanomena bei stereotakischen Hirnoperationen, 1966; ABC für Parkinsonkranke, 1966; also numerous articles. Research on brain, spinal cord and peripheral nerve surgery, especially pain and stereotaxic surgery, electrophysiol. and autonomic regulation of deep brain structures, epilepsy treatment, biochemistry and treatment of extrapyramidal disorders, vascular surgery. Home: 67 Richard Wagnerstrasse, Freiburg, Germany. Office: Neurochirurgische Klinik, Jung Stilling-Krakenhaus, 59, Siegen, Germany.*

UMBREIT, Wayne William, Am. bacteriologist; b. Markesan, Wis., May 1, 1913; s. William Traugott and Augusta (Abendroth) U.; B.A., U. Wis., 1934, M.S., 1936, Ph.D., 1939; m. Doris McQuade, July 31, 1937; children—Dorayne Loreda, Jay Nicholas, Thomas Hayden. Instr. soil microbiology Rutgers U., 1937-38; faculty U. Wis., Madison, 1938-44, asst. prof. bacteriology and chemistry, 1941-44; faculty Cornell U., 1944-47, prof. bacteriology, 1946-47; head dept. enzyme chemistry Merck Inst., Rahway, N.J., 1947-58; asso. dir., 1958; chmn. dept. bacteriology Rutgers U., New Brunswick, N.J., 1958-—. Recipient Biochem. Congress Symposium medal, Paris, France, 1952. Fellow Am. Acad. Microbiology, N.Y. Acad. Sci., A.A.A.S.; mem. Am. Soc. for Microbiology (Eli Lilly award in bacteriology 1947, Carski Found. award for distinguished teaching 1968), Soc. Biol. Chemists, Am. Chem. Soc., Theobald Smith Soc. (Waksman award in microbiology 1957, past pres.), Am. Assn. U. Profs., Sigma Xi. Author: (with Burris, Stauffer) Manometric Techniques, 1945; (with Oginsky) An Introduction to Bacterial Physiology, 1954; Metabolic Maps, 1960; Modern Microbiology, 1962; also numerous articles. Editor: Advances in Applied Microbiology, vols. 1-8; 1959-66. Research on chemistry of autotrophic bacteria, function of vitamins; action of antibiotics; metabolic effects of hormones. Home: 527 Prospect St., Westfield, N.J. 07090. Office: Dept. Bacteriology, Nelson Hall, Rutgers U., New Brunswick, N.J. 08903.*

UMEZAWA, Hiroomi, physicist; b. Saitama-ken, Japan, Sept. 20, 1924; s. Junichi and Takako (Sato) U.; grad. elec. engring. U. Nagoya, 1947; D.Sc., 1952; m. Tamae Yamagami, July 30, 1958; children—Rui, Ado. Assoc. prof. U. Nagoya, 1952-56; prof. U. Tokyo (Japan), 1956-65; prof. physics U. Naples (Italy), 1965-66; prof. physics U. Wis., Milw., 1966-—; vis. prof. U. Wash., 1956, U. Md., 1957, U. Ia., 1957, U. Marseille (France), 1959-60. Leader Naples Group Structure of Matter, Centera Nat. Research in Italy, 1964-66. Imperial Industry fellow U. Manchester (U.K.), 1953-55. Mem. Japan, Italy, Am. phys. socs. Author: Quantum Field Theory, 1956; also numerous other books, articles on quantum field theory or quantum mechanics in Japanese. Research on high energy particle physics, theory of quantum field and theory of many-body physics. Home: 4304 N. Murray Av., Shorewood, Wis. 53211.*

UNANUE, José Hipolito, physician; b. Arica, Chile, 1758; dep., minister of finance, pres. council of state of new republic of Peru; a founder El mercurio peruano, 1791; helped found anat. teaching amphitheater, Lima, also sch. of medicine, San Fernando; propagator of vaccination in Peru; pub. numerous bot. studies. Died 1833.

UNDERHILL, Anne Barbara, astrophysicist; b. Vancouver, B.C., Can., June 12, 1920; d. F. Clare and Irene A. (Creery) Underhill; B.A., U. B.C., Vancouver, 1942, M.A., 1944; postgrad. U. Toronto (Ont., Can.); Ph.D., U. Chgo., 1948. Staff, Dominion Astrophys. Obs., Victoria, B.C., 1949-62, sr. sci. officer, 1960-62; prof. astrophysics U. Utrecht (Netherlands), 1962-—; vis. lectr. in astrophysics Harvard, 1955-56. Fellow Royal Astron. Soc., Royal Astron. Soc. Can.; mem. Am. astron. Soc., Astron. Soc. Pacific, Netherlands Astron. Club. Author: The Early-type Stars, 1966; also numerous articles. Research on astrophysics especially study and interpretation of stellar spectra with application to study atmospheres of hot stars.*

UNDERKOFLER, Leland Alfred, Am. biochemist; b. Fairfield, Neb., Apr. 24, 1906; s. William Eugene and Laona (Hobson) U.; A.B., Neb. Wesleyan U., 1928, D.Sc., 1953; Ph.D., Ia. State U., 1934; m. Mabel Ann Horn, Sept. 1, 1932; children—William Leland, Leon Milton. Faculty, Ia. State U., 1935-55; research dir. Takamine Lab., dir. enzymology research lab. Miles Labs., Inc., Clifton, N.J., Elkhart, Ind., 1955-—. Mem. Am. Chem. Soc., Am. Soc. Biol. Chemists, Am. Assn. Cereal Chemists, Soc. Indsl. Microbiology (president), Am. Acad. Microbiology, Inst. Food Technologists, A.A.A.S., Sigma Xi. Author: Introduction to Organic Chemistry, 1953; also numerous articles. Editor: (with R. J. Hickey) Industrial Fermentations, 1954. Research in fermentation processes for chems. and enzymes, processes for producing indsl. enzymes, isolation and purification of enzymes, investigations on microbial nutrition. Home: 210 Larson Av. Office: 1127 Myrtle Av., Elkhart, Ind. 46514.*

UNDERWOOD, Michael, English physician; b. Surrey, Eng., 1736; ed. St. George's Hosp., also studied in Paris, France. Practiced medicine, specializing in surgery, London, then as male-midwife, Brit. Lying-In Hosp.; attended Princess of Wales, at birth of Princess Charlotte, 1796. Licentiate Royal Coll. Physicians, 1784; mem. Co. Surgeons. Author: A Treatise Upon Ulcers of the Legs, 1783; Surgical Tracts on Ulcers of Legs, 1788; A Treatise on the Diseases of Children, 1784, enlarged 2d edit., 1801. Wrote best English treatise of period on children's diseases, which contained first description of sclerema neonatorum (underwood's disease), a children's disease identified by hardening and tightening of skin and depression of respiration and pulse; also described aphthae or thrush. Died Knightsbridge, Eng., Mar. 14, 1820.

UNDERWOOD, Newton, Am. physicist; b. Atlanta, Nov. 19, 1906; s. Emory Marvin and Ruth E. (Newton) U.; B.S. in Math., Emory U., 1928; M.S. in Physics, Brown U., 1930, Ph.D., 1934; m. Hazel Briggs, June 18, 1934; children—Emory Marvin II, Marion Louise (Mrs. Timothy N. Taylor), Robert Gordon. Instr., Hood Coll., Frederick, Md., 1932-36; faculty Vanderbilt U., Nashville, 1936-50; prof. physics N.C. State U., Raleigh, 1950-62; prof. U. N.C., Chapel Hill, 1962-—; physicist Columbia, 1940-42. Mem. Am. Phys. Soc., Health Physics Soc., Am. Pub. Health Assn., Phi Beta Kappa, Sigma Xi. Patentee gaseous diffusion separation isotopes. Research, publs. on measurement bacterial growth instrumentation, measurements optical properties crystals, measurements stopping powers materials for charged particles, thermoluminescence, soil moisture, electron diffraction. Home: 111 Lone Pine Rd., Chapel Hill, N.C.*

UNDERWOOD, Ralph Sylvester, Am. mathematician; b. Anoka, Minn., Oct. 3, 1891; s. Charles Milford and Letta (Barrett) U.; B.A., U. Minn., 1916, M.A., 1917; Ph.D., U. Chgo., 1930; m. Zelda Wisdom Ray, Feb. 1, 1930. Faculty, Purdue U., 1919-22, Ala. Poly. Inst., 1922-27, U. Tex., summers 1943-45; faculty Tex. Technol. Coll., Lubbock, 1927-—, prof. math. and astronomy, 1930-61, emeritus, 1961-—. Mem. A.A.A.S., Math. Assn. Am., Am. Math. Soc., Am. Assn. U. Profs., Tex. Acad. Sci., Phi Beta Kappa, Sigma Xi. Author: Jaunts into Space, 1935; (with F. W. Sparks) Living Mathematics, 1940, rev., 1949, Analytic Geometry, 1947, rev., 1956, 61; (with Nelson, Selby) Intermediate Algebra, 1947; (with Woodward) Practical Trigonometry, 1956; Silhouette Mathematics, 1961; also articles. Originator extended analytic geometry field. Address: 2220 Broadway, Lubbock, Tex. 79401.*

UNGAR, Frank, Am. biochemist; b. Cleve., Apr. 30, 1922; s. Michael and Susan (Gelberger) U.; B.A., Ohio State U., 1943; M.S., Western Res. U., 1948; Ph.D., Tufts U., 1952; m. Shirley Ruth Katz, Sept. 26, 1948; children—Leanne Ruth, William Stephan, Barbara Louise, Joanne Sue. Research staff Cleve. clin., 1947-48, Worcester Found. for Exptl. Biology, Shrewsbury, Mass., 1951-58; faculty U. Minn. Med. Sch., Mpls., 1958-—, prof. biochemistry, 1966-—, dir. tng. program for steroid biochemistry, 1962-—; vis. asst. prof. Clark U., 1956-58. Mem. Am. Assn. Biol. Chemists, Am. Chem. Soc., Soc. for Exptl. Biology and Medicine, A.A.A.S., Endocrine Soc., Sigma Xi. Author: (with Ralph I. Dorfman) Metabolism of Steroid Hormones, 1953, 65; also articles. Research on metabolism of steroid hormones including reduction and hydroxylation reactions, periodicity, circadian rhythms in adrenal-pituitary-hypothalamus, effects of endocrine glands and growth hormone on enzymes, hormone action of progesterone. Home: 5232 Stevens Av., Mpls. 55419.*

UNGAR, Georges, physiologist, pharmacologist; b. Budapest, Hungary, Mar. 30, 1906; s. Alexandre and Julia (Barec) U.; D.Sc., U. Paris (France), 1934, M.D., 1939; m. Alberte Levillain, Oct. 2, 1937; 1 dau., Catherine Anne. Came to U. S., 1947, naturalized, 1954. Asst. prof., U. Paris, 1934-37; head Lab. Exptl. Medicine, Paris, 1937-39; mem. research unit Ministry Home Security, Oxford, Eng., 1941-44; lectr. U. London (Eng.), 1944-47; research asso. Northwestern U., Chgo., 1948-53; dir. dept. pharmacology U. S. Vitamin and Pharm. Corp., N.Y.C., 1954-62; prof. pharmacology, Baylor U. Coll. Medicine, Houston, 1963-—; Claude Bernard vis. prof. U. Montreal (Que., Can.), 1948; dir. Inst. for Comparative Biology, San Diego, 1962-63; vis. prof. Laval U., Quebec, Que., 1966. Sci. adviser to French Embassy, London, 1945-47. Fellow A.A.A.S., Royal Soc. Medicine London, N.Y. Acad. Medicine; mem. Am., French, Brit. physiol. socs., Soc. for Exptl. Biology and Medicine, Biochem. Soc., Soc. for Endocrinology, Histamine Club (sec. 1957-—). Author: Role Physiologique et pathologique de l'histamine, 1937; Excitation, 1963; also numerous articles. Research on role histamine in life processes and diseases, mechanism shock, inflammation and allergic reactions, chem. changes taking place in nervous system during excitation, chem. mechanism memory; co-discoverer antithistamine drugs, oral antidiabetic drug. Home: 1800 Holcombe Blvd., Houston 77025.*

UNGER, Franz, botanist, paleontologist; b. Arnthof, Styria, Nov. 30, 1800; grad. in medicine, Vienna, 1827; prof. botany, Graz, Austria, 1836-50; became prof. plant anatomy and physiology, Vienna, 1850-—. Author: Die Urwelt in ihren Verschiedenen Bildungsgraden, 1847; Synopsis plantarum fossilium, 1845; Genera et species plantarum fossilium, 1850; Iconographia plantarum fossilium, 1852; Anatomie und Physiologie der Pflanzen, 1855; Botànische Streifzüge auf der Gebiet der Kulturgesch, 1851; Sylloge plantarum fossilium, 3 vols., 1860-66. First to point out resemblance between plant protoplasm and animal sarcode, 1855; 1st description of male gametes in mosses and liverworts; one of 1st to demonstrate plant evolution, 1852; contributed to devel. of paleobotany; studies on cell formation in growing plants. Died Graz, Feb. 13, 1870.

UNGHVARY, László, Hungarian physician; b. Petrozsény, Hungary, May 29, 1904; s. Gy. and E. (Emmerth) U.; student U. Debrecen (Hungary), 1928-34, U. Budapest (Hungary), 1934-49, U. Zürich (Switzerland), 1937, U. Bad-Nauheim (Germany), 1938; m. I. Takách, Aug. 1, 1934. Staff, U. Debrecen, 1928-34; asst. prof. Inst. for Gen. Pathology, U. Budapest, 1934-49, asst. prof., chief physician Inst. for Rheumatic Diseases, 1949-—, leader various med. labs., 1938-—. Mem. German Cardiological Soc., Deutsche Gesellschaft für Kreislaufforschung. Author: Electrocardiography, 1958; Medical Syndromes, 1961; also numerous articles. Research on normal and path. significance of R-T angle-distance, various diseases of heart and electrocardiographic diagnosis. Home: 36 Rakoczi, Budapest, Hungary.*

UNKLESBAY, Athel Glyde, Am. geologist; b. Byesville, O., Feb. 11, 1914; s. Howard Ray and Madaline (Archer) U.; A.B., Marietta Coll., 1938; M.A., U. Ia., 1940, Ph.D., 1942; m. Wanda Strauch, Sept. 14, 1940; children—Kenneth, Marjorie, Carolyn, Allen. Geologist, U. S. Geol. Survey, Alaska, Fla., 1942-45, Ia. Geol. Survey, 1945-46; instr. Colgate U., 1946-47; faculty U. Mo., Columbia, 1947-—, prof., 1954-—, chmn. dept. geology, 1957-—, v.p. for adminstrn., 1966-—. Mem. Am. Assn. Petroleum Geologists, Geol. Soc. Am., Nat. Assn. Geology Tchrs., Paleontol. Soc., Sigma Xi. Author: Geology of Boone County, Missouri, 1952; Pennsylvanian Cephalopods of Oklahoma, 1961; also articles. Research on Paleozoic era cephalopods, water resources of central Fla. Home: 806 N. Valley View St., Columbia, Mo. 65201.*

UNNA, Klaus Robert Walter, pharmacologist; b. Hamburg, Germany, July 30, 1908; s. Karl and Marie (Boehm) U.; M.D., U. Freiburg, 1930; m. Dr. Maya Stromberg Grossmann, May 25, 1939; 1 son, Jan Erik. Came to U. S., 1937, naturalized, 1944. Intern Univ. Hosps., Berlin, Hamburg, Cologne, 1931; asst. pharmacology U. Hamburg, 1932, U. Vienna, 1933-37; sr. pharmacologist Merck Inst. Therapeutic Research, 1937-44; instr. pharmacology U. Pa., 1944-45; asst. prof. U. Ill., 1945-48, asso. prof., 1948-51, prof., 1951-—, head dept. pharmacology, 1954-—. Executive hon. sec. International Brain Research Organization, UNESCO, 1965-66. Mem. Am. Physiol. Soc., Am. Soc. Pharmacology and Exptl. Therapeutics, A.A.A.S., Am. Inst. Nutrition, Soc. Exptl. Biology Med., Sigma Xi. Mem. ed. bd. Jour. Pharmacology and Exptl. Therapeutics, 1953-57; asso. editor Proc. of Soc. Exptl. Biology and Medicine, 1959. Contbr. to various physiol., pharmacol., med. jours. Research on cardiovascular drugs; Vitamin B deficiencies; muscle paralysants; mode of action of drugs in the central nervous system. Home: Brook Pl., Hinsdale, Ill. Office: 901 S. Wolcott Av., Chgo.*

UNNA, Paul Gerson, German dermatologist; b. Hamburg, Germany, Sept. 8, 1850; ed. U. Strasbourg (now in France); became physician gen. hosp., Hamburg, 1878; founder pvt. clinic for skin diseases; prof. skin pathology, U. Hamburg from 1919. Author: Histopathologie der Hautkrankheiten, 1894; Histologischer Atlas zur Pathologie der Haute, 1897-1910; Biochemie der Haut, 1913. Research on biochemistry of skin; described stratum granulosum of skin, 1876, seborrheic eczema (Unna's disease), 1887; originated idea of using coated pills to provide for local absorption in the intestine, 1884; introduced resorcin and ichthyol in medicine, 1886; credited with introducing diascopy in skin examinations, 1894; credited with discovery of plasma cell; after A. Ducrey's discovery of bacillus of soft chancre, undertook important research on it; described what he considered the source of eczema; modified treatment of some skin diseases. Died Hamburg, Jan. 29, 1929.

UNNERUS, Carl-Erik, Finnish physician; b. Viborg, Finland, Aug. 25, 1911; s. Sven A. and Edit E. (Erving) U.; Med.lic., U. Helsinki, 1941, M.D., 1957; m. Agron Dora Maria, Dec. 22, 1941; children—Hans-Anders Herbert, Carola. Docent med. radiology U. Helsinki, 1958-66; prof. U. Oulu (Finland), 1966—; head radiol. dept. Womens U. Hosp., Helsinki, 1950—; chmn. Finnish Red Cross Health Dept., 1962—; head Cancer Outpatients Clinic, 1957—. Decorated Finnish Cross Freedom, Cancer medal in silver, 1961, Finnish Red Cross medal for merit, 1965; WHO research grant, 1959-60; Fulbright fellow, 1962, France, 1965. Mem. Finnish Radiol. Soc. (chmn. 1962—), Scandinavian Radiol. Soc. (sec. gen. 1964—), Congress of the Scandinavian Radiol. Socs. (pres. 1964—), Finnish Med. Assn. (past chmn., vice chmn. 1966—), Finnish Cancer Soc. (chmn. 1961—). Author: A Radiological and Obstetrical Survey of the Female Pelvis, 1957; Radiology on Obstetrics, 1956; Volume Dose and Gonad Dose in X-ray Diagnostics, 1963; Radiotherapy in Cancer of the Cervix, 1964; also numerous articles. Research on cancer, decreasing of X-ray dosages in X-ray diagnostics, new methods and apparatus in X-ray diagnostics in gynecology and obstetrics, urology, X-ray diagnostics of skeleton. Home: 5 Havsvindsvägen, Tapiola-Hagalund, Finland. Office: 2 Haartmansg., Helsinki, Finland.*

UNSÖLD, Albrecht Otto Johannes, German astrophysicist; b. Bolleim, Germany, Apr. 20, 1905; s. Johannes and Clara (Müller) U.; student U. Tübingen, U. Munich; Ph.D.; Dr.honoris causa U. Utrecht, 1961; m. Liselotte Kühnert, Sept. 26, 1934; children—Hans-Jürgen, Eberhard, Wolfgang, Annelotte. Became prof. agrégé physics and astrophysics U. Munich, 1929, U. Hamburg, 1930; became prof. theoretical physics, 1932; dir. Inst. and Obs., U. Kiel. Recipient Copernicus prize, 1943; Catherine Wolfe Bruce Gold medal Astron. Soc. Pacific, 1956; gold medal London Royal Astron. Soc., 1957. Hon. mem. Gottingen, Internat. acads. astronautics. Author: Physik der Sternatmosphären mit besonderer Berüchsichtigung der Sonne, 1938. Editor, Zeischrift für Astrophysik. Research in astrophysics, especially stellar atmospheres. Home: Sternwartenweg 17, 23. Office: Neue Universität, Olshausenstrasse 23, Kiel, West Germany.

UNTERSTENHÖFER, Günter, German biologist; b. Wipperfürth, Germany, July 8, 1914; ed. U. Bonn, U. Jena; m. Kardula Koppelberg, July 7, 1942; children—Christopher, Thomas. Became head biology dept. Farbenfabriken Bayer, 1955; named prof. U. Bonn, 1955. Research and publs. in phytopharmacy, econ. phytopathology and methodics. Home: Leichingerstrasse 13, Opladen, West Germany. Office: Leverkusen, Bayerwerke, West Germany.

UNVERDORBEN, Otto, German chemist; b. Dahme, nr. Potsdam, Germany, Oct. 13, 1806; studied chemistry; undertook career in commerce; 1st to prepare crystalline (renamed aniline by Fritzsche), 1826. Died Dahme, Dec. 28, 1873.

UNWIN, William Cawthorne, English engr.; b. Coggeshall, Eng., Dec. 12, 1838; s. William Jordan and Eliza (Tailer) U.; B.Sc., London, 1861; LL.D., Edinburgh, 1905; became sci. asst. to William Fairbairn, 1856; apptd. prof. hydraulics Royal Indian Engring. Coll., 1872; prof. engring 1st London U., from 1900. Recipient Kelvin medal, 1921. Author: Elements of Machine Design, 1877; The Testing of Materials of Construction, 1888; Treatise on Hydraulics, 1907. Conducted original research on strengths of materials, masonry dams; established principles of design of reinforced concrete; studied steam engines; pioneer in internal combustion engine; worked on Canadian Niagara Falls project. Died Kensington, Eng., Mar. 17, 1933.

UNZER, Johann August, German physiologist; b. Halle, Germany, 1727; m. Johanna Charlotte. Editor: Der Arzt med. jour.; author: Erste Gründe einer Physiologie der eigentlichen thierischen Natur thierischer Körper, 1771. Concluded that movement occurs when nerve ganglia reflect an external stimulus to given organ, and used theory to explain Hydra's movements; believed living beings were mech. systems, not requiring brain or soul. Died 1799.

UOTILA, Urho A(ntti) (Kalevi), geodesist; b. Pöytyä, Finland, Feb. 22, 1923; s. Antti Samuli and Vera Justina (Kytö) U.; came to U. S., 1951, naturalized, 1957; B.S., Finland's Inst. Tech., 1946, M.S., 1949; Ph.D., Ohio State U., 1959; m. Helena Vanhakartano, Aug. 8, 1949; children—Heidi, Kirsi, Elizabeth, Julie, Trina, Caroline. Surveyor, geodesist Finnish Govt., 1945-46, 48-51; geodesist Swedish Govt., 1947; research asst. Ohio State U., 1952-53, research asso., 1953-58, lectr. in geodesy, 1955-57, asso. research supr., 1958-61, research supr., 1961—, asst. prof., 1959-62, asso. prof., 1962-65, chairman of the department geodetic science, 1964—, professor, 1965—. Member advisory panel on geodesy for U. S. Coast and Geodetic Survey, Nat. Acad. Sci., 1964-66. Served with Finnish Army, 1942-44. Mem. Am. Geophys. Union (v.p. geodesy sect.), Am. Congress Surveying and Mapping, Am. Soc. Photogrammetry, Soc. Expln. Geophysicists, Canadian Inst. Surveying, Maanmittaustieteiden Seura, Sigma Xi. Contbr. articles profl. jours. Research in geometric geodesy, spheropotential and geopotential surfaces, gravity and gravitation, adjustment computation; computed several geoids using surface gravity material; these geoids compared with satellite observations; supervised world-wide collection of gravity material, put into uniform Potsdam system and accuracy evaluations made. Home: 4329 Shelbourne Lane, Columbus, O. 43221.*

UPHAM, Warren, Am. geologist, archaeologist; b. Amherst, N.H., Mar. 8, 1850; s. Jacob and Sarah (Hayward) U.; A.B., Dartmouth, 1871, A.M., 1894, D.Sc., 1906; m. Addie M. Bixby, Oct. 22, 1885; 1 dau., Pearl (dec.). Asst. geol. survey of N.H., 1875-78; on geol. survey of Minn., 1879-85; on U. S. Geol. Survey, 1885-95; sec. and librarian of Minn. Hist. Soc., St. Paul, 1895-1914, and archaeologist, 1914—. Fellow A.A.A.S. Author: The Glacial Lake Agassiz, 1895; Greenland Icefields and Life in the North Atlantic, with a New Discussion of the Causes of the Ice Age (with Prof. G. F. Wright), 1896; Minnesota in Three Centuries, Vol. I, 1908; Catalogue of the Flora of Minnesota, 1884; Minnesota Geographic Names, 1920; Congregational Work of Minnesota, 1832-1920; Stages of the Ice Age, 1922; Chapters of Minnesota and Its People, 1924; also many geol. reports and papers in scientific and hist. mags. Editor: Minn. Hist. etc. Collections, vols. 8-17, 1898-1920, contributing papers on Groscilliers and Radisson, the First White Men in Minnesota, and the Progress of Discovery of the Mississippi River. Research on glaciers and ice age; archaeology Minn. Died Jan. 29, 1934.

UPTON, Arthur Canfield, Am. pathologist; b. Ann Arbor, Mich., Feb. 27, 1923; s. Herbert Hawkes and Ellen (Canfield) U.; grad. Phillips Acad., Andover, Mass., 1941; B.A., U. Mich., 1944, M.D., 1946; m. Elizabeth Bache Perry, Mar. 1, 1946; children—Rebecca A., Melissa P., Bradley C. Intern Univ. Hosp., Ann Arbor, 1947, resident, 1948-49; instr. pathology U. Mich., 1950-51; pathologist Oak Ridge Nat. Lab., 1951-54, chief pathology-physiology sect., 1954—; mem. various committees nat. and internat. orgns. Served with AUS, 1943-46. Recipient Ernest Orlando Lawrence award for atomic field, 1965. Member Am. Assn. Pathologists and Bacteriologists, Internat. Acad. Pathology, Radiation Research Soc. (councilor 1963-64, v.p. and pres. elect. 1964), Am. Assn. Cancer Research (pres. 1963-64), Am. Soc. Exptl. Pathology (councilor 1965), A.A.A.S., Gerontological Soc., Sci. Research Soc. Am., Soc. Exptl. Biology and Medicine, Phi Beta Kappa. Mem. editorial bd. Lab. Investigation, Nuclear Medicine, Excerpta Medica, Internat. Jour. Cancer, Internat. Union Against Cancer. Research on effects of ionizing radiation on animals; pathology of endocrine glands, cancer, leukemia, carcinogenesis, radiation injury. Home: 637 Pennsylvania Av. Office: Biology Div., Oak Ridge Nat. Lab., P.O. Box Y, Oak Ridge.

URBACH, Franz, physicist; b. Vienna, Austria, June 28, 1902; s. Julius and Rosa (Fischer) U.; Ph.D., U. Vienna, 1926; m. Anna Federn, June 27, 1931; children—John Charles, Jane Muriel. Came to U. S., 1939, naturalized, 1945. Research physicist Inst. Radium Research, Vienna, 1926-31; head physics lab. Municipal Hosp. Vienna, 1931-34; work on infrared sensitive phosphor telescopes Radium Inst., Inst. Tech., Vienna, 1934-38; dir. NDRC phosphor lab. U. Rochester Inst. Optics, 1941-45; with Research labs. Eastman Kodak Co., Rochester, N.Y., 1945-67, sr. research asso., head solid state dept., sci. cons., 1968—. Recipient certificate merit U. S. War and Navy depts., 1947. Fellow Am. Phys. Soc., Optical Soc. Am.; mem. Sigma Xi. Research on absorption, photochem. reactions, luminescence and impurity effects in including alkali halides, silver halides, and phosphors; 1st use of thermoluminescences (glow-curves) for trap analysis; discovered and developed infrared-sensitive and non-linear temperature-sensitive phosphors; formulated Uhrbach's rule concerning shape and temperature dependence of absorption edges of solids; applied research on infrared detection, photog. latent image, electrophotography. Home: 153 E. Parkway, Rochester, N.Y. 14617. Office: Research Labs., Eastman Kodak Co., Kodak Park, Rochester, N.Y. 14604.*

URBAIN, Georges, French chemist; b. Paris, France, Apr. 12, 1872; ed. L'Ecole Municipale de Physique et Chimie, 1891-94; D.Sc., U. Paris, 1899. Lab. asst. Ecole de physique et Chimie, 1894-95, later prof. chemistry; prof. chemistry Ecole Centrale des Arts et Manufactures; asst. lectr. Faculté des Sciences, U. Paris, 1906-08, prof. chemistry, from 1908, dir. Inst. Chemistry, from 1928. Mem. French Acad. Scis., 1921. Author: Recherches sur la separation des terres rare, 1899; Introduction a l'etude de la Spectrochimie; (with Senechal) Introduction a la chimie complexe des mineraux, 1913; (with Senechal) Les Disciplines d'une science, 1921; Les Notions fondamentales d'élement chimique et d'atome, 1925. Research on rare earth metals; discovered lutecium, 1907, also law of optimum phosphorescence of binary systems; proved that several elements previously considered pure were really mixtures; research on complex inorganic salts; believed had discovered another element which he called celtium (true element, no. 72, was isolated by Hevesy and called hafnium). Died Paris, Nov. 5, 1938.

URBAN, Paul, Austrian physicist; b. Vienna, June 15, 1905; s. Moritz and Friederike (Suchonek) U.; Ph.D., engring. diploma. Elec. engr. Austrian railroads, 10 years; agrégé phys. theory U. Innsbruck (Austria), 2 years; dir. Theoretical Physics Inst., U. Gratz (Austria), 1947—. Mem. Austria Acad. Sci. (corr.), Italian Physics Soc., Am. Phys. Soc. Research and publs. in theoretical physics. Home: Goethestrasse 13, Graz, Austria.

URBANO OF BOLOGNA (or Urbanus Averroista), Italian philosopher; flourished 14th century; Servite; author: of a commentary on Ibn Rushd's commentary on Aristotle's physics, circa 1334. Early rep. of growing sch. of Averroism in Italy.

URE, Andrew, Scottish chemist; b. Glasgow, Scotland, May 18, 1778; student Edinburgh U.; M.D., Glasgow U., 1801; at least 1 son, Alexander. Became prof. chemistry and natural philosophy Anderson's Coll., Glasgow, 1804; helped found Glasgow Obs., 1809; went to London, 1830, comml. analytical chemist thereafter. Original mem. Royal Astron. Soc.; hon. mem. Geol. Soc. Fellow Royal Soc., 1821. Author: Dictionary of Chemistry, 2 vols., 1821; New System of Geology, 1829; Dictionary of Arts, Manufacturers, and Mines, 1837-39; also papers. Determined specific gravity of solutions of varying strengths of sulphuric acid; conducted elec. expts. on living subject. Died Jan. 2, 1857.

UREY, Harold Clayton, Am. chemist; b. Walkerton, Ind., Apr. 29, 1893; s. Samuel Clayton and Cora (Reinoehl) U.; B.Sc., U. Mont., 1917; Ph.D., U. Cal. at Berkeley, 1923, LL.D., 1955; D.Sc. (hon.) Princeton, 1935, Oxford (Eng.) U., 1946, Yale, 1951, U. Chgo., 1963, numerous others; m. Frieda Daum, June 12, 1926; children—Gertrude Elizabeth (Mrs. Michel Baranger), Frieda (Mrs. Joseph Brown), Mary Alice (Mrs. Emmett Lorey), John Clayton. Research chemist Barrett Chem. Co., Balt., 1918-19; instr. chemistry U. Mont., 1919-21; asso. in chemistry Johns Hopkins, 1924-29; faculty Columbia, 1929-45, exec. officer, dept. chemistry, 1939-42, dir. war research SAM Labs., 1940-45; faculty U. Chgo., 1945-58, Martin A. Ryerson Distinguished Service prof. chemistry Inst. for Nuclear Studies, 1952-58; George Eastman vis. prof. Oxford U., 1956-57; prof. chemistry-at-large U. Cal. at San Diego, 1958—. Cons. NASA. Recipient Nobel Prize in Chemistry, 1934; Medal for Merit, U. S. Govt., 1946; Cordoza award Tau Epsilon Rho, 1954; Honor Scroll award Chgo. chpt. Am. Inst. Chemists, 1954; Joseph Priestley award Dickinson Coll., 1955; Alexander Hamilton award Columbia, 1961; U. Paris medal, 1964; Nat. Medal Sci., 1964; Gold medal Royal Astron. Soc., 1966. Mem. Am. Chem. Soc. (Willard Gibbs medal 1934, Remsen Meml. award 1963), Am. Phys. Soc., Am. Philos. Soc., Am. Geophys. Union, Geol. Soc. Am., Am. Astron. Soc., Nat. Acad. Sci. (J. Lawrence Smith medal 1962), A.A.A.S., Am. Acad. Arts and Scis., Franklin Inst. (Franklin medal 1943), Royal Soc. London (Davy medal 1940), Am.-Scandinavian Found. (bd. dirs., v.p., fellow Copenhagen, Denmark 1923-24), Sigma Xi, Epsilon Chi, Phi Lambda Upsilon, Gamma Alpha, Phi Sigma, numerous fgn. socs. Author: (with A. Ruark) Atoms, Molecules and Quanta, 1930; The Planets, 1952. Editor: Jour. Chem. Physics, 1933-40; exec. editor: Geochimica et Cosmochimica Acta, 1963; asso. editor: Icarus, 1962—. Research, publs. on chem. properties relative to origin of the solar system; discovery of heavy hydrogen; separation of isotopes; entropy of gases; absorption spectra and structure of molecules; atomic structure; measurement of paleotemperatures. Home: 7890 Torrey Lane, La Jolla, Cal. 92037.

URI, Joseph, Hungarian pharmacologist; b. Balkány, Hungary, Feb. 21, 1920; s. L. and Ethel (Hetey) U.; M.D., U. Debrecen, (Hungary), 1944, postgrad., 1952-53; postgrad. U. Breslau (Germany), 1942, London (Eng.) Sch. Hygiene and Tropical Medicine, 1961; m. Antonia Kiss, May 25, 1954; children—András, János. Mem. faculty U. Debrecen, 1940—, asso. prof., dept. pharmacology and antibiotics, 1953—. Med. officer biol. standardization WHO, Geneva, Switzerland, 1955—. Recipient Markusovski award, 1963, First class decoration, 1965. Mem. Hungarian Physiol. Soc., Hungarian Microbiology Soc., Hungary Pharm. Soc., Internat. Soc. Human and Animal Mycology. Author: (with P. Adler), Local Anaesthetics, 1952; (with others) Medical Mycology, 1956, 66; (with L. Vaczi) Industrial Bacteriology. Research, publs. on mode of action of local anaesthetics; isolation of antimicrobial substances from plants; prodn. of drugs and chems., also penicillin and penicillin-like substances; discovered crystalline antibiotics, primycin, desertomycin, flavofungin). Home: 78 Nagyerdei Körut, Debrecen, Hungary.*

URIBE-VILLEGAS, Oscar, social scientist; b. Toluca, Mexico, Nov. 6, 1928; s. Rafael and Luz (Villegas) Uribe-Pichardo; B.Sc., Escuela Nacional Preparatoria, 1946; M.A. in Linguistics, Nat. Sch. Anthropology, 1952; postgrad. Nat. Sch. Polit. and Social Scis., Nat. U. Mexico. Research worker, fellow Inst. Social Research, Nat. U. Mexico, 1952—; tchr. social statistics Nat. Sch. Polit. and Social Scis., 1950,60. Mem. Mexican Sociol. Assn. Author: Suomi, 1955; Primeras Ideas sobre Vicariato y Sociopatologia, 1956; Técnicas Estadisticas para Investigadores Sociales, 1958; El A.B.C. de la Correlación. Causación Social y Vida Internacional, 1958; La Matemática la Estadistica y las Ciencias Sociales, 1963; 25 Conceptos de Uso Sociológico, 1965; also numerous articles. Critical approaches to some methods and basic concepts of sociology; research in description, comparison and reconstrn. Indian langs. Mexico, social and polit. aspects of "castellanización" of Indian groups Mexico. Home: 117, Palestina, México 16, D.F. Mexico. Office: Instituto de Investigaciones Sociales, Torre de Humanidades 50, Ciudad Universitaria, Mexico, D.F. Mexico.*

URONE, Paul, Am. chemist; b. Pueblo, Colo., Nov. 29, 1915; s. Pietro and Angelina (Vellino) U.; A.B., Western State Coll. of Colo., 1938; M.S., Ohio State U., 1947, Ph.D. (Research fellow), 1954; m. Florence Louise Genova, May 4, 1943; children—Paul Peter, Mayri Angeline. High sch. sci. tchr., 1938-41; chemist Colo. Fuel & Iron Corp., 1942-45, USAF Research Labs., Wright-Patterson Field, O., 1945-46; chief chemist State Indsl. Health Lab., Columbus, O., 1947-55; prof. chemistry U. Colo., Boulder, 1955—. Cons. on air pollution, instrumental analytical techniques, gas chromatography, 1955—. Colo. Research fellow to study air pollution U. Cal. at Los Angeles, 1961, U. Fla., 1968. Mem. Am. Chem. Soc., Am. Indsl. Hygiene Assn., Air Pollution Control Assn., Rocky Mountain Gas Chromatography Discussion Group, Sigma Xi, Phi Lambda Upsilon. Research, publs. on analysis of air contaminants, sulfur dioxide, halogenated hydrocarbons, chromium dusts and mists, peroxides and oxidants, research in support and liquid phase effects in gas chromatography. Home: 2891 20th St., Boulder, Colo. 80302.*

URSUS, Nicholas Reimarus (or Nicolai Reymers Bär), mathematician; b. Hemste de Ditmarschon, circa 1550; mathematician to Kaiser Rudolph II. Author: Fundamentum astronomician, 1588; De astronimicis hypothesibus, 1597. Set forth astron. system similar to Tycho's, but stated that earth rotates on its own axis. Died circa 1599.

URVANTSEV, Nikolai Nikoleevich, Russian geologist; b. Jan. 29, 1893; grad. Tomsk Tech. Inst., 1918. Mem. staff conducting geol. excavations in Taimyr Peninsula, particularly nr. Norilsk, for Tomsk Tech. Inst.; adj. geologist Siberian br. Geol. Com., 1920-24, geologist, 1924-30; sci. dir. expdn. to Novaia Zemlia, 1930-32; dep. dir. Arctic Inst., 1932-38; charge geology research region of Norilsk, 1943—; editor monographs on research, 1950—. Author: The Carboniferous Area of Norilsk, 1921; Quarternary Glaciation of the Taymyr, 1931; The Taymr Geological Expedition, 1929; Severnaya Zemlya, 1933; The Problem of Gas-Bearing Mineral Deposits in the Northwestern Part of the Siberian Plateau, 1959. Address: Vsesoyuzny Arktichesky Inst., Leningrad, USSR.

USACHEV, Lev Nikolaevich, Russian physicist; b. 1926; grad. Moscow U., 1948; candidate's degree in physico-math. scis., 1954. Mem. staff Physics Inst. of State Com. Council of Ministers for Utilization Atomic Energy, USSR, 1948—, div. head, 1956—. Recipient Lenin prize for participation in sci. experiments in physics of nuclear reactors on fast-moving neutrons. Author: The B.R.-2 Experimental Fast Neutron Reactor, 1957; Soviet Experimental Fast Reactor B.R.-2, 1957. Address: Acad. Scis. USSR, B. Kaluzhskaya 14, Moscow V-71, USSR.

USAMI, Shoichiro, Japanese biologist; b. Tokyo, Japan, Jan. 27, 1913; s. Yei and Kayo (Takusagawa) U.; grad. Tokyo U., 1936; D. Sc., Hokkaido U., Sapporo, Japan, 1946; m. Ayako Saito, Oct. 25, 1936; children—Kiwamu, Ron. Faculty, Hokkaido U., 1930—, prof. plant physiology and microbiology, 1953—. Mem. Japanese Biochem. Soc., Japanese Soc. Plant Physiology, Bot. Soc. Japan. Research, books, publs. on respiratory metabolism of plant and microorganisms, inhibition of cellular respiration, variation and regulation of metabolism during growth and by environmental change, vernalization, amino acid metabolism, comparative biochemistry of metabolism. Home: East 3 North 27th St., Sapporo, Japan.*

USHAKOV, Sergei Nikolaevich, Russian organic chemist; b. Sept. 16, 1893; grad. Petrograd Poly. Inst., 1921. Became prof. Leningrad Technol. Inst., 1930; staff Sci. Research Inst. Plastics, 1931-41, dir., 1931-38; dir. Sci. Research Inst. Polymerized Plastics, 1945-49; dir. Inst. High Molecular Compounds, USSR Acad. Sci., 1948-53. Recipient Stalin prize, 1942, 50, Order of Lenin, twice. Corr. mem. USSR Acad. Sci. Author: Artificial Resins and their Use in Varnish Industry, 1929; Plastics from Cellulose Esters, 1932; Cellulose Esters and Plastics Based on Them, 1941. Research and publs. on cellulose esters, phenolaldehyde condensation, polymers; prodn. of benzyl cellulose, phenolic resins poison, synthetic camphor, polyvinyl acetate, polyvinyl alcohol, ethyl cellulose. Office: Inst. High Molecular Compounds, USSR Acad. Scis., Birzhevoy, Prospekt, Leningrad, USSR.

USINGER, Robert L(eslie), Am. entomologist; b. Ft. Bragg. Cal., Oct. 24, 1912; s. Henry Clay and Edith (Johnson) U.; B.S., U. Cal. at Berkeley, 1935, Ph.D., 1939; m. Martha Boone Putnam, June 24, 1938; children—Roberta Christine (Mrs. Ronald Manuto), Richard Putnam. With Bishop Mus., Honolulu, 1935-36, Cal. Acad. Scis., 1936-39; faculty U. Cal. at Berkeley, 1939—, entomologist in expt. sta., 1953—, prof. entomology, 1953—, chmn. div. entomology and acarology, 1963—. NIH spl. research fellow Brit. Mus. Natural History, 1948-49; chmn. Pacific sci. bd. NRC-Nat. Acad. Sci., 1961-63, participant bd.'s Coral Atoll. study, Marshall Islands, 1950, Laysan expdn., 1961; chmn. biology div. Pacific Sci. Congress, Honolulu, 1961; mem. comite permanent Internat. Congress Entomology and Internat. Union Biol. Scis., 1953—; dir. Galapagos Internat. Sci. Project, 1964; participant Congo expdn. Institut pour la Recherche Scientifique en Afrique centrale, 1959. Decorated Gold medal King Frederick of Denmark, 1956; medal and award of merit Govt. of Ecuador, 1964. Fellow Royal Entomol. Soc. London, Linnean Soc. London; mem. Entomol. Soc. Am. (pres. 1966), Pacific Coast Entomol. Soc. (pres. 1952), Soc. Systematic Zoology (pres. 1967). Author: Elements of Zoology, 2d edit., 1961; General Zoology, 4th edit. 1965; Methods and Principles of Systematic Zoology, 1953; Classification of Aradidae, 1959; Aquatic Insects of California, 1956; Sierra Nevada Natural History, 1964, Cimicidae of the World, 1966. Editor Pan-Pacific Entomologist, 1939-49. New classification of Hemiptera; study of evolution of false chinch bugs of Hawaiian Islands; cross-breeding of different strains of bed bugs. Home: 4 Yale Circle, Berkeley, Cal. 94708.*

USKOKOVIC, Milan Radoje, chemist; b. Belgrad, Yugoslavia, July 14, 1924; s. Radoje and Milica (Stojisiljevic) U.; came to U. S., 1956, naturalized, 1962; Diplome Chem. Engr., U. Belgrad, 1950; Ph.D. in Organic Chemistry, Clark U., 1960; m. Nada Rakidzic, Aug. 4, 1952; children—Moira, Lila, Charles. Staff, Prolek, pharm. co., Belgrad, 1950-55, UN, Geneva, Switzerland, 1955-56; counsellor Worcester Found. for Exptl. Biology, Shrewsbury, Mass., 1956-60; research chemist research div. Hoffmann La Roche Co., Nutley, N.J., 1960—. Mem. Am. Chem. Soc., N.Y. Acad. Sci., Worcester Found. Research, publs., patents on syntheses of natural products including steroids and alkaloids, syntheses of steroid with unnatural configurations, aza- and oxa-steroids, novel heterocyclic systems. Home: 7 Windermere Rd., Upper Montclair, N.J. 07043. Office: Hoffmann La Roche Co., Nutley, N.J. 07110.*

UTTER, Merton Franklin, Am. biochemist; b. Westboro, Mo., Mar. 23, 1917; s. Merton Franklin and Gertrude (McMichael) U.; B.A., Simpson Coll., 1938; Ph.D., Ia. State U., 1942; m. Marjorie Fern Manifold, Sept. 2, 1939; 1 son, Douglas Max. Instr. bacteriology Ia. State U., 1942-44; asst. prof. physiol. chemistry U. Minn., 1944-46; asso. prof. biochemistry Western Res. U., 1946-56, prof., 1956—, dir. dept. biochemistry, 1965—. Mem. metabolic biology panel NSF; mem. biochem. study sect. NIH; mem. basic sci. adv. com. VA. Fulbright Research fellow, Australia, 1953-54; NSF Sr. research fellow, Oxford, 1960-61. Mem. Soc. Exptl. Biology and Medicine, Am. Soc. Biol. Chemists, Biochem. Soc. (Eng.), Am. Chem. Soc. (Paul-Lewis award in enzyme chemistry 1956), Soc. Am. Microbiologists, A.A.A.S., Sigma Xi. Biochemistry; bd. editors Jour. Biol. Chemistry. Research, numerous publs. on delineation of metabolic pathways in liver concerned with initial steps of carbohydrate synthesis from non-carbohydrate precursors and structure and mechanism of action of specific enzymes involved in this metabolic area. Home: 2569 Berkshire Rd., Cleveland Heights, O. 44106. Office: 2109 Adelbert Rd., Cleve. 44106.*

UTZ, John Philip, Am. physician; b. Rochester, Minn., June 9, 1922; s. Gilbert C. and Marion (Hoy) U.; student Notre Dame U., 1940-42; B.S., Northwestern U., 1943, M.D., 1946; M.S., Georgetown, U., 1949; m. Dorothy Mary Griffin, July 2, 1947; children—Margaret, Christopher, Charles, Jonathan, Stephen. Chief infectious disease service Lab. Clin. Investigation, Nat. Inst. Allergy and Infectious Diseases, NIH, Bethesda, Md., 1952-65; faculty Georgetown U. Sch. Medicine, 1952-65, asso. prof. medicine, 1962-65; prof. medicine, chief div. infectious disease and allergy Med. Coll. Va., Richmond, 1965—; lectr. preventive medicine Howard U., 1960—; vis. investigator Pasteur Inst., Paris, France, 1962-63. Cons. to fgn. govts., VA. Diplomate Am. Bd. Internal Medicine. Mem. Am. Fedn. for Clin. Research, A.C.P., Am. Coll. Chest Physicians, Soc. for Exptl. Biology and Medicine, Am. Thoracic Soc., Am. Assn. Immunologists, Soc. for Clin. Investigation, So. Soc. for Clin. Investigation, Infectious Disease Soc. Am., Am. Soc. for Microbiology, Am. Coll. Clin. Pharmacology and Chemotherapy, Internat. Soc. for Human and Animal Mycology, Richmond Acad. Medicine. Author: (with C. W. Emmons, C. H. Binford), Medical Mucology, 1963; also numerous articles. Research on diseases due to fungi, bacteria, viruses. Home: 4310 Sulgrave Rd., Richmond, Va. 23221.*

UVNÄS, Börje Karl Magnus, Swedish physiologist; b. Malmö, Sweden, June 20, 1913; s. Lorens and Debora U.; Med.Kan., U. Lund, 1934, Med.lic., 1938, Acad.thesis, 1942; children—Kerstin, Magnus, Anna. With U. Lund (Sweden), 1934-52, prof. physiology Med. Faculty, U. Lund, 1949-52; prof. pharmacology Karolinska Institutet, Stockholm, 1952—. Recipient KNO, LFS, V. A. Regnell's prize Swedish Med. Assn., 1958, 62; Björkén's prize U. Uppsala, 1963. Mem. Nobel com. Karolinska Institutet, pres. 1st Internat. Pharmacological Meeting, Stockholm, 1961. Fellow Deutsche Akademie der Naturforscher Leopoldina, A.A.A.S.; mem. Internat. Union Pharmacology (pres.), Swedish Pharmacological Soc. (pres.), Internat. Soc. for Biochem. Pharmacology (founding), European Soc. for Study Drug Toxicity (founding). Research and numerous publs. on mechanisms of release of biogenic substances, including histamine, gastrin; pharmacology and physiology of circulation; stimulatory and inhibitory mechanisms in gastric secretion. Home: 12 Karlaplan 12, Stockholm No, Sweden.*

UYEDA, Seiya, Japanese geophysicist; b. Tokyo, Japan, Nov. 27, 1929; s. Seiichi and Hatsuo (Okino) U.; B.Sc., U. Tokyo, 1951, D.Sc., 1958; m. Mutsuko Kosaka, July 6, 1951; children—Taro, Makiko, Naoko. Research asso. Earthquake Research Inst., U. Tokyo, 1955-63, asso. prof. Geophys. Inst., 1963—; vis. scholar dept. geophysics U. Cambridge (Eng.), 1958-59; geophysicist Scripps Instn. Oceanography, La Jolla, Cal., 1961-62; vis. prof. Stanford, 1964-65. Recipient Tanakadate prize Japanese Soc. Terrestrial Magnetism and Electricity, 1953; Okada prize Japanese Soc. Oceanography, 1968. Mem. Am. Geophys. Union, Royal Astron. Soc. (London, Eng.), Japan Seismol. Soc. Author: (with H. Takeuchi and H. Kanamori) Debate About the Earth, 1966; also numerous articles. Discovered reverse thermoremanent magnetism; established distbn. heat flow in and around Japanese Islands, also E. Pacific area. Home: 5-4-6 Minami Aoyama, Minato-ku, Tokyo, Japan.*

UYEHARA, Otto Arthur, Am. mech. engr. b. nr. Hanford, Cal., Sept. 9, 1916; s. Rikichi and Umi (Nakamura) U.; B.S., U. Wis., 1942, M.S., 1943, Ph.D., 1945; m. Chisako Suda, Aug. 12, 1945; children—Kenneth Otto, Susan Joy, Emi Ryu. Faculty, U. Wis., Madison, 1945—, prof. mech. engring., 1957—, chmn. div. thermosci. dept. mech. engring. Recipient Benjamin Smith Reynolds award U. Wis., 1967. Mem. Soc. Automotive Engrs. (Arch Colwel award 1966, Hornung Meml. award 1968; chmn. advanced power-plant com.), Am. Soc. M.E. (Oil and Gas div. award for meritorious paper 1952), Am. Soc. for Engring. Edn., Sigma Xi, Pi Tau Sigma, Phi Lambda Upsilon, Tau Beta Pi. Research, publs. on flame temperature measurements in both diesel and spark ignition engine as a function of time; combustion studies and research of internal combustion engine; factors involved in auto exhaust emission. Home: 1610 Waunona Way, Madison Wis. 53713.*

UYEO, Shojiro, Japanese chemist; b. Kyoto, Japan, Sept. 20, 1909; s. Shobel and Toyo (Inada) U.; grad. U. Tokyo, 1932, Ph.D., 1937; m. Haruko Yagi, Nov. 9, 1935; children—Hisako (Mrs. Satoshi Nakamura), Shoichiro, Toyoji, Keizo. Asst. dept. pharm. scis. U. Tokyo, 1937-39; faculty medicine Kyoto (Japan) U., 1939-51, prof. Faculty Pharm. Scis., 1960—, dean, 1964—; prof. Faculty Pharm. Scis., Osaka (Japan), U., 1951-60, dean, 1959-60. Mem. Pharm. Soc. Japan, Chem. Soc. Japan, Am., Swiss chem. socs, Chem. Soc. (London). Research, numerous publs. on devel. new synthetic drugs including vitamins, antimalarials, analgesics, anthelmintics, structure of alkaloids by degradation and synthesis, non-nitrogenous natural products. Home: Yanagiroku Nakagyo-ku, Kyoto, Japan.*

V

VAARAMA, Antero Otto, Finnish botanist; b. Kuopio, Finland, June 17, 1912; s. Edvard and Martha (Virtamo) V.; M.Sc., U. Helsinki, 1935, D.Sc., 1939; m. Thelma Teodora Nikula, Jan. 28, 1945; children—Heikki Antero, Martha Elina, Hannu Markus. Tchr. natural sci. and plant protection State Hort. Coll., Lepaa, Finland, 1939-45; sr. scientist State Hort. Exptl. Inst., Piikkiö, Finland, 1945-55; prof. botany, head dept. U. Turku (Finland), 1955—, dean Faculty Sci., 1960-63; research asso. U. Cal. at Berkeley, 1951-52; dir. U. Bot. Garden, Turku, 1945-65; vis. prof. U. Colo., Boulder, 1963-64. Mem. Nat. Sci. Bd., Finland, 1961-63. Decorated Commandeur Finland's Lion, 1963. Mem. Finnish Acad. Scis. Research, numerous publs. on spl. structural characteristics of communities formed by higher water plants; cytology of raspberry genus, Rubus; developed new berry, a hybrid between red raspberry and arctic berry; studies in spontaneous variation in chromosome number in plant tissues, chromosome structure in Bryophytes, induced mutations in birches. Home: 28B Yliopistonkatu, Turku. Office: Dept. Botany, U. Turku, Turku 2, Finland.*

VAARTAJA, Olavi Lauri, biologist; b. Viipuri, Finland, Mar. 19, 1917; s. Lauri and Ilme (Seppala) V.; B.A., Helsinki U., 1940, M.A., 1946, Ph.D., 1951. Asst. biologist Helsinki U., 1946-51; forest biologist Can. Dept. Agr., Saskatoon, Sask., Can., 1952-60; sr. lectr. Waite Agrl. Research Inst., Adelaide, South Australia, 1960-63; research scientist Can. Dept. Forestry, Ottawa, Ont., Can., 1963—. Fellow A.A.A.S.; mem. Am. Phytopath. Soc., Mycol. Soc. Am., Ecol. Soc. Am., N.Y. Acad. Scis., Internat. Soc. Biometeorology, Sigma Xi. Research, numerous publs. in micrometeorology, plant ecology and physiology, plant pathology, mycology and soil microbiology. Home: Saguenay Dr., R.R. 2, Aylmer-East, Que. Office: Dept. Forestry, Ottawa, Ont., Can.*

VACCAREZZA, Jorge Raul, Argentinian physician; b. Buenos Aires, Argentina, July 14, 1929; s. Silvio

A. and Sara Perez (Piacentini) V.; M.D., Sch. Medicine, Buenos Aires, 1954; D.Med. Scis., U. Buenos Aires, 1959; m. Maria Ines Peña, Oct. 14, 1959; 1 dau., Silvia. Fellow, Med. Sch. Madrid (Spain), 1954-55, Forlanini Inst., Rome, Italy, 1955; head asthma and respiratory diseases service dept. phthisiology Buenos Aires U., 1957—; fellow Med. Sch. Paris (France), 1959-61; head physiopath. research dept. phthisiology Nat. U. Buenos Aires, 1962—, docent, 1962—. Recipient award Infectious and Tb Diseases Soc., 1959; Nat. Acad. Medicine, 1961, 65, 66; award Argentine Union Against Cancer, 1965. Fellow Am. Coll. Chest Physicians; mem. Argentine Allergic Soc. (pres. 1962-64), French Pathologic Respiratory Soc. Author: Blood Proteins and Electrolytes Allergic Asthma, 1959; Adrenal Function in Asthma, 1960; also articles. Research on demonstration electrolytic disturbances in asthma, suprarenal hypofunction in asthma, stimulant effects ACTH and corticosteroids on cholinesterase activity, new test for presumptive diagnosis of cancer. Home: 1835 Guido St. Office: 405 Av. Velez Sarsfield, Buenos Aires, Argentina.*

VACEK, Zdenek, Czechoslovakian embryologist, histologist; b. Prague, Czechoslovakia, June 29, 1923; s. Ferdinand and Jarmila (Janska-Vosahlíkova) V.; M.D., Charles U., Prague, 1951, D.Sc., 1964; m. Vlasta Vacková-Navratilova, June 23, 1951; children—Zdenek, Vladimír. Faculty, Charles U., 1948—, prof. histology and embryology, 1965—, head Institut Embryology, Faculty Medicine, 1962—, vice dean Med. Faculty, 1964—. Recipient Medal Johannes Hus, Anat. Soc., 1965. Mem. Czechoslovak Anat. Soc., Assn. des Anatomistes (France). Author: Textbook of Histology, 1954; Histological Techniques, 1964; Electron Microscopy and Cytochemistry of Endometriusm and Placenta, 1966; also numerous articles. Research on peripheral nerve endings, electron microscopy of developing epithelium in small intestine especially its resorptive function, contrbn. of lungs to fat metabolism, blastogenesis and comparative placentation; discovered innervation of cutaneous covering of horns in Cervidae, erythrocyte resorption in paraplacenta in Felidae. Home: 5 Jungmannova, Prague-1, Czechoslovakia.*

VACHON, Alexandre, Canadian chemist; b. Aug. 16, 1885; s. J. Alexander and Mary (Davidson) V.; ed. Quebec Sem., Laval U., Harvard, Mass. Inst. Tech.; M.A., D.Sc., LL.D. Became priest, 1910; dir. Sta. Biologique du St. Laurent à Trois-Pistoles, dir. École Supérieure de Chimie, prof. analytical chemistry, univ. rector, dean faculty sci. Laval U.; vicar-gen. Archdiocese of Que.; asst. Pontifical Throne and Roman Count; Roman Cath. archbishop of Ottawa, from 1940. Mem. NRC Can. Fellow Royal Soc. Can.; pres. Canadian Chem. Assn., Canadian Inst. Chemistry. Author: Traité élémentaire de chimie; Minéralogie, géologie, botanique; L'étoile de mer: son utilité comme engrais; Hydrography of Passamaquoddy Bay; also articles. Died Mar. 30, 1953.

VACHTENHEIM, Julius, physician; b. Brod, USSR, Jan. 26, 1927; s. Adolf and Alexandra (Weiss) V.; M.D., Charles U., Prague, Czechoslovakia, 1952; specialist internal medicine Postgrad. Med. Sch., Prague, 1956, specialist rheumatic diseases, 1959; m. Irene Pospichal, Mar. 7, 1953; children—Jirí, Olga. Staff internal dept. Regional Hosp., Havlickuv Brod, 1952-54; staff internal dept. State Regional Hosp., Jihlava, Czechoslovakia, 1954—, chief dr. rheumatological div., 1959—; chief dr. Regional Rheumatological Center, Jihlava, 1956—. Mem. Czechoslovak Rheumatological Assn., Internal Assn., Gerontological Assn., Med. Assn. in Jihlava. Research, publs. in rheumatology especially connective tissues aspects, epidemiology of systemic lupus erythematosus. Home: 32 brí Capku. Office: State Regional Hosp., Jihlava, Czechoslovakia.*

VACZI, L., microbiologist; b. Komadi, May 2, 1917; s. L. and E. (Berei) V.; sub auspiciis Gubernatoris Med. Doktor, U. Med. Sch. Debrecen, 1942; m. Edith Petho, Jan. 20, 1945; children—Peter, Lajos, Gabor. Faculty, U. Med. Sch. Debrecen (Hungary), 1936-45, 1958—, prof. microbiology, dir. Microbiol. Inst., 1958—; head hosp. dept. Hungarian Ministry Health, 1945-51, head bacteria dept. State Inst. Hygiene, 1951-59 (both in Budapest). Mem. Hungarian Microbiol. Soc. Author: (with A. Jeney) Applied Bacteriology, 1966. Research, publs. on biol. significance of bacterial lipids, role of lipids in autibiotic resistance of bacteria, role of viruses in human carcinogenesis, biol. properties and multiplication of the Herpes simplex, varicella-zoster and cytomegalo viruses. Home: 36 Simonyi St., Debrecen, Hajdu Bihar, Hungary.*

VAGELOS, Pindaros Roy, Am. biochemist; b. Westfield, N.J., Oct. 8, 1929; s. Roy John and Marianthe (Lambrinides) V.; A.B., U. Pa., 1950; M.D., Columbia, 1954; m. Diana Touliatos, July 10, 1955; children—Randall, Cynthia. Physician, chemist Nat. Heart Inst., 1956-66, NIH, USPHS, 1956-64; prof., chmn. dept. biol. chemistry Washington U. Sch. Medicine, St. Louis, 1966—. Mem. Am. Soc. Biol. Chemistry, Am. Chem. Soc. Numerous publs. on fatty acid synthesis, enzymology of fatty acid synthesis, discovery of new cofactor acyl carrier protein necessary to anabolism. Home: 4 Rolling Rock Lane, St. Louis 63124.*

VAGLIO, Nicola, Italian physician, gynecologist; b. Naples, Italy, Aug. 7, 1914; s. Ruggero and Bianca (Pollio) V.; M.D., U. Naples (Italy), 1938, specialist in obstetrics and gynecology, 1945; m. Barbella Bruna, Aug. 28, 1960; children—Stefania, Claudio. Asst. obstetrics and gynecology dept. U. Naples, 1949-61; dir. obstet. and gynecol. dept. U. Perugia (Italy), 1961—. Mem. Italian Soc. Obstetrics and Gynecology, Italian Acad. Scis. (hon.). Co-Author: Italian Textbook of Gynecology, 1965; also numerous articles. Research on clin. and path. problems of sterility and infertility, gynecological endocrinology, therapy of gynecol. tumors. Home: 132 XX Settembre, Perugia, Italy.*

VAGUE, Jean Marie, French physician; b. Draguignan, France, Nov. 25, 1911; s. Victor François and Marie (Voiron) V.; M.D., Faculty Medicine, also Faculty Scis., Marseilles, France, 1935; m. Denise Marie Jouve, Sept. 3, 1936; children—Philippe, Thierry, Irène, Maurice. Physician hosps., 1943-46; asso. prof., 1946-52; dir. Lab. Endocrinology Pub. Assistance, 1952-57; prof. Endocrinologic Clinic, Hôpital de la Conception, Marseilles, 1957—; dir. Center Alimentary Hygiene and Prophylaxis of Nutrition Diseases in Nat. Rys., Mediterranean region, 1958—. Expert chronic degenerative diseases WHO, 1962—. Decorated knight Legion d'Honneur, Acad. palmes, knight Public Health, knight Mil. Merit. Mem. French Soc. Endocrinology, French Lang., European, Am., diabetes assns., Royal Soc. Medicine London. Author numerous books including: Human Sexual Differentiation, 1953; Notions of Endocrinology, 1965; also numerous articles. Research on relation of diabetogenic and atherogenic power of obesity with topographic distbn. fat in upper part of body, evolution of android diabetogenic obesity from 1st stage of efficacious hyperinsulinism to less efficacious hyperinsulinism and hypoinsulinism, neurogerminal degeneration, assn. of degenerative lesions of both germinal epithelium and nervous system. Home: 19 Rue Fontange, 13 Marseilles 6. Office: Clinique Endocrinologique, Hôpital de la Conception, 144 rue St.-Pierre, 13 Marseilles 6, Bouches du Rhone, France.*

VAHL, Martin, botanist; b. Bergen, Norway, Oct. 10, 1749; studied natural scis., Copenhagen, Denmark; pupil of Linné at Uppsala, Sweden; became lectr. botany, Copenhagen, 1779, 1st lectr., 1789, prof. botany, 1801; mem. Acad. Scis. Copenhagen. Author: Symbolae botanicae, 1790-94; Ecologae americanae, seu descriptiones plantarum praesertim Americae meridionalis nondum cognitarum, 1796-1807; Enumeratio plantarum vel ab aliis vel ab ipso observatarum, 1827; (with Müller) Zoologia danica. Made studies in Norway, Holland, France, Barbary coast. Died Copenhagen, Dec. 24, 1804.

VAIDJAN, astronomer; dir. obs., Bagdad, 988. Author: Commentaries on Euclid's Elements; On the Construction and Usage of the Astrolabe for Observations; Addition to the Second Book of Archimedes.

VAIL, Alfred Lewis, Am. inventor; b. Morristown, N.J., Sept. 25, 1807; s. Stephen and Bethiah (Young) V.; grad. U. City N.Y., 1836; m. Jane Elizabeth Cummings, July 23, 1839; m. 2d Amanda Eno, Dec. 17, 1855; 3 children. Bought an interest in Samuel F. B. Morse's telegraph, 1837, agreed to manufacture complete set of telegraphic instruments and to finance U. S. and fgn. patents; made 1st public exhbn. of telegraph, N.Y.C., 1838; demonstrated telegraph before Franklin Inst., Phila., also U. S. Congress, 1838; Phila. rep. Speedwell Iron Works of Morristown, 1839-43; became Morse's chief asst. after Congress provided funds for exptl. telegraph line between Washington, D.C. and Balt., 1843; received test message "What hath God wrought" at Balt., 1844; supt. telegraph lines at Phila., 1844-48. Author: The American Electro Magnetic Telegraph, 1845. Invented horizontal-lever motion for Telegraph; devised dot-dash alphabet; built grooved roller and automatic Telegraph lever. Died Morristown, Jan. 18, 1859.

VAIL, Derrick Tilton, ophthalmologist; b. Cincinnati, O., May 15, 1898; s. Derrick Tilton and Della (Harriss) V.; A.B., Yale, 1919; M.D., Harvard, 1923; grad. study Oxford U., Eng., 1927; m. Elizabeth Yeiser, Aug. 31, 1921; children—Derrick Tilton, III, David Jameson, Ann Elizabeth, Peter, Ophthalmic interne Mass. Eye and Ear Infirmary, Boston, 1923-24; instr. in ophthalmology, Coll. of Medicine, U. of Cincinnati, 1926-37, prof. of ophthalmology, 1937-45; dir. eye dept. Children's Hosp. and Cincinnati Gen. Hosp., 1937-45. Prof. ophthalmology, head dept. Northwestern U. Med. Sch., 1945-66, prof. of ophthalmology emeritus, since 1966. DeSchweinitz lecturer, 1945, Francis Proctor lecturer, 1947, Montgomery lecturer R.C.S. (Dublin, Ireland) 1952, Doyne lectr. and medallist, 1957. Served in S.A.-T.C., 1918; served as lt. col. to col., U. S. Army, 1942-45. Decorated Bronze Star, Legion of Merit (U. S.); Medaille de Reconnaisance (France); Officer Order Crown of Belgium. Mem. various civic health groups, former mem. council Nat. Institute Neurology & Blindness USPHS; vice president Ill. Soc. Prevention Blindness. Recipient Outstanding Contribution medal A.M.A. (sect. ophthalmology), 1956; Doyne lecture and medal Oxford Ophthalmological Congress (England), 1957; Leslie Dana gold medal, Nat. Soc. Prevention of Blind, 1959; Lucien Howe gold medal Am. Ophthal. Soc., 1960. Decorated Comdr. Knights St. John Jerusalem. Diplomate Am. Bd. Opthalmology (dir. 1946-54, pres. 1954). Fellow Royal Coll. Surgeons, Am. Coll. Surgs. 2d v.p.); hon. mem. several fgn. profl. socs.; mem. Am. Ophthal. Soc. (pres. 1958-59), Internat. Council Ophthalmology (pres. 1962-66); member other nat., state and local profl. med. socs., past officer of several. Author: Gifford's Ocular Therapeutic, 1946; Truth About Your Eyes, 1950. Editor-in-chief, gen. mgr. emeritus American Jour. of Ophthalmology; asso. editor Experta Medica, Ophthalmology; asso. editor Graduate Medicine. Editorial com. L'Année Thérapeutique en Ophthalmologie, Paris. Contbr. to med. jours. Research on anatomy blood supply optic nerve; retinal detachment, cataract and glaucoma surgery. Home: 2450 Lake View Av. Office: 700 N. Michigan Av., Chgo. 11.*

VAIL, Theodore Newton, Am. inventor; b. Carroll County, Ohio, July 16, 1845; s. Davis and Phoebe (Quinby) V.; ed. Morristown (N.J.) Acad.; studied medicine 2 yrs. with uncle; LL.D., Dartmouth, Middlebury Coll., Princeton, Harvard; D.Sc., U. of Vermont; m. 1st Emma Louise Righter, Aug. 3, 1869; 1 son, Davis Righter. m. 2d. Mabel Rutledge Sanderson, 1907. Asst. supt., 1873, asst. gen. supt., 1874, gen. supt., 1875-78, ry. mail service, Washington; in telephone business, 1878-87; traveled for health, 1887-93; Vt. farmer, 1893-96; in elec. enterprises in Argentine, S.A., 1896—; introduced Am. electric system street rys. in Buenos Aires and installed telephone systems in prin. cities; pres. Am. Telegraph & Telephone Co., 1907—; also pres. or dir. many corps. in U. S. and London. Mem. Am. Acad. Polit. and Social Science (Phila.), Acad. Polit. Science in City of New York, New York Academy Sciences, Nat. Inst. Social Sciences. Developed numerous improvements in telephone technology; switchboards, signalling apparatus, circuit-closing devices. Died Apr. 25, 1920.

VAILLANT, Sébastian, French botanist; b. Vigny, France, May 26, 1669; student of Tournefort; sec. to dir., later dir., prof. Jardin du Roi, Paris; mem. French Acad. Scis., 1716. Author: Discours sur la structure des fleurs, leurs différences et la structure de leurs parties, 1717; Sermo de structura florum, 1718; Etablissement de nouveaux caractères de trois familles ou classes de plantes à feuilles composées; Botanicum parisiense ou dénombrements par ordre alphabetique des plantes qui se trouvent aux environs de Paris (pub. posthumously by Boerhaave). Disseminated idea of plant sexuality; founder 1st hothouse in France. Died Paris, May 20, 1722.

VAINSHTEIN, Boris Konstantinovich, Russian physicist; b. D.Phys.-Math. Scis. Became dir. U Inst. Crystallography, USSR Acad. Sci., 1962. Recipient prize Presidium, USSR Acad. Scis., 1958. Corr. mem. USSR Acad. Sci. Research and publs. in molecular structures, crystallography, atomic structure of organic matter. Office: Inst. Crystallography, USSR Acad. Sci., Pyzhevskii, Pereulok, 3, Moscow, USSR.

VAJK, Raoul, geophysicist; b. Vajdahunyad, Hungary, Dec. 19, 1896; s. Joseph A. and Piroska (Malom) V.; B.M.E., U. Polytech. Budapest, Hungary, 1924; Ph.D. Franz Josef U. Sci., Hungary, 1922; Ph.D., U. Sci. Budapest, 1932; m. Mary Louise Gillespie, May 14, 1932; children—Antonia, William J., J. Peter. Came to U. S., 1946, naturalized, 1948. Geophys. interpreter Torsion Balance Exol. Co., Houston, 1928-31; chief geophysicist Hungarian-Am. Oil Indsl. Co., Budapest, 1933-46; research geophysicist Standard Oil Co. (N.J.), N.Y.C., 1946-50, sr. geophysicist, 1950-54, geophys. adviser, 1954-61; research asso. Lamont Geol. Obs. Columbia, 1962-65; research geophysicist Alpine Geophys. Assos., Norwood, N.J., 1965—; adj. asso. prof. geology N.Y. U., 1965-66; cons. geophysicist, 1962—. Mem. Am. Geophys. Union, Soc. Exploration Geophysicists (Medal award 1961), European Assn. Exptl. Geophysicists, Seismol. Soc. Am., Nat. Assn. Geology Tchrs., Am. Assn. Petroleum Geologists. Contbr. numerous articles to sci. jours. Developed new methods of interpretation of geophys. data especially in field of gravity; new methods for computation of gravity effect of topography on land and for gravity effect of submarine topography for gravity surveys at sea; new methods for separation of gravity effects of geol. structures of econ. importance from gravity effect of deep seated masses, an important problem in exploration for oil; invented mech. devices for rapid computation of path of refracted seismic waves. Home: P. O. Box 73, Princeton Junction, N.J. 08550. Office: 65 Oak St., Norwood, N.J. 07648.*

VALASEK, Joseph, physicist, educator; b. Cleve., Apr. 27, 1897; s. Jozef and Frantiska (Pytlikova) V.; B.S., Case Inst. Tech., 1917; M.A., U. Minn., 1920, Ph.D., 1921; m. Leila Elizabeth Munson, June 26, 1924; children—Frances Elizabeth, Marion Louise. Asst. physicist Nat. Bur. Standards, 1917-19; teaching asst. U. Minn., Mpls., 1919-20, instr. physics, 1920-21, asst. prof., 1922-27, asso. prof., 1927-39, prof., 1939-65. NRC fellow, 1921-22; dir. def. tng. in optical engring., 1941; research U. Upsala (Sweden), 1928-29. Fellow A.A.A.S., Am. Phys. Soc., Optical Soc. Am.; mem. Am. Assn. Physics Tchrs., Am. Assn. U. Profs., Sigma Xi, Tau Beta Pi, Gamma Alpha. Showed experimentally that Rochelle salt crystals have elec. properties similar to magnetic properties of iron; located upper and lower Curie temperature between which Rochelle salt is ferroelectric. Home: 300 Seymour Av. S.E., Mpls. 55414.*

VALDEN, Paval, see Walden, Paul.

VALENCIENNES, Achille, French zoologist; b. Paris, Aug. 9, 1794; asst. Mus. Natural History, prof. zoology, from 1844; mem. French Acad. Scis., 1844, also French Soc. Agr. Author: Histoire naturelle des poissons, 11 vols. (1st 6 with Georges Cuvier), 1828-49. Died Paris, Apr. 13, 1865.

VALENSI, Gabriel, phys. chemist; b. Tunis, Tunisia, Nov. 12, 1900; s. Raymond and Regina (Cardoso) V.; Licencié es Sciences, U. Paris, 1922, D.Phys. Scis., 1929; Chem. Engr., E.N.S.C.P., 1923; m. Alice Nevart Bedeyan, Sept. 28, 1928; children—Raymond, Christian Emile (dec. 1951), Régine Eugénie Marianne. Prof. phys. chemistry Istanbul (Turkey) U., 1929-34; with C.N.R.S., Paris, 1935-36; chief work Faculty Sci., Caen, France, 1937-38; prof. Faculty Sci., Montpellier, France, 1939-46; titular prof. Faculty Sci., Poitiers, France, 1947—. Vice pres. or sec. electrochem. commn. phys.-chemistry div. Internat. Union Pure and Applied Chemistry. Decorated chevalier la Légion d'Honneur; Commandeur des Palmes Académiques; officier du Nichan-Ifthikar (Tunisia). Mem. Comité International de Thermodynamique et de Cinétique électrochimiques (sec. French sect. 1949—; v.p. intermittently 1953-67), Société de Chimie-physique, Société Chimique de France, Société Francaise de Métallurgie. Contbg. author: L'Oxydation des Metaux, 1962; Atlas d'Equilibres Electrochimiques, 1963; Research, numerous publs. on thermodynamical equilibrium of chrome and manganese reactions, kinetics of nickel, cobalt and copper reactions; new electrochem. found. and new standards of pH potential pH diagrams of sulfur and halides; concentrated aqueous solutions of potash at low temperatures and high pressures: solubility and diffusion of oxygen, conductivity, viscosity. Home: Le Logis, Bignoux, par St.-Julien l'Ars, Vienne 86. Office: Faculté des Sciences, 40 Avenue du Recteur Pineau, Vienne 86, France.*

VALENSTEIN, Elliot S(piro), Am. physiol. psychologist; b. N.Y.C., Dec. 9, 1923; s. Louis and Helen (Spiro) V.; B.S., Coll. City N.Y., 1949; M.A., U. Kan., 1953, Ph.D., 1954; m. Thelma Lewis, June 15, 1947; children—Paul, Carl. Research fellow U. Kan., 1954-55; exptl. psychologist Walter Reed Army Med. Center, Washington, 1955-57, chief neuropsychology, 1957-61; lectr. psychology U. Md., 1956-57; lectr. grad. exptl. psychology Cath. U. Am., 1958; sr. research asso. Fells Research Inst., Yellow Springs, O., 1961—; faculty Antioch Coll., Yellow Springs, 1961—, prof. psychology, 1964—. Mem. nat. sci. adv. bd. exptl. psychology study sect. NIH, 1964—, chmn. exptl. psychology, 1966-67, chmn. neuropsychology research review committee, 1967-68. Fellow A.A.A.S.; Am. Psychol. Assn.; mem. N.Y. Acad. Scis., Phi Beta Kappa, Sigma Xi, Phi Sigma, Psi Chi. Research, publs. on role hormones in devel. behavior patterns, brain functioning, motivation and emotional behavior. Home: 1126 Livermore St. Office: Fels Research Inst., Yellow Springs, O. 45387.*

VALENTE, Frank Anthony, physicist; b. Padua, Italy, Jan. 22, 1899; s. Angelo and Rosina (de Stefano) V.; brought to U. S., 1902, naturalized, 1920; B.S. in Chem. Engring., N.Y. U., 1922, M.Sc., in Physics, 1924, Ph.D. in Nuclear Physics, 1939; postgrad. John Hopkins, 1929-30; m. Laura Edna Neske, June 26, 1926 (dec. Feb. 1964). Teaching fellow N.Y. U., 1922-24, instr., 1924-25, summers 1941-42; head sci. dept. Suffern (N.Y.) High Sch., 1925-26; physicist Ordnance Dept., U. S. Army, Picatinny Arsenal, 1926-28, Nat. Bur. Standards, 1928-30, Westinghouse Electric Co., 1930-31, Socony Mobil Oil Co., Bklyn., 1931-42; commd. Capt. AUS, 1942, advanced through grades to col., U. S. Army, 1949; operations officer Atomic Energy Devels., Manhattan Engr. Dist., U. Chgo., 1943, Oak Ridge, 1943-44, chief prodn. units, Hanford, Wash., 1944-47, mem. tech. staff sci. dir. atomic weapons tests, Los Alamos, operation sandstone, 1948, Operation Greenhouse, 1951 (both Eniwetok), Operation Ranger, Nev., 1951; prof physics Rensselaer Poly. Inst., Troy, N.Y., 1956-60, prof. nuclear engring. and sci., 1960—, dir. summer insts. in reactor physics NSF, 1960, 61, 63. Lectr. nuclear physics Georgetown U., Washington, 1948-56; vis. prof. NSF Summer Inst., Seattle U., 1963; expert cons. atomic energy U. S. Govt., 1956—; mem. nat. panel arbitrators Am. Arbitration Assn., 1964—. Fellow Am. Phys. Soc., A.A.A.S.; mem. Am. Nuclear Soc., Am. Assn. Physics Tchrs., Sigma Xi, Sigma Pi Sigma. Contributing editor: A Manual of Experiments in Reactor Physics, 1963. Contbr. articles on nuclear and reactor physics to profl. publs. Research on calorimetry propellants, heat transfer, electricity and magnetism, x-rays, spectroscopy, various aspects nuclear physics atomic energy.*

VALENTIN, Gabriel Gustav, physiologist; b. Breslau, Germany, July 8, 1810; apptd. prof. physiology, Bern, 1846; author: De phaenomeno motus vibratorii in membranis, 1835; De functionibus nervorum cerebralium, 1839; Lehrbuch der Physiologie des Menschen, 2 vols., 1844. Discovered diastatic action of pancreatic juice, also (with J. E. Purknite) ciliary epithelial motion; introduced term nucleolus, 1836; research on functions of cerebral nerves. Died Bern, Switzerland, May 24, 1883.

VALENTINE, David Henriques, English botanist; b. Salford, Eng., Feb. 16, 1912; s. Emmanuel and Dora (Besso) V.; M.A., St. John's Coll., Cambridge, 1936, Ph.D., 1937; m. Joan Winifred Todd, Apr. 21, 1938; children—Helen (Mrs. D. Parsons), Samuel Henriques, Susanna (Mrs. D. Jones), Rachel Mary, James Emmanuel. Demonstrator, U. Cambridge, 1936-45; fellow St. John's Coll., 1938-45; reader in botany Durham U., Eng., 1945-50, prof., 1950-66; prof. botany U. Manchester, Eng., 1966—. Fellow Inst. Biology, Linnean Soc. London. Author: Flora Europaea, vol. 1, 1964; also numerous articles. Research on plant classification, evolutionary problems at species level in flowering plants. Home: 4 Pine Rd., Didsbury, Manchester 20, Eng.*

VALENTINE, Frederick Albert, Am. mathematician; b. Portland, Ore., May 8, 1911; s. John August and Luella (Bruns) V.; A.B., Reed Coll., 1933; M.S., U. Chgo., 1934, Ph.D., 1937; m. Edith M. Johnson, Mar. 24, 1934; children—Virginia (Mrs. Robert Keyes), Judith Patricia. Faculty, U. Tenn., 1936-37; faculty U. Cal. at Los Angeles, 1937—, now prof. math. Mem. Math. Assn. Am. (chmn. So. Cal. sect. 1967—), Am. Math. Soc., Sigma Xi. Author: Convex Sets, 1964; also articles. Research on convex sets. Home: 733 Moreno Av., Los Angeles 90049.*

VALENTINI, Michael Bernhard, German physician; b. Giessen, Germany, Nov. 26, 1657; M.D. (under Strauss, Heiland, Tack), U. Giessen, 1686; attended med. lectures of Maets, Marggrav, mech. lectures of Muschenbroek, also lectures on botany, anatomy; student hosp., Leyden, Netherlands. Practiced medicine, Giessen, later also pvt. instrn., from 1682; extensive European travel, 1685-86; ordinary prof. physics U. Giessen, from 1687, extraordinary prof. medicine, 1696, ordinary prof., from 1697; physician to Landgräfin Elizabeth Dorothea, Buzbach, Germany. Fellow Royal Soc., 1717; mem. Academia Naturae Curiosorum (dir. Ephemerides), Italian Societe Recuperati, Royal Prussian Soc. Author: Museum Museorum . . . , 1704; Historia literaria S.R.I. Academiae Naturae Curiosorum, complectens recensionem et contenta librorum, a . . . Praesidibus, Adjunctis et Collegis, loco pensi Academici, ad norman et forman in Legibus praescriptam, editorum, seorsim olim et per partes continuata nunc vero auctior et emendatior conjunctim emissa, 1708. Wrote on medicine, natural history, physics; Museum Museorum describes properties and artistic and medicinal uses of minerals, plants and vegetable products, animals, discusses collections of natural history and art objects, rare plants, animals, birds, fossils, describes apparatus for natural philosophy demonstrations, also includes discussion on divining rod, 1704. Died Mar. 18, 1729.

VALERIO, Luca, Italian mathematician, physicist; b. Naples, Italy, 1552; prof. math. and physics, Rome. Author: De quadratura parabolae, 1606. Research on center of gravity of starry bodies. Died 1618.

VALERIUS CORDUS, see Cordus, Valerius.

VALLA, Giorgio, Italian mathematician, physician; b. Piacenza, Italy, circa 1430. Prof., Venice, Italy. Author: De expetendis et fugiendis rebus, 2 vols, 1501. Trans. sci. Greek texts into Latin including: Commentary on Archimedes (Eutocios); Commentary on the Lunes of Hippocrates (Simplicios); wrote book containing 3 sects. on arithmetic, 6 on geometry; 1st in West to take geometry as far as conic sects. Died Venice, 1499.

VALLENTYNE, John Reuben, Canadian zoologist; b. Toronto, Ont., Can., July 31, 1926; s. Harold James and Alice (Laurie) V.; B.A., with honors, Queens U., Can., 1949; Ph.D., Yale, 1953; m. Ann Vera Tracy, Aug. 30, 1947; children—Peter Lloyd, Stephen Way, Jane Leslie. Faculty, Queen's U., 1952-58, asso. prof., 1955-58; faculty Cornell U., 1958-66, prof. zoology, 1963-66; sci. leader Fisheries Research Bd., Can. Freshwater Inst., 1966—. Vis. fellow Carnegie Inst. Washington, 1956-57; Guggenheim fellow, 1964-65. Mem. A.A.A.S., Internat. Limnological Assn., Geochem. Soc., Ecol. Soc. Am. Research, publs. on biochemistry of sediments in freshwaters and organic matter in old fossils. Home: 190 Oxford St., Winnipeg 9. Office: Fisheries Research Bd., Freshwater Inst., 501 University Crescent, Winnipeg 19, Man., Can.*

VALLEY, George Edward, Jr., Am. physicist; b. N.Y.C., Sept. 5, 1913; s. George Edward and Edith (Cummins) V.; S.B., Mass. Inst. Tech., 1935; Ph.D., U. Rochester, 1939; m. Louisa King Williams, July 19, 1941 (div. Dec. 1960); children—George Cummins, John Williams, Katherine; m. 2d, Shea LaBonte. Optical engr. Bausch & Lomb Optical Co., 1935-36; teaching asst. U. Rochester, 1936-39; research asso. Harvard, 1939-41, NRC fellow nuclear physics, 1940-41; project supr., sr. staff radiation lab. Mass. Inst. Tech., 1941-45, editorial bd. Radiation Lab. Tech. Series, 1945, faculty, physics, 1946—, prof., 1957—, assisted founding Lincoln Lab., 1949, asso. dir., 1953-57. Dir. Capital for Tech. Industries, Inc., Santa Clara, Cal.; mem. sci. adv. bd. USAF, 1946-55, chmn. Air Def. System, engring. com. for chief of staff, 1950-51, chief scientist, 1957-58. Recipient certificate appreciation U. S. Army, Pres.'s certificate merit, Exceptional Civilian Service medal USAF. Fellow Am. Phys. Soc., I.E.E.E.; mem. Air Force Assn. (Sci. award), Sigma Xi. Co-editor: Vacuum Tube Amplifiers, 1948, Cathode Ray Displays, 1948. Home: 132 Main St., Concord, Mass. Office: Dept. Physics, Mass. Inst. Tech., Cambridge, Mass. 02139.*

VALLI, Eusebio, Italian physician; b. Casciana, Pisa, Italy, Dec. 16, 1755; s. Ponsacco Valli; M.D., U. Pisa, 1783; Ph.D., Coll. Knowledge, Pisa, 1783. Studied exotic diseases in Nr. East for 6 years; returned to Tuscany, 1789; staff physician, hosp., Center Clin. Medicine, Mantua, Italy, from 1800; studied yellow fever in Spain, from 1809. Head physician Italian Armed Forces, Mantua. Author: Dissertazione nella quale si examinano le teorie delle Acrimonie e sull'epidemie in genere, 1783; Memoire sulla peste di Smyrne nel 1784-1788; Experiments on Animal Electricity, 1794. Advocated mass inoculation, immunization in Italy to combat plagues, epidemics; studies of epidemics, yellow fever. Died of yellow fever, Sept. 24, 1816.

VALLISNIERI, Antonio, Italian physician, naturalist; b. Modena, Italy, May 3, 1661; student of Malpighi; prof. medicine, Padua, Italy; author treatises on ostrich, 1712, chameleon, 1715; studied sperm under microscope; studied reproductive systems of insects; described hymenoptera egg deposits which form gall on buds and leaves, thus destroying belief that gall insects reproduce spontaneously; studied fossils; 1st to recognize nature of geol. faults; repudiated theory of the flood; eponym of water weed Vallisneria spiralis. Died Padua, Jan. 28, 1730.

VALLOT, Antoine, French physician; b. Reims, or Montpellier, France, 1597; physician to Anne of Austria; named 1st physician to Louis XIV, 1652; supt. Jardin des Plantes, from 1658. Author: Hortus regius, 1665. Died Paris, 1671.

VALLOT, Joseph, French astronomer, geographer; b. Lodève, France, 1854; founder obs., Chamonix, France, also at Mont Blanc, France; research in bot. geography. Died Nice, France, 1925.

VALMONT DE BOMARE, Jacques-Christophe, French naturalist, mineralogist; b. Rouen, France, Nov. 17, 1731, Mem. French Acad. Scis., 1796, Société d'Agriculture. Author: Mineralogie, 1794; Dictionnaire raisonné universel d'histoire naturelle, 1800; Catalogue of a Cabinet of Natural History, 1758; New Exposition of the Mineral Kingdom, 2 vols., 1762. Collected natural history specimens from various regions of Europe. Died Paris, Aug. 24, 1807.

VALSALVA, Antonio Maria, Italian anatomist; b. Imola, Italy, Feb. 15, 1666; student of Malpighi at Bologna; prof., Bologna. Author: De aure humana tractatus (classic work on anatomy of hearing), 1704. Founder science of otology; originated term Eustachian tube; credited with classification of ear into 3 parts (internal, middle, external), 1704; described means whereby discharge in otitis media appears in auditory canal, 1704; described tympanic antrum (Valsalva's antrum), 1704; devised (Valsalva's) maneuver whereby intrapulmonic pressure is increased by forcible exhalation against closed glottis. Died Bologna, Feb. 2, 1723.

VALTER, Anton Karlovich, Russian physicist; b. Dec. 24, 1905; grad. Leningrad Poly. Inst., 1926. Asso. Leningrad Physicotech. Inst., 1924-30, Physicotech. Inst., Ukrainian (USSR) Acad. Sci., 1930-37; prof. Kharkov (USSR) U., 1937—. Mem. Ukrainian Acad. Sci. Co-author: A Study of the Reaction of He8 with Deuterons, 1956; Measuring the Effective Cross-Sections of the C12 (y.u) and C12 (an) Reactions in the Low Energy Region of Bombarding Particles, 1957; The 4-Mv Vertical Electrostatic Generator at the Physicotechnical Institute of the Ukrainian Academy of Sciences, 1957. Research on physics of dielectrics, semiconductors, atomic nucleus, solutions to problems of vacuum engring., constrn. electrostatic charged particle accelerators; devised method generating ultra-high tensions. Address: Gosudarstvenny Universitet, Universitetskaya 16, Kharkov, Ukraine, USSR.

VALUDE, Émile Marie, French ophthalmologist; b. Vierzon, France, 1857; head clinic Panas, 1886-89; became physician Quinze-Vingts, 1888; pres. Soc. Ophthalmology, 1900. Editor: (with Lagrange) L'encyclopédie d'ophthalmologie, 9 vols. Specialized in diseases of children. Died Paris, 1930.

VALVERDE DE AMUSCO, Juan, Spanish anatomist; Author: Historia de la composición del cuerpo humano (stimulated interest in anat. study in Spain), 1556. His work shows the influence of Vesalius.

VAMOS, Rezso, Hungarian soil biologist; b. Barcs, Hungary, Nov. 27, 1913; s. Rezso and Margit (Makk) V.; D.Agr., U. Szeged, 1938; m. Ibolya Kiss, Dec. 7, 1944; 1 dau., Katalin. Tchr. pub. schs., Sopron, Hungary, 1946-48; research fellow High Sch. Forestry, Sopron, 1949-52; with Plant Physiol. Inst., U. Szeged (Hungary), 1952-—. Mem. Internat. Soc. Soil Sci. Research, numerous publs. on microbiol. processes of waterlogged soils, reduction processes occurring in mud, biol. soda formation, injurious effect of hydrogen sulphide formed by sulphate reduction, root rot of rice plant and fish decay caused by hydrogen sulphide. Home: 23 Petofi. Office: 2 Tancsics, Szeged, Hungary.*

VAN ALLEN, James Alfred, Am. physicist; b. Mt. Pleasant, Ia., Sept. 7, 1914; s. Alfred Morris and Alma E. (Olney) Van A.; B.S., Ia. Wesleyan Coll., 1935, Sc.D., 1951; M.S., U. of Iowa, 1936, Ph.D., 1939; Sc.D., Grinnell College, 1957, Coe College, 1958, Sc.D., Cornell College, 1959; U. Dubuque, 1960, U. of Michigan, 1961, Northwestern University, 1961, Illinois College, 1963, Butler University, 1966, Boston College, 1966; married Abigail Fithian Halsey, October 13, 1945; children—Cynthia Olney, Margo Isham, Sarah Halsey, Thomas, Peter. Research fellow, physicist department terrestrial magnetism Carnegie Institution of Washington, District of Columbia, 1939-42; physicist, group and unit supervisor, applied physics lab. Johns Hopkins, 1942, 1946-50; organizer, leader sci. expdns. study cosmic radiation, Peru, 1949, Gulf of Alaska, 1950, Greenland, 1952, 57, Antarctica. 1957; prof. physics, head dept. U. of Iowa, 1951-—; research associate Princeton, 1953-54. Development radio proximity fuze National Defense Research Council, OSRD. Received C. N. Hickman medal for development Aerobee rocket, Am. Rocket Soc., 1949; physics award Wash. Acad. Sci., 1949; research fellow Guggenheim Meml. Found., 1951; space flight award Am. Astronautical Soc., 1958; Louis W. Hill space transp. award, Inst. Aero. Scis., 1959, Elliott Cresson medal Franklin Inst., 1961, John A. Fleming award Am. Geophys. Union, 1963, 64; Golden Omega Award Electrical Insulation Conf., 1963; Comdr. Order du Merite Pour la Recherche et L'-Invention, 1964; La. Broadcasters Assn. Award, 1964. Fell. Am. Rocket Society I.E.E.E., Am. Phys. Soc., Am. Geophys. Union (pres. planetary sciences section); member of Ia. Acad. Sci., Nat. Acad. Scis. (mem. space sci. bd.), Internat. Acad. Astronautics (founding mem.), Am. Philos. Soc., Royal Astron. Soc. (U.K.), Sigma Xi. Contbg. author: Physics and Med. of Upper Atmosphere, 1952; Rocket Exploration of the Upper Atmosphere. Editor: Scientific Uses of Earth Satellites, 1956; asso. editor Jour. Geophysical Research, 1959-—, Physics of Fluids, 1958-62. Contbr. numerous articles sci. jours. Discoverer radiation belts around earth; pioneer in high altitude research with rockets, satellites and space probes; research on primary cosmic radiation above atmosphere. Home: 5 Woodland Mounds Rd., R.F.D. 5, Iowa, City.*

VAN ALMELOUEEN, Theodorus Jannson, Dutch physician; b. Mydrecht, Holland, 1651-57; prof. Greek and medicine, Harderwyk; pub. an edit. of Hippocrates' Aphorisms, also of Celsus de medicina, 1687. Author: Theological and Philological Ameneties, 1694. Died 1712.

VAN ALPHEN, Jan, Dutch chemist; b. Mar. 8, 1900; s. H. J. and W. (Vlasblom) A.; D.Chemistry, U. Leyde; m. M. G. van den Bosch, 1924. Curator chemistry U. Leyde, 1925-46; collaborator Encyclopedia of Chemistry, Elsevier, 1946-50; dir. lab. Found. Rubber, Delft 1943-55; chief lab. research Unilever, Vlaardinge, 1955-57, apptd. dir. detergents div., 1958, coordinator of sci. publs., 1962. Author: Overzicht van de Geschiedenis der Organische Chemie, 1933; Organische Scheikunde, 1939; (with W. J. K. Schönlau and M. van den Temple) Rubber Chemicals, 1956; also numerous articles. Address: 302 Van Halewijnlaan, Voorburg, Netherlands.

VANAMEE, Parker, Am. physician; b. Portland, Me., Aug. 9, 1919; s. Talcott Ostrom and Eleanor (Wright) V.; B.S., Yale, 1942; M.D., Cornell U., 1945; m. Rosemary Brown, Aug. 1, 1953; children—Charles T, James P., Peter B., Helen F., Jessie W., Norman R. Staff, Meml. Hosp., N.Y.C., 1954-—, asso. attending physician, 1961-—; practice medicine specializing in renal physiology, N.Y.C., 1952; Am. Cancer Soc. fellow, vis. investigator Rockefeller Inst., 1948-51; with Sloan-Kettering Inst., N.Y.C., 1951-—, asso. mem., 1960-—, chief div. exptl. surgery and physiology, 1962-—; asso. prof. medicine Cornell U. Med. Coll., 1961-—. Mem. Harvey Soc., Am. Fedn. for Clin. Research, N.Y. Acad. Scis., A.A.A.S., Cornell Discussion Group, James Ewing Soc., Am. Soc. for Clin. Nutrition, A.M.A. Council on Foods and Nutrition, Sigma Xi. Author: (with Roberts, Poppell) Electrolyte Changes in Surgery, 1958; also articles. Research on malnutrition following gastrectomy, acid-base balance in surg. patients, in patients with kidney disease, in patients with cancer. Home: 345 E. 68th St. Office: 444 E. 68th St., N.Y.C. 10021.*

VAN ARKEL, Anton Eduard, Dutch chemist; b. The Hague, Nov. 19, 1893; s. Dirk and A. P. Ris (Lambers) Van A.; ed. U. Leyden, U. Utrecht; m. N. G. Adriani, July 21, 1920; children—Elizabeth Maria, Anna Petronella, Dirk, Anthony Ewoud. Asst. in chem. research N. V. Philips Labs., Eindhoven, Netherlands, 1920-21; prof. inorganic chemistry U. Leiden, 1934-—. Decorated Order Dutch Lion. Mem. Royal Acad. Sci., Royal Dutch Chem. Assn. (hon.), Batavia Soc. Author: Chemische binding als Electrostatisch Verschijnsel, 1930; Reine Metalle, 1939; Molecules and Crystals in Inorganic Chemistry, 1949; Moleculen en kristallen, 1961; Eerste nederlandse systematisch ingerichte encyclopedie, 1949. Home: Zoeterwoudsesingel 60, Leyden, Netherlands.

VAN ARSDEL, Eugene Parr, Am. forest pathologist, meteorologist; b. Emaus, Pa., Dec. 4, 1925; s. William C. and Mabel (Hedde) Van A.; B.S., Purdue U., 1947; M.S., U. Wis., 1952, Ph.D., 1954; m. Rose Price, Aug. 23, 1948; children—Elizabeth Rose, Jonathan Eugene. With Central States Forest Expt. Sta., 1947-48; Ind. supr. white pine blister rust control, 1948-51; research asst. U. Wis., 1951-54; biologist U. S. Army, 1954-56; plant pathologist Lake States Forest Expt. Sta., U. Wis., Madison, 1956-61, project leader No. conifer disease research, St. Paul, 1963-65; vis. prof. forest pathology Yale, 1965-66; prin. plant pathologist N.Central Forest Expt. Sta., St. Paul, 1965-—. Del. 4th Internat. Biometeorol. Conf. U. S. Dept. Agr., Pau, France, 1963. NSF grantee, 1962-63. Mem. Am. Phytopath. Soc., Am. Meteorol. Soc., Ecol. Soc. Am., Internat. Biometeorological Soc., Am. Inst. Biol. Sci., A.A.A.S., Sigma Xi. Research, publs. on meteorol. epidemiology of tree diseases. Home: 792 Redwood Lane, St. Paul 55112. Office: N.Central Forest Expt. Sta., Forest Service U. S. Dept. Agr., Folwell Av., St. Paul 55101.*

VAN ARSDOL, Maurice Donald, Jr., Am. sociologist; b. Seattle, May 4, 1928; s. Maurice Donald and Madge (Belts) Van A.; B.A., U. Wash., 1949, M.A., 1952, Ph.D., 1957; m. Marian C. Gatchell, Aug. 18, 1950. Acting instr. U. Wash., 1953-57; faculty sociology U. So. Cal., 1957-—, prof., 1965-—, research coordinator Youth Studies Center, 1958-59, co-dir. Population Research Lab., 1961-64, dir., 1964-—; vis. lectr. U. Hawaii, 1965. Fellow A.A.A.S., Am. Geog. Soc., Am. Sociol. Assn.; mem. Internat. Union for Sci. Study Population, Pacific Sociol. Assn. (mem. adv. council, 1965-—, v.p., 1964-65), Population Assn. Am. Author: (with C. F. Schmid, E. H. MacCannell) Mortality Trends in the State of Washington, 1955. Research, publs. on factor analyses of social structure of urban neighborhoods, reality and perception of air pollution in U. S. cities, residential mobility, suicide and juvenile delinquency, met. population growth. Office: Dept. Sociology and Anthropology U. So. Cal., Los Angeles 90007.*

VAN ATTA, Lester Clare, Am. physicist; b. Portland, Ore., Apr. 18, 1905; s. Lester C. and Martha (Murray) Van A.; B.A., Reed Coll., 1927; M.S. in Physics, Washington U., St. Louis, 1929, Ph.D., 1931; m. Elvene Winkleman, June 6, 1929; children—William Karl, John Reynolds, Martha (Mrs. B. Fred Mills). Research asst. Princeton, 1931-32; with Mass. Inst. Tech., 1932-45, asst. prof., 1937-40, head antenna group Radiation Lab., 1940-45; head antenna research br. Naval Research Lab., Washington, 1945-50; with Hughes Aircraft Co., Culver City, Cal., 1950-62, dir. Hughes Research Labs., 1961-62; chief scientist Lockheed Missiles and Space Co., Sunnyvale, Cal., 1962-64; asst. dir. electromagnetic research Electronics Research Center, NASA, Cambridge, Mass., 1965-—. Spl. asst. on arms control to dir. def. research and engring. Dept. Def., 1960-61. Fellow Am. Phys. Soc., I.E.E.E., Am. Inst. Aeros. and Astronautics (asso.); mem. Am. Geophys. Union. Research, publs. on critical potentials in gases by electron bombardment, high voltage and high vacuum engring., nuclear and X-ray bombardment studies, microwave antennas and circuits, radar systems, deep space communications. Patentee microwave field, retroreflective arrays. Home: 9 Hampshire Rd., Wellesley, Mass. 02181. Office: 575 Technology Sq., Cambridge, Mass. 02139.*

VAN ATTA, Robert Ernest, Am. chemist; b. Ada, O., Feb. 29, 1924; s. Ernest A. and Agnes (Klingler) Van A.; B.A., Ohio No. U., 1948; M.S., Purdue U., 1950; Ph.D., Pa. State U., 1952; m. Mary Ellen Koons, Jan. 20, 1946; children—John Robert, Richard Lewis, Matthew Ernest. Instr., Pa. State U., 1951-52; asso. prof., chmn. dept. chemistry Ohio No. U., 1952-54, dir. div. natural scis., 1953-54; asst. prof. So. Ill. U., Carbondale, 1954-59, professor, 1967-—; cons. Tech. Tape Corp., 1963-—. Mem. A.A.A.S., Am. Chem. Soc., Am. Assn. U. Profs., Sigma Xi, Phi Lambda Upsilon. Author: An Introduction to Analytical Chemistry, 1964. Research, publs. on devel. of analytical methods of chem. measurement, instruments for analytical measurements, inorganic complex ions, kinetics of organic reduction reactions, polarographic behavior of organic compounds, infrared spectroscopic techniques, devel. of expts. for use in instrn. in instrumental analytical measurements. Home: 1002 Skyline Dr., Carbondale, Ill. 62901.*

VAN BEBBER, Wilhelm Jacob, German meteorologist; b. Grieth, Prussia, July 10, 1841. Asso. with weather telegraphy for aiding sea travel from Hamburg, Germany. Author: Die Regenverhältnisse Deutschlands, 1877; Die Meteorologie im Dienste der Landwirtschaft (set forth importance of agrl. weather prognosis), 1877; Handbuch der ausübenden Witterungskunde, 1886; Lehrbuch der Meteorologie, 1890; Die Wettervorhersage, 2d edit., 1898; Anleitung zur aufstellung von wettervorhersagen, 1902; works on weather forecasting. Contbr. articles to Meteorologische Zeitschrift. Indicated routes of migration of barometric minima, 1881; gave detailed explanation of relationship between atmospheric conditions and man's health (called his works in this field hygienic meteorology), 1895. Died Altona, Germany, July 1, 1909.

VAN BEEK, Leendert Klaas Hellinga, Dutch chemist; b. Oostvoorne, Holland, Nov. 15, 1925; s. Krijn Hugo and Martha (Meijer) Van B.; Candidate Degree, U. Groningen (Holland), 1949, doctorandus Degree, 1954; Ph.D., Leiden (Holland), 1955; m. Joan Law Walters, June 27, 1959. Faculty U. Khartoum (Sudan), 1955-57; research asso. Cornell U., Ithaca, N.Y., 1958-59; chief dielectrics sect. Central Lab., TNO, Delft, Netherlands, 1959-64; research chemist Philips Research Labs., Eindhoven, Netherlands, 1964-—. Mem. Royal Netherlands Chem. Soc. Research, publs. on dielectric loss and conductivity in polymer solutions and in heterogeneous systems, in carbon black-loaded rubber vulcanizates, in protonic conductors, and in organic crystals, the purification and fractionization of gum arabic, thickness determination of adsorbed water layers from optical transmission, phys. chemistry of diazosulfonates. Home: 64 Le Sage ten Broeklaan. Office: Philips Research Labs., Eindhoven, Netherlands.*

VAN BEKKUM, Dirk Willem, Dutch radiobiologist; b. Batavia, Indonesia, July 30, 1925; s. Dirk Willem and Johanna Catharina (van Oordt) Van B.; M.D., Leyden U., The Netherlands, 1950, Ph.D., 1952; m. A. R. E. Kijlstra, May 21, 1950; children—Marion, Pleuntje, Joosje, Nienke. Asst. biochemistry Oxford U., Eng., 1947-48, State U. Hosp., Leyden. The Netherlands, 1950-52; mem. Med. biol. lab. TNO, Rijswijk (ZH), The Netherlands, 1952-60, head Radiobiol. Inst., 1960-—; also postgrad. tchr. in health physics and radiobiology. Prof. radiobiology Med. Faculty Rotterdam, 1966; prof. transplantation biology Med. Faculty State U. Leyden, 1967. Recipient award for med. research I. Korteweg and A. I. Overwater Founds., 1960. Mem. Netherlands Radiobiol. Soc. (sec.), Transplantation Soc. (councillor exec. bd.), Internat. Assn. Radiation Research (councillor). Author: Radiation Chimaeras, 1967. Research, publs. on bone marrow transplantation in rodents and primates, biochem. radiation effects, chem. radioprotective agts. Home: Meijerskade 9, Leyden, The Netherlands. Office: Radiobiol. Inst., TNO, 151 Lange, Kleiweg, Rijswijk (ZH), The Netherlands.*

VAN BEMMELEN, Jakob Maarten, Dutch chemist; b. Almelo, Overijssel, Nov. 3, 1830; Ph.D., Leyden, Netherlands, 1854; become prof. chemistry, Leyden, 1874. Research on adsorption and affinity; disposed toward a chem. absorption theory. Died Leyden, Mar. 13, 1911.

VAN BEMMELEN, Reinout W., Dutch geologist; b. Djakarta, Java, Apr. 14, 1904; s. W. and S. H. (de Iongh) Van. B.; Mining Engr., d.techn. scis. Tech. U. Delft (Netherlands), 1927; m. Lucie Clara Vanden Bos, Dec. 8, 1930; 1 son, Reinout J. Geol. survey Indonesia, 1927-46, chief volcanological survey, 1940-46; mem. staff Geol. Inst., State U. Utrecht (Netherlands), 1949-—, prof. applied geology, 1951-—. Mem. Soc. belgde geol. (hon.), Geol. Gesellschaft Vienna (corr.). Author: The Geology of Indonesia, 1949; Mountain Building, 1954; also numerous articles. Research on regional geology and undation theory of earth's crust. Home: 139 Hanedoesstr., The Hague, Netherlands. Office: 320 Oude Gracht, Utrecht, Netherlands.*

VAN BENEDEN, Edouard, Belgian cytologist; b. Louvain, Belgium, Mar. 5, 1846; s. Pierre Joseph van B. Beneden; engaged in research on reproduction; independently discovered the centrosome, 1876; determined that chromosome number is constant for each cell of a given body and characteristic of the species, 1887; demonstrated that chromosome number is halved during meiosls and restored during sexual process. Died Liege, Belgium, Apr. 28, 1910.

VAN BENEDEN, Pierre Joseph, Belgian naturalist; b. Mechlin, Belgium, Dec. 19, 1809; studied medicine; at least 1 son, Edouard; became curator Natural History Mus., Louvain, Belgium, 1831; prof. zoology, comparative anatomy and paleontology Catholic U., Louvain, 1836-94; pres. Royal Acad. Belgium, 1881; mem. French Acad. Scis., 1866. Specialist in marine invertebrates; investigated life history of parasitic worms; founder 1st lab. and aquarium for study of marine life, 1843. Died Louvain, Jan. 8, 1894.

VAN BERGEIJK, Willem Andre, biologist; b. Eindhoven, Holland, Dec. 20, 1929; s. Andre J. M. and Wilhelmina (Learbuch) van B.; B.S., Rijks-Universiteit Utrecht (Holland), 1951; M.S., U. Ia., 1955, Ph.D.,

1956; m. Betty Vacik, May 13, 1955; children—Andre W., Eric J. Came to U. S., 1953, naturalized, 1961. Mem. tech. staff Bell Tel. Labs., Inc., Murray Hill, N.J., 1956-66; prof. Ind. U. Center for Neural Scis., Bloomington, 1966—. Fellow Acoustical Soc. Am., A.A.A.S., Am. Inst. Biol. Sci., Am. Soc. Zoologists, Animal Behavior Soc., Psychonomic Soc., Sigma Xi. Author: (with J. R. Pierce, E. E. David) Waves and the Ear, 1960. Contbr. articles to sci. jours. Contbd. theories on evolution of hearing, binaural hearing, cochlear processes; contbd. toward understanding lateral line organ in fishes. Died Oct., 1967.

VAN BERKHEY, Johann Lefrancq, Dutch natural historian; b. Leyden, Netherlands, Jan. 25, 1729. Prof. natural history U. Leyden. Author: Hetverheelijke Leyden, 1774; Naturlijke Historie van Holland, 2 vols., 1769-79; Naturlijke Historie van Het Rundvee in Holland, 1805-11. Work in comparative anatomy. Died Mar. 13, 1812.

VAN BIESBROECK, George, astronomer; b. Ghent, Belgium, Jan. 21, 1880; s. Louis P. and Pharailde (Colpaert) Van B.; D.Sc., U. Ghent, 1902; D.Sc., U. Brussels, 1935; m. Julia Sterpin, Apr. 9, 1910; children—Simone (Mrs. John Titus), Edwin, Micheline (Mrs. Charles Wilson). Came to U. S., 1917, naturalized, 1922. Faculty, Yerkes Obs., U. Chgo., 1917—, prof. practical astronomy, 1923-45, prof. emeritus, 1945—; research asst. Lunar and Planetary Lab. U. Ariz., Tucson, 1963—. Mem. Eclipse Expdns. Nat. Bur. Standards, 1947, Nat. Geog. Soc. and Air Force, 1948, Nat. Geog. Soc., 1952. Recipient Donahoe Comet medal, 1926, 35; Burr prize Nat. Geog. Soc., 1952; Watson prize Acad. Sci., 1962. Mem. Am., Royal astron. socs., Soc. Astronomy France. Research in double stars and measures and orbit computing comets, measures of positions and orbit computations, measures of Einstein shift at solar eclipses. Home: 1640 E. 7th St., Tucson 85719.*

VAN BRUGGEN, John Timothy, Am. biochemist; b. Chgo., Aug. 12, 1913; s. Peter John and Anna (Vennema) Van. B.; B.A., Lenfield Coll., 1937; M.S., U. Ore. Med. Sch., 1939; Ph.D., St. Louis U. Sch. Medicine, 1944; m. Ruth Rosetta Hurt, July 12, 1936; children—Philip John, Anna Kaye (Mrs. David Quenelle), Linda Kaye. Instr., St. Louis U. Sch. Medicine, 1943-45; faculty U. Ore. Med. Sch., Portland, 1945—, prof. biochemistry, 1962—; biochemist AEC, Washington, 1967—. Mem. A.A.A.S., Am. Soc. Biol. Chemists, Am. Chem. Soc., Am. Heart Assn., Biophys. Soc. Author: (with E. S. West, W. R. Todd, H. S. Mason) Textbook of Biochemistry, 1967; also articles. Research on lipid metabolism and diabetes, mechanism of action of DMSO, membrane structure and function. Home: 991 S.W. Westwood Dr. Office: 3181 S.W. Sam Jackson Park Rd., Portland, Ore. 97201.*

VAN BURKALOW, Anastasia, Am. geologist, geographer; b. Buchanan, N.Y., Mar. 16, 1911; d. James Turley and Mabel Ritchie (Ramsay) Van Burkalow; B.A., Hunter Coll., 1931; M.A., Columbia, 1933, Ph.D., 1944. Research asst. Columbia, 1934-37, Am. Geog. Soc., 1944-47; mem. faculty Hunter Coll., 1938—, prof. geology and geography, chmn. dept., 1961—. Fellow Geol. Soc. Am., A.A.A.S., N.Y. Acad. Scis.; mem. Assn. Am. Geographers, Nat. Assn. Geology Tchrs. (editor Jour. Geol. Edn. 1954-56), Phi Beta Kappa, Sigma Xi. Contbg. editor Geog. Rev., 1949—. Research on factors affecting angle of repose and angle of sliding friction of loose material, distbn. and geol. origin of natural flourides in water supplies in U. S., geography of swimmers' itch. Home: 160 E. 95th St., N.Y.C. 10028.*

VANCE, Elbridge Putnam, Am. mathematician; b. Cin., Feb. 7, 1915; s. Selby Frame and Jeannie (Putnam) V.; student Haverford Coll., 1932-33; A.B., Coll. of Wooster, 1936; M.A., U. Mich., 1937, Ph.D., 1939; m. Margaret G. Stoffel, Aug. 5, 1939; children—Susan Margaret (Mrs. Robert Bezucha), Peter Selby, Douglas Putnam, Emily Louise. Instr., asst. prof. U. Nev., Reno, 1939-43; faculty Oberlin (O.) 1943—, prof. math., 1956—, chmn. dept., 1948—, acting dean Coll., 1966. NSF fellow Stanford, 1960-61; math. asso. U. Auckland (New Zealand), 1967. Mem. Math. Assn. Am., Am. Math. Soc., Phi Beta Kappa, Sigma Xi, Phi Kappa Phi. Author: Trigonometry, 1954; Unified Algebra and Trigonometry, 1955; Fundamentals of Mathematics, 1960; Modern College Algebra, 1962; Modern Algebra and Trigonometry, 1962; Introduction to Modern Mathematics, 1963. Research and numerous publs. on elementary math., topology. Home: 55 Glenhurst Dr., Oberlin, O. 44074.*

VANCE, John Edward, Am. chemist; b. Dayton. O., May 21, 1905; s. Edward Delmar and Kathleen (Phillips) V.; B.S., Yale, 1926; Ph.D., 1929; m. Esther Christensen, July 3, 1931; children—Joan (Mrs. Norman T. Rae), Philip. Faculty, Yale, 1931-42, 46-48; prof. N.Y. U., 1948—, head dept. chemistry, 1948-57. Mem. adv. groups Army Sci. Adv. Panel, 1956—, Army Research Office, 1958—, Army Test and Evaluation Command, 1964—, Army Missile Command, 1966—. Fellow N.Y. Acad. Scis.; mem. Am. Chem. Soc., Faraday Soc. Author: (with J. C. Warner) Manufacture of Uranium and Uranium Compounds, 1951. Research, numerous publs. on improved procedures in chem. analyses, studies of crystal growth of ionic crystals. Home: 29 Washington Sq., N.Y.C. 10011. Office: Washington Sq., N.Y.C. 10003.*

VANCE, Rupert Bayless, Am. demographer; b. Plumerville, Ark., Mar. 15, 1899; s. Walter J. and Lula M. (Bayless) V.; B.A., Henderson State Coll., 1920; M.A., Vanderbilt U., 1921; Ph.D., U. N.C., 1928; LL.D., Hendrix Coll., 1938, U. Ark., 1954; m. Rheb Cecile Usher, June 5, 1930; children—David Rupert, Donald Ernest, Victor Stuart. Faculty, U. N.C., Chapel Hill, 1928—, Kern prof. sociology, 1945—, mem. tech. adv. com. 17th Census U. S. Internat. Population Union, 1942—. With pub. health sect. NIH, 1950-58; cons. Social Security Bd. Recipient Thomas Jefferson award, 1965. Mem. Am. Sociol. Assn. (past pres.), So. Sociol. Soc. (past pres.), Population Assn. Am. (past pres.). Author: Human Factors in Cotton Culture, 1929; Human Geography of the South, 1932; All These People, 1945; also numerous articles. Editor: (with Guy B. Johson) Social Forces, 1958—. Research in regional sociology and demography analysis of So. U. S.; demonstrated that differential sex mortality between males and females is due to biology and not environment. Home: 501 Dogwood Dr., Chapel Hill, N.C. 27515.*

VAN CEULEN, Ludolph, Dutch mathematician; b. Hildesheim, Holy Roman Empire, Jan. 28, 1540; s. Jan (or Gerdt) and Hester (de Roode) van C.; m. 2d, Adriana Symons (or Simons), June 17, 1590; 12 children. Tchr. math. in Breda, Amsterdam, Delft, 1584-86; became tchr. in Arnheim, 1589; joined faculty at Leyden (Netherlands), 1595, became prof. fortification, 1600; taught fencing in Delft and Leyden. Author: Van den Circkel, 1596; Foundamenten, 1615. Using Archimedes' method and decimal notation, he worked out he value of pi to 20 decimal places in 1596, to 35 places (Ludolphian or Ludolf's number) in 1610. Died Leyden, Dec. 31, 1610.

VAN CITTERS, Robert Lee, Am. cardiovascular physiologist; b. Alton, Ia., Jan. 20, 1926; s. Charles and Wilhelmina (Heemstra) Van C.; A.B., U. Kan., 1949, M.D., 1953; m. Mary Ellen Barker, Apr. 9, 1949; children—Robert, Mary, David, Sarah Mary. Research fellow U. Kan., 1955-58; spl. research fellow U. Wash., 1958-61; asso. cardiopulmonary Inst., Scripps Clinic Research Found., La Jolla., Cal., 1961-62; exchange scientist U. S.-USSR Sci. Exchange Program, 1962; Robert L. King Chair cardiovascular research U. Wash., Seattle, 1963-65, asso. prof., 1965—, research staff Regional Primate Research Center, 1965—. Mem. A.A.A.S., Am. Physiol. Soc., Am. Heart Assn., N.Y. Acad. Scis. Research, publs. on devel. and implementation of systems for radio telemetry of blood flow and blood pressure from free-ranging exptl. animals, function and control of left ventricle performance and characteristics of arteries, distbn. blood flow. Home: 9131 184th St. S.W., Edmonds, Wash. 98020. Office: Dept. Physiology and Biophysics, U. Wash., Seattle 98105.*

VANCOUVER, George, Brit. explorer; b. 1758. Became seaman on Resolution, 1771; midshipman with Cook on Discovery, returned 1780; became lt. on sloop Martin, 1780; sailed with Rodney on Flame for W.I., 1781; participated in battle of Apr. 12, 1782; returned 1783; sailed under Gardner on Europa, 1784, to Jamaica, 1786; became mem. exploring expdn. under Capt. Robert on Discovery, 1789, but reassigned after dispute with Spain to Courageux under Gardiner until 1790; became comdr. of Discovery, 1790; assigned to receive Nootka Sound back from Spain and to survey coast from 30° northward; returned to Eng., 1795. Author: A Voyage of Discovery to the North Pacific Ocean and Round the World in the Years, 1790-95, in the Discovery Sloop of War and Armed Tender Chatham, under the Command of Captain George Vancouver, 3 vols., 1798. Surveyed S.W. coast of Australia, N.Am. coast from San Francisco north; circumnavigated Vancouver Island (named after him); charted coast line of Dusky Bay, New Zealand; discovered Gulf of Georgia. Died Petersham, Eng., May 10, 1798.

VAN CREVELD, Simon, Dutch pediatrician; b. Amsterdam, Netherlands, Aug. 21, 1894; s. A. and H. (Bos) van C.; M.B., U. Amsterdam, 1910, M.D., 1922; m. Elisabeth van Dam. Asst., physiol. lab. U. Groningen (Netherlands), 1919-23; asst., clinic U. Amsterdam, 1923-24, asst. pediatric clinic, 1924-26, asst. lab. gen. pathology, 1924-30, head dept. pediatrics, 1930-64; with Finkelstein's clinic, Berlin; practiced medicine, specializing in pediatrics, Amsterdam, 1926—; head baby clinic Univ. Hosp., also head health center for babies, dept. pub. health City of Amsterdam, 1930-38; founder clinic for convalescing children, 1950, hemophilia clinic, 1964. Recipient Gold medal U. Gröningen, 1920, medal Dutch Red Cross, 1960. Mem. Dutch Pediatric Soc. (pres. 1961-64), many other med. socs. Gave 1st clin. description of glycogen disease, 1927; discovered (with Bendien) clotting factor lacking in hemophilia, 1936. Address: 87 Joh. Vermeerstraat, Amsterdam, The Netherlands.*

VANDAM, Leroy David, Am. physician; b. N.Y.C., Jan. 19, 1914; s. Albert Herman and Esther (Cowan) V.; Ph.B., Brown U., 1934; M.D., N.Y. U., 1938; m. Regina Phyllis Rutherford, Nov. 30, 1939; children—Albert Rutherford, Samuel Whiting. Practice medicine, specializing in anesthesiology, Boston, 1954—; dir. dept. anesthesia Peter Bent Brigham Hosp., 1954—; faculty Med. Sch. Harvard, 1954—, prof. anesthesia, 1967—. Chmn. subcom. on anesthesia NRC, 1965—; mem. various coms. NIH, 1963—. Mem. A.M.A., Am. Assn. Anesthesiologists, Assn. U. Anesthetists, Halsted Soc., Horace Wells Soc., Phi Beta Kappa, Sigma Xi, Alpha Omega Alpha. Author: (with R. D. Dripps, J. E. Eckenhoff) Introduction to Anesthesia, 1957; The Physiology and Clinical Pharmacology of Anesthetic Administration, 1965. Editor-in-chief Anesthesiology, 1962—. Research, numerous publs. in surgery and anesthesiology, circulatory effects of anesthetics, neurol. effects of anesthetics, metabolic effects of anesthetics. Home: 11 Holly Rd., Waban, Mass. 02168. Office: 721 Huntington Av., Boston 02115.*

VAN DEENEN, Laurens L. M., Dutch biochemist; b. Maastricht, Netherlands, Aug. 14, 1928; s. L. A. A. and J. (Hendriks) van D.; Ph.D., U. Utrecht (Netherlands), 1957; m. M. L. Graaff, Sept. 11, 1956; children—Yolanda, Wouter. Faculty, U. Utrecht, 1950—, prof. biochemistry, 1961—. Research, numerous publs. on role of phospholipids as constituents of membranes of animal cells and bacteria, chem. synthesis of phospholipids in relation to model system for membranes, mode of action of phospholipases. Home: 233 Paltzerweg, Den Dolder, Netherlands. Office: Dept. Biochemistry, State U. Utrecht, Utrecht, Netherlands.*

VAN DE GRAAFF, Robert Jemison, Am. physicist; b. Tuscaloosa, Ala., Dec. 20, 1901; s. Adrian Sebastian and Minnie Cherokee (Hargrove) Van de G.; B.S., U. Ala., 1922, M.S., 1923, hon. D.Sc., 1941; postgrad. Sorbonne, 1924-25; B.Sc., Oxford, 1926, Ph.D., 1928; m. Catherine Boyden, Apr. 12, 1936; children—John Hargrove, William Boyden. Rhodes scholar, 1925-28; Internat. Edn. Bd. fellow, Oxford, 1928-29; Nat. Research fellow, Princeton, 1929-31. Research asso. Mass. Inst. Tech., 1931-34; asso. prof., 1934-60; dir. OSRD radiographic project, 1941-46; dir. High Voltage Engrng. Corp., 1946—, mem. exec. com. 1957—. Awarded Elliott Cresson medal, Duddell medal. Fellow Am. Phys. Soc.; mem. Am. Acad. Arts and Scis., Am. Indsl. Radium and X-ray Soc., Inc., Sigma Xi. Developed high-voltage electrostatic generator, 1933; research in nuclear physics. Died Jan. 16, 1967.

VAN DE HULST, Hendrik Christoffell, Dutch astronomer; b. Utrecht, Netherlands, Nov. 19, 1918; s. W. G. and Jeanette (Maan) van de H.; Ph.D., U. Utrecht, 1946; m. Wilhelmina Mengerink, June 1946; children—Elske, Erik, Henriette, Willem. Postdoctoral fellow U. Chgo., 1946-48; lectr. astronomy Leyden U., 1948-52, prof., 1952—. Vis. prof. Harvard, 1951, Cal. Inst. Tech., 1954, Columbia, 1962-63; pres. Com. on Space Research, 1958-62; chmn. Netherlands Com. for Geophysics and Space Research, 1959—. Recipient Eddington medal Royal Astron. Soc., 1955, Henry Draper medal Nat. Acad. Scis., 1956, Rumford medal Royal Soc. London, 1964. Mem. Internat. Astron. Union (pres. commn. 34 on interstellar matter 1952-58), Royal Dutch Acad. Scis., European Space Research Orgn. (v.p. prep. com. 1960-65, chmn. council 1968). Author: A Course of Radio Astronomy, 1951; (with C. A. van Peursen) Phaenomenologie en Natuurwetenschape, 1953; Light Scattering by Small Particles, 1957, also numerous articles. Discovered 21 cm. line in radio astronomy; research on interstellar matter, solar corona, light scattering in planetary atmospheres. Home: Sterrewacht 8. Office: Sterrewacht, Leyden, Netherlands.*

VAN DE KAMP, Peter, astronomer; b. Kampen, Netherlands, Dec. 26, 1901; s. Lubbertus and Engelina (van der Wal) van de K.; student U. Utrecht, 1918-22; Ph.D., U. Cal. at Berkeley, 1925; Ph.D., U. Groningen, 1926; came to U. S., 1923, naturalized, 1942. Asst., Kapteyn Astron. Lab., Groningen, Netherlands, 1922-23; research asso. McCormick Obs., U. Va., 1923-24, instr., 1925-28, asst. prof., 1928-37; Martin Kellogg fellow Lick Obs., U. Cal., 1924-25; faculty Swarthmore (Pa.) Coll., 1937—, prof., 1940—, dir. Sproul Obs., 1937—; vis. prof. Harvard, 1936, Wesleyan U., 1956—, New Sch. for Social Research, 1944-62, also univs. in Europe. Mem. Alsos mission OSRD, Paris, 1945; program dir. for astronomy NSF, 1954-55; dir.-at-large Assn. of Univs. for Research in Astronomy, 1958-62; pres. commn. on double stars Internat. Astron. Union, 1958-64, mem. U. S. nat. com., 1964—. Recipient Pres. and Visitors prize U. Va., 1927, 37, 38; Glover award Dickinson Coll., 1961; Nason award Swarthmore Coll., 1963; Rittenhouse Soc. Phila. medal, 1965. Mem. Am., Pacific astron. socs., Royal Dutch Acad. (corr.), Sigma Xi, Phi Beta Kappa. Author: Basic Astronomy, 1952; (with G. Honegger) Space, The Architecture of the Universe, 1961; Elements of Astromechanics, 1964; Principles of Astrometry, 1967; also numerous articles. Parallaxes of nearby stars, especially masses of double stars, search for unseen companions of nearby stars; discovered 1st extra-solar planet, 1963. Home: 15 Wellesley Rd., Swarthmore, Pa. 19081.*

VANDEMARK, Noland Leroy, Am. physiologist; b. Columbus Grove, O., July 6, 1919; s. Daniel Leroy and Mary Frances (Bogart) VanD.; B.S., Ohio State U., 1941, M.S., 1942; Ph.D., Cornell U., 1948; m. Beda Alta Basinger, Aug. 18, 1940; children—Gary Lee, Judy Beth, Linda Kay. Asst. animal husbandry Ohio State U., 1941-42; vitamin chemist, div. plant industry Ohio Dept. Agr., 1942; livestock specialist U. S. Allied Commn. Austria, 1946-47; asst. animal husbandry Cornell U., 1942-44, 48; from asst. prof. to prof. physiology, dept. dairy sci. U. Ill., 1948-64; prof. dairy sci., chmn. dept. Ohio State U. and Ohio Agrl. Expt. Sta., 1964—. Recipient Borden award in dairy sci., 1959; Italian Master Pioneer Gold Medal award 5th Internat. Congress Reprodn., Trento, 1964. Fellow A.A.A.S.; mem. Am. Physiol. Soc., Soc. Study Fertility, Am. Dairy Sci. Assn., Am. Soc. Animal Sci., Sigma Xi. Author: (with G. W. Salisbury) Physiology of Reproduction and Artificial Insemination of Cattle, 1961. Made fundamental contbns. to knowledge of reproductive physiology in cattle, in efficient use of artificial insemination in cattle breeding, semen preservation, combating problems of infertility. Home: 1930 Snouffer Rd., Worthington, O. 43085. Office: Dept. Dairy Sci., Ohio State Univ., Columbus, O. 43210.*

VAN DEMARK, Paul John, Am. microbiologist; b. Ithaca, N.Y., June 17, 1923; s. Legrand S. and Agnes (McAllister) Van D.; B.S., Cornell U., 1947, M.S., 1948, Ph.D., 1950; m. Eileen M. Maleski, June 2, 1946; children—Paul John, John, Michael, Stephen, Peter. Faculty, Cornell U., Ithaca, 1950—, prof., microbiology, 1957—. Mem. Am. Acad. Microbiologists, Am. Soc. Microbiologists, A.A.A.S., Sigma Xi. Author: Microbes in Action, 1962; also articles. Research on nutritional requirements and metabolic pathways in microorganisms. Home: South St., Trumansburg, N.Y. 14886. Office: Stocking Hall, Cornell U., Ithaca, N.Y. 14850.*

VAN DEN AKKER, Johannes Archibald, Am. physicist; b. Los Angeles, Dec. 5, 1904; s. John and Mabel (Freebairn) Van Den A.; B.S. in Physics, Cal. Inst. Tech., 1926, Ph.D., 1931; m. Carmen Laura Haberman Beck, June 9, 1958; 1 dau. (by previous marriage) Valerie (Mrs. George Wellington Emmert III). Instr. physics, Washington U., St. Louis, 1930-35, research asso., head physics dept., 1935-56; sr. research asso., chmn. dept. physics and math., Inst. Paper Chemistry, Lawrence U., Appleton, Wis., 1956—. Sr. Fulbright lectr., U. Manchester, Eng., 1961-62. Fellow Am. Phys. Soc., A.A.A.S., Optical Soc. Am.; mem. Am. Assn. Physics Tchrs., T.A.P.P.I., Am. Arbitration Assn., Sigma Xi, Tau Beta Pi, Phi Gamma Delta. Research, publs. on atomic physics, spectrophotometry, optical characteristics of light-scattering media, color measurement, physics of paper; inventor electron counter, analog computer. Home: 1 Brokaw Pl., Appleton, Wis. 54912.*

VAN DEN BERGH, Sidney, astronomer; b. Wassenaar, Holland, May 20, 1929; s. Sidney J. and S. M. (van den Berg) van den B.; student Leiden (Holland) U., 1947-48; A.B., Princeton, 1950; M.Sc., Ohio State U., 1954; Dr.rer.nat., Göttingen U. (Germany), 1956; m. Roswitha E. E. Koropp, Sept. 3, 1957; children—Peter, Mieke. Asst. prof. Ohio State U., 1956-58; faculty David Dunlap Obs., U. Toronto (Ont., Can.), 1958—, prof. astronomy, 1966—. Research and publs. on extragalactic nebulae, star-clusters. Home: 343 Sugar Maple Lane, Richmond Hill, Ont., Office: David Dunlap Obs., Richmond Hill, Ont., Can.*

VAN DEN BERGHE, Herman, Belgian geneticist; b. Overboelare, Belgium, June 12, 1933; s. Gustave and Maria (Steenhoudt) van den B.; Dr. Medecine, U. Louvain (Belgium), 1958, B.Basic Med. Scis., 1960, B.Electroradiology, 1966, Agregaat Hoger Onderwijs, 1966; m. Marie van Orshoven, Aug. 26, 1959; children—Tom, Ingrid, Godelieve, Lucas. Research asst. dept. virology Rega Inst., U. Louvain, 1958-63, faculty, 1963—, asso. prof. histology, 1966—, in charge dept. cytogentics, 1963-66, in charge dept. human genetics, 1966—; med. dir. registration cancer and genetic diseases ins. co., 1963—. Recipient 1st prize concours Nat. des Bourses de Voyage, Belgian Govt., 1961. Author: Het Zogenaamde Syndroom van Turner, 1966; also articles. Research on animal and human cytogenetics, dermatoglyphics, med. genetics. Office: Vesalius Inst., U. Louvain, 12 Minderbroedersstraat, Belgium.*

VAN DEN BOLD, Willem Aaldert, geologist; b. Amsterdam, Netherlands, Mar. 30, 1921; s. Hermanus Johannes and Willemina Alida (Grötendorst) van den B.; Ph.D., U. Utrecht, 1946; m. Antoinette Marie Jeanne van Doornik, June 11, 1958; children—Roberto Ramon, Lucia Eleanor, Marieke Saskia Yvette, Michiel Christiaan, Hans Adriaan. Paleontologist, Royal Dutch Shell Group, Venezuela, 1946-49, Colombia, 1949, Trinidad, 1950-56, New Zealand, 1957; asso. prof. La. State U., Baton Rouge, 1958-59, prof. geology, 1959—; vis. prof. U. Cologne (Germany), 1962. Mem. Geol. Soc. Am., Soc. Econ. Paleontology and Mineralogy, Am. Geophys. Union, Paleontol. Research Instn. Author: (with others) vol. Q of Treatise on Invertebrate Paleontology, 1961; also articles. Research on paleontology and stratigraphy of Tertiary of Caribbean region, taxonomy of ostracoda. Home: 413 Kimbro Dr., Baton Rouge 70808.*

VAN DEN BOSCH, Robert, Am. entomologist; b. Martinez, Cal., Mar. 31, 1922; s. Pieter and Aline (Schulthess) van den B.; A.B., U. Cal. at Berkeley, 1943, Ph.D., 1950; m. Margaret Irene Bryant, Mar. 11, 1944. Asst. entomologist U. Cal., Riverside, 1951-57, asso. entomologist, 1957-63, asso. entomologist, Berkeley, 1963-64, entomologist, 1964—, lectr., acting chmn. div. biol. control, 1966—, prof. entomology, 1967—. Guggenheim fellow, 1958-59. Fellow A.A.A.S.; mem. Entomol. Soc. Am., Ecol. Soc. Am., Pacific Coast Entomol. Soc., Am. Inst. Biol. Scis. Contbr. chpts. to books, numerous articles to profl. jours. Research in biol. control of pest insects, ecology of parasitic insects, integrated control of agrl. pests, biosystematics of aphids. Home: 4 Kensington Ct., Kensington, Cal. 94707. Office: 1050 San Pablo Av., Albany, Cal. 94706.*

VAN DEN BRENK, Hendrik Athos Sydney, radiobiologist; b. Sydney, Australia, June 22, 1921; s. Adriaanus Joannes and Anna (Faber) Van den B.; M.B., B.S., U. Melbourne (Australia), 1944, M.S., 1954; m. Alice Miriam Glasier, July 1, 1944; children—Christine Anna Agnes (Mrs. Warwick Stott), Judith Elise. Practice medicine, 1944-48; surgeon St. Vincent's Hosp., Melbourne, 1949-52; radiotherapist Royal Melbourne Hosp., 1952-54; sr. research fellow dept. physiology Royal Coll. Surgeons, London, Eng., 1954-56; chief scientist radiobiol. research unit, cons. radiotherapist Cancer Inst., Melbourne, 1956-67; hon. radiotherapist Royal Victorian Eye and Ear Hosp., 1955-67; lectr. radiobiology U. Melbourne, 1962-67; Richard Dimbleby fellow cancer research St. Thomas' Hosp., London, 1967—. Rouse Travelling fellow, 1967. Fellow Royal Coll. Surgeons; mem. Coll. Radiologists Australasia, Brit. Med. Assn., A.M.A., Australian Physiology Soc., Australian Radiation Research Soc., Cell Culture Soc. Victoria, Australian Assn. Med. Research. Editorial bd. Internat. Jour. Radiation Biology, 1961—, Australasian Radiology, 1960—. Research, numerous publs. on radiobiology and radiotherapy especially oxygen effect in radiation, radioprotective chem. action, physiology oxygen toxicity. Office: St. Thomas Hosp., London, Eng.*

VANDENHEUVEL, Frantz Aime, chemist, biophysicist; b. Brussels, Belgium, Apr. 30, 1913; s. Valere and Marthe (de Poplimont) V.; student Athenee Royal d'Ixelles, Brussels U.; Ph.D., Royal Coll. Scis., London, 1938; Dr., Imperial Coll., 1938; m. Denise Guillemine, Oct. 18, 1936. Asst. prof. U. Brussels, 1938-40; group leader Service de Recherches, Soc. Belge Carbochimique, 1940-47; head Fisheries Tech. Sta., Halifax, N.S., Can., 1947-60; sr. research officer biochemistry Animal Research Inst., Dept. Agr., Ottawa, Ont., 1960-66, prin. research officer, 1966—. Mem. Canadian Com. on Fats and Oils, 1951—; lectr. biology Ottawa U., part-time 1966—. Fellow Chem. Inst. Can.; mem. Am. Chem. Soc., N.Y. Acad. Scis., Am. Oil Chemists' Soc., Canadian Biochem. Soc., N.S. Inst. Scis. Research, publs. on molecular distillation, separation, identification, analysis of fatty acids, upgrading of fish protein and other fish products, biol. membrane structure and function, analysis, identification, and properties of steroids of natural origin, also others. Home: 268 MacLaren St., Ottawa 4. Office: Can. Dept. Agr., Research Br., Animal Research Inst., CEF, Ottawa, Ont., Can.*

VAN DEN SPIGELIUS, Adriaan, anatomist; b. Brussels, Belgium, 1578; student Louvain, Belgium, Padua, Italy; state physician Moravia until 1616; prof., Venice, Italy; became prof., Padua, 1616. Author: De semitertiana, 1624; De humani corporis fabrica, pub., 1627; De formatu foetu, liber singularis; De humani corporis fabrica. First account of malaria, 1624. Died Padua, Apr. 7, 1625.

VAN DEPOELE, Charles Joseph, scientist, inventor; b. Lichtervelde, Belgium, Apr. 27, 1846; s. Peter John and Maria (Algoed) Van D.; attended Imperial Lyceum, Lille, France; m. Ada Van Hoogstraten, Nov. 23, 1870, 7 children including Romaine Adeline. Came to U. S., settled in Detroit, 1869; became mfr. ch. furniture; exhibited arc lights, 1870, demonstrated feasibility of electric transp. by both overhead and underground conductors, 1874; worked on vibratory regulation for arc lights, demonstrated improved lights publicly, 1879; formed Van Depoele Electric Light Co., Inc. as Van Depoele Electric Light Co. of Chgo., 1881, Van Depoele Electric Mfg. Co., 1884; made 1st practical demonstration in world of a spring pressed under-running trolley at Chgo. Inter-State Indsl. Expn., 1883; successful with both the underground and overhead circuits in Toronto, Cnt., Can., 1884, 85; overhead system in operation in South Bend, Ind., 1885, adopted in Minneapolis, Minn., Montgomery, Ala., 1885-86; eight lines installed in U. S., Can., 1886; sold electric railway patents to Thomson-Houston Electric Co., Lynn, Mass., 1888; sold Van Depoele Electric Mfg. Co., 1889; made 444 patent applications, granted 249 in his name. Inventions include little "Giant" generator, 1880, 1st patent on electric rys., 1883, 1st on overhead conductor, 1885, patent for carbon contract brushes in electric motors, 1888, coal mining machine, 1891; experimented with electric refrigeration, 1886; made photographs in color, 1880-90. Died Lynn, Mass., Mar. 18, 1892.

VAN DER HOEVEN, Jan, Dutch naturalist; b. Rotterdam, Netherlands, Feb. 9, 1801; Ph.D., 1822; M.D., 1824. Prof. zoology U. Leyden (Netherlands), from 1835. Author: Handboek der Dierkunde, 2 vols., 1827-33; Over de natuurlyke Geschiedenis van het dierenrijk, 1851-57; Philosophia zoologica, 1864. Distinguished zoologist; research on natural history, anatomy of limuli, 1838; discourse, dissertations, 1846. Died Leyden, Mar. 10, 1868.

VAN DER LINDEN, Johann Antonides, Dutch anatomist, botanist; b. Enkhuysen, Holland, 1609; s. Antonides-Henrich Linden; studied medicine, Acad. Leiden; D.Physics, Franeker, 1630; named prof. medicine, Leiden, Netherlands, 1650. Author: De scriptes medicis, 1637; Medicina physiologica; wrote med. history, med. reference books; also edit. of Hippocrates in Greek. Died 1664.

VAN DER MENSBRUGGHE, Gustav, Belgian physicist; b. Ghent, Belgium, Feb. 13, 1835; prof. U. Ghent. Measured surface tension of liquid thus determining Laplacean constant, 1907. Died Ghent, Oct. 20, 1911.

VANDERMONDE, Alexandre (or Alexis) Théophile, French mathematician; b. Paris, France, Feb. 28, 1735. Co-founder Conservatoire des Arts et Métiers, dir., from 1782; helped inst. study polit. economy Ecole Normale. Mem. French Acad. Sci., 1771. Author: Sur la résolution des équations algébriques, 1771; Recherches analytiques sur les irrationnelles d'une nouvelle espèce, 1772; L'elimination des inconnues dans les quantités algébriques. Made 2 important studies on 2 gen. methods, substitution, combination, for solving lower order equations, 1770; obtained formulae for solving gen. quadratic, cubic, quartic equations: solved equation x to the 11th power-1=0; foresaw that equation x to the nth power-1=0 must have solution when n is a prime; (with LaGrange) originated combination method for solving equations; 1st to give logical exposition of theory of determinants; (with Monge and Berthollet) discovered difference between pig-iron, steel. Died Paris, Jan. 1, 1796.

VANDERSLICE, Joseph Thomas, Am. chem. physicist; b. Phila., Dec. 21, 1927; s. Joseph R. and Mae (Daly) V.; B.S. in Chemistry, Boston Coll., 1949; Ph.D. in Phys. Chemistry (Allied Chem. and Dye fellow), Mass. Inst. Tech., 1953; m. Patricia Mary Horstman, Nov. 20, 1954; children—Sharon, Joseph, Julie, Peter, John, Polly, Jeffrey. Faculty, Cath. U. Am., 1952-56, asst. prof., 1955-56; faculty U. Md., College Park, 1956—, prof. chem. physics, 1963—, dir. Inst. for Molecular Physics, 1967—, head of department of chemistry, since 1968—. Fellow Washington Acad. Scis., Am. Phys. Soc., mem. Am. Chem. Soc., Philos. Soc. Washington, Sigma Xi. Author: (with H. W. Schamp, E. A. Mason) Thermodynamics, 1966; also numerous articles. Determined interactions between atoms and molecules; calculated and measured intensities spectral lines. Home: Route 1, Box 582 B, Accokeek, Md. 20607. Office: Inst. for Molecular Physics, U. Md., College Park, Md. 20740.*

VAN DER STELT, Cornelus, Dutch chemist; b. Rotterdam, Netherlands, Mar. 27, 1921; s. Johan Machiel and Adriana (Middlehoek) van der S.; student Free U., Amsterdam, Netherlands, 1938-48, Ph.D., 1956; m. Willemina Fraukje Hofman, Aug. 4, 1948. Research chemist research dept. N. V. Koninklijke Pharmaceutische Fabrieken v/h Brocades-Stheeman en Pharmacia, 1948—, head organic chemistry sect., 1951—. Mem. Koninklijke Nederlandse Chemische Vereniging, Christelijke Vereniging van Natuur en Geneeskundigenin Nederland. Research, publs. and patents on synthetic compounds of medicinal interest, including tuberculostatic p-aminosalicyclic acid and derivatives, dibenzo cycloheptene derivatives, aromatic trifluoromethyl compounds. Home: 24 Laurens Reaellaan. Office: Brocades Research Lab., 26 Gonnetstraat, Haarlem, Netherlands.*

VANDERVOORT, Peter Oliver, Am. astronomer; b. Detroit, Apr. 25, 1935; s. William Bernard and Geraline (Case) V.; A.B., U. Chgo., 1954, B.S., 1955, M.S., 1956, Ph.D., 1960; m. Frances Sheridan, Sept. 23, 1956; children—William F., Dirk S. Research asso. Nat. Radio Astronomy Obs., Green Bank, W.Va., 1960; NSF postdoctoral fellow Princeton Obs., 1960-61; asst. prof. dept. astronomy U. Chgo., 1961-65, asso. prof., 1965—. Mem. Am. Astron. Soc., Am. Phys. Soc., Internat. Astron. Union. Research in hydrodynamic stability, gas dynamics, interstellar matter, stellar dynamics, structures of star clusters and galaxies. Home: 1413 E. 57th St. Office: 1100-14 E. 58th St., Chgo. 60637.*

VAN DER WAALS, Johannes Diderik, Dutch physicist; b. Leyden, Netherlands, Nov. 23, 1837; doctorate Leiden U.; prof., Debenter, also La Haye; prof. physics U. Amsterdam (Netherlands), 1877-1907. Recipient Nobel prize in physics (for equation of state for gasses and liquids), 1910. Mem. French Acad. Scis., Royal Acad. Scis., Amsterdam. Author: Over de continuiteit van den gasen vloestoftoestand, 1873. Research on gaseous and liquid phase; sought to determine why so-called perfect gas equation did not hold exactly for real gases; modified the gas equation which expressed deviation from laws of Boyle and Gay-Lussac, 1881; investigated electrolytic dissociation formula, 1891, surface tension, 1894, theory of mixtures, 1894, continuity of gaseous and liquid states, 1899, thermodynamic theory of capillarity, fluid statics. Died Amsterdam, Mar. 9, 1923.

VAN DER WAERDEN, Bartel Leendert, mathematician; b. Amsterdam, Netherlands, Feb. 2, 1903; s. Theodorus and Dorothea (Endt) van der W.; Ph.D.,

U. Amsterdam, 1926; Ph.D. (hon.), U. Athens; m. Camilla Rellich, Sept. 27, 1929; children—Helga (Mrs. Habicht), Ilse (Mrs. Peterschmitt), Hans. Prof., U. Groningen (Netherlands), 1928-31, U. Leipzig (Germany), 1931-45, Johns Hopkins, Balt., 1947-48, U. Amsterdam, 1948-51, U. Zürich (Switzerland), 1951-—. Mem. acads. of Amsterdam, Leipzig, Halle, Heidelberg, München, Göttingen. Author: Modern Algebra, 1931; Science Awakening, 1954; Sources of Quantum Mechanics, 1967; also numerous articles. Research in found. of algebraic geometry, 1925-62; algebra, geometry, number theory, history of sci., theoretical physics. Home: 18 Bionstrasse, Zürich, Switzerland.*

VANDERWERF, Calvin Anthony, Am. chemist; b. Friesland, Wis., Jan. 2, 1917; s. Anthony and Anna (Schaafsma) V.; A.B., Hope Coll., 1937, Sc.D., 1963; Ph.D., Ohio State U., 1941; LL.D., St. Benedict's Coll., 1966; Sc.D., Rose Poly. Inst., 1966; m. Rachel Anna Good, Aug. 22, 1942; children—Gretchen, Kasina, Julie, Lisa, Pieter, Marte. Instr., U. Kan., 1941-42, mem. faculty and staff, 1943-63, prof. chemistry, 1949-63, chmn. dept., 1961-63, prof. Hope Coll., 1963-—. Vis. scientist NSF, 1955-—; cons. various firms. Dir. Kativo Chem. Co., Ltd., Costa Rica, 1963-—. Grantee Petroleum Research Fund, 1962. Mem. Am. Chem. Soc., Chem. Soc. London, Sigma Xi. Author: (with Sisler and Davidson) General Chemistry, 2d edit., 1959, Korean transl., 1959, A Systematic Approach, 2d edit., 1961, Korean transl., 1964, Asian edit., 1964; (with Brewster and McEwen) Unitized Experiments in Organic Chemistry, 2d edit., 1964; Acids, Bases and the Chemistry of the Covalent Bond, 1961, also Japanese, French, Spanish and Italian transls., numerous articles. Mem. editorial bd. Jour. Chem. Edn. Research on synthesis of medicinals; mechanisms of organic reactions; aromatic fluorine compounds; unsaturated lactones; epoxides; phase diagrams; organo-boron and organo-phosphorus compounds. Home: 92 E. 10th St., Holland, Mich. 49423.

VANDERZEE, Cecil Edward, Am. chemist; b. Wetonka, S.D., Apr. 26, 1912; s. Cornelius E. and Clara (VanGronigen) V.; B.S., Jamestown Coll., 1938; Ph.D., State U. Ia., 1949; m. Helen May Thomas, July 16, 1944. Instr., Jamestown Coll., 1939, Hot Springs (S.D.) High Sch., 1939-42; faculty U. Neb., Lincoln, 1949-—, prof., 1958-—, vice chmn. dept. chemistry, 1965-—. Mem. Am. Chem. Soc., Neb. Acad. Scis., Calorimetry Conf. (chmn. elect 1967-—). Research and publs. on thermodynamics of electrolytes and some pure compounds, using calorimetric, electrometric and spectrophotometric techniques; determination of equilibrium constants, heats of formation, dilution, solution, reaction. Home: 3320 Melrose St., Lincoln, Neb. 68506.*

VAN DE VOORDE, Herman Pieter Jozef, Belgian microbiologist; b. Laeken, Belgian, Oct. 16, 1928; s. Urbain and Maria Theresia (Rutgerts) Van de V.; M.D., U. Louvain; agrégé; m. Godelieve Stoppie, Dec. 15, 1955; children—Pieter, Goedele, Machteld, Frieda. Dir. clin. microbiology and analysis labs. U. Louvain Clinic, St. Raphael, 1956-61; dir. Sch. Pub. Health, U. Louvain (Belgium), 1961-—, dir. Inst. Paramed. Sci., 1963-—. Mem. Belgian Hygiene Assn., Belgian Microbiol. Assn., Internat. Acad. Legal Medicine. Author: Staphylomycine, a new antibiotic; Een studie over potentieel pathogene mycobacteriën; Hygiene in het ziekenhuis. Home: Naamsestrasse 147, Louvain, Belgium.

VAN DE VOOREN, Adriaan Isak, Dutch aero. engr.; b. Hague, Netherlands, Mar. 1, 1919; s. William and Jacoba (van Veen) Van De V.; ed. Tech. U. Delft; Dr. ès. sc.; m. Catharina Barre, Aug. 4, 1943; children—Floris, Ernst, Paul, Juliette, Marleen. Engr., Nat. Inst. Aero. Research, 1941-52, engr.-in-chief 1952-58; asso. prof. aeros., Delft, Netherlands, 1956-58; prof. applied math. U. Groningen (Netherlands), 1958-—. Mem. Amsterdam Math. Soc., Royal Aero. Soc. London. Author: Unsteady Airfoil Theory; Flutter of Systems with Many Degrees of Freedom. Research and publs. on fluid mechanics, applied and numerical math. Home: Dilgtplein 6, Haren (Gr.), Netherlands. Office: Reitdiepskade 4, Groningen, Netherlands.

VANDE WIELE, Raymond Laurent, physician; b. Kortryk, Belgium, Oct. 2, 1922; s. Jan. B. and Marie L. (Lampe) Vande W.; M.D., U. Louvain, 1947; m. Beatrice Silides, Sept. 25, 1954; children—Barbara, Caroline, Margaret. Practice medicine, specializing in obstetrics, gynecology, N.Y.C., 1958-—; faculty Columbia, 1955-—, asso. prof., 1962-—; asst. obstetrician, gynecologist Presbyn. Med. Center, 1958-62, asso. attending, 1962-—; mem. endocrinology study sect. div. research grants NIH. Mem. Am. Gynecol. Soc., Assn. Profs. Gynecology and Obstetrics, Endocrine Soc., N.Y. Acad. Scis., Soc. Clin. Investigation, Soc. Gynecol. Investigation, Soc. Study Fertility. Contbr. numerous articles to profl. jours. Studies on biosynthesis and metabolism of steroid hormones of the adrenal, testis and ovary, induction of ovulation in the human. Home: Closter Dock Rd., Alpine, N.J. Office: 630 W. 168th St., N.Y.C. 10032.*

VAN DIEMERBROECK, Isbrand, Dutch physician; b. Montfort, Netherlands, 1609. Practiced at Nimeguen, Netherlands, several years; became asso. prof. U. Utrecht (Netherlands), 1642, prof., 1651, rector, twice. Author: Disputationes practicae; De Peste, 1646 (reprinted with other writing as Opera omnia, 1685); De variolis et morbillis; Observat et curationes medicae; Anatome corporis humani, 1671. Described plague at Nimeguen, 1635-37, including fumigation of bed-chambers with burning sulphur; studies in anatomy, measles, smallpox, diseases of lower abdomen. Died Nov. 17, 1674.

VAN DILLA, Marvin Albert, Am. physicist; b. N.Y.C., June 18, 1919; s. Albert and Hattie (Isenberg) Van D.; B.S., Coll. City N.Y., 1939; Ph.D., Mass. Inst. Tech., 1951; m. Jean Frances Noertker, Aug. 18, 1951; children—Teresa, Peter, Laura, Wendy. Instr., U. Utah, Salt Lake City, 1951-57; physicist Los Alamos Sci. Lab., 1957-—. Mem. Radiation Research Soc. Research, publs. on nuclear radiation detectors, metabolism radioisotopes, biol. effects radiation, radiation dosimetry, radioactive fallout, space radiation and meteorite radioactivity, biol. cell counting and sizing. Home: 154 Barranca Rd. Office: Los Alamos Sci. Lab., Los Alamos 87544.*

VAN DRIEST, Edward Reginald, Am. aero. engr.; b. Cleve., Sept. 16, 1913; s. Paulinus Marius and Regina (van Duffellen) van D.; B.S., Case Inst. Tech., 1936, Ph.D., 1940; M.S., U. Ia., 1938; Sc.D., Eidgen. Tech. Hoch., Zurich, Switzerland, 1948; m. Marie Claire Lang, Sept. 21, 1942; children—Edward Reginald, Marie Claire, James Paul. Aerodynamicist, N.Am. Aviation, Inc., 1948-56, dir. Space Scis. Lab., 1957-64, chief scientist ocean systems operations, Anaheim, Cal., 1965-—. Fellow Am. Astronautical Soc. (1st v.p. 1965-—), Am. Inst. Aeros. and Astronautics (asso.). Contbg. author: Handbook of Engineering, 1952; High-Speed Aerodynamics and Jet Propulsion, 1959. Research, publs. on devel. and formulation of theories of dimensional analysis for data interpretation; developed theory of supersonic turbulent boundary layer flow necessary for calculation of heat transfer and friction of high speed aircraft; developed fundamental theories and formulas for practical determination of region of transition of a laminar to a turbulent boundary layer; invented device using heat transfer as a means of controlling drag of supersonic aircraft. Home: 9732 El Arco Dr., Whittier, Cal. 90603. Office: 350 S. Magnolia Av., Long Beach, Cal. 90802.*

VAN DYKE, Harry Benjamin, Am. pharmacologist, physician; b. Des Moines, Jan. 31, 1895; s. Benjamin Isaac and Louise (Boody) van D.; B.S., U. Chgo., 1918, Ph.D., 1921, M.D., 1923; m. Elizabeth E. Allan, Apr. 14, 1920; children—Jane Elizabeth (Mrs. John H. Felber), Arthur Cushny (dec.). Faculty, U. Chgo., 1926-32, Peking Union Med. Coll., 1932-38; head div. pharmacology Squibb Inst. for Med. Research, 1938-44; David Hosack prof. pharmacology Columbia, N.Y.C., 1944-63, Hosack prof. emeritus, 1963-—; vis. prof. pharmacology Nat. Def. Med. Center, Taiwan, 1963-64, U. Malaya, Kuala Lumpur, Malaysia, 1965-—. Mem. numerous coms. NIH, NRC. Mem. Am. Soc. for Pharmacology and Exptl. Therapeutics (past pres.), Am. Physiol. Soc., Brit. Pharm. Soc., Am. Endocrine Soc., Assn. Am. Physicians, Soc. for Endocrinology (Gt. Britain), Biochem. Soc. (Gt. Britain). Author: The Physiology and Pharmacology of the Pituitary Body, Vol. I, 1936, Vol. II, 1939. Research, numerous publs. on physiology and pharmacology of pituitary secretions, especially gonadotropins and posterior pituitary hormones; comparative biochemistry of posterior pituitary hormones (vasotocin and others), pharmacology of autonomic (involuntary) nervous system, pharmacology of chemotherapeutic drugs (antibacterial and antimalarial). Home: 116 Pinehurst Av., N.Y.C. 10033. Office: 630 W. 168th St., N.Y.C. 10032.*

VAN DYKE, John Howard, Am. anatomist; b. Palisades Park, N.J., Sept. 13, 1911; s. Joseph Smith and Mabelle (Baird) Van D.; A.B., Colgate U., 1935; M.S., U. N.H., 1937; Ph.D., Cornell U., 1941; m. Margaret Mary Brubaker, Apr. 3, 1935; children—Jeanne (Mrs. Peter J. McKenna), John Howard, Virginia (Mrs. John R. Lattig), David, Susan, Kathleen, James. Research histologist U. S. Nutrition Lab., 1941-42; asst. prof. anatomy Washington U. Sch. Medicine, St. Louis, 1942-47; asso. prof. Ind. U. Sch. Medicine, Bloomington, Indpls., 1947-51; prof. anatomy Hahnemann Med. Coll., 1961-62. Fellow A.A.A.S., Gerontological Soc.; mem. N.Y. Acad. Scis., Cancer Research, Am. Assn. Anatomists, Am. Soc. Zoologists. Contbg. author: Comparative Endocrinology, 1959. Research, publs. on exptl. pathology abnormal devel. and carcinogenesis, endocrinology, biomagnetics. Home: Route 2, Box 246, Mounted Route, Telford, Pa. 18969. Office: 235 N. 15th St., Phila. 19102.*

VAN DYKEN, Alexander Robert, Am. chemist; b. Bozeman, Mont., Sept. 25, 1917; s. Alex and Anna (Van Dyke) Van D.; A.B., Calvin Coll., 1941; M.S., Purdue U., 1943; Ph.D., U. Chgo., 1949; m. Rose Dekker, July 19, 1946 (dec. 1965); children—Deborah Anne, Robert Webb; m. 2d, Nell DeMey, Feb. 4, 1966. Asso. chemist Argonne Nat. Lab., Chgo., 1948-56; chemist U. S. AEC, Washington, 1956-62, asst. dir. div. research for chemistry programs, 1962-—. Mem. Am. Chem. Soc., Radiation Research Soc. Research in organic fluorine chemistry, effect of fast neutrons on graphite, radiation chemistry, use of tracers carbon-14 and tritium in organic reactions. Home: 3913 Parsons Rd., Chevy Chase, Md. 20015. Office: Div. Research, U. S. AEC, Washington 20545.*

VAN DYNE, George Mason, Am. environmental biologist; b. Pueblo, Colo., Sept. 6, 1932; s. Riley Roy and Stella (Bentley) Van D.; Asso. in Arts and Scis., Pueblo Coll., 1952; B.S., Colo. State U., 1954; M.S., S.D. State U., 1956; Ph.D., U. Cal. at Davis, 1963; m. Shirley Mae Anderson, Apr. 8, 1955; children—Rita Rilene, Leslie Lenore, David Mason, Julie Janelle. Instr., Colo. State U., 1956-57; asst. prof. Mont. State U., 1957-61; asst. research nutritionist U. Cal. at Davis, 1961-64; ecologist radiation ecology Oak Ridge Nat. Lab., asso. prof. biology U. Tenn., 1964-66; asso. prof. biology Colo. State U., Coll. Forestry and Natural Resources, Ft. Collins, 1966-68, prof. biology, 1968-—. Mem. Ecol. Soc. Am., Am. Soc. Animal Sci., Am. Soc. Range Mgmt., Wildlife Soc., A.A.A.S., Am. Inst. Biol. Scis., Sigma Xi. Research, numerous publs. on nutrition, ecology range plants and animals, methods range measurement, application biometrical and computer methods to environmental problems. Home: Box 49, Bellvue, Colo. 80512. Office: Coll. Forestry and Natural Resources, Colo. State U., Ft. Collins, Colo. 80521.*

VAN EIJNATTER, Cornelis Leonardus Maria, Dutch plant breeder; b. Princenhage, Holland, Apr. 6, 1930; s. C. A. and C. (Brekoo) Van E.; Ph.D., Agrl. U., Wageningen, Holland, 1955; m. Jacqueline VanBoeckel, Apr. 4, 1956; children—Judith, Bastiaan, Lidwien, Joris, Wendy. Research officer W. African Maize Research Unit, Ibadan, Nigeria, 1956-59; sr. research officer Fed. Dept. Agrl. Research, Ibadan, 1959-63; prin. research officer Cocoa Research Inst., Ibadan, 1963-67; dir. Dutch Found. for Home Gardens Project, Amsterdam, 1967-—. Author: Towards the Improvement of Maize in Nigeria, 1965; The Cultivation of Kola, 1967; also articles. Introduced maize from Americas and bred new synthetic varieties; investigated methods of improving maize for Nigeria; studied breeding behavior of kola nut trees and selected clonal materials from local populations. Address: Tropical Inst., Mauritskade, Amsterdam, Holland.*

VAN ERMENGEM, Émile Pierre Marie, Belgian physician, bacteriologist; b. Louvain, Belgium, Aug. 15, 1851; M.D., U. Louvain, 1875; postgrad. Paris, London, Edinburgh, Vienna, 1876-78. Worked with Koch, Berlin, 1883; practiced medicine, Brussels, Belgium); also established pvt. bacteriological lab. (1st in Belgium); studied cholera inoculation methods of Ferran in Spain, 1885; prof. bacteriology U. Ghent (Belgium), 1888-1919. Mem. Académie Royale de Médecine de Belgique (perpetual sec. 1919-32). Discovered bacillus botulinus in cases food poisoning, 1897; research bacteriology, hygiene. Died Elsene, Belgium, Sept. 30, 1932.

VAN FLEET, Dick Scott, Am. botanist; b. Trenton, N.J., Mar. 23, 1912; s. Peter and Stella (Kise) Van F.; A.B., Ind. U., 1936, M.A., 1937, Ph.D., 1940; m. Clara Mootz, Sept. 4, 1939; children—Pete-Scott, Susan Jo. Asst. prof. biology Heidelberg Coll., 1940-43; Sterling Research fellow Yale, 1943-44; research asso. U. Mo., 1944-45, faculty, 1945-57, prof. botany, 1953-57; prof., chmn. dept. botany, U. Toronto (Ont., Can.), 1957-59; head dept. botany U. Mass., 1959-62; head dept. botany U. Ga., Athens, 1962-—. Mem. Bot. Soc. Am., Canadian Soc. Plant Physiology, Histochem. Soc., Plant Phenolics Group N.Am. Research and publs. on enzyme localization, microelectrophoresis of lipid-bound enzymes, distbn. and function of polyacetylene compounds in plants, evolution of polyene systems, origin of bioelectric currents in endodermis. Home: 530 Greencrest Dr., Athens, Ga. 30601.

VAN FOREEST, Pieter, Dutch physician; b. Alkmaar, Holland, 1522; ed. Louvain; M.D., Bologna, Italy; student of André Vésale, Padua, Jacques Dupois, Paris; practiced medicine, Alkmaar, 12 years, thereafter at Delft; declined chair of medicine U. Leiden; outstanding as clinician; recipient pension City of Delft for services during an epidemic. Author: De incerto, fallaci urinarum judicio, 1783; Observationum et curationum medicinalium, 1587. Died Alkmaar, 1597.

VANGHETTI, Giuliano, Italian surgeon; b. Greve, Italy, Oct. 8, 1861; s. Dario and Matilde (Rossi) V. Author: Plastiche e protesi cinematiche, 1906; Vitalizzazione delle membra artificiali, 1916; also numerous articles. Introduced kineplastic amputation in which stump is prepared for artificial limb, 1898. Died 1940.

VAN GITTERT, Pieter Hendrik, Dutch physicist, historian in physics; b. Gouda, Netherlands, May 30, 1889; s. Benjamin Pieter and Antonia Petrenolla (Huber) van G.; Dr. in Physics, Utrecht (Netherlands) U., 1919; m. Joha Geertruida Eymers, Apr. 11, 1938; children—Benjamin Pieter, Johanna Hermine Aleida. With Utrecht U., 1912-55, conservator, 1951, curator Mus., 1951-55; tchr. high sch., 1912-50. Decorated Order of Oranji Nassau. Author: Descriptive Catalogue of the Collection of Microscopes in the Possession of the University Museum, 1934; Astrolabes, 1952; Leerboek der Natuurkunde onder red.

van. R. Kronig, 1947; also numerous articles. Research on optics, both instrumental and theoretical, old phys. instruments; built double monochromator, 1926. Office: Utrecht U. Mus., Utrecht, Netherlands. Died Oct. 8, 1959.*

VAN GOORLE, David (Goarle, Gorlaeus), Dutch natural philosopher; b. Utrecht, Netherlands, Jan. 15, 1592 (or 1591); student Leyden, Netherlands, 1609-11. Author: Exercitationes philosophicae . . . , 1620. Classified, as 2d kind, properties which he saw as not inherent in atoms but appearing in aggregates wuch as fluidity; as 3d kind; properties which belong to atoms, such as largness; believed weight is not property but force in atoms, fire is an accident generated by heat. Died Cornjum, Apr. 21, 1612.

VAN HARREVELD, Anthonie, biologist; b. Haarlem, The Netherlands, Feb. 16, 1904; s. Anthonie and Elizabeth (de Tombe) Van H.; B.A., Amsterdam U., 1925, M.A., 1928, Ph.D., 1929, M.D., 1931; m. Traus Smits, Oct. 28, 1932; children—Anthonie, Frieda (Mrs. Oscar W. Ford). Asst., U. Utrecht, 1932-34; faculty Cal. Inst. Tech., Pasadena, 1934-—, prof. biology, 1947-—. Mem. Am. Physiol. Soc., other profl. orgns. Author: Brain Electrolytes, 1966; also numerous articles. Research on nerves and muscles, physiology of central nervous system, distbn. of water and electrolytes in central nervous tissue. Home: 764 S. Oakland Av., Pasadena, Cal. 91106.*

VAN HELMONT, see Helmont, van.

VAN HEURALT, Hendrik, Dutch mathematician; b. Haarlem, Netherlands, 1633. Author: Epistolae de curvarum linearum in rectas transmutatione, 1659. Noticed that problems of quadrature and rectification are identical, and that it is possible to reduce one to the other; rectified (independently of Fermat and Neile) semi-cubic parabola ay to the 2d power=x to the 3rd power. Died 1660.

VAN HEYNINGEN, William Edward, biochemist; b. Harrismith, South Africa, Dec. 24, 1911; s. George Phillipus Stephanus and Mabel Constance (Higgs) van H.; B.Sc. cum laude U. Stellen, South Africa, 1931, M.Sc. cum laude, 1932; Ph.D, U. Cambridge (Eng.), 1936, Sc.D., 1951; D.Sc., U. Oxford (Eng.), 1954; m. Ruth Eleanor Treverton, June 24, 1940; children—Simon, Joanna. Staff, Geol. and Geophys. Survey, Western Transvaal, 1932; jr. research officer Fuel Research Inst. S. Africa, 1932-34; Commonwealth Fund fellow Harvard, Coll. Phys. and Surg., Columbia U., 1936-38; Med. Research Council research grantee, 1938-43; staff Wellcome Labs., 1943-46; sr. research officer pathology Oxford U., 1946-66, reader bacterial chemistry, 1966-—, Master St. Cross Coll., 1965-—, curator Bodleian Library, 1963-—. Mem. Société Francaise de Microbiologie (corr.), Biochem. Soc., Soc. for Gen. Microbiology. Author: Bacterial Toxins, 1949; also numerous articles. Research on bacterial toxins including gas gangrene, dysentery, staphylococcal and tetanus toxins. Home: College Farm, Ferry Hinksey, Oxford. Office: St. Cross Coll., Oxford, Eng.*

VAN HISE, Charles Richard, Am. geologist; b. Fulton, Wis., May 29, 1857; s. William Henry and Mary (Goodrich) V.; B.M.E., U. of Wis., 1879, B.S., 1880, M.S., 1882, Ph.D., 1892; (LL.D., U. of Chicago, 1903, Yale, 1904, Harvard, 1908, Williams, 1908, Dartmouth, 1909); m. Alice Bushnell Ring, Dec. 22, 1881; 3 daus. Instr. metallurgy, 1879-83, asst. prof., 1883-86, prof. same, 1886-88, prof. mineralogy, 1888-90, prof. archaean and applied geology, 1890-92, prof. geology, 1892-1903, pres., 1903-—, U. of Wisconsin, Non-resident prof. structural geology, U. of Chicago, 1892-1903; mem. geologic br. U. S. Geol. Survey, 1883-—; geologist in charge of Div. of Pre-Cambrian and Metamorphic Geology, same, 1900-08, and cons. geologist, same, 1909-15; cons. geologist, Wis. Geol. and Natural History Survey, 1897-1903; mem. Nat. Conservation Commn., 1909; chmn. Wis. State Conservation Commn., 1908-15; chmn. Bd. of Arbitration in Controversy between Eastern Railroads and Brotherhood of Locomotive Engrs., 1912; trustee Carnegie Foundation for Advancement of Teaching, 1909-—. Author: Correlation Papers: Archean and Algonquin, 1892; A Treatise on Metamorphism, 1904; The Conservation of the Natural Resources of the United States, 1910; Concentration and Control—A Solution of the Trust Problem in the United States, 1912. Joint author: Penokee Iron Bearing Series of Michigan and Wisconsin; The Marquette Iron Bearing District of Michigan; The Menomonee Iron Bearing District of Michigan; Geology of the Lake Superior Region. Studied influence of mineral resources on civilization, mineral geology middlewestern U. S. Died Milw., Nov. 19, 1918.

VAN HORN, Frank Robertson, Am. geologist, mineralogist; b. Johnsonburg, N.J., Feb. 1872; s. George W. and Ellen J. (Robertson) V.; State Model Sch., Trenton, 1886-88; B.S., Rutgers, 1892, M.S., 1893, D.Sc., 1919; U. of Heidelberg, 1893-97, Ph.D., 1897; m. Myra Van Horn, June 8, 1898; children—Kent R., Hilda L. Instr. in mineralogy, Rutgers, 1892-93; instr. in geology and mineralogy, 1897-99, asst. prof., 1899-1902, prof., 1902-—, Case Sch. Applied Science, Cleveland. Fellow Geol. Soc. America (librarian 1913-18), Mineral Soc. America (sec. 1923-—), A.A.A.S. Author: Lecture Notes on Systematic Zoology, 1902; Lecture Notes on General and Special Mineralogy, 1903; Geology and Mineral Resources of the Cleveland District, Ohio, 1931. Research with X-ray techniques on pyrite or marcasite concentrations and tungsten minerals. Died Aug. 1, 1933.

VAN HORN, Kent Robertson, Am. metallurgist; b. Cleve., July 20, 1905; s. Frank Robertson and Myra (Van Horn) Van H.; B.S., Case Sch. Applied Sci., 1926; M.S., Yale, 1928, Ph.D. (Sterling Research fellow), 1929; D. Sc., Case Inst. Tech., 1955; married Estelle Yost, July 23, 1932; children—Karl Robertson, Neil Yost, Research metallurgist Aluminum Co. Am., 1929-62, dir. research, 1952-62, v.p., 1962-—. Lectr. metallurgy Case Inst. Tech., 1930-—. Mem. minerals and metals adv. bd. NRC. Recipient St. Claire Deville medal French Soc. Metallurgy, 1965. Mem. Am. Soc. Metals (pres. 1944-45, hon. mem.), Am. Inst. Mining and Metall. Engrs. (exec. com. Inst. of Metals div. 1940-43, Iron and Steel div. 1936-39), Brit. Inst. Metals, Am. Indsl. Radium and X-Ray Soc. (nat. dir. 1943-44, pres. 1944-45), Am. Soc. Testing Materials, Soc. for Non-Destructive Testing (honorary), Sigma Xi. Author: Practical Metallurgy (with George Sachs), 1940; Aluminum In Iron and Steel (with George Case), 1953. Contbr. to tech. jours. Pioneer in indsl. application of radiography and X-ray diffraction to research and comml. metall. problems, prodn. quality control; inventor aluminum alloys and fabricating processes in field of forgings; pioneer in orgn. of large scale indsl. research in U. S.; organized large light metal labs. Home: 373 Fox Chapel Rd., Pitts. 15238. Office: Alcoa Bldg., Mellon Sq., Pitts. 15219.*

VAN HOUTEN, Franklyn Bosworth, Am. geologist; b. N.Y.C., July 14, 1914; s. Charles Nicholas and Hessie (Bosworth) Van H.; B.S., Rutgers U., 1936; Ph.D., Princeton, 1941; m. Jean Oliver Sholes, Feb. 18, 1943; children—Jean, Bosworth, David. Instr., Williams Coll. 1939-42; faculty Princeton, 1946-—, now prof. geology; geologist U. S. Geol. Survey, 1947-—. Cons. to oil cos., 1954-—. Mem. Geol. Soc. Am., Am. Assn. Petroleum Geologists, Internat. Assn. Sedimentologists, Sigma Xi. Research, numerous publs. on clay minerals in soils and sedimentary rocks, problems origin nonmarine redbeds, iron oxides in sedimentary rocks, Cenozoic sedimentary rocks in Western U. S.; Cenozoic geology, Magdalena Valley, Colombia; interpreted history ancient saline lakes. Home: 176 Western Way, Princeton, N.J. 08540.*

VAN HOVE, Léon Charles Prudent, physicist; b. Brussels, Belgium, Feb. 10, 1924; s. Achille Romain and Alice (Spanoghe) Van H.; Ph.D. in Math., U. Libre Bruxelles, 1946; m. Jenny Jacquemain, July 20, 1946; 1 son, Michel André. Asst., Université libre de Bruxelles, 1945-49, 50-52; mem. Inst. for Advanced Study, 1949-50, 52-54; prof. theoretical physics, dir. Inst. for Theoretical Physics, U. Utrecht (Netherlands), 1954-61, asso. prof., 1964-—; dir. dept. theoretical physics European Orgn. for Nuclear Research, Geneva, Switzerland, 1961-—. Recipient Prix Francqui, Belgium, 1958; Heineman prize for math. physics Am. Phys. Soc./Am. Inst. Physics, 1962. Mem. Dutch Phys. Soc. Author: (with N. M. Hugenholtz, L. P. Howland) Problems in Quantum Theory of Many Particle Systems, 1961; also numerous articles. First proof that pressure of a thermodynamic system cannot decrease for increasing density; demonstrated that divergencies of quantum field theory imply occurrence of inequivalent representations of commutation relations, research on gen. theory of neutron scattering by solids and liquids, new derivation of approach of quantum systems to thermo dynamic equilibrium; research on high energy scattering. Office: CERN, 1211 Geneva 23, Geneva, Switzerland.*

VAN INGHEN, Marsilius, philosopher, astronomer, physicist; b. Ingen or Nijmegen, Netherlands; ed. English Nation of U. Paris. Rector, U. Paris, 1367-71, proctor English Nation, 1373-75; represented univ. at Ct. of Avignon, France. Discussed gravity; adopted theory of latitudes and longitudes by Oresme, mech. views of Buridan and Albert of Saxony; believed Aristotelian doctrine of impossibility of vacuum; derived his astron. theories from Bernard of Verdun, Giles of Rome. Died Heidelberg, Germany, Aug. 20, 1396.

VAN ITALLIE, Theodore Bertus, Am. physician; Hackensack, N.J., Nov. 8, 1919; s. Dorus Christian and Lucy M. (Pohle) Van I.; diploma Deerfield Acad., 1937; B.S., Harvard, 1941; M.D., Columbia, 1945; m. Barbara Cox, Sept. 25, 1948; children—Lucy M., Theodore Bertus, Christina M., Elizabeth B., Katharine R. Dir. Nutrition and Metabolism Research Lab. St. Luke's Hosp., N.Y.C., 1952-55; asso. Peter Bent Brigham Hosp., Boston, 1955-57; dir. medicine St. Luke's Hosp. Center, 1957-—; asst. prof. Sch. Pub. Health Harvard, 1955-57; asso. clin. prof. medicine Columbia, 1957-65, clin. prof., 1965-—. Diplomate Am. Bd. Internal Medicine. Fellow A.C.P.; mem. Am. Soc. Clin. Investigation, Am. Inst. Nutrition, Soc. Exptl. Biology and Medicine, Am. Clin. and Climatological Assn. Research, numerous publs. on pancreatic hormone, glucagon, mechanism of energy balance regulation, treatment of pruritus and hypercholesteremia in biliary cirrhosis, physiology and clin. use of medium chain triglyceride, cholesterol control with cholestyramine and polyunsaturated fats.*

VAN KRANENDONK, Jan, theoretical physicist; b. Delft, Netherlands, Feb. 8, 1924; s. Anthonius G. and Elisabeth (Dubois) Van K.; Ph.D., U. Amsterdam (Netherlands), 1952; m. Ruth Mechanicus, July 14, 1951; children—Olga, Daniel, Martin. Research fellow Harvard, 1953; lectr. Inst. Lorentz, U. Leiden (Netherlands), 1954-58; faculty U. Toronto (Ont.) Can., 1958-—, prof. physics, 1960-—. Recipient Steacie Meml. Fund prize for natural scis., 1964. Fellow Royal Soc. Can.; mem. Netherlands, Am. phys. socs., Canadian Assn. Physicists. Research, publs. on theory pressure-induced infrared absorption, theory Raman line broadening, theory nuclear spin-lattice relaxation, theory infrared and Raman spectrum solid hydrogen. Home: 51 Lower Links Rd., Willowdale, Ont. Office: Dept. Physics, U. Toronto, Toronto, Ont., Can.*

VAN LANSBERGEN, Philip, Dutch mathematician, astronomer; b. Goes, Holland, 1561; minister, Anvers, France, until 1585; pastor, Goës; lived at Middleburg, Holland, until his death. Author: Progymnasmata astronomiae restitutae; Uranometriae Libri III; Triangulorum geometricorum Libri IV; Commentationes in motum terrae diurnum et annum, . . . (supports Copernican system), 1630; also astron. tables. Died 1632.

VANLERENBERGHE, J., French physiologist; b. Lille, France, Nov. 10, 1921; s. Charles and Alice (Vanadruel) V.; Docteur en médecine, Faculté de Médecine, Lille, 1948; Docteur ès sciences, Faculté des Sciences, Lille, 1958; m. Francine Rotru, July 3, 1948; children—Brigitte, Edith. With Faculté de Médecine, Lille, 1953-—, titular prof. physiology, 1965-—; became dir. Institut Régional d'Education Physique de Lille, 1954; named chief service in exptl. and applied physiology Centre Anticancéreux de la Région du Nord, 1962. Decorated Chevalier de l'Ordre National du Mérite, Officier des Palmes Académiques, Chevalier de la Santé Publique. Mem. Assn. French-speaking Physiologists, Soc. Pharmacology and Therapeutics, Lille Biol. Soc. Research and numerous publs. on biliary functions, metastatic diffusion of malignant tumors in rat. Home: 21 Avenue Louise Michel, Lille 59, France.*

VAN LOOM, Joost (or Jodocus Lommius, Joost Lomm), Dutch physician; b. Buren, Holland, 1500. Practiced medicine at Tournai and Brussels. Author: Observationes Medicinales (an accurate description of many diseases), 1560; also other studies. Died circa 1562.

VAN LOPIK, Jack R(ichard), Am. geologist; b. Holland, Mich., Feb. 25, 1929; s. Guy M. and Minnie (Grunst) Van L.; B.S., Mich. State U., 1950; M.S., La. State U., 1953, Ph.D., 1955; m. Grace A. Manning, Aug. 10, 1952; 1 son, Charles Robert. Geologist, sect. chief, asst. chief, chief geology br. U. S. Army C.E. Waterways Expt. Sta., Vicksburg, Miss., 1954-61; chief area evaluation sect., tech. dir., mgr. Space and Environmental Sci. Programs, tech. requirements dir. geosciences operations Tex. Instruments, Inc., Dallas, 1961-—; guest lectr. NSF Inst. on Marine Geology Fla. State U., 1956; Ofcl. del. XX Congreso Internacional, Mexico City, 1956, XII Gen. Assembly Internat. Union Geodesy and Geophysics, Helsinki, 1960. Fellow Geol. Soc. Am., A.A.A.S.; mem. Am. Astronautical Soc., Am. Soc. Photogrammetry (dir. Tex.-La. sect., liaison rep. earth scis. div. NRC), Am. Geophys. Union, Am. Assn. Petroleum Geologists, Assn. Am. Geographers, Soc. Econ. Paleontologists and Mineralogists, Sigma Xi. Research in geomorphic/geologic problems in mil. engrng. programs, photogeology and remote sensing, deltaic and arid zone geomorphology and sedimentation, terrain analysis and quantification, lunar and earth-orbiting-satellite exploration. Home: 5752 Brookstown Dr., Dallas 75230. Office: P.O. Box 5621, Dallas 75222.*

VAN MAANEN, Adriaan, astronomer; b. Sneek, Holland, Mar. 31, 1884; s. Johan Willem Gerbrand and Catharina Adriana (Visser) Van M.; B.A., U. of Utrecht, 1906, M.A., 1909, Sc.D., 1911; U. of Groningen, 1909-10; came to U. S., 1911; unmarried. Volunteer asst., Yerkes Obs., Williams Bay, Wis., 1911-12; astronomer, Mt. Wilson Obs., since 1912. Mem. Astron. Soc. America, Astron. Soc. Pacific, Internat. Astron. Union, Soc. Astron. de France, Astr. Gesellschaft, Royal Astr. Soc., Amsterdam Acad., Utrecht Soc., Sigma Xi, etc. Specialized in study parallaxes and proper motions of stars and nebulae, gen. magnetic field of sun, 1st use large reflector to measure stellar parallax; determined motions of galactic clusters and variable stars; proved that nuclear matter flows along arms of spiral nebulae. Died Jan. 26, 1946.

VAN MARUM, Martin, Dutch botanist, physicist; b. Delft, Netherlands, Mar. 20, 1750; M.D., U. Groningen (Netherlands), Ph.D., 1773. Dir. Naturaliën-Kabinet der Hollandsche Maatschappij der Wetenschappen, from 1777; prof. physics, Haarlem, Netherlands. Corr. mem. French Acad. Scis., 1803. Author: Dissertatio de Motu Fluidorum in Plantis, 1773; A Treatise on Electricity, 1776. Studied circulation of sap; research on and inventor gigantic friction machine, 1785. Died Haarlem, Dec. 26, 1837.

VAN MONS, Jean-Baptiste, Belgian pomologist, chemist; b. Brussels, Belgium, 1765. Prof. chemistry École Centrale, Brussels, from 1797, U. Louvain (Belgium), 1817-30. Founder Journal de Chimie et Physique. Mem. French Acad. Scis., 1796, Soc. Agr. Author: Essai sur les principes de la chimie antiphlogistique, 1785; Théorie de la combustion, 1802; Principes d'électricité, 1803; Principes elementaires de chimie philosophique, 1818; Arbres fruitiers, leur culture en Belgique et leur propagation par la graine ou pomologie belge, 1835-36. Expts. on prodn. new varieties fruit, especially pear (produced Beurre Diel); theorized that seedlings of new varieties are more likely to improve than those of old varieties. Died Louvain, Sept. 6, 1842.

VAN MUSSCHENBROEK, Pieter, Dutch physicist; b. Leyden, Mar. 14, 1692; ed. U. Leyden; studied medicine, later physics; visited Eng., 1717, became acquainted with Newton. Prof. physics and math. Duisburg, 1719, later at Utrecht; prof. philosophy, Leyden, 1740-61. Fellow Royal Soc., 1734, French Acad. Sci., 1734; mem. prin. learned instns. of Europe. Author: Physicae experimentales et geometricae dissertationes, 1729; Elementa physicae, 1734; Introductio ad philosophiam naturalem, 1762. Made important discoveries, especially in magnetism and cohesion of bodies; invented pyrometer; discovered principle of Leyden jar, 1746; introduced Newton's ideas into Holland. Died Leyden, Sept. 19, 1761.

VAN NAME, F(rederick) W(arren), Jr., Am. physicist; b. Bklyn., Dec. 14, 1920; s. Frederick Warren and Gladys Evelyn (Griffen) Van N.; B.A., Swarthmore Coll., 1942; M.S., Yale, 1943, Ph.D., 1948; m. Barbara F. Leaper, Mar. 5, 1944; children—Frederick Warren 3d, Theodore Griffen, Patricia Francis; m. 2d, Jeanne D. Haight, Sept. 30, 1964. Mem. faculty Franklin and Marshall Coll., 1948-57, asso. prof. physics, 1953-57, chmn. dept., 1954-57; prof. physics, chmn. dept. U. Del., 1957-62, Pratt Inst., 1962-—. Cons. Harrisburg (Pa.) Polyclinic Hosp., 1953-57. Fellow Am. Phys. Soc., mem. Am. Assn. Physics Tchrs., Am. Inst. Physics, Sigma Xi. Author: Modern Physics, 2d edit., 1962, Polish transl., 1965; Analytical Mechanics, 1958; Elementary Physics, 1966; also papers and book revs. Died 1967.

VAN NIEL, Cornelis Bernardus, microbiologist; microbiology; b. Haarlem, Netherlands, Nov. 4, 1897; s. Jan Hendrik and Geertien Gesiena (Hagen) van N.; Chem.E., Tech. Univ., Delft. Holland, 1922, D.Sc., 1928; D.Sc., Princeton U., 1946, Rutgers U., 1954; LL.D., U. Cal. at Davis, 1968; m. Christina van Hemert, August 17, 1923; children—Ester, Ruth, Jan. Came to U. S., 1928. Asst. in microbiology Tech. U., Delft, 1922-23, conservator, 1923-28; faculty Stanford, 1929-63, prof., 1935-63, Herzstein prof. biology emeritus, 1963-—; vis. prof. microbiology U. Cal., Santa Cruz, 1964-68. Recipient Nat. Medal Science, 1964, Emil Christian Hansen Medal, 1964; Rumford medal, 1967. Fellow Am. Acad. Microbiol. (charter); mem. Deutsche Akademie Der Naturforscher Leopoldina, A.A.A.S., Royal Danish Sci. Soc., Royal Netherlands Acad. Scis., Göttingen Acad. Scis., Am. Soc. Microbiol. (hon.), West Soc., Am. Philos. Soc., Am. Acad. Arts and Scis., Soc. Gen. Microbiology (Gt. Britain) (hon.), Netherlands Soc. Microbiology, Nat., Cal. acads. sci., Am. Soc. Plant Physiol., Nat. Acad. Sci., Am. Soc. Naturalists (hon.), French Soc. de Microbiol. (hon.). Author: The Propionic Acid Bacteria, 1928; The Microbe's Contribution to Biology (with A. J. Kluyver), 1956. Contbr. to sci. jours. Research on photosynthesis; general microbiology; biochemistry of microorganisms. Home: P.O. Box 1833, Carmel, Cal. 93921. Office: Hopkins Marine Station, Pacific Grove, Cal. 93950.

VAN NORT, Leighton, Am. sociologist; b. Phila., Sept. 18, 1930; s. Joseph Sands and Isabelle (Fox) van N.; A.B. with honors and distinction, U. Pa., 1952; A.M., Princeton, 1954. Research asst. Behavioral Research Council, U. Pa., 1951-52; research fellow Office Population Research, Princeton, 1952-55, vis. lectr. sociology, 1959-60; faculty Bowdoin Coll., 1955-60; vis. lectr. Dartmouth, 1960-61, U. Cal. at Berkeley, 1963; demographer, sociologist Dept. State, Washington, 1961-62, officer charge FAO and population affairs, 1962-65, chief div. UN econ. affairs, 1965-—; lectr. Sch. Advanced Internat. Studies, Johns Hopkins, 1968-—. Mem. U. S. delegation UN Asian Population Conf., 1963, 3d-8th sessions World Food Program, 1963-—, Asian Industrialization Conf., 1965, UN Econ. Commn. for Asia and Far East, 1966. Mem. Am. Sociol. Assn., Am. Statis. Assn., Population Assn. Am., Internat. Union for Sci. Study Population. Contbr. articles to profl. jours. Research on role of values in population theory, relation of population changes to econ. devel. Home: 825 New Hampshire Av. N.W., Washington 20037; also Brunswick, Me. Office: Office Internat. Economic and Social Affairs, Dept. State, Washington.*

VAN PATTER, Douglas Macpherson, nuclear physicist; b. Montreal Que., Can., July 4, 1923; B.Sc., Queen's U., 1945; Ph.D. in Physics, Mass. Inst. Tech., 1949; m. 1950; 3 children. Jr. physicist NRC Can., 1945-46; asst. physics Mass. Inst. Tech., 1946-49, research asso., 1949-52; asst. prof. U. Minn., 1952-54; physicist Bartol Research Found., Franklin Inst., Swarthmore, Pa., 1954-—. Chmn. subcom. nuclear constants Nat. Acad. Scis.-NRC, 1959-64. Fellow Am. Phys. Soc. Research in nuclear reactions using electrostatic accelerators, radioactive decay schemes, systematics of level properties. Home: 97 W. Sproul Rd., Springfield, Pa. 19064. Office: Bartol Research Found., Franklin Inst., Whittier Pl., Swarthmore, Pa. 19081.*

VAN PILSUM, John Franklin, Am. biochemist; b. Prairie City, Ia., Jan. 28, 1922; s. John P. and Vera (Moore) Van P.; student Ia. State Coll., 1939-41; B.S., State U. Ia., 1943, Ph.D., 1949; m. Shirley Elaine Newsom, Oct. 14, 1958; children—John Robert, Patricia Mona, Barbara Joyce, Mary Ann, Elisabeth Joan. Instr. biochemistry L.I. Coll. Medicine, Bklyn., 1949-51; asst. prof. U. Utah Coll. Medicine, Salt Lake City, 1951-54; asst. prof. biochemistry U. Minn. Coll. Medicine, Mpls., 1954-63, asso. prof., 1963-—. Mem. Am. Chem. Soc., Am. Soc. Biol. Chemists, Sigma Xi. Research, publs. on metabolism of optical isomers of essential amino acids, Guanidinium compounds, hormonal regulation of Guanidinium compound metabolism. Home: 4356 Leander Lane, Mpls. 55421.*

VAN PRAAG, Herman Meir, Dutch psychiatrist; b. Schiedam, Netherlands, Oct. 17, 1929; s. M. M. and Ch. (Leverpoll) Van P.; ed. medicine State U. Leyden (Netherlands), 1948-56, Found. for Advanced Clin. Edn., Rotterdam, Netherlands, 1958-66; m. Cornelia Eikens, Nov. 17, 1956; children—Marinus Cornelius Gideon, Gideon Herman, Charlotte Adriana. Sr. sci. officer dept. psychiatry Dijkzigt Hosp., Rotterdam, 1963-66; head div. biol. psychiatry Psychiat. U. Clinic, State U. Groningen (Netherlands), 1966-—. Mem. Royal Netherlands Soc. for Advancement Medicine, Netherlands Soc. Psychiatry and Neurology (Ramaer medal 1965), Dutch Interdisciplinary Soc. Biol. Psychiatry (founder, 1st pres.), Soc. Biol. Psychiatry (U. S.). Author: A Critical Investigation of the Importance of Monoamine Oxidase Inhibition as a Therapeutic Principle in the Treatment of Depression, 1962; Psychopharmaca, 1966; also numerous articles. Research on biol. psychiatry especially in state of depression; clin. research for new psychotropic drugs. Home: 19 Maluslaan, Groningen, Netherlands.*

VAN REEN, Robert, Am. biochemist; b. Paterson, N.J., June 12, 1921; s. Cornelius and Martha (Miller) Van R.; A.B., N.J. State Coll., 1943; postgrad. N.Y. U.; Ph.D., Rutgers U., New Brunswick, N.J., 1949. Research fellow Rutgers U., 1947-49; asso. biochemist Brookhaven Nat. Lab., Upton, N.Y., 1949-51; research asso. Johns Hopkins, 1951-53, asst. prof., 1953-56; supervisory chemist Naval Med. Research Inst., Bethesda, Md., 1956-61, head nutritional biochemistry, 1961-—. Alternate mem. Interdepartmental Com. on Nutrition for Nat. Def., 1961-63, mem., 1963-65; lab. dir., Vietnam Survey, 1959; Navy liaison rep. to nutrition study sect., research grants div. NIH, 1964-—. Recipient McLester award Assn. Mil. Surgeons, 1959. Mem. Am. Soc. Biol. Chemists, Am. Inst. for Nutrition, Am. Chem. Soc., Internat. Assn. for Dental Research. Research, numerous publs. in normal and abnormal calcification, mineral metabolism, role of diet in vesical calculus disease in Thailand, Pakistan, Egypt. Home: 3900 Watson Pl. N.W., Washington 20016. Office: Navy Med. Research Inst., Bethesda, Md. 20014.*

VAN RHIJN, Pieter Johannes, Dutch astronomer, astrophysicist; b. Gouda, Netherlands, Mar. 24, 1886; Ph.D., U. Gröningen (Netherlands), 1915. Staff mem. Mt. Wilson Obs., 1912-14; asst. to Kapteyn, Astron. Lab., from 1914; dir. lab., prof. U. Gröningen, from 1921. (With Kapteyn) concluded elaborate investigation of number of stars (estimating total in stellar system at 30,000 million), 1925.

VAN ROOMEN, Adriaen, mathematician; b. Louvain, Brabant, Sept. 29, 1561; studied medicine and math. in Italy and Germany. Prof., U. Louvain; prof. math., U. Würzburg; royal mathematician, Poland. Author: Ideae Mathematicae pars prima, seu Methodus polygonium, 1593; Treatment of the Circle by Archimedes, 1597; Spherical Triangles, 1609. Gave value of pi to seventeen decimal places. Died Mainz, Germany, May 4, 1615.

VAN ROSSUM, Herman, Dutch mathematician; b. Amsterdam, Netherlands, July 16, 1917; s. Wilhelm Albertus and Antonetta (Gobel) Van R.; Doctor's degree in math. cum laude, Utrecht U., 1953; m. Julie Stoutjesdyk, Dec. 21, 1943; children—Paul Herman, Else Julie. Tchr. high sch., 1943-56; lectr. math. Amsterdam U., 1956-61, sr. lectr. applied math., 1964-—; vis. asso. prof. Michigan State U., 1961-62. Author: Mechanica, 1956; A Theory of Orthogonal Polynomials Based on the Padé Table; 1953; (with Boland and Oort) Lineaire Algebra Vraagstukken, 1965. Research, publs. on continued fraction theory, orthogonal polynomials, approximation theory, moment problems. Home: 92 Winkerlaan, Utrecht, Netherlands. Office: 30 Plantage Muidergracht, Amsterdam, Netherlands.*

VAN RYSSELBERGHE, Pierre, phys. chemist; b. Brussels, Belgium, May 18, 1905; s. Francois and Jeanne (Alvin) Van R.; Ingenieur, U. Brussels, 1927; M.A., Stanford, 1928, Ph.D., 1929; m. Lily Laura Chloupek, Sept. 2, 1930; children—Jane Frances (Mrs. Enrico Bernasconi), Pierre Louis, Jacqueline Grace. Came to U. S., 1927, naturalized, 1941. Instr., Stanford, 1929-31, asst. prof., 1931-41, lectr. in chemistry and chem. engring., 1956-—; faculty U. Ore., Eugene, 1941-56, prof., 1945-56. Vis. lectr. to Belgium, Belgian Am. Ednl. Found., 1935-36; Fulbright lectr., Italy, 1950-51; pres. Internat. Com. Electrochem. Thermodynamics and Kinetics, 1949-54, chmn. commn. on electrochem. nomenclature and definitions, 1950-—; mem. commn. on electrochemistry Internat. Union Pure and Applied Chemistry, 1953-61, chmn., 1961-67. Fellow A.A.A.S., Am. Inst. Chemists, N.Y. Acad. Scis.; mem. Am. Chem. Soc., Electrochem. Soc., Nat. Assn. Corrosion Engrs., U. Brussels Engrs. Author: (with T. De Donder) Thermodynamic Theory of Affinity, 1936; Electrochemical Affinity, 1955; Thermodynamics of Irreversible Processes, 1963; also many articles. Research on mixing effects in solutions of electrolytes, applications of polarography in corrosion, applications of thermodynamics to aspects of electrochemistry. Home: 551 Santa Rita Av., Palo Alto, Cal. 94301. Office: Stanford U., Stanford, Cal. 94305.*

VAN SANDE, Marc Henry, Belgian biochemist; b. Brasschaat, Belgium, May 30, 1927; s. Remy and Maria (Helskens) Van S.; Docteur ès Sciences, U. Poitiers (France), 1962; Chem. Engr., Institut Meurice-Chimie, Brussels, Belgium, 1950; m. Helena van Lier, Feb. 8, 1954. Biochemist, Inst. Tropical Medicine, Antwerp, Belgium, 1950-65, sci. collaborator Technicon, Inc., Brussels, 1965-—; neurochemist Institut Bunge, Berchem-Antwerp, part-time, 1952-—, cons. neurochemist Service de Neurologie, 1952-—. Named Chevalier de la Couronne, 1965; recipient Lauréat prix Matthis, 1963, Laureat prix Broden-Rodhain, 1965. Mem. Belgian Soc. Indsl. Chemistry, Belgian Soc. Tropical Medicine, Belgian Soc. Clin. Chemistry, Belgian Soc. Clin. Research, N.Y. Acad. Sci., Internat. Soc. Neurochemistry. Research, numerous articles on proteins in different physiol. fluids in normal and path. conditions, enzymes in human brain, amino-acids and lipids in nervous system; set up method for differential diagnosis of neurol. diseases by means of electrophoresis in agar-gel. Home: 40, Donkseinde, Brasschaat, Antwerp. Office: Institut Bunge, F. Williotstreet, Berchem, Antwerp, Belgium.*

VAN SCHOOTEN, Frans, Jr., Dutch mathematician; b. 1615; prof. math.; tchr. of Huygens, Hudde, Sluze; wrote on perspective; editor Vieta's works, also Latin edit. of La géométrie (Descartes). Died 1661.

VAN SCOTT, Eugene Joseph, Am. physician; b. Macedon, N.Y., May 27, 1922; s. Charles J. and Eleanor (DuPré) Van S.; student U. Mich., 1941-44; B.S., U. Chgo., 1945, M.D., 1948; m. Mary Louise Yox, Oct. 9, 1948; children—Christopher J., Stephen J., David J. Asso. dermatology U. Pa., Phila., 1952-53; head dermatology service gen. medicine br. Nat. Cancer Inst., NIH, Bethesda, Md., 1953-61, chief dermatology br., 1961-—, sci. dir. gen. labs. and clinics, 1965-—. Recipient Mr. and Mrs. J. N. Taub Internat. Meml. award for psoriasis research, 1964; James Clarke White award for achievement in dermatologic medicine, 1965; Pub. Health Service Meritorious Service award, 1966. Mem. Am. Acad. Dermatology, A.A.A.S., Am. Assn. for Cancer Research, Am. Dermatol. Assn., A.M.A., Am. Soc. for Clin. Investigation, Am. Soc. for Exptl. Pathology, Am. Soc. Dermatopathology, Soc. for Investigative Dermatology, Washington Dermatol. Soc., Washington Soc. Pathologists. Research, publs. on cell biology, dermatology, cancer including kinetics of cell populations in normal tissues, diseases of the epidermis and cancerous conditions of the skin; determined treatment benefits of anti-metabolic drugs on cancer and diseases of the skin. Home: 7009 Garrett Rd., Derwood, Md. 20855. Office: Nat. Cancer Inst., NIH, Bldg. 31, Bethesda, Md. 20014.*

VANSHIN, Aleksandr Leonidovich, Russian geologist; b. Mar. 28, 1911; grad. Moscow Geol. Survey Inst., 1923. Staff mining-geol. dept. Sci. Inst. for Fertilizer, 1929-36; staff Geol. Instr., USSR Acad. Sci., 1936-56, chmn. dept. on regional tectonics, 1956-—, bur. mem. dept. geology and geog. sci., 1960. Mem. USSR Acad. Sci. Recipient A.P. Karpinskii prize. Author: The Geology of the Northern Aral Region, 1953; The Geology, Geophysics and Geography of Siberia, 1960. Research and publs. in tectonics, stratigraphy, lithology, hydrogeology of Western Kazakhstan and So. Urals; assisted in compilation of tectonics maps of USSR; worked out stratigraphy of tertiary sedimentation of Avalo-Turgaisk depression, theory of correlation of plicate formations; discovered indsl. deposits of lignites, iron ores, bauxite, phosphorite, potassium salt, cement raw materials, series of Artesian basins. Home: Leninskii Prospekt 25. Office: Dept. Regional Tectonics, USSR Acad. Sci. Inst. Geology, Pyzhevskii, Pereulok 7, Moscow, USSR.

VAN SIEBOLD, Philipp Franz, German physician, biologist; b. Würzburg, Germany, 1796; grad. U. Würzburg, 1820. Practice medicine; commd. maj. Army of Dutch E. Indies; health officer Java and Japan, 1823-30; entered service of shogun, 1861; became mediator between Japan and European countries, 1859; became adviser to lt. gen. of Java, 1862. Author: Fauna japonica, 1833-51; Nippon, 1832-51; Flora japonica, 1835-70. Research and publs. on Japanese flora, fauna, lang.; his collections in natural history and ethnography are now in Mus. of Leyden; introduced tea plant to Europe. Died Munich, 1866.

VAN SLYKE, Donald D., Am. biol. chemist; b. Pike, N.Y., Mar. 29, 1883; s. Lucius L. and Lucy (Dexter) Van S.; A.B., U. of Michigan, 1905; Ph.D., 1907, Sc.D., 1935; studied U. of Berlin, 1911; Sc.D., Yale, 1925; M.D., U. of Oslo, 1938; Sc.D., Northwestern University, 1940, University Chicago, 1941, University of London, 1951; M.D. (honorary), University of Amsterdam, 1962; m. Rena Mosher, June 24, 1907 (dec.); children—Elsa, Karl Keller; m. 2d, Else von Bardenfleth Brock Aug. 1948. Research chemist, Rockefeller Inst., since 1907; chief chemist at hosp. for same, 1914-48, emeritus; asst. dir. in chge. research in biology and medicine, Brookhaven Nat. Lab., 1948-51, research chemist, 1951—; counselor Eli Lilly Research Grants, 1951-56; visiting professor at Peking Medical Sch., China, 1922. Pres. Am. Bureau for Med. Aie to China. John Phillips Meml. Award, Am. Coll. Physicians, 1954, 57, Donald D. Van Slyke award Am. Soc. Clin. Chemistry, 1957; Franklin medal 1965. Mem. of American Chem. Soc., Biol. Chemists (pres. 1921-22), Harvey Soc. (pres. 1927-28), N.Y. Acad. Medicine, Assn. Am. Physicians, Nat. Acad. Science, Acad. Science of India (honorary), Royal Danish Academy of Sciences and Letters, Brit. Physiol. Soc. (hon.), Swedish Royal Acad. Science, Accademia Medica Lombarda (hon.), Societa Italiana di Biologia Sperimentale (hon.), Société de Biologie Chimique (France), Société de Pathologie Renale, Sigma Xi, hon. member Renal Association (London), Physiological Society (Britain). Recipient Conné medal, Assn. Am. Physicians, 1937; Willard Gibbs Medal, 1939; Kober medal, 1942; Mickle fellowship, 1936; Order of Jade (China); Am. Chem. Soc. award, 1953; Sci. Achievement award, AMA, 1962. Contributor articles in American and foreign jours., Author: Factors affecting distribution of electrolytes, water, and gases in the animal body, 1926; Micromanometric Analyses, 1961; co-author (with J. P. Peters) Quantitative Clinical Chemistry (2 vol.), 1931-32; (with C. Lundsgaard) Cyanosis, 1923; (with E. Stillman and others) The Course of Bright's Disease. Developed diagnostic tests including test for acidosis in diabetes; research on gases and electrolytes of blood; chemistry of proteins and protein derivitives and their role in physiology and pathology, enzyme action, blood chemistry, and metabolic conditions of diabetes and nephritis. Home: Belle Terre, Port Jefferson, N.Y. Office: Brookhaven Nat. Lab., Upton, N.Y. 11973.*

VAN SLYKE, Lucius Lincoln, Am. chemist; b. Centerville, N.Y., Jan. 6, 1859; s. William J. and Katherine (Keller) V.; A.B., U. of Mich., 1879, A.M., 1881, Ph.D., 1882; student and fellow by courtesy in chem. lab., Johns Hopkins, 1889-90; m. Lucy W. Dexter, June 15, 1882 (died 1885); children—Donald Dexter, Carl Osborne; m. 2d, Julia Hanford Upson, Apr. 5, 1888 (died 1924); 1 son, Lawrence Prescott; m. 3d, Mrs. Hedwig Sheul, June 2, 1926. Asst. chem. lab., U. of Mich., 1882-85; prof. chemistry, Oahu Coll., Honolulu, and govt. chemist, H.I., 1885-88; lecturer on gen. chemistry, U. of Mich., 1888-89; chief research chemist, New York Agrl. Experiment Station, 1890-1929, also prof. dairy chemistry, N.Y. State Coll. Agr., Cornell Univ., 1920-29 (emeritus). Fellow A.A.A.S.; pres. Assn. Official Agrl. Chemists, 1900, N.Y. State Dairymen's Assn., 1897. Author: Modern Methods of Testing Milk and Milk Products, 1906; Science and Practice of Cheese Making (with C. A. Publow); Fertilizers and Crops, 1911; Cheese (with W. V. Price), 1927. Research on chemistry of dairy products, especially components of milk and their inter-relationships. Died Sept. 30, 1931.

VAN SOEST, Johannes Leendert, Dutch engr.; b. The Hague, Netherlands, Oct. 13, 1898; s. L. W. and J. (Smit) van S.; Electrotech. engr., Delft Inst. Tech., 1925; Hon. Dr. degree U. Utrecht (Netherlands), 1958; m. J. M. L. Wytman, June 6, 1950 (dec.); 1 dau., Viola J. (Mrs. J. Zeeman). Engr., Com. Phys. Warfare, Govt. of Netherlands, 1927-40; chief engr. Nat. Def. Research Orgn., 1945-48, dir. physics lab., 1948-57, dir. research, 1957-63, adviser, 1963—; asso. prof. information theory Delft Inst. Tech., 1949-64. Decorated knight Order Netherlands Lion, 1965; officer Order Orange Nassau, 1948. Fellow I.E.E.E., Internat. Acad. for Art and Sci.; hon. mem. Netherlands Electronic and Radio Soc., Royal Netherlands Bot. Soc.; mem. Royal Soc. Diligentia (pres. 1959—), Royal Inst. Engrs. Author: (with J. Doekse, P. Jansen, A. A. Kruyne, G. J. Vervelde) Grassen en Granen, 1951; also numerous articles. Research in directional hearing, illumination problems, information theory and cybernetics, noise phytogeography, flora of Netherlands, taxonomical studies. Home: van Soutelandelaau 35, The Hague, Netherlands.*

VAN STEENIS, Cornelius Gijsbert Gerrit, Dutch botanist; b. Utrecht, Holland, Oct. 31, 1901; s. Hendrik Jan and Louise (van Vuuren) Van S.; M.Sc., Utrecht U., 1925, Ph.D., 1927; hon. doctorate U. Montreal, 1959; m. Maria Johanna Kruseman, Nov. 8, 1927; children—Hein, Liesbet Marieke. First asst. Bot. Mus. and Herbarium, Utrecht U., 1926-27; botanist in herbarium Botanic Gardens, Bogor, Java, 1927-40; dir. Found. Flora Malesiana, Leyden, Holland, 1950—; extraordinary prof. tropical botany U. Amsterdam, 1951-62, U. Leyden, 1954-62; prof. systematic botany, dir. Rijksherbarium, Leyden, 1962—. Decorated Ridder in de Orde van de Nederlande Leeuw. Mem. Societas phytogeagrafika Suecica, Physiografiska Selskapet, Linnean Soc., Am. Bot. Soc. Author: Origin of the Malaysian Mountain Flora, 1934, 36; Maleise Vegetatieschetsen, 1934; Flora Malesiana, 1949; Flora voor de Scholen van Indonesië, 1949; Pacific Plant Areas, 1963; Specific and Infraspecific Delimitation, 1957; Landbridge Theory in Botany, 1962; also numerous articles. Research of affinities and composition of tropical flora of Indonesia and adjacent countries and its vegetation; prins. of vegetation description; devel. of flora in geol. time; modes of speciation. Home: 55 Marelaan, Oegstgeest, Holland. Office: 6 Schelpenkade, Leyden, Holland.*

VANSTRUM, Paul Ross, Am. chem. engr.; b. Mpls., Aug. 3, 1920; s. Paul August and Ruth (Ross) V.; B.Chem.Engring., U. Minn., 1942; m. Kathleen W. Johnson, Sept. 26, 1942; children—Marilyn K., Richard W., Paul W. Staff, Union Carbide Corp., 1942; chem. engr. Prest-O-Lite Co., 1942-44; design engr. SAM Labs., Columbia U., 1944; with Oak Ridge Gaseous Diffusion Plant, 1944—, supt. barrier devel. dept., 1953-54, tech. and adminstrv. supt. research and process devel. depts., 1954-60, dep. dir. research and devel., 1960—. Recipient Ernest Orlando Lawrence Meml. award U. S. AEC, 1966; U. Minn. Outstanding Achievement award, 1966. Mem. A.A.A.S., Sci. Research Soc. Am. Tech. adminstrn. in devel. isotopic separation processes for uranium including corrosion chemistry, gas-solids reactions, metallurgy, fluid mechanics, turbomachinery. Home: 104 Ogden Lane. Office: Oak Ridge Gaseous Diffusion Plant, P.O. Box P, Oak Ridge 37830.*

VAN SWIETEN, Gerard, physician; b. Leyden, Netherlands, May 7, 1700; ed. Leyden; pupil of Boerhaave; M.D., 1725; prof. medicine, Leiden, forced to resign because of Catholic faith; became 1st physician to Maria Theresa of Austria, 1745; apptd. supt. Imperial Library; reorganized Wiener Medizinische Fakultät (Vienna med. sch.), became dir.-gen., censor, perpetual pres., 1st to lecture on physiology and pathology, separated surgery from anatomy, emphasized study of anatomy, promoted exact observation and diagnosis, established bot. garden, chem. lab., improved teaching of barbers and midwives. Author: Commentaria in Hermanni Boernaave aphorismos de cognoscendis et curandis morbis, 4 vols., 1745-72, Index, 1776. Died Vienna, June 18, 1772.

VAN SWINDEN, Jan Hendrick, Dutch physicist, mathematician; b. The Hague, Netherlands, 1746; ed. U. Leiden (Netherlands); m. Sara Riboulot; 1 son, 3 daus. Prof. physics and astronomy, Amsterdam, from 1785; held several governmental offices; corr. mem. French Acad. Scis. Author: Observations on the Regular Variations of the Magnet Needle, 1777; Recueil des mémoires sur l'analogie de l'éléctricité et du magnetisme, 1784; Treatise on Weights and Measures, 1802. Died 1823.

VAN'T HOFF, Jacobus Hendricus, chemist; b. Rotterdam, Holland, Aug. 30, 1852; ed. Polytechnic, Delft, Netherlands, U. Leyden (Netherlands), Bonn, Germany, École de Médecine, Paris; doctorate, Utrecht, 1874; Became lectr. physics Vet. Sch., Utrecht, 1876; named prof. chemistry Mineralogy and geology Amsterdam U., 1878, Liepzig, Germany, 1887, Berlin, 1896; leader in founding Inst. Phys. Chemistry, Amsterdam, 1888. Recipient 1st Nobel prize in chemistry, 1901. Fellow Royal Soc. (Davy medal with LeBel 1893), 1897; corr. mem. French Acad. Scis., 1905. Author: La chimie dans l'espace, 1874; Ansichten über die organische Chemie, 2 vols., 1878-83; Etudes de dynamique, 1884; Lois de l'équilibre chimique dans l'étate dilué, 1885; Vorlesungen über theoretischen und physikalischen Chemie, 3 vols., 1898-1900; Leçons de chimie physique, 1918. A father of phys. chemistry; research on formation and decomposition of double salts; related (independently of Le Bel) optically active carbon compounds to 3 dimensional and asymmetrical molecular structure (led to devel. stereochemistry), 1874; 1st work relating thermodynamics to chem. reactions; developed law relating chem. equilibrium to temperature, method for determining order of reaction; discovered law of melting point and lowering of steam pressure in solutions; demonstrated that partial pressure of dissolved substances is identical to that in gas mixtures and related it to theory of electrolytic dissociation; introduced concept of chem. affinity as maximum work available as result of reaction and developed calculations for it. Died Berlin, Mar. 1, 1911.

VAN UDEN, Nicolaas Johannes, microbiologist; b. Venloo, Netherlands, Mar. 5, 1921; s. Adriaan and Cornelia (Baaijens) Van U.; Dr.Med., U. Vienna (Austria), 1948; m. Adelaide de Braganea, Oct. 17, 1945; children—Adiaan, Nuno, Francisco, Filipa, Miguel, Teresa. Free asst. Portuguese Inst. Cancer Research, 1949; free asst. Faculty Scis. U. Lisbon (Portugal), 1951, investigator, 1957-64; dir. Lab. Microbiology, Gulbenkian Inst., Sci., Oeiras, Portugal, 1964—; mem. Microbiol. Expdn., E. Africa, 1958-59; vis. scientist Inst. Marine Sci., Miami, Fla., 1960. Gulbenkian fellow Scripps Instn. Oceanography, La Jolla, Cal., 1960. Mem. Mycol. Soc. Am., Internat. Soc. Med. Mycology, Soc. for Gen. Microbiology. Research, publs. on yeast taxonomy including descriptions of new species, yeast ecology and their distbn. in wild mammals and birds, in Pacific and Atlantic waters and marine animals, kinetics of population growth, thermobiology of yeasts. Home: 8 Rua Vasco da Gama, Trafaria, Portugal. Office: Lab. Microbiology, Gulbenkian Inst. Sci., Oeiras, Portugal.*

VANUXEM, Lardner, Am. geologist; b. Phila., Pa., July 23, 1792; s. Jame and Rebecca (Clarke) V.; grad. École des Mines, Paris, France, 1819; m. Elizabeth Newbold, 1830. Prof. chemistry and mineralogy S.C. Coll. (now U. S.C.), 1819-27, made geol. surveys of N.C., S.C.; made geol. surveys of N.Y., Ohio, Ky., Tenn., Va., 1827-30; bought farm nr. Bristol, Pa., 1830; assigned to 3d and 4th dists. for geol. surveys of N.Y., 1836-42; founder Assn. Am. Geologists and Naturalists (now A.A.A.S.), 1840. Mem. group which established 1st uniform geol. nomenclature; held advanced views on creation, evolution, emancipation of women; studied phrenology. Died Bristol, Pa., Jan. 25, 1848.

VAN VALEN, Leigh, Am. evolutionist; b. Albany, N.Y., Aug. 12, 1935; s. Donald and Eleanor (Williams) Van V.; B.A., Miami U., Oxford, O., 1956; M.A., Columbia, 1957, Ph.D., 1961; m. Phebe Hoff, Sept. 12, 1959. Head, Grad. Genetics Lab., Columbia, 1957-59, Boese fellow, 1961-62; NATO fellow U. Coll., London, Eng., 1962-63; research fellow Am. Mus. Natural History N.Y.C., 1963-66; asst. prof. anatomy U. Chgo., 1963-66; research asso. Zoller Dental Clinic, 1967—. Mem. Soc. for Study Evolution, Soc. Vertebrate Paleontology, Genetics Soc. Am., Ecol. Soc. Am., Biometric Soc. Research, numerous publs. on early radiation placental mammals, with erection new order Deltatheridia and evidence for origins many orders; revision Insectivora; co-describer and discoverer earliest known primates and ungulates; developed new method for measurement natural selection; research on maintenance genetic variation, growth fields in mammalian dentition, evolution mammalian communities including extinction multituberculates and dinosaurs.*

VAN VLECK, John Hasbrouck, Am physicist; b. Middletown, Conn., Mar. 13, 1899; s. Edward Burr and Hester Laurence (Raymond) Van V.; A.B., University of Wisconsin, 1920, Sc.D., 1947; A.M., Harvard University, 1921, Ph.D., 1922; Sc.D., Wesleyan U., 1936, U. Md., 1955, Rockford Coll., 1961; Dr. honoris causa, U. Grenoble, 1950, U. Paris, 1960; Sc.D., honoris Causa, Oxford University, England, 1958; D.Sc. (hon.), University of Nancy, France, 1961; D.Sc. (honorary), Harvard University, 1966; married Abigail June Pearson, June 10, 1927. Instructor in Physics, Harvard, 1922-23; asst. prof. physics, University of Minn., 1923-26, asso. prof., 1926-27, prof., 1927-28; prof. theoretical physics, U. of Wis., 1928-34; asso. prof. mathematical physics, Harvard, 1934-35, professor, since 1935; chairman, physics department, 1945-49, Hollis professor of mathematics and natural philosophy, 1951—, dean engring. and applied physics, 1951-57; visiting lectr. Stanford, 1927, 34, 41, U. Mich., 1933, Columbia, 1934, Princeton, 1937; Lorentz prof., Leiden, 1960; Eastman prof., Oxford University, 1961-62. Head theory group Radio Research Laboratory, 1943-45; cons., Radiation Laboratory, 1942-45. Associate editor Phys. Rev., 1926-28, 32-34, Jour. Chem. Physics, 1935-43. Guggenheim Meml. Found. fellow, 1930. Recipient Albert A. Michelson award Case Inst. Tech., 1963; Irving Langmuir award in chem. physics Gen. Elec. Found., 1965; National Medal of Science, 1967. Fellow American Physical Soc. (councillor 1932-35, president 1952-53), A.A.A.S. (v.p. 1960), Am. Acad. Arts and Sciences (vice pres. 1956-57); mem. Academic des Sciences France (correspondent), National Academy of Science, Am. Philos. Soc., Am. Math. Soc., Holland Soc. N.Y., Netherlands, Japan Phys. Socs., Phi Beta Kappa, Sigma Xi; hon. mem. Societe Francaise de Physique; fgn. mem. Royal Netherlands Acad. Scis., Royal Soc. Sci. Uppsals, Royal Swedish Acad. Sci. Author: Quantum Principles and Line Spectra, 1926; The Theory of Electric and Magnetic Susceptibilities, 1932. Research on quantum theory of atomic structure and magnetism. Address: Harvard U., Cambridge, Mass.

VAN VLECK, Lloyd Dale, Am. geneticist; b. Clearwater, Neb., June 11, 1933; s. Harold F. and Patricia (Scott) Van V.; B.S., U. Neb., 1954, M.S. (Alpha Zeta Found. fellow), 1955; Ph.D. (NSF fellow), Cornell U., 1960; m. Dee O'Connor, June 28, 1958; children—Elizabeth S., John P. Research asso. dept. animal sci. Cornell U., Ithaca, N.Y., 1959-60, research geneticist, 1960-62, asst. prof., 1962-66, asso. prof., 1966—. Mem. Nat. Acad. Scis., Am. Soc. Animal Sci., Am. Dairy Sci. Assn., Biometrics Soc., Genetics Soc. Am., Am. Statis. Assn., Am. Inst. Biol. Scis., A.A.A.S., Sigma Xi, Gamma Sigma Delta, Alpha Zeta, Phi Kappa Phi. Research, numerous publs. on quantitative genetics of dairy cattle, breeding and selection plans to optimize rate of genetic improvement. Home: 322 Winthrop Dr., Ithaca, N.Y. 14850.*

VAN VOORTHUYSEN, Johannes Henricus, Dutch micropaleontologist; b. Amsterdam, Holland, Jan. 17, 1907; s. Gerrit and Anna Dorothea (Terpstra) Van V.; Dr.geology, U. Amsterdam, 1940; m. Jeanne Jacoba Lindemulder, June 2, 1909; children—Douwe, Maryke. Faculty, U. Amsterdam, 1934-41; with Shell Oil Co., Indonesia, 1929-34; micropaleontologist Netherlands Geol. Survey, Haarlem, 1941—. Publs. on Holocene, Pleistocene and Tertiary of North Sea

Basin. Home: 220 Lorentzkade. Office: 17 Spaarne, Haarlem, Netherlands.

VAN WAGTENDONK, Willem Johan, biochemist; b. Jacarta, Indonesia, Apr. 10, 1910; s. Andries and Maria Kathrina (Briede) van W.; A.B., Rijks Universiteit Utrecht (Netherlands), 1931, M.A., 1934, Ph.D., 1937; m. Roelina Jantina Lottering, Mar. 30, 1937; children—Anne Marie (Mrs. Robert Owen Petty), Jan Willem, Caroline Joyce. Chief chemist N.V. Polaks Frutal Works, Amersfoort, Netherlands, 1937-39; research asso. biology Stanford, 1939-41; faculty Ore. State Coll., Corvallis, 1941-46, asso. prof. biochemistry, 1945-46; asso. prof. zoology Ind. U., Bloomington, 1946-60; prof. biochemistry U. Miami (Fla.) Med. Sch., 1960—; chief basic research VA Hosp., Coral Gables, Fla., 1960—. Cons. health grants div. NIH, 1963—. Fellow A.A.A.S., N.Y. Acad. Sci.; mem. Am. Soc. Biol. Chemists, Soc. Protozoology, Soc. for Gen Microbiology, Biochem. Soc., Sigma Xi. Contbr. numerous articles to tech. jours. Research on growth requirements for microorganisms, metabolism protozoa, role endosymbiotes of protozoa in metabolism of their hosts. biochem. changes during aging in protozoa and chick; discovered role steriods as essential nutrilities for protozoa. Home: 9720 S.W. 114 St., Miami, Fla. 33156. Office: 120 Anastasia Av., Coral Gables, Fla. 33134.*

VAN WATERSCHOOT VAN DER GRACHT, Willem Antonius Josephus Maria, Dutch geologist; b. May 15, 1873; s. W.S.J. and Lady Jonkvrouwe M.C.A.J. (van der Does de Willebois) van W. van der G.; LL.D., U. Amsterdam (Netherlands), 1899; student geology and mining engring. Stoneyhearts Coll.; Mining Engr., Freiburg, 1903; m. J.R.F.G.M. Baroness Hammer Purgstall. Sec., Bd. Mines, Netherlands, 1903; became dir. Govt. Inst. for Exploration Mineral Resources, 1903; named mem. bd. Bd. Mines, Netherlands, 1903; cons. geologist in Rumania, Caucasus, Spain, Patagonia, Africa; const. to govt. on mining and geology, Sumatra and Java, 1913; cons. Shell Co., N.Am., 1915-17; worked for Morgan Group until 1927; became dir., pres. Roxana Petroleum Co., 1917; to N.S., 1931-32; returned to Netherlands, 1932; chief insp. Mining Dist. until 1940. Author: Permo-Carboniferous Orogeny in the South Central United States, 1931; also articles on geology of Netherlands and U. S. Died Roermond, Netherlands, Aug. 12, 1943.

VAN WEEL, Pieter Boudewijn, zoologist, physiologist; b. Ambon, Indonesia, Mar. 20, 1910; s. Kornelis M. and Elizabeth Tissot van Patot Van W.; Cand. Biol., State U. Utrecht (Netherlands), 1932, Dr. Biol., 1935, Ph.D. in Zoology, 1937; m. Emilia Elizabeth van Dijk, Mar. 21, 1941; 1 son, Kornelis Marinus. Came to U. S., 1950, naturalized, 1958. Asst. microbiology U. Amsterdam (Netherlands), 1938; Donders Found. fellow Cambridge (Eng.) U., 1939; instr. dept. biochemistry Med. Faculty, U. Indonesia, 1939-40, dept. histology, 1940-42, 45-47, asso. prof., 1947-50, head dept. histology, 1947-50; prof. zoology U. Hawaii, Honolulu, 1950—. Mem. A.A.A.S., Am. Soc. Zoologists, Soc. Gen. Physiology, Dutch Zool. Soc., N.Y., Hawaiian acads. sci., Sigma Xi. Research, numerous publs. on metabolism pancreas, histophysiol. studies glandular activity in intestines, liver and thyroid, digestion and resorption, respiration, activity smooth muscle, electrophysiol. studies on perception in crabs, vital staining during excretion and secretion, chemoreception in tuna. Home: 45001 Lilipuna Pl., Kaneohe, Hawaii 96744. Office: Dept. Zoology, 2538 The Mall, Honolulu, Hawaii 96822.*

VAN WINKLE, Walton, Am. biochemist; b. Seattle, July 16, 1910; s. Walton and Armantine (Monges) Van W.; A.B., Stanford, 1933, M.D., 1938; m. Frances A. Otwell, July 3, 1938; 1 dau., Elizabeth Armantine. Instr. pharmacology Stanford Sch. Medicine, San Francisco, 1938-41; asst. med. dir. FDA, Washington, 1941-46; sec. com. on research A.M.A., Chgo., 1946-51; v.p. research Ethicon, Inc., Somerville, N.J., 1951-65, v.p. med. affairs, 1965—. Professorial lectr. U. Ill. Sch. Medicine, Chgo., 1946-51. Mem. Soc. for Exptl. Biology and Medicine, Pharmacology Soc., Endocrine Soc., Am. Therapeutic Soc., Thoracic Soc., A.A.A.S., N.Y. Acad. Scis. Research, numerous publs. on pharmacology of bismuth compounds, pharmacology of glycols, hormonal effects in breast cancer, radiation sterilization of pharms., biochemistry of connective tissue, biochemistry of wound healing. Home: 68 Overbrook Dr., Princeton, N.J. 08540. Office: Ethicon, Inc., Somerville, N.J. 08876.*

VAN ZANDT, Thomas Edward, Am. physicist; b. Highland Park, Mich., July 10, 1929; s. Herman S. and Marguerite (Fortney) Van Z.; B.S., Duke, 1950; Ph.D., Yale, 1955; m. Ridi Stolker, Jan. 19, 1961; children—Tineke Renée, Saskia M. Physicist, Sandia Corp., Albuquerque, 1954-57; physicist Central Radio Propagation Lab., Nat. Bur. Standards, Boulder, Colo., 1957—. Lectr. dept. astrogeophysics U. Colo., 1960—. Recipient Gold medal U. S. Dept. Commerce, 1965. Mem. Am. Geophys. Union, Internat. Sci. Radio Union. Research, publs. on distbn. ionization in ionosphere and causative photochem. processes. Home: 2025 Alpine Dr. Office: Environmental Scis. Service Adminstrn., Boulder, Colo. 80302.*

VAQUEZ, Louis Henri, French physician; b. Paris, France, 1860. Staff physician, Paris; prof. clin. therapeutics Faculty Medicine, U. Paris. Mem. Académie de Médecine. Author works on heart diseases, bloodvessels, high blood pressure. Specialist in heart diseases; gave 1st description of erythremia (Vaquez's disease or Osler-Vaquez disease), 1892. Died Paris, 1936.

VARAHAMIHIRA, Hindu astronomer; b. nr. Ujjain, India; flourished circa 505. Author: Pañcasiddhantika; Brihatsamhita (Great Compendium, treatise on astrology with some interesting geog. details on India); Brhatjataka (Great Horoscopy); Laghujataka (Short Horoscopy). Wrote summary astronomy, astrology (historically valuable because contained 5 Siddhantas, summarized and in Karana form); showed advancement in math. astronomy; showed necessary computation for finding positions of planets; said earth is spherical.

VARAY, André Jean, French physician; b. Annecy, France, June 17, 1909; s. F. and M. (Fournier) V.; dr. U. Paris (France), 1937; m. Erika Kremenezuy, Oct. 29, 1958; children—Alexandre, François. Practice medicine specializing in gastroenterology, Paris, 1939—; chief medicine Beauson Hosp., Paris, 1956—; prof. Coll. Medicine, Paris, 1966—; chief Am. Hosp. Paris, 1963—. Named officier Légion d'Honneur. Mem. Société médicale des Hôpitaux de Paris, Société Française de Gastroenterologie. Author: Traite de Therapeutique, 1954; Précis de Gastroenterologie, 1966; also numerous articles. Research on diseases of digestive system. Address: 38 Av. Hoche, Paris, France.*

VARENIUS, Bernhardus (Varen, Bernhard), geographer; b. Hitzacker, Germany, 1622; student Königsberg (now Kaliningrad, USSR), 1943-45; med. degree Leyden (Holland) U., 1945-49. Practice medicine, Amsterdam, Holland. Author: Descriptio regni japoniae, 1649; Geographia generalis, 1650. Research in systematic and regional geography; applications of math. and physics to geography; set down gen. principles for study of geography. Died Leyden, circa 1650.

VARENTSOV, Mikhail Ivanovich, Russian geologist; b. Jan. 20, 1902; grad. Moscow Mining Acad., 1929. Dir., Inst. Geol. Scis., 1949-55; became chief lab. Inst. Oil, USSR Acad. Sci., 1956; geol. research, mem. various expdns., 1929-49; head expdns. in Georgia, Turkmen, Transcausia, Volga-Bash areas. Corr. mem. USSR Acad. Scis. Author: Geology of Oil and Gas Deposits of Tamanskii Peninsula, 1934; The Geology and Oil Deposits of Viennese Basin, 1948; The Geological Structure of Western Kura Depression, 1950; The New Oil-Bearing Region of Pannonian Basin in Southeastern Europe and Possible Analogies to It, 1950; (with V.T. Mordvoskii), Geological Structures of the Northern Edge of Gori-Mukhranskaya Depression, 1954. Research in petroleum, 1929-49, regional studies in tectonics, stratigraphy, oil geology, 1935-49. Office: Inst. Oil, USSR Acad. Scis., Moscow, USSR.

VARGA, Emil, Hungarian physiologist; b. Debrecen, Hungary, Nov. 30, 1921; s. Imre and Eszter (Bakos) V.; M.D., U. Debrecen, 1944; m. Sára Ember, July 2, 1949; children—Emil, Judit. Staff dept. physiology U. Debrecen, Med. Sch., 1942—, prof., chmn. physiology, 1966—; research staff dept. physiology and pharmacology Duke U., Durham, N.C., 1962-63. Mem. Hungarian Physiol. Soc., Hungarian Biophys. Soc., Hungarian Biochem. Soc. Contbr. articles to tech. jours. Showed (with S. Went) importance of moblzn. of biologically active substances exerting antagonistic effects on stimuli in regulation of circulation and heart function; discovered characteristic differences in structure and enzymatic properties of contractile proteins of different kinds of muscles. Home: 78 Nagyerdei krt., Debrecen, Hungary.*

VARGA, Magdolna, Hungarian plant physiologist; b. Szarvas, Hungary, Mar. 7, 1922; d. János and Teréz (Moravetz) Varga; student Tchrs. Tng. High Sch., Szeged, Hungary, 1941-45; Ph.D., U. Szeged, 1948, Candidate Biol. Sci., 1958; m. László Bertényi, Dec. 6, 1950; 1 dau., Magdolna. Staff bot. dept. U. Szeged, 1948-52, dept. plant physiology, 1952—, docent, 1960—. Mem. Hungarian Biol. Soc., Hungarian Biochem. Soc. Research, publs. on physiology of plant growth, effect and interrelation of naturally occurring plant growth regulators, phenolic inhibitors, mode of interaction of phenolic compounds with auxins and gibberellins. Home: Nr. 1 Dugonics Sq., Szeged I, Hungary.*

VARGA, Richard Steven, Am. mathematician; b. Cleve., Oct. 9, 1928; s. Steven and Ella (Krejcs) V.; B.S., Case Inst. Tech., Cleve., 1950; A.M., Harvard, 1951, Ph.D., 1954; m. Esther Marie Pfister, Sept. 22, 1951; 1 dau., Gretchen Marie. Adv. mathematician Bettis Atomic Power Lab., Westinghouse Elec. Corp., Pitts., 1954-60; prof. math. Case Inst. Tech., 1960—. Cons. Argonne Nat. Labs., Brookhaven Nat. Labs, Gulf Research & Devel. Co. Guggenheim fellow, 1963. Author: Matrix Iterative Analysis, 1962. Mem. editorial bds. Numerische Math., SIAM Jour., Math. of Computation. Research, numerous publs. on efficient numerical solution of partial differential equations on digital computers; such equations arise in design of neutron reactors as well as in study of conduction and convection in petroleum reservoirs. Home: 7065 Arcadia Dr., Cleve. 44129. Office: Math. Dept., Case Western Res. U., University Circle, Cleve. 44106.*

VARIAN, Russell Harrison, Am. physicist; b. Washington, D.C., Apr. 24, 1898; s. John Osborne and Agness (Dixon) V.; A.B., Stanford, 1925, M.A., 1927; D.Eng. (hon.), Poly. Inst. Bklyn., 1943; m. Dorothy Hill, 1947; children—George Russell, Charles John, Susan Aileen. Research physicist Humble Oil & Refining Co., 1929, Farnsworth Television Co., 1930-33; pvt. research 1934-35; research asso. Stanford, 1937-40. 46—; research engr. Sperry Gyroscope Co., 1940-46; pres. Varian Associates, mfrs. ultra-high frequency microwave tubes, 1948-56, chmn. bd. 1956—. Awarded John Price Wetherill medal, Franklin Inst., 1950. Fellow Inst. Radio Engrs., Cal. Acad. Sci., Am. Phys. Soc., A.A.A.S.; mem. Sigma Xi. Inventor klystron radio tube for prodn. waves in range of one to a few centimeters; patentee approximately 100 devices in microwave, applied physics field. Died July 28, 1959.

VARIGNON, Pierre, French mathematician; b. Caen, France, 1654. Became prof. math. Coll. Mazarin, 1688, Collège de France, 1704. Mem. French Acad. Scis., 1688, became asst. dir., 1710, dir., 1711-12, 19. Author: Projet d'une nouvelle mécanique, 1686; Nouvelles conjectures sur la pesanteur, 1690; Nouvelle mécanique, 1725; Traité du mouvement et de la mesure des eaux courantes, 1725. Studies in mechanics, pure math.; one of 1st to recognize value of calculus; originated term logarithmic spiral. Died Paris, Dec. 22, 1722.

VARLEY, Cornelius, English inventor, instrument maker; b. Eng., Nov. 21, 1781; s. Richard Varley; learned trade under his Uncle Samuel, watchmaker, jeweller, instrument maker; m. Elizabeth Straker, 1821; children include—Cromwell Fleetwood, Frederick Henry, Samuel Alfred, Theophilus. Worked with uncle until 1800; returned to art studies with his Brother John; both went to Norfolk and Suffolk, Eng., 1801; sketching tours in Wales, 1802-03; exhibited at Royal Acad., 1803. Recipient several medals including Isis Gold medal Soc. Arts, premium, Water Color Soc., 1819; prize medal Royal Instn., 1851. Mem. Watercolor Soc. (founder 1804, treas. 1815), Soc. Arts, Royal Instn. Author: Treatise on Optical Drawing Instruments; Etchings of Shipping Barges, Fishing Boats . . . , 1809; also other publs. Studies on atmospheric electricity; invented graphic telescope, 1811, composition for polishing lenses, lever microscope; contributed to constrn. 1st soda-water apparatus, large elec. machine with conductor 12 feet long; built lens 1/100 inch in focus; one of 1st (after F. Unger) to describe male gametes of mosses and liverworts. Died Stoke, Eng., Oct. 21, 1873.

VARLEY, Cromwell Fleetwood, English elec. engr.; b. London, Apr. 6, 1828; s. Cornelius and Elizabeth (Straker) V.; ed. St. Saviours, Southwark, Eng.; student telegraphy; m. Jesse Smith, Jan. 11, 1877; 2 sons, 2 daus. (from previous marriage). With Electric and Internat. Telegraph Co., 1846-68; independent inventor, from 1868; founder (with brother) chemistry class London Mechanics Inst. Mem. com. reporting on possibility of success for 2d Atlantic cable. Fellow Royal Soc., 1871; mem. Instn. Civil Engrs., Soc. Telegraph Engrs. (mem. council). Research and numerous publs. on electricity and telegraphic communication; invented double current key and relay, 1854, cymaphen (forerunner of telephone), 1870; developed (with brother) powder with variable resistance to currents of different tensions, 1856; observed electrocapillary phenomenon, 1870; proposed idea that cathode rays were composed of particles of matter projected by electricity; invented polarized relay, translating system for use in cables of Dutch lines; originated system of localizing faults in submarine cables, killing treatment of telegraphic wire; built device to simulate conditions of submarine cable. Died Bexleyheath, Eng., Sept. 2, 1883.

VARLEY, George Copley, English zoologist; b. Manchester, Eng., Nov. 19, 1910; s. George Percy and Elsie (Sanderson) V.; B.A., Sidney Sussex Coll., Cambridge (Eng.) U., 1932, M.A., Ph.D., 1938; m. Margaret Elizabeth Brown, June 28, 1955; children—Elizabeth, Robert. Supt., Entomol. Field Sta., Cambridge, Eng., 1933-36; research fellow Sidney Sussex Coll., Cambridge U., 1935-38, jr. curator insects zoology dept., 1938-40; reader entomology King's Coll., Newcastle on Tyne, Eng., 1945-48; prof. entomology U. Oxford (Eng.), 1948—; research fellow U. Cal. at Riverside, 1937-38. Fellow Royal Entomol. Soc. (past pres.), Zool. Soc. London (past v.p.), Brit. Ecol. Soc. (past pres.); mem. Brit. Trust for Entomology (pres. 1956—), Finnish Entomol. Soc. (corr.), Entomol. Soc. Can., Ecol. Soc. Am. Publs. on population studies of effects of parasites, predators and other factors on insects, pest control problems. Home: 18 Apsley Rd., Oxford, Eng.*

VARNAUSKAS, Edvardas, cardiologist; b. Zaidelai, Lithuania, Feb. 2, 1923; s. Jurgis and Ursule (Jackeviciute) V.; M.D., Karolinska Inst., Stockholm, 1951, Ph.D., 1955; m. Solveig Nordanfors, Jan. 5, 1954; children—Maria, Anna, Theresia, Caroline. Research fellow Johns Hopkins, Balt., 1955-57; asso. prof. medicine U. Göteborg (Sweden), 1957—; chief cardiology sect. med. clin. I, Sahlgren's Hosp., Göteborg, 1957—. Recipient Regnell's prize, 1954. Mem. Swedish Med. Assn., Swedish Cardiological Soc. Author: Studies in Hypertensive Cardiovascular Disease, 1955; also numerous articles. Research on hemodynamic evaluation of hypertensive cardiovascular disease and its treatment, hemodynamic basis for phys. tng. of coronary patients, coronary artery changes in obscure cardiomyopathy, regulation of pulmonary circulation in health and disease; evaluation of total and coronary circulation at rest and on exercise in coronary disease. Home: Box 489, Hovas, Sweden. Office: Med.clin. I, Sahlgren's Hosp., Göteborg, Sweden.*

VARNER, Joseph Elmer, Am. biochemist; b. Nashport, O., Oct. 7, 1921; s. George Ezra and Inez (Gladden) V.; B.Sc., Ohio State U., 1942, M.Sc., 1943, Ph.D., 1949; m. Carol Roberta Dewey, Apr. 29, 1945; children—Lee, Lynn, Karen, Beth. Chemist, Owens-Corning Fiberglas, Newark, O., 1943-44; chemist Battelle Meml. Inst., Columbus, O., 1946-47; faculty Ohio State U., 1949-61, prof. biochemistry, 1957-61; research fellow Cal. Inst. Tech., Pasadena, 1953-54; sr. scientist Research Inst. for Advanced Studies, Balt., 1961-65; staff AEC Plant Research Lab., prof. Mich. State U., East Lansing, 1965—. Sr. NSF fellow U. Cambridge (Eng.), 1959-60. Fellow A.A.A.S.; mem. Am. Chem. Soc., Am. Soc. Biol. Chemists, Am. Soc. Plant Physiologists. Author: (with J. Bonner) Plant Biochemistry, 1965; also numerous articles. Research on pathways nitrogen metabolism in plants, mechanisms enzyme action as controlled by plant hormones. Home: 4364 Wausau Rd., Okenos, Mich. 48864. Office: Biochemistry Dept., Mich. State U., East Lansing, Mich.*

VARNEY, Robert Nathan, Am. physicist; b. San Francisco, Nov. 7, 1910; s. Frank Hastings and Emily (Rhine) V.; A.B., U. Cal., Berkeley, 1931, M.A., 1932, Ph.D., 1935; m. Astrid M. Riffolt, June 19, 1948; children—Nils R., Natalie R. Instr. physics U. Cal., Berkeley, 1935-36, N.Y. U., 1936-38; faculty Washington U., St. Louis, 1938-64, prof., 1948-64; sr. cons. scientist, sr. mem. research lab. Lockheed Missiles and Space Co., Palo Alto, Cal., 1964—; mem. tech. staff Bell Telephone Labs., Murray Hill, N.J., 1951-52. Mem. Gov.'s Sci. Adv. Com., State of Mo., 1961-64; cons. McDonnell Aircraft Corp., 1960-63. Sr. NSF Postdoctoral fellow Royal Inst. Tech., Stockholm, Sweden, 1958-59. Mem. Phi Beta Kappa, Sigma Xi, Omicron Delta Kappa, Tau Beta Pi. Author: Lecture Notes in Engineering Physics, 1948; Lab Manual of Physics, 1951; also articles. Research on conduction of electricity in gases, on electron and ion collisions in gases, chemistry of gas ions. Home: 4156 Maybell Way, Palo Alto 94306. Office: 3251 Hanover St., Palo Alto, Cal. 94306.*

VAROLIO, Constanzo (Varoli), Italian physician, anatomist; b. Bologna, Italy, 1543; student U. Bologna (Italy); prof. Bologna, Rome; physician to Pope Gregory XIII, Rome. Author: De nervis opticis, 1573. Research in human brain, viscera; discovered new technique for brain dissection; described named mass of nerve fibers on underside of brain (pons Varolii). Died Rome, 1575.

VARRO, Marcus Terentius, Roman natural philosopher; b. Reate, Land of Sabines, Italy, 116, B.C.; pupil of Lucius Aelius Stilo (earliest Roman philogist), also Antiochus of Ascalon, Athens. Became praetor; studied under Pompey, Spain, 76 B.C.; became proquaestor to Pompey, Spain, served under him in war against pirates, 67 B.C.; became mem. Board of 20 assigning landgrants to veterans, Campania, 59 B.C.; fought for Pompey in Spain, 49 B.C.; pardoned by Caesar; apptd. to supervise collection and arrangement of books in pub. library in Rome, 47 B.C.; proscribed by Antony, 43 B.C.; apptd. supt. of library founded by Asinius Pollion by Augustus. Author numerous works including: De re rustica; 6 of 24 books of De lingua latina (only surviving works). Wrote on measurment, arithmetic, geometry, etymology, grammar, dialectics, rhetoric, astrology, music, medicine, botany, animal diseases, astronomy, geography, architecture, history, biography, edn., lit., drama, ancient agr. Died 27 B.C.

VARSANOFIEVA, Vera Aleksandrovna, Russian geologist; b. July 21, 1889; grad. Higher Women's Courses in Moscow, 1914; doctor's degree in geologic and mineral. scis. Faculty, Prechistensk Workers' Sch., Moscow, 1916-20; prof. 2d Moscow U., 1925-59, Lenin Pedagogical Inst., Moscow, 1930—. Mem. Moscow Soc. Explorers of Nature (became v.p., exec. editor geologic bulls. 1945). Research in geo-morphology of Quarternary deposits of No. Caucasus, stratigraphy of Paleozoid deposits of upper Pechora River basin.

VARSHNEY, Ishwar Prasad, Indian chemist; b. Aligarh, India, May 22, 1927; s. Gobind Prasad and Ranna (Devi) V.; Intermediate, Aligarh Muslim U., 1944, B.Sc., 1946, M.Sc., 1948, Ph.D., 1954; D.Sc., Sorbonne, U. Paris (France), 1956; m. Sharda Devi, July 12, 1951; children—Archana, Anil, Arvind. Research fellow Aligarh Muslim U., 1948-54, lectr., 1957-66; research fellow Mus. Nat. d'Histoire Naturelle, Paris, France, 1954-57; vis. scientist Organische Chemisches Institut, Bonn, Germany, 1966; prof., head chemistry dept. Shri Govindram Seksaria Technol. Inst., Indore, India, 1966—. Fellow Royal Inst. Chemistry (London), Indian Chem. Soc. (life); mem. Société Chimique de France. Research, publs. on chemistry of natural products from Indian medicinal plants especially saponins, sapogenins, steroids, triterpenes. Home: Mahabir Ganj, Aligarh (U.P.), India. Office: Dept. Chemistry, Shri Govindram Seksaria Technol. Inst., Indore-3 (M.P.), India.*

VASEY, George, botanist; b. Scarborough, Yorkshire, Eng., Feb. 28, 1822; brought to N.Y., 1823; attended Oneida (N.Y.) Inst., Berkshire Med. Inst., Pittsfield, Mass.; M. Miss Scott, 1846; m. 2d, Mrs. (Barber) Cameron, 1867; at least 6 children. Practiced medicine, Dexter, N.Y., 1846-48, Elgin and Ringwood, Ill., 1848-66; organizer, 1st pres. Ill. Natural History Soc., 1866; accompanied Maj. John Wesley Powell on Colo. bot. expdn., 1868; curator natural history mus. Ill. State Normal U., 1869; became co-editor Am. Entomologist and Botanist, 1870; botanist U. S. Dept. of Agr., 1872, also in charge U. S. Nat. Herbarium. Author: Catalogue of the Forest Trees of the United States Which Usually Attain a Height of Sixteen Feet or More, 1870; Agricultural Grasses of the United States, 1884; Grasses of the Southwest, 1890-91; Grasses of the Pacific Slope, 1892-93. Collected, classified and studied prairie flora, especially grasses. Died Washington, D.C., Mar. 4, 1893.

VASILEV, Leonid Leonidovich, Russian physiologist; b. 1891; D.Biol. Sci., St. Petersburg U., 1914. Instr. biology Ufa Gymnasia, 1915-20; head dept. physiology Bekhterev Brain Inst., Leningrad, 1921-48; head chair physiology Leningrad U., 1943—. Mem. editorial council Sechenov Physiol. Jour. of USSR. Corr. mem. USSR Acad, Med. Sci. Author: Electric Restoration of Physiological Functions, 1937; The Theory and Practice of Ionized Air Treatment, 1953; The Significance for Neuropathology of N. E. Vvedensky's Physiological Teachings, 1953; Mysterious Manifestations of The Human Psyche, 1959; Notes of a Physiologist, 1962. Co-editor physiology sect. Large Med. Ency., 2d edit. Research in parabiosis, nervous processes, ionized air; devel. theory of two kinds of inhibition, 1925-37; established law of threshold parabiosis, 1925; research elec. influence on tissue restoration, 1937; showed parabiotic nature of electronarcosis and anaphylaxis, 1944; found central nervous system emits parabiosic-eliminating influences; established 1st para-psychol. lab. in USSR. Address: Univ. n. 7-9, Gosudarstvenny Univ., Leningrad. USSR.

VASILEVSKIS, Stanislaus, astronomer; b. Laucesa, Latvia, July 20, 1907; s. Janis and Rozalija (Zakarauskis) V.; Mag.math., U. Latvia, Riga, 1932, Habilitation, 1939; m. Marija Rucevskis, May 18, 1929; children—Velta (Mrs. Arnis Zebergs), Janis. Came to U. S., 1949, naturalized, 1955. With U. Latvia, 1928-44, asst. prof., 1939-44; research asso. Leipzig (Germany), 1944-45; asso. prof. UNRRA U., Munich (Germany), 1946-48; staff Lick Obs., U. Cal. at Santa Cruz, 1949—, asso. astronomer, 1958-64, astronomer, 1964—. Mem. Internat. Astron. Union, Am Astron. Soc., Royal Astron. Soc., Astron. Soc. Pacific. Research and publs. on stellar proper motions and trigonometric parallaxes, cluster membership, automation of observing and measurement. Home: 156 La Canada Way, Santa Cruz, Cal. 95060.*

VASSALLI-EANDI, Antonio Maria, Italian physicist; b. Turin, Italy, Jan. 30, 1761. Prof. philosophy, Tortona, Italy; prof. physics U. Turin; dir. Natural History Mus., also Obs. of Turin. Corr. mem. French Acad. Scis., 1805; mem. Turin Acad. Scis. (life sec.). Author: Conjectures on the Art of Building Lightning Rods among the Ancient Romans, 1791; Letters on Galvanism, 1799; Annals of the Turin Observatory, from 1809-18; Memoires of the Academy of Sciences of Turin, from 1792-1809. Research and publs. on galvanism and lightning rods. Died Turin, July 5, 1825.

VASSAMILLET, Lawrence François, physicist; b. Elizabethville, Congo, Sept. 14, 1924; s. Emile G. and Laura (Wyatt) V.; B.Sc. in Physics, Mass. Inst. Tech., 1947, M.Sc., 1950; D.Sc., U. Liège (Belgium), 1952; Ph.D. in Physics, Carnegie Inst. Tech., 1957; m. Edith B. Pennoyer, Aug. 29, 1953; children—Laura Anne, Martha Louise. Jr. physicist Monsanto Chem. Co., 1947-48; instrument engr. Mass. Inst. Tech., 1948-50; X-ray diffractionist Nat. Carbon Co., 1953; project physicist Carnegie Inst. Tech., 1956-57; fellow Mellon Inst., Pitts., 1957-63, sr. fellow, 1963—. Mem. Am. Phys. Soc., Am. Crystallographic Assn., Pitts. Diffraction Soc., Electron Microscope Soc. Am., Sigma Xi. Research, publs. on solid state physics, crystallography, metallurgy and meteorites especially imperfections in metallic crystals using X-ray diffraction and electron microscopy; electron microprobe techniques to study non-equilibrium situation in meteorites. Home: 200 Fieldcrest Dr., Forest Hills, Pitts. 15221. Office: Mellon Inst., 4400 5th Av. Pitts. 15213.*

VASSEUR, Gaston, French geologist; b. Paris, Aug. 5, 1855. Prof. geology Faculty Scis., Aix-Marseille. Mem. French Acad. Scis., 1914. Prepared geol. map France, 1889. Died Marseille, Oct. 9, 1915.

VASSILIADIS, Peter, bacteriologist; b. Cairo, Egypt, May 17, 1907; M.D., U. Louvain, Belgium, 1930, Agrégó de l'Enseignement Superieur in Bacteriology, 1938; Diploma Tropical Medecine, Sch. Tropical Medecine, Antwerp, Belgium, 1958; m. Grammatiki Moutafi, May 28, 1938. Bacteriologist, pub. health dr. Quarantine Adminstrn. Egypt, 1933-39, 42-44; dir. Antirabies Inst. Alexandria, Egypt, 1947-49; dir. med. lab. Anglo-Swiss, Jewish. hosps., Alexandria, 1947-57, dir. dept. Central Labs. Leopoldville, Congo, 1958-60; dir. Enterobacteria Center Congo, Leopoldville, 1959-60; dir. toxoid dept. Greek Pasteur Inst., 1963-65; dir. Greek Nat. Shigello Center, 1963—, prof. bacteriology and immunology Athens Sch. Hygiene, 1965—. Recipient chevalier l'Ordre de la Santé Publique (France); médaille l' Institut de Médecine Tropicale Prince Léopold of Belgium. Fellow Royal Soc. Tropical Medicine and Hygiene; mem. Belgian, French bacteriological socs., Belgian and French Soc. Tropical Medicine, others. Discovered parasite of mice, eperythrorzoon dispar, also 4 new type of microorganisms of food poisoning group; contbns. to knowledge of cholera vibrios and colicintyping of shigella. Home: 119 Queen Sofias Av., Athens, Greece.*

VASSILIOU, Philon, Greek mathematician; b. Constantinople, Turkey, Nov. 5, 1904; s. Marcos and Argiro (Vitali) V.; ed. U. Leipzig, U. Athens, U. Hamburg; Dr.ès sc.; Instr., U. Salonika, 1933-35; asst. prof., 1935-37; prof. Tech. U. Athens, 1937—. Decorated Order Phoenix, Order of King George I. Mem. Am., French, German math. socs. Author: Varlesungen über höhere Mattematik, 1950-60; höhere Mattematik, 1962-64. Research and publs. on algebraic number theory, founds. of math., math. logic. Home: Odos Evans 23, Athens, Greece.

VASSY, Etienne Joseph, French physicist; b. Hauterive, France, Nov. 14, 1905; s. Louis Etienne and Marie Justine (Pellerin) V.; student Faculty Sci., Lyons, also Paris; Dr.ès.sc.; m. Arlette Tournoire, July 30, 1936. Successively asst., lectr., asso. prof., titular prof. atmospheric physics Faculty Sci. Paris. Pres., French Com. on Radio-Electricity, com. on ionospheric research Adv. Group for Aero. Research and Devel. Mem. Internat. Astronautics Soc., N.Y. Acad. Scis. (hon. life), I.E.E.E., Am. Rocket Soc., Phys. Soc. Research and numerous publs. in atmospheric physics. Home: 110, quai Louis Blériot, Paris 16, France.

VATER, Abraham, German anatomist; b. Wittenberg, Germany, Dec. 9, 1684. Prof. medicine, pathology, therapeutics, Wittenberg, Germany. Discovered papilla in duodenum (Vater's papilla); described ampulla hepatopancreatica (ampulla of Vater), corpuscula lamelossa (Vater's corpuscles). Died Nov. 18, 1751.

VATNSDAL, John Russell, Am. mathematician; b. Duxby, Minn., Sept. 10, 1901; s. Thomas and Anna (Johnson) V.; B.A., Reed Coll., 1921; postgrad. Yale; M.A., Wash. State Coll., 1930; Ph.D., U. Mich., 1940; m. Mildred Hunt, July 23, 1932; 1 dau., Mary (Mrs. G. Todd Cowan). Faculty, Wash. State U., Pullman, 1927-66, prof. math., 1949-66. Mem. Am. Math. Soc., Inst. Math. Statistics, Math. Assn. Am., Sigma Xi. Research, publs. on math. statistics. Died Apr. 14, 1966.

VAUBAN, Sébastien le Prestre de, see de Vauban, Sébastien le Prestre.

VAUCANSON, Jacques de, see de Vaucanson, Jacques.

VAUCHER, Jean Pierre Étienne, Swiss botanist; b. Geneva, Switzerland, 1763. Clergyman; prof. ecclesiastical history Acad. Geneva, became rector, 1818. Author: Histoire des conferves d'eau douce, 1803; Histoire physiologique des plantes de l'Europe, 1804; Monographie des orobanches, 1826; Souvenirs d'un pasteur genevois, 1842. Discovered function of spores; 1st to observe conjugation in algae (later named Vaucheria after him) and suggested it is a sexual act. Died 1841.

VAUGHAN, Benjamin, polit. economist; agriculturist; b. Jamaica, Apr. 30, 1751; s. Samuel and Sarah (Hallowell) V.; ed. Cambridge (Eng.) U.; read law Inner Temple, London, Eng.; studied medicine, Edinburgh, Scotland; hon. degrees Harvard, 1807, Bowdoin Coll., 1812; m. Sarah Manning, June 30, 1781, 3 sons, 4 daus. Propagandized for independence during Am. Revolution; joined father-in-law's mcht. firm; unofcl. mem. Brit. commn. which concluded Treaty of Paris, 1782; became mem. Parliament from Caine, 1792; escaped to France when he feared investigation of French Revolution supporters; imprisoned in France, 1794, soon released to Switzerland; came to Am., settled in Hallowell, Me., 1796; maintained wide correspondence with Am. polit. figures; founder Me. Hist. Soc.; mem. many lit. and sci. socs.; owned largest individually-owned library in New Eng. (divided after his death among Harvard, Bowdoin Coll. and Augusta Insane Hosp.). Author: Letters on the Subject of the Concert of Princes and the Dismemberment of Poland and France, 1793; the Rural Societies, 1800. Editor: Political, Miscellaneous and Philosophical Pieces . . . written by Benjamin Franklin, 1779. Supported free trade, Am. independence, French Revolution; conducted numerous agrl. expts. communicated to Mass. Agrl. Soc. under name of "Kennebec Farmer." Died Dec. 8, 1835.

VAUGHAN, Edward Kemp, Am. plant pathologist; b. East Las Vegas, N.M., Nov. 16, 1908; s. John Henry and Cora Lee (Spainhower) V.; B.S., N.M. State U., 1929; M.S., Ore. State U., 1932; Ph.D., U. Minn., 1942; m. Ruth May Bullis, July 10, 1932; children—Lawrence E., Rodney K. Prof. plant pathology Ore. State U., Corvallis, 1947—, acting head botany dept., 1966—. Guggenheim Found. fellow, Wageningen, Netherlands, 1954-55; New Zealand Dept. Sci. Indsl. Research sr. research fellow, Auckland, 1964-65. Mem. Am., Netherlands phytopath. socs., Bot. Soc. Am., Am. Soc. Plant Physiologists, Assn. Applied Biologists, Sigma Xi, Phi Sigma. Contbr. numerous articles to profl. jours. Developed method of freeing valuable strawberry breeding stocks from infection by Phytophthora fragariae, method for rapid removal of tannins from virus infected plant tissues without inactivation of the virus; work on interaction of pathogenic microorganisms. Home: 1606 Alta Vista Dr., Corvallis, Ore. 97330.*

VAUGHAN, John Heath, Am. physician; b. Richmond, Va., Nov. 7, 1921; s. Warren T. and Emma (Heath) V.; A.B. cum laude, Harvard, 1942, M.D., 1945; m. Margaret Lamb, May 25, 1946; children—John Heath, Nancy, David, Margaret. NRC fellow Columbia-Presbyn. Med. Center, N.Y.C., 1951-53; asst. prof. Med. Coll. Va., 1953-58; faculty U. Rochester (N.Y.), 1958—, prof. medicine, 1963—. Chmn. research com. Arthritis Found., N.Y.C., 1964—; cons. NIH, 1956-64. Mem. Am. Acad. Allergy (pres. 1967-68), Am. Assn. U. Profs., Soc. for Exptl. Biology and Medicine, Assn. Am. Physicians, Am. Soc. for Clin. Investigation, Am. Assn. Immunologists, Am. Rheumatic Assn. Editor: (with others) Immunological Diseases, 1966; editorial bd. Arthritis and Rheumatism, 1956—, Jour. Allergy, 1958, Jour. Clin. Investigation, 1965—. Research, numerous publs. on allergic, rheumatic and certain blood diseases as related to immunological mechanism. Home: 44 Golfside Pkwy., Rochester, N.Y. 14610.*

VAUGHAN, Thomas (or Eugenius Philalethes), Welsh alchemist; b. Newton, Wales, Apr. 17, 1622; s. Thomas Vaughan; B.A., Jesus Coll. Oxford (Eng.) U., 1642, studied medicine, chemistry, Oxford, London; m. Rebecca, Sept. 28, 1651. Received maintenance of St. Bridget's, Brecknockshire, Wales, from 1640-58 (deprived of this for supporting Royalists in civil wars). Author: Anthroposophia theomagica, 1650; Anima magica abscondita, 1650; The Man-Mouse Taken in a Trap, 1650; Magia adamica or the Antiquities of Magic, 1659; The Second Wash, 1651; Aqua vitae; Non Vitis or the Radical Humiditie of Nature Mechanically and Magically Dissected (Sloane manuscript, 1741); Lumen de Lumine, 1651; Aula lucis or the House of Light, 1652; Euphrates on the Waters of the East, 1655; The Chymists' Key to Shut or to Open or the True Doctrine of Corruption and Generation, 1657. As disciple of Cornelius Agrippa opposed philosophy of Aristotle and of Descartes. Died Feb. 27, 1666.

VAUGHAN, T(homas) Wayland, Am. geologist, oceanographer; b. Jonesville, Tex., Sept. 20, 1870; s. Dr. Samuel Floyd and Annie R. (Hope) V.; B.S., Tulane U., 1889, D.Sc., 1944; A.B., Harvard, 1893, A.M., 1894, Ph.D., 1903; LL.D., U. of B.C., 1933, U. of California, 1936; studied mus. in Europe; m. Dorothy Q. Upham, March 22, 1909 (deceased, Aug. 18, 1949); 1 dau., Caroline Ely (Mrs. James H. Fortune, Jr.). Engaged in geologic and paleontologic researches, with U. S. Geol. Survey, 1894-1923; geologist in charge Coastal Plain investigations, 1907-23, sr. geologist 1924-28, prin. scientist, 1928-39; retired. Custodian of Madreporarian corals, U. S. Nat. Museum, 1903-23; asso. in marine sediments, 1924-42, in paleontology since 1942; dir. Scripps Inst. of U. Calif., La Jolla, 1924-36, dir. emeritus, since 1936. Decorated Order of Rising Sun, 3d class, Japan, 1940; awarded Agassiz medal for research in oceanography, Nat. Acad. of Science, 1935; Mary Clark Thompson medal for geology and paleontology, 1945; Penrose Medal, Geol. Soc. of America, 1946. Fellow Am. Academy Arts and Sci., Am. Philos. Soc., Calif. Acad. Sciences, A.A.A.S. (pres. Pacific div. 1930-31), Geol. Soc. America (1st v.p. 1938, pres. 1939), Assn. Am. Geographers, Paleontol. Soc. (pres. 1923), Washington Acad. Science (pres. 1923); San Diego Soc. Natural History (pres. 1926); mem. Nat. Acad. Sciences, Am. Geophys. Union (chmn. oceanography sect. 1926-28), Geol. Soc. of Washington (pres. 1915); Oceanographic Soc. of the Pacific (pres. 1935-36); Philos. Soc. Tex.; corr. mem. Zoological Soc. of London, Acad. Natural Sciences Philadelphia, Konk. Nat. Ver. Nederl.-Ind., Soc. Geograf., Cuba, fgn. mem. Linnean Soc., fgn. fellow Geol. Soc., London; hon. mem. Geological Soc., Peru. U. S. del. 1st Pan-Pacific Science Conf., Honolulu, 1920 (chairman sect. geology); del. U. S., Nat. Acad. Sciences, etc., to 2d Pan-Pacific Science Congress, Melbourne and Sydney, Australia, 1923; 3d Pan-Pacific Science Congress, Japan, 1926; 4th Pacific Science Congress, Java, 1929, 5th Congress, Victoria and Vancouver, 1933; mem. div. geology and geography, Nat. Research Council, 1919-26; chmn. Internat. Com. Oceanography of Pacific, Pacific Science Assn., 1926-35; chmn. sect. geol. sciences, 8th Am. Sci. Congress, Washington, 1940. Author: The Eocene and Lower Oligocene Coral Faunas of the United States (Monograph 39, U. S. Geol. Survey); Recent Madreporaria of Hawaiian Islands and Laysan (Bull. 59, U. S. Nat. Museum); Contributions to the Geology and Paleontology of the Canal Zone, Panama (Bull. 103, U. S. Nat. Mus.); Geologic Reconnaissance of the Dominican Republic; International Aspects of Oceanography; and more than 300 other scientific papers. Specialist on tertiary geology, fossil and recent corals, larger Foraminifera, and marine sediments. Died Jan. 16, 1952.

VAUGHAN, Victor Clarence, Am. chemist, physician; b. Mt. Airy, Mo., Oct. 27, 1851; s. John and Adeline (Dameron) V.; B.S., Mt. Pleasant Coll., Mo., 1872; M.S., U. Mich., 1875 Ph.D., 1876, M.D., 1878, LL.D., 1900; hon. Sc.D., U. Western Pa., 1897; LL.D., Central Coll., 1910; Jefferson Med. Coll., Phila., 1915, U. Mo., 1923; m. Dora Catherine Taylor, August 21, 1877; children—Victor C., John Walter, Herbert Hunter, Henry Frieze, Warren Taylor. Assistant in chem. lab., 1875-83, lecturer med. chemistry, 1879-80, asst. prof., 1880-83, prof. physiol. and pathol. chemistry and asso. prof. therapeutics and materia medica, 1883-87, prof. hygiene and physiol. chemistry, and dir. Hygienic Lab., 1887-1909, dean dept. medicine and surgery, 1891-1921, U. Mich. in Santiago campaign, 1898, as maj. and surgeon 63d Mich. Vol. Inf.; apptd. div. surgeon, 1898; recommended by President for bvt. of lt. col.; col., M.C. U. S. A., in charge of communicable diseases, 1917-18. Chmn. div. of med. sciences, Nat. Research Council; member Typhoid Commn. Awarded D.S.M. Pres. Assn. Am. Physicians, 1908-09; pres. A.M.A., 1914-15. Knight Legion of Honor, France, 1923. Author: Osteology and Myology of the Domestic Fowl, 1876; Text-book of Physiological Chemistry (3 edits.), 1879-83; Ptomaines and Leucomaines and Cellular Toxins (with Dr. Novy); Protein Split Products (with Victor C. Vaughan, Jr., and J. Walter Vaughan), 1913; (with Henry F. Vaughan and George T. Palmer) Epidemiology and Public Health, 3 vols.; A Doctor's Memories. Mng. editor Jour. Lab. and Clin. Medicine, 1915-23. Research on bacteria and proteins; found poisonous groups in molecular structure of proteins; described reticulocytes. Died Richmond, Va., Nov. 21, 1929.

VAUGHAN, Victor Clarence III, Am. physician; b. Toledo, July 19, 1919; s. Warren Taylor and Emma Elizabeth (Heath) V.; A.B., Harvard, 1939, M.D., 1943; m. Deborah Cloud, Dec. 27, 1941; children—Jonathan, Judith (dec.), Sarah Fox, Joanna. Research fellow in pediatrics Harvard, Children's Hosp., Boston, 1947-49; instr., then asst. prof. pediatrics Yale Sch. Medicine, 1949-52; asso. prof. pediatrics Temple U. Sch. Medicine, 1952-57, prof. chmn. dept., 1964—; prof., chmn. dept. pediatrics Med. Coll. Ga., Augusta, 1957-64; med. dir. St. Christopher's Hosp. for Children, Phila., 1964—. Cons. Nat. Inst. Child Health and Human Devel., NIH, 1963—. Diplomate Am. Bd. Pediatrics (dir. 1960-65, sec. 1963-65). Mem. A.M.A., Am. Pediatric Soc., Soc. for Pediatric Research (v.p. 1964-65), Am. Acad. Pediatrics, Am. Acad. Allergy, Am. Coll. Allergists, Am. Soc. for Human Genetics, Med. Com. for Human Rights. Research on hemolytic disease of newborn, childhood allergies, growth and devel. Home: 125 W. Walnut Lane, Phila. 19144. Office: St. Christopher's Hosp. for Children, 2600 N. Lawrence St., Phila. 19133.*

VAUGHAN, Willard Stanley, Jr., Am. psychologist; b. Pitts., Apr. 6, 1929; s. Willard Stanley and Virginia (Evans) V.; B.S., Allegheny Coll., 1951; M.A., U. Md., 1954, Ph.D., 1956; m. Barbara Woodward, Sept. 6, 1952; children—Martha Grace, Andrew Evans, Roger Daniel, Emily Virginia. Instr. psychology Lehigh U., 1955-56; research asso. Psychol. Research Assts., Inc., Arlington, Va., 1956-57; dir. Human Scis. Research, Inc., McLean, Va., 1957-65; pres., treas. Whittenburg Vaughan Assos., Inc., Alexandria, Va., 1966—. Mem. Am., Eastern psychol. assns., Human Factors Soc., Sigma Xi. Psychol. surveys, studies, publs. on influences upon and aids to tactical mil. decision-making, requirements of certain mil. positions. Home: 3223 Alice Ct., Falls Church, Va. 22042. Office: 4810 Beauregard St., Alexandria, Va. 22312.*

VAUGHAN, Wyman Ristine, Am. organic chemist; b. Mpls., Oct. 28, 1916; s. James Albert and Katherine (Wyman) V.; A.B., Dartmouth, 1939, A.M. (Cramer fellow), 1941; M.A. (Austin fellow), Harvard, 1942, Ph.D., 1944; m. Hope Brutschy, June 19, 1943; children—Sarah Joan, Douglas Wyman. Faculty, Dartmouth, 1939-41, 44-46, Harvard, 1942-44; faculty U. Mich., Ann Arbor, 1946-66, prof. chemistry, 1959-66; prof., dept. head U. Conn., Storrs, 1966—. Mem. Am. Chem. Soc. (organic div.), N.Y. Acad. Scis., Phi Beta Kappa, Sigma Xi, Phi Lambda Upsilon, Chi Psi. Translator: Organische Chemie (K. Freudenberg, H. Plieninger), 1964. Research, numerous publs. on structural effects in organic chemistry, organic chem. reaction mechanisms using isotopic tracers and kinetics, synthesis candidate anti-cancer agts. and bicyclic terpenoids. Home: R.R. 1, Box 153, Storrs, Conn. 06268.*

VAUGHAN WILLIAMS, Edward Miles, physician; b. Bangalore, India, Aug. 8, 1918; s. Arthur and Stella (Pressy) V.W.; B.A., Oxford U., 1942, B.Sc., 1946, B.M. B.Ch., 1947, D.M., 1952, D.Sc., 1960; m. Marie De Lagarde, May 25, 1956; children—Penelope, Armelle, Roland. Univ. lectr. Oxford (Eng.) U., 1953-55, ofcl. fellow Hertford Coll., 1955—. Mem. Pharm. Soc. Gt. Britain, Physiol. Soc. Gt. Britain. Research, numerous publs. on intestinal movements, method for simultaneous measurement of mech. and elec. behavior of single cardiac cells, analysis of mode of action of antirhythmic agts., introduced double-shock method for increasing force of cardiac contraction, investigated anti-arrhythmic properties and discovered local anaesthetic activity of B-receptor blockers. Home: 153 Woodstock Rd., Oxford, Eng.*

VAUGHN, Reese Haskell, Am. microbiologist; b. Farragut, Ia., Oct. 1, 1908; s. Charles Arthur and Nellie (McAllister) V.; A.B., Simpson Coll., 1930; M.S., Ia. State Coll., 1932, Ph.D., 1935; m. Marjorie Helen Barton, June 23, 1935; children—Reese Barton, Mary Elizabeth (Mrs. Russell Lee Hunter), Charles Edward, John Arthur, Virginia Ann. Faculty, U. Cal., 1936—, prof., Davis, 1953—, bacteriologist, 1952-57, food technologist, 1957—, chmn. dept. food sci. and tech., 1963-66. Rockefeller Found. fellow, 1957; Fulbright lectr., 1967-68. Fellow Am. Acad. Microbiology, Royal Soc. Health; mem. Soc. Microbiology, Soc. Enology, Pub. Health Inst., Inst. Food Tech., Inst. Sanitation Mgmt. Research, numerous publs. on food fermentations and bacteria causing them, spoilage of foods, pectinolytic enzymes involved in softening of fruit and vegetable tissues, bacterial indices of sanitation, food plant waste disposal. Home: 739 N. Campus Way, Davis, Cal. 95616.*

VAUQUELIN, Louis Nicolas, French chemist; b. St. André D'Hébertot, Normandy, France, May 16, 1763; student pharmacy. Asst. to Fourcroy, 1783-91; prof. chemistry École des Mines, École Polytechnique, 1795, Collège de France, Paris, 1801, Muséum d'histoire naturelle, from 1804, Faculté de Medicine, Paris, from 1809; assayer gold, silver, 1802; dir. École de Pharmacie, 1803. Mem. Conseil des Arts et Manufactures. Legislative dep. from Calvados, French Assembly, from 1827. Mem. French Acad. Scis. (v.p. 1818, pres. 1819), 1795, Acad. Medicine, Soc. Agr. Author: Manuel de l'essayeur, 1812; Dictionnaire de Chimie et de Metallurgie, 6 vols., 1815. Contbd. more than 60 memoirs with Fourcroy. Discovered chromium and its compounds, 1798, beryllium compounds, 1798, quinic acid, (with P. J. Robiquet) asparagine (1st of amino acids to be discovered), 1806, camphoric acid, composition of alum; contbd. to studies animal chemistry. Died St. André D'Hébertot, Nov. 14, 1829.

VAVILOV, Sergei Ivanovich, Russian physicist; b. Mar. 12, 1891; ed. Moscow U. Asst. prof. physics Moscow Higher Tech. Sch., 1918-20, became prof., 1920; named lectr. Moscow U., 1920; named dir. P. N. Lebeden Phys. Inst., Moscow, 1932; dept. from Leningrad, Supreme Soviet of USSR; founder new br. phys. optics Physics Lab., USSR Acad. Scis. Recipient Order of Lenin, twice; Stalin prize, 1942. Mem. USSR Acad. Scis. (became pres. 1945), (chmn. organizational com. Soc. for Dissemination Polit. and Sci. Knowledge 1947). hon. mem. Ljubljana, Bulgarian, Belgrade, Croatian acads. Editor physics sect. Soviet Ency. Translated Newton from Latin into Russian. Research and publs. on optics, especially photoluminescent phenomena, relation of light to vision and optical physiology; studied fading of colors under heat action; discovered law for light luminescence, source of cold light by conversion of ultra-violet radiation. Died Jan. 25, 1951.

VAZQUEZ, Jacinto J., physician; b. Havana, Cuba, Aug. 17, 1923; s. Francisco and Ramona (Lopez) V.; M.D., U. Havana, 1948; M.S., Ohio State U., 1954; m. Martha D. Sierra, Aug. 19, 1948; children—Flavia I., Rachel M. Naturalized Am. citizen, 1955. Instr., asst. prof. pathology U. Pitts., 1955-59, asso. prof., 1959-61; asso. mem. Scripps Clinic and Research Found., La Jolla, Cal., 1961-63; practice medicine, specializing in pathology, Durahm, N.C.; asso. prof. pathology Duke, 1963-65, prof., 1965—. Cons. Oak Ridge Nat. Lab., NIH. Mem. Am. Assn. Pathologists and Bacteriologists, Am. Soc. Exptl. Pathology, Am. Assn. Immunologists, Internat. Acad. Pathology, A.A.A.S., Soc. Exptl. Biology and Medicine. Editorial bd. Am. Jour. Pathology. Research, numerous publs. in pathogenesis of immunologic diseases, immunopathology cellular aspects of antibody formation, cytokinetics of immune response. Home: 1707 Woodburn Rd., Durham 27705. Office: Duke U. Med. Center, Durham, N.C. 27706.*

VDOVENKO, Viktor Mihailovich, Russian chemist; b. Jan. 5, 1907; grad. Kiev Chemico-Technol. Inst. Food Industry, 1930. Staff, Kiev Chemico-Technol. Inst. Food Industry, until 1935; joined Inst. Chem. Physics, USSR Acad. Scis., 1935, staff Radium Inst., 1953, later dir., joined faculty Leningrad U., 1935, became prof., 1953. Corr. mem. USSR Acad. Sci. Research and publs. in radiochemistry, inorganic and phys. chemistry, radioactive element exchange between immiscible solvents; determined solubility of nonaqueous solutions of radioelements; established relation between structure and extractability of organic solvents. Office: V. G. Khlopin Radium Inst., USSR Acad. Sci., Ulitsa Roentgena 1, Leningrad., USSR.

VEATCH, Ralph Wilson, Am. mathematician; b. Ringwood, Okla., May 1, 1900; s. Henry W. and Alice E. (Keller) V.; A.B., Henry Kendall Coll., 1925; M.A., Northwestern U., 1927; postgrad. Columbia, U. Chgo., U. Wis.; m. Potter Lockwood, June 11, 1932; children—Annie Laurie (Mrs. Charles B. Stebbins), Ralph Wilson. Faculty, Ursinus Coll., 1927-30; faculty U. Tulsa, 1930—, prof. math., 1942—, head math. dept., 1946-65. Mem. Math. Assn. Am., Am. Soc. for Engring. Edn., Am. Assn. U. Profs., Sigma Xi, Kappa Mu Epsilon. Research in number theory, theory of equations, empirical forms and curve fitting. Home: 332 N. Santa Fe St., Tulsa. 74127.*

VEBLEN, Oswald, Am. mathematician; b. Decorah, Ia., June 24, 1880; s. Andrew A. and Kirsti (Hougen) V.; A.B., U. Ia., 1898; A.B., Harvard, 1900; Ph.D., U. Chgo., 1903; hon. D.Sc., Oxford, 1929; hon. Ph.D., U. Oslo, 1929, Hamburg, 1933; honorary D.Sc., U. Chgo., 1941, LL.D., Glasgow U., 1951; m. Elizabeth M. D. Richardson, 1908. Asso. in mathematics, U. Chgo., 1903-05; preceptor in mathematics, 1905-10, prof., 1910-32, Princeton; prof. Inst. for Advanced Study, Princeton, 1932-50, professor emeritus since 1950. Pres. International Congress Mathematicians held at Harvard, 1950. Chmn. phys. sciences, Nat. Research Council, 1923-34; Fellow Am. Acad. Arts and Sciences, Am. Phys. Soc., A.A.A.S.; mem. Nat. Acad. Sciences, Am. Philos. Soc., Am. Math. Soc. (pres. 1923-24), Math. Assn. of America; honorary mem. London Math. Soc., Circolo Matematico di Palermo, Société Mathématique de France (hon. mem. bureau); foreign corr., Academia Nacional de Ciencias Exactas, Lima, Peru; hon. fellow, Royal Soc., Edinburgh; member Royal Irish Academy (Department of Science); foreign member Royal Danish Acad. of Sciences; Polish Acad. Scis. and Letters, Accademia dei Lancei; Knight, 1st Class, Royal Order of St. Olav (Norway). Army-Navy Certificate of Merit, 1948. Author: Infinitesimal Analysis (with N. J. Lennes), 1907; Projective Geometry (Volume I, with J. W. Young), 1910, Vol. II, 1918, Cambridge Colloquium Lectures on Analysis-Situs, 1922; Invariants of Quadratic Differential Forms, 1927; Foundations of Differential Geometry (with J. H. C. Whitehead), 1932; Projektive Relativitätstheorie, 1933; Geometry of Complex Domains (with Wallace Givens), 1936. (With W. H. Bussey) founded finite projective geometry; worked on four-color mapping of topology; research on relation between differential geometry and topology, variable theory, axiom system in projective geometry, ordinal numbers. Died Aug. 10, 1960.

VEBLEN, Thorstein B., Am. economist, social theorist; b. Cato township, Manitowoc Co., Wis., July 30, 1857; s. Thomas Anderson and Kari (Bunde) V.; A.B., Carleton Coll., 1880; grad. student Johns Hopkins; Ph.D., Yale, 1884; fellow in economics and finance, Cornell, 1891-92; fellow U. of Chicago, 1892-93; m. Ellen May Rolfe, Apr. 10, 1888 (div.); m. 2d Anne Bradley Fessenden, June 17, 1914. Reader in polit. economy, 1893-94, asso., 1894-96, instr., 1896-1900, asst. prof. 1900-06, U. of Chicago; asso. prof. economics, Leland Stanford Jr. U., 1906-09; lecturer in economics, U. of Mo., 1911-18; teacher, New School for Social Research, New York, 1918-29. Mng. editor Journal of Political Economy, 1896-1905. Author: The Theory of the Leisure Class, 1899; The Theory of Business Enterprise, 1904; The Instinct of Workmanship, 1914; Imperial Germany and the Industrial Revolution, 1915; An Inquiry Into the Nature of Peace and the Terms of Its Perpetuation, 1917; The Higher Learning in America, 1918; The Vested Interests, 1919; The Place of Science in Modern Civilization and Other Papers, 1920; The Engineers and the Price System, 1921; Absentee Ownership and Business Enterprise in Recent Times, 1923; translator: The Laxdaela Saga, 1925. Exerted profound influence on Am. thought; drew parallel between barbarian rulers and modern aristocrats, whom he accused of conspicuous waste and conspicuous consumption; distinguished between production for profit and for use; believed prime duty of economic and social scholarship is to separate reality from abstractions. Died Palo Alto, Calif., Aug. 3, 1929.

VEDDER, Edward Bright, Am. physician; b. N.Y.C., June 28, 1878; s. Henry C. and Minnie (Lingham) V.; Ph.B., U. Rochester, N.Y., 1898, D.Sc., 1924; M.D., U. Pa., 1902, M.A., 1903; honor grad. Army Med. Sch., 1904; m. Lily S. Norton, June 22, 1903; children—Sibyl Norton, Henry C. 1st lt., asst. surgeon Med. Corps, U. S. Army, July 1903; advanced through grades to col., 1929. Served at various stas., U. S. and Pi., 1904-10; mem. U. S. Army Bd., for study tropical diseases, 1910-13; asst. prof. pathology, Army Med. Sch., Washington, 1913-19; charge So. Dept. Lab., Ft. Sam Houston, Tex., 1919-22; chief, med. research div., Edgewood Arsenal, Md., 1922-25; sr. mem. U. S. Army Bd. Med. Research, 1925-28; dir. Army Med. Sch., 1930-32; ret., 1933; prof. exptl. medicine, George Washington U., 1933-42; dir. med. edn., Alemeda County Hosp., 1942-47, Fellow A.C.P., A.C.S., A.M.A., A.A.A.S.; mem. Am. Soc. Tropical Medicine, Acad. Tropical Medicine, Washington Acad. Scis., Washington Acad. Medicine Assn. Mil. Surgeons U. S., Sigma Xi. Author: Beriberi 1913, Sanitation for Medical Officers, 1917; Syphilis and Public Health, 1918; The Medical Aspects of Chemical Warfare, 1925; Medicine—Its Contribution to Civilization, 1929; also many papers med. jours. Studied tropical medicine; developed use of emetine to treat amoebic dysentery. Died Jan. 30, 1952.

VEIBEL, Stig Erik, Danish chemist; b. Korsør, Denmark, Apr. 19, 1898; s. Bertel Christian and Vilhelmine (Petersen) V.; cand.polyt., Tech. U., Denmark, 1920; Dr.phil., U. Copenhagen (Denmark), 1929; Dr.h.c., U. Bordeaux (France), 1961; m. Ellen Kirk, Aug. 20, 1940. Asst. chem. lab. U. Copenhagen, 1920-44; Rockefeller grantee, Berlin, Germany, 1930-31; faculty Tech. U. Denmark, Lyngby, 1932-68, prof., head dept. organic chemistry, 1944-68. Mem. Danish Acad. Tech. Scis., Royal Danish Acad. Scis., Société Chimique de France (hon.), Polski Towarszystvo Chemiczne (hon.), Academie des Sciences Paris (corr.). Author: Vejledning i organiske stoffers identifikation, 1926; Kemiens historie i Danmark, 1939; Dansk Kemisk bibliografi 1800-1935, 1943; Identification of Organic Compounds, 1954; Organisk Kemi I-II, 1951; also numerous articles. Research on nitration on phenol, enzymic reaction (beta-glycosidases), heterocyclic compounds (pyrazolones, benzodiazephines), organic functional group analysis. Home: 4, Enighedsvej, Charlottenlund, DK2920, Denmark. Office: Bldg. 201, Anker Engelundsvej, Lyngby DK2800, Denmark.*

VEIRS, Everett Raymond, Am. physician; b. Vine Grove, Ky., Sept. 5, 1908; s. John Owen and Mary (Logsdon) V.; B.S., U. Ky., 1931; M.D., U. Louisville, 1935; m. Mary Stone, Sept. 2, 1942; children—Mary Ann, Susan Carol, Laura Elizabeth. Practice medicine specializing in ophthalmology; chief dept. ophthalmology Scott and White Clinic, Scott and White Meml. Hosp., 1945——; chief oculist Santa Fe R.R. Hosp., Temple, Tex., VA Hosp., Temple. Cons. ophthalmology Ft. Hood Sta. Hosp (Tex.); lectr. ophthalmology U. Tex. Postgrad. Sch. Medicine, Temple. Diplomate Am. Bd. Ophthalmology. Fellow Am. Acad. Ophthalmology and Otolaryngology; mem. Bell County Med. Soc., Tex., So. med. assns., A.M.A., Pan Pacific Surg. Assn., Tex. Ophthalmology and Otolaryngology Soc. Author: The Lacrimal System, Clinical Application, 1955; So You Have Glaucoma, 1958; The Lacrimal System in Clinical Practice, 1963; also articles. Research in ophthalmology especially lacrimal system and viral infection of eyes. Home: 505 W. Park St. Office: Scott and White Clinic, Temple, Tex. 76501.*

VEIS, Arthur, Am. chemist; b. Pitts., Dec. 23, 1925; s. Fred Max and Sara (Landis) V.; B.S. in Chem. Engring., U. Okla., 1947; Ph.D. in Phys. Chemistry, Northwestern U., 1951; m. Eve M. Zenner, June 24, 1951; children—Judith Hannah, Sharon Lynn, Deborah Jean. Instr., U. Okla., Norman, 1951-52; research chemist Armour & Co., Chgo., 1952-60, head phys. chemistry dept., 1959-60; spl. instr. Crane Jr. Coll., Chgo., 1955-56, Loyola U., 1957-58; faculty Northwestern U. Sch. Medicine, Chgo., 1960——, prof. dept. biochemistry, 1965——. Mem. adv. com. on animal products mil. personnel supplies NRC-Nat. Acad. Scis., 1965. Guggenheim fellow, 1967-68. Mem. Am. Chem. Soc. (dir. Chgo. sect. 1954——, past vice chmn., councilor 1963——), Am. Soc. Biol. Chemists, N.Y. Acad. Scis., Biophys. Soc. Author: The Macromolecular Chemistry of Gelatin, 1964; also articles. Research in chemistry collagen, formation prebiologic systems interacting polyelectrolytes. Home: 7633 Lowell St., Skokie, Ill. 60076. Office: 303 Chicago Av., Chgo. 60611.*

VEIS, George, rural and topographical engr.; b. Athens, Greece, Sept. 8, 1929; s. Konstantinos and Aspasia (Aslanisdou) V.; student Nat. Tech. U. Athens, Ecole Nationale des Sciences Geographique, Sorbonne, Paris; Ph.D., Ohio State U., 1957; m. Zoe Svolou, Oct. 6, 1955; children—Alexander, Konstantinos. Asst., Nat. Tech. U., Athens, 1952-55, asso. prof. of Surveying, 1960——; asst. Inst. Geodesy, Photogrammetry and Cartography, 1957; research asso. Harvark Coll. Obs., 1959——; dir. Astrophys. Observing Sta., Athens. Cons. Smithsonian Astrophys. Obs., 1958——. Decorated Golden Cross. Mem. Greek Aero. Union, Greek Geog. Soc., Am. Astron. Union. Author: The Uses of Artificial Satellites in Geodesy, 1963; Optical Tracking of Artificial Satellites, 1963; The Determination of Absolute Direction in Space, 1965. Research, numerous publs. on satellite geodesy. Home: 2 Lahitos St., Athens 601, Greece. Office: 60 Garden St., Cambridge, Mass. 02138.*

VEKSLER, Vladimir Iosifovich, Russian physicist; b. Mar. 4, 1907; grad. Moscow Energetics Inst., 1931. Staff, All Union Electro-Tech. Inst., 1930-36, Physics Inst., USSR Acad. Sci., United Inst. for Nuclear Problems, 1956——. Recipient Lenin prize for participation in creation synchro-cyclotron 10 billion electronic volts, 1959; Order of Lenin; (with E. M. McMillan) U. S. Atoms for Peace award, 1963. Author: Experimental Methods in Nuclear Physics, 1940; Ionization Methods in Irradiation Research (with L. Groshev, B. Isaev), 1949. Research and publs. on physics, X-ray machines, cosmic rays, particle accelerators; a founder of high-energy exptl. physics. Developed principle of particle outphasing. Died Sept. 22, 1966.

VEKUA, Ilya Nestorovich, Russian mathematician; b. Sheshelety, USSR, May 6, 1907; grad. Georgia State U., Thilisi, 1930; Dr.Phys.-Math. Sci. Prof., Tbilisi Stalin U., until 1952, Moscow U., 1952-57; named dep. dir. Math. Inst., USSR Acad. Sci., 1957, editorial bd., 1959——. Recipient Stalin prize, 1950. Mem. USSR Acad. Sci. (mem. presidium Siberian dept. 1958——), Georgia Acad. Sci. Author: New Methods of Solving Elliptical Equations, 1948; Stationary Singular Points of Generalized Analytical Functions, 1962. Research and publs. on various elliptical partial differential equations, general boundary problems, analytical functions of complex variables; solved equations of steady-state oscillations. Home: Novopeschanava, Korp. 25. Office: V.A. Stekkov Inst. Math., USSR Acad. Sci., 1-y Akademicheskii Proyezd, 28, Moscow, USSR.

VELARDO, Joseph Thomas, Am. physiologist, anatomist; b. Newark, Jan. 27, 1923; s. Michael A. and Antoinette (Iacullo) V.; A.B., Colo. State Coll., 1948; S.M., Miami U., Oxford, O., 1949; Ph.D., Harvard, 1952; m. Forresta M. Power, Aug. 12, 1948. Research fellow Harvard, 1952-53, research asso., Sch. Medicine, 1953-55; asst. in surgery Peter Bent Brigham Hosp., 1954-55; asst. prof. Yale Sch. Medicine, 1955-61; prof., chmn. dept. anatomy N.Y. Med. Coll., 1961-62; dir. Inst. for Study Human Reprodn., Cleve., 1962-67; prof. biology and endocrinology John Carroll U., Cleve., 1962-67; staff research and edn. St. Ann Hosp., Cleve., 1962-67; head dept. research, 1964-67; prof., chmn. dept. anatomy, Loyola U. Stritch School of Medicine, 1967——. Recipient Lederle Med. Faculty award 1955-58. Fellow A.A.A.S., N.Y. Acad. Scis.; mem. Am. Soc. for Study Sterility (Rubin award 1955), Am. Assn. Anatomists, Gerontological Soc., Am. Soc. Zoologists, Am. Physiol. Soc., Endocrine Soc., Soc. for Endocrinology, Soc. for Exptl. Biology and Medicine, Am. Soc. for Study Sterility, Internat. Fertility Soc., Sigma Xi, Kappa Delta Pi, Phi Sigma, Gamma Alpha, Alpha Epsilon Delta. Author: (with Charles G. Rosa, G. Fischer) Enzymes in Female Reproductive System, 1963; also numerous articles. Editor: Essentials of Human Reproduction, Clinical Aspects, 1958; Endocrinology of Reproduction, 1958; cons. editor: The Uterus, 1959. Editor, Continuing Med. Edn., 1963——. Research on physiology placenta, induction ovulation, hormonal influence on uterine growth, physiology of ovary. Address: Loyola Med. Center, Maywood, Ill.

VELEZ, Ismael, Am. botanist; b. Yauco, P.R., June 22, 1908; s. Carmelo and Maria (Irizarry) V.; A.B., Inter Am. U., 1930; M.S., La. State U., 1934, Ph.D., 1939; m. Lucia Irizarry, Dec. 27, 1935; children—Iris Alicia (Mrs. Kelvin Wheatley), Lester Rene. Faculty, Inter Am. U., San German, P.R., 1935——, prof., 1946——, dir. Summer Sch., acting dean, 1948-51, dir. Natural Sci. div., 1958-65, now head biology dept. Fulbright lectr. botany U. Ceylon, 1956-67; Fulbright lectr. plant pathology Universidad Nacional del Cuzco, Peru, 1960; adviser Banco Interamericano Desarollo and Ford Found. Program U. San Marcos, Lima, Peru, 1964; travelling lectr. NSF, 1962-65. Fellow A.A.A.S.; mem. Am. Soc. Agrl. Scis. (past pres. P.R. chpt.), Assn. Biology Tchrs., Sci. Tchrs. Assn. (past pres. P.R. chapt.), Ecol. Soc., Beta Beta Beta, Gamma Sigma Delta, Zigma Xi. Studies, publs. on various tropical plants; methods, programs of teaching of biology. Home and office: Inter Am. U. Campus, San German, P.R.*

VELICAN, Constantin Constantin, Rumanian physician; b. Murgeni, Iasi, Rumania, Apr. 18, 1919; s. Constantin A. and Eugenia (Lazaresco) V.; grad. Faculty Medicine, Bucharest, 1943; m. Doina Constantinesco, Dec. 24, 1957. Staff, Inst. Internal Medicine, R. S. Rumania Acad., Bucharest, 1949——, head lab., 1953-54, head dept., 1954——; asst. Faculty Medicine, Bucharest, 1945-50. Recipient Acad. Prize laureate, 1944, 64. Mem. Histochem. Soc. (chief Bucharest sect. since 1965——). Union Med. Sci. Socs. Author: (with N. G. Lupu) Role of the Nervouse System in the Pathologeny of Pneumoconiosis, 1953, Nonoccupational Pneumoconiosis, 1955; Nervous Regulation of Adrenal Gland, 1956; (with L. Kleinerman) Pulmonary Circulation, 1959; (with N. G. Lupu, C. Racoveanu, M. Gociu) Pulmonary Sclerosis, 1960; (with Doina Velican, N. Carp) Histochemistry and Pathophysiology of Mucopolysaccharides, 1963; (with Doina Velican) Histochemie des glucides en pathologie humaine, 1968; also numerous articles. Research in histochemistry, especially atherosclerosis, pulmonary fibrosis, chronic hepatitis and rheumatic diseases, role of muco-polysaccharides, glycoproteins and scleroproteins; studies on mast cells, basement membranes and arterial wall; promoted preclin. orientation in path. anatomy, theory of biopolymers' self-protection. Home: 4 Columb, Bucharest. Office: Inst. Internal Medicine, Sos. Stefan cel Mare Bucharest 10, Rumania.*

VELIKANOV, Mikhail Andreevich, Russian hydrologist, hydrodynamicist; b. Jan. 22, 1879; grad. Inst. Engrs. Lines Communication. Became engr. O'bt Yenisei Rivers; field studies on rivers, 1912-21; faculty Moscow Hydrometeorol. Inst., 1922-29, Central Asiatic U., Tashkent, 1942-43, Moscow U., 1945-54; organizer flow stas. Corr. mem. USSR Acad. Sci. Author: Hydrology of Continents, 1925. Research and publs. on water balance, river-bed hydrology, large hydraulic projects; founder labs. for hydrometeorol. service, also USSR Acad. Scis. Home: Leninskii Prospekt, 13. Office: USSR Acad. Sci., Leninskii Prospekt 14, Moscow, USSR.

VELIKOVSKY, Immanuel, cosmologist; b. Vitebsk, Russia, June 10, 1895; s. Simon-Yehiel and Beila-Rachel (Grodensky) V.; student Edinburgh U., Moscow Econ. Inst., Kharkov U.; M.D., U. Moscow, 1921; post grad. Charite, Berlin, Monakow Brain Inst., Zurich; studied psychoanalysis under Wilhelm Stekel, 1933 Vienna; m. Elisheva Kramer, Apr. 15, 1923; children: Shulamith, Ruhama-Ruth; practiced as M.D. then psychoanalyst, Palestine, 1924-39. Pub., organizer, ed. Scripta Universitatis atque Bibliothacae Hierosolymitanarum, 1921-24. Author: Worlds in Collision, 1950; Ages in Chaos, 1952; Earth in Upheaval, 1955; Oedipus and Akhnaton, myth and history, 1960; contbr. to scientific jours.; editorials on Middle East, New York Post, 1948-49. Suggested characteristic brain waves in epileptics, 1930. Cosomological postulations caused controversy in modern science; hypothesized Venus' atmosphere must be very hot, 1950 (confirmed by recent Venus space probe); posited that Jupiter emits radio waves, 1953 (confirmed experimentally by B. F. Burke and K. L. Franklin, 1955); suggested existence of terrestrial magnetosphere reaching to moon, 1956 (confirmed by Van Allen, 1958). Address: 78 Hartley Ave., Princeton, N.J.*

VELLE, Weiert, Norwegian physiologist; b. Lista, Norway, May 18, 1925; s. Olav Reinertsen and Severine Velle; student Vet. Coll. Norway, Oslo, 1947-53; Dr.med.vet., U. Oslo, 1959; m. Kristin Lunaas, Nov. 17, 1956; children—Ole Jacob, Hans Jorgen, Tone, Stine. Research fellow Vet. Coll., Oslo, 1956-60, Harvard, 1960-62; faculty Vet. Coll., Oslo, 1962—, prof. physiology, 1964—. Mem. Am. Endocrine Soc., Soc. for Reprodn. and Fertility, Scandinavian Physiol. Soc. Research, publs. on estrogenic hormones in domestic animals, control of ovulation, biosynthesis of fructose in various animals. Home: 15 Homans vei, Blommenholm, Norway. Office: Vet. Coll. Norway, Oslo, Norway.*

VELPEAU, Alfred Louis Armand Marie, French surgeon; b. Breches, France, May 18, 1795; M.D., U. Paris (France), 1823. Health officer, Paris; tchr. anatomy, surg. pathology, operatory medicine, after 1823; surgeon La Pitié, from 1830; prof. clin. surgery Faculté de Médecine, from 1835; chief surgeon Hopital de la Charité, from 1841. Mem. French Acad. Sci. (v.p. 1862, pres. 1863), 1843; Académie de Médecine. Author: Traité complet d'anatomie chirurgicale, 1825; Embryologie et ovologie humaine, 1833; De l'opération du trépan dans les plaies de tete, 1834; Traité complet de l'art des accouchements ou tocologie Théorique et practique, 1835; Nouveaux éléments de médecine opératoire, 1839; Leçons orales de clinique chirurgicale, 1840-41; Traité des maladies du sein et de la region mammaire, 1842; (with Beraud) Manual d'anatomie chirurgicale topographique, 1862; Leçons sur le diagnostic et le traitement des maladies chirurgicales, 1866. Well known surgeon 1st half 19th century; used anesthetic in childbirth; surg. handling aneurysm; wrote 1st textbook on diseases of female breast; invented bandage which supports arm in case of fractured clavicle (Velpeau's bandage). Died Paris, Aug. 24, 1867.

VELVOD, William, see Welwood, William.

VENABLE, Francis Preston, Am. chemist; b. Prince Edward County, Va., Nov. 17, 1856; s. Charles Scott and Margaret Cantey (McDowell) V.; grad. U. Va., 1879; studied at Bonn, 1879-80; A.M., Ph.D., Göttingen, 1881; Berlin, 1889; Sc.D., Lafayette, 1904; LL.D., U. Pa., U. Ala., U. S.C., Jefferson Med. Coll.; m. Sally Charlton Manning, Nov. 3, 1884; children—Louise M., Cantey McD., Charles S., John M., Frances P. Prof. chemistry, 1880-1900, pres., 1900-14, prof. chemistry, 1914-30, U. N.C. Mem. advisory bds. of Bur. Mines, 1917-23, and Chem. Warfare Service, 1918. Pres. Southern Ednl. Assn., 1903, Am. Chem. Soc., 1905, Southern Assn. Sch. and Colls., 1909; fellow Chem. Soc., London, A.A.A.S. Author: Manual of Qualitative Analysis, 1883; Short History of Chemistry, 1894; Development of Periodic Law, 1896; Inorganic Chemistry According to the Periodic Law (with James Lewis Howe), 1898; Study of the Atom, 1904; Radioactivity, 1917; Zirconium and Its Compounds, 1921. Determined atomic weight of zirconium, 1898; developed mfg. process for calcium carbide making comml. prodn. acetylene gas practical, 1895. Died Chapel Hill, N.C., Mar. 17, 1934.

VENABLES, Peter Henry, English psychologist; b. Ilfracombe, Devon, Eng., Apr. 3, 1923; s. Harry Francis and Lilian (Harris) V.; B.A. with 1st class honors, U. Coll. London, 1951; Ph.D., London Inst. Psychiatry, 1953; m. Agnes Hawkins, July 11, 1948; children—Peter Allan, Andrew John. Mem. sci. staff Med. Research Council Social Psychiatry Research Unit, Maudsley Hosp., London, 1951-63; reader in psychology Birkbeck Coll., U. London, 1964—. Fellow Brit. Psychol. Soc. (treas. 1963-65, dep. pres. 1965-66); mem. Exptl. Psychology Soc. (sec. 1964-), Soc. Psychophysiol. Research. Editor: A Manual of Psychophysiological Methods, 1966; asso. editor Psychophysiology, 1965—. Publs. on application of exptl. psychol. and psychophysiol. techniques to study of chronic schizophrenia; analysis of certain physiol. indices to psychol. state. Home: 62 Whyteleafe Rd., Caterham, Surrey, Eng. Office: Dept. Psychology, Birkbeck Coll., Malet St., London W.C. 1, Eng.*

VENCOVSKY, Eugen Josef, Czechoslovakian psychiatrist; b. Protejov, Czechoslovakia, Apr. 9, 1908; s. Frantisek Jan. and Marie (Stepánková) V.; Diploma, Faculty Medicine, Charles U., Prague, Czechoslovakia, 1933, D.Sc., Diploma, 1962; m. Blanka Fiedlerová, Mar. 19, 1915; children—Eva (Mrs. Valtr Bubenícek), Pavel. Asst. dr. dept. psychiatry Hosp. Prague, 1933-39; leading physician Psychiat. dept. Municipal Policlinic, Plzen, Czechoslovakia, 1939-42, primarius, 1942-45; dept. head Psychiat. Clinic, Faculty Medicine, Plzen, 1945-51, head, 1951—, faculty, 1947—, prof. psychiatry, 1957—; expert in psychiatry for W. Bohemia, 1960—. Mem. adv. bd. for psychiatry Ministry of Health, 1960—, mem. main com. for state research in psychiatry, 1962—; mem. Psychiat Chair, Inst. for Postgrad. Studies of Physicians in Prague, 1962—. Recipient medal of merit for devel. psychiatry in W. Bohemia, 1957, certificate of merit UNRRA, 1946. Mem. Czechoslovakian Psychiat. Soc. (pres. 1963—), J. E. Purkyne in Prague, Polish Psychiat. Soc. in Warsaw, Collegium Internationale Neuro-Psychopharmacologicum, Internat. Soc. for History Medicine, Internat. Assn. for Prevention Suicides, N.Y. Acad. Scis., World Psychiat. Assn., World Fedn. for Mental Health. Author: Paraphrenias, 1948; History of Czech Psychiatry, 1957; Historical Development of Psychiatry from Hippocrates to Pinel, 1963; also numerous articles. Research on rare sexual aberration lactophilia and colostrophilia, phenomenon of hydropsy, syndromollitical effect of psychopharmaca, anti psychotic effect of thiethylperazin and its use in psychopharmacotherapy; first Czech monographs on history of psychiatry. Home: Drevená ul. 3, Plzen. Office: 69 Dukelská ul., Plzen, Czechoslovakia.*

VEND, John, Brit. philosopher; b. Hull, Eng., Aug. 4, 1834; s. Henry and Martha (Sykes) V.; degree, 6th wrangler Gonville and Caius Coll., Cambridge, 1857; m. Susanna Carnegie Edmonstone, 1867; 1 son, John Archibald. Became deacon, 1858, priest, 1859; curate at Cheshunt, also Montlake, Eng.; became lectr. moral sci., Cambridge, 1862; became Hullsean lectr., 1869; tchr. logic until 1888, then univ. history; pres. Caius Coll., 1903-23. Fellow Royal Soc., 1883. Author: The Logic of Chance, 1866; Some Characteristics of Belief, Scientific and Religious, 1869; Symbolic Logic, 1881; The Principles of Empirical Logic, 1889; Biographical History of Gonville and Caius College, 1897; Matriculations and Degrees, 1544-1659, 1913; Venn Family Annals, 1904; Early Collegiate Life, 1913. Studies in logic and moral philosophy, univ. history. Died Cambridge, Eng., Apr. 4, 1923.

VENDITTI, John Michael, Am. pharmacologist; b. Balt., Feb. 19, 1927; s. John Anthony and Mary (Aiello) V.; B.S., U. Md., 1949, M.S., 1957; Ph.D., George Washington U., 1965; m. Nancy Orth, Mar. 24, 1951; children—Nancy May, John Michael, Mary Ruth. Biologist, Nat. Cancer Inst., Bethesda, Md., 1951-58, supervisory research pharmacologist, 1958-63, head screening sect. Drug Evaluation br. Cancer Chemotherapy Nat. Service Center, 1963-65, chief Drug Evaluation br., 1966—. Mem. A.A.A.S., Am. Assn. Cancer Research, Soc. Exptl. Biology and Medicine, Sigma Xi. Research, numerous publs. on chemotherapy of cancer including devel. assay systems for drug evaluation, combination chemotherapy, metabolite-antimetabolite relationships, studies of resistance and cross-resistance. Office: Wiscon Bldg., 7550 Wisconsin Av., Bethesda, Md. 20014.*

VENEL, Gabriel F., French chemist, surgeon; b. Combes, France, 1723; B.A., 1741; student Montpellier, France; later studied under Rouelle, Paris. Mem. Royal Soc. Montpellier. Author: Instruction sur l'usage de la houille, 1774; also publs. on physiology, pharmacology, chemistry, medicine. Developed method for teaching of medicine using philos. basis. Died 1775.

VENEL, Jean André, Swiss surgeon; b. Geneva, Switzerland, May 28, 1740; studied under Cabanis and Tronchin; 1 son, Jean-François-Henri. Founder 1st orthopedic Hosp., Orbe, Switzerland, 1790; founder sch. midwifery, Yuerdon. Contbr. articles to jours. A founder orthopedic medicine. Died 1791.

VENING MEINESZ, Felix Andries, Dutch geophysicist, geodesist; b. The Hague, Netherlands, July 30, 1887; s. Sjoerd Anne and Cornelia (den Tex) V. M.; Dr., Tech. U., Delft, Netherlands, 1915; hon. Dr. degree U. Liège (Belgium), 1947, U. Helsinki (Finland), 1949, U. Brussels (Belgium), 1949, U. Strassbourg (France), 1950, Columbia, 1955. Engr., Netherlands Geodetic Commn., 1927-66; prof. geophysics U. Utrecht (Netherlands), 1927-57; prof. phys. geodesy U. Delft, 1938-57. Named chevalier Order of Netherlands Lion, 1933; commdr. Order of Oranje Nassau, 1957, comdr. Belgian, Italian Orders; recipient Vetlesen prize, 1962; Great Cross Order of Queen Juliana, 1965; also medals from various profl. orgns. Mem. Royal Netherlands Acad. Scis., Royal Soc. (fgn.), Nat. Acad. Sci. (fgn. asso.). Author: Observations de Pendule dans les Pays Bas, 1923; Gravity Expeditions at Sea, vol. I, 1932, vol. II, 1936, vol. III, 1941, vol. IV, 1948; (with W. A. Heiskamen) The Earth and its Gravity Field, 1958; The Earth's Crust and Mantel, 1964; also numerous articles. Research and determination of gravity at sea, and applications, especially determination of shape of earth, forces causing surface features of earth. Died Aug. 10, 1966.

VENKATARAMAN, G.S., Indian microbiologist; b. Tenkasi, India, Feb. 1, 1930; s. G.R. Sitaraman and Lokanayaki; B.Sc., U. Madras, 1951; M.Sc., Banaras Hindu U., 1953, Ph.D. (Govt. India research fellow), 1960; m. Rukmini, July 8, 1953; 1 dau., Sarayu. Staff div. botany Indian Agrl. Research Inst., New Delhi, 1956-62, microbiologist div. microbiology, 1962—; tchr. postgrad. classes, 1958—. Recipient Gold medal. Fellow Nat. Acad. Scis. Phycological Soc., India, Nat. Geog. Soc. India; mem. Assn. Microbiologists India. Author: Vaucheriaceae, 1961; Charophyta (with V. S. Sundaralingam, B. C. Kundu), 1962; The Cultivation of Algae. Research, publs. on blue-green algae and bacteria with spl. reference to their biology and physiology, their utilization as a biol. manure to increase crop prodn., interactions between crop microbe assn. Home: 13/30 East Patel Nagar, New Delhi-8, India. Office: Div. Microbiology, Indian Agrl. Research Inst., New Delhi-12, India.*

VENKATESWARLU, K., Indian physicist; b. Nuzvid, India, June 3, 1916; s. K. Poorniah and K. (Annapurna) Sastry; student A.C. Coll., Guntur, India, 1931-33; B.Sc. with honors, Andhra U., 1936, M.Sc., 1939, D.Sc., 1946; m. Lakshminarasanna, June 6, 1934; 6 children. Faculty, Andhra U. Colls., Waltair, India, 1939-48, sr. lectr., 1946-48; Govt. of Madras Overseas scholar in geophysics, 1948-50; spl. officer for ground water resources Govt. of Madras, India, 1951-53; reader in physics Annamalai U., Annamalainagar, India, 1954-55, prof., head dept. physics, 1955-62; prof., head dept. physics Kerala U., Ernakulam, India, 1963—. Recipient Metcalfe Gold medal. Fellow Inst. Physics London, Indian Acad. Scis. Editor: Proc. Seminar on Raman and Infra Red Spectroscopy, 1964. Research, numerous publs. on theory of molecular vibrations, determination of molecular constants; X-ray diffraction studies in fibres, intensity studies in Raman effect. Home: 29-23-23 Tadepallivari St., Suryaraopet, Vjayawada, Andhra, India. Office: Dept. Physics, Kerala U., Ernakulam Centre, Kalamassery, Kerala, India.*

VENKITASUBRAMANIAN, T. A., Indian biochemist; b. Tathamangalam, India; s. T. P. Ananthanarayanan and T. R. Narayany; B.Sc., Maharaja's Coll., Ernakulam, India, 1947; M.Sc., Indian Inst. Sci. Bangalore, Ph.D., 1950; m. Lakshmanan Sarada, June 8, 1958; children—Viveka, Divya. Lectr., Indian Inst. Sci., Bangalore, 1950-52; postdoctoral fellow Columbia, N.Y.C., 1952-54; research asso. U. Wis., 1954-56; sr. research officer V. Patel Chest Inst., U. Delhi, 1956-59, asst. dir., head biochemistry dept., 1959—, univ. prof., head depts. biochemistry, pathology, anatomy, physiology, faculty med. scis., U. Delhi 1966—. Mem., fellow Royal Inst. Chemistry Gt. Britain; mem. Royal Inst. Chemistry (sec. North India br.), Am. Chem. Soc., Research, publs. on lipid metabolism of tubercle bacilli, carbohydrate metabolism in exptl. Tb, nucleic acid metabolism in bacteria. Home: 1, Patel Chest Flats, Maurice Nagar, Delhi, U., Delhi, -7, India.*

VENNESLAND, Birgit, biochemist; b. Kristiansand, Norway, Nov. 17, 1913; d. Gunnuf Olaf and Sigrid Kristine (Bandsborg) Vennesland; came to U. S., 1917, naturalized, 1925; B.S., U. Chgo., 1934, Ph.D., 1938; S.D., Mt. Holyoke Coll., 1960. Fellow dept. physiol. chemistry Harvard Med. Sch., 1939-41; mem. faculty U. Chgo., 1941—, prof. biochemistry, 1957—. Cons. molecular biology NSF; cons. physiol., chemistry USPHS. Mem. Am. Soc. Biol. Chemists, Am. Chem. Soc. (Garvan medal 1964), A.A.A.S., Am. Soc. Plant Physiologists (Stephen Hales award 1950). Research on enzyme mechanism and photosynthesis. Home: 6019 Ingleside St., Chgo. 60637.*

VENNING, Eleanor Hill, Canadian biochemist; b. Montreal, Que., Can., Mar. 16, 1900; d. George W. and Annette (Kent) Hill; B.A., McGill U., 1920, M.Sc., 1921, Ph.D. in Exptl. Medicine, 1933. Research fellow McGill U., Montreal, 1934-44, faculty, 1944—, prof. exptl. medicine 1960.—. Fellow Royal Soc. Can. Brit. Soc. Endocrinology; mem. Soc. Biol. Chemistry, Endocrine Soc. (Koch medal 1962), Canadian Physiol. Soc. (past pres.). Research, numerous publs. on reprodn. and adrenal physiology. Home: 250 Clarke Av., Westmount 6, Que. Office: McGill U., Montreal, Can.*

VENTENAT, Etienne Pierre, French botanist; b. Limoges, France, Mar. 1, 1757; entered monastery, student philosophy, theology. Renounced religious life; travelled in Eng.; prof. botany Lycée Républicain, Paris, France, 1796; adminstr. Bibliothèque du Panthéon, from 1799; with D'Entrecasteaux expdn. to rescue French explorer, La Perouse. Mem. French Acad. Scis., 1795. Author: Dissertation sur les parties des mousses qui ont été regardées comme fleurs males et comme fleurs femelles; Principes de botanique, 1794; Le Botaniste voyageur aux environs de Paris, 1803; Jardin de la Malmaison, 1803; Choix de plantes, 1803-08; Decas generum novum, 1808. Gave very detailed, exact descriptions of plants (some of most comprehensive of time). Died Paris, Aug. 13, 1808.

VENTRIS, Michael George Francis, English architect; b. Wheathampstead, Eng., July 12, 1922; ed. Gstaad, Switzerland, Stone Sch., Eng.; grad. Archtl. Assn. Sch., London, 1948; hon. Ph.D., U. Uppsala (Sweden); m., 2 children. Worked on archtl. team Ministry of Edn.; interested in epigraphy. Served as navigator RAF, World War II. Decorated Order Brit. Empire; named hon. research asso. U. Coll., London; Architects Jour. research fellow, 1956. Author: (with John Chadwick), Documents in Mycenaean Greek, 1956. Deciphered Minoan-Mycenaean Linear B script from tablets uncovered by Evans at Knossos, Crete, from 1899, and Blegen at Pylos, 1939; issued "Mid-Century Report", 1950, stating views of leading authorities on Linear B; in face of prevalent opinion that Linear B could not be Greek, hypothesized that tablets were in Greek and succeeded in assigning phonetic values to Linear B syllabary; Blegen's discovery of tripod tablet at Pylos, 1952, confirmed Ventris' decipherment; ability to read Linear B tablets found in Crete and at several sites on Greek mainland provided description of highly organized bureaucratic Bronze Age civilization and necessitated thorough reevaluation of accepted history of Aegean between 16th and 12th centuries B.C. Died in traffic accident, nr. Hatfield, Eng., Sept. 6, 1956.

VENTURI, Giovanni Battista, Italian physicist; b. Bibbiano, Italy, Sept. 11, 1746; s. Giovanni-Domenico and Domenica (Gallian) V.; became prof. philosophy, Modena, Italy, 1773; lived mostly in Paris after 1796. Author: Considérations sur la connaissance de l'etendue que nous donne la sens de l'avie, 1776; Indagine fiska sui calori, 1814; Précis de quelques experiences sur la section que des cylindres de camphre éprouvent á la surface de l'eau, 1905. Research on lateral movement in fluids (led to invention of Venturi tube by Clemens Herschel) 1797, frequency of audible tones, 1796, movement of camphor on water surfaces, 1805, colors, arty., 1815; invented tube with diverging nozzles. Died 1822.

VENZKE, Walter George, Am. vet. scientist; b. White Lake, S.D., June 18, 1912; s. Carl John and Katherine (Ebert) V.; student U. Ia., 1930-31; D.V.M., Ia. State U., 1935, Ph.D., 1942; M.Sc., U. Wis., 1937; m. Elaine King, Apr. 15, 1939; 1 dau., Ann Rees. Research asst. U. Wis., 1935-37; asst. prof. vet. anatomy Ia. State U., 1937-42, asst. prof. vet. physiology and pharmacology, 1942; faculty Ohio State U., Columbus, 1946—, prof., chmn. dept. vet. anatomy, 1954—, asst. dean, sec. Coll. Vet. Med., prof. animal sci., 1960—. Mem. Am., Ohio vet. med. assns., Am. Assn. Anatomists, World, Am. assns. vet. anatomists, Ohio Acad. Sci., Sigma Xi, Phi Zeta, Gamma Sigma Delta. Research, publs. on reproductive processes in cattle and relation to breeding efficiency, morphogenesis of endocrine glands of chick embryo. Home: 2535 Andover Rd., Columbus, O. 43221.*

VERBA, Sidney, Am. polit. scientist; b. N.Y.C., May 26, 1932; s. Morris Harry and Recci (Salman) V.; B.S., Harvard, 1953; M.A., Princeton, 1957, Ph.D., 1959; m. Esther Cynthia Winston, June 17, 1954; children—Margaret Lynn, Erica Kim. Faculty, Princeton, 1959-64; prof. polit. sci. Inst. Polit. Studies, Stanford, 1964-68; prof. polit. sci. U. Chgo., 1968—. Mem. com. comparative politics Social Sci. Research Council. Center For Advanced Study Behavioral Scis. fellow, 1963-64. Mem. Am. Polit. Sci. Assn., Am. Assn. U. Profs. Author: Small Groups and Political Behavior, 1961; (with Gabriel Almond) The Civic Culture, 1963; author, editor: (with Klaus Knorr) The International System, 1961; (with Lucian Pye) Political Culture and Political Development, 1965. Research, publs. on comparative polit. behavior. Home: 4908 S. Kimbark Av., Chgo.*

VERBIEST, Ferdinand, Flemish astronomer; b. Pitthem, Flanders, Oct. 9, 1623; student math. under André Tacquet. Joined Soc. Jesus, missionary to China, from 1659; dir. Chhin Thien Chien (Imperial Bur. Astronomy), China, 1669-88. Author numerous works in Latin and Chinese including: Liber organicus astronominae europaeae apud Sinas restitutae sub Imp Cam-Hi Appellato, 1633; Coeli phenomena, 1679. Determined prin. geog. points of Chinese empire; dir. negotiations that established frontiers of China and Russia; supervised constn. large bronze astron. instruments for Peking (China) Obs., 1674. Died Peking, Jan. 28, 1688.

VERCHÈRE DE REFFYE, Jean-Baptiste-Philippe-Dieudonné, French inventor; b. Strasbourg, France, 1821; weapons mfr. Meudon, Tarbes, France, 1853; inventor casting process and cannon machineing machine; built 1st rear loader with breech cartridge and reduced thickness of charge; 1st French mitrailleuses, 1870; steel cannon with rifled barrel; gun carriage with hydraulic brake, 1876; 1st steel sheet shells; investigated speed and pressure in barrel using Tresca's method but with device of own constrn. Died 1880.

VERDET, Marcel E., French physicist; b. Nimes, France, 1824. Prof. physics Ecole Normale; prof. Ecole Polytechnique. Research on induction phenomena by elec. discharges, 1848, using movement of magnetic or nonmagnetic metals, 1851, optical properties by magnetism in transparent objects, 1854-63. Died 1866.

VERDUIN, Jacob, Am. ecologist; b. Orange City, Ia., Nov. 19, 1913; s. Peter and Jennie (Lagestee) V.; student Northwestern Jr. Coll., 1935-37; B.S., Ia. State U., 1939, M.S., 1941, Ph.D., 1947; m. Bethy Albertha Andersen, July 3, 1942; children—Lans, Jan Christine, Charlotte Maria, Leslie Emilie, Bethy Andrea. Faculty, Ia. State U., 1941, 45, U. S.D., 1946-48, Ohio State U., 1948-55, Bowling Green State U., 1955-64; prof. botany So. Ill. U., Carbondale, 1964—. Fellow A.A.A.S., Ohio Acad. Sci.; mem. Am. Inst. Biol. Scis., Internat. Assn. Theoretical and Applied Limnology, Ecol. Soc. Am., Internat. Assn. Quaternary Research, Ill. Acad. Sci. Contbr. articles to profl. jours. Establisher interference law for diffusion through multiperforate septa, determined rates of photosynthesis and respiration in plant communities under natural conditions, investigator daily energy budgets in air and water of natural environment. Home: R.F.D. 4, Carbondale 62903. Office: So. Ill. U., Carbondale, Ill. 62903.*

VEREECKEN, Josephus Leonardus, Dutch psychiatrist; b. Amsterdam, Holland, June 3, 1929; s. Petrus and Johanna (Verpoort) V.; student St. Ignatius Coll., 1942-46; m. Theresia van Lith, Dec. 29, 1964. Practice medicine, 1954—; chief of psychiatric dept. St. Lucas Hosp., Amsterdam, 1966—. Mem. Commn. for Examinations of Gen. Practitioners, 1958-64. Mem. Royal Soc. Medicine, Dutch Soc. Psychotherapy, Dutch Soc. Neurology and Psychiatry. Author: Over ogencontact, 1957. Publs. on contbns. to psychotherapy in field of human relationships, aspects of bull-fighting, psychol. factors of depression, self-mutilation, suicide. Home: 29 Marquette. Office: St. Lucas Hosp., Amsterdam, Holland.*

VERESHEHAGIN, Leonid Fedorovich, Russian physicist; b. 1909; postgrad. student, 1930-32; Dr. Phys.-Math. Sci. Sr. engr. turbogenerator plant, 1932-34; engr., then chief engr. Physico-Tech. Inst., Khar'kov, 1934-39; lab. supr. Inst. Organic Chemistry, USSR Acad. Sci., 1939-54; chief lab. in ultra-high pressures, 1954-58, became dir. Inst. High Pressure Physics, 1958. Recipient Stalin Prize, 1952. Corr. mem. USSR Acad. Sci. Research and publs. on ultra-high pressures, pressure deformation of metals, high density gases at high temperature. Home: Dorogomilovsk nab. 9. Office: Inst. Physics High Pressures, USSR Acad. Sci., Leninskii Prospekt 31, Moscow, USSR.

VERHEYEN, Philippe, Flemish anatomist; b. E. Flanders, Apr. 23, 1648; student Louvain, Belgium, from 1677; med. license, 1681, doctorate U. Leyden (Netherlands), 1693. Prof. anatomy U. Louvain, from 1689. Author: Compendii theoriae practicae, 1683; De Febribus, 1692; Anatomia corporis humani, 1693. Described stellate plexus of veins located under capsule of kidney (called Verheyen's stars, venae stellatae renis and stellate veins), 1693. Died Jan. 28, 1710.

VERHOEK, Frank Henry, Am. chemist; b. Grand Rapids, Mich., Feb. 12, 1909; s. Henry and Cornelia (Benjamin) V.; B.S. cum laude, Harvard, 1929; M.S., U. Wis., 1930, Ph.D., 1933; D.Phil., Oxford U., 1935; m. Cordula Clara Thurow, Mar. 17, 1940; children—Susan Elizabeth, Helen Katherine and Louise Gretchen (twins). Teaching asst. chemistry U. Wis., 1929-33; faculty Ohio State U., Columbus, 1936—, prof. chemistry, 1953—. Chemist, Gen. Elec. Co., Schenectady, 1938; prin. chemist Argonne Nat. Lab., 1947; sr. chemist Olin Mathieson Chem. Corp., Niagara Falls, N.Y., 1955; cons. Liberty Mirror div. Libbey-Owens-Ford Glass Co., Brackenridge, Pa., 1943-52, U. S. Naval Ordnance Test Sta., China Lake, Cal., 1957-62; vis. prof. U. Fla., 1958-59; lectr. chem. bond approach project NSF, 1959-68. Mem. Am. Chem. Soc. (chmn. Columbus sect. 1948-49, nat. councilor 1950-52), Assn. Am. Rhodes Scholars. Author with others: Introductory Chemistry for the Laboratory, 1942, 2d edit., 1946; Textbook of Chemistry, 1949, 2d edit., 1956; Chemistry for the Laboratory, 1951, 2d edit., 1957; Chemistry, 1959; Experimental General Chemistry, 1963; Chemistry: A Study of Matter, 1968. Research, publs. on rates of chem. reactions, rate of decomposition of carboxylic acids in various solvents and solvent mixtures, oxidation of hydrocarbons in solution, gas phase, hydrocarbon derivatives of boron; rate of decomposition of solid explosive complexes of cobalt. Home: 37 E. Riverglen Dr., Worthington, O. 43085. Office: 88 W. 18th St., Columbus, O. 43210.*

VERHOOGEN, John, geophysicist; b. Brussels, Belgium, Feb. 1, 1912; s. Rene and Lucy (Vincotte) V.; Mining Engr., U. Brussels, 1933; Geol. Engr., U. Liege, 1934; Ph.D., Stanford, 1936; m. Ilse Goldschmidt, Nov. 28, 1938; children—Robert II., Alexis R., Therese, Sylvia. Came to U. S., naturalized 1953. Asst. U. Brussels, Brussels and Belgian Congo, 1936-39; Fonds National Rocherche Scientifique, Belgian Congo, 1939-40; Mines d'or de Kilo-Moto, Belgian Congo, 1940-43; dir. prodn. Miniere de Guerre, Belgian Congo, 1943-46; asso. prof. U. Cal., 1947-51, prof., 1952—. Guggenheim fellow, 1953-54, 60-61; recipient Day medal Geol. Soc. Am., 1958. Member Internat. Assn. Volcanology (v.p. 1951-54), Nat. Acad. Scis., Am. Geophys. Union. Geol. Soc. Am., Mineral. Soc. Am., Sigma Xi. Author: Igneous and Metamorphic Petrology (with F. J. Turner), 1951; articles on petrology, volcanology, geophysics. Research on temperature distribution within the earch; metamorphic reactions; magnetic properties of rocks; past history of earth's magnetic field; volcanology. Home: 2100 Marin Av., Berkeley 7, Cal.*

VERIGO, Aleksandr Bronislavovich, Russian physicist; b. Mar. 5, 1893; staff Radium Inst. Leningrad, 1918—; chief Geophys. Obs. for Study Cosmic Rays and Radioactive Phenomena, 1925-37; organizer expdn. for study cosmic radiation Eastern peak of Mt. Elbrus, 1928-30; sailed through Baltic in submarine, 1929-30; mem. polar expdn. to Franz Joseph Land, 1932. Made (with K. I. Zille, I. G. Prilutskii) flight into stratosphere; research on cosmic ray absorption in nuclei of atoms of various substances; dependence of absorption on atomic weight.

VERIGO, Bronislav Fortunatovich, Russian physiologist; b. 1860; prof. Odessa U., Perm U. Author: Fundamentals of Physiology of Men and Animals, 2 vols., 1905. Research in electrophysiology; discovered nerve composition, 1892; established law of Verigo. Died 1925.

VERIN, Pierre, archeologist; b. Niort, France, Apr. 6, 1934; s. Felix and Yvonne (Pineau) V.; licence en droit et lettres, 1955; diploma in oriental languages, Ecole France D'Outre Mer, 1956; M.A., Yale, 1964; doctorate prehistoire Sorbonne, 1965; m. Juliette Razanamasy, Aug. 6, 1956; children—Sonia, Emmanuel, Florence, Pierre. With mission to Society Islands, Bishop Mus.-Overseas Office Sci. Research, 1960-62; dir. Archeol. Center, U. Madagascar, Tananarive, 1962—, also asso. prof. anthropology. Mem. Malagesy Acad. Author: L'Ancienne Civilisation de Rurutu, 1968; also articles. Excavations in Austral Islands, Polynesian civilization; study of local Islamic cultures, origin of Malagasy people through archeol. study in Madagascar. Home: 46 Rue Dr. Besson, Tananarive, Madagascar.*

VERKSHINSKII, Sergei Arkadevich, Russian electronics physicist; b. Oct. 15, 1896; s. student Leningrad, Don poly. insts. Chief engr. Electrovacuum plant, Leningrad, 1922-28; became chief vacuum lab. Svetlana, 1928; chief engr., 1936-39, cons., 1939-41; named dir. Sci. Research Vacuum Inst., 1947; bur. mem. dept. tech. sci. USSR Acad. Scis. Recipient Stalin prize, 1946, Order of Lenin, 1956, Hero of Socialist Labor, 1956, A. S. Popov Gold medal, 1962. Mem. USSR Acad. Sci., Author: A New Method of Metallographic Study of Alliys, 1944. Research and publs. in electrovacuum engring.; metal alloys; designed electronic devices; directed devel. indsl. vacuum equipment. Office: USSR Acad. Scis., Sci. Research Inst., Moscow, USSR.

VERLINDE, Jacobus Dirk, Dutch bacteriologist; b. Spijkenisse, Netherlands, Oct. 11, 1910; s. Arie and Francina Margrietje (Dekker) V.; M.D., Vet. M.D., State U. Utrecht; m. Lydia Maria Sieling, June 8, 1939; children—Arie Dirk, Karel Hendrik, Frans Jacobus, Lydia Maria, Marianne. Asst. bacteriologist Dutch Prevention Inst., Leyden, 1935-40; bacteriologist Nat. Inst. Pub. Health, Utrecht, Netherlands, 1940-41; head med. microbiol. sect. Dutch Prevention Inst., Leyden, 1941-47; prof. med. microbiology Faculty Medicine, State U. Leyden, 1947—. Recipient medal of honor Free U. Brussels; van-Esveld medal. Mem. Royal Dutch, N.Y. acads. scis., Dutch Sci. Soc. Research and publs. on viral diseases, including smallpox, polio, encephalitis; exptl. and comparative pathology; epidemiology and immunology of infectious diseases, especially of nervous system. Home: van Slingelandtlaan 9, Leyden, Netherlands.

VERLY, Walter Georges, biochemist; b. Peronnes, Belgium, Apr. 10, 1923; s. Georges A. and Yvonne (Plaquet) V.; M.D., U. Liège (Belgium), 1947, agrégé, enseignement superieur en biochimie, 1956; postgrad. (Brit. Council scholar) U. Edinburgh, (Belgian Am. Ednl. Found. fellow) Cornell U.; m. Jacqueline Mairlot, Sept. 27, 1947; children—Pierre, François, Georges. Faculty, U. Liège, 1947-48, 57-64, Cornell U., 1949-52; research fellow biochemistry U. Edinburgh, 1948-49; asso. Belgian Nat. Fund for Sci. Research, 1953-63; prof., chmn. dept. biochemistry U. Montreal (Que., Can.), 1964—. Cons. Belgian Nuclear Center, 1957-64; mem., gen. sec. European Group cancer chemotherapy, 1962-64. Recipient Alumni prize U. Found. Belgium, 1954, Leon Fredericq prize, 1958. Mem. Am. Chem. Soc., Am. Soc. Biol. Chemists, Biochem. Soc., Canadian Soc. Biochemistry, Société belge de Biochimie, A.A.A.S., N.Y. Acad. Scis. Research, publs. on biochemistry, intermediary metabolism and tracer methodology; nucleic acids and molecular genetics. Home: 807 Wilder Av., Outremont, Que., Can.*

VERMEULEN, Alex, Belgian physician; b. St. Martens-Latem, Belgium, July 31, 1927; s. Maurice and Maria (Coussement) V.; M.D. with honors, U. Ghent (Belgium), 1951, Lic.Bioch. with honors, 1953, Agrégé, 1960; m. Anita Blanckaert, Sept. 4, 1954. Staff, U. Hosp., Ghent, 1953-60, chief, clinic, 1959-60; asso. prof. medicine U. Ghent, 1961—. Fellow Brit. Council, King's Coll. Hosp., London, also Roswell Park Meml. Inst., Buffalo, 1960; Fulbright fellow, 1960; Advanced C.R.B. fellow, 1964. Mem. Soc. Belge d'Endocrinologie (v.p. 1959—), Royal Flemish Acad. Medicine (award 1958), Soc. de Chimie Hormonologique (pres. 1959—). Contbg. author: Excerpta Medica, 1966; also numerous articles. Research on metabolism of natural and synthetic steroid-hormones, devel. methods for determination of steroids in biol. fluids. Home: 11 Paolalaan, Zwijnaarde, Belgium. Office: Med. Clinic, Akademische Ziekenhuis, Ghent, Belgium.*

VERMEULEN, Cornelius William, Am. physician; b. Paterson, N.J., Aug. 23, 1912; s. William and Ada (Ihrman) V.; A.B., Calvin Coll., 1933; M.D., U. Chgo., 1937; m. Ruth Carol Van Harn, Jan. 29, 1936; 1 son, Carl William. Asso. prof. urology U. Ill., 1946-53; prof. surgery U. Chgo., 1956—, asso. dean div. biol. sci., 1966—, chief of staff, 1967—. Mem. A.M.A., Am. Urol. Assn., Am. Assn. Genito-Urinary Surgery. Research, publs. on causation and mechanisms of urinary stone formation. Home: 1270 Oakmont St., Flossmoor, Ill. 60422. Office: 950 E. 59th St., Chgo. 60637.*

VERMUND, Halvor, radiologist; b. Solör, Norway, Aug. 5, 1916; s. Aarstein Posaas and Astrid (Wermund) V.; M.D., U. Oslo, 1943; Ph.D., U. Minn., 1951; m. Karen Kristine Bergfjord, June 23, 1943; children—Sten Halvor, Anita Torund. Came to U. S., 1947, naturalized, 1953. Asst. prof., asso. prof. U. Minn., 1953-57; prof. radiology, dir. radiotherapy U. Wis. Med. Sch., Madison, 1957—. Mem. sci. adv. com. on therapy of cancer Am. Cancer Soc., 1957-59; mem. radiation study sect. NIH, 1965—. Diplomate Am. Bd. Radiology. Fellow Am. Coll. Radiology; mem. Am. Radium Soc., Am. Soc. Therapeutic Radiology, Am. Roentgen Ray Soc., Radiol. Soc. N.Am., A.M.A. Research, publs. on chemotherapeutic adjuvants to radiotherapy of cancer, role of tumor bed in radiation-induced response of tumors. Home: 5013 Risser Rd., Madison, Wis. 53705.*

VERNANDSKI, Vladimir Ivanovich, Russian mineralogist; b. St. Petersburg, Russia, Mar. 12, 1863; grad. U. St. Petersburg, 1885. Prof. crystallography and mineralogy U. Moscow, 1890-1911; dir. State Radium Inst., Leningrad, USSR, 1926-38; founder biogeochem. lab. Acad. Scis., Leningrad, became dir., 1928; rector Tauride U., 1920-21. Mem. French Acad. Scis., 1936, USSR Acad. Scis., Ukrainian Acad. Sci. (pres. 1918-20). Author: History of Minerals, 1925; Selected Works, 2 vols., 1954. Developed method for calculation of rocks and minerals by their radioactive attrition; founder biogeochemistry; research on geochemistry, isomorphism of chem. elements, meteorites, microorganisms in biogeochemistry of earth's crust. Died Moscow, Jan. 6, 1945.

VERNE, C. M. Jean, French histologist; b. Saint-Julien, France, Oct. 1890; s. Verne Frederic and Marie (Hallberg) V.; Dr.en médecine, Paris, 1913, Dr.es sciences, 1921; agrégé in histologie, 1923; m. Louise Douin, Aug. 5, 1912; children—Jean-Marie, Christiane (Mrs. Baudelot), José. Prof., Faculté de Médecine, Paris, 1932-62, then hon. prof., dir. Institut d'Histochimie. Decorated Comdr. Palmes Acad. Mem. Academie nationale de Médecine, Association francaise pour l'avancement des sciences (sec.-gen.), Societé francaise d'histochimie. Author: Le protoplasma cellulaire, 1923; Les pigments dans l'organisme animal, 1926; La vie cellulaire hors de l'organism, 1930; Précis d'Histologie, 1953; L' Histologie, 1966. Research and numerous publs. on pigments and cultures of tissues; histochem. research on enzymes. Home: 38, rue de Varenne, Paris 7, France.*

VERNEAU, René, French physician, anthropologist; b. La Chapelle/Loire, France, 1852; M.D. Made study voyage to Canary Islands; prof. anthropology, Muséum d'histoire naturelle, Institut de Paléontologie Humaine; prof. ethnography Ecole coloniale; curator Musée d'ethnography du Trocadéro. Author: Le bassin dans les sexes et dans les races, 1875; Cinq Années de sejour aux iles Canaries, 1890; L'enfance de l'humanité, L'age de la pierre, 1890; Les Races Humaines, 1891; In Brehm's Merveilles de la nature: Les Ancien Patagons, 1903; (with Rivet) Ethnographie ancienne de l'Equateur, 1912; L'Homme, 1931. Research on comparative anatomy, anthropology. Died Paris, 1938.

VERNEUIL, Aristide Auguste Stanislaus, French surgeon, anatomist; b. Paris, Nov. 23, 1823; M.D., Faculty of Paris, 1852. Became chief surgery Hosp. Lourcine, 1862, Hosp. Midi, 1865, Hosp. Lariboisière, 1865, Hosp. Pitié, 1872; named prof. clin. surgery Faculty Medicine, Paris, 1867, prof. external pathology, 1868. Mem. French Acad. Scis., 1887, Acad. Medicine, Soc. Surgery. Author: Recherches sur la locomotion du coeur, 1852; Le Systeme veineux: Anatomie et physiologie, 1853; Découverte de la staphylorragie au XVIIIe siècle, 1861; Chirurgie réparatric, 1877; Mémoires de chirurgie, 1877-88; Etudes expérimentales et cliniques sur la tuberculose, 1887-90; Documents inédits, tirés des archives de l'ancienne Academie de chirurgie. Research on origin of tetanus; described syphilitic disease of bursae (Verneuil's bursitis), 1860; introduced forcipressure in surgery, 1875. Founder, L'Oevure de la Tuberculose. Died Maisons-Laffite, Seine-et-Oise, France, June 11, 1895.

VERNEUIL, Philippe Edouard Poullettier de, see de Verneuil, Philippe Edouard Poullettier.

VERNIER, Pierre, French mathematician, inventor; b. Ornans, France, 1580; s. Claude Vernier; studied under his father. Served King of Spain in Low Countries; dir. gen. monies for County of Bourgogne, France. Author: Construction usage et propriétés du quadrant nouveau de mathématiques, 1631. Invented aux. (vernier) scale for subdividing divisions on ordinary scale, thus allowing more accurate measurements of linear or angular magnitudes, 1631. Died Ornans, Sept. 14, 1637.

VERNON, Glenn Morley, Am. sociologist; b. Vernal, Utah, Apr. 6, 1920; s. William Morley and Roseltha (Bingham) V.; B.S., Brigham Young U., 1947, M.S. in Sociology, 1950; Ph.D., Wash. State U., 1953; m. June Andersen, Dec. 24, 1941; children—Gregory Glenn, Rebecca, Paul Bingham. Asst. prof. Auburn U., 1953-54; asst. prof., then asso. prof. Central Mich. U., 1954-59; asso. prof. Brigham Young U., 1959-63; prof. sociology, head dept. sociology and anthropology U. Me., 1963-68; prof. sociology U. Utah, 1968—. Fellow Am. Sociol. Assn.; mem. Eastern Sociol. Soc., Soc. Sci. Study Religion, Religious Research Assn., Nat. Council Family Relations, Sigma Xi, Phi Kappa Phi. Mem. Ch. of Jesus Christ of Latter Day Saints. Author: Sociology of Religion, 1962; (with others) Introduction to Sociology, 1961; Human Interaction, 1965; contbn. to International Yearbook on the Sociology of Religion, 1966; also articles. Research on sociology of religion, self definitions, attitudes toward death; contbd. to devel. of symbolic interactionism theory. Address: U. Utah, Salt Lake City.*

VERNON, Horace Middleton, English physician; b. Oct. 3, 1870; s. Thomas Heygate Vernon; grad. with 1st class honors in chemistry (scholar) Merton Coll., Oxford, 1891, in physiology, 1893; M.A., M.D., St. George's Hosp., London; m. Katharine Dorothea Ewart; 1 son, 3 daus. Became Radcliffe travelling fellow, 1897; univ. lectr. on chem. physiology, Oxford; fellow Magdalen Coll., Oxford, 1898-1920; investigator Indsl. Health Research Bd., 1919-32. Became Naples Biol. scholar, 1894, Tolleston Meml. prizeman, 1896, George Henry Lewes student, 1896. Author: Variation in Animals And Plants, 1903; Intracellular Enzymes, 1908; Industrial Fatigue and Efficiency, 1921; The Alcohol Problem, 1928; The Principles of Heating and Ventilation, 1934; The Shorter Working Week, 1934; Accidents and their Prevention, 1936; Health in Relation to Occupation, 1939; The Health and Efficiency of Munition Workers, 1940; also numerous articles. Contributed to devel. of indsl. medicine, med. aspects of social problems; studies in cytology and cell differentiation. Died Feb. 11, 1951.

VERNON, Jack Allen, Am. psychologist; b. Kingsport, Tenn., Apr. 6, 1922; s. John Allen and Mary Jane (Peters) V.; A.B., U. Va., 1948, M.A., 1950, Ph.D., 1952; m. Betty Jane Durbon, Dec. 16, 1945; children—Stephen Mark, Victoria Lynn. Faculty, Princeton, 1952-66, prof. psychology, 1965-66; prof. otolaryngology U. Ore. Med. Sch., Portland, 1966—; mem. study sect. Nat. Inst. Neurol. Diseases and Blindness. Mem. Acoustical Soc. Am., Am. Psychol. Assn., A.A.A.S. Author: Inside The Black Room, 1963. Editor: Self-Selection Text in Introductory Psychology, 1966. Research, numerous publs. on sensory deprivation where man's sensory stimuli are drastically reduced, electrophysiology of inner ear in many unusual animals such as bats, high frequency stimulation of ears of animals and quantification of high frequency stimuli. Home: 15556 N.W. Oak Hills Dr., Beaverton, Ore. 97005. Office: Kresge Hearing Research Labs., 3515 U. S. Veteran's Hospital Rd., Portland, Ore. 97201.*

VERNON, Leo Preston, Am. biochemist; b. Roosevelt, Utah, Oct. 10, 1925; s. William M. and Roseltha (Bingham) V., A.B., Brigham Young U., 1948; Ph.D., Ia. State Coll., 1951; m. Marion Fern Trunkey, Sept. 6, 1946; children—Ricky Leo, Marion Elise, Martin Preston, Jillaine, Eric Eugene. Trainee, Enzyme Inst., U. Wis., 1951-52; postdoctoral fellow Washington U., St. Louis, 1952-54; faculty Brigham Young U., 1954-60; fellow Nobel Inst., Stockholm, 1960-61; dir. Charles F. Kettering Research Lab., Yellow Springs, O., 1961—; prof. chemistry Antioch Coll., Yellow Springs, 1962—. Acting div. dir. life sci. div. Wright State U., Dayton, O., 1964-65; vis. prof. Purdue U., 1965. Mem. Am. Soc. Biol. Chemists, Am. Chem. Soc., Am. Soc. Plant Physiologists. Author: (with H. Gest and A. San Pietro) Bacterial Photosynthesis, 1964 (with G. R. Seely) The Chlorophylls, 1966; also numerous articles. Research on photosynthesis, expts. with radioactive isotopes to determine path of carbon fixation, path of electron transfer from photoexcited chlorophyll to carbon dioxide fixation, correlation of photo-chem. steps by chlorophyll in vivo and in vitro. Home: 202 Northwood Dr. Office: 150 E. South College St., Yellow Springs, O.*

VERNON, P. E., Brit. psychologist; b. Oxford, Eng., June 6, 1905; s. Horace Middleton and Katherine (Ewart) V.; B.A., Cambridge U., 1927, M.A., 1930, Ph.D., 1931; D.Sc., London U., 1954; m. Dorothy Anne Fairley Lawson, Sept. 30, 1947; 1 son, Philip Anthony. Fellow, univ. lectr. St. John Coll., Cambridge U., 1930-33; psychologist Child Guidance Clinic, Maudsley Hosp., London, 1933-35; head psychology dept. Londanhill Tng. Coll., Glasgow, Scotland, 1935-38, Glasgow U., 1938-47; psychol. research adviser Admiralty and War Office, 1942-45; prof. ednl. psychology U. London Inst. Edn., 1944-65, prof. psychology, 1965—. Mem. Am., Brit. psychol. assns., Brit. Assn. for Advancement Sci. Author books with G. W. Allport, 1933, with J. B. Pamy, 1949; also numerous articles. Research on psychology of musical appreciation, mental measurement, test of intelligence, selection and guidance, cultural and environmental factors in intellectual devel., assessment of personality. Home: 30 Shermardspark Rd., Wekoyn Garden City, Herts., Eng. Office: Inst. Edn., Malet St., London W.C.1., Eng.*

VERNON-HARCOURT, Augustus George, English chemist; b. London, Dec. 24, 1834; s. Frederick E. and Marcia (Tollemache) V.-H.; grad. 1st class in Natural sci. Harrow and Balliol Coll., Oxford, 1858; became sr. student Christ Church Coll., 1859; D.C.L., LL.D.; m. Rachel Mary Aberdare, 1872; 2 sons, 8 daus. Sci. tutor Christ Church, 1871-82; became Lee's reader in chemistry, 1869; met. gas referee. Fellow Royal Soc., 1868; mem. Chem. Soc. (v.p.). Author: (with H. G. Madan) Exercises in Practical Chemistry; also numerous articles. Invented a standard lamp of 1 candle-power and another of 10 candle-power which burned pentane, inhaler for adminstrn. chloroform mixed with air in variable proportions. Died Ryde, Isle of Wight, Aug. 23, 1919.

VERNOV, Sergei Nikolaevich, Russian physicist; b. July 11, 1910; grad. Leningrad Poly. Inst., 1931; Dr. Phys.-Math. Sci. Staff, Inst. Radium, USSR Acad. Sci., 1930-35, staff Physics Inst., 1935—; named dir. Sci. Research Inst. on Nuclear Physics, Moscow U., 1946, prof., 1943—. Recipient Stalin Prize, 1949, Lenin Prize, 1960. Corr. mem. USSR Acad. Sci. Author: Research on Cosmic Rays, 1948. Research and publs. on earth's external radiation belt, cosmic ray effects in stratosphere; devel. mechanisms for study cosmic rays, for rockets and satellites; determined origin of soft radiation component; demonstrated primary particle are protons. Home: Leninskiye gory sekt L., Moscow, USSR.

VÈRON, Marcel, French physician, engr.; b. Paris, Nov. 28, 1900; s. Alphonse and Marie (Roszak) V.; Ingenieur, Ecole Centrale de Paris, 1922; Docteure h.c., U. Louvain (Belgium); D'Ingenieur h.c. TH, Hanover, Germany; m. France Gouttefangea, July 30, 1930; children—Monique (Mrs. Jean Bilbille), Jean, Nicole (Mrs. Yves Lizoret). Became examiner Ecole Centrale, 1924, prof., 1930; became prof., dir. lab. Conservatoire national des Arts et Metier, Paris, 1935; Mem. Soc. Civil Engrs. France (became pres. 1947), French Assn. Regulation and Automation (pres. 1956-57), French Soc. Thermal Engrs. (pres. 1960-64), Internat. Conf. French Inst. Combustion Engrs. (pres. 1964-), French Soc. Physics; hon. mem. or laureate of French, Belgian, Italian, Swiss, German assns. Nouvelles études sur la chaleur, 1929; Traité de chauffage, 1941; Chuaffes thermiques flux, et colorifiques, 1951; also numerous articles. Founder, Gen. Rev. Thermics. Pioneered (with W. Margoules) French research in convection, (with Y. Rocard) convection laws; analytic studies of thermal conduction; theory of deflagration in motors. Home: 3 bis, rue du General Delanne, Neuilly (Seine), France. Office: 292, rue du St. Martin, Paris, France.*

VERONESE, Guiseppe, Italian mathematician; b. Chioggia, Italy, 1854; ed. U. Rome; Prof. analytic geometry U. Padua (Italy); dep. in Parliament; senator of Kingdom. Mem. Lincei. Described new properties of myscic hexagon; studies in finite fields, n-dimensional space; 1st conception of non-Archimedian geometry, 1891; demonstrated that if 4th dimension existed body could be removed from closed 3-dimensional volume without going through enclosing surface. Died 1917.

VERONESI, Umberto, Italian physician; b. Milan, Italy, Nov. 28, 1925; s. Francesco and Erminia (Verganti) V.; M.D., U. Milano, 1951; specialization in gen. surgery U. Pavia (Italy), 1956; m. Susy Razon, Apr. 13, 1961; children—Paolo, Marco, Alberto. Research worker exptl. oncology Consiglio Nazionale delle Ricerche, 1951-55; surgeon Nat. Cancer Institute, Milano, 1956-61; asso. dir. surg. div., 1961—. Mem. exper com. on cancer treatment WHO. Recipient Lustig prize for cancer research, 1958. Mem. Royal Soc. Medicine, Italian Soc. Cancerology, Italian Soc. Surgery, Internat. Coll. Surgeons, Med. Acad. Scis. Barcelona (hon.). Author: Handbook of Surgical Anatomy, 1961; (with P. Bucalossi,) Italian Contribution to Cancer Research, 1960; also numerous articles. Research on biology and pathology of tumors, especially breast and thyroid tumors; developed original surg. procedures, including enlarged and superradical mastectomy for breast cancer, hypophisectomy for advanced thyroid cancer. Home: 2 Via Vitali, Milano. Office: 22 Piazza Gorini, Milan, Italy.*

VERONIS, George, Am. oceanographer; b. New Brunswick, N.J., June 3, 1926; s. Nicholas Emmanuel and Angeliki (Eftimakis) V.; A.B., Lafayette Coll., 1950; Ph.D. in Applied Math., Brown U., 1954; M.A. (honorary), Yale University, 1966; m. to Catherine Elizabeth Berger, Jan. 29, 1949 (div. Nov. 1962); children—Melissa, Benjamin; m. 2d, Anna Margareta Olsson, Nov. 8, 1963. Staff meteorologist Inst. for Advanced Study, Princeton, N.J., 1953-56; staff mathematician Woods Hole (Mass.) Oceanographic Instn., 1956-64; asso. prof. oceanography Mass. Inst. Tech., Cambridge, 1961-64, research oceanographer, 1964-66; prof. geology and applied sci. Yale, 1966—. Fellow Am. Acad. Arts and Scis. Mem. Am. Phys. Soc., Phi Beta Kappa, Sigma Xi. Research, publs. on large scale ocean circulation and its response to wind systems over oceans, fluid motions which are generated by heat sources, electric fields, differential rotation of system, long wave motions in ocean and atmosphere. Home: 183 Colony Rd., New Haven 06511.*

VERRILL, Addison Emery, Am. zoologist, geologist; b. Greenwood, Me., Feb. 9, 1839; s. George Washington and Lucy (Hilborn) V.; B.S., Harvard, 1862; hon. A.M., Yale, 1867; m. Flora L. Smith, June 15, 1865; children—George Elliot, Evalina Flora (dec.), Alpheus Hyatt, Edith Barton, Clarence Sidney (dec.), Lucy Lavinia. Asst. in Museum Comparative Zoology, 1860-64, prof. zoology, 1864-1907, emeritus prof. 1907, curator Zoöl. Mus., 1865-1910, instr. geology, Sheffield Scientific School, 1870-94, Yale. Curator Boston Soc. Natural History, 1864-74; prof. comparative anatomy and entomology, U. of Wis., 1868-70; asso. editor Am. Jour. of Science, 1869-1920;

asst. in charge of scientific explorations by the U. S. Fish Commn., 1871-87. Mem. Nat. Acad. Sciences; pres. Conn. Acad. Arts and Sciences; fellow Am. Acad. Arts and Sciences. Contributed all the zool. matter to Webster's Internat. Dictionary, 1890, and Supplement, 1900. Author: Report upon the Invertebrate Animals of Vineyard Sound and Adjacent Waters, 1873; The Bermuda Islands, 1903; Zoology of the Bermuda Islands, Vol. I, 1903; Geology and Paleontology of the Bermudas, 1906; Coral Reefs of the Bermudas, 1907; Monograph of the Shallow Water Starfishes of the North Pacific Coast, 1914; Report on West Indian Starfishes, 1915; Reports on Alcyonaria and Actinaria of Canadian Arctic Expedition, 1921; Crustacea of Bermuda, 3 parts, 1923; Alcyonaria of the Blake Expedition, 1925. Investigated invertebrata of entire Atlantic and Pacific Coasts N. America, especially deep-sea fauna; research on marine faunas of Bermuda, W. Indies, Brazil, H.I. and Panama. Died Santa Barbara, Cal. Dec. 10, 1926.

VERRILL, Alpheus Hyatt, Am. naturalist, explorer; b. New Haven, Conn., July 23, 1871; s. Addison Emery and Flora L. (Smith) V.; ed. Hopkins Grammar Sch.; Yale Sch. Fine Arts; spl. course in zoology under father; m. Kathryn L. McCarthy, Jan. 21, 1892; children—Dorothy I. (Mrs. Russell Rhodes), Eric E. (dec.), Loyola K. (Mrs. F. Cintron, Jr.), Valerie G. (Mrs. P. A. C. Ellis) (dec.); m. 2d, Lida Ruth Shaw, Nov. 11, 1944. Illustrated natural history dept. Webster's Internat. Dictionary, 1896; Clarendon Dictionary; many sci. reports, and other publs. Extensive explorations in Bermuda, West Indies, Guiana, Central America, Panama, 1889-1920. Resided Dominica, British West Indies, 1903-06, British Guiana, 1913-17, Panama, 1917-21. Made enthnological expeditions to Panama, Peru, Bolivia, Chile and Surinam, 1916-28; archaeological explorations, Central America, 1924-27; carried on extensive excavations, Panama, 1924-27; engaged in making series of oil paintings of South and Central Am. Indians from life, 1926-28; archaeol. expdns., Peru and Bolivia, 1928-32; in charge of expdns., 1933, 34, salvaging Spanish galleon sunk in West Indies in 17th Century; expdn. to Brit. West Indies, 1948; expdn. to Mexico, 1953. In 1940 established the Anhiarka exptl. gardens and Natural Sci. Mus. at site of the ancient Indian village of Anhiarka where De Soto made his first settlement in Fla., 1944, established shell business, Lake Worth, Fla. Author adventure, natural history books. Invented autochrome photographic process, 1902; rediscovered supposedly extinct Solenodon parodoxus in Santo Domingo, 1907; discovered remains of unknown prehistoric culture in Panama. Died Nov. 14, 1954.

VERSCHUER, Otmar von, see von Verschuer, Otmar.

VERSTRAETE, Marc, Belgian physician; b. Brugge, Belgium, Apr. 1, 1925; s. Louis and Jeanne (Verstraete-Coppin) V.; M.D., U. Leuven (Belgium), 1951; Agrégé, U. Louvain (Belgium), 1955; m. Bernadette Moyersoen, July 10, 1955; children—Anneli, Benedicte, Luc, Frances, Beatrijs. Research staff Lab. Physiology, 1945-46, Lab. Gen. Pathology, 1945-51; fellow U. Basel (Switzerland), 1952; fellow Brit. Council Found. Radcliffe Infirmary, Oxford, Eng., 1955; Fulbright fellow U. S. Edn. Found., Cornell U. Med. Center Vascular Research Lab., 1955-56; faculty, head sect. vascular diseases and coagulation disorders, head lab. coagulation and proteolysis U. Louvain, 1957—, prof., 1963—. CRB Advanced fellow, U. S., 1966. Author: Het hemofilie syndroom, 1955; Hypertensie, 1966; also numerous articles. Coeditor: Thrombosis et Diathesis Haemorrhagica, 1957—, Medicina Experimentalis, 1959—. Research on clotting factors, characterization, in field of blood coagulation and proteolysis. Home: 29 Minderbroederstraat, Leuven, Belgium.*

VÉRTES, László, Hungarian archaeologist; b. Budapest, Hungary, Nov. 3, 1914; s. Isidor and Aranka (Reiner) V.; acad. cand. degree geol. scis., 1957; D.archaeology, 1965; m. Agnes Meller, June 3, 1950; 1 son, László. With Paleontol. dept. Hungarian Nat. Hist. Mus., 1946-50, dir., 1948-50; in charge Paleolithic Coll. archeol. dept. Hungarian Nat. Mus., Budapest, 1950—, sr. scientist, 1963—; faculty paleolithic period Eötvös Lóránd U., Budapest, 1949—, Author: (with others) Istállöskö-cave, 1955; Medveemberek krónikája, 1957; Untersuchungen an Höhlensedimenten, 1959; also numerous articles. Editor: Tata, 1964; Az öskökor és az átmeneti kökor Magrarországon, 1965. Research on Paleolithic period of Hungary and adjoining territories, including numerous excavations; most important find was lower Paleolithic living site at Vértesszöllös where 1st certain European Archantropus bones were found; research on evolution of culture. Home: 4/a Cserje, Budapest. Office: 14-16 Muzeum krt., Budapest, Hungary.*

VERWEY, Willard Foster, Am. microbiologist; b. Walkill, N.Y., July 14, 1913; s. Daniel George and Delora (Zimmerman) V.; B.Sc., Rutgers U., 1934; Sc.D., Johns Hopkins Sch. Hygiene, 1937; m. Selma Louise Stoner, June 6, 1940; children—Delora (Mrs. James McWilliams), James R. Asst., instr. N.Y. U. Coll. Medicine, 1937-40; dir. bacteriological research Sharp & Dohme, 1940-53; dir. bacteriology Merck, Sharp & Dohme, 1953-57; prof., chmn. dept. microbiology U. Tex. Med. Br., Galveston, 1957—. Mem. cholera panel U. S.-Japan Coop. Med. Sci. Program, 1965—; mem. cholera adv. com. NIH, 1967—; mem. com. Intersci. Conf. on Antimicrobial Agts. and Chemotherapy, 1964-68. Diplomate Am. Bd. Microbiology. Fellow Am. Acad. Microbiology; mem. Am. Soc. for Microbiology, A.A.A.S., Soc. for Exptl. Biology and Medicine, Sigma Xi. Mem. editorial bd. Antimicrobial Agts. and Chemotherapy, 1966-68. Research, numerous publs. on immunizing agts. against bacterial diseases, devel. of dehydrated Brucella vaccine, purified antigens for immunization against pertussis and cholera, dynamics of antibacterial chemotherapy. Home: 6810 Wayside Dr., Hitchcock, Tex. 77563. Office: U. Tex. Med. Br., Galveston, Tex. 77550.*

VERWORN, Max, German physiologist; b. Berlin, Germany, Nov. 4, 1863; Ph.D., 1887; M.D., 1889. Became asst. Inst. Physiology, Jena, Germany, 1891, prof., 1895; prof. Göttingen, also Bonn, Germany. Author: Etudes Psycho-Physiologiques sur les protistes, 1889; La signification physiologique du noyau cellulaire, 1891; le Mouvement de la substance vivante, 1892; Contribution a la physiologie du systeme nerveux central, 1898; La Physiologie générale, 1895; Allgemeine Physiologie, 1905; Naturwissenschaft Von Weltanschaung, 1905; Die mechanik des Geistes, 1906; Die Frage nach den Grenzen der Erkenntnis, 1908. Research on physiology of cell, relation of neurol. function and physiology to cognitation. Died Bonn, Nov. 23, 1923.

VERY, Frank Washington, Am. astronomer; b. Salem, Mass., Feb. 12, 1852; s. Washington and Martha Needham (Leach) V.; S.B. (in chemistry), Mass. Inst. Tech., 1873; m. Portia Mary Vickers, Apr. 11, 1893; children—Alice Needham (Mrs. Edmund R. Brown), Marjorie Vickers, Arthur Oldfield, Eleonora Virginia, Ronald Winthrop. Instr., physical lab., Mass. Inst. Tech., 1877; astronomer at Allegheny (Pa.) Obs., 1878-95; adj. prof. astronomy and instr. in geology, Western U. of Pa., 1890-95; acting dir. Ladd Obs. and prof. astronomy, Brown U., 1896-97, completing, at own expense, research on Atmospheric Radiation, published as bull. "G" by U. S. Weather Bur., 1900; spl. agt. U. S. Weather Bur., 1900; dir. Westwood Astrophys. Obs., 1906—. For 10 yrs. Prof. Langley's prin. asst.; made many original investigations. Swedenborgian. Fellow A.A.A.S.; mem. Am. Astron. Soc.; hon. mem. Acad. Arts and Sciences, Utrecht, Holland. Frequent contbr. to leading scientific mags. Demonstrated that the solar constant exceeds 3.5 c.g. min.; proof of light absorption by a medium filling all space, and theory of origin of matter; proof that the white nebulae are galaxies; confirmation of author's previous announcement of existence of water vapor and oxygen in the atmosphere of Mars, showing amount of oxygen to be about half of that in the earth's atmosphere; also zonal distribution of water vapor on Mars, etc. Invented instrument for making accurate quantitative measurements of intensities of Fraunhofer lines in the solar spectrum. Died Westwood, Mass., Nov. 24, 1927.

VERZAR, Frederiq, physiologist; b. Budapest, Hungary, Sept. 18, 1886; s. Gustav and Irene (Pfleiderer) V.; ed. U. Budapest, U. Tübingen (Germany), Cambridge U.; m. E. Jean McDougall, Dec. 28, 1938; children—Chrisine, Andrew. Prof. physiology U. Debrecen (Hungary), 1918-30, U. Basel (Switzerland, 1930-56; dir. Inst. Exptl. Gerontology, Basel, 1956—; dir. Hungarian Biol. Research Inst., 1926-38. Hon. dr., Freiburg, U. Derecen. Hon. mem. Acad. Medicine Switzerland. Author: (with Jean McDougall) Absorption from the Intestine, 1936; Physiology of Adrenal Cortex, 1939; Experimental Gernotology, 1963. Research on forces of selective absorption, adrenal cortical hormone activity, molecular aging of collagen and DNA, respiratory role of atmospheric-condensation nuclei. Home: 1 Gempenweg, Arlesheim Bld, Switzerland. Office: Nonkenweg 7, 4000, Basel, Switzerland.*

VERZELE, Maurice Julien Eduard, Belgian chemist; b. De Panne, Belgium, Jan. 21, 1923; s. Julien and Madeleyn (Stephanie) V.; Licentiate chemistry, U. Ghent (Belgium), 1945, D.Sci., 1949; m. Denise Careel, Nov. 10, 1945; children—John, Frank. With U. Ghent, 1945—, asso. prof., 1957-61, prof. chemistry, 1961—; asso. Nat. Fund for Sci. Research, 1954-57. Decorated officer Order Leopold. Fellow Chem. Soc.; mem. Inst. Brewing, Flemish Chem. Soc. (mem. central com. 1952—). Research, numerous publs. on chem. constn. of hops-derived substances, separation methods including preparative scale gas chromatography, counter current distbn., synthetic organic chemistry. Home: 13, Krijgslaan, Ghent, Belgium.*

VESALIUS, Andreas, Flemish anatomist, physician; b. Brussels, Flanders, probably Dec. 31, 1514; s. Andreas; studied medicine at Castle School of U. Louvain, 1530-33, 1536-37; U. Paris (under Sylvius and Guenther) 1533-36; M.D., U. Padua, 1537. Lectr. in surgery and anatomy, U. Padua, 1537-42; imperial physician to Holy Roman Emperor Charles V, 1543-56; probably in private practice, 1556-58; physician to Flemings at Court of Philip II in Spain, 1559-64; probably obtained appointment to old chair at Padua, 1564; but on return from pilgrimage to Holy Land died suddenly. Author: Tabulae anatomicae sex, 1538; De humani corporis fabrica libri septem, 1543; De humani corporis fabrica librorum epitome, 1543; others. Founder of modern anatomy; using Galenic exhortation to dissect and observe, overthrew traditional Galenic anatomy (based on nonhuman material) pointing out errors of Galen and describing what his own dissections of human cadavers revealed; made important contributions to osteology, myology, and cardiology; rejected Galenic pores in intraventricular septum of the heart in second edition of De fabrica, 1555; De fabrica contained magnificent woodcut illustrations (from Titian's studio) and was influential for 2 centuries; Vesalius also standardized form and meaning of anatomical nomenclature. Died Oct. 15, 1564, off island of Zante, west of Greece.

VESPUCCI, Amerigo, Italian explorer, navigator; b. Florence, Italy, Mar. 9, 1451; son of Nastogio Vespucci. Employed by comml. house of Medici, for a time; accompanied Alonso de Ojeda on expdn., 1499-1500, separated from Ojeda, and he alone discovered and explored mouth of Amazon River, then sailed along coast of No. South American; entered service of Portugal, 1501; made voyage on which he discovered mouth of Rie De La Plata, explored 6,000 miles of S.Am. coastline, 1501-02; on this trip calculated earth's equatorial circumference (only 50 miles short of correct figure); also proved that S.Am. was separate continent, not part of Asia; astronomer to King of Spain, until 1512; also pilot major for Spain (in which position he prepared and revised master chart of the Atlantic Ocean and Western lands); name America (in his honor) was 1st applied to New World by Martin Waldseemüller in Cosmographiae introductio, 1507. Died of malaria, Seville, Spain, Feb. 22, 1512.

VEST, Markus Friedrich, Swiss physician; b. Basel, Switzerland, Jan. 23, 1923; s. Carl Gottliev and Betty (Buri) V.; M.D., U. Basel, 1947; m. Annemarie Behn-Eschenburg, Feb. 12, 1948; children—Veronika, Eleonora, Hans. Staff, Swiss Inst. for Research in Tb, Davos, 1948, biochem. dept. U. Basel, 1949-50; staff Children's Hosp., U. Basel, 1950—, vice dir., 1963—, asso. prof. pediatrics, 1965—; staff Children's Med. Service, Mass. Gen. Hosp., Boston, 1958-59. Author: Vest Neonatal Jaundice, 1959; also numerous articles. Research on liver function, drug metabolism, physiology of neonatal period, bile pigment, metabolism somatotropin. Home: 4 Erlisacker, Bottmingen 4103, Switzerland. Office: 8 Romergasse, Basel, Switzerland.*

VEST, Marvin Lewis, Am. mathematician; b. Elkins, W.Va., May 17, 1906; s. Marvin J. and Margaret (Kley) V.; B.S., Davis and Elkins Coll., 1927; M.S., W.Va. U., 1932; A.M., U. Mich., 1942; Ph.D. in Math., 1948; m. Winifred Louise Buzzerd, Aug. 3, 1930; children—Marvin Lewis, Charles Marstiller. Prin., Franklin (W.Va.) High Sch., 1927-28; instr. Elkins (W.Va.) High Sch., 1928-31; Asso. prof. math. Davis and Elkins Coll., 1933-38; faculty W. Va. U., Morgantown, 1931-33, 38—; prof. math., 1955—. Cons. U. S. Bur. Mines, 1952-57. Mem. W.Va. Acad. Sci., Math. Assn. Am., Am. Math. Soc., Sigma Xi. Research, publs. on birational transformations asso. with configurations of lines in 3-space. Home: 417 Elm St., Morgantown, W.Va. 26505.*

VESTAL, Paul Anthony, Am. biologist; b. Roosevelt, Okla., Nov. 2, 1908; s. James Elroy and Florence (Anthony) V.; A.B., Colo. Coll., 1930; A.M., Harvard, 1932, Ph.D., 1935; m. Mary Roscoe, Aug. 21, 1936; children—Mary Fon, James Robie, Eleanor Anne. Faculty, Harvard, 1930-42; faculty Rollins Coll., Winter Park, Fla., 1942—, prof., 1949—, chmn. biology dept., 1949—, dir. Thomas R. Baker and Beal-Maltbie Shell Mus., 1943—. Vis. lectr., Uruguay, 1963. Mem. Fla. Acad. Scis. (pres., council), Fla. Audubon Soc. (v.p., council). Author: (with Oakes Ames) Outlines of Economic Botany, 1938; (with Richard E. Schultes) The Economic Botany of the Kiowa Indians, 1939; Ethnobotany of the Ramah Navajo, 1952; (with others) High School Biology Texts, 1960. Research on plant anatomy in relation to phylogeny, ethnobotany of Am. Indian tribes. Home: 1399 Richmond Rd., Winter Park, Fla. 32789.*

VESTLING, Carl Swensson, Am. biochemist; b. Northfield, Minn., May 6, 1913; s. Axel E. and Bertha (Swensson) V.; B.A., Carleton Coll., 1934; Ph.D., Johns Hopkins, 1938; m. Christina B. Meredith, Dec. 24, 1938; children—Martha M., Christina L. (Mrs. J. W. Copenhaver, Jr.), Anne S. Faculty, U. Ill., 1938-63; prof., head dept. biochemistry U. Ia., Iowa City, 1963—. Guggenheim fellow Med. Nobel Inst., Stockholm, Sweden, 1954. Mem. Am. Chem. Soc., Am. Soc. Biol. Chemists, Biochem. Soc. (Eng.), Soc. Exptl. Biology and Medicine, A.A.A.S., Phi Beta Kappa, Sigma Xi, Alpha Chi Sigma, Phi Lambda Upsilon. Editor: Biochemical Preparations, 1957. Numerous publs. on isolation and characterization of liver enzymes; flavoproteins; steroid hydroxylation. Home: Rural Route 1, Fairview Knoll, Iowa City, Ia. 52240.*

VETTE, James Ira, Am. physicist; b. Evanston, Ill., Mar. 4, 1927; s. Ira Charles and Alma (Ishmael) V.; student Northwestern U., 1944-45; B.A. in Physics, Rice U., 1952; Ph.D., Cal. Inst. Tech., 1958; m. Zoe Jane Cashmore, June 9, 1951; children—James Alden, Zoe Jane, Pamela Anne. Staff scientist Convair Sci. Research Lab., San Diego, 1958-62; sr. staff scientist Gen. Dynamics/Astronautics, San Diego, 1962; mgr. nuclear physics Vela satellite program Aerospace Corp., El Segundo, Cal., 1962-63; staff scientist space physics lab. Aerospace Corp., 1963-66; dir. Nat. Space Sci. Data Center, NASA Goddard Space Flight Center, Greenbelt, Md., 1967—. Dow Chem. Co. fellow, 1954-55; Gen. Electric Co. fellow, 1955-56; IBM fellow, 1956-57. Mem. Am. Phys. Soc., Am. Geophys. Union, A.A.A.S., Profl. Group Nuclear Sci., I.E.E.E. Research and publs. on prodn. sub-nuclear particles by high energy gamm rays (photo prodn. of mesons), detection of high energy X-rays from the sun during solar flares; measurements of particle, X-ray, gamma rays in space with satellites; constrn. of trapped radiation model environments utilizing measurements from many satellites. Home: 12502 White Dr., Silver Spring, Md. 20904. Office: Nat. Space Sci. Data Center, Greenbelt, Md. 20771.*

VETTER, Herbert, Austrian physician; b. Vienna, Austria, Aug. 24, 1920; s. Oskar and Paula (Antscherl) V.; student Technische Hochschule Vienna, 1938-39; student physics U. Vienna, 1939-40, M.D., 1948; m. Eleonore von Hacklaender, July 7, 1946; 1 dau., Barbara V. Physician, Allgemeines Krankenhaus, Vienna, 1948-60, head radioisotopes lab. and thyroid clinic, 1951-60; faculty U. Vienna, 1959—, prof. medicine, 1967—; chief med. sect. Internat. Atomic Energy Agy., 1958-67, asst. dir. div. life scis., 1963-67, sr. officer, 1967—, Chmn. planning bd. on medicine and biol. applications of radioactivity Internat. Commn. on Radiation Units and Measurements, 1962—. Brit. Council fellow, London, 1951; research fellow Sloan-Kettering Inst. for Cancer Research, N.Y.C., 1954; recipient prize for art and sci. Pres. Republic of Austria, 1956. Mem. Royal Soc. Medicine, Gesellschaft der Aerzte, Gesellschaft für Innere Medizin, Osterreichische Roentgen Gesellschaft. Author: (with N. Veall) Radioisotope Techniques in Clinical Research and Diagnosis (English), 1958, (German) 1960, (Spanish) 1964; also numerous articles. Editor: (with K. Fellinger) Radioaktiv Isotope in Klinik und Forschung, vol. I, 1955, vol. II, 1956, vol. III, 1958; Jour. Nuclear Medicine. Devel. or improvement of several med. applications of radioactive isotopes especially study of thyroid function and liver blood flow. Home: 18 Landesgerichtstrasse. Office: 11 Kärntnerning, A-1010 Vienna, Austria.*

VETTER, Klaus Jürgen, German chemist; b. Berlin, July 31, 1916; ed. U. Göttingen, U. Berlin; Dr. ès. sc.; m. Margarete Binder, July 2, 1955; children—Sabine, Andreas. Asst., U. Berlin, 1946-48; asst. Fritz Haber Inst. of Max Planck Soc., Berlin, 1948—, now collaborating scientist, chief div.; successively agrégé in chemistry Free U. Berlin, 1951, instr., 1955, asso. prof., 1961, titular prof., 1964. Mem. Deutsche Bunsengesellschaft, Gesellschaft deutscher Chemiker Comité internat. de thermodynamique et cinetique electrochimique, Electrochemistry Soc., Faraday Soc. Author: Elektrochemische Kinetik, 1961. Research and publs. on electrochemistry, especially kinetics of electrode processes, electrolytic passivity, corrosion. Home: Murellenweg 45, 1 Berlin 19. Office: Institut für Physikalische Chemie der Freien Universitat, Thielallee 63-67, 1 Berlin 33, West Germany.

VEVERS, Henry Gwynne, biologist; b. Girvan, Scotland, Nov. 13, 1916; s. Geoffrey Marr and Catherine (Andrews) V.; B.A., Magdalen Coll., Oxford, Eng., 1938, M.A., 1947, D.Phil., 1949; m. Joyce Brigstocke, Sept. 9, 1950; 1 son, Geoffrey William Gwynne. Bursar, zoologist Plymouth Marine Biol. Lab., 1946-55; curator aquarium Zool. Soc. London, 1955—, asst. dir. sci., 1966—; mem. expdns. to Greenland, Iceland, Faeroe Islands, Solomon Islands; hon. research fellow Bedford Coll., U. London, 1965—. Mem. Order Brit. Empire, 1942. Fellow Linnean Soc. London (zool. sec. 1960-67), Inst. Biology. Author: The British Seashore, 1954; (with H. Munro Fox) The Nature of Animal Colours, 1960; also articles. Research on influence of sex hormones on color and pattern in feathers, breeding population counts of N. Atlantic gannet, photography of sea floor with specially designed underwater camera, distbn. of carotenoid and porphyrin pigments in marine invertebrates. Home: Curator's Flat, Zool. Soc. Office: Zool. Soc., Regent's Park, London N.W.1, Eng.*

VIALOV, Oleg Stepanovich, Russian geologist; b. Jan. 23, 1904; grad. Leningrad U., 1928. Staff, Geologic Com., until 1933; staff All-Union Oil Inst., Leningrad, 1933-47; staff Geology Inst. Ukrainian Acad. Sci., Lvov, 1948—. Recipient Stalin prize, 1947. Research in geology of Central Asia, Western China, Antarctica, Carpathian Mountains, Caucasus, Kamchatka; stratigraphy and molluscs of Paleogene in Central Asia.

VIAUT, André Jules Armand, French meteorologist; b. Civry, France, Oct. 16, 1899; s. Armand and Berthe (Tavoillot) V.; licence ès sc., Collège Tonnerre; m. Marcelle Moreau, Sept. 22, 1921; 1 son, Michel. Meteorologist, 1921-28; head meteorologist, 1926-32; head warning services, 1932-39; head central management services, 1939-40; dir. Nat. Meteorol. Orgn., 1945—. Recipient Aeronautics medal; Léon-Grelaud prize French Acad. Scis., 1950. Mem. French Meteorology Soc. (sec.-gen.). Author: La Météorologie du navigant; La Météorologie du vol à voile; La mer et le vent; dir. of Monographies de météorlogie. Home: 1 avenue Victor-Hugo, Meudon (Seine and Oise), France. Office: 1 quai Branly, Paris, 7, France.

VICK, Francis Arthur, Brit. physicist; b. Solihull, Eng., June 5, 1911; s. Wallace Davenport and Clara (Taylor) V.; B.Sc. with honors in Physics, Birmingham (Eng.) U., 1933, Ph.D., 1936; m. Elizabeth Dorothy Story, Oct. 11, 1943; 1 dau., Christine Mary. Faculty, U. Coll., London, 1936-39; asst. dir. sci. research Ministry of Supply, London, 1939-44; faculty Manchester (Eng.) U., 1944-50; prof. physics U. Coll. N. Staffordshire (now Keele U.), 1950-59, acting prin., 1952-53; dep. dir. Atomic Energy Research Establishment, Harwell, Eng., 1959-60, dir., 1960-64; dir. research group U.K. Atomic Energy Authority, 1961-64, mem. for research, 1964-66; pres., vice chancellor Queen's U., Belfast, No. Ireland, 1966—. Hon. sec. Inst. Physics, 1956-60. Decorated Order Brit. Empire; knight comdr. Liberais Humane Order African Redemption. Fellow Inst. Physics, Instn. Elec. Engrs. Research, numerous publs. on proximity fuses, use of electron emission to study adsorption of gas layer on solids, structure and properties of complex thermionic cathodes, emission ions from heated solids, field emission of electrons from solids, field-ion microscope. Home: Vice Chancellor's Lodge, Lennoxvale, Belfast 9, No. Ireland.*

VICKERY, Robert Kingston, Jr., Am. geneticist; b. Saratoga, Cal., Sept. 18, 1922; s. Robert Kingston and Ruth (Bacon) V.; A.B., Stanford, 1944, M.A., 1948, Ph.D., 1952; m. Marcia Hoak, July 7, 1951; children—David K., Peter H. Instr., Pomona Coll., 1950-51; faculty U. Utah, Salt Lake City, 1952—, prof. genetics, 1962—, chmn. dept., 1962-65, prof. molecular and genetic biology, 1966—. Research asso. Cal. Inst. Tech., 1955; vis. prof. Harvard, 1961. Mem. Evolution Soc., Bot. Soc. Am., Genetics Soc. Am., Cal. Bot. Soc., Internat. Soc. Plant Taxonomy, Indian Genetics Soc., Am. Inst. Biol. Scis., A.A.A.S., Phytochem. Soc. N.Am., Sigma Xi. Research, publs. on evolution in Monkey flowers using chromosome numbers, characteristics of hybrids and biochemistry of pigments, geog. patterns. Home: 3376 Louise Av., Salt Lake City 84109.*

VICKREY, William Spencer, economist; b. Victoria, B.C., Can., June 21, 1914; s. Charles Vernon and Ada Eliza (Spencer) V.; grad. Phillips Acad., 1931; B.S., Yale, 1935; M.A., Columbia, 1937, Ph.D., 1947; m. Cecile Montez Thompson, July 21, 1951. Jr. economist Nat. Resources Com., Washington. 1937-38; economist 20th Century Fund Elec. Power Study, N.Y.C., 1939-40, OPA, Washington, 1940-41, tax research div. U. S. Treasury, Washington, 1941-43; tax cons. Gov. P.R., 1946; faculty Columbia, N.Y.C., 1947—, now prof. econs., chmn. dept., 1964-67. Fellow, Center for Study Behavioral Scis., Stanford, 1967-68. Mem. Shoup Tax Mission to Japan, 1949-50; instr. IBM Systems Research Inst., 1964; economist World Bank Coal Transport Study, New Delhi, 1963, also many others. Social Sci. Research Council fellow, 1938-39; Guggenheim fellow, 1955-56; Ford research prof., 1958-59. Mem. Am. Econ. Assn., Am. Statis. Assn., Econometric Soc., Royal Econ. Soc., Nat. Tax Assn., Tax Inst. Am. Author: Agenda for Progressive Taxation, 1947; The Revision of the Rapid Transit Fare Structure of the City of New York, 1953; Microstatics, 1964; Metastatics and Macroeconomics, 1964; also articles, revs. Research in tax field, pricing of urban transp., improved sorting procedures on digital computers, and methods of financing higher edn. Home: 162 Warburton Av., Hastings-on-Hudson, N.Y. 10706. Office: Fayerweather Hall, Columbia U., N.Y.C. 10027.*

VICO, Giovanni Battista (Giambattista), Italian social scientist; b. Naples, Italy, June 23, 1668; ed. Jesuit sch., U. Naples; m. 1699; 8 children. Tutor to nephews of G. B. Rocca, Bishop Ischia, Italy, for 9 years (devoted time to classical reading in family's library, nr. Salerno); prof. rhetoric U. Naples, from 1699; apptd. royal historiographer by Don Carlos of Bourbon (Charles II of Spain), 1735. Author: De nostri temporis studiorum ratione, 1709; De antiquissima italorum sapientia ex latinae linguae originibus eruenda, 1710; Diritto universale, 1720-22; Principi di scienze nuova d'intorno alla comune natura delle nazioni, 1725-44 (major theories developed in this work). Often regarded as father modern social history; developed systematic methods hist. research, drawing on study lang., linguistics, mythology, tradition; believed history to be account of birth and cyclical devel. human socs. and instns. believed each hist. period has distinct character, that similar periods recur, but with such modification that historians cannot be prophetic; influenced many areas modern social thought with views on class struggle, heroism, man's innate, divinely inspired sense justice, evolution of truth. Died Naples, Jan. 23 or 24, 1744.

VICQ-D'AYR, Félix, French physician, anatomist; b. Valognes, Basse, Normandy, France, Apr. 28, 1748; grad. medicine Paris, France; m. Mlle. Daubenton. Acting prof. anatomy Jardin du Roi; physician to Marie-Antoinette, also comte d'Artois. Mem. French Acad. Scis., 1774, Société d'agriculture, Académie francaise (sec.), 1788, Société Royale de Médecine (founder with Lassone 1776, perpetual sec.). Author: Médecine des Betes à Cornes, 1781; Traité d'anatomie et de physiologie, 1786; Système anatomique des quadrupèdes, 1792; Oeuvres complètes, 1805. Research, publs. on flexor, extensor muscles of animals, man, morphology of brain, vocal cords, anatomy of birds, quadrupeds; described bundle of Vicq d'Aryr (fibers extending from mammillary body to anterior nucleus of thalamus), 1781; leading comparative anatomist, 2d half 18th century. Died Paris, June 20, 1794.

VICTOR, Paul-Emile, French explorer; b. Geneva, Switzerland, June 28, 1907; D.Eng., Ph.D., Sc.D., Ecole Central, Lyons, France; student Inst. Ethnology, Paris; m. Elaine Decrais, July 30, 1946; children—Jean-Christophe, Stephanie Daphne. Mem. Greenland expdn., 1934-35; crossed Greenland by dog sleigh, 1936; wintered on E. coast of Greenland, 1936-37; made trans-Alpine crossing by dog sleigh Nice-Chamonix, 1938; mem. expdn. to Lapland, 1939; organizer, dir. expdns. Polaires Francaises, including missions Paul-Emile, Arctic and Antarctic expdns., 1947—; head internat. glaciological expdn. to Greenland. Pres., French Antarctic com. IGY. Recipient Gold medal Royal Geog. Soc. London, Vega Gold medal, Sweden, Médaille spéciale de l'Administrn. des Monnaies. Author: Boréal, 1938; Banquise, 1939; Apoutiek, 1947; Aventura esquimau, 1948; la Grande Faim, 1958; Pôle Sud, 1960; la Vole lactée, 1961; l'Homme à la conquête des pôles, 1962. Studies on Arctic and Antarctic regions, Eskimo life and culture. Office: Expéditions Polaires Francaises, 47 Avenue du Maréchal Fayolle, Paris 16e, France.

VICTORINUS, astronomer, mathematician; flourished 5th century. Commd. by Pope Hilarius to correct calendar, fix date of Easter, circa 465 A.D. Prepared arithmetical table; combined metonic cycle (19 years) with solar cycle (28 years) to form period of 532 Julian years (19 x 28), at end of this period, Easter moon would come in same month and on same day of week; prepared arithmetical tables from which desired products could be copied.

VICTORIUS OF AQUITANIA, see Victorinus.

VIDAL DE LA BLANCHE, Paul, French geographer; b. Pézenas, Herault, France, Jan. 22, 1845; ed. École Normale, École d'Athènes; doctorate, Sorbonne, Paris, 1871. Lectr., École Normale Supérieure, later asst. dir. for lit. dept.; prof. geography Faculté des Lettres, Paris, 1898-1909. Mem. Académie des sciences morales et politique. Author: Hérode Atticus; Etats et nations de l'Europe autour de la France, 1889; La Rivière Vincent Pinzon, 1902; Tableau de la géographie de la France, 1903; La France de l'Est, 1917; (with Lucien Gallois) Le Bassin de la Sarre 1919; Principes de geographie húmaine, 1922; Atlas général, Vidal-Lablache. Planned and prepared for publ. Géographie universelle, 15 vols. Founder, French sch. human geography; rejected theory of absolute geog. control; research on relation between man's environment and his activities. Died Tamaris, France, 1918.

VIDIUS, Guido, Italian anatomist; b. Florence, Italy, 1500; s. Giuliano Guidi and Costanza (De Domenico) V.; apptd. prof. Coll. Royal, Paris, France, 1542; went to Pisa, Italy, 1547. Author: De chirurgia, 1544; De anatomia, 1611. Contbd. to progress of anatomy by giving exact descriptions of vertebrae, cartilage, brain, heart and eye. Died Pisa, 1569.

VIER, Dwayne Trowbridge, Am. phys. chemist; b. Washington, Sept. 17, 1914; s. William Frederick and Charlotte (Trowbridge) V.; B.S. (Valentine Smith scholar), U. N.H., 1937, M.S., 1939; Ph.D., Columbia U., 1943; m. Marion Hazel Corson, June 30, 1951; children—David Corson, Alan Dwayne. Research chemist S.A.M. Labs., N.Y.C., 1943-45; asso. scientist Manhattan Dist., U. Cal., 1945-46; group leader Los Alamos Sci. Lab., 1946—. Recipient Cogswell prize U. N.H., 1936. Mem. Am. Chem. Soc., Sigma Xi, Phi Kappa Phi, Phi Lambda Upsilon. Research, publs. in phys. chemistry, inorganic chemistry of rare radioactive elements and lithium hydride, high temperature chemistry. Patentee in field. Home: 764 43d St. Office: Los Alamos Sci. Lab., Box 1663, Los Alamos 87544.*

VIERKANDT, Alfred, German sociologist; b. Hamburg, Germany, June 4, 1867. Prof., Berlin. Author: Die Stetigkeit im Kulturwandel, 1908; Familie, Volk und Staat, 1936. Editor: Handwörterbuch der Soziologie, 1931. Research on psychol. question of culture; analyzed community and its forms; Died Berlin, Germany, Apr. 24, 1953.

VIÈTE, François (or Vieta, Franciscus), Seigneur de la Bigotière, French mathematician; b. Fontenayle-Comte, Poitou, France, 1540; studied law at Poitiers; became councillor to parlements; maitre des requetes de l'hotel du roi, 1580; mem. privy council of Henry IV; broke cipher, decoded messages of Philip II to his soldiers in Netherlands for Henry IV. Author: Canon mathematicus seu ad triangula cum appendibus, 1579; In artem analyticam Isagoge, 1591; Supplementum geometriae, 1593; De numerosa potestatum purarum atque ad fectarum ad exegesin resolution e tractatus, 1600; Apollonius gallus, 1600; De aequationem recognitione et emendatione libri duo 1615; Greatest algebraist of 16th century and a founder of modern algebra; made many contributions to trigonometry including solutions of triangles and account of relations between functions of an angle and those of its submultiples; pioneered study of symbolic algebra and theory of equations; introduced letters of alphabet to denote known (consonants) and unknown (vowels) quantities (later he used symbol N for unknown); used words, not symbols, for equality and multiplication; knew and improved on Cardan's method of solving cubic equations and Ferrari's method of solving quartic; (invariably ignored negative roots); sought solution of quintic; enunciated 4 theorems of relations between roots of an equation and coefficients of its tens; gave method of approximating roots of an equation; calculated pi to 10 correct decimal places, the best value of his time, and expressed value of ratio pi by an infinite product; stressed superiority of decimal over sexagesimal fractions; made many other original contributions to mathematics; preferred Latin analysis to Arabic algebra. Died Paris, France, Dec. 13, 1603.

VIETEN, Heinz Wilhelm Josef, German radiologist; b. Bonn, Germany, Mar. 18, 1915; s. Wilhelm Franz and Mathilde (Ohmen) V.; med. state exam., U. Bonn, 1938; m. Marianne Dhein, Mar. 11, 1940; children—Birgitta (Mrs. Jörg Balau), Marianne (Mrs. Hans-Otmar Neher), Renate, Susanne. Faculty, U. Düsseldorf, 1949-—, prof. med. radiology, 1963-—, dir. med. radiol. clinic. Recipient Dr. Schleussner X-ray prize, 1950, 53. Mem. German, Rhenish-Westphalian X-ray socs., Cologne Bonn X-ray Assn., World Assn. Acad. Tchrs. Med. Radiology, Bonn Soc. Natural History and Medicine, Düsseldorf Med. Soc., Lower Rhenish-Westphalian Surgeons Assn., German Soc. Surgery, Am. Coll. Chest Physicians. Author: (with J. Schoenmackers) Atlas postmortaler Angiogramme, 1954; (with E. Stutz) Die Bronchographie, 1955; (with H. Oberdalhoff, H. Karcher) Klinische Röntgendiagnostik chirurgischer Erkrankungen, 1959; (with A. Gebauer, E. Muntean, E. Stutz) Das Röntgenschichtbild, 1959; also monographs, numerous contbns. to handbooks and profl. lit. Editor: (with others) Handbuch der Medizinischen Radiologie; Der Radiologe, 1961-—; adviser several sci. jours. Research in diagnostic film representation, contrast medium (heart, blood vessels, lungs), series exposures, cinematography; supervolt therapy. Address: 34 Schlossmannstrasse, Düsseldorf, West Germany.*

VIEUSSENS, Raymond, French physician, anatomist; b. Vigan, France, 1641. Physician, Hôpital St.-Eloy, Montpellier, France, from 1671. Mem. French Acad. Scis., 1708. Author: Neurologia universalis, 1685; Tractatus duo . . . , 1688; Deux dissertations . . . , 1698; Epistola de sanguinis humanis . . . , 1698; Epistola nova quaedam in corpore humano inventa exhibens, 1703; Traité sur la structure de l'oreille, 1714; Sur les Liqueurs du Corps Humain, 1715; De la structure et du mouvement naturel du coeur, 1715. Research on brain, spinal cord; gave 1st accurate descriptions of structure of left ventricle of heart, valve of large coronary vein, course of coronary vessels, all 1705, centrum ovale, 1685, aortic insufficiency; asso. various movements with different sects. of brain; probably discovered that blood contains acid. Died Montpellier, Aug. 16, 1715.

VIGANI, John Francis (Giovanni Francesco), chemist; b. Verona, Italy, circa 1650; studied mining, metallurgy and pharmacy in travels, Spain, France, Holland; m. Elizabeth, circa 1682; children—Frances, Jane. Came to Eng., circa 1682; became pvt. tutor chemistry and pharmacy, Cambridge, 1683, 1st prof. chemistry, Cambridge, 1703; became lectr. pharm. chemistry Queen's Coll, 1705. Author: Medulla Chymirae, 1682, enlarged, 1683, 85, 93, 1718-19. Developed method for purification of iron sulphate from copper, for making ammonium sulphate; proved that to make given salt, metallic base always needs same quantity of acid. Died Newark-on-Trent, Eng., Feb. 1712.

VIGENERE, Blaise de, French humanist; b. Saint-Puorcain, in Bourbonnais, France, 1523; ed. Paris, France; studied Greek under Turnebus and Dorat, also Hebrew; m. 1570. Offl. at French Ct.; present at diet of Worms; traveled to Rome, then returned to Paris; 1st sec. to Duke de Nevers, also Henry III. Author: Traité des Chiffres ou Secrets Manieres d'es crire, 1586; also wrote on fire and salt (pub. 1608), historical works. Translator: Icones of Philostratus; also works of Livy, Cicero, and Tasso. Work in cryptography. Died Paris, 1596.

VIGNOLES, Charles Blacker, English engr.; b. Woodbrook, Wexford, Eng., May 31, 1793; s. Charles Henry and Camilla (Hutton) V.; student Sandurst, from 1810; m. Mary Griffiths, July 13, 1817 (dec. 1834); m. 2d, Elizabeth, 1849; children—Charles Francis Ferdinando, Henry, Hutton, Olinthus John. With S.C. survey, 1816-23; ry. engr., 1825-65; engr. in chief Dublin & Kingston Ry (1st of Irish lines), 1834; 1st prof. C.E. civil engring. U. Coll., London, Eng. (1st in Eng.), from 1841. Fellow Royal Soc., 1855, Royal Astron. Soc. (council); mem. Instn. C.E.'s (pres. 1869), Royal Irish Acad., Royal Instn. Author: Observations on the Floridas, 1823. Contbr. articles to Ency. Metropolitana, 1817-45. (With J. Ericsson) patentee new method for ascension steep inclines on rys. (involved 3d rail), called center-rail system, 1830; inventor flat-footed Vignoles rail (widely used on continent), 1837; built suspension bridge over Dnieper River at Kiev, USSR (then longest of its kind in world), 1st ry. in western Switzerland, 1853-55. Died Hythe, Hampshire, Eng., Nov. 17, 1875.

VIGNON, Georges Eugène Paul, French physician; b. Lyons, France, Nov. 24, 1916; s. Paul and Madeleine (Mollade) V.; grad. Faculty Medecine, Lyon; m. Simone Piaton, Mar. 6, 1943; children—Florence (mrs. Jean-Pierre Boissel), Eric, Patricia. Physician, hosps. Lyons, 1950-—; faculty Faculty Medecine, Lyons, 1955-—, prof., 1965-—; head physician Hopital des Charpennes, Hospices Civils de Lyon, 1955-—. Named Chevalier des Palmes Académiques. Mem. French League Against Rheumatism (v.p.). Author: Lecons de Rhumatologie, 1964; (with P. P. Ravault) Rhumatologie clinique, 1956; also numerous articles. Research on rheumatic diseases and biol. studies of diseases of bone especially phospho-calcic metabolism using isotopes. Home: 13 Qual de Serbie, Lyon VI, France. Office: Hopital des Charpennes 31, Grde Rue des Charpennes, Velleurbanne, France.*

VIGO, Giovanni, Italian surgeon; b. Genoa, circa 1460. Practiced in Rome; physician to Pope Julius II. Author: Practica in Arte Chirurgica Copiosa, Continens Novem Libros, 1514; De Morbo Gallico, 1518. Author of most complete book on surgery (of the time); compiled already discovered med. knowledge; advocated use of warm oil in treatment of gunshot wounds. Died circa 1520.

VIGORS, Nicholas Aylward, Irish zoologist; b. Old Leighlin, Ireland, 1785; s. Nicholas Aylward and Catharine (Richards) V.; B.A., Trinity Coll., Oxford, 1817, M.A., 1818, (Hon.) D.C.L., 1832. Ensign, Peninsula War, 1809-11; became landowner in County Carlow, 1828; became Irish mem. Parliament, 1832. Fellow Royal Soc., 1826, Linnean Soc., Soc. Antiquaries, Geol. and Hist. Socs.; mem. Royal Irish Acad., Royal Instn., Zool. Soc. (charter, 1st sec. 1826-33). Contbg. author: Zoology of Captain Beechey's Voyage, 1839. Editor, (with others) Zool. Jour., vols. III, IV, 1828-35. Tried to apply quinary arrangment to class Aves. Died London, Oct. 26, 1840.

VIK, Rolf, Norwegian zoologist, parasitologist; b. Trondheim, Norway, Sept. 5, 1917; s. Vidar and Margit (Norgaard) V.; B.Sc., U. Oslo (Norway), 1942, M.Sc., 1950, Doctor's Degree, 1958; m. Marta Bergh, Sept. 8, 1945; children—Tore, Pal. Lectr. biology Oslo (Norway), Tng. Coll., 1951-60; faculty U. Oslo, 1960-—, prof. zoology, 1965-—, dir. Zool. Mus., 1965-—, v.p. Faculty Sci., 1965-—. Chmn. Norwegian Nat. Com., Internat. Biol. Programme, 1965-—; cons. parasitologist FAO, 1958-—. Mem. Am. Soc. Parasitologists, Wildlife Disease Assn., Brit. Soc. Parasitology. Author: Book of Norwegian Birds, 1953; (with Per Bergan A. S. Johansson, Jan Okland), Illustrated Zoological Dictionary, 1964; also articles. Research on occurrence and distbn. of parasites in fishes, birds and mammals of Norway; worked out life cycle of cestode Diphyllobothrium norvegicum and part of life histories of Eubothrium salvelini and Eubothrium crassum and nematod Philonema agubernaculum. Home: 21 Snaret, Eiksmarka, Oslo 7, Norway.*

VIKHERT, Amatoly Mihailovitch, Russian pathologist, physician; b. Moscow, USSR, Nov. 23, 1918; s. Mihail Osipovitch and Yadviga (Aurih) V.; student First Moscow Med. Inst., 1936-41, M.D., Ph.D., 1958; m. Tatyana Lebedeva, Jan. 24, 1959; children —Olga, Mihail. Mil. pathologist, 1941-46; mem. staff Inst. Morphology, Acad. Med. Sci. USSR, 1946-50; mem. staff dept. pathology Inst. Therapy, Acad. Med. Sci. USSR, 1950-—, chief dept., 1959-—, also chief Centre Studies Epidimiology and Pathology of Arterosclerosis; chief pathologist Ministry Pub. Health USSR. Mem. USSR Pathol. Soc. (v.p.), USSR Cardiology Soc. (bd. dirs.). Author numerous articles. Research in pathology and pathogenesis of urinary tract changes in spinal injuries, also sepsis leuta; Y.G. cells in difference condition of human pathology; epidemiology of arterosclerosis in USSR. Home: 11 Bourdenco. Office: 10 Petroverigsky, Moscow, USSR.*

VILARDEBO, A., entomologist, nematologist; b. Beira, Africa, Nov. 10, 1922; s. Ricardo and Suzanne (Cardozo) V.; Ingénieur Agronome, Institut Nat. Agronomique, 1942; Spécisation en Entomologie, Office Recherche Scientifieque Outre-Mer, 1944; m. Marthe Cauchy, May 11, 1946; children—Dominique, Christine Staff, Institut Francais de Recherches Fruitières Outre-Mer; staff Station Centrale à Kindia, Guinea, 1947-58; later head researcher, Ivory Coast until 1961; dir. Service Entomologie Nématologie au Siège de l'Institut Paris, 1962-—; tech. assistance, Mexico, Equador, Canary Islands, Somalie, 1960-—; faculty entomology culture of tropical fruit trees Centre de formation d'Inspectuers de la défense des Végétaux. Research and publs. on depredatory insects; nematodes of tropical fruit including bananas, citrus fruits, pineapples, avocadoes, mangoes and dates; biology, ecology and behavior of various parasites and methods of control using chemicals or biol. control. Home: 23, avenue Pasteur, Vanves, France 92. Office: 6, rue du Générale Clergerie, Paris, France.*

VILLA, Luigi, Italian physician; b. Milan, Italy, Sept. 7, 1896; s. Luigi and Emma (Romagnoli) V.; grad. in medicine, U. Pavia (Italy), 1920; m. Vincenza Cazzulani, Apr. 26, 1930. Faculty, U. Pavia, 1921-37, prof. dept. medicine, 1933-37; prof. U. Milan, 1937-50; dir., prof. Inst. Clin. Medicine, U. Milan, 1950-—, dean Faculty Medicine, since 1966-—, dir. Postgrad. Med. Sch., Sch. Internal Medicine, 1950-66, dir. Center Molecular Pathology, 1965-—. Recipient Vittorio Emanuele, Inst. Lombardo di Scienze e Lettere, 1925; Marzotto; Officier de la Santé publ. de France, 1957; Commendatore di San Gregorio Magno, 1963; Gold medal Italian Health Office, 1959, Ministry Edn., 1960. Mem. Italian Soc. Internal Medicine (pres.), Italian Soc. Metabolism (pres.), Internat. Soc. Medicine (Italian rep.). Author: Ricambio Idrico, 1932; Insufficienza Epatica, 1956; Anemia Perniciosa, 1961; also other books, numerous articles. Editor: Haematologica Latina, 1958; Enzumologia Biolog. et Clinica, 1960; Metabolism, 1963. Research on molecular pathology and enzymology of liver disease, blood diseases, water and electrolyte metabolism, endocrinology, cardiology, rheumatology; one of 1st to apply enzymological research to problem of pathology and clinic. Home and office: 19, via Fatebenefratelli, Milan, Italy.*

VILLANOVANUS, Arnaldus (Novicomensis), physician; b. nr. Valencia, Spain, circa 1235; student Paris, Montpellier, France, also in Italy. Charged with heresy for religious speculations; diplomat for kings of Sicily and Naples; physician to Pope Clement V; taught and practiced medicine, Montpellier, also Paris. Wrote several essays on medicine, alchemy, religion. Translator from Arabic to Latin: De tremore, palpitatione, regore et convulsione (Galenos); De medicinarum compositarum gradibus (Al-Kindi); De physicis litaturis (Qusta B. Luga); De viribus cordis (Ibn Sina); De conservatione corporis et regimine sanitatus (Abu al-Ala Zuhr); Albuzale de medicinis simplicibus (Abu al-Salt Umayya B. Abd al-Aziz). Studies in chemistry, medicine, essence of wine, turpentine. Died 1313.

VILLAR, German Eduardo, Uruguayan chemist; b. Montevideo, Uruguay, Apr. 29, 1900; s. German and Emma (Fernandez) V.; Civil Engr., U. Montevideo, 1925, hon. prof., 1954; m. Constance Eastman, Dec. 9, 1926; children—Susana (Mrs. Rolf Apolant), German, Mercedes (Mrs. Emilio Isla), Aileen (Mrs. Luis Serrano), Carlos. Prof. analytical chemistry U. Montevideo, 1926-31, prof. phys. chemistry, 1931-—, head chem. lab., 1932-35, dir. Inst. Tech. and Chemistry, 1935-—; prof. colloidal chemistry Inst. Advanced Studies of Montevideo, 1932-52. Mem. Assns. chemistry Chili, Peru, Uruguay, Royal Acad. Scis. Madrid, Nat. Acad. Engring. Uruguay, Assn. Engrs. Uruguay. Author: Elementos de Atomistica, 1939; Energia Atomica, 1959; Coloides, 1963; also numerous articles. Research on electronic configurations of atoms and position of elements in periodic chart. Home: 2924 21 de Setiembre, Montevideo, Uruguay.*

VILLARD, Paul-Ulrich, French physicist; b. Lyon, France, Sept. 28, 1860; ed. École normale; agrégé; Dr. ès sc., Prof., Conservatoire des Arts et Mètiers; mem. French Acad. Scis., 1908. Author: Les Rayons cathodiques, 1900. Studied hydrates; discovered silver hydrate; research in radiation; invented several radiological instruments, such as the osmoregulator and Villard's valve; observed for first time rays that he called gamma rays, whose penetrating power is superior to that of beta rays, (gamma rays are not deflected by electric or magnetic fields, have weak power of ionization, and are considered electromagnetic radiations), 1900; gave first phys. definition of unit of ionization, unit V. Died Bayonne, France, Jan. 13, 1934.

VILLARS, Dominique, French botanist; b. Haut-Dauphiné, France, Nov. 14, 1745; M.D., 1778. Prof. botany, Grenoble, France; prof. medicine and botany Faculté de Médecine, Strasbourg, France, 1805. Mem. French Acad. Scis., 1796. Author: Natural History of the Plants of Dauphiné, 4 vols., 1786; Principles of Medicine and Surgery. Genus Villarsia named in his honor. Died Strasbourg, June 20, 1814.

VILLARS, Donald Statler, Am. chemist; b. nr. Blanchester, O., Dec. 21, 1900; s. John Oscar and Lula (Statler) V.; A.B., Wilmington Coll., 1921; M.S., Ohio State U., 1922, Ph.D., 1924; m. Clarine Booth, Aug. 11, 1926; children—Dorothy Elizabeth (Mrs. Wares Ishaq), Robert John. NRC fellow, 1925-26; asso. in chemistry U. Ill., 1927-29; asst. prof. phys. chemistry U. Minn., 1929-30; physicist Gen. Electric Co., 1930; phys. chemist Standard Oil Co. (Ind.), 1931-34; phys. chemist planations U. S. Rubber Co., Sumatra, 1934-38, gen. labs., 1938-46; head sci. dept., prof. chemistry Jersey City Jr. Coll., 1946-49; research scientist U. S. Naval Ordnance Test Sta., China Lake, Cal., 1949-—. Recipient Alumni citation Wilmington Coll., 1948. Fellow Am. Phys. Soc.; mem. Am. Chem. Soc. (past chmn. profl. relation com. Mojave Desert sect., past chmn. Cal. coordinating com.), Sci. Research Soc. Am. (past pres., governing bd. China Lake chpt.), Sigma Xi, Phi Lambda Upsilon. Author: Statistical Design and Analysis of Experiments for Development, 1951. Research, patents, articles on thermodynamics, photochemistry, combustion, spectroscopy, quantum mechanics, statistics, phys. chemistry; developed 1st monochromator for photochem. photolyses; quantum-phys. explanation of electron jump; ultra-speed tensile measurement of elastomers. Home: 703

Saratoga Av., China Lake 93555. Office: Code 50503, U. S. Naval Ordnance Test Sta., China Lake, Cal. 93557.*

VILLAT, Henri-René-Pierre, French mathematician; b. Paris, France, Dec. 24, 1879; s. Louis Achille and Félicie (Lesper-Mont) V.; grad. Ecole Normale Supérieure, 1899; agrégé of Univ.; D.Sc.; m. Marie Le Saulnier, Nov. 14, 1904; children—Jacques, Philippe, Daniel; m. 2nd. Marian Arthur, Dec. 24, 1964. Prof., U. Caen (France), 1902; U. Montpellier, 1911; prof. theoretical mechanics Faculté des Sciences, U. Strasbourg, 1920; dir. Inst. Mechanics; prof., Faculté des Sciences, U. Paris, 1927. Mem., French Acad. Scis., 1932 (vice-pres. 1947, pres. 1948); decorated Commendeur de la Légion d'Honneur. Research and publs. on math., mechanics of fluids, friction on solid body moving in fluid, flexibility; contributed to modern aerodynamics. Home: 47, Boulevard Auguste-Blanqui, Paris 13e, France.*

VILLEE, Claude Alvin, Am. biologist; b. Lancaster, Pa., Feb. 9, 1917; s. Claude Alvin and Mary (Nestle) V.; B.S., Franklin and Marshall Coll., 1937; Ph.D., U. Cal. at Berkeley, 1941; A.M. (hon.) Harvard, 1957; m. Dorothy Theresa Balzer, Jan. 21, 1952; children—Claude Alvin III, Stephen Eric, Suzanne Claire, Charles Andrew. Asst. prof. Armstrong Coll., Berkeley, 1941-42, U. N.C., 1942-45; faculty Harvard Med. Sch., Boston, 1946—, prof., 1963-64, Andelot prof. biol. chemistry, 1964—. Cons. NIH, 1958—, NSF, 1958—. Recipient Rubin award Am. Soc. Fertility and Sterility, 1957; CIBA Found. (London) award, 1956. Mem. Am. Acad. Arts and Scis., Am. Soc. Biol. Chemistry, Endocrine Soc., Biochem. Soc. (Eng.), Am. Chem. Soc., Am. Inst. Biol. Sci. Author: Biology, 1950; (with Warren Walker, Fred Smith) General Zoology, 1958; The Placenta and Fetal Membranes, 1960; Gestation, 1955; Control of Ovulation, 1961; (with L. L. Engel) Mechanisms of Action of Steroid Hormones, 1961; also numerous articles. Research on mechanism genic control of devel. in fruit flies, sea urchins, and mammals; demonstrated effect hormone on purified enzyme system in vitro; discovered mechanism transport riboflavin and sugars across placenta. Home: 43 Cottage Farm Rd., Brookline, Mass. 02146. Office: 25 Shattuck St., Boston 02115.*

VILLEGAS, Alejandro H(ector), Argentine physician; b. Buenos Aires, Argentina, July 6, 1925; s. Desalín and María Beatriz (Scarzolo Harispe) V.; M.D., U. Buenos Aires, 1950, Dr.Medicine, 1952; m. Martha Josefina Siaz Lagar, Nov. 10, 1953; children— Maria Silvia, Alejandro Arturo. Asst. surgeon, instr. surgery Instituto de Cirugía Torácica de Buenos Aires, also Cátedra de Cirugía Torácica, 1950-57; fellow surgery Masst. Gen. Hosp., Boston, also Harvard Med. Sch., 1957-60; asst. surgeon, instr. surgery Hosp. Ramos Mejía, 7th Ward gen. surgery, also asso. prof. clin. surgery U. Buenos Aires, Med. Sch., 1961—. Cons. surgeon Centro de Rehabilitación Respiratoria de Buenos Aires, 1965—, Centro de Educación Médica e Investigaciones Clínicas, 1961—. Recipient Premio Ministro de Justicia e Instrucción Dr. Jorge E. Coll, 1939; Premio Facultad de Ciencias Medicas, U. Buenos Aires, 1952. Mem. Asociación Médica Argentina, Assn. Argentina de Circugía, Sociedad Argentina de Cirujanos, Soc. Argentina de Cirugía Torácica, Assn. Argéntina de Tisiología, Soc. de Cirugía de Buenos Aires, Soc. Argentina de Gastroenterología, Colegio Argentino de Cirujanos. Fellow A.C.S., Am. Thoracic Soc., Am. Coll. Chest Physicians. Research, publs. on thoracic, cardiovascular, vascular and gen. surgery, including recovery of ischemic heart, healing of arteries, bowel ischemia, local perfusion for treatment of malignant tumors. Home: 1565 Callao, Buenos Aires R. 17. Office: Clínica Bazterrica, Billinghurs 2084, Buenos Aires, Argentina.*

VILLEMIN, Jean Antoine, French physician, surgeon; b. Prey, France, Jan. 28, 1827; ed. Hosp. Instrn. Mil. Health, Strasbourg, France, from 1849, Val-de-Grace, from 1853, Sch. Strasbourg, from 1860; M.D., 1862. Phys. to Mil. Hosp., Strasbourg, from 1860; agrégé École du Val-de-Grace, from 1863, prof. hygiene, legal medicine, from 1867, prof. clin. medicine, Author: Du tubercule, au point de vue de son siege, Saulnier, Nov. 14, 1904; children—Jacques, Philippe, from 1873; med. insp. French Army, from 1885. Mem. Acad. Medicine. Author: Du tubercule, au point de vue de son siege, de son evolution et de sa nature, 1861; Role de la lesion organique dans les maladies, 1862; (with Morel) Traité d'histologie humaine, 1864; Recherches sur la vésicule pulmonaire et l'emphysème, 1866; Études sur la tuberculose, 1868. Demonstrated that origin of Tb is not spontaneous but is caused by virulent principle which is transmissible, 1868. Died Paris, France, Oct. 6, 1892.

VILLERMÉ, Louis René, French physician, statistician; b. Paris, France, 1782. Founded Annales D'Hygiene, 1829. Mem. Acad. Medicine, Académie des Sciences Morales et Politiques. Author: Sur la distribution de la population francaises par sexe et par etat civil, 1834; Tableau de l'etat physique et moral des ouvriers dans les fabriques de coton, de laine et de soie, 1840; Des Associations ouvrières, 1848. Study on condtions of poor for Académie des Sciences Morales et Politiques (as result of findings Act of 1841, regulating child labor, was passed in France), 1837; advocated reform in treatment prisoners; credited with being 1st to apply statistics to questions of hygiene. Died Paris, 1863.

VINACKE, William Edgar, Am. psychologist; b. Denver, July 26, 1917; s. Harold M. and Edna (Lewis) V.; A.B., U. Cin., 1939; Ph.D., Columbia U., 1942; m. Winifred Ross, Feb. 8, 1947; children —Susan K., Alan R., Edna M. Research asst., project supr. CAA-NRC, 1939-44; faculty U. Hawaii, Honolulu, 1946-63, prof. psychology, 1957-63; prof. psychology State U. N.Y. at Buffalo, 1963—; vis. prof. U. Cin., 1951, U. Colo., 1959, 63. Fellow, Fund for Advancement of Edn., 1955-56; Guggenheim fellow, 1960. Mem. Am. Psychol. Assn., Am. Sociol. Assn., A.A.A.S., Phi Beta Kappa, Sigma Xi. Author: The Psychology of Thinking, 1952; The Miniature Social Situation, 1954; (with W. Wilson and G. Meredith) The Dimensions of Social Psychology, 1964; also articles. Research in racial stereotypes and facial expression of emotion, systematic treatment of human thinking, expts. on strategy and formation of coalitions in intra-group competitive situations, intergroup competition, theory of relations between motivation and thinking. Home: 104 Saratoga Rd., Buffalo, N.Y. 14226.*

VINCE, Samuel, English mathematician, astronomer; b. Fressingfield, Eng., Apr. 6, 1749; s. John Vince; ed. Caius Coll., Cambridge (Eng.) U., M.A., Sidney-Sussex Coll., 1778; m. Mary Paris, 1780; 1 son, Samuel Berney. Prof. astronomy Cambridge U., 1796-1821; archdeacon Bedford, from 1809. Fellow Royal Soc. (Copley medal 1780 Bakerian lectr. 1794, 97, 99), 1786. Author: Sections, 1781; Treatise on Practical Astronomy, 1790; (with James Wood) The Principles of Mathematical and Natural Philosophy, 1793-99; A Complete System of Astronomy (held great reputation during his time), 3 quarto vols., 1797-1808; The Credibility of Christianity Vindicated, in answer to Mr. Hume's Objections, 1798, 2d edit., 1809; A Treatise on Plane and Spherical Trigonometry, 1800, 4th edit., 1821; Observations on the Hypotheses which have been assumed to account for the cause of Gravitation from Mechanical Principles, 1806; A Confutation of Atheism from the Laws of the Heavenly Bodies, 1807; Observations on Deism, 1845. Important mathematician of English synthetical sch.; many writings became univ. texts. Died Ramsgate, Eng., Nov. 28, 1821.

VINCENSINI, Paul Felix, French mathematician; b. Bastia, France, April 30, 1896; s. Dominique and Louise (Castelli) V.; ed. U. Toulouse, U. Strasbourg; agrégé math., 1921; D.Sc. 1927; m. Henriette Maréchal, Sept. 16, 1936; children—Dominque, Marie-Louise. Prof., Lycée de Bastia, 1921-34; head research, Paris, 1934-36; prof., Marseilles Lycée, 1936-58; prof. calculus, U. Besançon, 1938-49; prof. calculus, U. Marseilles, 1949-66; hon. prof., 1966. Recipient, Palms of Academy, Legion of Honor. Mem., French Math. Soc., Italian Math. Union; Austrian Math. Soc.; Am. Math. Soc., Circolo Matematico di Palermo; Royal Soc. Liège. Author: Corps convexes; séries linéaites; domaines vectoriels, 1938; many published articles. Studied differential geometry, deformation and stratification of ranges; ordinale geometry and multidimensional representations; congruences of laws of circles; spheres and their curvacity; theory of convex bodies. Home: 24 rue de la Garderie, Marseilles 4, France.*

VINCENT, Jean Hyacinthe, French bacteriologist, epidemiologist; b. Bordeaux, France, Dec. 22, 1862; ed. U. Bordeaux, Sch. Application, Val-de-Grace, France, 1888; hon. dr. U. Brussels, Belgium; m. Lucie Durand, 1897; 2 daus. Bacteriological asst., Ecole de Santé Militaire, from 1889; prof. agrégé, from 1896; dir. lab. Hôpital du Dey, Algiers, Algeria; prof. bacteriology, epidemiology, Val-de-Grace; prof. epidemiology Collège de France, Paris; gen. med. insp.; founder, became dir. Lab. Anti-Typhoid Vaccination, Val-de-Grace, 1911; dir. epidemiological lab. Collège de France, 1925. Mem., French Acad. Scis. (vice-pres. 1950, pres. 1941), 1922; Académie de Médecine; corr. mem. Am. Acad. Arts and Scis.; hon. fgn. mem. acads. medicine Brussels, Rome; Royal Soc., Soc. Epidemiology (London). Author books, articles. Isolated streptothrix madurae, 1894; discovered bacteria that causes trench mouth (Vincent's angina); developed vaccine for typhoid and paratyphoid fevers; (with Stoedel) developed serum for gas gangrene, another for streptococcus and colon bacillus; discovered cryptotoxines; used chlorine to treat infected wounds; showed madura foot due to actinomyces fungus. Died Paris, France, Nov. 23, 1950.

VINCENT, Walter Sampson, Jr., Am. biologist; b. Veneta, Ore., Aug. 6, 1921; s. Walter Sampson and Eva L. (Pickert) V.; student Ore. Coll. Edn., 1939-40; B.S., Ore. State U., 1946, M.S., 1948; Ph.D., U. Pa., 1952; postgrad. U. Libre Bruxelles, 1950-51; m. Helen Jane Sandberg, June 20, 1942; children—Walter Sampson, III, Elizabeth Ann, Jane Martha. Research asso. Ia. State Coll., 1951-52; faculty State U. N.Y. Upstate Med. Center, Syracuse, 1952-61, asst. prof. anatomy, 1956-61; asso. prof. anatomy and cell biology U. Pitts., 1961—; vis. research prof. U. Edinburgh (Scotland), 1964. Trustee Marine Biol. Lab. AEC fellow, 1949-50; Fulbright fellow, Belgium, 1950-51; Lalor fellow, 1954; USPHS Sr. Research fellow, 1958-63; recipient Ednl. Distinction award Syracuse Pub. Schs., 1960. Mem. A.A.A.S., Am. Soc. for Cell Biology, Internat. Inst. for Embryology, Genetics Soc. Am., Soc. Gen. Physiology. Asso. editor Jour. Morphology. Author: (with O. L. Miller, Jr., M. O. Drets, F. Saez) The Nucleolus, Its Structure and Function, 1966; also articles. Research on mechanisms genetic expression by biochem. techniques; pioneered isolation and chem. analysis nucleolus; demonstrated presence ribonucleic acid in nucleolus, presence messenger RNA in normal cells.*

VINCENT OF BEAUVAIS, French encyclopedist; b. circa 1190. Dominican, subprior Dominican monastery Beauvais, France, 1246; librarian, tutor Louis IX and his sons. Author: Speculum Maius (4-part encyclopedia representing knowledge accessible to ed. Westerners, 3d. quarter 13th century), printed for 1st time, 1473. 1st part, Speculum naturale, on alchemy, illustrates great interest in this field among well-known scholars, clerics of time; information derived from Latin translations Arabic works; work also dealt with meteorology, geography, geology, astronomy, botany, zoology, anatomy, physiology, psychology, husbandry, medicine, physics, math. Died, 1264.

VINES, Sydney Howard, English botanist; b. Ealing, Eng., Dec. 31, 1849; s. William Reynolds and Jessie (Robertson) V.; ed. Guy's Hosp., Christ's Coll., Cambridge, 1872; B.Sc., London, 1873, D.Sc., 1879; B.A., Cambridge, 1876, M.A., 1879, D.Sc., 1883; M.A., Oxford, 1888; m. Agnes Bertha Perry, 1884; 2 sons, including Walter Sherard; 1 dau. Fellow, lectr. Christ's Coll., Cambridge, 1876-88; reader botany Cambridge U., 1883-88; fellow Magdalen Coll., Sherardian prof., Oxford, 1888-1919; a founder Sch. Botany, Christ's Coll., Cambridge; became hon. fellow U. London, 1892, Cambridge, 1897. Fellow Royal Soc., 1885; mem. Linnean Soc. London (pres. 1900-04), Brit. Assn. (pres. bot. sect. 1900). Author: Lectures on the Physiology of Plants, 1886; A Student's Text-Book of Botany, 1895; also articles. Editor, a founder Annals of Botany, 1887-99. Introduced micros. methods to botany in Gt. Britain, 1876-1888; research and publs. on proteolytic enzymes in plants. Died Exmouth, Eng., Apr. 4, 1934.

VINEYARD, George Hoagland, Am. physicist; b. St. Joseph, Mo., Apr. 28, 1920; s. George Hoagland and Mildred M. (Barkley) V.; B.S., Mass. Inst. Tech., 1941, Ph.D., 1943; m. Phyllis Ainsworth Smith, Feb. 3, 1945; children—John H., Barbara Gale. Staff, Radiation Lab., Mass. Inst. Tech., 1943-45; asst. prof. to prof. U. Mo., Columbia, 1946-52; physicist to sr. physicist, dept. chmn., associate director, deputy director of Brookhaven Nat. Lab., Upton, N.Y., 1953—. Cons. Los Alamos Sci. Lab., 1964—. Fellow Am. Phys. Soc.; mem. Sigma Xi. Bd. editors Am. Jour. Physics, 1948-50, Phys. Rev., 1959-61; editor Documents in Physics series, 1964—. Research on theory of x-ray and neutron scattering; crystal growth; microwave devices; solid state structure of liquids. Home: 10 Brewster Lane, Bellport, N.Y. 11713. Office: Brookhaven Nat. Lab., Upton, N.Y. 11973.

VINGIELLO, Frank Anthony, Am. chemist; b. N.Y.C., Aug. 20, 1921; s. Frank Anthony and Madeline (Pelletiere) V.; B.S., Poly. Inst. Bklyn., 1942; Ph.D., Duke, 1947; postgrad. Northwestern U.; m. Ruth Reade McDonald, Oct. 11, 1947; children—Madeline, Arthur, Paul. Chemist, Gen. Chem. Co., 1942; research asst. Bklyn. Poly. Inst., 1943; faculty Duke, 1944-47, U. Pitts., 1947-48, Northwestern U., 1948; faculty Va. Poly. Inst., Blacksburg, 1948—, prof. organic chemistry, 1956—. Lectr. on air pollution local, nat., internat. sci. meetings, 1957—. Mem. Am. Chem. Soc., N.Y. Acad. Scis., Sigma Xi, Phi Lambda Upsilon, Sigma Pi Sigma. Research, numerous publs. on organic chemistry, organic synthesis, anti-cancer compounds, air pollution. Home: 107 Monte Vista Dr., Blacksburg, Va. 24060.*

VINOGRADE, Bernard, Am. mathematician; b. Chgo., Ill., May 7, 1915; s. James and Rose (Epstein) V.; B.S., City Coll. N.Y., 1937; M.A., U. Mich. 1940, Ph.D., 1942; m. Ann Costikyan, Dec. 30, 1942; children—Polly Jean, Peter James, Alice, Barbara. Instr. U. Wis., 1942-44, Tulane U., 1944-45; mem. staff radiation lab. Mass. Inst. Tech., 1945; mem. faculty Ia. State U., 1946—, prof. math., 1955—, chmn. dept., 1960-64; vis. prof. San Diego State Coll., 1959-60, City Coll. N.Y., 1964-65. Operations analyst standby unit USAF, 1951—. Mem. Math. Assn. Am. (bd. govs. 1957-59; Am. Math. Soc., Am. Assn. U. Profs.; Phi Beta Kappa. Author: Linear and Matrix Algebra, 1967; co-author: Laplace Transformation (with Maple and Holl), 1959. Research on structure of rings; commutative algebras; linear algebra. Home: 1206 Michigan Av., Ames, Ia. 50010.*

VINOGRADOV, Aleksandr Pavlovich, Russian chemist; b. Aug. 21, 1895; grad. Mil. Med. Acad. and Chemistry Faculty, Leningrad U., 1924. With USSR Acad. Scis., 1928—, now dir. Inst. Geochemistry and Analytical Chemistry, also dir. Siberian br. Inst. Geochemistry. Dep. chmn. peace fund Soviet Com. for Def. of Peace, 1961; presentated papers Internat. Conf. Peaceful Uses of Atom, 1955. Named Hero of Socialist Labor, 1949; recipient Stalin prize, 1949, 51; decorated Order of Lenin (3); Order Red Banner of Labor (2). Mem. USSR Acad. Sci., 1953. Author: Elementary Chemical Composition of Marine Organisms,

1935-44; Biogeochemical Regions, 1949; Geochemistry of Dispersed Elements in the Sea Water, 1944; Geochemistry of Dispersed Elements in Soils, 2d edit., 1957. Research in geochemistry, biogeochemistry and analytical chemistry; continued Vernadskii's work on distbn. chem. elements in earth's crust; studied ocean salts and primary rocks; described many rare soil elements; theories of biogeochem. regions, fertilization, sea water cations; devel. analytical methods. Home: 2-aya Filevskaya, Moscow 10. Office: V.I. Vernadskii Inst. Geochemistry, USSR Acad. Sci., Vorobevskoye Shosse 47-a, Moscow, USSR.

VINOGRADOV, Ivan Matveevich, Russian mathematician; b. Milolyub (now Velikie Luki Oblast), Sept. 14, 1891; grad. Petrograd U., 1914, D.Phys. Math. Scis., 1929; Ph.D. (hon.), U. Oslo (Norway), 1950. Lectr., then prof. Perm U., 1918-20; head chair theory of numbers Leningrad U., 1925-34; prof. Leningrad Poly. Inst., 1920-34; dir. Math. Inst., USSR Acad. Sci., 1932-—; prof. Moscow U., 1934-—. Recipient Stalin prize 1st class, 1941, Gold Hammer and Sickle medal, 1945; named Hero of Socialist Labor, 1945 decorated Order of Lenin (2). Mem. USSR Acad. Sci., 1929. Fellow Royal Soc., 1942; hon. mem. Moscow, London, Indian math. socs., Am., Hungarian acad. scis., Am. Philos. Soc., Moscow Naturalist Soc.; hon. fgn. mem. Am. Acad. Arts and Sci., Amsterdam Math. Soc.; Danish Acad. Sci., Dei Lincei Nat. Acad. (Rome); corr. mem. Paris, German acads. sci. Author: A New Method of Obtaining Asymptotic Expressions of Arithmetical Functions, 1917; New Method in Analytical Theory of Numbers, 1937; The Method of Trigonometrical Sums in the Theory of Numbers, 1947; Selected Works, 1952; Basis of Theory of Numbers, 6th edit., 1952; Trigonometrical Sums Containing the Value of the Polynomisal, 1957. Specialist in analytical theory of numbers; devel. formula for representations of odd number that led to his solution of the Goldbach problem; contbr. methods of solving additive problems. Home: al. Gorkogo 22-a, Moscow. Office: V. A. Steklov Math. Inst., USSR Acad. Sci., 1-y Akademicheski, Proyezd 28, Moscow, USSR.

VINOGRADSKI, Sergei Nikolaevich, microbiologist; b. Kiev, USSR, Sept. 1, 1856. Prof., St. Petersburg (now Leningrad, USSR), Pasteur Inst., Paris, France. Mem. French, USSR acads. scis. Author: Beiträge zur Morphologie und Physiologie der Bacterien; Sur l'assimilation de l'azote gaseux par les microbes. Leading researcher in soil microbiology; introduced selection cultures in microbiology, 1889; discovered closterium pasterianum, 1893. Died Brie-comte-Robert, France, Feb. 24, 1953.

VINSON, Porter Paisley, Am. physician; b. Davidson, N.C., Jan. 24, 1890; s. William Daniel and Lillie (Helper) V.; B.S., Davidson Coll., 1909, M.A., 1910; M.D., U. Md., 1914; m. Lenore Dunlap, May 14, 1919; children—Mary Lenore, Portia Ann (Mrs. John Duval Lawson), William Daniel. Intern, Trudeau Sanatorium, 1914-16, Montreal (Que., Can.) Gen. Hosp., 1916; fellow medicine Mayo Found., Rochester, Minn., 1916, then joined staff; became asso. prof. Mayo Found., 1932; prof. bronchoscopy, esophagoscopy, and gastroscopy U. Va.; chief div. Coll. Hosp., 1936-59; also pvt. practice, Ricmond, Va. Fellow A.C.P.; mem. Am. Bronchoscopic Soc. Author: The Diagnosis and Treatment of Diseases of the Esophagus, 1940; Diseases of the Esophagus, 1947; also articles, chpts. in books. Described dysphagia accompanied with glossitis and anemia and often splenomegaly and atrophy of oral and pharyngeal tissue (Plummer-Vinson syndrome or siderpenic dysphagia). Died Aug. 28, 1959.

VINTI, John Pascal, Am. physicist; b. Newport, R.I., Jan. 16, 1907; s. John Joseph and Anna (Sild) V.; S.B., Mass. Inst. Tech., 1927, Sc.D., 1932; m. Ella Keen Johnson, Feb. 17, 1951. Research fellow U. Pa., 1932-34; faculty Brown U., 1936-37, Citadel, 1937-38, Worcester Poly. Inst., 1939-41, N.C. State U., 1966; research physicist Ballistic Research Labs., Aberdeen Proving Ground, 1941-57, Nat. Bur. Standards, 1957-65; cons. Exptl. Astronomy Lab., Mass. Inst. Tech., Cambridge, 1967-—. Professorial lectr. celestial mechanics Georgetown U., 1963-64; adj. prof. Cath. U., 1965-66; cons. mem. commn. on celestial mechanics Internat. Astron. Union. Recipient Distinguished Service award Nat. Bur. Standards, 1961. Fellow Am. Phys. Soc., A.A.A.S., Brit. Interplanetary Soc., Royal Astron. Soc., Washington Acad. Scis., Am. Inst. Aeros. and Astronautics (asso.); mem. Am. Astron. Soc., Am. Geophys. Union, Philos. Soc. Washington, Sigma Xi. Research, publs. on wave functions of helium, quantum-mech. sum rules, theory of isotope shift in atomic spectra, chronographs, interior ballistics of guns and rockets, gamma ray scattering, multipath propagation of FM electromagnetic waves, Legendre and Bessel functions, theory of magnetically affected spin of artificial satellites, theory of satellite orbits. Home: Longwood Towers, 20 Chapel St., Brookline, Mass. 02146. Office: Exptl. Astronomy Lab., Mass. Inst. Tech., 265 Massachusetts Av., Cambridge, Mass. 02139.*

VIOLLE, (Louis) Jules (Gabriel), French physicist; b. Langres, France, Nov. 16, 1841; ed. École Normale Supérieure, Paris, France; D.Sc., 1870. Prof. physics sci. faculties Grenoble, Lyons (both France) univs., from 1883, École Normale Supérieure, 1890, Conservatoire des Arts et Métiers, 1891. Mem. French Acad. Scis., 1897, Académie d'agriculture. Author: Traité de physique. Research on temperature of sun, determination mech. equivalent of heat; measured solar constant, 1875; proposed photometric unit (Violle's standard), 1881; invented calorimeter (principle led to thermos bottle), 1882; research on geysers, origin of hail, exploration of atmosphere by sounding balloons. Died Fixin, Côte-D'Or, France, Sept. 12, 1923.

VIRCHOW, Rudolph Ludwig Karl, German pathologist; b. Schivelbein, Pomerania, Oct. 13, 1821; M.D., U. Berlin, 1843. Apptd. asst.-surgeon Charité Hosp., Berlin, 1843, pro-rector, from 1846; privatdozent U. Berlin, 1847, prof. path. anatomy, from 1856; mem. govt. commn., investigated outbreak of typhus, Upper Silesia, 1848; prof. U. Würzburg, from 1849; dir. Path. Inst., also founder center for research on diseases; elected mem. Prussian Lower House, 1862, entered Reichstag, 1880; joint-founder German Anthrop. Soc., 1869; presided over Berlin Anthrop. Soc., 1869; Croonian lectr., 1893. Recipient Copley medal Royal Soc. London, 1892, Grand Gold medal for Sci. from Kaiser Wilhelm II. Author: Die Cellularpathologie, 1858; Die Krankhaften Geschwülste, 1862-63. Regarded as founder of modern pathology; numerous contbns. to histology and morbid anatomy; studied tumors, phlebitis, leukemia, Tb, rickets, trichinosis, thrombosis and embolism; described pulmonary aspergillosis and septicemia; first to recognize that cell theory extended to diseased tissue; thought all cells arise from cells (implicit repudiation of theory of spontaneous generation); refused to accept Pasteur's germ theory of disease; viewed disease as civil war between cells; helped to instigate hosp. and sanitary reform. Died Berlin, Sept. 5, 1902.

VIREY, Jules Joseph, French biologist, physician; b. Haute-Marne, France, 1775. Became chief pharmaceutist Hosp. of Val de Grace, Paris, 1812. Author: Theoretical and Practical Treatise on Pharmacy, 1811; Ephemerides of Human Life, 1814; On Vital Power, 1922; Philosophical Hygiene, 2 vols., 1828; contbg. author: Dictionnaire des sciences naturelles; Dictionnaire des sciences médicales. Died 1847.

VIRGILI, Pedro, Spanish surgeon, anatomist; b. 1699; founder navy surg. colls. at Cadiz and Barcelona, Spain; performed tracheotomy for quinsy, Cadiz, 1743. Died 1776.

VIRGIN, Hemming Ivar, Swedish botanist; b. Stockholm, Sweden, Oct. 19, 1918; s. Ivar and Ida (Gadd) V.; Ph.D., Stockholm U., 1951. Asst. prof. Lund (Sweden) U., 1951-58; asso. prof. Royal Agr. Coll. Uppsala, 1958-62; prof. plant physiology Göteborg (Sweden) U., 1962-—, head Bot. Instn., 1965-—; vis. investigator Carnegie Instn. Washington, 1953-54. Mem. Swedish Natural Sci. Council, 1965-—. Mem. Royal Soc. Göteborg. Research and publs. on light effects on protoplasm, pigment formation in leaves, stomatal transpiration, mechanism of chlorophyll formation, methods for turgor measurements. Home: 58 Torild Wulffsgatan, Göteborg SV, Sweden.*

VIRIZGTSOV, Nikolai Nikolaevich, Russian organic chemist; b. June 5, 1907; s. N. N. Vorozhtsov; grad. Moscow Technol. Coll., 1928. Staff, Lab. Commn. on Study Natural Productive Forces, USSR Acad. Sci., Moscow, 1928-30; staff Leningrad Inst. High Pressures, 1930-38; faculty Kazakh State U., 1938-43, named prof., 1939; dir. Sci. Research Inst. Organic Semi-products and Dyes, 1943-47; chmn. dept. Moscow Mendeleev Chemico-Technol. Inst., 1945-58; dir. Inst. Organic Chemistry, Siberian Dept., USSR Acad. Sci., 1958-—. Recipient Stalin prize, 1952, Order Red Banner of Labor. Corr. mem. USSR Acad. Sci. Author: Chemistry of Natural Tanning Substances, 1932; Continuous Processes in Dye Industry, 1939; Principles in the Synthesis of Intermediary Products and Dyes. Research in organic dyes, intermediates; structure, synthesis of natural organic substances. Office: USSR, Novosibirsk, RSFSR Institut organicheskoy khimii Sibirskogo otdeleneiya AN SSSR.

VIRTAMA, Pekka Fero Juhani, Finnish physician; b. Helsinki, Finland, June 23, 1922; s. Eero and Anna (Hietanen) V.; candidate medicine, U. Helsinki, 1948, licentiate medicine, 1950, M.Med.Sci., 1958; m. Sinikka Nivala, June 27, 1949; children—Lisa, Eva, Rita. Registrar, U. Central Hosp., Helsinki, 1952-54; radiologist Kivela Town Hosp., Helsinki, 1954-56; cons. radiologist Finnish State Rys., 1954-62; cons. radiologist, med. dept. Hosp. of U. Helsinki, 1958-60, dir. radiodiagnostic dept., 1960-63, docent roentgen diagnostics, 1960-64; prof. Roentgenology, U. Turku, 1963-—, dir. Roentgen dept. U. Hosp., 1963-—, chmn. med. bd., 1965-—. Decorated Medal of Finnish Winter War, 1939-40. Fellow Internat. Coll. Angiology (sci. council); mem. Am. Coll. Chest Physicians, Scandinavian Radiological Soc. (mem. bd.). Author: Determination of Mineral Content of Human Finger Bones, 1957. Studies, publs. on radiol. determination of bone minerals; methodological improvements; a modified and improved method of arterial catherization; use of oxygen insufflation to visualize rectal wall. Home: 5 Kiinamyllynk, Turku 2, Finland.*

VIRTANEN, Artturi Ilmari, Finnish biochemist; b. Helsinki, Finland, Jan. 15, 1895; s. Kaarlo and Serafiina (Isotalo) V.; grad. Classical Lyceum, Viipuri, 1913; M.Sc., U. Helsinki, 1916, Ph.D., 1919; Dr. Medicine honoris causa, U. Lund, 1936; Dr. honoris causa, Royal Tech. Coll., Stockholm, 1949; Finland Inst. Tech., 1949, U. Paris, 1952; Dr. Agr., U. Justus-Liebig, Giessen, 1955; Dr. Agr and Forestry, U. of Helsinki, 1955; married Lilja Moisio, Feb. 29, 1920; children—Kaarlo Ilmari, Artturi Olavi. First asst. Central Lab. of Industry, 1916-17; chem. asst. state butter and cheese control sta., 1919; chemist lab. Butter Export Assn., Valio, 1919-20, dir. since 1921; dir. Biochem. Inst. (including lab. Valio and lab. Found. for Chem. Research) 1931-—; docent chemistry U. Helsinki, 1924-39; prof. biochemistry Finland Inst. Tech., 1931-39; professor biochemistry Univ. Helsinki, 1939-48; mem. and pres. Acad. Finland, 1948-63. Mem. Editorial staff Karjantuote since 1924, Suomen Kemistilehti since 1928, Enzymologia since 1936, Acta Physiologia Scandinavica since 1940, Annales Scientiarum Fennicae Series Chemica since 1945, Acta Chemica Scandinavica since 1946, Annales Medicinae Experimentalis et Biologiae Fenniae since 1947, Acta Endocrinologica since 1948, Plant and Soil since 1948. Mem. state com. on popular nutrition, 1936-40, state com. on prodn., 1940-42; rep. of Finland, League of Nations com. on nutrition, 1938; mem. com. chem. industry since 1942; expert of Ministry of Supply, 1942-44. Decorated Scheele medal, Swedish Chem. Soc., 1938; Adelsköld medal, Swedish Acad. Scis., 1943; Prize of Honor, Fund of Wihuri, 1943; Nobel Prize for chem., 1945; Golden Plaque of Mjolkpropagandan, Stockholm, 1946; Plaque of Finnish Tech. Soc., 1946; Great Medal of U. Ghent, 1946; Gadolin-Medal of Finnish Chem. Soc., 1949. Kairamo-Medal of Societas Zoologica Botanica Fennica Vanamo, 1950; comdr. 2d class Order White Rose of Finland, 1943; Cross of Liberty 2d class, 1945; Emanuele Paternó medal of Societa Chimica Italiana, 1957; medal of University of Pavia, 1957, U. Helsinki, 1960, Faber Foundation, 1964, Friesland Prize, Holland, 1967. Fellow A.A.A.S.; mem. European Nutritionists, Am. Society Microbiology, German Academy Leopoldina, Am. Soc. Biol. Chemists (hon.), German Soc. Nutrition, Finnish Acad. Sciences (chairman 1944-45), Swedish Acad. Agrl., Brit. Assn. Advancement, Finnish Acad. Agr., Royal Sci. Soc. Uppsala, Royal Sci. Acad. Sweden, Swedish Acad. Engring. Sci., Pontifical Acad. Sci.; hon. mem. Finnish Chem. Soc., Finnish Soc. Agronomists, Finnish Econ. Soc., Austrian Chem. Soc., Viipuri Student Union, Biochem. Soc. Stockholm, Finnish Med. Soc. Duodecim, Finnish Tech. Soc., Royal Soc. Edinburgh, Am. Institute of Nutrition, Academia Pugliese delle Scienze, Bari Italia, Higher Council Sci. Research, Madrid (hon. councillor). Author: Cattle Fodder and Human Nutrition, 1938; AIV System as the Basis of Cattle Feeding, 1943. Contbr. articles to tech. jours. Researches in agriculture and nutrition; discovered acidity prevents fodder sooilage; studied bacterial fermentations; nitrogen fixation in legume root nodules; enzyme proteins in young cells; nitrogen metabolism in plants. Address: Biochemical Research Institute, Kalevank 56b, Helsinki 18, Finland.

VISAPAA, Asko Edvard, Finnish chemist; b. Nurmes, Finland, Apr. 30, 1921; s. Edvard and Helmi (Wiik) Backberg; student U. Helsinki (Finland), 1944-48. Research asst. State Inst. for Tech. Research, Helsinki, 1948-62, research asso., 1962-—. Mem. Suomalaisten Kemistien Seura. Research, publs. on functional groups of cellulose, degree of order in cellulose, analytical methods based on infrared and X-ray spectrometry, X-ray diffraction; computer programming. Home: 12 Ulvilantie, Helsinki 35. Office: 37 Lönnr.K., Helsinki 18, Finland.*

VISCHER, Ernst Benedict, Swiss chemist; b. Basel, Switzerland, July 28, 1917; s. Ernst B. and Louise (Geigy) V.; Ph.D., U. Basel, 1944; m. Annemarie Wadler, May 2, 1949. Research asst. Inst. Biochemistry, U. Zürich (Switzerland), 1944-46; research asso. dept. biochemistry Columbia U., N.Y.C., 1946-48; research chemist Imperial Chem. Industries, Akers Research Labs., Welwyn, Gt. Britain, 1948-51; with research dept. CIBA Ltd, Basel, 1952-—, dir. chem. research pharm. div., 1963-—. Mem. Swiss, Am. chem. socs. Research and publs. on carbohydrate chemistry, synthesis of naturally occurring sugars, nucleic acids, including characterisation of nucleic acids of various origin using paper chromatography, isolation and identification of various metabolites, microbial transformations of steroids. Home: 3 Augustinergasse, Basel. Office: CIBA Ltd., Basel, Switzerland.*

VISCONTINI, Max, chemist; b. Mascara, Algeria, Feb. 1, 1913; s. Albert and Marie-Rose (Diliberto) V.; student Institut Nat. Agronomique, Paris, France, 1933-36; D. ès Scis. Physiques, Sorbonne, Paris, 1944; m. Maria-Pia Vladesco, Feb. 2, 1942; children—Marie-José (Mrs. Goetsch), François, Isabelle. Staff, Institut Pasteur Paris, 1936-47, chief lab., 1946-47; prof. organic chemistry Chem. Inst., U. Zurich (Switzerland), 1947-—. Recipient Bruylants medal U. Louvain (Belgium), 1960. Mem. Am., Swiss chem. socs. Research, numerous publs. on coenzymes, chemistry of insect pigments, chemistry and biochemistry of hydrogenated Pterins; determination of absolute configuration of steroids; synthesis of pyrrolizidines. Home: 43 Freiestrasse, 8032 Zurich, Switzerland.*

VISHNEVSKII, Aleksandr Aleksandrovich, Russian surgeon; b. 1906; s. Akesandra V. Vishnevskii; grad. Med. Faculty, Kazan U., 1929; Dr.Med. Sci., 1936. Intern, Kazan State U., 1929-31; instr. Leningrad Mil. Med. Acad., 1931-33; sr. asso. Surg. Clinic, All-Union Inst. Expt. Medicine, Leningrad, also Moscow, 1933-41; cons. surgeon Red Army, 1939-40; chief field surgeon Main Mil. San. Bd., Soviet Army, 1941-45; dir. Surg. Clinic, Inst. Neurology, 1946-48; chief surgeon Moscow Mil. Dist., 1946-56; dir. Surg. Clinic, Inst. Surgery, USSR Acad. Med. Sci., Moscow, 1948-50; dir. Inst. Surgery, USSR Acad. Med. Sci., 1950—; head 2d chair surgery Central Postgrad. Med. Inst., Moscow, 1950—; chief surgeon USSR Ministry Def., 1956—. Lectr. Soviet surgery Istanbul U., 1959. Recipient Internat. La Riche prize, 1953, World Med. and Surg. Assembly prize, 1957, World Peace Council's Scroll. of Honor, 1959, Lenin prize, 1960, Order of Lenin, Order of Red Banner, Order of Red Star; named Hon. Sci. Worker, RSFSR, 1957. Mem. USSR Acad. Med. Sci., All-Russian Soc. Surgeons (chmn. 1955—), Internat. Surg. Soc. (mem. presidium). Author: Notes of a Field Surgeon, 1943; Co-author: The Integration of the Animal Organism. Notes on the Correlation of the Endocrine Glands, 1931; Research and numerous publs. on devel. local anesthesia using creeping infiltration methods, neurotrophy in surgery; 1st to use cervical vagosympathetic blocking for prophylaxis, pleuropulmonary shock therapy for thoracic wounds, 1939; successfully performed valvotomy for mitral stenosis, 1953; replaced sect. of esophagus with polyvinyl pyrolidol tube, operated on dry heart, 1957; used novocain blocking of joint capsule in extremity. Home: Tsentralny Institut Usovershenstvovaniya Vrachey, Pl. Vosstaniya 1-2, Moscow, USSR, G-242.

VISHNIAC, Wolf, microbiologist; b. Berlin, Germany, May 20, 1922; s. Roman and Lea (Bagg) V.; A.B., Bklyn. Coll., 1945; M.S., Washington U., St. Louis, 1946; Ph.D., Stanford, 1949; m. Helen Frances Simpson, Aug. 18, 1951; children—David O. (dec.), Ethan T., Ephraim M. Came to U. S., 1940, naturalized, 1946. Fellow dept. biochemistry N.Y. U. Med. Sch., 1949-50, USPHS research fellow dept. pharmacology, 1950-52; faculty Bklyn. Coll., 1952, Yale, 1952-61; prof. dept. biology U. Rochester (N.Y.) 1961—, chmn. dept., 1965—. Vis. microbiologist Brookhaven Nat. Lab., 1959-60; cons. space sci. bd. Nat. Acad. Scis., 1961—; cons. bioscis. subcom. NASA, 1965—. Mem. Am. Soc. Microbiologists, Am. Soc. Biol. Chemists, Am. Chem. Soc., A.A.A.S., Am. Soc. Plant Physiologists, Am. Soc. Cell Biology, Harvey Soc., Soc. Gen. Microbiologists, Am. Acad. Microbiology, Am. Assn. U. Profs., Sigma Xi. Research, numerous publs. on electron transport in photosynthesis, metabolism of autotrophic hydrogen bacteria and sulfur bacteria, devel. techniques for biol. exploration of other planets. Home: 105 Brooklawn Dr., Rochester, N.Y. 14618.*

VISSCHER, Maurice B. Am. physiologist; b. Holland, Mich., Aug. 25, 1901; s. Johannes W. and Everdena (Bolks) V.; A.B., Hope Coll., Holland, Mich., 1922; Ph.D., University of Minn., 1925, M.D., 1931; grad. student University Coll., London, England, 1925-26; m. Janet Gertrude Pieters, Aug. 12, 1925; children—Barbara Ruth, (Mrs. Frederick Kahn), William Maurits, Janet Constance (Mrs. Allan Simpson), Pieter Bernard, Asst. in physiology, University of Minn., 1922-25, asst. prof. on leave, 1925-26; NRC fellow, University Coll., London, 1925-26, U. of Chicago, 1926-27; prof. of physiology, Coll. of Medicine, U. of Tenn., 1927-29; prof. of physiology and pharmacology, U. of Southern Calif., 1929-31; prof. of physiology and head of dept., U. of Ill., 1931-36, U. of Minnesota, 1936—, distinguished service professor, 1960—, regents prof., 1967. Co-director Medical Nutrition Study for UNRRA; chmn. NRC com. on UNESCO, 1950-52; pres. bd. trustees Biological Abstracts, 1949-53; dir. Am. Heart Assn., 1947-54; mem. nat. bd. dirs. Am. Cancer Society, 1952-54; v.p. Nat. Soc. Med. Research, 1952-55, pres., 1965—. Recipient Research Achievement award, Am. Heart Assn., 1962. Fellow N.Y. Acad. Scis.; member Am. Physiol. Soc. (pres. 1948-49), Soc. for Experimental Biology and Medicine (pres. 1967-69), American Chemical Society, A.M.A., (chmn. sect. on pathology and physiology 1938-39), Am. Assn. Univ. Profs, Acad. of Medicine of Chicago, A.A.A.S., Minn., Pathol. Soc., Minn. Acad. Medicine, Am. Assn. Scientific Workers (pres. 1947-48), secretary to Internat. Union Physiol. Scis.; Nat. Academy Sciences, American Academy of Arts and Sciences, Internat. Union Physiol. Scis. (chmn. U.S. com.), Sigma Xi, Alpha Omega Alpha. Author: Chemistry and Medicine, 1940; also numerous papers. Editor: (with F. Grande) Claude Bernard and Experimental Medicine, 1965. Analyzed mech. efficiency of heart and demonstrated a lowering in heart failure which is reversed by cardiotonic glycosides; initiated use of isotopes in study of bi-directional fluxes of water and ions across intestinal epithelia; studied conditions for prodn. and prevention pulmonary edema; analyzed physiol. mechanisms of endotoxic shock. Home: 1 Orlin St. S.E., Mpls. 55414.*

VITAL DU FOUR, see du Four, Vital.

VITALIANO, Charles Joseph, Am. geologist; b. N.Y.C., Apr. 2, 1910; s. Joseph and Catherine (DeBarberi) V.; B.S., Coll. City N.Y., 1936; A.M., Columbia, 1938, Ph.D., 1944; m. Dorothy A. Brauneck, Oct. 19, 1940; children—Judith E., Peter W. Instr. ceramic petrography Rutgers U., 1940-42; geologist U. S. Geol. Survey, 1942-47, part-time 1947-59; faculty Ind. U., Bloomington, 1947—, prof. geology, 1957—. J. F. Kemp fellow geology, 1939-40; Fulbright Sr. Research fellow, New Zealand, 1954-55; NSF grantee, 1957-60, 66—. Fellow Geol. Soc. Am., Mineral. Soc. Am.; mem. Soc. Econ. Geologists, Geochem. Soc., Am. Asssn. U. Profs., Sigma Xi. Research, publs. on ore deposits of Paradise Peak Quadrangle, Nev., igneous and metamorphic petrography of Western Nev., So. New Zealand and S.W. Mont., volcanic rocks of Western U. S., theories of volcanism and formation of volcanic rocks, principles governing the origin and paragenesis of metamorphic rocks. Home: 1114 Brooks Dr., Bloomington, Ind. 47401.*

VITO, Francesco Maria, Italian economist; b. Pignataro, Italy, Oct. 21, 1902; s. Federico and Rosa (De Vita) V.; LL.D., U. Naples (Italy), 1925, Pol. Sc.D., 1926, Ph.D., 1928. Research work in Germany, Eng., U. S. A., Scandinavian countries, 1928-35; prof. econs. Sacred Heart U., Milan, Italy, 1935—, dir. Inst. Econs., 1939—; vis. prof. U. Munich, U. Freiburg, U. Göttingen, U. Hamburg (all Germany), U. Paris, U. Strasbourg, U. Nancy, U. Grenoble (all France), U. Chgo., U. Que. (Can.), U. Delhi (India), U. Buenos Aires (Argentina), U. Rio De Janeiro (Brazil), U. Monterrey (Mexico), U. Montevideo (Uruguay), U. Santiago (Chile), U. Valparaiso (Chile). Italian Govt. del. to UNESCO gen. confs., 1948-66. Recipient Gold medal Ministry Pub. Edn. Mem. Italian Assn. Polit. Scis. (pres. 1951—), Sci. Com. on Productivity (pres. 1960—), Lincei Acad., Acad. Scis. and Lit., Pontifical Acad. St. Thomas. Editor: Money and Credit, 1964; Price and Distribution, 1965; Introduction to Economics, 1965; Internat. Rev. Social Scis., 1948—, Italian Bibliography Social Scis., 1956—. Research, numerous publs. on social scis., their need to be based on ethical conception of man and soc., prins. of internat. monetary system. Home: 2 Via Lanzone, Milan, Italy.*

VITRUVIUS (Marcus Vitruvius Pollio), Roman architect, engr., author; flourished 25 B.C.; little is known of his life; probably a native of Formiae in Campania; served as a military engr. in North Africa in the time of Julius Caesar; architect of basilica at Fanum Fortunae (now Fano, Italy); with three others, apptd. a superintendent of balistae and other military engines during the reign of Augustus. Author: De architectura (10 books). Remained authority on architecture for centuries; eventually closely studied by Bramante, Michelangelo, and Vignola; wished to preserve classical (Hellenistic) tradition in design of temples and public buildings; wrote on city planning and architecture in general; the four orders of architecture (Ionic, Doric, Corinthian, and Tuscan); interior decoration, hydraulics, clocks, mensuration, astronomy, geometry, and mech. engring.

VITVER, Ivan Aleksandrovich, Russian geographer; b. Feb. 25, 1891; grad. Moscow U., 1921. Prof. chair econ. and polit. geography of fgn countries, Moscow U., 1934—; also prof. Inst. Internat. Relations. Recipient Stalin prize, 1951. Author: South America, 1930; The Caribbean Countries, 1931; Germany, A Brief Economic and Geographical Outline, 1945; The French School of the Geography of Man, 1940; The Economic Geography of Germany, 1939; A Historico-Geographical Introduction to the Economic and Political Geography of the Capitalist World, 1945; Great Britain, An Economic and Geographical Outline, 1947; The Economic Geography of Foreign Countries, 1955; France, An Economic Geography, 1958; An Historico-Geographical Introduction to the Economic Geography of the Foreign World, 1963; also numerous articles. Author of one of most widely used geog. textbooks in USSR. Address: Leninski gory, Gosudarstvenny, Univ., Moscow, USSR.

VIVES, Juan Luis (Ludovicus Vives), humanist, philosopher; b. Valencia, Spain, Mar. 6, 1492; student, Paris, 1509-12. Became prof. humanities, Louvain, Belgium, 1519; went to Eng., 1523; tutor to Princess Mary; LL.D., Corpus Christi Coll., Oxford; lectr. philsophy Corpus Christi Coll., Oxford; lost favor because of his opposition to Henry VIII's divorce from Catherine of Aragon and went to Bruges, Belgium, 1527. Author: De Ratione studii puerilis, 1523; Introductio de ratione studii puerilis, 1523; Introductio ad sapientiam, 1524; Satellitium sive symbola, 1524; De institutione femina christianae, 1524; De disciplinis libri XX, 1531; Exercitationes linguae latinae, 1538; De anima et vita libri tres, 1538; De veritate fidei christianae, 1543. Editor: De civitate dei (St. Augustine), 1522. Introduced various ednl. methods, including use of vernacular, adaptation of teaching methods to student abilities, orgn. parent-tchr. assns., recreation in sch. curriculums; emphasized inductive methods for psychology and philosophy; established pub. schs. in each twp., Died Bruges, May 6, 1540.

VIVIAN, Donald Lindsay, Am. chemist; b. Boulder, Colo., Nov. 12, 1907; s. Alfred James and Helen (Smiley) V.; Ph.B. cum laude, Yale, 1929; Ph.D., George Washington U., 1934; m. Sallie Perrie Robinson, June 6, 1931. Jr. chemist U. S. Dept. Agr., 1935-36, asst. chemist, 1936-40; research asso. dept. pharmacology U. Md. Med. Sch., 1940-41; chemist WPB, 1941-43, sr. chemist, 1943-44; research chemist Gen. Aniline & Film Corp., 1944-46; research chemist Aromatic Products, 1946-47; chemist Nat. Cancer Inst., Bethesda, Md., 1947-49, sr. chemist, 1949-58, prin. chemist, 1958-60; prof. pharmacy U. Ariz. Coll. Pharmacy 1960—. Contbr. articles to sci. jours. Co-discovered new organic reaction widely applicable to a number of heterocyclic organic compounds containing nitrogen. Patentee in field. Home: 1237 E. Drachman St., Tucson 85719.*

VIVIANI, Vincenzo, Italian mathematician; b. Florence, Italy, Apr. 5, 1622; disciple of Galileo; student math. under Torricelli. Mathematician to Ferdinand II, Grand Duke Tuscany. Mem. Florentine Acad. Expts., French Acad. Scis., 1699, Fellow Royal Soc., 1696. Author: De Maximis et Minimis Geometrica Divinatio in Quintum Conicorum Apollonii Pergaei adhuc Desideratum, 1659; Quinto Libro degli Elementi D'Euclide, 1674; De Locis Solidis Secunda Divinatio Geometrica, 1673, 1701; Formazione e Misure di Tutti i Cieli con la Struttura e Quadratura Esatta Dell'Intero, E. . . . Uno degli Antichi delle Volte Regolari degli Architetti, 1692. Constructed 1st barometer (based on Torricelli's ideas), 1643; solved trisect. problem with aid of equilateral hyperbola; determined tangent to cycloid; wrote on conic sects.; leading geometer of his time. Died Florence, Sept. 22, 1703.

VIZIOLI, Raffaello, Italian neurologist; b. Naples, Italy, Apr. 18, 1926; s. Francesco and Maria Antoinetta (Basso) V.; M.D.; m. Maria Elisa Reduzzi, June 21, 1952; children—Francesco, Stefano, Maria Carola. Asst. dir. Clinic Nervous and Mental Diseases, Rome (Italy) U. Publns. on epilepsy and faculties of memory, electroencephalography and epilsepsy. Research on problem of conscience; 1st Western neurologist to establish sci. relations with East. Home: Largo Bradano 4. Office: viale Università 30, Rome, Italy.

VLACQ, Adriaan, Dutch mathematician; b. Gouda, Netherlands, circa 1600. Bookseller, pub., London, Eng., for 10 years. Author: (with E. de Decker) Eerste Deel van de Nieuwe Talkonst, 1626. Repub. Tables of Logarithms (Briggs), 1628. Logarithmic tables (1628, 33) filled gap between 20,000 and 90,000 left in Brigg's Arithmetica Logarithmica; issued 1st work on logarithms in Netherlands. Died The Hague, Netherlands, 1667.

VLADIGEROV, Theodor Andreev, Bulgarian economist; b. Sofia, Bulgaria, Aug. 28, 1898; s. Andrei Stainov and Vassilka (Radeva) V.; student U. Humboldt, Berlin, Germany, 1920-24; Dr.rer.pol. habil, U. Berlin, 1932. Asst., Institu Volkswirtschaft, Berlin, 1929-32; dozent U. Berlin, 1936-46; prof. polit. economy Svistov, Bulgaria, 1936-46; dir. Statis. Inst., U. Sofia, 1946-47, prof. Faculty Law, 1946-47; prof. High Econ. Inst. K. Marx, Sofia, 1947-61; vice dir. Inst. Econs., Bulgarian Acad. Scis., Sofia, 1953—. Recipient Award Dimitrove, 1959; decorated Civil merit, 1947, Red Banner of Work, 1961, Peoples Republic of Bulgaria 1st degree, 1963, Cyril and Methodius, 1st degree, 1963. Mem. Bulgarian Acad. Scis. (corr.), Soc. Sci. Workers, Soc. Economists. Author: Directed Economy, 1939; Agrar-Economic Theories, 1941; Fictiv capital, 1957; Problems of Agrarian Credit in the EEC, 1962; European Economic Community, 1967; also articles. Home: 39, Moskovska. Office: 3, Aksakov, Sofia, Bulgaria.*

VLADZIMIRSKAYA, Elena Vasil, Ukrainian chemist; b. Buhajiwka, Ukraine, Nov. 22, 1929; d. Basil and Ann (Martyniuk) V.; candidate sci. Med. Inst. at Lvov, 1956; dr. pharmacy Moscow Pharm. Inst., 1965. Faculty, Med. Inst. at Lvov, 1955—, prof. pharm. chemistry, 1967—. Research, publs., patents on chemistry of thiazolidones and thiazanones. Home: 37 Mayakovsky, Lvov, Ukraine, USSR.

VLAMIS, James, Am. plant physiologist; b. N.Y.C., June 8, 1914; s. Emanuel G. and Pauline (Haramis) V.; B.S., U. Cal. at Berkeley, 1935, Ph.D., 1941; m. Nancy Elise MacBride, May 2, 1941; children—Pauline Emily, Catherine Anne, Michael James, Barbara Annette. Faculty, U. Cal. at Berkeley, 1937-41, 46—, plant physiologist, dept. soils and plant nutrition, 1966—; research asst. Cabot Found., Harvard, 1941-43. Fulbright scholar to Portugal, 1962-63. Mem. Am. Soc. Plant Physiologists, Sigma Xi. Research, numerous publs. on relation of calcium in problem of serpentine soils using lettuce and barley as test plants, toxicity of aluminum and manganese in acid soils, oxygen relations in roots of rice, barley, tomato as they affect ion absorption and respiration, micronutrient deficiency study of rice, iron and manganese relations in rice, manganese and boron toxicity in barley, soil fertility of Cal. Home: 1029 Mariposa St., Berkeley, Cal. 94707.*

VLASOV, Kuzma Alekseivich, Russian geochemist, mineralogist; b. Nov. 14, 1905; grad. Timiryazev Moscow Agrl. Acad., 1931. Staff, Inst. Geol. Scis., USSR Acad. Scis., 1932-52; became chief Lab. Mineralogy and Geochemistry Rare Elements; named dir. Inst. Mineralogy, Geochemistry and Crystallography of Rare Elements, USSR Acad. Sci., 1956, bur. mem. dept. geology and geog. sci., 1960—. Corr. mem. USSR Acad. Scis. Research and publs. on genesis and classification of granite pegmatites, also deposits of other rare elements. Home: Lavrushinskii p. 17, Office: Inst. Mineralogy, Geochemistry and Crystallography of Rare Elements, USSR Acad. Sci., Ulitsa Kuybysheva, 8, Moscow, USSR.

VOCCI, Frank Joseph, Am. chemist; b. Balt., Aug. 13, 1924; s. Amilcare and Santa (Ambrosetti) V.; B.S., Loyola Coll., Balt., 1949; postgrad. U. Md., 1950-56; m. Jeanne Rose Gordon, Aug. 21, 1948; children—Francis J., Mark J., Donna P., Janet M., Mary C., David P., Stephen J., Jeanne R. Gen. chemist, group leader aerosol br., asst. chief basic toxicology br., toxicology div. Army Chem. Center, Md., 1949-61; chemist, chief basic toxicology br. Directorate Med. Research, Chem. Research and Devel. Labs., Edgewood Arsenal, Md., 1961—. Recipient Distinguished Service award State of Md., 1961. Mem. Am. Chem. Soc., Am. Indsl. Hygiene Assn., Sci. Research Soc. Am. Research, publs. on toxicological effects of drugs and poisons, analysis of body fluids and tissues in biol. assay, enzyme analysis by coulometric method, permeability of membranes; developer lab. apparatus for dissemination, collection and pulmonary retention of aerosols, apparatus for preparation of tissue homogenates. Home: 6009 Winthrope Av., Balt. 21206. Office: Bldg. 3220, Chem. Research and Devel. Labs., Edgewood, Md. 21010.*

VODAR, Boris, physicist; b. Petrograd, Russia, May 22, 1910; s. Vladimir and Eugenie (Kuenemann) V.; Ph.D., Sorbonne, Paris, 1944; m. Marie-Aimee Laurent, Jan. 15, 1952; 1 son, Michel. Founder high pressure lab. French Nat. Center for Sci. Research (CNRS), 1948; dir., 1948—, research prof., Bellevue, France, 1950—. Mem. French Adv. Comm. for Sci. Research, 1960-64, French Commn. on Molecular Physics and Optics., 1953-65. Decorated chevalier Legion d'Honneur; officer Merite Nat.; commandeur des Palmes Académiques; recipient Prix Hughes Academie des Sciences, 1962. Fellow Am. Phys. Soc.; mem. Internat. Union Pure and Applied Chemistry (commn. chem. thermodynamics 1958—), Internat. Council Sci. Unions. Research, numerous publs. on high pressure physics, including molecular collisions using spectroscopy; observed (with Coulon, H. Vu) new pressure induced spectra; 1st observation of inverse Brillouin effect and simulated inverse Raman effect (with Damartin, Oksengorn); invented source of line spectra on which is based a new method and apparatus for spectrochem. analysis (with G. Balloffet, J. Romand). Home: Virly, Jouaignes Aisne, France. Office: Laboratoire des Hautes Pressions, P.O. Box 30, Bellevue 92, France.*

VOEDODSKII, Vladislav Vladislavovich, Russian phys. chemist; b. July 25, 1917; grad. Leningrad Poly. Inst., 1940, also postgrad. work. Became sr. sci. research worker Inst. Chem. Physics, USSR Acad. Sci., 1944; faculty Moscow U., 1946-52; joined staff Moscow Physico-Tech. Inst., 1953, became prof., 1955. Recipient D.I. Mendeleev prize, 1952. Corr. mem. USSR Acad. Sci. Author: (with Y. B. Zel'dovich) Thermal Explosion and Velocity of Flames in Gases, 1947; (with A. B. Malbandyan) Mechanism of Oxidation and Combustion of Hydrogen, 1949. Research in chem. kinetics, free radicals, combustion theory, carbon oxidation; established details of hydrogen oxidation chain reaction; demonstrated (with others) possibility of radical chain in heterogenesis-catalytic processes. Office: Inst. Chem. Physics of USSR Acad. Sci., Vorobévskoye Shosse, 2, Moscow, USSR.

VOEGTLIN, Carl, pharmacologist; b. Basel, Switzerland, July 28, 1879; s. Carl V.; ed. univs. of Basel, Munich, Geneva and Freiburg: Ph.D., Freiburg, 1903; studied Victoria U., Manchester, Eng., 1903-04; D.Sc., Rochester, 1947; m. Lillian Kreuter, December 21, 1912; 1 son, Hugh Stewart. Instructor chemistry, University of Wisconsin, 1905; assistant in med.; associate and associate prof. pharmacology, Johns Hopkins Med. Sch., 1906-13; chief Div. of Pharmacology, Nat. Inst. of Health, U.S.P.H.S., 1913-39; chief Nat. Cancer Inst., 1938-42; lecturer in pharmacol. U. of Rochester Med. Sch. since 1943, now sr. consultant in cancer research, toxicology; cons. Manhattan Dist. Herter lecturer N.Y. U. Med. Coll., 1938, Barnard Hosp., 1940. Mem. com. on drug addiction Nat. Research Council. Mem. A.M.A., Am. Chem. Soc., A.A.A.S., Soc. Biol. Chemists, Am. Physiol. Soc., Soc. Pharmacology and Exptl. Therapeutics (pres. 1927-30), Am. Fed. for Exptl. Biology (chmn. 1928), Acad. Sciences (Washington), Clinico-Pathol. Soc. Washington, Harvey Soc., Phi Beta Kappa, Sigma Xi. Mem. 1st, 2d Internat. Conf. for Biol. Standardization of Drugs, Edinburgh and Geneva, Permanent Commn. on Biol. Standardization of League of Nations, Acad. of Med. of Washington, D.C. (pres. 1938-40), Soc. Exptl. Biology and Medicine, Am. Assn. Cancer Research (pres. 1941), Am. Soc. for the Control of Cancer (dir.). Retired. Research in physiology and pathology of nutrition, beri beri, pellagra, function of parathyroid gland, pharmacological action of serum preservatives, anaphylaxis, cancer, indsl. poisoning, chemotherapy, biol. oxidation-reduction. Home: 4700 Connecticut Av., Washington 8.

VOELCKER, John Christopher August, agrl. chemist; b. Frankfort/Main, Germany, Sept. 24, 1822; s. Frederick Adolphus Voelcker; ed. Göttingen U.; studied under Prof. Wöhler; Ph.D., 1846; student agrl. chemistry under Justus von Liebig, Giessen, Germany; m. Susanna Wilhelm, 1852; children—John Augustus, William, others. Pharmacist's asst., Frankfort/Main; pharmacist Schaffhausen; went to Edinburgh, Scotland, 1847; prof. chemistry Cirencester Agr. Coll., 1849-63; cons. agrl. chemist, 1855-84; lived in London, after 1863. Fellow, Royal Soc., 1870; mem. Inst. Chemistry Gt. Britain and Ireland (a founder 1877, 1st v.p.), Farmer's Club (became chmn. 1875). Research and numerous publs. on applications of chemistry to farming; feed, soil, artificial manures. Died Kensington, Eng., Dec. 5, 1884.

VOET, Andries, chemist; b. Amsterdam, Netherlands, Nov. 21, 1907; s. Herman I. and Maria (Boom) V.; B.Sc., U. Amsterdam, 1929, M.Sc., 1931, Ph.D., 1935; m. Henriette Mogendorff, Jan. 5, 1932 (dec. 1963); children—Louise E. (Mrs. Elliott J. Blumberg), Donald H., Marion J. (Mrs. Peter Goodman), Martin A.; m. 2d, Rebecca Cohen, July 3, 1964. Came to U. S., 1939, naturalized, 1944. Instr., U. Amsterdam, 1931-35; Netherlands Am. Found. fellow, 1935; chemist Van Son's Ink Works, Hilversum, Holland, 1935-40; research chemist Sun Chem. Corp., N.Y.C., 1940-43, J. M. Huber Corp., Borger, Tex., 1943—. Faculty, N.Y. U., 1948-50. Fellow N.Y. Acad. Scis.; mem. Soc. Rheology, Am. Phys. Soc., Am. Chem. Soc. Author: Ink and Paper in the Printing Process, 1952. Research, publs., numerous patents in colloid chemistry and physics, printing ink, carbon black; developed visco-elastic concepts on printing inks, processes for mfr. printing inks, process for masterbatching carbon blacks in rubber. Home: 901 Country Club Rd. Office: J. M. Huber Corp., Borger, Tex. 79007.*

VOGEL, Friedrich, German human geneticist; b. Berlin, Germany, Mar. 6, 1925; s. Reinhard and Ingeborg (Wahrenholz) V.; D.med. Freie Universität, Berlin, 1952; m. Adelheid Kurth, Aug. 6, 1951; children—Sebastian, Christiane, Rüdiger, Tilman. Staff, Max Planck Inst. Vergl. Erbbiologie, 1953-62; prof. anthropology and human genetics U. Heidelberg (Germany), dir. Inst. Human Genetics, 1962—. Author: Lehrbuch der Allg. Human-genetik, 1961; Über die Erblichkeit des Normalen EEG, 1958; also numerous articles. Research on problems of mutation and natural selection in man, genetics of electroencephalogram. Home: 13 Im Bubenwingert, Leimen o. Heidelberg, Germany.*

VOGEL, Heinrich August von, see von Vogel, Heinrich August.

VOGEL, Hermann Karl, German astronomer; b. Leipzig, Germany, Apr. 3, 1841; ed. U. Leipzig; founder, became dir. Astrophys. Obs., Potsdam, Germany, 1882. Corr. mem. French Acad. Scis., 1906. Author: Observations on Clouds and Stars, 1867; Observations Made at the Observatory of Bothcamp, 1872-73; Research on the Spectrum of Planets, 1878; also catalogue of 51 stars. Editor: Populärer Astronomie (Newcomb-Engelmann), 1905. Discovered spectroscopic binaries, circa 1890; proved variable stars of Type beta Lyrae are eclipsing stars; showed spectral lines of Algol are alternately displaced to red and violet during period of brightness variation, 1889; originated method for spectroscopic identification. Died Potsdam, Aug. 13, 1907.

VOGEL, Hermann Wilhelm, German photo-chemist; b. Dobrilugk, Brandenburg, Germany, Mar. 26, 1834. Prof. Acad. Arts and Crafts (tech. sch.) Charlottenburg, Berlin, Germany, from 1864. Author: Handbuch der Photographie, 1870. Invented orthochromatic photographic plate by adding certain dyes to emulsion to increase color sensitivity, 1873, also photometer, universal spectroscope; studies on spectroscopic photography. Died Berlin, Dec. 17, 1898.

VOGELSANG, Heinz Georg, German neuro-radiologist; b. Berlin, Feb. 13, 1928; s. Georg and Martha (Plump) V.; M.D., U. Würzburg, 1952; m. Marianne Greiner, Aug. 19, 1954; children—Ulrike, Andrea, Christine, Jens-Peter. With med. clinic U. Wurzburg, 1952-53, neurol. clinic U. Frankfort/Main, 1953-61; head physician, neurosurg. clinic U. Giessen/Lahn, 1961—. Research and publs. on vascular diseases of brain, vascular malfomations, infectious diseases; ossovenography. Home: 16 Gunthersgraben, Giessen/Lahn, Germany.*

VOGELSANG, T. M., Norwegian physician; b. Bergen, Norway, Aug. 1, 1896; s. M. P. and Elisa (Bonge) V.; M.D., U. Oslo, 1922; m. Kirsten Vogelsang, Sept. 12, 1923; 1 son, Sidsel. State epidemiologist Western Norway, 1929-35; supt. Gade Inst., 1935-48; prof. microbiology U. Bergen, 1948—; cons. physician U. Hosp. Bergen, 1948—. Author: Sero-Diagnostic de la Syphilis, 1939; Typhoid and Paratyphoid B Carriers and Their Treatment, 1950; also numerous articles. Research on typhoid and paratyphoid investigations, serological syphilis reactions, staphylococcal studies. Home: 1 Sydnes Plass. Office: 10 Haukelandsvn., Bergen, Norway.*

VOGL, Otto F(ranz), chemist; b. Traiskirchen, Austria, Nov. 6, 1927; s. Franz and Leopoldine (Scholz) V.; Ph.D., U. Vienna, 1950; m. Katherine Jane Cunningham, June 10, 1955; children—Eric, Yvonne. Came to U. S., 1953, naturalized, 1959. Instr., U. Vienna (Austria), 1948-53; research asso. U. Mich., 1953-55, Princeton, 1955-56; research chemist E. I. du Pont de Nemours & Co., Wilmington, Del., 1956—. Mem. Am. Chem. Soc. (past treas. polymer sect. Del. sect. 1964), A.A.A.S., Austrian Chem. Soc., Sigma Xi. Asso. editor Jour. Macromolecular Chemistry, 1965—. Contbr. numerous articles to tech. jours., chpts. to books. Research on isolation, structure, synthesis alkaloids, and purines, mechanisms chem. reactions, preparations, properties aldehyde polymers. Home: 2 Breeze Hill Rd., Wilmington 19807. Office: Central Research Dept., Exptl. Sta., E. I. du Pont de Nemours & Co., Wilmington, Del. 19898.*

VOGT, Marthe Louise, physiologist; b. Berlin, Germany, 1903; d. Oskar and Cécile (Mugnier) Vogt; M.D., Berlin (Germany) U., 1928, Ph.D., 1929; Ph.D., Cambridge (Eng.) U., 1938. Research asst. pharmacology dept. Berlin U., also head chemistry dept. Kaiser Wilhelm Inst. for Brain Research, Berlin-Buch, 1929-35; Rockefeller Travelling fellow London, also Cambridge U., 1935-36; Alfred Yarrow Research fellow Girton Coll., Cambridge, 1937-40; pharmacologist Pharm. Soc. Gt. Britain, 1941-46; faculty pharmacology dept. U. Edinburgh (Scotland), 1947-60; head pharmacology unit Inst. Animal Physiology, Agrl. Research Council, Babraham, Cambridge, 1960—; vis. asso. prof. Columbia Med. Sch., 1949; vis. prof. Sydney (Australia) U., 1965. Fellow Royal Soc., 1952; mem. Physiol. Soc., Brit. Pharmacological Soc., Endocrine Soc., Brit. Soc. for Endocrinology, German Physiol. Soc. Research, numerous publs. on regional distbn. drugs in brain tissue, control of secretion of adrenal cortex and medulla, occurrence and effect drugs on active substances in peripheral and central nervous systems. Home: 5, Marion Close, Cambridge. Office: Inst. Animal Physiology, Babraham, Cambridge, Eng.*

VOGT, Oskar, German neurologist; b. Husum, Prussia, Apr. 6, 1870; m. Cécile. Founder, dir. Brain Research Inst., Berlin-Buch, Neustadt, (both Germany). Author: Sitz und Wesen der Krankheiten, 2 vols., 1937-38; Thalamusstudien, 1941. (With wife) did detailed research on brain structure; developed new views on brain architectonics; founder scientifically based hypnotism; described condition marked by athetosis, hyperkinesis, due to lesion of corpus striatum (Vogt's syndrome), 1920. Died Neustadt/Schwarzwald, Germany, July 31, 1959.

VOGT, Walther, German pharmacologist; b. Dessau; July 5, 1918; s. Wolfram and Adelheid (Külz) V.; student U. Marburg; M.D., U. Munich, 1942; m. Erika Pfersdorff, Dec. 1, 1946; children—Albrecht, Reinhard. Asst., U. Frankfort/Main, 1945-53; with Max Planck Inst. Exptl. Medicine, Göttingen, 1953—, sci. mem., 1965—; with Nat. Inst. Med. Research, London, 1955-56; unscheduled prof. pharmacology, U. Göttingen, 1960—, Mem. German Pharmacological Soc., Soc. Biol. Chemistry, Leopoldina; asso. mem. Brit. Physiol. Soc. Research, numerous publs. on pharmacological, biochem. and physiol. aspects of biologically active substances originating from blood and tissues; discovery of pharmacologically active acidic phospholipids. Address: 3k Hermann Rein-Strasse, Göttingen, West Germany.*

VOGÜÉ, Charles Jean Melchoir, Marquis de, see de Vogüé, Charles Jean Melchoir, Marquis.

VOIGT, Adolf Frank, Am. chemist; b. Upland, Cal., Jan. 31, 1914; s. Adolf Frank and Marie (Hirschler) V.; B.A., Pomona Coll., 1935; M.A., Claremont Coll., 1936; Ph.D., U. Mich., 1942; m. Mary London, Dec. 26, 1941; children—Maryanne, Richard L. Instr., Smith Coll., 1941-42; chemist Manhattan Dist., Ia. State U., Ames, 1942-46, faculty, 1946—, prof. chemistry, 1955—, sr. chemist Ames Lab., AEC, 1955—, asst. dir., 1959—. Mem. Am. Chem. Soc., Am. Phys. Soc., Am. Nuclear Soc., A.A.A.S., Phi Beta Kappa, Sigma Xi, Phi Kappa Phi, Phi Lambda Upsilon. Contbr. numerous articles to profl. jours., also chpts. on tracer methodology, thorium toxicology to books. Research on application radioactive tracers to chemistry, equilibrium problems, kinetics, activation analysis, behavior of energetic carbon atoms produced in nuclear reactions, processing of nuclear fuel by high temperature methods. Home: 212 N. Russell Av., Ames, Ia. 50010.*

VOIGT, G. K., Am. soil scientist; b. Merrill, Wis., Jan. 17, 1923; s. M. W. and Hattie (Krause) V.; B.S., U. Wis., 1948, M.S., 1949, Ph.D., 1951; m. Jane Wurster, Jan. 12, 1946; children—Timothy, Valerie, Jeffrey. Faculty, U. Wis., 1950-54, asst. prof., 1953-54; faculty Yale, 1955—, prof. soil sci., 1965—. Collaborator U. S. Forest Service, 1952-55. Mem. Soil Sci. Soc. Am. (past divisional chmn., asso. editor Proc. 1964—), Am. Soc. Plant Physiologists, Wis. Acad. Arts, Letters and Sci., Sigma Xi, Alpha Zeta. Author: Analysis of Soil and Plants, 1955; also articles. Research on ion uptake by trees. Office: 370 Prospect St., New Haven.*

VOIGT, Peter Hans Ludwig, German rehab. specialist; b. Berlin, Germany, Aug. 5, 1908; s. Karl and Helene (Weber), June 29, 1938; ed. univs. Humboldt and Berlin; Dr.Edn.; m. Annemarie Ewerth, June 29, 1938; children—Helga, Wolf-Dietrich, Helmut. Formerly asst., chief asst., also instr., now prof., dir. Pub. Information Inst., Humboldt U., Berlin. Mem. German Rehab. Soc., Internat. Soc. Rehab. Handicapped, World Spl. Edn. Com. Author: Geschichte der Sprachheilschulen in Deutschland, 1954; Über den Begriffsschatz bei langfristig hospitalisierten Kindern, 1960; Rehabilitation geschädigter Kinder, 1961; Die Sonderschulwesen und die Rehabilitation geschädigter Kinder, Handbuch der Sonderpädagogik, 1964; also numerous articles. Home: Am Friedrichschain 15, Berlin NO 18. Office: Humboldt-Univ., Unter den Linden 6, Berlin, West Germany.

VOIGT, Woldemar, German physicist; b. Leipzig, Saxony, Sept. 2, 1850; prof. Leipzig, Königsberg, Göttingen univs.; corr. mem. French Acad. Scis., 1911; author: Kompensium der theoretetischen Physik, 2 vols., 1895-96; Thermodynamik, 2 vols., 1903-04; Lehrbuch der Kristallphysik, 1910. Research in crystallography, thermodynamics, magneto-optics, and electronics. Died Göttingen, Dec. 13, 1919.

VOINOV, Dimitrie, Rumanian biologist; b. 1867; mem. Acad. Socialist Republic Rumania. Author: Principles of Microscopy, 1906; Mitochondria, 1916; The Biological Problem of Sex Differentiation, 1929. A leading supporter of Darwinism in Rumania; studied cytology of Coleopterae, Lepidopterae, Orthopterae. Died 1951.

VOIPIO, Paavo Taneli, Finnish zoologist; b. Viipuri, June 19, 1913; s. Väinö Taneli and Elsi (Orkamo) V.; M.A., U. Helsinki (Finland) 1945, Ph.D., 1951; m. Meri Elisabeth Toivonen, Mar. 6, 1938; children—Mervi (Mrs. Erkki Häsänen), Ilkka (dec.), Arja (Mrs. Jan Höglund), Inari. Asst., Zool. Mus., U. Helsinki, 1945-53, curator vertebrates, 1954-58; prof. zoology U. Turku (Finland), 1958—; staff Finnish Game Research Inst., 1945-52, spl. research worker, 1951-52. Lectr. zoology U. Helsinki, 1952-62. Decorated comdr. Cross of Order of Finnish Lion. Mem. Finnish Ornithologists Union (past sec. gen.), Zool. and Bot. Soc. Turku (pres. 1966—). Translator into Finnish: Evolution in Action (Julian Huxley), 1956; Animals of Ancient Times (J. Augusta, Z. Burian), 1960. Research, publs. on evolutionary systematics with application of results of population genetics to phenomena of geog. variation, toi ecol. differentiation and population dynamics, to genodynamics of subspecific boundaries, to analysis of population of yellow-legged Herring gulls, to vertebrate polymorphism especially significance of morphs as ecol. indicators.*

VOISEY, Peter William, bioengr.; b. London, Eng., Mar. 10, 1934; s. William J. and Elizabeth (Kindler) V.; diploma Kingston Tech. Coll., 1955; m. Mary Frances Tindle, May 20, 1957; children—Peter William, Robin Elizabeth. With Canadair Ltd., Montreal, Que., Can., 1955-60; instrumentation engr. research br. Engring. Research Service, Can. Dept. Agr., Ottawa, Ont., 1960—. Mem. Soc. for Exptl. Stress Analysis (v.p. Ottawa Valley sect.), Instrument Soc. Am., Instn. Mech. Engrs. (asso.), Engring. Inst. Can., Assn. Profl. Engrs. Ont., Corp. Engrs. Que. Research, numerous publs. on phys. properties of foods and instrument devel. Home: 687 Gainsborough Av., Ottawa 13. Office: Engring. Research Service, Central Exptl. Farm, Ottawa, Ont., Can.*

VOISIN, Guy André, French immuno-pathologist; b. Paris, France, Dec. 11, 1920; s. Maurice Roger and Marguerite (Nommés) V.; Thesis in Medicine, Sorbonne, Paris, France, 1945, Thesis in Scis., 1958; m. Janine Terrassier, July 16, 1959; children—Jacques-André, Jean-Michel, Véronique. Research staff Pasteur Inst., Paris, 1947-50, tchr. immuno-pathology, 1953—; research fellow Johns Hopkins, 1951; asst. investigator Institut National d'Hygiene Paris, 1952-55; chief clinic Faculty Medicine Paris, 1952-54; asso. dir. Center Immuno-Pathology, Paris, 1958—; dir. research Claude Bernard Assn., Paris, 1964—; asso. biologist hosps. of Paris, 1962—; head dept. exptl. immuno-pathology I.N.S.E.R.M. Research Center, Saint-Antoine Hosp., Paris, 1966—. Invited investigator N.Y. U., 1956, 58, 60. Introduction of term immuno-pathology; research, publs. on auto-immunity, transplantation immunity, immunological enhancement, hypersensitivity reactions. Home: 40 rue Condorcet, Paris 9e. Office: Hôpital Saint-Antoine, Paris 12e, France.*

VOIT, Karl von, see von Voit, Karl.

VOKES, Harold Ernest, paleontologist, geologist; b. Windsor, Ont., Can., June 27, 1908; s. Albert John and Beatrice (Howlett) V.; B.A., Occidental Coll. 1931; Ph.D., U. Cal. at Berkeley, 1935; m. Gertrude Dutton Lawrence, Mar. 1, 1932 (div.); children—Gertrude Ann (Mrs. Peter E. Martin), Rosina Beatrice (Mrs. Robert Stephenson), Francis Elizabeth, Arthur Walter; m. 2d, Emily Anne Hoskins, Mar. 7, 1959. Came to U. S., 1922, naturalized, 1936. Hon. fellow Yale, 1935-36; staff Ill. State Geol. Survey, 1937; staff Am. Mus. Natural History, N.Y.C., 1937-43, curator geology and invertebrates, 1940-43; geologist U. S. Geol. Survey, 1943-46; faculty Johns Hopkins, 1946-56, prof. geology, 1947-56; prof. dept. geology Tulane U., New Orleans, 1956—, chmn. dept. geology, 1957-67. Mem. Internat. Commn. Zool. Nomenclature, 1942—. Mem. Paleontol. Soc. (past sec., past pres.), Geol. Soc. Am. (past v.p.), Am. Assn. Petroleum Geologists, Malacological Soc. London, Phi Beta Kappa, Sigma Xi. Research, numerous publs. on evolution and distbn. of fossils and recent Mollusca especially Bivalvia. Home: 2523½ Broadway, New Orleans 70125.*

VOLD, Marjorie Jean Young, chemist; b. Ottawa, Can., Oct. 25, 1913; d. Reynold Kenneth and Whilhelmine (Aitken) Young; came to U. S., 1918, naturalized, 1921; B.S., U. Cal. at Berkeley, 1934, Ph.D., 1936; m. Robert Donald Vold, June 14, 1936; children—Mary Louise (Mrs. Lynn Clifford Linman), Robert Lawrence, Wylda Bryan (Mrs. Stephen Cafferata. Spl. lectr. U. Cin., 1936; jr. research asso. Stanford U., 1937-41; mem. faculty U. So. Cal., Los Angeles, 1941—, now adj. prof. chemistry;

---

research chemist Union Oil Co., 1942-46; prof. phys. chemistry Indian Inst. Sci., Bangalore, 1955-57. Cons. industry, govt. agys. Guggenheim Found. fellow, 1953-54; recipient Western Regional 50th Anniversary research award (Iota Sigma Pi, 1952; Univ. medal U. Cal. at Berkeley, 1934. Mem. Am. Chem. Soc. (Garvan medal 1967; chmn. womens service com. 1945-52), Phi Beta Kappa, Sigma Xi. Author: (with R. D. Vold), An Introduction to the Physical Sciences, 1961, Colloid Chemistry, 1964; also chpts. in books, numerous articles. Research in structure of aqueous soap systems and lubricating greases, gels, liquid crystals, colloidal solids; factors controlling life time of fine suspensions, phys. form of polymers, digital computer simulation of colloid and polymer systems, relation of perfect vs. imperfect small crystals to useful characteristics, phys. properties of biol. colloids. Home: 8107 Billowvista Dr., Playa del Rey, Cal. 90291.*

VOLGER, Georg Heinrich Otto, German mineralogist, seismologist; b. Luneburg, Hanover, 1822. Taught at cantonal sch. of Zurich, 1851; prof. mineralogy and geology Senckenburg Mus. and Frankfort-am-Main, 1856-60. Author: Study on the Development of Mineralogy, 1854; Crystallography, 1855. Died Sulzbach, Germany, 1897.

VOLHARD, Jacob, German chemist; b. 1834; student (under Liebig) U. Giessen (Germany); prof. U. Munich (Germany), from 1869; U. Erlangen (Germany), from 1878, U. Halle (Germany), from 1882. Author: History of the Metals, 1897; Life of Hofmann, 1902; Life of Liebig, 2 vols., 1909. Synthesized sarcosine, 1862, creatine, 1869, thiophene, 1885; invented thiocyanate silver titration, 1874; research on guanidine, cyanamide; developed iodimetry. Died 1910.

VOLK, Wesley Aaron, Am. microbiologist; b. Mankato, Minn., Nov. 23, 1924; s. Albert Lee and Della (Buelow) V.; B.S., B.S. in Food Tech., U. Wash., 1948, M.S., 1949, Ph.D., 1951; m. Rose Marjorie Keller, Feb. 24, 1945 (dec. 1966); children—Pamela Lee, Bradley George. Faculty U. Va., Charlottesville, 1951—, professor microbiology, 1964—. Fellow Am. Acad. Microbiology; mem. Am. Soc. for Microbiology, Am. Soc. Biol. Chemists. Author: (with M. Wheeler) Basic Microbiology, 1964; also articles. Research on several intermediary pathways carbohydrate metabolism, purification and characterization enzymes on these pathways, cell wall components bacteria in genus Xanthomonas. Home: 1706 Yorktown Dr., Charlottesville, Va. 22901.*

VOLKERT, Mogens, Danish microbiologist; b. Kolding, Denmark, Sept. 2, 1913; s. Gustaf and Agnes (Nielsen) V.; M.D., U. Copenhagen (Denmark), 1939; m. Ellen Winkel, June 30, 1955. Asst. physician Municipal Hosp. Gentofte, 1939-40; asst. Biol. Inst., Carlsberg Found., 1940-47, bacteriol. dept. Statens Seruminstitut, 1943-46, Rockefeller Inst. for Med. Research, N.Y.C., 1946-47; head rickettsia and virus dept. Statens Seruminstitut, 1948-65; dir. Inst. Med. Microbiology, U. Copenhagen, 1965—, prof. med. microbiology, 1965—; head Inst. for Tumor Virus Research, 1965—. Recipient medal Norwegian Red Cross, 1945, Swedish Red Cross, 1945, Danish Red Cross, 1945. Research and numerous articles on gen. virology especially immunological non-responsiveness to viruses. Home: 10 Jagersvinget, Gentofte, Denmark. Office: 22 Juliane Mariesvej, Copenhagen, Denmark.*

VOLKMANN, Alfred Wilhelm, German physiologist; b. Leipzig, Germany, July 1, 1800; degree, Leipzig, 1826; postgrad., London, Paris. Became prof. zoology, 1834; became prof. physiology and pathology, Dorpat, Estonia, 1837; became prof. physiology, 1843, Halle, Germany, prof. anatomy, 1854. Author: Observatio biologica de magnetismo animali, 1826; Neue Beiträge zur physiologie des Gesichtssinnes, 1836; (with F. H. Bidder) Die Selbstandickeit des sympathetischen Neruensystems durch Anatomischen Untersuchungen Nachgewiessen, 1842; Anatomia animalium, 1831-33; Die Hämodynamik nach versuchen, 1850; Physiologische Untersuchungen im Gebiet der Optik, 1863-64; The Doctrine of the Corporeal Life of Man, 1837. Described canals transmitting blood vessels through bones (Volkmann's canals), 1863; studies on nervous system; Volkmann's membrane named after him. Died Halle, Apr. 21, 1877.

VOLKOVICH, Semen Isaakovich, Russian chemist; b. Oct. 23, 1896; grad. Inst. Nat. Economy, 1920. Prof., Moscow Higher Tech. Coll., 1929-32, Mil. Chem. Acad., later Acad. Chem. Def., 1932-47; dir. Research Inst. Fertilizers, Insecticides and Fungicides, All Union Lenin Acad. Agr. Sci., 1921—; prof. Moscow U., 1947—. Recipient Order of Lenin; Order of Red Banner of Labor; Stalin Prize, 1941. Mem. USSR Acad. Sci.; mem. Mendeleev All-Union Chem. Soc. (chmn.). Author: Production of Potassium Chloride, 1930; Technology of Nitrogenous Fertilizers, 1934; coauthor: General Chem. Technology, 1940-46. Research and publs. on processing of fertilizer and salts, organic chemistry; studies (with Zhukovskii) on electrothermal distillation of phosphorrous (led to 1st elec. furnace for phosphorous in USSR), 1922; devel. prodn. methods for flourides, nitrates, chlorides, sulfates, phosohides, 1945-50. Home: M. Bronnaya Moscow, USSR.

VOLLMAR, Joerg Friedrich, German surgeon; b. Plüderhausen, Germany, Sept. 22, 1923; s. Friedrich and Johanna (Trautmann) V.; student medicine U. Berlin, 1942-43, U. Würzburg, 1943-44, U. Heidelberg, 1945-48; staatsexamen, 1948. Surgeon to surg. clinic U. Heidelberg (Germany), 1948—; head physician, priv. dozent surgery, 1962—. Mem. Deutsche Gesellschaft Chirurgie, Deutsche Gesellschaft Angiology, Soc. Internat. de Chirurgie, Internat. Soc. for Cardiovascular Surgery. Research, publs. on blood banking, typical injuries in automobile accidents, reconstructive vascular surgery, new method for embolectomy. Home: 46, H. Thomas-str., Heidelberg, Germany.*

VOLPE, Erminio Peter, Am. biologist; b. N.Y.C., Apr. 7, 1927; s. Rocco and Rose (Ciano) V.; B.S., Coll. City N.Y., 1948; M.A., Columbia, 1949, Ph.D., 1952; m. Carolyn Thorne, Sept. 1, 1955; children—Laura Elizabeth, Lisa Lawton, John Peter. Instr., Coll. City N.Y., 1950-51; faculty Newcomb Coll., Tulane U., New Orleans, 1952—, prof. biology, 1960—, head dept. zoology, 1954-64, chmn. dept. biology Tulane U., 1964-66, asso. dean Grad. Sch., 1967—. Cons. Commn. on Undergrad. Edn. in Biol. Scis., 1964-65; steering com. Biol. Scis. Curriculum Study, 1966—. Recipient Newberry award Columbia, 1952. Fellow A.A.A.S.; mem. Genetics Soc. Am, Am. Soc. Zoologists, Soc. for Study Evolution, Soc. Systematic Zoologists, Ecol. Soc. Am., Am. Soc. Naturalists, Genetics Assn., Assn. Southeastern Biologists, Am. Assn. U. Profs., Am. Soc. Ichthyologists and Herpetologists, Phi Beta Kappa, Sigma Xi. Author: Understanding Evolution, 1967. Asso. editor Jour. Exptl. Zoology, 1968—. Research, numerous publs. on genetics, embryology, and evolution amphibians, mode inheritance traits, nature species formation, origin isolating mechanisms, nature immune response. Home: 7704 Sycamore St., New Orleans 70118.*

VOLPITTO, Perry Paul, anesthesiologist; b. Italy, July 9, 1905; s. George and Mary A. (Cavallo) V.; came to U.S., 1911, naturalized, 1921; B.S. Washington and Jefferson Coll., 1928; M.D., Western Res. U., 1933; m. Mary E. Stevens, Jan. 9, 1937; children—Maryann (Mrs. Simeon B. Fulcher), Perry Paul, George David. Research fellow Wis. Alumni Research Found. 1934-35; resident anesthesiology Bellevue Hosp., N.Y.C., 1936-37; faculty Med. Coll. Ga., Augusta, 1937—, prof. anestheology, chmn. dept., 1938—. Area cons. Atlanta area VA, 1948—; mem. Unitarian Service Com. Med. Mission to Columbia, 1948, Med. Mission to Japan, 1951. Diplomate Am. Bd. Anesthesiology. Mem. Am. (pres. 1965), So. (pres. 1949), Ga. (pres. 1948) socs. anesthesiologists, Assn. Univ. Anesthetists (pres. 1961), A.M.A., Am. Coll. Anesthesiologists, Richmond County Med. Soc. (pres. 1946). Contbr. numerous articles in field. Research on intravenous route for amnesic and analgesic drugs in obstetric labor; used stellate ganglion block for treatment and prognosis in cerebral vascular accidents and treatment of Parkinson's disease; made simultaneous use of alternating and direct currents for electronarcosis. Home: 3024 Bransford Rd., Augusta, Ga. 30902.*

VOLSKII, Anton Nikolaevich, Russian metallurgist; b. June 24, 1897; grad. Moscow Inst., 1924; Dr. Tech.Sci. Asso., State Research Inst. Non-Ferrous Metals, 1928-48; instr. Moscow Inst. Non-Ferrous Metals, 1929-34, prof., 1934—; prof., head dept. Kalinin Inst. Non-Ferrous Metals, Krasnoyarsk. Recipient Stalin prize, 1949, 53. Mem. USSR Acad. Sci. Author: The Theory of Metallurgical Processes, 1935; Principles of the Theory of Metallurgical Smelting, 1943; Research and publs. on non-ferrous metallurgy, chem. balance during smelting of metals. Office: Moscow Inst. Non-Ferrous Metals and Gold, Moscow, USSR.

VOLTA, Count Allessandro Giuseppe Antonio Anastasio, Italian physicist; b. Como, Italy, Feb. 18, 1745; s. Filippo and Maria Magdalen (Inzaghi) V.; m. Teresa de' Peregrini, 1794; 3 sons. Prof. physics Royal Sch., Como, from 1774; prof. natural philosophy U. Pavia (Italy), from 1779, dir. philos. faculty, 1815-19, ret., 1819; senator of the Kingdom of Lombardy, 1810; received the title of count and senator from Napoleon. Performed expts. on generation electric current before French Acad., Paris, 1801. Fellow Royal Soc. (Copley medal 1794 contbr. Transactions), 1791; fgn. asso. mem. French acad. Scis. (Gold medal), 1803. Author: De vi Attractiva Ignis Electrici, 1769; Lettere sull Aria Inflammabile Delle Paludi, 1776; Opere di Volta (collected works), 5 vols., 1816. Contbr. articles to profl. jours. Built Voltaic pile (1st electric battery), 1800, making it possible to decompose water by electrolysis, to electro plate precious metals, to form electromagnet; developed idea of electrochem. series; unit of electromotive force, volt, named in his honor, 1881; invented electrophorus, 1775 (improved 1777), electroscope, 1782, eudiometer, an electric pistol, lamp which burned inflammable air; isolated methane gas, 1778; research on atmospheric electricity, analysis marsh gas, 1776; investigations on heat. Died Como, Mar. 5, 1827.

VOLTAIRE (Jean François Arouet), French philosopher; b. Paris, France, Nov. 21, 1694; s. François Arouet and Marie Marguerite Daumart; ed. by Jesuits at Collège Louis-le-Grand; also studied law. Imprisoned in Bastille, 1717; acquired independent fortune through speculation; again in Bastille, 1726; to avoid difficulties with French authorities traveled in England, 1726-29; returned to France, 1729; engaged in literary projects in Paris, 1729-33; lived at Cirey (with Marquise du Chatelet), 1734-49; wrote tragedies: Brutus, 1730; Eriphile, 1732; Zaïre, 1732; L'Enfant prodigue, 1736; Mahomet, 1742; and Mérope, 1743; remained on shaky relations with French authorities; visited Berlin, 1743; appointed French royal-historiographer, 1745; after death of Marquise du Chatelet accepted offer of Frederick the Great to reside at his court in Potsdam, 1749; expelled from Prussia by Frederick the Great (after jewelry and bond fraud scandal), 1753; traveled about while writing, 1754-58; lived in Ferney, Switzerland, 1758-78; built church at Ferney, 1760-61; died a wealthy man. Author: Lettres philosophiques, 1734; Elements de la philosophie de Newton; Charles XII: Zadig, 1747; Le Siècle de Louis XIV, 1751; Micromégas, 1752; Diatribe du Docteur Akakia, 1752; Essai sur l'histoire generale et sur les moeurs, 1753, 1756; Poème sur la lois naturelle, 1756; Poème sur le désastre de Lisbonne, 1756; Candide (attacked Leibniz's optimism and best of all possible world concept), 1759; Dictionnaire philosophique, 1764; contributed to the Encyclopédie. A leading figure of French enlightenment; composed historical treatises with critical method and analytical clarity; disseminated Newtonian scientific thought on the continent through his writings; urged election of French encyclopedists to Berlin Academy of Sciences; opposed Leibnizo-Wolffian science; opposed Maupertuis in priority dispute for discovery of principle of least action. Died Paris, France, May 30, 1778.

VOLTERIS, Eduard Karl Gottfried, archeologist; b. Riga, Latvia, Mar. 19, 1856; s. Alexander and Mathilde (Wolter) V.; ed. U. Leipzig (Germany), St. Petersburg (now Leningrad), Moscow, Kharkov (all Russia) univs.; m. Alexandra Maslowski, 1886. Master, Russian lang., lit. St. Petersburg Acad. Sci., from 1883; lectr. Lithuanian-Baltic lang.; became citizen Lithuania, 1919; organized Lithuanian pub. central library, Kaunas; dir. Kaunas city libraries, from 1920, also city mus.; prof. Baltic archeology. Chmn. Archeol. Commn., Vilna, Lithuania; mem. reform group Vilna pub. libraries. Mem. Lithuanian Hist. Soc., German Archeol. Inst., Russian Geog., Archeol., Philos. Soc., Latvian-Lithuanian Reconciliation Attempts Soc. (hon. pres.). Author: Materialen und Reisenotizen über die Wohnsitze, Sprache und Sitten der Letigalier; Reiseberichte über littauische Diatekte und Altertümer; Wo lag Rimberts Apulia?; Archäologische Missverständnisse; Litauische Reformaten und Humanisten Des litauischen Grossfürsten Blutschwur und die ethnologische Bedentung von Blut und rote Farbe im litaurischen Folklor. Worked at excavations in Apulia-Skuodas, Castle Hill in Kaunas, firetombs in Rumshichki, also in Memel dist., Samogitia. Deceased.

VOLTERRA, Vito, Italian mathematician, physicist; b. Ancona, Italy, May 3, 1860; s. Abramo and Angelica (Almagia) V.; D.Sc., U. Rome, 1882; Hon. dr., Sorbonne, Cambridge, Oxford, Edinburgh, Oslo, Stockholm. Prof. mechanics U. Pisa, later at Turin, Italy; became prof. physics and math. U. Rome, 1900; senator from Riposo, 1905. Pres., Internat. Commn. Weights and Measures. Fellow Royal Soc., 1910; mem. Inst., French Acad. Scis., 1904, acads. of Edinburgh, Madrid, Göttingen, Washington, Leningrad, Author: Sulle apparenze elettrochimalla superficie di un cilindro, 1882; Sur la généralisation de la Théorie des fonctions d'une variable imaginaire, 1889; Sur les vibrations lumineuses dans les milieux buèfrigents, 1892; Sur les vibrations des corps élastiques isotropes, 1894; Lezioni di meccanica, 1896; Sur l'équilibre des corpes élastiques multiplement connexes, 1907; Leçons sur les équations différentielles de la phisique mathématique, 1912; Sur les équations intégrales, 1913; Leçons sur les fonctions des lignes, 1913; Saggi scientifici, 1920; Leçons sur la composition et les fonctions perrantables, 1924; Teoria de las funcionales, 1927; Theory of Functionals and of Integral and Integro-differential Equations, 1930; Leçons sur la théorie mathématique de la lutte pour la vie, 1930. Devel. of theory of intergo-differential equations, 1890; expanded theory of functional analysis. Died Rome, Oct. 11, 1940.

VOLTOLINI, Friedrich Edward Rudolph, otolaryngologist; b. Germany, 1819. First to use galvanocautery in laryngeal surgery, 1867; 1st to do laryngeal operation through mouth, 1889; described acute inflammation of internal ear (called Voltolini's disease), 1867. Died 1889.

VON ABICH, Otto Wilhelm Hermann, geologist; b. Berlin, Germany, Dec. 11, 1806; studied mineralogy in Berlin. Prof. mineralogy in Dorpat, Russia. Author: Geologische Forschungen in den Kaukasischen Ländern, 1878-87; also various reports, papers. Made valuable discoveries in geology, mineralogy, geography; explored extensively in Caucasus, Armenia, Persia; worked on spinels, fumaroles, volcanic problems; mineral clinoclasite (basic copper arsenate) named Abichite in his honor. Died Vienna, Austria, July 1, 1886.

VON AHORNRAIN, Joseph Georg Franz von Paula Ahorner, German physician; b. Augsburg, Germany, Apr. 1, 1764; s. Franz Jakob and Maria Katherina (Mages) A.; student theology Brixen (then Austria), medicine Innsbruck, Vienna (both Austria); m. Karolina Sophia Juliana v. Tromp, 1791; 1 son, Joseph Karl Andreas. Practiced medicine Augsburg, from 1793; dean Collegium Medicum, 1801-06. Mem. Mineral. Soc. Jena, Academia Latina (Rome). Author: Hippocrates Aphorismen, 1791; Bibliothek für Kinderärzte, 2 vols., 1792. Died Augsburg, Dec. 31, 1839.

VON ALBERTI, Friedrich, geologist; b. 1795; deriving divs. from strata sequences in central Germany, gave name Trias or Triassic system to lowest major div. of Mesozoic era. Died 1878.

VON AMMON, Friedrich August, German ophthalmologist; b. Göttingen, Germany, Sept. 10, 1799; s. Christoph Friedrich and Elisabetha (Breyer) von A.; ed. Leipzig, also Göttingen; m. Natalie Redlich, Dec. 1, 1824; m. 2d, Natalie Ernestine Bodelschwingh, Jan. 29, 1853. Physician to Eye Hosp., also Inst. for Blind, Dresden, Germany, from 1822; apptd. prof. gen. pathology, materia medica, 1828; dir. Surg.-Med. Acad.; physician to King Friedrich August II of Saxony. Author: Parallele der französischen und deutschen Chirurgie, 1823; Brunnendietätik, 1825; Die erste Unterpflichten, 1827; Klinische Darstellungen der Krankheiten und Bildungsfehler des menschlichen Auges, nebst Atlas (most comprehensive work of era preceding ophthalmoscopy), 4 vols., 1838-47; Die angeborenen chirurgischen Drankheiten des Menschen, 1839-42; Die plastische Chirurgie, 1842. Described cilia on inner surface of ciliary body, also pyriform scleral fissure of early fetal life, posterior scleral protuberance, 1838-47; introduced biepharoplasty, also dacryocystotomy. Died Dresden, May 18, 1861.

VON ANREP, Gleb Vassilievitch, physiologist; b. St. Petersburg, Russia, Sept. 23, 1891; s. Vassili von Anrep. M.D., Med. Acad. St. Petersburg, 1915; M.B., M.Sc., D.Sc., U. Coll., London; M.A., Cambridge; m. Olga Volkova; m. 2d, Dina; m. 3d, Ida; m. 4th Annie Weininger; a son, John. Served in Russian Army, 1914-17; with Counter-Revolutionary Army, 1918-20; moved to Eng., 1920; asst. U. Coll., London; became lectr., Cambridge, 1925; named prof. physiology Kasr el Aini, Cairo, Egypt, 1930; prof. physiology Faud I U., Cairo. Recipient Sharpey-Schafer prize, W. Mickel prize. Sydney Ringer Lectr.; Cooper Lane lectr., Stanford; fellow U. Coll. Fellow Royal Soc., 1928; mem. Acad. Medicine Roumania, Acad. Sci. Vienna. Research and publs. on diseases of coronary vessels, conditioned reflexes, circulatory system. Died Jan. 10, 1955.

VON ARCO, Georg Wilhelm Alexander Hans, German engr.; b. Grossgorschuetz, nr. Ratibon, Germany, Aug. 30, 1869; s. Alexander and Gertrude Arco; ed. Berlin U., Tech. High Sch., Charlottenburg; m. Elisabeth Annemarie Schwandt, Dec. 21, 1917; Served as officer Guards, Berlin, 1893-98, 1900-03; asst. to Prof. Siaby; engr. A.E.G. (Gen. Electric Co.), Berlin; dir. German radio system, 1903-30; originated and developed German radio sta., Nauen, nr. Berlin, also Telefunken Gesellschaft (German radio corp.). Studied high frequency radio senders, role of Ionosphere in radio communications; initiated wireless devels. and broadcasting in Germany within the reach of everyone; contbd. to establishment of overseas wireless telephone communication and telephotographic services in Germany. Died Berlin, May 5, 1940.

VON ARLT, Carl Ferdinand Ritter, ophthalmologist; b. Obergraupen, Bohemia, Apr. 18, 1812. Prof., Prague, Czechoslovakia, Vienna, Austria. Author several texts on eye surgery. Developed operation for distichia involving excision of ciliary bulbs and transplantation away from edge of lid (Arlt-Jaesche operation), 1854; described granular form trachoma (Arlt's trachoma), 1854-56, recess in lower part lacrymal sac; work on medico-legal aspects injuries of eye. Died Vienna, Mar. 7, 1887.

VON ARX, William Stelling, Am. oceanographer; b. Highland Mills, N.Y., Sept. 27, 1916; s. Arthur William and Helen (Stelling) von A.; B.A. magna cum laude in Geology, Brown U., 1942; B.S. (Dana fellow) Yale, 1943; Sc.D., Mass. Inst. Tech., 1955; m. Ruth Marie Lineback, Sept. 1, 1944; children—Frederick William, Katherine. Instr., Yale, 1943-45; phys. oceanographer Woods Hole (Mass.) Oceanographic Instn., 1945——; lectr. Harvard, 1948-49; faculty Mass. Inst. Tech., Cambridge, 1957——, prof. oceanography, 1959——. Mem. atmospheric scis. panel Pres.'s Sci. Adv. Com., 1957-64; mem. adv. com. U. S. Coast and Geodetic Survey, 1961-66, Geophys. Inst., U. Alaska, 1962——; mem. spl. commn. on weather modification NSF, 1964-65; Mem. exec. com. NRC, 1965-68. Fellow Am. Geophys. Union, Am. Acad. Arts and Scis.; mem. Am. Meteorol. Soc., Am. Soc. Limnology and Oceanography, Assn. Internat. d'Oceanographie Physique, Sigma Xi. Author: Introduction to Physical Oceanography, 1962; also numerous articles. Research on gen. circulation oceans by means rotating models and electromagnetic measurements, inertio-optical techniques for precise astron. nav. at sea, leveling at sea, marine phys. geodesy.*

VON ASTEN, Friedrich Emil, German astronomer; b. Cologne, Germany, Jan. 26, 1843; s. Alexander Friedrich and Auguste (Curtius) von A.; ed. Bonn, Germany; m. Berta Werner; astron. calculator Bonn Observatory, also computing sta., Berlin, Germany; contb. papers to astron. bulletins, jours. Made numerous calculations tracing comet number II, 1852, comet number III, 1864, Tempel, 1867, Encke, 1819-75, Uranus moonlets, Oberon, Titania, Terpsichod, Diana; also studied planets. Died Petrograd, Russia, Aug. 15, 1878.

VON AUENBRUGG, see Auenbrugger, Leopold Edler von Auenbrugg.

VON AUWERS, (Georg Frierich Julius) Arthur, German astronomer; b. Göttingen, Germany, Sept. 12, 1838; s. Gättfried Daniel and Emma Sophie Christiane (Borkenstein) A.; ed. U. Gottingen, 1857-59, U. Königsberg (now Kaliningrad, USSR), to 1862; m. Marie Henriette Jacobi, Nov. 1, 1862; 3 sons, including Karl Friedrich. Observer, Obs. Gotha; astronomer, from 1866. Mem. Acad. Scis. Berlin (perpetual sec.), Astron. Soc., French Acad. Scis. (corr.). Author: Geschichte des Fixsternhimmels (based on observations made 1743-1810). Made observations on transits of Venus, 1874, 82; completed J. F. W. Herschel's nebular observations; research on proper motion of fixed stars, positions of stars; made new reduction of Bradley's stars, 1888. Died Berlin, Jan. 24 or 25, 1915.

VON AUWERS, Karl Friedrich, German chemist; b. Gotha, Germany, Sept. 16, 1863; s. Arthur I.G.F. and Mane Henriette (Jacobi) von A.; ed. Heidelberg, also Berlin, Germany; degree, 1885; student A. W. Hofmann, 1887-89, also Victor Meyer; m. Elisabeth Pauline Cäcilie Koch, Apr. 8, 1893; 2 children, including Otto A. Asst. in Berlin, also Königsberg; became pvt. lectr., Heidelberg, 1890, asso. prof., 1894; apped. dir. Inst. Chemistry, Greifswald, Germany, 1900, Marburg, Germany, 1913-28. Mem. German Chemists' Assn. (v.p.). Contbd. more than 500 papers to sci. lit., 1884-1938. Formulated (with Victor Meyer) term stereochemistry; developed spectrochemistry of organic compounds, also naphthalene, anthracene and acridine derivatives of pyrazole; studied molecular refraction and dispersion; investigated isomerism of oximes and Beckmann rearrangement; discovered 2 modifications of benzil monoxime, 2 of benzil dioxime. Died Marburg, May 3, 1939.

VON BACH, Julius Carl, German mech. engr.; b. Stolberg, Saxony, Germany, Mar. 8, 1847; s. Heinrich Julius and Karoline Juliane (Keller) Bach; student Dresden (Germany) Poly. Sch., Technische Hochschule, Stuttgart, Germany; diploma Technische Hochschule, Karlsruhe, Germany, 1873; 4 hon. degrees. Chief engr. Maschinenfabrik W. Knaust, Berlin, Germany, 1874-76; dir. A.-G. Lausitzer Maschinenfabrik, Bautzen, Germany, 1876-78; prof. mech. engring. Technische Hochschule, Stuttgart, from 1878. Mem. Kuratorium der Physikalisch-Technischen Reichsanstalt, Württemberg Ersten Kammer (life). Author: Die Maschinenelemente, 1881; (with R. Baumann) Elastizität und Festigkeit, 1889-90; Milderung der Klassengegensätze, 1913; Mein Lebensweg und meine Tätigkeit, 1926. Originator of static elasticity, tensile strength theory in engring; worked for more practical teaching of engrs.; established engring. labs., Stuttgart, 1881, materials testing plant, 1883; aided in devel. zeppelin dirigible. Died Stuttgart, Oct. 10, 1931.

VON BACKSTRÖM, Johan Willem, South African geologist; b. Standerton, South Africa, Nov. 12, 1916; s. George and Aletta (Badenhorst) von B.; student U. Pretoria (S. Africa), 1935-37, D.Sc., 1961; m. Cornelia Müller, Mar. 9, 1943; children—Theodor, Aletta, Johan. Asst.-dir. Geol. Survey, Pretoria, head team govt. cons. geologists in engring. geology Orange River Scheme, 1960-63; head geology div. Atomic Energy Bd., Pelindaba, S. Africa, 1964——. Mem. Geol. Soc. S. Africa (Jubilee Gold medal 1949, mem. council 1966——). Author: The Geology of the Ottosdal Area, Transvaal, 1962; The Geology of the Keimoes Area, Cape Province, 1963; also articles. Research in regional, engring., and econ. geology and hydrology, Pre-Cambrian geology and charnockites of Namaqualand and N.W. Cape Province, Home: 155 Klaradyn St. Office: Private Bag 256, Pretoria, Transvaal, S.Africa.*

VON BAER, Karl Ernst, see Baer, Karl Ernst von.

VON BAESDOW, Karl Adolf, German physician; b. Dessau, Germany, Mar. 28, 1799; s. Luswig and Johanna (Krüger) von B.; studied medicine, Paris, France, Halle, Germany; M.D., 1821; m. Friderike Louise Scheuffelhuth, Apr. 23, 1823; 1 son, 3 daus. Settled in Merseburg, 1821; studied cholera epidemic, Magdeburg, Germany, 1830-31; became ofcl. physician Magdeburg, 1834. First to recognize and describe exophthalmic goiter (Basedow's disease), 1840. Died Merseburg, Apr. 11, 1854.

VON BAEYER, Adolf (Johann Friedrich Wilhelm), German chemist; b. Berlin, Oct. 31, 1835; s. Johann Jakob and Eugeine (Hitzig) B.; studied chemistry under R. W. Bunsen, F. A. Kekule; Ph.D., Berlin, 1858; m. Lida Bendemann, Aug. 8, 1868; 1 son, 1 dau. Became asst. prof., 1866; named prof. chemistry, Strasbourg (now in France, 1871, U. Munich, 1875. Recipient Davy medal Royal Soc. London for research on indigo, 1881, Nobel prize for work on organic dyes and hydroaromatic compounds, 1905. Elected to French Acad. Scis., 1886. Author: Gesammelte Werke, 2 vols., 1905. Discovered barbituric acid (forerunner of sleeping pill), 1963; distinguished photochem. reaction and dark reaction as 2 distinct phases of photosynthesis, 1870; discovered molecular structure of indigo; his research on various organic substances contbd. to devel. of chem. industry in Germany. Died Starnberg, Bavaria, Aug. 20, 1917.

VON BAEYER, Otto, German physicist; b. Riechenhall, Germany, Sept. 12, 1877; s. Adolf and Eugenie (Hitzig) von B.; ed. Munich, Leipzig, Germany; degree, 1905. Asst., U. Berlin, became lectr., 1908, prof., 1910; prof. physics Agr. U. Berlin, 1921-39. Contbr. papers to phys. jours. Research in spectroscopy, on cathode rays, beta ray spectra, ultra short wavelengths. Died Tutzing, Germany, Aug. 14, 1946.

VON BÄLZ, Erwin Otto Eduard, anthropologist, physician; b. Bietigheim, Germany, Jan. 13, 1849; s. Carl Gottlob F. and Wilhelmine Caroline (Essich) B.; ed. Tübingen, Leipzig, Germany; M.D., 1872; m. Hana Arai; 1 son. Prof. Tokyo (Japan) Med. Sch., 1875-92, lectr. on physiology, also on internal medicine and gynecology, from 1876; mem. staff U. Tokyo, to 1902; apptd. physician to Imperial Household, 1902; returned to Germany, 1905. Author: Über die Körperlichen Eigenschaften der Japaner, 1883; Lehrbuch der inneren Medizin (in Japanese), 3 vols., 1910. Promoted sci. cooperation between Japan and Germany; contbd. to understanding of beriberi by his report on an outbreak in Tokyo, 1881; studied parasitic diseases; gave classic description of disease characterized by painless papules on mucous membranes of lips (Baelz's disease), 1890; studied anthrop. characteristics of Japanese people. Died Stuttgart, Germany, Aug. 31, 1913.

VON BANDROWSKY, Ernst Titus, Polish chemist; b. Poland, Jan. 3, 1853; Ph.D., U. Lvov (Poland), 1874; postgrad. Bonn, Germany, 1875, Bern, Switzerland, 1876. Became asst., Lvov, 1877, prof., Cracow, Poland, 1879. Author: Inorganic Chemistry, 1891; Organic Chemistry, 1893. Discovered propiolic acid. Died Pisa, Italy, Dec. 13, 1891.

VON BARDELEBEN, Heinrich Adolf, German surgeon; b. Frankfort, Germany, Mar. 1, 1819; mil. surgeon, 1886, 70; engaged in gen. practice medicine, 1872; prof. Charité, Greifswald, Germany; became prof., Berlin, 1868. Author: Lehrbuch der Chirurgie, 1852. Early advocate of Lister's antiseptic methods; devised bandage dressing impregnated with bismuth and starch, for use in burns. Died Berlin, Sept. 24, 1895.

VON BASCH, (Samuel) Siegfried Carl, physician, pathologist; b. Prague, Czechoslovakia, Sept. 9, 1837; studied medicine, Prague; med. degree, Vienna, 1862; m. Adele Frankl; 2 daus. Employed in physiol. lab. of von Brücke, Vienna; asst. med. clinic, 1861-65; went to Mexico to become personal physician to emperor Maximillian, 1865; became pvt. docent for exptl. pathology, Vienna, 1870, prof., 1877. Author: Der Sphygmomanometer, 1887; Allgemeine Physiologie und Pathologie der Kreislaufes, 1892; Über Herzkrankheiten bei Arteriensklerose, 1900; Beiträge zur Entwicklungsgechichte der experiemntellen Pathologie, 1905. Devised 1st practical instrument for measurement of blood pressure, 1880. Died Vienna, Apr. 25, 1905.

VON BAUER, Karl Josef, German physician; b. Erlhammer, Germany, Oct. 1, 1845; s. Joseph Anton and Ernestine (Schlör) Bauer; student medicine, Munich; state exam, 1868; doctorate, 1870; m. Eva von Ziemssen; 6 children; m. 2d, Klara Rambaldi. Became lectr. internal medicine under Joseph von Lindwurm, Munich, 1873, asso. prof., 1876, prof., 1885, dir. Med. Clinic and Hosp., 1892. Author: Geschichte der Aderlasse, 1870; Über Krankenernährung und diätetische Therapie, 1883; Herz-und Gefässkrankheiten, 1895; Über den iodopathischen Herzhypertrophie, 1893; also articles. Applied Karl von Voit's metabolism physiology in clinics; founder dietetic therapy; helped introduce chem. direction of Munich sch.; studies on disturbances in heart function, liver. Died Munich, May 9, 1912.

VON BECKH, Harald J., physician, aeromed. scientist; b. Vienna, Austria, Nov. 17, 1917; s. Johannes and Elisabeth (Flach-Hillé) von B.; M.D., U. Vienna, 1940. Came to U. S., 1957, naturalized, 1963. Staff mem., lectr. Aeromed. Acad., Berlin, Germany, 1941-43; lectr. Nat. Inst. Aviation Medicine, Buenos Aires, Argentina, 1947-56; mem. sci. staff Aeromed. Research Lab., Holloman AFB, N.M., 1957-64, chief scientist, 1964-—. Prof. physiology N.M., State U., 1959-—; mem. Armed Forces-NRC Com. on Bio-Astronautics, 1958-61; mem. bio-astronautics com. Internat. Astronautical Fedn., 1961-—. Fellow Am. Inst. Aeros. and Astronautics (asso.), Brit. Interplanetary Soc.; mem. Aerospace Med. Assn., Internat. Acad. Aerospace Medicine, Internat. Acad. Astronautics, Assn. Mil. Surgeons U. S.; hon. mem. German Rocket Soc., Med. Assn. of Armed Forces of Argentina, Center Astronautical Studies of Portugal. Author: Physiology of Flight, 1955. Contbr. numerous articles on aerospace medicine to sci. jours. Conducted 1st weightlessness expts. with animals (turtles) in aircraft, proved temporary deterioration of neuromuscular coordination, 1953; gave 1st description of deconditioning effect of weightlessness in man, showing that exposure to weightlessness in jet aircraft lower tolerance to acceleration, 1958; inventor multidirectional anti-G device for protection against inflight accelerations by automatic positioning of the subject. Address: P.O. Box 696, Holloman AFB, N.M., 88330.*

VON BEETZ, Friedrich Wilhelm Hubert, German physicist; b. Berlin, Germany, Mar. 27, 1822; s. Friedrich Wilhelm and Dorothea Louisa (Schuchard) Beetz; student physics, chemistry and physiology Berlin U., 1840; m. Else Richter, May 6, 1848; 2 sons, 1 dau. Became asst. to H. G. Magnus, Berlin, 1843; became tchr. physics Berliner Kadettenhause, 1843, prof., 1850; joined faculty U. Berlin, 1849, Arty. and Engring. Sch., 1855; went to Bern, Switzerland, 1858; joined U. Erlangen, Germany, 1858; joined Technische Hochschule, Munich, 1868, dir., 1874-77. First pres. Internat. Electricity Exhbn., Munich, 1882. Mem. Phys. Soc. (a founder), Bavarian Acad. Scis. Author: Leitfaden der Physik, 1846; Grundzüge der Elektrizitätslehre, 1878. Research on electricity; improved various phys. apparatus; improved Wheatstone bridge. Died Munich, Jan. 22, 1886.

VON BEHRING, Emil Adolf, bacteriologist; b. Hansdorf, Prussia, Mar. 15, 1854; s. August G. and Augustine (Zech) Behring; ed. med. U. Pepinière, Berlin, doctorate, 1880; m. Else Spinola, 1896; 6 sons. Engaged in study of infectious diseases and bacteriology during service as mil. physician; became asst. to Robert Koch, Berlin, 1889; named prof. hygiene, Halle, Germany, 1893; left mil. service, 1894; apptd. prof., dir. Hygiene Inst., Marburg/Lahn, Germany, 1895. Recipient 1st Nobel prize in medicine, 1901; (with Pierre Roux) prize Paris Acad. Medicine, prize Inst. of France for work on diphtheria antitoxin. Author: Die praktische Ziele der Blutserumtherapie, 1892; Geschichte Abhandlungen zur ätiologischen Therapie von ansteckenden Krankheiten, 1893; Die Geschichte der Diphtherie, 1893; Diphtherie, 1901; Ätiologie und ätiologischen Therapie des Tetanus, 1904; Einfuhrung in der Lehre vond der Bekämpfung der Infektionskrankheiten, 1912. Founder of immunology as a science; discovered and improved (with Shibasaburo Kitazato) 1st effective immunizing serums for tetanus and diphtheria; developed bobobaccine (protective against Tb in cattle; introduced term antitoxin. Died Marburg/Lahn, Mar. 31, 1917.

VON BÉKÉSY, George, physicist; b. Budapest, Hungary, June 3, 1899; s. Alexander and Paula (Mazaly) Von B.; student U. Berne, 1916-20, M.D. (hon.), 1959; Ph.D., U. Budapest, 1923; M.D. (hon.), Wilhelm U., Munster, Germany, 1955, U. Padua, 1962; D.Sc. (hon.) U. Pa., 1965. Researcher, Hungarian Telephone System Lab., 1923-46, Central Lab. Siemens & Halske, Berlin, 1926-27; faculty U. Budapest, 1932-46, Karolinski Inst., Stockholm, 1946-49; sr. research fellow in psychophysics Harvard, Cambridge, Mass., 1949-66; prof. sensory scis. U. Hawaii, Honolulu, 1966-—. Recipient Denker prize in otology, 1931; Guyot prize for speech and otology Groningen U., 1939; Leibnitz medal Akad. Wiss, Berlin, 1937; Akad. award Acad. Sci., Budapest, 1946; Shambaugh prize in otology, 1950; Howard Crosby Warren medal Soc. Exptl. Psychologists, 1955; gold medal Deafness Research Found., 1961; Nobel prize in medicine, 1961. Fellow Acoustical Soc. Am. gold medal 1961); mem. Nat. Acad. Scis., Am. Acad. Arts and Scis., Am. Otol. Soc. (gold medal 1957, hon.), also several fgn. sci. socs. Author: Experiments in Hearing, 1960; Sensory Inhibition, 1967. Research on physics of sensory perception. Office: Lab. Sensory Scis., U. Hawaii, Honolulu 96822.*

VON BELL, Johann Adam Schall, astronomer, mathematician; b. Cologne, Germany, 1591; became mem. Soc. Jesus, 1611; sent as missionary to China, 1620; worked at Singa-fou, then called by emperor to Peking to correct Chinese calendar; in charge Imperial Bur. Astronomy under Ch'ing (Manchu) Dynasty (1644-69). Author: Hsi-Yang Msin Fa Li Shu (astron. ency.), 1645; also 150 treatises in Chinese on calendar. Died Peking, Aug. 15, 1661.

VON BENNEWITZ, Peter, see Apianus, Petrus.

VON BERGEN, Karl August, German anatomist; b. Frankfort/Oder, Germany, Aug. 11, 1704; s. Johann Goerg von Bergen; ed. univs. Leiden (Netherlands), Strasbourg (France); became asst. prof. U. Frankfort, 1732, prof. anatomy and botany, 1738, prof. therapeutics and pathology, 1744. Genus Bergena named in his honor by Adanson. Author: Elementa physiologie, 1749; Elementa anatomiae experimentalis, 1758. Research on arachnoid membrane. Died Frankfort/Oder, Oct. 7, 1759.

VON BERGMANN, Ernst, physician, surgeon; b. Riga, Latvia, Dec. 16, 1836; at least 1 son, Gustav. Became prof. Dorpat (now Tartu, Estonia) U., 1871, U. Würzburg (Germany), 1878; apptd. prof. surgery, dir. surg. clinic U. Berlin, 1882. Author: Die Behandlung der Schusswunden des Kniegelenkes im Krige, 1878; Die Lehre von den Kopfverletzungen, 1880. Noted for work in aseptic surgery; introduced steam sterilization in surgery, 1886. Died Mar. 25, 1907.

VON BERLEPSCH, Count Hans Hermann Carl Ludwig, German ornithologist; b. Fehrenbach, Germany, July 29, 1850; s. Karl and Johanna (Koch) V. B.; student zoology Leipzig, Halle (both Germany); m. Emma von Bülow, 1881; 3 sons, 5 daus. Expert S.Am. ornithology; developed large pvt. collection exotic birds, especially S.Am. varieties (collection, begun in 1872, eventually contained over 50,000 specimens, including 300 newly discovered and described ones), now in pub. mus., Frankfort/Main, Germany. Died Göttingen, Germany, Feb. 27, 1915.

VON BERTALANFFY, Ludwig, see Bertalanffy, Ludwig von.

VON BESANEZ, Baron Eugen Franz Gorup, chemist; b. Graz, Austria, Jan. 15, 1817; student Graz, Vienna, Padua, Munich. Became asso. prof. Erlangen, Germany, 1849, prof., 1855. Author: Organische chemie, 6th edit., 1881; Anleitung zur qualitativen und quantitativen zoochemischen Analyse . . . , 1850; Lehrbuch der Chemie für den Unterricht auf Universitäten, 3 vols., 1859-60; Anorganische chemie, 7th edit., 1885. Research and publs. on silica in feathers, analyses of mineral waters, physiol. chemistry; obtained thiamine, valine, impure mannose. Died Erlangen, Nov. 24, 1878.

VON BEZOLD, Albert, German physiologist; b. 1836; s. Johann Daniel Christoph von Bezold; became prof. physiology, Jena, Germany, 1861, Würzburg, Germany, 1865. Author: Untersuchung über die Innervation des Herzens, 1863. Research in elec. stimulation of nerves and muscles, blood circulation physiology; known for Bezold-Jarisch Effect; discovered accelerator nerves in heart and their origin in spinal cord, 1863, also demonstrated that when spinal nerves and vagi have been severed, heart could be stimulated to beat by pressure. Died 1868.

VON BEZOLD, (Johann Friedrich) Wilhelm, German meteorologist, physicist; b. Munich, Germany, June 21, 1837; s. Daniel Gustav and Sabine (Albrecht) von B.; student natural scis., Munich, Göttingen, 1856-60; m. Marie Hörmann, 1868; 1 son, 1 dau., Wilhelmine (Mrs. Ernst Hagen). Prof. tech. physics Polytechnikum, Munich, 1868-85; became dir. Bavarian Central Meteorol. Sta., 1878; placed in charge reorg., became dir. Prussian Meteorol. Inst., Berlin; became 1st German prof. meteorology, Berlin, 1885. Recipient numerous awards including Corthenius medal Leopoldinisch-Karolinsche Akademie Deutscher Naturforscher. Mem. Bavarian Acad. Scis., Prussian Acad. Scis., German Meteorol. Soc., German Assn. for Advancement Aviation. Author: Die Farbenlehre im Hinblick auf Kunst und Kunstgewerbe, 1874; Beobachtungen der meteorologischen Station en in Bayern, 5 vols., 1879-85; Zur Theorie des Erdmagnetismus, 1897; Geschichtlicher Abhandlungen aus dem Gebiet der Meteorologie und des Erdmagnetismus, 1906; also numerous articles. Built and expanded meteorol. instns.; contributed to devel. aeros.; creator thermodynamics of atmosphere; research in atmospheric electricity, terrestrial magnetism, cyclones. Died Berlin, Germany, Feb. 17, 1907.

VON BIBRA, Baron Ernst, German natural scientist; b. Schwebheim, Germany, June 9, 1806; s. Ferdinand and Lukretia von Bibra; m. Josephine Pickel, 1836; 3 sons, 1 dau. Adminstr. to family estate; moved to Nuremberg, Germany, 1846; made trip to S.Am., 2 years. Author: (with L. Geist) Über die Krankheiten der Arbeiter in den Zündholzfabriken, 1847; Beiträge zur Naturgeschichte von Chile, 1853; Reise in Südamerika, 1854; Die narkotischen Genussmittel und der Mensch, 1855; Reiseskizzen und Novellen, 1864; Getreidearten und Brot; Schwefeläther. Research in applied chemistry; indsl. hygiene including Phosphornecrosis; medicine including ethyl ether; nutrition chemistry; chem. studies of ancient coins. Died Nuremberg, June 5, 1878.

VON BIELA, Wilhelm, astronomer; b. Rossla, Austria, Mar. 19, 1782; served as Austrian officer; discovered Biella comet (made 4 returns, failed to reappear thereafter), 1826, meteorites which occupied its path after its disappearance provided 1st evidence of close connection between comets and meteors. Died Venice, Feb. 18, 1856.

VON BLUMENTHAL, (Ludwig) Otto, German mathematician; b. Frankfort/Main, Germany, July 20, 1876; s. Ernst and Eugenie (Posen) Blumenthal; student math., Göttingen, Germany, Munich, 1894-98, Paris, 1899-1900; doctorate under Hilbert, 1898; m. Mali Ebstein, 1908; 1 son, 1 dau. Prof., Marburg, Germany, 1904-05, Aachen, Germany, 1905-33; lectr. in Holland, Switzerland, Belgium, Bulgaria until 1935; sought refuge in Holland, 1939; evidently captured in German occupation of Holland; dir. Aussen-Institut, Aachen (Germany) Technische Hochschule. Mem. Deutsche Mathematiker Vereinigung (became chmn. 1924), Aachener Mathematische Gesellschaft, Förderungsverein für den mathematischen naturwissenschaftlichen Unterricht. Author several books including Principes de la théorie des fonctions entières d'ordre infini, 1910; also numerous articles. Editor, Mathematische Annalen, 1906-38. Studies in constrn. and representation of modular functions, whole functions of infinite order, applied math, generalized sphere function; contributed to theory of functions with complex variables; gave approximation used later in communications. Died Theresienstadt, Germany, 1944.

VON BODENSTEIN, Adam, physician, alchemist; b. Karlstadt, Habsburg Empire, (now Yugoslavia), 1528; s. Andreas and Anna (von Mochau) V.; M.D., U. Basel (Switzerland), 1548; m. Esther Wyss, 1549; m. 2d, Maria Jakoba Schenk, Aug. 13, 1565; 15 children. Med. faculty, coll. physicians U. Basel, expelled for heretical and scandalous books, 1564. Author: Onomasticon (commentary on Paracelsus' writings), 1574; Consilium philosophicum, 1576; De veritate alchemiae; Epistola ad Fugeros pro asserenda alchemia; also collection lesser writings, 1581. First to introduce teachings of Paracelsus at U. Basel; described theriac against plague; strong believer in alchemy. Died Basel, 1577.

VON BOGUSLAWSKI, Palm Heinrich Ludwig, German astronomer; b. Magdebourg, Germany, 1789; studied astronomy under Bode; became lt. in army,

1811; visited important observatories of Europe, 1812-15; ret. from mil. service after Waterloo; studied rural economy until 1829, then resumed astron. research; named curator Breslau (now Wroclaw, Poland) obs., 1831, dir., 1843; prof., Breslau, from 1836. Author: L'uranus, 3 vols., 1846-48. Observed comet of 1807; discovered comet named after him, Apr. 20, 1835. Died Breslau, 1851.

VON BOHNENBERGER, Johann Gottlieb, German geodesist, astronomer, physicist; b. Simmozheim, Germany, June 5, 1765; s. Gottlieb Christoph and Johanna Friederike (Schmid) Bohnenberger; student theology, Tübingen, Germany; m. Johanna Christine Luz, 1798; 2 sons, 2 daus. Became vicar, 1789; named asso. prof. Tübingen, 1798, prof. math. and astronomy, 1798; dir. surveying Wurtemberg, Germany, 1818-31. Corr. mem. French Acad. Scis., 1820. Author: Anleitung zur geographischen Ortsbestimmung, 1795; Lehrbuch der Astronomie, 1811; Anfangsgründe der höheren Analysis, 1812; De computandis dimensionibus trigonometicis . . . , 1826. Editor, Zeitschrift für Astronomie und verwandte Wissenschaften, 1816-18. Research and publs. on geographic positioning, influence of earth's rotation on free fall, instrumental error for improvement accuracy of astron. observations; introduced mercury level, more sci. methods of land survey; invented electrometer, a centrifuge, a reversion pendulum; observed simultaneous evaporation and freezing of water in vacuum, 1815. Died Tübingen, Apr. 19, 1831.

VON BONIN, Gerhardt, anatomist; b. Warendorf, Germany, July 17, 1890; s. Eberhard and Elema (Gerhardt) von B.; student Breslau (Germany) U., 1909-11; M.D., Freiburg (Germany) U., 1914; m. Lilian Ruzicka, Aug. 4, 1932; 1 son, Stephen. Came to U. S., 1930, naturalized, 1935. Practice medicine specializing in surgery, Heidelberg, China, 1921-27; anatomist Peking (China) Union Med. Coll., 1927-38; faculty U. Leyden (Netherlands), 1928-30; anatomist U. Ill. Med. Sch., Chgo., 1930-58, anatomist emeritus, 1958—; researcher dept. neurosurgery Mt. Zion Hosp., San Francisco, 1959. Author: Essay on Cerebral Cortex, 1950; (with Percival Bailey) Neocortex of Macac Mulatta, 1949, Isocortex of Man, 1950; (with Percival Bailey, Warren McCulloch) Isocortex of Chimpanzee, 1950; Evolution of Human Brain, 1961; also numerous articles. Research on comparative anatomy cerebral cortex. Home: 47 Upland St., Mill Valley, Cal. 94941.*

VON BORN, Ignaz, Austrian mineralogist, metallurgist; b. Karslburg, Germany, Dec. 26, 1742; s. Ludwig and Maria (von Deutis) von B.; student law, Prague, Czechoslovakia; m. Magdalena Montac, 1765; 2 daus. Joined Dept. Mines, Prague, 1770; studies in mining in Germany, France, Low Countries; became curator Imperial Mineral. Collections, Vienna, 1776; became Aulic councillor Dept. Mines and Mint, Vienna, 1779; founder a pvt. bus., (later became Gesellschaft der Wissenschaften, 1791, then Königlich Böttmische Gesellschaft), 1769. Mem. acads. Halle, St. Petersburg, Stockholm, Lund, Siena, Munich, London, Göttingen, Toulouse, Uppsala. Author: Lithophyllacium bornianum, 2 vols., 1772-75; Index rerum naturalium musei caes. vindobonnae, 1778; Physikalische Arbeiten der einträchtigen Freunde in Wien, 1738-88; Über das Anquicken der Gold- und Silberhaltigen Erze, 1786; Bergbaukunde, 2 vols., 1789; also articles. Improved and introduced to Austria amalgamation process for extraction of silver from its ores (used in Mexico, 1590); one of 1st to use chem. criteria to mineralogy, 1790; made improvements in mining, salt working and bleaching; invented bellows-blowpipe for lab. Died Vienna, July 24, 1791.

VON BÖTTIGER, Rudolph Christian, German physicist, chemist; b. Aschersleben, Germany, Apr. 28, 1806; s. Johann Christoph and Henrietta Eleonora Veronica (Walther) von B.; student theology, Halle, beginning 1824; doctorate, Jena, Germany, 1837; m. Sophie Elisabeth Christiane Harpke, 1841; 5 sons, including Oskar; 3 daus. Became tchr., 1835; joined faculty physics Physikalischer Verein, Franfort/Main, Germany, 1842; founder, dir. Polytechnische Notizblatt, 35 years. Author: Beiträge zur Physik und Chemie, 1838-46. Invented gun cotton, 1846, safety matches using red phosphorous, 1848, collodion (independently of Schönbein), galvanic process of copying etched copper plates, silver plating of glass, and nickel plating of iron; Böttger test for grape sugar named after him; studies in stypnic acid, amalgams, asbestos. Died Frankfort/Main, Apr. 29, 1881.

VON BÖVENTER, Edwin, German economist; b. Göttingen, Germany, Mar. 9, 1931; s. Edwin and Erna (Tonhose) v.B.; student U. Göttingen; M.A., U. Mich., 1954, Ph.D., 1956; Rockefeller fellow, U. Pa., Mass. Inst. Tech., Harvard, 1961-62; m. Berta Heidenhain, Apr. 14, 1956; children—Matthias, Axel. Instr., U. Mich., Ann Arbor, 1955; research, teaching asst. U. Münster (Germany), 1956-61; instr. U. Saarbrücken (Germany), 1962-63; prof. econs., dir. Alfred Weber Inst. Social and Polit. Scis., U. Heidelberg, 1963—. Mem. Regional Sci. Assn. (pres. German speaking sect. 1965—), Soc. Econ. and Social Scis., Am. Econ. Assn., Royal Econ. Soc., Econometric Soc. Author: Theorie des räumlichen Gleichgewichts, 1962; (with Bombach, Henn) Optimales Wachstum und optimale Standortverteilung, 1963; (with W. Mueller) Der Studienbeginn in den Wirtschaftswissenschaften, 1967; also articles in field. Research in econometrics: internat. impact of U. S. recessions, effects of German mark revaluation, analyses of demand for chem. products in Germany, on location theory: gen. equilibrium of space-economics; generalized models of spatial structure. Home: Am Schuetzenhaus, 6901 Schoenau. Office: 104 Bergheimer Strasse, 6900 Heidelberg, West Germany.*

VON BRAMBILLA, Johann Alexander, physician; b. St. Zenone nr. Pavia, Italy, Apr. 15, 1728; s. Giuseppe Brambilla; student medicine, Pavia, 1747-52; m. Therese Hann; 1 son, 2 daus. Became personal physician to later Emperor Joseph II; took charge of leadership of all mil. med. affairs in Austria, 1779; head Josephinum Inst., 1785-95. Author: Trattato chirurgico practico sopra il flemmone, 1777; Instrumentarium chirurgicum Viennense, 1780; Storia della scoperte fisico-mediche-anatomiche-chirurgiche, 1780-82; Verfassung und Statuten der Josephinum medizinisch-chirurgische Militär-Akademie, 1786. Suggested founding of Josephium Med-Surg. Acad., Vienna for improved treatment of soldiers. Died Padua, July 29, 1800.

VON BRAND, Theodor Curt, physiologist; b. Ortenberg, Germany, Sept. 22, 1899; s. Phillip Paul and Diane (von Hirsch) von B.; Ph.D., U. Munich, 1922; M.D., U. Erlangen, 1928; m. Margarethe Brandeis, Dec. 20, 1923; 1 son, Theodor. Came to U. S., 1936, naturalized, 1941. Asst., U. Erlangen, 1923-30, Inst. Tropical Diseases, Hamburg, Germany, 1931-33; fellow U. Copenhagen, 1934-35, Johns Hopkins Sch. Hygiene, 1936-37; prof. biology Barat Coll., Lake Forest, Ill., 1937-40, Cath. U., Washington, 1940-47; head sect. physiology and biochemistry Lab. Parasitic Diseases, USPHS, NIH, Bethesda, Md., 1947—. Fellow A.A.A.S.; mem. Mexican (hon.), Am. (past v.p.), socs. parasitologists, Helminthological Soc. Washington (pres. 1945), Sigma Xi. Author: Anaerobiosis in Invertebrates, 1946; Chemical Physiology of Endoparasitic Animals, 1952; Biochemistry of Parasites, 1966. Research, numerous publs. on life without oxygen in free-living and parasite animals, metabolism of protozoan and helminth parasites, nitrogen cycle in sea, effects of over dosage of vitamin D. Home: 8606 Hempstead Av., Bethesda 20034. Office: Lab. of Parasitic Diseases, NIH, Bethesda, Md. 20014.*

VON BRAUN, Wernher, space scientist; b. Wirsitz, Germany, Mar. 23, 1912; s. Magnus and Emmy (von Quistorp) von B.; B.S., Berlin (Germany) Inst. Tech., 1932; Ph.D., U. Berlin, 1934; numerous hon. degrees including D.Sci., St. Louis U., U. Pitts., Tech. U. (Berlin), Nat. U. Cordoba (Argentina), Emory U., D'Youville Coll; LL.D., Adelphi Coll., Iona Coll.; m. Maria Louise von Quistorp, Mar. 1, 1947; children—Iris Careen, Margrit Cecile, Peter Constantine. Came to U. S., 1945, naturalized, 1955. Rocket devel. engr. German Ordnance Dept., 1932-37; tech. dir. army exptl. sta. Peenemuende, Baltic Sea, 1937-45; project dir U. S. Army Ordnance Corps, Ft. Bliss, Tex., 1945-50; tech. dir. guided missile devel. group Redstone Arsenal, 1950-52, chief guided missile devel. div., 1952-56, dir. devel. operations div. Army Ballistic Missile Agy., 1956-60, dir. George C. Marshall Space Flight Center, NASA, 1960— (all Huntsville, Ala.). Recipient numerous awards including Distinguished Civilian Service award U. S. Dept. Def., 1957, Exceptional Civilian Service decoration U. S. Dept. Army, 1957; Dr. Robert H. Goddard Meml. trophy Clark U., 1958; Herman Oberth award, Am. Rocket Soc., 1961; Diesel medal in gold, German Soc. for Inventions, 1965; William Boelsche medal in gold, 1967; Galabert Internat. Astronautical prize, 1967; Langley medal Smithsonian Instn., 1967. Fellow Am. Inst. Aeros. and Astronautics (Astronautics award 1955), Am. Astronautical Soc. (Space Flight award 1957), Brit. Interplanetary Soc. (hon. fellow, Gold Medal award 1961); hon. mem. Deutsche Gesellschaft für Raketentechnik und Raumfahrt, Hellenic Astronautic Soc. (Greece) (Hon. fellow 1961) Hermann Oberth Soc., German Soc. Aviation and Space Medicine; mem. Nat. Acad. Engring., Norwegian Interplanetary Soc., Internat. Acad. Astronautics, Tau Beta Pi, others. Author: Across the Space Frontier, 1952; The Mars Project, 1953; Conquest of the Moon, 1953; The Exploration of Mars, 1956; Project Satellite, 1958; Start in den Weltraum, 1958; First Men to the Moon, 1960; A Journey Through Space and the Atom 1962; co-author History of Rocketry and Space Travel, 1966; Space Frontier, 1967. Pioneered devel. large liquid fueled rockets; responsible for devel. V-2 long-range missile, antiaircraft guided missile; devel. Redstone, Jupiter, Jupiter-C and Juno II missiles which successfully launched U. S. satellites and space probes including Explorer I (1st U. S. satellite); Redstone booster also employed in Mercury program and placed chimpanzee and 1st two U. S. astronauts in suborbital flights; responsible for Saturn I launch vehicles which successfully placed into orbit 3 Pegasus Meteroid Tech. satellites, and for Saturn V launch vehicle to be used for manned lunar landing of Apollo spacecraft. Home: P.O. Box 6822. Office: George C. Marshall Space Flight Center, NASA, Huntsville, Ala. 35812.*

VON BRILL, Alexander Wilhelm, German mathematician; b. Darmstadt, Germany, Sept. 20, 1842; s. Heinrich Konrad and Julie Henriette Simonetta (Wiener) Brill; student Karlsruhe, Giessen, Germany; passed exams in architecture and engring; doctorate in math., Giessen; postgrad., Berlin; m. Anna Schleiermacher, 1875; 3 sons, including Alexander, Eduard; 1 dau. Became asst. prof. Giessen, 1867; named prof. Polytechnische Schule, Darmstadt, 1869; joined Polytechnikum, Munich, 1875; prof., Tübingen, Germany, 1884-1919. Mem. Accademia dei Lincei Rome (fgn.), Deutschen Mathematiker Vereinigung (chmn.). Author: Über dem Relativitätsprinzip, 1912; Algebraische Kurven, 1925; Vorlesungen über allgemeiner Mechanik, 1928. Research and numerous publs. in theory of algebraic functions; a founder (with Max Noether) of algebraic geometry direction theory of algebraic functions; studies on Riemann-Roch's theorem and theory of residuation; supervised (with F. Klein) manufacture of geometric models for advanced students; defined conditions for application of principle of correspondence. Died Tübingen, June 18, 1935.

VON BRONK, Otto, German physicist, inventor; b. Danzig (now in Poland), Feb. 29, 1872; s. Franz and Modesta (Kalkstein) B.; m. Helene Zimmermeister, 1904. Concerned himself with electrophysics; (with F. Clausen) established laboratory for research in area of television; entered telephone business, 1911; took out several German and foreign patents; contributed to development of German radio industry. Died Berlin, Germany, Oct. 5, 1951.

VON BRÜCKE, Ernst Theodor, physiologist; b. Vienna, Austria, Oct. 8, 1880; s. Theodor and Emilie (Wittgenstein) Brücke; student Leipzig, Germany; m. Pauline Roelfs, 1905; 2 sons, 2 daus. Came to U. S., 1938. Asst. to E. Hering, Physiologisches Institut, Leipzig, joined faculty, 1907; became prof., head Physiol. Institut, Innsbruck, Austria, 1916; staff A. Forbes' Inst., Harvard. Research and numerous publs. in function of vegetative organs, comparative physiology, physiol. optics, physiology of nerves and muscles; developed process of simultaneous floating stimulation of reflex exciting and reflex inhibiting nerves with various frequencies. Died Boston, June 12, 1941.

VON BRÜCKE, Ernst Wilhelm, German physiologist; b. Berlin, Germany, June 6, 1819; s. Johann Gottfried and Christine (Müller) Von B.; student Berlin, Heidelberg (both Germany) univs.; studied under Johannes Müller; m. Dorette Brünslow, 1848; 2 sons. Became asso. prof. physiology, Königsberg (now Kaliningrad, USSR), 1848; prof. physiology U. Vienna (Austria), 1849-91, became rector, 1879; M. P. Decorated Order pour le mérite. Mem. Austrian Acad. Scis. Author: Über das Leuchten der menschalichen Auges, 1847; Grundzüge der Physiologie und Systematik der Sprachlaute, 1856; Die Physiologie der Farben für die Zwecke des Kunstgewerbes, 1866; Bruchstücke aus der Theorie der bildenden Künste, 1877; Vorlesungen über Physiologie, 2 vols., 1885-87; Schönheit und Fehler der Menschlichen Gestalt, 1891. Attempted (with others) to develope new physiology based on chemistry and physics; developed method for examining fundus of eye (led to devel. of ophthalmoscope), 1845; studies in optics, acoustics, rigor mortis, blood coagulation, protoplasm, mechanism of speech, skeletal muscle, nervous system, color change in chameleon, urine, protoplasm, effectiveness of pepsin, bile dyes, relation between aesthetics and vision. Died Vienna, Austria, Jan. 7, 1892.

VON BRUNCK, Heinrich, German chemist; b. Winterborn, Germany, Mar. 26, 1847; s. Friedrich Karl and Elisabeth (Ritter) B.; student chemistry, Zurich, Ghent, Tübingen; doctorate, 1867; hon. dr. engring. Tchnische Hochschule, Karlsruhe, Germany, 1905; m. Barbara Wilhelmine Emilie Fitting, 1871; 1 son, 2 daus. With chem. factory E. de Haën, Hannover, Germany, 2 years; became chemist, joined Badische Anilinund Sodafabrik, Ludwigshafen, Germany, 1869; named dir. Hochfeldbranch works, 1873; became dir. alizarin dept. of main plant, 1875, chief tech. dir., 1884, chmn. bd., 1907; organizer I. G. Farbenindustrie AG (merger Badische Anilin-und Sodafabrik, Bayer, and Höchst). Author: Über einige Abkömmlinge der Indigo-Fabrikation, 1901; also articles. Responsible for devel. of synthetic indigo, sulfuric acid contact process, indsl. prodn. anthraquinone, alizarin, electrolytic chlorine, process for fixation nitrogen; made various social and san. improvements. Died Ludwigshafen, Germany, Dec. 4, 1911.

VON BRUNN, Heinrich, German archaeologist; b. Wörlitz, Germany, Jan. 23, 1822; s. Heinrich and Albertine (Lindstedt) Brunn; studied under Friedrich Gottlieb Welker, also Friedrich Wilhelm Rietschel, Bonn, Germany, 1839-43; m. Ida Bürkner; 2 sons. Joined Archaeol. Inst., Rome, 1843; colleague of Theodor Mommsen, also Rietschel; became lectr., Bonn, 1854; became sec. Archaeol. Inst., Rome, 1856; named prof. archaeology, curator coin and vase collection, Munich, Germany, 1862; ministerial commr. for secondary sch. affairs; chmn. Commn. for Bavaria's Prehistory. Mem. Bavarian Acad. Scis. Author: Geschichte der griechsichen Künstler, 2 vols., 1853; I Relievi delle urne Etrusche, 1870; Griechische Götterideale, 1893; Griechische Kunstgeschichte, 2 vols., 1893-97. Revived classical archaeology; discoveries include Marsyas of Myron, 1858, votive offering of King Attalos, 1862, Eirene of Kephisodotus, 1868. Died Schliersee, Germany, July 23, 1894.

VON BRUNN, Walter Albert Ferdinand, surgeon, med. historian; b. Göttingen, Germany, Sept. 2, 1876; s. Albert and Fanny (Stelzner) von B.; student Göttingen, Germany, Rostock, Germany, until 1899; M.D.; m. Else Range, 1910; 1 son, 1 dau. Began practice surgery, Rostock, 1905; turned to med. history after losing right arm from infection in wartime operation;

worked under Karl Sudhoff, Leipzig, Germany; joined faculty, Rostock, 1919, became asso. prof., 1924, prof. med. history, 1934; named dir. Karl-Sudhoff-Institut für Geschichte der Medizin und der Naturwissenschaften, 1938. Mem. Leopoldina (v.p. 1947-51). Author: Die Stellung des Guy de Chauliac in der Chirurgie des Mittelalters, 1919; Kurze Geschichte der Chirurgie, 1928; Geschichte der Chirurgie, 1948; also numerous articles. Important in history of surgery. Died Leipzig, Dec. 21, 1952.

VON BRUNNER, Johann Conrad, anatomist, physician; b. Diessenhofen, Switzerland, Jan. 16, 1653. Anatomist, Diessenhofen; prof. medicine U. Heidelberg (Germany), from 1687. Recognized glandular nature of pancreas. Died Mannheim, Germany, Oct. 2, 1727.

VON BRUNS, Viktor, German surgeon; b. Helmstedt, Brunswick, Germany, Aug. 9, 1812; a son, Paul. Prof., U. Tübingen, Germany. Author: Lehrbuch der allgemeinen Anatomie, 1841; Handbuch der praktischen Chirurgie, 2 vols., 1854-60; Die Durchschneidung der Gesichtsnerven beim Gesichtsschmerz, 1859; Die Laryngoskopie, 1865. A founder modern laryngology; research in gen. anatomy, surgery, laryngoscopy; 1st bloodless enucleation of laryngeal polyp, 1862; opposed Lister's method of using carbolic acid spray during operations. Died Tübingen, Germany, Mar. 18, 1883.

VON BUBNOFF, Serge, geologist; b. St. Petersburg, Russia, July 15, 1888; ed. U. Freiburg (Germany); studied under Hans Cloos, Breslau (now Wroclaw, Poland), 1921; Prof., E. Europe Inst., Breslau; became prof. geology U. Greifswald (Germany), 1929; prof., dir. Geotectonic Inst., Humboldt U., Berlin. Author: Die Kohlenlagerstätten Russlands, 1923; Geologie von Europa, 3 vols., 1926-36; Grundproblem der Geologie, 1931; Geschichte und Bau des deutschen Bodens, 1936; Einführung in die Erdgeschichte, 1941. Research on structural and regional geology, complex fractioning of region around Basel, Switzerland, folded jura to Alpine thrusts, jura plateau, crystalline structure of uplifted massifs of Black Forest, Vosges, Oldenwald; structure of Silesian coal basins, geology of extra-Alpine Europe, crystal behavior, Precambrian formation in So. Sweden, flat-lying formation of Russian basement of Fenno-Scandia, and Urals. Died Nov. 16, 1957.

VON BUCH, Baron Christian Leopold, German geologist, paleontologist; b. Stolpe, Germany, Apr. 26, 1774; s. Adolf Friedrich and Charlotte Julie (von Arnim-Suchow) von B.; ed. Mining Acad., Freiburg, also Halle, Germany, Göttingen, Germany, 1793-96. Travelled in Austria, 1797-98, Italy (witnessed eruption of Mt. Vesuvius) 1798, 1805, Switzerland, 1799, Auvergne, 1802, Scandinavia 1806-08, Canary Islands, 1815. Corr. mem. French Acad. Scis., 1815; mem. Berlin Acad. Scis. Author: Geognotische Beobachtungen auf Reisen durch Deutschland und Italien, 2 vols., 1802-09; Physikalische Beschreibung der Kanarischen Inseln, 1825. Gesammelte Schriften, 4 vols., 1867-85; Research in tectonics, regional geology, history of earth, paleontology, volcanic processes, invertebrate paleontology; prepared 1st geol. map of Germany, 1832; originated class of Cystoids, 1845; 1st to notice gradual rising of Sweden; developed doctrine of slow rising of continents. Died Berlin, Mar. 4, 1853.

VON BUHL, Ludwig, German pathologist; b. Munich, Germany, Jan. 4, 1816; s. Friedrich and Sabina (Hönigst) von B.; m. Margarete Crescentia Burger, 1850; 2 sons, 1 dau. Became lectr. phys. diagnostics and path. anatomy U. Munich (Germany), 1847; became 1st prof. path. anatomy, Munich, 1859, named head Path. Inst., 1875. Asso. mem. Bavarian Acad. Scis. Author: Lungenentzündung, Tuberkulose und Schwindsucht, 1874; also numerous articles. Research on causes of disease; discovered origin of phthisic and tuberculor changes, significance of ground water flow in typhus; contributed to establishment of pathology on sci. basis; Buhl pneumonia named after him. Died Ebenhausen, Germany, July 30, 1880.

VON BULOW, Kurd, German geologist; b. Allenstein, Germany, July 20, 1899; s. Curt and Margarete (Sander) von B.; ed. U. Berlin, U. Breslau; Ph.D., U. Greifswald, 1920; m. Liselott Wendt, Aug. 6, 1927; children—Marianne (Mrs. Gerhart Ruickoldt), Werner, Kurd, Gerda. Ofcl. geologist, Berlin, 1920-35; prof. U. Rostock, 1935-67, dir. geol.-paleontol. inst.; dir. Mecklenburg State Geol. Inst., 1935-36, 45-52. Recipient German Nat. prize for Art and Sci., 1954. Mem. Deutsche Academie der Naturforscher (Leopoldina), Gesellschaft Deutscher Naturforscher und Arzte, Deutsche Gesellschaft fur Raketentechnik und Raumfahrt, many other geol. socs. Author numerous books, articles on geology, soils sci. including Allgemeine Küstendynamik, 1954; Tektonische Analyse der Morinde, 1957; Geología para Todos, 1958; Geologie voor jedereen, 1961; Die Entstehung der Kontinente und Meere, 1963; Geologie für Jedermann, 1964. Research on diluvial and alluvial geology, new system of soil types and pedology, lunar geology, history of geology; systematic bog (peat) geology. Home: 98 Karl Marxstrasse, Rostock, East Germany.*

VON BUNGE, Alexander, Russian botanist; b. Kiev, Russia, Sept. 24, 1803; M.D., Dorpat, Latvia, 1825; children—Gustav, Alexander. Traveled in Siberia and Eastern region of Altai Mountains, 1825; mem. mission to Peking, China, Acad. St. Petersburg; made 2d Asiatic journey, 1830; became dir. Bot. Gardens, Dorpat, 1836. Collected and published catalogues of plants in China and Altai Mountains. Died Dorpat, July 18, 1890.

VON CALW, Ulrich Rülein, German physician, metallurgist; b. circa 1465; student Leipzig, Germany; became physician, Freiberg, 1497. Author: Bergbüchlein, 1500. Studies on mining, ores, assaying; believed all metals are composed of sulfur and mercury. Died 1523.

VON COTTA, Bernhard, German geologist; b. Eisenach, Germany, Oct. 24, 1808; prof. geology Bergakademie, Freiburg. Author: Deutschlands Boden, 2 vols., 1854; Die Erzlagerstätten, 1855; Geologie der Gegenwart, 1866. Emphasized geol. modes of origin; clearly expressed modern theory that molten materials poured from volcanoes or formed within crust of earth may be crystallized into rock at any geol. epoch and therefore do not indicate age. Died Freiburg, Sept. 14, 1879.

VON CRANTZ, Heinrich Johann Nepomuk, obstetrician; b. Roodt, Luxembourg, Nov. 25, 1722; s. Pierre and Anne (Simon) Crantz; M.D., Vienna, 1750; one of 1st students of Van Swieten; studied obstetrics under André Levret, also Nicholas Puzos, Paris, London; m. Anna Susanne Petrasch; m. 2d, Magda Lena de Tremon; 2 sons, 1 dau. Became lectr. obstetrics St. Mary Hosp., Vienna, 1754; prof. physiology and materia medica U. Vienna, 1756-74. Author: Einleitung in eine Warhe und gegründete Hebammenkunst, 1756; Commentarius de rupto in partus doloribus a foetu utero, 1756; Commentatio de instrumentorum in arte obstetricia historia utilitate et recta ac praepostera applicatione, 1757; De systemate irritabilitatis, 1761; Materia medica et chirurgica, 3 vols., 1762; De aquis medicatis principatus Transsylvapiae, 1773; Die Gesundbrunnen der Österreichischen Monarchie, 1777. Promoted better edn. for midwives, use of Levret's forceps; research and publs. on tearing of uterus during birth, chemistry, botany, mineral springs. Died Judenburg, Austria, Jan. 18, 1797.

VON CRELL, Lorenz Florenz Friedrich, German chemist, mineralogist; b. Helmstädt, Jan. 21, 1744; Ph.D., M.D. Prof. chemistry and mineralogy Carolinum at Brunswick, 1771-73; prof. philosophy, Helmstädt, 1773-1810, prof. medicine, 1774-1810; ordinary prof. chemistry at Göttingen, 1810-16; Brunswick mining councillor, 1780; edited Crell's Annalen, Journal, Archiv and Entdeckungen, 1778-1803. Translated several chem. books into German including Black's Lectures and Crawford on Heat, also articles in jours. Most important contbn. was in editing of journals; did much chem. research. Died Göttingen, June 7, 1816.

VON CUBA, Johannes, physician; flourished Frankfort/Main, Germany, circa 1484-1503; m. 2d., June, 1500. Became citizen Frankfort/Main, 1503. Personal physician Count Friedrich I, Adolph II of Nassau, archbishop of Mainz, Count Eberhard II von Eppstein, other noblemen; city physician Frankfort/Main from 1484. Author: (supposed and now presumed translator from Latin and editor) Ortus Sanitatis uff Teutsch ein Gart der Gesundheit (first comprehensive pharmacopeia in German), 1485. Illustrations of plants he selected for book indicated trend to sci. bot. drawing. Died winter of 1503-04.

VON DARDEL, Guy F., Swedish physicist; b. Stockholm, Sweden, Aug. 26, 1919; s. Fredrik E. A. and Maj (Vising) Von D.; grad. Royal Inst. Tech., Stockholm, 1944; doctors degree, Stockholm, 1954; m. Matilda Jungstedt, Dec. 17, 1949; children—Louise, Marie. Physicist, SAAB, Linkoping, 1944-46, A.B. Atomenergi, Stockholm, 1946-54, CERN, Geneva, Switzerland, 1954-65; prof. physics U. Lund (Sweden), 1965—. Mem. Am. Phys. Soc. Research, publs. on neutron physics, including exptl. and theoretical founds. of pulsed neutron source method; high energy physics; cross section measurement at 2-20 GeV, prescision measurement of lifetime of neutral pion; dir. neutrino expts., photoprodn. Home: 8 Katedervagen, Lund, Sweden.*

VON DECHEN, Ernst Heinrich Karl, German geologist; b. Berlin, Germany, Mar. 25, 1800; s. Theodore and Elisabeth (Martinet) D.; ed. Berlin U., Berlin Mining Acad., 1818-21; m. Luise Gerhard, 1828; 1 son, 3 daus. In state service; travelled to Eng., Scotland, 1826-27; prof. mining engring., Berlin, from 1834; chief dist. dir. mines, Bonn, Germany, 1841-64. Corr. mem. French Acad. Scis., 1887. Author: Geognostische Umrisse der Rheinlande zwischen Basel und Mainz, 2 vols.; Geognostische Karte von Deutschland, England, Frankreich und den Nachbarländern, 1839; Sammlung der Hohenmessungen in der Rheinprovinz; Geognostische Beschreibung des Siebenjahriges; also several mineral. publs., geol. atlas (excellent) with 35 sheets, 2 vols. explanations. Wrote on mining, geognosy, mineralogy, metallurgy. Died Bonn, Feb. 15, 1889.

VON DELIUS, Heinrich Friedrich, German physician, chemist; b. Wernigerod, Germany, July 8, 1720; s. Jakob and Sophia Elisabeth (Schütz) Delius; student medicine, Berlin; M.D., Halle, 1743; m. Barbara Margarethe Besserer, 1752; 3 sons; 2 daus., including Margarethe Johanna Louisa (Mrs. August Schott). Practice medicine, Wernigerode, Germany; ct. physician, asst. state physician, Bayreuth, Germany; became prof. medicine, Erlangen, Germany, 1749. Mem. numerous socs., including Leopoldina (became pres. 1788). Research and numerous publs. in chemistry; introduced chemistry as academic subject to Erlangen. Died Erlangen, Oct. 22, 1791.

VON DELLINGHAUSEN, Baron Nicolas, Russian physicist; b. Kattenack, Russia, 1827; studied at Dorpat, Estonia, 1827, Leipzig, Germany, Heidelberg, Germany; served in Russian Army. Author: Versuch einer spekulativer Physik, 1851; Grundzüge der vibrationen theorie der natur, 1872; Beiträge mechanischen Wärmetheorie, 1874; Grundzüge der kinetischen naturlehre, 1898; wrote on vibration in nature. Originated a kinetic theory of matter, mech. theory of heat; research on gravitation. Died Riga, Latvia, 1896.

VON DIETZ, Johann Simon Jeremias, German surgeon, ophthalmologist; b. Nuremberg, Germany, Oct. 23, 1803; s. Johann Zacharias and Dorothea Cathrine (Raw) D.; student, Erlangen, Göttingen, Germany; M.D., Würzburg, Germany, 1825; m. Adelheid Lehmus, 1832; 1 son, Theodor. Studied hosps. and operations methods in Vienna, Paris, London, Edinburgh, Glasgow, Berlin; began practice medicine, specializing in surgery and ophthalmology, Nuremberg, 1828; named head surgeon Heiligen Geist-Spital, 1932; prof. surgery U. Erlangen, 1933-35; returned to Nuremberg, 1835, built its 1st modern hosp., dir. surg. dept., 32 years; also head Maximilians-Augenheilanstalt. Chmn. Arztlicher Reformcongress, Munich, 1848; pres. (with Ohm) 23d conv. of German scientists and physicians, Nuremberg, 1845; founder Nürnberger ärztlicher Lokalverein. Responsible for change from non-acad. to acad. surgery in Nuremberg. Died Nuremberg, July 8, 1877.

VON DITTEL, Leopold, surgeon, urologist; b. Fulnek (Mähren), May 29, 1815; s. Abraham and Marianne (Oppenheim) D.; studied medicine, Vienna, until 1840; m. Marie Girtler, 1860; 2 sons, including Leopold Gottfried, 1 dau. Asst., orthopedic inst., Vienna; physician, Trnecsin-Teplitz; gave up practice, 1848; asst. in forensic medicine to J. Dlauhy; surgeon J. F. von Dumreicher's clinic, became asst., 1853; named lectr. and chief surgeon Allgemeines Krankenhaus, Vienna, 1856, asso. prof., 1865; declined position as Dumreicher's successor, 1880; left teaching, 1891. Pres. Austrian Soc. Physicians. Contbr. to Austrian med. jours. Inventor new surg. instruments and methods of operation; 1st to perform intestine resection; founder (with Sir Henry Thompson and Felix Guyon) of urology; 1st to introduce rubber catheter into a suprapubic opening; contbd. to endoscopic diagnostics, performed surg. removal of bladder tumors; re-introduced lithotomy. Died Vienna, July 28, 1898.

VON DRASCHE, Anton, internist; b. Lobendau, Bohemia, July 2, 1826; s. Josef Drasche; ed. Prague, Leipzig, Vienna; M.D., Vienna, 1853; m. Marie Serkiss. Asst. physician Allgemeines Krankenhaus, Vienna became head physician, 1877; apptd. head provisional cholera dept., 1855; dir. cholera hosp., 3d dist. Vienna, 1866; became mem. faculty U. Vienna, 1858, asso. prof. epidemiology, 1874; ofcl. Austrian del. to internal. health conbress, Vienna, 1874. Author: Über den Harnstoff-Beschlag der Haut und Schleimhäute im Cholera-Typohoide, 1856; Die epidemische Cholera, 1960; Über den Einfluss der Hochquellenleitung auf die Salubrität der Bevölkerung in Wien (prizewinning monograph), 1883; Influenza, 1890; Über den gegenwärtigen Stand der bacillären Cholerafrage und über diesbezügliche Selbstinfektionsversuche, 1894, also articles. Co-editor Osterreichische Zeitschrift für praktishce Heilkunde, 1860-64; Bibliographie der gesamten medizinischen Wissenschaft, 11 vols. Contbd. to therapy and prevention of epidemics; basic epidemiological work in cholera, typhus, plague, influenza; doubted importance of cholera vibrios and diphtheria serum; worked in specialized areas of internal medicine, including heart, lungs, stomach. Died Vöslau, nr. Vienna, Aug. 23, 1904.

VON DREYSE, (Johann) Nikolaus, German inventor; b. Sömmerda, nr. Erfurt, Germany, Nov. 20, 1787; s. Johann Christian and Susanne (Fleischmann) D.; learned locksmith trade from father; m. Dorthea Luise Ramann, 1821; 1 son, Franz, also 2 daus. Employed by cousin Beck in Altenburg, then by locksmith in Dresden; joined Pauly weapons factory, Paris, 1809; returned to Sömmerda, 1814, enlarged (with Collenbusch) family workshop; formed percussion cap factory Dreyse & Collenbusch, 1824-43; received contract from Friedrich Wilhelm IV for 60,000 rifles, 1840, set up rifle and munitions factory financed by him, 1841. Holder 1st patent on muzzle loading rifle, 1828; inventor breech loading needle rifle, 1836, also (with pharmacists Baudius and Kahleys) an ignition composition; inventor a cartridge shell. Died Sömmerda, Dec. 9, 1867.

VON DRIGALSKI, (Karl Rudolf Arnold Artur) Wilhelm, German hygienist; b. Dresden, Germany, June 21, 1871; s. Arthur and Minna (Kuhn) von D.; student Kaiser-Wilhelm-Akademie for mil. physicians, Berlin, until 1895; m. Elisabet Dill, 1905; 1 son, Wolfgang, 1 dau. Health officer in army, 1895-1907, assigned to Koch's inst. for infectious diseases, 1901-04; named lectr. Tech. U. Hannover, 1906; apptd. hon. prof., city physician, Halle, Germany, 1908; anti-cholera worker, Serbia, 1913; epidemics officer, World War I.; became med. adviser on pub. health, Berlin, 1925; mem. Riech Health Bd., from 1927 until discharged by Nazis; ship's doctor several years; head pub. health dept. Hessian Ministry Interior,

1945-48; reestablished German Assn. for Pub. Health Oare and Politics, adminstr., from 1949. Author: Die Bekämpfung der Säuglingssterblichkeit in Halle a.S., 1908-09; Städtische Gesundheitspflege in Halle a.S., 1910-11; Schulgesundhietspflege, 1912; Der Aufsteig des Sanitätskorps, 1939; Im Wirkungsfelde Robert Kochs, 1948; Männer gegen Mikroben, 1951. One of Koch's most important students in bacteriology and epidemiology; concentrated on typhus and dysentery; contbd. proof of typhus bacilli from chronic carriers in pure culture (Drigalski culture plate). Died Wiesbaden, May 12, 1950.

VON DRYGALSKI, Erich Dagobert, geographer, geophysicist; b. Königsberg, Prussia, Feb. 9, 1865; s. Fridolin and Lydia (Siegfried) von D.; ed. univs. Königsberg, Bonn, Leipzig, Berlin; pupil of F. von Richthofen; doctorate, 1887; m. Clara Wallach, 1907; 4 daus. Leader preliminary expdn. of Gesellschaft für Erdkunde, Berlin, to W. Greenland, 1891, main expdn., 1892-93; leader 1st German S. Polar expdn., on the Gauss, 1901-03; became mem. faculty Berlin, 1898, prof., 1899; prof. Munich, 1906-35. Chmn. Geographische Gesellschaft, Munich, 29 years. Author: Expedition der Gesellschaft für Erdkunde zu Berlin, 2 vols., 18 -93; Deutsche Südpolar Expedition, 1901-03; Zum Kontinent des eisigen Südens, 1904; 20 text vols., 3 atlas vols., 1905-31; (with F. Machatschek) Gletscherkunde, 1942. Pioneer in polar research; student of polit. geography; influential on devel. of German geography. Died Munich, Jan. 10, 1949.

VON DUSCH, (Georg) Theodor, German physician; b. Karlsruhe, Germany, Sept. 17, 1824; s. Alexander Anton and Maria Anna (v. Weiler) von D.; studied law in Freiburg (Germany), then medicine in Heidelberg (Germany); doctorate from Heidelberg, 1846-47; studied in Paris, France until 1848; m. Auguste Gmelin, 1849; 2 children. Asst. to M. Chellius in surg. clinic, Heidelberg; mil. surgeon for a short period; practiced medicine in Mannheim, Germany, 1848-54; joined faculty of Heidelberg, 1854, became asso. prof. pathology, dir. med. polyclinic, 1856; surgeon, dir. typhus hosp., 1870-71; opened children's hosp. Luisenanstalt, 1885. Author: Beiträge zur Pathogenese der Icterus, 1854; Lehrbuch der Herzkrankheiten, 1868. First to perform tracheotomy for diphtheria in Heidelberg; after A. Hoche, performed first craniotomy. Died Heidelberg, Jan. 13, 1890.

VON DYCK, Walther Franz Anton, German mathematician; b. Munich, Germany, June 12, 1856; s. Hermann and Marie (Royko) D.; studied math. at Munich Polytechnikum under Alexander Brill and Felix Klein; received doctorate, 1879; hon. dr. from Hannover (Germany), Tübingen (Germany); m. Auguste Müller, 1886; 2 daus. including Gertrud (Mrs. Otto Hertwig). Became asst. to Felix Klein, 1879, and went with him to Leipzig, Germany; joined faculty of Leipzig, 1882; named prof. math. Munich Polytechnikum, 1884, dir. for 12 years, also rector; vice chmn. Deutsches Mus. Decorated Order of Maximilian for Sci. and art. Active in Mathematische Annalen, Encyklopädie der mathematischen Wissenschaften, Verbund der deutschen Hochschulen, also bavarian and Reichs sch. commns. Published an edit. of Opera omnia Johannes Keplers. Research on abstract group theory; graphic geometry, especially geometrical form of Kronecker's analytic, algebraic characteristic theory; discussed curve system defined by differential equations. Died Munich, Nov. 5, 1934.

VON EBERHARD, Otto (Eduard Hermann), German physicist; b. Frankfort/Main, Germany, Feb. 3, 1877; s. Wilhelm and Anna Margarete (Fuld) Eberhard; student tech. Mil. Acad., Vienna, 1893-96; trained as artilleryman, Austria, later Prussia; m. Charlotte Reiners, 1906; hon. dr. engring., Tech. U., Berlin, 1928; 1 son. Asst. to Carl Cranz, Mil. Tech. Acad., 1903-06; gave up mil. career because of accident in ballistic lab.; joined firm of Friedrich Krupp AG, Essen, Germany; became prof. honoris causa, 1919. Author: (with C. Cranz) Lehrbuch der Ballistik, 1925-36. Research in ballistics, theory of sighting equipment, 1908; measured air resistance of flying shell under various weather conditions (thus deriving step-wise calculation of trajectory used in constrn. of distance guns in World War I). Died Düsseldorf, Germany, Oct. 17, 1940.

VON ECK, Heinrich Adolf, German geologist; b. Gleiwitz, Germany, Jan. 13, 1837; s. Ludwig and Lina (Bönisch) Eck; student under F. Römer, Breslau (now Wroclaw, Poland), 1858-61; doctorate, 1865; m. Friederike Marie Kannabich, 1871. Staff, Geologische Landesanstalt, Berlin; active Prussian Geol. Survey Thuringia, Silesia; joined faculty Bergakademie, Berlin, 1866; became prof. mineralogy and geology Polytechnikum, Stuttgart, Germany, 1871; ret. because of blindness, 1900. Hon. mem. Deutsche Geologische Gesellschaft. Author: Über die Formationen des bunten Sandsteins und des Muschelkalks und ihre Versteinerungen in Oberschlesien, 1865; Zur Kenntnis des süddentschen Muschelkalks, 1880; Zur Gliederung des Buntsandsteins im Odenwalde, 1884; Geognostische Übersichtskarte des Schwarzwaldes, 1886-87; Übersicht über die wahrgenommenen Erdbeben in Württemberg und Hohenzollern von 1867-87, 1887, 1887-88, 1888; Geognostische Beschreibung der Gegend von Baden-Baden, Rotenfels, Gernsbach und Herrenalb, 1892; Verzeichnis der mineralogischen, geognostischen, urgeschiehtlichen und balneographischen Literatur von Baden, Württemberg und Hohenzollern von den Jahren 1513-1890, 2 vols., 1890/1901. Research on German Triassic deposits, S. Germany earthquake phenomena; cartographic-geol. studies of Black Forest. Died Stuttgart, Mar. 11, 1925.

VON ECONOMO, Baron Constantin Alexander (San Serff), physician, neurologist; b. Braila, Wallachia, Aug. 21, 1876; s. Johannes and Helene (Murati) Von E.; student Tech. U. Vienna (Austria); M.D., U. Vienna, 1901; postgrad. Paris, Strasbourg (both France), Berlin, Munich (both Germany) univs.; m. Karoline von Schönburg-Hartenstein, June 10, 1919. Asst. Vienna Psychiat. Clinic, from 1906, staff, until 1931, lectr., from 1913; field pilot Austrian Air Force, World War I; mil. physician Vienna Clinic. Author: Beiträge zur normalen Anatomie der Ganglienzelle, 1905; (with J. P. Karplus) Zur Physiologie und Anatomie des Mittelhirns, 1908; Die dissoziierte Empfindungslähmung bei Ponstumoren und über die zentralen Bahnen des sensiblen Trigeminus, 1913; Die hereditären Verhaltnisse bei Paronoia querulans, 1914; Die Encephalitis lethargica, 1917; Der striäre Symptomen komplex, 1923; Die Pathologie des Schlafes, 1925; (with G. Koskinas) Die Cytoarchitektonik der Grosshirnrinde des Menschen, 1925; Über den Schlat, 1925; Über den feineren Bau des Uncus, 1926; Die progressive Cerebration, ein Naturprinzip, 1928; Wie sollen wir Elite-Gehirne verarbeiten . . . , 1929; Die Encephalitis lethargica, ihre Nachkrankheiten und ihre Behandlung, 1929; Aufgabe der modernen Hirnforschung, 1931. Pioneer aviation devel.; balloonist; one of 1st in Austria to hold pilot's license; 1st to describe encephalitis lethargica (Von Economo's disease), 1917; research on sleep function, its localization problems; work on cytoarchitectonics of cerebrum cortex (beginning neurol. anthropology), elite brain, neuropathology, psychiatric genetics. Died Vienna, Oct. 21, 1931.

VON EHRENFELS, (Maria) Christian (Julius Leopold Karl), Austrian philosopher, psychologist; b. Rodaun, Austria, June 20, 1859; s. Leopold and Clothilde (von Coith) V. E.; student philosophy U. Vienna (Austria), 1882-85, under A. Meinong, Graz, Austria, 1885-88; Ph.D., 1885; m. Emma André, 1894; children—Rolf, Imma (Mrs. von Bodmershof). Lectr. philosophy, Vienna, from 1888; asso. prof. philosophy U. Prague (Czechoslovakia), 1896-99, prof., from 1899. Author: Über Gestaltqualitäten: in Vierteljahrschrift für Philosophie, 1890; Das Primzahlengesetz, 1922. Created concept of form-quality (gestaltqualitat); one of originators Gestalt psychology; wrote on philosophy math., law prime numbers. Died Lichtenau, Austria, Sept. 8, 1932.

VON EICHWALD, Karl Eduard, Russian naturalist; b. Mitau, Russia, July 16, 1795; s. Johann Christian and Charlotte Elisabeth (Louis) von E.; entered U. Berlin, 1814; M.D., Vilna, Russia, 1819; m. Sophie von Fincke, 1825; children include Eduard Georg. Practice medicine, Kurland, Russia; lectr. zoology, geology and paleontology, Dorpat, Estonia 1821; named prof. zoology and obstetrics, Kazan, Russia, 1823; became prof. U. Vilna, 1827; prof.-sec. St. Petersburg Med.-Surg. Acad., 1838-51; prof. paleontology St. Petersburg Mining Inst., also prof. mineralogy and geology St. Petersburg Engring. Sch., 1839-55. Author: Zoologia specialis quam expoditis animalibus tum vivis, tum fossilibus potissimum Rossiae in universum et Poloniae in specie, 3 vols., 1829-31; Naturhistorische Skizze von Lithauen, Volhynien und Podolien, in geognostisch-mineralogischer, botanischer und zoologischer Hinsicht, 1830; Reise auf dem Caspischen Meere und in der Caucasus, 2 vols., 1834-38; Sur le système silurien de l'Estonie, 1840; Die Urwelt Russlands, 4 vols., 1840-47; Fauna caspico-caucasica, 1841; Lethaea Rossica ou Paléontologie de la Russie, 1853-68. Explored many parts of Russia; research in archeology, geology, mineralogy, botany, ethnography; founder Russian paleontology; tried to find phylogenetic relationship in his zool. system. Died St. Petersburg, Nov. 22, 1876.

VON EISELSBERG, Baron Anton, Austrian surgeon; b. Schloss Steinhaus, Austria, July 31, 1860; s. Guido and Marie Pirquet (von Cesenatico) von E.; student medicine, Würzburg, Zurich, Paris; M.D., Vienna, 1884; student surgery under Billroth; hon. dr. univs. Athens, Budapest, Debrezin, Edinburgh, Geneva, Leyden, Paris, Vienna; m. Agnes von Pirquet, 1895; 1 son, 7 daus. Became asst. to Billroth, Vienna, 1887, lectr., 1890; named head surg. clinic, Utrecht, Netherlands, 1893; became prof., Königsberg (now Kaliningrad, USSR), 1896; named dir. I. surg. Univ. Clinic, Vienna, 1901-31. Mem. Vienna Acad. Scis., Deutsche Gesellschaft für Chirurgie (became chmn. 1918), Internat. Soc. Physicians (became pres. 1921). Author: (biography) Lebensweg eines Chirurgen, 1940; also articles. Contributed to devel. gastrointestinal surgery; studies in physiology and pathology of thyroid gland, jaw surgery, surgery of central nervous system; eliminated tetany cramps by means of aux. thyroid gland; 1st in Europe to remove spinal column tumor. Died nr. St. Valentin, Oct. 25, 1939.

VON EMPERGER, Friedrich Ignaz, Austrian engr.; b. Prague-Beraun, Bohemia, Jan. 11, 1862; s. Friedrich and Johanna Anna (Zdrahal) von E.; ed. German Tech. U., Prague, Vienna; tech. doctorate, 1903; hon. dr. Tech. univs. Prague, Dresden; m. Gabriele Seiche von Nordenheim, 1889; 2 daus. Asst. bridge bldg., Prague (Czechoslovakia) Tech. U.; staff firm of Ruston, Prague; made study trip to Paris, 1889-90; joined Jackson Archtl. Iron Works, N.Y., 1890; opened office for reinforced concrete, N.Y.C.; became rep. Austrian State R.R., N.Y.C. 1894; returned to Austria; lectr. Tech. U. Vienna, 1898-1902; rep. Austrian engrs. to World's Fair, Paris, 1900, Glasgow, 1901; mem. Austrian Patent Office; a founder Austrian Reinforced Concrete Com., pres., 1926-38. Recipient Mörsch medal, Goethe medal for arts and scis. Mem. Am. Inst. C.E. (hon.), Royal Inst. Dutch Engrs., Polish Acad. Scis. Author: Neue Bugenbrücken aus umschürtem Gusseisen, System Emperger, 1913; Vorspannung in Eisenbetonbau, 1940; Stahlbeton mit vorgespannten Zulagen aus höherwertigen Stahl, 1941; also numerous articles. Editor: Handbuch für den Eisenbetonbau, 4 vols., 1907-09. Founder, Beton und Eisen, 1902. Pioneered reinforced concrete engring.; designed bridges, skyscrapers, indsl. plants. Died Vienna, Austria, Feb. 7, 1942.

VON ERDMANN, Johann Friedrich, German physician; b. Wittenberg, Germany, July 18, 1778; prof. medicine, Wittenberg, also Kasan, Dorpat (now Tartu, Estonia); concluded from a quantitative expt. that hydrogen and oxygen are evolved in proportions in which they exist in water; adopted Fourinoy's theory; described an improvement of Cruickshank's trough battery. Died Wiesbaden, Germany, Jan. 28, 1846.

VON ERICHSEN, Lothar Carl Manuel, German chemist; b. Niedobschütz, Germany, May 15, 1915; s. Herbert B. E. and Erna (Wiosa) von E.; Dipl.-Ing., Tech. U., Breslau, Germany, 1939, D.Sc., 1942; m. Edelgard Kerber, Mar. 13, 1940; children—Frithjof Olaf, Sven Torsten. Chemist, Niederschl. Bergbau AG, 1942-44; mng. dir. Schwefelsäurefabrik Waldenburg, 1944-46; cons. engr., 1946-48; asst. prof. U. Bonn (Germany), 1948-55, prof. nuclear chemistry, 1957—; guest prof. Tech. U., Valparaíso, Chile, 1955-56. Sci. adviser IAEA, Iraq, 1962-63, to U. Kabul, 1965. Recipient award Heidelberg Acad. Sci., 1961. Mem. Gesellschaft Deutscher Chemiker, Deutsches Atomforum. Author: Friedliche Nutzung der Kernenergie, 1962; also articles. Research on correlations between solubility and molecular structure, isotope exchange reactions, parameters influencing exploitation of nuclear energy, neutron shielding materials for gen. use, nuclear fuel reprocessing and radioactive waste disposal. Home: 13 Am Hahnsberg, D 548, Oberwinter, W. Germany. Office: 12 Wegelerstrasse, D 53 Bonn, West Germany.*

VON ESMARCH, Johannes Friedrich August, German surgeon; b. Tönning, Germany, Jan. 9, 1823; s. Theophil Christian and Friederike (Homann) E.; ed. U. Göttingen; M.D., U. Kiel (Germany), 1848; m. Anna Stromeyer, 1854; 2 sons, including Erwin, 1 dau.; m. 2d Henriette of Schleswig-Holstein, 1872; 3 sons. Became asst. to Prof. von Longenbeck, 1848; mil. surgeon in campaign against Danes, 1848-50, 64; asst. to prof. Stromeyer, 1849; became prof. surgery U. Kiel (Germany), 1857; dir. surg. faculty in hosps., Berlin, during Austrian War, 1866; surgeon gen. during Franco-Prussian War; founder Samariter Verein, 1882. Author: Der erste Verband auf dem Schlachtfelde, 1870; Handbuch der Kriegschirugischen Technik; Die erste Hilfe bei plötzlichen Unglücksfallen, 1882. Invented rubber tourniquet used to expel blood from limb, 1869, rubber bandage for producing bloodless field in limb operations, 1873; introduced 1st-aid package for battlefields, 1869, antiseptic treatment in Germany. Died Kiel, Feb. 23, 1908.

VON ETTINGSHAUSEN, Baron (Johannes) Andreas (Jakob), German mathematician, physicist; b. Heidelberg, Germany, Nov. 25, 1796; s. Constantin and Anna Maria Franziska (Walther) V.E.; student U. Vienna (Austria), Bombardier Sch., Vienna; m. Antonie Skarnitzl, 1824; 1 son, Constantin. Asst. in math., physics U. Vienna, 1817-19, prof. higher math., from 1821, prof. physics, from 1835, dir. Phys. Inst., from 1853; prof. physics, Innsbruck, Austria, from 1819; dir. math. studies Engring. Acad., Vienna, from 1848; prof. higher engring. sci. Vienna Polytechnikum, from 1852. Mem. Göttingen (Germany), Vienna (co-founder, 1st gen. sec.) acads. scis. Author: Die combinatorische Analysis als Vorbereitungslehre zum Studium der höheren Mathematik, 1826; Vorlesungen über die höhere Mathematik, 1827; Anfangsgründe der Physik, 1860. Editor: (with A. Baumgartner) Zeitschrift für Physik und Mathematik, 10 vols., 1826-32. Pioneer math. physics; works on math. analysis, algebra, differential geometry, mechanics, ray optics, wave optics, electromagnetism; constructed magneto electric engine (considerable improvement on Pixii's engine), 1837. Died Vienna, May 25, 1878.

VON ETTINGSHAUSEN, Constantin, Austrian botanist, paleontologist; b. Vienna, Austria, 1826; studied medicinal plants, ferns, Austrian flora; made detailed study of floral fossils of Austria. Author: (with Pokorny) Physiotypa plantarum Austriacarum. Died Graz, Austria, 1897.

VON EULER, Ulf Svante, Swedish physiologist; b. Stockholm, Sweden, Feb. 7, 1905; s. Hans and Astrid (Cleve) Von E.; M.D., Karolinska Inst., 1930; M.D. h.c. U. Dijon, 1963, U. Ghent, 1963, U. Tübingen, 1964, U. Rio de Janeiro, 1953; m. Jane Sodenstierna, Apr. 12, 1930; children—Leo, Christopher, Ursula (Mrs. L. Sjöberg), Marie (Mrs. Anthony John); m. 2d, Dagmar Cronstedt, Aug. 20, 1958. Asst. prof. Karolinska Inst, Stockholm, 1930-

39, prof. physiology, 1939-—. Mem. Nobel Com. for Medicine, 1953-60, chmn., 1958-60, gen. sec., 1961-65, pres. Nobel Found., 1966-—; vice chmn. Med. Research Council Def. Medicine Sect., 1964. Decorated comdr. Order North Star, Sweden; Cruzeiro do Sul, Brazil; Gairdner award, 1961; Jahre award, 1965. Mem. Acads. sci. Sweden, Denmark, Leopoldina of East Germany. Author: Noradrenaline, 1956. Publs. on discovery of an active polypeptide, Substance P (with Gaddum); discovery of prostaglandin in human semen; identification of noradrenalin as the neurotransmitter of sympathetic nervous system. Home: 14 Karlaplan, 11522 Stockholm. Office: Dept. Physiology, Karolinska Inst., Stockholm, Sweden.*

VON EULER-CHELPIN, Hans, chemist; b. Augsburg Bavaria, Feb. 15, 1873; s. Rigas and Furtner (Von Mosham) Von Euler-C.; student Acad. of Painting, Munich, also univs. Berlin, Strassburg and Gottingen (Germany), Institut Pasteur (Paris); Dr. phil., univs. Zurich, Athens, M.D. (hon.), univs. Kiel, Bern, Torina; D.Sc., Rutgers U.; m. 2d, Beth af Ugglas, 1913; 9 children. Faculty U. Stockholm, 1898-—, became prof. chemistry, dir. inst. biochemistry, dir. Vitamin-Institut. Recipient (with Sir Arthur Harden), Nobel prize in chemistry, 1929, Gadolin medal, 1954. Fellow acads. sci. Bangalore, Berlin, Rome, Paris, Vienna, Copenhagen, Moscow, Halle, Gottingen; mem. Royal Swedish Acad. Sci., Royal Swedish Acad. Engring. Scis., Royal Inst. London, Finish Acad. Scis, Acad. New Delhi; fgn. mem. Max Planck Ges. Research on sugar fermentation and fermentative enzymes. Died Nov. 6, 1964.

VON EYTH, (Eduard Friedrich) Max(imilian), German engr.; b. Kirchheim und Teck, Germany, May 6, 1836; s. Eduard and Julie (Capoll) von E.; mech. engr. Polytechnikum, 1852-56; hon. dr. engring., Stuttgart, 1905. Mem. staff mech. factory of G. Kuhn, Stuttgart, Germany; joined steam plow factory of John Fowler, Leeds, Eng., 1861, resigned 1882; also chief engr. Prince Halim Pascha, Egypt; founder Deutsche Landwirtschaftsgesellschaft, 1885, adminstrv. dir., 11 years; returned to Ulm, Germany, 1896. Recipient Grashof Meml. medal, 1905; Max-Eyth Meml. medal in gold donated in his honor by Deutsche Landwirtschaftsgesellschaft. Author: (letters) Im Strom unserer Zeit, 1905; Hinter Pflug und Schraubstock, 1899; also novels. Numerous patents in field of locomobile, steam plow, steam engine; promoted cooperation of agriculture and technology. Died Ulm, Aug. 25, 1906.

VON FABER DU FAUR, Adolf Friedrich, engr., metallurgist; b. Wasseralfingen, Germany, Mar. 27, 1826; s. Wilhelm F. and Auguste (Gottlieb) F.; student engring., Freiberg (Germany) Mining Acad., 1846; 1 son, 4 daus. Emigrated to U. S., 1850. Capt., C.E., Civil War; later foundry engr., Newark, N.Y.; in later years in office with Franz Fohr, N.Y. Invented tilting furnace, in which crucible remained in furnace when pouring smelting still in use today (named after him); used to refine silver from lead, later to smelt gold-silver deposits of cyanide process. Died Newark, N.Y., Aug. 18, 1918.

VON FEHLING, Hermann Christian, German chemist; b. Lübeck, Germany, June 9, 1812; s. Hermann and Margarethe Elisabeth (Heitmann) Fehling; doctorate U. Heidelberg, 1837; student Giessen, Paris; studied under Liebig, Giessen, Germany; m. Sophie von Cless, 1844; 1 son, Hermann; 2 daus., including Clara (Mrs. Blohm). Pharmacist, Lübeck & Bremen; prof. chemistry Polytechnic Sch., Stuttgart, Germany, 1839-82, became prof. emeritus, 1882. Mem. jury for world exhbns. Mem. Deutsche Chemische Gesellschaft (v.p.). Author: Textbook of Organic Chemistry. Contbg. author: Pharmacopoea Germanica; Hanwörterbuch (Liebig, Wöhler, Poggendorf); Textbook of Chemistry (Graham-Otto). Editor: Glossary of Chemical Terms. Translator: Précis de chimie industrielle (Anseline Payen), 1849. Research in qualitative analysis, products of Württemberg salt industry, synthesis of aldehydes, derivatives of benzoic and succinic acids; isolated benzonitrile; introduced Fehling solution to detect presence of glucose, 1848; discovered succinosuccinic ester. Died Stuttgart, Germany, July 1, 1885.

VON FELLENBERG, (Ludwig) Rudolf, chemist; b. Bern, Switzerland, Mar. 17, 1809; s. Emanuel Rudolf and Henriette (Gruner) von F.; student tech., chemistry, Geneva, Bern, Paris; Ph.D., Giessen, Germany, 1841; m. Louise Reisse, 1836; 1 son, Edmund; m. 2d, Susanne Rivier, 1842; 3 sons, 2 daus. studied under H. F. Gaultier de Claubry, Paris. Returned to Bern, 1835; became prof. chemistry and mineralogy Acad., Lausanne, 1841-46; research in pvt. lab., Bern. Pioneered use of chemistry in archeology; publs. on analysis of Swiss mineral waters, various metals, ancient bronzes and glasses, nephrites and jades from lake-dwellings and Orient; determined prehistoric trade routes; Died Cannes, France, Feb. 13, 1878.

VON FICKER, Heinrich, physicist, meteorologist; b. Munich, Germany, Nov. 22, 1881; s. Julius and Maria (Tschafeller) V. F.; ed. Innsbruck, Vienna (both Austria), from 1901; Ph.D. in Meteorology, 1906; Ph.D. (hon.), Agrl. U. Vienna. Asst. to Trabert, Innsbruck, 1907-09, mem. faculty, from 1909; asso. prof. geophysics U. Graz (Austria), 1911-19, prof., from 1919; prof. meteorology U. Berlin (Germany), dir. Prussian Meteorol. Inst., from 1923; prof. geophysics U. Vienna, dir. Central Inst. Meteorology and Geodynamics, Vienna, 1937-53, ret., 1953. Pres. Internat. Climatology Commn., 1928-45. Mem. Austrian (pres. 1946-51, v.p. 1951-57), Russian (sec. 1932-37), Bavarian, USSR acads. scis., Leopoldina. Author: Wetter und Wetterentwicklung, 1932; (with B. de Rudder) Föln und Fohnwirkungen, 1943. Contbr. articles to sci. publs. Considered weather phys. phenomenon of atmosphere whose energy transformations he tried to explain; 1st to recognize importance of weather fronts; recognized that low pressure areas determined by higher pressure changes and that weather phenomena in gen. come from troposphere, stratosphere (prepared way for aerological research). Died Vienna, Apr. 29, 1957.

VON FISCHER, (Ludwig) Eduard, Swiss botanist; b. Bern, Switzerland, June 16, 1861; s. Ludwig and Mathilde (Berri) V. F; student mineralogy, geology, botany (crytogams), Bern, Strasbourg, France; Ph.D., 1883; m. Sophie Luise Johanna Gruner, 1899; 3 sons, 1 dau. Worked under Schwendener, Eichler, Ascherson, Berlin, Germany, 1884-85; botany faculty U. Bern, from 1885, asso. prof., 1893-97, prof, from 1897. Recipient Haller medal, Schläfli prize Swiss Natural Sci. Soc. Author: Die Uredineen der Schwiez (in Beiträge zur Kryptogamenflora der Schweiz 2, H.2, 1904; (with E. Gäumann) Biologie der pflanzenbewohnenden parasitischen Pilze, 1929. Research in determination of whole course of devel. of object, as a fungus, to show relationships, other factors; studies compatative morphology, biology; wrote several comprehensive standard works. Died Nov. 18, 1939.

VON FRANQUÉ, Otto Friedrich Wilhelm Paul, German gynecologist; b. Würzburg, Germany, Sept. 11, 1867; s. Otto and Marguerite (Pascault) Franqué; student Würzburg, Munich, Berlin; M.D., Würzburg, 1889; m. Erna Prym, 1902; 3 sons, including Otto; 3 daus. Asst. to E. von Rindlfleish, Path. Inst., Würzburg, then to M. Hofmeier, Women's Clinic; joined faculty U. Würzburg, 1894, became asso. prof. gynecology and obstetrics, 1901; became prof. German U. Prague, 1903, Giessen, 1907; succeeded H. Fritsch, Bonn, Germany, 1912, rector, 1922-23; ret. 1935; after retirement lived at Schloss Kalkum, Germany. Mem. German Soc. Gynecology (became pres. 1926). Research and numerous publs. on gynecology and obstetrics, including deformities, birth disturbances, use of radiation therapy, cancer of uterus, Tb of female sex organs. Died Schloss Kalkum, Apr. 11, 1937.

VON FRERICH, Friedrich Theodor, German physician, pathologist; b. Aurich, Germany, Mar. 24 1819; s. Jörg and Almuth (Rohden) von F.; m. Clara Offelsmeyer, 1868. Began practice medicine, Aurich, 1841; joined faculty U. Göttingen (Germany), 1846, became asso. prof., 1848; became dir. Internal Clinic, Kiel, Germany, 1850, Breslau (now Wroclaw, Poland), 1852, Berlin, 1859. Lecturing adviser in Prussian Ministry of Pub. Edn.; mem. sci. deputation for medicine. Author: Untersuchungen über die Galle, 1845; Die Bright'sche Nierenkrankheit und deren Behandlung, 1851; Klinik der Leberkrankheiten, 1858-61; Über den Diabetes mellitus, 1884. Research on metabolism of kidney and liver diseases, diabetes mellitus; 1st sci. description of multiple sclerosis, 1849; 1st description of progressive lenticular degeneration, 1861; described uremic syndrome, 1851; introduced sci. methods of chemistry and physics to med. practice; discovered leucine and tyrosine in urine of patients with acute yellow atrophy of liver, 1854. Died Berlin, Mar. 14, 1885.

VON FREY, Max(imilian) (Ruppert Franz), physiologist; b. Salzburg, Austria, Nov. 16, 1852; s. Carl and Anna (Gugg) von F.; student medicine, Vienna, Leipzig, Frieburg, Munich; m. Leonie von Parseval, 1888; 3 children. Became asst. to C. Ludwig, Leipzig, Germany, 1880, later lectr., asso. prof.; named prof. U. Zurich, 1898; became prof., Würzburg, Germany, 1899. Research and publs. on physiology of muscles, sense organs, circulation; devel. methods for measurement of threshold of skin sensitivity, including limits and characteristics of cold, heat, pain, pressure. Died Würzburg, Germany, Jan. 25, 1932.

VON FRIESEN, Sten, Swedish physicist; b. Uppsala, Sweden, Mar. 18, 1907; s. Otto and Vendla (Ohlsson) von F.; F.M., U. Uppsala, 1930, F.L., 1933, F.D., 1936. Faculty, U. Uppsala, 1935-37, 45-46; staff Nobel Inst. Physics, Stockholm, Sweden, 1937-39, Research Inst. Nat. Def., 1940-44; faculty U. Lund (Sweden), 1946-—, prof. physics, 1948-—. Mem. Swedish Atomic Energy Commn., 1949-59; mem. organizing comm. Lund Inst. Tech., 1960-—; mem. Physics Nobel Prize Com., 1965-—. Mem. Royal Physiographic Soc. Lund, Royal Swedish Acad. Sci., Royal Soc. Sci. Uppsala, Am. Phys. Soc. Research and publs. on precision measurement of atomic constants using electron waves; devel. new methods for determination of mass and charge of elementary particles and nuclei in cosmic radiation; liquid scintillators for study of nuclear radiation; designed Stockholm cyklotron. Home: 15 Per Henrik Lings väg, Lund, Sweden.*

VON FRISCH, Anton, Austrian surgeon, urologist, bacteriologist; b. Vienna, Austria, Feb. 16, 1849; s. Anton and Franziska (Heylmann) von F.; ed. U. Vienna; M.D., 1871; student surgery, Billroth's Clinic, 1874; m. Maria Exner, 1874; children—Hans, Otto, Ernst, Karl. Became demonstrator for J. Hyrtl, Anat. Inst., 1868; became asst. Billroth's Clinic, 1874; named prof. anatomy Acad. Graphic Arts, Vienna, 1874; also lectr. Austrian Mus.; dir. surg. dept. Crown Prince Rudolf Children's Hosp.; joined univ. faculty, 1882; later asso. prof., head surg. sect. Vienna Gen. Polyclinic, succeeded Ultzmann as head dept. urology, 1889; staff Pasteur Inst., Paris, 1886-87. Author: Studien über Tuberkulose, 1883; Handbuch der Urologie, 1903-05. Editor: (with O. Zuckerandl) Handbuch der Urologie, 1903. Research and numerous publs. on surg. treatment of prostate hypertrophy, bladder tumors and stones, bacteriology, including description of Klebsiella rhinoscleromatis; application of histological, bacteriological, anat., cytoscopic research to diagnosis; developed new disinfection method. Died Vienna, May 24, 1917.

VON FRISCH, Karl, zoologist; b. Vienna, Nov. 20, 1886; s. Anton and Marie (Exner) von F.; Dr.phil., U. Munich, Vienna, 1910; Dr.h.c., U. Bern, 1949, T. H. Zürich, 1955, U. Graz, 1957, Harvard, 1963, U. Tübingen, 1964; m. Margarete Mohr, July 20, 1917; children—Johanna (Mrs. Theo Schreiner), Maria, Helen (Mrs. E. Pflueger), Otto. Prof., dir. Zool. Inst., Rostock, 1921-23, Breslau, 1923-25, Munich, 1925-46, Graz, Austria, 1946-50, Munich, 1950-58. Recipient pour le mérite, Friedenskl., 1952; Magellanic prize Am. Philos. Soc., 1956; Kalinga prize UNESCO, 1959; Balzan price, 1963. Mem. acads. sci. Vienna, Munich, Washington, Stockholm, Copenhagen, Uppsala, Helsinki, Royal Soc. London, Royal Entomol. Soc. London. Author: The Dancing Bees, 1954; Man and the Living World, 1963; The Dance Language and Orientation of Bees, 1967; A Biologist Remembers, 1967; also numerous publs. Research on color sense in bees and fishes, sense of smell and taste in bees, hearing in fishes, lang. of bees, perception of polarized light in bees. Home: 10 Über der Klause, Munich 90, Germany.*

VON FRITSCH, Karl George Wilhlem, German geologist, paleontologist; b. Weimar, Germany, Nov. 11, 1838; s. Georg August and Hanci (von Rosenbach) von F.; student Eisenach Forestry Acad.; Ph.D., U. Göttingen, 1862; m. Elisabeth Kenngott, 1867; 3 sons; 4 daus., including Elisabeth (Mrs. Adolf Cluss). Traveled to Madiera and Canary Islands; mem. 1866 expdn. to E. Mediterranean; joined faculty U. Polytechnikum, Zürich, Switzerland, 1863; became lectr. mineralogy and geology Senckenberg Natural Sci. Soc., Frankfort/Main, 1867; traveled through Morocco, Atlas Mountains, 1872; named asso. prof. geology U. Halle, 1873, prof., 1876. Mem. Leopoldina (became pres. 1895), natural sci. assns. of Saxony and Thuringia. Author: (with G. Hartung, W. Reiss) Tenerife, geologisch und topographisch dargestellt, 1867; (with Reiss) Geologische Beschreibung der Insel Tenerife, 1868; Reisebilder von den Kanarischen Inseln, 1867; Das Gotthardgebiet, 1874; Allgemeine Geologie, 1888. Research on volcanic islands, especially Tenerife, geology of Gotthard area of Switzerland (basis for tunnel later bored through mountain), geol. formations of Saxony from Precambrium to Diluvium; described tephrite and basanite of Tenerife; proved modern volcanic rock is the same as lava from Tertiary period. Died Goddula, Germany, Jan. 9, 1906.

VON FRORIEP, August Friedrich, German anatomist; b. Weimar, Germany, Sept. 10, 1849; s. Robert and Wilhelmine (Ammermüller) F.; M.D., Leipzig, Germany, 1874; m. Caroline Elise Lenoir, 1875 (d 1880); m. 2d, Marie von Herman (auf Wain), 1890; 1 adopted son. Became asst. to Braune, Leipzig, 1874; tchr. anatomy Art Sch., Leipzig, for short period; became prosector, Tübingen, 1878; became asso. prof., Tübingen, 1884, prof., 1895, also dir. Anat. Inst.; ret. 1917. Corr. mem. Bavarian Acad. Scis. Author: Anatomie für Künstler, 1880; Die Lagebeziehung zwischen Grosshirn und Schädeldach, 1897; Der Schädel Friedrich Schillers und des Dichters Begräbnisstätte, 1913. Research in morphology, histology, embryology, skulls of various famous personalities; described a type of myositis (Froriep's induration, vestigial dorsal root ganglion of hypoglossal nerve; theorized infant's skull developed from 2 parts; believed devel. individual skull types connected to devel. of brain. Died Tübingen, Germany, Oct. 11, 1917.

VON FUCHS, Johann Nepomuk, German chemist, mineralogist; b. Mattenzell, Bavaria, May 15, 1774; studied at Vienna, Heidelberg, Freiburg, Berlin and Paris. Prof. of chemistry and mineralogy, Landshut, 1805; prof. Munich, 1823; conservator of Museum of Mineralogy, Munich, 1854. Ennobled by king of Bavaria. Mem., Munich Acad. Sci. Author: Gesammelte Schriften, 1856. Discovered water glass, 1823; applied it in stereochromy; fuchsite, a variety of muscovite, named in his honor; research on potassium silicate, cement, zeolites; introduced terms: amorphous and solidified liquids; wrote on dyes, sugar manufacture, crystals. Died Munich, Bavaria, Mar. 5, 1856.

VON FÜRTH, Otto, Austrian biochemist; b. Strakonitz, Bohemia, Nov. 18, 1867; s. Joseph and Wilhelmine (Forchheimer) von F.; student medicine and natural scis., Vienna, Prague, Heidelberg, Berlin;

M.D., Vienna, 1894; m. Margarete von Grünebaum, 1900; 1 son, 1 dau. Became venia legendi med. chemistry, Strasbourg, France, 1899; returned to Vienna, 1905; became asso. prof., dir. chemistry dept. Physiol. Inst., 1906, prof., 1917; became prof. med. chemistry U. Vienna, 1929; ret., 1938. Author: Probleme der physiologischen und pathologischen Chemie, 2 vols., 1912-13; Lehrbuch der physiologischen und pathologischen Chemie, 2 vols., 1925-28; also articles. Research in iodothyrine, choline, human bile, chemistry of muscle tissue, fermentative melanine formation, blood protein coagulation and its relation to rigor mortis; one of 1st to study hormones; isolated hormone of adrenal gland and named it supravenin. Died Vienna, June 7, 1938.

VON GIERKE, Edgar Otto Konrad, German pathologist; b. Breslau (now Wroclaw, Poland), Feb. 9, 1877; s. Otto Friedrich and Lili (Loening) von G.; ed. U. Heidelberg (Germany), U. Freiburg (Germany); m. Julie Braun, 1912; 3 sons, 1 dau. Joined faculty U. Heidelberg, 1905; joined staff Cancer Inst., London, 1907; later dept. head Charité Hosp., Berlin; became prosector Municipal Hosp., Karlsruhe, Germany, 1908; faculty bacteriology Tech. U. Karlsruhe, became asso. prof., 1911; field doctor, then mil. pathologist; World War I.; ret., 1938. Author: Taschenbuch der pathologischen Anatomie, 1911; Grundriss der Sektionstechik, 1911; also articles. Revised Technik der histologischen Untersuchung pathologischanatomischer Präparate, (C. von Kahlden), 1904. Research in thyroid gland structure, bone tumors, metabolic diseases; described abnormal storage of glycogen in liver in childhood (von Gierke's disease), 1931. Died Karlsruhe, Oct. 21, 1945.

VON GIERKE, Henning Edgar, biophysicist; b. Karlsruhe, Germany, May 22, 1917; s. Edgar and Julie (Braun) Von G.; student Tech. U., Munich, Germany, 1939-40; Diplom Ingineur, Tech. U., Karlsruhe, 1943, Doctor Engring., 1944; m. Hanlo Weil, Oct. 22, 1950; children—Karin, Susanne. Research asst. lectr. Tech. U. Karlsruhe, Inst. for Communication Engring., 1944-47; with Aerospace Med. Research Labs., USAF, Wright Patterson AFB, O., 1947—, sr. scientist, chief biodynamics and bionics div., 1956—; asso. prof. dept. preventive medicine Ohio State U., Columbus, 1963—. Recipient Distinguished Civilian Service award U. S. Dept. Def., 1963. Fellow Aerospace Med. Assn. (v.p. 1966), Acoustical Soc. Am.; mem. Internat. Acad. Aerospace Medicine, Biophys. Soc. Contbg. author: Handbook of Noise Control, 1957; Shock and Vibration Handbook, 1961; also numerous articles. Research on effects mech. energy on biol. systems, human tolerance for noise and vibration, effects of noise on hearing and vestibular organ. Home: 1325 Meadow Lane, Yellow Springs, O. 45387. Office: Biodynamics and Bionics div. Aerospace Med. Research Labs., Wright Patterson AFB, O. 45433.*

VON GLEICHEN, Friedrich Wilhelm (Russworm), German naturalist; b. Bayreuth, Germany, 1717. Lt.-col.; ret., 1756. Author: Microscopic Discoveries in Plants, Insects, 1777. Built microscope and used it to study semen, putrified vegetation; 1st to stain bacteria with indigo and carmine, 1778; 1st description of pollen tube (Asclepias). Died 1783.

VON GOEBEL, Karl Immanuel Eberhard, German botanist; b. Billigheim, Germany, Mar. 8, 1855; s. Michael and Luise Auguste Bruckmann) Goebel; student theology, philosophy, later botany Tübingen; Ph.D., Strasbourg, France, 1877; also numerous hon. doctorates; m. Charlotte Papellier, 1894; 1 son, 1 dau. Became asst. to J. von Sachs, Würzburg, Germany, 1878, joined faculty, 1880; asst. to A. Schenk, Leipzig; later asso. prof., Strasbourg; went to Rostock, Germany, 1882, became prof., 1883; became prof., Marburg, Germany, 1887; prof., Munich, 1891-31, rector, 1916-17; mem. expdns. to India, Ceylon, Java, 1885-86, Venezuela, Brit. Guiana, 1890-91, Australia, Tasmania, New Zealand, 1898-99; Planned Bot. Inst. and Garden, Nymphenburg, Germany; gen. dir. Bavarian sci. collections. Recipient Gold Citizen's medal, Munich, 1930. Fgn. mem. Royal Soc., 1926, Linnean Soc. (Gold medal 1931); mem. numerous acads., Bavarian Acad. Scis. (pres.). Author: Morphologische und biologische Studien, 2 vols., 1887-90; Pflanzenbiologische Schilderungen, 2 vols., 1889-91; Vergleichende Entwicklungs Geschichte der Pflanzenorgane, 1892; Grundzüge der Systematik und speziellen Pflanzenmorphologie, 1882; Organographie der Pflanzen, 2 vols., 1898-1901; Einleitung in die experimentelle Morphologie der Pflanzen, 1908; Die Entfaltungsbewegungen der Pflanzen, 1920. Research in evolution, botany, biology, physiology, morphology, nutritional and physiol. factors in flower formation, symmetry relationships; proved homology in devel. of sporangia. Died Munich, Oct. 9, 1932.

VON GRAEFE, Albrecht Friedrich Wilhelm Ernst, German ophthalmologist; b. Finkenheerd, Prussia, May 22, 1828; s. Karl Ferdinand and Auguste (Van Alten) V.G.; M.D., U. Berlin (Germany), 1847; also ed. Prague, Vienna, Paris, London, Glasgow, Dublin; m. Anna Knuth, 1862. Ophthalmologist, Berlin, from 1845; lectr., from 1850; asso. prof. ophthalmic surgery U. Berlin, 1857-65, prof., from 1865. Founder Berlin Ophthalmology Clinic, 1850, Archiv für Ophthalmologie, 1857. Mem. Deutsche Ophthalmologische Gesellschaft (founder 1863), Leopoldina. Author: Über die wirkung der Augenmuskeln, 1850; Handbuch der gesammten Augenheilkunde, 7 vols., 1874-80. Considered founder ophthalmology; introduced use Helmholtz' ophthalmoscope in diagnosis, 1851, (into Germany) iridectomy for treatment glaucoma, 1857; invented new operation for treatment strabismus, 1857; 1st to diagnose embolism of retinal artery as a cause of sudden blindness, 1859; differentiated between functional, organic loss vision; 1st to describe eye signs seen in exophthalmic goiter as lag of upper eyelid in following downward movement of pupil (Von Graefe's sign), 1864; described sympathetic ophthalmia, 1866, progressive ophthalmoplegia, 1868, keratoconus, 1868; used linear extraction in operation for cataract. Died Berlin, July 20, 1870.

VON GRAEFE, Karl Ferdinand, surgeon, ophthalmologist; b. Warsaw, Poland, Mar. 8, 1787; s. Karl and Christiane (Zschernig) von G.; student medicine, Dresden, Halle, Germany; M.D., Leipzig, Germany, 1804; m. Auguste van Alten, 1814; 3 sons, including Karl, Albrecht; 2 daus., Ottilie (Mrs. Hermann van Thile), Wanda (Mrs. Sigismudn von Dallwitz). Became personal physician to Duke Alexius of Anhalt-Bernburg, Ballenstedt, Germany, 1808; prof., surgery, dir. Ophthalmo-Surg. Clinic, U. Berlin, 1810-40; became 3d gen. staff physician Prussian Army, 1822; prof. surgery Mil. Acad. Medicine and Surgery; asst. dir. Friedrich Wilhelm Inst. Medicine and Surgery, also Mil. Acad.; a founder Berlin Surg. Clinic, also Polyclinic. Mem. Leopoldina. Author: Normen für Ablösung grösser Gliedmassen nach Erfahrungsgrundsätzen entworfen, 1812; Rhinoplastik oder die Kunst, den Verlust der Nase organisch zu ersetzen . . . , 1818. Father modern plastic surgery; developed 1st satisfactory method for rhinoplasty, 1818, operation for congenital cleft palate, 1820; invented numerous surg. and ophthal. instruments; first to apply ligature to arteria anonyma; 1st partial resection of lower jawbone. Died Hanover, Germany, July 4, 1840.

VON GRONK, Otto, German physicist, inventor; b. Danzig, Poland, Feb. 29, 1872; s. Franz and Modesta (von Kalkstein) B.; m. Helen Püschel, 1904; 1 son. Founder (with F. Clausen) lab. with expts. and lectures on wireless telegraphy, X-rays, prodn. selenium cells; became head patent dept. Telefunken, 1911. Research and publs. on TV; inventory high-frequency amplifications, reflex circuit; patentee solution for color TV, 1902. Died Berlin, Aug. 5, 1951.

VON GROTH, Paul Heinrich, German crystallographer, mineralogist; b. Magdeburg, Germany, June 23, 1843; s. Philipp and Marie (Steffen) G.; ed. Freiberg Mining Acad., Dresden Poly. Sch.; Ph.D., U. Berlin, 1868; m. Rosalie Levy, 1874; 11 children, including Otto, Alfred. Joined faculty U. Berlin, 1870; became lectr. Berlin Mining Acad., 1871; joined U. Strasbourg (France), 1872, founder Mineral. Inst.; prof. U. Munich, 1882-1923. Mem. Bavarian Acad. Scis. Author: Tabellarische Übersicht der (einfachen) Mineralien nach ihren krystallographischen-chemischen Beziehungen, 1874; Grundriss der Edelsteinkunde, 1887; Chemische Krystallographie, 5 vols., 1906-19; Elemente der Physikalischen und chemischen Krystallographie, 1921; Entwicklungsgeschichte der mineralogischen Wissenschaft, 1926; also articles. Founder, Zeitschrift für Krystallographie und Mineralogie, 1877, editor until 1920. Discovered yttrium-containing titanite (Grothit) nr. Dresden, 1866; studies on relation between crystal form and phys. properties of quartz, smaltite, halite; relation between form and chem. nature of crystals, crystal measurements; forerunner of crystal structure research based on later discovery of X-ray interference. Died Munich, Dec. 2, 1927.

VON GROTTHUSS, Christian Johann Dietrich (Theodor), see Grotthuss, Christian Johann Dietrich (Theodor) von.

VON GRUBER, Max(imilian) Franz Maria, hygienist, bacteriologist; b. Vienna, Austria, July 6, 1853; s. Ignaz and Gabriele (von Menninger) V.G.; M.D., U. Vienna, 1876; postgrad. U. Munich (Germany); m. Julie von Aichinger, 1878; 4 sons, 2 daus.; m. 2d, Alwine Friederike Ehrhardt, 1891. Faculty hygiene U. Vienna, 1882-84, asso. prof., dir. Hygiene Inst., 1887-91, prof., from 1891; asso. prof. hygiene U. Graz (Austria), from 1884; prof. U. Munich, 1902-23, emeritus, from 1923. Mem. Austrian State Bd. Health, from 1891. Mem. Bavarian (pres. from 1924), Vienna (corr.) acads. scis. Author: Die bauliche Neugestaltung der Wiener medizinischen Fakultat, 1895; Pasteurs Lebenswerk, 1896; Die Prostitution vom Standpunkt der Sozialhygiene aus betrachtet, 1900; Hygiene des Geschlechtslebens, 1903; (with E. Rüdin) Die Pflicht, gesund zu sein, 1909. Research, publs. on health laws Bavaria, Austria, combating typhus, cholera, other diseases; main work on bacteriology, immunity; considered founder serology for discovery agglutination (in practice known as Gruber-Widal reaction), 1896; work on problems resistance powers of organisms and their parasites. Died Berchtesgaden, Germany, Sept. 16, 1927.

VON GRUBER, Otto Heinrich Franz Anton, geodesist; b. Salzburg, Austria, Aug. 9, 1884; s. Max and Julie (von Aichinger) von. G.; student Tech. U. Munich, U. Berlin, U. Würzburg; Ph.D., U. Munich, 1911; m. Elsie Lennox, 1911; 1 son, 4 daus. Became asst. to H. Ebert, Tech. U. Munich, 1911; joined firm Stereographik GmbH, Vienna, 1913; leader photogrammetry troop, during World War I; joined faculty Tech. U. Munich, 1919; joined Zeiss, also Stereographik Gmbtl, Munich, other firms, 1922; became prof. geodesy Tech. U. Stuttgart (Germany), 1925; became sci. dir. dept. geodetic instruments Zeiss, 1930. Author: Einfache und Doppelpunkteinschaltung im Raum, 1924; also articles. Research in photogrammetry, including theory, devel. of instruments, applications; devel. Bauersfeld's stereoplanigraph, self-focussing rectifiers; contbr. to devel. radial triangulation, spatial aero triangulation; interpreted pictures from Zeppelin's Arctic Expdn., 1931, German Antarctic Expdn., 1938-39. Died Jena, Germany, May 3, 1942.

VON GRÜTZNER, Paul, physiologist; b. Festenberg, Germany, Apr. 30, 1847; s. Ferdinand and Emma (Staats) Grützner; student medicine, Breslau (now Wroclaw, Poland), Würzburg, Berlin; M.D., Berlin, 1879; m. Stephanie Ziegler, 1883; 2 sons, 1 dau. Asst. to R. Heidenhain, U. Breslau, became lectr., 1875, asso. prof., 1881; prof. physiology U. Bern (Switzerland), 1881-84; prof. physiology U. Tübingen, 1884-1916, became emeritus 1916, named rector, 1900; moved to Bern after retirement Mem. Leopoldina. Research and publs. in digestive physiology, digestive enzymes and their presence in urine, peristalsis of stomach and intestine, optical nerves, physiology of voice and speech, physiol. properties of red and white muscle fibers, muscle metabolism; originated colorimetric process for determination of pepsin; proved (with Heidenhain) heightening blood pressure and blood vessel constriction of visceral stimulation. Died Bern, July 29, 1919.

VON GUDDEN, (Johann) Bernhard Aloys, German psychiatrist; b. Cleves, Germany, June 7, 1824; s. Johann Jakob and Bernhardina (Firtzen) G.; student medicine, Bonn, Halle, Berlin, Germany; M.D., 1848; m. Clarissa Voigt, 1855; children include Clemens, Rudolf, Anna (Mrs. Hubert von Grashey). Became physician Sieburg Insane Asylum, 1849; joined Illenau under C. Roller, 1851; became dir. Werneck Insane Asylum (became most modern asylum in Germany), nr. Würzburg, Germany, 1855; named prof. psychiatry U. Zurich, also dir. Burghölzli Asylum, 1869; became prof. U Munich, also dir. Gabersee Asylum, 1873; drowned while escorting King Ludwig III of Bavaria to Schloss Berg for treatment. Author: Bernhard von Gudden's gesamte und hinterlassene Abhandlungen (editor H. Grashey), 1889; also articles. First German to introduce no restraint principle of Connolly; studies on mut. influence of ind. skull and brain growth, influence of trigeminal bisection on cornea, hematoma auris and rib fractures of insane brain anatomy; developed Gudden's method for extirpation of peripheral brain fibers of newborn animals which determined secondary degeneration of conduction routes and centers; research leading to discoveries in brain, especially thalamus; Gudden's commissure and ganglion named after him. Died Starnberger Lake, June 13, 1886.

VON GÜMBEL, (Carl) Wilhelm, German geologist; b. Dannenfels, Germany, Feb. 11, 1823; s. Johann Friedrich and Charlotte (Roos) Gümbel; student Munich, Heidelberg, Germany, 1842-48; state exam, Munich, 1848; m. Emma Wahl, 1855; 1 dau.; m. 2d, Katharina Labroisse, 1885. Mining asst., mining surveyor, for short period; became dir. Geognostic Study Inst., Bavaria, 1851, placed in charge all work, 1856; dir. State Mining Bur., 1879-98; hon. prof. geognosy and mining surveying U. Munich; chmn. mineral collection Poly. Sch. Mem. Bavarian Acad. Scis. Author: Geognostische Beschreibung des bayerischen Alpengebirges und seines Vorlandes, 1861; Geog. Beschr des ostbayerischen Grenzgebirges . . . , 1868; Geog Besch. des Fichtelgebirges mit dem Frankenwald, 1879; Geog. Beschr. der Fränkischen Alb mit dem Keupervorland, 1891; Geologie von Bayern, 2 vols., 1888-94. Analysis of rock and strata sequence. Died Munich, June 18, 1898.

VON HAAST, Sir Julius Johann Frank, geologist, paleontologist; b. Bonn, Germany, May 1, 1822; s. Matthias and Anna Eva Theodora Ruth Haast; student Bonn, for short period; Ph.D., Tübingen, Germany, 1862; hon. D.Sc., Cambridge, 1886; m. Antonia Schmitt, 1846; 1 son; m. 2d, Mary Dobson, 1863; 4 sons, including Heinrich Ferdinand; 1 dau. Salesman in Belgium, France, Frankfort/Main; went to New Zealand in contract to London shipping co., 1858; joined Austrian Novara expdn., 1859; named provincial geologist, 1859-68; founder Canterbury Mus., Christ Church, New Zealand, 1866. Became prof. geology Coll. at Christ Church, 1879; named German consul, 1880. Surveyor-gen. for Canterbury, New Zealand, 1861-71; founder Imperial Inst., London. Recipient Gold medal Royal Geog. Soc. London, 1884. Mem. Royal Soc., 1867, Leopoldina, Linnean Soc. Author: Geology of the Provinces of Canterbury and Westland, 1879; also articles. Determined Pleistocene glacial formations in New Zealand Alps unknown in So. hemisphere, 1861-76; discovered coal beds, gold in New Zealand, 1859, fossil and subfossil remains of moa and other non-flying birds; studies in botany and ethnology. Died Wellington, New Zealand, Aug. 15, 1887.

VON HACKER, Viktor, Austrian surgeon; b. Vienna, Austria, Oct. 21, 1852; s. Franz and Caroline (von Forazest) von H.; student Medicine, Vienna; Ph.D., 1878; hon. dr. U. Graz (Austria); m. Mary von Niebauer, 1890. Staff under Heschl, Path. Anat. Inst.; worked wunder Duchek med. clinic, Vienna; joined Billroth's 2d Surg. Clinic, 1881; joined faculty U. Vienna, 1888; also dir. surg. dept. Sophien Hosp., Vienna; became head physician Vienna Gen. Polyclinic, 1891; named asso. prof., 1894; apptd. prof. U. Innsbruck (Austria), 1895, U. Graz, 1903; ret., 1924. Named hon. citizen Graz. Hon. mem. German Soc. Surgery, Soc. Physicians in Vienna, Sci. Assn. Physicians in Styria. Author: Anleitung zur antiseptischen Wundbehandlung nach der an Prof. Billroth's Klinik gebräuchlichen Methode, 1883; Die Magenomerationen an Prof. Billroths Klinik, 1880-85, 86; Über die nach Verätzungen entstehenden Speiserohrverengungen, 1889; also articles. Developed gastro-intestinal surgery; pioneered esophragoscopy; studies in surgery of esophagus, gastrointestinal tract, surgery of pharynx, heart, and trachea; created artificial esophagus from transverse intestine; suggested triangular sling for upper arm breaks. Died Graz, Austria, May 20, 1933.

VON HAHN, Karl, explorer; b. Friedrichsthal, Germany, Apr. 29, 1848; s. Friedrich and Margarete Henrike Charlotte (von Gros) H.; student theology and philology, Tübingen, Germany; m. Helene von Franken, 1875; at least 1 son. Vicar, Brenz, also Reichenbach, Germany; med. aide in war of 1870-71; became ct. tchr. Gov. of Caucasia, Tiflis, USSR, 1872; became tchr. German-speaking secondary sch., also other instns., 1874; rep. Red Cross in Russian-Turkish War, 1877-78. Author: Nachrichten der alten griechischen und römischen Schriftsteller über den Kaukasus, 2 vols., 1884-90; Aus dem Kaukasus, Reisen und Studien, 1892; Kaukasische Reisen und Studien, Neue Beiträge zur Kenntnis des Kaukasischen Landes, 1896; Bilder aus dem Kaukasus, Neue Studien zur Kenntnis Kaukasiens, 1905; Neue Kaukasische Reisen und Studien, 1911; also 1st textbook on geography of Georgia, USSR, 1924. Gathered hist., ethnol., biol. and geol. data in Caucasus and Armenian highlands. Died Tiflis, Aug. 16, 1925.

VON HAIDINGER, Wilhelm Karl, Austrian mineralogist, geologist; b. Vienna, Austria, Feb. 5, 1795; s. Carl H. and Josepha (Schwab) von H.; studied under F. Mohs, Joanneum State Mus., Graz, Austria; m. Auguste Mohn; 2 daus., including Sidonie (Mrs. Eduard Döll). Followed Mohs to Freiburg, Germany, 1817; came to Edinburgh, Scotland, to organize mineral. collection of banker R. Allan; accompanied Allan's son on collection expdns. in Europe until 1827; assisted in mgmt. of brothers' procelain factory, Elbogen, Bohemia, beginning 1827; became dir. minerals collections of court monetary and mining chamber, 1840; founder (with F. von Thinnfeld) State Geol. Inst., Vienna, 1849, dir. until 1866. Haidingerite, also plant family, fossil fish and mountain chain in New Zealand named in his honor. Mem. Imperial Royal Geog. Soc. Vienna (co-founder), Werner Assn. for Geol. Research Moravia and Silesia (co-founder), Geol. Assn. for Hungary (co-founder), Società geologica Milan. (co-founder). Author: Anfangsgründe der Mineralogie, 1829; Bericht über die Mineralien-Sammlung der K. K. Hofkammer in Münzund Bergwesen, 1843; Geognotische Übersichtskarte der österreichischen Monarchie, 1845; Handbuch der bestimmenden Mineralogie, 1845; Bemerkungen über die Anordnung der kleinsten Theilchen in Christallen, 1853; Interferenzlinien am Glimmer, 1855; Vergleichungen von Augit und Amphibol, 1855. Editor: Jahrbuch of Vienna K. K. Geologische Reichsanstalt, 1850. Co-founder, editor Naturwissenschaftlich Abhandlungen, 4 vols., 1847-51; Berichte über Mitteilungen von Freunden der Naturwissenschaften in Wien, 7 vols., 1847-50. Pioneered in crystallography, especially mineral optical work; mineral studies and observations on meteorites; named numerous known minerals; studied and named numerous mineral types; recognized and interpreted pseudomorphosis; discovered phenomenon which enables detection of plane polarized light (Haidinger's brushes). Died Dornbach, Austria, Mar. 19, 1871.

VON HALBAN, Hans, chemist; b. Vienna, Austria, Oct. 21, 1877; s. Heinrich and Marie (Adler) Blumenstock; Ph.D., U. Zurich, Switzerland, 1901; m. Zora von Fialka, 1906; a son, Hans. Asst. to W. Ostwald and Le Blanc, U. Leipzig, 5 years; joined faculty U. Würzburg (Germany), 1906, became asso. prof. phys. chemistry, 1915; named dir. phys.-chem. research lab. Metall-Gesellschaft, Frankfort/Main, Germany 1923; named prof. U. Zurich, 1930. Research and publs. in dissociation of electrolytes and reaction kinetics, kinetics of fast reactions with photoelectric extinction measurements; developed optical method using alkali photocells filled with inert gas in spectral photometry used for study of dissociation of strong electrolytes and ion association. Died Zurich, Oct. 7, 1947.

VON HALBAN, Josef, Austrian gynecologist; b. Vienna, Austria, Oct. 10, 1870; s. Philipp and Anna (Damask) Blumenstock; M.D., Vienna, 1894; studied under Guyen and Péau, Paris; student bacteriology Institut Pasteur, Paris, gynecology, Vienna; m. Selma Kurz, 1910; 1 son, 1 dau., Desirée. Joined faculty U. Vienna, 1903, became asso. prof., 1909; named head gynecol. dept. Wieden Hosp., 1909. Author: (with J. Tandler) Anatomie und Aetiologie des Genitalprolapses beim Weibe, 1907; (with L. Seitz) Handbuch der Biologie und Pathologie des Weibes, 1924-29; Gynäkologische Operationslehre, 1932; also articles. Discovered trophic effect from ovaries and connection between ovaries and menstruation is a hormonal process rather than nerve impulse; 1st to mention internal secretion of placenta; pregnancy sign named after him; improved surg. techniques; studies in urology. Died Vienna, Apr. 23, 1937.

VON HALEM, Friedrich Wilhelm, German physician; b. Aurich, Germany, Nov. 13, 1762; s. Wilhelm and Dorothea (Schnedermann) von H.; student medicine, Halle, Göttingen, Berlin; m. Susanne, 1788; m. 2d, Anna Mencke, 1791. Began practice medicine, Emden, Germany, 1786; elected state physician of East Friesland, 1797. Author: Über die Seebadeanstalt auf der ostfriesischen Insel Norderney, 1801; Beschreibung der zum Fürstenthum Ostfriesland gehörigen Insel Norderney und ihrer Seebade-Anstalten, 1813; Die Insel Norderney und ihr Seebad nach den gegenwärtigen Standpunkte, 1822; also articles. Reformed midwifery sch., E. Freisland; founder 1st German ocean bathing healing establishment on island of Norderney in North Sea. Died Aurich, May 25, 1835.

VON HALLER, Albrecht, Swiss physician, physiologist, anatomist, botanist; b. Bern, Switzerland, Oct. 16, 1708; s. Niklaus Emanuel and Anna Marie (Engel) V. H.; student medicine (under Camerarius) U. Tübingen (Germany), 1723; M.D. (under Boerhaave), U. Leiden (Netherlands), 1727; postgrad. Paris (France), London, Oxford (both Eng.) univs., 1728; student math. U. Basel (Switzerland), 1928; m. Mariaanne Wyss, 1731; 2 sons, 1 dau; m. 2d, Elisabeth Bucher, 1739; 1 son; m. 3d, Sophie Teichmeyer, 1741; 4 sons, 3 daus. Collected flora of Alps, Basel, 1728; practiced medicine, also city librarian, Bern, from 1729; chair medicine, anatomy, surgery, botany U. Göttingen (Germany), 1736-53, established bot. garden, founded sch. midwifery, coll. surgery, theater, anat. mus.; physician to king Eng.; mem. bd. health, marriage magistrate, author, Bern, 1753-77. Helped reorganize Acad. Lausanne (Switzerland). Fellow Royal Soc., 1739; mem. Royal Soc. Göttingen (perpetual sec.), French, 1754, Swedish acads. scis.; others. Author: Historia stirpum Helvetiae indigenarum, 1742; Icones anatomicae, 1743-54; De respiratoine experimenta anatomica, 1747; Opuscula botanica, 1749; Opuscula pathologica, 1754; Elementa physiologiae corporis humani, 1757-66; Bibliotheca botanica, 2 vols., 1771-72; Bibliotheca anatomica, 2 vols., 1774-77; Bibliotheca chirurgica, 2 vols., 1774-75; Bibliotheca medicinae practice, 4 vols., 1776-88. Argued for classification based on natural relationships; showed distinction between and independence of sensibility (of nerves) and irritability (contractability of muscles), that nerves all lead to brain or spinal cord (centers of perception, response); gave results of stimulating or damaging parts of animal brain; clarified relationships of portal system (branching of coeliac axis named tripus Halleri); made distinction between sweet, foul, indifferent odors; 1st to show that arteries, capillaries transmit heart pulse; recognized mechanism of respiration, automatism of heart; admitted use of bile in digestion of fat; important work on anatomy of genitals, brain, heart, many arteries; described embryonic devel.; elevated physiology to separate br. sci.; suggested term graffian follicle, 1730; showed that lingual vein was blood vessel, not salivary duct; described circulus callosus halleri, 1747, circulus arteriosis halleri, 1756, circulus conosus halleri, 1757. Died Bern, Dec. 12, 1777.

VON HAMM, Wilhelm Philipp, agronomist; b. Darmstadt, Germany, July 5, 1820; s. Konrad and Wilhelmine (Emrich) Hamm; s. student Agr. and Forestry Acad., Hohenheim, Germany; Ph.D., Giessen, Germany, 1845; m. Nanny du Moissy, 1850; 6 children. Became tchr., P. E. von Fellenberg's Agrl. Inst., Hofwyl, Switzerland, 1844; founder 1st factory for agrl. machines, Leipzig, Germany; became ministerial advisers, dir. dept. agr. Ministry Commerce, 1867; joined Minnistry Agr., 1868. Author: Die landwirtschaftlichen Maschinen und Geräte Englands, 1845-49; Chemische Bilder aus dem täglichen Leben, 2 vols., 1854; Weinbuch, 1886; Das Wesen und die Ziele der Landwirtschaft, 1866; Gesammelte kleine Schriften, 2 vols., 1881; also numerous articles, monographs. Editor, Agronomische Zeitung, Leipzig, 1846-69. Assisted in founding Agr. U., Vienna, also exptl. stas.; promoted agrl. convs., exhbns., assns.; adapted works on plant cultivation, animal breeding, chem. topics, agrl. machines to popular form. Died Vienna, Austria, Nov. 8, 1880.

VON HAMMER, Ernst Hermann Heinrich, German geodesist, mathematician; b. Ludwigsburg, Würtemberg, Germany, Apr. 20, 1858; s. Andreas and Rosine Katherine (Benignus) V.H.; student math. Stuttgart (Germany) Poly. Sch., engring. sch., 1874-78; Ph.D., U. Leipzig (Germany), 1896; D.Eng. (hon.), Stuttgart, 1924; m. Wilhelmien Häberle, 1886; 1 dau. Topog. work for Black Forest Adminstrn; lectr. trigonometry, other math. Stuttgart Poly. Sch., from 1882, prof. lower, higher geodetics, plan drawing, methods of smallest squares, 1884-1925. Author: Lehr und Handbuch der ebenen und sphärischen Trigonometrie, 1885; Dei Berechnung der trigonometrischen Vermessungen mit Rücksicht auf die sphäroidische Gestalt der Erde, 1885; Die Hetzentwürfe geographischen Karten, 1887. Numerous transls. Significant part in Württemberg's work for Internat. Geodesy; numerous inventions; prepared new isogone chart, map projections, high altitude survey, self calculating tachymeter; wrote history math., several texts on practical geodetics. Died Stuttgart, Sept. 11, 1925.

VON HANDEL-MAZZETTI, Heinrich Raphael Eduard, Austrian botanist; b. Vienna, Austria, Feb. 19, 1882; s. Eduard and Friederika (de Mauro) V.H.-M.; Ph.D. in Botany, U. Vienna. Demonstrator Bot. Inst., U. Vienna, 1903-05, asst., 1905-25; with Vienna Natural History Mus., 1923-31, curator, 1925-31, ret., 1931. Journey to S.W. China, commd. by Austrian Acad. Scis., 1914-19. Author: Naturbilder aus Südwest-China, 1927. Contbr. monographs, articles to bot. publs. Specialist on flora of China, from 1919 (best known expert in this field); worked on collections of other botanist; gave numerous descriptions of new types, new plant geog. distbn. of China; interested in questions of phylogenic devel; traveled extensively in Switzerland and the Near East. Died Vienna, Feb. 1, 1940.

VON HANN, Julius Ferdinand, Austrian mineralogist; b. Schloss Haus, nr. Linz, Austria, Mar. 23, 1839; s. Joseph and Anna (Scheichenfellner) V.H.; ed. U. Vienna (Austria); m. Louise Weissmayr, 1878; 3 sons, 1 dau. Tchr. intermediate schs.; asst. Central Inst. for Meteorology, Vienna, 1867, dir., 1877-97; faculty U. Vienna, from 1868, asso. prof. phys. geography, from 1874, prof. geophysics, from 1877, emeritus, from 1910; prof. U. Graz (Austria). Lectr. on climatology Agrl. Inst. Vienna, 1872. Mem. numerous sci. acads., scholarly socs. Author: Handbuch der Klimatologie (1st comprehensive work in field), 1883, 3 vols., 1908-11; Atlas der Meteorologie, 1887; Verteilung des Luftdruckes über Mittel-und Südeuropa, 1887; Die Erde als Ganzes, ihre Atmosphäre und Hydrosphäre, 1896; Lehrbuch der Meteorologie, 1901, 3d edit., 1914; Klimatographie von Niederösterreich, 1904. Editor, contbr. Meteorologische Zeitschrift (leading organ of meteorol. world), for 55 years. Introduced thermodynamic principles into meteorology, marking beginning modern meteorology; developed new theory of mountain, valley winds, problems gen. circulation, tropical storms, connection between anomalies of weather in widely separated areas, daily air pressure variations; laid methods groundwork of climatology. Died Vienna, Oct. 1, 1921.

VON HANSEMANN, David Paul, German pathologist, anatomist; b. Eupen, Germany, Sept. 5, 1858; s. Gustav and Malthilde (Vorländer) von H.; student medicine, Berlin, Kiel, Leipzig, Germany; state exam., 1885; M.D., 1886; m. Elisabeth Walter, 1885; 1 son. Asst. to Virchow, U. Berlin, joined faculty path. anatomy, 1890; became prosector Friedrichshain Municipal Hosp., Berlin, 1895, Rudolf-Virchow Hosp., 1906; named titular prof., 1897, hon. prof., 1912; chief physician at a reserve hosp., later army pathologist, World War I. Author: Studien über die Spezifität, den Altruismus und die Anaplasie der Zellen, 1893; Die mikroskopische Diagnose der bösartigen Geschwülste, 1897; Der Aberglaube in der Medizin und seine Gefahr für Gesundheit und Leben, 1905; Descendenz und Pathologie, 1909; Atlas der bösartigen Geschwülste, 1910; Das konditionale Denken in der Medizin und seine Bedeutung für die Praxis, 1912. Research in path. anatomy, comparative pathology, cancer; opposed indiscriminate use of tuberculin in TB therapy. Died Berlin, Aug. 28, 1920.

VON HANSTEIN, Johannes Ludwig Emil Robert, German botanist; b. Potsdam, Germany, May 15, 1822; s. Ludwig and Emilie (Sello) H.; student horticulture Berlin, Potsdam univs., 1838-43; Ph.D. in Botany, U. Berlin, 1848; m. Helene Ehrenberg, 1857; 3 sons. Tchr., Berlin, from 1849; faculty U. Berlin, from 1855; 1st curator Berlin Bot. Mus., from 1861; prof., dir. Bonn (Germany) Bot. Garden, from 1865. Author: Untersuchung über den Bau und die Entwicklung der Baumrinde, 1853; Die Milchsaftgewächse und die verwandten Organe der Rinde, 1864; Übersicht des naturlichen Pflanzensystems, 1867; Untersuchungen über die Anordnung der Zellen in den Vegetationspunkten der Phanerogamen, 1868; Die Entwicklung des Keimes der Monokotylen und Dikotylen, 1870. Pioneer in microscopic anatomy, evolution of plants, devel. vascular tissue, devel. cortex and sap movement in dicotyledons; among 1st to describe fertilization process of ferns; created basis for evolutionary differentiation of phanerogams and higher cryptogams; denial of Darwin's principles left him behind in plant taxonomy. Died Bonn, Aug. 27, 1880.

VON HAUER, Franz, Austrian geologist, paleontologist; b. Vienna, Austria, Jan. 30, 1822; ed. U. Vienna, Schemmity Sch. Mines. Joined Montanistiche Mus. Vienna, 1843; began 1st course in paleontology, Vienna, 1844; became dir. Mus. Geologischen Reichsanstalt, 1867; dir. Hofmuseum. Mem. Acad. Scis. Vienna. Author: Die Cephalopoden des Salzkammergutes, 1846; (with F. Foetterle) Geologische Übersicht der Bergbaue der österreichischen Monarchie, 1855; Beiträge zur Palaeontographie von Österreich, 1858-59; (with Stache) Geologie Siebenburgens, 1863; Die Geologie und ihre Anwendung auf die Kenntniss der Bodenbeschaffenheit der österrungarische Monarchie, 1874; Die Cephalopoden des bosnichen Muschelkalkes von Han Bulog, 1887; Geologic Maps of Transylvania, 1861; Geologic Maps of Austria-Hungary, 1867-83. Research in geology of

Central and Western Alps, Carpatian Mountains; developed new classification for their terrains; paleontol. studies on cephalopods, and other Triassic and Liasic fossil formations. Died Vienna, Mar. 20, 1899.

VON HEBRA, Ferdinand Ritter, Austrian dermatologist; b. Brunn, Moravia, Sept. 7, 1816; M.D., U. Vienna (Austria); disciple of K. Rokitansky. Prof. U. Vienna. Author: Lehrbuch der Hautkrankheiten, 1860; Atlas of Skin Diseases, 1876. Founder modern dermatology; devised new nomenclature, classification for skin diseases; attributed most diseases of skin to result of local irritation instead of to condition of body fluids (as held by humoral pathology); described pityriasis ruba (Hebra's pityriasis), 1857, erythema multiforme exudativum (Hebra's disease), 1860, impetigo herpetiformis, 1872; originated compound ointment of sulfur for treatment of parasitic skin diseases, circa 1856. Died Vienna, Aug. 5, 1880.

VON HEFNER-ALTENECK, Friedrich, German elec. engr., inventor; b. Aschaffenburg, Apr. 27, 1845; s. Jakob Heinrich Von H.-A.; Ph.D. (hon.), U. Munich. Devised Hefner amyl acetate standard light source, 1884; build differential arc lamp, 1879; perfected Hughes' telegraph; invented drum armature which made possible building of longer life of D.C. motors, 1873; devised Hefner-Alteneck dynamometer. Died Berlin, Jan. 7, 1904.

VON HEINE, Jakob, German physician, orthopedist; b. 1800; practice medicine, Connstatt, Germany. Author: Lämungszustände der unteren Extremitäten und deren Behandlung, 1840. First description of deformities caused by poliomyelitis (Heine-Medin disease), 1840; described congenital cerebral spastic paralysis (Little's disease), 1840; suggested atrophy of anterior horns of spinal cord could occur in infantile poliomyelitis. Died 1879.

VON HERBERSTEIN, Baron Siegmund (Sigismund), zoologist; b. Wippach, Carnidla, 1468. Ambassador of Maximilien I of Denmark to Poland, 1516, to Russia, 1517, 26, also to Netherlands, Bohemia, Germany; envoy for Karl V, Ferdinand I. Author: Rerum Moscoviticarum Commentarii, 1549. Works contained some of oldest maps of Russia, also history early Russia; studied, wrote on animals of Muscovy. Died Vienna, Austria, 1566.

VON HERDER, Johann Gottfried, German philosopher, historian; b. Mohrungen, E. Prussia, Aug. 25, 1744; ed. U. Königsberg (Germany); m. Caroline Flachsland, 1773; 8 children. Ordained priest Roman Catholic Ch., 1765; tchr. cathedral sch., Riga, Latvia; traveling tutor to Prince Peter von Holstein-Gottorp, 1769-70; chief pastor, Bückeburg, Germany, 1771-76; gen. supt. ch. dist., ct. preacher Weimar, Germany, from 1776; tour Italy, 1788-89; v.p. Upper Consistory, Weimar, from 1793, pres., from 1801. Author: Abhandlung über den Ursprung der Sprache, 1772; Von deutscher Art und Kunst, 1773; Ideen zur Philosophie der Geschichte der Menschheit, 4 vols., 1784-91; Auch eine Philosophie der Geschichte zur Bildung der Menschheit, 1774; Gott, einige Gespräche, 1787; Cid, 1805; others. Lifelong attempt to formulate philosophy of human nature, culture, history; said that individual genius can be understood only as part of culture that formed it; conceived of Bible as hist. (as well as sacred) document to be understood in terms of cultural milieu which produced it; held that nature, history obey uniform set of laws; developed theory of opposing forces which are in equilibrium but which do erupt to create new forms and a new equilibrium; maintained that nations could be seen as developing organic entities which could be illuminated through hist. study of their culture (requiring study of folk lang., lit., religion). Died Weimar, Dec. 18, 1803.

VON HERTWIG, Richard Carl Wilhelm Theodor, German zoologist; b. Friedberg, Germany, Sept. 23, 1850; s. Carl and Elise (Trapp) V.H.; ed. U. Zurich (Switzerland), Jena, Bonn (both Germany) univs.; Ph.D.; M.D.; m. Jula Braun, 1887; children—Otto, Marianne. Lectr. in zoology U. Jena, from 1875, prof., from 1878; prof. zoology U. Königsburg (Germany), from 1881, U. Bonn, from 1883, U. Munich (Germany), 1885-1925; dir. Zool., Zootech. Staatssammlung. Author: Der Organisums der Radiolarien, 1879; Entwicklung des mittleren Keimblattes der Wirbeltiere, 1883; Über die Konjugation der Infusorien, 1889; Lehrbuch der Zoologie, 1891-92, 15th edit., 1931; Die Aktinien der Challenger-Expedition; Uber die Entwickelung des Unbefruchteten Seeieeleies, 1896; Die Protozoen und die Zelltheorie, 1902; Abstammungslehre und neuere Biologie, 1927; (with Oskar Hertwig) Die Cölomtheorie, 1881. First to describe parthenogenesis in sea urchin eggs; (with bro. Oskar) developed germ-layer theory, also worked on Haeckel's theory of coelom formation; proposed that conjugation in paramecium is process of meiosis and fertilization; studied nucleoplasmic ration in relaton to cell div.; originated chromidia theory; worked on sex determination in amphibians. Died Munich, Oct. 3, 1937.

VON HIPPEL, Arthur Robert, physicist; b. Rostock, Germany, Nov. 19, 1898; s. Robert and Emma (Bremer) von H.; Ph.D., U. Goettingen (Germany); m. Dagmar Franck, June 2, 1930; children—Peter H., Arndt R., Frank N., Eric A. Marianne M. Rockefeller fellow U. Cal., 1927-28; pvt. docent U. Jena, 1928-29. U. Goettingen, 1930-33; prof. U. Istanbul, 1933-34; guest prof., U. of Copenhagen, 1935-36; faculty Mass. Inst. Tech., 1936-—, prof., 1947-62, Inst. prof., 1962-64, emeritus, 1964-—, also lab. for insulation research. Chmn. conf. on electrical insulation, NRC, 1952; sci. adviser U. S. Naval Research Lab., 1965. Recipient Presidential Certificate of Merit, 1948. Fellow Am. Acad. of Arts and Scis., Am. Physics Soc., Washington Acad. Scis., Am. Assn. Advancement Sci., New York Acad. Sci. Author; Dielectrics and Waves, 1954; Molecular Science and Molecular Engineering, 1959. Editor: Dielectric Materials and Applications, 1954; The Molecular Designing of Materials and Devices, 1965. Contbr. articles to jours. Pioneer in modern materials research that ranges from atoms to complex systems and unites science and engrng. in molecular designing of materials and devices; contbd. to conduction and electric strength of materials, dielectric spectroscopy, crystal physics, ferroelectrics, magnetics, ceramics, polymers. Home: 265 Glen Road, Weston, Mass. 02193. Office: Mass. Inst. Tech., Cambridge, Mass.*

VON HOCHENEGG, Julius, Austrian surgeon; b. Penzing, nr. Vienna, Aug. 2, 1859. Prof. in Vienna. Author: Lehrbuch der Speziellen Chirurgie, 1907. Added to intestinal surgery, particularly by his method of operation on cancer of rectum. Died May 11, 1940.

VON HOCHSTETTER, Ferdinand, geologist; b. Esslingen, Germany, Apr. 30, 1829. Naturalist with Novara Expdn. around the world, 1857-59; became prof. mineralogy, geology U. Vienna, 1860; named curator Natural History Mus., dir. mineral. collections, Vienna, 1877. Author:Neuseeland, 1863; Reise der österreichischen Fregatte Novara um die Erde geologischer Teil, 3 vols., 1864-66. Research on geology of New Zealand; Hochstetter Peak in New Zealand named in his honor. Died Ober-Döbling, Austria, July 18, 1884.

VON HOFF, Karl Ernst Adolf, German geologist, seismologist; b. Gotha, Germany, 1771; ed. univs. Jena, Göttingen. Became sec. of legation, 1791, counselor of chancellery, 1813, pres. Superior Consistory, Gotha, 1828; named dir. sci. and art collections, Gotha, 1822. Author: Chronik der Erdbeben und Vulkan-Ausbrüche, 2 vols., 1840-41; Magazine für die gesamte Mineralogie, Geognosie, . . . , 1801; Geschichte der natürlichen Veränderungen der Erdoberfläche, 3 vols., 1822-34; Deutchland nach seiner natürlichen Beschaffenheit, 1838. Editor, Hofkalendar, 1801-16. First to issue annual lists of earthquakes; 1st compilation of gen. catalogue of earthquakes of world. Died 1837.

VON HOFMANN, August Wilhelm, German chemist; b. Giessen, Germany, Apr. 8, 1818; Ph.D. summa cum laude, U. Giessen, 1841; m. Helene Moldenhauer; m. 2d, Rosamund Wilson, 1856; m.3d, Elise Moldenhauer, 1866; m.4th, Bertha Tiemann, 1873; 11 children. Asst. to Liebig; privatdozent U. Bonn (Germany), until 1845, prof., dir. own lab., 1864-65; 1st dir. Royal Coll. Chemistry, London, Eng., 1845-63; prof. chemistry U. Berlin (Germany), 1865-92. Chemist to Royal Mint, 1856-65. Recipient Gold medal Société de Pharmacie de Paris, Grand Prix, Paris Exhbn., 1867. Fellow Royal Soc. (Copley medal 1875), 1851; mem. French, 1859, Prussian acads. sci., German (founder 1868, pres.), London (pres. 1861, fgn. sec.) chem. socs. Author: Handbook of Organic Analysis, 1853; Introduction to modern chemistry, 1865; Berliner Alchemisten und Chemiker. Ein Jahrhundert chemischen Forschung unter dem Schrim der Hohenzollern. Cheimsche Erinnerungen, 1882; Zur Erinnerung an vorangegangene Freunde, 1888. Editor: Annalen der Chemie, from 1873. Contbr. to sci. jours. Research on coal tar led to locating benzene in tar and to establishing nature of aniline, 1843; developed technique for preparation of aniline from benzene, 1845; discovered Hofmann's reduction reaction for converting an amide into an amine; perceived analogy between aniline and ammonia which led to work on amines, organic ammonium bases, allied phosphorus compounds; prepared rosaniline, 1858, violet dyes (Hofmann's violets) made from it, culmination in discovery of quinoline red, 1887; (with August Cahours) discovered allyl alcohol, formaldehyde, hydrazobenzene, isonitriles; developed method for determining molecular weights of liquids by means of vapor densities; synthesized esters of isothiocyanic acid; laid gt. stress on use of demonstrations in lectures, invented many standard ones. Died Berlin, May 5, 1892.

VON HOMEYER, Eugen Ferdinand, German ornithologist; b. Nerdin, Germany, Nov. 11, 1809; independent scholar, Stolp, Germany. Author: The Migration of Birds, 1881; Verzeichnis der Vögel Deutschlands, 1885. Owned largest collection of European birds. Died Stolp, May 31, 1889.

VON HÜFNER, Carl Gustav, German physiologist; b. Köstritz, Thüringen, Reuss-Schleiz, May 13, 1840; M.D., Leipzig, 1866; studied in Jena and Heidelberg. Prof. chemistry U. Tübingen, 1875; asst. physiology instr. U. Leipzig. Measured combining power of hemoglobin with oxygen (1 gm. hemo with 1.34 cc. O), 1894; improved spectrophotometer. Died Tübingen, Mar. 14, 1908.

VON IMHOF, Maximus, German physicist; b. Rissbach, Germany, July 26, 1758; student Landshut, Germany. Became Augustinian, 1780; tchr. monastery, Munich, Germany, 1786-91; became prof. physics and math. Electoral Lyceum, 1790, prior, 1798; left order, 1802; became canon, Munich. Publs. in physics; one of 1st Germans to support Lavoisier's chemistry. Died Munich, Apr. 11, 1817.

VON JACOBI, Moritz Hermann, physicist, chemist, engr.; b. Potsdam, Germany, Sept. 21, 1801. Practice architecture; went to Russia, 1818; apptd. to establish electric telegraph, 1832; became prof. U. Dorpat, 1834, U. St. Petersburg, 1835. Mem. Acad. Scis. St. Petersburg. Author books including Galvano Plastik, 1840. Research on electromagnetism, theory of electrochemical machines, galvanoplasty, application of electromagnetism to power machinery and boats, arc lamp; invented (independently of Spencer) galvanoplasty (electroplating), 1837; perfected voltmeter; built motor from electromagnets charged with cell battery (used to drive paddle steamer on Neva River), 1838. Died St. Petersburg, Mar. 10, 1874.

VON JAUREGG, see Wagner-Jauregg.

VON JOLLY, Philipp Johann Gustav, German physicist; b. Mannheim, Germany, Sept. 26, 1809; student U. Heidelberg, U. Vienna; children—Ludwig, Friedrich, Julius. Prof. math. and physics U. Heidelberg, 1834-54; became prof. physics, Munich, 1846. Author: Anleitung zur Differential-und Integral Rechnung, 1846; Die Principien der Mechanik, 1852; Über die Physik der Molecularkräfte, 1857; Die Veränderlichkeit in der Zusammensetzung der Atmosphärischen Luft, 1878; Die Anwendung der Waage, 1878-81. Invented air thermometer, spring balance (both named after him), eudiometer, pneumatic machine; determined mass and density of earth, 1881. Died Munic, Dec. 24, 1884.

VON KAISER, Kajetan George, German chemist; b. Kelheim, Bavaria, Jan. 3, 1803. Apptd. prof. tech., Munich, Bavaria 1851; tchr. applied chemistry Munich Tech. Hochschule, from 1878. Research, articles on chemistry of fermentation. Died Munich, Aug. 28, 1871.

VON KARMAN, Theodore, aeronautical engr.; b. Budapest, Hungary, May 11, 1881; s. Prof. Maurice and Helen (Konn) de Kármán; M.E., Royal Tech. U., Budapest, 1902; Ph.D., U. of Göttingen, Germany, 1908; hon. D.Eng., Tech., U. Berlin, 1929, U. Liege, Belgium, 1947, Princeton, 1947, Columbia, 1948, Yale, 1951, Hebrew Inst. Tech., Israel, 1952, Technische Hochschule, Aachen, 1953; D.Sc., Univ. Brussels, 1937, U. d'Aix, Marseille, 1948, U. Lille, 1953, U. Istanbul, 1953; Dr. Applied Scis., Technische U., Berlin, 1953; LL.D., U. Cal., 1942; Dokfor Honoris Causa, Istanbul Tech. U., 1955, U. Sevilla, Spain, 1958, U. Braunschweig, 1960, Politecnico d. Torino, 1960, U. Athens, 1961; D.Sc., Northwestern, 1956, U. So. Cal., 1958, New York University, 1960, Brown University, 1960. Came to U. S., 1930, naturalized citizen, 1936. Research engr. Ganz & Co., Germany, 1902; assistant prof. Royal Tech. U., Budapest, 1903-06; privat docent, U. Göttingen, 1909-12; dir. Aero. Inst., U. of Aachen, 1912-29; cons. Junkers Airplane Works, also Luftschiffbau Zepplin, Germany, 1924-28; research asso. Cal. Inst. Tech., 1928-30, dir. Guggenheim Aero. Labs., 1930-49, dir. jet. propulsion lab., 1942-45; founder, chief cons., Aerojet Engring. Corp., now Aerojet-General Corp., 1942-—; chmn. adv. group for Aero. Research and Development NATO. 1951-—; chmn. scientific advisory board to chief of staff USAF, 1944-55; cons. numerous cos. and govtl. agencies on aero. items, 1930-—; has held named lectureships in several univs. and confs. Wright Brothers lectr. Inst. Aeronautical Sciences, 1946. Awarded Am. Soc. of Mechanical Engineers medal, 1941; Sylvanus Reed award of Inst. Aeronautical Sciences, 1942; Medal for Merit, 1946; John Fritz Medal, 1948; Kelvin Gold Medal, 1950; officer Legion of Honor, 1947; also among others, Astronautics Award, Am. Rocket Soc., 1954. Wright Bros. Memorial Trophy, 1954, Daniel Guggenheim medal, 1955, Vincent Bendix award, 1957, Timoshenko medal, 1958; Lamme award, 1960; Robert H. Goddard Meml. award, 1960; Karl Friederich Gauss medal, 1960; Christopher Columbus gold medal, Genova, Italy, 1960. Hon. fellow Inst. Aeronautical Sciences, Royal Aeronaut. Soc.; fellow Am. Rocket Soc., Am. Acad. Arts and Scis., Am. Physical Soc.; mem. Nat. Acad. Scis. (Rome), Royal Soc. London, Academy of Sciences (Paris, France), 1946, Academy of Sciences (Torino, Italy), American Society M.E., Nat. Acad. Science, American Philos. Society, Academmia dei Lincei, Spanish Academy Sciences, A.A.A.S., Am. Soc. Civil Engrs., Franklin Inst., Am. Geophysical Society, Sigma Xi. Nominated by Pope Pius XII, Pontifical Acad. Sciences. Editor of books on aerodynamics in German lang. Author: (with J. M. Burgers) General Aero-dynamic Theory (2 vols), 1924; (with M. A. Biot) Mathematical Methods in Engineering (with M. A. Biot), 1940; Aerodynamics, 1954. Contbr. scientific articles to jours. Developed 1st theory of supersonic drag (called Kármán Vortex Trail), 1935; pioneer in devel. high speed aircraft and missiles; initiated research that developed 1st plane to break sound barrier; designed supersonic wind tunnels; research in math. analysis thermodynamics, aerodynamics, hydrodynamics. Died Aachen, Germany, May 6, 1963.

VON KERN, Vincenz Sebastien, Austrian surgeon; b. Grätz, Austria, Jan. 20, 1760; M.Surgery, Vienna, 1784. Prof. surgery, Laibach, 1797-1805; surgeon to Duke of Saxe-Hildburghausen; prof. surgery, Vienna, 1805-24, also founder library Clin. Sch.; became surgeon Establishment Deaf Mutes, Vienna, 1795; vice dir. Schs. Medicine, surgery and vet. medicine U. Vienna. Author: Manuel de Chirurgie, 1831; Beobachtungen und Bermerkungen aus dem Gebiete der Praktischen Chirurgie, 1828; Bermerkungen über den Gebrauch der Bader, 1802; also numerous articles. Contributed to devel. surgery and surg. teaching in Germany and No. Italy; publs. on smallpox innoculation; advocated simple aseptic treatment for wounds, water dressing instead of plasters and salves. Died Apr. 16, 1829.

VON KESZYCKI, Carl Heinrich, physicist; b. Utrecht, Holland, Sept. 4, 1922; s. Walter E. and Irene Countess (Dankelmann) von K.; Dr.rer.nat., Ph.D. in Physics, Ludwig Maximilians U., Munich, Germany, 1959; m. Meina R. Frucht, Sept. 24, 1956. Came to U. S., 1960, naturalized, 1965. Mem. Inst. for Theoretical Physics, Ludwig Maximilians U., 1956-59; scientist Lockheed-Cal. Co., 1960-61, head Space Sci. Lab., Burbank, Cal., 1961-63, asso. dir. research, Marietta, Ga., 1963—. Mem. Am. Phys. Soc., Am. Geophys. Union, Am. Inst. for Aeros. and Astronautics. Contbr. articles to tech. jours. Determined energy levels in heavy nuclei; research on solar-earth interactions; patentee electro-optical devices. Home: 3057 Pharr Ct. N., Atlanta 30305. Office: Dept. 72-14, Zone 400, Marietta, Ga. 30061.*

VON KETHAM, Johannes, physician; b. Kircheim, Württemburg; flourished circa 1480-95. Probably lived in Venice, circa 1492. Author: Fasciulus Medicinae, part I, part II, 1491, part III, 1495 (treatises on topics including uroscopy, pregnancy, blood-letting, regimen in epidemics, various surg. techniques; 1st illustrated, printed work of its kind in Venice).

VON KLEIST, E. Georg, German inventor; b. Pomerania. Dean cloister chpt. Kammin, Pomerania. (Independently of P. van Musschenbroek of U. Leyden) invented Leyden jar (also called Kleistian jar), earliest form of elec. condenser, 1745. Died 1748.

VON KOBELL, Franz (Ritter), German mineralogist; b. Munich, Germany, July 19, 1803; prof. mineralogy, Munich. Author: Tafeln zur Bestimmung der Mineralien mittels chemischen Versuch, 1883; Die Mineralnamen und die mineralogische Nomenklatur; Geschichte der Mineralogie, 1864. Invented galvanograph. Died Munich, Nov. 11, 1882.

VON KÖLLIKER, Rudolf Albert, anatomist, physiologist; b. Zurich, Switzerland, July 6, 1817; ed. Zurich, Bonn, Berlin (both Germany) univs.; M.D., U. Heidelberg (Germany). 1842; at least 1 son, Hans Theodor Alfons. Asst. to Henle, 1841-45; extraordinary prof. physiology, comparative anatomy U. Zurich, from 1844; extraordinary prof. U. Würzburg (Germany), from 1847-49, prof. anatomy, 1849-1902. Author: Entwicklungsgeschichte der Cephalopoden, 1844; Handbuch der Gewebelehre (1st formal histology text), 1852; Entwicklungsgeschichte des Menschen und der höhern Tiere, 1861; Icones histiologicae oder Atlas der Gewebelehre, 2 vols., 1864-65; Grundriss der vergleichenen Entwicklungsgeschichte des Menschen und der höhern Tieren, 1880; Mikroskopische Anatomie. Editor: (with von Siebold) Zeitschrift für wissenschaftlichen Zoologie, from 1849. Research on spermatozoa, refuted parasitic theory of their origin (proved origin to be in testicular cells), 1841; anticipated modern genetics in regarding nuclei of spermatozoon as carriers inherited characteristics; work in embryology; 1st to interpret developing embryo in terms of cell theory; work in cytology; viewed nerve fibers as portions of cell and anticipated Cajal's neuron theory; 1st to isolate smooth muscle cells, 1848; showed that muscle contraction produced electric current, 1856; contemporary of Darwin, noted that evolution could proceed by jumps as well as by gradual steps; (with F. G. J. Henle) introduced term Pacinian corpuscles, 1844. Died Wurtzburg, Nov. 2, 1905.

VON KRIES, Johannes, German physiologist; b. Roggenhausen, Germany, Oct. 6, 1853; ed. Halle, Leipzig (both Germany) univs., U. Zurich (Switzerland). Asso. with Helmholtz, Berlin, Germany, from 1876; asst. (with Carl Ludwig) Physiol. Inst. Leipzig, 1877-80; extraordinary prof. U. Freiburg (Germany), 1880-83, ordinary prof., 1883-1924. Author: Die Gesichtsempfindungen und ihre Analyse, 1882; Logik, 1916; Allgemeine Sinnesphysiologie, 1923; Immanuel Kant und sine Bedeutung für die Naturforschung der Gegenwart, 1924; Wer ist musikalisch, 1926. Research on physiology of vision; related retinal rods to twilight vision, retinal cones to daylight vision, 1894; determined differential thresholds for hue, 1882; worked on effect of tension of response of muscle to stimuli, 1880, color mixture with spectral lights, 1881; touched on problem of auditory localization, 1878-90; (with Carl Ludwig) measured blood-pressure in capillaries, 1875. Died Freiburg, Dec. 30, 1928.

VON KRUSENSTERN, Baron Adam Ivan, Russian navigator; b. Haggud, Estonia, Nov. 19, 1770. Joined Naval Cadet Corps.; participated in war with Sweden, 1788; served in Brit. fleet, 1793-99; commanded Russian fleet, 1803-06; mem. expdn. to N.W. coast of Am., E. coast of Asia, then round world; explored Bering Straits, 1815; dir. Russian Naval Sch., 1827-42; became adm., 1841. Mem. French Acad. Scis., 1810. Author: Voyage Round the World . . . , 3 vols. and atlas, 1809-13; Atlas de l'Océan Pacifique, 1824-27; Recueil de mémoires hydrographiques, 1824-35. First Russian navigator to sail around world; discovered Orloff Islands; studies in hydrography of Pacific; stimulated fur trade in N.Am. Died Reval, Estonia, Aug. 24, 1846.

VON LANGENSTEIN, Heinrich, German mathematician, astronomer, philosopher; b. Hainbuch, Germany, circa 1325; Licentiate in Theology, 1375; Doctor, 1376; in Paris 1363; mem. Sorbonne, became vice chancellor before 1381; a founder U. Vienna, Austria. Author: Commentary on Genesis I-3 (discussed astron. theories in 1st book); De improbatione epicyclorum et concentricorum; also others. Introduced math. to Vienna; his idea of the common nature of all beings and bodies led to an almost mechanistic concept of universe. Died Vienna, Feb. 11, 1397.

VON LANZ, Titus Ritter, German physician; b. Passau, Jan. 4, 1897; s. Tutus Ludwig and Luise (Leuze) L.; M.D., U. Munich; m. Ursula Schmidt, Mar. 12, 1949; children—Elizabeth, Titus, Herta, Ursula, Ulrich. Prof. at large, 1931; head. Reichsforschungsrat mission, 1938-45; became asso. prof., 1945, prof., 1947; dir. anatomy inst. U. Munich. Pres. Soc. Anatomy. Author: Praktische Anatomi; also articles. Home: Feichthofstrasse 161, 8 München-Obermenzing. Office: Pettenkoferstrasse 11, 8 München, W. Germany.

VON LASAULX, Arnold, German mineralogist, seismologist; b. Castellaun, nr. Coblenz, Germany, 1839; Ph.D., U. Berlin (Germany), 1868. Prof. mineralogy U. Breslau (now Wroclaw, Poland), from 1875, prof. mineralogy, geology U. Bonn (Germany), from 1880. Author: Elemente der Petrographie, 1875; Einführung in die Gesteinslehre, 1885; Precis de Petrographie, 1887. Editor: Der Aetna (W. Sartorius von Waltershausen), 1890. Noted for research on minerals, crystallography; one of earliest workers on microscopic petrography; described eruptive rocks of Saar, Moselle dists., 1878. Died 1886.

VON LEBENSTEIN, Gabriel, physician; flourished 14th century; doctor in medicina et magister parisiensis. Author: Von den Wassern die man prent aus den crentern und aus den Plumen (earliest known German book on brandies made from plants and herbs and how they were used in medicine). Described effect of convallaria majalis in heart disease.

VON LEBER, Ferdinand Joseph, Austrian surgeon; b. Vienna, Austria, 1727; M.D., 1751. Became prof. anatomy and surgery, 1761; knight counsellor, 1st surgeon to Emperor of Austria. Author: Traité d'anatomie, 1775. Died Vienna, Austria, 1808.

VON LEDEBOUR, Karl Friedrich, German botanist; b. Stralsund, Pomerania, July 8, 1785. Prof. botany Dorpat (now Tartu, Estonia), 1811-36; travelled to Altai Mountains, also Russia; prof. Heidelberg, Munich (both Germany) univs. Author: Icones plantarum Novarum Floram Rossicam, Imprimis Altaicam Illustrantes, 5 vols., 1829-34; Flora Altaica, 4 vols., 1829-34; Flora Rossica (one of best works on flora of Russia), 4 vols., 1842-53; Died Munich, Germany, July 4, 1851.

VON LEWENHEIMB (or Löwenheim), Philip Jacob Sachs, physician; b. Breslau (now Wroclaw), Silesia, Aug. 26, 1627; ed. Leipzig, Saxony; grad. Phil. Magister, 1648; studied medicine univs. Netherlands, Strasbourg (France), Paris (France), Montpellier (France), Padua (Italy); M.D., 1651; m., 1653. Practiced medicine Breslau, apptd. Stadt-physikus Breslau, 1670. Mem. Academia Naturae Cariesorum (which, as a result of Von L's influence, obtained patronage of Leopold I, Hapsburgs), 1677, and title Imperial Leopoldine Acad., 1687). Author: Ampelographia, Sive Vitis Viniferae Ejusque Partium Considerato Physico-Philogico-Historico-Medico-Chymica, 1661; Responsoria Dissertatio de Mirand Lapidum Natura, 1664; Oceanus Macro-Microcosmicus, seu Dissertatio Epistolica de Analogo Mota Aquarum ex et aD Oceanum Sanguinis ex et ad Cor, 1664; Gammarologia, id est, Gammarorum Sive Cancrorum Considerato, 1665. Editor Ephemerides, 1666. Died Jan. 7, 1672.

VON LEYDIG, Franz, German histologist; anatomist; b. Rothenburg, Bavaria, May 21, 1821; studied medicine and natural sci., Würzburg, Munich. Asst. prof. U. Tübingen, 1849, extraordinary prof., 1855, prof., 1857; prof. U. Bonn, 1875, ret., 1895. Author: Beiträge zur nukrostropischen Anatomie und Entwickelungsgeschichte des Rochen und Haie, 1952; Anatomisch-Histologische Untersuchungen über Fische und Reptilien, 1853; Lehrbuch der Histologie des Menschen und der Tiere, 1857; Naturgeschichte der Daphniden, 1860; Das Auge der Gliedestiere, 1864; Vombau des tierischen Körpers, 1864; Der Einstock und die Samentasche der Insekten, 1866; Über die Molche der württembergischen Fauna, 1868; Die in Deutschland lebenden Arte der Saurier, 1872; Die Hartdecke und Schale der Gashopoden, 1876; Die anuren Batrachier der deutschen Fauna, 1877; Die augenahnlichen Organe der Fische, 1881; Untersuschungen zur Anatomie und Histologie der Tiere, 1883; Zelle und Gewebe, 1885; Das Parietalorgan der Amphibien und Reptilien, 1890; Zur Kenntnis der Zeibel und der Parietalorgane, 1896; Horae zoologicae zur vaterländischen Naturkunde, 1902, also numerous articles. Demonstrated importance of and developed use of microscope in anat. demonstration; his work characterized by extreme precision; 1st to describe interstitial cells of testis believed to secrete male hormone, 1850 (called Leydig cells); founder of comparative histology. Died Rothenburg, Apr. 13, 1908.

VON LICHTENBERG, Alexander, urologist; b. Budapest, Hungary, Jan. 20, 1880; M.D., Budapest, 1902. Worked under Czerny and Narath, Heidelberg, Germany, 1902-08, then under Madelung, Strasbourg, France, 1908-18; apptd. asso. prof. surgery, Berlin, Germany, 1922. Author: Circulatory Disturbances in Peritonitis; Sodium Chloride Suprarenin Therapy, 1909; General Diagnosis and Symptomatology in Urology; General Diagnosis by Means of X-ray, 1929. Editor: Zeitschrift für Urologische Chirurgie; Jahresberichte für Urologie; Hanbuch für Urologie. Introduced (with M. Swick) Uroselection in radiography, 1929; recommended method of injection of Bismuth suspension, to make kidney, pelvis and ureter opaque to X-rays, in order to diagnose tumors; described operation for hydronephrosis, 1929; researches on Bottini's method of removing obstruction to urinary flow by thermogalvanic diuresis of prostate with modern instruments. Died 1949.

VON LIEBIG, Baron Justus, see Liebig, Baron Justus von.

VON LINDENAU, Baron Bernhard August, German astronomer; b. Saxe-Altenburg, Germany, June 11, 1780; served King of Saxony in various positions including councillor, minister of interior, 1826-43; ret. from polit. life, 1843; dir. Obs. Seeberg, later Obs. Gotha. Mem. French Acad. Scis. Publs. include: Tabulae Veneris, 1810; Tabulae Mortis, 1811; History of Astronomy during the First Decade of the Nineteenth Century, 1811. Died Saxe-Altenburg, May 21, 1854.

VON LITTROW, Joseph Johann, astronomer; b. Bischof-Teinitz, Bohemia, Mar. 13, 1781; a son, Karl Ludwig. Became prof. astronomy, Cracow, Poland, 1807, Kazan, Russia, 1810; dir. Obs. Vienna (Austria), 1819-40; prof. astronomy U. Vienna. Mem. French Acad. Scis., 1838, Acad. St. Petersburg. Author: Theoretical and Practical Astronomy, 1822-26; Diotropics, 1830; Die Wunder des Himmels, 1834-36; Atlas des Gestirnten Himmels, 1838. His work on telescope constrn. led to dialytic tlescope; studies on refraction of light; improved Obs. of Vienna. Died Vienna, Nov. 30, 1840.

VON LÖHNEYSS, Georg Engelhard, metallurgist, chemist; flourished mid to late 16th century. Entered service of Heinrich Julius of Brunswick-Wolfenbüttel, 1583. Author: Von Zeumen, circa 1588; Bericht vom Bergwerck, wie man dieselben bauen, circa 1617. Described refining of gold by antimony, preparation of sulphur, vitrols and alum, prodn. and refinement of saltpeter, aqua fortis; method for testing bismuth ore; 1st description of zinc with method for its extraction.

VON LUSCHKA, Hubert, German anatomist; b. Konstanz, Germany, 1820; student Freiberg, 1841-43, Heidelberg, 1843-44. Prof. anatomy, Tübingen, Germany, 1849-75. Author textbook of human anatomy in 3 vols.; also numerous articles. First description of polyposis of colon, 1861; described anatomy and physiology of larynx, especially arytenoid and asso. cartilages (one is Luschka's cartilage), 1873, glomus coccygens (Luschka's gland), Luschka's bursa in pharnyx of fetus or infant, 1873, lateral openings in 4th ventricle of brain (Luschka's foramen), atypical bile ducts in gall bladder wall; ileocolic fold named after him. Died 1875.

VON MÄDLER, Johann Heinrich, German astronomer; b. Berlin, Germany, May 29, 1794; ed. as tchr.; student astronomy, Berlin. Worked with student, Wilhelm Beer in obs. (built by Beer) for several years; prof., dir. Dorpat (now Tartu, Estonia) Obs., 1840-65. Author: Mappa selenographica, 1834-36; Der Mond nach seinen komischen und individuellen Verhaltnissen oder allgemeine, 2 vols.; Undersuchungen über der Fixsternsysteme, 2 vols., 1847-48; Die Eigenbewegung der Fixsterne, 1854; Der Fixsternhimmel, 1858; Gegenstände der Himmelskunde, 1870; Reden und Abhandlung über Geschichte der Himmelskunde, 2 vols., 1872-73, Drew (with Beer) 1st reliable map of moon which helped establish lunar nomenclature, 1837, 1st map of Mars, 1840; studies on determination of orbits; measured double stars, distances of fixed star systems; believed stars revolve around center of gravity of entire mass of stars which he placed in Pleiades nr. Alcyone. Died Hanover, Germany, Mar. 14, 1874.

VON MANGOLDT, Hans, German mathematician; b. Weimar, Germany, May 18, 1854; became prof. math. Tech. High Sch., Hannover, Germany, 1884, Aachen, Germany, 1886, Danzig, Poland, 1804. Author: Einfhrung üin die höhere Mathematik für Studierende und zum Gelbststudium, 1911. Research in number theory. Died Danzig, Oct. 27, 1925.

VON MANNERHEIM, Carl Gustav, entomologist; b. 1804. Gov. Finland. Pres. Kayserlichen Hofgerichtes, Viborg, Denmark. Decorated Cross of Order of St. Stanislaus, knight Order St. Vladimir. Mem. Entomol. Soc. France, others. Contbr. numerous articles on Coleoptera. Major work on insect collections in museums Dorpat (now Tartu, Estonia), St. Petersburg, Moscow (both Russia); described large number of beetles from Siberia, Alaska, Cal. Died Stockholm, Sweden, Oct. 9, 1854.

VON MARTIUS, Karl Friedrich Philipp, German naturalist; b. Erlangen, Bavaria, Apr. 17, 1794; studied medicine in Erlangen. Joined sci. expdn. to Brazil, 1817-20; prof. botany, Munich, 1826-64, also dir. botanic garden. Fellow Royal Soc., 1938, French Acad. Sci., 1826. Author: Historia naturalis palmarum, 3 vols., 1824-53; Brazilian Travels, 1824; Nova genera et species plantarum, 3 vols., 1827-32; The Plants and Animals of Tropical America, 1831; Flora Brasiliensis, 1840. Wrote on plants, animals, inhabitants and langs. of Brazil; discovered spiral tendency in plants, in particular studied palms. Died Munich, Dec. 13, 1868.

VON MAYER, Julius Robert, German physicist, physician; b. Heilbronn, Württemberg, Germany, Nov. 25, 1814; M.D., U. Tübingen (Germany); postgrad. Munich, Germany, Paris, France. Ship's surgeon, voyage to E. Indies, 1840; practice medicine Heilbronn, from 1841. Author: Bermerkungen über die Kräfte der unbelebten Natur, 1842; Die organische Bewegung im Zusammenhang mit dem Stoffwechsel, 1845; Beiträge zur Dynamik des Himmels, 1848; Bemergungen über das mechanische Äquivalent der Wärme, 1851. Determined quantitatively mech. equivalent of heat; studied independently principle of conservation of energy (1st law thermodynamics), also extended it to include living phenomena and applied it to many cosmic and terrestial phenomena. Died Heilbronn, Mar. 20, 1878.

VON MERING, Baron Joseph, German physician; b. Cologne, Germany, 1849; became prof., Strasbourg, France, 1886; named dir. Halle (Germany) Med. Polyclinic, 1891. Produced diabetes in animals by excision of pancreas, 1886; introduced (with E. Fischer) barbital, 1902. Died 1908.

VON MIDDELDORPF, Albrecht Theodor, German surgeon; b. Breslau, Prussia, 1824. Prof. at Breslau. Credited with performing 1st operation for tumor of esophagus, 1857, excision of cancer of tongue, 1854, relief of gastric fistula, 1859; also credited with introducing improved galvanocautery for use in major surgery, 1854; made various substantial contbns. to subject of fractures and dislocations. Died 1868.

VON MIKULICZ-RADECKI, Johann, surgeon; b. Czernowitz (now Ukrainian SSR), May 16, 1850; M.D., Vienna, 1875. Became asst. to Billroth, 1875; named prof. surgery, Krakow, Poland, 1882, Königsberg (now Kaliningrad, USSR), 1887, Breslau (now Wroclaw, Poland), 1890-1905. Research and publs. in goitre, cancer, blood loss replacement, anesthesia, asepsis; described chronic enlargement of lacrimal and salivary glands (Mikulicz's disease), 1892, certain large cells occurring in rhinoscleroma, 1876; introduced gauze drain for wound cavity, 1881; 1st to use electric esophagoscope, 1881; reported a resection and plastic reconstrn. of esophagus, 1886; originated surg. operation for removal of diseased segment of colon (Mikulicz's resection of intestine); invented surg. face mask and various surg. instruments. Died June 4, 1905.

VON MONAKOW, Constantin, neurologist; b. Bobrezowa, Vologda, Russia, Nov. 4, 1853. Prof. in Zurich. Research on structure and parts of brain; described rubospinal tract, 1909 (known as Monakow's bundle, fibers, tract); described function of red nucleus, 1909. Died Zurich, Switzerland, Oct. 19, 1930.

VON MÜLLER, Baron Ferdinand, botanist; b. Rostock, Germany, June 30, 1825; bot. studies in Schleswig and Holstein, 1840-47; went to Australia, 1847; traveled through South Australia, 1847-51; mem. expdn. exploring North Australia; botanist for Victoria; dir. Bot. Gardens, Melbourne, Australia, 1857-73. Fellow Royal Soc., 1861; corr. mem. French Acad. Scis., 1895. Author: Fragmenta phytographiae Australiae, 11 vols., 1858-81; Plants Indigenous to the Colony of Victoria, 2 vols., 1860-65; (with Bentham) Flora Australiensis, 7 vols., 1863-70; Eucalyptographia, 1879-82; Iconography of Australian Species of Acacia and Cognate Genera, 1887-88. Research (with Spallanzani) on fertilization of basil, hemp, spinach; discovered and described numerous plants from S. Australia and alpine Victoria; introduced blue-gum tree (Eucalyptus globulus) to So. Europe, and nontropical regions of S.Am. and Cal. Died Melbourne, Oct. 9, 1896.

VON MURALT, Johannes, Swiss physician; b. Zurich, Feb. 18, 1645. Practice medicine, Zurich; became prof. physics and math. U. Zurich, 1645; founder anat. collegium, Zurich, 1686. Author: Chirurgische Schriften, 1699; Dissertation Physicae experimentalis, 1705; Experimenta anatomica, 1670; Vademecum anatomicum, 1677. Tried to unite medicine and surgery; held 1st anat. lectures for practicing surgeons in vernacular. Died Feb. 12, 1733.

VONNEGUT, Bernard, Am. phys. chemist; b. Indpls., Aug. 29, 1914; s. Kurt and Edith (Lieber) V.; B.S., Mass. Inst. Tech., 1936, Ph.D., 1939; m. Lois Gloria Bowler, Dec. 25, 1943; children—Peter, Scott, Terrence, Kurt, Alex. With Hartford Empire Co., 1939-41; research asso. Mass. Inst. Tech., 1941-46; with Gen. Electric Co., Schenectady, N.Y., 1946-53; mem. staff Arthur D. Little, Inc., Cambridge, Mass., 1953—. Mem. Am. Meteorol. Soc., Am. Geophys. Union. Contbr. articles to profl. jours. Devised technique for studying nucleation of supercooled liquid metals; discovered silver iodide cloud seeding; research in origin and role of thunderstorm electricity. Home: 204 Merriam St., Weston, Mass. 02193. Office: Arthur D. Little, Inc., Acorn Park, Cambridge, Mass. 02140.*

VON NEUMANN, John, mathematician; b. Budapest, Hungary, Dec. 28, 1903; s. Max and Margaret (Kann) V.; student Berlin U., 1921-23, Zurich Inst., 1923-25; Ph.D., Budapest, 1926; D.Sc. Princeton, 1947, U. Pa., Harvard, 1950, Case Inst. Tech., U. Istanbul, 1952, U. Md., 1952, Munich Inst. Polytechnics, 1953, Columbia U., 1954; m. Mariette Kovesi, Jan. 1, 1930; 1 dau., Marina; m. 2d, Klara Dan, Dec. 18, 1938. Privatdozent mathematics, Berlin U., 1927; visiting prof. mathematical physics, Princeton U., 1930, prof., 1931-33, prof. Institute for Advanced Study, 1933. Mem. and consultant various Army, Navy, O.S.R.D., AEC, Committees, 1940—; appointed to membership AEC, October, 1954. Received Medal for Merit and Distinguished Civilian Service award, 1947; Medal of Freedom, 1956, Albert Einstein award, 1956, Enrico Fermi award, 1956. Fellow Am. Physical Society; mem. Am. Math. Soc. (pres. 1951-53), American Mathematical Association, National Academy of Scis., Am. Philos. Soc., Am. Acad. Arts and Scis.; corr. mem. Royal Dutch Acad. Scis. (The Hague), Istituto Lombardo (Milan, Italy); associate mem. Academia Nacional de Ciencias Exactas, Lima, Peru. Mem. Sigma Xi. Author: Mathematical Foundation of Quantum Mechanics, 1931; (with O. Morgenstern) The Theory of Games and Economic Behavior, 1944. Contbr. articles on math. subjects. Editor Annals of Mathematics (Princeton). Co-editor Compositio Mathematica (Amsterdam, Holland). Developed game theory, 1928; showed how game theory could be applied to social scis., esp. econs.; showed math. equivalence of Schrödinger's wave mechanics and Heisenberg's matrix mechanics, 1944; contbn. to operator and set theories, math. logic and theory continuous groups, periodic functions, ergodic theorem, computer theory and design. Died Washington, Feb. 8, 1957.

VON NEUMAYER, Georg, German meteorologist, hydrographer; b. Kirchheimbolanden, Bavaria, June 21, 1826; ed. U. Munich; went to Australia under auspices of Maximilian II; founder, dir. Flagstaff Obs., Melbourne, 1857-64; returned to Germany, 1864; promoted German-African Co.; organized several expdns. to N. and S. polar regions; with hydrographic bur. Berlin; became dir. marine obs., Hamburg, Germany, 1876-1903. Mem. German Polar Commn. Author: Results of the Observations at the Flagstaff Observatory, Melbourne 1860-64, 3 vols.; Die Erforschung des Südpolargebietes, 1872; (with Börgen) Die Beobachtungsergibnisse der deutschen Stationen im Systeme der internationalen Polarforschung, 1886; Atlas der Erdmagnetismus, 1891; Anemometer-Studien, 1897; Auf zum Südpol! 1901. Died Neustadt/Hardt, May 24, 1909.

VON NEUSSER, Edmund, physician; b. Galicia, Austria, 1852; prof. internal medicine, dir. med. clinic, Vienna, prof. clin. medicine, 1892-1912; author: works on pellagra and other diseases; described basophilic granules (Neusser's granules) around nucleus of some leukocytes, 1889. Died 1912.

VON NOORDEN, Carl, German internist; b. Bonn, Germany, Sept. 13, 1858; prof., Berlin, Frankfurt/Main, Germany, Vienna, Austria. Author: Lehrbuch der Pathologie des Stoffwechsels, 1893; Die Zuckerkrankheit und ihre Behandlung, 1895; Die Fettsucht, 1900; (with others) Handbuch der Pathologie des Stoffwechsels, 1906-97; (with H. Salomon) Allgemeine Diätetik, 1920, Die Die spezielle Diätetik der Magenkrankheiten, 1929. Founder dietetics as science; research on diabetes mellitus, nutrition and metabolism; introduced calculation of diabetic diet in white bread units; discovered effect of oats in diabetes, also of fruit diet for dehydration. Died Vienna, Oct. 26, 1944.

VON OETTINGEN, Wolfgang Felix, toxicologist; b. Marburg, Germany, Dec. 3, 1888; s. Wolfgang and Karoline (Wilmanns) von O.; Ph.D., U. Göttingen, 1912; M.D., U. Heidelberg, 1916; m. Helene Maria Bauer, Dec. 27, 1916; children—Elizabeth Leonore, Ursula Marianne. Came to U. S., 1924, naturalized, 1929. Faculty, Western Res. U. Sch. Medicine, 1925-34; dir. Haskell Lab. Indsl. Toxicology, E. I. Du Pont de Nemours Co., 1934-38; prin. indsl. toxicologist; div. indsl. hygiene NIH, Bethesda, Md., 1938-46, chief indsl. toxicologist, 1946-54, med. officer in toxicology, 1954-58, ret., 1958. Cons. to various govt. agys. Recipient Superior Service award Dept. Health, Edn. and Welfare, 1955. Fellow A.A.A.S.; mem. Am. Soc. Exptl. Pharmacology and Therapeutics (emeritus), Am. Indsl. Hygiene Assn. (hon. mem.), Soc. Toxicology (hon. mem.), Sigma Xi. Author several books including: Poisoning—Clinical Diagnosis and Treatment, 1952, 58; Handbook of toxicology, vol. I, 1956, The Halogenated Hydrocarbons of Industrial and Toxicological Importance, 1964; also monographs, numerous articles. Research on toxicology of indsl. poisons, clin. diagnosis and treatment of poisonings. Address: 5802 Wyngate Dr., Bethesda, Md. 20034.*

VON OLSHAUSEN, Robert, German gynecologist, obstetrician; b. Kiel, Germany, July 3, 1835; s. Justus Olshausen. Prof. gynecology and obstetrics Berlin U.; dir. Univ. Clinic for Women, Berlin. Introduced operations for excision of vagina, 1895, ventral fixation as independent procedure in displacements of uterus, 1886; made important use of curette in Germany. Died Feb. 1, 1915.

VON OPPOLZER, Theodor Egon, astronomer; b. Prague, Bohemia, Oct. 26, 1841; s. Johann Von Oppolzer. Prof. in Vienna. Author: Lehrbuch der Bahnbestimmung der Kometen und Planeten, 2 vols., 1870-80; Canon der Finsternisse, 1887; a table of lunar and solar eclipses from 1207 B.C. to 2163 A.D. Died Vienna, Dec. 26, 1886.

VON PAYER, Julius, Austrian explorer; b. Teplitz, Schönau, Sept. 1, 1842; explored Alps; leader expdn. on Tegetthoff to Arctic; discovered and named Franz Joseph Land. Author: Die Ortleralpen, 4 parts, from 1867; Die österreichisch-ungarische Nordpolexpedition in den Jahren 1872-74, 1876. Died Veldes, Aug. 30, 1915.

VON PECHMANN, Hans, German chemist; b. Nürnberg, Germany, Apr. 1, 1850; student U. Munich (Germany), 1869-71; doctorate U. Greifswald (Germany), 1875; prof. chemistry; joined U. Munich, 1877; with U. Tübingen (Germany), last 7 years of life. Established (independently of C. Loring Jackson) symmetrical structure of anthraquinon, 1879; synthesized (with Carl Duisberg) beta-methylumbelliferone; developed procedures of substituted coumarins (von Pechmann-Duisberg procedures); developed von Pechmann coumarin synthesis; discovered and named coumalinic acid, 1884; 1st preparation and study of acetonedicarboxylic acid; 1st preparation of diazomethan, 1894, 1,2,-diktones, methylglyoxal, diphenyltriketone; discovered osotriazoles and (independently of Bamberger) formazyl compounds, 1891. Died Tübingen, Apr. 19, 1902.

VON PECHMANN, Hubert Freiherr, German forester; b. Munich, Germany, July 19, 1905; s. Friedrich and Sophie Elisabeth (Dornier); Ph.D., U. Fribourg (Germany); m. Gertrud Maria Schnitzler, Oct. 12, 1933. Insp. water and forests; pres. Bayerisches Forstant, Tegernsee, Germany; instr. forestry sci. U. Munich, prof., 1948—. Mem. Verein zum Schutze der Alpenflanzen und Tiere. Author articles on sylviculture. Editor: Forstwissemschaftiches Centralblatt. Home: Munchnerstrasse 12, Tegernsee/Obb. Office: Universität München, Munich, West Germany.

VON PETTENKOFER, Max Joseph, German chemist; b. Lichtenheim, Bavaria, Dec. 3, 1818; ed. Royal Ct. Pharmacy, 1839, Würzburg and Giessen; M.D., U. Munich, 1843; m. Helen Pettenkofer; several children. Employed in mint, 1845-47; prof. med. chemistry U. Munich, 1847-50, ordinary prof., 1852, prof. hygiene, 1865; with physiol. Inst., Munich, 1852-78, Inst. Hygiene (his own), 1878; chief Ct. Pharmacy, peronal apothecary to ct., 1850-96; chmn. Royal Commn. on Cholera, 1854. Recipient Odre pour le Mérite in scis. and art, 1883. Mem. Bavarian Acad. Scis. (pres. 1899), Council Medicine. Author: Untersuchungen und Beobachtungen über die Verbrietungsart der Cholera, 1855; Das Kanal-oder Sielsystem in München, 1869; (with Ziemssen) Handbuch der Hygiene, 1882. Opposed Koch's view that subsoil water played main role in cholera epidemics (Boden theory); discovered influence of nourishment on composition of urine; discovered creatine; founded modern sci. hygiene; devised Pettenkofer test for bile in urine; detected arsenic by Marsch apparatus; separated arsenic from antimony; discovered how to use gas from wood for light; with his pupil (Carl von Voit) made 1st studies of metabolism in health and disease, 1st accurate studies of caloric values of numerous foodstuffs; Bavarian founder of exptl. hygiene and 1st hygienic inst. at Munich; studied home ventilation, recommended quarantine of travelers from plague areas, 1875. Died nr. Munich, Feb. 10, 1901.

VON PFAUNDLER, Leopold, Austrian physicist, chemist; b. Innsbruck, Austria, 1839; studied under Wurtz and Regnault in Munich and Paris; student U. Innsbruck (Austria); became prof. physics, Innsbruck, 1867; named dir. Inst. Physics, Graz, Austria, 1891. Author: (monograph with L. Barth) Die Stubaier Gebirgsgruppe, 1865. Research on molecules, flavones, flourine. Died 1920.

VON PFOLSPEUNDT, Heinrich, surgeon, physician; b. 1450-60; probably ed. under German and Italian tchrs.; surgeon Bavarian army; mem. Deutscher Orden. Author: Buch der Bündth-Ertznei 1460, pub. 1868. Wrote 1st surg. book in German, mainly on arrow wounds; made 1st reference to bullet extraction and facial plastic surgery; appears to have known Indian methods of plastic surgery. Died 1533.

VON PHILIPSBORN, Helmut, German mineralogist; b. Berlin, Germany, 1892; s. Ernst and Josepha (von Meibom) P.; ed. univs. Lausanne, Munich and Berlin; m. Annelene Michel, July 4, 1925; children—Rüdiger, Volker, Henning. Referendary, 1913-14; asst. U. Giessen, 1925-27, Berlin Tech. Coll., 1927-29; qualified at Berlin Tech. Coll., 1928; prof. mineralogy Friberg Sch. Mines, 1929, resigned, 1945; curator, later prof. applied mineralogy U. Bonn, 1950——. Mem. German Mineralogy Soc., French Soc. Mineralogy and Crystallography. Author: Tabellen zur Berechnung von Mineralund Gesteinsanalysen, 1933; Tafeln zum Bestimmen der Minerale bach äusseren Kennzeichen, 1954; Erzunde, 1964; also articles on petrology industry, biocrystallography, history of microscopy. Address: Mineralogisches Inst., Poppelsdorfer-Schloss, Bonn 53, West Germany.*

VON PIRQUET, Clemens (Freiherr), Austrian physician, pathologist; b. nr. Vienna, May 12, 1874; became prof. pediatrics Johns Hopkins, 1908, Breslau (now Wroclaw, Poland), 1910-11, Vienna, 1911; gen. commr. Am. Relief Adminstrn. for Austrian Children, 1919-23. Author: Die Serumkrankheit, 1905; Klin. Studien über Vakzination und vakzinale Allergien, 1907; Allergie des Lebansalters, die bösartigen Geschwülste, 1930. Developed diagnostic reaction for Tb; coined word allergy. Died Vienna, Feb. 28, 1829.

VON PLENCIZ, Marcus Antonius (Plenkiz), bacteriologist, physician; b. near Görz, Austria, Apr. 28, 1705; student medicine, Vienna, Austria; M.D. under Morgagni, Padua, Italy. Practiced medicine, Vienna. Author: Opera medico-physica; Tractatus de scarlatina. Suggested all diseases are caused by micro-organisms. Died 1786.

VON PLENCK, Joseph Jacob, Austrian physician; b. Vienna, Austria, Nov. 28, 1738; prof., Basel, Switzerland, Vienna. Author: Anfangsgründe der Geburtshülfe, 1769; Doctrina de morbis cutaneis, 1776. First classification of skin diseases, 1776; proposed induction of early childbirth to reduce number of Caesarean sections. Died Aug. 24, 1807.

VON PROWAZEK, Stanislaus, zoologist; b. Neuhaus, Nov. 12, 1875; prof. Hamburg Tropical Inst. Author: Einführung in die Physiologie der Einzelligen, 1910; Handbuch der pathogenen Protozoen, 3 vols., 1912-31; Herpetomonasmuscoendomestice. Studied disease-causing protozoa and problems of immunology; described analogous process of multiplication of individual male protozoa and coined term etheogenesis for the process; eponym of typhus bacillus Rickettsia Prowazekii. Died Cottbus, Germany, Feb. 17, 1915.

VON RECKLINGHAUSEN, Friedrich Daniel, German pathologist; b. Gütersloh, Germany, Dec. 2, 1833; M.D., Berlin, 1855. Asst. to Virchow, 1855-61; became prof. pathology U. Königsberg (now Kaliningrad, USSR), 1865; named prof. Würzburg, Germany, 1866, 1st prof. pathology U. Strasbourg (now France), 1872-1906. Author: Über die multiplen Fibrome der Haut, 1882. Described smallest lymph channels in connective tissue (Recklinghausen's canals), 1862, multiple neurofibromatosis (Recklinghausen's disease) and osteitis fibrosa cystica (Recklinghausen's bone disease), 1882, histiocytes; introduced term hemochromatosis, 1889. Died Strasbourg, Aug. 26, 1910.

VON REICHENBACH, Georg, German optician; b. Durlach, Germany, Aug. 24, 1772. Founder factory, Munich; founder (with Liebherr and Utzschneider) Optic-Mechanical Benedict-Beuren Inst., 1804. Mem. French Acad. Scis., 1815. Built telescopes and other astron. and math. instruments. Died Munich, May 21, 1826.

VON REICHENBACH, Baron Karl, German naturalist; b. Stuttgart, Germany, Feb. 12, 1788; Ph.D., U. Tübingen (Germany). Founder iron mfg. co. (1st large German firm of its kind), Villingen, also Hausach, Baden (both Germany), 1821-34; established several industries, Moravia. Author: Das Kreosot, 1833; Geologische Mittheilungen aus Marhen, 1834; Die Dynamide des Magnetismus, 1840; Odisch-Magnetische Briefe, 1852; Der sensitive Mensch und sein Verhalten zum Ode, 1854; Odische Erweiterungen, 1856; Köhierglaube und Afterwissenschaft, 1856; Aphorismen über Sensibilität und Od, 1866; Die Odische Lohe, 1867. Discovered paraffin, 1830, creosote, 1833; conducted studies on what he thought was new force or power (called od by him). Died Leipzig, Germany, Jan. 19, 1869.

VON RHEITA, Anton Maria Schyrlaus, astronomer; b. Bohemia, circa 1597. Capuchin monk. Invented 4-lens terrestrial telescope, 1645; created words ocular and objective to describe lens. Died Revenna (Italy), 1660.

VON RICHTHOFEN, Ferdinand Paul Wilhelm (Baron), German geographer, geologist; b. Karlsruhe, Baden-Württemberg, May 5, 1833; studied chemistry, physics and geology at Breslau, later at Berlin. Made investigations in Dolomites and Translyvania; invited on German econ. mission to Far East as geologist, 1860; visited Ceylon, Japan, Formosa, Celebes, Java, P.I., traveled Malay Peninsula from Bangkok to Moulmein; carried out geol. investigations in Cal., 1863-68; traveled in China, 1868-72; returned to Germany to prepare his writings, 1872; prof. at Bonn, 1875, began duties, 1879; prof. Leipzig, 1883, Berlin, 1886. Organizer Berlin Internat. Geog. Congress, 1899. Recipient Founders medal Royal Geog. Soc., 1878. Mem. Berlin Geog. Soc. (pres. 1873-78), Berlin Inst. Oceanology (founder), French Acad. Sci., 1894. Author: China, Ergebnisse eigener Reisen und darauf gegrundeter Studien, 5 vols., 1877-1912; Aufgaben und Methoden der heutigen Geographie, 1883; Füher für Forschungreisende, 1886; Triebkräfte und Richtungen der Erdkunde im neunzehnten Jahrhundert, 1903; Tagebücher aus China, 2 vols., 1907; Vorlesungen über allgemeine Siedlungs- und Verkehrsgeographie, 1908. Made major contbns. towards establishment of sci. of geomorphology, gave 1st systematic treatment of this area; also contbd. to devel. of geog. methodology in general; did important studies on geol. construction of China; his work in Cal. led to discovery of gold fields there. Died Berlin, Germany, Oct. 7, 1905.

VON ROKITANSKY, Baron Karl, pathologist, physician; b. Königgrätz, Bohemia, Feb. 19, 1804; student U. Prague (Czechoslovakia); M.D., U. Vienna (Austria), 1828. Asst., Path. Inst., Vienna; asst. prof. U. Vienna, 1833-44, prof. path. anatomy, 1844-75, rector. Mem. French, Vienna acads. sci. Author: Handbuch der pathologischen Anatomie, 3 vols., 1842-46; Lehrbuch der pathologischen Anatomie, 1855; Der selbstandige Wert des Wissens, 1867. First to describe spondylolisthesis and to differentiate lobar, bronchopneumonia; described appearance of lungs in emphysema; research on diseases of arteries; pioneer clin. pathology; performed thousands of autopsies. Died Vienna, July 23, 1878.

VON SACHS, Julius, German botanist; b. Breslau, Germany (now Wroclaw, Poland), Oct. 2, 1832; studied under Purkinje; grad. Ph.D., U. Prague, 1856. Privat dozent, U. Prague; asst. Agricultural Acad., Tharandt, 1859; dir., Polytechnic, Chemnitz, 1861; prof. botany, Freiburg, 1867; prof., U. Würzburg, 1868. Author: Handbuch der Experimetalphysiologie der Pflanzen, 1865; Lehrbuch der Botanik, 1868; Vorlesungen über Pflanzenphysiologie, 1882; Geschichte der Botanik vom 16 Jahrhundert bis 1860, 1875; Gesammelte Abhandlungen über Pflanzenphysiologie, 1892-93. Works mainly on nutrition of plants; influence of heat and light on plant growth; discovered that chlorophyll contained in bodies called chloroplasts and not dissufed in tissues; introduced clinostat, which he used in work on heliotropism and geotropism. Died Würzburg, Bavaria, May 29, 1897.

VON SCHAFHÄUTL, Karl Emil, German naturalist; b. Ingolstadt, Germany Feb. 16, 1803; prof. geognosy, Munich, Germany; music historian. Author: Geognostische Untersuchungen des sübayerischen Alpengebirges, 1851; Der Gregorianische Choral in seiner Entwicklung, 1869. Expert on mining and foundries; discovered nitrogen in iron, 1838. Died Munich, Feb. 25, 1890.

VON SCHELLING, Hermann, mathematician; b. Berlin, Germany, May 25, 1901; s. Ulrich and Lina (von Jagemann) von S.; Ph.D., Berlin U., 1931; m. Hildegard D. Bockhorn, Aug. 6, 1943; Came to U. S., 1947, naturalized, 1954. With Astron. Computational Lab., Berlin, 1928-36; biomathematician Med. Research Lab., U. S. Submarine Base, New London, Conn., 1947-56; applied mathematician Gen. Electric Co. Research and Devel. Center, Schenectady, 1956-65. Mem. Internat. Statist. Inst. Contbg. author: Bio-astronautics, 1964. Research and numerous publs. on concept of most frequent random walks, with applications to large scale weather phenomena as seen from weather satellites, meandering of rivers; discovered an affine invariant distance applied to color vision, depth perception, and to a perceptual theory of relativity, math. explanation for performance of our memory. Home: 143 Willow Lane, Scotia, N.Y. 12302. Office: Gen. Electric Co., P.O. Box 8, Schenectady 12301.*

VON SCHERER, Alexander Nicolaus, chemist; b. St. Petersburg (now Leningrad, USSR), Dec. 30, 1771; Ph.D. (under Göttling, Voigt), U. Jena (Germany), 1794; studies in chemistry, tech., Eng., Scotland. Tchr. pub. courses chemistry Weimar, Germany; ordinary prof. physics U. Halle (Germany), from 1800; mgr. stoneware factory of Baron von Eckarstein, Potsdam, Germany; prof. chemistry, Dorpat (now Tartu, Estonia), 1803; prof. Medico-Chirurg. Acad., St. Petersburg, from 1804. Named Bergrath by Grand Duke of Weimar. Founder, Naturforschende Gesellschaft Jena, Pharm. Soc. St. Petersburg (pres. until 1824); mem. St. Petersburg Acad. Scis., other European sci. socs. Author: Grundriss der Chemie zu Vorlesungen; Versuch einer populären Chemie, 1795; Grundzüge der neuen chemischen Theorie. Editor: Allgemeines Journal der Chemie, 1798-1803; Archiv für die theoretische Chemie; Nordische Blätter für die Chemie; Allegemeine Nordische Annalen der Chemie. Contbd. account of compilers of pharmacopoeias to Codex Medicamentarius Europaeus. Died St. Petersburg, Oct. 28, 1824.

VON SCHLECHTENDAL, Dietrich Franz Leonard, German botanist; b. Xanten, Germany, Nov. 27, 1794; prof., Halle, Germany. Author: Flora Berolinensis, 2 vols., 1823-24; (with others) Flora von Deutschland, 24 vols., 1840-75. Died Halle, Oct. 12, 1866.

VON SCHLOTHEIM, Ernst Friedrich Freiherr, German paleontologist; b. Almenhausen, Germany, Apr. 2, 1764; privy councillor, pres. chamber Ct. of Gotha, Germany. Author: Beschreibungmerkwürdiger Kräuterabdrücke und Pflanzenversteinerungen, 1804; Ein Beitrag zur Flora der Vorwelt, 1804; Die Petrefakterkunde, 1820. Developed a theory of evolution on basis of fossil remains; named fossils according to binominal system for 1st time in Germany. Died Gotha, Mar. 28, 1832.

VON SCHMOLLER, Gustav, German economist, historian; b. Heilbronn, Germany, June 25, 1838; prof., Halle, Germany, Strasbourg (now in France), Berlin; Prussian ct. historiographer. Author: Geschichte des deutsche Kleingewerbes im 19. Jahrhundert, 1870; Über einige Grundfragen des Rechts und der Volkswirtschaft, 1875; Die Strassburger Tucher- und Weberzunft, 1879; Zur Literaturgeschichte der Staats- und Sozialwissenschaften, 1888; Uber Wesen und Verfassung der Grossen Unternehmungen, 1889; Umrisse und Untersuchungen zur Verf.- und Wirtschaftsgeschichte besonders des preuss Staates im 17. und 18. Jahrhundert, 1898; Grundrisse der allgemaine Volkswirtschaftslehre, 2 vols., 1900-04; Deutsche Staatswesen in älterer Zeit, 1922; Deutsche Städtewesen in älterer Zeit, 1922. Founder Acta Borussica, Jahrbuch für Gesetzgebung, Verwaltung und Volkswirtschaft. Founder new hist. sch. of economics; attempted to interrelate economics and other social scis.; believed in regulation of economy by state as an ethical establishment. Died Bad Harzburg, Germany, June 27, 1917.

VON SCHRANK, Franz Paula, naturalist; b. Varnbach, Austria, Aug. 21, 1747; ed. Vienna, Oedenburg, Raab, Tyrnau; mem. Soc. Jesus; ordained, Vienna, 1774, doctorate in theology, 1776; became tchr. Coll., Linz, Austria, 1769; tchr. math., Amberg, tchr. rhetoric, Burghausen; apptd. prof. botany, Ingolstadt, 1784; dir. bot. garden, Munich, Germany, 1809-35: Mem. Munich Acad. Scis. Author: Enumeratio insectorum austriae, 1781; Naturhistoriche Briefe, 2 vols., 1785; Fauna boica, 1789-1803; Bayrische Flora, 2 vols., 1789; Primitiae florae Salisburgensis, 1792; Flora Monacensis, 8 vols., 1811-18; Plantae rariores horti academici Monacensis, 2 vols., 1819. Died Munich, Mar. 22, 1835.

VON SCHUBERT, Gotthilf Heinrich, German natural scientist; b. Hohenstein, Germany, Apr. 26, 1780; practiced medicine, Altenburg; prof., Erlangen, Germany, also Munich. Author: Handbuch der Naturgesch., 1813-23; Symbolik des Traums, 1814; Altes und Neues aus den Gebiete der inneren Seelenkunde, 5 vols., 1817-44; Die Krankheiten und störungen der mesnchlichen Seele, 1845; Vermischten Schriften, 2 vols., 1856-60. Died Laufzorn, Upper Bavaria, July 1, 1860.

VON SCHWEINITZ, Lewis David, Am. botanist; b. Bethlehem, Pa., Feb. 13, 1780; s. Baron Hans Christian Alexander and Anna Dorothea Elizabeth (de Watteville) von S.; entered Moravian Theol. Sem., Niesky, Silesia, Germany, 1798; m. Louisa Amelia Ledoux, 4 children, including Edmund Alexander de Schweinitz. Tchr., Moravian Theol. Sem., Niesky, 1800-07; pastor, Gnadenberg, Germany, 1807-14, Gnadou, Saxony, 1807-12; gen. agt. Moravian Ch., Salem, N.C., 1812-21, adminstr. northern province, 1821-34. Author: Fungi of Lusatia, 1805; Fungi of North Caroline, 1818; Narrative of an Expedition to the Source of St. Peter's River, 1824; A Synopsis of North American Fungi (description of 3,098 species belonging to 246 genera including 1,203 new species and 7 new genera), 1831; also pamphlet describing 76 Hepaticae, 1821, monograph on genus Viola, naming 5 new species, submitted to Am. Jour. Sci., 1821. First to describe fungi in N.C. and Pa.; discovered over 1000 new species; Schweinitzia Ororata (North Atlantic plant) named in his honor. Died Bethlehem, Feb. 8, 1834.

VON SEIDEL, Philipp Ludwig, German astronomer, mathematician; b. Zweibrücken, Germany, Oct. 24, 1821; prof. math., Munich. Author: Untersuchungen über die geneseitigen Helligkeiten der Fixsterne I. Grösse und über die Extinktion des Lichtes in der Atmosphäre, 1852; Untersuchungen über die Lichtstärke der Planeten Venus, Mars, Jupiter und Saturn, 1859; Resultate photometrischer Messungen an 208 der vorzüglichsten Fixsterne, 1862. Research on refraction and dispersion relationships of light in various media; 1st exact measurement (with Steinheil) of intensity of celestial bodies. Died Munich, Aug. 13, 1896.

VON SENGBUSCH, Reinhold, German biologist; b. Riga, Feb. 16, 1898; s. Reinhold and Hanna (Becker) von S.; Dr ès sc, U. Halle; m. Ursula Klein, Jan. 14, 1935; children—Kurt, Werner, Günter, Karin. Asst. div. head Kaiser Wilhelm Inst., Muncheberg; head Research Center, Luckenwalde; now dir. Max Planck Inst. Cultivated Plant Improvement, Hamburg-Volksdorf; hon. prof., 1958. Corr. mem. Berlin Agronomy Acad. Sci., Acad. nazionale di agricoltura. Author: Süslupine und Oellupine, 1942; Die Geschlechtsvererbung bei Hanf und die Züchtung eines monözischen Hanfes, 1943; SENGA-Erdbeersorten, 1951; Aktivmycelspickung von Champignonkulturen, 1959; Nierensteinauflösung beim Menschen durch Komplexbildung, 1960. Research on plant animal breeding, especially monoccious hemp, perennial rye, disease resistant tomatoes, boneless carp. Address: Waldredder 4, Hamburg-Volksdorf, West Germany.

VON SIEBOLD, Carl Theodor Ernst, German zoologist; b. Würzburg, Germany, Feb. 16, 1804; prof. physiology, comparative anatomy and vet. sci. U. Erlangen (Germany), U. Fribourg-en-Brisgau, 1845, U. Breslau (now Wroclaw, Poland), 1850, U. Munich (Germany), 1853; in charge zool. collection, Munich. Mem. French Acad. Scis., 1867. Author: Lehrbuch der vergleichenden Anatomie de wirbellosen Tiere, 1848. Infected dogs with Taenia echinococcus, 1854; 1st demonstration of complete life cycle of a parasite, using sheep staggers, 1854; rediscovered parthenogenesis in insects; studied intestines of worms and fish, devel. of jellyfish, protozoa, classification of invertebrates; showed unicellular organisms could use their hairlike projections for locomotion. Died Munich, Apr. 7, 1885.

VON SIEMENS, Wilhelm, German physicist, engr.; b. Berlin, Germany, 1855; s. Ernst Von Siemens; consolidated firms of Ernst Von Siemens & Halske. Studied theory of incandescent lamps, 1883, utilization of alternating current for electric rays. Died Arosa, 1919.

VON SOEMMERING, Samuel Thomas, naturalist, anatomist; b. Thorn, Prussia, Jan. 25, 1755; M.D., U. Göttingen (Germany), 1778; worked with John and William Hunter, Monro, Pieter Camper, in Eng., Scotland, Holland; prof. anatomy Carolinum Coll., Cassel, Germany; prof. anatomy, Mainz, Germany, 1784-95; practiced medicine, Frankfort/Main, Germany, 1797-1803; became asso. with U. Heidelberg, 1803; named royal physician of Bavaria, 1805; worked in Munich; mem. Munich Acad. Scis., Imperial Acad. Scis. St. Petersburg (hon.). Applied physiology to anatomy; descriptive anatomy said to have reached its height in his work; gave study of path. anatomy a precise sci. base; specialized in lesions and disorders of sense organs; demonstrated that human races differ somewhat in morphology; gave early description of achondroplasia; introduced classification of cranial nerves which superseded that of Thomas Willis, 1778; described long pudendal nerve (Soemmering's nerve), 1790; gave 1st full description of macula lutea (Soemmering's spot), 1795-98 (observed 1791); described adverse effects of tight corsets on internal organs (led to decrease in use), 1793; clearly described suspensory ligament of lacrimal gland (Soemmering's ligament), 1802; studied sun spots, meteors, fossils; inventor electric telegraph (1st practical system proposed for wireless communication), 1809. Died Frankfort/Main, Mar. 2, 1830.

VONSOVSKII, Sergéi Vasilievich, Russian physicist; b. Sept. 2, 1910; grad. Leningrad U., 1932. Joined staff Ural Phys.-Tech. Inst., Sverdlovsk, after graduation; staff Inst. for Physics Metals, Urals affiliate USSR Acad Sci., 1939—, later dep. dir.; prof. Urals U., 1944—; bur. mem. dept. physico-math. sci. USSR Acad. Sci., 1960—. Recipient Order of Red Banner Labor, 1960. Author: (with Y. S. Shur) Ferromagnetism, 1948; Contemporary Theory on Magnetism, 1953; Nuclear Research Methods in Solid Physics, 1963. Research and publs. on magnetism, theories of transition metals, exchange in ferrites; devel. (with Shubin) polar and intercharge theories; co-developer gen. ferro-magnetic theory; explained ferro-magnetic phenomena (divisibility of atomic moment). Office: USSR, Sverdlovsk, ul Belinskogo 71a, Gosudarstvenny universitet.

VON SPANHEIM, Ezechiel, scholar; b. Geneva, Switzerland, Dec. 18, 1629; prof., Geneva; ambassador of Brandenburg to Paris and London. Mem. Berlin Acad. Scis. (a founder). Wrote hist., numis. and cultural hist. works. Died London, Nov. 25, 1710.

VON STAUDT, Karl Georg Christian, German mathematician; b. Rothenburg, Germany, Jan. 24, 1798; prof., Erlangen, Germany; Author: Geometrie der Lage, 1847; Beiträge zur Geometrie der Lage, 1856-60. First complete theory of imaginary points lines and planes in projective geometry; made constrn. of regular inscribed polygon of 17 sides using only compasses. Died Erlangen, June 1, 1867.

VON STEINHEIL, Carl August, German physicist; b. Ribeauville, Alsace, Oct. 1801; student law U. Erlangen (Germany), astronomy Göttingen, Germany, Königsberg (now Kaliningrad, USSR). Prof. math. and Physics U. Munich; named head telegraph dept. Austrian Ministry Commerce, 1849; became mfr. optical, astron. and photog. materials, 1854, also made telescopes for obs. of Uppsala, Sweden, Mannheim, Germany, Leipzig, Germany. Steinheilite or iolite, a transparent mineral, was named in his honor. Devel. Austrian telegraph system; built 1st printing telegraph, 1836; discovered possibility of using earth as return conductor in telegraphy, 1838; invented electromagnetic telegraph, electric clock, pyroscope; 1st daguereotype picture in Germany; also optical instruments, electromagnetic motor. Died Munich, Sept. 12, 1870.

VON STERNBERG, Graf Kaspar Maria, botanist, paleontologist; b. Prague, Czechoslovakia, Jan. 6, 1761; dean cathedral of Regensburg, Germany. Author: Versuch einer geognotische-botanischen Darstellung der Flora der Vorwelt, 2 vols., 1820-32; Umriss einer Gesch. der böhmischen Bergwerke, das., 2 vols., 1836-38. Wrote on botany and paleontology; founder Bohemian Nat. Mus.; studied culture of Bohemia. Died Brezina, Raudnitz, Dec. 10, 1838.

VON STERNECK, Robert Daublebsky, see Daublebsky von Sterneck, Robert.

VON STORCK, Anton Baron, physician; b. 1731; succeeded van Swieten as tchr. and physician-in-ordinary to Viennese Ct. Research and publs. on hemlock, meadow-saffron, aconite, stramonium, henbane. Died 1803.

VON STRÜMPELL, Adolf, German internist; b. Neue-Autz, Germany, June 28, 1853; s. Ludwig Strümpell; became prof., Leipzig, Germany, 1882, Erlangen, Germany, 1886, Breslau (now Wroclaw, Poland), 1903, Vienna, 1909, Leipzig, 1910. Author: Lehrbuch der spezieuen Pathologie und Therpaie der inneren Krankheiten, 2 vols., 1883-84; Aus Demleben eines deutschen Kliniken. Research on in neurology; mentioned ankylosing spondylitis, 1884, described it, 1897; described a form of acute encephalitis of infants, 1885; gave 1st descriptions of Spondylitis deformans, hereditary spinal paralysis, 1886; described pseudosclerosis, 1898. Died Leipzig, Jan. 9, 1925.

VON SUCHTEN, Alexander, physician; ed. Cracow, Poland, Louvain, Belgium, Italy; became canon, Frauenberg 1539; apptd. librarian to Elector Otto Heinrich, 1549; became physician to King Sigismund Augustus, Cracow, 1554. Author: Liber unus de secretis antimonii . . ., 1570.

VON TRÖLTSCH, Baron Anton Friedrich, German otologist; b. Schwabach, Germany, Apr. 3, 1829; student law, Erlangen, Germany, 1947, natural sci., Munich, 1848, medicine Würzburg, 1849-53. Visited Dublin, London, Paris, beginning 1854; prof. Würzburg, Germany, 1864-90. Author: Die Anatomie des Ohres in ihrer Anwendung auf die Praxis und die Krankheiten des Gehörorganes, 1861; Gesammelte Beiträge zur pathologischen Anatomie des die Anatomie des Ohres in ihrer Anwendung auf die Praxisund die Krankheiten des Gehörorganes, 1861; Lehrbuch der Ohrenheilkunde, 1862; Die Chirurgischen Wundkrankheiten des Ohres, 1866; also articles. Founder, Archiv für Ohrenheilkunde, 1864. Originated 1st modern otoscope, 1860; 1st modern mastoid operation, 1861; 1st to spread use of centrally perforated concave mirror. Died Würzburg, Jan. 9, 1890.

VON VERSCHUER, (Frieherr) Otmar, German physician, anthropologist; b. Richelsdorfer, Germany, July 16, 1896; s. Hans and Charlotte (von Arnold) von V.; ed. univs. Marburg, Hamburg, Freiburg; M.D., Munich, 1923; m. Erika Flad, 1925. Became research asst., Munich, 1923; named privatdozent U. Tübingen (Germany), 1927; became dept. head, dir. Kaiser Wilhelm Inst. for Anthropology, Racial Biology of Mankind and Eugenics, 1942; named prof. U. Berlin, 1933; apptd. dir. Hereditary Biol. Inst., Frankfort U., 1935; joined Inst. Human Genetics, Münster (Germany) U., 1951; dir. Max Planck Inst. for Anthropology. Mem. acads. of Berlin, Mainz, Vienna, German Acad. Natural Sci.; hon. mem. Italian Soc. Med. Genetics, Japanese Soc. for Human Genetics, Anthropol. Soc. Vienna. Author: Zwillingstuberkose, 2 vols., 1932-36; Erbpathologie, 1934; Wirksame Faktoren im Leben des Menschen, 1954; Genetik den Menschen, 1959; Erblehre von Menschen, 1963; also numerous articles. Research on twins, hereditary pathology. Address: 27 Waldeyerstrasse, Münster, Germany.

VON VOGEL, Heinrich August, German chemist; b. Westerhof, Germany, 1778; prof. Paris, Munich. Author: Lehrbuch der Chemie, 2 vols., 1830-32. Research on natural substances; believed nature of light based on chem. reactions. Died Munich, 1867.

VON VOIT, Karl, German physiologist; b. Amberg, Germany, Oct. 31, 1831; student Munich (under Liebig), Würzburg, Göttingen (all Germany); M.D., 1854. Asst. Physiol. Inst., U. Munich, 1856-60, extraordinary prof. physiology, 1860-63, ordinary prof., dir. inst., 1863-1908. Co-founder Zeitschrift für Biologie, 1864. Author series standard texts on metabolism in humans, animals. Contbr. numerous articles to sci. publs. Worked with M. J. von Pettenkofer; devised calorimeter large enough to hold human being; made 1st determination of amounts of protein, fat, carbohydrates broken down by body, 1862; basic work on chem. processes undergone by foodstuffs after entering body; developed test for studying nitrogen intake-output; began work which led to discovery of essential amino acids; pioneer in studies on nutrition, metabolism. Died Munich, Jan. 31, 1908.

VON VOLKMANN, Richard, German surgeon; b. Leipzig, Germany, Aug. 17, 1830; student medicine Halle, Giessen, Berlin (all Germany). Prof. surgery, dir. clinic, Halle, from 1867. Author: Beiträge zur Churugie, 1873. Contbr. sect. on diseases locomotive organs Pitha-Billroth Handbuch der Churugie, 1865-72. Editor: Sammlung Klinischer Vorträge, 1870-89. Co-founder surgery on joints, extremities; pioneer in introduction of antiseptics into Germany; 1st to describe cancer from irritation of skin by coal tar, paraffin, 1873; 1st to excise rectum for cancer, 1878; described contracture of fingers or wrist as result of injury, pressure (Volkmann's contracture), 1881. Died Jena, Germany, Nov. 28, 1889.

VON WASSERMANN, August, German bacteriologist; b. Bamberg, Germany, Feb. 21, 1866; student at Erlangen, Munich, Vienna; M.D., Strasbourg, 1889; became dir. dept. exptl. therapy and biochemistry Koch Inst. for Infectious Diseases, 1906; dir. dept. exptl. therapy Kaiser Wilhelm Inst., Berlin-Dahlem, Germany, from 1913. Author: Hämolysine, Cytotoxine und Präzipitine, 1902; (with W. Kolle) Handbuch der pathogenen Mikroorganismen, 6 vols., 1903-09; also articles. A leading immunologist; extended serodiagnosis; developed (with Albert Neisser and Carl Bruck) specific blood test (Wassermann test) for diagnosis of syphilis, recorded in 1906. Died Berlin, Mar. 16, 1925.

VON WEISÄCKER, Carl Friedrich, German physicist; b. Kiel, Germany, June 28, 1912; ed. univs. Berlin, Göttingen and Leipzig. Asst., U. Leipzig, 1933-36, Kaiser Wilhelm Inst. für Physik, Berlin/Dahlem, 1936-41; prof. U. Strasbourg, 1942-45; dir. dept. Max-Planck Inst. für Physik, Göttingen, 1946-57; prof. philosophy U. Hamburg, 1957—. Recipient Goethe prize, Arnold-Raymond prize; decorated Orden Pour le Mérite. Author: Die Atomkerne, 1937; Zum Weltbilt der physik, 1944; Die Geschichte der Natur, 1948; Die Verantwortung der Wissenschaft im Atomzeitler, 1957; Atomenergie und Atomzeitler, 1957; (with J. Juilfs) Physik der Gegenwart, 1958; The Revelance of Science, 1964. Research in atomic physics; suggested long lifetime of excited nuclei due to difference in angular momentum between ground state and isomeric state, 1936; refuted collision theory of planetary cosmology and restated parthogenetic theory basing argument on gaseous composition of the universe. Address: Schwarzbuchenweg 40, Wellingsbuettel, Hamburg, German Federal Republic.

VON WEIZSACHER, Viktor (Freiherr), German neurologist; b. Stuttgart, Germany, Apr. 21, 1886; prof., Breslau (now Wroclaw, Poland), also Heidelberg, Germany. Author: Studien zur Pathogenese, 1935; Der Gestaltkreis, 1940; Körpergeschehen und Neurose, 1947; Natur und Geist, 1954; Der kranke Mensch, 1954; Pathosophie, 1956. Pioneer in unified psychosomatic medicine; creator Gestaltkreis doctrine; interpreted disease in man as a part of the self. Died Heidelberg, Jan. 9, 1957.

VON WEIZSÄCKER, Carl Friedrich Freiherr, German philosopher, physicist; b. Kiel, Germany, June 28, 1912; s. Ernst Freiherr and Marianne (von Graevenitz) von W.; student U. Berlin (Germany), 1929, Dr. phil., U. Leipzig (Germany), 1933, Habilitation, 1936; m. Gundalena Wille, Mar. 30, 1937; children —Carl-Christian, Ernest-Ulrich, Elisabeth (Mrs. Raiser), Heinrich. Asst., Kaiser-Wilhelm Inst., Berlin, 1936-42; prof. U. Strassburg, 1942-44; staff Max Planck Inst., Göttingen, 1946-57; prof. philosophy U. Hamburg (Germany), 1957—. Recipient Max-Planck medaille, 1958; Goethe-Prize, Frankfurt, 1958; Bundesverdienstkreuz, 1959; Friedensklasse Orden Pour-le mérite, 1961; Friedenspreis des deutschen Buchhandels, 1963. Mem. Max-Planck-Gesellschaft, Deutsche Akademie der Naturforscher, Deutsche Akademie für Sprache und Dichtung, Deutsches P.E.N. Zentrum der Bundesrepublik, Göttinger Akademie der Wissenschaften, Vereinigung deutscher Wissenschaftler. Author: Die Atomkerne; Atomenergie und Atomzeitalter, 1958; (with J. Juilfs) Physik der Gegenwart, 1958; Die Geschichte der Natur, 1962; Die Verantwortung der Wissenschaft in Atomzeitalter, 1963; Zum Weltbild der Physik, 1963; Die Tragweite der Wissenschaft, 1964. Research on astrophysics and cosmology, theory of origin of solar system, galactric systems and evolution of stars; studies in atomic physics, including axiomatic fqund. of quantum theory, quantum logic, and a unified theory of elementary-particle physics with cosmology. Home: 2 Hamburg 64, Schwarzbuchenweg 40, Germany.*

VON WELSBACH, Carl Auer (baron), Austrian chemist; b. Vienna, Sept. 1, 1858; s. Aloys Auer von W. Welsbach; ed. Vienna, Heidelberg; pupil of Bunsen; research on rare earths; discovered neodymium and praseodymium; decomposed ytterbium, 1906; inventor incandescent gas mantle (Welsbach mantle), 1885, osmium lamp, 1897, pyrophoric ferrocerium alloy (used in gas and cigarette lighters), 1903, also Welsbach burner. Died Treibach, Aug. 4, 1929.

VON WETTSTEIN, Dietrich Holger, geneticist; b. Göttingen, Germany, Sept. 20, 1929; s. Fritz and Elsa (Jesser) von W.; student U. Tübingen (Germany), 1947-49, Dr.rer.nat., 1953; student Fed. Inst. Tech., Zürich, 1949-50; Fil.lic., U. Stockholm (Sweden), 1953, Fil.dr., 1957; m. Penelope Margaret Knowles, Aug. 5, 1967. Research asst. dept. genetics Royal Coll. Forestry, Stockholm, 1951-57; asst. prof. Inst. Genetics, U. Stockholm, 1957-61; prof. genetics, head Inst. Genetics U. Copenhagen (Denmark), 1962—. Mem. bd. directing research in phytotron Royal Coll. Forestry, Stockholm, 1963—. Mem. Royal Danish Acad. Sci. Research, publs. on cell biology and mutation research. Home: 13 Asevej Ll., Vaerløse, Denmark. Office: 2 A Øster Farimagsgade, Copenhagen K, Denmark.*

VON WETTSTEIN, Richard, botanist; b. Vienna, Austria, June 30, 1863; prof., dir. bot. garden, Prague, Czechoslovakia, also Vienna. Author: Lehrbuch der Botanik, 1890; Grundzüge der geographische Methode der Pflanzensytematik, 1898; Handbuch der systematike Botanik, 3 vols., 1902-24. Developed a modern, influential classification system of plants. Died Aug. 10, 1931.

VON WIESER, Friedrich, economist; b. Vienna, Austria, July 10, 1851; prof., Prague, Czechoslovakia, also Vienna. Author: Theorie der gesellschaftlichen Wirtschaft, 1914; Das Gesetz der Macht, 1926. A leader of Austrian sch. of economics; developed system of economics related to sociology; analysed manifestation of power, regarded it the pivot of world history; formulated Wieser's theory that marginal value equals marginal cost. Died Brunwinkel/St. Gilgen am Wolfgangsee, July 22, 1926.

VON WIESNER, Julius Ritter, botanist; b. Tschechen/Brno, Moravia, Jan. 20, 1838; prof. anatomy and plant physiology U. Vienna; mem. French Acad.

Scis., 1909. Author: Die Rohstoffe des Pflanzenreichs, 1873; Die heliotropischen Erscheinungen im Pflanzenreich, 2 vols., 1878-80; Elemente der wissenschaftliche Botanik, 3 vols., 1881-89. Contbd. to knowledge and methodology of plant physiology. Died Vienna, Oct. 9, 1916.

VON WOLFF, Christian, see Wolff, Christian von.

VON WRANGEL, Ferdinand Petrovich, Baron (or Wrangell), Russian explorer; b. Pleskov, Russia, Dec. 29, 1796; apptd. comdr. expdn. to Arctic Sea, 1820, reached 72 degrees, 2 minutes north, on ice sledges; became gov. Russian possessions in northwestern Am.; made vice adm., 1847; mem. French Acad. Scis., 1856. Died Dorpat (now Tartu, Estonia), June 3, 1870.

VON WROBLEWSKI, Zygmunt Florenty, Polish physicist; b. Grodno, Poland, Oct. 28, 1845; student Kiev, Russia (banished to Siberia 1863-69 for participation in insurrection), then in Heidelberg and Berlin, Germany; degree, U. Munich (Germany); asst., phys. lab. U. Munich; lectr. univs. Strasbourg, Paris (both France) London, Oxford, Cambridge (Oil Eng.); apptd. prof. physics U. Cracow (Poland), 1882. First to liquify air on a large scale, 1885; transformed (with Karol Stanislaw Olszewski) large quantities of oxygen, nitrogen, carbon monoxide, also reduced nitrogen and carbon monoxide to solid state, thus creating low temperature tech. and making possible more accurate determination of specific heat. Died Cracow, Apr. 19, 1888.

VON WULFEN, Franz Xaver, botanist, mineralogist; b. Belgrade, Servia, Nov. 5, 1728; ed. Kaschau; became mem. Soc. Jesus, 1745; tchr. in Vienna, Laibach, then in Klagenfurt, Germany, from 1764. Genus Wulfenia and yellow lead ore wulfenite named in his honor. Mem. numerous sci. socs. Author: Abhandlungen vom karnthenschen pfauenschweifigen Helmintholith oder dem sogenannten opalisirend Muschelmarmor . . ., 1793; Flora norica phaneragama . . ., 1853. Discovered many new plant species of eastern Alps. Died Mar. 17, 1905.

VON WURZELBAU, Johann Philipp, German astronomer; b. Nuremberg, Germany, Sept. 28, 1651; staff obs. at Spitzenberg, Germany. Corr. mem. French Acad. Scis., 1699. Invented and improved various astron. instruments; astron. observations at Spitzenberg. Died Nuremberg, Mar. 22, 1725.

VON ZACH, Baron Franz Xaver, astronomer; b. Pressburg, Hungary, June 4, 1754; ed. Jesuit Coll. Lt.-col. in Austrian Army; went to Eng., 1783; entered service of Ernest II of Saxe-Coburg-Gotha; dir. Seeburg Obs.; accompanied Duchess of Saxe-Coburg-Gotha on travels in So. Europe, beginning 1806. Corr. mem. French Acad. Scis., 1805. Author: Novae et correctae tabulae motuum solis, 1792; Explicatio et usus tabellarum solis et catalogi stellarum fixarum, 1792; De vera latitudine et longitudine erfordiae, 1794; Nouveau calendrier séculaire français, 1797; also articles. Editor: Correspondance astronomique, géologique, hydrographique et statistiqué, 14 vols., 1818-26; Fixarum stellarum catalogus novus, 1804; Tabulae speciales aberrationis et nutationis, 2 vols., 1806-07; Monatliche Korrespondenz zur Beförderung der Erd- und Himmelskunde, 28 vols., 1800-13, Allgemeine geographische Ephermeriden, 4 vols., 1798-99. Rediscovered asteroid Ceres, 1801; determined geog. positions of several European cities and towns with sextant; 1st mention of Olbers technique of plotting cometary orbits, 1797; contributed to founding obs. at Lucca and Naples, Italy. Died Paris, Sept. 2, 1832.

VON ZENKER, Friedrich Albert, German pathologist; b. Dresden, Germany, Mar. 13, 1825. Prof., Dresden. Author: Über Trichinenkrankheit des Menschen . . ., 1860; Beiträge zur Normalen und Pathologishen Anatomie der Lungen, 1862. Described trichinosis as exactly defined disease; described colorless needle-like crystals of phosphate which were found in sputum of humans afflicted with bronchial asthma, other conditions (Charcot-Leyden-Zenker or Charcot-Leyden crystals), recorded 1851; 1st to describe pulmonary fat embolism, recorded 1862; invented fixing fluid made chiefly of mercury chloride and potassium dichromate, recorded 1894. Died Reppentin, Germany, June 13, 1898.

VON ZEPPELIN, Count Ferdinand, German inventor; b. Constance, Baden, Germany, July 8, 1838; ed. Ludwigsburg (Germany) Mil. Acad., U. Tübingen (Germany); D. Philosophy and Natural Sci. Commd. German Army, 1859, advanced through grades to gen., 1891; observer with Union Army, U. S. Civil War, 1863; served in 7 weeks war, 1866, Franco-Prussian War, 1870-71. Constructed 1st dirigible balloon of rigid type (aluminum frame), Friedrichshafen, Germany, 1900 (dirigible balloons often called zeppelins in his honor); zeppelins used for bombing raids on London, World War I (proved too vulnerable); Graf Zeppelin went around world, 1929; Hindenburg crashed, Lakehurst, N.J., 1937. Died Charlottenburg, Germany, Mar. 8, 1917.

VON ZIEMSSEN, Hugo Wilhelm, German physician; b. Griefswald, Germany, Dec. 13, 1829. Became prof. pathology, dir. Med. Clin., Erlangen, Germany, 1863; named prof., dir. Gen. Hosp., Munich, 1874. Editor: Handbuch der speziellen Pathologie und Therapie, 17 vols., 1874-84; Handbuch der allgemeinen Therapie, 1883-85; Klinische Vorträge, after 1887; co-editor: Handbuch der Hygiene und der Gewerbekrankheiten, 1882-86. Founder, editor (with F. A. von Zenker) Archiv für Klinische Medizin, 1865. Founder 1st specialized clin. lab. in Germany, Munich, 1885. Research in children's diseases, diseases of larynx and esophagus, electrotherapeutics; described therapeutic application of electricity to muscles, 1857, treatment for anemia using subcutaneous injections of defibrinated human blood, 1887, blood transfusion using a cannula, 1892; recommended baths of gradually lowering temperatures for hyperpyrexia; contributed to modernization of electrotherapy. Died Munich, Jan. 21, 1902.

VON ZITTEL, Karl Alfred (Ritter), German paleontologist, geologist; b. Bahlingen, Baden, Sept. 25, 1839; ed. in Heidelberg, Paris, Vienna. Asst. to royal mineral cabinet of Vienna, 1863; prof. mineralogy geognosy and paleontology Poly. Sch., Karlsruhe; prof. paleontology, Munich, 1866, later prof. geology and conservator geol. collections until his death; with F. G. Rohlfs's expdn. to Libyan dessert, 1873-74. Mem. Bavarian Acad., French Acad. Sci., 1900. Author: Beiträge zur Geologie und Palaeontologie der Libyschen Wüste, 2 vols., 1883-1906; Handbuch der Palaeontologie, 4 vols., 1876-93; Geschichte der Geologie und Paläontologie, 1899. Proved Sahara Desert had been land during Pleistocene Ice Age; accepted evolution, leader in applying it to paleontology, especially in his studies of ammonites; studied fossil sponges, established their classification, laid basis for that of modern forms, 1876; contbd. to vertebrate paleontology, dealt with turtles and pterodactyles of Bavarian lithographic limestones. Died Munich, Jan. 5, 1904.

VOORHORST, Reindert, Dutch physician; b. Wijhe, Netherlands, June 15, 1915; s. Aart and Geesje (Rensink) V.; student State U. Utrecht, 1932-39; doctorate State U. Groningen, 1940; m. Frieda Smeenk, Nov. 13, 1940; children—Lucius C. Frederick J., Catharinus D. Gen. practice medicine, Stadskanaal, 1940-45; with dept. internal medicine U. Groningen, 1945-46; physician Mil. Hosp., Dept Internal Medicine, U. Utrecht, 1946-50; internist Psychiat. Observation Clinic, Utrecht, 1947-56; staff mem. Microbiologic Lab., State U. Utrecht, 1950-56; allergist State U. Leiden, 1956—, lector allergology, 1961. Mem. med bd. Dutch Asthma Fund. Mem. Dutch Soc. for Allergology (pres.). Author: Basic Facts of Allergy, 1962; Het Atopisch Syndroom, de vreemde ziekte van Coca, 1966. Publs. on methods of counting vial tubercle bacteria; differences between effects of chemotherapeutics and disinfectants on tubercle bacteria; devel. of a new anti-hepatitis syringe; Thorn test (daily rhythm); clin. characteristics of Coca's atopic syndrome, of anaphylaxis, and of delayed-type allergy; house-dust allergen; eosinophilia. Home: 87 Plantsoen, Leiden, The Netherlands.*

VORIS, Frank Burkhart, Am. surgeon; b. East St. Louis, Ill., Apr. 27, 1909; s. Henry McMunn and Marjorie (Burkhart) V.; B.S., U. Ill., 1930, M.D., 1933; m. Shirley Ann Cannon, Mar. 11, 1944; children—Scarlett Francis (Mrs. Robert Dwyer), April Adelaide (Mrs. Thomas Schmoller), Henry McMunn II, Mary Frances, Georgia Cannon. Practice surgery, Miami Beach, Fla., 1936-41; commd. lt. (j.g.) USN, 1941, advanced through grades to capt., M.C., 1955; naval flight surgeon, 1942, naval aviator, 1948; sr. med officer Naval Air Sta., Barbers Point, Hawaii, 1948-49; asst. dir. Aerospace Crew Equipment Lab., 1949-50; chief spl. activities Bur. Medicine and Surgery, Washington, 1950-55; sr. med. officer U.S.S. Forrestal, 1955-57; asst. to chief naval operations for aviation medicine, 1957-61, dir. human research NASA, 1961-64; spl. asst. to chief naval material, life scis. and dir bio-astronautics Bur. Medicine and Surgery, Washington, 1964—. Mem. com. on vision Nat. Acad. Scis.-NRC, exec. council, 1962—, mem. com. on hearing, bioacoustics and biomechanics, mem. Man-in-Space com. NASA Space Sci. Bd.; mem. life scis. subpanel supporting space research and tech. NASA/DOD Aero. and Astronautics Coordination Bd: Recipient Aerospace Medicine Honor citation A.M.A., 1962. Diplomate Aviation Medicine, Am. Bd. Preventive Medicine. Fellow Aerospace Med. Assn. (exec. council 1961—, pres.-elect 1965); mem. A.M.A., A.C.S., Am. Coll. Preventive Medicine (v.p. aviation medicine 1960), Internat. Acad. Aviation and Space Medicine, Mil. Surgeons U. S. (Founder Medal 1952). Contbr. articles to sci. publs. Home: 10816 Stanmore Dr., Potomac, Md. Office: Bureau Medicine and Surgery, Navy Dept., Washington 20390.

VOROBIEV, Vladimir Petrovich, Russian anatomist; b. July 15, 1876; prof. U. Kharkov. Author: Atlas of Human Anatomy, 5 vols., 1938. Developed theory of macromicroscopy and stereomorphology. Died Nov. 31, 1937.

VORONOFF, Serge, physiologist; b. Voronezh, Russia, July 10, 1866; M.D., U. Paris; m. Gertrude Schwetz, 1934; became naturalized French citizen, 1897. Head exptl. surgery lab. Collège de France; surgeon in chief Russian Hosp., Paris; surgeon in chief Mil. Hosp. 197, World War I; dir. biology lab. l'Ecole des Hautes; in charge spl. sect. on bone and skin Grafting, French Army, Cannes, France, 1940; came to U. S. about 1940; named head mil. hosp., Cannes, 1939. Author: Greffes testiculaires, 1923; Rejuvenation by Grafting, 1925; Conquest of Life, 1928; Love and Thought in Animals and Men, 1937; (Contbr.) Sidelights from the Surgery, 1938; From Cretin to Genius, 1942; The Study of Old Age and my Method of Rejuvenation; co-author: Testicular Grafting from Ape to Man, 1933. Research on grafting of animal glands in humans, including testicle transplants in attempted rejuvenation, monkey bone graftings in wounded soldiers; advocated use of monkey thyroid glands in children with thyroid deficiency; proposed theory which related hormonal activity to aging; studies in cancer serum. Died Lausanne, Switzerland, Sept. 1, 1951.

VORONTSOV, Daniil Semenovich, Russian physiologist; b. Dec. 23, 1886; D.Biol.Sci., Physico-Math. Faculty, St. Petersburg U. Asso., Physiol. Lab., chair physiology St. Petersburg U., 1912-16; asst. chair physiology Bestuzhev Higher Women's Courses, Petrograd, 1914-16; asst., then lectr. Novorossiysk U., Odessa, 1916-22; prof. physiology Smolensk U., 1922-30, Kazan U. and Kazan Med. Inst., 1930-35, Kiev Med. Inst., 1935-41; prof. Kiev U., 1944—; dir. Lab. Electrophysiology, Inst. Physiology, Ukranian Acad. Sci., 1956—. Founder chair physiology Smolensk U., 1922; mem. editorial council Sechenov Physiol. Jour. of USSR. Mem. Ukranian Acad. Sci., 1957. Author: Electrogram of the Auricle, 1917; The Effects of Direct Current on a Nerve Treated with Water, 1924; The Electronic Reaction of the Cerebrospinal Roots, 1947; co-author The Physiology of Animals and Man, 1952; A Summary of the Electronic Reactions of the Cerebrospinal Roots, 1952; General Electrophysiology, 1961; The Effects of Strychnine Gamma-Aminobutric Acid, Acetyl Choline and Quinine on Development of the Physical Electron of the Nerve, 1963. Co-editor physiology sect. Large Med. Ency., 2d edit. Specialist on electrophysiology and nervous system; student of N.E. Vvedenskii; work in gen. physiology. Address: Ukranian SSR, Vladimirskaya 58, Gosudarstvenny Univ., Kiev, USSR.

VORONTSOV-VELIAMINOV, Boris Aleksandrovich, Russian astro-physicist; b. Feb. 14, 1904. Prof. Moscow U., 1934—. Recipient Bredikhin prize, 1962. Corr. mem. RSFSR Acad. Pedagogy Sci., 1947. Author: Catalogue of Planetary Nebulae and its Statistical Study, 1934; Course of Practical Astrophysics, 1940; The Blue-White Sequence on the Ressel Diagram, 1947; Gas Nelulae and Nova Stars, 1948; Outline History of Astronomy in Russia, 1956; Collected Problems and Exercises on Astronomy, 4th edit., 1957; Astronomy, 2d edit., 1957; (textbook) Astronomy, 1963. Research on gas nebulae and nova stars; compiled list of planetary nebulae, 1934; discovered blue-white star sequence of Hertzsprung-Ressel, 1947; catalogued and mapped 350 interacting galazies, 1959; work on morphological catalogue of 30,000 galaxies, 1962. Address: Leninski gory, Gosudarstvenny Univ., Moscow, USSR.

VOS, Isaac, see Vossius, Isaac.

VOSBURGH, Warren Chase, Am. chemist; b. Voorheesville, N.Y., June 14, 1894; s. Franklin Edward and Mary (Chase) V.; B.S., Union Coll., Schenectady, 1914, M.S., 1916; Ph.D., Columbia, 1919; m. Mary Marguerite Jones, June 5, 1927; children—Elizabeth (Mrs. Wallace B. Gordon), James R., Eleanor Rachel (Mrs. Lawrence M. Hettick). NRC fellow Columbia, 1919-21; dir. research Eppley Lab., Newport, R.I., 1921-25; instr. State U. Ia., Iowa City, 1925-28; faculty Duke, Durham, 1928—, prof. chemistry, 1934-59, prof. emeritus, 1959—; exchange tchr. U. Edinburgh (Scotland), 1932-33. Mem. Am. Chem. Soc., Electrochem. Soc., Sigma Xi, Phi Lambda Upsilon. Author: Qualitative Analysis, 1938; (with D. G. Hill, J. H. Saylor, R. N. Wilson) Elementary Chemistry, 1941; Quantitative analysis, 1941; also numerous articles. Research in methods analysis, complex salt formation in solution, voltaic cells and discharge mechanisms manganese oxide electrode. Home: Route 1, Box 377, Bahama, N.C. 27503.*

VOSKUYL, Roger John, Am. chemist; b. Cedar Grove, Wis., May 16, 1910; s. Anthony and Jennie (Baden) Voskuil; A.B., Hope Coll., 1932; M.A., Harvard, 1934, Ph.D., 1938; m. Gertrude Johanna Schaap, Aug. 15, 1935; children—Ruth (Mrs. J. Morey Ferguson), Nancy, Jane, Howard. Faculty, Wheaton (Ill.) Coll., 1938-50, prof. chemistry, dean, 1947-50; group leader Manhattan Project, 1943-45; pres. Westmont Coll., Santa Barbara, Cal., 1950—. Pres., Council for Advancement Small Colls., 1962-64. Mem. N.E.A., Sigma Xi. Contbr. sect. Modern Science and the Christian Faith, 1948. Developed a method of analyzing deuterium content in water of natural resources, method of analysis of heavy water; contbr. to determination of atomic weight of hydrogen. Home: 40 Cedar Lane, Santa Barbara, Cal. 93103.*

VOSSELER, Paul Christian, Swiss geologist; b. Gelterkinden, Mar. 20, 1890; s. Christian and Marie (Keiser) V.; ed. univs. Basel, Lausanne and Berlin; serum. Died Lausanne, Switzerland, Sept. 1, 1951. Ph.D.; m. Maris Zwicky, Oct. 16, 1944; children—Paul, Rosemarie, Beat, Martin. Prof. U. Basel. Mem. Schwizer Geographielehrer Verein, Verein Abstin Lehrer und Lehrerinnen; corr. mem. Neuchatel Geog. Soc.; hon. mem. Georg-ethnolog. Gesellschaft, Basel. Author: Murphologie des Aargauer Jurá, 1918; Einfürhing in die Geologie der Umgebung von Basel, 1938; Geomorphologische Forschung in der Schweiz, 1957. Address: 190 Bruderholzallee, Basel, Switzerland.

VOSSIUS, Isaac, physicist; b. Leyden, Netherlands, 1618; s. Gerard John and Elizabeth du Jon (Junius)

Vos; studied under his father and a pvt. tutor; D.C.L., Oxford, 1670. Traveled to Rome 1642; visited libraries in France and Eng.; went to Stockholm, Sweden by invitation of Queen Christiana, 1649; tutor to Queen Christiana, also collected Royal Library; left Sweden, 1652; visited Geneva, Paris; came to Eng., 1670; became prebend of Royal Chapel, Windsor, Eng., 1673, canon of Windsor. Author: De lucis natura et proprietate, 1662; Responsum, 1663; De motu mari[illegible] et ventorum, 1663; De nili et aliorum flum[illegible] origine, 1666; also numerous other publs. Pro[illegible] theory that light and heat are accidents; stu[illegible] capillarity, color, ocean currents; describe[illegible]scope for predicting storms at sea; attributed flooding of Nile to heavy rains in Ethiopia; attributed tides to effect of sun. Died Windsor, Feb. 21, 1689.

VOS VAN STEENWIJK, Baron Jacob E. de., Dutch astronomer; b. Zwolle, Holland, Apr. 30, 1889; s. Carel and Johanna (van Nes van Meerkerk), Vos van S.; Dr ès sc., U. Leyden; m. J. W. van Royen, May 4, 1916; children—Catharina, Carel, Johanna. Prof. secondary edn., head Inst. Intellectual Coop. UN; pres. U. Leyden, now curator. Burgermaster, Zwolle and Haarlem; commnr. of the Queen. Decorated Order Dutch Lion, Order Orange Nassau. Rotarian. Publs. on atronomy. Address: Duinweg 9, Aerdenhout, Netherlands.

VOTAVA, Zdenek, Czechoslovakian pharmacologist; b. Brno, Czechoslovakia, Jan. 11, 1914; s. Ludvik and Jaroslava (Kesslerova) V.; M.D., Charles U., Prague, 1945, Dr.Sc., 1959; m. Bozena Samesová, June 12, 1941; children—George, Zuzana (Mrs. George Jelínek), Zdenek. Head dept. pharmacology Research Inst. for Pharmacy and Biochemistry, Prague, 1946-59; prof., head dept. pharmacology Charles U. Sch. Medicine, Prague, 1955—; med. research asso. Thudichum Research Lab., State Research Hosp., Galesburg, Ill., 1966-67. Mem. Collegium Internat. Neuropsychopharmacologicum (v.p. 1958). Author (with H. Rasková) Textbook of Pharmacology, 1961; Psychopharmacologic Methods, 1962; also numerous articles. Research in field of mechanism of action of central cholinergic drugs; pharmacology in new drugs with spasmolytics in sulfonium group, antihistaminics, tricyclic antidepressives effects. Home: 21 Capku St., Praha Czechoslovakia. Office: 50 Srobarova St., Praha, Czechoslovakia.*

VOTH, Paul Dirks, Am. botanist; b. Gotebo, Okla., June 12, 1905; s. Peter Richert and Anna (Dirks) V.; A.B., Bethel Coll., 1929; M.S., U. Chgo., 1930, Ph.D., 1933; m. Selma J. Graber, June 12, 1930; children—Felice Ann (Mrs. Clyde Richard Goering), Pamela Kathy (Mrs. Albert Edward Dahlberg). Tchr., Buhler, Kan., 1924-25; fellow in botany U. Chgo., 1932-33; instr. biology Tex. Technol. Coll., 1930-32; faculty botany, U. Chgo., 1933—, prof., 1948—; exchange instr. botany Cornell U., 1938. Bot. cons. film strips Ency. Brit. Films, Inc., 1958-61; bot. cons. World Book Ency., 1963, 65, 67—. Mem. Bot. Soc. Am., Ecol. Soc. Am., Am. Soc. Plant Physiologists, A.A.A.S., Am. Assn. U. Profs., Am. Bryol. Soc., Sigma Xi, numerous others. Asso. editor Bot. Gazette, 1942-46, 59—. Research, publs. on patterns, distbn. of tissue systems in ferns, other plants; vegetative reprodn., growth responses in Marchantia. Home: 5725 Maryland Av., Chgo. 60637.*

VOUTILAINEN, Antero, Finnish physician; b. Finland, Jan. 6, 1914; s. Antti and Lahja (Vainie) V.; Med.cand., U. Helsinki (Finland), 1938, registered physician, 1945, M.D., 1954. Lectr., U. Helsinki, 1957-65, prof. roentgenology Oulu (Finland) U., 1965—; prof. radiotherapy Turku (Finland) U. Mem. Finnish Soc. Radiology (vice chmn. 1962-67), Duodecim, Finnish Physicians Union. Home: Mäntykallio A 5, Matinkylä, Finland. Office: Turku U., Radiotherapy Dept., Central Hosp., Turku, Finland.*

VOYACHEK, Vladimir Ignatevich, Russian physician; b. Dec. 7, 1876; ed. St. Petersburg Mil. Acad., 1899; M.D. Staff, Simonovskii's Clinic, 1899-1917; prof. otolaryngology Leningrad Mil. Med. Acad., Kirov, 1917, became v.p., asst. dir. acad., 1919, dir., 1925; founder sch. otolaryngology; founder dept. for treatment speech defects Leningrad Acad. Clinic; founder teaching methods mus. and new exptl. lab. Mem. USSR Acad. Med. Scis.; corr. mem. Otolarygologische Gesellschaft. Author: Practical Methods of Research into Labyrinth Functions, 1915; Ear, Nose and Throat Diseases, 2 vols., 1929; Problems of Practical Otolaryngology, 1930; Principles of Otolaryngology, 3d edit., 1939; Principles of Aviation Medicine, 1939; Military Otolaryngology, 3d edit., 1946. Improved and developed methods in eye, ear, nose, throat surgery; studies in inner ear; developed theory of air and sea sickness.

VRANCEANU, Gheorghe, Rumanian mathematician; b. 1900; prof. Bucharest (Rumania) U.; mem. Acad. Socialist Republic of Rumania. Author: Neolonomous Spaces, 1936; Sur la théorie des espaces à connexion affine, 1941; Lecons de géométrie différentielle, 3 vols., 1947-60. Founder sch. of global differential geometry in Bucharest; research on math. analysis, Riemann spaces, neolonomous spaces, theory of generalized relativity.

VROMAN, Leo, physiologist; b. Gouda, Netherlands, Apr. 10, 1915; s. Samuel Jacob and Anna (Vromen) V.; candidate biology Rijks Universiteit, Utrecht, Netherlands, 1937, Ph.D. in Animal Physiology 1958; doctorandus biology Med. Coll., Djakarata, Indonesia, 1941; m. Georgine Marie Sanders, Sept. 10, 1947; children—Geraldine Elizabeth, Peggy Ann. Came to U. S., 1945, naturalized, 1952. Research asso. St. Peter's Gen. Hosp., New Brunswick, N. J., 1947-55, Mt. Sinai Hosp., N.Y.C., 1955-58; sr. physiologist dept. animal behavior Am. Mus. Natural History, 1958-61; physiologist dept. med. service VA Hosp., Bklyn., 1961—. Recipient Netherlands State prize poetry, 1955. Mem. N. Y. Soc. for Study Blood, Poetry Soc. Am., A.A.A.S., Sigma Chi. Author numerous books including: Poems in English, 1953; Proza, 1960; 126 Gedichten, 1964; also articles. Research on behavior blood coagulation factors and other proteins at solid/liquid interfaces, adhesiveness blood platelets, theory blood clotting in relation to protein conformation and evolution. Home: 2365 E. 13th St., Bklyn. 11229. Office: VA Hosp., Bklyn. 11209.*

VUAGNAT, Marc Bernard, geologist; b. Annemasse, France, Jan. 30, 1922; s. Hubert and Louise (Rochat) V.; Lic. ès sc., U. Geneva (Switzerland), 1941, Dr. ès Sc., 1944; postgrad Swiss Fed. Inst. Tech., 1944-46; m. Anne-Marie Mermier, Mar. 22, 1954; children—Hubert Blaise, Bernard Bruno. Faculty, U. Geneva, 1946—, prof., head dept. mineralology, 1961—; asso. prof. U. Lausanne, 1956-61; acting asst. prof. Wash. State Coll., 1954-55. Mem. Societé suisse de Minéalogie et Pétrographie (past pres.), Mineral. Soc. Am., Am. Chem. Soc. Research and numerous publs. on petrology of basic and ultrabasic rocks, alpine graywackes. Home: 1249 Dardagny, Geneva, Switzerland.*

VUIA, Ovidiu, Rumanian neuropathologist; b. Arad, Rumania, Mar. 18, 1929; s. Tiberiu and Veturia (Bixa) V.; M.D., U. Cluj (Rumania), 1954. Chief, Neuropath. Lab., Cluj, 1955-62; researcher Neurol. Inst. Bucharest (Rumania), 1962—. Mem. Rumanian Neurol. Soc. Research, numerous publs. on path. aspects of neuroreticuloses, description of syndrome of neuromyositis, venous intracerebral system. Home: 55 Birsanesti, Et. VI, Sectorul V. Office: 42 Str. Povernei 30, Bucharest, Rumania.*

VUKOV, Konstantin, Hungarian chem. engr.; b. Törökbecse, Hungary, Mar. 5, 1920; s. Konstantin and Emilia (Kreuzberger) V.; Chem. engr. Budapest (Hungary) Tech. U., 1943, Doctor techn., 1960; D.Chem. Sci., Hungarian Acad. Sci., 1965; m. Irma Kovács, Aug. 7, 1944; children—Peter, Konstantin. Staff, Sugar Factory, Kaposvár, Hungary, 1943-51; chem. engr. Research Inst. Hungarian Sugar Industry, 1951—. Mem. sci. com. Commn. Internationale Technique de Sucrerie, Tirlemont, Belgium, 1964—. Research, numerous publs. on phys. and chem. properties of sugar-beet; phys. and chem. techniques in sugar-beet processing. Home: 19, Báthory utca, Budapest V. Office: 25, Tolnai Lajos utca, Budapest VIII, Hungary.*

VUL, Bentsion Moiseevich, Russian physicist; b. Belaya Tsericov, Russia, May 22, 1903; grad. Kiev Poly. Inst., 1928; completed postgrad. course Kiev Poly. Inst., 1929, USSR Acad. Sci., 1932; Dr. Phys.-Math. Sci., 1935. Joined Physics Inst., USSR Acad. Scis., 1932, dep. dir. Physics Inst., 1944—, in charge semiconductor research Physic Inst., 1948—, founder, head non-conductor lab. Lebedev Physics Inst. lectr. Zhukovsky Airforce Engring. Acad. 1939—. also staff Kiev Poly. Inst., later Leningrad Poly. Inst. Recipient Stalin prize, 1946, Order of Lenin, 3 times, Order Red Star. Corr. mem. USSR Acad. Sci. Research and publs. on physics of dielectrics and semi-conductors, end-effect dielectrics breakdown, titanium compounds, breakdown of compressed gases in heterogeneous fields; discovered ferro-electric-barium titanate, 1944; built quantum generator, 1962. Office: A. N. Lebedev Physics Inst., USSR Acad. Sci., Leninskii Prospekt, 53, Moscow, USSR.

VULF, Georgi Viktorovich, Russian crystallographer; b. Russia, June 10, 1863; prof. univs. Kazan, Warsaw, Moscow. Research and publs. in crystallography and crystallophysics; discovered Vulf-Breg law for X-ray analysis in crystallography. Died Dec. 25, 1925.

VULPIAN, Edme-Félix-Alfred, French physician; b. Paris, Jan. 5, 1826; Agrégé, Faculté de Médecine, 1860. Replaced Flourens at Muséum, 3 years; physician, Saltpetre Works; became prof. path. anatomy Faculté de Médecine, 1867, prof. comparative pathology, 1872, dean, 1875. Mem. French Acad. Scis. (became life sec. 1886), French Acad. Medicine. Author: Leçons sur la physiologie générale et comparée du système nerveux, 1866; Leçons sur l'appareil vasomoteur, 1875; Clinique médicale de l'Hôpital de la Charité, 1878; Cour de pathologie expérimentale, Maladies du système nerveux, 1879; Cours de pathologie expérimentale; Leçons sur l'action physiologique des substances toxiques et médicamenteuses, 1882. Research on nerve functions, pathology of nervous diseases, physiology of nervous system. Died Paris, May 18, 1887.

VUORELA, Lauri Antero, Finnish meteorologist; b. Kuopio, Finland, Feb. 27, 1913; s. Gustaf Adolf and Aina (Rossi) V.; Mag.Phil., U. Helsinki (Finland), 1945, Ph.D., 1951; m. Aini Lovisa Lax, Nov. 4, 1943. Forecaster, Finnish Meteorol. Office, Helsinki, 1941-46; meteorologist Swedish Metecrol. and Hydrological Inst., Stockholm, Sweden, 1946-50; research asst. Acad. Finland, Helsinki, 1950-58; faculty U. Helsinki, 1954—, prof. meteorology, head dept. 1958—, asso. dean sect. math. and natural scis. Phils. Faculty, 1960—, chmn. bd. Inst. Seismology, 1960—, chmn. bd. Computing Center, 1967—. Sec. Finnish nat. com. IGY, 1955-60; mem. Nat. Council for Sci. Research, 1961-65, vice chmn., 1964-65. Decorated comdr. Order Finland's Lion, 1965. Mem. ICSU, (sec. Finnish com. 1965—), IUGG (mem. Finnish com. 1959—, chmn. sect. meteorology 1961—), Finnish Geophys. Soc. (past pres.), Swedish Geophys. Soc., Am. Geophys. Unicn, A.A.A.S., Societas Scient. Fennica, Finnish Acad. Sci. and Letters (mem. bd., sec. sect. scis. 1964—). Exec. editor Geophysica, 1953-60, editor, 1960—. Research, publs. on gen. synoptics, tropical and arctic meteorology, gen. atmospheric circulation. Home: 15 A, Sinebrychoffinkatu, Helsinki 12, Finland.*

VVEDENSKII, Boris Alekseievich, Russian radio-physicist; b. Apr. 19, 1893; grad. Moscow U., 1915 Staff, All-Union Electro-Tech. Inst. 1927-35, named prof., 1929; staff Physics Inst., USSR Acad. Sci., 1941-44, mem. presidium, 1946-53, staff Inst. Radio Tech. and Electronics, 1953—; dir. constrn. ultra-short wave sta., 1929; chief editor Gt. Soviet Ency., 1953—. Dep. chmn. Com. for Stalin Prizes. Recipient Popov Gold medal, 1949, Order of Lenin, 3 times. Mem. USSR Acad. Sci. (acad. sec. 1946-51, chmn. sect. for processing problems of radio engring. 1944-53). Author: Physical Phenomena in Cathode Tubes, 4th edit., 1932; Basis of the Theory of Propagation of Radiowaves, 1934; (with A. G. Arenberg) Propagation of Ultra Short Waves, 1938; (with A. G. Arenberg) Charts for Calculating the Ultrashort Wave Field Beyond the Horizon, 1942; Problems of Ultrashort Wave Propagation, 1946; Radio Waves, 1946; Questions on Propagation of Ultra Short Waves, Part I, 1948. Research and publs. on ultrashort waves, magnetism, meter and decimeter sea waves; devel. diffraction formula; directed constrn. ultra-short wave sta., 1929. Address: USSR, Moskva, Pokrovsky b. 8, Red., Bolshaya Sovetskaya Entsiklopediya.

VYSHNEGRADSKI, Ivan Alekseievich, Russian engr.; b. Dec. 20, 1831; prof. St. Petersburg Technol. Inst.; minister finances 1888-92. Developed theory of automatic regulation; designed safer powder machines; invented regulator for direct current, 1877. Died Mar. 25, 1895.

VYSSOTSKY, Alexander N(icholas), astronomer, educator; b. Moscow, Russia, May 23, 1888; s. Nicholas G. and Alexandra A. (Atryganieva) V.; diploma 1st grade U. Moscow, 1912; Ph.D., U. Va., 1926; m. Emma T. R. Williams, Sept. 21, 1929; 1 son, Victor Alexander. Came to U. S., 1923, naturalized, 1929. Jr. astronomer Central Obs., Pulkovo, Russia, 1913-15; faculty U. Va., Charlottesville, 1923—, prof. astronomy, 1945-58, prof. emeritus, 1958—. Mem. Am. Astron. Soc., Internat. Astron. Union, Sigma Xi. Author: (with K. L. Bayeff) Atlas of Astronomical Illustrations, 1914; (with P. van de Kamp) A Study of Proper Motions, 1937; (with wife) An Investigation of Stellar Motions, 1948; also numerous articles. Research on stellar motions, structure of galaxy, spectra of faint stars, discovered faint red dwarf stars by their spectral characteristics on objective prism photographs. Address: 905 Garden Dr., Winter Park, Fla. 32789.*

W

WAAGE, Peter, Norweigian chemist; b. Flekkefjord, Norway, June 29, 1833; ed. Christiana (now Oslo, Norway) U.; prof. chemistry Christiana U., 1862-1900; developed (with Guildberg) fundamental law of mass action that rate of homogenous chem. reaction is proportional to products of molar concentrations of reactants, 1867. Died Christiana, Jan. 13, 1900.

WAAGEN, Wilhelm Heinrich, geologist; b. Munich, Germany, June 23, 1841; Ph.D., Munich; instr. paleontology, Munich; tutor to Princess Theresa and Prince Arnulf of Bavaria; with geol. surg survey of India, 1870-75; became instr., Vienna, 1877; apptd. prof. mineralogy and geology German tech. high sch., Prague, 1879; prof. paleontology U. Vienna, 1890—. Recipient Lyell medal Geol. Soc. London, 1898. Editor Geognostisch-paläontologische Beiträge; editor Beiträge zur Paläntologie, 1894-1900. Contbr. numerous papers on geology. Studied fossils from Lower Cambrian to Trias. Died Vienna, Mar. 24, 1900.

WAALER, Erik, Norwegian microbiologist; b. Hamar, Feb. 22, 1903; s. Per F. and Frederikke Amalie (Rynning) W.; M.D., U. Oslo; m. Esther Fasmer Dahl, Nov. 23, 1929; children—Julie, Per Erik, Alf, Gudmund, Anne Margrethe. Student microbiology and pathology U. Oslo, also N.Y.C.; asso. prof. pathology U. Oslo, 1938; pathologist-in-chief Bergen Hosp., 1941—; prof. pathology Bergen U., 1948—, dean Faculty Medicine, 1948-51, rector, 1954-59. Decorated Order St. Olav; recipient Fridtjof Nansen prize, 1964. Mem. Norwegian Acad. Sci.; hon. mem. Am. Rheumatology Soc., Finnish Med. Soc. Publns. on microbiology and pathology. Research in pathology and immunology of rheumatic disease and connective tissue disorders; immunopathology. Home: Kollen 2, Paradis. Office: Heukelandsveien 17, Bergen, Norway.

WABER, James Thomas, Am. physicist; b. Chgo., Apr. 8, 1920; s. James Warren and Anna (Cline) W.; B.S. in Chem. Engring., Ill. Inst. Tech., 1941, M.S., 1943, Ph.D., in Metallurgy, 1946; m. Santon Fotheringham, May 12, 1951; children—Lauriene, Sue, Gay, John. Research asst. prof. Ill. Inst. Tech., 1946-47; staff Los Alamos Sci. Lab., 1947-66; prof. materials sci. Northwestern U., Evanston, Ill., 1967—. Recipient Young Authors award Electrochem. Soc., 1945. Mem. Am. Inst. Mining, Metall. and Petroleum Engrs. (sec. nuclear metallurgy com. 1965—), Nat. Assn. Corrosion Engr. (Willis Rodney Whitney award 1962), Electrochem. Soc., Faraday Soc., Am. Soc. for Metals. Research, patentee, numerous publs. on metals especially theories of stress, corrosion cracking, applications of conformal mapping to study of corrosion, study and devel. alloys, calculation electronic band structure, research in atomic physics. Home: 2324 Hartzell St., Evanston, Ill. 60202.*

WACE, Alan John Bayard, English archaeologist; b. 1879; s. F. C. Wace; ed. Pembroke College, Cambridge U.; M.A., Litt.D.; hon. LL.D., Liverpool, hon. D.Litt. Amsterdam and Pennsylvania; m. Helen Pence, 1925; one daughter. Lectr., ancient history, archaeology, St. Andrews U., 1912-14; dir., Brit. School, Athens, 1914-23; Vanuxem Lectr., Princeton U., 1923; Norton Lectr., Am. Arch. Inst., 1923-24; deputy keeper, Victoria and Albert Museum, 1924-34; Armstrong Lectr., U. Toronto, 1939; Laurence Prof. Classical Arch., Cambridge U., 1934-44; prof. classics and arch., Farouk I U., Cairo, 1943-52. Fellow of British Academy; Fellow Soc. Antiquaries; fgn. mem. Swedish Acad., Am. Philos. Soc., German Arch. Inst.; hon. mem. Am. Arch. Inst., Greek Arch. Soc., Royal Northern Antiquaries. Author: Catalogue of Sparta Museum; Prehistoric Thessaly, 1912; Nomads of the Balkans; Excavations at Mycenae; Chamber Tombs at Mycenae; Cretan Statuette in Fitzwilliam Museum; Catalogue Algerian Embroideries; Sheldon Tapestries; Near Eastern and Mediterranean Embroideries; Approach to Greek Sculpture; Mycenae, an Archaeological History, 1949; numerous papers. Excavated at Mycenae, Sparta, Troy, Thessaly, Alexandria, Corinth; expanded categories of Neolithic pottery developed by Tsountas; excavated houses west of citadel of Mycenae containing Linear B tablets and graves of Middle Bronze through Iron Ages at Mycenae, 1952-54. Died 1957.

WACHMANN, Arthur Arno, German astronomer; b. Harburg, Mar. 8, 1902; s. Johannes and Radke Agnes Wachmann; ed. univs. Hamburg, Göttingen and Kiel; Ph.D.; m. Helmi Steffen, Apr. 4, 1933; children—Ekkehard, Inge. Asst., Bamberg Obs., 1926-27; asst. Hamburg Obs., 1927-41, observer, 1941-58, observer-in-chief, 1958—. Named hon. prof.; recipient A. Donnohoe medals (4) for discovery four new comets, Pacific Astronomy Soc. Mem. Astron. Soc. Germany, Am. Astron. Soc., Internat. Astron. Union. Author articles in field. Research on astronomy, especially variable stars. Address: Sternwarte, Hamburg-Bergedorf, West Germany.

WACHOWSKI, Theodore John, Am. radiologist; b. Chgo., Nov. 20, 1907; s. Albert and Constance (Korzeniewski) W.; B.S., U. Ill., 1929, M.D., 1931; m. Barbara Florence Benda, June 1, 1931; 1 son, Ted James. radiologist Copley Meml. Hosp., Aurora, Ill., 1935—; clin. prof. radiology U. Ill. Coll. Medicine, Chgo., 1949—. Fellow A.A.A.S.; mem. Radiol. Soc. N.Am. (dir., past pres.), Am. Coll. Radiology (chmn. bd. chancellors 1962-63, past pres.), A.M.A., Am. Roentgen Ray Soc., Ill., Kane County, Aurora (past pres.), Polish (past pres. Chgo.) med. socs. Publs. on gallbladder disease, tumors of paranasal sinuses, diagnosis intracranial lesions, treatment of brain tumors, diagnosis lung cancer, post traumatic hemorrhage in small bowel, effects on blood of alkylating agents. Home: 310 Ellis Av., Wheaton, Ill. 60187. Office: 502 S. Lincoln Av., Aurora, Ill. 60507.*

WACHSMANN, Felix, physicist; b. Banjaluka, Yugoslavia, Dec. 20, 1902; s. Wilhelm and Helene (Kollmann) W.; M.D., Tech. U. Munich, 1934; m. Eugenie Besold, Sept. 17, 1932. Engr., Rumania, 1930-39; staff Robert Koch Hosp., Berlin, 1939-43; faculty U. Erlangen, 1944-64, head Inst. for Radiation, 1958-64; head Inst. for Radiation Protection, Soc. for Radiation Research, Munich, 1964—; prof. Tech. U. Munich. Mem. Health Physics Assn., numerous German, fgn. med. socs. Author: Moving Field Radiation Therapy, 1962; also numerous articles. Research on radiotherapy methods, radiobiology, radiodosimetry, personnel-dosimetry. Home: 11 Pienzenauerstrasse, Munich. Office: 1 Ingolstadter Landstrasse, Neuherberg/Munich, Germany.*

WACHSMUTH, Charles, paleontologist; b. Hanover, Germany, Sept. 13, 1829; s. Christian Wachsmuth; m. Bernandina Lorenz, 1855. Came to N.Y.C. as agt. merc. house in Hamburg, 1852; went to Burlington, Ia., 1854; gathered large collection of rare crinoids and established spl. library on the subject which attracted attention of scientists; Louis Agassiz purchased the material for Mus. of Comparative Zoology, Cambridge, Mass., 1873; went with Agassiz to Cambridge, worked with him, until 187[illegible] engaged in study of N.Am. crinoids (with Frank Springer), after 1873. Author: (with Springer) N[illegible] American Crinoidea Camerata (monograph) 1[illegible] Died Burlington, Feb. 7, 1896.

WACHTER, Friedrich Ludwig, mathematician; b. 1792; became prof., Danzig, 1809. Author: Demonstratio axiomatis geometrici in Euclideis undecimi, 1817. Demonstrated that geometry on a sphere becomes identical with geometry of Euclid when radius is infinitely increased, also that limiting surface is not a plane. Died 1817.

WADA, Hiroshi, Japanese endocrinologist; b. Sobara, Japan, Feb. 11, 1918; s. Yutaka and Tora (Aoki) W.; student Morioka Agrl. Coll., 1937-40; grad. Coll. Agr., Kyoto U., 1942, Ph.D., 1962; m. Hiroko Michioka, May 5, 1944; children—Gosei, Setsuko, Momoko. Tech. ofcl. Ministry Agr. and Forestry, 1942-45; prof. Shizuoka Agrl. Coll., Shizuoka, Japan, 1948-52; faculty Okayama (Japan) U. Sch. Agr., 1952-62, prof., 1962—; research asso. U. Mo., Coll. Agr., Columbia, 1957-59, biomed. research fellow Population Council, 1958-59. Mem. Japanese Soc. Zootech. Sci., Japanese Soc. Animal Reproduction. Author: (with Shoji Vyesaka) The Japanese Cattle, 1956; Milking, 1967; Milkers and Milking Equipments, 1968; also articles. Research on effect of hyaluronidase in bull semen on artificial insemination, function of relaxin in physiology of reprodn.; discovered presence of relaxin in serum of pregnant cow; studies of estrogens in pasture plants, growth stimulants in cattle and poultry feeding. Home: Campus of Okyama U., Tsushima, Okayama, Okayama-prefecture, Japan.*

WADA, Juhn Atsushi, physician; b. Tokyo, Japan, Mar. 28, 1924; s. Teijuhn and Haruko (Miyakita) W.; M.D., Hokkaido Imperial U., Sapporo, Japan, 1946, D.Med.Sc., 1951; m. Mary Iwasaki, July 25, 1956; children—Kent, Eileen. Asst. prof. neurology and psychiatry, chief brain surgeon Hokkaido U. Hosp., 1952-56; asso. prof. neurology U. B.C., Vancouver, 1956—; attending neurologist, dir. seizure clinic and EEG labs. Vancouver Gen. Hosp. Mem. Am., Canadian neurol. and EEG socs., Canadian Soc. Electroencephalographers (councillor 1966—). Research and publs. on devel. Wada speech test to verify side of brain center for speech function, brain mechanism of human behavior, epilepsy. Home: 4729 W. 7th Av., Vancouver 8, B.C., Can.*

WADA, Juro, Japanese surgeon; b. Sapporo, Japan, Mar. 11, 1922; s. Teihuhn and Ysuko W.; student Kokkaido Imperial U., Sapporo, 1938-41; M.D., Hokkaido U., Sapporo, Japan, 1944, Ph.D. in Surgery, 1949; m. Christina Shuko, Sept. 18, 1958. Chief surgery Yakumo Nat. Hosp., Japan, 1949-50; fellow surgery U. Minn. Hosp., 1950-51; asst. in surgery Peter Bent Brigham Hosp., also Boston City Hosp., Harvard U. Med. Sch., 1953-54; faculty Sapporo Med. Coll., 1954—, prof. surgery, 1958—, chmn. dept. thoracic and cardiovascular surgery, 1968—. Cons. in thoracic and cardiovascular surgery Sapporo central Ry. Hosp., 1963—, Tomakomal Prefectural Sanatorium, 1955, Kushiro Prefectural Sanatorium, 1962—, Fellow Internat. Coll. Surgeons, Internat. Coll. Angiology, Am. Coll. Chest Physicians, Soc. Thoracic Surgeons, Academia Peruana de Chirugia; mem. Soc. Thoracic Surgeons (Founding), Am. Trudeau Soc., N.Y. Acad. Sci., Pan Pacific Surg. Assn. Author: Diagnosis and Treatment of Cardiac Diseases, 1964; also numerous articles. Research on introduction of thoracic and cardiovascular surgery in Japan; developed universal thermodisc oxygenator, Wada-Cutter hingeless heart valve, numerous new operative methods in thoracic and cardiovascular surgery. Home: 19 N. 21, W. 4, Sapporo, Hokkaido, Japan.*

WADA, Takeo, Japanese physician; b. Sapporo, Hokkaido, Japan, Nov. 15, 1914; s. Kango and Kin (Nagoya) W.; M.D., Hokkaido U., 1940, Ph.D., 1947; m. Yuriko Miyoshi, Oct. 5, 1942; children—Tatsuhiko, Satoko. Research fellow Hokkaido U. Sch. Medicine, 1946; lectr. internal medicine Hokkaido Women's Med Coll., 1946-50; faculty Sapporo Med. Coll., 1950—, prof. internal medicine, 1954—, staff Cancer Inst., 1961—. Mem. Japanese Assn. Internal Medicine (dir. 1965—). Author: Cachexia, 1966; also numerous articles. Biochem. analyses of metabolic changes in cancer; biochem. and biophys. studies on gastric secretion; analyses on serum protein in miscellaneous disease states; psychol. and pathophysiol. studies on patients with chronic diseases. Home: 711 W-9 S-15, Sapporo, Hokkaido, Japan.*

WADDELL, William Rhoads, Am. physician; b. Ft. Smith, Ark., Oct. 12, 1918; s. William Turner and Bonnie (Roper) W.; B.S., U. Ariz., 1940; M.D., Harvard, 1943; m. Barbara Ann Christie, July 31, 1944; children—Bonnie, Susan, Nancy, Sarah. Intern Mass. Gen. Hosp., Boston, 1944-45, surg. resident (USPHS fellow 1946, 50), 1945-51; mem. faculty Harvard, 1951-61, asso. clin. prof. surgery, 1960-61; prof. surgery, chmn. dept. U. Colo., 1961—; chief surgery Colo. Gen. Hosp.; cons. Denver, Grand Junction, Albuquerque VA hosps. Active Colo. region Nat. Conf. Christians and Jews. Served to lt. (j.g.), M.C., USNR, 1946. Diplomate Am. Bd. Surgery, Am. Bd. Thoracic Surgery. Fellow A.C.S.; mem. Am. Assn. Thoracic Surgery, Boston Surg. Soc., Am. Trudeau Soc., A.A.A.S., Am. Fedn. Clin. Research, Am., Western surg. assns., Soc. Univ. Surgeons, Southwestern Surg. Congress, Am. Gastroent. Assn. Contributions in field lipid metabolism, gastric physiology, transplantation and clinical surgery. Home: 5745 E. 6th Av., Denver 20.*

WADD[illegible]GTON, Cecil Jacob, physicist; b. Cam[illegible] E[illegible]g., July 6, 1929; s. Conrad Hal and Cecil [illegible] (Lascelles) W.; B.Sc. with 1st class honours, U. [illegible], 1952, Ph.D., 1955; m. Jean Constance Webb, Sept. 10, 1957. Royal Soc. McKinnon research student U. Bristol, 1955-59, lectr., 1959-62; research asso., lcetr. U. Minn., Mpls., 1957-58, asso. prof. physics, 1962—; Nat. Acad. Sci-NASA sr. postdoctoral fellow Goddard Space Flight Center, 1961. Fellow Am. Phys. Soc.; mem. A.A.A.S., Am. Geophys. Union, Am. Astron. Soc. Research, numerous publs. on nature and composition of primary cosmic radiation, influence of sun on these particles. Home: 2517 Thomas Av. S., Mpls. 55405.*

WADDINGTON, Conrad Hal, Brit. biologist; b. Evesham, Eng., Nov. 8, 1905; s. Hal and Mary (Warner) W.; B.A., Cambridge (Eng.) U., 1927, Sc.D., 1938; D.Sc., U. Montreal (Que., Can), 1958, U. Dublin (Ireland), 1965; D.Sc., Prague U., 1966; LL.D., Aberdeen U., 1966; m. Margaret Justin Blanco White, June 24, 1934; children—Caroline, Margaret Dusa. Lectr. embryology Cambridge U., also research worker Strangeways Lab., 1933-45; Buchanan prof. animal genetics Edinburgh (Scotland) U., 1946—; hon. dir. Agr. Research Council's unit animal genetics, 1946—, epigenetics research group Med. Research Council, 1962—. Mem. Adv. Council on Sci. Policy, 1961-64, Slater Com. on Environmental Scis., 1962; v.p. Internat. Biol. Programme, 1963-66. Decorated Comdr. Order Brit. Empire. Fellow Royal Soc., 1947, Internat. Inst. Embryology; fgn. mem. Am. Acad. Arts and Sci., Finnish Acad., Genetics Soc. Japan; mem. Orgn. Civil Sci., Internat. Union Biol. Scis. (pres. 1961—), World Soc. Ekistics (mem. exec. com. 1966—), Brit. Genetical Soc. (past pres.). Author numerous books including: New Patterns in Genetics, 1962; also numerous articles. Research on embryonic organizers, making several discoveries of their activity, inductive ability, competence; developmental genetics, evolution. Home: 15 Blacket Pl., Edinburgh 9, Scotland.*

WADDINGTON, Guy, phys. chemist; b. Victoria, B.C., Can., May 22, 1904; s. Frederick and Elinor (Naylor) W.; B.A., U. B.C., 1928, M.A., 1929; Ph.D., Cal. Inst. Tech., 1932; m. Winifred Martha Tervo, Aug. 21, 1930; 1 dau., Elizabeth Anne (Mrs. John Leslie Arrington, Jr.), Came to U.S., 1929, naturalized, 1941. Warren Fellow Cal. Inst. Tech., Pasadena, 1932-35; prof. chemistry, chmn. sci. div. Rollins Coll., Winter Park, Fla., 1935-43; chief thermodynamics br. U.S. Bur. Mines, Bartlesville, Okla., 1943-57; dir. office critical tables Nat. Acad. Scis.-NRC, Washington, 1957—. Chmn. commn. on symbols, terminology and units Internat. Union Pure and Applied Chemistry, 1961—, v. p. div. phys. chemistry, 1965—; mem. com. on phys. chemistry NRC, 1956—; exec. dir. com. on data for sci. and tech. Internat. Council Sci. Unions, 1966—. Recipient Rollins Coll. Decoration of Honor, 1943—, U.S. Dept. Interior Distinguished Service award 1953. Fellow A.A.A.S.; mem. Am. Chem. Soc. (S.W. Regional award 1957), Am. Inst. Chemists, Am. Soc. Testing Materials (mem. com. on numerical data 1962—), Am. Documentation Inst., N.Y. Acad. Scis. Author: (with others) Experimental Thermochemistry, 2 vols.; also numerous articles. Research on thermodynamics and thermochemistry of hydrocarbons and related substances, heat capacity vapors, heat vaporization, coordination critical tables programs, internat. relations in sci. Home: 2475 Virginia Av. N.W., Washington 20037. Office: 2101 Constitution Av. N.W., Washington 20418.*

WADDINGTON, William Henry, French archeologist, numismatist; b. St.-Remy-sur-Avre, France, Dec. 11, 1826; ed. Trinity Coll., Cambridge; m. Mary A. King; traveled in Asia Minor, Greece, Syria; elected rep. dept. of Aisne, France, 1871, senator for Aisne, 1876; French minister pub. instrn., 1873, 76-77, prime minister, 1879; ambassador to Eng., 1883-93. Author: Mélanges de numismatique de philologie, 1861; Voyage archéologique en Grèce et en Asie Mineure, 6 vols., 1847-77. Died Paris, Jan. 13, 1894.

WADE, Preston Allen, Am. surgeon; b. Helena, Mont., Mar. 22, 1901; s. John William and Claudia (Hilman) W.; A.B., Cornell U., 1922, M.D., 1925; m. Evangline Schreiter, Aug. 4, 1934. Faculty, Cornell U. Med. Coll., Ithaca, N.Y., 1927—, prof. clin. surgery, 1950—; attending surgeon N.Y. Hosp., 1950—; chief combined fracture service N.Y. Hosp.-Hosp. for Spl. Surgery, 1955—. Cons. surgeon to hosps.; 2d Watson-Jones lectr. Royal Coll. Surgeons, London, Eng., 1962; Sommer Meml. lectr., Portland, Ore., 1963; Stephen C. Meigher Meml. lectr., Schenectady, 1964. Diplomate Am. Bd. Surgery. Mem. A.C.S. (chmn. bd. regnts 1963—, pres. elect), N.Y. Acad. Medicine (pres. 1967—), Am. Surg. Assn., Société Internationale de Chirurgie, Assn. Internat. de Médicine des Accidents du Traffic, Am.

Orthopaedic Assn. (hon.), Am. Assn. for Surgery Trauma (past pres.), Greater N.Y. Safety Council (dir.), N.Y. Surg. Soc., Assn. Mil. Surgeons U.S., Soc. Med. Consultants to Armed Forces. Research on fractures, trauma. Home: 1035 Park Av., N.Y.C. 10028. Office: 898 Madison Av., N.Y.C. 10021.

WADE, Thomas Leonard, Jr., Am. [illegible]ian; b. Ridgeway, Va., Apr. 21, 1905; s. T[illegible] and Betty (Eggleston) W.; B.S., U. Va., [illegible] 1930, Ph.D., 1933; m. Margaret Millner [illegible] 24, 1931; children—Margaret Elizabeth [illegible]row McKay, Jr.), Ida Katherine (Mrs. [illegible]tt C. Adair), Thomas Leonard III, Mary Susan (Mrs. Michael J. Cuda). Asst. instr. U. Va., 1929-34; prof. math. Mercer U., 1934-39; asst. prof. math. U. Ala., 1939-43; prof. math. Fla. State U., Tallahassee, 1943-—, head dept., 1943-63. NSF fellow, 1962. Mem. Am. Math. Soc., Math. Assn. Am., Nat. Council Tchrs. Math., Sigma Xi, Pi Mu Epsilon. Author numerous books including: Introductory College Mathematics, 1959; (with H.E. Taylor) University Calculus, 1962, Subsets of the Plane: Plane Analytic Geometry, 1962, University Freshman Mathematics, 1963; also articles. Research on tensor algebra and invariant theory, expository texts on calculus, founds. math., matrix algebra. Home: 1003 Washington St., Tallahassee 32303.*

WADE, Walter, Irish botanist; M.D., Licentiate King's and Queen's Coll. Physicians; began practice medicine, Dublin, Ireland, 1790; lectr. botany to Dublin Soc.; collected plants in Ireland, 1796, 1801, 05; physician to Dublin Gen. Dispensary; lectr. botany to Royal Coll. Surgeons, Ireland. Recipient prize for discovery mosses new to Ireland, Dublin Soc., Asso. Linnean Soc. Author: Catalogus systematicus planetarium indigenarum in comitatu Dublinensi . . . pars prima, 1794; Syllabus of a Course of Lectures on Botany, 1802; Plantae rariores in Hibernia inventae, 1804; Sketch of Lectures on Meadow and Pasture Grasses Delivered in the Dublin Society's Botanical Garden, 1808; Salices, 1811; Sketch of Lectures on Artificial or Sown Grasses, 1808; also articles. First to arrange plants of Ireland systematically; 1st discovery of pipewort (Griocavlon) in Ireland; popularized botany in Ireland. Died Dublin, 1825.

WADEY, Walter Geoffrey, physicist; b. Whangarei, New Zealand, Sept. 9, 1918; s. Walter and Ada Isabel (Browne) W.; came to U.S., 1925, naturalized, 1940; B.S., U. Mich., 1941, M.A., 1942, Ph.D., 1947; m. Jean Mary Desjardins, Dec. 26, 1945; children—Susanna Marie Gordon Charles, Bryan Thomas. Research asso. Harvard, 1943-45; instr., then asst. prof. Yale, 1948-56; prof. physics So. Ill. U., 1956-57; sr. systems analyst Remington Rand Univac, 1957-59; tech. staff Hughes Aircraft Co., 1959-60; dept. mgr. Univac div. Sperry Rand Corp., Blue Bell, Pa. 1960-62; chief sci. Bowles Engring. Corp., Silver Spring, Md., 1962-63; chief scientist Washington Tech. Assos., Rockville, Md., 1963-64; sr. scientist Operations Research, Inc., Silver Spring, Md., 1964-—. Chmn. vacuum standards com. Am. Vacuum Soc. Mem. Am. Phys. Soc., Operations Research Soc. Am., Marine Tech. Soc., A.A.A.S., Phi Beta Kappa, Sigma Xi, Phi Kappa Phi. Research, publs., patents in beta and alpha-ray spectroscopy, nuclear particle accelerators, neutron spectroscopy, magnetic shielding, computer arithmetics; devel. electronic computers, fluid amplifiers, operations research. Home: 7505 Holiday Terrace, Bethesda, Md. 20034. Office: 1400 Spring St., Silver Spring, Md. 20910.*

WADIA, Darashaw, Nosherwan, Indian geologist; b. Surat, Bombay, India, Oct. 25, 1883; s. Nosherwan and Coonverbai Wadia; B.Sc., Delhi (India) U., 1905, M.A., 1906, D.Sc. (hon.), 1948; m. Meher Gustadji, Mar. 27, 1940. Prof. geology Kashmir State U., 1907-20; staff Geol. Survey India, 1921-38; geol. survey Kashmir Himalayas, Hazara, N.W. Punjab, 1921-38; mineral and geol. survey Ceylon Govt., 1938-44; geol. adviser Govt. of India, New Delhi, 1945-—, dir. Bur. Mines, 1947-48, mem. atomic energy commn., 1952-—, nat. prof. geology, 1962-—. Chmn., Indian Nat. Com. Oceanic Research, 1960-—; mem. Council Sci. and Indsl. Research, 1945-—. Recipient Royal Geog. Soc. London award, 1934; Padma Bhushna, 1957. Fellow Royal Soc., 1957; hon. fellow German, Am., Belgian geol. socs.; mem. Nat. Geophys. Research Inst. (chmn. council 1963-—, Indian Sci. Congress (past pres.), Nat. Inst. Sci. India (past pres.), Geol. and Mining Inst., India (past pres.), Geol. Soc. India (past pres., Lyell medal 1943), Indian Statis. Inst. (v.p. 1965-—). Author: Geology of India, 1919; also numerous articles. Research on gen. geology, stratigraphy and tectonics of N.W. Himalyas, their trendine and syntaxial bend, nat. mineral policy, econ., strategic and atomic minerals of India, rare metals, physiography, evolution of main phys. feature of India, evolution of deserts of Central Asia. Home: 10 King George Av. Office: South Block, Central Secretariat, Govt. India, New Delhi, India.*

WADSWORTH, Frank Lawton Olcott, Am. engr., inventor; b. Wellington, O., 1867; s. Francis Sage W.; grad. E.M. (mining engr.), Ohio State U., 1888 B.Sc., 1889, M.E. (mech. engr.), 1889; Clark U., 1889-92; m. Laura Poole, Sept. 1893; m. 2d, Mildred Schinneller, July 1914. Del. from Smithsonian Instn. to Internat. Bur. Weights and Meas., Paris, to assi[illegible]in establishing absolute length of standard m[illegible], 1892; sr. asst. in charge Astrophys. Obs., [illegible]shington, 1892-94; asst. prof. physics, U. Chgo., [illegible]94-96; asst. prof. astrophysics, Yerkes Obs., 1896-[illegible], asso. prof., 1897-98; dir. Allegheny Obs., 1900-04; spl. engr. and expert work, Pitts. and Washington, 1898-99; cons. expert John A. Brashear Co., 1901-04; gen. mgr. Pressed Prism Plate Glass Co., 1904-05; chief engr. Am. Window Glass Co., 1905-08; cons. engr., 1908-—. Pres. Miller Non-Corrosive Metal Co., 1909-—. Asst. editor Astrophys. Jour.; asso. editor Harper & Bros. Sci. Memoirs. papers and reports in sci. publs., and proceedings. Patentee of over 250 inventions relating to manufacture of glass, steel, electric lights, railway appliances, machine tools, engring. instruments, wire working machinery, apppliances, tires; originator, designer vertical tower telescope, curved plate camera, fixed deviation spectroscope, polar reflecting heliostat, precision interferometers, and other novel forms of phys. and astrophys. instruments; author of the hexaplex system of golf course design, testified as tech. expert in patent suits in U. S. and Canadian courts. Died Apr. 11, 1936.

WADSWORTH, Marshman Edward, Am. geologist; b. Livermore Falls, Me., May 6, 1847; s. Joseph and Nancy F. (Eaton) W.; A.B., Bowdoin Coll., 1869, A.M., 1872; A.B., Harvard, 1874, A.M., Ph.D., 1879; postgrad., U. Heidelberg, 1884-85; M.D., Nat. Med. Coll., 1894; E.M. (hon.), Pa. State Coll., 1917; Sc.D., (hon.), U. Pitts., 1919. Principal and supt. in Maine, N.H., Minn. and Wis., 1863-73; prof. chemistry Boston Dental Coll., 1873-74; instr. math. and mineralogy, Harvard, 1874-77, asst. in geology. Mus. Comparative Zoölogy, 1877-87; prof. mineralogy and geology; Colby, 1885-87; asst. geologist Minn. Geol. Survey, 1886-87; dir., prof. mining geology and petrography Mich. Coll. of Mines, 1887-97, pres., 1897-99; state geologist of Mich., 1888-93; geologist and mining expert Keweenawan Assn., 1898-1903; prof. mining and geology, and dean sch. mines and metallurgy Pa. State Coll. 1901-08; dean Sch. of Mines, prof. mining geol., U. Pitts., 1907-12, emeritus dean and prof., curator of mineral and petrographical collections, 1912-—. Geologist Pa. State Bd. Agr., 1902-05. Author: Geology of the Iron and Copper Districts of Lake Superior, 1880; Lithological Studies, 1884; Report of the Mich. Geol. Survey, 1893; The Azoic System (with Josiah Dwight Whitney), 1884; Crystallography, 1909; Michigan College of Mines in the Nineteenth Century, 1916. Taught 1st course in microscopic petrography in U. S.; research in pre-Cambrian geology, on meteorites. Died Apr. 21, 1921.

WADSWORTH, Milton Elliot, Am. metallurgist; b. Salt Lake City, Feb. 9, 1922; s. Thomas Guy and Agnes (Flockhart) W.; B.S., U. Utah, 1948, Ph.D., 1951; m. Mary Mirian Baily, Nov. 19, 1943; children—Mary Christine, Kathryn Ann, Jane Isabel, Amy Carol, Leslie Elizabeth, Margaret Allison. Faculty, U. Utah, Salt Lake City, 1948-—, prof. metallurgy, 1959-—. Cons. Oak Ridge Nat. Labs., Kennecott Copper, ESSO Research Labs., Baton Rouge, La. Mem. Am. Soc. for Metals, Am. Inst. Metall. Engrs., Sigma Xi, Phi Kappa Phi, Tau Beta Pi. Research on rate process in extractive metallurgy, surface chemistry in flotation, other separate processes. Home: 3437 S. 13th St. E., Salt Lake City 84106.*

WAELBROECK, Lucien, Swiss mathematician; b. Geneva, Switzerland, May 11, 1929; s. Pierre F. A. and Elizabeth (Varlez) W.; Licence en Science Mathématiques, U. Brussels (Belgium), 1950, D.Sc., 1953, Agrégé de l'Enseignement Supérieure, 1960; m. Christine Léonie Irminie Van Gaver, Aug. 5, 1957. Staff, Fonds National de La Recherche Scientifique, 1955-58; faculty U. Brussels, 1958-—, prof., 1965-—; mem. Inst. for Advanced Study, Princeton, N.J., 1957-58; research asso. Yale, 1962, Mass. Inst. Tech., 1967; vis. prof. U. Cal. at Los Angeles, 1967. Mem. Société Mathematique de Belgique, Société Mathematique de France, Am. Math. Soc., Soc. for Indsl. and Applied Math. Research and publs. in algebras in which a continuity structure is defined (Banach algebras, locally convex algebras, and generalizations); showed how it would be preferable, in many cases, to consider a bounded structure rather than a topology when studying an algebra. Home: 43 Ave. d'Italie, Brussels 5, Belgium.*

WAELSCH, Heinrich Benedict, biochemist; b. Brno, Czechoslovakia, Jan. 20, 1905; s. Emil B. and Marianne (Loew) W.; M.D., German U. Prague, 1929, Sc.D., 1930; m. Salome Gluecksohn, Jan. 8, 1942; children—Naomi Barbara, Peter Benedict. Came to U. S., 1938, naturalized, 1943. Research fellow, Graz, 1925; faculty German U., Prague, 1928-38; faculty Columbia, N.Y.C., 1943-—, prof. biochemistry, 1953-—, civilian OSRD, Washington, 1941-43; asso. research biochemist N.Y. State Psychiat. Inst. and Hosp., N.Y.C., 1943-51, chief, psychiat. research, 1958-—; dir. N.Y. State Research Inst. for Neurochemistry and Drug Addiction, Ward's Island, 1961-—. Mem. various coms. NIH, USPHS. Recipient Carl Neuberg medal, 1953. Govt. fellow, Dresden, Germany, 1930-31. Mem. Am. Soc. for Biol. Chemists, Am. Soc. Pharmacology and Exptl. Therapeutics, Biochem. Soc., Am. Chem. Soc., Am. Coll. Neuropharmacology, Collegium Internationale Neuropsychopharmacologicum, Am. Neurol. Assn., Harvey Soc., Soc. Exptl. Biology and Medicine, Assn. for Research in Nervous and Mental Diseases (award 1965), A.A.A.S. Editor: Biochemistry of the Developing Nervous System, 1955, Ultrastructure and Cellular Chemistry of Neural Tissue, 1957. Editor: Internat. Jour. Neuropharmacology, Jour. Neurochemistry, Life Scis.; Psychopharmacologia; Brain Research; Exptl. Brain Research. Research, numerous publs. on protein metabolism of central nervous system, hypothesis of existence of compartments of metabolism is significant contbn. to brain biochemistry. Home: 90 Morningside Dr., N.Y.C. 10027. Office: 722 W. 168th St., N.Y.C. 10032.*

WAELSCH, Salome Gluecksohn (Mrs. Heinrich Waelsch), geneticist; b. Danzig, Germany, Oct. 6, 1907; d. Ilja and Nadia (Pomeranz) Gluecksohn; Ph.D., U. Freiburg (Germany), 1932; m. Heinrich Waelsch, Jan. 8, 1943; children—Naomi Barbara, Peter Benedict. Came to U.S., 1933, naturalized, 1938. Asst., dept. exptl. cell research U. Berlin, 1932-33; research asso. Columbia, 1936-53, Coll. Phys. and Surg., 1953-55; faculty Albert Einstein Coll., N.Y.C., 1955-—, prof. anatomy, 1958-63, prof. genetics, acting chmn. dept., 1963-—. Mem. Am. Soc. Zoologists, Am. Assn. Anatomists, Genetics Soc., Growth Soc., Am. Soc. Naturalists, Am. Soc. Human Genetics. Research, numerous publs. on developmental genetics, role genes in normal and abnormal cell differentiation, genetically controlled congenital abnormalities nervous system, skeleton, musculature and other systems. Home: 90 Morningside Dr., N.Y.C. 10027.*

WAFA, Abul, see Abul Wafa.

WAGGONER, Paul Edward, Am. climatologist; b. Centerville, Ia., Mar. 29, 1923; s. Walter L. and Kathryn (Maring) W.; B.S., U. Chgo., 1946; M.S., Ia. State U., 1949, Ph.D., 1951; m. Barbara Ann Lockerbie, Nov. 3, 1945; children—Von L., Daniel M. Agt., U.S. Dept. Agr., Ames, Ia., 1948-51; asst. pathologist Conn. Agrl. Expt. Sta., New Haven, 1951-54, asso. pathologist, 1954-56, chief dept. soils and climatology, 1956-—. Guggenheim fellow, 1963. Mem. Am. Meteorol. Soc., Am. Soc. Agronomy, Am. Phytopath. Soc., Am. Soc. Plant Physiology. Author: (with others) Agricultural Meteorology, 1965; also numerous articles. Demonstrated role of stomata or plant leaf pores, in hydrological cycle; analyzed effect of plants upon energy budge of man shaded or sheltered by plants; demonstrated presence of toxin in vivo in prodn. of wilt disease; combined theory of turbulence transfer and probability theory in an epidemiology of plant diseases. Home: 314 Vineyard Rd., Guilford, Conn. 06437. Office: 123 Huntington St., New Haven 06504.*

WAGLEY, Charles Walter, Am. anthropologist; b. Clarksville, Tex., Nov. 9, 1913; s. Walter C. and Sally (Ridling) W.; A.B., Columbia, 1936, Ph.D., 1941; Hon.D., U. Bahia (Brazil), 1961; LL.D., U. Notre Dame, 1964; m. Cecilia Roxo, Sept. 12, 1942; children—Isabel Anne (Mrs. Conrad Kottak), Carlos (Dec.). Exec., Inst. Inter-Am. Affairs, U.S. Govt., 1942-46; faculty Columbia, 1946-—, prof. anthropology, 1951-—, dir. Inst. Latin Am. Studies, 1961-—, Franz Boas prof. anthropology, 1965-—. Mem. Am. Anthrop. Assn., Am. Ethnol. Soc., Am. Philos. Soc. Author: Amazon Town, 1949; Introduction to Brazil, 1963; also numerous articles. Research on Indian, peasant and urban socs. Latin Am., particularly Brazil. Home: 15 Claremont St., N.Y.C. 10027.*

WAGMAN, Irving Henry, Am. neurophysiologist; b. N.Y.C., Aug. 15, 1916; s. Jacob and Ida (Kramer) W.; B.S., Coll. City N.Y., 1936; M.A., U. Cal. at Berkeley, 1937, Ph.D., 1941; m. Dorothy E. Leach, Aug. 9, 1942; children—William J., Karen E., Daniel E., Ann A. Research asso. U. Cal. Med. Sch., San Francisco, 1941-43, research physiologist biomechanics lab. Med. Center, 1961-64; asst. physiologist USHPHS, 1943; fellow Eldridge Reeves Johnson Found. for Med. Physics, U. Pa., 1943-46; staff physiology dept. Jefferson Med. Coll., Phila., 1946-53, asso. prof., 1950-53; pvt. sci. cons., 1954; research asso. dept. neurology Mt. Sinai Hosp., N.Y.C., 1954-61; lectr. neurophysiology N.Y. U. Coll. Medicine, 1957-60; research physiologist Nat. Center for Primate Biology, U. Cal. at Davis, 1965-—, prof. physiology dept. animal physiology, 1965-—, prof. cipient Nat. Acad. Sci. Travel award, 1959, 62. Fellow A.A.A.S.; mem. Am. Physiol. Soc., Harvey Soc. N.Y., Am. Neurol. Assn., Am. Acad. Neurology, Assn. for Research in Nervous and Mental Diseases. Research, numerous publs. on properties of skeletal muscle fibers and neuromuscular junction, neural limitations of vision, patterned normal motor behavior. Home: Route 1, Box 2038, Davis, Cal. 95616.*

WAGNER, Adolph, German economist; b. Erlangen, Germany, Mar. 25, 1835; s. Rudolf Wagner; prof., Tartu, Estonia, also Freiburg im Breisgau, Berlin. Author: Die Abschaffung des privaten Grundiegentums, 1870; Grundlegung der politischen Ökonomie, 2 vols., 1876; Finanzwissenschaft, 4 vols., 1877-1901; Allgemeine und theoretische Volkswirtschaftslehre oder Sozialökonomik, 1901. Leading systematizer; favored social legislation, abolition of pvt. capital, distbn. of property after extensive nationalization. Died Berlin, Nov. 8, 1917.

WAGNER, Bernard Meyer, Am. pathologist; b. Phila., Jan. 17, 1928; s. John L. and Katherine (Levinson) W.; student U. Pa., 1944-45; M.D., Hahnemann Med. Coll., 1949; m. Patricia Hoffman, Mar. 23, 1951; children—Cynthia, Nancy, Robert. Dir. exptl. pathology lab. div. pathology Hahnemann Med. Coll., 1954-55; asst. prof. pathology U. Pa. Sch. Medicine, 1956-58; dir. pathology, pathologist Children's Hosp. Phila., 1955-58; asso. prof. dept. pathology, Robert L. King chmn. for cardiovascular research U. Wash. Sch. Medicine, Seattle, 1958-60; prof., chmn. dept. pathology N.Y. Med. Coll., N.Y.C., 1960-66; v.p. Warren-Lambert Research Inst., Morris Plains, N.J., 1966—. Spl. cons. to dir. Nat. Inst. Child Health and Human Devel., 1963—; mem. panel on neoplastic disease Health Research Council, City N.Y., 1965—; mem. sci. adv. bd. chief of staff USAF, 1960-65; cons. U.S. Vitamin and Pharm. Corp., 1964-66. Recipient Life Scis. award Am. Inst. Aeros. and Astronautics, 1963-65. Mem. Internat. Acad. Pathology, Am. Heart Assn., Nat., N.Y. acad. scis. Author: (with Black) Dynamic Pathology, 1964; also numerous articles. Research on connective tissue diseases. Home: 38 Mohawk Rd., Short Hills, N.J. Office: Warner-Lambert Research Inst., Morris Plains, N.J.*

WAGNER, Carl (Wilhelm), phys. chemist; b. Leipzig, Germany, May 25, 1901; s. Julius Eugen and Mathilde (Böse) W.; student U. München and Leipzig, 1920-24; Ph.D., U. Leipzig, 1924; Dr. rer. nat. (hon), Technische Hochschule Darmstadt, 1952. Asst. U. München, 1924-27; research fellow U. Berlin, 1927-28; asst. and privatdozent U. Jena, 1928-33; dep. prof. U. Hamburg, 1933-34; asso. prof. phys. chemistry Technische Hochschule Darmstadt, 1934-40, prof., 1940-45 (all in Germany); sci. advisor Ordnance Research and Development div. Suboffice Rocket, Fort Bliss, Tex., 1945-49; vis. prof. metallurgy Mass. Inst. Tech., 1950-55, prof. metallurgy, 1955-58; head of Max Planck Institut für physikalische chemie, Göttingen, West Germany, 1958-66, emeritus, 1967—. Mem. Nat. Academy Sciences (fgn. asso.) A.A.A.S., Am. Chem. Soc., Electrochem. Soc., Nat. Assn. Corrosion Engrs., Am. Phys. Soc., Am. Soc. Metals, Am. Inst. Mining and Metall. Engrs., American Academy of Arts and Sciences, Faraday Soc., Deutsche Bunsengesellschaft, Deutsche Gesellschaft für Metallkunde. Author: Thermodynamics of Alloys, 1952. Explored lattice defects in ionic crystals and semi-conductors, their role in heterogeneous catalysis and solid state reactions. Home: 63 Nikolausberger Weg., Gottingen. Office: Göttingen, Bunsenstr 10, Max Planck Institut für physikalische Chemie, West Germany.*

WAGNER, Christian Nikolaus Johann, metallurgist; b. Dudweiler, Germany, Mar. 6, 1927; s. Christian J.P. and Regina (Bungert) W.; student U. Poitiers (France) 1948-49; Licence ès Sciences Mathématiques, U. Saarbrücken (Germany), 1951, Dipl. Ing., 1954, Dr.rer.nat., 1957; m. Rosemarie A. Mayer, Apr. 5, 1952; children—Thomas, Karla R. Research asst. Institut für Metallforschung, Saarbrücken, 1953-55, research asso., 1957-58; vis. fellow Mass. Inst. Tech., 1955-56; faculty Yale, New Haven, 1959—, asso. prof. engring. and applied sci. 1962—. Mem. German, Am. phys. socs., Am. Inst. Mining Engrs., Am. Soc. for Metals, Am. Crystallographic Assn., Sigma Xi. Research and publs. on X-ray diffraction studies of structure of plastically deformed metals and alloys, atomic distbns. and electronic transport properties of liquid metals and alloys, structure of vapor-quenched alloy films. Home: 15 Wellsweep Rd., Branford, Conn. 06405. Office: Hammond Lab., Yale, New Haven 06520.*

WAGNER, Ernst Leberecht, German pathologist; b. Dehlitz, nr. Weissenfels, Germany, Mar. 12, 1829; ed. univs. Leipzig, Prague, Vienna; M.D., 1852; became lectr. path. anatomy, Leipzig, Germany, 1855, asso. prof., 1860, prof., 1862, prof. spl. pathology and therapy, 1877. Author: Handbuch der allgemeinen Pathologie, 1862. Gave 1st description of dermatomyositis, 1863, of colloid milium (Wagner's disease), 1866; found lymphoma of liver in a case of typhus; 1st to describe amyloid-degeneration of liver; discovered fat embolies in lung tissues; studied tuberculotic and syphilitic new growths (for which he coined word syphilom). Died Leipzig, Feb. 10, 1888.

WAGNER, Frederic H., animal ecologist; b. Corpus Christi, Tex., Sept. 26, 1926; s. Frederick Herman and Margaret (Burgess) W.; B.S., So. Meth. U., 1949; M.S., U. Wis., 1953, Ph.D., 1961; m. Marilyn Christensen, June 24, 1949; children—Gregory Alan, Jeffrey Hall. Research fellow Wildlife Mgmt. Inst., Madison, Wis., 1950-51; research biologist Wis. Conservation Dept., Madison, 1952-58, 59-61; faculty Utah State U., Logan, 1958-59, 61—, prof. wildlife resources. Participant planning session Internat. Biol. Program, 1966; staff short courses, in-service tng. schs. for state, fed., internat. conservation agys., 1962—; cons. research program Caesar Kleberg Found. for Conservation of Wildlife, Tex. A. and M. U. Mem. Ecol. Soc. Am., Wildlife Soc., Am. Soc. Mammalogists, Cooper Ornithol. Soc., Soil Conservation Soc. Am., Southwestern Assn. Naturalists, Sigma Xi. Author: (with C. D. Besadny, Cyril Kabat) Population Ecology and Management of Wisconsin Pheasants, 1965; also articles. Research on demographic patterns of populations and their meaning in terms of environmental influences, homeostatic mechanisms of bird and mammal populations, exploitation and manipulation of animal populations, conservation of renewable resources. Home: 1066 N. 1730 E., Logan, Utah 84321.*

WAGNER, Günther Albert, German pharmaceutist; b. Grünhainichen/Erzgebirge, Germany, Oct. 27, 1925; s. Albert Ernst and Ella (Franke) W.; pharmacist U. Greifswald (Germany), 1951, Dr.rer. nat., 1952, Dr.rer.nat.habil., 1955; m. Christa-Maria Claus, July 26, 1952; children—Detlef, Ulrich. Faculty, U. Greifswald, 1951-59, prof., 1959; prof. pharmacy Karl-Marx- U., Leipzig, Germany, 1959—, chair prof., 1964—, dir. Pharm. Inst., 1959—. Author: Pharmazeutische Chemie, 1962; Lehrbuch der Pharmazeutische Chemie, 1966; also numerous articles. Research on glycosides especially synthesis of nucleosides; pharm. analysis. Home 9 Störmthaler Strasse, Leipzig, East Germany.*

WAGNER, Heinrich Karl Albin, German biometrician; b. Buchholz (Saxony), Feb. 14, 1910; s. Reinhard and Elsa (Pfab) W.; student univs. Graz, Leipzig; Dr. U. Leipzig, 1936; m. Charlotte Roth, July 31, 1939; children—Dieter, Klaus, Michael. Meteorologist, 1937-46; with Bioclimatic Research Station, 1950-65, Research Inst. Spa Sci., 1966— (both Bad Elster). Mem. Soc. Biol. Rhythm Research, Biometric Soc. Author: Bad Elster—Heilbad und Landschaft; also numerous articles. Research on reactions of body temperature, basimetry—correlations of blood pressure measurements, analysis of standard deviations to characterize fluctuation of biol. measurements. Address: 11 Rossbacherstrasse, 9933 Bad Elster, East Germany.*

WAGNER, Helmut Rudolf, sociologist; b. Dresden, Germany, 1904; s. Rudolf Richard and Olga (Fischer) W.; came to U.S., 1941, naturalized, 1947; M.A., New Sch. Social Research, 1952, Ph.D., 1955; m. Hannelore Joseph, July 16, 1951; 1 dau., Claire Marianne. Lectr. grad. faculty New Sch. for Social Research, 1952-56; from asst. to prof. Bucknell U., 1956-64; prof. sociology, head dept. anthropology and sociology Hobart and William Smith Colls., Geneva, N.Y., 1964—. Mem. Am., Eastern, Ohio Valley sociol. socs., Soc. Sci. Study Religion. Research on advancement gen. theory of social relationships on basis of phenomenological considerations, application of this approach to selected problems of religious and ideological motivations for social behavior of various groups and categories of people. Home: 401 W. High St., Geneva, N.Y. 14456.*

WAGNER, Henry Nicholas, Jr., Am. physician, radiologist; b. Balt., May 12, 1927; s. Henry N. and Gertrude (Loane) W.; A.B., John Hopkins, 1948, M.D., 1952; m. Anne Barrett, Feb. 3, 1951; children—Henry Nicholas III, Mary Randall, John Mark, Anne Elizabeth. Faculty, Johns Hopkins Med. Instns., 1958—, asso. prof. medicine, radiology and radiol. scis., 1963—, head div. nuclear medicine, 1964—. Am. Fedn. for Clin Research (pres-elect 1967—), Soc. Nuclear Medicine (v.p. 1967—), Am. Soc. forClin. Investigation, A.C.P., Assn. Am. Physicians Research, numerous publs. on applications radioactive materials to clin. medicine and biomed. research; inventor radioactive pharms. Home: 3410 Guilford Terrace, Balt. 21218.*

WAGNER, Hildebert, German phytochemist; b. Laufen, Germany, Aug. 28, 1929; s. Karl and Kressenz (Bschorer) W.; state exam. in pharmaceutics U. Munich, 1954, Dr.rer.nat., 1956; m. Ursula Wagner, Sept. 19, 1958; children—Christine, Thomas, Michael. Faculty, U. Munich, 1960—, prof. pharmacognosy, 1965—, co-dir. Pharm. Inst., 1965—; prof. pharmacognosy, U. Vienna, 1964. Mem. Plant Chem. Group (England), German Assn. Fat Research. Research and numerous publs. in plant drug analysis, chemistry of medicinal plant drugs, fats and phospholipids of plants, synthesis of flavonoids and anthrachinons. Address: Editor: Hallucinogenic Drugs, 1967. Barlachstrasse 6/18, Munich, West Germany; also Breitbrunn/Chiemsee, Nelkenweg 5, Munich, West Germany.*

WAGNER, John Garnet, pharmacokineticist; b. Weston, Ont., Can., Mar. 28, 1921; s. Herbert William and Coral (Cates) W.; Phm.B., U. Toronto, 1947; B.S. in Pharmacy, U. Sask., 1948, B.A., 1949; Ph.D. in Pharm. Chemistry, Ohio State U., 1952; m. Eunice Winona Kelsey, July 4, 1946; children—Wendie Lynn, Linda Beth. Instr., asst. prof. pharm. chemistry Ohio State U., 1951-53; with Upjohn Co., Kalamazoo, 1953—, sr. research scientist Med. Research div., 1963-68; prof. pharmacy U. Mich., Ann Arbor, 1968—. Member Am. Pharm. Assn. (Ebert prize 1961), Am. Chem. Soc., Am. Fedn. for Clin. Research, Am. Coll. Clin. Pharmacology and Chemotherapy, Am. Soc. for Pharmacology and Exptl. Therapeutics. Research, numerous publs. on phytochemistry, methods of assay of drugs, in vitro and in vivo testing of enteric coated tablets and sustained action formulations of drugs, degrading and solubilization of antiinflammatory steroid, relationship between biol. activity and dosage form in which drug is administered to animals and man, absorption, distbn., metabolism and excretion of drugs, and kinetics of latter processes. Home: 3445 Andover Rd., Ann Arbor, Mich. 48105.*

WAGNER, Moritz, German zoologist; b. Bayreuth, Germany, Oct. 3, 1813; brother of Rudolf Wagner; prof., Munich. Author: Die Darwinische Theorie und das Migrationsgesetz der Organismen, 1868; Die Enstehung der Arten durch räumliche Sonderung, 1889. Originator migration theory that species are formed by adaptation to local environments. Died Munich, May 31, 1887.

WAGNER, Paul, German chemist; b. Liebenau/Hannover, Germany, Mar. 7, 1843; prof., dir. agronomic sta., Darmstadt, Germany; mem. French Acad. Scis. Author: Lehrbuch der Düngerfabrikation, 1877; Die Stickstoffdüngung, 1890; Anwendung küstlicher Düngemittel, 1900. Authority on agrl. chemistry and artificial fertilizers. Died Darmstadt, Aug. 25, 1930.

WAGNER, Richard, physician; b. Vienna, Austria, Oct. 30, 1887; s. Moriz and Gisela (Ratzersdorfer) W.; M.D., U. Vienna, 1912; m. Eleanor H. Culbert, Apr. 18, 1943. Came to U.S., 1938, naturalized, 1943. Biochemist, U. Strassbourg (France), 1912-14; asst. U. Childr. Hosp., Vienna, 1918-30; asst. prof. U. Vienna, 1924-38; faculty Tufts U. Sch. Medicine, Boston, 1939—, prof. pediatrics emeritus, 1958—. NIH grantee, 1947; Guggenheim fellow, 1960; AEC grantee, 1957-60. Mem. A.M.A., Am., New Eng. pediatric socs., Am. Diabetes Assn. Author: (with Pirquet) Die Ernährung des Diabetikers, 1928; (with Priesel) Die Zuckerkrankheit und ihre Behandlung in Kinderalter, 1937; also numerous articles. Discovered glycogen storage disease. Home: 104 Irving St., Cambridge, Mass. 02138. Office: 20 Ash St., Boston 02111.*

WAGNER, Richard Alfred Max, German physiologist; b. Augsburg, Germany, Oct. 23, 1893; s. Fritz and Babette Wagner; M.D.; M.D. honorius causa; m. Johanne Ingenrieth, Sept. 22, 1941. Successively agrégé, asso. prof., prof. univs. Graz, Breslau, Erlangen, Innsbruck and Munich. Decorated Order Bavarian Merit; silver decoration of honor Order Austrian Merit; Order Merit German Republic. Mem. Halle Acad., Bavarian, Austrian acad. scis. Publns. on muscular physiology, pulmonary hemorrhage, biol. regulation. Home: Lamontstrasse 3, Munich 27. Office: Pettenkoferstrasse 12, Munich 15, West Germany.

WAGNER, Robert Howard, Am. biochemist; b. Peru, Ind., Aug. 11, 1921; s. John E. and Ella (Stickel) W.; A.B., DePauw U., 1943; Ph.D., U. Cin., 1950; m. Jean Marjorie Linde, Jan. 27, 1945; children—Sibyl M., Lucy A., John A., Virginia E.L., Jean T., Elsbeth. Lab. instr. DePauw U., 1943-44; research fellow Children's Hosp., Cin., 1946-50; faculty U. N.C., Chapel Hill, 1950—, sr. research fellow NIH, 1959-64, asso. prof. pathology, chemistry, 1962—, research career devel. fellow, NIH, 1964—. Mem. Am. Chem. Soc., A.A.A.S., Soc. Exptl. Biology and Medicine, Sigma Xi. Research, publs. in antihemophilic factor, purification, use in treatment, study of properties. Home: 311 Burlage Dr., Chapel Hill, N.C. 27514.*

WAGNER, Robert Philip, Am. geneticist; b. N.Y.C., May 11, 1918; s. Philip Joseph and Annette (Pavelka) W.; B.S., Coll City N.Y., 1940; Ph.D., U. Tex., 1943; m. Margaret Lillian Campbell, June 12, 1947; children—Philip Campbell, James Robert, Ruth Annette. Faculty, U. Tex., Austin, 1940—, prof. zoology, 1956—. Research biologist Nat. Cotton Council, 1944-45, vis. prof. Ind. U., Bloomington, 1962; cons. NSF, NIH. NRC fellow Cal. Inst. Tech., 1946-47; Guggenheim fellow U. Cal. at Berkeley, 1957-58; Recipient Research Career award NIH, 1962. Mem. Genetics Soc. Am. (sec. 1965-67), Am. Soc. Naturalists, Am. Chem. Soc., Cell Biology Soc., Am. Soc. Human Genetics. Author: (with H.K. Mitchell) Genetics and Metabolism, 1955, 64; also numerous articles. Research on role of insect nutrition in evolution of insects, effect of gene mutation on metabolism of organism; demonstrated steps involved in synthesis of certain amino acids, showed metabolic machinery necessary to make these is highly organized in living cell. Home: 2410 Pemberton Pkwy., Austin, Tex. 78703.*

WAGNER, Robert Roderick, Am. physician, mirobiologist; b. N.Y.C., Jan. 5, 1923; s. Nathan and Mary (Mendelsohn) W.; A.B., Columbia, 1943; M.D., Yale, 1946. Research fellow Nat. Inst. Med. Research, London, Eng., 1950-51; instr., then asst. prof. medicine Yale, 1951-55; faculty Johns Hopkins, 1956—, 1957-63, prof. microbiology, 1964—. Mem. coms. USPHS, Dept. Def. Assn. Am. Med. Colls., A.M.A., Nat. Bd. Med. Examiners. Fellow A.A.A.S.; mem. Assn. Am. Physicians, Am. Soc. Clin. Investigation, Am. Assn. Immunologists, Soc. Exptl. Biology, Soc. Am. Bacteriologists. Research on relationship to pathogenesis of virus infection, host-parasite interactions, immunity, interferon. Office: 725 N. Wolfe St., Balt. 21205.*

WAGNER, Rudolf, German anatomist, physiologist; b. Bayreuth, Germany, July 30, 1805: ed. Erlangen, Würzburg, and Paris. Prof. Erlangen, Germany, 1832-40, Göttingen, Germany, from 1840. Author: Lehrbuch der vergleichenden Anatomie, 1834-35; Icones physiologicae, 1838-40; Lehrbuch der Physiologie, 1839; Handatlas der vergleichenden Anatomie, 1841; Handwörterbuch der Physiologie, 1842-53; Neurologische Untersuchungen, 1854; Der Kampf um die Seele vom Standpunkt der Wissenschaft, 1857; Vorstudien, 1860-62. Discovered germinal spot in human ovum (Wagner's spot), 1835; discovered (with Georg Meissner) tactile end organs (Meissner's corpuscles), 1852; opponent of materialism; attacked by C. Vogt for his irrational views; conducted research in anthropology and archeology. Died Göttingen, May 13, 1864.

WAGNER, Walter, chemist; b. Serajevo, Yugoslavia, Sept. 18, 1904; s. Stefen and Sophia (Altmann) W.; Ch.E., Vienna (Austria) Inst. Tech., 1927, D.Sc., 1928; m. Frances Weil, June 4, 1954; came to U.S., 1938, naturalized, 1944. With Indsl. Abrasives, Inc., Chgo., 1940-42; faculty Worcester (Mass.) Poly. Inst., 1942-43, Westbrook Jr. Coll., Portland, Me., 1943-44, Denison U., Granville, O., 1944-45; prof. analytical chemistry U. Detroit, 1945—. Mem. Am. Chem. Soc., Am. Assn. U. Profs., Assn. Analytical Chemistry. Author: (with Hull, Markle)) Advanced Analytical Chemistry, 1954; (with Shiple) Introduction to Quantitative Analyis, 1959; also articles. Asso. editor Ency. Chem. Reactions, 1946—; abstractor Chem. Abstracts, 1945—. Research in analytical chemistry including spectrophotometric analysis. Home: 17385 Warrington St., Detroit, 48221.*

WAGNER, Warren Herbert, Jr., Am. botanist; b. Washington, Aug. 29, 1920; s. Warren Herbert and Harriet (Claflin) W.; A.B., U. Pa., 1942; Ph.D., U. Cal. at Berkeley, 1950; postgrad. Harvard; m. Florence Signaigo, July 16, 1948; children—Warren Charles, Margaret Frances. Instr., Harvard, 1951; faculty U. Mich., Ann Arbor, 1951—, prof. botany 1962—, curator pteridophytes, 1961—, dir. bot. gardens, 1966—. Investigator project on evolutionary characters of ferns NSF, 1960—; dir. bot. research in aeroallergen project NIH, 1957-64. Fellow A.A.A.S. (sec. sect. bot. scis. 1964-67, v.p. 1968—); mem. Am. Fern oc. (past sec., curator, librarian 1957—, v.p. 1968), Am. Soc. Plant Taxonomists (past mem. council, pres. 1966-67), Soc. for Study Evolution (v.p. 1965-66, council 1968—), Am. Soc. Naturalists, Bot. Soc. Am., Torrey Bot. Club, New Eng. Bot. Club, Internat. Soc. Plant Morphologists, Internat. Assn. Plant Taxonomists, Sigma Xi, Phi Kappa Tau. Author: (with E. Steiner, A.S. Sussman) Botany Laboratory Manual, 1957, rev., 1965; The Fern Genus Diellia, 1952; also numerous articles. Research on higher plants, origin and evolution of ferns, methods of accurate deduction of phylogenetic relationships of fossil and living plants. Home: 2111 Melrose Av., Ann Arbor, Mich. 48104.*

WAGNER, William Frederick, Am. chemist; b. Canton, Mo., Sept. 13, 1916; s. William C. and Lilly (Giegerich) W.; A.B., Culver Stockton Coll., 1938; M.S., U. Chgo., 1940; Ph.D., U. Ill., 1947; m. Jean Rosselot, July 22, 1945; children—Jennie Ann, Lenore Elizabeth, Russell William. Research chemist Ill. Geol. Survey, Urbana, 1940-45; faculty Hanover (Ind.) Coll., 1947-49; faculty U. Ky., Lexington, 1949—, prof. chemistry, 1958—, chmn. dept., 1965—. Recipient Alumni award Culver-Stockton Coll., 1961. Mem. Am. Chem. Soc., Sigma Xi, Alpha Chi Sigma. Research, publs. in analytical chemistry, spectographic analysis, solvent extraction of metal chelates, precipitation processes. Home: 1252 Summit Dr., Lexington, Ky. 40502.*

WAGNER-FISCHER, Anne-Marie, German physician; b. Munich, Germany, Mar. 7, 1918; d. Georg and Nanine (Ritter) Fischer; student U. Munich, 1939-41; Dr.med. U. Würzburg, 1944; m. Erich Wagner, Mar. 6, 1962. Staff med. sect. Youth Office, Hamburg, Germany, 1947-51, Pub. Health Office, 1952-54; sci. researcher supr. German Red Cross, Bonn, Germany, 1957—. Mem. German Soc. for Rehab. Disabled (vice chmn. com. for housing and aides for daily life 1962—), Bundesausschuss für gesundheitliche, Volksbelehrung German Fed. Rep. (mem. bd. 1960—). Research, numerous publs. on devel. gadgets for disabled and old people especially kitchen- furniture for wheel-chair patients, practical dressing for disabled and older persons, health edn. Home: 8 Rheinblick 5486 Oberwinter. Office: 71 Friedrich Ebert Allee, 53, Bonn, Germany.*

WAGNER-JAUREGG, Julius, Austrian physician, psychiatrist; b. Wels, Austria, Mar. 7, 1857; M.D., U. Vienna (Austria), 1880; 1 dau., Julia. Asst. dept. exptl. pathology and internal medicine U. Vienna, 1881-83, asst. clinic psychiatry, 1883-89, dir. univ. hosp. for nervous and mental diseases, also prof., 1893-1928; prof. psychiatry and neurology U. Graz, 1889-93. Recipient Nobel prize (for work in use of malaria and other artificially induced fevers in treating syphilitic paralysis), 1927, Gold medal Am. Com. Research in Syphilis, 1937, also honored as Father of Fever Therapy, 1st Internat. Conf. on Fever Therapy, N.Y.C., 1937. Author: General Paralysis by Inoculation with Malaria. First to use (as early as 1887) fever therapy, or use of tertian malaria and other artificially induced fevers to treat dementia paralytica, a disease previously regarded as incurable; expert on cretinism and other thyroid diseases; noted for treatment of cretinism with thyroid gland preparations, treatment of goiter with iodine; also research on sleeping sickness, myxedema, forensic psychiatry, pharmacology. Died Vienna, Sept. 27, 1940.

WAHBA, Ahmed Hassan, bacteriologist; b. Berlin, Germany, Dec. 15, 1922; s. Hassan Ibrahim and Elfriede (Hauptstein) W.; M.B.,B.Ch., U. Cairo (Egypt), 1946; D.T.M. & H., U. London (Eng.), 1949, Ph.D., 1964; Diploma in Bacteriology, U. Manchester (eng.), 1950; m. Islah El-Falaky, July 23, 1959; children—Hassan, Mona. Bacteriologist, Egyptian Govt., 1946-49; research worker Statens Serum Inst., Copenhagen, Denmark, 1950, Pasteur Inst., Paris, France, 1951; dir. bacterial vaccines dept. Ministry Pub. Health, Cairo, 1952-58, dir. biol. control dept., 1959-60; research worker Cross-Infection Reference Lab., Central Pub. Health Lab., Colindale, London, 1961-63; lectr. London Sch. Hygiene and Tropical Medicine, 1963-65; dir. gen. Prodn. Labs. Adminstrn., Agouza, Cairo, 1966—. Recipient Cholera epidemic award, 1947. Research and publs. on application of vibriocine prodn. in cholera vibrios to diagnostics; discovered action of bacteriocines of Pseudomonas aeruginosa and its application by devising typing method of organism; developed new type of cholera vaccine, use of Lupinus seeds as media for vaccine prodn. studied various epidemic diseases, including measles, typhoid fever; modified various lab. tests, including hemolysis test of vibrios. Home: 27, Mohamed Hegab St., Heliopolis, Cairo, U.A.R. Office: Serum and Vaccine Inst., Agouza-Dokki, Giza, Egypt., U.A.R.

WAHL, Arthur Charles, Am. chemist; b. Des Moines, Ia., Sept. 8, 1917; s. Arthur C. and Mabel (Mussetter) W.; B.S., Ia. State Coll., 1939; Ph.D., U. Cal., 1942; m. Mary E. McCauley, Dec. 1, 1943; children—Kathryn Ann, Nancy Jean. Chemist, Manhattan Project, 1942-46; asso. prof. chemistry Washington U., 1946-53, Farr prof. radio-chemistry, 1953—; cons. Los Alamos Sci. Lab., 1950—. Mem. Am. Chem. Soc. (award for nuclear application in chemistry 1966). Editor, contbr.: Radioactivity Applied to Chemistry, 1951. Contbr. tech. articles. Co-discoverer of plutonium; research on oxidation-reduction reactions occurring by electron transfer, also nuclear charge distbn. in fission. Home: 7400 Teasdale, University City, Mo. 63130. Office: Washington U., St. Louis 63130.*

WAHL, Eberhard Wilhelm, meteorologist; b. Berlin, Germany, May 24, 1914; s. Edmund C. and Gertrud (George) W.; Ph.D., U. Berlin, 1937; m. Lore J. Rading, Mar. 12, 1946; children—Gisela A., Dorothee M. Came to U.S., 1949, naturalized, 1957. Research asso. Astron. Obs., Babelsberg, Germany, 1937-41; sci. asst. U. Berlin, 1938-45; councilor N.W. German Weather Service, 1946-49; with Air Force Cambridge Research Labs., Bedford, Mass., 1949-63, tech. dir. space track, 1961-63; vis. prof. meteorology U. Wis., Madison, 1962, prof., 1963—. Recipient Superior Performance awards USAF, 1960-63. Mem. Am. Astron. Soc., Am., German meteorol. socs., Am. Geophys. Union, Sigma Xi. Publs., systematic research and devel. on extended and long range forecasting; research in weather singularities and dynamic climatology; devel. space tracking system of USAF from ground up to an operational system. Home: 5510 Raymond Rd., Madison, Wis. 53711.*

WAHLEN, Friedrich Traugott, Swiss agriculturist; b. Gmeis, Germany, Apr. 10, 1899; s. Johannes and Katharina (Stucki) W.; Diploma as Agrl. Engr., Dr. Sc.Tech., ETH, Zurich; Dr. Med. honoris causa U. Zurich (Switzerland); Dr. honoris causa Agr. Sci., U. Göttingen (Germany); m. Helen Hopf, Oct. 11, 1923; Became asst. agronomy, Zurich, 1920; named supervising analyst Dominion Dept. Agr., Quebec, Can., 1922; chief analyst Seed br., Ottawa, Ont., 1924; became dir. Swiss Exptl. Sta. for Agr., Zurich, 1929; commr. war food prodn., 1942-49; mem. council states for Canton of Zurich; dir. agr. div., 1943-57; dep. dir. gen. FAO, UN, Rome, 1957-58; elected to Swiss Fed. Council; became pres. Swiss Confed, 1961. Recipient Marcel Benoist prize, 1943. Mem. Royal Swedish Acad. Agr. Author: Unser Boden heute und morgen, 1943; also numerous articles. Originated Wahlen plan for systematic utilization of Swiss soil, raw materials, man power to achieve self-sufficiency in food supplies, 1942; discovered new type of Bent grass used for lawns and fodder; agrl. devel. of under developed world. Home: Palais fédéral, Berne, Switzerland.

WAHLSTROM, Ernest Eugene, Am. geologist, mineralogist; b. Boulder, Colo., Dec. 30, 1909; s. Charles S. and Christine (Dahlin) W.; B.A., U. Colo., 1931, M.A., Harvard, 1936, Ph.D., 1939; m. Kathryn Kemp, Sept. 23, 1931; children—Karen J. (Mrs. Steven V. Guzak), David K. Faculty, U. Colo., Boulder, 1936—, prof. geology, 1947—, acting dean faculties, 1964; geologist U.S. Geol. Survey, intermittently, 1943—. Cons. in engring. and mining geology to firms, municipalities, City of Denver, 1939—. Recipient Brunton award U. Colo., 1931. Mem. Geol. Soc. Am., Mineral. Soc. Am., Geochem. Soc., Soc. Econ. Geologists, A.A.A.S., Am. Inst. Profl. Geologists, Assn. Engring. Geologists. Author: Optical Crystallography, 1943; Igneous Rocks and Minerals, 1947; Theoretical Igneous Petrology, 1950; Petrographic Mineralogy, 1955; also articles. Research in geology and mineralology. Home: 2875 Dover Dr., Boulder, Colo. 80302.*

WAHRHAFTIG, Austin Levy, Am. chemist; b. Sacramento, May 5, 1917; s. Moses Solomon and Irma (Levy) W.; student Sacramento Jr. Coll., 1934-36; A.B., U. Cal. at Berkeley, 1938; Ph.D., Cal. Inst. Tech., 1941; m. Ruby Martha Dixon, Aug. 24, 1957. Research fellow Cal. Inst. Tech., 1941-45; univ. fellow Ohio State U., 1946-47; faculty U. Utah, Salt Lake City, 1947—, prof. chemistry, 1960—. Mem. exec. com., com. on mass spectrometry Am. Soc. Testing Materials, 1962-64. Mem. Am. Chem. Soc., Am. Phys. Soc., A.A.A.S., Phi Beta Kappa, Sigma Xi. Research, publs. on devel. of theory of fragmentation of polyatomic molecules after ionization by electrons or ultraviolet radiation. Home: 2239 Logan Av., Salt Lake City, 84108.*

WAIBEL, Paul Edward, Am. nutritionist; b. Hawthorne, N.J., June 22, 1927; s. Paul and Emily (Scheibel) W.; B.S., Rutgers U., 1948; M.S., U. Wis., 1951, Ph.D., 1953; m. Joyce Avis Wickham, June 16, 1951; children—Gilbert Paul, Alan Carl, Ronald Mal. Research asso. Cornell U., Ithaca, N.Y., 1953-54; faculty U. Minn., St. Paul, 1954—, prof. animal sci., 1964—. Chmn., Minn. Nutrition Conf., 1963-66. Mem. Am. Inst. Nutrition, Soc. Exptl. Biology and Medicine, N.Y., Minn. acads., sci., A.A.A.S., Poultry Sci. Assn., Assn. Vitamin Chemists, Am. Chem. Soc. Research, numerous publs. on nutrition of turkeys and chickens including requirements of proteins, amino acids, vitamins, minerals, aortic rupture syndrome in turkeys, gnotobiology. Home: 2222 S. Rosewood Lane, St. Paul 55113.*

WAID, John Saville, English soil microbiologist; b. Gosport, Eng., Nov. 16, 1927; s. Saville Charles and Edith (Venelle) W.; B.Sc., London U., 1951; B.Sc., Oxford U., 1953, Ph.D., 1959; m. Sally Cecilia Taylor, May 29, 1954; children—Caroline Patricia, Giselle Alice, Saville Charles Sheridan, Susan Natasha Margaret. Soil microbiologist Grassland Research Sta., Hurley, Eng., 1954-56; soil microbiologist Nature Conservancy, Grange-over-Sands, Eng., 1956-58; biologist Fisons Fertilizers, Felixtowe, Eng., 1958-62; sr. lectr. botany dept. U. New Eng., Armidale, New S. Wales, Australia, 1962-64; research asso. dept. agronomy Cornell U., Ithaca, N.Y., 1964-65; lectr. dept. soil sci. Reading Berks., Eng. U., 1965—. Mem. Brit. Soc. Soil Sci. Author: Ecology of Soil Fungi—An International Symposium, 1960; also articles. Exec. editor Jour. Soil Biology and Biochemistry. Research on fungi of root systems of plant and in soil, urea decomposition in soil, pesticide decomposition in soil; discovered ultrasonic detection of forest tree pathogens and hydroxamate inhibition of urease activity in soil. Home: 18 Reeds Av., Earley, Reading, Berks, Eng.*

WAIN, Ralph Louis, English chemist; b. Hyde, Eng., May 29, 1911; s. George and Eliza (Hardy) W.; B.Sc. with 1st class honors in Chemistry, U. Sheffield, 1932, M.Sc., 1933, Ph.D., 1935; D.Sc., U. London (Eng.), 1950; m. Joan Bowker, Aug. 17, 1940; children—Rosemary Joan, Michael Louis. Research asst. U. Manchester (Eng.), 1935-37; lectr. chemistry Wye Coll., U. London, 1937-39, prof. agrl. chemistry, head dept. phys. scis., 1945—; research chemist U. Bristol (Eng.), 1939-45; dir. unit on plant growth substances and systemic fungicides Agrl. Research Council, 1953—. Recipient Ptuthivi gold medal for research, 1957; research medal Royal Agrl. Soc. Eng., 1960; John Scott medal and award, Phila., 1963. Fellow Royal Soc., 1960; mem. Royal Inst. Chemistry (past v.p.), Chem. Soc., London, Am. Chem. Soc. Editor (with F. Wightman) Planet Growth Substances, 1956. Research, numerous publs. on mode of action of chems. used for control plant pests, diseases and weeds, plant growth-regulating substances; new chems. for control weeds: Home: Staple Farm, Hastingleigh, Kent. Office: Wye Coll., Wye, Kent, Eng.*

WAINIO, Walter, Am. biochemist; b. Astoria, Ore., Sept. 8, 1914; s. Waldemar and Wilhelmina (Taipale) W.; B.S., U. Mass., 1936; M.S., Pa. State U., 1940; Ph.D., Cornell U., 1943; m. Betty Dent Hill, June 30 1949; 1 dau., Marguerite. Instr., Pa. State U., 1938-41, Cornell U. Med. Coll., 1941-43; faculty N.Y. U. Coll. Dentistry, 194348, asst. prof., 1944-48; with Rutgers The State U. N.J., New Brunswick, 1948—, prof. biochemistry, 1959—, chmn. dept. physiology and biochemistry, 1960-63, chmn. dept. biochemistry, 1966—. Fellow N.Y. Acad. Scis.; mem. Am. Soc. Biol. Chemists, A.A.A.S., Am. Chem. Soc., Biochem. Soc. (Eng.), Research, publs. on purification and characterization of enzymes of respiratory chain of mammalian heart muscle mitochondrion, link between oxidation and phosphorylation. Home: 477 Walnut Lane, Princeton, N.J. 08540. Office: Nelson Biol. Labs., Rutgers U., New Brunswick, N.J. 08903.*

WAIT, James Richard, physicist; b. Ottawa, Can., Jan. 23, 1924; s. George Enoch and Doris (Browne)

W.; B.A.Sc., U. Toronto, 1948, M.A.Sc., 1949, Ph.D., 1951; m. Gertrude Laura Norman, June 16, 1951; children—Laura, George. Jr. research engr. Ont. Hydro Electric Power Commn., Toronto, 1948-49; research engr. Nemmont Exploration Ltd., Jerome, Ariz., 1950-52; sci. officer Def. Research Telecommunications Establishment, Ottawa, 1952-55; cons., sr. research fellow Nat. Bur. Standards, Boulder, Colo., 1955-65; sr. research fellow Environmental Sci. Service Administrn., Boulder, 1965—. Research fellow Tech. U. Denmark, Copenhagen, 1960-61; adj. prof. elec. engring. U. Colo., 1961-66; mem. U.S. nat. com. Internat. Sci. Radio Union, 1963-66; vis. prof. Harvard, 1966-67. Recipient Gold medal Dept. Commerce, 1959; Stratton award Nat. Bur. Standards, 1962; Flemming award Washington C. of C., 1964. Fellow I.E.E.E.. (Diamond award 1964), A.A.A.S.; mem. Am. Geophys. Union, Can. Assn. Physics, Optical Soc. Am. Author: Electromagnetic Waves in Stratified Media, 1962 Electromagnetic Radiation from Cylindrical Structures, 1959; Overvoltage Research and Geophysical Applications, 1959; Electromagneotics and Plasmas, 1968. Editor, Radio Sci., 1959--61, 63-66, 67-68; U.S. editor Electronics Letters. Research in boundary value problems in electromagnetic theory; developed method for calculating radio fields at very low frequency; derived new solutions for electromagnetic prospecting problems. Home: 756 14th St. Office: Environmental Sci. Service Administrn., Boulder, Colo. 80302.*

WAIT, Samuel Charles, Jr., Am. chemist; b. Albany, N.Y., Jan. 26, 1932; s. Samuel Charles and Isabel (Cassedy) W.; B.S., Rensselaer Poly. Inst., 1953, M.S., 1955, Ph.D., 1956; m. Carol D. Petrie, June 6, 1957; children—Robert J. Alison R. Fulbright fellow U. Coll., London, Eng., 1956-57, asst. lectr., 1957-58; post-doctoral fellow U. Minn., 1958-59; asst. prof. Carnegie Inst. Tech., 1959-60; chemist Nat. Bur. Standards, Washington, 1960-61; asst. prof. Rensselaer Poly. Inst., Troy, 1961-64, asso. prof., 1964—. Mem. Am. Chem. Soc., Am. Optical Soc., Coblentz Soc., Sigma Xi, Phi Lambda Upsilon. Research, publs. on spectroscopy of heterocyclic organic molecules, electronic and vibrational spectroscopy, quantum chemistry of aromatic molecules, ionic interactions in liquids. Home: 909 Ash Tree Lane, Schnectady 12309. Office: Walker Lab., Rensselaer Poly. Inst., Troy, N.Y. 12181.*

WAITZ, Theodor, German anthropologist, philosopher; b. Gotha, Germany, Mar. 17, 1821; ed. Leipzig and Jena; lectr., Marburg, 1844-48; prof. philosophy, from 1848. Author: Anthropologie der Naturvölker, 1859. Among 1st to realize full scope of anthropology as a science; examined current views of anthrop. questions; analyzed anthropology as study of relation between man's mental and phys. composition and his social (historical) devel.; asserted that differences in degree of civilization among races are due to differences of opportunity rather than native capacity; follower of Herbert in philosophy; tried to base philosophy on psychology. Died Marburg, May 21, 1864.

WAJCHENBERG, Bernardo Léo, Brazilian physician; b. Sao Paulo, Brazil, Apr. 6, 1926; s. Max and Maria (Kanczuk) W.; student State Coll. Sao Paulo, 1938-44; M.D. summa cum laude, U. Sao Paulo, 1950, D.Medicine summa cum laude, 1961; m. Rosa Mohrer, May 24, 1951; children—Mauro, Silvia Eli Lilly fellow U. Minn. Hosps., 1954-55; W.K. Kellogg fellow U. Mich., U. Pa., 1955-57; asst. in medicine 1st Med. Clinics, Hosp. das Clinicas Sao Paulo, 1957—, physician-in charge diabetes and adrenal unit, 1958—, steroid lab., 1958—; lectr. U. Sao Paulo Med. Sch., 1957—. Recipient Rockefeller Found award, 1950; numerous awards including Nami Jafet, and prizes Brazilian Acad. Medicine. Mem. Sao Paulo Assn. Medicine (past pres.), Am. Fedn. Clin. Research, N.Y. Acad. Scis., Am. Diabetes Assn., Endocrine Soc., A.C.P. Numerous publs. on 1st clin. study renal lesion in rattlesnake poisoning; research on presence of specific globulin in fatal hepatitis, fatty acid metabolism, genetics of diabetes in children. Home: 2607 Gabriel Monteiro Silva. Office: Hosp. das Clinicas, Sao Paulo, Brazil.*

WAKE, Charles Staniland, anthropologist; b. Kingston-upon-Hull, Eng., Mar. 22, 1835; ed. Hull Coll., Eng. With Field Mus. of Natural History, Chgo., from 1895. Dir. Anthrop. Inst. of Gt. Britain and Ireland; mem. gen. com. British Assn. Advancement of Sci.; corr. mem. Bklyn. Ethical Soc.; asso. Soc. for Psychical Research, Eng. Author: Chapters on Man. 1862; The Evolution of Morality, 2 Vols., 1878; The Origin and Significance of the Great Pyramid, 1882; Serpent Worship, and Other Essays, 1888; The Development of Marriage and Kinship, 1889; Vortex Philosophy, or The Geometry of Science, 1907; System of Color and Musical-Tone Relations. Editor Memoirs Congress of Anthropology, Chgo., 1893-94. Died 1910.

WAKELY, Andrew (or Wakerly), mathematician; flourished 1631-65; tchr. applied math., maker of nautical instruments; author textbook on use of navigational instruments (in use more than a century), 1633.

WAKERLIN, George Earle, Am. physician; b. Chgo., July 1, 1901; s. George and Emma (Kenzig) W.; B.S., U. Chgo., 1923, Ph.D., 1926, M.D., 1929; M.S., U. Wis., 1924; m. Ruth Billings Coleman, Feb. 14, 1952; children—Susan Helen, George Earle. Research instr. U. Wis., 1923-25; asso. instr., fellow U. Chgo., 1925-28; practice medicine, Chgo., 1930-31; faculty U. Louisville Sch. Medicine, 1931-37, prof. physiology and pharmacology, 1934-37, head dept., 1932-37; prof. physiology, head dept. U. Ill. Coll. Medicine, 1937-58; med. dir. Am. Heart Assn., 1958-66; adj. prof. physiology Columbia Coll. Phys. and Surg., 1960-66; state coordinator heart disease, cancer and stroke, prof. medicine U. Mo. Sch. Medicine, Columbia, 1966—. Recipient medal for distinguished service in cancer control Am. Cancer Soc., 1957; Distinguished Service award Med. Alumni Assn. U. Chgo., 1962. Fellow A.C.P., Am. Pub. Health Assn.; mem. Soc. for Exptl. Biology and Medicine (past chmn. Ill. sect.), A.M.A. (past chmn. sect. on pathology and physiology), Am. Soc. for Study Arteriosclerosis (past dir.), Gerontological Soc. (past dir.), Am. Physiol. Soc. (past chmn. circulation group), Nat. Soc. for Med. Research (dir. 1965—), Am. Heart Assn. (past chmn. profl. edn. com.), Nat. Interagy. Council on Smoking and Health (exec. com. 1965-66), Am. Soc. for Pharmacology and Exptl. Therapeutics. Contbg. author: Laboratory Textbook of Human Physiology, 1937-58; also numerous articles. Discovered (with others) antirenin, new type exptl. hypertension may be due to some change in character blood flow through brain; demonstrated importance renin in hypertension, exptl. renal hypertension in dogs accelerates onset and accentuates severity exptl. hardening arteries. Home: 3002 Bray Av., Columbia, Mo. 65201.*

WAKIL, Salih Jawad, biochemist, educator b. Kerballa, Iraq, Aug. 16, 1927; s. Jawad Hamood and Millook (Atraqchi) W.; B.Sc., Am. U., Beirut, Lebanon, 1948; Ph.D. in Biochemistry, U. Wash., 1952; m. Fawzia Bahrani, Sept. 15, 1952; children—Sonya, Aida, Adil and Youssef (twins). Came to U.S., 1948, naturalized, 1964. Research fellow U. Wash., 1949-52; research asso., asst. prof. Inst. for Enzyme Research, U. Wis., 1952-59; faculty dept. biochemistry Duke, Durham, N.C., 1959—, prof., 1965—. Mem. Am. Chem. Soc., Am. Soc. Biol. Chemists, Sigma Xi, Phi Lambda Epsilon. Research, numerous publs. on discovery of mechanism of fatty acid biosynthesis, isolation of biotin enzymes, malonyl Co A, Acyl carried protein and elucidating control of fatty acid biosynthesis. Home: 2527 Sevier St., Durham, N.C. 27705.*

WAKIM, Kalil Georges, physiologist; b. Sidon, Lebanon, July 17, 1907; s. Georges Elias and Marian (Francis) W.; B.A., Am. U. Beirut, 1928, M.D., 1933; Ph.D., U. Minn., 1941. Came to U.S., 1938, naturalized, 1943. Instr. physiology Am. U. Beirut Med. Sch., 1933-38, dir. pub. health projects, rural areas Lebanon and Syria, 1937; research health conditions rural areas Palestine, Brit. Govt., 1934; practice medicine, research health conditions among Nomadic tribes Arabian Desert, 1936; fellow physiology, exptl. surgery Mayo Clinic and Found., Rochester, Minn., 1938-41; acting prof. physiology State U. Ia. Med. Coll., 1940; with Ind. U. Sch. Med., 1941-46, prof. physiology, 1943-46; prof. physiology Mayo Found., 1946—, cons., 1946—. Cons. Surgeon Gen.'s Office, Armed Forces Med. Center, 1948-54; hon. prof. medicine Loma Linda U., 1960—, asso. dir. Manger Research Found., 1964—. Recipient Gold medal Highest Order of Merit, Republic of Lebanon, 1950; Gold medal First Order of Merit, Republic of Syria, 1950. Fellow A.A.A.S.: mem. Am. Soc. Tropical Medicine, Am. Physiol. Soc. Am. Soc. Exptl. Biology and Medicine, Am. Soc. Pharmacology and Exptl. Therapeutics, A.M.A., Sigma Xi, Alpha Omega Alpha. Research, publs. on influence of circulation of blood on function of organs of body, body fluids and salts. Home: 522 21st St. N.E., Rochester, Minn. 55901.*

WAKLEY, Thomas, English med. reformer; b. Membury in Devonshire, Eng., July 11, 1795; s. Henry Wakley; apprentice to apothecary, 1810; pupil of surgeon Phelps, then Coulson; ed. Borough Hosps., London, Eng., 1815, pvt. anatomy sch. of Edward Grainger; m. Feb. 20, 1820; 3 sons, including Thomas Henry, James Goodchild. Pvt. practice medicine from 1818; founder med. publ. Lancet, 1823; M. P. from Finsbury, 1835-52; coroner West Middlesex, Eng. from 1839. Publicized through mag. Lancet med. information and details surg. performances hitherto kept for exclusive use mems. London hosps.; exposed nepotism, other unethical and undemocratic practices within London med. establishment; sought to make council of decision-making body of Coll. of Surgeons, London more widely representative of entire membership; as M. P. and coroner introduced higher standards of pub. health system, including legislation on uniform qualifications and registration standards for med. practitioners; investigation into adulterated foodstuffs eventually led to pure food and drugs legislation, 1860, 75, 79; reformed office of coroner by raising status of coroner's juries, improving inquest procedures, especially those on deaths in pub. instns.; proposed med. man rather than lawyer for coroner's position; credited with exposure charlatans and with bringing about end of flogging as army punishment. Died Deeping St. James, Eng., May 16, 1882.

WAKSMAN, Byron Halsted, Am. immunologist; b. N.Y.C., Sept. 15, 1919; s. Selman; A. and Bertha (Mitnik) W.; B.S., Swarthmore Coll., 1940; M.D., U. Pa., 1943; m. Joyce Ann Robertroy, Aug. 11, 1944; children—Nan, Peter. Fellow Mayo Found., 1946-48; NIH fellow Columbia Med. Sch., 1948-49; asso., then asst. prof. bacteriology and immunology Harvard Med. Sch., 1949-63; research fellow, then asso. bacteriologist (neurology) Mass. Gen. Hosp., 1949-63; prof. microbiology Yale, 1963—, chmn. dept., 1964—. Mem. expert panel immunology WHO, 1962—. Served as psychiatrist AUS, 1944-46. Mem. Am. Assn. Immunologists (councillor 1965—), British Soc. Immunology, Am. Soc. Microbiologists, Société Française d'Immunologie (councillor 1968—). Contbr. numerous articles in field. Editor Progress in Allergy, 1960—. Study of switching and automata theory; transduction of information. Home: Rock Hill Rd., Woodbridge, Conn. Office: 310 Cedar St., New Haven.

WAKSMAN, Selman Abraham, microbiologist; b. Priluka, Russia, July 22, 1888; s. Jacob and Fradia (London) W.; came to U.S., 1910, naturalized, 1916; B.S., Rutgers U., 1915, M.S., 1916, D.Sc. (hon.), 1943; Ph.D., U. Cal. at Berkeley, 1918; Ph.D. (hon.), U. Madrid, Strasbourg U. (France), Hebrew U.(Israel); M.D. (hon.) U. Liege, U. Athens, U. Pavia (Italy); D.Sc., Princeton, Brandeis U., U. Brazil, others; m. Deborah Mitnik, Aug. 5, 1916; 1 son, Byron H. Faculty, Rutgers U., New Brunswick, N.J., 1918-58, prof., 1930-50, head dept.. microbiology, 1940-58, dir. Inst. Microbiology, 1949-58, prof. and dir. emeritus, 1958, researcher, lectr., 1958—; marine bacteriologist Woods Hole Oceanographic Inst., 1931-42, now life trustee. Founder, pres. Found. for Microbiology, 1951—; cons. to indsl. labs., brs. of govt. and sci. instns. Recipient numerous awards including Passano Found. award, Emil Christian Hansen medal, award Carlsberg Labs., Denmark, 1948, Leeuwenhoek medal Netherland Acad. Scis., 1950, Albert and Mary Lasker award Am. Pub. Health Assn.; Nobel Prize in Physiology and Medicine, 1952; Order of Merit of Rising Sun, Emperor Japan, 1952; British Sholom Humanitarian award, 1952; St. Vincent award Acad. Medicine Turin (Italy), 1954; Great Cross Pub. Health, Spain, 1954; Instituto Carlo Forlanini medal, 1959; Am. Trudeau medal, 1961; commendatore Order of So. Cross Brazil, 1963; comdr. French Legion of Honor, 1950. Fellow A.A.A.S.; mem. Am. Soc. Microbiology, Nat. Acad. Scis., Am. Acad. Arts and Scis. (Amory award 1948), Internat. Soc. Soil Sci., Am. Soc. Agronomy, others fgn. mem. French Acad. Sci., also hon. mem. numerous sci. fgn. socs., Alpha Zeta, Phi Beta Kappa, Sigma Xi. Author numerous books including: Enzymes, 1926; Principles of Soil Microbiology, 1927; My Life With the Microbes, 1954; The Actinomycetes, vol. I, 1959, vol. II, 1961, vol. III, 1962; The Brilliant and Tragic Life of W.M.W. Haffkine, 1964; The Conquest of Tuberculosis, 1964; Jacob G. Lipman, Agricultural Scientist and Humanitarian, 1966. Editor: Streptomycin—Nature and Practical Applications, 1949; The Lit. on Streptomycin, 1948, 52; Neomycin, 1952, 56; Actinomycin, 1968, others. Research, publs. on soil microbiology and role of bacteria in decomposition, prodn. and nature of organic matter; isolation of new antibiotics, including actinomycin, streptomycin, neomycin and candicidin, all important in treatment of human disease. Home: 16 Logan Lane, Piscataway, N.J. 08804. Office: Inst. Microbiology, Rutgers U., New Brunswick, N.J. 08903.*

WALBORSKY, Harry M., chemist; b. Lodz, Poland, Dec. 25, 1923; s. Israel and Sarah (Miedowicz) W.; B.S., Coll. City N.Y., 1945; Ph.D., Ohio State U., 1949; children—Edwin, Eric, Lisa, Irene. Research asso. AEC project U. Cal., Los Angeles, 1949-50; Faculty chemistry Fla. State U., Tallahassee, 1950—. Cons. Dow Chem., 1956—. Mem. Am. Chem. Soc., Chem. Soc. London. Research, numerous publs. on mechanisms of organic reactions, organo-metallic chemistry, asymmetric syntheses. Home: 2223 Ruadh Ride, Tallahassee, Fla. 32303.*

WALCH, Johann Ernest Immanuel, German natural scientist; b. Jena, Germany, Aug. 29, 1725; prof., Jena. Author: Das Steinreich, 1761-64; Die Naturgeschichte der Versteinerungen, 1763-73. Died Dec. 1, 1778.

WALCOTT, Charles Doolittle, Am. paleontologist; b. New York Mills, N.Y., Mar. 31, 1850; s. Charles D. and Mary (Lane) W.; ed. pub. schs., Utica, N.Y.; LL.D., Hamilton, 1898, U. Chicago, 1901, Johns Hopkins, 1902, U. Pa., 1903, Yale 1910, St. Andrews, 1911, Pitts., 1912; Sc.D., U. Cambridge (Eng.), 1909, Harvard, 1913; Ph.D., Royal Fredericks U., Christiania, 1911; m. Helena B. Stevens, June 22, 1888 (dec. 1911); children—Charles D., Helen B. (Mrs. Cole B. Younger), Sidney S., Stuart B.; m. 2d, Mary Morris Vaux, June 30, 1914. Became asst. in N.Y. State Survey, 1876; asst. geologist U.S. Geol. Survey, 1879, paleontologist in charge invertebrate paleontology, 1888-93, geologist in general charge geology and paleontology, 1893-94, and dir.,

1894-1907; hon. curator dept. paleontology, 1892-97; at head of Nat. Mus., 1897-1898, with title of acting asst. sec. Smithsonian Instn., sec., 1907——; sec. Carnegie Instn., Washington, 1902-05; dir. U. S. Reclamation Service, 1905-07; dir. Research Corporation, N.Y.C. Fellow Christiana Sci. Soc., Am. Acad. Arts and Scis., Geol. Soc. Sci. (pres. 1901), Geol. Soc. London (Bigsby medal, and Wollaston medal, 1918), Imperial Soc. Naturalists (Moscow), Royal Geog. Soc. (London), Acad. Science Inst. Bologna; Soc. Géol. (Gaudry medal France), French Acad. Scis., Acad. Natural Sciences Phila. (Hayden medal), Nat. Acad. Scis. (Mary Clark Thompson medal 1921), Washington Soc. Archaeol. Inst. Am. (pres.), Royal Swedish Acad. Scis. Author: The Trilobite; Paleontology of the Eureka District; The Cambrian Faunas of North America; The Fauna of the Lower Cambrian or Olenellus Zone; Pre-Cambian Fossiliferous Formations; Correlation Papers; Cambrian Geology and Paleontology; Cambrian Brachiopoda; The Cambrian Faunas of China; The Cambrian and Its Problems in the Cordilleran Region; Pre-Cambrian Algonkian Algal Flora; Discovery of Algonkian Bacteria; Evidences of Primitive Life; Appendages of Trilobites. Research in geology and paleontology, especially Cambrian trilobites and stratigraphy. Died Washington, D.C., Feb. 9, 1927.

WALCOTT, William Welch, Am. physiologist; b. Englewood, N.J., Dec. 5, 1909; s. Frederic Collin and Mary (Guthrie) W.; A.B., Yale, 1933; Ph.D., Columbia, 1944; m. Martha Agusta Blake, Sept. 11, 1937; children—Alexandra (Mrs. A. Nicholas Wahl), Collin. Vis. fellow Rockefeller Inst. for Med. Research, 1938-39; faculty Columbia, 1941——, asso. prof. physiology, 1952——. Chmn., Jane Coffin Childs Meml. Fund for Med. Research, 1966. Mem. N.Y. Acad. Scis. (mem. council 1957-59), Am. Physiol. Soc., Soc. for Exptl. Biology and Medicine, Harvey Soc., A.A.A.S., Sigma Xi Research, publs. in cardiovascular physiology hemorrhagic and traumatic shock, regulation blood volume; devel. sci. instrumentation. Home: Norfolk, Conn. 06058.*

WALD, George, Am. biologist; b. N.Y.C., Nov. 18, 1906; s. Isaac and Ernestine (Rosenmann) W.; B.S., N.Y. U., 1927, D.Sc., 1965; Ph.D., Columbia, 1932; M.D. (hon.), U. Berne, 1957; D.Sc., Yale, 1958, Wesleyan U., 1962, McGill U., 1966; m. Ruth Hoffmann Hubbard, June 11, 1958; children—Michael, David, Elijah, Deborah. NRC fellow in biology, Berlin, Zurich, U. Chgo., 1932-34; faculty Harvard, Cambridge, Mass., 1935—, prof. biology, 1948—. Recipient Eli Lilly award Am. Chem. Soc., 1939, Lasker award Am. Pub. Health Assn., 1953, Proctor medal Assn. for research in Ophthalmology, 1955, Rumford medal Am. Acad. Arts and Scis., 1959, Ives medal Optical Soc. Am., 1966, (with Hartline and Granit) Nobel prize in physiology and medicine, 1967. Mem. Nat. Acad. Scis. Research, numerous publs. in biochemistry and physiology of vision and its evolution, biochem. evolution and origin of life, discovery of vitamin A and A2 in retina and their rules as precursors of visual pigments, mechanism of human color vision and color blindness. Home: 21 Lakeview Av., Cambridge, Mass. 02138.*

WALD, Niel, Am. physician; b. N.Y.C., Oct. 1, 1925; s. Albert and Rose (Fischel) W.; A.B., Columbia, 1945; M.D., N.Y. U., 1948; m. Lucienne Hill, May 22, 1953; children—David, Phillip. Practice medicine, specializing in radiation medicine and cytogenetics, Pitts., 1952——; sr. hematologist Atomic Bomb Casualty Commn., Hiroshima, Japan, 1954-57; head biologist Health Physics div. Oak Ridge Nat. Lab., 1957-58; faculty U. Pitts., 1958—, prof. radiation health, 1962——, prof. radiology, 1965——; cons. USPHS, Union Carbide Nuclear Co., Oak Ridge Nat. Lab., Westinghouse Electric Co., Bettis Atomic Power Lab. Recipient Nat. Def. Service medal, 1954. Mem. A.A.A.S., A.M.A., Internat., Am. socs. hematology, Radiation Research Soc., Health Physics Soc. (chpt. pres.), Am. Soc. Human Genetics, Indsl. Med. Assn., Am. Fedn. Clin. Research, Am. Pub. Health Assn., Soc. Nuclear Medicine. Research, numerous publs. on effects of radiation on blood, including prodn. of chromosome damage and leukemia; prevention and treatment of radiation-induced disease. Home: 5422 Normlee Pl., Pitts. 15217. Office: 130 DeSoto St., Pitts. 15213.*

WALDEN, Paul, chemist; b. Tsesis, Latvia, July 26, 1863; ed. Riga, Leipzig, Munich; prof. chemistry Poly. Inst., Riga; Latvia, before 1910, dir., after 1918; prof., St. Petersburg, Russia 1910-19; prof. phys. chemistry, dir., Chem. Inst., U. Rostock, Germany, 1919-34; vis. prof. Cornell U., Ithaca, N.Y., 1927-28; also prof., Tübingen, Germany, 1947. Del. Internat. Congress Applied Chemistry, N.Y., 1912. Mem. St. Petersburg Acad. Scis., French Acad. Scis. Author: Optische ("Waldensche") Umkehrung, 1918; Elektrochemie nichtwässigriger Lösungen, 1924; Goethe als Chemiker und Techniker, Geschichte der Chemie, 1947. Discovered bimolecular Walden inversion, 1895; studied elec. conductivity of aqueous solutions and of organic acids, the dielectric constants of liquids, also dissociation constants. Died Gammertingen (nr. Sigmaringen), Germany, Jan. 22, 1957.

WALDENSTROM, Jan, Swedish physician; b. Stockholm, Sweden, Apr. 17, 1906; s. Henning J. and Elsa (Laurin) W.; M.D., Uppsala (Sweden) U., 1937; post-grad. Cambridge (Eng.) U., 1927, U. Munich (Germany), 1934-35; M.D. (hon.), U. Oslo (Norway), 1961, Trinity Coll., Dublin, Ireland, 1962; M.Sc. Oxford (Eng.) U., 1964, M.D., U. Mainz (Germany), 1967; m. Elizabeth, 1932; children—Magnus, Johan Agnes (Mrs. Peter Nobel), Anders, Fanny; m. 2d, Karin Nordsjö, Jan. 26, 1957; children—Dag, Erik. Faculty, U. Uppsala, 1937-50, prof. theoretical medicine, 1947-50; prof. internal medicine Lund (Sweden) U., 1950——; head dept. medicine Gen. Hosp., Malmö, Sweden, since 1950——; vis. prof. univs. of Louvain, Belgium, Leyden, Holland, Oxford, Eng., Washington, St. Louis, Harvard, U. Cal. at Los Angeles. Recipient Jahre award, Oslo, 1964, Gairdner award, Toronto, Ont., Can., 1966. Fellow A.C.P., Royal Coll. Physicians; mem. Royal Soc. Medicine, London (hon.), Academia Leopolodina Halle (fgn.). Am. Acad. Arts and Scis., Swedish Acad. Scis. Contbr. chpts. to text books, numerous articles to tech. jours. Research on diseases where function of red blood pigment is disturbed (porphyrias), treatment of malignant diseases of blood and bone marrow, carcinoid syndrome; described disease macroglobulinemia; studied other globulins in blood. Home: Sanekullavägen 17, Malmö V. Office: Dept. Medicine, Gen. Hosp., Malmö, Sweden.*

WALDEYER, Heinrich Wilhelm Gottfried, German anatomist; b. Hehlen, Germany, Oct. 6, 1836; prof. path anatomy, Breslau (now Wroclaw, Poland), 1865-72; prof. normal human anatomy, Strasbourg (now in France), 1872-83; prof. anatomy, dir. anat. inst. U. Berlin, from 1883. Mem. French Acad. Scis. Author: Beiträge zur Entwicklung der Sexualorgane, 1870; Eierstock und Ei, 1870. Described the 2 duodenal fossae (Waldeyer's fossa) with reference to hernia, 1868; described germinal epithelium, also vascular layer of ovary, Waldeyer's line bounding insertion of mesovarium in overy at the hilus, 1870; credited with introducing term chromosome, 1888; proposed neuron theory of nervous system, 1891. Died Berlin, Jan. 23, 1921.

WALDHAUSEN, John Anton, Am. surgeon; b. N.Y.C., May 22, 1929; s. Max H. and Agnes (Stettner) W.; B.S. magna cum laude, Coll. Great Falls (Mont.), 1950; M.D., St. Louis U., 1954; m. Marian R. Trescher, June 4, 1957; children—John Henry Trescher, Robert Rodney, Anthony Gordon Scarlett. Sr. asst. surgeon Clinic Surgery Nat. Heart Inst. NIH, Bethesda, Md., 1957-59; instr. surgery, Ind. U. Med. Center, Indpls., 1962-63, asst. prof., 1963-66; asso. prof. surgery U. Pa., Phila. 1966——; practice medicine, specializing in surgery, Phila., 1966——. Recipient USPHS Career Devel. award, 1964. Diplomate Am. Bd. Surgery, Bd. Thoracic Surgery. Mem. A.C.S., Soc. U. Surgeons, Soc. Vascular Surgery, Am. Fedn. Clin. Research, Am. Acad. Pediatrics, Central Surg. Assn., Am. Soc. Artificial Internal Organs, Am. Heart Assn., A.A.A.S., Am. Physiol. Soc., Soc. Internationale de Chirurgie, Sigma Xi, Alpha Omega Alpha. Studies, numerous publs. on circulatory effects of arteriovenous fistulae, extra-corporeal circulation and digitalis preparations; fundamental problems of shock in the primate; methods of repair in coarctation of the aorta. Home: 1425 Colton Rd., Gladwyne, Pa. 19035. Office: U. Pa. Hosp., Phila. 19104.*

WALDI, Dieter, German chemist; b. Freiburg, Germany, Feb. 19, 1916; s. Karl Heinrich and Katharina (Wolff) W.; Diplomchemiker, U. Freiburg (Germany), 1942; Doctor phil II, U. Basel (Switzerland), 1948; m. Martha Bender, May 7, 1946. Control chemist in pharm. industry E. Merck AG KL., Darmstadt, Germany, 1951——. Mem. Soc. pharmaceutique. Author: Chromatography, 1960; also articles; contbg author: Thin Layer Chromatography, 1962. Research on paper chromatography of amino acids in biol. materials; systematic analysis of alkaloids using paper and thin layer chromatography; early pregnancy test by thin layer chromatography of steroids hormones; toxicological investigations by thin layer chromatography. Home: 7 Eichbergstr. Office: E. Merck AG KL Vit., 61 Darmstadt, Germany.*

WALDICHUK, Michael, oceanographer; b. Mitkau, Rumania, Oct. 23, 1923; s. Emanoil and Miranda (Macovichuk) W.; B.A., U. B.C., 1948, M.A., 1950; Ph.D., U. Wash., 1955; m. Shirley Isabelle Rabbage, Sept. 17, 1955; children—David Michael, Andrew John, Thomas Craig. Lab. asst. chemistry dept. U. B.C., Vancouver, Can., 1948-50; teaching fellow oceanographic labs., dept. oceanography U. Wash. Friday Harbor, Seattle, 1950-52; oceanographer Pacific Oceanographic Group Fisheries Research Bd. Can., Nanaimo, B.C., 1952-54, scientist-in-charge pollution investigation Biol. Sta., 1954-66, oceanographer-in-charge Pacific Oceanographic Group, 1966——. Mem. working com. on oceanography Nat. Acad. Scis.-NRC, U.S., 1958-62; mem. asso. com. water pollution research NRC, Can., 1965. Mem. Am. Chem. Soc., Am. Geophys. Union, Am. Soc. Limnology and Oceanography (past div. pres.), Chem. Inst. Can., Sigma Xi. Numerous publs. on investigations of phys. oceanography of coastal waters of B.C., evaluation of rates of renewal of contaminated waters, characteristics of various water-borne pollutants. Home: 2625 Lynburn Crescent. Office: Fisheries Research Bd. Can., Pacific Oceanographic Group, Nanaimo, B.C., Can.*

WALDMAN, Bernard, Am. physicist; b. N.Y.C., Oct. 12, 1913; s. Joseph and Mollie (Rothman) W.; A.B., N.Y. U., 1934, Ph.D., 1939; m. Florence Finley, June 13, 1942 (dec. June 61); children—John William, Nancy (Mrs. John Lee Melcher); m. 2d, Glenna Frank Ryan, Mar. 21, 1964; children—Tawny Ryan, Sharon Ryan. Research asso. U. Notre Dame, 1938-40, faculty 1940-43, 45—,prof., 1951——, asso. dean, prof. physics, 1964-67, dean Coll. Sci., 1967—; staff, group leader Los Alamos Sci. Lab., 1943-45. Mem. physics adv. panel NSF, 1965——; v. p. Midwestern Univs. Research Assn., Madison, Wis., 1959-60, 1965—, dir., 1960-65. Fellow Am. Phys. Soc.; mem. Am. Assn. Physics Tchrs., Am. Assn. U. Profs. Research, publs. on interaction of electrons and x-rays with matter; contbd. to improvement of electrostatic accelerators; constrn. of models for high energy accelerators. Home: 54728 Merrifield Dr., Mishawaka, Ind. 46544. Office: Coll. Scis., U. Notre Dame, Notre Dame, Ind. 46556.*

WALDMANN, Thomas Alexander, Am. physician; b. N.Y.C., Sept. 21, 1930; s. Charles and Elizabeth (Sipos) W.; A.B. with honors, U. Chgo., 1951; M.D., Harvard, 1955; m. Katherine Emory Spreng, Mar. 29, 1958; children—Richard Allen, Robert James, Carol Ann. Staff, Nat. Cancer Inst., NIH, 1956—, Am. Heart Assn. fellow, 1958-59, sr. investigator metabolism br., 1959-65, acting chief, 1965——. Med. cons. FDA, 1964——. Mem. Am. Physiol. Soc., Am. Soc. for Clin Investigation, Am. Fedn. for Clin. Research, Alpha Omega Alpha. Research, numerous publs. on red blood cells and serum protein; discovered intestinal lymphangiectasia, allergic gastroenteropathy and described new immunological disorders in patients with these diseases; devel. techniques for demonstrating gastrointestinal protein loss. Home: 3910 Rickover Rd., Silver Spring, Md. 20902. Office: Nat. Cancer Inst., NIH, Bethesda, Md. 20014.*

WALDO, C(lifford) D(wight), Am. polit. scientist; b. DeWitt, Neb., Sept. 28, 1913; s. Cliff Ford and Grace Gertrude (Lindley) W.; B.A., Neb. State Tchrs. Coll. (Peru), 1935; M.A., U. Neb., 1937; Ph.D. (Cowles fellow), Yale, 1942; m. Doris Gwendolyn Payne, Sept. 17, 1937; children—Mary Grace, Martha Gwen, Margaret Ann. Instr., Yale, 1941-42; asso. bus. econs. OPA, 1942-44; administrv. analyst U.S. Bur. Budget, Exec. Office of Pres., 1944-46; faculty U. Cal. at Berkeley 1946-67, prof. polit. sci., 1953—, dir. Inst. Govtl. Studies, 1958-67; prof. humanities Maxwell Sch., Syracuse (N.Y.) U., 1967—, Albert Schweitzer professor in the humanities. Fulbright fellow Oxford (Eng.) U., 1953; Fund for Advancement Edn. fellow, 1952-53. Recipient Silver medallion U. Bologna. Mem. Internat., Am. (past exec. com., v.p.) polit. sci. assns., Am. Soc. for Pub. Adminstrn. (mem. council 1963-66 past pres. San Francisco Bay area chpt.), Internat. Inst. Adminstrv. Sci. Author: The Administrative State: A Study of the Political Theory of American Public Administration, 1948; The Study of Public Administration, 1955; Perspectives on Administration, 1956; Political Science in the U.S.A.: A Trend Report, 1956; also articles. Editor: Ideas and Issues in Public Administration: A Book of Readings, 1953; Strengthening Management for Democratic Government, 1958; The Research Function of University Bureaus and Institutes for Government-Related Research, 1960; editor-in-chief Pub. Adminstrn. Rev. Survey, interpretation evaluation, analysis of Am. polit. sci. with emphasis on pub. adminstrn. Office: Dept. Humanities, Maxwell Sch., Syracuse U., Syracuse, N.Y.*

WALDO, Leonard, Am. metall. and elec. engr.; b. Cin., May 4, 1853; s. Frederic Augustus and Frances (Leonard) W.; B.S., Marietta College, 1872, A.M., 1877; student, Columbia School of Mines; Sc.D., Harvard, 1879; A.M. (hon.), Yale, 1880; m. Dora Fullerton, 1875 (dec.); m. 2d, Ada Louise Purdy, 1887. Asst. astronomer U. S. Transit of Venus expdn. to Tasmania, 1874; asst. Harvard Obs., 1875-80; astronomer in charge horological bur. Yale Obs. 1879-88. Cons., U. S. Steel Corp. in steel research; Cons. engr. War Dept. in prodn. of shells and illuminants in World War. Mem. fatigue of metals com. NRC; chmn. library bd. United Engring. Soc. Recipient medal Royal Soc. Arts. Inventor magnesium prodn. process. Died Jan. 25, 1929.

WALDSEEMÜLLER, Martin, German cartographer; b. Radolfzell or Freiburg, Baden Germany, circa 1470; studied theology U. Freiburg. Canon at St. Dié, Alsace (now in France). Became interested in cartography and geography as a youth; published map of world, Universalis cosmographia (contains 1st mention of name America, shows S.Am. as island; and Cosmographiae introductio (contains explanation of use of name America, suggests this name for New World), 1507 (a 1st edit. now in N.Y. Public Library); published Latin translation of 4 voyages of Amerigo Vespucci; produced Carta itineraria Europae (1st printed wall map of Europe), 1511; helped prepare 1513 edit. of Ptolemy's Geography (considered 1st modern atlas); apptd. canon of St. Dié, Lorraine, France, 1514; produced Carta marina navigatoria, 1516; often signed his maps with Greek spelling of his name, Illacomilus. Died St. Dié, Alsace, circa 1522.

WALES, William, English mathematician, astronomer; b. circa 1734; 1 dau. Astron. observer Hudson's

Bay transit of Venus expdn., 1769, James Cook's 2d voyage, 1772-74, 3d voyage, 1776-80; math. master Christ's Hops., London, circa 1781-98. Sec., Bd. Longitude. Fellow Royal Soc., 1776. Author: General Observations Made at Hudson's Bay, 1772; Nautical Almanac, 1773; The Two Books of Apollonius concerning Determinate Sections, 1772; The Method of Finding the Longitude by Timekeepers, 1794; Observations on a Voyage with Captain Cook, 1777; Remarks on Mr. Forster's Account of Captain Cook's Last Voyage, 1778; An Inquiry into the Present State of the Population in England and Wales, 1781. Editor: Astronomical Observations Made during the Voyages of Byron, Wallis, Carteret and Cook, 1778. Computed tables relating longitude to time, 1769. Died London, Dec. 29, 1798.

WALFORD, Roy L., Am. biologist; b. San Diego, June 29, 1924; s. Roy L. and Jean (Hyams) W.; student Cal. Inst. Tech., 1942-44; B.A., U. Chgo., 1946, M.D., 1948; m. Martha S. Schwalb, June 7, 1950; children—Lisa, Barney, Peter. Chief of lab. Chanute AFB, Ill., 1952-54; faculty U. Cal. at Los Angeles Med. Sch., 1954—, prof. pathology, 1966—, dir. Sch. Med. Tech., 1964-66. Mem. Am. Soc. Exptl. Pathology, Gerontol. Soc., Transplantation Soc. Author: Leukocyte Antigens and Antibodies, 1960; also numerous articles. Research on immunology of white blood cell, typing of lymphocytes in relation to organ transplantation; originator and immunologic theory of aging, prodn. of cancer by immunologic mechanisms. Home: 20 Latimer Rd., Santa Monica, Cal. 90402. Office: U. Cal. Sch. Medicine, Los Angeles 90024.*

WALK, Richard David, Am. psychologist; b. Ft. Dix, N.J., Sept. 25, 1920; s. Arthur Richard and Elsie (Roberts) W.; A.B., Princeton, 1942; M.A., State U. Ia., 1947; M.A., Harvard, 1949, Ph.D., 1951; m. Lois MacDonald, Apr. 1, 1950; children—Joan MacDonald, Elizabeth Howes, Richard David. Research asso. Human Resources Research Office, George Washington U., Ft. Benning, Ga., 1952-53; asst. prof. Cornell U., 1953-59; faculty George Washington U., Washington, 1959—, prof. psychology, 1964—. Vis. prof. Mass. Inst. Tech., 1965-66. Mem. Am., Eastern psychol. assns., Psychonomic Soc., Soc. for Research in Child Devel., A.A.A.S., Sigma Xi. Research, publs. on visual depth perception, psychol. stress with paratroopers, concept formation, perceptual learning; developer visual cliff (with E.J. Gibson). Home: 7100 Oakridge Av., Chevy Chase, Md. 20015. Office: Dept. Psychology, George Washington U., Washington 20006.*

WALKER, A. Earl, physician; b. Winnipeg, Manitoba, Can., Mar. 12, 1907; s. Arthur G. and Jennie B. (Shaw) W.; A.B., University Alberta, 1926, M.D., 1930. LL.D. (honorary) 1952; Rockefeller fellow Yale University and University of Amsterdam, 1935-37; married Betty D. Booth. Came to United States, 1931, naturalized, 1940. Intern, Toronto Western Hosp., 1930-31; resident U. of Chicago Clinics, 1931-34, instr. U. of Ia., 1934-35, instr. to prof. neurol. surgery, U. of Chicago, 1937-47; prof. neurol. surgery, Johns Hopkins since 1947, neurosurgeon-in-charge, Johns Hopkins Hosp. since 1947. Mem. A.M.A., Am. Neurol. Assn. (pres. 1965-66), Soc. Neurol. Surgeons (pres. 1966-67), Am. Assn. Neuropathol., Am. Electroencephalographic Soc., Soc. Neurol. Surgery, Am. Acad. Neurology, Am. Acad. Neurol. Surgery Harvey Cushing Soc., A.C.S., Eastern Assn. of Electroencephalographers, World Fedn. Neurosurg. Socs. (sec. fedn. affairs; pres.). Author: The Primate Thalmus, 1938; Pencillin in Neurology, 1947; Posttraumatic Epilepsy, 1949; Transtentorial. Herniations, 1962. Editor: A History of Neurological Surgery, 1951. Research on physiology of cerebral injuries; experimental physiology of cerebral cortex; cerebello-cerebral relationships; anatomy and physiology of thalmus; neurophysiological basis of epilepsy; neurosurgical pain therapy; visual mechanisms. Office: 601 N. Broadway, Balt. 5.

WALKER, Adam, Brit. physicist; b. Patterdale, Eng., 1731; studied math., Macclesfield, Scotland; children—William, Adam, John, Eliza; tchr., Eton, Winchester, Westminster; became writing master, accountant, free sch., Macclesfield, 1749; founder sem., Manchester, Eng. Author: Analysis of Course of Lectures on Natural and Experimental Philosophy, 1771; A System of Familiar Philosophy, 1779; also articles. Designed machines for pumping water, various wind and steam propelled vehicles, meteorol. apparatus, an ediouranion; improved the harpsichord; introduced a thermo-ventilation method. Died Richmond, Eng., Feb. 11, 1821.

WALKER, Burnham Sarle, Am. biochemist; b. Hamilton, N.Y., Oct. 22, 1901; s. Louis Abel and Edna (Sarle) W.; student Colgate U., 1919-21; A.B., Boston, 1923, A.M., 1924, Ph.D., 1926, M.D., 1934; m. Elisabeth Wellington French, July 23, 1927; dau., Ann Elisabeth (Mrs. Ann E. Massé). Faculty dept. biochemistry Boston U., 1926-56, chmn. div. med. scis. Grad. Sch., 1946-56; asso. pathologist in chemistry Burbank Hosp., Fitchburg, Mass., 1956—. Fellow A.A.A.S.; mem. Am. Chem. Soc., Am. Soc. Biol. Chemists, Soc. Exptl. Biology and Medicine, A.M.A., Mass. Med. Soc., Sigma Xi, Alpha Omega Alpha. Author: (with Boyd, Asimov) Biochemistry and Human Metabolism, 1952, 3d edit., 1957; (with Asimov, Nicholas) Chemistry and Human Health, 1956. Contbr. numerous articles to profl. jours. Devel. and application analytical methods for amino acids, carbon monoxide, alkaloids, sodium, iron, urethane, catalase, acid phosphatase applicable to body fluids, application chem. methods to kidney, adrenal and gastrointestinal physiology, therapeutic studies in urolithiasis, investigation enzymes of Staphylococcus aureus. Home: French Rd., Ashby, Mass. 01431. Office: Burbank Hospital, Fitchburg, Mass. 01420.*

WALKER, Charles Vincent, Brit. elec. engr.; b. 1812; ed. in engring. Electrician, South Eastern Ry., 1845-82. Fellow Royal Soc., 1855, Royal Astron. Soc.; mem. London Elec. Soc. (became sec. 1843), Soc. Telegraph Engrs. and Electricians (pres. 1869), Meteorol. Soc. (became pres. 1870). Author: Electrotype Manipulation, 1841; Electric Telegraph Manipulation, 1850; also treatises on electricity. Introduced improvements in telegraphy, 1848-49; designed device for communication of train passengers with guard, 1866, device for indicating trains on distant dial, 1876; contributed to introduction of system of time signals sent by telegraph from Greenwich, 1849. Died Tunbridge, Eng., Dec. 24, 1882.

WALKER, Darrell E., Am. genecist, plant breeder; b. Hubbell, Neb., June 7, 1920; s. Robert Stephen and Rosa Alma (Slagle) W.; B.S., Neb. State Tchrs. Coll., Kearney, 1941; Ph.D., U. Cal. at Berkeley, 1952; m. Margaret Louise Creswell, June 26, 1943; children—Ann Elizabeth, Stephen Creswell. Tchr., Hubbell High Sch., 1941-42; tchr., prin. Hebron (Neb.) High Sch., 1946-48; research asst. U. Cal. at Berkeley, 1948-50; research asso. Rod McLellan Co., orchid growers, 1950-53; mem. faculty Pa. State U., 1954—, prof. horticulture, head dept., 1963—. Mem. A.A.A.S., Am. Soc. Hort. Sci., Sigma Xi. Conducted genetic, cytological, plant breeding research on viburnam, snapdragons, petunias, geraniums, other ornamental species; supervised research on determination of chromosome numbers and relationships, sporogenesis, gametogenesis of poinsettia cultivars; introduced (with R. Craig) 1st true-breeding geranium cultivar propagated from seed for which All-American Selections gave spl. purpose recommendation. Home: 394 Park Lane, State College, Pa. 16801. Office: Tyson Bldg., University Park, Pa. 16802.*

WALKER, Edmund Murton, Canadian zoologist, entomologist; b. Windsor, Ont., Can., Oct. 5, 1877; s. Edmund and Mary (Alexander) W.; B.A., U. Toronto, 1900, M.B., 1903; student U. Berlin; m. Eleanora Walzel, 1909; children—Cynthia, Edmund Hugo, Elinor, Mary. Asst. dept. biology U. Toronto, 1904-05, successively lectr., asst. prof., asso. prof., 1906-26, prof. invertebrate zoology, 1926-48, head dept. zoology, 1934-48; asst. dir. Royal Ont. Mus. Zoology, 1918-31, hon. curator invertebrates, 1931—. Recipient Flavelle medal Royal Soc. Can., 1960. Fellow A.A.A.S., Entomol. Soc. Am. (became pres. 1939); mem. Royal Soc. Can. (became pres., sec. 1936), Entomol. Soc. Ont. (past pres.), Toronto Field Naturalists Club (founding, past pres.), Fedn. Ont. Naturalists (dir., pres. 1946-48), Royal Canadian Inst. (hon.), Acad. Natural Scis. Phila. (corr.). Author: The Odonata of Canada and Alaska, 2 vols., 1953-58. Research on dragonflies in Can.; morphology and phylogenetic relationships of Grylloblattodea; distbn. and taxonomy of crickets and grasshoppers; dipterous larvae; orthopteroid insects. Home: 120 Cheltenham Av., Toronto 12, Ont., Can.

WALKER, Edward Lewis, Am. psychologist; b. Connersville, Ind., June 18, 1914; s. Erle Lewis and Norma (Cloud) W.; A.B., Ind. U., 1938, M.A., 1940; postgrad. U. Ia.; Ph.D., Stanford, 1947; m. Alice Elizabeth Johnson, June 21, 1939; 1 son, Bruce Edward. Faculty, U. Mich., Ann Arbor, 1947—, prof. psychology, 1956—; vis. asso. prof. Stanford, 1951. Recipient Research Career award USPHS, 1964. Fellow Am. Psychol. Assn., A.A.A.S., Sigma Xi; mem. Midwestern, Eastern psychol. assns. Author: An Anatomy for Conformity, 1962; also numerous articles. Research on understanding learning process and role motivation or arousal in learning and performance. Home: 22 Harvard Pl., Ann Arbor, Mich. 48104.*

WALKER, Egbert Hamilton, Am. botanist; b. Chgo., June 12, 1899; s. William Hammersley and Ella (King) W.; B.A., U. Mich., 1922; M.S., U. Wis., 1928; Ph.D., Johns Hopkins, 1940; m. Dorothy J. Kemball, Apr. 10, 1936; children—William King, Jeanne Kemball (Mrs. Edward A. Houghton). Instr., Canton (China) Christian Coll., 1922-26; with Smithsonian Instn., Washington, 1928—, asso. curator dept. botany, ret., 1958, hon. research asso., 1965-67, botanist emeritus, since 1967—. Consultant spl. project Am. Inst. Biol. Scis., Washington, 1958-60, Pacific Sci. Bd., NRC-Nat. Acad. Sci., Washington, 1961—. Recipient Oberly prize Ala., 1938, 62. Fulbright grantee, 1948. Mem. Am. Soc. Plant Taxonomists, Internat. Assn. Plant Taxonomists, bot. socs. Washington, Korea, Japan, Sigma Xi. Author: A Bibliography of Eastern Asiatic Botany, 1938; (with Elmer D. Merrill) A Bibliography of Eastern Asiatic Botany, Supplement, 1960; Important Trees of the Ryukyu Islands, 1954; also numerous articles. Research on taxonomy of Eastern Asiatic botany especially family Myrsinaceae. Home: 7413 Holly Av., Takoma Park, Md. 20012. Office: Dept. Botany, Smithsonian Instn., Washington 20560.*

WALKER, Eric Arthur, engineer; b. Long Eaton, England, Apr. 29, 1910; s. Arthur and Violet Elizabeth (Haywood) W.; came to U. S., 1923, naturalized, 1937; B.S., Harvard, 1932, M.S, 1933, Sc.D., 1935; LL.D., Temple U., 1957, Lehigh U., 1957, Hofstra College, 1960, Lafayette College, 1960, U. Pa., 1960, U. of R.I., 1962, St. Vincent Coll., Latrobe, Pa.; L.H.D., Elizabethtown Coll., 1958; D.Litt., Jefferson Med. Coll., 1960; D.Sc., Wayne State U., 1965, U. Notre Dame; m. L. Josephine Schmeiser, Dec. 20, 1937; children—Gail, Brian. Asst. prof. elec. engring., Tufts Coll., 1933-35; asso. prof., 1935-38, head elec. engring. dept., 1938-40; head elec. engring. dept., 1938-40 ; head elec. engring. dept., Univ. of Conn., 1940-43; asso. dir. Harvard Underwater Sound Lab., 1942-45; dir. Ordnance Research Lab., Pa. State U., 1945-52, head elec. engring. dept. 1945-51, dean Sch. of Engring., 1951-56, v.p. Pa. State U., 1956, pres. univ., 1956—; executive secretary of the Research and Development Board, 1950-51; consultant National Research Council, 1949-50, mem. and past chmn. com. on undersea warfare; chmn. Pres. com. on tech. and distbn. research for benefit of small business, 1957; trustee Inst. for Def. Analysis; dir. Engring. Found. of United Engring. Trustees, Inc.; chmn. nat. sci. bd. Nat. Sci. Found.; chmn. Naval Research Adv. Com., Army Sci. Adv. Panel; vice chmn. Pres.'s Com. for Scientists and Engrs., 1956-58; mem. adv. panel on engring. and tech. manpower Pres.'s Sci. Adv. Com. Recipient Horatio Alger award, 1959, Tasker H. Bliss award, Am. Soc. Mil. Engrs., 1959; Golden Omega award Am. Inst. E.E. and Nat. Elec. Mfg. Assn., 1962. Registered profl. engr., Pa. Fellow I.E.E.E., Am. Acoustical Soc., Am. Phys. Soc.; mem. Am. Inst. Physics, Am. Soc. Engring. Edn. (pres. 1960-61), Pa. Assn. Colls. and Univs. (pres. 1960-61), Middle States Assn. Colls. and Secondary Schs. (pres. 1962; commn. higher edn. 1958-61), Engrs. Joint Council (pres. 1962-63), Am. Assn. Land-Grant Colls. and State Univs. (exec. com. 1958-60), Def. Sci. Bd., Nat. Acad. Engring. (pres.), Sigma Xi. Contbr. to tech. mags. Research on acoustic properties of fluids, electrostatics precipitation, electronic instrumentation, high voltage insulation; worked on devel. homing torpedoes. Home: West Campus, University Park, Pa.

WALKER, Gordon Northrop, Am. chemist; b. Bryn Mawr, Pa., Nov. 12, 1925; s. Joseph Godfrey and Leontine (Northrop) W.; B.S., U. Pa. 1947, M.S., 1949, Ph.D., 1951; m. Barbara Goodwin Jones, Dec. 6, 1952; 1 son, Alan Craig. Scientist, NIH, USPHS, Bethesda, Md. 1951-57; sr. research chemist CIBA Pharm. Co. subsidiary CIBA Corp., Summit, N.J., 1957—. Mem. American Association for Advancement Science. Am. Chem. Soc., Sigma Xi. Research, publs. on new organic synthetic methods especially cyclization reactions for polycyclic and heterocyclic ring systems, including those in podophyllotoxin, lysergic acid, morphine and colchicine; developed novel reduction methods. Home: Lake Trail W., Mt. Kemble Lake, Morristown, N.J. 07960. Office: CIBA Corp., Summit, N.J. 07901.*

WALKER, Harold Leroy, Am. metall. engr.; b. Benton, Ill., June 19, 1905; s. Roy Earl and Sophrona I. (Snider) W.; B.S., Mich. Coll. Mining and Tech., 1932, M.S., 1933, E. Metall., 1935; m. Violet Fraser Wood, Nov. 4, 1953. Research engr., asst. to supt. Munsing Paper Co., Mich., 1927-29; instr. metall. engring. Mich. Coll. Mining and Tech., 1932-37; asst. prof. metallurgy, metallography Washington State Coll., 1937-38; asst. prof., asso. prof., prof. metall. engring. U. Ill., 1938-54, acting head dept. mining and metal engring., 1940-42, head, 1942-54; with Internat. Cooperation Administrn. as vis. prof. metall. Indian Inst. Sci., Bangalore, India, 1954-56; spl. asst. for edn., office gen. mgr. AEC, Washington, 1956-57; dir. research U. N.M., Albuquerque, 1957—. Exec. council Rocky Mountain Sci. Council, 1959. Pres. Ill. Mining Inst., 1954. Fgn. duty with Fgn. Econ. Administrn., 1945; dir. dept. mines and minerals State Ill., 1947, 49. Recipient Ben M. Vallat Award, 1933. Mem. Am. Inst. Mining and Metall. Engrs., Am. Soc. for Metals, Ill. Mining Inst., Sigma Xi. Contbr. many articles to profl. jours. Spl. work in processes for heat treatment of armor piercing projectile steels and ordnance matériel; research on recrystallization phenomena in cold-worked metals and alloys. Home: 5101 Royene, N.E. Albuquerque 87110. Office: Adminstrn. Bldg., U. N.M., Albuquerque 78106.*

WALKER, Sir James, Scottish chemist; b. Dundee, Scotland, Apr. 6, 1863; s. James and Susan (Cairns) W.; apprentice to flax and jute spinner, 1879-82; B.Sc., U. Edinburgh, 1885, D.Sc., 1886, LL.D., 1929; at Univ. Coll., Dundee, before 1886; Ph.D., Leipzig, Germany, 1889; LL.D., U. St. Andrews, 1909; m. Annie Sedgwich, 1897; 1 son. Research asst. to Crum Brown, Edinburgh, 1889-92, succeeded him as prof. U. Edinburgh, 1908; entered lab. of William Ramsay, Univ. Coll., London, 1892, 2d asst. by 1893; named prof. chemistry Univ. Coll., Dundee, 1894; manufactured trinitrotoluene (T.N.T.), Edinburgh, during World War I; began planning new chem. labs. for Edinburgh, 1918; founder chem. research center; ret., 1928. Fellow Royal Soc. (Davy medal 1926); pres. Chem. Soc., 1921-23. Author: Introduction to Physical Chemistry, 1899, 10th edit., 1927. Conducted research in organic chemistry on electrolytic synthesis, amphoteric electrolytes, hydrolysis, ionization constants. Died Edinburgh, May 6, 1935.

WALKER, James Frederick, Am. biologist; b. Riverton, Ala., July 30, 1904; s. James Fred and Elizabeth D. (Ellis) W.; student Chicksaw Coll., 1922-23; B.A., U. Miss., 1927, M.S., 1931; Ph.D., U. Ia., 1935; postgrad. U. Cal. at Los Angeles, 1962-63; m. Mildred Beulah Lewis, Mar. 20, 1927; 1 son, James Frederick. Faculty, U. So. Miss., Hattiesburg, 1926-42, 45——, prof. biology, 1945——, head div. biology, 1946-57; faculty dept. anatomy Cal. Coll. Medicine, 1944-45. Fellow A.A.A.S.; mem. Am. Physiol. Soc. (asso.), Miss. Acad. Scis., Sigma Xi, Beta Beta Beta, Omicron Delta Kappa, Phi Eta Sigma, Alpha Epsilon Delta. Research, publs. on release of thyroid gland secretions in snakes, effects of carbon dioxide on embryo of Melanoplus differentialis marine life off coast of Miss., black spotting in shrimp, alcohol edn. Home: 200 S. 28th Av., Hattiesburg, Miss. 39401.*

WALKER, John, English chemist; b. Stockton-on-Tees, Eng., 1871; apprentice to Watson Alcock, surgeon in Stockton-on-Tees; studied at Durham and York; asst. surgeon to Alcock; turned to chemistry; began drug and chem. bus. in Stockton, 1818; inventor 1st friction match (composed of sulphur, potassium chlorate, rubber), 1826. Died Stockton, May 1, 1859.

WALKER, John Charles, Am. plant pathologist; b. nr. Racine, Wis., July 6, 1893; s. Samuel B. and Alice (Davis) W.; B.S., U. Wis., 1914, M.S., 1915, Ph.D., 1918; D.Sc., Göttingen U., 1960; m. Edna Merrill Dixon, Aug. 11, 1920; 1 son, John W.; m. Marian Dixon Tudor, Dec. 17, 1966. Plant pathologist U.S. Dept. Agr., Madison, Wis., 1917-45, collaborator, 1945——; faculty U. Wis., 1919——, prof., 1928-64, emeritus prof., 1964——. Mem. NRC, 1943-51. Recipient Man of Year award Vegetable Growers Assn. Am., 1954; Forty Niners Service award, 1955. Fellow A.A.A.S.; Am. Phytopath. Soc.; mem. Am. Soc. Naturalists, Bot. Soc. Am. (certificate of merit 1963.), Brit. Assn. Applied Biologists (hon.), Nat. Acad. Scis. Author: Plant Pathology, 1950; Diseases of Vegetable Crops, 1952; also numerous articles. Research on nature of disease resistance in plants and devel. numerous disease resistant varieties of crop plants. Home: 809 Oneida Pl., Madison 53711. Office: Russell Labs., Madison, Wis. 53706.*

WALKER, Norma Ford, Canadian biologist; b. St. Thomas, Ont., Can., Sept. 3 ,1893; d. Norman W. and Margaret (Dyke) Ford; B.A., U. Toronto (Ont.), 1918, Ph.D., 1923; m. Edmund Murton Walker; Aug. 2, 1943. With U. Toronto, 1918-58, asso. prof., 1943-58, prof. human genetics, 1958——; dir. dept. genetics Hosp. for Sick Children, Toronto, 1947-63; acting dean women Victoria Coll., U. Toronto, 1931-34. Mem. Am. Soc. Human Genetics, Ont. Soc. Biology, Fedn. Ont. Naturalists, A.A.A.S., N.Y. Acad. Scis., Canadian Genetic Soc. (v.p.). Author: A Comparative Study of the Musculature of Orthopteroid Insects, 1923; A Biological Study of the Dionne Quintuplets—an Identical Set, 1937; also numerous sci. articles. Pioneered in genetic and heredity counseling. Home: 120 Cheltenham Av., Toronto, Ont.

WALKER, Philip Leroy, Jr., Am. chemist; b. Balt., Jan. 10, 1924; s. Philip L. and Mary Belle (Heaps) W.; B.Engring., Johns Hopkins, 1947, M.S., 1948; Ph.D., Pa. State U., 1952; m. Virginia Lee Strobel, July 14, 1949; children—Page Elise, John, Lawrence. Control chemist Lever Bros. Co., 1948-49; faculty Pa. State U., University Park, 1952——, prof. fuel tech., 1955——, head dept., 1954-59, chmn. div. mineral tech., 1959-64, head dept materials sci., 1967——. Dir. C-Cor Electronics, Inc., Centre Video Corp., State College, Pa.; cons. AEC, carbon industry, 1955——; chmn. Am. Carbon Com., 1963——. Mem. Am. Chem. Soc., Am. Phys. Soc., Am. Vacuum Soc., Sigma Xi, Phi Lambda Upsilon. Author: Chemistry and Physics of Carbon, vols. 1, 2, 4, 1966, 68; numerous articles. Asso. editor Carbon, 1963——; editor Chemistry and Physics of Carbon, 1966-68. Research on chemistry and physics carbon and graphite, structure carbons and their graphitizability, their interaction with oxidizing gases, their adsorptive properties, their electronic properties; discovered more effective techniques for producing activated carbons and molecular sieve carbons. Home: 222 Hunter Av., State College, Pa. 16801. Office: Pa. State U., University Park, Pa. 16802.*

WALKER, Robert Milnes, English surgeon; b. Wakefield, Eng., Aug. 2, 1903; s. John William and Elizabeth (Holdsworth) W.; student U. London, U. Coll. Hosp.; m. Grace Anna McCormick, Sept. 5, 1931; children—Thomas Milnes, Edward Milnes, Ruth M. (Mrs. John Barker), Anna Guilford (Mrs. Peter McCormick), Susan Holdsworth (Mrs. Chander Colé), Selma Janis (Mrs. Jack Clay). Surgeon, Royal Hosp., Wolvehampton, Eng., 1931-46; prof. surgery U. Bristol (Eng.), 1946-64. Mem. Med. Research Council, 1959-63. Decorated comdr. Order Brit. Empire, 1964. Fellow U. Coll., London, 1953. Hon. fellow A.C.S.; mem. Royal Coll. Surgeons Eng. (mem. council 1953——, past v.p.), Assn. Surgeons Gt. Britain and Ireland (past pres.), Surg. Research Soc. (past pres.). Author: Medical Education in Britain, 1965; Pathology and Management of Portal Hypertension, 1959; also numerous articles. Research on liver function, especially in relation to surgery of liver disease, cardio-vascular surgery. Home: Wergs Copse, Kintbury, Newbury, Berkshire, Eng.*

WALKER, Robert Mowbray, Am. physicist; b. Phila., Feb. 6, 1929; s. Robert and Margaret (Seivwright) W.; B.S. in Physics, Union Coll., 1950; M.S., Yale, 1951, Ph.D., 1954 D.Sc. (hon.), 1967, m. Alice J. Agedal, Sept. 2, 1951; children—Eric, Mark. Physicist, Gen. Elec. Research Lab., Schenectady, 1954-62, 63-66; McDonnell prof. physics Washington U., St. Louis, 1966——. Vis. prof. U. Paris, 1962-63 adj. prof. physics Rensselaer Poly. Inst., 1965-66; pres. Vols. for Internat. Tech. Assistance, 1960-61, 65-66, founder, 1960. Recipient Distinguished Service award Am. Nuclear Soc., 1964; Yale Engring. Assn. award for contbn. to basic and applied sci., 1966; Indsl. Research awards, 1965, 66. NSF fellow, 1962-63. Fellow Am. Phys. Soc., Meteoritical Soc.; mem. Am. Geophys. Soc., Am. Astron. Soc., Am. Inst. Aeros. and Astronautics, Fedn. Am. Scientists, A.A.A.S. Research, publs. on cosmic rays, nuclear physics, radiation effects in solids, particularly devel. solid state track detectors and their applications to geophysics and nuclear physics problems; discovery of fossil particle tracks in terrestrial and extra-terrestrial materials and the fission track method of dating. Home: 6 Portland Pl., St. Louis 63108.*

WALKER, Roland, zoologist; b. Stellenbosch, South Africa, Feb. 8, 1907; s. Thomas and Ella (Dudley) W.; B.A., Oberlin Coll., 1928, M.A., 1929; Ph.D., Yale, 1934; m. Vivian V. Trombetta, May 1, 1942; children—David, Helen. Faculty, Oberlin Coll., 1931-32; faculty Rensselaer Poly. Inst., Troy, N.Y., 1934——, prof. biology, 1954——. Mem. Am. Soc. Zoologists, Am. Ornithologists Union, Wildlife Soc., Am. Micros. Soc., A.A.A.S., Am. Assn. U. Profs., Phi Beta Kappa, Sigma Xi. Research, publs. on neuroanatomy of arthropods, devel. sense organs and heart of vertebrates, old-age changes in lacrimal gland, tumors in fish and mammals, electron microscope studies of lymphocystis virus and fish-host cell reaction.*

WALKER, Sears Cook, Am. mathematician, astronomer; b. Wilmington, Mass., Mar. 23, 1805; s. Benjamin and Susanna (Cook) W.; grad. Harvard, 1825. Taught school, then became actuary Pa. Co. for Ins. on Lives and Granting Annuities, 1836; founded one of 1st astron. observatories in connection with Phila. High Sch., 1837; prepared parallactic tables which reduced time required to compute phases of occultation, 1834; mem. staff U. S. Naval Obs., Washington, 1845, discovered that planet Neptune was identical with star seen twice by Lalande in 1795, which had been referred to as star Number 26266, 1847; in charge of computations of geog. longitude in U. S. Coast Survey, 1847-53; originated telegraphing of transits of stars, (with Alexander D. Bache) method for determining differences of longitude by telegraph. Developed registry of time observations known as Am. method. Author: Researches relating to the Planet Neptune, 1850; Ephemeris of Planet Neptune for 1848-52, 1852, also articles. Died Cin., Jan. 30, 1853.

WALKER, Thomas Leonard, Canadian geologist; b. Brampton, Ont., Can., Dec. 31, 1867; M.A., Queen's U., Kingston, 1890; Ph.D. Leipzig (Germany) U., 1896; D.Sc., U. Toronto, 1938; m. Mary Augusta Woods, 1906; 2 sons, 1 dau. Asst. supt. Geol. Survey of India, 1897-1901; prof. mineralogy U. Toronto, 1901-37; apptd. dir. Royal Ont. Mus. Mineralogy and Petrography, 1913. Fellow Royal Soc. Can. (Flavelle medal 1941). Author: Crystallography, 1914. Founder series Contbns. to Canadian Mineralogy, 1921. Died Aug. 6, 1942.

WALKER, Warren Franklin, Am. biologist; b. Malden, Mass., Sept. 27, 1918; s. Warren Franklin and Aida (Miner) W.; S.B., Harvard, 1941, Ph.D., 1946; m. Hortense B. Allen, June 24, 1944; children—Edward A., Henry A., Susan Gay, Carol Ann. Instr. anatomy Boston U., 1945-47; faculty Oberlin (O.) Coll., 1947——, prof. biology, 1957——, chmn. department of biology, since 1917——. Member of Ohio Biology Tchrs. Conf. (past pres.), Ohio Acad. Sci. (past exec v.p. for zoology 1962-63), Am. soc. Zoologists (chmn. div. vertebrate morphology 1965-66), Phi Beta Kappa, Sigma Xi. Author: Vertebrate Dissection, 1954, 60, 65, 68; (with C.A. Villee, F.E. Smith) General Zoology 1968, 63, 68; Dissection of Fetal Pig, 1964; Dissection of Frog, 1967; A Study of Cat, 1967; also articles. Research on anatomy of locomotor apparatus of primitive terrestrial vertebrates, comparative studies of muscles and skeleton. Home: 160 Morgan St., Oberlin, O. 44074.*

WALKER, William Delaney, Am. physicist; b. Dallas, Nov. 1923; s. William Delany and Mildred (Ramsey) W.; B.A., Rice U., 1944; Ph.D., Cornell U., 1949; m. Suzanne Portor, Dec. 23, 1946; children—Nancy, Elizabeth, Samuel. Asst. prof. Rice U., 1949-52; lectr. U. Cal. at Berkeley, 1951-52; asst. prof. U. Rochester, 1952-54; prof. physics U. Wis., Madison, 1954——, chmn. dept., 1964-66; Chmn. Argonne Users Group, 1964-66; mem. physics panel NSF, 1964-66; mem. panel for high energy physics Nat. Acad. Scis., 1964. Fellow Am. Phys. Soc.; mem. Phi Beta Kappa, Sigma Xi. Editor: Proc. Rochester Conf., 1952. Research in elementary particle physics, especially strong interactions of particles; co-discoverer several fundamental particles. Home: 1811 Vilas Av., Madison, Wis. 53711.*

WALKER, William Hultz, Am. chem. engr.; b. Pitts., Apr. 7, 1869; s. David H. and Anna (Blair) W.; B.S., Pa. State Coll., 1890; A.M., Ph.D., U. Göttingen, 1892; Sc.D., U. Pitts., 1915; m. Isabelle Luther, Sept. 15, 1896. Prof. indsl. chemistry, then chem. engring. Mass. Inst. Tech., 1894-1921; lectr. indsl. chemistry Harvard, 1905-08; cons. chem. engr., 1900——; mem. Little & Walker, 1900-05; dir. research lab. applied chemistry Chem. Products Co. Recipient Nichols medal, 1908. Fellow Am. Acad. Arts and Scis., Am. Iron and Steel Inst., Am. Electrochem. Soc. (pres. 1910-11), Am. Chem. Soc. (pres. Eastern Sect. 1904), Am. Soc. for Testing Materials, Soc. Chem. Industry (London). Inventor indsl. processes; research on prodn. of art glass and of sterling silver, chemistry and indsl. uses of cellulose, technology of petroleum, corrosion of iron and steel. Died July 9, 1934.

WALL, Charles Terence Clegg, English mathematician; b. Bristol, Eng., Dec. 14, 1936; s. Charles and Ruth (Clegg) W.; student Marlborough Coll., 1949-54; B.A., Trinity Coll., Cambridge (Eng.) U., 1957, Ph.D., 1960; m. Alexandra Joy Hearnshow, Aug. 22, 1959; children—Nicholas C. T., Catherine E., Lucy R., Alexander T. fellow Trinity Coll., 1959-64, Lectr. Cambridge U., 1961-64; Harkness fellow Princeton, 1960-61; reader Oxford U., 1964-65; fellow St. Catherine's Coll., 1964-65; prof. Liverpool (Eng.) U., 1965——. Fellow Cambridge Philos. Soc. (editorial bd. Proc. 1963-64); mem. London, Am. math. socs. Mem. editorial bd. Topology, 1964-65. Research, numerous publs. on higher dimensional analog of surface called a manifold, problem of classifying manifolds especially differentiable ones, either up to complete differentiable equivalence, homotopy or cobordism. Home: 5 Kirby Park, West Kirby, Cheshire, Eng. Office: Dept. Pure Math., Liverpool U., Liverpool 3, Eng.*

WALL, Clifford Nathan, Am. physicist; b. Dayton, O., July 6, 1899; s. Winfield Scott and Dora (Kistler) W.; student N. Central Coll., 1918-20, Carnegie Inst. Tech., 1920-21; B.A., U. Ill., 1922, M.S., 1923, Ph.D., 1926; m. Mildred Ecki, Aug. 3, 1927; children—Gerald C., Robert E. Instr. physics U. Ill., 1926-28; prof. physics N. Central Coll., Naperville, Ill., 1929-42; faculty U. Minn., Mpls., 1942——, prof. physics, 1949-67, prof. emeritus, 1967——. Am. Field Service fellow, France, 1928-29; recipient Research Corp. award, 1947, Otustanding Alumnus award N. Central Coll., 1965. Mem. Am. Phys. Soc., Am. Assn. Physics Tchrs. (Oersted medal 1953), A.A.A.S., Sigma Xi. Author: (with R.B. Levine) Physics Laboratory Manual, 1951; (with H. Kruglak) Laboratory Performance Tests for General Physics, 1959; also articles. Research on energy of space lattices, atomic distbn. functions for liquids, statis. mechanics of condensation phenomena. Home: 2239 Como Av., St. Paul 55108.*

WALL, Frederick Theodore, Am. chemist; b. Chisholm, Minn., Dec. 14, 1912; s. Peter and Fanny (Rauhala) W.; B.Chemistry, U. Minn., 1933, Ph.D., 1937; m. Clara Elizabeth Vivian, June 5, 1940; children—Elizabeth, Jane. Mem. faculty U. Ill., Urbana, 1937-64, prof. chemistry, 1946-64, dean grad. coll., 1955-63; prof. chemistry, U. Cal. at Santa Barbara, 1964——, chmn. dept., 1964-66, vice-chancellor for grad. studies and research, 1966——. Bd. govs. Nat. Acad. Scis.-NRC, 1962——; pres. Assn. of Grad. Schs., 1961; trustee Inst. for Def. Analyses, 1962-64. Recipient Outstanding Achievement award U. Minn., 1959. Mem. Nat. Acad. Scis., Am. Acad. Arts and Scis., Am. Chem. Soc. (dir. 1962-64, Pure Chem. award 1945), Am. Phys. Soc., A.A.A.S., Faraday Soc. Editor, Jour. Phys. Chemistry, 1965——. Author: Chemical Thermodynamics, 1958, 65; also many articles. Research on theory of rubber-like electricity, configurations of rubber-like molecules, calculations of reaction probabilities, use of high speed digital computers for solution of phys. chemistry problems, statis. mechanics.

WALL, Hubert Stanley, Am. mathematician; b. Rockwell City, Ia., Dec. 2, 1902; s. Samuel Hugh and Gracia (Wright) W.; B.A., M.A., Cornell Coll., Mt. Vernon, Ia., 1924; Ph.D., U. Wis., 1927; postgrad. Inst. for Advanced Study; m. Mary Kate Parker, Oct. 18, 1947. Faculty, Northwestern U., 1927-44, prof., 1943-44; faculty Ill. Inst. Tech., 1944-46, prof., 1945-46; prof. math. U. Tex., Austin, 1946——. Mem. Am. Math. Soc., Circolo Mat di Palermo, Phi Beta Kappa, Sigma Xi. Author: Analytic Theory of Continued Fractions, 1948; Creative Mathematics, 1963; also numerous articles. Research in continued fractions, summability, differential and integral equations. Home: 2903 Breeze Terrace, Austin, Tex. 78722.*

WALL, John, English physician; b. Powick, Eng., 1708; s. John Wall; ed. Worcester Coll., Oxford, also Merton Coll.; B.A., 1730, M.A., M.B., 1736, M.D., 1759; m. Catherine Sandys, June 27, 1776;

1 son, Martin. Practiced medicine, Worcester, Eng., 1759-76; also a painter. Author: Medical Tracts, 1780. First med. writer to point out a condition in man similar to foot and mouth disease in cattle; gave early report of post-mortem in a case of angina pectoris; noticed calcification of aortic valves. Died 1776.

WALL, Leo Aloysius, Am. chemist; b. Washington, May 26, 1918; s. Aloysius Joseph and Mary Ellen (Costello) W.; B.S., Cath. U. Am., 1941, Ph.D., 1945; m. Leola Grace Ingalls, 1940; children—Julia Anne, Mary Ellen (Mrs. R.F. Beckham), Kathleen Cecilia (Mrs. Robert E. Prangley), Margaret Teresa (Mrs. David H. Biddle). Instr., Cath. U. Am., 1943-45; chemist mass spectrometry sect. Nat. Bur. Standards, Washington, 1946, with plastics sect., 1947-51, polymer chemistry sect., 1952-62, chief polymer chemistry sect., 1962—; research chemist Cal. Research Corp., Richmond, 1947. Recipient Dept. Commerce Silver medal, 1955, Gold medal, 1962; Fleming award, 1957; Cath. U. Am. Alumni Achievement award, 1964. Fulbright fellow U. Paris, 1951-52. Fellow A.A.A.S., Am. Inst. Chemists; mem. Washington Acad. Scis., Faraday Soc., Chem. Soc. (London), Philosoph. Soc. Washington, Albertus Magnus Guild, Am. Chem. Soc., Phi Beta Kappa, Sigma Xi. Research, numerous publs. on theories and mechanisms of polymer degradation, fluorine-containing and high temperatures polymers, free radical and radiation chemistry. Home: 1 Given St., Colonial Beach, Va. 22443. Office: Nat. Bur. Standards, Washington 20234.*

WALL, Martin, English physician; b. Worcester, Eng., 1747; s. John and Catherine (Sandys) W.; B.A., New Coll., Oxford, 1767, M.A., 1771, M.S. M.B., 1773, M.D., 1777; studied medicine St. Bartholomew's Hosp., London, also in Edinburgh, until 1774; at least 1 son, Martin Sandys. Began practice medicine, Oxford, 1774; became physician Radcliffe infirmary, 1775; apptd. lectr. and reader in chemistry Oxford U., 1781, Lichfield prof. clin. medicine, 1785-1824; Harveian orator, 1788. Fellow Royal Soc., 1788, also Royal Coll. Physicians. Author: Inaugural Dissertation, 1781; Syllabus of Lectures in Chemistry, 1782; Dissertations in Chemistry and Medicine, 1783; Clinical Observations on the Use of Opium in Low Fevers, with Remarks on the Epidemic fever at Oxford in 1785, 1786. Gave early discussion of surface chemistry with reference to use of oil to still waves. Died 1824.

WALL, Patrick David, neurophysiologist; b. Nottingham, Eng., Apr. 5, 1925; s. Thomas and Ruth (Cresswell) W.; M.A., B.M., B.Chem., Oxford U., 1948, D.M., 1960; m. Betty Tucker, Aug. 10, 1950. Came to U. S., 1948. Instr., Yale, 1948-50; asst. prof. U. Chgo., 1950-53; instr. Harvard, 1953-55; asso. prof. Mass. Inst. Tech., Cambridge, 1957-59, prof., 1959—, exec. officer dept. biology, 1957-64. Mem. neurology study sect. NIH. Mem. Am. Physiol. Soc., Am. Acad. Arts and Scis. Author: T.R.I.O., The Revolting Intellectuals Organization, 1965; also numerous articles. Editor; Exptl. Neurology. Research in various aspects of central and peripheral nervous system with emphasis on mechanisms of transmission over sensory pathways. Home: 205 Pleasant St., Arlington, Mass. 02174.

WALL, Robert LeRoy, Am. physician; b. Doylestown, Pa., July 7, 1921; s. Charles L. and May (Merkel) W.; A.B., Oberlin Coll, 1943; M.D., Temple U., 1946; m. Julia Allison, Nov. 24, 1952; children—Bruce, Kim, Robert, Wendy. Faculty, Ohio State U. Coll. Medicine, Columbus, O., 1952—, prof., 1965—; practice medicine specializing in internal medicine-hematology, Columbus, 1952—; mem. attending staff Hosp., Columbus, 1952—, pres. staff, 1964—. Mem. com. on plasma proteins NRC, 1961—; cons. Dayton (O.) VA, Wright-Patterson AFB hosps. Diplomate Am. Bd. Internal Medicine. Fellow A.C.P.; mem. A.M.A., Ohio, Columbus med. socs., A.A.A.S., Am. Heart Assn., Central Soc. for Clin. Research, Am., Internat. socs. hematology, Sigma Xi. Author: Immunohematology, 1952; also articles on hematology, chemotherapy, plasma proteins; research on Factor IX, deficiency in blood coagulation in Amish. Home: 2990 Riverside Dr. Office: University Hosp., Columbus, O. 43210.*

WALLACE, Alfred Russel, English naturalist; b. Usk, Monmouthshire, Jan. 8, 1823; s. Thomas and Mary (Greenell) W.; ed. Hertford Grammar Sch.; m. Annie Mitten, 1866; 1 son, 1 dau. Land surveyor, architect, 1838-44, 46-48; master at collegiate sch., Leicester, 1844-46; went (with Bates) to Amazon, 1848-52, to Malay Archipelago, 1854-62; then became interested in natural history, social sci., and sci. lit.; granted Civil List pension, 1881; lectr. in U. S., 1886-87. Recipient first Darwin medal, 1890. Fellow Royal Soc., 1893 (Royal medal, 1868). Author: Travels in the Amazon, 1853; Palm Trees of the Amazon, 1853; The Malay Archipelago, 1869; Natural Selection, 1870; Miracles and Modern Spiritualism, 1874; The Geographical Distribution of Animals, 1876; Tropical Nature, 1878; Australasia, 1879; Island Life, 1880; Land Nationalism, 1882; Bad Times, 1885; Darwinism, 1889; Vaccination, a Delusion, 1898; The Wonderful Century: its Successes and its Failures, 1898; Studies, Scientific and Social, 1900; Man's Place in the Universe, 1903; My Life, 1905; Is Mars Habitable?, 1907; The World of Life, 1910; social Environment and Moral Progress, 1912. Noted that deep-water strait between Bali and Lombok in Malay Archipelago divided the region into 2 very distinct zool. regions, with Oriental fauna on western side and Australasian fauna on eastern side (still called Wallace's line); collected 125,660 specimens during his East Indies journeys; reflecting on what he observed in travels, gradually became a convinced evolutionist; influenced by ideas of Malthus and Lyell; conceived idea of natural selection as method of evolution (1858) and sent views to Charles Darwin; wrote paper (with Charles Darwin) on theory of evolution, which was read to Linnaean Soc., 1858; stressed heredity and variation acted upon by natural selection to explain evolutionary change; called attention to colors and markings of animals by means of recognition at a distance; contbr. greatly to zoogeography, providing fundamentals for all subsequent investigations in field; mapped out a series of zoogeographic regions and discussed patterns of continuity and discontinuity in distbn. of animal types; criticized Darwin's theory of sexual selection and Lamarckian elements in Darwin's idea; became (with August Weisman) apostle of neo-Darwinian movement. Died Broadstone, Dorset, Nov. 7, 1913.

WALLACE, Arthur, Am. plant physiologist; b. Bear River City, Utah, Jan. 4, 1919; s. James Newton and Clara (Mackay) W.; B.S., Utah State U., 1943; Ph..D, Rutgers U., 1949; m. Elna Kemp, Apr. 16, 1943; children—Garn Arthur, Loretta, Marlin Scott, Val Kemp, Scott Kemp. Asso. prof. plant physiology U. Cal. at Los Angeles, 1956-62, prof., 1962—, div. chief environmental radiation Lab. Nuclear Medicine and Radiation Biology, 1966—. Mem. Am. Soc. Agronomy, Am. Soc. Plant Physiologists, Am. Chem. Soc., Am. Soc. Hort. Sci., Sigma Xi. Author: Decade of Synthetic Chelating Agents in Plant Nutrition, 1962; Solute Uptake by Intact Plants, 1963; Evidence in Science and in Religion, 1966; Current Topics in Plant Nutrition, 1966; also numerous articles. Research on physiology of trees, in acceptance of synthetic chelating agents as micronutrient sources for crops in alkaline and calcareous soils, salt and water transport in plants, biosynthesis of organic acids in citrus fruits, desert plant physiology and ecology. Home: 2278 Parnell Av., Los Angeles 90064.*

WALLACE, Franklin Gerhard, Am. parasitologist; b. Deer River, Minn., Apr. 3, 1909; s. William Rutledge and Mary (Gerhard) W.; B.A., Carleton Coll., 1928; Ph.D., U. Minn., 1933; m. Martha Hynes, June 19, 1935; children—Julia (Mrs. Robert M. Copeland), Lucy (Mrs. Harley Rankin, Jr.), John, Martha. Asst. prof. Lingnan U., Canton, China, 1933-37; faculty U. Minn., Mpls., 1937—, prof. zoology, 1963—. Mem. Am. Soc. Parasitologists, Am. Micros. Soc., Am. Soc. Tropical Medine Hygiene, Soc. Protozoologists. Studies, publs. on trypanosomatid parasites of insects; life cycles of various species of parasitic trematode.*

WALLACE, George John, Am. ornithologist; b. Waterbury, Vt., Dec. 9, 1906; s. James Moses and Florence (Richardson) W.; A.B., U. Mich., 1932, M.A., 1933, Ph.D., 1936; m. Martha Cooper, Sept. 12, 1934; children—Sylvia (Mrs. W.T. McGrath), Myra (Mrs. J. Rowley Connor). Biologist, Vt. Fish and Game Service, Montpelier, 1936-37; dir. Pleasant Valley Sanctuary, Lenox, Mass., 1937-42; staff dept zoology Mich. State U., East Lansing, 1942—, prof., 1954—. Recipient Placque for distinguished contbns. to ornithology and conservation Detroit Audubon Soc., 1959. Mem. Am., Brit. ornithologists unions, Am. Inst. Biol. Scis., A.A.A.S., Mass. (hon. v.p. 1940—), Mich. (dir. 1945—) Audubon socs. Author: An Introduction to Ornithology, 1955; also numerous articles. Research on N.Am. and neotropical thrushes, food habits predatory birds, effect DDT on birds. Home: 517 Ann St., East Lansing, Mich. 48823.*

WALLACE, James Merrill, Am. plant pathologist; b. Ripley, Miss., Oct. 13, 1902; s. John Chesterfield and Katherin (Hovis) W.; B.S., Miss. State U., 1923; M.S., U. Minn., 1927, Ph.D., 1929; m. Adeline Emilie Hoien, June 13, 1929; children—Jane (Mrs. Keith Le Bahn Nelson). Asso. botanist, plant pathologist Clemson Coll., 1928-29; asso. plant pathologist U.S. Dept. Agr., Ida., 1929-35, Cal., 1935-41; plant pathologist, prof. plant pathology U. Cal. at Riverside, 1942—. Cons., com. mem. European Plant Protection Orgn., 1956—. Mem. A.A.A.S., Internat. Orgn. Citrus Virologists (past chmn.), Am. Phytopath. Soc. (past pres. Pacific div.), Sigma Xi, Alpha Zeta, Gamma Alpha. Editor: Citrus Virus Diseases, 1959. Research and numerous publs. on immunological reactions of plants to virus infections; developed inoculation techniques for transmission and identification of citrus viruses; discovered and described 5 virus diseases of citrus. Home: 3261 Redwood Dr., Riverside, Cal. 92501.*

WALLACE, James Sim, Brit. dentist; b. Braehead, Scotland, June 29, 1869; s. James Wallace; ed. Glasgow U., Glasgow Dental Hosp., Nat. and Royal Dental hosps.; M.S., D.Sc.; m. Annie Alexander; 2 sons. Lectr. dental surgery and pathology London Hosp.; lectr. preventive dentistry King's Coll. Hosp.; lectr. in U. S. and Can. Recipient John Tomes prize, Cartwright medal and prize Royal Coll. Surgeons, 1920-25, Neech prize Soc. Med. Officers of Health. Fellow Soc. Med. Officers of Health (hon.), Am. Coll. Dentistry, Royal Coll. Surgeons (fellow in dental surgery); mem. Brit. Dental Assn. (hon.), Brit. Soc. for Study of Orthodontics (pres. 1910, hon. mem.), other dental socs.; v.p. Food Edn. Soc. Author: The Cause and Prevention of Decay in Teeth, 1900; The Irregularities of the Teeth, 1904; The Role of Modern Dietetics in the Causation of Disease, 1905; Supplementary Essays on the Cause and Prevention of Dental Caries, 1906; The Prevention of Dental Caries, 2d edit., 1912 (Swedish transl.); The Prevention of Common Diseases in Childhood, 1912; Dental Diseases in Relation to Public Health, 1914; Child Welfare, 1919; Oral Hygiene, 1923, 2d edit., 1929; The Teeth and Health, 1926; Variations in the Form of the Jaws, with special reference to their Etiology and their relation to the occlusion of the Dental Arches (Cartwright prize essay), 1926. Died July 13, 1951.

WALLACE, Philip Russell, Canadian physicist; b. Toronto, Ont., Can., Apr. 19, 1915; s. George Russell and Mildred (Stillwaugh) W.; B.A., U. Toronto, 1937, M.A., 1938, Ph.D., 1940; m. Jean Elizabeth Young, Aug. 15, 1940; children—Michael David, Kathryn Joan, Robert Philip. Fellow in applied math. U. Toronto, 1938-40; instr. math. U. Cin., 1940-42; instr. math. Mass. Inst. Tech., 1942-43; asso. research physicist Atomic Energy div. Nat. Research Council, Montreal Lab., 1943-46; asso. prof. math. McGill U., Montreal, 1946-49, prof., 1949-61, prof. physics, 1961—, dir. Inst. of Theoretical Physics, 1966—. Mem. Canadian Assn. Physicists (chmn. theoretical physics div. 1956-57), McGill Assn. U. Tchrs. (pres. 1964-65), Am. Phys. Soc., Canadian Assn. Physicists, Am. Assn. Physics Tchrs. Research, publs. on theory of electric and magnetic properties of solids; helicons; propagation of unusual electromagnetic waves in solids in the presence of a magnetic field; motion of charged particles in theory of relativity; nucleonics, theory of neutron diffusion; theory of scattering of electromagnetic waves from randomly distributed scatterers. Home: 125 Spartan Crescent, Pointe Claire, Que., Can. Office: McGill U., 805 Sherbrooke St. W., Montreal, Que., Can.*

WALLACE, Robert Charles, geologist; b. Orkney Isles, Scotland, June 15, 1881; s. James and Mary (Swanney) W.; M.A., U. Edinburgh (Scotland), 1901, B.Sc., 1907, D.Sc. (Carnegie scholar, 1907-08 fellow, 1910, Royal Exhbn. scholar, 1908-10), 1912; Ph.D., U. Göttingen (Germany), 1909; LL.D., univs. Scotland, Can., U. S.; m. Elizabeth Marcus Smith, 1912; 3 daus. Came to Can., 1910. Tchr. sci. secondary schs. U.K., 1901-04; research scholar crystallography, also demonstrator St. Andrews U., Scotland, 1910; lectr. geology, mineralogy U. Man. (Can.), 1910-12, prof., 1912-28; pres. U. Alta. (Can.), 1928-36; principal Queen's U., Kingston, Ont., Can., 1936-51; fellow Canadian Inst. Mining Metallurgy, 1913, pres., 1918-21; exec. dir. Arctic Inst. N. Am., from 1952. Pres. Man. Ednl. Assn., 1924-25; commnr. Mines and Natural Resources, Man., and for No. Man., 1926-28; chmn. Ont. (Can.) Research Commn., 1945-48; pres. Research Council Ont., 1948-51; pres. Canadian Assn. Adult Edn. from 1952. Fellow Royal Soc. Can. (pres. 1940-44), Geol. Soc. Author: The Burwash Lectures, 1932; A Liberal Education in a Modern World, 1932; Religion, Science, and the Modern World, 1952. Canadian adv. editor Ency. Americana, from 1950. Research, publs. on phys. chemistry of rock magmas, petrology, crystallography, and natural resources. Contbr. establishment schs. nursing, fine arts, phys. and health edn., depts. indsl. relations, biol. research in Ont. Died Kingston, Jan. 29, 1955.

WALLACE, Thomas Joseph, Am. chemist; b. Jamaica, N.Y., May 1, 1934; s. Thomas Joseph and Catherine (Reardon) W.; B.S. in Chemistry, St. Francis Coll., 1956; M.S., U. Conn., 1958, Ph.D., 1960; m. Anne Fuller Batton, June 3, 1961. With Esso Research and Engring. Co., Linden, N.J., 1960—, group leader, 1964-66, sect. head, 1966—, head petroleum products and chem. synthesis sect., 1966—. Mem. sci. adv. com. St. Francis Coll., 1965—. Named one of Outstanding Young Men of Am., Jr. C. of C., 1965. Fellow Royal Soc., 1953; mem. Am., Brit. chem. socs., Catalysis Soc., Am. Ordnance Assn., Sigma Xi. Research, numerous publs. and patents on sulfur chemistry, metal catalysis, autoxidation, base catalysis, solvent effects in organic reactions. Home: 10 Knollwood Rd., Whippany, N.J. 07981. Office: P.O. Box 121, Linden, N.J. 07036.*

WALLACE, William, Scottish mathematician; b. Dysart, Scotland, Sept. 23, 1768; student U. Edinburgh; LL.D., 1838. Bookbinder's apprentice; bookseller's shopman, Edinburgh; became asst. math. tchr., Perth, 1794; math. master Gt. Harlow Mil. Sch., Buchinghamshire, England, 1803; prof. math. U. Edinburgh, 1819-38; a founder obs. Calton Hill, Edinburgh. Fellow Royal Soc. Edinburgh. Author: A New Book of Interest, containing Aliquot Tables, truly proportioned to any given rate, 1794; New Series for the Quadrature of the Conic Sectons and the Computation of Logarithms, 1808; Account of the Invention of the Pantograph and Description of the Eidograph, 1831; Geometrical Theorems and Analytical Formulae, 1839; also articles. Invented eidograph for copying drawings, chorograph for drawing triangles. Died Edinburgh, Apr. 28, 1843.

WALLACE, William Edward, Am. chemist; b. Fayette, Miss., Mar. 11, 1917; s. James David and Matti Lee (Rogers) W.; student East Miss. Jr. Coll., 1932-34; B.A., Miss. Coll., 1936; Ph.D., U. Pitts., 1941; m. Helen M. Meyer, June 21, 1947; children—Richard G., Donald A., Marcia L. Post-doctoral fellow Carnegie Instn. of Washington and Buhl Found., 1940-42; lectr. physics U. Pitts., 1942-44; research asso. Manhattan Dist., Ohio State U., 1944-45; faculty U. Pitts., 1945-—, prof., chmn. dept. chemistry, 1963-—; cons. Am. Inst. for Research, Pitts., W. R. Grace Co., Westinghouse Elec. Corp., Wright-Patterson AFB; mem. edl. bd. Jour. Chem. Physics, 1962-65. Guggenheim fellow, 1954. Mem. Am. Chem. Soc. (dir. Pitts. sect. 1962-—), A.A.A.S., Sigma Xi. Research, numerous publs. on intermetallic compounds, particularly those containing rare earth metals; forming compounds, determining magnetic characteristics; structure determination and heat capacity; electronic states of rare earths and iron, cobalt and nickel in intermetallic compounds and factors responsible for compound formation. Home: 201 Pinecrest Dr., Pitts. 15237.*

WALLACH, Hans, exptl. psychologist; b. Berlin, Germany, Nov. 28, 1904; s. Albert and Thekla (Mayer) W.; Ph.D., U. Berlin, 1935. Research asso., mem. faculty Swarthmore (Pa.) Coll., 1936-—, prof. psychology, 1953-—; mem. Inst. Advanced Study, Princeton, N.J., 1954-55. Research, publs. on sound localization; perceptions of color, visual motion, tridimensional form, size; perceptual adaptation in stereo-vision and in visual direction. Home: 604 Elm Av., Swarthmore, Pa. 19081.*

WALLACH, Otto, German chemist; b. Königsberg, Prussia (now Kaliningrad, USSR), Mar. 27, 1847; Ph.D., Göttingen, Germany, 1869; pupil of Wöhler and Hofmann; became asst. to Kekulé, organic chemistry lab. U. Bonn., 1870, pvt. docent, 1873, asso. prof. chemistry, 1876; dir., instruction in pharmacy, 1879; prof., Göttingen, from 1889; dir. Chem. Inst., Göttingen, 1899-1915. Recipient Nobel prize in chemistry (for work in alicylic compounds), 1910. Fellow Chem. Soc. (hon.); mem. Verein Deutscher Chemiker (hon.), Deutsche Chemische Gesellschaft (pres.). Author: Terpene und Kampfer, 1909; also numerous papers. Studied pharmacy, 1879; began research on terpenes, 1884; studied chem. composition of camphors, essential oils; laid found. for research in aromatic chemicals, including perfumes and spices; studied azo and diazo compounds, studied thujone and fenchone, synthesized optically active compounds which lacked an asymmetric carbon atom; studied transformation of chloral into dichloric acetic acid; his research on terpenes led to work revealing their importance in vitamins and sex hormones and made possible indsl. use of ethereal oils. Died Göttingen, Feb. 29, 1931.

WALLEN, Irvin Eugene, Am. oceanographer; b. Afton, Okla., Oct. 4, 1921; s. Stuvie C. and Mittie (Hames) W.; B.S., Okla. State U., 1941, M.S., 1946; Ph.D., U. Mich., 1950; m. Dorothy Wiehe, Feb. 12, 1945; children—Karen Kaye, Gary Lee, Kathy Lynn, Shelley Jean, Mark Edward. Faculty, Okla. State U., 1948-56; asst. dir. Sci. Teaching Improvement Program A.A.A.S., 1956-57; fgn. tng. officer AEC, Washington, 1957-59, marine biologist, 1959-62; asst. dir. oceanography Mus. Natural History Smithsonian Instn., Washington, 1962-66, head Office of Oceanography and Limnology, 1966-—; prof. zoology Asia Found. Pakistan, 1960, 63. Chmn. research panel Interagy. Com. Oceanography, 1962-—; observer Presidents Sci. Adv. Com. Subpanel on Biol. Oceanography, 1965-66; U.S. rep. Internat. Council For Exploration of the Sea, 1965; mem. Pacific Sci. Bd., 1966-—; mem. Nat. Acad. Scis. Sino-Am. Com., 1966-—; mem. other state, fed., internat. commns. Investigator on grants Office Naval Research, NIH, Okla. Petroleum Refiners Waste Control Council, NSF, Link Found., Bur. Comml. Fisheries. Fellow A.A.A.S. (chmn. Council Com. coop. with developing countries); mem. Marine Tech. Soc., Am. Soc. Limnology and Oceanography, Soc. Systematic Zoology, Nat. Geog. Soc., Mediterranean Assn. Marine Biology and Oceanology, Internat. Assn. Limnology. Author: (with Roy W. Jones) Biological Science Notebook, 1950. Contbr. chpts. to Ocean Science, 1964, Radioecology, 1963; also numerous articles. Research on effects of clay turbidity and oil refinery wastes on fresh water fishes, limnological conditions in Okla. waters. Home: 6310 12th Rd. N., Arlington, Va. 22205. Office: Smithsonian Instn., Washington 20560.*

WALLER, Augustus Désiré, physiologist; b. Paris, France, July 12, 1856; s. Augustus Volney and Matilda (Walls) W.; ed. Coll. de Genève, Aberdeen, Edinburgh univs.; M.D., LL.D., M.A.; D.Sc., Perth; m. Alice Mary Palmer; 3 sons, 1 dau. Dir. physiol. lab. U. London. Named lauréat de l'Institut de France, 1889; recipient premio Aldini sul Galvanismo Royal Acad. Sci. of Inst. of Bologna, 1892. Fellow Royal Soc.; corr. mem. or asso. mem. Biol. Soc. Paris, Physiol. Soc. Moscow, royal acads. medicine of Rome and Belgium; hon. mem. council U. Tomsk. Author: Introduction to Human Physiology; Animal Electricity, 1897; Signs of Life, 1903; Physiology, the Servant of Medicine, 1910; The Psychology of Logic, 1912; also papers. First to use electrodes to study electric currents of beating heart, circa 1887, recorded them by electrocardiogram. Died Mar. 11, 1922.

WALLER, Augustus Volney, physiologist; b. Faversham, Eng., Dec. 21, 1816; s. William Waller; M.D., Paris, 1840; m. Matilda Walls, 1842; 1 son, Augustus, 2 daus. Practiced medicine, London, 1841-51; researcher, Bonn, Germany, 1851-56, Paris, 1856; invalid in Eng., 1856-57; prof. physiology Queen's Coll., Birmingham, Eng., 1858-60; went to Bruges, Belgium, 1860, then to Switzerland; began practice medicine, Geneva, Switzerland, 1868. Recipient Monthyon prize French Acad. Scis., 1852, 65. Licentiate Soc. Apothecaries, London. Fellow Royal Soc. (Royal medal 1860, Croonian lectr. 1869). mem. Soc. Physics and Natural History, Geneva. Contbr. to profl. jours. Demonstrated nerve degeneration after sect. of glosso-pharyngeal and hypoglossal nerves of frog, also that nerve fibers survive only when they maintain continuity with their cell bodies (Wallerian theory of degeneration), 1850. Died Geneva, Sept. 18, 1870.

WALLER, Cecil, English chemist; b. London, Sept. 27, 1907; s. Thomas and Ada (Girton) W.; B.Sc., U. Coll., London, 1928, M.Sc., 1929; D.Sc. (hon.), U. Bristol (Eng.), 1961; m. Margaret Willcocks, May 16, 1931; children—Anne (Mrs. Peter Gowlland), Alan, Hilary. With Ilford Ltd. (Eng.), 1930-—, chief emulsion chemist, 1950-60, research mgr., 1960-—. Recipient Duddell medal Phys. Soc. London, 1952; Longstreth medal Franklin Inst., 1955. Fellow U. Coll., London, 1960. Fellow Royal Inst. Chemistry, Royal Photog. Soc. (hon., Progress medal 1959); mem. Am. Chem. Soc. Research on sci. and tech. photog. manufacture, especially photog. emulsions, formation, growth, flocculation and sensitization of silver halide crystals, spl. photog. emulsions for nuclear research. Home: 41 Algers Rd., Loughton, Eng. Office: Ilford Ltd., Roden St., Ilford, Eng.*

WALLER, Frederic, Am. inventor; b. Bklyn., Mar. 10, 1886; s. Frederic and Katherine (Stearns) W.; ed. Bklyn. Poly. Inst.; m. Irene Seymour, Oct. 2, 1905 (div. 1919); m. 2d, Grace Fortescue Hubbard, Aug. 12, 1920 (dec. 1941); m. 3d, Doris Barber Caron, July 21, 1942; children—Muriel, Stuart. Short film producer Famous Players Lasky (now Paramount Pictures), 1919-22, Paramount Pictures, 1924-26, 29-36; made films for N.Y. World's Fair. Recipient U. S. Camera Achievement award, 1952; award for cinerama Acad. Motion Pictures Arts and Scis., 1954. Fellow Soc. Motion Picture, TV Engrs. Inventor cinerama, 1st automatic photographic printer, timer, 1st water skis, Waller gunnery trainer, gunnery analysis camera (used by USAF, World War II). Died May 18, 1954.

WALLER, Hans Dierck, German physician; b. Kiel, May 29, 1926; s. Hans and Annelise (Wessel) W.; M.D., U. Kiel, 1951; m. Friederike Scharre, Dec. 18, 1954; children—Hans Ulrich, Matthias, Bettina, Cornelius, Thomas. Faculty U. Marburg (Germany), 1952-62; faculty U. Tübingen, 1962-—, prof. internal medicine, 1966-—. Recipient Oehlecker prize, German Soc. Blood Transfusion, 1959, Theodor Frerichs prize, German Soc. Internal Medicine, 1960 Homburg prize, Regensburg Soc. Further Instrn. Physicians, 1964, Hufeland prize, Hufeland Found., 1965. Mem. German Soc. Internal Medicine, German Soc. Hematology, German Soc. Scientists and Physicians. Author: (with Löhr) Pharmakogenetik und Präventivmedizin, 1966; also numerous articles. Research on metabolism and enzymes of blood cells and their pathological alterations as cause of hematological disorders; screening tests of glycogenosis. Address: 28 Melanchthonstrasse, Tübingen, West Germany.*

WALLER, Ivar, Swedish physicist; b. Flen, Sweden, June 11, 1898; s. Erik and Signe (Frigell) W.; grad. U. Uppsala (Sweden), 1916, D.Sc., 1925; D.honoris causa, U. Leiden, 1965; m. Irène Julia Lydia Glucksmann, Apr. 7, 1932. Faculty, U. Uppsala (Sweden), 1925-64, prof. theoretical physics, 1934-64, emeritus, 1964-—. Mem. Swedish Atomic Com. and Council for Atomic Research, 1947-65; mem. bd. Swedish Atomic Energy Co., 1947-—, cons., 1964-—; del CERN, Geneva, Switzerland, 1952-65. Decorated comdr. 1st class Royal Swedish Order No. Star, 1967. Mem. Royal Swedish Acad. Sci. (mem. Nobel com. physics 1947-—), Norwegian Acad. Sci. Oslo, Royal Soc. Sci. Uppsala, Royal Physiographic Soc. Lund. Author: Theoretical Studies on Interference and Dispersion Theory of X-rays, 1925; also articles. Research on influence of heat motion in a crystal on intensity of X-rays scattered by crystal, diffuse scattering, Mössbauer effect, dynamics of lattices including impurities, quantum electrodynamics, slowing down of neutrons and paramagnetic relaxation. Home: 10 Tradgardsgatan, Uppsala, Sweden.*

WALLERANT, Frédéric-Félix-Auguste, French mineralogist; b. Trith-St.-Léger, France, July 25, 1858; student École normal supérieure, 1880-83; D.Sc., 1889. Prof. lycée de Montepellier (France), 1883-84; lectr. Faculté de Marseille (France), 1884-87; prof. Faculté de Rennes (France), 1887-91; lectr. École normale supérieure; named prof. mineralogy Sorbonne, 1903. Mem. French Acad. Scis. Author: Groupements cristallins, 1899. Research on crystalline groups, apparent symmetry of crystals, polymorphism, liquid crystals; made geol. studies of Esterel and Maures region. Died Paris, July 11, 1936.

WALLERIUS, Johann Gottschalk, Swedish physician; b. Nerike, Sweden, July 11, 1709; M.A., U. Uppsala (Sweden), 1730; studied math., philosophy, medicine; M.D., 1735. Elected adj. of med. faculty at Lund, 1732, lectr. physics, physiology, math., 1733; traveled to Copenhagen, Stockholm, Uppsala, Nyköping, Norköpping, Lindköpping; resigned appointment at Lund, moved to Uppsala, chosen sec. of med. faculty there; lectr. medicine; practiced medicine, Uppsala; elected assessor Royal Med. Coll., Stockholm, 1739; adj. of med. faculty, Uppsala; lectr. chemistry, metallurgy and materia medica; became physician to king; public prof. chemistry, metallurgy and pharmacy, Uppsala, 1760, also had seat in philos. faculty. Mem. Academia Naturae Curiosorum, Royal Swedish Acad. Scis. (pres. 1783), sci. socs. Uppsala, Lund, St. Petersburg. Introduced system of natural classification of minerals; research on composition of mineral, vegetable and animal substances; applied chemistry to agriculture research on artifical fertilizers. Died Uppsala, Nov. 16, 1785.

WALLERSTEIN, George, Am. astronomer; b. N.Y.C., Jan. 13, 1930; s. Leo and Dorothy (Calman) W.; B.A., Brown U., 1951; M.S., Cal. Inst. Tech., 1954, Ph.D., 1958; m. Marcia Lightbody, Apr. 10, 1965. Mem. faculty U. Cal. at Berkeley, 1958-65; prof., chmn. astronomy dept. U. Wash., Seattle, 1965-—. Mem. Am., Royal astron. socs., Astron. Soc. Pacific, Arctic Inst. N.M. Research and numerous publs. on chem. composition of stars, spectra of peculiar stars especially novae and related objects, color-magnitude diagrams of clusters and their interpretation in terms of stellar evolution and chem. composition.*

WALLGREN, Arvid J., Swedish physician; b. Norra Rada, Sweden, Oct. 5, 1889; s. Per J. and Signe (Hultgren) W.; Dr.Med., U. Uppsala (Sweden), 1918; Dr.Hon. Causa, U. Zürich, 1950, Sorbonne, Paris, 1947, Alger, Algeria, 1948, Santiago, Chile, 1949, Cardiff, Wales, 1953, Havanna, Cuba, 1953; m. Elsa Maria Carlborn, Oct. 5, 1922; children—Caral Göran, Gudrun Signe (Mrs. Joseph Merrill). Asso. prof. Uppsala U., 1921-22; head Children's Hosp., Göteborg, Sweden, 1922-43; prof. pediatrics, head pediatric clin. Caroline Hosp., Stockholm, Sweden, 1943-56; librarian Swedish Med. Soc., 1956-—; med. dir. Ins. Co. Svea, Göteborg, 1930-43, Sverige, Stockholm, 1943-60. Chmn. adv. bd. med. research WHO, Geneva, Switzerland, 1958-63. Decorated officier la Légion d'Honneur; Comdr. Ordre de la Santé Public, Paris; comdr. L'Ordre de Vasa, Stockholm; comdr. Finnish Ordre de la Rose Blanche; comdr. Ordre Polar Star. Fellow Royal Coll. Physicians; hon. mem. Am. Acad. Pediatrics, Am., Canadian, German, Argentine, Czechoslovakian, Finnish, Swiss, Italian, Polish pediatric socs.; mem. Göteborg (past chmn.), Swedish (past chmn.), med. socs., Swedish Pediatric Soc. (past chmn.), Internat. Pediatric Assn., Swedish Med. Research Council (vice chmn.). Author: Clinical and Experimental Pathology of Tuberculosis, 1918; Researches on Myeloma Disease, 1920; Splenomegalies, in Children, 1927; Iron in the Diet of Children, 1939; Epidemic Jaunidice, 1932; (with J. A. Miller) Pulmonary Tuberculosis in Children, 1938; Welfare of Mother and Child, 1939; (with G. Fanconi) Textbook of Pediatrics, 1950; Textbook of Pediatrics for Nurses, 1938; also numerous articles. Contbd. to global use of intradermal BCG vaccination against Tb; studies in Tb disease in childhood and role of erythema nodosum as sign of tuberculous infection, diagnostic criteria of rheumatic fever, reticulo-endothelial granuloma, pathogenesis of hypertrophic pyloric stenosis. Home: 53 Strandvägen. Office: Pediatric Clinic, Karolinska Sjukhuset, Stockholm, Sweden.*

WALLICH, Nathaniel, botanist; b. Copenhagen, Denmark, Jan. 28, 1786; M.D., Copenhagen, also U. Aberdeen; at least 1 son, George. Became surgeon Danish settlement, Serampore, India, 1807; entered English Med. Service, 1813; supt. Calcutta bot. garden, 1815-50. Obelisk erected in his honor by E. India Co, in bot. gardens, Calcutta. Fellow Royal Soc., 1829, Royal Asiatic Soc., Linnean Soc. (v.p.); mem. French Acad. Scis. Author: Tentamen floaee napalensis illustratae . . . , 1824, 26; A Numerical List of Dried Specimens of Plants in the East India Company's Museum, 1828; Plantae asiaticae rariores, or Descriptions and Figures of a Select Number of Unpublished East Indian Plants, 3 vols., 1830-32; also papers. Published (with William Carey) Flora India (by William Roxburgh), 1820. Collected plants in Nepal, Bengal, Burma. Died Apr. 28, 1854.

WALLING, Cheves (Thomson), Am. chemist; b. Evanston, Ill., Feb. 28, 1916; s. Willoughby George and Frederika Christina (Haskell) W.; A.B., Harvard, 1937; Ph.D., U. Chgo., 1939; m. Jane Ann Wilson, Sept. 17, 1940; children—Hazel (Mrs. David Nourse), Rosalind (Mrs. Stephen M. Holton), Cheves, Janie, Barbara. Research chemist E. I. duPont de Nemours & Co., 1939-43; U. S. Rubber Co., 1943-49; tech. aide OSRD, 1945-46; chief supvr. organic research Lever Bros. Co., 1949-52; prof. chemistry Columbia, 1952—, chmn. dept., 1963-66. Member National Academy Sciences. Am. Chemical Society, A.A.A.S., Am. Acad. Arts and Scis. Mem. editorial bd. Jour. Am. Chem. Soc., 1955-65. Author; Free Radicals in Solution, 1957. Research on phys. organic chemistry, in particular study of reactions involving free radials as transient intermediates; polymerization, halogenation, autoxidation; studied effect of high pressures on organic reactions. Home: 284 Upper Mountain Av., Upper Montclair, N.J. Office: Columbia U., N.Y.C. 27.*

WALLINGFORD, Richard, English mathematician; b. Berkshire on the Thames, Eng., circa 1292; s. William and Isabella Walingford; adopted upon father's death by William de Kirkeby; A.B., Oxford U., circa 1315, also B.D., license to lecture on Sentences; became mem. Benedictine order, St. Albans, Eng., circa 1315, abbot, 1326. Author: Quadripartitus de sinibus demonstratis (based on Greek and Arabic geometry and astronomy, includes account of Euclidean manner of fundamental theorems of plane trigonometry); Canones de instrumento, 1326; De arte componendi rectangulum, 1326, Ars operandi cum rectangulo, 1326 (both on his invention the rectangulus). Considered the greatest mathematician of his day; helped introduce trigonometry into Christian Western Europe; built astron. clock which showed motions of sun and moon, ebb and flow of tides. Died May 23, 1335.

WALLIS, John, English mathematician; b. Ashford, Kent, Eng., Nov. 23, 1616; s. Rev. John and Joanna (Chapman) W.; ed. Emmanuel College, Cambridge, B.A., 1637, M.A., 1640, ordained priest in Church of Eng., 1640, Fellow, Queen's College, Cambridge, 1644-45; Exeter College, Oxford, M.A., 1644, D.D., 1654 (confirmed by diploma, 1662); m. Susanna Glyde (d. 1687); 1 son, John W., 2 daus. Chaplain to Sir Richard Darley; to Mary, Baroness Vere, 1642-44; inherited estate in Kent on death of his mother, 1643; rector of St. Gabriel, London, 1643-47; St. Martin in Ironmonger Lane, 1647-49; Savilian prof. of geometry, Oxford, 1649-1703; Royal chaplain, 1660; on commission to revise Book of Common Prayer, 1661; Custos archivorum, Oxford U., 1657; Fellow Royal Soc., 1663. Author: Truth Tried, or Animadversions on the Lord Brooke's Treatise on the Nature of Truth, 1643; Grammatica linguae Anglicanae, 1652; Arithmetica infinitorum, 1655; Mathesis universalis, 1657; Mechanica sive tractatus de motu, 3 parts, 1670-71; De algebra tractatus, 1685; Institutio logica, 1687; Opera mathematica, 3 vols., 1693-99; Theological Writings, 1691; Sermons, 1691; plus numerous articles on grammar, logic, theology, and cryptology; ed. various works of Archimedes, Ptolemy, Pappus, Aristarchus, and postumous works of Jeremiah Horrocks, 1673. Served parliamentary cause as cryptographer by deciphering captured Royalist documents; associated with R. Boyle's scientific group, which met in Oxford and London, from 1649; one of founders of Royal Soc.; engaged in arguments with Hobbes (on his geometrical ideas); effected quadrature of many curves; established value of 4/pi by infinite approximation; introduced symbol for infinity; expanded range of higher algebra through his study of negative and fractional exponents; prepared way for binomial theorem and differential and integral calculus; probably introduced term mantissa in logarithms; studied cycloids; using indivisibles, determined center of gravity of cycloid; presented correct theory of impact of inelastic bodies (based on principle of conservation of motion) to Royal Society, 1668, insisted on mathematical foundation for physics, also on experimental verification; investigated momenta of forces; influential in delaying acceptance of Gregorian Calendar in Eng. Died Oxford, Eng., Oct. 28, 1703.

WALLIS, Richard Fisher, Am. physicist; b. Washington, May 14, 1924; s. William F. and Alberta (Sigelen) W.; B.S., George Washington U., 1945; M.S., 1948; Ph.D., Catholic U. Am. 1952; m. Mary Camilla Williams, Aug. 20, 1955; children—Maria F., Sylvia C. Postdoctoral fellow U. Md., College Park, Md., 1951-53; chemist Applied Physics Lab., Johns Hopkins, Silver Spring, Md., 1953-56; physicist U.S. Naval Research Lab., Washington, 1956-58, head semiconductors br., 1958-66, 67—; prof. physics U. Cal. at Irvine, 1966-67. Fellow Am. Phys. Soc.; mem. Am. Chem. Soc., Philos. Soc. Washington, Phi Beta Kappa, Sigma Xi. Research and publs. on theory of molecular structure and spectra; coloration of crystals by X-rays; theory of surface effects on vibrations of atoms in crystals, theory of magnetic field effects on solids. Home: 7615 Hamilton Spring Rd., Bethesda, Md. 20034. Office: U.S. Naval Research Lab., Washington 20390.*

WALLIS, Wilson Dallam, Am. anthropologist; b. Forest Hill, Md., Mar. 7, 1886; s. William Randall and Sarah (Kellogg) W.; A.B., Dickinson Coll., 1907, A.M., 1909; B.Sc., Diploma in Anthropology, Oxford U., 1910; Ph.D., U. Pa., 1915; m. Grace Steele Allen, Dec. 28, 1911; children—Wilson Allen, Virginia Dallam (Mrs. Raymond V. Bowers); m. 2d, Ruth Otis Sawtell, June 17, 1931. Faculty, U. Pa., 1911-15, U. Cal. at Berkeley, 1915-16, Fresno Jr. Coll., 1916-21, Reed Coll., 1921-23, U. Conn., 1955-56, Annhurst Coll., 1956; faculty U. Minn., Mpls., 1923—, prof. anthropology, 1926-54, emeritus, 1954—. Mem. Am. Anthrop. Assn., Am. Ethnol. Soc. Author: Culture and Progress, 1930; Culture Patterns in Christianity, 1964; also numerous articles. Field studies on Dakota, Malecite, Micmac tribes, changes in head form of children during growth, anatomic lag in correlated anatomic dimensions, critiques in inferring, relative ages of culture traits, independent origins of similar culture traits. Address: Box 66, South Woodstock, Conn. 06267.*

WALRAS, Leon, economist; b. Evreaux, France, 1834. Author: L'économie politique et la justice, examen critique et réfutation des doctrines économiques de P.-J. Proudhon, 1860; Théorie critique de l'impôt, 1861; Recherches de l'ideal social, 1868; Éléments d'économie pure, 1874. Mem. neo-classical sch. economics; worked on value theory, theory of gen. equilibrium; advocated cooperative assns.; used math. in analysis. Died Clarens, Switzerland, 1910.

WALSH, Benjamin Dann, entomologist; b. Clapton, London, Eng., Sept. 21, 1808; s. Benjamin Walsh; B.A., Trinity Coll., Cambridge (Eng.) U., 1831, M.A. (hon.), 1834; m. Rebecca Finn, 1837, Came to U. S., 1838; farmer, Henry County, Ill., 1838-51; in lumber bus., Rock Island, Ill., 1851-58. Founder and editor (with Charles V. Riley) Am. Entomologist, 1868; asso. editor, later editor Practical Entomologist, Phila. state entomologist Ill., published his only ofcl. report in Trans. Ill. State Hort. Soc., 1867. Author: The Comedies of Aristophanes, Translated into Corresponding English Metres, published in Blackwood's Mag., 1837; also agrl. articles to Proc. Boston Soc. Natural History, Trans Am. Entomol. Soc. Pioneer in demonstrating that Am. farmers aided multiplication of insects by improper planting of crops, also pioneered introduction of fgn. parasites and natural enemies of imported insect pests. Died Rock Island, Ill., Nov. 18, 1869.

WALSH, John, physicist; b. 1725; s. Joseph and Elizabeth (Maskelyne) W.; with mil. service; became sec. to Baron Clive, E. India Co., 1757; fellow Royal Soc., 1770, recipient Copley medal, 1774; experimented (with Jan. Ingenhousz) on torpedo fish in La Rochelle, found that one receives an elec. shock by grasping the fish on both sides; Walsh's letter to Benjamin Franklin (1773) explaining the fish expts. was the 1st instance its elec. nature was accepted as conclusive. Died 1795.

WALSH, John Edward, Am. mathematician; b. Grand Junction, Colo., May 28, 1919; s. Henry Edward and Lulu (Boren) W.; B.S. in Math., U. Notre Dame, 1941; M.A., U. Cal. at Los Angeles, 1944; Ph.D., Princeton, 1947; m. Pearl Ione Lane, Sept. 22, 1951; children—Luanne, John Edward. Research specialist Lockheed Aircraft Corp., Burbank, Cal., 1941-45, 54-58; design specialist Douglas Aircraft Co., Santa Monica, Cal., 1947-48; asso. mathematician RAND Corp., Santa Monica, 1948-51; cons. U.S. Census Bur., Washington, 1951; mathematician U.S. Naval Ordnance Test Sta., China Lake, Cal., 1951-54; sr. scientist System Devel. Corp., Santa Monica, 1958—. Asst. prof. U. Cal. at Los Angeles, 1949-51, lectr., 1952, 58-59; vis. prof. Stanford, 1962, U. Hawaii, 1964; adj. prof. U. So. Cal., 1965—. Fellow Am. Statis. Assn.; mem. Internat. Assn. for Statistics in Phys. Scis., Operations Research Soc. Am. (pres. 1967-68), Inst. Math. Statistics, Biometric Soc., Soc. Actuaries (asso.), Inst. Mgt. Scis., Research Soc. Am., Soc. Tech. Writers and Pubs., Operations Research Soc. Japan, Internat. Fedn. Operational Research Socs. (mem. council 1963, 66), Sigma Xi. Author: Handbook of Nonparametric Statistics, Vol. I, 1962, Vol. II, 1965; contbg. author: Contributions to Order Statistics, 1962; also numerous articles. Research on applied nonparametric statistics, operations research, biostatistics, math. statistics, actuarial fields. Home: 5019 Donna Av., Tarazana, Cal. 91356. Office: 2500 Colorado Av., Santa Monica, Cal. 90406.*

WALSH, Joseph Leonard, Am. mathematician; b. Washington, Sept. 21, 1895; s. John Leonard and Sallie (Jones) W.; student Balt. Poly. Inst., 1908-12, Columbia U., 1912-13; S.B., Harvard, 1916, Ph.D., 1920; postgrad. U. Chgo., 1916, U. Wis., 1917, U. Paris (France), 1920-21, U. Munich (Germany), 1925-26; m. Aline Natalie Burgess, July 15, 1931 (div. July 1941); children—Sallie Elizabeth (Mrs. Gordon Taylor Curtis), Elizabeth Hildegard (Mrs. David Wolfe); m. 2d, Elizabeth Cherry Strayhorn, Apr. 2, 1946. Faculty, Harvard, 1915-66, prof. math., 1935-66; prof. math. U. Md., College Park, 1965—. Mem. Am. Math. Soc. (past pres.), Nat. Acad. Sci., Math. Assn. Am., A.A.A.S., Phi Beta Kappa, Sigma Xi. Fellow Am. Acad. Arts and Sci. Author: Interpolation and Approximation, 1935; Location of Critical Points, 1950; Approximation by Polynomials, 1935; Approximation by Bounded Functions, 1960; (with A.H. Ahlberg and E.N. Nilson) Theory of Splines, 1967; also numerous articles. Research on approximation to functions, locations zeros polynomials, orthogonal functions. Home: 4105 Van Buren St., University Park, Md. 20782.*

WALSH, Michael Patrick, Am. biologist; b. Boston, Feb. 28, 1912; s. Coleman and Bridget (McDonough W.; A.B., Boston Coll., 1934, A.M., 1935; M.S., Fordham U., 1938, Ph.D., 1948; S.T.L., Weston Coll., 1942; LL.D., U. Mass., 1961, St. Anselm's Coll., 1963, Brandeis U., 1963; D.Ed., Suffolk U., 1963; D.Sc., Villanova, 1961, Stonehill Coll., 1964; Litt.D., Coll. Holy Cross, 1962; L.H.D., Northeastern U., 1962. Joined Soc. of Jesus, 1929, ordained priest Roman Cath. Ch., 1941; acting prin. Fairfield U. Prep., 1942; instr. biology Boston Coll., 1943-45, asso. prof., 1948-58, pres., 1958—. Mem. Pres.'s Commn. Presdl. Scholars, Northeastern Regional Com. Marshall Scholarships, Mass. Adv. Com. Racial Imbalance and Edn., Massachusetts Higher Edn. Facilities Commission; member of Boston Civic Progress Com., 1958—; chmn. ednl. div. United Fund, 1958; mem. commn. on legislation Assn. Am. Colls., 1958. Trustee of John F. Kennedy Memorial Library. Recipient Freedom Foundation Award, 1960. Mem. A.A.A.S., American Assn. Zoologists, Bot. Soc. Am., Am. Micros. Soc., Genetics Soc. Am., Am. Assn. Jesuit Scientists (pres. 1951), Nat. Cath. Ednl. Assn. (pres. coll. and univ. dept. 1966-68), Sigma Xi. Author publs. on cytology, genetics. Address: Boston Coll., Chestnut Hill, Mass. 02167.*

WALSH, Robert John, Australian biologist; b. Brisbane, Australia, Jan. 3, 1917; s. John James and Catherine (Ahern) W.; M.B., B.Surgery, U. Sydney, 1939; m. Kathleen Helen Tooth, June 5, 1944; children—Michael John, Richard George, Robert Mark and Rosemary Helen (twins). Dir. New South Wales chpt. Red Cross Blood Transfusion Service, 1946-66; vis. prof. human genetics U. New South Wales, 1962-66, chmn., 1967—; cons. hematologist Sydney Hosp., 1956—; vis. lectr. hematology U. Sydney, 1955—. Fellow Australian Acad. Sci. (sec. biol. sci. 1966—), Royal Australian Coll. Physicians; mem. Coll. Pathologists of Australia, Internat. Soc. Hematology, Internat. Biol. Programme. Research, numerous publs. on intermediary iron metabolism, hemochromatosis; skin pigmentation, genetics of blood proteins, blood groups in Pacific. Home: 237 Midson Rd., Epping, Sydney. Office: U. New South Wales, Sydney, Australia.*

WALSKE, M(ax) Carl, Jr., Am. physicist; b. Seattle, June 2, 1922; s. Max Carl and Margaret Ella (Fowler) W.; B.S. in Math. cum laude, U. Wash., 1944; Ph.D. in Theoretical Physics, Cornell U., 1951; m. E(lsa) Marjorie Nelson, Dec. 28, 1946; children—C. Susan, Steven C., Carol A. Staff mem., then asst. theoretical div. leader Los Alamos Sci. Lab., 1951-56, mem. staff, 1965-66; dep. research dir. Atomics Internat., Canoga Park, Cal., 1956-59; mem. U. S. delegation Conf. Suspension Nuclear Tests, Geneva, Switzerland, 1959-61; sci. rep. AEC in U.K., London, Eng., 1961-62; theoretical physicist RAND Corp., 1962-63; sci. attache U. S. missions to NATO and OECD, Paris, France, 1963-65; asst. to sec. def. for atomic energy, 1966—. Chmn. Dept. Def. mil. liaison com. to AEC, 1966—. Mem. Am. Phys. Soc., Am. Nuclear Soc., Phi Beta Kappa, Sigma Xi. Research on stopping power theory, nuclear weapons tech., reactor physics, nat. and internat. science policy, disarmament. Home: 1907 Windsor Rd., Alexandria, Va. 22307. Office: Dept. Def., The Pentagon, Washington 20301.*

WALT, Martin, Am. physicist; b. West Plains, Mo., June 1, 1926; s. Martin and Dorothy (Mantz) W.; B.S., Cal. Inst. Tech., 1950; M.S., U. Wis., 1951, Ph.D., 1953, m. Mary Estelle Thompson, Aug. 16, 1950; children—Susan Mary, Stephen Martin, Anne Elizabeth. Staff, Los Alamos Sci. Lab., 1953-56; research scientist Lockheed Missiles & Space Co., Palo Alto, Cal., 1956-60, cons. scientist, 1960-65, mgr. phys. scis. lab., since 1956—. Fellow Am. Phys. Soc.; mem. Am. Geophys. Union, Am. Inst. Aeros. and Astronautics (mem. space and atmospheric physics com. 1964—. Author: Auroral Phenomena, Experiments and Theory, 1965; also articles. Research on interactions fast neutrons with nuclei, total cross sections, differential elastic scattering cross sects., inelastic collision cross sects. for testing and evaluating optical model parameters, Van Allen radiation belts, polar aurora, cosmic rays. Home: 12650 Viscaino Ct., Los Altos Hills, Cal. 94022. Office: 3251 Hanover St., Palo Alto, Cal. 94304.*

WALTER, Albert G., surgeon; b. Germany, June 21, 1811; M.D., Konigsberg U.; postgrad. in Berlin; m. Frances Anne Butler, 1846; 1 son, 1 dau. Practiced medicine, Pitts., 1837-76; Pioneer in orthopedic surgery in U. S., also noted as oculist and as accident surgeon; performed 1st japarotomy for relief of ruptured bladder, 1859; early practitioner of antisepsis. 1st pres. Humane Sac. Pitts. Author: Conservative Surgery in its General and Successful Adaptation in Cases of Severe Traumatic Injuries of the Limbs, 1867. Died Pitts, Oct. 14, 1876.

WALTER, Edward Joseph, Am. seismologist; b. St. Louis, Dec. 6, 1914; s. Jacob and Catherine (Wallace) W.; B.S., St. Louis U., 1937, M.S., 1940, Ph.D., 1944; m. Constance Elms, June 7, 1939; children—Constance, Barbara, Edward, Mary, Francis X., Kevin, James, Thomas, John. Field seismologist Shell Oil Co., 1937-38, research seismologist, 1944-46; instr. St. Louis U., 1941-44; mem. faculty John Carroll U., Cleve., 1946—, prof. math., 1952—, dir.

Seismol. Obs., 1962—. Mem. Seismol. Soc. Am. (chmn. Eastern sect., editorial bd. Eastern sect., chmn. sect. vibrations com.), Am. Geophys. Union, Soc. Exploration Geophysicists, A.A.A.S., Ohio Acad. Sci., N. Ohio Geol. Soc., Research Soc. Am., Sigma Xi, Pi Mu Epsilon, Alpha Sigma Nu. Research and publs. on structure of earths crust, seismic wave velocities, attenuation of seismic waves, math. analysis and reduction of seismograms, devel. computer techniques, relationship between seismicity and volcanism, analysis of bldg. vibration problems. Home: 10306 Mayfield Rd., Chesterfield, O. 44026.*

WALTER, John Harris, Am. mathematician; b. Los Angeles, Dec. 14, 1927; s. Monroe Walter and Edith (Harris) W.; B.S., Cal. Inst. Tech., 1951; M.S., U. Mich., 1953., Ph.D., 1954; m. Jane Prashker, Mar. 20, 1955; children—Marjorie, Charles, Harry. Instr. U. Wash., 1954-56, asst. prof., 1956-61; asso. prof. U. Ill., Urbana, 1961-66, prof., 1966—; vis. asst. prof. U. Chgo., 1960-61, vis. asso. prof., 1966. NSF fellow, 1957-58. Mem. Am. Math. Soc., Sigma Xi. Contbr. articles in field to sci. jours. Characterized certain important classes of finite simple groups; studies on galois theory to infinite division ring extensions; automorphisms of certain classical groups. Home: 804 E. Mumford Dr., Urbana Ill. 61801.*

WALTER, Thomas, botanist; b. Hampshire, Eng., circa 1740; m. Anne Lesesne, Mar. 26, 1769; m. 2d, Ann Peyre, Mar. 20, 1777; 2 daus.; m. 3d, Dorothy Cooper, after 1780; 1 dau. Collected herbarium, presented to Linnean Soc. of London, 1849, acquired by Brit. Mus. Natural History, 1863; tried to introduce (with John Fraser) a native Carolina grass, Agrostis perennans, into gen. cultivation in Eng. Author: Flora Caroliniana (sole record of work, describing under binomial system of nomenclature approximately 1,000 species of flowering plants representing 435 genera, from specimens collected with Fraser in S.C.). Died Jan. 17, 1789.

WALTER, William Grey, physiologist; b. Kansas City, Mo., Feb. 19, 1910; s. Karl W. and Margaret (Hardy) W.; B.A., Cambridge U., 1931, M.A., 1935, Sc.D., 1947; M.D. (hon.), Aix-Marseille, 1949; m. Lorraine Donn, June, 1960; children—Nicolas Hardy, Jeremy Karl, Timothy Grey (by previous marriage). Rockefeller fellow Muadsley Hosp., London, 1935-39; head physiol. dept. Burden Neurol. Inst., Stoke Lane, Stapleton, Bristol, Eng., 1939—. Fellow Inst. Biology; mem. Internat. Fedn. Electroencepholograph Socs. (pres. 1953-57), Physiol. Soc. Author: The Living Brain, 1953; Further Outlook, The Curve of the Snowflake, 1956. Studies, publs. on mechanisms of elec. activity of brain; diagnosis of brain diseases; discovery of delta and theta brain waves; invented 1st artificial animal; discovery of expectancy wave as a sign of learning in human brain. Home: 35 Mariners Dr., Bristol. Office: Burden Inst., Stoke Lane, Stapleton, Bristol, Eng.*

WALTER OF EVESHAM, see Odington, Walter.

WALTERS, Alan Arthur, economist; b. Leicester, Eng., June 17, 1929; s. James Arthur and Clarabel (Heywood) W.; B.Sc. with 1st class honors in Econs., U. Coll., Leicester, Eng., 1951; postgrad. Nuffield Coll., Oxford, 1951-52; m. Audrey Elizabeth Claxton, Mar. 30, 1950; 1 dau., Louise. Faculty, U. Birmingham (Eng.), 1952—, prof. econometrics and statistics, 1961-67; prof. econs. Mass. Inst. Tech., Cambridge, Mass., 1967; Sir Ernest Cassel professor of economics London School of Economics, since 1968—. Consultant, Brookings Inst., 1964, World Bank, 1965—. Recipient Gertenberg prize, 1951. Fellow Royal Econ. Soc., Royal Statis. Soc. Author: (with R.W. Clover, G. Dalton) Growth Without Development, 1966; also articles. Research on theory and measurement of transport prices especially application of price theory to structure of road uses charges; measurement of prodn. relationships, effects of money on econ. variables. Home: 71 Eastern Rd., Birmingham 29, Eng.; also 76 Garfield St., Cambridge, Mass. 02138.*

WALTERS, G. King, Am. physicist; b. Baton Rouge, La., Aug. 23, 1931; s. Robert K. and Harriett (Fuller) W.; B.A., Rice U., 1953; Ph.D., Duke U., 1956; m. Jeanette Long, June 14, 1954; children—Terry Lynne, Jeffrey King, Gina Leslie. NSF Postdoctoral fellow Duke U., 1956-57; mem. tech. staff Texas Instruments, Inc., Dallas, 1957-62, sr. scientist, 1962, corporate research asso., 1962-63, cons., 1963—; prof. physics Rice U., Houston, 1963-64, prof. physics, space sci., 1964—. Asso. editor Jour. of Geophysical Research, 1966—. Mem. Am. Inst. Physics, Am. Geophys. Union, A.A.A.S. Contbr. articles in field to sci. jours. Pioneered application of magnetic resonance techniques to surface physics; co-discoverer optical pumping technique for polarizing gaseous and liquid He3 and successful application in polarized targets for nuclear structure studies; co-discoverer low temperature phase separation in He3-He4 solution; co-discoverer Fermi-Dirac degeneracy in liquid He3 at low temperatures. Home: 5102 Jackwood St., Houston 77035.*

WALTERS, Vladimir, Am. zoologist; b. N.Y.C., Dec. 18, 1927; s. Walter John and Sylvia Alice (Pottala) W.; B.S., Cornell U., 1947, M.S., 1948; Ph.D., N.Y. U., 1954; m. Lisa Hamilton, Jan. 30, 1960 (div. 1965); 1 dau., Sylvia Hamilton. Research asst. U.S. Naval Arctic Research Lab., Point Barrow, Alaska, 1948-49; research asso. in fishes Am. Mus. Natural History, N.Y.C., 1954-56, asst. curator fishes, 1956-60; faculty U. Cal. at Los Angeles, 1961—, asso. prof., 1964—. Mem. Am. Soc. Ichthyologists and Herpetologists (bd. govs. 1958-63, 65—), Arctic Inst. N.Am., Sigma Xi, Am. Soc. Zoologists, Am. Soc. Naturalists. Research, numerous publs. on metabolic physiology, systematics and zoogeography arctic, W. Indian and deep sea fishes, functional interpretation body structure deep sea and cave fishes in terms of conditions environment, swimming performances fishes by measuring power output tunas and marlins, and by characteristics water flow past swimming fishes. Home: 12332 Montana Av., Los Angeles 90049.*

WALTERS, Waltman, Am. surgeon; b. Cedar Rapids, Ia., July 29, 1895; s. Frank and Mellie (Phipps) W.; B.S., Dartmouth, 1917; M.D., U. Chgo., 1920; M.S. In Surgery, U. Minn., 1923; D.Sc., (hon.) Dartmouth, 1937; LL.D., Hahnemann Med. Coll., 1942; m. Phoebe Mayo, Feb. 5, 1921; children—Phoebe Mayo (Mrs. Alvin Marks), Waltman Mayo, James Mayo, Carolyn Damon (Mrs. Fred Brown). Head surgery sect. Mayo Clinic, Rochester, Minn., 1924-60; prof. surgery Mayo Grad. Sch., U. Minn., 1936-60. Recipient Miss. Valley Med. Soc. honor award, 1954, Am. Med. Writer's Assn. honor award, 1957, Dartmouth Coll. Alumni award, 1957, Shattuck Sch. Achievement award, 1957. Diplomate Am. Bd. Surgery (a founder), Am. Bd. Urology. Mem. A.C.S., Minn. State, So. Minn., Interstate Postgrad. med. assns., Zumbro Valley Med. Soc., Minn. Acad. Sci., Am. Urol. Assn., Am. Assn. for Cancer Research, Am. Gastroenterol. Assn., A.M.A., Am., Central So., Western, Pan-Pacific surg. assns., Assn. Mil. Surgeons U.S., Central Soc. for Clin. Research, Internat. Soc. Surgery, Pan-Am. Med. Assn., Soc. Clin. Surgery, German Surg. Soc., Leopoldina. Author: (with A.M. Snell) Diseases of the Gallbladder and Bile Ducts, 1940; (with Howard K. Gray, James T. Priestley) Carcinoma of the Stomach, 1942; also numerous articles. Chief editor A.M.A. Archives of Surgery, 1938-62, Lewis' Practice of Surgery, 1938—. Research on operations for postoperative stricture of common bile duct, path. lesions in duodenal ulcer; comparative studies on cancer of stomach lesions of bilary tract. ome: 304 8th Av., S.W., Office: 200 1st St., S.W., Rochester Minn. 55901.*

WALTHALL, Wilson Jones, Jr., Am. psychologist; b. San Antonio, July 7, 1918; s. Wilson Jones and Nina (Trimble) W.; B.A. with high honors, U. Tex., 1944, M.A., 1944, Ph.D., 1948; m. Annie Marie Joekel, Sept. 2, 1944; children—Anne, Wilson Jones 3d. Faculty U. Wyo., Laramie, 1947—, prof., 1957—, head dept. psychology, 1965—. Fullbright lectr. U. Coll., Madalay, Burma, 1954-55. Mem. Am., Rocky Mountain (pres. 1959-60), Wyo. (pres. 1961) psychol. assns., Phi Betta Kappa, Sigma Xi, Psi Chi. Research and publs. on factor analysis of maze environment effects on learning of rat, discovered that two different measures of perceptual field strength each equal to the same standard were equal to each other, that figural after-effects from real and apparent motion were quantitatively equal. Home: 1314 Steele St., Laramie, Wyo. 82070.*

WALTHER, Augustin Friedrich, German anatomist; b. 1688; Wittenberg, Germany, 1688; became prof., Leipzig, Germany, 1723; described Walther's ganglion on anterior surface of coccyx nr. the tip, circa 1722. Died 1746.

WALTHER, Bernard, astronomer; b. 1430; worked (at first with Regiomontanus) on nautical almanacs used by Spanish and Portuguese navigators; made astron. observations with reference to ray refraction and use of wheeled clocks. Died 1504.

WALTHER, Heinz, German dermatologist; b. Mullheim/Baden, Germany, Aug. 29, 1919; s. Adolf and Ida (Vollmar) W.; ed. U. Halle, U. Würzburg, U. Tübingen, U. Erlangen; M.D., U. Berlin, 1945; m. Elisabeth Hauck, Apr. 1, 1945; children—Gerhard, Inge, Peter. Asst. dermatol. clinic U. Erlangen, 1945-47, U. Tübingen, 1947-48; head physician Municipal Dermatol. Clinic, Regensburg, 1949-52; pvt. practice dermatology, also staff Municipal Hosp., Pforzheim, 1954—. Del. to Union européenne des Medecins specialistes; state chmn. dermatologists, Baden. Author: (with Leo Hauck) Behandlung der Geschlechtskrankheiten, 1948; Lexikon der aktuellen Therapie, 1960; (with Ernst Heinke) Taschenbuch der Hautund Geschlechtskrankeiten, 1966; also numerous articles. Research on veneral diseases, trichomonads, salve improvements, histology, determination of resistance to antibiotics, skin and internal medicine, skin cancer, occupational skin diseases. Home: 76 Alte Pforzheimerstrasse, Birkenfeld, Germany. Office: 41 Westliche, Pforzheim, Germany.*

WALTHER, Johannes, German geologist; b. Neustadt/Orla, Germany, July 20, 1860; became pvt. docent, Jena, Germany, 1866; apptd. asso. prof. U. Jena, 1890, prof. geology and paleontology, 1894. Author: Einleitung in der Geologie als historischen Wissenschaft, 3 vols., 1893-94; Das Gesetz der Wüstenbildung, 1900; Lehrbuch der geologie Deutschlands, 1910. Research on ocean bottom and its fauna, origins of coral reefs in Res Sea; made sci. expdns. to desert regions of Libya, India, N. Am., Urals, Turkestan. Died Berlin, May 10, 1937.

WALTHER, Paul, German physician; b. Delitzsch, nr. Leipzig, Germany, Aug. 7, 1849; s. Karl and Hermine (Gumprecht) W.; ed. Jena, Berlin; m. Rose Baumgärtel, 1878; 1 son, 1 dau.; m. 2d, Ella Faulhaber, 1908. Asst., Pathology Inst., Jena, Germany, 1872-74; asst. Clinics of Internal Medicine, Heidelberg; lecturing prof. pharmacology and med. chemistry, 1876, later also on dermatology, pediatrics, forensic medicine, hygiene; with govt. service, until 1921. Author: Zur Wirkung der Salizylsäure, 1875; Die gebräuchlichsten Rezeptformeln . . . , 1877; Über Spermatorrhoe und Prostatorrhoe, 1881; Die Störung der Geschlechtsfunktion des Mannes, 1899. Authority on indsl. medicine, accidents; on sexual pathology. Died Berlin, July 21, 1930.

WALTON, Arthur Calvin, Am. biologist; b. Meadville, Pa., Oct. 16, 1892; s. Calvin Levi and Minnie Belle (Stevens) W.; B.A., Northwestern U., 1914, M.A. in Cytology, 1915; Ph.D. in Parasitology, U. Ill., 1922-23; M.A., Harvard, 1917; m. Isyl Spiker, July 30, 1919; children—James Calvin, Robert Leslie, Richard Earl. Prof. zoology Knox Coll., Galesburg, Ill., 1924-58, chmn. dept., 1926-54, Abbott prof. biology, 1939-58, emeritus prof., 1958—, curator memorabilia, 1958—. Mem. agr. div. NRC, 1952-56. Mem. Ill. State Acad. Sci. (past sec.), A.A.A.S. (past pres. acad. div.), Am. Soc. Parasitologists (past mem. council, past pres.), Am. Micros. Soc. (past pres.), Am. Biol. Inst. (past mem. council), Soc. Systematic Zoologists, Am. Soc. Zoologists, Midwest Assn. Parasitologists (past pres.), Phi Beta Kappa, Sigma Xi, Beta Beta Beta, Pi Gamma Mu, Omega Beta Pi. Author: (with Chitwood and Chitwood) Introduction to Nematology, 1940; (with others) Biological Sciences Curriculum Studies High School Biology, 1960; also numerous articles. Editorial bd. Jour. Parasitology, 1947-50; abstractor Biol. Abstracts, 1926-63. Research on systematics, cytology, life history nematode parasites, amphibia. Died Apr. 23, 1967.

WALTON, Cyprian James, Am. biologist; b. Bklyn., July 1, 1909; s. James John and Katherine (Greevy) W.; B.A., Manhattan Coll., 1932, M.A. in English, 1936; M.S. in Biology, Fordham U., 1939, Ph.D. in Biology, 1944. Tchr. pub. schs., N.Y.C., Bklyn., 1929-34; prin. Hillside Sch., Troy, N.Y., 1934-35; asst. prin. St. Joseph High Sch., Manchester, N.H., 1935-36; faculty dept. biology Manhattan Coll., Bronx, N.Y., 1936—, prof., head biology dept., 1947—. Mem. A.A.A.S., Am. Inst. Biol. Scis., Entomol. Soc. Am., Am. Genetic Assn., Am. Assn. Med. Colls., N.Y. Acad. Sci., Sigma Xi. Author: One Hundred Great Scientists, 1964; Manual of Histology, 1964. Research on life cycle of plant lice causing tumors on shade trees in northeastern U.S. Address: Manhattan College, Bronx, N.Y. 10471.*

WALTON, Ernest Thomas Sinton, Irish physicist; b. nr. Waterford, Ireland, Oct. 6, 1903; s. John Arthur and Anna (Sinton) W.; B.A., U. Dublin, 1926, M.Sc., 1928, M.A., 1934; Ph.D. (Overseas Research scholar), Cambridge (Eng.) U., 1931, postgrad. (Clerk Maxwell scholar), 1932-34; D.Sc., Queen's U. Belfast, 1959; m. Winifred Isabel Wilson, Aug. 23, 1934; children—Alan John, Marian Elizabeth, Philip Wilson, Jean Margaret. Fellow, Trinity Coll., U. Dublin, 1934—, prof. natural and exptl. philosophy, 1946—; chmn. Sch. Cosmic Physics, Dublin Inst. for Advanced Studies, 1952-60. Mem. Scholarship Exchange Bd., Ireland and U.S., 1958—. Recipient Sr. Research award dept. sci. and indsl. research Cambridge U., 1930-34; Nobel prize for Physics (with Sir John Cockcroft), 1951; Hughes medal Royal Soc. London, 1938. Fellow Phys. Soc. London; mem. Royal Irish Acad. (past sec.). Cockcroft-Walton type particle accelerator used to produce fast atomic particles showing that lithium and other light elements could be disintegrated by bombardment with fast protons. Home: 26 St. Kevin's Park, Dartry Rd., Dublin 6, Ireland.*

WALTON, Harold Frederic, chemist; b. Aug. 25, 1912, Tregony, Eng.; s. James and Martha Florinda Jane (Harris) W.; B.A., Exeter Coll., Oxford U., 1934, D.Phil., 1937; student Princeton, 1937-38; m. Sadie Goodman, June 17, 1938; children—James, Elizabeth Louise, Daniel Goodman. Come to U. S., 1937, naturalized, 1948. Research chemist Permutit Co., 1938-40; instr., then asst. prof. Northwestern U., 1940-46; volunteer British Red Cross, 1946-47; mem. faculty U. Colo., 1947—, prof. chemistry, 1956—, chmn. dept., 1962-66; vis. prof. (Fulbright lectr.) U. Trujillo, Peru, 1966-67; honorary professor University Trujillo. Member of regional com. Woodrow Wilson Fellowships, 1957—. Mem. Am. Chem. Soc. (chmn. Colo. sect. 1957), A.A.A.S., Faraday Soc., Geochem. Soc., Colo.-Wyo. Acad. Scis., Am. Assn. of University Professors, Sigma Xi. Author: Inorganic Preparations, 1948; Principles and Methods of Chemical Analysis, 2d edit., 1964; Elementary Quantitative Analysis, 1958. Research on ion exchange and its use in chem. analysis, especially new technique of ligand exchange; separation of hydrogen isotopes by electrolysis, chemistry of uranium ore deposition, analytical methods in geochemistry. Home: 750 6th St., Boulder, Colo. 80302.*

WALTON, Henry John, S. African psychiatrist; b. Kuruman, S. Africa, Feb. 15, 1924; s. Dirk and Elizabeth (Henn) W.; M.B., Ch.B., U. Cape Town, 1946; M.D., 1954; D.P.M., U. London Inst. Psychiatry, 1957; Ph.D., U. Edinburgh, 1966; m. Sula Wolff, Jan. 5, 1959. Physician, dept. pediatrics, neurology and surgery U. Capetown Teaching Hosp., 1947-49, sr. registrar, depts. neurology and psychiatry, 1950-54, sr. lectr., dept. psychiatry, univ., 1958-59; registrar, sr. registrar Mandsley Hosp., London, 1955-57; sr. lectr. dept. psychiatry U. Edinburgh, 1962—; cons. professorial unit Royal Edinburgh Hosp., 1962—; in charge psychiatry dept. Western Gen. Hosp., Edinburgh, 1966—. Research fellow Columbia, 1960-61. Mem. Soc. for Research in Higher Edn. (mem. governing council 1966—), Royal Coll. Physicians, Royal Medico-Psychol. Assn. Author: Alcoholism, 1965. Research, publs on role of parental deprivation in causing suicidal behaviour, discrepancies in reporting by observers of same clin. phenomena, types med. students graduating from med. sch., effect of personality factors on drinking pattern and outcome in alcoholism, influence of mental pathology on social behaviour of old people. Home: 8 Abbotsford Crescent, Edinburgh 10, U.K.*

WALTON, Isaak, English biologist; b. Stafford, Eng., Aug. 9, 1593; s. Jervis Walton; apprenticed to ironmonger, London; m. Rachel Floud, Dec. 27, 1626; m. 2d, Anne Ken, circa 1646; children—Anne (Mrs. William Hawkins), Isaak, Isaac. Self-employed, London, 1614; became freeman of Ironmonger's Co., 1618; favoured royalists, 1942; lived with Bishop George Morley, Farnham, Eng., 1662-78; lived with son-in-law (canon of Winchester) Winchester, Eng., 1678-83. Author: The Compleat Angler (comments on natural philosophy, habits of fish); also published biographies of Dr. John Donne, 1640, Sir Henry Walton, 1651, Richard Hooker, 1565, George Herbert, 1670, Bishop Robert Sanderson, 1678; contributed verses to books, 1638-61. Died Dec. 15, 1683.

WALTON, James, English inventor; b. Yorkshire, Eng., 1802; s. Isaac Walton; a son, William. Began as cloth friezer; later machine mfr., nr. Halifax, Eng.; moved to Lancashire, Eng., before 1846; bought Dolfargan Hall, Montgomeryshire, Eng., 1870. Improved cotton-spinning machine, 1834-40, including India rubber card, 1834, endless sheet machine, machines for cutting and facing tappets, double twill wheels; developed 1st practical wire stop motion for machines, system for drawing wire; patent rolled angular wire. Died Nov. 5, 1883.

WALTON, Sir James, English surgeon; b. London, 1881; ed. Framlingham Coll., London Hosp. Med. Coll.; B.Sc., London U., 1906, M.B., B.S., 1908, M.S., 1909; m. Nancy Mary Trevett (dec. 1953); 1 son, 1 dau.; m. 2d, Renee Carrington, 1953. Various positions London Hosp., including asst. dir. path. inst., surg. registrar, demonstrator anatomy, surgeon; surgeon Poplar Hosp. for Accidents, Evelina Hosp. for Children, Seamen's Hosp., Greenwich; cons. surgeon Victoria Hosp., Kingston; surgeon to George V, Edward VIII, George VI; surgeon His Majesty's Household, 1930-36; extra surgeon to the Queen, from 1952, also to Queen Mary; Hunterian prof. surgery. Licentiate Royal Coll. Physicians. Fellow Royal Coll. Surgeons Eng. (v.p., mem. council), Royal Soc. Medicine (pres. surg. sect.); pres. Assn. of Surgeons, Med. Soc. London; mem. several med. socs. Author: Fractures and Separated Epiphyses; A Text-Book of the Surgical Dyspepsias; Physical Gemmology; also articles on surgery of upper abdomen, thyroid gland, central nervous system. Editor: A Text-Book of Surgical Diagnosis. Died Aug. 27, 1955.

WALTON, John, Brit. botanist; b. London, May 14, 1895; s. Edward Arthur and Helen (Henderson) W.; M.A., Sc.D., Cambridge U.; D.Sc., Manchester, Eng.; LL.D., McMaster U.; D.ès Sc., Montpellier, France, Lille, France; m. Dorothy, Aug. 10, 1918; children—Huon Seward, Camilla Marion (Mrs. John Douglas Uytman). Lectr., U. Manchester, 1924-30; Regius prof. botany U. Glasgow (Scotland, 1930-62, prof. emeritus, 1962—. Forestry commr., 1949-54; botanist Oxford U. Expdn. to Spitsbergen, 1921. Fellow Royal Soc. Edinburgh; hon. fellow Bot. Soc. Poland, Am. Bot. Soc., Geol. Soc. Belgium, Palaeobot. Soc. India. Author: Introduction to Study of Fossil Plants; also articles. Research on fossil plants of Palaeozoic Period; tech. methods used in study of fossil plants. Home: 9 Windsor, Dundee, Scotland.*

WALTON, John Nicholas, English neurologist; b. Rowlands Gill, Eng., Sept. 16, 1922; s. Herbert and Eleanor (Ward) W.; M.B., B.S. with 1st class honors, U. Durham, 1945, M.D., 1952; m. Mary Elizabeth Harrison, Aug. 31, 1946; children—Elisabeth Ann, Judith Mary, Christopher John. House physician Royal Victoria Infirmary, Newcastle upon Tyne, 1945-46, med. registrar, 1949-51; research asst. in medicine U. Durham, 1951-57; Nuffield Found. fellow Mass. Gen. Hosp. and Harvard Med. Sch., 1953-54; Kings Coll. fellow in neurology, Neurol. Research Unit, Nat. Hosp., London, 1954-55; 1st asst. in neurology Royal Victoria Infirmary, 1957-58; cons. neurologist Newcastle Gen. Hosp. and Royal Victoria Infirmary, 1958—; lectr. neurology Newcastle U., 1966. Vice chmn. Muscular Dystrophy Group of Great Britain. Terr. Decoration, 1964; Goulstonian lectr. Royal Coll. Physicians, London, 1964. Fellow Royal Soc. Medicine, Coll. Physicians London; mem. Assn. Brit. Neurologists. Author: Subarachnoid Haemorrhage, 1952; (with R.D. Adams) Polymyositis, 1956; Essentials of Neurology, 1961; Disorders of Voluntary Muscle, 1964; also numerous articles. Editor-in-chief Jour.. Neurol. Sci., 1966—. Research on clin. and genetic aspects of muscular dystrophy and other muscle diseases. Home: Holmwood, 9 Beechfield Rd., Gosforth, Newcastle upon Tyne 3, Eng. Office: Regional Neurol. Centre, Newcastle Gen. Hosp., Newcastle upon Tyne 4, Eng.*

WALTON, Matt Savage, Am. geologist; b. Lexington, Ky., Sept. 16, 1915; s. Matt Savage and Lillias (Wheeler) W.; B.A., U. Chgo., 1936; M.A., Columbia, 1946, Ph.D., 1951; m. Kathryn Ralston, Dec. 6, 1939; children—Matt Savage III, Kate Johns, Lisa Baar. Geologist, U. S. Geol. Survey, 1942-46; lectr. geology Columbia, 1947-48; geologist N.Y. State Sci. Service, 1948-56; faculty Yale, 1948-65, asso. prof., 1957-65; cons. geologist, Denver, 1965—. Cons. numerous maj. engring. projects, 1946—. Fellow Geol. Soc. Am., Am. Mineral. Soc.; mem. A.A.A.S. Research, publs. on diffusion in metamorphism; produced evidence showing Precambrian rocks of Eastern Adirondack mountains had complex history in which some maj. rock masses and units, previously regarded as late magnetic intrusion were probably elements of an older basement terrain remobilized during Grenville orogeny. Address: Brooks Towers, Denver 80202.*

WALTON, William Winfield, Am. chemist; b. Phila., Sept. 1, 1904; s. Thomas B. and Edith (Reazor) W.; B.S., Pa. State U., 1927; M.S., U. Md., 1941, Ph.D., 1947; m. Mabel S. Funk, Sept. 24, 1927; children—William Winfield, Theodore R., Nancy Lee (Mrs. Alan H. Singleton). Chemist, Bur. Engraving and Printing, 1927-28, U.S. Dept. Agr., 1929; chemist Nat. Bur. Standards, Washington, 1929—, chief surface chemistry sect., 1956-60, chief materials and composites sect., 1960—; faculty Am. U., 1948-49. Recipient Ludwig medal Pa. State U., 1927. Mem. Am. Chem. Soc., Chem. Soc. Washington (past pres.), Washington Acad. Scis. (past v.p.), A.A.A.S., Am. Forestry Assn., Alpha Chi Sigma. Research, publs. on precise conductivity measurements for pH standards, preparation pure substances for organic microanalytical standards, separation isotopes for Manhattan project, determination minute quanitites oxygen in organic compounds, preparation derivatives ascorbic acid. Home: 1705 Edgewater Pkwy., Silver Spring, Md. 20903. Office: Nat. Bur. Standards, Washington 20234.*

WANG, Chi Che, biochemist; b. Soochow, China, Oct. 30, 1894; d. Zone Wai and Jan Dan (Siah) Wang; came to U.S., 1907, naturalized, 1947. B.A., Wellesley Coll., 1914; M.S., U. Chgo., 1916, Ph.D., 1918; postgrad. Johns Hopkins, 1917. Asst. in nutrition U. Chgo., 1918-20; head dept. chemistry, charge clin. research Michael Reese Hosp., Chgo., 1920-30; faculty U. Cin., also Childrens Research Hosp., Cin., 1931-40, Northwestern U. Med. Sch., Chgo., 1943-46; research chemist Northwestern Yeast Co., Chgo., 1940-43, Mayo Clinic, Rochester, Minn., 1946-47; research biochemist charge Metabolic Lab., VA Hosp., Hines, Ill., 1947-54; biochemist charge research and clin. chemistry VA Hosp., Topeka, 1954-60; lectr. Washburn U., Topeka, 1955-60; ret., 1961. Fellow A.A.A.S., Soc. for Research in Child Devel.; mem. Am. Soc. Biol. Chemists, Am. Inst. Nutrition, Soc. Exptl. Biology and Medicine, Am. Chem. Soc., Am. Assn. Clin. Chemists, Inst. Medicine of Chgo., Phi Beta Kappa, Sigma Xi, Iota Sigma Pi, Sigma Delta Epsilon. Research and numerous publs. on chemistry of biol. fluids, food products, energy, mineral and protein metabolism of obese and undernourished children and adults. Address: 2817 Gage Blvd., Topeka 66614.*

WANG, Chih Hsing, chemist; b. Shanghai, China, Sept. 20, 1917; s. T.Y. and J.J. (Wan) W.; B.S., U. Shantung, China, 1937; M.S., Ore. State U., 1947, Ph.D., 1950; m. Louise Hagman, June 30, 1957. Came to U.S., 1946, naturalized, 1957. Faculty Ore. State U., Corvallis, 1950—, prof. chemistry, 1958—, dir. radiation center, 1964—, dir. Inst. Nuclear Sci. and Engring., 1964—. Mem. adv. com. on radiation health Ore. State Bd. Health, 1962—; cons. NSF, 1965—. Recipient Ore. Mus. Sci. and Industry Gov.'s award, 1966, Sigma Xi Research award, 1967. Mem. Am. Nuclear Soc., Am. Chem. Soc., Am. Soc. Plant Physiology, Am. Soc. Biol. Chemists, Sigma Xi, Phi Kappa Psi, Phi Lambda Upsilon, Sigma Pi Sigma. Author: Radiotracer Methodology in Biological Science, 1965; also numerous articles, chpts. in books. Devel. radiotracer methodology in biochem. research especially in radiorespirometric method for studying the metabolism of various biol. systems. Home: 3110 Chintimini St., Corvallis, Ore. 97330.*

WANG, Ging Hsi, neurophysiologist; b. Tsinan, Shantung, China, July 7, 1897; s. Mao Sheng and Cheng-Chi (Shen) W.; B.A., Nat. Peiping (China) U., 1919; Ph.D., Johns Hopkins, 1923; m. Chao-Ching Ho, Aug. 20, 1933. Asst. Johns Hopkins, 1923-24; prof. psychology Chunchow U., Kaifeng, China, 1924-25; instr. Johns Hopkins, 1925-27, lectr. otology, 1954-57; prof. Nat. Sunyatsen U., Canton, China, 1927-31 prof. psychology Nat. Peking U., Peiping, China, 1931-34, head dept. zoology, 1946-47; dir. Inst. Psychology, Acedmia Sinica, Nanking, China, 1934-49; guest U.S. State Dept., 1944-45; head div. internat. sci. cooperation UNESCO, Paris, France, 1947-53; research asso.. Lab. Neurophysiology U. Wis., Madison, 1957—. Fellow A.A.A.S.; mem. Chinese Physiol. Soc., Academia Sinica, Am. Physiol. Soc., N.Y. Acad. Scis., Wis. Acad. Scis. Author: The Neural Control of Sweating, 1964; also numerous articles. Research on four day activity cycles in female rats and its relation to ovarian secretion, differentiation striatal from thalamic animals after removal cerebral cortex, effects removal cerebellum in normal and striatal cats, cortical center for constriction pupils in cats, sweat centers especially inhibitory centers in cat's central nervous system. Home: 333 W. Washington Av., Madison, Wis. 53703.*

WANG, Harry (Hsi), embryologist; b. Kashan, Chekiang, China, Apr. 23, 1907; s. Sin Tsai and Shiu-mei (Tai) W.; B.S., U. Shanghai, 1930, M.S., 1931; Ph.D., U. Chgo, 1943; m. Lily Sze Chao, Aug. 24, 1932; children—Betty (Mrs. Herbert Izuno), Frank. Came to U.S., 1935, naturalized, 1954. Instr. U. Shanghai, 1931-33; asst. prof. Hangchow Coll., 1933-35; asst. prof. U. Shanghai, 1936-38; research asst. U. Chgo., 1938-45; asso. prof. Berea Coll., 1945-46; instr. U. Wis., 1946-47; asst. prof. U. Chgo. Coll., 1947-49, chief div. histology Am. Meat Inst. Found., U. Chgo., 1949-57; asso. prof. anatomy Loyola U. Stritch Sch. Medicine, Chgo., 1957-59, prof. anatomy, 1959—; vis. prof. Nat. Taiwan U., China, 1961. Mem. Am. Soc. Zoologists, Am. Assn. Anatomists, Soc. Exptl. Biology and Medicine, A.A.A.S. Contbr. numerous articles in field to sci. jours. Research in developmental biology; differentiation of neurofibrils; mechanism of feather regeneration; function of ovarian germinal epithelium; correlation between structure and organoleptic quality of meat; proteolysis in unicellular animals; analyses of cigarette toxicities using Paramecium as test organism. Home: 231 S. Clinton Av., Oak Park, Ill. 60302. Office: P.O. Box 1336, Hines, Ill. 60141.*

WANG, Hsien Chung, mathematician; b. Peiping, China, Apr. 18, 1919; s. Fu-Kuen and Luiyuan (Tao) W.; B.Sc., Southwest Associated U., China, 1941; Ph.D., Manchester U., Eng., 1948; m. Lung-shien Kuan, Mar. 20, 1956; children—Angela, Louise, Clara. Came to U.S., 1949, naturalized, 1958. Lectr. La. State U., 1949-51; mem. Inst. for Advanced Study, Princeton, N.J., 1951-52; research prof. Ala. Poly. Inst., 1952-54; mem. Inst. for Advanced Study, Princeton, N.J., 1954-55; vis. lectr. U. Wash., 1955-56; asso. prof. Columbia U., 1956-57; asso. prof. Northwestern U., 1957-59, prof., 1959-66; prof. Cornell U., Ithaca, N.Y., 1966—. Guggenheim fellow, 1962-63. Mem. London, Am. math. socs. Research and publs. on theory of Lie groups, differential geometry, totally discontinuous groups.*

WANG, Jui Hsin, chemist, biophysicist; b. Peiping, China, Mar. 16, 1921; s. Lieh and Sun Li (Sun) W.; B.Sc., Nat. Southwest Asso. U., Kumming, China, 1945; Ph.D. (Univ. fellow 1948-49), Washington U., St. Louis, 1949; M.A. (hon.), Yale, 1960; m. Yen Chan Yang, Apr. 2, 1949; children—Jane, Nancy. Came to U. S., 1946, naturalized, 1958. Postdoctoral fellow radiochemistry Washington U., 1949-51; mem. faculty Yale, 1951—, prof. chemistry, 1960-62, Eugene Higgins prof. chemistry, 1962-65, chemistry and molecular biophysics, 1965—; spl. research molecular structure and biochem. activity. Guggenheim fellow Cambridge (Eng.) U., 1960-61. Mem. Am. Chem. Soc., Chemical Society London, A.A.A.S., Yale Chemists Association, American Society Biological Chemists, Biophysical Society, Academia Sinica, Sigma Xi. Contbr. articles profl. jours., chpts. books. Elucidation of relationship between molecular structure and biol. function, especially for hemoglobin, cytochrome oxidase, urease, chymotrypsin, catalase, carbonic anhydrase, oxidative phosphorylation and photophosphorylation; diffusion in liquids; reaction mechanisms. Home: 641 Ridge Rd., Hamden 17, Conn. Office: Dept. Chemistry, Yale Univ., New Haven.*

WANG, Shih Chun, physiologist; b. Tientsin, China, Jan. 10, 1910; s. Yen Sun and Hsi (Han) W.; B.S., Yenching U., 1931; M.D., Peking Union Med. Coll., 1935; Ph.D., Northwestern U., 1940; m. Mamie Kwoh, Jan. 10, 1939; children—Phyllis M., Nancy E. Came to U.S., 1937, naturalized, 1949. Practice medicine, specializing in physiology and pharmacology, N.Y.C., 1940—; faculty Coll. Phys. and Surgs. Columbia, N.Y.C., 1941—, prof. physiology, 1954-56, prof. pharmacology, 1956—; vis. prof. China Med. Bd. Rockefeller Found., Taiwan, 1958. John Guggenheim fellow, 1951, Commonwealth fellow U. Goteborg, Sweden, 1966. Mem. Am. Physiol. Soc., Am. Pharmacol. Soc., N.Y. Acad. Scis., N.Y. Acad. Medicine, Assn. for Research on Nervous and Mental Diseases, Soc. Exptl. Biology and Medicine, Harvey Soc., A.A.A.S., N.Y. State Soc. Med. Research, Sigma Xi. Contbr. chpts. to Handbook of Physiology,

1965, Physiological Pharmacology, 1965. Research and numerous publs. primarily in field of physiology and pharmacology of automatic nervous system. Home: 18 Kent Rd., Tenafly, N.J. 07670. Office: 630 W. 168th St., N.Y.C. 10032.*

WANG, Shih Yi, chemist; b. Peking, China, June 15, 1923; s. Yen Tso and Li Hwa (Kung) W.; B.S., Nat. Peking U., 1944; Ph.D., U. Wash., 1952; m. Chun Lien Chi, Aug. 17, 1947; children—Robert C., Ethylin T. Research fellow Harvard, 1954-55; research asso. Tufts Med. Sch., 1955-56, asst. prof., 1956-61; faculty Johns Hopkins U., Balt., 1961—, prof. biochemistry, 1966—; NATO guest prof., Cologne, Germany, summer 1962. Mem. Am. Chem. Soc., Chem. Soc. London. Studies, publs. on interaction of deoxyribonucleic acid with ultra-violet radiation and chems.; proposed molecular aggregation-puddle formation hypothesis for frozen solutions and insight into understanding of photochemistry of nucleic acids. Home: 14 Windemere Pkwy., Phoenix, Md. 21131. Office: 615 N. Wolfe St., Balt. 21205.*

WANG, Yang, physician; b. Tangshan, China, May 12, 1923; s. Yü-Shen and Chun-jung (Po) W.; student Yenching U., Peiping China, 1940-41; M.B., Med. Coll. Shanghai, 1948; M.D. cum laude, Harvard, 1952; m. Helen Huang, June 18, 1966. Came to U.S., 1949, naturalized, 1954. Fellow physiology Mayo Found., 1958-59; asso. prof. medicine, dir. cardiac catherization lab. U. Minn., Mpls., 1959—, asst. dir. cardiovascular clin. research program project Gt. Plains Research Council of Am. Heart Assn., 1966—. Mem. Am. Fedn. for Clin. Research, Am. Heart Assn., A.M.A., Central Soc. for Clin. Research, A.C.P., Am. Coll. Sports Medicine, Am. Coll. Chest Physicians, Assn. U. Cardiologists, Soc. for Exptl. Biology and Medicine, A.A.A.S. Research and publs. on cardiology and cardiovascular physiology. Home: 18 Red Fox Rd., N. Oaks, St. Paul 55110. Office: U. Minn. Hosps., Mpls.*

WANGENSTEEN, Owen Harding, Am. surgeon; b. Lake Park, Minn., Sept. 21, 1898; s. Owen and Hannah (Hanson) W.; A.B., U. of Minn., 1919, M.D., 1922, Ph.D. (in surgery), 1925, LL.D., University of Buffalo, 1946; D.Sc. U. Chgo., 1956, St. Olaf Coll., 1958, Temple U., 1961, Hamline U., 1963; Docteur honoris causa, U. Paris (Sorbonne), 1962; married Helen Griffin, January 6, 1923; children—Mary Helen (Brink), Owen Griffin, Stephen Lightner; married second, Sarah Anne Davidson, Sept. 16, 1954. Fellow in medicine, U. of Minnesota, 1923; fellow in surgery, Mayo Clinic, 1924; resident surgeon, University Hosp., 1925; instr. in surgery, U. Minn., 1926, asst. prof., 1927, asso. prof., 1928, prof., 1931-67, prof. emeritus, 1967—, Regents prof., 1966, director department, 1930-67; asst. in Prof. F. de Quervain's Surgical Clinic, and Physiol. Inst. Prof. Leon Asher, Berne, Switzerland, 1927-28. Dir. Am. Cancer Soc.; mem. heart council USPHS, 1953-57, mem. council research faculty div., 1957-60. Served as pvt. S.A.T.C., World War I. Recipient Passano Award, 1961. Fellow Royal College Surgeons England (honorary), Hellenic Surgical Society Athens (hon.), A.C.S. (president 1959-60); mem. A.M.A. Minn. Academy Medicine, Minnesota, Minneapolis surgical societies, Society Experimental Biology and Medicine, Society Experimental Pathology, American Assn. Thoracic Surgery (pres. 1968), A.A.A.S., Minn. Pathol. Soc., Norweigian Acad. Sci. and Letters, Halsted Surg. Soc. (pres. 1957-59), French Acad. Surgery, Norwegian Acad. Sci. (hon.), Nat. Acad. Sci., Royal Coll. Surgeons of Edinburgh (Scotland) (hon.), German Surg. Congress, Nu Sigma Nu, Alpha Omega Alpha, Sigma Xi. Received John Scott award and medal in 1941; Alvarenza prize, Philadelphia College of Medicine, 1949; American Cancer Society (Minnesota division) award 1949; Spl. citation, Am. Cancer Soc., 1962. Received Samuel D. Gross prize award Phila, Acad. Surgery, 1935. Author: The Therapeutic Problem in Bowel Obstructions, 1937 (3d edit. 1955); Cancer, Esophagus and Stomach, 1956. Contbr. to med. jours. Co-editor of Surgery. Research on bowel obstruction, appendicitis, genesis of acid-peptic ulcer and its surgical management, etiology of gall stones; developed suction siphonage treatment of acute intestinal obstruction. Home: 2832 W. River Rd., Mpls. 55406. Office: Diehl Hall, University Med. Center, Mpls. 55455.*

WANGSNESS, Roald K., Am. physicist; b. Sleepy Eye, Minn., July 24, 1922; s. Gudbrand K. and Katherine (Fromm) W.; student St. Olaf Coll., 1940-41; B.A. summa cum laude, U. Minn., 1944; Ph.D. (AEC predoctoral fellow) Stanford, 1950; m. Cleo M. Abbott, July 29, 1944; children—Peter A.A., Steven J. Asst. prof. U. Md., College Park, 1950-51, prof., 1953-59; physicist U. S. Naval Ordnance Lab., Silver Spring, Md., 1950-59; prof. U. Ariz., Tucson, 1959—. Recipient, Distinguished Civilian Service award, USN, 1957. Fellow Am. Phys. Soc.; mem. Am. Assn. Physics Tchrs., A.A.A.S., Phi Beta Kappa, Sigma Xi. Author: Introduction to Theoretical Physics, 1963; Introductory Topics in Theoretical Physics, 1963. Research in nuclear induction, ferrimagnetism and statis. mechanics. Home: 5035 E. Scarlett St., Tucson 85711.*

WANLESS, Harold Rollin, Am. geologist; b. Chgo., Dec. 5, 1898; s. William Frank and Rhoda (Tanner) W.; B.Sc., Princeton, 1920, M.A., 1921, Ph.D., 1923; m. Grace Rogers, Aug. 7, 1926; 1 son, Harold Rogers. With dept geology U. Ill., Urbana, 1923-67, emeritus, 1967—, geologist U. S. Geol. Survey, 1954—. Sr. Fulbright fellow, Sydney, Australia, 1958-59; Geol. Soc. Am. fellow, 1935-46; NSF fellow, 1962-66. Mem. Ill. Acad. Sci. (pres. 1938), Geol. Soc. Am. (chmn. coal research div. 1967), Am. Assn. Petroleum Geologists, Am. Geophys. Union, A.A.A.S., Am. Soc. Photogrammetry, Soc. Econ. Paleontology and Mineralogy, Paleontol. Soc., Sigma Xi (pres. Ill. chpt. 1963-64). Author: Aerial Stereo Photographs, 1965. Research, numerous publs. on regional analysis of Pa. rocks of U. S., cyclic sedimentation, environmental mapping of ancient rocks, late Paleozoic glaciation, coastal changes as interpreted from aerial photographs, photogeology particularly developing sets of stereograms for use in teaching geology. Home: 704 S. McCullough St., Urbana, Ill. 61801.*

WANNAGAT, Ulrich, German chemist; b. Koenigsberg, May 31, 1923; s. Richard and Helene (Riemann) W.; diploma in chemistry U. Frankfurt, 1948, Dr.rer. nat., 1949; m. Inge Carduck, May 30, 1952; children—Antje, Elke, Ute, Gernot. Faculty, Tech. U., Aachen, Germany, 1952-61; prof., dir. Inst. Chemistry, Tech. U. Graz, Austria, 1961-66; prof., dir. Inst. Chemistry, Tech. U. Braunschweig, Germany, 1966—. Recipient F. S. Kipping award Am. chem. Soc., 1968. Author: (with others) Fischers Chemie Lexikon, 1967; also numerous articles. Research on devel. of silicon nitrogen chemistry, silylsubstituted metal amides, hydrazines; chemistry in electric discharges; preparation of polysulfurperoxides. Home: 1 Roentgenweg, D-334 Wolfenbuettel, Germany. Office: 4 Pockels strasse, D-33 Braunschweig, Germany.*

WANNAMAKER, Lewis William, Am. physician; b. St. Matthews, S.C., May 19, 1923; s. Whitfield Wesley and Lucille (Long) W.; student Emory U., 1940-43; M.D., Duke, 1946; m. Hallie Etheridge, Sept. 11, 1948; children—Julie Lucille, Hallie Ann, Susan Elizabeth, Whitfield Claude. Faculty U. Minn., Mpls., 1952—, prof., mem. grad. faculty dept. pediatrics, 1958—, dept. microbiology, 1965—; mem. staff Mpls. Gen. Hosp. Cons. to govt. agys.; mem. numerous panels NIH, USPHS.; career investigator Am. Heart Assn., 1958—. Helen Hay Whitney Found. fellow, NSF fellow, 1955-57, Guggenheim Meml. fellow 1966—. Diplomate Am. Bd. Pediatrics, Nat. Bd. Med. Examiners. Mem. A.A.A.S., Am. Assn. Immunologists, Am. Fedn. for Clin. Research, Am. Heart Assn., Am. Soc. for Clin. Investigation, Central Soc. for Clin. Research (councilor 1963-65), Infectious Disease Soc. Am., Midwest Soc. for Pediatric Research (pres. 1960-61), Minn. Heart Assn. (dir. 1962-64), N.W. Pediatric Soc., Soc. Am. Bacteriologists, Soc. for Exptl. Biology and Medicine (bd. editors 1959-62), Soc. for Pediatric Research (mem. council 1964—), Beta Kappa, Sigma Xi. Mem. editorial bd. Jour. Bacteriology, 1963—. Research in infectious diseases, particularly streptococcal infections and prevention of their complications, acute rheumatic fever and acute nephritis. Home: 1484 W. Hwy. 96, Arden Hills, St. Paul 55112. Office: Dept. Pediatrics, U. Hosp., Mpls. 55455.*

WANNIER, Gregory Hugh, physicist; b. Basel, Switzerland, Dec. 30, 1911; s. Eugen O. and Clara (Stachelin) W.; Ph.D., Louvain, Cambridge, Basel, 1935; m. Carol Goodman, Sept. 21, 1939; children—Milton E., Peter G. Came to U.S., 1936, naturalized, 1943. Exchange fellow Princeton U., 1936-37; instr. U. Pitts., 1937-38; lectr. U. Bristol, 1938-39; instr. U. Tex., 1939-41; lectr. U. Ia., 1941-46; mem. staff Socony-Vacuum Labs., Paulsboro, N.J., 1946-48; mem. tech. staff Bell Tel. Labs., Murray Hill, N.J., 1949-55, 1956-60; prof. U. Geneva, Switzerland, 1955-56; prof. physics U. Ore., Eugene, 1961—. Mem. Am. Phys. Soc. Author: Elements of Solid State Theory, 1959; Statistical Physics, 1966; also numerous articles. Harmonization of the localized and non-localized description of electrons in solids; first precise results on phase transitions from statis. mechanics; three-body treatment of ionization by electron impact. Home: 2000 University St., Eugene, Ore. 97403.*

WANTZEL, Pierre Laurent, French mathematician; b. 1814; répétiteur École Polytechnique, Paris; worked on solutions to equations higher than 2d degree; gave 1st rigorous proofs of impossibility of duplicating cube or trisecting angles by ruler and compasses; investigated (with Saint-Venant) laws of air flow. Died 1848.

WAPPAUS, Johann Eduard, German statistician, geographer; b. Hamburg, Germany, May 17, 1812; prof. Göttingen, Germany; leading population statistician of his day. Author: Allgemeine Bevölkerungsstatistik, 2 vols., 1859-61; Über den Begriff und die statistische Bedeutung der mittleren Lebensdauer, 1859. Died Göttingen, Dec. 16, 1879.

WAPSTRA, Aaldert Hendrik, Dutch physicist; b. Utrecht, Netherlands, Apr. 24, 1922; s. Hidde and Hendrika (Veld) W.; M.A., U. Utrecht, 1948; D.Sc., U. Amsterdam, 1953; m. Annemarie Van der Sluis, Nov. 23, 1967; children—Marlies, Hugo. Sci. collaborator Institut voor Kernfysisch Onderzoek, Amsterdam, 1949-62, sci. dir., 1962—; prof. nuclear physics Tech. U., Delft, 1956—; formerly with Nobel Inst., Stockholm, Cal. Inst. Tech., Argonne Nat. Lab., Oak Ridge Nat. Lab., U. Naples. Sec. com. nuclear masses and related constants Internat. Union Pure and Applied Physics; mem. com. atomic masses Internat. Union Pure and Applied Chemistry. Mem. Am., Dutch phys. socs., Internat. Astron. Union. Author: (with R. van Lieshout and G. J. Nygh) Nuclear Physics Tables, also numerous articles. Research on nuclear decay schemes, influence of nuclear size on conversion coefficients, calculation of nuclear masses from exptl. data, compilation of nuclear data of heavy elements. Home: 10 Voltaplein. Office: 18 Oosterringdijk, Amsterdam, Netherlands.*

WARBURG, Otto, German botanist; b. Hamburg, Germany, July 20, 1859; prof. Oriental Seminar, Berlin, from 1897. Fellow Royal Soc., 1934. Specialized in geog. and systematic botany, also culture of tropical plants. Died 1938.

WARBURG, Otto Heinrich, German biochemist; b. Freiburg, Baden, Oct. 8, 1883; s. Emil Warburg and Elizabeth Gertner; Dr. der Chemie, Berlin, 1906; M.D., Heidelberg U., 1911. Mem. Kaiser Wilhelm Gesellschaft, 1913—, titular prof., 1916—; dir. Max Planck Inst. Zellphysiologie, Berlin-Dahlem, 1931—. Served with Prussian Horse Guards, 1914-18. Decorated Ordre pour le Merite, Great Cross Star and Shoulder Ribbon, Bundes republik; recipient Nobel prize for medicine, 1931. Fgn. mem. Royal Soc., 1934. Author: Über den Stoffwechsel in der Tumoren, 1926; Wirkungsgruppen von Fermenten, 1946; Chemie der Photosynthese; Freedom of West Berlin, 1963. Discovered nature and action of respiratory enzyme; also that cancer cells can grow without oxygen; 1st to determine that cancer cells derive energy from lactic acid fermentation and can be damaged by radiation; studied photosynthesis mechanism; identified oxygen-activiting respiratory enzyme. Address: Garystrasse 18, Berlin-Dahlem, Germany.

WARD, Alger Luman, Am. chemist; b. Easthampton, Mass., May 4, 1890; s. Oscar and Ella Jeanette (Alexander) W.; B.S., Syracuse U., 1914, M.S., 1915, Sc.D., 1941; m. Emma Undritz, June 5, 1915. Research chemist E. I. du Pont de Nemours & Co. (discovered processes for synthesis of acetic acid, acetone, alkyl anilines, ethylene glycol and glycerine), 1915-20; research chemist United Gas Improvement Co., Phila. 1920-34; mgr. chem. labs. United Gas Improvement Co., 1934-45; dir. research lab. Pa. Indsl. Chem. Corp., 1946-55. Recipient Louis E. Levy medal Franklin Inst., 1938. Mem. Franklin Inst., Am. Chem. Soc., Soc. Chem. Industry. Co-author: Styrene, 1951; Styrene Monomer, Ency. Chemical Technology, 1954. Contbr. to, hon. asso. editor The Science of Petroleum, 1938. Contbr. tech. articles to jours. Specialist in correlation of relationship between density, refractive index and structure of hydocarbons; pioneer in lab. and comml. devel. of Am. petrochem. industry, particularly prodn., separation and recovery of unsaturated hydrocarbons from pyrelysis of petroleum; research on metal corrosion, cause and prevention of formation of liquid phase and vapor phase gums in gas distbn. system, synthetic resins, plasticizers for synthetic rubber, binders for asphaltic floor tile; discoverer and exponent (with S. S. Kurtz, Jr.) refractivity intercept. Address: 108 Colwyn Lane, Bala-Cynwyd, Pa. 19004.*

WARD, Arthur Allen, neurosurgeon; b. Manipay, Ceylon, Feb. 4, 1916; s. Arthur Allen and Alice (Bookwalter) W.; B.A., Yale, 1938, M.D., 1942; m. Janet L. Miller, Dec. 20, 1941; children—Sally Lawton, Linda Allen. Instr., Yale Med. Sch., 1945; research asst. Ill Neuropsychiat. Inst., 1945-46; instr. U. Louisville, 1946-47; faculty U. Wash. Sch. Medicine, Seattle, 1948—, prof., chmn. dept. neurosurgery, 1955—. Mem. neurophysiology panel Internat. Brain Research Orgn., 1962—; cons. NIH, 1959—. Mem. Am. Electroencephalograph Soc. (past pres.), Western Neurosurg. Soc. (pres. 1967—), Harvey Cushing Soc., Am. Neurol. Assn., Am. Physiol. Soc., Soc. Neurol. Surgeons. Editorial bd. Jour. Neurosurgery, 1965—, Rev. Surgery, 1955—. Research, numerous publs. on pathophysiol. mechanisms in epilepsy, function cerebral cortex brain, physiol. mechanisms in Parkinsonism, surg. therapy Parkinsonism. Home: 3922 N.E. Belvoir Pl., Seattle 98105.*

WARD, Darrell Neilsen, Am. biochemist; b. Logan, Utah, Jan. 22, 1924; s. Huburt C. and Phyllis (Nielson) W.; B.S., Utah State U., 1949; M.S., Stanford U., 1951, Ph.D., 1953; m. Afton Hall, Mar. 25, 1946; children—Kathleen, Pamela, Becky, Janeen, Alan Darrell, Melissa, Gregory Hall. Research asso. Cornell U. Med. Coll., 1952-54, instr., 1954-55; faculty and staff U. Tex. M.D. Anderson Inst., Houston, 1955—, head dept. biochemistry, 1963—, prof. biochemistry, 1961—, mem. com. on grad. studies Grad. Sch. Biomed. Scis., 1963—. Mem. Am. Soc. Biol. Chemistry, Am. Chem. Soc., Am. Assn. Cancer Research, A.A.A.S. Research, numerous publs. on posterior pituitary hormones; azo dye carcinogenesis; carcinogenesis by synthetic estrogens; chemistry and structure of luteinizing hormone of anterior pituitary. Home: 4617 Tonawanda St., Houston 77035.*

WARD, Harry Marshall, English botanist; b. Hereford, Eng., 1854; s. Francis Marshall Ward; attended Huxley's lectures, 1874-75; student Owens Coll., Manchester, 1875-76; B.A., Christ's Coll., Cambridge, 1879, M.A., Cambridge U., 1885, Sc.D., 1892; D.Sc., Victoria, 1902; m. Linda Kingdon, 1883; 1 son, 1 dau. Apptd. by Colonial office to study coffee-leaf disease in Ceylon, 1880-82; fellow Christ's Coll., Manchester, 1883-85; prof. botany Royal Indian Engring Coll., Coopers Hill., 1885-95; prof. botany Cambridge U., from 1895, also fellow Sidney Sussex Coll. Fellow Royal Soc., 1888, recipient Royal medal, 1893; fellow Linnean Soc. Author: Lectures on the Physiology of Plants, 1887; Timber and Some of its Diseases, 1889; Diseases of Plants, 1889; The Oak: a Popular Introduction to Forest-Botany, 1892; Grasses, 1901; Trees, 1904-05; also articles. Research on root tubercles of bean, also sources of nitrogen, 1887-88, symbiosis, 1892, bacteriology of water, 1892-00; also studied ferment action in color matter of Persian berries and in piercing of cell walls by fungal hyphae. Died Babbacombe, Torquay, Aug. 26, 1906.

WARD, Henry Augustus, Am. naturalist; b. Rochester, N.Y., Mar. 9, 1834; s. Henry Meigs and Eliza (Chapin) W.; A.M., Williams Coll.; LL.D., Rochester U.; to Louis Agassiz, Harvard Sci. Sch., 1854; studied at Paris, traveled through Europe and Orient, 1855-59; m. Phoebe A. Howell, Nov. 1860 (dec. 1880); m. 2d, Mrs. Lydia Avery Coonley, Mar. 18, 1897. Prof. natural scis. Rochester U., 1860-65; mgr., gold mines in Mont. and S.C., 1866-69; traveled in all countries making large and valuable cabinets of mineralogy and geology (Ward Cabinets), compiled at Rochester Ward's Natural Sci. Establishment, to be distributed to univs., colls. throughout U. S., 1870-1900; acting naturalist of U. S. Expdn. to Santo Domingo, 1871. Fellow A.A.A.S. Author: Notices of the Megatherium Cuvieri; Description of the Most Celebrated Fossil Animals in Royal Museums of Europe. Died 1906.

WARD, James, English psychologist; b. Hull, Eng., Jan. 27, 1843; s. James and Hannah (Aston) W.; B.A., Springhill Coll., Eng, 1869; theol. degree; student Liverpool (Eng.) Inst., U. Berlin, U. Göttingen (Germany); student Trinity Coll., Cambridge, 1873-74; LL.D., Edinburgh, Scotland, 1889, Cambridge, 1920; m. Mary Martin, 1884; 1 son, 2 daus. Became fellow Trinity Coll., Cambridge, 1875; prof. mental philosophy, Cambridge, 1897-1925; vis. lectr. Aberdeen U., 1895-98, St. Andrews, Scotland, 1907-10. Author: Naturalism and Agnosticism, 1899; Heredity and Memory, 1913; Psychological Principles, 1918; A Study in Kant, 1922; Essays in Philosophy (edited by Surley and Stout), 1927; also numerous articles. Anticipated emphasis of psychoanalysis on unconscious activities and opposed associationistic ideas; stressed importance of physiol. factors in psychology. Died Cambridge, Mar. 4, 1925.

WARD, James Wellington, Am. anatomist, parasitologist; b. Weathersby, Miss., Oct. 27, 1904; s. William Robert and Elizabeth (Stringer) W.; B.A., U. Ala., 1929, M.A., 1932; M.S., Miss. State U. Agr. and Applied Sci., 1931; B.S., U. Minn., 1945; Ph.D. in Basic Med. Scis., U. Miss., 1950; m. Mildred Daphne Brumfield, June 5, 1934; children—Mildred Daphne, Jane (Mrs. Cornelius B. Holcombe, Jr.). Faculty, Miss. State Coll., 1929-32; instr. zoology, U. Okla., 1932-33, ast. prof., 1933-36; asso. prof. U. Minn., 1938-44; parasitologist Miss. Expt. Sta., 1944-47; faculty Sch. Medicine, U. Miss., 1947—— prof. histology, 1953-62, acting chmn. dept. preventive medicine, 1953-55, acting dir. admissions, chmn. admissions and promotions, 1954——, prof. anatomy in charge neuroanatomy, 1962——; vis. prof. biology Belhaven Coll., 1957——; vis. prof. marine biology Gulf Coast Research Lab., 1947——. Cons. entomologist Redd Pest Control, 1955——, Miss. Game and Fish Commn., 1940——. Rockefeller Commonwealth scholar Tufts Grad. Med. Sch., 1943. Mem. Am. Assn. Anatomists, So. Assn. Anatomists, Am. Soc. Parasitologists, Soc. Exptl. Zoologists, Miss. Acad. Scis., Sigma Xi. Research and publs. on devel. corticospinal tract of oppossum, neuroembryology of marine catfish (Caleichthys felis), anaomalies in fish, neuromas in fish, occurence of Echinococcus granulosus in So. U. S., hemopoiesis in marine fish. Home: 2075 London Av., Jackson, Miss. 39211.*

WARD, Lester Frank, Am. geologist, sociologist; b. Joliet, Ill., June 18, 1841; s. Justus and Silence (Rolfe) W; A.B., Columbian (now George Washington) U., 1869, LL.B., 1871, A.M., 1873, LL.D., 1897; m. 1st Elizabeth Vought, Aug. 13, 1862; m. 2d Rosamond Simons, Mar. 6, 1873. Served in Civil War; in U. S. Treasury Dept., 1865-72; asst. geologist, 1881-88, geologist, 1888. U. S. Geol. Survey; prof. Brown U., 1900-13. Pres., Institute Int'nl. de Sociologie, 1900-03; pres., Am. Sociol. Soc., 1906-07, Mem. Am. Acad. Polit. and Social Science; fellow A.A.A.S. Author: Guide to the Flora of Washington and Vicinity, 1881; Dynamic Sociology, 1883; Sketch of Paleobotany, 1885; Synopsis of the Flora of the Laramie Group, 1886; Types of the Laramie Flora, 1887; Geographical Distribution of Fossil Plants, 1888; Psychic Factors of Civilization, 1893; Psychological Basis of Social Economics; Political Ethics of Spencer; Principles of Sociology; Outlines of Sociology, 1898; Sociology and Economics, 1899; Pure Sociology, 1903; Text-book of Sociology (with James Quayle Dealey), 1905; Glimpses of the Cosmos, 6 vol., 1913-18. Research in paleobotany and pioneer in Am. sociology; developed concept of telesis (conscious effort of society for progress); stressed education in aiding direction of social evolution. Died Washington, April 18, 1913.

WARD, Louis Emmerson, Am. physician; b. Mt. Vernon, Ill., Jan. 19, 1918; s. Henry Ben Pope and Aline (Emmerson) W.; A.B., U. Ill., 1939; M.D., Harvard, 1943; M.S. in Medicine (fellow), Mayo Grad. Sch. Medicine, U. Minn., 1949; m. Nan Talbot, June 5, 1942; children—Nancy, Louis, Robert, Mark. Cons. in medicine Mayo Clinic, Rochester, Minn., 1950——; prof. clin. medicine Mayo Grad. Sch. Medicine 1965——. Bd. govs. Mayo Clinic 1962——, chmn., 1964——; bd. trustees Mayo Found.; bd. dirs. Arthritis Found. Mem. A.M.A., A.C.P., Central Soc. for Clin. Research, Sigma Xi, Alpha Omega Alpha. Research, publs. on pathophysiology and treatment of rheumatic diseases. Home: 1110 10th St. S.W., Rochester 55901. Office: 200 1st St. S.W., Rochester, Minn. 55902.*

WARD, Nathaniel Bagshaw, Brit. botanist; b. London, 1791; s. Stephen Smith Ward; student London Hosp.; student at bot. demonstrations of Thomas Wheeler. Visited Jamaica, 1804; practice medicine, London; examiner botany Soc. of Apothecaries, 1836-54. Fellow Royal Soc., 1852, Linnean Soc.; mem. Bot. Soc. Edinburgh (charter mem., became sec. 1836), Soc. Apothecaries (became master 1854, later treas.), Micros. Soc. (now Royal Micros. Soc.) (co-founder 1839). Author: (pamphlet) Growth of Plants without open Exposure to Air, 1836. Inventor, Wardian case for transport of plants, 1829. Died June 4, 1868.

WARD, Robert De Courcy, Am. climatologist; b. Boston, Nov. 29, 1867; s. Henry Veazey and Anna Saltonstall (Merrill) W.; A.B., Harvard, 1889, A.M., 1893; m. Emma Lane, Apr. 28, 1897; children—Henry DeCourcy, Robert Saltonstall, Anna Saltonstall Magruder, Emma Lane. Asst. in physical geography, 1890-94, asst. in meteorology, 1894-95, instr., 1895-96, instr. climatology, 1896-1900. asst. prof., 1900-10. prof., 1910——. Harvard; exchange prof., Western colleges, 1927. Editor Am. Meteorol. Jour., 1892-96, contbg. editor Geog. Review. Member Shaler Memorial Expdn. to Brazil, 1908. Fellow Am. Acad. Arts and Sciences, Royal Meteorol. Soc., London, Assn. Am. Geographers (pres. 1917), Am. Meteorol. Soc. (pres. 1920, 21), Am. Assn. Advancement of Science; mem. Harvard Travellers Club (gold medalist, 1926); one of founders, 1894, Immigration Restriction League. Author: Practical Exercises in Elementary Meteorology, 1899; Climate Considered Especially in Relation to Man, 1908; The Climates of the United States, 1925. Translator: Hann's Handbuch der Klimatologie, Vol. 1, 1903. Contbr. to scientific jours. First professor of climatology in U. S.; formulated theories of relationships between climate and human life. Died Nov. 12, 1931.

WARD, Roland, chemist; b. Bishopton, Scotland, Aug. 22, 1902; s. Samuel and Jean (Barclay) W.; B.Sc., Paisley Tech. Coll., London, Eng., 1924; Ph.D., Western Res. U., 1931; m. Irene Belle Thurston, Apr. 1, 1933; 1 son, Barclay. Chemist Willard Storage Battery Co., Cleve., 1927-29; Jeavons fellow Western Res. U., 1929-31; instr. U. Ill., 1931-32; instr. Bklyn. Poly. Inst., 1932-36, asst. prof., 1936-43; prin. investigator OSRD, Bklyn., 1943-46; asso. prof. Bklyn. Poly. Inst., 1946-47, prof., 1947-49; prof. U. Conn., Storrs, 1949-67, research prof., 1967——. Mem. Am. Chem. Soc., Chem. Soc. of London, Faraday Soc., Electrochem. Soc., Sigma Xi, Phi Lambda Upsilon. Research, numerous publs. on solid state inorganic chemistry; preparation and properties of infrared-sensitive phosphors; transition metal ternary oxides; structure and magnetic properties. Home: Spring Hill, Storrs, Conn. 06268.*

WARD, Ronald Anthony, Am. entomologist; b. N.Y.C., Jan. 25, 1929; s. Milton M. and Marguerite (Foss) W.; B.Sc., Cornell U., 1950; Ph.D. (Chgo. Natural History Mus. fellow), U. Chgo., 1955; m. Harriet Arthur, July 24, 1950; children—Melinda Anne, Katherine B. Faculty, Gonzaga U., 1955-58; research entomologist U.S. Army Med. Research and Devel. Command, Walter Reed Army Inst. Research, Washington, 1958——, asst. chief dept. entomology, 1964——. U.S. Sec. Army Research fellow London Sch. Hygiene and Tropical Medicine, 1967. Mem. A.A.A.S., Am. Mosquito Control Assn., Am. Soc. Tropical Medicine and Hygiene Royal Soc. Tropical Medicine, Research, publs. on med. entomology with emphasis on transmission of malaria by mosquitoes. Home: 11220 Markwood Dr., Silver Spring, Md. 20902. Office: Dept. of Entomology, Walter Reed Army Institute of Research, Washington 20012.*

WARD, Seth, Brit. astronomer; b. Buntingford, Eng., 1617; s. John and Martha (Dalton) W.; B.A., Sidney-Sussex College, Cambridge, 1637, M.A., 1640, Fellow, 1640; M.A., Oxford U., 1649, D.D., 1654. Became math. lectr., Cambridge U., 1643; Savilian prof. astronomy, Oxford U., 1649-57; prof. Jesus College, 1657-59; pres. Trinity College, 1659-60; rector St. Laurence Jewry, 1660-61; St. Breock in Cornwall, 1662; consecrated bishop Ch. of Eng., 1662; fellow Sidney-Sussex Coll. Original mem., Royal Soc.; mem. Philos. Soc., Oxford; Author: Clavis mathematicae, rev. edit. with Sir Charles Scarburgh, 3d edit., 1652; In Ismaelis Bullialdi astronomiae philolaicae fundamenta inquisitio brevis, 1653; Astronomia geometrica, 1655; others. Claimed planets travel in elliptical orbits in such a way that a planet moves in uniform angular motion about empty focus of ellipse; engaged in philos. controversy with Thomas Hobbes; refuted John Webster's plan for the reform of the universities; opposed Solemn League and Covenant. Died Knightsbridge, Eng., Jan. 6, 1689.

WARD, Thomas Greydon, Am. physician; b. Athens, La., Sept. 25, 1911; s. Joe Miller and Alice (Finley) W.; M.D., Baylor U., 1935; Ph.D., Johns Hopkins, 1940, D.P.H., 1941; m. Opal Queen Doss, Sept. 11, 1934; children—Greydon Doss, Andra Moneen (Mrs. John Henry Green), Clayron. Gen. practice medicine, Coushatta, La., 1935-37; health officer, St. Joseph, La., 1937-39; asso. prof. microbiology Johns Hopkins, Balt., 1945-56; prof. virology U. Notre Dame (Ind.), 1956-60; sci. dir. Microbiol. Assos., Inc., Bethesda, Md., 1960——. Mem. Am. Epidemiological Soc., Am. Immunology Soc., Soc. for Exptl. Biology and Medicine. Research, numerous publs. on virology, viruses causing common cold, cancer viruses; co-discoverer of adenoviruses. Home: 6116 Robinwood St., Bethesda 20034. Office: 4813 Bethesda Av., Bethesda, Md. 20014.*

WARDER, Robert Bowne, Am. chemist; b. Cin., Mar. 28, 1848; s. John A. and Elizabeth Bowne W.; grad. Earlham Coll., 1866. A.M., 1873; grad. Lawrence Sci. Sch., Harvard, B.S. (in chemistry), 1874; postgrad. in Germany, 1874-75; m. Gulieima M. Dorland, Mar. 25, 1884. Prof. U. Cin., 1875-79, Haverford Coll., 1879-80; prof. Purdue U., Ind., also state chemist, 1883-87; prof. physics and chemistry Harvard, 1887——. Fellow A.A.A.S. (v.p. for chemistry, 1890). Contbr. papers on phys. chemistry. Died 1905.

WARDLAW, Claude Wilson, Brit. botanist; b. Glasgow, Scotland, Feb. 4, 1901; s. Johnston and Mary (Hood) W.; B.Sc., Glasgow U., 1921, Ph.D., 1925, D.Sc., 1928; M.Sc., Manchester U., 1943; D.Sc. (honorary), McGill University, 1959; married Jessie Connell, Dec. 1928; children—Alastair, Norman. Asst. lectr., lectr. botany dept. Glasgow U., 1921-28; pathologist, head low temperature research sta. Imperial Coll. Tropical Agr., Trinidad, B.W.I., 1928-40; prof. crytogramic botany Manchester U., Eng., 1940-58, prof. botany, 1958-66, now emeritus; Prather lectr. biology Harvard Univ., 1951. Served to lt. col., Territorial Army, 1940-52. Fellow Royal Soc. Edinburgh. Linnean Society (London); honorary foreign member of Agricultural Academy France, Am. Acad. Arts and Scis., Royal Acad. Belgium (hon. fgn. asso.), Am. Bot. Soc. (hon. fgn. corr.). Author: Diseases of the Banana, 1935; Green Havoc, 1935; Phylogeny and Morphogenesis, 1952; Morphogenesis in Plants, 1952; Embryogenesis in Plants, 1955; Banana Diseases, 1961; Organization and Evolution in Plants, 1965; Morphogenesis in Plants, 1967. Author sci. papers. Research in morphogenesis, embryogenesis, phylogeny, and diseases of plants. Home: 6 Robins Close, Bramhall, Cheshire, Eng.

WARDROP, James, Brit. surgeon; b. Torbane Hill, Scotland, Aug. 14, 1782; s. James and Marjory (Marjoribanks) W.; apprenticed to surgeon Andrew Wardrop, 1797; studied in London hosps., 1801-03; M.D. (hon.) St. Andrews U., 1934; m. Margaret Dalrymple, 1813; 4 sons, 1 dau. Asst. to John Barclay, anatomist, circa 1797 or 98; house surgeon Royal Infirmary, 1801; apptd. surgeon extraordinary to prince regent, 1818, surgeon in ordinary to king, 1828; founder (with William Willocks Sleigh) W. London Hosp. of Surgery, 1826. Fellow colls. surgeons Edinburgh, Eng., Royal Coll. Surgeons, London. Author: Essays on the Morbid Anatomy of the Human Eye, 2 vols., 1808-18; Observations on Fungus Haematodes, 1809; On Aneurysm and its Cure by a New Operation, 1828. First surgeon in Eng. to remove tumor of lower jaw by total vertical sect. of bone; credited with introducing term keratitis, circa 1808; 1st to use distal ligation in treating aneurysm of carotid artery, circa 1827. Died London, Feb. 13, 1869.

WARE, Arnold G., Am. chemist; b. Butler, Ill., June 1, 1915; s. Frank Scott and Nina (Grassel) W.; B.A., Carthage Coll., 1937; M.S., U. Colo., 1939, Ph.D., 1942; m. Freda Catherine Cowerthwaite, Mar. 7, 1941; children—Stephen Arnold, Randolph Howard, Mary Ann. Research asso. Wayne U., 1946-49; head clin. biochemist Los Angeles County Hosp., 1949——; faculty U. So. Cal., Los Angeles, 1949——, prof. biochemistry, nutrition 1956——; cons. San Gabriel (Cal.) Community Hosp.; mem. editorial adv. bd. Clin. Chemistry. Contbr. chpts. Clinical Anticoagulant Therapy, 1965; also numerous articles. Research in identification, isolation and measurement of blood clotting factors; research, devel. in gen. field clin. chemistry. Home: 1721 Bushnell St., South Pasadena, Cal. 91030. Office: 1200 N. State St., Los Angeles 90033.*

WARES, Gordon Webb, astrophysicist; b. Tynecastle, Sask., Can., Feb. 10, 1911; s. Arthur Wilbur and Cora Louise (Webb) W.; B.S. U. Wash., 1933; postgrad. U. Cal., 1933-36; S.M., U. Chgo., 1937, Ph.D., 1940; m. Mabel Zilla Martin, Sept. 8, 1936; children—John Paul, Elizabeth (Mrs. Roger Dowd), Virginia (Mrs. Andrew R. Martin), Vincent. Fellow, U. Chgo., Yerkes Obs., 1936-38, asst. 1938-49; prof. physics Brenau Coll., 1939-41; instr. Wis. State Tchrs. Coll., Milw. 1941-42; physicist, head computing sect. U. Cal. Radiation Lab., 1944-45; physicist Naval Ordnance Test Sta., China Lake, Cal., 1945-46, research physicist, 1946-51; with Air Force Cambridge Research Labs., Bedford, Mass., 1951——, research astrophysicist Space Physics Lab., 1963——. Mem. Am. Astron. Soc., Internat. Astron. Union. Contbr. articles to sci. jours. Extended theory of completely degenerate gas stars to intermediate case of partial degeneracy; developed open-range firing method of determining aeroballistic coefficients of air-launched spin-stabilized rockets; studies of nuclear fireballs; originator, dir. combined lab. and telescope astrophysics program for determination of chem. abundances of stellar atmospheres and galaxies. Home: 73 Perkins St., West Newton, Mass. 02165. Office: Air Force Cambridge Research Labs., Hanscom Field, Bedford, Mass. 01730.*

WARF, James Curren, Am. chemist; b. Nashville, Sept. 1, 1917; s. James Curren and Susie (McCord) W.; B.S., U. Tulsa, 1939; Ph.D., Ia. State U., 1947; m. Lee Walker, Apr. 14, 1940 (dec. 1959); children—Sandra, Curren Walter, Barnaby Louis; m. 2d, Kyoko Sato, Dec. 29, 1965. Instr., U. Tulsa, 1941-42; group leader Manhattan Dist., Ia. State U., 1942-47; Guggenheim fellow U. Berne, 1947-48; faculty U. So. Cal., Los Angeles, 1948—, asso. prof., 1952——. Vis. prof. U. Indonesia, 1957-59, Airlangga U., Surabaja, Indonesia, 1962-64; cons. Jet Propulsion Labs., 1965. Mem. Am. Chem. Soc., Fedn. Am. Scientists, Sigma Xi. Author: A Guide to Problem Solving in Quantitative Analysis, 1956; Physical Chemistry for Medical Students, 1964; also articles. Research on chemistry of rare earth elements, their carbides, refractory compounds, hydrides, solid state chemistry, thermodynamics and crystallography of rare earth compounds, liquid ammonia solutions, chemistry of copper hydride, its decomposition kinetics and thermodynamics and lattice energy, chemistry of uranium and thorium, methods of purification of uranium and thorium by extraction. Home: 3930 Franklin Av., Los Angeles 90027.*

WARFIELD, George, Am. physicist; b. Piombino, Italy, Apr. 21, 1919 (parents Am. citizens); s. Vincent and Ada (Donati) W.; B.S., Franklin and Marshall Coll., 1940; Ph.D., Cornell U., 1950; m. Lauraine Serra, July 14, 1945; children—Cheryl, Pamela, Richard. faculty Princeton (N.J.) U., 1950——, prof. elec. engring., 1966——; cons. RCA Labs., Princeton, 1950——. Mem. Am. Phys. Soc., Phi Beta Kappa, Sigma Xi. Research, publs. on devel. of the MOS transistor. Home: 19 Longview Dr., Princeton, N.J. 08540.*

WARGA, Mary Elizabeth, Am. physicist; b. Donora, Pa., Feb. 5, 1904; d. John and Mary (Obruba) W.; B.S., U. Pitts., M.S., 1928, Ph.D. 1937. Staff, Mellon Inst. Indsl. Research, Pitts., 1929-36; faculty U. Pitts., 1936—, prof. physics and applied spectroscopy, 1952-62, adj. prof. physics and engring., 1962——. Named Distinguished Dau. of Pa., 1954, Woman of Year in Sci. Research, Pitts., 1959; recipient Distinguished Service medal, N.Y. sect. Soc. Applied Spectroscopy, 1962. Fellow Optical Soc. Am. (exec. sec. 1959——), A.A.A.S., Am. Phys. Soc., Inst. Physics, Phys. Soc. (London); mem. Am. Inst. Physics (mem. governing bd. 1960——), Am. Chem. Soc., Washington Acad. Scis., Am. Assn. Physics Tchrs., Groupement pour l'avancement des méthodes spectrographiques (hon. Paris), Phi Beta Kappa, Sigma Xi, Sigma Pi Sigma, Iota Sigma Pi, Theta Phi Alpha. Research on optical absorption and upper atmosphere spectroscopy, spectrochem. analysis, ultraviolet, visible and infrared optical emission. Home: 1727 Massachusetts Av., Washington 20036. Office: 1155 16th St., N.W., Washington 20036.

WARGENTIN, Pehr Vilhelm, Swedish astronomer; b. Sunne Prästgard, Sweden, Sept. 11, 1717. Fellow Royal Soc., 1764; sec. Acad. Scis., Stockholm; mem. French Acad. Scis. Author: De satellitibus Iovis, 1741; Tabulae pro eclipsibus satellitum Iovis 1741. Observed opposition of Mars, 1751; investigated eclipses of Jupiter's moons, celestial mechanics, phys. geography. Died Stockholm, Dec. 13, 1783.

WARING, Charles Emmett, Am. chem. physicist; b. Phila., Jan. 24, 1909; s. Charles Wilson and Elizabeth (Way) W.; B.S., Muskingum Coll., 1931; M.Sc., Ohio State U., 1934, Ph.D., 1936; postgrad. (Lalor Found. fellow), Oxford U. (Eng.), 1940; m. Geraldine Howald, Dec. 19, 1936. Mem. faculty Poly. Inst. Bklyn., 1936-45; sci. editor div. war research Columbia, 1946; prof. chemistry U. Conn., Storrs, 1946——, head dept. 1946-66; v.p. research and devel. Raycon Corp., South Windsor, Conn. 1963——; dir. Convec Corp., Wethersfield, Conn. Sci. adviser to tech. dir. U.S. Naval Ordnance Test Sta. 1963——; also cons. to numerous govt. brs. Recipient Presidential Certificate of Merit, 1948. Mem. Am. Chem. Soc., A.A.A.S., Sigma Xi, Phi Kappa Phi, Sigma Pi Sigma. Research and publs. on establishing the mechanisms of free radical reactions in the gas phase

and the mechanisms of chemiluminescent reactions. Home: Separatist Rr., Storrs, Conn. 06208.*

WARING, Edward, English mathematician; b. Old Heath, Eng., 1734; s. John Waring; B.A., Magdalene Coll., Cambridge, 1757, M.A., 1769; M.D., Cambridge U., 1767; m. Mary Oswell, 1776. Practiced medicine London hosps., later at Addenbroke hosp., Cambridge, St. Ives, Huntingdonshire; gave up practice, 1770; Lucasian prof. math. Cambridge U., 1760-98. Fellow Royal Soc., 1763; recipient Copley medal, 1784; fellow royal socs. Göttingen (Germany), Bologna (Italy). Author: Miscellanea analytica . . . , 1762; Meditationes algebraicae algebraicae, 1770; Proprietates algebraicarum curvarum, 1772; Meditationes analyticae, 1776; On the Principle of translating Algebraic Quantities into Probable Relations and Annuities, 1792; An essay on the Principles of Human Knowledge, 1794. Gave Cauchy's ratio test, 1776; solution (in 1909) of Waring's problem of 1770 marked new era in analysis and made possible far-reaching theorems in arithmetic; 1st to give a process for approximation to values of imaginary roots; classified quartic curves into 12 main divisions and 84551 species; gave proofs of Descartes' rule of signs; presented theory of symmetrical functions of roots of algebraic equations, 1770, 82; presented series of propositions in number theory regarding decomposition of a number into sum of cubes, biquadratics; 1871; decomposed every even number into sum of 2 prime numbers. Died Aug. 15, 1798.

WARINGTON, Robert, English chemist; b. Sheerness, Eng., Sept. 7, 1807; s. Thomas Warington; student Mchts. Taylors' Sch., 1818-22; m. Elizabeth Jackson, 1835; 3 children, including Robert. Apprentice to mfg. chemist, 1822-27; became asst. to chemistry prof. London U., 1828; chemist to brewery, 1831-39; chemist London Apothecaries Soc., 1842-67; named chem. referee by four of met. gas cos. 1854. Fellow Royal Soc., 1864; mem. Cavendish Soc. (a founder, sec.), Chem. Soc. (a founder, hon. sec. until 1851), Royal Coll. Chemistry (founder). Revised the Translation of the Pharmacopoeia of the Royal College of Physicians, 1851. Contbr. articles to sci. jours. Editor (with Boverton Redwood) British Pharmacopoeia, 1867. Died Nov. 17, 1867.

WARINGTON, Robert, English agrl. chemist; b. London, Eng., Aug. 22, 1838; s. Robert Warington; student chemistry in father's lab.; attended lectures by Faraday, Brande, Hofmann; M.A. (hon.), Oxford, 1894; m. Helen Louisa Makins, 1884 (d. 1898); 5 daus.; m. 2d, Rosa Jane Spackman, 1902. Asst. Rothamsted Lab., 1859-60; research asst. to Sir Edward Frankland, London, Eng., 1860-62; asst. to Royal Agr. Coll., Cirencester, Eng., 1862-67; lectr. on Rothamsted expts., Cirencester, beginning 1864; chemist Millwall Lab., 1867-76; pvt. asst. to Sir John Bennet Lawes, Rothamsted, 1876-77; investigator Rothamsted Lab., 1876-91; lectr. in Am. under Lawes Trust, 1891; examiner in agr. sci. and art dept., also Sibthorpian prof. agr. U. Oxford, 1894-97. Fellow Chem. Soc., Royal Soc., 1886; Author: Chemistry of the Farm, 1881; Lectures on the Physical Properties of the Soil, 1900; Lectures on the Rothamsted Experiments, 1892; also articles. Research on ferric oxide and alumina in decomposing soluble phosphates and their retention in soil, citric and tartaric acids; demonstrated that 2 different organisms are responsible for nitrification. Died Harpenden, Eng., Mar. 20, 1907.

WARK, David Quentin, Am. meteorologist; b. Spokane, Mar. 25, 1918; s. Percival Damon and Clara Belle (Mackey) W.; B.A., U. Cal., Berkeley, 1941, Ph.D., 1959. Astronomer U.S. Naval Obs., Washington, 1942; weather forecaster Pan Am. Airways, Balboa, C.Z., 1942-43; research meteorologist Cal. Inst. Tech., 1943-44; weather forecaster U.S. Weather Bur., Marseilles, Munich, Cairo, 1946-49, San Francisco, 1949-58; meteorologist Nat. Environmental Satellite Center, Washington, 1958——. Mem. Am. Meteorol. Soc., Am. Astronom. Soc., Sigma Xi. Research, publs. on measurement of upper atmosphere temperatures from doppler broadening of airglow lines; interpretation and planning of meteorol. satellite expts. Home: Route 1, Box 226, Brandywine, Md. 20613. Office: Nat. Environmental Satellite Center, Washington 20233.*

WARKANY, Josef, physician; b. Vienna, Austria, Mar. 25, 1902; s. Jacob and Hermine (Buchwald) W.; student Real-Gymnasium, Vienna, 1912-20; M.D., U. Vienna, 1926; m. Suzanne Blanche Buhlman, Apr. 24, 1937; children—Joseph Henry, Stephen Fred. Came to U. S., 1932, naturalized, 1938. Intern Children's Clinic, Vienna, 1926-27; asst. Reichsanstalt for Mothers and Children, 1927-31; scholar Children's Hosp. Research Found., Cin., 1931-34, fellow, 1935——; asst. attending pediatrician Children's Hosp. Cin., 1934-35, attending pediatrician, 1935——; attending pediatrician pediatric and contagious divs. Cin. Gen. Hosp., 1935——; asst. prof. pediatrics U. Cin., 1932-45, asso., 1945——, prof. research pediatrics, 1953——. Mem. Sci. adv. bd. Nat. Assn. Retarded Children, Nat. Found.; drug research bd. Nat. Acad. Scis.; mem. study sect. human embryol. and devel. NIH. Recip. Mead-Johnson award, 1944, Borden award, 1950, award medal Am. Assn. Plastic Surgeons, 1962. Diplomate Am. Bd. Pediatrics, Ohio Med. Bd. Mem. Am. Pediatric Soc., Soc. Pediatric

Research, Ohio Med. Assn., Soc. Exptl. Biology and Medicine, Acad. Medicine Cin., Harvey Soc., Teratology Soc. (1st pres. 1960), Internat. Inst. Embryology, Sociéte de Pédiatrie, Paris. Research on causes of congenital malformations and mental retardation. Home: 3535 Biddle St. Office: Children's Hosp. Research Found., Cin. 29.

WARLTIRE, chemist; flourished 18th century. Believed that air releases its moisture when phlogisticated (confirmed by Priestley, 1781); exploded mixture of gases by elect. spark in closed copper globe and found small decrease in weight which he believed was caused by escape of heat.

WARMING, Johannes Eugenius Bülow, Danish botanist; b. Mano, Denmark, Nov. 3, 1841; ed. U. Copenhagen, Ph.D., 1871; taught at U. Copenhagen, 1873-82; botany, Royal Inst., Stockholm, Sweden, 1882-85; prof. botany, dir. bot. garden, Copenhagen, Denmark, 1886-1911; research expdns. to Brazil, 1863-66, Greenland, 1884, Norway, 1885, W.I., Venezuela, 1890-92. Author: Hanbog i den systematiske botanik, 1875, English edit., 1895; Den almindelige botanik, 1880; Om Skudbygniug, overvintring og foryngelse, 1884; Om Gronlands vegetation, 1888; Lagoa Santa, 1892; Plantesanfund (laid found. of modern plant ecology), 1895, English edit., Oecology of Plants, 1925; Dansk plantevaekst, 1906-19; Okologiens grundformer, 1923; also articles. Considered the founder of plant ecology; contbd. to knowledge of anture of ovules; studied relationship of plants to environmental factors, such as light, temperature, and precipitation; showed that plants, trees, and shrubs adapt to environmental changes through spontaneous modifications of structure and function; demonstrated that environmental conditions control internal plant processes; classified life forms of flowering plants; gave classic descriptions of Arctic and Brazilian flora; undertook research on problems of plant morphology. Died Copenhagen, Apr. 1, 1924.

WARNACA, Charles Edward, Am. geophysicist; b. Enid, Okla., Feb. 27, 1929; s. Albert August and Essie (Fuller) W.; A.B., Phillips U., 1951; m. La Donna June Froese, Oct. 13, 1949; children—Pamela Irene, Carl Albert. Party chief Seismograph Service Corp., Neb., Colo., Kan., Ill., Okla., 1951-54, Egypt, 1954-57, Libya, 1957-59, Miss., 1959-60; research, exploration Continental Oil Co., Ponca City, Okla. 1960-64, supervising geophysicist, 1964——. Mem. Soc. Exploration Geophysicists, Tulsa Geophys. Soc. Research on theory of wave fronts in anisotropic media, calculations of various gravity processing, migration of steep dip, data processing techniques and design of other method for computer operations, cross and auto correlation studies, data correction problems, evaluation of new techniques. Home: 2012 Shasta St. Office: Continental Oil Co., Box 1267, Ponca City, Okla. 74601.*

WARNER, Emory Dean, Am. physician; b. North English, Ia., July 5, 1905; s. Jasper B. and Minnie (Roller) W.; B.S., State U. Ia., 1927, M.D., 1929; m. Irene Ketchum, June 18, 1930; children—Carmen Marie (Mrs. Seymour Rafferty), Carol Cornelia, Helen Irene. Asst. pathologist U. Ia. Hosps., Iowa City, 1933-45, pathologist, 1945——; faculty U. Ia., Iowa City, 1933——, prof., head dept. pathology, 1945——. Diplomate Am. Bd. Pathology. Mem. A.M.A., Am. Soc. Exptl. Pathology, Am. Soc. Clin. Pathologists, Coll. Am. Pathologists, Soc. Exptl. Biology and Medicine, A.A.A.S., Internat. Acad. Pathology, Am. Assn. Pathologists and Bacteriologists, Central Soc. for Clin. Research, Ia., Johnson County med. socs., N.Y. Acad. Scis., Sigma Xi, Alpha Omega Alpha, Phi Chi. Research, numerous publs. on clotting of blood in health and disease; mechanisms of bleeding disorders and of thrombosis. Home: 1402 E. Court St., Iowa City 52240.*

WARNER, Francis James, neuropathologist; b. Warsaw, Poland, Oct. 3, 1897; s. Sam and Molly (Smith) W.; came to U.S., 1901, naturalized, 1925; M.D., Loyola U., Chgo., 1919; B.A., Ia. State U., 1922; M.A., U. Mich., 1925. Lectr. neuro-anatomy U. Utah, 1946-48; guest investigator neuro-anatomy, neuro-pathology U. Coll. Med. Sch., Nat. Hosp., London, Eng., 1949-50, 54-56; instr. clin. neurology Temple U., Phila., 1957——. Diplomate Am. Bd. Pathology. Fellow A.A.A.S., Royal Soc. Medicine (London), Zool. Soc. (London); mem. Am. Assn. Anatomists, Am. Soc. Zoologists, Am. Assn. Neuropathologists, Assn. Research in Nervous and Mental Disease, Am. Acad. Neurology (asso.), Am. Psychiatric Assn., Anat. Soc. Gt. Britain and Ireland, Am. Micros. Soc., Sigma Xi. Research, publs. on comparative neurology and neuro-embryology of reptiles; human neuropathology. Home: P.O. Box 523, Phila. 19105.*

WARNER, John Christian, Am. chemist; b. Goshen, Ind., May 28, 1897; s. Elias and Addie (Plank) W.; A.B., Ind. U., 1919, M.A., 1920, Ph.D., 1923, D.Sc., 1954; D.Sc., Northeastern U., 1951, Bucknell U., 1955, U. Md., 1955, Worcester Poly. U., 1956, Duquesne U., 1963, Grove City Coll., 1964, Carnegie Inst. Tech., 1965; L.H.D., Youngstown U., 1959, U. Geneva, 1964, Rose Poly. U., 1965; D.Engr., U. Toledo, 1960; LL.D., U. Pitts., 1953, Washington and Jefferson U., 1965; m. Louise Hamer, June 17,

1925; children—William H., Thomas P. Chemist Barrett Co., 1918, Cosden Oil Co., 1920-21, Wayne Chem. Corp., 1925-26; instr. chemistry Ind. U., 1919, 1921-24; faculty Carnegie Inst. Tech., Pitts., 1926—, prof., head dept. chemistry, 1938-49, dean grad. studies, 1945-60, v.p., 1949-50, pres., 1950-65, pres. emeritus, 1965—. Indsl., ednl. cons., 1965—; pres. Ednl. Projects, Inc., 1966—; research coordinator Manhattan Project, 1943-45; mem. gen. adv. com. AEC, 1952-64; trustee Carnegie Inst. Tech., Carnegie Inst., Mellon Inst., others dir. corps. Recipient Am. Inst. Chemists Gold medal; Pitts. Am. Chem. Soc. award, Award for Excellence Commonwealth of Pa. Mem. Nat. Acad. Scis., N.Y. Acad. Sci., Am. Chem. Soc. (pres. 1956), Electrochem. Soc. (pres. 1952-53), Am. Inst. Chemists, Am. Inst. Mining and Metall. Engrs. Author: (with T. P. McCutcheon, Harry Seltz) General Chemistry, 1939; (with Robert Leighou) Chemistry of Engring. Materials, 1942; (with Guido Stemple) General Chemistry - Problems and Experiments, 1929; also numerous articles. Research on kinetics of reactions in solutions including salt and medium effects; electrostatic contbn. to activation energies; molecular spectra; di-pole moments; thermodynamic properties of solutions; acid-base properties of mixed solvents; thermodynamics, rates and mechanism of corrosion reactions; phys. chemistry of metall. reactions; ednl. and research adminstrn. Home: 552 N. Neville St. Office: 155 N. Craig St., Pitts. 15213.*

WARNER, Joseph, surgeon; b. Antigua, W.I., 1717; s. Ashton Warner; ed. Westminster Sch.; apprenticed to Samuel Sharpe, 7 years. Qualified as surgeon, 1741. Became army surgeon, Scotland, 1745; surgeon Guys Hosp., London, 1746-80; practice medicine, London. Became mem. ct. assts. Corp. Surgeons, 1764, mem. ct. of examiners, 1771, master, 1780, 84. Author: Cases of surgery . . . to which is added an Account of the Preparation and Effects of the Agaric of the Oak in Stopping of Bleedings after some of the most capital Operations, 1754; A Description of the Human Eye and its adjacent parts, together with their Principal Diseaes, 1773; An Account of the Testicles . . . and the Diseaes to which they are liable, 1774. Fellow Royal Soc., 1750; mem. Coll. Surgeons (charter). First to tie common carotid artery, 1775. Died July 24, 1801.

WARNER, Robert Collett, Am. biochemist; b. Denver, Aug. 31, 1913; s. Hayward Dare and Grace (McKibben) W.; B.S., Cal. Inst. Tech., 1935, M.S., 1937; Ph.D., N.Y. U., 1941; children—Peter David, Caroline Dare, Victoria Anne. Teaching fellow N.Y. U., 1935-41; chemist Eastern Regional Lab., U.S. Dept. Agr., Phila., 1941-46; faculty N.Y. U. Sch. Medicine, N.Y.C., 1946—, prof. biochemistry, 1960—. Mem. study sect. on biophysics and biophys. chemistry USPHS, 1961-64; mem. editorial bd. Jour. Biol. Chemistry, 1960-65, asso. editor, 1965—. John Simon Guggenheim fellow, 1958. Mem Am. Soc. Biol. Chemists, Am. Chem. Soc., Biophys. Soc. Research, numerous publs. on chemistry of proteins, alkaline hydrolysis, fractionation of casein, crystal density, metal binding of conalbumin; chemistry of nucleic acids, interaction of synthetic polynucleotides, RNA polymerase, strand separation of DNA, RNA of bacterial viruses. ome: 3 Washington Sq. Village, N.Y.C. 10012.*

WARNER, Seth L., Am. mathematician; b. Muskegon, Mich., July 11, 1927; s. Seth Lemoine and Agnes (Brustad) W.; B.S., Yale, 1950; Ph.D., Harvard, 1955; postgrad. U. Nancy, 1951-52; m. Susan Emily Rose, June 16, 1962; 2 daus., Susan, Sarah. Faculty, Duke, Durham, N.C., 1955—, prof. math., 1965—. Mem. Am. Math. Soc., Math. Assn. Am., Phi Beta Kappa, Sigma Xi. Author: Modern Algebra, Vols. I and II, 1965; also articles. Research on theory of topological rings and algebras. Home: 2406 Wrightwood Av., Durham, N.C. 27705.*

WARNER, W. Lloyd, Am. anthropologist, sociologist; b. Redlands, Cal., Oct. 26, 1898; s. William Taylor and Clara (Carter) W.; B.A., U. Cal., 1925; postgrad. Harvard; M.A., Trinity Coll., Cambridge, Eng., 1954; m. Mildred Hall, Jan. 12, 1932; children—Ann Covington (Mrs. Michael Arlan), Caroline Ranghild (Mrs. John Hightower), William Taylor. Faculty, Harvard, Radcliff Coll., 1929-35; prof. sociology and anthropology U. Chgo., 1935-59; prof. social research Mich. State U., Lansing., 1959—. Sr. cons. Social Research, Inc., Chgo. Mem. Am. Anthrop. Assn., Am. Sociol. Soc., Am. Acad. Arts and Scis., Sigma Xi. Author: numerous books including The American Federal Executive, 1963; The Corporation in the Emergent American Society, 1961; The Family of God, 1961; Yankee City Series, 1941-59; A Black Civilization, 1937, rev., 1964. Research, publs of social and religious life of Australian aborigines, social class of communities, careers of bus. leaders and govt. ofcls. Home: 22 Summitt Dr., Dune Acres, Chesterton, Ind. Office: 432 Eppley Center, Mich. State U., East Lansing, Mich.*

WARNICK, Alvin Cropper, Am. animal physiologist; b. Hinckley, Utah, Nov. 15, 1920; s. Parley P. and Grace (Cropper) W.; B.S., Utah State U., 1942; M.S., U. Wis., 1947, Ph.D., 1950; m. Barbara A. Webster, Aug. 20, 1947; children—John A., Barbara Ann, Mary Louise. Faculty dept. animal husbandry Ore. State U., 1950-53; faculty dept. animal sci. U. Fla., Gainesville, 1953—, prof., 1962—; physiologist FAO, Balcarce, Argentina, 1963. Cons., Central U. Venezuela, 1960. Fellow A.A.A.S.; mem. Am. Soc. Animal Sci., Am. Genetics Assn., Sigma Xi. Author: (with T. J. Cunha, M. Koger) Crossbreeding Beef Cattle, 1963, Factors Affecting Calf Crop, 1967; also numerous articles. Research on low fertility in swine and beef cattle, genetic and environmental factors influencing pregnancy rates in beef cattle, effects of energy and protein deficiencies on male and female physiology in sheep and cattle. Home: 518 N.W. 36th St., Gainesville, Fla. 32601.*

WARNTZ, William, Am. geographer; b. Berwick, Pa., Oct. 10, 1922; s. Sterling Adrian and Lillian (Grey) W.; B.Sc., U. Pa., 1949, A.M. (Carnegie Found. fellow), 1951, Ph.D., 1955; m. A. Minerva Mosdell, June 19, 1947; children—Christopher William, Pamela Mary Elizabeth. Faculty U. Pa., 1949-56; research asso. Am. Geog. Soc., N.Y.C., 1956-66; vis. lectr. Hunter Coll., N.Y.C., 1956-66; research asso. dept. astrophys. sci. Princeton, 1957-65; vis. lectr. regional sci. dept. U. Pa., 1958-66; seminar asso. Columbia U., 1964—. Mem. adv. council Bur. Census, Washington, 1958-66; cons. NSF, 1958—; 1962-66; prof. theoretical geography and regional planning Harvard, 1966—, dir. Lab. for Computer Graphics and Spatial Analysis, 1968. Recipient Social Scis. Research Council Travel award, 1964, NSF Travel award, 1965. Fellow A.A.A.S. (councillor 1964—); mem. Regional Sci. Assn. (pres. 1965-66). Author: Toward a Geography of Price, 1959; Geography, Geometry and Graphics, 1963; Geographers and What They Do., 1964; Geography Now and Then, 1964; Macrogeography and Income Fronts, 1965; also articles. Application dimensional analysis to geog. problems; devel. relationships among geography, geometry and graphics; research in statis. methodology for areal distbns.; introduction and refinement concepts for macrogeography econ. phenomena. Home: 85 Argilla Rd., Andover, Mass. 01810; also 1054 Ohio Av., Cape May, N.J. 08204. Office: Meml. Hall, Harvard U., Cambridge, Mass. 02138.*

WARREN, Bertram Eugene, Am. physicist; b. Waltham, Mass., June 28, 1902; s. Willard and Grace (Adams) W.; S.B., Mass. Inst. Tech., 1924, S.M., 1925, Sc.D., 1929; m. Elna J. Peterson, July 26, 1934; children—Robert, Bruce. Faculty, Mass. Inst. Tech., Cambridge, 1929—, prof. physics, 1939—. Mem. Am. Phys. Soc., Am. Crystallographic Assn., Am. Acad. Arts and Scis. Research and numerous publs. on application of X-ray diffraction studies to imperfections in metals, structure of amorphous solids. Home: 71 Chester St., Arlington, Mass. 02174.*

WARREN, Cyrus Moors, Am. chemist, mfr.; b. Fox Hill, West Dedham, Mass., Jan. 15, 1824; s. Jesse and Betsey (Jackson) W.; B.S., Lawrence Sci. Sch., 1855; studied chemistry, Paris, France, Heidelberg, Germany; m. Lydia Ross, Sept. 12, 1849; 7 children. Partner (with brother Samuel) tarred roofing mfg. co., Cin., 1847; established lab., Boston, 1663; founder, pres., treas. Warren Chem. Mfg. Co.; Boston, Warren-Scharf Asphalt Paving Co., N.Y.C.; prof. organic chemistry Mass. Inst. Tech., 1866-68. Left bequests for promotion of science to Harvard and Am. Acad. Arts and Scis. Contbr. 13 papers to jours. or sci. publs. Developed improved process of fractional condensation, studied complex mixture of hydrocarbons in Pa. petroleum, invented process of purifying Trinidad asphalt. Died Manchester, Vt., Aug. 13, 1891.

WARREN, George Harry, Am. microbiologist; b. Morrisville, Pa., Sept. 7, 1916; s. Samuel and Ida (Konowitz) W.; A.B., Temple U., 1939; M.A., Princeton, 1940, Ph.D. (Class of 1877 fellow, John DeWitt Sterry fellow), 1944; m. Mildred Denker, Feb. 17, 1946; 1 son, Samuel Leslie. Head dept. bacteriology Wyeth Inst. Med. Research, Phila., 1947—, with dept. antibiotics research, 1954—, dept. virology, 1955-60; prof. microbiology Jefferson Med. Coll., Phila., 1966—. Mem. Soc. Exptl. Biology and Medicine, Am. Soc. Microbiology (past councilor), Am. Assn. Cancer Research, Reticuloendothelial Soc., N.Y. Acad. Scis., Sigma Xi. Research and numerous publs. on antigenic structure, antibiotics, tumor research, enzymes, mucopolysaccharides, viruses, chemotherapy, preparation of biols. Home: 349 Glen Gary Dr., Haverton, Pa. 19083. Office: Wyeth Lab., Box 8299, Phila. 19101.*

WARREN, Herbert Stetson, Am. anatomist, zoologist; b. N.Y.C., Aug. 24, 1887; s. George W. and Mary Alice (Flood) W.; B.S., City Coll. N.Y., 1911; M.A., Columbia U., 1920; Ph.D., Stanford U., 1926; m. Esther Christenson, June 2, 1924; 1 son, Leland Christenson. Tutor, City Coll. N.Y., 1914-18; instr. Columbia U., 1920-23; asst. prof. zoology U. Ida., 1926-30; asst. prof. zoology Temple U., 1930-37; asso. prof. anatomy Hahnemann Med. Coll., Phila., 1937-53; asso. prof. zoology Villanova U., 1955-61, ret., 1961. Recipient Pell medal City Coll. N.Y., 1911. Fellow A.A.A.S.; mem. Am. Genetic Assn., Am. Soc. Human Genetics, Biol. Photographic Assn., Genetics Soc. Am., Marine Biol. Lab., Nat. Geog. Soc., Am. Soc. Parasitologists. Research and publs. on similarities in nervous system and segmental excretory glands of primitive crustacean, Artemia, and Annelid worms showing evolution of Crustacea from Annelida; showed failure of human brain to develop causes absence of skull cap in Anencephalic monsters; described agenesis of external genitalia in humans. Home: 2768 Egypt Rd., Audubon, Pa. 19401.*

WARREN, Howard Crosby, Am. psychologist; b. Montclair, N.J., June 12, 1867; s. Dorman Theodore and Harriet (Crosby) W.; A.B., Princeton, 1889, A.M., 1891; postgrad. univs. of Leipzig, Berlin, Munich, 1891-93 Ph.D., Johns Hopkins, 1917; m. Catherine Campbell, Apr. 5, 1905. Instr. logic, Princeton, 1890-91, demonstrator psychology, 1893-96, asst. prof., 1896-1902, prof. exptl. psychology, 1902-14, head of psychol. lab., 1904-24, Stuart prof. psychology, 1914-22; 1st chmn., dept. psychology, 1920. Fellow A.A.A.S. Compiler, Psychol. Index, 1894-1907, 1910-14; asst. editor Am. Naturalists, 1896-97; asso. editor, 1900-04, co-editor, 1904-10, Psychol. Rev.; sr. editor Psychol. Rev. Publs., 1910—; pres. Psychol. Rev. Co., 1911-25. Contbr. Johnson's Cyclo., Baldwin's Dictionary of Philosophy and Psychology. Translator: Trade's Social Laws, 1899. Author: Human Psychology, 1919; History of the Association Psychology, 1921; Elements of Human Psychology, 1922. Asserted the method of introspection is indispensable in scientific psychology; supported double-aspect theory of the body-mind relationship; although he felt observations of overt behavior alone offered inadequate materials for research, he was sympathetic to behaviorism of John B. Watson. Died Princeton, Jan. 4, 1934.

WARREN, James V(aughn), Am. physician, educator; b. Columbus, O., July 1, 1915; s. James Halford and Lucile (Vaughn) W.; B.A., Ohio State U., 1935; M.D., Harvard, 1939, m. Gloria Kicklighter, May 27, 1954. Med. house officer Peter Bent Brigham Hosp., 1939-41, asst. resident medicine, 1941-42; research fellow medicine Harvard, 1941-42; med. investigator problems of shock and vascular injuries OSRD, 1942-46, faculty Emory U. Med. Sch., 1942-52, prof. physiology, chmn. dept., 1947-51, prof. medicine, 1951-52; asst. prof. medicine Yale Med. Sch., 1946-47; prof. medicine Duke Sch. Medicine, 1952-58; prof. medicine, chmn. dept. internal medicine U. Tex. Sch. Medicine, 1958-61; prof. medicine, chmn. dept. Ohio State U., 1961—. Mem. cardiovascular sect. USPHS. 1952-56 mem. tng. grant com. Nat. Heart Inst., 1960-64. Diplomate Am. Bd. Internal Medicine. Fellow A.C.P.; mem. A.M.A., Soc. Univ. Soc. Cardiologists, Am. Heart Assn. (pres. 1962-63), Am. Soc. Clin. Investigation Am. Fedn. Clin. Research (nat. pres. 1952-53), So. Soc. Clin. Research, Am. Physiol. Soc., Soc. Exptl. Biology and Medicine, Assn. Am. Physicians, Central Soc. Clin. Research, Sigma Xi. Editorial bd. Am. Heart Jour., Excerpta Medica, also Circulation (jour.). Contbg. author: Pre-Eclamptic and Eclamptic Toxemia of Pregnancy, 1941; Methods in Medical Research, Vol. VII; numerous articles profl. publs. Clin. med. and physiol. research on failure of heart and circulation; early investigator in technique of cardiac catheterization (used in diagnosis of congenital heart disease); proposed (with Eugene A. Stead, Jr.) new concepts in mechanism of congestive heart failure. Home: 5526 Ashford Rd., Dublin, O. 43017. Office: Ohio State U. Hosp., Columbus 43210.*

WARREN, John, Am. surgeon; b. Roxbury, Mass., July 27, 1753; s. Joseph and Mary (Stevens) W.; grad. Harvard, 1771, M.D. (hon.), 1786; m. Abigail Collins, Nov. 4, 1777; 17 children, including John Collins, Edward. Surgeon, Col. Pickering's Regt.; Mass. Militia, 1773; active in Boston Tea Party, 1773; sr. surgeon hosp. at Cambridge, Mass., 1775; went to N.Y., surgeon gen. hosp. on L.I., 1776; established hosp. for inoculation when smallpox was prevalent, Boston, 1778; gave pvt. course of anat. lectures at mil. hosp., Boston, 1780-81; an organizer Boston Med. Soc., 1780; established 1st school of medicine connected with Harvard, 1782, became prof. anatomy and surgery, 1783. A founder, pres. Mass. Humane Soc.; mem. Agrl. Soc., Am. Acad. Arts and Scis. Author: A View of the Mercurial Practice in Febrile Diseases, 1813. Pioneer in abdominal operations, amputation at shoulder joint; prominent in dealing with yellow fever epidemic, Boston, 1798. Died Boston, Apr. 4, 1815.

WARREN, J(ohn) Collins, Am. surgeon; b. Boston, May 4, 1842; s. Jonathan Mason and Annie (Crowninshield) W.: A.B., Harvard, 1863, M.D., 1866; LL.D., Jefferson Med. Coll., 1895, Harvard, 1906, McGill, 1911; m. Amy Shaw, May 27, 1873. Instr. surgery, Harvard, 1871-82, asst. prof., 1882-87, asso. prof., 1887-93, prof., 1893-99, Molesley prof. surgery, 1899-1907, prof. emeritus, 1907—, overseer, Harvard, 1908-14. Hon. fellow Royal Coll. Surgeons, Eng., Royal Coll. Surgeons, Edinburgh; fellow Am. Acad. Arts and Scis.; pres. Am. Surg. Assn., 1896. Editor Boston Med. and Surg. Jour., 1873-81. Author: The Anatomy and Development of Rodent Ulcer, 1872; Healing of Arteries in Man and Animals After Ligature, 1886; Surgical Pathology and Therapeutics, 1895. Editor, part author: international Text-Book of Surgery, 2 vols., by 1900. Died Nov. 2, 1927.

WARREN, John Collins, Am. surgeon; b. Boston, Aug. 1, 1778; s. John and Abigail (Collins) W.; grad. (valedictorian) Harvard, 1797, M.D. (hon.) 1819; studied medicine under his father in Europe, 1799-1802; m. Susan Powell, Nov. 17, 1803; m. 2d, Anne Winthrop, Oct. 1843; 6 children. Pres., Hasty Pudding Club, Harvard; partner in father's med. office, 1802; an original mem. Anthology Club; helped prepare Pharmacopeia for Mass. Med. Soc., 1808; adj. prof. anatomy and surgery Harvard Med. Sch., 1809-15, prof., 1815-47, emeritus, 1847, dean, 1816-19; surgeon Mass. Gen. Hosp., 1821; a founder New Eng. Jour. of Medicine and Surgery, 1821; active in temperance reform, 1827-56; pres. Mass. Temperance Soc., 1827-56; contributed $10,000 in his will to temperance cause; active mem. Mass. Agr. Soc.; pres. Boston Soc. Natural History; left many paleontol. and geol. specimens to Harvard Med. Sch., formed Warren Museum. Author: A Comparative View of the Sensorial and Nervous Systems in Men and Animals, 1822; Surgical Observations on Tumours with Cases and Operation, 1837; Physical Education and the Preservation of Health, 1845; Etherization; with Surgical Remarks, 1848; The Mastodon Giganteus of North America, 1852, 55; The Preservation of Health, 1854. First surgeon in U. S. to operate for strangulated hernia; introduced staphlorraphy into U. S., circa 1828; invited W. T. G. Morton to demonstrate ether anesthesia at Mass. Gen. Hosp.; with Morton, performed 1st surgical operation in which ether anesthesia was used, 1846; introduced John Hunter's operation for anerurysm into U. S.; interested in geology and paleontology, especially mastodon skeleton. Died Boston, May 4, 1856.

WARREN, Joshiah, Am. philosopher, inventor; b. Boston, circa 1798. Orchestra leader, tchr. of music, Cin., circa 1818; granted patent for lard-burning lamp, 1821; establshed lamp factory, Cin., became Owenite; formulated theory having basic principle of sovereignty of individual and cost the limit of price; invented speed press, 1830, not patented; started jour. The Peaceful Revolutionist, 1833; made own press type molds and stereotype plates: invented and perfected cylinder press self-inking and fed from continuius roll of paper, 1837-40; established town Modern Times, L.I., N.Y., early 1850's, became noted as gathering place for many eccentric people, lasted until 1862; at various times operated equity stores in which he neither received profit nor loss; founder of philos. anarchism in Am. Author: Equitable Commerce, 1846; Written Music Remodeled, and Invested with the Simplicity of an Exact Science, 1860; True Civilization and Immediate Necessity, 1863; True Civilization: a Subject of Vital and Serious Interest to all People, 1875. Died Charlestown, Mass., Apr. 14, 1874.

WARREN, Kenneth Wayne, Am. surgeon; b. Cartersville, Ga., July 16, 1911; s. George H. and Judia (Williams) W.; B.S., Harvard, 1934; M.D., Temple U., 1938; m. Ann Trimble, Aug. 19, 1940; children—Thomas, Sally, George. Staff, Lahey Clinic, Boston, 1945—, chmn. dept. gen. surgery Lahey Clinic Found., 1963—; surgeon New Eng. Baptist Hosp., 1945—, New Eng. Deaconess Hosp., 1945—, Fifth Surg. Service Harvard, 1966—, Boston City Hosp., 1966—; lectr. surgery Harvard, 1966—, dir. postgrad. med. inst., 1955—. Recipient Temple U. Alumni award, 1964. Fellow A.C.S., Am. Gastroenterological Assn.; mem. A.M.A., Mass. Med. Soc., Boston, New Eng. surg. socs., Am., So. surg. assns., Soc. for Surgery Alimentary Tract, Pan Am. Med. Assn., Soc Internat. de Chirugie, Am. Pancreatic Club. Author: (with R.B. Cattell) Surgery of the Pancreas, 1953; Liver, Pancreas and Billiary Tract, 1954; also numerous articles. Research on disease of liver, gallbladder, bile ducts and pancreas. Home: 59 Ledgeways , Wellesley, Hills, Mass. 02183. Office: 605 Commonwealth Av., Boston 02215.*

WARREN, Richard, Am. surgeon; b. Brookline, Mass., May 12, 1907; s. Joseph and Constance M. (Williams) W.; A.B., Harvard, 1929, M.D., 1934; m. Cora Lyman, Apr. 1, 1933; children—Janet (Mrs. George C. Buell), Constance (Mrs. Richard W. Warton), Richard A., John C. Practice surgery, Boston, 1938—; vis. surgeon Peter Bent Brigham Hosp., 1953—; staff West Roxbury VA Hosp., 1946—, chief cardiovascular surgery, 1963—; chief gen. surgery sect. VA Central Office, Washington, 1965—; dir. dept. surgery Cambridge City Hosp., 1966—; faculty Harvard Med. Sch., 1936—, clin. prof. surgery, 1953—. Mem. Am. Surg. Assn., Soc. Vascular Surgery, New Eng. Surg. Soc., A.M.A., Mass. Med. Soc., Soc. U. Surgeons. Author: Procedures in Vascular Surgery, 1960; also numerous articles. Editor: Surgery, 1963. Research in vascular surgery. Home: 200 Highland St., Dedham, Mass. Office: 319 Longwood Av., Boston.*

WARREN, S. Reid, Jr., Am. elec. engr.; b. Phila., Jan. 31, 1908; s. Reid and Lora (Chandler) W.; B.S. in Elec. Engring., U. Pa., 1928, M.S., 1929, Sc.D., 1937; m. Marian Stradling, Sept. 20, 1930; children—S. Reid, III, Alan Warren. Med. research com. Nat. Tb. Assn. grantee U. Pa., Phila., 1929-42, faculty, 1933—, prof., 1949—, radiol. physicist dept. radiology, 1938—. Cons. to hosps., govt. offices; sec. working group tech. com. 1 Internat. Electrotech. Comm., 1957—. Diplomate Am. Bd. Radiology. Registered profl. engr., Pa. Fellow I.E.E.E., A.A.A.S., Am. Coll. Radiology (ass.); mem. Am. Soc. Engring. Edn., Franklin Inst., History Sci. Soc., Soc. for History Tech., Radiation Research Soc., Am. Assn. Physicists in Medicine, Fellows for Am. Studies. Author: (with Charles Weyl) Apparatus and Techniques for Roentgenography of the Chest, 1935; (with Charles Weyl, Dallet B. O'Neill) Radiologic Physics, 1941, 2d edit., 1951; (with Harold Pender) Electric Circuits and Fields, 1943; also articles. Research on analysis phys. factors chest roentgenography and synthesis improved roentgenographic techniques, preparation definitions terms in radiology and radiol. physics. Home: 45 Server Lane, Springfield, Pa. 19064. Office: 220 S. 33d St., Phila. 19104.*

WARREN, Shields, Am. pathologist; b. Cambridge, Mass., Feb. 26, 1898; s. William Marshall and Sara Bainbridge (Shields) W.; A.B., Boston U., 1918; M.D., Harvard, 1923; D.Sc., Boston U., 1949, Western Res. U., 1952, Case Inst. Tech., 1956; LL.D., Tulane U., 1953; D.Sc. (hon.), Northwestern University, 1959; Doctor Honoris Causa, University of Brazil, 1964; married Alice Springfield, Aug. 11, 1923; children—Alice (Mrs. Guy C. McLeod), Patricia (Mrs. David R. Palmer). Instr. in pathology, Harvard Med. Sch., 1925-36, asst. prof., 1936-48, professor, 1948-65, prof. of pathology emeritus, 1965—; pathologist N.E. Deaconess Hospital, 1927-63, dir. labs., 1963—; cons. Nat. Cancer Inst.; dir. Mass. Tumor Diagnosis Service, 1928-55, now cons.; pathologist New Eng. Bapt. Hosp. to 1963. Dir. Mallinckrodt Chem. Works, dir. biol. and med. U. S. AEC, 1947-52, mem. adv. com., 1952-58, cons. AEC, 1959—; chmn. med. adv. com. U. S. AEC, P.R. Nuclear Center; spl. adviser, acting U. S. rep. to 2d Internat. Conf. on the Peaceful Uses of Atomic Energy, Geneva, 1958; mem. sci. adv. bd. consultants Armed Forces Inst. Pathology, 1952—; consultant to Nat. Adv. Cancer Council; U. S. rep. on UN Sci. Com. on Effects of Atomic Radiation; sr. statesman, sci. adv. bd. USAF. Chmn. corp. and bd. trustees Boston U. Recipient Ward Burdick award, Am. Soc. Clin. Pathology, 1949, American Cancer Society Medal, 1950, William Procter Award, Science Research Society of America, Modern Medicine Award, 1953, Banting Medal, American Diabetes Association, Charles V. Chapin medal City of Providence, 1958, Albert Einstein Medal and Award, 1962, Life trustee American Board Pathology. Mem. Nat. Acad. Scis., Am. Philos. Soc., Am. Cancer Soc. (hon. life mem., bd. dirs.; pres. Mass. div. 1958-60), A.A.A.S., A.M.A., American Academy of Arts and Sciences, American Association Pathologists and Bacteriologists (past president), American Soc. Exptl. Pathology (past pres.), Soc. Exptl. Biology and Medicine, Am. Assn. Cancer Research (ex-pres.), Phi Beta Kappa. Author: Medical Science for Everyday Use, 1927; Pathology of Diabetes Mellitus, 1930, 3d edit. 1952; (with O. Gates) A Handbook for the Diagnosis of Cancer of the Uterus, 1947, 48, 49; Introduction to Neuropathology (with S. P. Hicks); (with A. W. Oughterson) Medical Effects of the Atomic Bomb in Japan, 1956; The Pathology of Ionizing Radiation. 1961; (with P. M. LeCompte and M. A. Legg) Pathology of Diabetes Mellitus, 4th edit., 1966. Contbr. on sci. subjects. Research on effects of ionizing radiation on man; radiation on normal and neoplastic cells and mammals; pathology of tumors and diabetes. Home: 301 Otis, West Newton 65, Mass. Office: 194 Pilgram Rd., Boston 02215.

WARSCHAWSKI, Stefan Emanuel, mathematician; b. Lida, Russia, Apr. 8, 1904; s. Solomon and Olga (Lowenstein) W.; student U. Koenigsberg, U. Gottingen; Ph.D., U. Basel, 1930; m. Ilse Kayser, Mar. 24, 1947. Came to U.S., 1934, naturalized, 1941. Research asso. Columbia U., 1934-35, Cornell U., 1935-37; instr. math. U. Rochester, 1937-38, Brown U., 1938-39; faculty Washington U., St. Louis, 1939-45; sr. research mathematician, applied math. group Brown U., 1944-45; prof. math. U. Minn., Mpls., 1945-63, head dept. math. Inst. Tech., 1952-63; prof., chmn. dept. math. U. Cal. at San Diego, 1963—. Sr. mathematician Inst. Numerical Analysis, Nat. Bur. Standards, 1949; vis. prof. U. Cal. at Los Angeles, 1948-49. Mem. Am. Math. Soc., Math. Assn. Am. Research and publs. in complex analysis, particularly study of behavior of conformal mapping and other analytic functions at the boundary; constructive methods in conformal mapping; potential theory and differential equations. Home: 8902 Nottingham Pl., LaJolla, Cal. 92037.*

WARSI, Nazir A., mathematician; b. Gorakhpur, India, June 30, 1939; s. Akbar H. and Majeedun (Husain) W.; B.Sc., St. Andrew's Coll., Gorakhpur, 1957; M.Sc., U. Gorakhpur, 1959, Ph.D., 1961; m. Rashida Omer, Aug. 14, 1966. Asst. prof. U. Gorakhpur, 1959-63; faculty Savannah (Ga.) State Coll., 1963-66, prof., 1964-66; prof. math. Atlanta U., 1966—. Recipient Merit scholarships. Mem. Am. Math. Soc., Tensor Soc. Reviewer, Zentralblatt für Mathematik. Research and publs. on discovery of method to determine thermodynamic as well as flow parameters behind 3-dimensional unsteady shock wave of ordinary and magneto-gas-dyanic flows. Home: 600 Beckwith St., S.W., Atlanta 30314.*

WARTHIN, Aldred Scott, Am. pathologist; b. Greens burg, Ind., Oct. 21, 1866; s. Edward Mason and Eliza Margaret (Weist) W.; A.B., Ind. U., 1888, LL.D., 1928; A.M., U. Mich., 1890, M.D., 1891, Ph.D., 1893; music diploma (teacher), Cin. Conservatory Music, 1887; postgrad. in medicine, Vienna and Freiburg; m. Dr. Katharine Angell, June 27, 1900; children—Margaret, Aldred Scott, Virginia, Thomas Angell. Asst. in internal medicine, U. Mich. 1891-92, demonstrator internal medicine, 1892-95, demonstrator pathology, 1896, instr., 1897, asst. prof., 1899, jr. prof., 1902, prof. and dir. pathol. lab., 1903—. Mem. Am. Assn. Pathologists and Bacteriologists (pres.) Internat. Assn. Med. Museums (editor Bulletin, 1913-19, Annals of Clinical Medicine, 1924—), Assn. Am. Physicians (pres. 1927-28), Society Exptl. Medicine and Biology, Am. Assn. for Cancer Research (pres. 1927-28), Internat. Med. Hist. Soc. (vice-pres., 1929-30), Am. Heart Assn., Assn. Exptl. Pathology (v.p. 1922-23, pres. 1924), other socs. 1925-28. Author: Practical Pathology, 1896, 1911. Editor and translator of 10th edit. Ziegler's General Pathology, 1903; editor dept. Pathology, 2d and 3d edits., Wood's Reference Hand-Book of the Medical Science; Text-Book of General Pathology, 1914; Medical Aspects of Mustard Gas Poisoning, 1919; Old Age—The Major Involution, 1929; The Creed of a Biologist, 1930. Contbr. articles in med. jours. and textbooks; contbns. from the Pathological Laboratory of U. of Mich., 14 vols., 1896-1927. Most important researches on anatomy and pathology of haemolymph glands, pathology of diseases of blood and blood-forming organs, cardiac syphillis, latent syphillis, tuberculosis, toxic action of mustard gas, fat embolism, action of X-rays, heart in diphtheria, thymus, heredity in cancer, pathology of goiter; gave classic description of traumatic lipidemia and fatty embolism of lung, 1913. Died May 23, 1931.

WARTIK, Thomas, Am. chemist; b. Cin., Oct. 1, 1921; s. Abraham and Lena (Monnes) W.; A.B., U. Cin., 1943; Ph.D., U. Chgo., 1949; m. Louise Dreifus, Apr. 8 1952; children—Nancy, Steven Philip. Mem. faculty Pa. State U., State College, 1950—, prof., head dept. chemistry, 1960—; vis. scientist U. Cal. Radiation Lab., Livermore, 1957, 1959, 1961; cons. Callery Chem. Co., 1955-59; cons. Koppers Co., Inc., 1959—. Mem. Am. Chem. Soc. (councillor 1963-66), A.A.A.S., Phi Beta Kappa, Sigma Xi, Phi Lambda Upsilon. Research, publs. on chem. and phys. properties of boron and aluminum compounds; chemistry of organometallic compounds. Home: 736 W. Hamilton Av., State College, Pa. 16801.*

WARTIOVAARA, Veijo Kalervo, Finnish botanist; b. Helsinki, Finland, Nov. 25, 1907; s. Arvi and Tyra (Herlinn) W.; phil.kand., U. Helsinki (Finland), 1935, phil.doctor, 1942; m. Rakel Lekari, July 2, 1937; children—Liisa (Mrs. Pertti Ertama), Terttu (Mrs. Mauno Melvasalo), Jyrki, Leena. Biologist, Wihuri Research Inst., 1945-64; head fibre morphological dept. Research Inst. Finnish Pulp and Paper Industries, 1948-54; faculty U. Helsinki, 1948—, prof., 1962—. Research, publs. on cell permeability, restricted exchange of substances between living protoplasm and bathing fluid around a cell. Home: 8 A 24, Caloniuksenkatu, Helsinki 10, Finland.*

WARWICK, James Walter, Am. astronomer; b. Toledo, May 22, 1924; s. Walter and Alice (Mauk) W.; A.B., Harvard, 1947, M.A., 1948, Ph.D., 1951; m. Constance Sawyer, Sept. 6, 1947 (div. Mar. 1966); children— Sarah H., David I., Rachel J., Joel H.; m. June Pellillo, Dec. 9, 1966; 1 son, Joseph. Asst. prof. astronomy Wellesley Coll., 1950-51; research asso. Harvard Coll. Obs., 1951-55; staff mem. High Altitude Obs., Boulder, Colo., 1955-62; prof. astro-geophysics U. Colo., Boulder, 1962—. Mem. Am. Astron. Soc., Am. Geophys. Union, Am. Math. Soc. Research, publs. on analysis spectrum in stellar and solar magnetic fields, phenomenology solar flares, solar radio emission at low frequencies—its spectrum and place origin on the sun, radiophysics planet Jupiter, morphology Jupiter's magnetic field, ionospheric structure. Home: 295 Green Rock Dr., Boulder, Colo. 80302.*

WASER, Jurg, chemist; b. Zurich, Switzerland, Dec. 23, 1916; s. Ernest and Margrit (Ruttimann) W.; student U. Zurich, 1939; Ph.D., Cal. Inst. Tech., 1944; m. Grace Merritt, 1942; children—Peter, Nickolas, Katherine. Came to U.S., 1939, naturalized, 1949. With Rice Inst., 1948-58, prof. chemistry, 1956-58; prof. Cal. Inst. Tech., Pasadena, 1958—. Guggenheim fellow, 1963-64. Mem. Am. Crystallographic Assn. (pres. 1960), Am. Chem. Soc., Swiss Phys. Soc., Sigma Xi, Phi Lambda Upsilon. Author: Quantitative Chemistry, 1961, 2d edn., 1964; Basic Chemical Thermodynamics, 1966; also articles. Research on crystal structure by x-ray diffraction; chemical education.*

WASHBURN, A(lbert) L(incoln), Am. geologist; b. N.Y.C., June 15, 1911; s. Albert Henry and Florence Belle (Lincoln) W.; A.B., Dartmouth, 1935; Ph.D., Yale 1942; m. Barbara Tahoe Talbot, Sept. 14, 1935; children—Nuna Lincoln, Land Lincoln, Sila Talbot. Mem. Harvard-Dartmouth Mt. Crillon Expdn., Alaska, 1934; mem. Mt. McKinley Expdn., Alaska, 1936; mem. Nat. Geog. Mt. MsKinley Expdn., Alaska, 1936, Louise A. Boyd E. Greenland Expdn., 1937; mem. expdns. to Arctic N. Am., 1938, 39, 40, 41, 43, 47, 49, geomorphic investigations in Greenland, 1954, N.E. Greenland, 1955, 56, 57,

58, 60, 64, Antarctic, 1957-58, USNC IGY, 1957-58, Nat. Acad. Sci. Com. Polar Research, 1958-59, 63-—; exec. dir. Arctic Inst. of N.A., 1945-51; hon. lectr. geography, McGill Univ., 1948-51; dir. Snow, Ice, and Permafrost Reasarch Establishment, U. S. Army C.E., 1951-52; prof. No. geology Dartmouth Coll., 1953-59; prof. geology Yale, New Haven, 1960-—. Cons. several govt. bds. Fellow Am. (councillor), Canadian geog. socs., Geol. Assn. Can., Geol. Soc. Am, mem. Arctic Inst. N. Am. (hon.), N.H. Acad. Sci., Am. Geophys. Union, Sigma Xi. Author: Reconnaissance Geology of Portions of Victoria Island and Adjacent Regions, Arctic Canada, 1947; also papers. Specialist in geomorphology, glacial geology; research on cold climate geomorphic processes related to frost action. Home: 253 St. Ronan St., New Haven.*

WASHBURN, Edward Roger, Am. chemist; b. Big Rapids, Mich., Sept. 22, 1899; s. Edward Rush and Myrtilla (Rogers) W.; B.S., U. Mich., 1922, M.S., 1923, Ph.D., 1926; m. Dorothy May Adams, Aug. 23, 1926; children—Dorothy Elaine (Mrs. Robert Dudley Olney), Edward Roger, Robert Henry, Carolyn May. Faculty U. Neb., Lincoln, 1926-—, prof. chemistry, 1941-—, chmn. dept., 1955-64. Fellow A.A.A.S.; mem. Am. Chem. Soc., Neb. Acad. Sci., Sigma Xi, Phi Lambda Upsilon, Alpha Chi Sigma. Research, publs. in fields surface tension, spreading, solubility, phase rule. Home: 1681 Smith St., Lincoln, Neb. 68502. Office: Avery Lab., U. Neb., Lincoln, Neb.*

WASHBURN, Edward Wight, Am. chemist; b. Beatrice, Neb., May 10, 1881; s. William Gilmor and Flora Ella (Wight) W.; student U. Neb., 1899-1901; B.S., Mass. Inst. Tech., 1905, Ph.D., 1908; m. Sophie Wilhelmina de Veer, June 10, 1910; children—William de Veer, Janet, Roger D., Barbara. Research asso. in phys. chemistry Mass. Ins. Tech., 1906-08; asso. in chemistry, U. Ill., 1908-10, asst. prof., 1910-13, prof. phys. chemistry, 1913-16, prof. ceramic chemistry and head of dept. of ceramic engring., 1916-22; editor internat. Critical Tables (phy. chem. and engring. constants), 1922-30; chief chemist U. S. Bur. Standards, 1926-—. Vice-chmn. and setg. chmn. NRC, 1918-19, chmn., 1922-23, div. chemistry; del. Internat. Chem. Union, London, 1919, Lyons, 1922. Cambridge, 1923, and Internat. Research Council, Brussels, 1919 and 1922; Am. commr. Internat. commn. annual tables phys. and chem. constants, 1921-29; chmn. Internat. Commn. on Physico-Chemical Standards; mem. Internat. Com. Thermochemistry, 1929-—. Fellow A.A.A.S (chmn. Sect. C 1923-24), Royal Soc. Arts, Am. Ceramic Soc. (editor Jour. 1920-22). Author: Introduction to the Principles of Physical Chemistry, 1915 (French translation by Noyes and Weiss, 1925). Died Feb. 6, 1934.

WASHBURN, Henry Bradford, Jr., Am. geographer; b. Cambridge, Mass., June 7, 1910; s. Henry Bradford and Edith (Hall) W.; grad. Groton Sch., 1929; A.B., Harvard, 1933, A.M., 1960, grad. work, Inst. Geog. Exploration; hon. doctorates U. Alaska, 1951, Tufts U., 1957, Colby Coll., 1957, Northeastern U., 1958, Suffolk U., 1965; m. Barbara T. Polk, Apr. 27, 1940; children—Dorothy Polk, Edward Hall, Elizabeth Bradford. Instr. Inst. Geog. Exploration, Harvard, 1935-42; dir. Mus. of Sci., Boston, 1939-—. Mountaineer Alps, 1926-—; explorer Alaska Coast Range, 1930-40; leader first ascent, Mt. Crillon, Alaska, 1934, Nat. Geog. Soc. Yukon Expdn., 1935; leader 1st aerial photog. exploration Mt. McKinley, 1936, ascended its summit, 1942, 47, 51; leader aerial exploration St. Elias range, 1938; Mt. Lucania, Sanford , Mt. Marcus Baker, Alaska, 1938, Mt. Lucania, Alaskia, 1937, Mt. Bertha, Alaska, 1940, Mt. Hayes, Alaska, 1941; leader numerous mountain and Arctic area explorations; cons. various govtl. agys. on Alaska and cold climate equipment; leader spl. expdns. in investigating high altitude cosmic rays, Alaska, 1947; leader expdn. to Mt. Kennedy, 1965. Recipient Royal Geog. Soc. Cuthbert Peek award for Alaska Exploration and Glacier Studies, 1938; Burr prize Nat. Geog. Soc., 1940, 65; Gold medal Harvard Travellers Club, 1959; award established in his honor by Mus. of Sci., 1964. Fellow Royal Geog. Soc. London, Harvard Travelers Club, A.A.A.S., Am. Acad. Arts and Scis., Am. Geog. Soc. (hon.), Cal. Acad. Scis.; mem. Arctic Inst. N. Am., Mass. Audubon Soc. (dir.), French Alpine (Groupe de Haute Montague, Paris), others. Contbr. articles, photographs on Alaska, glaciers and mountains to mags., books Editor, pub. 1st large scale map Mt. McKinley, Am. Acad. Arts and Scis.-Swiss Found Alpine Research, Bern, 1960. Home: 76 Sparks St., Cambridge, Mass. 02138. Office: Science Park, Boston 02114.*

WASHBURN, Jack, Am. metallurgist; b. Mt. Vernon, N.Y., Apr. 30, 1921; s. Morgan and Marjory (Peck) W.; B.S., U. Cal. at Berkeley, 1949, M.S., 1950, Ph.D., 1954; m. Gertrude Gastelum, June 22, 1947; children—Jo Ann, Kenneth, Melissa. Research engr. Inst. Engring. Research, U. Cal. at Berkeley, 1949-52, faculty, 1952-—, prof. metallurgy, 1960-—, chmn. dept. mineral tech., 1966-—; research prof. Miller Inst. for Basic Research in Sci., Berkeley, 1962-63. NSF fellow Cambridge (Eng.) U., 1959-60, U. Paris Faculty Sci., Orsay, France, 1965; recipient Mathewson Gold medal Am. Inst. Mining and Metall. Engrs., 1956. Mem. A.A.A.S., Am. Inst. Mining and Metall. Engrs., Am. Soc. Metals, Sigma Xi, Tau Beta Pi. Editor: (with Gareth Thomas) Electron Microscopy and Strength of Crystals, 1963; also numerous articles. Research on stress induced motion small angle boundaries in zinc crystals proving existence dislocations, etch pit and transmission electron microscopy experiments for understanding behavior dislocations in crystals. Home: 2 Ajax Pl., Berkeley, Cal. 94708.*

WASHINGTON, Henry Stephens, Am. petrologist; b. Newark, Jan. 15, 1867; s. George and Eleanor P. (Stephens) W.; A.B., Yale, 1886; A.M., 1888; Ph.D., Leipzieg, 1893; postgrad. Yale, Leipzig, Am. Sch. Classical Studies (Athens); m. Martha Rose Beckwith, Oct. 25, 1893. Asst. in physics Yale, 1886-88, asst. in mineralogy, 1895-96; excavations in Greece, 1889-94; geol., volcanic, petrological investigations in Greece, Asia Minor, Italy, Spain, Brazil, Hawaiian Islands, and U. S. and chem. study of igneous rocks and minerals; cons. mining geologist, 1906-12; bd. mgrs. Geol. Survey N.J., 1909-14; with geophy. lab. Carnegie Instn. of Washington, 1912-34. Chem. asso. and Sci. attaché Am. Embassy, Rome, 1918-19; vice chmn. sect. volcanology Internat. Geophys. Union; chmn. Am. Geophys. Union, 1927-29. Author: Chemical Analysers of Igneous Rocks, 1902, 17; Manual of the Chemical Analysis of Igneous Rocks, 1904; The Roman Comagmatic Region, 1906. Joint Author: Quantitative Classification of Igneous Rocks. 1903. Contbr. of articles to sci., jours., 1887-—. Devised (with J. P. Iddings, L. V. Pirsson, Whitman Cross) systematic classification of igneous rocks on basis of chem. composition. Died Washington, D.C., Jan. 7, 1934.

WASICKY, Richard, pharmacologist; b. Tescen, Silesia, Feb. 6, 1884; s. Johann and Anna (Loubek) W.; M.D., U. Vienna (Austria); Dr. honoris causa, Sorbonne, U. Vienna; m. Marianne Joachimovits, Nov. 4, 1924; 1 son, Richard Robert. Prof., then prof. pharmacology U. Vienna; prof. pharmacology and pharmacognosy U. Sao Paulo, Brazil; prof. univ., dir. Inst. Biochem. Research, Santa Maria, Brazil. Recipient Hamburg medal for research and discovery in sci.; decorated Legion of Honor, Order Pub. Health, Cross of Honor for sci. and art (Austria). Publns. on pharmacognosy, pharmacology, toxicology, microchemistry, biol. methods, poison and medicinal plants. Address: rua Maranhao 600, Sao Paulo 4, Brazil.

WASILEWSKI, Ludwik, Polish chemist; b. Zamosc, Poland, Sept. 27, 1891; s. Konstanty and Marchwicka (Wasilewski; ed. Poly. U.; became asst., adj. lectr. Poly, U. Lvov; dir. prodn. sect. State Spirit Monopoly, Poland; chmn. sect. inorganic chem. industry Chem. Inst. for Researches. Research and publs. on clay, thermal isolation materials, aluminum, extraction of iron from aluminum salts and of zinc from ammonium solutions, stonebitumen mixtures, also various problems of applied chemistry.

WASMANN, Erich, zoologist; b. Meran, Tyrol, May 29, 1859. Ordained priest Roman Catholic Ch., joined Soc. Jesus; animal psychologist; entomologist. Author: Vergleiche studien über das Seelenleben der Ameisen und höhern Tiere, 1897; Instinkt und Intelligenz im Tierreich, 1897; Die Gastpflege der Ameisen, 1920; Ameisen, Termiten und ihre Gäste, 1934. Studied insects living in anthills; helped establish ecology at turn of century. Died Valkenburg, Netherlands, Feb. 17, 1931.

WASOW, Wolfgang Richard, mathematician; b. Vevey, Switzerland, July 25, 1909; s. Eduard and Alma (Thal) W.; came to U. S., 1939, naturalized, 1944; grad. Goettingen (Germany) U., 1933; Ph.D., N.Y.U., 1942; m. Mona Cantor, Oct. 1959; children—Bernard, Thomas, Robin Murie, David Murie, Oliver. Instr. math. N.Y.U., 1942-46; asst. prof. Swarthmore Coll., 1946-49; numerical analysis research U. Cal. at Los Angeles, 1949-56; prof. math. U. Wis., 1957-—. Fulbright fellow, Rome. 1954-55, Haifa, Israel, 1962. Mem. Am. Math. Soc., Math. Assn. Am., Am. Assn. U. Prof. Author: (with G. E. Forsythe) Finite Difference Methods for Partial Differential Equations, 1960; Asymptotic Expansions for Ordinary Differential Equations, 1966. Research on solution of differential equations by means of asymptotic series and finite difference approximations (applicable to study of relationships between viscous and non-viscous flows, quantum and classical mechanics, also computational techniques). Home: 2533 Branch St., Middleton, Wis. 53562. Office: Univ. Wisconsin, Madison, Wis. 53706.*

WASSERMAN, Günter, German physicist; b. Berlin, Germany, Sept. 19, 1902; s. Ewald and Margarete Wasserman; Ph.D., U. Berlin. Asst., Kaiser Wilhelm Inst. Metal Research, 1928-33, research on iron, 1933; prof. agrégé U. Darmstadt; work in industry; prof. sci. of metals Clausthal-Zellerfeld, also Acad. Mines Clausthal Tech. Sch., 1944-—. Mem. Brunswick Sci. Soc., German Metallography Soc., Verein deutscher Eisenhüttenleute, German Physics Soc. Author: Texturen metallischer Werkstoff, 1939; Zeitschrift für Metalkunde; Acta Metallurgica; also articles. Address: Grosser Bruch 23, 3392 Claudthal-Zellerfeld, West Germany.

WASSERMAN, Louis Robert, Am. hematologist; b. N.Y.C., July 11, 1910; s. Jacob and Ethel (Ballin) W.; A.B., Harvard, 1931; M.D., U. Chgo., 1935; m. Julia B. Wheeler, Feb. 20, 1957. Practice medicine, specializing in hematology, N.Y.C., 1940-—; hematologist, dir. dept. hematology Mt. Sinai Hosp., 1954-—, prof. medicine Mt. Sinai Sch. Medicine, 1966-—; asso. clin. prof. medicine Coll. Phys. and Surgs. Columbia, 1960-—; research collaborator Brookhaven Nat. Labs., Upton, N.Y., 1960-—. Diplomate Am. Bd. Internal Medicine. Fellow A.M.A., A.C.P.; mem. Soc. Study Blood, Internat., Am. socs. hematology, Harvey Soc., N.Y. Acad. Scis., Am. Soc. Clin. Pathology, Am. Fedn. Clin. Research, Soc. Exptl. Biology and Medicine, Reticulo-Endothelial Soc., Soc. Nuclear Medicine, N.Y. Acad. Medicine, Société des Hôpitaux de Paris, Am. Assn. Cancer Research, Leukemia Soc., Alpha Omega Alpha. Research, numerous publs. on blood composition, physiology, and diseases; radiation, antibiotic, nutritional treatment of blood disorders.*

WASSERMAN, Robert Harold, Am. phys. biologist; b. Schenectady, Feb. 11, 1926; s. Joseph and Sylvia (Rosenberg) W.; B.S., Cornell U., 1949, Ph.D., 1953; M.S., Mich. State Coll., 1951; m. Marilyn Mintz, June 11, 1950; children—Diane, Arlene, Judy. Research asso. U. Tenn.-Atomic Energy Program, 1953-55; sr. scientist Oak Ridge Inst. Nuclear Studies, 1955-57; faculty dept. phys. biology N.Y. State Vet. Coll., Cornell U., Ithaca, N.Y., 1957-—, prof. phys. biology, 1963-—, acting head dept. phys. biology, 1963-64; vis. scientist Inst. Biol. Chemistry, Copenhagen, Denmark, 1964-65. NSF-OECD fellow, 1964; John Simon Guggenheim fellow, 1965. Mem. Am. Physiol. Soc., Biophys. Soc., Soc. for Exptl. Biology and Medicine, Am. Inst. Nutrition, A.A.A.S., Sigma Xi. Phi Kappa Phi. Author: (with C.L. Comar) Annotated Bibliography of Strontium and Calcium Metabolism in Man and Animals, 1961; also numerous articles. Editor: Transfer of Calcium and Strontium Across Biological Membranes, 1963. Research on calcium, strontium and cesium metabolism, and factors affecting their behavior in animals, metabolic dissocoiation between cesium 137 and Barium 137m daughter, increased absorption calcium by amino acids, reduced retention strontium by certain substances, factor in intestinal tissue influencing calcium binding, mechanism lactose and vitamin D stimulation calcium absorption. Home: 207 Texas Lane, Ithaca, N.Y. 14850.*

WASSERMANN, Friedrich, anatomist; b. Munich, Germany, Aug. 13, 1884; s. Franz and Analie (Techheimer) W.; M.D., U. Munich, 1910; M.D. (hon.), U. Giessen, 1959; Ph.D. (hon.), U. Frankfurt, 1958; m. Margaret Schmidgall, Feb. 1917 (dec.); children—Gertrude (Mrs. Fetcher), Franz Walter. Faculty, U. Munich, 1914, 1919-36, prof. anatomy, head dept., 1931-36; prof. anatomy U. Chgo., 1937-49; sr. biologist Argonne (Ill.) Nat. Lab., 1949-—. Vis. prof. univs. U. S., Germany; mem. 1st Med. Mission to Germany, 1948. Decorated Comdrs. Cross, Order of Merit, German Fed. Republic. Fellow A.A.A.S.; mem. Am. Soc. Anatomists, German Anat. Soc. (hon), Am. Soc. Zoologists, Soc. Exptl. Biology and Medicine, others. Author: Growth and Reproduction of Living Matter (in German), 1929; also numerous articles. Research on mitosis, devel. and function of adipose tissue, histogenesis of dental tissues, formation of collagen fibers, fine structure of bone. Home: 7416 Webster St., Downers Grove, Ill. 60515. Office: 9700 S. Cass Av., Argonne, Ill. 60439.*

WASZ-HOCKERT, Ole, Finnish physician; b. Helsinki, Finland, Aug. 28, 1918; s. Hugo and Dagmar (Lucander) W.-H.; M.D., U. Helsinki, 1946, Dr.Med.Sc., 1950, specialist in pediatrics, 1951; m. Nina Gorbatow, Sept. 11, 1940; children—Nina-Lis, Barbara, Marina, Helena. Asst. prof. U. Helsinki, 1954-64, asst. chief Children's Hosp., U. Helsinki, 1957-64; prof. pediatrics, head dept. pediatrics U. Oulu (Finland), 1964-—; resident, research fellow dept. pediatrics Karolinska Institutet, Stockholm, Sweden, 1947, U. Paris (France), 1950, Cornell U., N.Y.C., 1954-55. Mem. Internat. Pediatric Assn., Scandinavian Pediatric Soc. (mem. bd. 1964-—). Contbg. author: Nordisk laroboki pediatrik. Research and numerous publs. on pediatrics, infectious problems, childhood Tb., immunology and social pediatrics. Home: 11 Kaupprienkatu, Oulu, Finland.*

WATANABE, Kenichi, Am. physicst; b. Honolulu, Dec. 17, 1910; s. Mataichi and Haru (Takata) W.; B.S., Cal. Inst. Tech., 1936, Ph.D. 1940; m. Betty Keiko, June 25, 1945; children—Mie Caroline, Mari phyllis, Kiyo Edith. Instr., then asst. prof. math. U. Hawaii, 1940-47; asst. prof. physics Wabash (Ind.) Coll., 1947-48; physicist Naval Research Lab., 1948-51; sect. head Air Force Cambridge (Mass.) Research Lab., 1951-54; prof. physics U. Hawaii, 1955-61, sr. prof., 1961-—. Mem. planetary atmospheres subcom. NASA, 1960-63. Fellow Am. Phys. Soc.; mem. Optical Soc. Am., Am. Geophys. Union, A.A.A.S., Sigma Xi. Mem. editorial adv. bd. Planetary and Space Sci. Jour., 1959-—. Research in spectroscopy and atmospheric physics; conducted exptl. study of extreme ultraviolet absorption and ionization of molecules, their applications to upper atmosphere. Home: 2460 Ferdinand Av., Honolulu 96822.*

WATANABE, Yoshio, Japanese oral surgeon; b. Tokyo, Japan, Nov. 6, 1914; s. Kenichi and Machiko W.; D.D.S., Tokyo Koto Shika Igakko, 1939, postgrad. oral surgery, 1939-41; M.D., Kanazawa U., 1944, D.Med. Sc., 1949; m. Tsuyako Kogusuri, June 11, 1945; children—Keiko, Michiko, Yoshiaki. Lectr., Tokyo Med. and Dental U., 1944-52, asst. prof. oral surgery, 1952-55; clin. asst. oral surgery Tokyo Imperial U. Med. Sch., 1945-46; prof. Okayama (Japan) U. Med. Sch., 1955—; research scholar oral exfoliative cytology Meml. Center for Cancer and Allied Diseases, N.Y.C., also research scholar plastic surgery Hosp. U. Pa., Phila., 1955-56; vis. prof. oral pathology U. Ill. Coll. Dentistry, 1964-65. Fulbright-Smith-Mundt grantee, 1955-56; Japan Soc. grantee, N.Y.C., 1956; Fulbright Travel grantee, 1964-65. Mem. Japanese Soc. Oral Surgeons, Japanese Stomatological Soc., Japanese Soc. Clin. Cytology, Japanese Soc. Oral Roentgenology, Japanese Soc. Plastic Reconstructive Surgery, Japanese Cancer Soc., Internat. Assn. Dental Research, Am. Soc. Cytology, Am. Coll. Dentists. Research and numerous publs. on oral exfoliative cytology, oral cancer and its dental etiologic factors, wound healing process in oral surg. operations from viewpoint of capillary vascularization; roentgenographic diagnosis of diseases of jaw-bones. Home: 934 Higashi Oizumi, Nevima-ku, Tokyo, Japan.*

WATASE, Yuzurn, Japanese physicist; b. Hyogo Prefecture, Japan, 1907; grad. Kyoto U., 1930; D.Sc., Tohoku U., 1933. Lectr., asst. prof. Osaka U., became prof., 1947; named prof. Osaka Municipal U., 1949. Author: Progress of Quantum Physics. Research on atomic nucleus, ultra-short waves, cosmic rays.

WATERFALL, Umaldy Theodore, Am. plant taxonomist; b. Frederick, Okla., Aug. 13, 1910; s. Fred C. and Thersia (Dean) W.; B.S., Okla. State U., 1935; M.S., U. Okla., 1942, Ph.D., 1956; m. Eudora LaClida Cotter, July 24, 1935; children—Patricia Lee (Mrs. Billy George Lorenz), Mary Kathleen, Richard Theodore. Tchr. pub. schs., 1934-44; range ecologist Soil Conservation Service, Sulphur, Okla., 1944-46; faculty U. Okla., 1946-49; faculty Okla. State U., Stillwater, 1949—, prof. dept. botany, 1960—, curator Herbarium, 1949—. Mem. Internat., Am. socs. plant taxonomists, Southwestern Assn. Naturalists (pres. 1965), New Eng. Bot. Club, Okla. Acad. Sci. (editor for botany 1958—), Sigma Xi, Phi Kappa Phi. Author: A Catalogue of the Flora of Oklahoma, 1952; Keys to the Flora of Oklahoma, 1962; also articles. Research on composition and distbn. of flora of Okla., Southwestern U.S., Mexico, genus Physalis, genus Eucnide. Home: 1917 W. Admiral Rd., Stillwater, Okla. 74074.*

WATERHOUSE, Benjamin, Am. physician; b. Newport, R.I., Mar. 4, 1754; s. Timothy and Hannah (Proud) W.; apprenticed to Dr. John Halliburton, surgeon, 1770; med. student at Edinburgh, 1775; grad. U. Leyden, 1780; attended med. dept. of Harvard, 1783; m. Elizabeth Oliver, June 1, 1788; m. 2d. Louisa Lee, Sept. 19, 1819; 6 children. Prof theory and practice of physic, med. dept. Harvard, 1783-1812 (forced to resign); performed vaccination against smallpox by Jennerian method in U. S.; upon receiving from Eng. vaccine in form of infected threads, immediately used it on his son, 1800; vaccinated others with cowpox with good results; sent vaccine to Pres. Jefferson (who had about 200 persons vaccinated with it), 1802; wrote many newspaper articles on vaccination; lectured on natural history, mineralogy and botany at R.I. Coll. (now Brown U.), 1784-86, at Cambridge, Mass., from 1788; drew up plans for Humane Soc. of Commonwealth of Mass., 1785; med. supt. of all mil. posts in New Eng., 1813-20; editor John B. Wyeth's Oregon (published to deter Western emigrations), 1833. Author: A Synopsis of a Course of Lectures, on the Theory and Practice of Medicine, 1786; A Prospect of Exterminating the Small Pox (his 1st report on smallpox), 1800. Cautions to Young Persons Concerning Health . . . Shewing the Evil Tendency of the Use of Tobacco . . . with Observations on the Use of Ardent and Vinous Spirits, 1805; Information respecting the origin, progress, and efficacy of the Kine Pock Inoculation, 1810. The Botanist, 1811; A Circular Letter, from Dr. Benjamin Waterhouse, to the Surgeons of the Different Post, 1817. First Am. physician to establish vaccination as a general practice. Died Cambridge, Oct. 2, 1846.

WATERHOUSE, George Robert, English naturalist; b. Somers Town, Eng., Mar. 6, 1810; s. James Edward and Mary (Newman) W.; sent to sch., Koekelberg, Belgium, 1821; articled to architect, Eng., 1824; m. Elizabeth Ann Griesbach, Dec. 21, 1834, Practice architecture, London; curator London Zool. Soc., 1836-43; assigned to description of mammals and Coleoptera, collected in voyage of Beagle by Darwin; keeper mineral. and geol. br. Brit. Mus., 1851-57, keeper dept. geology, 1857-80. Mem. Entomol. Soc. London, (co-founder 1833; hon. curator; pres. 1849-50). Author: Natural History of the Mammalia (including important descriptions of Marsupialia and Rodentia), 1846-48; Catalogue of British Coleoptera, 1861; also numerous articles. Died Jan. 21, 1888.

WATERHOUSE, J. B., paleontologist; b. Wellington, New Zealand, Apr. 5, 1932; s. John Henry and Nellie (Carr) W.; grad. with 1st class honors (Royal Exhbn. scholar), U. New Zealand, 1954; Ph.D., Cambridge (Eng.) U., 1958; m. Loris Vivienne Castle, Jan. 6, 1962; children—David Ray, Merrin Elizabeth, Rodney Bruce. Prin. sci. officer New Zealand Geol. Survey, 1954-66; asso. prof. paleontology U. Toronto (Ont., Can.), 1966—. Mem. Paleontol. Assn., Geol. Soc. New Zealand, Royal Soc. New Zealand, Geol. Soc. Australia. Author: Permian Stratigraphy and Faunas of New Zealand, 1964; Permian Brachiopods of New Zealand, 1964; also articles. Research on Permianrocks and fossils of New Zealand and related fossils of Australia and S.E. Asia resulting in discovery of 3 separate glacial episodes in Permian; studies of rock structure in So. Alps have been coupled with faunal studies to trace movement of continents around W. Pacific. Home: 2911 Bayview Av., Willowdale, Ont. Office: Dept. Geology, U. Toronto, Toronto 5, Ont., Can.*

WATERHOUSE, Rupert, English physician; b. Sheffield, Eng., Jan. 15, 1873; s. John Hodgson Waterhouse; ed. Wesley Coll., Sheffield, Eng., St. Bartholomew's Hosp., London; M.B., 1900; M.D., 1903. m. Mabel Dorothy Conor, 1919; 1 son, 1 dau. Became resident med. officer Royal United Hosp., 1901, Royal Mineral Water Hosp., 1903; named comdr. 4th So. Gen. Hosp., 1920; cons. physician, v.p. Royal Mineral Water Hosp.; cons. physician Royal United Hosp., Bath, Eng., Victoria Hosp., Frome Chippenham Cottage Hosp., Trowbridge and Dist. Hosp. Fellow Royal Coll. Physicians, Contbr. ariicles to med. jours. Described malignant form of meningococcus meningitis characterized by acute collapse with hemorrhages in adrenal glands (also described by C. Friderichsen, 1918), Waterhouse-Friderichsen Syndrome. Died Sept. 1, 1958.

WATERMAN, Alan Tower, Am. physicist; b. Cornwall-on-Hudson, N.Y., June 4, 1892; s. Frank Allan and Florence (Tower) W.; A.B., Princeton, 1913, M.A., 1914, Ph.D. in Physics, 1916; D.Sc., U. Akron, 1958, Northeastern U., 1953, Tufts Coll., 1952, U. Vt., 1955, U. Ariz., 1958, Bowdoin Coll., 1958, U. Notre Dame, 1960, Kenyon Coll., Norwich U., Bklyn. Poly. Inst., U. So. Cal., Loyola U., Chgo., 1962, U. Pitts., 1963; LL.D., Cornell Coll., Mt. Vernon, Ia., 1956, Am. U., U. Chattanooga, 1958, U. Mich., 1959, U. Cin., 1959, U. Cal. at Berkeley, 1960, Ill. Inst. Tech., 1962, Mich. State U., 1963, Rockefeller Inst., 1963; Dr. Pub. Affairs, Denison U., 1964; m. Mary Mallon, Aug. 28, 1917; children—Alan Tower, Neil John, Barbara (Mrs. Joseph C. Carney), Anne (Mrs. William C. Cooley), Guy. Faculty Yale, 1919-42, asso. prof., 1931-42; NRC fellow King's Coll., London, Eng., 1927-28; vice chmn. div. D, NRDC, 1942-43; dep. chief office of field service OSRD, 1943-45, chief, 1945-46; dep. chief, chief scientist Office Naval Research, U.S. Navy, 1946-51; dir. NSF, Washington, 1951-63, cons. to dir., 1964—; cons. to pres. Nat. Acad. Scis., to adminstr. NASA. Mem. numerous govt., profl. bds., coms., including sci. adv. com. Exec. Office Pres., 1950-56; mem. Nat. Aeros. and Space Council, 1958-60, Fed. Council Sci. and Tech., 1958-63, Com. Distinguished Service Awards, 1951-61, Def. Sci. bd. Dept. Def., 1956-63; cons. President's Sci. Adv. Com., 1957-63; mem. adv. bd. Georgetown U. Center for Strategic Studies, 1963—; research adv. council U.S. Office Edn., 1963—; adv. com. Pacific Sci. Center, Seattle, 1963—; liason com. sci. and tech. U.S. Library Congress, 1963. Bd. dirs. Center for Advanced Study Behavioral Scis., Palo Alto, Cal., 1952-58. Recipient Presidential medal for merit, 1948, Pub. Welfare medal Nat. Acad. Scis.-NRC, 1960, Conrad award U.S. Navy, 1957, Distinguished Service award Jacksonville U., 1958, Midwest Research Inst. citation, 1961, Presidential medal freedom, 1963. Fellow A.A.A.S. (bd. dirs. 1956-62, pres. 1963), Am. Phys. Soc., Am. Inst. Aeros. and Astronautics, Am. Acad. Arts and Scis., Am. Assn. Physics Tchrs.; mem. Am. Inst. E.E., Washington Acad. Aci., Philos. Soc. Washington, Washington Acad. Medicine, N.Y. Acad. Scis., Sci. Research Soc. Am. (Procter prize 1960), Phi Beta Kappa, Sigma Xi. Editor: (with Lincoln Thiesmyer) Combat Scientists, 1947; Kimball's College Physics, 1952. Editorial bd. Am. Jour. Sci., 1935-42. Research, numerous publs. on elec. properties of solids; positive elec. emission from various heated salts; formulated classical electron theory of elec. conduction in solids based on energy level difference between bound and free electrons. Home: 5306 Carvel Rd., Washington 20016. Office: 2101 Constitution Av. N.W., Washington 20418.*

WATERMAN, Alan Tower, Jr., Am. elec. engr.; b. Northampton, Mass., July 8, 1918; s. Alan T. and Mary (Mallon) W.; A.B., Princeton, 1939; B.S., Cal. Inst. Tech., 1940; postgrad. U. Minn., 1941-42; A.M., Harvard, 1949, Ph.D., 1952; m. Loretta A. Walker, Oct. 12, 1946; children—Linda E. (Mrs. Lester Sloan), Donna L. (Mrs. John Hickey), Alan D., Bruce E. Meteorologist, Am. Airlines, N.Y.C., 1940-41; research asso. Cal. Inst. Tech., Pasadena, 1942-45; research scientist Columbia, N.Y.C., 1945; chief meteorologist U. Tex., Austin, 1945-46; prof. elec. engring. Stanford (Cal.) U., 1952—. Cons. to govt. and industry. Fellow I.E.E.E.; mem. Am. Phys. Soc., Am. Meteorol. Soc., A.A.A.S., Am. Soc. for Engring. Edn., Internat. Sci. Radio Union, Sigma Xi. Contbr. to ency., textbooks. Analysis of structure of lower atmosphere in scattering, reflecting, and bending short radio waves, including their propagation to distances beyond the horizon. Home: 562 Gerona Rd., Stanford, Cal. 94305.*

WATERS, Aaron Clement, Am. geologist; b. Waterville, Wash., May 6, 1905; s. Richard Jackson and Hattie (Clement) W.; B.S., U. Wash., 1926, M.S., 1927; Ph.D., Yale U., 1930; m. Elizabeth Pearl von Hoene, June 10, 1940; children—Susan Margaret, Sharon Christine. Instr. Yale U., 1928-30; with Stanford U., 1930-42, 1945-50, prof., 1938-42, 1945-50; geologist, research geologist U.S. Geol. Survey, 1942-45, 1950-52; prof. Johns Hopkins U., 1952-63; prof. U. Cal., Santa Barbara, 1963-67, U. Cal., Santa Cruz, 1967—; geol. cons. various firms, govt. agys., 1932—. Guggenheim Meml. fellow, 1938; Columbia U. Bicentennial lectr., 1954; NSF Sr. Postdoctoral fellow, 1960, Condon lectr., 1964. Mem. Nat. Acad. Scis., Am. Acad. Arts and Scis., Geol. Soc. Am., Mineral. Soc. Am., Sigma Xi. Study of the geology of the Columbia River Plateau, and the Cascade Mountains, especially peterology of granites and basalts. Home: 308 Moore St., Santa Cruz, Cal. 950660.*

WATERS, John Richard, physicist; b. London, Eng., Apr. 26, 1930; s. George R. and Alice (Nowell) W.; B.A., Oxford (Eng.) U., 1952; M.S., U. Pitts., 1953; Ph.D., Cornell U., 1957; m. Mary Slater, Mar. 26, 1952; children—Joan M., Sandra M. Came to U.S., 1952, naturalized, 1965. Sr. scientist A.E.R.E., Harwell, Eng., 1957-59; research asso. Princeton, 1960-62; research specialist Jet. Propulsion Lab., Cal. Inst. Tech., Pasadena, 1962-63; project dir., sr. scientist Am. Sci. & Engring., Inc., Cambridge, Mass., 1963-67; v.p. research Johnston Labs., Inc., 1967—. Mem. Am. Phys. Soc., Am. Geophys. Union, Am. Inst. Aeros. and Astronautics, Sigma Xi. Contbr. articles to tech. jours. Measured nuclear resonance parameters using slow neutrons; developed scintillation chamber and used to measure meson decay and Cerenkov radiation; research in detection X-rays from celestial sources. Home: 801 Stags Head Rd., Towson, Md. 21204. Office: 3617 Woodland Av., Balt. 21215.*

WATERS, Ralph Milton, Am. anesthesiologist; b. North Bloomfield, O., Oct. 9, 1883; A.B., Adelbert Coll, 1907; M.D., Western Reserve U., 1912; m. 1913; 4 children. Became extern Emergency Hosp., 1908, intern German Hosp., 1909; practice medicine, 1916-27; asst. prof. surgery Med. Sch., U. Wis., 1927-28, asso. prof, 1928-33, prof. anesthesia, 1933-49, also chmn. dept., dir. anesthesia univ. hosps. Mem. subcom. anesthesia NRC, 1942-46. Recipient Hickman medal Royal Soc. Medicine, London, 1944, Cutter medal Phi Rho Sigma, 1946. Mem. A.A.A.S., Am. Soc. Anesthesiology, A.M.A., Am. Assn. for History Medicine, Internat. Anesthesis Research Soc., Wis. Acad. Scis. First clin. use (with others) of cyclopropane as anesthetic, 1934; studies on anesthetic drugs, anesthetic inhalation equipment, chloroform, nitrous oxide. Home: Rural Route 3, Box 433, Orlando, Fla.

WATERS, W. E., Am. horticulturist; b. Smithtown, Ky., Sept. 19, 1931; s. Abraham Lincoln and Mildred (Childers) W.; student Cumberland Coll., 1950-52; B.S., U. Ky., 1954, M.S., 1958; Ph.D., U. Fla., 1960; m. Mary Elizabeth Sumler, May 30, 1952; children—Marie Estelle, Alan Lee, Sheila Louise. Asst. county farm agt., Hazard, Ky., 1954; asst. horticulturist Gulf Coast Expt. Sta., Bradenton, Fla., 1960-66, asso. horticulturist, 1966—. Mem. Am., Fla. State socs. hort. sci., So. Weed Conf., Fla. Soil and Crop Soc. Research and publs. on nutritional requirements of chrysanthemums and gladiolus; photoperodic responses of Lillium longiflorum; effectiveness of chem. herbicides on floral crops. Home: 1418 37 St. Ct. W. Office: Gulf Coast Expt. Sta., Manatee Sta., Bradenton, Fla.*

WATERSON, Anthony Peter, English virologist; b. Hornsea, Eng., Dec. 23, 1923; s. Frederick and Frances (Cooper) W.; B.A., Emmanuel Coll., Cambridge (Eng.) U., 1944, M.D., 1954; postgrad. London Hosp. Med. Coll., 1944-47. Faculty, U. Cambridge, 1953-64, lectr., 1958-64; fellow, tutor Emmanuel Coll., 1954-64; prof. med. microbiology St. Thomas's Hosp. Med. Sch., U. London, 1964-67, prof. virology Royal Postgrad. Med. Sch., U. London, 1967—. Cons. virologist Central Vet. Lab., Weybridge, Eng., 1965—. Mem. Royal Coll. Physicians London, Soc. for Gen. Microbiology, Anglo-German Med. Soc. Author: Introduction to Animal Virology, 1961, 2d edit. 1967; also articles. Research on structure of viruses and its significance, structure influenza virus, antigenic components and measles virus. Home: Brackendene House, Woburn Hill, Weybridge, Surrey, Eng. Office: Royal Postgrad. Med. Sch., Ducane Rd., London, W.12, Eng.*

WATERTON, Charles, English naturalist; b. Walton Hall, Eng., June 3, 1782; s. Thomas and Anne (Bedingfeld) W.; ed. Arthur Storey Sch., Tudhoe, Eng., 1792, Stonyhurst Coll., Lancashire, Eng., 1796-180; m. Anne Edmonstone, 1829; 1 son, Edmund. Explorer, collector birds, reptiles, fauna in Brit. Guiana (now Guyana), 1804-13, 16-17, 20-21, Brazil, 1816-17,

Antilles, U. S., Can., 1824; established wildlife sanctuary home Walton Hall. Author: Wanderings in South America, the North-West of the United States, and the Antilles in the Years 1812, 1816, 1820, 1824 (pub. 1825); Essays in Natural History, 1838; Essays, 1844, 57; Autobiography. Studied habits of jacamars, red grosbeak, sunbird, tinamous, hummingbirds, vampires, sloths, and monkeys; investigated whether a vulture is attracted to food by sight (Audubon's contention) or by scent; studied tropical fauna; wrote articles taxidermy procedures birds. Died Walton Hall, May 27, 1865.

WATKINS, John Goodrich, Am. psychologist; b. Salmon, Ida., Mar. 17, 1913; s. John Thomas and Ethel (Goodrich) W.; student Coll. Ida., 1929-30, 31-32; B.S., U. Ida., 1933, M.S., 1936; Ph.D., Columbia, 1939; m. Doris Wade Tomlinson, June 8, 1946; children—John Dean, Jonette Alison, Richard Douglas, Gregory Keith, Rodney Philip. Instr. high sch., Ida., 1933-39; faculty Ithaca Coll., 1939-40, Ala. Poly. Inst., 1941-43, Wash. State Coll., 1946-49; clin. psychologist VA Hosp., American Lake, Wash., 1949-50; chief clin. psychologist VA Mental Hygiene Clinic, Chgo., 1950-53, VA Hosp., Portland, Ore., 1953-64; prof. psychology, dir. clin. tng. U. Mont., Missoula, 1964—. Lectr. numerous univs.; clin. asso. U. Ore. Med. Sch., 1957; pres. Am. Bd. Examiners in Psychol. Hypnosis, 1960-62. Mem. Internat. Soc. for Clin. and Exptl. Hypnosis (co-founder, pres. 1965-67, recipient awards 1960- 65), A.A.A.S., Am. Psychol. Assn., Acad. Psychosomatic Medicine, Am. Acad. Psychotherapists, Sigma Xi. Author: Objective Measurement of Instrumental Performance, 1942; Hypnotherapy of War Neuroses, 1949; General Psychotherapy, 1960; also articles. Developed and pub. standardized tests of performance for band instrument; devised new techniques of hypnotherapy and hypnoanalysis; formulated a theory of psychotherapy modifying and extending psychoanalytic, phenomenological and existential theories. Home: 631 Pattee Creek Dr., Missoula, Mont. 59801.*

WATSON, Adam, Scottish ecologist; b. Turriff, Scotland, Apr. 14, 1930; s. Adam and Margaret (Rae) W.; B.Sc., Aberdeen (Scotland) U., 1952, Ph.D., 1956; m. Jenny Raitt, Mar. 19, 1955; children—Jenny, Adam Christopher. Arctic Inst. scholar McGill U., Montreal, Que., Can., 1952; zoologist Arctic Inst. Expdn. to Baffin Island, 1953; asst. zoology dept. Aberdeen U., 1953-56, sr. research fellow, 1956-60; sr. sci. officer Nature Conservancy, Banchory, Scotland, 1960-65, prin. sci. officer, 1965—, officer-in-charge unit of grouse and moorland ecology, 1966—. Mem. Brit. Ecol. Soc., Arctic Inst. N. Am., Brit. Ornithologists Union. Research and publs. on population regulation particularly red grouse and ptarmigan; discovered that population density is regulated by aggressive behavior and social status of birds. Home: Clachnaben, Crathes, Aberdeenshire, Scotland. Office: Unit of Grouse and Moorland Ecology, Banchory, Scotland.*

WATSON, Alfred James, Australian chemist; b. Melbourne, Australia, June 20, 1916; s. James Ferrie and Elsie (Basterfield) W.; Asso. Diploma, Melbourne Tech. Coll., 1939; Fellowship Diploma, Royal Melbourne Inst. Tech., 1962; m. Margaret Lucille Miller, Jan. 3, 1942; children—Peter James, Judith Lucille, Pamela Margaret. With div. forest products Commonwealth Sci. and Indsl. Research Orgn., South Melbourne, Australia, 1933—, research officer, from 1942, prin. research scientist, 1963—. Fellow Royal Australian Chem. Inst.; mem. Australian and New Zealand Pulp and Paper Tech. Assn. (past pres., chmn. editorial com. 1958—). Research, publs. on chem. composition of wood, pulping and papermaking properties of tropical woods, eucalyptus, plantation grown softwoods, factors influencing pulping and papermaking operations, relationship between wood fibre properties and papermaking characteristics. Home: 24 Glendearg Av., Malvern, Victoria. Office: 69 Yarra Bank Rd., South Melbourne, Victoria, Australia.*

WATSON, Cecil James, Am. physician; b. Mpls., May 31, 1901; s. James Alfred and Lucia (Coghlan) W.; B.S., U. Minn., 1923, M.S., 1925, M.D., 1926, Ph.D., 1928; M.D. (hon.), U. Mainz (Germany), 1963; m. Joyce Petterson, Sept. 10, 1925. Pathologist, dir labs. Mpls. Gen. Hosp., 1926-28; instr. U. Minn., Mpls., 1925-28, fellow, 1932-33, faculty, 1934—, prof., head dept. medicine, 1942-66, Distinguished Service prof., 1961—, dir. med. unit for teaching and research at Northwestern Hosp., 1966—; NRC fellow, Munich, Germany, 1930-32; vis. prof. medicine, asso. dir. health div. Metall. Lab., U. Chgo., 1943-45. Mem. med. fellowship bd. NRC, 1957-62, chmn., 1960-62; bd. councillors for intramural programs Nat. Inst. Arthritis and Metabolic Disease, 1957-61, 64—, chmn., 1957-61, mem. council, 1951-54. Recipient Order of Merit, Republic of Chile, 1960, James F. Bell Distinguished Service award Minn. Med. Found., 1961, John Phillips medal A.C.P., 1957, Gordon Wilson medal Am. Clin. and Climatol. Assn., 1947; named Hon. Prof., San Marcos U., Lima, Peru, 1947, U. Chile, 1947. Mem. Nat. Acad. Sci., Assn. Am. Physicians, (past pres.), Assn. Prof. Medicine (past pres.), Am. Soc. Clin. Investigation (past pres.), Central Soc. Clin. Research (past pres.). Research and numerous publs. on discovery of histoplasmosis in U.S.; delineation (with Lerner) of cryoglobulinemia; isolation of crystalline stercobilin, crystalline naturally occurring porphyrins; basic classification of porphyria; bovine porphyria; excretion and metabolism of bile pigments and porphyrins in human disease; basic and clin. studies of jaundice and liver disease, especially hepatitis; urine Ehrlich reaction in relation to porphyria and liver disease. Home: 3318 Edmund Blvd., Mpls. 55406. Office: 810 E. 27th St., Mpls. 55407.*

WATSON, David Meredith Seares, English paleontologist; b. Manchester, Eng., June 18, 1886; s. David and Mary (Seares) W.; B.Sc., Manchester U., 1907; D.Sc. (hon.), U. Capetown (S. Africa), 1929, Manchester U., 1943, U. Reading (Eng.), 1948, D.Sc., U. Wales, 1948, U. Witwatersrand (S. Africa), 1948; LL.D., Aberdeen (Scotland) U., 1943; m. Katherine Margarite Parker, Dec. 21, 1917; children—Katherine Mary (Mrs. P. A. J. Powell), Janet Vida (Mrs. John Sutton). Hon. lectr. U. Coll., London, Eng., 1911-21; Jodrell prof. zoology and comparative anatomy U. London, 1921-51, emeritus prof., 1951—; Alexander Agassiz prof. Harvard, 1952. Mem. Agrl. Research Council, 1933-43. Hon. fellow U. Coll., 1948. Recipient Erzhertzog Rainer medal K.K. biol. gesellschaft zu Wien, 1928; Lyell medal Geol. Soc. London, 1935; Linnean Soc. London medal, 1949; Wollaston medal Geol. Soc., 1965. Fellow Royal Soc., 1922 (Darwin medal 1942); hon. mem. Acad. Sci. USSR, Am. Soc. Herpetologists and Ichtyhologists, N.Y., Bavarian, Indian, Cal., Royal Swedish acads. sci., Paleontol. Soc. Russia, Nat. Acad. Sci. (Mary Clarke Thompson medal 1941), Kaiserlich Leopold.-Carolinische Deutsche Acad. Naturf. in Halle, Am. Mus. Natural History, Paleontol. Soc. U. S. A., Geol. Soc. France, Royal Acad. Belgium, Royal Soc. Sci. Uppsala, Soc. Vertebrate Paleontologists, Royal Physiol. Soc. Lund, Royal Soc. Edinburgh (Makdougal Brisbane, prize 1939), Geol. Soc. Am., Am. Acad. Arts and Scis. Author: Paleontology and Modern Biology, 1951. Research, numerous publs. on fossil plants and vertebrates. Home: 2 The Knoll, Knoll Rd., Godalming, Surrey, Eng.*

WATSON, Dennis Wallace, microbiologist; b. Morpeth, Ont., Can., Apr. 29, 1914; s. William and Sara (Verity) W.; B.S.A., U. Toronto, 1934; M.Sc., Dalhousie U., 1937; Ph.D., U. Wis., 1941; m. Alicemay Whittier, June 15, 1941; children— Catherine W., William V. Came to U.S., 1938, naturalized, 1946. Research asso. U. Wis., 1942, faculty, 1946-49; vis. investigator Rockefeller Inst. for Med. Research, 1942; investigator Connaught Labs. for Med. Research, U. Toronto, 1942-44; faculty U. Minn., Mpls., 1949—, prof., head dept. microbiology, 1964—. Vis. prof. U. Washington Med. Sch., 1950; spl. research fellow USPHS, Dr. A. Wander Forshungsinstitut, Freiburg, Germany, 1960-61. USPHS career award prof., 1962-64. Mem. Am. Soc. Microbiology (pres. 1968-69), Am. Acad. Microbiology (gov.), Soc. for Exptl. Biology and Medicine (council), Am. Assn. Immunologists, Am. Chem. Soc., A.A.A.S., Minn. Path. Soc., Sigma Xi. Research, numerous publs. on immunology of bacterial toxins, hostparasite factors in Group A streptococcal infections,B. anthracis and Tb, specific factors in resistance to infection, mechanisms of antibody prodn. Home: 2106 Hendon Av., St. Paul 55108. Office: University of Minn., Mpls. 55455.*

WATSON, Donald Gordon, Am. biologist; b. Moscow, Ida., Mar. 12, 1921; s. Allan and Olive (Spitler) W.; student Mt. Vernon Jr. Coll., 1939-41; B.S., U. Wash., 1948; m. Madge Shardlow, Apr. 15, 1950; children—Elaine M., Marian J., Ian A. Biologist Fisheries Research Inst. U. Wash., 1947-48; biologist State of Wash. Dept. Game, Ford, Wash., 1949; biologist, sr. scientist Gen. Elec. Co., Richland, Wash., 1949-64; sr. scientist Pacific Northwest Lab. Battelle Meml. Inst., Richland, 1965—. Mem. Am. Fisheries Soc., Am. Soc. Limnology and Oceanography, Am. Inst. Fishery Research Biologists, Pacific Fishery Biologists. Studies, publs. on effects of radioactive isotopes on fish; distbn. of worldwide fallout in arctic ecosystems; mineral cycling in aquatic ecosystems; population dynamics of chinook salmon in the vicinity of nuclear reactors. Office: Pacific Northwest Lab., Battelle Meml. Inst., Richland, Wash. 99352.*

WATSON, Donald Stevenson, economist; b. Greenwood, B.C., Can., Oct. 28, 1909; s. James Livingstone and Roberta (Stevenson) W.; B.A., U. B.C., 1930; Ph.D., U. Cal., Berkeley, 1935; m. Liselotte Bunge, Oct. 4, 1935; children—Margot (Mrs. Thomas A. Zener), Wendy . Came to U.S., 1930, naturalized, 1938. Faculty George Washington U., Washington, 1935—, prof. econs., 1948—, chmn. dept., 1945-50, 56-60. Cons. govt. agys., bus. firms, research orgns., 1937—; prin. investigator govt. research projects George Washington U., 1960-61, 63-66. Mem. Am. Econ. Assn., Econometric Soc., A.A.A.S. Author: Economic Policy, 1960; Price Theory and Its Uses, 1963; (with A.E. Burns) Government Spending and Economic Expansion, 1940; (with A.E. Burns and A.C. Neal) Modern Economics, 1948; also articles. Editor: Price Theory in Action, A Book of Readings, 1965. Research on operation of fed. patent policies in fed.. contracts for research and devel. Home: 8540 Georgetown Pike, McLean, Va. 22101.*

WATSON, Earnest Charles, Am. physicist; b. Sullivan, Ill., June 18, 1892; s. Charles Grant and Alice (Smith) W.; Ph.B., Lafayette Coll., 1914, Sc.D., 1958; postgrad. U. Chgo., 1914-17; m. Elsa Jane Werner, Oct. 6, 1954. Faculty, Cal. Inst. Tech., Pasadena, 1919-62, prof., 1930-62, prof. emeritus, 1962—, dean faculty, 1945-60, chmn. div. math., astronomy, physics, elec. engring., 1946-49, acting pres., 1956. Sci. attaché Am. Embassy, New Delhi, India, 1960-62; U.S. del. sci. adv. com. Central Treaty Orgn., 1961, 62, mem. sci. survey team, 1963; cons. S. and S.E. Asia Program, Ford Found., 1964—. Recipient Presidential Certificate of Merit, 1945. Mem. Phi Beta Kappa, Sigma Xi, Tau Beta Pi, India Internat. Centre. Adminst. research on spectroscopy, x-rays, beta-ray, torpedoes, atomic bomb, also rocketry, World War II; adminstrn. of edn. in scis. Home: 930 Knollwood Dr., Santa Barbara, Cal. 93103.*

WATSON, George Elder, III, Am. ornithologist; b. N.Y.C., Aug. 13, 1931; s. George Elder and Forsyth (Patterson) W.; B.A., Yale, 1953, M.S., 1961, Ph.D., 1964; postgrad. Am. Sch. Classical Studies, Athens, Greece, 1953-54. Asst. curator birds U.S. Nat. Mus., Smithsonian Instn., Washington, 1962-64, asso. curator, 1964-65, curator, 1966—67, chmn. dept. of vertebrate zoology. 1967—. Fellow A.A.A.S.; mem. Am., Brit. ornithologists unions, Cooper, Wilson, German ornithol. socs., Soc. Study Evolution, Soc. Systematic Zoology, Biol. Soc. Washington. Author: (with Zusi and Storer) Preliminary Field Guide to the Birds of the Indian Ocean, 1963; Preliminary Smithsonian Identification Manual, Seabirds of the Tropical Atlantic Ocean, 1965, rev. 1966; also articles. Evolutionary and faunistic studies of birds of Aegean islands, moults, seabird identification and ecology. Home: 2621 O St. N.W., Washington 20007. Office: Dept. Vertebrate Zoology, U.S. Nat. Museum, Smithsonian Institution, Washington 20560.*

WATSON, George Mario, phys. chemist; b. Mexico City, Mexico, June 29, 1914; s. Jorge F. and Margarita (Peza) W.; B.S. in Chem. Engring., U. Tex., 1938, M.S., 1940, Ph.D., 1943; m. Margaret Alexander, Oct. 28, 1938; children—Mary Margaret (Mrs. Marco A. Solis), George Mario, Maribel. Came to U.S., 1930, naturalized, 1941. Plant mgr., gen. supt. Sochule S.A., A. Guayule Rubber Mfg. Co., Cuatro Cienegas, Coah, Mexico, 1943-48; faculty Tex. A. and M. Coll., College Station, 1948-55, prof., 1950-55; sr. chemist Oak Ridge Nat. Lab., 1955-60, asso. dir. reactor chemistry div., 1960—. Mem. Am. Chem. Soc., Am. Nuclear Soc. (chmn. Oak Ridge chpt. 1965-66). Research, numerous publs. on pressure-vol. temperature behavior hydrocarbons, chem. kinetic studies in aqueous solutions, solubility gases in molten fluoride solvents, chemistry fission products in molten fluoride systems, gas transport in porous media. Home: 824 Carrington Rd., Knoxville, Tenn. 37919. Office: P.O. Box X, Oak Ridge Nat. Lab. Bldg. 4500S, Oak Ridge 37831.*

WATSON, Hamish, Brit. cardiologist; b. Edinburgh, Scotland, June 26, 1923; s. John Thomas Richardson and Annie Ewing (Spence) W.; M.B., Ch.B., U. Edinburgh, 1945, M.D., 1962; m. Lesley Leigh Dick Wood, Feb. 8, 1951 children—Penelope Jane, Jillian Amanda. Cardiologist, Dundee (Scotland) Group Teaching Hosps.; lectr. cardiology and pediatric cardiology U. St. Andrews (Scotland), dep. dir. postgrad. med. edn. Recipient Territorial Efficiency Decoration. Fellow Royal Coll. physicians Edinburgh; mem. Royal Coll. Physicians London, Assn. European Paediatric Cardiologists (pres.), Internat. Soc. Cardiology (chmn. pediatric sect.), Assn. Physicians Gt. Britain and Ireland, Brit. Cardiac Soc. Author: A Textbook of Paediatric Cardiology, 1967; (with R. Walmsley) The Clinical Anatomy of the Heart, 1967; also articles. Asst. editor Brit. Heart Jour. Research on congenital heart disease especially in newborn, cine-angiocardiography, biplane image intensification, intracardial electrocardiography, pressure/flow relationships in right ventricular obstruction, clin. anatomy of heart. Home: Nethermains of Kinnaird, Inchture, Perthshire, Scotland. Office: Dept. Medicine, U. St. Andrews, Queen's Coll., Dundee, Scotland, U.K.*

WATSON, Herbert Edmeston, Brit. chemist; b. Kew, England, May 17, 1886; s. Arthur and Helen (Rumpff) W.; ed. Marlborough, also univs. London, Berlin, Geneva and Switzerland; Dr. ès sc.; m. Kathleen Rowson, Apr. 9, 1917; children—Bruce, Margaret. Prof. phys. and inorganic chemistry Indian Inst. Sci., Bangalore, 1911-34; prof. chemistry Univ. Coll., London, 1934-51; with Admiralty, 1939-45. Mem. Royal Inst. Chemistry, Inst. Chem. Engrs. Publns. on rare gases, dielectric constants, devel. Indian industries, distillation; creator incandescent neon lamp, 1911. Address: Westside, Knowl Hill, Woking, Eng.

WATSON, Hewett Cottrell, Brit. botanist; b. Yorkshire, Eng., May 9, 1804; s. Holland and Harriett (Powell) W.; student Phrenology and natural history, Edinburgh, Scotland, 1828-32. Inherited estate in Derbyshire, Eng., circa 1826; settled at Thames Ditton, Eng., 1853; visited Azores, 1842. Fellow Linnean Soc., mem. Royal Med. Soc. Edinburgh (elected sr. pres. 1831-32). Recipient Prof.'s Gold medal for bot. essay, 1831. Author: Outlines of the Geographical Distribution of British Plants, 1832; The New Botanists Guide to the Localities of the Rarer Plants of Britain, 1835-37; Cybele Britannica, 1870; Typographical Botany, 1873-74; Editor, Phrenological Jour., 1837-40, London Catalogue of British Plants, 1844-74. Classified Brit. plants according to local distbn. Died July 27, 1881.

WATSON, James Bennett, Am. anthropologist; b. Chgo., Aug. 10, 1918; s. James B. and Elizabeth (Thaxter) W.; student U. Me., 1936-37, N.J. State Coll., 1937-39; A.B., U. Chgo., 1941, A.M., 1945, Ph.D., 1948; m. Virginia Drew, Mar. 18, 1943; children—Anne Thaxter, James B. Asst. prof. Escola Livre de Sociologia e Politica, São Paulo, Brazil, 1944-45; faculty Beloit Coll., 1945-46, U. Okla., 1946-47, Washington U., St. Louis, 1947-55; prof. anthropology U. Wash., Seattle, 1955—. Field expdns. Ariz., 1942, Brazil, 1943, Colo., 1949, 50, Australian Papua, New Guinea, 1953-55, 59, 63-64. Mem. A.A.A.S., Am. Anthrop. Assn., Am. Ethnol. Soc., Soc. for Applied Anthropology, Polynesian Soc., Far Eastern Prehistory Assn., Sigma Xi. Editor, Contbr. New Guinea: the Central Highlands, 1964. Research, publs. on analysis, documentation small peripheral cultures in S.Am. and New Guinea and influence nearby civilizations upon them. Home: 2610 E. Interlaken Blvd., Seattle 98102.*

WATSON, James Craig, astronomer; b. Fingal, Ont., Can., Jan. 28, 1838; s. William and Rebecca (Bacon) W.; grad. U. Mich., 1857; Ph.D. (hon.), U. Leipzig (Germany), Yale; LL.D., Columbia; m. Annette Waite, May 1860. Came to U. S., 1850; mastered theoretical and practical astronomy studying under Francis Brunnow; while still a student, ground, polished and mounted 4 inch achromatic objective; published 15 astron. papers before age 21; prof. astronomy in charge obs. U. Mich., 1859, prof. physics, 1860-63, prof. astronomy, dir. obs., 1863-79; discovered asteroid Eurynome (1st of his 22 astron. discoveries), 1863; participated in eclipse expdns. to Ia., 1869, Sicily, 1870, Wyo., 1878; in charge of expdn. to observe transit of Venus in China, 1874; dir. Washburn Obs., U. Wis., Madison, 1879-80; mem. Nat. Acad. of Scis., 1868, Royal Acad. of Scis. in Italy, Am. Philos. Soc.; recipient Lalande prize French Acad. of Scis., 1870; Author: Popular Treatise on Comets, 1861; Theoretical Astronomy (became textbook in U. S., Germany, France, Eng.), 1868; Tables for the Calculation of Simple or Compound Interest, 1878; also papers. Discovered a number of asteroids and comets; wrote on the theory of motion of comets and planets. Died Madison, Wis., Nov. 22, 1880.

WATSON, James Dewey, Am. molecular biologist; b. Chgo., Apr. 6, 1928; s. James Dewey and Jean (Mitchell) W.; B.S., U. Chgo., 1947, D.Sc., 1961; Ph.D., Ind. U., 1950, D.Sc., 1963; LL.D., Notre Dame U., 1965; m. Elizabeth V. Lewis, Mar. 28, 1968. Merck fellow of NRC, U. Copenhagen, 1950-51; research Cambridge U., Cavendish Lab., 1951-53, 55-56; sr. research fellow in biology Cal. Inst. Tech., 1953-55; faculty Harvard, Cambridge, Mass., 1956—, prof. biology, 1961, dir. Cold Spring Harbor Labs. Quantitative Biology, 1968—. Trustee Cold Spring Biol. Labs., 1965—; cons. President's Sci. Adv. Com., 1962—; bd. sci. counselors Nat. Inst. Arthritis and Metabolic Diseases, 1967. Recipient Lasker prize, 1960, Eli Lilly award in biochemistry, 1959, Nobel prize in medicine and physiology (with Crick and Wilkins), 1962, Research Corp. prize (with Crick), 1962. Mem. Am. Soc. Biol. Chemists, Nat. Acad. Scis., Royal Danish, Am. acads. arts and scis. Author: The Molecular Biology of the Gene, 1965; The Double Helix, 1968; also articles. Elucidated (with Crick) double-helical structure of DNA molecule; research on role of RNA in protein synthesis. Home: 10 Appian Way, Cambridge, Mass. 02138.*

WATSON, John, surgeon; b. Londonderry, Ireland, Apr. 16, 1807; M.D., N.Y. Coll. Phys. and Surg., 1932. Mem. surg. staff N.Y. Hosp., 1832-33; physician N.Y. Dispensary, 1833-35; attending surgeon N.Y. Hosp., introducing many reforms and improvements, 1839-62; established (with Dr. Henry D. Bulkley) infirmary for cutaneaus diseases which became Broome St. Sch. Medicine where he held chair of surg. pathology. Instrumental in organizing N.Y. Med. and Surg. Soc., A.M.A., N.Y. Acad. Medicine (pres. 1859-60). Author: The True Physician, 1960; The Medical Profession in Ancient Times, 1856; A History of Medicine, 1862. Credited with performing 1st recorded operation of esophagotomy for relief of esophageal stricture, 1844. Died N.Y.C., June 3, 1863.

WATSON, John B(roadus), Am. psychologist; b. Greenville, S.C., Jan. 9, 1878; s. Pickens Butler and Emma K. (Roe) W.; A.M., Furman U., 1900; grad. student in psychology U. Chgo., 1900-03. Ph.D., 1903; LL.D., Furman U., 1919; m. Mary Ickes, Oct. 1, 1904; children—Mary I., John I.; m. 2d. Rosalie Rayner, Dec. 31, 1920; children—William Rayner, James Broadus. Prin. Batesburg Inst., 1899-1900; asst. in exptl. psychology U. Chgo., 1903-04, instr., 1904-08; prof. exptl. and comparative psychology, dir. psychol. lab., Johns Hopkins, 1908-20; became v.p. J. Walter Thompson Co., N.Y., 1924; v.p. William Esty & Co., N.Y., 1936-46. Editor Psychol. Rev., 1908-15, Jour. Exptl. Psychology, 1915-27. Commd. maj. Aviation Sect., Signal Corps, U.S.R., 1917; on duty, Washington, Mineola, and with A.E.F. Fellow Am. Acad. Arts and Scis.; mem. Am. Psychol. Assn. (pres. 1915), Am. Physiol. Soc., Sigma Xi, Phi Beta Kappa. Author: Animal Education, 1903; Behavior, 1914; Homing and Related Activities of Birds, 1915; Suggestions of Modern Science Concerning Education, 1917; Psychology from the Standpoint of the Behaviorist, 1919; Behaviorism, 1925. Ways of Behaviorism. 1928; Psychological Care of Infant and Child, 1928. Contbr. on neurology, animal and infant psychology; founded behaviorist sch. of psychology opposing introspective sch.; formulated stimulus-response pattern as basic unit of behavior; accepted Pavlov's conditioned response concept; from his research, concluded that there are 3 unlearned responses in newborn infants—love, fear, rage; stressed that psychology should be empirical study of what man observedly does and not study of conscious and unconscious mind. Died N.Y.C., Sept. 25, 1958.

WATSON, Kenneth Marshall, Am. physicist; b. Des Moines, Sept. 7, 1921; s. Louis Erwin and Irene (Marshall) W.; B.S., Ia. State Coll., 1943; Ph.D., U. Ia., 1948; m. Elaine C. Miller, Mar. 30, 1946; children—Ronald M., Mark Louis. Staff, Naval Research Lab., 1943-46; mem. Inst. for Advanced Study, 1948-49; asst. prof., U. Ind., 1951-53; asso. prof., U. Wis., Madison, 1953-57; prof. U. Cal., Berkeley, 1957—, staff Lawrence Radiation Lab., 1949-51, 1957—. Cons., Gen. Dynamics Corp., Lockheed Corp., Boeing Co., Rand Corp., NASA. Mem.., Am. Phys. Soc., Am. Geophys. Union. Author: (with M.L. Goldberger) Collision Theory, 1964; (with J.W. Bond and J.A. Welch) Atomic Theory of Gas Dynamics, 1964 (with J. Nuttall) Topics in Several Particle Dynamics, 1967. Research and publs. on theory of scattering bycomposite systems, charge independence of pi-meson phonomena, fluctuation phenomena in scattering e xperiments. Home: 635 Spruce St., Berkeley, Cal. 94707.*

WATSON, Sir Malcolm, Brit. physician; b. Aug. 24, 1873; s. George Watson; M.D., U. Glasgow (Scotland), 1903; student U. Coll., London; LL.D., U. Glasgow, 1924; M.B., C.M., 1895; D.P.H.; (hon.) LL.D., 1924; Hon. Dipl., Singapore, 1926; m. Jean Alice Gray (dec. 1935); 3 sons; m. 2d, Constance Evelyn Loring; 1 dau. House surgeon, house physician Glasgow Royal Infirmary; resident med. officer Smithston Asylum, Greenock Malayan Med. Service, 1900-08; pvt. and cons. practice, Malaya, 1900-28; prin. Malaria Dept., also dir. Inst., Putney, 1928-33, then hon. cons.; chmn. Kundong Rubber Co., Jas. Craig Ltd., Engrs., Malaya. Adviser to govts. of Nepal, Hyderabad, Baroda, Patiala, Bengal, So. Rhodesia, also rubber, tea and engring. cos. Recipient Stewart medal Rubber Grower's Assn., Stewart prize Brit. Med. Assn., 1927, Sir Wm. Jones Gold medal Asiatic Soc. Bengal, 1928, Mary Kingsley medal Liverpool Sch. Tropical Medicine, 1934, Albert medal Royal Soc. Arts, 1939. Fellow Geol. Soc., Inc. Soc. Planters (hon.), Royal Faculty Physicians and Surgeons (hon.); corr. mem. Soc. Path. Exotique Paris. Author: Rural Sanitation in the Tropics, 1915; Prevention of Malaria in the Federated Malay States, 2d edit. 1921; African Highway, 1952; also articles. Research on malaria in Klang, Federated Malaya State, 1901, Netherlands East Indies, 1911, India, 1923, Africa, 1929, Balkans, 1930; discovered origin of crescent malaria, 1901, quartan malaria and nephritis, 1904; developed methods for controlling malaria using sanitation, sub-drainage of ravines, 1909, larvicide for mosquitoes in running water, 1914; developed method of controlling dust in mines, for prevention of silicosis, for prevention of explosions in coal mines, 1948; rubber tapping process. Died Dec. 28, 1955.

WATSON, Richard, English chemist; b. Heversham, Aug. 1737; s. Thomas Watson; studied chemistry; D.D., 1771; m. Dorothy Wilson, Dec. 21, 1773; 6 children, including Richard. Became exhibitioner Trinity Coll., Cambridge, 1753, pvt. tutor, 1757, head tutor, 1767, fellow, 1764, prof. chemistry, 1764, Regius prof. div., 1771; became bishop of Llandaff, 1782. Fellow Royal Soc., 1769. Author: Institutiones metallurgica, 1768; An Essay on the Subjects of Chemistry . . . , 1771; Chemical Essays, 1781; also papers. Experimented on manufacture of gunpowder; suggested that air might be solidified by lowering its temperature sufficiently; anticipated Blagden by showing that resistance to congelation in salts of the same kind is in direct simple proportion to quantity of salt dissolved; studied smelting of lead ore; inventor black-bulb thermometer; credited with making coal gas from coal heated in a clay tobacco pipe. Died 1816.

WATSON, Samuel, Brit. clockmaker; flourished Eng., 1674-1709; mem. Clockmakers Co.; noted for astron. clocks; built a celestial orbitary for Queen Mary; gave solution to longitude problem by a timepiece, 1707. Died Eng., 1709.

WATSON, Sereno, Am. botanist; b. East Windsor Hill, Conn., Dec. 1, 1826; s. Henry and Julia (Reed) W.; grad. Yale, 1847, began study of chemistry and mineralogy Sheffield Scientific Sch., 1866; Ph.D. (hon.), Ia. Coll., 1878; never married. Permitted to join Clarence King exploration party in Cal. as volunteer aid, 1867, commissioned to collect plants and secure data regarding them; asst. Gray Herbarium, Cambridge, Mass., 1873, curator, 1874-92, revised Gray's Manual of Botany, 1889. Author: Botany, 1871; Botany of California, 1st vol., 1876, 2d vol., 1880; Bibliographical Index to North American Botany, 1878; Manual of the Mosses of North America, 1884. Died Cambridge, Mass., Mar. 9, 1892.

WATSON, Stanley Willard, Am. microbiologist; b. Seattle, Jan. 3, 1931; s. Harry Sheller and Maud (Nichols) W.; B.S., U. Wis., 1949, M.S., 1951; Ph.D., U. Wis., 1957; m. Margaret Ewald, Sept. 16, 1952. Fellow, Woods, Hole (Mass.) Oceanographic Inst., 1951-52, asso. scientist, 1957—; fisheries biologist Fish and Wildlife Service, 1952-54. Mem. Soc. Am. Microbiologists, Soc. Gen. Microbiologists, Soc. Limnologist and Oceanography. Research, publs. on ecology, ultra-structure, biochemistry and taxonomy of marine nitrifying bacteria, 1st to culture Labyrinthula in pure culture. Home: Box 224, Woods Hole. Office: Woods Hole Oceanographic Institute, Woods Hole, Mass. 02543.*

WATSON, Sir Thomas, English physician; b. Montrath, Eng., Mar. 7, 1792; s. Joseph and Mary (Catton) W.; B.A., St. John's Coll., Cambridge, 1815, M.A., 1818, also LL.D.; student medicine St. Bartholomew's Hosp.; M.D., Cambridge, 1825; m. Sarah Jones, Sept. 15, 1825; children—Sir Arthur Townley; 1 dau. Practice medicine, London, 1825-70; Gulstonian lectr., 1827; Lumelian lectr., 1831; censor, 1828, 37, 38; physician Middlesex Hosp., London, 1827-43; prof. U. Coll., London, 1823-31, King's Coll., London, 1831-40. Fellow Royal Soc., 1859, Royal Coll. Physicians. Author: Lectures on the Principles and Practice of Physic, 1843. Suggested use of rubber gloves in surgery, 1843.

WATSON, Thomas Leonard, Am. geologist; b. Chatham, Va., Sept. 5, 1871; s. Fletcher B. and Pattie Booker (Tredway) W.; B.Sc., Va. Agrl. and Mech. Coll. (now Va. Poly. Inst.), 1890, M.Sc., 1893; postgrad. U. of Va., 1891; Ph.D., Cornell U., 1897; m. Adelaide Stephenson, Feb. 8, 1899. Instr. geology and mineralogy Va. Agrl. and Mech. Coll., 1892-95; asst. chemist Va. Expt. Sta., 1890-95; mem. Cornell U. party of geologists on 6th Peary Arctic expdn. to N. Greenland, 1896; pvt. research worker on rock decay U. S. Nat. Mus., 1897-98; asst. state geologist of Ga., 1898-1901; prof. geology Denison U., 1901-04; geologist Ga. Geol. Survey, 1902, N.C. Geol. Survey, 1903; field asst. U. S. Geol. Survey, 1903-08; prof. geology Va. Poly. Inst., 1904-07; prof. economic geology U. Va., 1907—, head prof, schools of geology, 1910—. State geologist and dir. Va. Geol. Survey, 1908—. Mem. Com. of 100 on Sci. Research; mem. exec. com. Nat. Conservation Congress. Fellow Geol. Soc. Am. (councilor, 1915-17), A.A.A.S, Mineral. Soc. Am. (councilor 1922—; chmn. com. on nomenclature and classification of minerals), Soc. Econ. Geologists; mem. Am. Inst. Mining Engrs., NRC (sub-com.). Co-author: Ries and Watson's Engineering Geology; Elements of Engineering Geology Contbr. on theoretic and economic geology to various geol. publs. and reports of state and federal surveys. Died Nov. 10, 1924.

WATSON, Tully Franklin, Am. physicist; b. Boswell, Okla., Oct. 11, 1902; s. Oliver L. and Leona (Adkins) W.; B.A., U. Okla., 1928, M.S., 1930; Ph.D., U. Ill., 1935; m. Faye Williams, Jan. 12, 1924; 1 son, Robert Lee. Head physics dept. Northeastern State Coll., Tahlequah, Okla., 1935-43, Phillips U., Enid, Okla., 1944-47; prof. physics Wichita (Kan.) State U., 1947—. Mem. Am. Assn. Physics Tchrs., Phi Beta Kappa, Sigma Xi, Sigma Pi Sigma. Research, publs. on band spectra and molecular structure. Home: 4108 N. Oliver St., Wichita, Kan. 67220.*

WATSON, Sir William, English physician, botanist, sci. experimenter; b. London, Eng., Apr. 3, 1715; grad. Merchant Taylor's Sch., 1726, univs. Halle, Wittenberg (Germany); D. Physic, 1738; M.D. (hon.) U. Halle, 1757; 2 children. Apothecary, London, from 1738; physician Foundling Hosp., London, 1762-87. Recipient Copley medal, 1745. Licentiate, fellow Royal Coll. Physicians (censor, trustee 1785-86), Royal Soc. (v.p.), 1741; mem. Royal Acad. Madrid. Knighted, 1786. Author: Experiments on the Nature . . . of Electricity (theory similar to Benjamin Franklin's exptl. findings on transmission speed, conductivity of electricity, 1746-48), 1746; sequel, 1746; Account of a Series of Experiments Instituted with a View of Ascertaining the Most Successful Method of Inoculating the Smallpox, 1768. Research, publs. on physics and botany; credited with inclusion of platinum as metal; among earliest experimenters on electricity, 1st to investigate passage of current through rarefied gas and to discover that conductivity increases; attempted to improve output Leyden jar, 1750; discovered polygamous nature of holly, 1754; described nature of starpuff ball, 1744; largely responsible for intro. Linnean system bot. classification into England. Died London, May 10, 1787.

WATSON, William, social anthropologist; b. Clydebank, Scotland, Jan. 5, 1917; s. Thomas and Jessie (Brockett) W.; B.A., Cambridge (Eng.) U., 1947, M.Sc., 1949; Ph.D., Manchester (Eng.) U., 1953; m. Pamela Eugenie Matthews, Feb. 9, 1948; children—Hamish Brockett, Calum Macfarland, Angus

Clark. Research anthropologist Med. Research Council, U.K., 1948-51, cons., sociologist social medicine research unit, 1959-63; research fellow Rhodes-Livingstone Inst., Lusaka, No. Rhodesia, 1951-55; sr. lectr. sociology Manchester U., 1956-63; prof. sociology U. Va., Charlottesville, 1963—. Mem. Youth Service Devel. Council, U.K., 1960-63. Decorated D.F.C. Mem. Assn. Social Anthropologists, Am., Brit. sociol. assns., Internat. African Inst., African Studies Assn. Author: Tribal Cohesion in a Money Economy, 1959; (with M. W. Susser) Sociology in Medicine, 1962; also articles. Research on labor migration and its impact on tribal African life, labor migrants in indsl. societies and effects on social cohesion, problems of med. sociology. Home: 1006 Rugby Rd., Charlottesville, Va. 22903.*

WATSON, William Weldon, Am. physicist; b. Eveleth, Minn., Sept. 14, 1899; s. Thomas Tolman and Nellie Dewey (Ford) W.; B.S., U. Chgo., 1920, M.S., 1922, Ph.D., 1924; M.A. (hon.), Yale, 1940; m. Elizabeth Wells, Sept. 10, 1927; children—Ruth Pomeroy, Alice Daggett. Instr. physics U. Chgo., 1924-27, asst. prof., 1927-28, with metall. lab., 1943-45; faculty Yale, New Haven, 1928—, prof., 1940—, chmn. physics dept., 1940-61, dir. Sloane Physics Lab., 1941-46, 54-61. Mem. Conn. Adv. Com. on Atomic Energy, 1954-63; cons. Am. Inst. Physics, 1963—, Bridgeport Engring. Inst., 1964—. Chmn. bd. trustees Am.-Philippine Sci. Found., 1964—. Guggenheim fellow, 1928-29. Fellow A.A.A.S., Am. Phys. Soc.; mem. Phi Beta Kappa, Sigma Xi, Chi Psi. Author: (with Lloyd W. Taylor and Carl E. Howe) Physics for the Laboratory, 1929; (with Henry Margenau and Carol G. Montgomery) Physics: Principles and Applications, 1952. Cons. editor: McGraw-Hill Ency. Sci. and Tech., 1957—. Research, publs. on molecular spectroscopy, isotope separation by thermal diffusion, nuclear studies using separated isotopes. Home: 294 Livingston St., New Haven 06511.*

WATSON-MUNRO, Charles Norman, physicist; b. Dunedin, New Zealand, Aug. 1, 1915; s. Marshell Christopher and Ethel (Penny) W.-M.; M.Sc. with 1st class honors, U. New Zealand, 1937; m. Yvette Diamond, Oct. 16, 1947; 1 son, Timothy Machell. Geophysicist, Dept. Sci. and Indsl. Research, New Zealand, 1938-40; staff Mass. Inst. Tech., 1941; dir. Radar Lab., New Zealand, 1942-44; staff Atomic energy, Can., U.K., 1944-47; dep. head Dept. Sci. and Indsl. Research, New Zealand, 1947-49; prof. physics Victoria U., Wellington, New Zealand, 1950-54; chief scientist Australian AEC, 1955-59; prof. plasma physics U. Sydney (Australia), 1960—. Decorated Order Brit. Empire. Fellow Inst. Physicists, Australian Inst. Physics; mem. Instn. Elec. Engrs. (asso.). Research, publs. on devel. radar equipment used, 1939-44; built 1st atomic energy reactor, Can., 1945, U.K., 1947, Australia, 1958. Home: 66 Stuart, Sydney, N.S.W., Australia.*

WATSON-WATT, Sir Robert (Alexander), physicist; b. Brechin, Angus, Scotland, Apr. 13, 1892; s. Patrick and Mary (Matthew) W.W.; Student U. Coll., Dundee in U. of St. Andrews; LL.D., St. Andrews, 1943; D.Sc., U. Toronto, 1943; D.Sc., Laval U., 1952; m. Margaret Robertson, 1916 (div. 1952); m. 2d. Mrs. Jean Smith, 1952 (deceased December 1964); stepchildren—Anthony, Dennie (Mrs. Douglas Reburn). Assistant to prof. natural philosophy Univ. Coll., Dundee, 1912-21; various posts in meteorology, radio and radar in meteorol. office Dept. Sci. and Indsl. Research, Air Ministry, Ministries of Aircraft Prodn., Supply, Civil Aviation and Transport, 1915-52; dep. chmn. radio bd. of War Cabinet, 1943-46; advisory board Axe Science Corporation; consultant to the Sterling Forest Research Center. Created Knight, 1942. Decorated Companion of the Bath, 1941; U. S. Medal of Merit, Valdemar Poulsen medal Danish Acad. Tech. Scis., Hughes medal Royal Soc., Elliott Cresson medal, Franklin Institute. Fellow Royal Meteorological Society (past president), Royal Aero. Soc., Inst. Physics. Am Phys. Soc.; mem. Inst. Nav. (past pres.), I.R.E. (past v.p.). Author: Through the Weather House, 1935; Three Steps to Victory, 1958; The Pulse of Radar, 1959; Man's Means to His End (with others), 1961; The Cathode Ray Oscillograph in Radio Research, 1933. Worked on devel. radar; took out 1st patent on a radar instrument used in atmospheric study, 1919; patented new type of radiolocator, 1935; studied radio direction finders, telecommunications, electromagnetic radiation. Home: P.O. Box 653, Maple Brook, Tuxedo, N.Y.

WATT, James, Brit. engr., inventor; b. Greenock, Scotland, Jan. 19, 1736; s. James and Agnes (Muirhead) W.; LL.D., Glasgow U., 1806; m. Margaret Miller, 1763; (d. 1773); 2 sons, including James, 2 daus; m. 2d, Ann MacGregor, 1775; 1 son, Gregory; 1 dau. Apprenticed to instrument maker, London, at age of 19; went to Glasgow, Scotland at age of 21; apptd. math. instrument maker U. Glasgow; made surveys and reports on canals, rivers and harbors, 1769-73; adviser to Carron Foundry; became partner (with M. Boulton) Soho Engring. Works, nr. Birmingham, Eng., 1775; ret., 1800; visited Paris, 1786. Fellow Royal Soc. Edinburgh, 1784, Royal Soc. London, 1785; fgn. mem. French Acad. Scis., 1814; corr. mem. French Inst.; mem. Lunar Soc. Birmingham (a founder). Invented separate condenser for Newcomen's steam engine, 1765, also conversion of reciprocating motion to rotary motion using sun-and-planet gear, application of centrifugal governor to steam engines, invented throttle valve; patentee double-acting engine, improved combustion furnace; invented air-pump and cylinder steam jacket, indicator for drawing diagram steam pressure, spl. ink for copying letters, method for determining specific gravity of fluids, micrometer, locked-up automatic counter; originated term, horsepower; independently discovered chem. composition of water; proposed screw propeller be used for navigation; unit of power, Watt, named in his honor. Died Handsworth, Eng., Aug. 19, 1819.

WATT, James Jr., Brit. management scientist; b. Glasgow, Scotland, 1769; s. James and Margaret (Miller) W.; studied chemistry, mineralogy, natural philosophy; studied in Paris, but left for Italy, then England, 1789. With Boulton Jr., in charge of letter-copying press business, 1790; Boulton, Watt & Sons formed, 1794; opened Soho Foundry (Watt Jr. organized and administered foundry), 1796; Boulton, Watt and Company formed under management of Boulton Jr. and Watt Jr., 1800; formed banking firm, Matthew Robinson Boulton & James Watt & Company, London, 1802; contested steam engine patents of competitors, 1803; retired, 1840. Fellow Royal Soc., 1820. With Boulton Jr., pioneer in science of management; developed many approaches to industrial organization and control that foreshadowed practices today; on small scale applied exec. development plan, work study, statistical records and control records, cost accounting, market research, planned site location. Died 1848.

WATT, Richard Franklin, Am. forest tree physiologist; b. N.Y.C., Jan. 16, 1921; s. Harry C. and Grace (Ringwald) W.; B.S., N.Y. State Coll. Forestry, 1943; M.F., Yale, 1947; Ph.D., U. Minn., 1961; m. Rachel M. Gates, Dec. 30, 1944; children— Deborah H., Constance M. Jennifer G., Kathryn J. Instr., Pa. State Coll., 1947-48; research forester, silviculturist No. Rocky Mountain Forest and Range Expt. Sta., Spokane, Wash., 1948-54; with North Central Forest Expt. Sta (formerly Lake States Forest Expt. Sta.), St. Paul, 1954—, project leader, 1961—, plant physiologist, 1960—. Mem. Soc. Am. Foresters, Ecol. Soc. Am., Scandinavian Soc. Plant Physiologists, Sigma Xi, Gamma Sigma Delta, Xi Sigma Pi, Alpha Xi Sigma, Phi Kappa Phi. Research, publs. on growth western white pine and asso. species in no. Rocky Mountains, reaction eastern trees to length day, possibility growing No. species in Fla. with photoperiod artifically lengthened, role mineral element levels in growth and devel. No. forest species in Minn., Wis., Mich. Home: 905 Westport Dr. Office: U. Mo., Columbia, Mo. 65201.*

WATTENBERG, Albert, Am. physicist; b. N.Y.C., Apr. 13, 1917; s. Louis and Bella (Wolff) W.; B.S., Coll. City N.Y.; M.A., Columbia, 1939; Ph.D., U. Chgo., 1947; m. Shirley Hier, Sept. 5, 1943; children—Beth, Jill, Nina Diane. Spectroscopist, Schenley Distilleries, N.Y.C., 1939-42; physicist Manhattan Project, Metall. Lab., Chgo., 1942-46; group leader Argonne Nat. Lab., Chgo., 1946-50; asst. prof. U. Ill., Urbana, Ill., 1950-51, prof. physics, 1958—; research physicist Mass. Inst. Tech., 1951-58. Recipient award for 1st nuclear reactor Am. Nuclear Soc. NSF fellow U. Rome, 1962-63. Pioneered controlled nuclear reactor. Home: 701 W. Delaware St., Urbana, Ill. 61801.*

WATTERS, James Isaac, Am. chemist; b. Broadus, Mont., Apr. 4, 1908; s. James Oliver and Hilda (Erickson) W.; student U. Fla., 1926 28, U. Cal. at Berkeley, 1929; B.S., U. Minn., 1931, Ph.D., 1943; m. Louise Chambers, Aug. 28, 1938; children—Louise Emilie (Mrs. Robert Pflaum), Molly Marie, (Mrs. Joseph Sutherland), James Norman, Kathryn Verne. Instr., Cornell U., 1941-43; dir. analytical research div. metallurgy project U. Chgo., 1944-45; faculty U. Ky., 1945-48; faculty Ohio State U., Columbia, 1948—, prof., 1958—, head analytical div., 1963-66; vis. prof. Northwestern U., 1947. Mem. Am. Chem. Soc., A.A.A.S., Sigma Xi, Phi Lambda Upsilon. Contbg. author: Treatise of Analytical Chemistry, 1959; also articles. Editor: (with others) Analytical Chemistry Manhattan Project, 1948. Research on electrode reactions, spectrophotometry, complex ions in aqueous solution, new equations for calculating equilibrium constants, chemistry cobalt, copper, mercury, polyphosphates. Home: 1400 Ridgeview Rd., Columbus, O. 43221.*

WATTERSON, Ray Leighton, Am. biologist; b. Greene, Ia., Apr. 15, 1915; s. George Robert and Alice (Hesalroad) W.; A.B., Coe Coll., 1936, grad. student, 1936-37; Ph.D., U. Rochester, 1941; grad. student Johns Hopkins, 1940-41; m. Evelyn Lily Goddard, July 30, 1941; children—Richard Dean, James Robert, Donald Kent (dec.), Jean Marie. Instr. zoology Dartmouth, 1941-42; asst. prof. zoology U. Cal. at Berkeley, 1942-46; asst. prof., then asso. prof. U. Chgo., 1946-49; faculty Northwestern U., 1949-61, prof. biology, 1955-61, chmn. dept., 1958-61, chmn. adminstrv. com., 1956-58; prof. zoology, U. Ill. at Urbana 1961—; instr. Marine Biol. Lab. Woods Hole, Mass., summers 1942, 44, 46, Frank R. Lillie fellow exptl. embryology, summer 1952; asst. prof. Hopkins Marine Sta., Stanford, summers 1945-46; vis. prof. Friday Harbor Labs., U. Wash., Summer 1954. Fellow Royal Soc., 1941, A.A.A.S.; mem. Am. Assn. Anatomists, Am. Soc. Zoologists, Am. Soc. Naturalists, Soc. Study Devel. and Growth, Soc. Exptl. Biology and Medicine, Institut International d'Embryologie, Am. Soc. Cell Biology, Sigma Xi. Research on analysis of devel. pigment cells in chick embryos and tunicates; spinal cord, brain and their internal blood supply; vertebral column and ribs; endocrine glands and hormone dependent structures in avian embryos; effects of destructive agents on liver, kidney, heart devel.; cause faulty spinal cord and vertebral column devel. Home: 1717 Lincoln Rd., Champaign, Ill. 61820. Office: Dept. Zoology, U. Ill., Urbana, Ill.*

WATTS, Alva B., Am. animal scientist; b. Shreveport, La., Aug. 1, 1918; s. Alva Jackson and Julia (Thomas) W.; B.S., U. Southwestern La., 1939; M.S., La State U., 1941; Ph.D., Okla. State U., 1950; m. Katherine Corrine Wilbanks, Jan. 9, 1942; children—Katherine Gregg (Mrs. William S. Prescott), Corinne, Alva Burl II. Jr. biochemist So. Regional Research Lab., U.S. Dept. Agr., New Orleans, 1941-43; faculty La. State U., Baton Rouge, 1946—, prof., head dept. poultry sci., 1955—; dir. The Poultry Industries of La., Inc., 1955—. Fellow A.A.A.S.; mem. Am. Chem. Soc., Poultry Sci. Assn. (dir. 1958-59), Inst. Food Technologists (chmn. Gulf Coast sect. 1967), Am. Inst. Nutrition, Soc. Animal Sci., Soc. for Exptl. Biology and Medicine, Biometrics Soc., Am. Inst. Biol. Scis., World Poultry Sci. Assn. Research, numerous publs. on protein and amino acid requirements and metabolism in the chick; trace mineral requirements and chelates in poultry nutrition; digestibility of protein by chicks and hens; improvement of nutritional value of cottonseed meal; fat and energy utilization and requirements by the chick; calcium and phosphorus requirements and sources in poultry nutrition. Home: 1291 Lee Dr., Baton Rouge, La. 70808.*

WATTS, Betty Monaghan, Am. chemist; b. Johnstown, Pa., June 4, 1907; d. John H. and Anne (Young) Monaghan; B.S., Wilson Coll., 1928; Ph.D., Washington U., 1932; m. Hilary Watts, Dec. 19, 1936 (dec. Apr. 1956); 1 son, Jeremy Alan (dec.). Prof., Fla. State U., Tallahassee, 1950—, Distinguished prof., 1965—. Mem. adv. com. animal products, Q.M. Food Container Inst., 1960-63; mem. research adv. com. human nutrition, consumer use U.S. Dept. Agr., 1964—. Recipient Vibrans Sr. Scientist award Am. Meat Inst. Found., 1958, Borden award for research in food sci., 1964. Mem. Am. Chem. Soc., Am. Soc. Oil Chemists, Inst. Food Technologists, Home Econs. Assn., Phi Beta Kappa, Sigma Xi, Phi Kappa Phi, Omicron Nu. Contbr. chpts. to Advances in Food Research, 1954; Proceedings Flavor Chemistry Symposium, 1961; Lipids and Their Oxidation, 1962; also numerous articles. Research on chem. and enzymatic mechanisms of flavor and color deterioration in animal and vegetable tissues used as foods, particularly as related to decomposition of unsaturated fatty acids, control measures suitable for foods preserved by low or high temperatures, curing, freeze drying or irradiation. Home: 729 Monticello Dr., Tallahassee, Fla. 32303.*

WATTS, Henry, English chemist; b. London, Jan. 20, 1815; B.A., London, 1841; m. Sophie Hanhart, 1854; 3 sons, 2 daus. Asst. prof. chemistry U. Coll., London, 1846-57. Fellow Royal Soc., 1866, Chem. Soc. (editor Jour. 1849—), Phys. Soc.; hon. mem. Pharm. Soc. Editor: Dictionary of Chemistry, 1868; Manual of Chemistry, 10th, 11th, 12th and 13th edits. Translated, expanded: Handbuch der Chemie (Leopold Gmelins), vols., 1848-72. Died June 30, 1884.

WATTS, James Winston, Am. neuro-surgeon; b. Lynchburg, Va., Jan. 19, 1904; s. Thomas Ashby and Fannie (Cheatwood) W.; B.S., Va. Mil. Inst., 1924; M.D., U. of Va., 1928; m. Julia Meem Harrison, Oct. 24, 1931; children—James Winston, III, Randolph Harrison. House officer, neurology and neurosurgery, Mass. Gen. Hosp., Boston, 1928-29; interne surgical service, Long Island Coll. Hosp., Brooklyn, 1929-30; asst. resident neuro-surgeon, Chicago U. Clinics, 1930-31, vol. asst. in pathology to Prof. O. Foerster, Breslau, Germany, 1931-32; instr., research asst., dept. physiology, Yale, 1932-33; Fell neurosurgery and instr. neuro-surgery, Pa. U. Hosp., Phila., 1933-35; senior attending neurosurgery Children's Hosp.; chief neurological surgery D.C. Gen. Hosp., George Washington U. Hosp., consultant neurosurgery Walter Reed Hospital, VA Hospital, St. Elizabeth's Hosp. (Washington); cons. neurological surgery Doctors Hospital, National Institutes of Health Clinic; professor neurological surgery, chmn. dept. of neurology and neurological surgery George Washington U. Received John Horsley Memorial prize, U. of Va., 1942; decorated Comdr. Order of Merit of Duarte (Santa Domingo), 1956. Diplomate Nat. Bd. Med. Examiners, Am. Bd. Neurological Surgery. Hon. fellow or mem. several fgn. surg. and medical socs.; fellow A.C.S., Internat. Coll. Surgeons, Southeastern Surg. Congress; mem. Assn. Am. Med. Colls., Assn. Mil. Surgeons, A.M.A., Pan-Am. Med. Soc., Med. Soc. D.C. (pres. 1957-58, chmn. exec. bd. 1961-62), So. Med. Assn., Assn. Research Nervous and Mental Disease, Harvey Cushing Soc. (past v.p.), Am. Neurol. Assn., Wash. Acad. Surgery, Wash. Acad. Medicine, Phi Beta Kappa, Sigma Xi. Author: Psychosurgery—Intelligence, Emotion and Social Behavior Following Prefrontal Lobotomy for Mental Disorders (with Walter Freeman), 1942; Psychosurgery-Prefrontal Lobotomy for Mental Disorders and Pain (with Walter Freeman), 1950. Contbr. numerous articles to med. jours. Pioneer in devel. of prefrontal lobotomy in psychosurgical treatment of mental disorders. Home: 4661 Garfield St. N.W., Washington. Office: 1911 R St., Washington 9.

WAUD, Russell Amos, Canadian pharmacologist; b. Norwich, Ont., Can., Mar. 4, 1893; s. Edwin and Emily (Sims) W.; student Valparaiso U., 1916-18, Loyola U. at Chgo., 1918-19; M.D., U. Western Ont., 1921, M.Sc., 1925; Ph.D., U. Chgo., 1927; m. Mildred Adams, Apr. 19, 1922; children—Douglas Russell, Donald Adams. Gen. practice medicine, 1922-23; faculty U. Western Ont., London, 1923-60, prof., head dept. pharmacology, 1936-60; med. dir. Will Pharmaceuticals, London, Ont., 1960—. Fellow Am. Coll. Cardiology; mem. Can., Ont. med. assns., Pharmacol. Soc. Can. (pres. 1959-60), Can. Physiol. Soc., Am. Soc. Pharmacology and Exptl. Therapeutics, Am. Soc. Artifician Internal Organs. Author: Applied Pharmacology and Materia Medica, 1938; (with others) Pharmacology in Medicine, 1954, 58; also numerous articles. Research on viscosity of blood in relation to anaphylactic shock, use of transducer and thermionic valves in study of physiol. processes, action of plant alkaloids in circulation, artificial heart and lung, plant alkaloids having a cataleptic action. Home: 493 Baker St., London, Ont. Office: 1 Wilton Grove Rd., London, Ont., Can.*

WAUGH, John Lodovick Thomson, chemist; b. Avonhead, Scotland, Nov. 13, 1922; s. John Thomson and Joanna (Nicolson) W.; B.Sc. with 1st class honors, Glasgow U., 1943, Ph.D., 1949, A.R.I.C., 1943; m. June Mannering, June 28, 1949; children—Iain, Shena, Finlay, Malcolm. Plant supt. I.C.I. Ltd., 1943-46; faculty Glasgow U., 1946-49; Climax Molybdenum Research fellow Cal. Inst. Tech., 1949-50; asst. prof. chemistry U. Hawaii, 1950-51; Inst. Research fellow Cal. Inst. Tech., 1951-53; research chemist U.S. Borax Research Corp., Los Angeles, 1953-56; asso. prof. chemistry U. Hawaii, Honolulu, 1956—, program dir. NSF undergrad. research participation program, 1960-63. Asso. research officer Neutron Physics Br., Atomic Energy of Can. Ltd., 1962-63. Fellow Chem. Soc. London; mem. Royal Inst. Chemistry (asso.), N.Y., Hawaii acads. sci., Sigma Xi. Contbr. articles to profl. jours. Developed low density base for blasting explosives; discovered new series of heteropoly compounds; determined crystal structure of complex intermetallic compounds and heteropoly compounds by X-ray diffraction methods; discovered new borate mineral in Cal.; determined lattice dynamics of gallium arsenide by inelastic scattering of thermal neutrons. Home: 3181 Beaumont Woods Pl., Honolulu 96822.*

WAUGH, John S., Am. chemist; b. Willimantic, Conn., Apr. 25, 1929; s. Albert E. and Edith (Stewart) W.; A.B., Dartmouth Coll., 1949; Ph.D., Cal. Inst. Tech., 1953; m. Nancy Alicia Collier, July 24, 1954; children—Alice Collier, Frederick Pierce. Research fellow in physics Cal. Inst. Tech., 1952-53; faculty Mass. Inst. Tech., 1953—, prof., 1962—. Asso., Retina Found.; cons. to industry. Alfred P. Sloan Research fellow, 1958-62. Fellow Am. Acad. Arts and Sci., Am. Phys. Soc.; mem. Sigmi Xi, Phi Beta Kappa. Asso. editor Jour. Chem. Physics, Spectrochimica Acta; editor: Advances in Magnetic Resonance. Research on intermolecular forces and magnetic resonance. Home: 8 Pamela Dr., Arlington, Mass. 02174. Office: 77 Massachusetts Av., Cambridge, Mass. 02139.

WAWZONEK, Stanley, Am. chemist; b. Valley Falls, R.I., June 23, 1914;s. John and Sophie (Krol) W.; B.S., Brown U., 1935; Ph.D., U. Minn., 1939; m. Marian Lucile Matlock, Sept. 2, 1943; children—Ann Elizabeth, Lucile Kay, Mary Louise. Research fellow U. Minn., 1939-40; NRC fellow U. Ill., 1940-41, instr., 1941-43; instr. U. Tenn., 1943-44; with U. Ia., Iowa City, 1944—, prof. chemistry 1952—, chmn. dept. 1962—. Recipient Am. Chem. Soc. Iowa award, 1960. Mem. Am. Chem. Soc. (councilor, 1963—), Electrochem. Soc. (chmn. electroorganic div. 1957-59), Polarographic Soc., A.A.A.S., Ia. Acad. Sci., Sigma Xi, Phi Lambda Upsilon, Alpha Chi Sigma. Author: (with others) Laboratory Manual of Organic Chemistry, 2d edit., 1962; also numerous articles. Pioneered research on electrochemistry in non-aqueous solvents; developed new method for preparation of isocyanates which are important in polymer field. Home: 2014 Ridgeway Dr., Iowa City, Ia. 52240.*

WAY, E(dward) Leong, Am. pharmacologist; b. Watsonville, Cal., July 10, 1916; s. Leong M. and Lay-har (Shew) W.; B.S., U. Cal., Berkeley, San Francisco, 1938, M.S., 1940, Ph.D., 1942; m. Madeline Li, Aug. 11, 1944; children—Eric, Linette. Faculty, George Washington U., 1946-49; faculty U. Cal., San Francisco, 1949—, prof., 1957—, vice chmn. pharmacology, toxicology, 1957-66; USPHS Spl. Research fellow U. Berne, 1955-56; vis. prof. pharmacology U. Hong Kong, 1962-63, external examiner pharmacology, 1964-66. Mem. com. biochemical toxicology Drug Research Bd. Nat. Acad. Sci.-NRC, 1965—; mem. com. on abuse of depressant and stimulant drugs FDA; chmn. pharmacology study sect. NIH. Recipient Am. Pharm. Assn. Found. Research Achievement award in pharmacodynamics, 1962, Ebert prize Am. Pharm. Assn., 1962. Fellow A.A.A.S.; mem. Soc. Exptl. Biology and Medicine (vice chairman No. Cal. sect. 1966-67), American Society Pharmacology and Experimental Therapeutics, Am. Pharm. Assn., Am. Chem. Soc., A.M.A., Western Pharmacol. Soc. (past pres.), Sigma Xi. Studies, numerous publs on pharmacology of narcotic analgetic drugs, diagnosis of addiction, mechanisms of tolerance. Home: 185 Belgrave Av., San Francisco 94117.*

WAY, Harold Emory, Am. physicist; b. Colchester, Ill., Jan. 31, 1904; s. Clarence O. and Margaret (Gibbs) W.; B.S., Knox Coll., 1925; M.S., U. Pitts., 1928; Ph.D., U. Ia., 1937; m. Fern Robbins, Dec. 24, 1926; children—Mary Ann (Mrs. Fred Lauder), John H., Jane Ellen. Faculty, Knox Coll., 1927-47; prof., chmn. dept. physics Union Coll., Schenectady, 1948-67, dean Sci. and Engring., 1965-67. Cons., NSF, 1961—. Recipient Alumni Achievement award Knox Coll., 1941, Distinguished Service award Union Coll., 1965. Mem. Am. Phys. Soc., Am. Assn. Physics Tchrs., Sigma Xi. Exec. editor: Collected Works of Irving Langmuir, 12 vols., 1959-61. Studies of single crystals of zinc. Home: 5515 E. Covina Rd., Mesa, Ariz. 85201.

WAY, James Leong, Am. pharmacologist; b. Watsonville, Cal., Mar. 21 1927; s. Leong Man and Shee Shee (Leong) W.; B.A., U. Cal., Berkeley, 1950; Ph.D., George Washington U., 1955; m. Helen Wong, June 15, 1957; children—Lani, Jon, Lori. Instr. U. Wis., 1957-60, asst. prof., 1960-63; asso. prof. Marquette University, 1963-67; prof. Wash. State U., Pullman, 1967—. Greenwald scholar, 1947; Baxter N.Am. fellow, 1952; USPHS fellow, 1953, 56, 58, Career Devel. award, 1960. Mem. Am. Soc. Pharmacology and Exptl. Therapeutics, Am. Chem. Soc., A.A.A.S., Soc. Exptl. Biology and Medicine. Research, numerous publs. on anticancer drugs, effect and mechanism of action; antidote for alkylphosphate poisoning, molecular mechanisms for metabolism of 2-PAM; cyanide poisoning, improvement of antidote; methods to improve treatment of cyanide poisoning; molecular mechanism of action of new antidote. Home: 2000 N. State St., Pullman, Wash. 99163.*

WAY, Stanley Albert, English gynecol. surgeon; b. Portsmouth, Eng., Jan. 16, 1913; s. Albert Edward and Mary (Ellery) W.; ed. U. London, 1929-35; m. Ruth Noble, June 4, 1938; children—Bernard Gordon, Elizabeth Mary. Tutor, U. Durham, 1938-45; registrar obstetrics and gynecology Royal Victoria Infirmary, Newcastle, 1938-45, asso. surgeon, 1946—; surgeon, dir. gynecology research Queen Elizabeth Hosp., Gateshead, 1949—; lectr. in gynecol. oncology U. Newcastle upon Tyne. Vis. prof. U. Sydney, 1960; examiner U. New Zealand, 1960; sci. adv. com. Brit. Empire Cancer Campaign; chmn. Brit. Soc. Clin. Cytology. Blair Bell lectr. Royal Coll. Gynecologists, 1948; Hunterian Prof., Royal Coll. Surgeons, 1948; Ingleby lectr. U. Birmingham, 1953. Fellow Am. Assn. Obstetricians and Gynecologists (hon.); mem. Soc. Pelvic Surgeons, Am. Cytology Soc., North of Eng. Obstetrics and Gynecology Soc., Newcastle Obstetric Soc., Royal Soc. Medicine. Author: Malignant Disease of the Female Pelvis, 1951; Diagnosis of Early Cancer of the Cervix, 1963; also numerous articles. Pioneer studies of exfoliative cytology and radical surgery of female pelvis. Home: 33, Moor Crescent, Newcastle upon Tyne 3, Eng. Office: Queen Elizabeth Hosp., Gateshead 9, Eng.*

WAYGOOD, Ernest Roy, plant physiologist; b. Bramhall, Cheshire, Eng., Oct. 26, 1918; s. Edward Samuel and Alice (Harrison) W.; student Reaseheath Agrl. Coll., Cheshire, Eng., 1936; B.S. in Agr., Ont. Agrl. Coll., 1941; M.S.A. in Plant Physiology, U. Toronto, 1947, Ph.D., 1949; m. Adoree Magdalyn Woolf-LeBrooy, Dec. 30, 1950; 1 dau., Pamela Mimi. Mem. faculty McGill U., 1949-54; prof., head, dept. botany U. Man., Winnipeg, 1954—. Fellow Chem. Inst. Can.; mem. Canadian, Am. socs. plant physiologists, N.Y. Acad. Scis., Sigma Xi. Research, publs. on enzymatic control mechanisms in respiration and photosynthesis. Home: 163 St. Mary's Rd., St. Germain P.O., Man., Can.'*

WAYMAN, Cooper Harry, Am. chemist; b. Trenton, N.J., Jan. 29, 1927; s. Cooper O. and Helen (Univerzagt) W.; B.S., Rutgers U., 1951; M.S., U. Pitts., 1954; Ph.D., Mich. State U., 1959; J.D., U. Denver, 1967; m. Ruth Treier, June 16, 1951; children—Carol Beth, Andrea Lee. Mining engr. Am. Agrl. Chem. Co., Pierce, Fla., 1951-52; geologist Lone-Star Steel Co., LoneStar, Tex., 1952-53; research supr. U.S. Steel Corp., Monroeville, Pa., 1953-59; research chemist U.S. Geol. Survey, Denver, 1960—; patent law clk. Marathon Oil Co., Littleton, Colo., 1965-67; asst. prof. chemistry Colo. Sch. Mines, Golden, Colo., 1965—; cons. U.S. Fisheries and Wildlife, 1967—. Registered profl. engr., Colo. Mem. Am. Chem. Soc., Am. Soc. Agronomy, Water Pollution Control Fedn., A.A.A.S., Sigma Xi. Research, numerous publs. on biodegradation of detergents, which encouraged mfrs. to produce new detergent in attempt to abate water pollution. Home: 6936 Newcombe St., Arvada, Colo. 80002.*

WAYMOUTH, Charity, cell biologist; b. Blackheath, London, Eng., Apr. 19, 1915; d. Charles Sydney Herbert and Ada Curror Scott (Dalgleish) Waymouth; B.Sc., Bedford Coll., U. London, 1936; Ph.D., U. Aberdeen (Scotland), 1944. Beit Meml. fellow for med. research, U. Aberdeen, 1944-47, Nat. Inst. Med. Research, London, 1945-46, Carlsberg Fondets Biolgiske Institut, Copenhagen, Denmark, 1946, St. Thomas Hosp., London, 1946-47; staff Chester Beatty Research Inst., London, Eng., 1947-53; Brit. Empire Cancer Campaign-Am. Cancer Soc. exchange fellow The Jackson Lab., Bar Harbor, Me., 1952-53, staff scientist, sr. staff scientist, 1953—; hon. lectr. U. Me., Orono, 1964—. Mem. Am. Assn. U. Women (recipient State of Me. Achievement citation, 1962), A.A.A.S., N.Y. Acad. Scis., Tissue Culture Assn., Internat. Soc. for Cell Biology, Biochem. Soc., Physiol. Soc., Sigma Xi. Research and publs. on role of nucleic acids in growth, design and use of chemically defined nutrient media for growing cells. Home: 10 Atlantic Av., Office: Jackson Lab., Bar Harbor, Me. 04609.

WAYNER, Matthew John, Am. physiologist; b. Clifton, N.J., Sept. 7, 1927; s. Matthew John and Marie (Krabatz) W.; A.B., Dartmouth Coll., 1949; M.S., Tufts Coll., 1950; Ph.D., U. Ill., 1953; m. Therese Marie Desjardins, Oct. 22, 1949; children—Elizabeth Ann, Matthew John, Timothy John. Prof. Syracuse U., 1953—; vis. prof. Fla. State U., 1962-63. Mem. Am. Physiol. Soc., Am. Psychol. Assn., Gerontology Soc., A.A.A.S., N.Y. Acad. Sci. Editor: Thirst, 1964; editor-in-chief Physiology and Behavior, 1966—. Contbr. articles in field to sci. jours. Studies on thirst in the regulation of body water; how the brain works in controlling thirst. Home: Highbridge Terrace, Fayetteville, N.Y. 13066. Office: 601 University Av., Syracuse, N.Y. 13210.*

WEALE, Robert Alexander, physiologist; b. Prague, Czechoslovakia, Sept. 13, 1922; s. Frederick J. and Mary (Eksten) W.; B.Sc., U. London, 1944, M.Sc., 1947, Ph.D., 1953, D.Sc., 1959; m. Margaret Elizabeth Drury, July 31, 1952; children—Graham, Martin, Rosalind. Research asst. Bisra, 1945-47; lectr. physics S.W. Essex Tech. Coll., 1947-48; sci. staff, Med. Research Council Vision Research Unit, 1948-60; sr. lectr., head dept. physiol. optics Inst. Ophthalmology, U. London, 1960—, reader in physiol. optics, 1964—. Recipient Sr. award Spectacle Makers Co., 1962. Mem. Physiol. Soc., Colour Group, Photo-Biology Group, Royal Soc. Medicine, Illuminating Engring. Soc. Author: The Eye and Its Function, 1960; The Aging Eye, 1963; also articles. Research on photo-chem. reactions in living eyes using spectrophthalmoscope; properties of crystalline lens of eye; color vision in normal and color defective persons.*

WEATHERWAX, Paul, Am. botanist; b. Worthington, Ind., Apr. 4, 1888; s. Charles and Sarah Ellen (Newsom) W.; student Wabash Coll., 1909, DePauw U., 1910; A.B. magna cum laude Ind. U., 1914, A.M., 1915, Ph.D., 1918; Sc.D., Franklin Coll., 1963; m. Anna May Stanton, June 11, 1916; children—Helen (Mrs. J.B. Mosier), Robert, Charles. Tchr. elementary and high schs., Ind., 1907-13; faculty Ind. U., Bloomington, 1915-19, 21-59, prof. botany, 1935-59, prof. emeritus, 1959—; asso. prof. botany, U. Ga., 1919-21; vis. prof. State U. of Ia., 1936, Franklin Coll., 1960-63, Hanover Coll., 1966; tech. asst. sci. edn. Ind. U. Contract, Bangkok, Thailand, 1957-59. Waterman fellow, Ind. U., 1931-36; Guggenheim fellow, 1944-45. Mem. Ind. Acad. Sci., Bot. Soc. Am., Torry Bot. Club, A.A.A.S., Phi Beta Kappa, Sigma Xi. Author: Elementary Botany, 1942; The Story of the Maize Plant, 1923; Indian Corn in Old America, 1954; also numerous articles. Research on morphology grass family, morphology history and anthropology Indian corn plant. Home: 416 S. Dunn St., Bloomington, Ind. 47403.*

WEAVER, Albert B., Am. physicist, educator; b. Anaconda, Mont., May 27, 1917; s. John Bruce and Myrtle (Dragstedt) W.; student Mont. Sch. Mines, 1935-37; A.B., U. Mont., 1940; M.S., U. Ida., 1941; postgrad. U. Minn.; Ph.D., U. Chgo., 1951; m. Eva Adeline Okerberg, Sept. 20, 1945; children—Janet Lynn, Gail Lorraine, John Bruce. Physicist, Naval Ordnance Lab., 1942-45; research asso. U. Chgo., 1952-53, U. Wash., Seattle, 1953-54; asst. prof. physics U. Colo., Boulder, 1954-56, asso. prof., chmn. dept., 1956-58; prof. physics, head dept. U. Ariz., Tucson, 1958—, asso. dean Sch. Liberal Arts, 1961—. Fellow Am. Phys. Soc.; mem. Assn. Physics Tchrs. Research on cosmic rays. Home: 5726 E. Holmes St., Tucson.*

WEAVER, John Dodsworth, geologist; b. London, Eng., Nov. 5, 1914; s. Percy Albert and Theresa (Exall) W.; B.Sc. with 1st class honours in geology, U. London, 1949, Ph.D. in Geology, 1953; m. Suzanne Wiley, Sept. 6, 1941; children— Charlotte Theresa, Abbie Marion, Mary Elizabeth. Lectr. geology Columbia, 1949-52; faculty Mt. Holyoke Coll., 1952-55; prof. geology U. P.R., Mayaguez, 1955—, chmn. Inst. Caribbean Sci., 1960—. Fellow Geol. Soc. London, Geol. Soc. Am.; mem. A.A.A.S., Geol. Soc. London, Geol. Soc. Am.; mem. Geol. Soc. P.R. (past pres.). Research, publs. on geomorphological history of Caribbean region. Home: SU 47, Calle 20, Valle Hermosa, Mayaguez, P.R. 00708.*

WEAVER, John Ernst, Am. plant ecologist; b. Villisca, Ia., May 5, 1884; s. John and Amelia W.; B.S., U. Neb., 1909, A.M., 1911; Ph.D., U. Minn., 1916; m. Martha Hasse, Sept. 6, 1906; children—Cornelia (Mrs. Everett Hahne), Robert John. With Wash. State Coll., 1912-14, asst. prof., 1913-14; with U. Neb., 1914-52, prof. plant ecology, 1917-52, ret., 1952; research asso. Carnegie Inst., 1922-32. Mem. Bot. Soc. Am., Ecol. Soc., Ecol. Soc. Am. Author: North American Prairie, 1954; (with Albertson) Grasslands of the Great Plains, 1956; Native Vegetation of Nebraska, 1965; numerous others. Publs. on study of nutrition, growth of vegetation in relation to grazind and agriculture, ecology, devel. of roots. Home: 1636 S. 20th St., Lincoln, Neb. 68502. Died June 8, 1966.*

WEAVER, John Richard, Am. chemist; b. Goshen, Ind., Sept. 3, 1920; s. John E. and Elenora (Kauffman) W.; A.B., Bluffton Coll., 1942; M.S., U. Mich., 1949, Ph.D., 1953; m. Margaret Shelly, June 8, 1943; children—Gerald, Cynthia, Louise, Sally, Shelly, Paul, Marsha. Faculty, Bluffton (O.) Coll., 1950—, prof. chemistry, 1960—. Vis. prof. U. P.R., 1963-64; research cons. U. Mich., 1955—. Mem. Am. Chem. Soc. Contbr. articles to profl. jours. Research on new form of mercury electrode, theory of dipole moment measurements, measurement dipole moments of some boron compounds. Home: 136 N. Spring St., Bluffton, O. 45817.*

WEAVER, Lawrence Clayton, Am. pharmacologist; b. Bloomfield, Ia., Jan. 23, 1924; s. Wilber C. and Faye (Ballew) W.; B.S., Drake U., 1949; Ph.D. (Foxbilt fellow), U. Utah, 1953; m. Delores Hillman, Sept. 8, 1949; children— Karen Celeste, Kevin Neal, Gordon Douglas, Elizabeth. With Pitman-Moore Research (became div. Dow Chem. Co., 1961), Indpls., 1953-66, asst. gen. mgr., 1965-66; dean, prof. Coll. Pharmacy, U. Minn., Mpls., 1966—. Recipient Research Achievement award Am. Pharm. Assn. Found., 1963, Alumni Distinguished Service award Drake U., 1966. Fellow A.A.A.S.; mem. Am. Soc. Pharmacology and Exptl. Therapeutics, Soc. Exptl. Biology and Medicine, Biometric Soc., Am. Pharm. Assn., Soc. Toxicology, Royal Soc. Medicine, Research Soc. Am., Sigma Xi, Phi Delta Chi, Rho Chi, others. Research and publs. on effects of anticonvulsant and tranquilizing agts. Home: 703 Forest Dace Rd., New Brighton, Minn. 55112.*

WEAVER, Robert John, Am. botanist; b. Lincoln, Neb., Sept. 23, 1917; s. John Ernst and Martha (Hasse) W.; A.B., U. Neb., 1939, M.S., 1940; Ph.D., U. Chgo., 1946; m. Lucinda Jean Fraser, Oct. 7, 1951; children—Jeanne E. (Mrs. John Martin), Martha Helen. Plant physiologist, Ft. Detrick, Md., 1944-46; research asso. botany U. Chgo., 1946-48; with U. Cal. at Davis, 1948—, prof. viticulture, viticulturist Expt. Sta., 1958—; vis. research worker Forschungs Inst. für Rebenzüchtung, Landau, Germany, 1963-64. Sr. Fulbright Research scholar Superior Coll. Agr., Athens, Greece, 1955-56. Mem. Am. Bot. Soc., Am., Scandinavian, Japanese socs. plant physiologists, Am. Soc. Hort. Sci., Am. Soc. Enology and Viticulture (editorial bd. 1956—), A.A.A.S., Am. Inst. Biol. Sci., Phi Beta Kappa, Sigma Xi. Research and numerous publs. on plant hormones and regulators in plants, naturally occuring and manufactured regulators; adapted uses of hormones to comml. grape prodn. Home: 662 Elmwood Dr., Davis, Cal. 95616.*

WEAVER, Rufus B., Am. anatomist; b. Gettysburg, Pa., Jan. 10, 1841; A.B., Pa. Coll., 1862, A.M., 1865; M.D., Pa. Med. U., 1865, Hahnemann Med. Coll., 1891; Sc.D.; m. Madeleine Louise Bender, Dec. 21, 1869. Appt. demonstrator of anatomy Hahnemann Med. Coll., Phila., 1869, lectr. regional anatomy, 1876-96, prof. applied anatomy, 1896—, Dissected and mounted entire cerebro-spinal nervous system of a human body, 1888. Died July 15, 1936.

WEAVER, Warren, Am. mathematician; b. Reedsburg, Wis., July 17, 1894; s. Isaiah and Kittie Belle (Stupfell) W.; B.S., U. Wis., 1916, C.E., 1917, Ph.D., 1921, LL.D., 1948; Sc.D., U. Sao Paulo 1949, Drexel Inst. Tech., 1961, U. Pitts., 1964, N.Y. U. 1964; D.E., Rensselaer Poly. Inst., 1962; L.H.D., U. Rochester, 1963; m. Mary Hemenway, Sept. 4, 1919; 1 son, Warren. Helen Hemenway. Asst. prof. math., Thropp Coll., 1917-18, Cal. Inst. Tech., 1919-20; faculty U. Wis., 1920-32, prof. and chmn. dept. math., 1928-32; dir. div. of natural Scis., Gen. Edn. Rd., 1932-37; dir. div. natural Scis. Rockefeller Found., 1932-55, v.p. for natural and med. scis., 1955-59. chmn. basic research group, research and devel. bd. Dept. Defense, 1952-53; trustee Sloan-Kettering Inst., 1954—, mem. exec. com., 1956—, v.p. 1958-59; v.p. Alfred P. Sloan Found., 1959-64, cons, sci. affairs, 1964—, mem. exec. com., 1956—; chmn. bd., chmn. com. sci. policy Meml. Sloan-Kettering Cancer Center, 1960—: mem. exec. com., mgr. Meml. Hosp. Cancer and Allied Diseases, 1960—; mem., officer, cons. numerous other govt. and pvt. agys. Recipient King's medal for Service in Cause of Freedom, 1948; U. S. medal for Merit; Pub. Welfare medal Nat. Acad. Scis., 1957; Kalinga prize, 1964; Arches of Science award, 1965. Fellow Am. Acad. Arts and Scis., A.A.A.S. (past pres.), N.Y. Acad. Scis., Am. Phys. Soc.; mem. Am. Math. Soc., Math. Assn. Am., Am. Phiols. Soc. (councillor), Am. Soc. Naturalists, Am. Soc. for Symbolic Logic, NRC (Div. Phys. Scis., 1936-39, 1944-47), Chr. Michelsens Institutt. Bergen, Norway (corr.), Sigma Xi. Author: The Electromagnetic Field (with Max Mason), 1929; Elementary Mathematical Analysis, 1925; The Mathematical Theory of Communication (with C. E. Shannon) 1949; Lady Luck-The Theory of Probability, 1963; Alice in Many Tongues, 1964. Editor: The Scientists Speak, 1947. Contbr. papers on math. research in sci jours. and on gen. aspects of sci. in gen. jours. Study of theory of probability; electrodynamics. Home: Second Hill, R.F.D., New Milford, Conn. 06776. Office: Alfred P. Sloan Found. 630 Fifth Av., N.Y.C. 10020.*

WEBB, Albert Dinsmoor, Am. enologist; b. Victorville, Cal., Oct. 10, 1917; s. Ralph Hough and Vida (Dinsmoor) W.; B.S., U. Cal. at Berkeley, 1939, Ph.D., 1949; m. Nancy May Mathews, Sept. 5, 1943; children—Robert D., Bradford C. With Manhattan Dist. Project, U.S. Army Engrs., Oak Ridge, 1943-45; faculty U. Cal. at Davis, 1949—, prof. enology, 1960—. Fulbright research scholar Adelaide (Australia) U., 1958; NATO scholar U. Bordeaux (France), 1962. Mem. A.A.A.S., Am. Chem. Soc., Am. Soc. Enologists. Research and publs. on aroma and flavor of grapes and wine.*

WEBB, (Martha) Beatrice Potter, English economist, sociologist; b. Glouster, Eng., Jan. 22, 1858; d. Richard and Laurencia (Heyworth) Potter; pvt. edn.; D.Litt., U. Manchester, LL.D., U. Edinburgh, D.Polit. Economy, U. Munich; m. Sidney James Webb, 1892. Studied conditions of working classes for survey of Life and Labour of the People of London, 1891-1903; served on Royal Commn. of Poor Laws, 1905-09, World War I com. Aid of Distress, 1914-15, com. Statuatory War Pension, 1916-17, Reconstruction and Nat. Registration coms., 1917-18; war cabinet com. Women in Industry, 1918-19, Lord Chancellor's com. for Women Justices, 1919-20; justice of peace, London, 1919-27. Author: The Co-operative Movement in Great Britain, 1891; Men's and Women's Wages, 1919; My Apprenticeship, 1926; (with Sidney Webb) The History of Trade Unionism, 1894, Industrial Democracy, 1897, English Local Government, 7 vols., 1906-22, Consumer's Co-operative Movement, 1921, Soviet Communism: a New Civilization? 2 vols., 1935. Did pioneer work (with her husband) in study of British trade unionism; helped stimulate public interest in principles of social insurance; leading mem. of Fabian Soc. and Brit. Labour party, influential in guiding intellectual devel. of both. Died Liphook, Eng., Apr. 30, 1943.

WEBB, David Allardice, Irish botanist; b. Dublin, Ireland, Aug. 12, 1912; s. George Randolph and Isabella (Ovenden) W.; B.A., U. Dublin, 1935, Ph.D., 1937, M.A., 1938, Sc.D., 1951; student Trinity Coll., Dublin, 1931-36; M.A., U. Cambridge, 1938, Ph.D., 1939; Staff, Trinity Coll., Dublin, 1940—, prof. botany, 1949-65, prof. systematic botany, 1966—, fellow., 1949. Mem. Royal Irish Acad. (past mem. council). Fellow Linnean Soc. London; Author: An Irish Flora, 1943, 5th edit., 1967; also articles. Co-editor, contbg. author: Flora Europaea, vol. I, 1964. Research on biochemistry of marine invertebrates, especially interchange of chems. with sea-water, flora and vegetation of Ireland, integration of knowledge in flora of Europe, taxonomy of Saxifraga. Home: 23 Trinity Coll., Dublin, Ireland.*

WEBB, G(eorge) A(rthur), chem. engr.; b. Liverpool, Eng., July 7, 1910; s. George and Alice E. (Shields) W.; brought to U. S., 1920, naturalized, 1924; B.S. with highest honors, U. Pitts., 1934, Ph.D., 1941; postgrad. Carnegie Inst. Tech.; m. Sara R. Baumann, July 31, 1937; 1 dau., Barbara Jeanne. With Clairton By-Products Coke plant U. S. Steel Corp., 1934-37; indsl. fellow Mellon Inst., 1937-40, sr. fellow, 1941-43, dir. engring., 1956-57, dir. adminstrn.; research engr. Firestone Tire & Rubber Co., 1940-41; with Koppers Co., Inc., 1943-56, successively research engr., mgr. engring sect., asst. mgr. devel. dept., asst. to v.p. research, exec. sec. new products com., mgr. planning dept., 1945-56. Profl. engr., Pa. Mem. Am. Inst. Chem. Engrs. (chmn. Pitts. sect. 1949), Am. Chem. Soc. (dir. Pitts. sect. 1956), Am. Soc. M.E., A.A.A.S., Soc. Chem. Industry, Chemist Club, N.Y., Am. Soc. Engring. Edn., Nat. Soc. Profl. Engrs., Sigma Xi, Phi Lambda Upsilon, Tau Beta Pi, also Sigma Tau. Holder 19 U. S. and fgn. patents in field of dehydrogenation, hydrolysis, halogenation, polymerization. Home: 4822 Rolling Hills Rd., Pitts. 15236. Office: Mellon Inst., Pitts. 15213.*

WEBB, Harold Donivan, Am. physicist; b. Franklin, Ind., Sept. 23, 1909; s. Guilford and Bertha (Owens) W.; A.B., Franklin Coll., 1931; A.M., Ind. U., 1932, Ph.D., 1939; m. Mary Margaret Hougham, Aug. 15, 1937; children—Stephen R., Patricia L. (Mrs. James S. Muirhead), Sharon E. (Mrs. Philip E. Sticha), Mary Diana (Mrs. Richard E. Slavens). Tchr., Needham Jr. High Sch., Franklin, 1934-35, Baylor Sch., Chattanooga, 1935-36, Franklin High Sch., 1936-39; prof. math. and physics West Liberty (W.Va.) Coll., 1939-42; engr., physicist Evans Signal Lab., Belmar, N.J., 1942-47; faculty U. Ill., Urbana, 1947—, prof. elec. engring., 1958—, dir. radiodirection finding research, 1948-57, dir. ionospheric research, 1957—. Recipient Alumni Citation, Franklin Coll., 1961. Mem. I.E.E.E., Am. Geophys. Soc., A.A.A.S., Am. Mus. Natural History, Am. Forestry Assn., Ind. Acad. Sci., Sigma Xi, Tau Beta Pi, Eta Kappa Nu Alpha. Research and publs. on moon by radar, radio direction finding, electron content ionosphere measurements, relation of ionospheric disturbances to other natural phenomena. Home: 812 W. Delaware Av., Urbana, Ill. 61801.*

WEBB, Harold Worthington, Am. physicist; b. Ithaca, N.Y., July 27, 1884; s. John Burkibb and Mary (Gregory) W.; A.B., Columbia, 1905; Ph.D., 1909; m. Vivienne J. Mackenzie, Oct. 20, 1920; children—William M., Gregory W. Faculty, Columbia, N.Y.C., 1909—, prof. physics, 1929-53, prof. emeritus, 1953—. Fellow A.A.A.S., Am. Phys. Soc. (sec. 1923-28); mem. Optical Soc. Am. Research, publs. on radio waves, elec. excitation of spectra, theory of elec. arcs, electronic circuits. Home: 328 Westview Av., Leonia, N.J. 07605.*

WEBB, Philip Barker, Brit. botanist; b. Surrey, Eng., July 10, 1793; s. Philip Smith and Hannah (Barker) W.; ed. Christ Church, Oxford; entered Lincoln's Inn, 1812; B.A., 1815; studied geology under William Buckland. Re-discovered Scamander and Simois in travels in Italy, Greece, Troad, 1817-18; made natural history collections, Spain, 1826, Portugal and Morocco, 1827, Canary Islands, 1928-30, Italy, 1848-50, west of Ireland, 1851. Fellow Royal Soc., 1824; corr. mem. Acad. Scis. Madrid. Author: Osservazioni entorno allo stato antico e presente dellagro Trojano, 1820; Topographie de la Troade ancienne et moderne, 1844; Iter Hispaniense, 1838; Otia Hispanica, 1853; Histoire Naturelle des iles Canaries, 1836-50; Fragmenta floruiae aethiopico-aegypticae, 1845. Collected and studied (with M. Savin Berthelot) plants, birds, fish, shells, insects, rocks and waters of Canary Islands. Died Aug. 31, 1854.

WEBB, Raymon E., Am. plant pathologist; b. Truxno, La., Nov. 20, 1919; s. Abner and Miley (McCormick) W.; B.S., La. State U., 1942, M.S., 1948; Ph.D., U. Wis., 1952; m. Mary Virginia O'Brien, Feb. 19, 1943; children—Mary Ann, Raymon E. II, John Marshall. Horticulturist La. State U., 1946-50, 1952-53; plant pathologist U.S. Dept. Agr., Beltsville, Md., 1953—. Mem. Am. Phytopathology Soc. (pres. Potomac div. 1964-65, councilor Potomac div. 1965-67). Contbr. numerous articles in field to sci. jours. Identified new viruses infectious to potatoes and cucurbits; developed new potato, spinach, tomato varieties resistant to one or more diseases. Home: 10504 43d Av., Chestnut Hills, Beltsville, Md. 20705. Office: Crops Research Div., Plant Industry Sta., Beltsville, Md. 20705.*

WEBB, Robert Bradley, Am. microbiologist; b. Guy, Ark., Nov. 20, 1926; s. Raymond H. and Byrtie (Bradley) W.; B.S., Harding Coll., 1947; M.S., U. Okla., 1950, Ph.D., 1956; m. Mary Jane Rose, June 2, 1950; children—Rebecca Jane, Colleen Louise. Mem. faculty Harding Coll., Searcy, Ark., 1949-50, U. Tenn., 1950-52; NRC fellow Argonne Nat. Lab. (Ill.), 1956-58, asso. scientist, 1958—. Mem. A.A.A.S., Am. Soc. for Microbiology, Radiation Research Soc., Biophys. Soc., Genetics Soc. Research, publs. in life cycle and genetic recombination in Nocardia; evidence for direct involvement of organic peroxy-radicals in x-ray damage in biol. systems; role of water and glycerol as protective agts. in X-ray effects; demonstrated photodynamic mutagenesis with acridine dyes, visible light mutagenesis. Office: Argonne Nat. Lab., Argonne, Ill. 60439.*

WEBB, Robert Wallace, Am. geologist; b. Los Angeles, Nov. 2, 1909; s. Robert Remme and Sharlie Jeannette (Ward) W.; A.B., U. Cal., Los Angeles, 1931; postgrad. U. Wash., 1931, U. So. Cal., 1931-32; M.S., Cal. Inst. Tech., 1932, Ph.D., 1937; m. Evelyn Elaine Gourley, June 28, 1933; children—Robert Ian Arthur, Leland Frederick, Donald Gourley. Faculty U. Cal., Los Angeles, 1932-48; faculty U. Cal., Santa Barbara, 1948—, prof. geology, 1951—, chmn. dept. phys. scis., 1953-59; dir. Ford Found. Exptl. Program Instrs. for Colls., 1960-63. Coordinator vets. affairs univs. Cal., 1947-52; exec. sec. NRC div. geology and geography, 1953; dir. Am. Geol. Inst., 1953. Fellow Geol. Soc. Am., Mineral. Soc. Am., Meteoritical Soc. (sec. 1937-41); mem. Nat. Assn. Geology Tchrs. (pres. Far Western sect. 1959). Author: (with Joseph Murdoch) Minerals of California, 1966; also bulls. Cal. Div. Mines and Geology. Study of mineral occurrences in Cal.; interpreter of landscape features of U.S., especially So. Sierra Nevada of Cal. Home: 898 Via Campobello, Santa Barbara, Cal. 93105.*

WEBB, Sidney James, English economist, social reformer, historian; b. London, July 13, 1859; s. Charles and Elizabeth (Stacey) W.; ed. Switzerland, Mecklenburg-Schwerin, Birkbeck Inst., City of London Coll.; LL.B., London U., 1886; hon. degrees from London, Wales, Munich univs.; m. (Martha) Beatrice Potter, 1892. Clerk for firm of colonial brokers, 1875-78; 2d div. clerk War Office, 1878; with Surveyor of Texas office, 1879; 1st. div. clerk Colonial Office, 1881; called to bar by Gray's Inn, 1885; leader in Progressive campaign for London County Council: held seat in County Council, 1892-1910, chmn. Tech. Edn. Bd., 1912-27; hon. prof.

pub. adminstrn. London Sch. Economics and Polit. Sci.; served on Royal Commn. to review trade-unionism in light of Taff Vale judgement, 1903; mem. Labour Party's exec. com., 1915-25; mem. Parliament, from 1922; pres. Bd. Trade, 1924; sec. of state Dominion Affairs, 1929-30, Colonies, 1930-31. Author: Facts for Socialists, 1887; Facts for Londoners, 1889; Fabian Essays in Socialism, 1889; The Education Muddle and the Way Out, 1901; (with Beatrice Potter Webb) The History of Trade Unionism, 1894, Industrial Democracy, 1897, English Local Government, 7 vols., 1906-29, A Constitution for the Socialist Commonwealth of Great Britain, 1920, Soviet Communism: A New Civilization?, 2 vols., 1935. Worked (with his wife) for pub. sevrice, did pioneer work in study of trade unions, devoted time from the mid 30's to the study of Russian events, and greatly influenced trend of Brit. social thought of period; instrumental (as chmn. edn. bd.) in raising standards of London education. Died Passfield Corner, Liphook, Eng., Oct. 13, 1947.

WEBB, Sydney James, microbiologist; b. London, Eng., July 25, 1925; s. Arthur Watson and Kate (Goodall) W.; B.Sc., Regent Street Poly. U. London, 1951; M.Sc., Imperial Coll. Sci., U. London, 1953, D.I.C., 1955, Ph.D., 1959; m. Joyce Constance Carter, Apr. 3, 1948; children—John Sydney, Ian Andrew, Jane Elizabeth Anne, Kevin Watson. With Boots Pure Drug Co., 1955-56; with Def. Research Bd. Labs., Suffield, Alta., Can., 1956-60, cons., 1960; prof. microbiology U. Sask. (Can.), Saskatoon, 1960—. Mem. Soc. Gen. Microbiology, Soc. for Applied Microbiology, Canadian Soc. Microbiology, Canadian Biochemistry Soc., Canadian Soc. Cell Biology (dir.). Author: Bound Water in Biological Integrity, 1965; also articles. Research on role of water bound to macromolecules in their structure and function, discovered that loss of this water during partial desiccation leads to changes in structure of nucleo proteins and to death of mutation of cells, certain compounds can take over function of these water molecules and prevent desiccation and radiation damage. Home: 2616 Cascade St., Saskatoon, Sask., Can.*

WEBB, Watt Wetmore, Am. physicist; b. Kansas City, Mo., Aug. 27, 1927; s. Watt and Anna (Wetmore) W.; B.S., Mass. Inst. Tech., 1947, Sc.D., 1955; m. Page Chapman; children—Watt, III, Spahr Chapman, Bucknell Chapman. Staff Union Carbide Corp., Niagara Falls, N.Y., 1947-52, research scientist, asst. dir. research, 1955-61; faculty Cornell U., Ithaca, N.Y., 1961—, prof. engring. physics, 1965—. Cons. to industry, govt. adv. coms., 1953—. Mem. Am. Phys. Soc., Metall. Soc., A.A.A.S., Electrochem. Soc., Inst. for Metals, Am. Ceramic Soc., Am. Soc. for Metals, Sigma Xi. Research, publs. in high current electric arcs, welding engring., solid solutions, dislocations in crystals, whisker crystals, crystal growth, oxidation kinetics, fracture, superconductivity, critical phenomena, diffraction topography, ice, coop. phenomena.*

WEBB, Watts Rankin, Am. physician; b. Columbia, Ky., Sept. 8, 1922; s. Frank Elbert and Sue Josephine (Rankin) W.; B.A. with spl. distinction U. Miss., 1942; M.D., Johns Hopkins Sch. Medicine, 1945; m. Frances Luella Cooke, Aug. 19, 1944; children—Gordon Lewis, Harvey Elbert, Paul Alan, Andrew Michael. Faculty, U. Miss. Sch. Medicine, 1955-63, prof. surgery, 1963; prof., chmn. div. thoracic and cardiovascular surgery U. Tex. Southwestern Med. Sch., Dallas, 1964—. Cons. VA, U.S. Army, USAF, 1966; mem. surg. study sect. B, NIH, 1964. Decorated Silver medal, Shipley medal. Mem. Am. Thoracic Soc. (pres. So. chpt. 1964), Am. Coll. Chest Physicians (pres. So. chpt. 1965), So. Med. Assn. (chmn. surg. sect. 1965), Am. Assn. Thoracic Surgery, Am. Bd. Surgery, A.C.S., Am. Coll. Cardiology, Am. Fedn. Clin. Research, A.M.A., Am. Physiol. Soc., Am. Soc. for Artificial Internal Organs, Am. Surg. Assn., Bd. Thoracic Surgery, Soc. for Cryobiology, numerous others. Research, numerous publs. on devel. techniques and theory of freezing tissues and organs; heart and lung transplantation; use of metabolic inhibitors to prolong organ survival; surg. diagnosis of Tb; cardiac effects of adrenalectomy, alcohol, cigarettes and hypothermia. Home: 3500 Euclid Av., Dallas 75205. Office: 5323 Harry Hines Blvd., Dallas 75235.*

WEBBER, Robert Trumbull, Am. physicist; b. Newton, Mass., Aug. 22, 1921; s. Wolfert Gerson and Gertrude (Harris) W.; B.A., Amherst Coll., 1943; M.S., Yale, 1944, Ph.D., 1947; m. Clytie Frances Closky, Mar. 27, 1947. Instr. Yale U., 1947-49; sect. head. Naval Research Lab., Washington, 1949-53, branch head, 1953-57, sci. liason officer and dep. sci. dir. Office Naval Research, London Br., 1957-59; Tokyo rep. NSF, Tokyo, Japan, 1960-62; sci. attaché, Am. embassy, Tel Aviv, Israel, 1962-65, Tokyo, Japan, 1966—. Fellow Am. Phys. Soc.; mem. A.A.A.S., Washington Acad. Scis., Sigma Xi. Research, numerous publs. on low temperature physics, particularly superconductivity and the thermal and magnetic properties of metals. Address: Am. Embassy, APO, San Francisco 96503.*

WEBER, Alfred, German sociologist; b. Erfurt, Germany, July 30, 1868; brother of Max Weber; prof., Prague, Czechoslovakia, also Heidelberg, Germany. Author: Über den Standort der Industrien, 1909; Kulturgeschichte als Kultursoziologie, 1935; Abschied von den bischerigen Geschichte, 1946; Prinzipien der Geschichts und Kultursoziologie, 1952. Founder culture sociology based on society, civilization, culture. Died Heidelburg, May 2, 1958.

WEBER, Alfred Herman, Am. physicist; b. Phila., Jan. 15, 1906; s. Frank Curt and Anna Josephine (Kling) W.; A.B., St. Joseph's Coll., Phila., 1928, M.S., 1931; Ph.D., U. Pa., 1936; m. Frances Theresa Lever, Dec. 26, 1932; children—Constance Marie, Judith Ann, Christine Frances. Alfred Joseph, Mary Linda, Mark Frances, June Elizabeth. Mem. faculty St. Joseph's Coll., 1928-39, prof., head physics dept., 1936-39; asst. prof. physics St. Louis U., 1939-40, asso. prof., 1941-45, prof. since 1945, chmn. dept, since 1950; cons. physicist exptl. nuclear physics Argonne Nat. Lab., Chgo., 1947-60; cons. Army Ballistic Missile Agy., Huntsville, 1957-60, Marshal Space Flight Center for NASA, 1960—. Fellow Am. Phys. Soc.; mem. Physics Tchrs., Am. Assn. U. Profs., Am. Nuclear Soc., Am. Inst. Aeros. and Astronautics, Cath. Commn. Intellectual Cultural Affairs, Am. Crystallographic Soc., Sigma Xi. Author: How to Solve Problems in College Physics (with J. Harty), 1949; Vacuum Tube Characteristics and Design Sheets (with M. J. Schnurr and R. E. Schultz); College Laboratory Physics; (with J. F. McGee) Atomic, Nuclear and Electronic Physics Outline, 1960. Research, publs. on photoemission, photoconductivity thin metallic films, neutron diffraction in glasses, aluminum, slow neutron scattering in gases, thermionic, photoelectric (Schottky) deviations in simple W and TA crystals, atomic structure of viruses, organic compounds; space physics. Home: 1211 Du Bois, Kirkwood, Mo. 63122. Office: 221 N. Grand Blvd., St. Louis 63103.*

WEBER, Bernard Paul, French physician; b. Paris, France, Oct. 23, 1927; s. Jean-Paul and Micheline (Kastler) W.; M.D., U. Paris, 1960; m. Denise Gascuel, June 26, 1959; children—Benoit, Christian, Brigitte. Research on stress in animals and man (burns), 1958—. Mem. N.Y. Acad. Sci., Assn. French Anesthesiologists. Editor: Agressologie. Research on pharmacological and therapeutic action of aspartic acid, artificial hibernation, new hypnotics, pharmacological properties of new pyridazones, clin. therapy of burns. Home: 20 Boulevard de la Bastille, Paris XII. Office: 78 rue de la Convention, Paris 75, France.*

WEBER, Edouard Frederic, German physiologist; b. Wittenberg, Germany, 1806; s. Michael Weber; mem. faculty U. Leipzig (Germany); author several physiol. treatises; determined (with brother Ernst Heinrich) velocity of pulse wave, 1825, discovered inhibitory effect of vagus on heart, circa 1845. Died Leipzig, 1871.

WEBER, Ernst, elect. engr.; b. Vienna, Austria, Sept. 6, 1901; s. Hermann and Josefine (Swoboda) W.; E.E., Tech. U., Vienna, 1924. Ph.D., 1926. D.Sc., 1927; Doctor of Science, Pratt Institute, 1958; D.E., Newark Coll. Engring., 1959, U. Mich., 1964; m. Sonya C. Escherich, 1936. Came to U. S., 1931, naturalized, 1937. Adjunct prof. Tech. U. Berlin, 1929-31; vis. prof. Poly. Inst., Bklyn., 1930, research prof. elec. engring., 1931-41, prof. grad. elec. engring., head dept., 1942-45, head dept. elec. engring., 1945-57, v.p. research, 1957-61, pres., 1958; dir. Microwave Research Inst., 1945-57; dir., sec. Poly. Research Development Co., Bklyn., 1945-52, pres., 1952-60, cons., 1960—. Official investigator Office Sci. Research and Development, 1942-45. Awarded Presdl. Certificate of Merit. Fellow Am. Inst. E.E., Am. Phys. Soc., N.Y. Acad. Sci., Inst. Radio Engineers (president 1959), A.A.A.S.; mem. Am. Mathematical Society, American Standards Association, Am. Soc. Engring. Edn., Nat. Acad. Scis., Sigma Xi. Author: Electromagnetic Fields, 1950; Linear Transient Analysis (2 vol.), 1954-56. Contbd. to devel. of microwave theory and method; studied precision measurement of attenuation and tech. of metallized glass. Home: 159 Lorraine Av., Mt. Vernon, N.Y. Office: 333 Jay St., Bklyn.

WEBER, Ernst Heinrich, German anatomist, physiologist; b. Wittenberg, Saxony, June 24, 1795; s. Michael Weber; ed. Wittenberg. Prof. anatomy at Leipzig from 1818, prof. physiology, from 1840. Devised method for determining sensitivity of skin; with brother discovered inhibitory power of vagus nerve; made studies of acoustics and wave motion; discovered small bones (Weberian ossicles) between labyrinth and swim bladder of some fresh-water fish; formulated Weber's law of sensation that increase in stimulus necessary to produce increase in sensation depends on strength of preceding stimulus; law revised by Fechner and became known as Weber-Fechner Law; implications of law not now accepted but at time Weber's work both original and significant; sensations research marked beginning of exptl. psychology; pioneer exptl. studies of nervous impulses controlling heartbeat. Died Leipzig, Jan. 26, 1878.

WEBER, Frederick Parkes, English physician; b. 1863; s. Hermann Weber; M.A., M.D., Cambridge U.; student St. Bartholomew's Hosp., London, Paris, Vienna; m. Hedwig Unger-Laissle, 1921. Physician, Mt. Vernon Hosp. for Consumption; cons. diagnostician to German Hosp. Recipient Moxon medal Royal Coll. Physicians. Fellow Royal Numis. Soc. (hon.), Royal Soc. Medicine (hon.), Royal Coll. Physicians; mem. Danish Soc. Dermatologists, Am. Clin. and Climatol. Assn., Soc. Antiquaries, Brit. Med. Assn., Med. Soc. London, Assn. Physicians Gt. Britain and Irland (hon.), Brit. Assn. Dermatologists, German Soc. Dermatologists. Author: Aspects of Death and Correlated Aspects of Life in Art, Epigram and Poetry, 4th edit., 1922; Some Thoughts of a Doctor, 1935; Endocrine Tumours, 1936; Rare Diseases and Some Debateable Subjects, 2d edit. 1947; More Thoughts and Comments of a Doctor, 1952; On Naevi, 1952; Medical Teleology and Miscellaneous Subjects, 1958; Miscellaneous Notes, 1960; also articles. Described nodular nonsuppurative punniculitis, a skin disease with painful nodules under skin (Christian-Weber Disease), 1925, formation of certain kinds of tumours in skin and mucus membranes called hereditary hemorrhagic telangiectasia (Rendu-Osler-Weber disease). Died June 2, 1962.

WEBER, George, physician; b. Budapest, Hungary, Mar. 29, 1922; s. Salamon and Hajnalka (Arvai) W.; B.A., Queen's U., Kingston, Ont., Can., 1950, M.D., 1952; m. Catherine Elizabeth Forrest, June 30, 1958; children—Elizabeth Dolly Arvai, Julie Vibert Wallace. Came to U.S., 1959, naturalized, 1965. Postdoctorate research U. B.C., Can., 1952-53; research asso. Montreal (Can.) Cancer Inst. Research Labs., Notre Dame Hosp., U. Montreal, 1953-59; asso. prof. biochemistry and microbiology Ind. U. Sch. Medicine, Indpls., 1959-60, prof. pharmacology, 1960—, cancer coordinator of basic research, 1962—. Recipient Best Pre-Clin. Prof. award Student A.M.A., 1965-66. Fellow Royal Soc. Medicine; mem. Am. Soc. Pharmacology and Exptl. Therapeutics, Am. Physiol. Soc., Am. Assn. Cancer Research, Endocrine Soc., Canadian Physiol. Soc., Biochem. Soc., Sigma Xi. Editor: Advances in Enzyme Regulation, vols. 1-6, 1963—. Contbr. numerous articles to profl. jours. Discovered action of cortisone and insulin at enzyme biosynthetic level, correlation of growth rate with glycolysis and other metabolic parameters in liver tumors, biochem. alterations in diabetic liver, regulatory mechanisms of enzyme biosynthesis in mammalian tissues, feedback inhibition by free fatty acids of key enzymes of glycolysis and direct oxidation. Home: 7307 Lakeside Dr., Indpls. 46278.*

WEBER, George F(rederick), Am. plant pathologist; b. Alexandria, S.D., May 19, 1894; s. Lewis and Clara (Bashford) W.; B.S., S.D. State U., 1916; M.S., U. Wis., 1920, Ph.D., U. Wis., 1922; m. Kate Caldwell, Sept. 15, 1917. Asst., Barberry Eradication, Wis., 1919; agt. wheat smut investigation U. S. Dept. Agr., 1920, corn root rot investigation, 1921, collaborator, 1922—; asst. plant pathologist, U. Wis., 1922; faculty U. Fla., Gainesville, 1922-64, prof., 1937-64. Fellow A.A.A.S.; mem. Am. Phytopath. Soc. (past pres.), Fla. Acad. Scis. (past pres.), Bot. Soc. Am., Am. Mycol. Soc., Am. Assn. U. Profs., Sigma Xi, Phi Sigma, Gamma Alpha, Delta Tau Delta, Sigma Delta Psi, Phi Kappa Phi (pres. Fla. chpt. 1937—). Research, numerous publs. on control of plant diseases, descriptions of fungi. Home: 1122 S.W. 3d Av., Gainesville, Fla. 32601.*

WEBER, Gerhard, Swiss physician; b. Basel, Switzerland, Dec. 18, 1914; s. Anton and Marie (Schladerer) W.; Dr.med., U. Basel, 1939; m. Anne-Marie Schurpf, May 20, 1944. With Neurosurgery Clinic, Zurich, 1946-54; privatdozent for neurosurgery U. Zurich, 1954-62, prof., 1962—. Author: Hirnabszess, 1957; also articles. Research in neurol., neurosurg. topics. Home: 10 Hedwigstrasse, Zurich, Switzerland.*

WEBER, Heinrich, German mathematician; b. Heidelberg, Germany, Mar. 5, 1842; prof., Heidelberg, also Zurich, Switzerland, Königsberg (now Kaliningrad, USSR), Berlin, Marburg. Göttingen (all Germany), Strasbourg (now in France). Author: Lehrbuch der Algebra, 3 vols., 1891-96; Vorlesungen über partielle Differentialgleichungen der mathematische Physik, 2 vols., 1900-01; Enzyklopädie der Elementarmathematik, 3 vols., 1903-07. Study of partial differential equation; algebra. Died Strasbourg, May 17, 1913.

WEBER, Jon N., geochemist; b. Kitchener, Ont., Can., July 8, 1935; s. E.A. and E. (Esch) W.; B.Sc. with honors, McMaster U., Hamilton, Ont., 1958, M.Sc., 1959; Ph.D., U. Toronto, 1962; m. Patricia Carroll, Oct. 9, 1959; children—Duncan J., Alan S. Faculty, Pa. State U., University Park, 1961—, now asso. prof. geochemistry. Field studies in N. Africa, Central Am., E. Asia, Pacific Ocean, 1961—. Mem. Geochem. Soc., Geol. Soc. Am., Am. Assn. Petroleum Geologists. Research and publs. on stable isotope and trace element geochemistry of sedimentary rocks and minerals of carbonate sediments and of calcareous marine invertebrates. Office: 224 Mineral Sci., University Park, Pa. 16802.*

WEBER, Joseph, Am. physicist; b. Paterson, N.J., May 17, 1919; s. Jacob and Lena (Stein); B.S., U.S. Naval Acad. 1940; Ph.D., Cath. U. Am., 1951; postgrad. Lorentz Inst. for Theoretical Physics, Leiden, Holland, 1956; m. Anita Straus, Oct. 18, 1942; children—Jonathan, Paul, James, David. With U. Md., College Park, 1948—, prof. physics, 1958—; mem. Inst. for Advanced Study, Princeton, N.J.,

1955, 62. Recipient Sci. Achievement award, Wash. Acad. Scis., Gravity Research Found., 1959. Author: General Relativity and Gravitational Waves, 1961. Research on fluctuation theory; gravitational radiation theory and experiments discoverer of maser principle.*

WEBER, Max, German sociologist, polit. economist; b. Erfurt, Thuringia, Germany, Apr. 21, 1864; student law Heidelberg, Berlin, Göttingen univs.; m. Marianne Schnitger, 1892. Worked in Berlin criminal ct.; lectr. U. Berlin, 1892-94; apptd. prof. U. Freiburg, 1894, U. Heidelberg, 1896, U. Vienna, 1918, U. Munich, 1919. Founder Deutsche Gesellschaft für Soziologie; asso. editor Archiv für Sozialwissenschaft und Sozialpolitik (became Germany's leading social-sci. jour. until advent of Hitler), from 1903; served in adminstrn. of German army hosps., World War I; cons. to German Armistice Commn., 1918, also commn. to draft Weimar constn.; participated in Congress Arts and Scis., World's Fair, St. Louis. Author: Die römische Agrargeschichte, 1891 (survey of rural labor in Germany), 1892; The Protestant Ethic and the Spririt of Capitalism, 1904-05; Wissenschaft als Beruf, 1919; Politik als Beruf, 1919; Geschtliche Aufsuchungen zur Religionssoziologie, 3 vols., 1920-21; Wirtschaft und Gesellschaft, 1922; Methodology of the Social Sciences, 1922; Theory of Social and Economic Organization, 1924; General Economic History, 1924; Basic Concepts in Sociology; The City. Viewed sociology as empirical science based on comparative social history, including preliterate cultures, and based upon the elimination of every judgment of value; studied sociology of leadership, law, economics, the state, music, religion; research on reruralization and latifundia of later Roman Empire; critic of Marxist econ. determinism; strove to show that history is shaped by plurality of causes; perceived intimate relation between Calvinism and rise of capitalism; evolved concept of ideal types (working models) as tool for classifying and comparing social conduct and systems; his work has greatly influenced 20th century sociohistorical theories and methodology of sociology. Died Munich, June 14, 1920.

WEBER, Max Karl, Swiss geophysicist; b. Wohlen; s. Max and Stehli (Vreny) W.; Dr ès sc., Zurich Tect. Soc.; m. Kuhn Margrit, May 1, 1943. Sci. asst.; asso.; prof. agrégé and instr.; dir. Swiss Seismological Service; now prof. geophysics Eidgenossiche Tech. Hochschule. Mem. Swiss Natural Research Soc., European Assn. Exploration Geophysicists. Publns. on propagation time of seismic quakes; seismographs; vibration measurement and biophys. measuring devices. Home: Bremgarters rasse 6, Wohlen, Switzerland. Cffice: Eidgenossische Tech. Hochschule, Leonhardstrasse 33, Zurich, Switzerland.

WEBER, Morton M., Am. biochemist; b. N.Y.C., May 26, 1922; s. Morris and Mollie (Scherer) W.; B.S., Coll. City of N.Y., 1949; Sc.D., Johns Hopkins, 1953; m. Phyllis Levy, July 31, 1955; children—Stephen Abbott, Ethan Lenard. Instr. bacteriology Johns Hopkins, Balt., 1951-54, Am. Cancer Soc. postdoctoral fellow McCollum Pratt Inst., 1953-56; instr. Harvard, 1956-59; faculty St. Louis U. Sch. Medicine, 1959—, prof. microbiology, 1963—, chmn. dept., 1964—; cons. McDonnel Aircraft Corp., 1965—. Fellow A.A.A.S.; mem. Am. Soc. Microbiology (chmn. physiology div. 1966-67), Am. Soc. Biol. Chemists, Soc. Gen. Microbiology (Eng.), Sigma Xi. Research, publs. on physiology and biochemistry of microorganisms, mode of action of antibiotics and other antimicrobial agts., mechanisms of microbial pathogenicity with spl. reference to Mycoplasma; discovery of role of vitamin K in electron transport in micro-organisms. Home: 7068 Waterman St., St. Louis 63130.*

WEBER, Neal Albert, Am. biologist; b. Towner, N.D., Dec. 14, 1908; s. Albert and Kathryn (Boom) W.; B.A., U. N.D., 1930, M.A., 1932; M.A., Harvard, 1934, Ph.D., 1933; D.Sc., U. N.D., 1958; m. Jean Jeffery, May 29, 1940; children—Nancy Beth, Cornelius Jeffery, Peter Albert. Asso. prof. biology U. N.D., 1936-43, anatomist Med.Sch., 1943-47; collaborator Am. Mus. Natural History, 1947-50; faculty Swarthmore (Pa.) Coll., 1947—, prof. zoology, 1958—; prof., head dept. zoology Coll. Arts and Sci., Baghdad, Iraq, 1950-52; prof. anatomy Royal Coll. Medicine, Baghdad, 1950-52; vis. research prof. bacteriology U. Wis., 1955-56; sci. attaché Dept. State, Buenos Aires, Argentina, 1960-62. Mem. com. on polar research NRC-Nat. Acad. Scis., 1958-60, mem. postdoctoral fellowship panel, 1963-66; undergrad. research leader NSF, Trinidad, W.I., 1964-65. Harvard Traveling fellow, Cuba, 1933; NRC fellow Trinidad, 1934-35, Arctic Inst. N.Am. fellow, 1949. Fellow A.A.A.S.; mem. Am. Entomol. Soc. (past pres.), Ecol. Soc. Am., Am. Assn. Anatomists, Am. Soc. Ichthyologists and Herpetologists. Author: (with H.B. Glass, G. Moment) Laboratory Manual in Zoology, 1959; also numerous articles. Biol. exploration throughout the Ams., Central Africa, Mid-East, Arctic Alaska; research on biology fungus-growing ants and their fungi. Home: 1 Whittier Pl., Swarthmore, Pa. 19081.*

WEBER, Thomas Byrnes, Am. bioinstrumentation scientist; b. Oklahoma City, Sept. 1, 1925; s. Louis W. and Myrtle (Schmidt) W.; B.S., Okla. State U., 1949; M.S., La. State U., 1950, Ph.D., 1954; m. Marian L. Keller, June 14, 1958; children— Cynthia, Krista, Kathyreen, Kurt. Research scientist Animal Disease Inst., U.S. Dept. Agr., Washington, 1954-57; head biochem. dept. Dental Research Facility, Great Lakes, Ill., 1957-59; head atmospheric monitoring Sch. Aerospace Medicine USAF, San Antonio, 1960-62; mgr. advanced research Beckman Instruments, Inc., Fullerton, Cal., 1962-67; pres. Biosci. Planning, Inc., Anaheim, 1967—. Fellow Am. Inst. Chemists, Royal Soc. Medicine; mem. A.A.A.S., Am. Chem. Soc., Aerospace Med. Assn., Instrument Soc. Am., Am. Inst. Aeros. and Astronautics (past dir.). Author: Instrumentation Methods for Predictive Medicine, 1965; also numerous articles. Defined concepts of predictive medicine, including selection med. parameters, data reduction and correlation systems; developed continuous monitoring techniques for respiratory gases and micro-contaminants in environmental chambers. Home: 809 Larchwood St., Brea, Cal. Office: 1720 W. La Palma St., Anaheim, Cal. 92801.*

WEBER, Wilhelm Eduard, German physicist; b. Wittenberg, Germany, Oct. 24, 1804; ed. Halle, Göttingen, Germany; became prof. physics U. Göttingen, 1831-37; Leipzig, 1843-49; resumed Göttingen position, 1849. Fellow Royal Soc., 1840; mem. Göttinger Sieben, French Acad. Scis. Weber (magnetic unit) named in his honor. Author: Resultate aus den Beobachtungen des magnetischen Vereins, 6 vols., 1837-43; Elektrodynamische Massbestimmungen, 1846-77. Established (with Gauss) measuring system of electricity, patterned on Gauss' magnetic units (now in internat. use), developed new measuring instruments, made numerous measurements with precision never before attained; the weber, a magnetic unit, named for him; studied acoustical phenomena, the flute, temperature compensation of organ pipes, 1826-29; studied wave motion; studied (with Gauss) terrestrial magnetism, invented an electromagnetic telegraph, 1833; used Bunsen's ice calorimeter to determine specific heat of diamond; invented an electrodynamometer to counter effect of terrestrial magnetism in studies of elec. resistance and current strength, 1864; discovered connection between electric and magnetic power, found the velocity in play equal to speed of light; foreshadowed work of Maxwell. Died Göttingen, June 23, 1891.

WEBER, William Alfred, Am. botanist; b. N.Y.C., Nov. 16, 1918; s. Henry Paul and Emilie (Rilke) W.; B.S., Ia. State Coll., 1940; M.S., Wash. State Coll., 1942, Ph.D., 1946; m. Selma R. Herrmann, Aug. 5, 1940; children—Linna Louise (Mrs. Ludger Müller-Wille), Eunice Oma (Mrs. David Fulker), Erica Marion (Mrs. Philip Rice). Faculty dept. history, since 1962—, curator Herbarium, 1946—. Member of the British Lichen Society. Internat., Am. socs. plant taxonomists, Am. Bryological Soc., Swedish, Cal. bot. socs., Bot. Soc. Lund, Soc. for Study Evolution. Author: Handbook of Plants of the Colorado Front Range, 1952, rev., 1966; also numerous articles. Research on plant geography of So. Rocky Mountains, variation and taxonomy of lichens and bryophytes, mosses and lichens of Galapagos Islands, perennial sunflowers. Home: 1905 Bluff St., Boulder, Colo. 80302.*

WEBSTER, Arthur Gordon, Am. physicist; b. Brookline, Mass., Nov. 28, 1863; s. William Edward and Mary Shannon (Davis) W.; A.B., Harvard, 1885; studied Berlin, Paris, Stockholm, 1886-90; Ph.D., U. of Berlin, 1890; D.Sc., Tufts, 1905; LL.D., Hobart, 1908; m. Elizabeth Munroe Townsend, Oct. 8, 1889. Instr. mathematics, Harvard, 1885-86; docent in physics, 1890-92, asst. prof., 1892-1900, prof. and dir. of phys. lab., 1900—, Clark U. Awarded Thomson prize (5,000 francs), Paris, 1895, for exptl. research on the Period of Electrical Oscillations. Fellow A.A.A.S., Am. Acad. Arts and Sciences, Am. Inst. E.E., Inst. Radio Engrs. Del. U. S. Govt. to Internat. Radiotelegraphic Conf., London, 1912. Mem. U. S. Naval Consulting Bd., 1915—, Nat. Research Council, 1917—. Author: A Mathematical Treatise on the Theory of Electricity and Magnetism, 1897; Dynamics of Particles, and of Rigid, Elastic and Fluid Bodies, 1904; Lowell Institute Lectures on Electricity and Ether, 1897; Harrison Lectures on Sound, U. of Pa., 1911; also many papers. Research on physics, chiefly on mat. physics, mechanics, sound, electricity and ballistics. Died May 15, 1923.

WEBSTER, Bruce Peck, physician; b. Lansdowne, Ont., Can., Nov. 1, 1901; s. Thomas Amos and Bertha (Peck) W.; M.D. C.M., McGill U., 1925. Came to U. S., 1927, naturalized, 1932. Intern. asst. resident medicine Montreal Gen. Hosp., 1925-27; asst. resident medicine, Jacques Loeb fellow Johns Hopkins Hosp. and University, 1927-29; fellow medicine NRC, 1929-31; asst. prof. medicine Tulane U., 1931-32; asst. prof. medicine Cornell U., 1932-46. asso. prof. clin. medicine, 1949-66, clin. prof. medicine, 1966—; pvt. practice medicine, 1946—; attending physician N.Y. Hosp.; med. dir. Time, Inc., 1947—. Mem. pub. adv. com. to surgeon gen. USPHS; cons. Dept. Army; exec. com. Japanese Christian U. Found., St. Luke's Hosp., Tokyo; med. adv. bd. Am. Hosp., Paris. Past pres. the Internat. Union Against Trepanomatoses. Served from lt. col. to col., AUS, 1942-46. Decorated Legion of Merit. Mem. Soc. Exptl. Biology and Medicine, Soc. Clin. Investigation, Am. Inst. Nutrition, Harvey Soc. Soc. Research Child Development, N. Y. Acad. Medicine, Am. Soc. Hygiene (v.p., chmn. exec. com., also chmn. internat. com.), Soc. Med. Cons. to Armed Forces (past pres.). Clubs: University, Century Assn. (N.Y.C.); Maidstone (Easthampton, N.Y.). Research on pathological physiology of thyroid gland, growth and metabolism, cardiovascular syphilis; public health aspect of syphilis control. Home: 14 Sutton Pl. S., N.Y.C. 22; also Easthampton, N.Y. Office: 449 E. 68th St., N.Y.C. 21.*

WEBSTER, David Locke, Am. physicist; b. Boston, Nov. 6, 1888; s. Andrew Gerrish and Florence (Briggs) W.; A.B. Harvard, 1910, Ph.D., 1913; m. Anna Cutler Woodman, June 12, 1912; children—Nancy (Mrs. Anthony S. Felsovanyi), Helen (Mrs. Paul T. Feeley), David Locke III, Cutler; m. Olive Durbin Ross, Sept. 18, 1951. Asst., instr. math., physics Harvard, 1909-15; asst. prof. physics U. Mich., 1917, Mass. Inst. Tech., 1919-20; prof. physics Stanford U., 1920-54, exec. head physics dept., 1920-42, coordinator civil pilot tng., 1939-41, prof. emeritus, 1954—. Head physicist, later chief physicist Signal Service, Ordnance Dept., U.S. Army, 1942-45, Cons. CAA, 1941, Ordnance Dept., U.S. Army, 1945-50, Hawaii Inst. Geophysics, 1961-64, Ames Research Center, NASA, 1962. Mem. Nat. Acad. Scis., Am. Acad. Arts and Scis., Am. Philos. Soc., Am. Assn. Physics Tchrs. (pres. 1935-36), A.A.A.S. (v.-p., chmn. sect. B 1932), Am. Phys. Soc., Am. Geog. Soc., Am. Geophys. Union. Author: (with E.R. Drew and H.W. Farwell) General Physics for Colleges, 1923; revised editions of Civil Pilot Training Manual, and Flight Instructors Manual, 1941. Discovered critical potentials in X-ray excitation, 1915-17; research on ionization of X-ray shells 1920-36; participant early devel. of klystron, 1937-39. Address: 1830 Cowper St., Palo Alto, Cal. 94301.*

WEBSTER, Donald Robertson, Canadian surgeon; b. Pictou, N.S., Can., Apr. 1, 1902; s. Conrad Ogilvy Hall and Ella (Langille) W.; B.A., Dalhousie U., Halifax, N.S., 1922, M.D.C.M., 1925; M.Sc., McGill U., Montreal, Que., Can., 1930, Ph.D., 1932; m. Jane Kern Ross, Nov. 24, 1945; children—Janet, Donald Ross. With McGill U., 1936—, prof. surgery, 1953-62, prof. exptl. surgery, dir. lab., 1962—; staff Royal Victoria Hosp., Montreal, 1937—, surgeon-in-chief, 1953-62, attending staff, 1962—. Decorated Order Brit. Empire; named hon. surgeon to H.M. Queen Elizabeth. Fellow Royal Coll. Physicians and Surgeons Can., Royal Coll. Surgeons Eng. (hon.,; mem. A.C.S., Am. Surg. Assn., Am. Gastroent. Assn., N.Y. Acad. Scis., Internat. Surg. Group, Soc. U. Surgeons, Canadian Med. Assn., Soc. Clin. Surgeons Can., Canadian Soc. Gastroenterology, Internat. Soc. Surgery, Montreal Medico-Chirurg. Soc., Sigma Xi. Contbg. author: Textbook of Surgery, 2d edit., 1955, 3d edit., 1959. Home: 565 Roslyn St., Montreal, Que. 6, Can.

WEBSTER, George Calvin, Am. biochemist; b. South Haven, Mich., July 17, 1924; s. Eugene H. and Hazel (Empson) W.; B.S., Western Mich. U., 1948; M.S., U. Minn., 1949, Ph.D., 1952; m. Sandra Lee Whitman, Jan. 23, 1960; children—Jeffrey Calvin, Kimberley Ann. Research fellow Cal. Inst. Tech., 1952-55; prof. biochemistry Ohio State U., Columbus, O., 1955-61; vis. prof. enzyme chemistry U. Wis., Madison, 1961-65; chief chemist Cape Kennedy Missile Center, Cocoa, Fla., 1965—. Established investigator Am. Heart Assn., 1963—. Recipient Pub. Health Service Career Devel. award, 1963. Fellow A.A.A.S.; mem. Am. Soc. Biol. Chemists, Am. Chem. Soc., Biochem. Soc. (Eng.), Am. Soc. for Cell Biology, Sigma Xi. Author: Nitrogen Metabolism in Plants, 1959; also numerous articles. Research in clarification chem. pathways nitrogen metabolism in living cells; discovered several enzymes concerned with nitrogen metabolism and energy transfers; pioneered synthesis plant protein outside living cell. Home: 389 Chester Dr., Cocoa, Fla. 32922. Office: P.O. Box 563, Cocoa, Fla. 32922.*

WEBSTER, Harold Frank, Am. physicist; b. Buffalo, June 25, 1919; s. Stephen and Florence (Frank) W.; B.A., U. Buffalo, 1941, M.A., 1944; Ph.D., Cornell U., 1953; m. Helen Voorhis, Sept. 15, 1951; children—Sue Helen, Kenneth Harold, Jean Phyllis. Staff, Mass. Inst. Tech., Cambridge, 1943-45; teaching and research asst. Cornell U., 1945-51; physicist Gen. Electric Research Lab., Schenectady, 1951—. Mem. Internat. Sci. Radio Union, Am. Phys. Soc., I.E.E.E. (W.R.G. Baker award 1958). Contbg. author: Advances in Electronics, Vol. 17, 1962; also articles. Research on impedance matching techniques for microwave waveguide components; discovered vortex instability in electron sheet beams; derived current voltage characteristics thermionic diodes from space charge distbn.; measured importance crystal face variable on emission electrons from cesium covered surfaces. Home: 77 St. Stephens Lane W., Scotia, N.Y. 12302. Office: Gen. Electric Research and Devel. Center, P.O. Box 8, Schenectady 12301.*

WEBSTER, John, English chemist; b. Thornton, Craven, Eng., Feb. 3, 1610; probably student, Cambridge; studied under John Huniades. Ordained as minister; became curate of Kildwick, Craven, Eng., 1634; named master Clitheroe Grammar Sch., 1643; chaplain, surgeon in parliamentary army; vicar of Milton, Eng., circa 1649; became preacher, London, 1653; practiced medicine, Clitheroe, 1657-

82. Author: The Saint's Guide, 1653; The Judgement Set and the Books Opened, 1654; Academiarum examen, 1654; The Displaying of Supposed Witchcraft, 1647; Metallographia . . . , 1671. Suggested ednl. reform of Oxford and Cambridge Universities based on Christian principles; opponent of witchcraft. Died Clitheroe, Eng., July 18, 1682.

WEBSTER, Thomas, geologist; b. Orkneys, 1773; ed., Aberdeen, Scotland. Travelled in Eng. and France making archtl. sketches; practiced architecture, London; curator Geol. Soc.'s Mus.; prof. geology U. Coll., London, 1842-44. The rare mineral Websterite, also various fossils named in his honor. Author: On the Fresh-water Formation in the Isle of Wight with some Observations on the Strata over the Chalk in the South-east of England, 1814. Research and publs. on geology of Upper Secondary and Tertiary strata of S.E. Eng., also Isle of Wight. Died Dec. 6, 1844.

WEBSTER, Victor Stuart, Am. chemist, educator; b. Maquoketa, Ia., Nov. 4, 1908; s. Charles Orange and Frances (McComb) W.; B.S., State U. Ia., 1930, M.S., 1931, Ph.D., 1933; m. Delia Cornelia DeBoer, June 5, 1931; children—David Charles, Stuart John. Research chemist State U. Ia., 1933-34; chemist Ia. Planning Bd., 1934-35; faculty Moorhead (Minn.) State Coll., 1935-36; faculty S.D. State U., Brookings, 1936——; prof. chemistry, head dept., 1944——, acting dean Grad. Sch., 1962-63. Mem. Am. Chem. Soc., A.A.A.S., S.D. Acad. Sci., S.D. Edn. Assn.; Sigma Xi, Phi Kappa Phi, Phi Lambda Upsilon, Kappa Delta Pi. Research, publs. on oxidation of unsaturated acids, condensation of phenols with anhydrides, color test for amines, preparation alcohols by Grignard method. Home: 122 12th Av., Brookings, S.D. 57006.*

WECHSLER, David, psychologist; b. Lespede, Rumania, Jan. 12, 1896; s. Moses and Leah (Pascal) W.; A.B., Coll. City N.Y., 1916; M.A., Columbia, 1917, Ph.D., 1925; m. Ruth A. Halpern, July 21, 1939; children—Adam F., Leonard M. Chief psychologist Bellevue Psychiat. Hosp., 1932-66; clin. prof. N.Y. U. Coll. Medicine, 1942——; vis. prof. clin. psychology Hebrew U., Jerusalem, 1967. Spl. cons. sec. of war, 1941-43. Board govs. Hebrew U. 1965——; v.p., dir. Am. Friends of Hebrew U. Fellow Gerontological Soc.; mem. A.A.A.S., Am. Psychol. Assn. (award of year clin. div. 1960), Am. Psychopath. (pres. 1959-60), N.Y. State Psychol. Assn., N.Y. Acad. Scis., Am. Assn. Mental Deficiency, Soc. Gen. Systems Research Child Devel., Sigma Xi. Author: Wechsler-Bellevue Intelligence Scale, 1939; The Measurements of Adult Intelligence, 1944; Wechsler Intelligence Scale for Children, 1949; Wechsler Preschool & Primary Scale of Intelligence, 1967. Contributions in devel. and devising of scales of intelligence; psychological diagnosis; change of ability with age; inventor lie detector (psychogalvanograph). Home: 145 E. 92d St., N.Y.C., 10028.*

WECHSLER, Wolfgang, German neurologist; b. Ulm, Germany, June 20, 1930; s. Hermann and Gertrud (zum Tobel) W.; student univs. Freiburg, Bonn, Kiel, Munich, 1950-57; med. state exam., U. Munich, 1957; m. Doris Rotter, Sept. 20, 1958; children—Bettina, Isabel. Asst., German Research Inst. Psychiatry, Munich, 1957-61; chief asst., neurol. sect. Max Planck Inst. Brain Research, 1961——; docent exptl. neuropathology and neuroanatomy U. Cologne. Mem. Germans socs. Neuropathology and Neuroanatomy, Pathology, Electron Microscopy, Anatomy, Internat. Acad. Pathology, N.Y. Acad. Scis. Research, numerous publs. on brain hemorrhaging, infections of nervous system, muscular diseases, brain tumors, brain edema, regeneration and degeneration of nerve fibers, devel. of nervous system. Home: 2 Eichenhainallee, Bensberg-Frankenforst. Office: 200 Cologne-Merheimer-Strasse, Cologne-Merheim, West Germany.*

WECKEL, Kenneth Granville, Am. chemist; b. Canton, O., Oct. 9, 1905; s. John F. and Lena (Westmeyer) W.; B.S., U. Wis., 1931, M.S., 1932, Ph.D., 1935; m. Genevieve Brinkman, July 29, 1939. Faculty, U. Wis., Madison, 1935——, now prof. food sci. and industries, asso. dir. univ. industry research program. Recipient Borden award, 1938, Internat. Assn. Milk and Food Sanitarians citation, 1952. Mem. Am. Chem. Soc., Internat. Assn. Milk and Food Sanitarians, Am. Dairy Sci. Assn., Inst. Food Technologists, Am. Assn. Candy Technologists. Research, numerous publs. on food product processing, vitamin fortification of foods.*

WECKER, Johann Jacob, physician; b. Basel, Switzerland, 1528; qualified in medicine between 1560-66; became prof. logic, Basel, 1557, prof. Latin, 1560; became town physician, Colmar, Germany, from 1566. Author: Antidotarium speciale, 1577; De secretis libri XVII, 1582. Publs. on antidotes, chem. operations, distillation. Died Colmar, 1586.

WEDDELL, James, Brit. explorer, navigator; b. Ostend, Austrian Netherlands, Aug. 24, 1787 (parents Brit. citizens). Mcht. seaman; imprisoned on Rainbow frigate for mutiny, 1808; became master, 1810; became comdr. Firefly, 1810, Hope, 1812; comdr. Jane of Leight, sealing ship, Antarctic Ocean, 1819-24. Author: A Voyage towards the South Pole performed in the Years, 1822-24, 1825. Discovered and explored South Orkneys; explored Falkland Islands, South Shetlands, Cape Horn, South Georgia; surveyed South Shetlands; discovered Weddell Sea; Weddell quadrant, also type of seal found in area named in his honor. Died London, Sept. 9, 1834.

WEDDELL, James Blount, Am. physicist; b. Evanston, Ill., Apr. 29, 1927; s. Reid and Wilmirth (Gamble) W.; student Northwestern U., 1943-44, M.S., 1951, Ph.D., 1953; A.B., Drew U., 1949; m. Beatrice Lyerly, Oct. 15, 1960. Research engr. Westinghouse Research Labs., Forest Hills, Pa., 1953-57; sr. staff scientist Martin-Marietta Corp., Balt., 1957-62; staff scientist N. Am. Rockwell Corp., Downey, Cal., 1962——. Lectr., Drexel Inst. Tech., Phila., 1958-59. Mem. Am. Phys. Soc., Am. Geophys. Union, A.A.A.S., Am. Astron. Assn. (sr.), Sigma Xi. Contbr. articles to tech. jours. Measurement neutron inelastic scattering cross sects.; integrated radio-isotope heat source with thermoelectric semiconductor, long range predictability solar flares; developed hydromagnetic theory acceleration solar protons. Home: 936 S. Peregrine Pl., Anaheim, Cal. 92806. Office: 12214 Lakewood Blvd., Downey, Cal. 90241.*

WEDDERBURN, Joseph Henry Maclagan, mathematician; b. Forfar, Scotland, Feb. 26, 1882; s. Alexander Stormonth Maclagan and Anne (Ogilvie) W.; M.A., Edinburgh U., 1903, D.Sc., 1908; studied Leipzig, Berlin and U. of Chicago; unmarried. Asst. in mathematics, Edinburgh U., 1905-09; asst. prof. mathematics, Princeton University, 1909-21, associate professor, 1921-28, professor, 1928-45, professor emeritus since 1945. Fellow A.A.A.S., Royal Soc. (London), 1933, Royal Soc. (Edinburgh); mem. Am. Math. Soc., Circolo Matematico di Palermo. Author: Lectures on Matrices, 1934. Contbr. on scientific topcis. Worked on hypercomplex numbers and matrices, algebraic structural analysis. Died Oct. 3, 1948.

WEDDING, Randolph Townsend, Am. biochemist; b. St. Petersburg, Fla., Nov. 6, 1921; s. Randolph T. and Rowena (Reid) W.; A.A., St. Petersburg Jr. Coll., 1940; B.S., U. Fla., 1943, M.S., 1947; Ph.D., Cornell U., 1950; m. Mary Elizabeth Kennedy, Apr. 17, 1943; children—Sheila Tracy, Randolph Edward. Faculty, Cornell U., 1949-50; jr., asst., asso. plant physiologist U. Cal. at Riverside, 1950-60, faculty, 1960——, prof. biochemistry, biochemist, 1962——, chmn. dept., 1965——. Agrl. Research Council of Gt. Britain fellow, 1959-60; OECD sr. sci. fellow, 1964. Mem. Biochem. Soc., Soc. for Exptl. Biology, Am., Scandanavian, Japanese socs. plant physiologists, Am. Inst. Biol. Scis., Sigma Xi, Phi Kappa Phi. Research, numerous publs. on metabolic mechanisms in control plant growth, mechanisms of auxin toxicity to plants, high-energy sulfate compounds in sulfate uptake and metabolism by plants, clarification of ammonia toxicity to plants. Home: 5445 Brittany Av., Riverside, Cal. 92506.*

WEDEL, Georg Wolfgang, chemist, physician; b. Golzen, Niederlavsitz, Nov. 13, 1645; pupil of Rolfinck; sons include Ernst Heinrich, Johann Adolph; became prof. medicine, Jena, Germany, 1672. Author: Compendium chimiae theoreticae et practicae . . . , 1672; Physiologia medica, 1680; Pharmacia acroamatica, 1686; Physiologia reformata, 1688; Tabulae chimicae . . . , 1692; Schedaisma de sale volatili oleoso, 1711; also papers. Died Jena, Sept. 7, 1721.

WEDEL, Waldo Rudolph, Am. archeologist; b. Newton, Kan., Sept. 10, 1908; s. Peter John and Lena (Krehbeil) W.; student Bethel Coll., 1926-28; A.B., U. Ariz., 1930; M.A., U. Neb., 1931; Ph.D., U. Cal. at Berkeley, 1936; m. Mildred Ingram Mott, Aug. 12, 1939; children—Waldo Mott, Frank Peter, Linda Margaret. Archeologist, Neb. State Hist. Soc., 1936; with U.S. Nat. Mus., 1936——, head curator anthropology, 1962-65, sr. research scientist, 1965-——. Field dir. Missouri River Basin Survey, 1946-49. Recipient award in biol. scis. Washington Acad. Scis. 1947. Mem. Am. Anthrop. Assn., A.A.A.S., Soc. Am. Archaelogy (past pres.), Anthrop. Soc. Washington, (past pres.), Kan. Acad. Sci., Mo. Archeol. Soc., Nat. Acad. Scis. Author: Archeological Investigations Buena Vista Lake, Kern County, California, 1941; Archeological Investigation in Platte and Clay Counties, Missouri, 1943; An Introduction to Kansas Archeology, 1959; Prehistoric Man on the Great Plains, 1961; also numerous articles. Research on man's adjustments to life in interior grasslands Midwest and West. Home: 5305 Ridgefield Rd., Bethesda, Md. 20016. Office: Office of Anthropology, Smithsonian Instn., Washington 20560.*

WEDGWOOD, Josiah, English potter, inventor; b. Burslem, Eng., July 12, 1730; s. Thomas and Mary (Stringer) W.; ed. common schs., Eng.; m. Sarah Wedgwood, Jan. 25, 1764; 3 sons, 4 daus., including Susannah (Mrs. Robert Darwin), Thomas, Josiah, Emma (Mrs. Charles Darwin). Apprentice bro.'s potworks nr. Stoke-on-Trent, Eng., 1751-54; partner with Thomas Whieldon, Fenton, Eng., from 1754; opened own works Burslem, 1759, partner with Thomas Wedgwood as of 1766 and with Thomas Bentley from 1768; opened pot-works, Etruria, Eng., 1769; apptd. potter to Queen, 1762. Fellow Royal Soc., 1783, Soc. Antiquaries. Research, publs. on pottery craft; contbd. greatly to improvement of pottery and tech., artistic devel. Eng. ceramic industry; perfected cream-colored earthenware (later named Queen's ware) which became standard domestic pottery, patented 1763; also manufactured ornamental ware decorated in classical style; imparted finer grain and smoother surface to black basaltes or Egyptian Black; improved variegated or marbled ware; used sulphate of baryta to produce so-called Jasper ware; invented Wedgwood pyrometer, 1782. Died Etruria, Eng., Jan. 3, 1795.

WEDGWOOD, Thomas, English physicist, inventor; b. Etruria Hall, Eng., May 14, 1771; s. Josiah and Sarah (Wedgwood); ed. Edinburgh U., 1787-1789. Worked in family potteries for short period. Used light-sensitive properties of silver nitrate to produce image on glass or paper, but could not preserve image; 1st to suggest all bodies became red hot at same temperature, 1791-92. Died Eastbury, Eng., July 10, 1805.

WEED, Lewis Hill, Am. anatomist; born Cleveland, O., Nov. 15, 1886; s. Charles Henry and Mary Frances (Lewis) W.; A.B. Yale, 1908, A.M., 1909, M.D., John Hopkins, 1912; Sc.D., U. of Rochester, 1929; Univ. of Pennsylvania, 1942. Washington Univ. (St. Louis), Lafayette Coll., 1944, U. of Western Ont., 1947; LL.D Duke, 1938, Tufts College, 1943, Tulane University, 1944, Birmingham U., England, 1947; unmarried. Fellow in charge of Laboratory of Surgical Research, Harvard Medical School, 1912-13; Arthur Tracy Cabot fellow, same school, 1913-14; instructor, asso. and asso. prof. anatomy, Johns Hopkins U. 1914-19, professor, 1919-47, dean medical faculty, 1923-29, director School of Medicine 1929-46. Served as capt., Med. Corps, U. S. Army, Aug. 1917-May 1919; was mil. dir. Army Neuro-Surg. Lab. at Johns Hopkins Med. Sch. Research asso. Carnegie Instn. of Washington, 1922-35. Trustee Inst. for Advanced Study since 1930; trustee Carnegie Instn. of Washington since 1935; mem. Med. Fellowship Board, Nat. Research Council, 1935-39, chairman Division of Medical Sciences, same, 1939-49; successor fellow Yale Corp. Mem. Health and Med. Com., Council of Nat. Defense, 1940-41; mem. and v. chmn. Com. on Med. Research, Office of Scientific Research and Development, 1941-47. Chmn. med. and health advisory com. Am. Red Cross, 1943-44, chmn. advisory bd. for The Health Services, 1945-47. Mem. bd. of honorary consultants Army Med. Library since 1944. Mem. Am. Assn. Anatomists, A.A.A.S., Am. Physiol. Soc., Am. Philosophical Soc. Author of monographs and original articles on anatom. and neurology subjects. Research in experimental neurology. Died Dec. 21, 1952.

WEED, Lyle Alfred, Am. physician; b. Winterset, Ia., Jan. 15, 1904; s. Arthur Ray and Flora (Little) W.; B.A., Simpson Coll., 1927; M.S., Ia. State Coll., 1928; Ph.D., Western Res. U., 1932; M.D., U. Ia., 1939; m. Garnet Luella Holman, Aug. 29, 1928; children—Flora, Karl, David. Teaching fellow Ia. State Coll., 1927-28; bacteriologist Chgo. Sanitary Dist., 1928-29; spl. research asst. Western Res. U., 1929-32; fellow NRC, 1932-33; instr. bacteriology U. Ia., 1933-38; asst. prof. bacteriology Ind. U., 1939-42, asso. prof., 1942-45; cons. bacteriology Mayo Clinic, Rochester, Minn., 1945——, head of section on microbiology, Mayo, 1960; prof. bacteriology Mayo Found., 1957——. Pres. Rochester Bd. Pub. Health and Welfare, 1951-56. Diplomate Am. Pathology. Research on bacteria in air; diptheria Toxin; infectious granulomata; staphylococcal lipases; germicides; thyphoid serology; gas gangrene; toxicity of serum. Home: 1607 9th Av. N.E. Office: Mayo Clinic, Rochester, Minn.*

WEEKS, John Elmer, Am. physician; b. Painesville, O., Aug. 9, 1853; s. Seth and Deborah A. (Blydenburgh) W.; M.D., University of Michigan, 1881, D.Sc. (hon.), 1912; LL.D., New York University, 1923; interne Almshouse and Workhouse Hospital, resident phys. Emigrant Hosp. and interne, Ophthalmic and Aural Inst., New York, 5 ½ yrs.; (hon. degree Sc.D., Univ. of Mich., 1912); m. Jennie Post Parker, Apr. 29, 1890. Chief of Vanderbilt Clinic, New York, 1888-90; lecturer diseases of the eye, Bellevue Hosp. Med. Coll., 1890-1900; prof. same, Univ. and Bellevue Hosp. Med. Coll. (New York U.), 1900-1920, prof. emeritus; prof. diseases of the eye and ear, Woman's Med. Coll. of New York Infirmary for Women, 1893-99; cons. surgeon New York Eye and Ear Infirmary; 1st lt. U. S. A. Med. Res. Corps. 1911. Mem. A.M.A., N.Y. Acad. Med., etc. Author: Diseases of the Eye, 1910. Developed operation for inserting artificial eye; discovered bacillus responsible for pinkeye and Egyptian opthalmia, 1886. Died Feb. 2, 1949.

WEEKS, John Randel, IV, Am. metallurgist; b. Orange, N.J., Oct. 30, 1927; s. John Randel and Marion (Heberton) W.; Met.E., Colo. Sch. Mines, 1949; M.S., U. Utah, 1950, Ph.D. in Metallurgy, 1953; m. Barbara Ann Brewster, July 16, 1951; children—Ann Brewster, John Randel, V. Asso. metallurgist Brookhaven Nat. Lab., Upton, N.Y., 1953-59, metallurgist, 1959——, group leader liquid metal corrosion research, 1956——; adj. asso. prof. materials sci. State U. N.Y. at Stony Brook, 1962-63. Cons. to Aerojet Gen. Corp., Azusa, Cal., 1965——. Mem. Metall. Soc., Am. Inst. Mining, Metall. and Petroleum Engrs., Am. Soc. for Metals. Research and publs. on liquid metal corrosion mechanisms; developed and determined three and four component solubilities in liquid bismuth required for design of liquid uranium-bismuth fueled nuclear reactor. Home: 25 Acorn Lane, Stony Brook, N.Y. 11790. Office: 76 Cornell Av., Upton, N.Y. 11973.*

WEEKS, Lewis George, Am. geologist; b. Chilton, Wis., May 22, 1893; s. George T. and Katherine (Schneider) W.; m. Una Austin, Oct. 12, 1921 (dec.); 1 son, Lewis Austin; m. 2d, Anne Sutton, Dec. 14, 1957. Mining geologist Calumet & Ariz. Copper Mining Co., 1917, Cananea Consol. Copper Mfg. Co., Mexico, 1918-19; petroleum geologist Whitehall Petroleum Co. Ltd., London, Eng., India, 1920-24; with Standard Oil Co., 1924-58, chief geologist, N.Y.C., 1953-58; geol. cons., Westport, Conn., 1958—; founder Oil Basins, Inc., 1960, pres., dir., 1960-65; founder Offshore Royalties, Inc., 1965, pres., dir., 1965—. Adviser to cos., instns., govt. agys., fgn. govts.; mem. adv. bd. Internat. Oil and Gas Ednl. Center, Southwestern Legal Found., Dallas, 1962—. Mem. Am. Assn. Petroleum Geologists (Powers Meml. medal 1962; past pres.), N.Y. Acad. Scis. (past chmn. geol. scis. sect.), geol. socs. Am., Australia, Soc. Exploration Geophysicist, A.A.A.S., Am. Geophys. Union, Am. Inst. Profl. Geologists, Am. Inst. Mining Engrs. Author: Basin Development, Sedimentation and Oil Occurrence, 1948; also numerous articles. Contbg. author, editor: Habitat of Oil, 1958. Research on crustal architecture earth, devel. sedimentary basins, petroleum deposits. Home: 14 Bluewater Hill, Westport, Conn. 06880.*

WEERTMAN, Johannes, Am. physicist; b. Fairfield, Ala., May 11, 1925; s. Roelof and Christine (van Vlaardingen) W.; student Pa. State Coll., 1943-44; B.S., Carnegie Inst. Tech., 1948, D.Sc., 1951; postgrad. Ecole Normale Superieure, Paris, France, 1951-52; m. Julia Ann Randall, Feb. 10, 1950; children—Julia Ann, Bruce Randall. Solid state physicist U.S. Naval Research Lab., Washington, 1952-58, cons., 1960-67; sci. liaison officer U.S. Office Naval Research, Am. embassy, London, Eng., 1958-59; faculty Northwestern U., Evanston, Ill., 1960—, prof. materials sci. dept., 1961—, chmn. dept., 1964—, prof. geology dept., 1963—; vis. prof. geophysics Cal. Inst. Tech., 1964. Cons. U.S. Army Cold Regions Research and Engring. Lab., 1960—, Oak Ridge Nat. Lab., 1963—, Los Alamos National Laboratory, 1967—; co-editor materials sci. books MacMillan Co., 1962—. Mem. Am. Geophys. Union (Horton award 1962), Am. Soc. for Metals, Am. Inst. Mining, Metall. and Petroleum Engrs., Am. Phys. Soc., Am. Inst. Physics, Inst. Metals, Glaciological Soc., Geol. Soc. Am., A.A.A.S., Arctic Inst., Sigma Xi, Phi Kappa Phi, Pi Mu Epsilon. Author: (with wife) Elementary Dislocation Theory, 1964; also numerous arcticles. Developer theories for sliding of glaciers, flow of glaciers and ice sheets, stability of ice sheets, flow in earth's mantle, creep of metals and crystalline solids, internal friction of solids, dislocation phenomena in crystals, dislocation motion on earthquake faults, exptl. work on internal friction, creep, high velocity deformation. Home: 834 Lincoln St., Evanston, Ill. 60201.*

WEFA, Abul, see Abul Wafa.

WEGELIUS, Otto Carl, Finnish physician; b. Raumo, Finland, Oct. 21, 1920; s. Alfred Emil and Saga (Stackelberg) W.; M.D., U. Helsinki, 1947, specialist certificate, 1953; m. Maria Christina Lundqvist, Aug. 18, 1948; 1 son, Fred Otto Bertel. Asst. prof. internal medicine U. Helsinki, 1956-59, asst. tchr. 4th dept., 1959-64, asso. prof. internal medicine, 4th dept., 1964—, head connective tissue research unit, dept. pathology, 1962—; researcher Connective Tissue Research Lab., Copenhagen, 1955, 57; fellow Rockefeller Found. at Columbia U., 1958-59; head cons. Ins. Co. Elake-Varma, Helsinki. Mem. Finnish Med. Soc., Internat. Soc. Internal Medicine (mem. bd.), Finnish Soc. Sci., Endocrine Soc. Am. Publs., exptl. studies on elastic fibers of connective tissue; hormonal influence on connective tissue components, exophthalmos; clin. studies on connective tissue disorders. Home: Mynt gard, Koklaks, Finland. Office: 38 Unionsgatan, Helsinki, Finland.*

WEGENER, Alfred Lothar, German geophysicist, meteorologist; b. Berlin, Germany, Jan. 11, 1880; ed. Berlin, Heidelberg, also Innsbruck, Austria; prof. meteorology, Hamburg, Germany, 1919-24; Graz, Austria, 1924-30. Author: Thermodynamik der Atmosphäre, 1911; Die Entstehung der Kontinente und Ozeane, 1915. Studied thermodynamics of atmosphere on expdn. to Greenland, 1906-08, 12-17, 29, 30; originated hypothesis that continents have slowly drifted apart from a single original land area (Wegener hypothesis of continental drift), 1912; studied thickness of polar icecap and rate of drift of Greenland. Died Greenland, Nov. 1930.

WEGENER, Peter Paul, physicist; b. Berlin, Germany, Aug. 29, 1917; Sc.D., U. Berlin, 1943; m. Annette Schleiermacher, Aug. 14, 1961. Came to U. S., 1946, naturalized, 1953. Research supersonic wind tunnels, Kochel, Germany, 1943-45, in gasdynamics, hypersonic wind tunnels U. S. Naval Ordnance Lab., 1946-53, gasdynamic research Jet Propulsion Lab., Cal. Inst. Tech., 1953-60; prof. applied sci. Yale, 1960—. Mem. Am. Inst. Aero. and Astronautics, Am. Phys. Soc. Contbr. papers hypersonic research, condensation metastable state, chem. kinetics, flow systems real gases, design of continuous hypersonic wind tunnels. Home: Montgomery Pky., Branford, Conn. Office: Yale Univ., New Haven.*

WÉGRIA, René William Edward, physician; b. Fumal, Belgium, June 9, 1911; s. Edward and Odile (Vandensteen) W.; M.D. with great distinction, U. Liege, 1936; Laureat du Concours Interuniversitaire de Belgique, 1934-36; D.M.S., Columbia, 1945. Came to U. S., 1937, naturalized, 1947. Fellow Belgian Am. Ednl. Found., Vanderbilt U., Mayo Found. and Western Res. U., 1937-39; fellow, then instr. dept. physiology Western Res. U., 1939-42; asst. resident medicine Cleve. City Hosp., 1942-43; asst. resident cardiology, asst. medicine Presbyn. Hosp. and Columbia U. Coll. Phys. and surg., N.Y.C., 1943-46, instr. medicine, 1946-47, asso. medicine, 1947-49, asst. prof. medicine, 1949-56, asso. prof., 1958; cons. medicine St. Barnabas Hosp., 1948-58; vis. prof. Lovanium U., Leopoldville Belgian Congo, 1958; prof. medicine, dir. dept. St. Louis U., 1958-63, prof. of pharmacology and dir. dept., 1963—. Chmn. dean's com. VA Hosp., St. Louis, 1961-62; cons. Food and Drug Adminstrn., Dept. Health, Edn. and Welfare. Fellow A.A.A.S., New York Acad. Sciences; mem. Am. Physiol. Soc., Am. Heart Assn., Soc. Exptl. Biology and Medicine, Am. Soc. Clin. Investigation, Harvey Soc., Am. Pharmacological Soc., N.Y. Heart Assn., N.Y. County, N.Y. State med. assn., A.M.A. (cons. council drugs), Am. Assn. U. Profs., Am. Overseas Educators Assn. Research on physiology and pathological physiology of cardiovascular system and kidneys, coronary circulation and cardiac arrhythmias, rheumatic fever; pharmacology of salicylates; pathogenesis of cardiac, renal and nutritional edema. Address: 1402 S. Grand Blvd., St. Louis 63104.*

WEHMEYER, Lewis Edgar, Am. botanist; b. Quincy, Ill., Jan. 1, 1897; s. August H. and Emma (Grimm) W.; B.S. in Forestry, U. Mich., 1921, Ph.D. in Botany, 1925; m. Elaine F. Prince, Sept. 1, 1927. Nat. Research fellow Harvard, 1925-28; faculty U. Mich., Ann Arbor, 1928—, prof. botany, 1947—. Fellow A.A.A.S.; mem. Mycol. Soc. Am., Internat. Soc. Plant Taxonomists, Mich. Research Club, Sigma Xi, Gamma Alpha. Author: The Genus Diaporthe, 1933; Revision of Melanconis Pseudovalsa Prosthecium and Titania, 1941; The Fungi of New Brunswick, Nova Scotia and Prince Edward Island, 1950; A World Monograph of the Genus Pleospora and its Segregates, 1961; also numerous articles. Cultural and taxonomic study of pyrenomycetous fungi, studies of local flora of N.S., Wyo., Mt. Ranier. Home: 2721 N. Wagner Rd., Ann Arbor, Mich. 48103.*

WEHNELT, Artur Rudolph Berthold, physicist; b. Rio de Janeiro, Brazil, Apr. 4, 1871; became prof. U. Berlin, (Germany), 1908; named dir. Physics Inst., 1926. Research on electron emission; 1st to notice that metal covered with oxides of certain bases emits more electrons when heated (of use in radio and cathode tubes); developed a breaker which bears his name, also a cathode ray tube; devised hardness scale for X-rays by measurement of absorption; inventor high frequency electrolytic switch. Died Berlin, Feb. 15, 1944.

WEIBEL, Ewald Rudolf, Swiss cytologist, anatomist; b. Weggis, Switzerland, Mar. 5, 1929; s. Jakob and Berthe Méry (Pilet) W.; student med. schs. U. Paris (France), 1953, U. Göttingen (Germany), 1952-53; M.D., U. Zürich (Switzerland), 1955; m. Anna Verena Trachsler, Apr. 4, 1956. Staff dept. anatomy U. Zürich, 1955-58, asso. prof. dept. anatomy, 1963-66; research fellow dept. pathology Yale, 1958-59; research asso. dept. medicine Columbia U., N.Y.C., 1959-61, dept. cytology Rockefeller U., N.Y.C., 1961-62; prof., chmn. dept. anatomy U. Berne (Switzerland), 1966—. Mem. Internat. Soc. Stereology (pres. 1967—), Am. Soc. Cell Biology, Anatomische Gesellschaft, Royal Soc. Medicine London (affiliate), Schweiz. Naturforschende Gesellschaft. Author: Morphometry of the Human Lung, 1963; also articles. Editor: (with H. Elias) Quantitative Methods in Morphology, 1967. Devel. and refinement of methods for morphometric study of lung and other organs especially at cellular level; systematic quantitative study on structure of human and mammalian lung; research on oxygen toxicity; discovered specific cytoplasmic organelle of blood vessel endothelia. Home: 49 Halen, 3037 Stuckishaus BE, Switzerland. Office: 26 Bühlstrasse, 3000 Bern, Switzerland.*

WEIBULL, Ernst Hjalmar Waloddi, Swedish applied physicist; b. Wittskövle, Sweden, June 18, 1887; s. Richard and Gerda (Holmberg) W.; B.A., Stockholm (Sweden) U., 1912, M.A., 1924; Ph.D. honoris causa, Uppsala (Sweden) U., 1932; m. Karin Ewe, Oct. 8, 1912; children—Hans, Erik, Wiwica (Mrs. Qvitt Holmgren), Jan, Göran, Bengt, Torsten; m. 2d, Inga Brita Testad, Nov. 3, 1943; children—Helena (Mrs. Jonas Martin), Moana. Dir. research NKA Ball Bearing Co., Göteborg, Sweden, 1916-22; prof. mech. engring. Royal Inst. Tech., Stockholm, 1923-41, pro-rector, 1931-40, prof. applied physics, 1941-53; sci. adviser AB Bofors, Bofors, Sweden, 1941-52. Research on statistics aero. fatigue and reliability sponsored by Air Force Materials Lab., Research and Tech. Div. through European Office Aerospace Research, USAF, 1955—; vis. prof. Columbia, 1963-64. Mem. Royal Swedish Acad. Sci., Royal Swedish Acad. Engring. Sci., Royal Swedish Acad. Mil. Sci., Royal Swedish Acad. Naval Sci., Royal Soc. Uppsala. Author: Fatigue Testing and the Analysis of Results, 1961; also textbooks, articles. Proposed and developed statis. theory of strength of materials, including a statis. distbn. function (weibull distbn.) which is applied to problems of reliability and many other fields; inventions and patents on roller bearings, electric hammers, fatigue testing machines; oceanographic researches. Home: Bockamöllan, Brösarp, Sweden. Office: 15 Ch. Fontanettaz, 1012 Lausanne, Switzerland.*

WEICHERT, Charles Kipp, Am. zoologist; b. West Hoboken, N.J., Dec. 26, 1902; s. Maximilian Jacob and Ella (Kipp) W.; B.S., Rutgers U., 1924; M.S., U. Wis., 1926, Ph.D., 1928; m. Kathryn Hilberg, June 12, 1937; 1 dau., Kathryn Ann. Mem. faculty U. Cin., 1928—, prof., 1943-58, head dept. biol. scis., 1948-58, prof. zoology, dean McMicken Coll. Arts and Scis. 1958—. Fellow A.A.A.S.; mem. Am. Soc. Zoologists, Am. Soc. Anatomists, Endocrine Soc., Soc. for Exptl. Biology and Medicine, Phi Beta Kappa, Sigma Xi. Author: Laboratory Manual for Vertebrate Embryology, 1930; The Art and Science of Marriage, 1938; Laboratory Directions for Comparative Anatomy of Vertebrates, 1947; Anatomy of the Chordates, 1951, 1958, 1965; Elements of Chordate Anatomy, 1951, 1957, 1967; Representative Chordates, 1954, 1961. Research, numerous publs. on endocrine control of reproductive phenomena in rat. Home: 1150 Burney Lane, Cin. 45230.*

WEICHSELBAUM, Anton, Austrian pathologist; b. Schiltern, Langenlois, Feb. 8, 1845; mil. surgeon; later prof. path. anatomy U. Vienna; research on bacteriology of Tb; reported his discovery of meningococcus (causative agt. of cerebrospinal meningitis), 1887. Died Vienna, Oct. 22, 1920.

WEICKMANN, Helmut K., physicist; b. Munich, Germany, Mar. 10, 1915; s. Ludwig and Therese (Mayer) W.; student U. Leipzig, 1934-36; Ph.D., U. Frankfurt, 1939; m. Ruth Erika Burchardt, Apr. 28, 1945; children—Joachim, Klaus, Sabine, Peter. Came to U.S., 1949, naturalized, 1958. Research physicist Deutsche Forschungsanstalt fur Segelflug, 1939-45; dir. Observatory Hohenpiessen-Gerg, 1946-50; research physicist U.S. Army Signal Research and Devel. Lab., Ft. Monmouth, N.J., 1949-62; chief atmospheric physics br. U.S. Army Electronics Labs., Ft. Monmouth, 1962-65; dir. Atmospheric Physics and Chemistry Lab., Inst. of Environmental Research, Environmental Scis. Services Adminstrn., Dept. Commerce, Boulder, Colo., 1965—. Mem. panel on weather and climate control Nat. Acad. Sci. Recipient Achievement award Army Signal Research and Devel. Lab., 1961. Mem. Internat. Union of Geodesy and Geophysics (chmn. ad hoc com. on cloud physics and could modification), Am. Meteorol. Soc., Geophys. Union, Royal Meteorol. Soc. London, Deutsche Physikalische Gesellschaft. Editor: Artificial Stimulation of Rain, 1957; Physics of Precipitation; also author numerous articles, monograms. Research in physics of cirrus cloud crystals, size distbrn. of cloud droplets in various kinds of clouds, artificial dissipation of super-cooled clouds, formation of hail and precipitation, role of condensation nuclei in devel. of cloud physics processes. Home: 603 Wewoka Dr., Indian Hills, Boulder 80302. Office: 325 Broadway, Boulder, Colo. 80302.*

WEIDENREICH, Franz, anatomist, anthropologist; b. Edenkoben, Germany, June 7, 1873; ed. univs. Munich, Kiel, Gerlin (all Germany), Strasbourg (France); M.D., 1899. Anatomist U. Strasbourg, 1899-1901, 1902-18; prof. from 1904; prof. anatomy U. Heidelberg (Germany), 1921-24, Peking (China) Union Med. Coll., 1935-42; prof. anthropology Inst. Phys. Anthropology, Frankfurt/Main, Germany. Vis. prof. U. Chgo., 1934; with Am. Mus. Natural History, N.Y.C., 1941-48. Author: The Brain and Its Role in the Phytogenetic Transformation of the Human Skull, 1941; The Skull of Sinanthropus Pekinensis (1st classic description Peking Man), 1943; Apes, Giants and Man, 1946; Giant Early Man from Java and South China, 1946; Morphology of Solo Man, 1951. Research, publs. on Pithecanthropus (Java Man), Neanderthal skeletons Ehringsdorf, skeletal anatomy in relation problems primate evolution, Europe, Teshik Tash, Central Asia; participated excavation Sinanthropus fossils near Peking; major study (uncompleted) Solo Man. Died N.Y.C., July 11, 1948.

WEIDLEIN, Edward Ray, Am. chem. engr.; b. Augusta, Kan., July 14, 1887; s. Edward and Nettie (Lemon) W.; B.A., University of Kan., 1909, M.A., 1910; hon. Sc.D., Tufts Coll., 1924; LL.D., U. Pitts., 1930, Washington and Jefferson, 1947; Sc.D. (hon.), Rutgers, 1937, Waynesburg (Pa.) Coll., 1939. U. Wichita, 1945, Southwestern Coll., 1952, Phila. Coll. Pharmacy and Sci., 1952, U. Miami, 1953; Eng.D., Rensselaer Poly. Inst., 1949; m. Hazel I. Butts, April 24, 1915; children—Edward Ray, Robert Butts, John David, Industrial fellow, research on camphor, U. of Kan., 1909-10, research on ductless glands, 1910-12; sr. fellow, Mellon Inst. of Industrial Research, Pittsburgh, Pa., having charge of investigations on metallurgy and hydrometallurgy of copper, and also of the exptl. plant, at Thompson, Nev., 1912-16; asst. dir. of the Inst. at Pittsburgh, July-Oct. 1916, asso. dir., 1916-21, actg. dir., 1918-19, dir. 1921-51; dir. Mellon Inst., and v.p. bd. trustees, 1929-51, pres., 1951-56, chmn., 1951-55, ret; tech. adviser to Rubber Research Group, NSF, 1941—; past mem. wartime and emergency bds. on materials. President Regional Indsl. Devel. Corp. Fund Southwestern Pa.; bd. advisers Pitts. Indsl. Devel. Council; adv. com. of the Exposition Chemical Industries. Trustee of the Shadyside Academy (Pittsburgh); mem. bd. dirs. Western Pa. Hosp. Mem. Guild of Brackett Lecturers, Princeton U. Served as chemical expert War Industries Bd., 1918-19. Fellow A.A.A.S., Am. Inst. of Chemists, Royal Soc. of Arts (London); mem. nat., state and local profl. and scientific assns. and orgns., past pres. or other officer of several, mem. coop. coms. Recipient awards, State of Pa., 1939; Soc. of Chem. Industry, 1935; Pitts. award for 1939; Junior C. of C., 1941; U. of Kan., 1943; Naval Ordnance Development award, 1947; Priestley medal by Am. Chem. Soc., 1948; Richard Beatty Mellon award. air pollution control. 1956; William Proctor prize, A.A.A.S., 1961. Co-author books and articles chiefly relating to indsl. research. Research on metallurgy, hydro-metallurgy of copper; adminstrn. research. Home: Weidacres, P.O. Box 45, Rector, Pa. 15677. Office: care Mellon Inst., Carnegie-Mellon U., 4400 5th Av., Pitts. 15213.*

WEIDNER, Herbert, German zoologist; b. Hof/S., May 9, 1911; s. Albrecht and Marie (Klug) W.; Dr ès sc; m. Erna Rauh, July 7, 1938. Sci. asst. Nat. Zool. Inst., Hamburg (Germany) Zoology Museum, 1934-38, asst., 1938-41, asso. scientist, 1941-42, curator, 1942-55, sect. chief., 1955—; dir. entomology sect. U. Hamburg, 1950, agrégé, asso. prof., 1955—. Mem. German Zool. Soc., German Applied Entomology Assn., German Soc. Naturalists and Doctors, Internat. Soc. Study Social Insects, German, Viennese entomology socs., Internat. Entomology Union, Soc. Naturalists Berlin. Author: Bestimmungstabellen der Vorratsschädlinge under des Hausungeziefers Mitteleuropas, 1937; also articles on parasites and their process of destruction (zooecidia, orthoptera and isoptera). Editor: Ent. Mitt. zoologisches Staatsinstitut und zoologisches Museum, 1952—. Home: Uhland Strasse 18, 2 Hamburg 22. Office: Van Melle Park 10, 2 Hamburg 13, West Germany.

WEIDNER, Richard Tilghman, Am. physicist; b. Allentown, Pa., Mar. 31, 1921; s. Miles Percival and Mabel (Aichroth) W.; B.S. summa cum laude Muhlenberg Coll., 1943; M.S., Yale, 1943, Ph.D., 1948; m. Jean E. Fritsch, June 28, 1947; children—Christopher L., Allegra L., Timothy M. Instr., research asst. Yale, 1943-44, 1946-47; physicist Naval Research Lab., Washington, 1944-46; faculty Rutgers U., New Brunswick, N.J., 1947—, prof. physics, 1963—. Fellow Am. Phys. Soc., Sigma Xi. Author: (with R.L. Sells) Elementary Modern Physics, 1960; Elementary Classical Physics, 1965. Research on electron spin resonance, absorption of microwaves by magnetic materials. Home: 59 Hillcrest Rd., Martinsville, N.J.*

WEIER, Thomas Elliot, Am. botanist; b. N.Y.C., Jan. 4, 1903; s. John Edward and Edna (Hammill) W.; A.B., Hope Coll., 1926; Ph.D., U. Mich., 1929; m. Katherine Schmid, Aug. 18, 1927. Faculty, Ore. State Coll., 1934-36; faculty U. Cal. at Davis, 1936—, prof. botany, 1950—. Mem. A.A.A.S., Bot. Soc. Am., Sigma Xi. Author: (with W.W. Robbins, C.R. Stocking) Botany, an Introduction to Plant Science, 1950; also numerous articles. Discovered with electron microscope chlorplast is divided in three spaces, carbon cycle occurs in stroma, photo reaction occurs in array stacked membranes formed by 2 rows lipoprotein subunits, chlorophyll is bounded between these layers. Home: 631 Oak Av., Davis, Cal. 95616.*

WEIERSTRASS, Karl Theodor Wilhelm, German mathematician; b. Ostenfelde, Germany, Oct. 31, 1815; s. Wilhelm and (von der Forst) W.; student law, U. Bonn (Germany), 1834-38; studied math. under Gudermann, Münster Acad.; hon. doctorate, Königsberg (now Kaliningrad, USSR), 1854. Tchr. math. Munster Gymnasium, Deutschekrone, Germany, 1842-48, Collegium Hoseanum, Braunsberg, Germany, 1848-56; became asso. prof. math, Berlin, also lectr. Sch. Tech., 1856, prof., 1864. Mem. French Acad. Scis., 1868, Berlin Acad. Sci. Author: Abhandlungen aus der Funktionenlehre, 1886; Gesammelte Abhandlungen, 8 vols., 1894-1927; also numerous articles. Editor: Gesammelte Werke Abhandlungen aus der Funktionlehre (Steiner), 1886. Contributed to theory of functions; based his function concept on theory of exponential series; studies on Abelian, elliptic, and analytic functions, calculus of variations, complex variables, functions of real variables, converging infinite products, theory of bilinear and quadratic forms; developed tests for convergence of series; discovered function which is continuous over an interval but does not have derivative at any point on interval; constructed found. for arithmetization of math. analysis, making irrational numbers understandable in terms of sequences of rational numbers. Died Berlin, Feb. 19, 1897.

WEIGALL, Arthur Edward Pearse Brome, archaeologist, Egyptologist; b. Nov. 20, 1880; s. A. A. D. and Alice (Cowan) W.; ed. Wellington Coll., New Coll., Oxford (Eng.) U., from 1900; m. Hortense Schleiter; m. 2d, Muriel Frances Lillie. Asst. to prof. Flinders Petrie, staff Egypt Exploration Fund, 1901-02; archaeologist, Saqqara, Egypt, 1902-04; insp. gen. antiquities Egyptian govt., 1905-14. Author: A Report on the Antiquities of Lower Nubia, 1907; A Catalogue of the Weights and Balances in the Cairo Museum, 1908; Travels in the Upper Egyptian Deserts, 1909; A Guide to the Antiquities of Upper Egypt, 1910; The Life of Akhnaton, Pharaoh of Egypt, 1910; (with A. H. Gardiner) A Topographical Catalogue of the Tombs of Thebes, 1913; The Life of Cleopatra, Queen of Egypt, 1914; Egypt from 1798 to 1914, 1915; Ancient Egyptian Works of Art, 1924; A History of the Pharaohs, vol. I, 1925, vol. II, 1926; A Short History of Ancient Egypt, 1934. Excavated, preserved tombs of nobles at Thebes, 1907-12; excavated mortuary temple of Thothmes III, Thebes, 1905; investigated condition of antiquities in Nubia, 1906-07; explorations in eastern desert, Wady Hammamat, Kassir. Died Jan. 2, 1934.

WEIGEL, Christian Ehrenfried, German inventor; b. Straslund, Germany, May 24, 1748; prof., Greifswald, Germany. Author: Grundrisse der reinen und angewandten Chemie, 2 vols., 1777. Translated works of Lavoisier de Morveau and others. Inventor counter current cooler (Liebig cooler). Died Greifswald, Aug. 8, 1831.

WEIGEL, Erhard, German mathematician; b. Weiden/Naab, Germany, Dec. 16, 1675; prof., Leipzig, also Jena, Germany; tchr. of Leibniz and Pufendorf. Author: Philosophia mathematica, 1657; Astronomia sphaerica, 1657; Speculum uranicum, 1661; Miroir du temps, 1664. Introduced Gregorian calendar reform in protestant states; demanded math. found. for natural scis.; inventor several astron. instruments; opposed pedantic teaching methods, Died Jena, Mar. 21, 1699.

WEIGEL, Valentin, German astrologer, philosopher; b. Grossenhain, Germany, 1533; preacher, Zschopau, nr. Chemnitz, 1567-88. Author: Libellus de vita beata, 1609; Ein Schön Gebet Büchlein, 1612; Philosophia theologica, 1614; Principal und Haupttractat von der Gelassenheit, 1618; Soli deo gloria, 1618; Astrologie theologized. Works are of a mystical nature with similarities to those of Paracelsus. Died June 10, 1588.

WEIGERT, Carl, German pathologist, histologist; b. 1845; 1st to succeed in staining bacteria, circa 1871; brought attention to necrotizing effect of virus upon skin in explaining formation of smallpox lesions, resulting in intro. of term coagulation necrosis, 1874-75; developed a myelin sheath stain (Weiger's method or stain), 1882, also other stains; credited with emphasizing relation between myasthenia gravis and hypertrophy of thymus, 1901. Died 1904.

WEIGT, Ernst Felix Max, German geographer; b. Marbourg/Lahn, Aug. 12, 1907; s. Max and Else (Metzing) W.; Ph.D., U. Leipzig; m. Irene Bräuer, Aug. 28, 1938. Asst. Deutsche Mus. für Landeskunde, Leipzig; dir. German Sch., Lushoto, Tanganyika; asst., prof. agrégé univs. Hamburg and Cologne; now prof., dir. social and econ. geography U. Erlangen-Nuremberg. Mem. Deutsche Afrika Gesellschaft, Deutsche-Indische Gesellschaft, Verband Deutscher Hochschullehrer der Geographie. Author: Die Kolonisation Kenyas, 1932; Europäer in Ostafrika, 1955; Kenya-Uganda, 1956; Die Geographie, 1964; Beiträge zur Entwicklungspolitik, 1964. Editor: Nürnberger Wirtschaftsund Sozialgeographische Arbeiten. Home: Lohengrinstrasse 23. Office: Findelgasse 7, Nuremberg, West Germany.

WEIKEL, John Henry, Jr., Am. pharmacologist; b. Palmerton, Pa., June 14, 1929; s. John Henry and Anna (Mosey) W.; B.S., Trinity Coll., 1951; Ph.D., U. Rochester, 1954. Jr. scientist atomic energy project U. Rochester, 1954; pharmacologist FDA, 1954-56; pharmacologist Mead Johnson & Co., Evansville, Ind., 1956-62, dir. chem. pharmacology and safety evaluation Mead Johnson Research Center, 1962—. Mem. Am. Soc. Pharmacology and Exptl. Therapeutics, Soc. Exptl. Biology and Medicine, Soc. Toxicology, A.A.A.S., Am. Chem. Soc. Research, publs. on mechanism of calcium and phosphorus incorporation in bone mineral, use of radioisotopes in toxicology, effects of pesticides on cells, absorption and metabolic fate of drugs, safety evaluation of new drugs. Home: 1407 Howard St., Evansville 47713. Office: Mead Johnson Research Center, Evansville, Ind. 47721.*

WEIL, Adolf, internist; b. Germany, 1848; prof., Berlin, also Tartu, Estonia; gave classic description of leptospiral jaundice (Weil's disease), 1886; isolated (with Emil Abderhalden) norleucine from decomposition products of nervous tissue, circa 1913. Died 1916.

WEIL, Alfred Julius, physician; b. Mainz, Germany, Apr. 27, 1900; s. Edmund and Elisabeth A. (Hochheimer) W.; M.D., U. Munich, 1923; m. Elisabeth M. Seligmann, Mar. 29, 1925; children—Peter E., Eva G. (Mrs. B.L. Harris); m. 2d, Hilde H. Heymann. Came to U.S., 1936, naturalized, 1941. Mem. faculty various German univs., 1923-33; in charge immunol. research Lederle Labs., Am. Cyanamid Co., 1936-48; practice medicine specializing in allergies, Pearl River, N.Y., 1948—; attending microbiologist Bronx-Lebanon Hosp. Center, N.Y.C., 1948—. Cons. N.Y.C. Dept. Health, 1965—, mem. bd. examiners, 1964—. Fellow N.Y. Acad. Scis., N.Y. Acad. Medicine, Am. Coll. Allergists, Coll. Am. Pathologists, Am. Pub. Health Assn.; mem. A.M.A., Soc. Exptl. Biology and Medicine, Am. Assn. Immunologists (treas. 1946-47). Author: (with I. Saphra) Shigellae and Salmonellae, 1953; also numerous articles. Mem. editorial bd. Jour. Immunology, 1939-51, Annals of Allergy, 1941—, Rev. of Allergy, 1945-65. Research on antigenicity and biology of Wasserman reagins, classification and epidemiology of Enterobacteriaceae, antigenicity and biol. significance of substances of male genital tract, exptl. and clin. research in chemotherapy, especially antibiotics. Home: 143 Forest Av., Pearl River, N.Y. 10965. Office: Bronx-Lebanon Hosp. Center, Fulton Av., Bronx, N.Y. 10456.*

WEIL, Benjamin Henry, Am. chem. engr.; b. St. Joseph, Mo., July 8, 1916; s. Henry S. and Jeannette (Mayer) W.; Asso. Sci., St. Joseph Jr. Coll., 1937; B.S. in Chem. Engring., U. Mo., 1939; M.S., U. Wis., 1940; m. Carolyn R. Loeb, Mar. 11, 1944; children—Lewis H., Jeanne L., Carolyn R. Head information sect. chemistry div. Gulf Research and Devel., Pitts., 1940-45; head tech. information div. Ga. Tech. Engineering Experiment Station, 1945-50; Ethyl Corp. Research Labs., Detroit, 1950-57; chief Ga. Tech. Engring. Expt. Sta., Atlanta, 1945-50; mgr. Tech. information div. Ethyl Corp. Research Labs., Detroit, 1950-57; chief editor tech. information div. Esso Research and Engring. Co., Linden, N.J., 1957-60, head information processing sect., 1960-67, head tech. information section, 1967—. Fellow A.A.A.S. (councilor 1964-65), mem. Am. Chem. Soc. (chmn. div. chem. lit. 1958, councilor 1959-62, certificate merit 1956), Am. Documentation Inst. Spl. Libraries Assn., Soc. Tech. Writers and Pubs., Am. Petroleum Inst. (chmn. tech. information sub-com. 1959-61). Dir., Engring. Index, Inc., v.p., 1965-67. Author: (with V. Anhorn) Plastic Horizons, 1944; (with J.C. Lane) Synthetic Petroleum from the Synthine Process, 1948; also numerous articles. Editor: The Technical Report, 1953; Technical Editing, 1958. Developed quick-reading tech. abstracts; applied key word-in-context computer-based indexing to synthetic titles. Home: 4 Wells Lane, Warren, N.J. 07060. Office: Esso Research and Engring. Co., P.O. Box 51, Linden, N.J. 07036.*

WEIL, George Leon, Am. physicist; b. N.Y.C., Sept. 18, 1907; s. Leon and Elsie (Maiden) W.; B.A., Harvard, 1929; M.A., Columbia, 1931, Ph.D. in Physics, 1942; m. Veniette F. Caswell, 1950 (div. 1961); 1 son, Stephen G. Group leader Metall. Lab., Chgo., 1942-44; physicist E.I. duPont de Nemours & Co., Richland, Wash., 1944-45; group leader Los Alamos Sci. Lab., 1945; U.S. sci. rep. Chalk River Lab. (Ont.), 1945-46; research asso. Gen. Elec. Co., Schenectady, 1947; chief, reactors br. AEC, Washington, 1947-49, asst. dir. div. reactor devel., 1949-52; cons. indsl. orgns. and govt. agys., Washington, 1952—. Adviser, dir. numerous govt. agys. and confs. Recipient Certificates of Appreciation War Dept., 1945, AEC, 1955. Fellow Am. Phys. Soc., A.A.A.S.; mem. Am. Nuclear Soc. (dir. 1959-62), Sigma Xi. Research in nuclear physics including measurement of radioactive characteristics of emissions from arsenic and certain fission products, preliminary expts. (with Enrico Fermi) on feasibility of chain reaction using uranium and graphite; co-author (with Fermi and others) on uses of nuclear reactors to determine neutron absorption cross sects., and description of first exptl. prodn. of divergent chain reaction; patentee in method of utilizing chain reacting structure to determine neutron absorption of cross sect.; inventor measurement of deposition of icing on airplanes using fast and slow neutron techniques. Home: P.O. Box 35, Cabin John, Md. 20731. Office: 1101 17th St. N.W., Washington 20036.*

WEIL, Louise, French physicist; b. Selestat, France, Apr. 17, 1914; s. Paul and Camilla (Goldschmidt) W.; Agrége, Ecole Normale Supérieure de la rue d'Ulm, 1936, Docteur, 1941; m. Odette Marguerite Amidt, Mar. 26, 1940; children—Francois (Mrs. Pierre Rambaud), Antoine, Catherine, Bertrand. Asst., Faculty Scis., Strasbourg, France, 1938-45; lectr. phys. metallurgy, Grenoble, France, 1945-50, titular prof., 1950-57, prof. thermodynamics, 1957—, dean Faculty Scis., 1961—, dir. Research Center on Very Low Temperatures, 1961, dir. courses superior promotion of work, 1954-61; dir. Lab. Mech. Expts., Poly. Inst. Grenoble, 1954-64. Recipient Nat. Order of Merit; commandeur l'ordre du Palmes Academiques; officier l'ordre du Mérite pour la Recherche et l'Invention. Mem. Nat. Com. for Sci. Research, Commn. for Reform of Advanced Instrn., Internat. Union Physics (commn. on very low temperatures). Author: Thermodynamique; Eléments des Echanges thermiques; also numerous articles. Research on coercive field of fine grains; founder Indsl. Center Power Metallurgy; formed group for study very low temperatures, founder Center for Research on Very Low Temperatures, Nat. Center Sci. Research. Home: 149 ave. de l'Eygala, La Tronche (38), France. Office: CNRS chemin des Martyrs, Grenoble (38), France.*

WEIL, Max Harry, physiologist; b. Baden, Switzerland, Feb. 9, 1927; s. Marcel and Gretchen (Winter) W.; A.B., U. Mich., 1948; M.D., U. N.Y., 1952;

Ph.D., U. Minn., 1957; m. Marianne Judith Posner, Apr. 9, 1955; children—Susan Margot, Carol Juliet. Chief cardiology City of Hope Med. Center, Duarte, Cal., 1957-59; asst. prof. medicine U. So. Cal., 1959-63, asso. prof., 1963-—, dir. U. So. Cal.-Los Angeles County Hosp., Shock Research Unit, 1961-—, sr. attending cardiologist Children's div. Los Angeles County Hosp., 1965-—. Mem. com. on shock NRC, 1965-—; cons. medicine, clin. pharmacology Mead Johnson Research Center, 1965-—; cons. Cedars-Sinai Med. Center, Los Angeles, 1965-—; cons. Orange County (Cal.) Gen. Hosp., 1965-—; cons. IBM, Statham Instruments, 1965-—. Fellow A.C.P., Am. Coll. Cardiology; mem. Am. Physiol. Soc., Am. Fedn. Clin. Research, Western Soc. Clin. Research, Soc. for Exptl. Biology and Medicine, Sigma Xi. Editor: (with H. Shubin) Diagnosis and Treatment of Shock, 1967. Research, numerous publs. in clin. cardiology; cardiovascular pathophysiology, pharmacology; studies on shock; instrumentation for improving critical care and computer monitoring as applied to the critically ill; developed the first shock unit for direct study and management of patients in circulatory shock. Home: 6380 Drexel Av., Los Angeles 90048. Office: U. So. Cal. Sch. Medicine, 2025 Zonal Av., Los Angeles 90033.*

WEIL, Richard, Am. pathologist; b. N.Y.C., Oct. 15, 1876; s. Leopold and Matilda (Tanzer) W.; A.B., Columbia, 1896, M.D., 1900; m. Minnie Strauss, 1905; three children. Intern German Hosp., 1900-02; with clinics and labs. in Vienna and Strassburg, 1902-04; demonstrator pathology Cornell U. Med. Coll., 1905-08, asst. exptl. pathology, 1908-10, instr. exptl. therapeutics, 1910-11, asst. prof. exptl. therapeutics, 1911-15, asst. prof. exptl. pathology, 1915, head combined depts., 1916; adj. pathologist German Hosp., 1904-10; mem. staff Mt. Sinai Hosp., 1908-13; joined Huntington Fund for Cancer Research, 1906; studied serology of cancer and gen. problem of immunity at Loomis Lab. Mem. Am. Soc. Control Cancer, Am. Soc. Clin. Investigation, Assn. Am. Physicians, Am. Soc. Cancer Research (v.p.; established jour., editor 1915-17), Soc. Serology and Hematology (pres.), Am. Assn. Immunologists (pres.). Contbr. articles med. jours. Asso. editor Jour. Immunology, Am. Rev. Tb. Devised exceptionally delicate clin. test for luetic infection, based on blood hemolysis by cobra venom; worked upon anaphylaxis; promulgated theory that sensitization is essentially cellular in origin; contbn. to clin. medicine was method of transfusing citnated blood. Died Mason, Ga., 1917.

WEIL-MALHERBE, Hans, biochemist; b. Stuttgart, Germany, Dec. 29, 1905; s. Ludwig and Milly (Schweizer) Weil; M.D., U. Heidelberg, Germany, 1929; M.Sc., U. Durham, Eng., 1940, D.Sc., 1945; m. Rosanne Malherbe, Nov. 20, 1934; children—Claude, Jacqueline. Came to U.S., 1958, naturalized, 1963. Biochemist, Research Lab., North of Eng. Council of Brit. Empire Cancer Campaign, Newcastle-on-Tyne, 1935-45; lectr. in biochemistry King's Coll., Newcastle-on-Tyne, 1945-47; dir. research Runwell Hosp., Wickford, Essex, Eng., 1947-58; chief sect. neurochemistry div. Spl. Mental Health Research Programs, Nat. Inst. Mental Health, Washington, 1958-—; asso. prof. neurochemistry dept. psychiatry George Washington U. Med. Sch., 1965-—. Mem. Biochem. Soc., Am. Soc. Biol. Chemistry, Internat. Soc. for Neurochemistry, A.A.A.S. Research and numerous publs. on biochem. action of brain tissue, formation of various acids in brain; formation, transfer, chemistry of compounds in human, animal tissues. Home: 6213 E. Halbert Rd., Bethesda, Md. 20034. Office: St. Elizabeth's Hosp., Washington 20032.*

WEIMER, Bernal Robinson, Am. biologist; b. Port Royal, Pa., Dec. 4, 1894; George McCullough and Ada Ruth (Robinson) W.; A.B., A.M., W. Va. U.; Ph.D., U. Chgo., 1927; m. Margaret Grace Robinson, Aug. 31, 1918; children—John Robinson, Margaret Brown, George Alexander. Supervising prin. schs., Mifflintown, Pa., 1918-21; prof. biology Bethany College, 1921-—, distinguished prof. biology, 1966-—, chairman of the science and mathematics group, 1931-—, dean faculty, 1936-—, acting pres., 1952-—; instr. zoology W.Va. U., summers 1924, 32, U. Chgo., summer 1927. Served with inf., M.C., U. S. Army, World War I. Fellow A.A.A.S.; mem. Am. Soc. Zoologists, Am. Assn. Biology Tchrs., W.Va. Acad. Sci., W.Va. Biol. Survey, Phi Beta Kappa, Sigma Xi, Beta Beta Beta (nat. pres.). Democrat. Mem. Disciples of Christ Ch. Mason. Author: (with P. D. Strausbaugh) General Biology, 1938; A Manual for Biology Lab., 1938; Elements of Biology, 1944; (verse) Nature Smiles, 1940; (with E. M. Core) A New Manual for Biology Lab., 1944; Man and the Animal World, 1951; Of Things Bi-illogical, 1957. Contbr. articles sci., gen. mags. and jours. Research on hydra. Home: Ross St., Bethany, W.Va., 26032.*

WEINBERG, Alvin Martin, Am. physicist; b. Chgo., Ill., Apr. 20, 1915; s. J. L. and Emma (Levinson) W.; A.B., U. Chgo., 1935, A.M., 1936, Ph.D., 1939; Litt.D., U. Chattanooga; m. Margaret Despres, June 14, 1940; children—David, Richard. Research asso. math. biophysics U. Chgo., 1939-41, Metall. Lab., 1941-45; joined Oak Ridge (Tenn.) Nat. Lab., 1945, dir. physics div., 1947-48, research dir., 1948-55, dir. lab., 1955-—. Mem. Pres.'s Sci. Adv. Com., 1960-62, chmn. panel on sci. information, 1962. Recipient Atoms for Peace award, 1960; E. O. Lawrence award AEC, 1960. Fellow Am. Acad. Arts and Scis.; mem. Nat. Acad. Scis. (physics sect.), Am. Nuclear Soc. (pres. 1959-60). Author: (with E. P. Wigner) The Physical Theory of Neutron Chain Reactors, 1958; Reflections on Big Science, 1967. Contbd. to understanding of nerve activity, to underlying theory of nuclear chain reactors, to design and constrn. of nuclear chain reactors for research and for power; early work (with E. P. Wigner) led to pressurized water line of reactors. Home: 111 Moylan Lane, Oak Ridge 37832. Cffice: P.O. Box P, Oak Ridge, Tenn. 37830.*

WEINBERG, Eugene David, Am. microbiologist; b. Chgo., Mar. 4, 1922; s. Philip and Lenore (Bergman) W.; B.S. in Physiology, U. Chgo., 1942, A.M. in Microbiology, 1948, Ph.D., 1950; m. Frances Murl Izen, Sept. 5, 1949; children—Barbara Anne, Marjorie Jean, Geoffrey Alan, Michael Benjamin. Faculty Ind. U., Bloomington, 1950-—, prof. microbiology, 1961-—; vis. scientist NIH, 1957, Comml. Solvents Corp., 1960. Cons. U.S. Vitamin & Pharm. Corp., 1961-—. Mem. Am. Soc. Microbiology (past pres. Ind. br.), A.A.A.S., Am. Chem. Soc., Ind. Acad. Sci., Sigma Xi. Research, numerous publs. on antagonistic and synergistic effects trace metals and antimicrobial compounds, trace metal control specific biosynthetic processes mature cultures. Office: Jordan Hall, Ind. U., Bloomington, Ind. 47401.*

WEINBERG, Samuel Kirson, Am. sociologist; b. Chgo., Oct. 10, 1912; s. Herman Y. and Rebecca (Kirson) W.; A.B., U. Chgo., 1934, A.M., 1935; Ph.D., 1942; m. Rita C. Mohr, Sept. 15, 1946; children—Carol, Rebecca, Roger Maynard, Douglas Daniel. Co-dir. mental hosp. study Ohio State Mental Health Dept., 1946; faculty Roosevelt U., Chgo., 1947-—, research prof. sociology, chmn. dept., 1965-—; vis. prof. sociology U. Minn., 1951-52, U. Ghana, 1961. Research cons. dept. psychiatry Northwestern U. Med. Sch., 1964-—, Inst. for Juvenile Research, State of Ill., 1965-—; vis. scientist NSF, 1964-65. Fellow Am. Sociol. Assn.; mem. Am. Psychol. Assn., Soc. for Study Social Problems (chmn. psychiat. sociology 1962-65). Author: Society and Personality Disorders, 1952; Incest Behavior, 1955; Culture and Personality, 1958; Social Problems in Our Time, 1960; Culture of the State Mental Hospital (with H.W. Dunham), 1960; also numerous articles. Research on factors leading to onset transient schizophrenia, processes involved in incest, social intimacy, sociology, psychiatry of disordered behaviour. Home: 7420 S. Luella Av., Chgo. 60649. Office: 430 S. Michigan Av., Chgo. 60605.*

WEINBERG, Steven, Am. physicist; b. N.Y.C., May 3, 1933; s. Fred and Eva (Israel) W.; B.A., Cornell U., 1954; postgrad. Copenhagen Inst. for Theoretical Physics, 1954-55; Ph.D., Princeton, 1957; m. Louise Goldwasser, July 6, 1954; 1 dau., Elizabeth. Research asso., instr. Columbia U., 1957-59; research physicist Lawrence Radiation Lab., Berkeley, Cal., 1959-60; faculty U. Cal. at Berkeley, 1960-—, prof. physics, 1965-—. Cons., Inst. for Def. Analyses, Washington, 1960-—, Brookhaven Nat. Lab.; vis. prof. Mass. Inst. Tech., 1967-69. A.P. Sloan fellow, 1961-65; Loeb lectr. in physics Harvard, 1966-67. Mem. Am., Italian phys. socs., Am. Astron. Soc. Research, numerous publs. on elementary particles, quantum field theory, cosmology. Office: Physics Dept., U. Cal., Berkeley, Cal.*

WEINBERGER, Hans Felix, mathematician; b. Vienna, Austria, Sept. 27, 1928; s. Walter and Linda (Haas) W.; came to U.S., 1938, naturalized, 1944; B.S., Carnegie Inst. Tech., 1948, M.S., 1948; Sc.D., in Math., 1950; m. Laura C. Larrick, Apr. 20, 1957; children—Catherine, Sylvia, Ralph. Faculty, Inst. for Fluid Dynamics and Applied Math., U. Md., 1950-60, asso. research prof., 1957-60; mem. Math. Research Center, U.S. Army, U. Wis., 1957-58; asso. prof. U. Minn., Mpls., 1960-61, prof. math., 1961-—. Mem. Am. Math. Soc. Author: A First Course in Partial Differential Equations, 1965; (with M. H. Protter) Maximum Principles in Differential Equations, 1967; also articles. Research on lower bounds for eigenvalues, error estimate in approximation eigenvalues, error bounds in finite difference approximation, approximate solution linear elliptic partial differential equations. Home: 141 Orlin Av., Mpls. 55414.*

WEINER, Joseph Sidney, physiologist; b. Johannesburg, S. Africa, June 29, 1915; s. Robert and Fanny (Simon) W.; B.Sc., Witwatersrand U., 1934, B.Sc. with honors, 1936; student St. Georges Hosp., London, 1947; Ph.D., U. London, 1946; M.A., Oxford U., 1945; m. Marjorie Winifred Daw, June 23, 1943; children—Julia, Edmund. Research climate physiologist Rand Mines Ltd., 1935-37; grantee dept. applied physiology London Sch. Hygiene and Tropical Medicine, 1939-41; physiologist climatic unit Med. Research Council, Nat. Hosp., London, 1941-45; reader phys. anthropology Oxford U., 1945-63, hon. asst. dir. climate and working efficiency unit Med. Research Council, 1955-62, dir. unit, 1962-—; faculty London Sch. Hygiene, 1963-—; prof. environmental physiology London U., 1965-—. Internat. convener human adaptability sect. Internat. Biol Program, 1962-—. Recipient Vernon medal in indsl. physiology. L.R.C.P. Mem. Royal Coll. Surgeons, Physiol. Soc., Royal Anthrop. Inst. (past pres.), Ergonomics Research Soc. (past chmn.), Soc. for Study Human Biology (past sec.). Author: Piltdown Forgery, 1955; (with Harrison, Tanner, Barnicot) Human Biology, 1964; (with Pavi Baker) Biology of Human Adaptability, 1966; also numerous articles. Editor: Ergonomics, 1963-66. Research on adaptation of human body to hot climates, control of sweating, water and salt in body, human evolution using fossil materials; anthrop. studies of Bushmen and Hottentot in So. Africa; exposed Piltdown Man as forgery, 1953. Home: 95 Bedford Ct. Mansions, Bedford Av., London, Eng.*

WEINER, Murray, Am. physician; b. N.Y.C., Apr. 18, 1919; s. Samuel O. and Gussie (Begun) W.; B.S., Coll. City N.Y., 1939; M.S. in biochemistry N.Y. U., 1943, M.D., 1943; m. Marilyn R. Greenberg, Jan. 14, 1951; children—Eve, George, Joan. Fellow N.Y. U. Coll. Medicine, 1946-51, faculty, clin. medicine, 1951-—, asst. prof., 1956-—, asso. vis. physician N.Y. U. Research Service, Goldwater Meml. Hosp., 1958-—, in charge coagulation research lab., 1955-—, mem. vis. staff U. Hosp., 1951-—; asso. vis. physician Willard Parker Hosp. Chest Service, N.Y.C., 1951-56; cons. peripheral vascular disease L.I. Jewish Hosp., 1954-59; vis. staff North Shore Hosp., Manhasset, 1955-60; cons. Geigy Pharmaceuticals, 1953-—, research coordinator, 1958-60, asso. dir. medicinal research, 1965-67, dir. clin. pharmacology, 1959-67, v.p., dir. biol. research, 1967-—. Cons. NRC Com. on Biochem. Studies in Evaluating Drug Toxicity. Mem. Westchester County, N.Y. State med. socs., Am. Physiol. Soc., Am. Soc. Pharmacology and Exptl. Therapeutics, Soc. Exptl. Biology and Medicine, Am. Therapeutic Soc., A.C.P., Am. Soc. for Clin. Research, N.Y. Acad. Sci., Westchester Acad. Medicine, Soc. for Study of Blood, Am. Heart Assn., Sigma Xi. Author: (with Shepard Shapiro) Coagulation, Thrombosis and Dicumarol, 1949; also numerous articles. Developed improved methods for studying coagulation and clot dissolution; use of drugs to treat diseases of clotting and hemorrhage; study of fate of drugs in animals and man and mechanisms of their pharmacologic action. Home: 3 Thomas Way, White Plains, N.Y. 10607. Office: N.Y.U. University Hosp., N.Y.C. 10016.*

WEINER, Robert Franz Johann, German chemist; b. Horn, Apr. 1, 1902; s. Franz and Leopoldine (Niederleuthner) W.; Dr. eng., Vienna (Austria) Tech. Sch.; m. Hilde Fucik, Aug. 21, 1930; children—Helga, Ingrid, Gerda, Gudrun. Asst., Vienna Tech. Sch., 1924-32; engaged in chem. work in industry, 1932-—; sci. research Stuttgart and Dresden tech. schs., 1933-35; chemist Degussa Soc., Frankfort, 1938-46; dir. Clorat-Chemie, 1948-58; instr., then asso. prof. U. Innsbruck, 1949-—; dir. Electronic Microscopy Labs., Innsbruck, 1958-61; dir. Elektrolyse Soc., Munich, 1961-—. Mem. German Bunsen Soc. Phys. Chemistry, German Galvanotech. Soc. Publns. on phys. chemistry, electrochemistry, analytic chemistry, galvanotechnology, chem. technology. Home: Freisingstrasse 3, Innsbruck, Austria. Office: Chiemgastrasse 109, Munich 9, West Germany.

WEINER, Walter, hematologist, serologist; b. Prague, Czechoslovakia, June 14, 1899; s. Adolf and Pauline (Nossal) W.; M.D., U. Prague, 1924; m. Stephanie Dewidels, Jan. 22, 1928; 1 son, Robert Thomas. Specialist physician, 1929-39; dir. Regional Blood Transfusion Service, Birmingham, Gt. Britain, 1946-66; sr. research fellow U. Birmingham, 1966-—; cons., part-time lectr., Birmingham U., 1948-—. Fellow Coll. Pathologists; mem. Assn. Clin. Pathologists, Brit. Soc. for Immunology. Contbg. author: Clinical Aspects of Immunology 1963. Research and publs. on blood group serology, serology of hemolytic anemias of acquired auto-immune type, definition of anti-bodies in respect to their blood-group specificity; described 1st specific antibody occuring as an auto-antibody in auto-immune hemolytic anemia. Home: 26 Fellows Lane, Birmingham 17, Eng.*

WEINERT, Hans, German anthropologist; b. Bruanschweig, Germany, Apr. 14, 1887; prof., Berlin, also Kiel, Germany. Author: Menschen der Vorzeit, 1930; Ursprung der Menschheit, 1932; Rassen der Menschheit, 1934; Der geistige Aufstieg des Menschen, 1940; Stammesgeschichte der Menschheit, 1941; Die Reisenaffenmenschen und ihre stammesgeschichte Bedeutung, 1948.

WEINGÄRTNER, Lothar, German physician; b. Düren, Germany, Sept. 23, 1910; s. Julius and Elisabeth (Magnus) W.; M.D., U. Göttingen, 1937; m. Ilse Rolff, Oct. 3, 1939; children—Ingrid, Rolf, Tom. Faculty, Leipzig U. Clinic for Children, 1938-—, prof. pediatrics, 1956-—, dir. clinic, 1959-—. Mem. German Soc. Pediatrics and Tb, Bulgarian Soc. Pediatrics (hon.). Author: Lehrbuch die Therapie, 1958; Nebenwirkungen, 1959; Pädiatrie und Grenzgebiete, 1965; also numerous articles. Research on Tb of children, mechanisms of antibiotics, X-rays, side-effects of drugs, kidney diseases, vaccination with BCG. Home: Krokusweg 32, 402 Halle/Saale, Germany. Office: Univ. Clinic for Children, Halle-Wittenberg, Germany.*

WEINHOUSE, Sidney, Am. chemist; b. Chgo., May 21, 1909; s. Harry and Dora (Cutler) W.; B.S., U. Chgo., 1933, Ph.D., 1936; m. Sylvia Krawitz, Sept. 15, 1935, (dec. Aug. 1957); children—Doris Joan,

James Lester, Barbara May. Eli Lilly fellow U. Chgo., 1936-38, Coman fellow, 1941-44; head biochem. research Houdry Process Corp., 1944-47; research dir. Temple U. Research Inst., 1947-50; head dept. metabolic chemistry Lankenau Hosp. Research Inst. and Inst. Cancer Research, Phila., 1950-57; chmn. div. biochemistry Inst. Cancer Research, 1957-60; asso. dir. Fels Research Inst. Temple U. Sch. Medicine, 1961-63, dir., 1963—, also prof. biochemistry Temple U. Recipient Am. Chem. Soc. Phila. sect. award, 1965. Mem. Am. Chem. Soc., Am. Soc. Biol. Chemists, Am. Assn. Cancer Research, Am. Diabetes Assn., Soc. Exptl. Biology and Medicine. Research, numerous publs. on carbohydrate and fatty acid metabolism; effects of hormones on gluconeogenosis; comparative studies on control of respiration and glycolysis in liver and liver tumors. Home: 6419 Lawnton Av., Phila. 19126. Office: Fels Research Inst., Temple U. Sch. Medicine, 3420 N. Broad St., Phila. 19140.*

WEINLAND, James David, Am. psychologist; b. Banning, Cal., June 21, 1894; s. William Henry and Caroline (Yost) W.; A.B., Pomona Coll., 1917; A.M., Columbia, 1922, Ph.D., 1927. Faculty, Pratt Inst., 1922-27; faculty Lehigh U., 1922-25, Moravian Coll. for Women, 1922-24; instr. to prof. psychology N.Y.U., 1927-59. Mem. Am. Psychol. Assn. Author: General Psychology for Students of Business, 1940; (with Margaret Gross) Personal Interviewing, 1952; How To Improve your Memory, 1957; How To Think Straight, 1963; How To Study, 1965; also articles. Research on procedural thinking, memory and methods of study. Address: P.O. Box 962, Boulder City, Nev. 89005.*

WEINMAN, David II, Am. microbiologist; b. Albuquerque, Jan. 13, 1909; s. David and Belle (Lewinson) W.; B.A., Columbia, 1929; M.D., U. Paris, 1935, Medecin Colonial, 1935; postgrad. Institut Pasteur, Paris, 1935; m. Elizabeth Grover, July 28, 1946. Research fellow, instr. Harvard Med. Sch. and Sch. Pub. Health, 1936-44; faculty Columbia Coll. Phys. & Surg., 1944-46; faculty Yale Sch. Medicine and Grad. Sch., New Haven, 1947—, prof. microbiology, 1964—. Mem. Harvard expdn. to Peru, 1937, Liberia, 1944; div. chief SEATO Med. Research Lab., Bangkok, Thailand, 1964-65; mem. WHO Expert Panel on Parasitic Diseases, 1963—; cons. Yale-New Haven, VA hosps. Fulbright research fellow, E. Africa, 1956-57. Diplomate Am. Bd. Pathology. Mem. Royal, Am. socs. tropical medicine and hygiene, Belgian Soc. Tropical Medicine, Soc. Exptl. Biology and Medicine. Research, publs. in infectious diseases, particularly tropical; studies in microbiology and pathogenesis. Home: 91 Mill Rock Rd., New Haven 06511.*

WEINSTEIN, Alexander, mathematician; b. Saratov, Russia, Jan. 21, 1897; s. Julius and Alexandra (Levcowitz) W.; Ph.D., U. Zurich, 1921; Docteur es Sciences Mathematiques, Sorbonne, Paris, 1937; m. Marianne Ganz, Mar. 13, 1928. Came to U.S., 1940, naturalized, 1946. Research prof. U. Md., College Park, 1950—. Guggenheim fellow, 1954-56; Fulbright fellow, 1955-56. Mem. Accademia Nazionale dei Lincei, Rome, Nat. Acad. Sci. Peru, Am. Math. Soc. Author: Equations aux Derivees Partielles, 1937. Research, numerous publs. on theory of fluid jets, water waves, vibration and stability of plates, intermediate problems in eigen-values, generalized axially symmetric potential theory, Euler-Poisson-Darboux equation, generalized radiation problem, singular partial differential equations. Home: 9300 Piney Branch Rd., Silver Spring, Md. 20903. Office: U. Md., College Park, Md. 20740.*

WEINSTEIN, Leonard Harlan, Am. plant physiologist; b. Springfield, Mass., Apr. 11, 1926; s. Barney Willard and Ida (Feinberg) W.; B.S., Pa. State U., 1949; M.S., U. Mass., 1950; Ph.D., Rutgers U., 1953; m. Sylvia Jane Sherman, Oct. 15, 1950; children—Beth Rachel, David Harold. Fellow plant nutrition Rutgers U., 1953-55; plant physiologist, plant biochemist Boyce Thompson Inst. for Plant Research, Inc., Yonkers, N. Y., 1955-63, program dir. for plant chemistry, 1963—. Mem. Am. Soc. Plant Physiologists, Plant Phenolics Soc. N.Am., Harvey Soc., T.A.P.P.I., Am. Inst. for Biol. Scis., Sigma Xi. Research, numerous publs. on role trace elements in plant nutrition, effects air pollutants on plants. Home: 75 Lincoln Av., Ardsley, N.Y. 10502. Office: 1086 N. Broadway, Yonkers, N.Y. 10701.*

WEINSTEIN, Louis, Am. physician; b. Bridgeport, Conn., Feb. 26, 1909; s. Harry and Minnie (Rubenstein) W.; B.S., Yale, 1928, M.S., 1930, Ph.D., 1931; M.D., Boston U., 1943; m. Ethel Raport, Oct. 7, 1934; children—Allen Joseph, Carol Frances. Practice medicine, specializing in infectious disease and microbiology, Boston, 1944—; instr., asso. prof. medicine Sch. Medicine Boston U., 1943-57, lectr., 1957—; chief infectious disease service Mass. Meml. Hosps., 1947-57; instr., lectr. Med. Sch. Harvard, 1949—; asso. physician-in-chief, chief infectious disease service New Eng. Med. Center Hosps., 1957—; prof. medicine Sch. Medicine Tufts U., 1957—; asso. physician med. service Mass. Gen. Hosp., 1961—. Recipient Chapin medal City Providence, 1965. Mem. Mass. Med. Soc., A.C.P., A.A.A.S., N.Y. Acad. Scis., Am. Fedn. Clin. Research, A.M.A., Am. Soc. Clin. Investigation, Assn. Am. Physicians, Soc. For Exptl. Biology and Medicine, Am. Acad. Microbiology, Am. Acad. Arts and Scis. Author: (with Ehrenkranz) Streptomycin, 1957; The Practice of Infectious Disease, 1958; also numerous articles. Research and publs. primarily in field of microbiology, infectious diseases. Home: 26 Greylock Rd., Newtonville, Mass. 02160. Office: 171 Harrison Av., Boston 02111.*

WEINSTEIN, Roy, Am. physicist; b. N.Y.C., Apr. 21, 1927; s. Harry and Lillian (Ehrenberg) W.; S.B., Mass. Inst. Tech., 1951; Ph.D., 1954; m. Janet Elizabeth Spiller, Mar. 26, 1954; children—Lee Davis, Sara Lynn. Instr., Brandeis U., Waltham, Mass., 1954-56; fellow Inst. for Advanced Studies, Mass. Inst. Tech., 1954-56, asst. prof., 1956-59; NSF fellow Bohr Inst., Copenhagen, Denmark, 1959-60; faculty Northeastern U., Boston, 1960—, prof. physics, 1963—. Dir. IKOR, Inc., Burlington, Mass. Cons. to pvt. cos. Mem. Am. Phys. Soc., N.Y. Acad. Scis., Am. Assn. Physics Tchrs. Author: Atomic Physics, 1964; Nuclear Physics, 1964; Interaction of Radiation and Matter, 1964; also articles. Confirmed basic laws quantum electrodynamics at very small distances; discovered appearance to a single observer of objects moving at relativistic velocities; confirmed deuteron model nucleus for gamma ray interactions; research on positronium atom, annihilation electrons and positrons via single quantum emission. Home: 10 Daniels St., Lexington, Mass. 02173.*

WEINSTOCK, Bernard, Am. chemist; b. N.Y.C., Dec. 7, 1917; s. George and Fannie (Alexenburg) W.; A.B., Bklyn. Coll., 1938; postgrad. Columbia U., 1939-41; Ph.D., U. Chgo., 1948; M.A., U. Oxford (Eng.), 1958; m. Thelma Barkin, Dec. 19, 1947; children—George Matthew, Stephen Jay. Research asso. SAM Labs., Manhattan Project, N.Y.C., 1941-43; staff Los Alamos Sci. Lab., 1943-45; fellow Inst. for Nuclear Studies, U. Chgo., 1946-47; sr. scientist Argonne (Ill.) Nat. Lab., 1947-60; sr. staff scientist Sci. Lab., Ford Motor Co., Dearborn, Mich., 1960—. Cons., Argonne Nat. Lab., 1960—. Guggenheim fellow, 1957; recipient Award of Honor, Bklyn. Coll., 1966. Fellow Am. Phys. Soc. (past mem. exec. com. div. chem. physics); mem. Am. Chem. Soc., Research Soc. Am. (past local pres.), Combustion Inst. Mem. editorial adv. bd. Chem. and Engring. News, 1967—. Research and publs. on properties of liquid helium-3, properties of hexafluoride molecules, gas kinetics. Home: 4987 Malibu Dr., Bloomfield Hills, Mich. 48013. Office: P.O. Box 2053, Dearborn, Mich. 48121.*

WEIPPL, Gerald Theodor, Austrian pediatrician; b. Vienna, Austria, Aug. 4, 1927; s. Theodor and Ludmilla (Wagner) W.; M.D., U. Vienna, 1952. With Pediatric U. Clinic, Vienna, 1952—, head premature and newborn unit, 1962—; lectr. pediatric hematology U. Vienna, 1965—. Mem. European Club of Pediatric Research, Austrian and German Pediatric Soc., Soc. Physicians in Vienna. Editor: Pädiatrie und Pädologie, 1965. Research, numerous publs. on iron metabolism, anemia and polycythemia in infancy and childhood, haptoglobin, transferrin, hemopexin, lipoprotein, polymorphism in serum proteins, adult and fetal hemoglobin, diagnosis of prenatal dystrophia. Office: Spitalgasse 23, A-1090 Vienna, Austria.*

WEIR, Horace McColloch, Am. chem. engr.; b. Muncie, Ind., Nov. 19, 1894; s. Daniel Thomas and Lulu (McColloch) W.; B.Chem. Engr., Purdue U., 1917; postgrad. U. London, 1924-25; Ph.D., U. Munich, 1926; m. Helen C. Gillette, Feb. 18, 1922 (dec.); 1 son, Robert M. With Prest-O-Lite Co., 1919-20, Standard Oil Co., 1920-24, Atlantic Refining Co., 1927-34, Lummus Co., 1935-36, Rheinmetall-Borsig, Berlin, Germany, 1936-38, United Engring. & Constrn., Inc., 1938-43, Davison Chem. Co., 1943-44; cons. chem. engr. H.M. Weir Co., 1944—. Fellow A.A.A.S.; mem. Am. Chem. Soc., Am. Inst. Chem. Engrs., Franklin Inst., Sigma Xi, Tau Beta Pi, Phi Lambda Upsilon. Contbr. articles to profl. jours., also chpts. to books. Patentee fields of petroleum and chem. processing, metallurgy of gas sensitive metals and compounds. Home: Brynwood Apts., Wynnewood, Pa. 19096. Office: Land Title Bldg., Phila. 19110.*

WEIR, Sir William Douglas, Scottish engr.; b. Glasgow, Scotland, May 12, 1877; s. James and Mary (Douglas) W.; LL.D., Glasgow U.; m. Alice Blanche MacConnachie, 1904; children—James Kenneth, John William, Mrs. E. M. Indare. Apprentice engr. G. & J. Weir, Ltd., Holm Foundry Cathcart, Glasgow, then mng. dir., chmn., hon. pres.; Scottish dir. munitions, 1915-16; controller aero. supplies, mem. Air Bd., 1917-18; dir. gen. aircraft prodn. Ministry Munitions, became mem. council, 1918, sec. of state, pres. Air Council, 1917-18; chmn. Adv. Com. Civil Aviation, 1919; chmn. Com. on Amalgamation Depts. Common to Fighting Services, 1922-23; chmn. Nat. Rev. Com. on Electric Energy Supply, 1925; chmn. Com. on Electricification Main Line Rys. Gt. Britian, 1931; mem. com. on trading and financial positions Brit. Shipping Cos., 1932; indsl. adviser Empire Econ. Conf., Ottawa, Ont.; adviser on nat. def. Brit. Govt., 1936-37; named dir. gen. explosives Ministry Supply, 1939; became chmn. Tank Bd., 1942; dir. Cathart Investment Trust Ltd., Internat. Nickel Co. Can. Ltd. Knighted, 1917. Contributed to use of air power in World War I. Died July 2, 1959.

WEISBACH, Albin, German mineralogist; b. Germany, 1833; s. Julius Weisbach; student of Ferdinand Reich; prof. mineralogy Freiberg Sch. Mines; conducted research in crystallography and physics; 1st to identify argyrodite, 1855 (led to C. Winkler's discovery of element germanium 1886). Died 1901.

WEISBERGER, Austin S., Am. physician; b. Cleve., Oct. 1, 1913; s. Elias and Rachel (Schoenfeld) W.; A.B., Western Res. U., 1939, M.D., 1941; m. Eleanor Burt, Feb. 28, 1943; children—Susan, Betsy, Deborah. With Western Res. U. Sch. Medicine, Cleve., 1946—, prof. medicine, 1961-65, John H. Hord prof. medicine, chmn. dept. medicine, 1965—; staff Univ. Hosp., Cleve., 1956—, physician, 1965—, dir. dept. medicine, 1965—. Diplomate Am. Bd. Internal Medicine, mem. bd., 1967—. Fellow A.C.P.; mem. Soc. for Exptl. Biology and Medicine, Am. Soc. for Clin. Investigation, Am. Assn. Physicians, Central Soc. for Clin. Research (pres. 1967), Am. Soc. Hematology (mem. adv. council 1964-65), Am. Cancer Soc. (mem. study sect. 1963—), Cleve. Acad. Medicine (dir. 1965—). Research, numerous publs. in leukemia and lymphomas, iron metabolism, chloramphenicol toxicity, mechanisms action chloramphenicol, directed protein synthesis by RNA. Home: 3175 Laurel Rd., Shaker Heights, O. 44120. Office: 2065 Adelbert Rd., Cleve. 44106.*

WEISBURGER, John Hans, biochem. pharmacologist; b. Stuttgart, Germany, Sept. 15, 1921; s. William and Selma (Barth) W.; came to U.S., 1943, naturalized, 1944; student U. Brussels, 1939-40, U. Havana (Cuba), 1941-43; A.B., U. Cin., 1947, M.S., 1948, Ph.D., 1949; m. Elizabeth A. Kreiser, Apr. 7, 1947; children—William R., Diane S., Andrew J. Office Naval Research fellow U. Cin., 1949-50; fellow Nat. Cancer Inst., NIH, USPHS, Bethesda, Md., 1949-50, biochemist, 1951-61, chief carcinogen screening sect., Nat. Cancer Inst., scientist dir. USPHS, 1961—. Mem. interdepartmental Panel Tech. Experts, FDA-USPHS, 1961—; mem. biochem. and nutrition panel, div. research grants NIH, 1959-60. Mem. Am. Chem. Soc., Chem. Soc. Washington (past mgr.), Am. Assn. for Cancer Research, Am. Soc. Biol. Chemists, Biochem. Soc. (London), Soc. Toxicology, Nat. Cancer Inst. Assembly Scientists (past councilor), Sigma Xi, Alpha Chi Sigma. Mem. editorial bd. Cancer Research, 1960-64, Jour. Nat. Cancer Inst., 1960-62. Research and numerous publs. on chem. causes, prevention cancer, endocrine and other host factors in tumor devel., detection and bioassay of potentially hazardous materials. Home: 5309 McKinley St., Bethesda 20014. Office: Nat. Cancer Inst., NIH, Bethesda, Md. 20014.*

WEISE, Artur, German physicist; b. Berlin, May 29, 1904; s. Hans and Luise (Becker) W.; Dr. eng., Dresden Tech. Sch.; m. Ruth Amspach; m. 2d, Erike Schmerglatt; children—Monika, Volker. Asst. tech. labs. Dresden Tech. Sch., 1928-34; asst. exptl. sta. aerial navigation, dir. Inst. Gas Dynamics, Berlin-Adlershof, 1934-46; prof., dir. Aerodynamics Inst., Stuttgart Tech. Sch. Mem. Wissenschaftliche Gesellschaft fur Luft und Raumfahrt, Gesellschaft fur angewandte Math. und Mechan., Baden-Wurzburg Physics Soc. Author: Gegadbelter Verduchtungsstoss, Überschallaxialverdichter; Strömungslehre; Grundlagen des Maschinenbaus. Research on shock tube-wave tunnels; numerical computations of compressible flow fields; shock waves; gasdynamics turbo-machinery; shock tubes. Home: Bruno-Frank-Strasse 20, Stuttgart-Heumaden. Office: Pflaffenwaldstrasse, Stuttgart-Baihingen, West Germany.

WEISEL, Wilson, Am. physician; b. Alexandria, Neb., June 4, 1914; s. Lewis William and Ada (Wilson) W.; B.A., U. Wis., 1935, M.A., 1936; M.D., Harvard Med. Sch., 1938; M.S., U. Minn., 1942; m. Betty K. Amos, Sept. 3, 1938; children—Thomas Wilson, Richard Dale, Amy Kathleen. Clin. prof. surgery, chmn. dept. thoracic-cardiovascular surgery Marquette U. Sch. Medicine, Milw., 1946—; chmn. dept. surgery St. Joseph Hosp., Milw. Mem. Am. Assn. for Thoracic Surgery, Soc. for Thoracic Surgeons, Am. Thoracic Soc., Central, Wis. (sec.-treas. 1967) surg. socs., Am. Bd. Surgery, Bd. Thoracic Surgery. Research, numerous publs. on vascular prosthesis, pulmonary surgery, lung transplantation, bronchial surgery, esophogal carcinoma palliation, thoracic trauma. Home: 8395 N. Pelican Lane, Milw. 53209.*

WEISER, Jaroslav, Czeckoslovakian biologist, zoologist; b. Prague, Czechoslovakia, Jan. 13, 1920; s. Martin and Anna (Bekova) W.; R.N.Dr., Charles U., Prague, 1946; D.Sc., Prague Acad. Sci., 1961; m. Jaroslava Janacek, June 12, 1947; children—Jaroslav, Helena. Faculty, Charles U., 1947-51, Med. Faculty, Sarajevo, Yugoslavia, 1948-49; head dept. parasitology Acad. Scis., Prague, 1951-54, head dept. insect pathology, 1954—, dep. dir. Inst. Entomology, 1964—. Mem. expert com. on biol. control WHO, 1962—. Recipient State award for sci., 1964. Mem. Acad Sci. (corr.), Soc. Protozoologists, Soc. Invertebrate Pathologists, Entomol. Soc. Am., Parasitological Soc. France, Internat. Adv. Com. for Biol. Control. Author: Moderni insekticidy, 1951; Skudci lid. zdravi, 1952; Nemoci hmyzu (Diseases of Insects), 1966. Research, numerous publs. in insect protozoa, microsporidia, application in field control of pests, studies on insects attacking fungi and nematodes, theory of use of insect diseases. Home: 964, Herálecká St., Prague 4. Office: 2, Fleming Sq., Prague 6, Czechoslovakia.*

WEIS-FOGH, Torkel, zoologist; b. Aarhus, Denmark, Mar. 22, 1922; s. Svend and Dagmar Foldagen (Larsen) W.-F.; student U. Aarhus, 1940-41; Mag. scient., U. Copenhagen (Denmark), 1947, Dr.phil., 1952; m. Hanne Heckscher, Aug. 31, 1946. Asst. in research to Prof. August Krogh, Denmark, 1947-49, head of his pvt. lab., 1949-53; asso. lectr. Copenhagen U., 1953-54; fellow Rockefeller Found., 1954-55; Balfour student U. Cambridge, also mem. Trinity Coll., 1956-59; prof. zoophysiology Copenhagen U., 1959-66; prof. zoology, head dept. zoology, Cambridge U., 1966—, also fellow Christ's Coll. Became mem. Danish State Research Found., 1962; became chmn. Nat. Sci. Sect., 1963. Recipient Gold medal U. Copenhagen, 1944. Fellow Royal Danish Acad., Acad. Tech. Scis. Research and publs. on biophysics of insect flight, respiration, structural proteins and nervous coordination. Home: 7, Almoner's Av., Cambridge, Eng.

WEISMANN, August, German biologist; b. Frankfort/Main, Germany, Jan. 17, 1834; student medicine, Göttingen, 1852-56, then Paris, 1860, zoology, Giessen, 1861. Asst., Med. Clinic, Rostock, 1856; at Chem. Inst., 1857; practicing physician, Frankfort, 1858, 1860; joined med. faculty, U. Freiburg im Breisgau, 1863; prof. zoology, 1866-1912. Fellow Royal Soc., 1910. Author: Die Entwicklung der Dipteren, 1864; Studien zur Deszendenztheorie, 1875-76; Das Keimplasma, 1892; also others. Did extensive work in study of heredity; insisted on non-inheritance of acquired characteristics; devised germ-plasma theory of heredity; believed germ-plasm formed eggs and sperm, which is passed on from generation to generation (continuity of germ-plasm); among first to suggest chromosomes contain hereditary material; suggested that offspring receives half its germ-plasm from father and half from mother; attacked Lamarck's theories and favored those of Darwin; active proponent in Germany of evolution; studied insect embryology; investigated one-celled organisms. Died Freiburg-im-Breisgau, Baden, Nov. 6, 1914.

WEISNER, Louis, mathematician; b. N.Y.C., May 1, 1899; s. Jacob H. and Rose (Weber) W.; B.S., Coll. City N.Y., 1920; A.M., Columbia U., 1922, Ph.D., 1923; m. Celia Garris, June 3, 1923; 1 son, Allan Garris. Instr., U. Rochester, 1923-26; Nat. Research fellow Harvard, 1926-27; faculty Hunter Coll., 1927-54, asso. prof., 1936-54; prof. maths. U. New Brunswick (Can.), Fredericton, 1955—. Mem. Am. Math. Soc., Math. Assn. Am., Canadian Math. Congress, A.A.A.S. Author: Introduction to the Theory of Equations, 1938; also articles, book revs. Research in group theory, invariant theory, theory equations. Home: 335 Connaught St., Fredericton, N.B., Can.*

WEISS, Adolph Kurt, physiologist; b. Graz, Austria, Mar. 22, 1923; s. Alfred and Anna (Giesskann) W.; came to U.S., 1940, naturalized, 1943; B.S., Okla. Baptist U., 1948; M.S., U. Tenn., 1950; Ph.D., U. Rochester, 1953; m. Mary Davis, June 1, 1948; 1 son, Thomas Alfred. Faculty, U. Miami Sch. Medicine, 1953-61, Oklahoma City U., 1961-64; asso. prof dept physiology U. Okla. Med. Center, Oklahoma City, 1961-66, professor department of physiology, 1967—. Investigator, Howard Hughes Med. Inst., 1955-58; vis. prof. U. Coll. of W.I., 1960. NIH research career award, 1964. Mem. Am. Physiol. Soc., Am. Soc. Zoologists, Endocrine Soc., Gerontological Soc., Soc. for Exptl. Biology and Medicine, Sigma Xi. Contbr. chpt. to Behind the Dim Unknown, 1966. Research, publs. on oxygen consumption rates in rat tissues, discoverer tissues responding to thyroid hormone, cold exposure and aging influences. Home: 2224 N.W. 43d St., Oklahoma City 73112.*

WEISS, Armin, German chemist; b. Stefling, Germany, Nov. 5, 1927; s. Michael and Therese (König) W.; student U. München (Germany), 1948-51, U. Würzburg (Germany), 1947-48; Dr.rer.nat., Technische Hochschule, Darmstadt (Germany), 1953, Habilitation, 1955; m. Angela K. Rauh, Aug. 20, 1955; children—Michael Norbert, Eva Carola, Doris Angela, Ulrich Peter. With Technische Hochschule, Darmstadt, 1953-61, dozent, 1955-61; prof. inorganic chemistry U. Heidelberg (Germany), 1961-65, U. Munich, 1965—. Editor, Kolloid-Zeitschrift, Zeitschrift für Polymere, 1966—. Research and numerous publs. on chemistry of silican and silicates, uni- and two dimensional expanding lattices, reactions in 2-dimensional systems, surface chemistry, structural chemistry and X-ray crystallography of silicides and basic mercury compounds; discovered fibrous silica and intercalation compounds of kaolinite. Home: 4 Sanderplatz 8, München, Germany.*

WEISS, Bernard, Am. psychologist; b. Bklyn., May 27, 1925; s. Max and Sadie (Albert) W.; A.B., N.Y. U., 1949; Ph.D., U. Rochester, 1953; m. Ann J. Bartlett, Oct. 28, 1950; children—Wendy Bartlett, Thomas Wales. Research asso. U. Rochester, 1953-54; exptl., physiol. psychologist USAF Sch. Aerospace Medicine, Randolph Field, Tex., 1954-56; faculty Johns Hopkins Med. Sch., 1956-65, asst. prof. pharmacology and exptl. therapeutics, 1961-65; asso. prof. radiation biology, biophysics U. Rochester (N.Y.) Sch. Medicine, 1965-67, prof., 1967—. Mem. behavioral pharmacology com. NIH, 1965-67; field editor Jour. Pharmacology and Exptl. Therapeutics, 1965—. Mem. Am. Psychol. Assn. (mem. exec. bd.), Am. Soc. Pharmacology and Exptl. Therapeutics, Soc. for Exptl. Analysis of Behavior (dir.), Behavioral Pharmacology Soc. (pres. 1961-64), A.A.A.S., Psychonomic Soc. Research, numerous publs. in psychopharmacology with an emphasis on understanding how drugs affect the mechanisms of behavior. Home: 39 Ellison Hills Dr., Rochester, N.Y. 14625.*

WEISS, David Walter, immunologist, educator; b. Vienna, Austria, July 6, 1927; s. Hillel and Julie (Kohn) W.; student Yeshiva U., 1933; B.A. cum laude in Biology and Chemistry, Bklyn. Coll., 1949; Ph.D. in Microbiology and Biochemistry, Rutgers U., 1952; D.Phil. in Medicine, Oxford U., 1957; m. Judith Weintrob, Mar 25, 1951; children—Hillel, Joshua, Jeremy. Asst. prof. bacteriology U. Cal. at Berkeley, 1957-60, asso. prof., 1960—, asso. research immunologist Cancer Research Genetics Lab. Mem. Path. Soc. Gt. Britain and Ireland, Am. Thoracic Soc., Am. Assn. for Cancer Research, Acid Fast Club (Eng.). Research, publs. on tumor immunology, immunological responsiveness and host-parasite interactions in infectious diseases. Home: 215 Purdue Av., Berkeley, Cal.*

WEISS, Emilio, microbiologist; b. Pakrac, Yugoslavia, Oct. 4, 1918; s. Edoardo and Vanda (Schrenger) W.; student U. Rome, 1936-38; A.B., U. Kan., 1941; M.S., U. Chgo., 1942, Ph.D., 1948; m. Hilda Damick, June 23, 1943; children—Natalie Ann, Elizabeth Rae. Faculty, U. Chgo., 1948-50, Ind. U., 1950-53; br. chief VR div. Camp Detrick, Frederick, Md., 1953-54; with Naval Med. Research Inst. Nat. Naval Med. Center, Bethesda, Md., 1954—, now dep. dir. dept. microbiology. Mem. Am. Soc. Microbiology, Am. Acad. Microbiology, Soc. Exptl. Biology and Medicine, Am. Assn. Immunologists, Sigma Xi. Research, publs. on bacteria that have not yet been grown on lifeless media. Home: 3612 Raymond St., Chevy Chase, Md. 20015. Office: Naval Med. Research Inst., Bethesda, Md. 20014.*

WEISS, Harold Samuel, Am. physiologist; b. N.Y.C., Sept. 10, 1922; s. Seymour and Evelyn (Vormund) W.; B.Sc., Rutgers U., 1947, M.Sc., 1949, Ph.D., 1950; certificate meteorology N.Y. U., 1945; m. Ann Socolow, Aug. 27, 1948; children—Ronald, Karen, Pamela, Seymour. Research asst. Rutgers U., 1947-50, faculty, 1950-62, asso. prof., 1956-62; research officer USAF Aero-Med. Lab., Dayton, O., 1951-53; asso. prof. Ohio State U., 1963-67, prof., 1967—. Mem. Am. Physiol. Soc., A.A.A.S., N.Y. Acad. Sci., Soc. Exptl. Biology and Medicine. Research, numerous publs. on factors affecting spontaneous avian atherosclerosis; role of inert gases in artificial atmospheres; importance of diet, blood pressure, exercise in hardening of the arteries; importance and function of nitrogen in the air we breathe. Home: 1174 W. First Av., Columbus, Ohio 43212.*

WEISS, Herbert Vinik, Am. chemist; b. Bklyn., Nov. 16, 1921; s. Benjamin and Rebecca (Katz) V.; M.S., N.Y. U., 1949; Ph.D. (Ethicon fellow) U. Cin., 1952; m. Beatrice Hirth, May 29, 1955; children—Brett, Grant. Chemist, Chief Med. Examiners Office, N.Y.C., 1946-49; fellow U. Rochester (N.Y.), 1952-53; instr. N.Y. U., 1953-54; research fellow Columbia U., 1954-55; research chemist U.S. Naval Radiol. Def. Lab., San Francisco, 1956—. Cons., Radiation Detection Co., Mountain View, Cal., 1962—. Mem. Am. Chem. Soc., A.A.A.S., Sigma Xi. Research and publs. on deposition and accumulation of radioactivity formed in atomic detonations; discovered number radionuclides formed in fission uranium and applied to study charge distbn. in nuclear fission; devised mechanism for removal of minute traces of elements from solution. Home: 840 Buckland Av., San Carlos, Cal. 94070. Office: Hunter's Point, San Francisco 94135.*

WEISS, Leonard, pathologist; b. London, Eng., June 15, 1928; s. Alfred and Sophia (Barak) W.; B.A., U. Cambridge, 1950; M.A., M.B., B.Chir., Westminster Med. Sch., London, 1953, M.D., 1958, Ph.D., 1963; m. Maureen Anne Jones, Feb. 1951; children—Gregory, Simon, Emma. House physician, resident pathologist, research asst., hon. registrar in morbid anatomy Westminster Hosp., London, 1953-58; mem. sci. staff Med. Research Council-Nat. Inst. Med. Research, London, 1958-60, Strangeways Research Lab., Cambridge, 1960-64; practice medicine, specializing in pathology, Buffalo, 1964—; dir. dept. exptl. pathology Roswell Park Meml. Inst., Buffalo, 1964—; research prof., chmn. dept. exptl. pathology Roswell Park Grad. div. State U. N.Y. at Buffalo. Fellow Inst. Biology; mem. Path. Soc. Gt. Britain, U.K. Coll. Pathology, Internat. Soc. Cell Biology, Am. Soc. Cell Biology, Soc. Exptl. Pathology. Studies, numerous publs. on nature and function of the cell periphery, particularly in relation to cellular interactions; protective effects of deep hypothermia in mammals against X-irradiation. Home: 60 Willowbrook Dr., Williamsville, N.Y. 14221. Office: Roswell Park Meml. Inst., Buffalo 14203.*

WEISS, Lionel Ira, Am. indsl. engr.; b. N.Y.C., Sept. 5, 1923; s. Charles and Beatrice (Horn) W.; B.A., Columbia, 1943, M.A., 1945, Ph.D., 1953; m. Rhoda Storch, June 7, 1946; children—Robert Martin, David Lawrence, Paul Storch. Faculty U. Va., 1949-56, U. Ore., 1956-57; faculty Cornell U., Ithaca, N.Y., 1957—, prof. indsl. engring., 1961—. Mem. Inst. Math. Statistics, Phi Beta Kappa, Sigma Xi. Author: Statistical Decision Theory, 1961; also articles. Constructed sampling plans with optimal properties. Home: 124 Christopher Circle, Ithaca, N.Y. 14850.*

WEISS, Paul Alfred, biologist; b. Vienna, Austria, Mar. 21, 1898; s. Carl S. and Rosalie M. (Kohm) W.; Doctor of Philosophy, University Vienna, 1922; honorary M.D., University Frankfort, 1949; hon. D.Sc., U. Giessen, 1957; Dr. Med. and Surg. (hon.), University Helsinki, 1966; m. Marie Helen Blaschka, Jan. 30, 1926. Came to U. S., 1931, naturalized, 1939. Asst. dir. Biol. Research Inst., Acad. of Sciences, Vienna, 1922-29; Rockefeller fellow, 1927-30, 35, 37; Sterling fellow, Yale, 1931-33; asst. prof. to prof. zoology U. Chgo., 1933-54, chmn. div. biology master's program, 1947-54; head lab. of developmental biology and mem. Rockefeller Inst. Med. Research, N.Y., 1954-64; dean Grad. Sch. Biomed. Scis., U. Tex. at Houston, 1964-66; now prof. emeritus Rockefeller U., Univ. prof. U. Tex., Austin, Distinguished vis. prof. Tex. A &M U., Bryan; Mellon lecturer University of Pittsburgh, 1964; visiting scholar at Pratt Institute, New York, 1964; visiting professor Washington Univ., 1947, U. Frankfort, 1948, U. Neb., 1951, Massachusetts Inst. Tech., 1956-57, N.Y. U., 1960; Hopkins lecturer Stanford University, 1950, University Buffalo, 1967; ofcl. investigator charge Govt. War Research Project, 1942-45; spl. cons., U. S. High Commn., Germany, 1951. Chmn. U. S. delegation to Internat. Union Biol. Scis., 1953-55; chief sci. advisor U. S. commn. gen. Brussels Worlds' Fair, 1957-58; mem. adv. bd. Institute of Basic Research, Mass. Gen. Hosp. 1952-55; mem. sci. adv. bd. Pacific Sci. Center, 1963—; member International Commission on Cell Biology for UNESCO; chmn. div. biology and agr. Nat. Research Council, 1951-55, chmn. biol. council also Merck fellowship bd.; mem. Fulbright Com., 1949-52; trustee Fund for Neurobiology; mem. corp. Marine Biol. Lab., Woods Hole, Mass., Bermuda Biol. Sta.; mem. sci. adv. com. to Pres. U. S., 1958-60; council indsl. research and development Gov. N.Y. State; sci. adv. bd. Seattle 1961 World Fair. Decorated Grand Medal Geoffroy St. Hilaire (France). Mem. Harvey Soc. (president 1962-63), Am. Acad. Arts and Sciences, Am. Assn. for the Advancement of Sci (v.p. 1953), Internat. Inst. Embryology (v.p.), Internat. Soc. Cell Biology (pres. 1964—), Internat. Union Biol. Scis. (chmn. policy bd.), Tissue Culture Assn., Nat. Acad. Scis. (council), Am. Soc. Naturalists, Am. Assn., Anatomists, Am. Soc. Physiologists, Soc. Exptl. Biology and Medicine, Soc. for Study Growth and Devel. (pres. 1941-42), Am. Soc. Zoologists, Royal Swedish, Serbian acads. sci., German Acad. Sci. Leopoldina, Am. Soc. Cell Biology, Am. Assn. Electron-microscopists, Assn. Research Nervous and Mental Diseases, Nat. Acad. Scis. (hon.), Am. Philos. Soc., Phi Beta Kappa (hon.), Sigma Xi. Editor Devel. Biology, others. Research on electron microscopy; developmental biology; theoretical and experimental analysis of growth and differentiation in animals; tissue culture; nerve development; regeneration and wound healing; functional adaptation; coordination of nerve centers. Home: 201 E. 66th St. Office: Rockefeller U., N.Y.C. 10021.

WEISS, Pierre-Ernst, French physicist; b. Mulhouse, France, Mar. 25, 1865; ed. Poly. Inst. Zurich (Switzerland), École Normale Supérieure, Paris; became conf. master Faculty Scis., Rennes, France, 1895, Lyons, France, 1899; named prof. Poly. Inst. Zurich, 1902; apptd. prof. Faculty of Strasbourg (France), 1918, also dir. Physics Inst., Strasbourg; mem. French Acad. Scis., 1926. Author: (with G. Foëx) Le magnetisme, 1926. Research in magnetism; extended theory of paramagnetism to ferromagnetics; discovered heat producing capacity of magnetism; determined Weiss magneton (unit of magnetic moment); advanced theory of magneton of which all magnetic moments are multiples. Died Lyons, Oct. 24, 1940.

WEISS, Raymond Emile, French chemist; b. Bas-Rhin, France, Nov. 8, 1929; s. Louis Henri and Line (Loeffler) W.; M.Sc., U. Strasbourg (France), 1954; Ph.D., U. Nancy (France), 1959; m. Jacqueline Arlen, Aug. 19, 1954. Demonstrator, U. Nancy, 1955-60, asst. prof., 1960-61; faculty U. Strasbourg, 1961—, prof. inorganic chemistry, 1967—. Mem. Chem. Soc. France, French Soc. Mineralogy and Crystallography. Author: Lead—Traetise of Inorganic Chemistry of R. Pascal, 1963; also numerous articles. Research on structural chemistry of coordination compounds. Home: 13 Rue de la Paix, Mundolsheim (67), France. Office: Institute of Chemistry, BP 296/R8, Strasbourg (67), France.*

WEISS, Richard Jerome, Am. physicist; b. N.Y.C., Dec. 14, 1923; s. Morris A. and Anna (Harper) W.; B.S. in Physics, Coll. City N.Y., 1944; M.A. in Physics, U. Cal. at Berkeley, 1947; Ph.D., N.Y. U., 1951. Research fellow Brookhaven Nat. Lab., Upton, N.Y., 1949-54; physicist Army Materials Research Agy., Watertown, Mass., 1951—. Recipient Rockefeller Pub. Service award, 1956. Mem. Am. Phys. Soc. Author: Solid State Physics for Metallurgists,

1964; X-ray Determination of Electron Distributions, 1966; also articles. Determined electron distbn. in solids by X-ray techniques; thermodynamics of metals and alloys; magnetic properties of metals. Home: 4 Lawson St., Avon, Mass. 02322. Office: Materials Research Agy., Watertown, Mass. 02172.*

WEISS, Samuel Bernard, Am. biochemist; b. N.Y.C., May 18, 1926; s. Joseph and Helen (Mendelsohn) W.; B.S., Coll City N.Y., 1948; Ph.D., U. So. Cal., 1948; m. Maria Alda Tereza Cimoli, May 14, 1961; children—David Andre, Mark Adam. Fellow, U. Chgo., 1954-56, Mass. Gen. Hosp., Boston, 1956-57; prof. biochemistry Rockefeller U., N.Y.C., 1957-58; faculty U. Chgo., 1958—, prof. biochemistry, 1963—, dir. biochemistry lab., 1958—. Cons., NIH, 1963—. Recipient Theobold Smith award in med. research Eli Lily Co., 1961, Charles Pfizer award Am. Chem. Soc., 1965. Mem. Am. Soc. Biol. Chemists, Am. Chem. Soc., Sigma Xi. Contbr. articles to tech. jours. Co-discoverer cytidine cofactors for biosynthesis phospholipids; discovered RNA polymerase activity for RNA in mammalian and bacterial extracts, defined its mechanism action; demonstrated presence RNA thiol transferase activity in bacterial extracts which thiolate 5 RNA molecules. Home: 5527 S. Hyde Park Blvd., Chgo. 60637.*

WEISS, Tomás Radil, Czechoslovakian physiologist; b. Bratislava, Czechoslovakia, Nov. 8, 1930; s. Ernest and Margita (Büchlerovó) W.; ed. Faculty Medicine, Charles U., Prague, Czechoslovakia, 1949-55; m. Jirina Radilova, Aug. 14, 1956; children—Gabriela, Michaela. Clin. neurologist, 1956-57; postdoctoral fellow Inst. Physiology, Czechoslovak Acad. Scis., 1957-60; sci. worker Lab. Central Nervous System, Inst. Physiology, Acad. Scis., 1960-63; head Lab. Neurocybernetics, Inst. Physiology, Acad. Scis. 1963—; asst. research anatomist Brain Research Inst., U. Cal. at Los Angeles, 1964; faculty of philosophy Charles U., 1960—, docent psychophysiology, 1967—. Mem. Czechoslovak Med. Soc., Czechoslovak Cybernetic Soc. Author: (with others) New Data on the Brain, 1964; for Psychologists, 1963; also numerous articles, chpt. in textbook. Research on tetanotoxin intoxication, functional interrelationships of different structures of brain, neurophysiology of sleep and arousal, attention and vigility, neurocybernetics, neuronal networks, processing electrophysiol. data by computers. Home: 30 Vratislavova, Prague 2. Office: 1083 Budejovica, Prague 4, Czechoslovakia.*

WEISS, Ulrich, chemist; b. Prague, Czechoslovakia, Jan. 24, 1908; s. Gustav and Margarete (Perutz) W.; Dr.Rer.Nat., German U., Prague, 1930; m. Anna A. Loewi, May 23, 1937; children—Ruth A. (Mrs. Theodore T. Bolliger), Margaret R. Came to U.S., 1941, naturalized, 1946. Chemist Norgine Co., Czechoslovakia, 1930-38, Soc. An. Norgan, Paris, France, 1939-40, Endo Products, Inc., Richmond Hill, N.Y., 1941-51, Tb. Research Lab. USPHS, N.Y.C., 1951-53; research asso. dept. obstetrics and gynecology Columbia U., 1953-54; research chemist N.Y. Bot. Garden, N.Y.C., 1955-57; exec. sec. dental study sect., div. research grants NIH, Bethesda, Md., 1957-58, chemist Nat. Inst. Arthritis and Metabolic Diseases, 1958—. Mem. Am. Chem. Soc., N.Y. Acad. Scis., Chem. Soc. (London), Société Chimique de France. Research, publs. on optical rotary dispersion, pain -relieving drugs related to morphine, intermediates in biosynthesis of aromatic compounds in bacteria, antibiotics and pigments from molds, gold compounds used for treatment of rheumatoid arthritis. Home: 9411 Kingsley Av., Bethesda 20014. Office: Bldg. 2 NIH, Bethesda, Md. 20014.*

WEISS, Volker, metallurgist; b. Rottenmann, Austria, Sept. 2, 1930; s. Othmar and Pauline (Morianz) W.; student U. Tech., Graz, Austria, 1948-51; Dipl. Ing., U. Tech., Vienna, 1953; M.S., Syracuse U., 1955, Ph.D., 1957; m. Peg Hake, Sept. 14, 1957; children—Erick Volker, Christopher John. Came to U.S., 1953, naturalized, 1960. Research metallurgist Demka Steel Factory, Utrecht, Holland, 1952; with Syracuse (N.Y.) U., 1957—, prof. metallurgy, asso. chmn. for metallurgy dept. chem. engring. and metallurgy, chmn. solid state sci. tech. program, 1965—. Metall. cons. Moog Servocontrols, East Aurora,, N.Y., 1957—, Rollway Bearing Co., Syracuse, N.Y., 1958—, Wurttembergische Metallwarenfabrik, Geislingen, Germany, 1961—, United Aircraft Corp., 1966—. Fulbright scholar, 1953. Mem. Am. Soc. Metals, Am. Inst. Mining, Petroleum and Metall. Engrs., German Soc. Metallurgy, Am. Soc. M.E., Am. Soc. Testing Materials, Brit. Iron and Steel Inst., Sigma Xi. Research, numerous publs. on metal fatigue, low cycle fatigue, plastic flow of metals and alloys during phase changes; developed notch analysis of fracture and fatigue crack propagation. Home: 238 Scottholm Terrace, Syracuse, N.Y. 13210.*

WEISSBERGER, Arnold, chemist; b. Chemnitz, Germany, Oct. 19, 1898; s. Eduard and Clara (Ladewig) W.; Ph.D., U. Leipzig, 1924, docent, 1928; fellow Oxford U., 1933-36; m. Louise Harris, Mar. 16, 1938 (dec., 1962); children—Edward, Dorothy. Came to U.S., 1936, naturalized, 1940. Mem. research labs. Eastman Kodak Co., Rochester, N.Y., 1936-64, asst. div. head Color Photography div., 1955-61, asso. div. head, 1961-64, cons. research labs., 1964—; cons. Interscience Pubs. div. John Wiley & Sons, N.Y., London. Fellow Am. Acad. Arts and Scis.; mem. Am. Chem. Soc. (past sect. chmn.), Sigma Xi. Author: Grundriss der Organischen Chemie, 1926; (with E.S. Proskauer) Organic Solvents, 1935. Editor: Technique of Organic Chemistry, 1945—; Chemistry of Heterocyclic Compounds, 1950—; co-editor: Organic Analysis, 1953—, Physics and chemistry of the Organic Solid State, 1963—. Research, numerous patents and publs. in stereochemistry, reaction mechanisms, synthetic organic chemistry, relation of constitution to color and other properties, color photography. Home: 36 Southern Pkwy., Rochester 14618. Office: Research Labs., Eastman Kodak Co., Rochester, N.Y. 14650.*

WEISSBLUTH, Mitchel, physicist; b. Russia, Jan. 7, 1915; s. Elias and Mary (Saltzman) W.; came to U.S., 1922, naturalized, 1928; B.A., Bklyn. Coll., 1936; Ph.D., U. Cal. at Berkeley, 1950; m. Margaret Hochhauser, Feb. 25, 1940; children—Steven, Marc, Thomas. Radio engr. Crosley Radio Co., Cin., 1941-42; sr. research engr. Jet. Propulsion Lab., Pasadena, 1942-46; lectr. U. Cal. at Berkeley, 1950; staff Stanford (Cal.) U., 1950—, asso. prof. applied physics, 1967—, dir. biophysics lab., 1964—; liasion scientist Office Naval Research 1967-68. Recipient Fulbright award, 1960-61. Mem. Am. Phys. Soc., Biophysical Soc., Sigma Xi. Author: (with B. Pullman) Molecular Biophysics, 1965; Quantum Aspects of polypeptides and Polynucleotides, 1964; also articles. Research on physics hemoglobin, Mossbauer resonance in proteins, thermoluminescence in proteins, electron spin resonance, electronic states in metal-organic complexes, short-lived free radicals, radiation therapy, devel. linear accelerators for med. use, meson prodn., rocketry. Home: 820 Pine Hill Rd., Stanford, Cal. 94305.*

WEISSKOPF, Victor Frederick, physicist; b. Vienna, Austria, Sept. 19, 1908; s. Emil and Martha (Gut) W.; Ph.D., U. of Göttingen, Germany, 1931; m. Ellen Tvede, Sept. 5, 1934; children—Thomas Emil, Karen Louise; came to U. S., 1937, naturalized, 1942. Research asso. U. of Copenhagen, Denmark, 1932-34, Inst. of Tech., Zurich, Switzerland, 1934-37; asst. prof. physics, U. of Rochester, N.Y., 1937-43; with Manhattan Project, Los Alamos, N.M., 1943-46; prof. physics, Mass. Inst. Tech., Cambridge, Mass., 1946—; dir. gen. European Center Nuclear Research, Geneva, 1960-65. Recipient Max Planck medal (Germany), 1956. Fellow Am. Phys. Soc.; mem. Nat. Acad. Scis., French Academie des Scis. (corres.). Author: (with J. Blatt) Theoretical Nuclear Physics, 1952. Author articles in science jours. Research on theories of elementary particles, nuclear phenomena, quantum dynamics and electrodynamics, electron theory, nuclear physics. Home: 36 Arlington St., Cambridge, Mass.*

WEISSKOPF, Walter Albert, economist; b. Vienna, Austria, Nov. 14, 1904; s. Emil and Martha (Gut) W.; Dr.J., U. Vienna, 1927; m. Gertrude Rosenfeld, Apr. 29, 1937; 1 son, Martin Charles. Lawyer, Vienna, 1927-38; faculty U. Omaha, 1939-43; prof. Central YMCA Coll., Chgo., 1943-45; panel mem. War Labor Bd., Chgo., 1943-45; prof. econs. Roosevelt U., Chgo., 1945—; prof. Salzburg (Austria) Seminar for Am. Studies, 1952. Mem. Am. Econ. Assn., Assn. for Humanistic Psychology, Econ. History Assn. Author: The Psychology of Economics, 1955. Mem. bd. editors Jour. for Humanistic Psychology. Publs. dealing with the interrelations between economics, psychology, philosophy. Home: 5700 Blackstone Av., Chgo. 60637.*

WEISSLER, Arnold Mervin, Am. physician; b. Bklyn., May 13, 1927; s. Solomon F. and Dora (Hocheiser) W.; B.A. cum laude, N.Y. U., 1948; M.D., N.Y. State U., 1953; m. Gloria Diane Lazarus, June 22, 1953; children—Suzanne Robin, Mark Douglas, Leslie Ann, Jonathan Scott. Resident, Duke, 1954-59, asso. dept. medicine, Duke Hosp., 1959-60; asst. prof. dept. medicine Med. Br. U. Tex., 1960-61; asst. prof. dept. medicine Coll. Medicine, Ohio State U., Columbus 1961-63, asso. prof., 1963-—; USPHS Career Devel. awardee, 1961—, dir. div. cardiology, 1965—. Recipient Dudley Meml. award in surgery, 1953. Fellow A.C.P., Am. Coll. Cardiology; mem. Am., So. socs. clin. investigation, Am. Fedn. Clin. Research, Central Soc. Clin. Research, Assn. U. Cardiologists, Phi Beta Kappa, Sigma Xi, Alpha Omega Alpha. Studies, publs. on circulatory derangement in primary shock, physiology of upright posture on the cardiovascular system; abnormal heart sounds; actions of cardiac glycosides, cardiac function in man. Home: 2095 Sheringham Rd., Columbus O. 43221.*

WEISSLER, Gerhard Ludwig, physicist; b. Eilenburg, Germany, Feb. 20, 1918; s. Otto and Margret (Wendt) W.; B.S., Tech. U., Berlin, Germany, 1938; M.A., U. Cal., Berkeley, 1941, Ph.D., 1942; m. Claire B. Markoff, Aug. 15, 1953; children—Roderick A., Robert Eric, Mark Gregory (dec.). Came to U.S., 1939, naturalized, 1944. Instr. radiologic physics, Med. Center, U. Cal., San Francisco, 1942-44 faculty U. So. Cal., Los Angeles, 1944—, prof., 1952—, head dept. physics, 1951-56. Cons. in radiation physics, vacuum ultraviolet spectroscopy. Fellow Am. Phys. Soc. (chmn. div. electron physics 1964-65)); mem. Am. Bd. Radiology. Research, publs. on interaction of radiation with gases; interrelationships between properties of solid surfaces. Home: 5152 Veronica St., Los Angeles 90008.*

WEISSMAN, Sherman Morton, Am. physician; b. Chgo., Nov. 22, 1930; s. David Harry and Ann (Segal) W.; B.S., Northwestern U., 1950; M.S., U. Chgo., 1951; M.D., Harvard, 1955; m. Myrna Milgram, Feb. 28, 1959; children—Susan, Judith, Sharon, Jonathan. Clin. asso. Nat. Cancer Inst., NIH, Bethesda, Md., 1956-58, sr. investigator, 1960-67; asso. prof of medicine Yale University, 1967—; Nat. Cancer Inst. spl. research fellow U. Glasgow, 1959-60. Mem. Am. Physiol. Soc., Biochem. Soc. (U.K.). Research, publs. on human pyrimidine metabolism, in vivo and in vitro effect of anti-metabolites, abnormalites of hemoglobin synthesis, nucleic acid prodn. in cultured cells, plasma protein turnover in man. Home: 131 W. Park Av., New Haven.*

WEISZ, Paul Burg, physicist; b. Pilsen, Czechoslovakia, July 2, 1919; s. Alexander and Amalia (Sulc) W.; student Tech. U. Berlin, Germany, 1938-39; B.S., Ala. Poly. Inst., 1940; Sc.D., Swiss Fed. Inst. Tech., 1965; m. Rhoda A. M. Burg, Sept. 4, 1943; children—Ingrid B., P. Randall. Came to U.S., 1939, naturalized, 1946. Research asst. U. Berlin Inst. Cosmic Radiation Research, 1938; physicist Bartol Research Found., Swarthmore, Pa., 1940-46; instr. Swarthmore Coll., 1942-43; with Socony Mobil Oil Co. (name later changed to Mobil Research and Devel. Corp.,) 1946—, sr. scientist, Paulsboro, N.J., 1961—. Asso. editor Advances in Catalysis, 1956-58, editor, 1959—; mem. editorial bd. Jour. Catalysis, 1962—. Fellow Am. Phys. Soc.; mem. Am. Chem. Soc., Am. Inst. Chem. Engring., Faraday Soc., German Bunsen Soc., Sigma Xi. Research, numerous patents, publs. on mechanism of Geiger counter discharge action; patentee of Q-gas Geiger counting flow gas; relationship of solid state properties and catalytic properties of solids; diffusional phenomena in solids; polyfunctional catalysis; devel. of molecular shape selective catalysts. Home: 1253 Hunt Club Lane, Media, Pa. 19063. Office: Mobil Oil Corp., Paulsboro, N.J. 08066.*

WEISZFEILER, Jules Gyula, Hungarian microbiologist; b. Brassó, Hungary, July 18, 1902; s. Coloman and Josephine (Rezer) W.; student U. Paris (France), 1920, U. Brussels (Belgium) 1921, U. Jena (Germany), 1922-24; Sc.D., U. Geneva (Switzerland), 1925, M.D., 1928, hon. prof., 1943; m. Valentina Karasseva, May 2, 1952; children—Madeleine, (Mrs. Serge Choumoff). Jacqueline (Mrs. Hansjurg Ringger), Boris, Olga, Victor. Asst., Tb. Sanatory, Bad Lippspringe, 1928-30; asst. Hygiene Inst., Zurich, Switzerland, 1931-32; chief microbiol. dept. Central Tb. Inst., Moscow, USSR, 1932-52; chief microbiol. dept. Tb. Inst., Tashkent, USSR, 1952-57; chief microbiol. dept. Inst. Exptl. Medicine, Budapest, Hungary, 1958-63; dir. Microbiol. Research Group, Hungarian Acad. Scis., Budapest, 1963—; sci. dir. Tb. Inst., Sverdlowsk, 194-43, Inst. Microbiology and Epidemiology, Tashkent, 1954-57. Pres. microbiol., serum and vaccine commn. Sci. Health Council, Budapest, 1959—; mem. internat. com. on bacteriol. nomenclature IAMS. Decorated Order of Labour, Hungary, 1962. Mem. Assn. Microbiologists Hungary, Assn. Microbiologists of Cuba (fgn.), Internat. Soc. Chemotherapy. Editor: Control of Poliomyelitis by Live Poliovirus Vaccine, 1961; Proceedings of the Microbiological Research Group of the Hungarian Academy of Science, 1966. Research and numerous publs. on first demonstration of regeneration of brain and olfactory nerves in Amphibian Urodeles, mutational changes in mycobacteria, conservation of viability of dry BCG vaccine by liophylisation with Leshinskaya, mechanisms and relation of allergy and immunity in Tb.; devel. of new attenuated Tb. vaccine; discovered new facultative pathogenic mycobacteria M. simiae in Macacus monkeys. Home: 72/b Nemetvölgyi ut Budapest XII. Office: 1 Pihenö ut, Budapest XII, Hungary.*

WEITZ, Joseph, Am. psychologist; b. N.Y.C., Feb. 25, 1916; s. Abraham and Emma (Newman) W.; B.S., Carnegie Inst. Tech., 1937; M.A., U. Va., 1938, Ph.D., 1940; m. Anne Stainback, Sept. 29, 1951; 1 son, Wallace Roger. Mem. faculty Tulane, 1940-42, 46-48, Carnegie Inst. Tech., 1948-51; asso. dir. research Life Ins. Agy. Mgmgt. Assn., 1951-60; dir. research Richardson Bellows Henry, Inc., 1960-61; prof. psychology, N.Y. U., N.Y.C., 1961— also cons. Inst. Def. Analysis, Xerox. Mem. several profl. assns. Author: (with H. W. Karn) An Introduction to Psychology, 1955. Research, publs. on chem. basis of cutaneous sensitivity, approach to studying independent variables via criteria; job satisfaction concepts, including selection and tng. Home: 7 Washington News, N.Y.C., 10003.*

WEITZENHOFFER, André Muller, psychologist; b. Paris, France, Jan. 16, 1921; s. Henry Muller and Germaine (Muller) W.; B.S., Mass. Inst. Tech., 1943; M.S. in Math., Brown U., 1944, M.S. in Biology, 1949; Ph.D. (NSF, USPHS fellow) U. Mich., 1956; m. Geneva Ballinger, Aug. 26, 1950; children—Mark David, Janet; m. 2d Mildred Hinson, Mar. 13, 1964. Asst. prof. psychology Stanford, 1957-62; chief research psychologist VA Hosp., asso. prof. med. psychology U. Okla. Med. Center, Oklahoma City, 1962-—; cons. Office Naval Research. Recipient award for best contbn. to sci. hypnotism Soc. Clin. and Exptl. Hypnosis, 1960. Center For Advanced Study in Behavioral Scis. fellow. Fellow Soc. Clin. and Exptl. Hypnosis, Am. Soc. Clin. Hypnosis; mem. Am., Southwestern psychol. assns., N.Y. Acad. Scis., Sigma Xi, Phi Mu Sigma. Author: Hypnotism: An Objective Study in Suggestibility, 1953; General Techniques of Hypnosis, 1957; (with E.R. Hilgard) The Stanford Hypnotic Susceptibility Scale, 1959, The Stanford Profile Scales of Hypnotic Susceptibility, 1963. Publs. on establishing a sci. found. for hypnotic and suggestion phenomena; devel. of three major psychol. measuring instruments, studies in field of measurement, primarily in relation to logical and math. founds. Home: 500 N.W. 41st St., Oklahoma City 73118. Office: VA Hosp., 921 N.E. 13th St., Oklahoma City 73104.*

WEITZMAN, Ellis, Am. psychologist; b. Atlanta, Nov. 26, 1910; s. Nathan and Rachel (Stieffel) W.; A.B., Emory U., 1932; M.A., Creighton U., 1935; Ph.D., U. Neb., 1940; m. Ann Ruth Goldenberg, Aug. 23, 1934; children—Sandra Rae, Warren Ray. Social case investigator Douglas County Relief Adminstrn., Omaha, 1934-36; ednl. dir. Bellevue (Neb.) Vocational Sch., 1936-37; supr. spl. problems occupational analysis sect. USES, 1940-42; faculty Am. U., Washington, 1946-67, prof. psychology, 1950-67, chmn dept. psychology, div. social scis., 1960-67, dir. student personnel univ. exam., 1946-53; vis. prof. Mt. Vernon Jr. Coll., 1956-58; pres. Ellis Weitzman Assos., Inc., 1960-67. Psychol. stress research USPHS, 1961-62, cons. psychometrics heart disease control program, 1962-63; cons. D.C. Dept. Vocational Rehab., 1963-67. Diplomate Am. Bd. Examiners in Profl. Psychology. Fellow A.A.A.S., Am. Psychol. Assn.; mem. Am. Personnel and Guidance Assn., Am. Assn. U. Profs. (past chpt. pres.), D.C. Psychol. Assn., N.Y. Acad. Scis., Interam. Soc. Psychologists, Phi Delta Kappa, Psi Chi. Author: Growing Up Socially, 1949 (with W. J. McNamara) Constructing Classroom Examinations, 1949; Guiding Children's Social Growth, 1951; also articles. Research leading to 1st group test of social maturity; studies on psychol. precursors of heart disease, aptitude testing for industry and counseling of students. Home: 4534 Que Pl. N.W., Washington 20007. Died Aug. 17, 1967.*

WEITZMAN, Stanley Howard, Am. icthyologist; b. Mill Valley, Cal., Mar. 16, 1927; s. John Howard and Iva (Hager) W.; A.B., U. Cal. at Berkeley, 1951, M.A., 1953; Ph.D., Stanford, 1960; m. Marilyn Jean Sohner, Feb. 8, 1948; children—Earl David, Anna Lisa. Sr. lab. technician U. Cal. at Berkeley, 1950-56; instr. Stanford (Cal.) Sch. Medicine, 1957-63; asso. curator fishes U.S. Nat. Mus., Smithsonian Instn., Washington, 1963—. Mem. Am. Fisheries Soc., Am. Soc. Ichthyologists and Herpetologists (editorial bd. 1964—), Soc. for Study Evolution, Soc. Systematic Zoology. Author: (with others) Phyletic Studies Teleastean Fishes with a Provisional Classification of Living Forms, 1966; also numerous articles. Tech. editor Aquarium Jour., 1964-65. Research on anatomy and evolution teleostean fishes specializing in S. Am. fresh-water fishes and oceanic fishes.*

WEITZMANN, Kurt, archaeologist; b. Klein Almerode, Germany, Mar. 7, 1904; s. Wilhelm and Antonie (Keiper) W.; student Univs. Munster, Wurzburg, Vienna, Berlin, 1923-29; Ph.D., U. of Berlin, 1929; D. hon. causa, U. Heidelberg; m. Josepha Fiedler, Jan. 13, 1932. Came to U. S., 1935, naturalized, 1940. Stipend German Archaeolog. Institut, Greece, 1931; with Archaeol. Institut, Berlin, 1932-34; permanent mem. Inst. Advanced Study, Princeton since 1935, asso. prof. art and archaeology, 1945-50, prof. art and archaeology Princeton since 1950; vis. lectr. Yale, 1954-55; vis. prof. Univ. Alexandria, Egypt, 1960; guest prof. U. Bonn, 1962. Member of the bd. scholars Dumbarton Oaks Research Library and Collection, Harvard since 1949. Fellow Med. Acad. Am., German Archaeol. Institut Berlin, Academy of Sciences Goettingen (corr. member); mem. Brit. Acad. (corr.), American Philos. Society, College Art Association, Archaeological Inst. Am., Assn. Internat. des Etudes Byzantines (v.p.). Author: Die Byzantinischen Elfenbeinskulpturen (with Adolph Goldschmidt), 2 vols., 1930-34; Die Armenische Buchmalerei, 1933; Die Byzantinische Buchmalerei des 9 und 10 Jahrhunderts, 1935; Illustrations in Roll and Codex, 1947; The Joshua Roll, 1948; Greek Mythology in Byzantine Art, 1951; The Fresco Cycle of S. Maria Di Castelseprio, 1951; Ancient Book-illumination (Martin Classical Lectures Volume XVI), 1959; Geistige Grundlagen und Wesen der Makedonischen Renaissance, 1963. Editor: The Illustrations in the Manuscripts of the Septuagint 1941—; Studies in Honor of A. M. Friend, Jr., 1955; Expedition Mt. Sinai, 1956, 58, 60, 63, 65. Contbr. articles to profl. publs. Home: 30 Nassau St., Princeton, N.J. 08540.*

WEIZMANN, Chaim, biochemist; b. Motol, nr. Pinsk, Russia, Nov. 27, 1874; s. Reb Oizir and Rachel (Czemerinsky) W.; ed. Technische Hochschule, Darmstadt, also Berlin-Charlottenburg, Germany, U. Freiburg (Switzerland); doctorate, 1900; D.Sc., U. Manchester (Eng.), 1909, LL.D., 1919; m. Vera Chatzman, 1906; children—Benjamin, Michael. Went to Eng., 1904, naturalized, 1910. Lectr. organic chemistry U. Geneva, 1901-04; reader U. Manchester, from 1904; dir. Admiralty Labs., 1916-19; chmn. 1st Zionist Commn. (founded 1918); pres. World Zionist Orgn., 1920-31; 1st pres. of Israel, 1948-52. Author: Trial and Error (autobiography), 1949. Research on synthesis of polycylic substances, prodn. of acetone, butyl alcohol and their derivatives, also devel. of protein foodstuffs, including meat substitutes. Died Tel Aviv, Israel, Nov. 9, 1952.

WEIZSÄCKER, Carl Friedrich von, see von Weizsäcker, Carl Friedrich.

WELBER, Benjamin, physicist; b. Perecin, Czechoslovakia, Nov. 17, 1920; s. Isadore and Hannah (Friedman) W.; B.A., Yeshiva U., 1942; M.A., Columbia U., 1947, Ph.D., 1956; m. Eunice Sybil Scheinbok, June 26, 1949; children—Elihu R., Rachel J., C. Michael. Research staff NASA, Cleve., 1949-60; UNESCO expert Hebrew U., Jerusalem, Israel, 1960-61; sr. staff mem. IBM Research Center, Yorktown Heights, N.Y., 1961—; vis. scientist Clarendon Lab., Oxford (U.K.) U., 1958-59. UNESCO fellow phys. scis., 1960-61. Fellow Am. Phys. Soc.; mem. A.A.A.S., Research Soc. Am., Sigma Xi. Research and publs. on solid state physics, low temperature physics, electron paramagnetic resonance and optics. Home: 182 S. Bedford Rd., Chappaqua, N.Y. 10514. Office: IBM, P.O. Box 218, Yorktown Heights, N.Y. 10598.*

WELCH, Arnold DeMerritt, Am. pharmacologist, biochemist; b. Nottingham, N.H., Nov. 7, 1908; s. Lewis Hiram and Stella May (Batchelder) W.; B.S., U. of Fla., 1930, M.S., 1931; Ph.D., U. of Toronto, 1934; M.D., Washington U., 1939; m. Mary Grace Scott, June 15, 1933; children—Michael Scott, Stephen Anthony, Gwyneth Jeanne; m. 2d Erika Peter, March 15, 1966. Research assistant U. of Fla., 1929-31; fellow in pharmacology U. of Toronto, 1931-35; asst. in pharmacology Washington U., 1935-36, instr., 1936-40; dir. of pharmacol. research Sharp and Dohme, Inc., Phila., 1940-44, dir. of research, 1943-44; prof. pharmacology, dir. dept., sch. medicine Western Res. U., 1944-53; Fulbright sr. research scholar Oxford U., 1952; prof. pharmacology, sch. of medicine Yale, 1953-57, chmn. dept., 1953-66, Eugene Higgins prof. of pharmacology, 1957-66; dir. Squibb Inst. Med. Research, v.p. E. R. Squibb & Sons, Inc., 1966—. Member com. on growth Nat. Research Council (chmn. panel on mech. of action 1946-48, chmn. section on chemotherapy 1948-52, chmn. com., 1952-54); mem. sci. adv. bd. Leonard Wood Meml., 1947-53, Nat. Vitamin Foundation, 1953-56; member division committee biol. and medicine, National Science Foundation 1953-55; member study section on pharmacology and experimental therapeutics for USPHS, 1952-56, 1959-63, chmn., 1960-63, chmn. study sect. on chemotherapy, 1963-65; chairman of panel Pharmacology and Biochem., mem. coordinating com. cancer chemotherapy Nat. Service Center, US PHS, 1955-57; mem. research adv. council Am. Cancer Soc., 1956-59. Recipient alumni awards, U. Fla., 1953, Washington U., 1957; Commonwealth fellow U. Frankfurt, 1964-65; Torald Sollmann award Am. Soc. Pharm., 1966. Mem. Am. Soc. Microbiology, Association American Physicians, Association for Cancer Research. A.A.A.S., American Society Biological Chemistry, Central Soc. Clinical Research, Am. Soc. Hematology, Am. Chem. Soc., Am. Soc. Pharm. and Exptl. Therapeutics, Am. Therapeutic Soc., Soc. Exptl. Biology and Medicine, Biochem. Soc. of Great Britain, Phi Beta Kappa, Sigma Xi, Alpha Omega Alpha, Phi Kappa Phi. Author chapters in books, also papers sci. jours. Asso. editor Cancer Research, 1950-58; Pharm. Revs., 1962-66, Ann. Rev. Pharm., 1966—. Editor for Am. Cont., Biochemical Pharmacology, 1958-62, vice chmn. internat. bd. editors, 1962—. Research on inhibition of enzyme induction; biosynthesis of labile methyl group; cellular localization of pressor amines; utilization of nucleic acid precursors; choline-like compounds; sulfonamides; metabolic approaches to virus and cancer chemotherapy; filariasis. Home: 541 Lake Dr., Princeton, N.J. 08540. Office: Georges Rd., New Brunswick, N.J. 08903.*

WELCH, Billy Edward, Am. physiologist; b. West, Tex., Sept. 16, 1929; s. Perry S. and Elizabeth (Kelley) W.; B.S., Abilline Christian Coll., 1950; M.S., Tex. A. and M. Coll., 1952, Ph.D., 1954; m. Dorothy Poling, Mar. 10, 1956; children—William, Rebecca, Janet, Susan. Physiologist, U.S. Army Nutrition Lab., Fitzsimons Army Hosp., Denver, 1954-57; physiologist Northrop Aircraft Corp., Hawthorne, Cal., 1957-59; physiologist USAF Sch. Aerospace Medicine, Brooks AFB, Tex., 1959, chief Environmental Systems br., 1959—. Co-recipient Gen. Hoyt S. Vandenberg award Arnold Air Soc., 1964. Fellow Aerospace Med. Assn.; mem. A.A.A.S., Am. Physiol. Soc., Sigma Xi. Studies, publs. on effects of long-term exposure to selected spacecraft atmospheres; environmental physiology; selection of spacecraft atmospherse. Office: Environmental Systems Br., USAF Sch. Aerospace Medicine, Brooks AFB, Tex. 78235.*

WELCH, Claude Emerson, Am. physician; b. Stanton, Neb., Mar. 14, 1906; s. John Hayes and Lettie (Phelan) W.; A.B., Doane Coll., 1927; M.A., U. Mo., 1928; M.D., Harvard, 1932; Sc.D., Doane Coll., 1955; m. Phyllis Heath Paton, Apr. 14, 1937; children—Claude Emerson, John Paton. Practice surgery, Boston, 1937—; vis. surgeon Mass. Gen. Hosp., Boston, 1954—; faculty Harvard Med. Sch., Boston, 1937—, clin. prof. surgery, 1963—; cons. surgeon to various hosps. in New Eng. Diplomate Am. Bd. Surgery. Mem. A.C.S. (past gov.; regent), A.M.A., New Eng., Am. So. surg. assns., New Eng. Cancer Soc., Internat. Soc. Surgery (past treas. Am. chpt.), Soc. for Surgery Alimentary Tract (past pres.; trustee), Mass. Med. Soc. (past pres.), Internat. Cardiovascular Soc., James IV Assn. Surgeons. Author: Surgery of Stomach and Duodenum, 1951; Intestinal Obstruction, 1955; Polypoid Lesions of Colon and Rectum, 1965; Advances in Surgery, Vol. I, 1965, Vol. II, 1966; also numerous articles: Numerous clin. studies in surg. procedures and therapy for cancer. Home: 25 Rockmont Rd., Belmont, Mass. 02178. Office: 275 Charles St., Boston 02114.*

WELCH, Henry, Am. bacteriologist; b. Newburyport, Mass., July 16, 1902; s. John Henry and Nellie (McIntire) W.; Ph.B., Brown U., 1925, M.S., 1928, Sc.D., 1954; Ph.D., Western Res. U., 1930; m. Mary Hunt Brownell, Apr. 16, 1933; children—Laura Sumner (Mrs. Bruce R. Hamilton), David Dickey. Research mibrobiologist Conn. State Dept. Health, Hartford, 1930-38; with FDA, U.S. Dept. Health Edn. and Welfare, 1938-60, chief microanalytical div., 1942-45, dir. div. antibiotics, 1945-60; cons. pharmaceuticals, part time 1960—. Recipient Distinguished Service award U.S. Dept. Health Edn. and Welfare, 1955. Fellow Am. Pub. Health Assn.; mem. Am. Assn. Immunologists. Author: (with Charles E. Lewis), Antibiotic Therapy, 1954; Manual of Antibiotics, 1954; A Guide To Antibiotic Therapy, 1959; (with Felix Marti-Ibanex) The Antibiotic Saga, 1960; also numerous articles. Research on safety and efficacy antibiotics in man, pharmacological effects in animals standardization antibiotics and method assay. Home: 4201 N.E. 27th Av., Lighthouse Point, Fla. 33064.*

WELCH, Jack W., Am. physician; b. Lewistown, Ill., Nov. 2, 1918; s. Alan R. and Lee (Hinkle) W.; A.B. cum laude Bradley U., 1941; D.D.S., Northwestern U., 1944; M.D., U. Kan., 1951; m. Jane Chesky, Nov. 18, 1944; children—Susan, Sara, Alan, Joan. Chief oral surgery sect. Tripler Gen. Hosp., U. S. Army, Honolulu, 1946-47; instr. anatomy U. Kan. Med. Sch., 1947-48; staff surgeon Hertzler Clinic, Halstead (Kan.) Hosp., 1957—, dir., 1960-65, chmn. dirs., 1965—. Recipient U. Kan. Guffy award, 1951. Fellow Soc. Head and Neck Surgeons, Internat. Coll. Surgeons, Southwestern Surg. Congress, Acad. Internat. Medicine; mem. N.Y. Acad. Sci., Am. Thyroid Assn., Am. Med. Writers Assn., Am. Soc. Abdominal Surgeons, A.M.A., Harvey County Med. Soc., Kan. State Med. Assn., Am. Assn. Ry. Surgeons, Pi Gamma Mu. Research, numerous publs. on thyroidology, advances in thyroid surgery. Home: 326 Spruce St. Office Hertzler Clinic, 4th and Chestnut Sts., Halstead, Kan. 67056.*

WELCH, John S., Am. physician; b. Lincoln, Neb., Jan. 6, 1922; s. J. Stanley and Jessie (Graves) W.; student U. Neb., 1939-43; B.S., Northwestern U., 1943, B.M., 1946, M.D., 1947; m. Eleanor English, Sept. 19, 1947; children—John Michael, Jane Sarah, Judith Melinda. Fellow in surgery Mayo Found., Rochester, Minn., 1949-53; asst. to staff Mayo Clinic, Rochester, Minn., 1953-55, head sect. surgery, 1955—; instr. surgery U. Minn., 1958-67, asso. prof., 1967—, asst. dir. Mayo Grad. Sch., 1961—. Diplomate Am. Bd. Obstetrics and Gynecology, Am. Bd. Surgery. Mem. Central Assn. Obstetrics and Gynecology, Central Surg. Assn., A.M.A., A.C.S., Internat. Soc. Surgeons, Minn. State Med. Assn. Minn. Surg. Soc., Minn. Obstetrics and Gynecology Soc., Zumbro Valley County Med. Soc. Sigma Xi, Alpha Omega Alpha. Research and publs. on diseases of female pelvis. Home: 1520 5th St. S.W. Office: 200 1st St. S.W., Rochester, Minn. 55901.*

WELCH, William Henry, Am. pathologist; b. Norfolk, Conn., Apr. 8, 1850; s. William Wickham and Emeline (Collin) W.; A.B., Yale, 1870; M.D., Coll. Phys. and Surg. (Columbia), 1875; postgrad. univs. of Strassburg, Leipzig, Breslau, Berlin, 1876-78, 1884-85; numerous hon. degrees. Prof. pathology and anatomy and gen. pathology Bellevue Hosp. Med. Coll., 1879-84; Baxley prof. pathology Johns Hopkins, 1884-1916, dean med. faculty, 1893-98, dir. Sch. Hygiene and Pub. Health, 1916-26, prof. history medicine, 1926-30, emeritus, 1931; pathologist Johns Hopkins Hosp., 1889-1916. Pres. Med. State Bd. Health, 1898-1922, mem. to 1929; mem. Internat. Health Bd. and China Med. Bd., Rockefeller Found., 1900-32. Recipient Gold medal Nat. Inst. Social Scis.; medal of honor U. Vienna; Kober medal, 1927; Harbin Gold medal, 1931. Pres. Med. and Chirurg. Faculty of Md., 1891-92, Congress of Am. Phys. and Surg., 1897, Assn. Am. Physicians, 1901, A.A.-A.S., 1906-07, A.M.A., 1910-11, Nat. Tb Assn., 1910-11 (hon. pres.), Nat. Acad. Scis., 1913-16, Am. Social Hygiene Assn., 1916-19, Nat. Com. Mental Hygiene (hon. pres.), History of Science Soc., 1931; fellow Am. Acad. Arts and Scis., Coll. of Physicians, Phila.; hon. fellow Royal Soc. Medicine, Royal Sanitary Inst., London, Royal Coll. of Phys., Edinburgh, Soc. Med. Officers of Health (Eng.). Author: General Pathology of Fever, 1888; The Biology of Bacteria, Infection and Immunity, 1894; Bacteriology of Surgical Infections, 1895; Thrombosis and Embolism, 1899;

Papers and addresses by William Henry Welch, 3 vols., 1920. Discovered Staphylococcus epidermidis albus (or Micrococcus albus), studied its relation to wound infection, circa 1892; studied pathology of diphtheria; said to have brought scientific medicine to U. S. and to have established modern medical education by his part in the founding of Johns Hopkins Med. Sch., 1893. Discovered causative agt. of gas gangrene (known variously as Bacillus aerogenes capsulatus, Bacillus achalme, Welch bacillus, Bacillus perfringens, Clostridium welchii), 1892. Died Balt., Apr. 30, 1934.

WELCH, Winona Hazel, Am. botanist; b. Goodland, Ind., May 5, 1896; d. Charles Alfred and Carrie Eliza (Johnson) Welch; A.B., DePauw U., 1923; A.M., U. Ill., 1925; Ph.D., Ind. U., 1928. Faculty, DePauw U., Greencastle, Ind., 1930—, prof. botany, 1939-61, head dept. botany, bacteriology, 1956-61, prof. emeritus, 1961—, curator herbarium, 1964—; vis. prof. botany Ind. U., summers 1956, 59, 63; lectr., cons. cryptogamic botany Field Biology Program, Wis. State U., Pigeon Lake Field Sta., summer 1964. Recipient Coulter award Ind. Acad. Sci., 1951, First Outstanding Woman Tchr. award DePauw U., 1964. Fellow A.A.A.S., Ind. Acad. Sci. (pres.); mem. Am. Assn. U. Profs., Am. (pres.), Brit. bryological socs., Bot. Soc. Am., Am. Soc. Plant Taxonomists, Internat. Assn. for Plant Taxonomy, Phi Beta Kappa, Sigma Xi, Sigma Delta Epsilon (nat. pres.). Author: Mosses of Indiana, 1957; A Monograph of the Fontinalaceae, 1960; A Monograph of the Hookeriaceae of North America, 1962; Monograph of The Hookeriacea of Mexico, 1966. Home: 102 W. Poplar St., Greencastle, Ind. 46135.*

WELCKER, Hermann, German anatomist, anthropologist; b. Giessen, Germany, 1822; became prof. anatomy, Halle, Germany, 1859; named dir. anat. inst., 1876; devised method of counting blood globules which Vierordt had conceived; credited with 1st determination of total blood volume and volume of red blood cells, 1858. Died Winterstein, 1897.

WELDEN, Arthur Luna, Am. biologist; b. Birmingham, Ala., Jan. 27, 1927; s. Arthur L. and Marywoodson (Smith) W.; A.B., Birmingham-So. Coll., 1950; M.S., U. Tenn., 1951; Ph.D., State U. Ia., 1954; m. Frances Merkl Colvin, Aug. 19, 1950; children—Charles Woodson, Arthur Frederick. Faculty, Millikin U., 1954-55; faculty Tulane U., New Orleans, 1955—, asso. prof. biology, 1963—. Mem. Mycol. Soc. Am., Am. Soc. Plant Taxonomists, Internat. Assn. for Plant Taxonomy, Assn. for Tropical Biology, Bot. Soc. Mexico. Research and publs. on systematics and morphology of tropical and subtropical fungi (Basidiomycetes), especially wood-rotting forms in Gulf of Mexico-Carribean Sea area. Home: 7826 Willow St., New Orleans 70118.*

WELDON, Walter, English chemist; b. Loughborough, Eng., Oct. 31, 1832; s. Reuben and Esther (Fowke) W.; m. Anne Cotton, Mar. 14, 1854; 3 children. With father's mfg. bus. several years; journalist, contbr. Dial publs., London, from 1854; propr. Weldon's Register Facts, 1860-64; practice exptl. chemistry Walker Chem. Co. works, J. C. Gamble & Co., St. Helens, from 1866. Recipient gold medal Société D Encouragement pour L'Industrie Nationale, Paris, Grand prix Paris Exhbn., 1878. Fellow Royal Soc., 1882; mem. Soc. Chem. Industry (charter, pres. 1883). Developed extraction of copper from ore; devised recovery of manganese peroxide in manufacture chlorine with 1st comml. application, 1869; developed theorems of temperature-atomic volume relationship in compounds formation, exponential relationship of atomic weight of glucinum family are related to ratio of magnesium to glucinum. Died Burstow, Eng., Sept. 20, 1885.

WELDON, Walter Frank Raphael, English zoologist; b. Highgate, London, Mar. 15, 1860; s. Walter and Anne (Cotton) W.; student London U., 1876, Univ. Coll., London, 1876-77, Kind's Coll., London, 1877-78; commoner St. Johns Coll., Cambridge, 1878, exhibitioner, 1879, scholar, 1881, 1st class in natural sci. tripos, 1881; M.A.; m. Florence Tebb, Mar. 13, 1883. Research at zool. sta., Naples, Italy, 1881-82; demonstrator zoology, Cambridge, 1882-84; became fellow St. John's Coll., 1884, univ. lectr. in invertebrate morphology, 1884-91; with Lab. Marine Biol. Assn., Plymouth, 1888-91; Jordell prof. zoology Univ Coll., London, 1891-99; Linacre prof. comparative anatomy, Oxford, 1899-1906. Weldon Meml. prized established at Oxford in his honor, 1907. Fellow Royal Soc. (sec. com. for conducting statis. inquiries into measurable characteristics of plants and animals 1894), 1890. A founder, co-editor Biometrika, 1901-06. One of 1st to use statis. analysis to study variation, inheritance and natural selection; father of biometrics; demonstrated apparently useless characteristic of crabs in correlated with selective death rate and showed it is related to efficient filtration of water in gill chamber; believed Mendelian inheritance is not universal and proposed range with simple Mendelism at one end and blended inheritance at other, 1900-06. Died London, Apr. 13, 1906.

WELKER, Heinrich Johann, German physicist; b. Ingolstadt, Sept. 9, 1912; s. Karl and Bertha (Hecht) W.; Ph.D., U. Munich; dr. of eng. honoris causa, Karlsruhe Tech. Sci.; m. Elfriede Berthold, July 17, 1941; children—Sabina, Cornelia. Asst. to A. Sommerfeld, Munich, 1935-40; work at exptl. wireless telegraphy sta., Grafeling, nr. Munich, 1940-41, at K. Clusius, Munich, 1942-45; lab. chief Westinghouse Soc., Paris, France, 1947-51; sect. head exptl. labs. Siemens-Schuckertwerke, Erlangen, 1951-52, dir., 1963—; hon. prof. U. Munich. Mem. Max Planck Inst. Com., Heidelberg. Author: Lücke im Energiespektrum zur Erklärung der Supraleitung; Germanium als Detektormaterial, 1942; Magnetische Sperrschichten, 1951; III-V-Verbindungen als neue Halbleiter, 1951. Home: Föhrenweg 5. Office: Siemens-Schuckertwerke AG, W. v. Siemens-Strasse 50, Erlangen, West Germany.

WELLER, J(ames) Marvin, Am. geologist; b. Chgo., Aug. 1, 1899; s. Stuart and Harriet Ann (Marvin) W.; B.S., U. Chgo., 1923, Ph.D., 1927; m. Phyllis Vincent Gothwaite, Sept. 28, 1923; 1 dau., Harriet Vincent. Asst. geologist Ill. State Geol. Survey, Urbana, 1916-19, geologist, head dir. stratigraphy and paleontology, 1925-45; exploration in India, Whitehall Petroleum Corp. of London, 1920-22; geologist Chanute Spelter Co., Joplin, Mo., 1923; Ky. Geol. Survey, Frankfort, 1924-25; asst. prof. geology U. Ill., 1936-37; geol. exploration in China and Tibet. Standard Vacuum Oil Co., N.Y.C., 1937-38; prof. invertebrate paleontology U. Chgo., 1945-64, emeritus, 1964—; research asso. Field Mus. of Natural History, 1963—. Vis. prof. U. Tex., 1956; geologist U. S. Geol. Survey, attached to Mut. Security Agy., P.I., 1952-54; mem. com. on stratigraphy NRC, 1932-56, chmn. com. stratigraphic paleontology, 1951-52. Fellow Geol. Soc. Am., Paloentol. Soc.; mem. Am. Assn. Petroleum Geologists, Soc. Econ. Paleontologists and Mineralogists (pres. 1964-65), Phi Beta Kappa, Sigma Xi. Author: Stratigraphic Principles and Practice, 1960; The Record of Life, Fossils and the Course of Evolution, 1968; also tech. articles on geol. of Ill., Ky., and neighboring states, India, China, Philippine Islands, Carboniferous stratigraphy and paleontology, cycles of sedimentation, Carboniferous and Permian trilobites, flourspar deposits, coal geology. Editor Jour. Paleontology, 1942-51. Home: 5735 Blackstone Av. Office: Rosenwald Hall, Univ. of Chicago, Chicago 60637.*

WELLER, John Martin, Am. physician; b. Ann Arbor, Mar. 4, 1919; s. Carl Vernon and Elsie (Huckle) W.; A.B., U. Mich., 1940; M.D., Harvard, 1943; m. Virginia Brigham, Jan. 10, 1942; children—Wendy J., Mary M. Teaching and research fellow in biol. chemistry Harvard Med. Sch., Boston, 1948-50, Alfred Stengel Research fellow, 1948-49, NRC fellow, 1949-50, instr., 1950-52; asst., then jr. asso. Peter Bent Brigham Hosp., Boston, 1950-52; faculty U. Mich., Ann Arbor, 1955—, prof. internal medicine, 1963—. Cons., Ann Arbor VA Hosp., 1955—. Mem. sci. adv. bd. Mich. Kidney Found. Fellow A.C.P.; mem. Soc. Exptl. Biology and Medicine (sec.-treas. Mich. sect. 1965-67), Am. Physiol. Soc., Am. Fedn. Clin. Research, Am. Heart Assn., A.M.A., Central Soc. for Clin. Research. Author: (with James A. Greene, Jr.) Examination of the Urine, A Programmed Text, 1966; also numerous articles. Research on sodium and potassium metabolism, effects of drugs used to treat edema and high blood pressure, renal function, renal disease, treatment of renal failure. Home: 2508 Londonderry Rd., Ann Arbor, Mich. 48104.*

WELLER, Lowell Ernest, Am. chemist; b. Continental, O., Apr. 17, 1923; s. Everett Earl and Della F. (Miller) W.; B.S., Bowling Green State U., 1948; M.S., Mich. State U., 1951, Ph.D. in Chemistry, 1956; m. Eloise Leone Barrick, May 11, 1944; children—Ronald A., Donald E. Faculty, Mich. State U., 1948-57, asst. prof., 1956-57; asso. prof. U. Evansville, 1957-58, prof. chemistry, 1958—, head dept. chemistry, 1957—. Fellow Chem. Soc. (London), A.A.A.S.; mem. Am. Chem. Soc. (past chmn. Ind.-Ky. Border sect.), Ind. Acad. Sci., Sigma Xi. Research and publs. on synthesis, reactions, biosynthesis and biol. activity of indoles and pyrroles, plant growth regulators; organic synthesis and reaction mechanisms. Home: 1617 Bayard Park Dr., St., Evansville, Ind. 47714.*

WELLER, Thomas Huckle, Am. physician; b. Ann Arbor, Mich., June 15, 1915; s. Carl V. and Elsie A. (Huckle) W.; A.B., U. Mich., 1936, M.S., 1937, Doctor of Laws, 1956; Doctor of Medicine, Harvard University, 1940; m. Kathleen R. Fahey, Aug. 18, 1945; children—Peter Fahey, Nancy Kathleen, Robert Andrew, Janet Louise. Teaching fellow bacteriology Harvard, 1940-41, research fellow tropical medicine, pediatrics, 1947-48, instr. comparative pathology, tropical medicine, 1948-49, asst. prof. tropical pub. health, sch. pub. health, 1949-50, asso. prof., 1950-54, Richard Pearson Strong prof. tropical pub. health, head dept., 1954—; intern bacteriology and pathology Children's Hosp., Boston, 1941, intern medicine, 1942, asst. resident medicine, 1946, asst. dir. research div. infectious diseases, 1949-55; mem. commn. parasitic diseases Armed Forces Epilemiol. Bd., dir., 1953-59; in charge parasitology, bacteriology, virology sections Antilles Dept. Med. Lab., P.R. Recipient E. Mead Johnson award for development tissue culture procedures in study virus diseases, Am. Acad. Pediatrics, 1953; Kimble Methodology award, 1954; Nobel prize in physiology and medicine (with J. F. Enders and F. C. Robbins), 1954. Diplomate Am. Bd. Pediatrics. Fellow Am. Acad. Arts and Scis.; mem. Harvey Soc., A.M.A., Am. Soc. Parasitologists, Am. Soc. Tropical Medicine and Hygiene, Royal Soc. Tropical Medicine and Hygiene, Am. Pub. Health Assn., A.A.A.S., Am. Epidemiological Soc., National Academy Scis., Am. Pediatric Soc., Assn. of American Physicians, Society of Experimental Biology and Medicine, Am. Assn. Immunologists, Soc. Pediatric Research, Phi Beta Kappa, Sigma Xi. Author sci. papers. Specialist in study of viruses and parasites; demonstrated common etiology of chicken pox and shingles; (with Neva) 1st propagated German measles virus; developed tissue culture methods in studying viruses; (with Enders and Robbins) discovered that poliomyelitis viruses could grow in non-nervous tissue; studied congenital viruses. Home: 56 Winding River Rd., Needham 92, Mass. Office: 25 Shattuck St., Boston.

WELLES, Samuel Paul, Am. paleontologist; b. Gloucester, Mass., Nov. 9, 1907; s. Paul Irving and Anne L. (Krause) W.; A.B., U. Cal., 1930, Ph.D., 1940; m. Harriet Giles, Jan. 15, 1931 (dec. Jan. 1962); children—Samuel Paul, John Dunning, Ruth Anne; m. 2d, Lydia Ann Lynn, Jan. 25, 1963. Field and lab. asst. U. Cal. Mus. Paleontology, Berkeley, 1931-39, asst. curator reptiles and amphibians, 1939-42, sr. mus. curator, 1942-46, prin. mus. curator, 1946-47, lectr., prin. mus. paleontologist, 1946—, acting dir., 1947-48, mem. expdns., 1931-42, 45—. Fellow Geol. Soc. Am.; mem. Paleontol. Soc. Am., Cal. Acad. Sci., Soc. Vertebrate Paleontology (past pres.). Author: (with W. Fox) Bones to Bodies, 1959; (with others) Bibliographies of Fossil Vertebrates, 1942; also articles. Research on Labyrinthodont amphibians and pre-dinosaurian reptiles from Lower and Middle Triassic of No. Ariz.; elasmosaurid plesiosaurs world; ancestral carniverous dinosaur from Ariz. Jurassic. Home: 982 Santa Barbara Rd., Berkeley, Cal. 94707.*

WELLHAUSEN, Edwin John, Am. plant geneticist; b. Fairfax, Okla., Sept. 10, 1907; s. Henry Robert and Bertha (Rippe) W.; B.S., U. Ida., 1932; Ph.D. in Genetics, Ia. State Coll., 1936; postgrad. (Gen. Edn. Bd. fellow) Rockefeller Inst. for Med. Research-U. Cal. at Berkeley, 1936-37; m. Vivian Siemers, Aug. 8, 1937; 1 dau., Anita (Mrs. Roger C. Heiser). With Mont. State Coll. Agrl. Expt. Sta., 1937-39, W.Va. Agrl. Expt. Sta., 1939-43; with Rockefeller Found., 1943—, local dir. Mexican Agrl. Program, 1952—, in charge Inter-Am. Maize Improvement Program, 1959—, dir. Internat. Maize and Wheat Improvement Center, Mexico City, 1966—. Fellow Am. Soc. Agronomy; mem. A.A.A.S., Am. Phytopathol. Soc., Genetics Soc. Am., Latin Am. Assn. Plant Sci. Research and publs. on genetics of corn and its application to corn improvement through breeding; bacterial pathogens of corn. Home: Sierra Vertientes 614, Mexico 10, D.F., Mexico. Office: Rockefeller Foundation, 111 W. 50th St., N.Y.C. 10020.*

WELLINGS, Sefton Robert, Am. pathologist; b. Tacoma, Oct. 2, 1927; s. Donald and Norah (Atkins) W.; B.S., U. Wash., 1951, M.D., 1953; Ph.D. in Zoology, U. Cal. at Berkeley, 1961; m. Marjorie Helen Plumb, June 15, 1953; children—Anne Katherine, Elizabeth Norah, Julie Virginia, Mary Martha, James Sefton. Faculty, U. Cal. Med. Sch., San Francisco, 1957-61, asst. prof. pathology, 1960-61; faculty U. Ore. Med. Sch., Portland, 1961—, prof., chmn. dept. pathology, 1965—. Mem. Am. Soc. Zoologists, Am. Soc. Exptl. Pathologists, Am. Ornithologist's Union, Internat. Acad. Pathology (mem. council 1966—), Sigma Xi. Research, and numerous publs. on function and structure of mammary gland, skin neoplasia in marine fishes, natural history, etiology, gametogenesis in hydrozoa. Home: 5033 S.W. Humphrey Blvd., Portland, Ore. 97221.*

WELLISZ, Stanislaw, economist; b. Warsaw, Poland, Mar. 28, 1925; s. Leopold and Jadwiga (Landau) W.; B.A., Harvard, 1946, M.A., 1949, Ph.D., 1954; m. Izabella Gajewska, Apr. 28, 1955; children—Tadeusz, Krzysztof. Came to U.S., 1941, naturalized, 1947. With Standard Oil Co. (N.J.), 1946-47, McGraw-Hill Pub. Co., 1948-49, Center for Internat. Studies, Mass. Inst. Tech., 1953-55; faculty dept. bus. econs. U. Chgo. Grad. Sch. Bus., 1957-64; prof. econs. Columbia U., N.Y.C., 1964—. Vis. prof. Warsaw U., 1960; econ. cons. Calcutta Met. Planning Orgn., 1962-63; asso. Joel Dean Assos.,

N.Y.C., 1965—. Mem. Am. Econ. Assn., Econometric Soc., Polish Inst. Arts and Scis. in Am. Author: Economics of the Soviet Bloc, 1964. Research and publs. on econ. planning under diverse instnl. arrangements, capitalism, socialism, mixed econs. Home: 290 Riverside Dr., N.Y.C. 10024.*

WELLS, Bertram Whittier, Am. botanist; b. Troy, O., Mar. 5, 1884; s. Edward Thompson and Maria (Morehouse) W.; A.B., Ohio State U., 1911, M.A., 1916; Ph.D., U. Chgo., 1917; Sc.D. (hon.), N.C. State U., 1963; m. Maude Barnes, Feb. 25, 1941., Knox Coll., 1911-12, Conn. Agr. Coll., 1912-13, Kan. Agr. Coll., 1913-15; prof. botany U. Ark., 1918-19; faculty N.C. State U., Raleigh, 1919-54, prof., head dept. botany, prof. ecology, 1949-54, ret. 1954. Cons., NSF. Fellow A.A.A.S.; mem. N.C. Acad. Sci. (pres. 1933), So. Appalachian Bot. Club (pres. 1947), Bot. Soc. Am., Ecol. Soc. Am. Author: The Natural Gardens of North Carolina, 1932; also articles. Discovered (with I.V. Shunk) role of salt spray in determining form and zonation of seaside plants; discovered evidence to support catastropic meteorite theory of origin of Carolina Bays. Home: Route 1, Wake Forest, N.C. 27587.*

WELLS, Charles Prentiss, Am. mathematician; b. Tabor, Ia., Jan. 23, 1904; s. Ellis Robert and Anna (Clark) W.; B.A., Simpson Coll., 1930; M.S., Ia. State U., 1932, Ph.D., 1935; m. Dorothy Louise Hopson, June 20, 1941; children—Robert L., Michael C., Virginia F. Instr., N.D. State Coll., 1935-38; instr. Mich. State U., 1938-42, asst. prof., 1942-46, asso. prof., 1946-51, prof., 1951—, chmn. math. dept., 1960—; lectr. Brown U., 1948-49; research asso. Cal. Inst. Tech., 1957-58; research U.S. Army Office of Ordnance, 1954-59. Mem. Am. Math. Soc., Math. Assn. Am., Am. Assn. U. Profs., Sigma Xi, Phi Kappa Phi. Author: Differential Equations, 1950. Research and publs. in antenna radiation, especially of spheroidal and biconical antennas; diffraction of electro magnetic waves using Wiener-Hopf techniques; spl. functions especially parabolic functions and partial differential equations. Home: 1036 Beech St., East Lansing, Mich. 48823.*

WELLS, George Philip, English zoologist; b. Sandgate,Kent, Eng., July 17, 1901; s. Herbert George and Amy Catherine (Robbins) W.; B.A., Trinity Coll., Cambridge, 1923, M.A., 1930, Sc.D., 1951; m. Marjorie Stewart Craig (dec.); children—Catherine Anne (Mrs. John Walter Stoye), Oliver Craig. Mem. faculty Univ. Coll., London, 1928—, prof. zoology, 1954—. Fellow Royal Soc., 1955; mem. Soc. Exptl. Biology (hon.), Brit. Mus. (hon. asso.), Zool. Soc. London, Marine Biol. Assn. U.K., Freshwater Biol. Assn. Author: (with H.G. Wells and Julian Huxley) The Science of Life, 1929-30; also numerous articles. Research in invertebrate physiology; devised means of recording activity of marine worms under natural conditions for long periods of time, and demonstrated importance of spontaneously active rhythmic pacemakers in patterning their lives; importance of temperature in controlling geog. distbn. of marine worms. Home: Savile Club, 69 Brook St., London, W.1, Eng.*

WELLS, Harry Warren, Am. astronomer; b. Washington, Jan. 13, 1907; s. Eugene William and Mary Drucilla (Hughes) W.; B.S. in Elec. Engring., U. Md., 1928, E.E., 1937; m. Sarah Lewis Williamson, June 27, 1933; children—Harry Warren, Mary Louisa. Mem. research staff Carnegie Instn., specializing geophysics, upper atmospheric research, communications and radio astronomy, 1932-62; sci. attache Am. embassy, Rio de Janeiro, Brazil, 1960-62; exec. sec., com. polar research Nat. Acad. Scis., 1962—; pioneer research inosphere in Peru and U. S., also spectra of radio stars in Eng. and U. S. Fellow Inst. Radio Engrs.; mem. Internat. Sci. Radio Union, Am. Geophys. Union, Philos. Soc. Washington. Pioneer research inosphere in Peru and U. S., also spectra of radio stars in Eng. and U. S. Home: 3200 Leland St., Chevy Chase 15, Md. Office: Com. Polar Research, Nat. Academy Sciences, Washington 25.*

WELLS, Homer Douglas, Am. plant pathologist; b. Blaine, Ky., Nov. 11, 1923; s. Sam and Rachel (Wheeler) W.; student Ashland Jr. Coll., 1941-43; B.S., U. Ky., 1948, M.S., 1949; Ph.D. in Plant Pathology, N.C. State U., 1954; m. Shirley Marie Arrington, May 9, 1942; 1 dau., Lora Lynette. Asst. agronomist U. Ga. Coastal Plain Expt. Sta., Tifton, 1953—; research plant pathologist forage and range research br. Crops Research div. U.S. Dept. Agr., Tifton, 1953—. Mem. Am. Phytopath. Soc., Assn. So. Agrl. Workers, Sigma Xi. Research and numerous publs. on control of cottony blight disease on golf greens; developed resistance in Lupines to number of diseases. Home: R.F.D. 4, Box 208A. Office: Coastal Plain Experiment Station, Tifton, Ga. 31794.*

WELLS, Horace, Am. dentist, anesthetist; b. Hartford, Vt., Jan. 21, 1815; s. Horace and Betsy (Heath) W.; m. Elizabeth Wales, July 4, 1838; 1 child. Opened dentistry office, Hartford, Conn., 1836; became interested in narcotic effects of nitrous oxide inhalation, circa 1840, believed to have made 1st use of it as dental anesthetic, 1844; claimed to have used ether in extractions, and believed it might be used in major operations (also wrongly believed nitrous oxide superior to ether); 1st printed statement of his claims to discovery of anesthesia appeared in Hartford Courant, 1846; opened an office in N.Y.C.; taught dentistry to William T. G. Morton. Author: An Essay on Teeth; Comprising a Brief Description of Their Formation, Diseases and Proper Treatment, 1838; A History of the Discovery of the Application of Nitrous Oxide Gas, Ether, and Other Vapors, to Surgical Operations, 1847. Experimented with uses of chloroform. Died N.Y.C., Jan. 24, 1848.

WELLS, Ibert Clifton, Am. biochemist; b. Fayette, Mo., Apr. 12, 1921; s. Ira Clifton and Emma (Pulliam) W.; A.B., Central Meth. Coll., Fayette, Mo., 1942; Ph.D., St. Louis U., 1948; m. Katherine JoeAnn Haerle, Aug. 7, 1948; children—Mary Jeanne, Kevin Clifton, Bruce Anthony, Gary Vincent, Mary Leanne, Mary Katherine. Postdoctoral fellow chemistry Cal. Inst. Tech., Pasadena, 1948-50; faculty State U. N.Y., Syracuse, 1950-61; faculty Creighton U. Sch. Medicine, Omaha, 1961—, prof. biochemistry, chmn. dept., 1961—. Recipient Comml. Solvents Corp. award in antibiotics Am. Soc. Bacteriologists, 1953. Mem. Am. Soc. Biol. Chemists, Am. Chem. Soc., Soc. Exptl. Biology and Medicine, Sigma Xi. Research, publs. on antibiotics produced by Psuedomonas aeruginosa; action of choline in metabolism; work on hemoglobin of sickle cell anemia.*

WELLS, John West, Am. paleontologist; b. Phila., July 15, 1907; s. Raymond and Maida (West) W.; A.B., U. Pitts., 1928; M.A., Cornell U., 1930, Ph.D., 1933; m. Elizabeth Baker, Dec. 30, 1932; 1 dau., Ellen B. Faculty, U. Tex., 1929-31, Ohio State U., 1938-48; prof. geology Cornell U., Ithaca, N.Y., 1948—, chmn. dept., 1962-65. NRC fellow, 1931-32, 33-34. Prof., U. Queensland, 1954. Mem. Nat. Acad. Scis., Geol. Soc. Am., Paleontol. Soc. (pres. 1962), Paleontol. Research Inst. (pres. 1960-62), Paleontol. Assn. Numerous publs. on classification of fossil and living scleractinian corals, Cretaceous corals of U. S., recent corals of Marshall Islands, studies of fossil fish, corals and geochronometry. Home: 104 Brook Lane, Ithaca, N.Y. 14850.*

WELLS, Joseph Albert, Am. pharmacologist; b. Wellsville, Mo., Apr. 6, 1916; s. Clyde H. and Gladys (Retallack) W.; B.S., U. Denver, 1937, M.S., 1938; Ph.D., Northwestern U., 1942, M.D., 1946; m. Dorothy Hoar Spindle, Oct. 20, 1940; children—Daniel Clyce, Thomas B., James F., David W. From instr. to prof., chmn. dept. pharmacology Northwestern U. Med. Sch., 1938—, asso. dean, 1965—. Mem. Am. Soc. Pharmacology and Exptl. Therapeutics, A.A.A.S., Soc. Exptl. Biology and Medicine. Research and numerous publs. on autonomic pharmacology; treatment of shock; histamine and antihistaminics. Home: 540 Forest St., Evanston, Ill. 60602. Office: 303 E. Chicago Av., Chgo. 60611.*

WELLS, Patrick Harrington, Am. biologist; b. Palo Alto, Cal., June 19, 1926; s. Harrington and Doris V. (Lausten) W.; A.B., U. Cal. at Santa Barbara, 1948; Ph.D., Stanford, 1951; m. Pearl Marie Pernich, July 27, 1951; children—Harrington, Patricia Ann, John Thomas. Asst. prof. zoology U. Mo., 1951-57; faculty Occidental Coll., Los Angeles, 1957—, asso. prof., 1960—. Fellow A.A.A.S.; mem. Am. Soc. Zoologists, Am. Soc. Mammalogists, Bee Research Assn., Nat. Assn. Biology Tchrs., Western Soc. Naturalists, So. Cal. Acad. Sci., Am. Inst. Biol. Scis., Sigma Xi. Author: (with Harrington Wells) General Biology, 1956; (with H. Wells, W. Muller) Laboratory Exercises in Biology, 1956; also articles. Research on effects, photoreversal ultraviolet injury, physiology of cave animals, flatworms, honey bee communications system. Home: 1873 Campus Rd., Los Angeles 90041.*

WELLS, Roger Clark, Am. chemist; b. Peterboro, N.Y., Oct. 24, 1877; s. Byron Wells and Lucy (Clark) W.; A.B., Harvard, 1901, Ph.D., 1904; m. Etta May Card, Feb. 24, 1914; children—Arthur Byron, Roger Clark. Asst. in chemistry, Harvard, 1902-04, working on atomic weights of sodium and chlorine with T. W. Richards, also instructor, 1904-05; instructor physical chemistry, Univ. of Pennsylvania, 1905-07; research chemist, Gen. Electric Co., 1907-08; physical chemist, U. S. Geol. Survey, 1908-30, mineral resource specialist on sodium compounds, 1917-20, chief chemist since 1930. Examined potash deposits in Chile, 1916; del. 1st Pan-Pacific Scientific Congress, Honolulu, 1920. Mem. Chem. Soc. of Washington (pres. 1921), Am. Chem. Soc., Washington Acad. Sciences, Geol. Soc. Washington (pres. 1937), Am. Inst. Mining and Metal. Engrs., Am. Geophys. Union, Sigma Xi, fellow Am. Mineral Soc., A.A.A.S., Geol. Soc. America. Author: Electric Activity in Ore Deposits, Analyses of Rock and Minerals (U. S. Geol. Survey); Determination of the Common and Rare Alkalis in Mineral Analysis (with R. E. Stevens); over 100 other govt. publs. and scientific papers. Research on sodium compounds; chem. analysis of radioactive minerals; developed nephelometer. Died Apr. 19, 1944.

WELLS, Sir Thomas Spencer, English surgeon; b. St. Albans, Eng., Feb. 3, 1818; s. William and Harriet (Wright) W.; studied under Sadler, at Barnsley, Eng., student, Leeds, Eng., for year, Trinity Coll., Dublin, 1836, St. Thomas' Hosp., London, 1839-41; student pathology, Paris; m. Elizabeth Lucas Wright, 1853; 5 daus.; 1 son, Arthur Spencer. Asst. surgeon in Navy; staff Naval Hosp., Malta, 1841-47; civilian and naval practice medicine; he studied gunshot wounds, Paris, 1848, malaria, Egypt; set up pvt. practice, London, 1853; surgeon Samaritan Free Hosp., 1854-78, became cons. surgeon, 1878; became lectr. surgery Grosvenor Place Sch. Medicine, 1857; surgeon Smyrna, then Renkioi, during Crimean War; returned to Eng., 1856; surgeon to queen's household, 1863-96. Named hon. fellow King's and Queen's colls. Ireland. Fellow Royal Coll. Surgeons Ireland; mem. Royal Coll. Surgeons (became mem. council 1871, Hunterian prof. 1877, v.p. 1879, pres. 1883). Author: The Scale of Medicines with which Merchant Vessels are to be Furnished . . . with Observations on the Means of Preserving the Health and Increasing the Comforts of Seamen, 1851; Practical Observations on Gout and its Complications, 1854; Cancer Cures and Cancer Curers, 1860; Diseases of the Ovaries . . . , 1865; Notebook for Cases of Ovarian and other Abdominal Tumors, 1865; On Ovarian and Uterine Tumours . . . , 1882; Diagnosis and Surgical Treatment of Abdominal Tumours, 1885. Editor, Med. Times and Gazette. Contributed to devel. ovariotomy, surg. technique especially in care, discipline and order; described expression observed in patients with ovarian dsease, 1865; observations on malarial fever, 1850; advocated cremation. Died Cap de Antibes, France, Jan. 31, 1897.

WELLS, Webster, Am. mathematician; b. Boston, Sept. 4, 1851; s. Thomas F. and Sarah M. W.; B.S., Mass. Inst. Tech., 1873; studied civil engring.; m. Emily W. Langdon, June 21, 1876. Instr. math. Mass. Inst. Tech., 1873-80, 82-83, asst. prof., 1883-85, asso. prof., 1885-93, prof., 1893-1911. Author: Elementary Treatise on Logarithms; University Algebra; Plane and Spherical Trigonometry; Academic Algebra; Elements of Geometry; Higher Algebra; Essentials of Trigonometry; Four Place Tables; College Algebra; Academic Arithmetic; Revised Plane and Spherical Trigonometry; Six Place Tables; Essentials of Algebra; Essentials of Plane and Solid Geometry; New Higher Algebra; Complete Trigonometry; New Four Place Tables; Advanced Course in Algebra, 1904; Algebra for Secondary Schools, 1906; Text-Book in Algebra, 1906. Died May 23, 1916.

WELLS, William Charles, physician; b. Charleston, S.C., May 24, 1757; s. Robert and Mary Wells; ed. Dumfries, Scotland; apprentice to Dr. Alexander Garden, Charleston, 1771-75; M.D., U. Edinburgh, 1780; practiced medicine, London, Eng., from 1784; physician Finsbury dispensary, 1789-99; asst. physician St. Thomas's Hosp., 1795-1800, physician, 1800-17. Recipient Rumford medal Royal Soc. for 1st sci. explanation of phenomenon of dew. Fellow Royal Soc., 1793. Contbr. articles to jours. Credited with recognizing that erysipelas is contagious, circa 1800; described cardiac complications of rheumatism, 1812; gave 1st description of hematuria and albuminuria in dropsy, 1812; recognized principle of natural selection, 1913; described concept and significance of dew point, 1814. Died Surry, Eng., Sept. 18, 1817.

WELSH, Harry Lambert, Canadian physicist; b. Aurora, Ont., Can., Mar. 23, 1910; s. Israel and Harriet (Collingwood) W.; B.A., U. Toronto, 1930, M.A., 1931, Ph.D., 1936; postgrad. U. Goettingen (Germany), 1931-33; D.Sc., U. Windsor, 1964; m. Marguerite Hazel Ostrander, June 13, 1942. Faculty, U. Toronto (Ont., Can.), 1935—, prof., 1954—, chmn. dept., 1962—. Fellow Royal Soc. Can. (Henry Marshall Tory medal 1963), Royal Soc. (London), Am. Phys. Soc.; mem. Canadian Assn. Physicists (Gold medal 1961), Royal Astronom. Soc., Societe Francaise de Physique. Research in high resolution Raman effect of gases, pressure-induced infrared absorption, effect of intermolecular forces on spectra. Home: 8 Tally Lane, Willowdale, Ont., Can.*

WELSH, John, meteorologist; b. Boreland, Sept. 27, 1824; s. George Welsh; student Edinburgh, Scotland, 1839-42. Staff, Makerstown Obs., 1842-50, Kew Obs., 1850-59. Fellow Royal Soc., 1857. Contbr. articles to sci. jours. Improved self-recording magnetic instruments. Died May 11, 1859.

WELT, Louis Gordon, Am. physician; b. Elizabeth N.J., Sept. 6, 1913; s. Sigmund and Hattie (Gordon) W.; B.A., N.Y. U., 1934; M.D., Yale, 1938; m. Deborah Cushing Leary, Apr. 25, 1941 (dec. Oct. 1957); m. Mary MacRae Patton, Oct. 14, 1959; children—Robert Gray Sigmund, Frederick Gordon Patton. Asst. prof. Yale U. Sch. Medicine, 1949-52; faculty U. N.C. Sch. Medicine, Chapel Hill, 1952—, prof. medicine, 1954—, chmn. dept. medicine, 1965—. Chmn. medicine test com. Nat. Bd. Med. Examiners, 1964—; mem. adv. council Nat. Kidney Found., 1966—. Diplomate Am. Bd. Internal Medicine. Mem. Am. Heart assn. (exec. com. renal sect. 1962—), Am. Physiol. Soc., Am. Soc. for Clin. Investigation, Assn. Am. Physicians, Soc. Gen. Physiologists, So. Soc. for Clin. Investigation (past pres.), Sigma Xi (pres. N.C. chp. 1958-59). Author: Clinical Disorders of Hydration and Acid-Base Equilibrium, 1955; also numerous articles, chpts. in books, abstracts. Editor: Essays in Metabolism, 1957; (with Maurice B. Strauss) Diseases of the Kidney, 1963. Editorial bd. Annals of Internal Medicine, 1960—, Am. Jour.

Physiology, 1964-—, Jour. Applied Physiology, 1964-—. Research on normal and abnormal physiology kidney, internal and external exchanges electrolytes and water. Home: 614 Morgan Creek Rd., Chapel Hill, N.C. 27514.*

WELWITSCH, Friedrich Martin Joseph, botanist; b. Carinthia, Feb. 5, 1807; M.D., Vienna, Austria, 1836. Lived in Portugal, 1839-53; in charge bot. gardens, Lisbon, also Coimbra, Portugal; travelled in Portuguese W. Africa, 1853-60; met David Livingstone in interior, 1854; lived in London and compared his specimen with Brit. collections, 1863-72. Contbr. articles to sci. jours. Collected Portuguese plants, fungi, algae, molluscs, insects; developed herbarium of tropical plants; discovered numerous plants new to sci. Died Oct. 20, 1872.

WELWOOD, William (Welwod, or Velvod), mathematician; b. Scotland; flourished 1578-1622. Prof. math., St. Andrews, Scotland, 1578-87, prof. law, 1587-97; expelled by royal visitors, 1597; ordered to be replaced by James I, 1600. Author: Gullielmi Velvod aqua in altum per fistulas plumbeas facile exprimenda apologia demonstratina, 1582; Sea-Law of Scotland, 1590; Abridgement of all Sea Lawes, 1613; also 3 Latin treatise, 1594. Discovered principle of siphon, 1577.

WENDELBO, Per Erland Berg, botanist; b. Oslo, Norway, Sept. 19, 1927; s. Per Kr. Lund and Sigrun (Berg) W.; cand. real., U. Oslo, 1953; Dr. philos., U. Bergen (Norway), 1961; m. Ellen Schjölberg, Oct. 13, 1951; children—Rune, Gro, Frede. Curator, Bot. Garden, U. Bergen, 1953-65; dir. Bot. Garden, Gothenburg, Sweden, 1965-—; prof. systematic botany and plant geography U. Gothenburg (Sweden), 1965-—; mem. bot. collecting expdns., W. Pakistan, 1950, Iran, 1959, Afghanistan, 1962. Research and numerous publs. on systematics and plant geography of flora of S.W. Asia especially families Primulaceae and Liliaceae, flora of Norway. Address: 22 Frölundagatan,, Gothenburg SV, Sweden.*

WENDELIN, Godefroi, Flemish astronomer; b. Herken, nr. Liège, Belgium, 1580. Canon of Tournay. Author: Lunar Eclipses observed from 1573 to 1640. Credited with determining parallax of sun; studied moon and its eclipses. Died 1667.

WENDELL, Oliver Clinton, Am. astronomer; b. Dover, N.H., May 7, 1845; s. Oliver Ellsworth and Vienna (Willey) W.; A.B., Bates Coll., 1868, A.M., 1871, D.Sc., 1907; m. Sarah Butler, July 11, 1870. Asst. in Harvard Obs., 1868-69, 79-98, asst. prof., 1898-—civil and hydraulic engr., 1869-79. Fellow Am. Acad. Arts and Scis., A.A.A.S. Contbr. observations and reductions in various vols. Obs. Annals; also calculator of many orbits and ephemerides, and writer of papers of scientific, prose and poetic character. Died Nov. 6, 1912.

WENDEN, Henry Edward, Am. mineralogist; b. N.Y.C., Nov. 24, 1916; s. Henry Edward and Eileen (Crowley) W.; B.S., Yale, 1938; M.A., Harvard, 1950, Ph.D., 1958; m. Beatrice Gertrude Wilson, May 24, 1943; children—Eric Robert, Sylvia Louise, Mark Edward, Elaine Marcia, Rebecca Sue. Faculty, Boston U., 1949-53, Tufts U., 1953-57; prof. mineralogy Ohio State U., Columbus, 1957-—. Cons., Owens-Corning Fiberglas Co., Nat. Cash Register Co., E.I. du Pont de Nemours & Co., Inc., 1960-—. Recipient Alumni award for distinguished teaching Ohio State U., 1963. Fellow Mineral. Soc. Am., Ohio Acad. Sci.; mem. Mineral. Soc. Gt. Britain, Mineral. Soc. Can., Am. Crystallographic Soc., Nat. Assn. Geology Tchrs. (Neil Miner award 1962), Sigma Xi. Author: (with C.S. Hurlbut) The Changing Science of Mineralogy, 1964. Research and publs. on elec. properties, conductivity of crystals. Home: 52 E. South St., Worthington, O. 43085. Office: Dept. Mineralogy, Ohio State U. Columbus, O. 43210.*

WENDER, Mieczyslaw Bogumil, Polish neurologist, neurochemist, b. Poznan, Poland, July 8, 1926; s. Mieczystaw Boleslaw and Kamila (Danielec) W.; M.D., Med. Sch. Poznan, 1951; m. Helena Filipek, Aug. 8, 1958; 1 dau., Ewa. Mem. faculty Med. Sch. of Poznan, 1948-—, dir. dept. neurology, 1963-—; research fellow in neuropathology Inst. Bunge, Antwerp, Belgium, 1956; research asst. in chem. pathology Inst. Neurology, London, 1962. Mem. Polish Neurol. Soc. (sec.), Neurochem. Comm. World Fedn. Neurology (corr. mem.). Research, publs. on biochemistry of developing nervous system, especially problem of myelination in nervous fibers; late effect of prenatal X-irradiation; neurochemistry in relation to neuropath. problems such as exptl. allergic encephalomyelitis and multiple sclerosis. Home: 42 Przybyszewskiego, Poznan, Poland.*

WENDER, Simon Harold, Am. biochemist; b. Dalton, Ga., Sept. 4, 1913; s. Louis David and Ella (Mendel) W.; A.B., Emory U., 1934, M.S., 1935; Ph.D., U. Minn., 1938; m. Ruth Wisenberg, Sept. 6, 1942; children—Joseph Harris, Barbara Ray, Sherri Ann. Research chemist Tex. Agrl. Expt. Sta., College Station, Tex., 1939-41; faculty U. Ky., 1941-46; faculty U. Okla., Norman, 1946-—, prof., 1948-53, research prof. biochemistry, 1953-—, George L. Cross Research professor, 1968-—; visiting research asso. Argonne (Ill.) Nat. Lab., 1954-64; chmn. council Oak Ridge Inst. Nuclear Studies, 1961-64. Bd. dirs. Oak Ridge Asso. Univs., Inc., Mem. Plant Phenolics Group N.Am. (1st nat. pres. 1961-62), Am. Soc. Biol. Chemists, Am. Chem. Soc., A.A.A.S., Am. Soc. Plant Physiologists, Soc. for Exptl. Biology and Medicine, Phi Beta Kappa, Sigma Xi, Phi Lambda Upsilon. Research, numerous publs. and patents on elucidation of role of phenolic compounds in plant and animal metabolism; developed procedures for separation, identification and quantitative analysis of individual phenolic compounds in natural products. Home: 2614 Meadowbrook Dr., Norman, Okla. 73069.*

WENDLANDT, Wesley William, Am. chemist; b. Galesville, Wis., Nov. 20, 1927; s. William Walter and Bernice (Polzin) W.; B.S., Wis. State U., 1950; M.S., State U. Ia., 1952, Ph.D., 1954. Research asso. Argonne Nat. Lab., 1954-55; prof. Tex. Technol. Coll., 1954-66; vis. prof. N.M. Highlands U., 1961; prof., chmn. chemistry dept. U. Houston, 1966-—. Mem. adv. bd. John Wiley-Intersci. Pub. Co., 1965-—; cons. Thiokol, Tech. Equipment Corp. Mem. Am. Chem. Soc., Chem. Soc. London, A.A.A.S., Ia., Tex. Acad. scis., Sigma Xi, Alpha Chi Sigma, Phi Lambda Upsilon. Author: Thermal Methods of Analysis, 1964; (with H. G. Hecht) Reflectance Spectroscopy, 1966; (with J. P. Smith) Thermal Properties of Transition Metal Complexes, 1967. Research and numerous publs. in coordination compounds; thermoanalytical techniques; reflectance spectroscopy; chem. instrumentation; solid-state reactions; chemistry at high pressure and high temperatures. Home: 9900 Memorial St., Houston 77024.*

WENDT, Georg Gerhard, German geneticist; b. Rostock, Meckl, Germany, Apr. 10, 1921; s. Georg and Ilse (Timm) W.; student U. Rostock, 1940, U. Wurzburg, 1941-42, U. Berlin, 1942-44; Dr. Med., U. Prag, 1945; m. Ingeborg Nickel, Mar. 27, 1945; children—Thomas Stephan, Michael. With U. Marburg, Germany, 1952-59, prof., 1959-—, head dept. human genetics Med. Faculty, 1963-—. Mem. World Assn. Neurol. Commns. Research, numerous publs. in anatomy, anthropology and genetics of Papillary ridges, population genetics of Huntington's Chorea, blood and serum groups, diagnosis of paternity. Home: 3 Forsthausstrasse, Cappel üb Marburg 3554. Office: 7 Robert-Koch-Strasse, Marburg/Lahn 355, Germany.*

WENDT, G(eorge) R(ichard), Am. psychologist; b. Rochester, N.Y., Jan. 27, 1906; s. George and Gertrude (Baker) W.; A.B., Univ. of Rochester, 1927; A.M., Columbia, 1928, Ph.D., 1931; m. Ruth Frances Beecher, Aug. 8, 1928; children—Eleanor Beecher, Barbara Beecher, Peter Beecher. Fellow Nat. Research Council, Yale Inst. Human Relations, 1930-32; instr. in physiology, Yale Univ. Sch., of Medicine, 1932-35; instr. in physiology and psychology, Yale, 1935-36; asst. prof. psychology, Univ. of Va., 1936-37, Univ. of Pa., 1937-39; asso. prof. of psychology and chmn. psychology dept., Wesleyan Univ., 1939-45; prof. psychology, Univ. of Rochester, 1945-—, chmn. psychology dept., 1945-60; part time or summer teacher, Columbia Univ., 1935-36, New Haven Women's Coll., 1935-36, Rutgers Univ., 1938-39, Univ. of Calif. at Los Angeles, 1947. Dir. various mil. research projects on selection of aircraft pilots, prevention of motion sickness et al since 1939; consultant U.S.P.H.S. since 1946. Mem. exec. subcom. Nat. Research Council com. on selection and training of aircraft pilots, mem. and sec. subcom. on motion sickness com. on aviation medicine, World War II; mem. research study sect., Nat. Mental Health Council, 1946-52; mem. N.Y. adv. council in psychology, 1957-—; mem. VA area adv. council in psychology, 1957-—. Mem. Am. Assn. Univ. Profs., A.A.A.S., Am. Psychol. Assn., Nat. Inst. of Psychology (sec.-treas.), Soc. Exptl. Psychologists, Eastern Psychol. Assn., New York State Psychological Association, Sigma Xi. Asso. editor Journal of Experimental Psychology 1946-50, of Journal of Comparative and Physiological Psychology, 1946-50. Contbr. articles to psychol. and physiol. journals. Research on hearing, conditioned reflexes, motion sickness and physiological recording methods, effects of drugs on emological recording methods, effects of drugs on emotional social behavior. Home: East Lake Rd., Honeoye, N.Y. Office: U. Rochester, Rochester, N.Y. 14627.

WENDT, Gerald (Louis), Am. chemist; b. Davenport, Ia., Mar. 3, 1891; s. Johannes Heinrich and Dora (Albrecht) W.; grad. high sch., Davenport; A.B., Harvard U., 1913, A.M., 1914, Ph.D., 1916; Sc.D., MacMurray, Jacksonville, Ill., 1958; married Elsie Paula Lerch, Sept. 5, 1916 (div. 1938); 1 son, Robert Louis; m. 2d, Anne D. Powers, Feb. 22, 1947. Jr. chemist, U. S. Bur. Mines, 1916; instr. in chemistry, Rice Inst., Houston, Tex., 1916-17; instr., U. of Chicago, 1917-18, asst. prof. chemistry, 1918-21, asso. prof., 1921-22; asst. dir. research, Standard Oil Co. of Ind., 1922-24; dean of Sch. of Chemistry and Physics, Pa. State Coll., 1924-28; dir. Battelle Memorial Inst., Columbus, O., 1927-28; pres. Coffee Products Corp., 1929-35; dir. Am. Inst. of City of N.Y., 1936-38; dir. science and edn. New York World's Fair, 1938-40; science editor Time, Inc., 1942-45; editorial dir., Science Illustrated, 1946-49; head, div. of teaching and dissemination, Dept. of Natural Scis., United Nations Ednl., Scientific and Cultural Organization, Paris, 1952-54; pres. Nat. Agency for Internat. Publications, 1956-67. Editor Chem. Reviews, 1927-38. Served as captain Research Division, Chem. Warfare Service, U. S. Army, 1918. Fellow A.A.A.S.; mem. American Chemical Society, Phi Beta Kappa, Sigma Xi. Author: Matter and Energy, 1930; Science for the World of Tomorrow, 1939; Chemistry, 1942; The Atomic Age Opens, 1945; Atomic Energy and the Hydrogen Bomb, 1950; You and the Atom, 1955; Prospects of Nuclear Power and Technology, 1957; Atoms for Industry, World Survey, 1960. Editor: The Humanist, 1959-64; editor-in-chief United Nations Conference on Science and Technology, 1963. Known for interest in social consequences of modern sci. Home: 160 E. 48th St., N.Y.C. 10017. Office: 317 E. 34th St., N.Y.C. 10016.

WENGER, Rudolf, Austrian cardiologist; b. Linz, Austria, Sept. 4, 1915; s. Rudolf and Maria (Stallinger) W.; M.D., U. Vienna (Austria), 1938; m. Gisela Rieger, Dec. 24, 1943; children—Brigette, Sylvia, Thomas. Faculty, I Med. U. Clinic, U. Vienna, 1945-—, prof. internal medicine, 1965-—; chief cardiovascular lab. and dietetic dept. I, Med. U. Hosp., Vienna 1949-67; dir. 3d med. dept. Krankenanstalt Rudolfstiftung, Vienna, 1967-—. Mem. expert com. cardiovascular diseases WHO, 1965-—. WHO fellow Postgrad. Med. Sch., U. London (Eng.), 1948; Fulbright fellow Columbia U., N.Y.C., 1955. Mem. Italian Soc. Cardiology (corr.), Deutsche Gesellschaft Inn. Med., Deutsche Gesellschaft Kreislaufforschung. Author: Dietetics, 1955; Vektorkardiography, 1956; Endokardfibrosen, 1964; also numerous articles. Further introduction of vectorcardiography into cardiological clinics; systematic work on endocardial fibrosis; research on arteriosclerosis. Home: and office: 5 Esteplatz A1030, Vienna, Austria.*

WENGERD, Sherman Alexander, Am. geologist; b. Berlin, O., Feb. 17, 1915; s. Allen Stephen and Elizabeth (Miller) W.; A.B., Wooster Coll., 1936; A.M., Harvard, 1938, Ph.D., 1947; m. Florence Margaret Mather, June 12, 1940; children—Anne Marie (Mrs. Michael Riffey), Timothy Mather, Diana Elizabeth, Stephanie Katherine. Prof. geology U. N.M., 1947-—; cons. research geologist, Albuquerque, 1947-—. Recipient Am. Assn. Petroleum Geologists Presdl. award, 1948. Mem. Four Corners Geol. Soc. (pres. 1953), Geol. Soc. Am., Am. Assn. Petroleum Geologist, Am. Soc. Photogram., Soc. Econ Paleontologist and Mineralogist, Am. Assn. U. Profs., Internat. Assn. Sedimentology, Am. Inst. Profl. Geologists (charter; nat. editor 1965-66), Sigma Xi, Sigma Gamma Epsilon, Phi Kappa Phi. Contbr. numerous articles to nat. geol. bulls., chpts. to reference books and encys. Stratigraphic analysis and sedimentology in oil and gas discovery; oceanography; geog. and geol. exploration; hydrogeology and engring. geology. Home: 1044 Stanford Dr. N.E., Albuquerque, 87106.*

WENKERT, Ernest, chemist; b. Vienna, Austria, Oct. 16, 1925; s. Moses and Elcie (Seidmann) W.; came to U. S., 1941, naturalized, 1946; B.S., U. Wash., 1945, M.S., 1947; Ph.D., Harvard, 1951; m. Ann Davis, June 22, 1948; children—Naomi, David, Daniel, Deborah. Faculty, Lower Columbia Jr. Coll., Longview, Wash., 1947-48; faculty Ia. State Coll., 1951-61, prof., 1959-61; prof. Ind. U., Bloomington, 1961-—, Guggenheim fellow, 1965-66; vis. prof. U. Cal., Berkeley, 1961-62; Fulbright lectr. Institut de Chimie des Substances Naturelles, Gif-sur-Ivette, France, 1963; prof., acting head dept. organic chemistry Weizmann Inst. Sci., Rehovoth, Israel, 1964-65. Mem. Am. Chem. Soc., Chem. Soc. (London), Am. Assn. U. Profs., Sigma Xi. Research, numerous publs. in chemistry of natural products, synthesis and elucidation of structures of naturally occurring substances. Home: 711 S. Clifton St., Bloomington, Ind. 47401.*

WENNER, Herbert Allan, Am. physician; b. Drums, Pa., Nov. 14, 1912; s. Herbert C. and Verna (Walp) W.; B.S., Bucknell U., 1929-33; M.D., U. Rochester, 1939; m. Ruth I. Berger, June 27, 1942; children—Peter W., James M., Susan T., Thomas Hauze. Faculty, Yale Sch. Medicine, 1944-46; faculty U. Kan. Med. Sch., Kansas City, 1946-—, research prof. pediatrics, 1951-—. Cons., Kan. Bd. Health, USPHS, Ft. Riley Army Hosp., 1950-—; mem. panel for picornaviruses and panel for respiratory and related viruses NIH, 1960-64. Fellow Am. Pub. Health Assn., Royal Soc. Health, A.A.A.S.; mem. Soc. Pediatric Research, Am. Pediatric Soc., Soc. Exptl. Biology and Medicine, Biometrics Soc., Am. Assn. History Medicine, A.M.A., Am. Epidemiology Soc., Am. Bd. Microbiology,Infectous Diseases Soc. Am., Middle States Pub. Health Assn., N.Y. Acad. Scis., Kan. State, Wyandotte County med. socs., Sigma Xi, Phi Sigma. Author: (with others) Evaluation of the 1954 Field Trial of Poliomyelitis Vaccine Final Report, 1957. Research and numerous publs. on etiology, pathogenesis and epidemiology of infectious disease. Home: 9711 Johnson Dr., Merriam, Kan. 66102. Office: University of Kan. Medical Center, 39th and Rainbow Sts., Kansas City, Kan. 66103.*

WENRICH, David Henry, Am. zoologist; b. Lecompton, Kan., June 29, 1885; s. Christian K. and Frances C. (Lippy) W.; A.B., U. Kan., 1911, A.M., 1912; Ph.D., Harvard, 1915; m. Myra E. Spangler, June 19, 1916; children—Frances Anne (Mrs. Alvah L. Underwood), David Henry. Austin Teaching fellow Harvard, 1912-13, 14-15; teaching asst. Radcliffe

Coll., 1913-14; faculty U. Pa., Phila., 1915—, prof. zoology, 1928-55, prof. emeritus, 1955—. Fellow A.A.A.S., N.Y. Acad. Sci.; mem. Soc, Protozoologists (past pres., hon. mem. 1956), Nat. Acad. Sci. (hon. asso.). Author: Rambling Rhymes and Whimisical Jingles, 1954. Editor: (with Ivey F. Lewis, John R. Raper) Sex in Microorganisms, 1954. Contbr. numerous articles to tech. jours. Research on individuality and behavior chromosomes in spermatogenesis; structure, behavior, geog. distbn. certain ciliates; host relations between parasitic protoza and their hosts; morphology and host relations, intestinal protozoa. Home: 1900 Rittenhouse Sq., Phila. 19103.*

WENT, Arthur Edward James, Brit. zoologist; b. London, Eng., Jan. 21, 1910; s. Edward Benjamin and Louisa (Gurney) W.; B.Sc., London, 1934, Ph.D., 1941, D.Sc., 1948; m. Phyllis Nelly Howell, Apr. 2, 1938; children—Janice, David. With fisheries div. Irish Dept. Agr. and Fisheries, Dublin, 1936—, sci. adviser on fisheries, 1948—; dir. research Salmon Research Trust Ireland, 1955—. Pres., AnTaisce, Nat. Trust for Ireland, 1967—; pres. Internat. Council for Exploration Sea, 1966—. Fellow Inst. Biology; mem. Royal Irish Acad., Inst. Biology Ireland. Author: Irish Salmon and Salmon Fisheries, 1956; also numerous articles. Research in life history Irish Salmon, methods of fishing in Irish waters through ages. Home: Merton 46 Castlepark Rd., Sandycove, County Dublin, Ireland. Office: 3 Cathal Brugha St., Dublin 1, Ireland.*

WENT, Frits W(armolt), botanist; b. Utrecht, Holland, May 18, 1903; s. Friedrich August Ferdinand Christian and Catharina Jacomina (Tonckens) W.; student U. Utrecht, 1920-27 (A.B., A.M., Ph.D.); Ph.D., U. Paris; Sc.D. (hon.), McGill U., 1959; m. Catharina Helena van de Koppel, Nov. 1927; children—Hans Adriaan, Anna Catharina. Came to U. S., 1933. Asst. in plant taxonomy, U. Utrecht, 1923-25, asst. in gen. botany, 1925-27; botanist Bot. Gardens, Java, 1928-30, dir. Foreigners Lab., 1930-32; asst. prof. plant physiology Cal. Inst. Tech., 1933-35, prof., 1935-58; dir. Mo. Bot. Garden, 1958-63; prof. botany Washington U., St. Louis, 1958-65; Distinguished prof. botany, desert research inst. U. Nev., Reno, 1965—. Mem. Bot. Soc. Am., Ecol. Soc. Am., Am. Soc. Hort. Scis., Am. Soc. Plant Physiology, Western Soc. Naturalists. Nat. Acad. Scis. (Washington), Bot. Soc. Brazil (hon.), Deutsch Bot. Gesellschaft (hon.), Royal Dutch Acad. Sci., Amsterdam (corr.), French Acad. Scis. (corr.), Royal Belgian Acad. (corr.), Sigma Xi. Author: Phytohormones (with K. V. Thimann). 1937; Experimental Control of Plant Growth, 1957; The Plants, 1964. Contbr. profl. jours. Research on role of plant hormones in growth and tropisms, effects of climate on plants, ecology of desert plants, plant emanations as source of atmospheric hazes. Address: 3450 Lodestar Lane, Reno 89508.*

WENTORF, Robert Henry, Jr., Am. phys. chemist; b. West Bend, Wis., May 28, 1926; s. Robert Henry and Sophia (Rusch) W.; B.S., U. Wis., 1948, Ph.D., 1952; m. Vivian Marty, Aug. 20, 1949; children—Jill, Laine, Rolf. Research asso. Gen. Elec. Research Lab., Schenectady, 1952—. Gooch lectr. Yale, 1957; Brittingham vis. prof. U. Wis., 1966-67. Recipient Ipatieff prize, 1965. Mem. Am. Chem. Soc., Am. Geol. Soc., Sigma Xi, Alpha Chi Sigma, Tau Beta Pi. Editor: Modern Very High Pressure Research, 1962. Contbr. articles to sci. jours. Mem. team which synthesized diamonds by high pressure techniques; subsequent high-pressure work has yielded other materials including semiconducting diamonds and a new substance, cubic boron nitride, about as hard as diamond. Home: 2521 Hilltop Rd., Schenectady 12309. Office: Gen. Elec. Research and Devel. Center, Schenectady 12301.*

WENTWORTH, George Albert, Am. mathematician; b. Wakefield, N.H., July 31, 1835; s. Edmund and Eliza (Lang) W.; prep. Phillips Exeter Acad.; grad. Harvard, 1858; m. Emily J. Hatch, Aug. 8, 1864. Prof. math. Phillips Exeter Acad., 1858-91. Author: Wentworth Series of Mathematical Works (about 40 works on arithmetic, algebra, geometry, trigonometry); also (with G. A. Hill) of Wentworth & Hill Series of exercise books in algebra, arithmetic and geometry; also physics. Died 1906.

WENTZEL, Donat Gotthard, astrophysicist; b. Zürich, Switzerland, June 25, 1934; s. Gregor and Anna (Wielich) W.; B.A., U. Chgo., 1954, B.S., 1955, M.S., 1956; Ph.D., 1960; postgrad. U. Leiden (Holland), 1957-58; m. Maria Mayer, Mar. 21, 1959; 1 dau., Tania Maria. Came to U.S., 1948, naturalized, 1954. Faculty, U. Mich., Ann Arbor, 1960-67, asso. prof., 1964-67; asso. prof. U. Md., College Park, 1967—. Alfred P. Sloan Research fellow, 1962-66. Mem. Am. Astron. Soc., Internat. Astron. Union, Am. Geophys. Union, A.A.A.S., Am. Assn. U. Profs., Sigma Xi. Research and publs. on magnetic fields as they affect interstellar medium, sun, energetic charged particles. Office: Dept. Physics and Astronomy, U. Md., College Park, Md. 20742.*

WENTZEL, Gregor, physicist; b. Dusseldorf, Germany, Feb. 17, 1898; s. Josef and Anna (Joesten) W.; Ph.D., U. Munich (Germany), 1921; m. Anna Wielich, Dec. 5, 1929; 1 son, Donat G. Prof. physics U. Leipzig (Germany), 1926-28. U. Zurich (Switzerland), 1928-48, U. Chgo., 1948—. Fellow Am. Phys. Soc.; mem. Schweizerische Naturforschende Gesellschaft, Schweizerische Physikalische Gesellschaft, Nat. Acad. Scis. Author: Quantentheorie der Wellenfelder, 1943, English transl. (Quantum Theory of Fields), 1949. Research on theory of atomic spectra, wave mechanics, quantum electrodynamics, meson field theories, statis. mechanics of many-body problems, particularly superconductivity. Address: U. Chgo., Chgo. 60637.*

WENZEL, Carl, physician; b. Germany, 1769; Author: Über die Krankheiten des Utures, 1816; Allemeine geburtshülfliche Betrachtungen und über die künstliche Frühgeburt, 1818. Credited with being 1st to induce premature labor as part of ethical med. practice, 1804. Died 1827.

WENZEL, Carl Friedrich, German chemist; b. Dresden, Germany, circa 1740; studied surgery and chemistry, Amsterdam, Netherlands; studied chemistry, Leipzig, Germany, 1766; ship's surgeon in Dutch service; became chemist to Freiburg founderies, 1780; apptd. assesor to superintending bd. of founderies, 1785; metall. chemist, also in service of Meissner porcelain factory, 1786-93. Author: Einleitung zur höhern Chemie, 1774; Lehre von der Verwandtschaft der Körper, 3 vols., 1777-82; Chymische Untersuchung des Flussspaths, 1783. Discovered that degree of solubility of metals in acids is parallel to acid concentration, also that speed of chem. reactions is generally proportional to concentration of the substances (mass effect law); determined with great precision weights of numerous compounds, thus may be considered founder of stoichiometry as well as J. B. Richter; studied number relationships by union of acids and bases. Died Freiburg, Germany, Feb. 26, 1793.

WENZEL, Duane Greve, Am. pharmacologist; b. Wausau, Wis., Sept. 18, 1920; s. Rudolph Henry and Flora Leona (Greve) W.; B.S. in Pharmacy, U. Wis., 1945, Ph.D., 1948; m. Mary Ann Waltmire, Apr. 18, 1943; children—Bruce Duane, Richard Hyde, Scott Greve, Judd Bradford. Mem. faculty Sch. Pharmacy, U. Kan., 1948—, prof., 1956—, dean, 1964—. Vis. lectr. NSF-Am. Assn. Colls. Pharmacy, 1964. Mem Am. Assn. Coll. Pharmacy (exec. com. 1964—), Am., Kan. (librarian, exec. com.) pharm. assns., A.A.A.S., N.Y. Acad. Sci., Am. Heart Assn. Research on effects of nicotine on cardiovascular system, with spl. reference to heart disease. Home: 306 Parkhill Terrace, Lawrence, Kan. 66044.

WEPFER, Johann Jakob, Swiss physician; b. Schaffhausen, Dec. 23, 1620; city physician, Schaffhausen; personal physician to several princes. Author: Observationes anatomicae ex cadaveris eorum, quos sustulit apoplexia, 1658; Cicutae aquaticae historia et noxae commentario illustrata, 1679. Used autopsies and expts. on animals in med. research; devised method of injecting blood vessels with colored liquid; added to knowledge of brain and intestines; gave classical description of poison by water hemlock. Died Jan. 26, 1695.

WEPMAN, Joseph M., Am. psychologist; b. Copemish, Mich., Dec. 25, 1907; s. Benjamin and Jennie (Morris) W.; A.B., Western Mich. U., 1931; Ph.M., U. Wis., 1934; Ph.D., U. Chgo., 1948; m. Ruth P. Rolde, Jan. 3, 1933. Faculty, U. Chgo., 1936—, now prof. psychology. Fellow Am. Psychol. Assn., Am. Speech and Hearing Assn.; mem. Sigma Xi. Author: Recovery from Aphasia, 1951; Psychologist as a Witness, 1964. Co-editor: (with R. Heine), Concept of Personality, 1964. Developer psycholinguistic concept of aphasic lang. disorders, assessment of devices for lang. disorders and for detecting change in devel. of lang. Home: 300 N. State St., Chgo. 60610.*

WERBOFF, Jack, Am. psychologist; b. Bklyn., Apr. 24, 1928; s. Harry and Fannie (Davis) W.; B.A., Bklyn. Coll., 1949; M.A., Columbia U., 1950; postgrad. U. Minn., 1950-51; Ph.D., Washington U., St. Louis, 1957; m. Elaine Betz, Sept. 29, 1951; children—Judith, Susan, Barbara, Neal, Ann. Instr., Fla. State U., 1952-54; clin. psychologist Eglin AFB, Fla., 1952-54; research asst. Washington U., 1954-57; mem. faculty Wayne State U., Detroit, 1957-63, asso. prof. psychology, 1962-63; mem. faculty Lafayette Clinic, Detroit, 1957-63, head animal behavior lab., 1959-63; staff scientist Jackson Lab., Bar Harbor, Me., 1963—; lcetr. U. Me., 1964—. Fellow Am. Psychol. Assn.; mem. A.A.A.S., Teratology Soc., Soc. Research in Child Devel., Soc. Psychophysiol. Research, Radiation Research Soc., Animal Behavior Soc., Am. Care Panel, Psychonomic Soc., Behavorial Pharmacology Soc., Midwestern Psychol. Assn., Sigma Xi.. Research and publs. on devel. of field of behavioral teratology; studies of prenatal, genetic, and early experience factors affecting later behavior; studies of neonatal learning. Home: 11 Atlantic Av., Bar Harbor. Office: Jackson Lab., Bar Harbor, Me. 04609.*

WERGELAND, Harald Nicolai, Norwegian physicist; b. Norderhov, Norway, Mar. 14, 1912; s. Nicolai and Ebba (Wein) W.; Diploma in Chemistry, Trondheim (Norway) Inst. Tech., 1936; Ph.D. in Physics, Oslo (Norway), 1941; m. Hedvig Louise Erding, Aug. 21, 1937; children—Nicolai, Ebba. Tchr. various instns., 1936-41, 45-46; prof. physics Trondheim Inst. Tech., 1946-48, 49—; vis. prof., then asso. prof. Purdue U., Lafayette, Ind., 1948-49. Chmn., Seminar Theoretical Physics, Trondheim, 1964—. Mem. Royal Norwegian Acad. Scis. Trondheim (past pres.), Soc. for Social Responsibility in Scis. Author: (with D. ter Haar) Elements of Thermodynamics, 1966; also numerous articles. First to compute correctly pressure exerted by sound waves; contributed to theory of Brownian motion; studied radiation from fast electrons. Home: 5/Prost Castbergs Vei, Trondheim, Norway.*

WERLE, Eugen, German biochemist; b. Kaiserslautern, Oct. 10, 1902; s. Heinrich and Philippine (Wasem) W.; Ph.D., U. Munich, 1928, M.D., 1947; m. Anita Koenig, Mar. 19, 1943; children—Rolf, Martin, Klaus. Asst., Kaiser Wilhelm Inst. Work Physiology, Berlin, Dortmund, 1928-30; faculty Med. Acad., Düsseldorf, 1931-43; asso. prof. U. Munich, West Germany, 1947-59, prof. clin. chemistry, biochemistry, 1966—. Mem. Physiol. Clin. Soc., German Chem. Soc. Author: Kallikrein-Padutin, 1950; also numerous articles. Research on and discovery of enzymes in blood plasma during pregnancy, amino acid decarboxylases and aminoxydases, nicotine enzyme in liver, kallidin and kininases, colostrokinin; studies on kallikrein and trypsin, protease-inhibitors. Address: 6 Rümelinstrasse, Munich, West Germany.*

WERLHOF, Paul Gottlieb, German physician; b. Helmstedt, Germany, Mar. 24, 1699; practiced medicine, Hanover, Germany. Fellow Royal Soc., 1735. Author: Disquistio medica et philologica de variolis et anthracibus, 1735. Gave early description of thrombopenic purpura (essential thrombopenia, Werlhof's disease), 1735; described a bloodstain disease (morbus maculosus Werlhof). Died Hanover, July 26, 1767.

WERMAN, Robert, Am. neurologist; b. Bklyn., May 2, 1929; s. Sidney and Rose (Sutton) W.; A.B., N.Y. U., 1948, M.D., 1952; m. Golda Spiera, Feb. 14, 1954; children—Michael, Aaron, Rachel M. vis. scientist Columbia U., 1958-60; asst. prof., 1960-61; vis. scientist Cambridge (Eng.) U., 1960-61; faculty Ind. U. Med. Sch., Indpls., 1961—, prof. psychiatry, 1961—, prof. anatomy and physiology, 1964—. Recipient USPHS Career Devel. award, 1965. Research and publs. on pantopaque ventriculography, effects of intracarotid drugs in man, eye movement control, membrane physiology, synaptic psysiology. Home: 1123 E. 1st St., Bloomington, Ind. 47401. Office: 1100 W. Michigan St., Indpls. 46207.*

WERNER, Abraham Gottlob, German geologist; b. Wehrau, Silesia, Sept. 25, 1750; student mining, Freiberg, Germany, law and mineralogy, Leipzig, Germany. Lectr. mining and metallurgy Freiberg Sch. Mines, 1775-1817. Mem. French Acad. Scis., 1804. Author: Über die äussern Kennzeichen der Fossilien, 1774; Kurse Klassifikation und Beschreibung der Gebirgsarten, 1787; Neue Theorie über die Entstehung der Gänge, 1791. Established separation of geology from mineralogy; studies on classification, especially on rocks of Harz mountains; supported Neptunian sch. geol. theory which believed rocks and geol. formations were caused by sedimentation occuring in the past when the entire surface of the earth was covered with water and the only volcanic activity is of recent origin. Died Dresden, Germany, June 30, 1817.

WERNER, Alfred, Swiss chemist; b. Mülhausen, Alsace, Dec. 12, 1866; ed. univs. Karlsruch, Zurich; doctorate, 1890; worked with Berthelot in Paris; asst., Zurich Polytech. Sch., 1889; taught at Coll. de France, Paris; prof. chemistry U. Zurich, Switzerland, 1893-1915; head Chem. Inst., Zurich, 1909-15. Recipient Nobel prize in chemistry (for work on linkage of atoms in molecules), 1913. Author: Lehrbuch der Stereochemie, 1904; Neuere Anschauungen auf den Geibiet der anorganische Chemie, 1905; Über die Konstitution und Konfiguration von Verbindungen höhere Ordnung, 1914; and over 150 articles. Research on structure of complex inorganic compounds (cobalt, chromium, plantinum), stereochemistry of nitrogen compounds, isomeric forms of oximes; explained isomerism and made possible a classification of inorganic compounds by his coordination theory of valency, 1893; separated chlordiethylene ammines of cobalt into optically active components, thus introducing stereoisomerism into inorganic chemistry, 1911. Died Zurich, Nov. 15, 1919.

WERNER, Floyd Gerald, Am. entomologist; b. Ottawa, Ill., June 1, 1921; s. Frank and Edith (Mumper) W.; B.S., Harvard, 1943, Ph.D., 1950; m. Frances Ruth Watson, June 28, 1952; children—Susan Mae, William Edward, John David. Asst. prof. zoology U. Vt., 1950-53, asso. prof., 1953-54; asst. prof. entomology U. Ariz., Tucson, 1954-58, asso. prof., 1958-60, prof., 1960—, asst. entomologist, 1954-58, asso. entomologist, 1958-60, entomologist, 1960—. Mem. A.A.A.S., Entomol. Soc. Am., Entomol. Soc. Washington, Cambridge Entomol. Club, Soc. Systematic Zoology, Ariz. Acad. Sci. Studies, numerous publs. on classification of beetle families, Meloidae and Anthicidae, and econ. insects of Southwestern U.S.*

WERNER, Gerhard, pharmacologist; b. Vienna, Austria, Sept. 28, 1921; s. Rudolf and Elizabeth (Lukas) W.; M.D., U. Vienna, 1945; m. Marion E.

Hollander, July 24, 1958; children—Philip Ralph, Karen Nicole. Came to U.S., 1957, naturalized, 1965. With dept. pharmacology U. Vienna, 1945-50; prof. pharmacology, head dept. U. Calcutta (India) Sch. Tropical Medicine, 1952-54; prof. pharmacology, head dept. U. Sao Paulo, Med. Sch. of Ribeirao Preto, Brazil, 1955-57; asso. prof pharmacology Med. Coll., Cornell U., N.Y.C., 1957-61; asso. prof. pharmacology and physiology Johns Hopkins, Balt., 1963-65; prof. pharmacology, head dept. U. Pitts., 1965—. Mem. Am. Soc. Pharmacology and Exptl. Therapy, Harvey So. Mem. editorial bd. Jour. Clin. Pharmacology, Internat. Jour. Neuropharmacology. Research and publs. on drug actions on peripheral and central synapses and on sensory systems; information processing in sensory systems. Office: Dept. Pharmacology, U. Pitts., Pitts. 15213.*

WERNER, Henry James, Am. histo-cytologist, b. Hartford, Wis., Oct. 9, 1913; s. J. Fred and Emile (Beck) W.; B.S., Marquette U., 1939, M.S., 1941; postgrad. Harvard, 1944-45; Ph.D., U. Md., 1947; m. Sue Belle Hill, Apr. 5, 1947. Head dept. sci. Shenandoah Coll., 1941-44; biology teaching fellow Harvard, 1944-45; faculty U. Md., 1947-48, Otterbein Coll., 1948; faculty La. State U., Baton Rouge, 1949—, prof. histo-cytology, 1964—. Mem. Am. Soc. for Cell Biology, La Soc. Electron Microscopy (past pres.), La. Acad. Sci., Sigma Xi, Phi Kappa Phi, Phi Sigma, Alpha Epsilon Delta, Sigma Zeta. Author: Synopsis of Histology, 1961. Research and publs. on light and electron microscopic observations of various glands of body, electron microscopic aspects of human superficial pathogenic fungi. Home: 872 Nelson Dr., Baton Rouge 70803.*

WERNER, Johann, German mathematician; b. Germany, 1486; priest, Nuremberg, Germany. Author: Commentarius . . . quod cubi duplicatio dicitur, Libellus super vigintiduobus elementis conicis (1st original study of conics written in West), 1522. Studied conic sects. in relation to cone, from which he directly derived their properties; discovered prostapheresis (similar to logarithms); made astron. and meteorol. observations; recommended use of moon's eclipse for determinations of longitude. Died 1528.

WERNER, Rolfinck, German physician, chemist; b. Hamburg, Germany, Nov. 15, 1599; M.D., Padua, Italy, 1625. Became 1st prof. chemistry, Jena, Germany, 1641; prof. anatomy and surgery, Wittenburg, Germany, also Jena; held pub. dissections, twice 1629. Author: De objecto chymiae, 1637; De metallis imperfectis et mollibus, 1638; Dissertationes chymicae sex, 1660; Chymiae in artis forman reducta, 1661; Scrutimium chymicum vitrioli, 1666; De mimera martis, 1668; Ordo et methodus medicinae, 1669; De vegetabilibus, 1670. Founder chem. lab., Jena. Studies on cataracts in lens of eye; 1st to use cadavers in anat. lectures; discredited alchemical ideas. Died Jena, May 6, 1673.

WERNICKE, Carl, neuropsychiatrist; b. Tarnowitz, May 15, 1848; prof., Berlin, Breslau (now Wroclaw, Poland), Halle, Germany. Author: Der aphasische Symptomenkomplex, 1874; Lehrbuch der Gehirnkrankheiten, 2 vols., 1881-83; Atlas des Gehirns, 1897-1900; Discovered sensory speech center of brain; described sensory aphasia (Wernicke's aphasia), 1874; described spastic hemiplegia of extremities in which some muscle groups are affected more than others (Wernicke-Mann type of palsy or paralysis), 1889. Died June 13, 1905.

WERTHEIM, Gunther Klaus, physicist; b. Berlin, Germany, Feb. 26, 1927; s. Frederick M. and Elisabeth A. (Seiffert) W.; came to U.S., 1939, naturalized, 1945; M.E., Stevens Inst. Tech., 1951; A.M., Harvard, 1952, Ph.D., 1955; m. Lee Hilles, Sept. 8, 1956; children—Elisabeth Lee, Frederick Whiley, Charles Gunther. Research asso. Brookhaven Nat. Lab., 1955; staff Bell Telephone Labs., Murray Hill, N.J., 1955—, head crystal physics research dept., 1962—; adj. prof. physics Stevens Inst. Physics, Hoboken, N.J., 1965. Fellow Am. Phys. Soc.; mem. Research Soc. Am., A.A.A.S., Sigma Xi. Author: Mössbauer Effect, Principles and Applications, 1964; also numerous articles. Research on ocean currents, semi-conductor physics including radiation damage and properties of chem. impurities, Mössbauer effect in fields magnetism, chem. physics, and metal-organic chemistry. Home: 175 Woodland Av., Convent Station, N.J. 07961. Office: Bell Telephone Labs., Murray Hill, N.J. 07971.*

WERTHEIM, Willem Frederik, sociologist; b. Leningrad, USSR, Nov. 16, 1907; s. Jonas and Heintje (van Gelder) W.; M. Netherlands Law, Leiden (Netherland) U., 1928, M. Netherlands-Indies Law, 1930, D.Law, 1930; m. Annie Henriette Gijse Weenink, Dec. 30, 1930; children—Marijke (Mrs. The Siauw Giap), Anne Ruth (Mrs. Albert van Kammen), Hendrik Hugo. Judiciary, South Sumatra, 1931; with Batavia Dept. Justice, 1931-36; prof. ency. law and conflict of laws Law Sch., Batavia, 1936-42; prof. modern history and sociology of Indonesia, Amsterdam (Netherlands) U., 1947-55, prof. modern history and sociology of S. and S.E. Asia, 1955—; vis. prof. rural sociology Faculty Agr., U. Indonesia, Bogor, 1956-57. Cons. refresher course for Asian sociologists UNESCO, Delhi, India, 1965. Mem. Internat. Union for Sci. Study Population, Nederlandse Sociologische Vereniging, Oosters Genootschap, Verlond van Wetenschappelijke Onderzoekers. Author: In Gesprek met mijzelf, 1948; Het rassenprobleem, 1949; Indonesian Society in Transition, 1956, rev., 1964; East-West Parallels, 1964; also numerous articles. Analysis of influence of law upon soc. in terms of psychical causation; elaborated causality concept in terms of identity and novel creation; research on parallelism between social devels. in Asia and W.; application of evolutional concept of dialectics of progress as formulated by Jan Romein. Home: 12 Stadionkade, Amsterdam-Z, Netherlands.*

WERTHEIMER, Emile, French physiologist, physician; b. Rosheim, France, 1852; prof. physiology Faculty Medicine, Lille, France; corr. Acad. Medicine, Lille; research on pancreatic secretions, circulation of bile, glyco-secreting nerves. Died Sedan, France, 1924.

WERTHEIMER, Ernst, biochemist; b. Bühl, Germany, Aug. 24, 1893; s. Leo and Hermine (Wertheimer) W.; student medicine U. Kiel (Germany), 1913, U. Bonn (Germany), 1914; Dr.med., U. Heidelberg (Germany), 1920; m. Ruth Lehmann, Dec. 24, 1928; children—Ajalah (Mrs. Abraham Abrahamov), Noemi (Mrs. Eliser Kaplinski), Ada, Dorit. Staff, Physiol. Inst., U. Halle-Wittenberg, 1923-34; faculty Hebrew U., Jerusalem, Israel, 1934—, prof. human biochemistry, 1953-63, prof. emeritus, 1963—. Recipient Banting medal, Toronto, Ont., Can., 1964, Israel prize med. sci., 1956, Prize Bublic, 1964. Mem. Israel Acad. Scis. and Humanities. Research and publs. on metabolic regulation, physiology of adipose tissue, insulin effect thyroid glands. Home: Abarbanelstr. 26, Jerusalem, Israel.*

WERTHEIMER, Max, psychologist, philosopher; b. Prague, Apr. 15, 1880; ed. Prague and Berlin; Ph.D., U. Würzburg, 1904. Prof. at Frankfort and Berlin; came to U. S., became mem. grad. faculty New Sch. Social Research; vis. prof. Columbia, 1934. Author: Drei abhandlungen aur Gestalt-theorie, 1925; Productive Thinking, 1944. Founder (with Wolfgang Köhler and Kurt Koffka) Gestalt sch. of psychology, which introduced method of inquiry that maintains we are always presented with sensory field rather than with isolated sensations; insisted that in man's perception of objects there are characteristics not belonging to single sensation; stressed importance of wholes in learning and problem solving; pointed out relevance of Gestalt principles to teaching and attacked old emphasis on repetition and routine derived from associationist theory of learning; discovered phi phenomenen concerning illusion of motion in preception, 1910-12. Died N.Y., Oct. 12, 1943.

WERTZ, John Edward, Am. phys. chemist; b. Denver, Dec. 4, 1916; s. John and Amalie (Dreuth) W.; B.S. in Chem. Engring., U. Denver, 1937, M.S., 1938; Ph.D., U. Chgo., 1948; m. Florence Marie Carlson, June 5, 1943; children—John A., Byron A., Kristin M. Asst. prof. Augustana Coll., Rock Island, Ill., 1941-44; instr. Gustavus Adolphus Coll., St. Peter, Minn., 1944-45; faculty U. Minn., Mpls., 1945—, prof. phys. chemistry, 1957—. Sr. Fulbright scholar, Guggenheim fellow, Clarendon Lab., Oxford, Eng., 1957-58. Mem. Am. Phys. Soc., Am. Chem. Soc., Phys. Soc. London (Eng.), Faraday Soc., N.Y. Acad. Scis. Author: (with J. Bolton) Electron Spin Resonance, 1966; also numerous articles. Applications technique magnetic resonance to study structure molecules, study geometry of imperfections in insulating solids. Home: 1847 Ashland Av., St. Paul 55104.*

WERZ, Gunther Wolfram, German cell biologist; b. Ravensburg, Germany, Mar. 26, 1927; s. Stefan and Clara (Eisele) W.; Ph.D., U. Cologne, 1955; m. Gudrun Schillemeit, Oct. 14, 1958; children—Christian Stefan, Matthias Curt. Sci. asst. Max Planck Inst. for Marine Biology, Wilhelmshaven, Germany, 1955-65, chief electron microscope dept., 1965—. Mem. German Bot. Soc., A.A.A.S. Research and publs. on microscopic and sub-microscopic level of behavior of certain substances and structural constituents of cells in relation to realization of genetically determined morphogenesis, mainly Acetabularia and related species. Address: Max Planck Inst. for Marine Biology, Wilhelmshaven, Germany.*

WESLER, Oscar, Am. mathematician; b. Bklyn., July 12, 1921; s. Israel Edward and Sarah (Hartman) W.; B.S., Coll. City N.Y., 1942; M.S., N.Y. U., 1943; postgrad. Princeton, 1943-46; Ph.D., Stanford, 1955. Faculty Stanford 1952-56, U. Mich., 1956-64; prof. statistics and math. N.C. State U., Raleigh, 1964—. Cons., Inst. Sci. and Tech., U. Mich., 1957-64; vis. prof. statistics Stanford, 1962-63; vis. lectr. NSF Program vis. lectrs. in statistics, 1963—. Mem. Inst. Math. Statistics, Am. Math. Soc. Author: Solutions to Problems in Theory of Games and Statistical Decisions, 1954; also articles. Research in areas of math. statistics, statis. decision theory, probability theory, stochastic processes, convexity theory, linear algebra. Home: 1926 Smallwood Dr., Raleigh, N.C. 27605.*

WESOLOWSKI, Sigmund Adam, Am. surgeon; b. Saugus, Mass., Feb. 6, 1923; s. Joseph and Adamina (Ploharska) W.; M.D., Tufts Coll. Medicine, 1948, M.S., 1951; m. Wanda B. Kirbi, Oct. 7, 1945; children—Carl Adam, Paul David, Joan Marie, Adam John, Edward Alan. Charlton research fellow Ziskind Surg. Research Lab., New Eng. Med. Center, Boston, 1949-50, 1951-52; third research scholar A.C.S., London, Eng., Bklyn., 1956-59; instr. surgery State U. N.Y., 1957—, clin. prof. surgery, 1964—; chmn. dept. surgery Meadowbrook Hosp., East Meadow, 1964-66; dir. L.G.H. Lab., Mercy Hosp., Rockville Centre, N.Y., 1966—; cons. thoracic cardiovascular surgery Mercy Hosp., Rockville Centre, N.Y., Nassau Hosp., Mineola, N.Y., Meadowbrook Hosp., East Meadow, South Nassau Community Hosp., Oceanside, Mid Island Hosp., Bethpage, Providence Hosp., Seattle, Jay Hughes Meml. Med. Research Found., Wilmington, Del., Lutheran Med. Center, Bklyn., St. Johns Episcopalian Hosp., Bklyn. Mem. Am. Soc. for Artificial Internal Organs (pres. 1966-67), Soc. for Vascular Surgery, Internat. Cardiovascular Soc., Soc. University Surgeons, A.C.S., Am. Assn. Thoracic Surgeons, Soc. Thoracic Surgeons, Am. Soc. for Testing and Materials, A.A.A.S. Author: Evaluation of Tissue and Prosthetic Vascular Grafts, 1962. Editor (with C. Dennis) Fundamentals of Vascular Grafting, 1963; also numerous articles. Developed an improved arterial prosthesis now being used in humans. Home: 44 Roosevelt Av., East Rockaway, N.Y. 11518. Office: 165 N. Village Av., Rockville Centre, N.Y. 11570; also 520 Franklin Av., Garden City, N.Y. 11530.*

WESPI, Hans Jakob, Swiss physician; b. Schönenberg, Switzerland, Oct. 18, 1908; s. Jakob and Sophie (Hess) W.; ed. U. Montpellier (France); M.D., U. Zürich, 1934; m. Marie Adelheid Eggenberger, Apr. 22, 1937; children—Ruth (Mrs. Andreas Speich), Hans Heinrich, Markus Jakob, Peter Andreas, Anneliese. Asst. physician Krankenhaus Neumünster Zürich, also staff U. Frauenklinik, Zurich, 1934-43; chief physician obstet.-gynecol. dept. Krankenhaus Frauenfeld, 1944-46, Kantonsspital, Aarau, Switzerland, 1947—. Mem. Swiss Goiter Com. Recipient Prix quadriennal internat., 1946. Hon. Mem. Jugoslavian Gynecol. Soc., Sociedad argentina de patologia cervical y colposcopia, Sociedad medical de La Plata; mem. Société Francaise de Gynécologie (corr.), Schweiz Akademie Medizin. Wissenschaften (mem. fluorkommssion). Author: Entstehung und Früherfassung des Portiokarzinoms, 1946; Early Carcinoma of the Uterine Cervix, 1949; Fluor-Vollsaz zur Kropf und Kariesbekämpfung, 1956; (with G. Mestwerdt) Atlas der Kolposkopie, 1961; also numerous articles. Early diagnosis of cervical cancer with colposcopy; colpolhotography; goiter prevention with iodized salt; introduced fluroidated salt as alternative to water fluoridation in caries prevention. Home: Sonneckstr. 9, Suhr (Aargau) 5034. Office: Kantonsspital, Aarau (Aargau) 5000, Switzerland.*

WESSEL, Caspar, mathematician; b. Norway, 1745; worked in Denmark. Author: Essai sur le representation analytique de la direction (on vector theory), 1897. Produced 1st consistent and useful interpretation of complex numbers, in which he used Argand diagrams (properly called Wessel diagrams), 1897. Died 1818.

WEST, Charles Donald, Am. physician; b. Ogden, Utah, Oct. 25, 1920; s. Charles B. and Vera (Pickford) W.; B.A., U. Utah, 1941, M.D., 1944, Ph.D., 1950; m. Helen Hanigan, Sept. 20, 1946; children—Gregory John, Steven Charles, Richard Donald, Janet. With Cornell U., 1950-54, asst. prof. medicine, 1953-54; with Sloan-Kettering Inst., 1950-57, asso. mem., 1954-57; with U. Utah Coll. Medicine, 1957—, asso. prof. medicine, 1964—. Mem. Western Soc. Clin. Research, Endocrine Soc., Harvey Soc., Am. Assn. Cancer Research, A.A.A.S., Am. Fedn. Clin. Research, Soc. Exptl. Biology and Medicine, Am. Soc. Clin. Investigation, Western Assn. Physicians, Phi Beta Kappa, Sigma Xi, Phi Kappa Phi. Research, numerous publs. on fate and excretion of steroid hormones in body; developed and applied new methods for measuring hormones in the diagnosis of endocrine diseases; assessed therapeutic value of removal of ovaries, adrenals, and pituitaries in patients with breast cancer and other neoplasias. Home: 2623 E. 3820 South, Salt Lake City 84109. Office: 50 N. Medical Dr., Salt Lake City 84112.*

WEST, Charles Tyrrell, Am. mech. engr.; b. Crestline, O., July 1, 1915; s. LaFoy Allen and Mabelle (Hillson) W.; student U. S. Naval Acad., 1934-37; B.C.E., Ohio State U., 1939, M.S., 1946; postgrad. U. Mich.; Ph.D., Cornell U, 1951; m. Estella Marie Fahs, Apr. 5, 1947; children—Mary Bronwyn, Charles Tyrrell III. Bridge test analyst Assn. Am. Railroads, 1941-46; instr. engring. mechanics Ohio State U., 1946-53, prof., chmn. dept. engring. mechanics, 1953—, cons. Boeing Airplane Co., 1959-60. Research scientist Autometric Corp., 1959-63; cons. in dynamics aeronautics, aerospace and inertial systems to various corps.; 1964—. Mem. Am. Soc. C.E., Am. Math. Soc., Sigma Xi. Research and publs. on vibration and dynamics. Home: 1047 Glendale Av., Columbus, O. 43212.*

WEST, Clark Darwin, Am. physician; b. Jamestown, N.Y., July 4, 1918; s. Clark D. and Frances (Blanchard) W.; A.B., Coll. Wooster, 1940; M.D., U. Mich., 1943; m. Ruthann Asbury, Apr. 12, 1944; children—Charles Michael, John Clark, Lucy Frances. Faculty dept. pediatrics U. Cin., Coll. Medicine, 1951—, prof., 1962—; fellow Children's Hosp. Research Found., Cin., 1953—, asso. dir., 1963—; staff

Children's Hosp., Cin., 1952—, Cin. Gen. Hosp., 1953—. Mem. Soc. for Pediatric Research (past pres.), Midwest Soc. for Pediatric Research, Am. Pediatric Soc., Am. Physiol. Soc., Am. Assn. Immunologists, Alpha Omega Alpha. Editorial bd. Jour. Pediatrics, 1960—. Research and numerous publs. on glomerulnephritis in children, definition of types, role auto-antibody and complement in etiology and use of immunosuppressive agts. in treatment; developed means for measurement specific serum proteins employing immunoelectrophoretic analysis; research on immunoglobulin levels in children. Home: 8 Walsh Lane, Cin. 45208.*

WEST, Edward Staunton, Am. biochemist; b. Stuart, Va., Sept. 9, 1896; s. John Thrash and Mary (Jefferson) W.; A.B., Randolph-Macon Coll., 1917, LL.D., 1965; M.S., Kan. State Coll., 1920; Ph.D., U. Chgo., 1923; m. Ruth Evalyn Hurd, May 27, 1920. Faculty, Kan. State Coll., 1917-22, U. Chgo., 1922-23, Washington U. Sch. Medicine, St. Louis, 1923-34; prof. biochemistry, head dept. U. Ore. Med. Sch., Portland, 1934—, sr. scientist Ore. Regional Primate Research Center, 1961—. Mem. Am. Chem. Soc., Am. Soc. Biol. Chemists, A.A.A.S., Soc. Exptl. Biology and Medicine, Am. Assn. U. Profs., Ore. Acad. Sci., Portland Acad. Medicine, Phi Beta Kappa, Sigma Xi, Phi Beta Pi (hon.), Alpha Omega Alpha (hon.). Author: (with W.R. Todd) Textbook of Biochemistry, 1951; Textbook of Biophysical Chemistry, 1942. Research and numerous publs. on methods of biochem. analysis, carbohydrate metabolism, lipid, ketosis, muscle metabolism, carbohydrate chemistry. Inventor lab. condenser. Home: 18800 N.W. Holly Av., Beaverton, Ore. 97005. Office: University of Ore. Med. Sch., Portland, Ore. 97201.*

WEST, Geoffry Buckle, English pharmacologist; b. Surrey, Eng., Sept. 13, 1916; s. Alfred Victor and Lucy (Smith) W.; Student U. Coll., London, Eng., 1935-36, Sch. Pharmacy, London, 1936-38; m. Jean Lilian Bates, Aug. 1, 1942; children—Derek Buckle, Pauline Marina. Faculty, U. London, 1938-45, 46-50, reader pharmacology, 1956-66; research fellow U. Edinburgh (Scotland), 1945-46; sr. lectr. pharmacology U. St. Andrews (Scotland), 1950-56; cons. Brit. Indsl. Biol. Research Assn., 1966—. Mem. Brit. Pharmacological Soc., Physiol. Soc., N.Y. Acad. Scis. (A. Cressy Morrison award in natural scis. 1963). Author: Progress in Medicinal Chemistry, Vol. 1, 1961, Vol. 2, 1962, Vol. 3, 1963, Vol. 4, 1965, Vol. 5, 1967; also numerous articles. Editorial bd. Jour. Pharmacy and Pharmacology, 1958—, Internat. Archives of Allergy, 1960—. Discovered that young mammals have norepinephrine in their adrenal glands, that tissue mast cells contain histamine, that some animal reactions are genetically controlled. Home: 22 Burgh Heath Rd., Epsom, Surrey. Office: Brit. Indsl. Biol. Research Assn., Woodmansterne Rd., Carshalton, Surrey, Eng.*

WEST, Louis Jolyon, Am. psychiatrist; b. Bklyn., Oct. 6, 1924; s. Albert Jerome and Anna (Rosenberg) W.; B.S., U. Minn., 1946, M.B., 1948, M.D., 1949; m. Kathryn Louise Hopkirk, Apr. 29, 1944; children—Anne Kathryn, Mary Elizabeth, John Stuart. Faculty, Cornell, 1950-52; chief psychiatry service USAF Hosp., Lackland AFB, San Antonio, 1952-56; prof. psychiatry, head dept. psychiatry, neurology and behavioral scis. U. Okla. Sch. Medicine, Oklahoma City, 1954—; cons. psychiatry Oklahoma City VA, USAF hosps., Tinker AFB, Okla., USAF Aero-Space Med. Center, Bur. Social Research, others. Mem. adv. coms. USPHS, Nat. Assn. Mental Health; mem. Okla. Gov.'s Com. on Alcoholism; chief mental health sect. Okla. Med. Research Found., 1956—. Bd. dirs. Inst. Research in Hypnosis. Diplomate Nat. Bd. Med. Examiners, Am. Bd. Psychiatry and Neurology. Fellow A.A.-A.S., Am. Coll. Neuropsychopharmacology, Am. Psychiat. Assn.; mem. Acad. Psychoanalysis, Aerospace Med Assn., A.M.A., Am. Fedn. Clin. Research, Am Psychopath. Assn., Am. Psychosomatic Soc., Nat. Acad. Religion and Mental Health (founding), Nat. Council on Alcoholism, Alpha Omega Alpha, others. Author: Hallucinations, 1962. Adv. editor Internat. Jour. Clin. and Extl. Hypnosis, 1958—, mem. adv. editorial bd. Jour. Nervous and Mental Disease, 1961—. Research, publs. on prisoners of war and tactics of brainwashing, sleep deprivation, induction of hallucinations in normal persons, hypnosis, behavioral scis. in med. edn. Home: 4900 Willard Av., Oklahoma City 73105.*

WEST, Luther Shirley, Am. biologist; b. Utica, N.Y., Sept. 6, 1899; s. Norman Luther and Rena (Shirley) W.; B.S., Cornell U., 1921, Ph.D., 1925; m. Beatrice Emma Ryan, June 28, 1922; children—Ruth S. (Mrs. Donald S. MacVean), Richard L., William E., Alice B. (Mrs. M. Stuart Eldred), Betsey L. (Mrs. Robert A. Dahlgren), David J. Mem. faculty Cornell U., 1921-25, Battle Creek Coll., 1926-38; prof., head dept. biology No. Mich. U., Marquette, 1938-43, 1946-62, dean arts and sci., 1962-65; asso. mem. grad. faculty U. Mich., 1952-60. Sci. cons. WHO, 1952-53. Recipient Honorable Mention award Mich. Acad., 1950. Fellow A.A.A.S.; mem. Entomol. Soc. Am., Am. Micros. Soc., Mich. Acad. Sci. (pres. 1965-66), Am. Assn. U. Profs., Am. Inst. Biol. Scis., Mich. Assn. for Higher Edn., Am. Soc. Parasitologists, Sigma Xi, Phi Kappa Phi, others. Author: Eugenic Aspects of Race Betterment, 1932; (with others) Practical Malariology, 1946, 2d edit., 1963; The Housefly - Its Natural History, Medical Importance and Control, 1951; (with Francis C. Lundin) The Free-Living protozoa of the Upper Peninsula of Michigan. Research and publs. on various insects; world authority on the housefly (Musca domestica); studies in edn. and methods. Home: 137 W. Ridge St., Marquette, Mich. 49855.*

WEST, Philip William, Am. chemist; b. Crookston, Minn., Apr. 12, 1913; s. William Leonard and Annie Sylvia (Thompson) W.; B.S., U. N.D., 1935, M.S., 1936, D.Sc., 1958; Ph.D., U. Ia., 1939; m. Foymae S. Kelso, July 1, 1963; children—Dorothy (Mrs. Harry Bristol), Linda Kay, Patty Sue. Asst. chemist N.D. State Geol. Survey, 1933-36; asst. chemist, bacteriologist Ia. State Dept. Health, Iowa City, 1937-40; research microchemist Econs. Lab., St. Paul, 1940; with La. State U., 1940—. Dir. Inst. for Environmental Scis. Cons. WHO, USPHS. Recipient Charles E. Coates Meml. award, 1967. Mem. Internat. Union of Pure and Applied Chemistry (pres. analytical div. 1965-69), Austrian Chem. Soc. (hon.), A.A.A.S., Midlands Soc. for Analytical Chemistry (hon.). Author: Calculations of Quantitative Analysis, 1941; (with L. Vick) Qualitative Analysis and Analytical Chemical Separations, 1953; also numerous articles. Research on spot test methods of analyses and application of coordination chemistry and organic reagents to inorganic analysis; developed various instrumental techniques for analytical work; published numerous methods especially for air and water pollution studies as flame photometry, polarography, ring oven, reagent crayons, atomic absorption spectroscopy, polarized light microscopy, high frequency conductometric titrations and measurements. Home: 605 Nelson St., Baton Rouge, La. 70808.*

WEST, Robert, Am. chemist; b. Glen Ridge, N.J., Mar. 18, 1928; s. Robert C. and Constance (MacKinnon) W.; B.A., Cornell U., 1950; A.M., Harvard, 1952, Ph.D., 1954; m. Margaret F. Taylor, Nov. 5, 1950; children—David Russell, Arthur Scott. Asst. prof. Lehigh U., 1954-56; faculty U. Wis., Madison, 1956—, prof. chemistry, 1963—. Indsl. and govt. cons., 1961—; Abbott lectr. U. N.D., 1964; Fulbright lectr. Kyoto U., 1964-65. Mem. Am. Chem. Soc., Chem. Soc. (London), Japan Chem. Soc., A.A.A.S., Wis. Acad. Sci. Co-editor: Advances in Organometallic Chemistry, Vols. I-III, 1964-66. Research and numerous publs. on new types of organosilicon and organolithium compounds, perhalogenated compounds, theory of chem. bonding in organometallic and aromatic substances. Home: 5702 Old Sauk Rd., Madison, Wis. 53705.*

WEST, Theodore Clinton, Am. pharmacologist; b. Central, S.C., May 17, 1919; s. Henry A. and Aurena B. (Herington) W.; student Marion Coll., 1936-37; B.S. in Pharmacy, U. Wash., 1948; M.S. in Pharmacology, 1949, Ph.D., 1952; m. Juliann Orton, Jan. 16, 1942; children—David, Don, Lynne. Faculty, U. Wash., Seattle, 1949—, prof. pharmacology, 1964—, asst. chmn. dept. pharmacology, 1961—, asst. dean Sch. Medicine, 1966—, co-dir. pharmacology tng. grant, 1965—. Mem. Am. Soc. Pharmacology and Exptl. Therapeutics, Western Pharmacology Soc. Research and publs. on cardiac tissue of lab. animals, cardiac spontaneous rhythmicity including influence of drugs. Home: 1215 N.E. 135th St., Seattle 98125.*

WESTENBERG, Arthur Ayer, Am. chemist; b. Menomonie, Wis., Mar. 1, 1922; s. Arthur Enoch and Florence (Ayer) W.; B.A., Carleton Coll., 1943; M.A., Harvard, 1948, Ph.D., 1950; m. Jane Fairbank Burbach, Oct. 31, 1945; 1 son, Lee Arthur. Asst. prof. chemistry Lafayette Coll., Easton, Pa., 1950-52; sr. staff mem. Applied Physics Lab., Johns Hopkins, Silver Spring, Md., 1952-57, prin. profl. staff, supr. chem. physic research, 1957—. Cons. Project Squid, Office Naval Research 1959-64. Recipient Profl. Achievement award Washington Acad. Sci. and D.C. Soc. Engrs. and Architects, 1956. Mem. Am. Chem. Soc., Am. Phys. Soc., Combustion Inst. (Silver medal 1966), Washington Acad. Sci. Author: (with R.M. Fristrom) Flame Structure, 1965; also articles. Precise measurement of gas transport properties at high temperatures; interpretation of detailed flame structure data on temperature and chem. reactions in flames, application of electron spin resonance spectroscopy to study elementary chem. reaction rates. Home: 1805 Alcan Dr., Silver Spring, Md. 20902.*

WESTENDORP, Willem Fredrik, physicist; b. Amsterdam, Holland, May 7, 1905; s. Willem F. and Gesina (Stom) W.; E.E., Delft Technical U., 1928; advanced course engring., Gen. Electric Co; D.Eng., (hon.), Rensselaer Poly. Inst., 1947; m. Mary Andrews, Apr. 1, 1933; children—Mary Wilhelmina (Mrs. Philip Rex Kleitz), Ray (dec.). Came to U. S., 1928, naturalized, 1934. Radio engr. Gen. Electric Co., 1928-29, research engr. research lab., 1929-35, research high voltage x-rays and betatrons, 1935-54, physicist thermonuclear research, 1954—. Recipient John Price Wetherill medal Franklin Inst., 1944. Fellow Inst. Elec. and Electronic Engrs., Am. Phys. Soc. Inventor resonant transformer X-ray and cathode ray generator; research on light signals, 1932, gaseous discharges, 1934, million volt X-rays, 1939, 2 million volt X-rays, 1944, one hundred million electron volt betatron, 1945, electron gun performance in betatrons, 1949, ion pump, 1954, electron beam generators, 1957, ionization in gases for magnetohydrodynamics, 1961, thermonuclear fusion research stabilization effects, 1962, fusion neutron prodn., 1962, fusion in sustained magnetic field, 1965. Home: 17 Front St., Schenectady 12305. Office: Research and Devel. Center, Gen. Electric Co., P.O. Box 1088, Schenectady.*

WESTERFELD, Wilfred Wiedey, Am. biochemist; b. St. Charles, Mo., Dec. 13, 1913; s. August Henry and Caroline (Wiedey) W.; B.S. in Chem. Engring., Mo. Sch. Mines, 1934; Ph.D., St. Louis U. 1938; m. Nora McElroy, June 25, 1938; children—Mary, Margaret, William, Noel, John. NRC fellow Oxford (Eng.) U., 1938-39, Columbia U., 1939-40; faculty Harvard Med. Sch., 1940-45, asst. prof. biochemistry, 1940-45; prof., chmn. dept. biochemistry Syracuse (N.Y.) U. Med. Coll., 1945-50; prof., chmn. dept. biochemistry State U. N.Y., Syracuse, 1950—, acting dean Med. Sch., 1956-57, dean Grad. Studies, 1965. Recipient St. Louis U. Alumni award, 1959. Mem. Am. Soc. Biol. Chemists, Am. Chem. Soc., Soc. for Exptl. Biology and Medicine, Sigma Xi. Editor: Biochemical Preparations, Vol. 4, 1955. Research and numerous publs. on molybdenum as a trace metal in enzyme xanthine oxidase, biol. assays for molybenum, thyroxine, antithyrotixoic substances; discovered new intermediate in metabolism of alcohol. Home: 726 Ostrom Av., Syracuse, N.Y. 13210.*

WESTERFIELD, Holt Bradford, Am. polit. scientist; b. Rome, Italy, Mar. 7, 1928 (parents Am. citizens); s. Ray Bert and Beatrice (Putney) W.; B.A., Yale, 1947; M.A., Harvard, 1951, Ph.D., 1952; m. Carolyn Elizabeth Hess, Dec. 17, 1960; children—Pamela Bradford, Leland Avery. Teaching fellow Harvard, 1949-51, Sheldon Traveling fellow, 1951-52, instr. polit. sci., 1952-56; Congl. fellow Am. Polit. Sci. Assn., 1953-54; asst. prof. polit. sci. U. Chgo., 1956-57; asst. prof. polit. sci. Yale, 1957-63, asso. prof., 1963-65, prof., 1965—; research asso. Washington Center Fgn. Policy Research, Johns Hopkins Sch. Advanced Internat. Studies, 1965-66; vis. prof. Wesleyan U., 1967. Mem. Am. Polit. Sci. Assn., Internat. Studies Assn. Author: Foreign Policy and Party Politics: Pearl Harbor to Korea, 1955; The Instruments of America's Foreign Policy, 1963; also articles. Studies in bipartisanship in Am. fgn. relations, undercover techniques of internat. competition; comparisons of fgn. policy processes in Britain and Am. Home: 115 Rogers Rd., Hamden, Conn. 06517. Office: Dept. Polit. Sci., Yale New Haven 06520.*

WESTERGAARD, Mogens, Danish geneticist; b. Copenhagen, Denmark, Dec. 6, 1912; s. Henry and Mathilde (Wanning) W.; M.Sc., Copenhagen U., 1936, Ph.D., 1940; m. Ebba Caspesen, 1936; children—Kirsten, Poul, Susanne. Asst. prof. genetics lab. Royal Vet. and Agrl. Coll., Copenhagen, 1936-49; prof. genetics, chmn. Inst. Genetics, Copenhagen U., 1949-62; research asso. Carlsberg Found., Copenhagen, 1962—; vis. prof. Cornell U., 1955. Mem. adv. panel human genetics WHO; mem. adv. com. on radiation hazards Danish Pub. Health Orgn. Rockefeller fellow Cal. Inst. Tech., 1946-47; temp. fellow Carnegie Instn. Washington, 1958; Gosney fellow Cal. Inst. Tech., 1963-64. Fellow Arctic Inst. N. Am.; mem. Royal Danish Acad. Sci., Kungl. Fysiograf. Sälsk. Lund, Brit. Genetical Soc. (hon.), Nordic Genetical Soc. (hon.; Munksgaard award 1958). Author: Sex Determination in Plants, 1940; Cytotaxonomy of Arctic Plants, 1953; Arvelighedslren, 1953; Chemical Mutagenesis, Cytogenetics of Crossing Over, 1965. Home: 91 Abildgaardsvej, 2830 Virum, Denmark. Office: 10 Gl. Carlsbergvej, 2500 Copenhagen Valby, Denmark.*

WESTERMARCK, Edward Alexander, Finnish anthropologist, sociologist; b. Helsingfors, Finland' Nov. 20, 1862; s. N.C. and Constance (Blomqvist) W.; Ph.D., U. Helsingfors, 1886; LL.D. (hon.), univs. Aberdeen and Glasgow (Scotland). Docent, U. Helsingfors, 1890-93, then prof. moral philosophy; lectr. U. London (Eng.), 1904, prof. sociology, 1907-30; rector Turku U., 1918, prof. to 1935; prof. emeritus Acad. of Abo, 1935-39. Author: The Origin of Human Marriage, 1885; The History of Human Marriage, 1891; The Origin and Development of the Moral Ideas, 2 vols., 1906-08; Ritual and Belief in Morocco, 2 vols., 1926; A Short History of Marriage, 1926; The Goodness of Gods, 1926; Memories of my Life, 1927; Wit and Wisdom in Morocco, 1930; Ethical Relativity, 1932; Pagan Survivals in Mohammedan Civilization, 1933; Three Essays on Sex and Marriage, 1934; The Future of Marriage in Western Civilization, 1936 Christianity and Morals, 1939. Noted for Comparative studies of marriage, the family and morals; contradicting the social evolutionists, he asserted that monogamy and paternalistic family patterns existed in the earliest cultures of man; among 1st to give inductive character to sociology through use critical ethnological data. Died Lapinlahti, Sept. 3, 1939.

WESTERVELT, Peter Jocelyn, Am. physicist; b. Albany, N.Y., Dec. 16, 1919; s. William Irving and Dorothy (Jocelyn) W.; B.S., Mass. Inst. Tech., 1947, M.S., 1949, Ph.D., 1951; m. Alice Francis Brown, June 2, 1956; children—Dirck Edgell, Abby Brown. Faculty, Brown U., Providence, 1951—, prof. physics, 1963—. Cons., adviser to govt. agys., aircraft cos.; mem. com. on hearing and bioacoustics NRC, 1957——, chairman exec. council, 1968-69. Fellow of the Physical Society, also of the Acoustical Soc. Am. Research and publs. on theory of intense noise generation and propagation, gen. relativistic theory of gravitational waves. Home: 600 Spring Gree Rd., Warwick, R.I. 0288. Office: Dept. Physics, Brown U., Providence 02912.*

WESTFALL, Bertis Alfred, Am. pharmacologist; b. Halfway, Mo., June 4, 1907; s. Lawrence Lee and Nora (Davison) W.; A.B., U. Mo., 1933, M.A., 1934, Ph.D., 1938; postgrad. U. Toronto (Ont., Can.), 1948, U. Utah, 1949; m. Nell Frances Martin, June 7, 1933. Faculty, U. Mo., Columbia, 1936——, prof., 1946——, chmn. dept. physiology and pharmacology, 1953-65, chmn. dept. pharmacology, 1965——. Fellow A.A.A.S.; mem. Soc. for Exptl. Biology and Medicine; Am. Soc. for Pharmacology and Exptl. Therapeutics, Am. Physiol. Soc., Sigma Xi, Phi Beta Kappa. Author: (with M.M. Ellis, M.D. Ellis) Determination of Water Quality, 1946; also articles. Research on effects of barbiturates on oxygen consumption, effects of tumors on hypnotic action of barbiturates in mice, pharmacology of plant extractives, comparative pharmacology. Home: 1404 Anthony St., Columbia, Mo. 65201.*

WESTHEIMER, Gerald, neurophysiologist, optometrist; b. Berlin, Germany, May 13, 1924; s. Isaak and Ilse (Cohn) W.; optometry diploma, Sydney (Australia) Tech. Coll., 1943, fellowship diploma, 1950; B.Sc., U. Sydney, 1947; Ph.D., Ohio State U. 1953. Australian citizen, 1944; came to U. S. 1951. Practice optometry, Sydney, 1945-51; research fellow Ohio State U., 1951-53; prof. physiol. optics U. Houston, 1953-54; asst. prof., then asso. prof. physiol. optics Ohio State U., 1954-60; postdoctoral fellow neurophysiology Marine Biol. Lab., Woods Hole, Mass., 1957; vis. researcher Physiol. Lab., U. Cambridge (Eng.), 1958-59; mem. faculty U. Cal. at Berkeley, 1960——, prof. physiol. optics, 1963——, chmn. group physiol. optics, 1964——. Mem. com. vision NRC, 1957——; mem. visual scis., study sect. NIH, 1966-——. Fellow A.A.A.S., Optical Soc. Am., Am. Acad. Optometry; mem. Royal Soc. New So. Wales, Nat. Acad. Scis., Sigma Xi.; asso. mem. Physiol. Soc. Great Britain, Author research papers on servoanalytic study of movement and focusing responses of human eye; formation optical image on human retina; resolving capacity human visual system; psychophys. and electrophysiol. analyses of spatial interaction of human visual apparatus; optical and photoelectric instrumentation for testing correction of human vision. Home: 582 Santa Barbara Rd., Berkeley, Cal. 94707.*

WESTIN, Sverre, Norwegian physicist; b. Overhalla, Norway, Dec. 5, 1909; s. Nils Winther and Bereth (Hildrum) W.; student chemistry, Norges tekniske hogskole, 1929-33, Dr.Technicae in Physics, 1949; m. Karen Skjeseth, Nov. 28, 1937; children—Steinar, Gunnar Andreas. Research fellow physics dept. Norges tekniske hogskole, 1933-37; head phys.-chem. lab. Luma Stockholm, 1937-39; head chem.-phys. lab. Statens Teknologiske Institutt, 1940-48; prof. applied physics Norges tekniske hogskole, Trondheim, 1949-——. Mem. Det Kgl. Norske Videnskabers Selskab., Det Norske Videnskaps-Akademi i Oslo, Norges Tekniske Vitenskaps-Akademi. Research, publs. on electron scattering, phys. electronics and high vacuum technique. Home: 4 Sigurd Slembes vei, Trondheim, Norway.*

WESTINGHOUSE, George, Am. inventor; b. Central Bridge, N.Y., Oct. 6, 1846; s. George and Emeline (Vedder) W.; ed. Union Coll., hon. Ph.D. 1890; Dr. Eng., Königliche Technische Hochschule, Berlin, 1906; m. Marguerite Erskine Walker, Aug. 8, 1867. Asst. engr. U. S. Navy, after 1865; worked in his father's agrl. machinery shop, Schenectady; organized Westinghouse Air Brake Co., 1869, Union Switch and Signal Co., 1882; founder works at Wilmerding, East Pittsburgh, Swissvale, and Trafford City, Pa.; Hamilton, Can.; Manchester, and London, Eng.; Havre, France; Hanover, Germany; St. Petersburg, Russia; Vienna, Austria; Vado, Italy; pres. of 30 corps. Second recipient of John Fritz medal. Hon. mem. Am. Soc. M.E. (pres., 1909-10), Nat. Electric Light Assn. Invented rotary engine at 15; invented a device for replacing derailed steam cars, 1865; patented his invention of air brake, 1868; applied pneumatic devices to switching and signaling, greatly increasing efficiency; also utilized electricity in this connection, became interested in electric machinery; acquired Gaulard & Gibbs patents, 1885, introduced alternating current system of elec. distbn. for light and power; made many elec. inventions; backed Tesla financially and with shop facilities in developing induction motor which made possible utilization of alternating current for power purposes; built first 10 great dynamos for Niagara, dynamos for elevated and subway roads in N.Y., and for Met. Ry., London; devised a complete system for controlling natural gas and conveying it through pipe lines for long distances, thereby established practicability of utilizing natural gas as fuel in homes, mills and factories; took a foremost part in developing gas engines, and in adapting steam turbines to electric driving. Died N.Y.C., Mar. 12, 1914.

WESTLAKE, Donald William Speck, Canadian microbial biochemist; b. Woodstock, Ont., Can., Feb.

27, 1931; s. Frank and Irene Marybell (Yake) S.; B.Sc. in Agr., U. B.C., 1953; M.Sc. (B.C. Sugar Refinery scholar 1953, Agrl. Inst. Can. scholar 1954), U. Wis., 1955, Ph.D., 1958; m. Maureen Elizabeth Walsh, Aug. 28, 1954; children—Douglas McGregor, Jill Anne. Asst. research officer Prairie Regional Lab., Nat. Research Council Can., Saskatoon, Sask., 1958-63, asso. research officer, 1963-66; sessional lectr. dept. bacteriology Coll. Medicine, U. Sask., Saskatoon, 1964-66; asso. prof. Faculty Sci., U. Alta. (Can.), Edmonton, 1966-——. Mem. Soc. for Gen. Microbiology, Canadian Soc. Microbiology, Profl. Inst. Pub. Service Can. Contbr. articles to tech. jours. Research on metabolism of microorganisms. Office: Dept. Microbiology, Faculty Sci., U. Alta., Edmonton, Alta., Can.*

WESTLAND, Alan Duane, chemist; b. Toledo, Dec. 29, 1929; s. Courtis Owen and Bertha (Palmer) W.; B.A., U. Toronto, 1953, M.A., 1954, Ph.D., 1956; m. Lieselotte Schloemer, Sept. 30, 1956. Postdoctorate fellow Munster U., West Germany, 1956-58; asst. prof. U. Ottawa, 1958-64, asso. prof., 1964-——; examiner-in-chief for chemistry Dept. Edn., Province of Ont., 1966-67. Fellow Chem. Inst. Can., Chem. Soc., Am. Chem. Soc. Studies and publs. on analytical chemistry of platinum group elements and their minerals, chem. bonds formed by transition elements. Home: 971 Elsett Dr., Ottawa, Ont., Can.*

WESTOFF, Charles Francis, Am. sociologist, demographer; b. N.Y.C., July 23, 1927; s. Frank Barnett and Evelyn (Bales) W.; A.B., Syracuse U., 1949, M.A., 1950; Ph.D., U. Pa., 1953; m. Joan P. Uszynski, Sept. 11, 1948; children—David, Carol. Instr. U. Pa., 1950-52; research asso. Milbank Meml. Fund, N.Y.C., 1952-55; research asso. Office Population Research, Princeton, 1955-62, asso. dir., 1962-——, prof. sociology, 1962-——, chmn. dept. sociology, 1965-——; asso. prof. sociology N.Y. U., also chmn. dept. sociology Washington Sq. Coll., 1959-62. Fellow Am. Sociol. Assn.; mem. Population Assn. Am. (past dir.), Internat. Union for Sci. Study Population. Author: (with R.G. Potter, P.C. Sagi) Family Growth in Metropolitan America, 1961, The Third Child, 1963; (with R.H. Potvin) College Women and Fertility Values, 1967; also articles. Research on Social aspects of fertility. Home: 221 Herrontown Rd., Princeton, N.J. 08540.*

WESTON, Edward, electrician; b. England, May 9, 1850; ed. there and studied medicine; LL.D., McGill U., Can., 1903; Sc.D., Stevens Inst. Tech., 1904, Princeton, 1910; LL.D., U. of Pa., 1924. Came to U. S., 1870, and became chemist to Am. Nickel Plating Co.; and established in Newark, 1875, the first factory in America devoted exclusively to that class of machines; consol. business, 1881, with U. S. Electric Lighting Co., of which was electrician until 1888; formed Weston Elec. Instrument Co., Waverly Park, Newark, N.J., 1888, to mfr. the Weston measuring instruments; v.p. same co., 1888-1905, pres. 1905, later chmn. bd. Charter mem. Am. Inst. E.E. (pres., 1888). Fellow A.A.A.S. Awarded Franklin medal, 1924. Introduced improvements in nickel plating; later invented several dynamo-electric machines. Died Aug. 20, 1936.

WESTON, John Fred, Am. economist; b. Ft. Wayne, Ind., Feb. 6, 1916; s. David Thomas and Bertha (Schwartz) W.; B.A., U. Chgo., 1937, M.B.A. in Bus. Econs., 1942, Ph.D., in Finance, 1948; m. June Mildred Sherman, May 16, 1942; children—Kenneth F., Byron L., Ellen J. Instr., U. Chgo., 1940-42; econ. cons. to pres. Am. Bankers Assn., 1945-46; asst. prof. Sch. Bus., U. Chgo., 1947-48; prof. Grad. Sch. Bus., U. Cal., Los Angeles, 1949-——. Dir. Mgmt. Assistance, Inc.; Ford Found. Faculty Research fellow, 1961-62. Mem. Western Econ. Assn. (past pres.), Am. Finance Assn. (pres. 1966-——). Author: Competitive Bidding, 1942; Role of Mergers in Growth of Large Firms, 1953; Procurement and Profit Renegotiation, 1961; Managerial Finance, 1962; Editor: Defense-Space Market Research, 1964. Asso. editor Jour. Finance, 1948-55, editorial bd., 1957-59. Formulated basic concepts on nature of bus. profits; developed theories for guiding bus. firms in their diversification policies; research on bus. and social effects of corporate mergers. Home: 533 Spoleto Dr., Santa Monica, Cal. 90402. Office: Sch. Bus., U. Cal. at Los Angeles, 90024.*

WESTON, Ralph Emerson, Jr., Am. phys. chemist; b. San Francisco, Nov. 9, 1923; s. Ralph Emerson and Ruth (Fields) W.; student Sacramento Jr. Coll., 1940-42; B.S., U. Cal. at Berkeley, 1946; Ph.D., Stanford, 1950; m. Virginia Priest, May 26, 1951; children—Judith Fields, Joan Drew, Barbara Sears. Research fellow in chemistry Harvard, 1949-51; mem. staff Brookhaven Nat. Lab., Upton, N.Y., 1951-——, sr. chemist, 1965-——; lectr. Hunter Coll., N.Y.C., 1963; vis. scientist Center Nuclear Studies, Saclay, France, 1960-61. Mem. Am. Chem. Soc., Sigma Xi, Phi Lambda Upsilon. Research and publs. on kinetics and mechanisms of chem. reactions; infrared and Raman spectroscopy; effects of isotopic substitution on chem. equilibria and rates of chem. reactions. Home: Library Lane, Brookhaven, N.Y. 11719. Office: Brookhaven Nat. Lab., Upton, N.Y. 11973.*

WESTON, Sir Richard, English agriculturist; b. Sutton, Surrey, Eng., 1591; s. Richard Weston; m.

Grace Harper; 7 sons, 2 daus. Supt. works to make Wey River navigable to Guilford, Eng., completed 1653. Author: Discours of Husbandrie used in Brabant and Flanders (written while residing in Low Countries), circa 1645, printed, 1650. Introduced canal locks, irrigation to increase hay crops, rotation of crops. Died 1652.

WESTPHAL, Carl Friedrich Otto, German psychiatrist; b. Berlin, Mar. 23, 1833; prof., Berlin. Author: Geschichtliche Abhandlungen, 2 vols., 1892. Research on psychoses; introduced term paranoia; gave 1st description of agoraphobia; demonstrated (independently of Wilhelm Erb) diagnostic value of patellar tendon reflex, also its absence (Westphal's sign) in locomotor ataxia, 1875; described eye muscle paralization by alteration of 3d cranial nerve; described pseudosclerosis, 1883, also a gray nucleus under aquaduct of Sylvius giving rise to some fibers of trochlear nerve (Westphal's, or Edinger's, nucleus), 1887. Died Kreuzlingen, Germany, Jan. 27, 1890.

WESTPHAL, Otto Hermann Eduard, German biochemist; b. Berlin, Germany, Feb. 1, 1913; s. Wilhelm and Olga (Meyer-Delius) W.; student chemistry U. Freiburg (Germany), 1931-32, U. Berlin, 1932-34; D.Sc., U. Heidelberg (Germany), 1937, Habilitation, 1941; M.D. h.c., U. Giessen (Germany), 1967; m. Olga von Gayling, Apr. 5, 1941; children—Nikolaus, Katharina. Chief dept. biochemistry U. Göttingen, 1941-46, docent, 1942-52; dir. Wander Research Inst., Säckingen and Freiburg, 1947-61; dir. Max-Planck-Institute for Immunobiology, Freiburg, 1962-——; prof. chemistry U. Freiburg, 1952-——. Recipient Claude Bernard medal, Montréal, 1956, Pasteur medal, Paris, 1959, Emil Fischer medal, Munich, 1962, Emil von Behring Preis, Marburg, 1964, Aronson-Preis, Berlin, 1964, Carus medal Acad. Leopoldina, Halle, 1965, Paul Erlich Preis, Frankfurt 1968. Fellow N.Y. Acad. Scis.; mem. Acad. Wiss. Lit. Mainz, Leopoldina Halle, Gesellschaft für Biol. Chem. (pres. 1966-——), Akad. Wiss. Göttingen, Soc. franc. Microbiol. (corr.), Soc. Suisse Biochem. (corr.). Research and numerous publs. on syntheses of sugars and derivatives, prodn. of artificial antigens (vaccines); immunochem. and biochem. investigations of surface structures of bacteria. Home: 6 Eggstrasse, 78 Freiburg/Germany. Office: 51 Stübeweg 78, Freiburg/Germany.*

WESTPHAL, Ulrich Friedrich, biochemist; b. Göttingen, Germany, May 3, 1910; s. Emil and Mathilde (Klingeberg) W.; Ph.D., U. Göttingen, 1933; Dr.phil.habil., U. Berlin (Germany), 1941; m. Ilse Schlegelmilch, Mar. 20, 1940; children—Karin, Antje, Birgit. Came to U.S., 1949, naturalized, 1958. Liebig fellow, staff mem. Inst. for Organic Chemistry, Inst. Tech., Free City of Danzig, 1933-36; staff mem., asst. dir. Kaiser Wilhelm Institut für Biochemie, Berlin-Dahlem, 1936-45; chief chem. dept. Medizinische Klinik, U. Tuebingen (Germany), 1945-49; research scientist, chief protein and steroid sect. biochemistry dept. U.S. Army Med. Research Lab., Ft. Knox, Ky., 1949-58, chief biochemistry div., 1959-61; prof. biochemistry dept. U. Louisville Sch. Medicine, 1961-——; Rockefeller Found. Research fellow Columbia U., 1939. Mem. biochemistry study sect. div. research grants USPHS, 1961-65. Recipient Research Career awards USPHS, 1962-——. Mem. Am. Soc. Biol. Chemists, Am. Chem. Soc., Soc. for Exptl. Biology and Medicine, A.A.A.S., Deutsche Physiologisch-Chemische Gesellschaft, Sigma Xi. Research and numerous publs. on structure of estrogens; isolation, structure, partial synthesis; excretion of progesterone; steroid interactions with proteins; corticosteroid-binding globulins. Home: 10217 Hartley Dr., Louisville 40223.*

WESTRUM, Edgar Francis, Jr., Am. chemist; b. Albert Lea, Minn., Mar. 16, 1919; s. Edgar Francis and Nora (Kipp) W.; student Hamline U., 1936-38; student U. Minn. Inst. Tech., 1938-40; B. Chem. U. Cal., Berkeley, 1940, Ph.D., 1944; m. Florence Emily Barr, June 13, 1943; children—Ronald Mark, James Scott, Michael Lauren, Margaret Kristin. Scientist, U. Chgo. Metall. Lab., 1944-46; scientist U. Cal. Lawrence Radiation Lab., Berkeley, 1946; prof. dept. chemistry U. Mich., 1947-——. Cons., AEC Labs. Fellow Am. Phys. Soc., A.A.A.S.; mem. Am. Chem. Soc., Netherlands Phys. Soc., Sigma Xi. Author (with D.R. Stull, G.C. Sinke) The Thermodynamics of Organic Compounds, 1967. Editor, Bull. of Thermodynamics and Thermochemistry, 1964-——; Jour. Thermodynamics, 1967-——. Research, numerous publs. on chem. thermodynamics, solid state physics, thermal measurements in cryogenic, ambient, high temperature ranges. Home: 2019 Delaware Dr., Ann Arbor, Mich. 48103.*

WESTWATER, James William, Am. chem. engr.; b. Danville, Ill., Nov. 24, 1919; s. John and Lois (Maxwell) W.; B.S., U. Ill., 1941; Ph.D., U. Del., 1948; m. Elizabeth Jean Keener, June 9, 1942; children—Barbara, Judith, David, Beverly. Mem. faculty U. Ill., Urbana, 1948-——, prof., head dept. chem. engring., 1962-——. Papers chmn. 5th Nat. Heat Transfer Conf., Buffalo, 1960; chmn. 3d Internat. Heat Transfer Conf., Chgo., 1966. Recipient Conf. award 8th Nat. Heat Transfer Conf., 1965, William H. Walker award Am. Inst. Chem. Engrs. 1966. Mem. Am. Inst. Chem. Engrs. (chmn. energy conversion and transport division 1963, member board of directors 1968—), American Chemical Soc., Am. Soc. M.E., Am. Soc. Engring. Edn. Contbr. articles to profl. jours., textbooks. Developed specialized techniques for taking motion pictures at speeds up to 6,000 frames per second through a microscope at magnifications up to 66 times. Home: 116 W. Iowa St., Urbana, Ill. 61801.*

WESTWOOD, John Craig Nicholson, bacteriologist; b. Sonepur, India, May 27, 1917; s. John David and Anne M. (Paulin) W.; B.A., Cambridge U., 1939, M.B., B.Chir., 1942; diploma in Bacteriology, London Royal Coll., 1947. Dir. virus unit with rank of sr. prin. sci. officer Central Pub. Health Lab., Colindale, 1954-55; dep. chief sci. officer Microbiol. Research Establishment, Porton, Eng., 1955-63; faculty U. Ottawa, Ont., 1963—, prof., head dept. bacteriology, 1966—; cons. virology Ottawa Civic Hosp., 1963—; head bacteriology dept. Ottawa Gen. Hosp., 1966. Mem. Can. Soc. Microbiology, Soc. Gen. Microbiology, Canadian Assn. Med. Bacteriologists. Research, numerous publs. on tissue culture and virology, poxvirus and their antigenic structure, mechanisms by which they cause disease and possibilities of specific treatment. Home: 21 Briarcliff Dr., P.O. Box 124, R.R. 1, Ottawa. Office: 275 Nicholas St., Ottawa 2, Ont., Can.*

WESTWOOD, Melvin Neil, Am. pomologist; b. Hiawatha, Utah, Mar. 25, 1923; s. Neil and Ida (Blake) W.; student U. Utah, 1949; B.S., Utah State U., 1953; Ph.D., Wash. State U., 1956; m. Wanda Mae Shields, Oct. 12, 1946; children—Rose Dawn (Mrs. Richard Lee Summers), Nancy Gwen, Robert Melvin, Kathryn Mae. Research fellow Wash. State U., 1953-55; research pomologist U.S. Dept. Agr., Wenatchee, Wash., 1955-60; asso. prof. horticulture Ore. State U., Corvallis, 1960-67, professor of horticulture, since 1967—. Recipient of the Joseph H. Gourley award for research in pomology Am. Soc. Hort. Sci., 1958. Mem. Am. Soc. Hort. Sci., Am. Soc. Plant Physiology, Am. Pomology Soc., A.A.A.S., N.Y. Acad. Scis., Am. Sci. Affiliation. Contbg. author: Tree Fruit Nutrition, 1966. Research, numerous publs. on use of hormones and chem. regulators on plant devel., dynamics of seasonal growth of tree stems and fruits, physiology of winter dormancy in seeds and buds of trees, methods of propagating hard-to-root cuttings of woody plants, tree rootstock physiology as related to dwarfing, flowering and nutrition. Home: 2130 Elmwood Pl., Corvallis, Ore. 97330.*

WETHERBEE, David Kenneth, Am. biologist; b. Worcester, Mass., Jan. 29, 1927; s. Kenneth Brackett and Olive (Pike) W.; B.A., Clark U., 1950, M.A., 1952; Ph.D., U. Conn., 1955; m. Nancy Moore Schimpf, June 18, 1955; children—Rachel Olive, Sally Mills, Polly Stockwell, Peter Kenneth. Biologist, Neb. Game Commn., Lincoln, 1957-58; asst. prof. biology Fisk U., 1956-57; biologist N.H. Fish and Game Dept., Concord, 1957; biologist Patuxent Wildlife Research Refuge, Laurel, Md., 1958; biologist U.S. Fish and Wildlife Service, Amherst, Mass., 1959—; asso. prof. U. Mass., Amherst, 1959—. Author: Birds and Mammals of Worcester County, Mass., 1945. Research and publs. on relationships between infantile condition of birds and evolution; physiology and ecology of avian sterility. Home: Whitaker Rd., New Salem, Mass. 01355. Office: Holdsworth Hall, U. Mass., Amherst, Mass. 01003.*

WETHERILL, George West, Am. geophysicist; b. Phila., Aug. 12, 1925; s. George West and Leah (Hardwick) W.; Ph.B., U. Chgo., 1948, S.B., 1949, S.M., 1951, Ph.D., 1953; m. Phyllis May Steiss, June 17, 1950; children—Rachel, George, Sarah. Staff dept. terrestrial magnetism Carnegie Instn. Washington, 1953-60; vis. prof. geochemistry Cal. Inst. Tech., 1959; prof. geophysics and geology U. Cal. at Los Angeles, 1960—. Chmn. subcom. on nuclear geophysics Nat. Acad. Sci.-NRC, 1961—. Fellow Am. Geophys. Union; mem. Am. Phys. Soc., Geochem. Soc., Phi Beta Kappa, Sigma Xi. Research and numerous publs. on devel. new techniques for measurement ages of rocks and minerals by radioactive decay, applied techniques to problems of Precambrian correlation in N.Am., Europe. Home: 427 Denslow Av., Los Angeles 90049.*

WETMORE, (Frank) Alexander, Am. biologist; b. North Freedom, Wis., June 18, 1886; s. Nelson Franklin and Emma Amelia (Woodworth) W.; B.A., U. Kan., 1912; M.S., George Washington U., 1916, Ph.D., in Zoology, 1920, Sc.D., 1932; Sc.D., U. Wis., 1946, Centre Coll. Ky., 1947, Ripon Coll., 1959; m. Fay Holloway, Oct. 13, 1912; 1 dau., Margaret Fenwick (Mrs. John Graydon Harlan, Jr.); m. 2d, Beatrice Thielen, Dec. 16, 1953. Asst., Mus., U. Kan., 1905-08, 10; asst. Colo. Mus. Natural History, 1909; staff Bur. Biol. Survey, U.S. Dept. Agr., 1910-24, biologist, 1924; supt. Nat. Zool. Park, 1924-25; with Smithsonian Instn., Washington, 1925—, sec., 1945-52, research asso., 1953—. Chmn. com. selection Intern-Am. and Philippine fellowships John Simon Guggenheim Meml. Found., 1946-63, mem. adv. bd., 1953-63. Trustee Textile Mus. Washington. Recipient Isidore Geoffroy St. Hilaire medal Société Nationale d'Acclimitation de France, 1927, Otto Herman medal Hungarian Ornithol. Soc., 1931, Alumni award for achievement in sci. George Washington U., 1945, Orden de Mérito Carlos Manuel de Céspedes, Cuba, 1948, Brewster medal Am. Ornithologists' Union, 1959, Explorers Club medal, 1962; cited for distinguished service U. Kan., 1941. Fellow Am. Ornithologists Union; mem. Nat. Acad. Sci., Am. Philos. Soc., Nat. Geog. Soc. (trustee 1933—). Research and numerous publs. on taxonomy, classification of higher groups, ecology and geog. distbn. of birds, fossil birds. Home: 5901 Osceola Rd., Washington 20016. Office: Smithsonian Instn., Washington 20560.*

WETMORE, Ralph Hartley, botanist; b. Yarmouth, N.S., Can., April 27, 1892; s. Herman Augustus and Josephine Cordelia (Moses) W.; B.Sc., Acadia U., 1921; A.M., Harvard, 1922, Ph.D., 1924; D.Sc. (hon.); m. Marion Geraldine Silver, June 20, 1923 (dec.); children—Katherine Beryl, Evelyn Jean; m. 2d, Mary Olive (Hawkins) Smith, June 24, 1940. Naturalized, 1936. Nat. Research fellow, 1924-25; asst. prof. biology Acadia U., Wolfville, N.S., 1925-26; faculty Harvard, 1926-62, prof. botany, 1943-62, emeritus, 1962—, chmn. dept., 1932-34, dir. Bot. Labs., 1933-34, chmn. dept. biol., 1946-47, dir. Bio. Labs., 1953-56. Fellow Nat. Acad. Scis., Am. Acad. Arts and Scis., A.A.A.S., N.Y Acad Sci; mem Scandinavian Soc. Plant Physiology, Am. Soc. Plant Physiology, Bot. Soc. Am. (pres. 1953), Am. Soc. Naturalists, Torrey Bot. Club, N.E. Bot. Club (pres. 1949-52), Soc. Study of Devel. and Growth (pres. 1948-49), Biol. Stain Commn., Internat. Soc. Plant Morphologists, Internat. Assn. Wood Anatomists, French Soc. Plant Physiology, Sigma Xi. Contbr. numerous articles in sci. jours. Contbd. to understanding of influence of environment on hereditary pattern of orgn. in vascular plants. Home: 12 Francis Av., Cambridge, Mass. 02138.*

WETTERER, Erik Josef, German physiologist; b. Mannheim, Jan. 25, 1909; s. Josef and Elisabeth (Back) W.; M.D., U. Heidelberg; m. Erna Doesel, June 25, 1948. Asst., Munich Physiol. Inst., 1934; agrégé physiology U. Munich, 1940, prof. physiology, 1952. Mem. German Physiol. Soc., German Soc. Circulatory Research, Soc. Morphologists and Physiologists. Publns. on circulatory dynamics and application of electronics in physiology. Research on elec. methods in physiology; physiology of circulation. Address: Physiologisches Inst., Pettenkoferstrasse 12, Munich, West Germany.

WETTSTEIN, Albert, Swiss chemist; b. Frauenfeld, Switzerland, Mar. 3, 1907; s. Albert and Hedwig (Stutz) W.; Dr.phil., U. Zürich (Switzerland), 1930; Dr.med. (hon.) U. Basel (Switzerland), 1951; Dr.pharm. (hon.) U. Torino (Italy), 1961; hon. Prof., U. Freiburg, (Germany), 1965; m. Berthe Schmutz, July 3, 1941; children—Monique, Ulric. Chemist, CIBA A.G., Basel, 1931—, dir. pharm. research and prodn., 1955-64, exec. com., 1957-64, mem. bd. com., 1964—; prof., lectr. U. Freiburg, 1966—; dir. chem. cos. in Switzerland, Gt. Britain, U.S., Can., Italy, India. Recipient Royal Swedish Acad. Engring. Scis., prize, 1957, Marcel Benoist prize Switzerland, 1960. Mem. Swiss Chem. Soc. (pres. 1966-68), Swiss Pharmakopoea Com. Research and numerous publs. on chemistry of steroid hormones; 1st syntheses of natural male hormones and adrenocortical hormone aldosterone; isolation, synthetic biosynthetic and microbiol. work toward androgenic, progestative and corticoid hormones; isolation and research on antibiotics. Home: 40 Hellring, CH 4125 Riehen-Basel, Switzerland. Office: CIBA Ltd., CH 4000, Basel, Switzerland.*

WEVER, Ernest Glen, Am. psychologist; b. Benton, Ill., Oct. 16, 1902; s. Ernest Sylvester and Mary Jane (Shurtz) W.; A.B., Ill. Coll., 1922; A.M., Harvard, 1924, Ph.D., 1926; m. Suzanne Rinehart, 1928. Instr., U. of Calif., 1926-27; instr., Princeton U., 1927-29, asst. prof., 1929-31, asso. prof., 1931-41, Eugene Higgins professor of psychology, 1941—. Member of The American Psychological Association, Acoustical Soc. of America, Nat. Acad. of Sciences, Sigma Xi. Frequent contbr. to tech. and scientific jours. Research on physiological and experimental psychology; acoustics. Home: 29 Snowden Lane, Princeton, N.J.

WEXLER, Bernard Carl, Am. endocrinologist; b. Boston, Apr. 30, 1923; s. William and Dora (Gerson) W.; B.S., U. Ore., 1946; M.A., U. Cal. at Berkeley, 1947; Ph.D., Stanford, 1952; m. Jean A. Berkel, Sept. 3, 1946; children—Nancy Carol, Helen Jane, William Carl. Responsible investigator Stanford Research Inst., 1950-52; lectr. Dominican Coll., 1951-52; instr., Stanford, 1952; dir. endocrine and pathology labs. Baxter Labs., Morton Grove, Ill., 1952-55; research asso. May Inst. for Med. Research, Cin., 1955-58, asst. dir., 1958-62, dir., 1962—; faculty U. Cin. Coll. Medicine, 1955—, asso. prof., 1958—. Marrell scholar in gerontology, 1955-58; Advanced Research fellow Am. Heart Assn., 1959-63; recipient Research Career award Nat. Heart Inst., NIH. Mem. Endocrine Soc., Diabetes Soc., Gerontology Soc., Am. Soc. Pathologists and Bacteriologists, Am. Soc. for Exptl. Pathology, Soc. for Exptl. Biology and Medicine, Am. Assn. Med. Colls., A.A.A.S., N.Y. Acad. Medicine, Sigma Xi. Research and numerous publs. on hormonal aspects of heart attacks, strokes, and arteriosclerosis, particularly degenerative and aging effects of adrenal stress hormones. Home: 7640 De Mar Rd., Cin. 45243. Office: 421 Ridgeway Av., Cin. 45229.*

WEXLER, Harry, Am. meteorologist; born Fall River, Mass., Mar. 15, 1911; s. Samuel and Mamie (Starr) W.; S.B., Harvard, 1932; D.Sc., Mass. Inst. Tech., 1939; m. Hannah Paipert, Dec. 3, 1934; children—Susan Carol, Libby. Meteorologist U. S. Weather Bur., 1934-40, 41-42, chief sci. services, 1946-55; asst. prof. meteorology University Chicago, 1940-41; dir. meteorol. research, U. S. Weather Bureau, 1955—; chief scientist U. S. Expdn. to Antarctic for International Geophys. Year, 1955-58; chairman meteorol. panel of Nat. Acad. of Science, com. to study biological effects of atomic radiation; mem. Nat. Acad. to Sci. space sci. bd., com. polar research, also chmn. com. meteorological aspects of satellites, U. S. nat. com. IGY. Recipient Robert M. Losey award for outstanding services to aeronautics, Inst. Aeronautical Sci., 1945; Exceptional Service award, Dept. of Air Force, 1956, Dept. Commerce, 1958; Distinguished Pub. Service award, U. S. Navy, 1960; Nat. Civil Service League award, 1961. Fellow A.A.A.S., American Astronautical Society, also the American Acad. Arts and Scis.; mem. Am. Meteorol. Soc., Am. Geophys. Union, Royal Meteorol. Soc. Gt. Britain Am. Vets. Com. Author sci. articles. Devel. theory of ozone circulation explaining high concentration in Antarctic atmosphere; research on atmospheric and solar radiation; dynamic and synoptic meteorology; storms in upper atmosphere; uses of weather satellites; climatic variations; volcanic dust. Died Falls Church, Va., Aug. 11, 1962.

WEXLER, Solomon, Am. chemist; b. Milw., June 16, 1919; s. David and Rebecca (Leavitt) W.; S.B., U. Chgo., 1941, Ph.D., 1948; m. Ruth Weiss, June 17, 1951; children—Lauren Renee, Ronald Myron. Chemist, Metall. Lab., U. Chgo., 1942-44, Los Alamos Lab., U. Cal., 1944-46, Oak Ridge Nat. Lab., Monsanto Chem. Co., 1946; chemist Argonne (Ill.) Nat. Lab., 1948—, lectr. in radioactivity and neutron physics, 1950-51. Contbg. author: Actions Chimiques et Biologiques des Radiations, 1965. Research and publs. on primary effects nuclear transformations. Home: 5825 Eldon Pl., Downers Grove, Ill. 60515. Office: 9700 S. Cass Av., Argonne, Ill. 60440.*

WEYER, Edward Moffat, Jr., Am. anthropologist; b. Washington, Pa., July 12, 1904; s. Edward Moffat and Julia (Ross) W., Sr.; B.S., Washington and Jefferson Coll., 1925; Ph.D., Yale, 1930; anthrop. studies, Mato Grosso, Brazil, Chiapas, Mexico; m. Susann Lenore Moffat, May 29, 1935; children—Georgia Marodes, Andrew Moffat, Douglas Moffat. Archaeol. asst. on expdn. of Met. Mus. of Art to Luxor, Egypt, 1925-26; archaeologist on Stoll-McCracken Expdn. of Am. Mus. Natural History to Aleutian Islands, Bering Sea and Bering Straits, 1928; archaeol. and geog. research, Ariz., N.M. and Colo. for Am. Mus. Natural History, 1929; Peary Meml. Expdn. to N. Greenland, 1932; tchr. anthropology and geography, Washington and Jefferson Coll., 1931; photographer for Metropolis, The Story of New York City in pictures, pub. by Harper & Bros., 1934; editor Natural History Mag., 1935-57; acting curator and lectr. in anthropology Mus. Anthropology, U. Cal. at Berkeley, 1958; dir. Sch. Am. Research, Santa Fe, N.M., 1960-64; editor-in-chief N.Y. Acad. Scis., 1965-67. Fellow A.A.A.S., N.Y. Acad. Scis.; mem. Arctic Inst. N.Am., Am. Geog. Soc., Archeol. Soc. of N.M., Soc. Am. Archeol., Am. Anthropol. Assn., Phi Beta Kappa, Sigma Xi. Author: The Eskimos, 1932; An Aleutian Burial, 1929; Archaeological Material from Port Moller, 1930, Anthropological Papers, A.M.N.H.; Jungle Quest; Primitive Peoples Today, 1959; Two-Star Position Finding, 1968. Editor: Strangest Creatures on Earth, 1953; Illustrated Library of Natural Sciences, 1958. Contbr. mag. articles, book reviews, Encyl. Arctica. Spl. studies in seasonal variation in sunlight in different latitudes. Home: Grove Beach Rd., Westbrook 06498. Office: P.O. Box 175, Westbrook, Conn. 06498.*

WEYL, Hermann, mathematics; b. Elmshorn, Germany, Nov. 9, 1885; s. Ludwig and Anna (Dieck) W.; ed. univ. of Munich and Göttingen;; Ph.D., Göttingen, 1908; hon. Dr. Philosophy, University of Oslo, 1929; hon. Dr. Technology, Hochschule, Stuttgart, 1929; D.Sc., University of Pennsylvania, 1940. Columbia, 1954; Dr. Math. (honorary), Tech. Hochschule, Zürich, 1945; Dr., Université de Paris, 1952; married Helene Joseph, Sept. 1, 1913, (dec.); children—Fritz Joachim, Michael; m. 2d Ellen Baer Lohnstein, Jan. 7, 1950. Privatdozent U. of Göttingen, 1910-13; professor mathematics Tech. Hochschule, Zürich, 1913-30; Jones research prof. in mathematical physics, Princeton, 1928-29; prof. mathematics, U. of Göttingen, 1930-33; prof. mathematics Inst. for Advanced Study, Princeton, since Dec. 1, 1933. Mem. National Academy, American Academy Arts and Sciences, American Philosophical Society, American Math. Society, and mem. or hon. mem. various European socs. including Royal Soc. of London, Pontificial Acad. of Sciences. Award Lobatschefsky prize, Kazan, 1925. Author of 8 books pub. in Europe, and 7 books pub. in U. S.—the Open World, 1932; Mind and Nature, 1934; The Classical Groups, 1939; Algebraic Theory of Numbers, 1940; Meromorphic Functions and Analytic Curves, 1943; Philosophy of Mathematics and Natural Sci., 1949; Symmetry, 1952; also articles in mags. Research work in differential equations, topology, relativity, theory, infinitesimal geometry, group theory, philosophy of mathematics. Died Dec. 8, 1955.

WHALEY, W(illiam) Gordon, Am. biologist; b. N.Y.C., Jan. 16, 1914; s. Frank H. and Mae (Manson) W.; B.S., Mass. State Coll., 1936; Ph.D., Columbia, 1939; m. Clare Youngren, 1938; 1 dau., Patricia Anne. Lectr. Barnard Coll. Columbia, 1939-40; instr. Columbia, 1940-43; asso. agronomist to sr. geneticist, bur. plant industry, soils and agrl. engring. U.S.D.A., 1943-46, cons. geneticist agrl. research administration, 1946-48; asso. prof. botany U. Tex., 1946-48, chmn. dept., 1948-62, prof., 1948—, dir. plant research inst., 1948-63, dir. cell research inst. 1964—, asso. dean grad. sch., 1955-57, dean, 1957—; vis. prof. Rockefeller U., 1964—. Chmn. adv. com. for sci. edn. NSF, 1966—. Fellow A.A.A.S.; mem. Internat., Am. socs. cell biology, Am. Soc. Plant Physiologists, Soc. Developmental Biology, Nat. Assn. Land-Grant Colls. and State Univs. (chmn. commn. on grad. edn.), Bot. Soc. Am., Torrey Bot. Club, Sigma Xi. Club: Cosmos. Author: Biology for Everyone, 1948; Principles of Biology, 1954, rev., 1957, 64; also articles. Editor: Grad. Jour.; Jour. of Proc., Assn. Grad. Schs. Editorial bd. Jour. Ultrastructure Research. Contributions in analyses of gene action; physiological aspects of heterosis; developmental changes in cell ultrastructure; structure, differentiation, and functioning of Golgi apparatus. Home: 313 E. 35th St., Austin, Tex. 78705. Office: Cell Research Inst., U. Texas, Austin, Texas 78712.*

WHALING, Ward, Am. physicist; b. Dallas, Sept. 29, 1923; s. Horace Morland and Annie Byrd (Ward) W.; B.A., Rice U., 1944, M.A., 1947, Ph.D., 1949; m. Mary Lou Slighter, Sept. 1, 1955; children—Anne Buell, Carol Susan. With Cal. Inst. Tech., Pasadena, 1949—, Research fellow, 1949-52, prof. physics, 1962—. Mem. Am. Phys. Soc. (regional sec. for Western states 1964—). Research in nuclear spectroscopy, nuclear masses, range-energy relations. Home: 401 S. Parkwood Av., Pasadena 91107. Office: 1201 E. California St., Pasadena, Cal. 91109.*

WHAPLES, George William, Am. mathematician; b. Neponset, Ill., Nov. 27, 1914; s. George Warren and Harriet (Hill) W.; B.A., Knox Coll., 1935; M.A., U. Wis., 1937, Ph.D., 1939; m. Miriam Karpilow, Aug. 24, 1949; children—Rebecca Tamar, Barbara Caroline. Fellow Ind. U., 1939-41; asst. Inst. for Advanded Study, Princeton, N.J., 1941-42; instr. Johns Hopkins, 1942-43; asst. prof. U. Pa., 1943-46, U. Wis., Madison, 1946-47; faculty Ind. U., Bloomington, 1947—, prof. math., 1958—. Research and publs. on existence of local class fields. Home: 424 S. Dunn St., Bloomington, Ind. 47403.*

WHARTON, George Willard, Am. biologist; b. Belleville, N.J., Jan. 25, 1914; s. George Willard and Marguerite (Middleton) W.; B.S., Duke U., 1935, Ph.D., 1939; m. Camilla Richie, Mar. 28, 1938; children—Jane Middleton, Margaret Agnes, George Willard; m. 2d, Mildred Gerke. Faculty Duke, 1939-53; prof., head dept. zoology U. Md., Balt., 1953-61; prof., chmn. dept. zoology and entomology Ohio State U., Columbus, 1961—. Chmn. tropical medicine and parasitology study sect., spl. cons. USPHS, 1956-62. Guggenheim fellow, 1950-51. Mem. Nat. Allergy and Infectious Inst., Soc. Systematic Zoology (past pres., past mem. council), Am. Soc. Parasitologists (past editorial bd., past mem. council), A.A.A.S., Phi Beta Kappa, Sigma Xi. Author: An Introduction to Acarology, 1952; A Manual of Chiggers, 1952; A Manual of Mesostigmatid Mites Parasitis on Vertebrates, 1958. Research and numerous publs. on systematic zoology, acarology, parasitology, ecology. Home: 1535 Trentwood Rd., Columbus 43221.*

WHARTON, Thomas, English physician; b. Durham, Eng., Aug. 31, 1614; s. John and Elizabeth (Hodson) W.; grad. Pembroke Coll., Cambridge (Eng.) U., 1638; M.D., Trinity Coll., Oxford (Eng.) U., 1647; m. Jane Ashbridge; children—Thomas, Charles, William. Physician to St. Thomas' Hosp., London, Eng., 1659-73. Fellow Royal Coll. Physicians (censor 1658, 61, 66-68, 73). Author: Adenographia, 1656; Sive glandularum descriptio, 1656. Discovered duct of submaxillary gland for conveyance saliva into mouth (Wharton's duct); physician in London throughout plague, 1665-73. Died London, Nov. 15, 1673.

WHEAT, Myron William, Jr., Am. physician; b. Sapulpa, Okla., Mar. 24, 1924; s. Myron William and Mary (Hudiburg) W.; A.B., Wash. U., 1949, M.D. cum laude, 1951; m. Erlene Adele Plank, June 12, 1949; children—Penelope Louise, Myron William III, Pamela Lynn. Clin. fellow chest surgery Barnes Hosp., St. Louis, 1956-58; instr. Wash. U., 1956-58; with U. Fla., Gainesville, 1958—, prof. surgery, chief div. thoracic and cardiovascular surgery, 1965—, chmn. cardiovascular group J. Hillis Miller Health Center, 1964—, dir. profl. services, emergency room, 1958-64, 1965-66, dir. Cardiovascular Lab., 1966—, chmn. Preceptors U. Fla. Coll. Medicine, 1965—. Recipient George F. Gill Anatomy prize, 1948. Fellow Am. Coll. Cardiology (gov.), A.C.S.; mem. Alachua County Med. Soc., Am. Assn. Thoracic Surgery, Am. Heart Assn., A.M.A., Fla. Thoracic Soc., N.Y. Acad. Scis., Southeastern Surg. Congress, So. Thoracic Surg. Assn., Alpha Omega Alpha. Research on heart surgery; developed tng. program in thoracic and cardiovascular diseases; localized radioactive digitalis in heart muscle, established relationship between lysosomes and heart disease in humans; carried out first complete replacement of the aortic root and aortic valve with reimplantation of coronary arteries in a human being, 1962; co-originator drug therapy for dissecting aneurysms of aorta. Home: 2844 N.W. 4th Lane, Gainesville, Fla. 32601.*

WHEATSTONE, Sir Charles, physicist, inventor; b. Gloucester, Eng., Feb. 6, 1802; s. W. Wheatstone; D.C.L., Oxford U., 1862; LL.D., Cambridge U., 1864; m. Emma West, Feb. 12, 1847; 5 children. Maker of musical instruments, London, from 1823; apptd. prof. exptl. philosophy, King's Coll., London, 1834. Fellow Royal Soc., 1836; fgn. asso. mem. French Acad., 1873. Knighted, 1868. Author: Scientific Papers, 1879. Conducted expts. on sound and light, circa 1823; suggested stereoscope; demonstrated speed of electricity in conductor; designed (with William F. Cooke) electric telegraph, 1837, also single needle telegraph, 1845; invented concertina, 1829, rheostat kaleidophon, automatic telegraph, Wheatstone's bridge to measure elect. resistance; improved dynamo; worked on magnetoelectricity, submarine telegraphy, cryptography, and vision. Died Paris, France, Oct. 19, 1875.

WHEDON, George Donald, Am. med. investigator; b. Geneva, N.Y., July 4, 1915; s. George Dunton and Elizabeth (Crockett) W.; A.B., Hobart Coll., 1936; M.D., U. Rochester, 1941; Sc.D., Hobart Coll., 1967; m. Margaret Brunssen, May 12, 1942; children—Karen Anne, David Marshall. Asst. in medicine U. Rochester Sch. Medicine, 1942-44; asst. physician outpatient dept. N.Y. Hosp., N.Y.C., 1944-47, physician, 1947-52; instr. medicine Cornell U. Med. Coll., 1944-50, asst. prof., chief Metabolic Diseases br. Nat. Inst. Arthritis and Metabolic Diseases, NIH, 1950-52; 1952-65, asst. dir. Inst., 1956-62, dir., 1962—. Mem. sub.-com. on calcium, com. on dietary allowances Food and Nutrition Bd., NRC, 1959-64; cons. space medicine div. NASA. Diplomate Am. Bd. Internal Medicine. Mem. Endocrine Soc., Am. Physiol. Soc., Assn. Am. Physicians, Am. Diabetes Assn., Am. Inst. Nutrition, Am. Fedn. Clin. Research, Am. Rheumatism Assn., Gerontol. Soc., N.Y., Md. acads. scis., A.A.A.S., Am., Pan-Am. med. assns. Editorial bd. Jour. Clin. Endocrinology and Metabolism, 1960-67, Calcified Tissue Research, 1967—. Research, publs. in metabolic and physiol. aspects of convalescense and immobilization; techniques for continuous measurement of human energy metabolism; metabolic studies of mineral diseases with emphasis on interrelationships of nutritional and hormonal factors in osteoporosis; 1st studies of metabolic effects of space flight. Home: 5605 Sonoma Rd., Bethesda, Md. 20034. Office: Nat. Inst. Arthritis and Metabolic Diseases, NIH, Bethesda, Md. 20014.*

WHEELER, C(harles) Gilbert, chemist; b. London, Canada, July 23, 1836; s. William and Caroline M. W.; B.S., Harvard, 1858; studied in German univs.; m. Sarah Jenkins, May 10, 1863. Asst. state geologist of Mo., 1859-61; U. S. consul to Nuremberg, 1862-67; traveled in Europe and N. Africa, 1867-68; prof. chemistry U. Chgo., Chgo. Med. Coll., 1868; frequent visits to Mexico and C.Am., examining mines for Am. capitalists, 1868-1900; invented Babcock chem. fire extinguisher, 1869; sci. expert Bell Telephone Co. and other cos. in patent and other litigation; state commr. from Ill. to Vienna Expn., 1873; pres. Chicago Coll. Pharmacy, 1882; geologist and interpreter in commn. to examine route Nicaragua Canal, 1899; consul for Republic of Panama, Chgo. Author: Natural History Charts; Catalogue Polyglottus; Determinative Mineralogy; Chemistry of Building Materials; Medical Chemistry. Died 1912.

WHEELER, George Carlos, Am. biologist; b. Bonham, Tex., Apr. 10, 1897; s. Charles Allen and Elizabeth (Truss) W.; B.A., Rice U., 1918; M.Sc., Harvard, 1920, D.Sc., 1921; m. Esther Wadsworth Hall, Dec. 19, 1921 (dec. 1940); m. 2d, Jeanette Elizabeth Norris, Dec. 22, 1941; children—Walter Hall, Ralph Allen. Faculty, Syracuse U., 1921-26; faculty U. N.D., Grand Forks, 1926-67, prof. biology, 1926-65, univ. prof., 1965-67, emeritus, 1967—, head dept., 1926-63; research scientist Whittell Forest, U. Nev., 1967—. Mem. A.A.A.S., Am. Assn. U. Profs., Entomol. Soc. Am., Ecol. Soc. Am., Soc. Systematic Zoology. Author: (with Jeanette Wheeler), The Ants of North Dakota, 1963; also numerous articles. Research on ant larvae world, biogeography N.D.; pioneered studies on fauna N.D. Home: 3206 Gypsum Rd., Reno 89503.*

WHEELER, Harold Alden, Am. radio engr.; b. St. Paul, May 10, 1903; s. William Archie and Harriet Maria (Alden) W.; B.S., George Washington U., 1925; student Johns Hopkins, 1925-28; m. Ruth Gregory, Aug. 25, 1926; children—Dorothy, Caroline, Alden Gregory. Began with the Hazeltine Corp., Little Neck, N.Y., 1924-45, engr., 1924-39, v.p., chief cons. engineer, 1939-45, vice president, 1959-65, chmn., 1965—; cons. radio physicist, Great Neck, N.Y., since 1946; pres. Wheeler Labs., Inc., 1947-68, chief scientist, 1968—; expert consultant to Office Sec. Def. on guided missiles, 1950-53. Mem. Def. Sci. Bd., 1961-64. Recipient Medal of Honor I.E.-E.E., 1964; Armstrong medal Radio Club of Am., 1964. Fellow Radio Club of America, I.E.-E.E.; member of the Inst. E.E. Gt. Britain (asso.), Sigma Xi. Author of: Wheeler Monographs, volume I, 1953. Contributed articles sci., profl. jours. Holder 180 U. S. patents. Invented diode automatic volume control for radio receivers, 1926; designer of missile-tracking radar antennas and microwave circuits; work on submarine and subsurface radio systems. Home: 18 Melbourne Rd. Office: 122 Cutter Mill Rd., Great Neck, N.Y. 11021.

WHEELER, Harry Ernest, Am. botanist; b. West Charleston, Vt., Jan. 25, 1919; s. Gary Hollis and Hattie (Locke) W.; B.S., U. Vt., 1941; M.S., La. State U., 1947, Ph.D., 1949; m. Naomi Beatrice Lavo, Apr. 2, 1944. Asst. prof. botany La. State U., 1949-53, asso. prof., 1953-59, prof., 1959-67; prof. plant pathology U. Ky., 1967—; vis. investigator, research participant biology div. Oak Ridge Nat. Lab., 1949-50; research fellow Harvard, 1958. John Simon Guggenheim fellow, 1958. Fellow A.A.A.S.; mem. Am. Phytopath. Soc., Bot. Soc. Am., Mycol. Soc. Am., Sigma Xi, Phi Kappa Phi. Research and numerous publs. on antibiosis in relation to soil borne diseases; microflora of healthy plant tissues; genetics, cytology and sexuality of fungi; radiobiology, plant disease physiology; electron microscopy of diseased plants. Home: 3293 Bellefonte Dr., Lexington, Ky. 40502.*

WHEELER, Harry Eugene, Am. geologist; b. Pipestone, Minn., Feb. 1, 1907; s. Benjamin F. and Mary (Denhart) W.; B.S., U. Ore., 1930; M.A., Stanford, 1932, Ph.D., 1935; m. Loretta Rose Miller, Mar. 21, 1938; children—Eugene, Carolyn (Mrs. David S. Hirsch), David. Jordan Research fellow Stanford, 1930-33; faculty U. Nev., 1935-48, asso. prof. geology, 1942-48; faculty U. Wash., Seattle, 1948—, prof. geology, 1951—; geologist Nev. Bur. Mines, 1936-48; vis. prof. Ind. U., 1956-57, U. Tex., 1960, So. Meth. U., Dallas, 1966. Mem. Am. Stratigraphic Commn., 1958-60; vis. scientist USSR, Nat. Acad. Sci. and Soviet Acad. Sci., 1963. Fellow Geol. Soc. Am.; mem. Am. Assn. Petroleum Geologists, Soc. Econ. Paleontology and Mineralogists, N.W. Geol. Soc., Sigma Xi. Contbr. numerous articles to tech. jours. Research on stratigraphic principles and concepts and their implications in regional and interregional stratigraphic analysis as bases for hist. interpretation. Home: 5761 60th Av. N.E. Seattle 98105.*

WHEELER, John Archibald, Am. physicist; b. Jacksonville, Fla., July 9, 1911; s. Joseph Lewis and Mabel (Archibald) W.; Ph.D., Johns Hopkins, 1933; Sc.D., U. N.C., 1958, Western Res. U., 1959, U. Pa., 1968; m. Janette Hegner, June 10, 1935; children—Isabel Letitia (Mrs. Charles W. Ufford Jr.), James English, Alison Christie (Mrs. Beardley Ruml II). NRC fellow N.Y., Copenhagen, 1933-35; asst. prof. physics, U. N.C., 1935-38; faculty Princeton, 1938—, prof., 1947—, dir. Project Matterhorn, 1951-53; Lorentz prof. U. Leiden (Netherlands) 1956; Fulbright prof. Kyoto (Japan) U., 1962; vis. prof. U. Cal., Berkeley, 1960; 1st vis. fellow Clare Coll., Cambridge, 1964. Cons. physicist on atomic energy projects, Princeton, 1939-42, U. Chgo., 1942, E. I. du Pont de Nemours & Co., Wilmington, Del., Richland, Wash., 1943-45, Los Alamos, 1950-53; mem. adv. com. Oak Ridge Nat. Lab., 1957-65; mem. sci. adv. bd. USAF, 1961-62; chmn. advanced research projects agy. project Dept. Def., 1958; v.p. Internat. Union Physics, 1951-54. Trustee Battelle Meml. Inst. Recipient A. Cressy Morrison prize N.Y. Acad. Scis. for work on nuclear physics, 1947; Albert Einstein medal, 1965. Guggenheim fellow, Paris and Copenhagen, 1949-50. Fellow A.A.A.S. (dir. 1965-68), Am. Phys. Soc. (pres. 1966); mem. Am. Math. Soc., Am. Acad. Arts and Scis., Nat. Acad. Sci., Am. Philos. Soc. (councillor 1963—), Sigma Xi, Phi Beta Kappa. Author: Geometrodynamics, 1962; Gravitational Collapse, 1965; Spacetime Physics, 1966; also articles. Research in nuclear physics, control and peaceful uses of hydrogen bomb, Einstein unified field theory. Home: 30 Maxwell Lane, Princeton, N.J. 08540.*

WHEELER, Schuyler Skaats, Am. engr.; b. N.Y., May 17, 1860; s. James Edwin and Annie (Skaats) W.; ed. Columbia, M.Sc. (hon.), 1912; D.Sc., Hobart, 1894; m. Ella Adams Peterson, Apr. 1891; m. 2d, Amy Sutton, Oct. 1898. Asst. electrician Jablochkoff Elec. Light Co., 1881-82, next with U. S. Elec. Lighting Co.; soon after became one of Edison's engring. staff in charge of work at the first sta. at time of its starting, 1883, when incandescent light was introduced; contbd. many devices adopted; later erected sta. apparatus at Fall River, Mass., and Newburgh, N.Y.; electrician Herzog Teleseme Co.; electrician and mgr. C. & C. Electric Motor Co., 1886, first concern established for regular mfr. of electric motors; organized firm of Crocker & Wheeler, 1888; pres. of Crocker-Wheeler Co., Ampere, N.J., mfrs. elec. equipment, 1889—. Recipient John Scott medal Franklin Inst., 1904, for invention of electric buzz fan, 1886. Author: (with Francis B. Crocker) Practical Management of Dynamos and Motors, 1894. Prominent in devel. of electric motor and especially in direct application of electricity to driving tools; inventor numerous elec. and mech. devices, especially in the early days, such as electric elevator, electric fire engine, series multiple motor control, paralleling of dynamos; brought to U. S. the Latimer Clark library, the largest collection of rare elec. books in existence (catalogued, 2 vols. illus., as the Wheeler Gift) and presented it to Am. Inst. E.E., 1901, which led to erection of United Engring. Soc. Bldg. in N.Y., which he organized; author of code of profl. ethics for engrs. adopted by Am. Inst. E.E., 1912. Died Apr. 20, 1923.

WHEELER, Walter Hall, Am. geologist; b. Syracuse, N.Y., Dec. 21, 1923; s. George Carlos and Esther (Hall) W.; student U. N.D., 1940-42, 46; B.S., U. Mich., 1945, M.S., 1948; Ph.D., Yale, 1951; m. Eula Krueger, Jan. 16, 1945; children—Diana Esther, Roger Dana. Faculty, U. N.C., Chapel Hill, 1951—, asso. prof. geology, 1957—. Fellow Geol. Soc. Am.; mem. Am. Inst. Profl. Geologists, Am. Assn. Petroleum Geologists, Paleontol. Soc., Soc. Vertebrate Paleontology. Editor: Jour. Elisha Mitchell Sci. Soc., 1960-66. Research, publs. on uintatheres (extinct group fossil mammals), interpretations of Cretaceous stratigraphy of N.C. coastal plain. Home: 28 Mt. Bolus Rd., Chapel Hill, N.C. 27514.*

WHEELER, Wayne, Am. sociologist; b. Crete, Neb., July 21, 1922; s. Everett George and Eugenia (Merriman) W.; A.B., Doane Coll., 1944; M.A., U. Neb., 1948; postgrad. U. Stockholm, Sweden; Ph.D., U. Mo., 1959; m. Lola F. Everingham, Aug. 20, 1947; children—Alice Adelia de E., Brita Belinda de E. Instr. sociology Eastern N.M. U., 1947-49, U. Mo., 1949-51, Christian Coll., 1952-53, Bethany Coll., 1953-54; vis. asst. prof. sociology S.D. State U., 1953; vis. asst. prof. sociology Kan. State Tchrs. Coll., 1954-56; asso. prof. sociology, chmn. dept. Park Coll., 1956-61; dir. Kansas City Study of Adult Life, research asso. Com. on Human Devel. U. Chgo., 1960-63; prof. sociology, chmn. social scis. Tarkio Coll., 1963-65; prof. sociology Rice U., 1965-67; prof. sociology, chmn. dept., director of urban studies U. Neb., Omaha, 1967—. Swedish Govt. fellow, 1951-52. Fellow Am. Sociol. Assn., Soc. Applied Anthropology; mem. Midcontinent Am. Studies Assn. (pres. 1960-61), Midwest Sociol. Soc. (sec. 1961-62, editor Newsletter 1960-62), Soc. for Study Social Problems (chmn. new projects com. 1964-65), Am. Studies Assn., Norwegian-Am. Hist. Assn., Soc. Advancement Scandinavian Study. Author: Social Stratification in a Plains Community, 1949; The Other Directed Man; Concept and Reality, 1958; Social Change and Mental Health, 1960; Symposium on Max Weber, 1964; also articles. Research in community sociology, relation of social prestige and social class to social values, stress for basic social change theory, research on comparative study of pure instances with emphasis on non-quantitative techniques including historic, lit. and participation data; refinement of concepts frontier and romantic pluralism, use of perspective of continuity for prediction and explanation; research into contbns. of ethnic groups especially Scandinavians, Czechs, Acadians and their origins to unity and diversity in Am. Soc.; independence of thought and positions on issues confronting profl. socs. and intellectual circles. Home: 7610 Oakwood St., Ralston, Neb. 68127. Office: Dept. Sociology, U. Neb., Omaha 68101.*

WHEELER, William Morton, Am. zoologist; b. Milw., Mar. 19, 1865; s. Julius Morton and Caroline Georgiana (Anderson) W.; grad. German-Am. Normal Coll., Milw., 1884; Ph.D., Clark U., 1892; occupant Smithsonian table Naples Zool. Sta., 1893; univs. Würzburg and Liège, 1893; Sc.D., U. Chgo., 1916, Harvard, 1930, Columbia, 1933; LL.D., U. Cal., 1928; m. Dora Bay Emerson, June 28, 1898. Curator Milw. Pub. Mus., 1887-90; fellow and asst. in morphology Clark U., 1890-92; instr. embryology U. Chgo., 1893-97, asst. prof., 1897-99; prof. zoology U. Tex., 1899-1903; curator of invertebrate zoology Am. Mus. Natural History, 1903-08; prof. entomology Harvard, 1908-37, dean of Bussey Inst. for Research in Applied Biology, 1915-29. Research asso. Am. Mus. Natural History. Recipient Elliott medal; Leidy medal. Fellow Am. Acad. Arts and Natural History, Am. Acad. Arts and Sciences, A.A.A.S., Washington N.Y. acads. scis. Author: Ants, Their Structure, Development and Behavior, 1910; Social Life Among the Insects, 1923; The Social Insects, Their Origin and Evolution, 1928; Foibles of Insects and Man, 1928; Demons of the Dust, a Study in Insect Behavior, 1930; also numerous zool. publs. Asso. editor Biol. Bull., Jour. Morphology, Jour. Animal Behavior and Psyche. Died Cambridge, Mass., Apr. 19, 1937.

WHEELOCK, Earl Frederick, Am. physician, virologist; b. N.Y.C., Feb. 19, 1927; s. Morgan D. and Sylvia (Bender) W.; B.S., Mass. Inst. Tech., 1950; M.D., Columbia, 1955; Ph.D., Rockefeller U., 1961; m. Jean Lowery, Apr. 28, 1955; children—Lisa Lynn, Cynthia Anne, Scott Frederick. Postdoctorate fellow Rockefeller Inst., 1957-61; faculty dept. preventive medicine Western Res. U. Sch. Medicine, Cleve., 1961—, asso. prof. preventive medicine, 1964—. Mem. Soc. for Exptl. Biology and Medicine, Am. Soc. Immunologists, Harvey Soc., Soc. Am. Microbiologists, Am. Fedn. for Clin. Research, Central Soc. for Clin. Research. Research, publs. on mitosis and cell division in cells infected with respiratory tract viruses, host def. mechanisms in viral infections in man produced by live vaccines, the use and effectiveness non-tumor viruses for treatment human acute leukemia and virus-induced leukemia in mice. Home: 2690 Southington Rd., Cleve. 44120.*

WHELAN, Robert Ford, human physiologist; b. Belfast, No. Ireland, Dec. 22, 1922; s. Robert Henry and Dorothy Ivy (Whittington) W.; M.B., B.Ch., B.O.A., Queen's U., Belfast, 1946, M.D., 1952, Ph.D., 1955, D.Sc., 1960; m. Helen Elizabeth MacDonald Hepburn, July 31, 1952; children—Robert John, Elizabeth Janet, Jack Henry. Asst. lectr. physiology Queen's U., Belfast, 1948-52, lectr., 1953-57; research fellow Sherington Sch. Physiology, St. Thomas's Hosp., London, 1952-53; prof., head dept. human physiology and pharmacology U. Adelaide (Australia), 1958—, dean Faculty Medicine, 1964-65. Hancey cons. physiologist Royal Adelaide Hosp.; mem. Australian Drug Evaluation Com.; mem. research adv. com. Nat. Health and Med. Research Council. Fellow Royal Australian Coll. Physicians, Australian Acad. Sci.; mem. Brit. Pharmacological Soc., Brit., Australian physiol. socs., Australasian Soc. for Clin. and Exptl. Pharmacology, Australian Med. Assn., Royal Soc. S. Australia. Author: Control of the Peripheral Circulation in Man, 1966; also articles. Research on mode of action of various drugs and hormones on circulation in man, nature of nervous control of blood vessels, use of drugs in treatment of high blood pressure. Home: 108 Burton, North Adelaide. Office: Dept. Human Physiology and Pharmacology, U. Adelaide, Adelaide, Australia.*

WHELAND, George Willard, Am. chemist; b. Chattanooga, Apr. 21, 1907; s. Zenas Windsor and Lena (Willard) W.; B.S., Dartmouth Coll., 1928, D.Sc., 1954; A.M., Harvard, 1929, Ph.D., 1932; m. Elizabeth Brahson Clayton, Aug. 11, 1934; children—Margaret Elizabeth (Mrs. Leon W. Couch), Robert Clayton. Fellow in chemistry Cal. Inst. Tech., 1932-36; Guggenheim fellow U. Coll. London and Oxford U., Eng., 1936-37; faculty U. Chgo., 1937—, prof. chemistry, 1949—. Mem. Am. Phys. Soc., Am. Chem. Soc., A.A.A.S., Phi Beta Kappa, Sigma Xi. Author: The Theory of Resonance, 1944; Resonance in Organic Chemistry, 1955; Advanced Organic Chemistry, 3rd edition, 1960; also articles. Research in quantum mechanics of organic chemistry. Home: 9033 S. Bell Av., Chgo. 60620.*

WHETHAM, see Dampier, Sir William Cecil.

WHEWELL, William, English philosopher, mathematician; b. Lancaster, Eng., May 24, 1794; s. John and Elizabeth (Bennison) W.; ed. U. Cambridge, 1812-16; studied mineralogy under Prof. Mohs in Germany; ordained priest, 1825; m. 1st, Cordelia Marshall (d. 1855), Oct. 12, 1841; 2d, Everina Frances Affleck, July 1, 1858. Ass't. tutor, Trinity College, Cambridge, 1818-23; tutor, 1823-38; prof. mineralogy, 1828-32; Knightbridge prof. moral philosophy, 1838-55; master Trinity College, 1841-66; vice chancellor Cambridge, 1843, 56; founder tripos of moral sci. and tripos of natural sci., 1848. Fellow Royal Soc. 1820; gold medal, 1837; mem. Geological Soc. (pres., 1837). Author: Elementary Treatise on Mechanics, 1819; Essays on Mineralogical Classification and Nomenclature, 1828; Treatise on Dynamics, 1832; Astronomy and General Physics, 1833; History of the Inductive Sciences, 1837; Plurality of Worlds, 1853; plus numerous articles. Stimulated interest in philosophy at Cambridge; (with Airy) conducted experiments to determine the earth's density; studied geological nomenclature; supported introduction of analytical methods in mathematics at Cambridge; investigated tides; constructed map showing tide wave round earth; originated terms electrolysis, electrolyte, anode, cathode, anion, cation for Faraday; studied history of scis. Died Cambridge, Eng., Mar. 6, 1866.

WHIPPLE, Fred Lawrence, Am. astronomer; b. Red Oak, Ia., Nov. 5, 1906; s. Harry Lawrence and Celestia (MacFarland) W.; student Occidental Coll., 1923-24; A.B., U. of Calif. at Los Angeles, 1927, Ph.D., U. Cal. at Berkeley, 1931; M.A., (hon.), Harvard, 1945; Sc.D., Am. Internat. Coll., 1958; D. Litt. (hon.), Northeastern U., 1961; D.Sc. (hon.), Temple U., 1961; LL.D., C. W. Post Coll. of Long Island U., 1962; m. Dorothy Woods, 1928 (div. 1935); 1 son, Earle Raymond; m. 2d, Babette F. Samelson, Aug. 20, 1946; children—Dorothy Sandra, Laura. Teaching fellow U. of Calif. at Berkeley, 1927-29; Lick Observatory fellow, 1930-31; instr. Stanford summer 1929, U. of Calif. summer 1931; staff mem. Harvard Coll. Obs. since 1931, instr. Harvard, 1932-38, lectr., 1938-45, asso. prof. astronomy, 1945-50, professor astronomy, 1950—, Phillips prof. astronomy, 1968—, chairman department astron. 1949-56; dir. Smithsonian Astrophys. Observatory, 1955—. Research asso. Radio Research Lab., O.S.R.D., 1943-45. Member Rocket and Satellite Research Panel, U. S., since 1946; mem. U. S. subcom. NACA, 1946-52; mem. U. S. Research and Development Bd. Panel 1947-52; chmn. Tech. Panel on Rocketry, mem. Tech. Panel on Earth Satellite Program, other coms. Internat. Geophys. Year, 1955-59; mem., past mem., officer Internat. Astron. Union; dir. Optical Satellite Tracking Project, NASA, 1958—, project dir. Orbiting Astron. Obs., NASA, 1958—, dir. Meteorite Photography and Recovery Program, NASA, 1962—. Recipient Donohue medals, Presdl. Certificate of Merit, J. Lawrence Smith medal Nat. Acad. Scis. 1949; medal for astron. research, U. Liege, 1960; Space Flight award, Am. Astronautical Soc., 1961; Distinguished Fed. Civilian Service award, 1963. Fellow Am. Astronautical Soc. (v.p. 1962-64), Am. Rocket Soc., Am. Astron. Soc. (v.p. 1948-50), Am. Geophys. Union; mem. Am. Inst. Aeronautics and Astronautics (mem. Am. Inst. Aeronautics and Astronautics (mem. aerospace tech. panel for space physics 1960-63), Astronautical Soc. Pacific, Solar Assos.; mem. Internat. Sci. Radio Union (mem. U. S. A. nat. com. 1949-61), American Meteoritical Society, Am. Standards Assn., Am. Acad. Arts and Scis., Am. Philos. Soc., Internat. Acad. Astronautics (sci. adv. com. 1962—), com. on Space Research, Internat. Astronautical Fedn., Nat. Acad. Scis., A.A.A.S., Am. Meteorol. Soc., Phi Beta Kappa, Sigma Xi. Author: Earth. Moon and Planets, 1942, rev. 1963. Asso. editor Astron. Jour., 1954-56, 64—; editor Smithsonian Contbns. to Astrophysics, 1956—; regional editor Planetary and Space Sci., 1958—; editorial bd. Space Sci. Revs., 1961—. Contbr. sci. papers mags., other publs. Inventor tanometer, meteor bumper; independently discovered six new comets; developed theory of nature of comets and icy-comet model; research on spectrophotometry, meteors, planetary nebulae, stellar and solar system evolution, tracking of artificial earth satellites, study of earth's upper atmosphere. Home: 35 Elizabeth Bd., Belmont, Mass. 02178. Office: 60 Garden St., Cambridge, Mass. 02138.*

WHIPPLE, George Hoyt, Am. pathologist; b. Ashland, N.H., Aug. 28, 1878; s. Ashley Cooper and Frances Anna (Hoyt) W.; A.B., Yale, 1900, hon. M.A., 1927, hon. D.Sc., 1947; M.D., Johns Hopkins, 1905, hon. LL.D., 1947; hon. D.Sc., Colgate, 1927. Wesleyan U., 1935, Trinity Coll., Hartford, 1936, Western Reserve Medical School, 1943, University of Chicago, 1952; honorary LL.D., Tulane, 1935, University of Calif., 1935; U. of Buffalo, 1946, U. Rochester, 1950, University Glasgow, 1951; honorary doctorate, University of Athens, 1937; m. Katharine Ball Waring, June 24, 1914; children—George Hoyt, Barbara (Mrs. Grant Heilman). Assistant in pathology Johns Hopkins, 1905-06, instr., 1906-07; pathologist Ancon Hosp., Panama, 1907-08, also Bay View Hospital, Baltimore, 1908; associate in pathology, 1909-11, associate professor, 1911-14, Johns Hopkins; resident pathologist, Johns Hopkins Hosp., 1910-14; prof. research medicine, U. of Calif., and dir. Hooper foundation for med. research Med. Sch. U. of Calif., 1914-21; dean U. of Cal. Med. Sch., 1920-21; dean School of Medicine and Dentistry, U. Rochester, 1921-53, prof. pathology, 1921-55, professor pathology, emeritus, 1955. Served as a trustee of the Rockefeller Foundation. 1927-43; mem. bd. scientific dirs. Rockefeller Inst. for Med. Research, 1936-53, bd. trustees, 1939-60. Mem. and trustee Gen. Edn. Board, 1936-43. Recipient Gold-Headed Cane award, Am. Assn. Pathologists and Bacteriologists, 1961; Kovalenko medal, National Academy Scis., 1962. Fellow American Acad. of Oral Pathology (hon.); mem. International Assn. Dental Research (hon.); hon. member Pathol. Society Gt. Britain. and Ireland; mem. Am. Soc. Exptl. Pathology (pres. 1925), Assn. Am. Physicians, Am. Physiol. Soc., A.M.A., Am. Assn. Pathologists and Bacteriologists (pres. 1930), Nat. Acad. Sciences, 1929, Am. Philos. Soc., 1938, British Medical Association (fgn. corr.), Sigma Xi, Phi Beta Kappa. Nobel prize in physiology and medicine (with Minot and Murphy), 1934; William Wood Gerhard gold medal, 1934; Charles Mickle fellowship, U. of Toronto, 1938; Kober medal, Georgetown U., 1939; Rochester Civic medal, 1943. Research and over 200 articles on study of anemia, Tb, pancreatic lesions, black-water fever, hookworm; collaborated in liver treatment pernicious anemia; investigated bile pigments, pigment metabolism, blood plasma protein regeneration, liver injury and repair, hemoglobin and stroma, protein metabolism, iron metabolism. Home: 56-D Manor Pkwy., Rochester, N.Y. 14620.

WHIPPLE, George Mathews, English physicist; b. Middlesex, Eng., Sept. 15, 1842; s. George Whipple; B.Sc., King's Coll., London, Eng., 1871; Staff Kew (Eng.) Obs., from 1858, supt., from 1876; asst. examiner natural philosophy U. London, 1876-81. Fellow Royal Astron. Soc.; mem. Meteorol. Soc. (council 1876-77, fgn. sec. 1884-45). Contbr. sci. publs. Lifelong study wind pressure, velocity; drew plates for Researches in Solar Physics (Warren de la Rue), 1865-66; re-investigated cupanemometer (invented by Thomas Romney Robinson) at Crystal Palace, 1874; (with Richard Strackey) research in cloud-photography, 1890; improved Kew magnetic instruments; invented optical apparatus, device for testing dark shades of sextants; (with Captain Heaviside) made series of pendulum expts., 1873 (repeated with Colonel Herschel in 1881, with Gen. Walker, 1888) for determining constant of gravitation. Died Feb. 8, 1893.

WHIPPLE, Guy Montrose, Am. psychologist; b. Danvers, Mass., June 12, 1876; s. John Francis and Cornelia Eliza (Hood) W.; A.B., Brown U., 1897; scholar and asst. in psychology Clark U., 1897-98; Ph.D., Cornell U., 1900; m. Clarice Johnson Rogers, Sept. 4, 1901; children—Philip Montrose, Richard Randolph, Guy Montrose; m. 2d, Helen Davis, Oct. 13, 1925; 1 son, William Davis. Asst. in psychology Cornell U., 1898-1902, lectr. in science and art of edn., 1902-04, asst. prof., 1904-11, asst. prof. ednl. psychology, 1911-14; asso. prof. edn. U. Ill., 1914-15, prof. edn., 1915-18; acting dir. bur. of salesmanship research, 1917-19; prof. applied psychology, Carnegie Inst. of Tech., 1918-19; prof. exptl. edn., U. Mich., 1919-25. Acting prof. ednl. psychology, U. Mo., 1907-08. Fellow A.A.A.S. Mem. bd. editors Jour. Ednl. Research, Jour. Applied Psychology; editor Yearbooks of Nat. Soc. Study Edn., Edn. Problem Series. Author: Guide to High School Observation, 1908; Questions in General and Educational Psychology, 1908; Questions in School Hygiene, 1909; Manual of Mental and Physical Tests, 1914; How to Study Effectively, 1916, 27; Classes for Gifted Children, 1919; Problems in Educational Psychology, 1922; Problems in Mental Testing (with Helen D. Whipple), 1925. Translator:

Mental Fatigue (from German of M. Offner), 1911; The Psychological Methods of Testing Intelligence (from the German of W. Stern), 1914. Editor of elementary sch. books for D. C. Heath & Co., 1928-37. Died Aug. 1, 1941.

WHISTLER, Daniel, English physician; b. Eng., 1619; s. William Whistler; B.A., Merton Coll., Oxford, 1642; M.D., U. Leiden (Netherlands), 1645; M.A., Oxford U., 1644, M.D., 1647. Prof. geometry Gresham Coll., 1648—; also Linacre reader Oxford U.; attended wounded seamen in Dutch war, 1652; traveled to Sweden, 1653. Fellow Royal Soc., 1663, Royal Coll. Physicians; censor, registrar Coll. Physicians, London, 1674-82, treas., 1682, pres., 1683. Author: The Rickets (1st printed book on the subject, includes 1st description of rickets as clin. entity), 1645. Died 1684.

WHISTLER, Roy Lester, Am. chemist; b. Morgantown, W.Va., Mar. 21, 1912; s. Park Harris and Oneda (Mason) W.; B.S., Heidelberg Coll., 1934; M.S., Ohio State U., 1935; Ph.D., Ia. State U., 1938; m. Leila A.B. Kaufman, Sept. 6, 1935; 1 son, William Harris. Instr., Ia. State U., 1938; postdoctoral fellow Nat. Bur. Standards, 1938-40; head starch structure sect. U.S. Dept. Agr., Peoria, Ill., 1940-45; prof. chemistry Purdue U., West Lafayette, Ind., 1946—, asst. head biochemistry dept., 1948-61, chmn. Inst. for Utilization Research, 1962—. Indsl. cons., 1948—; vis. lectr. U. Witwatersrand, 1961, 65. Recipient Purdue U. Sigma Xi Research award, 1953, Hudson award, 1960. Mem. Am. Assn. Cereal Chemistry, Am. Chem. Soc., Am. Soc. Biol. Chemists, A.A.A.S., Phi Lambda Upsilon. Author: Polysaccharide Chemistry, 1953. Editor: Industrial Gums, 1959; Methods in Carbohydrate Chemistry, 5 vols., 1962-65; co-editor Starch, Chemistry and Technology, 2 vols., 1965, 67. Research, numerous articles on carbohydrates, sugars, starch, cellulose, gums. Home: 320 Laurel Dr., West Lafayette, Ind. 47906.*

WHISTON, William, English astronomer, mathematician; b. Norton, Eng., Dec. 9, 1667; s. Josiah and Catherine (Rosse) W.; B.A., Clare Coll., Cambridge, M.A., 1693; m. Ruth Antrobus, 1699; children—George, John. Fellow Clare Coll., 1691; chaplain to Bishop of Norwich; became vicar of Lowestoft, 1698; succeeded Isaac Newton as Lucasian prof. math. Cambridge U., 1703, expelled for Arianism, 1710; moved to London, became lectr.; Boyle lectr., 1707. Author: Theory of the Earth, 1696; Primitive Christianity Revived, 5 vols., 1711-12; Vindication of the Apostolic Constitutions, 1715; Sir Isaac Newton's Mathematical Philosophy Demonstrated, 1716; Life of Samuel Clarke, 1730; Primitive New Testament, 1745; transl. of Josephus, 1737. Devised method for longitude which engaged interest of Newton and of gen. public. Died London, Aug. 22, 1752.

WHITAKER, Ewen Adair, astronomer; b. London, Eng., June 22, 1922; s. George Frederick and Gladys (Johnstone) W.; Intermediate B.Sc. in Sci. Roan Sch. for Boys, Greenwich, London, 1940; Intermediate B.Sc. in Engring., Woolwich Poly. Inst., 1942, Higher Nat. Certificate in Mech. Engring., 1943; m. Beryl Joyce Georgina Horswell, June 22, 1946; children—Malcolm John, Graham David, Fiona Carolyn. Staff, Siemens Bros. & Co., Ltd., Woolwich, London, 1940-49; sci. asst. Royal Obs., Greenwich, 1949-50, asst. exptl. officer, 1951-55, exptl. officer, 1955-58; exptl. officer Royal Greenwich Obs., Herstmonceux Castle, Hailsham, Sussex, Eng., 1956-58; research asso. Yerkas Obs., U. Chgo., Williams Bay, Wis., 1958-60; research asso. Lunar and Planetary Lab., U. Ariz., Tucson, 1960—. Cons. Convair Astronautics, Chance-Vaught-Astronautics Space Sci. Div., Langley Field, Va., 1960-63; experimenter Ranger and Surveyor moonshot teams, 1963—. Fellow Royal Astron. Soc.; mem. Am. Astron. Soc., Internat. Astron. Union. Author: (with D. Arthur, E. Moore, J. Tapscott) Photographic Lunar Atlas, 1960; (with D. Arthur) Orthographic Atlas of the Moon, 1960; (with G.P. Kuiper, W. Hartmann, L. Spradley) Rectified Lunar Atlas, 1963; also articles. Pioneered use composite analytical photography on moon. Home: 5514 E. 1st St., Tucson 85711.*

WHITAKER, Thomas Wallace, Am. geneticist; b. Monrovia, Cal., Aug. 13, 1905; s. Walter R. and Mamie (Dunne) W.; B.S., U. Cal. at Davis, 1927; M.S., U. Va., 1929, Ph.D., 1931; m. Mary Beverley Somerville, Aug. 15, 1931; children—Thomas Wallace, Beverley W. (Mrs. Gilbert Rodgers). DuPont fellow U. Va., 1927-31; fellow Arnold Arboretum, Harvard, Cambridge, Mass., 1931-34; asso. prof. biology Agnes Scott Coll., 1934-36; with U.S. Dept. Agr., 1936—, prin. research geneticist, investigations leader, 1956—; asso. in marine biology U. Cal. at San Diego, 1964—. Recipient Andrew Flemming prize in biology U. Va., 1931; Guggenheim Found. fellow, 1946-47, 59. Fellow A.A.A.S.; mem. Fedn. Profl. Assns., Orgn. Profl. Employees Dept. Agr., Am. Soc. Naturalists, Torrey Bot. Club, Bot. Soc. Am., Am. Soc. Archeology, Soc. Hort. Sci., Am. Phytopath. Soc., Sigma Xi, Alpha Zeta, others. Author: (with G.N. Davis) The Cucurbits, 1962; also numerous articles. Developed several widely planted, disease resistant cultivars of lettuce and muskmelons; research on ethnobotany of cucurbits. Home: 2534 Ellentown Rd. Office: P.O. Box 150, La Jolla, Cal. 92037.*

WHITBY, G(eorge) Stafford, chemist; b. Hull, Eng., May 26, 1887; s. Stafford Beeston and Harriet (Smith) W.; A.R.C.S., Royal Coll. Sci., London, Eng., 1906, B.Sc., 1907; M.S., McGill U., Montreal, Que., Can., 1918, Ph.D., 1920, D.Sc., 1938; LL.D., Mt. Allison U., Sackville, N.B., Can., 1932; D.Sc., U. Akron, 1958; m. Wynne Atkinson, May 12, 1915 (dec. Nov.) 1955); children—Oliver, Phillida (Mrs. Harry Theodore Charly); m. 2d, Claire Newman, Aug. 3, 1964. Came to U. S., 1942, naturalized, 1956. Instr., Royal Coll. Sci., 1906-10; chemist Societe Financiere des Caoutchoucs, 1910-17; faculty McGill U., 1920-29, prof. organic chemistry, 1923-29; dir. div. chemistry Can. NRC, 1929-39; dir. Nat. Chem. Lab., Eng., 1939-42; prof. rubber chemistry, dir. rubber research U. Akron (O.), 1942-54, cons. on research, 1954—. Fellow Chem. Inst. Can.; mem. Am. Chem. Soc. (Charles Goodyear medal 1954), Instn. Rubber Industry (Colwyn Gold medal 1928), Canadian Chem. Inst. (past pres.), Canadian Chem. Assn. (past pres.). Author: Plantation Rubber and the Testing of Rubber, 1920; also numerous articles. Editor: Synthetic Rubber, 1954. Pioneered measurement variation of yield rubber from tree to tree; demonstrated that natural rubber contains fatty acid, that the addition basic substances enhances activity most accelerators vulcanization; developed process for low-temperature mfr. synthetic rubber. Home: 3715 Spring Valley Rd., Bath, O. 44210.*

WHITCOMB, John Grant, Am. physician; b. Detroit, July 12, 1926; s. Harold Gilbert and Irma (Kollar) W.; B.S., U. Mich., 1947, M.S., 1956; M.D., U. Cin., 1951; m. Mary Phyllis Hines, June 16, 1951. Staff surgeon Henry Ford Hosp., 1959-61; practice medicine, specializing in surgery, Albuquerque, 1961—; staff surgeon Lovelace Clinic, 1961—; clin. asso. dept. surgery Sch. Medicine U. N.M.; chmn. emergency center com. Bataan Meml. Meth. Hosp.; mem. courtesy staff S.W. Presbyn. Hosp. and Sanatorium; cons. surgeon VA Hosp. Diplomate Am. Bd. Surgery. Fellow A.C.S.; mem. A.M.A., Southwestern Surg. Congress, Pan-Pacific, Western surg. assns., Aerospace Med. Assn., Am. Diabetic Assn., Am. Coll. Chest Physicians, Am. Cancer Soc. (dir.-del. N.M. div.), Sigma Xi. Publs. on invention of intracaritary device for anesthesia; co-devel. ultra-clean, laminar air-flow operating room; origination of anteriomedial surg. approach to popliteal artery. Home: 1024 Alvarado Dr. S.E. Office. 5200 Gibson Blvd. S.E., Albuquerque 87108.*

WHITCOMB, Willard Hall, Am. entomologist; b. Manchester, N.H., July 2, 1915; s. Howard Clarkson and Nell (Hall) W.; B.S., Bates Coll., 1938; M.S., Tex. A. and M. U., 1942; Ph.D., Cornell U., 1947; student U. Rostock, Germany, 1938-39; m. Dorothy Elaine Goodwin, Jan. 7, 1943. Grad. asst. Cornell U., 1942, 1945-47; entomologist Ministerio de Agricultura y Cria, Maracay, Venezuela, 1947-52; entomologist Shell Oil Co. of Venezuela, Cagua, 1952-56; prof. entomology U. Ark., 1956-67; entomologist Big Bend Hort. Lab., U. Fla., Monticello, 1967—. Mem. Entomol. Soc. Am., Fla., Kan. entomol. socs., Entomol. Soc. Canada, Southwest Assn. Naturalists, Sigma Xi, Gamma Sigma Delta. Contbr. numerous articles in field to sci. jours. Study of cotton pests under tropical Am. conditions, especially in Venezuela and Colombia; study of predaceous insects of cotton field; devel. integrated control in cotton; life history and ecology of wolf spiders; insect pests of pecans and stone fruits. Office: Big Bend Hort. Lab., Box 539, Monticello, Fla. 32344.*

WHITE, Abraham, Am. biochemist, b. Cleveland, O., March 8, 1908; son of Morris and Lena (Garfinkle) W.; A.B., U. of Denver, 1927, A.M., 1928; Ph.D., U. of Mich., 1931; L.H.D. (honorary), Yeshiva University 1959. Sterling fellow, Yale, 1931-32, Porter fellow of Am. Physiol. Soc., 1932-33; instr. in physiol. chemistry, U. of Mich., summer 1932; instr. physiol. chemistry, Yale, 1933-37, asst. prof., 1937-43, associate prof., 1943-48; prof. and chmn. dept physiol. chemistry, Sch. of Medicine. University of Calif., 1948-51; vice pres., dir. research Chem. Specialties Co., Incorporated, 1951-53; professor and chairman of department biochemistry. Albert Einstein College of Medicine, Yeshiva University, 1953—, also associate dean. Awarded competitive fellowship (on basis of research work) to attend XVth Physiol. Congress (internat.), Leningrad and Moscow, 1935; Eli Lilly prize in biochemistry (for outstanding achievement in biochemistry), 1938. Mem. American Chemical Society, American Society for Cell. Biology, A.A.A.S., Am. Soc. Biol. Chemists, Am. Inst. Nutrition Soc. Exptl. Biol. and Med., N.Y. Acad. Scis., Endocrine Soc. Biochem. Soc., History Sci. Soc., Am. Acad. Arts and Scis., Sigma Xi, Phi Lambda Upsilon, Alpha Omega Alpha (hon.). Author: Principles of Biochemistry (with others). Contbr. numerous original research publs. on mode of action and functions of hormones; amino acid, protein chemistry and metabolism. Home: 200 E. 66th St., N.Y.C. 10021. Office: Eastchester Rd. and Morris Park Av., N.Y.C. 10461.*

WHITE, Addison Hughson, Am. chem. physicist; b. Clovis, Cal., Oct. 13, 1909; s. Arthur Thomas and Alice (Hughson) W.; A.B., Occidental Coll., 1930; m. Elizabeth C. McCusker, Oct. 12, 1934; 1 son, Arthur Thomas II. Mem. tech. staff Bell Telephone Labs., Inc., Murray Hill, N.J., 1930-47, research physicist in charge phys. electronics, 1947-53, dir. chem. physics, 1953-58, exec. dir. research, phys. scis. div., 1958—. Fellow Am. Phys. Soc.; mem. Am. Chem. Soc., Phi Beta Kappa. Research on dielectrics, electron diffraction, crystal structure, silicon rectifiers, thermionic emission, phys. electronics and metallurgy. Home: 15 Norwood Av., Summit, N.J. 07901. Office: Mountain Av., Murray Hill, N.J. 07971.

WHITE, Alan George Castle, Am. biochemist; b. Boston, Eng., Aug. 12, 1916; s. George Alec and Caroline (Castle) W.; came to U.S., 1925, naturalized, 1941; B.S., U. R.I., 1938; M.S., Pa. State U., 1940; Ph.D., Ia. State U., 1947; m. Juliette Marie Perotti, Sept. 13, 1945; children—David Edward, Margaret Marie. Research bacteriologist Joe E. Seagram & Son, Louisville, 1940-41; tech. asst. Rockefeller Inst., N.Y.C., 1941-43; instr. bacteriology Ia. State U., Ames, 1945-47; faculty Tulane U. Sch. Medicine, New Orleans, 1947-67, prof. biochemistry, 1960-67; prof., head dept. biology Va. Mil. Inst., 1967—; vis. prof. Osmania Med. Coll., Hyderabad, India, 1961-63. Fellow Ind. Sci. Research Inst., Ia. State U., 1943-45. Mem. Am. Soc. Biol. Chemists, Am. Soc. Microbiologists, Am. Chem. Soc., Soc. for Exptl. Biology and Medicine (sec. So. sect. 1965-67), A.A.A.S., Sigma Xi, Phi Sigma, Phi Lambda Upsilon. Research, publs. on carbohydrate metabolism hemolytic streptococci, fatty acid synthesis in yeast; determined arginine requirements streptocci and its metabolism; elucidated path glucose metabolism in entamoeba histolytica. Home: 613 Stonewall St., Lexington, Va. 24450.*

WHITE, Allen Ingolf, Am. pharm. chemist; b. Silverton, Ore., July 10, 1914; s. Anders O. and Ingeborg (Beaverson) W.; B.S., U. Minn., 1937, M.S., 1938, Ph.D., 1940; m. Edith Halverson, Sept. 24, 1938; children—Karen S. (Mrs. James H. Jordan), Connie L., Kathryn A. Faculty, Wash. State U., 1940—, prof., 1948—, dean, 1960—. Mem. Am. Pharm. Assn., Am. Chem. Soc., A.A.A.S., Am. Coll. Apothecaries, Am. Assn. U. Profs., Sigma Xi, Rho Chi, Phi Kappa Phi. Contbg. author: American Pharmacy, 1945, 5th edit., 1960; Textbook of Organic Medicinal and Pharmaceutical Chemistry, 4th edit. 1962, 5th edn., 1966. Research and publs. on relationship of chem. structure to physiol. mechanisms of action of potent analgesics; investigation of principal constituents in plant drugs used by American Indians. Home: 1908 Indiana St., Pullman, Wash. 99163.*

WHITE, Anthony, English surgeon; b. Durham, Eng., 1782; M.B., Emmanuel Coll., Cambridge (Eng.) U., 1804. Apprentice to Anthony Carlise; asst. surgeon Westminster Hosp., London, 1806-23, surgeon, 1823-46, cons. surgeon, after 1846. Hunterian orator, 1831. Mem. Royal Coll. Surgeons (council 1827, v.p. 1832, 40, pres. 1834, 42). Author: Treatise on the Plague, 1846; An Enquiry into the Proximate Cause of Gout and its Rational Treatment, 1848. First to excise head of femur for disease of hip joint, 1821. Died Mar. 9, 1849.

WHITE, Charles, English surgeon; b. Eng., 1728; s. Thomas and Rosamond W.; student medicine under father, also in London, Eng., Edinburgh, Scotland; m. Ann Bradshaw, Nov. 22, 1757; 8 children. Cofounder, surgeon Manchester (Eng.) Infirmary, from 1752; joint-founder Manchester Lying-in Hosp., 1790. Fellow Royal Soc., 1762; mem. Royal Coll. Surgeons, Manchester Literary and Philos. Soc. (co-founder, v.p. 1781). Author: Account of the Topical Application of the Spunge in the Stoppage of Haemorrhage, 1762; Cases in Surgery, 1770; Treatise on the Management of Pregnant and Lying-in Women, 1773; Inquiry into the Nature and Causes of that Swelling in one or both of the Lower Extremities which sometimes happens to Lying-in Women, 1784; Observations on Gangrenes and Mortifications, 1790; An Account of the Regular Gradation in Man and in different Animals and Vegetables, and from the former to the latter, 1799. Effected revolution practice of midwifery; 1st (since Hippocrates) to contribute substantially to understanding, treatment puerperal fever; performed 1st excision of head of humerus for caries, 1768; 1st excision of hip, 1769, of shoulder joint; gave 1st clin. description of phlegmasia alba dolens (milk leg), 1784. Died 1813.

WHITE, Charles Ablathar, Am. geologist; b. North Dighton, Mass., Jan. 26, 1826; s. Abiathar and Nancy (Corey) W.; M.D. Rush Med. Coll., 1864; A.M. (hon.), Ia. Coll., 1866; LL.D., State U. Ia., 1893; m. Charlotte R. Pilkington, Sept. 28, 1848. State geologist of Ia., 1866-70; prof. natural history State U. Ia., 1867-73, Bowdoin Coll., Me., 1873-75; geologist and palaeontologist to various U. S. Govt. surveys, 1874-92. Honorarily connected with Smithsonian Instn. and U. S. Nat. Mus., 1895—. Author: Manual of Physical Geography of Iowa, 1873; Report on Invert. Fossils (Geog. and Geol. Expl.) and Surveys West of 100th Meridian, 1875; Bibliography of North American Invertebrate Palaeontology, 1878; Contribçðes á Palaeontolgia do Brazil, 1887; The Relation of Biology to Geological Investigation, 1894. Died 1910.

WHITE, Colin, biometrician; b. Gympie, Australia, Aug. 25, 1913; s. William Robert and Susan (Smith) W.; B.Sc., Sydney (Australia) U., 1935, M.Sc., 1936, M.B., B.S., 1940; m. Jean Margaret Murphy, Dec. 12, 1943; children—Kenneth James, Allen Gregory. Med. officer Commonwealth Health Service, 1943-46; lectr. U. Birmingham (Eng.), 1946-48, 50-53; asst. prof. physiology U. Pa., 1948-50; faculty Yale, 1953—, prof. biometry, 1962—. Mem. adv. com. epidemiology and biometry NIH, 1960-65. Mem. Am. Statis. Assn., Biometric Soc., Royal Statis. Soc. Research, publs. on inheritance human twinning, statis. aspects cancer research, methods statis. analysis in med. research. Home: 107 Thornton St., Hamden, Conn. 06517. Office: 60 College St., New Haven 06510.*

WHITE, David, phys. chemist; b. Ukraine, Russia, Jan. 14, 1925; s. Maurice and Mania (Baron) W.; B.Sc. with honors in Chemistry, McGill U., 1944; Ph.D. in Phys. Chemistry, U. Toronto, 1947; m. Birdye E. Ruckenstein, Aug. 26, 1945; children—Sharon Miriam, Jacqueline Michele, Edward Irwin. Came to U. S., 1947, naturalized, 1956. Postdoctoral fellow Ohio State U., 1948-51, asst. dir. Cryogenic Lab., 1951-53; with Heat and Power div. Nat. Bur. Standards, 1953; faculty Syracuse U., 1953-54, Ohio State U., 1954-66; prof., chmn. dept. chemistry U. Pa., Phila., 1966—. Mem. task group on thermodynamics Internat. Union Pure and Applied Chemistry, 1963—; Fulbright prof. U. Kyoto, U. Tokyo, 1965; cons. Monsanto Research Corp., 1963—. Mem. Am. Phys. Soc., Am. Chem. Soc., Sigma Xi, Phi Lambda Upsilon. Contbr. numerous articles to sci. jours., chpts. to books. Research on thermodynamic properties at low temperatures, effect of hindered rotation and quenching of rotation on macroscopic and molecular properties of solids, high temperature chemistry, in particular stability and structure of inorganic molecules stable at elevated temperatures; developer miniature low temperature refrigeration devices, low temperature expanders. Home: 719 Clarendon Rd., Penn Valley, Narberth, Pa. 19072. Office: Dept. Chemistry, U. Pa., Phila. 19104.*

WHITE, Edith Grace, Am. ichthyologist; b. Boston, May 16, 1890; d. Joseph Adie and Elizabeth (Kelly) W.; A.B., Mt. Holyoke Coll., 1912; A.M., Columbia, 1913, Ph.D., 1918. Research asst. Princeton, 1915-16; lab asst. Western Res. U., 1917; faculty Heidelberg U., 1918-20, Shorter Coll., 1920-23; prof. biology, head dept. Wilson Coll., Chambersburg, Pa., 1923-58, prof. emeritus, 1958—. Research asso. ichthyology Am. Mus. Natural History, 1933-47; lectr. U. Tokyo, 1929. Fellow A.A.A.S.; mem. Am. Soc. Ichthyologists and Herpetologists, Am. Genetic Assn., Eugenics Soc., N.Y. Acad. Sci., Phi Beta Kappa. Author: Interrelationships of the Elasmobranch Fishes, 1937; Lab Manual of General Biology, 1934; Principles of Genetics, 1940; Genetics, 1962; also articles. Research and classification of sharks based on internal and external characters. Address: 1312 Edgar Av., Chambersburg, Pa. 17201.*

WHITE, Florence Roy, Am. chemist; b. New Castle, Pa., Mar. 6, 1909; d. Howard Meek and Florence (Burwash) Roy; A.B., U. Ill., 1930, A.M., 1931; Ph.D., U. Mich., 1935; m. Julius White, Mar. 28, 1932; children—Margaret Jane, Elizabeth Ann, William Jay, Richard Paul. Rockefeller fellow U. Mich., 1935-36, research asst., 1936-38; research asst. Yale, 1938-40; biochemist Bur. Home Econs., 1941; chemist NIH, Bethesda, Md., 1941-42, Nat. Cancer Inst., 1942-46, chemist Cancer Chemotherapy Nat. Service Center, 1958—, chemist head biochem. sect., 1964—; research asst. Nat. Acad. Sci., 1956-58. Mem. Am. Soc. Biochemists, Am. Assn. Cancer Research, Sigma Xi, Iota Sigma Pi, Omicron Nu. Research and publs. on metabolism of sulfur-containing amino acids and their relationship to carcinogenesis and tumor growth; search for exploitable biochem. differences between tumor and normal tissues; mechanisms of action and resistance to anti-cancer drugs. Home: 4217 Franklin St., Kensington, Md. 20795. Office: NIH, Bethesda, Md. 20014.*

WHITE, Francis Buchanan White, Scottish botanist, entomologist; b. Perth, Scotland, Mar. 20, 1842; s. Francis White; M.D., Edinburgh (Scotland) U., 1864; m. Margaret Juliet Corrie, 1866. Studied plants, animals; founded Scottish Naturalist mag., 1871. Mem. Cryptogamic Soc. Scotland (co-founder 1874), Entomol. Soc. London, Linnean Soc., Perthshire Soc. Natural Sci. (co-founder 1867, pres. 1867-72, sec. 1872-74, editor 1874-84, 92-94). Author: Fauna Perthensis, Lepidoptera, 1871; The Flora of Perthshire, 1898. Contbr. articles to sci. publs. Died Dec. 3, 1894.

WHITE, Frederick Elmer, Am. physicist, educator; b. Peabody, Mass., Jan. 21, 1909; s. Frederick W. and Catherine (Buckley) W.; A.B., Boston U., 1930; Sc.M., Brown U., 1932, Ph.D., 1934; m. Anna Mary Barron, July 10, 1943. Faculty, Williams Coll., 1935-36; faculty Boston Coll., 1936-42, 45—, prof., 1945—, acting chmn. dept. physics, 1963—; physicist div. phys. war research Duke, 1942-45. Fellow Acoustical Soc. Am., A.A.A.S.; mem. Am. Phys. Soc., Phys. Soc. Japan, Am. Assn. Physics Tchrs., Am. Assn. U. Profs., Phi Beta Kappa, Sigma Xi, Sigma Pi Sigma. Author: Acoustics, An Intermediate Text, 1954; Introduction to Analytical Mechanics, 1958. Asso. editor Jour. Acoustical Soc. Am., 1962—. Research and publs. on indeterminacy principle, acoustic filters, ultrasonics, atmospheric acoustics, mechanics, theoretical physics. Home: 12 Columbia Rd., Beverly, Mass. 01915. Office: Dept. Physics, Boston Coll., Chestnut Hill, Mass. 02167.

WHITE, Gilbert, English naturalist; b. Hampshire, Eng., July 18, 1720; s. John and Anne (Holt) W.; ed. under Thomas Warton; B.A., Oriel Coll. Oxford (Eng.) U., 1743, M.A., 1746. Curate Swarraton, 1751, Durley, 1757, Faringdon, 1758, West Deane (all Eng.); proctor Oxford U., dean Oriel Coll., from 1752. Author: Naturalist's Journal (diaries in Brit. Mus.), 1767; Natural History and Antiquities of Selbourne, 1789 (considered classic). Contbr. Transactions, Royal Soc. Died June 26, 1793.

WHITE, Harry James, Am. physicist, engr.; b. Fremont, Neb., July 29, 1905; s. Charles E. and Ines (Harris) W.; B.S., U. Cal. at Berkeley, 1928, M.S., 1931, Ph.D., 1933; m. Rhoda Palmer, Sept. 18, 1937. Instr. physics U. Cal. at Berkeley, 1934-35; research physicist Research Corp., N.Y.C., 1935-41; asso. group leader Mass. Inst. Tech. Radiation Lab., Cambridge, 1941-45; dir. research Research Corp., N.J., 1945-58; dir. research and devel. Research-Cottrell, Inc., Bound Brook, N.J., 1958-60; head dept. applied sci. Portland (Ore.) State Coll., 1960—; pres. Applied Math. Assos., Inc., 1961—. Vis. lectr. Harvard, 1957-60; cons. in electrostatic processes, air pollution control, systems sci. Bd. trustees, exec. com. Ore. Grad. Center. Fellow Am. Phys. Soc., I.E.E.E.; mem. A.A.A.S., Am. Math. Soc., Soc. Indsl. and Applied Math., Am. Soc. Engring. Edn., Am. Assn. Physics Tchrs., Air Pollution Control Assn., Soc. for Gen. Systems Research, Sigma Xi, Pi Mu Epsilon. Author: Industrial Electrostatic Precipitation, 1963. Research, publs., patents in fields of electrostatic processes, gaseous electronics, air pollution control, systems sci., radar. Home: 1717 S.W. Park Av., Portland, Ore. 97201.*

WHITE, Harvey Elliott, Am. physicist; b. Parkersburg, W.Va., Jan. 28, 1902; s. Elliott A. and Elizabeth (Wile) W.; A.B., Occidental Coll., 1925, D.Sc., 1961; Ph.D., Cornell U., 1929; m. Adeline I. Dally, Aug. 10, 1928; children—Donald H., Jerald P., Vernital (Mrs. Dennis Aigner). Internat. research fellow Physikalische Technische Reichsanstalt, Berlin, Germany, 1929-30; faculty U. Cal. at Berkeley, 1930—, prof. physics, 1942—, dir. Lawrence Hall Sci., 1960—. Tchr., NBC-TV physics course, 1958-59; with AEC-OSRD, 1942-45. Recipient War Dept. citation, 1946, Thomas A. Edison TV award 1959, Peabody-TV award, 1959, Parents Mag. medal, 1959, Sylvania award, 1959, Hans Christian Oersted medal, others; Guggenheim fellow, 1948. Fellow Am. Phys. Soc., Optical Soc. Am., Sigma Xi, Phi Beta Kappa, Phi Kappa Phi, Sigma Pi Sigma. Author: Introduction to Atomic Spectra, 1935; Classical and Modern Physics, 1940; Modern College Physics, 5 edits., 1947; Atomic and Nuclear Physics, 1964; Descriptive College Physics, 1955; Physics an Exact Science, 1959. Research and publs. on atomic spectra; discovered relations between electron shell structure of atoms adjacent to each other in chem. periodic table; determined the spin angular momentum of nucleus of a number of chem. elements; developed new type of demonstration expts. Home: 543 Spruce St., Berkeley, Cal. 94707.*

WHITE, Ian Gregory, Australian physiologist; b. Ariah Park, Australia, June 14, 1926; s. Arthur Thomas and Christina (Clarke) W.; B.Sc., U. Sydney, 1946, Ph.D., 1954, D.Sc., 1963; m. Hilary Irene Cornish, Aug. 20, 1949; 1 dau., Diana Susan. Faculty, U. Sydney, 1946—, acting prof., 1960, reader in vet. physiology, 1961—. Council for Sci. and Indsl. Research Orgn. scholar, 1955; Nuffield Dominion Travelling fellow, 1961. Mem. Biochem. Soc., Soc. for Study Fertility, Australian Physiol. Soc., Australian Biochem. Soc., Endocrine Soc. Australia, Australia-New Zealand Assn. Advancement Sci. Publs., research on toxicology of arsenic and chem. constituents of blood; physiology and biochemistry of reprodn.; studies on spermatozoa, seminal plasma; genitalia. Home: 48 Catalpa Crescent, Turramurra, New South Wales, Australia.*

WHITE, J. P., Am. polit. scientist; b. Cin., Mar. 7, 1924; s. Joseph H. and Anna (Cummings) W.; A.B., U. Cin., 1949; A.M., U. Chgo., 1952, Ph.D., 1953; m. Mary Louise Stewart, May 1, 1945; children—John P., Karen Ann. Faculty, U. Mich., Ann Arbor, 1954-62, asso. prof. polit. sci., 1962-63; prof. polit. sci. Ariz. State U., Tempe, 1963—, chmn. dept., 1964—. Research cons. Mich. Constl. Conv., 1961-62. Recipient Distinguished Service award U. Mich., 1962. Mem. Am., Western polit. sci. assns., Mich. Acad. Sci., Arts and Letters (chmn. history and polit. sci. sect. 1962-63). Author: Apportionment and Representative Institutions (with K. A. Lamb, W. J. Pierce), 1963; also articles. Research on legislative apportionment systems, polit. campaign finance, voting behavior. Home: 1724 N. McAllister Av., Tempe, Ariz. 85281.*

WHITE, Jack Lee, Am. metallurgist; b. Los Angeles, Oct. 29, 1925; s. John Maurice and Edith (Andrews) W.; B.S., Cal. Inst. Tech., 1949, Carnegie Inst. Tech., 1950; Ph.D., U. Cal. at Berkeley, 1955; D.I.C., Imperial Coll. Sci. and Tech., 1955; m. Elizabeth Ann Sherrard, June 30, 1950; 1 dau., Janette Ardith. NRC-Nat. Acad. Sci. postdoctoral research asso. U.S. Naval Research Lab., Washington, 1955-58; research staff mem. Gen. Atomic div. Gen. Dynamics Corp., San Diego, 1958-67; vis. metallurgist Petten (Netherlands) Lab. of EURATOM, 1967—. Mem. Am. Chem. Soc., Am. Ceramic Soc., Am. Soc. Metals, Inst. Metals (London), Am. Inst. Mining, Metall. and Petroleum Engrs., A.A.A.S., Phi Beta Kappa, Sigma Xi, Tau Beta Pi. Author: (with J. O'M. Bockris and J. D. Mackenzie) Physico-Chemical Measurements at High Temperatures, 1959. Research and publs. on new graphitic materials, mechanisms of graphitization, structure of high-temperature liquids, metals, glasses; developer differential calorimeter for stored energy measurements. Home: Dawson Nes 14, Bergen (N.H.). Office: EURATOM, Petten (N.H.), Netherlands.*

WHITE, James Edward, Am. physicist; b. Cherokee, Tex., May 10, 1918; s. William Cleburne and Willie (Carter) W.; B.A., U. Tex., 1940, M.A., 1946; Ph.D., Mass. Inst. Tech., 1949; m. Courtenay Brumby, Feb. 1, 1941; children—Rebecca, Peter McDuffie, Margaret Marie, Courtenay. Physicist, underwater sound lab. Mass. Inst. Tech., 1941-45, dir., 1944-45, Def. Research Lab., Austin, Tex., 1945-46; geophysics supr. field research lab. Mobil Oil Co., Dallas, 1949-55; mgr. physics dept. Denver Research Center, Marathon Oil Co., Littleton, Colo., 1955-60, research asso., 1960—. Spl. lectr. in geophysics Colo. Sch. Mines, 1964—. Fellow Acoustical Soc. Am.; mem. Soc. Exploration Geophysicists (pres. 1967-68). Am. Phys. Soc., Seismol. Soc. Am., A.A.A.S., Sigma Xi, Phi Beta Kappa, Phi Eta Sigma. Author: Seismic Waves: Radiation Transmission and Attenuation, 1965. Patentee directional underwater sound detector, also in field of seismic prospecting. Publs. on propagation and attenuation of seismic waves with emphasis on coupling to fluid-filled holes. Home: 330 W. Fair Av., Littleton 80120. Office: P.O. Box 269, Littleton, Colo. 80121.*

WHITE, James William, Am. surgeon; b. Phila., Nov. 2, 1850; s. James W. and Mary Anne (McClaranan) W.; M.D., Ph.D., U. Pa., 1871; LL.D., Aberdeen, 1906; m. Letitia Brown, June 22, 1888. On staff of Louis Agassiz during Hassler expdn. to W.I., Straits of Megellan, both coasts of S. Am., Galapagos Islands, 1871-72; resident phys. Phila. Hosp., 1873; surgeon Eastern State Penitentiary, 1874-76; surgeon First Troop Phila. City Cav., 1878-88; 1st prof. genito-urinary surgery, then prof. clin. surgery, then John Rhea Barton prof. surgery U. Pa.; surgeon Univ. Hosp.; cons. surgeon Phila. and Jewish hosps.; adv. surgeon Pa. R.R. Co. Author: American Text Book of Surgery (Keen and White), 1896; Genito-Urinary Surgery (White and Martin), 1897; Human Anatomy (Piersol), 1906; A Text Book of the War for Americans, 1915. Translator and editor. Cornil on Syphilis (Simes and White), 1875. An editor Annals of Surgery. Died Apr. 24, 1916.

WHITE, Julius, Am. chemist; b. Pitts., Mar. 23, 1904; s. Morris and Lena (Garland) W.; B.A., U. Denver, 1925, M.A., 1926; Ph.D., U. Ill., 1931; m. Florence Margaret Roy, Mar. 28, 1932; children—Margaret J., Elizabeth A., William J., Richard P. Research asst. prof. Yale Med. Sch., 1938-39; prin. biochemist NIH, Nat. Cancer Inst., Bethesda, Md., 1942-46, head metabolism sect., 1946-52, chief Lab. Physiology, 1952—. Cons. Am. Cancer Soc., 1961-64. Mem. Am. Chem. Soc., Am. Assn. Biol. Chemists, Am. Inst. Nutrition. Research, numerous publs. on metabolic changes in animals bearing malignant tumors, mechanism action carcinogenic agts., nutritional factors involved in controlling tumor growth. Home: 4217 Franklin St., Kensington, Md. 20795. Office: Nat. Inst. Health, Bethesda, Md. 20014.*

WHITE, Kerr Lachlan, physician b. Winnipeg, Man., Can., Jan. 23, 1917; s. John Alexander and Ruth (Preston); B.A. with honors (Alan Oliver Gold medal), McGill U., 1940, M.D., C.M., 1949; postgrad. Yale, (Commonwealth Fund Fellow) U. London (Eng.); m. Isabel Anne Pennefather, Nov. 26, 1943; children—Susan Isabel, Margot Edith. Came to U.S., 1948, naturalized, 1957. Hosmer research fellow McGill U., 1952-53; faculty U. N.C. Sch. Medicine, 1953-62, U. Vt., 1962-65; prof., chmn. dept. med. care and hosps. Sch. of Hygiene and Pub. Health, Johns Hopkins, Balt., 1965—; also cons., mem. various govtl. agys. Cons. surgeon gen. USPHS, 1957—, chmn. health services, research study sect. NIH, 1962-66. Diplomate Nat. Bd. Med. Examiners, Am. Bd. Internal Medicine. Fellow A.A.A.S., A.C.P., Am. Heart Assn., Am. Pub. Health Assn., Royal Soc. Medicine; mem. A.M.A., Am. Hosp. Assn., Am. Fedn. Clin. Research, Am. Sociol. Assn., Assn. Tchrs. Preventive Medicine, Assn. Am. Med. Colls., Group Health Assn. Am., Internat. Epidemiol. Assn., N.Y. Acad. Scis., Sigma Xi, Alpha Omega Alpha. Research, publs. on med. care, patient care and preventive medicine, also epidemiology of cardiovascular disease. Home: 1527 Park Av., Balt. 21217.*

WHITE, Leslie A(lvin), Am. anthropologist; b. Salida, Colo. Jan. 19, 1900; s. Alvin Lincoln and Mildred (Millard) W.; student La. State U., 1919-

21; A.B., Columbia, 1923, M.A., 1924; Ph.D., U. Chgo., 1927; Sc.D., U. Buffalo, 1962; m. Mary Augusta Pattison, Feb. 9, 1931 (dec.). Faculty, U. Buffalo, 1927-30; faculty U. Mich., Ann Arbor, 1930—, now prof. anthropology, chmn. dept., 1945-57; curator anthropology Buffalo Mus. Sci., 1927-30; vis. prof. Yale, 1947, Harvard, 1949, U. Cal. at Berkeley, 1957; ethnographic field work Pueblo Indians, intermittently, 1926-57. Lectr., Center for Advanced Study in Behavioral Scis., 1960-61. Recipient U. Mich. Distinguished Faculty Service award 1957; Viking medal Wenner-Grend Found., 1959. Mem. Am. Anthrop. Assn. (past pres.), A.A.-A.S. (v.p.), Am. Folk-Lore Soc., Am. Assn. U. Profs., Author: The Pueblo of Santa Ana, 1942; The Science of Culture, 1949; The Evolution of Culture, 1959; also numerous articles. Helped rehabilitate theory of evolution in cultural anthropology; introduced energy theory of cultural devel. into anthropology; developed sci. of culture introducing term culturology for it; research on ethnol. theory. Home: 819 Avon Rd., Ann Arbor, Mich. 48104.*

WHITE, Locke, Jr., Am. phys. chemist; b. South Boston, Va., Mar. 5, 1919; s. Locke and Emma Cabell (Edmunds) W.; B.S., Davidson Coll., 1939; Ph.D., U. N.C., 1943; m. Anne Stelzenmuller, June 12, 1948; children—Stephen Locke, Alan Edmunds, Diane Virginia, Brian Cabell, Helen Elizabeth. Chemist, U. S. Naval Research Lab., Washington, 1942-45; staff So. Research Inst., Birmingham, Ala., 1945-48, asst. dir., 48-61; prof. physics Davidson (N.C.) Coll., 1961—. Fellow A.A.A.S., Am. Phys. Soc., Am. Chem. Soc., Am. Assn. Phys. Tchrs., Am. Assn. U. Profs. Research, publs. on chem. warfare; adsorption, properties of aerosols; devel. electronic instrumentation. Home: 203 Lorimer Rd., Davidson, N.C. 28036.*

WHITE, Marsh William, Am. physicist; b. Claremont, N.C., Apr. 22, 1896; s. William Franklin and Dora (Marsh) W.; A.B., Park Coll., Parkville, Mo., 1917; M.S., Pa. State U., 1920, Ph.D., 1926; Sc.D. (honorary), Park College, 1958; married Stella Steele, Sept. 14, 1917; children—Laurence Marsh, Kenneth Steele, Malcolm Arthur. Instr. physics Pa. State U., 1918, asst. prof., 1920-26, asso. prof., 1926-42, prof. physics, 1942-60, prof. emeritus, 1960—; dir. Electronics Assos., Inc., Long Branch, New Jersey, C-COR, Centre Video, (both State College, Pa.). Special consultant with the OSRD, 1944-45; expert cons. sec. war, 1945-46; cons. research and development div. G-4, asst. chief of staff U. S. Army, 1946-55. Fellow A.A.A.S., Am. Phys. Soc.; mem. Am. Assn. Physics Tchrs. (pres. 1954), Am. Inst. Physics (governing bd. 1946-49), Am. Assn. U. Profs., Assn. of Coll. Honor Societies (pres. 1950, mem. national council), Omicron Delta Kappa, Sigma Xi (mem. exec. com. 1965—). Delta Chi (nat. treas., sec., pres.), Sigma Pi Sigma (nat. exec. sec. 1930—), Pi Mu Epsilon. Author: Practical Physics, 1943; College Technical Physics, 1946; College Physics, 1952; Physics for Science and Engineering, 1957. Contbr. sci. papers tech. jours. Research on high velocity electrons; reflection of electrons in x-ray tubes; wave-shape factor and meter. Home: 511 E. Prospect Av., State College, Pa. Office: Pennsylvania State University, University Park, Pa. 16802.*

WHITE, Michael James Denham, cytogeneticist; b. London, Eng., Aug. 20, 1910; s. James Kemp and Una (Chase) W.; B.Sc., U. Coll., U. London, 1931, M.Sc., 1932, D.Sc., 1940; M.Sc., U. Melbourne (Australia), 1958; m. Isobel Mary Lunn, Dec. 3, 1939; children—Nicholas James, Jonathan Charles, Charlotte Elizabeth. Asst. lectr. U. Coll. London, 1933-35, lectr. zoology, 1936-46; statistician, entomologist U.K. Ministry of Food, 1941-45; reader zoology U. London, 1947; guest investigator Carnegie Instn. Washington, Cold Spring Harbor, L.I., N.Y., 1947; prof. zoology U. Tex., 1947-53; sr. research fellow Commonwealth Sci. and Indsl. Research Orgn., Canberra, Australia, 1953-56; prof. U. Mo., 1957-58; prof. zoology U. Melbourne, 1958-64, prof. genetics, 1964—. Rockefeller Found. fellow Columbia, 1937-38; recipient Mueller medal Australian and New Zealand Assn. for Advancement Sci., 1965. Fellow Royal Soc., 1961, Australian Acad. Sci., U. Coll. London; mem. Am. Acad. Arts and Scis. (hon. fgn.), Genetics Soc. Am., Soc. for Study Evolution, Am. Soc. Naturalists, Sigma Xi. Author: The Chromosomes, 1937; Animal Cytology and Evolution, 1948; also numerous articles. Research on chromosomal differences between species and evolutionary significance, aberrant cytogenetic mechanisms, sex chromosomes. Home: 23 Leicester St., North Balwyn E.9., Victoria. Office: Dept. Genetics, U. Melbourne, Parkville N.2, Victoria, Australia.*

WHITE, Orland Emile, Am. geneticist, botanist; b. Sibley, Ia., Apr. 25, 1885; s. Henry Clay and Margaret Belle (Rush) W.; B.Sc., S.D. State U., 1909, M.S., 1911; M.Sc., Bussey Instn., Harvard, 1912, Sc.D. in Genetics, 1913; m. Loto Rollinstall Underwood, July 28, 1913; children—Margaret Underwood (Mrs. Dean Winchester), Prudence Wayne, (Mrs. James Quarles). With Bklyn. Bot. Garden, 1913-27, curator, 1916-27; botanist Mulford Biol. Explorations, Amazon Basin, 1921-22; prof. agr. and biology, dir. Blandy Expt. Farm, U. Va., 1927-55; research scholar U.S. Edn. Found., Burma, 1950-51; vis. prof. Sweetbriar Coll., 1955-57, Coll. William and Mary, 1961. Cited by U. San Simon, Cochabamba, Bolivia, 1958, Am. Horticulture Soc., 1963. Mem. A.A.A.S., Washington Acad. Sci., Bot. Soc. Am., Genetics Soc. Am., Explorers Club, Assn. Southeastern Biologists (Meritorious Teaching award 1953), Sigma Xi. Research, numerous publs. on genetics peas, castor beans, tobacco, on fascination, temperature tolerance in relation to mutation and geog. distbn. in plants and animals, origin cultivated plants. Home: 1708 Jefferson Park Av., Charlottesville, Va. 22903.*

WHITE, Paul Dudley, Am. physician; b. Boston, June 6, 1886; s. Herbert Warren and Elizabeth (Dudley) W.; A.B., Harvard, 1908, M.D., 1911; also numerous hon. degrees; m. Ina Reid, June 28, 1924; children—Penelope Dudley (Mrs. Peter Hofmann), Alexander Warren. Mem. faculty Harvard, 1920-50, clin. prof. medicine, 1940-50, emeritus, 1950—; staff Mass. Gen. Hosp., Boston, 1911—, vis. physician, 1930-50, cons., 1950—. Chmn. cardiovascular com. NRC, 1940-46; pres. Internat. Cardiology Found., 1957-67. Recipient numerous fgn. and domestic decorations and awards, 1919-66, including Albert Lasker award Am. Heart Assn., 1953, Presdl. Medal Freedom, 1964. Mem. Am. Heart Assn. (pres. 1940-41), Internat. Soc. Cardiology (pres. 1954-58), A.M.A., Am. Assn. Physicians, Aesculaptan Club, A.C.P., Am. Coll. Cardiology, Alpha Omega Alpha; also fellow many fgn. socs. Author numerous books including: Heart Disease, 1931; Hearts: Their Long Follow-up, 1967; also numerous articles. Pioneered research in field of cardiology; studied internal medicine, epidemiology, heart diseases. Home: 115 Juniper Rd., Belmont, Mass. 02116. Office: 264 Beacon St., Boston.*

WHITE, Philip Rodney, Am. physiologist; b. Chgo., July 25, 1901; s. Henry K. and Mary (Pattee) W.; A.B., U. Mont., 1922, D.Sc., 1956; postgrad. U. Wash., 1922-23; Ph.D., Johns Hopkins, 1928; m. Caroline D. M. Smith, Nov. 22, 1935; 1 son, Christopher J. Faculty, U. Mo., 1928-29; Nat. Research fellow Boyce Thompson Inst. for Plant Research, 1929-30; Rockefeller Found. fellow U. Berlin, 1930-31; fellow Rockefeller Inst. for Med. Research, 1932-34, asst., 1934-38, asso., 1938-45; sr. mem., dir. dept. physiology Inst. for Cancer Research, 1945-50; research asso. Jackson Lab., Bar Harbor, Me., 1951-57 sr. staff scientist, 1957—. Vis. prof. botany Yale, 1947-48; exchange prof. U. Paris, 1958-59; distinguished vis. prof. Pa. State U., 1963. Recipient Newcomb prize and medal A.A.A.S., 1937; Stephen Hales award Am. Soc. Plant Physiologists, 1940; Centennial medal Societe Botanique de France, 1954; medal of honor U. Liege, 1959. Fellow Am. Acad. Arts and Scis.; mem. Tissue Culture Assn. (past pres.), U. S., French, Scandinavian socs. plant physiology, Internat. Soc. Cell Biology, Internat. Soc. Plant Morphology. Author: A Handbook of Plant Tissue Culture, 1943; The Cultivation of Animal and Plant Cells, 1954, 2d edit., 1963; also numerous articles. Established cultivation of excised plant organs, and tissues, demonstrated existence of true malignancy in plants, importance of water secretion by roots as factor in rise of sap in trees, developed fully defined nutrients for animal cells. Home: 31 Hancock St. Office: Jackson Lab., Bar Harbor, Me. 04609.*

WHITE, Robert George, immunologist; b. Barwell, Eng., July 18, 1917; s. Thomas Percy and Robina (Fewkes) W.; student Oxford (Eng.) U., 1936-39, London Hosp., 1939-42; m. Joan Margaret Horsburgh, Sept. 15, 1952; children—Diana Margaret, Robert Anthony Horsburgh, Ruth Robina. Freedom Research fellow London Hosp., 1948-52; Med. Research Council Traveling fellow dept. bacteriology and immunology Harvard, 1952-53; reader London U., 1954-63; Gardiner prof., chmn. dept. bacteriology and immunology Glasgow (Scotland) U., 1963—. Mem. expert adv. panel on immunopathology WHO, 1962—. Mem. Brit. Soc. for Immunology (past meetings' sec.), Royal Soc. Medicine (pres. sect. clin. immunology and allergy), Med. Research Club, Path. Soc. Gt. Britain, Reticulo-Endothelial Soc. Author: Essentials of Bacteriology, 1963; (with J.H. Humphrey) Immunology for Students of Medicine, 1963; also articles. Research on cellular aspects of immunology, mechanism of adjuvant action on synthesis of antibody, histological tracer for antibody, fluorescent antibody method; characterized (with E. Lederer) chem. component of Freund adjuvant as peptidoglycolipid and elucidated (with John Gordon) its filamentous structure in electron microscope; discovered and named dendritic antigenreceptor cells of germinal centres of lymphoid tissue, 1963. Home: 47, Campbell St., Helensburgh, Scotland. Office: Western Infirmary, Glasgow W.2., Scotland.*

WHITE, Robert Stephen, Am. physicist; b. Ellsworth, Kan., Dec. 28, 1920; s. Bryon F. and Sebina (Leighty) W.; A.B., Southwestern Coll., Kan., 1942; M.S., U. Ill., 1943; Ph.D., U. Cal. at Berkeley, 1951; m. Freda M. Bridgewater, Aug. 30, 1942; children—Lynn, Peggy, John, David. Physicist, Lawrence Radiation Lab., U. Cal. at Berkeley and Livermore, Cal., 1951-61, also lectr.; physicist, head space particles and fields dept., space physics lab. Aerospace Corp., El Segundo, Cal., 1961-67; prof. physics, asso. dir. Inst. Geophysics and Planetary physics, U. Cal. at Riverside, 1967—. NSF sr. postdoctoral fellow Max-Planck Institüt für Physik und Astrophysik, Munich, Germany, 1961-62. Mem. Am. Geophys. Union, Am. Phys. Soc., A.A.A.S. Produced (with E.M. McMillian and J.M. Peterson) the first pi mesons from photon bombardment, 1949; measured their properties, studied their prodn. from hydrogen and deuterium; discovered (with S.C. Freden) protons in the earth's radiation belts, measured the energy and spatial distbns. of protons and electrons trapped in the earth's radiation belts, 1959; measured the spatial, time, and energy distbns. of low energy protons that diffuse inward from the solar wind through the earth's magnetic field, 1966. Home: 2808 Via Sola, Palos Verdes Estates, Cal. 90274. Office: 2535 Horace St., Riverside, Cal. 92506.*

WHITE, Sidney Edward, Am. geologist; b. Manchester, N.H., Mar. 14, 1916; s. Edward Noyes and Ruth (Putney) W.; B.Sc., Tufts Coll., 1939; M.A., Harvard, 1942; Ph.D., Syracuse, U., 1951; m. Patricia Lindsay Shearer, Jan. 16, 1946; 1 dau., Mary Lindsay. Staff, U.S. Geol. Survey, 1941, 48-49; instr. Tufts Coll., 1947-48, Syracuse U., 1949-51; faculty Ohio State U., Columbus, 1951—, asso. prof., 1958—; vis. asso. prof. U. Colo., 1957, 59, 61, 65-66; geologic field work, Ohio, Utah, Colo., Greenland, Mexico, Sweden; research asso. U. Colo. Inst. Arctic-Alpine Research, 1964—. Fellow Geol. Soc. Am.; mem. Geol. Soc. Mexico, Sigma Xi. Contbr. articles to tech. jours., book revs. Established glacial sequence for higher Mexican volcanoes; demonstrated turbulence and jerky flow Greenland glaciers by motion studies ice; research on amount talus debris and rate accumulation rock rubble in mountain valleys in Colo. Front Range for amount accumulated since last maj. substage glaciation, and effects present-day climate.*

WHITE, William Alanson, Am. physician; b. Bklyn., Jan. 24, 1870; s. Alanson and Harriet Augusta (Hawley) W.; student Cornell, 1885-89; M.D., L.I. Med. Coll., 1891; A.M. (hon.), Georgetown U., 1925; D.Sc., Washington U., 1932; m. Asst. physician Binghampton (N.Y.) State Hosp., 1892-1903; supt. St. Elizabeth Hosp. (making it a leading center for psychiat. care and tng.), Washington, 1904-37; prof. psychiatry Georgetown U., George Washington U.; lectr. insanity U. S. Naval and Army Med Sch. Mem. Nat. Com. for Mental Hygiene, Fed. Bd. Hospitalization; 1st pres. Internat. Congress Mental Hygiene, 1930; pres. Acad. of Medicine, Washington, D.C. Editor and Translator: (with Smith E. Jelliffe) The Psychic Treatment of Nervous Disorders, 1905. Editor: (with Dr. Jelliffe) Nervous and Mental Disease Monograph Series, also Modern Treatment of Nervous and Mental Diseases, 2 vols. the Psychoanalytic Rev. quar. (a founder 1913). Author: Outlines of Psychiatry; 1907; Mental Mechanisms; Mechanisms of Character Formation; Principles of Mental Hygiene; 1917; Diseases of the Nervous System (with Jelliffe); Mental Hygiene of Childhood; Thoughts of a Psychiatrist on the War and After; Foundations of Psychiatry, 1921; Insanity and the Criminal Law; Introduction to the Study of Mind; Essays in Psychopathology, 1925; The Meaning of Diseases, 1926; Lectures in Psychiatry; The Major Psychoses, 1928; Medical Psychology; Crimes and Criminals, 1933; Forty Years of Psychiatry. Abolished various forms of phys. restraint then in common use in Am. mental hosps.; emphasized humane treatment; followed concepts of Sigmund Freud; spread knowledge of psychoanalysis. Died Washington, Mar. 7, 1937.

WHITE, William North, Am. chemist; b. Walton, N.Y., Sept. 16, 1925; s. George F. and Frances (Peck) W.; A.B., Cornell U., 1950; M.A., Harvard, 1951, Ph.D., 1953; m. Hilda R. Sauter, Sept. 8, 1951; children—Carla A., Eric J. Fellow, Cal. Inst. Tech., Pasadena, Cal. 1953-54; faculty Ohio State U., Columbus, 1954-63, asso. prof., 1959-63; prof. chemistry U. Vt., Burlington, 1963—; chmn. dept. chemistry, 1963—; guest biochemist Brookhaven Nat. Lab., Upton, N.Y., 1963-64; research asso. Harvard, 1965. Mem. Am. Chem. Soc., Chem. Soc., N.E. Assn. Chemistry Tchrs., No. New Eng. Acad. Scis. Research, publs. on aromatic rearrangement, molecular complexes, and enzymology. Home: Pierson Dr., Shelburne, Vt. 05482. Office: Dept. Chemistry, U. Vt., Burlington, Vt. 05401.*

WHITEHAIR, Charles Kenneth, Am. nutritional pathologist; b. Abilene, Kan., Mar. 3, 1916; s. Charles William and Ceil (Hughes) W.; D.V.M., Kan. State U., 1940; M.S., U. Wis., 1943, Ph.D., 1947; postgrad. Rowett Research Inst. Aberdeen, Scotland; m. Margaret McKeegan, Feb. 15, 1958; children—Stanley Joseph, Robert Charles. Instr., U. Wis., 1940-47; faculty Okla. State U., Stillwater, 1947-52, prof., 1949-52; dir. vet. research, prof. physiology, 1953-54; asso. prof. U. Ill., Urbana, 1952-53; researcher Rowett Research Inst., Aberdeen, Scotland, 1955-56; prof. pathology Mich. State U., East Lansing, 1956—. Mem. adv. bd. Lobund Lab., U. Notre Dame, 1958—; mem. div. agr. and biology NRC, since 1956—. Mem. Am. Inst. Nutrition, Am. Soc. Exptl. Pathology, Soc. for Exptl. Biology and Medicine, Sigma Xi, Phi Kappa Phi. Contbg. author: Diseases of Cattle, 1963; Diseases of Swine, 1964; also numerous articles. Research on nutritional requirement during disease, interrelationship nutrition and infection, morphologic characteristics nutritional deficiencies, germfree research. Home: 304 Droste Circle, East Lansing, Mich. 48823.*

WHITEHEAD, Alfred North, philosopher; b. Ramsgate, Isle of Thanet, Kent, England, Feb. 15, 1861; s. Canon Alfred W.; B.A., Trinity Coll., Cambridge, 1884, M.A., 1887, D.Sc., 1905; D.Sc., U. of Manchester, 1920; LL.D., St. Andrews, 1921; D.Sc., Harvard and U. of Wis., 1925, Yale, 1926, McGill U., Montreal, Can., 1931; m. Evelyn Willoughby Wade, Dec. 16, 1890; children—T. North, Jessie Marie. Lecturer and later sr. lecturer on mathematics, Trinity Coll., 1885-1911; lecturer on applied mathematics and mechanics and later reader in geometry, University Coll., U. of London, 1911-14; prof. applied mathematics and later chief prof. of mathematics, Imperial Coll. of Science and Technology of same univ., 1914-24, senator, 1919, dean of faculty of science, 1921; prof. philosophy, Harvard, 1924-36, now emeritus prof. Fellow Royal Soc., British Acad.; mem. Math. Soc., British Assn. Advancement Science. Tarner lecturer, Trinity Coll., 1919. Awarded James Scott prize, Royal Soc., Edinburgh, 1922; Sylvester medal, Royal Soc., London, 1925. Author: A Treatise on Universal Algebra, 1898; (with Bertrand Russell) Principia Mathematica, 1910; An Introduction to Mathematics, 1910; The Organization of Thought, 1916; The Principles of Natural Knowledge, 1919; The Concept of Nature, 1920; The Principle of Relativity, 1922; Science and the Modern World, 1925; Religion in the Making, 1926; Symbolism: Its Meaning and Effect, 1927; The Aims of Education, 1928; Process and Reality (Gifford Lectures), 1929; The Function of Reason, 1929; Adventures of Ideas, 1933; Nature and Life, 1934; Modes of Thought, 1938. Recipient British Order of Merit, 1945. Made important contributions to study of foundations of mathematics, to mathematics, and to symbolic logic; formulated organic philosophy, which included critical examination of basic concepts of natural science. Died Dec. 30, 1947.

WHITEHEAD, George William, Am. mathematician; b. Bloomington, Ill., Aug. 2, 1918; s. George William and Mary (Gutschlag) W.; S.B., U. Chgo., 1937, S.M., 1938, Ph.D., 1941; m. Kathleen E. Butcher, June 7, 1947. Instr., Purdue U., 1941-45, Princeton, 1945-47; mathematician Aberdeen Proving Ground, 1945; faculty Brown U., 1947-49; faculty Mass. Inst. Tech., Cambridge, 1949—, prof. math., 1957—. Fellow Am. Acad. Arts and Scis.; mem. Am. Math. Soc. Author: Homotopy Theory, 1966. Research and publs. on homotopy theory. Home: 25 Bellevue Rd., Arlington, Mass. 02174. Office: Mass. Inst. Tech., Cambridge, Mass. 02139.*

WHITEHEAD, John Boswell, Am. elec. engr.; b. Norfolk, Va., Aug. 18, 1872; s. Henry Colgate and Margaret Walke (Taylor) W.; EE., Johns Hopkins, 1893, A.B., 1898, Ph.D. 1902; m. Mary Ellen Colston, Apr. 14, 1903; children—Clara (dec.), Margaret Walke, Joan Boswell. Elec. engr. Westinghouse Electric & Mfg. Co., 1893-96, Niagara Falls (N.Y.) Power Co., 1896-97; instr. applied electricity Johns Hopkins, 1897-1900, asso., 1901-04, asso. prof., 1904-10, prof., 1910—, dean, 1919-38, dir. Sch. Engring., 1938-42, prof. emeritus, 1942—; exchange prof. to France, 1926-27. Lab. asst. U. S. Bur. Standards, 1902; research asst. Carnegie Instn. Washington, 1902-05. Edison medalist, 1941. Fellow Am. Inst. E.E. (pres. 1933-34); Am. Phys. Soc., A.A.A.S.; mem. Nat. Acad. Scis., Nat. Research Council, Société Française des Electriciens, (hon.), Phi Beta Kappa. Author: Electric Operation of Steam Railways, 1909; Dielectric Theory and Insulation, 1927; Impregnated Paper Insulation, 1935; Electricity and Magnetism, 1939. Research in field of high voltage insulation; single phase railway systems; electric displacement; high alternating voltage. Died Nov. 16, 1954.

WHITEHEAD, Peter James Palmer, ichthyologist; b. Nairobi, Kenya, June 24, 1930; s. Philip Henry Rathbone and Gertrude (Palmer) W.; B.A., Cambridge (Eng.) U., 1953; m. Monica Fitzgerald-O'Dwyer, July 28, 1953; children—Amanda Oonagh, Paul James; m. 2d, Greta Ransom, Oct. 12, 1967; 1 dau., Victoria-Augusta. Fishery research officer Kenya Govt., 1953-60; sr. sci. officer Brit. Mus., London, 1961—. Mem. Am. Soc. Ichthyologists and Herpetologists, Inst. Biology. Research, publs. on biology of E. African river fishes and edible oyster in mangrove swamps, taxonomic studies of herring and anchovy-like fishes, culture of Tilapia in ponds, nomenclatural problems in fishes. Home: Pillar House, Harwell, Bershire, Eng. Office: Brit. Mus. Natural History, Cromwell Rd., London S.W. 7, Eng.*

WHITEHEAD, Robert, English mech. engr.; b. Mt. Pleasant, Bolton-Le-Moors, Eng., Jan. 3, 1823; s. James and Ellen (Swift) W.; ed. Mechanics' Inst., Manchester; m. Frances Maria Johnson, 1845; children—John, James Bethom, Robert Bovill, Alice, Countess Hoyos. Apprenticed to engring. firm, Manchester, Eng., 1837; with Philip Taylor & Sons, Marseilles, France, 1844-47; founder bus., Milan, Italy; with Austrian Lloyd Co., Trieste, 1848-50; mgr. Strudhoff Works, 1850-56; founder Stabilimento Tecnico Fiumano, 1856, bought co., 1872. Recipient Austrian Order Franz Joseph, 1868, orders from Prussia, Denmark, Portugal, Italy, Greece, Turkey. Inventor mobile torpedo (Whitehead torpedo), 1866; built compressed air motor in chamber to equalize pressure for diving, 1868; improved silk-weaving machinery; designed machinery for Austrian warships, machinery to drain Lombardy marshes. Died Beckett, Eng., Nov. 14, 1905.

WHITEHEAD, Thomas H., Am. chemist; b. Maysville, Ga., Sept. 5, 1904; s. Asa Hillyer and Clara (Comer) W.; B.S., U. Ga., 1925; M.A., Columbia, Ph.D., 1930; m. Dorothy Lou Simms, Dec. 19, 1931; children—Thomas H., John S. Mem. Faculty U. Ga., Athens, 1930-42, prof. chemistry, 46-51, 52—, also coordinator Instructional Instns., Grad. Sch., 1960—; with Chem. Warfare Service, U. S. Army, 1942 46, Army Chem. Center, 1951-52; cons. U. S. AEC, 1952—, Wright-Patterson AFB, 1964—. Decorated Legion of Merit. Fellow Ga. Acad. Sci.; mem. Am. Chem. Soc., A.A.A.S., Sigma Xi, Phi Kappa Phi, Phi Lambda Upsilon. Author: Theory of Elementary Chemical Analysis, 1950. Developer (with A. W. Thomas) complex coumpound theory of colloidal oxides; new process for bleaching and staining base ball bats; new methods for detecting casein glue in wood products, and determination of copper, also trace amounts of water in products. Home: 236 Henderson Av., Athens, Ga. 30601.*

WHITEHEAD, Walter D., Am. physicist; b. San Diego, Nov. 30, 1922; s. Walter D. and Lillian (Shepard) W.; B.S., U. Va., 1944, M.A., 1946, Ph.D. (Carnegie Instn. fellow, AEC fellow), 1949; m. Sophfe Louise Van Ketel, Apr. 6, 1949; children—Oliver Day, Thomas Kenyon. Physicist, Bartol Research Found., Swarthmore, Pa., 1949-53; asst. prof. N.C. State Coll., Raleigh, 1953-55, asso. prof., 1955-56; faculty U. Va., Charlottesville, 1956—, prof. physics, 1961—, dir. Center for Advanced Studies, 1965—, chmn. dept. physics, 1968—; guest physicist Instituut voor Kern Physisch Onderzoek, Amsterdam, The Netherlands, 1959-60. Recipient U. S. Naval Ordnance Devel. award, 1946. Fellow Am. Phys. Soc.; mem. Raven Soc., Va. Acad. Scis., Sigma Xi, Alpha Tau Omega, Sigma Pi Sigma. Research in nuclear physics, neutron physics, photo-nuclear reactions. Home: Box 345, Route 1, Earlysville, Va, Office: 1512 Jefferson Park Av., Charlottesville, Va. 22903.*

WHITEHILL, Alvin Richard, Am. bacteriologist; b. Groton, Vt., s. David L. and Abgail (Welch) W.; A.B., Dartmouth, 1934; Ph.D., Cornell U., 1942; m. Marie Josephine Battermann, May 30, 1943; children—Ency Anne, Wendy Lee, Pamela Nann, Austin Richard. Instr., Cornell U., 1938-41, Ill. Inst. Tech., 1941-44; nutritionist Am. Cyanamid Co., 1944-61; prof., head dept. bacteriology U. Me., Orono, 1961—. Mem. research council Inst. Am. Poultry Industries, 1958-61. Mem. Soc. Am. Microbiology, Am. Ind. Microbiology, Am. Food Technologists, N.Y. Acad. Sci. Contbr. articles to tech. jours. Research on bacterial and animal nutrition, antibiotics and plant tissue culture. Home: 76 Main St., Orono, Me. 04473.*

WHITEHOUSE, Michael Wellesley, English biochemist; b. London, Eng., Feb. 18, 1930; s. Stuart Wellesley and Ileene (Macbain) W.; M.A., Oxford (Eng.) U., 1955, B.Sc., 1953, D.Phil., 1955; m. Darling Jane Thorburn Boyd, Sept. 29, 1956; children—Hilary Louise, Dominic McBean Wellesley. Instr., asso. biochemistry U. Pa. Med. Sch., 1955-59; Arthritis and Rheumatism Council U.K. Geigy Travelling fellow Wenner-Gren Inst., U. Stockholm, 1960; Staines Med. Research fellow Exeter Coll., Oxford U., 1960-65, lectr. biochemistry, 1961—; vis. fellow Australian Nat. U., Canberra, 1966-67; NSF sr. fgn. scientist fellow Ohio State U. Coll. Pharmacy, 1967-68. Fellow Royal Inst. Chemistry London; mem. Am. Chem. Soc., Biochem. Soc., Sigma Xi. Author: (with P.W. Kent) Biochemistry of the Aminosugars, 1955; also articles. Editorial bd. Index Chemicus, 1960-66, Current Contests, 1958-66. Research on regulation of cholesterol oxidation in body by drugs, diet, and hormones; elucidating mode of action of antirheumatic drugs. Home: 6 Stanley Rd., Oxford, Eng.*

WHITEHOUSE, William John, physicist; b. Zoutpansberg, S. Africa, Mar. 9, 1910; s. Percy and Mildred (Fryer) W.; B.Sc. with 1st class honors, Manchester (Eng.) U., 1931, M.Sc., 1933, Diploma in Edn. with 1st class honors, 1934; m. Isabel Moffatt Brown, Apr. 14, 1941 (div.); children—Paul, Patrick David. Teacher, 1935-40; staff Telecommunications Research Establishment, Swanage, also Malvern, 1941-44; staff Joint U.K.-Canadian Mission, Montreal, Que., Can., 1944-46; staff Atomic Energy Research Establishment, Harwell, Eng., 1946-56, 58—; dir. Baghdad Pact Nuclear Centre, 1956-58. Mem. sci. adv. com. Oxford (Eng.) Coll. Tech., 1947-55; organizer advanced course in radiation protection Harwell Reactor Sch., 1959-64; lectr. Harwell Isotope Sch., intermittently 1951—. Fellow Inst. Physics and Phys. Soc. London; mem. Am. Phys. Soc. Author: (with J.L. Putman) Radioactive Isotopes, 1953; also articles. Research on X-ray crystalography; basic design of early microwave radar sets; a designer of 1st Brit. exptl. reactor; initiated prodn. radioactive isotopes from reactors in U.K.; research in nuclear physics especially spontaneous and induced fission, cosmic radiation, radiation protection and reactor safety. Home: 3, Pelham Ct., Westcote Rd., Reading, Eng. Office: Atomic Energy Research Establishment, Harwell, Didcot, Eng.*

WHITEMAN, Albert Leon, Am. mathematician; b. Phila., Feb. 15, 1915; s. Nathaniel and Clara (Brody) W.; A.B., U. Pa., 1936, A.M. (Harrison scholar), 1937, Ph.D., (Harrison fellow), 1940; m. Sally Seidelman, June 3, 1945. Mem. faculty Harvard, 1940-42, Purdue U., 1946; research mathematician Navy Dept., 1946-48; faculty U. So. Cal., Los Angeles, 1948—, prof. math., 1956—; mem. Inst. Advance Study, Princeton, 1952-54, 59-60, 67—, Inst. Def. Analyses, 1960-61. Chmn. number theory confs. NSF, 1955, 63. Mem. Am. Math. Soc., Math. Assn. Am., Am. Assn. U. Profs., Phi Beta Kappa, Sigma Xi, Pi Mu Epsilon. Editor, Pacific Jour. Math., 1957-62, Proceedings of Symposia in Pure Math, vol. 8, 1965; asso. editor Duke Math. Jour., 1957-62, spl. issue Jour. Math. Analysis and Applications, 1966. Research and publs. on number theory and combinatorial analysis. Home: 65 Einstein Dr., Princeton, N.J. 08540.*

WHITFIELD, Robert Parr, Am. geologist; b. New Harford, N.Y., May 27, 1828; s. William Fenton and Margaret (Parr) W.; self-ed.; hon. A.M., Wesleyan, 1882. Worked for his father, a spindlemaker, until 1848, when he entered employ of Samuel Chubbuck, instrument-maker, Utica, becoming mgr., 1849-56; asst. in palaeontology and geology. N.Y. State Natural History, 1856-76; U. S. Geol. Survey, 1872; tchr. Rensselaer Poly. Inst., Troy, N.Y. 1872-75, prof. geology, 1875-78; curator geol. dept., Am. Mus. Natural History, 1877—. Original fellow A.A.A.S.; fellow Geol. Soc. Am. contbr. paleontol. papers. U. S. Geol. Survey, Ohio Palaeontology, Palaeontology of New Jersey, 3 vols., also contbr. on geol. subjects. Died 1910.

WHITFORD, Albert E(dward), Am. astronomer, b. Milton, Wis., Oct. 22, 1905; s. Alfred Edward and Mary (Whitford) W.; A.B., Milton Coll., 1926; A.M., U. Wis., 1928, Ph.D., 1932; m. Eleanor Whitelaw, Oct. 23 1937; children—William Curtis, Mary, Martha. Fellow Nat. Research Council, Mt. Wilson Observatory, Cal. Inst. Tech., 1933-35; research asso. Washburn Observatory, U. Wis., 1935-38, asst. prof. astrophysics, 1938-46, asso. prof. astronomy, 1946-48, prof., dir. of observatory, 1948-58; on leave to Radiation Lab., Mass. Inst. Tech., 1941-46; dir. Lick Observatory, Univ. of Cal., 1958—. Member of the American Academy of Arts and Sciences, Internat. Astron. Union, Am. Assn. U. Profs., Nat. Acad. Sci., Am. Astron. Soc. (council 1950-53), Sigma Xi. Contbr. profl. jours. Contbn. to study atomic spectra, photoelectric photometry of stars and nebulae; worked on devel. instruments. Address: Lick Observatory, Mt. Hamilton, Cal.

WHITFORD, Larry Alston, Am. botanist, educator; b. Ernul, N.C., Apr. 11, 1902; s. David Purifoy and Grace (Price) W.; B.S., N.C. State Coll., 1925, M.S., 1928; Ph.D., Ohio State U., 1941; m. Duella Wade, June 2, 1928; children—Patricia (Mrs. A.T. Leary, Jr.), Bertram, David, Geoffrey. Faculty, N.C. State U., Raleigh, 1926—, prof. botany, 1962—. Lectr. hydrobiology U. Va., 1952; research asso. U. Fla., 1953. Fellow A.A.A.S.; mem. Phycological Soc. Am. (past pres.), Internat. Phycological Soc., N.C. Acad. Sci. (pres. 1964-65), Sigma Xi. Contbr. articles to profl. jours., chpt. to The Ecology of Algae. Compiler list of lower aquatic plants in N.C. with notes on habitat and distbn., research on habitat factors affecting distbn. of algae. Home: 3217 Oakgrove Circle, Raleigh, N.C. 27607.*

WHITFORD, Philip Burton, Am. plant ecologist; b. Argyle, Minn., Jan. 9, 1920; s. Manley C. and Ethel (Cadman) W.; B.E., No. Ill. State Coll., 1941; M.S., U. Ill., 1942; Ph.D., U. Wis., 1948, m. Kathryn Dana Rueber, Feb. 22, 1946; children—Mary Kathryn, Philip Clason. Edn. asst. Md. Bd. Natural Resources, 1948-49; faculty U. Wis.-Milw., 1949—, prof. botany, 1960—, chmn. dept. botany, 1966—. Fellow A.A.A.S.; mem. Ecol. Soc. Am., Am. Inst. Biol. Scis., Nature Conservancy. Author: (with M. Ackley) Chemistry of Photosynthesis, 1965; also articles. Research on edaphic control along prairie-forest border. Home: 4657 N. 117th St., Milw. 53225.*

WHITFORD, Walter George, Am. physiol. ecologist; b. Providence, June 12, 1936; s. Walter Albert and Mary-Helen (Gravier) W.; B.A., U. R.I., 1961, Ph.D., 1964; m. Ann Edith Rowley, Jan. 24, 1958 (div. Feb. 1968); children—William Brett, Eric Ian. Asst. prof biology N.M. State U., University Park, 1964-68, asso. prof. biology, 1968—. Mem. Am. Soc. Icthyologists and Herpetologists, Am. Soc. Mammalogists, A.A.A.S., Am. Soc. Zoologists, Herpetologists League, Sigma Xi, Phi Kappa Phi. Research, publs. on evaluation role of skin and lungs in oxygen consumption in amphibians, effect of environmental factors on gas exchanges in salamanders and frogs, including temperature, photoperiod, diurnal rhythms and body size, physiol. adaptations of desert amphibians and reptiles, bioenergetics of desert vertebrates. Home: 2712 Ridgeway Ct., Las Cruces, N.M. 88001. Office: Dept. Biology, N.M. State U., University Park, N.M. 88070.*

WHITHAM, Gerald Beresford, mathematician; b. Halifax, Eng., Dec. 13, 1927; s. Harry and Elizabeth (Howarth) W.; B.Sc., Manchester (Eng.) U., 1948, M.Sc., 1949, Ph.D., 1953; m. Nancy Lord, Sept. 1, 1951; children—Ruth H., Michael G., Susan C. Came to U. S., 1956. Lectr., Manchester U., 1953-56; asso. prof. N.Y. U., 1956-59; prof. math. Mass. Inst. Tech.,

1959-61; prof. aero. and math. Cal. Inst. Tech., 1961—. Fellow Royal Soc., Am. Acad. Arts and Scis.; mem. Am. Assn. U. Profs. Author research papers. Contributions in applied math., fluid dynamics. Home: 1689 E. Altadena Dr., Altadena, Cal. 91001. Office: Cal. Inst. Tech., Pasadena, Cal. 91109.

WHITING, Phineas Wescott, Am. geneticist; b. Lowell, Mass., Oct. 28, 1887; s. Henry Fairfax and Louise (Wescott) W.; A.B., Harvard, 1911, M.S., 1912; Ph.D., U. Pa., 1916; m. Anna Rachel Young, June 29, 1918. Prof., Franklin and Marshall Coll., Lancaster, Pa., 1918-20, St. Stephen's Coll., Annandale-on-Hudson, N.Y., 1920-21, U. Ia., Iowa City, 1921-24, U. Me., Orono, 1924-27, U. Pitts., 1927-33; prof., geneticist U. Pa., Phila., 1935-53, prof. emeritus, 1953—. Cons. biology div. Oak Ridge Nat. Lab., 1956—. Mem. Soc. Naturalists, Genetics Soc. (past pres.), Am. Soc. Zoologists, Genetics Assn. Am., Radiation Research Soc., Sigma Xi. Research, publs. on heredity bristles in blowflies, coat color in cats, color mutations in mealmoth Ephestia, cytology house mosquito and mud-dauber wasps, genetics parasitic wasps Habrobracon and Mormoniella, sex determination in animals with haploid males, compound genes. Home: 535 W. Vanderbilt Dr., Oak Ridge 37830. Office: Biology Div., Oak Ridge Nat. Lab., Oak Ridge 37830.*

WHITMAN, Charles Otis, Am. zoologist; b. Woodstock, Me., Dec. 14, 1842; s. Joseph and Marcia W.; A.B., Bowdoin, 1868, A.M., 1871; Ph.D., Leipzig, 1878; fellow Johns Hopkins, 1879; LL.D., U. Neb., 1894; Sc.D., Bowdoin, 1895; D. Biology, Clark U., 1909; m. Emily Nunn, Aug. 15, 1884. Prof. zoology Imperial U. Japan, 1879-81, Naples Zool. Sta., 1881-82; asst. in zoology Harvard, 1882-86; dir. Allis Lake Lab., 1886-89; prof. zoology Clark U., 1889-92; prof. and head dept. zoology and curator Zoo. Mus., 1892-1910. U. Chgo. Planner Marine Biol. Labs., Woods Hole, Mass., dir., 1888-1908. Author: The Embryology of Clepsine, 1878. Editor Jour. Morphology, 1887-1910, Biol. Bull., 1897-1910, Biol. Lectures, 1890-99. Research and numerous publs. on cell theory, eggs, devel. and anatomy of vertebrates and annelids, microscopic methods, animal behavior, genetics and evolution. Died Chgo., Dec. 6, 1910.

WHITMAN, Roy Milton, Am. physician; b. N.Y.C., June 16, 1925; s. Jack and Mae (Hoffman) W.; student Oberlin Coll., 1941-43; B.S., Ind. U., 1944, M.D., 1946 children—Joy, Bruce, Laura. Asst. chief psychiatry 17th Airborne div., Camp Pickett, Va., 1948-49, chief mental hygiene clinic, Ft. Riley, Kan., 1949-50; USPHS fellow in psychiatry U. Chgo. Clinics, 1950-52, instr., 1952-53, asst. prof. psychiatry, Northwestern U. Sch. Medicine, 1954-57; asso. prof. psychiatry U. Cin. Coll. Medicine, 1957-67, professor of psychiatry, since 1967—; cons. to VA Hosp., Cin., 1957—; clinician Central Clinic, Cin., 1957—; cons. Ill. State Psychiat. Inst., Chgo., 1962-66. Recipient Chgo. Psychoanalytic Inst. Alexander award, 1965. Mem. Am. Psychoanalytic Assn., Am. Psychiat. Assn., Group for Advancement of Psychiatry, Ohio Psychiat. Soc., Cin. Soc. Neurology and Psychiatry, Sigma Xi, Alpha Omega Alpha. Research, publs. in group dynamics and group therapy; research in the interrelationships of tranquilizing drugs and psychol. states; implantation of conflict by hypnosis; dream research, especially use of dream in psychiat. transactions; intrapsychic structure of negative ego ideal. Home: 4000 Ledgewood St., Cin. 45229.*

WHITMAN, William Tate, Am. economist; b. Boaz, Ala., Oct. 26, 1909; s. Edward Fenno and Jane Moore (Street) W.; A.B., Duke, 1929, M.A., 1933, Ph.D., 1943; m. Luisita Dye, Aug. 12, 1936; children—Melinda (Mrs. Phillip Robinson Certain), Ernestine. Accountant, Univ. Motors, Inc., Durham, N.C., 1934-36; mem. faculty The Citadel, Charleston, S.C., 1936-39, 40-47, Duke, 1939-40; faculty Emory U., Atlanta, 1947—, Charles Howard Candler prof. econs., 1960—. Mem. Am. So. (v.p. 1963-64) econ. assns. Indsl. Relations Research Assn., Beta Gamma Sigma, Alpha Kappa Psi, Delta Tau Delta. Author: (with C. Sidney Cottle) Investment Timing: The Formula Plan Approach, 1953; Corporate Earning Power and Its Market Valuation—1935-55, 1959; (with Frank J. Charvat) Marketing Management: A Quantitative Approach, 1964. Editor: (with Melvin L. Greenhut) Essays in Southern Economic Development, 1964. Developed method for testing formula plans investment timing; performance of different types of formula plans under varying market conditions; method for appraising comparative performance of publicly held corps.; appraisal of comparative performance of large sample of major Am. corps. over 20 year period. Home: 1533 Emory Rd. N.E., Atlanta 30306.*

WHITMER, Charles Austin, Am. physicist; b. Flat Rock, Ind., Feb. 10, 1907; s. Delbert Austin and Camoleon (Chales) W.; A.B., DePauw U., 1928; M.S., U. Ia., 1931; Ph.D., Ohio State U., 1936; m. Claire Ferris, Sept. 5, 1931 (dec. Feb. 1944); 1 son, John Charles; m. 2d Jeanette Keyes, July 13, 1945. Asst. prof. physics Olivet Coll., 1931-32; asst. prof. physics U. Okla., 1936-41, asso. prof., 1941-42; staff radiation lab. Mass. Inst. Tech., 1942-46; asso. prof. physics Rutgers U., 1946-51, prof., 1951—, chmn. dept., 1952-59, exec. officer, 1959-60; head student and curriculum improvement sect. NSF, 1960—. Fellow Am. Phys. Soc.; mem. Am. Assn. Physics Tchrs., A.A.A.S., Am. Soc. for Engring. Edn., Sigma Xi, Phi Beta Kappa. Author: Crystal Rectifiers (with H. C. Torrey), 1946. Contbr. articles on paramagnetic resonance. Early research in absorption of microwave radiation by paramagnetic solids through mechanism of electron spin resonance. Home: 612 S. Royal St., Alexandria, Va. 22314. Office: Nat. Sci. Found, Washington 20550.*

WHITNEY, Asa, Am. inventor, mfr.; b. Townsend, Mass., Dec. 1, 1791; s. Asa and Mary (Wallis) W.; m. Clarinda Williams Aug. 22, 1816; at least 5 children. Small scale mfr. of axles for horse-erecting machinery on inclined planes, Albany and Schenectady, N.Y., also bldg. railroad cars for Mohawk and Hudson R.R., 1830-39; canal commr. N.Y. State, 1839-42; obtained patent for locomotive steam engine, 1840, 2 patents for cast-iron car wheel with corrugated center web, and for method of manufacturing it, 1847; organized (with 3 sons) Asa Whitney & Sons, Phila., 1847; received patent for improved process of annealing and cooling cast iron wheels, 1848; pres. Reading R.R., 1860-61; bequeathed money to U. Pa. to establish chair of dynamic engring., also to Franklin Inst. Died Phila., June 4, 1874.

WHITNEY, Charles Allen, Am. astrophysicist; b. Milw., Jan. 31, 1929; s. Charles Smith and Gertrude (Schuyler) W.; B.S., Mass. Inst. Tech., 1951; A.M., Harvard, 1953, Ph.D. (Agassiz fellow, NSF fellow), 1955; m. Jane Ann Hall, Jan. 27, 1951; children—Elizabeth Ann, David Hall, Thomas Charles, Peter Schuyler, James Andrew. Physicist Smithsonian Astrophys. Obs., 1956—; research asso. Harvard Coll. Obs., 1956-62, faculty Harvard, 1958—, asso. prof. astronomy, 1963—. Mem. Am. Astron. Soc., Am. Acad. Arts and Scis., Internat. Astron. Union. Research and publs. on analysis of artificial satellite orbital data; structure of earth's upper atmosphere; contbns. to theory of stellar pulsation; shock waves in stellar atmospheres; oscillations of solar atmosphere and structure of stellar atmospheres. Home: 527 Boston Post Rd., Weston, Mass. 02193. Office: 60 Garden St., Cambridge, Mass. 02138.*

WHITNEY, Eli, Am. inventor; b. Westboro, Mass., Dec. 8, 1765; s. Eli and Elizabeth (Fay) W.; grad. Yale, 1792; m. Henrietta Frances Edwards, Jan. 6, 1817; 3 children. Made and repaired violins, manufactured nails in father's shop, 1780-82; designed cotton gin on Mrs. Nathanael Green's plantation with which one operator could clean 50 pounds of cotton daily, 1793; in partnership with Phineas Miller to patent and manufacture cotton gins, 1793; received patent for cotton gin, 1794, failed to benefit because of infringements and long litigation; received U. S. Govt. contract for 10,000 muskets to be delivered in 2 years, 1798; devised system of manufacturing interchangeable gun parts; purchased mill site for factory for firearms prodn., nr. New Haven, Conn (Whitneyville); built 1st successful milling machine; introduced div. of labor in his factories; original mem. Hall of Fame for Gt. Americans, 1900. Died New Haven, Jan. 8, 1825.

WHITNEY, Hassler, Am. mathematician; b. N.Y.C., Mar. 23, 1907; s. Edward Baldwin and Josepha A. (Newcomb) W.; Ph.B., Yale, 1928, Mus.B., 1929, Sc.D., 1947; Ph.D.; Harvard, 1932; m. Margaret R. Howell, May 30, 1930 (div. Nov. 1949); children—James Newcomb, Carol, Marian (Mrs. William Hugh Melhuish); m. 2d, Mary Barnett Garfield, Jan. 16, 1955; children—Sarah Newcomb, Emily Baldwin. Mem. faculty Harvard, 1933-52; prof. math. Inst. for Advanced Study, Princeton, 1952—. Mem. Am. Math. Soc., Math. Assn. Am., Nat. Acad. Scis., Am. Philos. Soc. Author: Geometric Integration Theory, 1957. Research and publs. on basic math. theory of networks and of differentiable functions and manifolds; set up theory of sphere bundles; defined tensor products, and extended algebraic topology; derived geometric and analytic properties of integration, singularities of smooth mappings, and analytic varieties. Home: 56 Maxwell Lane, Princeton, N.J. 08540.*

WHITNEY, John Edward, Am. physiologist; b. Casper, Wyo., July 6, 1926; s. Gary Myron and Harriet (Whitney) W.; A.B., U. Cal. at Berkeley, 1947, M.A., 1948, Ph.D., 1951; Ph.D., Cambridge U. (Eng.), 1956; m. Barbara Isabel Laundrie, June 11, 1949; children—Karen Lynn, Kathleen Diane, Eric Allen, Emily Allison, Brian Gary. Research fellow Cambridge U., 1954-56; faculty Sch. Medicine U. Ark., Little Rock, 1956—, prof., head dept., 1962—. Mem. Am. Physiol. Soc., Endocrine Soc., Soc. Exptl. Biology and Medicine, Sigma Xi. Studies, publs. on carbohydrate and lipid metabolism, especially in diabetes as result of adminstrn. of insulin and various pituitary hormones; insulin-like material found in non-pancreatic tissues and biosynthesis of insulin by pancreas. Home: 7016 Rockwood Rd., Little Rock 72207. Office: 4301 Markham St., Little Rock 72205.*

WHITNEY, Josiah Dwight, Am. geologist; b Northampton, Mass., Nov. 23, 1819; s. Josiah Dwight and Sarah (Williston) W.; grad. Yale, 1839; m. Louisa (Goodard) Howe, June 1854; 1 dau. Read law, 1841; travelled and studied in France, Germany, Italy, 1842-45; worked survey of mineral lands of No. peninsula of Mich., 1847-49; ind. cons. expert in mining, 1849-54; Ia. State chemist, also prof. mineralogy Ia. State U., 1855-58; mem. Ill. State Survey; geologist of Cal., 1860-74, becan survey of state; promoted Cal. Acad. Science; commr. Yosemite; apptd. to Harvard faculty, 1865, opened Harvard Sch. Mines, 1868, Sturgis-Hooper prof. geology, 1875-96; returned briefly to Cal. to continue survey; elected to Geol. Soc. London; mem. Am. Philos. Soc., Nat. Acad. Scis. Author: Metalic Wealth of the United States, 1854; Geological Survey of California, 1864-70; The Auriferous Gravela of the Sierra Nevada of California, 1880; Climatic Changes of Later Geological Times, 1882; Names and Places, 1888. Editorial staff Ency. Brit., 9th edit., also Century Dictionary and Cyclo. Discoverer Calaveras skull, Calaveras, Cal., 1886; Mt. Whitney (highest peak in U. S.) named in his honor. Died Lake Sunapee, N.H., Aug. 19, 1896.

WHITNEY, Mary Watson, Am. astronomer; b. Waltham, Mass., Sept. 11, 1847; d. Samuel B. and Mary W. (Crehore) W.; A.B., Vassar Coll., 1868, A.M., 1872 postgrad. U. Zürich, 1874-76. Became asst. to Prof. Maria Mitchell, 1881; prof. astronomy, dir. obs. Vassar Coll., 1889-1910. Research and papers on positions of asteroids and comets, on variable stars, on measurement of photog. plates. Died Jan. 20, 1921.

WHITNEY, Robert McLaughlin, Am. chemist; b. St. Paul, Sept. 28, 1911; s. John Cass and Anna Louise (Boyce) W.; B.A., Augustana Coll., 1936; Ph.D., U. Ill., 1944; m. Lola Elaine Hagan, Aug. 26, 1934; children—Mark Hagan, Robin Raulston; m. 2d, Ruth Ione Lemen Denison, Apr. 11, 1967. Faculty, Augustana Coll., 1936-37; head sci. dept. pub. schs., Harvey, N.D., 1937-38, Cairo, Ill., 1938-40; with Ill. Water Survey, 1940-42, Dean Milk Co., 1944-46; faculty U. Ill., Urbana, 1942-44, 46—, prof. dairy tech., 1959—. Mem. Am. Chem. Soc., Am. Dairy Sci. Assn. (Dairy Mfg. award 1961), Sigma Xi, Gamma Sigma Delta. Research, publs. on flavors of dried and evaporated milk, developer analytical tests for constituents of milk, especially milk salts; bactericidal effect of ultrasonic waves, composition of fat-globule membrane protein in milk, phys. state of milk protein in their natural environment and effect of heat on them. Home: 1706 Henry St., Champaign, Ill. 61820.*

WHITNEY, Simon Newcomb, Am. economist; b. N.Y.C., Apr. 5, 1903; s. Edward Baldwin and Josepha (Newcomb) W.; grad. Taft Sch., 1919; student Deep Springs (Cal.) Jr. Coll., 1919-21; B.A., Yale 1924, Ph.D., 1931; German-Am. exchange student Bonn U., Germany, 1925-26; m. Eunice Gilbert McIntosh, Aug. 23, 1941; children—Eunice Elizabeth, Simon Newcomb, Roger Sherman, Walter McIntosh. Instr. econs. Yale, 1926-28; economist antitrust div. Dept. Justice, 1928-29; asst. to economist Chase Nat. Bank of N.Y., 1931-34; sr. economist research and planning div. Nat. Recovery Adminstrn., 1934-36; staff div. research and satistics, bd. govs. Fed. Res. System, 1936; asso. Lionel D. Edie & Co., Investment counsellors, N.Y.C., 1936-40; economist O'Ryan Financial Commn. to Japan, 1940; prin. economist Office Export Control and Bd. Econ. Warfare, 1941-42; pres. Telluride Assn., endowed ednl. found., Ithaca, N.Y., 1930-31, dean, 1942-48, dir. Deep Springs (Cal.) Jr. Coll., 1942-48; cons. ECA, 1948; asso. prof. econs., coll. arts and sci. N.Y. U., 1948-49, prof., 1949-54, lectr., 1954-55 v.p. Econometric Inst., 1949; asso. economist Twentieth Century Fund, N.Y.C., 1949-54, chief research div., 1954-55; dir. Bur. Econs., FTC, 1956-61; prof. econs. Rutgers U., New Brunswick, N.J., 1961-67, New York University, N.Y.C., 1967—. Member Royal Economic Society, member American Econ. Assn., Phi Beta Kappa. Author: Trade Associations and Industrial Control: A Critique of the NRA, 1934; Antitrust Policies: American Experience in Twenty Industries, 1958. Research on effects of antitrust legislation. Home: 18 Circle Rd., Scarsdale, N.Y. 10583. Office: Dept. Econs., N.Y. U., N.Y.C. 10453.*

WHITNEY, Vincent Heath, Am. sociologist; b. Northwood, N.H., June 23, 1913; s. Harlan B. and Mary (Lowell) W.; A.B., U. N.C., 1936, M.A., 1937, Ph.D., 1944; m. Lucy Mansfield, June 14, 1941; children—Caroline Mansfield, Steven Lowell, Mary Starr. Project dir. N.C. State Planning Bd., Durham, 1938-39; instr. sociology U. Me., 1939-43; instr. econs. Wesleyan U., Conn., 1945-46; asst. prof. Brown U., 1946-49, asso. prof., 1949-52, prof., chmn. sociology, 1952-59; prof. sociology, chmn. dept. U. Pa., 1959—. dir. Population Studies Center, 1961—; sr. staff asso., rep. for Asia The Population Council, N.Y.C., 1957-58, 1962—. Mem. tech. adv. com. on population U.S. Census Bur., 1965—. Bd. dirs. Nat. Com. on Maternal Health, N.Y.C. Mem. Population Assn. Am. (dir. 1961-64), Eastern Sociol. Soc. (pres. 1957-58), Internat. Union for Sci. Study of Population, Sociol. Research Assn., Internat. Population Union, Am. Sociol. Assn., Regional Sci. Assn., Internat. Sociol. Assn. (mem. research com. on urban sociology). Author: (with W. Isard) Atomic Power: An Economic and Social Analysis, 1952-56. Contbr. numerous articles to sci. jours. Demonstrated urban character of much of U.S. population defined as rural and showed statistically other changes in the extent and nature of urbanization; applied principles of technol. change to predict peaceful uses of nuclear energy and showed interrelations with population change. Home: 3202 Huey Av., Drexel Hill, Pa. 19026. Office: Dietrich Hall, U. Pa., Phila. 19104.*

WHITROW, Gerald James, Brit. cosmologist; b. Kimmeridge, Dorset, Eng., June 9, 1912; s. George William and Emily (Watkins) W.; B.A., Oxford U., 1933, M.A., 1937, D.Phil., 1939; m. Annie Magda Mostel, Aug. 17, 1946. Lectr., Christ Church Coll., Oxford U., 1936-40; sci. officer Ministry of Supply, 1940-45; lectr. math. Imperial Coll., London, 1945-51; lectr. math. Imperial Coll., London, 1945-51, reader in applied math., 1951—. Fellow Royal Astron. Soc. (v.p. 1965-67); mem. Brit. Soc. for Philosophy of Sci. (pres. 1955-57), Internat. Soc. for Study of Time (pres. 1966—). Author: Structure of the Universe, 1949; Atoms and the Universe, 1956; Structure and Evolution of the Universe, 1959; The Natural Philosophy of Time, 1961; Einstein—The Man and His Achievement, 1967; also numerous articles. Research on theoretical cosmology, including theory of background radiation in different world-models; theory of relativity, including devel. of spl. relatiivty on basis of light-signalling and time measurement without appeal to rigid body concept; different theories of relativity with reference to astron. texts; gravitational collapse. Home: 6 Albert Mansions, Albert Bridge Rd., London S.W. 11, Eng.*

WHITTAKER, Edmund, economist; b. Radcliffe, Eng., Sept. 24, 1897; s. Edmund and Elizabeth Alice (Standring) W.; Nat. diploma in Agr. with honors, Harris Inst., 1923; B.Sc. with 1st class honors in Econs., U. Edinburgh, 1928, Ph.D., 1932; m. Katharine Wilmett Maxwell Cooper, July 30, 1929; children—Roger Standring, Margaret Katharine Maxwell (Mrs. Baker), Felicity Anne (Mrs. Henry Allyn Warner), Barry Edmund. Faculty, Edinburg and East of Scotland Coll. Agr., 1928-32, Natal U. Coll. (now U. Natal), 1932-37, U. Ill., 1937-42, Brown U., 1943-46, Ind. U., 1946-48; vis. prof. econs. and bus. adminstrn. U. Western Ont., 1942-43; faculty Colo. State U., 1948-63, emeritus prof., 1963—; faculty Hampden-Sydney Coll. (Va.), 1961-68, Patton prof. econs., 1966-68. Cons.; condr., supr. research projects, including 2 indsl. location studies Colo., 1956-59. Fellow Royal Econ. Soc. Author: A History of Economic Ideas, 1940; Elements of Economics, 1946; Economic Analysis, 1956; Schools and Streams of Economic Thought, 1960; also articles. Research, publs. on man as part of an evolving soc., influencing and being influenced by changing environment.*

WHITTAKER, Sir Edmund Taylor, English physicist, astronomer, mathematician; b. Lancashire, Eng., Oct. 24, 1873; s. John and Selina (Taylor) W.; ed. Trinity Coll., Cambridge, fellow, 1896-1907; LL.D., St. Andrews, Scotland, U. Cal.; Sc.D., Dublin, London, Manchester, Birmingham; m. Mary Boyd, 1901; 3 sons, 2 daus. Royal astronomer of Ireland, 1906-12; prof. math. Edinburgh U., 1912-46; became Hitchcock prof. U. Cal., 1934; hon. fellow Trinity Coll., Cambridge. Fellow Royal Soc. (Sylvester medal, 1931, Copley medal, 1954), 1905; hon. fellow Faculty Actuaries in Scotland; mem. Math. Assn. (pres. 1920-21), Brit. Assn. (pres. sect. A 1927), London Math. Soc. (pres. 1928-29, De Morgan medal 1935), Royal Soc. Edinburgh (pres. 1939-44); fgn. mem. Royal Accademia dei Lincei, Pontifical Acad. Sci., Royal Acad. Naples, Am. Philos. Soc., Indian Math. Soc. Author: Modern Analysis; Treatise on Analytical Dynamics, 1904; Theory of Optical Instruments, 1907; History of the Theories of Aether and Electricity, 2 vols., 1910 (with G. Robinson; The Calculus of Observations, 1924; The Beginning and End of the World; Space and Spirit; The Modern Approach to Descartes' Problem; From Euclid to Eddington; Eddington's Principle in the Philosophy of Science. Research on energy distbn. in continuous spectrum, 1907, partial differential equations, automorphic, Lamé and Mathieu functions; developed idea of movement of lines of force, 1916; gave relativistic basis to electrodynamics; anticipated a quantizing of field. Died Mar. 24, 1956.

WHITTAKER, James O., Am. social psychologist; b. Rochester, N.Y., Sept. 27, 1927; s. Roswell E. and Grace (Oliver) W.; B.A., Tex. Christian U., 1950; M.A., U. Pa., 1952; Ph.D., U. Okla., 1958; m. Sandra June Oshel, Aug. 16, 1957; children—Kellie Lynn, James O. II. Asst. prof. psychology Eastern Ill. State U., Charleston, 1955-56, U. N.D., Grand Forks, 1956-59; asso. prof., chmn. dept. psychology U. Bridgeport (Conn.), 1959-60, Gustavus Adolphus Coll., St. Peter, Minn., 1960-63; prof. N.D. State U., Fargo, 1963—. Mem. Red River Valley Psychol. Assn., Am. Assn. U. Profs., Sigma Xi. Author: Introduction to Psychology, 1965; also articles. Research on persuasion and attitude change, cultural factors in alcoholism and group dynamics. Home: 314 Forest Av. N., Fargo, N.D. 58102.*

WHITTAKER, John Macnaghten, English mathematician; b. Cambridge, Eng., Mar. 7, 1905; s. Edmund Taylor and Mary (Boyd) W.; M.A., Edinburgh (Scotland) U., 1924, D.Sc., 1929; M.A., Trinity Coll., Cambridge, 1931; m. Iona Elliot, Dec. 19, 1933; children—John Edmund, Richard James. Asst. lectr. Edinburg U., 1927-29; fellow, lectr. Pembroke Coll., Cambridge, 1929-33; prof. pure math. Liverpool (Eng.) U., 1933-52; vice chancellor Sheffield (Eng.) U., 1952-65. Dep. sci. adviser Brit. Army Council, 1944. Fellow Royal Soc., 1949. Recipient Freedom award City of Sheffield, 1965, Smith's prize, 1929, Adams prize, 1949. Author: Interpolatory Function Theory, 1935; Series de Base de Polynomes, 1949; also articles. Research on theory of functions of complex variable including a gen. theory of expansions. Home: 12 Endcliffe Crescent, Sheffield 10, Eng.*

WHITTAKER, Robert Harding, Am. ecologist; b. Wichita, Kan., Dec. 27, 1920; s. Clive Charles and Adeline (Harding) W.; A.B., Washburn Municipal U., 1942; Ph.D., U. Ill., 1948; m. Clara Caroline Buehl, Jan. 1, 1953; children—John Charles, Paul Louis, Carl Robert. Asst. prof. zoology Wash. State Coll., Pullman, 1948-51; sr. scientist biology dept. Hanford Atomic Products Operation, Richland, Wash., 1951-54; asso. prof. Bklyn. Coll., 1954-64; vis. scientist biology dept. Brookhaven Nat. Lab., Upton, L.I., N.Y., 1964-66; prof. biology U. Cal. at Irvine, 1966-68; prof. biology Cornell U., Ithaca, N.Y., 1968—. Mem. Ecol. Soc. Am., Am. Soc. Limnology and Oceanography, Brit. Ecol. Soc., Internat. Assn. Phytosociology, Torrey Bot. Club, Am. Soc. Zoologists. Author: Classification of Natural Communities, 1962; also articles. Research on theory of natural communities, cycling of nutrients; developed methods for analysing communities along gradients, measurement of net prodn. Home: 318 Winthrop Dr., Ithaca, N.Y. 14850.*

WHITTAKER, Roland Mapes, Am. chemist; b. Middletown, N.Y., Nov. 29, 1903; s. William and Jane C. (Collingwood) W.; B.Social Sci., Coll. City N.Y., 1927; A.M., Columbia, 1930; Ph.D., Fordham U., 1932; m. Mary R. LaDue, Apr. 4, 1931; 1 dau., Cynthia Jane (Mrs. Frank Frederick Holmberg, Jr.). Fellow, Coll. City N.Y., 1927-28; faculty Bklyn. Coll. 1928-37, Queens Coll., 1937-62; prof. chemistry Mt. St. Mary Coll., Newburg, N.Y., 1963—. Vis. prof. chemistry U. P.R., 1961-62. Mem. Am. Chem. Soc., New Eng. Assn. Chemistry Tchrs., Am. Assn. U. Profs., Sigma Xi. Author: A Laboratory Handbook and Syllabus for a 1st Course in Chemistry, 1938, 2d Course in Chemistry, 1939, 1st Year Course in Chemistry, 1940; (with A.P. Marion) A Laboratory Handbook for General Chemistry, 1947; Rudiments of Chemistry, The Chemist's View of the Nature of Matter, 1948; Laboratory Manual for General Chemistry, 1960; General Chemistry, 1949; Quimica General. Research on oxygen heterocyclics, organic insecticides. Home: 64 Wireless Rd., East Hampton, N.Y. 11937.*

WHITTAM, Ronald, English physiologist; b. Oldham, Eng., Mar. 21, 1925; s. Edward and May (Butterworth) W.; B.Sc. with 1st class honors, U. Manchester, 1951; Ph.D., U. Sheffield, 1954, Kings Coll., Cambridge U., 1958; m. Christine Patricia Lamb; children—Althea Monica, Harvey Stephen. Fellow dept. biochemistry U. Sheffield, 1953-55, physiol. lab. Cambridge U., 1955-58; mem. sci. staff Med. Research Council unit dept. biochemistry Oxford U., 1958-60, lectr. biochemistry, 1960-66; prof. gen. physiology U. Leicester (Eng.), 1966—. Bruno Mendel Traveling fellow Weizman Inst., Israel, 1965-66. Mem. Physiol. Soc., Biochem. Soc. Author: Transport and Diffusion in Red Blood Cells, 1964. Dep. chmn. editorial bd. Biochem. Jour. Research, numerous publs. on mechanism of active transport of inorganic ions across membranes of mammalian cells, energy supply is in form of adenosine triphosphate derived from cell metabolism, enzymes involved activated by ions in vectorial manner, energy utilization for active transport acts as pace-maker for energy prodn. Home: 9 Guilford Rd., Leicester, Eng.*

WHITTEMORE, Amos, Am. inventor; b. Cambridge, Mass., Apr. 19, 1759; s. Thomas and Anna (Cutter) W.; m. Helen Weston, June 18, 1781; at least 12 children. After public sch. edn. apprenticed to gunsmith; mfr. brushes for carding cotton and wool, 1795, supt. mech. equipment in 3 factories; received patents for machine which cut nails, loom for weaving duck, form of mech. ship's log, 1796; patented machine which eliminated hand labor in making cotton and wool cards, 1797, attempted unsuccessfully to introduce new machine into Eng., 1799-1800; in partnership (with brothers William and Robert Williams) to make card-making machines and cards themselves, 1800-12; renewed patent, 1808, sold patent rights and machinery to a N.Y.C. co., 1812, ret. Died West Cambridge (now Arlington), Mass., Mar. 27, 1828.

WHITTEN, Eric Harold Timothy, geologist; b. Ilford, Essex, Eng., July 26, 1927; s. Charles Alexander and Muriel (Smith) W.; B.Sc., Spl. in Geology U. London (Eng.), 1948, B.Sc. with gen. honors, 1948, Ph.D., 1952; m. Joan Margaret Thomas Aug. 4, 1953; 1 dau., Catherine Ann. Lectr. geology Queen Mary Coll., U. London, 1948-58; faculty Northwestern U., Evanston, Ill., 1958—, prof. geology, 1963—. Author: Structural Geology of Folded Rocks, 1966. Research, publs. in quantitative methodology applied to the nature and variability of granite rocks and to structural geology. Home: 2332 Ridge Av., Evanston, Ill. 60201.*

WHITTEN, Robert Craig, Jr., Am. physicist; b. Bristol, Va., Dec. 6, 1926; s. Robert Craig and Adeline (Gerke) W.; B.S., U.S. Mcht. Marine Acad., 1951; B.A., U. Buffalo, 1955; M.A., Duke, 1958, Ph.D., 1959; m. Sally Marie Kriz, Aug. 2, 1953; children—Robert Craig III, Lisa Marie. Research asso., instr. Duke, 1955-59; physicist Stanford Research Inst., Menlo Park, Cal., 1959-63, sr. physicist, 1963-67; research Scientist Ames Research Center NASA, Mountain View, Cal., 1967—; lectr. phys. sci. and physics Stanford, 1961-66; lectr. physics Santa Clara U., 1964. Mem. Am. Geophys. Union. Author: (with I.G. Poppoff) Physics of the Lower chem. processes in D-E-F regions of ionosphere, physics chem. processes in D-E-F regions of inosphere, physics of ionospheric perturbations and their relation to solar disturbances, planetary atmospheres, nuclear beta-decay. Home: 1117 Yorkshire Dr., Cupertino, Cal. 95014. Office: Ames Research Center, NASA, Moffett Field, Cal. 94035.*

WHITTINGHILL, Maurice, Am. biologist; b. St. Joseph, Mo., May 15, 1909; s. Robert Coleman and Melissa Hubbard (Polk) W.; A.B., Dartmouth, 1931; Ph.D., U. Mich., 1937; m. Doris E. Cummings, Sept. 3, 1932 (div. 1954); children—Diana (Mrs. H. Max Steele), Warren Cummings; m. 2d, Martha Harris Morrisson Speaks, June 4, 1955; 1 son, Alvin Dewey Speaks. Faculty dept. zoology U. N.C. Chapel Hill, 1942—, prof., 1953—; instr. Dartmouth, 1931-33, Bennington Coll., 1937-42. Mem. A.A.A.S., Am. Soc. Naturalists, Genetics Soc. Am., Am. Soc. Human Genetics, Am. Soc. Zoologists, Elisha Mitchell Sci. Soc. N.C. Acad. Sci. Author: Human Genetics and Its Foundations, 1965; also articles. Demonstrator action of some environmental agts., heat, gamma rays, chems., on chromosomal exchanges, genetic and environmental relationships of spondylitis, Still's disease, rheumatoid arthritis. Home: 24 S. Lake Shore Dr., Chapel Hill, N.C. 27514.*

WHITTINGTON, Harry Blackmore, geologist; b. Birmingham, Eng., Mar. 24, 1916; s. Harry and Edith M. (Blackmore) W.; B.Sc., U. Birmingham, 1936, Ph.D., 1938, D.Sc., 1949; m. Dorothy E. Arnold, Aug. 10, 1940. Came to U. S., 1949. Lectr. Judson Coll., U. Rangoon, 1940-41, U. Birmingham, 1945-49; asso. prof. geology Harvard, 1950-58, prof., 1958-66; Woodwardian prof. geology Cambridge (Eng.) U. 1966—. Recipient John Simon Guggenheim fellowship, 1957; Bigsby medal Geol. Soc. London, 1957. Mem. Am Acad. Arts and Scis. Research on stratigraphy and paleontology of Ordovician system; nature of evolutionary history of trilobites Home: 11 Causewayside, Fen Causeway, Cambridge, Eng.*

WHITTLE, Charles Edward, Jr., Am. physicist, educator, dean; b. Brownsville, Ky., Mar. 8, 1931; s. Charles Edward and Lillian (Skaggs) W.; A.B., Centre Coll. Ky., 1949; Ph.D., Washington U., St. Louis, 1953; postgrad. (Fulbright scholar) U. Leiden; m. Suzanne Lee Miller, June 22, 1952; children—Charles Edward III, Jane, Jessica, Gayl, Tamara, Thomas, David, Jeffrey, Mary. Research physicist Union Carbide Corp., 1954-56; faculty Western Ky. State Coll., 1956-62; prof. physics, coordinator research Centre Coll. Ky., Danville, 1962-64, aso. dean, 1964-65, dean faculty, 1965—. Ednl. cons. NSF, 1965. Mem. Am. Phys. Soc., Optical Soc. Am., A.A.A.S., Am. Assn. Physics Tchrs., Ky. Acad. Scis. (past pres.), Ky. Assn. Physics Tchrs. (past pres.), Sigma Xi. Research, publs. on low energy nuclear spectroscopy and nuclear structure, measurements of nuclear energy and spin states of intermediate and heavy mass nuclei, molecular structure. Home: 802 Hustonville Rd., Danville, Ky. 40422.*

WHITTLE, Sir Frank, English engr.; b. Coventry, Eng., June 1, 1907; B.A., Cambridge (Eng.) U., 1936, M.A., 1937, Sc.D. (hon.); D.Sc., Oxford Manchester, Leicester (all Eng.) univs.; LL.D., Edinburgh (Scotland) U.; D. Tech. (hon.), Trondheim (Norway) U.; m. Dorothy M. Lee, 1930; 2 sons. Served with Royal Air Force, 1923-48; tech. adviser Brit. Overseas Airways Corp., 1948-52; engr. Royal Dutch Shell Group, 1952-57, Bristol Siddeley Engines, 1957—. Recipient James Alfred Ewing medal Inst. C.E.'s, 1944, James Clayton prize Inst. Mech. Engrs., 1946, Daniel Guggenheim medal, 1946, Kelvin Gold medal, 1947, Melchott medal, 1949, Gold medal Internat. Aero. Fedn., 1951, Churchill Gold medal Soc. Engrs., 1952, Albert Gold medal Soc. Arts, Franklin Gold medal, 1956, John Scott award, 1957, Goddard award, 1965. Fellow Royal Soc. (Rumford medal 1950), 1947; mem. Royal Aero. Soc. (hon., Gold medal 1944). Author: Jet, The Story of a Pioneer, 1953. Developer jet engines, especially turbojet, turbofan, aftfan gas turbine engines; applied for 1st patent on gas turbine, 1930; 1st flight Gloster-Whittle aircraft, 1941.

WHITTLESEY, Charles R., Am. economist; b. Roseburg, Ore., Sept. 24, 1900; s. Charles Terrill and Penelope (Skinner) W.; A.B., Philomath Coll., 1921; A.M., AM. Univ. of Beirut (Syria), 1924; Ph.D., Princeton, 1928; Sc.D. (honorary), Univ. of Pennsylvania, 1967; m. Mary Weaver Fox, June 19, 1929; 1 dau., Margaret Terrill. Instr. Am. Univ. of Beirut, 1921-24; staff of Near East Relief, Athens, 1923; economist, U.S. Tariff Commn. 1928; instr. asst. prof., asso. prof. Princeton, 1928-40; mem. Hines-Kemmerer Econ. Mission to Turkey, 1934, fellow Soc. Science Research Council in Europe, 1935-36; prof. finance and economics. U. of Penn., 1940—; Ford Foundation faculty research fellow, 1957-58; visiting professor University of Bombay, 1959-60; economist, Penn Mutual Life Ins. Co., 1941-61, Fidelity-Philadelphia Trust Company (Pa.), 1948; member of staff,

Financial Research Program. National Bur. Econ. Research, 1942-44. Mem. Royal Econ. Society, American Econ. Assn. Author or co-author books relating to field; contbr. to econ. and social sci. publs. and to encys. Research on mometary policy; money and banking; trade and finance between nations. Home: 106 Avonbrook Rd., Wallingford, Pa. Office: Univ. of Pennsylvania, Phila.

WHITWORTH, Sir Joseph, English mech. engr., inventor; b. Stockport, Eng., Dec. 21, 1803; s. Charles and Sarah (Hulse) W.; LL.D., Trinity Coll., Dublin, Ireland, 1863; D.C.L., Oxford (Eng.) U., 1868; m. Fanny Ankers, 1825; m. 2d, Mary Louisa Broadhurst, 1871. Mechanic in Manchester and London; toolmaker, Manchester, from 1833; constructed measuring machine by which he elaborated his system of standard measures and gauges, 1840-50; made expts. relating to rifles; produced Whitworth steel for big guns, 1870; converted works at Manchester into limited liability co. which united with firm of Armstrong, Elswick, 1874. Recipient Grand Prix, Paris Exhbn., 1867. Fellow Royal Soc., 1857. Author: (with George Wallis) The Industry of the United States in Machinery, Manufacturers, and Useful and Ornamental Arts, 1854; Miscellaneous papers on Mechanical Subjects, 1858. Discovered method to produce plane surface of metal; introduced interchangeable machine parts, standardization of measures and screw threads (Whitworth thread), high precision calibrating apparatus; invented knitting machine, automatic drill, street cleaning machine, hexagonal bore for rifles, steel cannon with rifled barrel and moveable breech (adopted by War Office 1869); discovered new method to make ductile steel for guns, 1870. Died Monte Carlo, Monaco, Jan. 22, 1887.

WHYBURN, Gordon Thomas, Am. mathematician; b. Lewisville, Tex., Jan. 7, 1904; s. Thomas and Eugenia (McLeod) W.; B.A., U. Tex., 1925, M.A., 1926, Ph.D., 1927; Sc.D., Washington ahd Lee U., 1949; m. Lucille Enid Smith, Sept. 9, 1925; 1 son, Kenneth Gordon. Faculty, U. Tex., 1926-29; asso. in math. Johns Hopkins, 1929-34; prof. math. U. Va., Charlottesville, 1934-57, Alumni prof., 1957—, chmn. math. faculty, 1935-66, member Center for Advanced Studies, 1966—; summer vis prof. Stanford, 1929, 32, 38, 41, U. Cal. at Los Angeles, 1940, U. Cal. at Berkeley, 1947, U. Colo., 1956. Mem. War Policy Com. for Math., NRC Fellowship Bd., 1951-58. Mem. Nat. Acad. Scis., Am. Math. Soc. (pres. 1953-54, trustee 1948-64, mng. editor Transactions 1949-52), A.A.A.S. (v.p., chmn. sect. 1941), Math. Assn. Am. (Chauvenet prize 1938, gov.), Va. Acad. Sci., Phi Beta Kappa, Sigma Xi, Phi Lambda Upsilon. Author: Analytic Topology, 1942; Topological Analysis, 1958, rev., 1964; also numerous articles. Discovered and proved key theorems on structure of continua, including basic cut-point results and cyclic element theory; developed dynamic transformation topology clarifying preservation and evolution of structures under mappings; applied topological methods to solution of classical problems in analytic function theory. Home: West Lawn, U. Va., Charlottesville, Va. 22901.*

WHYBURN, William Marvin, Am. mathematician; b. Lewisville, Tex., Nov. 12, 1901; s. Thomas and Eugenia Elizabeth (McLeod) W.; B.A., U. Tex., 1922, M.A., 1923, Ph.D., 1927; LL.D., Tex. Technol. Coll., 1948; m. Aura Marie Barfield, Dec. 29, 1923; children—Willa Marie, Clifton Thomas. Faculty, South Park Coll., 1923-24, Tex. A. and M. Coll., 1924-25, Tex. Technol. Coll., 1925-27, U. Cal. at Los Angeles, 1928-44; pres. Tex. Technol. Coll., 1944-48; Kenan prof. math. U. N.C., Chapel Hill, 1948-67, v.p. grad. studies and research, 1955-60, chmn. dept., 1948-55, 60-65; Frensley prof. math. So. Meth. U., 1967—. Mem. Am. Math. Soc., Math. Assn. Am., A.A.A.S., Philos. Soc. Tex., Phi Beta Kappa, Sigma Xi. Author: (with Daus and Gleason) Basic Mathematics for War and Industry, 1944; (with P.H. Daus) First Year College Mathematics, 1949; Algebra for College Students, 1955; Introduction to Mathematical Analysis, 1958; Algebra with Applications, 1961. Research, publs. on boundary value problems for linear and non-linear differential systems, real variable theory. Home: Box 651, So. Meth. U., Dallas 75222.*

WHYTE, Henry Malcolm, physician; b. S. India, Oct. 26, 1920; s. Henry William and Ruby (Flower) W.; B.Sc., U. Queensland (New Zealand), 1942, M.B., B.S., 1944; D.Phil., Oxford (Eng.) U., 1951, m. Marguerite Mary Lamont, July 19, 1946; children—Bruce MacGregor, Christine. Sr. clin. research fellow Kanematsu Meml. Inst., Sydney (Australia) Hosp., 1952-56, dir. clin. research, 1956-60, dir. med. research, 1960-66; prof., head dept. clin. sci., John Curtin Sch. Med. Research, Australian Nat. U., Canberra, 1966—. Med. dir. Life Ins. Med. Research Fund Australia and New Zealand, 1960-66. Fellow Royal Australian Coll. Physicians; mem. Royal Coll. Physicians, Cardiac Soc. Australia and New Zealand, Australian Physiol. Soc., others. Author: The Fats of Life, 1961; also numerous articles. Research on body heating, blood, hypertension, renal function and disease, body composition, starvation, obesity, blood pressure, lipid metabolism, other factors relevant to coronary heart disease. Address: 25 Tasmania Circle, Canberra, Australia.*

WHYTE, William Foote, Am. sociologist, anthropologist; b. Springfield, Mass., June 27, 1914; s. John and Isabel (VanSickle) W.; A.B., Swarthmore Coll., 1936; mem. Soc. of Fellows, Harvard Univ., 1936-40; Ph.D., U. of Chicago, 1943; m. Kathleen King, May 28, 1938; children—Joyce, Martin, Lucy, John. Assistant prof. sociology, acting chmn. dept. anthropology, U. of Okla., 1942-43; asst., then asso. prof. sociology, U. of Chicago, 1944-48; exec. sec., Com. on Human Relations in Industry, U. of Chicago, 1946-48; prof. indsl. relations, N.Y. State Sch. of Indsl. and Labor Relations, Cornell U., 1948—, dir. Social Sci. Research Center, 1956-61; Fulbright fellow, Peru, 1961-62; career research award Nat. Inst. Mental Health. Trustee Found. Research Human Behavior, 1960-67. Fellow American Anthropological Assn., Am. Sociol. Assn; member Am. Acad. Arts and Scis., Soc. Applied Anthropology (pres. 1964), Indsl. Relations Research Assn. (pres. 1963). Rotarian. Author: Street Corner Society, 1943; Human Relations in the Restaurant Industry, 1948; Pattern for Industrial Peace, 1951; Money and Motivation, 1955, Man and Organization, 1959; Men at Work, 1961; Action Research for Management, 1965. Editor: Industry and Society, 1946; Human Organization, 1956-61, 62-63. Developed interviewing-observational research methods; studies of small group structures; application of interactional analysis to organizational research; cross cultural studies. Home: Trumansburg, N.Y. Office: Cornell University, Ithaca, N.Y.*

WHYTLAW-GRAY, Robert Whytlaw, English chemist; b. 1877; ed. Glasgow (Scotland) U., U. Coll., London, U. Bonn (Germany); Ph.D.; hon. D.Sc.; m. Doris Fortescue Carr, 1911; 2 daus. Began work under W. Ramsey, 1906; joined chem. staff U. Coll., London, 1906, became asst. prof., 1908; sci. master Eton Coll., 1914-23; civilian chem. adviser Chem. Warfare Com., 1914-18; prof. chemistry U. Leeds (Eng.), 1923-45, emeritus prof., 1945—. Became Brit. rep. comm. on atomic weights Internat. Union Chemists, 1940; cons., Imperial Chem. Industries. Fellow U. Coll., London. Fellow Royal Soc., 1928, Royal Inst. Chemistry. Research and publs. on determination of atomic weight of radium using micro-balance; determined atomic weights of nitrogen, chlorine, xenon, carbon, flourine, sulphur, silicon; studies on disperse systems in gases, smoke, dust, air pollution. Died Jan. 21, 1958.

WHYTT, Robert, Scottish physician; b. Edinburgh, Scotland, Sept. 6, 1714; s. Robert and Jean (Murray) W.; M.A., St. Andrews U., 1730; M.D., 1737; studied medicine, Edinburgh, London, Europe; M.D., Rheims, 1736; m. Helen Robertson; m. 2d, Louisa Balfour, 1743; 6 children by 2d wife. Prof. theory of medicine Edinburgh U., 1747; 1st physician to George III in Scotland, 1761. Fellow Royal Soc., 1752, Royal Coll. Physicians (pres. 1733-36). Author: On the Vital and Other Involuntary Motions of Animals, 1751 (threw aside doctrine of Stahl that rational soul is cause of all involuntary motions in animals, ascribed such movements to effect of stimulus acting on unconscious sentient principle); Physiological Essays, 1755; Nervous Hypochrondriac, or Hysteric Disease, 1764; Observations on the Dropsy of the Brain, 1768. Gave 1st demonstration that contraction of pupils in response to light is reflex action (known as Whytt's reflex), 1751. Died Apr. 15, 1766.

WIBERG, Egon Gustaf Martin, German chemist; b. Gustrow, (Mecklembourg), June 3, 1901; s. Gustaf and Gertrud (Holmstrom) W.; eng. diploma, Karlsruhe Tech. Sch.; Dr ès sc., honoris cause, Aix-la-Chapelle (France) Tech. Sch., 1959; m. Doris Schneider, Mar. 15, 1930; children—Nils, Birgit, Sven. Prof. agrégé, asso. prof. Karlsruhe Tech. Sch., 1931-38; asso. prof. U. Munich, 1938-51, prof., dir. Inorganic Chemistry Inst., 1951—. Recipient Alfred Stock gold medal; decorated Order Bavarian Merit. Mem. Soc. German Chemists (pres. 1959-61; Commn. prize 1950), Bavarian Acad. Sci., Leopoldina Soc. Halle, German Chem. Assn., German Bunsen Soc., Union Fed. Tech. Sci.; hon. mem. Austrian Chem. Assn. Author: Lehrbuch der Anorganischen chemie, 1964; also numerous articles. Home: Tiepolostrasse 1. Office: Meiserstrasse 1, Munich, West Germany.

WIBERG, Kenneth Berle, Am. chemist; b. Bklyn., Sept. 22, 1927; s. Dan and Solveig (Berle) W.; B.S., Mass. Inst. Tech., 1948; Ph.D., Columbia, 1950; m. Marguerite L. Koch, Mar. 17, 1951; children—Patricia, Robert, William. Mem. faculty U. Wash., Seattle, 1950-57, 58-62; vis. prof. Harvard, 1957-58; prof. chemistry Yale, 1962—. Mem. chemistry panel Air Force Office Sci. Research, 1960-66; mem. chemistry panel NSF, 1965-68, chmn., 1967-68; Boomer lectr. U. Alta., 1959. A.P. Sloan fellow, 1958-62; J.S. Guggenheim fellow, 1961-62. Mem. Nat. Acad. Scis., Am. Chem. Soc. (exec. com. organic div., 1961-67, Cal. sect. award, 1962), Chem. Soc., London, A.A.A.S., Sigma Xi. Author: Laboratory Technique in Organic Chemistry, 1960; (with B.J. Nist) Interpretation of NMR Spectra, 1962; Physical Organic Chemistry, 1964; Computer Programming for Chemists, 1965; also articles. Editor, Oxidation in Organic Chemistry, 1965. Mem. editorial bd. Organic Synthesis, 1964—. Syntheses of strained bicyclic compounds, and studies of properties of derivatives of these systems, particularly in solvolytic reactions; determined mechanisms of several oxidation reactions involving a variety of organic substrates. Home: 160 Carmalt Rd., Hamden, Conn. 06517. Office: 225 Prospect St., New Haven 06520.*

WIBERLEY, Stephen Edward, Am. chemist; b. Troy, N.Y., May 31, 1919; s. Irving Charles and Ruth (Stanley) W.; B.A., Williams Coll., 1941; M.S., Rensselaer Poly. Inst., 1948, Ph.D., 1950; m. Mary Elizabeth Bartle, Feb. 21, 1942; children—Stephen Edward, Sharon Elizabeth. Chemist, Congoleum Nairn, Inc., Kearney, N.J., 1941-44; prof. chemistry Rensselaer Poly. Inst., Troy, N.Y., 1946-63, dean Grad. Sch., 1964—; vis. sr. physicist Brookhaven Nat. Lab., Upton, N.Y., 1952; cons. to pvt. cos. Mem. Am. Chem. Soc., Am. Assn. U. Profs., Soc. Applied Spectroscopy, Sigma Xi, Phi Lambda Upsilon. Author: Instrumental Analysis, 1954; Laboratory Manual for General Chemistry, 1963; Introduction to Infrared and Raman Spectroscopy, 1964; also numerous articles. Developed analytical methods; determined structure by infrared absorption spectroscopy; research on reactions boron hydrides with oxygen, structure and gelling properties aluminum soaps in organic solvents. Home: 1676 Tibbits Av., Troy, N.Y. 12180.*

WIBORG, Kristian Fredrik, Norwegian marine biologist; b. Haegebostad, Norway, May 6, 1914; s. Andreas and Ida (Pedersen) W.; cand.real., U. Oslo (Norway), 1938; Dr.philosophy, Bergen (Norway) U., 1955; m. Kirsten Wiberg, July 16, 1949; children—Knut, Nina, Agnete. Staff, Directorate Fisheries, Inst. Marine Research (Fiskeridirektorates Havforskningsinstitutt), Bergen, 1939-63, 65—, leader plankton sect., 1946—; staff joint project Bur. Comml. Fisheries, George Washington U., 1964; lectr. zooplankton U. Bergen, 1961—. Mem. Selskabet til Videnskabenes Fremme, Class for Math. and Biology. Contbg. author: Havet og Vare Fisker (The Sea and our Fishes), 1960; Norges Dyreliv (Animal Life of Norway), 1950. Research publs. on life history, biology, biomass, taxonomy, comml. utilization of zooplankton, distbn., comml. utilization, culture of molluscs, fish eggs and larvae in relation to size of year classes, plakton instruments. Translator Soviet sci. lit. on fisheries biology. Home: 8 a Ankervegan Hop Fana, Norway. Office: 2 Nordnesparken, Bergen, Norway.*

WICHELHAUS, Carl Hermann, German chemist; b. Jan. 8, 1842; prof., Berlin; introduced alkaline neutralization of sulphuric acids into technology. Died Feb. 28, 1927.

WICHTERMAN, Ralph, Am. biologist; b. Phila. Sept. 8, 1907; s. John George and Selma E. (Zimmerman) W.; B.S., Temple U., 1930; M.A., U. Pa., 1932, Ph.D., 1936; m. Bessie Swift, Aug. 28, 1933; children—Robert John, Susan June (Mrs. Albert Szent-Gyorgi). Faculty, Temple U., Phila., 1918—, prof. biology, 1949—; vis. prof. U. Pa., 1950-51; guest investigator Dry Tortugas Lab., Carnegie Inst. Fla., 1939, NSF, Zool. Sta., Naples, Italy, 1962. Cons., referee govt. founds., nat. jours. Recipient 1st prize Darbaker award Pa. Acad. Sci., 1955. Fellow A.A.A.S.; mem. Am. Soc. Zoologists, Soc. Protozoologists, Am. Soc. Parasitologists, Am. Inst. Biol. Scis., Am. Micros. Soc., Pa. Acad. Sci., Marine Biol. Lab., Sigma Xi. Author: Biology of Paramecium, 1953; also numerous articles in protozoology and radiobiology; contbg. author to handbooks Nat. Acad. Sci. Research on fresh-water, marine, parastic protozoa; invented irradiation chambers to study biol. effects high dosage X and gamma irradiation on protozoa. Home: 7213 N. 20th St., Phila. 19138.*

WICK, Gian Carlo, physicist; b. Turin, Italy, Oct. 15, 1909; s. Federico Carlo and Barbara (Allason) W.; Dr. Physics, U. Turin, 1930; m. Antonella Civalleri, Dec. 4, 1943; children—Lionel Anthony, Julian Charles. Came to U. S., 1946, naturalized, 1955. Asso. prof. physics U. Palermo (Italy), 1937-38, U. Padova (Italy), 1938-40; prof. physics U. Rome, 1940-45, U. Notre Dame, 1946-48, U. Cal. at Berkeley, 1948-50; prof. Carnegie Inst. Tech. 1951-58; sr. physicist Brookhaven Nat. Lab., 1958-65; prof. physics Columbia, 1965—. Mem. Inst. for Advanced Study, Princeton, 1953. Fellow Am. Phys. Soc.; mem. Nat Acad. Scis. Research on atomic theory, structure of atomic nucleus, neutron scattering, elementary particles. Home: 3 Washington Sq. Village, N.Y.C. 10012.*

WICKE, Ewald, German chemist; b. Wuppertal-Elberfeld, Aug. 17, 1912; s. Gustav and Elisabeth (Rau) W.; Dr ès Sc., U. Göttingen; m. Elisabeth Gunther, Apr. 5, 1941; children—Angelika, Reinhard. Qualified as prof. U. Göttingen, 1943-44, prof. at large, 1949; prof., dir. Inst. Hamburg, 1954-56, Inst. Munster, 1959—. Laureate of Haber prize German Bunsen Soc., 1954; recipient Arnold Eucken V.D.1 medal, 1962. Mem. German Bunsen Soc., Soc. German Chemists, Dechema, German Engring. Assn., Combustion Inst. Author: Grundriss der physikalischen Chemie, 1959; Chemischer Apparatebau und Verfahrenstechnik, 1951; Porosität und katalytische Wirkung, 1956; Brennzonen in Kontaktschichten bei der katalytischen Co-oxydation, 1963; Zustandsdia-Chemischer Apparatebau und Verfahder Susteme Ph/H2 und Pd/D2 bei normalen Temperaturen; H/D Trenneffekte, 1964. Research on structure of liquids; chem. reaction engring.; thermodynamics; heterogenous reaction kinetics; chem. reaction engring. Home: Schluterstrasse 7. Office: Schlossplatz 4, Munster/Westphalia, West Germany.

WICKSELL, Johan Gustav Knut, Swedish economist; b. Stockholm, Dec. 20, 1851; s. provisions dealer; A.B., U. Uppsula, 1872, A.M., 1885, LL.B., 1899; also studied in France (Loren Scholar), Eng., Germany, Austria; m. Anna Brugge, 1889. Established reputation as social, polit. reformer with speech linking intoxication to poverty caused by over population, 1880; corr. for Swedish, Norwegian, Finnish newspapers in 1890's; became docent in polit. economy and fiscal law U. Uppsula; asso. prof. U. Lund, 1900-04, prof., 1904-16. Author: Value, Capital and Rent, 1893; Finanztheoretische Untersuchungen, 1896; Interest and Prices, 1898. Went beyond quantitative theory of money and traced complexities of nat. income; analyzed gen. price fluctuations in terms of supply and demand, savs. and investment; advocated planned variation in interest rates to maintain balance between savs. and investment; influenced J.M. Keynes and originated modern Swedish sch. of economists (includes G. Myrdal, among others. Died Stocksund, Sweden, May 3, 1926.

WIDAL, Fernand, physician; b. Dellys, Algeria, Mar. 9, 1862; student of Roux and Metchnikoff; physician of hosp., 1893; agrégé, 1895; prof. pathology and internal medicine, Paris; held chair, med. clinic Cochin Hosp.; recipient Osiris prize, also Gold medal of Paris interns (l'internat); mem. French Acad. Scis. (1919), French Acad. Medicine. Described (with Chantemesse) bacillus of epidemic dysentery, 1888; discovered (with Arthur Siczard) phenomenon of seroagglutination, also serodiagnosis for typhoid fever, (Widal test), 1896; pioneer in renal pathology; showed (with Lemierre and Javal) role of sodium chloride in manifestations of edema; studied shock; advanced theory of hemoclasie; discovered that temporary leukopenia occurs in anaphylactic shock after a meal of albuminoids in persons with malfunctioning liver. Died Paris, Jan. 14, 1929.

WIDDER, David Vernon, Am. mathematician; b. Harrisburg, Pa., Mar. 25, 1898; s. David Henry and Edith (Drabenstadt) W.; A.B., Harvard, 1920, M.A., 1923, Ph.D., 1924; m. Vera Adela Ames, June 12, 1939; children—David Charles, Edith Anne. Faculty, Bryn Mawr Coll., 1924-30, prof., chmn. math. dept., 1928-30; faculty Harvard, 1930—, prof. math., 1937—, chmn. dept., 1942-48, dir. Acad. Year Inst., 1963-66. Cons. Rand Corp., Santa Monica, Cal., 1949—, D.C. Heath & Co., Boston, 1960—. Guggenheim fellow, 1935-36; Fulbright grantee, Australia, 1962-63, Italy, 1955-56. Mem. Am. Math. Soc., Math. Assn. Am. Author: The Laplace Transform, 1946; Advanced Calculus, 1947; The Convolution Transform (with I.I. Hirschman), 1955; also numerous articles. Research on real inversion formula for Laplace transform, many representation theorems for Laplace transform; discovered reprentation theorems for Stielties and Convolution transform; research on representation for positive temperature function. Home: 30 Gould Rd., Arlington, Mass. 02174. Office: 572 Widener Library, Harvard, Cambridge, Mass. 02138.*

WIDGER, William Knowlton, Jr., Am. meteorologist; b. Lynn, Mass., Jan. 23, 1921; s. William K. and Dorothy (Harnden) W.; B.S., U. N.H., 1942; Sc.D., Mass. Inst. Tech., 1949; m. Constance Louise Keane, June 18, 1943; children—Dorothy Grace, Barbara Louise. Asst. prof. meteorology Cornell U., Ithaca, N.Y., 1949-51; chief satellite meteorology br. Air Force Cambridge Research Lab., Bedford, Mass., 1951-60; chief operational meteorol. satellites hdqrs. NASA, Washington, 1960-63; dir. satellite meteorology research Allied Research Assos., Inc., Concord, Mass., 1963-67; prof. physics Drexel Inst. Tech., Phila., 1967—. Sec., Joint Meteorol. Satellite Adv. Com., 1960-62; mem. com. on meteoral. aspects satellites Space Sci. Bd., 1958-62. Fellow A.A.A.S.; profl. mem. Am. Meteorol. Soc. (past chmn. com. for encouragement meteorol. research); mem. Am. Geophys. Union, Sci. Research Soc. Am., Sigma Xi, Phi Kappa Phi, Alpha Chi Sigma. Author: Meteorological Satellites, 1966; also articles. Research on satellite meteorology, systems analysis, operational applications, state-of-art summaries, gen. circulation, jet streams, mil applications of meteorology and climatology. Home: 136 Fox Rd., Media, Pa. 19063. Office: Dept. Physics, Drexel Inst. Tech., Phila. 19104.*

WIDGOFF, Mildred (Mrs. Anatole M. Shapiro), Am. physicist; b. Buffalo, Aug. 24, 1924; d. Leo and Rebecca (Shulimson) Widgoff; B.A., U. Buffalo, 1944; Ph.D., Cornell U., 1952; m. Anatole M. Shapiro, June 7, 1945; children—Eve, Jonathan. Asso. physicist Brookhaven Nat. Lab., 1952-55; research fellow Harvard, 1955-58; asst. prof. Brown U., Providence, 1958-66, asso. prof., 1966—; cons. Cambridge Electron Accelerator, 1958-60. Mem. Am. Phys. Soc., Phi Beta Kappa, Sigma Xi. Studies on interactions of elementary particles by means of nuclear emulsions, bubble chambers and spark chambers. Home: 10 Watson Av., Barrington, R.I. 02806. Office: Physics Dept., Brown U., Providence 02912.*

WIDHOLM, Olof Eric Bernhard, Finnish gynecologist; b. Helsinki, Finland, June 30, 1923; s. Eric and Saimi Widholm; B. Medicine, Helsinki U., 1947, licentiate of medicine, 1950, dr. medicine and surgery, 1953; m. Kerstin Stadigh, Feb. 2, 1952; children—Eric Mikael, Claes Olof. Docent gynecology and obstetrics Helsinki U., 1958; asst. lectr. U. Womens Hosp., Helsinki, 1958-63, asst. surgeon in chief, asst. prof., 1963—; prof. Turku (Finland) U., 1964, Oulu U., 1965—; gynecologist adolescent unit Folkhälsans, Helsinki, 1960—. Gynecol. expert for hosp. bldgs. in Ekenäs, Kotka and Esbo. Mem. Finnish Med. Assn., Finnish Gynecol. Assn., Scandianvian Assn. for Gynecologists and Obstetricians, Internat. Fertility Assn., Internat. Coll. Surgeons, Finnish Path. Assn. Research, publs. in serology and bacteriology, obstetrics, hosp. infections, cancer, adolescence endocrinology, exptl. medicine. Home: 11 Westendallé. Office: 10 Bulevard, Helsinki, Finland.*

WIDIMSKY, Jirí, Czechoslovakian physician, cardiologist; b. Pilsen, Czechoslovakia, Mar. 31, 1925; s. Bohus and Marie (Breiska) W.; M.D., Charles U., Prague, Czechoslovakia, 1950, C.Sc, candidatus scienciarum, D. Philosophy, 1957; m. Dagmar Petrovicka, July 30, 1953; children—Peter, Jirí. Physician, Karlsbad Hosp., 1950-51; research worker Inst. for Cardiovascular Diseases, Prague, 1951-61, chief cardiopulmonary group, 1962—; research fellow 1st med. dept. U. Gothenberg, Sweden, 1961-62; lectr. Postgrad. Med. Sch., Prague, 1962—; asso. prof. Charles U., Prague, 1968—. Mem. Czechoslovak Cardiological Soc. Author: Cor pulmonale bei der Lungentuberkulose, 1963; also numerous articles. Research on pulmonary circulation in health and diseases, particularly Tb, diagnosis, therapy, hemodynamics of pulmonary hypertension. Home: 934 Stallichova. Office: 800 Budejovicka, Prague 4, Czechoslovakia.*

WIDMANN, Johannes, physician; b. circa 1440; tchr. U. Tübingen (Germany). Author: Tractatus de pustulis (on the plague), circa 1497. Credited with introducing mercury for treatment of syphilis. Died 1524.

WIDMANN, Johannes, mathematician; b. Eger, Bohemia, circa 1460; credited with introducing modern signs for plus and minus. Author: Behende und hubsche Rechenung (influential in Germany), 1489. Died circa 1500.

WIED, George Ludwig, physician; b. Carlsbad, Czechoslovakia, Feb. 7, 1921; s. Ernst George and Anna (Traviczek) W.; M.D., Charles U., Prague, 1945; m. Daga M. Graaz, Mar. 19, 1949. Came to U. S., 1953, naturalized, 1960. Practiced medicine, specializing in obstetrics, gynecology, West Berlin, 1948-53, Chgo., 1954—; asst. obstetrics, gynecology Free U., West Berlin, 1948-52; asso. chmn. dept. obstetrics, gynecology Moabit Hosp. Free U., 1953; asst. prof., dir. cytology U. Chgo., 1954-59, asso. prof., 1959-65, prof., 1965—, mem. bd. adult edn., 1964—; hon. dir. Chgo. Cancer Prevention Center, 1959—; chmn. adv. com. cancer control Chgo. Bd. Health, 1961—; chmn. jury Maurice Goldblatt Cytology award, 1963. Recipient certificate merit U. S. Surgeon Gen., 1952, Maurice Goldblatt Cytology award, 1961. Mem. Am. (pres. 1965-66), Mexican (hon.), Spanish (hon.) Brazilian (fgn. corr.), Latin-Am. (hon.), Japanese (hon.) socs. cytology, N.Y. Acad. Sci., Internat. Acad. Cytology, Central Soc. Clin. Research, Internat. Corr. Soc. Obstetrics and Gynecology, Am. Soc. Cell Biology, German, Bavarian socs. obstetrics and gynecology, German Soc. Endocrinology, German Soc. Fertility and Sterility, German Soc. Applied Cytology, Royal Soc. Medicine, Sigma Xi. Contbr. articles profl. jours. Editor-in-chief: Acta Cytologica; Clinical Cytology; Introduction to Quantitative Cytochemistry. Research on cell morphology, cytochemistry in cancer; diagnosis of cancer; hormonal diagnostic and applied research. Home: 5000 Cornell Av., Chgo. 60615. Office: 5841 S. Maryland Av., Chgo. 60637.*

WIEDEMANN, Eilhard Ernst Gustav, German physicist; b. Berlin, Aug. 1, 1852; s. Gustav Heinrich Wiedemann; prof. physics, Erlangen, Germany; research on magnetism and electricity. Died 1928.

WIEDEMANN, Gustav Heinrich, German physicist; b. Berlin, Oct. 2, 1826; Ph.D., U. Berlin, 1847; children—Karl Alfred, Eilhard Ernst. Worked in lab of Magnus; prof. physics U. Basel (Switzerland), 1854-63; mem. faculty Brunswick (Germany) Polytechnic, 1863-66, Karlsruhe (Germany) Polytechnic, 1866-71; prof. phys. chemistry U. Leipzig (Germany), 1871-87, prof. physics, 1887-99. Fellow Royal Soc., 1884; mem. French Acad. Scis.; a founder Berlin Phys. Soc. Author: Die Lehre vom Galvanismus und Elektromagnetismus, 2 vols., 1861-63. Editor Annalen der Physik und Chemie (included all then known about electrotechnics), 1877-99. Investigated (with R. Franz) heat and electricity, 1853, relationship between torsion and magnetism, 1859; discovered that elec. and heat conductivity of metals are proportional to one another (Wiedemann-Franz law), 1853; discovered Wiedemann effect in electricity, research on elec. endosmosis and elec. resistance of electrolytes. Died Leipzig, Mar. 23, 1899.

WIEDEMANN, Hans Rudolf, German physician; b. Bremen, Germany, Feb. 16, 1915; s. Otto and Helene (Wilmanns) W.; ed. univs. Freiburg, Munchen, Hamburg, Lausanne, Jena; m. Gisela von Sybel, May 16, 1942; children—Jurgen, Gisela, Rainer, Ingeborg, Volker, Elisabeth. Physician, Children's Clinic, U. Jena, 1940-45; head dr. children's clinic, Bremen, 1945-46; head dr. Children's Clinic, U. Bonn, 1946-52, prof. medicine, 1950—; head children's Clinic, Krefeld, 1952-61; head Children's Clinic, U. Kiel, 1961—. Mem. Swiss Soc. Hematology and Pediatrics (corr. mem.) French (corr.), German socs. pediatrics, Internat. League Against Epilepsy, German Soc. Children's Psychiatry. Author: Der konstitutionelle, familiare, hamolytische Ikterus im Kindesalter, 1946; Die grossen Konstitutionskrankeiten des Skeletts, 1960; Dysostosen, 1966; Lehrbuch fur Kinderheilkunde, 1966. Publs. on contbns. to knowledge of congenital malformations, including first publ. of thalidomide disaster; hereditary diseases of skeleton and nervous system. Home: 26 Caprivistr., Kiel. Office: 15 Frobelstr., Kiel, Germany.*

WIEDMANN, Albert, Austrian dermato-venerologist; b. Magdeburg, Germany, Apr. 13, 1901; s. Johann M. and Johanna (Belani) W.; M.D., U. Vienna, 1928; m. Anny Konetzni, Mar. 12, 1941; children Johanna (Mrs. Robert Golda), Hans Jorg, Franz. Asst. physician Clinic for Venereal and Skin Diseases, Vienna, 1928-36; lectr. U. Vienna, 1936; head physician Rainer-Spital, Vienna, 1936—; faculty skin clinic U. Vienna, 1945—, prof., 1950—. Mem. Deutsche Dermatologische Gesellschaft (pres.), numerous other internat. profl. socs. Research, numerous publs. on syphilis, varicose symptom complex, neurovegetative care of skin. Home: 25 Alserstrasse, Vienna, Austria.*

WIEGAND, Clyde Edward, Am. physicist; b. Long Beach, Wash., May 23, 1915; s. Charles Lewis and Cecile (Compton) W.; A.B., Willamette U., 1940; Ph.D., U. Cal. at Berkeley, 1951; m. Della Willard, Oct. 26, 1946; children—Clyde Arthur, Carolyn Jeanne, Gary Edward. Research asst. U. Cal. Radiation Lab., Berkeley, 1941-43, physicist Los Alamos Sci. Lab., 1943-46; physicist Lawrence Radiation Lab., Berkeley, 1946. Fellow Am. Phys. Soc. Codiscoverer antiprotons, 1955. Home: 5964 Margarido Dr., Oakland, Cal. 94618. Office: Lawrence Radiation Lab., Berkeley, Cal. 94704.*

WIEGAND, Gunther Carl, economist; b. Marburg, Germany, May 25, 1906; s. Friedrich and Julia (von Zezschwitz) W.; B.A., U. N.M., 1944, M.A., 1945; Ph.D., Northwestern U., 1950; m. Elizabeth Kankel, June 14, 1947. With brokerage firms, N.Y.C., 1929-39; free-lance writer, 1931-41; asst. prof. econs. Loyola U., Chgo., 1946-50; asso. prof. U. Miss., 1950-55; vis. lectr. U. Ill., 1955-56; prof. econs. So. Ill. U., Carbondale, 1956—. Guest lectr. univs. Cape Town, Mysore, Bangalore. Mem. Am. Econ. Assn., Economists Nat. Com. on Monetary Policy, Mont Perlerin Soc. Author: A Preface to the Social Sciences, 1953; (with Robert Jacobs, F.G. Macomber) Developing Educational Resources to Assist with Educational Planning, 1964; (with Elizabeth Wiegand) History of Western Civilization—A Brief Guide, 1964; Economics: Its Nature and Importance, 1968. Numerous publs. on devel. internat. finance and effect on domestic econ. devels. Home: 701 Dixon St., Carbondale, Ill. 62901.*

WIEGLER, Johann Christian, German chemist; b. Lengensalza, Saxony, Dec. 21, 1732; s. Christian Ludwig Wiegleb; apothecary student under Sartorius, Dresden, Saxony. Supr. apothecaries, Langensalza, also senator, oberkämmerer of city. Mem. Academia Naturae Curiosorum; Churmaynizische Academie Nützlicher Wissenschaften. Author: Neuer Begrif von der Gährung und den ihr unterwürfigen Körpern, 1766; Vertheidigung der Meyerischen Lehre vom Acido pingui, 1770; Chemische Versuche über die Alcalischen Salze, 1774; (with Vogel) Lehrsatze der Chymie aus dem Lateinischen übersetzt und mit Anmerkungen Begleitet, 1775; Geschichte des Wachsthums und der Erfindungen in der Chemie in der Neuern Zeit, 1790; Handbuch der Allgemeinen Chemie, 1781; Historisch-Kritische Untersuchung der Alchemie, odor der Eingebildeten Goldmacherkunst, 1777; Apothekers in Langensalza, 1771; Die Natürliche Magie . . . , 1782; (with G. A. Hofmann) Anleitung zur Chemie für Künstler und Fabricanten (2d edit., together with remarks), 1779. Writer, translator, editor chem. books; as mem. older, conservative sch. chemistry, opposed Lavoisier, believed in Meyer's Acidum Pingue, but later changed views. Died Jan. 16, 1800.

WIELAND, G(eorge) R(eber), Am. paleontologist; b. Pa., Jan. 24, 1865; s. Washington Frederick and Margaret (Reber) W.; B.S., Pa. State Coll., 1893; studied U. Göttingen, 1894; U. Pa., 1896-97; Yale U., 1899, Ph.D., 1900; m. Edla Kristina Andersson, 1891 (deceased); 1 son, Hans Leonard. Asso. Carnegie Inst.; asso. prof. paleobotany, Yale. Mem. Wilderness Soc. Author: Polar Climate in Time the Major Factor in the Evolution of Plants and Animals, 1903; Osteology of Protostega, 1906; Armored Dinosauria, 1911; La Flora Liasica de la Mixteca Alta (in the Spanish), 1914; American Fossil Cycads, Vol. I, 1906, Vol. II, 1916; New North American Cycadedoids, 1921; Antiquity of the Anglosperms, 1926; A New Cycad from the Mariposa Slates, 1929; Raumeria of the Zwinger of Dresden, 1934; The Cerro Cuadrado Petrified Forest, 1935; Cycadeoid Types of the Kansas Cletaceous, 1942; The Carpathian-Black Hills Cycadeoid parallel, 1941; Fossil Cyad National Monument, 1944; The Yale Cycadeoids, 1945; also Chemistry of Petrifaction; Ancient Flower and cone-bearing Plants; The World's two greatest Petrified Forests; Land Types of the Trinity Beds; Mesaverde Cycadeoids; Wood Opalization Origin of Angiosperms; Ancient Climates; Dinosaur Extinction, etc. Awarded Archduke Rainer medal, Vienna, 1914; Bologna, 1907. Mem. N.Y. Acad. Sci.; hon. mem. Botanical Society of India, 1935. Proposed establishment "Fossil Cycad National Monument" of the Southern Black Hills, donating site of 320 acres including the world's finest flowering petrified forest, 1922; brought to light there, Nov. 1935, aided by the C.C.C. over one ton of in situ cycadeoids, proving a three-fold value—evolutionary, chemic, and stratigraphic. Discovered the giant fossil turtle Archelon in 1895, various paleontologic explorations in the West, Mexico and S. America. Died Jan. 18, 1953.

WIELAND, Heinrich Otto, German chemist; b. Pforzheim, Germany, June 4, 1877; ed. Berlin, Stuttgart, Germany; doctorate in chemistry U. Munich, 1901. Inst., U. Munich, 1904-13, prof. organic chemistry, 1913-17, prof., dir. chem. labs., 1925-52; prof. Kaiser Wilhelm Inst. for Chemistry, 1917-18; researcher and tchr., Freiburg, 1921-25. Recipient Nobel prize in chemistry, 1927. Fellow Royal Soc., 1931. Contbr. articles in many areas of organic chemistry and biochemistry to profl. jours. Research on bile acids, organic radicals, nitrogen compounds, highly toxic biol. substances, morphine, anesthetics, alkaloids of Strychnos; studied biol. and chem. oxidation, especially of iron compounds; discovered structure of cholesterol (with A. Windaus). Died Munich, Aug. 5, 1957.

WIELAND, Otto Heinrich, German clin. biochemist; b. Munich, Germany, May 21, 1920; s. Heinrich and Josefine (Bartmann) W.; Dr.med., U. Munich, 1944, Dr.med.Habil., 1951; m. Rosemarie Quilling, Nov. 11, 1944; children—Doris Schottenhamel, Felix, Isabella. Med. practice II. Med. U. Hosp., Munich, 1945-51, head dept. clin. chemistry and biochemistry, 1956-65; tng. in biochemistry Inst. Biochemistry, U. Munich, 1952-55; dir. Inst. Clin. Chemistry and diabetes research group Schwabing City Hosp., Munich, 1966—. Mem. German Assn. Clin. Chemistry (v.p.), German Assn. Biol. Chemists (v.p.), German Assn. Clin. Chemists, German Assn. Internal Medicine, European Assn. for study Diabetes. Research, publs. on enzymatic analytical methods in clin. chemistry and biochemistry, enzymatic and hormonal regulation of carbohydrate and fat metabolism in diabetes, cellular mechanisms of ketogenesis and gluconeogenesis, biochemistry of toxic liver injury, tumor metabolism. Home: 11 Alpenstrasse, Soecking, Bavaria 8135. Office: 1 Koelner Platz, Munich, Bavaria 8000 West Germany.*

WIELAND, Theodor Hermann Felix, German chemist; b. Munich, Germany, June 5, 1913; s. Heinrich and Josephine (Bartmann) W.; student Freiburg U.; doctorate U. Munich, 1937; m. Irmgard Porcher, Aug. 30, 1940; children—Sibylle Dunkel, Heinrich, Eberhard. Asst., Max Planck Inst. for Med. Research, Heidelberg, 1937-47; dozent U. Heidelberg, 1942; prof. U. Mainz, 1946-51; dir. Inst. Organic Chemistry, U. Frankfort, 1951—. Mem. Am., Swiss, German chem. socs., Max Planck Soc., Soc. Biol. Chemistry (Germany). Author: (with Gattermann) Die Praxis Des Organischen Chemikers, 1960; (with G. Pfleiderer) Molekularbiologie, 1966; also numerous articles. Coeditor Liebigs Annalen der Chemie, Analytical Biochemistry, Die Umschau; adviser European Jour. Biochemistry. Research on electrophoretic separation of mixtures on paper, synthesis of peptides, chemistry of toxins and antitoxins of mushroom Amanita Palloides; indole derivatives, amino acids, isozymes of lactic acid dehydrogenase; model reactions of oxidative phosphorylation. Home: 13 Am Rosengarten, Mainz, Germany. Office: 7-9 Robert Mayer Strasse, Frankfort am Main, Germany.*

WIEMANN, Joseph, French chemist; b. Colmar, France, Feb. 22, 1905; s. Joseph and Mathilde (Fuchs) W.; Licence es Sciences, Doctorat des Sciences, Faculté des Sciences, Paris; m. Anne Lucy Laporte, Oct. 15, 1937; children—Marie-Christine, Marie-Claude. Lectr., then titular prof. gen. chemistry and organic chemistry Faculté des Sciences, Lille, France, 1939-49; prof. Faculté des Sciences, Paris, 1949-55; titular prof. structural organic chemistry, U. Paris, 1955—; prof. Ecoles Normales Supérieures de St. Cloud et de Fontenay. Recipient Kuhlmann prize, Jecker prize; named officer des Palmes Académiques. Mem. Société Chimique de France (past pres. Lille sect, past v.p. Paris sect.), Am. Chem. Soc., Chem. Soc. London. Author: Relations entre la Structure et les Propriétés Physiques, 1965; also numerous articles. Numerous syntheses especially by means of electrochem. methods, of polyols in sugar series, Heterocyclic compounds and polyfunctional compounds; mechanism of these syntheses; detailed structure of compounds obtained by physico-chem. methods. Address: Faculté des Sciences, Lab. de Chimie Organique Structurale, 8 rue Cuvier, 75 Paris 5, France.*

WIEME, Roger Joseph, Belgium biochemist; b. Genth, Belgium, Feb. 15, 1923; s. E. E. and C. (Bruyneel) W.; M.D., U. Ghent, 1949; m. Diana Demol, May 22, 1963; 1 dau., Nathalie. Mem. staff med. dept. U. Ghent, 1950-59, asso. docent, 1959-65, asso. prof., 1965—. Asst. Nat. Found. Sci. Research in Belgium. Author: Agar Gel Electrophoresis, 1965. Research, publs. chpts. in books on electrophoretic separation of proteins; developed a highly resolving procedure using agar gel; discovered five lactate dehydrogenase isozymes in human serum and tissues; applied this observation to diagnosis of liver and muscular disease; investigations on genetic determination of human serum proteins. Home: 143 Martelarenlaan, Ghent. Office: Lab. for Protein Chemistry, Med. Dept., Acad. Hosp., De Pintelaan, Ghent, Belgium.*

WIEN, Max Carl, German physicist; b. Königsberg, Dec. 25, 1866; asst. to Rüntgen at Würzburg, Germany; prof., Aachen, Germany; became prof. physics, Danzig, Germany, 1904, Jena, Germany, 1911-35. Investigated high-frequency electromagnetic waves; credited with discovery of impulse excitation, 1906; studied behaviour of electrolytes in high voltage gradients; devised acoustical apparatus for measuring strength of tones. Died Jena, Feb. 24, 1938.

WIEN, Wilhelm, German physicist; b. Gaffken, Germany, Jan. 13, 1864; entered U. Göttingen (Germany), 1882, also studied at Heidelberg; doctorate Berlin U., 1886; adminstr. father's estate, 1886-90; asst. to Hermann Helmholtz, State Physico-tech. Inst., Berlin, 1890-02; lectr. Berlin U., 1892-96; lectr. physics Tech. High Sch., Aachen, Germany, 1896-99; prof. physics U. Giessen (Germany), 1899-1900, U. Würzburg (Germany), 1900-20, U. Munich (Germany), 1920-28. Recipient Nobel prize in physics (for discovery of (Wien's) displacement law that product of wavelength for maximum energy density and absolute temperature is constant 1893, (Wien's) distbn. law of radiation 1896), 1911. Authority on radiation from black bodies; studied diffraction of light aroung edges (doctoral thesis), thermodynamics, hydrodynamics, 1896-99, discharge of electricity through gases, magnetic and elec. deflection of canal rays, 1896; demonstrated that canal rays are residual gas atoms, 1902; made 1st rough determination of wave length of X-rays, 1905; studied cathode rays and optics. Died Munich, Aug. 30, 1928.

WIENER, Alexander Solomon, Am. immunohematologist; b. Bklyn., Mar. 16, 1907; s. George and Mollie (Zuckerman) W.; A.B., Cornell U., 1926; M.D., State U. N.Y. Coll. Medicine, 1930; m. Gertrude Rhoda Rodman, June 15, 1932; children—Jane (Mrs. Harold Einhorn), Barbara (Mrs. Theodore Krevit). Practice medicine specializing in hematology and immunohematology, Bklyn., 1934—; head blood transfusion div. Jewish Hosp. Bklyn., 1938-52, attending immunohematologist, 1952—; serologist to Office of Chief Med. Examiner of N.Y.C., 1938-50, sr. serologist, 1950—; faculty dept. forensic medicine N.Y. U. Sch. Medicine, 1949—, asso. prof. forensic medicine, 1955—; staff Adelphi Hosp., 1953—. Recipient Lasker award Am. Pub. Health Assn., 1946; Passano Found. award, 1951; Alumni medallion State U. N.Y. Coll. Medicine, 1955; Joseph P. Kennedy Jr. Found. Internat. award, 1966; named Hon. Citizen, Salerno, Italy, 1961. Fellow Coll. Am. Pathologists (founding), Internat. Soc. Hematologists, A.C.P., A.A.A.S., Am. Acad. for Forensic Sci.; life mem. Am. Soc. Clin. Pathology (Ward Burdick award 1946), N.Y. Acad. Medicine, N.Y. Acad. Sci.; mem. A.M.A., Am. Assn. Blood Banks (Karl Landsteiner award 1956), Am. Assn. Immunologists, Soc. for Study Blood (founder, 1st pres.), Royal Soc. Medicine (affiliate). Author: Blood Groups and Transfusion, 1932, 3d edit., 1943; Rh-Hr Blood Types, 1954; Advances in Blood Grouping, vol. I, 1961, vol. II, 1965; (with I.B. Wexler) Heredity of Blood Groups, 1958, Rh-Hr Syllabus, 1954, 2d edit., 63; also numerous articles. Co-discoverer Rh factor and its blocking antibodies; determined genetic mechanism of transmission of Rh-Hr blood types, also serology and nomenclature of Rh-Hr blood types; introduced treatment of erythroblastosis by exchange transfusion; discovered several other human blood factors and explained their roles in disease; comparative blood types in humans and other primates. Home: 90 Maple St., Bklyn. 11225. Office: 64 Rutland Rd., Bklyn. 11225.*

WIENER, Christian, physicist; b. 1826 (With E. Lemoine) defined coefficients of simplicity and accuracy of geometrical constrns.; introduced system named geometrographics, 1888. Died 1896.

WIENER, Norbert, Am. mathematician; b. Columbia, Mo., Nov. 26, 1894; s. Leo and Bertha (Kahn) W.; grad. Ayer (Mass.) High Sch., 1906; A.B., Tufts Coll., 1909; Harvard, 1909-10, 1911-13; Cornell, 1910-11; A.M., Harvard, 1912, Ph.D., 1913; studied Cambridge, England, 1913-14, 1914-15, Göttingen, 1914, Columbia U., 1915; m. Marguerite Engemann, Mar. 26, 1926; children—Barbara (Mrs. Gordon Raisbeck). and Margaret (Mrs. John T. Blake). Docent-lecturer, Harvard, 1915-16; instr. U. of Me., 1916-17; staff writer Ency. Americana, Albany, N.Y., 1917-18; newspaper work Boston Herald, 1919; instr. in mathematics, Mass. Inst. Tech., 1919-24, asst. prof., 1924-28, asso. prof., 1928-32, prof. since 1932; visiting prof. Tsing Hna U., Peiping, China, 1935-36. Lectured, Cambridge, England, 1931-32; Fullbright lecturer at College de France, 1951; guest prof. Indian Statis. Inst., Calcutta, 1955-56, Bowdoin prize, Harvard, 1914; Bocher prize, Am. Math. Soc., 1933; Guggenheim fellow, Göttingen and Copenhagen, 1926. Civilian computer, later pvt., Aberdeen Proving Ground, Md., 1918-19. Mem. Am. Math. Soc. (v.p.), Nat. Acad. Sciences, London Math. Soc. Author: The Fourier Integral and Certain of Its Applications; Harmonic Analysis in Complex Domain; Cybernetics, 1948; The Human Use of Human Beings, 1950; Ex-Prodigy, 1953; I Am a Mathematician, 1956; Extrapolation, Interpolation, and Smoothing of Stationary Time Series with Engineering Applications, 1949; Nonlinear Problems in Random Theory, 1958; God and Golem, Inc.; also articles in field and on epistemology. Worked in theories of probability and potential, theory of postulates, founds. of math., Fourier integrals and transforms, relativity and quantum theory; worked out math. theory of Brownian movement, 1920; research on math. physiology, harmonic analysis, vector and differential spaces; interest in automatic computers, feedback and statistical analysis of communication flow led to his foundation of cybernetics (science of communication and control). Died Stockholm, Sweden, Mar 18, 1964.

WIENER, Otto, German physicist; b. Karlsruhe, Germany, 1862; prof. Technische Hochschule, Aachen, Germany, also univs. Giessen, Leipzig (both Germany); proved stationary light waves occur between an incident and a reflected ray, by covering a mirror with fine grained photog. film. Died 1927.

WIENS, Herold Jacob, geographer; b. Shanghang, China, Dec. 26, 1912; s. Frank J. and Agnes (Harder) W.; m. Elizabeth Susong Burnett, Dec. 30, 1941; children—Thomas B., Linda C., Frank H., Judith C. Research analyst OSS 1941-46; faculty Yale, 1947-66, prof. geography, 1962-66, acting chmn. geography dept., 1965-66; prof. geography U. Hawaii, Honolulu, 1966—, director for Asian studies, since 1967—. Mem. Assn. Am. Geographers (councilor 1962-65), Assn. for Asian Studies, Am. Geog. Soc. Author: China's March Toward the Tropics, 1954; Atoll Environment and Ecology, 1962; Pacific Island Bastions of the U.S., 1962; (with Norton Ginsburg), Pattern of Asia, 1958, 67; (with M. Hertel) Asia with Focus on China, 1965; (with others) World Geography, 1965; Mongolia, 1967; also numerous articles. Synthesizing regional and hist. geography China, correlated and described environment and ecology coral reef islands, marine and terrestial. Office: Dept. Geography, U. Hawaii, Honolulu 96822.*

WIERSMA, Cornelis Adrianus Gerrit, physiologist; b. Naaldwyk, Holland, Oct. 10, 1905; s. Klaas and Cornelia (Rystenbil) W.; B.Sc., Leiden U., 1926; M.Sc., Utrecht U., 1929, Ph.D. (Dondersfonds research fellow), 1933; m. Jeanne J. Netten, Dec. 21, 1932. Came to U.S., 1934. Faculty, Utrecht U., Holland, 1929-34; faculty Cal. Inst. Tech., Pasadena, 1934—, prof. biology, 1947—; vis. prof. Cambridge (Eng.) U., 1957-58. Guggenheim fellow, 1957—. Mem. A.A.A.S., Am. Zool. Soc., Soc. Gen. Physiology. Research, publs. on neuromuscular transmission mechanisms in Crustacea, fast, slow, and inhibitory systems; functional connections of crayfish giant and other interneurons, including synaptic connectivity; stretch receptors in abdomen of crayfish; optical integrations and transformations in crustacean visual systems; electronarcosis in animals and man. Home: 350 S. Greenwood St., Pasadena, Cal. 91109.*

WIESE, Allen Franklin, Am. agronomist; b. Eyota, Minn., Dec. 16, 1925; s. Paul D. and Elizabeth (Hintgen) W.; B.S., U. Minn., 1949, M.S., 1951, Ph.D., 1953; m. Joan M. Hanson, Sept. 5, 1948; children—David P., Beth M., Ann M. Research asst. U. Minn., 1950-53; with Tex. A. and M. U., U.S. Dept. Agr. Southwestern Great Plains Research Center, Bushland, 1953—. Mem. Crop Sci. Soc. Am., Soil Sci. Soc. Am., Am. Soc. Agronomy, Weed Soc. Am., So. Weed Conf. Contbr. numerous articles to profl. jours. Developed chem. control methods for perennial weeds, ann. weeds in wheat, sorghum, cotton, castorbeans, soybeans and sugarbeets in southwestern U.S.; studied effect of weeds on yield and quality of wheat and sorghum; determined physiol. factors that affect persistence of weeds in semi-arid areas. Home: 6109 Hanson Rd., Amarillo, Tex. 79106. Office: P.O. Box 1, Bushland, Tex. 79012.*

WIESE, Alvin Carl, Am. biochemist; b. Milw., Aug. 13, 1913; s. Alvin John and Clara (Krause) W.; B.S., U. Wis., 1935, M.S., 1937, Ph.D., 1940; m. Hazel Marie Kuntz, Aug. 19, 1944; children—Jon Lee, Ray Alvin. Instr., Okla. State U., 1940-42; research asso. U. Ill., 1942-46; prof., head dept. agrl. biochemistry and soils U. Ida., Moscow, 1946—; partner Wiese-Jackson Assos., Consultants, Moscow, 1958—. Fellow A.A.A.S.: mem. Am. Chem. Soc., Am. Inst. Nutrition, Soc. for Exptl. Biology and Medicine, Poultry Sci. Assn., Sigma Xi. Research, publs. on biochem. and animal nutrition, vitamins, trace elements, deficiency symptoms and requirements vitamins for various animals, effect fluorides on plant enzyme systems, effects fluoride air pollutants on plants and animals. Home: 721 S. Lynn St., Moscow, Ida. 83843.*

WIESNER, Jerome Bert, Am. elec. engr.; b. Detroit, May 30, 1915; s. Joseph and Ida (Freedman) W.; B.S., U. Mich., 1937, M.S., Ph.D., 1950, D.Sc., 1962; Eng.D., Poly Inst. Bklyn., 1961; D.Sc., Lowell Technol. Inst., 1962, U. Mass., 1965, Lehigh U., 1965, Brandeis U., 1965, Northwestern U., 1966; m. Laya Wainger, Sept. 1, 1940; children—Stephen Jay, Zachary Kurt, Elizabeth Ann, Joshua Alec. Chief engr. Library Congress, 1940-42; staff Mass. Inst. Tech., 1942-45, Los Alamos Sci. Lab., 1945-46; prof. elec. engring., asso. dir., dir. research lab. electronics, chmn. dept. elec. engring. Mass. Inst. Tech., Cambridge, 1946-61, dean Sch. Sci., 1964-66, provost, 1966—. Spl. asst. to President for sci. and tech., dir. Office Sci. and Tech. White House, Washington, 1961-64; mem. sci. adv. com. Army, 1956-61; adv. bd. Television Fund, Inc.; tech. adv. com. Am. Found. For Blind. Recipient medal of honor Electronic Industries Assn., 1961. Fellow I.E.E.E.; mem. Am. Geophys. Union, Acoustical Soc. Am., Fedn. Am. Scientists, Am. Acad. Arts and Scis., Nat. Acad. Scis., Nat. Acad. Engring., A.A.A.S., Sigma Xi, Phi Kappa Phi, Eta Kappa Nu. Author: Where Science and Politics Meet, 1965; also numerous articles. Research on scatter communication techniques, anti-ballistic missile systems, information theory, communication theory, sci. mgmt. Home: 61 Shattuck Rd., Watertown, Mass. 02172.*

WIGAND, Albert, German botanist; b. Treysa, Apr. 21, 1821; prof., Marburg, Germany. Author: Lehrbuch der Pharmakognosie, 1863; Der Darwinismus und der Naturforschung Newtons und Cuviers, 3 vols., 1874-77. Opponent of Darwinism. Died Marburg, Oct. 22, 1886.

WIGGANS, Donald Sherman, Am. biochemist; b. Lincoln, Neb., July 14, 1925; s. Cleo Claude and Martha (Chinn) W.; B.Sc., U. Neb., 1949; Ph.D., U. Ill., 1952; m. Barbara Ellen Tarrant, June 9, 1951; children—David Frank, Richard Mason, John Sherman, Scott Donald. Research fellow in biochemistry U. Ill., 1949-51, U.S. AEC fellow 1951-52; instr. Yale, 1952-54; faculty U. Tex. Southwestern Med. Sch., Dallas, 1954—, prof., 1961—; vis. prof. biology So. Meth. U., Dallas, 1964-66. Mem. Am. Chem. Soc., Am. Soc. Biol. Chemists, Tex. Acad. Sci. (vis. scientist 1962—), Am. Inst. Nutrition, Soc. Exptl. Biology and Medicine, N.Y. Acad. Sci., A.M.A. Research, publs. on intermediatry metabolism proteins and amino acids, subcellular metabolism, comparative biochemistry. Home: 4129 Northview Lane, Dallas 75229.*

WIGGINS, Ira Loren, Am. botanist; b. Madison, Wis., Jan. 1, 1899; s. Edward T. and Minnie M. (Talbott) W.; A.B., Occidental Coll., Los Angeles, 1922, postgrad., 1922-23, D.Sc., 1965; M.A., Stanford, 1925, Ph.D., 1930; m. Dorothy Bruce, Aug. 20, 1923; children—Bruce Leon, Donnalie M. (Mrs. William H. McPherson). Tchr. high sch., Durham, Cal., 1923-24; faculty Occidental Coll., 1925-29; faculty Stanford (Cal.) U., 1929—, prof. biology, 1940-64, prof. emeritus, 1964—, dir. Natural History Mus., 1940-61; lectr. botany U. Fla., 1965, Cal. State Coll. at Fullerton, 1965-67. Dir., Belvedere Sci. Fund, 1961-64; dir. Arctic Research Lab., Point Barrow, Alaska, 1950-53, 56; mem. sci. team Galapagos Islands Sci. Project, 1964. Recipient Henry A. Gleason award N.Y. Bot. Garden, 1964. Mem. Am. Bot. Soc. (certificates of merit 1964, 65), Cal. Acad. Sci. (pres. 1953-57, research asso. 1966). Author: (with Forrest Shreve) Vegetation and Flora of the Sonoran Desert, 2 vols., 1964; (with John Hunter Thomas) A Flora of the Alaskan Arctic Slope, 1962. Research and publs. on the classification and geographic distbn. of plants of Am. Southwestern desert regions and adjacent Mexico, described about 30 new species of vascular plants from this region; collected numerous plants, preserved as herbarium specimens. Home: 111 Pope St., Menlo Park, Cal. 94025. Office: Div. Systematic Biology, Stanford U., Stanford, Cal. 94305.*

WIGGINS, Thomas Arthur, Am. physicist; b. Indiana, Pa., Feb. 24, 1921; s. Arthur William and Helen (Simpson) W.; B.S., Pa. State Coll., 1942; M.S., George Washington U., 1949; Ph.D., Pa. State U., 1953; m. Dorothy Blanchard, June 27, 1953; children—Bruce, James. Faculty, Pa. State U., University Park, 1953—, prof. physics, 1961—. Fellow Optical Soc. Am.; mem. Am. Phys. Soc., Sigma Xi. Research, publs. on molecular spectroscopy, phys. optics, atomic and molecular physics. Home: 352 Hillcrest Av., State College, Pa. 16801. Office: Dept. of Physics, Pa. State University, University Park, Pa. 16802.*

WIGGLESWORTH, Sir Vincent (Brian), English biologist; b. Kirkham, Eng., Apr. 7, 1899; s. Sidney Wigglesworth; M.A., Caius Coll., Cambridge (Eng.) U.; M.D., St. Thomas Hosp.; Ph.D., U. Bern (Switzerland); D.Sc., U. Paris (France); m. Mabel K. Semple, 1928; 3 sons, 1 dau. Lectr. med. entomology London Sch. Hygiene and Tropical Medicine, 1926; reader entomology U. London, 1936-44; reader entomology Cambridge U., 1945-52, dir. Agrl. Research Council Unit Insect Physiology, Quick prof. biology, 1952—. Lectr. various socs., univs. Fellow Royal Soc., 1939 (Royal medal 1955), 1939; mem. Am., Dutch, French, Egyptian, Finnish, Indian, German, Swiss socs. entomology, Royal Danish Acad. Sci., Am. Acad. Arts and Scis., Leopoldina, Zool. Soc. France, Indian Acad. Zoology, others. Author: Insect Physiology, 1934; The Principles of Insect Physiology, 1939; The Physiology of Insect Metamorphosis, 1954; The Control of Growth and Form, 1959; The Life of Insects, 1964. Research, publs. on comparative physiology; specialist insect physiology; 1st demonstrated function of neurosecretory cells; studied molting hormone in Rhodnius prolixus; developed theory of insect metamorphosis. Home: Chimney House, Lavenham, Suffolk, Eng.

WIGHTMAN, Arthur Strong, Am. physicist; b. Rochester, N.Y., Mar. 30, 1922; s. Eugene Pinckney and Edith (Stephenson) W.; B.A., Yale, 1942; Ph.D., Princeton, 1949; m. Anna-Greta Larsson, Apr. 28, 1945; 1 dau., Robin Letitia. Feculty, Princeton, 1949—, prof. math. physics, 1960—. NRC postdoctoral fellow Inst. Theoretical Physics, Copenhagen, 1951-52, 1956-57, Naples, 1957; vis. prof. Sorbonne, 1957; mem. Institut des Hautes Etudes Scientifiques, 1963-64. Fellow Am. Acad. Arts and Scis.; mem. Am. Phys. Soc., Am. Math. Soc., A.A.A.S. Author: (with R.F. Streater) PCT, Spin and Statistics, and All That, 1964. Studies in math. founds. of quantum mechanics, representations of commutation and anti-commutation relations, analytic functions and quantum field theory. Home: 30 The Western Way, Princeton, N.J. 08540.*

WIGNER, Eugene Paul, mathematical physicist; b. Budapest, Hungary, Nov. 17, 1902; s. Anthony and Elisabeth (Einhorn) W.; Chem. Engr. and Dr. Engring., Technische Hochschule, Berlin; hon. D.Sc., U. Wis., 1949, Washington University, 1950, Case Inst., 1956, U. Chgo., 1957, Colby Coll., 1959, U. Pa., 1961, Thiel Coll., 1965, University of Notre Dame, 1965, Technische Universität Berlin, 1966, Swarthmore College, 1966, Université de Louvain (Belgium), 1967; Dr. Jr., University Alberta; Doctor of Humane Letters, Yeshiva U., 1963; came to the U. S. 1930, naturalized, 1937; m. Amelia Z. Frank, Dec. 23, 1936 (died 1937); m. 2d, Mary Annette Wheeler, June 4, 1941; children—David Wheeler, Martha Faith. Lectr. Princeton U., 1930, halftime prof. mathematical physics, 1931-37, became Jones professor theoretical physics, 1938; professor of physics, U. of Wisconsin, 1937-38. On leave of absence, 1942-45, at Metall. Lab., U. of Chicago, 1946-47, director of research, development, Clinton Labs.; dir. Civil Defense Research Project, Oak Ridge, 1964-65. Lorentz lectr. Inst. Lorentz, Leiden, 1957. Mem. gen. advisory com. U. S. AEC, 1952-57, 59-64; math. sect. Nat. Research Council, 1952-54; physics panel National Science Foundation, 1953-64; visiting committee National Bureau Standards, 1948-52. Decorated Medal for Merit, 1946; Franklin medal of Franklin Inst., 1950; recipient Enrico Fermi Award U. S. AEC, 1958; Atoms for Peace award, 1960; Max Planck medal, German Phys. Society, 1961; Nobel prize for physics (with Mayer and Jensen) 1963; George Washington award Am. Hungarian Studies Found., 1964; Semmelweiss medal Am. Hungarian Med. Assn., 1965. Mem. Royal Netherlands Acad. Sci. and Letters, Am. Nuclear Soc. (dir. 1960-61), Am. Physical Society (vice president 1955, president 1956), American Math. Soc., Am. Acad. Arts and Scis., Am. Philos. Soc., Nat. Acad. Sciences, Sigma Xi; corr. mem. Acad. Sci., Göttingen, Germany. Author: Group Theory and its Application to the Quantum Mechanics of Atomic Spectra, 1959; Dispersion Relations and their Connection with Causality, 1964; Symmetries and Reflections, 1967. Research on role of invariance and symmetry, particularly in quantum mechanics; also on theory of rate of chem. reactions, structure of metals, theory of metallic cohesion, nuclear structure and collision theory; contbd. to establishing nuclear chain reaction and its exploitation. Home: 8 Ober Rd., Princeton, 08541. Office: Palmer Phys. Lab., Princeton U. N.J. 08541. Office: Palmer Phys. Lab., Princeton U. Princeton, N.J. 08540.*

WIIN-NIELSEN, Aksel Christopher, meteorologist; b. Juelsminde, Denmark, Dec. 17, 1924; s. Aage Nielsen and Marie Christophersen; B.S. in Math., U. Copenhagen (Denmark), 1947, M.S. in Math., 1950; Fil. Lic. in Meteorology, U. Stockholm (Sweden), 1957, Ph.D. in Meteorology, 1960; m. Bente Havsteen Zimsen, Dec. 5, 1953; children—Charlotte, Barbro Marianne, Karen Margrete. Came to U. S., 1959. Staff meteorologist Danish Meteorol. Inst., 1952-55; research meteorologist Internat. Meteorol. Inst., U. Stockholm, 1955-58, asst. prof., 1957-58, exec. editor publ. Tellus, 1957-58; research meteorologist Air Weather Service, USAF, also staff mem. joint numerical weather prediction unit and lectr. George Washington U., 1959-61; mem. research staff Nat. Center Atmospheric Research, 1961-62, asst. dir., 1962-63; prof., chmn. dept. meteorology and oceanography U. Mich., 1963—, rep. univ. council members Univ. Corp. Atmospheric Research, 1964—. Mem. Am. Geophys. Union, Swedish, Danish geophys. socs., Am. Meteorol. Soc., Tau Beta Pi. Observational studies of energy distbns. and energy conversions in atmosphere as function of horizontal scale; contbd. to understanding of dynamics and energetics of atmosphere, particularly atmospheric waves, also to foundation of phys. weather prediction. Home: 2171 Georgetown Blvd., Ann Arbor, Mich. 48105.

WIJSMAN, Robert Arthur, math. statistician; b. The Hague, Netherlands, Aug. 20, 1920; s. Johannes and Maria Alting (Mees) W.; Ingenieur in Physics Engring., Delft Inst. Tech., 1945; Ph.D. in Physics, U. Cal. at Berkeley, 1952; m. Gertrud Zierau, June 6, 1953; children—Ellen, Annette, Suzanne. Came to U.S., 1946. Lectr. med. physics, acting asst. prof. statistics U. Cal. at Berkeley, 1952-57; faculty U. Ill., Urbana, 1957—, prof. math., 1965—. Mem. Inst. Math. Statistics, Am. Math. Soc., Sigma Xi. Asso. editor Annals Math. Statistics, 1964-67. Research and publs. on theory of testing hypotheses, sequential analysis. Home: 507 S. Pine St., Champaign, Ill. 61820. Office: Dept. Math., U. Ill., Urbana, Ill. 61801.*

WILANSKY, Albert, mathematician; b. Newfoundland, Sept. 13, 1921; s. Isidore and Esther (Sidel) W.; Ph.D., Brown U., 1947; m. Ruth Paton, Feb. 19, 1946; children—Eleanor, Laura. Came to U.S., 1944, naturalized, 1954. Prof. math. Lehigh U., Bethlehem, Pa., 1957—. Mem. Math. Assn. Am. (vis. lectr. 1963—, past gov.), Am. Math. Soc., Nat. Council Tchrs. Math., Am. Assn. U. Profs. Author: Functional Analysis, 1964. Asso. editor Am. Math. Monthly, 1963—. Research, publs. on applications functional analysis to classical analysis. Home: 2450 Woodstock Dr., Bethlehem, Pa. 18017.*

WILCKE, Johan Carl, physicist; b. Wismar, Germany, Sept. 6, 1732; prof., Stockholm, Sweden. Fellow Royal Soc., 1789. Made numerous exptl. studies; investigated a theory of elec. dispersion and Leyden jar, 1757; worked with Aepinus; set up 1st chart of magnetic inclination, 1766; accepted theory of 2 fluids for electricity; formulated (independently of Joseph Black) theory of specific heat, 1772. Died Stockholm, Apr. 18, 1796.

WILCKE, Ortwin Eduard Emil, German neurosurgeon; b. Hohenbrueck, Pommern, Germany, Jan. 1, 1924; s. Kurt Friedrich Georg and Alice (Haenisch) W.; student U. Berlin (Germany); M.D., U. Heidelberg (Germany), 1948; m. Gisela Hochgeschurz, May 20, 1952; children—Andreas, Christa, Cornelia. Asst. in pathology U. Heidelberg, 1948, in internal medicine, 1949, in neurology, 1950, in surgery, 1951-53, in physiology, 1953-54; asst. in neuro-surgery U. Cologne (Germany), 1955—, med. specialist neurosurgery, 1962—, dozent in neurosurgery, 1965—. Editor: Handbuch der Neurochirurgie, 1963; Isotopendiagnostik in der Neurochirurgie, 1966. Research, publs. on histological examination of effect radiation on brain tissue, determination of cerebral blood circulation with isotopes.*

WILCOX, Calvin Hayden, Am. mathematician; b. Cicero, N.Y., Jan. 29, 1924; s. Calvin and Vera (Place) W.; student Syracuse U., 1947-48; A.B., Harvard, 1951, A.M., 1952, Ph.D., 1955; m. Frances Irma Rosekrans, May 29, 1947; children—Annette Faye, Victor Hayden, Christopher Grant. Mathematician, Air Force Cambridge Research Center, 1953-55; faculty Cal. Inst. Tech., 1955-61; vis. mem. Math. Research Center, U.S. Army, U. Wis., Madison, 1958-60, mem., prof. math., 1961-66; prof. math. U. Ariz., Tucson, 1966—. Cons. applied math. div. Argonne Nat. Lab., 1962—. Mem. adv. bd. sch. Math. Study Group, 1965—. Mem. Am. Math. Soc., Soc. for Indsl. and Applied Math. (mem.-at-large council). Editor: Asymptotic Solutions of Differential Equations and Their Applications, 1964; Perturbation Theory and its Application in Quantum Mechanics, 1966. Research, publs. on electromagnetic theory, analysis and design of radar antennas, waveform design, theory of shaping radar pulses (waveform design), transmission line theory, analysis of wave propagation on non--uniform transmission lines, gen. theory of wave propogation, contbr. to theory of wavepropagation in anisotropic media. Office: Math. Dept., U. Ariz., Tucson.*

WILCOX, John Marsh, Am. physicist; b. Iowa City, Jan. 31, 1925; s. Myron J. and Elizabeth (Schwabe) W.; B.S., Ia. State Coll., 1948; Ph.D., U. Cal. at Berkeley, 1954; m. Pauline Gan, Aug. 20, 1955; children—Sharon, David. Physicist, U. Cal. Lawrence Radiation Lab., Berkeley, 1952-64, U. Cal. Space Scis. Lab., Berkeley, 1964—. Mem. Am. Phys. Soc., Am. Astron. Soc., Am. Geophys. Union, A.A.A.S. Contbr. articles to profl. jours. Demonstrator, Alfvén hydromagnetic waves in lab. plasma, rotating plasma expts.; demonstrator extension of photospheric magnetic field in interplanetary space, observation longitudinal sector structure in interplanetary space and inner solar system. Home: 976 Grizzly Peak Blvd., Berkeley, Cal. 94708.*

WILCOX, Lee Roy, Am. mathematician; b. Chgo., June 8, 1912; s. Lee Alfred and Pauline (Miller) W.; S.B., U. Chgo., 1932, S.M., 1933, Ph.D., 1935; m. Virginia Marie Johnson, Dec. 21, 1940; children—Robert Harrison, Jean Marie. Mem., asst. Inst. for Advanced Study, Princeton, N.J., 1935-38; faculty U. Wis., Madison, 1938-40; faculty Ill. Inst. Tech., Chgo., 1940—, prof. math., 1958—; cons. in modern math., 1959—. Cons. editor Internat. Textbook Co., Scranton, Pa., 1948—. Mem. Am. Math. Soc., Math. Assn. Am., Am. Soc. for Engring. Edn., Am. Assn. U. Profs., Council for Basic Edn. Author: (with R.B. Kershner) The Anatomy of Mathematics, 1950; (with H.J. Curtis) Elementary Differential Equations, 1961; also articles. Discovered class of lattices; research on structure this class, theory sets, axiomatic devel. math. theories. Home: 1404 Forest Av., Wilmette, Ill. 60091. Office: 3300 S. Federal St., Chgo. 60616.*

WILCOXSON, Roy Dell, Am. plant pathologist; b. Columbia, Utah, Jan. 12, 1926; s. Roy Edwin and Bertha (Karren) W.; B.S., Utah State U., 1953; M.S., U. Minn., 1955, Ph.D., 1957; m. Iva Wall, Apr. 15, 1949; children—Bonnie, Paul Kevin, Karren Ray, John Wall. Faculty dept. plant pathology U. Minn., St. Paul, 1957—, now prof. Mem. Am., Indian phytopath. socs., Bot. Soc. Am., Am. Soc. Agronomy, Am. Inst. Biol. Scis., A.A.A.S., Minn. Acad. Sci., Sigma Xi. Research, numerous publs. on devel. of disease resistant varieties of crop plants, devel. of background information concerning effect of factors that alter disease severity, physiology of fungi, their reprodn. and translocation. Home: 1491 Raymond Av., St. Paul 55108.*

WILCZOK, Tadeusz Marian, Polish biochemist; b. Katowice, Poland, June 22, 1934; s. Pawel and Róza (Rózanska) W.; B.S. U. Wroclaw and Higher Pedagogic Acad., Katowice, Poland, 1955, M.S., 1956; Ph.D., U. Lódz (Poland), 1960, docent degree, 1963; m. Gizela Brzezina, Mar. 17, 1956; 1 son, Adam Tadeusz. Staff, Higher Pedagogic Acad., 1953-56; postgrad. Inst. Chemistry, Nat. Acad. Sci., Moscow, USSR, 1956-58; faculty dept. tumor biology Inst. Oncology, Gliwice, Poland, 1958-66; head dept. chemistry U. Med. Sch. Silesia, Katowice, 1966—. U. Mainz (Germany) fellow physiol. chemistry, 1960; Sloan-Kettering Inst. for Cancer Research fellow, N.Y.C., 1964-65. Recipient award Polish Acad. Scis., 1961, 64. Mem. Polish Biochem. Soc., Polish Oncological Soc., Polish Chem. Soc. Research, numerous publs. on changes in secondary structure of DNA and resultant biol. activity, incorporation of DNA into normal and neoplastic cells, repair of radiation damage by DNA. Home: 17 Brzozowa, Katowice 3. Office: 19 K. Marks St., Zabrze-Rokitnica, Poland.*

WILCZYNSKI, Ernest Julius, mathematician; b. Hamburg, Germany, Nov. 13, 1876; s. Max and Friederike (Hurwitz) W.; A.M., Ph.D., U. of Berlin, 1897; m. Countess Ines Macola, Aug. 9, 1906. Computer, Nautical Almanac Office; Washington, 1898; instr. math. summer session, Columbian (now George Washington) U., 1898, instr., asst. prof. and asso. prof. math., U. of Calif., 1898-1907; asso. prof. math., U. of Ill., 1907-10; asso. prof. mathematics, 1910-14, prof., 1914—, U. of Chicago. Research asst. and asso. Carnegie Instn., 1903-05; lecturer New Haven Colloquium, 1906. Fellow A.A.A.S.; laureate, Royal Belgian Acad. Sciences. Author: Projective Differential Geometry of Curves and Ruled Surfaces, 1906; The New Haven Colloquium (with E. H. Moore and Max Mason), 1910; Plane Trigonometry and Applications, 1913; College Algebra with applications, 1916. Important developer of projective differential geometry. Died Sept. 15, 1932.

WILD, Heinrich, physicist, meteorologist; b. Zurich, Switzerland, Dec. 17, 1883; dir. Berlin Obs., 1858-68, St. Petersburg Obs., 1868-94; dir. Central Meteorol. Bur., Bern, 1863-65; dir. Russian Meteorol. Bur., St. Petersburg, 1865-95. Established many meteorol. stas.; inventor topog. reprodn. device, also prism theodolite, 1918; inventor a barometer, weather vane, anemograf, polarization photometer, the polaristrobometer. Died Zurich, Sept. 5, 1902.

WILD, Robert Lee, Am. physicist; b. Sedalia, Mo., Oct. 9, 1921; s. Alwin B. and Nellie (Nowlin) W.; B.S., Central Mo. State Coll., 1943; M.A., U. Mo., 1948, Ph.D., 1950; m. Frances Elleta Wheeler, Oct. 7, 1943; children—James Robert, Janet Gayle, Margaret Nell. Faculty, U. N.D., 1950-53; faculty U. Cal., Riverside, 1953—, prof. physics, 1953—, chmn. dept., 1964—. NSF fellow, 1959-60. Mem. Am. Phys. Soc., Am. Assn. Physics Tchrs., Sigma Xi. Contbr. articles to sci. jours. Research in x-ray, solid state physics. Home: 5709 Durango Rd., Riverside, Cal. 92506.*

WILDE, Henry, Brit. elec. engr.; b. Manchester, Eng., 1833; D.Sc., D.C.L. Fellow Royal Soc., 1886. Contbr. articles to tech. jours. Discovered indefinite increase of magnetic and electric forces from extremely small quantities, 1864; invented dynamo-electric machine which generated powerful electric light for 1st time, 1864-65, and applied it to search-light (adopted by Brit. Navy 1875), to electroplating, 1868-80; invented magnetarium to reproduce terrestrial magnetism; discovered synchronizing property of alternating current to control rotations of dynamos, 1868, multiple relations among atomic weights (used to predict various elements and their properties), 1878; showed that moving force of celestial bodies is proportional to square of velocity and inversely proportional to square of distance, 1909. Died Mar. 29, 1919.

WILDER, Burt Green, Am. anatomist; b. Boston, Aug. 11, 1841; s. David and Celia Colton (Burt) W.; B.S. in anatomia summa cum laude, Lawrence Sci. Sch. (Harvard), 1862, M.D., 1866; m. Sarah Cowell Nichols, June 9, 1868 (dec. 1904); m. 2d, Mary Field, June 11, 1906. Asst. in comparative anatomy Mus. Comparative Zoology, 1866-68; curator herpetology Boston Soc. Natural History, 1867-68; prof. neurol. and vertebrate zoology Cornell U., Ithaca, N.Y., 1867-1910, emeritus, 1910—. Lectr. at various schs. The 1st Am. Festschrift was the Wilder Quarter-Century Book, comprising papers prepared by 15 former pupils, presented at 25th anniversary of Cornell U. Author: What Young People Should Know, 1874; Anatomical Technology (with S. A. Gage), 1882; Physiology Practicums; Emergencies, 1888; Health Notes for Students, 1890; The Brain of the Sheep, 1903; biographic articles on Jeffries Wyman, 1874, Louis Agassiz, 1907; also several songs, many sci. papers, mostly on the brain, numerous revs. and articles in mags. and in Reference Handbook of Med. Scis. Reeled 150 yards of silk from Nephila clavipes spider, nr. Charleston, S.C., 1863, woven into ribbon on steamloom, 1865, illus. account of spider's habits pub. in Atlantic, 1866; devised slip-system of notes, 1867, correspondence-slip, 1884; prepared nearly 2000 vertebrate brains, including 13 of educated persons; advocate of simplified anatomic nomenclature, also of objective study of the brain (beginning in primary schs.), use of chloroform in capital punishment, dissection of cat as prerequisite to that of man. Died Jan. 22, 1925.

WILDER, Harris Hawthorne, Am. Zoologist; b. Bangor, Me., Apr. 7, 1864; s. Solon and Sarah Watkins (Smith) W.; A.B., Amherst, 1886; Ph.D., U. Freiburg (Baden), 1891; m. Inez Luanne Whipple, July 26, 1906. Prof. zoology Smith Coll., 1892-28. Fellow Am. Acad. Arts and Scis. Author: Invertebrate Zoology, 1894; Synopsis of Animal Classification, 1902; The History of the Human Body, 1910, 2d edit., 1923; Personal Identification (with B. Wentworth), 1918; Manual of Anthropometry, 1920; Man's Prehistoric Past, 1923; The Pedigree of the Human Race, 1925; Early Years of a Zoologist, 1930; also numerous papers on comparative anatomy, vertebrate anatomy and physical anthropology, especially on amphibians, diploteratology and epidermic markings of human and simian palms and soles; investigated vertebrate evolution; laid anthrop. basis for identification through finger lines of hand. Died Feb. 1928.

WILDER, Oliver H.M., Am. nutritionist; b. Rose Lawn, Ind., s. Worden S. and Edith (Leeson) W.; B.S., Purdue U., 1928; M.S., Ohio State U., 1932, Ph.D., 1943; m. Caroline Nuckolls. With Ohio Agrl. Expt. Sta., 1928-43; chief animal feeds div. Am. Meat Inst. Found., 1943-63; dir. tech. services Nat. Renderers Assn., Des Plaines, Ill., 1963—. Mem. Am. Chem. Soc., Poultry Sci. Assn., A.A.A.S., Am. Soc. for Animal Sci., Inst. Nutrition. Contbr. numerous articles to tech. jours. Developed use animal fats, feather meal for use in animal feeds; research in plant sanitation practices, quality control, air pollution control. Home: 1395 Carol Lane, Des Plaines 60016. Office: 3150 Des Plaines Av., Des Plaines, Ill. 60018.*

WILDER, Raymond Louis, Am. mathematician; b. Palmer, Mass., Nov. 3, 1896; s. John Louis and Mary Jane (Shanley) W.; Ph.B., Brown U., 1920, Sc.M., 1921; Ph.D., U. Tex., 1923; Sc.D., Bucknell U., 1955, Brown U., 1958; m. Una Maude Greene, Jan. 1, 1921; children—Mary Jane (Mrs. Warren Hebard Jessop), Sally May (Mrs. Kermit Sparkman Watkins), Betty Ann (Mrs. Harry Clopton Dillingham), David Elliott. Asst. instr. Brown U., 1920-21; instr. U. Tex., 1921-24; asst. prof., Ohio State U., 1924-26; faculty U. Mich., Ann Arbor, 1926—, prof., 1935-47, research prof. math., 1947-67, professor emeritus, since 1967—. Member Institute for Advanced Study, Princeton, 1933-34, Guggenheim Meml. Found. fellow, 1940-41; research asso. Cal. Inst. Tech., 1949-50; vis. research prof. Fla. State U., 1961-62. Mem. Nat. Acad. Scis., A.A.A.S. (past v.p.), Am. Math. Soc. (past pres., past editor Colloquium), Assn. Symbolic Logic, Math. Assn. Am. (pres. 1965-67), Am. Anthropol. Assn. Author: Topology of Manifolds, 1949; Introduction to the Foundations of Mathematics, 1952, rev., 1965; also numerous articles. Editor: (with W. L. Ayres) Lectures in Topology, 1941. Research on topology; studied analog of Jordan curve theorem and its converse; unified set-theoretic and algebraic aspects of topology; investigated topology of manifolds; foundations of math. Home: 2427 Calle Montilla, Santa Barbara, Cal. 93105.

WILDER, Russell Morse, American physician; b. Cin., Nov. 24, 1885; B.S., U. Chgo., 1907, Ph.D., M.D., 1912; student U. Vienna (Austria), 1914; two children. Instr. anatomy and pathology U. Chgo., 1909-10; instr. internal medicine Rush Med. Coll., 1914-17; fellow Sprague Meml. Inst., 1915-17; mem. staff Mayo Clinic, 1919-29, head sect., div. medicine; asst. prof. medicine Mayo Found., 1919-22, asso. prof., 1922-29; prof. medicine, chmn. dept. U. Chgo. Med. Sch., 1929-31; prof., chief dept. Mayo Found., 1931-50, emeritus prof., 1950—; mem. staff Mayo Clinic, 1931-50. Dir. Nat. Inst. Arthritis and Metabolic Diseases, USPHS, 1951-53, cons., 1953—; mem. com. medicine NRC, 1940-46, chmn., 1940, chmn. food and nutrition bd., 1941, mem. exec. com., 1941-47; chief civilian foods requirements War Food Adminstrn., 1943; mem. biol. and nutritional study sect., office grants USPHS. Fellow A.M.A., A.C.P.; mem. Soc. Clin. Investigation, Physiol. Soc., Inst. Nutrition, Diabetes Assn. (pres. 1946). Research on nutrition, glucose metabolism, diabetes mellitus, typhus fever; differentiated between Rocky Mountain spotted fever and typhus fever, 1910. Address: Mayo Clinic, Rochester, Minn.

WILDEY, Robert LeRoy, Am. astronomer; b. Los Angeles, Aug. 22, 1934; s. Charles Dickens and Fern (Houts) W.; A.A., John Muir Coll., 1954; B.S., Cal. Inst. Tech., 1957, M.S., 1958, Ph.D., 1962; m. Diana Herberta Skolfield, June 27, 1959; children—Robert Bismarck, Wendy Carol. Research fellow in astronomy and geology Mt. Wilson and Polomar Observatories, also div. geol. scis. Cal. Inst. Tech., Pasadena, 1962-65; astrophysicist U.S. Geol. Survey Center of Astrogeology, Flagstaff, Ariz., 1965—; vis. prof. astronomy U. Cal. at Berkeley, 1966. Cons. United Electrodynamics Corp., 1962, Ford Motor Co., 1963, World Book Ency. Sci. News Service, 1963; lectr. Cal. Inst. Tech., 1964; guest lectr. Carnegie Inst. Tech., 1964, Inst. Geophysics U. Cal. at Los Angeles, 1963, Griffith Obs., 1963-64. Mem. Am. Astron. Soc., Am. Geophys. Union, Royal Astron. Soc., Internat. Astron. Union, Geol. Soc. Am., Sigma Xi. Research, publs. in stellar and galactic evolution; discovered 3 age groupings of stars in same star cluster; developed techniques in measuring thermal radiation from colder bodies in Solar system; discovered heat radiation from Jupiter sometimes increases when planet is shaded by solar eclipse from one of its satellites; detected and measured far-infrared radiation from stars; mapped heat radiation from Venus; discovered temporary hot spots on Venus and permanent hot spots on dark side of Moon; developed techniques for using lasers to measure shape of Moon. Home: 414 E. Cherry Av. Office: 601 E. Cedar Av., Flagstaff, Ariz. 86001.*

WILDHACK, William August, Am. physicist; b. Breckenridge, Colo., Sept. 24, 1908; s. Louis Amandus and Lizzie Frances (Smith) W.; B.S., U. Colo., 1931, M.S., 1932, postgrad., 1932-34; postgrad. George Washington U.; m. Martha Elizabeth Parks, Nov. 26, 1931; children—William August, Michael David. Physicist Nat. Bur. Standards, 1935—, chief missile instrument sect., 1948-50, chief office basic instrumentation, 1950-60, spl. asst. to dir., 1960-61, asso. dir., 1961-65, asso. dir. Inst. for Basic Standards, 1965—. Chmn. com. sci. equipment NRC, 1947-52; panel aircraft equipment Research and Devel. Bd., 1949-53; cons. sci. equipment for Europe, ECA, 1950; chief U. S. Meteorology Delegation to Russia, 1963. Fellow Instrument Soc. Am. (pres. 1954); mem. Am. Assn. Physics Tchrs., Am. Phys. Soc., Inst. Aero. Scis., A.A.A.S., Fedn. Am. Scientists, Sigma Xi, Tau Beta Pi, Sigma Pi Sigma. Author: Humidity and Moisture: Fundamentals and Standards, 1964. Invented and developed oxygen breathing equipment for aviation and space, pneumatic and elec. measuring instruments; research on optimum timing for parachute opening, oxygen concentration requirements, tilting of satellite orbits by atmospheric drag, nuclear energy in dense stars; contbd. to reliability and accuracy programs, fields of measurement standards and instrumentation. Home: 415 N. Oxford St., Arlington, Va. 22203. Office: Nat. Bur. Standards, Washington 20234.*

WILDMAN, Samuel Goodnow, Am. plant physiologist; b. Placerville Cal., May 26, 1912; s. Clifton Howe and Lucy (Goodnow) W.; B.A., Ore. State Coll., 1939; M.A., U. Mich., 1940, Ph.D., 1942. Asst. chemist Office Rubber Plant Investigations, U.S. Dept. Agr., Beltsville, Md., 1942-44; sr. research fellow Cal. Inst. Tech. Pasadena, 1944-50; faculty U. Cal. at Los Angeles, 1950— prof. plant physiology 1960—. Fulbright scholar, 1961; Guggenheim fellow, 1964. Mem. Am. Inst. Biol. Scis., A.A.A.S., Am. Soc. Plant Physiologists, Phytopath. Soc. Am., Japanese Soc. Plant Physiology. Research and publs. on structure of chloroplasts in relation to biochem. functions; protein and nucleic acid compositions of leaves and mechanisms of synthesis; mechanisms of tobacco mosaic virus reprodn. Home: 1024 Centinela Av., Santa Monica, Cal. Office: Dept. Bot. Scis., U. Cal. at Los Angeles 90024.*

WILDT, Rupert, astronomer; b. Munich, Germany, June 25, 1905; Ph.D., U. Berlin, 1927. Came to U. S., 1934, naturalized, 1942. Asst., Bonn. (Germany) Obs., 1928-29, Göttingen (Germany) Obs., 1930-34; Rockefeller fellow Mt. Wilson Obs., Carnegie Instn., 1935-36; vis. lectr. Harvard, 1938; mem. Inst. Advanced Study, Princeton, 1936-37; research asso. Princeton, 1937-42; asst. prof. astronomy U. Va., 1942-46; vis. prof., dir. Univ. Obs., Basle, Switzerland, 1947; asso. prof. astrophysics Yale, 1949-57, prof., 1957—, chmn. dept., 1966—; vis. prof. U. Hamburg (Germany), 1951, U. Cal., 1958, Nat. U., Mexico, 1963; emeritus prof. U. Göttingen, 1960. Fellow Royal Astron. Soc. (Eddington gold medal 1966); mem. Internat. Astron. Union, Assn. Univs. Research Astronomy (bd. dirs. 1958—, pres. 1965—). Research in theoretical astrophysics, geochemistry, stellar spectroscopy; atmospheric composition of outer planets. Address: Dept. Astronomy, Yale Univ., New Haven 06520.

WILES, Donald Roy, Canadian chemist; b. Truro, N.S., Can., Aug. 30, 1925; s. Neil D. and Hilda (Vaughan) W.; B.Sc. with honors, Mt. Allison U., 1946, B.Ed. 1947; M.Sc., McMaster U., 1950; Ph.D., Mass. Inst. Tech., 1953; m. Elisabeth Lilly, June 21, 1952; children—Anne, Karen, Peter. Research with A.C. Pappas, U. Oslo, Norway, 1953-55; research asso. in metallurgy U. B.C. (Can.) Vancouver, 1955-59; faculty Carleton U., Ottawa, Ont., 1959—, asso. prof. chemistry, 1964—. Dir. edn. Chem. Inst. Can., 1965-68. Mem. Chem. Inst. Can., Am. Chem. Soc., Chem. Soc. London, Norwegian Chem. Soc. Research and publs. on fission chemistry aqueous corrosion chemistry, nuclear recoil chemistry. Home: 93 Villa Cr., Ottawa, Ont., Can.*

WILETS, Lawrence, Am. physicist; b. Oconomowoc, Wis., Jan. 4, 1927; s. Edward and Sophia (Finger) W.; B.S., U. Wis., 1948; M.A., Pinceton, 1950, Ph.D., 1952; m. Dulcy Elaine Margoles, Dec. 21, 1947; children—Ileen Sue, Edward E., James D. Research asso. Project Matterhorn, Princeton, 1951-53; research asso. U. Cal. Radiation Lab., Livermore, 1953; NSF Postdoctoral fellow Inst. for Theoretical Physics, Copenhagen, Denmark, 1953-55; staff mem. Los Alamos Sci. Lab., 1955-58; mem. Inst. for Advanced Study, Princeton, N.J., 1957-58; faculty U. Wash., Seattle, 1958—, prof. physics, 1962—; NSF Sr. fellow Weizmann Inst. Sci., Rehovot, Israel, 1961-62. Cons. to pvt. and govt. labs. Fellow Am. Phys. Soc.; mem. Fedn. Am. Scientists, Phi Beta Kappa, Sigma Xi. Author: Theories of Nuclear Fission, 1964; also numerous articles. Research in theory nuclear structure and reactions, nuclear fission, atomic isotope shifts, atomic collisions, many body problem. Home: 8534 53rd Ct., N.E., Seattle 98115.*

WILEY, Richard Haven, Am. chemist, educator; b. Mattoon, Ill., May 10, 1913; s. John Frederick and Mary Frances (Moss) W.; A.B., U. Ill., 1934, M.S., 1935; Ph.D., U. Wis., 1937; LL.B., Temple U., 1943; m. Marybeth Signaigo, Dec. 28, 1940; children—Richard Haven, Frank Edmond. Research chemist E.I. duPont de Nemours & Co., Inc., 1937-45; faculty U. N.C., 1945-49, U. Louisville, 1949-65; prof. chemistry Hunter Coll., N.Y.C., 1965—; exec. officer doctoral program in chemistry City U. N.Y., 1965—. NSF sr. postdoctoral fellow Imperial Coll., U. London, 1957-58. Fellow A.A.A.S., N.Y. Acad. Sci.; mem. Am. Chem. Soc. (Midwest award 1965), Chem. Soc. (London), Phi Beta Kappa, Sigma Xi, Phi Lambda Upsilon Phi Eta Sigma, Phi Kappa Phi. Author: (with others) Five and Six Membered Heterocycles with Oxygen and Nitrogen, 1962; (with Paul F. Wiley) Pyrazolones, Pyrazolidones and Derivatives, 1964.Asst. editor Jour. Chem. and Engring. Data. Research, numerous publs. on synthesis and characterization or organic materials in pharmacological chemistry and devel. polymeric materials, behavior of organic materials under influence unusual radiation conditions, synthesis and characterization of new materials useful as plastic, fibers and elastomers. Home: 345 E. 56th St., N.Y.C. 10022.*

WILHELM, Harley A(lmey), Am. chemist, metallurgist; b. Ellston, Ia., Aug. 5, 1900; s. Bert C. and Anna B. (Glick) W.; A.B., Drake U., 1923, fellow in chemistry, 1924-25; Ph.D., Ia. State Coll., 1931; postdoctoral U. Mich., 1941; LL.D., Drake U., 1961; m. Orpha E. Lutton, May 29, 1923; children—Lorna (Mrs. Stewart B. Livingston), Max Gene, Myrna (Mrs. Charles W. Elliott), Gretchen. Tchr. Mapleton (Ia.) High Sch., 1923-24, Guthrie center (Ia.) High Sch., 1925-26; prof. and coach Intermountain Union Coll., Helena, Mont., 1926-27; grad. asst. Ia. State U., 1927-28, faculty, 1928—, prof. chemistry, 1945—, prof. metallurgy, 1963—. Asso. dir. Ames lab., U. S. AEC and Manhattan Dist. Engrs., 1942—. Recipient Alumni Merit award Ia. State Coll., 1949; Distinguished Service Award Drake U., 1959; William Hunt Eisenman award Am. Soc. Metals, 1962. Mem. A.A.A.S., Am. Chem. Soc. (award Ia. sect. 1959), Am. Inst. for Mining, Metall. and Petroleum Engrs., Am. Nuclear Soc., Am. Soc. Metals, Ia. Acad. Sci., Sigma Xi, Phi Beta Kappa. Research and inventions in phys. chemistry, spectrochemistry, metallurgy and chem. engring., particularly with regard to uranium, thorium, other materials and subjects related to atomic energy program of U. S. Home: 513 Hayward Av., Ames, Ia. 50010.*

WILHELMI, Alfred Ellis, Am. endocrinologist, biochemist; b. Lakewood, O., Sept. 28, 1910; s. Gustavus Carl and Helena Ethel (Looschen) W.; A.B. summa cum laude, Western Res. U., 1933; B.A. with 1st class honours in Animal Physiology (Rhodes scholar 1933), Oxford U., 1935, D.Phil. in Biochemistry, 1937; m. Jane Anne Russell, Aug. 26, 1940 (dec. Mar. 1967). Alexander Brown Coxe fellow physiol. chemistry Yale, 1939, from instr. to asso. prof., 1940-50; prof., chmn. dept. biochemistry Emory U., 1950—, Charles Howard Candler prof., 1960—. Cons. regulatory biology NSF, 1950-55; cons. endocrinology study sect. NIH, 1954-60, chmn., 1957-60; mem. Nat. Arthritis and Metabolic Diseases Council, 1962-67, Nat. Council for Humanities, 1967—. Fellow A.A.A.S., N.Y. Acad. Sci.; mem. Canadian Physiol. Soc. (hon.), Am. Soc. Biol. Chemists, Endocrine Soc. (Upjohn scholar 1960), Ga., Acad. Sci., Phi Beta Kappa, Sigma Xi, Omicron Delta Kappa. Contbr. articles in field. Research on metabolic changes in shock, effects of adrenal steroids and of growth hormone on fat metabolism, purification and chemistry of growth hormone of man and several other animal species. Home: 924 Castle Falls Dr. N.E., Atlanta 30329.*

WILHOFT, Daniel Charles, Am. zoologist; b. Newark, Nov. 16, 1930; s. Charles A. and Madeline (Fraser) W.; A.B., Rutgers U., 1956; M.A. U. Cal. at Berkeley, 1958, Ph.D., 1963; m. Dorothy Duxbury, Mar. 10, 1951; children—Daniel Charles, Linda Susan. Fulbright scholar, Australia, 1959-60; faculty Rutgers U. Newark, 1962—, asso.. prof. zoology, 1965—. Fellow Zool. Soc. London; mem. A.A.A.S., Am. Soc. Zoologists, Herpetologist League. Research, publs. on reproductive cycles of tropical and temperate zone lizards, effects of temperature on brain serotonin levels in lizards, thermoregulation in lizards, effects of thyroid gland and temperature on metabolism in lizards. Home: 240 Walton Av., South Orange, N.J. 07079. Office: 195 University Av., Newark 07102.*

WILKE, Charles Robert, Am. chem. engr.; b. Dayton, O., Feb. 4, 1917; s. Otto A. and Stella (Dodge) W.; B.S. in Chem. Engring., U. Dayton, 1940; M.S. in Phys. Chemistry, State Coll. Wash., 1942; Ph.D., U. Wis., 1944; m. Bernice Arnett, June 19, 1946. Chem. engr. Union Oil Co., 1944-45; instr. State Coll. Wash., 1945-46; faculty U. Cal. at Berkeley, 1946—, prof. chem. engring., 1953—, chmn. dept. chem. engring., 1953-63, chem. engr. Lawrence Radiation Lab., 1950—, chem. engr. Agr. Expt. Sta., Forest Products Research Lab., 1961—; vis. prof. Mass. Inst. Tech., 1960. Mem. Cal. Bd. Registration for Civil and Profl. Engrs., 1964—, v.p., 1966-67; indsl. cons., 1948—. Mem. Am. Inst. Chem. Engrs. (Alan P. Colburn award 1951, lectr. 1957, William H. Walker award, 1965), Am. Chem. Soc., N.Y. Acad. Sci., Am. Soc. Engring. Edn., Am. Soc. for Advancement Sci. Research, publs on methods for predicting diffuse coefficients in gases and liquid, viscosities and thermal conductivities gases, correlation flow capacities packed extraction columns, mass transfer studies in packed beds, tubes, electrolytic cells, co-current pipe flow, zone melting and thermal diffusion, sea water evaporation by direct contact heat transfer, sulfate reduction by bacteria, continuous culture of microorganisms, solvent seasoning wood. Home: 1327 Contra Costa Dr., El Cerrito, Cal. 94532. Office: Dept. Chem. Engring., U. Cal., Berkeley, Cal. 94720.*

WILKE, Sir David Percival Dalbreck, Scottish surgeon; b. Kirriemuir, Scotland, 1882; s. David Wilke; ed. Edinburgh (Scotland) U., Bonn (Germany) U., Bern (Switzerland) U., Vienna (Austria) U.; M.D., M.Ch.; m. Charlotte Erskine Middleton, 1911. Surgeon, Royal Infirmary, Edinburgh; dir. surgery Edinburgh Municipal Hosps.; cons. surgeon Leith (Scotland) Hosp., Falkirk (Scotland) Infirmary. Mem. Army Med. Adv. Bd., sci. adv. com. Scottish Bd. Health. Recipient Liston Victoria Jubilee prize, 1918. Fellow Royal Am. colls. surgeons, Royal Soc. Edinburgh; mem. French Acad. Surgery (corr.), Med. Research Council, Assn. Surgeons Gt. Britain and Ireland (pres. 1936-37). Research, publs. on abdominal surgery, acute intestinal obstruction, duodenal ulcer, cancer of rectum, diseases of appendix, biliary passages, spleen. Died Aug. 28, 1938.

WILKENING, Marvin Hubert, Am. physicist; b. Oak Ridge, Mo., Mar. 13, 1918; s. Theodore C. and Myrtle (Lang) W.; B.S. in Edn., S.E. Mo. State Coll., 1939; M.S. in Physics, Ill. Inst. Tech., 1943, Ph.D., 1949; m. Ruby A. Barks, Nov. 14, 1942; children—Laurel Lynn, J(ames) Wesley. Tchr. sci. Jackson (Mo.) High Sch., 1939-41; physicist Manhattan Project, Chgo., 1942-43, Oak Ridge, 1943-44, Richland, Wash., 1944-45, Los Alamos, 1945; instr Ill. Inst. Tech., Chgo., 1946-48; faculty N.M. Inst. Mining and Tech., Socorro, 1948—, prof. physics and geophysics, 1952—, dean of graduate studies, 1953—. Mem. Gov.'s Sci. Adv. Com., N.M., 1963—. Recipient Certificate of Service OSRD, 1945, Certificate of Service on Manhattan Project War Dept., 1945. Fellow A.A.A.S.; mem. Am. Phys. Soc., Am. Assn. Physics Tchrs., Am. Geophys. Union, Am. Meteorol. Soc., Am. Assn. U. Profs., Sigma Xi. Contbr. to Natural Radiation Environment, 1964. Constrn. of neutron monitors for nuclear reactors, including first nuclear reactor, Chgo., 1942; study of natural radioactivity in atmosphere; use of radon and dau. products to study convective and elec. environments of thunderstorms.*

WILKES, Alfred, zoologist; b. Birmingham, Eng., Jan. 7, 1909; s. Alfred Ernest and Mary (Rowe) W.; B.S.A., U. Toronto, 1935, M.S. in Agr., 1938, Ph.D., 1940; m. Reba Price Reach, Oct. 22, 1936; 1 dau., Willo Reba. Research officer Dominion Parasite Lab., Belleville, Ont., Can., 1935-50,, officer in charge, 1950-55; research scientist Entomology Research Inst., Ottawa, Ont., 1955—. Recipient Ramsay Wright award U. Toronto, 1939-40. Fellow A.A.A.S.; mem. Entomol. Soc. Ont. (dir. 1957—), Entomol. Soc. Can. (dir. 1960—), Entomol. Soc. Am., Gentics Soc. Can. (past sec.-treas.), Am. Genetic Assn., Que. Soc. Protection Plants. Research, publs. on sex determination and sex ratios in hymenopterous insects; developed selective breeding methods for prodn. parasites used in biol control pests. Home: 529 Fraser Av. Office: Entomology Research Inst., Central Exptl. Farm, Ottawa, Ont., Can.*

WILKIE, David, Brit. geneticist; b. Paisley, Scotland, Aug. 6, 1923; s. John Hay and Margaret (Murphy) W.; B.Sc., U. Glasgow, 1950, Ph.D., 1953; m. Moreen McGhee, Sept. 1, 1948; children—Alice Steen, Louise Margaret. Research asst. dept. genetics U. Glasgow, 1950-52, asst. lectr. dept. botany, 1952-54; lectr. dept. botany. U. Coll. London, 1954-65, reader dept. botany, 1965—; research fellow Rockefeller Found. dept., genetics U. Wash., 1959-60; vis. prof. dept. biochemistry Monash U., Australia, 1966-67; cons. Wm. Grant & Sons Ltd. Distillers, Scotland, 1961—. Mem. Inst. Biology, Brit. Genetical Soc. Author: Cytoplasm in Heredity, 1964; also articles. Discovered 1st case of genetic self-incompatibility in ferns; described a case of cytoplasmic inheritance in a fungus and a fern; developed a system for using monochromatic ultraviolet light in induction studies recombination, respiratory deficiency in yeast; pub. several papers on genetic and biochem. aspects of respiratory deficiency. Home: 7 Hill Close, Purley, Surrey, Eng. Office: U. Coll., Gower St., London W.C.1, Eng.*

WILKINS, Sir (George) Hubert, explorer; b. Mt. Bryan East, S. Australia, Oct. 31, 1888; s. H. Wilkins; ed. State Sch., Adelaide Sch. Mines; m. Suzanne Bennett, 1929. Photog. corr. with Turkish troops in Balkan War, 1912-13; 2d in command Stefansson's Party Canadian Arctic Expdn., 1913-17; served in war, 1917; navigator Blackburn Kangaroo Aeroplane, Eng.-Australia Flight, 1919; 2d in command Brit. Imperial Antarctic Expdn., 1920-21; naturalist Shackleton-Rowett Quest Expdn., 1921-22; leader Wilkins Australia and Islands Expdn. for Brit. Mus., 1923-25; comdr. Detroit Arctic Expdn., 1926-27, Wilkins Detroit-News Arctic Expdn., 1928, Wilkins Hearst Antarctic Expdn., 1928-29, Nautilus Arctic Submarine Expdn., 1931; mgr. Ellsworth Trans-Antarctic Expdn., 1933-39; cons. mil. planning div. U. S. Army, 1942-52, geographer research and devel. command, dept. def., from 1953. Recipient Patron's medal Royal Geog. Soc., 1928, numerous other medals. Fellow Royal Geog. Soc., Royal Meteorol. Soc.; hon. mem. many geog. socs. Author: Flying the Arctic, 1928; Undiscovered Australia, 1928; Under the North Pole, 1931; Thoughts Through Space, 1952; also naturalist's reports Proc. Linnean Soc., Ibis, articles in mags. and press. Died Dec. 1, 1958.

WILKINS, John, English natural philosopher; b. Eng., 1614; s. Walter and (Dod) W.; B.A., Magdalen Hall, Oxford (Eng.) U., 1631, M.A., 1634, B.D., 1648; D.D., Cambridge (Eng.) U., 1649; m. Oliver Cromwell's sister. Vicar of Fawsley, 1637; pvt. chaplain to prince palatine Charles Lewis, nephew of Charles I; adhered to parliamentary side during civil war, took covenant; warden Wadham Coll., Oxford U., 1648-59; master Trinity Coll., Cambridge U., from 1659 to Restoration; prebendary York, Eng., from 1660; vicar St. Lawrence Jewry, London, Eng., 1662; dean of Ripon, from 1663; prebendary, precentor of Exeter, from 1667; prebendary St. Paul's Cathedral, London, 1668; bishop of Chester, Eng., from 1668. Fellow Royal Soc. (co-founder, sec. 1662), 1663. Author: The Discovery of a World in the Moone, 1638; A Discourse tending to prove that Tis probable our Earth is one of the Planets, 1640; Mathematical Magick, 1648; An Essay towards a real character and a Philosophical Language, 1668. Died Nov. 19, 1672.

WILKINS, Maurice Hugh Frederick, biophysicist; b. Pongaroa, New Zealand, Dec. 15, 1916; s. Edgar Henry and Eveline (Whittaker) W.; Ph.D., St. John's Coll., Cambridge, 1940; m. Patricia Ann Chidgey, Mar. 12, 1959; children—Sarah Fenella, George Hugh, Emily Lucy Una, William Henry. Research with Manhattan Project U. Cal., Berkeley, 1944; lectr. St. Andrews U., 1945; faculty Kings Coll., London, Eng., 1946—, dep. dir. Med. Research Council, 1955—, prof. molecular biology, 1962—. Decorated Comdr. Brit. Empire; recipient Albert Lasker award Am. Pub. Health Assn., 1960, Nobel prize for physiology and medicine (with F. H. C. Crick and J. D. Watson), 1962. Fellow Royal Soc., 1959; mem. Brit. Biophys. Soc. (past chmn.). Research, publs. on X-ray diffraction analysis of structure of DNA; devel. of electron trap theory of phosphorescence and thermo-luminescence; light microscopy techniques for cyto-chemical research; including use of interference microscope for dry mass determination in cells. Home: 30 St. John's Park, London, S.E.3. Office: 26-29 Drury Lane, London, W.C.2, Eng.*

WILKINS, Robert Wallace, Am. physician; b. Chattanooga, Tenn., Dec. 4, 1906; s. James Drewry and Kate (Fristoe) W.; A.B., U. N.C., 1928; M.D., Harvard, 1933; m. Margaret Gayden Morrill, Aug. 30, 1941; children—Margaret Dodge, Mary Gayden, Robert Wallace. Staff Boston City Hosp., Harvard Med. Sch., 1933-37, Nat. Hosp., London, Eng., 1937-38, Johns Hopkins Hosp. and Med. Sch., Balt., 1938-40; asst. prof. medicine Boston U., 1940-42, asso. prof., 1942-55, prof., 1955—, chmn. dept. medicine, 1960—; staff Evans Meml. Hosp., 1940-55, asso. dir., 1955-60, dir., 1960—; vis. physician Massachusetts Memorial Hospitals, 1940-55, asso. physician-in-chief, 1955-60, physician-in-chief, 1960—. Secretary of subcommittee on acceleration, com. on aviation med. NRC, 1942-46. Recipient Lasker-Am. Pub. Health Assn. Award. Fellow A.C.P., A.M.A., N.Y. Acad. Scis.; mem. Am. Soc. Clin. Investigation, Soc. Exptl. Biology and Medicine, Assn. Am. Physicians, Am. (former pres.; chmn. central com. med. and community programs; dir.), Mass. (past pres.) heart assns., Am. Clin. and Climatol. Assn., A.A.A.S., Phi Beta Kappa. Research on cardiovascular physiology, arteroscleresis, peripheral circulation, hypertension; introduced to U. S. tranquilizing drug, reserpine, 1952. Home: 299 High St., Newburyport, Mass. Office: 65 E. Newton St., Boston 18.

WILKINS, Thomas Russell, physicist; b. Toronto, Ont., Can., June 6, 1891; s. Thomas and Annie (Cornell) W.; B.A., McMaster U., 1912; Ph.D., U. Chgo., 1921; postgrad. Cambridge (Eng.), 1925-26; m. Olive Cross, June 17, 1913. Science master Woodstock Coll. (Can.), 1913-14; instr. physics U. Chgo., 1916-17; prof. physics Brandon (Can.) Coll., 1918-25, U. Rochester, 1925—; dir. Inst. of Optics, Rochester, 1928—. Editor (with others) Orientation Course in Natural Science, 1938. Secured photographic recordings of cosmic rays and of successive distintegrations of radium atoms. Died Dec. 10, 1940.

WILKINSON, Denys Haigh, English physicist; b. Leeds, Eng., Sept. 5, 1922; s. Charles and Hilda (Haigh) W.; B.A., Cambridge (Eng.) U., 1943, M.A., Ph.D., 1947, Sc.D., 1961; hon. D.Sc., U. Sask., Saskatoon, Can., 1964; m. Christiane Andrée Clavier, June 25, 1947; children—Gabrielle-Anne, Judith Penelope, Erica Mary. With Brit. and Canadian atomic energy projects, 1943-47; faculty U. Cambridge, 1947-57, reader, 1956-57; prof exptl. physics, head nuclear physics lab. U. Oxford (Eng.), 1957—. Fellow, Jesus Coll., Cambridge, 1945-59, hon. fellow, 1960. Rutherford Meml. lectr. Brit. Phys. Soc., 1962. Recipient Holweck medal Brit. and French phys. socs., 1957, Hughes medal Royal Soc., 1965. Fellow Royal Soc., 1956; mem. Brit. Ornithologists Union. Author: Ionization Chambers and Counters, 1950; also numerous articles. Exptl. and theoretical research on nuclear structure and on mechanism of bird navigation.*

WILKINSON, John, English inventor; b. Clifton, Cumberland, Eng., 1728; s. Isaac Wilkinson; ed. acad. of Dr. Caleb Rotherham, Kendal, Eng.; m. Anne Mawdsley, 1755; m. 2d, Miss Lee; 3 illegitimate sons. Established blast furnace, Bilston, Staffordshire, Eng., circa 1748; operated cylinder boring plant, Bersham, Eng., from circa 1756; opened large bar rolling mill, 1812, mfr. wrought iron, Bosely, Eng. Fellow Royal Soc., 1764. Inventor 1st blast furnace, 1748; improvements cylinder boring beneficial to Watt in bldg. Soho engines; erected 1st large working steam-engine in France (asso. with Paris waterworks); patentee lead pipe, 1750, turning lathe, 1778; built 1st ship of iron (70 foot barge). Died July 14, 1808.

WILKINSON, Joseph Ridley, Am. chemist; b. Palatka, Fla., Sept. 24, 1917; s. Robert Ridley and Lucille (Kennerly) W.; B.S., The Citadel, 1941; M.S., U. Ga., 1949; Ph.D., Fla. State U., 1955; m. Mary Connally Wallis, Oct. 26, 1941; children—Mary Leila, Sally Ann. Mem. faculty The Citadel, Charleston, S.C., 1946—, prof. chemistry, 1964—. Cons., U.S. Army Chem. Corps., 1963-66. Mem. Am. Chem. Soc., Sigma Xi, Sigma Pi Sigma. Research and publs. primarily in fields of analytical and nuclear chemistry; determined decay scheme of chromium - 48; discoverer nuclide manganese - 53. Address: The Citadel, Charleston, S.C. 29409.*

WILKINSON, Walter D., Am. metallurgist; b. Chgo., Oct. 22, 1908; s. Walter D. and Mary (Stewart) W.; B.S., Cal. Inst. Tech., 1930; M.S., Mont. Sch. Mines, 1932; Dr.Ing., Freiberger Bergakademie, Germany, 1934; m. Grace Csepke, June 10, 1940; 1 dau., Beverly. With Linde Air Products Co., 1934-39, Electro Metall. Co., 1939-42, Hindustan Aircraft, Ltd., Bangalore, India, 1943-44, U.S. Vanadium Corp., Cal., Nev., 1945-47; sr. metallurgist Argonne (Ill.) Nat. Lab., 1947—, prof. Internat. Inst. Nuclear Sci., 1955-63, metallurgy div. editor, 1936—; vis. prof. nuclear reactor materials U. Wis., 1965-66. Mem. Am. Chem. Soc., Am. Inst. Mining and Metall. Engrs., Am. Soc. Metals. Author: (with W. Murphy) Nuclear Reactor Metallurgy, 1958; Uranium Metallurgy, 2 vols., 1962-63; also articles. Editor, contbg. author Extractive and Physical Metallurgy of Plutonium, 1960. Research in corrosion behavior metals in liquid state when contained in solid metal or ceramic containers; determined some properties U-Pu alloys. Patentee Uranium- Gallium solid-liquid dispersion nuclear fuel, process for making oxide fuel. Home: 609 S. 7th Av., Maywood, Ill. 60153. Office: 6700 S. Cass Av., Argonne, Ill. 60439.*

WILLAN, Robert, English physician, dermatologist; b. Yorkshire, Eng., Nov. 12, 1757; s. Robert William Willan; M.D., U. Edinburgh (Scotland), 1780. Practiced medicine Darlington, London (both Eng.); physician to Pub. Dispensary, London, 1783-1803. Recipient Fothergillian medal Med. Soc. London, 1790. Fellow Royal Soc., 1809, Soc. Antiquaries; licentiate Royal Coll. Physicians. Author: Observations on the Sulphur Waters of Croft, 1782; Description and Treatment of Cutaneous Diseases, Part I, 1798, Part II, 1808; Reports on the Diseases of London, 1801; First English physician to arrange diseases of skin in clear, intelligible manner, and to fix their nomenclature on satisfactory and classical basis. Died Apr. 12, 1812.

WILLARD, Bradford, Am. geologist; b. N.Y.C., May 17, 1894; s. Gates and Katrina (Haff) W.; B.A. with honors, Lehigh U., 1921; A.M., Harvard, 1922, Ph.D., 1923; m. Elise Krassa, June 22, 1923; children—Gates, Elizabeth Bradford (Mrs. Wesley A. Rusnell). Mem. faculty Brown U., 1923-30; staff Pa. Geol. Survey, 1930-39; faculty Lehigh U., Bethlehem, Pa., 1939-58, prof., head dept. geol. scis., 1938-58; emeritus, 1958—. Mem. Pa. Acad. Sci. (pres. 1945-46), Geol. Soc. Am., A.A.A.S., Sigma Xi. Author: (with A.B. Cleaves, F.M. Swartz) Devonian of Pennsylvania, 1939; (with others) Geology and Mineral Resources, Bucks County, Pennsylvania, 1959; Harvey Bassler Collection of Peruvian Fossils, 1966. Home: 145 Aurora St., Bethlehem, Pa. 18018.*

WILLARD, Harvey Bradford, Am. physicist; b. Worcester, Mass., Aug. 9, 1925; s. George Howard and Virginia (Wohlbruck) W.; S.B., Mass. Inst. Tech., 1948, Ph.D., 1950; m. Isabella Rallis, Nov. 2, 1950; children—Karl Philip, Karen Elizabeth. Physicist, Oak Ridge Nat. Lab., 1950-51, group leader 5.5 Mv Van de Graaff, 1951-57, co-dir. High Voltage Lab., 1957-63, asso. dir. physics div., 1963-67; prof. physics U. Tenn., 1963-67; prof., chmn. physics dept. Case Western Res. U., 1967—. Nuclear cross sect. adv. group AEC, 1961—. Fellow Am. Phys. Soc.; mem. A.A.A.S., Tenn. Acad. Sci., Research Soc. Am., Sigma Xi. Contbg. author: Fast Neutron Physics, 1963; Progress in Fast Neutron Physics, 1963. Research in nuclear structure, nuclear reactions, polarizations of neutrons, beta decay, light nuclei. Home: 15 N. Strawberry Lane, Moreland Hills, O. 44022. Office: Physics Dept., Case Western Res. U., Cleve. 44106.*

WILLARD, Hobart Hurd, Am. chemist; b. Erie, Pa., June 3, 1881; s. James Richard and Julia (Hobart) W.; A.B., U. Mich., 1903, A.M., 1905; Ph.D., Harvard, 1909; m. Margaret Sheppard, June 27, 1927; children—Ann Hobart (Mrs. Robert Roy Rorfhage), Nancy Margaret (Mrs. Eric Lindbloom). Faculty chemistry dept. U. Mich., Ann Arbor, 1909—, prof., 1923-51, emeritus, 1951—. Mem. Am. Chem. Soc., Electrochem. Soc., Am. Inst. Chemists, N.Y. Acad. Sci., Sigma Xi, Alpha Chi Sigma. Author: Instrumental Methods of Analysis, 4th edit., 1965; (with others) Elements of Quantitative Analysis, 4th edit., 1956; (with others) A Short Course in Quantitative Analysis, 1957; also numerous articles. Research in methods for separation and determination of many elements; precipitation from homogeneous solution. Address: 1926 Norway Rd., Ann Arbor, Mich. 48104.*

WILLARD, John Ela, Am. chemist; b. Oak Park, Ill., Oct. 31, 1908; s. Wallace Watson and Mary (Ela) W.; student Pomona Coll., 1926-27, Carleton Coll., 1927-28; B.S., Harvard, 1930; Ph.D., U. Wis., 1935; m. Adelaide Ela, June 12, 1937; children—Ann (Mrs. Kirkor Bozdogan), Mark, David, Robert. Faculty, Avon (Conn.) Old Farms Sch., 1930-32, Haverford (Pa.) Coll., 1935-37; faculty U. Wis., Madison, 1937-42, 46—, prof. chemistry, 1948—, dean Grad. Sch., 1958-63, Vilas prof., 1963—; asst. sect. chief Manhattan Project, U. Chgo., 1942-44, div. dir., 1945-46, area supr. Hanford Engr. Works (Wash.), 1944-45. Chmn. exec. bd. Council Participating Instns. Argonne Nat. Lab., 1952-53; mem. adv. coms. Argonne Nat. Lab., AEC. Mem. Am. Chem. Soc. (chmn. div. phys. and inorganic chemistry 1957 award for Nuclear Applications in Chemistry 1959), Am. Phys. Soc., Radiation Research Soc., A.A.A.S., Am. Assn. U. Profs., Sigma Xi, Phi Beta Kappa. Asso. editor Chem. Revs., 1955-58, Radiation Research, 1965-68. Research, publs. on chem. effects of nuclear transformations, radiation chemistry, photochemistry; mechanism of chemical reactions of highly energetic atoms, especially in process of neutron-capture; reactions and nature of trapped intermediates. Home: 2306 Hollister Av., Madison, Wis. 53705.

WILLARD, John Wesley, Am. chemist; b. Davenport, Ia., June 23, 1907; s. Herbert Hodson and Margaret Rose W.; B.S., Purdue U., 1929, M.S., Ph.D., 1942; m. Gwenethe E. Drake, Nov. 29, 1928; children—John Wesley, William J. Chemist, Anderson-Pricharol Oil Corp., Oklahoma City, 1930-32, Cleaner Corp. Am., 1932-34; asst. prof. Va. Mil. Inst., 1940-46; prof. chemistry S.D. Sch. Mines and Tech., Rapid City, 1946—. Cons. to armed services, 1940—. Fellow Am. Inst. Chemists; mem. Am. Chem. Soc., Sigma Xi, Phi Lambda Upsilon, Theta Tau, Sigma Pi Sigma. Author: (with Middleton) Semimicro Qualitative Analysis, 1940; (with others) General Chemistry, 1945, Systematic General Chemistry, 1947. Research and publs. on separation of carbon, carbonaceous materials from surfaces, oil shales, coals, bituminous sands, new menomic to show electron configuration of atomic particles, new periodic table. Patentee on removing paraffin sludge from oil well. Home: 1314 S. 7th St., Rapid City, S.D. 57701.*

WILLARD, Mary Louisa, Am. chemist; b. State College, Pa., May 19, 1898; d. Joseph Moody and Henrietta (Nunn) Willard; B.S., Pa. State U., 1921, M.S., 1923; Ph.D., Cornell U., 1927. Mem. faculty Pa. State U., University Park, 1920—, prof. chemistry, 1948-64, prof. emeritus, cons. chem. microscopy, 1964—. Fellow Am. Inst. Chemists (Honor award 1955); mem. A.A.A.S., Am. Crystallographic Soc., Am. Microscopic Soc., Am. Microchem. Soc., Am. Forensic Soc. Author: Introduction to Chemical Microscopy, 1961; (with Nevin Grenninger) Photomicrogrpahy, 1963; Drug Microscopy, 1960; Textile Microscopy, 1962. Asso. editor Mikrochemie, 1942—. Research and publs. on advancement of field of chem. microscopy and allied fields. Home: 363 Ridge Av., State Coll., Pa. 16801.*

WILLCOX, Richard R(obert) English venerologist; b. London, Eng., May 10, 1912; s. Robert Day and Winifred George (Elderkin) W.; MRCS (Eng.), LRCP (London); 1936; M.B., B.S., U. London, 1937, M.D., 1951; m. Alma Caffery, Sept. 28, 1938; children—Jane (Mrs. Charles Winzer Madge), Jeremy. Obstetric officer, house physician, med. supt. St. Mary's Hosp., 1936-38; cons. venerologist; St. Mary's Hosp., 1946—, King Edward VII Hosp., Windsor, Eng., 1947—, Govt. of So. Rhodesia, 1949. Periodic cons. in veneral diseases and treponematoses WHO, 1949—, mem. expert panel on venereal infections and treponematoses, 1959—; chmn. vd com. N.W. Met. Regional Hosp. Bd., 1965—. Recipient Cheadle Gold medal St. Mary's Hosp., 1937. Mem. Med. Soc. for Study Venereal Diseases (pres. 1965-67), Inst. Technicians in Venerealogy (pres. 1965-67), Brit. Med. Assn. (chmn. venerologists group 1965-67), Société Belge et Méd. Tropicale, Société de Derm. et Vener. Iran. Author: Textbook of Venereal Diseases, 1950; Progress in Venereology, 1952; (with T. Guthe) Treonematoses: a World Problem, 1954; Textbook of Venereal Diseases and Treponematoses, 1964; also numerous articles. Research on etiology and treatment non-genococcal urethritis, venereal diseases in So. Rhodesia and other countries, discovered njovera, an epidemic non-venereal syphilis of So. Rhodesia. Home: Tideway, Lonsdale Rd., Barnes, London S.W. 13. Office: 142 Harley St., London W.I., Eng.*

WILLDENOW, Carl Ludwig, German naturalist; b. Berlin, 1765; studied medicine, Halle, Germany; practiced medicine, Berlin; became prof. natural history, Berlin, 1798, also supt. botanic garden. Author: Species plantarum (incomplete); Grundriss der Kräuter-Kunde; Prodromus flora berolinenis; Catalogue of Butterflies in the Mark of Brandenburg; other treatises. Held that plant geography reveals the history of plants; observed similarities between trees and shrubs in no. areas of Asia and Am., between shrubs in Cape of Good Hope and Australia; studied protective mechanisms of plants; compared variations in wild and cultivated plants; defined 5 chief floristic areas of Europe, indicated their dispersal centers and boundaries, also observed that intermingling at boundaries enriches variety of forms there; 1st to appreciate essential unity of Hydropterideae (Filicineae). Died 1812.

WILLEMET, Pierre Remi, French naturalist; b. Norroy-sur-Moselle, 1735; lived at Nancy, France. Author: Flora of Lorraine, 3 vols., 1805; other works. Studied behavior of etiolated plants. Died 1807.

WILLENBROCK, Frederick Karl, Am. elec. engr.; b. N.Y.C., July 19, 1920; s. Berthold Daniel and Anna Marie (Koniger) W.; S.B., Brown U., 1942; A.M., Harvard, 1947, Ph.D., 1950; m. Mildred G. White, Dec. 20, 1944. Research fellow, lectr. Harvard, 1950-55, asso. dir. labs. div. engring, and applied physics, 1955-60, dir. labs., 1960-63, asso. dean div. engring. and applied physics, 1960-67; provost, engring. and applied scis. State U. N.Y. at Buffalo, 1967—. Cons. to research dir. Office Naval Research, 1961—. Recipient Distinguished Engring. Service award Brown U., 1962. Fellow I.E.E.E. (editor publs. 1965-66, v.p. publs. activities 1966—); mem. Am. Phys. Soc., Am. Soc. for Engring. Edn., Sigma Xi, Tau Beta Pi. Research, publs. in electronics and engring. edn. Home: 57 Tillinghast Pl., Buffalo 14216.*

WILLETT, Hurd Curtis, Am. meteorologist; b. Providence, Jan. 1, 1903; s. Allan Herbert and Mable (Hurd) W.; B.S., Princeton, 1924; Ph.D., George Washington U., 1929; m. Cynthia Jump, Sept. 5, 1936; children—Joyce (Mrs. Paul M. Bradley), Allan Brock, Steven Curtis; m. 2d, Dorothy Lloyd, Oct. 28, 1949; children—David Hurd, Dorothy Emily. With U. S. Weather Bur., 1924-28; faculty Mass. Inst. Tech., Cambridge, 1929—, prof. meteorology, 1945—. Expert cons. weather div. USAF, 1942-45. Mem. Am. Meteorol. Soc., A.A.A.S., Am. Geophys. Union, Assn. Am. Geographers, Royal Meteorol. Soc. (London), Am. Acad. Arts and Scis. Author: Descriptive Meteorology, 1944, 2d edit., 1959; also numerous articles. Research on climatic fluctuations, solar-weather relationships, long-range forecasting. Home: 74 Harvard Rd., Littleton, Mass. 01460. Office: 77 Massachusetts Av. Cambridge, Mass. 02139.*

WILLETT, Richard Wright, New Zealand geologist; b. Dunedin, New Zealand, Oct. 31, 1912; s. Wright and Evelyn (Michie) W.; D.Sc. (hon.) U. Otago; m. Kathleen Gilbert Fairbairn, Dec. 28, 1939; children—Shirley, Susan, Keith. Staff, New Zealand Geol. Survey, 1936-51, geologist, 1945-51, sr. geologist 1954-56, dir., Lower Hutt, 1956—; geol. liaison officer Commonwealth of Britain, London, 1951-54; Recipient Silver medal Royal Soc. Arts, London, 1954. Fellow Royal Soc. New Zealand (v.p.); mem. Australasian Inst. Mining and Metallurgy (v.p. New Zealand br.), Geol. Soc. New Zealand, New Zealland Assn. Scientists. Author book (with B. J. Gamier); 1948; also articles. Research on econ. geology of New Zealand; (with H. W. Wellman) structure of Alpine fault; Pleistocene studies and effect on biology. Home: 10 Walden, Wellington, New Zealand. Office:

New Zealand Geol. Survey, Andrews Av., Lower Hutt, New Zealand.*

WILLEY, Gordon Randolph, Am. anthropologist; b. Chariton, Ia., Mar. 7, 1913; s. Frank and Agnes (Wilson) W.; A.B., U. Ariz., 1935, M.A., 1936; Ph.D., Columbia U., 1942; m. Katherine Whaley, Sept. 17, 1938; children—Alexandra (Mrs. Peter Guralnick), Jeanne Winston. Archaeol. asst. Nat. Park Service, Macon, Ga., 1936-38; archaeologist research La. State U., 1938-39; archaeologist Columbia Grad. Sch., 1939-42, instr. Columbia U., 1942-43; anthropologist Smithsonian Instn., 1943-50; Bowditch prof. archaeology Harvard, 1950—; vis. prof. Cambridge (Eng.) U., 1962-63. Recipient Fiking Fund medalist, 1954. Mem. Nat. Acad. Scis., Am. Acad. Arts and Scis., Am. Anthropol. Assn. (pres. 1961), Soc. Am. Archaeology (pres. 1967). Author: Excavations in the Chancay Valley, Peru, 1943; Archaeology of the Florida Gulf Coast, 1949; Prehistoric Settlement Patterns in the Viru Valley, Peru, 1953; The Monagrillo Culture of Panama, 1954; (with Philip Phillips) Method and Theory in American Archaeology, 1958; Prehistoric Maya Settlement Patterns in the Belize Valley, 1965; An Introduction to American Archaeology, 1966. Editor: Prehistoric Settlement Patterns of the New World, 1956. Research and numerous publs. in settlement patterns of prehistoric peoples throwing light on rise of urban life in Precolumbian America; general syntheses of new world Precolumbian cultures, their history and evolution. Address: Peabody Museum, Harvard U., Cambridge, Mass. 02138

WILLIAM IV, LANDGRAVE OF HESSE-CASSEL, astronomer; b. 1532; s. Philip the Magnanimous, Landgrave of Hesse and Christina; a son, Maurice the Learned. Became landgrave after death of father, 1567. Built obs. at Cassel which was 1st with revolving roof; observed (with Christian Rothmann, Yoost Bürgi) fixed stars; used clock to record time of observations, to measure motion of celestial sphere; compiled star catalogue. Died 1592.

WILLIAM OF CONCHES (or Guilelmus de Conchis, William Shelley Anglice), French philosopher; b. Conches, nr. Evreux, Normandy, circa 1080; may have taught in Paris; tchr., Chartres, France, about 20 years; pupils included John of Salisbury; entered service of Geoffrey the Fair, count of Anjou, possibly tutor to his son (the future Henry II). Author: De philosophia mundi, rev. edit., Dragmaticon, circa 1144-50; De honesto et utili (or Summa moralium philosophorum), before 1150; commentary on Plato's Timaeus; commentary on Boetius' Consolation (popular for 2 centuries, pdagiarized by Nicholas Trivet of Norwich). Developed an atomic theory combined with theory of 4 elements and neo-Platonic ideas; noticed that density and temperature of air decrease with increasing altitude; attempted to explain gen. circulation of air and to relate it to oceanic circulation; distinguished astrology as the study of appearances from astronomy as a discipline dealing with realities and laws.

WILLIAM OF LUNIS (or Wilhelmus de Lunis apud Neapolim), Italian translator; b. Italy; flourished 13th century; translator book on algebra, also some of Ibn Rushd's commentaries on Aristotelian logic and on Porphyry's interpretation of it.

WILLIAM OF MOERBEKE (or Guilielmus Moerbecanus), neo-Platonic philosopher, translator; b. Moerbeke, E. Flanders, circa 1215; Dominican monk; penitentiary and chaplain to many popers; participant council of Lyons, 1273-74; archbishop of Cornith, 1378 86; in Vitorbo and Orvieto most of 1266-78. Author: De arte et scientia geomantiae. Translator or supr. of transl. (from Greek into Latin): Prognostics (Hippocrates); Politics (made at request of St. Thomas Aquinas, introduced a work unknown to western Christians and Muslims, marked beginning of new devel. of social philosophy in Christian West), 1260, History and the Generation of Animals, 1260, Rhetoric, 1281, 4th book of Meteorology, Economics, 11th book of Metaphysics (all Aristotle); De iis quae in humido vehuntur (Archimedes), 1269; Catoptrics (Hero); De alimentis (Galen), 1277; Commentaries on the Meteorology, 1260, on De sensu et sensibili (Alexander of Aphrodisias), 1260; Commentary on the Prior analytics and on the De anima (Themistios); Theorolgical Introduction (at St. Thomas' request), 1268, Commentaries on the Timaeos and the Parmenides (Proclos); Commentary on the 3d book of De anima (Philoponos); Commentaries on the De coelo et mundo (Simplicious), 1271; on the Catagories, 1266 (both Simplicious). Died 1286.

WILLIAM OF OCCAM (or Ockham), philosopher, b. Ockham, Surrey, Eng., circa 1285; studied at Oxford (said to have been pupil of Duns Scotus), B.D. U. Paris, D.D.; Franciscan. Lectured on theology, philosophy, Oxford, 1315-19; went to Avignon to answer charge of heresy, 1324, not condemned; involved in controversy on evangelical property, fled to Bavaria, 1328; excommunicated for defense of Emperor Louis of Bavaria in his struggle with pope on relationship between secular and ecclesiastical powers; held seal of Franciscan order after death of Michael of Cesena, 1342-49; called himself vicar of the order; attempted reconciliation with the pope, 1347; uncertain whether he obtained absolution. Author: Quaestiones in quatuor libros sententiarum; Quaestiones in octo libros physicorum; Quodlibeta septem; Centilogium theologicum; Opus nonaginta Dierum, circa 1330; De Dogmatibus Papae Johannis XXII; Dialogus; Epistola ad Fratres minores in capitulo apud Assisium congregatos, written 1334; Compendum errorum papae, written before 1338; Defensorium contra Johannem papam; Tractatus ostendens quod Benedictus papa XII nonnullas Johannis XXII haereses amplexus est et defendit, 7 books; Tractatus de potestate imperiali; Octo quaestiones super potestate ac dignitate papali, between 1339 and 1342; Tractatus de jurisdictione imperatoris in causis matrimonialibus, 1342; Tractatus de electione Caroli IV, 1348 or 1349; Summa logices, publ. 1488; Commentaries on Porphyry's Introduction to Aristotle's Organon. Greatly influenced schools of logic and philosophy of his and next 2 centuries; broke with previous medieval philosophy; believed natural reason was incompetent in matters of faith, that faith should be based on probability; basically an empiricist; his philosophical doctrine described as nominalism or terminalism; believed universal concepts to be natural signs based on similarity discovered by experience. Died Munich, Bavaria, circa 1349.

WILLIAM OF SAINT CLOUD (or Guilelmus de Sancto Clodoaldo, sometimes called William the Englishman), astronomer; possibly of English origin; probably flourished in Paris, circa 1292-96; magister and canon St. Cloud, Paris; may have been a founder astron. sch. of Paris; credited with making a calendar for Queen Elizabeth of Eng.; indirectly observed sun by means of camera obscura, 1285; compiled almanac giving positions of planets for 1292-1311; compiled perpetual calendar dedicated in 1296 to Queen Marie of France; inventor directorium (instrument of unknown description); only medieval Christian astronomer to make direct determination of obliquity of ecliptic at 23 degrees, 34 minutes, 1290.

WILLIAMS, A(rthur) O(lney), Jr., Am. physicist; b. Providence, Apr. 7, 1913; s. Arthur Olney and Elizabeth Caecilia (Gillespie) W.; S.B., Mass. Inst. Tech., 1934; Sc.M., Brown U., 1936; Ph.D., 1937; m. Jean Frances White, Mar. 26, 1938; 1 dau., Frances Jean. Instr. physics U. Me., 1937-40, asst. prof., 1940-42; faculty Brown 1942—, prof. physics, 1951—, chmn. exec. com. dept. physics, 1954-56, chmn. dept., 1956-60, 62-63. Cons. NRC, 1947—, Office of Naval Research, 1950—; mem. R.I. AEC, 1955—. Fellow Am. Phys. Soc., Acoustical Soc. Am.; mem. Am. Assn. U. Profs., Sigma Xi, Kappa Sigma. Author: Electronics, 1953. Co-author: College Physics, 1957. Contbr. to tech. jours. Research on phys. acoustics and underwater sound, atomic wave functions. Home: 7 Lee Rd., Barrington, R.I. 02806. Office: Brown University, Providence 02912.*

WILLIAMS, Carrington Bonsor, Brit. entomologist; b. Liverpool, Eng., Oct. 7, 1889; s. Alfred and Lillian (Kirkland) W.; B.A., Cambridge U., 1911, M.A., Sc.D., 1930; m. Ellen Margaret Bain, Sept. 25, 1920; children—John, Basil, Miles. Research scholar John Innes Hort. Inst., London, 1911-16; sugar cane entomologist Dept. Agr., Trinidad W.I., 1916-21; sect. dir., dir. dept. entomology Ministry Agr., Egypt, 1921-27; entomologist E. African Agr. Research Sta., Amani Tongaurike, 1927-29; lectr. agr. and forest entomology U. Edinburgh (Scotland), 1929-32; head dept. entomology Rothamsted Exptl. Sta., Harpendon, Eng., 1932-55; guest prof. U. Minn., 1932-58. Mem. adv. units to W. Africa, Brit. Colonial Office, 1952-53. Fellow Royal Soc., 1954, Inst. Biology; hon. mem., post pres. Brit. Ecol. Soc., Royal Entomol. Soc. London, Assn. Applied Biologists; others. Author: Migration of Butterflies, 1930; Insect Migration, 1958; Patterns in the Balance of Nature, 1964; also numerous articles. Research on environmental influence on balance of insect populations, insect migration. Home: Elm Park Lodge, Selkirk, Scotland.*

WILLIAMS, Carroll Milton, Am. biologist; b. Richmond, Va., Dec. 2, 1916; s. George Leslie and Jessie (Hendrick) W.; B.S., U. Richmond, 1937, D.Sc., 1960; A.M., Harvard, 1938, Ph.D., 1941, M.D., 1946; m. Muriel Anne Voter, June 26, 1941; children—John Leslie (dec.), Wesley Conant, Peter Glenn, Roger Lee. Faculty, Harvard, Cambridge, Mass., 1946—, prof. biology, 1953—, chmn. dept., 1959-62, Bussey prof., 1966—. Cons. NSF, 1955-60. Recipient Boylston prize and gold medal Harvard Med. Sch., 1961. Fellow A.A.A.S., Entomol. Soc. Am., Am. Acad. Arts and Scis.; mem. Am. Physiol. Soc., Harvey Soc., Soc. Exptl. Biology, Soc. Comparative Physiology, Nat. Acad. Scis., Phi Beta Kappa, Sigma Xi. Research, numerous publs. on physiology and biochemistry of insects with spl. reference to endocrine control of growth and metamorphosis. Home: 27 Eliot Rd., Lexington, Mass. 02173. Office: Harvard Biology Lab., Divinity Av., Cambridge, Mass. 02138.*

WILLIAMS, Charles Hanson Greville, English chemist; b. Cheltenham, Eng., Sept. 22, 1829; s. S. Hanson and Sophia (Billings) W.; ed. at Prestbury nr. Cheltenham, Eng.; m. Henrietta Bosher, Nov. 25, 1852; 4 sons, 4 daus. Cons. and analytical chemist Oxford Ct. Cannont St., London, 1852-53; asst. to prof. T. Anderson, Glasgow (Scotland), 1853-56; lectr. chemistry Normal Coll., Swansed, 1857-58; worked under L. Playfair, Edinburgh (Scotland) U.; chemist Geo. Miller & Co., Glasgow; asst. to W. H. Perkin, Greenford Green, Eng., 1863-68; partner E. Thomas & J. Dower, Star Chem. Works, 1868-77; chemist, photometric supr. Gas Light & Coke Co., London, 1877-1901. Fellow Royal Soc., 1862; mem. Chem. Soc. Author: A Handbook of Chemical Manipulation, 1857; Manual of Chemical Analysis for Schools, 1858; Contbg. author: Treatise on Coal Gas (King); Dictionary of Chemistry (Watt); also articles. Research on volatile bases produced by destructive distillation of shales, cinchonine, and some hydrocarbons, chemistry of coal gas, including determination of its specific gravity; discovered cyanine (quinoline-blue) 1857, lepione and numerous other alkaloids; isolated isoprene, 1880; developed method of making artificial emeralds; demonstrated that emeralds lost 9% of weight on fusion and specific gravity reduced to 2.4. Died Smallfields, Eng., June 15, 1910.

WILLIAMS, Charles James Blasues, English physician; b. Wiltshire, Eng., Feb. 3, 1805; s. David Williams; ed. Edinburgh (Scotland) U., 1820; M.D., 1824; studied at Laennec's clinique, Paris, France; m. Harriet Williams Jenkins, 1830. Prof. medicine, physician U. Coll., London, Eng., from 1839; physician extraordinary to Queen Victoria, from 1874. Lumleian lectr., 1862. Fellow Royal Soc., 1835, Royal Coll. Physicians (censor 1846-47); mem. Path. Soc. (original mem., 1st pres. 1846), Royal Med. and Chirurg. Soc. (pres. 1873-75). Author: Rational Exposition of the Physical Signs of the Diseases of the Lungs and Pleura, 1828; Principles of Medicine, 1843. Contbr. articles Cyclopaedia of Practical Medicine, 1833. Helped found Consumption Hosp., Brompton, Eng., 1841. Died Mar. 24, 1889.

WILLIAMS, Charles Melville, Canadian agrologist; b. Regina, Sask., Can., Mar. 18, 1925; s. Edward Percy and Charlotte Alice (MacDonald) W.; B.S.A., U. B.C., 1949; M.S. in Agr., U. B.C., 1952; Ph.D., Ore. State U., 1955; m. Patricia Jean Begbie, Mar. 24, 1953; children—Allan Begbie, Catherine Ann, Charlotte Lucy Elaine. Dist. agriculturist B.C. Dept. Agr., 1949-51; prof. animal sci. U. Sask., Saskatoon, 1954—, sr. extension specialist, 1964. FAO animal prodn. officer Bur. Animal Industry, Philippines, 1964. Mem. Canadian Soc. Animal Prodn., Agrl. Inst. Can., Am. Soc. Animal Sci., Internat. Soc. Bioclimatology. Research, publs. on effect low fluctuating temperatures on prodn. domestic animals. Home: 1 Moxon St., Saskatoon, Sask., Can.*

WILLIAMS, Clarke, Am. physicist; b. N.Y.C., May 4, 1902; s. William Robert and Flora (Nabersberg) W.; A.B., Williams Coll., 1922, Sc.D., 1964; B.S., Mass. Inst. Tech., 1924; Ph.D., Columbia, 1935; m. Lindsay Clement Field June 30, 1927 (div. Apr. 1933); 1 son, David Field; m. 2d, Margaret S. Button, Aug. 12, 1933; children—Evan Thomas, Thomas Button. With Duke Price Power Co., 1924-25, N.Y.C. R.R., 1925-26, Columbia, 1926-30, Coll. City N.Y., 1930-49; with Brookhaven Nat. Lab., Upton, N.Y., 1947-67, chmn. nuclear engrng. dept., 1952-62, dep. dir., 1962-67, emeritus, 1967—; research adminstr. Marine Resources Council, Nassau-Suffolk Regional Planning Bd., 1967—. Tech. adviser, paper coordinator 2d Internat. Conf. on Peaceful Uses Atomic Energy, 1958; chmn. Nuclear Congress, 1960; adviser U. S. Delegation to 3d Geneva Conf., 1964. Fellow Am. Nuclear Soc. (past pres., dir. publs.), A.A.A.S. (council); mem. Am. Phys. Soc., Fedn. Am. Scientists, Assn. For Scientists Atomic Edn., Phi Beta Kappa, Sigma Psi, Epsilon Chi. Research, publs. on properties of the solid state of matter, neutron physics; reactor tech. and design, resource mgmt. planning. Home: 200 S. Country Rd., Bellport, L.I., N.Y. 11713. Office: Marine Resources Council, County Center, Hauppauge, N.Y. 11787.*

WILLIAMS, Clyde Michael, Jr., Am. physician; b. Marlow, Okla., Oct. 8, 1928; s. Clyde Michael and Betty (Black) W.; B.A., Rice Inst., 1948; M.D., Baylor U., 1952; D.Phil. (Rhodes scholar), Oxford (Eng.) U., 1954; m. Martha Adelaide Smither, Jan. 29, 1953; children—Edith Gibbs, Clyde Michael III, Kathleen, John Thomas. Tech. dir. radioisotope Lab., VA Hosp., Pitts., 1958-60; faculty U. Fla. Coll. Medicine, Gainesville, 1963—, prof. medicine, chmn. dept. radiology, 1965—. Diplomate Am. Bd. Radiology; mem. Am. Physiol. Soc., Am. Coll. Radiology, Radiol. Soc. N.Am., Soc. Nuclear Medicine. Research, numerous publs. on theory thermal sensations, gas chromatography aromatic acids, computer aided diagnostic programs in radiology. Home: 1604 N.W. 8th Av., Gainesville, Fla. 32601.*

WILLIAMS, Daniel H., Am. physician; b. Hollidaysburg, Pa., Jan. 28, 1858; s. Daniel and Sarah Ann (Price) W.; grad. Janesville Classical Acad., 1878; M.D., Chicago Med. Coll., 1883; LL.D., Wilberforce U., 1908; m. Alice D. Johnson, Apr. 2, 1898. Surgeon to South Side Dispensary, Chicago, 1884-92; founded, 1891, and surgeon Provident Hosp.; phys., Protestant Orphan Asylum, 1884-93; surgeon-in-chief, Freedmen's Hosp., Washington, 1893-98; asso. on staff, St. Luke's Hosp., Chicago. Mem. Ill. State Bd. of Health, 1889, reapptd., 1891. Fellow Am. Coll. Surgeons. Prof. clin. surgery, Meharry Med. Coll., Nashville, Tenn., 1899—. Credited with 1st successful closing of wound of heart and pericardium. Died Aug. 4, 1931.

WILLIAMS, David Tyndale, Am. physicist; b. Manilla, P.I., Feb. 24, 1907; s. Hermon Porter and Beulah (MacFarland) W.; student Rutgers U., 1925-27, Montclair State Tchrs. Coll., 1927-29; A.B., Columbia, 1930, A.M., 1933; postgrad. U. Mich., 1933-36; Ph.D., N.Y. U., 1938; m. Charlotte Marie Rigg, Feb. 20, 1934; children—David Gerald, James

Hermon, Christopher Rigg. Mem. faculty Carleton Coll., Northfield, Minn., 1938-39; physicist NACA, 1939-45; staff Battelle Meml. Inst., Columbus, O., 1945-46, 47-57; faculty U. Mich., Ann Arbor, 1946-47; prof. aerospace engring. U. Fla., Gainesville, 1957—. Asso. fellow Am. Inst. Aeros. and Astronautics; mem. Am. Phys. Soc., Sigma Xi. Author: (with others) Essentials of Biological and Medical Physics, 1955. Research and publs. in aircraft engine performance, spectroscopy, med. research, aerodynamics, combustion; applied research in light propagation. Home: 3100 N.W. 1st Av., Gainesville, Fla. 32601.*

WILLIAMS, Dudley, Am. physicist; b. Covington, Ga., Apr. 12, 1912; s. Arthur Dudley and Ethel (Turner) W.; A.B., U. N.C., 1933, M.A., 1934, Ph.D., 1936; m. Loraine Decherd, June 12, 1937; children—Francis Dudley, Harriet Thomson. From instr. to asst. prof. U. Fla., Gainesville, 1936-41; staff mem. Radiation Lab., Mass. Inst. Tech., 1941-43; asst. prof. U. Okla., 1943-44; staff mem. Manhattan Project, U. Cal., Los Alamos, N.M., 1944-46; asso. prof. Ohio State U., Columbus, 1946-50, prof., 1950-63, acting chmn., 1952-53, 58-59; head dept. physics N.C. State U., Raleigh, 1963-64; Regents' Distinguished prof. Kan. State U., Manhattan, 1964—; Guggenheim fellow U. Amsterdam also Oxford U., 1956; NSF Sr. Postdoctoral fellow Institut d'Astrophysique, U. Liège, 1961-62. Fellow Am. Phys. Soc., Am. Optical Soc., Ohio Acad. Sci.; mem. Am. Assn. Physics Tchrs., Phi Beta Kappa, Sigma Xi (pres. Ohio chpt. 1962-63). Author several books including: (with G. Shortley) Elements of Physics, 1953, 55, 61, 65; Molecular Physics, 1962. Asso. editor Jour. Optical Soc. Am., 1959—. Contbr. articles on work in infrared spectroscopy, microwave transmission, magnetic-resonance phenomena, mass spectroscopy to tech. jours. Home: 120 Longview Dr., Manhattan, Kan. 66502.*

WILLIAMS, Elkanah, Am. ophthalmologist; b. Lawrence County, Ind., Dec. 19, 1822; s. Isaac and Amelia (Gibson) W.; grad. Ind. Asbury U. (now DePauw U.), 1847; M.D., U. Louisville, 1850; m. Sarah L. Farmer, Dec. 1847; m. 2d, Sarah B. McGrew, Apr. 7, 1857. Practiced medicine, Cin., 1855, specialized in diseases of eye and ear (one of 1st in country to limit his practice to this speciality); established charity eye clinic similar to European instns. in connection with Miami Med. Coll., 1855; prof. ophthalmology and aural surgery Miami Med. Coll. (1st chair devoted to this specialty in U. S.) 1865-68; asst. surgeon U. S. Marine Hosp., Cin., during Civil War, mem. staff Cin. Hosp., 1862-73; one of 1st in Am. to use ophthalmoscope; published article The Ophthalmoscope in London Med. Times and Gazette, 1854; co-editor Cin. Lancet and Observer, 1867-73. Mem. Am. Ophthal. Soc., pres., 1876; mem. Am. Otological Soc.; hon. mem. Ophthal. Soc. Gt. Britain. Died Oct. 5, 1888.

WILLIAMS, Ernest Edward, Am. zoologist; b. Easton, Pa., Jan. 7, 1914; s. Edward and Alma (Miller) W.; B.S., Lafayette Coll., 1933; Ph.D., Columbia, 1949. Faculty, Harvard, 1949-57, asst. prof., 1952-57, curator reptiles and amphibians Mus. Camparative Zoology, Cambridge, Mass., 1957—. Mem. A.A.A.S., Am. Soc. Ichthyologists and Herpetologists, Soc. Study Evolution, Am. Soc. Naturalists, Am. Soc. Zoology. Research, publs. in lizard classifications, behavior and ecology. Home: 28 Wendell St. Office: Mus. Comparative Zoology, Harvard, Cambridge, Mass. 02138.*

WILLIAMS, Ferd Elton, Am. physicist; b. Erie, Pa., June 9, 1920; s. Earl V. and Osa (Riddle) W.; B.S., U. Pitts., 1942; M.A., Princeton, 1945, Ph.D., 1946; m. Anne Katherine Lindberg, Nov. 25, 1948; children—Karen, Gareth, Geoffrey, Kevin, Quentin. Research engr. RCA Labs., Princeton, 1942-46; asst. prof. U. N.C., 1946-49; theoretical physicist, mgr. luminescence research Gen. Electric Research Lab., Schenectady, 1948-61; chmn. dept., H. Fletcher Brown prof. physics U. Del., Newark, 1961—. Cons. Am. Cyanamid Co., Stamford, Conn., 1962—; mem. adv. com. for solid state Oak Ridge Nat. Lab., 1964-67; cons. Picatinny Arsenal, Dover, N.J., 1966—; cons. basic luminescence program U. S. Army; Del. regional counselor in physics Am. Inst. Physics. Fellow Am. Phys. Soc., A.A.A.S., Optical Soc. Am.; mem. Am. Chem. Soc., Sigma Xi. Research and publs. on luminescence and semiconductivity, theoretical calculation of luminescent spectra of alkali halide phosphors; co-discoverer of electroluminescence in single crystals of zinc sulfide; prediction and theory of photoelectroluminescence; co-discoverer of donor-acceptor pairs in compound semiconductors and theory of their optical properties; mechanisms of infrared to visible conversion in crystals; patentee in field. Home: 1008 Dixon Dr., Christine Manor, Newark, Del. 19711.*

WILLIAMS, Francis Henry, Am. physician; b. Uxbridge, Mass., Apr. 15, 1852; s. Henry Willard and Elizabeth (Dewe) W.; B.S., Mass. Inst. Tech., 1873; M.D., Harvard, 1877; European study, 1877-79; m. Anna Dunn Phillips, Sept. 15, 1891. Asst. U. S. Transit of Venus Expdn. to Japan, 1874; tour around world, 1874-75; practicing phys. at Boston, 1879—. Instr. in materia medica, 1884-85, materia medica and therapeutics, 1885-86, asst. prof. materia medica and therapeutics, 1886-88, asst. prof. therapeutics, 1888-91, Harvard Med. Sch.; visiting phys., 1896-1913; sr. phys., 1913—, Boston City Hosp. Life member Corp. of Mass. Inst. Tech. (mem. exec. com. its 1st 25 yrs.), 1882—. Fellow A.A.A.S., American Acad. Arts and Sciences; pres. Assn. Am. Physicians, 1917-18. Author: The Roentgen Rays in Medicine and Surgery, 1901-03. Initiated bacteriol. examinations in diphtheria, Boston City Hosp., 1892 (first in community to use antitoxin, 1894); developed original methods of treatment of various diseases with the beta rays from radium, especially diseased tonsils (instead of operation) and many diseases of the eye; first paper on subject (published Med. News Feb. 6, 1904) "Some Physical Properties and Medical Uses of Radium Salts," with report of forty-two cases treated with pure radium bromide. Devised clinical method of measuring the X-rays and the beta rays from radium. Died June 22, 1926.

WILLIAMS, G. Rainey, Am. physician; b. Atlanta, Oct. 25, 1926; s. George Rainey and Hildred (Russell) W.; student U. Tex., 1944-46; B.S., Northwestern U., 1947, B.Med., 1950, M.D., 1951; m. Martha Alden Vose, June 16, 1950; children—Bruce, Alden, Ellen, George Rainey. William Stewart Halsted fellow in surgery Johns Hopkins Sch. Medicine, 1951-52; with U. Okla. Sch. Medicine, Oklahoma City, 1958—, prof. surgery, vice-chmn. dept. surgery, 1963—, chief div. thoracic surgery, 1965—; mem. Presbyn. Med. Center, Oklahoma City. Markle scholar in med. sci., 1960-65. Mem. Am. Assn. Thoracic Surgery, Am., So. surg. assns., Soc. for Vascular Surgery, Soc. U. Surgeons. Research, numerous publs. on cardiac arrest, open heart surgery, replantation of amputated extremities. Home: 6512 Hillcrest St., Oklahoma City 73116.*

WILLIAMS, Harold, Am. psychologist; b. Yankton, S.D., Mar. 29, 1921; s. C. Earl and Alice (Thompson) W.; B.A., U. Neb., 1944; Ph.D., U. Minn., 1951; m. Bette Sherwood, Oct. 18, 1942; children—Loxie Lee (Mrs. William A. Devin III), Cindy Lou, Candy Jane, Thomas Charles, Lucy Alice. Clin. psychologist, asst. chmn. Letterman Army Hosp., San Francisco, 1952-54; chief clin. and social psychology, neuropsychiat. div. Walter Reed Army Inst. Research, 1954-63, dep. dir. neuropsychiat. div., 1963-64; research prof. psychology U. Okla. Sch. Medicine, Oklahoma City, 1964—. Mem. com. on clin. drug evaluation Psychopharmacology Service Center, 1963—; mem. com. on hearing, bioacoustics and biomechanics Nat. Acad. Scis., 1965—. Mem. Am. Psychol. Assn., Psychonomic Soc., A.A.A.S., Assn. for Psychophysiol. Study Sleep, Soc. for Psychophysiol. Research, N.Y. Acad. Scis. Cons. editor Psychophysiology, Jour. Abnormal Psychology. Research, publs. on effects of stress on human performance and physiol. functioning, behavioral and physiol. properties of sleep and altered states of consciousness. Home: 3905 Coltrane Rd., Oklahoma City 73111.*

WILLIAMS, Harold Henderson, Am. biochemist; b. Blanchard, Pa., Aug. 29, 1907; s. Shuman S. and Bertha E. (Johnston) W.; B.S., Pa. State U., 1929; Ph.D., Cornell U., 1933; Sterling fellow Yale, 1933-35; m. Agnes T. Gainey, Nov. 28, 1935; children—Patricia M., Margaret H. (Mrs. Merlin L. Puck), Kathleen E. Mem. staff Children's Fund Mich., Detroit, 1935-45, asso. dir. research lab., 1942-45; prof. biochemistry Cornell U., Ithaca, N.Y., 1945—, head dept. biochemistry, 1955-64. Mem. A.A.A.S., Am. Soc. Biol. Chemistry, Am. Chem. Soc., Soc. Exptl. Biology and Medicine, Am. Inst. Nutrition (Borden award 1953). Author: (with Icie Macy Hoobler) Hidden Hunger, 1945. Research and publs. on nutrient constituents of human and cow's milk; lipid constituents of human and animal cells; amino acid content of animal feeds; protein and amino acid metabolism in animals; selenium metabolism in microorganisms. Home: 1060 Highland Rd., Ithaca, N.Y. 14850.*

WILLIAMS, Henry Shaler, Am. geologist; b. Ithaca, N.Y., Mar. 6, 1847; s. Josiah B. and Mary H. (Hardy) W.; Ph.B., Yale, 1868, Ph.D., 1871; m. Harriet H. Wilcox, 1871. Asst. in paleontology Yale, 1868-70, Silliman prof. geology, 1892-1904; prof. natural science Ky. U., 1871-72; asst. prof. geology, 1879-92; prof., Cornell U., 1892, prof. geology and head geol. dept., dir. mus., 1904-12, emeritus prof. geology, 1912—; also in charge Devonian Lab. of U. S. Geol. Survey. Am. commr. Internat. Congress Geology. Fellow London Geol. Soc., Soc. Geology du Nord, Geol. Soc. Am. A.A.A.S. Author: Geological Biology; Correlation Papers, Devonian and Carboniferous, 1891; Geological Biology, 1895; On the Theory of Organic Variation, 1911; also numerous papers on Devonian geology and palaeontology. Asso. editor Am. Jour. Sci., Jour. Geology. Research on Devonian systems of Ohio, Pa., N.Y., aided in confirming Darwinian theories; pioneered photog. technique of fossil illustration; originated stratigraphical method of geol. study. Died Havana, Cuba, July 31, 1918.

WILLIAMS, Henry Willard, Am. ophthalmologist; b. Boston, Dec. 11, 1821; s. Willard and Elizabeth (Osgood) W.; grad. Harvard Med. Sch., 1849; m. Elizabeth Dewe, 1848; m. 2d, Elizabeth Adeline Law, 1860; 6 children. Organized voluntary class of Harvard students for ophthalmology lectures, 1850; lectr. ophthalmology Harvard Med. Sch., 1866-71, 1st prof. ophthalmology, 1871-95; ophthal. surgeon Boston City Hosp., 1864-91; a founder Am. Ophthal. Soc., 1864, pres., 1868-75; wrote article on cataract operation for Boston Med. and Surg. Jour., 1850; pres. Mass. Med. Soc., 1880-82. Author: A Practical Guide to the Study of the Diseases of the Eye, 1862; Our Eyes, and How To Take Care of Them, 1871; The Diagnosis and Treatment of the Diseases of the Eye, 1881. Died Boston, June 13, 1895.

WILLIAMS, Hulen Brown, Am. chemist, b. Lauratown, Ark., Oct. 8, 1920; s. Ernest Burdett and Ann Jeanette (Miller) W.; B.A., Hendrix Coll., 1941; M.S., La. State U., 1943, Ph.D., 1948; postdoctoral student Cornell U., summer 1950, U. Cal. at Berkeley, 1953; m. Virginia Anne Rice, June 20, 1942; children—James Browning, Virginia Jean. Instr. to asso. prof. La. State U., 1947-56, prof., head chemistry dept., 1956—, mgr. Laboratory Stores, 1948-55. Served as lt. (j.g.) USNR, 1944-46. Mem. Am. Chem. Soc., Sigma Xi, Sigma Pi Sigma, Phi Lambda Upsilon, Phi Kappa Phi, Kiwanian. Contbr. research articles profl. jours. Research in organic, phys. and colloid chemistry. Home: 470 Castle Kirk Dr., Baton Rouge 70808.*

WILLIAMS, James Michel, Am. sociologist; b. Sangerfield, N.Y., June 22, 1876; s. Riley Walter and Mary Turnbull (Mickel) W.; A.B., Brown U., 1898; B.D., Union Theol. Sem., 1901; Ph.D., Columbia, 1906; LL.D., Hobart, 1941; m. Lucinda Chamberlain Noyes, Jan. 22, 1913; 1 son, Henry Noyes. Lecturer in economics at Vassar College, 1907-08; professor of sociology, Hobart Coll., 1908-41, prof. emeritus since 1941. Author: The American Town, 1906; The Foundations of Social Science, 1920; Principles of Social Psychology, 1922; Our Rural Heritage, 1925; The Expansion of Rural Life, 1926; Children and the Depression, 1932; Human Aspects of Unemployment and Relief, 1933; The Police School—An Experiment in Adult Education, 1936. Pioneer in study of U. S. rural life; known for field studies and description of isolated community of nineteenth century Am. farmers. Home: 169 St. Clair St., Geneva, N.Y.

WILLIAMS, Jesse Noah, Jr., Am. biochemist; b. Greenville, N.C., Apr. 1, 1924; s. Jesse Noah and Eulalie (Waldrep) W.; B.S. in Chemistry, U. N.C., 1945; M.S. In Chemistry, U. Ill., 1946, Ph.D., 1948; m. Henrietta Ver Meer, Sept. 6, 1946; 1 dau., Marylie C. Faculty, U. Wis., Madison, 1948-59, asso. prof. biochemistry, 1952-59; chemist NIH, Bethesda, Md., 1959—. Mem. Am. Soc. Biol. Chemists, Am. Inst. Nutrition, Phi Beta Kappa, Sigma Xi. Research, publs. in intermediary metabolism, nutritional biochemistry, enzymology. Office: Bldg. 10, NIH, Bethesda, Md. 20014.*

WILLIAMS, John Warren, Am. chemist; b. Woburn, Mass., Feb. 10, 1898; s. Charles Sampson and Genevieve (Allen) W.; B.S., Worcester Poly. Inst., 1921; M.S., U. Wis., 1922, Ph.D., 1925; m. Lois Mary Andrews, Sept. 5, 1925; 1 dau., Janet Andrews (Mrs. Philip Coussens). Faculty dept. chemistry U. Wis., Madison, 1925—, prof., 1938—. NRC fellow, Copenhagen-Leipzig, 1927-28; IEB fellow, Uppsala, 1934-35; Guggenheim Found. fellow, Copenhagen, Oxford, Pasadena, 1956-57. Mem. Am. Chem. Soc. (sec., vice chmn. and chmn. div. phys. and inorganic chemistry 1934-36, vice chmn. and chmn. div. colloid chemistry 1937-38, chmn. nat. colloid symposium 1947-52, Kendall award 1955), Nat. Acad. Scis., Am. Soc. Biol. Chemists, Sigma Xi. Author: (with others) Experimental Physical Chemistry, 1929, 6th edit., 1962. Asso. editor: Jour. Phys. and Colloid Chemistry, 1947-53, Ann. Reviews Phys. Chemistry, 1949-55; editor Ultracentrifugal Analysis in Theory and Experiment, 1963. Research and publs. on phys. chemistry of proteins, use of ultracentrifugal analysis in determination of number of combining sites on antibody molecule; size distbn. analyses of certain plasma extenders and fragmentation by enzymatic hydrolysis of gamma globulin molecules; patentee in recovery of gamma globulins from blood plasma. Home: 4129 Mandan Crescent, Madison 53711. Office: 1112 W. Johnson St., Madison, Wis. 53706.*

WILLIAMS, K(enneth) Lloyd, English surgeon; b. Leamore, Eng., Apr. 2, 1923; s. Reginald Lloyd and Helena (Powell) W.; M.A., M.B.B.Chir., Cambridge and St. Thomas Hosp., 1947, M.Chir., 1961, M.D., 1964; m. Frances Josephine Keegan, Feb. 8, 1958; children—Mark Julian, Julia. Sr. registrar Central Middlesex Hosp., London, Eng., 1957-59; sr. registrar Middlesex Hosp., London, 1959-65; cons. surgeon Bath Clin. Area, 1965—. Comyns Berkeley travelling fellow, 1962, WHO travelling fellow, 1963. Recipient Cressy Morrison award N.Y. Acad. Scis., 1964. Mem. Royal Soc. Medicine, N.Y. Acad. Scis., Assn. Surgeons of Gt. Britain and N. Ireland. Described necrotizing colitis; studies, publs. on infrared measurement of skin temperature, med. thermography. Home: 5 Gay St., Bath, Eng. Office: Royal United Hosp., Bath, Somerset, Eng.*

WILLIAMS, Lansing Earl, Am. plant pathologist; b. Spencer, W.Va., Aug. 8, 1921; s. Harry Earl and Nora (Hickman) W.; B.Sc., Morris Harvey Coll., 1950; M.Sc., Ohio State U., 1952, Ph.D., 1954; m. Mildred Juanita Hill, May 9, 1946; children—Patricia Marie, Lansing Earl. Grad. asst. Ohio State U., 1950, research asst., 1950-52, research fellow, 1952-54, faculty Ohio Agrl. Research & Devel.

Center, Wooster, 1954-—, prof., 1966-—; instr. Mansfield Campus Ohio State U., 1959-65. Mem. Am. Phytopathol. Soc., Ohio Acad. Sci., Am. Inst. Biol. Sci., Sigma Xi. Research, numerous publs. on effects of agronomic practices on soil-borne plant pathogens and soil fungal populations; factors affecting corn stalk rot; isolation and study of maize dwarf mosaic virus; effects of wheat streak mosaic virus on corn prodn. Home: 2580 Christmas Run Blvd., Wooster, 44691. Office: Ohio Agrl. Research & Devel. Center, Wooster, Ohio 44691.*

WILLIAMS, Louis Otho, Am. botanist; b. Jackson, Wyo., Dec. 16, 1908; s. Frank Richard and Emma (Olcott) W.; B.S., U. Wyo., 1932, M.S., 1933; Ph.D., Washington U., St. Louis, 1936; m. Terua Pierson, Sept. 16, 1934. Research asso. Harvard, 1936-47; prof., subdir. Escuela Agricola Panamericana, Tegucigalpa, Honduras, 1947-57; botanist U.S. Dept. Agr., Beltsville, Md., 1957-60; curator Central Am. botany, chief curator botany Field Mus. Natural History, Chgo., 1960-—; research asso. Mass. Inst. Tech., 1947-50. Author: (with Paul C. Standley) Flora Guatemala, 1960; Orchidaceae of Mexico, 1951; Orchidaceae of Central America, 1956. Research and publs. on flora of neotropics. Address: Field Mus. Natural History, Chgo. 60605.*

WILLIAMS, Marvin Martin Dixon, Am. radiol. physicist; b. Walla Walla, Wash., Aug. 24, 1902; s. Luther N. and Elizabeth (Dixon) W.; B.S., Whitman Coll., 1925; postgrad. U. Minn., U. Wash.; M.S., U. Pa., 1929; Ph.D., U. Minn., 1931; m. Orpha Stewart, May 25, 1931; children—Martha, Lila, Kenneth, Jay, Hester. Instr. physics Whitman Coll., 1926-27; asst. radiotherapeutic physics U. Pa., 1927-29; X-ray and radium technician Phila. Gen. Hosp., 1927-29; radiol. physicist Peiping Union Med. Coll., China, 1931-35; faculty Mayo Grad. Sch., U. Minn., Rochester, 1936-—, prof., 1950-—; research asso. Argonne Nat. Labs., 1952-60. Mem. adv. com. Edn. Visually Handicapped, Minn. Bd. Edn., 1958-61; various subcoms. Nat. Com. Radiation Protection; mem. Am. Bd. Health Physics, 1966-—. Diplomate Am. Bd. Radiology (mem. coms.). Fellow Am. Coll. Radiology (mem. commn. on cancer 1966-—), Health Physics Soc., Assn. Midwest Univs. (council), A.A.A.S., Am. Assn. Physicists in Medicine (pres. 1963), Radiol. Soc. N.Am. (coms., recipient Gold medal 1965), Am. Roentgen Ray Soc. (coms.), Am. Radium Soc. (2d v.p. 1964, coms.), Am. Phys. Soc., Am. Assn. Physics Tchrs., Radiation Research Soc., Am. Nuclear Soc., Fedn. Am. Scis., Phi Beta Kappa, Sigma Xi. Contbr. articles to profl. jours. Home: 1013 N.W. 2d St., Rochester, Minn. 55901.*

WILLIAMS, Melvin J(ohn), Am. sociologist; b. Stovall, N.C., Feb. 13, 1915; s. John Presley and Mary Jenera (Wilkerson) W.; A.B., Duke, 1932, B.D., 1936, Ph.D., 1941; m. Frances Clark, Oct. 15, 1936; children—Kay Frances (Mrs. Bradley Yont), Dorothy Virginia (dec.), Melvin John, Deboeah Susan, Steven Clark, Eric Stanton. From instr. to asst. prof. sociology Albion Coll., 1941-44; prof. sociology, head dept. Wesleyan Coll., Macon, Ga., 1944-47; asso. prof. sociology Fla. State U., 1947-52; prof. sociology, head dept. Stetson U., 1952-60, prof. sociology, dir. social work, 1960-63; prof. sociology, chmn. dept. E. Carolina Coll., Greenville, N.C., 1963-—. Co-dir. Addison project Kresge Found., 1942-44; dir. YWCA Community Survey, Macon, Ga., 1944; dir., family counselor Children's Center, Macon, 1946-47; dir. Adolescent Research project Wesleyan Coll, 1944-46, E. Carolina Coll., 1963; research and sociol. cons., 1946-—. Fellow Am. Sociol. Assn.; mem. A.A.A.S., Am. Assn. U. Profs., Am. Acad. Polit. and Social Sci., So. Sociol. Soc. (sec. treas. 1953-55, 1st v.p. 1955-56), Groves Conf., Nat. Council Family Relations. Author: (with others) Contemporary Social Theory, 1940; Catholic Social Thought, Its Approach to Contemporary Problems, 1950; Social Norms of Adolescents, A Study in Social Guidance, 1959; Moral and Spiritual Values in Education, 1955; also numerous articles. Research on models for analysis of theoretical systems, social apperception tests and survey methods for study of comparative behavioral systems and sub-systems at community, organizational and group levels, small groups sociometric analysis, behavioral case studies, social science and atomic power; digital computation research on value orientations and foci, particularly among adolescent structures; action research in social guidance, preventive and therapeutic counseling, parent-child relations. Home: 803 E. 3d St., Greenville, N.C. 27834.*

WILLIAMS, Merton Yarwood, Canadian geologist; b. Bloomfield, Ont., Can., June 21, 1883; s. Edwin Allison and Caroline E. (Bowerman) W.; B.Sc. in Mining Engring., Queen's U., 1909; Ph.D. in Geology, Yale, 1912; m. Lula M. Philp, Dec. 23, 1915; children—Edwin Philp, Marion Margaret, Kathleen Mavis (Mrs. W.R. Carruthers). Tchr., Prince Edward County, Ont., 1903-05; field asst. Geol. Survey Can., 1908-11, geologist, 1912-21; geologist, Hong Kong, 1924-25; prof. paleontology and stratigraphy U. B.C., Vancouver, Can., 1921-36, head dept. geology and geography, 1936-50, prof. emeritus, 1950-—. Cons. Shell Oil Co. of Can., 1950-54. Fellow Geol. Soc. Am. (past 2d v.p.), Royal Soc. Can. (past pres.). Author: Geological Survey Canada Memoirs; also numerous articles. Research on Silurian paleontology Can., structure oil and gas fields Western Ont., Peace River B.C., So. Plains Alta. Home: 2376 W. 5th Av., Vancouver 9, B.C., Can.*

WILLIAMS, Ralph Chester, Jr., Am. physician, educator; b. Washington, Feb. 17, 1928; s. Ralph Chester and Annie (Perry) W.; A.B., Cornell U., 1950, M.D., 1954; m. Mary Elizabeth Adams, June 23, 1951; children—Cathy, Frederick, John, Michael, Ann. Guest investigator Rockefeller Inst., 1961-63; faculty dept. medicine U. Minn., Mpls., 1963-65, asso. prof., 1965-—. Mem. Am. Rheumatism Assn., Assn. Am. Immunologists, Am. Soc. Clin. Investigation, Phi Beta Kappa, Alpha Omega Alpha. Research, publs. on autoantibodies and relationship to rheumatic diseases. Home: 3158 Arthur St. N.W., Mpls. 55418. Office: Box 91, University Hospital, Mpls. 55455.*

WILLIAMS, Richard Hays, Am. sociologist; b. Pomona, Cal., Feb. 19, 1912; s. Robert Day and Jessie (Hays) W.; A.B., Pomona Coll., 1933; student U. Bordeaux, France, 1933-34; M.A., Harvard, 1936, Ph.D., 1938; m. Donna Campbell, June 23, 1953; children—Robert Harold, Frank Rendler. Faculty, U. Buffalo, 1936-45, asso. prof., 1942-45; chief surveys ICD Opinion Surveys War Dept., Regional Mil. Govt., Stuttgart, Germany, 1945-46; dir. Bur. Indsl. Psychology Commn. Generale d'Organization Scientifique, Paris, France, 1947-48; asso. dir. Coll. Study in Intergroup Relations Am. Council on Edn., Detroit, 1948-49, also asso. prof. sociology Wayne State U., 1948-49; sociologist Operations Research Office, Johns Hopkins, 1949-51; social sci. research cons., br. chief, spl. asst. to dir. Nat. Inst. Mental Health, NIH, Bethesda, Md., 1951-—. William Lincoln Honnold fellow, Franco-Am. Exchange fellow, Bordeaux, 1933-34, Robert Treat Paine fellow, Harvard, 1935-36. Mem. Am. Sociol. Assn., Phi Beta Kappa. Author: (with M. Greenblatt and D. Levinson) The Patient and the Mental Hospital, 1957; (with Denise Kandel) Psychiatric Rehabilitation: Some Problems of Research, 1964; (with Claudine G. Wirths) Lives Through the Years: Styles of Life and Successful Aging, 1965. Home: 6017 Cheshire Dr., Bethesda, Md. 20014. Office: 5454 Wisconsin Av., Chevy Chase, Md. 20203.*

WILLIAMS, Richard Tecwyn, Brit. biochemist; b. Abertillery, Mon., South Wales, Feb. 20, 1909; s. Richard and Mary (Jones) W.; B.Sc., U. Wales, 1929, Ph.D., 1932; D.Sc., U. Birmingham, 1939; Docteur Honoris Causa U. Paris, 1966; m. Josephine Teresa Sullivan, Apr. 3, 1937; children—Peter, Stephen, Josephine, Maria, Clare. Research asst. Physiology Inst. Cardiff, 1930-34; lectr. biochemistry U. Birmingham, 1934-42; sr. lectr. biochemistry U. Liverpool, 1942-48; prof. biochemistry St. Mary's Hosp. Med. Sch., U. London, 1949-—; vis. prof. toxicology N.Y. U. Sch. Medicine, 1965-66; Fellow Royal Soc., 1967; mem. food additives and contaminants com. Ministry Agr. Food and Fisheries, 1964-—. Mem. Biochem. Soc., Chem. Soc., Physiol. Soc., Pharmacol. Soc., N.Y. Acad. Sci., Toxicol. Soc. (Merit award 1968), (hon.), Pan Am. Med. Assn. (hon. life). Author: Detoxication Mechanisms, 1947, 2d edit., 1959; also numerous articles. Pioneer in study of drug metabolism and fate of toxic substances in body. Home: 95 Vernon Dr., Stanmore, Middlesex, Eng. Office: St. Mary's Hosp. Med. Sch., London W.2, Eng.*

WILLIAMS, Robert, English physician; b. London, Eng., circa 1787; M.D., Trinity Coll., Cambridge (Eng.) U., 1816. Asst. physician St. Thomas' Hosp., London, 1816, physician, 1817-45. Fellow Royal Coll. Physicians (censor 1831). Author: Elements of Medicine, 1836-41. Discovered curative power of iodide of potassium in later stages of syphilis; introduced bromide of potassium into English practice. Died Nov. 24, 1845.

WILLIAMS, Robert Evan Owen, English bacteriologist; b. London, Eng., June 30, 1916; s. Gwynne Evan Owen and Cicely (Innes) W.; M.B., B.S., U. Coll. London, 1940, M.D., 1945; m. Margaret Lumsden, Jan. 30, 1944; children—Rosamund Jean, Gwyneth Jane, Peter Evan Owen. Bacteriologist, Med. Research Council Wound Infection unit, Birmingham, Eng., 1942-46; dir. Staphylococcus, Streptococcus, Air Hygiene Lab. Central Pub. Health Lab., Colindale, Eng., 1946-60; prof. bacteriology U. London, St. Mary's Hosp. Med. Sch., 1960-—. Author: with Garrod, Shooter Hospital Infection with Blowers, 1960, 2d edit., 1966. Research, numerous publs. on causes and prevention of infection spread in hosps., air-borne infection in schs. and offices, epidemiology of streptococcal infections, bacteriology of staphylococci. Home: 26 Brampton Grove, London N.W.4. Office: St. Mary's Hosp. Med. Sch., London W.2, Eng.*

WILLIAMS, Robert Fones, Am. mathematician; b. Bessemer, Ala., July 27, 1928; s. Elgin Bow and Anne (Brown) W.; B.A. in Math., U. Tex., 1948; Ph.D. in Math., U. Va., 1954; m. Catherine Gordon Barnes, Sept. 8, 1952; 1 dau., Ellen Lindsay. Instr., U. Va., 1949-54; asst. prof. Fla. State U., Tallahassee, 1954-55; lectr. U. Wis., Madison, 1955-56; asst. prof. Purdue U., Lafayette, Ind., 1956-59; mem. Inst. for Advanced Study, Princeton, N.J., 1959-61; asst. prof. U. Chgo., 1961-63; faculty Northwestern U., Evanston, Ill., 1963-—, prof. math., 1967-—. NSF fellow, 1959-61. Mem. Am. Math. Soc. Research and publs. on continuous mappings, dimension theory, area, transformation groups, differential dynamical systems and zeta-function. Home: 2315 Sherman St., Evanston, Ill. 60201.*

WILLIAMS, Robert Hardin, Am. physician; b. Savannah, Tenn., Sept. 27, 1909; s. Fayette C. and Ella (Walker) W.; B.A., Washington and Lee U., 1929; M.D., Johns Hopkins, 1934; m. Andrea Brown, Jan. 31, 1942; children—Lee, Hugh, Alan. Practiced medicine, Seattle, 1948-—; prof. medicine U. Wash., Seattle, 1948-—, chmn. dept. medicine, 1948-63; mem. council Nat. Inst. Arthritis and Metabolic Diseases. Mem. Am. Soc. Clin. Investigation (pres.), Assn. Am. Physicians (mem. council), A.C.P., Assn. Profs. Medicine (pres.), Western Assn. Physicians (pres.), Am. Diabetes Assn., Am. Fedn. Clin. Research, Endocrine Soc. Author: Textbook of Endocrinology, 1950; Diabetes, 1960; also numerous articles. Participant in pioneering investigations of mechanisms of actions and therapeutic uses of thioureas, sulfonylureas and biguanides; co-discoverer glutathione-insulin transhydrogenase, lecithin-cholesterol acyl transferase, glycogen synthetase, certain adipose lipases. Home: 4907 N.E. 39th St., Seattle 98105.*

WILLIAMS, Robert L(eon), Am. psychiatrist; b. Buffalo, July 22, 1922; s. Leon R. and L. Paulyne (Ingraham) W.; B.A., Alfred U., 1944; M.D., Union U., 1946; m. Shirley Glynn Miller, Feb. 5, 1949; children—Karen, Kevin. Chief neurology and psychology Lackland AFB Hosp., USAF, San Antonio, 1952-55; cons. neurology and psychiatry to USAF Surgeon Gen., 1955-58; Faculty Coll. Medicine, U. Fla., Gainesville, 1958-—, prof., chmn. dept. psychiatry, 1964-—; mem. faculty various univs. part time 1949-58, including Albany Med. Coll. at Union U., Coll. Phys. and Surg. at Columbia U., Boston U., U. Tex., Georgetown U. Mem. profl. adv. council Nat. Assn. for Mental Health. Recipient USAF Certificate Profl. Achievement, USAF Surgeon Gen., 1967. Mem. Am. Psychiat. Assn. (program com.), Am. Electroencephalographic Soc., Am. Coll. Psychiatry (constn. and by-laws com.), Am. Coll. Psychiatrists, Am. Acad. Neurology (com. on fed. services), A.M.A., Group for Advancement of Psychiatry, others. Author: (with W.B. Webb) Sleep Therapy: A Bibliography and Commentary, 1966. Research and publs. on basic psychophysiology of human sleep. Home: 32 S.W. 42d St., Gainesville, Fla. 32601.*

WILLIAMS, Robert Pierce, Am. microbiologist; b. Chgo., Oct. 27, 1920; s. Gross Taylor and Cornelia (Pierce) W.; A.B., Dartmouth, 1942; S.M., U. Chgo., 1947, Ph.D., 1949; m. Elizabeth Ranstead, Dec. 20, 1944; children—Robert Eskridge, Scott Ranstead. Instr., U. So. Cal., 1949-51; faculty Baylor U. Coll. Medicine, Houston, 1951-—, prof. microbiology, 1963-—, acting chmn. dept., 1961-66. Cons. in bacteriology VA Hosp., M.D. Anderson Hosp., Houston. Fellow Am. Acad. Microbiology (charter fellow); mem. Am. Soc. Microbiology (councilor 1967-—, pres. Tex. 1956, mem. coms.), A.A.A.S., Am. Chem. Soc., Soc. Gen. Microbiology Gt. Britain. Author: (with K.K. Burdon) Microbiology, 1968; also articles. Research on biosynthesis of bacterial pigment prodigiosin; effects of anerobiosis, antibiotics, temperature on biosynthesis of prodigiosin; methods for investigating host-parasite relationship of bacterial infection with electron microscope. Home: 5122 Brae Burn Dr., Bellaire, Tex. 77401. Office: Dept. Microbiology, Baylor U. Coll. Medicine, Houston 77025.*

WILLIAMS, Robert R., Am. chemist; b. Nellore, India, of Am. parents, Feb. 16, 1886; s. Robert Runnels and Alice Evelyn (Mills) W.; std. Ottawa (Kan.) U., 1905; B.S., U. Chicago, 1907, M.S., 1908; post grad. study, 1911-12; Sc.D., Ottawa U., 1935, Ohio Wesleyan U., 1938, U. of Chicago, 1941, Columbia, 1942, Yale, 1942, Stevens Institute Tech., 1950, U. Denver, 1952; LL.D., Washington U., 1956; m. Augusta C. Parrish, Mar. 27, 1912; children—Robert Reynolds, Elizabeth Alice, Jean Parrish, June Augusta. Began as teacher in Philippines, 1908; chemist Bur. of Science, Manila, 1909-15, Bureau of Chemistry, Washington, D.C., 1915-18, Western Electric Co., 1919-24; chem. dir. Bell Telephone Labs., N.Y. City, 1925-45, dir. grants, Research Corp., 1945-51, asst. to pres., 1946-56, ret., dir., 1951-—; dir. General Aniline & Film Corp.; research asso. Tchrs. College Columbia and Carnegie Inst., 1923-34. Engaged in C.W.S. and Air Service research, Washington, D.C., World War I. Chmn. Cereal Com., Food and Nutrition Bd. Nat. Research Council since 1940. Founder Williams-Waterman Fund, Research Corp. Fellow A.A.A.S. Mem. Am. Chem. Soc., Soc. Exptl. Biology and Medicine, Soc. Biol. Chemistry, Am. Philosophical Soc., Nat. Acad. Scis., Contributor technical articles to magazines. Awarded Willard Gibbs medal, 1938; Eliott Cresson medal, 1940; designated Modern Pioneer by National Association Mfrs., 1940; 8th Annual Award of Asso. Grocery Mfrs., 1941; John Scott medal of City of Philadelphia, 1941; Charles Frederick Chandler medal by Columbia U., 1942; Perkin Medal, 1947; Carlos Manuel de Cespedes medal (Cuba), 1950. Co-author: (with T. D. Spies) Vitamin B1 and Its Use in Medicine; Toward Conquest of Beriberi, 1961. Identified thiamine (vitamin B1) as substance whose lack causes beriberi; isolated, synthesized and determined structure of thiamine, 1936; worked for enrichment of flour, bread and rice; invented methods of submarine and textile insulation; research on synthetic rubber. Died Summit, N.J., Oct. 2, 1965.

WILLIAMS, Robert Walter, Am. physicist; b. Palo Alto, Cal., June 3, 1920; s. Philip S. and Louise (Brown) W.; A.B., Stanford, 1941; M.A., Princeton, 1943; Ph.D., Mass. Inst. Tech., 1948; m. Heide Kraus, June 16, 1958; children—Paul, David, Eric. Asso. physicist Los Alamos Lab., 1943-46; research asso. Mass. Inst. Tech., 1946-48, from asst. prof. to asso. prof., 1948-59; prof. physics, U. Wash., Seattle, 1959—. Cons. Boeing Co., Seattle, 1961—. Fellow Am. Phys. Soc., Am. Acad. Arts and Scis. Research, publs. on cosmic-ray air showers and cosmic radiation, analysis of nuclear radii, measurement of properties of fundamental particles. Home: 13006 42d Av. N.E., Seattle 98125.*

WILLIAMS, Robin Murphy, Jr., Am. sociologist; b. Hillsborough, N.C., Oct. 11, 1914; s. Robin Murphy and Mabel (Strayhorn) W.; B.S., N.C. State Coll., 1933, M.S., 1935; postgrad. Cornell U., 1935-36; M.A., Harvard, 1939, Ph.D., 1943; m. Marguerite York, July 29, 1939; children—Robin Murphy III, Nancy E., Susan Y. Asst. rural sociologist N.C. State Coll., 1935-36; instr., research asst. U. Ky., 1939-42; statistician U.S. War Dept., Washington, 1942-43; sr. statistician, analyst European Theater of Operations, 1943-45; asso. prof. sociology Cornell U., 1946-48, prof. sociology, 1948-67, Henry Scarborough prof. social sci., 1967—, dir. Social Sci. Research Center, 1949-54. Fulbright lectr. U. Oslo, 1954-55; mem. sci. adv. bd. USAF, 1950-54; nat. adv. bd. Mental Health Council, 1963-67. NSF fellow, 1961-62. Mem. Am. Sociol. Assn. (pres. 1958), Sociol. Research Assn. (pres. 1958), Eastern Sociol. Soc. (pres. 1966-67), Am. Philos. Soc., Rural Sociol. Soc., Soc. for Study of Social Problems. Author: The Reduction of Intergroup Tensions, 1947; The American Soldier, Vols. I-II, 1949; American Society, 1951, 2d edit., 1960; (with Margaret Ryan) Schools in Transition, 1954; (with others) What College Students Think, 1960; Strangers Next Door, 1964; also numerous articles. Developed partial synthesis of research on social integration; analyzed sources and consequences of intergroup conflict and cooperation, especially in Am. race relations; showed importance of group factors in military morale; described major instns. and organs. and their interrelations in a nat. soc. Home: 414 Oak Av., Ithaca, N.Y. 14850.*

WILLIAMS, Robley Cook, Am. biophysicist; b. Santa Rosa, Cal., Oct. 13, 1908; s. William Claude and Anna (Cook) W.; A.B., Cornell U., 1931, Ph.D., 1935; m. Margery Ufford, June 20, 1931; children—Robley Cook, Grace W. (Mrs. Cornelio Garnica). Faculty, U. Mich., 1935-50; prof. biophysics U. Cal. at Berkeley, 1950-64, prof. molecular biology, chmn. dept., 1964—, asso. dir. Virus Lab., 1959—. Cons. USPHS, 1955-61; pres. Commn. on Molecular Biophysics, Internat. Orgn. Pure and Applied Biophysics, 1961—. Recipient Longstreth medal Franklin Inst. Phila., 1939; John Scott award, 1954. Mem. Biophys. Soc. (past pres.), Electron Microscope Soc. Am. (past pres.), Nat. Acad. Scis., Am. Acad. Arts and Scis. Research on stellar spectrophotometry, devel. aluminizing process for astron. mirrors, biophys. investigations of viruses, electron microscopy, devel. shadowing process for electron microscopy. Home: 1 Arlington Ct., Berkeley, Cal. 94707.*

WILLIAMS, Roger John, biochemist; b. Ootacumund, India, Aug. 14, 1893; s. Robert Runnels and Alice Evelyn (Mills) W.; B.S., U. Redlands, 1914; M.S., U. Chgo., 1918, Ph.D., 1919; m. Hazel Elizabeth Wood, Aug. 1, 1916 (dec. Feb. 1952); children—Roger John, Janet Elizabeth, Arnold Eugene; m. 2d Mabel Phyllis Foote Hobson, May 9, 1953; 1 son, John Wallace Hobson (foster child). Prof. chemistry U. Tex., Austin, 1939—, dir. Clayton Found. Biochem. Inst., 1941-63, cons., 1963—. Mem. Nat. Polio Bd., 1951-53; mem. med. adv. bd. Muscular Distrophy Assn. Am., 1952—. Recipient Mead Johnson award Am. Inst. Nutrition, 1941; Chandler medal Columbia, 1942. Mem. Am. Chem. Soc. (past pres.), Nat. Acad. Scis. Author: The Human Frontier, 1946; (with others) Biochemistry of the B Vitamins, 1950; Nutrition and Alcoholism, 1951; Free and Unequal, 1953; Biochemical Individuality, 1956; Alcoholism: The Nutritional Approach, 1959; Nutrition in a Nutshell, 1962; You Are Extraordinary, 1967. Research in organic chemistry, biochemistry, nutrition; discovered, isolated and synthesized pantothenic acid, a B vitamin; pioneer biochem. investigations of alcoholism; studied individual metabolic patterns. Home: 1604 Gaston Av., Austin, Tex. 78703.*

WILLIAMS, Roger Wright, Am. med. entomologist; b. Gt. Falls, Mont., Jan. 24, 1918; s. Elmer Howard and Mary (Davidson) W.; B.S., U. Ill., 1939, M.S., 1941; postgrad. Cornell U., U. N.C. Sch. Pub. Health; Ph.D., Columbia Sch. Pub. Health, 1947; postgrad. U. London Sch. Hygiene and Tropical Medicine; m. Marjorie Madeline Jones, May 9, 1943; children—Barbara, Stuart Roger. Faculty, Sch. Pub. Health and Adminstrv. Medicine, Columbia, N.Y.C., 1947—, prof. med. entomology, 1966—. Recipient numerous fellowships. Fellow A.A.A.S.; mem. Entomol. Soc. Am., Am. Mosquito Control Assn., Am. Soc. Parasitologists, Am., Royal socs. tropical medicine and hygiene, Soc. Systematic Zoology, Elisha Mitchel Sci. Soc., N.Y. Acad. Sci., Entomol. Soc. Washington, N.Y. Soc. Tropical Medicine, Corp. Bermuda Biol. Sta. for Research, Phi Sigma. Research, numerous publs. on arthropods of med. importance and diseases they transmit. Home: 16 Virginia St., Tenafly, N.J. 07670. Office: 630 W. 168th St., N.Y.C. 10032.*

WILLIAMS, Stanley Burdg, Am. psychologist; b. Superior, Wis., July 10, 1913; s. Joseph Abraham and Beulah (Burdg) W.; B.A., U. Cal. at Los Angeles, 1934, M.A., 1937; Ph.D., Yale, 1940; m. Mary Elizabeth Young, Oct. 13, 1939; children—Carol Frances, Christopher Burdg. Instr., U. Me., 1940-42, Brown U., 1942-43; asst. prof. Johns Hopkins, 1946-48; prof. psychology Coll. William and Mary, Williamsburg, Va., 1948—. Fellow Am. Psychol. Assn.; mem. Va. Acad. Sci. (pres. 1966-67). Research, publs. on habit strength and transfer of learning in animals; tests for selection of Marine Corps Officers, visual detection of signals on radar scopes. Home: 208 Kingswood Dr., Williamsburg, Va. 23185.*

WILLIAMS, Thomas Rhys, Am. anthropologist; b. Martins Ferry, O., June 13, 1928; s. Harold K. and Dorothy J. (Lehew) W.; B.A., Miami U. (O.), 1951; M.A., U. Ariz., 1956; Ph.D., Syracuse U., 1956; m. Margaret Martin, July 12, 1952; children—Rhys, Ian. Mem. faculty Syracuse (N.Y.) U., 1955-56, Sacramento State Coll., 1956-65, U. Cal., Berkeley, 1962; prof. Ohio State U., Columbus, 1965—, chmn. dept. anthropology, 1967—. Fellow Am. Anthrop. Assn.; mem. A.A.A.S., Sigma Xi. Author: The Dusun: A North Borneo Society, 1965; Field Methods in the Study of Culture, 1967. Studies of Borneo culture, soc., and personality; contbns. to theory and method in studies of relations between culture and personality; study of methods for field studies of culture.*

WILLIAMS, Virginia Rice (Mrs. Hulen Brown Williams), Am. biochemist; b. North Little Rock, Ark., Oct. 27, 1919; d. Roderick John and Mattie (Thurman) Rice; A.B., Hendrix Coll., 1940; B.S. in Home Econs., U. Ark., 1942; M.S., La. State U., 1944, Ph.D., 1947; m. Hulen Brown Williams, June 20, 1942; children—James Browning, Virginia Jean. Faculty, La. State U., Baton Rouge, 1942—, prof. biochemistry; vis. prof. U. Cal. at Berkeley, 1953. Mem. Am. Chem. Soc., Am. Soc. Biol. Chemistry, Am. Soc. Microbiology, Sigma Xi, Phi Kappa Phi, Phi Upsilon Omicron, Iota Sigma Pi, Zeta Tau Alpha. Research, publs. on food composition, coenzymes, mechanism of enzyme action, alpha deaminases. Home: 470 Castle Kirk, Baton Rouge 70808.*

WILLIAMS, William Lane, Am. anatomist; b. Rock Hill, S.C., Dec. 23, 1914; s. Oscar Kell and Cora (Mobley) W.; B.S., Wofford Coll., 1936; M.A., Duke, 1939; Ph.D., Yale, 1941. Asst. in research, University scholar, research fellow Donner Found., instr. anatomy Sch. Medicine, Yale, 1938-41, 42-43; instr. anatomy U. Rochester Sch. Medicine and Dentistry, 1941-42; asst. prof. anatomy La. State U. Sch. Medicine, 1943-45; asst. prof. anatomy U. Minn., 1945-48, asso. prof., 1948-58, dental faculty, 1955-58; prof., chmn. dept. anatomy U. Miss., 1958—. Fellow A.A.A.S.; mem. Am. Assn. Anatomists, Am. Physiol. Soc., Am. Soc. Exptl. Pathology, Endocrine Soc., Histochem. Soc., Am. Assn. Cancer Research, Am. Fedn. for Clin. Research, Soc. for Exptl. Biology and Medicine, Am. Med. Writers Assn., Minn. Med. Found., Am. Assn. for Study Liver Diseases, Am. Soc. for Cell Biology, N.Y. Acad. Scis., Sigma Xi. Asso. editor The Anat. Record, 1958—. Research, numerous publs. on genetics in dietary prodn. of cirrhosis of liver; adrenal corticoids and sex hormones as causes of cardiovascular disease. Home: 1315 N. Jefferson St., Jackson 39202. Office: 2500 N. State St., Jackson, Miss. 39216.*

WILLIAMS, William Lawrence, Am. biochemist; b. St. Cloud, Minn., June 14, 1919; s. Will L. and Edith (Mottram) W.; B.S., U. Minn., 1942; M.S., U. Wis., 1947, Ph.D., 1949; m. Opal Fay Romedy, Apr. 4, 1946; children—Lawrence K., Jan R. Biochemist, Fleischmann Lab., Stamford, Conn., 1942-44, Naval Med. Research Inst., Bethesda, Md., 1945-46; faculty N.C. State Coll., Raleigh, 1949-50; group leader Lederle Labs., Pearl River, N.Y., 1950-59; research prof. biochemistry U. Ga., Athens, 1959—. Recipient Career Research award NIH, 1962—. Mem. Am. Chem. Soc., Soc. Am. Bacteriologists, A.A.A.S., Animal Nutrition Research Council (treas.), Soc. for Study Reprodn. (dir.), Am. Soc. Biol. Chemists, Mark L. Morris Animal Found. (hon. dir. research), Sigma Xi. Research and publs. on discovery of new bacterial vitamins which proved of importance in metabolic biochemistry; developed assays for known vitamins; described biol. effects of female reproductive tract on sperm; also described properties and purified sperm antifertility factor which prevents fertilization of ova. Home: Route 3, Athens, Ga. 30601.*

WILLIAMSON, Alexander William, English chemist; b. Wandsworth, Eng., May 1, 1824; s. Alexander and Antonia (McAndrew) W.; student medicine Heidelberg (Germany) U., 1840-41; Ph.D. in Chemistry under Liebig, 1846; student math., Paris, France, 1846-49; D.C. L. (hon.), U. Durham (Eng.), 1889; LL.D., univs. Dublin (Ireland), 1878, Edinburgh, Glasgow (Scotland), 1881; m. Emma Catherine Kay, 1855; 2 children. Prof. practical chemistry Coll. U. London (Eng.), 1849-87, prof. gen. chemistry, 1855-87, examiner chemistry, mem. senate, 1874, prof. emeritus, from 1887. Largely responsible for intro. degrees Sci., formation teaching univ. London; mem. 1st elec. standards com. 1861; chief gas examiner London Bd. Trade, 1876-1901. Fellow Royal Soc., 1855 (mem. council 1859-61, 69-71, fgn. sec. 1873-89, recipient medal 1862), (hon.) Royal Soc. Edinburgh, Berlin (Germany) Chem. Soc.; mem. French Acad. Sci., 1873, Berlin Acad. Sci., Reale Accademia dei Lincei, Rome, Italy, (hon.) Royal Irish Acad., Manchester (Eng.) Literary and Philos. Soc., Soc. Pub. Analysts, (charter) Soc. Telegraph Engrs., Soc. Chem. Industry, Brit. Assn. Advancement Sci. (pres., treas. 1874), Chem. Soc. (pres., council mem.), Royal Soc. Sci. Göttingen (Germany), Reale Accademia della Scienza di Torino (Italy). Author: (with Key) Invasion Invited by the Defenceless State of England, 1858; Chemistry for Students, 1865; contbd. Dictionary of Chemistry, 1863-66. Research, publs. on organic chemistry; expts. on relationship ether and alcohol led to clarification of organic molecular structure including structure alcohols, ethers; introduced classification organic compounds based on relationship molecular structure to that of water molecule; explicated reversible reaction which contbd. understanding chem. dynamics and chem. equilibrium also devel. theory catalytic action (led to clarification by others of valence and ionization phenomena); worked on dynamics of galvanic battery, analysis of gases; 1st to produce mixed ether (chem. reaction known as Williamson Synthesis); obtained correct formulae acetone, aldehyde, 1852. Died Hindhead Surrey, Eng., May 6, 1904.

WILLIAMSON, James, Brit. biochemist; b. Mid-Yell, Scotland, Sept. 20, 1918; s. James and Anne (Nisbet) W.; B.Sc., U. Aberdeen (Scotland), 1941; M.Sc., U. London (Eng.), 1946; Ph.D., U. Liverpool (Eng.), 1949; m. Eileen Quayle Wood, July 29, 1944; children—Elizabeth Anne, Laurence James. Exptl. officer armament research dept. Woolwich Arsenal, 1942-45; research asst., May and Baker Research fellow chemotherapy dept. Liverpool Sch. Tropical Medicine, 1945-49; Imperial Chem. Industries Research fellow, Wandsworth scholar parasitology dept. London Sch. Hygiene and Tropical Medicine, 1949-54; biochemist, sr. prin. sci. officer W. African Inst. for Tryanosoniasis Research, 1954-60; mem. sci. staff Med. Research Council, Nat. Inst. for Med. Research, London, 1960—. Fellow Royal Soc. Tropical Medicine and Hygiene; mem. Biochem. Soc., Brit. Soc. for Parasitology, Protozool. Soc. Research, publs. on chemotherapy of tropical disease and basis of drug resistance in these infections, chem. and immunological analysis of trypanosomes, malaria parasites and schistomes; devel. suramin complex depot prophylaxis for animal trypanosomiasis. Home: 41, Eltham Palace Rd., London S.E.9., U.K. Office: Nat. Inst. for Med. Research, London, N.W.7., U.K.*

WILLIAMSON, Martin Bernard, Am. biochemist; b. N.Y.C., Nov. 26, 1914; s. M. Nason and Rose (Wall) W.; B.S., Bklyn. Coll., 1938; A.M., U. Mo., 1940, Ph.D., 1942; m. Marian B. Fickes, Oct. 2, 1942; children—Arthur H., David A. Biochemist, Wyeth Inst., Phila., 1942-46; instr. Harvard, 1946-48; faculty Loyola U. Sch. Med., 1948—, prof. 1954—. Mem. Am. Soc. Biologists and Chemists, A.A.A.S., Am. Chem. Soc., Soc. Exptl. Biology and Medicine, Société de Dermo-Chimie (France, hon.), Sigma Xi (pres. Loyola chpt. 1954-55). Author: Healing of Wounds, 1957. Publs. on research into biochemistry of protein metabolism during healing of wounds. Home: Box 266, Villa Park, Ill. 60181. Office: Dept. Biochemistry, Stritch Sch. Medicine, Loyola U., Chgo. 60141.*

WILLIAMSON, Ralph Elmore, Am. physicist; b. Tulsa, July 8, 1917; s. Ralph Elmore and Mary (Hutchison) W.; A.B., Phillips U., 1938; M.A., Drake U., 1939; Ph.D., U. Chgo., 1943; m. Dorothy Elinor Stuart, July 3, 1953; 1 son, Gregory Stuart. Instr., U. Chgo., 1943-44, Cornell U., 1944-46; faculty U. Toronto (Ont., Can.), 1946-53, prof. astronomy, 1948-53; astronomer David Dunlap Obs., Richmond Hill, Ont., 1946-53; staff Los Alamos Sci. Lab., 1953—. Cons. radio astronomy project Cornell U., 1946-53. Fellow Royal Astron. Soc.; mem. Am. Astron. Soc. Research in stellar interiors, radio astronomy, nuclear weapons, design and physics. Home: 661 43d St. Office: P.O. Box 1663, Los Alamos 87544.*

WILLIAMSON, William Crawford, English naturalist; b. Scarborough, Eng., Nov. 24, 1816; s. John and Elizabeth (Crawford) W.; ed. Manchester and University Coll., London; LL.D., Edinburgh, 1883. m. Sophia Wood, 1842 (dec. 1871); children—Robert Bateson, Edith; m. 2d, Annie C. Heaton, 1874; 1 son, Herbert. Apprenticed to apothecary, Scarborough, 1832; surgeon to Charlton-on-Medlock Dispensary, 1841-68; contbd. to founding Manchester (Eng.) Inst. for Diseases of Ear, 1855, surgeon, 1855-70; became 1st prof. natural history, anatomy and physiology Owens Coll., Manchester, 1851, prof. natural history, 1872, prof. botany, 1880-92, prof. emeritus, 1892; popular sci. lectr. Recipient Wallaston medal Geol. Soc., 1890. Licentiate Soc. Apothecaries. Fellow Royal Soc. (Royal medal 1874), 1854; mem. Royal Coll. Surgeons, Lit. and Philos. Soc. Manchester, 1851. Author: Melicerta, 1853; On Recent Foraminifera, 1858; Reminiscences of a Yorkshire Naturalist, 1896; also monographs on histology of Teeth, fish, scales and bone, 1840-50; Research on plants of coal-measures; 1st to point out importance of plant life forms in coal; a founder palaeobotany. Died Clapham, London, June 23, 1895.

WILLIE, Charles Vert, Am. sociologist; b. Dallas, Oct. 8, 1927; s. Louis James and Carrie (Sykes) W.; A.B., Morehouse Coll., 1948; M.A., Atlanta U., 1949; Ph.D., Syracuse U., 1957; m. Mary Sue Conklin, Mar. 31, 1962; children—Sarah Susannah, Martin Charles, James Theodore. Faculty, Syracuse (N.Y.) U., 1952-55, 60-62, 64-66, prof., chmn. dept. sociology, 1967—; faculty State U. N.Y. Med. Sch., Syracuse, 1955-60; dir. research Washington Project Pres.'s Com. Juvenile Delinquency and Youth Crime, 1962-64; vis. lectr. dept. psychiatry Med. Sch. Harvard, Boston, 1966; evaluation cons. Joint Urban Program Episcopal Ch. U. S., 1964—; lectr. in ch. and soc. Episcopal Theol. Sch., Cambridge, Mass., 1966—; cons. Ford Found. Social Sci. Research Council Summer fellow Nat. Opinion Research Center U. Chgo., 1956. Fellow Am. Sociol. Assn.; mem. Soc. Study Social Problems, Am. Pub. Health Assn. Research on community decision making, epidemiology of infant mortality and mental illness, sch. integration, juvenile delinquency and poverty. Home: 215 Edgemont Dr., Syracuse, N.Y. 13214.*

WILLIER, Benjamin Harrison, Am. embryologist, biologist; b. Weston, O., Nov. 2, 1890; s. David and Alice (Rickard) W.; B.S., Coll. Wooster, 1915, D.Sc. (hon.), 1941; Ph.D. magna cum laude, U. Chgo., 1920; m. Helen Beatrice Shipman, Sept. 11, 1919; children—Helen Kathryn (Mrs. William C. Disser), Louise (Mrs. James J. Kehoe). Faculty, Coll. Wooster, 1915-16, U. Chgo., 1920-33, U. Rochester, 1933-40; faculty Johns Hopkins, Balt., 1940—, prof. emeritus biology, 1958—. Investigator, Marine Biology Sta. Carnegie Instn. Washington (Tortugas), 1936; vis. prof. U. Fla. Coll. Medicine, 1964. Fellow A.A.A.S.; mem. Nat. Acad. Scis., Am. Philos. Soc., Institut International d'Embryologie, Internat. Soc. Cell Biology, Soc. Exptl. Biology and Medicine, L.I. Biol. Assn. (dir. 1942-60), Phi Beta Kappa, Sigma Xi. Author: (with Edgar Allen) Sex and Internal Secretions, 1932; (with R. Courrier) La Différenciation Sexuelle chez les Vertébrés, 1951; (with Jane M. Oppenheimer) Foundations of Experimental Embryology, 1964. Editor, co-author: Analysis of Development, 1955; Quar. Rev. Biology, 1941-57. Research, publs. on sexual differentiation vertebrates, embryonic origin chick germ cells, genes and plumage, pigment patterns in birds. Home: 6002 Roland Av., Balt. 21210.*

WILLIS, Edwin Roy, Am. entomologist; b. N.Y.C., Feb. 23, 1911; s. Charles William and Jennie (Stark) W.; B.S., La. State U., 1938, M.S. (fellow), 1939; Ph.D. (fellow), Ohio State U., 1947; m. Kathryn Rebecca Ruka, June 20, 1943; children—Kathryn Ann. Asst. zoology Ohio State U., 1939-41; entomologist Q.M. Research and Devel. Center, U.S. Army, Phila., 1947-54, Q.M. Research and Devel. Command, Natick, Mass., 1954-59, United Fruit Co., Norwood, Mass., also La Lima, Honduras, 1959-62; prof. entomology Ill. State U., Normal, 1962—. Mem. Am. Soc. Zoologists, Animal Behavior Soc., Entomol. Soc. Am., A.A.A.S., Am. Inst. Biol. Scis., Ill. Acad. Sci. Author: (with L.M. Roth) The Biotic Associations of Cockroaches, 1960. Research and publs. on respiratory physiology of the honey bee; behavior of malaria and yellow-fever mosquitoes, flour beetles, cockroaches, insect pests of banana plants. Home: 209 Orleans Dr., Bloomington, Ill. 61701. Office: Ill. State U., Normal, Ill. 61761.*

WILLIS, Robert, English mechanician, archeologist; b. London, Eng., Feb. 27, 1800; s. Robert Darling Willis; B.A., Gonville and Caius Coll., Cambridge (Eng.) U., 1826; m. Mary Anne Humfrey, July 26, 1832. Frankland fellow, 1826, Found. fellow, 1829; ordained priest Roman Catholic Ch., 1827; Jacksonian prof. natural exptl. philosophy Cambridge U., 1837-75. Mem. commn. to inquire into application of iron to ry. structures, 1849; lectr. applied mechanics Sch. Mines, 1853. Fellow Royal Soc., 1830; mem. Archeol. Inst. Author: Remarks on Architecture of Middle Ages, 1835; Principles of Mechanism, 1841; Architectural Nomenclature of Middle Ages, 1843; numerous works elucidating mech. constn. English cathedrals. Inventor odontograph, 1837, cymagraph, 1841. Died Feb. 28, 1875.

WILLIS, Thomas, English physician; b. Gt. Bedwin, Wiltshire, Eng., Jan. 27, 1621; s. Thomas and Rachel (Howell) W.; M.A., Christ Ch., Oxford (Eng.) U., 1642, M.B., 1646, M.D., 1660; m. Mary Fell, Apr. 7, 1657; m. 2d, Elizabeth Nicholas, Sept. 1, 1672; children—Richard, Thomas. Sedleian prof. natural philosophy Oxford U., from 1660; practiced medicine, London, Eng., from 1666. Fellow Royal Soc., 1663, Royal Coll. Physicians. Author: Cerebri anatome nervorumque descriptio et usus, 1664 (most exact account of nervous system then extant) Pathologiae cerebri et nervosi generis specimen, 1667; De anima brutorum, 1672. First to distinguish form of diabetis known as diabetes mellitus; 1st to note sugar in urine of diabetics; 1st set forth anat. relations of main cerebral arteries, in doing so discovered circle of Willis (circle of arteries at base of brain); authority on brain and nervous system; described hysteria and several varieties of nervous affliction; discovered 11th cranial nerve. Died London, Nov. 11, 1675.

WILLISTON, Samuel Wendell, Am. paleontologist, b. Boston, July 10, 1852; s. Samuel and Jane A. (Turner) W.; B.S., Kan. Agrl. Coll., 1872, A.M., 1875; M.D., Yale, 1880, Ph.D., 1885; Sc.D., Yale, 1913; m. Annie I. Hathaway, Dec. 20, 1880. Asst. in paleontology and osteology, Yale, 1876-85, demonstrator of anatomy, 1885-86, prof., 1886-90; prof. hist. geology and anatomy and dean Med. Sch., U. of Kan., 1890-1902; prof. paleontology, U. Chgo., 1902—. Asst. paleontology, U. S. Geol. Survey, 1882-85; health officer, New Haven, Conn., 1888-90; mem. Kan. State Bd. Health, 1898-1900, Bd. Med. Examiners, 1900-02. Fellow Am. Acad. Arts and Scis.; mem. Nat. Acad. Scis. Author: Manual of North American Diptera, 1896, 1908; Reports University Geological Survey of Kansas, Vols. IV, VI, 1898, 1900; American Permian Vertebrates, 1911; Water Reptiles of the Past and Present, 1914; also about 250 scientific papers on entomology, zoology, sanitation, comparative anatomy and paleontology of reptiles and amphibians. Asst. editor Science, 1885-86. Died Aug. 30, 1918.

WILLMAN, Harold Bowen, Am. geologist; b. Newcastle, Ind., July 30, 1901; s. Earnest Floyd and Gay (Bowen) W.; A.B., U. Ill., 1926, A.M., 1928, Ph.D., 1931; m. Martha Evangeline Righter, Aug. 15, 1931 (dec.); children—Martha Gay (Mrs. Morris K. Holland), Virginia Anna. With Ill. State Geol. Survey, 1926—, geologist, head stratigraphy and areal geology sect., 1945—. Mem. Geol. Soc. Am., A.A.A.S., Sigma Xi. Research, publs. on glacial geology, Ordovician, Silurian, Pennsylvanian stratigraphy, geologic mapping of Ill. Home: 411 W. Indiana St. Office: State Geol. Survey, Urbana, Ill. 61801.*

WILLMAN, Vallee Louis, Am. physician; b. Greenville, Ill., May 4, 1925; s. Philip Louis and Marie (Doll) W.; student U. Ill., 1942-43, 45-47; M.D., St. Louis, U., 1951; m. Melba L. Carr, Feb. 2, 1952; children—Philip, Elizabeth, Susan, Stephan, Mark, Timothy, Jane, Sarah, Vallee. Ellen McBride fellow St. Louis U., 1956-57, faculty, 1957—, prof. surgery, 1964—, staff surgeon St. Louis U. Hosps., 1957—. Cons. John Cochran VA Hosps., 1952—. Mem. Am. Surg. Assn., Am. Physiol. Soc., Soc. U. Surgeons, Am. Assn. Thoracic Surgery, Internat. Cardiovascular Soc., Phi Beta Kappa, Alpha Omega Alpha. Research, numerous publs. on physiology of heart, transplantation of heart, circulatory physiology, extracorporal circulation. Home: 450 Sherwood St., Webster Groves, Mo. 63119. Office: 1325 S. Grand Av., St. Louis 63104.*

WILLMER, Edward Nevill, English histologist; b. Birkenhead, Eng., Aug. 15, 1902; s. Arthur Washington and Janet Mary (Cooper) W.; B.A., Oxford U., 1924; M.Sc., Manchester, 1928, M.A., Cambridge, 1930, Sc.D., 1945; m. Henrietta Noreen Rowlatt, Mar. 11, 1939; children—Arthur Patrick, Frances Janet, Erica Tickell, Hugh Richard Nevill. Demonstrator, Manchester U., 1924-29; lectr. Cambridge U., 1930-48, reader, 1948-66, prof. histology, 1966—. fellow Clare Coll., 1936—. Fellow Royal Soc., 1960. Mem. Brit. Assn. for Advancement Sci. (council 1960-65), Cambridge Philos. Soc. (pres. 1964-65), Physiol. Soc., Soc. for Exptl. Biology, Soc. Cell Biology. Author: Tissue Culture, 1934; Retinal Structure and Color Vision, 1946; Cytology and Evolution, 1960. Editor: Cells and Tissues in Culture, 1966. Publs. on studies of growth and differentiation of cells in culture; respiration of tropical fishes; color vision in central fovea; differentiation in an amoeba. Home: 11 Park Terrace, Cambridge, Eng.*

WILLOUGHBY, Francis, English naturalist, biologist; b. Middleton, Eng., 1635; s. Francis and Cassandra (Ridgeway) W.; B.A., Trinity Coll., Cambridge (Eng.) U., 1656; M.A., 1659; pupil of James Duport; also studied at Oxford U.; m. Emma Bernard, 1668; children—Francis, Cassandra (Mrs. James Brydges), Thomas. Accompanied John Ray bot. expdn. through No. midland counties Eng., 1662, also Europe, collecting natural history specimens. Original fellow Royal Soc., 1663. Author: Ornithologiae libritres, 1676, English edit., 1678; De Historia Piscium, 1686. First naturalist who treated study birds as sci., also 1st to make rational classification (system is basis of ornithol. classification of Linnaeus). Died Middleton, July 3, 1672.

WILLS, James Henry, Am. pharmacologist; b. Richmond, Va., Aug. 7, 1912; s. James Henry and Sarah Macy (Baughman) W.; B.S., Va. Poly. Inst., 1934; M.S., Med. Coll. Va., 1936; Ph.D., U. Rochester, 1941; m. Ruth Elizabeth Cass, June 12, 1948. Faculty, U. Rochester, 1941-46, U. Tenn. Med. Sch., 1946-47; asst. chief pharmacology br. Chem. Corps Med. Labs., Army Chem. Center, Md., 1947-49, chief, 1949-59, asst. chief physiology div. U.S. Army Chem. Research and Devel. Labs., 1956-60, chief, 1960-65; asst. dir. Inst. Exptl. Pathology and Toxicology, Albany (N.Y.) Med. Coll., 1965—, prof. pharmacology, 1965—. Exec. sec. Pesticide Control Bd., N.Y. State Dept. Health, 1965—; lectr. U. Md., 1950-65. Recipient Exceptional Civilian Service award U.S. Army, 1965. Sec. of Army fellow, 1958. Mem. Am. Chem. Soc., Am. Physiol. Soc. (Porter fellow 1941), Am. Soc. for Pharmacology and Exptl. Therapeutics, Research Soc. Am., Soc. for Exptl. Biology and Medicine (pres. Md. sect. 1961-63), A.A.A.S., N.Y. Acad. Scis., Mental Health Soc. Harford County, Md. (pres. 1962-63), Sigma Xi, Pi Delta Epsilon, Phi Lambda Upsilon, Phi Kappa Phi. Contbr. articles to sci. jours., textbooks. Home: 23 W. Bayberry Rd., Glenmont, N.Y. 12077.*

WILLS, Lawrence Alfred, Am. physicist; b. Portland, Ore., Feb. 16, 1908; s. Alfred N. and Louise (Rich) W.; B.A., Reed Coll., 1929; Ph.D., N.Y. U., 1933; m. Frances Nichols, July 12, 1931. Mem. faculty Coll. City N.Y., 1933—, prof., 1963—. Cons. United Nuclear Corp. Am., White Plains, N.Y., 1951-65, Radioptics, Inc., 1966—. Research in nuclear reactor physics, radiation transport shielding and heating. Home: 89 Summit Dr., Hastings-on-Hudson, N.Y. 10706.*

WILLSTÄTTER, Richard, German chemist; b. Karlsruhe, Germany, Aug. 13, 1872; Ph.D., Munich, 1894; taught at U. Munich, 1894-1905; asso. prof. Tech. Coll., Zurich, 1905-12; prof. chemistry, Berlin, 1912-16, dir. Kaiser Wilhelm Inst.; prof. organic chemistry, dir. chem. lab. U. Munich, 1916-25; worked privately at Munich, then Zurich, Switzerland. Recipient Nobel prize in chemistry (for work on plant pigments), 1915; Davy medal 1932, Gibbs medal, 1933. Fellow Royal Soc., 1928. Began study of chlorophyll and assimilation of carbon dioxide by plants, 1907; proved existence of 2 chlorophylls; showed relationship of flower pigments to chlorophyll, also relation of blue of cornflower to red of rose; studied alkaloids and their derivatives, including cocaine (led to comml. prodn. of local anesthetics), tropine, atropine, 1903; synthesized lecithin, 1904, chlorophyll, 1906; studied plant and animal enzymes, carotinoids, structure of anthocyanins; perfected technique called partition chromatography. Died Locarno, Switzerland, Aug. 3, 1942.

WILMER, Harry Aron, Am. physician; b. New Orleans, Mar. 5, 1917; s. Harry A. and Leona (Schlenker) W.; B.S., U. Minn., 1938, M.B., 1940, M.S., 1940, M.D., 1941; Ph.D., 1944; m. Jane Harris, Oct. 31, 1945; children—Harry, John, Thomas, James, Mary. Practice medicine, specializing in psychiatry, Palo Alto, Cal., 1951-55, 58-64, San Francisco, 1964—; faculty Med. Sch. Stanford, 1949-56, 58-64; mem. staff Mayo Clinic, Rochester, Minn., 1957-58; clin. prof. psychiatry Sch. Medicine U. Cal., San Francisco, Langley Porter Neuropsychiatric Inst., 1964—. Cons. USPHS, 1949-55, Cal. Dept. Mental Hygiene, 1949—, mental health study sect. Nat. Inst. Mental Health, 1956-57, USIA, 1958-60, Cal. Dept. Corrections, San Quentin Prison, 1961—. Recipient Rollin E. Cutts prize in surgery U. Minn., 1940. NRC fellow Johns Hopkins, 1944-45. Mem. A.M.A., Am. Psychiat. Assn., A.A.A.S., Sigma Xi, Alpha Omega Alpha. Author: Huber the Tuber, 1941; Corky the Killer, 1945; This is Your World, 1952; Social Psychiatry in Action, 1958; The Mind, First Steps, 1963; also numerous articles. Research on group psychotherapy, therapeutic community, social psychiatry, correctional psychiatry, mental hosp. milieu therapy, television research in teaching. Home: 1317 Forest Av., Palo Alto, Cal. 94301. Office: U. Cal. Sch. Medicine, San Francisco 94122.*

WILMS, Max, German surgeon; b. Hünshoven/Geilenkirchen, Germany, Nov. 5, 1867; prof., Leipzig, Germany, also Basel, Switzerland, Heidelberg, Germany. Author: Lehrbuch der Chirurgie, 2 vols., 1908-09. Introduced limited paravertebral rib resection into surg. treatment of Tb. Died Heidelberg, May 14, 1918.

WILSDORF, Doris Kuhlman (Mrs. Heinz Gerhard Friedrich Wilsdorf), physicist; b. Bremen, Germany, Feb. 15, 1922; d. Adolf Friedrich and Elsa (Dreyer) Kuhlmann; B.Sc., Gottingen U., 1944, M.Sc., 1946, Ph.D., 1947; D.Sc., U. Witwatersrand, 1954; m. Heinz Gerhard Friedrich Wilsdorf, Jan. 4, 1950; children—Gabriele, Michael. Came to U.S., 1956, naturalized, 1961. Postdoctoral asst. Gottingen U., 1948, U. Bristol, Eng., 1949; lectr. U. Witwatersrand, Johannesburg, S. Africa, 1950-56; faculty U. Pa., 1957-64; prof. engring. physics U. Va., Charlottesville, 1964-66; univ. prof. applied sci., 1966—. Recipient medal S.E. sect. Am. Soc. for Engring. Edn., 1965, 66; J. Shelton Horseley award Va. Acad. Sci., 1965. Fellow Am. Phys. Soc., Am. Soc. Engring. Edn.; mem. S. African Phys. Soc. (founder), Delta Kappa Gamma. Research solid state physics, particularly theory crystal defects.*

WILSDORF, Heinz Gerhard Friedrich, physicist; b. Pennekow, Germany, June 25, 1917; s. Moritz F. and Therese (Papenfuss) W.; M.S., Berlin U., 1944; Ph.D., Gottingen U., 1947; D.Sc., U. Witwatersrand, Johannesburg, S. Africa, 1954; m. C.J. Doris Kuhlmann, Jan. 4, 1950; children—Gabriele, Michael. Came to U.S., 1956, naturalized, 1961. Postdoctoral fellow in physics, Gottingen, 1947-49; research officer Nat. Phys. Lab., Pretoria, S. Africa, 1950-54, prin. research officer, 1954-56; sr. research metallurgist Franklin Inst. Labs., Phila., 1956-58, mgr. metal physics lab., 1958-60, tech. dir. solid state scis. div., 1960-63; prof. materials sci., chmn. dept. U. Va., Charlottesville, 1963—, Wills Johnson prof. materials sci., 1966—. Research in phys. metallurgy, electron microscopy. Home: 1855 Westview Rd., Charlottesville, Va. 22903.*

WILSON, A. T., New Zealand chemist; b. Feb. 8, 1930; s. Percy Frederick and Dorothy (Evans) W.; B.Sc., Victoria U. (New Zealand), 1950, M.Sc. with 1st class honors, 1951; Ph.D., U. Cal. at Berkeley, 1954; m. Joyce Dunlop, Aug. 22, 1960. With Standard Oil Co. Ind., 1954-56, New Zealand Inst. Nuclear Sci., 1957-60, Victoria U., Wellington

(New Zealand), 1960—. Recipient Easterfield medal. Research, numerous publs. on photosynthesis, application of radioisotopes to biochem. and geol. problems. Address: Victoria U. of Wellington, Wellington, New Zealand.*

WILSON, Sir **Alan Herries**, English physicist; b. Wallasey, Eng., July 2, 1906; s. Herries and Anne (Bridges) W.; student Emmanuel Coll., Cambridge, Eng., 1923-26; m. Margaret Constance Monks, July 28, 1934 (dec. 1961); children—Peter Robert, John Richard. Lectr. math. U. Cambridge, 1933-45; with Courtaulds, Ltd., 1945-62, mng. dir., 1954-57, dep. chmn., 1957-62; chmn. Glaxo Group, Ltd., London, 1963-J; dir. Internat. Computers & Tabulators Ltd., Created Knight Bachelor, 1961. Fellow Royal Soc., 1942. Author: Theory of Metals, 1936; Semi-Conductors and Metals, 1939; Thermodynamics and Statistical Mechanics, 1957. Discovered fundamental principles of semi-conductors, 1931. Home: 65 Oakleigh Park S., London N.20. Office: 6/12 Clarges St., London, W.1., Eng.*

WILSON, Albert George, Am. astronomer; b. Houston, July 28, 1918; s. Arthur Rector and Hazel (Straw) W.; B.S.E.E., Rice Inst., 1941; M.S., Cal. Inst. Tech., 1942, Ph.D., 1947; m. Donna Scott Kirby, May 31, 1961. Staff astronomer Mt. Wilson and Palomar Observatories, 1947-53; astronomer Lowell Obs., Flagstaff, Ariz., 1953-57, dir., 1954-57; sr. sci. staff Rand Corp., Santa Monica, Cal., 1957-65; asso. dir. Douglas Advanced Research Lab., Huntington Beach, Cal., 1966—. Cons. Indian Govt., U. Grants Commn.; mem. NRC of Peace Research Inst., State Dept. Am. Specialist Program. Fellow Royal Astron. Soc.; mem. Am. Astron Soc. (council 1955-58), Internat. Acad. Astronautics, Astronom. Soc. Pacific (dir. 1955-58), Internat. Astronom. Union, Phi Beta Kappa, Sigma Xi, Tau Beta Pi. Editor: ICARUS, Internat. Jour. of Solar System. Research, publs. on galaxies, clusters, observational cosmology, atmosphere of Mars, epistemological methodologies. Home: 19936 Grandview St., Topanga, Cal. 90290. Office: 5251 Bolsa Av., Huntington Beach, Cal. 92646.*

WILSON, Alexander, Scottish astronomer; b. St. Andrews, Scotland, 1714; s. Patrick Wilson; M.A., St. Andrews U., 1733, M.D. (hon.), 1763; several sons, including Patrick. Asst. to surgeon and apothecary, London, Eng., 1737-39; established typefoundry, St. Andrews, 1742, moved to Camlachie, nr. Glasgow, Scotland, 1744; 1st prof. practical astronomy U. Glasgow, 1760-84. Original mem. Royal Soc. Edinburgh. Author: Thoughts on General Gravitation and Views Thence arising as to the State of the Universe; A Specimen of some of the Printing Types Cast in the Foundry of Alexander Wilson & Sons, 1772. Contbr. articles to sci publs. Discovered that sun-spots are cavities in luminous matter surrounding sun, 1769; improved and produced printing types. Died Edinburgh, Oct. 18, 1786.

WILSON, Alexander, ornithologist; b. Seed Hills of Paisley, Scotland, July 6, 1766; s. Alexander and Mary (McNab) W. Published volume Poems, 1760; came to Am., 1794; took over sch. at Gray's Ferry on Schuylkill River, Phila., 1802; asst. editor Abraham Rees's Cyclopaedia; visited ornithological wilderness West of Alleghanies, 1810. Author: The Foresters, 1805; American Ornithology, 9 vols., 1808-14; Poems; Chiefly in the Scottish Dialect, 1816. Founder Am. ornithology; discovered 39 species of birds; wrote on nature and travel in Am. Died Phila., Pa., Aug. 23, 1813.

WILSON, Archie Spencer, Am. chemist; b. Tekoa, Wash., Jan. 19, 1921; s. Andrew Hamilton and Viola (Sledge) W.; B.S., Ia. State U., 1946; M.S., U. Chgo., 1950, Ph.D., 1951; m. Ivon Marie Teeter, June 11, 1944; children—Andrea Lou, Ronald Spencer, Steven Daniel. Faculty, U. Chgo., 1948-50, U. Neb., 1950-51; sr. scientist Hanford Labs., Richland, Wash., 1951—. Lectr., Center for Grad. Study, Richland, 1954—; sci. adviser 2d Internat. Conf. Peaceful Uses Atomic Energy, Geneva, Switzerland, 1958. Mem. Am. Chem. Soc., Sigma Xi. Contbr. articles to profl. lit. Research in x-ray structure determination of uranium and thorium alloys and compounds; basic ruthenium chemistry; theory and practice of amine extraction applied to nuclear fuel processing; patentee nuclear fuel processing field. Home: 2009 Van Giesen St. Office: P.O. Box 999, Richland, Wash. 99352.*

WILSON, Arthur James Cochran, physicist; b. Springhill, N.S., Can., Nov. 28, 1914; s. Arthur A.C. and Hildegarde (Geldert) W.; B.Sc., Dalhousie U., Halifax, Can., 1934, M.Sc., 1936; Ph.D., Mass. Inst. Tech., 1938, U. Cambridge (Eng.), 1942; m. Harriett Charlotte Friedeberg, June 24, 1946; children—John Henry Veit, Howard James Cochran, Mary Jean Alexandra. Faculty, U. Coll., Cardiff, Wales, 1945-65, prof. physics, head dept., 1954-65; prof. crystallography U., Birmingham, Eng., 1965—. Fellow Royal Soc., 1963, Inst. Physics, Instn. Metallurgists; Internat. Union Crystallography (past mem. exec. com.). Author: X-ray Optics, 1949; Mathematical Theory of X-ray Powder Diffractometry, 1963; also numerous articles. Introduced statis. methods in X-ray crystallography, research on design and aberrations of instruments, especially powder diffractometers. Office: The University, P.O. Box 363, Birmingham 15, Eng.*

WILSON, Basil Wrigley, civil engr.; b. Cape Town, S.Africa, June 16, 1909; s. George Hough and Sarah (Hearn) W.; B.S., U. Cape Town, 1931, D.Sc., 1953; M.S., U. Ill., 1939, C.E., 1940; m. Elizabeth Mary Davenport, Feb. 27, 1941; children—Mary D. (Mrs. Robert A. Holmstrom), Richard L., Gerald H., Derek W. Engr., S.African Rys. and Harbors, 1932-52; faculty Tex. A. and M. U., 1953-61, prof. engring. oceanography, 1959-61; sr. engr. Nat. Engring. Sci. Co., Pasadena, Cal., 1961-64; dir. engring. oceanography Sci. Engring. Assos., San Marino, Cal., 1964—. Commonwealth (Harkness) Fund Service fellow, 1938-39. Fellow Am. Soc. C.E. (Arthur M. Wellington prize 1952); mem. Inst. C.E. (London), S.African Inst. C.E. (award 1959), Am. Geophys. Union, Internat. Assn. Hydraulic Research, A.A.A.S., Permanent Internat. Assn. Nav. Congress, Seismol. Soc. Am., Sigma Xi. Contbr. sect. to Ency. Brit., 1961, Studies on Oceanography, 1964. Research, numerous publs. on nature long period waves in bays and harbors; developed numerical computer technique for prediction ocean waves generated by moving storms, system for numerical prediction hurricane-storm tides, theory dynamics mooring cable, moored ship motion in ocean waves. Home: 529 S. Winston Av., Pasadena, Cal. 91107. Office: 2450 Mission St., San Marino, Cal. 91108.*

WILSON, Benjamin, English physicist; b. Leeds, Eng., 1721; s. Major and Elizabeth (Gates) W.; ed. Leeds Grammar Sch.; student painting under Thomas Hudson; m. Miss Hetherington, circa 1771; 7 children. Clk. registry of prerogative ct. Doctors' Commons, to registrar Charter house; portrait painter, Dublin, Ireland, 1748-50, London, from 1750; mgr. pvt. theater for Duke of York, Westminster, Eng.; serjeant-painter Eng., from 1764; James Worsdale painter to Bd. Ordinance, from 1767. Fellow Royal Soc. (Gold medal for elec. expts. 1760); mem. several fgn. socs. Author: An Essay towards an Explication of the Phaenomena of Electricity deduced from the Aether of Sir Isaac Newton, 1746; A Treatise on Electricity, 1750; Observations on a Series of Electrical Experiments, 1756. Made important elec. expts. especially on Leyden jars and lightning conductors, 1760. Died June 6, 1788.

WILSON, Brian Graham, physicist; b. Belfast, North Ireland, Apr. 9, 1930; s. Charles Wesley and Isobel (Ferguson) W.; B.Sc. (first class honors), Queen's U., Belfast, 1952; Ph.D., Nat. U. Ireland, 1956; m. Barbara Elizabeth Wilkie, May 27, 1959; children—Bronwen Moira, Patrick Duncan, Brendan Charles. Postdoctoral fellow NRC, Ottawa, Ont., Can., 1955-57; officer-in-charge Sulphur Mountain Lab., Banff, Can., 1957-60; asso. prof. physics U. Calgary, Alta., Can., 1960-65, prof. physics, 1965—, dean faculty arts and sci., 1967—. Mem. asso. com. on space research NRC, 1959—. Mem. Am. Geophys. Union, Canadian Assn. Physicists, Inst. Physics London (asso.), Phys. Soc. London. Research, publs. in solar modulation of earth's environment, cosmic x-rays. Home: 640 Crescent Rd., Calgary, Alta, Can.*

WILSON, Carl Louis, Am. botanist; b. Springfield, O., Aug. 8, 1897; s. Charles Albert and Emma (Roettig) W.; A.B., U. Denver, 1919; M.A., Cornell U., 1921, Ph.D., 1924; m. Ellen Margaret Sands, Aug. 19, 1923; 1 dau., Margaret Joanne (Mrs. David E. Kinney). Asst. botany Cornell U., 1919-24; faculty Dartmouth, Hanover, N.H., 1924—, prof. botany, 1935-63, emeritus prof., 1963—. Fellow A.A.A.S.; mem. Sigma Xi, Phi Sigma, Lambda Chi Alpha. Author: (with W.E. Loomis) Botany, 1952. Contbr. articles on floral antomy and morphology, anatomy of vascular plants to profl. jours. Home: 6 Brewster Rd., Hanover, N.H. 03755.*

WILSON, Cedric William Malcolm, Brit. pharmacologist; b. Edinburgh, Scotland, Nov. 23, 1925; s. Malcolm and Maud (Stone) W.; B.Sc., U. Edinburgh, 1947, M.B. Ch.B., 1949, Ph.D., 1954, M.D., 1958; m. Joan Mowbray Russell, Sept. 18, 1951; children—Brennan, Kenneth, Helen, Colum. Med. Research Council fellow, 1951-54, with dept. pharmacology U. Edinburgh, 1951-53, Nat. Inst. for Med. Research, London, 1953-54; Wellcome Research fellow Middlesex Hosp. Med. Sch., London, 1954-55; lectr. pharmacology and gen. therapeutics U. Liverpool (Eng.), 1955-64; prof., chmn. dept. pharmacology Trinity Coll., U. Dublin (Ireland), 1965—. Lilly Research fellow Nat. Heart Inst., Bethesda, Md., 1958-59. Mem. Am. Coll. Clin. Pharmacology, and Chemotherapy, Collegium Internationale Neuro Psychopharmacologium, Brit. Pharmacological Soc., Physiol. Soc. Research, publs. on mechanisms of drug action and their clin. evaluation, patient use of prescribed drugs, control of allergies, drug dependence. 77 Orwell Rd., Rathgar, Dublin 6, Ireland.*

WILSON, Charles B., Am. physician; b. Neosho, Mo., Aug. 31, 1929; s. Byron Sanders and Maggie (Polson) W.; B.S., Tulane U., 1951, M.D., 1954; m. Mary Barksdale Craig, Jan. 14, 1956; children—Rebecca, Robert Craig, Byron Sanders. Intern, Charity Hosp., New Orleans, 1954-55, resident pathology, 1955-56; resident neurosurgery Ochsner Found. and Tulane U., 1956-60; faculty Tulane U., 1960-61, La. State U., 1961-63; practice medicine, specializing in neurosurgery, Lexington, Ky.; asst. prof., chmn. div. neurosurgery U. Ky., 1963—, asso. prof., 1965—; chief neurosurgery U. Ky. Hosp., 1963—; cons., chief neurosurgery VA Hosp., 1963—; cons. Pub. Health Service Hosp., 1966—. Recipient Borden Undergrad. Research award; Isadore Dyer Scholarship award. Diplomate Am. Bd. Neurol. Surgery. Mem. A.C.S., Am., So. neurosurg. socs., Harvey Cushing Soc., Am. Assn. Neuropathologists. Research, publs. in chemotherapy of malignant brain tumors, tissue culture of brain tumors and their sensitivity to chemotherapeutic agts., heterotransplantation of human brain tumors from tissue culture. Home: 137 S. Hanover St. Office: U. Ky. Med. Center, Lexington, Ky. 40506.*

WILSON, Charles Owens, Jr., Am. pharmacologist; b. Bothell, Wash., Jan. 25, 1911; s. Charles Owens and Aura Belle (Steele) W.; Ph.G., U. Wash., 1932, Ph.C., 1933, B.S., 1934, M.S., 1935, Ph.D., 1938; m. Vaughn Claire Howell, June 16, 1936; children—Charles Owens III, Heather Claire, Bonnie Claire, Felix Daniel. Teaching asst. Coll. Pharmacy, U. Wash., 1934-38; asst. prof. pharm. chemistry Coll. Pharmacy, George Washington U., 1938-40; faculty in pharm. chemistry Coll. Pharmacy, U. Minn., 1940-48; prof., 1947-48; prof. pharm. chemistry Sch. Pharmacy, U. Tex., 1948-59; dean, prof. pharm. chemistry Sch. Pharmacy, Ore. State U., 1959—. Engaged in malarial research OSRD, 1944-45. Mem. Am. Pharm. Assn. (chmn. sci. sect. 1940-42), Am. Chem. Soc., A.A.A.S., Sigma Xi, Kappa Psi. Phi Kappa Phi, Phi Sigma, Rho Chi. Author: (with Ole Gisvold) Textbook of Organic Medicinal and Pharmaceutical Chemistry, 5th edit., 1966; (with T. O. Soine) Rogers' Inorganic Pharmaceutical Chemistry, 8th edit., 1966; (with Tony E. Jones). The American Durg Index, 14th edit., 1968. Contbr. articles. Research on synthesis of organic med. compounds, hypnotics, local anesthetics. Patentee barbituates. Home: 4010 S.W. Fairhaven Dr., Corvallis, Ore. 97330.*

WILSON, Charles Thomson Rees, Scottish physicist; b. Glencorse, Scotland, Feb. 14, 1869; s. John and Annie Clerk (Harper) W.; ed. Owen's Coll., Manchester; M.A., Sidney Sussex Coll., Cambridge, 1892; m. Jessie Fraser, 1908; 2 sons, 2 daus. With Cavendish Lab., Cambridge, after 1892; Clerk Maxwell student, 1896-99; fellow Sidney Sussex Coll.; lectr. and demonstrator, Cambridge U., 1900-18, reader elec. meteorology, 1918-25, Jacksonian prof. natural philosophy, 1925-34, ret. Apptd. observed meteorol. physics Solar Physics Obs., 1913. Recipient Hopkins prize Cambridge Philos. Soc., 1920, Gunning prize Royal Soc. Edinburgh, 1921, Howard Potts medal Franklin Inst., 1925, (with A. H. Compton) Nobel prize in physics, 1927. Fellow Royal Soc., 1900, recipient Hughes medal, 1911, Royal medal, 1922, Copley medal, 1935. Research on condensation nuclei, ions, atmospheric electricity, from 1895; discovered that ionized particles can act as nuclei for water drop formation even in absence of dust; developed method of photographing activity of ionized particles by inventing Wilson cloud chamber, 1896, applied it to radiation and particle physics, 1923; developed method for protection of barrage balloons from lightning during World War II. published a theory of thunderstorm electricity, 1956. Died Carlops, Scotland, Nov. 15, 1959.

WILSON, Sir **Charles William**, English archaeologist; b. Liverpool, Eng., Mar. 14, 1836; s. Edward and Frances (Stokes) W.; student Liverpool Coll., 7 years, Cheltenham Coll., 2 years, Bonn U., 1 year; LL.D., Edinburgh U., 1886, hon. master, 1893; m. Olivia Duffin, Jan. 22, 1867; 4 sons, 1 dau. Commd. in Royal Engrs., 1855; surveyed boundary between B.C. and U. S. from Lake of Woods west to Pacific Ocean; surveyed Jerusalem, after 1864, Scotland, 1866-68; helped demarcate new frontier under treaty of Berlin, after 1873; began work dividing Anatolia in Asia Minor into 4 consulates, 1879. Recipient Silver medal Soc. Arts, 1891. Fellow Royal Soc., 1874. Author: Report on the Survey of Jerusalem, 1866; Report on the Survey of Sinai, 1869; From Korti to Khartoum, 1885; Lord Clive, 1890; Handbooks for Asia Minor and Constantinople, 1892, 95. Studied geography, topology, culture of countries visited during his career. Died 1905.

WILSON, Christopher L(umley), chemist; b. Leeds, Eng., Aug. 31, 1909; s. William Fletcher and Mary Elizabeth (Lumsden) W.; B.S., Leeds U., 1930; Ph.D., U. Coll., London, 1932, D.Sc., 1939, fellow Royal Inst. Chemistry, 1941; m. Evelyn Martha Rye, Mar. 25, 1937; children—Jeremy David and Elizabeth Ann Hamilton (twins). Came to U. S., 1946, naturalized citizen. Asst. lectr. Univ. Coll., London, 1932-36, lectr., 1936-46; research chemist Imperial Chem. Industries, Manchester, Eng., 1939-41, academic relations officer, 1941-42; dir. research Revertex Limited, London, 1942-46; tech. officer Ministry of Supply, 1945-46; prof. chemistry U. of Notre Dame, 1946-48, 54-61, Ohio State U., 1948-54; prof. High Point Coll., 1961—. Pres., dir. Unipoint Labs. and Unipoint Industries, Inc.; officer, dir. various corps. Fellow of Chem. Soc. London; mem. Soc. Plastics Industry, Soc. Chem. Industry (London), Fedn. Socs. Paint Tech., Am. Chem. Soc., Am. Electro. Chem. Soc., Assn. Cons. Chemists and Chem. Engrs., Sigma Xi, Phi Lambda Upsilon. Holder many chem. patents; inventor polyvinyl synthetic sponge, 1947; research and publs. in phys. and organic chemistry, especially in relation to mechanism of organic reactions also in organic electrochemistry, furan chemistry mechanisms of reactions, spectroscopy; contbr. indsl. inventions in plastic foams, polyurethanes. Home: 207 Shadow Valley. Office: 416 S. Elm St., High Point, N.C. 27260.*

WILSON, David Wright, Am. biochemist; b. Knoxville, Ia., Jan. 4, 1889; s. James Ewing and Katherine (Wright) W.; B.S., Grinnell (Ia.) Coll., 1910; M.S., U. Ill., 1912; Ph.D., Yale, 1914; m. Helene Connet, Nov. 24, 1921; children—John Ewing (dec.), Thomas Hastings, Juliet Connet (Mrs. William John Welch). Began career with Johns Hopkins Med. Sch., 1914-22, successively as asst., asso. and asso. prof. of physiol. chemistry; Benjamin Rush prof. of physiol. chemistry, U. Pa. Med. Sch., Phila., 1922-57, prof. emeritus biol. chemistry, 1957—, head dept., 1922-57, ret. Mem. Am. Soc. Biol. Chemists (pres. 1953-54), Am. Physiol. Soc., Soc. for Exptl. Biology and Medicine, A.A.A.S., Physiol. Soc. Phila., Am. Chem. Soc., Nat. Acad. Scis., Phi Beta Kappa, Sigma Xi. Author: A Laboratory Manual of Physiological Chemistry, 1928; (with Mier and Reimann) Preparation and Measurement of Isotopic Tracers, 1946. Discovered that carbon dioxide could be incorporated into position 2 of pyrimidines. Died July 13, 1965.

WILSON, E(dgar) Bright, Jr., Am. chemist; b. Gallatin, Tenn., Dec. 18, 1908; s. Edgar Bright and Alma (Lackey) W.; B.S., Princeton, 1930, A.M., 1931; Ph.D. (teaching fellow), Cal. Inst. Tech., 1933; A.M. (hon.), Harvard; m. Emily Buckingham, June 15, 1935 (dec. Sept. 1954); children—Kenneth, David, Nina (Mrs. E.K. Cornell); m. 2d, Thérèse Bremer, July 25, 1955; children—Anne, Paul, Steven. Jr. fellow Harvard, Cambridge, Mass., 1934-36, faculty, 1936—, Theodore William Richards prof., 1947—; Carothers research prof., 1951-52; research dir., dep. dir., weapons systems evaluation group Dept. Def., 1952-53. Guggenheim fellow, also Fulbright grantee Oxford, Eng., 1949-50; recipient Medal for Merit, U.S. Govt., 1948. Mem. Am. Chem. Soc. (prize 1937, Debye award 1962, J.F. Norris award 1966), Nat. Acad. Sci., Am. Phys. Soc., Am. Philos. Soc., Operations Research Soc. Am., Am. Acad. Arts and Scis. Author: (with L. Pauling) Introduction to Quantum Mechanics, 1934; Introduction to Scientific Research, 1952; (with J.C. Decius, P.C. Cross) Molecular Vibrations, 1955. Research and publs. on theory of vibrational and rotational spectra of molecules; exptl. infrared and microwave spectroscopy; molecular structure studies, application of quantum mechanics to chemistry. Home: 7 Martha's Point Rd., Concord, Mass. 01742. Office: 12 Oxford St., Cambridge, Mass. 02138.*

WILSON, Edmund Beecher, Am. zoologist; b. Geneva, Ill., Oct. 19, 1856; s. Judge Isaac C. and Caroline (Clark) W.; Ph.B., Yale, 1878, LL.D., 1901; Ph.D., Johns Hopkins, 1881; LL.D., 1902; postgrad. univs. of Cambridge, and Leipzig, 1881-82; LL.D., U. Chgo., 1901; Sc.D., U. Cambridge, 1909, Harvard, 1924; M.D., U. Leipzig, 1909; Ph.D., U. of Lwów, 1926; D.H.C., U. Louvain, 1927; Sc.D., Columbia, 1929; m. Anne Maynard Kidder, Dec. 27, 1904; 1 dau., Nancy. Lectr. biology, Williams, 1883-84. Mass. Inst. Tech., 1884-85; asso. professor and prof. biology Bryn Mawr, 1885-91; adj. prof. biology, Columbia, 1891-94, prof. invertebrate zoology, 1894-97, prof. zoology, 1897-1909, Da Costa prof. zoology, 1909-28, emeritus prof. in residence, 1928-39, also dean of faculty of pure science, 1905-06. Fellow Royal Soc., 1921, Am. Acad. Arts and Scis., A.A.A.S. (pres. 1913), N.Y. Acad. Scis. (pres. 1904); mem. numerous Am. and fgn. sci. socs., including French Acad. Scis. Author: General Biology (with W. T. Sedgwick), 1887; Atlas of Karyokinesis and Fertilization, 1895; The Cell in Development and Heredity, 1896, 1900, 1925; The Physical Basis of Life, 1923. Specialist in embryology, and experimental morphology; studied chromosomes (including significance of sex chromosomes) and function of cell in heredity; promoted theories of Gregor Mendel; pioneer in field of cytology. Died N.Y.C., Mar. 3, 1939.

WILSON, Edward Adrian, English physician, explorer; b. Cheltenham, Eng., July 23, 1872; s. Edward Thomas and Mary Agnes (Whishaw) W.; ed. Gonville and Caius Coll., Cambridge; B.A., Cambridge, M.B., B.C., 1900; m. Oriana Fanny, 1901. Mem. Antarctic expdn., 1901, studied habits and breeding of Emperor penguins; embarked on 2d expdn. to Antarctic, 1910, obtained chicks of Emperor penguin. Recipient Polar medal, 1904; Parton's medal Royal Geog. Soc., 1913. Perished with entire party of explorers on return from S. Pole, 1912.

WILSON, Edward Osborne, Am. zoologist; b. Birmingham, Ala., June 10, 1929; s. Edward Osborne and Inez (Freeman) W.; B.S., U. Ala., 1949, M.S., 1950, Ph.D., 1955; m. Irene Gertrude Kelley, Oct. 30, 1955; 1 dau., Catherine Irene. Faculty, Harvard, Cambridge, Mass., 1953—, prof. zoology, 1964—. Fellow Am. Acad. Arts and Scis., A.A.A.S. Research, numerous publs. on classification of the ants, evolution at species level, theory of distbn. of plants and animals, orgn. of insect socs., communication in animals by means of chem. secretions. Home: 5 Thoreau Rd., Lexington, Mass. 02173. Office: Biol. Labs., Harvard, Cambridge, Mass. 02138.*

WILSON, Ernest Henry, botanist; b. Chipping Campden, Eng., Feb. 15, 1876; s. Henry and Annie (Curtis) W.; ed. Royal Coll. Sci., London; M.A. (hon.), Harvard, 1916; m. Ellen Ganderton, June 8, 1902; 1 dau., Muriel Primrose. Came to U. S., 1906. Asst. dir. Arnold Arboretum, Harvard, 1919-27, keeper, 1927; traveled in China 11 yrs., 3 yrs. in Japanese Empire, also in Australasia, India, South Africa. Recipient Victoria medal of honor, George Robert White medal, Veitch meml. medal, Geoffrey St. Hilaire Gold medal. Fellow Am. Acad. Arts and Scis. Author: Naturalist in West China, 1913; Cherries of Japan; Conifers and Taxads of Japan, 1915; Aristocrats of the Garden, 1916; The Romance of Our Trees, 1920; A Monograph of Azaleas (Wilson and Rehder), 1921; Lilies of Eastern Asia, 1925; America's Greatest Garden, 1925; Plant Hunting, 1927; More Aristocrats of the Garden, 1928; China, Mother of Gardens, 1929. Introduced more than 1000 plant species from China into western world. Died 1930.

WILSON, George, Scottish physician, chemist; b. Edinburgh, 1818; s. Archibald and Janet Wilson; Ph.D., Univ. Coll., 1837; M.D., Royal Coll. Surgeons Edinburgh, 1839; apptd. lectr. Edinburgh Vet. Coll., Sch. of Arts, Scottish Instn., 1843; named dir. Scottish Indsl. Mus., 1855. fellow Royal Soc. Edinburgh. Author: Chemistry, 1850; Electricity and the Electric Telegraph, 1856; The Five Gate Ways of Knowledge, 1856; Memoir of Edward Forbes (pub. posthumously); Life of Cavendish (paper in which he disputed priority of James Watt's research on composition of water over that of Cavendish); other articles. Begin research on distbn. of fluorides, 1845; contbd. to understanding of color blindness. Died 1859.

WILSON, George Watkins, Am. chemist; b. N.Y.C., June 22, 1890; s. James and Dorothea (Becker) W.; B.S., Coll. City N.Y., 1910; M.S., N.Y. U., 1915, Ph.D., 1921; m. Josephine Van Pelt Spencer, Aug. 26, 1937; children—George Spencer, Susan Van Pelt, Richard William. Tchr. high schs., N.Y., 1910-15, 20-28; chemist U.S. Bur. Mines, Washington, 1917-18; aide Smithsonian Instn., Washington, 1919-20; faculty Coll. City N.Y., 1928—, prof. chemistry, 1951-60, prof. emeritus, 1960—. Mem. div. chemistry, chem. tech. NRC, 1940. Mem. Am. Chem. Soc., Am. Inst. Chemists, A.A.A.S., Am. Assn. U. Profs., Chemistry Tchrs. Club N.Y. (past pres.), Frat. Emile, N.Y. Acad. Sci. Research on gas-masks, use war gases, explosives, 1917-43. Died Sept. 25, 1967.

WILSON, George Wilton, economist; b. Winnipeg, Man., Can., Feb. 15, 1928; s. Walter and Ida (Wilton) W.; came to U. S., 1952; B.Commerce, Carleton U., Ottawa, Can., 1947-50; M.A., U. Ky., 1951; Ph.D., Cornell U., 1955; m. I. Marie McKinney, Sept. 6, 1952; children—Ronald Leslie, Douglas Scott, Suzanne Rita. Economist, Bd. Transp. Commrs., Ottawa, 1951-52; teaching fellow Cornell U., 1952-55; asst. prof. econs. Middlebury (Vt.) Coll., 1955-57; prof. transp. Ind. U., 1957—, also prof., chmn. dept. econs., Collaborator study S. Asia, 20th Century Fund, 1962-65, dir. research Canada's needs and resources, 1964-65; dir. case studies role transp. in econ. devel. Brookings Instn., 1964; mem. Presdl. Task Force Transp., 1964; mem. survey team on econ. future of Leeward, Windward islands, Barbadoes, 1966. Mem. Am. Econ. Assn., Author: Essays on Some Unsettled Questions in the Economics of Transportation, 1962; Mathematical Models and Methods in Marketing, 1961 (co author); Canada: An Appraisal of Its Needs and Resources, 1965; Growth and Change at Indiana University, 1966; Highway Investment and Economic Growth, 1966. Editor: Classics of Economic Theory, 1964. Research on applicability of contemporary econ. analysis to rate regulation in transport and to planning for devel. of underdeveloped countries, also on relationship of transport investment to econ. growth, econ. potential of Can., econ. and social prospects for S. Asia. Home: 2325 Woodstock Pl., Bloomington, Ind. 47401.*

WILSON, Harwell, Am. surgeon; b. Lincoln, Ala., May 23, 1908; s. Joseph Harwell and Linda (Harrison) W.; A.B., Vanderbilt U., 1928, M.D., 1932; postgrad. U. Chgo.; m. Helen Cobb, Jan. 18, 1941; children—Linda Helen, Margaret Marie, William Harwell. Faculty, U. Tenn., Memphis, 1939—, prof., chmn. dept. surgery, 1948—; surgeon-in-chief City of Memphis Hosps., 1948—; sr. staff Bapt. Hosp., Memphis. Cons. to hosps.; lectr. univs. Recipient Distinguished Service award U. Chgo. Alumni Assn., 1952. Mem. A.M.A., Am., So. surg. assns., A.C.S. (treas. 1963—) Southeastern Surg. Congress (past pres), Soc U. Surgeons, Soc. Clin. Surgery, Soc. Vascular Surgery, Soc. for Surgery Alimentary Tract (past pres.), Excelsior Surg. Soc., Internat. Soc. Surgery, So. Med. Assn. (past v.p.), Soc. Surg. Chmn. Am. Med. Schs. Research, numerous publs. on surg. diseases circulatory system, surg. treatment certain types cancer, surg. shock. Home: 521 Colonial Rd., Memphis 38117.*

WILSON, Henry Parke Custis, Am. gynecologist; b. Workington, Md., Mar. 5, 1827; s. Henry Parke Custis and Susan E. (Savage) W.; B.A., Coll. of N.J. (now Princeton), 1848; studied medicine U. Va.; grad. U. Md., 1851; m. Alicia Griffith, 1858; 5 children. Surgeon in charge Balt. City Almshouse Infirmary, 1857-58; cons. surgeon St. Agnes Hosp., 1879, Johns Hopkins Hosp., 1889; with William T. Howard founded Hosp. for Women of Md., 1882; 1st physician in Md. to remove uterine appendages by abdominal section, 2d in Md. to perform successful ovariotomy, 1866; 2d in world to remove intrauterine tumor filling whole pelvis by marcellation; devised instruments for gynecol. surgery; pres. Med. and Chirurg. Faculty of Md., 1880-81; founder Balt. Obstet. and Gynecol. Soc., also Am. Gynecol. Soc., 1880-81; mem. Brit. Gynecol. Assn.; hon. fellow Edinburgh (Scotland) Obstet. Soc. Died Balt., Dec. 27, 1897.

WILSON, Herbert Couper, Am. astronomer; b. Lewiston, Minn., Oct. 24, 1858; s. Thomas and Ann W.; A.B., Carleton Coll., 1879, A.M., 1882; Ph.D., U. Cin., 1886; studied astronomy, Cin. Obs., 1880-81; m. Mary B. Nichols, Dec. 20, 1882 (dec. 1924); children—Ralph Elmer, Ruth Edna (Mrs. William H. Geer), Mary Helen (Mrs. George M. Constans), Lois Norma (Mrs. W. Harlan Pye); m. 2d, Florence E. Rice, Aug. 26, 1926. Asst. astronomer, Cin. Obs., 1881-82, 84-86, astronomer in charge, 1882-84; computer to U. S. Transit of Venus Commn. U. S. Naval Obs., Washington, 1886-87; asst. prof. astronomy, Carleton Coll., 1887-1900, asso. prof. math. and astronomy, 1900-08, prof. and dir. Goodsell Obs., 1908-26, also dean faculty, 1906-10; ret. on Carnegie Found., 1926. Asst. astronomer Lick Obs., 1910-11, Harvard Obs., 1916, Mt. Wilson Obs., 1920-21. Asso. editor, Astronomy and Astrophysics, 1892-94, Popular Astronomy, 1894-1909; editor Popular Astronomy, 1909-26. Devoted much attention to double and variable star observations and astronomical photography. Died Mar. 1940.

WILSON, Irene Mossom, botanist; b. Broach, India, May 10, 1904; d. Robert and Mossom (Thomson) Wilson; B.Sc., U. London, 1928, M.Sc., 1931, Ph.D., 1937, D.Sc., 1963. Faculty, U. Wales, Aberystwyth, 1935—, reader in botany, 1965—. Mem. Brit. Mycol. Soc. (v.p. 1964). Research, publs. on cytology and devel. of fungi; rust diseases. Home: Brynheulog, North Rd., Aberystwyth, Eng.*

WILSON, Irwin B., Am. biochemist; b. Yonkers, N.Y., May 8, 1921; s. Meyer H. and Pauline (Blank) W.; B.S., Coll. City N.Y., 1941; Ph.D., Columbia, 1948; m. Norma Blum, Feb. 10, 1953; children—Susan H., Michael H. Faculty, Coll. City N.Y., 1946-49; faculty Columbia Coll. Phys. & Surg., N.Y.C., 1949-66, prof. biochemistry, 1964-66; prof. chemistry U. Colo., Boulder, 1966—. Recipient 1st Ann. award Am. Assn. for Research in Neurology and Psychiatry, 1958. Mem. Am. Chem. Soc., Fedn. for Exptl. Biology, Harvey Soc., Am. Acad. Neurology. Research on how living cells carry out chem. reactions; invented an antidote for nerve gases and certain insecticides; explained action of certain anticholinesterase drugs. Home: 2550 Stanford Av., Boulder, Colo. 80302.*

WILSON, James Tinley, Am. geophysicist; b. Claremont, Cal., Nov. 13, 1914; s. Raleigh and Mary (Brooks) W.; A.A., Porterville Jr. Coll., 1933; A.B., U. Cal. at Berkeley, 1935, Ph.D., 1939; m. Martha Wheeler, July 17, 1943; 1 dau., Deborah Mary. Research asst. Harvard, 1939-40; sect. supr. Airborne Instruments Lab. div. war research Columbia, 1943-45; faculty U. Mich., Ann Arbor, 1940-43, 45—, chmn. dept. geology, 1955-61, dir. Inst. Sci. and Tech., 1962—. Mem. com. on seismology Nat. Acad. Sci.-NRC, 1960—, chmn., 1960-65, chmn. com. on remote sensing of environment, 1964—; mem. contractors research and evaluation panel Air Force Office Sci. Research, 1961—, chmn., 1961-62; cons. adv. com. on reactor safeguards AEC, 1967—. Fellow Am. Geophys. Union, Geol. Soc. Am., Royal Astron. Soc.; mem. Am. Geol. Inst., Am. Inst. Mining, Metall. and Petroleum Engrs., Seismol. Soc. Am. (past pres.), Soc. Exploration Geophysicists, Soc. Metall. Engrs., Sigma Xi (past pres. Mich. chpt.). Research, publs. on structure of earth, local earthquakes, elastic waves. Home: 40 Underdown Rd., Ann Arbor, Mich. 48105.*

WILSON, Joe Bransford, Am. microbiologist; educator; b. Dallas, June 29, 1914; s. William Ringgold and Kate (Bransford) W.; B.A., U. Tex., 1939; M.S., U. Wis., 1941, Ph.D., 1947; m. Dorothy Louise Simmons, Oct. 16, 1944; children—Joe Bransford, Kathleen Scott. Instr., U. Tex., 1939; faculty U. Wis., Madison, 1939-42, 46—, prof. microbiology, 1955—, asso. dean Grad. Sch., 1965—. Mem. adv. council Chem. Corps., 1955-65; mem. tech. adv. com. Office Sec. Def., 1954-63; mem. bacteriology and mycology study sect. NIH, USPHS, 1965—. Diplomate Am. Bd. Microbiology. Fellow A.A.A.S., Am. Acad. Microbiology; mem. Am. Soc. Microbiology, Am. Assn. Immunologists, Soc. Exptl. Biology and Medicine, Soc. Gen. Microbiology (Gt. Britain), Phi Kappa Psi. Author: (with others) Microbiology, 1951. Mem. editorial bd. Jour. Bacteriology, 1963—. Contbr. numerous articles to profl. jours. Research on pathogenesis and metabolism of Brucella, Vibrio, Leptospira, and Cocci. Home: 3427 Sunset Dr., Madison, Wis. 53705.

WILSON, John Tuzo, Canadian geologist; b. Ottawa, Ont., Can., Oct. 24, 1908; s. John A. and Henrietta (Tuzo) W.; B.A., U. Toronto, 1930; B.A., U.

Cambridge, 1932, M.A., 1942, Sc.D., 1958; Ph.D., Princeton, 1936; D.Sc., U. Western Ont., 1958; LL.D., Carleton U., 1958; m. Isabel Jean Dickson, Oct. 29, 1938; children—Patricia, Susan. Geologist, Geol. Survey Can., 1936-39; prof. geophysics U. Toronto (Ont., Can.), 1946—, prin. Erindale Coll. Recipient Order Brit. Empire, 1945; Legion Merit; Civic award merit City of Toronto, 1961. Fellow Royal Soc., 1909, Royal Soc. Can., World Acad. Art and Sci.; mem. Geol. Soc. London, Geol. Soc. Am., Geol. Soc. Can., Canadian Assn. Physicists, Internat. Union Geodesy and Geophysics (pres. 1957-60), Internat. Council Sci. Unions, (exec. com. 1963-66), Nat. Acad. Scis. (fgn. asso.). Author: One Chinese Moon, 1959; (with Jacobs, Russell) Physics and Geology, 1959; Year of the New Moons, 1961; also numerous articles. Research on structure and growth continents, theory continental drift, geology ocean basins; discovered new class geol. faults, 1965. Home: 3359 Mississauga Rd., Mississauga, Ont. Office: Erindale Coll., U. Toronto, Toronto 5, Ont., Can.*

WILSON, Katherine Woods, Am. chemist; b. Cal., Feb. 8, 1923; d. Harry L. and Sallie (Woods) Wilson; B.S., U. Cal. at Berkeley, 1944, M.S., 1945; Ph.D., U. Cal. at Los Angeles, 1948. Faculty, W.Va. U., 1948-53, Pepperdine Coll., 1953-55; with Los Angeles County Air Pollution Control Dist., 1954-55; asso. research chemist, lectr. dept. engring. U. Cal. at Los Angeles, 1955-62; phys. chemist Stanford Research Inst., South Pasadena, Cal., 1962—. Mem. Am. Chem. Soc., Air Pollution Control Assn., Phi Beta Kappa, Sigma Xi. Research, publs. on chem. analysis, air pollution, textile chemistry. Home: 780 Monterey Rd. Office: 820 Mission St., South Pasadena, Cal. 91030.*

WILSON, Mathew Kent, Am. chemist; b. Salt Lake City, Dec. 22, 1920; s. Mathew B. and Helen (Marsh) W.; B.S., U. Utah, 1943; Ph.D., Cal. Inst. Tech., 1949; m. Gladys Accredo, Oct. 25, 1944; 1 son, Mathew Kent. Mem. faculty Harvard, 1948-56; prof., chmn. chemistry dept. Tufts U., 1956—; head chemistry sect. NSF, 1966—. Mem. Adv. Council Coll. Chemistry, 1963-66; dir. Chem. Bond Approach Project, 1959-65. Guggenheim fellow, also Fulbright scholar, 1954-55. Fellow Am. Acad. Arts and Scis.; mem. Am. Chem. Soc., Optical Soc. Am. Author: (with E.G. Rochow) General Chemistry: A Topical Approach, 1954. Editor: Spectrochimica Acta, 1964—. Research and publs. on determination of geometrical structure of molecules. Home: 9 Longfellow Rd., Winchester, Mass. 01890. Office: 1800 G St., Washington 20550.*

WILSON, Michael F(riend), Am. physiologist, biophysicist; b. Morgantown, W.Va., Jan. 13, 1927; s. Minter Lowther and Mildred (Friend) W.; B.A., W.Va. U., 1949; M.D., U. Pa., 1953; m. Clare Braddock Schofield, Apr. 10, 1954; children—Helen Horton, Rebecca Friend, Michael Lowther, Elizabeth Schofield, Sarah Anderson. Practice internal medicine and cardiology, Phila., 1954-58; research fellow physiology and biophysics U. Wash., Seattle, 1958-60; faculty U. Ky. Med. Center, Lexington, 1960-65, asso. prof. physiology and biophysics, 1963-65; prof., chmn. physiology W.Va. U. Med. Center, Morgantown, 1965—. Fellow Am. Coll. Cardiologists, A.C.P.; mem. A.A.A.S., Am. Assn. U. Profs., I.E.E.E., Am., W.Va. heart assns., Am. Physiol. Soc., Aerospace Med. Assn., N.Y. Acad. Sci. Pavlovian Soc. Am. Research, publs. on bodily function and regulation heart, lungs and brain. Inventor electromagnetic cardiometer.*

WILSON, Morley Evans, Canadian geologist; b. Bright, Ont., Can., Feb. 8, 1882; B.A., U. Toronto (Ont.), 1907; postgrad. U. Wis., U. Chgo.; Ph.D., Yale, 1921; m. Edna Eileen Wiggle, 1912; m. 2d, Lillian Agnes Salt, 1949. Asst. geologist Geol. Survey Can., 1907-15, asso. geologist, 1915-25, geologist, 1925-48. Fellow Geol. Soc. Am. (Penrose medal 1950); mem. Royal Soc. Can. (pres. sect. IV 1938, Miller Meml. medal 1945), Canadian Inst. Mining and Metallurgy (council 1924-25, chmn. Ottawa br. 1926-27, Blaylock medal 1950). Research, publs. Canadian geology; reconstructed geol. history Archean-type Precambrian rock; developed method of studying stratification of areas where lava flow and folding had occurred. Died 1965.

WILSON, Nixon Albert, Am. acarologist; b. Litchfield, Ill., May 20, 1930; s. Benezette Nixon and Adrian (Barnes) W.; A.B., Earlham Coll., 1952; M.S., U. Mich., 1954; postgrad. U. Md.; Ph.D., Purdue U., 1961; m. Flora Nell Hines, June 15, 1963; 2 sons, Stephen, David. Animal ecologist Hawaii Dept. Health, Honokaa, 1961-62; acarologist B.P. Bishop Mus., Honolulu, 1962—; leader sci. expdns. to West New Guinea, 1962, Philippines, Hong Kong, Taiwan, 1964, Laos, Indonesia, 1965, East New Guinea, 1966; affiliate grad. faculty U. Hawaii, 1962—. Mem. Am. Soc. Mammalogists, Am. Soc. Parasitologists, Wildlife Disease Assn., Sigma Xi. Mem. editorial bd. Jour. Med. Entomology, 1964—. Research, publs. in systematics of acarines, ecology of ectoparasites and ecology of bats. Office: B.P. Bishop Mus., Honolulu 96819.*

WILSON, Olin Chaddock, Am. astronomer; b. San Francisco, Jan. 13, 1909; s. Olin C. and Sophie (Clarey) W.; A.B., U. Cal., 1930; Ph.D., Cal. Inst. Tech., 1934; m. Katherine E. Johnson, Sept. 3, 1943; children—Nicole, Randall. Asst. astronomer Mt. Wilson Obs., 1936-50, astronomer Mt. Wilson and Palomar obs., 1950—. Mem. Nat. Acad. Scis., Am. Astron. Soc., A.A.A.S. Investigation of planetary nebulae; Wolf-Rayet stars; interstellar matter; stellar atmospheres; radial velocity programs.

WILSON, Perry William, Am. biochemist; b. Bonanza, Ark., Nov. 25, 1902; s. Comodore Lawson and Frankie Ellen (Smith) W.; student Rose Poly. Inst., 1922-23; B.S., U. of Wis., 1928, M.S., 1929, Ph.D., 1932; m. Helen Evelyn Hansel, Sept. 4, 1929; children—Gwenn Dellen, Richard Hansel. Instr., asst. prof., asso. prof., department bacteriology, U. of Wis., 1932-45, prof. since 1945; Guggenheim Fellow, 1936. Mem. Soc. Am. Bacteriologists, (pres. 1957), Am. Chem. Soc., Soc. Biol. Chemists, Biochemical Soc. (Eng.), Soc. of Experimental Biology and Medicine, National Acad. Sci., Sigma Xi. Editor: Bacteriological Revs., 1952-57. Author: Biochemistry of Symbiotic Nitrogen Fixation, 1940; Respiratory Enzymes (editor and co-author), 1939; Symposium on Respiratory Enzymes (editor), 1942; Symposium on Use of Isotopes in Biology and Medicine (editor), 1948; Bacterial Physiology, 1951. Studied bacterial enzyme systems, with Dr. Marjorie Stephenson, Cambridge U., Eng. and biol. nitrogen fixation, with Nobel prize winner A. I. Virtanen, Helsinki, Finland; now engaged in research dealing with use of isotopes in study of enzyme mechanisms (fermentation and biol. nitrogen fixation. Home: 3830 Cherokee Dr., Madison 5, Wis.*

WILSON, Raymond Edgar, Am. physicist; b. Salem, Ore., Aug. 20, 1915; s. John B. and Alice (Lewis) W.; B.A. in Physics, Reed Coll., 1937; Ph.D. in Physics, U. Wash., 1942; m. Alice P. Weil, June 13, 1939; children—Nancy Ruth, Bruce Henry. Asso. prof. physics George Washington U., Washington, 1946-47; asst. chief heat and power div. Nat. Bur. Standards, Washington, 1947-53, chief office engring. standard, 1963-64; chief physics sect. Emerson Research Lab., Washington, 1953-55; mgr. Engring. Services Lab., Hughes Aircraft Co., Tucson, 1955-63; asst. project mgr. Jet Propulsion Lab., Cal. Inst. Tech., Pasadena, 1964-66, Hughes Aircraft Co., Tucson, 1966—. Fellow Am. Phys. Soc., A.A.A.S.; mem. Washington Acad. Scis., Washington Philos. Soc., Sigma Xi, Sigma Pi Sigma. Contbg. author: Process Instruments and Controls Handbook 1957; Handbook of Physics, 1958; Am. Inst. Physics Handbook, 1963. Contbr. articles to tech. jours. Research on underwater ordnance, precision temperature measurements, system engring. Home: Route 2, Box 761-L, Tucson 85715. Office: Hughes Aircraft Co., Tucson.*

WILSON, Raymond Hiram, Jr., Am. astronomer; b. Gap, Pa., Feb. 14, 1911; s. Raymond Hiram and Agnes (Wright) W.; A.B., Swarthmore Coll., 1931; A.M., U. Pa., 1933, Ph.D., 1935, postgrad.; postgrad. Harvard; m. Irene Gladys Louise Hansing, Aug. 21, 1940; 1 dau., Kristin Marie. Research asso., instr. math. and astronomy various colls., 1929-40; astronomer Naval Obs., 1940-42; research on contracts with Office Naval Research, 1949-52; faculty Temple U., 1946-51, U. Louisville, 1951-54; physicist Naval Research Lab., Project Vanguard, 1954-58; applied mathematician Goddard Space Flight Center, 1958-62; chief applied math. br. NASA Hdqrs., Washington, 1962—. Profl. lectr. astronomy Georgetown U., 1962—. Fellow A.A.A.S.; mem. Am. Inst. Aeros. and Astronautics (sr.), Am., Rittenhouse (pres. 1949) astron. socs., Am. Astronautical Soc., Math. Assn. Am., Soc. Indsl. and Applied Math., Nat. Acad. Sci., Sigma Xi, Sigma Pi Sigma. Contbr. articles to profl. jours. Research on interferometric instrumentation applied to binary stars, applied math. Patentee method and apparatus for magnetic steering. Home: 3937 1st St. S.W., Washington 20032. Office: Code RRA, NASA Hdqrs., Washington 20546.*

WILSON, Robert, Brit. engr.; b. Dunbar, Haddingtonshire, Scotland, 1803; m. twice; 4 sons, 4 daus. Engr., Edinburgh, Scotland; later moved to Manchester, Eng.; mgr. J. Nasmyth's Bridgwater Foundry, Patricroft, Eng.; engr. Low Moor Ironworks, Bradford, Eng.; became mng. partner Nasmyth, Wilson & Co., 1856. War Dept. grantee, 1880. Fellow Royal Soc. Edinburgh; mem. Royal Scottish Soc. Arts (Silver medal 1832). Author: The Screw Propeller: Who Invented it?, 1880. Patentee self-acting motion for steamhammer, 1843, double-acting hammer, 1861, hydraulic packing-press, (later improved), 1856, also valves, pistons, propellers, hydraulic machiner, 1842-80; invented screw-propeller for ships. Died Matlock, Eng., July 28, 1882.

WILSON, Robert Lee, Am. mathematician; b. Champaign, Ill., Mar. 7, 1917; s. William Harold and Esther (Barney) Wilson; A.B., U. Fla., 1938; M.A., U. Wis., 1940, Ph.D., 1947; m. Anna Katherine Fulton, Sept. 11, 1940; children—Robert Lee, William F., Anna Katherine, Steven K. Faculty, U. Tenn., Knoxville, 1947-56, asst. prof. math., 1948-56; sr. aerophysics engr. Gen. Dynamics, Convair div., Ft. Worth, 1956-58; adj. prof. math. Tex. Christian U., Ft. Worth, 1956-58; prof. math., chmn. dept. Ohio Wesleyan U., Delaware, 1958—, dir. Computing Center, 1963—, asst. to pres. acad. affairs, 1962-63; dir. Computing Centre, vis. prof. math. U. Ibadon, Nigeria, 1966-68. Mem. Am. Math. Soc., Math. Assn. Am., Nat. Council Tchrs. Math., A.A.A.S., Ohio Council Tchrs. Math., Assn. for Computing Machinery, Phi Beta Kappa, Sigma Xi. Author: (with E.D. Eaves) Introductory Mathematical Analysis, 1958; also articles. Research in numerical analysis. Home: 149 Oak Hill Av., Delaware, O. 43015.*

WILSON, Robert Rathbun, Am. physicist; b. Frontier, Wyo., Mar. 4, 1914; s. Platt Elvin and Edith (Rathbun) W.; A.B., U. of Calif., 1936, Ph.D., 1940; m. Jane Inez Scheyer, Aug. 20, 1940; children—Daniel, Jonathan, and Rand. Engaged as instructor Princeton University, 1940-42, assistant professor, 1942-45; head exptl. nuclear physics division Los Alamos Lab., Atomic Bomb Project, 1944-46; asso. prof. physics Harvard, 1946-47, prof. physics and dir. Lab. Nuclear Studies, Cornell U., 1947-67; prof. physics U. Chgo., 1967—; dir. of Nat. Accelerator Lab., Weston, Illinois, 1967—. Recipient of the Elliott Cresson medal Franklin Inst., 1964. Fellow Am. Phys. Soc.; mem. Fedn. Am. Scientists (chmn. 1946), Nat. Acad. Sci., Sigma Xi. Contbr. articles to Physics Rev. Research on nuclear and particle physics. Home: 5825 S. Dorchester Av., Chgo.

WILSON, Robert Warren, Am. paleontologist; b. Oakland, Cal., July 26, 1909; s. Robert Gordon and Lillie (Silcott) W.; B.S., Cal. Inst. Tech., 1930, M.S., 1932, Ph.D., 1936; m. Geraldine Lee, Mar. 20, 1941; children—Robert C., Margaret L. Faculty, U. Colo., 1939-47, U. Kan., 1947-61; paleontologist, U.S. Geol. Survey, 1956—; prof. paleontology, dir. Mus. Geology, S.D. Sch. Mines and Tech., Rapid City, 1961—. NRC fellow Cal. Inst. Tech., 1946-47; Guggenheim fellow, 1956-57. Fellow Geol. Soc. Am.; mem. Soc. Vertebrate Paleontology (past pres.), Sigma Xi, Tau Beta Pi. Contbr. numerous articles to profl. jours. Research on geologic history of rodents. Home: 209 Franklin St., Rapid City, S.D. 57701.*

WILSON, Samuel Alexander Kinnier, neurologist; b. Cedarville, N.J., Dec. 6, 1877; grad. Edinburgh (Scotland) U., 1903; student Paris, 1903-04, Leipzig, Germany. Resident med. officer Nat. Hosp. for Paralized and Epileptic, London; became jr. neurologist King's Coll. Hosp., 1919, sr. neurologist, 1928. Author: Aphasia, 1926; Modern Problems in Neurology, 1928; Neurology (edited by Alexander Ninian Bruce), 2 vols., 1940. Described progressive lenticular degeneration (Wilson's disease), 1912. Died London, May 12, 1937.

WILSON, Thomas Hastings, Am. physiologist; b. Phila., Jan. 31, 1925; s. David Wright and Helene (Connet) W.; student Oberlin Coll.; M.D., U. Pa., 1948; Ph.D. (Am. Cancer Soc. exchange student), U. Sheffield, Eng., 1953; m. Dorothy Marguerite Blanchard, Dec. 26, 1952; children—Katherine, Peter, Susan, Elizabeth. Faculty, U. Pa., 1949-50, Washington U., St. Louis, 1956-57; faculty Harvard Med. Sch., Boston, 1957—, asso. prof. physiology, 1961—. Author: Intestinal Absorption. Research in permeability of living cells; developed methods for study of intestinal absorption in vitro. Home: 21 Hillside Terrace, Belmont, Mass. Office: 15 Shattuck S., Boston 15.*

WILSON, Volney Colvin, Am. physicist; b. Evanston, Ill., Feb. 27, 1910; s. Robert Edward and Ruth (Colvin) W.; B.S., Northwestern U., 1932; M.A., Ohio State U., 1934; Ph.D., U. Chgo., 1938; m. Virginia Dodd, June 11, 1938; children—Robert Edward, Lee Colvin. Instr., U. Chgo., 1938-40, dir. instrumentation and control Manhattan Project, 1942-45; radar devel. Radiation Lab., Mass. Inst. Tech., 1941; physicist Gen. Electric Research Lab., Schenectady, 1945—. Fellow Am. Phys. Soc.; mem. Am. Nuclear Soc., Am. Inst. Aeros. and Astronautics, A.A.A.S., Sigma Xi, Sigma Chi. Editorial bd. Advanced Energy Conversion. Research on cosmic rays, nuclear instrumentation, reactor control, magnetic materials, neutron spectroscopy, computer circuitry, energy conversion; inventor thermionic converter. Home: 1 Indian Kill Rd., Scotia, N.Y. 12302. Office: P.O. Box 1088, Schenectady 12301.*

WILSON, William, Brit. botanist; b. Warrington, Eng., June 7, 1799; s. Thomas Wilson; studied under Dr. Reynolds, Dissenter's Acad., Manchester, Eng.; m. Mrs. Lane, 1836. Articled to firm of solicitors, Manchester. Author: Bryologia Britannica, 1855; also numerous articles. Studies in bryology; discovered Cotoneaster on Great Ormes' Head, 1821, also species (new to Britain) of rose, fern, several mosses. Died Paddington, Eng., Apr. 3, 1871.

WILSON, Sir William James Erasmus, Brit. surgeon; b. Marylebone, Eng., Nov. 25, 1809; s. William Wilson; studied under George Lanstaff (surgeon to Cripplegate Dispensary), anatomy under John Abernethy, St. Bartholomew's Hosp.; student Aldersgate St. Sch. Medicine; LL.D., U. Aberdeen (Scotland); m. Miss Doherty, 1841. Demonstrator anatomy to Richard Quain, U. Coll., 1831-36; founder unsuccessful Sch. Anatomy, Sydenham Coll., 1836; became lectr. anatomy and physiology Middlesex Hosp., 1840; cons. surgeon to St. Pancras Infirmary; founder professorship dermatology Royal Coll. Surgeons, 1869, prof. dermatology, 1869-77; founder chair pathology U. Aberdeen, 1881. Licenciate Soc. Apothecaries. Fellow Royal Soc. (hon. gold medal, 1884), 1845. Royal Coll. Surgeons. Knighted, 1881. Author: Practical and Surgical Anatomy, 1838; The Anatomist's Vade mecum, 1840; A Practical and Theoretical Treatise

. . . on Diseases of the Skin; The Eastern or Turkish Bath: its History, 1861; The Vessels of the Human Body in a Series of Plates (with J. Quain), 1837. Editor: Jour. Cutaneous Medicine and Diseases Skin, 1867-70. Research in skin diseases; 1st to describe dermatitis exfoliativa (Wilson's disease), 1870. Died West Gate-on-Sea, Eng., Aug. 7, 1884.

WILSON, William Preston, Am. physician; b. Fayetteville, N.C., Nov. 6, 1922; s. Preston Puckett and Rosa Mae (Van Hook) W.; student Campbell Jr. Coll., 1939; B.S., Duke, 1943, M.D., 1947; m. Dorothy Elizabeth Taylor, Aug. 21, 1950; children—William Preston, Ben V., Karen E., Tammy, Robert E. Staff psychiatrist State Hosp., Raleigh, N.C., 1948-49; asst. dir. psychosurg. sect. State Hosp., Butner, N.C., 1949-52; staff VA Hosp., Durham, N.C., 1955-58; faculty Duke Sch. Medicine, 1949-58, 61—, prof. psychiatry, 1964—, dir. EEG Lab., 1961—; asst. prof., dir. psychiat. research U. Tex. Med. Br., 1958-60; chief Neuropsychist. Lab., VA Hosp., Durham, 1961—. Mem. A.M.A., Am. Psychiat. Assn., Am. EEG Soc., Am. Psychopath. Assn., So. Med. Assn., A.A.A.S., Assn. for Research in Nervous and Mental Diseases, Am. Acad. Neurology, Sigma Xi. Research, numerous publs. on brain function and behavior, biochemistry, neurophysiology, clin. observations mental illness. Home: 1209 Virginia Av., Durham, N.C. 27705.*

WILSON, William Solomon, Am. chemist; b. Reynoldsburg, O., Aug. 9, 1908; s. Luda and Lena (Metzger) W.; B.Sc., Brown U., 1931, M.Sc., 1934; Ph.D., Yale, 1936. Research asso.. Shell Devel. Co., Emeryville, Cal., 1937; instr. chemistry Blue Ridge Coll., 1939-40; faculty Georgetown Coll., 1940-47, asso. prof. phys. sci., 1945-47; faculty U. Alaska, College, 1947—, acting and asso. dir. Geophys. Inst., 1950-53, prof., head gen. sci., 1964—; asst. prof. chemistry U. Wyo., 1948-49. Fellow A.A.A.S., Am. Inst. Chemists; mem. Am. Phys. Soc., Am. Chem. Soc., Franklin Inst., N.E.A., N.Y. Acad. Sci., Am. Geophys. Union, Am. Meteorol. Soc., Nat. Sci. Tchrs. Assn., Sigma Xi, Phi Delta Kappa. Research, publs. on application of basic theories to nature of matter, application of spectroscopy to chem. reactions, adminstrv. devel. of depts. and programs, sci. edn. Home: Box 13, College, Alaska 99701.*

WILT, James William, Am. chemist; b. Chgo., Aug. 28, 1930; s. Edward Frank and Mary (Manarik) W.; A.B., U. Chgo., 1949, M.Sc., 1953, Ph.D. (NSF fellow), 1954; m. Catherine R. McAndrews, Jan. 10, 1953; children—Donna M., Susan A., Gregory E., Catherine M., Maureen T. Instr., U. Conn., 1955; faculty Loyola U., Chgo., 1955—, prof. chemistry, 1966—. Mem. Am. Chem. Soc., Phi Beta Kappa, Sigma Xi. Research and publs. on molecular rearrangements (chem. changes in which new compound has same number and type of atoms attached to one another differently). Home: 2745 Pauline St., Glenview, Ill. 60025. Office: 6525 N. Sheridan Rd., Chgo. 60626.*

WIMSATT, William Abell, Am. zoologist; b. Washington, July 28, 1917; s. William Church and Alma (Engelbretson) W.; A.B., Cornell U., 1939, Ph.D., 1943; m. Ruth Claire Peterson, June 18, 1940; children—William C., Michael H., Mary S., John L., Jeffry M., Ruth A. Instr., Harvard, 1943-45; faculty Cornell U., Ithaca, N.Y., 1945—, prof. zoology, 1951—, chmn. dept., 1959-64. Research collaborator Brookhaven Nat. Lab., 1954-59. Fellow A.A.A.S.; mem. Am. Assn. Anatomists, Am. Soc. Zoologists, Soc. Mammalogists, Histochem. Soc., Inst. Internat. d'embryologie, Am. Inst. Biol. Sci., Soc. Developmental Biology; Sigma Xi, Phi Kappa Phi, Phi Gamma Mu. Publs. on vertebrate reproductive physiology, hibernation, thermoregulation, histochemistry, ultrastructure. Home: 121 Cayuga Park Rd., Ithaca, N.Y. 14850.*

WIMSHURST, James, English engr., inventor; b. Poplar, Eng., Apr. 13, 1832; s. Henry Wimshurst; ed. Stearonheath House, London; apprenticed at Thames Ironworks to J. Mare, 1853; m. Clara Tubb, 1864; 2 sons, 1 dau. Surveyor of Lloyds, 1853-65; became chief Liverpool Underwriters' Registry, 1865; chief shipwright surveyor consultative dept. Bd. Trade, 1874-99; Bd. of Trade rep. to Internat. Conf., Washington, 1890; named assessor to City of London Ct., 1902. Fellow Royal Soc., 1898; mem. Inst. Naval Architects, Instn. Elec. Engrs., Phys. Soc. (became mem. council 1880), Röntgen Soc. (became mem. council 1898), Royal Inst. (became mem. com. mgmt. 1902). Author: The Particulars of Several Forms of Electrical Machines since 1880; A Book of Rules for the Construction of Steam Vessels, 1898. Contributed to devel. and design of static electricity generators; invented improved vacuum pump, improved method for electrically connecting light-ships with shore, instrument for measurement of stability of ships. Died Clapham, Eng., Jan. 3, 1903.

WINAND, Leon Jules François, Belgian physicist; b. Liège, Belgium, Jan. 29, 1911; s. Leon J. and Guillelmine (Galoppin) W.; Sc.D., U. Liège, 1932, postgrad., 1940; postgrad. Inst. Radium, Paris, 1932-34, 36-38; m. Anne Rideau, Apr. 21, 1938; children—Joëlle (Mrs. N. Verjus), François, Jean-Pierre, Chantal, Marie-France (dec.), Denis, Patrick. Faculty, State U., Liège, 1935—, prof. physics, 1961—, dir. Nuclear Physics Lab., 1952—, dean Faculty Sci., 1966-68. Decorated Commandeur Ordre Couronne; chevalier Ordre Leopold; Croix Civique le Classe. Mem. Royal Sci. Soc. Liège, Am. Phys. Soc., I.E.E.E. Author: Nuclear Physics Course, 1960, 65; also articles. Research on neutron physics, cosmic rays, high energy reactions, nuclear emulsions, low energy accelerators, low energy nuclear physics. Home: 7B Place St. Paul, Liège, Belgium. Office: 1A Quai Roosevelt, Liège, Belgium.*

WINANS, Thomas DeKay, Am. engr., inventor; b. Vernon, N.J., Dec. 6, 1820; s. Ross and Julia (De Kay) W.; m. Celeste Revillon, Aug. 23, 1847; 4 children. Engr. with Harrison, Winans & Eastwick (firm organized to handle Russian railroad constrn. venture), went to Russia to take charge of mech. dept. of railroad from St. Petersburg to Moscow, 1843, contracted to equip railroad with locomotives and other rolling stock in 5 years, established shops at Alexandrovsky; returned to U. S. (left brother in charge in Russia), 1851, recalled to Russia for new constrn. contract, 1866; business interests taken over by Russian Govt. with payment of large bonus, 1868; dir. B. & O. R.R.; established soup station opposite his home during Civil War; devised (with his father) cigar-shaped hull for trans-Atlantic steamers, 1859; invented device which made organ as easy of touch as piano, invented glass feeding vessels for fish (adopted by Md. Fish Commn.); used undulation of waves to pump water of a spring to reservoir at top of his villa, Newport, R.I. Died Newport, June 10, 1878.

WINBURY, Martin Maurice, Am. pharmacologist; b. N.Y.C., Aug. 4, 1918; s. Ervin and Helen (Stein) W.; B.S., L.I.U., 1940; M.S., U. Md., 1942, postgrad.; Ph.D., N.Y. U. 1951; m. Blanche M. Simons, July 11, 1942; children—Nancy Ellen, Gail Elizabeth. Mineral economist U.S. Bur. Mines, College Park, Md., 1942-44; biochemist, pharmacologist Merck Inst. for Therapeutic Research, Rahway, N.J., 1944-47; pharmacologist G.D. Searle & Co., Chgo., 1947-55; sr. pharmacologist Schering Corp., 1955-58, dir. pharmacology, 1958-61, asso. dir. biol. research, 1960-61; dir. pharmacology Warner-Lambert Research Inst., Morris Plains, N.J., 1961—, Vice chmn Gordon Research Conf. on Medicinal Chemistry, 1965, chmn., 1966. Mem. A.A.A.S., N.Y. Acad. Scis., Am. Heart Assn., Am. Coll. Cardiology, Soc. Exptl. Biol. Medicine, Am. Soc. Pharmacological and Exptl. Therapy, Cardiac Muscle Club. Contbg. author: Advances in Pharmacology, 1965; contbg. author, editor Lab. Evaluation of Antianginal Agents, 1967; also numerous articles. Research on regulation nutritional circulation heart; developed drugs for treatment ulcer, heart failure, edema, cardiac arrhythmias, coronary disease; developed methods for study new drugs; research on use isotopes in pharmacology tissue circulation, cardiovascular pharmacology. Home: 27 Baker Rd., Livingston, N.J. 07039. Office: 170 Tabor Rd., Morris Plains, N.J. 07950.*

WINCH, Ralph Philip, Am. physicist; b. Milton Junction, Wis., Oct. 19, 1905; s. Philip George and Eva (Thompson) W.; A.B., Milton Coll., 1927; M.A., U. Wis., 1929, Ph.D., 1931; m. Mary Elizabeth Johnson, June 8, 1930; children—Martha Elizabeth (Mrs. Anil Asher), Roger Holmes, Katherine Jean (Mrs. Charles Y. Wrigley). Faculty, Williams Coll., Williamstown, Mass., 1931—, prof. physics, 1945—, Barclay Jermain prof natural philosophy, 1950—, acting chmn. physics dept., 1943-45, 53-54, chmn., 1960—; vis. asst. prof. Princeton, 1942; vis. prof. Brown, 1953, Wesleyan U., summers 1959-65, 68. Fellow A.A.A.S.; mem. Am. Assn. Physics Tchrs. (sec. 1961-66, Distinguished Service citation 1960, 66), Am. Inst. Physics, Am. Phys. Soc. (chmn. N.E. sect. 1957-58), Am. Assn. U. Profs. Author: Electricity and Magnetism, 1955, 2d edit., 1963; also articles. Research on photoelectric effects in single crystals and polycrystalline metals. Home: 24 Bingham St., Williamstown, Mass. 01267.*

WINCH, Robert Francis, Am. sociologist; b. Lakewood, O., Aug. 31, 1911; A.B., Western Res. U., 1935; A.M., U. Chgo., 1939, Ph.D., 1942; m. Martha Brashares, Sept. 3, 1938. Faculty, U. Chgo., 1941-42, Vanderbilt U., 1946-48; faculty Northwestern U., Evanston, Ill., 1948—, prof. sociology, 1955—, chmn. dept. sociology, 1967—. Social Sci. Research Council fellow, 1945-46; Guggenheim fellow, 1955-56. Fellow A.A.A.S., Am. Sociol. Assn., Am. Psychol. Assn.; mem. Sociol. Research Assn., Am. Statis. Assn. Asso. editor Sociometry, 1956-61, Jour. Marriage and Family, 1965—. Research structural-functional theory of identification and family. Home: 942 Harvard Ct., Highland Park, Ill. 60035. Office: Centennial Hall, Northwestern University, Evanston, Ill. 60201.*

WINCHELL, Alexander, Am. geologist; b. Northeast, Dutchess County, N.Y., Dec. 31, 1824; s. Horace and Caroline (McAllister) W.; grad. Wesleyan U., Middletown, Conn., 1847, LL.D., 1867; m. Julia F. Lines, Dec. 5, 1849; 6 children. Sch. tchr., South Lee, Mass., 1841-42, Pennington (N.J.) Male Sem., N.J., 1847-49; prof. natural history Amenia Sem., N.Y., 1849; in charge of acad., Newbern, Ala., 1850; opened Mesopotamia Female Sem., Eutaw, Ala., 1851; pres. Masonic U., Selma, Ala., 1853; prof. physics and engring. U. Mich., Ann Arbor, 1853-55, prof. geology, zoology and botany, 1855-73, prof. geology and paleontology, 1879-91; directed state geol. survey Mich., 1859-61, 69-71, located salt beds of Saginaw Valley; chancellor Syracuse (N.Y.) U., 1872-74; prof. geology and zoology Vanderbilt U., 1875-78; chmn. com. to organize Geol. Soc. of Am., pres., 1891. Author: The Doctrine of Evolution, 1874; Sketches of Creation, 1870; Preadamites, 1880; World Life, 1883; (textbook) Geological Studies, 1886; also published bibliography of over 250 titles. Sci. popularizer; research on evolution, paleontology and geology; discovered many new fossilgenera and species. Died Ann Arbor, Feb. 19, 1891.

WINCHELL, Newton Horace, Am. geologist, archeologist; b. North East N.Y., Dec. 17, 1839; s. Horace and Caroline (McAllister) W.; A.B., U. Mich., 1866, A.M., 1869; m. Charlotte Sophia Imus, Aug. 24, 1864; children— Horace Vaughn, Alexander Newton Winchell. Supt. pub. schs., Adrian Mich., 1866-69; asst. state geologist Mich., 1869-70; asst. geol. survey, Ohio, 1870-72; state geologist of Minn., 1872-1900; prof. geology and mineralogy U. Minn., 1873-1900; archeologist Minn. Hist. Soc., 1906—. Member U. S. Assay Commn., 1887, founder and editor, Am. Geologist, 1888-1905. Author: Catalogue of the Plants of the State of Michigan, 1861; Geology of Ohio and Minnesota, 1872-1900; The Iron Ores of Minnesota (with Horace V. Winchell), 1891; Elements of Optical Mineralogy (with Alex. N. Winchell), 1909; The Aborigines of Minnesota; The Paleoliths of Kansas; also papers in sci. jours. Theorized that man inhabited Am. continent during late Ice Age. Died Mpls., May 2, 1914.

WINCHESTER, Albert McCombs, Am. biologist; b. Waco, Tex., Apr. 20, 1908; s. Robert S. and Katheryn (Moore) W.; A.B., Baylor U., 1929; M.A., U. Tex., 1931, Ph.D., 1934; postgrad. U. Chgo., Harvard, U. Mich., U. Munich (Germany); m. Josephine Milam Walker, Dec. 23, 1934; 1 dau., Betty Jo. Head biology dept. Okla. Bapt. U., 1936-42; prof. biology Baylor U., 1942-45; head biol. dept. Stetson U., 1942-45; Biol. Sci. Curriculum Study cons. U. Colo., 1960-61; prof. biology Colo. State Coll., Greeley, 1961—, head biology dept., 1968—; vis. prof. biology Emory U., 1945, U. Va., 1951-52, U. Vt., 1959. Mem. Colo. Bd. Examiners in the Basic Sciences, since 1967—. Fellow A.A.A.S.; mem. Genetics Soc. Am., Am. Soc. Human Genetics, Colo. Wyo. Acad. Sci., Am. Biology Tchrs. Assn. Am. Inst. Biol. Scis. (vis. lectr. 1964—). Author: Zoology, the Science of Animal Life, 1947, rev., 1956; Biology and Its Relation to Mankind, 1949, rev., 1957, 64, 68; Genetics, 1951, rev., 1958, 66; Heredity and Your Life, 1956, rev., 1960; Heredity, 1961; Modern Biological Principles, 1965; Human Heredity, 1966; also articles. Home: 2127 14th St. Rd., Greeley, Colo. 80631.*

WINCKELMANN, Johann Joachim, German archaeologist; b. Stendal, Prussia, Dec. 9, 1717; s. Ferdinand and Dorothea (Krützfeldt) W.; ed. univs. Halle and Jena, 1738-40; Pvt. tutor; asst. master grammar sch., Seehausen, 1743-48; librarian to Count von Bünar at Nöthnitz, 1748-54, to Cardinal Rassionei, Rome, 1754-58, to Cardinal Albani, 1758-63; chief antiquities in The Vatican and all Rome, 1763-68, Author: Gedanken über die Nachahmung der griech. Werke in der Malerei und Bildhauerkunst, 1755; Anmerkungen über die Baukunst der Alten, 1761; Abhandlung von der Fahigkeit der Empfindung des Schonen in der Kunst, 1763; Geschichte der Kunst des Altertums, 1764; Monumenti antichi inediti, 2 vols., 1767-68. Founder of sci. archaeology; formulated 1st system for dating classical statuary; made 1st careful descriptions of ancient articles including some excavated from Pompeii and Herculaneum; in 1st systematic study of classical art, he developed periods of classification still used; work contb. to devel. of neo-classic art in Europe and to appreciation and understanding of classical art. Murdered in Trieste, Austria (now Italy), June 8, 1768.

WINCKLER, George, anatomist; b. France, Apr. 2, 1901; s. Emile and Marcelle (Forster) W.; Dr en médecine, Faculté de Médecine de Strasbourg, France, 1926; m. Marie-Jeanne Forster, Nov. 27, 1928; children—Geneviève (Mrs. Lavillaureix), Monique, Marie-Thérèse (Mrs. Baille), Cécile (Mrs. Alméras), Véronique. Head normal anat. research, Strasbourg, 1928-46, prof., 1946-51; prof. normal anatomy, dir. Institut d'Anatomie, Lausanne, Switzerland, 1951—. Decorated Legion of Honor. Mem. Assn. Anatomists. Author: Manuel d'Anatomie topographique et fonctionelle, 1964; also numerous articles. Dir. Archives d'anatomie, d'Histologie et d'Embryologie. Research in muscles of back and abdomen, innervation of muscles. Home: 39 Bethusy. Office: Institut Anatomie normale, 9 Bugnon, Lausanne, Switzerland.*

WINCKLER, Hugo, German archaeologist; b. Grafenhainichen, July 4, 1863. Prof., U. Berlin; helped excavate Phoenician city of Sidon, also Hittite capital, Boghazkeui. Author: Die Keilschrifttexte Sargons, 2 vols., 1889; Geschichte Israels in Einzeldarstellungen, 2 vols., 1895-1900; Die Gesetze Hammurabis, 1902; Die babylon. Geisteskultue in ihren Beziehungen zur Kulturentwicklung der Menschheit, 1907; Vorderasien im 2. Jahrstausend auf Grund archivalischer Studien, 1913. Most important contbn.

discovery at Boghazkeui of cuneiform tablets, 1906-07; these tablets yielded much information on Hittite civilization and are one of few outside lit. references to Greek Bronze Age civilization; primarily an Assyriologist; with Puchstein and Curtius founded Hittitology. Died Berlin, Apr. 19, 1913.

WINCKLER, Johann Heinrich, German physicist, inventor; b. Wigendorf, Germany, 1703; prof., Leipzig, Germany. Fellow Royal Soc., 1746. Author: Thoughts on the Properties and Effects of Electricity; other works. Produced a perfected friction machine, 1744, a perfected Leyden jar, 1746; credited with influencing Franklin. Died 1770.

WINDAUS, Adolf Otto Reinhold, German chemist; b. Berlin, Germany, Dec. 25, 1876; s. Adolf and Marguerite (Elster) W.; studied medicine, Freiburg, with Emil Fischer, Berlin; Ph.D., Freiburg, 1899; m. Elizabeth Resau, Sept. 29, 1915; 2 sons, 1 dau. Became pvt. docent U. Freiburg, 1903; prof. applied med. chemistry, Innsbruck, Austria, 1913-15; prof. U. Göttingen (Germany), 1915-44, also dir. chem. lab. Recipient Nobel prize in chemistry, 1928; Pasteur medal, 1938; Goethe medal. Mem. Göttingen, Berlin, Prussian (fng.), Bavarian (corr.) acads. scis., Chem. Soc. London (hon.). Author: Abbau- und Aufbauersucke im Gebiet der Sterine. Research on steroids; discovered that ultraviolet light activates ergosterol and gives vitamin D2; investigated digitalis, colchicin, histidin, histamin; elucidated structure of animal and vegetable sterines, especially cholesterins, bile acids, antineuritic vitamins. Died Göttingen, June 9, 1959.

WINDELBAND, Wilhelm, German philosopher; b. Potsdam, Germany, May 11, 1848; student under Hermann Lotze. Prof. Zürich (Switzerland), Freiburg (Germany), Strasbourg (France), Heidelberg (Germany) univs.; founder Baden (S.W. German Sch. Philosophy). Author: Die Geschichte der Neueren Philosophie, 1878-80; Praludien, 1884; Geschichte und Naturwissenschaft, 1894; Geshichte der Philosophie, 1892. Prominent in neo-Kantian movement late 19th century; approach distinguishes between natural sci., scis. of culture (natural are nomothetic, that is, formulate gen. laws by combining facts empirically obtained, cultural are idiographic, that is, deal descriptively with particular in its value for human spirit). Died Heidelberg, Oct. 22, 1915.

WINDER, Charles Gordon, Canadian geologist; b. Ottawa, Ont., Can., June 13, 1922; s. Louis David and Irene (Stevenson) W.; B.Sc., U. Western Ont., 1949; M.S., Cornell U., 1951, Ph.D., 1953; m. Frieda Jean Clark, Sept. 25, 1948; children—Nancy, Paula. Faculty, U. Western Ont., London, 1953—, prof., head dept. geology, 1965—. Fellow Geol. Soc. Am.; mem. Paleontol. Soc., Soc. Econ Paleontologists and Mineralogists, Alberta Society Petroleum Geologists, Geol. Assn. Can., Sigma Xi. Author: Lexicon of Paleozoic Names in Southwestern Ont., 1961; also articles. Research in paleozoic geology southwestern Ont. Home: 288 Neville Dr., London, Ont., Can.*

WINDER, Claude Veyne, Am. pharmacologist; b. Fallon, Nev., May 26, 1909; s. Elbridge Lee and Charlotte (Sifford) W.; B.S., U. Nev., 1931; M.S., U. Mich., 1932, Sc.D., 1936; m. Harriet Owen, Aug. 19, 1932 (dec. Sept. 1959); children—Robert O., Ann (Mrs. Richard Taylor), Allan W., Thomas L., Carl H., Paul S., Mary J., Norma C.; m. 2d, Margaret Draughon Francis, Nov. 23, 1961. Faculty, U. Mich. Med. Sch., 1931-43; sr. research pharmacologist Parke, Davis & Co., Ann Arbor, Mich., 1943-66, lab. dir. in pharmacodynamics, 1966—. Fellow A.A.A.S.; mem. Mich. Acad. Sci., Arts and Letters, Soc. for Exptl. Biology and Medicine (mem. Mich. council 1963-64), Am. Statis. Assn., Biometric Soc., Am. Physiol. Soc., Am. Soc. Pharmacology and Exptl. Therapeutics, Sigma Xi. Research, publs. in central and chemoreflex actions of drugs and chems. on breathing, stimulation of breathing by oxygen want, control of breathing, circulation and heart rate via reflexes from arteries, pain in animals and evaluations of pain-relieving agts., acute tolerance to drugs, antihistamines, dilators of blood vessels, cough in animals and evaluation of cough controlling drugs, inflammation in animals and evaluation of drugs to manage inflammatory debilitation and pain; co-discoverer fenamic acids. Home: 1035 Martin Pl. Office: 2800 Plymouth Rd., Ann Arbor, Mich. 48106.*

WINDLE, Sir Bertram Coghill Alan, anthropologist, physician; b. May 8, 1858; s. S. A. and Syndey Katharine (Coghill) W.; M.A., B.A., (1st class honours in natural sci.), M.D., M.B., (1st Place), B.Ch., D.Sc., U. Dublin; M.S., LL.D., U. Birmingham (Eng.); LL.D., U. Boston; Ph.D. (hon.), Rome; D.Sc. (hon.), Marquette U.; m. Madeline Hudson, 1886; m. 2d, Edith Mary Nazer, 1901; 1 dau. Became prof. U. Coll., Cork, Ireland, 1904, pres., 1904-19; organizer Nat. U. Ireland; became prof. cosmology and anthropology St. Michael's Coll., Toronto (Ont., Can.) U., 1919, also spl. lectr. ethnology; dean med. faculty, prof. anatomy and anthropology U. Birmingham; became commr. under Irish U. Act. of 1908; examiner anatomy Cambridge, Aberdeen, Glasgow, Durham Univs., also Royal Coll. Physicians, London, Royal Coll. Surgeons, Ireland. Mem. Gen. Council Med. Edn. Fellow Royal Soc., 1899; mem. Irish Tech. Assn. (pres.). Knighted, 1909. Author: A Manual of Surface Anatomy, 3d edit., 1902; The Proportions of the Human Body, 1892; Tyson's Pygmies of the Ancients, 1894; Life in Early Britain, 1897; Shakespeare's Country, 1899; The Malvern Country, 1900; The Wessex of Thomas Hardy, 1901; Chester, 1903; The Prehistoric Age, 1904; A School History of Warwickshire, 1906; What is Life? . . . , 1908; Facts and Theories, 2d edit., 1923; A Century of Scientific Thought, 1915; The Church and Science, 1917; Science and Morals, 1919; Scholasticism and Vitalism, 1920; The Romans in Britain, 2d edit., 1923; On Miracles . . . , 1924; The Who's Who of the Oxford Movement, 1926; The Catholic Church and . . . Science, 1927; Religions Past and Present, 1928. Tried to demonstrate that there is no conflict between sci. progress and ch. teachings. Died Feb. 14, 1929.

WINDMUELLER, Herbert George, biochemist; b. Beckum, Westphalia, Germany, July 5, 1931; s. Alfred and Ida (Stein) W.; came to U.S., 1937, naturalized, 1942; B.S., Va. Poly. Inst., 1952, M.S., 1956, Ph.D., 1958; m. Ray Yordinsky, Oct. 11, 1958; children—Andrew Solomon, Dena Margaret. Research fellow in biochemistry Brandeis U., 1958-61; biochemist NIH, Bethesda, Md., 1961—. Mem. A.A.A.S., Am. Inst. Biol. Chemists, Am. Inst. Nutrition, Am. Chem. Soc. Showed mechanism of ethylene oxide toxicity; purified brain enzyme; synthesized and studies, pyridine nucleotide coenzyme analogues; studied hepatic lipogenesis and lipid secretion. Home: 5924 Wilmett Rd., Bethesda 20034. Office: NIH, Bethesda, Md. 20014.*

WINDORFER, Adolf, German pediatrician; b. Falkenstein, Germany, Feb. 7, 1909; s. Hugo and Ida (Raum) W.; M.D., Erlangen-Nürnberg U.; m. Liselotte Poppe, July 4, 1936; children—Liselotte (Mrs. Hans Truckenbrodt), Adolf. With orthopedic clinic U. Munich, 1934-36, U. Children's Clinic, Frankfort/Main, 1936-50; head physician Children's Clinic, Stuttgart, 1950-56; dir. U. Children's Clinic, Erlangen-Nürenberg, Germany, 1956—. Mem. Deutsche Gesellschaft fur Kinderheilkunde, Österreichische Gessellschaft fur Kinderheilkunde, Deutsche Gessellschaft der Naturforscher und Ärzte, Deutsche Gesellschaft zur Bekampfung der Mucoviscidose (chmn.). Author: Notfallfibel, 1965, 67; also articles. Research on poliomyelitis, epidemology of infectious diseases, cystic-fibrosis, and virological diseases. Home: 10 Gustav Specht Strasse, Erlangen, Germany.* Office: 15 Loschgestrasse, Erlangen, Germany.*

WING, G(eorge) Milton, Am. mathematician; b. Rochester, N.Y., Jan. 21, 1923; s. George O. and Louise (Weiss) W.; B.A., U. Rochester, 1944, M.S., 1947; Ph.D., Cornell U., 1949; m. Nina A. John, June 7, 1958. Staff mem. Los Alamos Sci. Lab., 1945-46, 51-58, Sandia Lab., Albuquerque, 1959-64; faculty U. Cal. at Los Angeles, 1949-52, U. N.M., 1958-59; prof. applied math. U. Colo., Boulder, 1964-66; prof. math. U. N.M., Albuquerque, 1966—. Cons. Rand Corp., 1958-65, Sandia Corp., 1958-59, E.H. Plesset & Assos., 1958-59, 64—, Los Alamos Sci. Lab., 1964—. Mem. Am. Math. Soc., Math. Assn. Am., Soc. Indsl. and Applied Math. (vis. lectr. 1961-62, 64—), Phi Beta Kappa, Sigma Xi. Author: An Introduction to Transport Theory, 1962. Research on transport theory, radiative transfer, applied math. Home: 912 McDuffie Circle N.E., Albuquerque.

WING, John Kenneth, English physician; b. London, Eng., Oct. 22, 1923; M.B., U. London, 1952, D.P.M., 1956, Ph.D., 1959, M.D., 1960. Sci. staff Med. Research Council, 1957—, dir. social psychiatry research unit, London, 1965—; organiser psychiat. teaching London Sch. Hygiene, 1963—. Hon. cons. psychaitrist Maudsley Hosp., London, 1959—. Author: (with D.H. Bennett, J. Denham) The Industrial Rehabilitation of Longstay Schizophrenic Patients, 1964; (with G.W. Bronw, M. Bone, B. Dalison) Schizophrenia and Social Care, 1966; also articles. Editor: Early Childhood Autism, 1966; Jour. Social Psychiatry, 1966—. Research on influence of social events on course of psychiat. disorders, social precipitants and social treatments of mental diseases. Home 46 Dacres Rd., Forest Hill, London, S.E. 23, Office: Med. Research Council, Social Psychiatry Research Unit, Maudsley Hosp., London, S.E. 5, Eng.*

WINGATE, Edmund, English mathematician; b. Flamborough, Eng., 1596; s. Roger and Jane (Birch) W.; B.A., Queens Coll., Oxford, 1614; m. Elizabeth Button, July 28, 1628; 5 sons, including Button; 2 daus. Entered Gray's Inn, 1614; became tchr. English to Princess Henrietta Maria, Paris, circa 1624; returned to Eng. where he sided with Parliament in civil war; became M.P. for Bedfordshire, 1655. Author numerous works including: L'usage de la règle de proportion en arithmétique, 1624; Arithmétique logarithmetique, 1626; The Construction and Use of the Line of Proportion, 1628; Of Natural and Artificiall Arithmetique, 1630; Ludus mathematicus, 1654; The Body of the Common Law of England, 1655; The Clarks Tutor for Arithmetick and Writing, . . . , 1671. Editor works of Samuel Foster. Wrote textbooks on elementary arithmetic; introduced rule of proportion to France; contributed to spread of logarithms in France. Died London, Dec. 1656.

WINGE, Ojvind, Danish geneticist, mycologist; b. Aarhus, Denmark, May 19, 1886; s. Sigfred and Petra (Rian) W.; M.Sc., U. Copenhagen, 1910, Ph.D., 1917; postgrad. U. Stockholm (Sweden), U. Paris, U. Chgo.; m. Julie Begtrup Moller, May 8, 1913; 1 son, Per. Asst., Carlsberg Lab., 1911-21; prof. genetics Royal Vet. and Agrl. Coll., Copenhagen, 1921-33; prof., dir. physiol. dept. Carlsberg Laboratorium, 1933-56. Lectr. genetics U. Copenhagen, 1923-35; Hitchcock prof. U. Cal., 1957; leader Carlsberg Breweries' Expt. Farm, 1938-62. Fellow Royal Soc., 1947; mem. Royal Danish Acad. Sci.; fgn. mem. all Scandinavian acads., Nat. Acad. Sci. Author: Textbook in Genetics, 1928; Mycological Excursion Manual; Inheritance in Dogs, 1950; also numerous articles. Research on cytology, sex-linked inheritance and sex determination in plants, chromosomes of malignant tumors, yeast genetics; plant species studies. Died Apr. 5, 1964.*

WINITZ, Milton, Am. chemist; b. Bklyn., July 20, 1924; s. Max and Sophie (Starr) W.; B.A., N.Y. U., 1948; Ph.D., Ia. State Coll., 1951; m. Charlotte Metalsky, Sept. 3, 1948; chidren—Mark, Jan. supervisory chemist Nat. Cancer Inst., Bethesda, Md., 1953-62; head sect. on amino acids, peptides and metabolic nutrition City of Hope Med. Center, Duarte, Cal., 1962-63; head life scis. lab. United Tech. Center, Sunnyvale, Cal., 1964—. Mem. subcom. on amino acids of com. biol. chemistry NRC, 1957-64, chmn., 1962-64. Recipient Superior Service medal Dept. Health, Edn. and Welfare, 1956. Mem. Fedn. Am. Socs. for Exptl. Biology. Author (with Jesse P. Greenstein): Chemistry of the Amino Acids, 3 vols., 1961. Research, publs. in amino acid and peptide chemistry, metabolic nutrition, enzymology, and chemically defined diets. Home: 844 Ross Ct., Palo Alto, Cal. 94303. Office: 1050 E. Arques Av., Sunnyvale, Cal. 94088.*

WINKELMANN, Richard Knisley, Am. dermatologist; b. Akron, O., July 12, 1924; s. H.A. and Ilene (Knisley) W.; student U. Akron, 1942-43, Mich State Normal Coll., 1943-44; B.S., U. Mich., 1947;M.D., Marquette U., 1948; postgrad. (fellow), Mayo Found.; Ph.D. in Dermatology, U. Minn., 1956; m. Anne Mary Robertson, Apr. 26, 1952; children—Richard Lowell, John Curtis, Lisa Anne, Susan Elizabeth. AEC fellow, 1949-51; research asso. Washington U., 1950; faculty Med. Coll. Ala., 1952-54; faculty U. Minn., 1955—, asso. prof. Mayo Found. faculty, 1962-65, prof., 1965—, asso. prof. anatomy, 1965—; practice medicine specializing in dermatology, Rochester, Minn., 1956—. Cons. sect. dermatology Mayo Clinic, 1956—; mem. med. sci. com. on skin NRC, 1961—. Fellow A.C.P.; mem. Am. Acad. Dermatology and Syphilogy, A.M.A., N.Y. Acad. Scis., Soc. for Investigative Dermatology, Research Com. Am., Am. Assn. Dermal Pathologists (mem. program com.), Am. Assn. Anatomists, Am. Fedn. for Clin. Research, A.A.A.S., Am. Assn. Phys. Anthropologists, Minn. Soc. for Clin. Pathology., Am. Israeli (corr.) dermatol. assns., Minn., Chgo., Brasilian (hon.), Danish (fgn. hon.) dermatol. assns. Author: Nerve Endings in Normal and Pathologic Skin, 1960; also numerous articles. Research on cutaneous nerve, comparative histochemistry in neuroanatomy skin, clin. disease skin, microscopic study diseases skin. Home: Oakledge, Mounted Route 72, Rochester, Minn. 55901.*

WINKLER, Clemens, German chemist; b. Freiberg, Germany, Dec. 26, 1838; mgr. cobalt-glass works, 1860-73; prof. Bergakademie, Freiberg from 1873. Author: Anleitung zur chemischen Untersuchung der Industriegase, 1876-79; Die Massanalyse nach neuem titrimetrischem System, 1883; Lehrbuch der technischen Gasanalyse, 1884. Leading inorganic chemist of his day; established gas analysis; contbd. new methods to gravimetric analysis and volumetric analysis; devised perforated rotating cathode, also separation of nickel, zinc, bismuth; introduced term normal weight into volumetric analysis; devised test for carbon monoxide by palladium chloride; determined sodium hydroxide in presence of alkali carbonates; reduced oxides with magnesium; prepared hybrides of metals; developed contact process for sulfuric acid, 1878; studied element Indium; discovered element Germanium (predicted by Mendeleeff 1871), 1886; inventor 3-way stopcock. Died Dresden, Germany, Oct. 8, 1904.

WINKLER, Kjeld, Danish physician; b. Odense, Denmark, Jan. 23, 1925; s. Gustav and Karla (Olsen) W.; M.D., U. Copenhagen (Denmark), 1952, dr.med., 1966; m. Anne-Birthe Jorgensen, Apr. 7, 1963; children—Thomas, Soren. With dept. internal medicine, cardiologic lab. Rigshospitalet, 1953-62; sr. registrar dept. internal medicine III Kommunehospitalet, Copenhagen, 1962-67, head dept. clin. physiology, 1967—. Mem. Danish Soc. Internal Medicine, Danish Soc. Biology, Danish Soc. Clin. Chemistry and Physiology. Research, publs. on elimination of bromsulfalein in man, liver diseases and methods in diagnosis, metabolism of alcohol and different carbohydrates in human liver, cardiac resuccitation, treatment of hypercholesterolaemia, blood flow in kidney. Home: 104. V.Voldgade, Copenhagen, Denmark.*

WINKLER, Klaas Christiaan, Dutch bacteriologist; b. Amsterdam, Netherlands, May 22, 1908; s. Cornelis and Maria (Junius) W.; student chemistry Utrecht (Netherlands), 1925-38, degree in medicine, 1933; m. Johanna Kruyt, July 17,, 1933; children—

Yolanda (Mrs. H. Peschar), Maria (Mrs. A.C. Noordanus), Peter, Wilhelmine (Mrs. Th. Winkel), Paul,, Madeleen. Practice gen. medicine, Leiden, Netherlands, 1933-38, reader, 1939-47, prof. bacteriology, 1947—. Recipient Dr. Dentz medaile, 1961, 1961. Mem. Soc. Gen. Microbiology, Internat. Soc. for Cell Biology, Royal Soc. Medicine, Internat. Assn. for Dental Research, N.Y. Acad. Sci. Contbg. author: Leerboek der Microbiologie en Immunologie, 1957. Research, numerous publs. on mechanism of action of sulfonamide, intermediate metabolism of bacteria, bacteriophage, bacterial genetics, survival of bacteria in air, properdin and natural immunity, oral bacteriology, dental caries and water fluoridation. Home: 413 Oude Gracht. Office: 59 Catharijnesingel, Utrecht, Netherlands.*

WINKLER, Walter Theodor, German psychiatrist; b. Pearadja, Sumatra, Dec. 16, 1914; s. Johannes and Luise (Metzler) W.; ed. univs. Tübingen, Rostock and Marburg; M.D.; m. Marie Seim, Feb. 8, 1941; children—Ursula, Karl-Rudolf, Susanne, Brigitte. Asst., Univ. Clinic Neurology, Marburg, 1941-46, Tübingen Neurology Clinic, 1946-48; dr.-in-chief Tübingen Clinic, 1948-61; agrégé in neurology and psychiatry; prof. at large, 1955; dir. Westfalisches Landerkrankenhaus, Gutersloh, 1961—. Mem. Soc. Psychotherapy, German Soc. Neurology and Psychiatry, German Constl. Research Soc. Author: Psychologie der modernen Kunst, 1949; also articles on psychiatry, psychotherapy, social psychiatry, especially psychodanics of endogenic psychoses. Home: Hermann Simonstrasse 2. Office: Hermann Simonstrasse, 483 Gutersloh, West Germany.

WINLOCK, Joseph, Am. astronomer; b. Shelby County, Ky., Feb. 6, 1826; s. Fielding and Nancy (Peyton) W.; grad. Shelby Coll., 1845; A.M. (hon.), Harvard, 1868; m. Mary Isabella Lane, Dec. 10, 1856; 6 children. Prof. math. and astronomy Shelby Coll., Shelbyville, Ky., 1845-52; mem. staff Am. Ephemeris and Nautical Almanac, Cambridge, Mass., 1852-57, supt., 1858-59, circa 1862-66; prof. math. U. S. Naval Obs., 1857; head dept. math. U. S. Naval Acad. 1859; Phillips prof. astronomy, also dir. obs. Harvard, 1866-75. Mem. Nat. Acad. Scis. (corporate), Am. Acad. Arts. and Scis. First astronomer to obtain photograph of corona during solar eclipse, also 1st to adapt to photographic purposes telescope of long focus, fixed horizontally and used without eye piece. Died Cambridge, June 11, 1875.

WINN, Alfred Vernon, Am. chemist; b. Galt, Cal., Mar. 15, 1915; s. Henry A. and Irene (Moon) W.; B.A., Pacific Union Coll., 1938; M.S., U. Wash., 1950; Ph.D., Stanford, 1958; m. Helen M. Corey, June 3, 1940; children—Alfred Vernon, Kenneth G. Tchr. high schs., 1938-49; faculty Canadian Union Coll., Lacombe, Alta., 1950-54; faculty Pacific Union Coll., Angwin, Cal., 1954-56, 58—, prof. chemistry, chmn. chemistry dept., 1960—. Lectr. Stanford, 1957-58. Mem. Am. Chem Soc., A.A.A.S., Cal. Assn. Chemistry Tchrs., Sigma Xi. Research on chemistry of terpens, mechanisms of organic chem. reactions. Home: 235 Cold Springs Rd., Angwin, Cal. 94508.*

WINNE, William Thomas, Am. botanist; b. Schenectady, May 25, 1913; s. Matthew and Alice (Gaffers) W.; A.B., Union Coll., Schenectady, 1934; M.S., Cornell U., 1935, Ph.D., 1947. Faculty, Union Coll., 1946—, prof. botany, 1955—, chmn. dept. biol. scis., 1965-66. Mem. Torrey Bot. Club, New Eng., Am. bot. socs., Am. Bryological Soc. Taxonomic research on plants of W. Africa, No. Mexico, taxonomy of brophytes. Contbr. articles to tech. jours. Home: 2517 Troy Rd., Schenectady 12309.*

WINNECKE, Friedrich August Theodore, German astronomer; b. 1835; dir. obs., Strasbourg (now in France), 1872-86; determined solar parallax; studied paths of comets and double stars. Died 1897.

WINOGRADOW, Leon, Polish chem. ceramist; b. Warsaw, Poland, Apr. 15, 1899; s. Abraham and Rachel (Nemzer) V.; Doctorate, Free City of Gdansk Poly., 1924; diploma Teplice-Sanov, Czechoslovakia, 1925; m. Elisabeth Tovbin, June 2, 1923; 1 dau., Helen (Mrs. Boghdan Matuszewski). Research asst. Gdansk Poly., 1923-24; lab. dir. J. Inwald's Glass Factory, Teplice-Sanov, 1924-27; tech. dir China Factory, Nowy Dwór, Poland, 1928-39, Ceramics Factory, Dermanka, Poland, 1935-40; sci. dir. Ukranian Porcelain Trust Labs., Kiev, USSR, 1940-41; chief technologist Abrasives and Porcelain Factory, Hajta, Siberia, USSR, 1941-47; dir. fine ceramics and bldg. materials br. Central Adminstrn. Polish Ceramic Industry, 1947-49; prof. fine and spl. ceramics Acad. Mining and Metallurgy, Cracow, Poland, 1952—; sci. dir. Inst. Glass and Ceramics, Warsaw, 1950-55, sci. adviser, 1955—. Expert, UNO Tech. Assistance, 1958-59. Decorated Polish Golden Merit Cros, Medal for Sacrifical War Service of USSR, Officer's Cross of Order of Polonia Restituta, Hon. Gold badge Polish Gen. Tech. Assn. Mem. Inst. Glass and Ceramics (mem. council), Inst. Electronics Polish Acad. Sci. (mem. council), Com. Ceramic Sci., Polish Gen. Tech. Orgn. (vice chmn. com. on bldg. materials), Brit. Ceramic Soc., Deutsche Keramische Gesellschaft, Internat. Comm. on Glass. Research, publs. on purification and use of ceramic raw materials especially kaolin; extraction of high quality feldspar from granites and pegmatities; strain hardening of glass; drying of ceramics; large pore abrasive tools; sintered alumina; electronic ceramic bodies; grinding and its effect on properties of some ceramic bodies; patentee sponge insulating glass, melting, casting and recrystallization of basalts. Home: 9 ul. Wiejska, Warsaw, Poland. Office: 30 al. Mickiewicza, Cracow, Poland.*

WINOGRADSKY, Sergei Nikolaievich, microbiologist; b. Kiev, Russia, Sept. 1, 1856; became mem. Imperial Inst. Exptl. Medicine, St. Petersburg, 1890, dir., 1902; dir. dept. plant microbiology Pasteur Inst., from 1924. Fellow Royal Soc., 1919; mem. French Acad. Scis., 1902, French Acad. Agr. Studied soil microbiology, bacteriology of cellulose decomposition, autotrophic bacteria; established existence of microorganism which with nitrobacteria transform ammonia into nitrous acid. Died Brie-Comte-Robert, France, Feb. 24, 1953.

WINOGRADZKI, Judith, theoretical physicist; b. Petrograd, USSR, Dec. 5, 1916; d. Michel and Ida (Guinsburg) Wittenberg; Licence ès sciences, U. Paris (France), 1939, doctorat ès sciences physiques, 1953; m. Anatole Winogradzki, Oct. 30, 1945. Tchr. math., physics, pvt. schs., Paris, 1940-45; research Institute Henri Poincaré, Centre National de la Recherche Scientifique, Paris, 1951-63, dir. seminar on theoretical physics, 1957—; lectr. U. Paris, 1958—; prof. U. Rouen (France), 1963—. Mem. French, Am. phys. socs., French Math. Soc. Mem. editorial bd. Cahiers de physique, 1966—. Research, publs. on determination of whole set of 2d rank spinors with Lorentz-invariant components (number is 8), constn. of tensorial and spinorial formalism including parities, applications in relativistic quantum mechanics; non-phenomenological establishment of Einstein's last field equations. Home: 3, rue du Bocage, 91-Orsay, France. Office: Faculte des sciences de Rouen, 76-Mont-St-Aignan, France.*

WINOKUR, George, Am. psychiatrist; b. Phila., Feb. 10, 1925; s. Louis and Vera (Pallay) W.; A.B., Johns Hopkins, 1944; M.D., U. Md., 1947; m. Betty Stricklin, Sept. 15, 1951; children—Thomas S., Kenneth L., Patricia L. Faculty, Washington U. Sch. Medicine, St. Louis, 1951—, prof. psychiatry, 1966—; asso. psychiatrist Barnes Hosp., Renard and affiliated hosps., St. Louis, 1963—; cons. med. staff Malcolm Bliss Mental Health Center, St. Louis, 1966—. Fellow Am. Psychiat. Assn.; mem. Assn. for Research in Nervous and Mental Diseases, Am. Psychopath. Soc., Psychiat. Research Soc., Sigma Xi. Editor: Determinants of Human Sexual Behavior, 1964. Contbr. numerous articles to profl. jours. Found assn. ability to develop conditioned response with degree of reactivity in human; research on epidemiology affective disorder with respect to normal prevalence, genetic factors, clin. assn. with alcoholism, sex differential, genetic and clin. differentiation types. Home: 6983 Cornell St., St. Louis 63130. Office: Renard Hosp., 4940 Audubon Av., St. Louis 63110.*

WINSBERG, Lester, physicist; b. Montreal, Que., Can., Jan. 31, 1921; s. Paul Henry and Hattie (Adler) W.; came to U.S., 1923, naturalized, 1948; B.S., U. Chgo., 1942, Ph.D., 1947; m. Sandra Schwid, Mar. 19, 1949; children—Elana, Paul, Daniel, Devora. Research scientist U. Chgo., 1942-47, 53-54; with Argonne (Ill.) Nat. Lab., 1947-50, 60-64, cons., 1964—; research scientist Weizmann Inst. Sci., Israel, 1950-52, 54-55, Lawrence Radiation Lab., U. Cal. at Berkeley, 1955-60; prof. physics U. Ill. at Chgo., 1961—, head physics dept., 1964-67. Mem. Am. Phys. Soc., Am. Assn. Physics Tchrs. Research and publs. in nuclear fission, neutron diffraction, natural radioactivity in Israel, nuclear reactions induced by high-energy particles and pions and muons, stopping of nuclear fragments, others. Home: 5508 S. Harper Av., Chgo. 60637.*

WINSLOW, E(arle) M(icajah), Am. economist; b. Marshalltown, Ia., Jan. 26, 1896; s. Calvin Starbuck and Laura Christina (Bogue) W.; A.B., Penn Coll., 1920; A.M., U. Ia., 1921; postgrad. Yale; Ph.D., Harvard, 1929; m. Blanche Mitchell, July 14, 1922. Asso. economist U. Ia., 1922-26; instr., tutor, Harvard, 1927-29; Social Sci. Research Council fellow, Europe, 1930-31; faculty Tufts U., 1929-36, prof. econs., head dept. econs. and sociology, 1931-36; economist U.S. Tariff Commn., 1937-62. Mem. Can.-U.S. Joint Econ. Commn., 1941-42; lectr. instr. internat. relations Haverford Coll., 1935, Wellesley Coll., 1936. Dir. pub. forums for U.S. Dept. Edn., 1936-37; mem. U.S. delegation Gen. Agreement on Tariffs and Trade Conf., Torquay, Eng., 1950, Geneva, Switzerland, 1961. Recipient Ford Hall Forum Gold medal, 1936. Author: The Pattern of Imperialism, 1948; The League and Concerted Economic Action, 1931; A Libation to the Gods, 1963. Research on internat. econs., trade agreements programs, theories of imperialism, Roman aqueducts. Home: 2333 N. Vernon St., Arlington, Va. 22207. Died Mar. 9, 1966.

WINSLOW, Jacques Bénigne, physician; b. Odense, Fyn Island, Denmark, Apr. 2, 1669; ed. Holland, Paris; tchr. anatomy and surgery Jardin du Roi, Pai Paris; directed constrn. of new anatomy dissection room, Paris, 1774; apptd. prof. anatomy U. Paris, 1743; mem. French Acad. Scis. Author: La Disposition des fibres musculaires du coeur, 1711; Description d'une valvule singulière de la veine cave, 1717; Exposition anatomique de la structure du corps humain, 1732; Sur l'incertitude des signes de la mort, 1742. Gave 1st exact description of foramen epiploicum, recognized it as a normal formation; demonstrated importance of synchronized movement of antagonistic muscles; discovered great sympathetic nerve and connections between it and branches of vagus nerve. Died Paris, Apr. 3, 1760.

WINSOR, Frederick Albert, inventor; b. Brunswick, Eng., 1763; s. Friedrich Albrecht Winzer; ed. in Hamburg, Germany; a son, Frederick Albert. Lived in Eng. before 1799; studied Lebon's thermo lamps, Paris, 1802; returned to Eng., 1803; lighted part of Pall Mall, London with gas, 1806; assisted in obtaining charter for Westminster Gas Light and Coke Co., 1810; founder gas lighting co., Paris, 1815. Author: Description of Thermo-lamp invented by Lebon of Paris, pub. with remarks by F.A.W. of London, 1802; The Superiority of the New Patent Coke over the use of Coals in Family Concerns, displayed every Evening, at the Large Theatre, Lyceum Strand, by the New Imperial Patent Light Stove, 1806; Analogy between Animal and Vegetable Life Demonstrating the beneficial application of the Patent Light Stoves to all Green and Hot Houses, 1807. Pioneered gas-lighting; patentee oven for manufacture of gas, 1804, new gas furnace and purifiers, 1807, 08, 09. Died Paris, May 11, 1830.

WINSTEAD, Nash Nicks, Am. phytopathologist; b. Durham, N.C., June 12, 1925; s. Nash L. and Lizzy (Featherston) W.; B.S., N.C. State U., 1948, M.S., 1951; Ph.D., U. Wis., 1953; m. Geraldine Larkin Kelly, Sept. 17, 1949; 1 dau., Karen Jewell. Faculty, N.C. State U., Raleigh, 1953—, prof. plant pathology, 1961—, dir. Inst. Biol. Scis., 1965—, asst. dir. Agrl. Exptl. Sta., 1965—. Chmn. So. Regional Edn. Bd. Com. Interaction Between Protoplasm and Toxicants. Recipient Sigma Xi Research award, 1960. Phillip Found. Intern Academic Adminstrn. Ind. U., 1965-66. Mem. Am. Phytopathological Soc. (com. chmn.), Am. Soc. Hort. Sci., Am. Inst. Biol. Sci., Nat. Council U. Research Adminstrs., So. Regional Agrl. Exptl. Sta. Dirs. (sec.), Sigma Xi, Phi Kappa Phi. Research, numerous publs. on phytopathology, primarily with physiology of parasitism and breeding for disease resistance, involving fungal, bacterial and nematode pathogens. Home: 1109 Glendale Dr., Raleigh, N.C. 27609.*

WINSTEIN, Saul, chemist; b. Montreal, Que., Can., Oct. 8, 1912; s. Louis and Anne (Dickovati) W.; came to U.S., 1924, naturalized, 1929; A.B., U. Cal. at Los Angeles, 1934, M.A., 1935; Ph.D., Cal. Inst. Tech., 1938; Hon.D., U. Montpellier (France), 1962; m. Sylvia V. Levin, Sept. 3, 1937; children—Bruce, Carolee. Research fellow Cal. Inst. Tech., 1938-39; NRC fellow organic chemistry Harvard, 1939-40; instr. Ill. Inst. Tech., 1940-41; faculty U. Cal. at Los Angeles, 1941—, prof. chemistry, 1947—. Recipient Cal. Inst. Tech. Alumni Distinguished Service award, 1966; Edward A. Dickson Achievement award as U. Cal. at Los Angeles Alumnus of Year, 1958, Distinguished Teaching award in grad. and postdoctoral edn. U. Cal. at Los Angeles, 1963, Herbert Newby McCoy award, 1966. Mem. Am. Chem. Soc. (award in pure chemistry 1948, Theodore William Richards medal 1962, James Flack Norris award 1967), Nat. Acad. Scis., Am. Acad. Arts and Scis., Bayerische Akademie der Wissenschaften, Brit. Chem. Soc., Sigma Xi, Alpha Chi Sigma, Phi Lambda Upsilon. Editorial bd. Jour. Am. Chem. Soc., 1960—, Internat. Jour. Chem. Kinetics. Mem. hon. editorial adv. bd. Tetrahedron, 1957—. Research, publs. on organic reaction mechanisms and phys. organic chemistry, ion pairs, homoconjugation and homoaromaticity. Home: 201 Ashdale Pl., Los Angeles 90049.*

WINSTEN, Seymour, Am. chemist; b. Jersey City, June 14, 1926; s. Louis and Esther (Kroll) W.; student Newark Coll. Engring., 1943-44, Hamilton Coll., 1944; A.B., Rutgers U., 1948, Ph.D., 1956; M.Sc., N.Y. U., 1950; m. Madeline Dashow, Apr. 10, 1949; children—Helene, Beth, Amy. Research asst. Merck Inst. for Therapeutic Research, Rahway, N.J., 1950-56; asso. in microbiology U. Pa. Sch. Vet. Med., 1956-57; chief div. biochemistry Albert Einstein Med. Center, Phila., 1957—. Cons. Moss Rehab. Hosp., Atlantic City Hosp., 1961—; cons. clin. chemistry to Surg. Gen. U.S., 1965—; vis. lectr. Temple U., 1963—. Mem. Am. Assn. Clin. Chemists. Research on natural resistance in tumors, diagnostic enzymes in cancer chemotherapy, instrumentation for clin. chemistry lab. Home: 218 Preston Rd., Flourtown, Pa. 19031. Office: Albert Einstein Med. Center, York and Tabor Rd., Phila. 19141.*

WINSTON, Sanford Richard, Am. sociologist; b. Jamaica, N.Y., Nov. 27, 1897; s. Ian and Mary (Franklin) W.; A.B., Western Res. U., 1925; Ph.D., U. Minn., 1929; postgrad. U. Chgo., U. N.C., Columbia; m. Ellen Black, Aug. 30, 1928. Tchr. U. Minn., 1925-26; faculty U. N.C., Raleigh, 1926-65, prof. sociology, 1932-65, head dept. sociology and anthropology, 1933-63. Tchr., Cornell U., summer 1944; lectr. Wake Forest Coll., 1949-50, Meredith Coll., 1941-45, Peace Coll., 1951-66; lectr. Ft. Bragg, 1956-66; cons., 1945—. Research fellow Laura

Spelman Rockefeller Found., 1925-26; grantee Gen. Edn. Bd., 1943-46, Purnell Fund, 1941-42, N.C. Council Housing Authorities, 1945-48. Fellow Am. Sociol. Assn.; past pres. N.C. Archaeol. Soc. Author: Culture and Human Behavior, 1933; Illiteracy in the United States, 1930; Social Aspects of Public Housing, 1947; Leadership in War and Peace, 1947; also articles. Research on interrelationships of societal life, cultural patterns, human behavior, processes involved in accumulation and diffusion of cultural traits and their impact on personality and behavior. Home: 1712 Piccadilly Lane, Raleigh, N.C 27608.*

WINTER, Chester Caldwell, Am. urologist; b. Cazenovia, N.Y., June 2, 1922; s. Chester Caldwell and Cora (Martin) W.; B.A., U. Ia., 1943, M.D., 1946; m. Cary Margaret Jones, June 18, 1945; children—Paul, Ann, Jane. Faculty, U. Cal. at Los Angeles, 1954-60; prof., dir. urology Ohio State U. Coll. Medicine, Columbus, 1960—; dir. urology Univ., Children's hosps. (both Columbus); cons. Vets. Hosp., Air Force Hosp. (both Dayton, O.). Diplomate Am. Bd. Urology. Mem. A.C.S., A.A.A.S., Internat. Soc. Urology, Internat. Soc. Nephrology, Am. Assn. Genito-Urinary Surgeons, Am. Urol. Assn., A.M.A., Soc. Nuclear Medicine, U. Surgeons, Soc. Pediatric Urology, Sigma Xi. Research, publs. on surg. correction, correctable renal hypertension, urologic radioisotope tests; co-inventor, radioisotope renography. Home: 176 Tucker Dr., Worthington, O. 43085. Office: 410 W. 10th Av., Columbus, O. 43210.*

WINTER, Edward Henry, Am. anthropologist; b. Poughkeepsie, N.Y., Aug. 21, 1923; s. Wallace A. and Agnes (Tinnins) W.; A.B., Harvard, 1944, M.A., 1949, Ph.D., 1953; student Balliol Coll., Oxford U., 1946-47, London Sch. Econs., 1949-50; m. Elisabeth Lee Stoffel, July 10, 1947; children—S. Anthony, Nicholas Savalle, Christopher Powell. Research fellow Brit. Social Sci. Research Council, Uganda, 1950-52; sr. research fellow E. African Inst. Social Research, Tanganyika, 1953-55; asst. prof. anthropology U. Ill., 1955-57, asso. prof., 1957-59; prof. anthropology, chmn. dept. sociology, anthropology U. Va., Charlottesville, 1959—. Fellow Am. Anthrop. Assn., Assn. Social Anthropologists. Author: Bwamba Economy, 1955; Bwamba, 1956; Beyond the Mountains of the Moon, 1959; Witchcraft and Sorcery in East Africa, 1963; Tanganyika, 1967. Contbns. to solution of problems of social structure, nature of African societies, and underdeveloped economics. Home: 620 Preston Pl., Charlottesville, Va.*

WINTER, Rolf Gerhard, physicist; b. Dusseldorf, Germany, June 30, 1928; s. Julius and Erika (Wolff) W.; came to U.S., 1938, naturalized, 1944; B.S., Carnegie Inst. Tech., 1948, M.S., 1951, D.Sc., 1952; m. Patricia Mae Saibel, Jan. 31, 1951; children—Erika Louise, Edward Paul Merritt, James Frederick. Faculty, Western Res. U., Cleve., 1951-54, Pa. State U., University Park, 1954-64; prof. physics Coll. William and Mary, Williamsburg, Va., 1964—, chmn. dept., 1966—. Vis. physicist and lectr. Carnegie Inst. Tech., 1955-56, Oxford U., 1961-62, U. Wis., 1963. Fellow Am. Phys. Soc.; mem. Am. Assn. U. Profs., Va. Acad. Sci. Contbr. articles to sci. jours. Research on nuclear beta decay, nuclear structure and reactions, quantum theory. Home: 702 College Terrace, Williamsburg, Va. 23185.*

WINTERBOTTOM, John Miall, S.African ornithologist; b. Capetown, S. Africa, Aug. 26, 1903; s. John Herbert and Nora (Miall) W.; B.Sc., U. London, 1926, Ph.D., 1931; m. Marjorie Grace Mash, Aug. 2, 1929; children—David Miall, Judith Cardinall, Richard. Publs. editor Edn. Dept., Gold Coast, Ghana, 1929-31; edn. officer No. Rhodesia, 1931-50; ednl. adviser Longmans, Green & Co., Cape Town, 1950-56; sr. bursar S. African Council for Sci. and Indsl. Research, 1957-60; dir. Percy Fitzpatrick Inst. African Ornithology, U. Cape Town, 1960—. Mem. S. African (hon. life, hon. sec. 1951—, Gill Meml. medal 1960), Am. (corr. fellow), Brit. (corr.) ornithologists unions, Asociacion Ornitologica del Plata ()corr.), Zool. Soc. London, Inst. Biology. Author: (with C.M.N. White) Check List of the Birds of Northern Rhodesia, 1949; Common Birds of the Bush, 1951; The Bird and its Environment, 1965; also numerous articles. Research on distbn., environment, numbers and classification of African birds. Home: 9 Alexandra Av., Oranjezicht, Cape Town, S. Africa.*

WINTERFELD, Karl, German chemist; b. Cologne, Germany, Dec. 17, 1891; s. Gustav and Gertrud (Butzler) W.; Dr.Phil., U. Marburg, 1921; m. Rosita Fonck, May 19, 1926; children—Karl Dietrich, Margarete (Mrs. Gerwalt Zinner), Lieselotte. Faculty, Freiburg U., 1927-49, dir. inst., prof., 1938-49; prod., dir. Inst. Pharm. Chemistry, Bonn (Germany) U., 1949—. Mem. Fed. Health Council, 1953—, Commn. of Pharmacopeia, 1953—. Recipient Fluckiger medal, 1962; Mannich medal, 1966; Great Fed. Distinguished Service Cross, 1961. Mem. Pharm. Soc. (hon.), German Chem. Soc., German Soc. Sci. Investigators and Physicians. Author: Preparative Pharmaceutica, 1965; also numerous articles. Research on chemistry of alkaloids, chemistry of heterocycles, active materials of mistletoe. Home: 5 Endenicher Allee, Bonn, Germany.*

WINTERGERST, Erich, German physicist; b. Munich, Germany, July 6, 1905; s. Hans and Mathilde (Weber) W.; eng. diploma dr., Munich Tech. Sch.; m. Hildegard Rietz, Dec. 11, 1943; children—Monika, Beate. with Chemische Fabrik Dr. A. Wacker, Burghausen, Germany, 1929-30; asst. tech. phys. lab. Munich Tech. Sch., 1930-35; with Heereswaffenamt, Berlin, Germany, 1935-38; dir., later tech. dir. Firma Armaturen und Apparatefabrik Preschona, Berlin, 1938-45; prof. Munich Tech. Sch., 1945—; dir., mem. adminstrv. council I.C. Eckart A.G., Stuttgart-Bad Catstannt, 1952—. Mem. Verein Deutscher Ingenieure. Author: Uber die Schmelzzeit von Schmelzisicherungen, 1950; Eigenschaften under Anwendungen von Metallmembranen, 1959; Die technische physik des Kraftwagens, 1961; Fernmessing in der Verfahrenstechnik, 1961; Methoden der unmittelbaren Durchflussmessung, 1961. Home: Leibnizstrasse, 47, Stuttgart, Germany. Office: Pragstrasse 80, Stuttgard-Bad Cannstat, West Germany.

WINTERINGHAM, Francis Peter Worsley, English biochemist; b. Grimsby, Eng., May 16, 1918; s. Francis and Anne (Coombe) W.; A.R.I.C., U. Bath, 1940, D.Sc., 1967; Sc.D. (hon.), U. N.C., 1967; m. Jeanne Farrar-Davis, June 8, 1940; children—Hilary, Rosemary (Mrs. Christopher Petrie), Timothy. Head biochemistry dept., Agrl. Research Council, Pest Infestation Lab., Slough, Eng., 1947-67; pesticide specialist Plant Prodn. and Protection Div., UN FAO, Rome, Italy, 1967—; vis. scholar U. Cal., 1951-52; vis. prof. U. N.C., 1964-65. Cons. Internat. Cocoa Office. Harkness fellow Commonwealth Fund of N.Y., 1951; sr. fgn. sci. fellow NSF, 1963. Fellow Royal Inst. Chemistry, Inst. Biology, Royal Soc. Arts; mem. Biochem. Soc. U.K., Soc. of Chem. Industry, Entomol. Soc. Am. Research, numerous publs. on devel. of radioactive tracer methods for insect biochemistry and insecticide toxicology. Home: Via Luigi Lifio, 59, Rome 00143. Office: Plant Protection Div., UN FAO, Rome, Italy.*

WINTERKORN, Hans Friedrich, engr.; b. Mannheim, Germany, Nov. 24, 1905; s. Valentin Guénémé and Elisabeth Christine (de Gedde) W.; student Ruperto-Carola U., Heidelberg, Germany, 1924-31; D. Natural Philosophy, U. Heidelberg, 1931; postgrad. U. Mo., Mass. Inst. Tech., U. Mich.; m. Hazel Zumwalt, Oct. 13, 1940; children—Hans Friedrich Zumwalt, Erik Guinemer de Gedde. Came to U. S., 1931, naturalized, 1938. Head div. chemistry Sch. Adult End., Mannheim, 1929-31; research chemist, engr. Mo. Hwy. Dept., 1931-32, research cons., 1932-42; research asso. prof. soil mechanics U. Mo., 1940-42, asso. prof. civil engring., 1942-43; asso. prof. civil engring. Princeton, 1943-55, prof., 1955—. Cons. U.S. Dept. Interior, 1954—, Gen. Electric Co. (lunar soils), 1961—, Jet Propulsion Lab. Cal. Inst. Tech., USAF, 1963—; cons. other agys. Fellow Am. Inst. Chemists, Am. Soc. C.E.; mem. Soil Sci. Soc. Am. (chmn. soil tech. sect. 1943-45), Am. Rd. Builders Assn. (cons. Pan Am. div.), Am. Geophys. Union, Am. Assn. U. Profs., Assn. Asphalt Paving Technologists, Hwy. Research Bd. (asso.; chmn. com. physico-chem. phenomena in soils 1936-66). Sigma Xi, Tau Beta Pi, Alpha Chi Sigma, Gamma Alpha. Author: Grundlagen Der Bodenstabilisierung, 1960. Contbr. articles profl. jours. Applied chem. physics to terrestrial and lunar soil engring.; developed soil stblzn. and water-proofing methods for fighter fields, also Navy method for stblzn. invasion beach heads, World War II; discovered thermo-electric effect in moist soils; also osmotic swelling; also demonstrated existence of water in solid state above its normal melting point. Home: 116 Prospect Av., Princeton, N.J. 08540.*

WINTERS, Robert Wayne, Am. physician; b. Evansville, Ind., May 23, 1926; s. Frank and Clara (Flentke) W.; B.A. with highest honors, Ind. U., 1948; M.D. cum laude, Yale, 1952; m. Madoris Seiler, Sept. 5, 1948; children—Henry Nicholas, R. George. Research fellow U. Cal. Hosp., 1952-54, U. N.C. Sch. Medicine, 1954-58; asst. prof. physiology U. Pa. Sch. Medicine, 1958-61; prof. pediatrics, Columbia Coll. Physicians and Surgeons, 1961—; career scientist Health Research Council, N.Y.C., 1963—. Mem. Am. Physiol. Soc., Am. Soc. for Clin. Investigation, Soc. for Pediatric Research, Am. Pediatric Soc., Phi Beta Kappa, Alpha Omega Alpha, Sigma Xi. Contbr. numerous articles on renal, electrolyte and acid-base physiology to tech. jours. Home: 110 E. End Av., N.Y.C. 10028.

WINTERSTEINER, Oskar Paul, chemist; b. Bruck a/d Mur, Austria, Nov. 15, 1898; s. Carl and Eva (Torkar) W.; Ph.D., U. of Graz, Austria, 1921; m. Margaret Ralston Prest, Sept. 21, 1934; children—Peter, Susanne. Instr. med. chemistry and organic microanalysis U. of Graz, Austria, 1921-26; fellow Internat. Edn. Bd., Johns Hopkins Med. Sch., Rockefeller Inst., 1926-27; instr. pharmacology Johns Hopkins Med. Sch., 1927-29; asst. prof. biochemistry Coll. Phys. and Surgs., Columbia U., 1929-39, asso. prof., 1939-41; dir. div. organic chemistry, Squibb Inst. Med. Research, 1941-59, director biological chemistry, 1959-63, sci. adviser, 1961-63, cons., 1964—; hon. professor biochemistry Rutgers U. since 1942; cons. and mem. antibiotic study sect. National Inst. Health, 1946-49; member board of scientific advisers National Inst. Arthritis and Metabolic Diseases, 1957-59. Awarded Presdl. Certificate of Merit, 1948; Nichols medal, 1950. Mem. Nat. Acad. Scis., Am. Chem. Soc. (chmn. North Jersey sect. 1957), A.A.A.S., Harvey Soc., Soc. Exptl. Biology and Medicine, Swiss Chem. Soc., Sigma Xi. Editor: Proceedings Soc. Exptl. Biology and Medicine, 1938-43, 1948-50; mem. editorial bd. Jour. Biol. Chemistry, 1952-62, Jour. Am. Chem. Soc., 1959—. Research in organic microanalysis; chemistry of insulin; organic arsenic compounds; steroids and steroid hormones; penicillin; streptomycin; curare; alkaloids. Home: 6 Dewey Dr. Office: Squibb Inst. Med. Research, New Brunswick, N.J.

WINTHER OF ANDERNACH, Johann, see Johannes Guinterius.

WINTHROP, Henry, Am. psychologist; b. N.Y.C., July 20, 1910; s. Charles and Rose (Sugar) W.; B.S., Coll. City N.Y., 1935; M.S., George Washington U., 1940, postgrad.; Ph.D., New Sch. For Social Research, 1953; m. Gussie Munk, Dec. 5, 1952; 1 son, Kenneth Robert. With U.S. Civil Service, 1936-42, 48-52, N.Y. State Civil Service, 1946-47; faculty Coll. William and Mary, 1953-54, Hollins Coll., 1956-57, U. Wichita, 1957-60; chmn. dept. interdisciplinary social sci., prof. grad. program in social studies U. South Fla., Tampa, 1960—. Cons., lectr. to various govtl. agys. and instns.; mem. sci. adv. bd. Inst. for Cybercultural Research, 1965. Mem. Am. Anthrop. Assn., Am. Assn. Existential Psychology and Psychiatry, Am. Assn. Humanistic Psychology, Am. Polit. Sci. Assn., Am. Psychol. Assn., Am. Sociol. Assn., Artoga, Inst. Mgmt. Sci., others. Editorial bd. various prof. jours. Research, publs. on devel. of first test for measurement of attitude consistency; formulation of theories in psychology; math. laws governing propagation of human behavior; constn. and personality, psychol. effects of automation. Home: 1816 Marvy Av., Tampa 33612.*

WINTHROP, John Am. astronomer; b. Boston, Dec. 19, 1714; s. Adam and Anne (Wainwright) W.; A.B., Harvard, 1732, LL.D. (1st hon. LL.D. given by Harvard), 1773; LL.D., U. Edinburgh (Scotland), 1771; m. Rebecca Townesend, July 1, 1746; m. 2d, Hannah (Fayerweather) Tolman, Mar. 24, 1756; several children, including James. Hollis prof. math and natural philosophy Harvard, 1738-79; research in astronomy, published results in Philos. Trans. Royal Soc.; made series of sun-spot observations, Apr. 19-22, 1739 (1st set observations sun-spots in Mass. Colony); made study of transit of Mercury over sun, 1740, also reported transits, 1743, 69; established 1st lab. of exptl. physics in Am. at Harvard, 1746, demonstrated laws of mechanics, light, heat, movement of celestial bodies according to Newtonian system, introduced into math, curriculum elements of fluxions (now known as differential and integral calculs), 1751; reported on earthquake that shook New Eng., 1755; delivered lecture on return of Halley's Comet of 1682, 1759 (1st predicted return of a comet); made preparations for transits of Venus, 1761, 69, dir. 1st astron. expdn. sent by Harvard, to St. John's, (Newfoundland) for 1761 transit; elected fellow Royal Soc., 1766; mem. Am. Philos. Soc., 1769; a founder Am. Acad. Arts and Scis., 1769; 1st astronomer in Am.; adviser to George Washington during Am. Revolution. Author: Relation of a Voyage from Boston to Newfoundland for the Observation of the Transit of Venus, 1761; Two Lectures on the Parallax and Distance of the Sun, 1769. Sometimes considered seismology; studied electricity. Died Cambridge, Mass., May 3, 1779.

WINTREBERT, Paul Marie Joseph, French zoologist, physician; b. Lille, France, July 28, 1867; M.D., D.Sc.; m. Eugénie Brise. Prof. anatomy, comparative histology Faculty Scis., Paris, France. Mem. French Acad. Scis., 1938. Author: le Vivant, créateur de son évolution, 1962; le Vivant, maitre de son développement, 1963. Home: rue des Orangers, Banyuls-sur-Mer, France.

WINTTER, John Ernest, Am. pharm. chemist; b. Birmingham, Ala., Oct. 31, 1924; s. Charles A. and Lona (Galloway) W.; B.S. in Pharmacy, Howard Coll., 1949; M.S. in Pharmacy, U. Fla., 1950, Ph.D., 1952; m. Mildred Irene Hogue, Oct. 25, 1944; children—Beverly Jeanne, John Michael. Asso. prof. Howard Coll., 1952-58, prof. Samford U. (formerly Howard Coll.), Birmingham, Ala., 1958—, chmn. dept. pharm. chemistry, 1952—. Mem. Am., Ala. pharm. assns., Am. Inst. History Pharmacy, A.A.A.S., Ala. Acad. Sci., Gamma Sigma Epsilon, Rho Chi, Kappa Psi. Research, publs. in synthetic and natural product chemistry. Home: 420 Redwood St., Birmingham, Ala. 35210.*

WINZLER, Richard John, Am. biochemist; b. San Francisco, Sept. 29, 1914; s. Joseph and Esther (Hoppe) B.; student San Mateo Jr. Coll., 1932-34; B.S., Stanford, 1936, Ph.D., 1938; m. Georganna E. Martin, June 17, 1939; children—Joan Adele (Mrs. M. Donald Cave), Natalie Jane (Mrs. Peter F. Rusch), Lee Martin. Sterling fellow Yale, 1938-39; NRC fellow Wenner Grens. Inst. Stockholm (Sweden), 1939-40, Cornell U. Med. Sch., N.Y.C., 1940-41; research fellow Nat. Cancer Inst., NIH, Bethesda, 1941-43; faculty U. So. Cal. at Los Angeles, 1943-52, prof. biochemistry, 1949-52; prof., head dept. biochemistry U. Ill. Coll. Medicine, Chgo., 1952-65; prof., head dept. biochemistry State U. N.Y. at Buffalo, 1965—;

vis. prof. U. Wis., Madison, 1941. Cons. in med. edn. U. Chiengmai (Thailand), 1962. Commonwealth fellow U. Freiberg (Germany), 1958. Research, publs. on studies on glycoproteins. Home: 220 Lakewood Pkwy., Snyder, N.Y. 14226. Office: Capen Hall, State U. N.Y., Buffalo 14214.*

WIRSUNG, Johann Georg, anatomist; b. Augsburg, Germany; prof., Padua, Italy. Author: Figura ductus in pancreate, 1642. Discovered excretory duct of pancreas. Died Padua, 1643.

WIRTSCHAFTER, Zolton Tillson, Am. physician; b. Cleve., Oct. 10, 1899; s. Adolph and Tillie (Gutman) W.; B.S., Case Sch. Applied Sci., 1921; M.D., U. Cin., 1926, M.D., 1927; m. Reitza Dine, Oct. 15, 1933; children—Jonathan Dine, David Daniel. Med. cons. Brotherhood Locomotive Fireman and Enginemen, 1932-42; sr. instr. Western Res. U. Med. Sch., 1939-42; staff Mt. Sinai Hosp., Cleve., 1939-42; practice medicine specializing in metabolis and diabetes, Cleve., 1928-46; asso. prof. clin. medicine U. So. Cal., Los Angeles, 1947-49; chief med. service VA Hosp., Portland, Ore., 1953-56, dir. research, 1956—; asso. prof. medicine U. Ore., Portland, 1956—. Diplomate Am. Bd. Internal Medicine. Fellow A.C.P., Am. Pub. Health Assn.; mem. Am. Diabetic Assn., A.A.A.S., A.M.A. Author: Diabetes Mellitus, 1941; Minerals and Nutrition, 1941; Genesis of the Mouse Skeleton, 1960; (with Walker, Wirtschafter) Genesis of the Rat Skeleton, 1957; also numerous articles. Research on connective tissue disorders, other metabolic problems in field internal medicine. Address: VA Hosp., Sam Jackson Park, Portland, Ore. 97207.*

WIRTZ, Karl W. J., German nuclear physicist; b. Cologne, Germany, Apr. 24, 1910; s. Carl and Hildegard (Krebs) W.; student Bonn, Fribourg, Breslau (Wroclaw, Poland); Ph.D.; agrege U. Berlin; m. Ottenie von Ziegner; a child Christiane. Asst., U. Leipzig, (Germany), 1935-37; asst., div. head Max Planck Physics Inst., 1937-57; became asso. prof. U. Göttingen; named prof. Karlsruhe Tech., 1957; also dir. Nuclear and Reactor Tech. Physics Inst., Karlsurhe (Germany) Atomic Center. Decorated Order Alfonso él Sabio, 1958; Order of Belgian Crown, 1958. Mem. Deutscher Atomforum, Deutsche Physikalische Gesellschaft, Verein Deutscher Ingenieure, Bunsen Soc., Am. Nuclear Soc. Author: (with K. H. Beckurts) Elementare Neutronenphysik 1958; Co-editor: Nuckleonik, Reactor Sci. and Engring., Atomwirtschaft. Research in reactor physics and design. Home: Siebenmannstrasse 10, 7501 Wolfartsweier, Karlsruhe. Office: Postfach 947, 75 Karlsruhe, West Germany.

WISCHMEYER, Carl Riehle, Am. elec. engr.; b. Terre Haute, Ind., Oct. 2, 1916; s. Carl and Grace (Riehle) W.; B.S., Rose Poly. Inst., 1937, E.E., 1942; M.E.E., Yale, 1939; m. Mary DeFord Sumners, Mar. 3, 1945; children—Mary Grace, Carl Edward, Margaret Kay. Faculty, Rice Univ., Houston, 1939—, master Baker Coll., 1956-68, director of continuing studies, 1968—, prof. elec. engring. Rice U., 1959—. Mem. tech. staff, resident vis. Bell Telephone Labs., Murray Hill, N.J., 1945-47, 63-64; cons. various indsl. firms. Mem. I E.E.E. (dir. 1964-67), Am. Soc. Engring. Edn., Soc. Exploration Geophysicists, European Assn. Exploration Geophysicists, Sigma Xi, Tau Beta Pi. Demonstrated highly localized nature of irregularities in migration of magnetic flux through hard supercondrs. Address: Rice U., Houston 77001.*

WISCHNITZER, Saul, Am. biologist; b. N.Y.C., Apr. 10, 1930; s. Solomon and Ray (Weiss) W.; student Bklyn. Coll., 1952-53, U. Basel, 1953-54; B.A., Yeshiva U., 1951; M.S., U. Notre Dame, 1954, Ph.D., 1956. Faculty, N.Y. Med. Coll., 1957-64, asst. prof., 1960-64; asso. prof. dept. biology Yeshiva U., 1964-68, professor department of biology, since 1968—, asst. to dean, 1964-66, asst. dean Yeshiva Coll., 1966—. Fellow Royal Micros. Soc.; mem. A.A.A.S., Am. Assn. Anatomists, Electron Microscope Soc. Am., Am. Soc. Cell Biology, Am. Assn. U. Profs., Sigma Xi, Alpha Epsilon Delta. Author: Introduction to Electron Microscopy, 1962; Outline of Human Anatomy, 1963; (with S. J. Piliero and M. S. Jacobs) Atlas of Histology, 1965; Atlas and Dissection Guide for Comparative Anatomy, 1967. Research on fine structure of developing vertebrate egg using electron microscope. Address: 186th St. and Amsterdam Av., N.Y.C. 10033.*

WISE, George Herman, Am. nutritionist; b. Saluda County, S.C., July 7, 1908; s. Herman Michael and Etta Mae (West) W.; B.S., Clemson Coll., 1930; M.S., U. Minn., 1932, Ph.D., 1937; m. Marie Larson, Aug. 8, 1937; children—George Michael, Nancy Marie (Mrs. John W. Cameron), Julia Melverda (Mrs. C. W. Friend), James Patrick. Grad. asst. U. Minn., 1933-36; asso. prof. Clemson Coll., 1937-44; asso. prof., then prof. Kan. State Coll., Manhattan, 1944-47; asso. prof. Ia. State Coll., Ames, 1947-49; mem. faculty N.C. State U., Raleigh, 1949—, prof. animal sci., 1949-51, head nutrition sect., dept. animal sci., 1949-66, William Neal Reynolds distinguished prof., 1951—. Mem. com. animal nutrition NRC, 1950-54; spl. cons. Dept. Agr., 1957-61. Recipient Am. Feed Mfrs. Assn. award, 1948; Borden award dairy prodn., 1949. Fellow A.A.A.S. (council 1955-56); mem. Am. Diary Sci. Assn. (exec. bd. 1959-62, pres. 1965, award of honor 1966), Am. Soc. Animal Sci., Am. Chem. Soc., Am. Inst. Biol. Sci., Am. Inst. Nutrition, Soc. Exptl. Biology and Medicine, N.C. Acad. Sci., Am. Assn. U. Profs., Sigma Xi. Contbr. tech. jours. Editorial bd. Jour. Dairy Sci., 1947-52, Jour. Animal Sci., 1947-50. First to report value of sulfa drugs in treating a deadly diarrhea of calves; discovered fat-disintegrating enzyme in mouth secretions of young calves (now used in cheese curing); showed that vitamin A transfer from mother to fetus can be increased by feeding massive doses, also that nursing produces reflex stimulus that directs flow of milk into stomach of calf. Home: 229 Woodburn Rd., Raleigh, N.C. 27605.*

WISE, Keith Arthur John, New Zealand entomologist; b. Wellington, New Zealand, June 1, 1926; s. Victor John and Grace (Chapman) W.; m. Rosemary Musgrave, 1951; children—Gillian, Jennifer, Vicki, Martin. Entomol. techician, plant diseases div. Dept. Sci. and Indsl. Research, Auckland, Australia, 1948-61; fellow in entomology Bernie P. Bishop Mus. Honolulu, Christchurch, 1961-65; entomologist Auckland War Meml. Mus., 1965—. Fellow Royal Entomol. Soc. London; mem. Royal Soc. New Zealand, Entomol. Soc. New Zealand, New Zealand Assn. Scientists, Art Galleries and Museums Assn. New Zealand. Research, publs. on taxonomy of moths, caddisflies, lacewings, springtails, aquatic insects; distbrn. and ecology of free-living insects, springtails, mites; discovered seveeral nw sepecies of mites, springtails in Antarctica. Home: 77 Portage Rd., Papatoetoe, New Zealand. Office: Auckland Mus., Auckland, New Zealand.*

WISE, M. J., English geographer; b. Stafford, Eng., Aug. 17, 1918; s. Harry Cuthbert and Evelyn (Lawton) W.; B.A., U. Birmingham, 1939, Ph.D., 1951; m. Barbara Mary Hodgetts, May 4, 1942; children—Barbara Janet, John Charles Michael. Lectr., U. Birmingham, 1946-51; lectr. London Sch. Econs., 1951—, Sir Ernest Cassel reader econ. geography, 1954-58, prof. geography, 1958—. Chmn. com. on statutary small holdings Ministry Agr., 1963-67. Fellow Royal Geog. Soc. (hon. sec. 1963—), Geog. Assn. (hon. treas. 1967), Brit. Assn. for Advancement Sci. (past pres. sect. E), Inst. Brit. Geographers. Editor: Birmingham and its Regional Setting, 1950; also articles. Research on econ. geography, especially relating to indsl. location, process of regional growth and change. Home: 45 Oakleigh Av., London N.20, Eng.*

WISELOGLE, Frederick Yager, Am. chemist; b. Albion, Mich., May 18, 1912; s. Andrew J. and Florence (Yager) W.; B.S., U. Mich., 1932, M.S. in Chemistry, 1934, Sc.D., 1936; m. Louise Ryder, Oct. 13, 1945 (dec. Jan. 1959); children—Elizabeth, Ann Dee, William, Margaret; m. 2d, Charlotte Pike, Mar. 2, 1963. Staff chemistry dept. Johns Hopkins, 1936-46; staff Squibb Inst. for Med. Research, New Brunswick, N.J., 1946—, asso. dir. in charge chem. research, 1949-61, dir. devel., 1962—. Recipient Presdl. Certificate of Merit for survey antimalarial drugs, 1948. Fellow N.Y. Acad. Scis. (past mem. sci. council, past pres., trustee); mem. Am. Chem. Soc., A.A.A.S., Soc. Chem. Industry. Research and publs. on antimalarial drugs, isoniazid treatment of Tb. Home: 1163 Revere Rd., Colonial Gardens, North Brunswick, N.J. 08902. Office: 5 Georges Rd., New Brunswick, N.J. 08903.*

WISELY, Hunter Baughan, zoologist; b. Ashburton, New Zealand, Mar. 12, 1924; s. John and Alberta (Watson) W.; B.Sc., Canterbury U. Coll., Christ Church, New Zealand, 1948, M.Sc., 1950, D.Sc., 1966; m. Valmai Mary Hanham, Dec. 17, 1953; children—Virginia Gae, Christopher David Bruce. Biologist wildlife div. New Zealand Dept. Internal Affairs, Wellington, 1949; jr. lectr. zoology Canterbury U. Coll., 1950-55; research officer div. fisheries and oceanography CSIRO, Cronulla, New South Wales, Australia, 1956-58, sr. research scientist, 1959—. Research publs. on gen. biology of mayflies and stoneflies, rearing, devel. and settling behavior of marine invertebrate, larvae, antifouling and anticorrosion protection of ships' hulls; developer protection system of solventless epoxy resin followed by multiple coats of soluble matrix cuprous oxide antifouling paint. Home: 146 B Woolooware Rd. Office: CSIRO Marine Lab., Cronulla, New South Wales, Australia.*

WISEMAN, Ralph Franklin, Am. microbiologist; b. Washington, Sept. 1, 1921; s. Isidore David and Gertrude (Mincosky) W.; B.S., U. Md., 1949; M.S., U. Hawaii, 1953; Ph.D., U. Wis., 1956; m. Millicent Becker, Apr. 22, 1951; children—Dvorah Ann, Eve Jessica. Bacteriologist, Nat. Inst. Dental Research, USPHS, 1949-51; research asst. U. Hawaii, 1951-53, U. Wis., 1953-56; faculty U. Ky., Lexington, 1956—, prof. microbiology, 1966—. Fellow Am. Acad. Microbiology; mem. Am. Soc. Microbiology, Soc. for Gen. Microbiology (Eng.), A.A.A.S., Am. Inst. Biol. Scis., Soc. for Indsl. Microbiology, Assn. for Applied Gnotobiotics, Sigma Xi, Phi Sigma, Phi Epsilon Phi. Research on physiology of intestinal microorganisms, uric acid-intestinal bacteria interrelationships, anomalies of germfree life. Home: 988 Maywick Dr., Lexington, Ky. 40504.*

WISEMAN, Richard, English surgeon; b. London, Eng., between 1621-23; apprenticed to Richard Smith at Barber-Surgeons Hall; m. Dorothy (d. 1674); m. 2d, Mary Mauleverer; 1 son. Served in Dutch Navy; joined Royalist Army of West, circa 1644; became med. attendant to Prince Charles, 1645; captured at Worcester, Eng., 1651; began practice medicine in London, 1652; freeman Barker-Surgeons Co, became master, 1665; imprisoned for assisting escape of royalist from Tower of London, 1654; became surgeon in ordinary for the person, 1660; named prin. surgeon and sgt.-surgeon to Charles II, 1672. Author: A Treatise of Wounds, 1672; Severall Chirurgical Treatises, 1672. Contbd. to raising of status of physicians in Eng. Died Bath, Eng., circa Aug. 20, 1676.

WISHARD, William Niles, Jr., Am. physician; b. Indpls., July 29, 1898; s. William Niles and Frances Cornelia (Scoville) W.; B.A., Williams Coll., 1921; M.D. cum laude, Harvard, 1925; m. Carolyn Louise Davis, Oct. 10, 1938; children—Susan (Mrs. David S. Poston), William Niles III, Gordon Davis. E. surg. house officer Mass. Gen. Hosp., Boston, 1926-27; pvt. practice genito urinary surgery, Indpls., 1928—; mem. staff Methodist Hosp., 1928—, chmn. urol. staff, 1938, chmn. med. adv. com., 1938-40, pres. med. staff, 1940-41; mem. staff St. Vincents, Ind. U. Med. Center hosps.; cons. staff Marion County Gen., Community, Sunnyside hosps.; chmn. unrol. staff Marion County Gen. Hosp., 1950-53; member faculty Ind. U. Sch. Medicine, 1928—, clin. prof. urology, 1953—. Mem. Bd., Med. Registration and Exam. Ind., 1945-57. Diplomate Nat. Bd. Med. Examiners, Am. Bd. Urology (sec.-treas. 1955—, chmn. credentials com. 1955—). Mem. Am. Urol. Assn. (pres. N. central sect. 1951), Am. Assn. Genito Urinary Surgeons (council 1963—, pres. 1965-66), A.M.A., Indpls. (pres. council 1946), Ind. (chmn. surg. sect. 1948) med. assns., Am. Assn. Med. History, Assn. Am. Med. Colls., A.C.S., Internat. Urol. Soc. Contbr. articles on urologic subjects to med. jours., 1935—. Home: 5720 N. Pennsylvania St., Indpls. 46220. Office: 1711 N. Capitol Av., Indpls. 46207.*

WISHART, George Macfeat, Scottish physiol. chemist; b. Aug. 18, 1895; s. George Wishart; grad. in medicine Glasgow U., 1918; B.Sc., M.D.; m. Elizabeth Mary Bedale, 1926; 2 daus., Asst. to Regius prof. physiology Glasgow U., 1919-21, Grieve lectr. physiol. chemistry, 1921-35, Gradiner prof. physiol. chemistry, 1935-47, dean faculty medicine, dir. port-grad. med. edn., 1947—. Mem. Gen. Med. Council; mem. Gen. Nursing Council for Scotland. Fellow Royal Soc. Edinburgh, Royal Faculty Physicians and Surgeons Glasgow. Author: Groundwork of Biophysics, 1931; also papers on physiology and biochemistry. Died July 18, 1959.

WISHNER, Julius, psychologist; b. Lomza, Poland, July 7, 1921; s. David and Pearl (Podbilevich) Wisnia; came to U. S., 1927, naturalized, 1929; B.A., Bklyn. Coll., 1946; M.A., Northwestern U., 1947, Ph.D., 1950; m. Dorothy Busker, Nov. 1, 1941; children—Karen Frances, Amy Rose. Faculty dept. psychology U. Pa., Phila., 1949—, prof., 1962—, asso. prof. Law Sch., 1959-62, dir. clin. program, dir. research program in personality and exptl. psychopathology, 1959—. Vis. lectr. psychology Harvard, 1958-59. Fellow Am. Psychol. Assn., A.A.A.S.; mem. Internat. Union Sci. Psychology, Internat. Assn. Applied Psychology (exec. com., asst. editor Bull.). Research on concept of efficiency as definition of continuum of psychol. health and psychopathology, impressions of personality as problem in information processing. Home: 5 Hathaway Circle, Wynnewood, Pa. 19096. Office: Dept. Psychology, U. Pa., Phila. 19104.*

WISLICENUS, Johannes, chemist; b. Kleineichstedt/Querfurt, Germany, June 24, 1835; s. Gustav Adolf Wislicenus; became prof., Zurich, Switzerland, also Würzburg, Germany, 1872, Leipzig, Germany, from 1885. Fellow Royal Soc., 1897 (Davy medal 1898). Author: Théories des types chimiques, 1859. Proposed term geometric isomerism to describe space relations in lactic acids, 1869; demonstrated 2 chemically alike but physically different natural lactic acids (contbd. to formulation of postulate of tetrahedral carbon atom by van't Hoff and LeBel 1875), 1872-73; fundamental research on cis-trans isomerism, 1888-1901; studied acetoaceticester as a synthetic agt., esterification of ketons, ring formations, activity of alkyl halides; introduced term alkyl. Died Leipzig, Dec. 5, 1902.

WISSLER, Clark Am. anthropologist; born, Wayne County, Ind., Sept. 18, 1870; s. Benjamin Franklin and Sylvania (Needler) W.; A.B., Ind. U., 1897, A.M., 1899; Ph.D., Columbia University, 1901; LL.D. from Indiana University, 1929; m. Etta Viola Gebhart, June 14, 1899; children—Stanley Gebhart, Mary Viola. Instr. edn., New York U., 1901-02; asst. in anthropology, 1903-05; lecturer, 1903-09, Columbia; asst. in anthropology, 1903-06, curator, 1906-41, Am. Museum Natural History; professor anthropology, Yale, 1924-40. Division chmn. Nat. Research Council, 1920-21; consulting anthropologist Bishop Mus., Honolulu since 1920. Fellow N.Y. Acad. Sciences, Am. Ethnol. Soc., Am. Geog. Soc.; mem. Nat. Acad. Sciences, Sigma Xi, Phi Beta Kappa. Author: North American Indians of the Plains, 1912; Man and Culture, 1922; The Relation of Nature to Man in Aboriginal America, 1926; Social Anthropology, 1929; The American Indian, 1938; Indian Cavalcade, 1938; The Indian in the United States, 1940. Developed culture-area and age-area theories in anthropology. Died Aug. 25, 1947.

WISSLER, Robert William, Am. pathologist; b. Richmond, Ind., Mar. 1, 1917; s. William O. and Muriel (Thomas) W.; A.B., Earlham Coll., 1939, D.Sc. (hon.), 1959; Ph.D., U. Chgo., 1946, M.D. with honors, 1948; m. Elizabeth Anne Polk, 1940; children—Barbara, Mary, David, John. Intern USPHS, Chgo. Marine Hosp., 1949-50; instr. pathology U. Chgo., 1943-47, ass. prof., 1947-52, asso. prof., 1952-57, prof., chmn. dept. pathology, 1957-—, cancer coordinator, 1962-—. Nutritional and immunologic research Army and Navy, 1941-45; cons. pathology study sect. Surg. Gen. USPHS, 1957-61; com. pathology Nat. Acad. Scis-NRC, 1958, chmn. com., 1962-—; cons. Armed Forces Inst. Pathology, 1961, chmn. sci. adv. com., 1966-67; adv. com. Ill. Bd. Higher Edn.; mem. pathol. tng. com. NIH, 1963-67. Recipient David Worth Dennis award for excellence in chem. Earlham Coll., 1939, Howard Taylor Ricketts grad. student award for outstanding research pathology and bacteriology, 1947. Joseph A. Capps prize Chgo. Inst. Medicine, 1951. Diplomate Nat. Bd., Med. Examiners (chmn. path. test com. 1962-64), Am. Bd. Pathology (trustee 1968). Mem. Am. Soc. Exptl. Pathology (pres. 1961-62), Am. Soc. Cell Biology, Reticuloendothelial Soc., Soc. Exptl. Biology and Med., Am. (council arteriosclerosis sec. 1964, chmn. 1966) Chgo. (trustee) heart assns., Am. Cancer Society (dir. Ill. div.), European Group Atherosclerosis, Am. Assn. Pathologists and Bacteriologist (council 1963-—, pres. 1968), Am. Soc. Clin. Pathology, Internat. Acad. Path., Am. Assn. Immunologists, A.M.A., Am. Assn. Cancer Research. Coll. Am. Pathologists (sec. acad. sect. 1960, mem. assembly 1966-—), N.Y. Acad. of Scis., Am. Assn. of Chmn. Med. Sch. Depts. Pathology (pres. 1967), Univs. Asso. for Research and Edn. in Pathology (v.p., dir.), Am. Assn. Accreditation on Lab. Animal Care (vice chmn. trustees 1967-—), Sigma Xi (pres. Chgo. 1961), Alpha Omega Alpha, Gamma Alpha. Unitarian (chmn. bd. trustees 1962-63). Author numerous sci. papers in field. Asso. editor Metabolism, 1953-58; Nutrition Reviews, 1954-63; editorial bd. Recent Results in Cancer Research. Editor: Endogenous Factors Influencing Host-Tumor Balance, 1967. Developed new models demonstrating relationship between protein malnutrition and decreased resistance to infection; research on pathology of serum sickness; study of artherosclerosis; tumor immunity using ascites tumor systems; new concepts based on experimental study cellular mechanisms of antibody formation. Home: 5521 Woodlawn Av., Chgo. 60637.*

WISWELL, Ozro Brandt, Am. anatomist, comparative endocrinologist; b. Seattle, May 12, 1914; s. Ozro Newcome and Anna Brandt; A.B., U. So. Cal., 1939, M.Sc., 1939; D.M.S., Shinshu U. Sch. Medicine, Matsumoto, Japan, 1960; m. Viola Lenore Olson, Nov. 27, 1942; children—Eric, Albert, Christina, Swanhild, Barnabas. Mem. faculty U. So. Cal., 1939-42, U. Cal. Med. Center, San Francisco, 1946-50; faculty U. Tex. Dental Br., Houston, 1950-—, asso. prof. anatomy, 1962-—. Lectr., Sacred Heart Dominican Coll., 1956-—; with various projects USAF, USN. Fellow Tex. Acad. Sci., A.A.A.S.; mem. State Anat. Bd. Tex., Royal Soc. Health, Eng., Am. Assn. U. Profs., Am. Assn. Anatomists, Endocrine Soc., Am. Soc. Zoologists; Kappa Alpha Theta, Rho Chi, Sigma Xi. Research and publs. on tissue transplantation, comparative endocrinology, hormone control of endocrines in subhuman primates, terrestrial gastropod response to environmental and hormonal stimuli. Home: 4430 Creekbend Dr., Houston 77035.*

WIT, Daniel, Am. polit. scientist; b. N.Y.C., June 8, 1923; s. Benjamin and Stella (Bloom) W.; A.B., Union Coll., 1943; A.M., Princeton, 1948, Ph.D., 1950; m. Phyllis Jo Citron, June 15, 1947; children—Pamela S., Frederick W. Instr., Ohio State U., 1948-50; asst. prof. U. Cin., 1950-54; vis. prof. U. Mich., 1954-56; vis. prof. pub. adminstrn. Indiana U.-Thamassat U., Bangkok, Thailand, 1956-58; vis. asso. prof. govt. Ind. U., 1958; dir. internat. studies Govtl. Affairs Inst., Washington, 1959-61; prof., head dept. polit. sci. No. Ill. U., 1961-—; prof. fgn. affairs Nat. War Coll., Washington, 1963-64; Fulbright prof. Inst. Social Studies, The Hague, Netherlands, 1966-67; profl. lectr. pub. adminstrn. George Washington U., 1959-61; cons. on Thai labor law, practice U. S. Dept. Labor, 1963; pub. adminstrn. adviser Thailand, 1956-58. Mem. Am. Polit. Sci. Assn., Midwest Polit. Sci. Conf., Am. Soc. for Pub. Adminstrn., Eastern Regional Orgn. for Pub. Adminstrn. Author: Comparative Political Institutions, 1953; (with M. L. Thomas, D. Grant) A Comparative Survey of Local Government and Administration, 1961; Labor Law and Practice in Thailand, 1964; Thailand: Another Vietnam?, 1968; also articles. Pioneer in studies in comparative govt. after World War II which included comparative socio-political and geo-polit. analysis of functional categories; early discussion of polit. implications of atomic power. Home: 222 Fairmont Dr., De Kalb, Ill. 60115.*

WITEBSKY, Ernest, physician; b. Frankfurt, Germany, Sept. 3, 1901; s. Michael and Hermine (Neuberger) W.; student U. Frankfurt Med. Sch., 1920-25; M.D., U. Heidelberg (Germany), 1926; Doctor of Medicine honoris causa, U. Freiburg (Germany), 1958; m. Ruth Mueller-Erkelenz, June 23, 1935; children—Frank G., Grace E. Came to U. S., 1934, naturalized, 1939. Asst. research div. Cancer Inst., U. Heidelberg Med. Sch., 1925-29, pvt. dozent, 1929-33; fellow Mt. Sinai Hosp., N.Y.C., 1934-35; faculty State U. N.Y. at Buffalo Sch. Medicine, 1936-—, prof. bacteriology and immunology, 1940-54, distinguished prof., 1954-—, head dept., 1941-67, acting dean, 1958-59, dean, 1959-60, dir. Center for Immunology, 1967-—; bacteriologist, serologist Buffalo Gen. Hosp., 1936-—, dir. Blood Bank, 1941-—. Cons. VA Hosp., 1951-—. Recipient Chancellor's gold medal U. Buffalo, 1950, Mem. Am. Assn. Blood Banks (Karl Landsteiner Meml. award 1959), Internat., Brit. (hon.), socs. hematology, A.M.A., Am. Assn. Immunologists, Am. Soc. Microbiologists, Soc. for Exptl. Biology and Medicine, Am. Assn. Pathologists and Bacteriologists, A.A.A.S., Am. Pub. Health Assn., Am. Acad. Microbiology, N.Y. Acad. Scis., Pan Am. Med. Assn. Research, numerous publs. in blood groups, transfusion problems, organ and tissue specificity, autosensitization. Home: 607 Ashland Av., Buffalo 14222.*

WITELO (Vitellius), physicist, philosopher; b. Silesia, circa 1220; ed. Paris, 1253; studied canon law, Padua, circa 1262-68; student at Viterbo, 1269; held views which offer parallels to modern psychology; significant as link in transmission of Greco-Arabic optics; his optics derive largely from Ibn al-Haitham; his theory of rainbow was inferior to those of Muslum contemporaries but superior to that of Aristotle. Author: De natura daemonum; De intelligentiis, circa 1225; Optics, or Perspectivae, circa 1270-78; De primaria causa paenitentiae. Died Premonstratensian monastery of Witow, nr. Piotrkow, Poland.

WITH, Torben Kiorboe, Danish physician; b. Copenhagen, Denmark, Aug. 18, 1910; s. Carl Johannes and Inge (Kiorboe) W.; M.D., Copenhagen U., 1941; m. Kamma Bentsen, Jan. 18, 1938; children—Jette, Kamma-Hanne (Mrs. Wilhelm Johannsen), Frants-Ulrik, Carl, Nete Marianne. Mem. staff Rigshospital, Copenhagen, 1939-45, 48-53; asst. physician German Refugee Adminstrn., 1945-48; acting head physician med. dept. State Hosp., Sonderborg, 1946; head physician lab. and blood bank County Hosp., Svendborg, 1954-—. Lectr. European profl. socs. Mem. Danish Soc. Internal Medicine and Clin. Chemistry, Biochem. Soc. London, Nutritional Soc. London, Assn. Clin. Pathologists London. Author: Absorption, Metabolism and Storage of Vitamin A and Carotene, 1940; Biology of Bile Pigments, 1954; Biologie der Gallen Frabstoffe, 1960; Bile Pigments, 1968. Numerous publs. on bone changes in blood diseases, vitamin A and carotenoids, recognition of uveoparotid fever as a manifestation of sarcoidosis; studies on bile pigments; methods for large-scale preparation of purified porphyrins from excreta. Home: 53 St. Jorgensvej, Svendborg, Denmark.*

WITHERING, William, Brit. physician, botanist, mineralogist; b. Wellington, Eng., Mar. 1741; s. Edmund and Sarah (Hector) W.; M.D., U. Edinburgh (Scotland), 1766; m. Helena Cookes, Sept. 12, 1772; children—William, Charlotte. Chief physician Birmingham Gen. Hosp. Fellow Royal Soc., 1784. Author: Dissertatio Inauguralis de Angina Gangraenosa, 1766; A Botanical Arrangement of all the Vegetables naturally growing in Great Britain, according to the System of the celebrated Linnaeus . . . , 1776; An Account of the Foxglove and some of its Medical Uses, 1785; Outlines of Mineralogy, 1783; An Account of the Scarlet Fever and Sore Throat, or Scarlatina Anginosa, 1778; Miscellaneous Tracts, 1822 Translated Sciagraphia regni mineralis (with additional notes) (Bergmann). Witheringia, a genus of Solanaceae, named in his honor; introduced use of digitalis in cardiac disease; 1st to point out connection between dropsy and heart disease. Died Isle of Wight, Oct. 6, 1799.

WITHNER, Carl Leslie, Jr., Am. botanist; b. Indpls., Mar. 3, 1918; s. Carl L. and Martha (Mevers) W.; A.B., U. Ill., 1941; M.A., Yale, 1943, Ph.D., 1948; m. Patricia Maxwell, June 4, 1941; children—Dennis Maxwell, Rika D'Almeida, Holly Nichols. Faculty, Bklyn. Coll., 1948-—, prof. botany, 1963-—; research asso. charge orchid collection Bklyn. Botanic Garden, 1948-—. Vis. asso. prof. biology Stanford, 1963. Guggenheim fellow, 1962. Mem. Am. Soc. Plant Physiologists, Bot. Soc. Am., Am. Orchid Soc., Torrey Botanical Club, Phi Beta Kappa, Sigma Xi, Delta Phi Alpha, Alpha Phi Omega, Phi Kappa Phi. Editor: The Orchids, A Scientific Survey, 1959. Research, numerous publs. on growth and nutritional requirements of orchid seedlings, origin and use of ovule culture in growing orchid seedlings, vanilla seed germination, evolution and speciation of orchids, wood anatomy, biochem. problems correlated with plant devel. Home: 2338 Royce St., Bklyn. 11234.*

WITKIN, Herman A(lan), Am. psychologist; b. N.Y.C., Aug. 2, 1916; s. Abraham and Anna (Baer) W.; student Cornell U., 1932-35; A.B., N.Y. U., 1935, A.M.A., 1936, Ph.D., 1939; m. Evelyn Maisel, July 9, 1943; children—Joseph, Andrew. Research asso. Swarthmore (Pa.) Coll., 1939-40; faculty Bklyn. Coll., 1940-52; prof., dir. psychology lab State U. N.Y. Downstate Med. Center, 1952-—. Mem. Am. Psychol. Assn., Soc. for Psychol. Study of Social Issues. Author: (with others) Personality through Perception, 1954; (with others) Psychological Differentiation, 1962; (with others) Experimental Studies of Dreaming. also articles. Research on cognitive styles, relation between cognitive style and personal orgn., cognitive devel. as function family experiences, cognitive patterning in deficit children, relation between presleep experiences and dreams. Home: 662 E. 26th St., Bklyn. N.Y. 11210.*

WITKOP, Carl Jacob, Am. geneticist; b. East Grand Rapids, Mich., Dec. 27, 1920; s. Carl J. and Frances (Miller) W.; B.S., Mich. State Coll., 1944; D.D.S., U. Mich., 1949, M.S., 1954; m. Mary Worcester, Sept. 3, 1946; children—Carl Gray, Mary Margaret, Martha Frances; m. 2d, Mary Ann Jacobson, Oct. 6, 1966; children—Andrea Jean, Steven Carl. Commd. asst. dental surgeon USPHS, 1949, advanced through grades to dental dir., 1961; dental officer Nat. Inst. Dental Research, 1950-54, chief human genetic br., Bethesda, Md., 1954-66; instr. med. genetics Georgetown U., 1959-65; chmn. div. genetics, prof. U. Minn. Dental Sch., 1966-—. Cons. Childrens Hosp., Washington, 1955-—. Diplomate Am. Bd. Oral Pathology. Fellow Am. Acad. Oral Pathology, A.A.A.S.; mem. Am. Dental Assn., Am. Soc. Human Genetics, Am. Soc. Cytology, Washington, Md. socs. pathologists, Sigma Xi. Author: Genetics and Dental Health, 1962. Research on human genetics, population isolates, nutritional and hereditary defects in perception, oral sensation and speech, congenital malformations, exfoliative cytology, oral nutrition and epidemiology, oral pathology. Home: 9 Manitoba Rd., Hopkins, Minn. Office: Sch. Dentistry, U. Minn., Mpls.

WITMER, R(obert) B(onner), Am. physicist; b. Bathgate, N.D., Jan. 30, 1900; s. Samuel T. and Charlotte (Fizette) W.; B.S. U. N.D., 1922, M.S., 1926; Ph.D., U. Mich., 1935; postgrad. U. Wash., Columbia; m. Lillian Leith, Aug. 15, 1927; 1 dau., Jean Leith. Grad. asst. U. N.D., 1922-24, faculty, 1924-—, dean freshmen, 1939-42, prof. physics, 1938-65, dean jr. div., 1946-49, acting dean College of Sci., Lit and Arts, 1948-49, dean, 1949-65, dean emeritus, hon. univ. prof., 1965-—. Mem. Am. Assn. Univ. Profs., N.D. Acad. Sci., N.E.A., Am. Phys. Soc., Am. Assn. Physics Tchrs., Sigma Xi, Sigma Tau, Phi Beta Kappa, Phi Delta Kappa, Phi Eta Sigma, Sigma Nu. Specialist in measurement of long X-rays by means of ruled grating; resistivity of N.D. clays. Home: 2920 Fifth Av. N., Grand Forks, N.D. 58201.*

WITNEY, Fred, Am. educator; b. Chgo., Nov. 25, 1917; s. Morris and Tilly (Sweet) W.; B.A., U. Ill., 1940, M.A., 1941, Ph.D., 1947; m. Judith Clein, Sept. 4, 1943; children—Eileen, Frank. Instr., Ind. U., 1947-48, asst. prof., 1948-53, asso. prof., 1953-60, prof., 1960-—. Arbitrator, labor disputes. Mem. Am., Midwest econs. assns., Nat. Acad. Arbitrators, Indsl. Relations Research Assn., Phi Beta Kappa, Beta Gamma Sigma, Phi Eta Sigma. Author: Wartime Experiences of the N.L.R.B., 1949; Government and Collective Bargaining, 1951; Indiana's Economic Resources and Potential, 1955; The Collective Bargaining Contract Agreement: Its Negotiation and Administration, 1957; Indiana Labor Relations Law, 1960; Labor Policy and Practices in Spain, 1965; (with Arthur Sloane) Labor Relations, 1967; also articles. Analysis of concepts and principles of indsl. relations and their application to contemporary problems; forecast of future issues and their solution in terms of alternatives; indsl. relations viewed as an interdisciplinary science within contemporary society. Home: 1212 E. Southdowns Dr., Bloomington, Ind. 47401.*

WITSCHI, Emil, biologist; b. Bern, Switzerland, Feb. 18, 1890; s. Johann and Elise (Blank) W.; M.S., U. Bern, 1911; Ph.D., U. Munich, 1913; M.D., (hon.), U. Basel, 1960; m. Ida Martha Muehlestein, July 14, 1914; children—Marianne (Mrs. Dwight J. Potter), Hans Walter. Mem. faculty U. Basel, Switzerland, 1921-27; prof. embryology and endocrinology U. Ia., Iowa City, 1927-58, prof. emeritus, 1958; guest prof. U. Tubingen, Germany, 1948-49, Sorbonne, U. Paris, 1959; vis. prof. Yale, 1961; China Found. vis. prof. Taiwan U., 1962; program specialist for reproductive physiology Ford. Found., 1962-67; sr. scientist Population Council, Biomed. Div., Rockefeller U., N.Y.C., 1967-—. Recipient Fred Conrad Koch medal Endocrine Soc., 1960; Jubile Scientifique dans les Salons de la Sorbonne et Dedication d'un Volume Jubilaire, 1959. Mem. Am. Soc. Zoology (pres. 1959-60), Endocrine Soc., A.A.-A.S., Leopolina, German Acad. Naturalists, N.Y. Acad. Sci., Schweiz. Nat. Ges., Am. Genetics Soc., Soc. Exptl. Biology and Medicine, Sigma Xi (nat. lectr. Pacific region 1960). Author: Development of Vertebrates, 1956; also numerous articles. Research on induction theory of embryonic sex differentiation, genetics of sex differentiation, evolution of sex chromosomes, hormonal mechanisms of seasonal sexuality in birds, breeding of sex-reversed amphibians, morphology and physiology of underwater hearing and evolution of vertebrate ear, overripeness of eggs as a cause of embryonic malformations, gonadotropins. Home: 435 E. 70th St., N.Y.C. 10021. Office: Population Council, Biomed. Div., Rockefeller U., N.Y.C. 10021.*

WITT, Otto Nikolaus, chemist; b. St. Petersburg, Russia, 1853. Became prof. Technische Hochschule, Charlottenburg, Germany, 1891; became indsl. chemist in Eng., 1875; named commr. Universal Expn., Chgo., 1893. Author: Die Chemische Industrie auf der internationalen Weltanstellung zu Paris; Chemische Homologie und Isomerie in Ihren Einfluss auf Erfindungen; also articles on dyes, dyeing. Founder, editor Prometheus; also Die chemische Industrie. Discovered dyes chrysoidine, tropaeoline; introduced theory of chromophores and chromogens. Died 1915.

WITT, Peter Nikolaus, pharmacologist; b. Berlin, German, Oct. 20, 1918; s. Felix H. and Emma (Von Mendelssohn) W.; student U. Berlin, 1941-44, U. Graz (Germany), 1939-41; M.D., U. Tubingen (Germany), 1946; m. Ingeborg Feiler, Mar. 29, 1949; children—Elise M., Mary E. Came to U.S., 1956, naturalized, 1962. Asst., U. Bern (Switzerland), 1949-56, privat-docent, 1956; faculty Coll. Medicine, State U. N.Y. Upstate Med. Center, Syracuse, 1956—, asso. prof. pharmacology, 1959-66; dir. div. research N.C. Mental Health Dept., Raleigh, 1966—. Rockefeller fellow Harvard Med. Sch., 1952-53; recipient Lederle Med. Faculty award, 1957-59; Buergi award N.J. Bur. Research Neruology and Psychiatry, 1956. Mem. Soc. Pharmocology, German Pharmacology Soc., Swiss Pharmacological Soc. Author: Die Wirkung von Substanzen auf den Netzbau der Spinne al biologischer Test, 1956; also numerous articles. Research on effects drugs on ion movements in heart muscle, effect behavioral drugs on web geometry of spider. Home: Route 1, Box 447, Knightdale, N.C. 27545. Office: N.C. Dept. Mental Health, P.O. Box 7532, Sta. B, Raleigh, N.C. 27602.*

WITTE, Siegfried Kurt Helmut, German physician; b. Garlitz, Germany, Sept. 22, 1922; s. Helmut and Hertha (Rahmel) W.; student univs. Berlin, Würzburg, Erlangen; Dr.med., 1948; m. Elisabeth Beck, Oct. 2, 1949; children—Ilsabe, Helmut, Reinhard. Asst., med. polyclinic U. Wurzburg, 1950-53; asst. med. clinic U. Erlangen, 1953-59, chief physician, 1959—, docent, 1957-62, unscheduled prof., 1962—. Author: (with N. Henning) Atlas of Gastroenterological Cytodiagnosis, 1957, Internationales Symposion über klinische Cytodiagnostik, 1958; Blutgerinnung und Blutgefässe, 1960; Hämostase, Thrombogenese, Pharmakologisch wirksame Gerinnungsprodukte, 1963; Knochenmarktransfusion, 1963; also numerous articles. Research on clin. hematology mainly cytodiagnosis and blood coagulation, relationship of coagulation to vascular permeability, cin. oncology, chemotherapy, bone marrow transfusion, heterotransplantation of human tumors. Address: 58 Möhrendorfer Strasse, 852 Erlangen, West Germany.*

WITTEN, Victor Herbert, Am. physician; b. Jacksonville, Fla., Aug. 8, 1916; s. Morris H. and Cecilia (Starr) W.; student Washington and Lee U., 1934-35; B.S., Tulane U., 1938, M.D., 1941; postgrad. N.Y. U. Med. Sch., 1946-49; m. Joan Adrienne Kalmine, June 22, 1956; stepchildren—Richard K. Paradies, Jed C. Paradies. Preceptee, Office of Marion B. Sulzberger, 1946-49; practice medicine specializing in dermatology, N.Y.C., 1949-62; faculty N.Y. U. Med. Sch., 1950-62, asso. clin. prof. dermatology, 1957-62; prof. dermatology U. Miami (Fla.) Sch. Medicine, 1962-68, clin. prof. dermatology, 1968—. Mem. adv. panel on med. and biol. scis. Office Research and Engrng., Dept. Def., 1962—; mem. adv. com. on gen. medicine to surgeon gen., chmn. subcom. on dermatology U. S. Army Med. Research and Devel. Command, 1962—. Fellow Am. Acad. Dermatology, A.C.P., N.Y. Acad. Medicine, N.Y. Acad. Scis.; mem. Am., Israeli dermatol. assns., Venezuelan Soc. Dermatology, Venereology, and Leprology, Danish Dermatology Soc., Soc. for Investigative Dermatology. Author: (with Marion B. Sulzberger, Jack Wolf) Essentials of Diagnosis and Treatment, 1961; (with Rudolf L. Baer) Yearbook of Dermatology, 1955-62; also numerous articles. Internat. editorial bd. Excerpta Medica, 1955—. Research on biologic effects of alpha and beta radiation on human skin, use and effectiveness of topical corticosteroids in diseases of skin, in vitro studies of metabolism of corticosteroids in human skin, effects of ionizing radiation to gonads and methods of protection, use of ionizing radiation in dermatology and mgmt. dermatol. disorders. Home: 480 Casuarina Concourse, Coral Gables, Fla. 33143. Office: 1150 N.W. 14th St., Miami 33136.*

WITTER, Robert Frank, Am. chemist; b. Chgo., Nov. 19, 1918; s. Frank and Julia (Gibson) W.; B.S., U. Ill., 1940; M.S., U. Mich., 1941, Ph.D., 1944; m. Lynette A. Ward, June 6, 1948; children—Frank R., Lester J. Research asso. N.Y. State Agrl. Expt. Sta., Geneva, 1944-47; faculty U. Rochester (N.Y.), 1947-57; chief biochemist toxicology sect. Communicable Disease Center, USPHS, Savannah, Ga., 1957-62, chief lipid standardization lab. Communicable Disease Center, Atlanta, 1962—. Mem. program com. Am. Assn. Biol. Chemists, 1950-57. Mem. Am. Chem. Soc., Am. Oil Chemists Soc., Phi Lambda Upsilon, Sigma Xi, Phi Sigma, Gamma Alpha. Asso. editor Lipid, 1965. Research in metabolism of lipids, methods for determinations of blood lipids, biochemistry of organic phosphorus insecticides. Home: 584 N. Superior Av., Decatur, Ga. Office: Lab. Br., Communicable Disease Center, USPHS, Atlanta.*

WITTGENSTEIN, Ludwig Josef Johann, philosopher; b. Vienna, Austria, Apr. 26, 1889; s. Karl W.; ed. Technische Hochschule, Berlin-Charlottenburg, 1908, U. Manchester, 1908-11, Trinity Coll., Cambridge (studied math. under B. Russell), 1912-14, Ph.D., 1929; Vienna Coll. for Elementary Tchrs., 1919-20. Served with Austrian Army, 1914-18; schoolmaster in Austrian dists. Schneeberg and Semmering, 1920-26; Gardiner's asst. Hütteldorf monastery, nr. Vienna, 1926; archtl. worker, Vienna, 1926-28; fellow Trinity Coll., Cambridge, 1929, lectr., 1930-36, 37-39, prof. philosophy, 1939-47. Visited Norway, 1936, Ireland, 1948, U. S., 1949, Norway and Vienna, 1950. Author: Tractatus Logico-Philosophicus (English transl.), 1922; (posthumous) Philosophical Investigations (transl. G. E. M. Anscombe), 1953; Remarks on Foundations of Mathematics (edited by G. H. von Wright, transl. Anscombe), 1956; The Blue and Brown Books, 1958; Notebooks 1914-16 (edited by von Wright and Anscombe), 1961; Lectures and Conversations on Aesthetics, Psychology and Religious Belief (edited Cyril Barrett), 1966. Influential in founding of philosophy of Logical Positivism, 1920's; later approached philosophy as an analysis of proper use of lang. Died Cambridge, Eng., Apr. 29, 1951.

WITTHAUS, Rudolph August, Am. toxicologist; b. N.Y., Aug. 30, 1846; s. Rudolph A. and Marie A. (Dunbar) W.; A.B., Columbia, 1867, A.M., 1870; M.D., Univ. Med. Coll. (New York U.), 1875; postgrad. Sorbonne and Collège de France, Paris, 1873-74. Asso. prof. chemistry and physiology N.Y. U., 1876-78; prof. chemistry and toxicology U. Vt., 1878-98; prof. physiol. chemistry Univ. Med. Coll., 1882-86, chemistry and physics, 1886-98; prof. chemistry and toxicology U. Buffalo, 1882-88; prof. chemistry and physics Cornell U. Med. College, 1898-1911, emeritus, 1911—. Toxicol. expert in Carlyle Harris, Buchanan, Mayer, Fleming, Benham, Molineux and many other cases. Fellow A.A.A.S. Author: Essentials of Chemistry, 1879; General Medical Chemistry, 1881; Manual of Chemistry, 1879, 1908; Laboratory Guide in Urinalysis and Toxicology, 1886. Editor: Witthaus and Becker's Medical Jurisprudence (4 vols.), in which contributed introduction and Vol. 4 on Toxicology, 1894, 1906. Contbr. articles on poisoning by hydrocyanic acid, oxalic acid, opium and strychnine, ptomaines, Wood's Handbook of Medical Sciences; on homicide by morphine; detection of quinine; postmortem imbition of poisons; Researches Loomis Laboratory. Died Dec. 23, 1915.

WITTKOWER, Eric David, psychiatrist; b. Berlin, Germany, Apr. 4, 1899; s. Louis and Bertha (Katz) W.; M.D., U. Berlin, 1924; postgrad. U. Edinburgh (Scotland), U. Glasgow (Scotland); m. Claire Francesca Weil, July 29, 1931; children—Andrew, Sylvia (Mrs. Benjamin Jefferson Adkins). Asst., Charite, Berlin, 1925-33; pvt. dozent psychosomatic medicine U. Berlin, 1932-33; research fellow Maudsley Hosp., U. London (Eng.), 1933-35; Sir Halley Stewart Research fellow Tavistock Clinic, London, 1935-40, physician, 1935-40; research psychiatrist dermatol. dept. St. Bartholomew's Hosp., London, 1948-51; psychiat. research fellow Nat. Assn. for Prevention Tb, London, 1945-48; lectr. psychiatry Maudsley Hosp., U. London, 1949-51; faculty McGill U., Montreal, Que., Can., 1951—, prof. psychiatry, 1963—; asso. psychiatrist Montreal Gen. Hosp., 1951-61, cons. psychiatrist, 1961—; dir. sect. transcultural psychiat. studies dept. psychiatry McGill U., 1956—. mem. WHO Expert Adv. Panel on Mental Health, 1964—. Licentiate Royal Coll. Physicians, Royal Coll. Surgeons, Royal Faculty Physicians and Surgeons. Fellow Psychiat. Assn., Brit. Psychol. Soc., Royal Soc. Medicine (Eng.), World Fedn. Mental Health, Acad. Psychoanalysis; mem. Am. (past pres.), Argentinian (hon.) psychosomatic socs., Brit. (asso.), Canadian (pres. 1965—, life mem.) psychoanalytic socs., Canadian, Que., Polish (corr.) psychiat. assns., Canadian Assn. Occupational Therapy (hon. adv. council), Pan Am. Med. Assn. (life), Internat. Inst. for Mental Health Research (mem. profl. bd. 1963—), Internat. Council Psychologists (profl. bd. 1963—), Internat. Council Psychologists (profl. affiliate), N.Y. Acad. Scis., Hellenic Soc. Neurology and Psychiatry (hon.). Contbr. chpts. to The Psychological Basis of Medical Practice, 1963; Marriage Counselling in Medical Practice, 1964; Psychological and Allergic Aspects of Asthma, 1965. Research, numerous publs. in biochemistry, hematology, internal medicine, high-altitude physiology and pathology, psychosomatic, transcultural psychiatry. Home: 363 Clarke Av., Montreal, Que. Office: Allan Meml. Inst., 1025 Pine Av., W., Montreal, Que., Can.*

WITTMANN, Heinz Günter, German molecular biologist; b. Stürlack, Jan. 16, 1927; s. Paul Michael and Margarete (Burbiel) W.; Diplom., Landw. Hochschule Hohenheim, 1951; postgrad. Techn. Hochschule Stuttgart (Germany), 1951-52; Dr.rer.nat., U. Tübingen (Germany), 1956; m. Brigitte Liebold, Mar. 28, 1961; 1 dau., Beate. Research fellow U. Cal. at Berkeley, 1956-57; scientist Max-Planck Inst. für Biologie, Tübingen, 1957-64; dir. Max-Planck Institut für Molekulare Genetik, Berlin-Dahlem, Germany, 1964—; dozent U. Tübingen, 1962-68; hon. prof. U. Berlin-Dahlem, 1968—. Mem. European Molecular Biology Orgn., Leopoldina. Research, numerous publs. on structure tobacco mosaic virus and its mutants and strains, structure of ribosomes especially their proteins; contbr. to deciphering genetic code. Home: 17 Meisenstrasse. Office: 23 Harnackstrasse, Berlin-Dahlem, Germany.*

WITTNER, Murray, Am. physician; b. N.Y.C., Apr. 23, 1927; s. Albert and Aranka (Reisch) W.; B.S., U. Ill., 1948, M.S., 1949; Ph.D. (Priscilla Clarke Hodges scholar), Harvard, 1955; M.D., Yale, 1961; m. Phyllis Gillman, June 19, 1955; children—Lisa Ann, David Gillman. Research fellow Harvard, 1954; instr. Albert Einstein Coll. Medicine, Bronx, N.Y., 1955, faculty, 1962—, asst. prof. pathology and parasitology, 1965—; clin., and teaching fellow pathology Harvard, also Mass. Gen. Hosp., 1961; cons. parasitologist to air surgeon Far East Air Forces USAF, 1951-53; attending pathologist Bronx Municipal Hosp. Center, Albert Einstein Hosp. Center, 1965—; dir. parasitology labs. Einstein Med. Center, 1965—; guest faculty State U. N.Y. at Buffalo, 1965. Career scientist Health Research Council, N.Y.C., 1965—. Mem. Soc. Protozoology, Am. Soc. Zoologists, Am. Microscopical Soc., Am. Fedn. Clin. Research, Internat. Acad. Pathology, Sigma Xi, Alpha Omega Alpha, Gamma Alpha. Research, publs. on pathophysiology protozoan parasites, physiology and biochemistry feeding mechanisms free-living protozoa, effects gaseous atmospheres at high pressures on cells and tissues. Home: 6 Pheasant Run, Larchmont, N.Y. 10538. Office: 1300 Morris Park Av., Bronx, N.Y. 10461.*

WITTWER, Sylvan Harold, Am. agriculturist; b. Hurricane, Utah, Jan. 17, 1917; s. Joseph and Mary Ellen (Stucki) W.; B.S., Utah State U., 1940; Ph.D., U. Mo., 1943; m. Maurine Cottle, July 27, 1938; children—LaRee (Mrs. Reed Farrer), Alice (Mrs. Mack Sowards), Arthur John, Carl Thomas. With U. Mo., 1940-46, instr. horticulture, 1943-46; faculty Mich. State U., East Lansing, 1946—, prof. horticulture, 1950—, dir. agrl. expt. sta., 1965—, asst. dean Coll. Agr. Recipient Vaughn award, 1953; Campbell award, 1957. Mem. Am. Inst. Biol. Sci., Am. Soc. Hort. Sci. (mem. edl. com.), A.A.A.S., Internat. Hort. Soc., Am. Soc. Plant Physiology, Bot. Soc. Am., Soc. Econ. Horticulture, Soc. for Study of Growth and Devel., Sigma Xi, Alpha Zeta. Contbr. numerous articles to sci. jours. Pioneer in use of radioisotopes in studies of foliar nutrition of agrl. crops; effects of plant growth regulators and environmental factors on control of flowering, fruiting and senescence in hort. crops; discovered use of malic hydroxide for control of storage sprouting in onions by pre-harvest foliar applications; conducted first studies demonstrating wide applicability of carbon dioxide enrichment of atmospheres for promoting growth and accelerating maturity of greenhouse vegetables. Home: 1767 Hitching Post Rd., East Lansing, Mich. 48823.*

WITZLEB, Erich, German physiologist; b. Bad Gandersheim, Germany, Apr. 5, 1924; s. Otto and Anna (Toedt) W.; student univs. Tübingen, Strasbourg; med. state exam., U. Hamburg, 1949; m. Ingeborg Oldach, Sept. 22, 1950; children—Work, Katrin, Madeleine. With Inst. Exptl. Pathology and Balneology U. Hamburg, 1950-55; with Gollwitzer-Meier Inst., Oeynhausen state spa, U. Münster/Westphalia, 1955—, docent applied physiology, 1958-64, head dept. physiology, 1959—, unscheduled prof., 1964—. Author: Bad Oeynhausener Gespräche, vols. 1-6, 1956-64; also numerous articles. Research on excitation mechanism of chemopresso-and mechano-receptors; regulation of coronary blood flow and myocardial oxygen consumption; regulation of venous tone. Home: 39 An der Südbahn, 4871 Oberbecksen. Office: 42 Herforderstrasse, Bad Oeynhausen 497, West Germany.*

WIZINGER, Robert Charles, Swiss Chemist; b. Vic-Seille, Apr. 28, 1896; s. Michael and Elizabeth (Marchal) W.; ed. univs. Strasbourg, Stuttgart Tübingen, Karlsruhe and Bonn; Dr. Chemistry; m. Gerty Aust, Mar. 4, 1924; 1 son, Hans. Agrégé, U. Bonn, 1727, asso. prof., 1947; prof. agrégé U. Zurich, 1938, asso. prof., 1942; now prof. U. Basel; prof. honoris causa univs. Bonn and Friburg/Breisgau. Recipient Werner prize, 1943, Broylants medal, 1955. Mem. Swiss Chem. So., Soc. German Chemists. Author: Organische Farbstaffe, 1933; Chemische Plaudereien, 1933; Kohle, Luft und Wasser, 1938. Home: Laupenring 145, Basel. Office: Inst. fur Farbenchemie Basel, St. Johanns-Vorstadt 10/12, Switzerland.

WODSEDALEK, Jerry Edward, Am. zoologist; b. Kewaunee, Wis., Aug. 5, 1884; s. Frank and Marie (Posepney) W.; Ph.B., U. Wis., 1910, M.Ph., 1911, Ph.D., 1913; m. Hazel Mae Phillips, Aug. 28, 1914; children—Helen Marie (Mrs. Stanley M. Schlosser), Stanley (dec.). Faculty, U. Ida., 1913-28; prof. zoology U. Minn., Mpls., 1928-50, emeritus, 1950—. Fellow A.A.A.S.; mem. Am. Soc. Zoologists (life), Am. Inst. Biol. Scis. (life), Am. Micros. Soc., Minn. Acad. Scis., Sigma Xi, Phi Gamma Mu, Sigma Alpha Epsilon. Author: (with H. L. Dean) General Biology Laboratory Guide, 1955; General Zoology, 1963; General Zoology Laboratory Guide, 1963; also articles. Research, publs. on cytology, genetics, cytogenetics, insect physiology, entomology, animal behavior. Died Jan. 5, 1967.

WOELFEL, Julian Bradford, Am. dentist; b. Balt., Dec. 17, 1925; s. Norman and Florence (Carpenter) W.; D.D.S., Ohio State U., 1948; m. Marcile Cottrell, May 1, 1948; children—Bradford, Barry, Jay Bryan. Faculty, Ohio State U., Columbus, 1948—, prof. dentistry, 1963—; pvt. practice dentistry, Columbus, 1956—; research asso., prosthedontic cons. Nat. Bur. Standards, Washington, 1957—; cons. VA, Dayton, O., 1965—. Fellow Internat. Coll. Dentists, Am.

Coll. Dentists; mem. Internat. Assn. for Dental Research (award for meritorious research in dental prosthetics 1967), Am. Dental Assn. (cons. council on dental trade and lab. relations, mem. council on materials and devices), Am. Prosthodontic Soc., Acad. Denture Prosthetics, Federation Dentaire Internationale, Sigma Xi. Author: (with others) Resins and Technics Used in Constructing Dentures, 1965. Research and publs. on varied posterior teeth on dentures, impression materials and techniques, electromyography, denture base materials, hard reliners and human chewing cycle using face-bow, denture home-reliners, swallowing pressures with dentures, functional 3 dimensional movements of jaw joints. Home: 4345 Brookie Ct., Columbus, O. 43214.*

WÖHLER, Friedrich, German chemist; b. Eschersheim, nr. Frankfort/Main, Germany, July 31, 1800; ed. U. Marburg, H. Heidelberg (under Gmelin); doctorate in medicine, surgery, obstetrics, 1823; lab. asst. to Berzelius, Stockholm; became tchr. chemistry and mineralogy Sch. of Arts et Métiers, Berlin, 1825, organized similar sch., Cassel, Germany, 1831; named. prof. chemistry U. Göttingen (Germany), 1836; dir. Inst. Chemistry, Göttingen; insp.-gen. pharmacies of Hanover. Fellow Royal Soc. (Copley medal 1872), 1854; mem. French Acad. Scis., other leading European sci. acads. and socs. Author: Grundriss der unorganischen Chemie, 1831; Grundriss der organischen Chemie, 1840; also numerous artilces. In charge of publ. of Berzelius' Ann. Reports, from 1825; helped with publ. of Liebig's Annalen, from 1832. Pioneer in study of organic chemistry and isomerism; isolated aluminum, 1827, beryllium, 1828; discovered calcium carbide, also preparation of acetylene from it; developed commonly used method for preparing phosphorus; synthesized urea from ammonium cyanate, 1828; asso. with Liebig in work on benzoyl radicals; discovered (with Liebig) amygdalin, 1837, hydroquinone, 1848, calcium carbide, 1862, also showed analogy between carbon and silicon, 1863; stated new method for tech. extraction of phosphorus through reduction of phosphates; research in chemistry of metabolism; studied carbazotic (picric) acid; experimented on tech. prodn. of nickel; studied narcotine, quinone, cocaine; constructed new galvanic elements with iron. Died Göttingen, Sept. 23, 1882.

WOHLSCHLAG, Donald Eugene, Am. ecologist; b. Bucyrus, O., Nov. 6, 1918; s. Herman and Agnes (Canode) W.; B.S. magna cum laude, Heidelberg Coll., O., 1940; Ph.D. (Ind. Conservation Dept. fellow, Nat. Acad. Scis.-NRC fellow), Ind. U., 1949; m. Elsie Marjorie Baker, June 5, 1943; children—William Eugene, Nancy Sue, Sarah Ann. Asst., Ind. Lake and Stream Survey, 1939-40; research asso. U. Wis., 1948-49, faculty Stanford, 1949-64, prof. biol. scis., 1964; dir. Inst. Marine Sci., prof. zoology U. Tex., Port Aransas, 1965—. Project leader Antarctic Biol. Lab. Support Program, U. S. Antarctic Research Program, 1961-65; sci. adviser San Francisco Aquatic Resources Com., 1957-64; adviser-cons. NSF Antarctic Research, 1961-65; cons. Antarctic N. Star Research and Devel. Corp., Mpls., 1965—. Fellow A.A.A.S., Arctic Inst. N. Am., Am. Inst. Fisheries Research Biologists; mem. Ecol. Soc. Am. (past asso. editor, chmn. Western sect. 1964-65), Am. Soc. Limnology and Oceanography, Am. Soc. Zoologists, Am. Inst. Biol. Scis., Biometric Soc., Am. Fisheries Soc. Research, publs. on ecology and energetics of fish populations, fish population dynamics, metabolic characteristics of polar fishes, metabolism, growth and energetics of Gulf fishes. Home: Box 607, Port Aransas, Tex. 78373.*

WOHNUS, John Frederick, Am. biologist; b. Chester, Pa., July 15, 1913; s. William Henry and Cora Charlotte (Haupt) W.; B.A., Williams Coll., 1935, M.A., 1937; Ph.D., U. Cal. at Los Angeles, 1941; m. Margaret Page, July 28, 1939; 1 son, William Henry. Faculty, Williams Coll., 1935-37, U. Cal. at Los Angeles, 1937-40; fellow Scripps Instn. Oceanography, U. Cal. at La Jolla, 1941-43; faculty dept. biology Bennington (Vt.) Coll., 1946—. Mem. A.A.-A.S., Am. Soc. Zoologists, Sigma Xi. Research, publs. in fields of embryology, cytology, chromosomes of amphibians and mammals. Home: Murphy Rd., Bennington, Vt. 05201.

WOJEWSKI, Alfons, Polish physician; b. Goscicino, Poland, May 19, 1912; s. Joseph and Mary (Willa) W.; Dyploma physician U. Poznan (Poland), 1939, D. degree, 1946, 57; m. Hildegard Tutkowski, Oct. 24, 1941; children—Elisabeth, Joanne. Faculty, Pomerania Med. Acad., Szczecin, Poland, 1962—, dean Med. Faculty, prof. urology, 1966—. Decorated Golden Cross of Merits. Mem. Polish, Internat. urol. assns. Author: On the Diagnosis of Prostatic Carcinoma, 1961; Carcinoma Prostate, 1966; also numerous articles. Research on carcinoma of prostate using original methods of diagnosis and therapy. Home: 3 Prusa, Szczecin, Poland.*

WOKER, Gertrude Jan., Swiss biochemist; b. Berne, Switzerland, Dec. 16, 1878; d. Philip and Elisabeth (Müller) Woker; ed. U. Berne; Ph.D.; prof. biophysics and biochemistry, dir. lab. phys. chemistry and biology U. Berne; mem. Chemische und biochemische Gesellschaft. Author: General Part, 1910; Anorganic Catalyse, 1916; Biochemical Catalyse, 1924, 1st part, Hydrolitcal Enzymes, 1931; Methoden zum Studium der Wirkung der einzelnen Verdauungssäfte (handbook of biol. methods), 1928; Der kommende Gift- und Brandkreig (6th and 9th edits. burned in Germany during Hitler era.)

WOLCOTT, Edson Ray, Am. physicist, chemist, metallurgist; b. Sharon, Wis., Apr. 22, 1877; s. John Lawrence and Adela Jane (Callender) W.; B.S., U. Wis., 1900; Ph.D., U. Berlin, 1903; m. Sallie Dingle, Oct. 15, 1907; children—Sallie Jane (Mrs. Hubert D. Eller), Edray (Mrs. Miles R. Minton). Faculty, U. Wis., 1902, Colo. Sch. Mines, 1903-07; operator elec. furnace, Chgo., 1907-13; dir. devel. Cottrell process for metal recovery from smelters, 1913-20; constrn. plant for mfg. anhydrous aluminum chloride Texaco, Port Arthur, Tex., 1920-23; constrn. lightning protection equipment major oil cos. Cal., 1923-28; constrn. oil pump Union Oil Co., Los Angeles, 1928-29; owner, operator lab. for prodn. spl. motion picture equipment, Los Angeles, 1923-66. Fellow Am. Phys. Soc., A.A.A.S.; mem. Am. Chem. Soc., Phi Beta Kappa, Sigma Xi, Sigma Alpha Epsilon. Patentee in field. Home: 917 Crenshaw Blvd., Los Angeles 90017. Died Feb. 24, 1966.

WOLCOTT, Robert Henry, Am. zoologist; b. Alton, Ill., Oct. 11, 1868; s. Robert N. and Agnes (Swain) W.; B.L., U. Mich., 1890, B.S., 1892, M.D., 1893; M.A., U. Neb., 1895; m. Clara Buckstaff, June 2, 1897; children—Robert Allen, Emily Agnes. Asst. in zoölogy U. Neb., 1894-95, instr., 1895-98, adj. prof., 1898-1902, asst. prof., 1902-03, asso. prof., 1903-05, prof. anatomy, 1905-09, head prof. zoölogy, 1909—, acting dean Coll. Medicine, 1909-13, jr. dean, 1913-15. On Mich. Fish Commn. biol. survey of waters of the state, 1893-94; engaged in faunal work in Neb. Fellow A.A.A.S. Contbr. to sci. jours. on ornithology, entomology, fresh water biology, fauna of Neb., and especially on Am. watermites. Died Jan. 23, 1934.

WOLD, Finn, biochemist; b. Stavanger, Norway, Feb. 3, 1928; s. Sverre Marcelius and Herdis (Rasmussen) W.; student Oslo (Norway) U., 1947-50; M.S. in Chemistry, Okla. State U., 1953; Ph.D. in Biochemistry, U. Cal., at Berkeley, 1956; m. Bernadine Moe, June 13, 1953; children—Eric Robert, Marc Sverre. Came to U. S., 1950, naturalized, 1957. Research asso. U. Cal. at Berkeley, 1956-57; faculty U. Ill., Urbana, 1957-67, asso. prof. dept. chemistry and chem. engring., 1962-67; prof. biochemistry U. Minn. Med. Sch., Mpls., 1967—. Fulbright Travel grantee, 1950; Guggenheim fellow, 1960-61; recipient Lalor Research award, 1958. Mem. Am. Soc. Biol. Chemists, Biochem. Soc. (London, Eng.), Am. Chem. Soc., A.A.A.S., Sigma Xi. Research, publs. on chemistry and biochemistry biol. important phosphate esters, phys., chem. and biochem. studies on proteins. 2161 Folwell St., St. Paul 55108. Office: U. Minn. Med. Sch., Mpls. 55455.*

WOLF, A(rnold) V(eryl), Am. physiologist; b. N.Y.C., Dec. 3, 1916; s. Paul and Anna E. (Greenfield) W.; B.S., City Coll. N.Y., 1938; Ph.D. in Physiology, U. Rochester, 1942; m. Orietta L. Behrens, May 18, 1944; children—Paula D., Robyn L., Rodney A. Asst. physiology expdn. in desert of S.W. U. S., U. Rochester, 1942; from instr. to asso. prof. physiology Albany (N.Y.) Med. Coll., 1942-52; chief renal sect. dept. cardiorespiratory diseases Walter Reed Army Inst. Research, 1952-58; prof. physiology, head dept. U. Ill. Coll. Medicine. 1958—. Mem. adv. com. gen. medicine, Office Surgeon Gen., cons. Am. Optical Co. Mem. Am. Physiol. Soc., A.A.A.S., Am. Soc. Artificial Internal Organs, Soc. Exptl. Biology and Medicine, Sigma Xi. Author: The Urinary Function of the Kidney, 1950; Thirst: Physiology of the Urge to Drink and Problems of Water Lack, 1958; (with N. A. Crowder) An Introduction to Body Fluid Metabolism, 1964; Aqueous Solutions and Body Fluids: Their Concentrative Properties and Conversion Tables, 1966; also papers in field water metabolism. Research on kidney function, thirst, problems of castaway at sea, theory of sea water drinking in man and animals, concentrative properties of solutions in body fluids; developed TS meter and concentrimeter for refractometric determination of water and solute in aqueous solutions and body fluids. Home: 2148 Magnolia Lane, Highland Park, Ill. 60035. Office: 901 S. Wolcott, Chgo. 60612.*

WOLF, Bernard Saul, Am. radiologist; b. N.Y.C., Sept. 8, 1912; s. Alex and Mildred (Levy) W.; B.S., N.Y. U., 1932, M.D., 1936; m. Isabel Masket, May 4, 1939; children—Ellen, Susan. Med. dir. N.Y. operations office AEC, 1947-49, cons., 1949—; dir. dept. radiology Mt. Sinai Hosp., N.Y.C., 1949—, prof., chmn. dept. radiology Mt. Sinai Sch. Medicine, 1965—; asso. clin. prof. radiology Columbia, 1950-67. Mem. Mayor's Tech. Adv. Com. on Radiation, 1960—. Recipient Samuel Morse medal for Physics, N.Y. U.; Luigi Celano medal for pathology N.Y. U. Med. Coll. Fellow Am. Coll. Radiology; mem. Am. Roentgen Ray Soc., A.M.A., Am. Gastroent. Soc. (editorial bd.), Am. Coll. Neuroradiologists. Research, numerous publs. on diagnostic radiology, radium dosimetry, radiation protections. Home: 1215 Fifth Av., N.Y.C. 10029. Office: 11 E. 100th St., N.Y.C. 10029.*

WOLF, Charles, Jr., Am. economist; b. N.Y.C., Aug. 1, 1924; s. Charles and Rosalie (Zeamans) W.; S.B., Harvard, 1943, M.P.A., 1948, Ph.D. magna cum laude, 1949; m. Theresa van de Wint, Mar. 1, 1947; children—Charles Theodore, Timothy van de Wint. With Tech. Cooperation Adminstrn., Dept. State, 1949-53; faculty Cornell U., 1953-54, U. Cal. at Berkeley, 1954-55; sr. economist RAND Corp., Santa Monica, Cal., 1955-67, head econs. dept., 1967—. Lectr. econs. U. Cal. at Los Angeles, 1960—; cons. Asian Productivity Orgn., 1963—; mem. S.E. Asia devel. adv. group AID, 1966—. Mem. Am. Econ. Assn., Econometric Soc., Am. Polit. Sci. Assn., Inst. Strategic Studies (London), Assn. for Asian Studies, Council on Fgn. Relations, Phi Beta Kappa. Author: Foreign Aid: Theory and Practice in Southern Asia, 1960; (with R. Gangadharan and Kee Chung Han) Industrial Productivity and Economic Growth, 1964; United States Policy and the Third World: Problems and Analysis, 1967. Research and publs. on quantitative methods in interdisciplinary scholarship and policy research in social scis. dealing with econ. devel., mil. problems and fgn. aid. Home: 440 Skyewiay, Los Angeles 90049. Office: RAND Corp., 1700 Main St., Santa Monica, Cal. 90406.*

WOLF, Charles Joseph Étienne, French astronomer; b. Borges, France, Nov. 9, 1827; ed. École normale supérieure; D.Sc., 1856; prof. physics, lycée of Metz, Faculty Scis. Montpellier (France); astronomer Paris Obs.; tchr. phys. astronomy Sorbonne; mem. French Acad. Scis., 1883. Author: Hypothèses cosmogoniques, 1886; Astronomie et géodésie, 1891; Histoire de l'Observatoire de Paris, 1902. Discovered spectroscopically (with Georges Rayet) the 3 Wolf-Rayet stars, 1867. Died St.-Servan, July 4, 1918.

WOLF, Emil, physicist; b. Prague, Czechoslovakia, July 30, 1922; B.Sc., U. Bristol (Eng.), 1945, Ph.D., 1948; D.Sc., U. Edinburgh (Scotland), 1955. Research asst. optics Cambridge (Eng.) U., 1948-51; research asst., lectr. math. and physics U. Edinburgh, 1951-54; research fellow theoretical physics U. Manchester (Eng.), 1954-59; asso. prof. optics U. Rochester (N.Y.), 1959-61, prof. physics, 1961—; Guggenheim fellow, vis. prof. U. Cal. at Berkeley, 1966-67. Fellow Optical Soc., Am. Phys. Soc., Brit. Inst. Physics and Phys. Soc. Author: (with M. Born) Principles of Optics; Progress in Optics; also numerous articles. Research in theory of partial coherence, optical physics, electromagnetic theory. Address: Dept. Physics and Astronomy, Univ. Rochester, Rochester, N.Y. 14627.

WOLF, Eric Robert, anthropologist; b. Vienna, Austria, Feb. 1, 1923; s. Arthur George and Maria (Ossinovski) W.; came to U. S., 1940, naturalized, 1943; B.A., Queens Coll., 1946; Ph.D. (Viking Fund fellow), Columbia U., 1951; m. Kathleen Louise Bakeman, Sept. 24, 1943; children—John David, Daniel Jacob. Asst. prof. U. Ill., 1952-55, U. Va., 1955-58, Yale, 1958-59; asso. prof. U. Chgo., 1959-61; prof. anthropology U. Mich., Ann Arbor, 1961—. Guggenheim fellow, 1960-61. Mem. Am. Anthrop. Assn., Am. Ethnol. Soc., Royal Anthrop. Inst., Soc. for Comparative Studies in Soc. and History (pres.). Author: Sons of the Shaking Earth, 1958; Anthropology, 1964; Peasants, 1966. Research and publs. on social orgn. among peasant populations and their bearing on integration of these populations into larger socs. Home: 1102 S. Forest Av., Ann Arbor, Mich. 48104.*

WOLF, Frank Louis, Am. mathematician; b. St. Louis, Apr. 18, 1924; s. Louis and Helen (Below) W.; B.S. in Chem. Engring., Washington U., St. Louis, 1944, M.A., 1948; Ph.D., U. Minn., 1955; m. Joyce E. Gifford, Aug. 16, 1947; children—Joan, Allison, Barbara, Jonathan. Faculty, St. Cloud (Minn.) State Coll., 1949-51; faculty Carleton Coll., Northfield, Minn., 1952—, professor of math., 1967—. Mem. Am. Math. Soc., Math. Assn. Am., Am. Assn. U. Profs., Nat. Council Tchrs. Math., Sigma Xi, Pi Mu Epsilon. Author: Elements of Probability and Statistics, 1962. Home: 12 Bundy Ct., Northfield, Minn. 55057.*

WOLF, Frantisek, mathematician; b. Prostejov, Czechoslovakia, Nov. 30, 1904; s. Karel and Josefa (Kytka) W.; Ph.D., Marsaryk U., Brno, Czechoslovakia, 1928; postgrad. U. Cambridge (Eng.), U. Stockholm (Sweden); m. Myrtle O. Richey, Jan. 13, 1945; 1 son, Thomas Francis. Came to U. S., 1941, naturalized, 1945. Docent, U. Prague (Czechoslovakia), 1937-39; instr. Macalester Coll., 1941-42; faculty U. Cal. at Berkeley, 1942—, prof. math., 1952—; traveling lectr., Europe, 1955-56, 66; vis. prof. Technische Hochschule, Aachen, Germany, 1963. Fulbright grantee, 1959-60; Miller Research grantee, 1963-64. Mem. Am. Math. Soc., Am. Math. Assn., Math. Council in Cal., Am. Czechoslovak Soc. for Art and Scis. Research, numerous publs. on differential operator, generalized spectral decomposition, areas supporting math. research in quantum physics and engring. Home: 181 Stonewall Rd., Berkeley, Cal. 94705.*

WOLF, Frederick Taylor, Am. botanist; b. Auburn, Ala., July 11, 1915; s. Frederick A. and Wynnette (Taylor) W.; student U. N.C., 1931-33; A.B., Harvard, 1935; M.A., U. Wis., 1936, Ph.D., 1938; m. Virginia Hudson Martin, June 10, 1945; children—(Virginia) Ann, Catherine Lee. NRC fellow Harvard, 1938-39; faculty Vanderbilt U., Nashville, 1939—, prof. botany, 1956—. Mem. Soc. Am. Bacteriologists (past pres. Ky.-Tenn. br.), Tenn. Acad. Sci. (past pres.), Assn. Southeastern Biologists (past v.p.), Am. Soc. Plant Physiologists (past chmn. So. sect.), Bot. Soc. Am. (past chmn. Southeastern sect.). Research, numerous papers on fungi, plant pigments, photosynthesis. Home: 2503 Kensington Pl., Nashville 37212.*

WOLF, George, biochemist; b. Vienna, Austria, June 16, 1922; s. Hugh H. and Greta (Schoenberger) W.; B.Sc., London (Eng.) U., 1944; D.Phil., Oxford (Eng.) U., 1947; m. Patricia A. Nicol, Nov. 3, 1948; children—Roger D., Camilla M., Lucas E. Came to U. S., 1948, naturalized, 1955. Postdoctoral fellow Harvard, 1948-50, U. Wis., 1950-51; asst. prof. U. Ill., 1951-55, asso. prof., 1955-62; asso. prof. Mass. Inst. Tech., 1962-—. Guggenheim fellow, 1958-59. Mem. Am. Assn. U. Profs., Am. Soc. Biol. Chemists, Sigma Xi. Author: Chemical Induction of Cancer, 1953; Isotopes in Biology, 1964; also articles. Editor: Recent research on Carnitine, 1964. Research on metabolism of amino acids and vitamins, their function in mammalian organism. Home: 38 Peacock Farm Rd., Lexington, Mass. 02173. Office: 56-235 Mass. Inst. Tech., Cambridge, Mass. 02139.*

WOLF, Helmut Walter, German pediatrician; b. Freiburg im Breisgau, Oct. 13, 1925; s. Walter Karl and Else (Schoeck) W.; M.D., U. Marburg/Lahn, 1952; m. Ingeburg Schneider, June 7, 1952; children—Werner, Otfried, Dietlind, Norgard. With children's hosp., U. Göttingen, 1954-—, head outpatients dept., head physician, faculty, 1960-—, unscheduled prof., 1966-—. Mem. Biochem. Soc., German Soc. Pediatrics (A. Cherny prize 1964. Co-author book, 1961. Publs. on vitamin B, nephrology, differential diagnosis of internal diseases. Address: 2 Schlegelweg, Göttingen, West Germany.*

WOLF, Hermann Gottfried, Austrian pediatrician; b. Vienna, Jan. 3, 1926; s. Hermann and Viktoria (Bruner) W.; M.D., U. Vienna, 1951; m. Gudrun Carl, Aug. 1, 1958; children—Hermann, Georg, Peter. Sr. physician Children's Clinic, U. Vienna. 1951-65; chmn. Mautner Markhofsches Children's Hosp., Vienna, 1965-—; prof. 1967-—. Mem. European Soc. Pediatric Radiology. Author: Rontgendiagnostik beim Neugeborenen und Saugling, 1959. Research on pediatric radiology, pediatric hematology. Home: 67 Cottagegasse. Office: 75 Baumgasse, Vienna, Austria.*

WOLF, Irvin Simon, Am. psychologist; b. Wabash, Ind., Dec. 23, 1914; s. Eugene and Rae (Simon) W.; A.B., Manchester Coll., 1937; M.A., Ind. U., 1939, Ph.D., 1948; m. Jeannette Hawk, Aug. 28, 1942; children—Thomas Gene, Carol Sue. Instr. psychology U. Ind., 1946-47; asst. prof. psychology U. Buffalo, 1947-48; asso. prof. psychology U. N.C., 1948-54; prof. psychology Denison U., 1954-—, chmn., 1954-61, 64-—, mng. editor Jour. Sci. Labs., 1964-—, editor Psychol. Record, 1959-—. Mem. Am., Ohio, Midwestern psychol. assns., Am. Assn. U. Profs., Sigma Xi. Contbr. articles to profl. jours.; jour. editor; book reviewer. Home: Mt. Parnassus, Granville, Ohio 43023.*

WOLF, Leonard Nichols, Am. biologist; b. McKeesport, Pa., Jan. 3, 1911; s. Leonard and Elizabeth (Kroneberger) W.; B.S. in Biology, St. Vincent Coll., 1933; M.S., U. Pitts., 1934, Ph.D. in Biology, 1939; m. Anne Marie Breslin, May 1, 1943; children—Gregory T., Leonarda A., Elizabeth J., Maria I., Andrea T., John P., Anitra C. Instr., St. Thomas Coll., Scranton, Pa., 1937-39; faculty U. Scranton, 1939-—, prof. biology, chmn. dept., 1941-—. Chmn., Commn. on Acad. Facilities, Commonwealth of Pa.; mem. Pa. Bd. Edn., Council Higher Edn. Mem. A.A.A.S., Am. Inst. Biol. Scis., Am. Soc. Zoologists, Nat. Assn. Biology Tchrs., N.Y. Acad. Scis., Pa. Acad. Sci., Sigma Xi, Phi Sigma, Alpha Sigma Nu. Contbr. articles to sci. jours. Research on effects of sex hormones in mammals. Home: 1132 Grandview St., Scranton, Pa. 18509.*

WOLF, Maxmilian Franz Joseph Cornelius, German astronomer; b. Heidelberg, Germany, June 21, 1863; ed. Heidelberg, Stockholm; Ph.D., 1888; erected small obs.; dir. Königstuhl astrophys. obs., from 1896; prof. geophysics U. Heidelberg, 1901-32. Author: Über die Bestimmung der Lage des Zodiakallichts und der Gegenschein, 1900; Die Entdeckung und Katalogisierung von kleinern Nebelflecken durch die Photographie, 1901; Stereoskopische Bestimmung der Eigenbewegung von Fixsternen, 1906; Die Milchstrasse, 1908. Inventor photog. method of discovering asteroids; discovered 1st photog. asteroid (No. 323) as well as 100 more; discovered numerous new nebulae, including N. Am. Nebula; studied Milky Way; confirmed (with Barnard) presence of dark nebulae; discovered many planetoids; discovered periodical comet (which bears his name) with course of 7.7 years, 1883; made studies in spectrum analysis; 1st to use stereo-comparator for scanning celestial photographs. Died Heidelberg, Oct. 3, 1932.

WOLF, Paul Leon, Am. physician; b. Detroit, Oct. 4, 1928; s. Meyer and Molly (Kammer) W.; B.S., Wayne State U., 1948; M.D., U. Mich., 1952; m. Florence Freedman, Dec. 21, 1952; children—Cheryl, Elliot, Robin. Faculty, Wayne State U., Detroit, 1956-—, prof. pathology, 1965-—, chief clin. pathology, 1960-—, dir. med. tech., 1964-—; clin. pathologist Detroit Gen. Hosp., 1960-—. Mem. A.M.A., Am. Soc. Clin. Pathologists, Coll. Am. Pathologists, Am. Assn. Pathologists and Bacteriologists, Detroit Physiol. Soc., Detroit Surg. Soc., Mich. Path. Soc., Mich. State Med. Soc., Wayne County Med. Soc., Reticuloendothelial Soc., Sigma Xi. Research, numerous publs. on histochemistry, cancer immunology. Home: 12750 Burton St., Oak Park, Mich. 48237. Office: 1400 Chrysler Expressway, Detroit 48207.*

WOLF, Rudolf, Swiss astronomer; b. Fällanden, nr. Zurich, Switzerland, July 7, 1816; dir. obs., Bern, Switzerland, 1847; prof. École polytechnic, also U. Zurich; dir. Zurich Obs., from 1855. Mem. French Acad. Scis. Author: Taschenbuch für Mathematik, Physik, Geodäsie und Astronomie, 1852; Geschichte der Astronomie, 1877; Handbuch der Astronomie, 2 vols., 1890-93. Discovered relationship between sunspots and terrestrial magnetic variations, 1852; believed aurorae also related to sunspot cycle. Died Zurich, Dec. 6, 1893.

WOLF, Stewart George, Jr., Am. physician; b. Balt., Jan. 12, 1914; s. Stewart George and Angeline (Griffing) W.; student Phillips Acad., 1927-31, Yale, 1931-33; A.B., Johns Hopkins, 1934, M.D., 1938; m. Virginia Danforth, Aug. 1, 1942; children—Stewart George III, Angeline Griffing, Thomas Danforth. Intern N.Y. Hosp., 1938-39, resident medicine, 1939-42, NRC fellow, 1941-42; research fellow Bellevue Hosp., 1939-42, clin. asso. vis. neuropsychiatrist, 1946-52; research head injury and motion sickness Harvard neurol. unit Boston City Hosp., 1942-43; asst., then asso. prof. medicine Cornell U., 1946-52; prof., head dept. medicine U. Okla., 1952-—, supr. clin. activities Okla. Med. Research Found., 1953-55, now head psychosomatic and neuromuscular sect.; cons. internal medicine VA Hosp., Oklahoma City; consultant European Office (Paris), Office International Research, Nat. Insts. Health, 1963-64. Mem. edn. and supply panel Nat. Adv. Commn. on Health Manpower, 1966-—; mem. U. S. Pharamacopeia Scope Panel on Gastroeuterology. Regent Nat. Library Medicine, mem. bd. Internat. Cardiology Found. Served as major M.C., AUS, World War II. Recipient award Am. Gastroenterological Assn., 1943, Hofheimer prize for research, 1952, Award of Merit, American Heart Assn., 1965. Mem. Am. Soc. Clin. Investigation, Am. Clin. and Climatological Assn., Assn. Am. Physicians, Am. Psychosomatic Society (pres. 1961-62), A.M.A. (Council mental health 1960-—), National Advisory Heart Council, Am. Heart Assn. (chmn. com. on internat. program), Oklahoma City Symphony Soc. (pres. 1956-61). Club: Cosmos (Washington). Author: Human Gastric Function, 1943; The Stomach, 1965; others. Studies on gastric fistula; quantitative approach to sensory and psychosomatic phenomena; documentation physical effects of placebos; discovered (with Caputto, Trucco and Tang) proteolytic enzyme, gastricsin; relation of psychosocial forces to myocardial infarction, sudden death. Home 644 N.E. 14th St. Office: 800 N.E. 13th St. Oklahoma City. 73104.*

WOLF, Walter, chemist; b. Frankfort am Main, Germany, May 25, 1931; s. Willy and Thea (Schmidt) W.; B.S., U. Montevideo (Uruguay), 1949, M.S., 1952; Ph.D., U. Paris (France), 1956; m. Gladys D. Wohl, July 6, 1955; children—Jeanette Ingrid, David Eric. Came to U. S., 1958, naturalized, 1965. Attache, Centre Nat. de la Recherche Scientifique, 1955-56; asso. prof. organic, biol. chemistry U. Concepcion, Chile, 1956-58; travelling fellow McGill U., 1958; research asso. Amherst Coll., 1958-59; research asso. U. So. Cal., Los Angeles, 1959-62, faculty, 1959-—, asso. prof. pharm. chemistry, 1965-—; cons. Atomics Internat., N.Am. Aviation Co. First ann. lectr. U. So. Cal. Chemistry Alumni Assn. Mem. Am., Swiss chem. socs., Societe de Chimie Biologique (France), Sigma Xi. Research, publs. on iodine compounds, particularly thyroid hormones and heterocyclic polyvalent iodine compounds, ESR radiobiology. Home: 10239 Babbitt Av., Northridge, Cal. 91324. Office: Sch. Pharmacy, U. So. Cal., Los Angeles 90007.*

WOLFE, Bertram, Am. physicist; b. N.Y.C., June 26, 1927; s. Paul and Sally (Greenberg) W.; A.B. in Physics, Princeton, 1950; Ph.D. in Nuclear Physics, Cornell U., 1954; m. Leila Ann Katz, Dec. 23, 1950; children—Sarah Elizabeth, Donald Stewart. Physicist, Eastman Kodak Co., Rochester, N.Y., 1954-55; with nuclear energy div. Gen. Electric Co., San Jose, Cal., 1955-—, mgr. plant engring. and projects sect. advanced products operation, 1965-—. NSF fellow, 1953. Mem. Am. Phys. Soc., Am. Nuclear Soc. (past chmn. No. Cal. sec.. 1965), Phi Beta Kappa, Sigma Xi, Phi Kappa Phi. Research, publs. on nuclear reactor physics and tech.; designer nuclear reactors. Home: 2416 Whitethorne Dr., San Jose, Cal. 95128. Office: 310 De Guigne Dr., Sunnyvale, Cal. 94086.*

WOLFE, Gustav, biologist, psychologist; b. Karlsruhe, Germany, Mar. 18, 1865; prof. psychiatry, Basel, Switzerland. Author: Der gegenwärtige Stand des Darwinismus, 1896; Mechanismus und Vitalismus, 1902; Die Bergründung der Abstammungslehre, 1907; Leben und Erkennen, 1933. Proponent of nevoitalism as means of fighting materialism; opposed Darwin's theory of natural selection; believed in purposeful behavior of life, which also finds expression in phylogenesis. Died 1941.

WOLFE, Hugh Campbell, Am. physicist; b. Parkville, Mo., Dec. 18, 1905; s. Arthur Lester and Gertrude Remington (Snow) W.; A.B., Park Coll., 1926, Sc.D. (honorary), 1962; M.S., University of Michigan, 1927, Ph.D., 1929; m. to Ithmer M. Coffman, Sept. 3, 1929; children—Arthur, Elizabeth, James, Helen. Nat. Research fellow Calif. Tech. and U. of Calif., 1929-31; fellow Lorentz Found., Utrecht, Holland, and Heckscher research asst. Cornell U., 1931-32; instr. Ohio State, 1932-33; instr. City Coll. of N.Y., 1934-42, asst. prof. physics, 1942-48, asso. prof. physics, 1949; prof., head dept. physics The Cooper Union Sch. of Engring., N.Y.C., 1949-60; dir. publs. American Institute of Physics, 1960-—; dir. and chmn. bd.'s management com. Eastern Coops., Inc., 1949-51, dir., 1952-59; pres. Mid-Eastern Coops., Inc., 1952-59. Tech. aide O.S.R.D. and Nat. Defense Research Com., 1944-46; exec. com. Nat. Com. for Sane Nuclear Policy, 1958-62; chmn. interdivisional com. on symbols units and terminology Nat. Acad. Scis.-NRC, 1962-—. Recipient Naval Ordnance Award, 1945. Fellow Am. Physical Soc.; mem. Internat. Union Pure and Applied Physics (pres. commn. on publ. 1966-—), Fedn. Am. Scientists (chmn. 1949, v. chmn. 1950), Am. Assn. Physics Tchrs. Research in quantum theory applications; computation of gun sights; information systems devel. Home: 30 Lawrence Pkwy., Tenafly, N.J. 07670. Office: 335 E. 45th St., N.Y.C. 10017.

WOLFENSBERGER, Wolf, psychologist; b. Mannheim, Germany, July 26, 1934; s. Friedrich and Helen (Loewit) W.; B.S., Siena Coll., 1955; M.A., St. Louis U., 1957; Ph.D., George Peabody Coll., 1962; m. Nancy Evelyn Artz, Feb. 13, 1960; children—Margaret Elisabeth, Joan Angela. Came to U. S., 1950, naturalized, 1956. Dir. human devel. Greene Valley Hosp. and Sch., Greeneville, Tenn., 1960-62; NIH Research fellow Maudsley Hosp., London, Eng., 1962-63; dir. research Plymouth State Home and Tng. Sch., Northville, Mich., 1963-64; mental retardation research scientist U. Neb. Coll. Medicine, Omaha, 1964-—. Mem. Neb. Gov.'s Com. on Mental Retardation, 1965. Fellow Am. Assn. on Mental Deficiency; mem. A.A.A.S., Am. Psychol. Assn., Am. Montessori Soc., Nat. Assn. for Retarded Children, Council for Exceptional Children. Research, publs. on clin. and theoretical aspects mental retardation and mgmt. mentally ill. Home: 2327 S. 103d St., Omaha 68124.*

WOLFENSTEIN, Lincoln, Am. physicist; b. Cleve., Feb. 10, 1923; s. Leo and Anna (Koppel) W.; S.B., U. Chgo., 1943, S.M., 1944, Ph.D., 1949; m. Wilma Caplin, Feb. 3, 1957; children—Frances Anne, Leonard, Miriam. Physicist, NACA, 1944-46; faculty Carnegie Inst. Tech., Pitts., 1948-—, prof. physics, 1960-—; fellow European Center for Nuclear Research, 1957-58, 64-65. Fellow Am. Phys. Soc. Research and publs. on theoretical analysis of elementary particle reactions, in particular, the use of polarized proton beams and weak interaction processes. Home: 5853 Marlborough Av., Pitts. 15217.*

WOLFF, Christian, Baron von, mathematician, philosopher, physicist, psychologist; b. Breslau (now Wroclaw, Poland), Jan. 24, 1679; s. a tanner; ed. in math. and natural philosophy U. Jena, Germany. Became privatdocent, U. Leipzig, Germany, 1703; aptd. (on recommendation of Leibniz) Prof. math., U. Halle, Prussia, 1706; banished for his sinophilism in theological disputes with Pietists, 1723; prof. math. and philosophy, U. Marburg, Hesse, 1723-40; sci. adviser to Peter the Great, 1716-25; helped found St. Petersburg Acad. Scis., through his recommended programs and members, 1723-25; refused vice-presidency, 1723; recalled to U. Halle by Frederick II, 1740; prof. U. Halle, 1741-54; Chancellor of U. Halle, 1743-54; refused Frederick II's offer of pres. of Berlin Acad. Scis., 1740. Mem., French, Berlin, and St. Petersburg Acads. Sci. Author: De philosophia practica universali, 1703; Aerometriae elementa, 1709; Vernünfftige Gedanken von Gott, der Welt und der Seele der Menschen, auch allen Dingen überhaupt, 1719; Vernünfftige Gedanken von der Würkung der Natur, 1723; Oratio der Sinarum philosophia practica, 1726; Discursus praeliminaris de philosophia in genere, 1728; Philosophia rationalis, 1728; Cosmologica generalis, 1731; Psychologica empirica, 1732; Psychologica rationalis, 1734; Theologica naturalis, 1736; Philosophia practica, 1738-39; Jus naturae methodo scientifica pertractatum, 8 vols., 1740-48; Philosophia moralis sive ethica, 5 vols., 1750-53. Collaborator Acta Eruditorum, Leipzig; Commentarii, St. Petersburg. Leader in German enlightenment; supplemented his thought with that of Descartes and Leibniz to develop Leibnizo-Wolffian philosophical system (Germany's 1st national philosophy); by 1740's this system displaced P. Melanchthon's peripatetic philosophy of 16th century in German speaking Europe; Wolff's philosophy provided rational dogmatism, strict math. method, and vague idealism; it centered around Leibnizian monad theory, and thus the elastic monad replaced hard Greek atom as basic building block of universe; methodology was basically rationalistic, although he advocated use of observation and experiment for control and completion of deduction. Wolffian philosophy was main opponent of Newtonian thought in mid and late 18th century, as it held dominant position in St. Petersburg and Berlin Acads. Sci.; Wolff underestimated Newton and abhorred much of his sci. thought; Wolff introduced several textbooks written in German vernacular; supported Leibnizian doctrine of conservation of vis viva; tried to make it a basis for different scis., including psychology; supported concept of energetics; supported S. Koenig in principle of least action priority dispute against Maupertuis and Euler in Berlin, 1750-52; friend of the Bernoullis; taught Lomonosov, 1736-39; opposed by Euler, particularly for monad theory; worked on math. series; studied meteorology; described anemometer with vanes and propellers which indicated wind strength through deflection of lever balance; supported Malpighi's views on necessity of air for plant life; carried out observations and wrote on plant physiology. Died Halle, Apr. 9, 1754.

WOLFF, Etienne, French embryologist; b. Auxerre, France, Feb. 12, 1904; s. Armand and Gabrielle (Cahen) W.; Ph.D., U. Strasbourg, 1936; m. Emilienne Hennig, July 13, 1927. Asst., Biol. Lab. of Wimereux, U. Paris, 1927-28; faculty U. Strasbourg, 1933-55, prof. zoology and embryology, 1942-55; prof. Collège de France, 1955—, pres. adminstr., 1965—. Dir., Inst. Embryology and Teratology, Centre Nat. de la Recherche Sci., 1946—; dir. Lab. of Embryology, Ecole Pratique des Hautes Etudes, 1946—. Decorated officer Légion d'Honneur; commandeur L'Order du Mérite; Commander des Palmes Académiques; prize l'Institut de France. Mem. Acad. Scis., Académie Royale des Sciences de Suède, Académie Royale de Belgique. Author: La Science des Monstres, 1948; Les Changements de Sexe, 1946; Les Chemins de la Vie, 1964; also numerous articles. Research on exptl. embryology and teratology, intersexuality, in vitro culture of embryonic organs, in vitro culture of human organized cancers. Address: 49 bis Avenue Belle Gabrielle, Nogent, France.*

WOLFF, Frederick William, clin. pharmacologist; b. Berlin, Germany, Aug. 21, 1920; s. Bruno and Ella (Landau) W.; M.B., B.S., Durham U., King's Coll., Newcastle upon Tyne, Eng., 1946, M.D., 1957. Practice medicine, specializing in clin. pharmacology, Balt., 1959-63, N.Y.C., 1963-65, Washington, 1965—; faculty, physician Johns Hopkins Hosp., 1959-63; asso. prof. medicine, pharmacology N.Y. Med. Coll., 1963-65; prof. medicine Sch. Medicine George Washington U., 1965—; dir. research Washington Hosp. Center, 1965—. Recipient Research Career Scientist award Health Research Council N.Y., 1963. Fellow Am. Pharm. Mfrs. Assn.; mem. Am. Fedn. Clin. Research, Am. Diabetes Assn., Brit. Pharmacol. Soc., Royal Soc. Medicine, N.Y. Acad. Scis. Author: (with W. A. Broom and F. G. Young) Mode of Action of Insulin, 1959. Research, numerous publs. on mode of action of oral hypoglycemic agts., metabolic effects of diuretic agts., utilization of antidiuretic hyperglycemic agt. diazoxide in human hypoglycemia, effects of treatment in hypertension. Home: 7710 Maple Av., Takoma Park, Md. 20012. Office: G. Hyman Research Bldg., Washington Hosp. Center, 110 Irving St. N.W., Washington 20010.*

WOLFF, George Louis, biologist; b. Hamburg, Germany, Aug. 24, 1928; s. Adolf and Eva (Nathan) W.; came to U. S., 1940, naturalized, 1946; B.S. cum laude, Ohio State U., 1950; Ph.D., U. Chgo., 1954; m. Eleanor Herstein, Aug. 30, 1953; children—David Bernard, Adrienne Ann. USPHS fellow Nat. Cancer Inst., Bethesda, Md., 1954-56, biologist, 1956-58; research asso. Inst. for Cancer Research, Phila., 1958-63, asst. mem., 1963—, supr. animal colony, 1958—. Mem. A.A.A.S., Am. Assn. for Cancer Research, Genetics Soc. Am., Animal Care Panel, Soc. for Exptl. Biology and Medicine, Sigma Xi. Contbr. articles to tech. jours. Research, publs. on genetic control of growth and metabolism as it affects cancer devel. Office: 7701 Burholm Av., Phila. 19111.*

WOLFF, Irving, Am. physicist; b. N.Y. City, July 6, 1894; s. Max and Julie (Gutman) W.; B.S., Dartmouth, 1916; Ph.D., Cornell, 1923; Hecksher research fellow, 1923-24; m. Consuelo Hope Hughes, July 30, 1941; 1 dau., Margaret Elizabeth. Instr. physics Ia. State Coll., 1919-20, Cornell, 1920-23; research in electronics, Radio Corp. of Am., N.Y. City, Camden and Princeton, N.J., 1924-59, dir. radio tube research lab. RCA Labs., Princeton, 1945-52, dir. research, 1952-54, v.p. research, 1954-59, retired, now a consultant. Member of the advisory com. dept. physics Princeton University. Recipient Modern Pioneer award Nat. Mfrs. Assn., 1937; Distinguished Public Service award, U. S. Navy, for early radar work, 1948; Cresson medal Franklin Institute, 1959. Fellow A.A.A.S., Institute Radio Engrs. (chairman of Philadelphia section 1935, dir. 1952-53; chmn. awards com. 1955), Acoustical Soc. Am.; Phys. Soc. Am., Fedn. Am. Scientists, Sigma Xi. Research in electronics, radar, directional antennas, air navigation, microwaves, sound measurement and sound equipment, polarization capacity, measurements of alternating current bridge. Address: 111 Red Hill Rd., Princeton, N.J.

WOLFF, Kaspar Friedrich, physiologist; b. Berlin, Germany, Jan. 18, 1733; pupil of J. F. Meckel, Berlin; M.D., Halle, Germany, 1759; prof. anatomy and physiology Acad., St. Petersburg, Russia, 1764-94. Author: Theoria generationis, 1759; Über die Bildung des Darmkanals in bebrüteten Hühnchen, 1812. Founder of modern embryology; gave 1st clear description of mesonephros (Wolffian body), 1759; described twisting tubules (Wolffian tubules) which serve as functional units of mesonephros, 1759; substituted layer theory for embryonic preformation theory, thus laying found. for germ-layer theory of Heinrich C. Pander and Carl Ernst von Baer, 1759; described lateral thickening of urogenital ridge (Wolffian ridge) from which mesonephros rises, circa 1765; eponym of Wolffian cyst; proved his theory of epigenesis by demonstrating that rolling of a blastoderm layer forms the chick's intestine, 1768. Died St. Petersburg, Feb. 22, 1794.

WOLFF, Kurt H(einrich), sociologist; b. Darmstadt, Germany, May 20, 1912; s. Oscar Louis and Ida (Kohn) W.; student U. Frankfurt (Germany), 1930-33, U. Munich (Germany), 1931-32; laurea in filosofia U. Florence (Italy), 1945; postgrad. U. Chgo., Harvard; m. Carla Elisabeth Bruck, June 11, 1936; 1 son, Carlo T. Came to U. S., 1939, naturalized, 1945. Tchr., Schule am Mittelmeer, Recco, Genoa, Italy, 1934-36, Istituti Mare-Monte, Ruta, Genoa, 1936-38; research asst. So. Meth. U., 1939-43; asst. prof. Earlham Coll., 1944-45; faculty Ohio State U., 1945-59, asso. prof., 1952-59; prof., chmn. dept. sociology Brandeis U., Waltham, Mass., 1959-62, prof., 1962—. Fulbright Sr. lectr. U. Rome (Italy), 1963-64; guest prof. sociology U. Freiburg, 1966; U. Frankfurt, 1966-67, University of Paris (Nanterre), 1967; research asso. personnel, research bd. Ohio State U., 1951-52, 58-59; U. S. specialist Dept. State, 1952-53. Wenner-Gren Found. for Anthropol. Research grantee, 1947, 49; Social Sci. Research Council Post-doctoral Research Tng. fellow, 1943-44; Fund for Advancement Edn. Faculty fellow, 1955-56. Mem. Am. Sociol. Assn., Eastern Sociol. Soc. Editor, translator: The Sociology of Georg Simmel, 1950; editor Emile Durkheim, 1959, Georg Simmel, 1960; Wissenssoziologie (Karl Mannheim), 1965. Research, publs. on analysis relations between society and intellectual life, analysis of conceptions of these relations, idea surrender and catch in its bearing on social scis., concepts of student's relation to world. Home: 58 Lombard St., Newton, Mass. 02158. Office: Dept. Sociology, Brandeis U., Waltham, Mass. 02154.*

WOLFF, Kurt Julius, physician; b. Mi Friedland, Germany, July 17, 1907; s. Oscar and Flora (Mochol) W.; M.D., Frederick William's U., Berlin, Germany, 1932; m. Maria De Capitani, May 8, 1937. Dir. geriatric treatment and research unit Osawatomie (Kan.) State Hosp., 1957-58; clin. dir. Galesburg (Ill.) State Research Hosp., 1958-59; asso. chief staff, dir. profl. edn. and residency tng. Coatesville (Pa.) VA Hosp., 1959—; asso. psychiatry U. Pa. Med. Sch., Phila., 1959—. Cons. to hosps. Diplomate Pan Am. Med. Assn. Fellow Am. Psychiat. Assn. (mem. com. on aging 1961-64), Am. Geriatric Soc., Gerontological Soc.; mem. A.M.A., Pa. Psychiat. Soc. (mem. com. on aging 1965—), Royal Soc. Health (Gt. Britain), A.A.A.S., Am. Group Psychotherapy Assn., Am. Soc. Group Psychotherapy and Psychodrama. Author: Therapia Vitalistico-Costitutionale, 1936; The Biological Sociological and Psychological Aspects of Aging, 1959; Geriatric Psychiatry, 1963; also numerous articles. Research on limitation and results group psychotherapy and individual psychotherapy with aged, drug studies improving emotional conditions and memory defects elderly patients, importance coordinated approach to problem aging. Home: VA Hosp., Coatesville, Pa. 19320.*

WOLFF, Manfred Ernst, medicinal chemist; b. Berlin, Germany, Feb. 14, 1930; s. Adolph Abraham and Kate (Fraenkel) W.; B.S., U. Cal. at Berkeley, 1951, M.S., 1953, Ph.D., 1955; m. Helen S. Scandalis, Aug. 1, 1953; children—Stephen Andrew, David James, Edward Allen. Research fellow U. Va., 1955-57; sr. medicinal chemist Smith, Kline & French Labs., Phila., 1957-60; faculty U. Cal. at San Francisco, 1960—, prof. medicinal chemistry, 1965—. Mem. Am. Chem. Soc., Am. Pharm. Assn., A.A.A.S. Sigma Xi. Research, patentee, publs. on synthesis steroids, relationship between chem. structure and biol. activity in these compounds, chem. routes for prodn. aldosterone and 19-hydroxyandrostenedione, synthesis cardiac aglycones and glycosides. Office: Dept. Pharm. Chemistry, Sch. Pharmacy, U. Cal., San Francisco, 94122.*

WOLFF, Paul Martin, Am. oceanographer; b. Newcomerstown, O., July 3, 1921; s. Arthur R. and Elva (Martin) W.; A.B., Wittenberg U., 1942; M.S., U. S. Naval Postgrad. Sch., 1949; M.A., U. Chgo., 1952; m. Margaret Corbett, June 27, 1943; children—Jeffrey C., William Alan, Michael Paul. Commd. ensign USN, 1942, advanced through grades to capt., 1963; meteorol. computer research Project Arowa, 1952-55; mem. Joint Numerical Weather Prediction Group, 1957-59; officer in charge Fleet Numerical Weather Facility, Monterey, Cal., 1959—. Mem. Am. Meteorol. Soc., Am. Geophys. Union, Sigma Xi. Contbr. chpts. to Computer Applications, 1960; Environmental Science, 1965; Scientific Computing Symposium, 1966. Research on computer pattern separation at 500 mb, long wave stblzn. model, computer produced surface pressure map, synoptic hemispheric ocean-atmosphere energy exchange computations. Home: 1 Gillespie Lane, Monterey 93940. Office: U. S. Naval Postgrad. Sch., Monterey, Cal. 93940.*

WOLFF, Torben Lunn, Danish zoologist; b. Copenhagen, Denmark, July 21, 1919; s. Jorgen de Lichtenberg and Karen (Lunn) W.; B.A., U. Copenhagen, 1943, M.A., 1945, Ph.D., 1947; D.Sc., Copenhagen U., 1963; m. Lisbeth Christensen, Dec. 5, 1954; children—Vibeke, Marianne. Mem. Danish Atlantide Expdn., tropical W. Africa, 1945-46; faculty Royal Agrl. and Vet. High Sch., 1947-49; preparation Danish Galathea Deep-Sea Expdn. Round the World, 1949-50, souschef, 1950-52; sec. XIV Internat. Congress Zoology, Copenhagen, 1952-53; asst. curator U. Zool. Mus., Copenhagen, 1953-57, curator, 1957-66, chief curator, 1966—; leader last sect. Danish Noona Dan Expdn., Bismarck and Solomon Islands, 1962; mem. Am. Te Vega Expdn., Central Pacific, 1965; lectr. systematic zoology U. Copenhagen, 1961—. Recipient Royal Galathea medal, 1955. Mem. Danish Natural History Soc. (pres. 1963—). Author: The Year in Nature, 1944; The Systematics and Biology of Bathyal and Abyssal Isopoda Asellota, 1962; also numerous sci. papers, articles. Editor: The Natural History of Rennell Island, 1955—; Galathea Report, 1959—; (with Knudsen) Atlantide Report, 1959—. Research on composition and ecology of fauna of deep-sea trenches, systematics and biology of various groups of deep-sea crustaceans, evolution of deep-sea hermit-crabs, technique in investigating great depths, history of Danish marine expdns., fauna of isolated Rennell Island, Brit. Solomon Islands. Home: 12 Hesseltoften, Hellerup, Copenhagen. Office: Zool. Mus. 15, Universitetsparken, Copenhagen Ö, Denmark.*

WOLFGANG, Marvin Eugene, Am. sociologist; b. Millersburg, Pa., Nov. 14, 1924; s. Charles T. and Pauline (Sweigart) W.; student Pa. State U., 1942-43; B.A., Dickinson Coll., 1947; postgrad. U. Oslo (Norway), 1948; M.A., U. Pa., 1950, Ph.D., 1955; m. Lenora Davida Poden, June 1, 1957; children—Karen Eleanor, Nina Victoria. Asst. prof., chmn. dept. sociology Lebanon Valley Coll., 1948-52; faculty dept. sociology U. Pa., Phila., 1952—, prof., grad. chmn., 1964—, dir. Center Criminological Research, 1961—; cons. Inst. Criminal Anthropology U. Rome, 1963—, President's Commn. on Law Enforcement and Adminstrn. Justice, 1965-66. Social Sci. Research Council grantee, 1955; Guggenheim fellow, Fulbright Research fellow, 1957-58; Guggenheim fellow, 1968-69. Member American Sociol. Assn., Am. Soc. Criminology (Research award 1960, pres.), Pa. Prison Soc. (pres.), Phi Beta Kappa. Author: Patterns in Criminal Homicide, 1958; The Sociology of Crime and Delinquency, 1962; The Sociology of Punishment and Correction, 1962; The Measurement of Delinquency, 1964. Research, numerous publs. primarily in field of criminology, theoretical integration of sociology and psychology. Home: 4628 Pine St., Phila. 19143.*

WOLFGANG, Richard Leopold, chemist; b. Frankfurt, Germany, July 24, 1928; s. Julius Jakob and Wilhelmine (Moser) W.; came to U. S., 1945, naturalized, 1951; Ph.B., U. Chgo., 1947, S.B., 1948, Ph.D. in Phys. Radiochemistry, 1951; m. Anne O'Neill-Butler, July 26, 1952 (div. Jan. 1966); children—William John, Michael Butler, Katherine Ellen. Jr. chemist Air Force radiochem. research project U. Chgo., 1949; chemist Brookhaven Nat. Lab., Upton, N.H., 1951-56; asso. prof. chemistry Fla. State U., Tallahassee, 1955-56; faculty Yale, 1956—, prof. chemistry, 1962—. Cons. Westinghouse Electric Corp., 1957-60, Pratt & Whitney Aircraft Corp., 1958-61. Mem. Am. Chem. Soc., Am. Phys. Soc., Phi Beta Kappa, Sigma Xi. Contbr. numerous articles to tech. jours. Developed field reactions high kinetic energy atoms; research on chemistry free carbon atoms, nuclear reactions and characterizations new-radio-isotopes; co-discoverer tritium in natural waters.*

WOLFLE, Dael, Am. psychologist; b. Puyallup, Wash., Mar. 5, 1906; s. David Henry and Elizabeth (Pauly) W.; B.S., U. Wash., 1927, M.A., 1928; Ph.D., Ohio State U., 1931, D.Sc., 1957; D.Sc., Drexel Inst. Tech., 1956, Western Mich. U., 1960; m. Helen Morrill, Dec. 28, 1929; children—Janet Helen (Mrs. William Densmore Chapman), Lee Morrill, John Morrill. Prof. psychology U. Miss., 1932-36; examiner biol. scis. U. Chgo., 1936-39, faculty, 1939-45, asso. prof., 1943-45; exec. sec. Am. Psychol. Assn., Washington, 1946-50; dir. Commn. on Human Resources, Washington, 1950-54; exec. officer A.A.A.S., Washington, 1954—. Cons. to govt. agys. OECD UNESCO. Recipient Presidential certificate merit, 1948, U. S. Air Force Execeptional Service award, 1957. Mem. A.A.A.S., Am. Psychol. Assn., Sigma Xi. Author: Factor Analysis to 1940, 1941; Human Factors in Military Efficiency, 1946; America's Resources of Specialized Talent, 1954; Science and Public Policy, 1959; also numerous articles. Editor: Symposium on Basic Research, 1959. Research on animal and human learning; pioneered analysis edn. and utilization Am.'s resources specialized talent. Home: 2801 Rittenhouse St., N.W., Washington 20015. Office: 1515 Massachusetts Av. N.W., Washington 20005.*

WOLFOWITZ, Jacob, mathematician, statistician; b. Warsaw, Poland, Mar. 19, 1910; s. Samuel and Helen (Pearlman) W.; B.S., Coll. City N.Y., 1931; M.A., Columbia, 1933; Ph.D., N.Y. U., 1942; m. Lillian Dundes, Mar. 3, 1934; children—Laura M. (Mrs. Tsvi Sachs), Paul D. Asso. prof. U. N.C., 1945-46; asso. prof. Columbia, 1946-49, prof., 1949-51;

prof. math. Cornell U., Ithaca, N.Y., 1951——; vis. prof. U. Cal. at Los Angeles, 1952-53, U. Ill., 1953, Israel Inst. Tech., Haifa, 1957-58, University of Paris, 1967. Fellow Institute Mathematical Statistics (past pres., Rietz lectr. 1957), Econometric Soc.; mem. Am. Math. Soc., I.E.E.E. Author: Coding Theorems of Information Theory, 1961; also numerous articles. Asso. editor Annals Math. Statistics, 1953-57. Research on math. statistics, information theory, probability theory, measure theory. Home: 241 Valley Rd., Ithaca, N.Y. 14850.*

WOLFRAM, J., mathematician; b. circa 1778; asst. to Flamsteed at English Royal Obs.; lt. of arty., Netherlands; Author: Zusätze zu den logarithmischen und trigonometrischen Tabellen; also tables pub. in Sammlung (J. C. Schulze). Computed common logarithms of numbers 1-100, also of primes from 100-1100, to 61 places.

WOLFROM, Melville Lawrence, Am. chemist; b. Bellevue, O., Apr. 2, 1900; s. Fred and Maria (Sutter) W.; A.B., Ohio State U., 1924; M.Sc., Northwestern U., 1925, Ph.D., 1927; m. Agnes Louise Thompson, June 1, 1926; children—Frederick L. (dec.), Eva M. (Mrs. David Frank), Betty J. (Mrs. Gale Hixson), Anne M. (Mrs. Wilson Fleming Jr.), Carl T. Faculty Ohio State U., Columbus, 1929——, Regents' prof., 1965——. Mem. coms. NDRC, NRC; chmn. symposium Internat. Union Biochemistry, Vienna, Austria, 1958; mem. com. book award in sci. Phi Beta Kappa, 1961-63, chmn., 1963. Recipient citation Austrian Govt., 1959. NRC fellow, 1927-29. Fellow Nat. Acad. Scis., Am. Acad. Arts and Scis., N.Y. Acad. Scis., Ohio Acad. Sci. A.A.A.S., Chem. Soc. London; mem. Am. Chem. Soc. (chmn. com. carbohydrate nomenclature; Honor award div. carbohydrate chemistry 1952), Am. Assn. Cereal Chemists, Am. Soc. Biol. Chemists, T.A.P.P.I., Phi Beta Kappa, Sigma Xi, Phi Lambda Upsilon, Pi Mu Epsilon, Alpha Chi Sigma. Editor (with R. L. Whistler): Methods in Carbohydrate Chemistry, 2 vols., 1962-63, cons, editor vols. III-V. Editor Advances in Carbohydrate Chemistry, 1945-49, 52——; sect. 43 Chem. Abstracts, 1959——, adv. bd., 1964——; bd. editors Science, P.R., 1963——. Research, publs. on organic chemistry of carbohydrates and other natural products. Home: 168 Fallis Rd., Columbus, O. 43214.*

WOLFSBERG, Max, chemist; b. Hamburg, Germany, May 28, 1928; s. Gustav and Ida (Engelman) W.; came to U. S., 1939, naturalized, 1945; A.B., Washington U., St. Louis, 1948, Ph.D., 1951; m. Marilyn L. Fleischer, June 23, 1957; 1 dau., Tyra G. With chemistry dept. Brookhaven Nat. Lab., Upton, N.Y., 1951——, sr. chemist, 1963——; prof. chemistry State U. N.Y., Stony Brook, 1966——; vis. prof. chemistry Cornell U., 1963, Ind. U., 1965. Mem. NRC Com. on Kinetics of Chem. Reactions, 1965——. NSF sr. post-doctoral fellow, 1958-59. Mem. Am. Chem. Soc., Phi Beta Kappa. Research, publs. on molecular orbital calculations of properties of inorganic compounds (Wolfsberg-Helmholz approximations); theory of isotope effects on equilibria and on rate processes, theory of fragmentation patterns observed in mass spectra, theory of chem. reactions. Home: 25 Barbara Ct., Sayville, N.Y. 11782. Office: Brookhaven Nat. Lab., Upton, N.Y.*

WOLFSON, Albert, Am. biologist; b. N.Y.C., Feb. 3, 1917; s. Sigmund and Pauline (Segall) W.; B.S., Cornell U., 1937; Ph.D., U. Cal. at Berkeley, 1942; m. Dorothy Duke, June 19, 1937; children—Linda Jean (Mrs Francis J. Brule), Robert Neal, Instr., U. Cal. at Berkeley, 1942-44; mem. faculty Northwestern U., 1944——, prof., 1957——; research asso. Field Mus., 1966——. Abraham Rosenberg research fellow zoology U. Cal, 1940-42; lectr. Am. Inst. Biol. Scis., 1959-61. Mem. Biol. Scis. Curriculum Study Com., 1962-65; cons. div. biol. and med. scis. NSF, 1965-67. Recipient Phi Sigma medal in biology U. Cal., 1941; Brewster Meml. medal and award Am. Ornithologists Union, 1962; sr. postdoctoral fellow NSF of Zool. Inst. Tokyo, 1961-62. Fellow A.A.A.S., Am. Ornithologists Union (sec. 1951-53); mem. Am. Soc. Naturalists, Am. Soc. Zoologists, Endocrine Soc., Internat. Platform Assn., Brit. Ornithologists' Union, Inst. Elec. and Electronics Engrs., Chgo. Acad. Sci. (gov. 1966——), Phi Beta Kappa, Sigma Xi, Phi Kappa Phi. Author: The Human, The Earthworm, The Frog (all 1955). Editor: Avian Biology, 1955. Contbr. articles to profl. jours., books, encys. Showed how day length times bird migration in spring and regulated lates ann. breeding cycle; proposed theory for evolution of migration based on geol. concept of continental drift, demonstrated importance of this concept in understanding worldwide distbns. of plants and animals. Home: 800 Becker Rd., Glenview, Ill. 60025.*

WOLFSON, Kenneth Graham, Am. mathematician; b. N.Y.C., Nov. 21, 1924; s. Roderick and Selma (Simon) W.; B.A., Bklyn. Coll., 1947; M.A., Johns Hopkins, 1948; Ph.D., U. Ill., 1952; m. Rosalind Rephen, Dec. 21, 1947; children—Douglas K., Glenn S. Mem. faculty Rutgers—The State U., 1952——. prof. math., 1960——, chmn. dept., 1961——. Mem. Am. Math Soc., Math. Assn. Am., Assn. Math Tchrs. N.J. Contbr. articles in field. Research on structure theory of rings, particularly endomorphism rings of various types of modules. Home: 446 Cedar Av., Highland Park, N.J. 08904. Office: Dept. Math., Rutgers—The State U., New Brunswick, N.J. 08903.*

WOLK, Elliot, Samuel, Am. mathematician; b. Springfield, Mass., Aug. 5, 1919; s. Max and Jennie (Robinson) W.; A.B., Clark U., 1940; Sc.M., Brown U., 1947, Ph.D., 1954; m. Eleanor F. Lisniansky, Dec. 24, 1950; children—Joel L., Daniel P., Sara F. Faculty, U. Conn., Storrs, 1950——, prof. math., 1964——. Cons. Electric Boat div. Gen. Dynamics Corp., 1957-58. Mem. Am. Math. Soc., Math. Assn. Am. Research, publs. on theory of partially ordered sets, topologies in partially ordered sets. Home: Birchwood Heights, Storrs, Conn. 06268.*

WOLKEN, Jerome Jay, Am. biophysicist; b. Pitts., Mar. 28, 1917; s. Abraham I. and Dina (Landa) W.; B.S., U. Pitts., 1946, M.S., 1948, Ph.D., 1949; m. Dorothy O. Mallinger, June 19, 1945 (dec. 1954); children—Ann Alexandra, A. Jonathon; m. 2d, Tobey J. Holstein, Jan. 25, 1956; children—H. Johanna, Erik Andrew. Research fellow Mellon Inst., Pitts., 1943-47, Rockefeller Inst., 1950-52; faculty U. Pitts., 1953——, prof. biophysics physiology, 1962——; dir. biophys. research lab. Eye and Ear Hosp., Pitts., 1953-64; dir. biophys. research lab. Carnegie Inst. Tech., Pitts., 1964——; vis. prof. biophysics Pa. State U., 1963, Center for Theoretical Studies, U. Miami, Coral Gables, 1966-67. Fellow A.A.A.S., Optical Soc. Am., N.Y. Acad. Scis.; mem. Am. Chem. Soc., Optical Soc. Am., Assn. for Research on Ophthalmology, Biophys. Soc., Pa. Acad. Sci., Soc. Gen. Physiology, Soc. Protozoology, Biophys. Soc., Author; Euglena: An Experimental Organism for Biochemical and Biophysical Studies, 1961; Vision: Biochemistry and Biophysics of the Retina, 1966; also numerous articles. Editor: Photoreception, 1958. Research on photobiology, photosynthesis and vision. Home: 5817 Elmer St., Pitts. 15232.*

WOLLAN, Ernest Omar, Am. physicist, govt. ofcl.; b. Glenwood, Minn., Nov. 6, 1902; s. T. C. and Caroline (Gjerset) W.; B.A., Concordia Coll., 1923; M.S., U. Chgo., 1927, Ph.D., 1929; m. Adelaide Hov, Sept. 11, 1930 (dec.); children—Thomas C., Kathryn (Mrs. R. L. Aagard), John J. Instr., N.D. State Coll., 1926-27; instr. U. Chgo., 1930-32, research asso., 1933-34, 39-42, physicist Manhattan Project, 1942-44; NRC fellow, Zurich, Switzerland, 1932-33; asst. prof. Washington U., St. Louis, 1934-38; physicist Chgo. Tumor Inst., 1938-39; with Oak Ridge Nat. Lab., 1944——, asso. dir. physics div. Fellow Am. Phys. Soc., Peruvian Chem. Soc. Contbr. numerous articles on x-ray scattering, cosmic rays, radiation physics, nuclear physics, neutron diffraction, solid state physics with emphasis on study magnetism to profl. publs.; invented pocket radiation meter, film badge for radiation monitoring. Home: 107 Oneida Lane, Oak Ridge 37831. Office: P.O. Box X, Oak Ridge 37831.*

WOLLASTON, William Hyde, English chemist, physicist; b. East Dereham, Norfolk, Eng., Aug. 6, 1766; s. Francis and Althea (Hyde) W.; ed. Charterhouse and Caius Coll., Cambridge U.; M.D., 1793. Set up pvt. research lab Royal Soc., 1793; practiced medicine, London, 1797-1800; became partially blind, 1800. Established Wollaston medal for research in mineralogy. Fellow Royal Soc. (Copley medal 1802, sec., 1804), 1793; fgn. asso. French Acad. Scis. Contbr. papers to jours. Discovered urates in gouty concretions, circa 1797; developed method of making platinum malleable (income from process enabled him to retire from medicine); investigated voltaic cell and showed it generated static electricity, 1801, thus demonstrated experimentally that frictional and current electricity are identical; studied emission and absorption spectra; 1st used method of total reflection to measure refractivity; discovered dark (Fraunhofer) lines in solar spectrum and ultraviolet rays, 1802; discovered palladium, 1803, rhodium, 1804; inventor camera lucida, a cryophorus, double image prism for splitting single ray into 2 divergent rays, reflecting goniometer to measure crystal angles, Wollaston lens; placed crystallography on firm quantitative base; contbd. to advance of electrochemistry through use of equivalents based on oxygen as ten; supported Dalton's atomic theory; studied theory of multiple proportions in chemistry; mineral compound Wollastonite named in his honor. Died London, Dec. 22, 1828.

WOLLENBERGER, Albert, German pharmacologist; b. Freiburg, Germany, May 21, 1912; s. Sigmund and Seline (Dukas) W.; student U. Berlin (Germany) Med. Sch., 1931-33; B.S., Springfield Coll., 1939; A.M., Harvard, 1942, Ph.D., 1946; m. Gertrude E. Basse, Nov. 14, 1951; children—Knud, Birgit, Ulla, Esther, Leah, Hannah, Ellen. Research asst., asso. pharmacology Harvard Med. Sch., 1945-51; guest investigator Carlsberg Lab., Copenhagen, Denmark, Inst. Biochemistry U. Uppsala (Sweden), dept. biochemistry Maudsley Hosp., London, Eng., Inst. Neurophysiology, U. Copenhagen, 1951-54; asso. prof. pharmacology Humboldt U., Berlin, 1954-55; head lab. Circulation Research, 1954-64; dir. Inst. Circulation Research, German Acad. Scis., Berlin-Buch, Germany, 1965——. Recipient Deutsche Nationalpreis, 1964. Mem. E. German Assn. for Heart and Circulation (pres. 1965——), Am. Soc. for Pharmacology and Exptl. Therapeutics, Deutsche Pharmakologische Gesellschaft, European Soc. for Biochem. Pharmacology, A.A.A.S., Royal Society of Medicine (London), Gesellschaft der Deutschen Naturforscher and Ärzte. Research, numerous publs. on biochem. changes underlying heart failure, elucidation factors and mechanisms in control myocardial energy metabolism, action cardio-active drugs using spontaneously beating single heart cell in culture; localization and differentiation ATP-splitting enzymes systems at fine structural level. Home: 114-116 Pölnitz Weg, Office: 70 Lindenberger Weg, Berlin-Buch, DDR, Germany.*

WOLLIN, Goesta, oceanographer; b. Ystad, Sweden, Oct. 4, 1922; s. Hans Oskar and Anna (Anderson) W.; B.S., Hermods Coll., Malmo, Sweden, 1939; postgrad. U. Lund, 1940; M.S., Columbia, 1953; m. Janet Abel, Oct. 21, 1950; 1 dau., Karen. Came to U. S., 1942, naturalized, 1943. Research, adminstrv. asst. Lamont Geol. Obs. Columbia, N.Y.C., 1950-56, research cons., 1956——; dir. Hydroelectrics, Inc., 1966——. Lectr. U. Rome (Italy), 1954; del. White House Conf. on Children and Youth, 1960; cons. on textbooks on sci. Ginn & Co., 1965——; del. UN Congress, Stockholm, 1965. Mem. A.A.A.S., N.Y. Acad. Scis., Brit. Glacial Soc. Author: Denna Var Skuld, 1946; Flugsmallan, 1948; (with David B. Ericson) The Deep and the Past, 1964. Studies, publs. on world's largest collection deep sea sediment cores from all the oceans. Home and office: Sneden's Landing, Palisades, N.Y. 10964.*

WOLLNY, Edwald, German agronomist; b. Berlin, Germany, 1846; student Acad. Agr., Proscau, 1866-68; doctor, Leipzig, Germany; became asst. Agronomy Inst., Berlin, 1870; named prof. agrl. Tech. Inst., Munich, 1872. Author: De l'application de l'electricité en agriculture, 1883; Soins à donner aux semis et à la culture des plantes, 1885; Culture des céréales, 1887; Décomposition des matières organiques et formation des humus du sol, 1897. A founder of agrl. physics; studied effects of light, heat, humidity, electricity on plant prodn. Died Munich, 1901.

WOLMAN, Abel, Am. civil engr.; b. Balt., June 10, 1892; s. Morris and Rose (Wachsman) W.; A.B., Johns Hopkins, 1913, B.S.E., 1915, D.Eng., 1937; m. Anna Gordon, June 10, 1919; 1 son, M. Gordon. Chief engr. Md. Dept. Health, 1922-39; faculty Johns Hopkins, Balt., 1938——, now prof. emeritus san. engring., environment. Chemn., Nat. Water Resources Com., 1935-41, Md. Water Resources Commn., 1935-45, Md. Planning Commn., 1934-45; cons. engr. USPHS, NRC, also various cities, countries, brs. armed forces. Mem. Nat. Acad. Scis., Nat. Acad. Engring. Research in control of environment for reduction of disease and welfare of man; devel. of community water service. Home: 3213 N. Charles St., Balt. 21218.*

WOLMAN, M., pathologist; b. Warsaw, Poland, Oct. 19, 1914; s. Yehuda Leib and Leah (Cukier) W.; student U. Florence, 1932-35; M.D., U. Rome, 1938; m. Brigitte Koebbel, Jan. 25, 1939; children—Dan, Ruth, Naomi, Amnon Yehuda. Chief physician dept. pathology Hadassah U. Hosp., Jerusalem, Israel, 1946-49; asso. prof. Hebrew U.-Hadassah Med. Sch., 1949-59; clin. asso. prof. pathology Govt. Hosp. Tel-Hashomer, Israel, 1959——; head dept. div, pathology Tel-Aviv (Israel) U. Med. Sch., 1959——, chmn. prof. pathology, 1964——. Mem. Israel Med. Assn., Israel Soc. Pathology, Israel Soc. Clin. Pathology, Israel. Biochem. Soc., Israel Soc. Electron Microscopy, Histochem. Soc., Arbeitsgemeinschaft f. Histochemie, Soc. Exptl. Biology and Medicine, Am. Acad. Neurology, Am. Neuropath. Soc., Internat. Soc. Neurochemistry. Research, publs. on devel. of histochem. procedures; study of molecular structure of myelin and demyelination. Home: 15 ido St., Ramat Gan, Israel. Office: Govt. Hosp., Tel-Hashomer, Israel.*

WOLPE, Joseph, b. Johannesburg, S. Africa, Apr. 20, 1915; s. Michel Salmon and Sarah (Millner) W.; M.B., B.Ch., Witwatersrand (S. Africa), U., 1939, M.D., 1948; m. Stella Ettmann, May 27, 1948; children—Allan, David. Practice gen. medicine, Roodepoort, S. Africa, 1940-42, specializing in psychiatry, Johannesburg, 1948-59; lectr. dept. psychiatry U. Witwatersrand, 1948-59; prof. dept. psychiatry U. Va. Sch. Medicine, Charlottesville, Va., 1960-65; prof. dept. behavioral sci. Temple U. Med. Sch., Phila., 1965——. Fellow Center for Advanced Study in Behavioral Scis., Stanford, Cal., 1956-57. Mem. Am. Psychol. Assn., Am. Psychiat. Assn., Royal Medico-Psychol. Soc. Author: Psychotherapy by Reciprocal Inhibition, 1958; (with A. Salter, L. J. Reyna) Conditioning Therapies, 1964; (with A. A. Lazarus) Behavior Therapy Techniques, 1966; also numerous articles. Research on devel. behavior therapy based on prins. learning. Home: 315 Baird Rd., Merion, Pa. Office: Temple U. Med. Sch., Eastern Pa. Psychiat. Inst., Henry Av. and Abbottsford, Rd., Phila. 19129.*

WOLSKY, Alexander Albert, biologist; b. Budapest, Hungary, Aug. 12, 1902; s. Alexander Bela and Margaret (Devecis) W.; Dr.Phil., U. Budapest, 1928, Private Docent, 1935; m. Maria de Issekutz, Mar. 9, 1940; children—Catherine, Thomas. Came to U. S., 1954, naturalized, 1960. Lectr., U. Budapest, 1925-29; research asso. Hungarian Biol. Research Inst., Tihany, 1929-39, dir., 1939-45; prof. zoology, chmn. dept. U. Budapest, 1945-48; prin. sci. officer, dir. sci. cooperation offices UNESCO, S.

and S.E. Asia, 1948-54; prof. exptl. embryology Fordham U., 1954-67; prof. biology, chmn. sci. depts. Marymount Coll., 1967—. Fellow N.Y. Acad. Scis.; mem. Am. Soc. Zoologists, Soc. Developmental Biology, Soc. Exptl. Biology (Gt. Britain), Indian Zool. Soc., Deutsche Zoologische Gesellschaft, Internat. Soc. Cell Biology, Internat. Inst. Embryology. Research, numerous publs. on physiology insect devel., effect chems. on devel. and regeneration, genetic and environmental effects on devel.; discovered changes in respiratory enzyme system during insect devel. Home: Bedford Rd., Greenwich, Conn. 06833. Office: Dept. Biology, Marymount Coll., Tarrytown, N.Y. 10591.*

WOLSTEIN, Benjamin, Am. psychoanalyst; b. Woodbine, N.J., Sept. 11, 1922; s. David and Reba (Rudnick) W.; B.A., Yeshiva Coll., 1944; M.A., Columbia, 1947, Ph.D., 1949; m. Irma June Holland, Jan. 12, 1952. Lectr., Columbia, 1947-49, State U. N.Y. Downstate Med. Center, 1950, New Sch., 1958—; faculty W. A. White Inst. Psychiatry, 1960—, clin. supr., 1964—; clin. prof. psychology Adelphi U., 1964—; pvt. practice psychodiagnostics and psychotherapy, 1950-51, psychoanalysis, 1951—. Mem. Am., N.Y. State psychol. assns., Am. Philos. Assn., A.A.A.S., W. A. White Psychoanalytic Soc. Author: Experience and Valuation, 1949; Transference, 1954; Countertransference, 1959; Irrational Despair, 1962; Freedom to Experience, 1964; Theory of Psychoanalytic Therapy, 1967; also articles. Defined structure of psychoanalytic inquiry in terms of personalities of analyst and patient, distinguishing between metapsychology and psychology. Address: 2 W. 67th St., N.Y.C. 10023.*

WOLTER, J(urgen) Reimer, physician; b. Halstenbek, Germany, May 9, 1924; s. C. Henry and W. Maria (Peters) W.; M.D., U. Hamburg, 1949; m. Lieselotte Schlote, Oct. 10, 1952; children—Klaus H., A. Maria, Jan D., Niels R. Came to U. S., 1953, naturalized, 1959. Instr., U. Hamburg Med. Sch., 1953; research asso. neuropathology U. Mich., Ann Arbor, 1953-55, faculty, 1955—, prof. ophthalmology, 1964—; chief eye service VA Hosp., Ann Arbor, 1962—. Mem. A.M.A., Am., German ophthal. socs., Am. Acad. Ophthalmology, Otolaryngology, Assn. Research Ophthalmology, Mich. Triological Soc., Am. Ophthalmol. Soc., Sigma Xi. Editor: Jour. Pediatric Ophthalmology. Research, numerous publs. in eye pathology and clin. ophthalmology, especially in neuropathology and neuroanatomy eye. Home: 2565 Bedford St., Ann Arbor, Mich. 48104.

WOLTERINK, Lester Floyd, Am. physiologist; b. Marion, N.Y., July 28, 1915; s. John and Ruth (Voorhorst) W.; A.B., Hope Coll., 1936; M.A., U. Minn., 1940, Ph.D., 1943; m. Lillian Ruth Nichols, June 30, 1938; children—Charles Paul, Timothy John. Faculty, Mich. State U., 1941-65, prof. physiology, 1952-65; project scientist biosatellite project NASA/Ames Research Center, Moffett Field, Cal., 1965—, cons., 1963-64; instr. pre-flight physics Mich. State U.-Air Force, Mich. Agrl. Expt. Sta., 1945-65; asso. scientist Argonne (Ill.) Nat. Lab., 1948. Mem. Am. Physiol. Soc., Am. Soc. Zoologist, Biophys. Soc., Radiation Research Soc., Soc. for Exptl. Biology and Medicine, Am. Astro. Soc., Instrument Soc. Am. Research, publs. on comparative physiology of energy metabolism kinetics of radio-isotope turnover as indices of metabolic activity, space biology. Home: 290 Avalon Dr., Los Altos, Cal. 94022. Office: NASA, Ames Research Center, Moffet Field, Cal. 94035.*

WONG, Harry Yuen Chee, Am. physiologist, endocrinologist; b. Kapaa, Kauai, Hawaii, Oct. 23, 1917; s. Sam and Sain (Fong) W.; B.S., Okla. State U., 1942; M.S., U. So. Cal., 1947; Ph.D., 1950; m. Mabel M. K. Liu, June 24, 1943; children—William James, Donald, Carol Jean. Asso. prof. biology Andrews U., Berrien Springs, Mich., 1949-51; instr. Howard U., Washington, 1951-52, faculty physiology Coll. Medicine, 1952—, prof., 1967—, dir. basic endocrinology and metabolism, 1953—. Cons. aviation physiologist Office Surgeon Gen., USAF, 1958-64; dir. gastroenterology research clinic Freedmen's Hosp., Washington, 1953-56. Mem. Endocrine Soc., Am. Physiol. Soc., Council of Arteriosclerosis, Am., Washington heart assns., Internat. Congresses Endocrinology, Physiology and Pharmacology. Research, publs. in atherosclerosis and endocrinology. Home: 7403 Aspen Av., Takoma Park, Md. 20012. Office: Dept. Physiology, Howard U. Coll. Medicine, Washington 20001.*

WOOD, Albert E(lmer), Am. vertebrate paleontologist; b. Cape May Court House, N.J., Sept. 22, 1910; s. Albert Norton and Edith (Elmer) W.; B.S., Princeton, 1930; M.A., Columbia, 1932, Ph.D., 1935; M.A., Amherst Coll., 1954; m. Frances Wright, Jan. 17, 1937; children—Albert Frederick, Roger Conant, Daniel Nixon. Cutting Travelling fellow Columbia, 1934-35; staff U. S. Army C.E., Binghamton, N.Y., 1936-46, geologist, 1941-46; faculty biology dept. Amherst (Mass.) Coll., 1946—, prof., 1954—. Sr. postdoctoral fellow NSF, Basel, Switzerland, 1966-67. Fellow Geol. Soc. Am., A.A.A.S.; mem. Paleontol. Soc., Am. Soc. Mammalogists, Soc. for Study Evolution, Soc. Systematic Zoologists, Soc. Vertebrate Paleontology (past pres.), Sigma Xi. Research, numerous publs. on evolution, origin, classification of rodent; origin and relationships of rabbits. Home: 11 Wildwood Lane, Amherst, Mass. 01002.*

WOOD, Alex James, Canadian nutritionist; b. Vancouver, B.C., Can., Sept. 3, 1914; s. Robert Wood and Elizabeth (Stirling) W.; B.S., U. B.C., 1935, M.S., 1938; Ph.D., Cornell U., 1940; m. Eileen Emma Ford, Feb. 14, 1942; children—Sandra Eileen, Janet Marion. Bacteriologist Fish Research Bd., Halifax, N.S., Can., 1940-43; bacteriologist Def. Research Bd. Can., Suffield, Alta., 1944-50; prof. nutrition U. B.C., 1946-65; prof. biochemistry, dean arts and sci. U. Victoria, Victoria, B.C., Can., 1965—. Mem. Am. Chem. Soc., N.Y. Acad. Sci., A.A.A.S. Contbr. numerous articles in field to sci. jours. Microbiology of fish spoilage; microbiol. def. studies; animal growth, body composition; nutritional requirements. Home: 3775 Haro St., Victoria, B.C., Can.*

WOOD, Alexander, Brit. physician; b. Cupar, Scotland, Dec. 10, 1817; s. James and Mary Wood; M.D., U. Edinburgh, 1839; m. Rebecca Massey, June 15, 1842. Med. officer Stockbridge Dispensary, Royal Pub. Dispensary, New Town; became extramural lectr. on medicine, 1841; became assesor Univ. Ct., Edinburgh, 1864. Fellow Royal Coll. Physicians (council 1846, sec. 1850, pres. 1858-61). Author: On the Pathology and Treatment of Leucorrhoea, 1844; What is Mesmerism, 1851; Smallpox in Scotland, 1860; Preliminary Education, 1868; Introduced use of hypodermic syringe for administering drugs, 1855. Died Feb. 26, 1884.

WOOD, Arthur Lewis, Am. sociologist; b. Portland, Ore., Sept. 19, 1912; s. Arthur Evans and Julia (Bishop) W.; A.B., Dartmouth, 1935; M.A., U. Mich., 1936; Ph.D., U. Wis., 1941; m. Helen Holden Wolfe, Aug. 25, 1939. Instr., U. Buffalo, 1939-44; asst. prof. Bucknell U., 1944-49; faculty U. Conn., Storrs, 1949—, prof., 1961—, head dept. sociology and anthropology, 1962-67. Fulbright Research fellow, Ceylon, 1956-57. Mem. Am. Sociol. Assn. Author: Crime and Aggression in Changing Ceylon, 1961; Criminal Lawyer, 1967. Research on deviant behavior. Home: Bundy Lane, Storrs, Conn. 06268.*

WOOD, Earl Howard, Am. physiologist; b. Mankato, Minn., Jan. 1, 1912; s. William C. and Inez (Goff) W.; B.A. summa cum laude, Macalester U., 1934; B.S. with distinction, U. Minn., 1940, M.S., 1939, M.D., Ph.D., 1941; m. Ada C. Peterson, Dec. 21, 1936; children—Phoebe, Mark, Guy, Andy. Instr. dept. physiology U. Minn., 1939-40; NRC fellow dept. pharmacology U. Pa. Med. Sch., 1941; instr. pharmacology Harvard Med. Sch., 1942; asst. in physiology Acceleration Lab. Mayo Aeromed. Unit Mayo Found., Rochester, Minn., 1942-46; sci. cons. to surgeon gen. USAF Aeromed. Center, Heidelberg, Germany, 1946; career investigator Am. Heart Assn., 1962—; prof. physiology Mayo Found., Rochester, Minn., 1951—, staff mem. Mayo Clinic, cons. physiology, 1942—; Commonwealth fellow vis. prof. Physiologic Inst., U. Bern (Switzerland), 1965-66. Member American Physiological Soc., Am. Heart Assn., Am. Soc. Exptl. Pharmacology and Therapy, Am. Soc. for Clin. Investigations, Am. Soc. Exptl. Biology and Medicine, A.A.A.S., Aerospace Med. Assn., Nat. Acad. Sci., Nat. Inst. Gen. Med. Scis., others. Research, numerous publs. on devel. of instrumental technics and procedures for study of heart and circulation in health and disease; applications of these procedures to detection and quantitation of various types of acquired and congenital heart disease, study of effects and compensatory reactions of heart and circulation to various types of circulatory stress. Home: 1147 2d St. N.W., Rochester 55901. Office: Mayo Clinic, Rochester, Minn. 55902.*

WOOD, Edward James Ferguson, marine microbiologist; b. Brisbane, Australia, June 23, 1904; s. James Boyne and Maud (Barrymore) W.; B.Sc., U. Queensland, Australia, 1927, M.Sc., 1929, B.A., 1934, D.Sc., 1965; m. Hazel Jessie Fisher, Apr. 29, 1931; children—Janet (Mrs. Brian K. Perry), Heather (Mrs. Murray Branch), Alison Ferguson. Asst. pathologist Bur. Sugar Expt. Stas., 1927-33; biochemist Commonwealth Serum Labs., Melbourne, Australia, 1933-34; asst. bacteriologist U. Melbourne, 1934-37; research officer div. fisheries and oceanography Commonwealth Sci. and Indsl. Research Orgn., Cronulla, Australia, 1937-63; prof. marine microbiology Inst. Marine Sci., head microbiology dept. U. Miami (Fla.), 1963—. Recipient Galthea medal King of Denmark, 1952. Mem. A.A.A.S., Am. Inst. Biol. Scis., Am. Soc. for Microbiology, Soc. for Gen. Microbiology (Gt. Britain), Australian and New Zealand Assn. for Advancement Sci. Author: Microbiology of the Australasian Region, 1964; Marine Microbial Ecology, 1965; Microbiology of Oceans and Estuaries, 1967; also numerous articles. Research on phytoplankton distbn. in oceans in relation to environmental factors, distbn. and significance of pigmented phytoplankton in depths of oceans and in deep sediments. Home: 3601 Royal Palm Av., Coconut Grove, Fla. 33133. Office: Inst. Marine Sci., 1 Rickenbacker Causeway, Miami, Fla. 33149.*

WOOD, Elizabeth Armstrong (Mrs. Ira Eaton Wood), Am. crystallographer; b. N.Y.C., Oct. 19, 1912; d. Herbert Ralph and Winona May (Hull) Armstrong; A.B., Barnard Coll., 1933; M.A., Bryn Mawr Coll., 1934, Ph.D., 1939; D.Sc., Wheaton Coll., Mass., 1963, Western Coll. for Women, Ohio, 1965; m. Ira Eaton Wood, May 13, 1947. Instr. geology Bryn Mawr Coll., 1934-35, 37-38; instr. geology and mineralogy Barnard Coll., 1935-37, 38-41; research asst. Columbia, 1941-42; fellow NRC, 1942-43; mem. tech. staff crystal research Bell Telephone Labs., Murray Hill, N.J., 1943—. Fellow Mineral. Soc. Am., Am. Phys. Soc.; mem. Am. Inst. Physics (dir.), Am. Crystallographic Assn. (pres. 1957), Am. Iris Soc., Sci. Research Soc. Am., Phi Beta Kappa. Author: Crystal Orientation Manual, 1963; Crystals and Light, An Introduction to Optical Crystallography, 1964. Research on x-ray crystallography, phys. properties of crystals, geology and petrology of metamorphic and igneous rocks, optical mineralogy. Home: 37 Pine Ct., New Providence, N.J. 07974. Office: Bell Telephone Labs., Murray Hill, N.J. 07971.

WOOD, Frank Bradshaw, Am. astronomer; b. Jackson, Tenn., Dec. 21, 1915; s. Thomas Frank and Mary (Bradshaw) W.; B.S., U. Fla., 1936; M.S., Princeton, 1940, Ph.D. in Astronomy, 1941; postgrad. U. Ariz., 1938-39; m. Elizabeth Hoar Pepper, Oct. 5, 1945; children—Ellen, Eunice (Mrs. Hilary Roche Davis), Mary Elizabeth, Stephen Bradshaw. Research asso. Princeton U., 1946; asst. prof., asst. astronomer Steward Obs., U. Ariz., 1947-50; asso. prof., exec. dir. Flower and Cook Astron. Obs., U. Pa., Phila., 1950-54, prof. astronomy, dir. obs., 1954—, Flower prof. astronomy, 1958—. Nat. Research fellow, 1946-47, Fulbright fellow, 1957-58. Mem. Am. Astron. Soc. (council 1958-61), A.A.A.S. (sec. sect. D 1958—), Royal Astro. Soc., Internat. Astro. Union, Sigma Xi. Editor: Astronomical Photoelectric Photometry, 1952; Present and Future of the Telescope of Moderate Size, 1956; Photoelectric Astronomy for Amateurs, 1963. Research and publs. on photometry, eclipsing variable stars. Home: Providencs Rd., R.D. 2, Malvern, Pa. Office: Dept. Astronomy, U. Pa., Phila. 19104.*

WOOD, Harland Goff, Am. biochemist; b. Delavan, Minn., Sept. 2, 1907; s. William Clark and Inez (Goff) W.; B.A., Macalester Coll., 1931, Sc.D., 1946; Ph.D., Ia. State Coll., 1935; m. Mildred Lenora Davis, Sept. 14, 1929; children—Donna (Mrs. Thomas McCutcheon), Beverly (Mrs. Ronald Abram), Louise (Mrs. Glenn Scott). NRC fellow U. Wis., 1935-36; faculty Ia. State Coll., 1936-43, U. Minn., 1943-46; faculty Western Res. U., Cleve., 1946—, prof. 1965—. Cons. AEC, 1949-62. Recipient Eli Lilly award in bacteriology, 1942, Carl Neuberg medal, 1952. Fulbright scholar, 1955; Guggenheim fellow, 1962-63. Mem. Nat. Acad. Scis., Am. Soc. Biol. Chemists (past pres.), Biochem. Soc. Gt. Britain, Am. Chem. Soc., Soc. Exptl. Biology and Medicine, Soc. Am. Microbiologists, N.Y. Acad. Sci., Am. Acad. Arts and Scis., Bayerischen Akademie. Editor: Jour. Biol. Chemistry, 1949-54. Editor: Biochemistry, 1966. Discoverer of utilization of CO_2 by heterotrophic organisms, use of carbon isotopes to study intermediary metabolism of animals and bacteria, estimation of pathways of carbohydrate metabolism, purification and properties of enzymes, mechanism of reactions by which CO_2 is utilized. Home: 16259 Oakhill Rd., Cleve. 44106.*

WOOD, James Edwin, III, Am. physician; b. Charlottesville, Va., Feb. 5, 1925; s. James Edwin and Emily (Battle) W.; student Davidson Coll., 1941-42, Duke, 1942-44; postgrad. U. Va.; M.D., Harvard, 1949; m. Ann Sherwood Jones, July 10, 1948; children—Edwin Duncan, James Barker, Emily Battle, Ann Jones. Faculty, Boston U. Sch. Medicine, 1953-58, Med. Coll. Ga., 1958-64; prof. medicine Va. Heart Assn.; research prof. cardiology U. Va. Sch. Medicine, Charlottesville, 1964—, acting chmn. dept. physiology, 1965-66. Mem. gen. com. on environmental research U. S. Army Surgeon Gen., 1964—. Mem. Am. Soc. for Clin. Investigation, Am. Physiol. Soc., Am. Clin. and Climatol. Assn., Am. Fedn. for Clin. Research (past pres. So. sect.), Am. Heart Assn. (chmn. council on circulation). Author: The Veins—Normal and Abnormal Function, 1965. Research, publs. on phys. characteristics of human veins, abnormal responses of veins, acclimatization to heat and cold in man. Home: 1408 Foxbrook Lane, Charlottesville 22901. Office: U. Va. Hospital, Box 146, Charlottesville, Va. 22902.*

WOOD, James J., engr., inventor; b. Kinsale, Ireland, Mar. 25, 1856; s. Paul H. and Elizabeth (Shine) W.; brought to U. S., 1864; grad. as mech. engr. and draftsman, Poly. Inst. Bklyn., 1878; m. Nellie B. Scott, Jan. 20, 1916; children—Venie Elizabeth (Mrs. Joseph H. Appel), Alexander Paul, Ella May (Mrs. William S. Savage). Began as employ of Branford (Conn.) Lock Co. at age of 11, and at 16 designed a horizontal steam engine; with Brady Mfg. Co., Bklyn., 1874-80, becoming supt. and chief engr.; partner Fuller-Wood Co., 1880-85; with Thomson-Houston and Gen. Electric cos. as inventor and cons. engr., 1885—. Fellow Am. Inst. E.E. Inventor of Wood system, and father of closed coil, constant current high tension, self-regulating series arc dyna-current, high tension, self-regulating series arc dynamo; Brayton oil engine, installed in 1st Holland submarine; built 1st lamps for Sir Hiram Maxim; built machines for constructing main cables, original Bklyn. Bridge; holder of 240 patents covering elec. and mech. devices, including dynamo for flood lighting, 1st used on Statue of Liberty. Died Apr. 20, 1928.

WOOD, James Rushmore, Am. surgeon; b. Mamaroneck, L.I., N.Y., Sept. 14, 1816; s. Elkanah and

Mary (Rushmore) W.; grad. Vt. Acad. Medicine, Castleton, 1834; m. Emma Rowe, 1853. Instr. anatomy Vt. Acad. Medicine, 1834; practiced medicine, N.Y.C., 1837; a founder Bellevue Hosp., N.Y.C., 1847, mem. bd.; an organizer Bellevue Hosp. Med. Coll., 1856, prof. operative surgery and surg. pathology, 1886, created 1st hosp. ambulance service in U. S., 1869; opened 1st tng. sch. for nurses in U. S. at Bellevue, 1873; pioneer in cure of aneurism by pressure; perfected bisector for rapid operation for vesical calculus; wrote paper Early History of the Operation of Ligature of the Primitive Carotid Artery; established one of largest collections of postmortem and pathological material in world; twice pres. N.Y. Path. Soc.; mem. N.Y. Acad. Medicine, N.Y. State, Mass. med. socs., mem. A.M.A., N.Y. Surg. Soc., Med. Jour. Assn. Died N.Y.C., May 4, 1882.

WOOD, Jesse Hermon, Am. chemist; b. Pine Knot, Ky., Apr. 29, 1903; s. William Thomas and Mary (Parker) W.; B.S., Eastern Ky. State Coll., 1928; M.S., U. Tenn., 1929; Ph.D., U. N.C., 1937; m. Lena Lovett, Aug. 29, 1930; children—Thomas Hermon, Robert Stuart. Faculty, U. Tenn., Knoxville, 1937—, prof. chemistry, 1946—, dir. gen. chemistry, 1942—. Mem. Am. Chem. Soc., Tenn. Acad. Sci., A.A.A.S., Tenn. Ednl. Assn., Sigma Xi. Author: (with C. W. Keenan) General College Chemistry, 1957, 66; Fundamentals of College Chemistry, 1963; (with Keenan, Bowman, Bull) Laboratory Manual for College Chemistry, 1957, 66; also articles. Research on thioketones and thioaldehydes, Sommelet reaction, chlormeythylation, sympathomimetic amines. Home: 2028 E. Velmetta Circle, Knoxville, Tenn. 37916.*

WOOD, John Karl, Am. physicist; b. Logan, Utah, July 8, 1919; s. John Karl and Phebe (Ricks) W.; B.S., Utah State U., 1941; M.S., Pa. State U., 1942, Ph.D., 1946; m. Margaret Wilson, Mar. 15, 1947; children—James, Robert, Donald, Elizabeth. Optical engr. Bausch & Lomb Optical Co., Rochester, N.Y., 1946-48; prof. U. Wyo., 1948-56; prof. Utah State U., 1956—. Mem. Am. Phys. Soc., Am. Assn. Physics Tchrs., Sigma Xi. Research, publs. on X-ray spectroscopy, optical interferometry. Home: 1359 Juniper Dr., Logan, Utah 84321.*

WOOD, John Lewis, Am. biochemist; b. Homer, Ill., Aug. 7, 1912; s. Walter Horton and Ellen (Palmer) W.; student Blackburn Coll., 1930-32, D.Sc., 1955; B.S., U. Ill., 1934; Ph.D., U. Va., 1937—; m. Amy Josephine Berks, Oct. 4, 1941; children—Linda Ellen, Philip Berks. With George Washington U. Med. Sch., 1937-38; research asso. Cornell U. Med. Sch., 1938-39; Finney-Howell fellow Harvard, 1940-41; asso. chemist U. S. Dept. Agr., Phila., 1941-42; research asso. Cornell U., 1942-44, asst. prof., 1944-46; faculty U. Tenn., Memphis, 1946—, prof. biochemistry, 1950, chmn. dept. biochemistry, 1954-68. Oak Ridge Inst. fellow, 1949; Guggenheim fellow, 1954; USPHS Spl. Research fellow, 1965. Fellow Tenn. Acad. Scis.; mem. Am. Chem. Soc., Harvey Soc., Assn. Cancer Research, Am. Assn. U. Profs., Soc. for Exptl. Biology and Medicine, A.A.A.S., Soc. Biol. Chemists, Sigma Xi, Alpha Chi Sigma, Phi Lambda Upsilon. Research, publs. on organic and bio chemistry. Home: 49 Sevier, Memphis 38111.*

WOOD, Lawrence Arnell, Am. physicist; b. Peekskill, N.Y., Jan. 15, 1904; s. Eben Albert and Elizabeth (Lawrence) W.; A.B., Hamilton Coll., 1925; Ph.D., Cornell U., 1932; m. Margaret Kreitzer Buoy, June 23, 1951; children—Margaret Elizabeth, Robert Eben. Asst. physics Cornell U., 1927-28, instr., 1928-35; physicist rubber sect. Nat. Bur. Standards, Washington, 1935—, chief rubber sect., 1943-62, cons. on rubber, 1962—. U. S. del. Internat. Rubber Tech. Confs., London, Eng., 1938, 48, 62. Elihu Root fellow Hamilton Coll., 1925-27. Recipient U. S. Dept. Commerce Meritorious Service award, 1958. Fellow Am. Phys. Soc. (chmn. div. high polymer physics 1947); mem. Washington Acad. Scis. (pres. 1960, award for achievement in phys. scis. 1943), Philos. Soc. Washington (pres. 1955), Emerson Lit. Soc., Phi Beta Kappa, Sigma Xi, Phi Kappa Phi, Gamma Alpha. Research, publs. on the physics and chemistry of rubber and other polymers. Home: 7014 Beechwood Dr., Chevy Chase, Md., 20015. Office: Nat. Bur. Standards, Washington 20234.*

WOOD, Richard Dawson, Am. marine biologist; b. Toledo, Apr. 28, 1918; s. Harold Lewis and Hadji (Dawson) W.; B.A., B.S., Ohio State U., 1940; Ph.D., Northwestern U., 1947; m. Urda Katherine Larson, Sept. 8, 1946; children—Kenneth Dawson, Steven Lewis. Faculty, U. R.I., Kingston, 1947—, prof. botany, 1959—; research asso. Narragansett Marine Lab., Kingston, 1955. Fulbright Research scholar, Australia, 1960; Darbaker prize in phycology, 1966. Mem. Ecol. Soc. Am., Am. Soc. Limnology and Oceanography, Internat. Am. phycological socs., Bot. Soc. Am., Am. Soc. Plant Taxonomy, Internat. Assn. Plant Taxonomists, Phi Beta Kappa, Sigma Xi, Phi Kappa Phi, Phi Epsilon Phi, Phi Eta Sigma. Author: Monograph of the Characeae, 1965; Benthic algal Ecology by Scuba, 1962; also articles. Research on re-orgn. of classification of Characeae; geog. distbn. and floristics of Characeae; ecology of aquatic plants and algae; systematics and distbn. marine algae. Home: Stonehenge Rd., Kingston, R.I. 02881.*

WOOD, Robert Williams, Am. physicist; b. Concord, Mass., May 2, 1868; s. Dr. Robert Williams and Lucy J. (Davis) W.; A.B., Harvard, 1891; student Johns Hopkins, 1891-92, U. Chgo., 1892-94, U. Berlin, 1894-96; LL.D., Clark U., 1909, U. Birmingham, 1913, U. Edinburgh, 1921; Ph.D., honoris causa, U. Berlin, 1931; D.Sc., honoris causa, Oxford U., 1948; m. Gertrude Ames, Apr. 19, 1892; children—Margaret, Robert Williams, Elizabeth, Bradford (dec.). Instr. physics U. Wis., 1897-99, asst. prof., 1899-1901; prof. exptl. physics, Johns Hopkins, 1901-38, prof. emeritus, 1938, later reappointed research professor. Awarded John Scott Legacy premium and medal Franklin Inst., Phila., for color-photography; Rumford premium, a gold and a silver medal, by Am. Acad., 1909, for researches on theory of light; silver medal London Soc. Arts for color-photography process; gold medal for physics for 1918, Societa Italiana delle Scienze (delta dei XL), Rome; Ive's medal Optical Soc. of Am., 1933; Rumford gold medal Royal Soc., London, 1938; Henry Draper gold medal Nat. Acad. of Sci. for distinguished contbns. to astronomy, 1940. Serced as cons. in development of atom bomb with Manhattan group in New York during World War II; served at Aberdeen proving ground, and as cons. in Navy Experiments on shock waves under water. Commd. major, Signal O.R.C., Aug. 1917; with A.E.F.; developed methods for secret signaling. Fgn. mem. Royal Soc., London, Acad. Sci., Leningrad; Royal Swedish Acad. (1932); hon. fellow Royal Micros. Soc., London, London Phys. Soc. (1933); hon. mem. London Optical Soc., Royal Instn., London, Am. Optical Soc. (1945); corr. mem. Königliche Akademie der Wissenschaften du Göttingen; fgn. asso. Academia dei Lincei, Rome; fellow Am. Acad. Arts and Sciences; mem. Nat. Acad. Sciences, Am. Philos. Soc., Am. Phys. Soc. (pres. 1935); hon. fgn. mem. Indian Assn. for Cultivation of Science, Calcutta, 1931. Author: Physical Optics, 1905, rev. edits., 1911, 34; Researches in Physical Optics, 2 vols.; (fiction) The Man Who Rocked the Earth, and the Moon-Maker (with Arthur Train), 1915. Illustrated nonsense verses, How to Tell the Birds from the Flowers and other wood-cuts. Researches in optics, spectroscopy, atomic and molecular radiation, supersonics improvements in diffraction gratings. In 1898 originated method now in general use of thawing frozen street mains and service pipes by passing an electric current through them. Died Aug. 11, 1955.

WOOD, Scott Emerson, Am. chemist; born Ft. Collins, Colo., Apr. 9, 1910; s. Jesse R. and M. Louise (Scott) W.; B.S., U. Denver, 1930, M.S., 1931; Ph.D., U. Cal. at Berkeley, 1935; m. Marie Simmons, July 15, 1936; 1 son, Edward S. Research asso. Mass. Inst. Tech., 1935-40; with Yale, 1940-48, asst. prof., 1943-48; faculty Ill. Inst. Tech., Chgo., 1948—, prof., 1954—, acting asso. dean research, 1962-64. Cons., Chem. Engring. div. Argonne Nat. Lab., 1960—, resident research asso., 1952, 65; vis. prof. U. Coll., Dublin, Ireland, 1966-67. Research, publs. on thermodynamics, properties of liquid solutions, chromatography and properties of fused salt mixtures. Home: 2507 Burr Oak Av., North Riverside, Ill. 60546. Office: Ill. Inst. Tech., Chgo. 60616.

WOOD, Searles Valentine (the elder), English geologist; b. Woodbridge, Eng., Feb. 14, 1798; s. John and Mary Ann (Baker) W.; m. Elizabeth Taylor, 1821; a son, Searles Valentine. Officer, E. India Co., 1811-25; partner in bank, Hasketon, Eng.; ret., 1835; went to London; curator Geol. Soc.'s Mus. Recipient Wollaston medal, 1860. Fellow Geol. Soc.; mem. Palaeontographical Soc. Author: Catalogue of Grag Shells, 1840-42; Grag Univalves; Bivalves, 1850-55; Grag Mollusca; also geol. articles. Presented his collection of fossils to Brit. Mus. Natural History. Died Martlesham, Eng., Oct. 26, 1880.

WOOD, William Barry, Jr., Am. physician; b. Milton, Mass., May 4, 1910; s. William Barry and Emily Niles, Lockwood) W.; A.B., Harvard, 1932; M.D., Johns Hopkins, 1936; m. Mary Lee Hutchins, July 2, 1932; children—William Barry III, Margaret, Peter, Jonathan, Jean. Med. house officer, Johns Hopkins Hosp., 1936-39; NRC fellow in bacteriology, Harvard Med. Sch., 1939-40; asst., dept. medicine Johns Hopkins Med. Sch., 1937-39, asso. 1940-42; prof. medicine Sch. Medicine, Washington U., St. Louis, 1942-55; v.p. Johns Hopkins U. and Hosp., 1955-59; prof. microbiology Johns Hopkins Schs. Medicine, Hygiene and Pub. Health, 1955—, dir. depts. microbiology, 1959—. Trustee Rockefeller Found. Recipient Bristol award Am. Soc. Infectious Diseases, 1965. Fellow A.C.P.; mem. Am. Soc. Clin. Investigation (pres. 1952), Soc. Exptl. Biology and Med., Central Soc. Clin. Research (pres. 1952), Soc. Am. Bacteriologists, Assn. Am. Physicians (pres. 1962-63), Nat. Acad. Scis (council 1962-65), Sigma Xi, Phi Beta Kappa. Research on mechanisms of recovery in acute bacterial infections, also on mechanism of fever. Home: Chittenden Lane, Owings Mills, Md. 21117. Office: 725 N. Wolfe St., Balt. 21205*

WOOD, William Wayne, Am. chem. physicist; b. Terry, Mont., Nov. 1, 1924; s. Richard Orville and Hope (Baltimore) W.; B.S., Mont. State Coll., 1947; student Carroll Coll., 1943-44; Ph.D., Cal Inst. Tech., 1951; m. Betty Jean King, June 28, 1946; children—Bernard E., William G., Elisabeth J., Richard L., John F. With Los Alamos (N.M.) Sci. Lab., 1950—, group leader, 1957—. Fellow Am. Phys. Soc. Contbr. articles to sci. jours. Research in statis. mechanics and theory of detonations. Home: 933 Tewa Loop. Office: P. O. Box 1663, Los Alamos, N.M. 87544.*

WOODALL, John, Brit. surgeon; b. circa 1556; s. Richard and Mary (Ithell) W.; m. Sara Henchpole; 3 sons, 1 dau. Became mil. surgeon Lord Willoughby's regt., 1591; lived in Germany, 8 years; became mem. Barber Surgeon's Co., 1599, warden, 1627, master, 1633; surgeon St. Bartholomew's Hosp., 1616-43; became 1st surgeon gen. E. India Co., 1612; apptd. examiner surgeons, 1641. Author: The Surgeon's Mate or Military and Domestique Surgery, with a Treatise for the Cure of the Plague, 1639; The Surgeon's Mate, or a Treatise discovering faithfully the due contents of the Surgeons Chest, 1617; Viaticum, being the Pathway to the Surgeon's Chest, 1628. Described surg. instruments, drugs including their preparations, various diseases, injuries and operations; earliest description of treatment for scurvy using lime juice, 1639; had secret remedy called aurum vitae for plague. Died Sept. 1643.

WOODBURN, Henry Milton, Am. chemist; b. Lockport, N.Y., May 30, 1902; s. David and Margaret Jane (Shaw) W.; A.C., U. of Buffalo, 1922, B.S., 1923; M.S., Northwestern U. (fellow Public Health Inst., 1924-25; fellow Am. Petroleum Inst., 1928-29), 1925; Ph.D., Pa. State Coll., 1931; m. Dorothea Williamson Brown, June 3, 1933; children—James Milton, Eleanor Margaret, Stephen Brown. Instr., U. of Buffalo, 1923-24, 1926-35, asst. prof., chemistry, 1935-43, asso. prof., 1943-45, prof. chemistry since 1945, acting head, chemistry dept., 1943-45, head, 1945-56, dean of the graduate sch. 1953-66. Mem. Am. Chem. Soc. (councillor 1945-51, chmn. Western N.Y. sect. 1952-53), Phi Beta Kappa, Sigma Xi, Phi Lambda Upsilon, Alpha Chi Sigma. Author: Laboratory Exercises in Inorganic Chemistry, 1935. Contbr. articles in chem. jours. Systematic study of cyanogen with organic compounds containing reactive hydrogen; chemistry of the principal products of these reactions (oxamidines, cyanoformamidines, oxalimidates, cyanoformimidates). Home: 93 Lehn Springs Dr., Williamsville, N.Y. 14221. Office: State U. N.Y., Buffalo 14.*

WOODBURY, Dixon Miles, Am. pharmacologist; b. St. George, Utah, Aug. 6, 1921; s. Angus M. and Grace (Atkin) W.; B.S., U. Utah, 1942, M.S., 1945; Ph.D., U. Cal. at Berkeley, 1948; m. Donna Budge, June 26, 1945; children—Bruce D., Marcia, Mark B. Faculty, U. Utah, Salt Lake City, 1947—, prof. pharmacology, 1961—. Mem. coms. NIH, 1961—. Lennox lectr. Am. Epilepsy Soc., 1966. Recipient Research Career award NIH, 1962—. Mem. Am. Soc. Pharmacology and Exptl. Therapeutics (John Jacob Abel medal 1957). Contbg. author: Pharmacological Basis of Therapeutics (L. S. Goodman and A. Gilman), 1965; Physiology and Biophysics (T. C. Ruch and H. D. Patton), 1965. Bd. editors Jour. Pharmacology and Exptl. Therapeutics, 1958-62, Endocrinology, 1960-64. Contbr. articles, revs. to sci. jours. Home: 3118 Crestview Circle, Bountiful, Utah 84010.*

WOODBURY, John Walter, Am. physiologist; b. St. George, Utah, Aug. 7, 1923; s. Angus M. and Grace (Atkin) W.; B.S., U. Utah, 1943, M.S., 1947, Ph.D., 1950; m. Betty E. Gunderson, June 10, 1949; children—David, Carolee, Dixon, Michele. Staff Mass. Inst. Tech. Radiation Lab., 1943-45; with U. Wash., Seattle, 1950—, asso. prof., 1957-62, prof. physiology, biophysics, 1962—. Mem. Am. Physiol. Soc., Biophysics Soc., I.E.E.E., A.A.A.S. Author: (with others) Neurophysiology, 1965; also numerous articles. Editorial bds. Am. Jour. Physiology, Jour. Applied Physiology, 1960-66, Physiol. Revs., 1967—. Recording of voltages across surface membranes of nerve, muscle and heart cells; mechanism of generation and conduction of action potential in heart; hydrogen ion permeability of surface membranes; theory of ion penetration of cell membranes. Home: 1213 N.E. 107th St., Seattle 98125.*

WOODBURY, Max Atkin, Am. biomathematician; b. St. George, Utah, Apr. 30, 1917; s. Angus Munn and Grace (Atkin) W.; B.S. in Math., U. Utah, 1939; M.S., U. Mich., 1941, Ph.D., 1948; m. Lida Gottsch, May 30, 1947; children—Carolyn, Max Ten Eyck, Christopher, Gregory. Faculty U. Mich., 1947-49, Princeton, 1949-52, U. Pa., 1952-54; prin. investigator logistics research project Office Naval Research, George Washington U., 1954-56; faculty N.Y. U., 1956-65; prof. math. Duke U., prof. biomath. Duke Med. Center, Durham, N.C., 1966—. Pres. Biomed. Information-processing Orgn., 1961-62, Inst. for Biomed. Computer Research, 1961—; cons. WHO, sci. orgns., univs., govt. agys., corps. Fellow A.A.A.S., Am. Statis. Assn., Inst. Math. Statistics, N.Y. Acad. Scis.; mem. numerous sci., profl. socs., Phi Beta Kappa, Sigma Xi, Phi Kappa Phi. Research, publs. on application math., statis., computer techniques to meteorology, medicine. Address: P.O. Box 3200, Durham, N.C. 27706.*

WOODBURY, Richard Benjamin, Am. anthropologist; b. West Lafayette, Ind., May 16, 1917; s. Charles G. and Marion (Benjamin) W.; student Oberlin Coll., 1934-36; B.S., Harvard, 1939, M.A., 1942, Ph.D., 1949; postgrad. Columbia; m. Nathalie Ferris Sampson, Sept. 18, 1948. Archeol. research, Ariz., 1938, 39, Fla., 1940, Guatemala, 1947-49, El Morro Nat. Monument, N.M., 1953-56; archaeologist United

Fruit Co. Zaculeu Project, Guatemala, 1947-50; asso. prof. anthropology U. Ky., 1950-52, Columbia, 1952-58, U. Ariz., 1959-63; curator archeology U. S. Nat. Mus., Washington, 1963-65, asst. dir., 1965-67; Smithsonian Instn., chmn. office anthropology, 1965-67. Mem. div. anthropology and psychology NRC, 1954-57. Fellow Am. Anthrop. Assn. (exec. bd. 1964-67); mem. Archeol. Inst. Am. (gov. Washington soc.), A.A.A.S. (council), Soc. Am. Archeology (treas. 1953-54, pres. 1958-59), Sigma Xi. Author: (with A. S. Trik) The Ruins of Zaculeu, Guatemala (2 vols.), 1953; Prehistoric Stone Implements of Northeastern Arizona, 1954. Editor: (with I. A. Sanders) Societies Around the World (2 vols.), 1953. Editor Am. Antiquity, 1954-58; Abstracts of New World Archaeology. Mem. editorial bd. Am. Jour. Archeology, also Archeology. Research on prehistoric agr. and water supply problems in Am. Southwest and Mexico. Office: Office of Anthropology, Smithsonian Instn., Washington 20560.*

WOODBURY, Robert Arthur, Am. pharmacologist, physician; b. Pittsburg, Kan., Sept. 1, 1904; s. Foust H. and Kathryn (Jernigan) W.; B.S., U. Kan., 1924, M.S., 1928, Ph.D., 1931; M.D., U. Chgo., 1934; m. Helen Blakesley, Dec. 28, 1931; children—George Robert, Joan Kathryn, Martha Alice. Faculty, U. Ga. Coll. Medicine, 1935-47, prof., chmn. dept. pharmacology, 1940-47; prof. pharmacology, chmn. dept. U. Tenn. Sch. Medicine, Dentistry and Pharmacy, Grad. Sch. Med. Scis., Memphis, 1947—. Cons. NIH, 1965—, FDA, 1946—. Mem. Am. Soc. for Pharmacology and Exptl. Medicine (past mem. council), Am. Physiol. Soc., Am. Soc. for Pharmacology and Exptl. Therapeutics, Am. Heart Soc., Schweizerischen Vereins der Physiologen and Pharmacen. Contbg. author: Pharmacology in Medicine; also numerous articles. Research on pharmacology cardiovascular system, reproductive system, posterior pituitary, autonomic nervous system. Home: 532 W. Clover Dr., Memphis.*

WOODBURY, Robert Morse, Am. economist; b. Worcester, Mass., July 15, 1889; s. John Charles and Jennie (Morse) W.; A.B., Clark U., Worcester, 1910, A.M., 1912; Pres. White traveling fellow Cornell U., 1913-14, Ph.D., 1915; postgrad. U. Berlin, U. Munich (Germany) 1914; LL.D., Clark U., 1955; m. Helen L. Sumner, Nov. 25, 1918; m. 2d, Mildred Fairchild, July 24, 1947. Instr. econs. Cornell U., 1912-13, 1915-16; asst. prof. econs. U. Kan. 1916-17; dir. statis. research Children's Bur., Dept. Labor, 1918-24; mem. staff Inst. Economics, Washington, 1924-28; asso. editor Social Science Abstracts, 1928-33; labor adviser NRA, 1933-35, economist, labor research sect., 1935-36; tech. expert Com. on Population, Nat. Resources Com., 1936; became mem. statis. sect. Internat. Labor Office, 1936, statis. adviser, 1941-46, chief statistician, 1946-53; sec.-gen. 7th Internat. Conf. Labor Statisticians, 1949; mem. Internat. Statis. Inst., 1949-53. Fellow A.A.A.S., Am. Pub. Health Assn., Am. Sociol. Soc., Royal Econ. Soc., Royal Statis. Soc.; mem. Am. Econ. Assn., Internat. Population Union, Population Assn., Am. Statis. Assn. Author: Social Insurance, 1917; Industrial Instability of Child Workers, 1920; Statures and Weights of Children Under Six Years of Age, 1921; Infant Mortality and Preventive Work in New Zealand, 1922; Causal Factors in Infant Mortality, 1925; Maternal Mortality, 1926; Infant Mortality and Its Causes, 1926; Workers' Health and Safety, 1927; Methods of Family Living Studies, 1940; International Comparisons of Food Costs, 1941; Food Consumption and Dietary Surveys in the Americas, 1942. Contbr. to Jour. Am. Statis. Assn. Developed Westergaard's method of expected cases for application to analyzing causes, separating out and measuring individual causes in their effects on ratios, with spl. reference to infant mortality. Address: 323 Caversham Rd., Bryn Mawr, Pa. 19010.*

WOODBURY, Walter Bentley, English inventor; b. Manchester, Eng., June 26, 1834; ed. in sci.; m. Studied engineering; went to Australian goldfields, 1852; migrated to Batavia, Java, and worked on collodion process of photography; returned to England, 1863; settled at Birmingham; patented many photographic devices. Contributed numerous papers to journals. Inventor of Woodbury-type process in photography, 1866; invented method of water-marking called filigrane. Died Margate, Eng., Sept. 5, 1885.

WOODCOCK, Karl Stanley, Am. physicist; b. Thomaston, Me., May 11, 1895; s. Elmer Forrest and Jessie (Killeran) W.; B.S., Bates Coll., 1918; M.S., U. Chgo., 1922, Ph.D., 1932; m. Hazel Luce, Sept. 5, 1922; children—Eugene Luce, Richard Forrest, Carol (Mrs. Horace Atwood Record). Faculty, Bates Coll., Lewiston, Me., 1918—, prof. physics and astronomy, 1935—, head dept., 1943-65. Fellow A.A.A.S.; mem. Am. Phys. Soc., Am. Assn. Physics Tchrs. (Distinguished Service citation 1966), Phi Beta Kappa, Sigma Xi. Research on secondary emission, mass spectroscopy, demonstration apparatus. Home: 86 Russell St., Lewiston, Me. 04240.*

WOODHALL, Barnes, Am. physician; b. Rockport, Me., Jan. 22, 1905; s. Charles Henry and Florence (Barnes) W.; A.B., Williams Coll.; M.D., Johns Hopkins; m. Frances Colman, Aug. 25, 1928; children—Colman, Elizabeth (Mrs. Charles Rackley). Instr. surgery and neurosurgery Johns Hopkins Hosp., 1930-37;

prof. neurosurgery, chmn. dept. Duke U. Med. Center, Durham, N.C., 1937-60, dean, 1960-63, asso. provost, 1967—. Cons. NIH, 1955—, VA., 1947—. Regent, Nat. Library Medicine, 1965—. Mem. Cushing Soc., Am. Acad. Neurol. Surgery, Soc. Neurol. Surgeons, Am., So. surg. assns., A.A.A.S. Author: (with Webb Haymaker) Peripheral Nerve Injuries, 1945—; (with William Lyons) Neuropathology of Peripheral Nerve Injury, 1947; (with B. S. Gilbert) Followup Peripheral Nerve Injuries World War II, 1950. also numerous articles. Research on diagnosis and treatment of peripheral nerve injuries, chemotherapy of brain tumors, biochemistry and immunology of brain tumors. Home: 4006 Dover Rd., Hope Valley, Durham, N.C. 27707.*

WOODHOUSE, Robert, English mathematician, astronomer; b. Norwich, Eng., Apr. 28, 1773; s. Robert and (Alderson) W.; B.A., Cauis Coll., Cambridge, Eng., 1795, M.A., 1798; m. Harriet Wilkins, 1823; a son, Robert. Fellow, Cambridge U., 1798-1823, became Lucasian prof. math., 1820, Plumian prof. astronomy and exptl. philosophy, 1822, supt. Cambridge Obs., 1824. Recipient 1st Smith prize, 1795. Fellow Royal Soc., 1802. Author: The Principles of Analytical Calculation, 1803; Elements of Trigonometry, 1809; A Treatise on Isoperimetrical Problems and the Calculus of Variations, 1810; A Treatise on Astronomy, 1812; Physical Astronomy, 1818. First to explain and advocate notation and methods of calculus in Eng. Died Cambridge, Dec. 28, 1827.

WOODRIFF, Ray Alan, Am. chemist; b. Pueblo, Colo., Jan. 22, 1909; s. Alan and Violet (Young) W.; student Colo. State Coll., 1929-30; B.S., U. Ore., 1933; M.S., Ore. State Coll., 1935; Ph.D., U. Colo. 1942; m. Margaret R. Oswald, Aug. 1, 1938; children—Daniel Thomas, Roger Lee, Susan Kay. Faculty, Mont. State U. 1939—, prof. chemistry, 1953—. Mem. Am. Chem. Soc., Soc. for Applied Spectroscopy, N.W. Soc. Natural and Earth Scis., Sigma Xi. Author: (with others) Introduction to Modern Chemistry, 1951, Experiments in General Chemistry, 1951. Research, publs. on resolution of dilemma concerning nature of light by explaining photoelectric effect on basis of waves; proposed new theory of origin of forces responsible for mountain bldg. Home: 521 W. Grant St., Bozeman, Mont. 59715.*

WOODRING, Wendell Phillips, Am. geologist; b. Reading, Pa., June 13, 1891; s. James Daniel and Margaret (Hurst) W.; A.B., Albright Coll., 1910, Sc.D., 1952; Ph.D., Johns Hopkins, 1916; m. Josephine Jamison, Feb. 9, 1918 (dec.); children—Jane Hurst (dec.), Judy Worth (Mrs. Robert Milton Armagast); m. 2d, Merle Crisler Foshag, Apr. 19, 1965. Geologist, Sinclair C.Am. Oil Corp., 1917-18, U. S. Geol. Survey, 1919-27, 30-61; paleontologist Tropical Oil Co., 1922-23; prof. invertebrate paleontology Cal. Inst. Tech., 1927-30; hon. research asso. Smithsonian Instn., Washington, 1961—. Mem. Geol. Soc. Am. (Penrose medal 1949, pres. 1953), Paleontol. Soc. (pres. 1948), Nat. Acad. Sci., Am. Philos. Soc. Author: (with J. S. Brown, W. S. Burbank) Geology of the Republic of Haiti, 1924; Miocene Mollusks from Bowden, Jamaica, 1925-28; (with R. Stewart, R. W. Richards) Geology of the Kettleman Hills Oil Field, Cal., 1941; (with M. N. Bramlette, W. S. W. Kew) Geology and Paleontology of Palos Verdes Hills, Cal., 1946; (with Bramlette) Geology and Paleontology of the Santa Maria dist., California, 1950; Geology and Paleontology of Canal Zone and Adjoining Parts of Panama, 1957; also articles. Research on geology and paleontology of Cal., Cuba, Jamaica, Haiti, Panama. Home: 5202 Westwood Dr., Westmoreland Hills, Md. 20016. Office: U. S. Nat. Museum, Washington 20560.*

WOODRUFF, Calvin Watts, Am. physician; b. New Haven, July 16, 1920; s. William W. and Myra (Kilborn) W.; B.A., Yale, 1941, M.D., 1944; m. Betty C. Perry, June 17, 1950; children—Virginia, Carl, William. Faculty, Vanderbilt U. Med. Sch., Nashville, 1948-60, asso. prof. pediatrics, 1958-60; porf. pediatrics Am. U. Beirut (Lebanon), 1960-63; prof. nutrition and pediatrics, U. Mich., 1963-65; prof. pediatrics U. Mo., Columbia, 1965—. Markle scholar in med. sci., 1952-58. Mem. Am. Pediatric Soc., Soc. for Pediatric Research, Am. Inst. Nutrition, Sigma Xi, Alpha Omega Alpha. Research, publs. in nutritional anemias and metabolic disturbances of growth and devel. Home: 1021 Lakeside Dr., Columbia, Mo. 65201.*

WOODRUFF, H(arold) Boyd, Am. microbiologist; b. Bridgeton, N.J., July 22, 1917; s. Harold E. and Velma (Smith) W.; B.S., Rutgers U., 1939, Ph.D., 1942; m. Jeanette I. Whitner, July 25, 1942; children—Brian W., Hugh B. Asst. in soil microbiology Rutgers U., 1939-42; staff Merck Sharp & Dohme Research Labs., Merck & Co., Inc., Rahway, N.J., 1942—, asst. dir. microbiology dept., 1949-52, dir. microbiol. and natural product dept., 1952—. Mem. Soc. for Indsl. Microbiology (past pres.), Theobald Smith Soc. (past pres.), Am. Soc. for Microbiology (treas. 1964—), Soc. for Gen. Microbiology, Soc. for Applied Bacteriology, Am. Chem. Soc., Soc. Chem. Industry, N.Y. Acad. Scis., Am. Acad. Microbiology, A.A.A.S., Phi Beta Kappa, Sigma Xi. Editor-in-chief Applied Microbiology. Research, publs. on isolation of new antibiotics including actinomycin, streptothricin, cycloserine, and novobiocin; developed fermentation processes for penicillin and streptomycin and vitamin B12; research on amino acid fermentations and antiviral agts. Home:

797 Valley Rd., Watchung, N.J. 07060. Office: Lincoln Av., Rahway, N.J. 07065.*

WOODRUFF, J(ames) Donald, Am. physician; b. Sparrows Point, Md., June 20, 1912; s. Trowbridge Blanchard and Luella Lloyd (Redding) W.; B.S., Dickinson Coll., 1933; M.D., Johns Hopkins, 1937; m. Bettye Muriel Gardner, Apr. 15, 1939; children—James Donald, David Crane, Gary B.M. Practice medicine, specializing in gynecology, Balt., 1942—; faculty Sch. Medicine Johns Hopkins, 1942—, professor obstetrics, gynecology, 1968—, asso. prof. pathology, 1961—; chief gynecologist Md. Gen. Hosp., 1951-57; chief gynecologist pathology lab. Johns Hopkins Hosp., 1953—; chief gynecologist Woman's Hosp., 1957-62. Mem. Am. Gynecol. Soc., Am. Assn. Obstetrics and Gynecology, Am. Coll. Obstetrics and Gynecology, Continental Gynecol. Soc. (past pres.). Author: (with Novak, Woodruff) Obstetric and Gynecologic Pathology, 1962. Research, numerous publs. on endocrine activity of common ovarian tumors, use of radioisotopes in study of differential metabolic activity of benign and malignant tissues. Home: 107 Cotswold Rd., Balt. 21210. Office: Johns Hopkins Hosp., Balt. 21205.*

WOODRUFF, Michael Francis Addison, Brit. physician; b. London, Eng., Apr. 3, 1911; s. Harold Addison and Margaret (Cooper) W.; M.B., B.S. with 1st class honors in Surgery, U. Melbourne (Australia), 1937, M.D., M.S., 1941, D.Sc., 1962; m. Hazel Gwenyth Ashby, June 12, 1946; children—Keith, Geoffrey, Margaret. Sr. lectr. in surgery U. Aberdeen (Scotland), also cons. surgeon Aberdeen Royal Infirmary and Hosp. for Sick Children, 1948-52; prof. surgery U. Otago, Dunedin, New Zealand, also surgeon Dunedin Hosp., 1953-56; prof. surg. sci. U. Edinburgh (Scotland), 1957—; surgeon Edinburgh Royal Infirmary, 1957—. Mem. clin. research bd. Med. Research Council, 1964—. Fellow Royal Coll. Surgeons Eng., Royal Soc. Medicine, Assn. Surgeons Gt. Britain and Ireland, Royal Soc.; fellow Academie de Chirurgie (associé étranger); mem. Am. Surg. Assn. (hon.), Deutsche Gesellschaft für Chirurgie (corr.), Surg. Research Soc. Author: Surgery for Dental Students, 1954; The Transplantation of Tissues and Organs, 1960; also numerous articles. Research on immunology of organs especially kidney. Home: The Bield, 506 Lanark Rd., Juniper Green, Midlothian, Scotland. Office: Dept. Surg. Sci., U. Edinburgh, Med. Sch., Teviot Pl., Edinburgh 8, Scotland.*

WOODS, Frank Wilson, Am. ecologist; b. Covington, Va., Apr. 1, 1924; s. Frank Edward and Margaret (Gilliam) W.; B.S., N.C. State Coll. 1949; M.S., U. Tenn., 1951, Ph.D., 1957; m. Sarah Ellen Moore, Dec. 13, 1948; children—Mary, Edward, Peter, Alice, James. Research forester U. S. Forest Service, Marianna, Fla., 1953-58; faculty Duke, Durham, N.C., 1958—, asso. prof. forest ecology, 1963—. Mem. Ecol. Soc. Am., Soc. Am. Foresters, A.A.A.S., Assn. Tropical Forestry, N.C. Acad. Sci., Am. Soc. Biologists. Author (with J. Janick, R. Schery, V. Ruttan) Plant Science, 1968. Studies and numerous publs. of root systems of plants from all aspects; ecol. relationships. Home: 1509 Pinecrest Rd., Durham, N.C. 27705.*

WOODS, Henry, Brit. paleontologist; b. Dec. 18, 1868; s. Francis Woods; ed. St. John's Coll., Cambridge; M.A.; m. Ethel Gertrude Skeat, 1910. Demonstrator paleontology Cambridge, 1892-00, univ. lectr. paleozoology, 1900-34. Fellow Royal Soc., 1916, Geol. Soc. (Lyell medal 1918, Wollaston medal 1940); hon. mem. Royal Soc. New Zealand. Author: A Monograph of the Cetaceous Lamellibranchia of England, 2 vols., 1899-1913; Catalogue of Type Fossils in the Woodwardian Museum, Cambridge, 1891; Palaeontology-invertebrate, 8th edit., 1948; The Cretaceous Fauna of Pondoland, 1906; The Palaeontology of the Upper Cretaceous Deposits of Northern Nigeria, 1911; The Cretaceous Fauna of the South Island of New Zealand, 1917; Monograph of the Fossila Mascrurous Crustacea of England, 1925-31; also material on tertiary molusca and crustacea in Geology of N.W. Peru (T. O. Bosworth), 1922. Died Apr. 4, 1952.

WOODS, Joe Darst, Am. chemist; b. Knoxville, Ia., Jan. 2, 1923; s. James Isaac and Mary (Darst) W.; B.S. cum laude, Central Coll., Ia., 1947; M.S., Ia. State U. 1950, Ph.D., 1954; m. Nadine Reed, Nov. 14, 1945; children—Miriam, John. High sch. sci. tchr., Albia High Sch., 1947-48; faculty Drake U., Des Moines, 1952—, prof. chemistry, 1964—. Chmn., Ia. sci. talent search Ia. Acad. Sci., 1960—. Mem. A.A.A.S., Am. Chem. Soc., Am. Assn. U. Profs., Sigma Xi, Phi Lambda Upsilon. Research, publs. on oxidation of metals, isotropic tracer reaction mechanisms. Home: 4107 Ardmore Rd., Des Moines 50310.*

WOODS, Joseph, English botanist; b. Stoke Newington, Eng., Aug. 24, 1776; s. Joseph and Margaret (Hoare) W.; self-educated; Joined office Daniel Asher Alexander, architect; practiced in London, 1819-33; ret., 1833. Fellow Linnean Soc., Geol. Soc., Antiquaries Soc.; mem. London Archtl. Soc. (founder, became 1st pres. 1806). Woodsea, genus of Brit. fern, named in his honor. Author: Tourists' Flora, 1850. Published catalogue of plants and ferns of Britain and various other European countries; also articles. Editor: Antiquities of Athens (James Stuart), vol. 4, 1816. Described several species new to Britain. Died Southover Crescent, Eng., Jan. 9, 1864.

WOODS, Kenneth B(rady), Am. civil engr.; b. Sunnyside, Wash., Aug. 14, 1905; s. Clarence Rockwell and Lucy (Brady) W:; B.C.E., Ohio State U., 1932, C. E., 1937; m. Vivian G. Trautman, Feb. 14, 1930; children—Ronald Edward Jerry Lee, Norman Kenneth. Asst. county engr., Greenville, O., summers 1929-32; part-time instr. Ohio State U., 1930-33; asst. engr. in charge of soils work. Ohio Dept. Hwys, Columbus, 1933-39; asso. dir. joint hwy research project, prof. highway engring. Purdue U., 1939-54, head sch. civil engring. and dir. joint hwy. research project, 1954-65, Gass Distinguished prof. engring. 1965—. Mem. Hwy. Research Bd. (chmn. exec. com. 1956—, com. on frost heave and action in soil 1943-52, other bodies). Recipient Hwy. Research Bd. award for paper of outstanding merit on hwy. materials, 1945, for outstanding contbn. to hwy. research, 1949; Distinguished Alumnus award Ohio State U., 1961; named Engr. of Year in Ind., 1962. Mem. Am. Soc. for Testing Materials (pres. 1958-59; Frank E. Richart award 1963), Am. Concrete Inst. (chmn. subcom. on subgrades and bases), Assn. Asphalt Paving Technologists, Nat., Ind., Lafayette socs. profl. engrs., Am. Soc. C.E. (sec.-treas., Ind. sect., 1948-49, pres. 1952); U. S. Council Soil Mechanics and Found. Engring., Arctic Inst., Am. Geog. Soc., Am. Soc. Engring. Edn., Ind. Acad. Sci., Am. Ry. Engring. Assn., Am. Assn. State Hwy. Ofcls. (chmn. adv. com), Sigma Xi (pres. Purdue chapt. 1949-51), Chi Upsilon, Tau Beta Pi. Contbr. numerous papers and bulls. Editor-in-chief: Highway Engring. Handbook, 1959-60. Research on surficial and military geology; materials of highway and airport construction; air photo interpretation of soils. Home: 902 N. Chauncey, West Lafayette, Ind. 47906.*

WOODS, Mark Winton, Am. biologist; b. Washington, Oct. 15, 1908; s. Albert Frederick and Bertha (Davis) W.; B.S., U.Md., 1931, B.S., 1933, Ph.D., 1936; m. Vera Lorraine Klein, Dec. 23, 1933; children—Judith Patricia, Mark Alvin. Faculty, U. Md., 1936-47; research biologist Nat. Cancer Inst., Bethesda, Md., 1947—. Mem. Am. Soc. Biol. Chemists, Am. Assn. Cancer Research, Am. Phytopath. Soc., N.Y. Acad. Sci., Washington Acad. Sci., A.A.A.S., Bot. Soc. Washington, Sigma Xi, Phi Kappa Phi, Alpha Zeta. Research, numerous publs. on metabolism of normal and cancer tissues; hereditary properties of mitochondria and role of mutant mitochondria in neoplasia in plants; mechanism of action of certain chemotherapeutic agts. in normal and cancer cells; plant virus diseases, plant nuclear cytology. Home: 4003 Quintana St., Hyattsville, Md. 20782. Office: Nat. Cancer Inst., Bldg. 6-B-29, Bethesda, Md. 20014.*

WOODS, Philip Sargent, Am. biologist; b. Concord, N.H., Nov. 25, 1921; s. Arthur Page and Beulah (Sargent) W.; B.S., Mich. State U., 1947; Ph.D., U. Wis., 1952; m. Dorothy Barrett, Mar. 15, 1946; children—Thomas B., Karen L. Radiobiologist, Army Med. Research Lab., Ft. Knox, Ky., 1952-53; USPHS fellow Columbia, 1953-55; cytologist Brookhaven Nat. Lab., Upton, L.I., N.Y., 1955-62; asso. prof. cytology dept. biol. sci. U. Del., Newark, 1962-67; prof. cytology dept. biology Queens Coll. of City U. N.Y., Flushing, 1967—. Research on duplication of chromosomes, ribonucleic acid metabolism. Office: Dept. Biology, Queens Coll. of City U. N.Y., Flushing, N.Y. 11367.*

WOODSIDE, Gilbert Llewellyn, Am. biologist; b. Curwensville, Pa., Feb. 9, 1909; s. William W. and Minnie (Kester) W.; A.B., De Pauw U., 1932; A.M., Harvard, 1933; Ph.D., 1936; m. Mary Calhoun Livingston, Sept. 1, 1934; children—Kenneth Hall, Richard Livingston, Asst. prof. biology U. Mass., 1936-46, prof. biology, 1946-64, head dept. zoology, 1948-61, dean grad. sch., 1950-61, provost, 1961-63, asst. to dir. for sci. program planning and devel. Nat. Inst. Child Health and Human Devel., Bethesda, Md., 1964-66, asso. dir., 1966—. Mem. Am. Soc. Zoologists, Soc. for Study Growth and Devel., Phi Beta Kappa, Sigma Xi, Phi Kappa Phi. Contbr. articles on exptl. embryology and cancer research. Research on embryonic induction in the chick, hyperthyroidism and auxin in chick embryos, chemotherapy of mouse tumors using 8-Azaguanine, electron microscopy of embryo mice lungs. Home: 9706 Parkwood Dr. Office: Nat. Inst. Child Health and Human Devel., Bethesda, Md. 20014.*

WOODSON, Robert Everard, Jr., Am. biologist; b. St. Louis, Apr. 28, 1904; s. Robert Everard and Ada Lee (Cowan) W.; A.B., Washington U., St. Louis, 1926, M.S., 1927, Ph.D., 1929; A.M., Harvard, 1928. Prof. botany Washington U., from 1945; curator herbarium Mo. Bot. Garden, from 1958. Member Batan Soc. Am., Am. Soc. Plant Taxonomists, St. Louis Acad. Sci., Phi Beta Kappa, Sigma Xi. Contbr. articles on evolution, plant geog., taxonomy, floristics. Died St. Louis, Nov. 6, 1963.

WOODSTOCK, Lowell Willard, Am. plant physiologist; b. Harvey, Ill., July 6, 1931; s. Willard Howard and Rhoda Wyatt (Salter) W.; B.S., U. Ill., 1954; Ph.D., U. Wis., 1959; m. Agnes Jean Winifred Smith, Dec. 30, 1961; 1 dau., Elizabeth Mary. Am. Cancer Inst. fellow Royal Botanic Garden, Edinburgh, Scotland, 1959-61, Madison, Wis., 1961-62; research asso. Argonne (Ill.) Nat. Lab., 1962-63; research plant physiologist Agr. Research Services, Beltsville, Md., 1963—. Mem. Am. Soc. Plant Physiologists, Bot. Soc. Am., Scandinavian Soc. for Plant Physiology, Am. Inst. Biol. Scis., A.A.A.S., N.Y. Acad. Scis., Washington Bot. Soc., Sigma Xi. Research on quantitative relationships between ribosenucleic acids and plant growth, respiration, protein synthesis, germination seeds; discovered relationship between rates of respiration and rapidity seed germination. Home: 10503 Royal Rd., Silver Spring, Md. 20903. Office: Plant Industry Sta., Beltsville, Md. 20705.*

WOODVILLE, William, English physician, botanist; b. Cumberland, Eng., 1752; ed. Edinburgh (Scotland) U.; studied under William Cullen; M.D., 1775; Became physician Middlesex Dispensary, London, 1782, to smallpox and inoculation hosps. at St. Pancras, London, 1791. Licenciate Royal Coll. Physicians. Fellow Linnean Soc. Author: A Comparative Statement of Facts and Observations relative to the Cowpox, published by Doctors Jenner and Woodville, 1800; Medical Botany (described medicinal plants from Royal Coll. Physicians catalogues), 1790-94; A History of the Inoculation of the Smallpox in Great Britain, 1796; Reports of a Series of Inoculations for the Variolae Vaccinae or Cow Pox, 1799. Adopted and advocated vaccination which had been introduced by Edward Jenner, Died London, Mar. 26, 1805.

WOODWARD, Arthur Eugene, Am. chemist; b. Los Angeles, Oct. 16, 1925; s. Arthur August and Mabel (Christopher) W.; B.A., Occidental Coll., 1949, M.A., 1950; Ph.D., Bklyn. Poly. Inst., 1953; m. Doris Elaine Bingham, June 20, 1952; 1 dau., Pamela Louise. Fulbright fellow U. Louvain, Belgium, 1953-54; research fellow Harvard, 1954-55; asst. prof. chemistry Pa. State U., 1955-59, asso. prof. physics, 1960-64; Guggenheim fellow Queen Mary Coll., U. London, England, 1962-63; asso. prof. chemistry City Coll., U. City N.Y., 1964-66, prof., 1967—. Mem. Am. Chem. Soc., Am. Phys. Soc., Am. Assn. U. Prof., N.Y. Acad. Scis. Research and numerous publs. on preparation, modification, phys. properties of large synthetic molecules such as polyethylene, nylon, polystyrene; study of thermal motion in long chain compounds in solid state; preparation and study of properties of crystals of such molecules. Office: 138th and Convent Av., N.Y.C. 10031.*

WOODWARD, Edward Roy, Am. physician; b. Chgo., Sept. 6, 1916; s. Lee Roy and Lynn (Webster) W.; B.A., Grinnell Coll., 1938; M.D., U. Chgo., 1942; m. Dorothy Louise Furry, Sept. 9, 1939; children—James Roy, Suzanne Louise. Instr., U. Chgo., 1949-53; faculty U. Cal. Med. Center, Los Angeles, 1953-57, asso. prof., 1957; prof., chmn. dept. surgery U. Fla. Coll. Medicine, Gainesville, 1957—. Cons. on extramural programs AEC, 1958-60; mem. surgery study sect. B, NIH, 1965—. John and Mary R. Markle scholar in med. sci., 1952-57. Fellow A.C.S.; mem. Am. Cancer Soc. (mem. com. on personnel for research 1954-61), Am. Physiol. Soc., Am. Surg. Assn., Am. Gastroent. Assn., Internat. Soc. Surgery. Author: The Postgastrectomy Syndromes, 1963; also numerous articles. Research on gastrointestinal physiology and pathology particularly peptic ulcer. Home: 601 N.W. 23d St., Gainesville, Fla. 32601.*

WOODWARD, John, English geologist, physician; b. Derbyshire, Eng., May 1, 1665; studied under Peter Barwick; created M.D. by Archbishop Thomas Tenison, 1695; M.D., Cambridge, 1695. Became prof. physic Gresham Coll., London, 1692; Gulstonian lectr., 1710-11. Fellow Royal Soc., 1693, served on council, expelled for insulting Sir Hans Sloane, 1710; fellow Royal Coll. Physicians. Author: An Essay toward a Natural History of the Earth, 1695; Brief Instructions for making Observations in all Parts of the World and sending over Natural Things, 1696; An Account of some Roman Urns . . . , 1713; Naturalis historia telluris illustrata et aucta, 3 parts, 1714; The State of Physick and of Diseases, 1718; An attempt towards a Natural History of the Fossils of England, 2 vols., 1728-29; Fossils of all Kinds digested into a Method, 1728; Select Cases and Consultations in Physic . . . , 1757; also other publs. A founder exptl. plant physiology; one of 1st to use water culture; discovered transpiration; recognized existence of strata in earth's crust, although he overlooked their fossils. Died Gresham Coll., Apr. 25, 1728.

WOODWARD, Richard Lewis, Am. civil engr.; b. Kansas City, Mo., Dec. 11, 1913; s. John Bennet and Grace (Baker) W.; B.S. in Civil Engring., Washington U., 1935; M.S., Harvard, 1948; Ph.D., Ohio State U., 1952; m. Helen Beal, June 17, 1939; children—Brenda, Melinda, Richard, Amy. Engr., W. W. Horner, St. Louis, 1935-36, Nat. Resources Com., Washington, 1937; research, investigations USPHS, 1937-63; sr. research asso., lectr. Harvard, 1963-67; cons. engr. Camp, Dresser & McKee, 1966—. Recipient Meritorious Service medal USPHS, 1963. Mem. Am. Soc. C.E., Am. Acad. San. Engrs. Research, publs. on water supply and treatment, effects water pollutants on health, control water pollution, water resources planning operations research in water quality mgmt. Home: 40 Carl St., Newton Highlands, Mass. 02161. Office: 1 Center Plaza, Boston 02108.*

WOODWARD, Robert Burns, Am. chemist; b. Boston, Apr. 10, 1917; s. Arthur Chester and Margaret (Burns) W.; B.S., Mass. Inst. Tech., 1936, Ph.D., 1937; D.Sc., Wesleyan U., 1945, U. Manchester (Eng.), 1954, Bucknell U., 1955. U. New Brunswick (Can.), 1956, Yale, 1956, Harvard, 1957, U. So. Cal., 1959, U. Chgo., 1961, New Eng. Coll. Pharmacy, 1961, Colby Coll., 1963, U. Cambridge (Eng.), 1964, Brandeis U., 1965, Stonehill Coll., 1966, U. Sheffield (Eng.), 1966, Israel Inst. Tech., 1966; LL.D., U. Glasgow, Scotland, 1966; A.M. (hon.) Harvard, 1946; m. Irja Pullman, July 30, 1938; children—Siiri Anne, Jean Kristen; m. 2d, Eudoxia M. M. Muller, Sept. 14, 1946; children—Crystal Elisabeth, Eric Richard Arthur. Mem. Soc. Fellow, Harvard, 1938-40, faculty, 1941—, prof., 1950-53, Morris Loeb prof. chemistry, 1953-60, Donner prof. sci, 1960—; dir. Woodward Research Inst., Basel, Switzerland, 1963—. Cons. Polaroid Corp., 1942—, Charles Pfizer & Co., Inc., 1951—; hon. lectr. numerous univs. Recipient numerous award including Ledlie prize Harvard, 1955, Research Corp. award, 1955, Davy medal Royal Soc., 1959, Pius XI Gold medal Pontifical Acad. Scis., 1961, Priestley medallion Dickinson Coll., 1962, Nat. medal Sci. U. S. A., 1964, Nobel prize in chemistry, 1965. Fellow Royal Sci., 1956, Am. Acad. Arts and Scis., Chem. Soc. (hon.), Indian Acad. Scis. (hon.); mem. Nat. Acad. Scis., German Chem. Soc. (hon.); Royal Irish Acad. (hon.), Austrian Acad. Scis. (corr.), Am. Philos. Soc., Belgian Chem. Soc. (hon.), others. Synthesized quinine, 1944, patulin, 1950, cholesterol, 1951, cortisone, 1951, strychnine, 1954, lysergic acid, 1954, lanosterol, 1954, reserpine, 1956, chlorophyll, 1960, tetracycline, 1962; deduced structure of penicillin, 1945, strychnine, 1947, terramycin and aureomycin, 1952, ferrocene, 1952, cevine, 1954, magnamycin, 1956, oleandomycin, 1960. Office: Dept. Chemistry, Harvard U., 12 Oxford St., Cambridge, Mass. 02138.*

WOODWARD, Samuel Pickworth, English naturalist; b. Norwich, Eng., Sept. 17, 1821; s. Samuel and Elizabeth (Bolingbroke) W.; studied under William Brooke, Priory Sch., Greyfriars; Ph.D. (hon.), U. Göttingen, 1864. Collected bot. specimens for Dawson Turner; staff library Brit. Mus., 1838; named subcurator Geol. Soc. London, Somerset House, 1839; named prof. geology and natural history Royal Agrl. Coll., Cirencester, Eng., 1845; 1st class asst. dept. geology and minerology Brit. Mus., 1846-65; examiner in natural sci. Council Mil. Edn., Sandhurst; examiner geology and paleontology U. London. Fellow Geol. Soc.; mem. Linnean Soc. (asso.), Bot. Soc. London, Cotteswold Naturalists Field Club (a founder). Author: Manual of Mollusca, 3 parts, 1851-56; also articles. Research on recent and fossil shells; named and described new genus, Echinothuria, from fossil form. Died Herne Bay, Eng., July 11, 1865.

WOODWARD, Val Waddoups, Am. geneticist; b. Preston, Ida., July 26, 1927; s. Rollo W. and Hazel (Waddoups) W.; B.S., Utah State U., 1950; M.S., Kan. State U., 1950; Ph.D., Cornell U., 1953; children—Dean, Jim, Kathryn. Postdoctoral fellow Brookhaven Nat. Lab., 1953-55; faculty Kan. State U., 1955-58, Wichita U., 1958-61; asso. prof. biology Rice U., Houston, 1961-67; prof. genetics U. Minn., St. Paul, 1967—. NSF fellow, Eng., 1962-63. Mem. A.A.A.S., Genetics Soc. Am., Am. Inst. Biol. Scis., Sigma Xi. Contbr. articles to sci. jours. Home: 2235 Hillside St., St. Paul 55108. Office: Genetics Dept., U. Minn., St. Paul.*

WOODWELL, George Masters, Am. ecologist; b. Cambridge, Mass., Oct. 23, 1928; s. Philip McIntire and Virginia (Sellers) W.; A.B., Dartmouth, 1950; A.M., Duke, 1956, Ph.D., 1958; m. Alice Katharine Rondthaler, June 23, 1955; children—Caroline Alice, Marjorie Virginia, Jane Katharine, John Christopher. Faculty, U. Me., 1957-61, asso. prof. botany, 1960-61; vis. asst. ecologist biol. dept. Brookhaven Nat. Lab., Upton, N.Y., 1961-62, ecologist, 1965—.; lectr. Yale U. Sch. Forestry, 1967—. Mem. adv. bd. mem. N.Y. State Pesticides Control Bd. Fellow A.A.A.S.; mem. Ecol. Soc. Am. (v.p. 1966-67), Bot. Soc. Am., Radiation Research Soc., Sigma Xi. Editor: Ecological Effects of Nuclear War, 1965; mem. editorial adv. bd. Radiation Botany. Research, publs. in structure and function of natural communities, especially ecol. effects of ionizing radiation and effects of persistent pesticides.*

WOODWORTH, Robert Sessions, Am. psychologist; b. Belchertown, Mass., Oct. 17, 1869; s. William Walter and Lydia Ames (Sessions) W.; A.B., Amherst Coll., 1891, Sc.D., 1951; A.B., Harvard, 1896, A.M., 1897; Ph.D., Columbia, 1899, Sc.D., 1929; LL.D., Lake Erie Coll., 1934; Sc.D., U. of N.C., 1946, U. of Pa., 1946; grad. study. univs. of Edinburgh, Liverpool, Bonn.; m. Gabrielle Marie Schjöth, Apr. 23, 1903; children—Svenssen, Greta Sessions, William, Mary Virginia. Teacher sciences, Watertown (N.Y.) High Sch., 1891-93; instr. mathematics, Washburn Coll., Kan., 1893-95; asst. in physiology, Harvard Med. Sch., 1897-98; instr. physiology, Univ. and Bellevue Hosp., Med. Coll., N.Y.C., 1899-1902; demonstrator in physiology, U. of Liverpool, Eng., 1902-03; instr. psychology, Columbia U., 1903-05, adj. prof., 1905-09, prof., 1909-42, prof. emeritus, 1942—. Supt. of sections anthropometry and psychometry, St. Louis Expdn., 1904. Recipient Gold Medal Am. Psychol. Found., 1956. Fellow A.A.A.S. (v.p. 1909, 1924), N.Y. Acad. Scis., British Psychol. Soc., Am. Acad. Arts and Scis.; mem. Am. Philos. Soc., Nat. Acad. Scis., Nat. Research Council (chmn. div. anthropology and psychology, 1924-25), Am. Psychol. Assn. (pres.

1914), Social Science Research Council (treas. 1927-29; pres. 1931-32), Psychol. Corp. (pres. 1929), Nat. Inst. Psychology. Author: Le Mouvement, 1903; Elements of Physiological Psychology (with George T. Ladd), 1911; The Care of the Body, 1912; Dynamic Psychology, 1917; Psychology, 1921, 5th edit. (with D. G. Marquis), 1947; Contemporary Schools of Psychology, 1931; Adjustment and Mastery, 1933; Experimental Psychology, 1938; Psychological Issues, 1939; Heredity and Environment, 1941; First Course in Psychology (with Mary R. Sheehan), 1944; also author many scientific papers in psychol. and physiol. jours. Editor: Archives of Psychology, 1906-45. Works became standard and popular textbooks; early critic of transfer-of-training theories (thinking this happens only when identical factors between old and new learning are perceived); stressed study of both physiol. and mental activities; used functional and dynamic approaches. Died N.Y.C., July 4, 1962.

WOOL, Ira Goodwin, Am. biologist; b. Newark, Aug. 22, 1925; s. Abraham Herman and Edith (Jacobs) W.; A.B., Syracuse U., 1949; M.D., U. Chgo., 1953, Ph.D., 1954; m. Glorye Needlman, Sept. 14, 1950; children—Christopher Douglas, Jonathan Andrew. USPHS fellow, U. Chgo., 1953-54, Commonwealth fellow med. scis., 1956-57, faculty, 1957—, prof. physiology and biochemistry, 1965—; vis. prof. Wayne State U., 1964, 66, Fla. State U., 1966; vis. faculty mem. Mayo Grad. Sch. Medicine, 1966. Eugene Bernstein lectr. Beth Israel Hosp., Harvard Med. Sch., 1964; Lederle Sci. lectr. Lederle Labs., 1966. Recipient Ginsburg award in physiology U. Chgo., 1952. Mem. Endocrine Soc., Am. Diabetes Assn., A.A.A.S. (mem. council 1962—), Am. Physiol. Soc., Biochem. Soc. (Eng.), Soc. for Exptl. Biology and Medicine, Sigma Xi., Phi Beta Kappa, Alpha Omega Alpha. Editor, Vitamins and Hormones, 1960—, Endocrinology, 1965—. Research and publs. on regulation of protein synthesis and mechanisms of hormone action especially mechanism of regulation of protein synthesis by insulin. Home: 1351 E. 55th Pl., Chgo. 60637.*

WOOLF, Arthur, Brit. engr.; b. Cornwall, Eng., Nov. 4, 1766; s. Arthur and Jane (Newton) W.; apprenticed to carpenter at Pool, Eng. Became millwright under Joseph Bramah, Pimlico, Eng.; became master engr., 1795; resident engr. Neux's brewery, 1796-1806; became partner in steam engine factory, Lambeth, Eng., 1806; engaged in mining, Cornwall, Eng.; supt. Harvey & Co.'s engine mfg., Hayle, Eng., until 1833. Patentee a compound engine (Woolf engine), 1810, an improved engine boiler, 1803, system for facilitating use of steam at high pressure; introduced various improvements to mining which were later superseded by high pressure single-cylinder engine of Richard Trevithick; built 1st fire damp engine with double expansion, engine with cylinder of 2.23 meters in diameter and 3 meter stroke (largest of its time), bell valve, 1830. Died The Strand, Guernsey, Eng., Oct. 26, 1837.

WOOLFENDEN, Glen Everett, Am. ornithologist; b. Elizabeth, N.J., Jan. 23, 1930; s. Lester Bancroft and Ethel (Lester) W.; B.S., Cornell U., 1953; M.A., U. Kan., 1956; Ph.D., U. Fla., 1960; m. Gwendolyn Sindle, Dec. 27, 1954; children—Kim Ellen, Scot Lester, Lisa Ruth. Instr., U. Fla., Gainesville, 1959-60; faculty U. S. Fla., Tampa, 1960—, asso. prof., 1965—. Cons., Encephalitis Research Center, Tampa, 1963—. Mem. Am., Brit. ornithologists unions, Cooper, Wilson ornithol. socs., Bird-Banding Assn., Fla. Acad. Scis. Editor, Recent Literature of the Auk, 1965—. Research and publs. on ecology, systematics, paleontology of birds. Home: 8011 Cardinal Dr., Tampa, Fla. 33610.*

WOOLLAM, David Henry Morgan, English anatomist; b. Caldscote, Eng., Aug. 12, 1921; s. Harry Morgan and May Duncan (Smith) W.; M.D., Cambridge (Eng.) U., 1953, Sc.D., 1964; m. Margaret Leonora Harries, Oct. 16, 1943; children—Christopher, Leonora, Katharine, Victoria. Lectr. anatomy U. Cambridge, 1953—, fellow Emmanuel Coll., 1959—, dir. med. studies, 1960—. Mem. Royal Coll. Physicians. Author: (with J. W. Millen) Anatomy of the Cerebrospinal Fluid, 1963; also numerous articles. Editor series Advances in Teratology, 1966—. Research in circulation of cerebrospinal fluid and teratology, especially drug-induced malformations, anatomy of normal and abnormal chromosomes. Home: 17 Cavendish Av., Cambridge, Eng.*

WOOLLARD, George Prior, Am. geophysicist; b. Savannah, Ga., Dec. 20, 1908; s. Alfred and Anne (Jevons) W.; B.S. in Civil Engring., Ga. Inst. Tech., 1932; M.S., 1934; A.M., Princeton, 1935, Ph.D., 1937; m. Eleanor Monroe McClintock, May 17, 1937; children—Bruce, Monroe, Nancy-Light, George, Matilda, Frank, Belinda, Elizabeth, Harold N. Pierson (foster son). Instr., Ga. Inst. Tech., 1933-34; William Pierson Field lectr. Princeton, 1940-48; research group leader Woods Hole Oceanographic Inst., 1941-48; asso. prof. geophysics and engring. geology U. Wis., 1949-53, prof., 1953-63, dir. Geophys. and Polar Research Center, 1938-63; dir. Hawaii Inst. Geophysics, Honolulu, 1963—. Chmn. solid earth geophys. panel U. S. Nat. Com. for Internat. Geophys. Year, 1957-61. Fellow Geol. Soc. Am., Am. Geophys. Union, Royal Astron. Soc.; mem. Internat. Union Geodesy and Geophysics (U. S. mem. earth mantle commn.), Soc. Exploration Geophysicist, European Assn. Exploration Geophysicists, Am. Assn. Petroleum Geologists, A.A.A.S. Author: (with J. C. Rose) International Gravity Measurements, 1963; also numerous articles. Standardization of gravity, world gravity control net, relation of gravity anomalies to coastal structure, relation gravity and magnetic ananalies to geology, sound propagation in oceans, acoustic structure of oceans, effect bottom geology on sound propogation in oceans, prediction of gravity from geology and topography. Home: 18 Palione Pl., Kailua, Oahu, Hawaii. Office: Inst. Geophysics, U. Hawaii, Honolulu.*

WOOLLEY, Sir Charles Leonard, English archaeologist; b. London, Eng., Apr. 17, 1880; s. George Herbert and Sarah Woolley; M.A., New Coll., Oxford method for preparing synthesizing urea from am- (Eng.) U.; D.Litt., U. Dublin (Ireland); LL.D. (hon.), U. St. Andrews (Scotland); m. Katharine Elizabeth Keeling. Asst. keeper Ashmolean Museum, 1905-07; with Coxe expdn. in Nubia, 1907-11, Oxford Nubian expdn., 1912, British Mus. excavations at Carchemish, 1912-14, excavations in Sinai, 1914, excavations at Carchemish, 1919, at Tel el Amarna, Egypt, 1921-22, at Ur, 1922-34, at Al Mina, Syria, 1936-37, At Atchana, Hatay, 1937-39, 46-49; Huxley meml. lectr., 1942; Myres meml. lectr., 1960. Served as capt. Royal F.A., 1914-16; prisoner in Turkey, 1916-18; maj. polit. dept., N. Syria, 1919; served to lt. col. Brit. Army, 1939-46. Hon. fellow Oxford U., also Turk Tarih Kurumu; hon. asso. Royal Inst. Brit. Architects; fellow Soc. Antiquities. Author: (with T. E. Lawrence) The Wilderness of Sin; (with T. G. Peet) The City of Aten; Dead Towns and Living Men; Carchemish, 3 vols., 1914-54; Ur Excavations; (with others) Al 'Ubaid, 5 vols., 1927; The Royal Cemeteries, 1934; The Early Periods, 1944; The Symerians, 1928; Ur of the Chaldees, 1929; Digging Up the Past, 1930; The Development of Sumerian Art, 1935; Abraham, 1936; The Protection of the Treasures of Art and History in War Areas, 1947; A Forgotten Kingdom, 1953; Spadework, 1953; Excavations at Ur, 1954; Alalakh, 1955; The Ziggurat and Its Surroundings, 1955; History Unearthed, 1958; The Art of the Middle East, 1961. Best known for excavations at Ur, the legendary home of Abraham; discovered mass ritual sacrifice of servants of kings of Ur; established Sumerian civilization was highly developed; major contbr. to knowledge of Sumerians. Died London, Feb. 20, 1960.

WOOLLEY, Dilworth Wayne, biochemist; b. Raymond, Alta., Can., July 20, 1914 (parents Am. citizens); s. Andrew Dilworth and Henrietta (Schonfeld) W.; B.Sc., U. Alta., 1935, LL.D., 1958; M.S., U. Wis., 1936, Ph.D., 1938; M.D. (hon.), U. Amsterdam, 1949; m. Janet McCarter, June 24, 1945. Research asso. U. Wis., 1938-39; prof. biochemistry Rockefeller Inst. for Med. Research, N.Y.C., 1939-65, Rockefeller U., N.Y.C., 1965-66. Cons., OSRD, 1944, NIH, 1945-53. Mem. Am. Chem. Soc. (Eli Lilly award 1948), Am. Soc. Biol. Chemists, Am. Soc. Pharmacology, Am. Inst. Nutrition (pres., Mead Johnson award 1945, councillor), Soc. Am. Bacteriologists (Eli Lilly award 1940), Soc. Exptl. Biology and Medicine (councillor), Am. Acad. Arts and Scis., N.Y. Acad. Medicine, Harvey Soc., Nat. Acad. Scis., German Pharm. Soc. Author: A Study of Antimetabolites, 1952; Biochemical Bases of Psychoses, 1962. First chem. isolation and proof of nicotinamide; contributed to characterization and establishment as vitamins, pantothenic acid and inositol. 1st demonstration of an anti-vitamin (pyrithiamine) effective in higher animals, and elucidated principles of antimetabolites; made 1st antimetabolites of serotonin and discovered it is involved in mental processes; research on receptor for serotonin. Died July 23, 1966.

WOOLLEY, George W(alter), Am. biologist; b. Osborne, Kan., Nov. 9, 1904; s. George Aitcheson and Nora Belle (Jackson) W.; B.S., Ia. State Coll., 1930; M.S., U. Wis., 1931, Ph.D., 1935; m. Anne Geneva Collins, Nov. 2, 1936; children—George Aitcheson, Margaret Anne, Lawrence Jackson. Fellow U. Wis., 1935-36; mem. staff Jackson Meml. Lab., Bar Harbor Me., 1936-49, bd. dirs., 1937-49, v.p. bd., 1943-47, asst. dir. and sci. adminstr., 1947-49, vis. research asso., 1949—; mem., chief div. steroid biology Sloan-Kettering Inst., N.Y.C., 1949-58; prof. biology Sloan-Kettering div. Cornell U. Med. Coll., 1951—, chief div. human tumor exptl. chemotherapy, 1958-61, chief, div. tumor biology, 1961—, asso. scientist Sloan-Kettering Inst. Cancer Research, 1966—; head biol. scis. sect. Nat. Inst. Gen. Med. Scis., NIH, 1966—. Cons. Nat. Edn. Service of U. S., Washington, 1961—; spl. cons. Nat. Cancer Inst., NIH, 1956—; mem. Expert Panel on Carcinogenicity, unio intern. contra cancerum, 1962—; mem. panel com. on growth NRC, 1945-51. Fellow A.A.A.S., N.Y. Acad. Sci.; mem. Am. Mus. Natural History, Nat. Sci. Tchrs. Assn. (cons. 1961—), Am. Assn. Cancer Research (dir. 1951-54), Am. Soc. Human Genetics, Mt. Desert Island Biol. Lab., Soc. Exptl. Biology and Medicine, Am. Inst. Biol. Scis., Am. Assn. Anatomists, Am. Genetic Assn., Wis. Acad. Arts Sci. and Letters, Jackson Lab. Assn., Genetics Soc. Am., Sigma Xi. Contbr. chpts. med. books. Mem. editorial bd. Jour. Nat. Cancer Inst., 1947-50. Research in endocrinology and on cancer. Home: 5300 Westbard Av., Washington 20016. Office: 5333 Westbard Av., Bethesda, Md. 20014.*

WOOLLEY, Richard van der Riet, English astronomer; b. Weymouth, Dorset, Eng., Apr. 24, 1906; s. Charles Edward Allen and Julia M. (van der Riet) W.; B.Sc., U. Cape Town, 1924, M.Sc., 1925; B.A., U. Cambridge, 1928, Ph.D., 1931, M.A., 1937, Sc.D., 1950; m. Gwyneth Jane Margaret Meyler, Mar. 2, 1932. Commonwealth Fund fellow Mt. Wilson Obs., Cal., 1929-31; Isaac Newton student Cambridge U., 1931-33; chief asst. Royal Obs., Greenwich, Eng., 1933-37; John Couch Adams astronomer U. Cambridge, 1937-39; Commonwealth astronomer Mt. Stromlo Obs., Canberra, Australia, 1939-55; astronomer royal, 1956—, Royal Greenwich Obs. Fellow Royal Soc., 1953, Royal Astron. Soc.; mem. Internat. Astron. Union. Author: (with F. W. Dyson) Eclipses of the Sun and Moon, 1937; (with D. W. N. Stibbs), The Outer Layers of a Star, 1953; also numerous articles. Research on dynamics of galaxy, observations of stars whose motions shed light on problem. Home: Herstmonceux Castle, nr. Hailsham, Sussex, Eng. Office: Royal Greenwich Obs., Herstmonceux Castle, nr. Hailsham, Sussex, Eng.

WOOLLEY, Tyler Anderson, Am. acarologist, zoologist; b. Los Angeles, Apr. 3, 1918; s. George Edwin and Elizabeth Amelia (Anderson) W.; B.S., U. Utah, 1939, M.S., 1941; Ph.D., Ohio U., 1948; m. Lucile Green, July 7, 1943; children—Stephen T., Spencer G., Lee G. Grad. Teaching fellow U. Utah, 1939-41, 46; grad. teaching asst. Ohio State U., 1946-47, asst. instr., 1947-48; asst. prof. zoology Colo. State U., Ft. Collins, 1948-52, asso. prof., 1952-58, prof., 1958—; dir. Acarology-Parasitology Tng. Grant, NIH, 1964—. Mem. Am. Micros. Soc., Soc. Systematic Zoology, Entomol. Soc. Am., Kan. Entomol. Soc., Sigma Xi, Phi Kappa Phi. Author: Invertebrate Zoology, 1953. Laboratory Directions for invertebrate Zoology, 1963. Research and numerous publs. on oribatid mites; taxonomy; ecology. Home: 1813 Crestmore Pl., Ft. Collins, Colo. 80521.*

WOOLNOUGH, Walter George, Australian geologist; b. Brushgrove, Australia, Jan. 15, 1876; s. James W. and E. P. Woolnough; ed. Newlington Coll., U. Sydney (Australia); B.Sc., 1898, D.Sc., 1904; m. Margaret Ilma Wilson, 1903; 2 sons, 1 dau. Demonstrator geology U. Sydney, 1898-1901, lectr. geology, 1905-10, asst. prof. 1911-13; lectr. mineralogy and petrology U. Adelaide (Australia), 1902-04; prof. geology U. Western Australia, 1913-19; geologist Brunner, Mond & Co., Ltd., Northwich, 1919-27; geol. adviser Australian Commonwealth Govt., 1927-41; translator, bibliographica, officer Ministry of Post-War Reconstrn., later Ministry Nat. Devel., Melbourne, 1942-51; ret., 1951. First chmn. Commonwealth Govt. Oil Adv. Com.; mem. Prof. David's Funafuti expdn., 1897; leader expdns. to Fiji, 1901, 04; mem. Commonwealth Preliminary Sci. Expdn. to No. Ty., 1911; toured oil fields of U. S. and Argentina for Commonwealth govt., 1930; leader Commonwealth Aerial Giol. Survey Expdn. in Australia, 1932. Fellow Geol. Soc., Royal Soc. New S. Wales (pres. 1926-27, medal 1956); hon. mem. Geol. Soc. Australia. Author: Direction Finding by Sun, Moon and Stars without Mathematical Calculation, 1943; also papers, ofcl. reports, transls. of articles. Made extensive geol. studies throughout Australia, Paupa, mandated ty. of New Guinea. Died Sept. 28, 1958.

WOOLRIDGE, Dean E(verett), Am. electronics exec.; b. Chickasha, Okla., May 30, 1913; s. Auttie Noonan and Irene Amanda (Kerr) W.; A.B., U. Okla., 1932, M.S., 1933; Ph.D., Cal. Inst. Tech., 1936; m. M. Helene Detweiler, Sept. 1936; children—Dean Edgar, Anna Lou, James Allan. Mem. tech. staff Bell Telephone Labs., N.Y.C., 1936-46; co-dir. research and devel. labs Hughes Aircraft Co., Culver City, Cal., 1946-51, dir., 1951-52, v.p. research and development, 1952-53; pres., dir. Ramo-Wooldridge Corp., Los Angeles, 1953-58, Thompson Ramo Wooldridge, Inc., Los Angeles, Cleve., 1958-62; dir. Thompson Ramo Wooldridge, Inc.; research asso. Cal. Inst. Tech., 1962—. Recipient Citation of Honor, Air Force Assn., 1950; Raymond E. Hackett award, 1955; Distinguished Service citation U. Okla., 1960; Westinghouse Sci. Writing award A.A.A.S., 1963. Fellow Am. Phys. Soc., I.E.E.E., Am. Inst. Aeros. and Astronautics; mem. Stanford Assos., Cal. Inst. Assos., Am. Inst. Physics, A.A.A.S., Phi Beta Kappa, Sigma Xi, Tau Beta Pi, Phi Eta Sigma, Eta Kappa Nu. Author: The Machinery of the Brain, 1963; The Machinery of Life, 1966; also articles. Research on phys. electronics, devel. of magnetic recording, electronic computers, guided missiles. Address: 4545 Via Esperanza, Santa Barbara, Cal. 93105.*

WOOLSEY, Clinton Nathan, Am. neurophysiologist; b. Bklyn., Nov. 30, 1904; s. Joseph Woodhull and Matilda (Aichholz) W.; A.B., Union Coll., Schenectady, 1928; M.D., Johns Hopkins, 1933; Rockefeller fellow med. sci., U. Pa., 1938-39; m. Harriet Runion, May 24, 1942; children—Thomas Allen, John David, Edward Alexander. Asst. to asso. prof. physiology Johns Hopkins Sch. Medicine, 1933-48; Charles Sumner Slichter prof. neurophysiology U. Wis., 1948—, dir. Lab. Neurophysiology, 1960—. Mem. mem. council Neurol. Diseases and Blindness Inst., NIH, 1958-62, mem. neurology program project com., 1963—, mem. bd. sci. counsellors Neurol. Diseases and Blindness Inst., 1965—; research adv. panel Nat. Multiple Sclerosis Soc., 1959-64; hon. mem. faculty U. Chile, Santiago, 1964—. Fellow A.A.A.S., Am. Physiol. Soc., Am. Neurol. Assn., Am. Acad. Neurology, Assn. Research Nervous and Mental Diseases; mem. Nat. Acad. Scis., Phi Beta Kappa, Sigma Xi,

Alpha Omega Alpha, Co-editor, contbr.: Biological and Biochemical Bases of Behavior, 1958; Cerebral Localization and Organization, 1964. Contbr. articles profl. jours. Comparative studies on cerebral localization in mammals, using evoked potential and elec. stimulation methods, with spl. attention to patterns of orgn. in somatic sensory and motor systems, auditory, visual and autonomic systems, cerebral cortex; research on thalamo-cortical relations (anat. and physiol.), localization of sensory, motor and inhibitory functions in cerebellum. Home: 106 Virginia Terrace, Madison, Wis. 53705.*

WOONTON, Garnet Alexander, Canadian physicist; b. London, Ont., Can., July 9, 1906; s. Alexander LeBreton and Margaret (Bass) W.; B.S. in Comml. Econs., U. Western Ont., 1925, M.A. in Physics, 1931, D.Sc. (hon.), 1955; m. Isabel Davenport, Aug. 4, 1934; 1 dau., Elizabeth LeBreton (Mrs. Raymond W. Riley). Faculty, U. Western Ont., 1931-48; faculty McGill U., Montreal, Que., Can., 1948—, dir. Eaton Electronics Research Lab., 1950—, chmn. dept. physics, 1955-68. Dir. Varian Assos. Can., Ltd. Recipient Gold medal for Achievement in Physics, Canadian Assn. Physicists, 1963. Fellow I.E.E.E. (past dir.); mem. Canadian Assn. Physicists (past pres.), Internat. Radio Sci. Union (past v.p.), Royal Soc. Can., Am. Phys. Soc. Research, publs. physiology of elec. impulses, microwave optics, paramagnetic resonance and relaxation. Home: 7400 de Tilly, Montreal 16, Que., Can.*

WOOSTER, William Alfred, English physicist; b. London, Aug. 18, 1903; s. Ernest Alfred and Rachel (Johnson) W.; M.A., U. Cambridge (Eng.), 1925, Ph.D., 1927, Sc.D., 1950; B.Sc., U. London (Eng.), 1924; m. Nora Anna Martin, July 28, 1928; children—Antony Martin, Geoffrey Archer, Anna Shirley (Mrs. Fabio Pasti). Faculty, Cambridge U., 1928-60, lectr., 1935-60; dir. Crystal Structures Ltd., Bottisham, Cambridge, 1960—. Fellow Inst. Physics; mem. Assn. Sci. Workers (pres. 1965—), World Fedn. Sci. Workers (treas. 1947—), Mineral. Soc. Author: Crystal Physics, 1938; (with Henry, Lipson) Interpretation of X-ray Diffraction Photographs, 1951; Experimental Crystal Physics, 1957; Diffuse X-ray Reflections from Crystals, 1962; also numerous articles. Research on natural radioactivity, phys. properties of crystals and its presentation to student, diffuse scattering of X-rays from crystals especially diamond; design of an automatic X-ray diffractometer. Home: 339 Cherry Hinton Rd., Cambridge. Office: Crystal Structures Ltd., Bottisham, Cambridge, Eng.*

WORCESTER, Edward Somerset (marquis), Brit. inventor; b. Eng., 1601; s. Edward Somerset Worcester; expended much money on cause of Charles I and on sci. expts.; impoverished after 1660; enabled by act of Parliament (1663) to receive profits from water commanding engine he had invented; built a simple to operate machine for lifting water; built apparatus which contained alternating vessels filled with cold and heated water and shot a stream of water 40 feet high, patented by Parliament, 1663; thus credited with building 1st steam engine. Author: Century of the Names and Scantlings of Inventions. Died 1667.

WORDEN, Edward Chauncey, I, Am. chemist; b. Ypsilanti, Mich., Apr. 17, 1875; s. Chauncey Perry and Elvira Mabel (Brainerd) W.; Ph.C., U. of Mich., 1896; B.S., N.Y. U., 1907, M.A., 1909, D.Sc., 1921; m. Anna Wilhelmina Breitsman, Sept. 25, 1901; children—Marian Alice (Mrs. De Witt Bell), Edward Chauncey II, Anna Lois, Waite Warren, Loanna. Served as chemist at N.Y. Agrl. Expt. Sta., Geneva, 1896-97; mem. Crane & Worden, chemists, N.Y., 1899-1900; chemist Celluloid Zapon Co., Springfield, N.J, 1900-02, Clark Thread Co., Newark, 1902-14, Worden Lab., Milburn, N.J., 1914—. Chmn. com. on airplane coatings NRC, 1916; edition Report Aviation Chemistry, 1914-18, prepared for U. S. Army Air Service; chief of airplane wing coating sect. Bur. of Aeronautics, Washington, 1916-18, crossing Atlantic 14 times for U. S. Govt. Fellow Chem. Soc., London, and French Acad., Paris. Author: Nitrocellulose Industry, 2 vols., 1911; Cellulose Acetate, 1915; Technology of Cellulose Esters, Vol. I, 1921; Chemical Patents Index (United States), 1915-1924, 5 vols., 1927; (with Edward C. Worden, II) Technical Dict. of Chemistry (containing over 400,000 separate headings in alphabetical arrangement). Asso. editor Kunstoffe, also of La Coutchoue et la Guttapercha. Died Sept. 22, 1940.

WORK, Harold Knowlton, Am. chem. engr.; b. Hartford, Conn., May 21, 1901; s. Norman Porter and Hattie (Lincoln) W.; A.B., Columbia U., 1923, Chem.E., 1925; Ph.D., U. Pitts., 1929; m. Margaret Virginia Leal, Dec. 28, 1925; children—Mary Jane (Mrs. David Sanford Laity), Harriet Lincoln (Mrs. Robert Leonard Luce), Harold Knowlton, Virginia Leal (Mrs. John David Foulke). Research fellow Mellon Inst., 1925-29; with Aluminum Co. Am., 1929-36, head div. aluminum research labs., 1929-36; mgr. research div., dir. research Jones & Laughlin Steel Corp., 1936-49; faculty N.Y. U., N.Y.C., 1949-66, asso. dean Coll. Engring., 1957-66, dir. Engring. Found., 1959-65. Sec., Nat. Acad. Engring., Washington, 1964—. Recipient Illig medal Columbia U., 1925, Robert W. Hunt prize Am. Inst. Metall. Engrs., 1942. Fellow N.Y. Acad. Scis.; (mem. standing com. engring. mgmt.), Am. Soc. Elec. Engrs. (mem. engring. coll. research council), Am. Soc. Metallurgists, Am. Soc. for Testing and Materials, Indsl. Research Inst., Metals Sci. Club, Nat. Conf. on Adminstrn. Research, Am. Inst. Chemists, Engring. Fedn. Am. Chem. Soc., Am. Soc. Chem. Engrs., Am. Inst. Mining, Metall. and Petroleum Engrs., Phi Beta Kappa, Sigma Xi, Phi Lambda Upsilon, Tau Beta Pi, Epsilon Chi, Pi Tau Sigma, Theta Tau. Contbr. articles to tech. jours. Invented oxalic-sulfuric electrolyte for anodizing, double anodic treat, oxalic acid electrolyte, coating aluminum pistons, oxide coating aluminum, white oxide coating aluminum, self-lubricating pistons, controlling Bessemer converter, steelmaking. Home: 9106 Courtley Ct., Fairfax, Va. 22030. Office: 2101 Constitution Av., Washington 20418.*

WORK, Lincoln T(honas), Am. engr.; b. Hartford, Conn., Dec. 28, 1898; s. Norman Porter and Hattie (Lincoln) W.; A.B., Columbia, 1918, Chem.E., 1921, A.M., 1924, Ph.D., 1929; m. Clara Radcliff, Sept. 7, 1922; children—Lillian Radcliff (Mrs. David Acker), Dorothy Lincoln (Mrs. Willard Beauchamp). Instr., then asso. prof. Columbia, 1921-40; also cons. chem. engr.; dir. research and devel. Metal & Thermit Corp., Rahway, N.J., 1940-49; cons. engr., N.Y. C., 1949—. Mem. Am. Ceramic Soc., Am. Chem. Soc., Am. Inst. Chem. Engrs., Am. Inst. Mining and Metall. and Petroleum Engrs., Am. Soc. for Metals, Am. Soc. Testing and Materials, Assn. Cons. Chemists and Chem. Engrs. (pres. 1957-58), Electrochem. Soc., Institute Metals, Soc. Chem. Industry (chmn. Am. Sect. 1941), Am. Inst. Chemists (prses. 1952-54, hon. mem.), N.Y. Chem. Club (pres. 1960-62). Author chem. publs.; contbr. sect. Perry Chemical Engineers Handbook, 4th edit. Holder patents. Research on heterogeneous systems in chemical and metallurgical engineering; paints, pigments, cement, surface chemistry. Home: 15 N. Crescent, Maplewood, N.J. 07040. Office: 50 E. 41st St., N.Y.C. 10017.*

WORK, Telford Hindley, Am. physician; b. Selma, Cal., July 11, 1921; s. Telford and Ada (Hindley) W.; A.B., Stanford, 1942, M.D., 1945; D.T.M. & H. London Sch. Trop. Medicine, 1949; M.P.H., Johns Hopkins, 1952; m. Ellen Trauman Fox, Sept. 17, 1949; children—Clemens Paul, Amrit Louise. Physician, Fiji Colonial War Meml. Hosp., Suva, Fiji Islands, 1949-51; mem. Staff Rockefeller Found., N.Y.C., 1952-61, arbovirus researcher, N.Y.C., 1952, virology dept. Cairo, Egypt, 1952-54, N.Y. Labs., 1958-60; arbovirus research Virus Research Center, Poona, India, 1954-58, dir., 1955-58; chief virology sect. USPHS Communicable Disease Center, Atlanta, 1960-66; prof. infectious and tropical diseases U. Cal., Los Angeles, 1966—. Cons. WHO, 1959—, regional arbovirus lab. dir., 1961-66; cons. Inter-Oceanic Canal; cons. Fla. State Bd. Health Encephalitis Research Center. Mem. Am. Soc. Trop. Medicine, Am. Pub. Health Assn., Am. Epidemiol. Soc., Royal Soc. Tropical Medicine, Wildlife Diseases Assn., Am. Bd. Microbiology, Am. Ornithologists Union, Cooper Ornithol. Soc., A.A.A.S. Contbr. numerous articles to sci. jours. Discovered and described Kyasanur Forest disease, a new hemorrhagic fever caused by a tickborne virus of the Russian spring-summer virus complex; devel. virological facilities and staff for India; discovered human infection with Venezuelan equine encephalitis virus, other neotropic arboviruses and encephalitis in U. S.; investigation of migratory birds as reservoirs and disseminators of arboviruses; initiation regional resources for detection of arbovirus activity in U. S. Home: 30461 Pacific Coast Hwy., Malibu, Cal. 90265. Office: Center for Health Scis., U. Cal., Los Angeles 90024.*

WORKING, Elmer Joseph, Am. economist; b. nr. Denver, May 29, 1900; s. Daniel W. and Grace (Booth) W.; student U. Denver, 1916-17, 18-19, George Washington U., 1917-18; B.S., U. Ariz., 1921; M.S., Ia. State Coll., 1922; postgrad. U. Minn.; postgrad. Harvard, Ph.D., 1952; m. M. Gertrude Brown, June 11, 1927; children—Nancy Joan (Mrs. Gerald A. Mendelsohn), Robert Daniel. Staff, Inst. Econs., 1923-26; asso. prof. U. Minn., 1927-28; sr. economist Bur. Agrl. Econs., 1928-35; asso. prof., prof. U. Ill., 1935-54; prof. agrl. econs. Wash. State U., 1954-66, chmn. 1954-65. Fellow Econometric Soc.; mem. Am. (v.p. 1945), Western (pres. 1959-60) farm econs. assns., Am. Econ. Assn., Am. Statis. Assn., Sigma Xi, Phi Kappa Phi, Gamma Sigma Delta. Author: Demand for Meat, 1954; also articles. Contbd. to understanding of relationships of econ. theory and statis. methods in estimation of demand and supply curves, including 1st recognition of identification problem and 1st statis. evidence of differences in short-run and long-run elasticities of demand. Home: 5917 E. 4th St., Tuscon 85711.*

WORKMAN, Everly John, Am. physicist; b. Loudonville, O., July 2, 1899; s. William Truman and Rosella (Boner) W.; B.S., Whitman Coll., 1924, D.Sc., 1955; Leland Stanford U., 1926-27; Ph.D., U. Va., 1930; m. Frances Emerson Cox, Sept. 29, 1968. Instr. physics Whitman Coll., 1924-26, N.D. Coll., 1927-28; Nat. Research Fellow, Bartol Research Foundation of the Franklin Inst., 1930-31; at Calif. Inst. Tech., 1931-32; research asso. Reed Coll., 1932-33; asso. prof. of physics and head physics dept., U. of N.M., 1932-42, prof. of physics and head of dept., 1942-46; pres. N.M. Inst. Mining and Technology, 1946-65; meteorologist, dir. Cloud Physics Obs., U. of Hawaii, Inst. Geophysics, Hilo, Hawaii, 1965—; specializing atmospheric physics, ordnance research and earth sciences. Fellow Am. Phys. Soc., American Geophys. Union, A.A.A.S.; member of New York Academy of Science, also mem. Sigma Xi, Phi Beta Kappa. Author publs. Research in fields of thermodynamics; penetrating radiation measurements; many types of instrumentation; thunderstorms, atmospheric electricity, precipitation phenomena, and lightning. Address: 17 Kuikahi St., Hilo, Hawaii. 96720.*

WORM, Ole (or Vorm), Danish anatomist; b. Jutland, 1588; physician to Christian IV, also to Fredrick III. Author: Fasti Danici, 1626; Lexicon runicum et appendix ad monumenta danica, 1650; numerous treatises on medicine and natural history. Described Wormian bones in suture lines of skull. Died 1654.

WORSHAM, James Essex, Jr., Am. chemist; b. Newport News, Va., Apr. 19, 1925; .s. James Essex and Thelma (Fox) W.; B.S., U. Richmond, 1947; M.S., Vanderbilt U., 1949; Ph.D., Duke, 1953; m. Esther Muriel Tappmeyer, Oct. 17, 1960; 1 son, Russell Mark. Faculty, Hampden Sydney Coll., 1953-54; faculty U. Richmond (Va.), 1954-60, asso. prof. chemistry, 1960—; research asst. Mass. Inst. Tech., 1959-60. Vis. prof. cardiopulmonary research group dept. medicine Med. Coll. Va., 1963—. Recipient J. Shelton Horseley Research award Va. Acad. Sci., 1958. Mem. Am. Chem. Soc., Am. Crystallographic Assn., Am. Assn. U. Profs., Sigma Xi, Omicron Delta Kappa. Research on structure of amides; diffraction studies of urea and palladium-hydrogen system. Home: 11 Bostwick Lane, Richmond, Va. 23173.*

WORTELS, see Ortelius Abraham.

WORTHEN, Amos Henry, Am. geologist; b. Bradford, Vt., Oct. 31, 1813; s. Thomas and Susannah (Adams) W.; m. Sarah B. Kimball, Jan. 14, 1834; 7 children. Moved to Warsaw, Ill., 1836, entered dry-goods bus.; moved to Boston, 1842; ret. from business, became geologist assisting in Ill. Geol. Survey, 1844; became geologist Ia. Geol. Survey, 1855; apptd. state geologist Ill., 1858; mem. Am. Philos. Soc., Nat. Acad. Scis. Author: Geological Survey of Illinois, 8 vols., 1866-90. Described over 1600 fossil species; specialist in classification of lower Carboniferous strata. Died Warsaw, May 6, 1888.

WORTIS, Joseph, Am. psychiatrist; b. Bklyn., Oct. 2, 1906; s. Harry and Selina (Brunswick) W.; B.A., N.Y. U., 1927; M.D., Vienna (Austria) Med. Coll., 1932; m. Helen Zunser, Jan. 13, 1934; children—Henry H., Emily (Mrs. William Leider), Edward Irving. Practice medicine specializing in psychiatry, N.Y.C., 1935—; asst. psychiatry Johns Hopkins Med. Sch., 1938-39; physiol. investigator Aviation Research Lab., Columbia U., 1942-43; asst. clin. prof. psychiatry N.Y. U. Med. Coll., 1945-52; asso. clin. prof. psychiatry N.Y. State Coll. Medicine, Bklyn., 1958—; prof. psychiatry N.Y. Sch. Psychiatry; dir. div. pediatric psychiatry Jewish Hosp. Bklyn., 1950-68; dir. devel. studies Maimonides Community Mental Health Center, 1968—. Fgn. Travel fellow, 1934-41. Diplomate Am. Bd. Psychiatry and Neurology. Fellow N.Y. Acad. Scis., Am. Psychiat. Assn., N.Y. Acad. Medicine, Am. Pub. Health Assn., Am. Orthopsychiat. Assn.; mem. A.A.A.S., Soc. for Biol. Psychiatry (past pres.), Am. Acad. Mental Retardation, Am. Assn. Mental Deficiency (v.p.), Author: Soviet Psychiatry, 1950; Fragments of an Analysis with Freud, 1954; also numerous publs. Editor: Basic Problems of Psychiatry, 1953. Editor, Recent Advances Biol. Psychiatry, 1958—, Mental Retardation Ann. Rev. Introduced shock treatment to U. S.; research on social and physiol. psychiatry, mental retardation. Home: 152 Hicks St., Bklyn. 11201. Office: 1082 10th Av., Bklyn. 11219.*

WORTIS, Sam Bernard, Am. physician; b. Bklyn., Mar. 25, 1904; s. Morris and Rose (Switzer) W.; A.B., N.Y. U., 1923; M.D., Cornell U. Med. Coll., 1927; m. Ethel Emerson, Sept. 21, 1934; children—Michael, Ethan. Instr. exptl. neurology N.Y. U. Med. Coll., 1930-32, asso. prof. neurology, 1934-44; asso. in psychiatry Johns Hopkins U. and Hosp., 1935-36; asst. prof. clin. medicine Cornell U. Med. Coll., 1938-42; prof. psychiatry and chmn. dept. of psychiatry N.Y. U. Med. Coll. 1942—, prof. neurology and chmn. dept., 1946—, dean Coll. Medicine, Postgrad. Med. Sch., 1960-64; cons. mental health WHO, 1960—; cons. Montefiore, Goldwater Meml. Polyclinic, Bellevue, St. Vincents hosps. (N.Y.C.), Danbury (Conn.) Hosp., VA, U. S. Marine hosps., Ellis Island and Stapleton, S.I.; dir. psychiatric and neurolog. services of Univ. Hosp., N.Y. Univ., Bellevue Med. Center; cons. Central Suffolk Hosp., Riverhead, N.Y., Eastern L.I. Hosp., Greenport, N.Y., U. S. Naval Hosp., St. Albans, N.Y.; cons. other govt. and pvt. agys. Diplomate Am. Bd. Psychiatry and Neurology (examiner, pres.), Am. Bd. Internal Medicine. Fellow A.C.P., Am. Pub. Health Assn.; mem. N.Y. Soc. Clin. Psychiatry (pres. 1960—), Am. Psychiat. Assn. (pres. N.Y. County dist. br. 1960—), also nat., state, local med., profl., sci. orgns. Contbr. to Am. and fgn. jours. Asso. editor: Diseases of the Nervous System, Am. Jour. Psychiatry. Editor Yearbook Neurology, Psychiatry and Neuro-Surgery. Research on metabolic, neurologic and psychodynamic aspects of nervous and mental disease. Home: 330 E. 33d St., N.Y.C. 10016. Office: 410 E. 57th St., N.Y.C. 10022.*

WORZEL, John Lamar, Am. geophysicist; b. West Brighton, N.Y., Feb. 21, 1919; s. Howard Henry and

Marie Alma (Wilson) W.; B.S., Lehigh U., 1940; M.A., Columbia U., 1948, Ph.D., 1949; m. Dorothy Crary, Nov. 22, 1941; Sandra Lee (Mrs. Adrian Anthony John Browne), Howard, Richard, William. Research asso. Woods Hole Oceanographic Instn., 1940-46; mem. faculty Columbia U., 1947-—, asst. dir. Lamont Geol. Observatory, 1951-64, prof., 1957-—, acting dir. Lamont Geol. Observatory, 1964-65, asso. dir., 1965-—; geophys. cons. Office Naval Research 1950; Guggenheim fellow Cambridge U., 1963-64. Recipient Meritorious Pub. Service citation U. S. Navy, 1964. Fellow Geol. Soc. Am., Am. Geophys. Union; mem. A.A.A.S., Am. Phys. Soc., Internat. Union Geodesy and Geophysics, Seismological Soc. Am., Soc. Exploration Geophysicists, Sigma Xi, Tau Beta Pi, Pi Mu Epsilon. Author: Pendulum Gravity Measurements at Sea, 1936-59, 1965; also articles. Research on normal mode of propagation of sound in shallow water, in SOFAR sound propagation, in seismic refraction measurements at sea, submarine photography, gravity measurements at sea with pendulums and on surface ships with gravimeters. Home: Ludlow Lane, Palisades, N.Y. 10964. Office: Lamont Geol. Observatory, Palisades, N.Y. 10964.*

WOTIZ, Herbert Henry, biochemist; b. Vienna, Austria, Oct. 8, 1922; s. Edward and Irene (Politzer) Wottitz; B.Sc., Providence Coll., 1944; Ph.D., Yale, 1951; m. Miriam S. Rose, June 15, 1947; children—Harriet-Sue, Robert P., Richard A. Came to U. S., 1938, naturalized, 1944. Faculty, Boston U. Sch. Medicine, 1951-—, prof. biochemistry, 1963-—; research fellow Mass. Meml. Hosps., Boston, 1951-55, research asso., 1955-—; v.p. Consilium, Inc., Milton, Mass., 1965-—. USPHS Research career Devel. fellow, 1960-64, 65-—. Mem. Am. Assn. Cancer Research, Am. Chem. Soc., Endocrine Soc., Am. Soc. Biol. Chemists, Soc. for Exptl. Biology and Medicine, Am. Assn. U. Profs., Sigma Xi. Author: (with Clark) Gas Chromatography in the Analysis of Steroid Hormones, 1966; also numerous articles, chpts. in books. Established final step in female sex hormone biosynthesis, extra-hepatic male sex hormone metabolism in man, rapid and ultramicro hormone analysis by gaschromatography. Home: 9 Cape Cod Lane, Milton, Mass. 02187. Office: 15 Stoughton St., Boston 02118.*

WOTTON, Edward, English zoologist, naturalist, physician; b. Oxford, Eng., 1492; med. degree Oxford U., 1555; physician to Henry VIII; fellow Coll. Physicians. Author: De differentis animalium. A founder of entomology; developed a new and critical method of animal classifications. Died 1555.

WOULFE, Peter, Brit. chemist, mineralogist; b. circa 1727; fellow Royal Soc., 1767, Copley medal, 1768. Contbr. articles to profl. jours. Discovered native tin, Cornwall, Eng., 1766; invented Woulfe's bottle which passed gases through liquids (1st convenient method for producing concentrated solutions of soluble gases, for purifying insoluble gases from soluble impurities); studies on structure of stannic sulphide; produced picric acid by treating indigo cochineal and other coloring matters with strong nitric acid. Died 1803.

WRAY, James Bailey, Am. physician; b. Knoxville, Tenn., Apr. 1, 1926; s. James Bailey and Elizabeth (Hobson) W.; B.S., U. Tenn., 1947, M.D., 1950; m. Margaret Mustoe, Apr. 9, 1950; children—Nancy Elizabeth, David James, Jonathan Lee. Faculty, Duke, 1955, Bowman Gray Sch. Medicine, 1957-60; faculty State U. N.Y. Upstate Med. Center, Syracuse, 1960-66, prof. orthopedica, 1962-66, chmn. dept. orthopedic surgery, until 1966; prof. orthopedic surgery, chmn. dept. orthopedics Ind. U. Med. Sch., Indpls., 1966-—. Recipient Kappa Delta award, 1962, Angiology Research Found. award, 1964, Nicolas Andry award, 1966. Diplomate Am. Bd. Orthopedic Surgery (examiner 1962-—). Mem. Am. Orthopaedic Assn., Am. Acad. Orthopedic Surgery, Am. Soc. Surgery of Hand, Orthopedic Research Soc., A.C.S. Research and publs. on human circulation, reaction of vascular system to bone fracture, post traumatic metabolism, effects of hormones on fracture healing. Home: 4707 Briar Patch Ct., Indpls. 46250. Office: 1100 W. Michigan St., Indpls. 46207.*

WREN, Sir Christopher, English architect and scientist; b. East Knoyle, nr. Tisbury, Eng., Oct. 20, 1632; s. Christopher and Mary (Cox) W.; ed. Oxford B.A., 1650/51; M.A., 1653; D.C.L., 1661; Cambridge, LL.D., 1661; fellow of All Souls' College, 1652-1660/61; m. 1st Faith Coghill, Dec. 1669; 2 sons, Gilbert and Christopher; m. 2d Jane Fitz William, 1676; 2 children, Jane and William. Prof. of astronomy, Gresham College, London, 1657-60; Savilian prof. astronomy, Oxford, 1660/61-1673; designed chapel of Pembroke college, Cambridge, 1663-65; built new "theatre", Oxford, 1664-69; journeyed to France, 1665; on commission to rebuild London after Great Fire of 1666; designed St. Paul's Cathedral; surveyor general, 1669-1718; planned 50 London churches, 1670-1711; Greenwich Observatory, 1675; Cambridge Library, 1676-84; built Chelsea hospital, 1682-91; Hampton court, 1689-94; Kensington Palace, 1689-1702; Marlborough house, 1709-11. A founder and Fellow of the Royal Society (pres. 1681-83); knighted, 1673. Friend of Evelyn, Boyle, Wallis, Isaac Barrow, Halley, Newton, Hooke, and Flamstead; exerted considerable influence upon architecture in Eng. and abroad; displayed clarity, order, and dignity in his work; rarely sought plastic effects through modeling; had baroque criteria for his architecture; evolved hypothesis concerning comets; made model of the moon; studied Saturn and its rings; discovered a graphic method of computing eclipses; studied variations of the barometer; developed method of finding longitude at sea; devoted time to anatomical and medical studies; devised invention for purifying and fumigating sick rooms. Died London, Eng., Feb. 25, 1723.

WRENN, T(homas) Randall, Am. reproductive physiologist; b. Herndon, Va., Sept. 12, 1924; s. Raymond N. and Winnie (Holden) W.; student Randolph-Macon Coll., 1941-43; B.S., U. Richmond, 1944; postgrad. U. Md., 1947-52; m. Almeda McMurray, Aug. 10, 1945; children—Roger, Leah, Douglas, Wendy. Research biologist research div. Bur. Dairy Industry and Animal Husbandry, U. S. Dept. Agr., Agrl. Research Center, Beltsville, Md., 1946-—. Mem. Am. Dairy Sci. Assn. Research and publs. on ovarian and pituitary hormones in mammary gland growth, histochem. characterization of female reproductive system in cattle, decidual reaction of uterus during implantation, placental influences in mammary gland devel., body temperature and hormone interrelationships in cows, histamine excretion of cattle, mechanisms of estrogen action. Home: 3225 Powder Mill Rd., Adelphi, Md. 20783. Office: Animal Husbandry Research Div., U. S. Dept. Agr., Agrl. Research Center, Beltsville, Md. 20705.*

WRENSHALL, Gerald Alfred, Canadian biophysicist; b. North Battleford, Sask., Can., Apr. 3, 1912; s. Alfred Davenport and Mabel (Gardiner) W.; B.Sc., U. Sask., 1933, M.Sc. in Physics, 1935; Ph.D. in Physics, Yale, 1939; M.A., U. Toronto, 1947, Ph.D. in Physiology, 1951; m. Constance Blake Luscombe, June 29, 1942. Radon pumpman Sask. Cancer Commn., 1935-36; lectr. physics McMaster U., 1939-43; faculty Banting and Best dept. med. research U. Toronto (Ont., Can.), 1943-—, prof., 1959-—, prof. dept. physiology, 1961-—. Joint chmn. com. on sci. program and publ. proc. Internat. Diabetes Fedn., 1964. Mem. Am. Inst. Physics, Canadian Assn. Physicists, Biophys. Soc., Am. Diabetes Assn., N.Y. Acad. Scis. Author: The Story of Insulin, 1962; (with G. Hetenyi, Jr., W. Feasby) On the Nature and Treatment of Diabetes, 1965; also numerous articles. Research absorption X-rays, range energy relationship for atomic nuclei, res. insulin stores in pancreas of diabetic man, tracer methodology. Home: 586 Hillsdale Av. E., Toronto 7, Ont., Can.*

WRIGHT, Sir Almroth Edward, English physician, pathologist; b. Middleton Tyas, Eng., Aug. 10, 1861; ed. Dublin (Ireland), Leipzig, Marburg (both Germany), Strasbourg (now France) univs. Prof. pathology Army Med. Sch., Netley, Eng., 1892-1902; prof. exptl. pathology U. London (Eng.), 1902-46; principal Inst. Pathology and Research St. Mary's Hosp., London. Fellow Royal Soc., 1906, (corr.) mem. French Acad. Scis., 1918. Author: Studies in Immunization (basis for modern immunology) 1909, Technique of the Teat and Capillary Glass Tube, 2d edit. 1921, Pathology and Treatment of War Wounds, 1942, Researches in Clinical Physiology, 1943, Alethetropic Logic, publ. posthumously, 1953. Research parasitic diseases and immunology; introduced antityphoid inoculation, 1896; developed (with D. Semple) agglutination test Malta Feber (Brucella melitensis) 1897; introduced vaccines enteric Tb and pneumonia; proved blood contains opsonins rendering bacteria subject to destruction by phagocytosis, 1903; introduced term opsonic index. Died Farnham Common, Eng., Apr. 30, 1947.

WRIGHT, Arthur Williams, Am. physicist; b. Lebanon, Conn., Sept. 8, 1836; s. Jesse and Harriet (Williams) W.; A.B., Yale, 1859, Ph.D., 1861; postgrad. univs. of Heidelberg and Berlin, 1868-69; m. Susan Forbes Silliman, Oct. 6, 1875. Assisted in preparing revised edit. Webster's Dictionary, 1862-63, 90; tutor in Latin, Yale, 1863-66, in physics, 1866-68, instr. physics, Sheffield Sci. Sch., 1867-68, prof. molecular physics and chemistry Yale 1871-87, prof. exptl. physics, 1887-1906, emeritus, 1906-—, in charge of Sloane Phys. Lab., built from his plans under his supervision, 1883-1906; prof. physics and chemistry Williams Coll., 1869-72. Cons. specialist U. S. Signal Service, 1881-86. Fellow Royal Astron. Soc. Gt. Britain. First to observe the electric shadow in air; discovered and analyzed gases in stony meteorites; measured polarization of zodiacal light, also of light of moon, solar corona and comets; applied cathode discharge in vacuo to form metallic films for mirrors. Died Dec. 19, 1915.

WRIGHT, Barbara Evelyn, Am. biochemist; b. Pasadena, Cal., Apr. 6, 1926; d. Gilbert M. and Leta L. (Browne) Wright; A.B., Stanford, 1947, Ph.D., 1951. Biologist, NIH, Bethesda, Md., 1953-61; research dir. dept. developmental biology Retinal Found., Boston, 1967-—. Tutor biochem. sciences, assistant professor bacteriology and immunology Medical School, Harvard University, since 1967-—. Mem. Am. Soc. Microbiologists, Am. Soc. Biol. Chemists, N.Y. Acad. Scis. Author: Control Mechanisms in Respiration and Fermentation, 1963; also numerous articles, chpts. in books. Research on biochem. basis differentiation in cellular slime mold Dictyostelium discoideum. Home: 106 Dudley Lane, Milton, Mass. 02186. Office: Retina Found., 20 Staniford St., Boston 02114.*

WRIGHT, Benjamin Fletcher, Am. polit. scientist; b. Austin, Tex., Feb. 8, 1900; s. Benjamin Fletcher and Mary (Blandford) W.; B.A., M.A., U. Tex., 1921; Ph.D., Harvard U., 1925; LL.D., Amherst Coll., 1950, Mt. Holyoke Coll., 1958, U. Pitts., 1962; Litt.D., Am. Internat. Coll., 1958; m. Alexa Foote Rhea, July 9, 1926; children—David Herndon, Janet (Mrs. Paul E. Jones, Jr.). Instr. govt. U. Tex., 1922-24, adj. prof., 1925-26; instr. govt. Harvard, 1926-28, asst. prof., 1928-40, asso. prof., 1940-45, prof., 1945-49, chmn. dept. govt., 1942-46, vis. prof., 1966-67; pres. Smith Coll., 1949-59; fellow Center for Advanced Study in Behavioral Scis. Stanford, 1959-60; prof. govt. U. Tex., 1960-—. Mem. Am. Acad. Arts and Scis., Am. Polit. Sci. Assn., So. Polit. Sci. Assn. Author: American Interpretations of Natural Law, 1931; Contract Clause, 1938; Growth American Constitutional Law, 1942; Consensus and Continuity, 1958; also articles. Editor: The Federalist, 1961. Interpretations of Am. polit. and constl. thought and institutions in relation to environmental, traditional, philos. and other factors. Home: 1415 Wathen Av., Austin, Tex. 78703.*

WRIGHT, Byron T., Am. physicist; b. Waco, Tex., Oct. 19, 1917; s. Wilbur and Dora (Thompson) W.; B.A., Rice U., 1938; Ph.D., U. Cal. at Berkeley, 1941; m. Lorna Doone Bloemers, Oct. 21, 1944 (dec. 1964); children—Carol Ann, Susan Lee, Gail Elizabeth. Physicist, Naval Electronics Lab., 1941-42, Manhattan Project, 1942-46; mem. faculty U. Cal. at Los Angeles, 1946-—, prof. physics, 1956-—; Ford Found. vis. scientist CERN, 1963-64. Fulbright research scholar, 1956-57; Guggenheim fellow, 1963-64. Fellow Am. Phys. Soc. Spl. research accelerator devel., nuclear structure physics. Home: 1225 Chickory Lane, Los Angeles 90049.

WRIGHT, Charles, Am. botanist; b. Wethersfield, Conn., Oct. 29, 1811; s. James and Mary (Goodrich) W.; grad. Yale, 1835. Taught sch., Miss., also East Tex., 1835-44; tchr. Rutersville (Tex.) Coll., 1845; accompanied battalion of U. S. troops from San Antonio to El Paso, Tex., summer 1849, collected plants and sent many specimens to Asa Gray at Harvard; botanist U. S.-Mexican boundary survey, 1851-52; became botanist N. Pacific Exploring and Surveying Expdn., 1852; made collections of plants at Cape of Good Hope, Hong Kong, Loo Choo Islands, Japan, 1853-56; conducted bot. explorations of Cuba, 1856-67; curator herbarium, Cambridge, Mass., 1868; librarian Bussey Instn., 1875-76; new specimens discovered by him pub. in Gray's Plantae Wrightianae, Part I, Smithsonian Contbns. to Knowledge, Vol. 3, 1852, Vol. 5, 1853, Botany of the Mexican Boundary Survey, 1859. Died Wethersfield, Aug. 11, 1885.

WRIGHT, Charles Robert, Am. sociologist; b. Pennsauken, N.J., Feb. 26, 1927; s. Frank Watson and Elizabeth (Price Demme) W.; A.B., Columbia U., 1949, M.A., 1950, Ph.D., 1954; m. Anne Marie Krefft, July 1, 1950. With Columbia U., 1953-56, instr., 1954-56; faculty U. Cal. at Los Angeles, 1956-—, prof. sociology, 1965-—; OAS vis. prof. to Chile, 1963. Cons., UN Research Inst. for Social Devel., 1965-66, USPHS, 1965-—; research div. for sociology and social psychology NSF, 1967-—. Fellow Am. Sociol. Assn., A.A.A.S.; mem. Am. Assn. for Pub. Opinion Research, Soc. for Study Social Problems, Pacific Sociol. Assn. Author: (with H. Hyman, G. Levine) Inducing Social Change in Developing Communities, 1967; Mass Communication: a Sociological Perspective, 1959; (with W. Bell, R. Hill) Public Leadership, 1961; (with H. Hyman, T. Hopkins) Applications of Methods of Evaluation, 1962; also articles. Editorial staff Am. Sociol. Rev., 1956-57; asso. editor Pacific Sociol. Rev., 1962-64, Sociol. Inquiry, 1964-67, Jour. Health and Social Behavior, 1966-—. Research on applications sociol. theory to communication, methodological problem of evaluation research on social action programs, innovations in tng. grad. students for sci. sociol. research. Home: 928 Stanford St., Santa Monica, Cal. 90403. Office: Dept. Sociology, U. Cal., Los Angeles 90024.*

WRIGHT, Claude-Starr, Am. physician; b. Laurens, S.C., Mar. 7, 1917; s. Wallace Howell and Inez Elizabeth (Starr) W.; B.S., U. S.C., 1939; M.D., Med. Coll. S.C., 1942; m. Louise Markwalter, Apr. 26, 1958; children—Claude Starr, Mary Louise, Henry Winston, Caroline Starr, James Markwalter. Faculty dept. medicine, Ohio State U., 1947-55, asso. prof., 1953-55; faculty dept. medicine Med. Coll. Ga., Augusta, 1955-—, prof., 1958-—, head div. hematology, 1955-—, dir. dept. continuing edn., 1958-64. Diplomate Am. Bd. Internal Medicine. Fellow A.A.A.S., A.C.P., Internat. Soc. Hematology; mem. Am. Assn. Immunologists, Am. Fedn. for Clin. Research (past. chmn. Midwestern sect.), Am. Soc. for Clin. Investigation, Am. Soc. Hematology, Central Soc. for Clin. Research, Alpha Omega Alpha, Phi Chi. Research and numerous publs. on humoral and cellular def., spleen, leukemias, cancer chemotherapy, erythropoietin, red blood cell destruction. Home: 1120 Glenn Av., Augusta 30904.*

WRIGHT, David McCord, Am. economist; b. Savannah, Ga., Aug. 1, 1909; s. Anton Pope and Hannah McCord (Smythe) W.; student The Citadel, Charleston, S.C., 1926-27, U. Pa. 1927-30; LL.B., U. Va., 1935; M.A., Harvard, 1939, Ph.D., 1940; m. Caroline Noble Jones, June 27, 1940; children—Antony Pope, Peter M., Anna H. Admitted to Ga. bar, 1935; atty. R.F.C., 1936-37; lectr. U. of Va. Law Sch. 1940, 47-55; Planning Board, 1943; faculty U. Va., 1939-55, prof., 1946-55; William Dow prof. econs. and polit. sci. McGill U., 1955-62; prof. econs. U. Ga., Athens, 1962—. Econ. adviser Fed. Res. Bank of Atlanta, 1963-66; lectr. various univs.; Fulbright lectr. Oxford U., 1953-54. Rockefeller fellow, Earhart fellow, 1954-55. Mem. Royal Econ. Soc., Am. (exec. com. 1952-55), So. (pres. 1950) econ. assns. Author: The Creation of Purchasing Power (David Wells prize Harvard), 1941; The Economics of Disturbance, 1947; Democracy and Progress, 1948; Capitalism, 1951; A Key to Modern Economics, 1954; The Keynesian System, 1962; Economics and Economic Growth-Elementary Economics, 1964; Growth and the Economy, 1965. Co-author econ. symposia and reports. Editor: The Impact of the Labor Union, 1951. Editorial com. UNESCO Dictionary of Social Scis., 1959-63. Contbr. articles to econ., legal, other jours. Died Jan. 7, 1968.

WRIGHT, Douglas Tyndall, Canadian civil engr.; b. Toronto, Ont., Can., Oct. 4, 1927; s. George Charles and Etta (Tyndall) W.; B.A.Sc., U. Toronto, 1949; M.S., U. Ill., 1952; Ph.D., Cambridge (Eng.) U., 1954; m. Margaret Anne Maxwell, May 21, 1955; children—William Maxwell, Matthew Clyde, Robert Tyndall, Sarah Jane, Anne Marie. From lectr. to asso. prof. civil engring. Queen's U., Kingston, Ont., 1954-58; prof. civil engring. U. Waterloo (Ont.), 1958-67, chmn. civil engring. dept., 1958-63, dean engring., 1959-66; mem. Ont. Com. on U. Affairs, Toronto, 1964—, chmn., 1967—; vis. prof. Universidad Nacional Autonoma de Mexico, 1964, 66, U. Sherbrook (Que., Can.), 1966-67; Cons. engr. on spl. problems in structural engring., Can., Mexico, U. S., Eng., Australia, 1954—; chmn. standing com. on structures Nat. Bldg. Code Can., 1963—. Athlone fellow Cambridge U., 1952-54. Mem. Engring. Inst. Can., Assn. Profl. Engrs. Ont., Am. Soc. C.E., Internat. Assn. for Bridge and Structural Engring. Research and publs. in structural engring. leading to design methods for problems in plastic design of structures, reticulated space frame shell and plate structures; research in methodology and rationale of engring. design. Home: 6B Wychwood Park, Toronto. Office: 481 University Av., Toronto 2, Ont., Can.*

WRIGHT, Edward, English mathematician, hydrographer; b. Graveston, Norfolk, Eng., circa 1558; s. Henry Wright; B.A., Caius Coll., Cambridge, 1580-81, M.A., 1584; 1 son, Samuel; fellow Caius Coll., 1587-96; accompanied Earl of Cumberland on voyage to Azores, 1589; lects. on nav. (with support from E. India Co.); tutor to Prince Henry, son of James I. Author: Voyages to Azores; Certain Errors in Navigation arising either of the ordinary erroneous making or using of the sea chart, compass, cross staff, and tables of declination of the sun and fixed stars (most influential book on nav. of 16th century), 1599; The Havenfinding Art . . . (adaptation of Simon Stevin's De Havenvinding), 1599; The Description and Use of the Sphaere, 1613; A Short Treatise of Dialling, 1614; A description of Napier's Table of Logarithms, translated 1616; A description of the admirable table of logarithms, 1616; A laudatory address concerning the books on magnets. Published 1st table of meridional parts; assisted in Gilbert's work on magnetism; devised chart in which degrees of latitude were lengthened in same proportions as degrees of longitude (basis for projection of Mercator's charts now in use); associated with Briggs and Napier; pub. method for determining magnitude of earth, 1610; produced table of latitude for dividing meridian into minutes; constructed an orrery for use in predicting eclipses; devised instrument for correcting variation of compass; explained phenomena of dip, parallax, refraction. Died London, 1615.

WRIGHT, Edward Maitland, Brit. mathematician; b. Yorkshire, Eng., Feb. 13, 1906; s. Maitland Turner and Kate (Owen) W.; M.A., D.phil., Oxford (Eng.) U., 1932; LL.D., St. Andrews (Scotland), 1963; m. Elizabeth Phyllis Harris, Aug. 15, 1934; 1 son, John Maitland. Lectr., Christ Church, Oxford, Eng., 1933-35; prof. U. Aberdeen (Scotland), 1936-62, prin., vice chancellor, 1962—. Mem. Anderson Com. on Student Grants, 1959-60, Hale Com. on U. Teaching, 1961-64, Royal Commn. on Med. Edn., 1965-67. Fellow Royal Soc. Edinburgh; mem. London, Am. math. socs. Author: (with G. H. Hardy) Theory of Numbers, 1938; also numerous articles. Research in theory of numbers, spl. functions, asymptotic expansions, delay-differential equations. Home: Chanonry Lodge, Chanonry, Aberdeen, U.K.*

WRIGHT, Frances Woodworth, Am. astronomer; b. Providence, Apr. 30, 1897; d. George William and Nellie (Woodworth) Wright; A.B., Pembroke Coll., 1919; A.M., Brown U., 1920; postgrad. Radcliffe Coll., 1927-30, Ph.D., 1958; postgrad. Harvard. Mem. faculty Elmira Coll., 1920-27; research asst. Harvard, 1928-61, exec. sec. dept. astronomy, 1946-63, lectr. on astronomy, 1958—; astronomer Smithsonian Astrophys. Obs., Cambridge, Mass., 1961—. Fellow Meteoritical Soc., A.A.A.S.; mem. Am. Astron. Soc., Inst. Navigation (mem. editorial bd. 1950-53), Phi Beta Kappa, Sigma Xi. Author: (with B. J. Bok) Basic Marine Navigation, 1944, rev. 1952. Research and publs. on variable stars, marine navigation, meteors, interplanetary dust, large Magellanic cloud. Home: 56 Concord Av. Office: Smithsonian Astrophys. Obs., Cambridge, Mass. 02138.*

WRIGHT, Fred Boyer, Am. mathematician; b. Roanoke, Va., Dec. 14, 1925; s. Fred Boyer and Irene (Straley) W.; student Va. Mil. Inst., 1943-44; B.A., U. N.C., 1947, M.A., 1948; Ph.D., U. Chgo., 1953; m. Katherine Virginia Settle, Sept. 4, 1948; children—Amanda Lois, Katherine Irene. Mem. faculty U. N.C., 1948-49; mathematician adv. bd. on simulation U. Chgo., 1953-54; faculty Tulane U., New Orleans, 1954—, prof. math., 1962—, dir. Internat. Symposium on Ergodic Theory, 1961. Vis. prof. Cambridge U., 1958-59; mem. com. on regional devel. math. Nat. Acad. Scis.-NRC, 1964-67. Alfred P. Sloan research fellow, 1958-62. Mem. Am. Math. Soc., Math. Assn. Am. (cons. com. undergrad. program math. 1965—, chmn. com. on ednl. media 1966), London Math. Soc., A.A.A.S., Am. Assn. U. Profs., Phi Beta Kappa, Sigma Xi, Pi Mu Epsilon. Mem. editorial com. Carus Monographs, 1957-60, Duke Math. Jour., 1962-64. Characterized absolute valued algebras; contbns. to ergodic theory, Boolean algebras, rings of operators. Home: 7725 Plum St., New Orleans 70118.*

WRIGHT, George F., Am. chemist; b. Council Bluffs, Ia., Feb. 23, 1904; s. Franklin Prentice and Mary (Griffith) W.; B.S., Ia. State U., 1929, Ph.D., 1932; m. Renee Alice Grunwald, June 21, 1937; children—George, John, Jill. Prof., U. Toronto (Ont.), 1937—; pres. Empire Labs., 1965-67. Recipient U. S. medal of Freedom, 1945; decorated Order Brit. Empire. Mem. Chem. Inst. Can. (past pres.), Royal Soc. Can., Am. Chem. Soc., N.Y. Acad. Scis., A.A.-A.S. Research and numerous publs. on chemistries of furans, explosives, organometallic compounds, tobacco; dipole monents of organic compounds; infra red spectroscopy; organic microanalyses. Home: 36 Bracondale Hill, Toronto, Ont., Can.*

WRIGHT, Helen, Am. astronomer; b. Washington, Dec. 20, 1914; d. Frederick E. and Kathleen E. (Finley) Wright; grad. Bennett Jr. Coll., Millbrook, N.Y., 1934; A.B., Vassar Coll., 1937, M.A., 1939; grad. student U. Cal.; m. John F. Hawkins, 1946 (div.). Assistant astronomy dept. Vassar Coll., 1937-39; Mt. Wilson Observatory, 1937, jr. astronomer U. S. Naval Observatory, 1942-43. Mem. Internat. Astron. Union, Am. Astron. Soc., History of Sci. Soc. Author: Sweeper in the Sky, The First Woman Astronomer in America, 1949; Palomar, The World's Largest Telescope, 1952; A New Treasury of Science, 1965; Anthropology, 1966; Biology, 1966; Great Underseas Adventures, pub. 1966; Explorer of the Universe, A Biography of George Ellery Hale, published in 1966. Editor of: A Treasury of Science (with Harlow Shapley, Samuel Rapport), 1943, latest, 1963; Readings in the Physical Sciences (with H. Shapley, S. Rapport) 1948; (Great Adventures in Medicine (with S. Rapport), 1952, The Crust of the Earth, 1955, Great Adventures in Science, 1956, The Great Explorers, 1957, Great Adventures in Nursing, 1960, The Amazing World of Medicine, 1961; Mathematics, 1963; Archaeology, 1963; (with S. Rapport) Science, 1963; Engineering, 1963; Physics, 1964; Astronomy, 1964. Studies in history sci., history telescopes, sci. in 19th century U. S. Home: 175 W. 12th St., N.Y.C. 10011.

WRIGHT, Herbert Fletcher, Am. psychologist; b. Muskegon, Mich., Dec. 11, 1907; s. William Henry and Jennie (Raymond) W.; B.A., Neb. Wesleyan U., 1930; M.A., U. Cin., 1931; Ph.D., Duke, 1934; m. Lorene Nell Graver, Sept. 22, 1929; children—William Henry, Katharine Lenore (Mrs. Patrick Dolan). Acting prof. psychology Mercer U., 1934-35; Gen. Edn. Bd. postdoctoral fellow State U. Ia., 1935-36; instr., asst. prof. psychology Carleton Coll., 1936-43; asst. prof. Northwestern U., 1945-47; asso. prof. U. Kan., 1947-49, prof. psychology, 1949—, chmn. dept., 1962—. Cons. Nat. Inst. Mental Health; co-dir. Midwest Field Sta., 1947-54; dir. City-Town Project, 1955—. Fellow Ford Found., 1953-54. Mem. Am., Midwestern, Kan. psychol. assns., A.A.A.S., Am. Assn. U. Profs., Phi Beta Kappa, Sigma Xi. Author: (with R. G. Barker) One Boy's Day, 1951; Midwest and its Children, 1955; Methods in Psychological Ecology, 1949; Recording and Analyzing Child Behavior, 1966. Author, editor (with R. G. Barker, J. S. Kounin) Child Behavior and Development, 1943. Participant early studies in psychol. ecology; helped develop methods for systematic description of naturally occurring conditions and behavior of children in communities; explored relationships between community size and child behavior. Home: 2536 Arkansas St., Lawrence, Kan. 66044.*

WRIGHT, Irving Sherwood, Am. physician; b. N.Y.C., Oct. 27, 1901; s. Harry J. and Cora Ann (Hassett) W.; A.B., Cornell, 1923; M.D., 1926; m. Grace Mansfield Demarest, Oct. 15, 1927; children—Barbara Mansfield, Alison Sherwood; m. 2d, Lois Elliman Findlay, Oct. 31, 1953. Asst. and asso. vis physician N.Y. Post Grad. Med. Sch. and Hosp. of Columbia U., 1929-39, prof. clin. medicine, dir. and exec. officer dept. medicine, 1939-46; asst. physician Bellevue Hosp. (Cornell Div.), 1931-34, asso., 1934-37; faculty Cornell Med. Coll., 1946—, prof. clin. medicine, 1948—; staff N.Y. Hosp., 1946—; physician Met. Opera, 1935-62; dir. medicine Cornell div. Welfare Hosp. Chronic Disease, 1937-40, 1st pres. med. bd. 1937-39, cons., 1940-46; cons. Orange Meml. Hosp., East Orange, N.J., Monmouth Meml. Hosp., Long Branch, N.J. Hackensack (N.J.) Hosp., Mt. Vernon (N.Y.) Hosp.; mem. staff or cons. navy and army agys. Chmn. Josiah Macy, Jr. Found. Conf. blood clotting, 1947-52; chmn. Internat. Com. on Haemostasis and Thrombosis, 1954-63, sec. gen., 1963—; chmn. com. on cerebral vascular diseases NIH, 1961-64. Recipient Albert and Mary Lasker award Am. Heart Assn., 1960. Diplomate Am. Bd. Internal Medicine. Fellow A.C.P. (pres. 1966-67), N.Y. Acad. Medicine; mem. Am. Asso. Physicians, A.M.A. (sect. exptl. med. and therapeutics, chmn., 1939-40), Am. Heart Assn. (sect. for study peripheral circulation, chmn., 1939-40; mem. exec. com., bd. dirs. 1935-57, pres. 1952-53; mem. Nat. Adv. Heart Council, 1954-58), Cornell U. Med. Coll. Alumni Assn. (pres. 1953), N.Y. Acad. Medicine, NRC (subcom. cardiovascular diseases), Am. Soc. Clin. Investigation Soc. Exptl. Biology and Medicine, N.Y. Acad. Scis., Harvey Soc., Acad. Columbian de Ciencas Exactas Fisico-Quimicas Y Naturales (corr., Bogota), Sigma Xi; hon. mem. Royal Soc. Medicine of London, med. socs. in Chile, Peru, Argentina, Cuba, Sweden and USSR, faculty U. Chile. Author numerous books on cardiovascular diseases. Editor-in-chief Modern Medical Monographs. Contbr. to jours. Pioneer in investigation of peripheral circulation by modern techniques; among first to study use of ascorbic acid, many types of anticoagulants in man; described hyperabduction syndrome; contbd. to standardization of blood pressure techniques and nomenclature of blood clotting factors; research on nature and treatment of strokes. Home: 25 East End Av. Office: 450 E. 69th St., N.Y.C.*

WRIGHT, Kenneth Elmer, Am. botanist; b. Logan, O., July 6, 1902; s. Val. H. and Mary (Sunderland) W.; B.S., Ohio State U., 1925, M.S., 1929, Ph.D., 1935; m. Martha R. Rockhold, Aug. 30, 1930; 1 son, Richard E. Instr., Logan High Sch., 1927-28; faculty U. R.I., 1930-46, asso. prof. botany, 1942-46; faculty Smith Coll., Northampton, Mass., 1946—, prof., 1963—. Mem. Am. Soc. plant Physiologists, Am. Soc. Plant Taxonomists, Bot. Soc. Am., Am. Assn. U. Profs., Sigma Xi. Research and publs. on water relations of plants, mineral nutrition of plants and local flora. Home: 91 Woodlawn Av., Northampton, Mass. 01060.*

WRIGHT, Kenneth Osborne, Canadian astrophysicist; b. Ft. George, B.C., Can., Nov. 1, 1911; s. Charles Melville and Agnes Pearl (Osborne) W.; B.A., U. Toronto, 1933, M.A., 1934; Ph.D., U. Mich., 1940; m. Margaret Lindsay Sharp, Sept. 25, 1937; 1 dau., Nora Louise (Mrs. Reymond Osborne). With Dominion Astrophys. Obs., Victoria, B.C., 1936—, asst. dir., 1961-66; dir., 1966—; vis. prof. dept. astronomy U. Toronto, 1960-61; research asso. Mt. Wilson and Palomar Obs., 1962; vis. fgn. prof. dept. astronomy Amherst-Mt. Holyoke Colls., 1963. Fellow Royal Soc. Can.; mem. Royal Astron. Soc. Can. (pres.), Am., Royal astron. socs., Internat. Astron. Union, Astron. Soc. Pacific. Research and numerous publs. on stellar spectra, especially curves of growth and spectra of giant eclipsing stars; research on solar chromosphere. Home: 3050 Devon Rd. Office: Dominion Astrophys. Obs., Rural Route 7, Victoria, B.C., Can.*

WRIGHT, Lemuel Dary, Am. biochemist; b. Nashua, N.H., Mar. 1, 1913; s. Clarence Herman and Avis (Dary) W.; B.S., U. N.H., 1935, M.S., 1936; postgrad., Pa. State U., 1936-37; Ph.D., Ore. State U., 1940; m. Dorthol Ernestine Quarles, May 3, 1941; children—Carolyn Wright (Mrs. David W. Corson), Martha Susan, Priscilla Dee, Barbara Ellen, Nancy Alice. Fellow, U. Tex., 1940-41; instr. U. W.Va. Sch. Medicine, 1941-42; from research biochemist to dir. microbiol. research Merck Sharp & Dohme Research Labs. (formerly Sharp and Dohme Labs.), 1942-56; prof. nutrition and biochemistry Grad. Sch. Nutrition, Cornell U., Ithaca, N.Y., 1956—. Mem. study sect. NIH. Mem. Am. Chem. Soc., Am. Soc. Biol. Chemists, Am. Inst. Nutrition (Borden award 1958). Contbr. articles to tech. jours., chpts. to books, revs. Editorial bd. Jour. Nutrition, Proc. Soc. for Exptl. Biology and Medicine. Research on mevalonic acid, biotin vitamers, microbiol. methods of assay, biosynthesis of pyrimidines and cholesterol. Home: 1035 Hanshaw Rd., Ithaca, N.Y. 14850.*

WRIGHT, Norman, Am. physicist; b. North Baltimore, O., Oct. 26, 1906; s. Neil B. and Clara (Withrow) W.; A.B., Phillips U., 1927; M.S., U. Okla., 1929; Ph.D. in Physics, U. Mich., 1933; m. Erma M. Brooks, Dec. 29, 1934; children—Peter N.C., Virginia D. Research asso. biophysics U. Mich., 1934-37; research physicist Dow Chem. Co., Midland, Mich., 1937-41, dir. chem. physics research lab., 1941—. Recipient Pitts. Spectroscopy award Spectroscopy Soc. Pitts., 1958. Fellow Am. Phys. Soc., A.A.A.S., Optical Soc. Am.; mem. Am. Chem. Soc., Soc. for Applied Spectroscopy (medal 1955), Coblentz Soc., Phi Beta Kappa, Sigma Xi, Gamma Alpha. Research and publs. in molecular spectroscopy, especially devel. instrumentation for application of infrared spectroscopy to chem. analysis and molecular structure determination. Home: 4201 Sherwood Ct., Midland 48640. Office: Chem. Physics Research Lab., Dow Chem. Co., Midland, Mich. 48642.*

WRIGHT, Orville, Am. inventor; b. Dayton, O., Aug. 19, 1871; s. Milton and Susan Catharine (Koerner) W.; ed. pub. and high schs. to 1890; hon. B.S., Earlham Coll., Ind., 1909, LL.D., 1931; Dr. Tech. Sci., Royal Tech. Coll., Munich, 1909; LL.D., Oberlin, 1910, Harvard Univ., 1930, Huntington (Ind.) Coll., 1935; Sc.D., Trinity, 1915, Cincinnati, 1917, Ohio State U., 1930; M.A., Yale, 1919; Dr. Engring., Univ. of Michigan, 1924; D.Sc., Otterbein Coll., Westville, Ohio, 1947; Doctor of Engineering, University of Dayton, 1943; unmarried, Dir. Wright Aeronautical Laboratory, Dayton, O. Awarded the Collier trophy, 1913, for development of the automatic stabilizer; gold medals: Aero Club of France, 1908; Aero Club of United Kingdom, 1908; Acad. of Sports of France, 1908; Aeronautical Soc. Gt. Britain, 1908; Congress of U. S, 1909; State of Ohio, 1909; City of Dayton, 1909; Aero Club America, 1909; French Acad. Sciences, 1909; Cross of Chevalier of Legion of Honor, French, 1909; Cross of Officer of Legion of Honor, 1924; Langley medal, Smithsonian Inst., 1910; Elliott Cresson medal, Franklin Inst., 1914; Albert medal, Royal Soc. Arts, 1917; The John Fritz medal, 1920; bronze medal, International Peace Society; The John Scott medal, 1925; Washington award, 1927; Distinguished Flying' Cross awarded, Feb., 1929; Daniel Guggenheim medal, 1930; Franklin medal, 1933; Medal for Merit, 1947; hon. mem. Aero Club of Sarthe, France, Aeronautical Society, Great Britain, Aero Club of United Kingdom, Osterreichischen Flugtechnischen Vereines, Vienna, Verein Deutscher Flugtechniker, Berlin, American Society Mech. Engrs., Aeronautical Soc. America, Nat. Aeronautic Assn. (gov. at large, 1929-39), Nat. Exchange Club, Ohio Society of New York, Inst. of Aeronautical Science, 1932, Franklin Inst., Nat. Fedn. Post Office Clerks, Inst. of Mech. Engrs., London, Air Line Pilots Assn., Inc.; hon. fellow Royal Aeronautical Soc.; mem. Nat Inventors Council, Nat. Acad. Sciences, Nat. Museum Engring. and Industry (v.p., 1924), Nat. Adv. Com. for Aeronautics, A.A.A.S., Franklin Inst., S.A.E., Engineers' Club of Dayton (hon. life); hon.. scout Boy Scouts America. Mem. Daniel Guggenheim Fund for Promotion of Aeronautics; chmn. advisory com., Daniel Guggenheim Sch. of Aeronautics, New York Univ.; hon. Aircraft Pilot Certificate No. 1, issued by Civil Aeronautics Authority, 1940. With his brother, Wilbur, was the first to fly with a heavier-than-air machine, Dec. 17, 1903, and with him the inventór of the system of control used in flying machines of today. Died Jan. 30, 1948.

WRIGHT, Paul Albert, Am. zoologist; b. Nashua, N.H., June 15, 1920; s. Clarence H. and Avis C. (Dary) W.; S.B., Bates Coll., 1941; A.M., Harvard, 1942, Ph.D., 1944; m. Claire Wilson, June 26, 1943; children—Loren Wilson, Darryl Price, Barton Day. Faculty, Harvard, 1944-45, U. Wash., 1945-46, Boston U., 1946-47, U. Mich., 1947-58; faculty U. N.H., Durham, 1958-—, prof. zoology, 1962-—, chmn. dept., 1963-—. Fellow A.A.A.S.; mem. Am. Soc. Zoologists (treas. 1965-—), Am. Physiol. Soc., Soc. for Exptl. Biology and Medicine, Corp. Marine Biol. Lab., Sigma Xi. Research in ovulation, carbohydrate metabolism, skin color change in cold-blooded vertebrates, particularly Amphibia. Home: 3 Lundy Lane, Durham, N.H. 03824.*

WRIGHT, Paul McCoy, Am. chemist; b. Helena, Okla., Sept. 11, 1904; s. James B. and Della (Maddox) W.; B.S., Wheaton Coll., 1926; M.S., Ohio State U., 1928, Ph.D., 1930; m. Mabel Katherine Davidson, Aug. 11, 1930; children—Marjorie Evelyn (dec.), Eugene Paul, Roger Lauren. Chemistry asst. Ohio State U., 1926-29; with Wheaton Coll., 1929-—, prof., chmn. dept. chemistry, 1940-—; research asso., cons. Argonne Nat. Lab., 1957-—; dir., chmn. Evangel-Air, Inc., 1962-—. Mem. Am. Chem. Soc., A.A.A.S., Am. Sci. Affiliation, Ill. Acad. Sci. Research on half life samarium-147, color centers in rose quartz, comparison of energies of alpha and beta particles in liquid scintillator. Home: 717 N. Washington St., Office: 501 Seminary Av., Wheaton, Ill. 60187.*

WRIGHT, Philip Lincoln, Am. zoologist; b. Nashua, N.H., July 9, 1914; s. Clarence Herman and Avis (Dary) W.; B.S., U. N.H., 1935, M.S., 1937; Ph.D., U. Wis., 1940; m. Margaret Ann Halbert, July 21, 1939; children—Alden, Philip Lincoln, Ann. With U. Mont., 1939-—, prof. zoology, 1951-—, chmn. dept., 1956-—. Fellow A.A.A.S.; mem. Mont. Acad. Scis. (pres. 1957), Am. Soc. Mammalogists, Am. Soc. Zoologists, Am. Assn. Anatomists, Wildlife Soc., Am. Ornithologists Union, Sigma Xi, Gamma Alpha, Phi Sigma. Research and publs. on growth and reprodn. in weasel family, including species of weasel, pine martew, fisher, wolverine, and badger; supervised studies on grouse, pronghorn, elk, spotted skunk. Home: 635 E. Sussex Av., Missoula, Mont. 59801.*

WRIGHT, Raymond William Henry, physicist; b. Ottery St. Mary, Devon, U.K., Sept. 2, 1926; s. William and Gladys (Baker) W.; B.Sc. Spl. Physics with 1st class honors, Queen Mary Coll., U. London (Eng.), 1947, Ph.D., 1949; m. Gladys Margaret Jarrett, Aug. 4, 1950. Faculty, U. Ibadan (Nigeria), 1949-57, sr. lectr., 1954-57; prof., head dept. physics U. Ghana, 1957-63, pro-vice chancellor, 1960-63; prof., head dept. physics, U. W.I., Kingston, Jamaica, 1963-—, dean faculty natural scis., 1965-—. Fellow Inst. Physics, Royal Astron. Soc., Ghana Acad. Scis. (past chmn. phys. scis. sect.), Instn. Elec. Engrs. (asso. mem.), Jamaican Inst. Engrs. Research and publs. on earths atmosphere at low latitudes, radio propagation and properties of ionosphere especially effect of irregularities, density of earth's atmosphere at various heights using laser techniques. Address: Dept. Physics, U. W.I., Kingston, Jamaica.*

WRIGHT, Sewall, Am. geneticist; b. Melrose, Mass., Dec. 21, 1889; s. Philip Green and Elizabeth Quincy (Sewall) W.; B.S., Lombard Coll., 1911; M.S., U. Ill., 1912, Sc.D., 1961; Sc.D., Harvard, 1915, Sc.D. (hon.), 1951; Sc.D. (hon.), U. Rochester, 1942, Yale, 1949, Knox Coll., 1957, Western Res. U., 1958, U. Chgo., 1959, U. Ill., 1961, U. Wis., 1965; LL.D., Mich. State U., 1955; m. Louise Lane Williams, Sept. 10, 1921; children—Richard, Robert, Elizabeth Quincy (Mrs. Phillip Cagan). Sr. animal husbandman in animal genetics Bur. Animal Industry, U. S. Dept. Agr., 1915-25; asso. prof. zoology U. Chgo., 1926-29, prof. 1930-37, Ernest D. Burton distinguished service prof., 1938-54; Leon J. Cole prof. genetics U. Wis., 1955-60, emeritus, 1960-—. Hitchcock prof. U. Cal., spring 1943; Fulbright prof. U. Edinburgh, 1949-50. Pres. X Internat. Congress Genetics, 1958. Recipient Daniel Giraud Elliott medal, Kimber genetics award, Weldon Meml. medal (Oxford), Lewis prize. Mem. Nat. Acad. Sci., Am. Philos. Soc., Am. Statis. Assn., A.A.-A.S., Am. Soc. Zoologists (pres. 1944), Am. Soc. Naturalists (pres. 1952), Am. Acad. Arts and Scis., Am. Genetic Assn., Soc. Animal Prodn., Genetics Soc. Am. (pres. 1934), Inst. Math. Statistics, Soc. for Study Evolution (pres. 1955), Royal Soc., 1963, Am. Soc. Human Genetics, Biometric Soc., Zool. Soc. (corr., London), Royal Soc. Edinburgh (hon.), Genetical Soc. London (hon.). Contbr. to jours. Worked out genetics of many characters of guinea pig with spl. regard to interaction effects; conducted exptl. studies which led to formulation of roles of inbreeding, crossbreeding and selection in livestock improvement, also to gen. theory of evolution in which steps are increasingly favorable interaction systems rather than mutations which are good in themselves. Home: 3905 Council Crest, Madison, Wis. 53711.*

WRIGHT, Sydney Edward, Australian pharm. chemist; b. Sydney, Australia, June 3, 1914; s. William Alfred and Emily Jane (Hayes) W.; Dip.Pharm., U. Sydney, 1935, Ph.D., 1955; B.Sc., U. Queensland, 1943, M.Sc., 1944, D.Sc., 1963; m. Phyllis May Edwards, Jan. 16, 1943; 1 son, William Haldane. Lectr. organic chemistry U. Queensland, 1944-45; prin. New Zealand Coll. Pharmacy, 1946-50; faculty U. Sydney, 1951-—, prof. pharm. chemistry, head dept. pharmacy, 1960-—. Mem. expert com. on food additives WHO; mem. Australian Commonwealth and state coms. on food additives, pesticides, drugs and poisons legislation. Fellow Royal Australian Chem. Inst.; mem. Australian and New Zealand Assn. for Advancement Sci. (chmn. New S. Wales div.). Author: Metabolism of Cardiac Glycosides, 1960; also articles. Research in chemistry of natural products especially constituents of Australian flora, metabolism of plant glycosides with cardiotonic action in man and other animal species, metabolism of drugs, pesticides and insecticides, food additives and other materials important in human environment. Home: 14 Elgin, Gordon, N.S.W. Office: Pharmacy Dept., U. Sydney, N.S.W., Australia. Died Oct. 7, 1966.*

WRIGHT, Thomas, English natural philosopher; b. Byer's Green, Eng., 1711. Instrument maker; declined professorship math. at Imperial Acad. St. Petersburg. Author: Louthiana, or an Introduction to the Antiquities of Ireland, 1748; Original theory . . . of the Universe, 1750; The Use of the Globes or the General Doctrine of the Sphere, 1740; Clavis Celestis being the Explication of a Diagram entitled a Synopsis of the Universe or the Visible World epitomized, 1742; The Universal Vicissitude of Seasons, 1737. Anticipated modern phys.-philosophical theory of material universe; gave theory of Milky Way, 1750, grindstone theory of universe; predicted ultimate resolution of rings of Saturn into congeries of small satellites. Died Byer's Green, Eng., 1786.

WRIGHT, Thomas, Brit. physician, paleontologist, geologist; b. Paisley, Scotland, Nov. 9, 1809; student Royal Coll. Surgeons, Dublin, Peter Street Anat. and Surg. Sch.; qualified, 1832; M.D., St. Andrews, Scotland, 1846; m. Elizabeth May; m. 2d, Mary Ricketts, 1845; children—Thomas Lawrence, Mrs. Edward Bestbridge Wethers, Mrs. Charles Byron Wilcox. Articled to surgeon, Paisley; practiced medicine, Cheltenham; surgeon to Gen. Hosp. Recipient Wollaston medal, 1878. Fellow Royal Soc., 1879, Geol. Soc., Royal Soc. Edinburgh, Brit. Assn. (pres. geol. sect. 1875). Author: Lias Ammonites 4 vols., 1878-84; also articles. Collected Jurassic fossils; described seaurchins and starfishes of Jurassic and Cretaceous formations, 1855-82. Died Nov. 17, 1884.

WRIGHT, Wilbur, Am. aeronaut; b. nr. Millville, Ind., Apr. 16, 1867; s. Milton and Susan Catharine (Koerner) W.; brother of Orville W.; ed. high schs., 4 yrs., Richmond, Ind., and Dayton, O.; (hon. B.S., Earlham College Ind., 1909; LL.D., Oberlin College, Ohio, 1910); unmarried. From 1903, with his brother, Orville, devoted time to heavier-than-air flying machine, patented by Wright Bros. in leading countries of world. Has made numerous flights in U. S. and abroad; sold a machine to U. S. Govt. for $30,000. Awarded gold medal by French Academy Sciences, 1909; also many others. Mem. Aero Club of America. With his brother, 1st to fly with heavier-than-air machine, Dec. 17, 1903. Died May 30, 1912.

WRIGHT, Willard H(ull), Am. med. parasitologist; b. Findlay, O., Dec. 9, 1894; s. William Henry and Ella Mary (Wright) W.; D.V.M., George Washington U., 1917, grad. work, 1929-30, Ph.D., 1935; U. of Del., 1926-27; M.S., Am. U., 1931; grad. work, Sch of Tropical Medicine, San Juan, Puerto Rico, 1935-36; m. Dorothy Walton, June 28, 1919; children—Mary Louise (Mrs. James T. Campbell), Robert Edward. Veterinarian U. S. Dept. Agr., 1919-25, animal disease control in N.C., Md., Tex.; veterinarian Md. State Bd. Agr., 1925-28; parasitol. U. S. Bur. of Animal Industry, 1928-36; sr. zoologist Nat. Inst. Health, U.S.P.H.S., 1936-38, chief div. of zoölogy, 1938-47, scientist dir. since 1945, chief div. tropical diseases, 1947-58; profl. asso. Nat. Academy Sciences-Nat. Research Council, 1958-62; lecturer Harvard School of Public Health, 1950-60; consultant Pan Am. Sanitary Bureau, 1962-63, Gorgas Meml. Inst., 1967-—. Mem. expert com. on schistosomiasis and expert com. on onchocerciasis, also adviser WHO, 1963, 1965, 1966. Professional lecturer medical zoology, George Washington University, 1939-58, member graduate council. Asst. sec. gen. 4th Internat. Congresses on Tropical Med. and Malaria, Washington, 1948, U. S. del. 5th internat. conf., Istanbul, 1953, alternate del. 6th internat. conf., Lisbon, 1958; U. S. del. 11th Pacific Sci. Congress, 1966; mem. commission on parasitic diseases Army Epidemiological Bd. Served as 2d lt., F.A., 32d div., A.E.F., 1917-19; col., U.S.P.H.S., assigned to War Dept. as spl. cons. to surgeon gen., 1945-46, field dir. commn. on schistosomiasis, Army Epidemiol. Bd., Southwest Pacific and Japan, 1945-46. Decorated Legion of Merit, 1946; Billings Medal, A.M.A., 1940; Walter Reed medal, 1958. Member Am. Soc. Tropical Medicine and Hygiene (vice pres. 1956, pres. 1959-60); mem. Acad. Medicine (Washington), also American Society Parasitol. (president, 1950), Société Belge de Médecine Tropicale (honorary foreign member), Sigma Xi. Author numerous science publs.; also co-author of Tropical Health. Contbr. to Ency. Brit., various textbooks on tropical med. and chemotherapy. Research in epidemiology and chemotherapy of tropical diseases. Home: 6624 32d Pl., N.W., Washington 20015. Office: Gorgas Meml. Inst., 2007 Eye St. N.W., Washington 20006.*

WRIGHT, William, Scottish physician, botanist; b. Crieff, Scotland, Mar. 1735; ed. U. Edinburgh (Scotland); M.D., St. Andrews, Scotland. Apprenticed to surgeon, Falkirk, Scotland; surgeon's mate Intrepid, Danae, also in W.I., 1760; became asst. to Dr. Gray, Jamaica, 1764; became partner (with Thomas Steel), Hampden, Trelawny, Jamaica, 1765; named hon. surgeon-gen., Jamaica, 1774; returned to Eng., 1777; sailed as regtl. surgeon to Jamaica regt., 1779; captured by French; sailed to Jamaica, 1782; named physician-gen., Jamaica, 1783; returned to Edinburgh, Scotland, 1785; physician to expdn. under Sir Ralph Abercromby, W.I., 1796-98. Fellow Royal Soc., 1778; mem. Linnean Soc. (asso.), Royal Coll. Physicians Edinburgh (became pres. 1801), Philos. Soc. (later Royal Soc.) Edinburgh (charter), Wernerian Soc. (charter, became v.p. 1808). Contbr. articles to profl. jours. Made natural history collections in Jamaica; verified 760 species of Jamaica plants; genus Wrightea named in his honor. Died Edinburgh, Sept. 19, 1819.

WRIGHT, William, Scottish structural eng.; b. Stewarton, Scotland Dec. 3, 1918; s. James and Jean (Robb) W.: student George Watson's Coll., U. Glasgow, U. Aberdeen; M.A., Sc.D., Ph.D., B.Sc.; m Mildred Anderson Robertson, May 4, 1944. Instr. polytech. teaching U. Aberdeen; head civil engring. studies dept. U. Southampton; prof., head Poly. Sch., Trinity Coll., Dublin; dir. Grad Sch. Engring. Studies, Trinity Coll., Dublin; fellow Trinity Coll. Dublin. Fellow Royal Soc. Edinburgh; mem. Instn. Civil Engrs., Instn. Civil Engrs. Ireland. Author: Beams on Elastic Foundation, Solution by Relaxation Methods, 1952; Design of Curved Beams, 1953. Research and publs. on structures and elasticity, hydraulic problems using models. Home: Gaile, Ballyedmonduff, Stepaside, Dublin, Ireland.

WRIGHT, William Hammond, Am. astronomer; b. San Francisco, Calif., Nov. 4, 1871; s. Selden Stuart and Joanna Maynard (Shaw) W.; B.S., U. Cal., 1893, grad. student, 1894-96, U. Chgo. (Yerkes Obs.), 1896-97; D.Sc., Northwestern U., 1929; m. Elna Warren Leib, Oct. 8, 1901. Asst. astronomer Lick Obs., 1897-1908, astronomer, 1908-44, emeritus, 1941-—, dir., 1935-42, in charge obs.' expdn. to So. Hemisphere, 1903-06. Fgn. asso. Royal Astron. Soc. London; mem. Nat. Acad. Sci.; Am. Philos. Soc., Am. Astron. Soc. Contbr. astron. publs. Photographed Mars; showed that the planets disk appears considerably larger in ultraviolet than in infrared light; attributed this difference to scattering of ultraviolet light by the atmosphere: led to calculation of. thickness of atmosphere. Died May 16, 1959.

WRINCH, Dorothy, biochemist; b. Rosario, Argentina, S.A.; d. Hugh Edward Hart and Ada Minnie (Souter) Wrinch; scholar Girton Coll., Cambridge, Eng., 1913, wrangler, math. tripos, 1916, moral scis. tripos, 1917; B.A., M.A., Cambridge U.; M.A., D.Sc., Oxford (Eng.) U.; M.Sc., D.Sc., London (Eng.) U.;

student univs. Vienna, Paris; m. J. W. Nicholson, 1922 (div. 1938); 1 dau., Pamela; m. 2d, Otto Charles Glaser, 1941 (dec. 1951). Lectr. pure math. Univ. Coll., London; 1918-20, hon. research asso., dept. biochemistry, 1962; research fellow Girton Coll., 1920-24, 30-34, Oxford U., 1931-33, 39-41, Rockefeller Found. fellow, 1935-41; mem. faculty phys. scis. Oxford, 1923-39; lectr. chemistry Johns Hopkins, 1939-41; vis. prof. Amherst, Mt. Holyoke, Smith colls., 1941-42; lectr. physics Smith Coll., 1942-54, vis. prof. physics, 1954-59, vis. research prof., 1959—. Mem. Internat. Commn. Tchrs. Math., 1932-39; mem. corp. Marine Biology Lab., Woods Hole, Mass., 1943. Fellow Am. Phys. Soc.; Mem. Am. Chem. Soc., Am. Crystallographic Assn., Soc. Vis. Scientists, British Assn. (sec. math. sect. 1932-38). Author: Fourier Transforms and Structure Factor, 1946; Chemical Aspects of the Structure of Small Peptides: an Introduction, 1960; Chemical Aspects of Polypeptide Chain Structures and the Cyclol Theory, 1965; also papers. Research and publs. on structural issues in applied math., sci. method, theoretical crystallography, molecular biology; devised cyclol theory of peptide structure, 1936; following Stoll's discovery of cyclol bonds in ergot tripeptides, applied theory to antibiotic and other small peptides, 1960, then after total syntheses in Basel and Moscau of ergot tripeptides and model cyclols and birth of cyclol chemistry, extended theory to polypeptide chains, drawing attention to impact of this updated picture on probable nature of active sites on enzymes and chem. events during peptide synthesis and degradation, other problems of molecular biology, 1965. Home: Greenhaven, Harbor Hill, Woods Hole, Mass. 02543. Office: Dept. Physics, Smith Coll., Northampton, Mass. 01060.*

WRISBERG, Heinrich August, German anatomist; b. St. Andressberg in Harz, Jan. 20, 1739. Held chair of obstetrics, then anatomy, Göttingen. Author: Descriptio anatomica embryonis, 1764; Satura observationum de animalculis infusoriis,1765; De testiculorum ex abdomine in scrotum descensu ad illustrandam in churugia de hernia congenitis utriusque sexus doctrinam, 1779; De utero gravido, tubis, overiis, et corpore luteo quorundam animalium aim usden partibus in homine coliatis, 1782; Observationum anatomicarum de nervis viscertum abdominalium particula prima quae de ganglio plexuque semilunari agit, 3 parts, 1780-1803; Observationum anatomicae de quinto pare nervorum encephali et de nervis qui ex edum duram ingredi falso discuntur, 1777; Observationes anatomicae de vena azyga duplici, aliisque huisus venae rerietatibus, 1778; Sylloge commentationum anatomicarum, 1786; Commentatio de secundarum humanarum varietate, 1773. Research on abdominal nerves; described cuneiform cartilages of larynx (Wrisberg's corpuscles), 1764, lateral meniscus ligament (Wrisberg's ligament), 1764, ganglion of superficial cardiac plexus (Wrisberg's ganglion), 1764, delicate fibers which connect motor and sensory roots of trigeminal nerve (Wrisberg's lines), 1764, med. cutaneous nerve of arm (Wrisberg's nerve), 1777, pars intermedia of facial nerve (Wrisberg's pars intermedia), 1777. Died Göttingen, Hanover, Mar. 29, 1808.

WRONG, Dennis Hume, sociologist; b. Toronto, Ont., Can., Nov. 22, 1923; s. Humphrey Hume and Mary Joyce (Hutton) W.; B.A., U. Toronto, 1945; Ph.D., Columbia, 1956; m. Elaine L. Glickstein, Nov. 24, 1949 (div.); 1 son, Terence Hume; m. 2d, Jacquelin C. Mehta, Mar. 26, 1966. Tchr., Princeton, 1949-50, Rutgers U., 1950-51, U. Toronto, 1954-56, Brown U., 1956-61; mem. grad. faculty New Sch. Sociol. Research, 1961-63; prof. sociology, chmn. dept. Univ. Coll., N.Y. U., 1963—. Mem. Am. Sociol. Assn., Eastern Sociol. Soc., Population Assn. Am., Canadian Polit. Sci. Assn. Author: American and Canadian Viewpoints, 1955; Population and Society, 1961. Editor Social Research, 1962-64. Contbr. articles of theoretical nature on social theory, polit. sociology, stratification, culture and personality, also articles on empirical research on population (fertility) and voting in Canadian politics. Home: Drakes Corner Re., Rd., Princeton, N.J. Office: Dept. Sociology, New York Univ., University Heights, N.Y.C. 10453.*

WRONSKI, Jozef Maria Hoene, mathematician; b. Poznan, Poland, Aug. 24, 1778; studied philosophy and math., Germany; served in Polish and Russian armies, 1793-98; lived in France, from 1801. Author: Prospectus Messianism, 1831; Messianism, Final Union of Philosophy and Religion, 1831-39; Oeuvres mathématiques, 4 vols., 1925. A combinatorialist; studied 4 spl. forms of determinants; noted for Wronskian determinant of a set of functions; made incisive criticisms of philosophy of math.; contbd. to knowledge of celestial mechanics and physics; attempted to find the basis for Leibniz' calculus in Kant's philosophy; held that absolute truth might be reached by math. formulae; applied his system of philosophy to math. analysis, theory of numbers, theory of probabilities. Died Paris, France, Aug. 8, 1853.

WU, Chia-Chu, Chinese physician; b. Taipei, Taiwan, June 30, 1917; s. Shia H and Apu (Chen) W.; M.D., Taipei Imperial U., 1937; m. Kuitu Man, Mar. 23, 1940; children—I-Tsu, I-Lee, I-Len, I-Jen. Prof. obstetrics and gynecology Coll. Medicine, Nat. Taiwan U., 1956—. Fellow Internat. Coll. Surgeons; mem. Assn. Obstetrics and Gynecology Republic of China (sec. gen. since 1961—). Research and publs. on Diplococcus tropicus, fertility and sterility, anaerobic bacteria in obstetrics and gynecology; discovered causative microorganism of tropical ulcer. Home: 173 Kun Ming St., Taipei, Taiwan, Republic of China.*

WU, Chien-Shiung, physicist; b. Shanghai, China, May 31, 1913; d. Zong-Ye and F. H. (Fan) W.; B.S., Nat. Central U., China, 1934; Ph.D., U. Cal. at Berkeley, 1940; D.Sc., Princeton, 1958; D.Sc., Smith Coll., 1959, Goucher Coll., 1960; m. Dr. Luke Chia-liu Yuan, May 30, 1942; 1 son, Vincent W. C. Yuan. Taught Smith Coll. and Princeton; staff mem., War Research, Columbia, 1944, tchr., 1952-57; prof. physics Columbia, 1957—. Recipient research award Research Coop. Am., 1958; Woman of Year award, Am. Assn. U. Women, 1962; Comstock award Nat. Acad. Sci., 1964. Member National Acad. Sci., Am. Phys. Soc., Acadamie Sinica of China. Author: Beta Decay (with Moszkowski), 1965. Specialist in nuclear physics; experimentally established non-conservation of parity in beta-decay, 1957, also conservation of vector current in beta decay, 1963. Home: 15 Claremont Av., N.Y.C. 27.

WU, Ta-You, physicist; b. Canton, China, Aug. 22, 1907; s. Kuo-Chi and Kia-Er (Kuan) W.; B.Sc., Nankai U., Tientsin, China, 1929; M.A., U. Mich., 1932, Ph.D. 1933; m. Kuan-Shih Yuan, Sept. 6, 1936; 1 son, Po Chi. Prof. physics Nat. Peking (China) U., 1934-46; vis. prof. U. Mich., 1947; research asso. Columbia U., 1947-49; vis. prof. N.Y. U., 1948-49; sr. and prin. research officer, head theoretical physics sect. NRC. Can., 1949-63; mem. Inst. for Advanced Study, Princeton, 1958-59; vis. prof. Nat. Taiwan U., 1956-57, U. Lausanne (Switzerland), 1960-61; prof. Poly Inst. Bklyn., 1963-65; prof. physics State U. N.Y., Buffalo, 1965—. Recipient Ting Meml. prize Academia Sinica, 1939, 1st prize Ministry Edn., Republic of China, 1943, medal, 1956. Fellow Am. Phys. Soc., Royal Soc. Can.; mem. Academia Sinica, Phi Beta Kappa, Sigma Xi. Author: Vibrational Spectra and Structure of Polyatomic Molecules, 1939; (with T. Ohmura) Quantum Theory of Scattering, 1962; Kinetic Equations of Gases and Plasmas, 1966; also numerous articles. Research in fields atomic, molecular and nuclear physics, scattering theory, kinetic theory plasmas. Home: 500 LeBrun Rd., Buffalo 14226.*

WU, Tai Tsun, physicist; b. Shanghai, China, Dec. 1, 1933; s. King Ching and Wei Van (Tsang) W.; B.S., U. Minn., 1953; S.M., Harvard, 1954, Ph.D., 1956.; m. Sau Lan Yu, June 18, 1967. Came to U. S., 1950, naturalized, 1964. Mem. Inst. for Advanced Study, Princeton, N.J., 1958-59, 60-61, 62-63; jr. fellow Harvard, 1956-59, faculty, 1959—, Gordon McKay prof. applied physics, 1966—. Author: (with R. W. P. King) Scattering and Diffraction of Waves, 1959. Research on statis. mechanics, electromagnetic theory, elementary particles. Home: 14 Fernald Dr. Office: Pierce Hall, Harvard, Cambridge, Mass. 02138.*

WUCHERER, Otto Eduard Heinrich, physician; b. Oporto, Portugal, July 7, 1820; physician St. Bartholomew's Hosp., London, also Lisbon, Portugal, Nazareth, Cacheoita, Bahia (all Brazil). Author: De mutationibus quals syphilis ejusque medendae natio subit. diss., 1841; also articles. First to discover presence of filarial worms (Wuchereria Bancrofti) in chylous urine, 1866; discovered hookworms in Brazil, 1866; eponym of filarial worm genus Wuchereria da Silva Araujo (phasmid nematode parasites of man), 1877. Died May 7, 1874.

WULFF, John, Am. physicist, metallurgist; b. N.Y.C., Apr. 13, 1903; s. Carl Louis and Dorothea (Gaude) W.; E.M., Cole Sch. Mines, 1924; M.S. in Metallurgy, Yale; D.Sc., Tuebingen U.; m. Eva Thomac, Dec. 19, 1929; children—Angelca (Mrs. A. T. Sawyer), Felicitas, Monica (Mrs. Alan Steinert, Jr.). Nat. Research fellow U. Tuebingen, 1929-30, U. Tuebingen and Munich, 1930-31; faculty Mass. Inst. Tech., 1931—, prof. metallurgy Class of 1922, 1946—. Mem. Am. Phys. Soc., Am. Inst. Mining Engrs. (Distinguished Teaching award), Am. Soc. Metals (Albert Easton White award), Sigma Xi, Theta Xi, Tau Beta Pi. Author: Friction and Surface Finish; Powder Metallurgy; Metallurgy for Engineers; Structure and Properties of Materials. Phys. and metall. research, including electron theory of passivity, surface tension of solid metals, theory and practice of sintering in powder metallurgy, superconductivity of solid solutions and intermetallic compounds. Home: 17 Hillside Av., Cambridge, Mass.*

WÜLLNER, Friedrich Hugo Anton Adolf, German physicist; b. Düsseldorf, June 13, 1835; prof., Bonn, also Aachen, Germany. Author: Über den Wechsel und die Erhaltung der Kraft, 1860; Lehrbuch der Experimentalphysik, 2 vols., 1862-65. Contbd. to theory of heat; investigated vapor pressure of salt solutions, also spectral analysis of gases under pressure. Died Aachen, Oct. 6, 1908.

WUNDERLICH, Bernhard, chemist; b. Brandenburg, Germany, May 28, 1931; s. Richard Otto and Johanne (Wohlgefahrt) W.; came to U. S., 1954, naturalized, 1960; student Humboldt U., Berlin, Germany, 1949-53, Goethe U., Frankfurt, Germany, 1953-54, Hastings Coll., 1954-55; Ph.D., Northwestern U., 1957; m. Adelheid Christa Felix, Dec. 28, 1953; children—Caryn Cornelia, Brent Bernhard. Instr. chemistry Northwestern U., 1957-58; faculty Cornell U., Ithaca, N.Y., 1958-63; faculty chemistry Rensselaer Poly. Inst., Troy, N.Y., 1963—, prof. chemistry, 1965—. Cons., Du Pont Co., 1962—. Mem. Am. Chem. Soc., Am. Phys. Soc., N.Y. Acad. Scis., A.A.A.S. Author: The Crystalline Solid State of Linear High Polymers, 1967; also numerous articles. Research on solid state of linear high polymers. Home: 211 Winter St. Extension, Troy, N.Y. 12180.*

WUNDERLICH, Carl August, German physician; b. Sulz/Neckar, Germany, Aug. 4, 1815; student, Tübingen, Germany, 1833-37, then in Paris, France; became asst. St. Catherine's Hosp., Tübingen, 1837; apptd. inst., Tübingen, 1839; named asst. Tübingen clinic, 1841, provisional chief, 1843, chief, 1846; prof. medicine, Leipzig, Germany, 1850-77; head Leipzig Med. Clinic, from 1850. Author: Ein Beitrag zur Geschichte und Beurtheilung der gegen wartigen Heilkunde in Deutschland und Frankreich, 1841; Vernoch einer pathologischen Physiologie das Blutes, 1845; Handbuch der Pathollogie und Therapie, 1846-56; Geschichte der Medizin, 1859; Das Verhalten der Eigen Warme in Krankeiten (laid found. for modern clin. thermometry), 1868, 2d edit., 1870, English transl. On the Temperature in Diseases: A Manual of Medical Thermometry (W. D. Woodman), 1871; (with Edward Seguin) Medical Thermometry and Human Temperature, 1871; (article) Ueber spontane und primare Pyamie, 1857. Editor Archif fur physiologische Heilkunde (founded with Roser to present medicine as br. of natural sci. and to establish pathology on physiol. basis), 1842-59. Recognized that fever is a symptom of disease rather than a disease itself; introduced regular checking of temperature and fever curve; described course of pyrexia in many conditions. Died Leipzig, Sept. 25, 1877.

WUNDT, Wilhelm Max, German psychologist; b. Neckarau, Baden, Aug. 16, 1832; ed. univs. Heidelberg and Tübingen; Ph.D., M.D.; student Johannes Muller's Inst. Physiology, Berlin, 1856. Dozent physiology U. Heidelberg, 1857-64, asst. to Helmholtz in Physiol. Inst., 1858-64, asst. prof. physiololgy, 1864-74; prof. inductive philosophy U. Zurich (Switzerland), 1874-75; prof. philosophy U. Leipzig, 1875-1917. Author: Lehre von den Musketbewegungen, 1858; Beiträge zue Theorie der Sinneswahrnehmung, 1858; Vorlesungen über die Menschenund Thierseele, 1863; Grundzüge der physiologischen Psychologie, 2 vols., 1873-74; Logik, 1880; Ethik, 1886; System der Philosophie, 1889; Grundriss der Psychologie, 1896; Völkerpsychologie, 10 vols., 1900-20; Einleiting in die Philosophie, 1901; Einführung in die Psychologie, 1911. Founder sci. psychology; founder 1st exptl. psychology lab., 1879; founder 1st psychol. jour. Philosophische Studien, 1881; offered 1st univ. course in sci. psychology, 1862; believed psychology should be concerned with the contents of the conscious mind; studied higher mental processes and their social manifestations; led in gaining acceptance of psychology as an independent discipline, experimental research on perception, feeling, and apperception. Died Grossbathen, nr. Leipzig, Aug. 31, 1920.

WÜNSCH, Erich, German chemist; b. Reichenberg, Germany, Mar. 17, 1923; s. Gustav and Adele (Hübner) W.; diplom, U. Munich (Germany), 1951, doctorate, 1956; m. Ilse Rumler, Mar. 29, 1952 (dec. Feb. 1963); children—Rolf Michael, Birgit Renate, Sabine; m. 2d, Edith Helmes, June 15, 1964. Sci. asst. organic chemistry inst. Hochschule Regensburg, 1951-53; colleague Max-Planck-Institut für Eiweissund Lederforschung, Munich, 1953-60, head dept. peptid chemistry, 1960—. Mem. Gesellschaft Deutscher Chemiker. Author: (with W. Grassman) Fortschritte der Chemie organischer Naturstoffe: Synthesen von Peptiden, vol. XIII, 1956; also articles. Syntheses in steroid field; ground work research in field peptide synthesis; devel. enzyme substrata; synthesis natural substances of peptides, total synthesis of pancreatic hormone glucagon. Home: 2 Midgardstrasse, 8132 Tutsing-Obb, Germany. Office: 46 Schillerstrasse 8, Munich, Germany.*

WURDACK, John J., Am. botanist; b. Pitts., Apr. 28, 1921; s. John H. and Mary (McMahon) W.; B.S., U. Pitts., 1942; B.S. in San Engrng. U. Ill., 1949; Ph.D., Columbia, 1952; m. Marie Louise Solt, Nov. 29, 1959; children—Douglas A., Kenneth J. Asst., asso. curator N.Y. Bot. Garden, N.Y.C., 1949-59; curator dept. botany Smithsonian Instn., Washington, 1960—. Mem. Am. Soc. Plant Taxonomists, Internat. Soc. for Plant Taxonomy, Bot. Soc. Washington, Assn. for Tropical Biology, Torrey Bot. Club. Contbr. numerous articles to sci. jours. Classification of S. Am. flowering plants. Home: 4400 Samar St., Beltsville, Md. 20705. Office: Smithsonian Instn., Dept. Botany, Washington 20560.*

WURTZ, Charles Adolphe, French chemist; b. Wolfsheim/Strasbourg, France, Nov. 26, 1817; student of Balard, Dumas, Liebig; asst. to J. B. Dumas, Sorbonne, 1845, prof. chemistry and toxicology, 1852-75; dean Faculty Medicine, 1866-75, became prof. organic chemistry Paris Faculty Scis., from 1875. Mem. French Acad. Scis., French Acad. Medicine. Author: Traité élémentaire de chimie médicale, 2 vols., 1864-65; Dictionnaire de chimie pure et appliquée, 5 vols., 1868-78, supplement, 2 vols., 1880-86; La theorie anatomique, 1879; Traité de chimie biologique, 1885. Editor Annales de chimie et de physique. Studied organic chemistry; supported atomic theory; research on acids of phosphorus, alkylamine glycol; determined constitution of hypophosphorous acid; discovered phosphorous oxychloride, 1847, primary aliphatic amines, 1848; synthesized hydrocarbons by action of sodium on alkylhalides, 1855; discovered aldol while studying polymerization products of acetaldehyde, 1872; developed synthesis for making paraffin from alkylhalogens through metallic sodium (named after him); established makeup of glycerine; prepared glycol and its derivatives, including lactic acid from propylene glycol, 1856; Wurtz column, Wurtz-Fittig reaction for synthesizing organic halogen compounds named for him, also mineral wurtzite. Died Paris, May 12, 1884.

WURZELBAU, Johann Philipp von, see von Wurzelbau, Johann, Philipp.

WUST, Carl John, Am. immunologist; b. Providence, July 2, 1928; s. Louis August and Ida (Jauernig) W.; B.Sc., Providence Coll., 1950; M.Sc., Brown U., 1953; Ph.D., Ind. U., 1957; m. Barbara Marion Russin, Sept. 5, 1951; children—Carl John, Stephen Louis, Catherine Joanne, Gregory Harold, Elizabeth Diane. USPHS-NIH-Nat. Cancer Inst. Postdoctorate fellow Yale, 1957-59; biochemist biology div. Oak Ridge Nat. Lab., 1959—. Mem. Am. Soc. for Microbiology, Reticuloendothelial Soc., Am. Assn. Immunologists, A.A.A.S. Research on protein synthesis, antibody biosynthesis, enzyme protection from antibody neutralization. Home: 132 Iroquois Rd., Oak Ridge 37830. Office: Biology Div., Oak Ridge Nat. Lab., P.O. Box Y, Oak Ridge 37831.*

WUST, Georg Adolf Otto, German phys. oceanographer; b. Posen, Germany, June 15, 1890; s. Max and Clara (Ebel) W.; Dr.phil., Berlin U., 1919; m. Marth Weyringer, Oct. 22, 1927 (dec. 1941); children—Ilse (Mrs. Emil Hohl), Luise (Mrs. Franz-Joseph Schmidt); m. 2d, Maria Volmer, Dec. 31, 1943. Asst., Institüt fur Meereskunde U. Berlin, 1919-28; chief oceanographer Meteor., Atlantic Expdn., 1925-27; asso. prof. Berlin U., 1929-45; chief oceanographer ALTAIR Gulf Stream Expdn., 1938; oceanographical adviser Navy Nautical Dept., 1939-45; prof. oceanography, dir. Inst. for Marine Research, Kiel (Germany) U., 1946-59, emeritus prof., 1959; vis. prof. oceanography Columbia U., N.Y.C., 1960-64; guest prof. U. Bonn (Germany), 1965—. Recipient Karl Ritter medal German Geog. Soc. Berlin, 1928. Hon. mem. Geog. Soc. Amsterdam, Geog. Soc. Frankfurt, Am. Geog. Soc. Author: Results of German Atlantic Expedition METEOR, 4 vols., 1925-27; also numerous articles. Research on ocean evaporation, stratification and circulation, bottom temperature and currents of Atlantic, Indian, Pacific oceans, interaction of sea and atmosphere. Home: 4 Endenicher Alee, Bonn, West Germany.*

WYATT, Gerard Robert, Am. biologist; b. Palo Alto, Cal., Sept. 3, 1925; s. Horace Graham and Aimée (Strickland) W.; B.A., U. B.C., 1945; postgrad. U. Cal. at Berkeley, 1946-47; Ph.D., U. Cambridge (Eng.), 1950; m. Sarah Silver Morton, Dec. 19, 1951; children—Eve M., Graham S., Diana S. Agrl. research officer Can. Insect Pathology Lab., Sault Ste. Marie, Ont., Can., 1950-54; faculty Yale, New Haven, 1954—, prof. biology, 1964—. Mem. Am. Soc. Biol. Chemists, Am. Soc. Zoologists, Biochem. Soc. (Gt. Britain), Soc. Exptl. Biology, (Gt. Britain), Sigma Xi. Research and publs. on chem. composition of nucleic acids, composition of insect blood, biochemistry of insects including carbohydrate metabolism, nucleic acids, hormone action and devel.; discovered 5-methylcytosine and 5-hydroxymethylcytosine as nucleic acid components, trehalose in insect blood. Home: 892 Indian Hill Rd., Orange, Conn. 06477. Office: Kline Biology Tower, Yale, New Haven 06520.*

WYATT, John, English inventor; b. Thickbroom, Eng Apr. 1700; s. John and Jane (Jackson) W; ed. Lichfield Sch.; m. twice; children (by second marriage)—Charles, John, 4 daus. Began as carpenter, Thickbroom; then worked at Soho Foundry. Writer various articles, plans, designs. Invented (with L. Psvel), spinning-machine, circa 1738; invented and perfected compound lever weighing machine, circa 1744. Died Nov. 29, 1766.

WYATT, Stanley Porter, Jr., Am. astronomer; b. Medford, Mass., Apr. 20, 1921; s. Stanley Porter and Maud Scott (Norton) W.; A.B., Dartmouth, 1942; A.M., Harvard, 1948, Ph.D., 1950; m. Catherine Jeannette Barber, Dec. 18, 1948; children—Stanley Porter III, Christopher Monroe, Scott Emery. Instr., U. Mich., 1950-53; faculty U. Ill., Urbana, 1953—, prof. astronomy, 1961—. Mem. Am. Astron. Soc., Internat. Astron. Union. Author: (with J. M. Atkin) Charting the Universe, 1961, The Universe in Motion, 1962, Gravitation, 1962. The Message of Starlight, 1964, The Life Story of a Star, 1965, Galaxies and the Universe, 1965; Principles of Astronomy, 1964. Research and publs. on galactic structure, external galaxies and interplanetary physics. Home: 510 W. Michigan Av., Urbana, Ill. 61801.*

WYCKOFF, Delaphine Grace Rosa, Am. microbiologist; b. Beloit, Wis., Sept. 11, 1906; d. Charles Darwin and Grace (Chamberlin) Rosa; Ph.B., U. Wis., 1927, Ph.M., 1928, Ph.D., 1938; m. John Franklin Wyckoff, Aug. 17, 1942. Faculty, N.D. Agrl. Coll., Fargo, 1928-37, asst. prof. bacteriology, 1936-37; faculty Wellesley Coll. (Mass.), 1938—, prof., 1957—. Cons., Biol. Scis. Curriculum Study, U. Colo., Boulder, 1959—. Mem. A.A.A.S., Am. Soc. Microbiology, Soc. Gen. Microbiology, Nat. Assn. Biology Tchrs., Sigma Xi, Sigma Delta Epsilon (nat. pres. 1961), Alpha Gamma Delta, Phi Delta Gamma. Research and publs. on yeast metabolism, antibacterial agts. Home: 78 Cedar St., Newington, Conn. 06111. Office: Wellesley Coll., Wellesley, Mass. 02181.*

WYCKOFF, Harold Orville, Am. physicist; b. Traverse City, Mich., Apr. 26, 1910; s. Joseph Nelson and Jennie (Bare) W.; B.S., U. Wash., 1934, Ph.D., 1940; m. Mildred Marie Haymond, June 29, 1940; children—Jeannette Marie, Harold Orville. Physicist, Nat. Bur. Standards, Washington, 1941-43, 45-66, chief x-ray standards sect., 1964-66; dep. dir., sci. Armed Forces Radiobiology Research Institute, Bethesda, Maryland, 1966—; operations analyst 9th Air Force, Europe, 1943-45. Sec., mem. Internat. Com. on Radiol. Units and Measurements, 1956-65; mem. bd. Nat. Council on Radiation Protection and Measurement; Dept. Commerce, rep. on working group Fed. Radiation Council, 1964—. Decorated Bronze Star; recipient Gold medal Dept. Commerce, 1960. Fellow Am. Phys. Soc.; asso. fellow Am. Coll. Radiology (vice chmn. commn. on radiol. units, standards and protection 1961—); mem. Radiol. Soc. N.Am. (Gold medal 1963), Health Physics Soc., Radiation Research Soc. Author: (with C. B. Braestrup) Radiation Protection, 1958; contbr. to Handbook on Radiation Hygiene, 1958. Studied radiation measurement and protection, radiation physics. Home: 9208 Whitney St., Silver Spring, Md. 20901. Office: Armed Forces Radiobiology Research Inst., Bethesda, Md. 20014.*

WYCKOFF, Ralph Walter Graystone, Am. crystallographer; b. Geneva, N.Y., Aug. 9, 1897; s. Abram Ralph and Ethel Agnes (Catchpole) W.; B.S., Hobart Coll., 1916; Ph.D., Cornell U., 1919; M.D. (hon.), Masaryk U., Czechoslovakia, 1947; Sc.D. U. Strasbourg, 1952; m. Laura Laidlaw; children—Ralph Walter Graystone, Anne Elias, Grietje Slaon. Instr. analytical chemistry Cornell U., 1917-19; phys. chemist Geophys. Lab., Carnegie Instn. of Washington, 1919-27; research asso. Cal. Inst. Tech., 1921-22; asso. mem. Rockefeller Inst. Med. Research, 1927-37; scientist Lederle Labs., Inc., 1937-40, asso. dir. in charge of virus research, 1940-42; tech. dir. Reichel Labs., Inc., 1942-43; resident lectr. epidemiology U. Mich., 1943-45; sr. scientist NIH, 1945, scientist dir., 1946-52, biophysicist, 1954-59; sci. attache Am. Embassy, London, 1952-54; pres. Internat. Union Crystallography, 1954-57; prof. physics, bacteriology U. Ariz. Exchange prof. Sorbonne, Paris, France, 1965. Fellow Am. Phys. Soc., A.A.A.S., Indian Acad. Sci.; mem. Am. Acad. Scis., Nat. Acad. Sci., Fr. Soc. Mineralogy and Crystallography (hon.), Royal Micros. Soc. (fgn.), Royal Netherlands Acad. Sci. and Lit. (hon.), French Soc. Microbiology (hon.), Royal Soc., 1951, French Acad. Scis. (corr.). Author: An Analytical Expression of the Results of the Theory of Space Groups, 1922, 30; The Structure of Crystals, 1924, 31, 35; Crystal Structures, 1947, 51, 53, 56, 58-60, 63-65; Electron Microscopy, 1950; The World of the Electron Microscope, 1958. Pioneer in application of X-rays to determination of atomic positions in crystals, in devel. of ultracentrifuge, in application of electron microscope to study and visualization of viruses and large molecules. Home: 4161 Camino Arco, Tucson 85718.*

WYGODZINSKY, Pedro, entomologist; b. Bonn, Germany, Oct. 5, 1916; s. Willy and Adele (Sperling) W.; Ph.D., U. Basle (Switzerland), 1961; m. Betty Martin. Researcher, Nat. Malaria Service, Rio de Janeiro, Brazil, 1941, Inst. Agrl. Experimentation, Ministry Agr., Rio de Janeiro, Brazil, 1941-48, Inst. Regional Medicine, Tucumán, Argentina, 1948-54; prof. Miguel Lillo Inst., Tucumán, 1954-58, Nat. U. Buenos Aires, Argentina, 1958-62; asso. curator Am. Mus. Natural History, N.Y.C., 1962—. Mem. Entomol. Soc. Am., N.Y. Entomol. Soc., Soc. Systematic Zoology. Research, numerous publs. on morphology, classification and geog. distbn. of insect families Thysaunra, Reduviidae, Simuliidae of world.*

WYLIE, Charles Murray, physician; b. Dunfermline, Scotland, Sept. 20, 1924; s. Thomas Murray and Mary (Symon) W.; M.B., Ch.B., U. Glasgow, 1947, M.D., 1957; M.P.H., Johns Hopkins, 1952, Dr.Pub. Health, 1956; D.T.M. and H., London Sch. Hygiene, 1958; m. Dorothy Hennessy, Jan. 31, 1959; children—Christine, Sheila, Adrian. Came to U. S., 1949, naturalized, 1955. Practice medicine, specializing in preventive medicine, Balt., 1958-68; faculty Sch. Hygiene and Pub. Health, Johns Hopkins U., Balt., 1958-68, dir. research Pub. Health Adminstrn. Project, 1958-68; prof. pub. health adminstrn. U. Mich, 1968—. Diplomate Am. Bd. Preventive Medicine. Fellow Am. Pub. Health Assn.; mem. Brit. Med. Assn., Am. Geriatrics Soc., Assn. Tchrs. Preventive Medicine, Am. Med. Writers Assn., Assn. Mgmt. Pub. Health, Soc. Med. Officers Health. Contbr. chpt. to Rypins Medical Licensure Examinations, 1966. Studies, numerous publs. on rehab. of patients disabled by strokes, epidemiology and acute med. care of strokes. Home: 1607 Dicken Dr., Ann Arbor, Mich. 48103.*

WYLIE, Sir James, physician; b. Kincardine-on-Forth, Scotland, 1768; s. William and Janet (Meiklejohn) W.; M.D., King's Coll., Aberdeen, Scotland, 1794. Became sr. surgeon in Eletsky Regt. of Russian service, 1790; became physician to Imperial Ct. at St. Petersburg, 1798; named surgeon-in-ordinary to Tsar Paul, also physician to heir-apparent Grandduke Alexander, 1799; named insp.-gen. Army Bd. Health, 1806; dir. med. dept. Ministry War, 1812; became physician in ordinary Tsar Alexander I, 1814, and accompanied him to Eng.; privy councillor under Tsar Nicholas. Author: On the American Yellow Fever, 1805; Pharmacopoeia castrensis Ruthenica, 1808; Practical Observations on the Plague, 1829; Rapport officiel à Sa Majesté Impériale sur la valeur comparée des méthodes thérapeutiques appliquées dans les hôpitaux militaires et à Saint-Pétersbourg aux sujets atteints de la maladie épidémique dite le choléra morbus, avec des observations pratiques sur la nature du fléau et sur ce que l'on apprend par l'ouverture des cadavres, 1831; Description de l'ophthalmie qui a sévi parmi les troupes, 1835. Founder, Med. Acad. St. Petersburg and Moscow, 1804, pres., 1804-34. Contributed to improvement of Russian hosp. system. Died St.Petersburg, Mar. 2, 1854.

WYLIE, Lloyd Ritchie, Am. astronomer; b. Idana, Kan., Apr. 12, 1892; s. Oliver Moses and Mary Ida (Ritchie) W.; A.B., Park Coll., Parkville, Mo., 1914; postgrad. U. Chgo., 1915-17; M.A., Northwestern U., 1922, Ph.D., 1932; m. Eula Sauls Treadway, Oct. 7, 1941. With U. S. Bur. Standards, 1917-20; mem. faculty Northwestern U., 1920-24, Ohio Wesleyan U., 1924-26, Lake Forest Coll., 1928-34, Eastern Ill. State Tchrs. Coll., 1934-36; with U. S. Naval Obs., 1936-46; prof. astronomy. Wittenberg U., 1946-60, head earth sci. dept., 1952-60, prof. emeritus astronomy, 1960-64; asso. prof. math. No. Ill. U., 1965—; cons. USAF, 1956-64. Fellow A.A.A.S., Ohio Acad. Sci.; mem. Am. Phys. Soc., Am. Astron. Soc., Sigma Xi, Sigma Pi Sigma. Investigation on orbits of Neptune and Uranus; determination of mass of Pluto; application of light amplification to astron. work. Home: 709 3d St., Waverly, O. 45690.*

WYLY, Lemuel David, Am. physicist; b. Seneca, S.C.; s. Lemuel David and Mary Julia (Reid) W.; B.S., The Citadel, 1938; M.A., U. N.C., 1939; Ph.D., Yale U., 1949; m. Martha Estelle Bruggemann, Dec. 28, 1938; children—Lemuel David III, Martha Jeannette. With Ga. Inst. Tech., Atlanta, 1939—, Regents' prof., 1958—; cons. Union Carbide Nuclear Co., Oak Ridge, Tenn., 1952—. Fellow Am. Phys. Soc.; mem. Am. Physics Tchrs., Sigma Xi. Research, publs. on fundamentals of beta decay. Home: 962 Canter Rd., Atlanta 30324.*

WYMAN, Donald, Am. horticulturist; b. Templeton, Cal., Sept. 18, 1903; s. Paul and Lillian (Kennedy) W.; B.S., Pa. State U., 1926; M.S., Cornell U., 1931, Ph.D., 1935; m. Florence Mary Dorward, Oct. 15, 1927; children—Donald, Mary Dorothea (Mrs. Roger M. Thomas), John Kennedy, Barbara Ann (Mrs. Dana A. Skiff). Horticulturist, Arnold Arboretum Harvard, Jamaica Plain, Mass., 1936—; hon. mem. Internat. Shade Tree Conf. Recipient Norman Jay Colman award Am. Assn. Nurserymen, 1949, 51, Distinguished Service award N.Y. Hort. Soc., 1960, medal of honor Garden Club Am., 1965. Mem. Am. Hort. Soc. (past pres., dir.), Am. Soc. Hort. Sci., Am. Assn. Bot. Gardens and Arboretums (past pres.), Am. Hort. Council, Sigma Xi, Phi Kappa Phi, Pi Alpha Xi. Author: Shrubs and Vines for American Gardens, 1949; Arnold Arboretum Garden Book, 1954; Ground Cover Plants, 1956; Trees for American Gardens, 1956, 65; The Saturday Morning Gardener, 1962; also numerous articles. Introduced several hundred new ornamental woody plants to Am. horticulture. Home: 102 Wellesley St., Weston, Mass. 02193. Office: Arnold Arboretum, Jamaica Plain, Mass. 02130.*

WYMAN, Jeffries, Am. biophysicist; b. West Newton, Mass., June 21, 1901; s. Jeffries and Helen (MacKay) W.; A.B., Harvard, 1923, postgrad., 1924; postgrad. Cambridge U., 1924; Ph.D., U. Coll. London (Eng.), 1927; m. Anne Codman Cabot, 1928 (dec.); children—Anne Cabot, Jeffries; m. 2d, Rosamond Forbes, 1946 (div.); m. 3d, Olga Lodigensky, June, 1954; Mem. staff biology dept. Harvard, 1928-51; sci. adviser U. S. Embassy, Paris, France; dir. Middle East regional office UNESCO; mem. staff Inst. Biol. Chemistry, U. Rome (Italy), also Instituto Regina Elena, Rome. Sec. gen., mem. council European Molecular Biology Orgn. Author: (with J. T. Edsall) Biophysical Chemistry, 1958. Research, publs. on molecular biology and related subjects; research on conformational changes and allosteric effects as basis of control and regulation in enzymes and working proteins generally. Home: 44 Piazza Farnese, Rome, Italy.*

WYMAN, Leland Clifton, Am. physiologist, anthropologist; b. Livermore Falls, Me., Feb. 20, 1897; s. Foscari A. and Hannah (Huff) W.; A.B., Bowdoin Coll., 1918; Ph.D., Harvard, 1922; m. Paula Berg, May 30, 1920. Faculty, Coll. Liberal Arts Boston U., 1922—, prof., 1948-62, prof. emeritus, 1962—; curator archives Mus. No. Ariz., Flagstaff, 1960—. Univ. lectr. Boston U., 1958-59. Fellow Am. Anthrop. Assn., Am. Acad. Arts and Scis.; mem. Am. Physiol. Soc., Am. Inst. Archaeology, Phi Beta Kappa. Author: Beautyway; A Navaho Ceremonial, 1957; The Windways of the Navaho, 1962; The Red Antway of the

Navaho, 1965; also numerous articles. Research on transplantation of the adrenal cortex, analysis of secretions of transplants of the adrenal cortex; Navaho Indian religion; Navaho Indian ceremonialism, ethnobiology. Home: 5 Furnival Rd., Jamaica Plain, Mass. 02130. Office: 2 Cummington St., Boston 02215.*

WYMAN, Marvin Eugene, Am. nuclear engr., b. North Branch, Minn., Apr. 9, 1921; s. Einar O. and Clara (Swanson) W.; B.A., St. Olaf Coll., 1942; M.S., U. Ill., 1943, Ph.D., 1950; m. Betty L. Sexton, Mar. 12, 1955; children—Christina, Dorcas. Asst. prof. physics St. Olaf Coll., Northfield, Minn., 1949-52, asso. prof. physics, 1952-53; staff mem. Los Alamos Sci. Lab, 1953-58; prof. nuclear engring. and physics U. Ill., Urbana, 1958—. Mem. subcom. on research reactors Nat. Acad. Sci—NRC, 1961—. Fellow Am. Phys. Soc.; mem., Am. Nuclear Soc., Am. Soc. E.E., Am. Assn. Physics Tchrs., Sigma Psi. Home: 605 Harding Dr., Urbana, Ill. 61801.*

WYMAN, Morrill, Am. physician; b. Chelmsford, Mass., July 25, 1812; s. Rufus and Ann (Morrill) W.; grad. Harvard, 1833, M.D., 1837; m. Elizabeth Aspinwall Pulsifer, Aug. 14, 1839; 2 children including Morrill. Practiced medicine, Cambridge, Mass.; adjunct Hersey prof. theory and practice of medicine Harvard Med. Sch., 1853-56; organized (with 3 others) pvt. med. sch., Cambridge 1857; U. S. insp. hosps. during Civil War; overseer Harvard, 1875-87; founder Cambridge Hosp., 1886. Author: A Practical Treatise on Ventilation, 1846; Autumnal Catarrh (Hay Fever), 1872. Performed 2 successful operations in treatment of pleurisy (1st in U. S.) 1850. Died Jan. 30, 1903.

WYMAN, Walter, Am. surgeon; b. St. Louis, Mo., Aug. 17, 1848; s. Edward and Elizabeth (Hadley) W.; grad. City U. of St. Louis, 1866; A.B., Amherst, 1870 A.M., 1889; M.D., St. Louis Med. Coll., 1873; LL.D., Western U. Pa., 1897, U. Md., 1907. In Marine Hosp. Service, 1876—, serving successively at St. Louis, Cin., Balt., N.Y., Washington; gave spl. attention to phys. conditions affecting seamen of merchant marine, instrumental in having laws passed for their benefit; also brought to notice cruelties imposed on deck-hands on Western rivers and on crews of oyster vessels in Chesapeake Bay; established hosp. for latter; had charge of Govt. measures to ward off cholera, 1893; supervising surgeon gen. U. S. Marine Hosp. Service, 1891-1902; surgeon gen. U. S. Pub. Health and Marine Hosp. Service, 1902—; adminstr. Nat. quarantine law and establishments founder 1st govt. sanatorium for consumptives, Ft. Stanton, N.M.; instrumental in securing enactment of laws relating to quarantine, quarters and food for seamen, govt. regulation of mfr. and sale of viruses, serums, toxins, the establishment of a leprosy investigation sta. in Hawaii, the creation of a hygienic lab. at Washington, establishment of a bur. of pub. health. Mem. bd. visitors Govt. Hosp. for Insane; chmn. Internat. San., Bur. of Am. Republics; pres. Am. Pub. Health Assn., 1902-03, Assn. Mil. Surgeons, 1904-05; hon. fellow Am. Assn. Obstetricians and Gynecologists; sec. sect. pub. and internat. hygiene, 9th Internat. Med. Congress, Washington, 1887; chmn. com. internat. quarantine, Pan-Am. Med. Congress, Mex., 1896; chmn. sect. pub. health Internat. Congress Arts and Scis. St. Louis, 1904; pres. sect. state and municipal control, Internat. Congress on Tb. Washington, 1908; mem. many med. and other socs. Author pamphlets on pub. health. Died 1911.

WYNDER, Ernest Ludwig, physician; b. Herford, Germany, Apr. 30, 1922; s. Alfred W. and Theresa (Godfrey) W.; B.A., N.Y. U., 1943; B.S., Wash. U., 1950, M.D., 1950. Came to U. S., 1938, naturalized, 1943. Asso. prof. preventive medicine Sloan-Kettering div. Cornell U. Med. Coll. Grad. Sch. Med. Scis., 1956—, asso. mem. Sloan-Kettering Inst. for Cancer Research, 1960—; asst. attending physician dept. medicine Meml. Hosp. for Cancer and Allied Diseases, N.Y.C., 1964—; asst. vis. physician dept. medicine James Ewing Hosp. of City N.Y., 1959—. recipient Borden Undergrad. Research award in medicine Wash. U., 1950. Mem. A.M.A., Am. Cancer Research Assn., Indsl. Med. Assn., Med. Soc. County N.Y., N.Y. Acad. Scis. Research, numerous publs. in etiology of major types of cancer, including lung, upper alimentary tract, stomach, colon, breast, cervix, endometrium and bladder, tobacco and air pollution carcinogenesis, possible effect of nutritional deficiencies on certain types of human cancer. Home: 180 East End Av., N.Y.C. 10028. Office: 410 E. 68th St., N.Y.C. 10021.*

WYNNE, Elmer Staten, Am. bacteriologist; b. El Paso, Tex., Oct. 23, 1917; s. P. D. and Mary (Durnell) W.; B.A., U. Tex., 1938, M.A., 1944, Ph.D., 1948; m. Regina Edith Everett, Mar. 4, 1938; children—Edith Jane, and Frank Staten. Member of the faculty Univ. of Okla., Norman, 1948-50, U. Tex., Houston, 1950-59; chief, microbiology USAF Sch. Aerospace Medicine, Brooks AFB, Tex., 1959-67. Diplomate Am. Bd. Microbiology. Charter fellow Am. Acad. Microbiology; mem. Am. Soc. Microbiology, Am. Inst. Biol. Sci., A.A.A.S. Research, publs. on physiology bacterial spore germination, cancer research, aerospace microbiology. Home: 4826 Hershey Dr., San Antonio, Tex. 78220. Office: Box 4328, USAF Sch. Aerospace Medicine, Brooks AFB, Tex. 78235.*

WYNNE, Lyman Carroll, Am. research psychiatrist; b. Tyler, Minn., Sept. 17, 1923; s. Nels Wind and Ella C. (Pultz) W.; war certificate, Harvard, 1943, M.D., 1947, Ph.D., 1958; m. Adele Rogerson, Dec. 22, 1947; children—Christine, Randall, Sara, Barry, Jonathan. Med. intern Peter Bent Brigham Hosp., Boston, 1947-48; intern neurology Queen Square Hosp., London, Eng., 1950; resident psychiatry Mass. Gen. Hosp., Boston, St. Elizabeth's Hosp., Washington, also Nat. Inst. Mental Health, 1951-54; psychoanalytic tng. Washington Psychoanalytic Inst., 1954-60, teaching analyst; cons., investigator WHO, 1965—; staff Nat. Inst. Mental Health, 1954—, chief sect. family studies, 1957—, chief adult psychiatry br., 1961—; mem. faculty Washington Sch. Psychiatry, also Washington Psychoanalytic Inst.; collaborating investigator WHO; vis. lectr. Am. U., Beirut, Lebanon, 1963-64. Med. dir. USPHS, 1961—. Recipient Commendation medal USPHS, 1965, Meritorious Service medal, 1966. Hofheimer Prize Am. Psychiat. Assn., 1966; Frieda Fromm-Reichmann award Am. Acad. Psychoanalysis, 1966. Fellow Am. Psychiat. Assn., Am. Orthopsychiat. Assn.; mem. Am. Psychosomatic Soc., Washington Psychiat. Soc., Washington Psychoanalytic Soc., Am. Psychoanalytic Assn. Club: Cosmos. Study of family relations in mental illness; cognitive styles and thinking disorders, psychodynamics and development of schizophrenic disorders; cross-cultural studies of families; psychoanalytic ego psychology; microlinguistic analysis of speech of psychiatric patients. Home: 5868 Marbury Rd., Bethesda 20034. Office: Clin. Center, Nat. Inst. Mental Health, Bethesda, Md. 20014.*

WYNNE-EDWARDS, Vero Copner, English zoologist; b. Leeds, Eng., July 4, 1906; s. John Rosindale and Lilian (Streatfeild) W.-E.; B.A., Oxford (Eng.) U., 1927, M.A., 1930; m. Jeannie Campbell Morris, Dec. 19, 1929; children—Janet (Mrs. Charles Sorbie), Hugh Robert. Asst. lectr. Bristol U., 1929-30; faculty McGill U., Montreal, Que., Can.,1930-46; Regius prof. natural history Aberdeen (Scotland) U., 1946—. Vis. Tom Wallace prof. Conservation, U. Louisville, 1959; vis. fellow New Zealand U., 1962; dir. Research unit grouse and moorland ecology Nature Conservancy, Banchory, Scotland, 1960—; mem. Nat. Environment Research Council U.K., 1965—. Fellow Royal Soc. Can., Royal Soc. Edinburgh, Am. Ornithologists Union (corr.); mem. Brit. Ornithologists' Union (pres. 1965—), Societas Scientiarum Fennica (fgn.). Author: Animal Dispersion in Relation to Social Behaviour, 1962; also numerous articles. Research on ecology No. sea-birds, birds, fish and plants in boreal and arctic Can., population ecology animals. Home: 70 High St., Old Aberdeen, Scotland.*

WYRTKI, Klaus, phys. oceanographer; b. Tarnowitz, Germany, Feb. 7, 1925; s. Wilhelm and Margarete (Pachazina) W.; student U. Marburg; Ph.D., U. Kiel, 1950; m. Helga Koecher, June 6, 1953; children—Undine, Oliver. With German Hydrographic Inst., Hamburg, 1950-51; research fellow German Research Council, U. Kiel, 1951-54; head Inst. Marine Research, Djakarta, Indonesia, 1954-57; sr. research officer Commonwealth Sci. and Indsl. Research Orgn., Div. Fisheries and Oceanography, Sydney, Australia, 1958-61; asso. research oceanographer, then research oceanographer Scripps Instn. Oceanography, LaJolla, Cal., 1961-64; prof. oceanography U. Hawaii, Honolulu, 1964—. Mem. Am. Geophys. Union, A.A.A.S., Soc. Limnology and Oceanography, Sigma Xi. Author: Physical Oceanography of the Southeast Asian Waters, 1961. Editor, Atlas on Phys. Oceanography, Internat. Indian Ocean Expedition. Research and publs on large-scale ocean circulation, on energy exchange between ocean and atmosphere, on thermal structure of ocean especially Baltic and North Sea, Antarctic Ocean, S.E. Asian waters, Eastern Tropical Pacific. Home: 635 Ahakea St., Honolulu, Hawaii 96816.*

WYSS, Orville, Am. microbiologist; b. Medford, Wis., Sept. 10, 1912; s. John and Gertrude (Walther) W.; B.S., U. Wis., 1937, M.S., 1938, Ph.D., 1941; m. Margaret Bess Bedell, June 26, 1915; children—Alice Ann (Mrs. Kenneth E. Hawker, Jr.), Jane Adair, Patti Bess. Research microbiologist Wallace & Tiernan, Inc., Belleville, N.J., 1941-45; faculty U. Tex., Austin, 1945—, prof. microbiology, 1948—, chmn. dept. microbiology, 1959—. Cons., NIH, 1959—. Mem. Am. Soc. for Microbiology (past pres.), Am. Chem. Soc., Am. Soc. Biol. Chemists, Am. Acad. Microbiology. Research and numerous publs. on microbial physiology and genetics. Home: 4606 Madrona Dr., Austin, Tex. 78731.*

X

XANTHAKIS, John Nikita, Greek astronomer; b. Gythion, Greece, Nov. 21, 1904; s. Nikitas Petros and Georgia (Patrinakos) X.; Ph.D., U. Athens (Greece), 1930; m., Oct. 10, 1942; 1 dau. Asst. astronomer Athens Obs., 1924-29; astronomer, chief asst. Athens U., 1929-31; research asso. Obsevatoire de Strasburg, 1931-34; prof. math. Mil. Sch. Greece, 1938-39; prof. astronomy U. Tessalouike, 1939-55. Pres., Greek nat. comms. Astronomy, Math, from 1957, Space Research, from 1964. Mem. Internat. Astron. Union, Athens Acad. (pres. 1964); pres. adminstr. council Greek Oceanographic Inst., from 1965. Author: Astronomy, 4 vols., 1949-55; General Mathematics, vol. 1, 1947; Probability Calculus, 1951; also articles. Research on math. analysis, positonal astronomy, geophysics. Address: 44 Panepistimione, Athens, Greece.*

XENOCRATES OF APHRODISIAS, Greek physician; flourished Rome, Italy, circa 70. Wrote on food derived from aquatic animals; also numerous works on food and drugs (interesting to history of ancient superstition), and work which served as source for Pliny's Natural History.

XENOCRATES OF CHALCEDON, Greek philosopher, mathematician; b. Chalcedon, Bithynia, circa 397 B.C.; pupil of Plato. Head of Platonic Acad., Athens, 339-314 B.C.; accompanied Plato to Syracuse; sent on embassies to Philip of Macedon; tchr., Epicurus was one of his pupils. Author over 75 treatises. Wrote history of geometry in 5 books; solved problem of combinatorial analysis; did work on theory of indivisible lines; 1st to work on theory of primary numbers; introduced into Acad. mystic Pythagorean doctrine of numbers. Died Athens, circa 314 B.C.

XENOPHANES (of Colophon), Greek philosopher; b. Colophon, Ionia, circa 570 B.C. Exile or fugitive from Colophon at age 25; lived mostly in So. Italy and Sicily; resided for time at Zancle (Messina) and Catana (Catania), then settled at Elea in So. Italy. Founder or forerunner of Eleatic sch.; invented several paradoxes to prove that things do not really move, that we only think they do; correctly interpreted fossils and saw them as demonstrating periodical submergences of dry land; his philosophy was monistic; believed that all things are made of earth and water; maintained that human knowledge was only opinion, not absolute truth which it could only approximate at best. Died probably in Sicily, circa 478 B.C.

XENOPHON, Greek natural philosopher and historian; b. Athens, Greece, circa 430 B.C.; student of Plato. Joined Greek mercenaries to dethrone Artaxerxes II Mnemon, King of Persia, after failure of cause, directed successful retreat of Greeks from Cunaxa to Chrysopolis (at Chalcedon, now Turkey); also aided Spartans against Persian satraps, Asia Minor, 399-394 B.C.; returned to Greece after Elean recovery Scillus from Sparta, circa 371 B.C., moved to Corinth. Author: Kyrou Anabasis, 7 vols.; Hel'enika, 7 vols., Kyrou Paideia, 8 vols., numerous other books. His Apomnemoneumata Socratous defends Socrates' life, teachings; also wrote on Greek affairs through 1st 40 yrs. 4th century, B.C. (411 B.C.), and on Persian life. Inventor retreat tactics still used in mountain warfare; known as Attic Bee because of style of writing. Died Corinth, circa 354 B.C.

XENOPOL, Alexandru, Rumanian sociologist, historian; b. Jassy, Moldavia, Mar. 23, 1847. Prof. Rumanian history U. Jassy. Corr. mem. Académie des sciences morales et politiques de France, 1901. Author: Les Roumains au moyen age, 1855; Etudes historiques sur le peuple roumain, 1887; Histoire des Roumains de la Dacie trajane, 6 vols., 1888-93; Les Guerres entre les Russes et les Turcs, et leurs conséquences sur les pays roumains, 2 vols., 1882; Les principes fondamentaux de l'histoire, 1899; Théorie de l'histoire, 1908. Attempted to derive regularities from recurrence of events at various times. Died Bucharest, Rumania, Feb. 27, 1920.

XYLANDER, Guilielmus, mathematician, translator; flourished mid 16th century. Prof. Greek, Heidelberg, Germany. Author: Opuscula mathematica, 1577. Translated 1st 6 books of Euclid's Elements into German, 1562; also various works from Greek to Latin, including Arithmetica (Diophantus), 1575, work of Psellus, 1556. Studies in astronomy, arithmetic, algebra, geometry.

Y

YAALON, Dan Hardy, geochemist, pedologist; b. Hradište, Czechoslovakia, May 11, 1924; s. Hugo and Elsa (Jellinek) Berger; B.Sc., Royal Agr. and Vet. Coll., Copenhagen, Denmark, 1947; Ph.D., Hebrew U. Jerusalem, Israel, 1954; m. Rita Singer, Mar. 17, 1952; children—David Zvi, Uri Michael. With Hebrew U. Jerusalem, 1956—, sr. lectr., 1961-67, asso. prof. pedology, 1967—. Contbr. articles to sci. jours. Application of clay mineral. and soil morphol. methods to study of soils of Israel, which led to clearer understanding of factors and processes in their formation and distbn.; instrumental in devising new soil classification and soil map of Israel. Office: Dept. Geology, Hebrew U., Jerusalem, Israel.*

YABUTA, Teijiro, Japanese agriculturist; b. Shiga Prefecture, Japan, Dec. 16, 1888; grad. Agrl. Chemistry, Tokyo U., 1911, then postgrad.; Agr.D., 1921. Technician, Agrl. Research Inst.; lectr. Tokyo U.; became prof. Tokyo U., 1921, now emeritus prof. Mem. Acad. (prize for research in filamentous fungi). Patentee method for manufacture of vitamin B, others.

YACOWITZ, Harold, Am. nutritionist, biochemist; b. N.Y.C., Feb. 17, 1922; s. Louis and Clara (Kurtzberg) Y.; B.S., Cornell U., 1947, M.S., 1948, Ph.D., 1950; m. Anne Ruth Barnett, Dec. 31, 1941; children—Caryn, Richard, Susan. Asso. research biochemist Parke, Davis & Co., Detroit, 1950-51; asst. prof. Ohio State U., 1951-53, asso. prof., 1953-55; head nutrition research dept. Squibb Inst. for Med. Research, New Brunswick, N.J., 1955-59; dir. applied research Nopco Chem. Co., Harrison, N.J., 1959-61; research asso. Health Research Inst., Fairleigh Dickinson U., Madison, N.J., 1961—; nutrition cons., 1961—. Robert Gould Research fellow 1948. Mem. Am. Inst. Nutrition, Am. Chem. Soc., Inst. Food Technologists, Poultry Sci. Assn., Animal Sci. Assn., Dairy Sci. Assn., N.Y. Acad. Scis., Animal Nutrition Research Council, Sigma Xi, Phi Kappa Phi. Patentee growth promoting action of antifungal antibiotics; research and numerous publs. in vitamin B12 interrelationships, micro-biol. assay for vitamin B12 and pseudo vitamin B12, effects of dietary fats on growth and feed efficiency of poultry, hemorrhagic syndrome in chicks, atherosclerosis in chicks, lipid metabolism in chicks, rats, and man, antifungal antibiotics for poultry diseases. Home: 221 2d Av., Piscataway, N.J. 08854.*

YAFET, Yako, physicist; b. Istanbul, Turkey, Jan. 2, 1923; s. Isak and Milka (Yafet) Y.; B.M.E., Tech. U. Istanbul, 1945; Ph.D., U. Cal., Berkeley, 1952; m. Judith L. Zublin, Dec. 2, 1949; children—Ellen Z., Steven B., Daniel. Came to U. S., 1946, naturalized, 1956. Research asso. U. Ill., 1952-54; physicist Westinghouse Research Lab., Pitts., 1954-60; lectr. Carnegie Inst. Tech., 1957-58; mem. tech. staff Bell Telephone Labs., Inc., Murray Hill, N.J., 1960—. Fellow Am. Phys. Soc. Research and publs. on theory of solid state physics, including electronic properties of semiconductors and metals, especially spin resonance and magnetic properties of conduction electrons; nuclear magnetic resonance; impurity states in semi-conductors. Home: 47 Curtiss Pl., Maplewood, N.J. 07040. Office: Bell Telephone Labs., Murray Hill, N.J. 07971.*

YAFFE, Leo, chemist; b. Devil's Lake, N.D., July 6, 1916; s. Samuel and Mary (Cohen) Y.; B.Sc. with honours, U. Man., 1940, M.Sc., 1941; Ph.D., McGill U., 1943; m. Betty Workman, Mar. 18, 1945; children—Carla Joy, Mark John. With Atomic Energy of Can. Ltd., 1945-52; faculty McGill U., Montreal, Que., Can., 1952—, Macdonald prof. chemistry, 1956—, chmn. dept. chemistry, 1965—; dir. research and labs. IAEA, 1963-65. Research collaborator Brookhaven Nat. Labs., Upton, L.I., N.Y., 1952—; cons. Pulp and Paper Research Inst. Can., 1954—. Fellow Chem. Inst. Can. (past chmn. bd. dirs.), Royal Soc. Can., Am. Phys. Soc.; mem. Sigma Xi. Research and numerous publs. on elucidation of nuclear fission and nuclear spallation processes, determination of absolute disintegration rates, discoverer new isotopes. Home: 5777 McAlear Av., Montreal, Que., Can.*

YAGER, William Alfred, Am. phys. chemist; b. Schenectady, Jan. 16, 1907; s. Hugo A. and Rose (Walther) Y.; B.S. in Chemistry, Union Coll., Schenectady, 1928; m. Ruth Wesley, July 12, 1930; children—Nancy Ann (Mrs. John C. Weller), William Alfred. Mem. tech. staff Bell Telephone Labs., Murray Hill, N.J., 1928—. Fellow Am. Phys. Soc.; mem. Sigma Xi. Research and numerous publs. on dielectrics, electron spin resonance. Patentee in field. Home: 41 Kendrick Rd., Summit, N.J. Office: Bell Telephone Labs., Murray Hill, N.J. 07974.*

YAGI, Kunio, Japanese biochemist; b. Yokohama, Japan, June 24, 1919; s. C. and N. (Wakabayashi) Y.; grad. Faculty Medicine, U. Nagoya (Japan), 1942, D.Med. Sci., 1947, Ph.D., 1958; m. K. Usami, Apr. 18, 1948; children—Keiko, Yasuyuki, Junji, Utako. With Faculty of Medicine, U. Nagoya, 1942—, prof. biochemistry, 1962—. Recipient Chunichi Cultural prize 1954, Japan Vitamin Soc. prize, 1961. Mem. Japanese Biochem. Soc., Japan Vitamin Soc., Chem. Soc. France. Author: Biochemistry of Flavins, 1957; also numerous articles. Devised fractional determination method for flavin compounds, preparation method of flavin adenine dinucleotide from Eremothecium Ashbyii; isolated by crystallization enzyme-substrate complex of D-amino acid oxidase; synthesized riboflavin tetrabutyrate. Home: 2-21 Nishizato-cho, Chikusa-ku, Nagoya, Japan.*

YAGI, Yasuo, biochemist; b. Kagoshima, Japan, Nov. 24, 1921; s. Bunya and Naka (Ito) Y.; B.Sc., Tokyo Imperial U., 1943; postgrad. Nagoya U., 1943-46; D.Sc., U. Paris, 1952; m. Michiko Ozawa, Apr. 11, 1947; children—Jean Koichiro, Mayumi, Kenji. Came to U. S., 1956. Asst., Nagoya U., Japan, 1946-52, instr., 1952-54, asst. prof. chemistry, 1954-56; sr. research scientist Roswell Park Meml. Inst., Buffalo, 1956-59, asso. cancer research scientist, 1959-63, prin. cancer research scientist 1963—; research asst. prof. chemistry State U. N.Y., Buffalo, 1956-68, research professor of chemistry, since 1968—. French Govt. Exchange scholar, 1950-52; Sloan Found. research fellow, 1952-54. Mem. Am. Assn. Immunologists, Am. Chem. Soc., Am. Assn. Cancer Research, Soc. Chim. Biol. (France), N.Y. Acad. Scis. Author: (with F. Egami) Immunochemistry, 1949; Toxic Proteins, 1951; also numerous articles. Immunochem. studies on antibodies, especially characterization, isolation, chem. and biol. properties; chem. identification of antigenic determinant regions in protein antigens; immunochem. studies on antigens of nonmalignant and malignant tissues and on corresponding antibodies. Home: 65 Campbell Rd., Buffalo 14215. Office: Roswell Park Meml. Inst., Buffalo 14203.*

YAHR, Melvin David, Am. physician; b. N.Y.C., Nov. 18, 1917; s. Isaac and Sarah (Riegelhauupt) Y.; A.B., N.Y. U., 1939, M.D., 1943; m. Felice Turtz, May 9, 1948; children—Carol, Nina, Laura, Barbara. Asst. dean Columbia U. Coll. Phys. and Surg., N.Y.C., 1959—, prof. neurology, 1962—; attending neurologist Neurol. Inst., 1962—; med. dir. Parkinson's Disease Found., 1958—. Mem. A.M.A., Am. Neurol. Assn. (sec.-treas.), Am. Heart Assn. (dir.), Am. Acad. Neurology, A.C.P., Assn. Research in Nervous and Mental Disease, N.Y. Neurol. Soc. Author: (with Dominick Purpura) The Thalamus, 1966. Research and publs. in Parkinson's disease, cerebro-vascular disease, epilepsy. Home: 23 Seely Pl., Scarsdale, N.Y. Office: Neurological Institute, 710 W. 168th St., N.Y.C. 10032.*

YAKOVLEV, Aleksandr Sergeevich, Russian aircraft designer; b. Moscow, USSR, Apr. 1, 1906; grad. Mil. Air Engring. Acad., Moscow, 1931. Served to col.-gen. Engring. Tech. Service, USSR; chief Exptl.-Designing Bur., from 1934, gen. designer, from 1957; dep. People's Commissariat, later dep. minister aviation industry. Recipient Stalin prize 1941, 42, 43, 46, 47, 48, also Order of Lenin. Corr. Mem. USSR Acad. Scis. Author: Stories of an Aircraft Designer, 1957. Designed many planes for various purposes; sports, tng., passenger, fighter, bomber, helicopter, also combat, piston and jet airplanes (under him 1st jet fighter, YAK-15, designed), series jet supersonic fighters; designed Flying Car. Home: Metrostroevskaya 1, Moscow. Office: USSR Acad. Scis., Leninskii Prospekt 14, Moscow, USSR.

YAKOVLEV, N. N., Russian geologist, paleontologist; b. Apr. 27, 1870. Mem. Geol. Com. (now All-Union Sci. Research Geol. Inst.), Leningrad, 1895, dir., 1923-26; prof. Petersburg Mining Inst., 1900-30. Recipient A. P. Karpinskii prize, 1948. Corr. mem. USSR Acad. Scis. Author: Studies on the Coral Rugosa, 1914; Fauna of the upper part of the Paleozoic deposits in the Donetz Basin, I-III, 1903-12; Attachment of Brachiopods as a Basis of Their Species and Genus, 1908; Extinction of Animals and Plants and Reasons according to Geologic Data, 1922; Crinoids and Blastoids of Carboniferous and Permian Deposits of the U.S.S.R., 1956; The Organism and the Environment, 1956. Conducted geol. expdns. in Donbas, Urals, Caucasus and beyond; worked out stratigraphy of lower Permain sediments; studied mineral sources; 1st to make paleoecological studies of invertebrates in Russia. Home: u/. Marksa i Engelsa 16, Moscow. Office: USSR Acad. Scis., Leninskii Prospekt 14, Moscow, USSR.

YAMAGATA, Noboru, Japanese chemist; b. Tokyo, Japan, Mar. 9, 1920; s. Yoshio and Haruyo (Takasu) Y.; B.Sc., Tokyo U., 1943, postgrad., 1949-50, D.Sc., 1953; m. Toshiko Tokunaga, Mar. 28, 1953; children—Ryoichi, Kenji, Miho. Faculty, Kiriu (Japan) Coll. Tech., Gunma U., 1950-60, prof., 1958-60; chief dept. radiol. health Inst. Pub. Health, Tokyo, 1959—. Mem. expert com. Radiation Council Japan, 1958—. Mem. Geochem. Soc., Health Physics Soc., Japan Chem. Soc., Japan Geochem. Soc., Japan Radiation Research Soc. Research and numerous publs. on radioanalytical and geochemistry of rare alkali elements especially cesium, environmental contamination with radioisotopes from viewpoint of pub. health. Home: 863 Oizumi-gakuen-cho, Nerima-ku, Tokyo. Office: Inst. Pub. Health, Shiba-shirokane-daimachi, Minatoku, Tokyo, Japan.*

YAMAGUCHI, Kazuo, Japanese economist; b. Tochigi Prefecture, Japan, 1907; grad. Tokyo U., 1932; entered Toyo Keizai Shimpo Sha, 1932. Became asst. prof. Hokkaido (Japan) U., 1947, prof., 1948, vice gov. Shizuoka Prefecture, Japan. Author: History of Foreign Trade of the Late Feudal Age; History of Japanese Fishing Industry; Study of the Japanese Fishing Economy. Research on fgn. trade history of Japan during later period of feudal age.

YAMAGUCHI, Nariyoshi, Japanese neuropsychiatrist; b. Nanao, Japan, Jan. 20, 1929; s. Narihiro and Kiyo (Hiraba) Y.; M.D., Kanazawa (Japan) U., 1952, D.Med.Sci., 1957; m. Masae Sugimori, Oct. 30, 1955; children—Narihito, Machiko. Faculty dept. neuro-psychiatry Kanazawa U. Sch. Medicine, 1957—, instr., 1959—; dir. Ishikawa Prefectural Mental Hygiene Center, Kanazawa, 1966—. Founds. Fund for Research in Psychiatry fellow dept. anatomy Brain Research Inst., U. Cal. at Los Angeles, 1961-63. Mem. Soc. Biol. Psychiatry, Japanese Soc. Psychiatry and Neurology, Japan EEG Soc. Research, publs. on recruiting responses observed during wakefullness and sleep in unanesthetized chronically ill cats, effects of chem. stimulation of preoptic region, nucleus centralis medialis or brain stem reticular formation with regard to sleep and wakefulness. Home: 1-21-8, Kodatsuno, Kanazawa, Ishikawa, Japan.*

YAMAGUCHI, Shigeto, Japanese physicist; b. Saga, Japan, Mar. 16, 1914; s. Keihachi and Mine (Ohtani) Y.; grad. Tokyo U., 1939, D.Sc., 1943. Prof. physics, mem. research staff Inst. for Catalysis, Hokkaido U., 1943-48; chief Applied Electron Beam Lab., Inst. Phys. and Chem. Research, Tokyo, Japan, 1953—. Research, numerous publs. on electron diffraction technique, magnetic analysis by electron diffraction, dielectric structure analysis by electron beam technique, study of solid catalysts by electron diffraction, discoverer new iron sulfide, electron diffraction and microscopy. Address: Institute of Physical and Chemical Research, 2-28 Hon-Komagome, Bunkyo-ku, Tokyo, Japan.*

YAMAGUCHI, Yoshio, Japanese physicist; b. Takefu, Japan, Jan. 29, 1926; s. Kazo and Teruko (Kutsuki) Y.; B.Sc., U. Tokyo (Japan), 1947, D.Sc., 1953; m. Yoriko Komiya, Mar. 5, 1953; children—Yukio, Kazuo, Akiko. Faculty, Osaka City U., 1956-62, prof. physics dept., 1957-62; prof. Inst. for Nuclear Study, U. Tokyo, 1962-68, prof. physics dept., 1968—. Mem. Phys. Soc. Japan. Author several books, numerous articles. Research on theoretical physics, particle physics, nuclear physics, cosmic rays; predicted asso. prodn. of strange particles; proposed SU (3) symmetry scheme of elementary particles. Home: 2-chome 22-11 Hikawadai, Kurume, Kitatamagun, Tokyo, Japan.*

YAMAGUTI, Satyu, parasitologist; b. Nagano Prefecture, Japan, Apr. 21, 1894; grad. Okayama Med. Coll., 1917; M.D., Tokyo U., 1925; D.Sc., Kyoto U., 1935; m. Ikuko Yamaguti, Nov. 22, 1923; children—Junko (Mrs. Kenneth E. Fletcher), Toshikazu, Noboru, Hideo. Lectr. parasitology Kyoto U., 1926-43; parasitologist Naval. Inst. Tropical Hygiene, Macassar, Celebes, 1943-44; spl. cons. 207th Malaria Survey Detachment, U. S. 8th Army, 1945-50; prof. parasitology Okayama U. Med. Sch., 1950-59; research staff Meguro Parasitological Mus., Tokyo, 1967—; research staff Beltsville, (Md.) Parasitological Lab., 1955—; parasitologist U. Hawaii, 1965—; faculty Tulane U. Grad. Sch., 1966—. Fulbright-Smith-Mundt grantee, 1955-56; NSF grantee, 1956-57, 1960-69. Mem. Japanese (hon.), Am. (hon.) socs. parasitologists, Helminthological Soc. India (hon.). Author: Keys to Adult Culicine Mosquitoes of America North of Mexico, 1952; Systema Helminthum, Vol. I-V, 1958-62; Parasitic Copepoda and Branchiura of Fishes, 1963; also numerous articles. Research on helminth fauna of Japan, Celebes and Borneo, mosquito fauna of Japan, Korea, Guam and N.Am., monogenetic and digenetic trematodes of Hawaiian fishes, parasitic copepods of Japanese fishes. Home: Tonodan, Kyoto, Japan; also 2612 Kirkwood Pl., Hyattsville, Md. 20782. Office: Beltsville Parasitological Lab., Beltsville, Md. 20705.*

YAMAMOTO, Giichi, Japanese meteorologist; b. Kanazawa, Japan, July 4, 1909; s. Kichitaro and Yoshi (Hata) Y.; B.Sc., Tohoku U., 1933, D.Sc., 1946; m. Fumi Usui, Dec. 27, 1935; children—Hiroyoshi, Tatsuo, Kozo, Kazuko. Meteorologist, Central Meteorol. Obs. Japan, 1933-42; research Central Aero. Inst. Japan, 1942-46; prof. meteorology Tohoku U., Sendai, Japan, 1946—, dean faculty sci., 1965—. Mem. Meteorol. Soc. Japan (Fujiwara prize 1965), Am., Royal meteorol. socs. Author: Atmospheric Radiation, 1954; also numerous articles. Research on transfer atmospheric radiation, radiative equilibrium earth's atmosphere, atmospheric turbulence. Home: 23 Yagiyama-yayoicho, Sendai, Japan.*

YAMAMOTO, Shutaro, Japanese vet. pathologist; b. Himeji City, Hyogo, Japan, Oct. 27, 1909; s. Saichiro and Yasu (Tanaka) Y.; Ph.D., U. Tokyo, 1944; m. Yasuko Fukumoto, Apr. 10, 1940; children—Atsuko, Keiko. Asst., Faculty of Agr., U. Tokyo, 1933-40, asst. prof., 1940-46, prof., head vet. pathology, 1946—, head div. vet. sci. Faculty Agr., 1957-59, head Lab. Animal Physiology Phys. and Chem. Inst., 1963—; mem. U. Senate, 1966—. Mem. Academie Veteriare de France, 1957—. Recipient Japan Agr. prize, 1946; Japan Acad. prize, 1959. Mem. Japanese Soc. Vet. Sci. (adv. com. 1946—), Japanese Soc. Pathology (mem. adv. com. 1946—), Japanese Soc. Bacteriology (adv. com. 1946—), Japanese Soc. Virology, Japanese Soc. Mycology, Japanese Soc. Parasitology. Author: Animal Diseases and Public Health, 1938; also numerous articles. Pathology of infectious diseases of animals, leptospirosis, avian-type Tb, toxoplasmosis, Japanese B encephalitis, necrobacillosis, cryptococcosis. Home: Omiyamae 5-221, Suginamiku, Tokyo, Japan.*

YAMAMOTO, Sukeyasu Steven, physicist; b. Tokyo, Japan, Aug. 11, 1931; s. Yuji and Makiko (Toyoda) Y.; came to U. S., 1950; B.S., Yale, 1955, M.S., 1957, Ph.D., 1959; m. Keiko Sato, Nov. 30, 1963; 1 son, Yuji Andrew. Mem. staff Brookhaven Nat. Lab., Upton, N.Y., 1959-65; asso. prof. physics U. Mass., Amherst, 1965—, also co-dir. high energy physics group. Mem. Am. Phys. Soc., Phi Beta Kappa, Sigma Xi. Research, publs. on high energy physics using bubble chambers to study nature of fundamental particles; took part in discovery of omega minus particle.*

YAMAMOTO, Yuroku, Japanese chemist; b. Fukui, Japan, Sept. 28, 1921; s. Yukichi and Shizu Yamamoto; B.Sc., Kyoto U., 1945, Ph.D. in Chemistry, 1957; m. Yoko Kinoshita, Sept. 1945; children—Hirofumi, Masafumi, Yoshifumi. With Kyoto U., 1945-61, lectr., 1957-61; prof. Hiroshima (Japan) U., 1961—. Mem. Chem. Soc. Japan, Japan Soc. for Analytical Chemistry. Author: Colorimetry, 1967; also articles. Research in analytical chemistry, especially spectrophotmetry and atomic absorption spectroscopy; developed (with others) idea for determination of anions by atomic absorption spectroscopy. Home: 1488 Midori-machi, Hiroshima, Japan.*

YAMAMURA, Hideo, Japanese physician; b. Kumagaya, Japan, Jan. 23, 1920; s. Hisachi and Kimiko Y.; M.D., Tokyo (Japan) U., 1943, M.Med. Sci., 1948; m. Miyoko, Oct. 26, 1942; children—Akio, Nobuko, Shiseru. Faculty, Tokyo U. Sch. Medicine, 1952—, prof. anesthesiology, 1956—. Mem. Japan Soc. Anesthesiology (past pres.), World Fedn. Socs. Anesthesiologists (pres. 1960-66, exec. com. 1966—), Japan Soc. Surgery, Japan Soc. Blood Transfusion. Author: Clinical Anesthesia, 1956; Conquest over the Pain, 1966; also numerous articles. Research on mode of action, uptake, and elimination of inhalation anesthetics, pulmonary function during artificial respiration. Home: 3-55 Hamacho, Nibonbashi, Chouku, Tokyo, Japan.*

YAMAMURO, Sadayuki, mathematician; b. Kanazawa, Japan, June 18, 1925; s. Sadao and Tokiwa (Tsukamoto) Y.; B.Sc., Tohoku U., Sendai, Japan, 1950; Ph.D., Hokkaido U., Sapporo, Japan, 1954; m. Akiko Maeda, June 28, 1953; 1 dau., Yuriko. Staff dept. math. Hokkaido U., 1951-56, asst. profl., 1955; mem. Inst. for Advanced Study, Princeton, N.J., 1956-58; asst. prof. Yokohama (Japan) City U., 1961-64; sr. research fellow Inst. Advanced Studies, Australian Nat. U., Canberra, 1964—. Mem. Japan Math. Soc., Am., Australian math. socs. Research and publs. on theory of functional analysis, especially theory of vector lattices, theory of non-linear functional analysis. Home: 5 Warren Pl., Chifley, A.C.T., Australia.*

YAMANAKA, Chiyoe, Japanese elec. engr.; b. Osaka, Japan, Dec. 14, 1923; s. Kichibei and Mitsuko (Kakudo) Y.; grad. dept. elec. engring. Faculty Engring., Osaka U., 1948, Ph.D., 1960; postgrad. Mass. Inst. Tech.; m. Tamiko Kashida, May 6, 1955; children—Kaoru, Chihiro. Faculty, Osaka U., 1955—, prof. elec. engring., 1961—. Sr. Mem. I.E.E.E. (U. S.); mem. Inst. Elec. Engrs. Japan, Elec. Communication Soc. Japan, Illuminating Soc., Phys. Soc., Applied Phys. Soc. Author: Plasma and its Application, 1966. Research, and publs. on advance of plasma diagnostics by micro and mm wave and laser; shock wave structures; quantum electronics. Home: 23 Nishiyama, Ashiya, Japan. Office: Osaka U., Osaka, Japan.*

YAMASAKI, Fumio, Japanese physicist; b. Tokyo, Japan, Aug. 23, 1907; s. Naomasa and Miyu (Tadokoro) Y.; B.S., U. Tokyo, 1928; D.Sc., Tokyo U. Lit. and Sci., 1951; m. Midori Iwanami, Apr. 9, 1938; children—Kazuo, Sumiko (Mrs. Harahiko Sakata). Asst., Hokkaido U., Sapporo, Japan, 1931, lectr., 1932-35; researcher Inst. Phys. and Chem. Research, Tokyo, 1935-51, chief Radiation Lab., 1951—. Recipient Japanese Gov. Blue Ribbon medal. Mem. Japan Radioisotope Assn. (exec. dir. 1954), Japan Health Physics Soc. (dir. 1963-65), Japan, Am. phys. socs. Research, numerous publs. on application of Wilson chamber to study of spark discharge, radioactivity of cyclotron produced isotopes, neutron capture, application of radioisotope, radiocarbon dating, dose measurement. Home: 8-23, Komachi 1-Chome, Kamakura, Japan. Office: 28-8 Honkomagome, Bunkyoku, Tokyo, Japan.*

YAMASAKI, Hoshito, Japanese geneticist; b. Tottori-ken, Japan, Jan. 14, 1905; s. Muneo and Takako (Nakagaki) Y.; M.Agr. Sci., Kyushu U., 1931, Dr.Agr. Sci., 1948; m. Haruko Kishida, Apr. 8, 1928; children—Kazumi (Mrs. Shuzo Takata), Naomi (Mrs. Hiroshi Ikeda), Sanae. Researcher, Nat. Agrl. Expt. Sta., Tokyo, 1931-32; chief researcher Nara Agrl. Expt. Sta., Nara-ken, 1932-37; chief researcher Nagano Agrl. Expt. Sta., Nagno-ken, 1937-50; head div. genetics Nat. Inst. Agrl. Scis., Tokyo, 1950-61, head dept. physiology and genetics, 1961-64; prof. Nihon U., Tokyo, 1964—; lectr. Tokyo Met. U., 1952—, Tokyo U., 1962-64, Iwate U., Morioka, 1965—. Recipient Agrl. Sci. prize Japanese Soc. Agrl. Sci. Mem. Genetics Soc. Japan, Japanese Soc. Breeding, Phytopath. Soc. Japan, Bot. Soc. Japan. Author: (in Japanese) Corn Breeding, 1954; also numerous articles. Editor: Ency. Agrl. Science, 1966. Devised method of embryotransplantation for genetico-physiol. study in cereals; studies on role of endosperm in low temperature vernalization and seed dormancy of cereals; discovered occurrence of parasexual cycle and transformation as a cause of variation in two phytopathogenic species. Home: 7045 Ikuta, Kawasaki, Kanagawa, Japan. Office: 3-49 Shimoumа, Setagaya-ku, Tokyo, Japan.*

YAMASHITA, Tadayoshi, Japanese physicist; b. Hiroshima, Japan, Jan. 10, 1919; s. Shotaro and Tatsuko (Moriwake) Y.; M.S. in Sci., Hiroshima U., 1941, D.Sc., 1955; m. Satoko Omoto, Jan. 15, 1946; children—Reiko, Manabu. Asst. prof. physics Nagoya (Japan) U., 1947-52; prof. physics Def. Acad., Yokosuka, Japan, 1953—; prof. physics Tokyo Coll., Sci., 1956—. Research fellow Cal. Inst. Tech., 1955-56. Mem. Soc. Japan Electron Microscopy, Phys. Soc. Japan. Author: General Physics, 1967; also articles. Observations on dislocations in deformed single crystals of iron by electron microscope and explanation of their work-hardening. Home: 2-4-59 Ikego Zushi, Japan. Office: Dept. Physics, Def. Acad., Yokosuka, Japan.*

YANDERS, Armon Frederick, Am. geneticist; b. Lincoln, Neb., Apr. 12, 1928; s. Fred W. and Beatrice (Pate) Y.; A.B., Neb. State Coll., 1948; M.S., U. Neb., 1950, Ph.D., 1953; m. Evelyn Louise Gatz, Aug. 1, 1948; children—Mark Frederick, Kent Michael. Research asso. Oak Ridge Nat. Lab. and Northwestern U., 1953-54; biophysicist U. S. Naval Radiol. Def. Lab., San Francisco, 1955-58; asso. geneticist Argonne (Ill.) Nat. Lab., 1958-59; with dept. zoology Mich. State U., 1959—, prof., asst. dean Coll. Natural Sci., 1963—. Mem. A.A.A.S., Am. Inst. Biol. Sci., Am. Soc. Naturalists, Genetics Soc. Am., Am. Soc. Zoologists. Research and publs. on effects of radiation on genes and chromosomes, fertilization phenomena, abnormal genetic ratios. Home: 1311 Bayshore Dr., Haslett, Mich. 48840.*

YANG, Chen Ning, physicist; b. Hofei, Anhwei, China, Sept. 22, 1922; s. Ke Chuan Yang and Meng Hwa Lo; B.Sc., Nat. S.W. Asso. U., China, 1942; Ph.D., U. Chgo., 1948; D.Sc. (honorary), Princeton University, 1958, Bklyn. Polytechnic Institute, 1965; m. Chih Li Tu, Aug. 26, 1950; children—Franklin, Gilbert, Eulee. Instr. U. Chgo., 1948-49; mem. Inst. Advanced Study, Princeton, 1949-55, prof., 1955-66; Albert Einstein prof. State U. N.Y., Stony Brook, 1966—. Recipient Albert Einstein Commemorative award in sci., 1957. Recipient (with Tsing-Dao Lee) Nobel Prize for Physics, 1957. Mem. Am. Phys. Soc., Sigma Xi. Research in theoretical physics; statistical mechanics; (with Lee) established the nonconservation of parity in weak interactions, 1956; investigated the behavior of elementary particles. Office: State U. N.Y., Stony Brook, N.Y.

YANG, Chung-Tao, mathematician; b. Pingyang, China, May 4, 1923; s. Jone Nieh and Yulin (Su) Y.; B.S., Chekiang U., 1946; Ph.D., Tulane U., 1952; m. Ages Kang, Feb. 2, 1957; children—Deane, Lynne, Jeanne. Came to U. S., 1950. Asst., Chekiang U., 1946-48, Academia Sinica, 1948-49; instr. Taiwan U., 1949-50; teaching asst. Tulane U., 1950-52; research asso. U. Ill., 1952-54; mem. Inst. for Advanced Study, 1954-56; faculty U. Pa., Phila., 1956—, prof. math., 1961—. Mem. Am. Math. Soc., Math. Assn. Am. Contbr. articles to math. jours. Home: 311 Hidden River Rd., Narberth, Pa. 19072. Office: Dept. Math., U. Pa., Phila. 19104.*

YANG HUI (Chien Kuang), Chinese mathematician; b. Chien-Tang, China; flourished circa 1261-75; pupil of Liu I. Author: Hsian-chieh chiu-chang suan-shu (The Analysis of the Arithmetic Rules), 9 sects., 1261; Yang Huis suan-fa, 7 books, circa 1275. Analyzed arithmetical rules; illustrated summation of an arithmetical series and listed rules for summing series; studies on proportions and compound proportions, linear equations with 4 or 5 unknowns, computations using decimal fractions, graphic explanations of quadratic equations, solution of numerical equations.

YANKWICH, Peter Ewald, Am. chemist; b. Los Angeles, Oct. 20, 1923; s. Leon Rene and Helen (Werner) Y.; B.S., U. Cal. at Berkeley, 1943, Ph.D., 1945; m. Elizabeth Pope Ingram, July 14, 1945; children—Alexandra Helen, Leon Rene, II, Richard Ingram. Mem. sci. staff Radiation Lab., U. Cal. at Berkeley, 1944-48, faculty, 1947-48; faculty U. Ill., Urbana, 1948—, prof. chemistry, 1957—, head div. phys. chemistry, 1962-67. Mem. Adv. Council on Coll. Chemistry, 1961-68. NSF Sr. Postdoctoral fellow, 1960-61. Mem. Phi Beta Kappa, Sigma Xi. Research on isotope kinetics, chem. effects of nuclear transformations. Home: 604 W. Washington St., Urbana, Ill. 61801.*

YANO, Kentaro, Japanese mathematician; b. Tokyo, Japan, Mar. 1, 1912; s. Seiichi and Chie (Nakayama) Y.; D.Sc., Tokyo U., 1939; D.Sc., U. Paris, 1938; m. Yukako Ogawa, Oct. 30, 1941; children—Teiichi, Junji. Asst. prof. math. Tokyo U., 1940-58; prof. math. Tokyo Inst. Tech., 1958—. Mem. Math. Soc. Japan, Am., French math. socs. Author: (with S. Bochner) Curvature and Betti Numbers, 1953; Differential Geometry on Complex and Almost Complex Spaces, 1965. Research and numerous publs. in Riemannian geometry using integral formulas and applications to unified field theory of gravitation and electro magnetism. Home: 31-10, 1-chome, Kaminakazato, Kita-ku, Tokyo. Office: Oh-okayama, Meguro-ku, Tokyo, Japan.*

YANOFSKY, Charles, Am. biologist; b. N.Y.C., Apr. 17, 1925; s. Frank and Jennie (Kopatz) Y.; B.S., Coll. City N.Y., 1948; M.S., Yale, 1950, Ph.D., 1951; m. Carol Cohen, June 19, 1949; children—Stephen, Robert, Martin. Research asst. microbiology Yale, 1951-53; asst. prof. microbiology Western Res. U., 1954-58; asso. prof. dept. biol. scis. Stanford, 1958-61, prof., 1961—. Recipient Lederle Med. Faculty award, 1955-57; Eli Lilly award in bacteriology, 1959; U. S. Steel award in molecular biology, 1964; Howard Taylor Ricketts award, 1966. Mem. Am. Acad. Arts and Scis., Nat. Acad. Scis., Am. Soc. Microbiologists, Genetics Soc. Am., A.A.A.S., Soc. Biol. Chemists, Sigma Xi. Contbr. numerous articles in field to sci. jours. Demonstrated linear correspondence between structure of a gene and structure of a protein; amino acid changes in proteins resulting from mutational changes. Home: 725 Mayfield Av., Stanford, Cal. 94305.*

YANOWSKI, Leo Kasimir, Am. chemist; b. Paterson, N.J., Nov. 27, 1901; s. John K. and Clara (Bender) Y.; B.S., Fordham U., 1927, M.S., 1928, Ph.D., 1930; m. Eleanor R. Chadwick, June 25, 1930; children—Barbara (Mrs. Howard L. Ryder), Brenda (Mrs. Warren S. Witters). Instr., Fordham U., 1927-31, asst. prof., 1931-42, asso. prof., 1942-60, prof., 1960-67, prof. emeritus, 1967—, adminstrv. asst. chemistry dept., 1967. Recipient Bene Merenti medal. Mem. Am. Chem. Soc. (chmn. N.Y. sect. radio com. 1953-65), Am. Microchem. Soc. (chmn. 1945-46), Phi Beta Kappa, Sigma Xi, Phi Lambda Upsilon. Research and publs. on use of complexes as analytical reagents; producer ednl. radio series Everybody's Chemistry, 1953-65. Home: 279 E. 203d St., N.Y.C. 10458.*

YAQUB, Adil Mohamed, mathematician; b. Anabta, Jordan, Jan. 19, 1928; s. Mohamed and Fawzieh (Gazala) Y.; Came to U. S., 1947, naturalized, 1954; A.B., U. Cal. at Berkeley, 1950, M.A., 1951, Ph.D., 1955; m. Nancy Shiddell, Apr. 20, 1951; children—Charles, Hanan. Faculty, Purdue U., 1955-60, asst. prof., 1957-60; vis. asso. prof. U. Cal. at Santa Barbara, 1960-61, asso. prof. math., 1961-67, now professor of mathematics, since 1967—. XL Research grantee, summer 1957; NSF Faculty Fellow, summers 1959, 60, 61; Purdue Research Found.-NSF Research grantee, summer 1965. Mem. Am. Math. Soc., Math. Assn. Am., Am. Assn. U. Profs., Sigma Xi. Research and publs. on ring theory, abstract algebra, gen. math. systems. Home: 602 Litchfield Lane, Santa Barbara, Cal. 93105.*

YA'QUB IBN ISHAQ AS'AD AL-DIN, see al-Mahalli.

YARINSKY, Allen, Am. parasitologist; b. Bklyn., May 6, 1929; s. William and Celia (Zaretsky) Y.; B.S., Coll. City N.Y., 1951; M.S., Columbia U., 1953; M.S. in Pub. Health, U. N.C., 1957, Ph.D., 1961; m. Estelle Kessler, Dec. 24, 1952; children—Steven, Carolyn, Adam. Research scientist, chief parasitology sect. Bur. Labs., N.Y.C. Dept. Health, 1961-65; instr. dept. parasitology Columbia U. Sch. Pub. Health and Adminstrv. Medicine, 1961-65; research biologist Sterling-Winthrop Research Inst., 1966—. Cons. in parasitology div. labs. Hackensack (N.J.) Hosp., 1964—. Mem. Am. Soc. Parasitologists, Am. Soc. Tropical Medicine and Hygiene, A.A.A.S., Am. Inst. Biol. Sci., N.Y. Soc. Tropical Medicine, Sigma Xi, Delta Omega. Research on Clostridium botulinum type E toxin, irradiation effects on immunity mice to infection with Trichinella spiralis, diagnosis intestinal protozoan infections of man. Home: 420 Wellington Rd., Delmar, N.Y. 12054. Office: Sterling-Winthrop Research Inst., Rensselaer, N.Y. 12144.*

YARMOLINSKY, Michael Bezalel, Am. molecular biologist; b. N.Y.C., Jan. 18, 1929; s. Avrahm and Babette (Deutsch) Y.; A.B., Harvard, 1950; Ph.D., Johns Hopkins, 1954; m. Sirpa Helvi Irmeli Tuhkanen, June 22, 1962; 1 dau., Miriam Laura. Instr., N.Y. U. Coll. Medicine, N.Y.C., 1954-55; research asso. Johns Hopkins, Balt., 1958-61, asst. prof. dept. biology, McCollum-Pratt Inst., 1961-63; NSF fellow Service de Génétique Microbienne, Institut Pasteur, Paris, France, 1963-64; research chemist Lab. Molecular Biology, Nat. Inst. Arthritis and Metabolic Diseases, NIH, Bethesda, Md., 1964—. Mem. Am. Soc. Biol. Chemists. Research on protein synthesis, biology of temperate bacteriophage. Home: 5506 Roosevelt St., Bethesda 20034. Office: Nat. Inst. Arthritis and Metabolic Diseases, NIH, Bethesda, Md. 20014.*

YARNELL, John Leonard, Am. physicist; b. Topeka, Mar. 1, 1922; s. Ray and Ruth (Leonard) Y.; A.B., U. Kan., 1947, A.M., 1949; Ph.D., U. Minn., 1952; m. Hazel Helen Heidtke, Sept. 28, 1952; children—Barbara, Ann, Thomas, Martha. Staff mem. physics div. Los Alamos Sci. Lab., 1952-65, group leader, 1965—. Fellow Am. Phys. Soc.; mem. Phi Beta Kappa, Sigma Xi. Research and publs. on lattice vibrations in solids, properties of liquid helium, low energy nuclear reactions, design and operation of research reactors. Home: 205 El Conejo St. Office: Los Alamos Sci. Lab., Box 1663, Los Alamos 87544.*

YARRELL, William, English zoologist; b. St. James, London, Eng., June 3, 1784; s. Francis and Sarah (Blane) Y.; ed. Dr. Nicholson's Sch., Ealing, London. Clerk, banking firm Messrs. Herries, Farquhar & Co., 1802-03; newspaper agt., bookseller, London. Fellow Linnean Soc. (treas. 1849-56, v.p.); mem. Royal Inst., charter mem. Zool. Soc. Author: History of British Fishes, 2 vols., 1835-36; History of British Birds, 3 vols., 1837-43. Research, publs. on zoology; pub. 2 works which became standard authorities; added new species of swan to European avifauna, identified notochord in the Amphioxus, 1836, proved the white bait is distinct species of fish. Died Great Yarmouth, Eng., Sept. 1, 1856.

YARROW, Sir Alfred Fernandez, Brit. engr.; b. London, Eng., Jan. 13, 1842; s. Edgar William and Esther (Lindo) Y.; ed. U. Coll. Sch., London; LL.D., U. Glasgow (Scotland); m. Minnie Florence Franklin, 1875 (d. 1922); 3 sons, including Harold Edgar; 2 daus. Apprenticed to marine engine bldg. firm, 1857; installed (with James Hilditch) probably 1st pvt. telegraph line in Eng., 1857; became London rep. Coleman & Sons, 1862; became partner with Hedley,

1866; worked for Admiralty during World War I; pres. Yarrow & Co. Ltd., Glasgow, Scotland. Fellow Royal Soc.; mem. Instn. C.E.'s (hon. life), Instn. Naval Architects (mem. council 1887, v.p., 1896). Developed new type of steam plow, 1860-62, Yarrow straight-tube boiler for ships; improved steam launches; built numerous small vessels using prefabrication methods, 1869-75; increased speed and power of torpedo boats; designed destroyers and gunboats for Brit. Navy, World War I; one of 1st designers to use high tensile steel and aluminum, to perform systematic expts. and speed trials; contbd. to devel. Yarrow-Schlick-Tweedy system for reduction of machinery vibration. Died London, Jan. 24, 1932.

YASUKOCHI, Ko, Japanese physicist; b. Matsuyama, Japan, Jan. 2, 1924; s. Yutaka and Umeko (Nakamura) Y.; Ph.D., Tokyo U., 1962. Asst. lectr. dept. physics Tokyo U., 1946-59; prof. magnetism, dept. physics Coll. Sci. and Tech., Nihon U., Tokyo, Japan, 1959-—. Mem. Phys. Soc. Japan. Research and publs. on magnetic properties of intermetallic compound between 3d elements and Ge, Sn, magnetism of rare earth compounds, non-ideal Type II superconductors. Home: 1-24-9 Matsugaoka, Nakano-ku, Tokyo, Japan.*

YASUNOBU, Kerry Tsuyoshi, Am. biochemist; b. Seattle, Nov. 21, 1925; s. Seiji and Chiyono (Kobayashi) Y.; B.S., U. Wash., 1950, Ph.D., 1954; m. Kikue Itami, Nov. 15, 1952; children—Steven Earl, Chrissie Lea. Research scientist U. Tex., 1954-55; research asso. U. Ore. Med. Sch., 1955-58; asst. prof. U. Hawaii, Honolulu, 1958-62, asso. prof., 1962-65, prof., 1965-—. Mem. Am. Chem. Soc. (sec. Hawaii chpt. 1961-62), Sigma Xi (councillor Hawaii chpt. 1962). Research and numerous publs. in enzyme and protein chemistry. Home: 3270 Melemele Pl., Honolulu 96814.*

YATES, Francis Eugene, Am. physiologist; b. Pasadena, Cal., Feb. 26, 1927; s. Francis Eugene and Helen (Smith) Y.; student U. Tex., 1944-45, U. Cal. at Los Angeles, 1945-46; B.A., Stanford, 1947, M.D., 1951; m. Margaret Adelle Barnett, Dec. 16, 1949; children—Katherine Kingsley, Gregory Barnett, Peter Franklin, Eugene Barnett, Anna Scott. Faculty, Harvard, 1953-60, asst. prof., 1959-60; asso. prof. physiology, Stanford (Cal.) U., 1960-—, acting exec. head dept. physiology, 1964-—. Cons., NIH, 1964-—. Markle scholar in med. sci., 1960-65; recipient Upjohn award Endocrine Soc., 1962. Mem. N.Y. Acad. Scis., A.A.A.S., Endocrine Soc., Am. Physiol. Soc., Am. Inst. Biol. Scis., Phi Beta Kappa, Sigma Xi. Research, publs. on application control theory to endocrine feedback systems. Home: 662 Mirada Av., Stanford, Cal. 94305.*

YATES, Frank, English research statistician; b. Manchester, Eng., May 12, 1902; s. Percy and Edith (Wright) Y.; B.A., St. John's Coll., Cambridge, Eng., 1924, Sc.D., 1938; m. Pauline Shoubersky, July 14, 1939. Research officer, math. adviser Gold Coast Geodetic Survey, 1927-31; with Rothamsted Exptl. Sta., Harpenden, Eng., 1931-—, head dept. statistics, 1933-68, head agrl. research statis. service, 1947-68, dep. dir., 1958-68. Sci. adviser to various ministries FAO; mem. sub-commn. on statis. sampling UN, 1947-52. Fellow Royal Soc. (Royal medal 1966), 1948; mem. Brit. Computer Soc. (pres. 1960-61), Royal Statis. Soc. (pres. 1967-68). Author: (with R. A. Fisher) Statistical Tables for Biological, Agricultural and Medical Research, 1938; Sampling Methods for Censuses and Surveys, 1949; also numerous articles. Research on math. statistics, especially design and analysis of expts. and sampling survey theory and practice; devel. use of electronic computers in research statistics. Home: Stackyard, Rothamsted, Harpenden. Office: Rothamsted Exptl. Sta., Harpenden, Herts, Eng.*

YATES, Keith, chemist; b. Preston, Eng., Oct. 22, 1928; s. Harold and Elizabeth (Wilson) Y.; B.A., U. B.C., Vancouver, Can., 1956, M.Sc., 1957; Ph.D., Oxford (Eng.) U., 1959, D.Phil., 1961; m. Norma June Charter, Aug. 21, 1953; children—Alison, Robyn, Nicola. Faculty, U. Toronto (Ont., Can.), 1961-—, asso. prof. chemistry, 1964-—, exec. asst. to chmn. dept., 1963-65, asst. dean Sch. Grad. Studies, 1967-—. Mem. Chem. Inst. Can. Research and publs. on phys. and theoretical organic chemistry, especially acidity. Home: 23 Broadlands, Don Mills, Ont. Office: Dept. Chemistry, U. Toronto, Toronto, Ont., Can.*

YATES, Peter, chemist; b. Wanstead, Eng., Aug. 26, 1924; s. Harold Andrew and Kathryn (Yexley) Y.; B.Sc., U. London, 1946; M.Sc., Dalhousie U., 1948; Ph.D., Yale, 1951; m. Mary Ann Palmer, Sept. 9, 1950; 1 son, John Anthony; stepchildren—William Palmer Franklin, Thomas Jay Franklin. Post-doctoral fellow Harvard, 1950-51, faculty, 1952-60; instr. Yale, 1951-52, vis. prof., 1966; prof. chemistry U. Toronto (Can.), 1960-—. Fellow Chem. Inst. Can. (Merck, Sharp & Dohme lectr. 1963); mem. Am. Chem. Soc., Chem. Soc. London. Contbr. articles to sci. jours. Research, both structural and synthetic, on natural products, photochem. products, heterocyclic compounds, and aliphatic diazo compounds.*

YATZIDIS, Hippocrates, Greek physician; b. Athens, Greece, Apr. 23, 1923; s. Alexandre Lazarus and Anastasia (Courtidis) Y.; M.D., Med. Sch., U. Athens (Greece), 1950, Ph.D., 1954; m. Catherine Maleas, Sept. 12, 1955. Head research program 2d dept. internal medicine Athens U., 1958-61, chief sect. renal diseases Hippocratean Hosp., 1961-—, sr. lectr., 1961-—, sr. cons. in internal medicine, 1961-—, asso. prof. medicine, 1962-—. Mem. Hellenic Hematological Soc., European Soc. Hematology, Soc. Nephrology, Internat. Soc. Nephrology, Internal. Coll. Angiology, European Dialysis and Transplant Assn., Med. Soc. Athens (past sec.). Asso. editor Nephron, 1964-—. Research and publs. in renal diseases; introduced effective method for replacement of renal function in treatment of endogenous and exogenous intoxications known as hemo-carbo perfusion or Yatzidis' charcoal artifical kidney. Home and office: 31 Academias St., Athens 135, Greece.*

YEAGER, Ernest Bill, Am. phys. chemist; b. Orange, N.J., Sept. 25, 1924; s. Ernest Frederick and Olga (Wittwer) Y.; B.A., Montclair State Coll., 1945; M.S., Western Res. U., 1946, Ph.D., 1948; Mem. faculty Western Res. U., 1948-—, prof. chemistry, 1958-—, acting chmn. dept., 1964-65; tech. dir. Ultrasonics Research Lab., 1951-62, dir., 1962-, co-dir. Condensed State Center, 1965-—; vis. prof. chemistry U. Southampton (Eng.), 1968; NATO sr. fellow, 1968. Cons., Union Carbide Corp. Mem. com. on undersea warfare Nat. Acad. Sci., 1962-—; mem. NRC, 1968-—. Recipient Ann. Tech. award Cleve. Tech. Socs. Council, 1954; Biennial award Acoustical Soc. Am., 1956; Certificate of Merit, Cleve. Chem. Professions, 1959. Fellow A.A.A.S.; mem. Electrochem. Soc. (v.p. 1962-65, pres. 1965-66), Acoustical Soc. Am. (exec. council 1964-66, v.p. 1967-68), Am. Chem. Soc., Optical Soc. Am., Soc. for Applied Spectroscopy, Faraday Soc., Internat. Com. on Electrochem. Thermodynamics and Kinetics (v.p. 1967-68), Am. Assn. U. Profs. Editor: Transactions of the Symposium on Electrode Processes, 1961, 66. Contbr. numerous articles to profl. jours. Research on electrochem. processes, phys. and chem. properties of ionic solutions, applications for ultrasonic techniques in chemistry; co-inventor sodium amalgam-oxygen fuel cell. Home: 3071 Ashwood Rd., Cleve. 44120.*

YEAGER, George Herschel, Am. surgeon; b. Davis, W.Va., Oct. 19, 1905; s. Louis G. and Susan Lambie (Osborn) Y.; B.S., U. W.Va., 1925; M.D., U. Md., 1929; m. Dorothy Stone, Nov. 26, 1936; children—Anne Stone, Barbara B., George Harvey. Practice medicine, specializing in surgery, Balt., 1933-—; faculty U. Md. Sch. Medicine, Balt., 1934-—, prof. clin. surgery, 1946-—; dir. profl. and supporting services U. Hosp., Balt., 1965-—. Diplomate Am. Bd. Surgery. Mem. Am., So. (sec. 1964-—) surg. assns., A.C.S., Southeastern Surg. Congress (pres. 1967-68), Soc. for Vascular Surgery, A.M.A. (mem. conf. com. on grad. tng. in surgery 1963-—), World, So. (councilor for Md. 1966-—) med. assns., So. Surgeons Club, Balt. City Med. Soc., Med. and Chirurg. Faculty Md., Soc. Med. Consultants to Armed Forces. Editor, Am. Surgeon, 1954-—. Research and publs. on vascular surgery. Home: 212 Ridgewood Rd., Balt. 21210.*

YEATER, Max Laverne, Am. physicist; b. Detroit, Feb. 18, 1917; s. Paul O. and Bessie (German) Y.; B.S. in Elec. Engring., U. Mo., 1939; M.S., Washington U., St. Louis, 1941, Ph.D., 1943; m. Virginia Thorman, June 22, 1943; children—Donald P., David A., Carol E. Research asso. Mass. Inst. Tech. Radiation Lab., 1943-45, Gen. Electric Research Lab., 1945-51; mgr. neutron physics group Knolls Atomic Power Lab., Schenectady, 1951-58; prof., asso. chmn. dept. nuclear engring. and sci. Rensselaer Poly. Inst., Troy, N.Y., 1958-—. Mem. Am. Nuclear Soc., Am. Phys. Soc., A.A.A.S., Am. Soc. Engring. Edn., N.Y. Acad. Scis., Am. Assn. U. Profs. Research and publs. in photonuclear disintegration, neutron cross sects., reactor physics, nuclear instrumentation. Home: 1391 Valencia Rd., Schenectady 12309. Office: Rensselaer Poly. Inst., Troy, N.Y. 12181.*

YEATMAN, Harry Clay, Am. zoologist; b. Ashwood, Tenn., June 22, 1916; s. Trezevant Player and Mary (Wharton) Y.; A.B., U. N.C., 1939, M.A., 1942, Ph.D., 1953; m. Jean Hansford Anderson, Nov. 24, 1949; children—Henry Clay III, Jean Hansford. Zoology asst. U. N.C., Chapel Hill., 1939-42, teaching fellow, 1946-47, instr., 1947-50; faculty U. South, Sewanee, Tenn., 1950-—, prof. zoology, 1960-—. Cons. div. crustacea U. S. Nat. Mus., Washington, 1942-—, Woods Hole (Mass.) Oceanographic Instn., 1960-—. Gen. Edn. Bd. fellow, 1940-41. Fellow A.A.A.S.; mem. Soc. Systematic Zoology (charter), Soc. Limnology and Oceanography (charter), Soc. Ichthyology and Herpetology, Tenn. Acad. Sci., Ornithol. Union. Contbg. author: Freshwater Biology, 1959; Ency. Sci. and Tech., 1960; also articles. Home: Mississippi Av., Sewanee, Tenn. 37375.*

YEH, Kung Chie, elec. engr.; b. Hangchow, China, Aug. 3, 1930; s. S. J. and A. S. (Shu) Y.; B.E.E., U. Ill., 1953; M.E.E., Stanford, 1954, Ph.D., 1958; m. Margaret Young; children—Joanna, Lisa, David. Teaching asst. elec. engring. Stanford, 1953-54, research asst. 1954-58; research asso. elec. engring. U. Ill., Urbana, 1958-59, asst. prof., 1959-62, asso. prof., 1966-—; vis. prof. dept. elec. engring. Nat. Taiwan U., China, 1966; vis. fellow elec. engring. U. Hawaii, 1967. Mem. I.E.E.E., Am. Phys. Soc., Am. Geophys. Union, URSI (mem. U. S. Commn. 3). Research, publs. on ionospheric propagation problems and ionospheric physics. Home: 618 Richards Lane, Champaign, Ill. 61820. Office: U. Ill., Urbana, Ill. 61801.*

YEH, Si-Jung, Chinese chemist; b. Taiwan, China, May 24, 1927; s. Ping-Cheih and Yu-Pao (Hsue) Y.; B.S., Nat. Taiwan U., 1950; M.S., U. So. Cal., 1956; Ph.D., U. Cin., 1958; m. Pi Chiu Huang, Feb. 11, 1951; children—Chih-chao, Hui-chu, Li-ju. Research asso. U. Ore., 1958-59; asst. prof. Nat. Tsing Hua U., 1959-63, prof., 1963-—; research prof. Nat. Council on Sci. Devel., 1964-—; dep. dir. Inst. Nuclear Sci., 1964-—. Recipient Hsieh Chih research award, 1964. Mem. Chinese Chem. Soc., Japan Atomic Energy Soc., Phi Tau Phi. Editor, Nuclear Sci. Jour. Research, publs. on techniques for prodn. of short-lived radioisotopes for practical application hot atom chemistry, radiation chemistry. Address: 397 Kuang Fu Rd., Hsinchu, Taiwan, China.*

YEH, Thomas Jui-ting, surgeon; b. Tainan, Formosa, Feb. 6, 1929; s. Tsuo-jou and Tsu-ai (Lin) Y.; M.D. suprema cum laude, Nat. Taiwan U., Taipei, Formosa, 1951; m. Doris McKibben, June 25, 1957; children—Thomas Jui-ting, Karen Anne. Came to U. S., 1951, naturalized, 1963. Resident surgery U. Louisville Hosps., 1953-57; research fellow cardiovascular surgery Med. Coll. Ga., Augusta, 1957-58, resident thoracic surgery, 1959-60, instr., asso. prof. thoracic surgery, 1960-—, also asst. dir. Hemodynamic Research Lab.; resident thoracic surgery McGill U., 1958-59; practice medicine, specializing in thoracic surgery, Augusta, 1960-—; attending staff Eugene Talmadge Meml. Hosp.; cons. VA Hosp. Fellow A.C.S.; mem. A.M.A., Am. Heart Assn., Am. Thoracic Soc., Am. Assn. Thoracic Surgery, Am. Soc. Artificial Internal Organs, Soc. Thoracic Surgeons, So. Thoracic Surg. Assn. Contbg. author: Current Therapy, 1966; Biochemical Clinics, The Lung, 1965. Research and publs. primarily in field of lung transplantation in dogs, extra corporal circulation. Home: 8308 Kent Dr. Office: Meml. Med. Center, P.O. Box 6268, Sta. C, Savannah, Ga.*

YEN, John Teng Chien, geologist; b. Canton, China, Feb. 15, 1903; s. Ker-Chuen and Yuan-Chen (Chang) Y.; B.A., St. John's U., 1927; B.S., Nat. U., Nanking, 1929; Humboldt scholar, Berlin U., 1937-39; D.Sc., Glasgow U., 1948; m. Marie Louise Henry, Apr. 7, 1945. Cons., research geologist, Phila., Washington, 1953-54; prof. geology Villanova (Pa.) U., 1957-—, chmn. dept. geology, 1958-—. Ind. field investigations in Scandinavian countries, Germany, Austria, Switzerland, France, 1955-56, Switzerland, France, 1960, Netherlands, Germany, France, 1963, Austria, France, Netherlands, Belgium, Spain, 1965. Fellow Geol. Soc. Am., Geol. Soc. London, A.A.A.S.; mem. Paleontol. Soc., Am. Geophys. Union, Am. Geochem. Soc., Am. Ecol. Soc., Malacological Soc. London, Malakozool. Soc. Germany. Research, numerous publs. on marine and land gastropods of China, fresh-water mollusks of the western U. S.; certain European fauna. Home: 4 Lowry's Lane, Rosemont, Pa. 19010. Office: Dept. Geology, Villanova U., Villanova, Pa. 19015.*

YEN, Kuo Tai, fluid physicist; b. Hankow, China, Aug. 1, 1919; s. Mon Kai and Fon Wen (Chiang) Y.; B.S., Nat. Tsing Hwa U., Peking, China, 1941; M.A.E., Bklyn. Poly. Inst., 1948; Ph.D., Brown U., 1952; m. Irene Lee, Aug. 11, 1951; children—Daniel, Ronald. Came to U. S., 1946, naturalized, 1961. Asst. aeros. Nat. Tsing Hwa U., 1941-42; engr. Second Aircraft Co., China, 1942-46; research asst. Bklyn. Poly. Inst., 1947-49; asst. prof. Rensselaer Poly. Inst., 1951-55, asso. prof., 1955-58, prof., 1958-60; research specialist Space Scis. Lab., Missile and Space div. Gen. Electric Co., Phila., 1960-66, prin. scientist, 1967-—. Mem. Am. Inst. Aeros. and Astro., Am. Geophys. Union, A.A.A.S., Sigma Xi. Research and publs. in compressible and viscous flow theories as applied to aeros. and lubrication, electromagnetic wave propagation, plasma physics, magnetogasdynamics, solar physics. Home: 436 Louella Av., Wayne, Pa. 19087. Office: P.O. Box 8555, Phila. 19101.*

YEN, Teh Fu, chemist; b. Kunming, Yunnan, China, Jan. 9, 1927; s. Kuang Pu and Ren Chen (Liu) Y.; B.S., Central China U., 1947; M.S., W. Va. U., 1953; Ph.D., Va. Poly. Inst., 1956; m. Shiao-Ping Siao, May 31, 1959. Came to U. S., 1949, naturalized, 1963. Sr. research chemist Goodyear Tire and Rubber Co., Akron, O., 1955-59; fellow Mellon Inst., Pitts., 1959-64, sr. fellow, 1965-—; vis. lectr. Nat. Taiwan U., 1965-—, Tsing-Hua U., 1965. Recipient Sigma Xi Research award, 1955. Fellow Inst. Petroleum; mem. Am. Chem. Soc., Am. Phys. Soc., Chem. Soc. (London), Sigma Xi, Phi Lambda Upsilon. Research on structural evaluation complex molecules, polynuclear aromatic hydrocarbons. Home: 304 Sharon Dr., Pitts. 15221. Office: 4400 5th Av., Pitts. 15213.*

YERGANIAN, George, Am. biologist; b. N.Y.C., June 14, 1923; s. Charles and Miriam (Krishjian) Y.; B.S., Mich. State U., 1947; Ph.D., Harvard, 1950; m. Sona Arzomanian, Sept. 3, 1950; children—Arra George, Athena Zee. Instr., U. Minn., Mpls., 1950-51; AEC fellow Brookhaven Nat. Lab., 1951-52; AEC fellow Boston U., 1952-53; USPHS fellow Nat. Cancer Inst., Children's Cancer Research Inst., Boston, 1953-

54, biologist, chief labs. cytogenetics, 1954—; research asso. pathology Harvard Med. Sch., 1956—. Mem. Am. Assn. for Cancer Research, Tissue Culture Assn., Am. Soc. for Cell Biology. Research and publs. on chromosomes and genetic features of normal, malignant and virus-transformed cells of Chinese and Armenian hamsters. Home: 89 Bellevue Hill Rd., West Roxbury, Mass. 02132. Office: 35 Binney St., Boston 02115.*

YERGIN, Paul Flohr, Am. physicist; b. N.Y.C., Apr. 21, 1923; s. Howard V. and Ida (Flohr) Y.; student Coll. City N.Y.; B.S., Union Coll., Schenectady, 1944; M.A., Columbia U., 1949, Ph.D., 1953; m. Eunice Carlson, June 14, 1947; children—Ann, Susan. Staff mem. div. war research Columbia U., 1944-46; faculty U. Pa., 1952-56, asst. prof., 1955-56; faculty Rensselaer Poly. Inst., Troy, N.Y., 1956—, asso. prof., 1957—. Chmn., Gordon Research Conf. on Photonuclear Reactions, 1967. Mem. Am. Phys. Soc., Am. Assn. Physics Tchrs., Fedn. Am. Scientist, Am. Assn. U. Profs., Sigma Xi. Research on fine structure of singly ionized helium, photonuclear reactions, neutron total cross sections for MeV neutrons. Home: 10 Crescent Terrace, Troy, N.Y. 12180.*

YERKES, Robert Mearns, Am. psychobiologist; b. Breadysville, Pa., May 26, 1876; s. Silas Marshall and Susanna Addis (Carrell) Y.; A.B., Ursinus, 1897; A.B., Harvard, 1898, A.M., 1899, Ph.D., 1902; LL.D., Ursinus, 1923; D.Sc., Wesleyan U., 1923; hon. M.A., Yale, 1931; m. Ada Watterson, 1905; children—Roberta Watterson, David Norton. Began as teacher and investigator at Harvard, 1901, asst. prof. comparative psychology, 1908-17, psychologist to the Psychopathic Hosp., Boston, 1913-17; prof. psychology and dir. Psychol. Lab., U. of Minn., 1917-19 (absent on mil. duty); chmn. research information service, Nat. Research Council, 1919-24 (chmn. com. for research in problems of sex, 1921-47); prof. psychology, Inst. Psychology, Yale, 1924-29; prof. psychobiology, Yale, 1929-44, emeritus, 1944—; organized and dir. Yale Labs. of Primate Biology, Orange Park, Fla., 1929-41, named Yerkes Labs. of Primate Biology, 1942. Chief, divison of psychology, office Surg. Gen., A.U.S., 1917-18. Consulting services War Dept. and Nat. Research Council, World War II. Fellow Am. Acad. of Arts and Sciences, A.A.A.S.; mem. Nat. Acad. Sciences, Am. Philos. Soc., Am. Psychol. Assn. (pres., 1916-17), American Physiological Society, American Society Naturalists (pres. 1938), Soc. of Mammalogists. Author: The Dancing Mouse, A Study in Animal Behavior, 1907; Introduction to Psychology, 1911; Methods of Studying Vision in Animals (with J. B. Watson), 1911; Outline of a Study of the Self (with D. W. La Rue), 1914; A Point Scale for Measuring Mental Ability (with R. S. Hardwick and J. W. Bridges), 1915; The Mental Life of Monkeys and Apes—A Study of Ideational Behavior, 1916; Psychological Examining in the U. S. Army (with others), 1921; Chimpanzee Intelligence and Its Vocal Expressions (with B. W. Learned), 1925; Almost Human, 1925; The Mind of a Gorilla, 1927; The Great Apes (with A. W. Yerkes), 1929; Chimpanzees: A Laboratory Colony, 1943. Research on physiology of the nervous system, animal behavior, comparative psychology and mental measurement. Especially interested in psychobiological research, mental engineering, and problems of population. Died Feb. 3, 1956.

YERSIN, Alexandre Emile John, bacteriologist; b. Canton Vaud, Switzerland, Sept. 22, 1863; ed. Lausanne, Switzerland, Marburg, Germany, Paris. Became naturalized French citizen, 1889. Research with Roux on diphtheria antitoxin Pasteur Inst., Paris; research in China and French Indochina; staff Colonial Health Service; founder, dir. Pasteur insts., Canton, China, Nha Trang, French Indochina; insp. gen. Pasteur insts., Indochina. Corr. mem. French Acad. Scis., 1916. Prepared (with Roux) anti-plague serum, 1894, anti-diphtheria serum, 1898; discovered plague bacillus, 1894; introduced rubber tree to Indochina. Died Nha Trang, Annam, French Indochina, Mar. 2, 1943.

YERUSHALMY, Jacob, bio-statistician; b. Vienna, Austria, Aug. 5, 1904; s. Zev and Anna (Ziev)) Y.; B.A., Johns Hopkins, 1927, M.A., 1929, Ph.D., 1930; m. Eva R. Zemil, Sept. 4, 1934; children—Zeva Anne, Paul Zev. Came to U. S., 1924, naturalized, 1931. Instr. mathematics Johns Hopkins, 1932-35; statistician, N.Y. State Dept. Health, 1934-38, Nat. Insts. Health, 1938-41; dir. div. statis. research Children's Bur., Dept. Labor, 1941-43; prin. statistician USPHS, 1943-48; prof. biostatistics U. Cal. at Berkeley, 1948—; cons. WHO, USPHS, Cal. Dept. Pub. Health. NRC fellow mathematics, 1930-32. Mem. Nat. Adv. Neurol. Diseases and Blindness Council, Am. Pub. Health Assn., Am. Statis. Assn., Population Assn. Am., Internat. Union Sci. Study Population, Phi Beta Kappa, Sigma Xi. Contbr. articles sci. jours. Research on infant mortality relating sex, birth order and parents' order and parents' ages to mortality, 1938; promoted establishment of improved X-ray techniques; correlated fat and heart disease mortality; stressed use of "age-specific" rates in studying group mortality. Home: 526 Santa Barbara Rd., Berkeley 7, Cal.

YEVICK, George Johannus, Am. physicist; b. Berwick, Pa., Apr. 24, 1922; s. John and Teresia (Powell) Y.; B.Sc., Mass. Inst. Tech., 1942, D.Sc., 1947; M.Engring. (hon.), Stevens Inst. Tech., 1958; m. Miriam Amalie Lipschutz, May 15, 1945; 1 son, David Owen. Staff radiation lab. Mass. Inst. Tech., 1944-46; faculty Stevens Inst. Tech., Hoboken, N.J., 1947—, prof., 1957—. Dir. Electrokinetics, Inc., Florham Park, N.J., Hydroelectrics Internat. Inc., Sparkill, N.Y. Mem. Phys. Soc., Sigma Xi. Research and publs. on elementary particles, dynamical theory of many interacting particles, causal theory of quantum mechanics, control of thermonuclear fusion. Patentee in field. Home: 536 Nordhoff Dr., Leonia, N.J. 07605. Office: Stevens Inst. Tech., Hoboken, N.J.*

YIH, Chia-Shun, hydrodynamicist; b. Kweiyang, China, July 25, 1918; s. Ting-Chien and Wan-Lan (Shiao) Y.; B.S., Nat. Central U., Chungking, China, 1942; M.S., State U., Ia., 1947, Ph.D., 1948; m. Shirley Gladys Ashman, Feb. 17, 1949; children—Yuen-Ming David, Weiling Katherine. Came to U. S., 1945, naturalized, 1954. Instr. math. U. Wis., 1948-49; lectr. math. U. B.C., 1949-50; asso. prof. civil engring. Colo. State U., 1950-52; research engr., asso. prof. fluid mechanics State U. Ia., Iowa City, 1952-56; faculty U. Mich., Ann Arbor, 1956—, prof. fluid mechanics, 1958-68, Stephan P. Timoshenko Univ. prof. engring., 1968—. attaché de recherché U. Nancy (France), 1951-52. Referee for fluid mechanics jours.; reviewer sci. proposals U. S. Govt., 1957—; cons. Hwyck Felt Co., Rensselaer, 1960-64. NSF Sr. postdoctoral fellow U. Cambridge, 1959-60; Guggenheim fellow, 1964. Fellow Am. Phys. Soc.; mem. Internat. Assn. for Hydraulic Research, Sigma Xi, Pi Mu Epsilon, Tau Beta Pi. Author: Dynamics of Nonhomogeneous Fluids, 1965. Contbg. author: Handbook of Fluid Mechanics, 1961; Advanced Fluid Mechanics, 1958. Reviewer, Math. Revs., 1958—. Research and publs. on hydrodynamics, hydrodynamic stability, gravity effects in fluid flow, rotating fluid, non-homogenous fluids, magnetohydrodynamics, geophys. fluid mechanics. Home: 3530 W. Huron River Dr. Ann Arbor, Mich. 48103.*

YNTEMA, Jan Lambertus, physicist; b. Leeuwarden, Netherlands, Oct. 5, 1920; s. Lambertus and Grietje (Vandermolen) Y.; Ph.D., U. Amsterdam, 1952; m. Ida G. Hoekstra, July 15, 1948; children—Grietje I., L. James, Richard J., David O. Came to U. S., 1949, naturalized, 1964. Fellow Canadian Nat. Research Lab., 1948-49; research asso. Princeton, 1949-52; asst. prof. U. Pitts., 1952-55; asso. physicist Argonne (Ill.) Nat. Lab., 1955—. Fellow Am. Phys. Soc.; mem. Sigma Xi. Research in intramolecular potential of helium, proton-proton scattering, nuclear reaction processes, structure of nuclei; instrumentation for nuclear research. Home: 5125 Grand Av., Western Springs, Ill. 60558. Office: Argonne Nat. Lab., Argonne, Ill. 60440.*

YOCUM, Lawson Edwin, Am. plant physiologist; b. nr. Catawissa, Pa., Aug. 10, 1890; s. Ambrose Heart and Ida M. (Cherrington) Y.; B.S., Pa. State Coll., 1916; tchr.'s certificate Bloomsburg Tchrs. Coll., 1911; M.S., Ia. State Coll., 1919, Ph.D., 1924; m. Mildred Hicks, Sept. 6, 1917; children—Mary Jean, (Mrs. Jack R. Harlan). Instr., Pa. State Coll., 1918-19, Ia. State Coll., 1916-18, 19-20, 23-24; faculty George Washington U., Washington, 1931-56, prof., dept. head, 1945-56; faculty N.C. Coll., 1921-30. Fellow Washington Acad. Scis.; mem. Washington Bot. Soc. (past pres.), Bot. Soc. Am., Phi Epsilon Phi. Research and publs. on translocation of foods in wheat seedling, stomata and transpiration of oaks. Author: Students Outline of Botany, 1937; Plant Growth, 1945. Home: 1257 Drew St., Clearwater, Fla. 33515.*

YODA, Naoya, Japanese chemist; b. Tokyo, Japan, June 11, 1931; s. Shuichi and Teruko (Ozaki) Y.; A.B. in chemistry magna cum laude, Nagoya (Japan) U., 1954, Ph.D., 1960; M.A., Harvard, 1956; m. Kazuko Hironaka, Apr. 7, 1960; children—Naohisa, Nobuhisa. Fulbright Research fellow Harvard, 1954-56; research asso. U. Ariz., 1962-63; research chemist Toyo Rayon Co., Ltd., Central Research Labs., 1957-61, sr. research chemist Basic Research Labs., 1964-65, research assos., 1966—. Inspection mem. Chem. Industry and Chem. Fibers, U. S., 1961, U. S. and Europe, 1962-63. Mem. Chem. Soc. Japan, High Polymer Soc. Japan, Am. Chem. Soc. Author: (with Y. Ishii) Physical Organic Chemistry, 1959; Introduction to the Chemistry of High Polymers, 1960; (with T. Hoshino) Preparative Methods of Polymer Chemistry, 1961; also articles. Editor: High Polymers Japan, 1964—. Research on new thermally stable polymers, aromatic polyanhydrides, aromatic polyamides, polymerizations of ring compounds, proton-transfer polymerization of aromatic compounds, nonbezenoid aromatic systems. Home: 937-1 Koshigoe, Kamakura. Office Toyo Rayon Co. Ltd., 1,111 Tebiro, Kamakura, Kanagawa, Japan.*

YODER, Hatten Schuyler, Jr., Am. petrologist; b. Cleve., Mar. 20, 1921; s. Hatten Schuyler and Elizabeth Katherine (Knieling) Y.; A.A., U. Chgo., 1940, S.B., 1941; student U. Minn., summer 1941; Ph.D., Massachusetts Institute of Technology, 1948; m. Elizabeth Marie Bruffey, Aug. 1, 1959; children—Hatten Schuyler III, Karen. Petrologist, Geophys. Lab., Carnegie Instn. of Washington, 1948—. Recipient award Mineral. Soc. Am., 1954, bicentennial medal Columbia, 1954; Arthur L. Day medal, Geol. Soc. America, 1962. Fellow Geological Soc. of America (member council 1966-68), Mineralogical Soc. Am. (council 1962-64), Am. Geophys. Union (pres. volcanology, geochemistry and petrology sect. 1962-64); mem. Mineral. Soc. London, Geol. Soc. Edinburgh (corresponding), Geol. Soc. Finland, Geochem. Soc. (council 1956-58), Am. Chem. Soc., Mineral. Assn. Can., Nat., Washington acads. scis., New York Academy of Sciences, Geological Society of Washington. Chemical Society of Washington. Sigma Xi. contbr. profl. papers sci. jours. Co-editor: Journal of Petrology, 1959—. Study of experimental petrology; piezochemistry; phase equilibria in mineral systems; hydrothermal mineral synthesis; properties of minerals at high pressure and high temperature. Address: 2801 Upton St., Washington 8.

YOE, John Howe, Am. chemist; b. Oxford, Ala., July 19, 1892; s. Rev. Alfred Moore and Tulliola May (Howe) Y.; student Birmingham Coll., 1908-10; B.S., Vanderbilt U., 1913, M.S., 1914; grad. student U. of Chicago, summer 1913; M.A., Princeton, 1917, Ph.D., 1923; m. Françoise Alexander Cheely, Aug. 18, 1919; children—Elizabeth Howe (Mrs. Richard W. Gordon), Françoise Cheely (Mrs. Chas. F. Schneider), Jane Randolph (Mrs. R. Warner Wood, Jr.), and Ann Everett (Mrs. John A. Hinckley). Assistant in chemistry Vanderbilt University, 1911-13, instr. in biol., 1913-14, in biol. and chemistry, 1914-15; instructor (summers in zoology, George Peabody College for Teachers, 1914, in chemistry, 1915-17; assistant in chemistry, Princeton, 1915-17; chemical engineer (gas research) U. S. Bur. Mines, Am. Univ. Expt. Sta., Washington, D.C., 1917-18; 1st lt. research div., C.W.S., U. S. Army, Apr. 1918-Jan. 1919, and chief of Canister Research Unit, June 1918-Jan. 1919; research chemist, Pa. Trojan Powder Co., Allentown, Pa., Jan.-Aug. 1919; with U. of Va. since 1919 as asst. prof., asso. prof., professor of chemistry, 1927-63, professor emeritus chemistry, 1963—, chairman of department, 1953-57; chemist Virginia Geology Survey, 1925-63; cons. chemist to Md., D.C. and Va. Laundry-owners Assn., 1927-31; cons. U. S. Bur. Mines, 1942, Philip Morris, Gliden Co., Universal Match Co.; sect. mem. Nat. Defense Research Com., 1940-41; ofcl. investigator O.S.R.D., 1941-45, Chem. Warfare Service, 1945-46; scientific observer at Atomic Bomb tests, Bikini, summer, 1946. Commd. col. Tennessee Nat. Guard and appointed aide-de-camp on govs. staff, 1941; mem. Internat. Commn. on analytical reactions of Union Internationale de Chimie, 1946-60; mem. Nat. Research Council Commn. Analytical Chemistry, 1947-53. Winner of Research Prize, Virginia Acad. Science, 1928; Grant-in-aid, 1933, 36, 37; research grant, Carnegie Corp., 1938; Jefferson award (with L. G. Overholser), Va. Acad. Science, 1940; Inter-Acad. award and Jefferson gold medal (with same), Va., N.C., S.C., Ga. and Fla. acads. science, 1941; Herty gold medal, 1943; Fisher award, American Chemical Society, 1957; on the Richmond Times-Dispatch "Va. Honor Roll of 1943." Fellow Am. Association for Advancement of Science, Am. Inst. Chemists (counselor, 1944-47); mem. Am. Chemical Soc. (editor Bulletin of Va. sect. 1933-34; vice chairman 1935-36, chairman 1936-37; counselor 1939, 42-45, 48-63. Distinguished Service award 1956), Southern Association of Science and Industry, Va. Acad. Sci. (chmn. chem. sect. 1934-35, counselor 1942-44), American Assn. University Profs., Sigma Xi, Phi Beta Kappa. Author or co-author books relating to field since 1920; contbr. to other publications, including scientific journals of United States and Europe. Asst. editor Apalytica Chimica Acta, 1948—; contbg. editor Michiochemical Journal, 1957—. Inventor of a photoelectric colorimeter and a color comparator. Home: Wayside Place, Charlottesville, Va.*

YOELI, Meir, physician; b. Kaunas, Lithuania, Aug. 20, 1912; s. Michael and Hanna Ethel (Karnovsky) Y.; student U. Kaunas, 1928-34; M.D., U. Basel, (Switzerland), 1939; m. Ketty Benardout, May 21, 1946; children—Michael, Gideon and Edith (twins). Came to U. S., 1956, naturalized, 1962. Asst., Malaria Research Sta., Rosh Pina, Palestine, 1934-38; lectr. tropical medicine and parasitology Hebrew U., Jerusalem, Palestine, 1946-56; lt. col. commanding Malaria Field Lab., 1941-46; officer commanding dept. hygiene Israel Def. Forces, 1948-49; vis. investigator London Sch. Tropical Medicine, also Rockefeller Inst., N.Y.C., 1951-53; prof. N.Y. U., Sch. Medicine, 1956—. Recipient Wallach prize Govt. of Israel, 1955, Laveran medal Soc. de Pathologie Exotique, Paris, France, 1965, Maimonides award Micheal Reese Med. Center, Chgo., 1966. Mem. A.M.A., A.A.-A.S., Am., Belgian socs. tropical medicine, Societe de Pathologie Exotique, Israel Soc. History Medicine and Sci. Numerous publs. on malaria, tropical and infectious diseases. Home: 215 W. 88th St., N.Y.C. 10024.*

YOFFE, Abraham David, physicist, chemist; b. Jerusalem, Nov. 26, 1919; s. Haim and Leah (Lorberbaum) Y.; B.Sc., U. Melbourne, 1939, M.Sc. with 1st class honors, 1941; Ph.D., Cambridge U., 1948, Sc.D., 1961; m. Elizabeth Hebden Mann, June 1949; children—Deborah, Jonathan, Gideon, Susan. With Council for Sci. and Indsl. Research, Australia, 1941-44, Weizmann Inst. Sci., Israel, 1950-54; faculty Cambridge (Eng.) U., 1955—, asst. dir. research, 1957—, fellow Darwin Coll. Mem. Faraday Soc., Cambridge Philos. Soc., Accademia Teatina (corr.). Author: (with F. P. Bowden) Initiation and Growth of Explosion in Liquids and Solids, 1952; Fast Reactions in Solids, 1958. Research and numerous publs. into factors which determine initiation and growth of explosions in gases, liquids, solids, stability of variety of solids to heat, light and ionizing radiations, electron energy levels in solids, behavior of electrons in very thin crystals of unit cell dimensions, properties of solids at very high pressures. Home: 6 Alwyn Rd. Office: Cavendish Lab., Cambridge, Eng.*

YOFFEY, J(oseph) M(endel), anatomist; b. Manchester, Eng., July 10, 1902; s. Israel Jacob and Pere (Jaffee) Y.; M.B., Ch.B., U. Manchester (Eng.), 1924, M.D., 1932, D.Sc., 1945; m. Betty Gillis, Aug. 28, 1940; children—Judith Rachel (Mrs. Adrian Boxer), Ruth Deborah, Naomi Sarah. Leech Research fellow U. Manchester, 1926-27, demonstrator anatomy, 1927-29; house surgeon, asst. lectr. Royal Infirmary, 1930-32; sr. lectr. U. S. Wales, 1932-37; Rockefeller Found. fellow Harvard, 1937-39; prof. anatomy U. Bristol (Eng.), 1940—; vis. prof. U. Wash., Seattle, 1960—, dean faculty medicine, 1959-62; Hunterian prof. Royal Coll. Surgeons Eng., 1943, 46. Decorated Knight of Dannebrog 1st Class, 1960. Fellow Royal Coll. Surgeons Eng.; Mem. Anat. Soc., Physiol. Soc., Am. Assn. Anatomists, Soc. Endocrinology. Author: (with F. C. Courtice) Lymphatics, Lymph and Lymphoid Tissue, 1956; Quantitacice Cellular Haematology, 1960; Bone Marrow Reactions, 1966; also numerous articles. Quantitative study of way various blood cells are produced in blood-forming tissues, and factors controlling such prodn., especially deficiency and excess of oxygen; studies on life history of lymphocyte and way which it migrates through blood between various parts of body. Home: 1 Rehov Degania, Beth Hakerem, Jerusalem, Israel.*

YONETANI, Takashi, biochemist; Kagawa-ken, Japan, Aug. 6, 1930; s. Yoshitaka and Kimiko (Kishii) Y.; B.Sc. in Biology, U. Osaka (Japan), 1953, Ph.D. in Biochemistry, 1961; m. Taeko Shiomi, Mar. 3, 1958. Came to U. S., 1958. Research fellow Johnson Found., U. Pa., Phila., 1958-61, asst. prof. 1964-66, asso. prof., 1966—; vis. fellow Nobel Med. Inst., Stockholm, Sweden, 1962-64; Mem. Am. Soc. Biol. Chemists, Am. Chem. Soc. Research and numerous publs. on purification, crystallization, characterization and mechanism of action of oxidation-reduction enzymes such as cytochrome oxidase, alcohol dehydrogenase, and cytochrome c peroxidase.*

YONEZAWA, Toshiyuki, Japanese geophysicist; b. Bunkyo-ku, Tokyo, Japan, Feb. 14, 1915; s. Seiji and Toyo (Sato) Y.; grad. U. Tokyo, 1942, D.Sc., 1955; m. Masako Horisawa, Mar. 21, 1945; children—Yukiko, Tatsuji. With Radio Research Labs. (formerly Phys. Inst. for Radio Waves), Ministry of Posts and Telecommunications, Tokyo, 1942—, chief subsect. radiophysics, 1948-56, chief research sect. radiophysics, 1956—; lectr. U. Tokyo, 1953—. Mem. ionosphere research com. Sci. Council Japan, 1954—; mem. working group IV, Com. on Space Research, 1962—. Recipient Tanakadate prize Soc. Terrestrial Magnetism and Electricity Japan, 1948. Mem. Phys. Soc. Japan, Am. Geophys. Union, Soc. Terrestrial Magnetism and Electricity Japan. Author: The Earth's Upper Atmosphere, 1958; also articles. Research on physics of upper atmosphere of earth especially disclosure of mechanisms of formation of F2 layer of ionosphere and constrn. theory of formation of F2 layer. Home: 666 Chofu-otsuka-machi, Ota-ku, Tokyo. Office: Radio Research Labs., Kokubunji, P.O., Koganei-shi, Tokyo, Japan.*

YONGE, Sir Charles Maurice, Brit. zoologist; b. Wakefield, Eng., Dec. 9, 1899; s. John Arthur and Edith (Carr) Y.; B.Sc., U. Edinburgh (Scotland), 1922, Ph.D., 1924, D.Sc., 1927; D.Sc., U. Bristol (Eng.), 1959; m. Martha Jane Lennox, June 30, 1927 (dec. Jan. 1945); children—Elspeth (Mrs. Bruno Touschek), Robin; m. 2d, Phyllis Greenlaw Fraser, Mar. 26, 1954; 1 son, Christopher. Asst. naturalist Plymouth Marine Labs., 1924-27, physiologist, 1930-32; Balfour student U. Cambridge (Eng.), 1927-30; leader Gt. Barrier Reef Expdn., 1928-29; prof zoology U. Bristol, 1933-44; Regius prof. zoology U. Glasgow (Scotland), 1944-64, research fellow, 1965—. Mem. Natural Environment Research Council, 1965—. Fellow Royal Soc. London, 1946, Royal Soc. Edinburgh; mem. Scottish Marine Biol. Assn. (pres., chmn. council 1944—). Author: (with F. S. Russell) The Seas, 1928; A Year on the Great Barrier Reef, 1930; The Sea Shore, 1949; Oysters, 1959; also numerous articles. Editor: (with K. M. M. Wilbur) Physiology of Mollusca, 1964. Research on marine biology especially, coral reefs, feeding and digestion in invertebrates, form and adaption in bivalve Mollusca to trace structural features of greatest significance in their evolutionary history. Home: 13 Cumin Pl., Edinburgh 9, Scotland. Office: Dept. Zoology, Univ., Glasgow, W2, Scotland.*

YORK, Carl Monroe, Jr., Am. physicist; b. Macon, Ga., July 2, 1925; s. Carl Monroe and Eugenia (Anderson) Y.; A.B., U. Cal., Berkeley, 1946, M.A., 1950, Ph.D., 1951; m. Nancy Sutton, Mar. 22, 1952; children—Christopher Anderson, Paul William, Leila Eugenia. Research fellow Cal. Inst. Tech., 1952-54; faculty U. Chgo., 1954-59; faculty U. Cal., Los Angeles, 1960—, prof. physics, 1963—, assistant chancellor, 1965—. Cons., Argonne Nat. Lab., Lemont, Ill., 1957-60, TRW Systems (formerly STL), Redondo Beach, Cal., 1961—, Film Assos. Cal., Los Angeles, 1962—. Fulbright fellow, 1951-52, Ford fellow, 1959-60, Guggenheim fellow, 1959-60. Mem. Am. Phys. Soc., A.A.A.S., Sigma Xi. Author: Handbuch der Physik, 1957. Co-discoverer positive sigma hyperon, 1953. Home: 908 Malcolm Av., Los Angeles 90024.*

YORK, Herbert, Am. physicist; b. Rochester, N.Y., Nov. 24, 1921; s. Herbert Frank and Nellie Elizabeth (Lang) Y.; A.B., U. Rochester, 1942, M.S., 1943; Ph.D., U. Cal., Berkeley, D.Sc. (hon.), Case Inst. Tech., 1960; LL.D. (hon.) U. San Diego, 1964; m. (Marie) Sybil Dunford, Sept. 28, 1947; children—David Winters, Cynthia, Rachel. Mem. research and teaching staffs Lawrence Radiation Lab., U. Cal., Livermore, 1943-58, dir., 1952-58; research adminstr. Inst. Def. Analyses, Washington, 1958, now trustee; dir. def., research and engring. Dept. Def., Washington, 1958-61; chancellor U. Cal. at San Diego, La Jolla, Cal., 1961-64, now prof. Mem. Pres.'s Scientific Advisory Committee, 1957-58, 64-67, vice chairman, 1965-66. Cons. govt. and industry. Trustee Aerospace Corp., El Segundo, Cal. Recipient Ernest Orlando Lawrence Meml. award AEC, 1962. Mem. Am. Phys. Soc., Internat. Acad. Astronautics, Phi Beta Kappa, Sigma Xi. Research and publs. in application of atomic energy to nat. def., elementary particles; high energy physics; nuclear weapons and power def. research. Home: 6110 Camino de la Costa, La Jolla, Cal. 92037.*

YOSHIDA, Zenichi, Japanese chemist; b. Kagawa-ken, Japan, Aug. 19, 1925; s. Takaichi and Masa (Yamada) Y.; B.Engring., Kyoto (Japan) U., 1949, Ph.D., 1954; m. Tetsuko Takeuchi, Feb. 3, 1958; children—Masahiko, Takako, Noriko. Faculty, Kyoto U., 1951—, prof. chemistry, 1963—. Mem. Chem. Soc. Japan (past exec. sec., award for advancement chemistry, 1958; editorial bd. Jour. 1964-65), Japan Soc. for Promotion of Sci., Am. Chem. Soc., Optical Soc. Am., Soc. Polymer Sci. Japan, Soc. Dyers, Colourists Eng. Author: (with K. Yagi, T. Tabata) Florescence, 1958; also numerous articles. Editor: (with S. Nagakura, K. Fukui) Organic Quantum Chemistry, 1966. Originated new theory on fluorescence and chem. structure of organic compounds, intermolecular hydrogen bond, chemistry of new aromatic systems, structure-reactivity relationship. Home: Higashi-hanaike-cho 1, kita-ku, Kyoto, Japan.*

YOSHII, Naosaburo, Japanese physiologist; b. Osaka, Japan, Feb. 20, 1911; s. Naraji and Masa (Yoshimura) Y.; M.D., Imperial U. Osaka, 1934, Dr. Med. Sci., 1939; m. Takako Izeki, Mar. 28, 1938. Research fellow Imperial U. Osaka, 1934-38, surgeon U. Hosp., 1938-43, lectr. surgery, 1941-43; prof. physiology Tokushima Med. Coll., 1943-45; asso. prof. Osaka U. Med. Sch., 1945-48, prof. physiology, 1948—. Mem. central com. Internat. Brain Research Orgn., UNESCO, 1960-64, central com. Japan, 1963—. Mem. Physiol. Soc. Japan, Japan EEG Soc., Japan EMG Soc., Japanese Soc. for Psychiatry and Neurology, Japan Med. Electronic Soc. Author: Clinical Physiology, 1953; also numerous articles. Editor in chief Jour. Clin. EEG Japan; editorial bd. Physiology and Behavior, Jour. Physiol. Soc. Japan, Internat. Jour. EMG, Brain Research jours. Research on heat prodn. of pancreas, stenosis and insufficiency of cardiac valves induced by ventriculotomy on dog, electrophysiol. studies on conditioning, stepwise involvement of brainstem reticular, thalamo-reticular, hippocampal systems. Home: 495 Komuro, Fujiidera-City, Osaka-fu, Japan.*

YOSHIKAWA, Haruhisa, Japanese biochemist; b. Kanagawa-ken, Japan, Jan. 25, 1909; s. Harujiro and Kiku (Tsujimura) Y.; grad. U. Tokyo (formerly Tokyo Imperial U.), 1931, Dr.Med.Sci., 1939; m. Haruko Murai, Nov. 17, 1943; children—Shoichi, Mitsuko (Mrs. Tetsuro Suzuki), Kenji, Yoko. Asst. prof. biochemistry Inst. Pub. Health, Tokyo, 1938-45; Faculty Medicine, U. Tokyo, 1945—, prof. physiol. chemistry and nutrition, 1952—, dean Faculty Medicine, 1965—. Mem. Japanese Biochem. Soc. (past dir.), Japan Radioisotope Assn. (dir. 1952), Clin. Chemistry Japan (dir. 1960—). Author: Textbook of Biochemistry for Medical Students, 1962; also numerous articles. Demonstrated effect of copper on formation of respiratory enzyme (Cytochrome oxidase); demonstrated spl. respiratory pigment (cytochrome b5) in mammalian liver cells; research on enzyme catalyzing combination of iron and protoporphyrin in avian red blood cells and liver mitochondria, metabolism of human red blood cells; discovered rejuvenation of long-stored red cells by incubation with adenine and inosine. Address: 12-1 Ohtuka 6-Chome, Bunkyo-Ku, Tokyo, Japan.*

YOSHIMOTO, Chiyoshi, Japanese med. electronic scientist; b. Sapporo, Japan, May 31, 1916; s. Senzo and Tsune (Watanabe) Y.; B.Engring., Hokkaido U., Sapporo, 1939, M.D., 1953; m. Haruko Mitomi, Apr. 12, 1941; children—Hideaki, Kuniyasu. Mem. research staff Research Lab., Tokyo (Japan)—Shibaura Electric Co., 1939-45; faculty Hokkaido U., 1945—, prof. med. electronics, Research Inst. Applied Electricity, 1962—, dir. dept. med. electronics, 1962—. Mem. Japan Assn. Med. and Biol. Engring. Research and numerous publs. on devel. med. transducers, analog simulations of respiratory system in man, analysis of electrocardiographic potential distbns., devel. med. analog data process and transmission. Home: W. 12, S. 17, Sapporo, Hokkaido, Japan.*

YOSHIMURA, Hisato, Japanese physiologist; b. Hyogo Prefecture, Japan, Feb. 9, 1907; s. Eiji and Chie (Kawai) Y.; M.D., Kyoto U., 1930, Dr.Med. Sci., 1936; m. Hisae Nagoka, June 29, 1933; children—Eiko (Mrs. Shiro Hosoda), Manabu, Seishi, Satoko. Faculty, Kyoto (Japan) U., 1930-39, 45-46, Kobe Med. Sch., 1946-47; prof. physiology Kyoto Prefecture U. Medicine, 1947—. Mem. Japanese Council Physiol. Scis., 1966—. Recipient Takeda award Japanese Soc. Food and Nutrition, 1962. Mem. Japanese Biometrical Soc. (pres. 1965), Japanese Soc. Biophysics, Internat. Biometrics Soc. (v.p. 1966), Kinki Nutritional Soc. (pres. 1966), Physiol. Soc. Japan (sec. 1966), Japanese Soc. Phys. Fitness (sec. 1964). Author: Theory and Practice of pH-Determination; (with R. Shoji) Medical Physiology, 1952; Theory and Practice of Protein Nutrition, 1964. Research, numerous publs. on optimal protein requirement of adults; inventor new method of frostbite treatment, device to estimate peripheral tolerance to cold, discoverer reabsorption nerve of salivary gland, discoverer of sportsanemia. Home: 6 Shodencho Matsugasaki, Kyoto, Japan.*

YOSHIMURA, Shouji, Japanese physician; b. Osaka, June 3, 1924; s. Masayuki and Yukiko Yoshimura; M.D., Tokyo U., 1949, Ph.D., 1950; m. Maruko Saito, Mar. 10, 1953; 1 dau., Sonoko. Asst. prof. Med. Sch., Tokyo U., 1952-60; asso. prof. internal medicine, Nihon Med. Sch., Tokyo, 1960—. Chief cons. Internat. Yokohama Hosp., 1964—. Mem. Japanese Med. Assn. Internal Medicine, Japanese Circulation Soc. Author: Rinshō-Seikagaku (Clinical Biophysics), 1966; also numerous articles. Devel. new method for physiol. determination of pulmonary surface tension, new technique for measurement of cerebral blood volume with I 131-albumin; studies of distal effects of cardiac beat actions. Home: 15 31-2 Niski-Ikebukuro. Office: 5-3-3- Iidabashi Chiyoda-Ku, Tokyo, Japan.*

YOSHINORI, see Kittoku, Isomura.

YOSHIO (Eisho), Kogyu, Japanese physician; b. 1724; s. Kozaemon family; studied medicine while serving as interpreter Dutch lang. to Shogunate; 3 sons, including Joen. Tchr. of physicians, Ryotaku Maemo and Gempaku Sugita. Author numerous med. works including In-eki-hatsubi (urine analysis). Famous physician of Dutch sch.; 1st to introduce methods for examining urine. Died 1800.

YOSIM, Samuel Jack, Am. chemist; b. St. Petersburg, Fla., Apr. 21, 1920; s. Phillip and Beatrice (Solomon) Y.; B.S. in Chemistry, U. Fla., 1948; M.S., U. Chgo., 1949, Ph.D. in Phys. Chemistry, 1952; m. Esther Jane Gilbert, Aug. 30, 1949; children—Paul, Robin. Asso. chemist Argonne (Ill.) Nat. Lab., 1951-52; group leader phys. chemistry group Atomics Internat., div. N.Am. Aviation, Canoga Park, Cal., 1952—. Mem. Am. Chem. Soc., Phi Beta Kappa, Sigma Xi. Research and publs. on systems of molten salts, theory of liquids, nuclear transformation in solids; calculation of thermodynamic properties of liquids. Home: 23812 Killion St., Woodland Hills, Cal. 91364. Office: P.O. Box 309, Canoga Park, Cal. 91304.*

YOST, Don Merlin Lee, Am. inorganic chemist; b. Tedrow, O., Oct. 30, 1893; s. William Nicholas and Viola Lorena (Lee) Y.; B.S., U. Cal., Berkeley, 1923; Ph.D. (duPont fellow), Cal. Inst. Tech., 1926; m. Susie Marguerite Sims, Mar. 7, 1917; children—Helen Marguerite, Max Cayley. Teaching fellow U. Utah, 1923-24; mem. faculty Cal. Inst. Tech., Pasadena, 1926—, prof. inorganic chemistry, 1948—. Sect. chmn. NDRC, also OSRD, World War II. Fellow Am. Phys. Soc., A.A.A.S.; mem. Nat. Acad. Scis., Am. Acad. Arts and Scis., Sigma Xi, Lambda Chi Alpha. Author: Systematic Inorganic Chemistry, 1944; Rare Earth Elements and their Compounds, 1946; also articles, revs., short stories. Research in inorganic and phys. chemistry, physics; linear algebras, elliptic and hyperelliptic functions; radioactivity; chem. kinetics, low and high temperature thermodynamics. Home: 1270 Cordora St., Pasadena, Cal. 91106.*

YOST, Henry Thomas, Jr., Am. biologist; b. Balt., Jan. 22, 1925; s. Henry Thomas and Martha (Minsker) Y.; A.B., Johns Hopkins, 1947, Ph.D., 1951; M.A. (hon.) Amherst Coll., 1965; m. Martha Jean Thomas, June 12, 1948; 1 dau, Brenna Duncan. Faculty Amherst (Mass.) Coll., 1951—, prof., 1965—. Mem. sci. adv. bd. Consumers Union, N.Y.C., 1959—. Adam T. Bruce fellow, 1951. Fellow A.A.A.S.; mem. Am. Assn. U. Profs. (nat. council 1965-66, 68—, pres. Amherst chpt. 1965—), Radiation Research Soc., Genetics Soc., Soc. for Developmental Biology, Bot. Soc., Sigma Xi. Research on effects of combined radiation on chromosome breakage, genetic recombination, mutation and energy generation in mitochondria, comparison of nucleic acids of normal and tumor tissues of plants. Home: 75 N. East St., Amherst, Mass. 01002.*

YOUNG, Allan Charles, biophysicist; b. Saskatoon, Sask., Can., May 23, 1911; s. Samuel George and Marie (Doe) Y.; B.A., U. B.C., 1930, M.A., 1932;

Ph.D., U. Toronto, 1934; m. Barbara Agnes Breeton, Dec. 28, 1947; children—Carolyn, Christopher. Came to U. S., 1949, naturalized, 1955. Instr., U. Rochester Med. Sch., 1936-38; Rockefeller Found. fellow, 1938-40; with NRC Can., 1940-45, B.C. Research Council, 1945-48; faculty dept. physiology and biophysics U. Wash. Sch. Medicine, Seattle, 1949-—, prof., 1960-—, coordinator biophysics tng. program, 1957-—. Mem. Am. Physiol. Soc., Contbg. author: Medical Physiology, 1960; Physiology and Biophysics, 1965; Medical Physics. Research, publs. on physiol. control systems. Home: 7621 N.E. 112th St., Kirkland, Wash. 98033. Office: Dept. Physiology and Biophysics, U. Wash. Sch. Medicine, Seattle 98105.*

YOUNG, Arthur, Brit. agriculturalist; b. London, Eng., Sept. 11, 1741; s. Arthur and Anna Lucretia (Goussmaker) Y.; m. Martha Allen, 1765; children—Mary, Elizabeth (Mrs. John Hoole), Arthur, Martha Ann. Apprenticed to merc. firm at Lynn, 1758; went to London, 1761, started mag. The Universal Museum, 1762; farmer at Bradfield, 1763-66, then at North Mimms, Herfordshire, 1768; agt. to Lord Kingsborough in County Cork, 1777-79; founded Annals of Agriculture, 1784; sec. to Bd. of Agr., 1793. Fellow Royal Soc., 1774. Author: Farmer's Letters to the People of England, 1767; A Six Weeks' Tour through the Southern Counties of England and Wales, 1768; A Six Month's Tour through the North of England, 1770; The Farmer's Guide in Hiring and Stocking Farms, 1770; Rural Economy, 1770; A Course of Experimental Agriculture, 1770; Farmer's Tour through the East of England, 1771; The Farmers Calendar, 1771; Observations on the Present State of the Waste Lands of Great Britain, 1773; Tour in Ireland, 1780; Travels in France during 1787-90, 1792; General View of the Agriculture of the County of Lincoln, 1799; The Question of Scarcity Plainly Stated, 1800; Inquiry into the Propriety of Applying Waste Lands to the Better Maintenance and Support of the Poor, 1801; Essay on Manures, 1804; General View of the Agriculture of Hertfordshire, 1804, of Norfolk, 1804, of the County of Essex, 2 vols., 1807, of Oxfordshire, 1809; On the Husbandry of the Three Celebrated Farmers, Bakewell, Arbuthnot, and Ducket, 1811. One of the greatest English writers on agriculture; his works helped to educate farmers and to raise standards of farming; advocated use of seed drill and horse hoe, improved crop rotation, better stockbreeding methods. Died London, Apr. 20, 1820.

YOUNG, Barton R., Am. radiologist; b. Spring City, Pa., Aug. 23, 1903; s. John B. and Eva (Rogers) Y.; M.D., Temple U., 1929, M.S. in Radiology, 1939; m. Mildred Hartman, July 20, 1949; children—Barbara Hartman, Barton R., Beverly Hartman. Faculty, Temple U., Phila., 1930-—, asso. prof. radiology, 1936-45, prof., 1945-—; dir. dept. radiology Germantown Dispensary and Hosp., Phila., 1948-—. Vis. lectr. U. Pa., Phila. Recipient Annual award Temple U., 1950, gold medal Germantown Hosp., 1958. Mem. Am. Coll. Radiology (bd. chancellors), Am. Roentgen Ray Soc. (pres. 1958). Author: Skull, Sinuses, Mastoids, 1948; also articles; contbg. author: Progress in Neuropsychiatry, 1964, 65. Chem. investigation of skull, sinuses, mastoid, acute abdominal disorders. Home: 1847 Lambert Rd., Jenkintown, Pa. 19046. Office: Germantown Hosp., Phila. 19144.*

YOUNG, Charles Augustus, Am. astronomer; b. Hanover, N.H., Dec. 15, 1834; grad. Dartmouth, 1853; Ph.D., U. of Pa., 1870, Hamilton Coll., New York, 1871; LL.D., Wesleyan, Conn., 1876, Columbia, 1897, Western Reserve, 1893, Dartmouth, 1903; m. Augusta S. Mixer, Aug. 26, 1857. Served as capt. Co. B, 85th Regiment, Ohio vols., 4 months, 1862. Prof. mathematics, natural philosophy and astronomy, Western Reserve Coll., 1857-66; prof. natural philosophy and astronomy, Dartmouth, 1866-77; prof. astronomy, Princeton U., 1877-1905. Mem. of Nat. Acad. Sciences and many other Am. and foreign learned socs. Author: The Sun (in Internat. Scientific Series), 1882; A General Astronomy, 1889; Elements of Astronomy, 1890; Lessons in Astronomy, 1891; Manual of Astronomy, 1902; Uranography. Specialist in solar physics; made spectroscopic studies of sun's atmosphere and protuberances; discovered greenline of solar corona, 1869, solar reversing layer, 1870.

YOUNG, Charles J(acob), Am. elect. engr.; b. Cambridge, Mass., Dec. 17, 1899; s. Owen D. and Josephine (Edmonds) Y.; A.B., Harvard, 1921; m. Eleanor L. Whitman, 1923; children—John Peter, David Whitman; m. 2d, Esther M. Christensen, 1929; children—Neils Owen, Esther Van Horne. Radio engr. Gen. Electric Co., Schenectady, N.Y., 1923-29; electronic development engr. RCA Mfg. Co., Camden, N.J., 1929-42; asso. lab. dir. RCA Labs., Princeton, 1942-—. Recipient Modern Pioneer award N.A.M., 1946. Fellow Inst. Radio Engrs., Sigma Xi. Research on develop. of facsimile equipment, electronic printing and counters; electrostatic photography, electrolytic recording; frequency control. Home: 78 Stockton St. Office: David Sarnoff Research Center, Princeton, N.J.

YOUNG, Charlotte Marie, Am. nutritionist; b. Mpls., Aug. 19, 1910; d. Will Morris and Charlotte (Webster) Young; B.S. with high distinction, U. Minn., 1935; M.S., Ia. State U., 1937, Ph.D., 1940. Instr., Mich. State U., East Lansing, 1940-42; mem. faculty Grad. Sch. Nutrition, Cornell U., Ithaca, N.Y., 1942-—, prof. med. nutrition, sec. Grad. Sch. Nutrition, 1952-—. Mem. Am. Bd. Nutrition, 1962-—; mem. U. S. Nat. Com., Internat. Union Nutrition Scis., 1962-—, cons. to WHO, AID, USOM. Recipient Centennial award Ia. State U., 1958; Outstanding Achievement award U. Minn., 1959; Borden award Am. Home Econs. Assn., 1963. Mem. Am. Dietetics Assn. (mem. council, speaker exec. bd. 1962-64), Am. Inst. Nutrition, Am. Soc. for Clin. Nutrition, Am. Pub. Health Assn., Am. Home Econs. Assn., Sigma Xi, Phi Kappa Phi, Omicron Nu (nat. pres. 1953-55), Iota Sigma Pi. Research and numerous publs. in applied human nutrition in fields of body composition of women, obesity and weight reduction, dietary methodology, nutritional status studies, dietary studies, food habit determinants. Home: 110 Warren Rd., Ithaca, N.Y. 14850.*

YOUNG, Elrid Gordon, Canadian biochemist; b. Quebec, Que., Can., Jan. 5, 1897; s. James and Jane (Douglas) Y.; B.A. with honors in Chemistry and Biology, McGill U., Montreal, Que., 1916, M.Sc., 1919; Ph.D., Cambridge (Eng.) U., 1921; D.Sc., Acadia U., 1957; LL.D., Dalhousie U., 1965; m. Madge L. Musgrave, May 15, 1926. Ramsay Meml. fellow for Can., Cambridge U., 1919-21; asso. prof. biochemistry U. Western Ont., 1921-24; prof. biochemistry Dalhousie U., Halifax, N.S., Can., 1924-50; dir. Atlantic Regional Lab., Nat. Research Council Can., Halifax, 1950-62. Mem. Canadian Council on Nutrition, 1938-—. Fellow Chem. Inst. Can. (past pres.), Royal Soc. Can. (past pres. sect. V); mem. Am. Chem. Soc., Am. Soc. Biol. Chemists, Biochem. Soc. (Gt. Britain), Canadian Biochem. Soc. (past pres.), Canadian Physiol. Soc. (past pres.), Nutrition Soc. Can., A.A.A.S. Research and numerous publs. on chemistry of proteins, purine metabolism, human nutrition, chemistry of marine algae. Home: 6262 Oakland Rd. Office: Nat. Research Council, 1411 Oxford St., Halifax, N.S., Can.*

YOUNG, Francis Allan, Am. psychologist; b. Utica, N.Y., Dec. 29, 1918; s. Francis Allan and Julia (McOwen) Y.; B.S., U. Tampa, 1941; M.A., Western Res. U., 1945; Ph.D., Ohio State U., 1949; m. Judith W. Wright, Dec. 21, 1945; children—Francis Allan, Thomas Robert. Faculty, Wash. State U., Pullman, 1948-—, prof., 1961-—, dir. Primate Research Center, 1957-—. Mem. Am. Psychol. Assn., A.A.A.S., Assn. Research in Ophthalmology, Ecol. Soc. Am., Am. Acad. Optometry. Studies, numerous publs. on pupillary responses to various types of stimulation; refractive characteristics of the primate eye; devel. of near and farsightedness. Home: 224 Webb St., Pullman, Wash. 99163.*

YOUNG, Gale Jay, Am. physicist; b. Baroda, Mich., Mar. 5, 1912; s. Otto Conrad and Alta (Houser) Y.; B.S. in Elec. Engring., Milw. Sch. Engring., 1933; B.S., U. Chgo., 1934, M.S. in Math., 1936; m. Margaret Casselman, June 20, 1949; children—Linda Adelma, Wendy Margaret. Research asst. math. biophysics U. Chgo., 1936-40, staff Metall. Lab. Manhattan Project, 1942-46; head dept. physics and math. Olivet Coll., 1940-42; tchr. reactor tech. Clinton Lab., 1946-48; staff Lexington Project, Mass. Inst. Tech., summer 1948; sr. v.p. Nuclear Devel. Corp. Am., White Plains, N.Y., 1948-61; v.p. United Nuclear Corp., White Plains, 1961-62; asst. dir. Oak Ridge Nat. Lab., 1962-—. Mem. sci. adv. bd. USAF, 1954-57; cons. Manhattan Coll., N.Y.C., 1960-62. Fellow Am. Phys. Soc., Am. Nuclear Soc. (dir. 1964-67). Research in math. biophysics; math. analysis of psychol. tests; nuclear reactor theory; desalination. Home: 110 Wiltshire Dr., Oak Ridge 37830. Office: P.O. Box X, Oak Ridge 37830.*

YOUNG, Harold Edwin, Australian agrl. scientist; b. Wellcamp, Australia, Apr. 29, 1907; s. James Edgar and Ruby (Hope-Johnstone) Y.; B.Sc. in Agr. with Honors, U. Queensland, 1932, M.Sc. in Agr., 1936, D.Sc.Agr., 1941; m. Hazel Rose Constance Sinclair, Dec. 16, 1937; children—Edwin S., Rosemary Hazel (Mrs. John D. Ward), Frances Susan, Erica, Margaret. Asst. pathologist Royal Brisbane (Australia) Hosp., 1932-34; staff Dept. Agr. and Stock, Brisbane, 1934-49, forest pathologist, 1946-48, sr. weed officer, 1948-49; spl. research officer Rubber Research Inst. Ceylon, 1949-52; dir. Rubber Research Inst. Ceylon, sr. physiologist Bur. Sugar Expt. Stas., Brisbane. Mem. Australian Inst. Agrl. Scis., Inst. Rubber Industry (London), Field Naturalists Club, Queensland Orchid Soc., Ceylon Orchid Circle. Author: Fused Needle Disease and its Relation to Nutrition of Pines, 1940; Weed Control, 1962; also articles. Discovered cause and cure of fused needle diseases of slash and loblolly pine in Australia; worked out fertilizer requirements of phosphorous of soils and leaf index of these requirements; worked out control for rubber mildew in Ceylon, cause and remedy for several diseases of forest trees in Queensland. Home: 25 McCaul, Taringa, Queensland. Office: 99 Gregory Terrace, Brisbane, Queensland, Australia.*

YOUNG, Hugh Hampton, Am. surgeon; b. San Antonio, Tex., Sept. 18, 1870; s. Gen. William Hugh and Frances Michie (Kemper) Y.; A.B., A.M., U. of Va., 1893, M.D., 1894; Johns Hopkins, 1894-95; D.Sc., Queen's U., Belfast, 1933; m. Bessy Mason Colston, June 4, 1901 (died May 21, 1928); children—Frances Kemper (Mrs. Wm. Francis Rienhoff), Frederick Colston, Helen Hampton (Mrs. Bennett Crain), Elizabeth Campbell (Mrs. Warren Russell Starr). Pathologist to Thomas Wilson Sanitarium, 1895; successively asst. resident surgeon, 1895-98, head of dept. urol. surgery, and asso. surgeon Johns Hopkins Hospital, and clinical professor of urology, Johns Hopkins U. Pres. Md. State Lunacy Commn. Pres. Am. Assn. Genito-Urinary Surgeons, 1909, Am. Urol. Assn., 1909, Medico-Chirurgical Faculty of Maryland, 1912; Chmn. Bd. of Mental Hygiene for Md.; chmn. Md. Aviation Commn. Awarded Keyes medal, 1936; Francis Amory Septennial Prize 1941. Mem. Internat. Assn. Congres Internationale d'Urologie (pres. 1927); corr. mem. Association Francaise d'Urologie, Deutsche Gesellschaft für Urologie, Sociedad de Cirujia de Buenos Aires, Societa Italiana di Urologia, R. Romanae Medicorum Academie Praeses, fellow Royal College Surgeons of Ireland; honorary fellow Royal Society of Medicine; fellow American College of Surgeons. Author: Studies in Urological Surgery (Vol. XIII, Johns Hopkins Hosp. Repts.), 1906; Hypertrophy and Cancer of the Prostate (Vol. XIV, Johns Hopkins Reports), 1906; Young Practice of Urology (2 vols.); Urological Roentgenology; Genital Abnormalities, Hermaphroditism and Related Adrenal Diseases; Hugh Young, A Surgeon's Autobiography, 1940. Founder and editor Jour. of Urology. Has contributed over 350 papers to Am. and foreign med. jours. Specialist in urology; research on carcinoma of prostate. Died Aug. 23, 1945.

YOUNG, J. Lowell, Am. soil chemist; b. Perry, Utah, Dec. 13, 1925; s. Isaac A. and Elzada (Nelson) Y.; B.S. in Chemistry with honors, Brigham Young U., 1953; Ph.D. in Soils, Ohio State U., 1956; m. Ruth Ann Jones, Sept. 15, 1950; children—Gordon, LoAnn, Colene, Kathryn. Research asso. dept. agrl. biochemistry Ohio State U., 1956-57; faculty Ore. State U., Corvallis, 1957-—, asso. prof. soils, 1963-—; chemist soil and water conservation research div. Agr. Research Service, U. S. Dept. Agr., Corvallis, 1957-61, research chemist, 1961-—. Mem. A.A.A.S., Am. Soc. Agronomy, Soil Sci. Soc. Am., Western, Internat. socs. soil sci. Research and publs. on inorganic and organic nitrogenous constituents of soils, especially amino acids of soil humic materials; pioneered demonstration of de novo synthesis of enzymes in storage organs of germinating seeds. Home: 1230 Lincoln St., Corvallis, Ore. 97330.*

YOUNG, James, Scottish chemist; b. Glasgow, Scotland, July 13, 1811; s. John and Jean (Wilson) Y.; studied chemistry Anderson's U., Glasgow, 1830, LL.D., St. Andrews U., 1879; m. Mary Young, Aug. 21, 1838; 3 sons, 4 daus. Lecture asst. to Thomas Graham, accompanied him in Univ. Coll., London, 1837; mgr. chem. works, nr. Liverpool, 1839, nr. Manchester, 1843; chemist James Muspratt's alkali works, Lancashire, 1839; set up movement for establishing Manchester Examiner (newspaper), 1st pub. 1846; manufactured oils from petroleum spring, Alfreton, Derbyshire, 1848-51; partner (with Edward Meldrum and Edward William Binney) to manufacture oils from Torbane Hill mineral or Boghead coal, Bathgate, 1850; began sale of paraffin, 1856, took over whole bus. from partners, 1865, sold it to Young's Paraffin Light & Mineral Oil Co., 1866; pres. Anderson's Coll., 1868-77, founder Young chair tech. chemistry, 1870. Fellow Royal Soc., 1873; mem. Chem. Soc. (v.p. 1879-81). Discovered cheaper methods to produce sodium stannate and potassium chlorate, his expts. led to manufacture of paraffin-oil and solid paraffin on large scale, 1847-50; helped establish Scottish mineral oil industry; suggested use of caustic lime to prevent corrosion of iron ships by bilge water; began determination of velocity of white and colored light by modification of H. L. Fizeau's method (with George Forbes), 1878. Died Edinburgh, May 14, 1883.

YOUNG, John Parke, Am. economist; b. Los Angeles, Oct. 24, 1895; s. William Stewart and Adele (Nichols) Y.; A.B., Occidental Coll., 1917; M.A., Columbia U., 1919; M.A., Princeton, 1920, Ph.D., 1922; m. Florence Hensel, Sept. 7, 1927 (dec. 1949); children—Douglas Parke (dec.), Richard Parke, Roger Hensel, Catherine Jean (Mrs. William R. Selleck); m. 2d, Marie Louise Smith, June 24, 1952. Examiner, FTC, 1917-18; instr. Princeton, 1921-23; dir. U. S. Senate Fgn. Currency and Exchange Investigation, 1923-25; prof., chmn. econs. Occidental Coll., 1926-41; economist U. S. Dept. Commerce, 1941-42, Bd. Econ. Warfare, 1942; economist U. S. Dept. State, Washington, 1943-65, chief div. internat. finance, 1954-62; vis. prof. econs. Claremont Grad. Sch., 1966-67. Mem. Commn. Financial Advisers to Govt. China, 1929-30; pres. Financial Cons. Corp., 1932-42. Financial adviser C.Am. Common Market, 1964-65; econ. adviser Govt. of Chile, 1952. Mem. Phi Beta Kappa, Phi Gamma Delta. Author: Central American Currency and Finance, 1925; European Currency and Finance, 2 vols., 1925; International Trade and Finance, 1938; The International Economy, 1942. Home: 1303 Wentworth Av., Pasadena, Cal. 91106.*

YOUNG, John Radford, Brit. mathematician; b. London, Apr. 1799; self-educated. Prof. math. Belfast (Ireland) Coll., 1833-49; Author: Elements of Geometry, 1827; Elements of Analytical Geometry, 1830; An Elementary Essay on the Computation of Logarithms, 1830; The Elements of the Differential Calculus, 1831; The Elements of the Integral Calculus, 1831; The Elements of Mechanics comprehending Statics and Dynamics, 1832; Elements of

Plane and Spherical Trigonometry, 1833; On the Theory and Solution of Algebraical Equations, 1835; Mathematical Dissertations for the Use of Students in the Modern Analysis, 1841; On the General Principles of Analysis . . . , 1850; An Introductory Treatise on Mensuration, 1850; An Introduction to Algebra and to the Solution of Numerical Equations, 1851; Rudimentary Treatise on Arithmetic, 1858; A Compendious Course of Mathematics, 1855; The Theory and Practice of Navigation and Nautical Astronomy, 1856; The Mosaic Cosmogony not adverse to Modern Science, 1861; Science Elucidative of Scripture and not Antagonistic to it, 1863. Discovered proof of Newton's rule for determining number of imaginary roots in an equation, 1844. Died Peckham, London, Mar. 5, 1885.

YOUNG, John Richardson, Am. biologist, physician; b. Elizabethtown, Md., 1782; s. Dr. Samuel and Ann (Richardson) Y.; grad. Coll. of N.J. (now Princeton), 1799; M.D., U. Pa. at Phila., 1803. Practiced medicine with his father, 1803-04. Author: An Experimental Inquiry into the Principles of Nutrition and the Digestive Process (written as a student, published in his inaugural thesis for med. degree; work later influenced William Beaumont's studies in digestion), 1803. Research on digestion in frogs showed that living matter is not digested; discovered that gastric juice is itself acid and that acidity is not result of fermentation. Died Hagerstown, Md., June 8, 1804.

YOUNG, John Wesley, Am. mathematician; b. Columbus, O., Nov. 17, 1879; s. William Henry and Marie Louise (Widenhorn) Y.; ed. Karlsruhe, Germany, and Columbus, O., 1885-89; Gymnasium Baden-Baden, Germany, 1889-95; Ph.B., Ohio State U., 1899, A.M., Cornell U., 1901, Ph.D., 1904; m. Mary Louise Aston, July 20, 1907; 1 dau., Mary Elizabeth. Instr. mathematics, Northwestern U., 1903-05; preceptor mathematics, Princeton, 1905-08; asst. prof. mathematics, U. of Ill., 1908-10; prof. and head of dept. of mathematics, U. of Kan., 1910-11; prof. mathematics, U. of Chicago, summer quarter, 1911; prof. of mathematics, Dartmouth Coll., 1911——. Chief examiner in geometry, Coll. Entrance Exam. Board, 1915-17. Editor, Bulletin Am. Math. Soc., 1907-25. Was pres. Math. Assn., America and mem. many other socs.; chmn. Nat. Com. on Math. Requirements, 1916-23. Author: Projective Geometry, Vol. I (with Oswald Veblen), 1910; Lectures on Fundamental Concepts of Algebra and Geometry, 1911 (Italian transl., 1919); Plane Geometry (with A. J. Schwartz), 1915, 2d edit., 1922; Elementary Mathematical Analysis (with F. M. Morgan), 1917; Plane Trigonometry (with F. M. Morgan), 1919; Projective Geometry, 1929; also papers in various math. jours. Editor, for the Houghton Mifflin Co., of a series of math. texts. Investigated founds. of math.; employed analytical techniques in math. research. Died Feb. 17, 1932.

YOUNG, Kimball, Am. sociologist; b. Provo, Utah, Oct. 26, 1893; s. Oscar Brigham and Anna Marie (Roseberry) Y.; A.B., Brigham Young U., 1915; A.M., U. of Chicago, 1918; Ph.D., Stanford, 1921; m. 2d, Lillian D. Jackson, April 6, 1940; 1 dau. Helen Anderson (Mrs. R. D. Willey) (by first marriage). Began as teacher high school, 1915; asst., later asso. professor psychology, Univ. of Oregon, 1920-22, 1923-26; asst. prof. psychology, Clark U., Worcester, Mass., 1922-23; asso. prof. sociology, U. of Wis., 1926-30, prof. social psychology, 1930-40; chmn. dept. of sociology, Queens Coll., Flushing, N.Y., 1940-47; chmn. dept. sociology, Northwestern U. since 1947. Expert consultant War Dept., 1944; head sociology branch Shrivenham Am. Univ. (U. S. Army), 1945. Mem. Social Science Research Council, Am. Psychol. Assn., Am. Sociol. Soc. (pres. 1945), Sigma Xi. Author: Mental Differences in Certain Immigrant Groups, 1922; Source Book for Social Psychology, 1927; Social Psychology, 1930 (2d edit., '44); The Madison Community (with others), 1934; Bibliography on Censorship and Propaganda (with R. D. Lawrence), 1928; Social Attitudes (with others), 1931; An Introductory Sociology, 1934; Source Book for Sociology, 1935; Personality and Problems of Adjustment, 1940; Sociology, A Study of Society and Culture, 1942. Gen. editor of American Sociology Series (Am. Book Co.). Contbr. to books and jours. of social science. Mem. bd. editors Jour. of Social Psychology, Am. Journal of Sociology. Wrote widely used textbooks in sociology and social psychology. Address: 1725 Orrington Av., Evanston, Ill.

YOUNG, Laurence Chisholm, mathematician; b. Göttingen, Germany, July 14, 1905; s. William Henry and Grace Emily (Chisholm) Y.; B.A. with 1st class honours, Trinity Coll., Cambridge (Eng.) U., 1928, M.A., 1931, Sc.D., 1938; m. Joan Elizabeth Mary Dunnett, June 9, 1934; children—Francis Edward, Rosalind Elizabeth (Mrs. Kenneth A. Ford), David L. (dec.), Sylvia Margaret Christabel (Mrs. Roger A. Wiegand), Angela Celia Grace, Beatrice Virginia. Came to U. S., 1949. Prof., head dept. pure math. U. Capetown (S. Africa), 1939-49; prof. math. U. Wis., 1949——, Distinguished prof., 1968——, chmn. dept., 1962-64. Mem. London, Am. math. socs., Cambridge Philos. Soc., Cirolo matematico di Palermo, Math. Assn. Am., Royal Astron. Soc., Royal Soc. S. Africa. Author articles analysis, geometry, topology. Research on theory of the integral; prime ends; calculus of variations; inequalities. Home: 5532 Lake Mendota Dr., Madison, Wis. 53705.

YOUNG, Leona Esther, Am. chemist; b. Alameda, Cal., Apr. 4, 1893; d. John Nelles and M. Josephine (Hamilton) Y.; B.S., U. Cal. at Berkeley, 1915, M.S., 1916, Ph.D., 1929. Research chemist El Dorado Oil Co., 1917-19, Berkeley (Cal.) Cons. Lab., Berkeley, Cal., 1920-21; tchr. Anna Head Sch., Berkeley, 1922-27; prof. chemistry Mills. Coll., Oakland, 1927-58, head dept. chemistry, 1935-58. Mem. Am. Chem. Soc., Sigma Xi, Alpha Chi Omega, Iota Sigma Pi. Author: (with Porter) General Chemistry, 1940; (with Petty) Chemistry for Progress, 1957; also articles. Research on elec. cells, effect of ultrasonic waves on many chem. reactions. Home: 2510½ Etna St., Berkeley, Cal. 94704.*

YOUNG, Leslie, Brit. chemist; b. Sunderland, Feb. 27, 1911; s. John and Ethel Herring; ed. Imperial Coll., Univ. Coll. London; Ph.D., D.Sc.; m. Ruth Elliott, Aug. 17, 1939; 1 son, Anthony. Prof. biochemistry U. Toronto; biochemist U. London, now prof. biochemistry. Mem. Biochem. Soc. (hon. sec. 1950-53), Royal Inst. Chemistry (v.p.). Author: The Metabolism of Sulphur Compounds, also articles. Research metabolism sulphur compounds, biochemistry toxic agts. Home: 23 Oaklands Av., Esher, Surrey. Office: St. Thomas' Hosp. Med. Sch., London S.E.1, Eng.

YOUNG, Otis Bigelow, Am. physicist; b. Dodge Center, Minn., Nov. 17, 1899; s. George Allen and Delia (Bigelow) Y.; A.B., Wabash Coll., 1921; A.M., U. Ill., 1923, Ph.D., 1928; m. Olive Eleanor Patmore, June 1929; children—Ruth Eleanor (Mrs. Russell W. King), Dorothy Jane (Mrs. Billy Turner). Head physics dept. McKendree Coll., Lebanon, Ill., 1928-29; faculty physics dept. So. Ill. U., Carbondale, 1929——, prof. physics and astronomy, 1945——, chmn. dept., 1938-53, coordinator fed. aeros. tng., 1939-43, dir. vet. information service bur., 1945-46, dir. atomic and capacitor research, 1953——; commr.-sec. So. Ill. Airport Authority, Jackson County, 1946——. Mem. civil effects test group AEC-FCDA, Nev. Proving Ground, 1957. Mem. Ill. Acad. Sci. (past pres.), Am. Assn. Physics Tchrs., Am. Inst. Physics, Am. Phys. Soc., Am. Fedn. Am. Scientists, Synton Radio Soc., Sigma Xi, Epsilon Chi, Kappa Phi Kappa, Pi Kappa Delta, Sigma Pi Sigma. Research and publs. on electric discharges in gases, dielectric constants, capacitors, cosmic rays, nuclear radiation from nuclear weapons for civil def., low energy elementary and fundamental particles, magnetic monopole. Home: Route 1, Heritage Hills, Carbondale, Ill. 62901.*

YOUNG, Paul Thomas, Am. psychologist; b. Los Angeles, May 26, 1892; s. William Stewart and Cynthia Adele (Nichols) Y.; A.B., Occidental Coll., 1914, D.Sc., 1961; A.M., Princeton, 1915; Ph.D., Cornell, 1918; m. Josephine Kennedy, July 27, 1929; children—Rosemary Adele (Mrs. Lawrence Lee Mitchell), Stewart Adams. Instr., U. Minn., 1919-21; faculty U. Ill., Urbana, 1921——, prof. psychology, 1934-60, prof. emeritus, 1960——; vis. prof. U. Cal. at Los Angeles, 1947, San Diego State Coll., 1950; vis. fellow Yale, 1939-40; research fellow Harvard, 1947-48. NRC fellow for study in Berlin, 1926-27; recipient Distinguished Alumnus award Occidental Coll., 1956, Distinguished Sci. Contribution award Am. Psychol. Assn., 1965. Mem. Phi Beta Kappa, Sigma Xi, Psi Chi, Phi Delta Kappa. Author: Motivation of Behavior, 1936; Emotion in Man and Animal, 1943; Motivation and Emotion, 1961. Address: 336 Notre Dame Rd., Claremont, Cal. 91711.*

YOUNG, Richard Stuart, Am. biologist; b. Southampton, N.Y., Mar. 6, 1927; s. P. Stuart and Myrtle (Terrell) Y.; A.B., Gettysburg Coll., 1948; Ph.D., Fla. State U., 1955; Sc.D., Gettysburg Coll., 1966; m. Nancy J. Mayer, June 7, 1955; children—Dee Ann, Sandra Lee, Mark Stuart. With cancer research div. Lederle Labs., Inc., Pearl River, N.Y., 1948-49; cancer researcher pharmacology div. FDA, Washington, 1956-58; chief spl. br. Army Ballistic Missile Agy., Huntsville, Ala., 1958-60; chief flight biology Office Life Scis. Programs, NASA, Washington, 1960-61; chief exobiology div. Ames Research Center, Mt. View, Cal., 1961-67; chief exobiology program OSSA, NASA, 1967——. Mem. A.A.A.S., Soc. for Exptl. Biology and Medicine, Sigma Xi, Phi Sigma. Research and numerous publs. on detection of extraterrestrial life, effect of environmental extremes, such as on Mars and in space on living organisms; analysis of extraterrestrial samples for evidence of life. Home: 7927 Falstaff Rd., McLean, Va. 22101. Office: NASA Hdqrs., Washington.*

YOUNG, Richard Wain, Am. anatomist; b. Albany, N.Y., Dec. 15, 1929; s. F. Eugene and Dorothy (Little) Y.; B.A. in Biology, Antioch Coll., 1956; Ph.D. in Anatomy (USPHS fellow), Columbia U., 1959; postgrad. Oak Ridge Inst. Nuclear Studies, 1963; m. Jennifer Watanabe, Dec. 26, 1955; children—Kevin Eugene, Michael David, James Blanchard, Richard Andrew. Research asst. Fels Research Inst., Yellow Springs, O., 1952-56; NSF fellow U. Bari (Italy), also Karolinska Inst., Stockholm, Sweden, 1959-60; asso. prof. anatomy U. Cal. at Los Angeles, 1960——; vis. investigator dept. biology Comm. à l'Énergie Atomique, Saclay, France, 1966-67. Markle scholar in med. sci., 1962-67. Mem. Am. Assn. Anatomists, Am. Soc. for Cell Biology, Assn. for Research in Ophthalmology, Jules Stein Eye Inst. Research and publs. on cranial bones, bone cells, ocular tissues especially lens and retinal photoreceptors using radioisotope techniques. Home: 1123 S. Carmelina Av., Los Angeles 90049.*

YOUNG, Robert Allen, Am. physicist; b. N.Y.C., June 8, 1929; s. Harry Allen and Betty (Hammer) Y.; student Lehigh U., 1947-48; B.S. in Physics, U. Wash., 1951, Ph.D., 1959; m. Nancy Field, 1951 (div. 1953); m. 2d, Muriel Ann Reese, 1958 (div. 1961); 1 dau., Carol Ann. Engr., Boeing Airplane Co., Seattle, 1959-60; physicist Stanford (Cal.) Research Inst., 1960-62, sr. physicist, 1962-67, chmn. dept. atmospheric chem. physics, 1967——. Vis. fellow Joint Inst. for Lab. Astrophysics, U. Colo., 1966-67. Mem. Am. Phys. Soc., Am. Geophys. Union. Contbr. articles to profl. lit. Evaluated, by lab. techniques, chemiluminescent reactions produced in earth's airglow; measured rates of same ionic procedure occurring in earth's ionosphere; studies of various processes of energy transition involving excited nitrogen and oxygen. Home: 3130 Euclid St., Boulder, Colo. 80302.*

YOUNG, Robert Thompson, Jr., Am. physicist; b. Grand Forks, N.D., June 8, 1908; s. Robert Thompson and Ellen (Pierce) Y.; student U. N.D., 1925-27; B.A., U. Mont., 1930; M.A., U. Ill., 1932; Ph.D., Harvard, 1936; m. Elvira Ogden, June 8, 1940; children—Ellen (Mrs. Edward Pischedda), Robert, Richard, Lucy. Instr., Worcester Poly. Inst., 1935-40, asst. prof., 1940-41; research asso. Scripps Inst. Oceanography, La Jolla, Cal., 1936-40; mem. staff Mass. Inst. Tech. Radiation Lab., 1941-45; physicist Naval Research Lab., Washington, 1945-48; sect. chief Nat. Bur. Standards, Washington, 1948-53; br. chief, cons. Harry Diamond Labs., Washington, 1953——; asso. mem. adv. group on electron devices Dept. Def., 1946-60. Fellow Am. Phys. Soc., Washington Acad. Scis.; mem. Sigma Xi. Research and publs. on cosmic rays, penetration of light in sea water, magnetrons, lasers. Home: 4123 Woodbine St., Chevy Chase, Md. 20015. Office: Harry Diamond Labs., Washington 20438.*

YOUNG, Roland Arnold, Am. polit. scientist; b. Loveland, Colo., May 27, 1910; s. Isaac and Anna (Selfors) Y.; A.B., Baylor U., 1932; Ph.D., Harvard, 1940; m. Kathleen Westlake, Jan. 17, 1952; 1 son, Nicholas. Sec., Senate Com. on Fgn. Relations, Washington, 1941-42; prof. polit. sci. Northwestern U., Evanston, Ill., 1948——, acting dir. program African studies, 1962-63. Vis. prof. U. Chgo., Harvard, Columbia U., Johns Hopkins. Fulbright fellow, 1958-59; Guggenheim fellow, London, 1959-60; Liberal Arts fellow in Law and Polit. Sci., Harvard Law Sch., 1965. Mem. Internat. Acad. Law and Sci., Internat. African Law Assn., African Law Assn. Am., Am. Polit. Sci. Assn., Am. Soc. Internat. Law, African Studies Assn., Lincoln's Inn. Author: This is Congress, 1943; Congressional Politics in the Second World War, 1956; The American Congress, 1958; (with H. Fosbrooke) Smoke in the Hills: Political Conflict in the Morogoro District of Tanganyika, 1960; The British Parliament, 1962; American Law and Politics, 1967. Editor: Approaches to the Study of Politics, 1957; Through Masailand with Joseph Thomson, 1963. Research in African polit. and legal systems; legislative process in U.K. and U. S. Home: Route 1, Box 71, Unionville, Va. 22567.*

YOUNG, Sydney, English chemist; b. Farnworth, Lancashire, Eng., Dec. 29, 1857; s. Edward and Anna Eliza (Gannery) Y.; B.Sc., London (Eng.) U., 1880, D.Sc., 1883; Sc.D. (hon.), U. Dublin (Ireland), 1905, U. Bristol (Eng.), 1921; m. Grace Martha Kimmins, 1896; children—Sydney Vernon Kimmins, Charles Edgar Kimmins. Lectr. chemistry (worked with William Ramsay) U. Coll., Bristol, Eng., from 1882, prof. chemistry, from 1887; prof. chemistry U. Dublin, 1904-28, ret., 1928. Mem. adv. council Dept. Sci. and Indsl. Research, 1920-25; external examiner in chemistry Victoria U., 1893. Fellow Royal Soc., 1893, London Chem. Soc. (council 1884), Inst. Physics (founding); mem. Berlin Chem. Soc., London Phys. Soc. (council 1894), Inst. Chemistry, Chem. Soc. (v.p. 1917-20), Royal Irish Acad. (pres. 1921-26). Author: Questions on Physics; Fractional Distillation, 1903; Stoichiometry, 1908, 2d edit., 1918; Distillation, Principles and Processes, 1922. Contbr. articles to Dictionary of Applied Chemistry (Thorpe). Research (with Ramsay) on vapor pressures of solids and liquids and thermodynamical relations; research on pure substances and systematic study of behavior of mixed liquids when distilled; further work on critical constants (with G. Thomas); also studies on ethyl-valerolactone (studied hydrocarbons from Am. petroleum), alcoholic thiorides; devised bubbling still-head. Died Bristol, Apr. 8, 1937.

YOUNG, Thomas, English physician, physicist, Egyptologist; b. Milverton, Somerset, Eng., June 13, 1773; s. Thomas and Sarah (Davis) Y.; student medicine London, Eng., Edinburgh, Scotland; D. Physic, U. Göttingen (Germany), 1796; M.B., Cambridge (Eng.) U., 1803, M.D., 1808; m. Eliza Maxwell, 1804. Practice medicine, London, from 1799; prof. natural philosophy Royal Instn., 1801-03; physician St. George's Hosp., London, 1811-29; insp. calculations Palladium Ins. Co., 1814; sec. commn. established to ascertain length of seconds pendulum, 1816, Bd. On Longitude, 1818; supt.

Naut. Almanac, 1818. Fellow Royal Soc. (fgn. sec. 1802-29, Bakerian lectr. 1801, 03, Croonian lectr. 1808), 1794, Royal Coll. Physicians (censor 1813, Croonian lectr. 1822-23); mem. French Acad. Scis. Author: A Course of Lectures on Natural Philosophy and the Mechanical Arts, 1807; An Introduction to Medical Literature, including a system of Practical Nosology, 1813; A Practical and Historical Treatise on Consumptive Disease, 1815; Elementary Illustrations of the Celestial Mechanics of Laplace, 1821; Account of Some Recent Discoveries in Hieroglyphic Literature and Egyptian Antiquities, 1823. As physician, interested in sense perception; called founder physiol. optics; discovered mechanism by which lens of eye changes shape (accommodation) in focusing on objects at differing distances; described, measured astigmatism; discovered that cause of astigmatism is irregularities in curvature of cornea, 1801; hypothesized that all color perception depends on 3 kinds nerve fibers in retina corr. to colors red, green, violet; investigated interference phenomena and revived wave theory of light as opposed to corpuscular theory; held radiant light consist of undulations in luminiferous ether; 1st to use word energy in its modern sense (product of mass of body into square of its velocity); introduced absolute measurements into elasticity by defining modulus as weight which would double length of rod of unit cross-sect. to which it was hung (Young's modulus); contbd. to understanding capillary action and surface tension of liquids; studied Egyptian hieroglyphic inscriptions, worked out beginning of an hieroglyphic alphabet; instrumental in deciphering Rosetta stone; important contbns. field haemodynamics. Died London, May 10, 1829.

YOUNG, Wesley Andrew, Am. veterinarian; b. Polk City, Ia., Mar. 16, 1898; s. Walter W. and Minnie L. (Kobi) Y.; D.V.M., Ia. State U., 1919; m. Violet Ackerman, Dec. 16, 1961; children—Loie D. (Mrs. Joseph Brooks), Robert H. Individual practice, Des Moines, 1919, field veterinarian, then chief veterinarian in livestock ins. work in Ia., 1919-25; chief veterinarian Animal Rescue League, Boston, 1925-36; mng. dir. Anti-Cruelty Soc., Chgo., 1936-52; Western regional dir. Am. Humane Assn., 1952-58; supr. Griffith Park Zoo, Los Angeles, 1958-63; dir. Los Angeles Zoo, 1963-68; animal cons., 1968—; cons. veterinarian Lincoln Park Zoo, Chgo., 1938-52. Asst. sec. Am. Humane Assn., 1945-52; pres., dir. Nat. Livestock Conservation, Inc.; pres. Am. Cat Assn., 1940-50; chmn. Nat. Dog Weeks, 1940-50. Recipient Alumni Service award Ia. State U., 1963; Fido Human award Gaines Dog Food Co., 1948; award outstanding service to dogs Dog World mag., 1951. Mem. N.G., 1926-40. Mem. Am. (treas. 1945-52), So. Cal. (treas. 1957—, parliamentarian 1955-57) vet. med. assns., Am. Assn. Zool. Parks and Aquariums. Columnist., Animal Care, Boston Herald Traveler, 1935; established radio program, Animals in the News, 1930; conducted television show, Animal Clinic, 1950. Research on new pharm. preparation for domestic and zoological animals; surgical procedure for control flight in birds. Home: 1917 Parnell Av., Los Angeles 90025.*

YOUNG, William Caldwell, Am. endocrinologist; b. Chgo., Sept. 8, 1899; s. William Henry and Mary (Ca'dwell) Y.; A.B., Amherst Coll., 1921, M.A., 1925; Ph.D., U. Chgo., 1927; m. Ruth Annis Hobby, June 19, 1934; children—Deborah Louise (Mrs. Floyd John Detering), Malcolm Caldwell. Asst. zoology U. Chgo., 1927-20; with Brown U., 1928-39, asst prof., 1932-39; asso. prof. primate biology Yale, 1939-44; prof. biology Cedar Crest Coll., 1944-46; asso. prof. anatomy U. Kan., 1946-48, prof., 1948-63; prof. anatomy, chmn. dept. reproductive physiology and behavior Ore. Reg. Primate Research Center U. Ore. Med. Sch., 1963—; vis. prof. Harvard, 1956-57. NRC fellow, Freiburg, Germany 1931-32; Population Council fellow, 1956-58; recipient Howard Crosby Warren medal, 1965. Mem. A.A.A.S., Am. Soc. Zoology, Endocrine Soc., Am. Assn. Anatomy. Editor: Sex and Internal Secretions, 3d edit., 1961. Research, publs. in gonadol hormones in establishment and regulation of reproductive capacities in mammals; problem of fetal loss during pregnancy and function of epididymis; study of relations of gonadal hormones to reproductive behavior; discovery that in most mammals the gonadal hormones have a qualitatively different action on the embryo than on the adult. Died Aug. 30, 1965.*

YOUNG, William Glenn, Jr., Am. physician; b. Washington, Feb. 26, 1925; s. William Glenn and Molly (Weaver) Y.; M.D., Duke, 1948; m. Frances Shields, Feb. 28, 1952; children—William Glenn III, Sarah Lee, Ellen Shields, John Weaver. Asst. prof. surgery Duke, 1957-60, asso. prof., 1960-63, prof., 1963—; cons. thoracic surgeon VA, Watts hosps., (both Durham, N.C.), Eastern N.C. Tb Sanitarium, Wilson, N.C. Contbr. numerous articles to sci., med. jours. Research in thoracic, cardiovascular surgery. Home: 3718 Eton Rd., Durham, N.C.*

YOUNG, William Gould, Am. chemist; b. Colorado Springs, Colo., July 30, 1902; s. Henry A. and Mary Ella (Salisbury) Y.; A.B., Colo. Coll., 1924, M.A., 1925, D.Sc., 1962; Ph.D., Cal. Inst. Tech., 1929; LL.D., U. Pacific, 1966; m. Helen M. Graybeal, June 4, 1926. Research asst. Coastal Lab. Carnegie Inst. of Washington, 1925-27; research asst. Am. Petroleum Inst., Cal. Inst. Tech., 1927-29; NRC fellow Stanford, 1929-30; faculty U. Cal., Los Angeles, 1930—, prof. chemistry, 1943—, chmn. dept., 1940-48, dean div. phys. sci., 1946-57, vice chancellor, 1957—; cons. NDRC, 1941-45. Recipient Tolman medal Am. Chem. Soc., 1961, award in chem. edn., 1962, Priestley medal, 1968. Mem. Am. Chem. Soc., Nat. Acad. Sci., Sigma Xi. Research and numerous publs. on allylic rearrangements; displacement reactions of allylic compounds; allylic Grignard reagents. Home: 955 Harvard St., Santa Monica, Cal. 90403.*

YOUNG, William Henry, mathematician; b. London, Eng., Oct. 20, 1863; s. Henry and Hephzibah (Jeal) Y.; M.A., Sc.D., Cambridge (Eng.) U.; D.Sc. (hon.) in Math., U. Geneva, (Switzerland); hon. degrees univs. Calcutta (India), Strasbourg (France); m. Grace Emily Chisholm, 1896; 3 sons incl. Laurence Chisholm Young; 3 daus. including Rosalind Cecily Young, Lectr. Girton Coll. several years; chief examiner Central Welsh Bd., 1902-05; examiner univs. Cambridge, London, Wales; 1st Hardinge prof. math. U. Calcutta, 1913-16; prof. philosophy, hist. math., Liverpool (Eng.) U., 1913-19; prof. pure math. U. Wales, 1919-23; traveled twice around world and throughout South America, 1936-37; guest chief univs., other ednl. insts. Europe, U. S., Can., Japan, Korea, Manchuria, China, India, Ceylon, Middle East, Brit.-ruled South Africa. Mem. exec. com. Internat. Research Council, also fgn. sec. Nat. Com. Mathematics. Hon. fellow Peterhouse, Cambridge U., 1939. Fellow Royal Soc., 1907 (recipient Sylvester medal, 1928); mem. London Math. Soc. (awarded De Morgan medal, 1917, v.p. 1922-24, (hon.) Bur. Societe Mathematique de France, (corr.) Inst. Coimbra (Portugal). Author: The First Book of Geometry, 1905. (with wife) The Theory of Sets of Points, 1906; The Fundamental Theorems of the Differential Calculus, 1910; Research, publs. on ednl., acad. topics; developed theory of integration, especially evident in treatment of Stieltjes integral, method of monotone sequences; worked on theory of Fourier series, other orthogonal series, and on differential calculus of functions of more than 1 variable. Died Lausanne, Switzerland, July 7, 1942.

YOUNGGREN, Newell A., Am. biologist; b. River Falls, Wis., Mar. 15, 1915; s. Nels A. and Nancy (Weberg) Y.; B.S., River Falls State Coll., 1937; M.P.H., U. Wis., 1941; Ph.D., U. Colo., 1956; m. Beth I. Hoveland, May 9, 1941; children—Stephen, Jeffrey. Faculty, Northland (Wis.) Coll., 1946-48, Bradley (Ill.) U., 1948-54; faculty U. Colo., Boulder, 1954-61, asst. prof., chmn. dept., 1958-61; prof. biology U. Ariz., Tucson, 1961—. Mem. Am. Soc. Zoologists, A.A.A.S., Nat. Assn. Biology Tchrs., Sigma Xi, Phi Sigma. Author: (with William Bond) General Biology Laboratory Manual, 1963; (with Cockrum, McCauley) Biology, 1966. Research on slime molds. Home: 5770 Vista Valverde St., Tucson 85718.*

YOUNGHUSBAND, Sir Francis Edward, Brit. explorer; b. Muree, India, May 31, 1863; s. John William Younghusband; ed. Clifton, Royal Mil. Coll., Sandhurst; LL.D., U. Endiburgh, D.Sc., U. Cambridge; m. Helen Augusta Magniac, 1897; 1 dau. Joined 1st King's Dragon Guards, 1882, apptd. capt., 1889; explorations in Manchuria, 1886, Peking to India, 1887, Pamirs and Hunza, 1889-91; with Indian Polit. Dept., 1890, became polit. officer, Hunza, 1892, Chitral, 1893-94; correspondent London Times, 1895-97; polit. agent, Haraoti and Tonk, 1898-1902; resident, Indore, 1902-03; headed mission to Tibet, 1903-04, resident, Kashmir. 1906-09; pres. Royal Geog. Soc., 1919; founder World Congress of Faiths, 1936. Author: Heart of a Continent, 1898; Relief of Chitral; South Africa of Today, 1898; Kashmir, 1909; India and Tibet, 1912; The Heart of Nature, 1921; The Gleam, 1923; Wonders of the Himalaya, 1924; But in Our Lives, 1926; The Epic of Everest, 1927; The Light of Experience, 1927; Life in the Stars, 1927; The Coming Country, 1928; Dawn in India, 1930; The Living Universe, 1933; Modern Mystics, 1935; Everest: The Challenge, 1936; A Venture in Faith, 1937; The Sum of Things, 1939. Arranged treaty that opened Tibet to British trade, 1904; explored unknown Pamirs region and other parts of Asia; studied oasis settlements; best known for 3 unsuccessful attempts on Mt. Everest and impetus he gave to interest in scaling it. Died Lytchett Minister, Eng., July 31, 1942.

YOUNGQUIST, Walter, Am. geologist; b. Mpls., May 5, 1921; s. Walter R. and Selma (Knock) Y.; B.A., Gustavus Adolphus Coll., 1942; M.S., U. Ia., 1943, Ph.D., 1948; m. Elizabeth S. Pearson, Dec. 11, 1943; children—John, Karen, Louise, Robert. Faculty, Coll. Mines, U. Ida., 1948-51, U. Kans., 1954-57; geologist I, Internat. Petroleum Co., Talara, Peru, 1951-52, sr. geologist, 1952-53, chief, spl. studies sect., 1953-54; prof. geology U. Ore., Eugene, 1957—. Fellow Geol. Soc. Am., A.A.A.S.; mem. Am. Assn. Petroleum Geologists, Am. Inst. Profl. Geologists, Paleontol. Soc., Nat. Planning Assn. Author: (with A. K. Miller) American Permian Nautiloids, 1949; (with A. K. Miller and C. Collinson) Ordovician Cephalopod Fauna of Baffin Island, 1954. Study, publs. on Paleozoic fossils, chiefly cephalopods and conodonts; geol. factors which control location of oil deposits, nat. resource econs. Home: 780 W. 40th Av. Eugene, Ore. 97405.*

YOUNGS, Edward George, English physicist; b. Dovercourt, Eng., June 3, 1932; s. Richard George and Bertha (Laflin) Y.; B.Sc., King's Coll., U. London (Eng.), 1953; Ph.D., Cambridge (Eng.) U., 1956; m. Betty Joan Bough, Apr. 12, 1958; 1 son, Richard Edward. Staff unit soil physics Agrl. Research Council, Cambridge, 1955—, sr. sci. officer, 1959-66, prin. sci. officer, 1966—; research asso., Hatley Found. grantee dept. physics U. Wis., Madison, 1960-61; supr. research students. Cambridge U., 1961-64, 65—. Mem. Inst. Physics (asso.). Research and publs. on physics of water movement through saturated and unsaturated porous materials especially as related to drainage and irrigation practices, analysis of horizontal groundwater seepage. Home: Hilary, 12, Cambridge Rd., Girton, Cambridgeshire, Eng. Office: Agrl. Research Council, Unit of Soil Physics, Huntingdon Rd., Cambridge, Eng.*

YOURASSOWSKY, Eugene, Belgian physician, bacteriologist; b. Brussels, Belgium, Jan. 8, 1929; s. Denis and Claire (Fievet) Y.; M.D., U. Brussels, 1955; postgrad. in mycology Tropical Institut Prince Leopold, Antwerp, Belgium, 1956; m. Francine Combaz, Feb. 25, 1959; children—Evelyne, Catherine, Nadine. Hosp. dir., Belgian Congo, 1956-59; bacteriologist, physician Brughmann Hosp., U. Brussels, Fondation Medicale Reine Elisabeth, Clin. Lab., 1959—. Mem. Société Belge de Medecine Interne, Société Belge de Medecine Tropicale, Société Belge de Biologie Clinique, Société Belge de Mycologie. Research and publs. on kinetics of antibiotic action, antibiotics in clin. field, bacterial taxonomy. Home: 12 Foestraets Av., Brussels, Belgium.*

YPERMAN, Jan (Jehan) (John of Ypres), Flemish surgeon; b. nr. Ypres, West Flanders, circa 1260; ed. Ecole de Médecine de Paris; m. 1285. Became surgeon Belle Hosp., Ypres, circa 1304; surgeon in militia. Author: Die chirurgie von Meister Jan Yperman; Desen Boec spreekt van Medicynen. Introduced ligature; early studies on treatment of sounds; understood diagnostic importance of anesthesia in leprosy; advocated use of anesthesia. Died circa 1331.

YU, Ts'ai Fan, physician; b. Shanghai, China, Oct. 24, 1911; s. I-Kang and Siu-Chen (Chou) Y.; B.A., Ginling Coll., Nanking, China, 1932; M.D., Peiping Union Med. Coll., Peiping, China, 1936. Came to U. S., 1947, naturalized, 1955. Faculty, Columbia U. Coll. Phys. & Surg., N.Y.C., 1947-65, asst. clin. prof., 1959-65; asso. prof. dept. medicine Mt. Sinai Sch. Medicine, N.Y.C., 1966—. Mem. Am. Physiol. Soc., Am. Soc. for Pharmacology and Exptl. Therapeutics, Am. Rheumatism Assn., Harvey Soc., A.M.A. Research and numerous publs. on pathogenesis of gout and uric acid metabolism. Home: 90 La Salle St., N.Y.C. 10027.*

YUAN, Luke Chia-Liu, physicist; b. Changtefu, China, Apr. 5, 1912; s. Ke Wen and Tan (Liu) Y.; B.S., Yenching U., 1932, M.S., 1934; postgrad. U. Cal. at Berkeley, 1936-37; Ph.D., Cal. Inst. Tech., 1940; m. Chien-Shiung Wu, May 30, 1942; 1 son, Vincent. Came to U. S., 1936, naturalized, 1954. Teaching asst. in physics Yenching U., 1932-34; grad. asst. in physics Cal. Inst. Tech., 1937-40, research fellow, 1940-42; research physicist labs. RCA, 1942-46; research asso. Princeton, 1946-49; sr. physicist Brookhaven Nat. Lab., Upton, L.I., N.Y., 1949—. Recipient Sci. Achievement award Chinese Inst. Engrs., 1961, medal for sci. achievement Ministry of Edn., Republic of China, 1959. Fellow American Physical Soc.; member of New York Academy of Sciences, Academia Sinica, Sigma Xi. Author: Methods of Experimental Physics (Nuclear Physics, Vols. A and B), 1961; Nature of Matter, Purposes of High Energy Physics, 1964; also articles. Established existence of 1st pion-nucleon resonance in high energy interactions, shrinkage effect in elastic scattering of elementary particles at very high energies and anti-shrinkage effect in anti proton-proton interactions, existence of real part of forward scattering amplitude in pion-proton and proton-proton interactions at very high energies, existence of maximum in cosmic ray neutron intensity spectrum. Home: 194 Bay Av., Patchogue, N.Y. Office: Brookhaven Nat. Lab., Upton, L.I., N.Y.*

YUKAWA, Hidekl, Japanese physicist; b. Tokyo, Japan, Jan. 23, 1907; s. Takuji and Koyuki Ogawa; grad. Third High Sch., Kyoto, Japan, 1926; M.S., Kyoto U., 1929; D.S., Osaka U., 1938; m. Sumiko Yukawa, April 3, 1932; children—Harumi, Takaaki. Lectr., Kyoto U., 1932, prof. since 1939; lectr., Osaka U., 1933-36, asst. prof., 1936-39, prof. emeritus, since 1950; vis. prof. Inst. Advanced Study, Princeton, 1948, Columbia, 1949-53; dir. Research Inst. Fundamental Physics, Kyoto University. Awarded, Imperial Prize of Japan Acad., 1940, Order of Decoration of Japan, 1943, Nobel Prize for Physics, 1949. Fellow Am. Physical Soc.; Fgn. Asso., Nat. Acad. Scis.; mem. Japan Acad.; mem. Japan Physical Soc. Author: Introduction to Quantum Mechanics, 1946; Introduction to the Theory of Elementary Particles, 1948. Pub. papers on Meson Theory since 1935. Predicted existence of new fundamental particle (meson) which acted as quantum for interactions within atomic nucleus, 1935; also predicted process of "k capture", 1936; research on theory of elementary particles and fields of force. Address: Kyoto Univ., Kyoto, Japan.

YULE, George Udny, Brit. statistician; b. Morham, Scotland, Feb. 18, 1871; s. George Udny and Henrietta (Peach) Y.; student U. Coll., London, 1887-90, U. Bonn (Germany), 1892-93; M.A., Cambridge, 1913; Demonstrator U. Coll., London, 1894-96, asst. prof. applied math., 1896-99, fellow, 1926; asst. dept. tech. City and Guilds of London Inst., 1899-1912; Newmarch lectr. statistics U. Coll., 1902-09; lectr., reader statistics Cambridge U., 1912-31, fellow St. John's Coll., 1922. Recipient Guy Gold medal, 1911. Fellow Royal Soc., 1921, Royal Anthrop. Inst. (mem. council). Author: An Introduction to the Theory of Statistics, 1911; The Statistical Study of Literary Vocabulary, 1944; also articles. Mem. Brit. probabilistic sch. statistics; developed math. theory of evolution based on work of J. C. Willis, 1924; other statis. studies. Died June 26, 1951.

YUNIS, Jorge J., physician; b. Sincelejo, Colombia, S.Am., Oct. 5, 1933; s. José J. and Victoria (Turbay) Y.; M.D., Central U., Madrid, Spain, 1956; Dr. Degree, Central U., Madrid, 1957; m. Olga M. Kretschmer, June 30, 1962; children—George Joseph, Olga Maria, Karl George. Fellow dept. lab. medicine U. Minn. Med. Sch., Mpls., 1959-62, dir. Med. Genetics Lab., 1962—, faculty, 1962—, asso. prof. lab. medicine, 1965—, dir. tng. program and postgrad. edn. in clin. pathology, 1966—. Mem. Am. Soc. Human Genetics, Am. Soc. Hematology, Am. Assn. Blood Banks, Central Soc. for Clin. Research. Author: Human Chromosome Methodology, 1965; Biochemical Methods in Red Cell Genetics, 1968; also articles. Research in cytogenetics and biochemistry of mammalian condensed chromatin in devel. and evolution. Office: 412 S.E. Union St., Mpls. 55455.*

YUNKER, Conrad E., Am. biologist; b. Matawan, N.J., Dec. 22, 1927; s. Conrad E. and Helen (Merrill) Y.; B.S., U. Md., 1952, M.S., 1954, Ph.D., 1958; m. Samira L. Abozeid, Aug. 1958; children—Conrad E., Dina L., Samira E. Staff, U. S. Naval Med. Research Unit 3, Cairo, Egypt, 1955-57; research asso. U. Md., 1958-59, asst. prof., 1959; entomologist Can. Dept. Agr., 1959-60; scientist Middle Am. Research Unit, C.Z., 1960-62; sr. scientist Rocky Mountain Lab., USPHS, Hamilton, Mont., 1962—. Mem. Am. Soc. Zoologists, Am. Soc. Parasitologists, A.A.A.S., Am. Soc. Tropical Medicine and Hygiene, Soc. Systematic Zoology, Helminthological Soc. Washington, Entomol. Soc. Am. Author: (with others) A Guide to the Families of Mites, 1958; also articles. Identification and systematic study of mites especially those parasitic on animals and important in human diseases, tick and insect borne viruses, adapted 1st insect cell line to media without insect hemolymph and established that it would support viruses.*

YUNUSOV, Sabir Yunushovich, Russian organic chemist; b. Tashkent, USSR, Nov. 11, 1909; grad. Chem. Faculty Central Asia U., 1935; Dr.Chem. Scis., 1948. Chief lab. alkaloid chemistry Inst. Plant Chemistry, Uzbek Acad. Sci., 1943—, v.p., 1952-62, now dir. Mem. Uzbek Acad. Sci. Research and publs. on alkaloid chemistry, alkaloid storage in various plant parts for determination of laws governing this accumulation; isolated new alkaloid compounds from plants; established structure of several alkaloids. Office: Uzbek SSR Acad. Sci., Inst. Plant Chemistry, Ulitsa Kuibysheva, 14, Tashkent, Uzbek SSR.

YVON VILLARCEAU, Antoine-Joseph-François, French mathematician; b. Vendôme, France, Jan. 15, 1813; prof. Paris; astronomer Paris Obs. Mem. French Acad. Scis., 1867, Bureau des longitudes. Originated theory on sectioning cylinder; contributed to astrophysics. Died Paris, Dec. 23, 1883.

Z

ZABARELLA, Jacobus, Italian logician, astrologer; b. Padua, Italy, Sept. 5, 1533; grad. as master in logic, math., physics, ethics, U. Padua, 1553. Prof. logic U. Padua, from 1563; several times ambassador to Venice. Author: De Rebus naturalibus libri XXX; De natura Logicae libri II; De Methodis libri IV; De propositionibus necessariis; Comment. in libros physicorum Aristotelis. One of greatest Aristotelians and logicians of his time; work includes discussions on constn. of natural sci., nature of heavens, movement of fire, of heavy and light objects, constn. of individual, generation and death, reaction, mixture, elementary qualities, regions of air, celestial heat, compounds, the soul, accretion and nutrition, mind and senses, others. Died Oct. 1589.

ZABIN, Irving, Am. biochemist, molecular biologist; b. Chgo., Nov. 13, 1919; s. Morris and Alexia (Saks) Z.; B.S., U. Ill., 1940; Ph.D., U. Chgo., 1949; m. Esther Marshall, Mar. 15, 1942; children—Lee Barbara, Fredric Marshall, Carol Ann. Research asso. dept. biochemistry U. Chgo., 1949-50; research asso. dept. biol. chemistry U. Cal. at Los Angeles, 1950-51, faculty, 1951—, prof., 1964—. Nat. Multiple Sclerosis Soc. scholar, vis. scientist Pasteur Inst., Paris, France, 1959-60; Guggenheim fellow, Pasteur Institute, 1967. Mem. Am. Soc. Biol. Chemists, A.A.A.S., Sigma Xi. Author: (with D. J. Hanahan, F. R. N. Gurd) Lipide Chemistry, 1960; also numerous articles. Research on pathway formation of cholesterol and other lipids, induced enzymes. Home:

937 Centinela St., Santa Monica, Cal. 90040. Office: Dept. Biol. Chemistry, U. Cal., Los Angeles, 90024.*

ZABLOCKI, Bernard, immunologist; b. Nowogródek, Russia, Jan. 1, 1907; s. Euphemius and Anna (Ginz) Z.; M.Chemistry, U. Vilna, 1928; M.D., U. Warsaw (Poland), 1939; m. Victoria Putrament, May 15, 1940; children—Andrew-Christopher, Eva-Sylvia. Research asst. State Hygiene Inst. Warsaw, 1929-39; faculty U. Lodz (Poland), 1946—, prof. immunology, 1950—, chmn. dept. microbiology, 1946—, vice rector sci., 1950-65; research asso. Yale, New Haven, 1960-61. Decorated Cavalier and Commandery Order Polonia Restituta; recipient Sci. award City of Lodz, 1964, Sci. award 1st degree Ministry Edn., 1965. Mem. Polish Acad. Scis. (corr.), Societas Sciantiarum Lodziensis, Polish Acad. Scis. (chmn. immunology sect. 1965—). Author: Fundamentals of Chemical Bacteriology, 1955; Outline of Immunology, 1959; Theoretical Bases of Immunopathology, 1963; also numerous articles. Physiochem. investigations on serological test for syphilis; isolation of typhoid and dysentery endotoxins and their chem. structure; research on bacterial and testicular hyaluronidase and hyaluronic acid, antigen O, protection of gram positive bacteria against penicillin; bioenergetic investigations of bacterial cultures, microcalorimetry, immunopathology. Office: 18 Nowotki, Lódz, Poland.*

ZABOLOTNYJ, Danylo, Russian microbiologist; b. Ukrainia, 1866; prof. in St. Petersburg, Russia, Kiev, Odessa (all Russia). Made expdns. for research and control of plague and cholera to epidemic areas of India and Mongolia. Died 1929.

ZABOROVSKII, Aleksandr Ignatievich, Russian geophysicist; b. May 24, 1894. Prof., Moscow Geol. Survey Inst., 1930-54; prof. Moscow U., 1954—. Author: Terrestrial Magnetism, 1932; Special Functions for Geophysical Surveyors, 1939; Electrical Surveying, 1943. First systematic magnetic survey of Kursk magnetic anomaly; geophys. surveys of petroleum deposits in Caucasus and Urals. Home: Gosudarstvenny universitet, Leninskie gory, Moscow, USSR.

ZACCHIAS, Paolo, Italian physician; b. Rome, 1584; physician to Pope Innocent X; author works on med. jurisprudence, book on hypochondriacal ailments. Put forensic medicine on more sci. basis. Died 1659.

ZACHARIAS, Jerrold Reinach, Am. physicist; b. Jacksonville, Fla., Jan. 23, 1905; s. Isidore A. and Irma (Kaufman) Z.; A.B., Columbia, 1926, A.M., 1927, Ph.D., 1933; m. Leona Hurwitz, June 23, 1927; children—Susan, Johanna. Asst. prof. Hunter Coll., New York City, Oct. 1931-Nov. 1940; staff mem. Radiation Lab. Mass. Inst. of Technology, 1940-45, professor of physics, 1946-66, Institute professor, 1966—, director Lab. for Nuclear Science and Engring., 1946-56; div. head U. Cal., Los Alamos Lab. 1945; dir. Sprague Electric Co.; mem. sci. bd. Itek Corporation. Research on atomic clocks; magnetic and electric shapes of atomic nuclei. Home: 32 Clifton St., Belmont, Mass.

ZACHARIASEN, Fredrik, Am. physicist; b. Chgo., June 14, 1931; s. William Houlder and Ragni (Durban-Hansen) Z.; B.S., U. Chgo., 1951; Ph.D., Cal. Inst. Tech., 1956; m. Nancy J. Walker, Jan. 27, 1957; children—Kerry Ellen, Judith Ann. Instr., Mass. Inst. Tech., 1956-57; asst. prof. Stanford, 1958-60; faculty Cal. Inst. Tech., 1960—, prof. physics, 1966—. Cons. to Rand Corp., 1956—, Los Alamos Sci. Lab., 1960—, Inst. for Def. Analyses, 1960—. Alfred P. Sloan fellow, 1960-64. Author: (with S. D. Drell) Electromagnetic Structure of Nucleons, 1960; also articles. Research in high energy physics of elementary particles. Home: 2235 N. Villa Heights Rd., Pasadena, Cal. 91007.*

ZACKS, Sumner Irwin, Am. pathologist; b. Boston, June 29, 1929; s. David and Rose (Krivitsky) Z.; B.A. cum laude, Harvard, 1951, M.D. magna cum laude, 1955; m. Marilyn Garfinkel, June 28, 1953; children—Nancy Alice, Charles Matthew, Susan Esther. Asst. pathologist Pa. Hosp., Phila., 1960-61, asso. pathologist, 1961—; instr. Harvard Med. Sch., 1956; staff U. Pa. Sch. Medicine, 1961—, asso. prof. pathology, 1966—. Cons., Children's Hosp. Phila., 1962—; mem. corp. Marine Biol. Lab., Woods Hole, Mass., 1965—. Recipient Hektoen Bronze medal, A.M.A., 1961. Diplomate Am. Bd. Pathology. Fellow Am. Coll. Pathologists; mem. Histochem. Soc. (sec. 1965—), Am. Soc. Exptl. Pathologists, Am. Soc. Neuropathologists, A.M.A., Pa. State, Phila. City med. socs., Sigma Xi, Alpha Omega Alpha. Author: The Motor Endplate, 1964; also articles. Research on structure of neuromuscular junctions in normal and disease states, ultra structure pathology of muscle, mode of action of bacterial neurotoxins. Office: Pa. Hosp., 8th and Spruce Sts., Phila. 19010.*

ZAHL, Harold Adelbert, Am. physicist; b. Chatsworth, Ill., Aug. 24, 1904; s. Arthur Herman and Bertha (Dieber) Z.; B.A. in Physics, N. Central Coll., Naperville, Ill., 1927; M.S. in Physics, U. Ia., 1929, Ph.D., 1931; m. Vera Virginia Hiller, Dec. 17, 1948; children—James F., Christopher Allen, Harold Alexander. Physicist, U. S. Army Labs., Ft. Monmouth,

N.J., 1931-66, dir. research, Electronics Labs., 1948-65, dir. Atmospheric Scis. Lab., U. S. Electronics Command, 1965-66; cons. U. S. Electronics Command, part-time 1966—. Recipient Dept. Army Decoration for Exceptional Civilian Service, 1962, Sci. Achievement award Service Clubs Long Branch (N.J.), 1963, Fed. Bus. Assn. N.Y. Outstanding Civilian award, 1964, Distinguished Alumnus award N. Central Coll., 1964. Fellow Am. Phys. Soc., I.E.E.E.; mem. I.R.E. (Harry Diamond Meml. award 1954), Sigma Xi, Gamma Alpha. Author: Electrons Away . . . or Tales of a Government Scientist, 1968. Research and publs. on exptl. verification of wave-particle dualism of athoms, propagation of sound through ocean, radar, electron tubes; developed infra-red detecting pneumatic cell, tubes used in radar, especially the Zahl tube and radar switching tubes. Home: Box 164-B, R.F.D.-1, Holmdel, N.J. 07733.*

ZAHM, John Augustine, naturalist, explorer; b. New Lexington, O., June 14, 1851; s. Jacob M. and Mary (Braddock) Z.; A.B., Notre Dame U., 1871; hon. Ph.D. from Pope Leo XIII, 1895. Entered Order of Holy Cross, 1871; apptd. to charge of scientific dept., 1874, dir. same, 1875, later pres. bd. trustees, Notre Dame U.; also for yrs. curator Notre Dame Mus. Lecturer at Plattsburg, N.Y., and Western (Madison, Wis.) summer schs., and New Orleans winter sch.; also lectured at Catholic U. of America. Author: Evolution and Dogma; Bible Science and Faith: Sound and Music; Catholic Science and Catholic Scientists; Scientific Theory and Catholic Doctrine; Science and the Church; Evolution and Theology; Souvenirs of Travel; Alaska—the Country and the People; Hawaii and the Hawaiians. Devoted many yrs. to study of S. America, and wrote under pseudonyms: Following the Conquistadores Up the Orinoco and Down the Magdalena; Following the Conquistadores Along the Andes and Down the Amazon; Women in Science. Wrote also under the name of J. A. Zahm: The Quest of El Dorado; Following the Conquistadores Through South America's Southland; The Great Inspirers, 1917. Known as advanced evolutionist; with Roosevelt expdn. to S. Am., 1913-14; contbd. to study of sci. and religion. Died Munich, Germany, Nov. 11, 1921.

ZAHN, Helmut Gustav, German chemist; b. Erlangen, Germany, June 13, 1916; s. Hermann Wolfgang and Irma (Brand) Z.; Dipl.-Ing., Karlsruhe Inst. Tech., 1939, Dr.-Ing., 1940; m. Roswitha Schmidt-Lorenzen, July 14, 1945; m. 2d, Ingrid Fricke, Oct. 5, 1961; children—Thomas, Manuel, Leopold, Alexandra. Asst., Karlsruhe Inst. Tech., U. Heidelberg, 1940-53; unscheduled prof. Chem. Inst., U. Heidelberg, 1953-57; asso. prof. Aachen Inst. Tech., 1957-60, prof. textile chemistry, 1960—; dir. German Wool Research Inst., 1952—. Research and numerous publs. on cross-linking proteins, amino acids; cross-linking theory in tanning reactions; synthesis of oligomers and pleionomers with nylon and polyester structure; total synthesis of insulin. Address: 20 St. Vitherstrasse, 51 Aachen, West Germany.*

ZAHN, Rudolf K., German biochemist; b. Bad Orb, Feb. 6, 1920; s. Jakob and Maria (Noll) Z.; grad. in medicine summa cum laude, U. Frankfort/Main, 1948; Rockefeller fellow Harvard, 1949-50; m. Gertrud Daimler, Jan. 17, 1942; children—Isabel, Matthias. With dept. pharmacology U. Pa., Phila., 1950; asst. U. Frankfort/Main (Germany), 1950-56, docent, 1956-61, asso. prof. physiology, biochemistry, 1961-67; prof. dir. inst. biochemistry U. Mainz (Germany), 1967—. Mem. Soc. Biol. Chemistry, Soc. Biographics, Soc. Electron Microscopy, Soc. Clin. Chemistry, A.A.A.S., N. Y. Acad. Sci. Research, numerous publs. on glomerular punctures and submicro analysis for kidney function; devel. methods for continuous extraction in protein analysis, for preparation, purification, stabilization of DNA; electron microscopy of DNA molecules; biol. activities of cross-linked DNA. Home: 12 Oderstrasse, 62 Wiesbaden-Schierstein. Office: 15 Johann Joachim Becher-Weg, 65 Mainz, West Germany.*

ZAHOR, Zdenek, Czechoslovakian physician, pathologist; b. Prague, Czechoslovakia, May 30, 1920; s. Zdenek and Marie (Nemcová) Z.; M.D., Charles U., Prague, 1949; m. Vlasta Boušková, Apr. 9, 1943; 1 son, Martin. Asst., II dept. pathology Med. Faculty, Charles U., 1949-60, asst. prof., 1960—. Mem. Czechoslovak Soc. Pathologists. Discovery of selective toxicity of cadmium salts on testicular tissue, postlathyric fibroelastosis of aorta in rats, pathogenesis of spontaneous arteriosclerosis in female breeder rats. Home: 7 Kourimská, Prague 3, Czechoslovakia.*

ZAHORSKY, John, Am. pediatrician; b. Mereny, Hungary, Oct. 13, 1871 (came to U. S., 1872, naturalized by father's citizenship); s. John and Amalia (Cura) Z.; A.B., Steelville (Mo.) Inst., 1892; M.D., Mo. Med. Coll., St. Louis, Mo., 1895; post-grad. Johns Hopkins, 1899; m. Elizabeth Silverwood, June 27, 1900; children—Theodore S., Elizabeth (Mrs. Joseph W. Cushing). Began practice 1895; resident physician, Bethesda Hosp., 1896-98; editor St. Louis Courier of Medicine, 1900-1905; lecturer pediatrics, Washington U., 1900-05, clin. prof., 1905-11; chief Children's Clinic, Wash. U. Hosp. Clinic, 1905-11; attending physician, St. Louis Children's Hosp., 1910-12; pediatrician St. Louis City Hosp., 1920-23, St.

John's Hosp., 1920-27; prof. pediatrics, St. L. Univ., 1912-48, emeritus, chmn. dept., 1928-33; dir. of department since 1933; pediatrician-in-chief, St. Mary's Group Hosp., 1924-48; Licentiate Am. Bd. of Pediatrics, 1936; pres. (emeritus) bd. dirs., Bethesda Gen. Hosp. Received awards from St. Louis U. and St. Louis Pediatric Soc. Fellow Am Coll. Physicians; mem. A.M.A., Southern Med. Assn., Acad. Pediatrics, St. Louis Med. Soc., Acad. Science, Mo. Hort. Soc., Mo. Hist. Soc., Sigma Xi. Author: Baby Incubators, 1905; Golden Rules of Pediatrics, 1913, Synopsis Pediatrics, 6th edit., 1953; The Infant and Child, 1939; From the Hills, Autobiography, 1950; also about 125 med. articles. Described roseola infantilis, 1910; first to describe herpangina, 1920. Home: Steelville, Mo.

ZAHRINGER, Joseph, German physicist; b. Schonenbach, Germany, Mar. 15, 1929; s. Johann and Maria (Straub) Z.; doctorate U. Freiburg, 1956; postgrad. fellow Brookhaven Nat. Lab., 1956-58; m. Elisabeth Kraaibeck, Aug. 30, 1955; children—Eva-Maria, Ulrike, Klaus-Peter. Sci. asst. U. Freiburg, 1956; sci. asst. Max Planck Inst., Heidelberg, 1958-64, asso. dir., 1965—; lectr. on geophysics at various univs. Mem. German Phys. Soc. (recipient award 1964), Max Planck Soc. Author: (with O. A. Schaeffer) K-Ar Dating, 1966; also articles. Research on origin of meteorites, detection of helium in space, mass spectrometer. Home: 4 Zur Forstquelle, Heidelberg, Germany.*

ZAIMIS, Eleanor Cristides, pharmacologist; b. Greece, June 16, 1915; d. Jean and Helen Cristides; ed. univs. Bucharest, Hungary, Athens, Greece; M.B., 1938; M.D., 1941; B.Sc. in Chemistry, Athens, 1947; m. Evanghelos Chrysafis, 1938; m. 2d, John Zaimis, 1943 (d. 1957). Asst. to prof. pharmacology U. Athens, 1938-47; head dept. health Youth Center, Athens, 1940-45; mem. Greek Penicillin and Streptomycin Com., 1945-47; Brit. Council Scholar, 1947-48; Med. Research Council fellow, 1948-50; research worker dept. pharmacology Bristol (Eng.) U., 1947; with depts. chemistry and physiology Nat. Inst. for Med. Research, London, 1948; staff dept. pharmacology Sch. Pharmacy, London U., 1948-50; lectr. pharmacology London U., 1950-54, reader London U. Royal Free Hosp. Sch. Medicine, 1954-58; prof., 1958—. Recipient Cameron prize, Edinburgh, Scotland, 1956; Gairdner Found. Internat. award, Toronto, Ont., 1958. Mem. Acad. Medicine Rome, Physiol. Soc. (mem. com.), Brit. Pharmacol. Soc., Société Francaise de thérapeutique et de pharmacodynamie. Author: Textbook on Hygiene (Greek Acad. prize) 1948; also articles. Research in physiology and pharmacology of skeletal, smooth and cardiac muscles; introduced (with W. D. M. Paton) hexamethonium bromide, 1949. Home: 4 Cambridge Gate, Regent's Park, N.W.1, London. Office: Royal Free Hosp. Sch. Medicine, Hunter St., London W.C.1, U.K.

ZAJDELA, Francois Engelbert, biologist; b. Ljubljana, Yugoslavia, Dec. 28, 1920; s. Franc and Rosa (Stary) Z.; med. dr. U. Zagreb (Yugoslavia); postgrad. U. Padova, Italy, U. Ljubljana, 19——; m. Renée Le Roy, July 2, 1951; children—Francois, Catherine, Nicolas, Pierre, Jean, Marie. Staff, Cancer Inst. Ljubljana, 1940-42, Lab. Path. Anatomy, Padova, 1943-47, Karolinska Institutet, Stockholm, Sweden, 1947; chief lab. exptl. cytology Radium Inst., Paris, 1947-65, dir. unit cellular physiology nat. health service, Orsay, France, 1965—. Recipient E.S.S.E.C. prize; Ordre Nat. du Merite. Mem. Societé francaise d'Histochimie, Société francaise de Cancérologie. Research, numerous publs. on mechanism of chem. carcinogenesis at cellular level, role of cellular nuclei in differentiated tissues. Home: 48, rue Velpeau, Antony, Paris. Office: Batiment 110, Institut du Radium, 91, Orsay, Paris, France.*

ZAJIC, Jirí, Czechoslovakian chemist; b. Horni Cerekev, Czechoslovakia, Mar. 29, 1926; s. Jan and Julie (Urbanová) Z.; engr. Inst. Chem. Tech., Prague, Czechoslovakia, 1951; D., Inst. Chem. Tech., Prague, 1961; m. Kveta Pesková, June 23, 1951. Staff, Research Lab. Fat and Milk Industry, 1951-53; staff dept. tech. of milk and fats Inst. Chem. Tech., Faculty Food Tech., Prague, 1953—, asst. prof. fats technology, 1964—. Cons. to fat industry. Mem. Internat. Soc. for Fat Research. Author: Fats Technology, 1965; also articles. Research on glycerol polymerization, refining of fats, adsorption of fatty acids on ionex, reaction of sucrose with fatty acids, hydrogenation of fats. Home: 15 Malinová, Prague 10, Czechoslovakia.*

ZAK, Bennie, Am. clin. chemist; b. Detroit, Sept. 29, 1919; s. Morris and Lena (Sneider) Z.; B.S. in Chemistry, Wayne U., 1948, Ph.D., 1952; m. Doris Kitty Selby, Sept. 7, 1946; children—Steven Dennis, Deborah Lise, Marsha Gale. Med. lab. analyst, jr. asso. pathology Detroit Gen. Hosp., 1951-59; faculty Wayne State U. Sch. Medicine, 1959—, prof. pathology, 1965—; free lance cons. clin. chemistry, 1952—. Research and numerous publs. on analytical biochemistry, visible and ultraviolet spectrophotometry, electrophoresis, automation and chromatography; determined numerous substances in biol. materials. Home: 25435 Southwood Dr., Southfield, Mich. 48075. Office: 1400 Chrysler Expressway, Detroit 48207.*

ZAKHARIN, Antonovich Grigorii, Russian elec. engr.; b. Batum, USSR, 1904; grad. Tiflis (USSR) Poly. Inst., 1932; D. Tech. Scis., 1954. Staff All-Union Agrl. Electrification Inst., 1932-43; head agrl. electrification group Gen. Power Engring. Dept., Krzhizhanovsky Inst. Power Engring., USSR Acad. Scis., from 1943; also with USSR Ministry Agr. Mem. All-Union Agrl. Exhbn. Orgn., 1954-56. Research on power supply apportionment for agrl. zones USSR.

ZAKIROV, Kadyr Zakirovich, Russian botanist; b. July 25, 1906; grad. Uzbek Pedagogical Acad., Samarkand, 1933. Staff Uzbek U., Samarkand, USSR, 1937-41, Central Asian U., Tashkent, USSR, 1941-43, Pedagogic Inst., Tashkent, 1943-52; dir. Inst. Botany, Uzbek Acad. Scis., 1952-56; rector Uzbek U., 1956-57. Named Hon. Sci. Worker Uzbek SSR, 1956. Mem. Uzbek Acad Scis., Uzbek Acad. Agrl. Sci. (pres.). Authr: Data on the Flora of the Zeravshan River, 1941; The Flora and Vegetation of the Zeravshan Valley, 1955. Made expdns. to study flora of Uzbek; studies in geography, biology and taxonomy of plants, flora of Zeravshan River; originated theoretical explanation of division of Central Asia into zones and belts. Office: Akademiya Selskokhozyaystvennsykh nauk. ul. Navbi 8, Tashkent, Uzbek SSR, USSR.

ZAKRZEWSKA, Marie Elizabeth, physician; b. Berlin, Prussia, Sept. 29, 1829; when 18 yrs. old began study of midwifery in Royal Hosp. Charité, Berlin; graduated in spring, 1852, as teacher of midwives; became accoucheuse-in-chief of that hosp.; but, learning that in the U. S., women could become full doctors of medicine, resigned that position and emigrated in 1853. Grad. Western Reserve Coll. Med. School, Cleveland, O., 1856. With Dr. Elizabeth Blackwell established, 1857, New York Infirmary for Indigent Women and Children; was resident physician until May 1859; prof. obstetrics, New England Female Med. Coll., Boston, 1859-61. In 1861 inaugurated New England Hosp. for Women and Children, of which she was a dir. and advisory physician. Died 1902.

ZAKRZEWSKI, Aleksander, otolaryngologist; b. Oleksiniec, USSR, Apr. 14, 1909; s. Nikolai and Felicia (Oranska) Z.; M.D., U. Poznan (Poland), 1932, M.A., 1935; m. Aleksandra Durska, Aug. 21, 1937; children—George, Anna (Mrs. Wasiewicz). Asst. to head of otolaryn. dept. U. Poznan, 1929-39, asst. prof. in charge otolaryn. dept., 1945-48, prof., dir. dept., 1948—; asst. U. Warsaw, 1939-45, habilitated and docent, 1945. Prorector, Med. Acad. Poznan, 1954-62. Named officer Polonia Restituta Order. Mem. Polish (pres. 1954-58), Poznan (chmn. 1946—) otolaryn. socs., Collegium Otolarnygologicum Amititiae Sacrum, French Otolaryn. Soc., Royal Soc. Medicine, London, Eng., Author: (with another Manual of Otolaryngology for Specialist, 1953. Research, publs., contbns. to encys. on quantitative hearing examinations, method of sound localization by measuring directional hearing acuity angle, own original method of otogenic brain abscess treatment, others. Home: 49 Przybyszewski Str., Poznan, Poland.*

ZALKIN, Allan, chemist; b. Haliburton, Ont., Can., Aug. 16, 1926; s. Samual M. and Bella (Fisher) Z.; brought to U. S., 1937, naturalized, 1944; B.S., U. Cal. at Los Angeles, 1948; student Stanford, 1945; Ph.D., U. Cal. at Berkeley, 1951; m. Maxine Palma Johnson, Dec. 20, 1965; 1 dau., Natasha. With Lawrence Radiation Lab., Livermore, Cal., 1951-60, Berkeley, 1960—, sr. research chemist, 1960—. Mem. Am. Chem. Soc., Am. Crystallographic Soc., A.A.A.S. Research and numerous publs. on crystal and molecular structure determinations of inorganic, organic, metallo-organic, biochem. and intermetallic materials by X-ray diffraction crystallography; developer computing programs and techniques for X-ray crystallographic uses. Home: 81 Edgecroft Rd., Berkeley 94707. Office: Chemistry Dept., Lawrence Radiation Lab., Berkeley, Cal. 94720.*

ZALOKAR, Marko, biologist; b. Ljubljana, Yugoslavia, July 14, 1918; s. Alojz and Ana (Kos) Z.; came to U. S., 1947, naturalized, 1952; Dipl. Phil., U. Ljubljana, 1940; D.Sc., U. Geneva, 1944; m. Julia Gay Ballantine, Dec. 31, 1951; children—Catherine Nadja, Mira Elizabeth. Research fellow Cal. Inst. Tech., 1947-49; asst. prof. U. Wash., 1949-51; vis. scientist NIH, 1951-54; research asso. Wesleyan U., Middletown, Conn., 1954-55, Yale, 1955-60, U. Cal. at San Diego, 1961-66, U. Cal. Davis, 1966-67, Cal. Inst. Tech., 1967—. Mem. A.A.A.S., Genetics Soc. Am., Bot. Soc. Am., Soc. Developmental Biology. Contbr. numerous articles to profl. jours., also chpts. to books. Research in genetics; demonstrated nuclear origin of ribonucleic acid and transfer to cytoplasm. Home: 6064 Av. Chamnez, La Jolla, Cal. 92038.*

ZAMBECCARI, Giuseppe, Italian surgeon, anatomist; b. Florence, Italy, 1655. Worked under Redi at Ospedale di Santa Maria Nuova, Florence, Italy; became prof. anatomy, Pisa, Italy, 1681. Wrote account of his expts., 1680. Research on splenectomy in dogs, demonstrating that spleen is not necessary for life, 1680; also studied exptl. removal of other organs. Died 1728.

ZAMBONI, Giuseppe, Italian physicist; b. Venice, Italy, June 1776; ed. Verona Sem. Prof. physics Verona, Italy. Invented dry cell (Zamboni pile), 1812, elec. clock. Died Venice, July 25, 1846.

ZAMECNIK, Paul Charles, Am. physician; b. Cleve., O., Nov. 22, 1912; s. John C. and Mary (McCarthy) Z; A.B., Dartmouth, 1933; M.D., Harvard, 1936; m. Mary Connor, Oct. 10, 1936; children—Karen, John, Elizabeth. Resident medicine Huntington Hosp., Harvard University, 1936-37, Collis P. Huntington professor oncologic medicine, 1956—; intern medicine U. Hosps. of Cleve., 1938-39; fellow Carlsberg Lab., Copenhagen, Denmark, 1939-40, Rockefeller Inst., N.Y.C., 1941-42; physician Mass. Gen. Hosp., 1956—; dir. J. C. Warren Labs., Harvard and Mass. Gen. Hosp. Mem. staff OSRD, World War II. Mem. Am. Acad. Arts and Scis., Assn. Am. Physicians, Am. Assn. Biol. Chemists, Am. Assn. Cancer Research. Research on cancer; protein metabolism of normal and malignant tissues; radioactive isotope tracers. Home: 101 Chestnut St. Office: Mass. Gen. Hosp., Fruit St., Boston.

ZANARDINI, Giovanni, Italian physician, botanist; b. Venice, Italy, June 12, 1804; s. Angelo and Anna Maria (Traffico) Z.; M.D., U. Padua (Italy), 1831; doctorate in surgery and obstetrics U. Pavia (Italy). Became med. judge Venetian Delegation, 1834; head physician Royal Hosp. Padua, 1834-47; asst. head physician Royal Hosp. Venice until 1869; surgeon Venetian Conservatory of Zitelle. Mem. Coll. Physicians. Research and publs. on algae of Adriatic, Red Sea, Indian Ocean, Mediterranean. Died Apr. 24, 1878.

ZANCHETTI, Alberto, Italian physician; b. Parma, Italy, July 27, 1926; s. Mario and Amelia (Leal) Z.; M.D., U. Parma (Italy), 1950; Ph.D. in Physiology, U. Pisa (Italy) Sch. Medicine, 1956; m. Carla de Renzi, June 6, 1959; children—Silvia, Mario, Giorgio. Asst. prof. physiology U. Pisa Sch. Medicine, 1950-56; asso. prof. medicine U. Siena (Italy) Sch. Medicine, 1956-66; asso. prof. medicine U. Milan (Italy) Sch. Medicine, 1966—. Rockefeller Found. fellow U. Ore. Sch. Medicine, 1953; recipient Marzotto award in medicine, 1959. Mem. A.A.A.S., Internat. Brain Research Orgn., Royal Soc. Medicine London, European Soc. for Clin. Investigation. Author: (with G. F. Rossi) The Brain Stem Reticular Formation: Anatomy and Physiology, 1957; also numerous articles. Research on nervous mechanisms of sleep and wakefulness, mechanisms of emotional behavior, nervous control of cardiovascular system, exptl. and human arterial hypertension, clin. pharmacology and treatment of human essential hypertension. Home: 7, via Caradosso, Milan, Italy.*

ZANDER, Alvin Frederick, Am. psychologist; b. Detroit, Oct. 13, 1913; s. Hugo Helmuth and Frieda (Meisler) Z.; B.S., U. Mich., 1936, M.S., 1937, Ph.D., 1942; m. Patience Dorothea Clare, Dec. 19, 1939; children—Constance Gwen (Mrs. Myron Howard Nadel), Christopher Alvin, Judith Ann. Research fellow U. Ia., 1942; asst. dir. research Boy Scouts Am., 1942-44; asst. prof. psychology Springfield Coll., 1946-47; program dir. Research Center for Group Dynamics, U. Mich., Ann Arbor, 1948-59, dir., 1959—. Fulbright Research award, Oslo Norway, 1957-58; fellow for grad. study U. Ia., 1942, U. Chgo., 1938-39. Mem. Am. Psychol. Assn. (sec. treas div. personality and social psychology 1964-66, mem. council reps. 1964—), Soc. for Psychol. Study Social Issues (past pres.), Mich. Psychol. Assn., Phi Kappa Phi, Phi Delta Kappa. Author: (with Dorwin Cartwright) Group Dynamics Research and Theory, 1953, rev., 1960; (with Arthur Cohen and Ezra Stotland) Role Relations, 1957; also numerous articles. Research on the hierarchy of groups, conditions determining group goals. Home: 3 Harvard Pl., Ann Arbor, Mich. 48104.*

ZANEVELD, Jacques Simon, biologist; b. Netherlands, Dec. 9, 1909; s. Cornelus Dirk and Dirkje (Van der Giessen) Z.; Dr. ès sc., U. Leyden (Netherlands); m. Engelina Van de Water, Feb. 3, 1939; children—Lourens Jan Dirk, Jacques Ronald Victor. Asst. in research services Rijksherbarium, Leyden, 1938-42; prof. biology Haganum Gymnasium, the Hague, 1942-48, 51-54; phycologist marine research labs., Jakarta, Indonesia, 1948-51; dir. Marine Biol. Inst. of Caraibes, Curacao, 1954-59; prof. biology, pres. biol. dept. Old Dominion Coll., Norfolk, Va., 1959—. Mem. Am., Dutch bot. socs., Internat. Oceanographic Found., Va. Acad. Sci., Phycological Soc. Am. Home: 1334 Upper Brandon Pl., Norfolk, Va.

ZANGERL, Rainer, paleontologist; b. Winterthur, Switzerland, Nov. 19, 1912; s. Hermann and Hedwig (Widmer) Z.; Ph.D., U. Zurich, 1936; m. Anna J. Kurz, Nov. 6, 1937; children—Carl H. E., Arthur R. Came to U. S., 1937, naturalized, 1943. Prof. vet. anatomy Middlesex U., Waltham, Mass., 1938-39; instr. zoology U. Detroit, 1939-43; asst. prof. comparative anatomy U. Notre Dame, 1943-45; curator fossil reptiles Field Mus. Natural History, Chgo., 1945—, chief curator of geology, 1962—; lectr. dept. geophys. scis. U. Chgo., 1948—. Mem. Soc. Vertebrate Paleontology, Geol. Soc. Am., Schweizerische Palaeontologische Gesellschaft, Am. Soc. Zoologists, A.A.A.S. Publs. on research leading to understanding of fossil reptiles, especially turtles; paleoecology of Pennsylvanian black shales; Pennsylvanian sharks. Home: 3100 Longfellow Dr., Hazelcrest, Ill. 60429. Office: Field Museum of Natural History, Chgo. 60605.*

ZANGHERI, Pietro, Italian botanist; b. Forli, Italy, July 23, 1889; s. Francesco and Geltrude (Mazzotti) Z.; ed. in natural sci. and botany; m. Maria Ragazzinni, Oct. 13, 1921; children—Vilfredo, Sergio, Miranda. Co-dir. Archivio Botanico e Biogeografico; dir. Zangheri Mus. Natural History, Rome. Recipient Gold medal for teaching culture and art. Order of Merit of Italian Republic. Mem. Botanica italiana, Italian Acad. Forestry Sci. Author: Romagna fitogeografica, 5 vols.; also numerous articles. Address: corso Diaz 182, Forli, Italy.

ZANOTTI, Eustachio, Italian mathematician, astronomer; b. Bologna, Italy, Nov. 27, 1709; ed. by Jesuits; student math. under Francesco Maria Zanotti. Prof. astronomy U. Bologna. Fellow Royal Soc., 1760; mem. Berlin, Kassel acads. Work with astron. ephemerides, perspectius, elastic force, light refraction; determined lunar parallax; observed meridian altitudes of Mars, moon during opposition of Mars, Bologna, 1751. Died May 15, 1782.

ZANOTTI, Francesco Maria, Italian mathematician, philosopher; b. Bologna, Italy, Jan. 6, 1692; ed. by Jesuits; student philosophy under Canon de St. Sauveur, algebra under Victor Stancoir. Prof. philosophy U. Bologna, from 1718, librarian, pres., from 1766. Fellow Royal Soc., 1740; mem. French Acad. Scis. (corr.), 1750. Author several treatises on physics, math., art. Wrote on vital forces, 1752, central forces, 1762, roots of cubics; advocated Descartes, later Newton. Died Bologna, Dec. 24, 1777.

ZANTEDESCHI, Francesco, Italian physicist; b. 1797. Priest, taught physics at Venice Lyceum and Padua. Author over 350 articles. Used magnet to produce electric currents in closed circuit, 1820-30; detected magnetic action on steel needles by ultraviolet light, 1838; studied repulsion of flames by strong magnetic field. Died Padua, Italy, Mar. 29, 1773.

ZAPFFE, Carl Andrew, Am. metallurgist; b. Brainerd, Minn., July 25, 1912; s. Carl and Ethel (Moberg) Z.; B.S., 1933, hon. D.Eng., Michigan Tech. U.; M.S., Lehigh U., 1934; Sc.D., Harvard U., 1939; post-grad. Johns Hopkins U.; m. Adelaid Camille Denise duPont, May 22, 1937; children—Denise (Mrs. Robert Digges), Carl Moberg, Jessie (Mrs. Richard Morast), Carlotta Karen, Barbara Ann, Augusta Camille, Isabel, Christina Ethel. Metallurgist, duPont Exptl. Sta., Wilmington, Del., 1934-36; research assoc., Batelle Mem. Inst., Columbus, O., 1938-40; research engr., Columbus, 1940-43; asst. tech. dir., Rustless Iron Steel Corp., Baltimore, Md., 1943-45; owner, C. A. Zapffe and Assoc., Baltimore, 1945-52; self-employed, Baltimore, 1952—. Mem., Am. Soc. Metals, A.A.A.S., Am. Chem. Soc., Am. Geophysical Union; Am. Inst. Mining Metal., Petroleum Engrs., Am. Phys. Soc., Am. Soc. Testing Materials, Am. Welding Soc., Electrochem. Soc. Nat., Assn. Corrosion Engrs., Brit. Iron and Steel Inst., Brit. Inst. Metals, others. Author: Stainless Steels, 1949; articles, movies. Inventor of fractography, fractocrystallography as research tools; wire bend, bar bend tests; planar-pressure theory, hydrogen embrittlement of metals; micellar theory for imperfect structure of solid state; submarine vulcanism theory for Pleistocene Ice Ages; discoverer of hydrogen as cause of defects in weld metals, castings and forgings, various coatings for metals, some archaeological findings. Home and office: 6410 Murray Hill Road, Baltimore, Md., 21212.*

ZARAFONETIS, Chris John Dimiter, Am. physician; b. Hillsboro, Tex., Jan. 6, 1914; s. James and Helen (Skouras) Z.; B.A., U. Mich., 1936, M.S., 1937, M.D., 1941; m. Sophia Levathes, Mar. 27, 1943; 1 son, John Christopher. Research fellow internal medicine U. Mich., 1946-47, faculty, 1947-50, prof. internal medicine, dir. Simpson Meml. Inst. Med. Sch., 1960—, coordinator med. edn. for nat. def., 1966—; faculty Temple U., Phila., 1950-60, prof. clin. and research medicine, 1957-60, chief hematology sect. univ. hosp., 1950-60. Recipient Sternberg medal U. Mich., 1941, Henry Russel award, 1950; U. S. Typhus Commn. medal U. S. Army, 1945; Order Ismail, Egypt, 1946. Diplomate Am. Bd. Internal Medicine. Fellow Internat. Soc. Hematology, A.C.P.; mem. Am. Micros. Soc. (editorial bd. Trans. 1956-65), A.M.A., Am. Soc. Tropical Medicine, Central Soc. for Clin. Research, Am. Fedn. for Clin. Research, Coll. Physicians Phila., Am. Trudeau Soc., N.Y. Acad. Scis., Am. Soc. Hematology, Am. Med. Writer's Assn., Am. Soc. Internal Medicine, Soc. Exptl. Biology and Medicine, Soc. Med. Cons. to Armed Forces, A.A.A.S., Internat. Soc. Internal Medicine, Aerospace Med. Assn., Internat. Coll. Angiology, Midwest Blood Club, Asociacion Medica Argentina, Sociedad de Farmacologic Y Terapeutica (hon.), Assn. Mil Surgeons U. S., Am. Therapeutic Soc., Internat. Soc. Toxinology, Am. Thoracic Soc., Am. Coll. Clin. Pharmacology and Chemotherapy (regent 1966—), Internat. Soc. Blood Transfusion, Mich. Soc. Internal Medicine (trustee), Drug Information Assn., Med. Mycol. Soc. Ams. Research and numerous publs. on histoplasmosis, infectious mononucleosis, lymphogranuloma and herpes viruses, rickettsial diseases, potassium para-aminobenzoate acid in collagen and bullous disorders and conditions with excess fibrosis, blood dyscrasias, lipid mobilizer hormone and dyscrasias. Home: 2721 Bedford Rd., Ann Arbor, Mich. 48104.*

ZARISKI, Oscar, mathematician; b. Kobrin, USSR, Apr. 24, 1899; s. Bezalel and Anna (Tannenbaum) Z.; came to U. S., 1927, naturalized, 1936; Dottore in Matematica, U. Rome (Italy), 1924; M.A. (hon.), Harvard 1947; D.Sc., Coll. Holy Cross, 1959, Brandeis U., 1965; m. Yole E. Cagli, Sept. 11, 1924; children—Raphael, Vera Letitia. Internat. Edn. Bd. fellow U. Rome, 1925-27; Johnston scholar Johns Hopkins, 1927-29, asso. math., 1929-32, faculty, 1932-45, prof., 1937-45; research prof. U. Ill., 1946-47; faculty Harvard, 1947—, prof. math., 1947-61, Dwight Parker Robinson prof., 1961—; vis. mem. Inst. for Advanced Study, Princeton, 1935-36, 39, 60-61, Hautes Etudes Scientifiques, Paris, France, 1961, 67; chmn. dept. math. Harvard, 1958-60. Guggenheim fellow, 1939-40; recipient Nat. medal of Science, 1965. Mem. Am. Math. Soc. (Cole prize 1944; president elect 1968), Mathematical Assn. Am., Société Math. de France, London Math. Soc. (hon.), Nat. Acad. Scis., Am. Philos. Soc., Am. Acad. Arts and Scis., Accademia Nazionale dei Lincei, Acad. Sci. Brazil, Acad. Sci. Peru. Author: Algebraic Surfaces, 1935; (with Pierre Samuel) Commutative Algebra, vol. I, 1958, vol. II, 1960; also numerous articles. Translator: Was sind und was sollen die Zahlen; Stetigkeit and irrazional Zahlen (R. Dedekind), 1926. Editor: Ill. Jour. Math., 1957-60. Research on founds. of algebraic geometry using methods modern algebra, theory of normal varieties, local uniformization and reduction of singularities of algebraic varieties, theory of birational transformations, abstract holomorphic functions, linear systems, minimal models. Home: 27 Lancaster St., Cambridge, Mass. 02140.*

ZARROW, M. X., Am. biologist; b. Worcester, Mass., Nov. 1, 1913; s. Max and Ida (Medlinsky) Z.; A.B., Clark U., 1934, A.M., 1936; postgrad. N.Y. U., 1939-41; Ph.D., Harvard, 1947; m. Irma Gue, Apr. 6, 1946; 1 son, Peter Gue. Asst. pharmacologists Hoffman-LaRoche, Nutley, N.J., 1942-46; research asst., instr. Yale Sch. Medicine, 1947-48; instr. Biol. Labs., Harvard, 1948-49; faculty Purdue U., Lafayette, Ind., 1949—, prof. zoology, 1958—; staff Worcester Fedn. for Exptl. Biology, 1955-56. Mem. endocrinology study sect. NIH, 1963-67. mem. subcom. on role of hormones in livestock prodn. div. biology and agr. Nat. Acad. Sci.-NRC, 1961—; cons. Eli Lilly & Co., Indpls., 1964—. Recipient Sigma Xi award Purdue chpt., 1955; NSF Sr. Postdoctoral fellow, 1962-63. Fellow A.A.A.S., N.Y. Acad. Scis.; mem. Am. Endocrine Soc., Endocrine Soc. (Eng.), Am. Soc. Zoologists, Am. Physiol. Soc., Soc. for Study Fertility, Soc. Gen. Physiologists, Soc. for Exptl. Biology and Medicine, Sigma Xi, Phi Beta Kappa (honorary member). Author: (with J. M. Yochim, J. L. McCarthy) Experimental Endocrinology, 1964; also numerous articles. Editor: Growth and Living Systems, 1961. Research on regulation ovulation, role hormones in maternal behavior, factors involved in determination of onset of puberty, factors involved in parturition and role adrenal cortex in early experience. Home: 1917 Indian Trail Dr., West Lafayette, Ind. 47906. Office: Dept. Biol. Scis., Purdue U., Lafayette, Ind. 47907.*

ZATZKIS, Henry, mathematician; b. Holzminden, Germany, Apr. 7, 1915; s. Markus and Lifscha (Eber) Z.; student U. Heidelberg, 1934-37; came to U. S., 1940, naturalized, 1946. B.S., Ohio State U., 1942; M.S., Ind. U., 1944; Ph.D., Syracuse U., 1950; m. Natalie Serlin, July 1, 1951; children—Mark, David. Instr., Ind. U., 1942-44, U. N.C., 1944-46, Syracuse U., 1946-51, U. Conn., 1951-53; faculty Newark Coll. Engring., 1953—, prof. math., 1959—, chmn. dept. math., 1959—. Mem. Am. Physics Tchrs. Assn., N.J. Assn. Math. Tchrs., Phi Beta Kappa, Sigma Xi. Author: (with others) Fundamental Formulas of Physics, 1955, International Dictionary of Applied Mathematics, 1960; co-author Handbook of the Engineering Sciences, vol. I, 1967; also articles. Research on quantum mechanics, relativity theory, heat conduction. Home: 5 Elliott Pl., West Orange, N.J. 07052.*

ZAUMEYER, William John, Am. plant pathologist; b. Milw., Dec. 10, 1903; s. Anthony J. and Margaret (Mueller) Z.; B.S., U. Wis., 1925, M.S., 1926, Ph.D., 1928; m. Ivy Rabbitt, Aug. 31, 1931; children—Margaret (Mrs. Paul Flood), Carol (Mrs. Bernard McCarthy). With U. S. Dept. Agr., 1928—, sr. pathologist crops research div. Agr. Research Service, Plant Industry Sta., Beltsville, Md., 1945-52, prin. plant pathologist, 1952—, investigations leader bean and pea investigations, 1945—. Cons., AID, Brazil, 1963, El Salvador, 1964—. Recipient U. S. Dept. Agr. Superior Service award, 1957, Gold medal award for release of Topcrop bean All Am. Selections, 1950. Mem. A.A.A.S., Am. Phytopath. Soc. (past pres.), Am. Inst. Biol. Scis., Bot. Soc. Washington. Research and numerous publs. on diseases of beans and peas; devel. numerous mosaic- and rust-resistant snap and dry beans; described several new virus diseases of beans and peas, numerous strains of bean rust fungus; research on action of antibiotics for control plant diseases; described inheritance of genetic resistance to various bean diseases. Home: 3804 Thornapple St., Chevy Chase, Md. 20015. Office: Crops Research Div., Agrl. Research Service, U. S. Dept. Agr., Plant Industry Sta., Beltsville, Md. 20705.*

ZAURIRSKI, Sigismond, Polish mathematician, sci. philosopher; b. Berezowica Mala, Poland, Sept. 29, 1882; s. Joseph and Stronska Z.; ed Lvov (formerly Poland), Berlin (Germany), Paris (France) univs.; Ph.D., 1905; m. Kamile Galotzy, 1908. Tchr. math., philos. propedeutics, grammar sch., from 1906; lectr. Poly. High Sch., from 1922; agrl. prof. Lvov, from 1924; prof. methodology of scis. U. Poznan (Poland), from 1929. Recipient Rignanto prize of Scientia, 1934. Mem. Sci. Soc. Poznan. Author: Causality and Functional Relation, 1912; Modal Propositions, 1914; Natural Science and Axiometical Method, 1924; Relation between Logic and Mathematics, 1927; Historical and Critical Enquiry of the Doctrine of Eternal Return, 1927; Indeterminism of the Physics of Quanta, 1931; The New Logics and their Application in Modern Science, 1932; The Evolution of the Concept of Time, 1934; The Relation between the Logic with Several Values to the Calculating of Probability. Office: Poznan Marsalkowska Foch 64, Poland.

ZAVALISHIN, Dmitrii Aleksandrovich, Russian electrotechnologist; b. 1900; grad. Leningrad Poly. Inst., 1925. Staff, Leningrad Poly. Inst. until 1939; chmn. dept. electric machines S. M. Buden Mil. Electrotech. Acad., 1939-41; prof. faculty spl. electrotech. Armed Forces Advanced Sch. Engring. and Tech., 1941-46; chmn. dept. elec. machines Leningrad Inst. Aero. Instrument Constrn., 1946-59; became chief lab. on sci. fundamentals of automatized elec. apparatus, USSR Acad. Scis. Inst. Electromechanics 1959. Named Honored Scientist and Technologists of R.S.F.S.R., 1957. Corr. mem. USSR Acad. Sci. Research and publs. in elec. machines, electron-ionic and semiconductor equipment. Office: Inst. Electromechanics, USSR Acad. Scis., Dvortsovaya Naberezhnaya 18, Leningrad, USSR.

ZAVOISKII, Evgenii Konstantinovich, Russian physicist; b. Sept. 28, 1907; grad. U. Kazan, 1930; D.Physico-Math. Scis., 1945. Faculty, U. Kazan, 1933—, prof., 1945—; asso. research insts. USSR Acad. Scis.; has worked on devel. of use of image convertors for scintillation chamber and investigation of means of comparatively short duration, 1947—. Recipient Lenin prize, 1957. Mem. USSR Acad. Scis. (corr.). Author: (with S. A. Al'tshuller, and B. M. Kozyrev) New Method of Investigating Paramagnetic Absorption, Paramagnetic Relaxation in Liquid Solutions with Perpendicular Fields; Paramagnetic Absorption in Solutions with Parallel Fields; Paramagnetic Abscription in Some Salts in Perpendicular Magnetic Fields, 1946; Spin Magnetic Resonance in the Decimeterwave Region, 1946; (with others) Scintillation Chamber, 1955. Discovered electronic paramagnetic resonance, 1944; (with S. A. Al'tschuller and B. M. Kozyrev) established series of connections between form of resonant lines. Address: Physics Dept., University of Kazan, Kazan, Tatar, ASSR, USSR.

ZAWADZKI, Wlodzimierz, Polish physicist; b. Warsaw, Poland, Jan. 4, 1939; s. Stanislaw and Zofia (Kwiatkowska) Z.; Master's Degree in physics, Warsaw U., 1961; Ph.D., Inst. Physics, Polish Acad. Scis., Warsaw, 1964. Research asst. Inst. Physics, Polish Acad. Scis., 1961-65, 67—; staff mem. Nat. Magnet Lab., Mass. Inst. Tech., Cambridge, 1965-67. Research, publs. on theory of transport, magnetic and thermodynamic properties of electrons in semiconductors (especially III-V intermetallic compounds) for which relation between energy and momentum is not quadratic; properties of electrons in solids in presence of crossed magnetic and electric fields; nonlinear optical properties of semiconductors. Office: 37 Zielna Str., Warsaw, Poland.*

ZAWISTOWSKI, Stanislaw, Polish histologist, physician; b. Wilno, Poland, Aug. 24, 1919; s. Leon and Maria (Biengo) Z.; grad. Med. Acad., Gdansk, Poland, 1954, M.D., 1959; m. Helena Hajdukiewicz, Mar. 15, 1952; 1 son, Wladyslaw. Staff, Inst. Histology and Embryology, Med. Acad. Gdansk, 1954—, dir., 1965—; lectr. histology Analytical Technik Med. Sch., Gdansk, 1951—. Mem. Polish Anat. Soc., Polish Histochem. and Cythochem. Soc., Polish Med. Soc., Polish Soc. Otolaryngology. Author: Technika mikroskopowa, 1956; Technika histologiczna, histologia oraz podstawy histopatologii, 1965; also articles. Research on Golgi complex and enzymes; histochemistry of kidney during hypoglycemia; histochemistry of nasal mucosa; cytological and histochem. study of parathyroids; histochem. investigation of cell cultures infected with variola virus, vaccinia virus and adenovirus. Home: 4 Zakopianska, Gdansk, Poland.*

ZAYED, Salah Abdel Dayem, Egyptian chemist; b. Shebin El-Kom, Egypt, Feb. 13, 1932; s. Mohamed and Zakia (Zayed) Z.; B.Sc., Cairo U., 1951, M.Sc., 1954, Ph.D., 1957; m. Soheir Mahmoud Amer, July 25, 1957. Demonstrator, Faculty Sci., Cairo (Egypt) U., 1951-57; fellow Max Planck Inst. for Biochemistry, Munich, Germany, 1957-61; research Nat. Research Center, Cairo, 1961-63, asso. prof., head chemistry of insecticides unit, 1963—. Recipient State prize in chemistry, 1966. Research, publs. on insect chemistry; fate, metabolism, detoxification mechanisms of insecticides in insects, plant, mammals, microorganisms; pesticides. Home: 48 Ansar, Cairo. Office: Nat. Research Center, Tahrir, Dokki, Cairo, Egypt.*

ZDRODOVSKII, Pavel Felikosovich, Russian microbiologist, immunologist; b. May 16, 1890; grad. Med. Faculty, Kazan (USSR) U. Asst. dept. microbiology Don U., USSR, head Cholera, Typhus and Vaccine Labs., Rostov (USSR) Bacteriological Inst., 1918-20; epidemiologist 11th Army in Caucasus, USSR, 1920-22; asst. dept. microbiology Azerbaidzhan (USSR) U., lectr., 1922-26, prof., chair microbiology, 1926-30, dir. Inst. Microbiology and Hygiene, Baku, 1922-30; head epidemiology sect. Leningrad (USSR) Inst. Exptl. Medicine, 1930-34; head epidemiology dept. All-Union Inst. Exptl. Medicine, Moscow, 1934-37; head epidemiology dept. All-Union State Pub. Health Inspectorate, Moscow, 1935-37; head dept. exptl. pathology and immunology of infections Inst. Epidemiology and Microbiology, USSR Acad. Med. Sci., 1945-47, founder Rickettsiosis dept., 1958. Founder Baku Inst. Microbiology and Hygiene, 1920-22; mem. preparatory com. 9th Internat. Microbiology Congress, Moscow, 1963, 66. Recipient Stalin prize, 1949, Lenin prize, 1959. Mem. USSR Acad. Med. Sci. Author: An Experimental Assessment of Abderhalden's Reaction, 1920; Malaria on the Mugan, 1926; Ancylostomiasis, 1929; Brucellosis Theory, 1933; The Modern Problem of Specific prophylaxix of Diphtheria and Diphtheria Vaccination Experience in Leningrad, 1933; Brucellosis Current Theory with Reference to Human Pathology, 1948; Rickettsia and Rickettsiosis, 1948; Problems of Reactivity in the Study of Infection and Immunity, 1950; The Physiological Principles of Immunity, 1950; Brucellosis, 1935; The Present State of Experimental Immunology and its Immediate Tasks, 1956; Rickettsia and Rickettsiosis Theory, 1956; Problems of Infectious Pathology and Immunology, 1958; Problems of Infection and Immunity, 1961; others. Made several expdns. studying infectious diseases in Central Asia, Caucasus, also regional pathology Azerbaidzhan, 1918-20; conducted campaign against malaria in Azerbaidzhan; pioneered diphtheria toxoid prodn. and mass immunization of children in USSR; dir. devel. prophylactic vaccination against brucellosis with live weakened vaccine, other preventive vaccinations; discovered phenomenon of non-specific allergy, hemorrhagic parallergy.

ZEA, Francisco Antonio, Colombian naturalist; b. Medellin, New Granada, (Colombia) 1770. Made sci. explorations with José Mutis; prof. natural scis., dir. Royal Bot. Garden, Madrid, 1805; minister of interior, gov. of Malaga under Joseph Bonaparte; went to S. Am., 1814, fought with Bolivar against Spainiards; prof. Angostura congress, 1819; v.p. Gran Colombia, 1819; Colombian minister to Eng., 1820. Author: Historia de Colombia, 1821, also numerous sci. treatises. Died Bath, Eng., 1822.

ZECH, Jakob, German clock maker; b. Germany, 16th century. Constructed 1st clock with balance wheel (made clock regular), 1525.

ZECHMEISTER, László Károly E., organic chemist; b. Györ, Hungary, May 14, 1889; s. Charles and Irene (Mocsáry) Z.; diploma chemist Eidg. Technische Hochschule, Zurich, Switzerland, 1910, Dr.-ing., 1913; m. K. Elizabeth Sulzer, Mar. 16, 1949. Prof. med. chemistry U. Pécs (Hungary), 1923-40; prof. organic chemistry Cal. Inst. Tech., Pasadena, 1940-59, prof. emeritus, 1959—. Recipient Pasteur medal, Paris, France, 1935, Claude Bernard medal, Paris, 1949, Labline award Am. Chem. Soc., Grand Prix, Hungarian Acad. Sci., 1937; Guggenheim fellow, 1949. Harvey lectr., 1951. Mem. Am. Chem. Soc., A.A.A.S., Danish Acad. Sci. (fgn. mem.). Author: (with L. Cholnoky) Principles and Practice of Chromatography, 1943; Progress in Chromatography, 1948; Cis-trans isomeric Carotenoids, Vitamins A and Arylpolyenes, 1962; also numerous articles. Editor, Progress in Chemistry Organic Natural Products, 1938—. Research on chemistry polysaccharides, enzymes, carotenoids, pigments, stereochemistry polyenes, fluorescent natural products, carcinogens, chem. conversions in ultrasonic field. Home: 370 S. Allen Av., Pasadena, Cal. 91106.*

ZECKWER, Isolde Therese, Am. pathologist; b. Phila., Oct. 10, 1892; d. Richard and Marie T. (d'-Invilliers) Z.; A.B., Bryn Mawr Coll., 1915; M.D., Woman's Med. Coll. Pa., 1919. Fellow path. anatomy Mayo Clinic, Rochester, Minn., 1921-22; staff research dept. pathology Harvard Med. Sch., 1922-26; pathologist L.I. Hosp., Boston, 1922-26; researcher physiology U. Coll., U. London (Eng.), 1926-27; faculty dept. pathology U. Pa. Sch. Medicine, Phila., 1927-58, prof. pathology, 1954-58, emeritus prof., 1958—. Diplomate Am. Bd. Pathology. Mem. Am. Soc. for Expt. Pathology, Am. Assn. Pathologists and Bacteriologists, Am. Physiol. Soc., Am. Assn. for Cancer Research. Research and publs. on functional and structural changes in endocrine organs under physiol. and path. conditions in animals, exptl. tumors in animals. Address: 2100 Walnut St., Phila. 19103.*

ZEEMAN, Pieter, Dutch physicist; b. Zonnemaire, Zeeland, Netherlands, May 25, 1865; student U. Leyden, from 1885, dr.'s degree, 1893. Faculty, U. Leyden, 1890-1900; prof. at Amsterdam, 1900; dir. Phys. Inst. Amsterdam, 1908. Recipient Baumgartner prize, Vienna, Wilde prize, Paris, Nobel prize in physics (with Lorentz), 1902. Fellow Royal Soc., 1921 (Rumford medal); mem. French Acad. Scis. Author: Messungen über das kerrsche magnetooptische Phänomen, 1893; Experimentaluntersuchungen über Teile, die kleiner als Atome sind, 1900; Magnetooptische Untersuchungen, 1914; Verhandelingen . . . over magnetooptische versuchijnselen, 1921. Discovered Zeeman effect while at Leyden, 1896 (when ray of light from source placed in magnetic field is examined spectroscopically spectral line is widened or occasionally doubled); worked on propagation of light in vibrating media, made observations in water, quartz and flint; determined Fresnel coefficient for various colors, 1914-16. Died Amsterdam, Oct. 9, 1943.

ZEIDMAN, Irving, Am. physician; b. Camden, N.J., Mar. 17, 1918; s. Neil and Sylvia (Lichtenstein) Z.; A.B., U. Pa., 1937, M.D., 1941; m. Elinor Sleeper, July 2, 1953; children—Thomas and Robert (twins). Faculty, U. Pa. Med. Sch., Phila., 1946—, prof., 1963—. Mem. Am. Soc. Pathologists and Bacteriologists, Assn. Cancer Research. Research and publs. on cancer spread in blood and lymph stream. Home: 449 Larchwood Rd., Springfield, Pa. 19064. Office: University of Pa. Med. Sch. Phila. 19104.*

ZEIGER, Herbert Jack, Am. physicist; b. Bronx, N.Y., Mar. 16, 1925; s. Isidore and Goldie (Orgel) Z.; B.S. Coll. City N.Y., 1944; M.A., Columbia U., 1947, Ph.D., 1951; m. Hanna Bloom, June 13, 1954; children—Joel, Susan, Judith. Postdoctoral fellow Columbia U., 1950-51; staff mem. Lincoln Lab., Mass. Inst. Tech., Lexington, 1952—. Mem. Am. Phys. Soc. Research and publs. on microwave masers, semiconducting lasers, stimulated Raman scattering processes, cyclotron resonance in semiconductors, gen. solid state band structure. Home: 167 Pond Brook Rd., Chestnut Hill, Mass. 02167. Office: Lincoln Lab., Lexington, Mass. 02173.*

ZEISEL, Hans, law sociologist; b. Kaaden, Czechoslovakia, Dec. 1, 1905; s. Otto and Elsa (Frank) Z.; Dr.Juris, U. Vienna (Austria), 1927, Dr.Pol.Sc., 1928; m. Eva Striker, July 30, 1938; children—Jean, John. Came to U. S., 1938, naturalized, 1944. Research analyst Market Research Co. Am., 1939-41; instr. statistics Rutgers U., 1942-43; dir. research devel. McCann-Erickson, 1943-50; expert cons. War Dept., also lectr. New Sch. Social Research and Grad. Sch., Columbia U., 1945-52; dir. research Tea Council, U. S. A., 1951-52; prof. law and sociology U. Chgo. Law Sch., 1953—. Chmn. bd. Marplan, 1964—. Mem. Am. Statis. Assn., Am. Assn. for Pub. Opinion Research, Am. Market Assn. Author: (with M. Jahoda, P. F. Lazarsfeld) Marienthal, 1932; Say it with Figures, 1947; (with H. Kalven, Jr., B. Buchholz) Delay in the Court, 1959; (with H. Kalven, Jr.) The American Jury, 1966; also numerous articles. Introduced large scale social sci. research of legal instns. Home: 1155 E. 57th St., Chgo. 60637.*

ZELDITCH, Morris, Jr., Am. sociologist; b. Pitts., Feb. 29, 1928; s. Morris and Anne (Hankin) Z.; B.A., Oberlin Coll., 1951; Ph.D., Harvard, 1955; m. Bernice Osmola, June 12, 1950; children—Miriam Lea, Steven. Instr., Columbia U., 1955-57, asst. prof., 1957-61, dir. program in social scis. in medicine, dept. psychiatry, Coll. Phys. and Surg., 1958-61; asso. prof., Stanford (Cal.), 1960—, exec. head dept. sociology, 1964—. Served with AUS, 1946-47. Author: Sociological Statistics, 1959; (with Berger, Cohen, Snell) Types of Formalization in Small Group Research, 1962. Asso. editor: Am. Sociol. Rev., 1964—. Contbr. articles to tech. jours.*

ZELDOVICH, Yakov Borisovich, Russian physicist; b. Mar. 18, 1914; grad. Leningrad U., 1931. Joined staff Inst. Chem. Physics, USSR Acad. Scis., 1931. Recipient Stalin prize, 1943. Mem. USSR Acad. Sci. Author: Theory of Burning and Detonation of Gases, 1944; Theory of Shock Waves and Introduction to Gas Dynamics, 1946; (with P.Y. Sadovnikov, D. A. Frank-Kamenetskii) Oxidation of Nitrogen during Combustion, 1947; (with A. S. Kompaneets) Theory of Detonation, 1955; also articles. Determined (with others) mechanism of nitrogen oxidation in an explosion, 1935-39; calculated (with Y. B. Khariton) chain reaction in uranium fission, 1939-40; proposed device for chem. reaction in shock wave; developed (with D. A. Frank-Kamenetskii) theory of flame propagation. Office: Inst. Chem. Physics, USSR Acad. Scis., Vorob'evskoye Shosse 2, Moscow, USSR.

ZELINSKI, Nikolai Dimitrievich, Russian chemist; b. Tiraspol, Moldavia, Feb. 6, 1860. Head dept. organic chemistry Moscow (USSR) State U.; prof. Moscow Soc. Nature Experimentalists; dir. organic anaysis dept. Inst. Organic Chemistry USSR. Recipient Stalin prizes, 1942, 46; named scientist of merit, 1926. Mem. USSR Acad. Scis., chem. socs. London (hon.), France, USSR Soc. for Sci. Knowledge (hon.). Research, publs. in theoretical and applied sci.; identified with study of petroleum, devel. oil industry; studies on chemistry of albumin; inventor 1st carbon gas mask; established sch. for systematic research on organic catalysis; works on naphtha, protein, active carbon, also synthetic rubber. Died 1953.

ZELINSKY, Daniel, Am. mathematician; b. Chgo., Nov. 22, 1922; s. Isaac and Ann (Ruttenberg) Z.; S.B., U. Chgo., 1941, S.M., 1943, Ph.D. in Math, 1946; m. Zelda Oser, Sept. 23, 1945; children—Mara, Paul, David. Instr. premeteorology U. Chgo., 1943-44; mem. applied math. group Columbia, 1944-45; instr. math. U. Chgo., 1946-47; NRC fellow Inst. Advanced Study, 1947-49; Guggenheim fellow, 1956-57; mem. faculty Northwestern U., 1949—, prof. math., 1960—; Fulbright fellow, Kyoto, Japan, 1955-56. Mem. div. math. NRC, 1962-65, exec. com., 1963-65. Mem. Am. Math. Soc. (editor trans. 1961-66). Research in homological algebra and rings. Home: 613 Hunter Rd., Wilmette, Ill. 60091. Office: Dept. Mathematics, Northwestern Univ., Evanston, Ill. 60201.

ZELLER, Alfred, Austrian chemist; b. Vienna, Austria, May 13, 1908; s. Ignaz and Stephanie (Gamper) Z.; Ph.D., U. Vienna; m. Erika Rausch, Aug. 7, 1938; children—Gertrud, Monika. Asst., Plant Physiol. Inst., U. Vienna, 1938-41, Exptl. Horticulture Inst. Pollnitz, 1938-41; staff Plant Physiol. Inst., U. Vienna, 1945-46; dir. Fed. Alpine Agronomy Center, Admont, Austria, 1947-51; staff Chesterford Park Research Sta., nr. Cambridge, 1951-57; dir. Fed. Exptl. Agronomy Chem. Center, Vienna, 1958—. Mem. Österreichische Biochemische Gesellschaft, Österreichischer Chemiker Verein, Verband Deutscher Landwirtschaft Untersuchungs und Forschungsanstalten, Chemphysikalische Gesellschaft Vienna. Home: Konradgasse 3, Vienna 2, Austria. Office: Trunnerstrasse 1, Vienna 27, Austria.

ZELLER, Edward Jacob, Am. geochemist; b. Peoria, Ill., Nov. 6, 1925; s. John George and Mabel (Singer) Z.; A.B., U. Ill., 1946; M.A., U. Kan., 1948; Ph.D., U. Wis., 1951; m. Anke M. Neumann, July 16, 1965. Project asso. U. Wis., 1951-55; faculty U. Kan., Lawrence, 1956—, prof. geology, 1962—; guest geologist physics dept. Brookhaven Nat. Lab., Upton, L.I., N.Y., 1965-66. Cons. to Advanced Research Project Adminstrn. Contractor, 1964—. Sr. Postdoctoral fellow NSF, Psykalisches Inst., U. Bern (Switzerland), 1961. Mem. Geol. Soc. Am., Geochem. Soc., Soc. Econ. Paleontologists and Mineralogists, Am. Assn. Petroleum Geologists, Sigma Xi. Author: (with L. B. Ronca) Geologic Dating, 1963; also articles; contbg. author: Nuclear Geology, 1952; Earth Science and Meteoritics, 1963. Research in field radiation damage in natural crystals, chem. reactions produced by high energy protons, thermoluminescence used to determine geologic age and for measurement of paleoclimate in Antarctica and N. polar region. Office: Dept. Geology, U. Kan., Lawrence, Kan.*

ZELLWEGER, Hans Ulrich, pediatrician; b. Lugano, Switzerland, June 19, 1909; s. Leopold and Helena (Abys) Z.; student Med. Sch. Zurich (Switzerland), 1928-34, Med. Sch. Rome (Italy), 1933, Med. Sch. Berlin (Germany), 1931-32, Med. Sch. Hamburg (Germany), 1932; M.D., U. Zurich, 1934; m. Margaret Olgiati, July 16, 1940; children—Andres, Judy. Came to U. S., 1957, naturalized, 1964. Coworker with Albert Schweitzer, Lambaréné, Republic of Gabon, 1937-39; asst. prof. pediatrics U. Zurich, 1944-51; chmn. dept., prof. pediatrics Am. U., Beirut, Lebanon, 1951-59; prof. pediatrics U. Ia., Iowa City, 1959—. Mem. A.M.A., Am. Pediatric Soc., Am. Acad. Cerebral Palsy, Am. Acad. Pediatrics, Am. Soc. Human Genetics, N.Y. Acad. Scis. Author: (with Fanconi, Botstejn) Die Poliomyelitis and ihre Grenzgebiete, 1945; Krampfe in Kindesalter, 1948; (with Adolph) Vitamine und Vitaminkrankheiten, 1954; Downs Syndrome, 1965; also numerous articles. Research on poliomyelitis, neuropediatrics, genetics, cytogenetics. Home: 630 Park Rd., Iowa City 52240.*

ZEMAN, Wolfgang, physician; b. Stuttgart, Germany, Apr. 14, 1921; s. Hans Gustav Franz and Irma (Schmidt) Z.; M.D., U. Tübingen (Germany), 1945; postgrad. in neuropathology Deutsche Forschungsanstalt für Psychiatrie, Max Planck-Inst., 1947-49, Neuropsychiat. Inst., U. Mich., 1951-52; m. Ruth Elizabeth Taenzer, Jan. 16, 1952; children—Pamela C. Cornelia A., Michael W. Came to U. S., 1951, naturalized, 1959. Instr., U. Hamburg (Germany), 1950-51, U. Mich., 1951-52, U. Ark., 1953-54; asst. prof. U. Pitts., 1955-56; faculty Ohio State U., 1956-60, asso. prof., 1958-60; faculty Ind. U. Med. Center, 1960-65, prof. pathology, 1965—; acting chief neuropathology br. Armed Forces Inst. Pathology, Washington, 1957; guest radiobiologist dept. biology Brookhaven Nat. Lab., Upton, N.Y., 1958—. Author: (with J. R. M. Innes), Craigie's Neuroanatomy of the Rat, 1963; also numerous articles. Research on pathogenesis of radiation necrosis in nervous tissue; genetic delineation of certain diseases of brain. Home: 1100 W. Michigan St., Indpls. 46202.*

ZEMANSKY, Mark Waldo, Am. physicist; b. N.Y.C., May 5, 1900; s. Abraham Philip and Rebecah (Cohen) Z.; B.S., City Coll. N.Y., 1921; A.M., Columbia U., 1922, Ph.D., 1927; m. Adele Cohen, Sept. 25, 1932; children—Philip L., Herbert D. From instr. to prof. physics, City Coll. N.Y., 1921-67; lectr. New School, N.Y., 1947-57. Served with U. S. Army, 1917-18. Recipient Townsend Harris medal, City Coll. N.Y. Alumni, 1960. Mem. Am. Phys. Soc. (pres. met. sect., 1948), Am. Assn. Physics Tchrs. (pres. 1951, awarded Oersted medal, 1957); exec. sec. 1967—), Am. Inst. Physics (treas. 1956), Phi Beta Kappa, Sigma Xi, Zeta Beta Tau. Author: (with A. Mitchell)

Resonance Radiation and Excited Atoms, 1934; Heat and Thermodynamics, 1937; (with F. W. Sears) College Physics, 1947, University Physics, 1949; Temperatures Very Low and Very High, 1964. Home: 736 Rutland Av., Teaneck, N.J. 07666.*

ZEMPLÉNYI, Tibor Karol, Czechoslovakian physician; b. Part. Lupca, Czechoslovakia, July 16, 1916; s. David and Irene (Pollak) Z.; M.D., Charles U., Prague, 1946; C.Sc., Czechoslovak Acad. Sci., 1960; D.Sc., 1964; m. Hana Bendová, Aug. 13, 1952; 1 son, Jan. Mem. staff, clin. asst. Dept. Medicine Prague Motol Clinic, 1946-52; head atherosclerosis research Inst. for Cardiovascular Research, Prague, 1952——; asst. prof. medicine Med. Faculty, Charles U., Prague. Recipient Czechoslovak Cardiol. Soc. award, 1959-60. Mem. Czechoslovak Cardiol. Soc., Czechoslovak Physiol. Soc., Czechoslovak Biochem. Soc., Italian Soc. Atherosclerosis (hon.), European Atherosclerosis Group, Czechoslovak Acad. Sci. Author: Enzyme Biochemistry of the Arterial Wall as Related to Atherosclerosis, 1968. Research, numerous publs. in electrophysiology, clin. cardiology, especially electrocardiography; unipolar leads and vectors; lipid metabolism and particularly arterial enzymes, indicating basic role of vascular metabolism in pathogenesis of atherosclerosis. Office: 809, Budejovická, Prague 4, Czechoslovakia.

ZENISEK, Ladislav, Czechoslovakian physician; b. Prague, Czechoslovakia, Oct. 17, 1918; s. Jaroslav and Anna (Hermanova) Z.; student Charles U., Prague, 1937-39, M.D., 1946, Cand.Scient., 1961; m. Renata Faltinova, July 8, 1946; children—Jan., Eva. Asst., Inst. Normal Anatomy, 1946; asst. dept. obstetrics and gynecology Sch. Medicine, Hradec Králové, Czechoslovakia, 1946-52; asst. dept. gynecology and obstetrics Sch. Medicine, Olomouc, Czechoslovakia, 1956-66, asst. prof., 1966——. Mem. Soc. Med. Jan. Ev. Purkyne. Author: (with Marsálek) Roentgen Diagnosis in Gynecology, 1961; also articles. Discovered female genital tuberculosis can be healed by antituberculotics, that mucous of cervix uteri has antimicrobial properties; diagnosis of all path. states of uterine tube using X-ray. Home: 363 Svermova Hradec Králové, Czechoslovakia. Office: Dept. Obstet. Gynecology, U. Olomouc, Czechoslovakia.*

ZENKER, Friedrich Albert von, see von Zenker, Friedrich Albert.

ZENKEVICH, Lev Alexandrovich, Russian oceanographer; b. June 17, 1889; grad. Law Faculty, Moscow U., 1912, Physico-Math. Faculty, 1916. Became asso. Inst. Oceanography, USSR Acad. Scis., 1947; faculty Moscow U., 1916——, prof., 1930——; staff State Oceanographic Inst., 1921-30; head Vityaz expdn., 1949-52; mem. various expdns. for studies of No. seas, Caspain, Far Eastern seas, Pacific. Became mem. adv. com. on marine sci. UNESCO, 1955. Corr. mem. USSR Acad. Scis. (chmn. oceanographic com.). Recipient Albert I medal French Inst. Oceanography, 1959. Mem. Internat. Council Sci. Unions (v.p. spl. com. on oceanographic research). Author: Fauna and Biological Productivity of the Sea, 2 vols., 1947-51; The Seas of the USSR, their Fauna and Flora, 1956; Life of the Ocean Depths, 1959; coauthor: Geography of Animals, 1946. Introduced quantitative system for marine fauna; formulated laws of distbn. for organisms in No. seas; studies in fauna of Russian seas, evolution of motility in invertebrates. Home: Lomonosovskii Prospekt, 14, Moscow, USSR. Office: Chmn., Oceanographic Com., USSR Acad. Sci., Leninskii Prospekt, 14, Moscow, USSR.*

ZENODODOROS, mathematician; flourished 2d century B.C. Wrote Treatise on Isoperimetry (14 of his propositions on isoperimetry preserved by Pappos and Theon of Alexandria); wrote on isoperimetric plane surfaces, showed that of all solid figures of which surfaces are equal the sphere is greatest in volume.

ZENO OF CITIUM, Greek philosopher; b. circa 335 B.C.; s. Mnaseas of Citium; attended lectures of Polemon, Diodorus, Crates of Thebes, Athens, Greece, from 313; taught in the Stoa Athens. Author: On Life According to Nature; On Human Nature; On Emotions; On Duty; On Law; On Vision; On the Whole World; On Signs; On Varieties of Style; Ethics; Pythagorean Questions; Rhetoric Universals; The State; The Republic. Founder of Stoic sch.; created complete philos. system consisting of logic, theory of knowledge, physics, ethics; taught that virtue is necessarily good, vice necessarily evil, and that most things in life are morally indifferent; also taught that wisdom is prime virtue, absolute law governs nature, man's essential nature is reason. Died Athens, circa 263 B.C.

ZENO OF ELEA, Greek mathematician, philosopher; b. Elea, Italy, circa 495, B.C.; s. Teleutagoras; pupil of Parmenides. Accompanied Parmenides to Athens, circa 450 B.C., propounded principles of Eleatic sch.; returned to Elea, later died in attempt to oust city's tyrant; few fragments of his writings remain. In argument regarded as inventor of dialectic; used paradoxes to illustrate his philos. arguments and until devel. of limits in calculus and Cantor's concepts of infinity they could not be adequately explained; attacked especially reality of motion and analyzed geometric continuity; his paradoxes on time, space and number contbd. to increase of logical and math. rigor; 4 best-known paradoxes are Achilles and Tortoise, Flying Arrow, Stadium, and Row of Solids. Died Elea, circa 430 B.C.

ZENO OF SIDON, Greek mathematician; b. Sidon, Phoenicia, circa 150 B.C. Directed Epicurean sch., Athens, after 100 B.C. Held that happiness lies in enjoyment of present pleasure and assurance that such enjoyment will last throughout life or most of life; critized deficiency of assumptions of Euclid's 5 postulates. Died Athens, after 73 B.C.

ZENTMYER, George Aubrey, Jr., Am. plant pathologist; b. North Platte, Neb., Aug. 9, 1913; s. George Aubrey and Mary (Strahorn) Z.; A.B., U. Cal. at Los Angeles, 1935; M.S., U. Cal. at Berkeley, 1936, Ph.D., 1938; m. Dorothy Anne Dudley, May 24, 1941; children—Elizabeth, Jane, Susan. Asst. forest pathologist U. S. Dept. Agr., San Francisco, 1937-40; asst. pathologist Conn. Agr. Expt. Sta., New Haven, 1940-44; staff U. Cal. at Riverside, 1944——, plant pathologist, 1955——, prof. plant pathology, 1963——. Mem. NRC; mem. coms. Nat. Acad. Sci.-NRC; cons. Trust Ter. Pacific Islands, 1964, 66, U. S. Dept. Agr., 1955, Pineapple Research Inst., 1961, Hawaiian Sugar Planters Assn., 1961. Guggenheim fellow, 1964-65; recipient award of honor Cal. Avocado Soc., 1954. Fellow A.A.A.S. (exec. com. Pacific div. 1964——); mem. Am. Phytopath. Soc. (pres. 1965-66), Mycol. Soc., Am. Inst. Biol. Scis., Philippine, Indian phytopath. socs., Assn. Tropical Biology, Sigma Xi. Contbg. author: Advances in Pest Control Research, Vol. 1, 1957; Annual Review of Phytopathology, Vol. 1, 1963. Editorial com. Ann. Rev. Phytopathology, 1967——. Research and numerous publs. on chemotherapy for control of vascular diseases of plants with devel. new chems., techniques; discovered chelation as mechanism of fungicidal action; devel. information on cause, life cycle, and control of diseases of avocado, macadamia, cacao, coffee, chemotaxis of zoospores in genus of plant pathogens, root diseases. Home: 3892 Chapman Pl., Riverside, Cal. 92506.*

ZEPPELIN, Count Ferdinand von, see von Zeppelin, Count Ferdinand.

ZERNIKE, Frits, Dutch physicist; b. Amsterdam, July 16, 1888; s. C. F. A. and Anne (Dieperink) Z.; Sc.D., U. Amsterdam, M.D., 1952; Sc.D., U. Poitiers, 1955; m. Dora van Bommel van Vloten, 1929; 2 children. Asst. astron. lab. U. Groningen, 1913, lectr. theoretical physics, 1915, prof. theoretical and tech. physics and theoretical mechanics, 1920-58, ret.; vis. prof. Johns Hopkins, 1948. Decorated Officer Legion of Honor (France), 1954; recipient gold medal U. Groningen, 1908, Dutch Soc. Scis., Haarlem, 1912; Rumford medal Brit. Royal Soc., 1952; Nobel prize for physics, 1953. Mem. Royal Netherlands Acad. Scis.; hon. mem. Royal Micros. Soc. of London Optical Soc. Am.; fgn. mem. Royal Soc. Contbr. articles learned jours., Holland, Germany, U. S., Great Britain. Developed phase-contrast microscope, making possible exam. internal structure living tissue without stain. Died Naarden, Netherlands, Mar. 10, 1966.

ZERNOV, Dmitrii Vladimirovich, Russian electronic expert; b. Mar. 20, 1907; grad. Moscow U., 1930. Staff, All-Union Electro-Tech. Inst., 1930-34; faculty Moscow Inst. Transport Engrs., 1932-38; staff Sci. Research Inst. Cinematography and Photography, 1936-39; joined Inst. Automation and Telemechanics, USSR Acad. Scis., 1939, Inst. Radiotech. and Electronics, 1953. Recipient Order of Red Banner of Labor. Corr. mem. USSR Acad. Scis. Author: The Mechanism of Electrical Sparkover in Solid Dielectrics, 1950; Vacuum Contact Tubes, 1960. Research and publs. on electron-beam devices of commutator type; 1st to build sodium vapor fluorescent lamps in Russia; developed TV system. Office: Inst. Radio Engring. and Electronics, USSR Acad. Scis., Mokhovaya Ulitsa 11, K-9, Moscow, USSR.

ZERRAS, Leon Grotius, Am. physician, biochemist; b. nr. Frankfort, Ind., Mar. 7, 1897; s. William Henry and Bertha Elizabeth (Maish) Z.; B.S., Ind. U., 1920, M.D., 1922; Ph.D., Cambridge (Eng.) U., 1939; postgrad. Karolinska Inst., Stockholm, Sweden, 1939, U. Chgo. Law Sch., 1940-41; m. Helen L. Lesh, June 15, 1922; children—Cherler Perry, Allen. Practice medicine specializing in clin. research, Indpls., 1926-36; dir. Lilly Lab. for clin. Research, Indpls., 1926-36; faculty Ind. U. Med. Sch., 1926-36, asso. prof., 1931-36; vis. physician, physician-in-chief Indpls. City Hosp., 1926-36, dir. med. teaching, 1930-36, dir. meningococcic meningitis epidemic, 1929-33. Fellow Ind. Acad. Sci., A.C.P. (life); mem. Central Soc. Clin. Research (charter), Am. Soc. Clin. Investigation (emeritus). Author: Oxidation-Reduction Mechanisms in Enzyme Systems; also numerous articles. Path. evaluation of yeast-like fungi, pernicious anemia, evaluation of barbituric acid derivatives; isolation and purification of lactic acid enzyme of yeast, oxidation-reduction mechanisms in enzyme systems; improved cell for measurement of oxidation-reduction potential. Home and office: Box 96, Rural Route 1, Camby, Ind. 46113.*

ZEUNER, Gustav, German physicist, engr.; b. Chemnitz, Saxony, Nov. 30, 1828. Prof. mechanics Poly. Sch., Zurich; dir. Mining Acad., Freiberg, 1871-75, Poly. Sch., Dresden, 1873-97; editor-in-chief L'Ingénieur civil. Mem. French Acad. Sci., 1901. Author several books and publs. including: Théorie mecanique de la chaleur, 1860; Etudes sur la statistique mathématique, 1869; Thermodynamique technique, 1905. Discovered polar diagram of steam engine, wrote treatise on valve gear; did work on hydraulics, thermodynamics, statis. math. Died Dresden, Oct. 17, 1907.

ZEUTHEN, Erik, Danish cell biologist; b. Hareskovby, Denmark, Nov. 15, 1914; s. Otto Ludvig and Ida Mathilde (Brondsted) Z.; Magister scientarum, U. Copenhagen, 1939, Dr.phil., 1947; m. Elisabeth Engberg, Aug. 5, 1945; children—Jesper, Morten, Elisabeth. Research asso. comparative physiology and cell biology Carlsberg Lab., 1941-46; research asso. U. Copenhagen, 1946-56 lectr., 1956-57, vis. lectr., 1964——; dir. Biol. Inst., Carlsberg Found., 1957——, prof., 1961——. Vis. prof. U. Cal. at Berkeley, 1949, research asso. 1956. Recipient Thunberg medal Kungl. Fysiografiska Sällskapet, Lund, Sweden. Mem. Royal Danish Acad. Sci. and Letters. Editor: Synchrony in Cell Division and Growth, 1964. Research, numerous publs. in comparative and cellular physiology, cell cycle in artificially synchronized mass systems of cells; micromanometric methods (Cartesian divers) for measurements of respiration and enzymic activities of single cells and for measuring submerged weight of single live cells. Home: 71, Vangeleddet, Virum 2830, Denmark. Office: 16, Tagensvej, Copenhagen N 2200, Denmark.*

ZHAVORONKOV, Nikolai Mikhailovich, Russian chem. technologist; b. Aug. 7, 1907; grad. Moscow Chem.-Technol. Inst., 1930. Faculty, Moscow Chem. Technol. Inst., 1930-48, prof., 1942-48; prof. Karpov Physicochem. Inst., 1944——; dir. Mendeleev Chem. Technol. Inst., Moscow, 1948——; mem. bur. dept. chem. sci. USSR Acad. Scis., 1960——. Recipient D. I. Mendeleev prize, 1950. Mem. USSR Acad Scis. Author: (with V. A. Malyusov) Molecular Distillation, 1950; Nitrogen in Nature and Engineering, 1951; Chemical Research in the Soviet Union, 1956. Research and publs. on prodn. and purification of hydrogen, nitrogen-hydrogen mixture from carbon dioxide and carbon monoxide, separation of liquid and gas mixtures; isolated isotopes of hydrogen, carbon, nitrogen, oxygen, boron and other light elements; studies (with others) on theory of concentration processes for stable isotopes. Home: N. Basmannaya 16, Moscow, USSR. Office: Inst. Gen. and Inorganic Chemistry, USSR Acad. Scis., Leninskii Prospekt 31, Moscow, USSR.

ZHDANOV, Yuri Andreevich, Russian chemist; b. Kalinin, USSR, Aug. 20, 1919; s. Andrew and Zinaida (Condratyeva) Z.; Candidate Philosoph.Sci., Moscow (USSR) State U., 1948, D. Chem.Sci., 1960; m. Thaissia Pogosskaya, June 30, 1958; children—Catherine, Andrew. Asst. chemistry dept. Moscow State U., 1946-53; faculty Rostov-on-Don (USSR) State U., 1953——, prof., 1960——, rector, 1957——. Decorated Order of Lenin, Order of Red Star, Badge of Honor. Mem. Mendeleev's Chem. Soc. USSR, Biochem. Soc. USSR. Author: Homology in Organic Chemistry, 1950; (with G. Dorofeenko) Chemical Transformation of Carbon Chain of Carbohydrates, 1962; Methodology of Organic Chemistry, 1960; (with V. Minkin) Correlation Analysis in Organic Chemistry, 1966; also numerous articles. Research on chemistry of natural organic compounds, electronic properties of azomethyns and their stereoisomerism; application of information theory to organic reactions mechanisms; devel. new methods of synthesis of C-glycosides, O-glycosides, new chem. mutagenes. Home: Souvorov Str., 1, Rostov-on-Don, USSR.*

ZHURKOV, Serafim Nikolaevich, Russian physicist; b. May 16, 1905; grad. Voronezh Inst., 1929. Became prof. Leningrad Phys.-Tech. Inst., 1930. Corr. mem. USSR Acad. Scis. Author: Molecular Mechanism of Polymer Hardening, 1945; The Relation of Strength to Time in Solids, 1953. Research and publs. on physics of solids and polymers, strength of brittle materials, mech. disintegration in solids, vitrification in amorphous substances and polymers; developed theory of polymer plastification. Office: Leningrad Phys.-Tech. Inst., Leningrad, USSR.

ZIEGLER, Hubert, German botanist; b. Regensburg, Sept. 28, 1924; s. Max and Laura (Sohler) Z.; Dr.rer.nat., U. Munich, 1950; m. Irmgard Günder, July 31, 1955; 1 son, Lothar. Asst. U. Munich, 1950-57, docent, 1957-59; asso. prof. Darmstadt Inst. Tech., 1959-62, dir. Bot. Inst. and Garden, 1959——, prof., 1962——. Mem. German, Bavarian, Swiss bot. socs., Zool.-Bot. Soc. Vienna. Editor Planta, Fortschritte der Botanik, 1967——. Contbr. articles in field. Research in translocation in plants, tree physiology, enzyme regulation in photosynthesis. Address: 3 Breslau-Place, Darmstadt, West Germany.*

ZIEGLER, Karl, German chemist; b. Helsa nr. Kassel, Germany, Nov. 26, 1898; s. Karl and Luise (Rall) Ziegler; student Marburg U.; m. Maria Kurtz, 1922. Lectr. Marburg U., 1923; prof. Heidelberg U., 1927; prof. chemistry, dir. Chem. Inst. U. Halle, 1936, dir. Max Planck Inst. Coal Research, Mülheim-Ruhr, 1943——; guest prof. U. Chgo., 1936;

Carl Folkers lectr., Madison, Wis., also Urbana, Ill., 1952. Recipient Liebig medal German Chemists' Soc., 1935, Carl Duisberg plaque, 1953; Lavoisier medal French Chem. Soc.; (with G. Natta) Nobel prize for chemistry. 1963; Swinburne medal Plastics Inst., London, 1964; also numerous other awards for discoveries in chemistry of carbon compounds, devel. of plastics. Mem. German Chemists' Soc., other sci acads. Research on free radicals; many membered ring systems; organometallic compounds. Home: 14 Lembkerstr. Office: 1, Kaiser-Wilhelm-Platz, 433 Mülheim-Ruhr, Germany.*

ZIELEN, Albin John, Am. chemist; b. Chgo., Dec. 22, 1925; s. Albin Paul and Ellen (Longridge) Z.; A.B., Miami U., Oxford, O., 1950; Ph.D. in Chemistry, U. Cal. at Berkeley, 1953. Chemist, Radiation Lab., U. Cal. at Berkeley, 1950-53; asso. chemist Argonne (Ill.) Nat. Lab., 1953——. Mem. Am. Chem. Soc., Phi Beta Kappa, Sigma Xi. Research and publs. on thermodynamics and kinetics of aqueous chemistry of neptunium, thorium, uranium and zirconium, precision electromotive force measurements with glass electrode. Home: 18 W. 008 Standish Lane, Villa Park, Ill. 60181. Office: 9700 S. Cass Ave., Argonne, Ill. 60439.*

ZIEMSSEN, Hugo Wilhelm von, see von Ziemssen, Hugo Wilhelm.

ZIERSKI, Marian, Polish physician; b. Lvov, Poland, May 1, 1906; s. Filip and Eugenia (Szaniawska) Z.; M.D., U. Prague, Lowow, 1933; d.med.sci., U. Lwów. 1938; m. Lina Unger, Oct. 2, 1933; 1 son, Jan. Tomasz. Asst. dept. Tb., Chest Clinic, Lwów, 1933-41; dir. Central Municipal Tb. Dispensary, Lodz, 1946-50, chief, supt., 1947-60; lectr. Med. Sch., Lodz, 1949-53; head, chmn. dept. Tb. and respiratory diseases Postgrad. Med. Sch., Lódz, 1954——; adviser cons. Ministry of Health for Respiratory Diseases. Decorated Gold Cross of Merit, Cross of Polonia Restituta, Medal of Honor, City of Lódz, Gold medal of Polish Students Assn. Mem. Nat. Tb. Inst. (mem. research council 1952——, chmn., 1965——), Polish Tb. Assn. (past v.p., mem. council 1961——), Internat. Union a Tb., Brit. Tb. Assn., Am. Thoracic Soc. Author: Tuberculosis and Pregnancy, 1952; Epidemiology of Tuberculosis, 1958; The Pneumonias, 1964; Modern Chemotherapy of Tuberculosis; also numerous articles. Research on method of chemotherapy in Tb, treatment of cases with resitant tubercle bacilli, epidemetric indices in Tb. Home: 30 Buczka, Zódz, Poland.*

ZIEVE, Leslie, Am. physician; b. Mpls., Aug. 6, 1915; s. Joseph and Hannah (Bulgats) Z.; B.A., U. Minn., 1937, M.A., 1939, M.D., 1943, Ph.D., 1952; m. Bernice Larson, June 26, 1941; 1 son, Franklin. Chief radioisotope service Mpls. Veterans Hosp., 1950——, asso. chief of staff, 1960——, dir. Cancer Lab., 1961——; prof. medicine U. Minn., 1962——, mem. exec. com. Grad. Sch., 1965——. Recipient VA Middleton award. Mem. A.C.P., Central Soc. for Clin. Research. Contbr. numerous articles to sci., med. jours. Editor Jour. Lab. and Clin. Medicine. Developed or analyzed methods for clinical evaluation of liver and of pancreatic function; described several clinical syndromes that focused attention on clinical abnormalities previously unrecognized; studied enzyme acting on phospholipids and applied technique in study of diseases. Address: VA Hosp., 13N, Mpls. 55417.*

ZIFF, Morris, Am. physician, chemist; b. N.Y.C., Nov. 19, 1913; s. Benjamin and Ethel (Seldowitz) Z.; B.S., N.Y. U., 1934, Ph.D. in Chemistry, 1937; M.D., N.Y. U., 1948; m. Ruth Rawson, Oct. 20, 1940; children—Edward, David. Research asst. Columbia U., 1939-41, vis. scholar, 1941-44; instr. Coll. City N.Y., 1941-44; faculty N.Y. U., N.Y.C., 1945-48, 50-58, asso. prof. medicine, 1957-58; prof. internal medicine U. Tex. Southwestern Med. Sch., Dallas, 1958——. Cons., USPHS, 1954-59, mem. arthritis tng. grants com., 1959-63. Recipient Heberden medal Heberden Soc. London, 1964, Research Career award USPHS, 1962——. Mem. Assn. Am. Physicians, Am. Soc. Clin. Investigation, Am. Assn. Immunologists, A.C.P., Am. Rheumatism Assn., Harvey Soc., Phi Beta Kappa, Sigma Xi, Alpha Omega Alpha. Research and numerous publs. on rheumatoid factor, study of immunology and biochemistry as related to rheumatoid arthritis and similar disease conditions. Home: 11116 Pinocchio Dr., Dallas 75229.*

ZILBER, Lew Alexander, Russian virologist; b. Novgorod, Russia, Mar. 28, 1894; s. Alexander Abram and Anne (Desson) Z.; ed. Physico-Math. Faculty, U. Petrograd, 1915; M.D., Ph.D., Moscow U.; m. Valeria Peter Kisseleva, Nov. 5, 1931; children—Lew, Pheodor. Mil. physician, 1919-21; asst. Microbiol. Inst., Moscow, 1921-30; privatreader Moscow U., 1923-29; prof. Advanced Tng. Inst. Doctors, 1900-39; head dept. virology and immunology of cancer Gamaleja Inst., Moscow, 1939——. Mem. Acad. Med. Sci. USSR; hon. life mem. N.Y. Acad. Scis.; mem. Microbiol. Soc. USSR, Oncological Soc. USSR Am., French assns. cancer research. Decorated Order Red Banner, Order Lenin. Author: (with G. Abelev) Virology and Immunology of Cancer, Research on hereditary serological transformation of proteus; nature of antibodies and complement; symbiosis of bacteria and viruses; discovery of Far-East tick encephalitis virus; virogenic theory cancer origin; discovery of oncogenicity of Rous virus for mammals; study human gastric cancer antigen. Home 2819 Givopisnaja, Moscow D-182. Office: 2 Gamaleya, Moscow D-182, USSR.*

ZILBOORG, Gregory, psychiatrist; b. Kiev, Russia, Dec. 25, 1890; s. Moses and Anne (Braun) Z.; grad. Realschule, Kiev, Russia, 1911; M.D., Psychoneurological Inst., St. Petersburg, Russia, 1917; M.D., Coll. Phys. and Surg., Columbia, 1926; m. Ray Leibow, Dec. 14, 1919 (div. 1946); children—Gregory, Nancy; m. 2d, Margaret Stone, August 19, 1946; children—Caroline Crawford, John Talcott, Matthew Stone (stepson). Came to U. S., 1919, naturalized, 1925. Physician in Russian Army, 1915-16; participated in first revolution in Petrograd, 1917; sec. to ministry of Labor in Cabinets of Lvov and Kerensky, 1917; editor daily paper, Kiev, until Germans occupied So. Russia, in 1918; forced to leave Russia and came to U. S., 1919; engaged in lecturing, journalism, and theater, 1919-22: Grad. Coll. Phys. and Surgeons, N.Y.C., 1926; staff Bloomingdale Hosp., 1926-31; asst. Psychoanalytic Inst., Berlin, 1929-30; pvt. practice in psychiatry and psychoanalysis, N.Y.C., 1931——; sec., dir. of research, com. for the Study of Suicide; Noguchi lectr. history of medicine, Johns Hopkins U., 1935; asso. in psychiatry, Catholic U. of Am., 1944-46; Gimbel lecturer, U. Cal., 1947; cons. in research and psychotherapy, Butler Hosp., Providence; asst. prof. clin. psychiatry, N.Y. Med. Coll., Flower and Fifth Av. Hosp., N.Y. Chmn. Consulting Delegation on Criminology to United Nations; asso. vis. psychiatrist, Kings Co. Hosp., N.Y.; clin. asso. prof. psychiatry college of medicine State U. of N.Y. Mem. A.A.A.S., Am. Assn. of History of Medicine, Am. Bd. of Psychiatry and Neurology (diplomate), A.M.A., Am. Orthopsychiatric Assn., Am. Psychiatric Assn., Am. Sociological Soc., American Psychoanalytic Assn., American Soc. for Research in Psychosomatic Problems, Assn. for Research in Nervous and Mental Diseases, Research Council on Problems of Alcohol, History Sci. Soc., Internat. Psychoanalytic Soc., Medical Correctional Assn., Med. Soc. County of N.Y., N.Y. Acad. Medicine, N.Y. Med. Soc., N.Y. Neurol. Soc., N.Y. Psychoanalytic Soc. (sec. 1933), N.Y. Soc. Clin. Psychiatry, N.Y. Soc. for Med. History (pres. 1944-45), Pan-Am. Med. Assn.; corresponding mem. Argentine Soc. of Neurology and Psychiatry Argentine Society of the History of Medicine, Argentine Psychoanalytic Society, Brazilian Institute of the History of Medicine. Author: The Medical Man and the Witch During the Renaissance, 1935; A History of Medical Psychology (co-author with George Henry), 1941; Mind, Medicine and Man, 1943; Sigmund Freud, 1951; The Psychology of the Criminal Act and Punishment, 1954; Freud and Religion, 1959. Translator: He, the One Who Gets Slapped (by Andreiev), 1921; We (by Zamiatin), 1924; The Criminal, the Judge and the Public (by Alexander and Staub, from the German), 1931; Outline of Clinical Psychoanalysis (by Fenichel), 1934. Asso. editor: One Hundred Years of American Psychiatry, 1944. Contbr. numerous articles to med. jours. and drama mags. Research on mental disorders associated with childbirth; legal aspects of psychiatry; relationship between psychiatry and religion. Died Sept. 17, 1959.

ZILLIKEN, Frederick William, German biochemist; b. Bonn, Germany, Oct. 28, 1920; s. Frits Willi and Christina (Huck) Z.; B.S., U. Bonn, 1939, Ph.D., 1943; M.S., U. Praha (Czechoslovakia), 1941; M.D., U. Heidelberg (Germany), 1948; m. Anneliese Heesen, Aug. 17, 1945; children—Claudia, Ingeborg, Rosemary. Faculty, U. Bonn., 1945-47; staff Max Planck Inst. for Med. Research, Heidelberg, 1947-50; faculty U. Pa., Phila., 1950-60, asst. prof. biochemistry, 1955-60; prof. biochemistry R.K. U., Nijmegen, Netherlands, 1960-65; prof. Philipps U., Marburg, Germany, 1965——. Mem. Fedn. Am. Biol. Chemists, Brit. Biochem. Soc., Chem. Soc. Holland. Research and numerous publs. on chemistry and biochemistry of human milk, amino sugars, neuraminic acid, cell differentiation, blood-coagulation and cell contact. Home: 2 Walter Voss, Marburg 355, Germany.*

ZILSEL, Paul Rudolph, physicist; b. Vienna, Austria, May 6, 1923; s. Edgar and Ella (Breuer) Z.; B.S., Coll. Charleston, 1943; M.A., U. Wis., 1945; Ph.D., Yale, 1948; m. Dorice Tentchoff, Aug. 26, 1947 (div.); children—Carrie, Joanna. Came to U. S., 1939, naturalized, 1945. Postdoctoral fellow Duke, 1948-49; faculty Colo. State U., 1949-50, U. Conn., 1950-54, Israel Inst. Tech., 1954-56, McMaster U., 1956-58; faculty dept. physics Case Western Res. U., Cleve., 1958——, prof., 1963——. NSF vis. fellow Princeton, 1964-65. Mem. Am. Phys. Soc., Am. Assn. Physics Tchrs., Am. Assn. U. Profs., Sigma Xi. Research on theory of quantum liquids, superfluid helium. Home: 2300 Overlook Rd., Cleveland Heights, O. 44106. Office: Dept. Physics, Case Western Res. University, Cleve. 44106.*

ZILVERSMIT, Donald Berthold, biochemist; b. Hengelo, Holland, July 11, 1919; s. Herman and Elizabeth (DeWinter) Z.; student U. Utrecht, Holland, 1936-39; B.S., U. Cal. at Berkeley, 1940, Ph.D., 1948; m. Kitty Fonteyn, June 28, 1945; children—Elizabeth Ann, Dorothy Suzan, Sarah Jo. Faculty, U. Tenn. Med. Coll., 1948-66, prof., 1956-66; prof., Grad. Sch. Nutrition, Cornell U., Ithaca, N.Y., 1966——; vis. prof. dept. phys. chemistry U. Leiden, Holland, 1961-62; vis. fellow Australian Nat. U., Canberra, Australia, 1966; career investigator Am. Heart Assn., 1959——. Mem. adv. bd. Circulation Research, 1962-67, Circulation, 1960-65; cons. NIH. Mem. A.A.A.S., Am. Assn. U. Profs., Am. Chem. Soc., Am. Physiol. Soc., Council for Study of Arteriosclerosis, Soc. Explt. Biology and Medicine, Reticulo Endothelial Soc. Editor, Jour. of Lipid Research, 1958-61. Research and numerous publs. on fat transport in blood; mechanisms of fat absorption; use of radioactive isotopes in biochem. research; abnormalities in fat metabolism in atherosclerosis. Home: 1030 Hanshaw Rd., Ithaca, N.Y. 14850.*

ZIMEL, Herman Iosef, Rumanian physician; b. Craiova, Rumania, May 9, 1919; s. Iosef H. and Sabina (Meltzer) Z.; M.D., Inst. for Human Medicine, Bucharest, Rumania, 1950, Dr. in Med. Scis., 1964; m. Stavrica Rodica, Dec. 29, 1949; 1 son, Dan Gabriel. Chief endocrinological dept. Endocrinological Inst. C. I. Parhon, Bucharest; asst. prof. Med. Inst., Bucharest, 1950-57. Fellow Rumanian Endocrinological Soc., Rumanian Oncological Soc. Author: (with O. Costakel) Clinical and Experimental Oncological Researcher, 1958; also numerous articles. Research on hormonocytostatic compounds synthesized by linking hormones and alkylating derivatives (indicated in hormonal dependent tumor mgmt.); role of hypothalamus in devel. exptl. tumors and their reactivity to cytostatics using stereotaxic method of stimulation or destruction. Home: 3 Baba Dochia. Office: 34 Bd. Aviatorilor, Bucharest, Rumania.*

ZIMEN, Karl Erik, Swedish-German chemist; b. Berlin, Feb. 5, 1912; s. Even Olof and Ella (Otto) Z.; Dr.phil., U. Berlin, 1937; m. Eva Haberlandt, Aug. 17, 1938; children—Erik, Ralf, Monica. Head, Inst. Nuclear Chemistry, Chalmers U. Tech., Göteborg (Sweden), 1946-56; prof. Berlin Inst. Tech., 1957——; dir., div. nuclear chemistry, Hahn-Meitner Inst. Nuclear Research, Berlin, 1957——. Mem. German AEC., German Reactor Safeguards Commn. Mem. Soc. German Chemists, Svenska Kemistamfundet. Author: Angewandte Radioaktivität, 1952. Research, publs. on radionuclides, especially rare gases, in solids (nuclear fuels, geo- and cosmochem. material. Home: 19 Hittorfstrasse, Berlin 33. Office: 100 Glienicker Strasse, Berlin 39, West Germany.*

ZIMM, Bruno H(asbrouck), Am. physical chemist; b. Woodstock, N.Y., Oct. 31, 1920; s. Bruno L. and Louise S. (Hasbrouck) Z.; grad. Kent (Conn.) Sch., 1938; A.B., Columbia, 1941, M.S., 1943, Ph.D., 1944; m. Georgianna S. Grevatt, June 17, 1944; children—Louis H., Carl B. Research asso. Columbia, 1944; research asso., instr. Polytech. Inst. Bklyn., 1944-46; instr. chemistry U. Cal. at Berkeley, 1946-47, asst. prof., 1947-50, asso. prof., 1950-51; vis. lectr. Harvard, 1950-51; research asso. research lab., Gen. Electric Co., 1951-60; professor chemistry U. Cal., La Jolla, Cal., 1960——. Recipient Bingham Medal, Soc. Rheology, 1960; also the High Polymer Physics prize from American Physical Society, 1963. Mem. Biophysical Society, Am. Chem. Soc. (Baekeland award 1957), Nat. Acad. Scis., Am. Phys. Soc., Fedn. Am. Scientists (exec. com. 1956). Asso. editor Jour. Chem. Physics, 1947-49; adv. bd. Jour. Polymer Sci., 1953-62, Jour. Bio-Rheology, 1962——, Jour. Biopolymers, 1963——, Jour. Phys. Chemistry, 1963——. Investigation of the structure and properties of high polymers and biological macromolecules; theromodynamics of solutions. Home: 2605 Ellentown Rd., La Jolla, Cal. 92037.

ZIMMER, Basil George, Am. sociologist; b. Smith Creek, Mich., June 29, 1920; s. Walter Nicholas and Mary Alice (Martin) Z.; A.A., Port Huron Jr. Coll., 1941; student Shrivenham Am. U., Eng., 1946; B.A., U. Mich., 1947, M.A., 1949, Ph.D., 1954; m. Janet M. Jackson, Nov. 3, 1942; children—Basil George II, Linda Jean. Teaching fellow U. Mich., 1948-50, research asst. Mich. Dept. Corrections, 1948; fellow Human Resources Research Inst., U. S. Dept. Air Force, 1952-53; instr. Eastern Mich. U., 1950-51, U. Mich., 1951-52; asst. prof. Fla. State U., 1952-53; lectr. sociology, resident dir. Social Sci. Research Project, U. Mich., 1953-59, asso. prof. Flint Coll., 1956-59; asso. prof. Brown U., 1959-61, prof., 1962——. Cons. met. area problems, Flint, Mich., 1956-59, Demographic and Tng. Center, U. Kerala, India, 1964; cons. Fertility Research, Med. Research Council, U. Aberdeen, Scotland, 1966. Fellow Am. Sociol. Assn.; mem. Population Assn. Am., Eastern Sociol. Soc., Internat. Union for Sci. Study Population. Author: Rebuilding Cities: The Effects of Displacement and Relocation on Small Businesses, 1964; (with A. H. Hawley) Resistance to Reorganization of School Districts in Metropolitan Areas, 1966. Research, publs. on urban problems facing 20th Century America; consequences of suburbanization movement of population; the effects of displacement and relocation on small businesses; examination and appraisal of consequences of federally sponsored public improvement programs; examined factor affecting fertility in U. S., Scotland and India; studied adjustment of migrants in urban areas; studies of socio-econ. status of U. S. population. Home: 10 Bridgham Farm Rd., Rumford, R.I. 02916. Office: 100 Maxcy Hall, Brown U., Providence 02912.*

ZIMMER, Hans Willi, chemist; b. Berlin, Germany, Feb. 5, 1921; s. Wilhelm J. F. and Martha (Schindler) Z.; B.S., Tech. U., Berlin, 1946, Dipl.Ing., 1949,

Dr.Ing., 1950; m. Marlies Wuensch, Oct. 26, 1946; 1 son, Hans Martin. Came to U. S., 1953. Instr., Technische U. Berlin, 1950-53; research asso. U. Ill., Urbana, 1953-54; faculty U. Cin., 1954—, prof. chemistry, 1961—; vis. prof. U. Mainz (Germany), 1966-67; vis. prof. University Bonn (Germany), 1967. Cons., Carlisle Chem. Works, Cowles Chem. Co., Cin., Cleve., 1960-65; Sigma Xi Distinguished Research prof. U. Cin., 1965. Mem. Am. Chem. Soc., German Chem. Soc., Sigma Xi. Contbg. author: Landolt-Boernstein Handbook, 1964. Research and numerous publs. on organic and organometallic synthesis including chemistry of butyro-lactons, chemistry Group IV and V elements. Home: 2910 Scioto St., Cin. 45219.*

ZIMMER, Herbert, psychologist; b. Fuerth, Bavaria, Germany, July 13, 1924; s. Sigmund and Paula (Kohn) Z.; came to U. S., 1940, naturalized, 1943; A.B. summa cum laude, Western Res. U., 1948; Ph.D., U. Rochester, 1952. Research psychologist Air Research and Devel. Command, Maxwell AFB, Ala., 1952-55; faculty Georgetown U. Med. Center, 1955-61, U. Ga., 1961-66; prof., dir. bioelectronic research lab. City U. N.Y., 1966—. Cons. various govtl. brs. and corps.; lectr. Mem. Psychonomic Soc., Southeastern Psychol. Assn., I.E.E.E., Phi Beta Kappa, Sigma Xi. Author: Computers in Psychophysiology, 1966; Psychophysiologic Components of Human Behaviour, 1966; (with others) Manipulation of Human Behaviour, 1961. Research, publs. on conditioning of automatic nervous system, neural timing mechanisms, feedback control of stimuli. Home: 2880 S. Moreland Blvd., Cleve. 44120. Office: 695 Park Av., N.Y.C. 10021.*

ZIMMER, Karl Guenter, German biophysicist; b. Breslau, Germany (now Wroclaw, Poland), July 12, 1911; s. Arthur K. Elsa (Geipel) Z.; Ph.D., M.A., U. Berlin (Germany), 1934; m. Elisabeth Cron, July 12, 1940. Research asso. genetics dept. Kaiser Wilhelm Institut, Berlin, 1934-45; head lab. Ministry Medium Machines, Moskva, USSR, 1945-55; asst. prof. Royal U. Stockholm, Sweden, 1955-56; prof. radiobiology U. Heidelberg (Germany), 1957—; dir. Inst. Radiobiology, Nuclear Research Center, Karlsruhe, Germany, 1957—. Mem. Assn. for Radiation Research, European Soc. for Radiobiology. Author several books; also numerous articles. Research on action of nuclear radiations on living material especially hereditary material. Home: 4 Ortelsburger, Karlsruhe. Office: P.O. Box 3640, Karlsruhe, Germany 75.*

ZIMMERMAN, Carle Clark, Am. sociologist; b. nr. Pleasant Hill, Mo., Apr. 10, 1897; s. Charles Payne and Lucinda (Payton) Z.; student Westminster Coll., 1914-16; A.B., U. Mo., 1920; S.M., N.C. State Coll., 1923, Sc.D., 1942; postgrad. U. Chgo., 1923; Ph.D., U. Minn., 1925; M.A., Harvard, 1940; m. Madeleine Andrist, June 16, 1925; children—Constance, Charlotte, Carle Clark. Asso. prof. sociology U. Minn., 1925-31; asso. prof. sociology, chmn. tutors in sociology Harvard, 1932-63; Distinguished prof. sociology N.D. State U., Fargo, 1964-67; Distinguished prof. sociology U. Sask., 1968—; vis. research prof. U. Rome (Italy), 1952; vis. prof. U. Mexico, 1957, Peruvian univs. 1960-61; vis. Fulbright prof. U. Istanbul (Turkey), also U. Ankara (Turkey), 1963-64. Mem. numerous commns. in U. S. and Can. Fellow A.A.A.S., Am. Sociol. Assn., Am. Acad. Arts and Scis.; mem. Institut Internat. de Sociologie (v.p. 1963—), Rural Sociol. Soc. (pres. 1958—). Author: (with P. A. Sorokin) Principles of Rural Sociology, 1929, Systematic Source Book Rural Sociology, 3 vols., 1933; Siam, Rural Survey, 1934; Standards of Living, 1935; Changing Community, 1939; Family and Civilization; 1947; Successful American Families, 1960; (with Seth Russell) Symposium Great Plains, 1967; also numerous articles, revs. Research on time in sociology and social change. Home: RFD 2, Box 180, Laconia, N.H. 03246. Office: Dept. Sociology, U. Sask., Saskatoon, Sask., Can.*

ZIMMERMAN, Elwood Curtin, Am. entomologist; b. Spokane, Wash., Dec. 8, 1912; s. Ernest W. and Ethel (Lingle) Z.; B.S., U. Cal. at Berkeley, 1936; student U. Hawaii, 1935; Ph.D., Diploma of Imperial Coll., U. London (Eng.), 1956; m. Hannah Louise Bond, Oct. 11, 1941. With Bishop Mus., Honolulu, 1934—, curator entomology, 1937-45, research asso., 1946-61, entomologist, 1961—; lectr. U. Hawaii, Honolulu, 1936-37, entomologist, 1958-61; asso. entomologist expt. sta. Hawaiian Sugar Planters' Assn., 1946-54; hon. asso. Brit. Mus., 1951—. Recipient Friends of Library Lit. award, 1948. Mem. A.A.A.S., Am., Pacific Coast, Cambridge, Hawaiian entomol. socs., Entomol. Soc. Am., Entomol. Soc. Washington, Coleopterists' Soc., Royal Entomol. Soc. London, Linnean Soc., Zool. Soc. London, Ray Soc., Honolulu Acad., Hawaiian Bot. Soc., Systematics Soc., Soc. Systematic Zoology, Soc. for Study Evolution, Sigma Xi. Author: Insects of Hawaii, vols. 1-8, 1948-58; also numerous articles. Exploration of Polynesian islands with studies of their insects; geog. distbn. of life in Oceania; taxonomic studies of Curculonidae (Weevils) of Pacific Islands; insects of sugarcane and their biol. control. Address: Bishop Mus., Honolulu 96819.*

ZIMMERMAN, Harry Martin, pathologist; b. nr. Vilna, Russia, Sept. 28, 1901; s. Jacob and Anna (Kaplan) Z.; B.S., Sheffield Sci. Sch., Yale, 1924, M.D., 1927; postgrad. Deutsche Forschungsanstalt für Psychiatrie, Munich, Germany, 1929; L.H.D., Yeshiva, U., 1957; m. Miriam Gordon, Sept. 2, 1930. Came to U. S., 1909, naturalized, 1924. Charles Linnaeus Ives fellow Yale Sch. Medicine, 1927-29, faculty 1930-43, asso. prof. pathology, 1933-43; prof. Coll. Phys. and Surg., Columbia U., 1947-64; prof. pathology Albert Einstein Coll. Medicine, N.Y.C., 1964—; chief lab. div. Montefiore Hosp. and Med. Center, N.Y.C., 1946—. Cons. to hosps., govt. agys. Recipient Golden Hope Chest award Nat. Multiple Sclerosis Soc., 1965. Mem. Am. Neurol. Assn., Am. Assn. Neuropathologists, Acad. Neurology, Am. Assn. Pathologists and Bacteriologists, Am. Soc. Exptl. Pathologists, Am. Soc. Clin. Pathologists, Harvey Soc., N.Y. Acad. Medicine, N.Y. Acad. Scis., N.Y. Path. Soc., A.A.A.S., A.M.A. Author: (with Martin G. Netsky, Leo M. Davidoff), Atlas of Tumors of the Nervous System, 1956; (with L. M. Davidoff, H. G. Jacobson) Neuroradiology Workshop, Vol. I, 1961, Vol. II, 1963, Vol. III, 1968; also numerous articles. Research on demyelinating diseases of nervous system, vitamin deficiencies, viruses in exptl. brain tumors, electron microscopy of brain. Home: 200 Cabrini Blvd., N.Y.C. 10033. Office: 111 E. 210th St., Bronx, N.Y. 10467.*

ZIMMERMAN, Howard Elliott, Am. chemist; b. N.Y.C., July 5, 1926; s. Charles and May (Cohen) Z.; B.S., Yale, 1950, Ph.D., 1953; m. Jane Kirschenheiter, June 3, 1950; children—Robert, Steven, James. NRC fellow Harvard, 1953-54; faculty Northwestern U., 1954-60, asst. prof., 1955-60; asso. prof. U. Wis., Madison, 1960-61, prof. chemistry, 1961—. Mem. Am. Chem. Soc., Chem. Soc. (London), German Chem. Soc., Phi Beta Kappa, Sigma Xi. Editorial bd. Jour. Organic Chemistry, Molecular Photochemistry. Research and publs. on phys. organic chemistry. Home: 1 Oconto Ct., Madison, Wis. 53705.*

ZIMMERMAN, Hyman Joseph, Am. physician; b. Rochester, N.Y., July 19, 1914; s. Philip and Rachel (Marine) Z.; A.B., U. Rochester, 1936; M.A. Stanford U., 1938, M.D., 1942; m. Kathrin Jewel Jones, Feb. 28, 1943; children—Philip M., David J., Robert L., Diane E. Instr. medicine George Washington U. Sch. Medicine, 1948-51; asst. chief med. service VA Hosp., Washington, 1948-51; asst. prof. medicine U. Neb. Coll. Medicine, 1951-53; chief med. service VA Hosp., Omaha, 1951-53; asso. prof. medicine U. Ill. Coll. Medicine, 1953-57; chief med. service VA Hosp., Chgo., 1953-57; prof., chmn. dept. medicine Chgo. Med. Sch., 1957-65; chmn. dept. medicine Mt. Sinai Hosp., Chgo., 1957-65; prof. medicine George Washington U. Sch. Medicine, 1965—; chief Liver and Metabolism Research Lab. VA Hosp., Washington, 1965—; cons. Chgo. Cook County Hosp., 1957-65. Mem. Am. Soc. Clin. Investigation, Central Soc. Clin. Research, Am. Assn. Study Liver Disease, Soc. Exptl. Biology and Medicine, Endocrine Soc., A.A.A.S., Endocrine Soc., Am. Diabetes Assn., Am. Soc. Pharm. and Exptl. Therapeutics, A.M.A., A.C.P., Am. Fedn. Clin. Research, Sigma Xi, Alpha Omega Alpha. Research, publs. clin. and exptl. liver disease; mechanism hepatic toxicity; clin. enzymology. Home: 7913 Charleston Ct., Bethesda, Md. 20034. Office: 50 Irving St. N.W., Washington 20422.*

ZIMMERMAN, Leonard Norman, Am. microbiologist; b. N.Y.C., Sept. 13, 1923; s. Harry and Mae (Coplin) Z.; B.S., Cornell U., 1948, Ph.D., 1951; m. Rima Grossman, Nov. 26, 1946; children—Erik E., Raul L., Leda E. With Pa. State U., 1951—, prof. bacteriology, 1961—. Mem. Am. Soc. Microbiology, A.A.A.S., Sigma Xi. Research and publs. on mechanism by which streptococci regulate synthesis of their enzymes. Home: 306 W. Ridge Av., State College, Pa. 16801.*

ZIMMERMAN, Lorenz Eugene, Am. pathologist; b. Washington, Nov. 15, 1920; s. Lorenz and Louise M. (Desgalier) Z.; A.B., George Washington U., 1943, M.D., 1945; m. Anastasia P. Urbaniak, Nov. 28, 1959; children—Spencer E., Patricia A., Brian L., Mary Louise, Barbara Anne. Commd. 1t M.C., U. S. Army, 1946, advanced through grades to lt. col., 1951; comdg. officer 8217 U. S. Mobile Med. Lab., Korea, 1950-51; staff Armed Forces Inst. Pathology, Washington, 1952—, chief ophthalmic pathology br., 1954—; asso. in pathology George Washington U. Sch. Medicine, Washington, 1952-63, prof. ophthalmic pathology, 1963—; lectr. ophthalmology Johns Hopkins Sch. Medicine, Balt., 1958—. Cons. pathology Am. Acad. Ophthalmology and Otolaryngology, 1958—. Decorated Legion of Merit; recipient Dept. Army Decoration for meritorious civilian service, 1963; commended by VA., 1958, Dept. Army, 1965. Author: (with Littman) Cryptococcosis, 1956; (with Hogan) Ophthalmic Pathology, 1962; Tumors of Eye and Adnexa, 1962; also numerous articles. Research in pathologic anatomy with prin. interest in pathology of eye. Home: 10016 E. Bexhill Dr., Kensington, Md. 20795. Office: Armed Forces Inst., Pathology, Washington 20305.*

ZIMMERMAN, Stanley W(illiam), Am. elec. engr.; b. Detroit, July 30, 1907; s. William Richard and Martha (Gebhardt) Z.; B.S. and M.S. in Elec. Engring., U. Mich., 1930; m. Evelyn G. Raney, Oct. 1, 1932; children—Dorothy (Mrs. Bynack), JoAnne (Mrs. Busch), William S., Richard L. Started as journeyman electrician, City of Detroit, 1926-30; asst., dept. physics, Wayne U., 1926-28; asst., dept. engring. research, U. Mich., 1928-30; asst., Detroit Edison research dept., 1929-30, Gen. Elec. Test., Schenectady and Pittsfield, Mass., 1930-32, Pittsfield works lab., Gen. Elec., 1932-34, High Voltage Lab. at Pittsfield, 1934, researcher devel. and design, Lightning Arrester Dept., Pittsfield, 1935-45; in charge high voltage research lab., Cornell U., 1945—, prof. elec. engring. 1945—, on leave industry engring. dept. Westinghouse, E. Pitts., 1952, electronic countermeasures reliability spl. studies Ramo Wooldridge, Los Angeles, 1959-60, Lawrence Radiation Labs., U. Cal. at Livermore, 1961, Argonne Nat. Lab., U. Chgo., 1962, high voltage lab. Nat. Bur. Standards, Washington, 1963, 64; cons. circuit interruption devices, transformer and substa. design, high voltage measurements, elec. insulation. Sec. conf. insulation NRC; sec. conf. elec. insulation Nat. Acad. Scis. Registered profl. engr., N.Y., Mass. Fellow I.E.E.E.; mem. Am. Soc. Engring. Edn., Internat. Conf. on Large Elec. Systems, N.Y. State Soc. Profl. Engrs. (v.p. Ithaca), Eta Kappa Nu. Contbr. various confidential and pub. reports and discussions. Patentee thyrite station arrestor component. Home: 102 Valley Rd., Ithaca, N.Y. 14850.*

ZIMMERMANN, Bernard, Am. surgeon; b. St. Paul, June 26, 1921; s. Harry Bernard and Mary Robertson (Prince) Z.; M.D., Harvard, 1945; Ph.D., U. Minn., 1953; m. Elizabeth Caldwell, June 18, 1949; children—Bernard III, Andrew Caldwell. Faculty, U. Minn., 1953-60, prof. surgery, 1959-60, cancer coordinator, 1953-60; prof., chmn. dept. surgery W.Va. U. Med. Center, 1960—. Mem. program project rev. com. Nat. Inst. Metabolism and Arthritis, 1964—. Diplomate Am. Bd. Surgery, Bd. Thoracic Surgery. Mem. A.C.S., Am. Soc. Exptl. Pathology, Am., Central surg. assns., A.A.A.S., Am. Cancer Soc. (pres. W.Va. dis. 1964—), Am. Trudeau Soc., Endocrine Soc., Soc. Exptl. Biology and Medicine, Halsted Soc. (sec.-treas. 1963-66), Southeastern Surg. Congress, So. Thoracic Surg. Assn., Soc. Surgery Alimentary Tract, Surg. Biology Club II, Soc. U. Surgeons, Sigma Xi, Alpha Omega Alpha. Editorial bd. Jour. Surg. Research, 1960—. Publs. on studies of electrolyte balance of post-operative patients; surg. considerations of adrenal gland and its hormones; various surg. techniques. Home: Route 5, Box 139, Morgantown, W.Va. 26506.*

ZIMMERMANN, Walter M., German botanist; b. Walldürn, Germany, May 9, 1892; s. Emil and Maria (Welte) Z.; student univs. Freiburg, Berlin, Munich, Karlsruhe Inst. Tech.; Dr.rer.nat., 1921; m. Anna Schleier, 1921; children—Gerda (Mrs. Zimmermann), Reinhard, Ute (Mrs. Stämpfli); m. 2d, Karin Krause, 1960. Asst., Bot. Inst., U. Freiburg, 1919-25; faculty U. Tübingen, (Germany), 1925—, prof., 1930—, dir. inst. applied botany, 1939—, emeritus, 1960—. Baden-Württemberg state nature conservation ofcl., 1946-62. Hon. mem. numerous sci. socs.; mem. Assn. Biologists (dep. chmn.). Author: Phylogenie der Pflanzen, 1930; Vererbung erworbener Eigenschaften, 1938; Grundfragen der Evolution, 1948; Geschichte der Pflanzen, 1949; Evolution, 1953; Telomtheorie, 1963; also numerous articles. Research on phylogeny of plants, evolution questions, paleobotany, algology, plant sociology, development physiology, taxonomy (splty. Ranunculacaea). Address: 2 Autenriethstrasse, Tübingen, West Germany.*

ZIMMERMANN, Wilhelm, German physician, chemist; b. Cologne, Germany, Aug. 19, 1910; s. Emil and Luise (Popp) Z.; student univs. Bonn, Vienna, Hamburg, Breslau; Dr.phil., U. Bonn, 1936; Dr.-med., U. Breslau, 1939; m. Maria Werz, Sept. 8, 1937; children—Anneliese, Bernd, Gerd (dec. 1959). Asst., docent hygiene inst. U. Breslau, 1936-45; 1st asst. Govt. Lab., Coblenz, Germany, 1946-49; dir. Govt. Lab., Trier, 1949-58; prof. head hygiene inst. U. Saar, Homburg, Germany, 1958—. Recipient Honor cross German Red Cross. Mem. N.Y. Acad. Scis., Royal Soc. Medicine London, several German socs. Author: Chemical Methods for Estimation of Steroid Hormones, 1955; also numerous articles. Discovered chem. reaction for 17-ketosteroids (Zimmermann reaction); research on hormones and infection, hormones and meteorbiology, problems of hygiene. Address: 665 Medical Faculty, Homburg/Saar, West Germany.*

ZINDER, Norton David, Am. geneticist; b. N.Y.C., Nov. 7, 1928; s. Harry and Jean (Gottesman) Z.; A.B., Columbia U., 1947; M.S., U. Wis., 1949, Ph.D., 1952; m. Marilyn Estreicher, Dec. 24, 1949; children—Stephen, Michael. Faculty, Rockefeller U., N.Y.C., 1952—, prof. genetics, 1964—. Cons. NSF, 1962—. Recipient U. S. Steel Found. award in molecular biology Nat. Acad. Scis., 1966. Mem. Soc. Am. Bacteriologists (Eli Lilly award in microbiology 1962), Harvey Soc., Genetics Soc. Am. Contbr. articles to sci. jours. Co-discovery of bacterial tranduction, RNA-containing bacteriophages. Home: 194-10L 64th Circle, Fresh Meadows, N.Y. 11365. Office: Rockefeller U., 66th St. and York Av., N.Y.C.*

ZINGG, Walter, surgeon; b. Kloten, Switzerland, Mar. 29, 1924; s. Ernst J. and Ida (Haab) Z.; M.D., U. Zurich, 1948; M.S., U. Man., 1952; m. Regula L.

Zollinger, June 25, 1949; children—Claudia E., Jeannette R., Esther A., David W. Practice medicine, specializing in surgery, Winnipeg, Man., Can., 1958-64; lectr. U. Man., 1956-64; asst. prof. surgery U. Toronto (Ont., Can.), 1964-68, asso. prof., 1968—; asso. scientist Research Inst. Hosp. for Sick Children, Toronto, 1964-68, sr. scientist., 1968. Fellow A.C.S., Royal Coll. Surgeons Can. Research and publs. on exptl. surgery, extracorporeal circulation, hypothermia, rheology of blood, shock. Home: 92 Highbourne Rd., Toronto 7. Office: 555 University Av., Toronto 2, Ont., Can.*

ZINK, Robert Edwin, Am. mathematician; b. Mpls., Nov. 16, 1928; s. Walter Leo and Blanche (Wisehaupt) Z.; B.A., U. Minn., 1949, M.A., 1951, Ph.D., 1953; m. Gloria M(ae) Brownell, Sept. 8, 1950; children—David M(atthew), Richard J(ames), William T(homas). Faculty George Washington U., Washington, 1955-56, Wabash Coll., Crawfordsville, Ind., 1961-62; faculty Purdue U., West Lafayette, Ind., 1953-54, 56—, asso. prof. math., 1960-66, prof., 1966—, asst. head dept., 1965—. Mem. Am. Math. Soc., Math. Assn. Am., Phi Beta Kappa, Sigma Xi. Research and publs. in measure theory. Home: 307 Sharon Rd., West Lafayette, Ind. 47906.*

ZINN, Donald Joseph, Am. zoologist; b. N.Y.C., Apr. 19, 1911; s. Arthur and Florence (Cohn) Z.; B.S., Harvard, 1933; M.S., U. R.I., 1937; Ph.D., Yale, 1942; m. Eleanor Louise Blevins, July 19, 1941; children—Donald Blevins, Jeffrey Arthur. Dir., Bass Biol. Lab., Englewood, Fla., 1933-35; research asst. Ro-Lab., Newtown, Conn., 1938-39; naturalist Marine Biol. Lab., Woods Hole, Mass., 1945-46; faculty U. R.I., Kingston, 1946—, prof., 1960—, chmn. dept. zoology, 1960-65; sr. vis. investigator systematic-ecology program Marine Biol. lab., 1965—; research asso. Narragansett Marine Lab., Kingston, 1950—. Cons., Green Plastics, Bradford, R.I., 1955-58. Mem. Am. Soc. Limnology and Oceanography (mng. editor 1955-57), Nat. Wildlife Fedn. (pres. 1966—), Phi Gamma Delta. Co-editor Psammonalia, 1966—. Contbr. numerous articles to tech. jours. Co-discoverer sub-class Mystacocarida; research on mesopsammon. Home: 1045 Kingstown Rd., Peace Dale, R.I. 02879. Office: Dept. Zoology, U. R.I., Kingston, R.I. 02881.*

ZINN, Johann Gottfried, German anatomist, botanist; b. Ansbach, Bavaria, Dec. 4, 1727. Prof. medicine, dir. bot. gardens U. Göttingen (Germany), from 1753. Author: Descriptio Anatomica Oculi Humani (1st book on anatomy of eye), 1755. Made 1st anat. description of eye, ear; well-known for research on anatomy of eye; name asso. with several parts of eye (Zinn's ligament annulus, zonule, central artery); Linnaeus named plant zinna after him for his work in botany. Died Göttingen, Apr. 6, 1759.

ZINN, Walter H., physicist; b. Kitchener, Can., Dec. 10, 1906; s. John and Maria Anna (Stoskopf) Z.; B.A., Queen's U., 1927, M.A., 1930, D.Sc. (hon.), 1957; Ph.D., Columbia, 1934; m. Jennie A. Smith, Mar. 4, 1933 (dec. Jan. 1964); children—John, Robert. Came to U. S., 1930, naturalized, 1938. Tchr. Queen's U., 1927-28; faculty Columbia, 1931-32, research asso., 1941-42; faculty Coll. City, N.Y., 1932-41; physicist metall. lab. U. Chgo. (Manhattan Project), 1942-46; dir. Argonne Nat. Lab., AEC, 1946-56; pres. Gen. Nuclear Engring. Corp., 1956—; v-p. Combustion Engring., Inc., 1959—. Mem. Pres.'s Sci. Adv. Com. Recipient certificate of merit Am. Power Conf., 1947; Atoms-For-Peace award, 1960. Fellow Am. Nuclear Soc. (pres. 1955), Am. Phys. Soc., A.A.A.S., Sigma Xi; mem. Nat. Acad. Scis., I.E.E.E. (sr.), Am. Inst. Chem. Engrs. (mem. exec. com. nuclear engring. div.). Author: Neutron Generator Utilizing the D-D Reaction, 1937; Scattering Cross Sections of Various Elements for D-D Neutrons, 1939; Emission of Neutrons by Uranium, 1939; Method for Measure Neutron-Absorption Cross Sections by the Effect on the Reactivity of a Chain Reactor, 1946; Diffraction of Neutrons by a Single Crystal, 1946; Basic Problems in Central-Station Nuclear Power, 1952; Engineering and Technical Problems of Atomic Power, 1953; and numerous other profl. publs. in field. Research in nuclear physics and development of reactors. Office: Combustion Engring., Inc., Windsor, Conn.

ZINSSER, Hans, Am. bacteriologist; b. New York, Nov. 1878; s. August and Marie Theresia (Schmidt) Z., A.B., Columbia U., 1899, A.M., 1903, and hon. D.Sc. in 1929; M.D., College of Phys. and Surg. (Columbia U.), 1903; hon. D.Sc., Western Reserve U., 1931, Lehigh U., 1933, Yale, 1939, Harvard, 1939; m. Ruby Handoforth Kunz, June 1905. Intern, 1903-05, bacteriologist, 1905-06, Roosevelt Hospital; asst. bacteriologist Coll. Phys. and Surg., 1905-06, instr. bacteriology and hygiene, same, 1907-10; asst. pathologist St. Luke's Hosp., 1906-10; asso. prof. bacteriology, Stanford U., Calif., 1910-11, prof., 1911-13; prof. bacteriology, Columbia, 1913-23; prof. bacteriology and immunity, Harvard Med. Sch., Boston, 1923—. Bacteriologist Presbyn. Hosp., 1913-23; consulting bacteriologist Children's Hosp., 1924-34, chief bacteriol. service, same, 1934—, Infant's Hosp., 1935—; consulting bacteriologist Peter Bent Brigham Hosp., 1924—. Mem. Am. Red Cross Sanitary Commn. to Serbia, 1915. Sanitary insp. 1st Army Corps, and 2d Field Army A.E.F., and as asst. dir. labs. and infectious diseases, A.E.F.; sanitary commr. in Russia for League of Nations, health sect., summer 1923; exchange prof. to France, 1935. Trustee Mass. Gen. Hosp. Decorated D.S.M., French Legion of Honor, Serbian Order of St. Sava. Mem. Nat. Acad. Sciences, A.M.A., Am. Acad. Arts and Sciences, Am. Acad. Tropical Medicine, and many other med. socs. Author: Text-Book of Bacteriology, 1911; Infection and Resistance; Resistance to Infectious Diseases; Rats, Lice and History, 1935. With others, developed immunization against typhus fever, 1930; research on cholera, bacteriological diseases. Died Sept. 4, 1940.

ZIPPIN, Leo, Am. mathematician; b. N.Y.C., Jan. 25, 1905; s. Max and Bella (Salwen) Z.; A.B., U. Pa., 1925, M.A., 1926, Ph.D., 1929; m. Frances Levinson, July 1, 1932; children—Nina (Mrs. Gordon Baym), Vivian (Mrs. Kenneth Narehood). Instr., Pa. State Coll., 1929; Nat. Research fellow, 1930-32; research asst. Inst. for Advanced Study, Princeton, 1932-36; instr. N.Y. U., 1936-38; instr. Queens Coll., 1938-41, asst. prof., 1941-45, asso. prof., 1945-51, prof., 1951—; exec. officer math. doctoral program City U. N.Y., 1963-68; sci. dir. Phila. computing br. Ballistic Research Lab., Aberdeen Proving Grounds, Md., 1942-45. Mem. Math. Assn. Am., Am. Math. Soc., Phi Beta Kappa, Sigma Xi. Author: (with Deane Montgomery) Topological Transformation Groups, 1954; Uses of Infinity, 1962. Research and publs. in topology and topological groups with emphasis on nature of low-dimensional intuitive spaces and groups of transformations of these spaces. Home: 340 Riverside Dr., N.Y.C. 10025. Office: 33 W. 42d St., N.Y.C. 10036.*

ZIRKEL, Ferdinand, German geologist, mineralogist; b. Bonn, Germany, May 20, 1838. Prof., Lemberg, Kiel, Leipzig (all Germany), 1870-1909. Author: Lehrbuch der Petrogaphie, 2 vols., 1866; Die Mikroskopische Beschaffenheit der Mineralien und Gesteine, 1873. One of 1st to investigate micros. insights of rocks; mineral zirkelite named for him. Died June 11, 1912.

ZIRKLE, Conway, Am. biologist; b. Richmond, Va., Oct. 28, 1895; s. Charles Milton and Mary Louise (Timberlake) Z.; B.S., U. Va., 1921, M.S. 1921; Ph.D., Johns Hopkins U., 1925; postgrad. U. London, 1923, U. Geneva, 1924; m. Helen Emily Kingsbury, Oct. 4, 1923. NRC fellow Harvard, 1925-28, research fellow, 1928-30; asso. prof. biology U. Pa., Phila., 1930-37, prof., 1937—. Pres., Biol. Stain Commn., 1964—. Recipient Silver medal Czechoslovak Acad. Sci., 1965. Mem. A.A.A.S. (v.p. 1951-52), Bot. Soc. Am., Genetics Soc., Am. Soc. Naturalists, History of Sci. Soc., Acad. Internat. History Sci. Author: The Beginning of Plant Hybridization, 1935; Death of a Science in Russia, 1949; Evolution, Marxian Biology and the Social Scene, 1959; (with M. J. Sirks) The Evolution of Biology, 1964; also numerous articles. Research in chemistry of cytological fixation, plant vacuoles, mitochondria, permeability of plant cells, history of biology. Home: 2307 Secand Rd., Secane, Pa. 19018.*

ZIRKLE, Raymond Elliott, Am. experimental biologist; b. Springfield, Ill., Jan. 9, 1902; s. Charles Peter and Lena May (Wettengel) Z.; A.B., U. of Mo., 1928, Ph.D., 1932; m. Mary Evelyn Ramsey, Apr. 26, 1924; children—Raymond Elliott, Thomas Edward. Asst. botany, U. of Mo., 1928-30, instr., 1930-32; National Research Fellow, U. of Pa., 1932-34, fellow med. physics, 1934-38, instr. exptl. radiology, 1937-40; asst. prof. biology, Bryn Mawr Coll., 1938-40; prof. botany, Ind. U., 1940-44; principal biologist, Metall. Project, Manhattan Dist., 1942-46; prof. botany, U. of Chicago, 1944-48; dir. Inst. Radiobiology and Biophysics, 1945-48; prof. radiobiology, 1948-59, prof. biophysics, 1959—, chmn. dept. biophysics, 1964-66; Hitchcock prof. U. Cal., 1951. Fellow A.A.A.S.; mem. Nat. Acad. Sci., Radiation Research Soc. (past pres.), Am. Philos. Soc., Botanical Society of America, American Soc. Zoologists, Am. Roentgen Ray Soc., Am. Society Naturalists, American Society of Plant Physiologists, Biophysical Society, Phi Beta Kappa, Sigma Xi. Contributor articles to profl. journals. Editor: Biological Effects of External Beta Radiation, 1951; Biological Effects of External X and Gamma Radiation. Part I, 1954, Part II, 1956; A Symposium on Molecular Biology, 1959. Research on effects of ultraviolet and ionizing radiations; mechanisms of mitosis; irradiation of individual cells. Home: Olympia Fields, Ill. Office: U. Chgo., Chgo. 60637.

ZISMAN, William Albert, Am. chemist; b. Albany, N.Y., Aug. 21, 1905; s. Leonard and Helena (Bernstein) Z.; B.S., Mass. Inst. Tech., 1927, M.S. (Malcolm Cotton Brown fellow), 1928; Ph.D. in Physics, Harvard, 1932; D.Sc., Clarkson Coll. Tech., 1965; m. Esther Baitz, Oct. 20, 1935; 1 dau., Sandra Ruth. With Harvard Geophysics Research Com., 1931-33, PWA, 1933-35, Resettlement Adminstrn., 1935-37, Geophys. Lab., Carnegie Inst., Washington, 1938; head lubrication br. chemistry div. Naval Research Lab., Washington, 1939-46, head surface chemistry br., 1946-56, supt. chem. div., 1956—. Lectr. colloid and surface chemistry Georgetown U., 1952-58, Am. U., 1959-63; mem. com. on colloid and surface chemistry NRC-Nat. Acad. Sci., 1961-67; mem. research adv. com. on dental materials Nat. Inst. Dental Research, NIH, 1962-67. Recipient Navy Distinguished Service award, 1954, Dept. Def. Distinguished Civilian Service award, 1964. Member American Chem. Society (recipient of Carbide and Carbon award 1955, Kendall award 1963), Am. Phys. Soc., Am. Soc. Lubrication Engrs. (Nat. award 1961), Sci. Research Soc. Am. (1st Ann. Applied Sci. award 1955), Washington, N.Y. acad. scis., Chem. Soc. Washington (Hillebrand award 1954), Internat. Union Pure and Applied Chemistry (sec. commn. on colloid and surface chemistry), Sigma Xi. Mem. adv. bd. Jour. Colloid and Interface Sci., Advances in Chemistry Series. Research and numerous articles on surface chemistry, lubrication, corrosion inhibition, effect of constitution on adhesion of solids, mechanisms of displacement of aqueous and organic liquids from solid surfaces by surface-chem. methods, surface-chem. fundamentals and principles. Patentee in field. Home: 200 E. Melbourne Av., Silver Spring, Md. 20390. Office: Chemistry Div., Naval Research Lab., Washington 20390.*

ZISSIS, George John, Am. physicist; b. Lebanon, Ind., Dec. 31, 1922; s. John Frank and Georgia (Antonakous) Z.; B.S., Purdue U., 1946, M.S., 1950, Ph.D., 1954; m. Wanda H. Evans, Jan. 16, 1954; children—Maida Anne, John George, Christopher George, Maria Cecelia. Research fellow, instr. Purdue U., West Lafayette, Ind., 1946-54; sr. scientist Westinghouse Atomic Power Div., Pitts., 1954-55; research physicist U. Mich., Ann Arbor, 1955-62, 64—, head Infrared Physics Lab., Inst. Sci. and Tech., 1964—; physicist Inst. for Def. Analyses, Washington, 1962-64, cons., 1964—. Mem. Optical Soc. Am., A.A.A.S., Am. Astronautical Soc., Sigma Xi, Sigma Phi Sigma. Author: (with M. Holter, G. Suits, S. Nudleman, W. Wolfe) Fundamentals of Infrared Technology, 1962; also numerous reports and articles. Research in ballistic missile def. and infrared tech. Home: 1549 Stonehaven Rd., Ann Arbor, Mich. 48104.*

ZITRIN, Arthur, Am. physician; b. Bklyn., Apr. 10, 1918; s. William and Lillian (Elbaum) Z.; B.S., Coll. City N.Y., 1938; M.S., N.Y.U., 1941, M.D., 1945; certificate psychoanalytic medicine, Columbia, 1955; m. Charlotte Marker, Oct. 4, 1942; children—Richard Alan, Elizabeth Ann. Research fellow animal behaviour Am. Mus. Natural History, 1939-42; instr. physiology Hunter Coll., N.Y.C., 1948-49; mem. faculty N.Y.U. Sch. Medicine, 1949—, prof. psychiatry, 1967—; mem. staff Bellevue Hops., N.Y.C., 1951—, dir. psychiatry, 1955—; dir. psychiatry N.Y.C. Dept. Hosps., 1962-64; individual practice, 1949—; attending psychiatrist Univ. Hosp., N.Y.C.; cons. psychiatrist Manhattan VA Hosp. Diplomate Am. Bd. Psychiatry and Neurology. Fellow Am. Psychiat. Assn., N.Y. Acad. Medicine: mem. N.Y. Soc. Clin. Psychiatry (pres. 1966-67), Am. Psychoanalytic Assn., A.M.A., Sigma Xi, Alpha Omega Alpha. Contbr. papers in field. Research on neural and hormonal bases for social and reproductive behaviour patterns in chickens and cats; clin. psychiat. problems including drug effects, suicide, addiction. Home: 56 Ruxton Rd., Great Neck, N.Y. 11023. Office: 550 1st Av., N.Y.C. 10016.*

ZMUDA, Alfred Joseph, Am. physicist; b. Shenandoah, Pa., Mar. 15, 1921; s. Frank Larence and Alice (Wasilewski) Z.; B.S. in Math., St. Francis Coll., 1942; Ph.D., Catholic U., 1951; m. Margaret Martha Koval, Feb. 23, 1946; children—Mary Alice, Carol Ann. Equipment engr. Western Electric Co., Kearney, N.J., 1946; physicist Johns Hopkins Applied Physics Lab., Balt., 1951—. Cons. to geophysics panel Sci. Adv. Bd., USAF, 1961-64; mem. panel on world magnetic survey geophysics research bd. Nat. Acad. Scis., 1962—; research chmn. Met. Citizens Council for Rapid Transit, 1964—. Fellow Washington Acad. Sci., Am. Phys. Soc.; mem. Am. Geophys. Union, Philos. Soc. (past chmn. com analysis earth's main magnetic field), Internat. Assn. Geomagneticism, Am. Geophys. Union (vis. scientist 1963), Internat. Sci. Radio Union (mem. com. 3). Research and publs. on geomagnetism, ionospheric physics, space physics. Home: 1421 Crestridge Dr., Office: 8621 Georgia Av., Silver Spring, Md. 20910.*

ZNANIECKI, Florian Witold, sociologist; b. Swiatniki, Poland, Jan. 15, 1882; s. Leon and Amelia (Holtz) Z.; B.A., Univ. of Warsaw; M.A., U. of Geneva; student Univ. of Zurich, Univ. of Paris; Ph.D., Univ. of Cracow, 1909; m. Emilia Szwejkowska, Sept. 1, 1906; one son, Julius; m. 2d. Eileen Markley, April 26, 1916; one dau., Helena (Mrs. Richard S. Lopata). Dir. Polish Emigrants Protective Assn., 1911-14; lecturer, Univ. of Chicago. 1917-19; prof. sociology, Univ. of Poznan, Poland, 1920-39; vis. prof., Columbia, 1931-38, summer 1939, Julius Beer lecturer, 1939-40; vis. prof., Univ. of Ill., 1939-40, prof. of sociology, 1941—. Decorated Commander Polonia Restituta. Founder, Polish Sociol. Inst. and Polish Sociol. Review. Fellow A.A. A.S.; mem. Am. Social. Soc. (pres. 1953-54). Internat. Sociol. Inst., London Sociol. Soc. (vice pres.). Author: The Problem of Values; Humanism and Knowledge; The Fall of Western Civilization; Introduction to Sociology; Sociology of Education, 2 vols.; The City Viewed by Its Inhabitants; The Men of Today and the Civilization of the Future; (all in Polish); (in English) The Polish Peasant, 5 vols. (with W. I. Thomas), 1918-20; Cultural Reality, 1919; Laws of Social Psychology, 1925; The Method of Sociology, 1934; Social Actions, 1936; The Social Role of the Man of Knowledge, 1940; Las sociedades de cultura nacional, 1944; Cultural Sciences. Modern Nationalities, 1952; about 30 articles in various languages. With Thomas, 1st to use personal data as sources for sociol. investigations and developed concept social disorgn. as perspective from which to study social phenomena; developed idea that individually determined attitudes and socially determined values were components of human behavior. Died Champaign, Ill., Mar. 23, 1958.

ZOBELL, Claude Ephraem, Am. marine microbiologist; b. Provo, Utah, Aug. 22, 1904; s. Ephraim A. and Stella (Davis) Z.; student So. Ida. Coll. Edn., 1922-24; B.S., Utah State U., 1927, M.S., 1929; Ph.D. (Thompson scholar), U. Cal. at Berkeley, 1931; m. Margaret Harding, June 5, 1930 (div. Apr. 1944); children—Karl M., Dean H.; m. Jean E. Switzer, May 30, 1946. Prin., Rigby (Ida.) Pub. Sch., 1924-26; instr. bacteriology Utah State U., 1927-28; research asst. Hooper Found., San Francisco, 1929-31; instr. microbiology Scripps Inst., U. Cal. at La Jolla, 1932-36, asst. prof., 1936-42, asso. prof., 1942-48, prof., 1948-—; research asso. U. Wis., 1937-38; vis. investigator Woods Hole Oceanographic Instn., 1939; Distinguished Overseas lectr. Australian New Zealand Assn. for Advancement Sci., 1956. Lectr. Meet the Scientist series La. Jolla Theatre and Arts Found., 1960-66. Spl. Rockefeller Found. fellow, 1947-48; recipient King Frederick IX (Denmark) medal, 1952, Utah State U. Distinguished Service award, 1961. Mem. Am. Petroleum Inst. (dir. research project 43A 1940-50), San Diego Zool. Soc. (mem. research council 1932-—), Microbiology Soc. (Thailand hon.), Surtsey Research Soc. (asso. founding mem., Iceland), Am. Soc. Microbiology, Soc. for Indsl. Microbiology, Am. Soc. Limnology and Oceanography, Soc. for Gen. Microbiology, Sigma Xi, Phi Kappa Phi. Author: Marine Microbiology, 1946; also numerous articles. Research on importance of bacteria as geochem. agts. in sea, and in origin of oil; pioneered discovery and cultivation of living organisms from ocean depths. Home: 2404 Ellentown Rd., La Jolla, Cal. 92037.*

ZOLCINSKI, Adam Emil, Polish physician; b. Bochnia, Poland, Nov. 20, 1907; s. Jan and Wiktoria (Godzisz) Z.; Med. Diploma, Lwów (Poland) U., 1932, Med. Doctors' degree, 1935; Docent degree Wroclaw Med. Acad., 1955; m. Irena Serkes, Feb. 1, 1949; children—Ewa, Malgorzata. Sr. asst. Clinic Gynecology and Obstetrics, Lwów U., 1933-40; staff Med. Acad., Wroclaw, 1947-—, asso. prof., 1955-—; chief dir. Hosp. Gynecology and Obstetrics, Walbrzych, Silesia, Poland, 1951-52; dean med. dept. Med. Acad. Wroclaw, 1958-66, lectr., 1955-—, med. adviser in all clinics. Recipient Gold Cross of Merit, Nat. Health Service award for outstanding work, 1959, medal to 15th Anniversary of People's Poland, 1955, medal to 15th Anniversary of Lower Silesia, 1960. Fellow Polish Med. Soc., Polish Gynecologic Soc., (head Wroclaw chpt. 1957-—, mem. bd. 1947-—). Research and publs. on terratogenic and toxic effects of various medicine on devel. of fetus, treatment and relationships of leucoplakia with carcinoma of cervix uteri, female sterility. Home: 105 Jednosci Narodowej, Wroclaw, Poland.*

ZOLLINGER, Heinrich, Swiss chemist; b. Aarau, Switzerland, Nov. 29, 1919; s. Fritz and Helene (Prior) Z.; Diploma Chem.Eng., Swiss Fed. Inst. Tech., Zürich, Switzerland, 1943, Ph.D., 1944; m. Heidi Frick, Oct. 9, 1949; children—Fritz, Ruedi, Hansjürg. With research dept. dyes CIBA Co., Basle, Switzerland, 1945-51; with chem. dept. Mass. Inst. Tech., Cambridge, 1951-52; privat-dozent U. Basle, 1952-60; prof. chemistry Swiss Fed. Inst. Tech., 1960-—. Recipient Alfred Werner prize, 1959; Ruzicka Medal, 1960; Herbert Levinstein award, 1965. Mem. Internat. Union Pure and Applied Chemistry (mem. bd. organic div.), Swiss Chem. Soc., Internat. Fedn. Textile Chemists and Colourists, Soc. Dyers and Colorists. Author: Chemie der Azofarbstoffe, 1958; Diazo and Azo Chemistry, 1961; also numerous articles. Research on mechanism of electrophilic and nucleophilic aromatic substitution, chemistry of metastable intermediates, reactive dye chemistry, azo dye chemistry, dyeing mechanisms, structure of cross-linked polymers, especially textile fibres, cotton. Home: 45 Boglerenstrasse, 8700 Küsnacht, Switzerland. Office: 6, Universitätstrasse, 8006 Zürich, Switzerland.*

ZÖLLNER, Fritz, German physician; b. Vienna, Austria, July 14, 1901; s. Josef and Karoline (Sager) Z.; Doctor med. U. Vienna, 1926; m. Franzi Uhrmann, Aug. 27, 1937; children—Brigitte, Christoff, Michael, Sibylle, Susanne. Fellow ENT Clinic, U. Vienna, 1929-31; staff U. Graz (Austria), 1931-45, U. Jena (Germany), 1931; head ENT-Clinic, U. Freiburg (Germany), 1947-—. Mem. Academia Leopoldina. Author: Tuba Eustachi, 1942; Audiology, 1954; (with Berendes, Link) Handbuch der Hals-Nasen-Ohrenheilkunde, 1963; also numerous articles. Research on function of Eustachian tube and middle ear mechanics, histology of tonsils and ear, surgery of ear, Tympanoplasty. Home: 17 in der Röte, Freiburg i. Breisgau, Bad.-Württberg, Germany.*

ZÖLLNER, Johann Karl Friedrich, German astrophysicist, astronomer; b. Berlin, Prussia, Nov. 8, 1834; ed. univs. Berlin, Basel (Switzerland). Prof. astron. physics, Leipzig, Germany, 1872. Author: Wissenschaftliche Adhandlungen, 4 vols., 1878-81; Photometrische Untersuchungen, 1865. Research, publs. on photometry, comets, electro-dynamic theory of matter; improved spectroscope, invented astrophotometer (still most common one in use) with which he made studies in sensory physiology (i.e., mental delusions, especially-optical ones). Died Leipzig, Apr. 25, 1882.

ZÖLLNER, Nepomuk Johann, German physician; b. Marktredwitz, Germany, Feb. 21, 1923; s. Otto Wilhelm and Else (Beauvais) Z.; ed. U. Munich (Germany), M.D., 1945; m. Marga L. Benker, Oct. 3, 1951; children—Michael S., Andreas O., Susanne M. Research asst. U. Munich, 1945-48, faculty, 1954-—, asso. prof. medicine, 1960-—; research asst. Tufts Coll. Med. Sch., 1948-50, 52-53, Bellevue Hosp., N.Y.C., 1953. Mem. German Nutrition Soc. (dir. dept. sci. 1966-—, also several European socs. for internal medicine, nutrition, biol. chemistry. Editor: Thannhausers Lehrbuch des Stoffwechsels, 1957; Untersuchung und Bestimmung der Lipoide im Blut, 1966; also numerous articles. Research on nutritional aspects of arteriosclerosis, drugs influencing lipid metabolism, new methods for analysis of lipids; discoveries in heparin metabolism; evaluation of sugars in parenteral nutrition. Home: 148 Tegernseer Landstrasse, Munich 9, West Germany.*

ZOLOG, Nicolae, Rumanian ophthalmologist; b. Coroi, Roumania, Dec. 19, 1910; s. Ion and Ann (Farcas) Z.; M.D., U. Cluj (Roumania), 1935; m. Florica Bucur, July 10, 1937; 1 son, Alexander. Staff, U. Eye Clinic Cluj, 1937-48, asst., 1939-48; faculty Med. Sch. Timisoara (Roumania), 1948-—, prof. ophthalmology, 1963-—. Cons. ophthalmologist Eye Hosp. Timisoara. Mem. Rumanian, French socs. ophthalmology, Internat. Assn. for Prevention Blindness. Contbg. author: Textbook of Ophthalmologie, 1958. Research and publs. on ocular tumors, water metabolism, auditive trouble, capillar permeability in glaucoma, therapy of retinal vascular diseases, ocular congenital anomalies; described first lacrimal glomustumor. Home: 1 Asanesti, Timisoara, Rumania.*

ZOLTAI, Tibor Zoltan, crystallographer, mineralogist; b. Györ, Hungary, Oct. 17, 1925; s. Nicholas Z. and Elizabeth (Schmidt) Z.; student Poly. U., Budapest, Hungary, 1944, U. Graz (Austria), 1946-48, Sorbonne, Paris (France), 1948-50; B.A.Sc., U. Toronto (Ont., Can.), 1955; Ph.D., Mass. Inst. Tech., 1959; m. Olga M. Wagner, Nov. 18, 1950; children—Peter, Katharina. Came to U. S., 1955, naturalized, 1966. Faculty, dept. geology and geophysics, U. Minn., Mpls., 1959-—, prof., 1964-—, chmn. dept., 1963-—. Fellow Mineral. Soc. Am.; mem. Geol. Soc. Am., Am. Crystallographic Assn., Am. Inst. Profl. Geologists, Mineral. Assn. Can., Sigma Xi. Research and publs. on crystal structures, crystal chemistry of silicates and other minerals. Home: 686 Heinel Dr., St. Paul 55113.*

ZONDEK, Herman, physician; b. Wronke, Germany, Sept. 4, 1887; s. Abraham and Sarah (Hollaender) Z.; student U. Goettingen, Berlin, 1907-12; Dr. med., U. Berlin, 1912; m. Gerda Wolfson, Feb. 20, 1950; children—Birgit, Bernd. Faculty Friedrich Wilhelm U., Berlin, 1918-33, prof. medicine, 1934; vis. prof. Hebrew U., Jerusalem, Israel, 1951; dir. Municipal Hosp. am Urban, 1926-33; hon. physician Victoria Meml. Jewish Hosp., Manchester, Eng., 1933-34; dir. med. div. Bicur Cholim Hosp., Jerusalem, 1934-57. Mem. Israel Med. Assn. (hon. pres. sci. council), Jerusalem Acad. Medicine (pres. 1963-—), Endocrine Soc. Israel (pres. 1963-66), Israel Acad. Scis. and Humanities, Royal Soc. Medicine (affiliate), Jerusalem Med. Assn. (hon.); hon. mem. fgn. endocrine socs. Author: (with C. Mase), Das Hungeroedem, 1920; The Diseases of the Endocrine Glands, 1923, rev. edits., 1926-58; also numerous articles. Research on heart and kidney pathology, especially endocrine function in health and disease; 1st description of endocrine disorders such as myxedema heart, pituitary hyper- and hypothyroidism, and various hypothalamic-endocrine syndromes; iodine therapy of hyperthyroidism; elucidation of mechanisms of hormonal action and regulation. Home: 8, Maimon St., Jerusalem, Israel.*

ZONNEVELD, Jan Izaak Samuel, Dutch phys. geographer; b. Kampen, Netherlands, Aug. 31, 1918; s. Jan and Lidewey Z.; drs. geology State U. Leiden (Netherlands), 1941; Doctor, State U. Leiden, 1947; m. Barendina Johanna Lablans, July 17, 1945; children—Elise Catherine, Jan Mathys, Annelies. Geologist, Geol. Bur., Mining Dist. Heerlen, Netherlands, 1945-48; acting head Central Bur. for Aerial Surveys, Suriname, S.Am., 1948-52; geologist Geol. Service, Haarlem, Netherlands, 1952-58; prof. geol. dept., State U., Utrecht, Netherlands, 1958-—. Mem. Kon. Ned. Academkskundig Genootschep., Assn. for Internat. Quaternary Research. Author: De Levende Aarde; Tussende Bergen an de Zee; also articles. Used differences between Rhine sands and Maas sand to study geol. history of S.E. Netherlands; studies on geomorphology of Netherlands Antilles and Suriname. Home: 24 Graaf Jaulaan, Zeist, Netherlands. Office: 21 Drift, Utrecht, Netherlands.*

ZOOK, Harry David, Am. chemist; b. Milroy, Pa., Feb. 8, 1916; s. Samuel Milton and Emma (Smucker) Z.; B.S., Pa. State U., 1938, Ph.D., 1942; M.S., Northwestern U., 1939; m. Margaret Olsen, Dec. 20, 1947; children—Stephen Michael, Terri. Faculty, Pa. State U., University Park, 1941-—, prof., 1960-—; vis. lectr. Stanford, 1962-63. Mem. Am. Chem. Soc., Chem. Soc. London. Author: (with R. B. Wagner) Synthetic Organic Chemistry, 1953. Research in nature of carbanions, kinetics and orientation of enolate alkylation. Home: 828 N. Thomas St., State College, Pa. 16801. Office: 207 Old Main, University Park, Pa. 16802.*

ZORBACH, William Werner, Am. chemist; b. Sandusky, O., June 15, 1916; s. William and Helen (Werner) Z.; B.S., Bowling Green State U., 1947; Ph.D., McGill U., 1951; m. Betty Canfield, June 28, 1946; 1 dau., Judy Lynne. Instr., Bowling Green (O.) State U., 1951-52; asst. prof. chemistry Georgetown U., Washington, 1952-56, asso. prof., 1958-64, prof., 1964-67; dir. bio-organic chemistry Gulf South Research Inst., New Iberia, La., 1957-—; research chemist NIH, Bethesda, Md., 1957-58. Cons. FTC, 1958-59, div. research grants, NIH, 1962-63. Fellow Am. Inst. chemists; mem. Am. Chem. Soc., N.Y. Acad. Sci., Nat. Wildlife Fedn. (asso.), Am. Amaryllis Soc., Sigma Xi, Sigma Alpha Epsilon. Co-editor: Synthetic Procedures in Nucleic Acid Chemistry. Research on biologically important glycosides. Home: 4109 Walnut Dr. Office: Gulf South Research Inst., New Iberia, La. 70560.*

ZOSIMOS OF PANOPOLIS, Greek alchemist; b. Panopolis (now Akhmon, Egypt), flourished end of 3d century and possibly 4th. Wrote alchem., magical, mystical works, also ency. of chem. arts in 28 vols.; earliest identifiable writer on alchemy of whom we have genuine writings.

ZSCHEILE, Frederick Paul, Jr., Am. biochemist; b. Burlington, Kan., May 11, 1907; s. Frederick Paul and Esther (Young) Z.; B.S. U. Cal. at Berkeley, 1928, Ph.D., 1931; m. Emily DeSylvester, Oct. 14, 1933; children—Frederick Paul, III, Richard Eugene, Elizabeth Therese (Mrs. Robert Earl Partrick). Nat. Research fellow dept. chemistry U. Chgo., 1931-33, research asso. 1934-37, 44-46; staff Purdue U., Lafayette, Ind., 1937-42, asso. chemist, asso. prof. agrl. chemistry, 1942-44; asso. prof., asso. biochemist dept. agronomy, U. Cal. at Davis, 1946-52, prof., biochemist, 1952-—. Guggenheim fellow, 1958-59. Member American Chemical Society (chairman Sacramento sect. 1954-55). Research on spectrophometric analysis biologically active substances, biochem. nature of disease resistance in plants, phytotron design for controlled environment for plant research. Home: 236 B St., Davis, Cal. 95616.*

ZSEBOK, Z., Hungarian physician; b. Budapest, Hungary, June 19, 1908; med. certificate Pazmany Pter U., 1940, specialist for radiology, 1947, hon. lectr. radiology, 1957, Dr. Med. Scis., 1962; m. Susanne H. Nagy, 1938; 1 son, Stephan. Sr. doctor X-ray dept. Radium Hosp., Budapest, 1934-39; head radiologist, hosps. at Munkacs and Beregszasa; 1939-44; asst. sec. state Ministry Pub. Welfare, 1945-48; asst. prof. Med. U., Budapest, 1948-62; dir. Radiol. Clinic, Med. U., Budapest, 1962-—. Dir. medico-radiol. research team Hungarian Acad. Sci. Recipient Kossuth prize, 1955. Mem. Soc. Hungarian Radiologists (pres.). Author: X-ray Anatomy of the Chest, 1953, subsequent edits.; Textbook of Radiology, 1966. Mem. Soc. German Radiologists (hon.), Soc. Finnish Radiologists (hon.), Soc. East German Radiologists (corr. mem.). Editor, Magyar Radiologia; editorial bd. Atompraxis, Der Rontgenologe, Investigative Radiology, Radiologia Diagnostica. Radiol. studies of chest, particularly lungs of newborn; methods of radiol. exam. Home: 54 Biro Lajos. Office: 78 Ulloi, Budapest, Hungary.*

ZSIGMONDY, Richard Adolf, chemist; b. Vienna, Austria, Apr. 1, 1865; Ph.D. in organic chemistry, U. Munich, 1889; student U. Vienna, U. Berlin. Asst. to physicist A. Kundt, Berlin; prof. U. Graz; with glass works of Schott, Jena, 1897-1900; worked in his pvt. lab. with Siedentopf, 1900-03; dir. Inst. for Inorganic Chemistry, U. Göttingen, 1908. Recipient Nobel prize in chemistry for work on colloids, 1925. Author: Zur Erkenntnis der Kolloide; Kolloidchemie, 1912. Explained precipitation of colloids by electrolytes, protective colloidal action and speed of coagulation; synthesized purple of Cassius; developed membrane and ultrafilter with uniform pores of 4 double MU diameter, 1922; produced several types of colored glasses (one variety called milk glass); co-inventor of ultra-microscope, 1903; discovered sols of gold and variables effecting color; did basic research on colloidal suspensions and on their phys.-chem. nature. Died Göttingen, Germany, Sept. 29, 1929.

ZSOLDOS, F., Hungarian plant physiologist; b. Sarkad, Hungary, Mar. 24, 1927; s. Ference and Margit (Nagy) Z.; student U. Eötvös Lórand, Budapest, Hungary, 1958; m. Ildikó Jeremiás, Aug. 28, 1962; 1 dau., Gábor. Staff, Plant Physiol. Inst., U. Eötvös Loránd, Morphological Inst.; technician Plant Physiol. Inst., József Attila U., Szeged, Hungary, 1958-—, lab. IAEA, Vienna, Austria, 1962-64. Research, publs. on use of radioactive and stable isotopes in nutrition of plants, plant physiol. exams. on effect of environmental factors (temperature, pH), affecting nutrient uptake, synthetic substances influencing growth and yield, cold-shock and nitrogen utilization by rice plants. Home: 26 Batthyány, Szeged, Hungary.*

ZSOLT, János, Hungarian microbiologist; b. Debrecen, Hungary, Oct. 4, 1920; s. János and Margit (Csuthy) Z.; Ph.D., U. Budapest (Hungary), 1943; Candidate biol. scis., Hungarian Acad. Sci., 1962; m. Edith Diószeghy, July 22, 1957. Asst., Inst. for Plant Physiology, U. Budapest, 1941-43; chemist Ampelogical Inst., Budapest, 1943-47; 1st asst. Chem. Inst., High Sch. for Agronomy, Budapest, 1947-50; researcher Research Inst. for Biology, Hungarian Acad. Sci., Tihany, 1950-53, Research Inst. for Botany, Vácrátót, 1953-57; researcher Inst. for Plant Physiology, U. Szeged (Hungary), 1957-—. Research and numerous publs. on biochemistry and taxonomy of yeasts, antifungal compounds of higher plants. Home: 6 Ságvári, Szeged, Hungary.*

ZUBECKIS, Edgar, food chemist and technologist; b. Jelgava, Latvia, Dec. 27, 1902; s. Ludwig and Lucija (Willert) Z.; Agriculturist, U. Riga (Latvia), 1929, Venia legendi, 1938; Dr.Agr., Agr. Acad., Jelgava, Latvia, 1944. Instr., U. Riga., 1931-36, privatdocent, 1936-41; asso. prof. Agr. Acad., Jelgava, 1941-44; asso. prof. U. Bonn (Germany), 1947-49; research scientist Ont. Research Council, 1953-55; research scientist Hort. Research Inst. Ont., Vineland, Can. Mem. N.Y. Acad. Scis., Agr. Inst. Can., Canadian Inst. Food Tech., Internat. Soc. Hort. Sci. Author: Fruit Juices, 1939; Chemical Analysis of Agricultural Products, 1940; Biological Preservation of Vegetables, 1943; also numerous articles. Research on chemistry and tech. agrl. products, chem. composition and nutritive value of raw materials and consumers' goods; developed new lines of fruit beverage. Address: Horticultural Research Inst. Ont., Vineland, Can.*

ZUBEK, John P(eter) physiol. psychologist; b. Trnovec, Czechoslovakia, Mar. 10, 1925; s. John Joseph and Mary (Hrubos) Z.; B.A., U. B.C., 1946; M.A., U. Toronto (Ont., Can.), 1948; Ph.D., Johns Hopkins, 1950; m. Sparling Mary, July 1, 1961. Asst. prof. psychology McGill U., Montreal, Que., Can., 1950-53; prof., head dept. psychology U. Man. (Can.), Winnipeg, 1953-61, research prof. psychology, 1961-—, dir. isolation lab., 1959-—. Mem. div. exptl. psychology NRC Can., 1954-61; mem. human resources sci. adv. com. Def. Research Bd. Can., 1959-65. Mem. Canadian, Am. psychol. assns., A.A.A.S., Sigma Xi, Phi Beta Kappa. Author: (with P. A. Solberg) Human Development, 1954; also numerous articles. Determined behavioral, physiol. and biochem. effects of prolonged sensory and social isolation, determined age changes in sensory processes and intellectual abilities; research on cerebral mechanisms in cutaneous sensitivity. Home: 21 Mayfair Pl., Winnipeg, Man., Can.*

ZUBIN, Joseph, psychologist; b. Rossenai, Lithuania, Oct. 9, 1900; s. Jacob Moses and Hannah Rachel (Brodie) Z.; naturalized Am. citizen, 1929; A.B., Johns Hopkins, 1921; Ph.D., Columbia U., 1932; m. Winifred M. P. Anderson, Oct. 12, 1934; children—Jonathan Arthur, David Anderson, Winifred Anne. Asst. ednl. psychologist Tchrs. Coll., Columbia U., 1930-31, instr. psychometrics Coll. Phys. and Surg., 1932-33; fellow Union Am. Hebrew Congregation, 1929-32; instr. ednl. psychology Coll. City N.Y., 1934-36; asst. psychologist Mental Hosp. Survey Com., Nat. Com. for Mental Hygiene, 1936-38; asso. research psychology N.Y. State Psychiat. Inst. and Hosp., 1938-56; chief psychiat. research biometrics N.Y. State Dept. Mental Hygiene, 1960-—; faculty Columbia U., N.Y.C., 1939-—, prof. psychology, 1956-—. Fellow Am. Psychol. Assn.; mem. Am. Psychopath. Assn. (past pres.), Am. Statis. Assn., Psychonomic Soc., Eastern Psychol. Assn., Psychometric Soc., Inst. Math. Statistics, Am. Pub. Health Assn., Sigma Xi. Author: (with Florence Schumer, Leonard Eron) An Experimental Approach to Projective Techniques, 1965; also numerous articles. Editor: Field Studies in the Mental Disorders, 1961; also co-editor numerous publs. Am. Psychopath. Assn., 1944-65. Home: 190 Highwood Av., Leonia, N.J. 07605. Office: 722 W. 168th St., N.Y.C. 10032.*

ZUBROD, Charles Gordon, Am. physician; b. N.Y.C., Jan. 22, 1914; s. Charles Augustus and Anna (Beyer) Z.; A.B., Holy Cross Coll., 1936; M.D., Columbia U., 1940; m. Christina Catherine Mullins, June 15, 1940; children—Christine, Gordon, Justin, Margaret, Stephen. Instr., Johns Hopkins Med. Sch., 1946-49, Roche Research fellow, 1946-49, asst. prof., 1949-53; asso. prof. medicine St. Louis U., 1953-54; chief gen. med. br. Nat. Cancer Inst., NIH, 1954-59, clin. dir., 1955-61, dir. intramural research, 1961-65, sci. dir. for chemotherapy, 1965-—. Mem. adv. com. Burroughs Welcome Fund., 1959-—; mem. com. drug efficacy Nat. Acad. Sci.-NRC, 1966-—; mem. cooperative studies evaluation com. VA, 1966-—; mem. research com. Am. Cancer Soc., 1956-59. Recipient Superior Service award U. S. Dept. Health Edn. and Welfare, 1964, Distinguished Service award, 1965. Mem. Am. Soc. for Pharmacology and Exptl. Therapeutics, Am. Soc. for Clin. Investigation, Assn. Am. Physicians, Am. Assn. for Cancer Research (dir. 1965-—, A.C.P. (mem. cancer com. 1961-—). Research and numerous publs. in chemotherapy of bacterial diseases, malaria and cancer, pharmacology of drug transport into central nervous system particularly elasmobronchs. Home: 5813 Marengo Rd., Bethesda (P.O. Washington 20016). Office: Nat. Cancer Inst., NIH, Bethesda, Md. 20014.

ZUCKER, Alexander, physicist; b. Zagreb, Yugoslavia, Aug. 1, 1924; s. William and Bertha (Klopfer); came to U. S., 1939, naturalized 1943; B.A., U. Vt., 1947; M.S., Yale, 1948, Ph.D., 1950; m. Joan-Ellen Jamieson, Nov. 28, 1953; children—Rebecca Marie, Claire Louise. Research asst. Yale, 1948-50; physicist Oak Ridge Nat. Lab., 1950-—, asso. div. dir. electronuclear div., 1960-—. Mem., U. S. Delegation to USSR on peaceful uses of atomic energy, 1963. Mem. Am. Phys. Soc., Phi Beta Kappa, Sigma Xi. Research in medium energy nuclear physics; nuclear reactions and scattering of heavy ions; elastic and inelastic scattering of polarized protons; particle accelerators, especially cyclotrons. Home: 103 Orange Lane, Oak Ridge, Tenn. 37830.*

ZUCKER, Marjorie Bass, Am. physiologist; b. N.Y.C., June 10, 1919; d. Murray H. and Agnes (Naumburg) Bass; A.B., Vassar Coll., 1939; Ph.D., Columbia U., 1944; m. Howard D. Zucker, June 25, 1938; children—Andrew, Ellen, Joan, Barbara. Instr., Columbia U., Coll. Phys. and Surg., 1945-49; with N.Y. U. Coll. Dentistry, 1949-55, asso. prof., 1954-55; asso. mem. Sloan-Kettering Inst., N.Y.C., 1955-63; asso. prof. physiology Cornell U. Med. Coll., 1955-63; sr. research asso. Am. Nat. Red Cross Research Lab., asso. prof. pathology N.Y. U. Sch. Medicine, N.Y.C., 1963-—. Mem. Am. Physiol. Soc., Soc. Exptl. Biology and Medicine, Am. Soc. Hematology, Internat. Soc. Hematology. Author: (with A. J. Marcus) Physiology of Blood Platelets, 1965. Editor-in-hematology Proc. Soc. Exptl. Biology and Medicine, 1960-65. Research and numerous publs. on relationship of platelet clumping to blood clotting and platelet metabolism and enhancement and inhibition of aggregation caused by various chemicals. Home: 333 Central Park West, N.Y.C. 10025. Office: 550 1st Av., N.Y.C. 10016.*

ZUCKER, Milton Lawrence, Am. physiologist; b. St. Louis, Aug. 19, 1928; s. Samuel and Mary (Goldberg) Z.; A.B., Washington U., St. Louis, 1948, Ph.D., 1952; m. Monica Beth Ribstein, Sept. 12, 1948; children—Jeffrey, Frank, Jonathan, Daniel. McCollum-Pratt fellow Johns Hopkins, 1952-54; asst. plant physiologist Conn. Agrl. Expt. Sta., New Haven, 1954-57, asso. plant physiologist, 1957-60, plant physiologist, 1960-—. NSF Sr. Postdoctoral fellow, 1963-64. Mem. Am. Soc. Plant Physiologists, Am. Soc. Biol. Chemists. Asso. editor Plant Physiology, 1966-—. Studies, publs. of nitrogen metabolism in plants; biosynthesis of phenolic compounds and their role in flowering and disease resistance; regulation of enzyme synthesis in plants. Home: 15 Homeland Terrace, Hamden, Conn. 06517. Office: Box 1106, New Haven 06504.*

ZUCKERMAN, Bert Merton, Am. biologist; b. N.Y.C., Mar. 26, 1924; s. Harry Louis and Pearl (Fine) Z.; B.S. in Forestry, N.C. State Coll., 1948; M.S. in Forest Pathology, N.Y. State Coll. Forestry, 1949; Ph.D., U. Ill., 1954; m. Harriette Rodetsky, June 18, 1950; children—Myra Sue, Linda, Jonathan. Asst. plant pathologist Ill. Natural History Survey, Urbana, 1951-54; faculty U. Mass. Cranberry Expt. Sta., East Wareham, 1955-—, prof. biology, 1963-—. Mem. Am. Soc. Nematologists, Am. Phytopath. Soc., European Soc. Nematologists. Co-translator: Selected East European Papers in Nematology, 1964. Research and numerous publs. on methods in exptl. plant nematology, pesticide degradative pathways asso. with microbial metabolism. Home: 98 Maple St., New Bedford, Mass. 02740. Office: U. Mass. Cranberry Expt. Sta., East Wareham, Mass.*

ZUCKERMAN, Sir Solly, anatomist; b. Cape Town, South Africa, 1904; ed. S. African Coll. Schl.; U. Cape Town; U. Coll. Hosp., London, M.A., M.D., D.Sc.; Dr. h.c. U. Bordeaux, 1961, hon. D.Sc., Us. Sussex, 1963, Jacksonville, 1964; m. Lady Joan Rufus Isaacs, 1939; 1 son, 1 dau. Demonstrator anatomy, U. Cape Town, 1923-25; Union Research Scholar, 1925; res. anatomist, Zoological Soc. London; demonstrator anatomy, U. Coll., London, 1928-32; res. assoc. and Rockefeller Research Fellow, Yale U., 1933-34; U. demonstrator, lectr. human anatomy, 1934-45; Wm. Julius Mickle Fellow, U. London, 1935; Beit Fellow, 1934; Hunterian prof., Royal Coll. Surgeons, 1937; Agricultural Research Council, 1949-59; Chmn., Defense Research Policy Cttee., 1960-64; chmn. Cttee. Sci. Manpower, 1950-64; dep. chmn., Advisory Counc. on Sci. Policy, 1948-64; Gregynog Lectr., U. Coll. Wales, 1956; Mason Lectr. U. Birmingham, 1957; Sand Cox prof. anatomy, U. Birmingham; other advisory positions. Recipient Medal of Freedom (U. S.), Chevalier de la Légion d'Honeur (France); Companion d'Honneur (France); Companion of Bath, 1946; Knight Commander of the Bath, 1964. Fellow Royal Soc., 1943; Mem. Royal Coll. Surgeons (hon. fellow, 1964); Fellow Royal Coll. Physicians; hon. mem. Academia das Ciencias de Lisboa; fgn. mem., Am. Philos. Soc. Author: The Social Life of Monkeys and Apes; Functional Affinities of Man, Monkeys and Apes; A New System of Anatomy. Ed. The Ovary, vols. 1 & 2; articles in various sci. journals. Research on anatomy, zoology, endocrinology. Home: 16 Ampton Road, Edgbaston, Birmingham, Eng.

ZUFFANTI, Saverio, Am. chemist; b. Boston, May 4, 1908; s. Vincent and Josephine (Ditavi) Z.; B.Chem. Engring., Northeastern U., 1930; A.M., Boston U., 1932; B.S., Northeastern U., 1934; m. Anne Butera, Apr. 19, 1933; 1 dau., Dorothy Anne (Mrs. William K. Perkins). Faculty, Northeastern U., Boston, 1930-—, prof. chemistry, 1948-—; cons. Salem Grease and Oil Co. (Mass.), 1941-—. Cons. to pvt. cos.; staff mem. Handbook of Chemistry and Physics, 1938-—; whole jour. abstractor Chem. Abstracts, 1937-—. Fellow Am. Inst. Chemists; mem. Am. Chem. Soc., New Eng. Assn. Chem. Tchrs., Sigma Xi. Author: General Chemistry, 3d edit. 1965; Laboratory Manual, 1959; Electronic Theory of Acids and Bases, 1946; also articles. Research on synthesis of organic sulfur compounds. Home: 112 Quincy Shore Dr., Quincy, Mass. 02171. Office: Northeastern U., Boston 02115.*

ZUHDI, Nazih, surgeon; b. Beirut, Lebanon, May 19, 1925; s. Omar and Lutfiye (Atef) Z.; B.A., Am. U., Beirut, 1946, M.D., 1950; m. Lamya Mujahed, Sept. 17, 1947; children—Omar, Nabil. Came to U. S., 1950, naturalized, 1963. Research fellow State U. N.Y., 1953-54, asst. instr. surgery, 1955-56; resident cardiac surgery U. Hosp., Mpls., 1956, chief resident thoracic surgery U. Hosp., Oklahoma City, 1957-58; practice medicine specializing in cardiovascular and thoracic surgery, Oklahoma City, 1958-—; chmn. research com. Mercy Hosp., Oklahoma City, 1959-—, mem. intensive care com., infection com., 1965-—; mem. staff Bapt. Meml. Hosp., St. Anthony Hosp. (both Oklahoma City). Mem. com. 8 Internat. Congress Diseases of Chest, Mexico City, 1964. Diplomate Am. Bd. Surgery, Am. Bd. Thoracic Surgery. Mem. A.M.A., Am. Thoracic Soc., Osler Soc., Internat. Coll. Angiology, Internat. Coll. Surgeons, Am. Coll. Angiology, Am. Coll. Chest Physicians, Am. Coll. Cardiology, A.C.S., Am. Soc. Artifical Internal Organs, Soc. Thoracic Surgeons (founding), Internat. Cardiovascular Soc., Am. Assn. Thoracic Surgery, others. Research and publs. on physiology of extracorpeal circulation, internal hypothermia, coronary perfusion, assisted circulation; open-heart surgery; devel. heart-lung machines; design, exptl. implantation in chest of plastic bypass hearts; originator total definitive priming of heart-lung machines with blood substitutes; other cardiovascular studies. Home: 1607 Brighton St., Oklahoma City 73120. Office: 1211 N. Shartel, Oklahoma City 73103.*

ZUIDEMA, George Dale, Am. surgeon; b. Holland, Mich., Mar. 8, 1928; s. Jacob and Reka (Dalman) Z.; A.B., Hope Coll., 1949; M.D., Johns Hopkins, 1953; m. Joan K. Houtman, June 2, 1953; children—Karen Sue, David Jay, Nancy Ruth, Sarah Kay. Intern, Mass. Gen. Hosp., Boston, 1953-54, asst. resident surgeon, 1954, 57-58, chief resident, 1959; practice medicine, specializing in surgery, Ann Arbor, Mich., 1960-64, Balt., 1964-—; asst. prof. surgery U. Mich., 1960-63, asso. prof., 1963-64; prof., dir. dept. surgery Sch. Medicine, Johns Hopkins, 1964-—; surgeon-in-chief Johns Hopkins Hosp., 1964-—; mem. sci., tech. adv. com. NASA, 1965-—; cons. Walter Reed Army Med. Center, NIH Clin. Center, Balt. City Hosp., Sinai Hosp. Recipient Henry Russel award U. Mich., 1963. John and Mary R. Markle scholar in academic medicine, 1961-—. Diplomate Am. Bd. Surgery. Fellow A.C.S.; mem. Assn. Am. Med. Colls., Central Soc. Clin. Research, Soc. U. Surgeons, Am., So. surg. assns., Soc. Clin. Surgery, Soc. Vascular Surgery, Internat. Cardiovascular Soc., Halsted Soc., Phi Beta Kappa, Tri Beta, Alpha Omega Alpha. Editor: (with O. H. Gauer) Gravitational Stress in Aerospace Medicine, 1961; (with R. D. Judge) Physical Diagnosis, 1963; (with G. L. Nardi) Surgery-A Concise Guide to Clinical Practice, 1961. Editor, Jour. of Surg. Research, 1966-—. Research, numerous publs. on liver diseases and accompanying nervous system alterations due to ammonia toxicity, defects in salt and water metabolism and kidney function. Home: Box 55-D, Chapel Ct., Timonium, Md. 21093. Office: Johns Hopkins Hosp., 601 N. Broadway, Balt. 21205.*

ZÜLICK, K. J., German neurologist; b. Allenstein, Germany, Nov. 4, 1910; s. Georg and Lilli (v. Brincken) Z.; m. Marie-Luise Neven, Jan. 9, 1947; children—Anne-Kathrin, Christiane-Maria, Johann-Christoph. Dir. dept. neurology Max Planck Inst. Brain Research; dir. neurol. unit Köln (Germany) City Hosp. Decorated Knight Brasilian Order of Merit, Cruzeiro Do Sul. Author: Die Hirngeschwülste, 1959; also numerous articles. Description and classification of brain tumors and reactions of brain to tumor growth; analysis of cerebrovascular diseases, especially stroke and studies in treatment; analysis of disturbances of brain function, particularly motility after brain traumatism and other brain diseases. Address: 200 Ostmerhei-merstr., Cologne, West Germany.*

ZULLO, Victor August, Am. marine biologist; b. San Francisco, July 24, 1936; s. Albino J. and Marie A. (Gius) Z.; B.A., U. Cal. at Berkeley, 1958, M.A., 1960, Ph.D., 1963; m. Janet E. Lewis, Aug. 13, 1960. Fellow in systematics systematics-ecology program Marine Biol. Lab., Woods Hole, Mass., 1962-63, resident systematist, 1963-67, asst. dir., 1964-66; asso. curator dept. geology Cal. Acad. Scis., 1967-—. Mem. Soc. Systematic Zoology, Paleontol. Soc., Soc. Vertebrate Paleontology, Am. Soc. Zoologists. Research and publs. on classification and evolution of Cirripedia, paleontology of marine invertebrates of Pacific Basin, systematics of fossil molluscs and echinoids. Home: 82 Graceland Dr., San Rafael, Cal. 94901. Office: Cal. Acad. Scis., San Francisco 94118.*

ZUMBERGE, James Herbert, Am. geologist; b. Mpls., Dec. 27, 1923; s. Herbert Samuel and Helen (Reich) Z.; student Duke, 1943-44; B.A., U. Minn., 1946, Ph.D., 1950; m. Marilyn Edwards, June 21, 1947; children—John Edward, Jo Ellen, James Frederick, Mark Andrew. Faculty dept. geology U. Mich., Ann Arbor, 1950-62, prof., 1960-62; pres. Grand Valley State Coll., Allendale, Mich., 1962-68; dir. Sch. Earth Sci., prof. geology U. Ariz., Tucson, 1968—; instr. Duke, Durham, N.C., 1946-47. Cons. geologist on ground water and non-metallic minerals, 1950-62. Mm. Geol. Soc. Am., Am. Geophys. Union, Soc. Econ. Geologists, Glaciological Soc., A.A.A.S., Sigma Xi. Author: The Lakes of Minnesota, 1952; Laboratory Manual for Physical Geology, 1967; Elements of Geology, 1963. Research and numerous publs. on movement of Ross Ice Shelf in Antarctic during IGY; studies in Lake Superior with new evidence bearing on its origin; glacial geology of Gt. Lakes region. Home: 2740 Avenida de Posada, Tucson 85718.*

ZUMPT, Fritz Konrad Ernst, German entomologist; b. Germany, May 11, 1908; s. Konrad and Olga (Gebauer) Z.; Staatsexamen, Humboldt U., Berlin, Germany, 1932, Ph.D. magna cum laude, 1931; m. Gertrude Elsner, June 3, 1935; children—Ingolf Fritz, Gisbert Fritz. With pvt. pest control firm, 1933-34; asst., head dept. Tropical Inst., Hamburg, Germany, 1934-48; with dept. entomology and parasitology S. African Inst. for Med. Research, Johannesburg, S.Africa, 1948—, head dept., 1963—. Lectr. med. entomology U. Witwatersrand. Fellow Royal Entomol. Soc. London; mem. Société Royal d'Entomologie de Belgique (hon.), Société Belge Medicine Tropicale (corr.), Mus. Natural History (corr.), S. African Entoml. Soc., Deutsche Entomologisches Gessellschaft Berlin, Mitteilungen Entomologisches Gessellschaft, Deutsche Gesellschaft für Parasitologic, Inst. Man., Author: Die Tsetsefliegen, 1936; Grundriss der medizinischen Entomologie (with F. Weyer), 1940; Gesundheitsschädliche Insekten und Spinnentiere der warmen Länder (with F. Weyer), 1942; Insekten als Krankheitserremer und Krankheitsüberträger, 1956; Calliphoridae (Diptera cyclorrhapha) part I, 1956, part II, 1958, part III, 1961; The Arthropod Parasites of Veretebrates in Africa South of the Sahara, vol. I (with Andy Uaud, Lawrence, Theiter, Till Vercammen, Grandjean), 1961, vol. III (with Haeselbarth, Segerman), 1966; Myiasis in Man and Animals in the Old World, 1965; also numerous articles. Research on arthoropods (ticks, mites, scorpions, insects) which cause or transmit diseases, of humans, domestic and wild animals. Home: 33, 16th St., Parkhurst, Johannesburg, S. Africa.*

ZUM WINKEL, Karl, German radiologist; b. Weida, Germany, May 15, 1920; s. Karl and Marta (Geber) zum W.; Student U. Jena (Germany), 1939, U. Munchen, 1940, U. Berlin, 1942-43, U. Königsberg, 1943-44, U. Breslau, 1944-45; Dr.med. U. Göttingen, 1945; m. Gonda Podszuweit, Dec. 21, 1946; children—Detlef, Albrecht. Asst., Roentgen-Inst. Gera, 1951-57; head X-ray dept. Hosp. Weida (Germany), 1957—; chief physician Universitats Strahlenklinik Heidelberg (Germany), 1962— lectr. Heidelberg U., 1962—. Mem. Deutsche Roentgensellschaft, Gesellschaft fuer Nuclearmedizin. Author: Nierendiagnostik mit Radioisotopen, 1964; also numerous articles. Clin. and exptl. research on kidney examination with isotopes as renography, conventional and camera scintigraphy, and autoradiography, nuclear med. and roentgenologic diagnostics lymphatic system, radiotherapy of malignant thyroid, facial tumors, and the malignant systemic diseases. Home: 10 Lud.-Krehl-Str., Heidelberg, Germany.*

ZÜTT, Jürg, German psychiatrist; b. Karlsruhe, Germany, June 28, 1893; s. Adolf and Ida (Müller) Z.; M.D., U. Berlin, 1920; m. Ilse Renate Braun-Wogau, Apr. 15, 1937. Faculty, U. Berlin, 1932-37; dir. pvt. neurol. clinic, Berlin, 1937-46; prof. psychiatry and neurology U. Würzburg, 1946-50, U. Frankfurt/Main, 1950-64. Pres. com. for sci. research World Orgn. Psychiatry. Author: Auf dem Wege zu einer anthropologischen Psychiatrie, 1963; also numerout articles. Research on advancement of cultural sci. anthropology, psychiatry. Home: 57 Holbeinstrasse, Frankfort/Main, Germany.*

ZVEREV, Mitrofan Stepanovich, Russian astronomer; b. Apr. 16, 1903; grad. Moscow Conservatory, 1929, Moscow U., 1931. Staff, Shternberg State Astron. Inst., Moscow, 1931-51; became dep. dir. Main Astron. Obs., USSR Acad. Scis., Pulkovo, 1951; faculty Moscow U., 1938-52, prof., 1948-52. Corr. mem. USSR Acad. Scis. Author: On the Catalogue of Faint Stars, 1940; Fundamental Astronomy, 1950, vol. 6, 1954. Research and publs. in meridional astronomy, faint stars, system of coordinates, time service, gravimetry, variable stars; compiled star catalogs. Office: Main Astron. Obs., USSR Acad. Sci., Leningrad M-140, Pulkovo, USSR.

ZVONKOV, Vasilii Vasilevich, Russian transport engr.; b. Borovichi (now Novgorod Oblast), USSR, Jan. 6, 1891; grad. Moscow Inst. Lines Communication Engrs., 1917; Dr.Tech. Sci. Staff various transport orgns. until 1929; with Moscow Inst. Lines Communication Engrs., 1929-33; prof. Mil. Transport Acad., 1935-55; staff sect. on sci. solution transp. problems USSR Acad. Scis., 1939-55, joined staff Inst. Complex Transp. Problems, 1955. Chmn. sci. and tech. council Ministry Mcht. Marine; chmn. council for water mgmt. USSR Acad. Scis., editorial bd. tech. sect. News; mem. Commn. for Compiling Prospective Plan, USSR Gosplan; chmn. commn. for future transp. equipment State Sci. and Econ. Council, USSR Council Ministers. Named Honored Scientist, R.S.F.S.R., 1948; recipient Order of Lenin, 4 times; Order Red Star; Badge of Honor. Corr. mem. USSR Acad. Scis. Author: The Organization of Shipping Enterprises. Calculations, 1929; Marine Trade Calculation Problems with Examples of Practical Solutions, 1932; The Dispatching System in Water Transport, 1932; Comprehensive Categories of Technical Resources in Inland Water Transport, 1948. Studies in planning and devel. of water transportation in USSR, 1922—. Home: Kote'nicheskaya nab. Y15. Office: Inst. Complex Transp. Problems, USSR Acad. Scis., Moscow, USSR.

ZWEIFEL, Paul Frederick, Am. physicist; b. N.Y.C., June 21, 1929; s. Fritz and Dorothy Mary (Alterskye) Z.; B.S., in Physics, Carnegie Inst. Tech., 1948; Ph.D., Duke, 1954; m. Constance Reed Bailey, June 11, 1960; children—Frederick Feza, Evan Rudolph. Research asso., mgr. theoretical physics, cons. physicist Knolls Atomic Power Lab., Gen. Electric Corp., Schenectady, 1953-58; asso. prof. nuclear engring. U. Mich., Ann Arbor, 1958-60, prof., 1960-68; prof. physics Va. Poly. Inst., Blacksburg, 1968—; vis. lectr. Union Coll., Schenectady, 1954-55, U. Wis., 1964; vis. prof. theoretical physics Middle E. Tech. U., Ankara, Turkey, 1964-65. Cons. to labs. and atomic energy cos.; mem. adv. com. reactorphysics U. S. AEC, 1958-64. Fellow Am. Phys. Soc., Am. Nuclear Soc. (past chmn. Mich. sect., dir. 1967—); mem. Fedn. Am. Scientists (past sec.). Research and numerous publs. in beta-decay theory, nuclear reactor theory, neutron transport theory, slow neutron scattering.*

ZWELFER (or Zwelffer, Swölfer), Johann, German physician; b. Palatinate, 1618; studied medicine; grad. at Padua. Apothecary for 16 years, then practiced medicine in Vienna; prof. medicine, Vienna. Author: Pharmacopoeia Augustana et eius mantissa cum animadversionibus, 1652; Pharmacopeia Regia sive Dispensatorium Novum locupletatum et absolutum cum annexa Mantissa Spagyrica et gemino discursu apologetico contra Ott. Jachenium et Franc. Vernis, 1668; Discursus Apologeticus . . . , 1675. Claimed to be 1st to reduce pharmacy to system; pub. pharm. works, including Augsburg Pharmacopoeia, 1652; criticized Tachenius for confusing lyes and alkalis with salts of minerals and animals and for his method of fixing volatile salt of vipers; used chem. remedies; a solar diaphoretic antimony, mercury precipitates, mercury sublimate, turpethum minerale, and turpethum minerale rubrum. Died 1668.

ZWEMER, Raymond Lull, Am. biologist; b. Bahrein Islands, Mar. 30, 1902 (parents Am. citizens); s. Samuel Marinus and Amy Elizabeth (Wilkes) Z.; A.B., Hope Coll., Mich., 1923; Ph.D., Yale, 1926; Nat. Research fellow Harvard, 1926-28; m. Dorothy Ingeborg Bornn, Sept. 13, 1929; children—Raymund Wilkes, Suzanne (Mrs. R. A. Visser), Theodore Lestrup, Jane Karen (Mrs. R. L. Koeser). Mem. faculty Columbia, 1928-46; exec. sec. Nat. Acad. Scis. and NRC, 1947-50; chief of sci. div. Library of Congress, 1950-55; with UNESCO, 1956-58; asst. sci. adviser Dept. of State, 1958-61; asso. editor jours. Am. Physiol. Soc., 1961-65; exec. editor fedn. preceedings Fedn. Am. Socs. for Exptl. Biology, Bethesda, Md., 1966—. Mem. advisery com. Insts. on Research Adminstrn., Am. U., 1959—; bd. dirs., sec. Council on Biol. Scis. Information; also cons., lectr. to univs. and brs. of govt. Guggenheim fellow, 1941. Fellow A.A.A.S.; mem. Am. Assn. Anatomists (com. on internat. anat. nomenclature), Am. Assn. Phys. Anthropologists (life), Am. Physiol. Soc., Nat. Geographic Soc., N.Y. Acad. Sci. (v.p., A. Cressy Morrison prize 1937), Conf. Biol. Editors (dir. 1963—), Soc. Exptl. Biology and Medicine, Washington Acad. Scis., Sigma Xi, Gamma Alpha, others; hon. mem. Socidad de Bologia Argentina, Sociedad de Medicina Uruguay, others. Research and publs. on adrenal gland structure and function; focused ultrasound; movement and effects of potassium under various physiol. and path. conditions. Home: 5008 Benton Av., Bethesda. Office: 9650 Rockville Pike, Bethesda, Md. 20014.*

ZWICKY, Fritz, astrophysicist; b. Varna, Bulgaria (citizen of Switzerland), Feb. 14, 1898; s. Fridolin and Franziska (Wrcek) Z.; B.S. in Physics, Fed. Inst. of Tech., Zurich, Switzerland, 1920, Ph.D., 1922; m. Margaritha Anna Zuercher, Oct. 15, 1947; children —Margarit, Franziska, Barbara. Came to the U. S., 1925. Began as research asst. Fed. Inst. Tech., Zurich, Switzerland, 1920-25; internat. research fellow, Cal. Inst. Tech., Rockefeller Found., 1925-27, asst. prof., theoretical physics, 1927-29, asso. prof., 1929-41, prof. astrophysics 1942—; dir. research Aerojet Engring. Corp., Azusa, Cal., 1943-49, tech. adviser, chief research cons., 1949-61; astronomer Mount Wilson and Palomar Observatories 1948—. Chmn. bd. trustees Pestaloozi Found. Am., 1958. Recipient Presdl. Medal of Freedom, 1949. Mem. Am. and Swiss phys. socs., Am. Astron. Soc., Internat. Acad. Astronautics (v.p. 1965), Soc. Morphological Research (founder, pres. 1961). Author: Morphological Astronomy, 1957; Morphology of Propulsive Power, 1962. Contbr. numerous articles to Am., Swiss and German sci. mags. Inventor aeropulse, hydropulse, hydroturbojet, mono-propellants, coruscatives discovered supernovae, dwarf galaxies, Humason-Zwicky stars, intergalactic matter; compact galaxies; theory of neutron stars, pygmy stars; inventor of morphological method. Patentee in field. Home: 2065 Oakdale St., Pasadena, Cal. 91107.*

ZWIKKER, Cornelis, Dutch physicist; b. Zaandam, Netherlands, Aug. 19, 1900; s. Klaas and Klaartje (Dil) Z.; Master, U. Amsterdam, 1923, Sc.Dr., 1925; m. Johanna Dorothea Theinert, Apr. 17, 1924; children—J. D. Koenen, Adriana Cornelia Yonker, Kees. Research asso. Philips Co., Eindhoven, Holland, 1923-29, dir. light div., 1945-52; prof. applied physics Delft Inst. Tech., 1929-45; dir. Nat. Aero. Lab., Amsterdam, 1952-56; prof. sci. materials Eindhoven Inst. Tech., 1956—. Mem. Netherlands Phys. Soc. (past pres.), Netherlands Acoustical Soc. (past pres.), Netherlands Inst. Engrs. (past pres. sect. research), Netherlands Math. Soc., Netherlands Chem. Soc. Author: Properties of Tungsten, 1925; Short Treatise on Illumination, 1932; Optics, 1932; Solid Materials, 1940; Fluorescent Lighting, 1954; Acoustical Materials, 1950; also numerous articles. Editor, Physica, 1930—, Netherlands Jour. Physics, 1932—. Research on refractory metals, reverberation in auditoriums, acoustic materials, ultrasonic delay lines, ferrites; devel. in radio-statics bldg., lighting, aircraft constrn. Home: 80 Helmerslaan, Eindhoven, Holland.*

ZWINGER, Theodor (the Elder), Swiss physician; b. Basel, Switzerland, Aug. 2, 1533; student (under Ramus), Paris, France, of medicine, Padua, Venice (both Italy). Prof. Greek, Basel, from 1559, prof. theoretic medicine, from 1571; ch. historian, tchr. moral philosophy. Author: Theatrum vitae humane, 1565; Methodus apodemica, 1577; Physiologia medica, 1610. Supporter of Paracelsus; med. work well-known (still considered scholarly). Died Basel, Mar. 10, 1588.

ZWISLOCKI, Jozef John, elec. engr.; b. Lwow, Poland, Mar. 19, 1922; s. Tadeus and Helena (Moscicki) Z.; diploma in Elec. Engring., Fed. Tech. Inst., Zurich, 1944, Sc.D., 1949; m. Sylvia Goldman, July 11, 1954. Came to U. S., 1952, naturalized, 1960. Research asst., head, electroacoustic lab. dept. otolaryngology U. Basel, Switzerland, 1945-51; research fellow psychoacoustics lab. Harvard, 1951-57; faculty Syracuse U. (N.Y.), 1957—, prof. elec. engring., 1962—, dir. Lab. Sensory Communication, 1963—. Research faculty N.Y. State Upstate Med. Center, Syracuse, 1961—, research prof., 1967—. Mem. review panel communicative scis. NIH, 1966—; chmn. exec. com. Com. on Hearing, Bioacoustics and Biomechanics, NRC-Nat. Acad. Scis., 1966-67. Fellow Acoustical Soc. Am., Am. Speech and Hearing Assn.; mem. Internat. Soc. Audiology (v.p. 1967, exec. com. 1967—), I.E.E.E. (sr.), N.Y. Speech & Hearing Assn., Psychonomic Soc., A.A.A.S., Sigma Xi. Research, publs. in hearing field, including mechanics of middle and inner ear, bone conduction, psycho-physiol. analysis, theory of temporal summation, central masking, adaptation, loudness, differential sensitivity, acoustic method ear examination, narrow-band masking for audiometry; theory and design ear protectors; patentee. Home: 6 Lakeview Rd., Fayetteville, N.Y. 13066.*

ZWOLINSKI, Bruno John, Am. phys. chemist; b. Buffalo, Nov. 4, 1919; s. Bronislaus John and Elizabeth (Glowska) Z.; B.S. in Chemistry, Canisius Coll., 1941; M.S. in Chemistry, Purdue U., 1943; Ph.D. in Chemistry, Princeton, 1947; m. Margery Williams, Aug. 16, 1952; children—Jan, Jeffrey, Maria. Research scientist Manhattan Project, Columbia U., 1944-46; asst. prof. U. Utah, 1948-53; sr. physicist Stanford Research Inst., 1953-54; lectr. chemistry Carnegie Inst. Tech., 1957-61; dir. Thermodynamics Research Center, Tex. A. and M. U., College Station, 1961—, head dept. chemistry, 1964-65, prof., 1961—. Fellow Am. Inst. Chemists, N.Y. Acad. Scis., Tex. Acad. Sci.; mem. Am. Chem. Soc., Am. Phys. Soc., A.A.A.S., Sigma Xi. Research and numerous publs. on theoretical phys. chemistry. Home: 903 Francis Dr., College Station, Tex. 77840.*

ZWORYKIN, Vladimir Kosma, physicist; b. Mourom, Russia, July 30, 1889; s. Kosma A. and Elena Z.; Elec. Engring., Petrograd. Inst. Tech., 1912; post-grad. in Physics, Coll. France, 1912-14; D.Sc., Poly. Inst. Bklyn., 1938; m. Katherine Polevitsky, Nov. 14, 1951; children (from previous marriage)—Nina, Elaine. Came to U. S., 1919, naturalized, 1924. Mem. research staff Westinghouse Corp., 1920-29; dir. Electronic Research Lab., RCA, Princeton, N.J., 1929-54, dir. Gen. Research Lab., Labs. Div., 1954—, hon. v.p., 1954—. Recipient De Forest Audion award, 1966; Nat. Medal of Sci., 1966; numerous other awards. Fellow I.E.E.E., Television Soc. Eng. (hon.), Inst. Internazionale delle Communicazoni (Italy; hon.); Nat. Acad. Engring., Nat. Acad. Scis., Am. Acad. Arts and Scis., Am. Inst. Physics, Am. Phys. Soc., A.A.A.S., Electron Microscope Soc. Am. (charter), Brit. Instn. Radio Engrs. (hon.), Television Engrs. Japan (hon.), Sigma Xi, Eta Kappa Nu (eminent mem.), others. Invention of iconoscope, kinescope, emission multiplier, image tube, electron microscope. Home: 103 Battle Rd. Circle. Office: RCA Labs., David Sarnoff Research Center, Princeton, N.J. 08540.*

ZYGMUND, Antoni, mathematician; b. Warsaw, Poland, Dec. 26, 1900; s. Vincent and Antonina (Perkowska) Z.; Ph.D., U. Warsaw, 1923; m. Irena Parnowska, Feb. 25, 1925; 1 son, George. Came to U. S., 1940, naturalized, 1947. Prof., U. Chgo., 1947—, Distinguished Service prof., 1963—. Mem. U. S., Polish, Argentina nat. acads., Am., Polish, London math. socs. Author: Trigonometric Series, 1935; (with S. Saks) Analytic Functions, 1938; also numerous articles. Home: 5420 East View Park, Chgo. 60615.*

ADDENDUM

ADAMS, John Bertram, English nuclear engr.; b. Kingston, Eng., May 24, 1920; s. John and Emily (Searles) A.; ed. Eltham Coll., London, Eng.; D.Sc. (hon.), U. Geneva (Switzerland), 1960, U. Birmingham (Eng.), 1961, U. Surrey (Eng.), 1966, M.A. Oxford 1967; m. Renie Warburton, 1943; children—Josephine, Katharine, Christopher John. Staff, Telecommunication Research Establishment, Swanage, Malvern, Eng., 1940-45, Atomic Energy Research Establishment, Harwell, Eng., 1945-53; dir. proton-synchrotron div. European Orgn. for Nuclear Research, Geneva, 1954-60, dir. gen. Europ. Orgn. for Nuclear Research, Geneva 60-61; dir. Culham. Lab., United Kingdom Atomic Energy Authority, Abingdon, Eng., 1960-67, mem. for research, 1966—; controller Ministry Tech., London, 1965-66. Mem. Council for Sci. Policy, 1965—, Adv. Council on Tech., 1965—. Recipient Röntgen prize U. Giessen, 1960, Duddell medal Phys. Soc., 1961; decorated Companion of St. Michael and St. George, 1962. Fellow Royal Instn. Elec. Engrs., Royal Soc. Design and constrn. of large accelerating machines for high energy nuclear physics; dir. nuclear fusion and plasma physics research. Home: Grey House, Lincombe Lane, Boar's Hill, Oxford, Eng. Office: Office Mem. for Research, UKAEA, Harwell, Eng.

ARNDT, Helmut, German economist; b. Königsberg, Germany, May 11, 1911; s. Adolf and Louise (Zabeler) A.; ed. Berlin, Marburg and Munich univs.; Ph.D. in Law and Economics; m. Elfriede, Jan. 25, 1947; children—Claudia, Rolf. Prof. agrégé U. Marburg, 1946; vis. prof. U. Syracuse (N.Y.), 1950-51; prof. U. Marburg, 1952, U. Istanbul, 1953, Technische Hochschule Darmstadt, 1954, Free U. Berlin, 1957; mem.-consul Commn. of Inquiry, 1961-62; vis. prof. U. Heidelberg, 1963; dir. Inst. Econs.; dir. Inst. for Research of Economic Concentration. Mem. Soc. Econ. and Social Sci. (pres. 1966); German Soc. Sociology; Am. Econ. Asso.; Internat. Polit. Sci. Asso.; Royal Econ. Soc.; Econometric Soc.; List Soc.; Internat. Econ. Asso. (council mem. 1966). Author: Voraussetzungen des Marktautomatismus, 1947; Schöpferischer Wettbew und Klassenlose Gesellschaft, 1952; Ausbeutung und Marktform, 1959; Die Konzentration in der Wirtsch (ed.), 1960; Marktmechanismus und heterogene Konkurrenz, 1964; Mikroökonomische Theorie, 2 vols., 1966; Die Konzentration der westdeutschen Wirtschaft, 1966; Die Konzentration in der Presse und die Problematik des Verleger-Fernsehens, 1967; Konzentration und Konzentrationspolitik (ed.), 1967. Office: 1 Berlin 33, Garystrasse 21, West Germany.* (replaces sketch page 64)

ASCENZI, Antonio, pathologic anatomist; b. Boulogne-sur-Mer, France, May 4, 1915; s. Armando and Adreina (Allevy) A.; M.D.; m. Roberta Graziani, May 15, 1929; children—Paolo, Maria-Grazia. Prof. pathologic anatomy U. Pisa (Italy). Recipient Feltrinelli prize, Nat. Italian Acad. of Lincei, 1964. Mem. Acad. Medicine Rome, Italian Soc. of Pathology, Am. Assn. Anatomists. Research, publs. on relation of structure, ultrastructure and micro-mechanical properties of bone; discoverer of mandible of Neanderthal man "Circeo III b". Home: 21 via R. Pereira, Rome. Office: Istituto di Anatomia Patologica, via Roma 57, Pisa, Italy.* (replaces sketch page 69)

ASTAUROV, Boris Lvovich, Russian biologist; b. Kazan, Russia, Oct. 27, 1904; s. Lev Mikhailovich and Olga Andreyevna (Teeckenko) A.; grad. Moscow U., 1927, Cand. Biol. Sci., 1936, D.Biol. Sci., 1938; m. Tatiana Michilovna Yakovleva, Jan. 15, 1936; 1 dau., Olga; m. 2d, Natalia Sergaeevna Skadovskaya, Jan. 8, 1944; 1 dau., Natalia. Asst. genetics lab. Inst. Exptl. Biology, Moscow, 1924-26; sci. worker genetics dept. Commn. for Investigation Natural Resources, USSR Acad. Scis., Moscow, 1926-30; sci. worker dept. genetics and breeding Middleasian Inst. Sericulture, Tashkent, 1930-36; sci. worker lab. developmental mechanics Inst. Exptl. Biology, 1939—; Inst. Cytology, Histology and Embryology, 1935-47; head Filatov Lab. Developmental Mechanics, 1947-48; sci. worker, 1948-55; head Filatov Lab. Exptl. Embryology, Severtzov Inst. Animal Morphology, 1955-67; head lab. developmental cytogenetics, Inst. Developmental Biology, 1967, dir. 1967—. Recipient Silver medal for invention heat shock method for thermic cure of silkworm nosema disease All-Union Exhbn. Achievements of Nation, 1963; G. Mendel 100th Anniversary Meml. medal Czechoslovak Acad. Sci., 1965. Fellow Internat. Inst. Embryology; mem. Internat. Soc. Cell Biology, Moscow Soc. Naturalists (chmn. genetics sect.), Am. Soc. Zoologists (corr.), USSR Acad. Scis. (academician), W. J. Vavilov Genetics and Breeding Soc. (pres. 1966—). Mem. editorial bd. Cytology, 1959—, div. biology Bull. Moscow Soc. Naturalists, 1962—, Genetics, 1965—, Priroda, 1965—. Research and numerous publs. on cytogenetics and developmental biology of Drosophila and silkworms; developer method of thermal artificial parthenogenesis in unfertilized silkworm eggs, 1934; complete androgenesis in animals especially for 1st time; interspecific androgenesis; devel. of exclusively adult male progeny from enucleated egg cytoplasm taken from one species with nucleus taken from another species; proved by means of androgenesis the nuclear versus cytoplasmic control of specific differences and nuclear versus cytoplasmic localization of injuries caused by ionizing irradiation, 1947, prodn. of exptl. tetraploid bisexual strain for 1st time in animal (silkworm), 1957-66. Home: 3 Demetr. Ulyanov ulitsa, Moscow B-333. Office: Inst. Developmental Biology, USSR Acad. Scis., Vavilov ulitsa 26, Moscow B-133, USSR.* (replaces sketch page 72)

ASTUNI, Enrico, Italian engr.; b. Avellino, Italy, Feb. 28, 1914; s. Enrico and Ida (Caretti) A.; studied classics, indsl. and electro-tech. engring.; Dr. in engring.; m. Milena Liguori; children—Enrico, Domenico, Giulio, Ida. Dir. electro-tech. Inst. U. Genoa (Italy). Mem. AEI Genoa (p. pres.), Tech. Counsel of Min. Telecommunications, Sci. Counsel of Int. Inst. Telec., Lig. Academy. Author: 4 vols. on foundations of electrical engring. and electric machines; also study on type of commutator machines and methods on a.c. networks analysis. Home: viale Gambaro 15, Genoa (16146), Italy. Office: via Montallegro 1, Genoa (16145), Italy.* (replaces sketch page 73)

AZIZBEKOV, Shamil Abduragimovich, Azerbaijan geologist; b. 1906; grad. Azerbaijan Polytechnical Inst., 1930; candidate, 1934, Dr. geological-mineralogical sci., 1943. Prof. of petrography and mineral resources Azerbaijan Industrial Inst., 1944—. Mem. Academician, Acad. of Sci. Azerbaijan SSR, 1945— (sec. of earth's sci. dept., mem. of presidium); Honored sci. worker of Azerbaijan, 1959—. Chief editor: Geology of Azerbaijan, 7 vols.; author over 230 publs. in geology and petrography. Made geological investigations of the Caucasus, since 1930. Home: Lenin prospect, 3, fl. 29, Baku, USSR. Office: Academy of Sciences of Azerbaijan SSR, Baku, USSR.* (replaces sketch page 80)

BALCELLS ROCAMORA, Enrique, Spanish zoologist; b. Barcelona, Spain, Mar. 31, 1922; s. Francisco and Asunción (Rocamora) B.; Licenciado en Ciencias Naturales, U. Barcelona, 1943; D.Naturals Scis., U. Madrid (Spain), 1950. Scientist, Consejo Superior de Investigaciones Cientificas, Barcelona, also prof. Barcelona U., 1944-64; dir. Centro Pirenaico de Biología Experimental, Jaca, Spain, 1963—. Research, publs. on insects and vertebrates. Home: 87 Vía Layetana, Barcelona, Spain. Office: 64 Apartado, Jaca, Spain.*

BARUK, Henri Marc, French physician; b. Saint-Ave (Morbihan), France, Aug. 8, 1897; s. Jacques and Marie (Brechon) B.; ed. Lycée at Angers, Faculté de médecine at Paris; M.D.; m. Suzanne Sorano, Dec. 29, 1947. Specialist in neuropsychiatry; intern. Hosps. of Paris, 1921-26; head clinic Faculté de médecine at Paris, 1926-30; now prof. agrégé; chief physician Nat. House of Charenton, 1932—, dir. Sch. Advanced Studies. Recipient award Com. of Pub. Health. Mem. Nat. Acad. Medicine, Am. Internat. Acad. (hon.), also numerous socs. in France and other countries. Author 13 books and over 500 articles on neuropsychiatry, experimental neuropsychiatry (experimental catatonia), social and moral psychiatry; studies about Hebraic civilization. Address: 5 quai de la Republique, Saint Maurice (Seine), France.*

BILO, Julien Eugène Charles, Belgian mathematician; b. Ostend, Belgium, May 21, 1914; s. Charles Joseph and Alida (Poppe) B.; B.S., State U. Ghent (Belgium), 1934, Lic.Sc., 1936, D.Sc., 1945; agrégé Ens. Sup., 1948; m. Lydie Josephine Ponjaert, Sept. 9, 1939; children—Frank, Rita. Asst., State U. Ghent, 1936, faculty, 1948—, prof. math., dir. research dept. higher geometry, 1952—; faculty State Secondary Normal Sch., Ghent, 1936-48. Mem. Royal Flemish Acad. Scis., Lit. and Fine Arts Belgium (ann. prize 1947), Belgian Math. Soc., Math. Assn. Am. Author: Bijdrage tot de Grondslagenleer van de gewone complexe projectieve meetkunde, 1949; Onderzoekingen betreffende de meetkundige grondslagen van de projectieve quaternionenmeetkunde, 1949; also papers. Editor jour. Simon Stevin, Wis- en Natuurkundig Tijdschrift. Research on founds. of synthetic complex projective geometry and of quaternion geometry, constrn. of theory by synthetic methods. Home: 164 Oude Brusselse Weg, Gentbrugge, Belgium. Office: Univ. of Ghent 22, J. Plateaustraat, Ghent, Belgium.*

BITTEL, Kurt, German archaeologist; b. Heidenheim, Germany, July 5, 1907; s. Emil B.; Ph.D., ed. Us. Heidelberg, Berlin, Vienna, Marburg; m. Maria Riediger, 1951. Fellow German Archaeological Inst., 1930; asst., Archaeological Inst., Istanbul, 1933, dir. 1938; prof. U. Tübingen, 1946; visiting prof. U. Istanbul, 1951; dir. German Archaeological Inst., Istanbul, prof. U. Istanbul, from 1953. Hon. mem. Royal Irish Acad.; Vienna, Mainz, Munich Acads.; Prehistorical Soc. (Eng.); Fellow Brit. Acad.; mem. Austrian, Bulgarian Archaeological Insts.; pres. German Archaeological Inst., 1960. Author: Prähistorische Forschung in Kleinasien, 1934; Die Kelten in Württemberg, 1934; (with H. G. Guterbock) Neue Untersungen in der hethitischen Hauptstadt, 1935; Die Ruinen von Boghazköy, der Hauptstadt des Hethiterreiches, 1937; (with R. Naumann) Neue Untersungen hethitische Architektur, 1938; (with R. Naumann and H. Otto) Yazilikaja, 1941; Kleinasiatische Studien, 1942; Grundzüge der Vor- und Frühgeschichte Kleinasiens, 1945; (with R. Naumann) Boghazköy-Hattusa, Ergebnisse der Ausgrabungen des Deutschen Archäologischen Instituts und der Deutschen Orient-Gesellschaft in den Jahren 1931-39, 1952. Archaeological excavations in Germany, Egypt, Turkey; directed excavations of Bogasköy, capital of Hittite Empire, uncovering over 800 inscribed clay tablets, fortified castle, and temple. Office: Peter-Lenné Strasse 28, 1 Berlin 33, Germany. Home: 45, Baseler Strasse 112, Berlin.

BOSSERT, Helmuth Th., German archaeologist; b. Landau, Germany, Sept. 11, 1889; s. Theodor and Emma (Seyb) B.; Ph.D., ed. Us. Heidelberg, Strasbourg, Freiburg, Munich; m. Hürmüz Aslivar, 1950. Scholar and publisher, Berlin, 1919-24; prof. and dir. inst. ancient Near East, U. Istanbul, from 1934. Mem. Austrian and German Archaeology Insts.; German Oriental Soc.; German Turkish Soc. Author: Volkskunst in Europa, 1926; Geschichte des Kunstgewerbes aller Völker und Zeiten, 6 vols., 1928-39; Altanatolien, 1942; Asia, 1946; (with H. Cambel) Karatepe, 1946; (with U. Bahadir Alkim) Karatepe, 1947; Altsyrien, 1951. Research on Hittite Empire Anatolia; directed excavations of Karatepe, from 1947; and Misis, from 1955; discovered tablet containing both Phoenician and Hittite hieroglyphics; with others, deciphered Hittite script. Address: University of Istanbul, Istanbul, Turkey.

BRUHNS, Carl Christian, German astronomer; b. Plön, Holstein, Germany, Nov. 22, 1830; doctorate, Berlin, 1856. Asst. Berlin Observatory, 1852; observer, 1854; instr. U. Berlin, 1859; prof. astronomy, dir. observatory, Leipzig, 1860. Author: Die astronomische Strahlenbrechung in ihrer historischen Entwickelung, 1861; Geschichte und Beschreibung . . . , 1861; Astronomische und geodätische Arbeiten . . . , 1864-74; Nouvelle Manuel de logarithmes à sept décimales, 1869; Biographie von J. F. Encke, 1869; Atlas der Astronomie, 1872; (with others) Al. von Humboldt, eine wissenschaftliche Biographie, 3 vols., 1872. Discovered 6 comets; computed cometary and planetary orbits; important work in connection with European triangulation; set up meteorological stations in Saxony, 1863; established weather forecasting bureau, Leipzig, 1878. Died Leipzig, Germany, July 25, 1881.

BURNS, Louisa, Am. osteopathic physician; b. Saltilloville, Indiana, Mar. 18, 1869; B.S., M.S., Borden Inst., 1906; Pacific College of Osteopathy, D.O. 1903, Sc.D. 1906. Taught physiology, Pacific College of Osteopathy, 1906-14; head, Still Research Inst., 1914-36; head, Louisa Burns Osteopathic Research Lab., 1936-57. Recipient Distinguished Service Certificate, Am. Osteopathic Assn., 1929. Mem. Am. Osteopathic Assn.; A.A.A.S. Author: Basic Principles, 1907; The Nerve Centers, 1911; The Physiology of Consciousness, 1911; Cells of the Blood, 1931; (editor with others) Pathogenesis of Visceral Disease Following Vertebral Lesions, 1948; numerous articles. Research on determination of physiologic and anatomic effects of structural abnormalities in man and exptl. animals. Died Whittier, Cal., Jan. 19, 1958.

CAIANIELLO, Eduardo Renato, Italian physicist; b. Naples, Italy, June 25, 1921; s. Giuseppe and Sammartino (Lidia) C.; Ph.D., Rochester, N.Y.; m. Persico Carla, 1946; children—Dora, Eva, Orietta, Silvia. Dir. Laboratorio di Cibernetica del C.N.R.; prof. theoretical physics U. Naples. Mem. Am. Phys. Soc., Accademia Pontaniana, N.Y. Acad. Sci. Home: Via Posillipo 102, Naples. Office: Laboratorio di Cibernetica del C.N.R., Arco Felice, Naples, Italy.* (replaces sketch page 288)

CARAPANCEA, Mihai Titus, Rumanian ophthalmologist; b. Bucharest, Rumania, Apr. 20, 1920; s. Titus G. and Eleonora (Zissu) C.; M.D., Faculty Medicine, Bucharest, 1944; m. Efterpi Cristescu, July 11, 1964. Asst. lectr. Bucharest U., 1944-52, cons. ophthal. clinics, 1946-52; sci. research worker Inst. for Normal and Path. Physiology, Acad. of Rumania, Bucharest, 1951-53, head lab. for clin. and exptl. physiology and physiopathology of eye, 1953—. Decorated Star of Rumania. Mem. Acad. Rumania (mem. astro. commn.), Rumanian Soc. Norm. Path. Physiology, Internat. Soc. Ophthalmology, Internat. Soc. Clin. ERG, Internat. Soc. Biometeorology, Internat. Union Physiol. Sci. Research, numerous publs. on corneal bio-architectonic structure, neurosis of hypermetropia, exophthalmia through hypertonia of striated peri-ocular musculature, pathognomonic ERG indications in gen. fatigue phenomenon, recurrent alternating viral uveitis, transient viral retinal spasm and sclerosis, retinal and accommodative modifications at high altitude, non-specific action of drugs on eye, syndrome of ciliary plexus of orbit. Home: 21, Prof. I. Bogdan St., Bucharest III. Office: 11, Bul. 1 Mai. Bucharest II, Rumania.*

CARTER, Harry Nelson, Am. mathematician; b. Haileyville, Okla., Mar. 7, 1912; s. Ed and Cora

(Baldwin) C.; B.S., Northeastern Okla. State Coll., 1940; postgrad. U. Tulsa, 1945-46; M.S., U. Colo., 1950; m. Bonnie Lucille Jackson, Oct. 27, 1939. Tchr. pub. schs. Ark., Okla., 1936-41; math. instr. Spartan Sch. Aeronautics, Tulsa, Oklahoma, 1941-45; from instr. to prof. math. U. Tulsa, 1945—, asst. dean engring., 1958-59, acting dean, 1959-60, men's counselor, 1960-62, dean of students, 1962—. Mem. Am. Soc. Engring. Edn., Am. Math. Soc., Okla. Acad. Sci., Math. Assn. Am. (contest chmn. Okla., Ark. sect.), A.A.A.S. Research in summation of divergent series, and numerical solution of differential equations; derived general formula for determining sum of certain infinite series. Home: 3739 South Fulton, Tulsa 74135. Office: 600 South College, Tulsa, Okla. 74104.*

CAZENEUVE, Jean, French sociologist; b. Ussel, France, May 17, 1915; s. Charles and Yvonne (Renoul) C.; Agrégé, Docteur ès lettres, Ecole Normale Supérieure; m. Germaine de Paraize, Sept. 10, 1963; children—Sabine, Emmanuel. Prof., U. Alexandria (Egypt), 1948-50; Rockefeller Found. scholar, 1954-55; instr., research chief Centre National de la Recherche Scientifique, 1956-64, dir. research, 1964-66; prof. sociology Sorbonne, Paris, 1966—; adminstr. French Radio and TV (ORTF), 1964—. Mem. Internat. Assn. French-speaking Sociologists (sec. gen. 1962—), Americanists Soc., French Sociology Soc. Author numerous books including: Les rites; Sociologie de la Radio-Télévision; La mentalité archaique; also articles. Research on ethnology and anthropology of Zuni Indians; sociology of communications. Home: 4 rue Pierre Corneille, St. Germain-en-Laye 78, France. Office: Sorbonne, 17 rue de la Sorbonne, Paris, France.*

CERNATESCU, Radu, Rumanian chemist; b. 1894; prof. Iasi U.; mem. Acad. Romanian People's Republic. Author: (with E. Papafil) Sur l'influence des sels sur la solubilité de l'eau dans le phénol, 1937; (with M. Poni) Elementary Treatise of Inorganic Chemistry, 3 vols., 1950-51. Research in physical and analytical chemistry; extended Dalton's law to concentrated solutions; studied, synthetized complex sulphocyanates, animes of cobalt hydrop hydrosulphites and bivalent nickel. Died 1958.

CESARO, Angelo Nunziante, Italian physician; b. Naples, Italy, May 7, 1913; s. Carlo Nunziante and Stella (Esposito) C.; grad. U. Naples, 1938; m. Franca Barbara, Apr. 14, 1946; 1 dau., Stelluccia Nunziante. Vol. asst. inst. occupational medicine U. Naples, 1938-40; asst., inst. occupational medicine U. Siena, 1941, asst., inst. pathology, 1941-44; asst., inst. occupational medicine, univs. Milan, Padova, 1944-53; prof. occupational medicine, dir. inst. occupational medicine, U. Messina (Italy), 1954—. Exchange lectr., USSR; lectr., univs. Milan, Perugia, Rome. Mem. Internat. Commn. Occupational Medicine, Geneva, Italian Soc. Occupational Medicine (directory), Italian Soc. Histochemistry (co-founder). Contbg. author: Trattato Italiano di medicina interna (P. Introzzi); also papers in periodicals. Research on percutaneous absorption of toxic agts. of indsl. use, histopathology of silicotic lung, illnesses of sulfur-mine workers, sulfocarbonic chronic intoxication (primitive encephalopathic arteriosclerosis), blood cyto- and histochemistry, mechanism of action of anticryptogamic products in man, effects of aniline and derivatives on isolated cells (cancrologic research), pneumoconiosis from pumice (liparosis). Home: 22, S. Agostino, Messina, Sicily, Italy.*

CESI, Federigo, Prince (Duke of Acqua Sparta), Italian naturalist, b. Rome, Italy, 1585. Founder-supporter Accademia dei Lincei, Rome (1 of earliest modern sci. socs. for advancement of natural history). Author: Apiarium (contains oldest known drawings of objects seen through microscope, done by Francesco Stelluti); Tabulae Phytosophicae, 1630; prepared illustrated edit. F. Hernandez's Natural History of Mexico, 1651. Helped spread use of microscope and telescope; devised system classification plants and animals according to genera and species (influenced Linnaeus); 1st discoverer spores of ferns. Died Rome, 1630.

CHAMBERLIN, Thomas Chrowder, Am. geologist; b. Mattoon, Ill., Sept. 25, 1843; A.B., Beloit Coll., 1866, A.M., 1869; grad. science, U. of Mich., 1868-69 (Ph.D., univs. of Mich. and Wis., 1882; LL.D., U. of Mich., Beloit Coll., Columbian U., 1887, U. of Wis., 1904, Toronto U., 1913; Sc.D., U. of Ill., 1905, U. of Wis., 1920); m. Alma Isabel Wilson, 1867. Prof. natural science, State Normal Sch., Whitewater, Wis., 1869-72; prof. geology, Beloit, 1873-82, Columbian, 1885-87; pres. U. of Wis., 1887-92; prof. and head dept. of geology and dir. Walker Mus., U. of Chicago, 1892-1919 (prof. emeritus). Asst. state geologist, Wis., 1873-76, chief geologist, 1876-82; studied glaciers of Switzerland, 1878; U. S. geologist in charge of glacial div., 1882-1907; geologist Peary Relief Expdn., 1894; cons. geologist, Wis. Geol. Survey; commr. Ill. Geol. Survey; cons. geologist U. S. Geol. Survey; investigator fundamental problems of geology, Carnegie Instn., 1902-09; research asso., same instn., 1909—; mem. commn. for Oriental Ednl. Investigation, 1909. Fellow Am. Acad. Arts and Sciences. Author: Geology of Wisconsin; General Treatise on Geology (with R. D. Salisbury), 1906; The Origin of the Earth, 1916. Research on fundamental geology of solar system; (with F. R. Moulton) formulated planetesimal, or spiral nebula hypothesis to account for origin of earth; studied glacial deposits and their evidence as to climatic conditions in past ages. Died Chicago, Ill., Nov. 15, 1928.

CHATTERJEE, Asima Mukherjee, Indian chemist; b. Calcutta, India; d. Indra Narayan and Kamala Devi (Ghosal) Mukherjee; B.Sc., Scottish Church Coll., 1936; M.Sc., Calcutta U., 1938, D.Sc., 1944; m. Barondananda Chatterjee, Aug. 1945; 1 dau., Julie. Prof., head dept. chemistry Lady Brabourne Coll., Calcutta, 1940-54; reader U. Coll. Sci., Calcutta U., 1954-61, Khaira prof. chemistry Calcutta U., 1962—. Expert reviewing com. on chemistry Indian University Grants Commn. Eli Lilly & Co. grantee, 1955-57; Smith Kline and French Labs. grantee, 1956-59; Govt. of West Bengal grantee 1951—; Indian Council Med. Research grantee, 1963—; Council Sci. and Indsl. Research grantee, 1960—; Premehand Roychand scholar, 1942; recipient Mouat medal, 1944, Nagarjuna Gold medal and prize, 1949; Shati Swarup Bhainager Meml. award Council Sci. and Indsl. Research, 1961. Fellow Asiatic Soc., Nat. Inst. Scis. India; mem. Indian Chem. Soc., Indian Assn. for Cultivation Sci., Indian Sci. Congress Assn. (treas.), Indian Sci. News Assn., Council Sci. and Indsl. Research (mem. council), Indian Council Med. Research, Indian Inst. Petroleum, Sigma Xi. Author: Bharater Banaushadhi; Saral Madhyamic Rasayan, 3 vols.; also numerous articles, chpts. in books. Research in chemistry plant products, synthetic organic chemistry, mechanism of organic reactions, organic analytical chemistry, stereochemistry, chromatographical resolution of biol. products; discovered hypotensive, sedative and cardiovascular drugs of rauwolfia series, antiepileptic and antileukodermic drugs. Address: Bengal Engring. Coll., Qr. No. 42, Mowrah-3, India; also 92, Acharya P.C. Rd., Calcutta-9, West Bengal, India.*

CHRYSOCOCCES, Georgios, Byzantine astronomer, physician; flourished 1335-46 at Trebizond, Byzantine Empire; ed. by priest named Manuel, Academy of Trebizond. Practiced medicine in Trebizond. Author: Treatise on the astronomy of the Persians (including tables), 1346; Determination of moon's orbit; Determination of the sun's orbit; Table of longitudes and latitudes of remarkable cities. His astronomical and geographical tables have been used in various modern works.

CLARKE, Samuel, English mathematician, philosopher, theologian; b. Norwich, Eng., Oct. 11, 1675; s. Edward and Hannah (Parmeter) C.; B.A. Caius College, Cambridge, 1695; D.D. Cambridge; m. Katherine Lockwood; seven children, including Samuel. Chaplain to Bishop Moore of Norwich, 1698, later executor; began study divinity; publ. several theological works; rector of Drayton, nr. Norwich; gave Boyle lectures, 1704, 1705; rector St. Benet's Paul's Wharf, 1706; apptd. chaplain in ordinary to Queen Anne; rector St. James's Westminster, 1709; master Wigston's Hosp., Leicester, circa 1718; corresponded with Leibnitz on nature of time and space and on free will (publ. 1717). Author: Jacobi Rohaulti Rhysica, 1697; Three Practical Essays upon Baptism, Confirmation, and Repentance, 1699; Reflections on part of a Book called "Amyntor," 1699; Paraphrases on the Four Gospels, 1701-2; Boyle Lectures in 1704 and 1705 (publ. in 2 vols., 1705-06; later publ. as A Discourse concerning the Beings and Attributes of God . . .); Discourse concerning the Connection of Prophecies, 1725; Letter to Mr. Dodwell, 1706; Is. Newtoni Optice . . . , 1706; C. Julii Caesaris quae estant . . . , 1712; The Scripture Doctrine of the Trinity, 1712; A Collection of Papers which Passed between Dr. Clarke and Mr. Leibnitz . . . , 1717; Seventeen Sermons, 1724; Letter to B. Hoadly on Velocity and Force; Homeri Ilias Graece et Latine, 1729; Exposition of the Church Catechism, 1729; Ten volumes of Sermons, 1730-31; translated Rohault's Traité de Physique into Latin; translated Newton's Optics. Follower of Newton; defended Newtonian principles; considered 1st English metaphysician; philosophy opposed to Locke; founder of "intellectual" school (Wollaston and Price its chief English followers); involved in controversies with orthodox divines and deists because of his theological doctrine. Died May 17, 1729.

COLLEE, Robert F. J. M. J., metallurgical engr.; b. Liège, Belgium, June 18, 1924; s. Robert and Marie Leontine Collee; C E. in Mining, U. Liège, 1948, C.E. in Metallurgy, 1949. spl. doctorate in applied sci., diploma in nuclear engring.; m. Marie-Therèse Polet, Aug. 2, 1952; children—Pierre-Emmanuel, Marie-Christine, Sabine-Claire, Jean-Benoit. Mem. Faculty Applied Sci., U. Liège, 1948—, asso., 1959-65, prof., 1965—. Author: Un aspect nouveau de la surtension de l'hydrogene, 1954; L'influence electrochimique du zinc et hydrometallurgie du cobalt, 1962; La corrosion eng. nucleaire, 1966; Analyse cobalt par spectrometre gamma, 1960. Research on electrochem. aspects of metallurgy of zinc and cobalt, corrosion in nuclear engring., gaseous evolution in metall. electrochemistry, spectrometry gamma applied to analysis of cobalt in metallurgy. Home: 19 Ponson, Jupille (Liège). Office: 2 Stevart, Liège, Belgium.*

CONTÉ, Nicolas Jacques, French chemist; b. St.-Ceneri, France, Aug. 4, 1755; student of Grueze, Paris; m. Mlle. de Brossard; at least 1 dau.; abandoned art for science; organizer sch. and workshops of mil. airship sta., Meudon, 1794; served in French Army in Egypt, 1798; dir. work on Egypt pub. by Egyptian Commn., France; comdr. of Aerostatiers (after battle of Fleurus); founder factory for prodn. of Conté crayons; a founder Conservatory of Arts and Crafts. Inventor hydraulic machine; conceived of use of aerostats in mil. operations; discovered artificial graphite, 1795; invented substitute for plumbago for use in Conté crayons or pencils; conducted expts. for inflation of mil. balloons; devised metal-covered barometer for measuring heights; as mem. Egyptian expdn., built mills, optical telegraphs on balloons, hydraulic machines, including one operated by tides; inventor machines for textiles, agr. and powder manufacture, also artificial brood hens; inventor machine for engraving large copper plates, 1802. Died Paris, Dec. 6, 1805.

CONTU, Paolo, anatomist; b. Orani, Italy, Aug. 17, 1921; s. Priamo and Maria Angela (Zichi) C.; M.D. U. Cagliari (Italy); M.D. U. Recife, 1957, M.D. U. Rio Grande do Sul, 1959 (Brazil); m. Sônia Araujo De Carvalho, July 3, 1959; 1 dau., Daniela. Asst., asso. prof. anatomy, U. Bologna (Italy), 1947-54; prof. anatomy, U. Recife (Brazil), 1954-58; prof., head of dept. neuro-anatomy, Sch. Medicine, U. Rio Grande do Sul, 1958; asso. prof. anatomy, U. Rio Grande do Sul. Mem. Italian, Am., Brazilian Anatomy Socs.; Italian Soc. Exptl. Biology; Brazilian Genetics Soc.; A.A.A.S.; Nat. Geog. Soc. Author: (with P. Osorio, in Portuguese) Book of NeuroAnatomy, 1967; 104 publ. articles. Research on glial architecture of man and vertebrates; electrolytic decalcification; cochlear innervation in cats; regeneration of nervous fibers through cord graft in dogs; exptl. study on pio-glial membrane during endovenous injection; studies of nervous cells on peripheral arteriopathies. Home: 523 Apt. 81, Joao Telles, Porto Alegre, Rio Grande do Sul, Brazil. Office: School of Medicine, Sarmento Leite, Porto Alegre, Rio Grande do Sul, Brazil.*

COTTON, Aimé Auguste, French physicist; b. Bourg, France, Oct. 9, 1869; s. Eugène and Nathalie (Vincent) C.; ed. École Normale Supérieure; aggregate in phys. scis., 1893; m. Eugenie Feytis, Aug. 9, 1913; 2 children. Prof. gen. physics Faculty Scis., Toulouse, France, 1895-1900; faculty École Normale Superieur, Paris, 1900-20; prof., dir. lab. phys. research Sorbonne, 1920-24; dir. laboratoire des basses temperatures et du grand electro-aimant Nat. Sci. Research Center. Mem. French Acad. Scis., 1923, pres., 1938; mem. Bur. Longitudes, French Soc. Physics, French Soc. Chemistry, Internat. Phys. Union (v.p.). Devised (with H. Mouton) apparatus for observing ultramicroscopic objects, discovered double refraction of liquids in a magnetic field, 1914; studies (with Pierre Weiss) Zeeman effect, deduced ratio (e divided by m) of charge of electron to its mass; installed large electromagnet at Bellvue Labs.; inventor scale for weighing and measuring magnetic fields; research on magnetism, optics, magno-optics, anomalous dispersion and dichroism. Died Apr. 15, 1951.

COUCHET, Gérard Henri, French mathematician; b. Montpellier, France, Jan. 2, 1909; s. Severin and Marie (Bessiere) C.; agrégé in math. scis., U. Paris, 1938, D.Sc. in Math., 1948; m. Yvone Bessiere, Aug. 14, 1934; children—Mireille (Mrs. Colin), Irène (Mrs. Temple), Yves. Pro. math. numerous secondary schs. in Paris, also the provinces; mem. Faculty Scis., U. Montpellier, 1948—, titular prof., 1956—. Outside collaborator Nat. Office Aero. Studies and Research, 1946-56; examiner, pres. jury admissions for Poly. Tech., 1949-58, v.p. jury aggregation in math., 1961-65. Decorated Chevalier Legion of Honor; Officier Palmes Acad.; laureate French Acad. Scis. Mem. Math. Soc. France. Author: Mouvements plans d'un fluide en présence d'un profil mobile, 1956; also numerous articles. Research on plane mechanics of incompressible fluids, also large generalization of theory of Joukowsky concerning non-stationary movements of a fluid, engendered by anticipated displacement of a sect.; studies on aerodynamic forces which act on a sect. in presence of whirlwind, also stability of certain movements of a sect.; critical speed of thin wing. Home: 14 rue du Collège, Montpellier (Hérault) France. Office: Faculty Scis., Pl. Eugene Bataillon, Montpellier (Herault) France.*

COUDERC, Paul, French astronomer; b. Nevers, France, July 15, 1899; s. Jean and Margarite (Chastang) C.; ed. Paris Faculty Sci. and École Normale Supérieure; agrégé ès sciences mathématiques, docteur ès sciences mathématiques; m. Blanche Jurus, July 6, 1926. Prof. math., Chartres lycée, 1926-29; prof., lycées Montaigne, Charlemagne, Janson-de-Sally, Paris, 1930-44; astronomer, Paris Observatory, 1944—; maitre de conférences, astronomy, Ecole Polytechnique, 1945—; Officer of French Legion of Honor, Palmes académiques. Author: l'Architecture de l'Univers, 1930; Parmi les étoiles, 1938; la Relativité, 1942; l'Expansion de l'Univers, 1952. Office: 61, avenue de l'Observatoire, Paris 14e, France.

COURTOIS, Jean Emile, French biochemist; b. Paris, Mar. 6, 1907; s. Léon Germain and Renée

(Thurissey) C.; Baccalaureat, Coll. de Saulieu, 1924; Docteur ès-Scis., Faculté des Scis. de Paris, 1938; Docteur en Pharmacie, Faculté de Pharmacie de Paris, 1931; m. Gilberte Quinque, Aug. 6, 1934; children—Michelle (Mme. Jean Marc Cheron), Marielle, Chantal, Marie-Aleth, Isabelle. Pharmacist, Paris hosp., 1932-—; with Faculté de Pharmacie de Paris, 1931-—, titular prof. biochemistry, 1955-—. Sec., French Nat. Com. Biochemistry, 1958-—. Mem. Acad. Pharmacy Paris, Royal Acad. Pharmacy Madrid (corr.), Norske Videns Kaps Akademi i Oslo (fgn.), Soc. Biol. Chemistry (sec. gen. 1955-—), Internat. Union Biochemistry (council 1949-64), Internat. Fedn. Clin. Chemists (pres. 1963-—). Author: (with P. Fleury) Les diastases, 1947; (with R. Perles) Précis de chimie biologique, 2 vols., 1959-60; also papers. Research on properties and purification of phosphatases, phytic acid, phytase, use of periodic acid as selective reagent for oxidation of carbohydrates, structure of polysaccharides and their degradation by enzymes, glycosidases, clin. chemistry; discovered and determined structure of many new naturally occurring oligosaccharides. Home: 1, rue Chardon Lagache, Paris XVI, France.*

CULLEN, William, Scottish physician; b. Hamilton, Scotland, Apr. 15, 1710; ed. Glasgow U., M.D., 1740; m. Anna Johnstone, 1741; 7 sons, 4 daus. Surgeon on merchant ship which sailed to West Indies, 1729; asst. to apothecary, London, to 1731; practiced as surgeon, Hamilton, 1736-44; chief magistrate of Hamilton, 1739-40; founded med. sch., Glasgow, lectr. there, 1744-50; prof. medicine, Glasgow U., 1751-55; prof. chem., Edinburgh U., 1756-66; prof. of physic (physiology), 1766-89. Fellow Royal Soc., 1777; fellow Royal Soc. Edinburgh; mem. Edinburgh College Physicians (pres. 1773-75), fgn. assoc. Royal Soc. Medicine, Paris. Author: Synopsis Nosologiae Methodicae, 1769; Institutions of Medicine, Part 1, 1772; A Treatise of Materia Medica, 2 vols., 1789; Clinical Lectures (pub. posth., 1797). Experimentally used and introduced many valuable drugs, including henbane, tartar, tartar emetic, and James's powder; stressed importance of influence of nervous system in diseases; noted distinctness of motor and sensory nerves; classified diseases as (1) pyrexiae, or febrile diseases, (2) neuroses, or nervous diseases, (3) cachexiae, or diseases resulting from poor hygiene, and (4) locales, or local diseases. Died Scotland, Feb. 5, 1790.

DE CLAVE, Estienne, physician, chemist; flourished 17th century Paris. Gave chemical courses, publ. chemical books, Paris; condemned by Sorbonne, 1624. Author: Paradoxes des pierres et pierreries, 1635; le Cours de chemie, 1646; (attributed to him) Principes de nature, 1635; Nouvelle lumiere philosophiques des vrays principes . . . de la nature (possibly an early edition of Cours de Chemie, circa 1641). Favored use of chemical remedies; believed chemistry to be most essential part of medicine; thought stones were generated from seed, produced by action of subterranean fire; held that earth, water, sulphur, salt and mercury were elements.

DE HAAS, Wander Johannes, Dutch physicist; b. Lisse, Netherlands, Mar. 21, 1878; s. Albertus and Maria (Efting) de H.; doctorate in physics, U. Leyden, 1912; m. Geertruida Luberta Lorentz, Dec. 22, 1910; 2 sons, 2 daus. Asst., Kamerlingh Onnes Lab., 1895-1911; private asst., Bosscha Lab., Berlin, 1911-13; physics teacher, Royal Acad., 1913-15; tchr., Deventer Secondary Sch., 1915-16; conservator, Teyler Inst., Haarlem, 1916-17; prof. physics, Polytechnical Sch., Delft, 1917-22; prof., U. Groningen, 1922-24; prof. physics, dir. cryogenic lab., U. Leyden, 1924-48; dir., Kamerlingh Onnes Lab.; prof. emeritus, from 1948. Recipient (with Einstein) Baümgartner Prize, Vienna Acad., 1916; Rumford medal, Royal Soc., 1934. Mem. Koninklyke academie van wetenschappen, Amsterdam; Warsaw Acad. Technical Sci.; fgn. assoc., French Acad. Sci.; hon. mem., French Physics Soc.; Fellow, Royal Soc. Arts, London. Author papers in Dutch, English, French and German physics publications. Research on effect of low temperature on superconductability of magnetism; achieved record temperature of 0,000 2° K; demagnetized paramagnetic salts, 1947. Died Bilthoven, Netherlands, Apr. 26, 1960.

DEHN, Max Wilhelm, mathematician; b. Hamburg, Germany, Nov. 13, 1878; s. Maximilian Moses and Bertha (Rat) D.; grad. U. Göttingen, 1900. Asst. lectr., Münster, 1901; assoc. prof., Kiel, 1911; assoc. prof. Tech. U. Breslau, 1913; ordinary prof., Frankfort, 1921; traveled in Europe then to U. S., 1935; prof. Black Mountain College (North Carolina). Author: Die Legendreschen Sätze über die Winkelsumme im Dreck, 1900; (with Pasch) Vorlesungen über neuere Geometrie, 1926; Über die Geistige Eigenart des Mathematikers, 1928. Research in fundamental geometry group theory, topology; demonstrated that proof of Euclid's theorem for triangular pyramids depends on continuity, 1900; developed non-Archimedian geometry linking similar triangles of ancient Babylonians to 19th century non-Euclidean geometry; defined group of knot, 1910. Died Black Mountain, North Carolina, June 27, 1952.

DE SIEBENTHAL, Jean Emmanuel, mathematician; b. Lausanne, Switzerland, June 26, 1917; s. Emmanuel and Rosa (Seiter) de S.; ed. École Normale, Lausanne, 1934-38; licencié sc. Math. faculty sci. U. Lausanne, 1942; postgrad. École Polytechnique Fédérale, Lausanne, 1944-46; Sc.D. in Math., 1951; m. Lucie Favre, Sept. 21, 1946; children—Jean-Luc, Bruno, Hugues, Francois, Marie-Luce. Lectr. spl. math. Lausanne Polytech. Sch., 1946-54, prof. geometry, 1954-—, full prof., 1964-—; prof. admissions dept., 1956-63. Mem. Swiss Math. Soc. (pres. 1964-66), Lausanne Math. Club (pres. 1965-—). Author: Cours d'Algèbre du C.M.S., 1950; Cours de Géometrie Descriptive, 1960. Research, publs. (with Armand Borel) on theory of subgroups in the maximal range of related groups, 1948, theory principal sub-group, 1951, theory unconnected related groups, 1955, gradations semi-simple algebras, 1966, renovation of descriptive geometry, 1961, stem of linear algebra, 1968. Home: 25 Grand Vennes, 1010 Lausanne, Switzerland.

DIDIER, André, French physicist; b. Paris, France, Mar. 10, 1914; s. Louis and Blanche (Deslandres) D.; ed. Nat. Conservatory Arts and Trades, Paris, engring. degree; m. Mabire, Dec. 7, 1965; children—Béatrice, Claude, Francois. Chef de travaux, titular prof., chair of acoustics and electro-acoustics, Nat. Conservatory Arts and Trades, Paris. Mem. Legion of Honor; Order of Nat. Merit; French Electrotechnicians Soc.; French Math. Soc. Author: Acoustique-Electro-acoustique; about 200 published articles. Specialist in acoustical engring.; patentee in field; research on sound mechanisms; fabrication of magnetic bonds; electronic medical research. Home: 4H rue de Paris, Boissy-St. Leger 94, France. Office: 292 rue St. Martin, Paris (75) 3, France.*

EGLINUS ICONIUS, Raphael (or Nicolas Niger Hapelius), alchemist; b. Götz, Münchhof, 1559. Went to Geneva and Basel; promoted to a sch. at Sonders in Veltelin, but left 1586; "Paedagogus alumnorum", prof. new testament and deacon, Cathedral of Zurich, 1592; 1st to introduce public discussions at Zurich Cathedral; archdeacon, 1st to introduce church songs, 1596; fled from Zurich to Marburg because of debts, 1601; became doctor and prof. of theology. Author: Disquisito de Helia Artium ad illustrissimum principem Mauritium, Hassiae Landgravium, 1606 (included in 2d work); Cheiragogia Heliana, 1612. Perhaps 1st to use word "phlogiston", 1606. Died Marburg, Hesse-Nassau, Aug. 20, 1622.

GHIORSO, Albert, Am. physicist; b. Vallejo, Cal., July 15, 1915; B.S., U. Cal. 1937; m. 1942; 2 children. Mem. staff metall. lab. U. Chicago, 1942-46; physicist Lawrence Radiation Lab., U. Cal., Berkeley, 1946-—. Research in transuranium elements, nuclear properties of heavy element isotopes, reactions induced by heavy ions, systematics of radioactive decay. Office: Heavy Ion Linear Accelerator, Lawrence Radiation Laboratory, University of California, Berkeley, Calif. 94720.

HARTKEMEIER, Harry P(elle), Am. statistician; b. Louisville, May 23, 1904; s. John Fred and Mathilda (Reichenbacher) H.; B.S., U. of Louisville (awarded univ. scholarship, James B. Speed scholarship), 1927; A.M., Harvard (awarded univ. scholarship), 1928; Ph.D., U. of Chicago, 1930; m. Mona McKittrick, Dec. 23, 1930; 1 son, Leonard Douglas. Instr. in econ. statistics U. of Chicago, 1929; asst. prof. economics U. of Mo., 1930, asso. prof. accounting and statistics, 1934, asso prof. bus. statistics, 1935, prof. bus. statistics, 1941-57, dir. Statis. Lab., 1946-57, principle research statistician Nat. Safety Council, Jan.-Aug. 1956; engr. AA, Data Reduction Analysis & Quality Control, Missile Test Project, Patrick Air Force Base, Aug. 56-Jan. 57 (while on leave absence U. Mo.); research asso. Research Center, Stanford University, Stanford, Cal., 1957-58; staff member, electronic data processing dept. Sandia Corporation, Sandia Base, Albuquerque, N.M., 1958-59; staff engr., information processing Lockheed Aircraft Corp., Sunnyvale, Cal., 1959-62; sr. staff engr. Lockheed Missiles & Space Co., 1962-—; prof. statis. analysis and electronic computer data processing Santa Clara (Cal.) U., 1960-—. Served as spl. agt. U. S. Census Bureau; sr. statistician U. S. Dept. of Agr., cons. to bus. orgns. With U. S. Office of Edn. and Office of Prodn. Research and Development, W.P.B., giving special instruction on use of statis. methods and sampling to control quality of war prodn., 1944. Fulbright lectr., Coll. Commerce & Econs., Baghdad, 1953-54. Mem. Am. Statis. Assn., Econometric Soc., Am. Econ. Assn., Am. Assn. U. Profs., Legion of Honor. Author books including: Business Statistics, 1946, rev. edit., 1947; Elementary Statistical Analysis, 1952; Punch Card Methods, 1952; Data Processing, 1966; Fortran Programming of Electronic Computers, 1966; Introduction to Statistical Analysis, 1967. Contbr. articles to prof. jours. Research in statistics using computers; applications to business and military uses. Home: 638 Tomi Lea St., Los Altos, Cal. 94022. Office: Lockheed Missiles and Space Co., P.O. Box 504, Sunnyvale, Cal. 94088.

HEATH, Robert Galbraith, Am. physician, psychiatrist; b. Pitts., May 9, 1915; s. Robert Malcolm and Minnie Coleman (Galbraith) H.; B.S., U. Pitts., 1937, M.D., 1938; D.M.Sc., Columbia, 1949; m. Eleanor Bugher Wright, Sept. 7, 1940; children—Anne, Shari, Barbara, Carol, Robert Galbraith. Intern Mercy Hosp., Pitts., 1939; instr. medicine U. Pitts., 1939-40; asst. chief resident neurology Neurol. Inst. N.Y., 1940-42, asst. attending neurologist, 1946-49; psychiatrist Pa. Hosp. Nervous and Mental Diseases, also demonstrator neurology Jefferson Med. Coll., 1942-43; instr. neurology Columbia, 1946-49, attending psychoanalyst Psychoanalytic Clinic, 1951-56; prof., chmn. dept. psychiatry and neurology Tulane U., 1949-—; cons. psychiatrist Knickerbocker Hosp., N.Y., also N.J. State Hosp., 1947-49; sr. vis. physician Charity Hosp.; cons. VA Hosp., Southeast La. State Hospital, East Louisiana State Hospital; med. staff DePaul Sanitarium; vis. staff Touro Infirmary; chief med. officer Mcht. Marine Rest Center, U. S. Marine Hosp.; chief psychiatrist U. S. Penitentiary, Lewisburg, Pa.; visiting professor N.Y. School of Psychiatry. Pres. Institute Mental Hygiene of New Orleans. Med. adviser to La. dir. SSS; USPHS coms. psychosurgery and tng. grants; com. med. edn. Am. Psychiat. Assn. Fellow Am., So. psychiat. assns., American Coll. Neuropsychopharmacology (charter), Am. Acad. Neurology, N.Y. Acad. Medicine, A.C.P.; mem. Am. Neurol. Assn., Am. Assn. Psychoanalytic Medicine, A.M.A., Soc. Research Psychosomatic Problems, Society of Biological Psychiatry (vice pres.), Assn. Research Nervous and Mental Disease, A.A.A.S., Group for Advancement of Psychiatry, New Orleans Soc. Neurology and Psychiatry, Orleans Parish Med. Soc., So. Soc. for Clinical Research, Acad. Psychoanalysis (trustee), Sigma Xi. Author: Selective Partial Ablation of the Frontal Cortex (ed. F. A. Mettler), 1949. Editor: Studies in Schizophrenia, 1954; Serological Fractions in Schizophrenia, 1963; The Role of Pleasure in Behavior, 1964. Research on brain function and behavior; biochem. basis of schizophrenia; pain and pleasure responses. Address: 1430 Tulane Av., New Orleans 70112.

HUTCHINSON, G(eorge) Evelyn, biologist; b. Cambridge, Eng., Jan. 30, 1903; s. Arthur and Evaline D. (Shipley) H.; ed. Greshams Sch., Holt Norfolk, and U. of Cambridge. Came to U. S., 1928, naturalized, 1941. Sr. lectr. zoology U. Witwatersrand Union, South Africa, 1926-28; instr., advancing to prof. Sterling prof. zoology, Yale, since 1928. Author: The Clear Mirror, 1936; The Itinerant Ivory Tower, 1953; A Treatise on Limnology, vol. 1, 1957, vol. 2, published 1967; A Preliminary List of the Writings of Rebecca West, 1912-51, 1957; The Enchanged Voyage, 1962; The Ecological Theater and the Evolutionary Play, published 1965; also numerous sci. papers on aquatic insects, limnology, biogeochemistry. Other research on aquatic Hemiptera; psychoanalytic aspects of evolution; theoretical aspects of ecology. Home: 269 Canner St., New Haven 11, Conn. Office: Osborn Zoological Lab., Yale Univ., New Haven, Conn.

ISSACSON, Robert Lee, Am. psychologist; b. Detroit, Sept. 26, 1928; s. Emil Alfred and Evelyn (Johnson) I.; A.B., U. Mich., 1950, M.A., 1954, Ph.D., 1958; m. Susan Doherty, Dec. 29, 1956; children—Gunnar, Lars, Mary Ingrid, Mary Christina. Faculty, U. Mich., Ann Arbor, 1956-—, prof. dept. psychology, 1967-—. Mem. Am., Mich. psychol. assns., N.Y. Acad. Scis., A.A.A.S., Psychonomic Soc., Sigma Xi, Phi Kappa Phi, Phi Sigma. Author: (with M. L. Hutt and M. L. Blum) Psychology: The Science of Behavior, 1965, Psychology: The Science of Interpersonal Behavior, 1966. Editor: Basic Readings in Neuropsychology, 1965. Research, publs. on ways structures of oldest parts of brain (limbic system) interact with each other and with neocortex to control and regulate behavior, effects of brain lesions on behavior and developing nervous system. Home: 810 Oxford St., Ann Arbor, Mich. 48104.*

KELLY, Brian Thomas, Brit. physicist; b. Cardiff, S. Wales, Aug. 13, 1934; s. James and Winifred (Knott) K.; B.Sc. with honors, U. Coll. S. Wales and Monmouthshire, Cardiff, 1955; m. Kathryn Lord, Nov. 30, 1957; children—Annette, Michael, Richard (dec.), Gareth. Sci. officer U.K. Atomic Energy Authority, Windscale Works, 1955-60, sr. sci. officer reactor materials lab., Culcketh, Eng., 1960-66, prin. sci. officer, 1966-—. Asso. mem. Inst. Physics. Author: Irradiation Damage to Solids, 1966; also articles. Research on effects of neutron irradiation in nuclear reactors on phys. and mech. properties of graphite used as reactor moderator (led to devel. improved nuclear graphites for use in reactors), effects of fast neutron irradiation on properties of zirconium and zincology, concrete. Home: 14 Fowley Common Lane, Glazebury, nr. Warrington. Office: U.K. Atomic Energy Authority, Reactor Materials Lab., Culcketh, nr. Warrington, Lancashire, Eng.*

KEREKES, Medard Francis, Rumanian biochemist; b. Cluj, Rumania. Oct. 9, 1925; s. Francis and Gisela (Urbansky) K.; ed. Babes-Bolyai U., Cluj, diploma in chemistry, 1949; dr. chemistry, Academy of Socialist Republic Rumania, 1965; m. Elizabeth Toth, June 14, 1958; children—Medard Francis, Ildiko. Demonstrator, Babes-Bolyai U., 1949; asst., Medical Sch. Tirgu Mures U., chair of biochemistry, 1949-58; principal researcher, Research Station Tirgu Mures, 1958-—. Recipient several public service decorations. Author: 70 publ. articles. Research in protein chemistry (denaturation of proteins); sensibility against trypsin of denatured proteins; protection of proteins against denaturation; studies on proteolytic enzymes (cathepsins) of the brain in normal animals and animals with exptl. allergic encephalomyelitis. Home: 4, str. Cringului, Tirgu Mures, Rumania. Office: 38, str. Gh. Marinescu, Tirgu Mures, Rumania.*

KORTE, Friedhelm, German chemist; b. Bielefeld, Nov. 24, 1923; s. Wilhelm and Luise (Schilling) K.; student univs. Freiburg, Marburg; Ph.D., U. Göttingen, 1948; m. Ingeborg Vogler, Sept. 18, 1948; children—Stephan, Gisela. Faculty, U. Hamburg, 1954-55, U. Bonn, 1955-—, asso. prof. organic chemistry, 1964-67, prof. organic chemistry, 1967-—; mgr., research dir. Shell Grundlagenforschung GmbH, Schloss Birlinghoven, 1959-64. Mem. Soc. German Chemists (chmn. Bonn sect.), Am. Chem. Soc. Regional Editor: Tetrahedron, Tetrahedron Letters. Numerous articles in field. Research on chem. classification of plants; isolation of bitter principles; heterocycles in living cells; acyllacton rearrangement; paraffins; ecological chemistry of pesticides, active compounds of hashish; chemistry of high pressure. Home: 5204 Hangelar-Niederberg, Im Erlengrund 6. Office: Organisch-Chemisches Institut der Universität Bonn, Meckenheimer Allee 168, 53 Bonn, West Germany.*

KUCKUCK, Hermann Adolf Friedrich, German plant geneticist; b. Berlin, Sept. 7, 1903; s. Hermann and Else (Vockrodt) K.; Dipl. agron., U. Berlin, 1928, Dr. agr., 1929, Dr. agr. habil., 1942; m. Erika Matthie, Mar. 28, 1931; children—Ingrid (Mrs. Horst Albert), Gisela (Mrs. Günther Alleweldt), Elke (Mrs. Dietrich Hoppenstedt), Holger. Prof. plantbreeding U. Halle/Saale, 1946-48; dir. Central Research Inst. Plantbreeding, Müncheberg, 1948-50; guest researcher Inst. Plantbreeding, Svalöf, Sweden, 1950-52; plantbreeding expert FAO, Iran, 1952-54; prof., dir. Inst. Applied Genetics, Horticulture Faculty, Hanover Inst. Tech., 1954-—. Cons. FAO, Rome. Mem. Eucarpia. Author: (with Mudra) Lehrbuch der Allgemeinen Pflanzenzüchtung, 1950; Pflanzenzüchtung I: Grundzüge der Pflanzenzüchtung, 1952; Pflanzenzüchtung II: Spezielle Gartenbauliche Pflanzenzüchtung, 1957; also articles. Research on chromosome mapping of antirrhinum, genetics and breeding of barley, evolution of cultivated wheats. Home: 3001 Thönse, Germany. Office: 2 Herrenhäuser-Strasse, Hanover, West Germany.*

LOESCHCKE, Hans Hermann, German physiologist; b. Cologne, Germany, Oct. 20, 1912; s. Hermann and Thekla (Freytag) L.; student univs. Heidelberg, Freiburg; state exam., U. Greifswald, 1935, Dr.med., 1937; m. Gertrud Wilckens, May 26, 1942; children—Gerhard, Henning, Volker, Gesine. Asst., Physiol. Inst., U. Göttingen, 1938-50, unscheduled prof., 1950-63, prof. clin. physiology, head dept., 1963-64; prof. physiology, chmn. dept. Ruhr U., Bochum (Germany), 1964-—; guest prof., Physiol. Inst., Bern, Switzerland, 1954-55; guest scientist Cardiovascular Research Inst., San Francisco, 1960. Rockefeller fellow, 1949-50. Mem. German Soc. Circulation Research, Soc. Lung and Respiration Research, N.Y. Acad. Scis. Research and numerous publs. on physiology of blood, circulation and respiration including problems of gas diffusion, effects of oxygen deficiency, effects of steroid hormones on breathing, regulation of ionic body constituents. Address: 18 Paracelsusweg, Bochum, West Germany.*

LOZACH, Noël, French chemist; b. Nantes, France, Aug. 2, 1915; s. Leon and Marguerite (Aubry) L.; ed. Ecole Normale Supérieure, 1935-45, doctor ès sciences, Paris, 1945; m. Odette Marie, Dec. 11, 1943; children—Rene, Mireille, Yves, Sylvie, Annie, Herve. Asst., 1941-47, lectr., 1947-49, Faculty Sci., Paris; reader, 1949-52, prof., 1952-—, Faculty Sci., Caen; dean, Faculty Sci., Caen, 1956-—; dir. Ecole Nat. Sup. Chimie, Caen, 1952-—. Mem. organic chem., C.N.R.S., 1967-—; IUPAC, 1954-—; Comite Consultatif des Universities, 1967-—. Decorated Chevalier, Legion of Honor, Commander, Palmes académiques. Mem. French Chem. Soc.; Am. Chem. Soc. Author: (with Vavon, Dulou) Manipulations, 1946; Nomenclature en chimie organique, 1967; sixty publ. articles, contributor to other books. Research in organo-sulphur chemistry, mainly heterocycles; organic nomenclature. Home: 17, rue Docteur Maugeais 14-Caen, France. Office: Faculte des Sciences, 14-Caen, France.*

MARKOWITZ, William, astronomer; b. Milicz, Poland, Feb. 8, 1907; s. Hyman and Rebecca (Baumstein) M.; came to U. S., 1911, naturalized, 1923; M.S., U. Chgo., 1929, Ph.D., 1931; m. Rosalyn Bessie Shulemson, Jan. 28, 1943; 1 son, Harold Toby. Instr. mathematics, astronomy, Pa. State U., 1931-32; with U. S. Naval Obs., Washington, 1936-—, dir. time service div., 1953-66; instr. math. and astronomy Catholic U., Washington, 1947-50, U. Md., 1965-66; prof. physics, Marquette U., 1966-—. Cons. NASA, Nat. Acad. Sci. Mem. Am. Astron. Soc., Am. Geophys. Union., Internat. Astron. Union (pres. com. on time, 1955-61), Internat. Union Geod., Geophys. (pres. sect. geod. astron. 1960), Commn. on Definition of the Second. Important work includes design of dual-rate moon camera. Office: Dept. of Physics, Marquette University, Milwaukee, Wis. 53233.*

MELLORS, Robert Charles, Am. pathologist; b. Dayton, O., June 18, 1916; s. Bert S. and Clementine (Steinmetz) M.; A.B., Western Res. U., 1937, M.A., 1938, Ph.D., 1940; M.D., Johns Hopkins, 1944; m. Jane K. Winternitz, Mar. 25, 1944; children—Alice J., Robert Charles, William K., John W. Instr. biochemistry Western Res. U., 1940-42; research asso. epidemiology Johns Hopkins, 1942-44; with Meml. Center for Cancer and Allied Diseases, N.Y.C., 1946-58, asso. attending pathologist, 1957-58; faculty Cornell Med. Coll., 1953-—, prof. pathology, 1961-—; pathologist, dir. labs., asso. dir. research Hosp. for Spl. Surgery, N.Y.C., 1961-—; asso. attending pathologist N.Y. Hosp., 1961-—; mem. research adv. com. NIH and Am. Cancer Soc. Mem. Coll. Am. Pathologists (mem. com. on nomenclature and classification of disease), Am. Acad. Orthopedic Surgeons (Kappa Delta award), Am. Soc. Exptl. Pathology, Am. Soc. Biol. Chemists, Am. Assn. Immunologists, Am. Assn. Pathologists and Bacteriologists, Am. Soc. Clin. Pathologists, Coll. Pathologists (Gt. Britain), Internat. Soc. Cell Biology, A.M.A., Am. Rheumatism Assn., Phi Beta Kappa, Sigma Xi. Research, publs. on viruses as agts. of cancer, hypersensitivity diseases of man and mouse. Home: 47 Oakwood Av., Rye, N.Y. 10580. Office: 535 E. 70th St., N.Y.C. 10021.*

MELTZER, Allan Harold, Am. economist; b. Boston, Feb. 6, 1928; s. George B. and Minerva I. (Simons) M.; A.B., Duke, 1948; M.A., U. Cal. at Los Angeles, 1955, Ph.D., 1958; m. Marilyn Ginsburg, Aug. 27, 1950; children—Bruce Michael, Eric Charles, Beth Denise. Lectr. dept. econs. U. Pa., Phila., 1956-57; faculty Carnegie Inst. Tech. Grad. Sch. Indsl. Adminstrn., Pitts., 1956-—, prof. econs., 1964-—; Ford Found. vis. prof. U. Chgo., 1964-65. Cons. U. S. Treasury, 1961-62, joint econ. com. U. S. Congress, 1960, com. on banking and currency U. S. Ho. of Reps., 1963-64; bd. govs. Fed. Res. System, Fed. Deposit Ins. Corp. Social Sci. Research Council fellow, 1955-56; Ford Found. fellow, 1962-63. Mem. Am. Econ. Assn., Am. Finance Assn., Econometric Soc. Author: (with G. von der Linde) A Study of the Dealer Market, 1960; (with Karl Brunner) An Analysis of Federal Reserve Monetary Policymaking, 1964; also articles. Research on relation of money and changes in monetary policy to nat. income, employment and prices. Home: 5830 Marlborough Av., Pitts. 15217.*

MORRIS, John Wesley, Am. geographer; b. Billings, Okla., Nov. 14, 1907; s. Henry L. and Lillian (Knowles) M.; B.S., U. Okla., 1930; M.S., Okla. State U., 1934; Ph.D., George Peabody Coll. for Tchrs., 1941; m. Mary E. Russell, Feb. 19, 1932; children—Carole (Mrs. Harvey Wilson), Russell A. Faculty, Seminole (Okla.) Jr. Coll., 1931-38, Southeastern (Okla.) State Coll., 1939-42, 46-48; prof. geography U. Okla., Norman, 1948-—, chmn. dept., 1965-—. Asso. dir. Okla. Center Urban and Regional Studies, 1951-—. Mem. Assn. Am. Geographers, Nat. Council Geog. Edn. (pres. 1960, Distinguished Service award 1966), Southwestern Social Sci. Assn. (pres. 1965). Author: (with Freeman) World Geography, 1958; Oklahoma Geography, 1960; Historical Atlas of Oklahoma, 1965. Devel. vegetation keys for boreal fringe areas by photo interpretation. Home: 833 McCall Drive, Norman, Okla. 73069.*

MORRISON, Philip, Am. physicist; b. Somerville, N.J., Nov. 7, 1915; B.S. Carnegie Inst., 1936; Ph.D. U. California, 1940; m. 1938. Instr. physics, San Francisco State College (Calif.), 1941; U. Illinois, 1941-42; physicist, metallurgy lab., Chicago, 1943-44; physicist, group leader, Los Alamos Sci. Lab., 1944-46; asso. prof. physics, Cornell U., 1946-56; prof. Cornell, 1956-65; prof. Mass. Inst. Technology, 1965-—. Recipient Pregel Prize, 1955, Babson Prize, 1957, Cersted medal, 1965. Mem. Physics Soc. Research on applications of nuclear physics in astronomy; theory of nuclei. Office: Physics Dept., Massachusetts Institute of Technology, Cambridge, Mass. 02139.

NEELS, James Van Gundia, Am. geneticist; b. Hamilton, O., Mar. 22, 1915; s. Hiram A. and Elizabeth (Van Gundia) N.; A.B., Coll. Wooster, 1935, D.Sc., 1959; Ph.D., U. Rochester, 1939, M.D., 1944; m. Priscilla Baxter, May 6, 1943; children—Frances, James Van Gundia, Alexander. Instr. zoology Dartmouth Coll., 1939-41; fellow in zoology NRC, 1941-42; faculty U. Mich., Ann Arbor, 1948-—, prof. human genetics, chmn. dept., 1956-—, prof. internal medicine, 1957-—, Russel lectr., 1966. Cons. USPHS, 1956-—, NRC, 1949-—, WHO, 1957-—. Recipient Albert Lasker award Am. Pub. Health Assn., 1960. Mem. Nat. Acad. Sci., Am. Soc. Human Genetics (pres. 1953-54 Allen awards, 1965), Assn. Am. Physicians, Am. Philos. Soc., Genetics Soc. Am., Japanese Soc. Human Genetics, Brazilian Soc. Genetics, Phi Beta Kappa, Sigma Xi, Alpha Omega Alpha. Human Genetic studies: Mutation Rates, Blood Abnormalities, Radiation and Inbreeding Effects, and Population Genetics of Primitive Man. Home: 2235 Belmont St., Ann Arbor, Mich. 48104.*

NIEMI, Mikko, Finnish anatomist; b. Turku, Finland, Sept. 13, 1929; s. Aimo and Ester (Lehto) N.; Dr.Med., U. Helsinki (Finland), 1958; m. Irmeli Kuusisto, Apr. 16, 1953; children—Jaana, Tiina, Veli-Mikko. Faculty, U. Helsinki, 1956-—, sr. lectr. anatomy, 1960-64, sr. lectr. pathology, 1964-—; prof. anatomy U. Turku, 1965-—. Research, numerous publs. on applied histochemistry, devel. and function of male reproductive organs. Home: 12 Piispankatu, Turku, Finland.*

NORTHUP, George Warren, Am. osteopathic physician; b. Syracuse, N.Y., Oct. 9, 1915; s. Thomas Larkham and Anna Laura (Carleton) N.; student Brothers Coll., Drew U., 1933-35; D.O., Phila. Coll. Osteopathy, 1939, postgrad. student, 1939-41; student Seton Hall Coll., 1949-50; licensed osteopathic physician, N.J., Pa., 1939, licensed to practice medicine and surgery, N.J., 1941; certified Am. Osteopathic Bd. Phys. Medicine and Rehabilitation; m. Elsie May Webster, Sept. 3, 1941; 1 son, Jeffrey Carleton. Mem. staff West Essex Gen. Hosp.; now chmn. dept. phys. medicine. Mem. 4-man com. rep. osteopathic profession to NRC; del. Am. Council Edn., 1953, 54, 59, Nat. Health Council, 1959-66 (also mem. bd. dirs.); cons. Surgeon General's Conf. on Asian flu, 1957; cons. Sec. HEW conf. on food additives, 1958; del. White House Conf. Children and Youth, 1960; treas. Nat. Coordinating Council on Drug Abuse Edn. and Information, 1968; pres. Livingston Profl. Labs., 1951-59. Named Physician of Year, N.J. Assn. Osteopathic Physicians and Surgeons, 1958; recipient Drew Alumni Achievement Award in Scis., 1962. Fellow N.Y. Acad. Osteopathy, Acad. Applied Osteopathy (pres. 1950-51), N.Y. Acad. Scis., Am. Coll. Osteopathic Internists (hon.), Am. Coll. Osteopathic Surgeons (hon.); mem. Am. Osteopathic Assn. (gen. program chmn. convention, Boston, 1948; mem. ho. dels. 1951, 53, 54, 55, 56, trustee 1954-59, pres. 1958-59, editor 1961-—), Am. Council Edn., N.J. (speaker ho. dels. 1964), Nat. Health Council (treas. 1964-—), A.A.A.S., Essex County Assn. Osteopathic Physicians and Surgeons. Author: Osteopathic Medicine: An American Revolution; also numerous articles, editorials on osteopathic medicine. Editor: Jour. Am. Osteopathic Assn., The D.O., Health, 1961-—. Home: 81 Woodland Rd., Madison, N.J. 07940. Office: Am. Osteopathic Assn., 212 E. Ohio St., Chgo. 60611; 104 S. Livingston Av., Livingston, N.J. 07039.*

PINNEY, Edmund Joy, Am. mathematician; b. Seattle, Aug. 19, 1917; s. Henry Lewis and Alice (Joy) P.; B.S., Cal. Inst. Tech., 1939, Ph.D., 1942; m. Eleanor Russell, Mar. 10, 1945; children—Henry Russell, Gail Shiela. Teaching fellow math. Cal. Inst. Tech., 1939-42; research asso. Mass. Inst. Tech., 1942-43; research analyst Consol. Vultee Aircraft Co., San Diego, 1943-45; instr. math. Ore. State Coll., 1945-46; lectr. U. Cal., 1946-47, asst. prof., 1947-52, asso. prof., 1952-59, prof. 1959-—; dir. contract Office Naval Research, 1953-—. Mem. Am. Math. Soc., Am. Phys. Soc., A.A.A.S. Research and publs. on electromagnetic theory, integral equations, hydrodynamics, mechanics, electricity nonlinear mechanics, calculus of variations in abstract spaces. Home: 66 Scenic Dr., Orinda, Cal. 94563. Office: Dept. Math., U. Cal. at Berkeley, Cal.*

PLATT, Robert Baxter, Am. ecologist; b. Knoxville, Tenn., Jan. 19, 1913; A.B. Emory & Henry College, 1933; M.A. Peabody College, 1935; Ph.D. U. Pennsylvania, 1948; m. Deanie Sherrod, 1941; children: Carolee (Mrs. Arthur Frost), Rosalind. Biology instr., Roanoke Sr. High, Roanoke Va., 1936-39; instr., Armstrong Jr. College, Savannah, Ga., 1939-41; lab. instr., U. Pennsylvania, 1945-48; asst. prof., Emory U., 1948-52; assoc. prof., 1952-57; prof., 1957-—; chmn. dept. biology, Emory U., 1966-—. Visiting prof. UCLA, 1966; consultant, radiation biology, UCLA, 1965-—. Mem. Ecological Soc. of Am. (vice-pres. 1966); fellow, AAAS, Internat. Soc. Biometeorology; Am. Inst. Biological Sci.; Ga. Acad. Sci.; Radiation Research Soc. Author: (with John F. Griffiths) Environmental Measurement and Interpretation, 1964; (with George K. Rein) Bioscience, 1967. Research on nature and quality of man's environment, especially physiological and ecosystems ecology, and environmental effects of ionizing radiation including nuclear events. Home: 1811 E. Clifton Road N.E., Atlanta, Ga. 30307. Office: 203 Biology Bldg., Emory University, Atlanta, Ga. 30322.*

PROSSER, Francis Ware, Jr., Am. physicist; b. Wichita, Kan., June 30, 1927; s. Francis Ware and Harriet (Osborne) P.; B.S., U. Kan., 1950, M.S., 1954, Ph.D., 1955; m. Nancy Lou Baugh, May 31, 1952; children—David Francis, Rebecca Ann, Martha Lou. Research asso. Rice Inst., 1955-57; faculty U. Kan., Lawrence, 1957-—, asso. prof., 1963-—, prof., 1967-—. NSF fellow, 1935-55. Mem. Am. Phys. Soc., Am. Assn. Physics Tchrs., Sigma Xi, Sigma Pi Sigma, Sigma Tau, Tau Beta Pi. Exptl. studies, publs. on properties of excited states of nuclei, primarily nuclei lighter than calcium, measurement of energy, angular momentum, parity and decay properties of these states; contbns. to theory and application of gamma-ray spectroscopy to such measurements. Home: 1622 Cambridge Rd., Lawrence, Kan. 66044.*

ROSENKRANZ, Alfred, Austrian physician; b. Vienna, Austria, Sept. 21, 1924; s. Emil and Maria (Cochlar) R.; M.D., U. Vienna, 1948; m. Evelyne

Schreder, Apr. 21, 1960; children—Clemens Johannes, Isabella Maria, Lukas Rupertus. Sci. worker Path. Anat. Inst. U. Vienna, 1948-51, pediatrician, 1951-64; dir. Kinderklinik der Stadt Wien Glanzing, Vienna, 1964——; U. prof., 1968——. Author: Diabetes Mellitus in Childhood, 1967; also numerous articles. Med. studies in nephrology, metabolism, diabetes in childhood; exams. of biochemistry of premature babies. Office: 37 Glanzing G.A. 1190, Vienna, Austria.*

RZHANOV, Anatolii Vasilevich, Russian engr.; staff P.N. Lebedev Inst. Physics, USSR Acad. Scis., to 1962; dir. Siberian br. Inst. Solid State Physics and Semiconductor Electronics (now Semiconductor Physics Inst. of Siberian Branch, Acad. Sci., USSR), 1962——. Corr. mem. USSR Acad. Scis. Research and publs. on metalloids, especially germanium. Home: Woewodsky 14, Novosibirsk, U.S.S.R. Office: Semiconductor Physics Institute, Novosibirsk 90, USSR.

SALVETTI, Carlo, Italian nuclear physicist; b. Milan, Italy, Dec. 30, 1918; s. Adriano and Irma (Smerzi) S.; D.Physics, U. Milano, 1940; m. Piera Pinto, June 16, 1951; children—Silvia-Sherin, Adriana. Head reactor group, a founder CISE Labs., 1946-57; dir. gen. Ispra Research Centre, 1957-59; dir. IAEA Research and Labs., 1959-62; mem. Italian Nuclear Energy Commn. (CNEN), 1960-64, vice chmn., 1964——. Gov. for Italy, IAEA, 1963, chmn. bd. govs., 1964; v.p. Italian Nuclear Energy Forum, 1964——. Mem. Atomic Indsl. Forum, Forum Italiano Energia Nucleare (v.p.), Società Italiana di Fisica, Associazione Elettrotecnica Italiana. Author: Lectures on Reactor Theory, 1951; Lezioni di Fisica Sperimentale, 1967; also articles. Research in operational methods in reactor theory, time dependent neutron density problems; contbr. to devel. of nuclear models (Shell model and collective models); design and calculation of criteria for heavy water reactor. Home: 38 Via Gramsci. Office: 15 Via Belisario, Rome, Italy.*

SRINIVASAN, Sumangali Kidambi, Indian mathematician; b. Madras, India, Dec. 16, 1930; s. Sumangali Kidambi Govindachari and Rukmani; student Voorhees Coll., Vellore, 1948-50; M.A., Madras U., 1953, M.Sc., 1955, Ph.D. (Govt. sr. research fellow), 1957; postdoctoral fellow U. Sydney (Australia), 1957-58; m. Vijaya Lakshmi, Nov. 25, 1959; children—Ramanujan, Venkatesan, Sridhar. Nat. Inst. Scis. sr. research fellow U. Madras, 1958-59; faculty dept. applied math. Indian Inst. Tech., Madras, 1959-67, asst. prof., 1961-67; prof., 1967——. Mem. Indian, Am. math. socs. Reviewer: Zentralblatt für Mathematik. Research, publs. on stochastic methods in kinetic theory, theory of multiple prodn. of mesons in photo nucleon reactions. Home: 2, XIII Cross Rd., Indian Inst. Tech., Madras, 36, India.*

STREET, Kenneth, Jr., Am. chemist; b. Berkeley, Cal., Jan. 30, 1920; A.B., U. Cal., 1943; Ph.D., U. Cal., 1949; m. 1944; three children. Instr. chem., U. Cal., 1949-51; chemist California Research and Development Co., 1951-52; Lawrence Radiation Laboratory, U. Cal., Berkeley, 1952-—; prof. chem. U. Cal. 1959——. Mem. Am. Chem. Soc. Research on transuranium elements; nuclear reactions; radioactive isotopes; molecular beam spectroscopy; chromatography of inorganic ion exchange. Office: Lawrence Radiation Laboratory, University of California, Berkeley, Cal. 94720.

TALBOTT, John Harold, Am. physician; b. Grinnell, Ia., July 10, 1902; s. Arthur D. and Caroline (Flook) T.; A.B., Grinnell Coll., 1924, D.Sc., 1946; M.D., Harvard, 1929; m. Mildred Cherry, Sept. 3, 1932; children—John A., Cherry (Mrs. John C. Trowbridge). Intern, Presbyn. Hosp., City N.Y., 1929-31; faculty Harvard Med. Sch., 1931-46; med. staff Mass. Gen. Hosp., Boston, 1931-46; prof. medicine U. Buffalo, 1946-59; physician-in-chief Buffalo Gen. Hosp., 1946-59; dir. div. sci. publs. A.M.A., 1959——, editor Jour. A.M.A., 1959——. Dir. U. S. Army Frostbite Commn. to Korea, 1951; chmn. Josiah Macy, Jr., Conf. on Cold Injury, 1952-58; mem. sci. adv. panel Dept. of Army, 1958-60; mem. sci. com. Nat. Arthritis and Rheumatism Found. Served to col. AUS, 1941-46; comdg. officer Climatic Research Lab. Decorated Legion of Merit. Mem. A.C.P. (gov. Western N.Y. 1955-59), A.M.A., Assn. Am. Physicians, Am. Soc. Clin. Investigation. Author: Gout, 3d edit., 1967; (with R. Ferrandis) Collagen Diseases, 1956; (with L. M. Lockie) Progress in Arthritis, 1958. Editor: Medicine, 1947-59. Research on responses of normal man to stresses of prolonged exercise, high and low temperatures, high altitude; studies on chemical abnormalities of uric acid metabolism; original findings in gouty arthritis; fundamental aspects of elevated serum uric acid values; integration of various aspects of pathological findings of connective tissue disorders such as polyarthritis, polymyositis, systemic scleroderma, systemic lupus erythematosus. Home: 900 Lake Shore Dr., Chgo. 60611. Office: 535 N. Dearborn St., Chgo. 60610.*

WRIGHT, Albert Hazen, Am. zoölogist; b. Hilton, N.Y., Aug. 15, 1879; s. Delos C. and Emily A. (Hazen) W.; grad. State Normal and Tng. Sch., Brockport, N.Y., 1899; A.B., Cornell U., 1904, A.M., 1905, Ph.D., 1908; m. Anna Allen, June 25, 1910. Asst. in vertebrate zoölogy Cornell U., 1905-08, faculty, 1908-46, prof., 1925-46, emeritus. Recipient Eminent Ecologist award, 1955. Fellow A.A.A.S., Am. Geog. Soc., Acad. Zoology India; mem. Am. Fish Soc. (hon.), Am. Soc. Zoologists, Biol. Soc. Washington, Ecol. Soc. Am., Am. Soc. Mammalogists, Am. Soc. Ichthyologists and Herpetologists, Am. Ornithologists Union, Am. Fisheries Soc., Sigma Xi. Author: North American Anura, 1914; Biol. Reconnaissance of the Okefinokee Swamp, Ga.; Old Northampton, N.Y., 1928; Life Histories of Okefinokee Frogs, 1931; Handbook of Frogs and Toads, 1933, 1942; History and Cartography of Okefinokee Swamp, Vol. I, 1945; also Studies in History, Nos. 1-34; Handbook of U. S. A. Snakes, 2 vols., 1958; Sullivan Expedition of 1779, Regimental Roster, 1965. Home: 113 E. Upland Rd., Ithaca, N.Y.*

NOBEL PRIZES

YEAR	PHYSICS	CHEMISTRY	PHYSIOLOGY OR MEDICINE
1901	Wilhelm Röntgen	Jacobus van't Hoff	Emil von Behring
1902	Hendrik Lorentz Pieter Zeeman	Emil Fischer	Sir Ronald Ross
1903	Antoine Henri Becquerel Pierre Curie Marie Curie	Svante Arrhenius	Niels R. Finsen
1904	Lord Rayleigh	Sir William Ramsay	Ivan P. Pavlov
1905	Philipp Lenard	Adolf von Baeyer	Robert Koch
1906	Sir Joseph Thomson	Henri Moissan	Camillo Golgi Santiago Ramon y Cajal
1907	Albert A. Michelson	Eduard Buchner	Alphonse Laveran
1908	Gabriel Lippman	Sir Ernest Rutherford	Paul Ehrlich Élie Metchnikoff
1909	Guglielmo Marconi Karl F. Braun	Wilhelm Ostwald	(Emil) Theodor Kocher
1910	Johannes van der Waals	Otto Wallach	Albrecht Kossel
1911	Wilhelm Wien	Marie Curie	Allvar Gullstrand
1912	Nils Gustaf Dalén	Victor Grignard Paul Sabatier	Alexis Carrel
1913	Heike Kamerlingh Onnes	Alfred Werner	Charles Richet
1914	Max von Laue	Theodore W. Richards	Robert Barany
1915	Sir William Bragg Sir Lawrence Bragg	Richard Willstätter	(No award)
1916	(No award)	(No award)	(No award)
1917	Charles G. Barkla	(No award)	(No award)
1918	Max Planck	Fritz Haber	(No award)
1919	Johannes Stark	(No award)	Jules J. Bordet
1920	Charles Guillaume	Walther Nernst	August Krogh
1921	Albert Einstein	Frederick Soddy	(No award)
1922	Niels Bohr	Francis Aston	Archibald V. Hill Otto Meyerhof
1923	Robert A. Millikan	Fritz Pregl	Sir Frederick G. Banting John J. Macleod
1924	Manne Karl Siegbahn	(No award)	Willem Einthoven
1925	James Franck Gustav L. Hertz	Richard Zsigmondy	(No award)
1926	Jean Baptiste Perrin	Theodor Svedberg	Johannes A. Fibiger
1927	Arthur (Holly) Compton Charles T. Wilson	Heinrich Wieland	Julius Wagner-Jauregg
1928	Sir Owen W. Richardson	Adolf O. Windaus	Charles J. Nicolle
1929	Prince Louis de Broglie	Sir Arthur Harden Hans von Euler-Chelpin	Christiaan Eijkman Sir Frederick Hopkins
1930	Sir Chandrasekhara Raman	Hans Fischer	Karl Landsteiner
1931	(No award)	Carl Bosch Friedrich K. Bergius	Otto. H. Warburg
1932	Werner Heisenberg	Irving Langmuir	Edgar D. Adrian Sir Charles Sherrington
1933	Paul A. Dirac Erwin Schrödinger	(No award)	Thomas Hunt Morgan
1934	(No award)	Harold C. Urey	George H. Whipple George R. Minot William P. Murphy
1935	Sir James Chadwick	Frédéric Joliot-Curie Irene Joliot-Curie	Hans Spemann
1936	Carl D. Anderson Victor F. Hess	Peter J. Debye	Sir Henry H. Dale Otto Loewi
1937	Clinton J. Davisson Sir George Thomson	Sir Walther Haworth Paul Karrer	Albert Szent-Györgyi
1938	Enrico Fermi	Richard Kuhn	Corneille Heymans
1939	Ernest O. Lawrence	Adolf F. Butenandt Leopold Ruzicka	Gerhard Domagk